SCIENTIFIC ENCYCLOPEDIA

Ninth Edition

VOLUME 2

VAN NOSTRAND'S

SCIENTIFIC ENCYCLOPEDIA

Ninth Edition

VOLUME 2

Glenn D. Considine
Editor

Peter H. Kulik
Associate Editor

Animal Life
Biosciences
Chemistry
Earth and Atmospheric Sciences
Energy Sources and Power Technology
Mathematics and Engineering Sciences
Medicine, Anatomy, and Physiology
Physics
Plant Sciences
Space and Planetary Sciences

WILEY-INTERSCIENCE

A John Wiley & Sons, Inc., Publication

This book is printed on acid-free paper. ⊗

Copyright © 2002 by John Wiley & Sons Inc., New York. All rights reserved.

Published simultaneously in Canada.

No part of this publication may be reproduced, stored in a retrieval system or transmitted in any form or by any means, electronic, mechanical, photocopying, recording, scanning or otherwise, except as permitted under Sections 107 or 108 of the 1976 United States Copyright Act, without either the prior written permission of the Publisher, or authorization through payment of the appropriate per-copy fee to the Copyright Clearance Center, 222 Rosewood Drive, Danvers, MA 01923, (978) 750-8400, fax (978) 750-4744. Requests to the Publisher for permission should be addressed to the Permissions Department, John Wiley & Sons, Inc., 605 Third Avenue, New York, NY 10158-0012, (212) 850-6011, fax (212) 850-6008, E-Mail: PERMREQ@WILEY.COM.

For ordering and customer service, call 1-800-CALL-WILEY.

Library of Congress Cataloging in Publication Data is available.

Considine, Glenn D., editor; Kulik, Peter H., associate editor
    Van Nostrand's Scientific Encyclopedia, Ninth Edition

ISBN 0-471-33230-5 (v. 2)

Printed in the United States of America.

10 9 8 7 6 5 4 3 2

REPRESENTATIVE TOPICAL COVERAGE

ANIMAL LIFE

Amphibians
Annelida
Arthropods
Birds
Coelenterates
Echinoderms
Fishes
Insects
Mammals
Mesozoa
Mollusks
Paleontology
Protozoa
Reptiles
Rotifers
Zoology

BIOSCIENCES

Amino Acids
Bacteriology
Biochemistry
Biology
Biophysics
Cytology
Enzymes
Fermentation
Genetics
Hormones
Microbiology
Molecular Biology
Proteins
Recombinant DNA
Viruses
Vitamins

CHEMISTRY

Acids and Bases
Catalysts
Chemical Elements
Colloid Systems
Corrosion
Crystals
Electrochemistry
Free Radicals
Inorganic Chemistry
Ions
Macromolecular Science
Organic Chemistry
Oxidation-Reduction
Photochemistry
Physical Chemistry
Solutions and Sales

EARTH AND ATMOSPHERIC SCIENCES

Climatology
Ecology
Geochemistry
Geodynamics
Geology
Geophysics
Hydrology
Meteorology
Oceanography
Tectonics
Seismology
Volcanology

ENERGY SOURCES AND POWER TECHNOLOGY

Batteries
Biomass and Wastes
Coal
Combustion
Electric Power
Geothermal Energy
Hydroelectric Power
Natural Gas
Nuclear Energy
Ocean Energy Resources
Petroleum
Solar Energy
Steam Generation
Tidal Energy
Turbines
Wind Power

MATHEMATICS AND INFORMATION SCIENCES

Automatic Control
Communications
Computing
Data Processing
Measurements
Navigation and Guidance
Statistics
Units and Standards

MATERIALS AND ENGINEERING SCIENCES

Chemical Engineering
Civil Engineering
Glass and Ceramics
Laser Technology
Mechanical Engineering
Metallurgy
Mining
Microelectronics
Plastics and Fibers
Process Engineering
Structural Engineering
Transportation

MEDICINE, ANATOMY, AND PHYSIOLOGY

Brain and Nervous System
Cancer and Oncology
Cardiovascular System
Chemotherapy
Dermatology
Diagnostics
Digestive System
Endocrine System
Genetic Disorders
Gerontology
Hematology
Immunology
Infectious Diseases
Kidney and Urinary Tract
Mental Illness
Muscular System
Ophthalmology
Otorhinolaryngology/Dental
Parasitology
Pharmacology
Reproductive System
Respiratory System
Rheumatology
Skeletal System

PHYSICS

Atoms and Molecules
Electricity
Electronics
Fluid State
Gravitation
Magnetism
Mechanics
Motion
Optics
Radiation
Solid State
Sound
Subatomic Particles
Surfaces
Theoretical Physics
Waves

PLANT SCIENCES

Agriculture
Algae
Botany
Diseases and Pests
Fruits
Fungi
Growth Modifiers
Nutritional Values
Plant Breeding
Seeds and Germ Plasm
Trees
Yeasts and Molds

SPACE AND PLANETARY SCIENCES

Astrochemistry
Astrodynamics
Astronautics
Astronomy
Astrophysics
Cosmology
Probes and Satellites
Solar Systems

PREFACE

The editors are pleased to introduce *Van Nostrand's Scientific Encyclopedia*, Ninth Edition, thus continuing a proud tradition of excellence that dates back some eight decades to the First Edition, published in 1938. Born before the Atomic Age and updated at intervals ever since, the book now finds itself in the Information Age, and much has changed. Indeed, so much has changed for this edition that, in answer to the first question of what is new, we might just as well ask: What isn't new?

The essence of *VNSE* is enduring, and it remains a fine, concise, comprehensive, and accessible general science work. Its intellectual scope ranges from the introductory to the highly technical in a vast and ever-expanding array of topical coverage in the sciences, engineering, mathematics, medicine, and more. As has long been the case, the editors have designed the book to be approachable by students of many ages. An important feature continued in this work, therefore, is the progressive development of the discussion of each topic, beginning with a simple definition expressed in plain terms, developing into a more detailed treatment, and augmented by often-extensive Additional Reading suggestions.

Contemporary readers can now turn to *VNSE* for information about how their daily lives are increasingly affected by the sophistication of today's science and the complexity of modern technology. They will be reminded that knowledge and discovery exist in a continuum, and that often, but not always, what is new depends entirely on what came before. As our esteemed, late editor of more than 30 years, Douglas M. Considine, was wont to say, "Science is history". With that mantra in mind, and noting that it has already been eight years since the Eighth Edition in 1994, it is time to ask: What's new?

First, the way that the editors wrote, gathered and assembled articles for this book is fundamentally different from the way it was done for previous editions. Douglas Considine could sit down after reading a dozen or more scientific journal articles and, from the print versions of those journals, proceed to type a synopsis of those works at blazing speed on a manual Underwood typewriter. Cutting and pasting were manual jobs that entailed scissors and glue. Articles were stored in a physical filing cabinet. If he needed expert help on a given topic, he wrote to someone, or he called on the telephone. Not so for the new generation of editors, for whom research was and will continue to be largely an electronic event, either via Internet or Email query letter, and only occasionally via phone. The amount of information available to them is so staggering that to call it merely overwhelming is to err on the side of paucity, and therein lies one of their greatest challenges—separating the wheat from the chaff, as it were. Typing is now called *keyboarding*, and it can be done via voice-activated software. Word processing software has rendered cutting and pasting mere perfunctory chores, and whole file cabinets can now fit on a pocket-sized diskette. In that sense, technology is history, too.

Second, we have entered a new Age of Discovery, as witnessed by the scores of new entries on topics that were in their nascent stages in 1994. The Ninth Edition features entirely new or completely rewritten home articles or whole families of articles on the full array of topical coverage, including but by no means limited to: Genetics Engineering, Human Genome Project (The), and Cloning; Bioprocess Engineering (Biotechnology); Space Shuttle, Space Stations, Spacecraft Missions, Satellites (Communications and Navigation), Cosmology, X-ray Astronomy (family of articles), Astrobiology and The Universe; Artificial Intelligence (family of articles); Medicine, Diseases, Vaccines, Vision (family of articles), AIDS, and STDs; Climate, Global Warming, and Acid Rain; Gerontology and Biochemical Theories of Aging; Computer Sciences and The Internet; and Flat Panel Display Technology (family of articles).

Third, the suggested readings at the end of articles now contain both updated print and Internet references. One has only to engage a typical search engine, on any server, on a large topic, say Artificial Intelligence, to realize the value of these sources. Instead of beginning with "hits" that number in the thousands, *VNSE* readers will have the luxury of having much of the culling already done for them, as they will be offered good "first places" to go for more information. These thousands of references throughout the Ninth Edition will, one hopes, provide a bridge to further and deeper knowledge on literally scores of topics.

Fourth, the editors have added detailed Time Lines and Glossaries to some of the large home articles (Bioprocess Engineering, Artificial Intelligence, Vision and the Eye, Optical Fiber Systems, The Internet and many others) to offer "at a glance" information and historical perspective.

Fifth, and this relates again to historical perspective, the editors have added brief biographies of scores of scientists whose work is alluded to in the text of the book. A history of their times is not complete without mention of their works. Science *is* history.

A statistical summary of the Ninth Edition would include (1) more than 8,000 entries; (2) more than 9,000 cross-references for convenient retrieval of information; (3) an alphabetical index with more than 19,500 lines; and (4) 4,378 diagrams, graphs and photographs, and in excess of 550 tables. The interior references in the book, where one article refers to another article that offers augmented or corollary coverage, and the visual aids, as well as the index, have been entirely overhauled; this will result in much greater ease in "navigating" the book.

Last, but certainly not least, the editors, faced with daunting amounts of information in highly specialized fields, have relied increasingly on the generous contributions of industry experts and scholars from all over North America and Europe. Special thanks go to two individuals in particular: to Michael Ladisch of Purdue University, for his home article on Bioprocess Engineering (Biotechnology); and to David Leake of Indiana University, both for his home article on Artificial Intelligence, and for quarterbacking the entire family of twelve AI "sidebar" articles. It has always been in the best tradition of the history of science to share knowledge. It is therefore no mere coincidence that so many contributors are teachers at the university level, for they not only have deep knowledge in their respective fields, but they also can communicate that knowledge effectively. The great improvements to the substance of this book would not have been possible without them, and the editors have preserved the individual styles of the authors in keeping with the tradition of *VNSE* as an eminently personal, and, one hopes, more accessible work of general science.

This "personality" of the book, if you will, was engendered and fostered by my father Douglas M. Considine. To him this book is fondly and respectfully dedicated, whose genius, insatiable curiosity, passion for knowledge, and idiosyncratic style kept the Fifth through Eighth editions not only on the map, but on thousands of shelves. For the Ninth Edition and ensuing editions, the torch has been passed. Thanks, Dad.

GLENN D. CONSIDINE, Editor
PETER H. KULIK, Associate Editor

CONTRIBUTORS

Several hundred scientists, engineers, and educators, located worldwide, made this Ninth Edition of the *Van Nostrand's Scientific Encyclopedia* a reality. Their inputs ranged from detailed information, graphics, and editorial guidance to the creation of comprehensive manuscripts on complex subjects. The editors and staff of this encyclopedia gratefully acknowledge their excellent cooperation and stress that the following abridged list of over 300 individuals and groups could be much longer.

Special appreciation must be extended for the efforts of Jeanne Maree Iacono, who authored and rendered invaluable assistance toward creating brief biographies on scores of scientists.—Ramon A. Mata-Toledo, *James Madison University*, who reviewed the Computer Sciences and authored several articles.—Joseph Castellano, President and CEO of *Stanford Resources*, who prepared numerous entries on Flat Panel Display Technology.—Dr. Thomas J. Harrison, who prepared numerous articles on computers and digital technology.—Dr. Steven N. Shore, who authored and arranged several entries dealing with astronomy and related sciences.—Dr. Ann C. DeBaldo, University of South Florida, who prepared numerous entries in the areas of immunology, oncology, and infectious diseases.—Drs. M. L. and W. L. Dilling, who skillfully summarized the complex world of organic chemistry, its nomenclature and equations.—Richard Q. Hofacker Jr., who authored articles on microelectronics and telephony and who rendered invaluable assistance toward creating comprehensive, yet concise, inputs concerning the broad field of telecommunications.—Peter E. Kraght, who not only authored several articles, but who also prepared the foundation for other descriptions in the spheres of meteorology and climatology.—Elmer Rowley, who made the coverage of mineralogy and crystallography in this encyclopedia truly outstanding.—*VisionRx*, Elmsford, NY., for the numerous entries on Vision and eye related disorders. Without exaggeration, the list of such very special efforts could be extended by several additional paragraphs.

NOTE: In the cases of relatively short articles, the authors' initials may be used instead of their full names. In the following list, an asterisk indicates such authors. For example: *Jeanne Maree Iacono (J. M. I.).

Mark Adams, *Fisher Controls International, Inc., Marshalltown, IA.*

O. J. Adlhart, *Engelhard Corporation, Iselin, NJ.*

H. J. Albert, *Parr Instrument Company, Moline, IL,* Calorimetry. *http://www.parrinst.com/*

P. S. Albright, *Wichita, KS.*

D. Allen, *NCR Corporation, Fort Collins, CO.*

American Forests, *Washington, DC. http://www.americanforests.org/*

American Gas Association (The), *Washington, DC. http://www.aga.org*

Ames Research Center, *National Aeronautics and Space Administration Moffett Field, CA. http://www.arc.nasa.gov/*

R. C. Anderson, *Jet Propulsion Laboratory, Pasedena, CA,* Pathfinder Mission to Mars.

Lorella Angelini, *NASA/Goddard Space Flight Center, Greenbelt, MD,* BeppoSAX (Satellite). *http://www.gsfc.nasa.gov/*

F. Arnold, *Kollmorgen Corporation, Northampton, MA.*

H. R. Arum, *Designatronics, Inc., New Hyde Park, NY.*

P. Auvray, *Levallois-Perret-Cedex, France*

J. Bakos, *J. H. Fletcher & Company, Huntington, WV.*

M. S. Baldwin, *Westinghouse Electric Corporation, East Pittsburgh, PA.*

D. Bane, *Jet Propulsion Laboratory/California Institute of Technology, Pasadena, CA.*

R. Q. Barr, *Climax Molybdenum Company, (A subsidiary of the Phelps Dodge Corporation), Phoenix, AZ. http://www.climaxmolybdenum.com/*

W. T. Barrett, *Foote Mineral Company, Exton, PA.*

E. Bendel, *McDonnell Douglas Corporation, Long Beach, CA.*

R. J. Benke, *Westinghouse Electric Corporation, Pittsburgh, PA. http://www.westinghouse.com/*

W. O. Bennett, *American Time Products, Woodside, NY.*

M. S. Bernath, *Gould, Inc., Andover, MA.*

J. Blackwell, *Department of Macromolecular Science, Case Western Reserve University, Cleveland, OH,* Macromolecular Science.

J. A. Blaeser, *Gould, Inc., Andover, MA.*

Robert E. Bodenheimer, Jr., *Georgia Institute of Technology, Atlanta, GA,* Computer Animation.

BorgWarner Chemicals, *Engineering Staff, Washington, WV.*

G. Bouissières, *University of Paris, Orsay, France.*

R. S. Boulton, *Ministry of Works, Wellington, New Zealand.*

C. O. Bounds, *St. Joe Minerals Corporation, Monaca, PA.*

R. G. Bowen, *Consulting Geologist, Portland, OR.*

Patricia T. Boyd, Ph.D., *U. Maryland Baltimore County,* and *NASA's Goddard Space Flight Center, Greenbelt, MD,* Rossi X-Ray Timing Explorer (RXTE).

J. Boyle, *Giddings & Lewis Electronics Company, Fond Du Lac, WI.*

J. M. Breen, *Adaptive Intelligence Corporation, Milpitas, CA.*

E. H. Bristol, *The Foxboro Company, Foxboro, MA.*

P. M. Brown, *Foote Mineral Company, Exton, PA.*

N. W. Browne, *Davy McKee (Oil & Chemicals) Ltd., London, UK.*

R. Brunner, *Semiconductor Products Sector, Motorola Inc., Phoenix, AZ, Bureau International de l'Heure, Paris, France*

Bruce G. Buchanan, *University of Pittsburgh, Pittsburgh, PA,* Artificial Intelligence Timeline.

B. M. Burns, *Coal Technology Association, Gaithersburg, MA. http://www.coaltechnologies.com/*

L. H. Busker, *Beloit Corporation, Beloit, WI.*

E. R. Caianiello, *Instituto di Fisica Teorica, Università di Napoli, Naples, Italy*

Canadian Association of Petroleum Producers, *Calgary, Alberta. http://www.capp.ca/*

J. Caraceni, *International Fuel Cells, Inc., South Windsor, CT.*

S. C. Carapella, Jr., *ASARCO Inc., Denver, CO. http://www.asarco.com/globe/net.html*

J. J. Carpenter, *American Time Products, Woodside, NY.*

M. S. Carrigy, *Alberta Oil Sands Technology and Research Authority, Edmonton, Alberta.*

R. T. Carson, *Eaton Corporation, Milwaukee, WI.*

Joseph Castellano, *Stanford Resources, Inc., San Jose, CA,* Display Technologies (Other); Electroluminescent Displays; Field Emission Displays; Flat Panel Display Technology; Inorganic Light-Emitting Diode Displays; Liquid Crystal Display Technology; Microdisplays; Organic Light-Emitting Diodes; (OLEDs); Plasma Display Panels; Vacuum Fluorescent Displays.

Centers for Disease Control and Prevention (CDC), *Atlanta, GA. http://www.cdc.gov/health/diseases.htm*

Centre National de la Recherche Scientifique, *Solar Energy Laboratory, Font Romeau, France.*

C. G. Chaggaris, *ORS Automation, Inc., Princeton, NJ. http://www.orsautomation.com/contact.html*

Vinton G. Cherf, *Internet Architecture and Technology, at MCI World-Com.*

R. H. Cherry, *Consultant, Huntington Valley, PA.*

A. Chiavello, *Satellite Communications, Denver, CO.*

W. Chow, *Electric Power Research Institute, Palo Alto, CA. http://www.epri.com/*

Henrik I. Christensen, *Royal Institute of Technology, Stockholm, Sweden,* Artificial Intelligence: Machine Vision.

David D. Clark, *MIT Laboratory for Computer Science, Cambridge, MA.*

D. L. Clark, *Department of Geology and Geophysics, University of Wisconsin, Madison, WI.*

J. Cobb, *Cognex Corporation, Needham, MA. http://www.cognex.com/*

R. L. Colona, *General Scanning Inc., Watertown, MA.*

R. K. Conolly, *American Petroleum Institute, Washington, DC. http://api-ec.api.org/frontpage.cfm*

P. J. Constantino, *Jervis B. Webb Company, Farmington Hills, MI. http://www.jervisbwebb.com/jbw/jerviswebbhomepage_def.htm*

Jimmy G. Converse, *Sterling Chemicals Inc., Texas City, TX.*

C. S. Cook, *University of Texas, El Paso, TX.*

P. H. Cook, *The Dow Chemical Company, Freeport, TX.*

T. E. Cook, *The Procter & Gamble Company, Cincinnati, OH. http://www.pg.com/main.jhtml*

A. B. Coon, *University of Illinois, Urbana, IL.*

G. R. Cooper, *School of Electrical Engineering, Purdue University, West Lafayette, IN.*

D. A. Corrigan, *Handy & Harman, Fairfield, CT.*

A. T. Coscia, *American Cyanamid Company, Stamford, CT.*

David L. Crawford, Ph.D., *International Dark-Sky Association, (Emeritus Astronomer at National Optical Astronomy Observatories/Kitt Peak National Observatory), Tuscon, AZ,* Light Pollution. *http://www.darksky.org/ida/index.html*

J. H. Cronin, *Westinghouse Electric Corporation, East Pittsburgh, PA.*

A. B. Crossman, *Brown & Root, Inc., Houston, TX.*

W. J. Culhane, *Mead Corporation, Chillicothe, OH. http://www.meadwestvaco.com/*

V. Cullen, *Woods Hole Oceanographic Institution, Woods Hole, MA. http://www.whoi.edu/*

Robert A. Daene, *Beloit Corporation, Beloit, WI.*

R. M. Dahlgren, *The Procter & Gamble Company, Cincinnati, OH,*

E. E. David, Jr., *Exxon Research and Engineering Company, Annandale, NJ.*

R. Davis, *NCR Corporation, Fort Collins, CO.*

R. Dean, *GA Technologies, Inc. San Diego, CA.*

Ann. C. DeBaldo, Ph.D., *College of Public Health, University of South Florida, Tampa, FL.*

D. F. DeCraene, *Chemetals Corporation, Baltimore, MD.*

W. E. Degenhard, *Carl Zeiss, Inc., New York, NY. http://www.zeiss.de/us/micro/home.nsf*

Ramon López de Mántaras, *Artificial Intelligence Research Institute, Spanish Council for Scientific Research,* Artificial Intelligence: Fuzzy Reasoning.

W. F. Dennen, *University of Kentucky, Lexington, KY.*

S. E. Desai, *Davy McKee Iron & Steel, Stockton-on-Tees, UK.*

Marie desJardins, *Department of Computer Science and Electrical Engineering Department, University of Maryland, Baltimore, MD,* Artificial Intelligence: Machine Learning.

D. L. Dexter, *University of Rochester, Rochester, NY.*

B. Dickie, *Ministry of Mines and Minerals, Edmonton, Alberta.*

J. Dietl, *Wacker Chemie, GMBH, Munich, Germany.*

E. D. Dietz, *Consultant, Toledo, OH.*

W. Dietz, *Wacker Chemie, GMBH, Munich, Germany.*

M. L., and W. L. Dilling, *The Dow Chemical Company Midland, MI.*

Z. C. Dobrowolski, *Kinney Vacuum Company, Cannon, MA.*

V. J. Dobson, *Dynapath System Inc., Detroit, MI.*

F. Dostal, *American Time Products, Woodside, NY.*

Jim Douglas, *Dammeron Valley, UT,* Electrical Ground Fault Circuit Interrupters (GFCI); Electrical Power Quality.

R. G. Douglas, *University of New York, Stony Brook, NY.*

E. A. Draeger, *McNally Pittsburg Mfg. Corp., Pittsburg, KA.*

H. Dressler, *Koppers Company, Inc., Monroeville, PA. http://www.koppers.com/about.htm*

R. M. Durham, *Infrared Industries, Inc., Santa Barbara, CA.*

C. J. Easton, *Sensotec, Inc., Columbus, OH. http://www.sensotec.com/index.html*

Jan-Olof Eklundh, *Royal Institute of Technology, Stockholm, Sweden,* Artificial Intelligence; Machine Vision

R. A. Elliott, *Qualiplus USA, Inc., Stamford, CT.*

Eurotunnel Exhibition Centre, *Victoria Plaza, 111 Buckingham Palace Road, London SW1W OST, UK.*

Eurotunnel Information Centre, *St. Martin's Plain, Cheriton High Street, Folkstone, Kent CT19 4QD, UK. http://ww1.eurotunnel.com/rcs/etun/pb_english/en_wp_corp/index.jsp*

B. Evans, *Rare-earth Information Center, Institute for Physical Research and Technology. Iowa State University, Ames, IA. http://www.external.ameslab.gov/RIC/index.html*

J. J. Faran, Jr., (retired), *Lincoln, MA.*

H. Fenninger, *Wacher Chemie, GMBH Munich, Germany.*

J. File, *Plasma Physics Laboratory, Princeton University, Princeton, NJ.*

T. Flack, *Westinghouse Electric Corporation, Madison Heights, MI.*

R. Fletcher, *J. H Fletcher & Company. Huntington, WV.*

P. A. Flinn, *GMF Robotics Corporation, Troy, MI.*

Kevin Flurkey, Ph.D., *The Jackson Laboratory, Bar Harbor, ME,* Geriatrics; and Gerontology. *http://www.jax.org*

Susan Eileen Fox, *Macalester College, St. Paul, MN,* Artificial Intelligence: Case-Based Reasoning.

J. A. Garman, *Great Lakes Chemical Corporation, West Lafayette, IN.*

Gas Research Institute, *DesPlaines, IL. http://www.gri.org/*

R. E. Gebelein, *Moore Products Company, Spring House, PA.*

F. B. Gerhard, Jr., *GTE Laboratories Incorporated, Waltham, MA.*

H. P. Gerrish, *National Hurricane Center, Coral Gables, FL. http://www.nhc.noaa.gov/*

I. Gilmour, *Polaroid Corporation, Cambridge, MA.*

K. F. Glasser, *Consolidated Edison Company of New York, Inc., New York, NY.*

Goddard Institute for Space Studies, *Columbia University, New York, NY. http://www.giss.nasa.gov/*

J. Golden, *National Oceanic and Atmospheric Administration, Boulder, CO. http://www.noaa.gov/*

D. T. Goldman, *National Bureau of Standards, Washington, DC.*

Avelino, J. Gonzalez, *University of Central Florida, Department of Electrical Engineering and Computer Science, Orlando, FL,* Artificial Intelligence: Expert Systems.

D. L. Gregory, *Boeing Aerospace Company, Seattle, WA. http://www.boeing.com/flash.html*

E. A. Groh, *Geologist, Portland, OR.*

L. Groszek, *Technical Center, Ford Motor Company Dearborn, MI.*

K. A. Gschneidner, Jr., *Rare-earth Information Center, Institute for Physical Research and Technology. Iowa State University, Ames, IA. http://www.external.ameslab.gov/RIC/index.html*

G. A. Hall, Jr., *Westinghouse Electric Corporation, Pittsburgh, PA.*

R. C. Hamilton, (retired), *Cornell University, Ithaca, NY.*

William Hankley, *Department of Computing and Information Science, Kansas State University, Manhattan, KS,* Software Engineering.

P. S. Hansen, *The Foxboro Company, Invensys Process Systems, Foxboro, MA. http://www.foxboro.com/*

P. S. Hansen, *Iowa State University, Ames, Iowa*

A. O. Hanson, *University of Illinois, Urbana, IL.*

P. W. Harland, *Ametek, Inc., Paoli, PA. http://www.ametek.com/*

*Thomas J. Harrison, (T.J.H) (retired), *IBM Corporation, Boca Raton, FL.*

W. Havemann, *Carl Zeiss, Inc., New York, NY. http://www.zeiss.de/us/micro/home.nsf*

B. W. Heinemeyer, *The Dow Chemical Company, Freeport, TX.*

E. W. Hewson, *Oregon State University, Corvallis, OR.*

S. P. Higgins, Jr., *Honeywell, Inc., Phoenix, AZ. http://www.honeywell.com/*

D. Hines, *New Mexico Institute of Mining and Technology, Socorro, NM. http://www.nmt.edu/*

Geoffrey Hinton, *Department of Computer Science, University of Toronto, Toronto, Canada,* Artificial Intelligence: Neural Networks.

S. E. Hluchan, *Pfizer, Inc., Wallingford, CT. . http://www.pfizer.com/main.html*

D. R. Hodge, *Alexandria, VA.*

Jessica K. Hodgins, *Georgia Institute of Technology, Atlanta, GA,* Computer Animation.

D. M. Hoelzl, *GTE Laboratories, Incorporated, Waltham, MA.*

Richard Q. Hofacker, Jr., (retired), *Bell Laboratories, Short Hills, NJ,* Satellites (Communications and Navigation); Telephony (Telecommunications).

K. Honchell, *Cincinnati Milacron, Lebanon, OH.*

J. C. Hoogendorn, *South African Coal, Oil and Gas Corp., Ltd., Sasolburg, Republic of South Africa.*

L. Hoover, *American Geological Institute (AGI), Washington, DC.* *http://www.agiweb.org/*

H. S. Hopkins, (retired), *Olin Corporation, Norwalk, CT.* *http://www.olin.com/*

David W. Howard, *Brookfield Engineering Laboratories, Inc. Stoughton, MA.*

Samuel C. Hsieh, *Department of Computer Science, Ball State University, Muncie, IN,* Programming Language.

Patrick Hughes, *Earth Observatory, NASA, Washington, DC.* *http://earthobservatory.nasa.gov/*

G. C. Humphreys, *Davy McKee (Oil & Chemicals) Ltd., London, UK.*

T. N. Hurst, *Hewlett-Packard Company, Boise, ID.*

***Jeanne Maree Iacono, (J. M. I.),** *Dammeron Valley, UT,* Brief biographies of scores of scientists.

R. P. Iacono, M.D., F.A.C.S., *Redlands, CA,* Parkinson's Disease. *http://Pallidotomy.com/index.html*

J. Ingle, *Caterpillar, Inc., Peoria, IL. http://www.caterpillar.com/*

Institute of Gas Technology, *Chicago, IL.*

Jyrki Jaakkola, *Valmet Corporation, Charlotte, NC.*

R. B. Jacques, *Black Mesa Pipeline, Inc., Flagstaff, AZ.*

Fred Jansen, *Space Science Department, ESA Directorate of Scientific Programmes, ESTEC, Noordwijk, The Netherlands,* XMM-Newton.

A. Jayaraman, *AT&T Bell Laboratories, Murray Hill, NJ.*

W. D. Jensen, *GTE Laboratories Incorporated, Waltham, MA.*

Robert E. Kahn, *Corporation for National Research Initiatives, Reston VA.*

D. Kaiser, *Parker Hannifin Corporation, Richmond, CA.* *http://www.parker.com/*

G. J. Kaminski, *The Procter & Gamble Company, Cincinnati, OH.* *http://www.pg.com/main.jhtml*

J. N. Karlberg, *The Procter & Gamble Company, Cincinnati, OH.*

Sir Maurice Kendall, *International Statistical Institute, London, UK.*

E. W. Kent, *National Bureau of Standards, Washington, DC.*

R. W. Keyes, *IBM Corporation, Yorktown Heights, NY.*

K. E. Kimball, *Siemans Capital Corporation, Iselin, NJ.*

J. P. King, *The Foxboro Company, Rahway, NJ.*

Leonard Kleinrock, *Professor of Computer Science, University of California, Los Angeles, CA,* Internet (The History). *http://www.lk.cs.ucla.edu/*

D. M. Koffman, *GTE Laboratories Incorporated, Waltham, MA.*

Michael Kohlhase, *Department of Computer Science, Carnegie Mellon University, Pittsburgh, PA,* Artificial Intelligence: Automated Reasoning. *http://www-2.cs.cmu.edu/~kohlhase/*

Peter E. Kraght, (retired), *Consulting Meteorologist, Mabank, TX.*

P. A. Kraska, *Pattern Processing Technologies, Inc., Minneapolis, MN.*

T. W. Krauss, *Intec Controls Corporation, Foxboro, MA.*

G. Kuebler, *GLI International, Inc., (formerly Great Lakes Instruments), Milwaukee, WI. http://www.gliint.com/*

I. A. Kunasz, *Foote Mineral Company, Exton, PA.*

W. Kupper, *Mettler Instrument Corporation, Hightstown, NJ.*

Michael R. Ladisch, *Director, Laboratory of Renewable Resources Engineering http://fairway.ecn.purdue.edu/IIES/LORRE/index and Department of Agricultural and Biological Engineering; http://abe.www.ecn.purdue.edu/ABE/Fac_Staff/ladisch Purdue University, West Lafayette, IN,* Bioprocess Engineering (Biotechnology).

A. H. Lalas, *Chrysler Corporation, Detroit, MI. http://www.chrysler.com/*

G. G. Lauer, (retired), *Koppers Company, Inc., Monroeville, PA.*

R. F. Lawrence, (retired), *Westinghouse Electric Corporation, East Pittsburgh, PA.*

W. W. Lawrence, Jr., *Ethyl Corporation, Baton Rouge, LA.*

David B. Leake, *Computer Science Department, Indiana University, Bloomington, IN,* Artificial Intelligence. *http://www.cs.indiana.edu/~leake/*

C. Lebarbier, *Électricité de France, Paris, France.*

J. M. Lee, *The M. W. Kellogg Company, Houston, TX.*

Barry M. Leiner, *Research Institute for Advanced Computer Science, Moffett Field, CA.*

L. Libby, *Simmons Refining Company, Chicago, IL.*

B. Lindal, *Virkir Consulting Group Ltd., Reykjavik, Iceland.*

N. C. Liston, *U. S. Department of Army Cold Regions Research and Engineering Laboratory, Hanover, NH.*

Jamie Love, *Science Explained,* Cloning (Mammals); and Cloning (The Story of Dolly the Sheep). *www.synapses.co.uk/science/index.html.*

S. Lovejoy, *McGill University, Montreal, Quebec.*

B. A. Loyer, *Motorola, Inc., Phoenix, AZ.*

Lucent Technologies, *Optical Fiber Solutions, Norcross, GA.*

David C. Lynch, *CyberCash Inc., New York, NY.*

Steven L. Lytinen, *School of Computer Science, Telecommunications, and Information Systems, DePaul University, Chicago, IL,* Artificial Intelligence: Natural Language Processing.

John B. Macauley, Ph.D., *The Jackson Laboratory, Bar Harbor, ME,* Biochemical Theories of Aging. *http://www.jax.org*

Ralph E. Mackiewicz, *Sisco, Inc., Sterling Heights, MI,* Manufacturing Message Specification (MMS). *http://www.sisconet.com/*

E. C. Magison, *Consulting Engineer, Ambler, PA.*

C. L. Mamzic, *Siemens Energy & Automation Inc., (formally Moore Products Company, Spring House, PA. http://www.mooreproducts.com/*

John Marafino, *Department of Mathematics, James Madison University, Harrisonburg, VA,* Calculators.

J. R. Masson, *Davy McKee (Oil and Chemicals) Ltd., London, UK.*

Ramon A. Mata-Toledo, *James Madison University, Harrisonburg, VA,* Algorithm; Database Management Systems; Translators.

H. L. Mayer, *Hydro-Quebec, Montreal, Quebec.*

J. Mazurkiewicz, *Pacific Scientific, Rockford, IL.*

W. R. McCown, *Westinghouse Electric Corporation, Pittsburgh, PA.*

W. F. McIlhenny, *The Dow Chemical Company, Midland, MI.*

Lisa Meeden, *Associate Professor and Director, Computer Science Program, Swarthmore College, Swarthmore, PA,* Artificial Intelligence: Robotics.

R. W. Miller, *Consultant, Foxboro, MA.*

E. D. Mohr, *Unimation (Westinghouse Electric Corporation), Danbury, CT.*

S. M. Moore, *Lawrence Berkeley Laboratory, Berkeley, CA.* *http://www.lbl.gov/*

J. A. Morgan, *North American Electric Reliability Council, Princeton, NJ.*

Kevin Mulrooney, *Newark, DE. Index*

T. Murphy, *IBM Corporation, Yorktown Heights, NY.*

J. Nagy, *Beckman Industrial Corporation, Cedar Grove, NJ.*

NASA Astrobiology Institute (NAI), *Washington DC.* *http://nai.arc.nasa.gov/*

NASA's Jet Propulsion Laboratory/California Institute of Technology, *Pasadena, CA.*

National Indoor Environmental Institute, *Plymouth Meeting, PA.*

National Institutes of Health (NIH), *Bethesda, MD. http://www.nih.gov/*

M. M. Nelson, *Honeywell Inc., Billerica, MA.*

L. R. Newitt, *Geological Survey of Canada, Ottawa, Ontario.*

E. R. Niblett, *Geological Survey of Canada, Ottawa, Ontario.*

S. Nojiima, *Japan Gasoline Company, Ltd., Tokyo, Japan.*

Northeastern Forest Experiment Station, *U.S. Department of Agriculture, Darby, PA.*

Oak Ridge National Laboratory, *Oak Ridge, TN. http://www.ornl.gov/ornlhome/index.htm*

James F. O'Brien, *Georgia Institute of Technology, Atlanta, GA.* Computer Animation.

H. Oeda, *Ojinomoto Co., Inc., Kawaski, Japan.*

R. L. Osborne, *Honeywell Inc., Billerica, MA.*

R. H. Osman, *Robicon Corporation, (A Subsidiary of High Voltage Engineering Corporation), New Kensington, PA. http://www.robicon.com/*

V. C. Oxley, *GTE Laboratories Incorporated, Waltham, MA.*

S. T. Oyama, *Lawrence Berkeley Laboratory, Berkeley, CA.* *http://www.lbl.gov/*

Pacific Gas and Electric Company, *(a subsidiary of PG&E Corporation), San Francisco, CA. http://www.pge.com/*

Panel on Mathematical Sciences, *Commission on Physical Sciences, Mathematics, and Resources, National Research Council, Washington, DC.*

J. M. Pasachoff, *Hopkins Observatory, Williams College, Williamstown, MA.*

R. Peacock, *LTV Steel Company, Inc. Independence, OH,* Radiation Thermometry.

P. Pesch, *Astronomy Department, Case Western Reserve University Cleveland, OH.*

L. V. Pfaender, *Owens-Illinois, Toledo, OH.*

Sir David Phillips, *University of Oxford, Oxford, UK.*

A. K. Pierce, *Kitt Peak National Observatory (a division of the National Optical Astronomy Observatories which is operated by the Association of Universities for Research in Astronomy (AURA), Inc. under cooperative agreement with the National Science Foundation, Tucson, AZ. http://www.noao.edu/kpno/*

D. Postma, *General Motors Corporation, Detroit, MI.*

D. B. Priddy, *The Dow Chemical Company, Midland, MI.*

J. H. Purnell, *Department of Chemistry, University of Swansea, Swansea, UK.*

N. Razo, *National Center for Atmospheric Research, Boulder, CO. http://www.ncar.ucar.edu/ncar/*

R. D. Reincke, *Caterpillar Inc., Peoria, IL. http://www.caterpillar.com/*

R. G. Reip, *Consulting Engineer, Sawyer, MI.*

R. P. Rich, *Eastman Chemical Company, Kingsport, TN. http://www.eastman.com/Markets/Textiles/Textiles_intro.asp*

E. H. Richardson, *Herzberg Institute of Astrophysics Dominion Astrophysical Observatory, Victoria, British Columbia, Canada. http://www.hia.nrc.ca/facilities/dao/*

J. A. Riddick, *Baton Rouge, LA.*

J. C. Riley, *Metrologist, Portland, OR.*

G. G. Robert, *University of Oxford, Oxford, UK.*

Lawrence G. Roberts, *Caspian Networks, San Jose, CA.*

T. H. Rogers, (retired), *Elastomers Consultant, Clearwater, FL.*

G. R. Romovacek, *Koppers Company, Inc.,Monroeville, PA.*

B. A. Ross, *General Motors Corporation, Indianapolis, IN.*

D. M. Ross, *Propellants Consultant, Lancaster, CA.*

Elmer B. Rowley, (retired), *Union College, Schenectady, NY.*

P. F. H. Rudolph, *Lurgi Mineralotechnik, GMBH, Frankfurt (Main), West Germany.*

L. Russell, *MTS Systems Corporation, Eden Prairie, MN. http://www.mts.com/*

S. J. Sansonetti, *Consultant, Reynolds Metals Company (ALCOA), Richmond. VA. http://www.alcoa.com/*

R. P. Santandrea, *Los Alamos National Laboratory, Los Alamos, NM. http://www.lanl.gov/worldview/*

E. J. Sare, *PPG Industries Inc., Barberton, OH.*

W. L. W. Sargent, *Royal Greenwich Laboratory, Sussex, UK.*

Jonathan Schaeffer, Ph.D., *Department of Computer Science, University of Alberta, Edmonton, Alberta Canada,* Artificial Intelligence: Game Playing Systems.

D. Schertzer, *Météorologie Nationale, Paris, France.*

W. R. Schiller, *Wacher Chemie, GMBH, Munich, Germany.*

M. Schussler, *Fansteel, North Chicago, IL.*

M. Sekino, *Toyobo Co., Ltd., Iwakuni, Yamaguch-Pref., Japan.*

W. G. Shequen, (retired), *Bausch & Lomb, Sunland, CA. http://www.bausch.com/*

*Steven N. Shore, (S.N.S), *University of Indiana South Bend, South Bend, IN.*

E. C. Shuman, *Consulting Engineering, State College, PA.*

Siemans Aktiengesselschaft Engineering Staff, *Erlangen, Germany.*

L. E. Simmons, *Simmons Refining Company, Chicago, IL.*

D. C. Sleeman, *Davy McKee (Oil & Chemicals) Ltd., London, UK.*

L. F. Small, *Oregon State University, Corvallis, OR.*

James S. Sochacki, *James Madison University, Harrisonburg, VA,* Differential Equations.

G. A. Somorjai, *Lawrence Berkeley Laboratory, Berkeley, CA. http://www.lbl.gov/*

E. Sperry, *Beckman Industrial Corporation, Cedar Grove, NJ.*

M. A. Stadtherr, *Department of Chemical Engineering, University of Illinois, Urbana, IL.*

S. Stamas, *Exxon Corporation, New York, NY. http://www.exxon.com/index_flash.html*

C. B. Steffenson, *Department of Astronomy, Case Western Reserve University, Cleveland, OH.*

J. Stevenson, *West Instruments, East Greenwich, RI.*

S. Stoddard, *Wough Controls Corp., Chatsworth, CA.*

T. S. Storer, *Hewlett-Packard Company, Palo Alto, CA. http://www.hp.com/*

E. Sulzer, *Siemens Energy & Automation, Inc., Peabody, MA.*

J. C. Summers, *Automotive Catalyst Company, Tulsa, OK.*

H. F. Szepan, (retired), *Ingersoll-Rand, Nashua, NH. http://www.ingersoll-rand.com/*

D. G. Terry, (retired), *Ingersoll-Rand, Nashua, NH.*

Tokyo Electric Power Company, *Tokyo, Japan.*

Wesley F. Tree, *The College of Wooster, Wooster, OH.*

W. A. Troeger, *Weston (Sangamo-Weston, Inc.), Newark, NJ.*

Joachim Truemper, Ph.D., *Professor, Max Planck Institute (MPE), Germany,* ROSAT (Roentgen Satellite).

Karen Tucker, *Chandra X-ray Observatory Center, Harvard-Smithsonian Center for Astrophysics, Cambridge, MA,* X-Ray Astronomy. *http://cfa-www.harvard.edu/*

Wallace Tucker, *Chandra X-ray Observatory Center, Harvard-Smithsonian Center for Astrophysics, Cambridge, MA,* X-Ray Astronomy. *http://cfa-www.harvard.edu/*

S. Turner, *National Bureau of Standards, Gaithersburg, MD.*

L. F. Urry, *Eveready Battery Company, Ltd., Westlake, OH. http://www.eveready.com/*

U. S. Department of Energy, *Office of Health and Environmental Research, Oak Ridge, TN,* Human Genome Project (The) "To Know Ourselves."

G. V. Van denBerg, *Shell Internationale Petroleum Mastschappij B. V., The Hague, Netherlands.*

O. Vandermarcq, *Ambassade de France aux Etats-Unis Services de la Mission Scientifique, Houston, TX.*

E. Van Haaften, *American Time Products, Woodside, NY.*

J. A. Vegeasis, *Shell Development Company, Houston, TX.*

M. G. Venegas, *The Procter & Gamble Company, Cincinnati, OH.*

*R. C. Vickery, (R.C.V), *Blanton/Dade City, FL.*

Video Logic Corporation, *Sunnyvale, CA.*

Ray Villard, *Space Telescope Science Institute, Baltimore, MD,* Hubble Space Telescope. *http://www.stsci.edu/resources/*

VisionRx, Inc., *Elmsford, NY. http://visionrx.com/*

G. T. Volpe, *University of Bridgeport, Bridgeport, CT.*

Kyle Wagner, Ph.D., NIH Fellow, *University Maryland at Baltimore, Baltimore, MD and University of Maryland Institute for Advanced Computer Studies, College Park, MD,* Artificial Intelligence: Artificial Life.

J. Walker, *Ontario Hydro, Toronto, Ontario.*

K. A. Walsh, *Brush Wellman Inc., Elmore, OH.*

J. D. Warnock, *Siemens Energy & Automation Inc., (formally Moore Products Company), Spring House, PA. http://www.mooreproducts.com/*

Martin C. Weisskopf, *Marshall Space Flight Center, Huntsville, AL,* Chandra X-Ray Observatory. *http://www.msfc.nasa.gov/*

J. Wells, *Edison International, parent company of (Southern California Edison Company), Rosemead, CA. http://www.edisonx.com/*

J. Y. Welsh, *Chemetals Corporation, Baltimore, MD.*

L. Werth, *Pattern Processing Technologies, Inc., Minneapolis, MN.*

J. R. Whiteway, *Ontario Hydro, Toronto, Ontario.*

Darrell Whitley, *Department of Computer Science, Colorado State University, Fort Collins, CO,* Artificial Intelligence: Genetic Algorithms and Evolutionary Computing.

R. M. Whittier, *Endevco Corporation, San Juan Capistrano, CA.*

P. R. Wiederhold, *General Eastern Instruments Corporation, Watertown, MA.*

R. N. Wilkinson, *The Procter & Gamble Company, Cincinnati, OH.*

E. Williams, *Cobalt Information Centre, London, UK.*

R. L. Wilson, *Honeywell, Inc., Fort Washington, PA.*

A. T. Winfree, *Professor Ecology and Evolutionary Biology, University of Arizona, Tucson, AZ,* Biological Rhythms; Circadian Clock; Fibrillation; and Jet-Lag.

Stephen Wolff, *Cisco Systems, Inc., San Jose, CA.*

A. S. Wood, *Jet Propulsion Laboratory/California Institute of Technology, Pasadena, CA.*

G. R. Woodcock, *Boeing Aerospace Company, Seattle, WA. http://www.boeing.com/flash.html*

Edward L. (Ned) Wright, *Professor of Physics and Astronomy, UCLA, Westwood, CA. http://www.astro.ucla.edu/%7Ewright/intro.html* Cosmology.

Mike Wright, *Marshall Space Flight Center, Huntsville, AL,* Rocketry.

G. Yazbak, *MetriCor, Inc.,Monument Beach, MA.*

C. K. Zimmerman, *E. I. DuPont de Nemours & Company, Inc., Wilmington, DE. http://www.dupont.com/*

VAN NOSTRAND'S

SCIENTIFIC
ENCYCLOPEDIA

Ninth Edition

VOLUME 2

L

LABARIA (*Reptilia, Sauria*). A poisonous South American snake belonging to the pit vipers. It ranges from eastern Brazil north into the Guianas. Related to the jararaca.

LABRADOR CURRENT. An ocean current that flows southward from Baffin Bay, through the Davis Strait, thence southeastward past Labrador and Newfoundland. East of the Grand Banks, the Labrador current meets the Gulf Stream, and the two flow east separated by the cold wall.

LABYRINTH FISHES (*Osteichthyes*). Of the suborder *Anabantoidea*, and family *Anabantidae*, this is a group of tropical freshwater fishes, quite small in size, and usually found in Africa and southeast Asia. The pelvic fins bear a long slender filament, apparently sensory. They are named because of their highly specialized breathing apparatus. In using this labyrinth-type anatomical mechanism, the fish draws in a bubble of air from which it extracts the oxygen, and when the fish next surfaces, it expels old air out of the gill covers. This is an accessory apparatus that enables these fishes to tolerate water deficient in oxygen. One member of this family is the walking fish (sometimes called climbing perch) (*Anabas testudineus*) which attains a length of about 10 inches (25 centimeters). It is found in Malaya, the Philippines, and India. Its name stems from the fact that it apparently can walk for considerable distances on land, possibly seeking another body of water. In walking, the fish uses its gill plates, which fortunately are equipped with spiny edges, serving as "feet." The fish can cover about 10 feet (3 meters) in a minute by this method of locomotion.

Another labyrinth fish is the *Betta splendens* (Siamese fighting fish), found in Thailand. The males of this species are renowned for their ability to fight other males of their species. They have a rather dull color and achieve a length of about 2 inches (5 centimeters). The so-called "gourami" (*Osphronemus goramy*) is the largest of the labyrinths, attaining a length

of 2 feet (0.6 meter). The fish is used as food in the Orient. Because of its unusual habits, the kissing gourami (*Helostoma temmincki*) is popular among tropical-fish hobbyists. There are several varieties of gourami. Favorites among fanciers are the genus *Colisia*, paradise fishes (genus *Macropodus*), and the croaking gourami (*Trichopsis vittatus*), the males of which make an odd croaking noise, particularly at night, when they come to surface for air. See Fig. 1.

LACCOLITH. An intrusive type of igneous rock. Studies of the forms of igneous rock masses have shown that the openings followed by volcanic lavas in their upward journey toward the surface are of two dominant types, *tubular* and *tabular*. Tubular openings are approximately circular in outline. These openings may vary from a few centimeters to many meters. Dikes are thin, tabular, parallel-walled masses of igneous rocks. Essentially, they have a vertical position and are formed by the injection of lava into fissures and joints in rocks. Sills are another form, but they are flat and essentially horizontal. Laccoliths, although somewhat similar to sills, differ from them in that the overlying beds are arched. The horizontal area occupied by a laccolith usually is smaller than that occupied by a sill. It is believed that the lava forming a laccolith was too viscous to flow far between the beds of the rock and thus pushed them up to form a dome. A laccolith generally is fed through a conduit. Laccoliths may merge into sills, and they are also commonly associated with dikes. Several laccoliths may occur in the same area. Laccoliths form numerous buttes and mountains in the western United States. See Figs. 1 and 2.

The formal definition (American Geological Institute) is "A concordant igneous intrusion with a known or assumed flat floor and a postulated dike-like feeder somewhere beneath its thickest point. It is generally lens-like

Fig. 1. Species of labyrinth fishes.

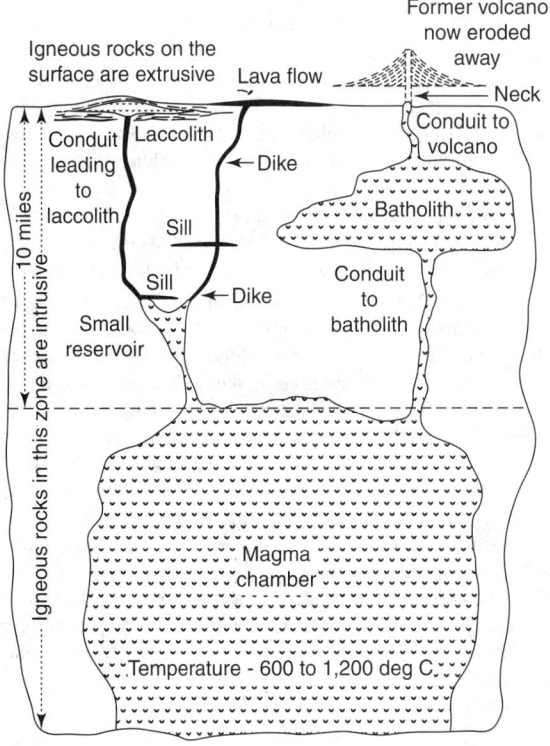

Fig. 1. An idealized magma, showing the various forms of igneous rocks, such as dikes, sills, batholiths, and laccoliths.

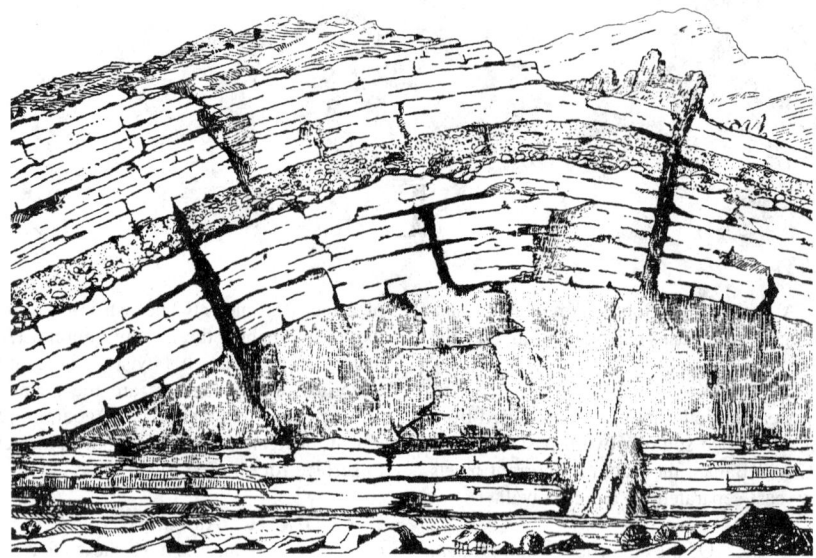

Fig. 2.   An idealized laccolith, with associated dikes and sills.

in form and roughly circular in plan, less than 5 miles in diameter and from a few meters to several hundred meters in thickness."

LACHRYMATOR.   A chemical substance that causes tears to form in the eyes. See also **Chlorinated Organics**.

LACQUER.   See Paints and Coatings; and Resins (Natural).

LACRIMAL SYSTEM.   Orbital structures of the eye responsible for tear production and drainage. Tears are produced in the lacrimal gland above the outer corner of the eye. They flow across the eye surface and drain into the upper and lower puncta, which are openings at inner eyelid margins. They then go through the upper and lower canaliculi to the common canaliculus, into the tear sac. From there, tears flow through the nasolacrimal duct, down into the nose.

Vision Rx, Inc., Elmsford, NY.

LACTIC ACID.   Alpha-hydroxy-propionic acid, $H-C_3H_5O_3$, formula weight 90.05, colorless liquid, mp $18\,^\circ C$, bp $122\,^\circ C$, sp gr 1.248, miscible with water, alcohol, or ether in all proportions. The substance exists in two forms: (1) *dextro* lactic acid, which rotates the plane of polarized light to the right; and (2) *levo* lactic acid which rotates the plane of polarized light to the left. A mixture of these two forms is ordinary lactic acid, which does not rotate the plane of polarized light. Ordinary lactic acid is termed dextrolevo lactic acid. Lactic acid is a product of corn refining.

Lactic acid was one of the first biological substances to be investigated from the standpoint of the existence of the two optically active forms.

Lactic Acidosis. Lactic acid is the cause of one of many possible disorders in human acid-base metabolism. Lactic acidosis represents an accumulation of lactic acid in the blood and tissues. This condition gradually depletes the natural buffers in the body and there is a consequent lowering of pH. As described in the entry on **Glycolysis**, lactic acid is the end product of that process. Lactic acid blood levels are determined by at least four factors. The rate of generation of lactic acid; the rate of transport from tissues to plasma and from plasma to the liver (point of utilization of lactic acid); the rate of utilization; and excretion of lactic acid by the kidneys. Normally, all of these functions are maintained in balance to give a normal blood lactate concentration of about 1 mEq/l.

On the generation side, three factors are involved. (1) The availability of oxygen is a major controlling determinant of lactic acid generation because, as adenosine triphosphate (ATP) generation from oxidative phosphorylation diminishes, the cells naturally respond with a greater rate of glycolysis. This increases tissue lactate levels and ultimately lactate blood levels. See also **Phosphorylation (Oxidative)**. (2) If, as may be caused by various factors, there is an increase in pH, the activity of phosphofructokinase will increase (this is the rate-limiting enzyme of glycolysis). With increases in pH, the enzyme is more active and more lactate is formed. (3) Factors that

affect the biological oxidation-reduction potentials also influence the rate at which glucose is metabolized to lactate.

Fundamental predisposing conditions causing an increased generation of lactate include: decreased tissue perfusion associated with shock, which may occur in cardiac arrest; increased skeletal muscle activity (the rate of glycolysis increases with exercise; this also may be associated with convulsive states that may follow severe exercise — brought about by increased blood lactate concentrations); large tumors, since tumors (leukemias, lymphomas, etc.) may have an increased rate of glycolysis even in the presence of a sufficient supply of oxygen; and both cyanide and carbon monoxide poisoning, which can increase lactate levels because of insufficient oxygen supply.

On the utilization side, there are a number of influencing factors. The liver is the principal lactic acid utilization center. In liver failure, a surplus of lactate builds up, a condition which may be associated with reduced hepatic perfusion, hepatocyte failure, and hepatocytes replaced by tumor. Blood lactate concentrations are elevated in persons with diabetic ketoacidosis. The observed elevation of lactic acid levels in cases of alcoholism is not fully understood: the condition may increase generation or by decreasing utilization. The latter effect is now favored by many authorities, the theory being that ethanol completes for electrons in the liver, thus decreasing utilization of lactic acid in that organ.

Additional Reading

Bozoglu, T., Faruk, and B. Ray (Editors): "Lactic Acid Bacteria: Current Advances in Genetics, Metabolism, and Application of Lactic Acid Bacteria," Springer-Verlag New York, Inc., New York, NY, 1996.
Wood, B.: "The Lactic Acid Bacteria in Health and Disease," Aspen Publishers, Inc., Gaithersburg, MD, 1999.
Wood, B. and W. Holzapfel: "The Genera of Lactic Acid Bacteria," Blackie Academic & Professional, UK, 1999.

LACTOSE.   See Carbohydrates.

LACTIC ACIDOSIS.   See Lactic Acid.

LACTULOSE.   See Sweeteners.

LADY BEETLE *(Insecta, Coleoptera).*   Small oval beetles, strongly convex and with relatively small legs. The common name and also the name lady bug apply chiefly to the more common red species, marked with black and white, but the family *Coccinellidae* to which they belong contains many others.

Lady beetles are found on plants and trees and deposit their eggs on the underside of leaves. They are practically round in shape and quite small, approximately $\frac{1}{8}$ inch (3 millimeters) across. The color varies with the species. They are harmless and often carried around by children as "pets."

The worm-like maggots eat plant lice. Aphids are a favorite food and thus the presence of lady beetles helps with gardening and growing flowers. See also **Beneficial Insects**.

LAG (Angle of). When two related quantities, such as an alternating voltage and an alternating current, vary sinusoidally with time and have the same frequency, they may be expressed as

$$Q_1 = A \left\{ \begin{array}{c} \sin \\ \cos \end{array} \right\} (\omega t + \phi)$$

$$Q_2 = B \left\{ \begin{array}{c} \sin \\ \cos \end{array} \right\} \omega t$$

where A, B, and ω are constants. It is then said that Q_2 lags (behind) Q_1 and ϕ is known as the angle of lag if it is positive. If ϕ is negative its magnitude is the angle of lead and Q_2 is said to lead Q_1.

LAGOON. See **Estuary**.

LAGRANGE FORMULA FOR INTERPOLATION. Used when $(n + 1)$-pairs of values are given for $y = f(x)$, but not necessarily at equally spaced increments of x or y. Let the given number pairs be (x_0, y_0), $(x_1, y_1), \ldots, (x_n, y_n)$. Then for any desired value of x within this interval,

$$y = y_0 L_0^{(n)}(x) + y_1 L_1^{(n)}(x) + \cdots + y_n L_n^{(n)}(x)$$

where

$$L_i^{(n)}(x) = \frac{(x - x_0) \cdots (x - x_{i-1})(x - x_{i+1}) \cdots (x - x_n)}{(x_i - x_0) \cdots (x_i - x_{i-1})(x_i - x_{i+1}) \cdots (x_i - x_n)}.$$

The quantities L_i, known as Lagrange coefficients, are independent of y; hence they may be calculated once for a given set of x values and used unchanged to obtain results for varying y. Moreover, it will be found that they remain unchanged with a change of variable to $u = (x - a)/h$, where h and a are constants.

Because of the symmetry in the equation, x and y may also be interchanged so that inverse interpolation may be effected.

See also **Interpolation**.

LAGRANGIAN COORDINATES. Sometimes called material coordinates. A system of coordinates by which fluid parcels are identified for all time by assigning them coordinates which do not vary with time. Examples of such coordinates are (a) the values of any properties of the fluid conserved in the motion; or (b) more generally, the positions in space of the parcels at some arbitrarily selected moment. Subsequent positions in space of the parcels are then the dependent variables, functions of time and of the Lagrangian coordinates.

Few observations in meteorology are Lagrangian: this would require successive observations in time of the same air parcel. Exceptions are the constant-pressure balloon observation, which attempts to follow a parcel under the assumption that its pressure is conserved, and certain small-scale observations of diffusion particles. See also **Eulerian Coordinates**.

LAGRANGIAN FUNCTION. Also called kinetic potential, the difference between the kinetic energy and the potential energy of a dynamic system. It is generally symbolized by L.

LAGRANGIAN POINT. One of the five solutions by Lagrange to the three-body problem in which three bodies will move as a stable configuration. In three of the solutions the bodies are in line; in the other two the bodies are at the vortices of equilateral triangles.

Lagrange predicted in 1772 that if the three bodies form an equilateral triangle revolving about one of the bodies, the system would be stable. This prediction was fulfilled in 1908 with the discovery of the asteroid Achilles approximately 60° ahead of Jupiter in Jupiter's orbit. Since then other asteroids have been discovered 60° ahead and 60° behind Jupiter.

LAGUERRE DIFFERENTIAL EQUATION. The linear equation $xy'' + (1 - x)y' + ny = 0$, having a simple pole at the origin. Its solutions are the *Laguerre polynomials*. Differentiation of the equation k times and replacement of the kth derivative by y gives

$$xy'' + (k + 1 - x)y' + (n - k)y = 0$$

which is the associated Laguerre equation with solutions as associated Laguerre polynomials. These functions occur in the quantum mechanical problem of the hydrogen atom.

The associated polynomials may be defined by the equivalent expressions

$$L_n^{(k)}(x) = \frac{e^x x^{-k}}{n!} \frac{d^n}{dx^n} (e^{-x} x^{n+k})$$

$$= \sum_{i=0}^{n} \binom{n+k}{n-i} \frac{(-x)^i}{i!}$$

The special case of $k = 0$ gives the Laguerre polynomials

$$L_n(x) = 1 - \binom{n}{1} x + \binom{n}{2} \frac{x^2}{2!} - \binom{n}{3} \frac{x^3}{3!} + \cdots$$

Both kinds may also be expressed in terms of the Gauss hypergeometric series and by generating functions. They are also related to the Hermite polynomials and the Bessel functions.

LAKE. See **Earth**; and **Limnology**.

LAKES (Colors). See **Colorants (Foods)**; and **Dyes (Textile)**.

LAMBDA PARTICLE. A hyperon with a rest-mass energy of 1115.6 MeV, an isospin quantum number zero, an angular momentum spin quantum number $\frac{1}{2}$, and a strangeness quantum number 1. Symbol, λ.

LAMBERT. See **Units and Standards**.

LAMBERT PROJECTION. The Lambert modified conformal conic projection (commonly known as the Lambert, and sometimes as the Gauss conformal) has been used for many years in the construction of maps. During the twentieth century, this projection gained favor rapidly for use in constructing charts for air navigators.

The projection is actually a mathematical type, but may be quite accurately described as a conical projection. It differs from the simple conical in that the cone is not tangent to the earth's surface, but cuts it on two latitude parallels known as standard parallels. This type of projection is particularly valuable for portraying large longitudinal areas, e.g., the entire United States. The graticule of the Lambert chart shows parallels of latitude as concentric circles centered at the nearer pole, and the meridians of longitude as straight lines converging on this pole. From simple geometric considerations, it is obvious that the meridians and parallels must be perpendicular. The angle of convergence of the meridians, and therefore the radii of the parallels, depends upon the distance of the nearer pole from the center of the area.

The great advantage of the Lambert projection is that the scale of distance is uniform, for all practical purposes, all over the chart. To indicate the accuracy of this statement, consider the Lambert projection on which the series of aeronautical charts of the United States are constructed by the Coast and Geodetic Survey. The standard parallels for these charts are N 45° and N 33°. The scale of distance, if considered as unity on the standard parallels, is 0.994 at the central parallel, expands to 1.010 at the extreme north boundary of the United States, and to 1.023 at the tip of Florida. Comparing the Lambert with the mercator distance scales, we find that if we consider the mercator scale as unity at N 39°, the scale at the northern limits of the United States is 1.154, whereas, at the tip of Florida, it is 0.846.

The nonorthogonal graticule of the Lambert chart is a distinct disadvantage for general navigational problems, since neither the rhumb line nor the great circle is straight. However, the uniformity of scale is of such great advantage that air navigators have begun to use this type of chart even in preference to the mercator, particularly when navigating in good visibility over land, or where good radio aids are available.

A straight line between two points on a Lambert chart is referred to as a Lambert line. Since the meridians on the graticule are convergent, this will not be a rhumb line. However, the distance measured along this line will be less than that along the rhumb line, and only slightly greater than the great-circle distance between the two points. This line is frequently used by aviators during conditions of good visibility. The standard procedure in using the Lambert line is to draw a straight line on the chart and pick out conspicuous landmarks separated by about 25 miles. The rhumb line indicated by measurement of the angle between the Lambert line and the meridian nearest the starting point is then followed. This will lead to a point at some distance from the first landmark but within visibility. From this first landmark, a new heading is adopted for the second, and so on to destination. If a rhumb line is desired for the entire route, the Lambert line is drawn as before, and the course is measured from the meridian halfway between the point just left behind and the point to be arrived at. This rhumb line will appear as a curve on the Lambert chart, but can be laid down by any one of a number of standard methods with sufficient accuracy for the selection of landmarks.

The convergence of the meridians prevents the use of the Lambert chart for graphical solution of dead-reckoning problems, and introduces difficulties in plotting lines of position obtained either by radio bearings or by celestial observation. The ease with which such problems can be solved on the mercator chart seems to cast doubt on any statement to the effect that "the Lambert Chart will completely supersede the mercator for all navigational purposes."

See also **Course**; **Great-Circle Course**; **Line of Position**; **Mercator Sailing**; **Navigation**; and **Rhumb Line**.

LAMBERT'S COSINE LAW.   The intensity from a surface element of a perfectly diffuse radiator is proportional to the cosine of the angle between the direction of emission and the normal to the surface. An element of a surface that obeys this law will appear equally bright when observed from any direction.

LAMB SHIFT.   The displacement between the $2S_{1/2}$ and $2P_{1/2}$ levels of hydrogen, which in the absence of radiative corrections would be zero due to the Coulomb degeneracy. The experimental value obtained by Lamb and Rutherford

$$E_{2S_{1/2}} - E_{2P_{1/2}} = 1057.8 \pm 0.1 \text{ megahertz}$$

is in agreement with the theoretical value. Of this, 27 MHz arises from vacuum polarization, the rest from self-energy corrections. The term is now used to indicate the displacement of any bound state level due to radiative corrections. See also **Field Theory**.

LAMÉ EQUATION (Generalized).   The most general second-order (linear) differential equation with five regular singularities, one of them being the point at infinity, with preassigned exponents differing from each other by $\frac{1}{2}$ at each singularity, all other points of the complex plane being ordinary points. This equation is remarkable because of the large number of important equations (Legendre, Bessel, etc.) obtainable from it by confluence. Letting a_1, a_2, a_3, a_4 and ∞ be the singular points, with exponents α_1, $\alpha_1 + \frac{1}{2}$, $\cdots \alpha_4$, $\alpha_4 + \frac{1}{2}$, μ_1, $\mu_1 + \frac{1}{2}$, the equation has the form

$$\frac{d^2w}{dz^2} + P\frac{dw}{dz} + Qw = 0$$

where

$$P = \sum_{i=1}^{4} \frac{\frac{1}{2} - 2\alpha_i}{z - a_i},$$

$$Q = \sum \frac{\alpha_i\left(\alpha_i + \frac{1}{2}\right)}{(z - a_i)^2} + \frac{Az^2 + 2Bz + C}{(z - a_1)\cdots(z - a_4)}.$$

A is expressible in terms of the α_i, and B, C are arbitrary constants. For example if the confluence takes the form $a_1 = a_2 = 0$, $a_3 = a_4 = \infty$, then choosing all $\alpha_i = 0$ and setting $z = \zeta^2$, with proper choice of B and C, we get

$$\zeta^2 \frac{d^2w}{d\zeta^2} + \zeta \frac{dw}{d\zeta} + (\zeta^2 - n^2)w = 0,$$

which is Bessel's equation.

LAMELLA (Botany).   The middle lamella is the compound layer composed of the primary walls and the cement-like intercellular substance that occurs between the primary walls of two cells. This composite layer is usually made up of pectic materials (colloids which have a great affinity for water), one of which is calcium pectate. The function of the middle lamella is to hold adjoining cells together. Sometimes, particularly in

mature fruits, the middle lamella substance breaks down. As a result, the cells of the fruit separate easily, giving to the fruit a meal-like character.

The middle lamella is the common source of pectin, which is added to concentrated fruit juices in the preparation of jellies. The term lamella is also applied to each of the concentric growth layers in large starch grains.

LAMELLA (Zoology).   1. A thin leaf or plate, such as a lamella of bone. See also **Bone**.

2. A flat plate formed by the fusion of ctenidial filaments in the bivalve mollusks. Two lamellae united by bridges of tissue form a gill through which water circulates under the influence of ciliary action in the persisting open spaces. This form of gill is the source of the name *Lamellibranchiata* applied to the class containing these animals.

LAMELLAE.   See **Bone**.

LAMELLIBRANCHIATA.   The bivalve mollusks, a class of the phylum *Mollusca* including the clams, mussels, oysters, scallops and related species. Many of these animals are valuable for food, and they produce pearls and mother-of-pearl. The class is also named *Pelecypoda*.

Bivalve mollusks differ from other members of the phylum in the following characters: (1) The body is bilaterally symmetrical and transversely compressed. (2) The mantle forms two lobes extending down along the sides of the body. In most species these lobes unite at the posterior end to form two passages, an upper excurrent and a lower incurrent siphon. Currents of water carry food and oxygen into the mantle cavity through the lower opening and a current bearing wastes passes out of the upper. (3) Each mantle fold secretes a valve of the shell formed of calcareous matter covered outside by a horny periostracum and inside by nacre, commonly called mother-of-pearl. The two valves of the shell are joined by a hinge ligament and the articulation is strengthened in some species by interlocking teeth. They are closed ventrally by the contraction of one or two adductor muscles. (4) The gills are thin plates on each side of the body in most species. They are formed of united ctenidial filaments. (5) The foot is a muscular wedge-shaped protuberance at the anterior end of the body. The head is rudimentary.

All bivalves are aquatic. Most species creep slowly by thrusting the foot into the muddy or sandy bottom but some propel themselves by jets of water squirted from the siphons or forced from the mantle cavity by rapidly closing the valves. The species vary from freshwater forms about $\frac{1}{8}$ inch long to giant marine shells more than a yard long.

The class is divided into four orders:

Order *Protobranchiata*. Gills in the form of small leaflets, two rows on each side of the body. Marine species.

Order *Filibranchiata*. Marine mussels, scallops, etc. Gills composed of filaments united only by ciliary junctions.

Order *Eulamellibranchiata*. Gill filaments united to form continuous plates. Freshwater clams or mussels, marine clams, oysters, shipworms, etc.

Order *Septibranchiata*. Gills replaced by a horizontal partition between the upper and lower divisions of the mantle chamber. A few marine species.

See also **Clam**; **Mollusca**; **Mussel**; and **Pearl**.

LAMINAR FLOW.   A condition of fluid flow in a closed conduit in which the fluid particles or "streams" tend to move parallel to the flow axis and not mix. This behavior is characteristic of low flow rates and high viscosity fluid flows. As the flow rate increases (or viscosity significantly decreases), the streams continue to flow parallel until a velocity is reached where the streams waver and suddenly break into a diffused pattern. This point is called the *critical velocity*. See also **Turbulent Flow**.

Laminar flow is characterized by a parabolic flow profile where the maximum velocity at or near the center of the conduit is approximately twice the average velocity in the profile. Laminar flow often is referred to as *viscous flow*, streamline flows, and *low-Reynolds number flow*. Special attention must be paid to the constancy of coefficient of most flowmeters in the region of laminar flow. See also **Reynolds Number**.

Laminar Sublayer. When a fluid is in turbulent flow past a rigid surface, fluctuations of velocity in the direction normal to the surface are inhibited, and very close to the surface they may be negligible. Then the Reynolds shear stress is small compared with the viscous stresses, and it has been common to describe the region as a laminar sublayer. In fact, turbulent fluctuations of velocity in planes parallel to the wall are considerable in comparison with the mean velocity.

See also **Aerodynamics and Aerostatics**; and **Fluid and Fluid Flow**.

LAMP. See **Illumination**.

LAMPBLACK. See **Carbon Black**.

LAMPREYS *(Agnatha).* A jawless fish of the family *Petromyzontidae*, the lamprey appears much like an eel. However, it is not an eel. Normally, various species of lampreys occur on both sides of the Atlantic. Like its close relative, the hagfish, the lamprey is characterized by the primitive features of jawless fishes—no scales, no sympathetic nervous system, a cartilage skeleton, and single nostril. Since about the mid-1800s, lampreys have not been considered of commercial value. However, they were eaten during the middle ages. Probably the parasitic landlocked lamprey *Pretomyzon marinus* is best known because of the very extensive damage it has done to many freshwater fish species in the Great Lakes. The lamprey attaches itself to a host and literally sucks life-giving juices from it. In the saliva of the lamprey, there is an anticoagulating material that continually dilutes the blood of its victim. Once the host is drained of vital juices, the lamprey moves on to another victim.

As an example of this damage, just a few decades ago, the Great Lakes yielded an annual catch of nearly 12 million pounds (5.4 million kilograms) of lake trout. Within a period of about 30 years, lake trout practically disappeared from the lakes. Initially, the lampreys migrated from their normal marine-water habitat to fresh water for spawning. Inasmuch as the Great Lakes are interconnected to the sea through various waterways, including the Welland Canal and the New York State Barge Canal, the lampreys ultimately invaded the lakes. During the latter 1970s, considerable progress was made toward specifically combating the lampreys without harming other species. Population of trout in the lakes is again rising.

LANDING GEAR. The apparatus comprising those components of an aircraft or spacecraft that support and provide mobility for the craft on land, water, or other surface. The landing gear consists of wheels, floats, skis, bogies, and treads, or other devices, together will all associated struts, bracing shock absorbers, etc.

LANDING SKID. A skid or runner used in the main landing gear of an aerodynamic vehicle, upon which the vehicle slides over the ground.

LANGLEY, SAMUEL PIERPONT (1834–1906). Born in the Boston suburb of Roxbury, Massachusetts. Samuel Langley was one of America's most accomplished scientists. His work as an astronomy, physics, and aeronautics pioneer was highly regarded by the international science community. Ironically though, Langley's formal education ended at the high school level, but he managed to continue his scientific education in Boston's numerous libraries.

Langley began his career as a civil engineer in Chicago, continuing later in St. Louis, before returning to Boston to accept an assistantship at the Harvard Observatory. Heading south once again, Langley later taught mathematics at the U.S. Naval Academy in Annapolis, Maryland. Then, from 1867–1887, he served as professor of physics and astronomy as well as director of the Allegheny Observatory at the Western University of Pennsylvania (now known as the University of Pittsburgh). After 1887, Langley was appointed Secretary of the Smithsonian Institution in Washington DC.

Langley's chief scientific interest was the sun and its effect on the weather, and believed that all life and activity on the Earth were made possible by the sun's radiation. In 1878, he invented the bolometer, a radiant-heat detector that is sensitive to differences in temperature of one hundred-thousandth of a degree Celsius (0.00001 °C). Composed of two thin strips of metal, a Wheatstone bridge, a battery, and a galvanometer (an electrical current measuring device), this instrument enabled him to study solar irradiance (light rays from the sun) far into its infrared region and to measure the intensity of solar radiation at various wavelengths. See Fig. 1.

Bolometers have been flown on numerous NASA missions including the Earth's Radiation Budget Experiment (ERBE) and the Clouds and Earth's

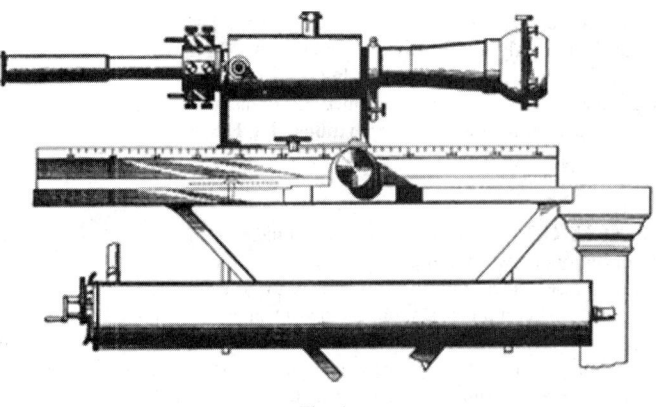

Fig. 1

Radiant Energy System (CERES), which provided accurate regional and global measurements of the components of the Earth's radiation budget. Langley's highly original and innovative research earned him honorary doctorates, awards, and medals from universities and scientific societies around the world.

In addition to his solar interests, Langley was the only professional scientist of his day who believed that the human race was destined to fly. While at the Allegheny Observatory, he made important experiments on the lift and drag of an aircraft moving through the air at a measured speed. Backed by these experiments, he was the first to offer a clear explanation of the way birds soar and glide without appreciable wing movement.

In 1886, he undertook a series of experiments on a rotating rig to measure the power needed to propel objects through the air. Encouraged by his findings, Langley set out to build a series of large working models of steam-powered flying machines he called "aerodromes," and, in 1896, became the first to build heavier-than-air machines capable of sustained (although uncontrolled) flight. Langley built two unmanned craft, each of which had two sets of 14-foot (4.3-meter) wings, weighed 26 pounds (11.8 kilograms), and were powered by steam engines.

Langley's first manned aircraft, powered by a five-cylinder air-cooled gasoline engine designed by Charles M. Manly did not fair as well as his unmanned craft. Piloted by Manly, the aircraft snagged upon launching from a catapult, and crashed into the Potomac River for the second and last time on Dec. 8, 1903, just nine days before the successful flights of the Wright brothers near Kitty Hawk, NC. This aircraft had a wingspan of 48 feet (14.6 meters) and a total weight (with pilot) of 850 pounds (386 kilograms). Some authorities believe that if his catapult had not failed, Langley would have been the first to achieve sustained flight in a manned heavier-than-air machine.

Langley's memory lives on in the names of the NASA Langley Research Center, the adjacent Air Force base, and several place names across the country. Our nation's first aircraft carrier, CV-1, built at the Norfolk Navy Yard in the early 1920's, was also named after Langley.

See also **Bolometer**.

Web References

Icarus on the Mall: *http://www.150.si.edu/chap6/six.htm*
Langley's Feat—and Folly (from Smithsonian Magazine): *http://smithsonian-mag.com/smithsonian/issues97/nov97/object_nov97.html*

LANGMUIR, IRVING (1881–1957). Langmuir was an American scientist whose fields of contribution include chemistry, physics, and technology. He graduated as a metallurgical engineer from the School of Mines at Columbia University in 1903. Postgraduate work in Physical Chemistry under Nernst in Göttingen earned him the degrees of M.A. and Ph.D. in 1906.

Returning to America, Dr. Langmuir became Instructor in Chemistry at Stevens Institute of Technology, Hoboken, New Jersey, where he taught until July 1909. In 1909, Langmuir began working for the General Electric Company in Schenectady, New York where he eventually became Associate Director.

His work on filaments in gases led directly to the invention of the gas filled incandescent lamp and to the discovery of atomic hydrogen. He later used the latter in the development of the atomic hydrogen welding process.

He was the first to observe the very stable adsorbed monatomic films on tungsten and platinum filaments, and was able, after experiments with oil films on water, to formulate a general theory of adsorbed films. He also studied the catalytic properties of such films.

Dr. Langmuir received twenty-three scientific medals and prizes, including the Nobel Prize in Chemistry in 1932 for his work on surface chemistry.

See also **Molecular and Supermolecular Electronics**.

J. M. I.

LANGUAGE (Computer). A communications means for transmitting information between human operators and computers. The human programmer describes how the problem is to be solved using the computer language. A computer language consists of a well-defined set of characters and words, coupled with a series of rules (termed syntax) for combining them into computer instructions or statements. There is a wide variety of computer languages, particularly in terms of flexibility and ease of use. There are three levels in the hierarchy of computer languages: (1) machine languages; (2) procedure-oriented languages; and (3) problem-oriented languages.

Machine Language (1) A language designed for interpretation and use by a machine without translation. (2) A system for expressing information which is intelligible to a specific machine; e.g., a computer or class of computers. Such a language may include instructions that define and direct machine operations, and information to be recorded by or acted upon by these machine operations. (3) The set of instructions expressed in the number system basic to a computer, together with symbolic operation codes with absolute addresses, relative addresses, or symbolic addresses. In this case, it is known as an Assembler Language. See also **Assembler (Computer System)**.

Procedure-oriented Language. A machine-independent language which describes how the process of solving the problem is to be carried out. For example, FORTRAN, ALGOL, PL/I, and COBOL.

Problem-oriented Language. A language designed for convenience of program specification in a general problem area. The components of such a language may bear little resemblance to machine instructions and often incorporate terminology and functions unique to an application. Also known as Applications Language.

Other computer languages include:

Algorithmic language. An arithmetic language by which numerical procedures may be precisely presented to a computer in a standard form. The language is intended not only as a means of directly presenting any numerical procedure to any appropriate computer for which a compiler exists, but also as a means of communicating numerical procedures among individuals.

Artificial language. A language specifically designed for ease of communication in a particular area of endeavor, but one that is not yet "natural" to that area. This is contrasted with a natural language that has evolved through long usage.

Common machine language. A machine-sensible information representation, common to a related group of data processing machines.

Common business-oriented language. A specific language by which business data processing procedures may be precisely described in a standard form. The language is intended not only as a means for directly presenting any business program to any appropriate computer for which a compiler exists, but also as a means of communicating such procedures among individuals.

Object language. A language which is the output of an automatic coding routine. Usually, object language and machine language are the same; however, a series of steps in an automatic coding system may involve the object language of one step serving as a source language for the next step and so forth.

THOMAS J. HARRISON, IBM Corporation, Boca Raton, FL.

LANGUR. See Monkeys and Baboons.

LA NIÑA. The coupled atmosphere–ocean phenomenon known as El Niño is frequently followed by a period of normal conditions in the equatorial Pacific Ocean. Sometimes, but not always, El Niño conditions give way to the other extreme of the El Niño-Southern Oscillation (ENSO)

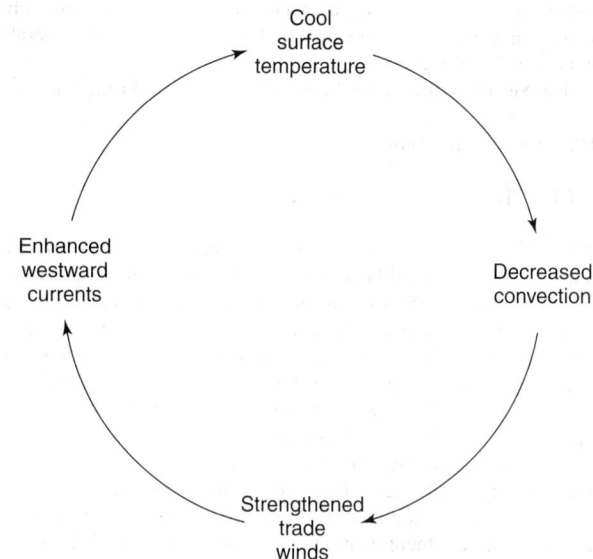

Fig. 1. Processes that affect La Niña. The cycle of La Niña is completed as the strengthened trade winds allow enhanced westward currents to flow, carrying cool water from the eastern Pacific.

cycle. This cold counterpart to El Niño is known as La Niña, Spanish for "the girl child." See Fig. 1.

The Southern Oscillation

While researching the collapse of the rainy phase of the monsoon system and resulting drought that occurred in India during the early years of the 20th century, Sir Gilbert Walker discovered a seesaw variation in pressure between the eastern and western Pacific Ocean. Walker found that when air pressure was high at Darwin, Australia (western Pacific) it was low at Tahiti, French Polynesia (eastern Pacific), and when air pressure was low at Darwin, it was high at Tahiti. Walker, however, failed to make the connection between this oscillating pressure pattern and El Niño. This link was made convincingly in the 1960s by the Norwegian meteorologist Jacob Bjerknes, who was also researching the anomalous drought in India.

How La Niña Forms

Researchers discovered that during non-El Niño years, surface pressures tend to be low over the warm waters of the equatorial western Pacific as overlying warm moist air rises and then diverges aloft. Over the colder waters of the eastern equatorial Pacific, surface pressures tend to be higher as converging winds aloft contribute to the sinking of cool air. In much the same way as a ball rolls down a hill, air flows from high pressure in the east to low pressure in the west along this equatorial pressure gradient. This contrast in pressure is what drives the trade winds, the prevailing large-scale surface winds that blow from east to west. As these winds blow along the surface of the equatorial waters, there is a net transport of ocean water in a westward direction. As this occurs, cold, nutrient-rich water rises up (or upwells) along the coast of South America to replace the westward-moving surface water. This upwelling brings nutrients to the surface waters off the coast allowing the fish population living in these upper waters to thrive.

During La Niña years, the trade winds are unusually strong due to an enhanced pressure gradient between the eastern and western Pacific. As a result, upwelling is enhanced. The cycle of La Niña is completed as the strengthened trade winds allow enhanced westward currents to flow, carrying cool water from the eastern Pacific along the coast of South America, contributing to colder than normal surface waters over the eastern tropical Pacific and warmer than normal surface waters in the western tropical Pacific. See Figs. 2 and 3.

The Effects of La Niña

Changes in global atmospheric circulation patterns accompany La Niña and are responsible for weather extremes in various parts of the world that are typically opposite to those associated with El Niño. These patterns result from colder than normal ocean temperatures inhibiting the formation of rain-producing clouds over the eastern equatorial Pacific region while at

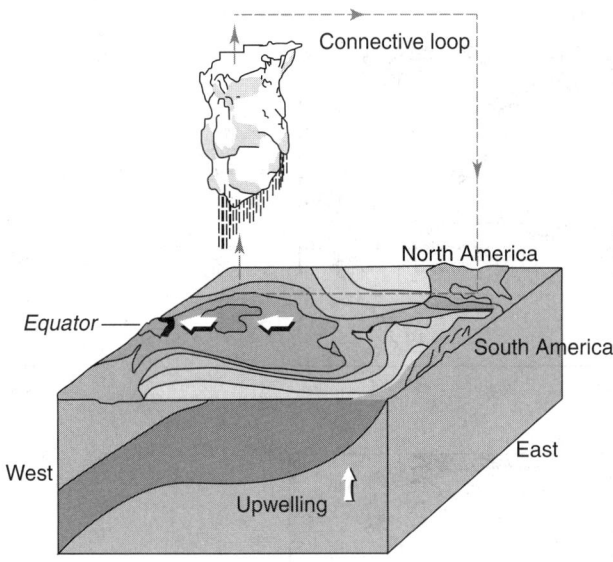

Fig. 2.   Normal conditions over the Pacific basin.

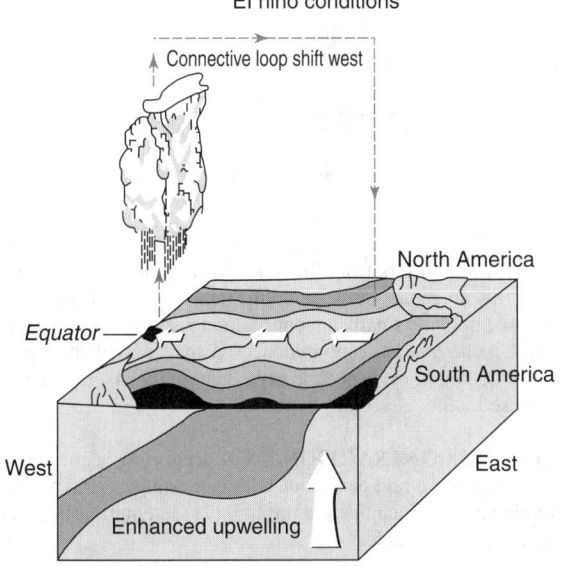

Fig. 3.   Disturbed conditions over the Pacific basin during a La Niña.

the same time enhancing rainfall over the western equatorial Pacific region (Indonesia, Malaysia, and northern Australia.) These patterns affect the position and intensity (weakening) of jet streams and the behavior of storms outside of the tropics in both the Northern and Southern hemispheres. Inside the tropics, ENSO strongly affects tropical cyclone activity around the world. During La Niña, weakened jet streams contribute to an increase in the number of Atlantic tropical storms and hurricanes. During El Niño, strengthened jet streams contribute to a decrease in tropical cyclone activity in the Atlantic and Australian basins.

U.S. La Niña Impacts

The first three months of the year during a La Niña typically feature below normal precipitation in the Southwest, the central and southern sections of the Rockies and Great Plains, and Florida. Meanwhile, the odds of surplus precipitation increase across the Pacific Northwest, in the northern Intermountain West, and over scattered sections of the north-central states, Ohio Valley, and upper Southeast. La Niña features unusually cold weather in the Northwest and (to a lesser extent) northern California, the northern Intermountain West, and the north-central states. Farther south, higher than normal temperatures are slightly favored in a broad area covering the southern Rockies and Great Plains, the Ohio Valley, the Southeast, and the mid-Atlantic states.

Global La Niña Impacts

Globally, La Niña is characterized by wetter than normal conditions west of the equatorial central Pacific over northern Australia and Indonesia during the northern hemisphere winter, and over the Philippines during the northern hemisphere summer. Wetter than normal conditions are also observed over southeastern Africa and northern Brazil, during the northern hemisphere winter season. During the northern hemisphere summer season, the Indian monsoon rainfall tends to be greater than normal, especially in north-west India. Drier than normal conditions are observed along the west coast of tropical South America, and at subtropical latitudes of North America (Gulf Coast) and South America (southern Brazil to central Argentina) during their respective winter seasons. See Fig. 4.[1]

NASA and NOAA Missions to Study La Niña

Over the years, several NASA missions have studied the effects associated with La Niña and El Niño, such as changes in sea-surface temperature (SST) and cloud cover. These studies are augmented by data from operational satellites of the National Oceanic and Atmospheric Administration (NOAA).

Initial efforts at mapping SST and cloud cover were conducted using data from NASA's Nimbus series of satellites. The four-channel Advanced Very High Resolution Radiometer (AVHRR), flown on NOAA's TIROS-N weather satellite in 1978 and on the NOAA-6 satellite in 1979, greatly increased the accurate measurements of El Niño effects. ("Four channel" means that the instrument views in four different parts of the electromagnetic visible and infrared spectrum.) Still further increases in accuracy resulted when a fifth channel was added to the AVHRR instrument flown on NOAA-7 in 1981, and on subsequent NOAA satellites. The fifth channel improved the measurement of SST by providing corrections for atmospheric water vapor that otherwise would have interfered with the temperature measurements.

The joint U.S.-French TOPEX/Poseidon mission was launched in 1992 and is providing global determinations of changes in ocean surface currents that are related to the La Niña and El Niño phenomena. Data retrieved from TOPEX/Poseidon are important because they provide measurements of the depth to which the cold or warm anomaly extends.

A NASA scatterometer called NSCAT flew on the Japanese Advanced Earth Observing System (ADEOS) spacecraft, which was launched in August 1996. NSCAT provided very high quality data on the speed and direction of ocean-surface winds worldwide. Unfortunately, after nine months in orbit, a spacecraft failure brought to an end the stream of NSCAT data. Recognizing the important contributions to Earth science made by NSCAT, NASA launched the QuikSCAT satellite in June 1999 to bridge the gap remaining before launch of the Japanese spacecraft designated ADEOS II (planned for 2000). The SeaWinds instrument onboard QuikSCAT and ADEOS II will provide detailed measurements of the winds above the oceans.

In addition to the scatterometer measurements, which use active microwave radar systems to determine surface wind speeds and directions over the ocean, surface wind speeds are also being obtained from the Special Sensor Microwave Imager (SSM/I), a passive microwave sensor onboard a Department of Defense spacecraft.

Key sources of data related to El Niño have been retrieved from the five-channel AVHRRs flown on NOAA-7, 9, and 11. These historic data sets cover the period 1981 through 1992 and beyond and will permit more accurate SST determinations than were previously available. These data are important to the development and testing of a new generation of computer models in which the interacting processes of the land, the atmosphere, and the oceans are coupled. These coupled models will lead the way to an increased understanding of phenomenon such as La Niña and the teleconnections that link La Niña with changes in weather patterns throughout the world.

NASA's SeaWiFS (Sea-viewing Wide Field of View Sensor) was launched on the OrbView-2 satellite in August 1997. SeaWiFS is designed to detect ocean color, which is an indicator of microscopic plant life in the ocean. The growth of such plants (called phytoplankton) is affected by the changes in sea surface temperature that are related to La Niña and El Niño. SeaWiFS data enable scientists to compare and contrast El Niño's impacts on the marine biosphere with those of La Niña.

---

[1] National Centers for Environmental Prediction-Climate Prediction Center (NOAA).

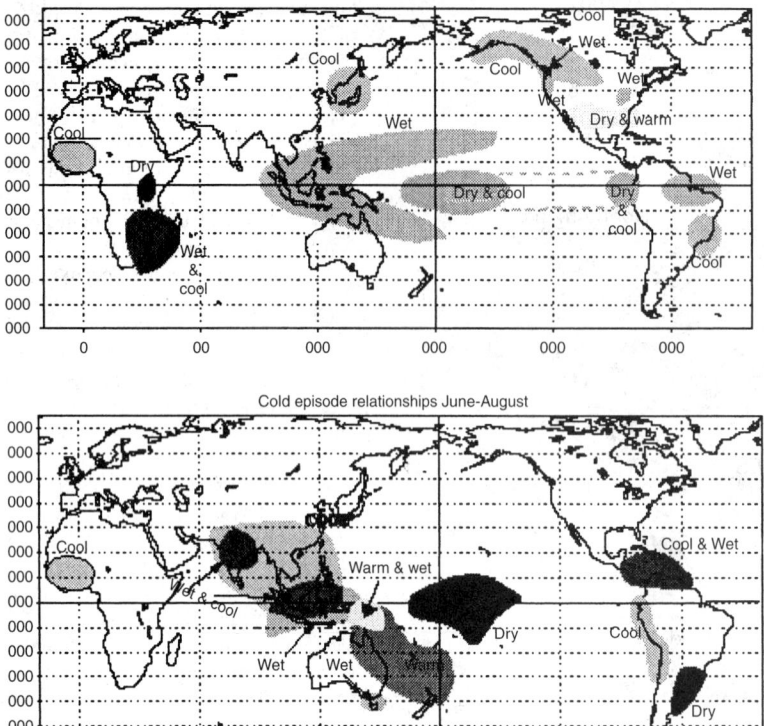

Fig. 4.   Anomalous precipitation and temperature patterns associated with La Niña. (*National Oceanic and Atmospheric Administration.*)

The Tropical Atmosphere Ocean (TAO) Array consists of nearly 70 moored buoys in the tropical Pacific designed by the National Oceanic and Atmospheric Administration (NOAA). These floating devices take real-time measurements of air temperature, relative humidity, surface winds, sea surface temperatures, and subsurface temperatures down to a depth of 500 meters. Data from these moored buoys are processed by NOAA and then made available to scientists for collaborative research studies.

The joint U.S.-Japanese Tropical Rainfall Measuring Mission (TRMM), launched in November 1997, uses for the first time both active (radar) and passive microwave detectors from space to provide measurements of precipitation, clouds, and radiation processes in lower latitudes, including those portions of the Pacific Ocean where El Niño and La Niña occur. TRMM research team members have successfully retrieved sea-surface temperature data from the TRMM Microwave Imager (TMI) instrument onboard the spacecraft, giving them new insight into the complex evolution of the La Niña event. TMI is an all-weather measuring instrument that can see through clouds to measure sea-surface temperature in the tropics. Similar observations will be continued by the Advanced Microwave Scanning Radiometer (AMSR) to be flown on ADEOS-II and the AMSR-E instrument to be flown onboard EOS PM-1, both of which will be launched in the year 2000.

With the launch of the EOS satellites, we will have the means to collect and analyze the most comprehensive data set ever acquired for the development of coupled models. This data set will increase markedly our understanding of the causes and effects of such large-scale ocean–atmosphere phenomena as La Niña and El Niño.

See also **El Niño**.

Additional Reading

Glantz, M.H.: "Currents of Change: El Nino and La Nina Impacts on Climate and Society," Cambridge University Press, New York, NY, 2000.
Philander, G.S.: "El Nino, La Nina, and the Southern Oscillation," Academic Press, Inc., San Diego, CA, 1989.

Web References

Climate Prediction Center. *http://www.cpc.ncep.noaa.gov/pacdir/*
Jet Propulsion Laboratory. *http://www.jpl.nasa.gov/elnino/990310.html*

LANTERN FISHES.   See **Iniomous Fishes**.

LANTERN FLY (*Insecta, Homoptera*).   Any member of the family *Fulgoridae*, which differs from the related leaf hoppers and other families in having the antennae inserted at the sides of the head. The North American species are small but one giant Brazilian species has a wing spread of 6 inches. A large prominence on the head of this species was once said to be luminous, hence the name lantern fly has persisted although none of these insects is actually luminous.

LANTHANIDE CONTRACTION.   The decreasing sequence of crystal radii of the tripositive rare-earth ions with increasing atomic number in the group of elements (57) lanthanum through (71) lutetium of the Lanthanide Series in the periodic table.

LANTHANIDE SERIES.   The chemical elements with atomic numbers 58 to 71 inclusive, commencing with cerium (58) and through lutetium (71) frequently are termed collectively, the Lanthanide Series. Lanthanum, the anchor element of the series, appears in group 3b of the periodic table. Some authorities consider lanthanum a part of the series. Members of the series, along with lanthanum and yttrium, are described under **Rare-Earth Elements and Metals**. See also **Actinide Series**.

LANTHANUM.   Chemical element, symbol La, at. no. 57, at. wt. 138.91, periodic table group 3, homolog of the Lanthanide Series of elements, mp 918°C, bp 3464°C, density 6.146 g/cm^3 (20°C). Elemental lanthanum has a double close-packed hexagonal crystal structure at 25°C. The pure metallic lanthanum is silver-gray in color, but with a luster that remains only briefly upon exposure to air, rapidly oxidizing to a white powder. The oxide is hygroscopic and tends to spall, thus exposing fresh surfaces of the metal for oxidation. Thus, the metal must be handled in an inert atmosphere. Chips and powdered lanthanum are quite pyrophoric. Under required inert atmospheric conditions, the metal is easy to work with normal tools, paralleling tin in its workability. There are two natural isotopes ^{139}La and ^{138}La. The latter is mildly radioactive with a half-life of 10^{10}–10^{15} years. The element becomes a superconductor below 6 K. There are 19 known artificial isotopes, all radioactive. Of the light (or cerium-group) rare-earth metals, lanthanum is the second most plentiful and ranks 57th in abundance of elements in the earth's crust, exceeding gold, tantalum, platinum, mercury, bismuth, and several other commonly-used elements. The element was first identified by C.G. Mosander in 1839. Electronic configuration

$$1s^2 2s^2 2p^6 3s^2 3p^6 3d^{10} 4s^2 4p^6 4d^{10} 5s^2 5p^6 5d^1 6s^2.$$

Ionic radius La^{3+} 1.045 Å. Metallic radius 1.879 Å. First ionization potential 5.577 eV; second, 11.06 eV. Oxidation potentials $La \rightarrow La^{3+} + 3e^-$, 237 V; $La + 3OH^- \rightarrow La(OH)_3 + 3e^-$, 2.76 V.

Other important physical properties of lanthanum are given under **Rare-Earth Elements and Metals**.

Much of the commercial lanthanum production uses bastnasite, a rare-earth fluorocarbonate found in Inner Mongolia and Southern California, as the source. See also **Bastnasite**. The element is separated from other rare-earth elements by a liquid solvent extraction or an ion-exchange process after acid leaching of bastnasite (or monazite) minerals. Pure lanthanum is obtained by (1) electrowinning from the oxide La_2O_3 in a molten fluoride electrolyte, (2) electrolysis of fused anhydrous $LaCl_3$, or (3) metallothermic reduction of LaF_3 by calcium in a reactor under an inert atmosphere.

Lanthanum metal dissolves readily in dilute mineral acids. The oxide is dissolved by concentrated mineral acids and acetic and formic acids. The metal is a component of mischmetal used for lighter "flints" and in the cores of carbon electrodes for high-intensity lighting. See also **Cerium**. Several of the best grades of optical glass require pure lanthanum oxide as an ingredient for lowering the dispersion of light and for improving the index of refraction. The oxide melts at $2,310\,^\circ C$. The oxide ranks eleventh among the most refractory metal oxides, but finds limited use because of its highly hygroscopic nature. The oxide also is used as a host matrix for fluorescent phosphors and in thermistors and capacitors and other elements of electronic circuitry. Lanthanum oxide combined with transition metal (e. g. Mn, Co, Cr) oxides are being used in solid oxide fuel cells (SOFC) as electrodes. By far, the largest use of lanthanum (mixed with other rare-earths) is for molecular-sieve catalysts for cracking crude petroleum.

As an alloying metal, lanthanum finds broad use. Although lacking mechanical strength, lanthanum has a high affinity for oxygen, sulfur, nitrogen, and hydrogen and thus makes an effective component for scavenging gases from molten metals. Cobalt-base alloys containing lanthanum have shown increased resistance to oxidation and hot corrosion. One of the intermetallic compounds of lanthanum $LaCo_5$ possesses excellent magnetic properties—well in excess of those of alnico and platinum cobalt permanent magnets. The intermetallic compound $LaNI.C$."eye"$_5$ shows exceptional properties for absorbing and desorbing large amounts of hydrogen at room temperature. It is the major component of rechargeable metal hydride batteries, which is a rapidly expanding major application.

See references listed at ends of entries on **Chemical Elements**; and **Rare-Earth Elements and Metals**.

K.A. GSCHNEIDNER, Jr. and B. EVANS
Iowa State University, Ames, IA.

LAPIS LAZULI. See **Lazurite**.

LAPLACE EQUATION. A second-order partial differential equation of elliptic type which, in vector form, is $\nabla^2 \phi = 0$. It is the homogeneous case of Poisson's equation. Its solutions, the scalar quantity ϕ, occur in problems involving steady-state temperatures, gravitational and electric potentials, hydrodynamics of ideal fluids, and many other physical phenomena. The equation is usually solved by the method of separation of variables in a suitable curvilinear coordinate system and with boundary conditions imposed by physical requirements. Such solutions are called harmonic functions. In two dimensions, an analytic function of the complex variable must satisfy Laplace's equation. Thus, $w = u + iv$ is an analytic function of $z = x + iy$ for $u_{xx} + u_{yy} = 0$ and $v_{xx} + v_{yy} = 0$.

We shall here cite two familiar physical examples to which Laplace's equation applies; there are many others.

1. Consider a region of space in which there is an electric field (due to electric charges in the vicinity) but no free electricity, that is, no such space charge as exists in a vacuum tube in operation. At any point in this space there is an electric potential, which varies with the position of the point and is therefore a function of its coordinates. It is shown in electrostatic theory that if the potential satisfies Laplace's equation, it is possible to trace the lines of force and equipotential surfaces in the region. This is done by means of special solutions of Laplace's equation, called harmonic functions, which satisfy the "boundary conditions" imposed by the

arrangement of the neighboring charges. As to which form of the equation is to be used, this depends upon which system of coordinates is most conveniently adapted to the shape and arrangement of the charged bodies.

2. If specified parts of the surface of a solid thermal conductor (such as a block of copper) are kept at different specified constant temperatures, heat will flow from the warmer toward the colder boundaries within the conductor. When this flow has become steady, the temperature of the conductor takes on a definite constant value at each point, dependent upon the location of the point. If the temperature is designated by u, it now satisfies Laplace's equation, and the lines of flow and isothermal surfaces may be determined accordingly. A similar procedure is applied to potential in the steady flow of electricity through a metallic conductor.

LAPLACE THEOREM. This limit theorem, of which the Bernoulli theorem is a corollary, states that if there are n independent trials, in each of which the probability of an event is p, and if this event occurs k times, then

$$P\left\{ z_1 \leq \frac{k - np}{\sqrt{npq}} \leq z_2 \right\} \longrightarrow \frac{1}{\sqrt{2\pi}} \int_{z1}^{z2} e^{-1/2z^2} \, dz$$

as $n \rightarrow \infty$, whatever the numbers z_1 and z_2. Roughly speaking, the theorem states that the number of successes k in n trials is normally distributed for large n.

LAPLACE TRANSFORM. This class of integral transform has so many applications in modern technology that it justifies the more extended discussion in this special entry.

The behavior of ac electric circuits is readily expressed in terms of integro-differential equations, as is understood by recalling that the voltage drop across an inductance is a differential function of the current, whereas the voltage across a capacitance is an integral function of the current. Moreover, many types of nonelectrical systems—mechanical, acoustical, hydraulic—lend themselves conveniently to analysis by analogy to electrical systems, so providing a convenient means of generalizing and unifying the analysis of complex systems involving subsystems of various kinds. This fact, together with the adaptability of the Laplace transform method to the solution of differential and integral equations, and the analysis of discontinuous functions, has brought about its wide application to control problems.

The Laplace transform of a function $f(t)$ is defined as the function given by

$$\mathcal{L}[f(t)] = \int_0^\infty f(t) e^{-st} \, dt = F(s) \qquad (1)$$

Here the double equation means that the Laplace transform may be written as $\mathcal{L}[f(t)]$ or as $F(s)$, since the transformation process results in a new function $F(s)$ of a variable s instead of the original function f of time, t. The variable s may be real but is usually complex. The transformation is effected by integrating the function $f(t)$ by the use of the factor e^{-st}, where e is the natural logarithmic base (2.71828 ...). The integral is, as shown, in the improper form, its upper value being considered to increase indefinitely, approaching ∞ as a limit.

Evaluating Laplace Transforms. One of the simplest discontinuous functions is the *unit function*, which has only two values, 1 for positive or zero values of t, and 0 for negative values of t; it is often expressed as u (t). Thus for values of $t \geq 0$,

$$\mathcal{L}[u(t)] \; \mathcal{L}(1) = \int_0^\infty 1 e^{-st} \, dt = -\frac{1}{s} e^{-st} \Big]_0^\infty$$

$$= -\frac{1}{s} 0 - \left(-\frac{1}{s} 1 \right) = \frac{1}{s} \qquad (2)$$

The *unit ramp function* is a linear and continuous function of t, thus expressed as $f(t)$ for positive or zero values of t, so that

$$\mathcal{L}[f(t)] = \int_0^\infty t e^{-st} dt$$

$$= -\frac{t e^{-st}}{s} \Big]_0^\infty + \frac{1}{s} \int_0^\infty e^{-st} \, dt$$

$$= 0 + \left[\frac{1}{s} \cdot \frac{-1}{s} e^{-st} \right]_0^\infty = \frac{1}{s^2} \qquad (3)$$

The Laplace transform of a *power of t* is found by a change of the variable of integration, as from t to $w = st$, where s is real and positive:

$$\mathcal{L}(t^p) = \int_0^\infty w^p e^{-st}\, dt = \frac{1}{s^{p+1}} \int_0^\infty w^p e^{-w}\, dw \tag{4}$$

The last integral at the right is essentially the gamma function, which is given in tables, and which is a generalization of the factorial $n!$, that is defined only for integers and zero. The gamma function may be written as

$$\Gamma(p) = \int_0^\infty w^{p-1} e^{-w}$$

and when $p = n$, the value of the function reduces to $n!$. Therefore, we can write for Equation (5),

$$\mathcal{L}(t^p) = \frac{1}{s^{p+1}} \Gamma(p+1) \tag{5}$$

To find the Laplace transform of the *exponential function* e^{-ct} (for positive values of t) where c is a real number:

$$\mathcal{L}(e^{-ct}) = \int_0^\infty e^{-ct} e^{-st}\, dt = \int_0^\infty e^{-(s+c)t}\, dt$$

$$= -\frac{1}{s+c} e^{-st} \Big]_0^\infty = \frac{1}{s+c} \tag{6}$$

To find the Laplace transform of the *trigonometric function* $\sin at$, where a is a real, positive number:

$$\mathcal{L}(\sin at) = \int_0^\infty \sin at\; e^{-st}\, dt$$

Now in the preceding examples, it was found that the evaluation of the Laplace transform of an exponential was quite simple because e^{-st} is also an exponential. So in the present example we replace $\sin at$ by its exponential expression $(e^{iat} - e^{-iat})/2i$, where $i = \sqrt{-1}$, giving

$$\mathcal{L}(\sin at) = \frac{1}{2i} \int_0^\infty (e^{iat} - e^{-iat}) e^{-st}\, dt$$

$$= \frac{1}{2i} \left(\frac{1}{s - ia} - \frac{1}{s + ia} \right)$$

$$= \frac{a}{s^2 + a^2} \tag{7}$$

A similar substitution is used to find the Laplace transform of $\cos at$. Since $\cos at = (e^{iat} + e^{-iat})/2i$, we have

$$\mathcal{L}(\cos at) = \frac{1}{2i} \int_0^\infty (e^{iat} + e^{-iat}) e^{-st}\, dt$$

$$= \frac{s}{s^2 + a^2} \tag{8}$$

Properties of the Laplace Transform. An important property of the Laplace transform is its linearity, so that the Laplace transform of the sum of two functions is equal to the sum of the individual Laplace transforms

$$\mathcal{L}[f_1(t) \pm f_2(t)] = F_1(s) \pm F_2(s)$$

Here the original functions are denoted by $f_1(t)$ and $f_2(t)$, and their Laplace transforms by $F_1(s)$ and $F_2(s)$.

(I) The full statement of the *linearity* (also called *superposition*) *theorem* introduces two constants (a and b below):

$$\mathcal{L}[af_1(t) \pm bf_2(t)] = aF_1(s) + bF_2(s) \tag{10}$$

conveying the further information that constant factors of functions appear unchanged in their Laplace transforms.

Since by Equation (2) the Laplace transform of $t = 1$ was found to be $1/s$, then it follows directly that the Laplace transform of any constant is expressed by

$$\mathcal{L}(c) = c\mathcal{L}(1) = \frac{c}{s} \tag{11}$$

It also follows from the linearity theorem that the Laplace transform of any polynomial is equal to the sum of the Laplace transforms of its terms:

$$\mathcal{L}(a_0 + a_1 t + a_2 t^2 + \cdots + a_n t^n) = a_0 \mathcal{L}(1) + a_1 \mathcal{L}(t)$$
$$+ a_2 \mathcal{L}(t^2) + \cdots + a_n \mathcal{L}(t^n) \tag{12}$$

The general expression for this series of terms is, therefore,

$$\sum^n a_i \frac{\Gamma(i+1)}{s^{i+1}} = \frac{a_0}{s} + \frac{a_1}{s^2} + \frac{2!a_2}{s^3} + \frac{3!a_3}{s^4} \cdots \frac{n!a_n}{s^{n+1}} \tag{12a}$$

Note that, as derived, Equation (12a) is valid only for positive and real values of s. In fact, in all operations with Laplace transforms constant attention must be paid to (1) whether the transform of a function exists at all, and, if it does, (2) for what values of the variables it has corresponding real values and for what values it becomes zero or indefinitely great.

(II) The *real differentiation theorem* asserts that if a function and its first derivative have Laplace transforms and if $\mathcal{L}[f(t)] = F(s)$, then

$$\mathcal{L} \frac{df(t)}{dt} = sF(s) - f(t \to 0+) \tag{13}$$

which states that the Laplace transform of the first time derivative of a function is equal to the product by s of the Laplace transform of the function itself minus the value of the function as t approaches 0 from the positive side.

(III) The *real integration theorem* asserts that if a function of t has a Laplace transform and if $\mathcal{L}[f(t)] = F(s)$, then

$$\mathcal{L} \left[\int f(t)\, dt \right] = \frac{1}{s} \left[F(s) + \int_{0+} f(t)\, dt \right] \tag{14}$$

which states that the Laplace transform of the integral of the function is equal to the product by $1/s$ of the Laplace transform of the function plus the value of the integral of the function as t approaches zero from the positive side.

The usefulness of these two theorems is due to the fact, as stated earlier, that so many of the circuit and control equations contain differentials or integrals or both.

(IV) There are two translation theorems used with Laplace transforms. The first of these is the *time-delay theorem*, which is written

$$\mathcal{L}[f(t - t_0)] = e^{-t_0 s} F(s) \tag{15}$$

where t_0 is real and ≥ 0, provided $f(r) = 0$ for $t < 0$ and, in the case of the $0 +$ transform, $f(t)$ in addition contains no impulse functions at $t = 0$. A direct corollary of the time-delay theorem is the *unit step function*. In Equation (2) it was found that for values of t of unity, the Laplace transform was $1/s$. The unit step function is $u(t - t_0)$, where t has a value of 1 for all positive and zero values of the function, i.e., for all values of t greater than or equal to t_0, while t is zero for all its other values. This function expresses the commencement of a steady-state operation or process of unit value at time t_0.

Equations (2) and (15) give the unit step function directly, for by (2) the Laplace transform of unity is $1/s$, which corresponds to $F(s)$ in (15), giving

$$\mathcal{L}u(t - t_0) = e^{-t_0 s} \cdot \frac{1}{s} = \frac{e^{-t_0 s}}{s}$$

(V) The *time advance theorem* is written

$$\mathcal{L}[f(t + t_0)] = e^{t_0 s} F(s) \tag{16}$$

Note the change of sign of t_0 on both sides of the equation from those in the time delay theorem of Equation (15). The same conditions apply: that t_0 is real and ≥ 0, provided $f(t) = 0$ for $t < t_0$ and, in the case of the $0 +$ transform, $f(t)$ in addition contains no impulse functions at $t = t_0$.

(VI) The *frequency shift theorem* is written

$$\mathcal{L}[e^{at} f(t)] = F(s - a) \tag{17}$$

where a is any finite constant, real or complex.

(VII) The *final value theorem* assets that if the function $f(t)$ and its first derivative have Laplace transforms and if $\mathcal{L}[f(t)] = F(s)$, then

$$\lim_{s \to 0} sF(s) = \lim_{t \to \infty} f(t) \tag{18}$$

provided that $sF(s)$ is analytic on the imaginary axis and on the right half-plane.

(VIII) The corresponding *initial value theorem* asserts that if a function and its first derivative have Laplace transforms, and if $\mathcal{L}\phi f(t) = F(s)$, then

$$\lim_{s \to \infty} sF(s) = \lim_{t \to \infty} f(t) \qquad (19)$$

provided that the $\lim sF(s)$ as s approaches infinity exists.

(IX) The last of these Laplace theorems that is widely used in elementary system calculations is the *convolution theorem*. The convolution of two functions $f_1(t)$ and $f_2(t)$, denoted by $f_1^* f_2 = f_2^* f_1$ is defined by the integral

$$f_1^* f_2 = \int_0^t f_1(t - \tau) f_2(\tau) \, d\tau$$

If $f_1 = t^n$ and $f_2 = t^m$ (n and m, two positive integers), it can be verified by direct calculation that

$$\mathcal{L}\{f_1^* f_2\} = F_1(s) F_2(s)$$

where $F_1(s)$ is the Laplace transform of $f_1(t)$ and $F_2(s)$ is the Laplace transform of $f_2(t)$. Thus, if $f_1 = t^2$ and $f_2 = t^3$, then

$$f_1^* f_2 = \int_0^t (t - \tau)^2 \cdot \tau^3 \, dt = \frac{t^6}{60}$$

On the other hand, if

$$\mathcal{L}(t^2) = 2/s^3 \quad \text{and} \quad \mathcal{L}(t^3) = 6/s^4$$

then

$$\mathcal{L}(t^2) \cdot \mathcal{L}(t^3) = \frac{12}{s^7} = \mathcal{L}\left(\frac{t^6}{60}\right)$$

From this result it follows immediately, if $f_1(t)$ and $f_2(t)$ can be expanded in two convergent power series, that

$$(f_1^* f_2) = F_1(s) F_2(s) \quad \text{(convolution theorem)} \qquad (20)$$

In fact, the theorem is true in more general cases. The convolution theorem can be used to derive additional Laplace transforms and also to solve so-called "integral equations," i.e., functional equations where the unknown function appears behind an integral sign.

The Inverse Laplace Transform. In solving problems by means of the Laplace transform, the inverse operation of finding functions which correspond to transforms is, of course, as important as the direct one of finding the transforms. For this purpose, it is convenient to have values of transform pairs, which are given in Table 1. The justification for the inverse operation rests, of course, upon the assumption of uniqueness, that for each transform there is only one function. This is true, although its proof is not given here.

TABLE 1. LAPLACE FUNCTION-
TRANSFORM PAIRS

	$f(t)$	$F(s)$
1.	1	$\dfrac{1}{s}$
2.	c	$\dfrac{c}{s}$
3.	t	$\dfrac{1}{s^2}$
4.	t^p	$\dfrac{\Gamma(p+1)}{s^{p+1}}$
5.	$\sin at$	$\dfrac{a}{s^2 + a^2}$
6.	$\cos at$	$\dfrac{s}{s^2 + a^2}$
7.	e^{-et}	$\dfrac{i}{s + c}$
8.	$e^{at} t^n$	$\dfrac{n!}{(s-a)^{n+1}}$
9.	$t \sin at$	$\dfrac{2as}{(s^2 + a^2)^2}$
10.	$t \cos at$	$\dfrac{s^2 - a^2}{(s^2 + a^2)^2}$

The symbol for the operation of inverse (Laplace) transformation is $\mathcal{L}^{-1}$.

There is one kind of inverse transform, however, that is not readily tabulated. It is the inverse transform of a fraction. Since fractions occur quite frequently in the problems solved by these methods, the following discussion of one case of finding their inverse is important.

Given a Laplace transform in the form of a rational algebraic fraction,

$$F(s) = \frac{A(s)}{B(s)} = \frac{a_m s^m + a_{m-1} s^{m-1} + \cdots + a_1 s + a_0}{b_n s^n + b_{n-1} s^{n-1} + \cdots + b_1 s + b_0} \qquad (21)$$

where the a_i and b_i are all real constants and n and m are positive integers. Where the roots of the denominator are all real or zero and different, a direct method is available to find $\mathcal{L}^{-1}$.

Let the roots of the denominator be $s_1, s_2, \ldots, s_n$. The $F(s)$ takes the form

$$F(s) = \frac{a_m s^m + a_{m-1} s^{m-1} + \cdots + a_1 s + a_0}{(s - s_1)(s - s_2) \cdots (s - s_n)}$$

This equation may be expressed as the sum of partial fractions

$$F(s) = \frac{k_1}{s - s_1} + \frac{k_2}{s - s_2} + \cdots + \frac{k_n}{s - s_n}$$

where the k_i are coefficients to be determined. The procedure is to multiply both sides of the equation by $(s - s_i)$, which is then equated to 0. Thus expressions for each coefficient are obtained of the general form

$$k_i = \left[(s - s_i) \frac{A(s)}{B(s)} \right]_{s = s_i}$$

where the term $A(s)/B(s)$ represents the transforms whose inverse is to be found.

Then using for each coefficient transform pair number 8 in the table of transforms, we have

$$\mathcal{L}^{-1}\left[\frac{k_i}{s - s_i} \right] = k_i e^{s_i t} \qquad (22)$$

which can be found for each fraction, so that the inverse transform of $F(s)$ that is sought is the sum of all i values

$$\mathcal{L}^{-1}[F(s)] = \sum^i k_i e^{s_i t}, \quad 0 \leq t \qquad (23)$$

Applications of the Laplace Transform. (a) *Simple resistance-capacitance circuit.* A simple example of the application of the Laplace transform in solving equations is provided by the RC-circuit. Let us find, for example, the behavior of the instantaneous current $i(t)$ from the time that switch S in Fig. 1 is closed. The current produced by a given electromotive forces (E) varies inversely as the resistance $(i(t) = E/R)$ or $E = i(t)R$, and the integrated current (with respect to time) varies directly as the capacitance $(\int i(t) \, dt = CE)$ or $(E = (1/C) \int i(t) \, dt)$ so that the equation for a circuit having both capacitance and resistance is

$$Ri(t) \frac{1}{C} \int i(t) \, dt = E \qquad (24)$$

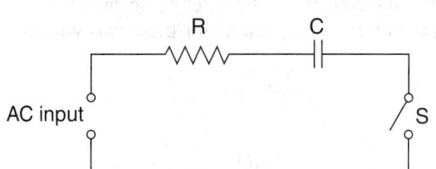

Fig. 1.   Resistance-capacitance (RC) circuit.

Taking the Laplace transform of both sides, we have

$$\mathcal{L}\left[Ri(t) + \mathcal{L}\left[\frac{1}{C} \int i(t) \, dt \right] \right] = \mathcal{L}(E) \qquad (25)$$

By theorem I, and the definition of the Laplace transform, $\mathcal{L}[Ri(t)] = RI(s)$, where $I(s)$ is the effective value of the current, by theorem II,

$$\mathcal{L}\left[\frac{1}{C} \int i(t) \, dt \right] = \frac{1}{sC} \left[I(s) + \int i \, dt(0+) \right]$$

Therefore the equation becomes

$$RI(s) + \frac{1}{sC}\left[I(s) + \int i\, dt(0+)\right] = \frac{E}{s} \tag{26}$$

by using transform pair number 2 from the table to show that $\mathcal{L}(E) = E/s$. If it is assumed that the initial condition of the circuit is with the switch open, then

$$\frac{1}{sC}\int i\, dt(0+) = 0$$

so that

$$RI(s) + \frac{1}{sC}I(s) = \frac{E}{s}$$

Transposing, we have

$$I(s) = \frac{E/s}{R + \dfrac{1}{sC}} = \frac{E}{s\left(R + \dfrac{1}{sC}\right)}$$

$$= \frac{E/R}{s + \dfrac{1}{RC}}$$

Substitute P for RC, so that

$$I(s) = \frac{E}{R} \cdot \frac{1}{s + 1/P} \tag{27}$$

Taking the inverse transform, we have

$$\mathcal{L}^{-1}[I(s)] = i(t) = \frac{E}{R}\mathcal{L}^{-1}\frac{1}{s + 1/P}$$

Then by inverse transform number 7 in the table,

$$i(t) = \frac{E}{R}e^{-t/P} = \frac{E}{R}e^{-t/RC} \tag{28}$$

which is the instantaneous value of the current for given values of E, R, and C, at time t.

(b) *Generalized resistance-capacitance* circuit. In investigating the behavior of control and servo systems, an effective approach is from the point of view of the ac electrical circuit. Thus, voltage or electromotive force in the electrical system has as its analogs force in a mechanical system, and pressure in an acoustical or hydraulic system. Current in an electrical system is analogous to velocity in a mechanical system or an acoustical system, and to rate of flow in a hydraulic system. Electrical resistance has as its analogs friction in a mechanical system or a hydraulic system, and the real component of acoustic impedance in an acoustical system. These analogies can be extended to capacitance, inductance, impedance, etc.

Because of these analogies between systems, the use of generalized concepts in control problems is particularly valuable. Let us take the first step in this direction by using again the simple resistance-capacitance circuit of the previous application, but denoting the input voltage by $\theta_i(t)$ to differentiate it from the voltage available to an external load by connecting the latter across the condenser, the latter being called the output, $\theta_0(t)$. See Fig. 2. Then the electrical equations for these two voltages are

$$\theta_i(t)Ri(t) + \frac{1}{C}\int i(t)\, dt \tag{29}$$

$$\theta_0(t) = \frac{1}{C}\int i(t)\, dt \tag{30}$$

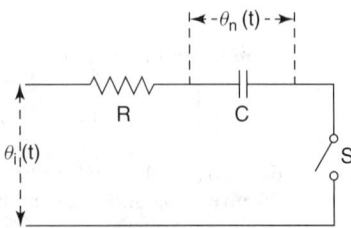

Fig. 2.   Resistance-capacitance (*RC*) ac circuit showing input and output connections.

Assuming, as in the previous application, that there is no initial charge on C (switch open), then the Laplace transforms of $\theta_i(t)$ and $\theta_0(t)$ are

$$\mathcal{L}[\theta_i(t)] = \theta_i(s) = RI(s) + \frac{1}{sC}I(s) \tag{31}$$

$$\mathcal{L}[\theta_0(t)] = \theta_0(s) = \frac{1}{sC}I(s) \tag{32}$$

We now generalize these terms; $\theta_i(t)$ is called the driving function of the system; $\theta_0(t)$ is called the response function of the system; $\theta_i(s)$ is called the driving transform of the system; and $\theta_0(s)$ is its response transform. The ratio of the response transform to the driving transform, $\theta_0(s)/\theta_i(s)$ is called the transfer function, and is a characteristic of the system independent of its particular input or output: it is often denoted by $G(s)$.

Then we have, from the foregoing equations,

$$G(s) = \frac{\theta_0(s)}{\theta_i(s)} = \frac{\dfrac{1}{sC}I(s)}{RI(s) + \dfrac{1}{sC}I(s)} = \frac{1}{RCs + 1} \tag{33}$$

(c) *Unit step driving function.* The first Laplace transform evaluated in this section (Equation (2)) was that of the unit function $u(t)$ for the two values 1 and $t \geq 0$, and 0 for $t < 0$. Its Laplace transform was found to be $1/s$.

Therefore, if the driving function, $\theta_i(t)$, of a system is of the unit kind, as shown in Fig. 3, its driving transform is

$$\theta_i(s) = 1/s \tag{34}$$

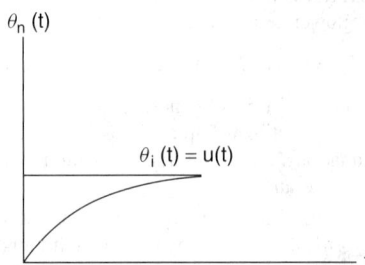

Fig. 3.   Response of *RC* circuit to unit input.

and its response transform is, from Equation (33),

$$\theta_0(s) = G(s)\theta_i(s) = \frac{1}{RCs + 1} \cdot \frac{1}{s} + \frac{1}{s(RCs + 1)} \tag{35}$$

Then applying the partial fraction method expressed in Equations (22) and (23), we have

$$\theta_0(s) = \frac{k_1}{s} + \frac{k_2}{RCs + 1} = \frac{1}{s} + \frac{-RC}{RCs + 1} \tag{36}$$

Therefore,

$$\theta_0(s) = \frac{1}{s} - \frac{1}{s - 1/RC}$$

Then using transform paris numbers 1 and 7 in the table,

$$\mathcal{L}^{-1}[\theta_0(s)] = \theta_0(t) = 1 - e^{-t/RC} \tag{37}$$

Figure 3 shows how the value of $\theta_0(t)$ approaches the unit value asymptotically as t increases, as is evident from the equation.

(d) *Unit ramp driving function.* The second Laplace transform evaluated in this treatment, Equation (3), was that of the unit ramp function $f(t)$ for values of $t \geq 0$. Its Laplace transform was found to be $1/s^2$.

Therefore, if a driving function $\theta_i(t)$ is of the Heaviside unit ramp kind, as shown in Fig. 4, its driving transform is

$$\theta_i(s) = 1/s^2$$

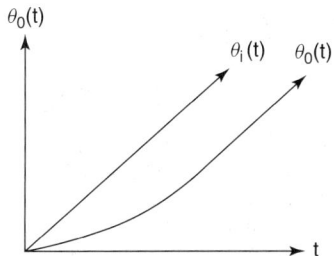

Fig. 4.   Response of *RC* circuit to Heaviside unit ramp input.

and its response transform is, from Equation (35),

$$\theta_0(s) = \frac{1}{s^2(RCs + 1)} \tag{38}$$

Then applying the partial fraction method expressed in Equations (22) and (23), we have

$$\theta_0(s) = \frac{k_1}{s^2} + \frac{k_2}{s} + \frac{k_3}{RC + 1}$$

$$= \frac{1}{s^2} - \frac{RC}{s} + \frac{(RC)^2}{RCs + 1}$$

$$= \frac{1}{s^2} - \frac{RC}{s} + \frac{RC}{s + 1/RC}$$

Then by use of transform pairs numbers 1, 2, and 7 in the table, we have so

$$\mathcal{L}^{-1}\theta_0(s) = \mathcal{L}^{-1}\left[\frac{1}{s^2} - \frac{RC}{s} + \frac{RC}{s + 1/RC}\right]$$

$$\mathcal{L}^{-1}\theta_0(s) = \theta_0(t) = t - RC(1 - e^{-t/RC}) \tag{39}$$

Figure 4 shows how the value of $\theta_0(t)$ lags the unit ramp function as t increases. See also **Time Constant**.

LAPLACIAN.   The vector operator

$$div \ \text{grad} = \nabla \cdot \nabla = \nabla^2$$

In rectangular coordinates its components are $\partial^2/\partial x^2$, $\partial^2/\partial y^2$, $\partial^2/\partial z^2$; in spherical polar coordinates

$$\frac{1}{r^2}\frac{\partial}{\partial r}\left(r^2\frac{\partial}{\partial r}\right), \ \frac{1}{r^2 \sin\theta}\frac{\partial}{\partial\theta}\left(\sin\theta\frac{\partial}{\partial\theta}\right), \ \frac{1}{r^2\sin^2\theta}\frac{\partial^2}{\partial\phi^2}$$

and in cylindrical coordinates

$$\frac{1}{\rho}\frac{\partial}{\partial\rho}\left(\rho\frac{\partial}{\partial\rho}\right), \ \frac{1}{\rho^2}\frac{\partial^2}{\partial\phi^2}, \ \frac{\partial^2}{\partial z^2}$$

See also **Del**; **Divergence (Mathematics)**; and **Gradient (Mathematics)**.

LAPPING.   The process of producing an extremely accurate, highly finished surface by means of a lap, which is a block charged with abrasive. Lapping reduces the possibility of wear on close-fitting running parts or on the surfaces of measuring equipment, by reducing the minute ridges and serrations left by machining and grinding operations to a more uniform bearing surface. Lapping may be done by hand or by machine. If a part is to be hand lapped to a final accurate dimension, a lap or mating form is made from a metal somewhat softer than the part to be finished. The surface of the lap is charged with a fine abrasive, or a small amount of abrasive mixed with grease, oil, or alcohol. A flat lap has a carefully trued surface with a series of grooves in it. The lapping compound is smeared on the face of the lap and the work is rubbed over the face along an ever-changing path. The grooves in the face of the lap act as channels for any excess abrasive and oil. Very little pressure is used in order to eliminate the danger of scoring the work or stripping the lap. Hand lapping requires skill and time. The amount of material removed by lapping should not exceed 0.0002 to 0.0005 inch (0.005 to 0.013 millimeter).

LAPSE.   See **Atmosphere (Earth)**.

LAPSE RATE.   The decrease of an atmospheric variable with height, the variable being temperature, unless otherwise specified.

LAPWING.   See **Waders, Shorebirds, and Gulls**.

LARCH TREES.   Members of the family *Pinaceae* (pine family), these trees are of the genus *Larix*. The larches are different among the conifers in that they are deciduous, turning a beautiful yellow color in the fall. They prefer a lot of sunlight and do well in most soils except dry, yellow, chalky soils. The principal species, in addition to those listed on accompanying table, include:

Dahurian larch	*Larix gmelini*
Dunkeld hybrid larch	*L. eurolepsis*
Japanese larch	*L. kaempferi*
Red larch	See Japanese larch
Siberian larch	See Dahurian larch
Siberian larch (red)	*L. olgensis*
Sikkim larch	*L. griffithiana*

The term *tamarack* frequently refers to the eastern larch, but is also used in connection with certain other larches, including the western larch.

The eastern larch is found through Alaska, Canada, south to Minnesota and through the northern parts of the midwestern United States, and south and eastward through West Virginia, Pennsylvania, and New York. In Canada, the tree follows the Mackenzie River northward into Yukon-Northwest Territory. The tree bark is thin, bright red-brown. The wood weighs about 38 pounds per cubic foot (609 kilograms per cubic meter). It is strong, durable, hard, and close-grained. In the spring, the flowers are of bright yellow. The needles are bright green from 1/2 to 1 inch (1.3 to 2.5 centimeters) long. The cone is winged, chestnut brown, and from 1/2 to 3/4 inch (1.3 to 1.9 centimeters) long. The eastern larch or tamarack of the far north does not grow so large as those found in more southern climes, but it grows very straight and is ideal for telephone poles and railroad ties. However, it is not as extensively used for timber as the western larch. Other names sometimes used for the tree are American larch, black larch, and eastern hackmatack. See Table 1.

The western larch is found principally from British Columbia to Oregon, Montana, and into Idaho. It grows at elevations up to about 7000 feet (2100 meters). The bark is thick (3 to 6 inches) (7.6 to 15.2 centimeters) and is reddish-brown with deep wide ridges. The twig is stout and brittle. The leaf

TABLE 1. RECORD LARCH TREES IN THE UNITED STATES[1]

Specimen	Circumference[2]		Height		Spread		Location
	Inches	Centimeters	Feet	Meters	Feet	Meters	
European larch (1996) (*Larix decidua*)	183	465	92	28	72	21.9	Vermont
Subalpine larch (1993) (*Larix lyallii*)	236	599	94	28.7	56	17.1	Washington
Western larch (1993) (*Larix accidentalis*)	230	584	189	57.6	35	10.7	Washington
Western larch (1995) (*Larix accidentalis*)	264	671	153	46.6	34	10.4	Montana

[1]From the "National Register of Big Trees," American Forests (by permission).
[2]At 4.5 feet (1.4 meters).

is pale green, 1e1/2 to 2 inches (3.8 to 5 centimeters) in length and grows in clusters of 20 to 30. The wood is heavy, durable, strong, and close-grained. The heart wood is red-brown in color; the sap wood is thin and nearly white. In the green condition, the moisture content of western larch is 58%, with a weight of 48 pounds per cubic foot (769 kilograms per cubic meter). When air-dried to 12% moisture content, the weight is 36 pounds per cubic foot (577 kilograms per cubic meter), and 1,000 board-feet (2.36 cubic meters) weigh about 3,000 pounds (1360 kilograms). The compressive or crushing strength parallel to the grain of the green wood is 3,990 pounds per square inch (27.5 MPa); of the air-dried wood, 8,110 pounds per square inch (56 MPa). The tensile strength perpendicular to the green of green wood is 330 pounds per square inch (2.3 MPa); of the air-dried wood, 430 pounds per square inch (3 MPa). The wood is valuable for many uses, including interior finished products, furniture, and cabinetry, but its major value lies in its use for heavy-duty construction, rough timbers for mining, railroad ties, and telephone poles. The tree usually grows very straight. In some areas, the tree has been extensively cut and is now under government protection. The tree frequently is planted for ornamental purposes where the size is appropriate. Douglas fir timber frequently is shipped along with the western larch timbers. Other commercial names for the wood include mountain larch, Montana larch, and hackmatack.

As noted from the list of species, larches also occur widely in parts of Europe and Asia. The European larch is an important source of wood in Russia, and other eastern European countries.

See also **Conifers**.

LARGE ION. An atmospheric ion of relatively large mass and low mobility which is produced by the attachment of a small ion to an Aitken nucleus. Also called *slow ion, heavy ion*, or *Langevin ion*. The ion density of large ions varies widely, depending upon the degree of atmospheric pollution. Representative low-altitude values might be 1000 per cubic centimeter in clean country air, 10,000 per cubic centimeter in an industrial area, and 100 per cubic centimeter over the oceans.

LARGEMOUTH BLACK BASS.   See **Sunfishes**.

LARGE NUMBERS (Law of).   See **Law of Large Numbers**.

LARK (*Aves, Passeriformes*).   Song birds of many species, confined to the northern hemisphere. The skylarks of Europe and Asia are the most famous members of the group because of the quality of their song. The common European species, *Alauda arvensis*, is established in Oregon and North America also has a native species, the horned lark, *Otocoris alpestris*. The meadowlark is more closely related to the blackbirds and orioles. See also **Meadowlark**.

LARMOR'S THEOREM.   This theorem concerns the motion which a charged particle or system of charged particles subject to a central force directed toward a common point experiences when under the influence of a small uniform magnetic field. If a coordinate system is chosen which rotates about the direction of the magnetic field with an angular velocity,

$$\omega = -\frac{q}{2mc}H$$

where ω is the angular velocity, q is electric charge in electrostatic units, m is mass, c is the velocity of light, H is magnetic field strength in electromagnetic units, then the motion in this system of coordinates is the same as the motion referred to a coordinate system fixed in space without a magnetic field. Application of this principle arises almost exclusively in the description of the motion of atoms and electrons in magnetic fields.

LARVA.   An immature form of animals, that undergoes metamorphosis between emergence from the egg and the attainment of adult life. The larva is often very different from the adult.

The larvae of many invertebrates of sessile habit, such as the sponges and some coelenterates, are ciliated (cilia) organisms, which swim about for a time before attaching themselves to the permanent support where they are to develop. In the two phyla mentioned the larva is little more than a ciliated gastrula. It is filled with solid endoderm in the coelenterates and is called a planula.

The flukes also begin life as a ciliated larva known as a miracidium. This form gives rise to more complex larvae called rediae and these in turn produce tailed larvae called cercariae. The cercaria is transformed into the

adult. In the same phylum the tapeworms hatch as 6-spined hexacanth larvae and pass through a bladder-worm stage before becoming adults.

Roundworms of many species pass through one or more larval stages in which they are worm-like but differ in habits and in some structural details from the adults.

Some of the segmented worms, mollusks, echinoderms, and chordates, hatch as complex larvae with localized zones of cilia for locomotion. The trochophore or trochosphere larva of annelids and mollusks has a ciliated alimentary tract with mouth and anus 90 degrees apart and a belt of cilia around the middle of the body. Larvae of echinoderms also have a bent alimentary tract but the cilia are arranged in one or more bands, sometimes of intricate form. These larvae bear various names: bipinnaria, auricularia, pluteus, of the starfishes, sea cucumbers and sea urchins and brittle stars, respectively, or collectively the dipleurula. The bipinnaria resembles the tornaria larva of Balanoglossus.

Among the arthropods the high development of metamorphosis is accompanied by great diversity of larval forms. In the class Crustacea these forms seem to represent previous evolutionary stages and are named after groups of the class whose adults they resemble. Among them are the nauplius, cypris, and cyclops larvae and many others. A single individual may pass through several of these stages in the course of its development. Insects present an entirely different type of metamorphosis in which the larval characteristics appear to have been acquired later in the course of evolution than those of the adult, as an adaptation to special conditions (Cenogenesis). In species with complete metamorphosis, only the first immature stage is called the larva. In this stage, butterflies and moths are called caterpillars; some of the beetles, grubs; and many flies, maggots. Species with less complex metamorphosis are called nymphs or naiads during development (in gradual and incomplete metamorphosis respectively).

See also **Metamorphosis**; and **Pupa**.

LARYNX.   Sometimes called voice box, the larynx is a cartilaginous organ of the throat that contains the vocal cords and produces most of the sound in phonation. The larynx is shaped fundamentally like a tube, wide at the top and narrow at its lower portion. Generally, it appears somewhat like a three-cornered tube with a prominent ridge on the front side. It is made up of several pieces of firm elastic tissue (cartilage), which are held together by muscles and ligaments. The thyroid cartilage is the largest cartilage of the larynx and consists of two plates standing on end, which meet in the front of the neck and form a ridge (Adam's apple).

The interior of the larynx extends from the pharynx above to the windpipe (trachea) below. The inner tube of the larynx is divided horizontally into two parts by the projection of the muscular *vocal folds*, which contain the two *vocal cords*. The cords are really folds in the lining of the larynx rather than true cords. These cords produce the sound that is converted into speech by the movements of the mouth and tongue. When the vocal cords are tightened, the air being exhaled causes the cords to vibrate, and sounds are produced. The tighter the cords, the higher the tone. These sounds are made into words by the tongue, teeth, and lips. The degree to which a person can tighten and relax his vocal cords determines his tone range in singing and speaking. When the vocal cords are fully relaxed, no sound is made.

Laryngitis is an inflammation of the mucous membrane of the voice box. Common microbial flora found in the larynx include *Haemophilus influenzae*, influenza virus, parainfluenza virus, adenovirus, coxsackievirus, rhinovirus, and respiratory syncytial virus. Uncommonly present in the larynx are pneumococci, Group A streptococci, *Mycobacterium tuberculosis*, and *Mycoplasma pneumoniae*. Viral laryngitis in children is commonly called *croup*.

Cancer of the larynx makes up only about 2% of all human cancers. It occurs most often in men between ages 40 and 60 years, although all age groups and both sexes may be affected. The main symptom is hoarseness. Any voice change that persists more than two weeks should be investigated by a physician. There may be a tickling sensation in the throat, or simply a discomfort of the throat. There may be difficulty in swallowing and pain on speaking. Other ensuing developments may be a cough, shortness of breath, wheezing, and halitosis. Lymph nodes in the neck may become enlarged. Occasionally the patient may expectorate blood.

In the United States, new cases of laryngeal cancer number about 12,000 per year, of which 3,700 are ultimately fatal. Epidemiologic studies from many parts of the world have demonstrated an increase in the relative risk of laryngeal cancer in smokers as compared with nonsmokers, ranging from 2.0 to 27.5, with a strong dose-response relation. Laryngeal cancer, in accordance with recent statistics on smoking, may be increasing among women and somewhat decreasing among men. As with lung cancer, the time necessary after smoking cessation for the risk levels for laryngeal cancer to approach those in nonsmokers is about 10 to 15 years.

Radiation and/or surgical therapy may be used in connection with laryngeal cancers. Surgery is usually indicated if the disease involves a large portion of the larynx. When the larynx is removed, the patient's windpipe is attached to the skin of the neck. Thus, the patient no longer breathes through the nose, but rather through a hole in the neck. The patient must take precaution to avoid breathing in foreign matter because of the lack of protection normally afforded by the nose. As soon as healing permits, the patient will normally commence speech lessons to learn to speak by application of one of two natural procedures, the esophageal method or the pharyngeal method.

Where sustained effort to learn either of these methods does not lead to success, the patient may use an artificial larynx, a device that is held against the side of the throat. When activated, the device transmits vibrations to the pharynx and mouth, providing intelligible but unnatural sounds.

Alternate Treatment. As pointed out in a 1991 study by the Department of Veterans Affairs, a Laryngeal Cancer Study Group reports, "Because laryngectomy results in substantial functional morbidity, including the loss of the natural voice, alterations in deglutition (swallowing), and the creation of a permanent tracheostoma (hole) in the neck, alternative forms of treatment have been developed. In selected patients with moderately advanced cancers, partial laryngeal resections that spare vocal function or primary radiation therapy achieve survival rates comparable to those obtained with total laryngectomy, permitting preservation of the larynx in 40 to 70 percent of patients. For patients with more advanced disease, however, treatment by radiation alone, with salvage by surgery, if necessary, results in lower survival rates."

The Veterans Affairs Group made a study of 352 patients, who were randomly assigned to receive either three cycles of chemotherapy (cisplatin and fluorouracil) and radiation therapy or surgery and radiation therapy. After two cycles of chemotherapy, the clinical tumor response was complete in 31% of patients and partial in 54% of patients. The report concludes: "These preliminary results suggest a new role for chemotherapy in patients with advanced laryngeal cancer and indicate that a treatment strategy involving induction chemotherapy and definitive radiation therapy can be effective in preserving the larynx in a high percentage of patients, without compromising overall survival."

Injuries. Freak, uncommon accidents can affect the functioning of the larynx and related organs. In 1991, T.C. James, T.C. Li (Mayo Clinic), and D. Gunderson (Park Nicollet Medical Center, Minneapolis) report that a young man suffering from asthma awoke in the middle of the night and reached for a metered-dose inhaler that he had been using. Unfortunately, still groggy from sleep, the man did not remove the cap of the inhaler and aspirated it into the larynx. He suffered some vocal cord damage, but recovered within a period of about 2 months. In another instance, a middle-aged man had acute respiratory distress after using an inhaler. A laryngoscopy revealed that he had inhaled a loose coin that he kept in his pocket along with the inhaler. Radiology revealed a coin in the right mainstem bronchus. This was removed, and the patient recovered without sequelae. This latter case was reported by Carol Shultz (Henry Ford Hospital) and associates in 1991.

Obviously, persons who are given inhalers should be instructed to check the mouthpiece before each use and to cap the device after each use.

Myasthenic laryngitis signifies a weakness and exhaustion of the muscles in the larynx resulting from overuse of the voice box. Resting the voice part of each day is important, as well as refraining from shouting, talking in a loud voice, and all unnecessary uses of the voice. Usually the voice returns to normal when these precautions are observed.

Additional Reading

Bailey, B.J. and H.F. Biller: "Surgery of the Larynx," W.B. Saunders Company, Philadelphia, PA, 1998.
Ferlito, A.: "Diseases of the Larynx," Arnold Publishing, London, UK, 2000.
Ferlito, A. and W. Arnold: "Cancer of the Larynx: Current Concepts in the Treatment of the Neck," S. Karger Publishers, Inc., Farmington, CT, 2000.
Fried, M.P. and R. Hurley: "The Larynx: A Multidisciplinary Approach," Mosby-Year Book, Inc., St. Louis, MO, 1995.
Hawkshaw, M., R.T. Sataloff, and J.R. Speigel: "Atlas of Laryngology," Singular Publishing Group, Inc., San Diego, CA, 1999.
Hirano, M. and K. Sato: "Histological Color Atlas of the Human Larynx," Singular Publishing Group, Inc., San Diego, CA, 1993.
Kavuru, M.S. and A.C. Merta: "Treatment of Recurrent Respiratory Papillomatosis (Juvenile largyngotracheobronchial papillomatosis)," N. Eng. J. Med., 204 (January 16, 1992).
Lang, J.: "Clinical Anatomy of the Oral Cavity, Pharynx, and Larynx," Thieme Medical Publishers, Inc., New York, NY, 1998.
Li, T.C., T.C. James, and D. Gunderson: 'Inhalation of the Cap of a Metered-Dose Inhaler," 'N. Eng. J. Med., 431 (August 8, 1991).
Schultz, C.H., S.W. Hargarten, and J. Babbitt: "Inhalation of a Coin and a Capsule from Metered-Dose Inhalers," N. Eng. J. Med., 431 (August 8, 1991).
Veterans Affairs Laryngeal Cancer Study Group: "Induction Chemotherapy Plus Radiation Compared with Surgery Plus Radiation in Patients with Advanced Laryngeal Cancer," N. Eng. J. Med., 1685 (June 13, 1991).
Stafford, N.: "Cancer of the Pharynx and Larynx," Blackwell Science, Inc., Malden, MA, 1999.
Tucker, H.M.: "The Larynx," Thieme Medical Publishers, Inc., New York, NY, 1992.

LASER. An acronym for *light amplification by stimulated emission of radiation*. The device is identical in theory of operation to the maser except that it operates at frequencies in the optical region of the electromagnetic spectrum, rather than in the microwave. See also **Maser**. By common usage, these devices are all called lasers, although more precise terminology would aptly use such terms as ultraviolent maser, optical maser, infrared maser, etc. Although the original microwave maser offers an extremely stable frequency source, its main use has been as an amplifier with very low noise output. In contrast, one of the principal attributes of the laser is its ability to produce a single frequency at high intensity in the optical region. Not only may the output be a single monochromative wave, but the wave may be coherent, or in phase, over the whole face of the radiator. In this mode of operation, the laser is an oscillator whose output depends upon the selective amplification of one of the single-frequency modes of the resonant cavity containing the active laser medium.

The decade of the 1980s represented the period when laser technology received profound acceptance, to the point where lasers, like semiconductors and digital computers before them, are now considered important components (hardware) of larger systems. The development of laser technology continued apace during the early 1990s, with numerous new and practical applications occurring at an accelerated rate. lasers now range from microlasers that measure but a few millionths of a meter, for use in optical communications and information processing, to the building-sized x-ray research laser that can deliver 100,000 joules of energy in less than one-billionth of a second — that is, a 10^{14}-watt pulse. See Fig. 1. One other feat, accomplished in the mid 90's, is a veritable tour de force. A group at MIT managed to coax single excited barium atoms crossing a miniature

Fig. 1.   A very powerful optical laser installed at the Lawrence Livermore National Laboratory. Known as NOVA, two of its ten beams serve as the energy source for x-ray lasers. The latter need about 1000 times the pump energy of optical lasers, delivered about 10,000 times faster. Note size of person in the foreground of this sketch.

resonator to serve as a laser. The crucial part is the extreme precision of the mirror arrangement that keeps the photons emitted by the atoms resonating. So far the main goal of this work has been pure physics: to test the theory of quantum electrodynamics (QED).

Historical Perspective

The fundamental theoretical concepts of the maser and laser date back many years, to the early workers in quantum mechanics who appreciated that an incident electromagnetic beam of an appropriate resonant frequency, passing through a medium, might stimulate molecules in an upper quantum energy state to return to a lower quantum energy state, and thus reinforce the primary beam by negative absorption. As early as 1940, Fabrikant (Russia) suggested that experiments be made to prove negative absorption. Fabrikant was the first scientist to introduce the term "collisions of the second kind" (later to prove of importance in laser patent litigation). By this he meant a collision in which some of the kinetic energy of motion of colliding particles is converted to internal energy (or a change in energy state of at least one of the colliding particles).

It is also interesting to note that, as early as 1950, researchers Lamb and Retherford (Columbia University), observed that, if an upper quantum energy state could be caused to be more highly populated (as compared with a lower quantum energy state), the result would be a net induced emission on an incident beam. They further indicated that such a population inversion would probably occur between the $2p$ and $2s$ levels in hydrogen. There shortly followed experiments by Purcell and Pound (Harvard University) who used magnetic techniques to invert the population of a pair of nuclear spin states in lithium fluoride and thus were the first researchers known to directly observe negative absorption of an applied pulse, a phenomenon which they called negative temperature.

Townes (Columbia University) is generally credited with first recognizing that stimulated emission could be utilized in the making of practical hardware. In 1951, Townes described an approach wherein an ammonia beam would be divided into two portions along the lines of experimentation carried out in Germany, wherein a quadrupolar focusing technique was used for separating a beam of molecules into two portions, one of which contained molecules predominantly in the upper of two energy states. Townes proposed to pass the high-energy portion through a cavity resonant at the frequency corresponding to the energy separation of the two states and so reported this proposal. Instead of a millimeter wave generator, a microwave oscillator was used. The latter was given the name maser. This led to the award of U.S. Patent No. 2,879,439, which covered the use of stimulated emission for the amplification and/ or generation of oscillatory electromagnetic energy. This patent subsequently was licensed to laser manufacturers.

Following the development of the microwave maser, Schawlow (AT&T Bell Laboratories) and Townes in 1958 proposed that optical maser action could be obtained by placing an active medium in an optical cavity. The medium would be a gas or a solid excited electrically or by light in such a manner that any optical wave present would be amplified as it moved through the material. This work led to the ultimate awarding of the basic laser patent in March 1960 (U.S. Patent No. 2,929,922). During the 1960s, laser research was carried on at a rapid pace at AT&T Bell Laboratories, among others. Considering the work at AT&T Bell Laboratories, in 1960 a laser capable of emitting a continuous beam of coherent light (using helium-neon gas) was developed; in 1961, the continuous-wave solid-state laser (neodymium-doped calcium tungstate) was developed. As a refinement of the helium-neon laser, in 1962 the basic visible light helium-neon laser was developed, of which several hundred thousand are in use as of the early 1980s. In 1964, the carbon dioxide laser (highest continuous-wave power output system known to date) was developed. Other developments during 1964 included the neodymium-doped yttrium aluminum garnet laser, the continuously operating argon ion laser, the tunable optical parametric Oscillator, and the synchronous mode-locking technique, a basic means for generating short and ultrashort pulses. In 1967, the continuous wave helium-cadmium laser (utilizing the Penning ionization effect for high efficiency) was developed. These lasers find use in high-speed graphics and biological and medical applications. In 1969, the magnetically tunable spin-flip Raman infrared laser, used in high-resolution spectroscopy, as well as in pollution detection in both the atmosphere and the stratosphere, was developed. Laser developments continued and, in 1970, semiconductor heterostructure lasers capable of continuous operation at room temperature were introduced. The distributed

feedback laser, a mirror-free laser structure compatible with integrated optics, was introduced, in 1971. These were followed by the tunable, continuous-wave color-center laser (1973), techniques for creating optical pulses of less than one-trillionth second duration (1974), and, in the late 1970s, long-life semiconductor lasers for lightwave communications.

One of the first uses of a laser in an astronomical setting was the placement of a laser retroreflector in the Sea of Tranquility on the moon during the landing of Apollo 11 on July 20, 1969. The retroreflector is passive, requiring no power. Two additional reflectors were mounted, to form a triangular arrangement, by Apollo 14 and 15 at later dates. The information continues to be checked periodically, when the atmosphere is clear by the McDonald Observatory in Texas, the Lure Observatory in Hawaii, and the Calhern Observatory in southern France. The reflectors are pulsed with 259-million joule neodymium-YAG (yttrium-aluminum-garnet) lasers. Because the retroreflectors were left on featureless land (not mountainous), telescope pointing is described as being akin to targeting a moving dime with a rifle 2 miles (3.7 km) away. The ranging experiments have contributed much to the knowledge of the moon's orbits and other geometric data. It has been noted that the moon recedes from Earth at a rate of about 1.5 inches (3.8 cm) per year. Minor variations in the moon's rotation also have been noted. These measurements indicate variations as small as 1/1000 of an arc-second over the course of a year. See also **Moon (Earth's)**.

Laser Principles

The basic concept of the laser may be described in general terms as follows. Optical maser action can be obtained by placing an active medium in an optical cavity. The medium may be a gas, solid, or liquid which can be excited electrically, by light, chemically, or thermally in such a manner that any optical wave present will be amplified as it moves through the material. One of the first cavities proposed, for example, is a Fabry-Perot resonator — two plane, parallel reflecting plates with a small transmission through which the radiation can escape. Upon excitation of the material, light is emitted with a band of frequencies determined by the particular material. In addition, the direction of emission is nominally random. In the presence of the cavity, some of the waves escape after several back-and-forth reflections from the parallel plates, "walking off" the edge of the reflectors, so to speak. Those waves that travel normal to the walls remain in the cavity and are amplified, provided that they reinforce each other after each round-trip reflection at the two surfaces. this reinforcement or resonance is only satisfied if the spacing of the plates is an integral multiple of one-half the wavelength in the medium. Thus, after a short time, only that frequency which satisfies the resonant condition and those waves traveling normal to the reflector will build up an appreciable intensity. The resultant light which is partially transmitted through one of the reflectors will thus be a single frequency, or several discrete frequencies if there is more than one cavity resonance within the band of frequencies emitted by the laser material. In addition, the wave front will be in phase across the surface of the reflector since waves striking the surface at normal incidence are amplified most strongly. The resultant beam will then be diffraction limited, i.e., the beam will spread by an angle in radians given approximately by the ratio of the wavelength to the diameter of the beam. In actual practice, single-mode operation is obtained only under special conditions. Several frequency modes may be present because of the multiple resonances of the cavity and numerous "off-axis" modes may be found that correspond to resonant waves, which travel at small angles from the normal to reflectors. These waves "walk off" so slowly that they still are amplified appreciably. Refinements of the simple cavity may consist of concave reflectors that decrease the diffraction losses, or several parallel reflectors, which limit the oscillation to a frequency common to each pair in the set, among other approaches.

Basic Requirements of Lasing Media. The key to successful laser operation is the active medium that amplifies the wave. Qualitatively, a material that fluoresces or exhibits luminescence is an obvious candidate. In fluorescence, electrons are excited to an upper-energy state by short-wavelength light, such as ultra-violet, while luminescence is produced by passing an electron current through the medium, such as in a gaseous discharge. In either process, stimulated emission can occur only if more electrons are produced in the upper-energy state than in the lower or terminal state for the radiating transition. In this case, an incident photon will stimulate further transitions and amplification will result. If the final state were more heavily populated, then the photon would cause more upward or absorbing transitions and the net effect would be absorption.

Laser source requirements vary widely among specific applications. Practical systems require a large range of wavelengths, output powers, spatial and temporal beam characteristics, among other features. In many applications, the following factors apply: (1) an optimum wavelength exists; (2) a specific minimum power level is required; (3) capital and operating costs of the laser must be minimized; (4) size and weight constraints must be met; (5) the laser should be capable of operating for extended periods with little maintenance; (6) the laser output should have a specific temporal and spatial characteristics; and (7) operator safety must be considered in design and use of the overall laser system.

Wavelength characteristics of available laser types are given in Table 1.

TABLE 1. WAVELENGTH RANGE OF SOME LASERS

Type of Laser	Wavelength (Micrometers)
Solid state	1.06
Ion	0.514, 0.488
Carbon dioxide	10.6
Diode	0.65–1.8
Dye	Visible (tunable)
Helium-neon	0.63
Rare gas halide	0.35, 0.25, 0.19
Helium-cadmium	0.422

Types of Lasers

From the above list of media requirements, it is obvious that a majority of materials are not candidates for making an effective laser. When the laser was first conceived, the possibility of finding numerous materials, as has turned out to be the case, initially was not envisioned by most researchers in the field.

Early Ruby Laser. It is recorded that the first optical laser was demonstrated by Maiman (Hughes Aircraft Research Laboratories) in 1960. Maiman used a ruby single crystal of aluminum oxide doped with chromium impurities. By applying semitransparent reflective coatings on the ends of a rod about 2 inches (5 centimeters) long, Maiman made the cavity and the crystal an integral unit. See Fig. 2. Exposure to an intense exciting light from a xenon flashtube was found to invert the population between the red-emitting level and the ground or lowest-energy state of the electrons. The result was a burst of intense red light emanating in a beam through the end reflectors. This was a powerful laser.

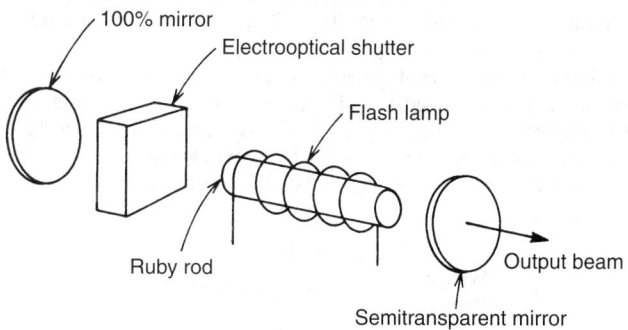

Fig. 2.   Schematic diagram of early giant pulsed ruby laser.

Because of off-axis modes and multiple resonances, the output is not a single frequency, single plane-wave mode, but generally consists of the order of 100 separate modes. The beam is still quite narrow, being of the order of 1 milliradian or 0.05 degree. As a comparison with conventional light sources, the energy radiated from 1 square centimeter of the brightest flash lamp is less than 10 kW and is distributed over the entire visible spectrum. In addition, the radiation is incoherent and is spread out uniformly in all angles from the source. Thus, the directivity and spectral purity of the laser source are many orders of magnitude superior to that of an incandescent source. The ruby laser suffers from low efficiency, about 1%, and except with elaborate cooling systems, only operates on a pulsed basis. The ruby laser was first reported in the literature (Maiman, 1960)

and, in 1967, Maiman was awarded U.S. Patent No. 3,353,115 for the optically pumped ruby laser.

Other crystalline or glass systems with impurity ions have been developed, which yield wavelengths from the ultraviolet to approximately 3 micrometers wavelength in the infrared. Some, such as neodymium-doped yttrium aluminum garnet (YAG), operate in a continuous mode at the one-watt level, while peak powers have reached values as high as 10^{14} W, or greater, in pulses of the order of 10^{-12} second. These ultrahigh powers are obtained in neodymium-doped glass systems, using several stages of amplification and novel pulse-forming techniques.

Gas Lasers. In 1961, Javian, Bennet, and Harriott demonstrated laser action in a gaseous discharge of helium and neon. The parallel-plate reflector cavity was used, but with much greater spacing. Later, concave mirrors were used to decrease the loss of energy out the sides of the cavity. See Fig. 3. The gas laser operated continuously and delivered power up to about one watt. Pulsing the gas discharge yielded peak power as high as 100 watts. The first laser radiated at 1.15 micrometers in the infrared, while further development with different gases yielded output from the ultraviolet to 330 micrometers or 0.33 millimeters in the far infrared. In contrast to the ruby laser, the gaseous laser beam may be diffraction limited and the frequency is pure, i.e., oscillation may be limited to one mode. By careful design, the frequency may be stabilized to within a few thousand cycles per second, or approximately one part in 10^{13} or better. Although the original gas laser utilized electrical excitation of electronic transitions, later versions use vibrational transitions in molecules, such as carbon dioxide, and the excitation may be electrical, chemical, or thermal. The helium-neon laser is commonly used in supermarket bar-code scanners.

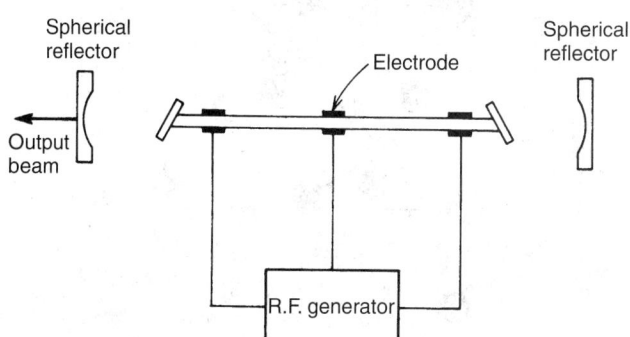

Fig. 3.   Schematic diagram of early helium-neon laser with external spherical reflectors. Curvature of reflectors is exaggerated.

In the chemical laser, atomic species such as hydrogen and fluorine can be reacted to produce molecules in an excited vibrational state, which, in turn, yields amplification or oscillation. Electrically excited lasers, particularly those using carbon dioxide at 10 micrometers, can operate at atmospheric pressure, using spark discharges or pre-ionization voltages in the 100-kV range. The high pressure and the powerful electrical excitation result in peak powers in the 10 to 100 MW region. For continuous laser operation, the gas may be circulated rapidly to avoid excessive heating.

In a gas dynamic laser, an appropriate fuel is burned to produce carbon dioxide and nitrogen at high temperature and pressure. When released through a nozzle into the optical resonator region, the gas cools rapidly in terms of its kinetic or translational energy, but the population of the vibrational energy levels of the carbon dioxide molecules becomes inverted since the lower level of the laser transition relaxes much more rapidly. In addition, the vibrationally excited nitrogen molecules are in near resonance with the upper laser state of the carbon dioxide and transfer energy with high efficiency to maintain the inversion. Lasers of this type are capable of producing continuous power at a relatively high level.

Semiconductor Laser. In the semiconductor laser, a solid (semiconductor) material is used. The electron current flowing across a junction between p- and n-type material produces extra electrons in the conduction band. These radiate upon making a transition back to the valence-band or lower-energy states. If the junction current is large enough, there will be more electrons near the edge of the conduction band than there are at the edge of the valence band and a population inversion may occur. To utilize this effect, the semiconductor crystal is polished with two parallel faces perpendicular to the junction plane. The amplified waves may then

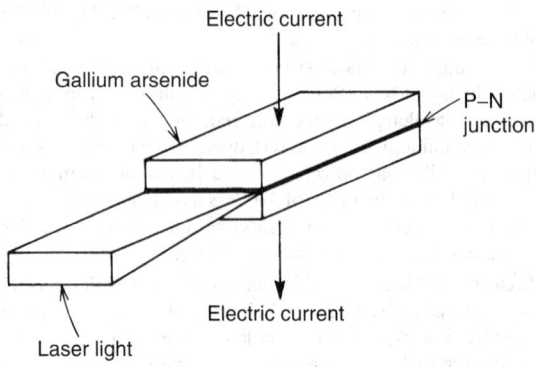

Fig. 4.   Schematic diagram of early gallium arsenide laser.

propagate along the plane of the junction and are reflected back and forth at the surfaces. See Fig. 4.

A major advantage of the gallium arsenide (GaAs) laser is that it has the electron distribution of a semiconductor. The main difference between electrons in semiconductors and electrons in other laser media is that in semiconductors all of the electrons occupy and thus share the entire crystal volume. Although all semiconductors possess this property, not all of them can be used as lasers. See Fig. 5.

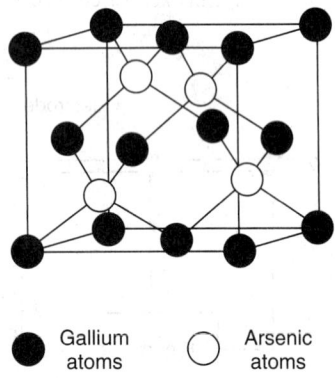

Fig. 5.   Schematic diagram of a gallium arsenide (GaAs) crystal, which is commonly referred to as a zinc-blende structure. The structure consists of a face-centered cubic lattice of gallium atoms with arsenic atoms positioned on the body diagonals. The arsenic atoms also lie on a face-centered cubic lattice displaced relative to the gallium lattice by one-fourth the body diagonal of the cube.

The gain in the material is high enough so that the reflection at the semiconductor-air interface is sufficient to produce oscillation without special reflective coatings. The first such device used gallium arsenide and radiated at 8400 angstroms, or just beyond the visible region in the

infrared. The efficiency of this laser, first demonstrated in 1962, was high (about 40%) and the power source was low-voltage direct current.

The compactness and efficiency of the semiconductor laser make it particularly attractive for systems use. Other substances used have included indium arsenide, indium phosphide, indium antimonide, and alloys such as gallium-arsenide-phosphide. These lasers may be tuned over several percent of their normal frequency of operation by varying the current flow through the device. The tuning results from the variations in temperature with current, which, in turn, changes the index of refraction and the resultant resonant frequency of the cavity.

Free Electron Laser. Lasers of this type depart markedly from conventional lasers. Free electron lasers employ an electron beam and a magnetic field. A "free" electron may be defined as an electron that is not bound into atoms or molecules. Traditional lasers use bound electrons. Thus, the conventional laser is limited to producing light (radiation) that is consistent with those frequencies that are specific to the vibration of a given atom or molecule. Carbon dioxide and helium lasers, for example, are so inhibited.

Free electrons are caused to vibrate by passing them through an alternating magnetic field. By changing (tuning) the apparatus, a broader range of frequencies is obtainable. Coherent radiation can range from the far-infrared to the far-ultraviolet regions of the spectrum. See Fig. 6.

Currently, free electron lasers are large and costly, but are ideal for certain kinds of research. Developers believe that ultimately the free electron laser will find numerous applications beyond research. Kim and Sessler suggest that applications may include surgery, fixing polymers, pharmaceutical manufacture, and lithography. It is predicted that the free electron laser will continue to expand in research usage, including condensed matter studies, nonlinear plasma studies, nonlinear quantum electrodynamics, nonlinear optics, and nonlinear microwaves, as well as microscopy, DNA studies, and cell response research in biology. One study is in progress to determine the feasibility of adapting the free electron laser to perform precision radar measurements in space. Researchers will study the potential of the laser as a compact space radar transmitter for discriminating objects in space, as would be required, for example, in connection with the Strategic Defense Initiative (SDI) program. The program will take advantage of the electron laser's inherent tunability, high power and efficiency, and ability to operate in frequency bands of 100 GHz and higher. The program's ultimate goal is a space-based, multiband, adaptive laser capable of operating efficiently at randomly chosen, stable frequencies.

Researchers have found that electrostatic accelerators are well suited for the far-infrared spectrum. An early device of this type was built at the University of California at Santa Barbara for the main objective of free-electron research and studies in solid-state physics and biophysics. The accelerator has an operating range of 2–6 MV, corresponding to wavelengths in the range from 100–800 micrometers. Pulse duration is from 3–30 microseconds.

Solid-State Laser Development. Over the years, the progress of solid-state lasers has depended heavily on the improvement of old and the establishment of new pump sources. The helical lamp used by the early ruby laser was replaced by the linear flash lamp and arc discharge lamps. The next step was that of using a diode laser to pump another solid-state

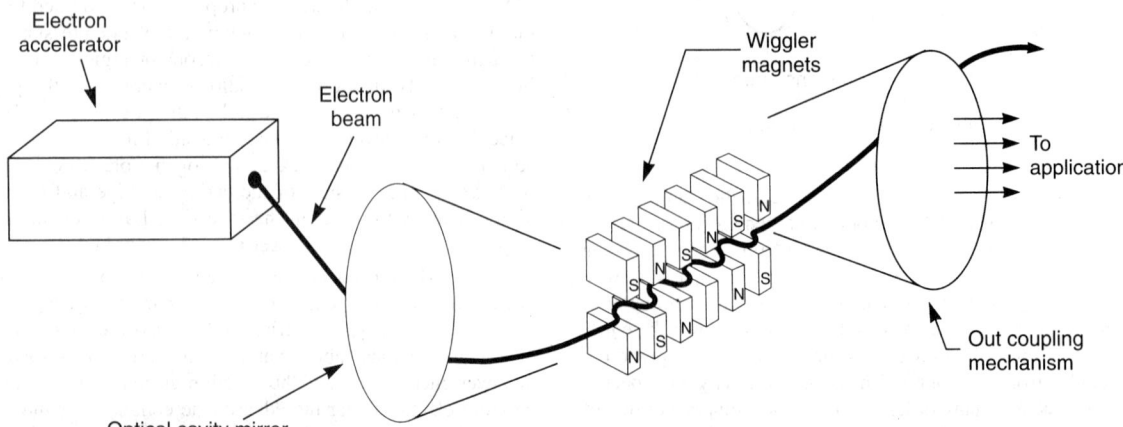

Fig. 6.   Schematic diagram of a free-electron laser. Beam of accelerated electrons passes through a field of alternating magnetism (wiggler magnets). The coherent light is generated and contained in an optical cavity defined by mirrors. (*Kim and Sessler.*)

laser. This latter approach is advantageous because the diode laser emits optical radiation into a narrow spectral band. If the emission of the wavelength of the diode laser lies within the absorption band of the ion-doped solid-state laser medium, diode laser optical pumping can be very efficient and accompanied by little excess heat generation. By contrast, flash lamp pumping is limited by the broad emission spectrum and by excess heat production.

As pointed out by R.L. Byer (Stanford University), "The diode laser is essentially a continuous wave device with low energy storage capability, whereas the solid-state laser can store energy in the long-lived metastable ion levels." Stored energy can be extracted by Q-switching (rapid switching) to provide peak power levels that are orders of magnitude greater than obtainable from the diode laser per se. Important, too, is the fact that a solid-state laser can collect output from several diode lasers and thus furnish greater average power than is obtainable from a single diode laser. Furthermore, the line width of the diode laser-pumped solid-state laser is many times less than that of the diode laser source. Finally, the solid-state laser source emits optical radiation in a diffraction-limited spiral beam that is easily focused into a fiber or small space.

As early as 1982, a diode laser-pumped miniature Nd:YAG laser with a linewidth of less than 10 kHz was demonstrated. The research in this area continued apace at Stanford University and by a number of commercial electronics firms, with emphasis placed on the development of three-level lasers, Q-switched and mode-locked operation, single-frequency operation (monolithic nonplanar ring oscillator), visible radiation by harmonic generation, and array-pumped solid-state lasers. See Fig. 7.

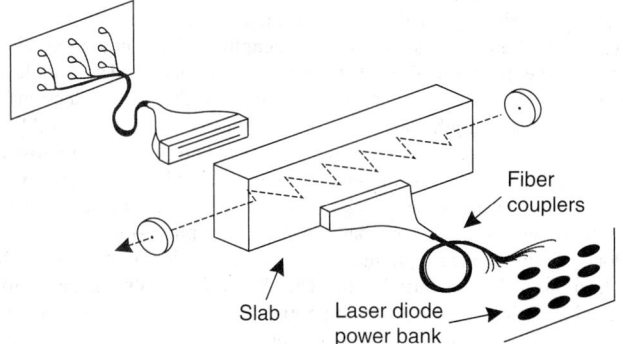

Fig. 7. Diagram of high-average power slab laser oscillator pumped by an array of diode lasers. Such an arrangement offers lower cost, ease of power scaling, and long-term reliability. (*After Byer.*)

A new class of lasers, so called vibronic solid state lasers, can emit — in contrast to other solid state lasers — a comparatively broad range of wavelengths. The lower level in these lasers is a band of energy levels that is caused by interaction between the electron motion and lattice vibrations. With the help of the usual tools (filters, etalons etc.) a narrow and tunable frequency bandwidth is selected for the laser output. The most popular materials are Ti-sapphire, i.e., titanium doped Al_2O_3 with output range from 600 to 1180 nm and Alexandrite, i.e., chromium doped $BeAl_2O_4$ lasing between about 700 and 825 nm.

Laser Microminiaturization — Target Optical Computer. For many years, scientists have accepted the fact that optical computers would not become a reality until optical components of micro size and exceptional performance equivalent to the already existing electronic switches and circuits could be developed. Thus, the optical computer became a major driving force toward the development of optical components. The problem was extraordinarily complex because size reductions of several orders of magnitude were mandatory.

As early as 1989, researchers at AT&T Bell Laboratories fabricated more than a million micron-size lasers (microlasers) on a single semiconductor chip, about 7 mm wide by 8 mm long. Individual devices ranged in size from 1 to 5 microns. Thus, these devices were two orders of magnitude smaller than conventional diode lasers. Researchers feel that it may be feasible to manufacture such devices that measure only between $\frac{1}{2}$ and $\frac{1}{4}$ micron. Traditional devices that measure a few microns wide by several hundred microns long have been well established for use in compact-disk players and fiber-optic communications. Thus, the *much, much smaller*

devices coming out of research comprise a major step toward achieving the needs of optical computing. Much competitive research continues during the early 1990s because the market demand for microminiature lasers can be immense.

As pointed out by Jewell and associates (Bellcore), "The principles of operation underlying a diode laser are the same as those for any laser. Atoms in a part of the laser called the *amplifying medium* — typically a solid, liquid, or gas- are pumped, or energized, either electrically or with a source of electromagnetic radiation. When a light wave of a specific wavelength traveling through the amplifying medium encounters a pumped atom, it can induce the atom to release its energy in the form of a light wave at the same wavelength. The process is coherent, which is to say that the crests and troughs of the waves match up, and the intensity of the light increases. Mirrors on each end of the amplifying medium form a cavity, and they force the light to bounce back and forth many times through the medium, maximizing the increase in intensity.

The differences in construction of a microlaser from a conventional gas laser or conventional diode laser are shown in Fig. 8.

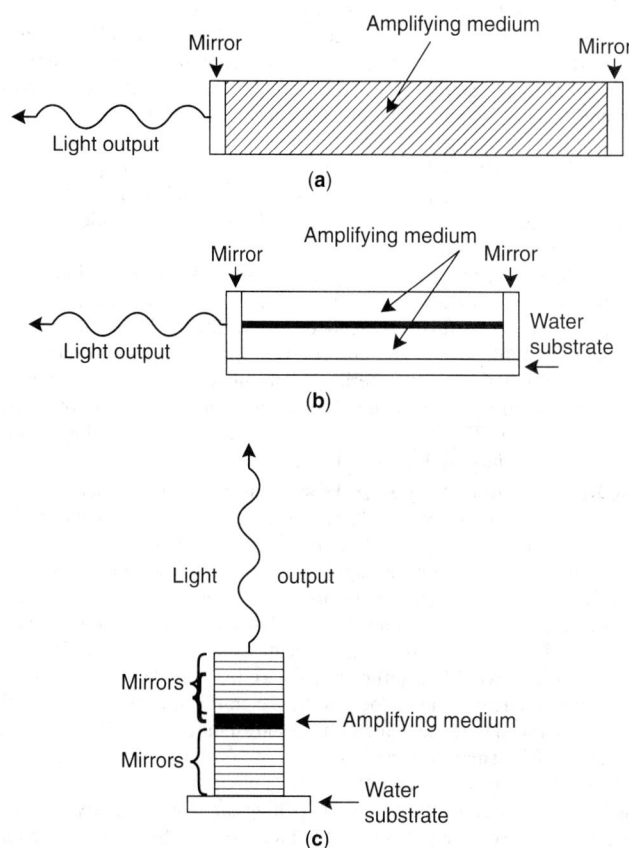

Fig. 8. Comparison of constructional configurations of (**a**) conventional helium-neon laser, (**b**) traditional diode laser, and (**c**) microlaser. Several orders of magnitude in size difference cannot be depicted here. The helium-neon laser ranges from 100 to 1,000 times longer than a traditional diode laser. The latter is some 100 times longer than a microlaser. (*After Jewell, Harbison, and Scherer.*)

To date, manufacture of the very tiny microlasers depend upon critical production techniques. For example, molecular beam epitaxy allows the basic material of each laser to be built up from layers of semiconducting materials. A typical microlaser may comprise 500 or more individual layers.

It was reported in late 1991 that a semiconductor laser that emits a blue-green light had been developed (3 M Company). Manufacturers of optical disks and other consumer electronic devices have been seeking a laser that would produce light in this range of the spectrum. The detailed development procedures required are beyond the scope of this article. However, it has been reported that the layered device is comprised of a gold electrode, a *p*-type zinc selenide, a quantum well (cadmium zinc selenide), an *n*-type zinc selenide, a substrate (gallium arsenide), and an indium electrode. By the end of the decade, progress in semiconductor research has led to reliable diode lasers with output in the blue and green part of the spectrum. In particular, SiC and GaN based systems promise to yield successful applications (See further reading by Fasol and by Fasol, Nakamura and Davies).

Liquid Lasers. Lasers operating in liquid media utilize rare earth ions in such organic hosts as chelates. Laser action is obtained in liquids using a flash tube or another laser as the pump. Early versions used rare earths in an organic liquid, while organic dyes, introduced somewhat later, were found to be more efficient, but required a separate laser for the exciting radiation. The dye laser has the special attraction that one laser may be tuned over a significant fraction of the visible spectrum by using a reflection grating as one of the cavity mirrors. One of the major areas of applications for dye lasers is the field of spectroscopy—for both basic research and applications such as combustion diagnostics or trace element analysis (See section on **Laser Spectroscopy**).

Another type of liquid laser utilizes a different principle and depends upon stimulated Raman scattering. Raman laser action was discovered by Woodbury in 1962, using a ruby laser and nitrobenzene. Here the laser excites the nitrobenzene, which in turn shows amplification at a frequency displaced from the ruby line by the vibrational frequency of the molecule. There is not true inverted population in this case. The incident photon is scattered by the molecule, which absorbs an amount of energy determined by its vibrational energy. The molecule is left in an excited state and the scattered photon is frequency-shifted by the energy loss. The process may be stimulated inasmuch as the rate at which the scattered photons are produced is proportional to the number of photons already present in the cavity at the scattering wavelength. As in the normal stimulated emission case, the frequency and phase of the output wave are identical with the wave that stimulates the scattering.

The Raman laser normally operates using the Stokes line, or the wavelength corresponding to the loss of one vibrational quantum. Other modes of operation utilize the second or third Stokes lines corresponding to double or triple vibrational absorptions. Similarly, higher-order effects in the medium may produce a series of anti-Stokes lines, which correspond to vibrational energy being added to the initial energy of the photons from the driving laser. The wavelength range of Raman lasers using different liquids is from the visible to the near-infrared.

X-Ray and Other Very High-Power Lasers. Almost since the inception of laser technology, the laser has been of interest to the military in connection with a variety of applications—weapons, radars, illuminators, rangers, etc. Among the highly sophisticated laser applications long envisioned by the military is the use of a laser to propel a spacecraft. In early concepts, a laser beam produced at ground level would vaporize an appropriate fuel, which could be water, and the supersonic jet caused by vaporization would be sufficient to place the vehicle into orbit without any chemical energy being expended by the craft itself. In the late nineties, experiments with tethered vehicles have shown initial albeit limited success in this quest. A small test craft without any on-board fuel was propelled 75 feet upward by 450 joule pulses from a powerful carbon dioxide (CO_2) laser. The small craft operates by expelling air that is rapidly heated by the absorbed laser energy. No longer motivated by SDI, the stated long-term goal is to loft small "picosatellites" into orbit (See further reading by Appell). Within the last few years, a major interest of the military has been directed toward the development of an x-ray laser for possible application in connection with the SDI (Strategic Defense Initiative) program. There are, of course, also strictly scientific interests in tunable coherent x-rays.

It is generally understood that for a laser to serve as a weapon it must have appropriate wavelength and brightness characteristics. The wavelengths produced by most lasers are absorbed by the atmosphere. Laser wavelengths between 0.3 and 1 micrometer are generally the most easily transmitted. Investigators have estimated that if a laser firing over a 3000 km engagement distance is to burn through a missile skin in 1 second, it must deliver 10,000 joules of energy per centimeter. (These requirements correspond to a brightness of 10^{21} watts per steradian, or unit of solid angle.) It is further estimated that a laser having the desired wavelength and brightness would require a beam power of approximately 100 MW. By comparison, a typical nuclear power plant has an output of about 1000 MW.

European Superlaser. In mid-1990, five European countries agreed to fund ($200 to $500 million) construction of a European High Performance Laser Facility. Sponsored by France, Germany, Italy, Spain, and the United Kingdom, the new laser would be three to four times more powerful than the Lawrence Livermore National Laboratory's NOVA, currently the world leader. The program was to progress in two principal stages, commencing with two intermediate power lasers—one neodymium glass and one KrF laser, built side by side. The major goal was to free European laser scientists from the dependence on high-powered machines in the United States and Japan. A superlaser of this type can be used to investigate some fundamental problems of physics. The intense pulse of the proposed laser would create conditions even hotter than the core of a burning star.

As of the late 1980s and early 1990s, four kinds of laser were under development. (1) *Chemical lasers* utilize chemical reactions between two gases to generate radiation. This technology is probably the most mature of the four kinds of lasers. It has been reported that the brightest of the chemical lasers is the MIRACL (mid-infrared advanced chemical laser), which in a demonstration at the White Sands Missile Test Range (New Mexico) destroyed a mock-up of a missile standing about a half-mile distant from the laser. The MIRACL was estimated to have a brightness of about 10^{17} watts per steradian, short of the SDI goal by a factor of about 10,000. (2) The *Excimer* (meaning excited dimer) consists of an unstable compound made up of two molecules. An electric discharge excites the molecules to form the dimer, and in breaking down, the dimer emits radiation. In some way, this radiation triggers a cascade of reactions that produce a laser beam. Radiation in the beam is estimated between 0.2 and 0.4 micrometer. A krypton-fluoride laser, tested at the Los Alamos National Laboratory, is estimated to operate at a wavelength of about 0.25 micrometer, delivering 10,000 joules of energy in a 380 nanosecond pulse. (One nanosecond = One billionth of a second.) It is reported that while the energy produced meets SDI goals, the pulse duration is off by a factor of about 3 million. (3) In the *free-electron laser*, a beam of electrons passes by a series of so-called wiggler magnets, which cause the electrons to vibrate and emit radiation. The wavelength can be tuned, theoretically, to any value between about 0.1 and 20 micrometers. The smaller the wavelength, the greater the energy. It is reported that a free-electron laser operated at wavelengths down to 10 micrometers in tests at Los Alamos. More recent research targets a 1-micrometer radiation of 100 microsecond pulses, containing 30 kW of power. As of the mid-1990s, both excimer and free-electron lasers were relatively poor converters of electric energy into beam energy, requiring massive power supplies—hence difficulties in locating the needed equipment in space. See prior description of free-electron laser in this article. (4) The *X-ray laser* also appears to be plagued with heavy and costly support equipment. Essentially, the x-ray laser consists of a nuclear explosive surrounded by an array (cylindrical configuration) of metal fibers. The emission of x-rays during the nuclear explosion stimulates the emission of a beam of x-rays from the fibers. This occurs within a microsecond prior to immolation of the device per se. For obvious reasons, further details remain sparse. It has been reported that to make an effective weapon, a particle beam would require energy of 250 MeV. If one assumes an acceleration gradient of 10 MeV per meter, it follows that the structure must be 25 meters long. When accounting is made of the mass of the power supply and its fuel, the weight of the weapon is found to be 50 to 100 tons. (Current typical payloads weigh a comparatively few tons.) Much additional work of a guarded nature continues.

Soft X-Ray Laser. A modern 1- to 2-billion-eV synchrotron radiation facility, based on high-brightness-electron beams and magnetic undulators, would generate coherent, laserlike soft x-rays of wavelengths as short as 10 angstroms. This radiation would be broadly tunable and subject to full polarization control. Radiation with these properties could be used for phase- and element-sensitive microprobing of biological assemblies and material interfaces as well as research on the production of electronic microstructures with features smaller than 1000 angstroms. These short-wavelength capabilities, which extend to the K-absorption edges of C, N, and O, are neither available nor projected for laboratory XUV (soft x-ray and ultraviolet radiation) lasers. Higher-energy storage rings (5 to 6 million eV) would generate significantly less coherent radiation and would be further compromised by additional x-ray thermal loading of optical components. Synchrotron radiation is discussed further in the article on **Particles (Subatomic)**.

To extend scientific and technological opportunities, authorities suggest that a bright source of tunable, partially coherent, XUV radiation is needed. Coherence, in the limited sense used here, refers to the ability to form interference patterns when wave fronts are separated and recombined. The availability of a tunable source of coherent soft x-rays, combined with other developments in x-ray optical techniques, would make it possible to construct an x-ray microprobe of sufficient intensity to permit fundamentally new, phase-sensitive experimentation in a number of scientific and technological fields. Various imaging and scattering techniques would be enhanced by the greatly increased photon flux available to study small samples, as well as providing the capability of tuning the radiation to the wavelength of interest. For example, with soft X-rays well matched to the absorption edges of elements, such as carbon (284), nitrogen (400), and oxygen (532), as well as other elements of relatively low atomic numbers (Na, P, S, K, and Ca), it should be possible to study elemental distributions and motion within biological specimens without the need for dehydration, fixing, or staining. Three-dimensional imaging, made possible by combining partially coherent undulator radiation and x-ray microholographic techniques, would complement the information available from electron microscopes.

Extensive development and experimental designs of soft x-ray lasers have been underway at the Lawrence Livermore National Laboratory and the Princeton Plasma Physics Laboratory, among other research institutions.

Researchers Suckewer and Skinner (Princeton University Plasma Physics Laboratory), in an early 1990 paper, observe, "Most of what is known about the internal structure of cells has been learned by the development and application of the techniques of electron microscopy. This knowledge rests on the premise that the intensive procedures necessary to prepare a specimen for electron microscopy do not significantly influence the structure, form, and high-resolution detail observed. Nonetheless, unanswered questions remain about the fidelity of the image of a cell that has been fixed, stained with heavy metals, and sectioned to the original living cell. X-ray microscopy offers a new way to look at unaltered cells in their natural state." X-ray laser microscopy can offer numerous advantages in this regard.

In summarizing the current status of soft x-ray laser research, the aforementioned researchers comment, "The general impact soft X-ray lasers will have in science and technology will depend on improvements in their performance and cost. It is necessary for their successful commercialization that these devices operate routinely at high gain-lengths ($GL > 4$), with the use of a low-cost driver laser, and this needs more system development and engineering. Most applications of visible-wavelength lasers are based on the fact that the brightness of these lasers is several orders of magnitude greater than that of conventional spontaneous emission sources, and this is achieved principally by the laser cavity mirrors. This technology is significantly more difficult in the x-ray region because of intrinsic limitations of x-ray absorption in materials and present limits in the soft x-ray laser pulse lengths. Nevertheless, a 'revolution' in x-ray optics is under way and the precedent of visible-wavelength lasers illustrates the potential benefits awaiting the creative inventor of applications of this technology to novel fields."

Laser Applications

During the early phases of laser development (late 1950s to early 1970s), there was a high tempo of research activity and confidence in the ultimate potential for practical applications. Some of the early suggestions for laser use included instrumental applications in metrology and spectroscopy, as well as working tools for industry, such as cutting, welding, and annealing. But during that period it was also observed by some researchers that the laser was an invention for nonexistent needs. The decade of the 1980s removed all such doubts, and as science entered the 1990s, the laser had become established as an essential component in numerous laboratory research programs and industrial and medical applications. Just a cursory inspection of the literature during the late 1980s and early 1990s is indicative of the wide scientific interest in the laser.

A number of representative uses are described here; many other uses are covered elsewhere in this encyclopedia. Check the alphabetical index. The use of lasers in fiber optic communications is described in the article on **Telephony**. Among medical applications for laser technology, photocoagulation procedures appear to predominate. Among other medical entries in this book, check the article on **Vision and the Eye**. Laser fusion is described under **Fusion Power**. Lasers are also described in the article on **Light**. The use of lasers in metrology is further described under **Interferometer** and in surveying and leveling large plots of land under **Irrigation**.

Atomic Cooling and Trapping. During the last few years, the ability to control the position and velocity of isolated atoms and microscopic

particles has progressed markedly. By the end of 1998 molecules have also been laser cooled and several groups have obtained sufficiently low temperatures and high densities to achieve Bose-Einstein condensation of trapped atoms. In such a state of matter—predicted by Bose and Einstein—the quantum nature of the atoms causes them to lose their individual existence and to coalesce into one collective system. The tremendous success of these efforts has been recognized by the award of the 1997 Nobel Prize in physics to Steven Chu, Claude Cohen-Tanoudji and William Phillips for development of methods to cool and trap atoms with laser light. As pointed out by S. Chu (Stanford University) in a late 1991 paper, "Light can exert forces on an atom because photons carry momentum. The exchange of photon momentum with an atom can occur *incoherently*, as in the absorption and reemission of photons, or *coherently*, as in the redistribution (or lensing) of the incident field by the atom."

Coherent interaction is called the *dipole force*. The incoherent interaction that alters the momentum of an atom is called the *scattering force*.

Successful atom manipulation, however, often depends more upon cooling the atoms than upon exciting the aforementioned forces. Dramatic cooling of atoms to extremely low temperatures is accomplished by employing counterpropagating laser beams, arranged along x, y, and z axes—in essence, creating three-dimensional cooling. As pointed out by Chu, "Because the cooling force is viscous (linearly proportional to the velocities of the atom for low velocities), we named the laser beams that generate the drag force, 'optical molasses'."

In 1991, a research team (Ecole Normale Superieure, Paris) reported the cooling of a sample of cesium atoms to 2.5 μK. At about the same time, a research group (Joint Institute for Laboratory Astrophysics, Boulder, Colorado) reported the achievement of 5 μK. The aforementioned "optical molasses" technique was used in both cases.

Laser cooling, trapping, and related techniques are finding numerous research and practical applications. For example, practical laser-cooled atomic clocks are now possible, constituting a major improvement in accuracy over present atomic clocks. As mentioned by Chu, "A cesium time standard based on a sealed design for which the cooling, manipulation, and detection of the atoms are all done with diode lasers should exceed the stability of the best present-day time standards."

In an excellent paper, Chu (See reference) observes, "Perhaps the most exciting applications in the field of laser cooling and trapping will come out of the ability to study problems in polymer physics and biology on a single molecular basis. Normally one examines the behavior of a large number of molecules, and the fundamental chemistry of the molecules must be inferred from the average behavior of the entire ensemble. On the other hand, the processes that govern the behavior of a single molecule are important: for example, the nucleus of a cell has a single molecular copy of its genetic blueprint, and its chemistry depends in part on the chemistry of single molecules."

In a 1990 paper, Zewail (See reference) describes how atoms can collide, interact, and give birth to molecules in less than a trillionth of a second. As an example of how high-speed imagery has improved over the years, he compares photos of a galloping horse (10 meters per second) taken in 1887 with quantitative observations (made in 5 trillionths of a second) of hydrogen iodide colliding with carbon dioxide to form carbon monoxide, hydroxide, and iodine.

Lasers as Mini-Manipulators. As scientists continue to probe the very minute aspects of natural organs and substances (nanotechnology), small lasers have been found to possess "manipulative" abilities of a kind not envisioned in the early years of laser technology. Scientists (Massachusetts Institute of Technology) in 1990 reported of how lasers can be used effectively as manipulators at the microscopic level. In a study of "mechanoenzymes," which are responsible for the rotary motions of flagella, laser light was used to lift up, move, and position microscopic objects with the "pressure" of the laser light itself, a phenomenon that has been described by Amato as "akin to a blast of air levitating a plastic ball."

Laser mini-tweezers also have been used to clip off regions of chromosomes, moving organelles around inside cells, pushing molecules tiny distances within crystals, and, when used as tiny scalpels or scissors, to catch, trap, puncture, and splice subcellular structures. Recently, measurements have been made of the elastic properties of DNA. Also, it has been found that bacteria can be moved around in a water solution without apparent damage to the organism. Medical applications are described later.

Laser Spectroscopy. In the early years of laser technology, spectroscopy was one of its major uses, an application that has expanded markedly during the past few decades. The review by Gupta provides a good starting point for further reading since it includes an—at the time—up-to date resource letter featuring a large number of annotated references. The techniques of laser spectroscopy parallel those of microwave or radio-frequency spectroscopy, but because lasers are imbued with high spectral purity, they permit vastly improved resolution of fine detail. Early lasers were limited to molecular lines that were coincident with the laser wavelengths. Then lasers using fluorescent organic dyes appeared. These instruments had relatively wide emission bands, offering a tuning capability. Both continuous-wave and pulsing dye lasers have been widely used in most of the visible and near-visible ranges of the spectrum. During the interim, much progress has been made, particularly in providing tunability to lasers. For example, a methyl fluoride molecular gas laser is continuously tunable over broad portions of the far infrared, a region that previously had been difficult. A highly schematic diagram of the operating principle used in early laser-probe emission spectrography is given in Fig. 9.

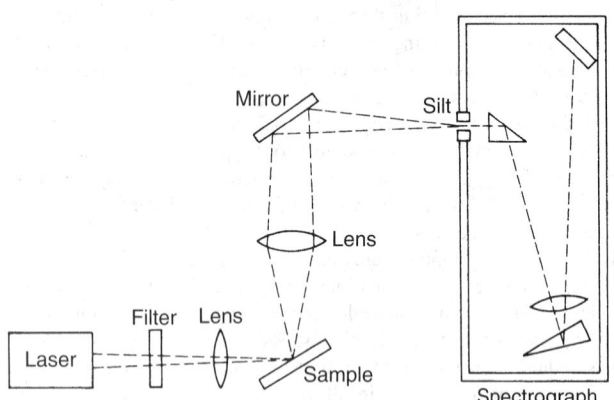

Fig. 9.   Operating principle of laser-probe emission spectrography.

A particularly interesting development is that of the so-called "atomic fountain." As noted by Chu in 1991, "The precision of a spectroscopic measurement depends on both the high Q (Q = quality factor of the resonance defined by $Q = V/\Delta V$) and the signal-to-noise ratio of the signal. Thus, it is important to create a high-flux source of cold atoms. Also, many applications would benefit from a continuous beam of atoms instead of the pulsed sources." An extreme limit of a slow beam is an "atomic fountain," which first was envisioned by Zacharias in the early 1950s. A group of Stanford University scientists has constructed an atomic fountain by first trapping atoms from a thermal beam in a magneto-optic trap and then pushing the atoms upward with a pulse of light from a continuous-wave laser. See Fig. 10.

One reason for the rapid advancement of chemical reaction dynamics research has been the availability of tunable laser sources that operate throughout the infrared, visible, ultraviolet, and vacuum ultraviolet regions of the spectrum. By using nonlinear optical techniques, the outputs from high-power, pulsed visible dye lasers can be summed and mixed to yield useful tunable ultraviolet and vacuum ultraviolet light, with wavelengths as short as 100 nm. Techniques have been developed to probe almost any kind of atomic or molecular state, quite often with sensitivities approaching number densities of 10^5 cm^{-3}, and, in special situations, with detection sensitivity for single atoms.

In a detailed reference, Grant and Cooks (Purdue University) explain in considerable detail the combining of the latest advances in mass spectrometry with laser spectrometers. This technique is contributing in a major way to studies of chemical dynamics, cluster structures, and reactivity, and to the elucidation of the properties of highly excited molecules and ions.

Laser Remote Sensing of Atmospheric Properties

LIDAR (an acronym for light detecting and ranging) is analogous to radar. In lidar, the projection of a short laser pulse is followed by reception of a portion of the radiation reflected from a distant target or from atmospheric

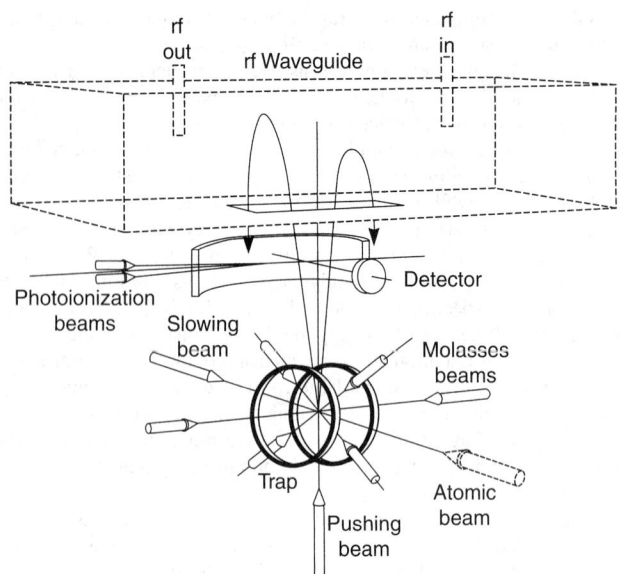

Fig. 10.   The atomic fountain makes it possible to determine precisely the energy states of atoms. Upon injection, the atoms in question are slowed down by a laser beam. Then, the atoms are captured and cooled by means of a magnetic field and several light beams. The cooled atoms follow a ballistic trajectory through a radio frequency (rf) waveguide and a resonant photoionization detection region. (*After Kasevich, Riis, Chu, and DeVoe.*)

constituents, such as molecules, aerosols, clouds, or dust. As explained by Killinger and Menyuk (See reference), the incident laser radiation interacts with the aforementioned constituents to cause alteration in the intensity and wavelength in accordance with the strength of the optical interaction and the concentration of the interacting species in the atmosphere. Information on both composition and physical state of the atmosphere can be deduced from lidar data. The range of the interacting species can be determined from the temporal delay of the backscattered radiation. See Fig. 11.

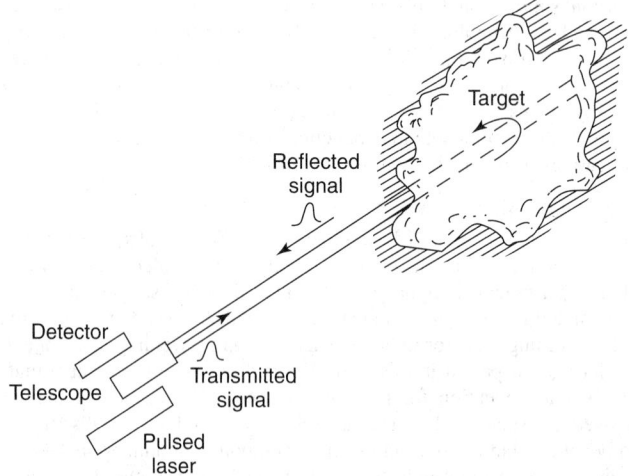

Fig. 11.   Basic components of lidar system used for remote sensing of the atmosphere. Backscattered information sometimes will contain spectral information useful for determining composition and physical characteristics of the cloud or of the intervening atmosphere.

Among specific uses of lidar have been: (1) measurement of movement and concentration of urban air pollution; (2) determinations of chemical emissions from and in the vicinity of industrial plants; (3) determination of atmospheric trace chemicals in the atmosphere; (4) measurement of the velocity and direction of winds near storms and airports, including windshear and gust fronts; and (5) determination of the global circulation of volcanic ash emitted into the atmosphere, relatively recent examples including Mount Pinatubo and Kilauea; among several other applications.

Shortly after the discovery of lasers (early 1960s), Fiocco and Smullin bounced a laser beam off the moon (1962). These researchers also investigated the turbid layers in the upper atmosphere. As early as 1963, Ligda used a ruby laser to obtain the first lidar measurements of cloud heights and tropospheric aerosols. In 1964, Scotland used a temperature-tuned ruby laser to detect water vapor in the atmosphere. Lidar, in recent years, has been greatly improved because of the availability of several kinds of laser sources and improvements made in optical instrumentation and data processing.

As summarized by Killinger and Kenyuk, the future of laser remote sensing is promising and will depend upon several factors, including: (1) development of practical, eye-safe laser sources that cover certain spectral gaps where lidar is currently weak; (2) a further simplification of lidar systems, including lowering size and cost of equipment needed; and (3) more experience to be gained from promising new applications. Among these new applications are: (a) detection of methane gas leaks in coal mines, using a diode laser lidar system; (b) detection of methane and natural gas leaks in industrial plants, using a laser coupled to a low-loss optical fiber network; (c) measurement of global wind fields through the use of Doppler lidar systems mounted in a satellite as a means for improving weather forecasting; and (d) the planned use of lidar on the NASA space-borne Earth Observing System for measurements of global temperature, water vapor, and pressure.

Classification of Lidar System. Lidar systems can be classified on the basis of particular optical interactions which they utilize. Classes of lidar include:

1. *Atmospheric backscatter lidar*, wherein the lidar system transmits one laser wavelength and detects changes in the backscatter due to the aerosols or dust in the atmosphere. This is the most common type of lidar and consists of a nontunable, high-power, pulsed laser. Atmospheric constituents having comparatively large optical scattering cross sections are relatively easy to detect. These systems are used in tracking turbid effluent and gas plumes from factories as well as for mapping rain, snow, ice crystals, and dense clouds in the atmosphere. This type of system was used for checking volcanic ash in the atmosphere.

2. *Differential-absorption lidar* (DIAL), a system which measures the concentration of a molecular species in the atmosphere. This is accomplished by transmitting two wavelengths, only one of which is absorbed. The difference in the intensity of the returns at the two wavelengths is measured. Backscatter in DIAL may come from a hard target or aerosols and dust. One wavelength will be absorbed by the target molecules; the other wavelength will not be absorbed. Many DIAL studies have been carried out in the infrared (IR) range, where almost all molecules of interest have extensive absorption bands. Molecules so far studied include SO_2, NH_3, O_3, CO, CO_2, HCl, NO, N_2H_4, N_2O, and SF_6.

3. *Fluorescence lidar* uses two wavelengths (as in DIAL) plus spectrometric techniques for separating the wavelength-shifted fluorescence signal from the strong Rayleigh backscatter in the atmosphere. The laser is tuned to an absorption line of the species to be measured. Reradiated fluorescence is detected by selective spectral filtering of the returned radiation. The fluorescence radiation may be at the same wavelength as the excitation wavelength, or it may have a longer wavelength because of the red-shift. The backscatter coefficient for fluorescence is greater in the ultraviolet (UV) than in the IR—this due to combined effects of absorption cross section, which is greater in the UV than in the IR. For some applications, fluorescence lidar is limited for remote sensing because of detector sensitivity coupled with solar background radiation. The latter tends to confine fluorescence measurements to nighttime studies and to wavelengths shorter than 1 micrometer, where photomultiplier detection can be used. Nevertheless, some investigators have been quite successful in using the method, particularly in the study of alkali metal (Na, K, Li, and Ca) profiles at altitudes of 80 to 100 km. The method also has been useful for studying the hydroxyl free radical (OH). This radical is of principal interest because of the catalytic role which it exerts in atmospheric chemistry. The OH radical, along with chlorine and nitrogen oxides, is involved in the ozone destruction cycle.

4. *Raman lidar*, a method that is limited by the small optical interaction strength for Raman scattering. High-energy pulsed lasers are employed in this method. The method is limited to the UV or visible regions to permit the use of sensitive photomultiplier tubes for detection. Raman lidar has been used effectively for species that either are at close range or in high concentration, such as N_2, O_2, and H_2O. As mentioned by

Killinger and Menyuk, this method does have some attractive features, the most noteworthy of which is that the laser wavelength need not be tuned across an absorption line because the spectral information is given by the frequency shift of the emission (independent of laser wavelength).

5. *Doppler lidar*, a method that detects only a very narrow spectral range ($\sim10^{-5}$ nm) that encompasses the Doppler-shifted backscatter lidar return. Doppler shifts in the return lidar signals have been used to measure wind velocities and to differentiate between molecular and aerosol returns in the atmosphere. Optical heterodyne techniques are used to detect the shifts which are very small—for example, a fractional change in frequency of about 10^{-8} for a velocity of 1 m/sec at a wavelength of 10 micrometers. Carbon dioxide lasers which provide high power and stable single-frequency operations, are commonly used in Doppler lidar systems. These are in the 10-micrometer range. The system has provided information on boundary layer flow near storm gust fronts and wind shears near airports. They also have been used to measure aircraft vortices and clear air turbulence. See Fig. 12.

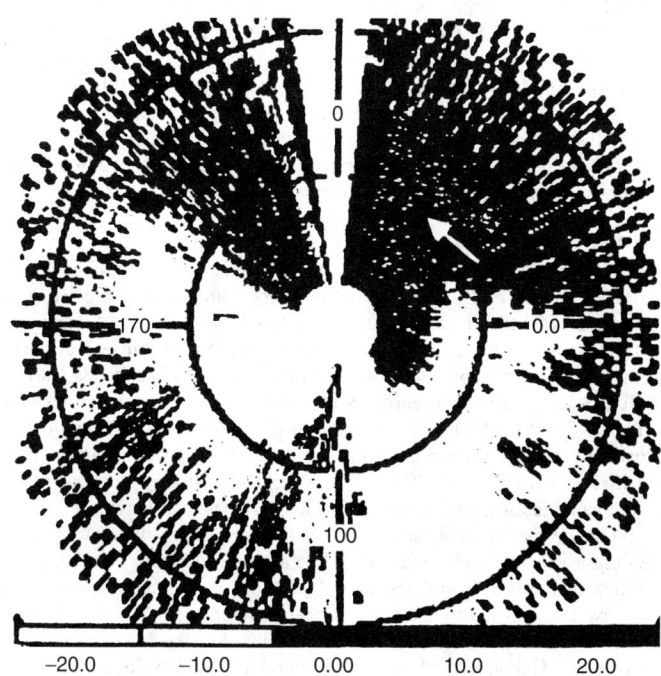

Fig. 12.   Doppler lidar measurement of wind direction and velocity near an airport during a storm. White arrow points to presence of a strong, localized downburst-gustfront.

Doppler broadening effects also have been used to separate backscattered lidar signals into molecular and aerosol components. Characteristics of some lidar systems are summarized in Table 2.

Laser Techniques in High-Pressure Geophysics

The experimental establishment of high pressures and temperatures in the laboratory comparable to conditions found in the Earth's interior is described in the article on the **Diamond Anvil High Pressure Cell**. Laser techniques used in conjunction with the diamond cell make it possible to study the high-pressure properties of material that heretofore had to be inferred from samples found on the surface. Spontaneous Raman scattering of crystalline and amorphous solids at high pressure demonstrates that dramatic changes in structure and bonding occur on compression. High-pressure Brillouin scattering is sensitive to the pressure variations of single-crystal elastic moduli and acoustic velocities. Laser heating techniques with the diamond anvil cell can be used to study phase transitions, including melting, under deep-earth conditions. Laser-induced ruby fluorescence has been essential for the development of techniques for generating maximum pressures now possible with the diamond anvil cell, and currently provides a calibrated in situ measure of pressure well above 100 gigapascals. Hemley, Bell, and Mao (Carnegie Institution of Washington) point out that applications of new spectroscopic techniques, such as double resonance, ultrafast kinetics, Fourier-transform, Raman, and nonlinear optical methods, are likely prospects in future

TABLE 2. SUMMARY OF LIDAR SYSTEM CHARACTERISTICS

Type of Lidar	Type of Laser Used	Nominal Accuracy	Range (km)	Atmospheric Targets
Atmospheric backscatter	Ruby, Nd:YAG	1–10%	10–50	Dust, clouds, volcanic ash, smoke plumes
DIAL, Raman	Dye, CO_2, optical parametric amplifier, $CO:MgF_2$	1 ppb–100 ppm	1–5	H_2O, O_3, SO_2, NO, NO_2, N_2O_3, C_2H_4, CH_4, HCl, CO_2, CO, Hg, SF_6, NH_3
Fluorescence	Dye	10^2–10^7 atoms/cm	1–90	OH, Na, K, Li, Ca, Ca^+
DIAL, Raman	Dye, Nd:YAG	1 K, 5 mbar	1–30	Temperature, pressure
Doppler	CO_2	0.5 m/sec	15	Wind speed

After Leone (1987).

work on geophysical problems with the diamond anvil cell. Recent high-pressure studies involving the use of picosecond spectroscopy and hyper-Raman scattering of perovskites may be representative of this trend. Time-resolved studies may permit the detailed investigation of the kinetics of high-pressure phase transitions and the rheology of minerals under in situ deep-earth conditions. The combination of spectroscopic and x-ray diffraction probes with laser-heating techniques may yield detailed structural information on earth materials at high temperatures and pressures, thus advancing an understanding of the connection between atomic-scale properties and global deep-earth processes.

Laser Metrology

Lasers are widely used for making precision measurements of geometric variables. In the early 1960s, laser pioneers demonstrated precise measurements with the device. Lasers introduced the concept of frequency metrology as contrasted with wavelength metrology. In an early experiment with a super-stabilized laser, scientists at the Massachusetts Institute of Technology during the early 1960s worked out a laser version of the famous Michelson-Morley experiment at Case Institute of Technology in Cleveland. The MIT scientists concluded that an advance in measurement sensitivity by a factor of 1000 over the Michelson-Morley data was potentially available through the use of frequency rather than length metrology. In 1962, the first laser measurement of the speed of light (c) was made, yielding a value of 299, 792, 462 ± 18 meters per second.

Since that time, several more sophisticated determinations have been made by leading metrology laboratories, including the National Institute of Standards and Technology (Boulder, Colorado), the National Physical Laboratory (United Kingdom), the Laboratoire de Physique des Lasers (Villetaneuse, France), and the Laboratory for Spectroscopy (Russia), among others. Based on these measurements, c is now given as precisely 299,792,458 meters per second. Furthermore, the length unit has been abandoned as a fundamental unit and is now derived from the time unit and the above quoted value for the speed of light.

At the practical manufacturing level of metrology, laser guidance systems can be used. Mergler (Case Western Reserve University) introduced a machine in 1978 along these lines. In a conventional machining operation, a part is cut, then measured, then remachined until the required dimensions are obtained. Manual measuring methods are tedious, time consuming, and somewhat limited in accuracy. In Mergler's system, a small modulated gas laser beam follows the surface of the part being machined and, within a precision of 1/5000 inch (0.005 millimeter) measures the piece as it is being cut. In later systems, the gas laser was replaced by a solid-state laser which occupies less space. See also laser interferometer described in entry on **Interferometer**; and **Electron Beam Lithography**.

Laser Doppler Flowmeter

As shown in Fig. 13 fluid flow can be determined by measuring the doppler shift in laser radiation scattered from particles in the moving fluid stream. No sensor is required in the moving stream. The laser radiation focal point can be moved across the flow tube to measure velocity profiles. Fluid linear flows from 0.01 to 5000 inches (0.03 centimeter to 127 meters) per second have been measured. Contaminants, such as smoke, may have to be added to gases to provide scattering centers for the laser beam.

Laser Gyroscope

As early as the beginning of the 20th century, some investigators suggested that light will exhibit gyroscopic behavior, that is, the time required by light

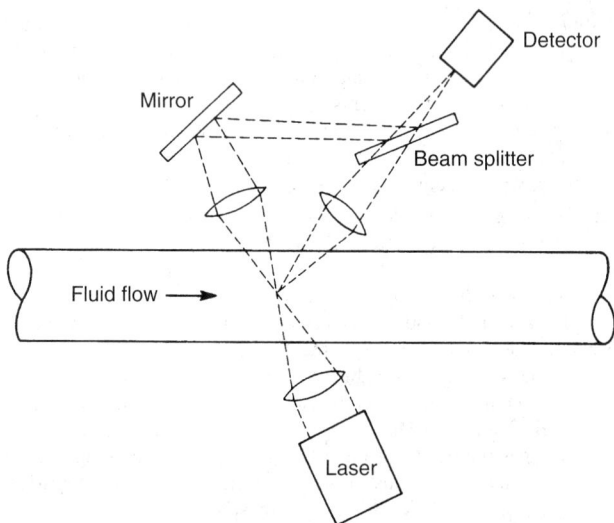

Fig. 13.   Operating principle of laser doppler flowmeter.

to traverse a circular pathway depends on whether the pathway is stationary or rotating. Thus the time difference can be used as a measure of the amount of rotation. The practical application of this observation, however, had to await vast improvements in optical systems, including the discovery of the laser, advances in fiber-optics, and better reflective mirrors. Within recent years, this principle has been applied in two configurations — fiber gyroscopes and ring-laser gyroscopes. The latter is described briefly here. As of the late 1980s, several aircraft depend upon ring-laser gyroscopes instead of their mechanical counterparts. The ring-laser gyroscope is more sensitive, has virtually no moving parts, and is as accurate as the best mechanical instruments. The rotation-induced difference in length of light path traversed is called the Sagnac effect, after the researcher who first demonstrated the phenomenon in 1913.

As previously mentioned in this article, a laser usually incorporates a resonant cavity. C.V. Heer (Ohio State University) in 1958 proposed that a resonant cavity could be used to measure rotation rates. In such an instrument, light circulates many times around a given path, not just back and forth between two mirrors. The first gyroscopes of this kind were constructed on a large scale, consisting of four glass tubes, each a meter long and arranged in a square. Light was made to travel around the device by placing a mirror in each corner. Over the last several years, the device has been markedly reduced in size (fits in the palm of the hand). Contemporary gyroscopes of this type are made from a single block of glass, into which a square channel is drilled. The channel is filled with a mixture of helium and neon. The laser is completed by attaching a small number of electrodes and four mirrors. As explained by Anderson, some ring-laser gyroscopes have a triangular channel and three mirrors; others have a hexagonal channel and six mirrors.

Beyond the scope of this article, Anderson explains the operation of the gyroscope in intimate detail and describes two problems that have proved most vexing to manufacturers of the ring-laser gyroscope, namely, frequency locking at low rotation rates and the bias effect. Improvements in this instrument are expected in the relatively near future because of what scientists have recently learned pertaining to the phenomenon of optical phase conjugation. See also **Light**.

Lasers in Manufacturing Operations. A principal advantage of the laser in manufacturing operations is its ability to apply an extremely high flux of energy to the surface of a workpiece, as compared with traditional heat sources, such as flames, torches, electric arcs, and plasma jets. For manufacturing operations, lasers are usually placed in two categories. (1) *Light-duty lasers* range from a few tens of watts to a few hundred watts. Typical applications include cutting and drilling ceramic substrates in the electronics industry, drilling gems (for example, rubies in watchmaking), and cutting light-gauge metals as well as cloth, plastics, wood, and a variety of materials. Light-duty lasers that have been used include ruby lasers, neodymium-doped glass lasers, and neodymium-doped yttrium aluminum garnet lasers, among others. Depending upon the particular laser selected, the laser may operate in a pulsed or continuous mode. Argon and CO_2 lasers usually are operated in the continuous-wave mode. (2) *Heavy-duty lasers* range from a few kilowatts to a few tens of kilowatts. Typical applications include pipeline welding, automobile part welding, surface heat-treating of engine and other parts, with the applications expanding as experience is gained.

The high flux of electromagnetic energy applied to the surface of a workpiece by a laser is absorbed in an outer layer only about 10 nanometers thick. Thus the heat source is confined essentially to a thin film. Through careful design of equipment, the heat energy required is maintained in a comparatively small region, thus preventing or reducing thermal damage to the rest of a given part, and achieving a very high energy efficiency, estimated to range from 10 to 1000 times greater than can be achieved with conventional energy sources.

The electronics industry utilizes laser welders for joining dissimilar materials, fixing electrodes to batteries and connectors to a host of devices. A whole new area of laser technology, sometimes called laser microchemistry, has been exploited in the microstructure engineering of semiconductors. Lasers are used to initiate chemical reactions that result in deposition of material at a surface, for removing materials, and for alloying or diffusively mixing two or more solids on microscopic spatial scales. Lasers thus have played a major role in establishing new dimensions in microfabrication technology. It is possible to use a single laser to produce both gas-phase photolysis and surface heating. As described by Christensen (See reference), solar cells have been fabricated by using a UV laser to photodissociate trimethylboron, $B(CH_3)_3$, over a silicon surface in the manufacture of solar cells. The laser also heats the surface so that the boron atoms absorbed on the surface after the photolytic step rapidly diffuse into the bulk of the material. After irradiation, the silicon is heavily doped with boron near the surface, and the p-n junction thus formed functions as a photovoltaic cell. In some other applications, it has proved advantageous to use two lasers of different wavelengths to separately achieve photolysis and heating.

Perspective. The industrial applications for lasers developed comparatively slowly. As previously mentioned, lasers depend upon raising active molecules of the lasing medium to what might be called an *upper laser level* of energy, after which they relax to a *lower laser level*. Energy is given up during this process. Part of this energy is represented by photons of which the laser beam is composed. The other part is waste heat, which raises the temperature of the lasing medium. Thus, an excess of waste heat be removed so that the upper-level population can be maintained, a significant problem in the case of a continuously emitting high-power laser. Higher packets of energy in pulses can be attained, but waste heat must be removed by conduction between pulses. The end result is a pulsed high-power laser, but one that has a comparatively low average energy level simply because of the pauses in between.

In early laser designs, the quantity of waste heat generated was a limiting factor and consequently the average power output was low. Such lasers were excited by diffuse longitudinal electric discharges in long tubes with relatively large diameters. Heat generated at the center of the tube diffused to the side walls essentially by conduction, and the rate of heat transfer varied inversely with the tube radius and essentially directly with the length of the tube. Thus, the length of laser tubes increases as greater output power was sought.

Various component cooling schemes were proposed and used, but a major improvement was made when the concept of cooling a flowing laser medium was proposed, thereby taking advantage of the far more effective cooling by convection than by conduction. Gas lasers were considered the

most apt for application of this concept and this led to the gas dynamic laser.[1]

With this concept, gas dynamic lasers increased in power outputs from less than 10 kilowatts by a factor of 13 to 14 within less than a decade (by the late 1960s). Success with the early gas lasers in this respect catalyzed a number of other refinements and improvements. However, the problem of maintaining a high-pressure glow discharge remained. Population inversion can be produced when electrons in an ionized gas are at a temperature relatively high as compared with the kinetic temperature of ions or molecules. This is a condition referred to as *glow discharge*. But an arc may form when the discharge is destabilized as the result of greatly increased gas pressure. Overheating of the molecules and ions destroys the population inversion. This problem was overcome by the concept of the ionizer/sustainer.

With the availability of high-power continuous electric discharge lasers capable of operating at up to 20 kilowatts output, a number of the previously predicted applications for lasers became practical. One of the first uses of a laser beam strictly for its power in cutting (exploding) a material was the fabric cutting system developed by Hughes Aircraft Company (circa 1966-1967) for which U.S. Patent No. 3,761,675 was awarded to W.J. Mason, D.W. Wilson, D.M. Considine, F.J. Viosca, and J.P. Wade on September 25, 1973. See Fig. 14. In this system cloth is carried in a single layer into a cutting area where a laser beam focused on the cloth is directed by computer commands to travel within the cutting area so as to cut many patterns in the cloth rapidly and accurately. The cut produced by the focused laser beam is sharp and narrow, leaving the fabric unfrayed. With synthetic materials, such as nylons and Dacrons®, the laser beam also serves to seal the cut edges by melting them during the cutting process. Unlike a mechanical blade, the laser beam does not dull; its cut remains uniform and is effective in cutting a wide range of materials, even those having metallized threads.

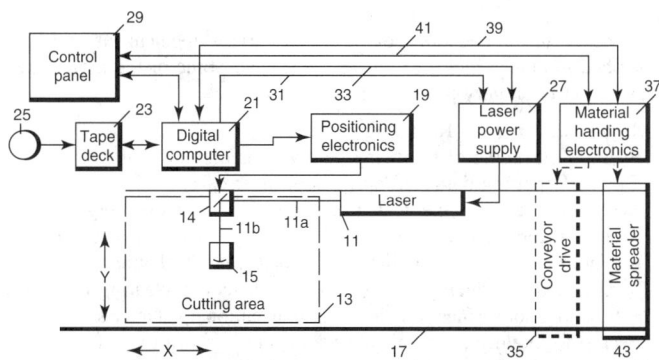

Fig. 14.   A cloth cutting system wherein cloth is carried in a single layer into a cutting area, where a laser beam is focused on the cloth and is directed by computer commands to travel within the cutting area so as to cut a plurality of patterns through the cloth rapidly and accurately. Invented by W.J. Mason, D.W. Wilson, D.M. Considine, and J.P. Wade (*Hughes Aircraft Company*) in 1973, this was one of the very early and successful industrial applications of the laser. Diagram is part of U.S. Patent 3,761,675.

In the early 1980s, a helium-neon laser was used in a scanner system for inspecting textiles. The system uses laser output split into three beams, each of which scans the fabric independently in a pattern covering the entire surface. The system, moving at a rate of four meters per second, detects flaws through changes in reflected light and flags these areas for elimination or repair. In terms of economics, one laser system working

---

[1] Invented by Kantrowitz in the late 1960s. In essence, the device had two compartments separated by a nozzle. In the first compartment, gas was held at a temperature of about 1400 K and pressure of 17 atmospheres. This high-pressure compartment held about 10% of the active CO_2 molecules in the total system. Expansion of this gas through an orifice caused cooling. Because of the cooling, the lower-level population essentially vanished a few centimeters downstream from the nozzle. This occurred before the upper-level population had an opportunity to decline significantly. The population "inversion" resulting was adequate for effecting a laser beam of considerable power.

one shift performs the same function as human inspectors at two plants working two shifts.

High-power lasers can perform many metalworking operations, including welding, cutting, surface hardening, and surface alloying. For small devices, the laser can perform much as a conventional electron beam, but without requiring the need for operation under a vacuum. High power densities can be achieved—up to 10^6 watts per square centimeter. It has been shown that a 16-kilowatt laser can make a 0.75-inch (1.9-centimeter) penetration weld in stainless steel at a rate of about 30 inches (76 centimeters) per minute. The laser beam can be directed by mirrors, thus making it effective for welding pipe from the inside. It also has been shown that a continuous-wave carbon dioxide device (15 kilowatts) can be used for welding half-inch (1.2-centimeter) thick steel plates at the rate of about 50 inches (127 centimeters) per minute. If the laser is focused to a spot size of about 0.03 inches (0.08 centimeter) in diameter, power densities of some 2200 kilowatts per square centimeter are produced.

Laser Recording. For many years, it has been known that lasers can be used to encode information on materials that respond in an irreversible manner to exposure to high-intensity light. However, it is only comparatively recently that the concept has been reduced to commercial practice—with the almost sudden appearance of optical disk recording (compact disk) in the entertainment field. It is because of the coherence and relatively short wavelength of laser radiation that such large volumes of information can be written onto a very small space of the recording medium. The potential for microlasers in this field is discussed earlier in this article.

Lasers in Medicine

Medical applications for lasers have developed at a very rapid rate and one that is continuing to expand today in an exponential fashion. Much of the laser equipment for the manipulations required in surgical procedures is available. Training professionals in the effective use of the equipment and equipment costs are the only factors retarding even faster growth of the field.

Some of the applications of lasers in various medical situations are described in other articles in this encyclopedia. Some of the more widely accepted laser surgery procedures include:

Abdominal cavity — Repairing hernias.

Brain — Shrinking or removing benign or malignant tumors.

Ears — Repairing damaged portions of inner ear.

Eyes — The earliest medical use of lasers was in eye surgery. Reattaching torn retinas; opening blocked tear ducts that cause dry-eye syndrome; coagulating bleeding blood vessels that cause diabetic retinopathy; reducing fluid buildup in glaucoma; clearing the cloudiness that sometimes remains after conventional cataract surgery.

Feet — Removing corns, plantar warts, and ingrown toenails.

Gastrointestinal tract — Removing hemorrhoids and intestinal polyps.

Genital-reproductive tract — Vaporizing venereal warts, fibroids, external genital growths, and tumors in the cervix, vagina, and perineum; vaporizing excess uterine lining that causes bleeding.

Mouth — Removing or vaporizing superficial tumors.

Nose — Vaporizing polyps; removing adenoids and the excess tissue that causes sinusitis.

Skin — Erasing port-wine stains, birthmarks, tattoos, age spots, and freckles; removing warts, broken facial capillaries, and precancerous patches.

Urinary tract — Vaporizing tumors; opening a narrow urethra.

Additional Reading

Adams, C.S. and E. Riis: "Laser Cooling And Trapping Of Neutral Atoms," *Progress in Quantum Electronics*, 1–79 (1997).

Amato, J.: "Moving Tiny Things by Optical Tweezers," *Science News*, 148 (March 10, 1990).

Anderson, D.Z.: "Optical Gyroscopes," *Sci. Amer.*, 94–99 (April 1986).

Appell, D.: "High-Power Laser Beam Launches Fuel-Less Craft," *Laser Focus*, 90 (March 1998).

Attwood, D., K. Halbach, and K. Kwang-Je, Kim: "Tunable Coherent X-rays," *Science*, **228**, 1265–1272 (1985).

Byer, R.L.: "Diode-Laser-Pumped Solid-State Lasers," *Science*, 742 (February 12, 1988).

Cherfas, J.: "A European Superlaser?" *Science*, 1073 (June 1, 1990).

Christensen, C.P.: "New Laser Source Technology," *Science*, **224**, 117–123 (1984).

Chu, S.: "Laser Manipulation of Atoms and Particles," *Science*, 861 (August 23, 1991).

Chu, S.: "Laser Trapping of Neutral Particles," *Sci. Amer.*, 71 (February 1992).

Corcoran, E.: "Diminishing Dimensions," *Sci. Amer.*, 122 (November 1990).

Corcoran, E.: "True Blue (Laser)," *Sci. Amer.*, 171 (September 1991).

Corcoran, E.: "Tacky Lasers Are the Tiniest Yet," *Sci. Amer.*, 28 (January 1992).

Duley, W. and K. Shibata: "1996 International Congress on Applied Lasers and Electro-Optics Proceedings," Laser Institute of America, Orlando, FL, 1997.

Fasol, G.: "Room-Temperature Blue Gallium Nitride Laser Diode," *Science*, 1751 (June 21, 1996).

Fasol, G., S. Nakamura, and I. Davies: "The Blue Laser Diode: GaN Based Light Emitters and Lasers", Springer-Verlag New York, Inc., New York, NY, 1997.

Feld, M.S. and K. An: "The Single Atom Laser," *Scientific American*, 56–63 (July 1998).

Feng, S. and P.A. Lees: "Mesoscopic Conductors and Correlations in Laser Spackle Patterns," *Science*, 633 (February 8, 1991).

Freund, H.P. and R.K. Parker: "Free-Electron Lasers," *Sci. Amer.*, 84 (April 1989).

Grant, E.R. and R.G. Cooks: "Mass Spectrometry and Its Use in Tandem with Laser Spectroscopy," *Science*, 61 (October 5, 1990).

Hemley, R.J., P.M. Bell, and H.K. Mao: "Laser techniques in High-Pressure Geophysics," *Science*, **237**, 605–612 (1987).

Hirschfelder, J.O., R.E. Wyatt, and R.D. Coalson: "Lasers, Molecules, and Methods," John Wiley & Sons, Inc., New York, NY, 1989.

Jewell, J.L., J.P. Harbison, and A. Scherer: "Microlasers," *Sci. Amer.*, 86 (November 1991).

Killinger, D.K. and N. Menyuk: "Laser Remote Sensing of the Atmosphere," *Science*, **235**, 37–45 (1987).

Kinoshita, J.: "Atomic Fountain: Laser Light Slows Atom Beam to a Trickle," *Sci. Amer.*, 26 (June 1990).

Kim, K. Kwang-Je, and A. Sessler: "Free-Electron Lasers: Present Status and Future Prospects," *Science*, 88 (October 5, 1990).

Lamb, W.E., Jr. and R.C. Retherford: *Phys. Rev.*, **79**, 549 (1950).

Langreth, R.N.: "Laser Cooling Made Simpler, Cheaper," *Science News*, 216 (October 6, 1990).

Maddox, John: "The Wonders Of The Microlaser," *Nature*, 101 (January 12, 1995).

Matthews, D.L. and M.D. Rosen: "Soft X-Ray Lasers," *Sci. Amer.*, 86 (December 1988).

Maiman, T.H.: *Br. Commun. Electron.*, **7**, 674 (1960).

Meyers, R. (Editor): "Encyclopedia of Lasers and Optical Technology," Academic Press, Inc., San Diego, CA, 1990.

Misaelides, P.: "Application of Particle and Laser Beams in Materials Technology," Kluwer Academic Publishers, New York, NY, 1995.

Morrison, D.C.: "An Unsung Legacy of the First Lunar Landing (Laser)," *Science*, 447 (October 27, 1989).

Murname, M.M. et al.: "Ultrafast X-ray Pulses from Laser-Produced Plasmas," *Science*, 531 (February 1, 1991).

Narayan, J.: "Surfaces, Interfaces, and Films: New Tools (Lasers) Aid Engineering," *Adv. Materials & Processes*, 51 (January 1988).

Pepper, D.M., J. Feinberg, and N.V. Kukhtarev: "The Photorefractive Effect," *Sci. Amer.*, 62 (October 1990).

Phillips, W.D., P.L. Gould, and P.D. Lett: "Cooling, Stopping, and Trapping Atoms," *Science*, 877 (February 19, 1988).

Pool, R.: "Making Atoms Jump Through Hoops," *Science*, 1076 (June 1, 1990).

Pool, R.: "Laser Cooling Hits New Low," *Science*, 1077 (June 1, 1990).

Purcell, E.M. and V. Pound: *Phys. Rev.*, **81**, 279 (1951).

Ruthen, R.: "Surfing Photons," *Sci. Amer.*, 12D (August 1989).

Svelto, O. and D.C. Hanna: "Principles of Lasers," 4th Edition, Perseus Publishing, Boulder, CO, 1998.

Staff: "Miniature Lasers Reach Mass Production," *Chem. Eng. Progress*, 15 (August 1991).

Taylor, N.: "Laser: The Inventor, the Nobel Laureate, and the Thirty-Year Patent War," Simon & Schuster Trade, New York, NY, 2000.

Suckewer, S. and C.H. Skinner: "Soft X-Ray Lasers and Their Applications," *Science*, 1553 (March 30, 1990).

Townes, C.H.: "Harnessing Light," *Science* **84**, 153–155 (November 1984).

Vander Been, M.R.: "Gallium Arsenide Sandwich Lasers," *Adv. Materials & Processes*, 39 (May 1988).

Waterbury, R.C.: "Catalysts Enable Sealed Carbon Dioxide Laser," *Instrumentation Technology*, 80 (April 1990).

Yamamoto, Y., M. Susumu, and W.H. Richardson: "Photon Number Squeezed States in Semiconductor Lasers," *Science*, 1219 (March 6, 1992).

Zewail, A.H.: "The Birth of Molecules," *Sci. Amer.*, 76 (December 1990).

LASER ALTIMETER.   See **Moon (Earth's)**.

LASER DOPPLER FLOWMETER.   See **Flow Measurement (Liquids and Gases)**.

LASER GLASS.   See **Glass**.

LASER IN-SITU KERATOMILEUSIS (LASIK).   LASIK is now the most popular of all laser vision correction procedures. It was estimated that

$1\frac{1}{2}$ to 2 million procedures would be completed in the United States in the year 2000. This highly successful procedure combines the minimal postoperative discomfort and rapid visual recovery of the Automated Lamellar Keratoplasty (ALK) procedure with the computer-controlled precision of the Photorefractive Keratectomy (PRK) procedure.

The first step in the LASIK procedure is the creation of a flap of tissue from the outer layer of the central zone of the cornea using the microkeratome. The flap is then folded back out of the way, but it is held ready for replacement upon completion of the procedure. The Excimer laser is then used to sculpt the remaining central zone in accordance with predetermined data that have been entered into the laser system's computer. Under this precise control, the laser reshapes the curvature of the cornea to correct for nearsightedness, farsightedness or astigmatism. This part of the procedure takes only 30 to 60 seconds, after which the corneal flap is replaced. No sutures are used, and the surface of the eye will normally heal itself. Most patients can see quite well within 24 hours or less. Complete healing of the cornea takes about one month.

The LASIK procedure is performed on an outpatient basis. Although the actual laser procedure takes only a few seconds per eye, the procedure requires a couple of hours at the surgery center. Some of this time is spent preparing the patient for the procedure. A few minutes are required afterwards for postoperative instructions and departure preparation.

Surgical Procedure

Step 1: Eye preparation. Before the procedure begins, a nurse or technician talks to the patient about any immediate health problems that may affect readiness for the procedure. Antibiotic and anesthetic eye drops are then placed in the eye to numb it and prevent infection. The eye is swabbed with a sterile solution. The eyelid is then propped open with a lid retainer, and a paper or plastic "mask" is placed over the eye to keep eyelashes out of the way. Then the cornea is marked with a blue "dye ring," which serves as a reference point for the surgeon throughout the procedure. Because the cornea is numb, most patients experience little if any discomfort during these pre-operative preparations.

Step 2: Creating the flap. Next, the doctor creates a flap from the central zone of the cornea using the microkeratome. This precision instrument works much like a miniature carpenter's plane. It contains a disposable cutting blade that is preset according to the thickness of the cornea, usually about 160 to180 microns or 1/3 the depth of the cornea. The microkeratome operates in conjunction with a suction ring that holds the eye perfectly still, and when activated by a vacuum tube, it raises and flattens the cornea so it can be reached easily for cutting the flap.

Step 3: The Excimer laser. After the flap has been folded back from the center of the eye, the doctor dries the underlying cornea with a sponge-tipped swab and aligns the Excimer laser's microscope with the central corneal area in order to monitor the laser's sculpting pulsations. The patient is asked to focus on a fuzzy red light inside the laser. As the doctor activates the laser, there is a "popping" or "tacking" sound, and there is a slight odor similar to that of hair burning, but no discomfort for the patient. The number of laser pulsations will depend on the nature of the refractive vision problem that is being corrected. This phase of the procedure takes only a minute or so. The doctor then carefully folds the flap back in place and irrigates the eye with a sterile saline solution. The corneal area may be dried with a gentle blower, which helps seal the flap. In addition, a contact lens may be placed in the eye.

Step 4: Post-operative measures. When the procedure is complete, additional antibiotic drops are placed in the eye, and it may be covered with a plastic shield. For a short while after the procedure, the eye is numb from the anesthetic drops. As the numbness wears off, the patient may experience some light sensitivity and a scratchy or dry sensation as though something is in the eye. This feeling usually goes away within a few hours. Patients must not drive themselves home following the procedure.

The patient returns to the doctor's office the next day for a post-operative examination. The doctor checks the flap to see if it is healing properly. If a contact lens was placed after surgery, it will be removed at this time. Vision is checked and, for most patients will range from 20/20 to 20/40 depending on the number of laser pulsations received. For some patients, vision may continue to improve for several weeks before totally stabilizing.

The LASIK procedure has several advantages over both the ALK and the PRK procedures on which it is based. Although it employs the Excimer laser precision control and accuracy of the PRK procedure, the LASIK procedure does not remove any part of the epithelium, the thin, filmlike protective outer layer of the cornea, as does PRK, and there is less chance of scarring. Thus, the primary healing process is the resealing of the corneal flap, which usually happens within 24 hours and with little postoperative discomfort. The LASIK procedure can also handle successfully higher degrees of myopia than PRK and can be used to treat cases of farsightedness and astigmatism.

See also **Automated Lamellar Keratoplasty (ALK)**; **Lasers (Eye Surgery)**; **Photorefractive Keratectomy (PRK)**; and **Refractive Eye Surgery**.

Vision Rx, Inc., Elmsford, NY.

LASER IONIZATION. See **Mass Spectrometry**.

LASERS (EYE SURGERY). The word "laser" is an acronym for light amplification by stimulated emission of radiation. In most lasers used in ophthalmology, an electric current is passed through a tube that contains an amplifying medium, usually a gas or solid material, which serves to intensify the energy. This energy is emitted as a narrow light beam, which when focused through a microscope, will either cut, burn, or dissolve various tissues.

Different types of lasers emit specific colors of light and are used to treat various eye problems. The lasers are usually named for the amplification materials used. For instance, the carbon dioxide laser is called a CO_2 laser, while the YAG laser contains a solid material made up of yttrium, aluminum, and garnet.

Ophthalmic lasers allow precise treatment of a variety of eye problems without risk of infection. Most laser procedures are also relatively painless and can be done on an outpatient basis. This combination of safety, precision, convenience, and reduced cost make lasers one of the most successful medical tools available to physicians.

Types of Lasers and Their Uses

Excimer laser. The Excimer laser is perhaps the best known of all lasers because of its use in laser vision correction surgery such as laser *in-situ* keratomileusis (LASIK) and photorefractive keratectomy (PRK). The Excimer or pulsed gas laser, emits an ultraviolet light beam, vaporizing tissue by breaking down molecular tissue bonds in a minute targeted area. It is called a *cold laser* because it does not produce heat that could have harmful effects to the surrounding tissue.

The most important feature of the Excimer laser for surgical applications is its ability to focus powerful energy on a microscopic target without affecting the surrounding area. Each pulse of the laser removes about 1/500 of the thickness of a human hair, which is about 125 microns in diameter. These two factors of precise depth and area control are of particular significance in surgical applications such as refractive vision correction.

YAG laser. An acronym for yttrium–aluminum–garnet, the YAG laser produces short-pulsed, high-energy light beams to cut, perforate, or fragment tissue. This laser may also be called a neodymium–YAG or ND–YAG laser.

Cataract patients often have the misconception that a YAG laser is used to remove their cataracts, but no lasers are used in cataract surgery. This misconception occurs because up to 75% of cataract patients develop a condition known as posterior capsular opacification, a clouding of the residual lens capsule left in place after cataract surgery. This gradual loss of vision resembles the symptoms of cataract development, making some people believe that their cataracts have grown back.

The YAG laser is commonly used to vaporize a portion of the capsule, allowing light to pass through to the retina. The procedure is completely painless, takes only a few minutes in the office, and is effective in eliminating the cloudy condition.

Holmium laser. Also known as the infrared holmium YAG laser, this laser is used in a refractive surgery procedure called laser thermal keratoplasty (LTK) to correct mild to moderate cases of farsightedness and some cases of astigmatism. Unlike the Excimer laser, which reshapes the cornea by removing or ablating tissue, the Holmium laser produces infrared light that reshapes the cornea by causing tissue to constrict. The pulsations from the Holmium laser are computer-controlled to produce a pattern of 8 to 16 tiny beams in concentric rings around the periphery of the cornea. The heated fluid in the spots where these beams hit the cornea creates a series of tiny craters. The subsequent shrinkage pulls in the periphery of the cornea, causing the center to bulge, much like tightening a belt, and thus correcting farsightedness.

CO_2 laser. The CO_2 laser is a specialized laser that is filled with carbon dioxide gas and uses an infrared emission for cutting tissue through heat absorption. It is one of the most common lasers used in surgery and is good for precise cutting and vaporization of tissue, such as that needed in the treatment of superficial lesions or removing small volumes of tissue.

The CO_2 laser is used by ophthalmic plastic surgeons to remove fine wrinkles from around the eyes. This laser precisely removes the outermost layer of skin and the underlying dermis, allowing the regrowth of wrinkle-free new skin.

Erbium laser. The Erbium laser, or erbium–YAG laser, is also used in skin resurfacing and is considered to be more precise and accurate than the CO_2 laser. It is able to remove finer wrinkles with less damage to the skin. The depth of penetration is about 5 microns compared with the 20 microns typical of the CO_2 laser. The Erbium laser also causes less irregular skin pigmentation in darker skinned individuals, because it produces a thinner laser area and less heat. Because the Erbium laser produces minimal thermal scatter, the healing time is less than the healing time with the CO_2 laser.

The Erbium laser is also being used in a promising new clinical procedure to emulsify the eye's natural lens during cataract surgery. Most cataract surgeons currently use a piece of equipment called a phacoemulsifier to break up and remove the cloudy lens. The Erbium laser was chosen for the new technique because of its high absorption rate in water, a primary component of the eye's natural crystalline lens.

Argon laser. The argon laser is filled with argon gas that produces blue/green wavelengths. These particular wavelengths are absorbed by the cells that lie under the retina and by the red hemoglobin in blood, but the blue-green wavelengths can pass through the fluid inside the eye without damage. For this reason, the argon laser is used extensively in the treatment of diabetic retinopathy, a severe disorder of the retina that causes blood vessels to leak. The argon laser can burn and seal these blood vessels.

Retinal detachment is another serious eye problem that can be treated by the argon laser. The laser is used to weld the detached retina to the underlying choroid layer of the eye.

Several forms of glaucoma, which is a leading cause of blindness, are also treated with argon lasers. The very serious angle closure glaucoma, for instance, is sometimes treated by using the laser to create a tiny opening in the iris, allowing excess fluid inside the eye to drain to reduce pressure.

Macular degeneration, a severe condition that affects central vision in older adults, is sometimes treated with an argon or krypton laser. In this treatment, the laser is used to destroy abnormal blood vessels so that hemorrhage or scarring will not damage central vision. See also **Laser In-Situ Keratomileusis (LASIK); Laser Thermal Keratoplasty (LTK); Photorefractive Keratectomy (PRK)**; and **Refractive Eye Surgery**.

Vision Rx, Inc., Elmsford, NY.

LASER (Telephony).   See **Telephony (Telecommunications)**.

LASER THERMAL KERATOPLASTY (LTK).   LTK is a refractive surgery procedure that uses a Holmium laser to reshape the cornea for correction of low ranges of hyperopia (farsightedness). The Holmium laser is an infrared (thermal) laser that uses heat to shrink corneal tissue.

On the other hand, the Excimer laser uses a cool beam to vaporize corneal tissue. The Excimer laser is used in the laser *in-situ* keratomileusis (LASIK) and photorefractive keratectomy (PRK) procedures.

In LTK, the Holmium laser is used to gently heat stromal collagen in a ring around the outside of the pupil. The heat causes the tissue to shrink, thereby creating an effect like tightening a belt. The periphery of the cornea is pulled, causing the center to bulge. Because the cornea of a farsighted eye is too flat, this bulging effect, when carefully controlled, corrects the problem.

People with mild hyperopia, +0.75 to +2.75, are prime candidates for LTK. It is also being used to treat presbyopia (age-related loss of focus) and overcorrection from radial keratotomy (RK), PRK, and LASIK procedures. LTK is currently under investigation by the FDA for approval in the United States, and patients can only be treated as part of the investigational protocol.

To understand how LTK works, it is first necessary to understand the visual function of the eye. See also **Visual Function (Eye)**.

If the cornea is too flat or the eye is too short from front to back, light rays are theoretically focused behind the retina, resulting in hyperopia (farsightedness). The objective of LTK is to make the cornea steeper in order to correct for farsightedness.

The LTK procedure is performed on an outpatient basis with topical anesthetic eye drops to numb the eye. Based on the patient's prescription, the laser's computer is calculated to deliver the number of pulses and the diameters of the circles needed to provide the proper amount of correction. After aligning the pupil with the use of a slit-lamp microscope, the surgeon activates the laser, and it transmits tiny beams of infrared light in two concentric rings around the periphery of the cornea. Because moisture in the cornea absorbs the energy in the laser pulses, the tissue shrinks slightly creating tiny craters, which tighten the cornea and result in a steeper surface. The laser never touches the eye, and the entire process takes just a few seconds per eye.

The LTK procedure is painless, although the patient may have blurry vision and a mild scratchy sensation for a couple days. Antibiotic eye drops are normally used for about a week, and, if needed, Tylenol and ice packs can be used to relieve discomfort. Most eyes are fully healed in three days, and, although many patients report almost instant vision improvement, vision stabilization usually occurs within two weeks. See also **Hyperopia (Farsightedness)**; **Laser In-Situ Keratomileusis (LASIK)**; **Photorefractive Keratectomy (PRK)**; **Presbyopia**; and **Refractive Eye Surgery**.

Vision Rx, Inc., Elmsford, NY.

LASSA FEVER (African Hemorrhagic Fever).   Lassa fever is an acute viral illness that occurs in West Africa. The illness was discovered in 1969 when two missionary nurses died in Nigeria, West Africa. The cause of the illness was found to be Lassa virus, named after the town in Nigeria where the first cases originated. The virus is a member of the virus family Arenaviridae. It is a single-stranded RNA virus and is zoonotic or animal-borne.

In areas of Africa where the disease is endemic, Lassa fever is a significant cause of morbidity and mortality. Although the disease is mild or has no observable symptoms in about 80% of people infected with the virus, the remaining 20% have a severe multisystem disease. Lassa fever is also associated with occasional epidemics, during which the case-fatality rate can reach 50%.

Lassa fever is an endemic disease in portions of West Africa. It is recognized in Guinea, Liberia, Sierra Leone, as well as Nigeria. However, because the rodent species that carry the virus are found in other regions outside of West Africa, the actual geographic range of the disease may extend to other portions of Africa. The number of Lassa virus infections per year in West Africa is estimated at 1,00,000 to 300,000, with approximately 5,000 deaths. Unfortunately, such estimates are crude because surveillance for cases of the disease are not uniformly performed. In some areas of Sierra Leone and Liberia, it is known that 10–16% of people admitted to hospitals have Lassa fever, which indicates the serious impact of the disease on the population of this region.

The reservoir, or host, of Lassa virus is a rodent known as the "multimammate rat" of the genus *Mastomys natalensis*. It is not certain which species of Mastomys are associated with Lassa. At least two species carry the virus in Sierra Leone: *M. huberti* and *M. erythroleucus*. Mastomys rodents breed very frequently, produce large numbers of offspring, and are numerous in the savannas and forests of West, Central, and East Africa. In addition, some species, like *M. huberti*, prefer to live in human homes. All these factors together contribute to the relatively efficient spread of Lassa virus from infected rodents to humans.

There are a number of ways in which the virus may be transmitted or spread to humans. The Mastomys rodents shed the virus in urine and droppings. Therefore, the virus can be transmitted through direct contact with these materials, through touching objects or eating food contaminated with these materials, or through cuts or sores. Because Mastomys rodents often live in and around homes and scavenge on human food remains or poorly stored food, transmission of this sort is common. Contact with the virus also occurs when a person inhales tiny particles in the air contaminated with rodent excretions. This is called aerosol or airborne transmission. Finally, because Mastomys rodents are sometimes used as a food source, infection may occur via direct contact when they are caught and prepared for food.

Lassa fever may also spread through person-to-person contact. This type of transmission occurs when a person comes into contact with virus in the blood, tissue, secretions, or excretions of an individual infected with the Lassa virus. A person may also become infected by breathing in small airborne particles which an already infected person may produce by actions like coughing. The virus cannot be spread through casual contact (including skin-to-skin contact without exchange of body fluids). Person-to-person transmission is common in both village settings and in health care settings, where, along with the above-mentioned modes of transmission, the virus also may be spread in contaminated medical equipment, such as reused needles (this is called nosocomial transmission).

Symptoms of Lassa fever typically occur 1–3 weeks after the patient comes into contact with the virus. These include fever, retrosternal pain (pain behind the chest wall), sore throat, back pain, cough, abdominal pain, vomiting, diarrhea, conjunctivitis, facial swelling, proteinuria (protein in the urine), and mucosal bleeding. Neurological symptoms have also been described, including hearing loss, tremors, and encephalitis. Because the symptoms of Lassa fever are so varied and nonspecific, clinical diagnosis is often difficult.

Lassa fever is most often diagnosed by using enzyme-linked immunosorbent serologic assays (ELISA), which detect IgM and IgG antibodies as well as Lassa antigen. The virus itself may be cultured in 7 to 10 days. Immunohistochemistry performed on tissue specimens can be used to make a postmortem diagnosis. The virus can also be detected by reverse transcription-polymerase chain reaction (RT-PCR); however, this method is primarily a research tool.

The most common complication of Lassa fever is deafness. Various degrees of deafness occur in approximately one-third of cases, and in many cases hearing loss is permanent. As far as it is known, severity of the disease does not affect this complication: deafness may develop in mild as well as in severe cases. Spontaneous abortion is another serious complication.

Approximately 15–20% of patients hospitalized for Lassa fever die from the illness. However, overall only about 1% of infection with the Lassa virus result in death. The death rates are particularly high for women in the third trimester of pregnancy, and for fetuses, about 95% of which die in the uterus of infected pregnant mothers.

Ribavirin, an antiviral drug, has been used with success in Lassa fever patients. It has been shown to be most effective when given early in the course of the illness. Patients should also receive supportive care consisting of maintenance of appropriate fluid and electrolyte balance, oxygenation and blood pressure, as well as treatment of any other complicating infections.

Individuals at risk are those who live or visit areas with a high population of Mastomys rodents infected with Lassa virus or are exposed to infected humans. Hospital staff are not at great risk for infection as long as protective measures are taken.

Primary transmission of the Lassa virus from its host to humans can be prevented by avoiding contact with Mastomys rodents, especially in the geographic regions where outbreaks occur. Putting food away in rodent-proof containers and keeping the home clean help to discourage rodents from entering homes. Using these rodents as a food source is not recommended. Trapping in and around homes can help reduce rodent populations. However, the wide distribution of Mastomys in Africa makes complete control of this rodent reservoir impractical.

When caring for patients with Lassa fever, further transmission of the disease through person-to-person contact or nosocomial routes can be avoided by taking preventive precautions against contact with patient secretions (together called VHF isolation precautions or barrier nursing methods). Such precautions include wearing protective clothing, such as masks, gloves, gowns, and goggles; using infection control measures, such as complete equipment sterilization; and isolating infected patients from contact with unprotected persons until the disease has run its course.

Further educating people in high-risk areas about ways to decrease rodent populations in their homes will aid in the control and prevention of Lassa fever. Other challenges include developing more rapid diagnostic tests and increasing the availability of the only known drug treatment, ribavirin. Research is presently under way to develop a vaccine for Lassa fever.

See also **Viral Hemorrhagic Fevers**.

Additional Reading

Buckley, S.M. and J. Casals: "Pathobiology of Lassa Fever," *Int. Rev. Exp. Path.* **18**, 97 (1978).

Holmes, G.P. et al.: "Lassa Fever in the United States," *N. Eng. J. Med.* 1120 (October 18, 1990).

Jarling, P.B. et al.: "Lassa Virus Infection," *J. Inf. Dis.* **141**, 580 (1980).

Johnson, K.M. and T.P. Monath: "Imported Lassa Fever—Reexamining the Algorithms," *N. Eng. J. Med.* 1139 (October 18, 1990).

Walker, D.H.: "Lassa Fever in Man," *Am. J. Path.* **107**, 349 (1982).

Centers for Disease Control and Prevention, (CDC), Atlanta, GA.

LASTUS RECTUM.   See **Parabola**.

LATENT HEAT.   Heat gained by a substance or system without an accompanying rise in temperature during a change of state. As examples, the latent heat of fusion is the amount of heat necessary to convert a unit mass of a substance from the solid state to the liquid state at the same temperature, the pressure being that to allow coexistence of the two phases. A considerable part of the latent heat arises from the entropy increase consequent on the greater disorder of the liquid state. The latent heat of sublimation is the amount of heat necessary to convert a unit mass of a substance from the solid state to the gaseous state at the same temperature, the pressure being that to allow coexistence of the two phases.

LATERAL.   A force that acts on a structure or a structural member in a transverse direction is sometimes called a lateral load. The wind blowing upon the exposed surface of a bridge or building at right angles to its length or upon the stationary or moving traffic using the bridge constitutes one type of lateral load. The sway of a moving train on a bridge or the centrifugal force transmitted if the bridge is on a curve is a type of lateral loading. A moving crane supported on girders exerts a side thrust on the girders that may also be included in this classification.

Trusses and girders, which constitute the main load-carrying members of bridges, are not ordinarily designed to carry side loads of this nature, and consequently have very little strength in that direction. For this reason, the trusses of girders of a bridge are joined together in a horizontal plane by a system of lateral bracing composed of struts and diagonals. These members are often referred to as the laterals or the lateral system. This lateral bracing stiffens the whole bridge and opposes any sidewise deflection or vibration.

The term lateral is also used in connection with sewerage systems. Any sewer which serves the abutting property owners and in which each owner has an equal right is a common sewer. A lateral sewer is one that has no other common sewer flowing into it.

LATERAL LINE.   See **Fishes**.

LATERITE.   The sub-aerial decay of rocks in tropical regions, having a distinctly moist or rainy climate, results in the development of a residual, reddish, and usually sticky soil frequently containing concretions. The principal products of laterization are the hydrated oxides of aluminum and iron either in the crystalline or amorphous form. If the concentration of iron oxide is sufficiently high the laterite may be valuable as an iron ore. If, on the other hand, the concentration of alumina is high the laterite may be valuable as an ore of that metal.

LATEX.   Latex is a milky substance found in many plants. It is a complex emulsion in which such substances as proteins, alkaloids, starches, sugars, oils, tannins, resins, and gums are found. In most plants the latex is white; but in some it is yellow; in others, orange or scarlet.

The cells or vessels in which latex is found make up the laticiferous system. There are two very different ways in which this system may be formed. In many plants the laticiferous system is formed from cells laid down in the meristematic region of the stem or root. Rows of these cells are formed. The cell walls separating them are dissolved, so that continuous tubes, called latex vessels, are formed. This method of formation is found in the poppy family; in the rubber plant, *Hevea brasiliensis*; and in the *Cichorieae*, a section of the composite family distinguished by the presence of latex in its members. Dandelion, lettuce, hawkweed, and salsify are members of the *Cichorieae*. See also **Rubber (Natural)**.

LATIN SQUARE.   An experimental design based on a $p \times p$ array of p letters such that each letter occurs once and only once in each row and column; e.g., for $p = 4$ and the letters, A, B, C, D:

```
A  B  C  D
B  D  A  C
C  A  D  B
D  C  B  A
```

A layout of this type, for example, may correspond to 16 plots and four treatments represented by the letters.

The design may be generalized to allow for a further treatment represented by Greek letters and it is then known as a Graeco-Latin square, e.g.,

$$
\begin{array}{cccc}
A\alpha & B\beta & C\gamma & D\delta \\
B\gamma & A\delta & D\alpha & C\beta \\
C\delta & D\gamma & A\beta & B\alpha \\
D\beta & C\alpha & B\delta & A\gamma
\end{array}
$$

In this case, no combination of Roman and Greek letters occurs more than once. More general designs are sometimes known as Hyper-Graeco-Latin squares. The purpose of all these designs is to provide independent comparisons of row, column, and treatment effects. See also **Orthogonal Squares**.

LATITUDE. The celestial latitude of a point on the celestial sphere is the spherical coordinate measured from the plane of the ecliptic along a great circle passing through the object and the poles of the ecliptic.

Because of the fact that the earth is not a perfect sphere, there are several different sorts of terrestrial latitude in use. In Fig. 1, we have an ellipse $PEP'E'$ representing a section of the earth in the plane of a meridian. C is the geometric center of the earth, and the line COZ' is the line to the geocentric zenith of the point. O. The angle $ECO(\phi')$ is the geocentric latitude of the point O.

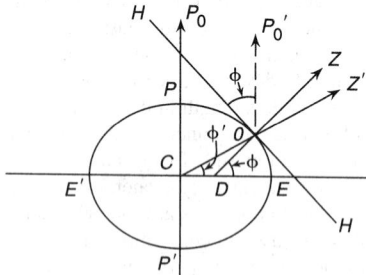

Fig. 1.   Ellipse representing a section of the earth in the plane of a meridian.

The line DOZ represents the direction of gravity at the point O and extends to the astronomic zenith of O. The angle $EDZ(\phi)$ is the astronomic latitude of the point O. The difference between the astronomic and geocentric latitude of a point, the angle $COD = \phi - \phi'$, is defined as the reduction of latitude for the point O.

Because of local influences, such as massive mountains in the vicinity, the direction of the plumb line may not be strictly perpendicular to the surface of the earth. The geographical latitude of a point is the angle, measured in the plane of the local meridian, between the equator and a line drawn perpendicular to the theoretical geoid (surface of the earth) through the point in question. The difference between astronomic and geographic latitude is always relatively small, but by no means an inappreciable angle, and is known as station error. Station error is commonly between 4 and 6 seconds of arc, but occasionally amounts to 30 or 40 seconds.

In Fig. 1, CP represents the axis of rotation of the earth, which, if extended, will pierce the celestial sphere in its pole of rotation. The parallel line OP_0' is the line from the observer at O to the pole of rotation of the celestial sphere, and the line HOH represents the plane of the astronomic horizon at O. HOP_0' is the altitude of the pole of rotation at O. Inspection of the figure will indicate that this is equivalent to the angle EDZ. This gives rise to the common definition of the astronomic latitude of a point as the altitude of the pole of rotation of the celestial sphere at the point.

Astronomic latitude may be determined in a variety of ways by observation of the celestial objects. The most direct method is to observe the altitude of some object on the meridian whose declination is known. In Fig. 2, we have a representation of the celestial sphere drawn in the plane of the local meridian of the point O. In the figure, HOH' represents the plane of the horizon; $HPZQH'$ represents the local meridian; OP the direction of the pole of rotation; OQ the direction of the equator. $HOP = \phi$ (the astronomic latitude of O), and $H'OQ = 90 - \phi$. Since $H'S$ represents the altitude of a celestial object, S, which is on the meridian, and QS represents the declination, d, of the object, we have at once the relation: $\phi = \delta + 90 - \text{altitude}$. This is the method of determination of latitude most

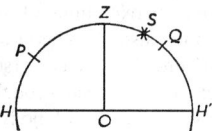

Fig. 2.   Representation of the celestial sphere drawn in the plane of the local meridian of the point O.

commonly used at sea, and it presents two fundamental difficulties to the navigator. The instant that the object is on the meridian must be accurately known and, also, the declination of the object must be observed. If both the Greenwich time and the longitude are known, the instant that the object should reach the meridian may be calculated in advance from the right ascension of the object; and the observation of altitude is taken at the predetermined instant. (Before chronometers came into use, it was necessary to watch the object very carefully and to record the maximum altitude attained by the object. If the object was the sun, the time that the maximum altitude was obtained became the local apparent noon, and was used by the navigating officer for setting the watch time for the ship.) If the observed object is a star, the declination may be immediately obtained from star catalogues; but if the sun, whose declination is changing rapidly, is the observed object, the Greenwich time of observation must be used to obtain the declination from the ephemeris.

Should the meridian observation be missed, because of cloud cover or for any other reason, the astronomical triangle may be solved to obtain the latitude if the local time of observation and the declination of the object are both known. If the object is observed very close to the meridian, and if the approximate latitude as well as the local time is known, the observation may be "reduced to the meridian" by tables. See also **Celestial Sphere and Astronomical Triangle**.

Modern navigational methods for determining latitude are discussed elsewhere. See also **Navigation**.

A meridian altitude of an object is always effected by the correction for astronomical refraction, which is always subject to error unless the object observed is close to the zenith. For accurate determination of latitude for purposes of geodetic surveying, the zenith telescope is used. See also **Earth**.

LATITUDE (Geomagnetic).   See **Geomagnetic Latitude**.

LATTICE COMPOUNDS.   Chemical compounds formed between definite stoichiometric amounts of two molecular species that owe their stability to packing in the crystal lattice, and not to ordinary valence forces.

LATTICE CONSTANT.   A length representing the size of the unit cell in a crystal lattice. In a cubic crystal, this is just the length of the side of the unit cell, but such a simple definition is not in general possible, and the lattice constant must be chosen according to the geometry of the structure in each case.

LATTICE DESIGNS.   Lattice designs form a class of experimental designs enabling a large number of unrelated treatments to be compared in randomized blocks of a reasonable size. If there are n treatments where $n = p \times q$, the treatments are thought of as generated by the combinations of two pseudofactors A and B, one at p, the other at q levels (see also **Factorial Experiment**). Two types of replicates are then laid down, confounding the main effect of A in one type and the main effect of B in the other. The most useful case is that in which $p = q$, which gives rise to blocks of equal size.

LATTICE DIMENSIONS.   According to the Bragg formula the spacing of the atomic planes can be deduced from the X-ray diffraction pattern and a knowledge of the X-ray wavelength, which can itself be measured by diffraction from a ruled grating.

LATTICE ENERGY OF CRYSTAL.   The decrease in energy accompanying the process of bringing the ions, when separated from each other by an infinite distance, to the positions they occupy in the stable lattice. It is made up of contributions from the electrostatic forces between the ions, from the repulsive forces associated with the overlap of electron shells, from the van der Waals forces, and from the zero-point energy.

LATTICE (Mathematics). A set S of elements $a, b, \ldots$, is *partially ordered* if a binary relation often denoted by the symbol $\leq$, which is reflexive, antisymmetric and transitive, is defined for certain of its elements. For example, let $a, b, \ldots$ denote the subsets of S, and let a stand in the given relation to b if the subset a is included in the subset b. A partially ordered set is a lattice if for any two elements a, b there exists an element c which is a least upper bound for a, b; that is, such that $a \leq c$, $b \leq c$ and if $a \leq e$, $b \leq e$, then $c \leq e$, and also an element d which is a greatest lower bound for a, b; that is, such that $d \leq a$, $d \leq b$ and if $f \leq a$, $f \leq b$ then $f \leq d$. These elements c and d are called the *join* (or *union*) and the meet (or *intersection*), respectively, of a and b, and are denoted by $c = a \cup b$ and $d = a \cap b$. The terms cup and cap are also used, and it is common to write $a + b$ for $a \cup b$ and ab or $a \times b$ for $a \cap b$. See also **Boolean Algebra**.

LATTICE WATER. See **Hydrate**.

LAUAN TREE. See **Mahogany Trees**.

LAUNCH WINDOW. The postulated opening in the continuum of time or space, through which a spacecraft or missile must be launched in order to achieve a desired encounter, rendezvous, impact, or the like.

LAUREL FAMILY (*Lauraceae*). Approximately one thousand species make up this family of trees and shrubs. They are characterized by alternate, simple, often evergreen leaves, and by panicles or umbels of flowers with one-Seeded drupes or berries. Some of the more familiar and economically important members of *Lauraceae* are described here. See Table 1.

Avocado. This tree, sometimes called the alligator pear, is a native of the lowlands of tropical America, but has been extensively cultivated in tropical and subtropical regions. The tree was introduced into California and Florida many years ago. The avocado is now of considerable economic importance in California. The avocado tree is attractive, with large oval to elliptical leaves and small yellowish flowers. The large green-to-brown fruit varies in shape from nearly spherical to that of a pear. The fruit is very nutritious and is rich in oil. The flavor is quite subtle and often is garnished with salt, vinegar, or salad oil. The Guatemalan avocado has an oil content up to 25%. The tree can withstand temperatures as low as $25\,°F$ ($-4\,°C$.) without damage. The Mexican species is the hardiest and of excellent quality. It can withstand temperatures as low as $20\,°F$ ($-6.7\,°C$.) if not prolonged. California growers bud this variety, using patch-a-bud technique. In Florida, the side graft is commonly used. The pulp is about 69% water, 20% oil, and contains close to 2.5% protein. The Western Avocado of the West Indies has the most tender fruit and is of a low oil content, ranging from 4 to 7%. The peel is smooth and purple. The tree cannot withstand temperatures below $28\,°F$ ($-2.2\,°C$.).

J.M. Haller (American Forests magazine, p. 29, May 1982) observes some of the unusual characteristics of the avocado. "Flowers, which in other species are certifiably and consistently male, female, or hermaphroditic, may on the avocado be male in the morning and female in the afternoon! Other species are rigorously grafted to prevent reversion to a primitive type bearing inferior fruit. But the avocado, though it may be and often is similarly grafted, will produce an astonishing variety of viable types from seed, most of them equal to any given grafted line and many of them superior. Other trees are either deciduous (leaf-shedding) or evergreen. The avocado manages both at the same time, shedding its leaves regularly each spring but not until the new season's crop is ready as a replacement (hence always green)."

Bay. This tree is native to the West Indies, but is found in France, Germany, and the coastal areas of the Americas. It is a small-to-medium-size tree, attaining a height of from 35 to 40 feet (10.5 to 12 meters), with a trunk of about 5 to 12 inches (12.7 to 30.5 centimeters) in diameter. The tree is related to the allspice and sassafras trees. The fruit is a berry. The oil from the fruit is yellow and aromatic and is the basis of bay rum. The bay tree sometimes is called the wax myrtle tree.

Cinnamon. The spice is obtained from a small tree native to Sri Lanka and India, where it is cultivated extensively. The tree grows to a height of from 25 to 40 feet (7.5 to 12 meters) and has shiny dark green, leathery leaves, small whitish flowers which have a rather disagreeable odor, and dark purple fruits. The bark of young twigs is smooth and somewhat mottled; in the older branches and the main stem, the bark becomes thick, rough and of little value. To insure the desideratum of many young branches, the limbs are severed so that many slender branches will form, a practice known as coppicing. From these slender stems, the bark is removed by lengthwise splitting and partial loosening from the stem. As it dries, it rolls back. It is then removed from the stem, the dry useless periderm scraped off, and the inner bark remaining allowed to dry completely. During drying, its color changes from pale yellow to deep brown. The tight rolls of dried bark are packed together in bundles, called pipes, and are ready for marketing as cinnamon. The bark contains considerable amounts of a powerful drug, which in large doses is a dangerous poison. The principal use of cinnamon is as a spice for pastries. By distillation of cinnamon stems and leaves, oil of cinnamon is obtained, used in flavoring candy and in scenting soaps.

Laurel. The laurel tree grows along the coastal mountains and in the Sierra Nevada mountains of California — at an altitude of about 4,000 feet (1220 meters). The California laurel and Oregon myrtle are essentially the same tree. The tree attains a height of from 50 to 80 feet (15 to 24 meters), with a trunk of 2 to 3 feet (0.6 to 0.9 meter) in diameter in mature trees.

TABLE 1. RECORD LAUREL TREES IN THE UNITED STATES[1]

Specimen	Circumference[2]		Height		Spread		Location
	Inches	Centimeters	Feet	Meters	Feet	Meters	
Avocado laurel (1999) (*Persea americana*)	172	437	75	22.9	60	18.3	California
Californialaurel (1997) (*Umbellularia californica*)	546	1387	108	32.9	118	36	California
Loblolly bay laurel(1993) (*Gordonia lasianthus*)	164	417	95	29	60	18.3	Florida
Mountain laurel (1999) (*Kalmia latifolia*)	56	142	20	6.1	19	5.8	Georgia
Redbay laurel (typ.) (1993) (*Persea borbonia var. borbonia*)	152	386	77	23.5	52	15.8	Florida
Sassafras laurel (1995) (*Sassafras albidum*)	262	665	78	23.8	69	21	Kentucky
Swampbay laurel (1999) (*Persea borbonia var. pubescens*)	63	160	36	11	47	14.3	North Carolina
PAWPAWS Common pawpaw (1986) (*Asimina triloba*)	26	66	63	19.2	29	8.8	Virginia
Smallflower pawpaw (1993) (*Asimina parviflora*)	21	53	24	7.3	17	5.2	Florida

[1]From the "National Register of Big Trees," American Forests (by permission).
[2]At 4.5 feet (1.4 meters).

The branches are erect, long, and thick. The bark is thin, scaly, and dark brown. The leaves are about 2 to 5 inches (5 to 12.7 centimeters) in length and one-half to one or more inches (1.3 to 2.5 centimeters) wide. The underside is light green; the top side is leathery, glossy, and thick. The flowers are in clusters, pale yellow and small. The fruit of approximately 1 inch in length hangs in clusters of two and three. It is about the size of an olive and is a yellow-green color, containing one Seed. Laurel wood, used for cabinet work, veneers, and garden tool handles, is fine-grained and hard and weighs approximately 40.5 pounds per cubic foot when dry (649 kilograms per cubic meter).

Pawpaw. This tree is found in the southern and midwestern states of the United States. It is related to the banana plant. The tree is small, usually grows wild, and is found in woodland areas. The fruit is exotic in appearance with a rich golden color. The fruit ranges from 3 to 5 inches (7.6 to 12.7 centimeters) in length and hangs from the tree in clusters. Pawpaw wood is spongy and weak and of no commercial value. The flower is purple and fragrant.

P. Stevenson presents an interesting portrait of the pawpaw in the March/April 1990 issue of *Amer. Forests,* page 46.

Sassafras. Frequently more of a shrub than a tree, the sassafras plant is found in the New England states, west to Wisconsin, and south to the Gulf coast. In many areas, the sassafras grows into a large tree, ranging up to 30 or 40 feet (9 to 12 meters) in height, with a trunk measuring from 8 inches to 2 feet (20 centimeters to 0.6 meter) in diameter. The record sassafras tree in the United States, as reported by American Forests Association, is located in Owensboro, Kentucky. See table.

All parts of the sassafras tree have a characteristic fragrance. The branch grows horizontal. The leaf is bright green and glossy and well known for its "mitten" shape. In the autumn, the tree turns a golden red and is quite showy.

The flower is small, yellow, and occurs in clusters. The flower appears before the leaf and is staminate with a six-lobed calyx, orange stalked glands, and nine stamens. The fruit is a dark blue, thin and fleshy berry of oblong shape. Although eaten readily by birds, the fruit is not enjoyed by humans. The bark is thick, scaly, and gray with longitudinal ridges. The wood is brittle and coarse grained and, although it resists moisture decay well, it is seldom considered of commercial value. Sassafras oil is sometimes used in soaps and toiletries.

H. Clepper elucidates further details of the sassafras tree in the March/April 1989 issue of *Amer. Forests,* page 33.

LAURIC ACID. Also called dodecanoic acid, formula $CH_3(CH_2)_{10}COOH$. A fatty acid that occurs in many vegetable oils and fats as the glyceride, especially in coconut oil and laurel oil. See also **Vegetable Oils (Edible).** Combustible. It takes the form of colorless needles at room temperature. Specific gravity 0.833; mp 44 °C; bp 225 °C (100 millimeters pressure). Insoluble in water; soluble in alcohol and ether. It is derived by the fractional distillation of coconut oil. Lauric acid is used in alkyd resins; wetting agents; soaps; detergents; cosmetics; insecticides; food additives.

LAVA. Molten material that has poured out on the surface of the earth and, due to relief of pressure, may have lost much of its original gas and water content during its relatively rapid consolidation. The term lava is used for both the liquid and the consolidated state of the igneous material. Lava may be erupted either by volcanoes or from fissures. Flowing lava is shown in Fig. 1. The most extensive lava flows are fissure eruptions, such as the Columbia Plateau basalts in Oregon or the plateau basalts of the Deccan, India, which are derived from basic magma. Had this magma, either basic or acid, cooled slowly beneath the surface of the earth under great pressure and with all its original gases, the resulting rock would have had a coarser texture and somewhat different mineral content.

See also **Earth; Ocean;** and **Volcano.**

LAVAGE. See **Empyema.**

LAVOISIER, ANTOINE (1743–1794). Lavoisier was a French chemist who is often referred to as the father of modern chemistry. He is remembered for showing that air is not an element but a mixture of gases. He is responsible for founding the system of naming compounds. Lavoisier recognized and named oxygen (1778). He explained combustion. While working for the government as a tax collector, he developed a metric system so as to have uniformity of weights and measures throughout France.

J. M. I.

LAW OF AREAS. See **Kepler's Laws of Planetary Motion.**

LAW OF COSINES. See **Direction Cosine;** and **Pythagorean Theorem.**

LAW OF LARGE NUMBERS. There are various laws of large numbers but the essential idea is exactly the same in each case. If the size of a sample is increased indefinitely or becomes very large, good sample estimates of population parameters will tend to concentrate more and more closely about the true value. Bernoulli's theorem is perhaps the simplest illustration of a law of large numbers.

Put another way, such laws state conditions under which random variables converge in probability to constants as some parameter n(usually a sample number) tends to infinity. *Strong laws* are concerned with showing that, for example, a variable $x$converges to a value μ with probability unity. Weak laws consider conditions under which the probability that $|x - \mu|$ is greater than some given $\in$, tends to zero.

LAWRENCIUM. Chemical element, symbol Lr, at. no. 103, at. wt. 257 (mass number of known isotope), radioactive metal of the Actinide series, also one of the Transuranium elements. ^{103}Lr was identified in 1961 by A. Ghiorso, T. Sikkeland, A. Larsh, and R. Latimer at the University of California at Berkeley.

This method used to produce and identify lawrencium was similar to that used in the later, direct-counting experiments performed in connection with the production of nobelium at Berkeley. About 3 micrograms of a mixture of californium isotopes were bombarded with boron ions accelerated in the heavy-ion linear accelerator. The atoms of lawrencium recoiled from the target into an atmosphere of helium, where they were electrostatically collected on a copper conveyor tape. This tape was then periodically pulled into place before radiation detectors to measure the emission rate and the energy of the alpha particles being emitted. By this means, it was possible to identify the lawrencium isotope ^{257}Lr, with a half-life of 8 seconds. At present, because of the short half-life and the lack of a suitable daughter isotope, available in the case of nobelium, it has not been possible to perform a chemical identification.

Another isotope, ^{256}Lr, half-life about 45 seconds, was reported by the Soviet Union in 1965. It was produced by impact of oxygen atoms (^{18}O) on americium (^{243}Am). It decayed by alpha-particle emission and electron capture to form ^{252}Fm. See also **Chemical Elements.**

Lawrencium has been found to behave quite differently from dipositive nobelium and, in fact, it is comparable to the tripositive elements that appear earlier in the Actinide series.

Additional Reading

Eskola, K., Eskola, P., Nurmia, M., and A. Ghiorso: "Studies of Lawrencium Isotopes with Mass Numbers 255 through 260," *Phys. Rev.,* **4,** 2, 632–642 (1971). (A classic reference.)

Fuger, J. and L.R. Morss: "Transuranium Elements: A Half Century," American Chemical Society, Washington, DC, 1992.

Fig. 1.   Lava flow.

Ghiorso, A., T. Sikkeland, Larsh, A.E., and R.M. Latimer: "New Element, Lawrencium, Atomic Number 103," *Phys. Rev., Lett.*, **6**, 9, 473–475 (1961). (A classic reference.)

Greenwood, N.N. and A. Earnshaw: "Chemistry of the Elements," 2nd Edition, Butterworth-Heinemann, Inc., Woburn, MA, 1997.

Lide, D.R.: "CRC Handbook of Chemistry and Physics 2000–2001," 81st Edition, CRC Press, LLC., Boca Raton, FL, 2000.

Seaborg, G.T. and W.D. Loveland: "The Elements beyond Uranium," John Wiley & Sons, Inc., New York, NY, 1990.

LAWSON CRITERION. See **Nuclear Reactor**.

LAWSONITE. This calcium aluminum silicate mineral, $CaAl_2(Si_2O_7)$ $(OH)_2 \cdot H_2O$, is found as grains and veins within the metamorphic rocks, gneisses, and schists. It was found originally on the Tiburon Peninsula, San Francisco Bay, California, but also occurs in schistose rocks in France and New Caledonia. The mineral has a hardness of 7; specific gravity of 3.09. It is colorless, pale blue to bluish gray, translucent, with vitreous to greasy luster. The mineral crystallizes in the orthorhombic system.

LAXATIVE. See **Constipation**.

LAZULITE. This mineral crystallizes within the monoclinic system, a basic phosphate of magnesium and aluminum, $MgAl_2(OH)_2(PO_4)_2$. Ferrous iron can substitute for the magnesium and the isomorphous mineral scorzalite is the product. Usually occurs massive but acute pyramidal crystals are not uncommon. Color is azure-blue to bluish-green, usually translucent (rarely transparent), with vitreous luster. It has a hardness of 5.5–6, with specific gravity of 3–3.1.

Lazulite is a rare mineral found principally within high-grade metamorphic rocks. Notable world crystal occurrences are Salzburg, Austria; Syria; Hörnsjöberg, Sweden; Madagascar; Brazil; and Graves Mountain, Georgia. When transparent, the mineral can be cut into gem stones.

LAZURITE. The mineral lazurite or lapis lazuli has been used since ancient times for jewelry and other ornamental purposes. Ground to powder it forms the pigment ultramarine, now, however, largely superseded by artificial preparations. Lapis lazuli is a mixture of minerals, lazurite being the chief component. This mineral is isometric, and chemically a sodium, calcium, aluminum sulfo-chlorosilicate. A general formula is $(Na, Ca)_8(Al, Si)_{12}O_{24}(S, SO_4)$. Lapis lazuli has a hardness of 5–5.5; specific gravity, 2.4; color, various shades of blue; luster, vitreous to greasy; translucent to opaque. Localities are Afghanistan, Siberia, Chile, and California.

LAZY EYE. See **Vision and the Eye**.

L-BAND. A frequency band used in radar extending approximately from 0.390 gigacycles per second to 1.55 gigacycles per second.

L-DISPLAY. In radar, a display in which a target appears as two horizontal blips, one extending to the right and one to the left, from a central vertical time base. When the radar antenna is aligned in azimuth at the target both blips are of equal amplitude. When not correctly pointed the relative blip amplitude indicates the pointing error. The position of the signal along the baseline indicates target distance. The display may be rotated 90 degrees when used for elevation instead of azimuth aiming. Also called *L-scan, L-scope*, or *L-indicator*.

L-DOPA. See **Parkinson's Disease**.

LEAD. Chemical element, symbol Pb. at. no. 82, at. wt. 207.2, periodic table group 14, mp 327.5°C. bp 1740°C, density 11.35 g/cm³. (20°C). Elemental lead has a face-centered cubic structure with an edge length of 4.950 Å.

Lead is a white to bluish-gray metal, soft, malleable, and slightly ductile; tarnishes in air, forming a film of oxide, forms oxide scum upon heating the molten metal in air; soluble in dilute HNO_3; HCP or H_2SO_4 attack lead only slightly, the extent depending markedly upon the concentration and the temperature; slowly dissolves in H_2O and consequently the use of lead constitutes a health hazard due to its toxic effect; attacked by solutions of organic acids or sodium hydroxide. Lead is one of the four most largely produced and utilized metals, and considerable scrap metal is recovered. Used (1) in construction and apparatus where workability is demanded, and definite resistance to corrosion is supplied by the metal, (2) as a constituent of various alloys, especially solder, type metal, pewter, and fusible alloys, (3) for storage battery plates, (4) for shot and bullets, and (5) as a protective coating for iron and steel.

Lead has four naturally occurring isotopes. In order of abundance, these are ^{208}Pb, ^{206}Pb, ^{207}Pb, and ^{204}Pb. There are ten unstable isotopes, 200–203, 205, and 209–214. See also **Radioactivity**. In terms of abundance, lead is scarcely represented in the earth's crust, the average composition of igneous rocks containing only 0.002% Pb by weight. In terms of cosmic abundance, an estimate made by Harold C. Urey in 1952, using silicon as a basis with the figure of 10,000, lead had an abundance figure of less than 0.02. In terms of presence in seawater, lead is 27th among the elements, with an estimated 14 tons per cubic mile (3 metric tons per cubic kilometer) of seawater. In this regard, it is comparable to tin, copper, arsenic, protactinium, and selenium.

The atomic weight varies because of natural variations in the isotopic composition of the element, caused by the various isotopes having different origins: ^{208}Pb is the end product of the thorium decay series, while ^{207}Pb and ^{206}Pb arise from uranium as end products of the actinium and radium series respectively. Lead-204 has no existing natural radioactive precursors. Electronic configuration $1s^2 2s; \& 22p^6 3s^2 3p^6 3d^{10} 4s^2 4p^6 4d^{10} 4f^{14} 5s^2 5p^6 5d^{10} 6s^2 6p^2$. Ionic radius Ph^{2+} 1.18 Å. Pb^{44} 0.70 Å. Metallic radius 1.7502 Å. Covalent radius (sp^3) 1.44 Å. First ionization potential 7.415 eV; second, 14.97 eV. Oxidation potentials $Pb \rightarrow Pb^{2+} + 2e^-$, 0.126 V; $Pb^{2+} + 2H_2O \rightarrow PbO_2 + 4H^+ + 2e^-$, -1.456 V; $Pb + 2OH^- \rightarrow PbO + H_2O + 2e^-$, 0.576 V; $Pb + 3OH^- \rightarrow HPbO_2^- + H_2O + 2e^-$, 0.54 V. Other physical properties are given under **Chemical Elements**.

Lead is of interest as being the terminal product of radioactive decay. Thus, while ordinary lead has the atomic weight 207.19 (being composed of 1.37% ^{204}Pb, 26.26% ^{206}Pb, 20.8% ^{207}Pb and 51.55% ^{208}Pb), the isotopic composition, and hence the atomic weight, varies somewhat in lead from meteorites, from deep-seated rocks and from uranium ores (the last being somewhat less dense, as would be expected from the fact that ^{206}Pb is the end-product of the uranium series). These variations in isotopic composition of lead permit of calculations of the age of the earth (and the meteorites).

Lead Melting Point as a Standard. Melting, defined as the equilibrium transition between crystalline and liquid states, is of large concern in the development of the physical and materials sciences. To date, some of the purest crystals of silicon, diamond, and other technologically important materials have been produced from melts. Studies of melts also are of significance in understanding the interiors of terrestrial plants and, in fact, of Earth. In research at the University of California (Berkeley), studies of the effects of high pressure on the fusion temperature of lead have been underway. The advantages of studying lead are outlined by the investigators as: (1) the melting temperature of lead at ambient pressures is low and well determined, (2) lead is highly compressible and therefore should show the effects of pressure, (3) the behavior of lead under pressure is relatively simple, involving only one known polymorphic transition (from face-centered cube to hexagonal close-packed crystal structure), and (4) shock-wave experiments have been carried out previously to document the compression of both crystalline and molten lead at simultaneously high pressures and temperatures.

Occurrence and Processing. Galena, PbS, lead sulfide, is the source of over 95% of the lead currently produced. Bodies containing galena range from 3% to 30% lead. One of the most widely distributed sulfide minerals, galena frequently occurs along with sphalerite, ZnS. The lead-zinc ores processed usually contain recoverable quantities of copper, silver, antimony, and bismuth. Principal sources being worked are in Australia's Broken Hill area in New South Wales, the western United States, Canada, Mexico, Peru, former Yugoslav Republics, and the former Soviet Union. When groundwater reacts with galena, cerussite, $PbCO_3$, is formed; when galena is in contact with sulfate solutions generated by the oxidation of sulfide minerals, anglesite, $PbSO_4$, may be formed. See also **Anglesite**; **Cerussite**; and **Galena**.

In processing, the ore first is crushed, wet-ground, and classified to a point where it is at least 90% less than 200 mesh. Separation of the sulfide ore from the gangue is aided by flotation agents. The resulting concentrates contain from 45% to 60% lead, from zero to 15% zinc, and often a few ounces (~50 grams) of gold and up to 50 ounces (1.4 kilograms) of silver per ton. Copper content may be as much as 3%, arsenic, 0.4%, and antimony, 2%. The sulfur content (10 to 30%) is reduced by roasting in a Dwight-Lloyd sintering machine. This sulfur reduction is necessary because PbS is not reduced by carbon or carbon monoxide at blast-furnace temperatures. Once formed, the sinter, together with limestone and coke, is fed into a blast furnace. Further oxidation and electrolytic methods may be used to refine the lead. Lead is commercially produced to standards of very high purity. The minimum lead content permitted by specifications for Pig Lead (7 classifications) is 99.73%. Fully refined lead averaging 99.99% lead is obtainable. Large quantities are used for production of chemicals. At one time, primary uses for lead chemicals were in the production of paint pigments and lead tetraethyl gasoline additive.

Lead Metals and Alloys. Lead is soft and ductile and is readily worked by common methods, predominantly by rolling and extruding. Lead is easily formed and readily joined by welding (burning), or by soldering and can be bonded to steel, or used as a liner for steel, wood, concrete, and other materials. Lead is widely used in this manner because of its excellent resistant to atmospheric and soil corrosion, and attack by sulfuric and phosphoric acids. Lead generally does not resist the action of the organic acids, nor the oxidizing mineral acids, such as HNO_3. Lead is attacked by alkalies.

Due to its low melting point, pure lead will very gradually flow or creep at room temperature. Thus, lead sheeting used as a roofing material on old buildings will usually be thicker at the lower edge than at the upper edge. Other examples of creep occur under low sustained stresses due to the oil pressure in lead-covered power conducting cable, for example, or due to the weight in the case of a deep tank lined with sheet lead. To counter the effects of creep, lead containing 0.06% copper (*chemical lead* or *acid lead*) is preferred.

The addition of antimony in amounts up to 12% greatly improves the casting properties and increases the hardness very materially. These properties make possible the casting of intricately shaped antimonial lead storage-battery grids which, including the weight of the lead oxide paste applied to them, constitute the largest single use for the metal.

Tin and lead in various proportions form a highly useful series of alloys generally known as the soft solders which are used for joining copper, iron, nickel, lead, zinc and even glass. The solders can be applied by means of a soldering tool, by wiping, by hot-dipping, or by special machines as in the tin-can industry. Numerous compositions are used, the most popular of which are listed in the accompanying table.

Further additions of bismuth, cadmium, and antimony to the tin-lead alloys result in the low melting or "fusible" alloys widely used as safety devices, the melting points of which can be varied to suit a wide range of requirements. The type metals of the printing industry are lead-tin-antimony alloys having the requisite hardness and good casting properties needed for high-fidelity reproduction.

Babbitt metals (white-metal bearing alloys) are generally classified as either tin-base or lead-base. The true tin-base Babbitts contain only tin, antimony and copper, and have been used for many years. The practice of adding up to 25% lead to the tin Babbitts to reduce their cost is to be avoided since the net result is an expensive series of alloys with inferior properties to the inexpensive lead-base Babbitts. The lead-base bearing alloys of the older type usually contain lead, antimony and tin, and while not considered the equal of the tin-base alloys for severe service have been widely employed due to their low cost. The lead-base alloy containing arsenic has found extensive use and has come to the fore of this group since it has successfully met many automotive and other severe service requirements. All of these alloys render their most efficient service when used in the form of a thin lining bonded to a bronze or steel shell. See Table 1.

Lead Eliminated from Free-cutting Alloys. Among the numerous efforts being made to eliminate lead from the environment, including the potable water plumbing systems, free-cutting copper alloys that contain no lead have been developed. As reported in late 1991, bismuth, as a replacement, has significant potential as a nontoxic alternative to lead to enhance the machinability of copper. When bismuth is used alone, however, the element embrittles copper because of its tendency to "set" grain boundaries. J.T. Plewes (see reference) ascribes this characteristic to the large difference in surface tension between copper and bismuth. It has been found that adding a third element in modest amounts removes this limitation of bismuth. Such elements include phosphorus, indium, and tin.

TABLE 1. REPRESENTATIVE LEAD AND TIN ALLOYS

Name	Pb	Sn	Sb	Cu	Bi	Ag	Cd	Typical Application
Lead Alloys								
Chemical or acid lead	99.9			.06				Tank linings, coils, etc., power cable sheath.
Cable sheath	98.9		1.0					Telephone cable sheath.
Hard lead	96–92		4–8					Cast shapes, wrought sheet and pipe.
Battery grid metal	92–88	.25	8–12					Cast battery grids.
Solders								
Soft solder	50	50						General purposes, most popular solder.
Wiping solder	60	40						For wiping joints in cables, lead pipes, etc.
	60	37.5	2.5					
"Fine solder"	40	60						For making joints at low temperature.
Solder	95–97.5					5–2.5		High temperature solder.
Fusible Alloys								
Wood's metal	25	12.5			50		12.5	Melts in hot water at 154°F. Wets glass. Wide range of melting points possible with changes in composition for automatic sprinkler systems and other safety devices.
Matrix metal	28.5	14.5	9		48			For anchoring punches, etc., in jigs and fixtures. Expands on freezing.
Bending alloy	26.5	13.5			50		10	Filler for tubes, etc., during bending. Melts out in hot water.
Type Metals								
Electrotype	93	3	4					
Linotype	84	4	12					
Stereotype	80.5	5.75	13.75					
Monotype	76	8	16					Single type.
Tin Base Babbitts								
		89	7.5	3.5				General usage.
		83.3	8.3	8.3				Hard Babbitt.
Lead Base Babbitts								
	82.5	1.0	15	.5	1.0 As			General usage.
	80	5	15					General usage.
	75	10	15					General usage.

Notes: Figures given in percent. Wood's metal melts at ~68°C in water.

Chemistry of Lead. A number of oxides of lead are known, but not all are daltonide compounds. Thus, lead(I) oxide, Pb_2O, made by heating lead(II) oxalate, has been shown by x-ray analysis to be a mixture of the metal and lead(II) oxide, PbO. The latter is obtained by heating lead in air, which yields a yellow, rhombic material, which has a peculiar layer structure having each lead atom attached to four oxygen atoms all lying on the same side of it, forming a square pyramid with the lead at the apex. Each oxygen atom is surrounded tetrahedrally by four lead atoms. Another form of PbO, somewhat more stable and soluble in water, red in color, and tetragonal in structure, may be obtained along with the yellow form by alkaline dehydration of $Pb(OH)_2 \cdot$ PbO is amphiprotic, but only weakly acidic. Lead(IV) oxide, Pb_2O, is obtained by action of chlorine on alkaline solutions of lead(II) oxide or acetate. The reaction is $Pb(OH)_3^- + ClO^- \rightarrow PbO_2 + Cl^- + OH^- + H_2O \cdot PbO_2$ can also be produced on a lead or platinum anode by electrolysis in acidic solution. Like the lower elements of main group 4, lead(IV) forms tetrahedral bonds exhibiting sp^3 hybridization. In its relatively more stable salts, however, the $6s^2$ electrons are unused, and Pb^{2+} ions are formed by loss of the $6p^2$ electrons. These facts explain the marked difference between the essentially covalent character of many of the tetravalent compounds and the essentially electrovalent character of the divalent compounds, as well as the peculiar structure of PbO and many other Pb(II) compounds.

The dioxide, Pb_2O, has rutile structure, and the compound is a strong oxidizing agent. It is also amphiprotic, giving unstable lead(IV) salts with acids, and orthoplumbates, $M_4^1PbO_4$, or metaplumbates, $M_2^1PbO_3$, upon fusion with alkalies. Lead dioxide dissolves in aqueous alkali with formation of the ion $Pb(OH)_6{}^{2-}$, the alkali salts of which are isomorphous with the corresponding stannates and platinates. Lead sesquioxide, Pb_2O_3, has been shown not to exist as a stable phase.

Lead orthoplumbate, Pb_2PbO_4, red lead, is similarly described as a salt, in this case an orthoplumbate of divalent lead, Pb_2PbO_4, because on treatment with nitric acid, two-thirds of the lead dissolves and one-third remains as PbO_2. It is prepared in the red form by atmospheric heating of PbO, and in a black form by reaction of PbO with pure oxygen. Red lead is formed of PbO_6 octahedra (with one common edge) linked by lead atoms covalently bonded to three oxygen atoms.

The lead dihalides are known for all four of the common halogens. They are not strictly ionic in the anhydrous state, but they dissolve in (hot) water to give Pb^{2+} ions, more or less hydrated. They are much less soluble in cold water. They also form complex compounds such as M_2PbCl_4, MPb_2Cl_5, M_4PbF_6, and $MPbF_3$, where M is an alkali metal. The compound formed, especially of the fluoroplumbates(II) depends somewhat on the alkali metal, some of which form/nondaltonide (berthollide) compounds. Of the lead tetrahalides, only PbF_4 and $PbCl_4$ are known, the fluoride being prepared by fluorination of PbF_2. The chloride, which easily loses chlorine, is made by careful acidification of a hexachloroplumbate(IV). $PbCl_4$ forms the complex compound ammonium hexachloroplumbate, $(NH_4)_2PbCl_6$, upon addition to its solution of solid ammonium chloride.

Lead(II) inorganic compounds and salts of organic acids are far more numerous than those of lead(IV), as is to be expected from the essentially covalent character of the latter. In addition to the oxides and halides already discussed, there are lead(II) compounds of essentially all of the common anions, including many basic compounds. Thus lead(II) chloride forms such basic compounds as $PbCl_2 \cdot Pb(OH)_2$, $PbCl_2 \cdot PbCl_2 \cdot 2PbO$, $PbCl_2 \cdot 3PbO$, and $PbCl_2 \cdot 7PbO$. In fact, a whole series of lead salts are derived from the hydroxide, some of which are double compounds, such as $PbX_2 \cdot 2Pb(OH)_2$ and some of which, of composition $\cdot Pb(OH)X$, have been shown to be dimeric of the general formula

$$\left[Pb \binom{HO}{HO} Pb \right] X_2$$

Other lead compounds include the following:

Acetates. Lead acetate, "sugar of lead" $Pb(C_2H_3O_2)_2 \cdot 3H_2O$, white crystals, soluble, formed by reaction of lead oxide and acetic acid, and then crystallization. Used (1) to furnish a soluble lead salt, (2) as a mordant in dyeing and printing textiles, (3) as a paint and varnish drier, basic lead acetate, white crystals, soluble, formed by reaction of lead acetate solution and lead oxide, and then crystallization. Used as a coagulating, clarifying, and deacidifying agent for many organic solutions.

Arsenate. Lead arsenate, arsenate of lead $Pb_3(AsO_4)_2$, white precipitate, formed by reaction of soluble lead salt solution and sodium arsenate solution. Used as an insecticide. Banned or tightly controlled in some countries.

Azide. Lead azide PbN_6, white precipitate, formed by reaction of soluble lead salt solution and sodium azide solution (white solid, formed by reaction of sodamide $NaNH_2$ upon heating in nitrous oxide N_2O gas). Used as a detonator.

Borate. Lead borate $Pb(BO_2)_2$, white crystals, insoluble, by reaction of lead oxide and boric acid solution. Used in preparing special types of glass.

Carbonates. Lead carbonate $PbCO_3$, white precipitate, formed by reaction of soluble lead salt solution and sodium carbonate solution in the cold; basic lead carbonate, formed by reaction of (1) soluble lead salt solution and hot sodium carbonate solution, and (2) lead sheets, carbon dioxide and acetic acid, and pigment, the quality depending largely upon the conditions of the reaction.

Chromates. Lead chromate, "chrome yellow" $PbCrO_4$, yellow precipitate, by reaction of soluble lead salt solution and sodium dichromate or chromate solution, melting point of lead chromate 844 °C. Used as a pigment; basic lead chromate, red solid, insoluble, formed by heating lead chromate and sodium hydroxide solution.

Nitrates. Lead nitrate $Pb(NO_3)_2$, white crystals, soluble, formed by reaction of lead oxide and nitric acid, and then crystallization, decomposes on heating leaving lead oxide residue. Used to furnish a soluble lead salt; basic lead nitrate, formed by reaction of lead nitrate solution and lead oxide.

Oxalate. Lead oxalate PbC_2O_4, white precipitate, formed by reaction of soluble lead salt solution and ammonium oxalate solution, yields lead suboxide on heating at 300 °C out of contact with air.

Phosphate. Lead phosphate $Pb_3(PO_4)_2$, white precipitate, by reaction of soluble lead salt solution and sodium phosphate solution.

Sulfates. Lead sulfate $PbSO_4$, white precipitate, formed by reaction of soluble lead salt solution and sulfuric acid or sodium sulfate solution; basic lead sulfate, "sublimed white lead," white solid, formed (1) by reaction of lead sulfate and lead hydroxide in water (slow reaction), and (2) by roasting galenite in a current of air.

Sulfide. Lead sulfide PbS, brownish-black precipitate, formed by reaction of soluble lead salt solution and hydrogen sulfide or sodium or ammonium sulfide, soluble in dilute nitric acid.

In the great majority of organometallic compounds of lead, the metal is tetravalent and covalently bonded, although the organolead group includes many compounds with both organic radicals and halogen atoms attached to Pb which are not to be described merely as covalent compounds. More than five hundred organometallic compounds of lead have been reported, many of which are named as substituted plumbanes, although PbH_4 is not a starting point in their production. Tetraethyl lead, $Pb(C_2H_5)_4$, is made from a sodium-lead alloy and ethyl chloride.

Like carbon and silicon, and to a lesser extent, germanium and tin, lead forms binary compounds with metals, such as Na_4Pb_7 and Na_4Pb_9. These materials are essentially salt-like, and contain polyplumbide anions. They are of theoretical interest, because they are intermediate in character between stoichiometric compounds (daltonide compounds) and intermediate phases. The two compounds cited dissolve in liquid ammonia, electrolyze in such solutions to give the metals, and apparently form ions such as $[Pb_7]^{4-}$ and $[Pb_9]^{4-}$ which readily form amine complexes.

Lead in Biological Systems — Toxicity

Lead has been identified as a biological system deterrent for decades. It is only recently, however, that studies pertaining to low-dosage exposures of lead have been published despite the fact that probably millions of words have appeared in various publications on the overall topic of lead poisoning.

From a qualitative standpoint, exposure to lead results in a clinical picture of hypertensive encephalopathy, neuropathy, and hemolytic anemia characterized by coarse basophilic stippling in red blood cells. The mechanism of lead's action on human tissue is complex. For one thing, lead blocks heme synthesis. This leads to a build-up of red blood cell protoporphyrin. Lead interferes with cell metabolism by causing a deficiency of pyrimidine $5'$-nucleotidase. Lead attacks erythrocyte membrane phospholipids with resultant loss of potassium and interference with the sodium-potassium balance. Diagnosing lead poisoning may involve a determination of the free erythrocyte protoporphyrin level as well as determination of blood and urine levels. Once confirmed, further exposure to lead must be stopped immediately. Chelating compounds, such as $CaNa_2EDTA$, may be administered intravenously over an 8-hour period for several days. This may be followed by treatment with oral penicillamine for several days.

Lead poisoning can lead to chronic renal failure. In its effect on kidney function, lead acts much like cadmium. Chronic exposure to or ingestion of practically any heavy metal, such as lead, is the most common path to polyneuropathy. Where effects of heavy metals on the peripheral nervous system are suspected, many physicians will require testing for metal in hair, fingernails, serum, and urine of the patient. Habitual sniffing of leaded gasolines can lead to lead poisoning. Scientists have compared the lead concentration in the diets of present Americans (0.2 part per million) with the diets of prehistoric peoples (estimated to be less than 0.002 part per million). Some investigators believe that the presence of "natural" lead contamination has been grossly overestimated and that what has appeared to be natural has been the result mainly of a gradual build-up of lead pollution in the air derived from anthropogenic sources. The principal sources of atmospheric lead contamination include (1) natural sources, such as wind-blow volcanic dust, sea spray, forest foliage, and volcanic sulfur compounds; and (2) anthropogenic sources, such as lead alkyls (present in fuels), iron smelting, lead smelting, zinc and copper smelting, and the burning of coal. Much remains by way of research into the sources of lead contamination, including the contributions of atmospheric pollution, of food containers, and of food processing equipment.

In 1990, H.I. Needleman and co-researchers (University of Pittsburgh, Boston University, and Harvard University) reported their findings on the long-term effects of exposure to low doses of lead in children. An abstract of the report is as follows:

> To determine whether the effects of low-level lead exposure persist, we reexamined 132 of 270 young adults who had initially been studied as primary school-children in 1975 through 1978. In the earlier study, neurobehavioral functioning was found to be inversely related to dentin lead levels. As compared with those we restudied, the other 138 subjects had somewhat higher lead levels on earlier analysis, as well as significantly lower IQ scores and poorer teachers' ratings of classroom behavior.
>
> When the 132 subjects were reexamined in 1988, impairment in neurobehavioral function was still found to be related to the lead content of teeth shed at the ages of six and seven. The young people with dentin lead levels >20 ppm had a markedly higher risk of dropping out of high school (adjusted odds ratio, 7.4; 95 percent confidence interval, 1.4 to 40.7) and of having a reading disability (odds ratio, 5.8; 95 percent confidence interval, 1.7 to 19.7) as compared with those with dentin lead levels <10 ppm. Higher lead levels in childhood were also significantly associated with lower class standing in high school, increased absenteeism, lower vocabulary and grammatical-reasoning scores, poorer hand—eye coordination, longer reaction times, and slower finger tapping. No significant associations were found with the results of 10 other tests of neurobehavioral functioning. Lead levels were inversely related to self-reports of minor delinquent activity.
>
> We conclude that exposure to lead in childhood is associated with deficits in central nervous system functioning that persist into young adulthood.

An interesting professional critique of the Needleman report is summarized by J. Palca (reference listed).

The lead elimination and clean-up problem has numerous parallels with the asbestos pollution problem. The main problem is not one of finding substitutes for these substances, because lead-free paints, for example, have been available for several years, just as substitute insulating materials for asbestos have been found. As pointed out in an excellent article by Pollack (reference listed), the problem (or dilemma) lies with the cleanup of old structures that have such materials installed; and to what limits must one go, within the limitations of financial resources, to remove such materials; and, once removed, how to dispose of them safely. While removing all lead-painted surfaces from schools, for example, ultimately can provide assurance that children will not be exposed to the long-term effects of lead, a great deal of new exposure to workmen and the immediate neighborhoods of such removal projects can occur. Modern technology has designed equipment to protect the safety of restoration or demolition crews, but there remains the enforcement of their using such equipment. One crux

of the problem is that posed by the apparent effects of lead in very low dosages. Such pollutants of low concentration, as may be typified by the creation of dust and windborne aerosols, definitely exacerbates the problem. Obviously, the problem becomes one more hampered by socioeconomic measures than by technology.

Additional Reading

Bodwal, B.K. et al.: "Ultralight-Pressure Melting of Lead: A Multidisciplinary Study," *Science*, 462 (April 27, 1990).

Considine, D.M. and G.D. Considine: "Van Nostrand Reinhold Encyclopedia of Chemistry," 4th Edition, Van Nostrand Reinhold Company, Inc., New York, NY, 1984. (A classic reference).

Davis, J.R.: "Metals Handbook," 2nd Edition, ASM International, Materials Park, OH, 1998.

Greenwood, N.N. and A. Earnshaw: "Chemistry of the Elements," 2nd Edition, Butterworth-Heinemann, Inc., Woburn, MA, 1997.

Holden, C.: "Resurrected Lead," *Science*, 192 (October 11, 1991).

Holden, C.: "Toxic Waste Program Lacks Science Base," *Science*, 797 (November 8, 1991).

Krebs, R.E.: "The History and Use of Our Earth's Chemical Elements: A Reference Guide," Greenwood Publishing Group, Inc., Westport, CT, 1998.

Lide, D.R.: "CRC Handbook of Chemistry and Physics 2000-2001," 81st Edition, CRC Press, LLC., Boca Raton, FL, 2000.

Needleman, H.L. et al.: "The Long-Term Effects of Exposure to Low Doses of Lead in Childhood," *N. Eng. J. Med.*, 83 (January 11, 1990).

Palca, J.: "Get-the-Lead-Out Guru Challenged," *Science*, 842 (August 23, 1991).

Plewes, J.T. and D.N. Loiacono: "Free-Cutting Copper Alloys Contain No Lead," *Advanced Materials & Processes*, 23 (October 1991).

Pollack, S.: "Solving the Lead Dilemma," *Technology Review (MIT)*, 22 (October 1989).

Raloff, J.: "Beverages Intoxicated by Lead in Crystal," *Science News*, 54 (January 26, 1991).

Stwertka, A. and E. Stwertka: "A Guide to the Elements," Oxford University Press, Inc., New York, NY, 1998.

LEAD-ACID BATTERY.   See **Battery**; and **Electric Car**.

LEAD SCREW.   The screw that controls the longitudinal motion of a tool on a lathe or other machine tool. Also, the screw that drives the cutting head across a recording disk in the initial recording process.

LEAD SULFIDE.   See **Galena**.

LEAF.   The food-manufacturing organ of a plant. Typically, leaves consist of a broad thin blade borne on a slender stalk. The important chemistry taking place within a leaf is described under **Photosynthesis**.

The leaf originates as a small protuberance from the surface of the growing tip of the stem. Numerous divisions of the cells of this protuberance produce a structure from five to eight cells thick. Many of these leaf primordia are borne together on the stem tip, and, together with any protecting scales that may cover them, form the buds of the stem. At first, all the cells of these leaf primordia are alike. Very early in their existence, however, certain cells become distinct by their somewhat elongated shape. These cells are the beginnings of the vascular elements. Cells divisions continue in these small bodies until there are present in the bud recognizable but very small leaves, which are folded in various ways. In woody plants, this development takes place in the year previous to that in which the leaf will unfold. With the advent of the new growing season, growth of the many minute cells of these tiny leaves is very rapid, so that within a few days' time the leaf has unfolded and grown to its mature size. During this enlargement, many changes have taken place in the cells of the leaf.

The mature leaf is commonly composed of two distinct parts, the broadly expanded, thin green blade, and the petiole or stalk which supports it and connects it with the stem. In many plants there is formed at the base of the petioles a pair of outgrowths called stipules, which in some plants may take the form of a complete sheath. See Fig. 1. This sheath is well developed in members of the Umbelliferae. Sometimes the petiole is completely lacking, the blade being attached directly to the stem; leaves of this kind are called sessile leaves. Less frequently the blade of the leaf is lacking, the petiole being expanded into a flattened object looking much like a blade. Certain Australian trees, species of *Acacia* and *Eucalyptus*, exhibit this peculiarity. Leaves of such plants often show progressive changes from those having well-developed blades to those in which the blade is completely lacking, showing clearly that the flattened portion present is a modified petiole.

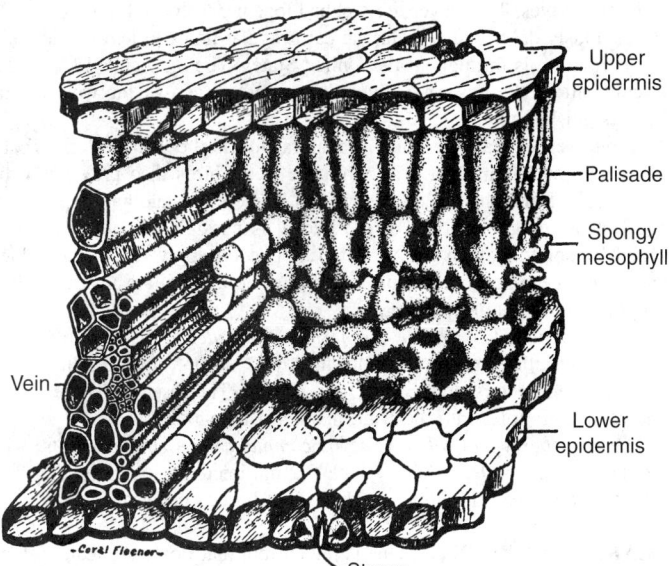

Fig. 1.   Leaf of apple, illustrating all parts — blade, petiole, and stipules.

Such flattened petioles are not uncommon, but usually the blade is present, as is the case in the lemon tree. Leaves may be deciduous, falling off at the end of a single growing season, or evergreen and persistent through several seasons. In nearly all cases the leaf fall is brought about by the development of a definite abscission layer. In many plants, such a layer is formed not only at the base of the petiole, but also at the point where the petiole joins the blade.

The shape of the blade is extremely varied, ranging from very slender linear leaves to those that are broader than they are long. The margin of the leaf may be entire, that is, without indentations of any sort, or toothed or lobed in various ways, until some are incised nearly to the midrib. If the leaf is completely divided into separate segments, it is said to be a compound leaf, in contrast with the undivided leaves, which are simple leaves, no matter how deeply they may be lobed. If the sections of a compound leaf all come from a common point, the leaf is said to be palmately compound; if they are borne along a central axis, the leaf is pinnately compound. While such infinite variations do exist, the leaves of any single species of plant are recognizably constant in shape.

The blade of the leaf is supported by a framework of veins, which are also very characteristically arranged. In many leaves, especially in dicotyledons, one vein, usually extending through the center of the blade, is more prominent than the others. This is called the main vein or midrib. The others are lateral veins. In most dicotyledons, the veins branch abundantly to form an intricately anastomosing network, which reaches all parts of the leaf. In most monocotyledons, the midrib and lateral veins extend in parallel lines from base to apex of the leaf. Between these, many minute veinlets exist, too small to be readily seen, reaching all parts of the leaf.

The cellular structure in leaves is very constant. See Fig. 2. Covering the entire surface of the leaf is the epidermis, a layer of tabular cells. On the upper surface of the leaf, the epidermal cells are frequently covered with a layer of cutin, a waxy substance that is impervious to water and so greatly reduces the loss of water by evaporation from the leaf surface.

Epidermal cells contain a scant peripheral cytoplasm, and a large central vacuole full of cellsap. Usually, there are no chloroplasts present in the epidermal cells. The cells of the epidermis of the lower surface are similar to those of the upper, but with a less evident cuticle. In the epidermis of the leaf, particularly that of the lower surface, there are many minute openings, called stomata, which permit a ready exchange of gases between the interior of the leaf and the external air. Each stoma is surrounded by a pair of guard cells containing chloroplasts. These cells close the stoma by collapsing and open it by expanding. All cells occurring between the upper and lower epidermal layers are called mesophyll cells. Beneath the upper epidermis the mesophyll cells form a very distinct layer, called the palisade mesophyll. These are elongated cells with their long axis perpendicular to the surface of the leaf. They contain large numbers of chloroplasts. In them, furthermore, active photosynthesis takes place. Occupying all the rest of the leaf is a loose tissue composed of irregularly arranged rounded cells known as the spongy mesophyll. Numerous intercellular spaces separate these cells from one another. Ramifying through the leaf just below the palisade cells are the veins. Each vein is composed of three types of cells. Some of them are thick-walled xylem cells, which carry water and dissolved mineral matter to all parts of the leaf. Others are phloem cells, which carry food away from the green cells of the leaf where they are elaborated. The xylem cells are toward the top of the leaf, the phloem cells toward the bottom. Outside these and often forming a conspicuous tissue are the masses of fibers, or collenchyma, thick-walled cells that give support to the leaf.

Leaves are often greatly modified. See Fig. 3. In many plants, they become greatly enlarged and fleshy, and serve as organs of storage of water and food. Many rock garden plants, such as species of *Sedum*, have leaves of this type. Of similar nature are the scale-like leaves that form the greater part of many bulbs, such as those of many lilies. The common onion is composed of the closely enwrapped bases of leaves, swollen with food material. In other plants, modification of the leaves becomes extreme, as, for example, in the pitcher plants and bladderworts. See also **Insectivorous Plants**.

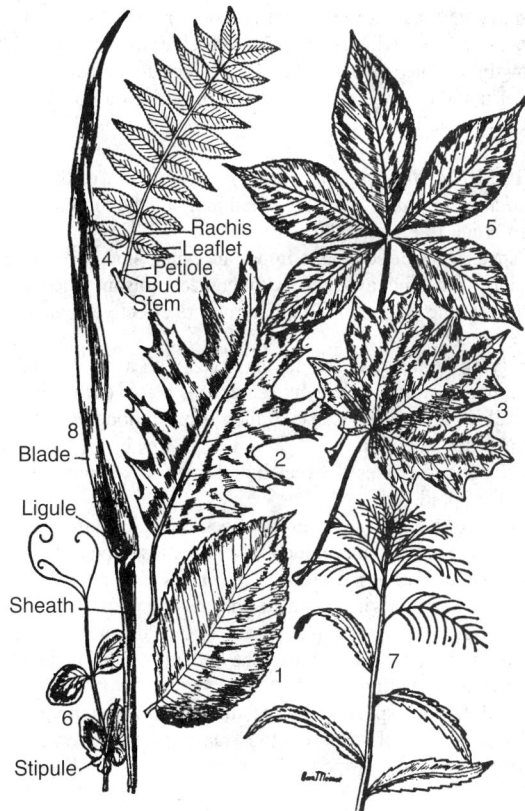

Fig. 2.   A portion of the blade of a leaf cut so as to show the internal structure. The cell contents are not shown.

Fig. 3.   Types of leaves: (1,2) elm leaf and oak leaf, both pinnately netted veined; (3) maple leaf, palmately netted veined; (4) black walnut leaf, pinnately compound; (5) buckeye leaf, palmately compound; (6) a pea leaf, with stipules, tendrils, and two unmodified leaflets; (7) portion of a plant of the water mermaid, *Proserpinaca*, with upper leaves modified by immersion in water; (8) grass leaf.

In other plants, such as the common barberry, the leaf is reduced to sharp-pointed branched spines; in many cases all gradations between these spines and typical leaves may be found on a single branch. In some plants, as the Locust, *Robinia pseudoacacia*, only the stipules are modified to short sharp spines. Many plants have leaves modified into tendrils, slender thread-like objects, which twine tightly around any suitable object with which they may come in contact. Sometimes only the tip of the blade functions in this way, and sometimes only the stipules are thus modified, as in the Carrion flower, *Smilax herbacea*. Many plants of the legume family have pinnately compound leaves, some of the segments of which are changed into tendrils. Weirdest of all are the leaves of species of *Nepenthes*, one of the pitcher plants. (See **Insectivorous Plants**, where this leaf is described.)

In a few plants, the leaf becomes a vegetative reproductive body, having in the notches of its margin, at its tip, or less commonly on its surface, groups of meristematic cells, which, when the leaf is mature, give rise to tiny plants that remain attached to the parent leaf for some time. Among the plants in which reproduction of this type occurs are species of *Bryophyllum* and *Kalanchoë*.

The principal function of the leaf is to carry on photosynthesis. To do this, the leaf must receive adequate light. Leaves are not distributed haphazardly on the stem, but in a very definite way which assures them the maximum of light. In many plants, the leaves are in pairs on opposite sides of the stem. Each successive pair usually grows out at right angles to the pair beneath it, thus preventing overshadowing. Leaves may occur in whorls, in which case there will be three or more leaves growing from each node of the stem. In many plants the leaves are alternate, each node bearing a single leaf. In every case, alternate leaves arise from the stem in such a way that a line passing around the stem and through the junction of the petiole with the stem forms a regular spiral. Examination of this spiral shows that the leaves are distributed on it in a very exact mathematical arrangement. In the simplest case, the leaves are in two longitudinal rows along the stem, every third leaf being directly above the first; in the next arrangement there are three longitudinal rows, the fourth leaf of the spiral being above the first. Other more complicated arrangements are found. The arrangement of leaves on a stem is called phyllotaxy.

Sometimes the exact arrangement is more or less obscured by twisting of the stem during growth. The leaves themselves turn considerably during their growth, petioles twisting to one side or the other, or elongating unequally, in such a way as to bring the blade into a position to receive the most favorable light.

LEAF HOPPER *(Insecta, Homoptera)*.   Any insect of the large family *Cicadellidae* or *Jassidae*. They are small to moderate jumping insects, and they often come freely to light.

Many species are of economic importance and, since they have sucking mouths, they must be attacked with contact poisons such as nicotine sulfate or kerosene emulsion. These sprays are effective against the tender immature insects.

The body of the insect is slender and long, with a round head. The four legs are stout and strong. A hair-like antenna is below each eye. Sometimes they are called "dodgers" because of their habit of dodging around various objects to escape attention.

The leaf hopper exudes a sweet liquid from its abdomen in a fashion similar to the aphid.

In Australia, the *Eurymela* group lives on eucalyptus leaves. These insects are attended by ants, much as the aphids. The potato and apple leaf hopper is the *Empoasca fabae*. The species *Eutettix tenellus* is a pest to gardens. It inserts a virus with its tiny mouth parts, causing leaves to curl. The species *Nephotettix* is a pest to rice fields and has caused much damage in India. The insect stunts the growth of the plant by robbing juices and causing wilting.

Other leaf hopper species include the *apple leaf hopper (Empoasca mali)*. The adults are of several colors, ranging from green to brown and yellow; or they may be striped. The insects are wedge-shaped and attain a length of about $\frac{1}{8}$ inch (3 millimeters). The nymphs are similar to the adults, but smaller, and crawl sideways like crabs. Action of the insects causes leaves to curl and turn yellow or reddish brown. The apple leaf hopper is found throughout the United States.

Young trees are the most seriously infested by leaf hoppers. The insects usually attack the underside of the leaves. Control should be directed toward the young nymphs; adult leaf hoppers often escape by flying away when disturbed. To control young trees infested by leaf hoppers, the tips of affected branches can be dipped into a container of soap solution, using about 1 pound of soap per 8 gallons of water (about 250 grams of soap per 15 liters of water). Dipping, which kills some of the young leaf hoppers, should be done in the latter part of June and again one month later. This is the period when the maximum number of nymphs will be found on the trees. Many adult leaf hoppers can be captured as they fly away by placing a shield covered with a sticky substance close to the tree.

Four generations of leaf hoppers are produced each year. The eggs, laid in blisters under the bark of the tree, winter over. Eggs laid in the summer are placed in leaf veins and petioles.

The *grape-vine leaf hopper (Typhlocyba comes)* is a small yellow-colored insect, sometimes mistakenly called thrips. This pest is most prevalent in the western United States. It sucks sap from the underside of leaves, causing the leaves to become brown and brittle. In treating for this hopper, care must be taken to thoroughly reach the underside of the foliage.

LEAF INSECT *(Insecta, Orthoptera)*.   Large insects of the Old World tropics related to the walking-stick insects. They have leaf-like wings and in some species the body and legs are extended in flat processes which also resemble leaves.

LEAF MINER *(Insecta, Lepidoptera)*.   Larval insects that work in the soft tissue of leaves between the upper and lower epidermis. They are necessarily small and are sometimes able to complete their development on a very small part of the food available in a single leaf. The burrow or mine shows as a brownish or transparent patch in the leaf and its form is characteristic of the insect making it. The larvae of many of the smallest moths and of some sawflies are leaf miners.

LEAF ROLLER *(Insecta, Lepidoptera)*.   Also sometimes referred to as the *leaf tyer* or leaf sewer, these small moths are usually members of the family *Tortricidae*. Their descriptive name derives from their practice (in caterpillar stage) of rolling all or part of a leaf into a cylindrically shaped case, tying the case with natural gum threads, and then lining the case with silk and thus forming a cocoon wherein the insect transforms into the pupa stage. Researchers have observed that several larvae may work cooperatively to form a common "nest." There are several species, each of which builds a characteristic nest.

The adults of the *apple leaf roller (Archips argyrospila)* are brown moths with light markings on the wings and having a wingspan of about $\frac{3}{4}$ inch (18–19 millimeters). The larvae are from pale yellow to a dirty green in color, with brown or black heads, and ranging up to $\frac{3}{4}$ inch (18–19 millimeters) in length. Light yellow, green, or grayish eggs are laid on branches in masses of 10–19. The red-banded leaf roller has a broad, reddish-brown band across the wings. The larvae feed on buds, fruit, and leaves. The leaves are webbed together to form a tent or cocoon as previously described. The larvae eat irregular holes in leaves and fruit.

Distribution is throughout the United States. The red-banded leaf roller is confined to the eastern United States and ranges as far west as the Mississippi Valley.

A number of parasites and predators attack the leaf rollers. Toads eat many caterpillars that drop from the trees; birds also prey upon the caterpillars. Since the insect overwinters in the egg stage and deposits its eggs on the twigs and bark of the tree, it is possible to control the first brood by spraying with a dormant fruit tree oil spray. Also, the folded leaves can be pinched by hand to destroy the caterpillars inside. Debris should be burned after picking.

The *avocado leaf roller (Amorbia emipratella)* is a yellowish-green caterpillar with a pinkish-brown stripe approximately an inch (2.5 centimeters) long when fully mature. The insect rolls the leaves and eats small holes into the fruit, making it unmarketable.

The *strawberry leaf roller (Ancylis comptana)* has similar habits, with the larva, usually less than $\frac{1}{2}$ inch (12 millimeters) long, feeding and folding the leaves. The insect produces two broods per year.

LEAKAGE CURRENT.   This is the current that flows or "leaks" along the surface or through the body of an insulator. Except under abnormal conditions such as dirty or moist surfaces or in electronic circuits having very minute currents the leakage is usually negligible.

LEAKAGE REACTANCE. This is the inductive reactance caused by the flux that links only one coil of a transformer. The useful flux, of course, links both windings and is the medium of transfer of energy between them. Leakage reactance is one of the major internal impedance components of the transformer.

LEAK DETECTION. See **Mass Spectrometry**.

LEAPFROGGING. The process of phasing, or delaying the ranging pulse of a tracking radar in order to move, or shift (on the radarscope presentation) the target blip past the target blip from another radar.

LEAPFROG TEST. In computer operation, a check routine which eventually occupies every possible position in the memory.

LEAPING MAMMALS. See **Rabbits and Hares**.

LEARNING DIFFICULTY. See **Dyslexia**.

LEAST ENERGY PRINCIPLE. A principle relating to stable equilibrium, and having very wide application. If a system is in stable equilibrium, any slight change in its condition or configuration, requiring the performance of work, will put it out of equilibrium, so that, if the system is now left to itself, it will return to its former state, and in so doing it will give up the energy imparted when it was disturbed. Consider, for example, a block of wood floating in a pail of water. If the block is lifted slightly, work is done and the center of mass of the wood-water system as a whole is raised, so that it now has more potential energy. The same would be true if the block were pushed a little farther into the water. In either case, when the block is released, it resumes its former level and the potential energy of the system diminishes to its former minimum value. This illustrates the general principle, which is that a system is in stable equilibrium only under those conditions for which its potential energy is at a minimum.

The principle of least energy is one aspect of the principle of virtual work.

LEAST SQUARES. Suppose that it is required to fit an equation of functional form $y = f(x_1, x_2, \ldots, x_p)$ to a series of observations on y and the x's. If the number n of observations exceeds the number of constants in the functional form, no exact fit is, in general, possible. The method of least squares determines a good fit by minimizing the sum of squares of residuals $\Sigma(y - f)^2$ over the observations.

The method is clearly reasonable in all cases. In some it has optimal properties; for example, if f is linear and the model is of the type $y = \beta_0 + \beta_1 x_1 + \cdots + \beta_p x_p + \varepsilon$, where ε is a random residual normally distributed, the estimators of the β's derived by least squares are unbiased and have minimum variance. See also **Regression**.

LEATHER-JACKET. 1. *Insecta, Diptera*. The tough-skinned larvae of some species of crane flies. They live in the ground in pastures, hay fields, and grain fields and are sometimes serious pests. Since they come to the surface at night they can be destroyed by the use of poison baits.

2. *Pisces*. File fish related to trigger fish. Coastal; frequenting reefs and rocks, mostly poisonous but two or three Australian species said to be good as food.

LEAVENING AGENTS. The generation of carbon dioxide for use as dough leavening is produced by reacting sodium carbonate (baking soda) with one of several leavening acids. In the case of an acidic phosphate salt (with two replaceable hydrogen atoms), the reaction is:

$$MH_2PO_4 + 2\,NaHCO_3 \longrightarrow MNa_2PO_4 + 2\,H_2O + 2\,CO_2$$

where M can be a hydrogen or an alkali metal ion. Claims for use of acidic phosphate salts, in addition to formation of carbon dioxide, are the buffering effects for providing an optimal pH for the baked product, as well as interactions with protein constituents of flour, with resulting optimal elastic and viscosity properties of the dough batter.

Other leavening acids used in modern bakeries include sodium aluminum sulfate, $Na_2SO_4 \cdot Al_2(SO_4)_3$; sodium aluminum phosphate hydrate (and anhydrous); potassium acid tartrate, $KHC_4H_4O_6$ (cream of tartar); and glucono-delta-lactone. The baker is concerned with (1) *dough rate of reaction* (DRR), a measure of the rate at which the leavening acid reacts with the baking soda during both the mixing stage and the holding period after mixing (bench action); and (2) *neutralizing value* or neutralizing strength, i.e., the weight of leavening acid required to neutralize a given weight of sodium bicarbonate. This value is used to compute the amount of leavening acid required to yield the needed amounts of leavening gas as well as its effect upon the pH of the baked goods.

Properties of the principal leavening acids are given in Table 1, which shows the most appropriate baking applications for each.

TABLE 1. PROPERTIES OF LEAVENING ACIDS

Chemical Name and Formula	Abbreviation	Relative Speed at Room Temperature	Neutralizing Value[1]
Sodium aluminum phosphate (anhydrous)	SALP	Medium	110
Sodium aluminum sulfate, $Na_2SO_4 \cdot Al_2(SO_4)_3$	SAS	Slow	100
Monocalcium phosphate (anhydrous), $CaH_4(PO_4)$	MCP	Slow	83
Monocalcium phosphate (monohydrate), $CaH_4(PO_4) \cdot H_2O$	$MCP \cdot H_2O$	Quite fast	80
Sodium acid pyrophosphate, $Na_2H_2P_2O_7$	SAPP	Medium	72
Glucono-delta-lactone, $C_6H_{10}O_6$	GDL	Slow	55
Potassium acid tartrate, $KHC_4H_4O_6$	—	Medium to fast	50
Dicalcium phosphate dihydrate, $CaHPO_4 \cdot 2H_2O$	DCP	Very slow	33
Sodium aluminum phosphate hydrate	$SALP \cdot H_2O$	Slow	100

[1] Values in this column indicate the parts of sodium bicarbonate that will be neutralized by 100 parts of the leavening acid under nominal conditions. Values vary with composition of dough.

Baking powders, as prepared for the home baker and for use in premixes, usually incorporate, along with sodium bicarbonate, one of the following leavening acids: (1) potassium hydrogen tartrate (2 parts for 1 part sodium bicarbonate); (2) tartaric acid (infrequent), 1 part; (3) calcium hydrogen phosphate (crystallized), 1.5 parts; or (4) sodium aluminum sulfate or ammonium aluminum sulfate, 1.8 parts. With 7 parts by weight of this finely powdered mixture, there is usually mixed about 3 parts by weight of starch to diminish the effects of moisture in storage. In some cases, dry powdered egg albumin is added to decrease the loss of carbon dioxide upon wetting the flour and baking powder mixture when used. For some purposes, ammonium carbonate can be used alone, since upon heating this material furnishes both ammonia and carbon dioxide gases to make the product light. These gases escape from the product during the baking process. In selecting a baking powder, one must keep in mind the speed with which the components react at room temperature: alum-containing baking powders act slowly; phosphate baking powders have a medium speed; and tartrate baking powders act quickly to produce carbon dioxide. Hence, when using the latter type, it is necessary to bake quickly after mixing to eliminate the loss of too much gas.

Within the last several years, advantage has been taken of mixing different leavening acids in premixes and household baking powders. Because the use of emulsifiers in most cake mixes reduces the need for early leavening action, it is common practice to use combinations of slow-acting leavening acids that retain much of their leavening reaction for the baking stage. In mixes, the leavening process must be regarded as a system because, in addition to gas generation, the leavening system controls the pH of the finished product and thus affects crumb and crust color, the intensity of flavor, as well as other properties. For various cakes, the optimum pH values are: white cakes, 6.9–7.2; yellow cakes, 7.2–7.5; chocolate or devil's food cakes, 7.1–8.0. Monocalcium phosphate (anhydrous) and sodium aluminum phosphate are frequently used together in white and yellow cake mixes; monocalcium phosphate and sodium acid pyrophosphate or dicalcium phosphate dihydrate are used in chocolate cake mixes. Generally, the combination will be comprised of 10–20% fast-acting leavening acid and 80–90% slow-acting leavening acid.

For pancake and waffle mixes, a common blend of leavening acids is 20–30% monocalcium phosphate monohydrate or monocalcium phosphate (anhydrous), combined with 70–80% sodium aluminum phosphate. A batter of this type can be prepared several hours in advance if retained under refrigeration. It has been observed that such a batter will sour before a serious loss of leavening power occurs.

Prepared biscuit mixes made of flour, shortening, and salt usually contain 30–50% monocalcium phosphate (anhydrous) and 50–70% sodium aluminum phosphate or sodium acid pyrophosphate. Self-rising flours and corn meals usually contain flour or corn meal, salt, soda, and leavening acid. Usually used in these products are combinations of sodium aluminum phosphate and monocalcium phosphate (anhydrous).

Refrigerated doughs available for preparation of biscuits, dinner rolls, and various sweet rolls, usually contain flour, water, shortening, nonfat milk solids (or dried whey solids), sugar (or corn sugar), salt, soda, and a leavening acid. Long-term refrigerated storage requires that only slow-acting leavening acids be used, frequently the sodium acid pyrophosphates. The latter have the disadvantage of possibly producing orthophosphates under certain conditions. The orthophosphates have a rather disagreeable, astringent flavor.

Unleavened Products. The principal unleavened bakery product is pie crust, which is low in moisture and high in fat content. The ingredients and method of preparation prevent the formation of a continuous gluten network through the dough mass. The porosity associated with leavened products is not desirable because the crust literally acts as a container and requires some strength.

LE CHÂTELIER'S PRINCIPLE. Let us perturb a system that is initially in stable equilibrium to a neighboring nonequilibrium state. Since the initial equilibrium is supposed to be stable, the system will return to an equilibrium state.

Theorems governing the behavior of perturbed systems are often known as *theorems of constraint or theorems of moderation*. The best known thermodynamic theorem of moderation is that of Le Châtelier-Braun, which in the form stated by Le Châtelier is:

"Any system in chemical equilibrium undergoes, as a result of a variation in one of the factors governing the equilibrium, a *compensating* change in a direction such that, had this change occurred alone it would have produced a variation of the factor considered in the *opposite direction*."

However, this principle suffers from a number of important exceptions. It is therefore preferable to study the "moderation" starting from the usual thermodynamic formalism without invoking a special principle.

LEEK. Of the family *Amaryllidaceae* (amarylis family), the leek (*Allium porrum*) is related to a great number of other species of the genus and of

Fig. 1.   Leek (*Allium porrum*), closely related to the onion. (*USDA photo.*)

similar odor and taste. Closely related species are chive, garlic, onion, shallot, and Welsh or Japanese onion. The leek resembles the onion in its adaptability and cultural requirements. Instead of forming a bulb, the leek produces a thick, fleshy cylinder that has the characteristics of a large, green onion. See Fig. 1.

Usually, the seeds are sown in a shallow trench so that the plants can be more easily hilled up as growth proceeds. Leeks are ready for use any time they reach a proper size. Under favorable conditions, they will grow to 1.5 inches (4 cm) in diameter or more, with white parts from 6 to 8 inches (15 to 20 cm) in length. They may be lifted in the fall and stored like celery in a dry, cool place.

LEE, TSUNG-DAO (1926).   Lee was born in Shanghai, China. He began his science studies at the National Southwest Associated University but in 1946 came to the University of Chicago. Here, he became friends with another student named Chen Ning (Frank) Yang. Here, Lee also completed his PhD. on white dwarf stars under Enrico Fermi. At this point, Lee starting researching with Chandrasekhar for about a year, and then went to UC, Berkeley and researched with Yang.

The rest of his scientific career, Yang spent at Columbia University. There he taught and did research in statistical mechanics. He is known for the "Lee model." Later, he again began working with Yang again on the nonconservation of parity. For this research, Lee and Chen Ning Yang shared the Nobel Prize in Physics in 1957.

J. M. I.

LEEWAY.   The difference between the actual direction in which a ship is moving relative to the surface of the water and the direction in which the keel of the ship is pointing. Leeway is usually produced by the pressure of the wind against the side of the vessel and is much more pronounced in the case of sailing vessels than in internally powered ships. The amount of leeway can best be determined by observing the angle between the wake of the vessel and the line of the keel.

In determining the true course of the vessel, the leeway is treated in the same manner as a compass correction. If the wind is blowing against the left side of the vessel, the vessel is said to be on the port tack, and the true course will be to the right of the course indicated by the keel. Hence, for a ship on the port tack, leeway has the same effect as an east or positive compass correction; on the starboard tack, leeway is applied as a west or negative correction.

See also **Course**; and **Navigation**.

LEFT-HANDEDNESS (System).   See **Handedness (Right- and Left-)**.

LEGENDRE DIFFERENTIAL EQUATION.   A second-order equation

$$(1 - x^2)y'' - 2xy' + n(n + 1)y = 0$$

It is a special case of the associated Legendre equation

$$(1 - x^2)y^n - 2xy' + \left[n(n + 1) - \frac{m^2}{1 - x^2} \right] y = 0$$

which in turn is a special case of the Gauss hypergeometric equation. Both Legendre equations have singular points at $x = \pm 1$, ∞ and if m, n are integers the solutions are Legendre polynomials. For nonintegral values of these parameters, the solutions are Legendre functions. These differential equations occur in the quantum mechanical problems of the rigid rotator and the hydrogen atom.

The associated polynomials may be defined by the expression

$$P_n(x) = (-1)^m (1 - x^2)^{m/2} \frac{d^m P_n(x)}{dx^m}$$

and the special case $m = 0$ is the Legendre polynomial of degree n

$$P_n(x) = \frac{1}{2^n n!} \frac{d^n}{dx^n} (x^2 - 1)^n$$

$$= \frac{(2n)!}{2^n (n!)^2} \left[x^n - \frac{n(n - 1)}{2(2n - 1)} x^{n-2} \right.$$

$$\left. + \frac{n(n - 1)(n - 2)(n - 3)}{2 \cdot 4(2n - 1)(2n - 3)} x^{n-4} \pm \cdots \right]$$

The first definition, in terms of the nth derivative, is called the *Rodrigues formula*. A second set of polynomials, linearly independent of P_n, is composed of polynomials of the second kind, Q_n. The general solution of the Legendre equation is then $y = AP_n(x) + BQ_n(x)$, where A, B are arbitrary constants. Many other definitions and relations for these polynomials are known.

One integral representation is the *Schläfli formula*

$$P_n(z) = \frac{1}{2\pi i} \int_c \frac{(t^2-1)^n}{2^n(t-z)^{n+1}} dt$$

where the contour encircles the point z in the counterclockwise direction in the complex plane. Another such representation is the *Heine formula*, which for the associated polynomial becomes

$$P_n^m(x) = (n+1)(n+2)\cdots(n+m)(-1)^{m/2}$$
$$\times \frac{1}{\pi} \int_0^\pi [x + \sqrt{x^2-1}\cos\phi]^n \cos m\phi \, d\phi$$

See also **Generating Function**.

LEGENDRE SYMBOLS. See **Number Theory**

LEGIONELLOSIS. Legionellosis is an infection caused by the bacterium *Legionella pneumophila*. This disease has two distinct forms:

- Legionnaires' disease, (LD), is the more severe form of legionellosis and is characterized by pneumonia.
- Pontiac fever, is an acute-onset, flu-like, nonpneumonic illness.

Legionnaires' disease acquired its name in 1976 when an outbreak of pneumonia occurred among persons attending a convention of the American Legion in Philadelphia. Later, the bacterium causing the illness was named *Legionella*.

An estimated 8,000 to 18,000 cases of Legionnaires' disease occur each year in the United States; 23% are nosocomial. Most LD cases are sporadic with 10–20% being linked to outbreaks. Pontiac fever has been recognized only during outbreaks. Some people can be infected with the *Legionella* bacterium and have mild symptoms or no illness at all.

Outbreaks of Legionnaires' disease have received significant media attention. However, this disease usually occurs, as a single, isolated case not associated with any recognized outbreak. When outbreaks do occur, they are usually recognized in the summer and early fall, but cases may occur year-round. About 5 to 15% of people who have Legionnaires' disease die. A substantially, higher proportion of fatal cases occur during noscomial outbreaks. Pontiac fever is a self-limited disease that requires no treatment.

Patients with Legionnaires' disease usually have fever, chills, and a cough, which may be dry or may produce sputum. Some patients also have muscle aches, headache, tiredness, loss of appetite, and occasionally, diarrhea. Laboratory tests may show that these patients' kidneys are not functioning properly. Chest X-rays often show pneumonia. It is difficult to distinguish Legionnaires' disease from other types of pneumonia by symptoms alone; other tests are required for diagnosis. Persons with Pontiac fever experience fever and muscle aches and do not have pneumonia. They generally recover in 2 to 5 days without treatment. The time between the patient's exposure to the bacterium and the onset of illness for Legionnaires' disease is 2 to 10 days; for Pontiac fever, it is shorter, generally a few hours to 2 days.

The diagnosis of legionellosis requires special tests not routinely performed on persons with fever or pneumonia. Therefore, a physician must consider the possibility of legionellosis in order to obtain the right tests. Several types of tests are available. The most useful tests detect the bacteria in sputum, find *Legionella* antigens in urine samples, or compare antibody levels to *Legionella* in two blood samples obtained 3 to 6 weeks apart.

People of any age may get Legionnaires' disease, but the illness most often affects middle-aged and older persons, particularly those who smoke cigarettes or have chronic lung disease. Also at increased risk are persons whose immune system is suppressed by diseases such as cancer, kidney failure requiring dialysis, diabetes, or AIDS. Those that take drugs that suppress the immune system are also at higher risk. Pontiac fever most commonly occurs in persons who are otherwise healthy.

Erythromycin is the antibiotic currently recommended for treating persons with Legionnaires' disease. In severe cases, a second drug, rifampin, may be used in addition. Other drugs are available for patients unable to tolerate erythromycin. Pontiac fever requires no specific treatment.

Outbreaks of legionellosis have occurred after persons have breathed mists that come from a water source (e.g., air conditioning cooling towers, whirlpool spas, showers) contaminated with *Legionella* bacteria. Persons may be exposed to these mists in homes, workplaces, hospitals, or public places. Legionellosis is not passed from person to person and there is no evidence of persons becoming infected from automobile air conditioners or household window air-conditioning units.

Legionella organisms can be found in many types of water systems. However, the bacteria reproduces to high numbers in warm, stagnant water (90–105 °F), such as that found in certain plumbing systems and hot water tanks, cooling towers and evaporative condensers of large air-conditioning systems, and whirlpool spas. Cases of legionellosis have been identified throughout the United States and in several foreign countries. It is believed to occur worldwide.

During outbreaks, the CDC and health department investigators seek to identify the source of disease transmission and recommend appropriate prevention and control measures, such as decontamination of the water source. The improved design and maintenance of cooling towers and plumbing systems will also limit the growth and spread of *Legionella* organisms. Current research will likely identify additional prevention strategies. See also **Bacterial Diseases**.

Additional Reading

Addis, D.G. and J.P. Davis: "Sporadic Cases of Legionnaires' Disease," *N. Eng. J. Med.* 1699 (June 18, 1992).

Bhopal, R.S., R.J., Fallon, and M.R.C. Path: "Sporadic Cases of Legionnaires' Disease," *N. Eng. J. Med.* 1699 (June 18, 1992).

Breiman, R.F. and J.C. Butler: "Legionnaires' Disease: Clinical, Epidemiological, and Public Health Perspectives," *Semin in Respir Infects* **13**, 84–89 (1998).

Fiore, A.E. and J.C. Butler: "Detecting Nosocomial Legionnaires' Disease," *Infections in Medicine* **15**, 625–635 (1998).

Fiore, A.E., J.P. Nuorti, O.S. Levine, et al.: "Epidemic Legionnaires' Disease two Decades Later: Old Sources, New Diagnostic Methods," *Clin. Infect. Dis.* **26**, 426–433 (1998).

Hoge, C.W.: "Sporadic Cases of Legionnaires' Disease," *N. Eng. J. Med.* 1700 (June 18, 1992).

Kool, J.L., J.C. Carpenter, and B.S. Fields: "Effect of Monochloramine Disinfection of Municipal Drinking Water on Risk of Nosocomial Legionnaires' Disease," *Lancet* **353**(9149), 272–277 (1999).

Kool, J.L., A.E. Fiore et al.: "More than Ten Years of Unrecognized Nosocomial Transmission of Legionnaires' Disease Among Transplant Patients: Difficulties of Legionella Control in a Complex Water System," *Infect. Contr. Hosp. Epidemiol.* **19**, 898–904, 7 (1998).

Lowry, P.W. et al.: "A Cluster of Legionella Sternal-Wound Infections Due to Postoperative Topical Exposure to Contaminated Tap Water," *N. Eng. J. Med.* 109 (January 10, 1991).

Marston, B.J., H.B. Lipman, and R.F. Breiman: "Surveillance for Legionnaires' Disease: Risk Factors for Mortality and Morbidity," *Arch. Intern. Med.* **154**, 2417–2422 (1994).

Morse, D.L., Birkhead, G.S., and S. Kondracki: "Sporadic Cases of Legionnaires' Disease," *N. Eng. J. Med.* 1700 (June 18, 1992).

Straus, W.L., J.F. Plouffe, T.M. File Jr. et al.: "Risk Factors for Domestic Acquisition of Legionnaires' Disease: Ohio Legionnaires' Diseases Group," *Arch. Intern. Med.* **156**(15), 1685–1692 (1996).

Yu, V.L. and J.E. Stout: "Sporadic Cases of Legionnaires' Disease," *N. Eng. J. Med.* 1701 (June 18, 1992).

LEGUMINOSAE. The Pea Family is second only to the Composite Family among the dicotyledons, with respect to the number of species it includes. Of its more than 10,000 species in early 500 genera, many are trees or shrubs, especially those in tropical regions. Herbaceous species are numerous in temperate regions. Many climbing plants, also, are found in the family. Leguminous plants are found in all sorts of environments and climates.

Nearly all the plants in this family have pinnately compound leaves. The stipules present in the leaves are sometimes modified to persistent spines. The flowers are either regular or irregular. When regular, the flowers have five sepals, commonly more or less united, five petals, a varying number of stamens, and a single pistil. Irregular flowers are of the type known as papilionaceous, a name given because of the fancied resemblance of the flower to a butterfly. In flowers of this type, the calyx has five unequal, more or less united sepals, which frequently persist during development of the fruit, five separate petals, showing very constant difference in form. The upper one, called the standard, is large and showy; the two lateral to this, called wings, are smaller in size; and the two lower ones are more or less united into one unit, called the keel or carina. Within this keel are the ten stamens, which may be separate but in many genera are united in groups, the nine lower ones having their filaments more or less completely joined, while the tenth stamen remains free. The pistil has a somewhat flattened ovary, a long style, and a terminal stigma. The ovary contains several ovules. The mature fruit is called a pod or legume, which when mature often splits open with sufficient force to eject the seeds to considerable distances. The seeds in most cases have large food reserves stored in the thick cotyledons.

There are three subfamilies in the *Leguminoseae*. The *Mimosoideae* have regular flowers and valvate corolla. The *Caesalpinioideae* have irregular (zygomorphic) flowers. These two subfamilies are essentially tropical. The *Papilionoideae* have irregular (papilionaceous) flowers, and include most of the important cultivated forms.

Many members of this family supply man with important foodstuffs, such as beans, peas, and peanuts, while others are important forage crops for domestic animals. The high protein content of the plant is the principal reason for its importance as a food source. Clovers and alfalfa are not only valuable as forage plants, but furnish an excellent hay. Legumes are also of immense value because of their association with nodule-forming bacteria, resulting in a considerable accumulation of nitrogenous substance, which is later liberated into the soil, greatly enriching it. Other members of the family yield valuable dyes, gums and resins, and oils; many are sources of timber.

The leaves of many genera of legumes are interesting because of their ability to move. In many of them, the leaflets fold together at night, so that the blade of the leaflet is vertical. Of particular interest in this connection is the Sensitive plant, *Mimosa pudica*, the leaves of which respond very quickly to external stimuli. A light blow will cause the many leaflets to fold together, and the whole section of the compound leaf to bend down. A sudden breeze or change of temperature will produce the same result. Recovery from the shock is gradual. When stimulated by a series of successive shocks, the plant recovers more and more slowly each time.

See also **Locust Trees**.

LEISHMANIASIS. There are two major subtypes of this disease — *visceral* and *cutaneous* leishmaniasis. Visceral leishmaniasis (kala-azar; tropical splenomegaly) is a chronic, systemic disease caused by the flagellate protozoan *Leishmania donovani*. Discrete foci are distributed in South and Central America, Africa, Asia, the former U.S.S.R., and Europe. Kala-azar is basically a domestic disease where poor hygiene, moderate heat and vegetation favor the activities of the sand-fly (*Phelbotomus*) vectors. There are marked geographic variations in epidemiological patterns: in India only about 40% of cases are infants and the disease is largely urban, but no extra-human reservoirs are known. In China the disease is both urban and rural, children are most affected, and dogs are an important reservoir. Clinical disease usually appears 2–4 months after the infectious sandfly bite. The tiny (2–3 mm) amastigote stage (LD body) multiplies as an intracellular parasite within reticuloendothelial cells, leading to cell destruction. In established infections there is irregular, undulant fever, emaciation and diarrhea, anemia and leukopenia, hepatosplenomegaly, and lymph node enlargement. Untreated disease shows high mortality, from 1 week to 3 years following onset. Recovery is followed by lasting immunity. Diagnosis is based on a suggestive clinical picture coupled with demonstration of LD bodies in biopsies of bone marrow or lymph nodes. Inoculation of culture media or hamsters may be necessary. After adequate treatment with antimony (Pentostam®) or diamidine compounds, prognosis is excellent. Control measures should include residual spraying and personal protection measures against sandflies, clearing of decaying vegetation, and reduction of the dog population.

Cutaneous leishmaniasis, also spread by sandflies, involves several *Leishmaniae* and takes a number of clinical forms. The amastigotes are restricted to the area of the small skin ulcers, producing a cutaneous rather than systemic disease. *Leishmania tropica* produces Old World cutaneous leishmaniasis, also known as Oriental sore, Baghdad boil, or Delhi boil. It occurs around the Mediterranean, in the Middle East, southern former U.S.S.R., northwest India, and parts of Africa. There are two subtypes of Old World cutaneous leishmaniasis: (1) The moist or rural form (*L. tropica major*) is a zoonosis of desert gerbils. In humans, it has an incubation period of 6–10 weeks and produces cutaneous ulcers, which may spread. (2) *L. tropica minor* produces the dry or urban form, harbored by both humans and dogs. The incubation period is longer (6–10 months), the dry skin lesion does not spread, and usually heals spontaneously within a year. New World cutaneous leishmaniasis likewise includes (1) *L. tropica mexicana* (Chiclero ulcer) and (2) *L. braziliensis* (mucocutaneous leishmaniasis or espundia). The first of these is a zoonosis of rodents in Central America, producing spontaneously healing ulcers on the pinna of the ear in humans. *Espundia* is a zoonosis, especially prevalent where jungle has recently been cleared. After inoculation the parasite spreads, especially to the oro-nasal areas, where it may produce progressive, destructive, disfiguring lesions, Espundia may persist for years, followed by death from septicemia or bronchopneumonia.

Diagnosis of cutaneous leishmaniasis is, by demonstrating the amastigotes in smears from ulcers or in cultures on NNN medium, or by the Montenegro skin test. Effective drugs include pentavalent antimonials (Pentostam®), pyrimethamine (Daraprim®), and amphotericin B (Funizone®). The prognosis is good, except in untreated mucocutaneous leishmaniasis. Infection often confers lifelong immunity. Control measures include covering individual lesions, reduction of natural reservoirs, and personal protection against sandflies.

Additional Reading

Berman, J.D.: "Human Leishmaniasis: Clinical Diagnostic and Chemotherapeutic Developments in the Last 10 Years," *Clin. Infect. Dis.*, **24**, 684–703 (1997).

Desjeux, R.: "Leishmaniasis: Public Health Aspects and Control," *Clin. Dermatol*, **14**, 417–423 (1996).

Hart, D.: "Leishmaniasis: The Current States and New Strategies for Control," Perseus Books, Boulder, CO, 1989.

Herwaldt, B.L.: "Leishmaniasis," *Lancet*, **354**, 1191–1199 (1999).

Herwaldt, B.L., S.L. Stokes, and D.D. Juranek: "American Cutaneous Leishmaniasis in U.S. Travelers," *Ann. Inter. Med.*, **118**, 779–784 (1993).

Hide, G., et al.: "Trypanosomiasis and Leishmaniasis," CAB International, New York, NY, 1997.

Tapia, F., et al.: "Molecular and Immune Mechanisms of Pathogenesis of Cutaneous Leishmaniasis," Landes Bioscience, Georgetown, TX, 1999.

R. C. V.

LEMMA. A proposition which is stated for the purpose of later use in the proof of another proposition. Strictly, the proposition is assumed. In current use, a short proof is often given.

LEMMING. See **Rodentia**.

LEMNISCATE OF BERNOULLI. A special case of a general type of higher plane curves known as lemniscates (*L. lemniscus*, loop) and also of the oval of Cassini, when $k = a$. Its equation in rectangular coordinates is

$$(x^2 + y^2)^2 + 2a^2(y^2 - x^2) = 0$$

or, in polar coordinates,

$$r^2 = 2a^2 \cos 2\theta$$

The curve is symmetrical about both X- and Y-axes and the node at the origin has tangents given by $(x^2 - y^2) = 0$. See Fig. 1. If $\cos 2\theta$ is replaced by $\sin 2\theta$ in the polar equation for the curve it is rotated about the origin by $45°$ and is then called a two-leaved rose lemniscate. (See also **Rose Curve**). A more general type of curve, known as Booth's lemniscate, has the equation

$$(x^2 + y^2)^2 = a^2x^2 \pm b^2y^2$$

The curve $x^4 = a^2(x^2 - y^2)$ is known as the lemniscate of Gerono.

See also **Curve (Higher Plane)**.

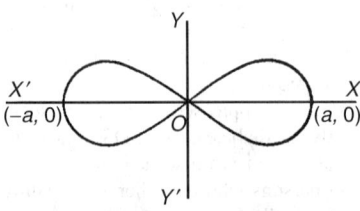

Fig. 1.   Lemniscate.

LEMON TREE. See **Citrus Trees**.

LEMUR *(Mammalia)*. Lemurs are the most primitive animals of the order of Primates. They have a well-developed thumb and great toe, like other primates, but the second toe bears a sharp claw instead of a nail. The body is generally more like that of a squirrel than like the apes and monkeys and the face in many species is peculiarly expressionless, with large staring eyes. Lemurs constitute a family *Lemuroids*, containing, in addition to the species named as lemurs, the indri, the sifakas or propitheques, the galagos, the awantibo, the pottos, the lorises or slow lemurs, and the avahi. They center in Madagascar, but a few species occur in eastern Africa, southern India, and on the islands of the Oriental region. The tarsiers and the aye-aye are closely related to the true lemurs. A ring-tailed lemur is illustrated in Fig. 1.

Fig. 1. Ring-tailed lemur. (*New York Zoological Society*.)

The African slow lemur is named the *potto* and resembles the lorises of Asia in many ways. The potto is remarkable for the rudimentary index finger, which is a stub without a nail or joints, and for the very short tail. The two species are Bosman's potto and the awantibo, *Perodicticus calabarensis*. The sifaka is the genus *Propithecus*, with several species found in Madagascar. They are related to the indri lemur, but have long tails and shorter muzzles.

The galago is an African lemur, also sometimes called bush-baby. The several species have long bushy tails, large ears, large staring eyes, their thumbs and big toes opposed, and thick woolly soft fur. Their nearest relatives are the mouse lemurs of Madagascar.

Additional Reading

Gould, E. and G. McKay: "Encyclopedia of Mammals," 2nd Edition, Academic Press, Inc., San Diego, CA, 1998.
Jolly, A.: "Madagascar's Lemurs," *National Geographic*, 132 (August, 1988).
Staff: "Encyclopedia of Mammals," Marshall Cavendish Inc., Tarrytown, NY, 1997.
Tattersall, I.: "Madagascar's Lemurs," *Sci. Amer.*, 110 (January, 1993).

LEMUR (Flying). See **Dermaptera**.

LENARD EFFECT. The separation of electric charges accompanying the aerodynamic breakup of water drops, first studied systematically by the German physicist P. Lenard. Also called *spray electrification*, or *waterfall effect*. Experiments have shown that the degree of charge separation in spray processes depends upon the drop temperature, presence of dissolved impurities, speed of the impinging airblast, and contact with foreign surfaces. The largest fragments of the broken drops are observed to carry positive charges and the fine spray of drops carried off in the impinging air current carries a net negative charge.

LENGTH OF A CURVE. The length of a straight line is interpreted experimentally to mean the number of times another straight line of unit length can be superimposed on the given line. Since this operation cannot be conveniently applied to a curve the concept is generalized to the limit of the sum of chords to the curve. As the number of chords increases without limit each chord separately approaches zero as a limit.

In rectangular coordinates, if the curve is described by $f(x, y) = 0$, its length from the point (a, c) to the point (b, d) is given by either one of the definite integrals

$$s = \int_a^b \sqrt{1 + y'^2}\,dx = \int_c^d \sqrt{1 + x'^2}\,dy$$

where $y' = dy/dx$ and $x' = dx/dy$. If t is a parameter, $x = f_1(t)$, $y = f_2(t)$, the arc length between $t = a$ and $t = b$ is

$$s = \int_a^b \sqrt{x_t^2 + y_t^2}\,dt$$

where $x_t = dx/dt$ and $y_t = dy/dt$. In polar coordinates, if $r = f(\theta)$, $r' = dr/d\theta$ and $\theta' = d'/dr$,

$$s = \int_{\theta_1}^{\theta_2} \sqrt{r^2 + r'^2}\,d\theta = \int_{r_1}^{r_2} \sqrt{r^2\theta'^2 + 1}\,dr$$

See also **Circular Curves**; and **Coordinate System**.

LENSES. See **Mirrors and Lenses**.

LENS (Eye). See **Vision and the Eye**.

LENTICELS. The young stems of plants are covered with a single layer of cells known as the epidermis. As the stem grows older, this epidermis is lost and replaced by a thicker protective tissue known as cork. The cork cells have walls that are suberized and impervious to gases. The living cells within the stem require an exchange of gases with the outside atmosphere. This exchange occurs through lenticels. A lenticel is a mass of thin-walled parenchyma cells loosely arranged so that air spaces are numerous. Through the lenticel gases pass readily. Lenticels appear on the surface of the stem as rough masses, usually protruding somewhat, and either circular or somewhat elongate in shape. They are very irregularly distributed.

LENTICULARIS. See **Clouds and Cloud Formation**.

LENZ' LAW. A general law of electromagnetic induction, stated by H.F.E. Lenz in 1833. It points out that the electromotive force induced by the variation of magnetic flux (with reference to a conductor, in the manner discovered by Faraday) is always in such direction that, if it produces a current, the magnetic effect of that current opposes the flux variation responsible for both electromotive force and current. An outstanding illustration is the drag on a generator armature; if the armature circuit is closed, the rotation is opposed by a torque arising from the reaction between the field and the current in the armature conductors. Power must therefore be applied to drive the machine; and the greater the armature current, the more power is required. The effect known as magnetic damping also depends upon Lenz' Law. A copper disk, when spun between the poles of a strong magnet, quickly comes to rest because of the opposing torque. This arrangement serves as a speed regulator in watt-hour meters.

LEO (the lion). One of the most easily distinguished of all the zodiacal constellations. The "sickle" of Leo, the fifth sign of the zodiac, is known to all watchers of the spring and early summer skies. The brightest star in the group, Regulus, is a double star, but cannot be resolved with telescopes having smaller than a 3-inch aperture because of the fact that, in small instruments, the bright star masks the fainter one. Gamma Leonis is one of the finest of the double stars, having two components of approximately the same magnitude, one yellow and the other orange in color.

This constellation is also noted for the location of the radiant point of the Leonids, one of the best known meteor showers. (See map accompanying entry on **Constellations**.)

LEONIDS. A name applied to a meteor shower that has probably attracted more attention than any other. Each year, about the 12th of November, a number of meteors are observed coming from a radiant point in the constellation of Leo. Records of the appearance of this shower are found back as far as 585 A.D. The Leonid shower is one in which meteors are distributed all along the orbit, so that a radiant point may be determined practically every year. There is, also, a very strong condensation of the meteors into a swarm through which the earth used to pass every 33 years. Probably the greatest display was in November 1833. In Silliman's Journal of that year, we find: "To form some idea of the phenomenon, the reader may imagine a constant succession of fireballs, resembling rockets, radiating in all directions from a point in the heavens." One observer counted 650 during 15 minutes. During the interval between 1833 and 1866, a great deal of computing was done on the Leonid shower, and November 13 was predicted as the date of passage of the earth through the main swarm. The prediction was fulfilled, and at Greenwich, England, eight observers actually counted 8000 meteors, 4860 of them being counted between one and two o'clock in the morning. However, brilliant as the shower was at that time, it apparently was not as striking as the display in 1833. In 1899, there was a moderately good display of the Leonids, but nothing comparable to the showers of 1866 and 1833. The newspapers had promised so much to the general public that the failure of the shower to come up to the expectations proved a rather serious blow to astronomy. The explanation for the failure of the shower to live up to its prediction is to be found in the fact that Jupiter passed very close to the swarm during 1899 and deflected it from the earth's orbit. Further perturbations have so deflected the orbit that no real shower was observed in 1932, 1933, and 1934, although enough meteors were seen during each November to permit a determination of the radiant point. In November 1966, however, perturbations again brought the main swarm into such a position that the earth passed through it, giving rise to showers comparable to the one of 1833. See Fig. 1.

Fig. 1.   Leonid shower as seen from Kitt Peak near Tucson, Arizona.

LEOPARD-CATS.   See **Cats**.

LEPIDOLITE.   This member of the mica group of minerals is a silicate of potassium, lithium and aluminum, sometimes with sodium, fluorine, or rarely rubidium. A general formula is $K(Li, Al)_3(SiAl)_4O_{10}(F, OH)_2$. Crystals of lepidolite are monoclinic but often pseudo-hexagonal; cleavage, basal and perfect, being susceptible of splitting into thin laminae; hardness, 2.5–4; specific gravity, 2.8–3.3; luster, pearly; color, reddish to violet, grayish-blue, gray to white. A variety carrying rubidium is yellowish-gray; translucent. It usually is found as granular to scaly masses, in short stocky prisms or less often in easily cleavable sheets. Lepidolite is characteristic of pegmatite veins, frequently being associated with other lithium-bearing minerals such as tourmaline, spodumene, amblygonite, and others. It occurs in the Ural Mountains, the Czech Republic and Slovakia, the Island of Elba, and Madagascar, where it is often found in large sheets. In the United States, it is found in the pegmatites of New England, California, South Dakota, and New Mexico. The name lepidolite is derived from the Greek, meaning scale. See also **Lithium**.

LEPIDOPTERA.   The butterflies, skippers, and moths. An order of insects characterized by sucking mouths in the adult stage with two pairs of wings covered at least in part with a vesture of flattened scales, complete metamorphosis and a larva with biting mouth-parts. The second order of insects in size, with about 90,000 known species.

Butterflies and moths are widely distributed and because of their bright colors are among the animals known to everyone. The butterflies are diurnal and so are readily observed, but most moths are nocturnal, hence many beautiful species are rarely seen unless they are sought. Skippers are an intermediate group more nearly like the butterflies. The most magnificent species of all three forms are tropical, but representatives are found even in the Arctic regions. See also **Butterfly**.

The adults visit flowers for nectar or take no food. In the larval stage most species are plant feeders, but a few carnivorous forms are known and some are scavengers. The order includes many economic species, among them the clothes moths, the bee moth, and the cut worms. See Fig. 1.

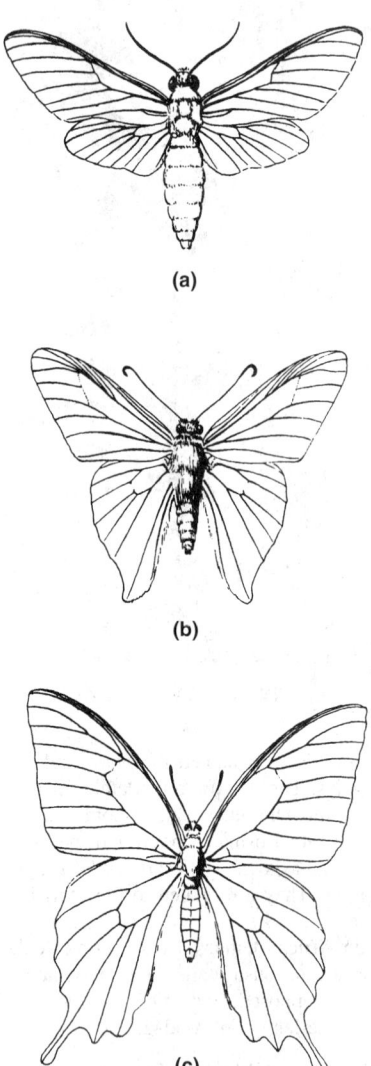

(a)

(b)

(c)

Fig. 1.   Typical members of Lepidoptera: (**a**) moth; (**b**) skipper; (**c**) butterfly. (*USDA*.)

The main families of *Lepidoptera* include:

Aegeriidae (also called *Sesiidae*)	Clear-winged moths
Arctiidae	Tiger moths
Citheroniidae (also called *Ceratocampidae*)	Royal moths
Coleophoridae (also called *Haploptiliidae*)	Casebearers
Cossidae	Carpenter moths
Eucleidae (also called *Limacodidae* or *Cochidiidae*)	Slug-caterpillar moths

Gelechiidae	Gelechid moths
Geometridae	Measuring-worm moths
Gracilariidae	Leaf blotch miners
Hepialidae	Swifts (swift-flying)
Hesperiidae	Common skippers
Incurvariidae	Yucca moths
Lasiocampidae	Lappet moths, tent caterpillars
Lycaenidae	Blues, hairstreaks, gossamer wings
Micropterygidae	Mandibulate moths
Nepticulidae	Nepticulid moths
Noctuidae (also called Phalaenidae)	Owlet or cutworm moths
Notodontidae	The prominents
Oecophoridae	Parsnip webworm
Olethreutidae (also called Eucosmidae)	Codling moths
Papilionidae	Swallowtail butterflies
Pieridae	White and sulfur butterflies
Psychidae	Bagworm moths
Pyralididae	Pyralid or snout moths
Saturnidae	Giant silkworm moths
Sphingidae	Hawk or sphinx moths
Tineidae	Clothes moths
Tortricidae	Leaf-roller moths
Yponomeutidae	Miscellaneous grouping, including diamondback moth and apple fruit moth.

LEPROSY. A chronic infectious but only mildly contagious disease caused by *Mycobacterium leprae*. The disease presents a great variety of signs and symptoms, depending on what tissue or organ of the body is involved.

Leprosy is a disease of antiquity, and there is evidence that it has existed at least since 2000 B.C. References to the disease are found in the Old Testament. Essentially, leprosy is a tropical disease. While it has been common in the Orient for several thousands of years, it appeared as a scourge in Europe in the eleventh and twelfth centuries, and did not subside until the sixteenth century when segregation of the victims was carried out on a large scale. At present the disease occurs endemically and sporadically, chiefly in the Orient, Australia, Asia, in Mediterranean countries, and in Central and South America. There are various other foci of sporadic cases such as some parts of northern and central Europe, the West Indies, Louisiana, Minnesota, and South Carolina in the United States, and several in Canada. Occasional cases are encountered in the larger seaports, both Atlantic and Pacific.

As of the early to mid-1990s, some 12 million cases existed in the world, of which 3.5 million were found in India. The approximate number of 25,000 cases seen now in Europe and the United States have been imported since World War II from tropical countries by way of immigrants.

The mode of infection by the causative agent is not definitely known. The nasal mucosa, the gastrointestinal tract, and the skin have been considered as possible portals of entry. The disease is definitely contagious, but years of exposure and contact Seem to be necessary for its transmission.

The organism causing leprosy is similar to the tubercle bacillus in appearance and staining characteristics. It is found in great numbers in the nodules occurring under the skin, in discharges from the nose and throat, and in discharges from ulcers. It was first discovered in 1873 by Hansen.

Much difficulty has been encountered in culturing, or rather maintaining, the organism, since no artificial medium has been found which will sustain it. Instead, mouse foot-pad inoculations have been the major source of organisms for study. More recently use of the immunodeficient mouse and the armadillo have provided almost adequate supplies of *M. Leprae*. To a great extent, leprosy may be considered an immunological disease. The leprosy bacillus is virtually nontoxic and most symptoms of the disease are due to immune reactions against antigenic constituents of *M. leprae*.

The period of incubation is variable. It may be from a few months to 20 to 30 years. There are two general types of the disease: (1) nodular leprosy in which the skin is primarily involved; and (2) maculoanaesthetic

leprosy, in which there is an involvement of nervous tissue. A mixed form also occurs showing symptoms of both forms.

In the first stages of nodular leprosy, brownish-red spots appear on the skin, usually on the limbs and face, covering large and small areas of the skin. Later nodular thickenings appear at these sites. The face may show the so-called leonine appearance, due to the thickening of the skin in the region of the forehead, eyes, lobes of ears and around the nose and mouth. The entire skin assumes an unhealthy, dusky appearance. Some of the thickened areas ulcerate and fingers and toes may atrophy. Ulceration also appears in the nose and throat and the voice becomes hoarse. The eyes are affected similarly, and blindness may result. This form of the disease may last 10, 20 years, or longer without treatment. Many of the patients die of complicating disorders such as pneumonia, nephritis, tuberculosis and malnutrition.

Maculo-anaesthetic leprosy is characterized by flat, red to brown lesions on the skin, distributed symmetrically. These lesions gradually become insensitive to pain (anaesthetic); trauma, even burns, may occur without the patients feeling pain. Ulceration of the area and contractions will produce deformities.

At one time, chaulmoogra oil was widely used in the treatment of leprosy. Cortisone drugs have been used successfully in the treatment of certain manifestations and for certain phases of the disease, notably for relief of acute lesions of the eye. Derivatives of diaminodiphenylsulfone (Dapsone), administered orally or by injection, have brought about marked improvements, particularly in advanced cases, and for ridding the body of *Mycobacterium leprae* in from two to four years. Para-aminosalicylate and streptomycin also have been used successfully.

Most recently, Dapsone resistance has been seen in the disease organisms so that emphasis is being placed upon the development of a vaccine. This is proving to be a difficult problem, but meanwhile, what appears to be a successful intermediate mode of treatment, has been devised in Venezuela. Purified *M. leprae* bacillus is combined with *Bacillus Calmette-Guerin* (BCG). When this is used to vaccinate leprosy victims, the bacilli were cleared from the patients' bodies in 18 months and progress of the disease was halted. Further work is being done to confirm the value of this form of prophylaxis.

Additional Reading

Antia, N. and V. Shetty: "The Peripheral Nerve in Leprosy and Other Neuropathies," Oxford University Press, Inc., New York, NY, 1998.

Courtright, P. and S. Lewallen: "Guide to Ocular Leprosy for Health Workers," World Scientific Publishing, Inc., Riveredge, NJ, 1993.

Hastings, R.C.: "Leprosy: Medicine in the Tropics," 2nd Edition, Churchill Livingstone, Inc., Philadelphia, PA, 1994.

Staff: "Chemotherapy of Leprosy: Report of a WHO Study Group," World Health Organization, Geneva, Switzerland, 1994.

Web References

American Leprosy Foundation: *http://users.erols.com/lwm-alf/*

Centers for Disease Control and Prevention: *http://www.cdc.gov/ncidod/dbmd/diseaseinfo/hansens_t.htm*

Global Alliance for Leprosy Elimination: *http://www.foundation.novartis.com/leprosy/global_alliance/index.htm*

Introduction to Leprosy: *http://web.raex.com/~bbeechy/introduction.html World Health Organization: http://www.who.int/lep/*

R.C. VICKERY, M.D., D.Sci.; PhD, Blanton/Dade City, FL.

LEPTONS. The electron, muon, and two kinds of neutrino are collectively called *leptons*. The leptons are considered to be point-like particles without structure and thus truly elementary. Leptons can interact with other particles through the weak interactions. Electrons and muons also can interact through electromagnetic and gravitational forces, but they appear to be without capability of interaction through the strong (nuclear) forces. The neutral members, the electron neutrino, the muon neutrino, and their antiparticles have extremely weak interaction with matter and do not participate in electromagnetic interactions. Leptons make excellent probes in particle physics experiments. The other major family of subatomic particles is referred to as *hadrons*. See also **Electron**; **Muon**; **Neutrino**; and **Particles (Subatomic)**.

The name *lepton* from its derivation means "light," referring to the fact that the masses of the leptons are all lighter than that of the lightest meson. The properties of the electron are discussed in that entry; here it will merely be noted that the term *electron* is used to denote the negative electron (often called the "negatron" when ambiguity might arise). Its antiparticle is the positron (also called positive electron).

LEPTOSPIROSIS. Caused by pathogenic spirochetes (*Leptospira interrogans*), leptospirosis may take a number of forms. The serotype *icterohemorrhagiae* causes the syndrome known as Weil's disease; the *autumnalis* serotype causes pretibial fever (Fort Bragg fever); serotypes *canicola*, *icterohemorrhagiae*, and *pomona* may cause aseptic meningitis (leptospiral meningitis), which differs from viral meningitis. In recent years, infections with the *canicola* serotype have occurred most frequently in the United States.

Humans may be infected through contact with the urine or affected tissues of an infected animal and the entry portal may be the skin (hands, etc.), the conjuctiva, or oral mucous membrane. Contaminated soil and water also can be infective. The 1964 outbreaks (approximately 150 cases) in the United States largely reflected water-related infections. Leptospiras can survive for several weeks in a warm, neutral or alkaline medium. Transmission of the disease between humans is rare. At one time, it was believed that leptospirosis was mainly associated with farming, husbandry, veterinary work, meat packing, and sewage facility work, but in recent years outbreaks have occurred among persons engaged in recreational activities, particularly involving contaminated water. The disease is usually seen during the summer months and shows no geographical patterns. Since 1950, the average number of cases per year has been rising slowly, but in a cyclic manner. A peak of 76 cases was reported in 1964. Over the last few years, average cases per year have numbered about ninety.

The incubation period ranges from 1 to 3 weeks. Frequently, the disease is self-limiting. The more serious cases involve secondary developments during the course of the disease. Symptoms usually include moderate to high fever, chills, headache, and prostration. Onset is usually quite abrupt. Fever and chills may persist for about one week. In some patients, cough and chest pain are also present. Muscle tenderness (calves, thighs) and stiff neck are common complaints. Conjunctivitis is present in some patients. This initial phase of the disease abates in about one week, followed by a period of a few days of apparent recovery, during which period the leptospira disappear from the blood. Then the second phase commences, usually with somewhat milder symptoms. In very serious cases, this period of apparent remission may not occur. Generalizations are difficult because of the numerous courses the disease can take.

In the Weil's disease syndrome, the second phase features high fever; the liver enlarges and becomes tender. There may be purpura and gastrointestinal bleeding. Renal damage may occur, with accompanying jaundice. Oliguria (reduction in urine flow) may be serious. Weil's disease is the most serious of the leptospirosis infections, with a mortality of 15–40 percent.

The use of antimicrobials is usually not effective unless their administration is commenced within the first few days after onset. Both penicillin G and tetracycline antibiotics have been used. Further treatment depends upon the course of the disease, with each feature addressed specifically. In cases of renal damage, hemodialysis or peritoneal dialysis may be indicated.

The vaccination of dogs, sometimes implicated in leptospirosis, is not fully protective because even healthy animals can shed the leptospiras in their urine.

Additional Reading

Staff: "Leptospirosis Annual Summary," in Center for Disease Control, Atlanta, Georgia (Issued annually).

Tappero, J.W., D.A. Ashford, and B.A. Perkins: "Leptospirosis," in: G.L. Mandell, J.E. Bennet, and R. Dolin, Editors: Principles and Practice of Infectious Diseases, 5th Edition, Churchill Livingstone, Inc., Philadelphia, PA, 1999.

Weyant, R.S., S.L. Bragg, and A.F. Kaufmann: "Leptospira and Leptonema," in: P.R. Murray et al.: Manual of Clinical Microbiology, 7th Edition, ASM Press, Washington, DC, 1999.

Web References

Centers for Disease Control and Prevention: *http://www.cdc.gov/ncidod/dbmd/ diseaseinfo/leptospirosis_t.htm*

University of Leicester, Department of Biology: *http://www.leicester.ac.uk/biology/ gat/virtualfc/weil.html*

New York State Department of Health: *http://www.medhelp.org/*

R.C. VICKERY, M.D., D.Sci.; PhD, Blanton/Dade City, FL.

LEPUS (the hare). A southern constellation located near Canis Major and Orion.

LESCH-NYHAN SYNDROME. A sex-linked genetic disorder characterized by random, uncontrollable movements, mental retardation, and a self-destructive psychotic behavior. An enzyme deficiency results from the absence of a gene (or presence of a defective gene) on the X chromosome. The disorder affects only males because males have only one X chromosome. Because females have two X chromosomes, the presence of a normal gene on one of these suffices and there is no resulting disorder.

LESION. An injured or otherwise impaired tissue of the body.

LESSER CATS. See **Cats**.

LETTUCE. See **Rose Family**.

LEUCITE. The mineral leucite is a metasilicate of potassium and aluminum corresponding to the formula $KAlSi_2O_6$. It is isometric at a temperature of about 600 °C (1112 °F) and pseudoisometric at lower temperatures, at which leucite is tetragonal but retains an external isometric crystal form, usually trapezohedral. It has a conchoidal fracture; is brittle; hardness, 5.5–6; specific gravity, 2.47–2.50; luster, vitreous; color, white or some shade of gray; translucent to opaque. It is commonly found in the more recent lavas of high alkali content. Leucite is seldom reported from plutonic rock types. It is a relatively rare mineral. It is found plentifully at Vesuvius and Monte Somma and elsewhere in Italy, and Germany in the Tertiary volcanic district of the Eifel. In the United States, leucite has been found in the Leucite Hills of Wyoming, the Highwood Mountains of Montana, and as pseudomorphs (pseudoleucites) representing a mixture of nepheline, orthoclase, analcime, and aegerine from New Jersey, Arkansas, and Montana. Its name is derived from the Greek word leukos, referring to its white color.

LEUCOCYTES. See **Blood**.

LEUCOPLASTS. See **Plastics**.

LEUKEMIAS. The leukemias, like the anemias, represent a group of disorders of the blood cells rather than a single disease and they rank as a significant cause of death in children.

Normally, precursor (hematopoietic) blood cells follow a fixed sequence in their development and differentiation into erythrocytes (red blood cells) and leukocytes (white blood cells) and produce the complement of blood cells required by the human body. In the leukemias, however, the development and differentiation do not cease and the group of diseases is characterized by the undifferentiated proliferation of malignant cells derived from the hematopoietic precursors with resultant replacement of the normal bone marrow and often infiltration of other organs. The cell clone has (1) poor responsiveness to normal regulatory mechanisms, (2) a tendency to have a diminished capacity for normal cell differentiation, (3) an ability to expand at the expense of normal myeloid or lymphatic tissue, and (4) an ability to suppress or impair normal myeloid cell growth.

The cause of this erratic behavior of hematopoietic cells is unclear. Many factors have been suggested in the etiology; these include RNA viruses, ionizing radiation, and a number of drugs and chemicals which have been shown to have adverse effects upon the body. Genetic effects also appear to be involved, especially those conditions associated with chromosomal damage.

About 1–2 kg of leukemic cells in the body ($1–2 \times 10^{12}$ cells) appear to be sufficient to cause death from leukemia. This number would be about the total marrow volume of an average adult. Because the diagnosis of leukemia cannot usually be made unless the cell burden is 10^9 cells, it is apparent that only ten tumor cell doublings separate the smallest detectable number from a potentially lethal cell number. When a patient is said to be in remission, it means only that the leukemic cell number is not clinically detectable and is, therefore, less than 10^9.

From a standpoint of terminology, the traditional basic classification of leukemia has been the acute and chronic forms of the disease. In acute leukemia, the predominant hematopoietic cell is a primitive precursor, which may be undifferentiated or have the features of lymphoblast, myeloblast, monocyte, or erythroblast. In the chronic disease there is more obvious differentiation into myeloid or lymphoid cells. Apart from this general classification, the leukemias are classed according to the kind of cell primarily involved.

The several different kinds of leukemia differ in symptoms and in life expectancy of the patient. The acute form occurs most frequently in young children; the chronic form usually in persons over 35 years old. *Lymphocytic leukemia* principally involves the white blood cells, which arise from the lymph nodes and spleen. In *granulocytic leukemia*, one or more of the three types of granulocytes, originating in the bone marrow, are involved. *Monocytic leukemia* is typified by the appearance of excessive numbers of monocytes derived from connective tissue. In lymphocytic leukemia, an early symptom is enlargement of the lymph nodes and acute leukemia is often first detected by prolonged hemorrhage following a minor surgical procedure, such as tooth extraction. Fever, arthralgia and anemia are also early symptoms. In some acute leukemias the total white cell count may be normal or even below normal, although the particular type of cell is found in excessive numbers in other tissues.

Three conditions characterize chronic leukemias: (1) enlargement of the spleen; (2) increased numbers of white cells, and (3) anemia. Because of the numerous symptoms and variation in degree, the diagnosis of leukemia is confirmed only by microscopic examination of blood and bone marrow which exemplifies severe anemia and reduced blood platelet counts.

Some of the clinical manifestations of acute leukemias can be related to bone marrow failure, leading to infection, anemia, and bleeding, together with the metabolic effects of a large tumor mass and infiltration of various organs by malignancy.

Treatment of leukemia dates back several decades. In 1948, Farber initiated the folic acid antagonistic route involving daily injections of drugs, such as prednisone, l-aspariginase, vincristine, methotrexate, cytabarine, or cyclophosphamide. Often, within a few weeks, the patient was free of measurable signs of the disease, but when therapy was discontinued, relapse occurred and symptoms recurred within three months. Modern drug treatment follows along similar lines. It is a long-term maintenance procedure, requiring from 3 to 5 years and continued maintenance as required if death is to be averted. Probably less than 50 percent of patients survive after stopping all therapy beyond the initial period of 5 years or longer. Ironically, the chemotherapy used to arrest other cancers has become an important causation of leukemia.

Treatment-Related Leukemias. The treatment methodologies used for primary cancers and the resulting effect upon the incidence of leukemia have been established, but require further elucidation. Much remains to be researched pertaining to the biologic nature of chemotherapy and radiation treatment effects that bring about leukemia. Past studies have indicated cytogenetic abnormalities, including the loss of entire chromosomes 5 and 7, produced by treatment of cancers as well as from a variety of environmental toxins, such as benzene. C.A. Coltman, Jr. and S. Dahlberg observe, "This consistent picture of resistant leukemia associated with defects of chromosomes 5 and 7, regardless of the underlying disease process and the type of carcinogen, implies a cause-and-effect relation. The deletion of these genes, which are critical to the proliferation and differentiation of hematopoietic (blood forming) tissues, must have a central role in the basic biology of treatment-related leukemia. The exact relation remains to be elucidated, but it is clearly a common pathway of carcinogenesis for ionizing radiation, chemotherapy, and environmental toxins in a wide variety of benign and malignant conditions."

It has been demonstrated that patients who have received chemotherapy with alkylating agents (cyclophosphamide, chlorambucil, melphalan, and other leukemogenic substances) for ovarian cancer have an increased risk of acute leukemia, particularly of the myeloid (resembling bone marrow) type. J.M. Kaldor (International Agency for Research on Cancer, Lyon, France) reported on an extensive study in 1990 and concluded, "The extent to which the relative risks of leukemia are offset by differences in chemotherapeutic effectiveness is not known." A further observation of the Kaldor report: "The trend in treatment for ovarian cancer is toward more use of chemotherapy and in particular an increasing reliance on combinations of drugs. Despite some promising recent findings (Tropé reference), there is limited evidence so far that the effectiveness of such chemotherapy improves with the intensity or number of agents, and our study clearly shows dramatic increases in the long-term risk of leukemia at high dosages."

The Kaldor group also investigated the effect of different treatments for Hodgkin's disease on the risk of leukemia. The conclusion: "We conclude that chemotherapy for Hodgkin's disease greatly increases the risk of leukemia and unaffected by concomitant radio therapy. In addition, the risk is greater for patients with more advanced stages of Hodgkin's disease

and for those who undergo splenectomy." However, in another observation, "In the case of Hodgkin's disease, a substantial risk of leukemia following combination chemotherapy is clearly offset by enormous gains in survival. However, after ovarian cancer, the extent to which the increased risk of leukemia is offset is still unclear."

Other Approaches to Leukemia Treatment. Distinct from chemotherapy, red cell transfusion has been used for severe anemia at the outset of treatment, and although granulocyte transfer is not clearly understood, bone marrow transplantation is recognized as a definitive modality for aplastic anemia, acute leukemia, and chronic myeloid leukemia. The effect of such treatment becomes apparent after 2 to 4 weeks, and it is marked by a rise in circulating granulocytes and later by an increase in the platelet count. However, because allogenic grafts of marrow contain immunocompetent cells as well as marrow stem cells, the engrafted cells may mount an attack against the host. This becomes apparent in approximately half of the cases so treated.

A major goal in the treatment of acute leukemia is to reduce the amount of therapy and to minimize the adverse side effects without cutting the effectiveness of established treatments. For related topics, See also **Blood**; **Bone**; and **Cancer**.

Additional Reading

Abraham, N.G., R. Haas, G. Brittinger et al.: "Molecular Biology of Hematopoiesis and Treatment of Leukemias and Lymphomas," S. Karger Publishers, Inc., Farmington, CT, 1998.

Berkow, R. and M.H. Beers: "The Merck Manual," 17th Edition, Merck & Company, Inc., Whitehouse Station, NJ, 1999.

Coltman, C.A., Jr. and S. Dahlberg: "Treatment-Related Leukemia," *N. Eng. J. Med.*, 52 (January 4, 1990).

Erickson, D.: "Molecular Trickster: Antisense RNA Pulls A Fast One on the Leukemia Virus," *Sci. Amer.*, 26 (July, 1991).

Freireich, E. and H. Kantarjian: "Leukemia: Advances in Research and Treatment," Kluwer Academic Publishers, Norwell, MA, 1993.

Goldman, J.M.: "Understanding Leukemia and Related Cancers," Blackwell Science, Inc., Malden, MA, 1999.

Henderson, E.S. and T.A. Lister: "Leukemia," 6th Edition, W.B. Saunders Company, Philadelphia, PA, 1996.

Juliusson, G. et al.: "Response to 2-Chlorodeoxyadenosine in Patients with B-Cell Chronic Lymphocytic Leukemia Resistant to Fludarabine," *N. Eng. J. Med.*, 1056 (October 8, 1992).

Kaldor, J.M. et al.: "Leukemia Following Chemotherapy for Ovarian Cancer," *N. Eng. J. Med.*, 1 (January 4, 1990).

Kaldor, J.M. et al.: "Leukemia Following Hodgkin's Disease," *N. Eng. J. Med.*, 7 (January 4, 1990).

Kantarjian, H. and M. Talpaz: "Medical Management of Chronic Myelogenous Leukemia," Vol. 16, Marcel Dekker, Inc., New York, NY, 1998.

Polliack, A.: "Leukemia and Lymphoma Reviews," Vol. 6, Gordon & Breach Publishing Group, Newark, NJ, 1999.

Pui, C.: "Childhood Leukemias," Cambridge University Press, New York, NY, 1999.

Schumacher, H.R.: "Acute Leukemia," Lippincott Williams & Wilkins, Philadelphia, PA, 1997.

Stass, S.: "Acute Leukemias: Biologic, Diagnostic and Therapeutic Determinants," Marcel Dekker, Inc., New York, NY, 1987.

Tropé, C.: "Melphalan With and Without Doxorubicin in Advanced Ovarian Cancer," *Obset. Gynecol.*, 582, 70 (1987).

Weeks, J.C. et al.: "Cost Effectiveness of Prophylactic Intravenous Immune Globulin in Chronic Lymphocytic Leukemia," *N. Eng. J. Med.*, 81 (July 11, 1991).

Weirsma, S.R. et al.: "Clinical Importance of Myeloid-Antigen Expression in Acute Lymphoblastic Leukemia of Childhood," *N. Eng. J. Med.*, 800 (March 21, 1991).

Web References

Leukemia Information Library: *http://www.meds.com/leukemia/leukemia.html*
Leukemia Resource Center and Links: *http://rodneyporter.com/leukaemia/*
The Leukemia & Lymphoma Society, Formerly Leukemia Society of America: *http://www.leukemia.org/*
The Myeloproliferative Disorders: *http://www.acor.org/diseases/hematology/MPD/*

R.C. VICKERY, M.D.; D.Sc.; Ph.D., Blanton/Dade City, FL.

LEUKOCYTOSIS. An abnormally high level of leukocytes (white cells) in the blood, a common manifestation of infection.

LEUKOPENIA. A decrease in the normal number of leucocytes in the bloodstream. This is a usual accompaniment of certain stages of some infectious diseases, while in other infections it is of grave prognostic significance, and indicates a failure of one of the lines of defense of the body. In agranulocytosis, leukopenia is the chief manifestation of the disease.

Drugs and chemicals sometimes implicated in leukopenia include arsenic, chloramphenicol, cimetidine, 5-fluorocytosine, phenytoin, sulfopyridine, thiazides, and trimethoprim. Leukopenia is sometimes associated with Colorado tick fever, meningococcemia, pneumococcal pneumonia, and salmonelolosis. Leukopenia is also a result of starvation.

LEVEL OF ESCAPE.   Sometimes called *critical level of escape*, that level, in the atmosphere, at which a particle moving rapidly upwards will have a probability of $1/e$ of colliding with another particle on its way out of the atmosphere (where e is the number **e**, the base of natural logarithms). It is also the level at which the horizontal mean free path of an atmospheric particle equals the scale height of the atmosphere. The critical level of escape is the base of the exosphere, which is the outermost, or topmost, portion of the atmosphere. See also **Atmosphere (Earth)**.

LEVEL (Surveyor's).   An instrument for determining differences of elevation. It is of use in obtaining comparative elevations of two points, or in defining the profile of a certain path, such as a roadway, drainage ditch, etc. A level of this type has an accurately made bubble level attached to and made exactly parallel with a telescope. In the wye level this telescope rests in Y-shaped supports. These supports are held in turn by the instrument base, which may be adjusted by hand, and which is attached, usually by screwing, to the top of a tripod, upon which the instrument rests when in use. The bubble tube being parallel to the center of the telescope, the latter will automatically be leveled, ready for a horizontal sight, when the bubble tube is level. The dumpy level has the telescope and supports cast in one piece or rigidly connected. Lasers are also used in leveling operations, particularly for grading irrigated areas. See also **Irrigation**.

LEVEL WIDTH (Excitation Energy).   A measure of the spread in excitation energy of an unstable state of a quantized system. In emission or absorption spectra, variations in the intrinsic line widths show that some energy levels in atomic or nuclear systems are broad and others are narrow. In nuclear physics, level widths have been observed chiefly in connection with neutron and charged-particle resonances, which are found to have nonuniform breadths in energy. The level width is related to the mean life of the level by the expression

$$\Gamma = \hbar/\tau$$

where Γ is the level width, h is $\hbar/2\pi$ (h is Planck's constant), and τ is the mean life. Level widths usually show themselves as the widths of resonance peaks observed when the cross section for the particular reaction is plotted as a function of the energy of the incident particle. The quantitative value of the level width is usually taken as the full width at half maximum of the resonance peak.

If a system at a given level has several alternate modes of disintegration, there is associated with each a partial level width proportional to the probability of disintegration by the particular mode. The total level width is the sum of the partial level widths. See also **Broadening of Spectral Lines**.

LEVERS.   See **Machine (Simple)**.

L'HOSPITAL RULE.   If two functions $f(x)$ and $g(x)$ together with their derivatives up to order $(n-1)$ vanish at $x = a$, and if their derivatives

$$\lim_{x \to a} \frac{f(x)}{g(x)} = \frac{f^{(n)}(a)}{g^{(n)}(a)}$$

of nth order do not both vanish there or both become infinite, then
See also **Indeterminate Form**.

LIBRA.   (the scales or the balances). A small constellation, best known as the seventh sign of the zodiac; its symbol is taken for the autumnal equinox (i.e., the point where the sun apparently crosses the celestial equator from north to south). The brightest star in Libra is a wide double star, which can easily be resolved with a field glass. This star carries the Arabic name Zuben el Genubi, the southern scale. (See map accompanying entry on **Constellations**.)

LIBRATIONS.   A term applied in astronomy to many periodic oscillations; in particular, to slight apparent oscillations of the moon, whereby

observes on the earth are enabled to observe somewhat more than 50% of the moon's surface. There are three principal librations of the moon: a libration in lunar latitude, a libration in lunar longitude, and a diurnal (or daily) libration.

The libration in lunar latitude arises from the fact that the orbit plane of the moon is inclined at about 6.5° to the plane of the moon's equator. This produces an effect relative to the earth similar to the terrestrial effects relative to the sun, which produce the seasons. During one-half of the month, the north lunar pole is directed slightly toward the earth, and during the remainder of the month the south lunar pole is toward the earth.

The libration in lunar longitude is due to the fact that, although the rotation of the moon is quite uniform, with a period equal to the period of revolution about the earth, the orbital motion is not uniform, but is in accordance with the second Keplerian law of planetary motion. Suppose that a certain point on the surface of the moon is directly toward the earth at a time when the moon is in perigee. Since the moon is moving more rapidly in its ellipse at perigee than at any other part of its orbit, by the time the elongation has increased to 90%, the selected point has not completed one-quarter of a rotation and is not directly toward the earth. The result is that the observer will see a slightly different hemisphere of the moon than he did at perigee. By the time the moon has reached apogee, the selected point will again be toward the observer, and he will have the same view of the moon as at perigee. At this point, the moon's orbital motion is a minimum; the selected point will complete a quarter rotation before a quarter revolution is completed, and another slightly different hemisphere of the moon will be visible.

The diurnal libration is due to the fact that the observer sees slightly "over the top" of the moon when the moon is rising, and slightly under the bottom when the moon is setting. This is really a libration of the observer rather than a libration of the moon, but is classed with the latter.

The combined result of the librations is that about 41% of the moon is always visible from the earth (or would be if the sun were shining upon it), 41% is never visible, and the remaining 18% is either visible or invisible, depending upon the particular position of the moon relative to the earth.

The term libration is also applied to certain periodic perturbations in the orbits of members of the solar system.

See also **Perturbation (Astronomy)**.

LICHEN.   Perennial plants, which are a combination of two plants growing together in an association so intimate that they appear as one. Either of the component plants alone possesses none of the characteristics shown by the two in combination. Lichens often are cited as examples of symbiosis, i.e., two organisms living together of benefit to each other. The components of a lichen are always an alga and a fungus. See also **Algae**; and **Fungus**. The algal constituent is usually one of the simple green algae, or, more rarely, a blue green one. The alga can live perfectly well by itself, and is often found growing free on rocks or tree trunks in regions where the lichen would exist. The fungal component is usually a member of the ascomycetes. In lichens growing in cooler regions it is always one of this group. There are certain lichens found in the tropics, however, in which the fungal component is a basidiomycete. In the lichen, the fungus alone is capable of fruiting, although the algal cells do divide and so increase in number. It is very difficult to see how such an association came about, and to determine whether it is really a case of symbiosis or whether it is not parasitism, one of the plants living on the other. Unquestionably the fungus benefits from the presence of the alga, since the latter carries on photosynthesis, making food materials which are used by the fungus. The latter, lacking chlorophyll, cannot manufacture its own food. It is possible that the alga benefits by the added moisture gathered by the fungus, that the latter protects the alga against desiccation. Certainly the association is well established, and seemingly has been so for a long period of time. Lichens have been "made" artificially, that is, the two components have been grown separately in pure cultures, and when brought together have produced a lichen. So lichens can be formed anew, always with a very constant appearance characterizing the particular form considered.

Lichens are found nearly everywhere where civilization has not killed them. They are found on the surface of rocks and soil; they occur on the bark of trees; in the tropics they may be found on the surface of thick evergreen leaves of trees; a few species even grow on rocks submerged by the tides.

The shape of the lichen body or thallus is very diverse. Some species are flat crusts growing on or even in the surface of the substratum, whether the latter be trunk of tree or barren rock. Lichens of this type are called crustose. In others the thallus is split up into many radiating divisions, and is called a foliose lichen. Many others have an erect, often much-branched thallus and are called fruticose lichens. The color of the thallus may be yellow, orange, brown, gray, or black.

The greater part of the lichen thallus is composed of fungus hyphae, which form a compactly tangled mass. The algal cells occur in an irregular loosely arranged layer near the outer surface of the thallus. Short irregular branches from the fungus hyphae grow tightly around each algal cell, sending into it short, absorbing structures called haustoria. The surface of the lichen is composed of enlarged thick-walled fungus cells, which form a compact layer over the more loosely arranged central portions. In many of the crustose and foliose lichens there are many rhizoids which anchor the plant firmly.

One of the ways in which a lichen reproduces is by means of soredia. These are minute bits of lichen, formed on the surface, and composed of one or more of the algal cells together with a small mass of closely associated hyphae. Often these soredia are so numerous as to give to the lichen a powdery appearance. Either through disintegration of the lichen body, or because the continuity of the hyphae breaks down, soredia become free from the thallus. They are then easily spread by wind or by water. They may even be carried about unintentionally by the many small insects and other animals that feed on lichens. Lichens also reproduce by means of spores; that is, the fungus component forms special reproductive structures very similar to those formed by similar fungi not forming lichen thalli. In the lichens composed of ascomycetes these reproductive structures are open cups or mounds, called apothecia. Commonly these apothecia are of a different color from that of the thallus.

Reindeer moss, *Cladonia rangiferina*, is the principal food of the reindeer, and may also be used as fodder for other animals. Reindeer moss, not a moss at all, is an erect much-branded lichen which grows abundantly over wide stretches of barren soil. The dense grayish-green tufts grow continuously at the tops, becoming 6–10 inches (15 to 25 centimeters) tall and attaining great age. Another lichen, *Cetraria islandica*, or Iceland moss, may be used as stock food. In habit it resembles reindeer moss but is coarser and less branched. See Fig. 1.

Fig. 1.   Lichen (reindeer moss). (*A.M. Winchester.*)

Litmus paper is prepared from *Lecanora tartarea*, a common lichen found in the Netherlands.

LICORICE.   This herbaceous native plant of southern Europe (*Glycyrrhiza glabra* L.) of the family *Leguminosae* is the source of popular flavorings used in the food industry. Stolons or roots (at least 2 years old) are extracted with hot water. The taste is sweet and rich; the essence is sometimes described as slightly spicy. The principal constituent of licorice is the potassium-calcium-magnesium salt of glycyrrhizic acid, which upon hydrolysis produces glycyrrhetic acid and 2 moles of glucuronic acid. Licorice roots also contain triterpene, flavonoids, and B vitamins. The powerful sweetening power of licorice (estimated at 50 times that of sucrose) is derived from the glycyrrhizin. In food and drug applications, licorice is used both to enhance and also to mask or subdue flavors, particularly of a bitter nature.

The licorice plant ranges from 4 to 5 feet (1.2 to 1.5 meters) tall and has many pinnately compound, pale-green leaves and purplish flowers resembling those of the perennial pea. Cultivated in southern Europe, the plant also is found wild in eastern Europe. Licorice root and extract are used in nonalcoholic beverages (up to 130 parts per million); in ice cream (200 ppm); in candies (as high as 2500 ppm); in baked goods (200 ppm); in gelatins and puddings (5 ppm); in syrups (50 ppm); and in some chewing gums (as high as 22,000 ppm). Licorice is also frequently used in tobacco products. A synthetic licorice flavoring is made and is the ammonium salt of glycyrrhizic acid.

LIDAR.   Acronym for Light Detection And Ranging. Like RADAR, except lidar uses light (laser) instead of radio waves.

LIE GROUP.   A group that is also an analytic manifold in which the group operations, multiplication and formation of inverse, are analytic. Historically, the first Lie groups were continuous transformations of the points of a manifold. Thus, consider the set of transformations of the points of a Euclidean plane,

$$x_1 = \phi(x, y, a), \quad y_1 = \psi(x, y, a)$$

for some range of values of the parameter a, given by functions ϕ and Ψ such that, if two transformations of the set are carried out in succession, the result is again a transformation of the set. An easily visualized example is the group of rotations

$$x_1 = x \cos a - y \sin a, \quad y_1 = x \sin a + y \cos a$$

Lie groups of transformations owe their practical importance partly to their usefulness in systematizing the solutions of differential equations, both ordinary and partial, and partly to the fact that many of the standard groups of linear transformations are Lie groups. For example.

The *full linear group* is the group of all nonsingular matrices with complex numbers as elements.

The *real linear group* is the group of all nonsingular n × n matrices with real numbers as elements.

The *unimodular group* is the group of all complex matrices with determinant equal to unity.

The *real unimodular* group is the group of all real matrices with determinant equal to unity.

The *unitary group* is the group of all unitary matrices.

The *unitary modular group* is the group of all unitary matrices with determinant equal to unity.

The *real orthogonal group* is the group of all real orthogonal matrices.

The *rotation group* is the group of all real orthogonal matrices with determinant equal to plus one.

See also **Group**.

LIFE SCIENCES.   The field of scientific disciplines encompassing biology, physiology, psychology, medicine, sociology, and other related areas.

LIGAMENT.   See **Tendon**.

LIGAND.   Any atom, radical, ion, or molecule in a complex (poly-atomic group) which is bound to the central atom. Thus, the ammonia molecules in $[Co(NH_3)_6]^{3+}$, and the chlorine atoms in $[PtCl_6]^{2-}$ are ligands. Ligands are also complexing agents, as for example, EDTA, ammonia, etc. See also **Chelates and Chelation**.

Ligand field theory incorporates elements from the valence bond theory of Pauling and the molecular orbital method of Hund, Mullikan, and others. As pointed out by Mortimer, the chemists of the late nineteenth century had difficulty in understanding how "molecular compounds" or "compounds of higher order" are bonded. The formation of a compound such as $CoCl_3 \cdot 6NH_3$, was baffling, particularly in this case since simple $CoCl_3$ does not exist. In 1893, Alfred Werner proposed a theory to account for compounds of this type. Werner wrote the formula of the cobalt compound as $[Co(NH_3)_6]Cl_3$. Werner assumed that the six ammonia molecules are symmetrically coordinated to the central cobalt atom by "subsidiary valencies" of cobalt, while the "principal valencies" of cobalt are satisfied by the chloride ions. Werner devoted over 20 years preparing and studying coordination compounds and perfecting and proving his theory. Although modern work has amplified his theory, it has required relatively little modification.

In ligand field theory, one is concerned with the origin and the consequences of splitting the inner orbitals of the central metal by the surrounding ligands. The most satisfactory correlations have been demonstrated with the first transition series, in which the $3d$-orbitals are split into different energy levels. To appreciate the effect of a ligand field, imagine that a symmetrical group of ligands is brought up to a charged ion from a distance. First, the electrostatic repulsions between the ligand electrons and those in the d-orbitals of the metal will raise the energy of all five d-orbitals equally. Then, as the ligands approach to within bonding distances, the repulsion interactions will take on a directional character that will vary with the particular d-orbitals under consideration. This arises because of the different shapes and orientations of the five d-orbitals in space along a Cartesian coordinate system. The splitting of the orbitals for a given central metal ion is dependent on the set of ligands.

Applications of ligand field theory to many transition metal complexes have played an important role in the interpretation of visible absorption spectra, magnetism, luminescence, and paramagnetic resonance spectra.

LIGASES.   See **Enzyme.**

LIGATURE.   A threadlike material or wire used for tying off blood vessels or other structures of the body during surgical operations. The material may be absorbable (catgut) or nonabsorbable (silk, nylon or linen) or metal wire. Ligatures are made in various grades of thickness and tensile strength.

LIGHT.   Although the use of the word *light* has been broadened over the decades, general usage still refers to the visible portion of the electromagnetic spectrum. The wavelengths of visible light extend approximately from 4000 to 7000 Å (1 Å = 10^{-8} centimeter). The speed of light in vacuum (symbol c) is generally published as 299,792,500 meters per second (~186,292 miles per second). The direct determination of the velocity of light, usually performed in air, conventionally has been based upon the measurement of the time for a light pulse to cover a known distance. (See reference to Michelson-Morley experiment in entry on **Laser**.) Such a pulse means an increase, followed by a decrease, of the amplitude of the light vibrations. What is observed is energy exchange associated with the amplitude changes and, as a result, the propagation velocity of light is obtained. A change of amplitude is, however, equivalent to an interference among a series of adjacent wavelengths, inasmuch as that change is created by just such an interference. The light pulse, therefore, consists of a whole group of adjacent wavelengths interfering with each other. Interference is a sum-product. If the participating waves have different velocities, one can find by simple addition of two sine oscillations that the group formed has a velocity different from those of the waves creating the group. An interesting experiment in the use of a laser to measure the speed of light was made in 1962. This experiment was later followed by a number of other, even more sophisticated laser measurements.

A review of the numerous experiments to measure the velocity of light makes fascinating reading for the student of science. An excellent starting review of the experiments of Galileo (1676), Bradley (1725), Foucault (1850), Kohlrausch (1856), Blondlot (1891), Michelson (1879 and 1926), Karolus and Mittelstaedt (1928), Anderson (1940), Jones and Conford (1947), Alaskon (1949), Essen and Gordon-Smith (1947–1950), Bergstrand (1950), Froome (1950–1958), Rank, Bennet and Bennet (1955), Mackenzie (1953), Kolibayev (1958–1963), Mockler (1961), and Cohen (1972) is given in "The Encyclopedia of Physics," (R.M. Besançon, editor), Van Nostrand Reinhold, New York.

Light has a physical character similar to that of radio waves. However, the frequency of light waves is almost a billion times higher and the wavelengths a billion times shorter, than the waves of standard radio broadcast bands. The perception of color depends upon the distribution of the electromagnetic energy over the visible wavelengths. White light is a superposition of waves at many frequencies. It can be decomposed into its monochromatic spectral components by a prism or other spectral apparatus. The violet end of the spectrum is near 4000 Å. The red end is near 7000 Å. Whereas light, in its narrow definition, should be confined to this relatively narrow portion of the electromagnetic spectrum, in recent years it has become customary to extend the definition to the ultraviolet and infrared portions of the spectrum. One sometimes speaks loosely of ultraviolet and infrared light, although electromagnetic waves at these frequencies are not detectable by the human eye. Instruments and photographic films, however, can be made sensitive to both the shorter and much longer wavelengths. See also **Infrared Radiation**; **Photography and Imagery**; **Spectrochemical Analysis (Visible)**; **Spectro Instruments**; **Spectroscope**; and **Ultraviolet Radiation**.

The study of the human eye as a detector of light is the task of *physiological optics*. The impression of light is not necessarily always connected with the simultaneous presence of electromagnetic energy at the retina. The eye is capable of creating false images, as when one "sees stars" from a heavy mechanical blow in the dark. The impression of light is retained for about 0.1 second after the light source is shut off. This fact is made use of in motion pictures to create the impression of motion through use of a series of still images. The eye is a detector with a relatively long response time. Photoelectric cells can react more than a million times faster. Color vision is also subject to physiological peculiarities that are quite complex. See also **Vision and the Eye**.

The quantum theory of radiation applies to light, the energy quanta of which are called photons. Some of the optical phenomena so readily interpreted on the wave theory, such as reflection, refraction, interference, diffraction, and polarized light, offer difficulties when studied in terms of quanta. The laws of photoelectric phenomena, photoconductivity, and the spectrum, on the other hand, appear more readily understandable when studied in terms of quanta.

The property of light most immediately accessible to observation is its propagation along straight lines. If light rays pass from one medium to another, their direction is changed according to the law of refraction. See also **Mirrors and Lenses**; **Refraction**; and **Refractive Index**. If light in medium I propagates with velocity v_1 and makes an angle v with the normal to the boundary between media I and II, the direction v in medium II, with a velocity of propagation v_2 is given by Snell's law, $\sin v_1 / \sin v_2 = v_1/v_2 = n$. The constant n is called the relative index of refraction of medium II with respect to medium I. These laws are the basis of *geometrical optics*. This branch of the science of light describes the paths of light rays, the formation of images by mirrors and lenses, the action of telescopes, microscopes, prisms, and numerous other optical instruments.

The wave character of light becomes apparent by more refined observations. The phenomena of diffraction, interference and polarization are the subjects of *physical optics*. Diffraction describes how waves are bent around obstacles. They represent corrections to the deviations from the laws of geometrical optics. These effects become pronounced only when the material has a characteristic dimension comparable to the wavelength of the wave. When light waves reach the same point along different paths, the resulting intensity may be smaller than that produced by each individual wave separately. The relative phases of the waves may be such that they interfere destructively, when the arrival of one wave with maximum positive deflection coincides with that of another wave with maximum negative deflection. Observations of light in crystals of calcite (Iceland spar) first showed that there are two different modes of vibrations for each direction of propagation. These are called the two transverse modes of polarization.

All phenomena of geometrical and physical optics are described consistently by Maxwell's equations of electromagnetic theory. Optical phenomena are, therefore, closely related to other electric and magnetic phenomena. Very early in the present century, the prevailing opinion was that the wave character of light was unambiguously established and the nature of light well understood.

There was, however, a mathematical difficulty with the intensity of radiation of ultraviolet and higher frequencies. The photoelectric effect could also be interpreted only by considering light to have a quality of particles. The number of electrons emitted from a photosensitive surface is proportional to the intensity of the light. The energy of the individual electrons is, however, determined by the light frequency. This led to the postulate of *light quanta* with energy hv, where h is Planck's constant. This duality in nature, in which wavelike and particle-like properties are combined, is described without internal contradiction by quantum mechanics. The combined particle-and-wave character of light is revealed by the combination of properties of the light sources, the electromagnetic field describing the light waves, and the detectors.

The combination of the laws of quantum mechanics and electromagnetic theory gives a consistent description of the generation, propagation and detection of light. Since these same laws also describe many other properties of matter, such as electronic structure, chemical binding, electricity, and magnetism, it may be said that the nature of light is well understood. In this context, it is not necessary and not even desirable to pose the question, "What is it, precisely, that vibrates in a light wave in vacuum?" The electromagnetic fields acquire meaning only through their relationships with detectors and sources. Human knowledge or understanding is here used in the operational sense that a relatively simple framework of physical concepts and mathematical relationships exist, which gives an accurate description of the wide variety of optical phenomena at present accessible to observation or verification in experimental situations.

The study of the interaction of light waves with matter in the sources and detectors is the subject of *spectroscopy*. This is a wide field, which encompasses atomic and molecular spectroscopy, parts of solid-state physics, and photochemistry. The quantum theory was largely developed on the basis of spectroscopic data. A light quantum is emitted by an excited atom, molecule, or other material system when an electron in such a particle makes a transition or "quantum jump" from a state with higher energy to a state with lower energy. The energy difference between these states is equal to the quantum energy hv. Similarly, the absorption of light quanta is accompanied by an electronic transition from a state with a lower energy to a state with an energy higher by an amount hv. In this manner, the frequencies of spectral lines are characteristic for the electronic energy levels in each material. The frequency of the light may be said to correspond to the frequencies of the vibrating charges or oscillators, which are represented by electrons.

Light sources are thus bodies with a sizeable population of electrons in excited states. This may be accomplished by raising the temperature of the material. The most important source of light is the sun. The moon and other planets are visible only because they reflect sunlight, just as all objects on earth which we can see by daylight, but not at night.

Human-engineered light sources range from primitive fire, candles, and oil and kerosene lamps to electric light bulbs, fluorescent-gas discharge tubes, arcs, and many others. See also **Illumination**. In early sources, the material particles of smoke or wick are heated by the chemical reaction of oxidation or burning; in incandescent lamps, a wire is heated to a very high temperature by an electric current. There are so many energy levels, in these luminous solid materials or gases at high pressures, that the emitted light is essentially white and contains all frequencies. The higher the temperature, the more radiation is emitted and the higher the average frequency of radiation. It should be realized that most of the energy is emitted as invisible (infrared) radiation, even in the better incandescent lamps. Hot gases in flames may also emit sharp spectral lines characteristic of the atoms occurring in the flame. The yellow color, which arises when sodium chloride is sprinkled in a flame, is due to the characteristic yellow spectral line of sodium atoms.

In gas discharge tubes, atoms or molecules are excited by collisions with electrons in ionized gas. The energy is provided by the generator, which provides the voltage necessary to maintain the discharge current. An arc is a discharge in air or in a high-pressure vapor. Mercury and sodium discharges are used for street lighting. Fluorescent tubes use a gas discharge with a substantial ultraviolet component. This ultraviolet light excites electrons in fluorescent centers on the walls of the tube. The electrons drop immediately from the highly excited state to an intermediate state with a lower energy. From this state, they finally drop down to the original ground energy level, with emission of visible light. Gas discharges at relatively low pressure may serve as spectroscopic sources to study the emission spectra of atoms, ions, and molecules. From the relationship between the energy levels and the frequency of radiation, it follows that a material, when heated, can emit precisely those frequencies that it absorbs when it is in the lower energy level at low temperature.

All of these light sources are incoherent in the sense that there is no phase relationship between the light waves emitted by the different atoms in the source. This is quite different from the property of the usual sources of electromagnetic radiation at lower frequencies. In electronic oscillators used in radio and microwave transmitters, all electrons move and vibrate in step with each other. Unlike the light sources previously mentioned, lasers emit a coherent beam of light. In lasers, the original, spontaneously emitted light forces the other excited atoms to emit their radiation in step, or coherently. If stimulated emission thus dominates the spontaneous emission, a laser results. This requires a high concentration of excited atoms and a sufficient feedback mechanism of light by mirrors. In its simplest form, a laser consists of a gas discharge in a tube of suitably chosen dimensions and gas pressure between a set of parallel mirrors. Because the atoms in the laser source all act constructively in step, these sources provide a more efficient means to transmit light energy.

The high light intensities available in focused laser beams have led to the development of the branch of *nonlinear optics*. The optical properties of materials are different at high intensities, because the electronic oscillators are driven so hard that enharmonic properties become evident. A typical effect is the harmonic generation of light in which red laser light is converted into ultraviolet light at exactly twice the frequency when the high-intensity beam traverses a suitable crystal, such as quartz. It thus becomes feasible to duplicate at light frequencies all nonlinear effects known from the field of radio communications, such as modulation, demodulation, frequencing mixing, among others. It is no longer correct to say that the propagation of a light wave is independent of the presence of other light waves. At high intensities, there is a noticeable interaction between light waves of different frequencies.

Optical Phase Conjugation

Even though the laser enjoys very wide application (see also **Laser**), its full potential has not been realized because of distortion that occurs in light waves when they pass through optical systems where inhomogeneities are present. There is a real need for means to reduce or compensate for static and dynamic distortions (noise) which frequently occur. High-power lasers, tracking systems, atmospheric communication networks, and photolithographic systems, among others, are degraded by such noise. A relatively new area of optical system research, known as *optical phase conjugation*, promises to provide at least a partial solution to this problem.

That light beams possess the property of reversible propagation has been known for many years. For example, assuming a *perfect* optical system, a coherent light beam introduced at point A in a system and traveling through a complex of components will exit at point B undistorted. If a light beam of exactly the same characteristics were introduced at point B, it would travel the same exact path and exit, undistorted, at point A. In other words, under perfect conditions, the propagation is reversible. What was not known concerning reverse propagation until the early 1970s was that a distorted (noisy) light beam, if propagated in reverse through a given optical system, will during the course of its backward track remove all the distortions introduced into it during its prior forward track, and thus exit at the point of origin as a "clean" beam, free of distortion. Thus, the concept of reversible propagation holds not just for a theoretically perfect system, but for a practical imperfect system as well.[1] The backward-traveling wave, in essence, is a "time-reversed" replica of the original incident wave.[2]

[1] In an experiment at the P.N. Labedev Physical Institute (Moscow) in 1972. Boeia Ya Zel'dovich and coworkers observed a curious phenomenon while doing an experiment. The researchers intentionally distorted an intense beam of red light from a pulsed ruby laser by directing it through a frosted glass plate. The degraded beam was directed down a long tube filled with methane gas under high pressure. Interactions occurred between the beam and the molecules of the gas (stimulated Brillouin scattering) and, acting as a mirror, the gas reflected the beam backward. The investigators were surprised to find that once the reflected wave passed back through the same piece of frosted glass, a nearly perfect, undistorted optical beam emerged. Thus, they found that the distortions introduced by the glass during the forward passage were, in essence, canceled out during the backward passage. (The phase difference between any two points of the reversed beam has a sign opposite to that of the phase difference between the same points of the original beam. As described by Shkunov and Zel'dovich, the mathematical operation of changing a phase sign is known as conjugation, and thus the coining of the term optical phase conjugation.)

[2] During the early phases of explaining this unexpected phenomenon, a number of homely analogies were developed, the most common being that of comparing the retracing of light waves back through the distorting media as "making a film run backward." As described by Shkunov and Zel'dovich, "The relation between the wave fronts of two mutually reversed waves is analogous to the relation between the positions of two opposing armies on a military map. The front line of each army coincides with that of the other, and the directions of desirable movement are opposite. One can say that the front lines are mutually reversed: a convex part of one army's front corresponds to a concave part of the other."

Producing a Phase-Conjugated Wave. Two general approaches have been investigated thus far for implementing the concept of optical phase conjugation in a practical way for the purposes previously mentioned. These two approaches employ different physical principles, but both rely on the laser light itself to interact with the nonlinear optical properties of a specific medium to initiate phase conjugation or a turnabout of the distorted light waves: (1) *stimulated Brillouin scattering* (SBS) and (2) *degenerate four-wave mixing* (DFWM). When any material — gas, liquid, or solid — is penetrated by light of intensity great enough to compete with the atomic forces that bind the material together, the material is modified, as is also the light penetrating it. This nonlinear interaction generates the SBS or DFWM time-reversed waves. The success in the reversal of wave motion is due to an extreme simplification of the problem: the quantum-mechanical and thermal motions of atoms and electrons that radiate light do not need to be reversed. It is sufficient for practical purposes, as observed by Shkunov and Zel'dovich, to reverse the temporal behavior of macroscopic parameters describing the averaged motion of a large number of particles.

In SBS, the modified material generates sound waves that serve as an appropriate reflective surface to produce the time-reversed waves.[3] In DFWM, the interaction uses a holographic process in a nonlinear material to generate the conjugate, or reversed light waves.

Examples of nonlinear mediums include semiconductors, crystals, liquids, plasmas, liquid crystals, aerosols (as in the atmosphere), and atomic vapors. The term *nonlinear*, as used here, pertains to media that are altered or affected by light. Linear materials, in contrast, are not so affected.

Prospective Uses of Optical Phase Conjugation. Currently, a number of scientific laboratories, in addition to those in Russia, have been conducting intensive research in this area in attempts to refine the technology and to find practical applications. Among the active laboratories are the California Institute of Technology, the AT&T Bell Laboratories, Philips Research Laboratories (the Netherlands), the University of Southern California, the University of Waikato (New Zealand), the University of Arizona, the National Institute of Standards and Technology, IBM, and Hughes Aircraft Research Laboratories. A number of scientists envision several important applications within the next decade. Admittedly, the field remains in an investigative state. Some potential applications as reviewed by Pepper (see reference) and others are reviewed briefly here.

Some scientists have suggested that phase-conjugated systems may be used for image transfer in photolithography. Researchers at IBM have demonstrated image transfer. Light from a laser passes through the mask pattern, a semitransparent mirror, and then an amplifier. The intensity of the beam increases, but at the cost of introducing distortions into the beam. When the image is returned through an amplifier by a phase-conjugating mirror, the "time-reversed" beam is both powerful and free of distortions. Conventional methods, such as compensating for optical aberrations, are no longer required. Another way in which lensless-imaging schemes may be used include fiber-optic communications and associated memory as used in pattern-recognition devices. As stressed by Pepper (see reference), the emergence of optical phase conjugation has unified many areas of applied and fundamental optical physics. Spectroscopy, the study of the interaction of matter and radiation, has particularly benefited. The concepts, techniques, and basic applications of optical phase conjugation can, in principle, be applied to most other areas of the electromagnetic spectrum. Microwave phase conjugation would have major applications in radar, millimeter-wave imaging systems, and high-frequency temporal signal processing, as well as microwave spectroscopy. Researchers are planning experiments in the acoustic-wave area. Acoustic signal-processing devices and sonar may benefit from such research.

[3] Traditionally, the frequency of scattered light has been regarded as identical to that of the incident light. In actual practice, as first predicted by Brillouin in 1914, a slight line broadening occurs due to motion of the scatterers (Doppler effect) and also due to variations in the directions or magnitudes of their polarizability tensors (due to chemical reactions). The Brillouin effect, simply stated, is as follows: upon the scattering of monochromatic radiation, a doublet is produced, in which the frequency of each of the two lines differs from the frequency of the original line by the same amount, one line having a higher frequency, and the other having a lower frequency.

Squeezed Light. Although research has been proceeding for over a decade, the study of "squeezed" light is still essentially in an experimental stage. For a number of years, scientists have recognized that a beam of light is not free from random fluctuations, but in electronic terms it is noisy. Light is used frequently to make measurements and to observe numerous instrumental phenomena. Thus, faulty light contributes to errors, if ever so small. This situation is explained by the quantum theory, which implies that light must be accompanied by a certain minimal amount of light fluctuation.

A beam of light has been defined as an oscillating electromagnetic field, which can be viewed as a smooth wave. The shape of the wave can be foretold with absolute certainty, but its slope must fit within a particular "envelope" of uncertainty. Even in darkness, quantum physicists would allow that the wave exists but is flat, with some degree of uncertainty.

Researchers Slusher and Yurke (AT&T Bell Laboratories) have observed, "In practical terms, this means that even in a vacuum, with no external light sources, there must still be small fluctuations in the electromagnetic field. Noise limits the precision of spectroscopy, in which the frequency and intensity of the radiation emitted by atoms or molecules yield information about their properties." It also is envisioned that quantum noise will also limit those technologies concerned with optical computing and communications.

In early investigations, researchers at several laboratories had experimented with so-called *squeezed light*. This field of study involves quantum fluctuations, and largely stemmed from studies of the coherent light generated by a laser. A number of theoretical physicists more than 30 years ago studied the statistical properties of coherent light. Even the presumed perfectly coherent light of an ideal laser was found to have a Poisson distribution of photons rather than a single, well-defined number. Thus, although not so noisy as a conventional incoherent light beam, the laser is also noisy and, as lasers are employed more frequently in very sophisticated research and practical applications, optical devices could ultimately be limited by such noise.

As related by Robinson (see reference), for squeezed states, the statistical variance of the photon number is not so important as the variances of the amplitude of the electric field, which is related to the photon number, and of its phase. Researchers at AT&T Bell Laboratories have successfully "squeezed" noise or unwanted fluctuations in *phase* at the cost of increasing it in amplitude and, conversely, they have also squeezed noise in *amplitude* at the cost of increasing it in phase. In essence, these researchers generated the squeezed light by using a technique known as *four-wave mixing*, previously mentioned under optical phase conjugation. The researchers directed two laser beams at each other from opposite directions. The laser beams met in a material (nonlinear medium). In their experiment, the nonlinear medium was a beam of sodium atoms oriented at right angles to the laser beams. The research team found that the laser beams interacted with the Na atoms in such a way that two output beams emerged in opposite directions along an axis tilted at a slight angle with respect to the axis of the two input laser beams. Attributed to the properties of the nonlinear medium, when the output beams were combined by mirrors, the result was a single beam of squeezed light. By varying the position of the mirrors that direct the beams emerging from the nonlinear medium, it was found possible to reduce the noise in either the phase or the amplitude of the squeezed light. The degree of success of the experiment was relatively small — about a 7% drop in noise. More recently, other researchers (University of Texas) have reported a 42% reduction in noise.

Practical applications of squeezed light must await further research findings.

Subwavelength Illumination. Superresolution light microscopy permits researchers to optically study specimens without being limited by the diffraction properties of visible light. This technique, however, requires the efficient emission of light from *subwavelength* light sources. In an experimental setup, an electromagnetic wave is generated to emerge from an aperture. First, the wave is highly collimated to the aperture dimension. As pointed out by K. Lieberman and coworkers (Hebrew University, Jerusalem), "It is only after the wave has propagated a finite distance from the aperture that the diffraction that limits classical optical imaging takes effect. Thus, in the near-field region, a beam of light is present that is largely independent of the wavelength and is determined solely by the size and shape of the aperture." In a 1990 paper (see reference listed), an approach for producing sources of light with subwavelength dimensions is described.

The methodology is based on the packaging of photons as molecular excitons, effectively reducing the volume of the light beam by 109 and making possible propagation through dimensions of 1 nanometer. The researchers further observe, "Molecular microcrystals are grown in the tips of micropipettes that have inner diameters of 100 nanometers or less. Measurements are presented that demonstrate this improvement in transmission for pipettes of various diameters. The ultrasmall dimensions of these light sources, the wavelength range (UV to IR) of their emission, their ease of production, and their expected unique abilities for high efficiency excitation-imaging of surfaces portend significant applications for this methodology."

Light Reflections in Computer Graphics. The technology of computer graphics has enabled the production of objects in realistic three dimensions from two dimensions, a technique that frequently has replaced the need for clay models in a number of fields, notably in the automotive and appliance fields and for creating realistic backgrounds in simulators as, for example, in aircraft flight training. To achieve realism of painted, metal, glass, and other reflective surfaces, "reflection" algorithms can be created.

Reflection patterns vary markedly from one type of product surface to the next. These patterns are unique for given materials or material families, including, for example, steel, chromium, rubber, and plastics, because of the wide range of diffuse and specular reflections for given materials. Through careful experimentation, appropriate algorithms can be developed. Considerable detail pertaining to these methodologies is given in the D.P. Greenberg reference listed. See also **Hurlbert Poggio reference listed**.

Extending the Light-Sensing Range. Since so-called "night vision" sensors were developed several decades ago, the practical use of the infrared (IR) portion of the light spectrum has continued in its importance, particularly in military operations such as tank and helicopter maneuvering. Traditionally, these sensors have depended upon mercury cadmium telluride (*mer-cal*) semiconductors. These sensors pick up wavelengths between 3 and 5 and 8 and 12 microns. A greater sensitivity over the whole IR range has been sought. The materials for these sensors also have been difficult to process. After some years of research, a group at AT&T Bell Laboratories has developed ways to produce gallium arsenide optical crystals that are claimed to approach atomic perfection. The new detectors, which have greatly increased sensitivity, feature what is known as a quantum well, which precisely controls the flow of electrons. Experimental arrays with more than 16,000 pixels have been built and tested. Such detectors can sense temperature differences as small as a ten-thousandth of a degree. Cool objects appear dark, with warmer objects ranging over a gray scale. The quantum-well infrared photodetectors (QUIPs) now are being studied in terms of mass producing them.

Sky Colors. As early as 1899, Rayleigh claimed that the sky is blue on a clear day because of scattering and color separation by the molecules contained in the air of the atmosphere. Rayleigh viewed the phenomenon as a composite of all the colors in the visible spectrum. See also **Scattering**. He proposed that short-wavelength light (blue) is scattered more than red light, which has a longer wavelength. Based upon the ratio of the blue and red wavelengths ($\sim$1.68), Rayleigh reasoned that the blue-scattered light is approximately eight times that of the red-scattered light.

Sunsets are reddened because the light reaching the observer has passed through a much longer path, allowing domination of the red end of the spectrum in contrast with the situation when the sun is viewed high in the sky.

Other investigators have attributed sky color to layers of ozone in the upper atmosphere. It has been postulated that, in these layers, molecules have absorption bands that favor the absorption of light at the red end of the spectrum. A number of postulations have emerged pertaining to the development of several colors, including the purple sky coloration. These are explored in more detail in the Meinel reference listed.

Some correlation of sunset color with volcanic eruptions has been made. As examples, the explosion of Krakatau (near Java) in 1883 produced brilliant sunsets over a span of about 5 years; the explosion of Agung on Bali in 1963 affected sunset coloration for about 3 years.

Optical Computer

The pace of research to develop an optical computer based upon beams of light rather than electric currents have increased markedly during the past decade. An optical analogue of the transistor was demonstrated in the early 1980s. An optical computer that might operate a thousand times

faster than contemporary electronic computers is indeed a powerful incentive. Theoretically, the operations carried out by a computer (logical and arithmetic) could be effected by numerous means and, in fact, this already has been demonstrated when one considers that early computers depended exclusively on vacuum tube switches. Although the transistor that replaced the tube was a great step upward in electronics, the transistor was a significant hardware departure — thus indicating that computer designers will quickly adapt to new hardware when there are significant demonstrable advantages. Thus, among some scientists, there is an attitude that tends to regard the optical computer as an inevitable development for the future. Some forecasts indicate that an optical computer may be capable of a trillion operations per second. Even though the current electronic computers are undergoing significant changes in the way they are organized to process information, such reorganization today probably would be easier to achieve if the system were basically optical rather than electronic — as considered by some contemporary scientists. See also **Digital Computer Systems**.

Additional Reading

Andreev, A.V. et al.: "Quantum Optics," SPIE International Society for Optical Engineering, Bellingham, WA, 1999.

Born, M. and E. Wolf: "Principles of Optics: Electromagnetic Theory of Propagation, Interference and Diffraction of Light," 7th Edition, Cambridge University Press, New York, NY, 1999.

Buchwald, J.D.: "The Rise of the Wave Theory of Light," University of Chicago Press, Chicago, IL, 1990.

Corcoran, E.: "Body Heat — QWIPs Offer a New Way to See in the Dark," *Sci. Amer.*, 123 (October, 1991).

Feynman, R.P.: "QED: The Strange Theory of Light and Matter," Princeton University Press, Princeton, NJ, 1988.

Flam, F.: "Through a Glass — Darkly (Light Scattering)," *Science*, 29 (April 5, 1991).

Flam, F.: "Scopes with a Light," *Science*, 30 (April 5, 1991).

Gower, M. and D. Proch: "Optical Phase Conjugation," Springer-Verlag, Inc., New York, NY, 1994.

Greenberg, D.P.: "Light Reflection Models for Computer Graphics," *Science*, 166 (April 14, 1989).

Hakfoort, C.: "Optics in the Age of Euler: Conceptions of the Nature of Light, 1700–1795," Cambridge University Press, New York, NY, 1994.

Harris, N.: "Lighting and Its Uses," *Science*, 543 (August 4, 1989).

Hirota, O.: "Squeezed Light," Elsevier Science, New York, NY, 1992.

Hurlbert, A.C. and T.A. Poggio: "Synthesizing a Color Algorithm from Examples," *Science*, 482 (January 29, 1988).

Leonhardt, U.: "Measuring the Quantum State of Light," Cambridge University Press, New York, NY, 1997.

Lieberman, K. et al.: "A Light Source Smaller than the Optical Wavelength," *Science*, 59 (January 5, 1990).

Loudon, R.: "The Quantum Theory of Light," 3rd Edition, Oxford University Press, Inc., New York, NY, 2000.

Marshall, E.: "Science Beyond the Pale," *Science*, 14 (July 6, 1990).

Maxwell, J.C. and P.M. Harman: "Scientific Letters and Papers, 1846–1862," Vol. 1, Cambridge University Press, New York, NY, 1990.

Maxwell, J.C. and P.M. Harman: "The Scientific Letters and Papers of James Clerk Maxwell: 1862–1873," Vol. 2, Cambridge University Press, New York, NY, 1995.

Meinel, A. and M. Meinel: "Sunsets, Twilights, and Evening Skies," Cambridge University Press, Cambridge, Massachusetts, 1983.

Pepper, D.M.: "Application of Optical Phase Conjugation," *Sci. Amer.*, 74–83 (January, 1986).

Peterson, I.: "Putting a Far Finer Point on Visible Light," *Science News*, 7 (January 6, 1990).

Robinson, A.L.: "Bell Labs Generates Squeezed Light," *Science*, **230**, 927–929 (1985).

Ruthen, R.: "Surfing Photons," *Sci. Amer.*, 12D (August, 1989).

Saki, Jun-Ichi: "Phase Conjugate Optics," The McGraw-Hill Companies, Inc., New York, NY, 1992.

Schivelbusch, W.: "Disenchanted Night: The Industrialization of Light in the Nineteenth Century," University of California Press, Berkeley, CA, 1988.

Shkunov, V.V. and B.Y. Zel'dovich: "Optical Phase Conjugation," *Sci. Amer.*, 54–59 (December, 1985).

Siegel, D.M.: "Innovation in Maxwell's Electromagnetic Theory: Molecular Vortices, Displacement Current, and Light," Cambridge University Press, New York, NY, 1991.

Slusher, R.E. and B. Yurke: "Squeezed Light," *Sci. Amer.*, 50 (May, 1988).

Staff: "Optics Source Book," McGraw-Hill, New York, NY, 1988.

Walker, J.: "The Colors Seen in the Sky Offer Lessons in Optical Scattering," *Sci. Amer.*, 102 (January, 1989).

Weber, M.J., Ed.: "CRC Handbook of Laser Science and Technology," CRC Press, Boca Raton, FL, 1987.

Weiss, P.L.: "Reflections on Refraction," *Science News*, 236 (October 13, 1990).

LIGHT (Effects on Plants).   See **Photoperiodism**.

LIGHT (Polarized).   See **Polarized Light**.

LIGHT-ACTUATED CELL.   See **Photoelectric Effect**.

LIGHT-COUPLED SWITCH.   A switch in which the switching signal is transmitted to the device in the form of light energy. The switching element may be a phototransistor, a photodiode, or field-effect transistor (FET). The receipt of light energy by such a device changes the transmission characteristics of the device, permitting conduction between two terminals. When no light is present, the resistance of the device is high. When excited by photon energy, the resistance drops to a much lower value. Various light sources are used in connection with light-coupled switches. Although common incandescent sources may be used, gas-discharge sources, such as neon lamps and solid-state devices, such as gallium arsenide light-emitting diodes, are more commonly used in instrumentation systems. As compared with incandescent sources, the other sources produce less heat and possess higher speed and reliability.

A major advantage of the light-coupled switch is isolation that can be obtained between signal and drive source. Signals can be controlled without introducing errors due to the drive source. Further, isolation helps in maintaining a high common-mode rejection ratio in analog signal-handling equipment.

See also **Analog Switch**.

LIGHT CURVE (Astronomical).   In the study of variable stars, and in kindred problems in astronomical research, it is desirable to represent graphically the variation of radiation intensity with time. A diagram in which light intensity, on any convenient scale, is plotted as ordinate against time as abscissa is known as a light curve. As the number of observations increases, it frequently becomes possible to detect a periodic variation in the light intensity. After a provisional period has been determined, some convenient epoch is selected, and all the observations are reduced to the cycle of variation embracing the selected epoch by the use of the provisional period. In order that the resulting points may fall on a regular curve, it is frequently necessary to apply a number of corrections to the provisional period. The curve drawn through the plotted points, all reduced to the selected epoch by means of the repeatedly corrected period, is known as the mean light curve. Examples of light curves will be found in the articles on **Cepheids**; **Eclipsing Binary**; and **Variable Star**.

LIGHT-EMITTING DIODE (LED).   See **Automotive Electronics**; **Luminescence**; and **Telephony (Telecommunications)**.

LIGHTING.   See **Illumination**.

LIGHTNING.   Under favorable circumstances, large electrical potential differences that are generated in the earth's lower atmosphere may be neutralized within an extremely short span of time in the form of lightning. The lightning may occur fully within the atmosphere, as by intra- or intercloud lightning, or between the atmosphere and the earth's surface, particularly the higher surfaces (mountains) or objects such as buildings and trees which extend upward from the ground for short distances. Favorable paths of electrical conductance predominantly determine the geometry of lightning. It has been observed that conductivities and air-earth current densities may be as much as ten times greater in the vicinities of high mountains than at sea level. This correlates well with other observations which indicate that lightning at sea occurs with a frequency only about 10% of the frequency over land. Conductivity decreases when humidity increases. Considering the earth as a whole, lightning is very common, occurring with a frequency of about 100 lightning events per second. Scores, hundreds, and even thousands of lightning events may occur in connection with a given thunderstorm. The frequency and variation of lightning characteristics ranges widely — with location, season of the year, and local time of day. Lightning over land occurs with the greatest frequency during late afternoon and early evening; and over the seas and oceans during late evening to an hour or two after midnight local time. There is, however, no time of day or night when lightning may not occur. Lightning can strike the same object or location not just once, but many times even during a given storm. This is particularly true of high structures. Trees do not provide protection against lightning, but represent

an unsafe location for persons to be during a lightning-productive storm. Some authorities have suggested that a metal-bodied automobile may be one of the safest places to be during a storm of this type if other hazards (falling trees, flying debris, etc.) are minimal. For persons caught in open fields, golf courses, etc., lying prone on the ground is considered best in the interest of protection from lightning.

Although lightning is an extremely common phenomenon of the lower atmosphere, it remains rather poorly understood, particularly in quantitative terms. In recent years, lightning has been found to be much more complex and variable than previously believed. High-speed camera techniques, notably with improvements that permit the analysis of lightning during daytime hours, have produced much new evidence both of a qualitative and quantitative nature. The unaided eye or still camera, incapable of breaking down the individual events which occur during what appears to be a single stroke of lightning, at best can provide very rough data — estimates of patterns, brilliance, and distances and heights. In recent years, concerted programs of research in this field have contributed much toward expanding the knowledge of lightning. Such efforts include work done in connection with the Thunderstorm Research International Program (TRIP), in which several organizations such as the Kennedy Space Center (NASA), the University of Florida, the University of Arizona, the New Mexico Institute of Mining and Technology, Rice University, and the State University of New York at Albany, among others, have participated. Uman et al. (1978) reported on the physics and meteorology of a lightning flash at Kennedy Space Center; Orville and Lala (1978) reported on the development and use of a streaking camera for producing daylight time-resolved photographs of lightning.

Characteristics of Lightning

From the standpoint of geometry, the fundamental types of lightning are: (1) cloud-to-cloud (intercloud); (2) between two portions of the same cloud (intracloud); (3) cloud-to-earth; and (4) earth-to-cloud. The last two types often occur as part of what visually appears to be a single event. Lightning also may be classified as (a) long time span and low current; and (b) short time span and high current. The first of these is generally the more damaging to objects, such as igniting forest and structural fires.

Rarely do the long-time-span lightning events persist up to or over one second, although it was reported by Godionton in 1896 that a single lightning flash lasted up to 15 or 20 seconds. If this observer was accurate, it is indeed an exceedingly rare case. Observations since then have not included events of this time magnitude. Most lightning events persist for fractions of a second, ranging from 10^{-2} second through an average of about 0.2 second, up to 1 second (unusual).

In the usual cloud-to-ground lightning event, the phenomenon commences with what is called a *stepped leader*. High-speed camera techniques have been invaluable in confirming this characteristic. See the very schematic sketch in Fig. 1. The stepped leader may be of relatively low

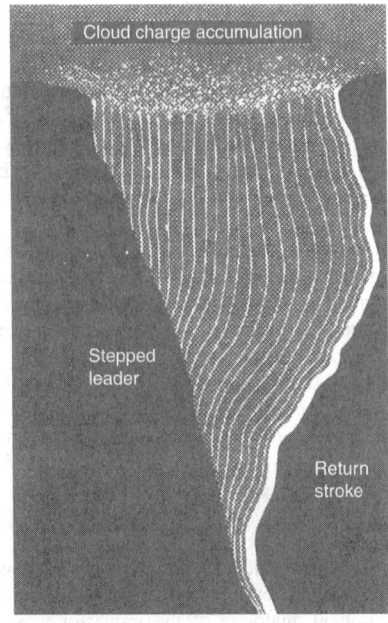

Fig. 1.   Stepped leader and return stroke in cloud-to-earth lightning event.

luminosity and possibly one to two meters in diameter. The leader sets up the conditions for the much more dramatic *return stroke* from earth to cloud. The time interval between steps of the stepped leader may range from a minimum of 30 microseconds to 50 microseconds (typical) to a maximum of 125 microseconds. The minimum length of a step may be as low as about 3 meters, ranging up to 50 meters (typical) to a maximum of about 200 meters. The average velocity of propagation of the stepped leader, at a minimum, may be 1.0×10^5 meters/second, ranging up to 1.5×10^5 meters/second (typical) to a maximum of about 2.6×10^6 meters/second. The charge deposited on a stepped-leader channel may be as low as 3 coulombs, ranging up to 5 coulombs (typical) to a maximum of about 20 coulombs. The return stroke may have a diameter measurable in terms of centimeters rather than meters, with a peak current ranging from 10 to 20 (typical), but up to a maximum of 110 kiloamperes. The channel length may range from 2 kilometers up to 5 kilometers (typical) to a maximum of about 14 kilometers. The velocity of propagation may range from 2.0×10^7 meters/second to 5.0×10^7 meters/second (typical) up to a maximum of about 1.4×10^8 meters/second. Temperatures within the return stroke range up to 30,000 K at pressures up to one MN/m^2 stroke.

Upon completion of the return stroke, a dart leader may again move downward because of residual potential differences, causing a second return stroke. As many as 40 secondary discharges (return strokes) may occur in what is called a multiple stroke flash. See various lightning configurations in Figs. 2, 3, and 4.

Fig. 2. Dramatic display of several lightning strokes in vicinity of Kitt Peak, Arizona. Buildings of the National Optical Astronomical Observatories are shown in the light of the lightning. *Nikkormat* FTn with 28 mm lens with tripod and cable release; *Kodachrome II* film; *f* 3.5; exposure, about 1 minute. (*Copyright Gary Ladd, 1972.*)

Fig. 3. Massive display of lightning over city in western United States. (*Electric Power Research Institute.*)

Energy Sources of Lightning

Numerous theories have been advanced pertaining to the accumulation of electric charge in the lower atmosphere. Advanced as early as 1885, the influence theory suggested that the earth's field is usually negative with relation to positive cloud charges, thus setting up the right conditions for lightning. This concept was probably in keeping with the general scientific knowledge of that time. It was also generally proposed that particles (hydrometeors, such as rain, snow, hail, fog, etc., associated with storms) developed a charge as the result of frictional forces with the atmosphere. In the free ionization theory, it was suggested that droplets of different sizes selectively collect available ions, thus establishing large potential differences. It is now generally believed that electric conduction current in the atmosphere is carried almost exclusively by fast ions. Positive ions include $(H_3O)^+(H_2O)_n$ and negative ions include $O_2^-(H_2O)_n$ or $CO_4^-(H_2O)_n$, where the probable values of n for positive ions in the troposphere are 6 or 7; in the stratosphere, 4; and in the mesosphere, 3 or 2. The value of n is dependent upon the water vapor pressure and temperature of the atmosphere, as well as the lifetime of the ion.

A few highlights pertaining to the energy in lightning would include:

1. Typical voltage drop in ground or other conductors after lightning impact in the neighborhood of 10 kV (dangerous!).
2. Energy delivered to an average stroke of lightning, about 100 kJ/m. Intracloud lightnings have been observed up to 100 km in length.

Fig. 4. Lightning.

Height from which a lightning stroke points at a target estimated several decameters, depending on conductivity distribution in ground.

3. Long-lasting, low-current (hundreds of amperes) flashes are more dangerous to people and objects (for example, forest fires) than short, high-current flashes.

Damage Wrought by Lightning

The natural high-voltage phenomena occurring during lightning are not only of general scientific interest and value. A greater knowledge of these phenomena is of practical value as well, in particular because of the danger from lightning to life and to susceptible structures, notably electric power systems and communications equipment. For many years, the direct and induced effects of lightning discharges have been simulated in industrial laboratories by means of high-voltage impulse generators. These frequently use the Marx method of first slowly charging a number of condensers in parallel and then suddenly connecting them in series by spark-gap switches which, at the same instant, impress the multiplied voltage upon the test circuit. A typical voltage wave produced by such impulse generators rises to its peak value of several million volts in one microsecond and then diminishes exponentially, reaching half voltage in about ten microseconds.

The Electric Power Research Institute (Palo Alto, California) has for a number of years conducted research on the characteristics of lightning, means of predicting lightning, and means of protecting electric utility equipment from damage, particularly overhead transmission lines. Although improvements in calculating lightning performance of transmission lines have been achieved, little information has been available for accurately predicting the lightning flashover performance of multiple-circuit transmission lines. See Fig. 5. In most double-circuit lines, when one circuit flashes over, the second circuit will simultaneously flash over 40–60% of the time. EPRI has generated a computer program, known as MULTI-FLASH and available to electric utility operators, that enables the transmission line designer to accurately predict the lightning performance of lines containing up to 12 ac phases, 12 dc poles, or any combination thereof on the same transmission tower, with a variety of tower shapes and insulator strings. All transmission voltage and significant corona effects are included, as well as the statistical distribution of footing resistance. The output includes an analysis of expected shielding failure performance, followed by a detailed tabulation of the expected flashover frequency of each of the phases for dc poles that are involved. Using this computer analysis, the designer can explore new and innovative shapes of materials and accurately predict the degree of improvement that may be achieved.

Fig. 5.   Much research in recent years has been conducted to understand and to reduce the effects of lightning flashover in multiple-circuit transmission lines. (*Electric Power Research Institute.*)

Research also has been conducted pertaining to arresters. These devices play a vital role in protecting substation equipment from high-voltage surges that originate from either substation switching equipment or lightning strikes. Through the proper selection of arresters, undesirable overvoltages can be limited. New materials having an extremely nonlinear voltage/current characteristic have been used in the construction of gapless

surge arresters. Materials include zinc oxide (ZnO). The new designs have been under evaluation, including some at Tennessee Valley Authority substations. See Fig. 6.

Fig. 6.   New materials, including metal oxides (ZnO), have been used in innovative surge arresters to protect electric power substation equipment from high-voltage surges that originate either from substation switching equipment or lightning strokes. The new materials have extremely nonlinear voltage/current characteristics that lend themselves well to gapless surge arrester designs. (*Electric Power Research Institute.*)

A few years ago, some utility system designers concluded that surge protection devices could not be further optimized without an improved database of lightning stroke characteristics. The EPRI thus established a project along these lines. In this study, lightning stroke current and energy magnitude were estimated by comparing the gaps of 2800 surge arresters that accumulated 32,000 arrester years of service on utility lines with gaps that had discharged known currents. The stroke energy statistics obtained are of particular interest to surge protection engineers because little information along these lines has been available. Under present standards, arrester durability is determined in part by tests using 65-kA and 8×20-microsecond current discharges that have a charge of 1.25 coulomb. The study showed that arresters in service are more frequently exposed to longer-duration, lower-current waves having charges of up to 4.2 coulombs, than to higher-current, lower-energy waves.

In another, associated study, the geographic density of lightning flashes was probed. As a starter, EPRI has mapped the location of lightning flashes in the eastern United States. Such data will provide surge protection engineers with information on the number of lightning flashes that can be expected to strike distribution and transmission lines directly or nearby. Data on charge polarity, number of strokes per flash, and peak field strength radiated from the first stroke are being gathered.

Additional Reading

Bazelkilan, E.M.: "Lightning Physics and Lightning Protection," Iop Publishing, Philadelphia, PA, 2000.

Franz, R.C., R.J. Nemzek, and J.R. Winckler: "Television Image of a Large Upward Electrical Discharge Above a Thunderstorm System," *Science*, 48 (July 6, 1990).

Marks, J.A., Ed.: "Electrical Systems," Electric Power Research Institute, Palo Alto, CA, 1985. *http://www.epri.com/*

Orville, R.E. and G.G. Lala: "Daylight Time-Resolved Photographs of Lightning," *Science*, **201**, 59–61 (1978).

Tahiliani, V.: "Metal Oxide Surge Arresters for Gas-Insulated Systems," Electr Power Research Institute, Palo Alto, California, 1983.

Uman, M.A.: "The Lightning Discharge," Academic Press, San Diego, California, 1987.

Uman, M.A. and E.P. Krider: "Natural and Artificially Initiated Lightning," *Science*, 457 (October 27, 1989).

Waterbury, R.C.: "Safir Forecasts Lightning Strikes," *Instrumentation Technology*, 72 (July, 1990).

LIGHTNING BUG.   See **Firefly**.

LIGHT POLLUTION.   Today, people who live near large cities have lost much of their view of the universe. The spectacular view of the night sky that their ancestors had above them on clear dark nights no longer exists. The great increase in urban population has caused an ensuing rapid increase in urban sky glow due to outdoor lighting. This has brightened the heavens to such an extent that the only view most people have of the Milky Way or most stars is when they are well away from cities. This excess light in the sky has had an adverse impact on the environment and seriously threatens to remove forever one of mankind's natural wonders—the dark sky.

The extent of vast illuminated cosmopolitan areas of the United States is dramatically illustrated by the "map" of Fig. 1.

Effects on Professional Astronomy.   While this increased urban sky glow brightens the night sky for the general public and for amateur

Fig. 1.   Satellite view of the United States at night, on a night when most of the country was clear. The upward light produced by the urban areas is evident in the photograph. Some of the light is direct-up light; the rest is light reflected from the ground. (*United States Air Force.*)

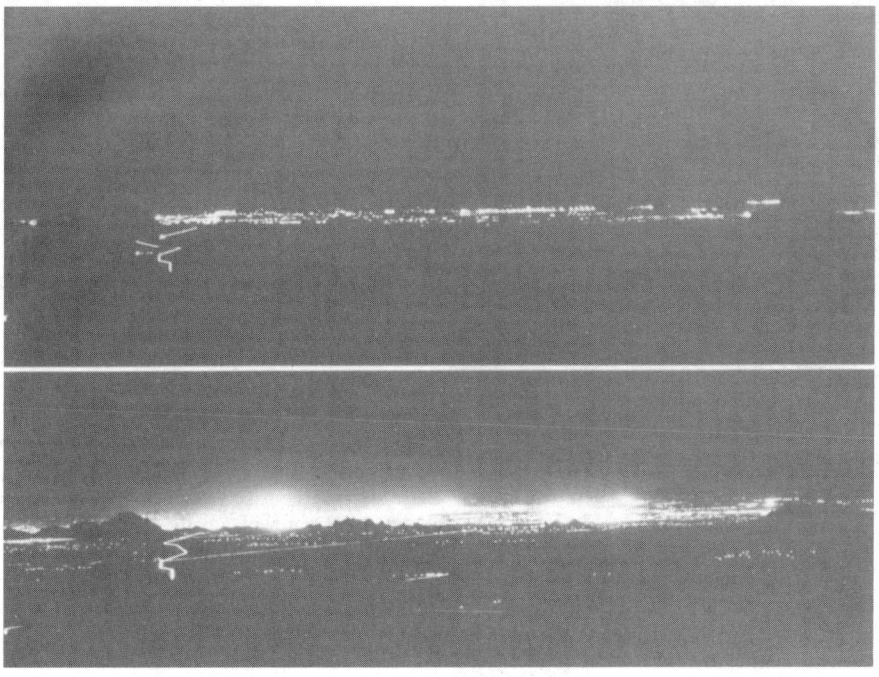

Fig. 2.   Two photographs of Tucson, Arizona as Seen from Kitt Peak, (about 35 miles 56 kilometers) west of the city. The growth of the city over 21 years (top view, 1959; bottom view, 1980) is evident. It is this type of growth, with its associated growth in outdoor lighting, that is comprising the research done at most of the major observatories located near cosmopolitan areas. (*National Optical Astronomy Observatories.*)

astronomers, it is a special threat to professional astronomy. Advances in frontier astronomy require observations of very faint objects that can be studied only with large telescopes located on prime observing sites, well away from sources of air pollution and from urban nighttime sky glow. For example, most observations of cosmological interest deal with extremely remote sources — galaxies or quasars at such distances that the light has traveled several billion years, sometimes twice the age of our solar system, before reaching us. This light is then often lost in the glare of anthropogenic sky glow.

Observations of such extremely faint extragalactic sources, and even of many objects of interest in our own galaxy, can be done only during the dark moon period. The sky background is much too bright when moonlight is present. Artificial lighting of the sky due to cities has, unfortunately, the same adverse effect on nighttime sky brightness in limiting observations. An interesting contrast in the situation at Tucson, Arizona, which is located near the Kitt Peak National Observatory, is given in Fig. 2. The worsening of the problem over a 20-year period is demonstrated.

This increased sky glow which adversely affects the environment and compromises astronomical research is called *light pollution*, for it is wasted light that does nothing to increase nighttime safety, utility, or security. It only serves to waste energy and money. An example of an improperly designed billboard that pollutes the night sky is given in Fig. 3.

Fig. 3.   Photograph of a billboard, showing a typical example of poor lighting. Most of the light output by the fixture Seems to miss the billboard. Better quality fixtures would put most of the light onto the board, thus less waste and less light pollution, all at less cost. To minimize light pollution, the quality fixtures should be mounted at the top of the board, thus minimizing the direct and reflected upward light. Also, the lights should be time-clocked to go off at 10 or 11 PM. There are few potential viewers after then, and the sky would be darker after those hours. Energy and money would be saved. (*National Optical Astronomy Observatories.*)

Ground-Based Astronomical Observations Still Much Needed. The argument that all astronomy can be done from space is not correct; the largest telescopes will continue to be ground-based for a long time because of cost factors. It does not make sense to do in space, at much higher cost, what can be done from the ground. There are many things that can only be done in space and the demand for that type of research is severe. The experience of more than two decades of space astronomy, however, has greatly increased the demand for ground-based facilities. Planning for several ground-based telescopes much larger than anything now in existence is already underway. There are exciting times ahead for astronomers, using present and future ground-based telescopes, which *complement* the telescopes in space. See also **Telescope (Astronomical-Optical)**.

Solutions to the Light Pollution Problem. Control programs are underway now in a number of communities to reduce the effects of light pollution of the night sky. Programs like these are critical to the long-term success of astronomical research, and to preserve people's view of the universe. Unlike dealing with the problems of water and air pollution, vast sums of money are not required to alleviate light pollution.

At present, the lack of awareness rather than resistance is generally the biggest problem in controlling light pollution. Educating the public, government officials and staff, and lighting professionals has been the major thrust of the current programs. These efforts have helped. The increase of light pollution near major observing sites is moderating. More can and must be done. Astronomers, amateurs and professionals, and many others are urging such cooperation.

Astronomers are *not* against night lighting. They have the same needs for quality lighting as everyone else. They advocate the best possible lighting for the task, with lighting designs that allow for all the relevant factors, such as glare control, efficiency, and the need for dark skies. An important added advantage is that everything that is done to minimize light pollution also saves energy by improving the efficiency and utility of the nighttime lighting.

There are other adverse effects of poor quality lighting. Light that comes out of a fixture essentially horizontally does little or nothing to light the ground, but it does cause glare. Such glare is annoying to the eye of the beholder, and it even can cause discomfort or disability. Its blinding effects have often even caused accidents. Glare never helps; it is always a sign of poor-quality lighting.

Confusion or clutter is another adverse effect of poor lighting. Nighttime lighting should provide guidance, providing help and safety, not confusion. Some installations mislead a driver, for example, leading to accidents. In addition, the clutter of outdoor lights that we see today in most cities is often just as trashy a sign of poorly controlled urban growth as is the litter of garbage we See.

Light trespass is another adverse effect of poor outdoor lighting. Light from a fixture that falls in someone else's yard or in through one's window is usually unwanted. It is indeed "trespass." This results in black paint on one side of the fixture, irate phone calls to the owner of the light or to the police or to the sports park owners. Such trespass is the sign of a poorly shielded light fixture, not of a quality lighting design or installation. There is no need for such an adverse effect.

It is a sad state that many people are not aware of quality lighting. Many even think that lighting does not exist if it does not exhibit the adverse effects, for they are so used to the associated glare and light trespass of the all too common poor lighting. Quality lighting does exist. It should always be used. There are many examples of it being used, and it should be used for *all* installations. Professional lighting designers are well aware of quality lighting, and use it whenever possible. Unfortunately, so much of today's outdoor lighting is not done by lighting professionals, or by people aware of or sensitive to quality lighting. That is what must be changed.

Specific Corrective Measures. What aspects of quality lighting can be used to help solve the light pollution problem? Following are some solutions that will greatly minimize light pollution without in any way compromising nighttime safety, security, or utility:

1.   Use night lighting only when necessary. Turn off lights when they are not needed. Timers can be very effective. Use the correct amount of light for the need, not overkill. Until recently, when energy conservation became an important issue, it often Seemed that the only "design" criterion used for outdoor lighting was: "If a certain amount of light is good, double it and things will be better." That is not good design, as any professional lighting designer will agree.

2. Direct the light downward, where it is needed. The use and effective placement of well-designed lighting fixtures will achieve excellent control. Shielding the light source to avoid the upward light helps. When possible, retrofit present poor fixtures, ones that spray light everywhere, especially directly up into the sky. In all cases, the goal is to use fixtures that control the light well, minimizing glare, light trespass, and light pollution. All of this also minimizes the energy waste. Light is used when and where needed, and not wasted.

3. Use low pressure sodium (LPS) light sources whenever possible. This is the best possible light source to minimize adverse sky glow effects on astronomy. LPS lamps are the most energy efficient light sources that exist. Areas where LPS is especially good are street lighting, parking lot lighting, security lighting. There are some applications where LPS should not be the sole lighting source, for applications where color rendering is critical. But, for most applications, LPS should be considered. It is an excellent, low-cost light source and helps greatly to minimize the adverse sky glow.

4. Avoid growth near the observatories, and apply rigid controls on nighttime lighting when such growth is unavoidable. Such controls do not compromise safety, security, or utility. Lighting ordinances have been enacted by many communities to enforce quality, effective lighting. These communities have found that the quality of lighting has improved, usually at lower cost.

All of these solutions to the problem of adverse nighttime lighting say, really: "Do the best possible professional lighting design for the task. Include all relevant factors, such as glare, light trespass, and light pollution." All the solutions needed for protecting astronomy have positive side benefits of maximizing the quality of the lighting, and of saving energy. See also **Illumination**.

The American Astronomical Society has a Committee on Light Pollution. The Illuminating Engineering Society of North America has a Committee on Light Trespass and another one on the Environmental Impact of Outdoor Lighting. Other groups are responding in a similar matter. Fortunately, lighting technology is advancing and there is an increasing interest in quality lighting. Such advances in technology, and increasing awareness of the problems of light pollution, will help greatly in promoting quality outdoor lighting, and thus minimizing light pollution.

DAVID L. CRAWFORD, Ph.D., Kitt Peak National Observatory, Tucson, AZ.

LIGHT TIME. The elapsed time taken by electromagnetic radiation to travel from a celestial body to the observer at the time of observation. The *American Ephemeris and Nautical Almanac* uses a light time of 498.8 seconds for 1 astronomical unit.

LIGHT WATER REACTOR. See **Nuclear Power Technology**.

LIGHT-YEAR. A popular method of expressing large distances; specifically, the distance that light will travel in the course of one year. The velocity of light is established (International Astronomical Union (Hamburg 1964)) at 299,792.5 kilometers per second, or about 186,282 statute miles per second. There are approximately 31,560,000 seconds in a mean solar year. Accordingly, a light-year represents a distance of approximately 9.454×10^{12} kilometers; 5.875×10^{12} miles. See also **Astronomical Unit**; and **Parsec**.

LIGNIN. Approximately 25% of the content of most woods is lignin. Lignin concentration in wood substance is greatest in the middle lamella (the zone around each individual fiber cell), decreasing in concentration through the cross section of the fiber, reaching a concentration of about 12% at the inner layer of the fiber adjacent to the fiber cavity, or lumen. Lignin and hemicellulose cement the fiber cells together, providing rigidity to the fibrous wood structure. In the destructive distillation of wood, the methanol produced is derived from the lignin. In the manufacture of paper pulp, it is necessary to remove the lignin, usually accomplished by treatment of the wood fibers with such agents as sulfur dioxide, calcium bisulfite, and sodium sulfate/sodium sulfide solutions. Sodium hydroxide is sometimes used. An important byproduct of the paper pulp industry is dimethyl sulfoxide, $(CH_3)_2SO$ which is produced from the lignin released during wood pulping by the Kraft process. Dimethyl sulfoxide has a number of industrial uses—as an intermediate in organic syntheses, as a solvent in spinning synthetic fibers, and in some pharmaceuticals.

The wall material of plant cells is one of their distinguishing characteristics. As a result, lignin, cellulose, and other wall constituents have been studied in many plant tissue cultures. Phenylpropanoids, for example, have been shown to be precursors of lignin formation in white pine, *Sequoia*, lilac, rose, carrot, and geranium tissue cultures. Moreover, the biosynthesis of lignin has been shown to be affected by kinetin, boron, and major elements, such as calcium.

Lignin is a major source of vanillin.

LIGNOCELLULOSE. See **Cellulose**.

LIGNUM VITAE *(Guaiacum).* The heartwood of a tree native to the West Indies. The tree also is found in other regions of moderate climate. It is a valuable, tough, resinous wood and very heavy, being the heaviest of all commercial woods; a cubic foot weighs 76 pounds (34.5 kilograms). Lignum vitae has been used in the making of bowling balls, pulley sheaves, and mallet heads.

The tree attains a height of about 40 feet (12 meters); the trunk may range from 2 to 4 feet (0.6 to 1.2 meters) in diameter and normally grows quite crooked. The branches are knotty.

The record *G. sanctum* (roughbark) tree growing in the United States is located in Biscayne National Park, Florida. See Table 1.

The record *G. angustifolium* (Texas lignum vitae) tree growing in the United States is located at Alamo, Texas. See Table 1.

LIKELIHOOD. If $P(x, \theta)$ denotes a probability function depending on one or more parameters collectively denoted by θ, the likelihood of a sample $x_1, x_2 \cdots x_n$ is defined as $L = P(x_1, \theta) \cdot P(x_2, \theta) \cdots P(x_n, \theta)$.

The method of maximum likelihood consists in estimating the parameters θ by choosing those values which maximize L (or log L). Under general conditions, maximum likelihood estimates are consistent and efficient and tend to be normally distributed in large samples; further, that they are sufficient if sufficient statistics exist. The large sample variances or covariances of the maximum likelihood estimates are given by the elements of the inverse matrix to

$$\left[nE \left(\frac{\partial^2 \log P}{\partial \theta_i \, \partial \theta_j} \right) \right]$$

where E denotes the expected value.

LILIACEAE. The Lily Family has representatives in all parts of the world, and more especially in the drier regions of the temperate zone. Several members of the family are important vegetables, notably asparagus and onions, while a great many more are cultivated for ornament. Among the latter are the true lilies (the genus *Lilium*), tulips, and hyacinths.

Most members of the Lily Family are herbaceous plants with a shallow fibrous root system. A few species of *Aloe* and *Dracaena* are shrubby or

TABLE 1. RECORD LIGNUMVITAE TREES IN THE UNITED STATES[1]

Specimen	Circumference[2]		Height		Spread		Location
	Inches	Centimeters	Feet	Meters	Feet	Meters	
Roughbark lignumvitae (1995) (*Guaiacum sanctum*)	37	94	31	9.4	39	11.9	Florida
Texas lignumvitae (1974) (*Guaiacum angustifolium*)	32	81	26	7.9	22	6.7	Texas

[1]From the "National Register of Big Trees," American Forests (by permission).
[2]At 4.5 feet (1.4 meters).

even small trees. Characteristic of the family are underground rhizomes or bulbs, storage organs which enable the plant to survive in regions where protracted dry seasons occur. As a rule, these plants have linear undivided leaves which do not show division into petiole and blade. The inflorescences of the family are very diverse. In some genera the flowers are solitary, in others they occur in racemes, while umbels occur in still others. The perianth of the flower has six separate members in two whorls of three, which are very much alike in size, shape, and color. The stamens have conspicuous anthers. The ovary is superior, 3-celled, and bears a single style with a 3-lobed stigma. The fruit is a capsule or a berry.

LIMACON. A higher plane curve, also known as Pascal's snail (named for Stefan Pascal, the father of Blaise Pascal, 1623–62, the famous French philosopher, mathematician, and physicist). Its equation in rectangular coordinates is $(x^2 + y^2 - 2ax)^2 = k^2(x^2 + y^2)$ and in polar coordinates, $r = 2a\cos\theta \pm k$. The curve is closed and symmetric to the X-axis. When $k < 2a$, there is an internal node at the origin and the limaçon cedilla is called hyperbolic. The loop disappears when $k > 2a$, so that a conjugate point exists and the limaçon is now elliptic. The case of $k = 2a$ is the cardioid, with a cusp at the origin. See Fig. 1.

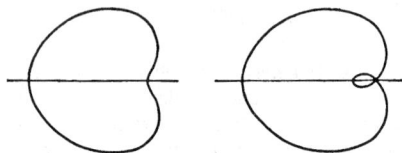

Fig. 1. Limaçon. In case of diagram at left, $k > 2a$; right, $k < 2a$.

The limaçon (its name comes from Latin, *limax*, snail) can be generated as follows. Let ODM be a circle with diameter $OD = 2a$ on the X-axis. A radius vector OM meets the circle at M and to it is added and subtracted a fixed length $MP = -MP' = k$. The locus of P and P' is the curve.

The various types of limaçons are special cases of the Cartesian oval. See also **Curve (Higher Plane)**.

LIMB DARKENING. The darkening of the limb of the sun or stars is due to the line of sight passing through greater thicknesses of cooler gases at the edge. Limb darkening follows the simple relation

$$I = I_0(1 - u + u\cos\theta)$$

where I is the observed intensity at a point, making the angle θ between the observer, the center of the star, and the point in question; I_0 is the intensity at the center of the disk; and u is the coefficient of limb darkening. See also **Sun (The)**.

LIME AND LIMESTONE. The term *lime* includes a variety of chemicals manufactured from limestone or derived from chemical processes that utilize calcium compounds. According to the composition of the parent limestone, lime may be designated as *high calcium lime* or *dolomitic lime*. Both *quicklimes*, CaO and CaO · MgO, and *hydrated limes*, CaO · H$_2$O, Ca(OH)$_2$ · MgO, and CaO · MgO · 2H$_2$O, are conventionally called lime. Precise terminology requires complex wording, e.g., *dolomitic quicklime* to denote CaO · MgO. The various lime oxides and hydroxides are among the lowest-cost and most widely used sources of alkali for the chemical and metallurgical industries. About 80% of the lime used in the United States is used by the chemical and related industries, mostly as quicklime. About 10% is dead-burned dolomite, and less than 10% goes into construction uses, mostly as hydrate. Very little lime is imported into or exported by the United States.

Limestone. This is a rock containing chiefly calcium carbonate and variable quantities of magnesium carbonate. Limestone is classified along the lines of lime as previously mentioned. *High-calcium limestone* contains 5% or less of MgCO$_3$ and occurs in two mineral forms, calcite and aragonite. See also **Aragonite**; and **Calcite**. *Dolomitic limestone* usually contains over 35% MgCO$_3$, with the remainder CaCO$_3$. See also Dolomite. *Magnesian limestone* is predominantly CaCO$_3$, but contains from 5 to 35% MgCO$_3$. All limestones evolve carbon dioxide and bubble in dilute hydrochloric acid. Dolomite reacts only with dilute HCl, while calcite will decompose in cold dilute HCl.

Limestones vary greatly in color and texture, the latter ranging from dense and hard limestone, e.g., marble or travertine, which can be sawed and polished, to soft, friable forms, e.g., chalk and marl. Chalk is a very fine-grained white limestone, while marl is an impure deposition product that contains clay and sand. Texture, hardness, and porosity appear to be functions of the degree of cementation and consolidation during the formation of these materials. Color variations arise from the presence of impurities. Some impurities, such as sulfur and phosphorus, make limestone unattractive for metallurgical uses.

A high percentage of all limestone is quarried; the balance is mined underground. Although limestone occurs widely, good chemical- and metallurgical-grade limestone is less plentiful. Along the seacoasts, oyster or clam shells are dredged as a source of CaCO$_3$. Limestone is normally processed through a series of crushing, screening, and grinding operations. Because of transportation costs, the proximity of limestone sources to points of use is highly desirable. The major uses of limestone are in construction (asphalt filler, road stone, riprap, and bituminous aggregate); in Portland cement; in agriculture; and in metallurgy.

Precipitated CaCO$_3$ is produced in a number of chemical processes. Sometimes it is economical to dry and calcine the byproduct to regenerate CaO or Ca(OH)$_2$. Some precipitated CaCO$_3$ is made to specific particle size and shape, whiteness, and purity for use as functional filler for paper coatings, paint, and polymers. These products command a premium price as compared with pulverized limestone fillers.

Manufacture of Lime. The basic processes are calcination and hydration. Commencing with high-calcium limestone, the reactions are:

$$CaCO_3 + heat \rightleftharpoons CaO + CO_2 \qquad (1)$$

$$CaO + H_2O \rightleftharpoons Ca(OH)_2 + Heat \qquad (2)$$

If dolomitic limestone is used, the reactions are:

$$CaCO_3 \cdot MgCO_3 + heat \rightleftharpoons CaO \cdot MgO + 2\,CO_2 \qquad (3)$$

$$CaO \cdot MgO + H_2O_{(liq)} \rightleftharpoons Ca(OH)_2 \cdot MgO + heat \qquad (4a)$$

or

$$CaO \cdot MgO + 2\,H_2O_{(gas)} + pres \rightleftharpoons Ca(OH)_2 \cdot Mg(OH)_2 + heat \qquad (4b)$$

High-calcium limestone dissociates at 900 °C (1650 °F) in 100% carbon dioxide atmosphere at 1 atm pressure. Under similar conditions, dolomitic limestone dissociates over 727–900 °C (1340–1650 °F). The heat of reaction required to convert CaCO$_3$ to CaO is about 2.8 million Btu per ton (0.64 million kg-Cal/metric ton) of CaO. In practice, heat input may vary from 4 to 10 million Btu/ton (0.9 to 2.3 million kg-Cal/metric ton) of lime. Calcination of limestone particles proceeds by a receding-surface mechanism. To attain reasonable rates of heat transfer into the center of the rock or pebble-sized stone, operating temperatures in lime kilns are 980–1260 °C (1800–2300 °F). Reaction rate is increased and opportunity for recarbonation of the oxide is decreased by rapid removal of CO$_2$ from the kiln.

Except for very old mixed-feed vertical kilns, lime kilns operate with countercurrent flow of raw material and heat. Modern lime kilns utilize coolers to preheat air by recuperating heat from the hot quicklime. Lime kilns may be fired directly with coal, oil, or gas.

Two major types of lime calciners are the rotary (Fig. 1) and the vertical kiln. In North America, rotary kilns are widely used for lime calcination, whereas in Europe vertical kilns are most popular. Rotary kilns typically have higher output [up to 600 tons (545 metric tons)/day] and lower labor cost. Vertical kilns can be designed for higher fuel efficiency and lower capital investment. They handle down to about $\frac{3}{4}$-in. (19-cm) stone, but normally the rock feed is at least 3×6 in. (7.5 × 15 cm). Rotary kilns can handle down to $\frac{1}{4}$-in. (0.6 cm) stone.

The long-established use of lime is as a structural material in masonry mortars, wall plasters, sand-lime brick, and for soil stabilization. Double-hydrated dolomitic lime or specially processed high-calcium lime mixed with gypsum plaster is troweled on interior walls or ceilings to provide a hard, white, finished surface. It is mixed with cement and sand to make exterior plaster or stucco. Mason's mortar used to lay up bricks or blocks usually contains lime. Lime provides plasticity, water retention, and easy troweling. Sand-lime bricks are more popular in Europe than in the United States. About 10% hydrated lime is mixed with graded sand and water, pressed into shape, and put into autoclaves for 4–8 hours at 150–205 °C

Fig. 1.   Large rotary lime kiln designed to operate 24 hours per day year around. Rugged construction is required for handling abrasive limestone rock at very high temperature.

(300–400 °F). The reaction product, calcium silicate, results in a strong, white brick. Dead-burned dolomite, formed by calcining dolomite at about 1650 °C (300 °F) to convert MgO to periclase, is used as a *refractory*. Lime is used in the sulfate process for making paper. In the seawater process for producing magnesium metal, lime reacts with $MgCl_2$ to precipitate $Mg(OH)_2$. Calcium metal is made from lime by reducing CaO with coke. In water treatment, hydrated lime can be added to remove temporary hardness, or in the lime-soda process to remove permanent hardness. Dolomitic lime removes silica from boiler feedwater due to silica absorption by $Mg(OH)_2$. For acid neutralization of industrial wastes, lime is widely used.

Additional Reading

Boynton, R.: "Chemistry and Technology of Lime and Limestone," 2nd Edition, John Wiley & Sons, Inc., New York, NY, 1980.
Gerhartz, W.: "Ullmann's Encyclopedia of Industrial Chemistry," John Wiley & Sons, Inc., New York, NY, 1987.
Oates, J.: "Lime and Limestone: Chemistry and Technology, Production and Uses," John Wiley & Sons, Inc., New York, NY, 1999.

LIME (Citrus).   See **Citrus Trees**.

LIMIT.   A finite number s approached by a sequence $\{s_n\}$ if, for every positive number $\in < 0$, there exists a number N for all $n > N$, where $|s_n - s| < \in$. The sequence is then said to converge, or symbolically,

$$s_n \rightarrow s; \quad n \rightarrow \infty$$

or

$$\lim_{n \rightarrow \infty} s_n = s$$

If the limit does not exist, the sequence diverges, but see also **Series**.

The limit of a variable or a function can be defined in a similar way. If the limit is zero, the function is called an infinitesimal. If the function increases without limit, it remains greater than any assigned number however large and it is said to be come infinite or approach infinity. A variable or function that decreases without limit approaches minus infinity.

The following definition is sometimes needed. If $\in$ is a positive number, no matter how small, δ is defined by $0 < (x - a) < \delta$, and $|f(x) - A| < \in$, then A, called the right-hand limit at $x = a$, is symbolized by

$$\lim_{x \longrightarrow a+} f(x)$$

A left-hand limit, $\lim_{x \rightarrow a+} f(x)$ can be defined in a similar way.

See also **Sequence**.

LIMIT SWITCH.   An enclosed electromechanical device which makes or breaks electrical circuits when actuated by an external force, such as by a machine member (lever, cam, or dog), or other object when a preselected position of travel or limit is reached. Circuits switched may be for safety stop, position indicating, or be part of a total sequence of automatic

or semiautomatic operations. The limit switch is important in many automated systems, such as machine tools, conveyors, and other materials handling equipment. Unlike photoelectric and proximity switches, the limit switch requires physical contact for actuation. These switches are designed to operate reliably for tens of millions of actuations in industrial environments. The switches are obtainable in numerous current and voltage ratings, ranging from the switching needs of solid-state devices to motor-controlling relays and large solenoids. Voltage requirements may be as high as 600 volts.

Limit switches are actuated in several ways, but the two principal methods are (1) the roller lever switch, and (2) the plunger switch. Low-force actuators, termed *cat-whisker* or *wobblestick*, also are available. See Fig. 1.

See also **Proximity and Object Detectors**.

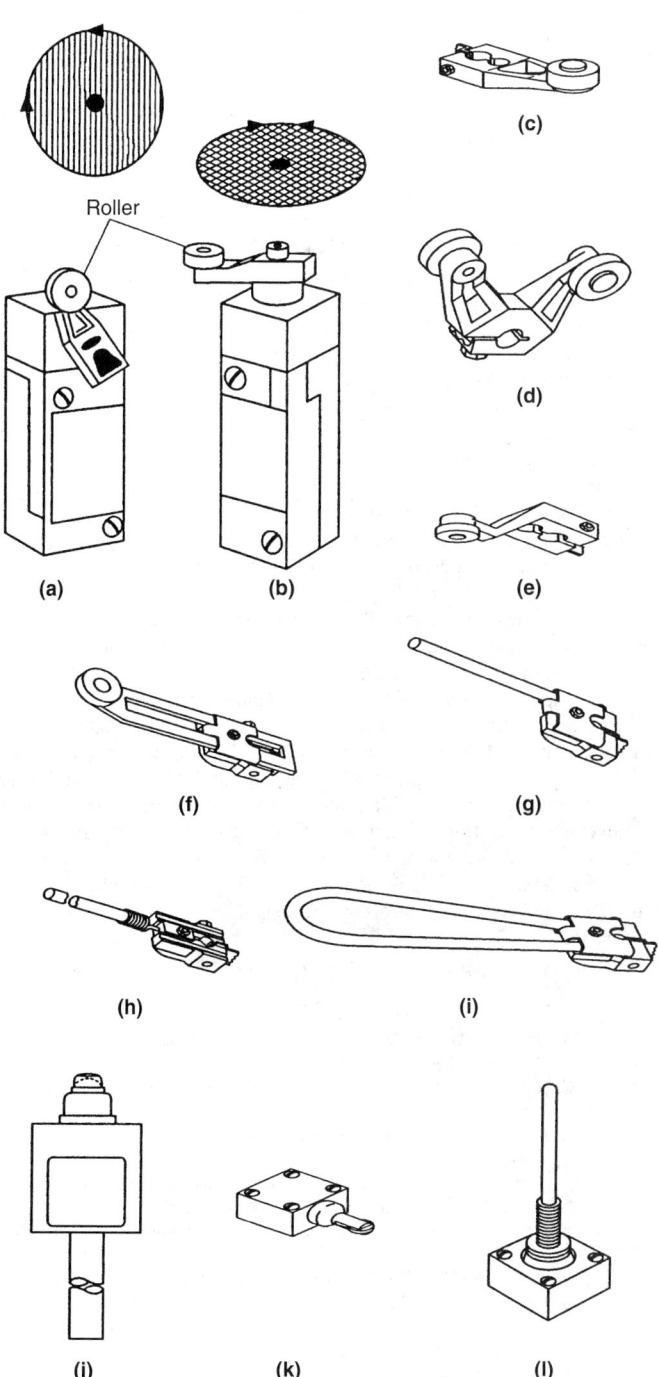

Fig. 1.   Various configurations of electromechanical limit switches. (**a**) Side-mounted roller. (**b**) Top-mounted roller. (**c**) Standard roller lever. (**d**) Yoke roller lever. (**e**) Offset roller lever. (**f**) Adjustable-radius roller lever. (**g**) Rod lever. (**h**) Spring-rod lever. (**i**) Flexible loop lever. (**j**) Top plunger. (**k**) Side roller plunger. (**l**) Wobbler switch. (*MICRO SWITCH.*)

LIMNOLOGY. The science of lakes, a synthesis of many disciplines, drawing its specialists from various scientific fields. The field includes the study of inland waters, although it is principally concerned with the physicochemical nature of lakes, their flora and fauna. Stream study has lagged behind lake investigation, although the ecological approach to rivers falls within the realm of limnology.

Lake basins owe their origins to diverse causes, many of which are geologic accidents, or catastrophic. Geologically, lakes are temporary phenomena in geomorphic evolution. Tectonic events have created some of the oldest and deepest lakes of the world: the African rift lakes and Lake Baikal of Siberia, with an ancient, largely endemic fauna. Vulcanism, glacial activities, solution of calcareous substrates, aeolian forces, and even meteoritic impact have created lake basins. In 1957, Hutchinson summarized 76 major categories and 8 subdivisions of these events, which have resulted in lake genesis.

No matter what its origin, a lake is doomed to eventual extinction because of its concave nature and accumulation of autochthonous and allochthonous materials,[1] which gradually obliterate the depression. Thus, a lake passes from a youthful stage to maturity, senility, and extinction. The rate of succession depends upon various factors. For example, introduction of domestic sewage enriches the lake and accelerates the aging process. The youthful lake may be described as *oligotrophic*, the mature lake as *eutrophic*. Many intermediate stages between extreme oligotrophy and extreme eutrophy occur, and the term *mesotrophy* can be applied to them. Senility is characterized by much shallow water and by the conspicuous encroachment of large aquatic plants upon the open water. Extinction often involves a marshy meadow, which is later colonized by plants typical to terrestrial situations. If drainage is poor and the lake is protected from wind, a floating bog mat may close over and eventually obliterate the open water. Bog lakes are acid, or at least circumneutral, and are typified by characteristic marginal vegetation contributing to the floating mat. When calcium content is low in bog-lake waters, decay of organic matter is reduced greatly. Plant fragments from the bog mat accumulate in flocculent layers, and the water may become tea-colored from humic matter. Flocculated humic colloids contribute to bog sediments to form a characteristic deposit termed *dy* by Scandinavian researchers. Under such conditions, nutrients are not recycled by decay, and the lake approaches extinction as a *dystrophic lake*.

In recent years, pollution of lakes has been of major concern. The effect of acid rain on certain lakes, usually at relatively high elevation (alpine lakes) is described in the entry on **Water Pollution**. Accelerated eutrophication, resulting primarily from phosphorus additions due to anthropogenic activities, is generally regarded as one of the major causes of the deterioration of the Great Lakes water quality. A mathematical model of the Great Lakes total phosphorus budgets indicates that a milligram per liter effluent restriction for point sources would result in significant improvement in the trophic status of most of the system. However, because large areas of their drainage basins are devoted to agriculture or are urbanized, western Lake Erie, lower Green Bay, and Saginaw Bay may require non-point source controls to effect significant improvements in their trophic status.

Volcanic lakes are discussed under **Volcano**.

Additional Reading

Bronmark, C. and Lars-Anders Hansson: "The Biology of Lakes and Ponds," Oxford University Press, Inc., New York, NY, 1998.

Brown, A.C. and A. McLachlan: "Ecology of Sandy Shores," Elsevier, Amsterdam, The Netherlands, 1990.

Ellis, W.S. and D.C. Turnley: "The Aral Sea Lies Dying," *National Geographic*, 73 (February, 1990).

George, D.G.: "Management of Lakes and Reservoirs during Global Climate Change," Kluwer Academic Publishers, Norwell, MA, 1998.

Imberger, J.: "Physical Limnology," Kluwer Academic Publishers, Norwell, MA, 2001.

Lampert, W.: "Limnoecology: The Ecology of Lakes and Streams," Oxford University Press, Inc., New York, NY, 1996.

Lerman, A., D.M. Imboden, and J.R. Gat: "Physics and Chemistry of Lakes," Springer-Verlag, Inc., New York, NY, 1995.

Niemi, T.M., J.R. Gat, and Z. Ben-Avraham: "The Dead Sea: The Lake and Its Setting," Oxford University Press, Inc., New York, NY, 1996.

Talling, J.F. and J. Lemoalle: "Ecological Dynamics of Tropical Inland Waters," Cambridge University Press, New York, NY, 1998.

[1] Defined in separate entries in this encyclopedia.

Wetzel, R.G.: "Limnology," 3rd Edition, Academic Press, Inc., San Diego, CA, 2001.

Classic References

Cole, G.A.: "Limnology" in The Encyclopedia of Geochemistry and Environmental Sciences (R.W. Fairbridge, Ed.), Van Nostrand Reinhold, New York, NY, 1972.

Hutchinson, G.E., published by Wiley, New York: "A Treatise on Limnology." 1957. "Chemistry of Lakes," 1957. "Introduction to Lake Biology and the Limnoplankton," 1966. "Limnological Botany," 1975.

LIMONITE. The mineral limonite, hydrated oxide of iron, corresponds to the formula $Fe_2O_3 \cdot nH_2O$, but is often very impure due to the admixture of sand and clay. It is not found crystallized but grades from loose porous material to compact masses. Its hardness is variable but pure material is 5–5.5; specific gravity 3.6–4; usual luster, dull to earthy but may be silky to submetallic; color, various shades of yellowish-brown, sometimes nearly black; streak, yellowish-brown; opaque. Limonite is a secondary mineral from the alteration of various other iron-bearing ores or minerals; it is of widespread occurrence and used both as an ore of iron and as a pigment. Limonite has been formed in marshy and boggy areas and is frequently called bog iron ore. Limonite is an important ore of iron in Lorraine, Luxemburg, Bavaria and Sweden. It is found in Saxony, Austria and England. In the United States, limonite is found particularly in Connecticut, Massachusetts, Pennsylvania, New York, Virginia, Tennessee, Georgia and Alabama, but these deposits are of little economic importance at the present time.

LIMPET (*Mollusca, Gasteropoda*). Marine and fresh-water animals related to the snails, with a low conical shell, not spirally twisted. In the common limpets the shell is solid and in the keyhole limpets it is either notched in front or perforated between that point and the apex. Mollusks of the family *Capulidae*, more closely related to some of the species with coiled shells than to the true limpets, also have shells which are not spiral and are called limpets. One form, *Crucibulum*, is called the cup and saucer limpet and another, *Crepidula*, the boat limpet or slipper shell.

LINEAR. An adjective often used in mathematics to describe certain properties. Given the quantities $x_1, x_2, x_3, \ldots, x_n$, a linear combination of them is $a_1x_1 + a_2x_2 + \cdots + a_nx_n = 0$. The quantities x_i are linearly dependent provided all a_i are not zero; otherwise they are linearly independent. The test for such dependence may be made by means of the Gram determinant or the Wronskian.

A linear function is a polynomial of the first degree in its variables and it usually means a polynomial in one variable. Thus, with the linear function, $y = mx + b$, a plot of it would be a straight line of slope m and intercept b on the Y-axis. The general case of a linear algebraic equation would be $a_1x_1 + a_2x_2 + \cdots + a_nx_n = a_0$, where the x_i are variables and the a_i are constants.

A set of simultaneous linear equations is

$$\sum_{j=1}^{n} a_{ij}x_j = b_i$$

$i = 1, 2, \ldots, n$ and a_{ij}, b_i are constants.

A linear differential equation is $A_0(x)y + A_1(x)y' + A_2(x)y'' + \cdots + A_n(x)y^{(n)} = f(x)$ where the $A_i(x)$ are functions of the independent variable only and y', y'', $\ldots$ are the first, second, etc., derivatives. These equations are also inhomogeneous. If the right-hand side is zero in any case, the equation is still linear but homogeneous. Similarly, if all b_i are zero in the case of simultaneous algebraic equations, they are also homogeneous.

See also **Transformation (Mathematics)**.

LINEAR ACCELERATOR. See **Particles (Subatomic)**.

LINEAR ENERGY TRANSFER. The linear energy transfer of charged particles passing through a medium was defined by the ICRU in 1962 as dE_L/dl, where dE_L is the average energy locally imparted to the medium by a charged particle of specified energy in traversing a distance dl. The term locally imparted may refer either to a maximum distance from the track or to a maximum value of discrete energy loss by the particle beyond which losses are no longer considered as local. In either case the limits chosen should be specified. The concept of linear energy transfer is different from that of stopping power. The former refers to energy imparted within a limited volume, the latter to loss of energy regardless of where this energy is absorbed.

LINEAR GRAPH.   See **Graph (Mathematics)**.

LINEAR HYPOTHESIS.   In statistics, generally, any hypothesis which is linear in the parameters entering into it. More specifically, the term has been used to refer to hypotheses that are linear in the means of a set of normal distributions, many of the simpler statistical hypotheses being capable of being thrown into such a form.

LINEAR INEQUALITIES.   A system of relations among variables x_i, possibly including linear equations among them, but also including at least one inequality of the form

$$\sum a_i x_i \geq b$$

(In practice a strict inequality is seldom required.) Such a system may be incompatible (e.g., $x_1 \geq 0, x_2 \geq 0, -x_1 - x_2 - 1 \geq 0$), may define a unique point (e.g., $x_1 \geq 0, x_2 \geq 0, -x_1 - x_2 \geq 0$), or else will define a region in space, not necessarily bounded (e.g., $x_1 \geq 0, x_2 \geq 0, -x_1 - x_2 + 1 \geq 0$ define a bounded region, $x_1 \geq 0, x_2 \geq 0, x_1 + x_2 - 1 \geq 0$ an unbounded region).

The inequality written above can be replaced by the equivalent pair

$$x_0 + \sum a_i x_i = b_i, \quad x_0 \geq 0$$

and, in general, it is possible to replace a system of inequalities by a system in the special form

$$\mathbf{Ax} = \mathbf{b}, \quad \mathbf{x} \geq 0$$

where $\mathbf{A}$ is a rectangular matrix, $\mathbf{x}$ and $\mathbf{b}$ are vectors.

LINEARITY.   With reference to industrial and scientific instruments, the Scientific Apparatus Makers Association defines linearity as the closeness to which a curve approximates a straight line. Linearity is usually measured as a *nonlinearity* and expressed as *linearity*; e.g., a maximum deviation between an average curve and a straight line. The average curve is determined after making two or more full range traverses in each direction. The value of linearity is referred to the output unless otherwise stated. As a performance specification, linearity should be expressed as *independent linearity, terminal-based linearity*, or *zero-based linearity*. When expressed simply as linearity, it is assumed to be independent linearity. See also **Conformity**.

Independent Linearity. The maximum deviation of the actual characteristic (average of upscale and downscale readings) from a straight line so positioned as to minimize the maximum deviation. See Fig. 1(a).

Terminal-Based Linearity. The maximum deviation of the actual characteristic (average of upscale and downscale readings) from a straight line coinciding with the actual characteristics at upper and lower range-values. See Fig. 1(b).

Zero-Based Linearity. The maximum deviation of the actual characteristic (average of upscale and downscale readings) from a straight line so positioned as to coincide with the actual characteristic at the lower range-value and to minimize the maximum deviation. See Fig. 1(c).

LINEARITY CONTROL.   1. In cathode-ray tube equipment, an adjustment that tends to correct any distortion in the sawtooth current or voltage waves used for deflection.

2. In television, a control which varies the distribution of scanning speed throughout the trace interval.

LINEARITY THEOREM.   See **Laplace Transform**.

LINEARLY INDEPENDENT VECTORS.   The vectors $A_1, A_2, \ldots, A_m$ are linearly independent if the equation $c_1 A_1 + c_2 A_2 + c_2 A_2 + \cdots + c_m A_m = 0$ implies that $c_1 = c_2 = \cdots = c_m = 0$. Any n-dimensional space contains n linearly independent vectors. They are a base of the vector space. Any other vector can be written as a linear combination of these base vectors. It is convenient to choose the base vectors mutually perpendicular or orthogonal and of unit length. In ordinary three-dimensional space, these vectors are denoted by i, j, k.

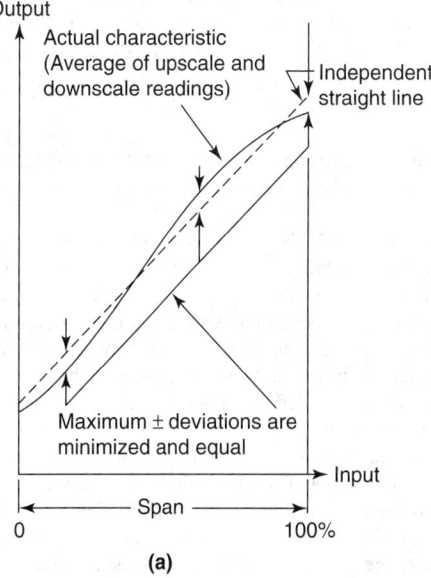

(a)

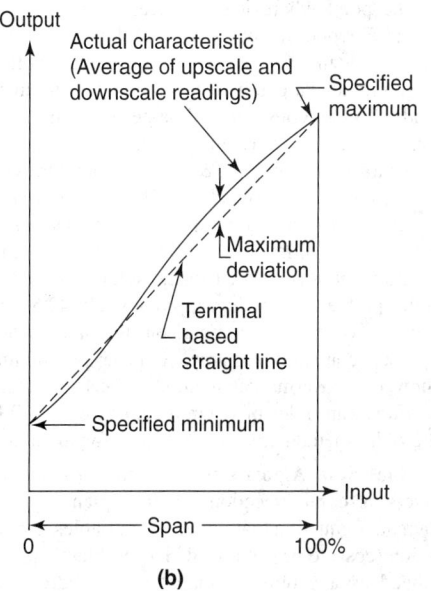

(b)

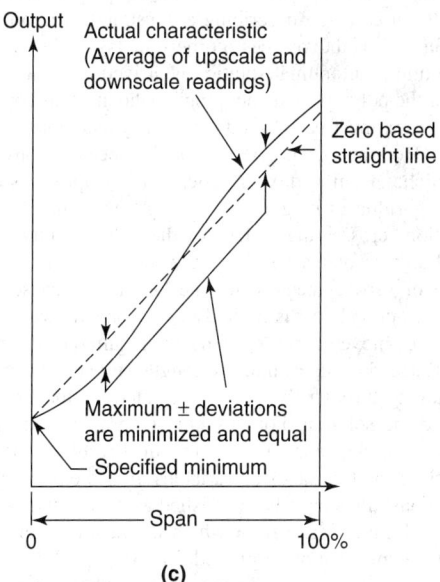

(c)

Fig. 1.   Fundamental relationships pertaining to linearity: (**a**) independent linearity; (**b**) terminal-based linearity; (**c**) zero-based linearity.

LINEAR MAGNIFICATION. For each optical surface in a system, the linear size of the image, I_i, is to the linear size of the object, O_i, as is the distance of the image, Q_i, to the distance of the object, P_i. Then

$$I_i/O_i = Q_i/P_i = m_i$$

The total linear magnification of the system is then the sum of the products of the magnifications of the parts.

LINEAR PROGRAMMING. A technique of mathematics and operations research for solving certain kinds of problems involving many variables where a best value or set of best values is to be found. This technique is not to be confused with computer programming, although problems using the technique may be programmed on a computer. Linear programming is most likely to be feasible when the quantity to be optimized, sometimes called the objective function, can be stated as a mathematical expression in terms of the various activities within the system, and when this expression is simply proportional to the measure of the activities; i.e., is linear, and when all the restrictions are also linear.

In 1947, mathematician George B. Dantzig devised a linear programming algorithm (the *simplex algorithm*). Long before the advent of modern computers, the method enabled machines to handle complex problems with hundreds of constraints. Over the years, the method has been improved—to the point where it can process problems with 15,000 to 20,000 constraints. Beyond this point, the simplex algorithm may become prohibitively slow and cumbersome. Many of today's problems, especially those of the telecommunications industry, are larger and may reach many tens of thousands or more constraints. A large problem may require 12 to 24 hours or longer of computer running time.

In 1984, Narendra Karmarkar (AT&T Bell Laboratories) demonstrated a new algorithm (called the *Karmarkar* or *AT&T algorithm*) that greatly reduces computation times for many practical problems and it is still being explored to determine its ultimate capabilities and limitations. In an early test of the algorithm, it was applied to the solution of a communication network planning problem involving approximately 42,500 variables and 15,000 constraints. For sake of comparison, the same problem was run with a widely used conventional simplex linear programming package. The results showed the Karmarkar method to be faster than the simplex method by more than an order of magnitude, i.e., every 10 hours of time required by the older method was cut to 1 hour for the new method.

Essence of the New Algorithm. A linear programming problem, such as an overseas communication facilities planning project, can be assembled as a matrix equation that represents variables, constraints, and an objective function (cost) to be optimized. The problem's possible solutions can be envisioned as a geometric shape (a polyhedron, referred to by mathematicians as a *polytope*). The boundary of the polytope is formed from multisided, irregular, flat planes called polygons. Each polygon corresponds to an equation describing a constraint. The optimal solution always lies on one of the vertices (corner points) of the polytope. In the simplex algorithm, the optimal solution is arrived at by starting at one of the vertices of the polytope and "hopping" to the most appropriate adjacent vertex. This vertex must be selected from many available, and must lie in the direction of optimal cost. The process is repeated many times as the computer searches from vertex to vertex for the optimal solution, using the simplex algorithm as its guide ("steering" the computer). The zig-zag path of iterations crosses the surface of the polytope until it ends at the final vertex. Large problems are difficult to solve using the simplex method because a lot of vertices need to be checked, and because movement in the direction of optimal cost is limited to only one vertex at a time. Many thousands of such movements (iterations) are required to solve the problem.

With the Karmarkar algorithm, the lengthy process is cut short, so to speak, by moving through the polytope's interior, finding a much more direct route to the solution. For this kind of problem, the solution does not consist of a single, short answer. The final result defines the optimal value for each of the thousands of variables. It also ensures that the many thousands of constraints have been satisfied as well. This massive solution is typically read onto a disk, from which the facility planner can extract all or part of the information in printed form.

Details and examples of the Karmarkar algorithm can be found in the "Record," 4–13 (March 1986), published by AT&T Bell Laboratories, 150 John F. Kennedy Parkway, Short Hills, New Jersey 07078.

LINEAR SPACE (Vector Space). A generalization of certain algebraic aspects of addition in Euclidean space.

A linear space is a set V with a binary relation called addition making it a group and a multiplication of the elements of V by scalars. The elements of V are called *vectors*. The scalars are usually the real or complex numbers.

The subset $x_1, x_2, \ldots, x_n$ of vectors in V is said to be *linearly independent* if the only scalars $\lambda_1, \lambda_2, \ldots, \lambda_n$ for which $\lambda_1 x_1 + \lambda_2 x_2 + \ldots + \lambda_n x_n = 0$ are $\lambda_1 = \lambda_2 = \cdots = \lambda_n = 0$. A subset $x_2, x_2, \ldots x_n$ is said to be a *basis* for V if every vector x in V can be expressed $x = \lambda_1 x_1 + \lambda_2 x_2 + \cdots + \lambda_n x_n$ for scalars $\lambda_1, \lambda_2, \ldots, \lambda_n$. The number of vectors in a linearly independent basis for V is said to be the (linear) dimension of V. The linear dimension of three-dimensional space is three.

If $e_1, e_2 \ldots, e_n$ is a linearly independent basis for V, then each vector in V corresponds to the n-tuple $(\lambda_1, \lambda_2, \ldots, \lambda_n)$ of scalars for which $x = \lambda_1 e_1 + \lambda_2 e_2 + \cdots + \lambda_n e_n$. Conversely, the set of n-tuples of scalars form a linear space.

A mapping L from V to the scalars is said to be a *linear functional* on V if it satisfies: $L(\lambda_1 x_1 + \lambda_2 x_2) = \lambda_1 L(x_1) + \lambda_2 L(x_2)$. The set of linear functionals on V can be made into a linear space in a natural way and this linear space is called the *adjoint* or *conjugate space* of V.

Since many systems describing physical situations (or a first-order approximation) are linear, that is, satisfy the principle of superposition, the space of possible states of the system is a linear space. Also linear spaces are important in discussing linear partial differential equations such as the wave equation or the heat equation.

See also **Euclidean Space**.

LINEAR SYSTEMS. Systems such that the interrelated quantities comprising the system are related by linear differential or differentio-integral equations. Such equations and therefore such systems obey the principle of superposition, namely, the combined effect of a number of causes acting together is the sum of the effects of the several causes acting separately.

LINEAR TOPOLOGICAL SPACE. A generalization of the algebraic and topological aspects of Euclidean space to infinite dimensional linear spaces.

A linear topological space is a linear space X with scalars either the real or complex numbers along with a topology $\mathscr{T}$ so that addition of vectors and scalar multiplication are continuous. Elementary examples of linear topological spaces are the Euclidean spaces with the usual topology. More representative examples are Hilbert space and the space of infinitely differentiable functions on some Euclidean space with the topology of uniform convergence of each derivative on closed bounded subsets.

A function $x \rightarrow \|x\|$ from a linear space V to the nonnegative reals is said to be a *norm* if (1) $\|x\| = 0$ implies $x = 0$, (2) $\|\lambda x\| = |\lambda|\|x\|$, and (3) $\|x + y\| \leq \|x\| + \|y\|$ and V is said to be a *normed linear space*. The "metric topology" defined on V by the metric $\rho(x, y) = \|x - y\|$ makes V into a linear topological space. V is said to be a *Banach space* if it is complete relative to the metric $\rho(x, y)$.

The *conjugate space* V^* of a linear topological space V is the set of all continuous linear functionals on V.

A continuous linear operator ϕ is a mapping between linear topological spaces V and W that is continuous and satisfies $\phi(\lambda_1 x_1 + \lambda_2 x_2) = \lambda_1 \phi(x_1) + \lambda_2 \phi(x_2)$.

The study of linear topological spaces provides the theoretical underpinning for the modern theory of distributions or generalized functions. Although used by engineers and physicists during most of this century, generalized functions were first defined in a mathematically rigorous way by L. Schwartz in the late forties to be the linear functionals on the space of infinitely differentiable functions.

R.G. Douglas, State University of New York at Stony Brook.

LINE (Mathematics). The path described by a moving point. If, as is generally meant, the line is straight its equation in a plane and in rectangular coordinates is $Ax + By + C = 0$, which is a degenerate conic section. Other forms of its equation are: (a) $y = mx + b$, where m is the slope and b the y-intercept; (b) $x/a + y/b = 1$, where a, b are the x-, y-intercepts, respectively; (c) $y - y_1 = m(x - x_1)$, where (x_1, y_1) is a point on the line; (d) $(y - y_1)/(y_2 - y_1) = (x - x_1)/(x_2 - x_1)$, where (x_2, y_2) is another point on the line; (e) $x \cos\theta + y \sin\theta = p$, where the perpendicular from the origin to the line has length p and makes an angle θ with the

X-axis; (f) $rr_1 \sin(\theta - \theta_1) + r_1 r_2 \sin(\theta_1 - \theta_2) + rr_2 \sin(\theta_2 - \theta) = 0$ where (r, θ) are polar coordinates of any point on a straight line passing through two points (r_1, θ_1) and (r_2, θ_2); (g) $x = x_0 + at$, $y = y_0 + bt$, where t is a parameter for the line of slope b/a which passes through the point (x_0, y_0).

If the straight line is located in a three-dimensional rectangular coordinate system, its equation may be taken as

$$(x - x_1)/\lambda = (y - y_1)/\mu = (z - z_1)/v$$

or

$$(x - x_2)/(x_2 - x_1) = (y - y_1)/(y_2 - y_1) = (z - z_1)/(z_2 - z_1)$$

where (x_1, y_1, z_1) and (x_2, y_2, z_2) are two points on the line and λ, μ, v are its direction cosines or numbers proportional to them. A straight line in space is also determined by the equations of two planes, intersecting to form the given line. Thus, such a line can be defined by the simultaneous equations for two planes, $Ax + By + Cz + D = 0$; $A'x + B'y + C'z + D' = 0$.

See also **Coordinate System**; and **Direction Cosine**.

LINE OF APSIDES. A line that contains the major axis of an ellipse is known as the line of apsides of the ellipse. In astronomy, the term is used to indicate the line joining perihelion and aphelion points in an orbit and extending to infinity to cut the celestial sphere. See also **Orbit (Astronomy)**.

LINE OF NODES. The astronomical term applied to a line of intersection of any two fundamental planes. The line of nodes for the moon is the line of intersection of the plane containing the moon's orbit with the plane of the ecliptic. The line of nodes for any member of the solar system, other than satellites, is the line of intersection of the plane of the orbit of the object with the plane of the ecliptic. The line of nodes for the earth is the line of intersection of the plane of the earth's equator with the plane of the ecliptic. For binary stars, the line of nodes is the line caused by the intersection of the plane of the orbit with the plane perpendicular to the line of sight and containing the center of gravity of the system.

LINE OF POSITION. In navigation any line on the surface of the earth upon which a ship is known to be located. If two or more lines of position are known, the ships must be at their point of intersection. A position determined by the intersection of lines of position is known as a fix.

Lines of position are usually circles, either great or small. In most cases, the distance of the center is so great in comparison with the length of the plotted line that the curvature is not apparent in the plot. In some cases, however, where both points of intersection of two circles of position appear on the plot, a dead-reckoning position will indicate which one is the desired fix. Lines of position are obtained by several methods. See also **Course**; **Navigation**; and **Pilotage (or Piloting)**.

LINE PRINTER. A printer, often used in conjunction with a computer, which is capable of printing an entire line of characters at one time.

LINE-REVERSAL PYROMETER. A thermometer for high-temperature gases in which the temperature of a calibrated radiator is adjusted until the spectral areal radiant intensity of its continuum radiation is equal to the intensity of radiation from some suitable characteristic spectral line emitted by the gas. The comparison is made at the wavelength of the spectral line. Seeding is often used to create such a line.

LINE WIDTH. A measure of the spread in wavelength (or energy) of radiation that is normally characterized by a single wavelength (or energy) value. In practice, the width is usually measured at one-half the maximum intensity of the line. The three phenomena that contribute to line broadening are Doppler broadening, pressure broadening, and the intrinsic level width. Due to the finite resolution of measuring apparatus, a broadening due to the characteristics of the instrument also must be considered.

LINKE SCALE. A type of cyanometer; an instrument used to measure the blueness of the sky. The Linke scale is simply a set of eight cards are numbered 2 to 16, the odd numbers to be used by the observer if he judges the sky color to lie between any of the given shades. Also called *blue-sky scale*. Sky-blueness study, or cyanometry, is a means of studying atmospheric turbidity.

LINOLEIC ACID. Also called linolic acid, formula $CH_3(CH_2)_4$ $HC{:}CHCH_2CH{:}CH(CH_2)_7COOH$. This is a polyunsaturated fatty acid (two double bonds) existing in both conjugated and unconjugated forms. It is a plant glyceride essential to the human diet. It is found in linseed oil, safflower oil, and tall oil. See also **Vegetable Oils (Edible)**. At room temperature linoleic acid is a colorless to straw-colored liquid. Specific gravity 0.905 (15/4 °C); mp -5 °C; bp 228 °C (at 14 millimeters pressure). Insoluble in water; soluble in most organic solvents. Combustible. Sources are the oils previously mentioned. Linoleic acid is used in soaps; special driers for protective coatings; emulsifying agents; pharmaceuticals; livestock feeds; and margarine.

LINOLENIC ACID. Also called 9,12,15-octadecatrienoic acid, formula $CH_3CH_2CH{:}CHCH_2CH{:}CHCH_2CH{:}CH(CH_2)_7COOH$. This is a polyunsaturated fatty acid (three double bonds). It occurs as the glyceride in many seed fats. It is an essential fatty acid in the diet. See also **Vegetable Oils (Edible)**. At room temperature, linolenic acid is a colorless liquid; soluble in most organic solvents; insoluble in water. Specific gravity 0.916 (20/4 °C); mp -11 °C; bp 230 °C. Combustible. Linolenic acid finds use in various pharmaceuticals and drying oils.

LION. See **Cats**.

LIOUVILLE EQUATION. In the statistical mechanics of an ensemble of systems, each containing N particles of mass m, it is useful to introduce a density or probability function $P_N(q_N, p_N)$, representing the probability that a system of the ensemble will have its point in phase space fall within the volume bounded by $q_1, q_2, \ldots, q_N, p_1, p_2 \ldots, p_N$, and $q_1 + \delta q_1, q_2 + \delta q_2, \ldots, q_N + \delta q_N, p_1 + \delta p_1, p_2 + \delta p_2, \ldots, p_N + \delta p_N$. If no new systems are created or destroyed, P_N will satisfy the Liouville equation:

$$\frac{\delta P_N}{\delta t} + \sum_{j=1}^{N} \left(\nabla_j \frac{P_N p_j}{N} + \delta_j P_N \dot{p}_j \right) = 0$$

where

$$\nabla_j = \partial/\partial q_j \quad \text{and} \quad \delta_j = \partial/\partial p_j$$

The p-terms are values in generalized coordinates for the positions of the systems, the δ-terms are their momenta, the δ-terms in p and q denote small changes, $\partial P_N/\partial t$ is the partial derivative, and ∇_j and δ_j are vector differential operators. The dot over p_j denotes its first derivative with respect to time.

LIOUVILLE-NEUMANN SERIES. An infinite series

$$\phi(x) = \sum_{n=0}^{\infty} \lambda^n \phi_n(x)$$

which is a unique, continuous solution of a Fredholm integral equation of the second kind. If the nth iterated kernel is defined as

$$K_n(x, z) = \int \int \cdots \int K(x, y_1) K(y_1, y_2) \cdots K(y_{n-1}, z) dy_1 dy_2 \cdots dy_{n-1}$$

then

$$\phi_n(x) = \int K_n(x, z) f(z) \, dz$$

The resolvent or solving kernel is given by

$$K(x, z; \lambda) = \sum_{n=0}^{\infty} \lambda^n K_{n+1}(x, z)$$

hence the solution of the integral equation becomes

$$\phi(x) = f(x) + \lambda \int K(x, z; \lambda) f(z) \, dz$$

Similar methods may be used to solve the Volterra equations.

LIPIDOSES. Disturbances of lipid metabolism in which abnormal deposits of lipids are found in various groups of cells in the body. Primary lipidoses, i.e., those due to an inborn error of metabolism include: (1) *Gaucher's disease*, in which the product accumulated is ceramide glucoside, caused by a beta-glucosidase enzyme defect. In the infantile form, there is mental retardation. In the adult form, there are hepatosplenomegaly and bone changes. (2) *Niemann-Pick disease*, in which the product accumulated is sphingomyelin, caused by a sphingomyelinase enzyme defect. In this disease, there is mental retardation and hepatosplenomegaly. (3) *Tay-Sachs disease*, in which the product accumulated is ganglioside G_{M2}, caused by a hexosaminidase A enzyme defect. In this disease, there is mental retardation and sometimes blindness. (4) *Generalized gangliosidosis*, in which the product accumulated is ganglioside G_{M1}, caused by a beta-galatctosidase enzyme defect. In this disease, there is mental retardation and hepatosplenomegaly (enlargement of liver and spleen). In secondary lipidoses there may be abnormal overgrowth of reticulum cells, in which lipoids are present. These include the Letterer-Siwe syndrome (destructive lesions chiefly in bone, fatal in infancy), Hand-Schäuller-Christian syndrome (bone lesions with fibrosis of the lungs, diabetes and stunted growth) and the eosinophilic granuloma of bone, a solitary, benign and often self-limiting tumor of children and young adults. See also **Gaucher's Disease**.

Additional Reading

Berkow, R. and M.H. Beers: "The Merck Manual," 17th Edition, Merck & Company, Inc., Whitehouse Station, NJ, 1999.
Bruyn, G.W. and H.W. Moser: "Neurodystrophies and Neurolipidoses," Elsevier Science, New York, NY, 1996.
Figueroa, M.L. et al.: "A Less Costly Regimen of Alglucerase to Treat Gaucher's Disease," *N. Eng. J. Med.*, 1632 (December 5, 1992).
Garber, A.M.: "No Price Too High?" *N. Eng. J. Med.*, 1676 (December 3, 1992).
Gatt, S., L. Douste-Blazy, and R. Salvayre: "Lipid Storage Disorders: Biological and Medical Aspects," Perseus Books, Boulder, CO, 1988.
Schwandt, P.: "Risk Prevention of Arterial Lipidoses," Warren H. Green, Inc., St. Louis, MO, 1990.

LIPIDS. A heterogenous group of substances which occur ubiquitously in biological materials. They may be categorized as a group by their extractability in nonpolar organic solvents, such as chloroform, carbon tetrachloride, benzene, ether, carbon disulfide, and petroleum ether. Structural types within the group range from simple, straight-chain hydrocarbon molecules to complex ring structures with varying side chains. A useful classification of the lipids is: (1) fatty acids; (2) neutral fats; (3) phosphatides; (4) glycolipids; (5) aliphatic alcohols and waxes; (6) terpenes; and (7) steroids. See also **Steroids**.

Many lipids, especially the phospholipids, have a strong tendency to form complexes with each other and with various substances. Complex formation is due to the electrostatic attraction of polar groups and to the mutual solubility of the long hydrocarbon chains. Thus, the lipoproteins and proteolipids are complexes of proteins and a variety of lipids, such as cholesterol, phospholipids, glycerides, and glycolipids. The lipids are linked to the proteins by several types of forces. Electrostatic forces, Van der Waals forces, hydrogen bonding, and hydrophobic bonding hold these complexes together. Because of their attraction for water, the polar groups of the protein and phospholipid arrange themselves on the outside of the complex, while the hydrocarbon groups of the lipids are folded into the center. Thus, there is presented to the aqueous phase those groups which have an affinity for water. This arrangement accounts for the solubility of the complexes in water. The phospholipids, owing to their polar groups, act as water solubilizers for the nonpolar lipids. The arrangement may be different in the proteolipids of the brain and nerves, since they are not soluble in water. In these complexes, the lipids may completely envelop the protein.

Knowledge of lipid metabolism has increased at an accelerated rate during the past few decades. The detailed biochemical reactions whereby the fatty acids are synthesized and oxidized; how phospholipids, glycolipids, and cholesterol are synthesized; and how lipids are absorbed and transported have been elucidated. Fatty acids are synthesized from acetyl coenzyme A and malonyl coenzyme A thiol esters. The vitamin biotin plays a vital part in the fixing of carbon dioxide to form malonyl coenzyme A, an important intermediate in fatty acid synthesis. The hormone insulin also favors fatty acid synthesis. The oxidation of fatty acids occurs as their coenzyme A esters in the Krebs cycle of the mitochondria. Cholesterol is biosynthesized from acetyl coenzyme A. Cholesterol in humans is converted to bile acids, fecal sterols, and to steroid hormones. The synthesis of lecithin is mediated via phosphatidic acid and diglyceride precursors. Cytidine nucleotides play a role in the transfer of choline (as phosphoryl-choline) to a diglyceride to form lecithin. Uridine nucleotides act to transfer sugar residues in the synthesis of glycolipids.

The transport of lipids in the blood plasma is effected by complex formation with proteins to yield lipoproteins. The liver is the major organ for the synthesis of the lipoproteins. Analysis of serum lipoprotein patterns is important in the understanding of vascular disease (atherosclerosis). The clearing of lipemic blood, such as may occur after a heavy fat meal, is brought about by enzyme known as lipoprotein lipase. This enzyme yields free fatty acids which combine immediately with the plasma albumin to form complexes known as NEFA (nonesterified fatty acids). NEFA act as important transport vehicles for transport of triglycerides and the levels of blood NEFA are very sensitive to hormonal control and neural control. Certain hormones, such as epinephrine, stimulate the membrane-bound adenyl cyclase which converts ATP (adenosine triphosphate) to cyclic AMP (adenosine monophosphate). The latter stimulates adipose tissue lipase and mobilizes depot fat.

The excess utilization of lipids and excess oxidation of fatty acids causes an increase in acetoacetic acid in the body. This condition is known as *ketosis* and can lead to *acidosis*. This situation is common in severe diabetes and can occur whenever carbohydrate utilization is severely decreased.

Research continues in an effort to gain a more thorough understanding of how lipoproteins are synthesized; how lipids are arranged and combined with proteins to form cell membranes; what specific role lipids play in transport across cell membranes; how hormones act to regulate lipid metabolism; the biochemical basis of such abnormal lipid metabolic states as Gaucher's disease, Niemann-Pick's disease, etc.; how lipids per se permeate cell membranes; and how many phenotypic lipoproteins occur in serum.

See also **Cholesterol**.

Additional Reading

Akoh, C.C.: "Food Lipids: Chemistry, Nutrition, and Biotechnology," Marcel Dekker, Inc., New York, NY, 1998.
Bornscheuer, U.T.: "Enzymes in Lipid Modification," John Wiley & Sons, Inc., New York, NY, 2000.
Chang, T.Y. and D.A. Freeman: "Intracellular Cholesterol Trafficking," Kluwer Academic Publishers, Norwell, MA, 1999.
Feher, M.D. and W. Richmond: "Lipids and Lipid Disorders," Mosby-Year Book, Inc., St. Louis, MO, 1997.
Fox, P.: "Advanced Dairy Chemistry: Lipids," Vol. 2, Chapman and Hall, New York, NY, 1999.
Gatt, S., L. Douste-Blazy, and R. Salvayre: "Lipid Storage Disorders: Biological and Medical Aspects," Perseus Books, Boulder, CO, 1988.
Gotto, A.M. and H.J. Pownall: "Manual of Lipid Disorders: Reducing the Risk for Coronary Heart Disease," 2nd Edition, Lippincott Williams & Wilkins, Philadelphia, PA, 1999.
Gurr, J., J. Harwood, K.N. Frayn: "Lipid Chemistry," 5th Edition, Blackwell Science, Inc., Malden, MA, 2001.
Hainik, T. and V. Passechnik: "Bilayer Lipid Membranes: Structures and Mechanical Properties," Kluwer Academic Publishers, Norwell, MA, 1995.
Katsaros, J. and T. Gutberlet: "Lipid Bilayers: Structure and Interactions," Springer-Verlag, Inc., New York, NY, 2000.
Keane, W.F. and B.L. Kasiske: "Lipids and the Kidney," S. Karger Publishers, Inc., Farmington, CT, 1997.
Vance, D. and J. Vance: "Biochemistry of Lipids Lipoproteins and Membranes, 3rd Edition," Elsevier Science, New York, NY, 1997.

LIPOMA. A benign tumor made up of fat cells. Lipomas occur commonly in the subcutaneous tissues about the head and neck. They cause no symptoms.

LIQUATION. 1. A process of magmatic differentiation believed to take place as a result of the separation of two immiscible liquids from the parent magma.

2. The separation of a more readily fusible substance from a less fusible one by controlled heating.

LIQUEFACTION. See **Natural Gas**.

LIQUID CRYSTAL DISPLAY TECHNOLOGY. Liquid crystal materials are organic compounds that have dual properties over some temperature range: They flow like viscous fluids, but at the same time, they have many of the optical characteristics of solid crystals. Liquid crystals were first discovered in 1888 by Reinitzer, who noted their dual liquid and crystalline nature.

The liquid crystalline state is called a *mesophase*. It exists between the normal liquid state (also known as the *isotropic phase*), which has no positional or orientational order, and the highly ordered solid crystalline state. That is, the liquid crystal phase exists above T_m, the melting point of the solid crystal, and below T_c, the "clearing point" at which the liquid becomes isotropic. Compounds that exhibit liquid crystallinity are called *mesomorphic materials*, and the liquid crystalline state is often referred to by the general term *mesomorphism*. In the mesophase, the molecules spontaneously adopt an ordered arrangement (a process called *self-alignment*), but some movement of the molecules is allowed, which is essential for the functioning of the liquid crystal display (LCD).

These thermal qualities make the operation of LCDs highly temperature dependent. Extremes of temperature render the display inoperative. Even within the temperature range that defines the mesophase, the liquid crystal material becomes more viscous when cooled. This greatly reduces the response time at 0 °C or below, making for a sluggish display, because it is more difficult for the liquid crystal molecules to rotate under the influence of an electric field physically. Furthermore, temperature fluctuations change the dielectric and optical anisotropy of the liquid crystal material. Generally, careful combinations of different liquid crystal compounds produce samples with more-stable temperature characteristics.

Liquid crystal molecules tend to be rod- or disc-shaped. The most important ones for display technologies are the rod-shaped molecules in mesophases called *nematic, smectic*, and *cholesteric*, distinguished by their molecular order. Nearly all of the LCDs made today use materials that exhibit the nematic mesophase, which is covered here in the greatest detail. A bulk sample of a nematic liquid crystal material in its mesophase is a white, opaque liquid that flows easily; it appears cloudy, because the ordering of the molecules occurs only in domains on the order of tens of microns.

The nematic liquid crystal is typically a mixture of various individual compounds, which consist of elongated molecules held together at their ends by polar forces (similar to those that hold the two helical strands of molecules together in DNA (deoxyribonucleic acid, the basic building block of all biological structures). Under a microscope, the liquid crystal material appears to consist of long strands, or threads, that move together like a series of flexible molecular chains. See Fig. 1. This gives liquid crystals the unique property of *cooperative alignment*, which means that the direction of alignment of one molecule influences the alignment of the others in its vicinity. Such preferential alignment gives liquid crystals their *anisotropic* properties; that is, various physical properties, such as the dielectric constant, refractive index, conductivity, viscosity, and magnetic susceptibility, have different values in the directions parallel and perpendicular to the molecular axis. The long axis direction that the molecular chains adopt is primarily determined by the structural nature of the surface on which they lie. However, the alignment of the molecular chains is also influenced by applied electric (or magnetic) fields.

These are the features that enable liquid crystals to be useful in display devices. Liquid crystal materials are highly tailored to have a wide operating temperature range, a fast response time, stability, and an appropriate response to an applied electric field. See also **Liquid Crystals**.

Liquid Crystal Displays

Table 1 summarizes the main types of LCD technology. The categorization is not designed to be comprehensive, but to highlight the key applications and indicate where they fit into the overall structure. Some categorizations are made on the basis of liquid crystal type, others on addressing technique, depending on which is more relevant to highlight the key applications.

The linear presentation of all of the types of LCDs is complicated by the fact that they can be grouped by addressing technique or by liquid crystal type, and the two groups are partially, but not completely, overlapping. This discussion adheres to a structure based primarily on addressing scheme. All liquid crystal types are covered, but not within one linear section. The liquid crystal types not presented in "LCD Addressing Schemes" are discussed in "Other Types of LCDs."

Basis of LCD Operation

This section outlines the fundamental properties of LCD operation that are common to all types of LCDs. The descriptions are very general because the particular type of liquid crystal or addressing scheme has a major effect on the overall functioning of the display. The most common type of liquid crystal used in displays is the *twisted-nematic* (TN) type, whose operation is described in detail under "LCD Addressing Schemes."

In its simplest form, an LCD is made by creating a sandwich of liquid crystal material several microns thick between two glass plates; this sandwich is sometimes called a *cell*.

A thin, transparent conductive coating, typically indium–tin oxide (ITO), is patterned to form an array of electrodes on each plate. A surface aligning agent, made from a solid polymer material (usually polyimide), is then deposited over the ITO electrodes and buffed so that when the liquid crystal is flowed between the plates, the molecules align in the proper way to react to an applied field. Although still the subject of research, one theory holds that the buffing action produces enough localized heating to realign the polyimide chains with their long axes in the direction of the buffing wheel. When the liquid crystal is applied to such a buffed polymer surface, a single layer of liquid crystal molecular chains becomes anchored to that surface with the chain axis parallel to the direction of alignment in the polymer film due to the strong polar forces between the benzene rings that make up the liquid crystal molecules and those of the polyimide. The space between the plates is sealed off, and the ITO leads are connected to control electronics. The display is bright because of a backlight or because

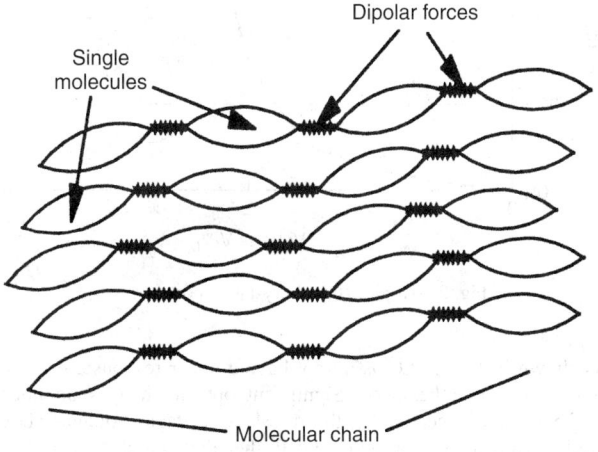

Fig. 1.   Microscopic structure of liquid crystals.

TABLE 1. MAIN TYPES OF LCD TOCHNOLOGY

Liquid Crystal Displays						
Active Matrix				Passive Matrix		
Thin-Film Transistor (TFT)			Liquid-Crystalon-Silicon (LCOS) (projector)	Supertwisted Nematic (STN)		Twisted Nematic (TN) (watch, calculator)
Amorphous Silicon (notebook computer, desktop monitor)	Polysilicon		CdSe, continuous grain silicon, single crystal silicon	Multiplexed (notebook computer)	Other addressing schemes and alignments (handheld)	
	HTPS (projector, view-finder)	LTPS (hand-held)				

of reflected light shone from the front of the cell; liquid crystals do not give off their own light (see "LCDs for Portable and Handheld Devices" for more detail on the difference between transmissive, transflective, and other types of light modulation in LCDs).

Without an electric field, the liquid crystal molecular chains at the surface of the glass plates align themselves in a direction determined by the direction of the polyimide chains. There are two alignment modes: *homeotropic alignment* means the molecules orient perpendicular to the alignment layer, and *homogeneous alignment* means the molecules are parallel to the alignment layer. The vast majority of LCDs use homogeneous alignment. Varying degrees of tilt between the liquid crystal and the polyimide can also be achieved; a small tilt angle ($1°\angle5°$) is required to prevent the formation of *reverse tilt* domains, which give the display a patchy appearance.

The top and bottom plates may have different alignment directions. For example, in a twisted-nematic cell the surface orientation of the polyimide on the bottom plate is rotated $90°$ from that of the top plate, so through the width of cell, the liquid crystal molecules twist continuously through $90°$.

When an external electric field of sufficient strength is applied to the cell, it overrides the surface alignment, and the molecules align uniformly over the entire sample in the direction of the field. The "turn-on" time is 10–15 milliseconds for TN liquid crystals and as short as microseconds for ferroelectric liquid crystals. This alignment occurs because liquid crystal molecules have different dielectric constants, e, along the long and short axes. This property is called *dielectric anisotropy* ($\Delta\varepsilon$) and is represented by

$$\Delta\varepsilon = \varepsilon_{\text{parallel}} - \varepsilon_{\text{perpendicular}}$$

Under this electric field excitation, the liquid crystal molecules orient themselves with the axis of greatest dielectric constant parallel to the field in order to be in the lowest (most stable) energy state. Thus, samples with positive dielectric anisotropy, where the dielectric constant is greatest along the long axis, will align parallel to the field (perpendicular to the plates), and samples with negative dielectric anisotropy align perpendicular to the field. When the field is removed, the sample returns to its previous semiordered state over the turn-off time, which is generally slightly longer than the turn-on time.

When the liquid crystal molecules are oriented either by a surface alignment layer or an electric field, the bulk sample has all the optical properties of an anisotropic crystal. Anisotropic crystals have two distinct optical states, which can be viewed using polarized light. Thus, LCDs have a polarizer associated with each glass plate; the second polarizer is often termed the *analyzer*. Light passes through the first polarizer, is modulated by the optical properties of the liquid crystal molecules, then encounters the analyzer. By changing the liquid crystal molecules' optical properties through an applied field, the light can be passed or blocked in a controllable way.

In the (more common) *normally white* (NW) TN cell with positive dielectric anisotropy, the polarizers have their polarization direction parallel to the polyimide alignment direction (recall that the top and bottom plates differ in this direction by $90°$). When there is no applied field, light from the backlight rotates its own polarization direction through $90°$ as it encounters the continuously twisting liquid crystal molecules. In this way, the light passes through the crossed polarizer-analyzer pair when there is no field, so the display looks bright. Under the action of the field, when the molecules are all aligned in a single direction, the light passes straight through the cell and is blocked by the crossed polarizers; the display looks dark. The *normally black* (NB) cell blocks light when the field is off and passes light when the field is on.

This light-modulating effect is the basis for a display because it is a controllable way of producing visible *contrast*. The contrast of a display is an important quality and is defined as the difference in brightness between a fully *on* (bright) state and a fully *off* (dark) state. The process of electronically modulating the cell between the two distinct optical states described above is called optical rotation. There are other optical effects in anisotropic media; for instance, double refraction (birefringence) and optical scattering. They are useful for displays. Optical scattering works by modulating the cell between a transparent state and a scattered or cloudy state (instead of a totally dark state). Birefringence uses the refractive anisotropy of liquid crystals and is the basis for many enhanced LCDs.

It is important to note that the liquid crystal cell rotates the light because of its anisotropic properties, not because it acts as a polarizer. Optically anisotropic materials have the property of creating elliptically polarized light from incident linearly polarized light (from the front polarizer) that is tilted with respect to one of the refractive axes. The different indices of refraction create a time delay (phase shift) in light propagating along the two orthogonal axes, resulting in elliptically polarized light. This light has a component perpendicular to the original incident polarized light, and so the light appears to have rotated. The amount of rotation is determined by the optical path difference; effectively, light polarized parallel to the incident direction travels a different distance through the material than light polarized perpendicularly. The optical path difference is written $\Delta n \times d$, where Δn, the optical anisotropy, is the difference between the index of refraction parallel to the axis of incoming light (n_{parallel}) and the index of refraction perpendicular to the axis of incoming light ($n_{\text{perpendicular}}$), and where d is the cell thickness. Typical TN cell spacing, d, is 7 to 10 microns; typical $\Delta n \times d$ values are 0.5 to 1.5 microns.

The liquid crystal layer in a TN cell cannot be made extremely thin because the refractive index of the liquid crystal material is wavelength dependent; i.e., red light is modulated differently than blue light. This can produce undesirable colors on the display. Cells with a thick liquid crystal layer show less of this effect, but it is noticeable if the cell spacing is thinner than 3 to 4 microns.

Figure 2 shows how liquid crystals respond to the applied electric field. Figure 2 illustrates the voltage/brightness (brightness of light transmitted or reflected) relationship for a typical TN-LCD cell.

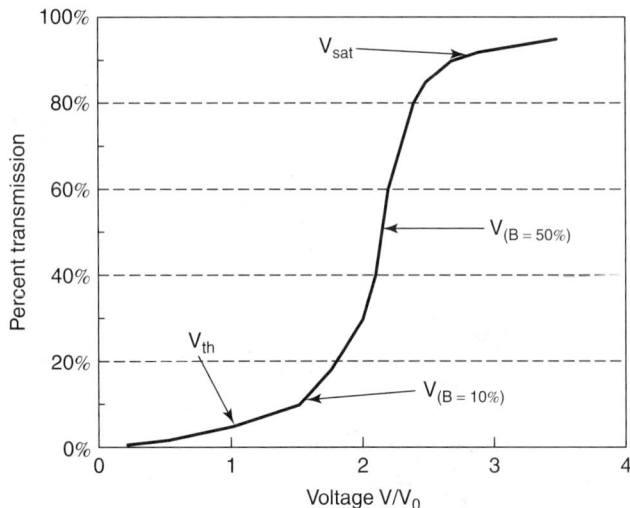

Fig. 2.   Typical liquid crystal response curve.

As shown in Fig. 2, LCDs do not have a linear response; instead, they exhibit a threshold behavior. Significant optical changes do not occur until the voltage exceeds V_{th}, the threshold voltage. Voltages above the saturation voltage, V_{sat}, produce no further darkening. A typical sample has V_{th} of 1 volt and V_{sat} of 2.5 volts. An important evaluation criterion is the performance coefficient, P, which is a measure of the nonlinearity of the display. P is usually defined as

$$P = \frac{V_{(B=50\%)} - V_{(B=10\%)}}{V_{(B=10\%)}},$$

where

$V_{(B=50\%)}$ = the driving voltage required to produce 50% of the maximum display brightness

$V_{(B=10\%)}$ = the driving voltage required to produce 10% of the maximum display brightness

The more nonlinear the response, the smaller the P value will be; a perfect step function will have a P value of zero. P is not simply the slope of the active region, however, but a relationship involving both the slope and the closeness of the transition to the origin. Although increased nonlinearity means better contrast, there is a tradeoff with how much gray scale can be achieved. A display with a step-function response curve can only be black or white; in fact, limited grayscale is a problem for the highly nonlinear ferroelectric liquid crystal displays.

During the 1920s and 1930s, work on liquid crystal materials and the electro-optic effects they produced was conducted in France, Germany, the USSR, and Great Britain. Perhaps the first patent on a light-value device (that is, one controlling the bright and dark states by passing or blocking a backlight) that used liquid crystals was awarded to the Marconi Wireless Telegraph Company (now part of GEC) in 1936. In the mid-1950s researchers at the Westinghouse Research Laboratories discovered that cholesteric liquid crystals could be used as temperature sensors. It was not until the 1960s, however, that serious studies on the materials and the effects of electric fields on them were carried out.

The idea of using liquid crystal materials for display applications was probably first conceived by Richard Williams and George Heilmeier at the David Sarnoff Research Center (then the central research arm of RCA Corporation) in Princeton, New Jersey, in 1963. Later, a larger group, headed by George Heilmeier and including Louis Zanoni, Joel Goldmacher, Lucian Barton, and Joseph Castellano, spearheaded the work to develop LCDs for the fabled "television on a wall" concept, a dream of the late television pioneer David Sarnoff. Between 1964 and 1968, this group discovered many effects, including dynamic scattering, dichroic dye LCDs, and phase change displays.

One of the major breakthroughs was the discovery that by mixing pure nematic liquid crystalline compounds together, it was possible to produce stable, homogeneous liquid crystal solutions that could operate over a broad temperature range. These early materials were composed of Schiff base and ester compounds. Although the Schiff bases worked well in all-glass sealed packages, they were susceptible to breakdown by hydrolysis (from moisture penetration) when less-expensive, nonhermetic plastic seals were used. To eliminate the problem, cyanobiphenyl materials were developed by Gray, Harrison, and Nash in 1973. These materials had improved properties and even broader temperature ranges. Later, in 1977, cyanophenylcyclohexanes, which showed improved operating properties in displays, were developed by researchers at E. Merck & Company in Darmstadt, Germany.

Once the principle of passing and blocking light through the action of liquid crystals was established, the next step was to fabricate a display with many areas whose brightness and darkness could be individually controlled so that words and pictures could be displayed. These areas are of course the *pixels* of modern displays, named for "picture elements."

Twisted-nematic field-effect displays were the first LCDs to undergo widespread commercial development, due to their inherent simplicity and ease of construction. Figure 3 illustrates the construction of a single pixel in a TN-LCD with positive dielectric anisotropy. The alignment direction on the upper glass plate is rotated at an angle of 90° to that on the lower plate; it is a standard normally white pixel as described above (most displays are

normally white; the normally black mode is more difficult to control and tends to have lower contrast).

LCD Addressing Schemes

The challenge in making a display with many pixels lies in finding a way to address all of the pixels. The addressing technique affects important image qualities such as grayscale (and hence color), visual contrast, and speed.

The simplest way to address an LCD is by direct addressing, in which each "pixel" (which could actually be a large area) is addressed by sending a direct command to its particular electrodes. Examples of this type of display are low information content segmented displays, such as the 7-electrode numeric displays common in watches and calculators.

To achieve acceptable contrast over a broad viewing angle in applications requiring high information content, one must go beyond direct addressing. There are two general approaches for accomplishing this in LCDs: passive matrix and active matrix. *Passive matrix* refers to improvements in LCD geometry and the use of different physical/optical LCD effects to produce a cell with improved contrast characteristics. Examples include multiplexed displays, as well as liquid crystal devices with memory, such as ferroelectric-smectic and cholesteric devices and cells that incorporate a greater-than-90°-twist (supertwist). Virtually all enhancements use the optical anisotropy of liquid crystal molecules to produce visible change. *Active matrix* refers to the addition of some type of nonlinear switching device, such as a diode or transistor, at the pixel site to improve the nonlinear response of the liquid crystal material. With active matrix devices, the pixel structure itself is relatively unchanged, but the devices can be complex.

The quest for more-efficient passive and active matrix displays has come about because of the limitations inherent in the performance of multiplexed displays. The fundamental property required for a large-area display with a large number of characters or pixels is that it can be written very quickly and produce a high-contrast image over a wide viewing angle. Cathode ray tubes (CRTs) excel in this regard, as the electron beam can address and activate any pixel on the faceplate within milliseconds. The main drawbacks to CRTs are their considerable weight, bulky size, and high power consumption. Passive matrix and, to a greater extent, active matrix LCDs (AMLCDs) are rapidly achieving performance comparable to CRTs, without these undesirable features.

Twisted-Nematic (TN) LCD Operation

There are three main methods to obtain grayscale in TN-LCDs. One is voltage modulation, whereby the cell receives a partial *on* voltage that produces only partial rotation of the liquid crystal molecules, resulting in partial light transmission. Another is spatial modulation, also known as *halftoning*, whereby the density of *on* to *off* pixels in a certain area is changed to create the illusion of a gray level over that area. The third, most common, method is time modulation, or pulse-width modulation. Because of the liquid crystal's finite response time (milliseconds to tens of milliseconds), it responds not to instantaneous voltage but to the time-averaged voltage over approximately its response period. Thus, applying a 10-volt, 1 kHz frequency with a duty cycle of 70% (70% at 10 volts, 30% at 0 volts) means the effective voltage is 7 volts. Applying such pulse-width modulation to individual pixels is an efficient way to produce grayscale.

For a very different reason, time modulation over the entire display is important. The time-averaged voltage across the liquid crystal cell must be zero in order to prevent deterioration by electrolytic effects over time. If the liquid crystal material and aligning layer were perfectly pure, this voltage inversion would not be necessary, but inevitable amounts of trace impurities move toward the plates when a net DC field is applied, ultimately degrading the performance. In all LCD cells, the write voltage is positive one cycle and negative the next, so the time-averaged control voltage is kept at zero. It is the voltage difference, not the polarity, that produces a visible effect, so reversing the voltage does not change any display characteristics (−7 volts produces the same visual effect as +7 volts). Current displays, all of which use this polarity-inverting scheme, have been tested with lifetimes of approximately 50,000 hours.

Twisted-nematic displays have good angular viewability when directly driven, although there is some contrast reduction when viewed at oblique angles because the liquid crystal molecules are being observed along a different propagation axis, which changes Δn and affects the optical rotation.

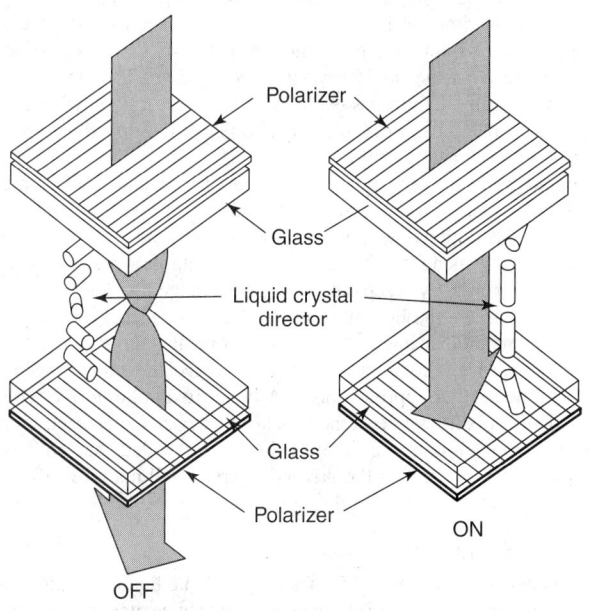

Fig. 3.   Single-pixel construction in a TN-LCD.

Directly driven TN-LCDs are ideal for low-cost, low-power devices. They offer good contrast and an acceptable response time for simple readout operations, where each segment of the cell can be individually addressed. However, for more complex graphic display functions, a dot matrix array of display pixels is needed; for this type of display, a more complex addressing scheme is required.

Historically, multiplexing of the basic direct-drive TN-LCD was the "next step up." In this case, pixels are formatted as a grid and handled in a more efficient way than communicating directly with each pixel. The display is scanned row by row, from top to bottom, at 60 to 100 Hz. The liquid crystal material reacts to the average of the voltage over time. When the proper voltage difference is generated across a row and a column, the liquid crystal material at the intersection of these electrodes, which is known as the *selected pixel*, is activated. A result of this arrangement is that the nonselected pixels immediately adjacent to the selected pixels also receive some fraction of the voltage. Thus, the liquid crystal molecules in nonselected pixels are partially aligned by the electric field, reducing the contrast between the *off* and *on* elements of the display. Highly multiplexed TN-LCDs suffer from contrast inversion at certain oblique angles, where the on state pixels become dimmer than the *off* state pixels. This is especially troublesome with grayscale displays, because the different gray levels fade together.

Multiplexed TN-LCD Operation

The most common multiplexing scheme is shown schematically in Figure 4. The horizontal row lines are the *address lines*, and the vertical column lines are the *data lines*. The row and column bus lines are on opposite glass plates to eliminate the possibility of shorts due to manufacturing defects. The driver circuitry sequentially selects the address lines by giving them a nonzero voltage, while the data lines simultaneously send a nonzero select voltage to all pixels in the selected row.

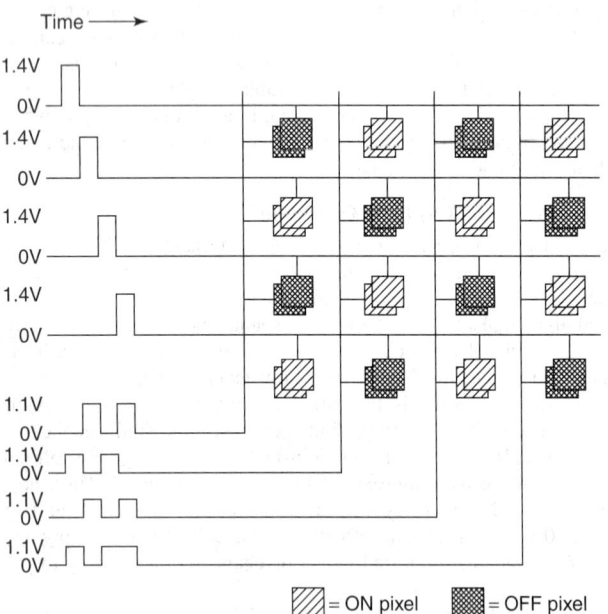

Fig. 4.   Standard Multiplexing Scheme ($V_{th} = 1.5$ volts).

The select voltages from the row and column lines are individually below the threshold voltage, but the sum of the two is greater than the saturation (on) voltage. Ideally, for a pixel to be on, both its address and data lines must have a select voltage. If either the data line or the address line is in its nonselect state, the pixel voltage is insufficient to make a change in contrast.

With multiplexing, the turn-*on* and turn-*off* times become even more important than in direct addressing. In the TN-LCD shown in Figure 4, each *on* pixel has a select voltage applied to it for only one-quarter of the frame write cycle, meaning it spends the majority of its time having no applied voltage. There is only one-quarter of a cycle for an *on* pixel to distinguish itself from one in the *off* state, not an easy task when the turn-on time is tens of milliseconds. Furthermore, a low temperature will slow the response time even more.

This type of response, in which the molecules respond to the time-averaged voltage (as opposed to the instantaneous or maximum voltage) is known as a *root-mean-square* (RMS) response. A selected pixel, which spends most of its time with its driving voltage below V_{th}, has only a slightly higher RMS voltage than a nonselected pixel. All contrast in the display needs to be produced with only this slight difference in voltage. A highly nonlinear liquid crystal material is required to produce viewable images with adequate contrast; a small difference in $V_{off} - V_{on}$ at the pixel needs to produce a large difference in contrast. The more lines a display has, the smaller $V_{on} - V_{off}$ at the pixel will be, thus the more nonlinear the liquid crystal sample needs to be.

As mentioned earlier, nonselected pixels in the same row or column of a selected pixel will receive partial voltages not intended for them. Simply raising the driving voltages to the rows and columns to the pixel response time will only turn the entire display on. The row and column select voltages must both be under V_{th}, but when added together, they must be above V_{th}. For example, if V_{th} of a liquid crystal material is 1.5 volts, the selected row may be placed at 1.4 volts, while all nonselected rows are kept at zero volts. The selected data lines are supplied with 1.1 volts, while nonselected data lines are kept at zero volts. Thus, every pixel in the selected row, regardless of whether it should be *on* or *off*, has a voltage of at least 1.4 volts. Likewise, every pixel in a selected column has a voltage of at least 1.1 volts. Because these voltages are below threshold, they produce no noticeable change in contrast. Only pixels where both the row and column are selected to produce a voltage difference of 2.5 volts exhibit a change in contrast. But a selected pixel spends only a fraction of the frame time at 2.5 volts; its RMS value will not be significantly higher than that of the nonselect state.

The values for V_{on} and V_{off} are chosen using the results of Alt and Plesko. They showed that the efficiency of multiplexed (and passive matrix) addressing is optimized by choosing

$$V_{on}/V_{off} = [(N^{1/2} + 1)/(N^{1/2} - 1)]^{1/2}$$

where N is the number of rows. Thus, in a multiplexed display with 100 lines, the ratio of V_{on}/V_{off} is only 1.11. This means that if the liquid crystal sample being used has $V_{(B=10\%)} = 2$ volts, then for a contrast ratio of 12:1, the sample must have $V_{(B=50\%)} = 2.22$ volts.

As N increases, the ratio of voltages approaches 1, meaning the display has zero contrast and is not viewable. Hence, this addressing scheme is inherently limited as displays are developed with more and more rows. Liquid crystal cells with low P values (high nonlinearity) are better suited for highly multiplexed displays because they are better able to produce acceptable contrast with only a small voltage change, but even these have increasingly worse performance as N increases.

To achieve a video rate with this addressing scheme, LCDs with 525 lines operating at 30 frames per second must produce adequate contrast with a V_{on} to V_{off} ratio of 1.045. A high definition television, with a line count of 1,100, must produce its contrast with a V_{on} to V_{off} ratio of 1.031 and a line scan time of $1/(1,100 \times 30)$, or 30 microseconds. (Note that the frame update frequency is a function of the IC driver speed and is a separate issue from the liquid crystal response time. The slower of these two factors usually determines overall display speed). Even with advanced liquid crystal materials, however, TN-LCDs have multiplexing capabilities limited to approximately 100 lines.

There are some ways to improve upon the above equation's limitations through display design. For instance, the number of pixels in a multiplexed display can be doubled without increasing the liquid crystal material's nonlinearity. By using a split-screen bus driving scheme, the display is essentially driven as two separate displays, arranged with one half on the top and one half on the bottom. The disadvantage of this approach is that a defect in manufacturing will be much more damaging to final display performance. Also, this approach requires twice the number of column drivers as the regular approach does. A break in a column line will blank out all of the pixels between the break and the end of the column line, while in a conventional multiplexed display (provided that both ends of the columns are attached to the drivers), a break will not affect the display unless two breaks occurred in the same bus line.

Passive Matrix LCD Operation

Supertwisted-nematic (STN) LCDs use available liquid crystal materials, fabrication processes, and drive electronics to achieve wider viewing angles and increased contrast ratios than TN-LCDs, but they require more precise fabrication. Particularly when combined with film compensation, STN-LCDs can produce superior contrast with a neutral color scheme. The drawbacks of relatively slow response time and low color quality have limited the use of STN-LCDs in video applications, although research is continuing to improve this technology.

An STN-LCD cell is drawn schematically in Figure 5. *Supertwist* refers to the fact that the aligning layers are rotated more than 90° (typically between 180 and 270°) relative to each other. The increased twist angle is facilitated by a combination of a high pre-tilt, as much as 35°, and doping of the liquid crystal mixture with a chiral additive of the correct pitch and handedness such that the larger rotation is possible. Care must be taken in the fabrication; if the pre-tilt is insufficient, or if the wrong chiral additive is used, the molecules will rotate in the opposite direction and the display will not function. Increasing the twist angle has the effect of steepening the voltage-contrast response curve shown in Figure 2.

STN-LCDs operate much like TN-LCDs, modulating light through liquid crystal anisotropy such that the effects of the modulation can be observed with two polarizers. Increasing the voltage across the cell rotates the liquid crystal molecules so they are more perpendicular to the plates. By changing the twist angle or the anisotropy of the liquid crystal, and/or by placing the front and rear polarizer at angles other than 90° to each other, different *on* and *off* colors can be attained. By placing the analyzer parallel to the Y-axis in the figure, each pixel will have a faint bluish appearance in the select state and a brighter greenish appearance in the nonselect state. Other common color modes for STN-LCDs include yellow/green, purple/green, and yellow-green/purple-pink. STN-LCDs operating in these modes have attained contrast ratios of 50:1 multiplexed at 100 lines, a result of the highly nonlinear response curve.

The type of STN-LCD described so far suffers from poor viewing angle dependence. An observer viewing the display at oblique angles sees both a different Δn (optical path difference) and a different d (cell spacing), as with a TN-LCD, but the effect is more dramatic because of the greater nonlinearity. Minimizing oblique-angle distortions requires manufacturing a very uniform gap that is slightly narrower than for a TN-LCD, approximately 4 to 5 microns. Such narrow cell gaps can be difficult to manufacture reliably. Other ideal technical specifications for STN-LCDs include the following:

- Small optical path difference and narrow cell spacing
- Large *bend/splay* ratio (terms related to the material)
- Small twist/splay ratio
- Small $D\varepsilon/\varepsilon_{perpendicular}$
- Large pre-tilt angle (20–35 degrees)
- Uniform cell thickness of 4 to 5 microns

Early STN-LCDs were superior to TN-LCDs in terms of multiplexability and contrast, but they were inferior in that they were not truly black and white. In addition to being aesthetically unpleasant, STN-LCDs increased the likelihood of eyestrain. All the current designs for large-area STN-LCDs employ some form of optical compensation in order to obtain color neutralization. The first acceptable black-and-white displays used an optical compensator, which consisted of an identical liquid crystal cell, except with a twist in the opposite direction of the active cell. However, the liquid crystal compensator cells increased the expense and weight of the display and increased the cell thickness and susceptibility to damage. As a result, it has been replaced by the film compensation type.

Film-Compensated STN (FSTN) LCD. Optical film compensation uses anisotropic organic retardation films (RF) or one or more polymer compensation layers laminated to the display. See Fig. 6. The polymer retardation film is typically a monoaxially stretched polycarbonate or poly(vinyl alcohol) film, approximately 80 microns thick. The film imitates the function of the liquid crystal compensator cell, but without the cost, bulk, weight, and fragility. One difficulty with film compensation is matching the thermal response and oblique viewing characteristics to the active cell. The compensating film needs to be formulated carefully to compensate for the active STN cell at as many temperatures and viewing angles as possible.

With film-compensated STN-LCDs, issues of alignment of optical axes between the polarizers and film become very complex. Optimizing the relation of a film's optical axis relative to the crossed polarizers is the subject of continuing research. The film is usually sandwiched between the front polarizer and the liquid crystal cell; alternate layerings are possible and will change display performance. These placement issues are even more unwieldy in the double-layer, film-compensated STN.

Color STN-LCDs. The sophisticated compensation techniques for STN-LCDs have resulted in displays with excellent black-and-white quality. This has allowed for the development of color STN-LCDs through the addition of color filters. Pixels are made to switch between a dark *off* state and a colored *on* state, and if the pixels are small enough, the eye perceives full color (through the mixing of primary colors). Light transmission is around 20%, and the contrast ratio is 20:1 or 30:1. Most

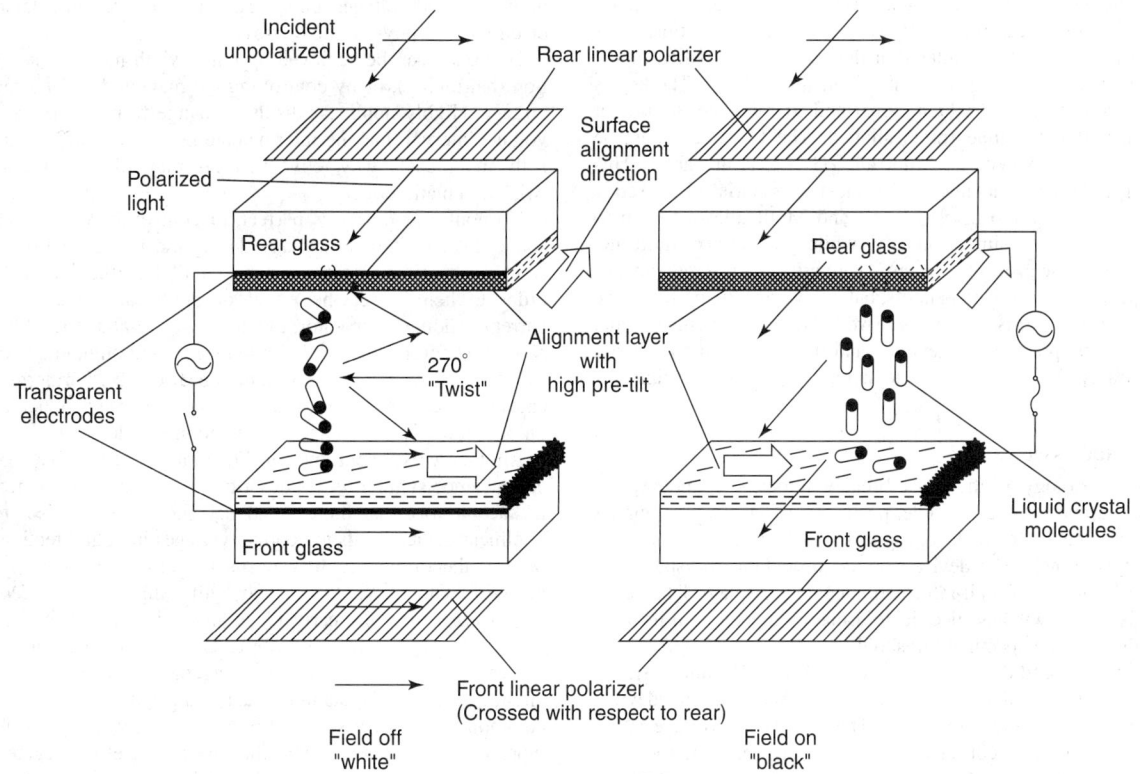

Fig. 5.   Supertwisted-nematic field effect LCD operation.

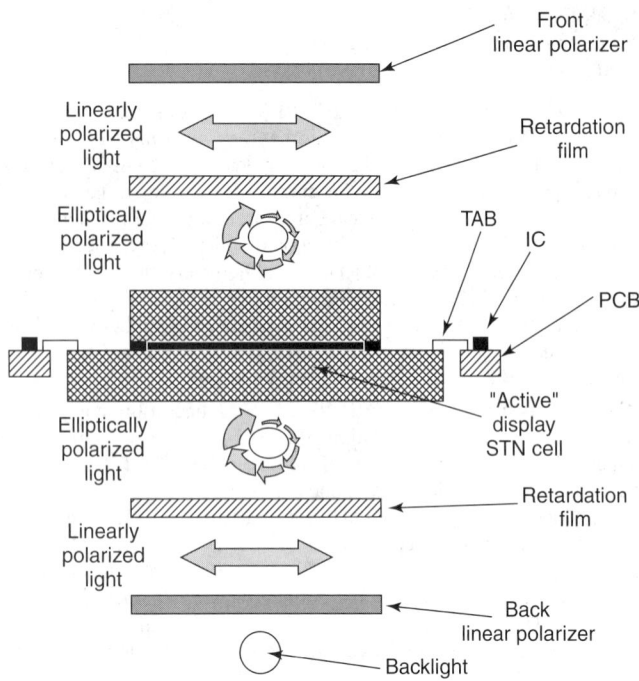

Fig. 6.   Operation and structure of a film-compensated STN-LCD.

current STN-LCDs use double-layer film compensation and a color-filter system.

Enhanced Driving Schemes for STN-LCDs. Another limitation of STN-LCDs is the tradeoff between contrast and switching speed: increasing one decreases the other when using the standard multiplexing scheme outlined above. Over the past few years, several companies have developed enhanced driving schemes for STN-LCDs that promise high-contrast STN-LCDs capable of displaying video images. The earliest prototypes of these displays demonstrated full-motion video with excellent contrast. The latest focus is to increase color saturation and viewing angle. The idea is to use the simple matrix structure of an STN-LCD, optimize it for video speed, and adjust the performance through sophisticated driving.

Instead of sending a high-voltage pulse down the rows as in standard multiplexing, a set of predetermined signals is applied to multiple rows simultaneously and a unique calculated analog voltage is applied to the columns, depending on the state of the pixels in that column. The key is to determine which rows need which signal and to calculate the proper column voltage within the time frame allotted. The LCD cell must be capable of switching at video rates (not all STN-LCDs are able). This means decreasing the cell thickness and using fast material. The Active Addressing approach developed by InFocus and Motif allows for up to 255 lines to be addressed simultaneously, while the Optrex Multi-line Selection (MLS) scheme uses 5 to 7 lines. There are certain screen images for which this technique offers no benefits, but on average the performance over standard multiplexing is noticeable. Two basic requirements are that the signals should be practical to generate and that the technique should produce no artifacts. Also, the entire calculation cycle must fit within the frame cycle.

Active Matrix Addressed LCDs

Active matrix technology is the most broadly pursued technology for increasing the number of addressable pixels in LCDs. Active matrix addressing leaves the basic mode of optical transformation unchanged, but uses some type of nonlinear device external to the liquid crystal cell to enhance the performance of high-information-content displays. The active matrix elements act as switches, electrically isolating a pixel until its input voltage exceeds a certain specified threshold.

Active matrix addressed devices do not respond to RMS time-averaged voltage, but rather to instantaneous peak voltage. Active matrix devices are driven at duty cycles close to 100%. This means the voltage at the pixel remains more or less constant during each frame, as if the pixel were directly driven. During nonaddress times, the nonlinear element has very high impedance, which electrically isolates the pixel from the drive circuitry. The pixel itself acts as a capacitor, storing charge and maintaining voltage throughout the nonaddress period. An additional storage capacitor can be added at the pixel site, in parallel to the pixel, to help maintain charge during nonselect periods.

Active matrix devices consist of two-terminal or three-terminal gain-producing devices. The currently researched two-terminal devices are primarily thin-film diode (TFD) networks (a single diode is unsuitable because of its rectification behavior) and metal-insulator-metal (MIM) devices. Three-terminal devices are almost exclusively thinfilm transistors (TFTs).

Active matrix devices perform much better than almost all passive technologies, but at the cost of increased manufacturing complexity and lower production yields.

Three-terminal devices are generally higher quality than two-terminal types, but they are usually more expensive to manufacture. In a price/performance comparison between passive matrix, two-terminal, and three-terminal types, there is no clear-cut winner. Lower-quality, text-only applications usually can use passive matrix STN-LCDs. High-quality, full-color video applications generally require AMLCDs.

Two-Terminal, Nonlinear Devices. The terms for the two primary types of two-terminal nonlinear devices used in AMLCDs diodes and MIM devices are sometimes used interchangeably, although they are different devices. In this article, a *diode* is a device that passes current in one direction only (rectification), and a *MIM* is a device that passes current in both directions. Diodes are usually made of amorphous silicon (a-Si), while MIMs are usually made of silicon nitride (SiN_x) or tantalum pentoxide (Ta_2O_5).

Diodes do not appreciably conduct a current until the voltage drop across the two terminals exceeds a certain threshold, $V_{th(NLD)}$, which ranges from 0.2 to 15 volts in practical applications (the subscript, NLD [nonlinear device] is used to prevent confusion with the liquid crystal optical switching threshold, which is called $V_{th(LC)}$). Once the threshold is met, only the voltage above the threshold is passed. For example, if $V_{th(NLD)}$ is 5 volts and the source voltage is 7.2 volts, then V_{lc}, the voltage across the liquid crystal cell, will only be 2.2 volts. Furthermore, if the source voltage dips below 5 volts, no current can flow through the nonlinear device, thereby electrically isolating the pixel from the source. The nonlinear device acts as an open switch such that the pixel retains its previous charge. Higher nonlinear device threshold voltages can be achieved by wiring several devices in series; three, 4.2-volt V_{th} nonlinear devices in series will have a cumulative threshold of 12.6 volts. Two or more nonlinear devices in series are sometimes preferable, because high-threshold voltages are spread over more than one device, minimizing electrical and physical field stress.

Grayscale can be achieved by pulse-width modulation, spatial modulation (halftoning), or by controlling the magnitude of the row (data) select voltage. Highly nonlinear displays change from off to on in a very short voltage interval, making intermediate gray levels difficult to achieve. This is the main reason why active matrix displays almost always use the simple twisted-nematic effect.

No nonlinear device is perfect, of course; the diode will always allow some leakage of pixel voltage during the nonselect time. An additional storage capacitor is often placed parallel to the liquid crystal pixel in order to help offset the effect of nonlinear device *off* state leakage current. Adding a capacitor, though, has several undesirable side effects. A capacitor requires extra fabrication steps, and pinhole defects between the capacitor plates can short out a nonlinear device. Furthermore, pixels with capacitors require higher *on* state charging currents, placing a greater load on both the driver circuitry and the nonlinear device *on* state conductance.

Another very important consideration for the active matrix display is conductance symmetry. As described previously, LCDs need to be driven in such a way that there is no net D.C. voltage. Therefore, either the nonlinear device needs to conduct voltages in both directions equally well, or else the driver circuitry needs to compensate for any asymmetrical behavior. Otherwise, the display will exhibit *image sticking* (the same as *ghosting* from the early black-and- white television era), flicker, and rapid aging from electrolytic effects. Much of the current two-terminal device research is aimed at refining the nonlinear device and/or driver circuitry such that it has more balanced response characteristics. The most common means of assuring driving symmetry is to wire together two or more nonlinear devices such that the asymmetries negate each other. In one of the few examples of TFT-LCDs being simpler than two-terminal devices, this symmetry balance is easier to achieve in a TFT-LCD.

Diodes. Two types of thin-film diodes (TFDs) are used in LCDs: Schottky and PIN (positive-doped/insulator/negative-doped). All current TFD networks use a-Si. Much of what is known about a-Si diodes is the result of research into thin-film photovoltaic solar cells, which were first heavily investigated in the 1970s. Although the physics and fabrication techniques were studied and understood during the brief push to develop lowcost solar cells to replace fossil fuel plants, interest dwindled after 1981. It was around this time when research into flat panel displays with thin-film diodes began.

Roughly speaking, diodes act by passing voltage signals of one sign and blocking those of the opposite sign. Because a single diode passes voltage only of one sign, it will convert an A.C. signal into a series of pulses of only one sign, a process called *rectification*. In practice, however, diodes do not abruptly turn on with an applied forward bias or turn off under reverse bias. Rather, they exhibit well-defined transition regions, which do not have sharp corners and are not located precisely at zero volts. It is this threshold behavior, not rectification, that is applied in thin-film devices for LCDs. In addition to forward rectification, many diodes also exhibit a reverse breakdown voltage under higher reverse biases.

A single diode is unsuitable for use in a pixel element because rectification would quickly lead to a net D.C. buildup, damaging the cell. Therefore, diodes are always found in geometries of two or more, such that they cancel out each other's asymmetrical response. The diode networks for each pixel are generally confined to a small physical space to minimize the effects of gradual manufacturing variation over the entire display surface. Diodes that are located far from each other can have different performance characteristics due to variations in doping levels or deposition thickness. State-of-the-art diode networks not only offer superior voltage symmetry but also optimize threshold voltage, *off* pixel current leakage, and *on* state resistance.

A major advantage of using a-Si is that thin-film materials can be deposited at a maximum substrate temperature of less than $300°C$. The economic implications of low-temperature fabrication are substantial. Instead of using polished fused quartz substrates or high-temperature glass, common soda-lime glass or relatively inexpensive alkali-free glass, such as electronics-grade Corning 7059, can be used.

Schottky diodes are formed as a simple junction between an intrinsic (pure) semiconductor and a metal. The capacitance of the diodes must be made much smaller than the capacitance of the liquid crystal cell in order to maximize both the number of addressable lines and the contrast ratio; the leakage of current through the capacitor formed by the diodes creates the limitation on addressability. Schottky diodes also have a reverse breakdown voltage, which means at large reverse voltages (approximately 15 volts) they show a threshold behavior and quickly become conductors. This reverse breakdown voltage is the principle of operation in the back-to-back diode configuration discussed below.

The PIN diodes consist of layers of conductive positive-doped semiconductor, an insulating layer, and a layer of conductive negative-doped semiconductor. The PIN diode is the basis of the a-Si solar cell. However, in a solar cell, the object is to maximize the photon-generated current, while in a display, the photocurrent must be reduced and the reverse current must remain very low. PIN diodes are formed by plasma-enhanced chemical vapor deposition (PECVD), usually with a Cr electrode above and below. The insulating layer is typically 5,000 angstroms thick, while the P and N layers are 500 angstroms thick. The on/off ratio in such a device can exceed 100:1.

As mentioned previously, a single diode is unsuitable for display applications because of its strongly asymmetrical response. Therefore, diodes must be arranged in a balancing network. The simplest networks consist of two diodes. There are a number of diode geometry variations:

- Back-to-back (BTB) diode structures
- Ring diodes
- Stacked-ring diodes
- Stacked-ring diode equivalent circuits
- Two-diode switches
- Double diode plus reset (D2R)

While some of these have faded into obscurity after some initial effort, Phillips Research Laboratories developed a D^2R circuit that eliminates the extra row electrodes while maintaining the benefits of the two-diode addressing scheme. The number of external connections is the number of rows plus the number of columns plus one. Current D^2R displays show

no flicker or crosstalk and a contrast ratio higher than 100:1. The essential feature is that the precision of the two-diode switch is accomplished with a similar level of fabrication complexity (four photomask steps), but without the extra bus lines. However, this bus line reduction comes at the price of a more complicated addressing scheme, and drive voltages must be precisely regulated to avoid net DC current. This technology has been used in production by Philips Flat Display Systems. See also **Organic Light-Emitting Diodes (OLEDs)**; and **Inorganic Light-Emitting Diodes Displays**.

Metal-Insulator-Metal (MIM) Devices. The MIM device as applied to LCDs has a longer history of development than other types of two-terminal devices and also has a simple fabrication process. The MIM device was first investigated as a method of providing threshold control for displays in the late 1970s at Bell Northern Research in Ottawa, Canada. Most this technology is used in production of 2- and 3-inch pocket television displays.

A MIM-addressed display relies on the extremely nonlinear current-voltage characteristics of a very thin layer of semiconducting material placed between two metal plates. Like a diode, the MIM allows current to flow only when a threshold voltage is exceeded. Conduction in this material occurs by field-assisted ionization of localized impurity states in the band gap. It does not conduct at low voltages because the impurity states are filled, neutral, and far from the conduction band. The application of a high field distorts the potential well around the impurity and reduces the barrier to electron flow. The structure, doping, and base material chosen influence the conduction properties significantly. Presently, the two most common MIM materials are anodized tantalum pentoxide (Ta_2O_5) and nonstoichiometric silicon nitride (SiN_x).

One of the most pressing issues in MIM device development is the reduction of parasitic capacitance (unwanted capacitance that occurs as a by product of device geometry). Because MIMs are constructed by sandwiching a nonconductor between two metal electrodes, they produce a very efficient capacitor. This is exacerbated if the insulator has a high dielectric constant. Large MIM capacitance slows response time, because charge that is intended for the liquid crystal pixel capacitor is shared by the MIM capacitor, reducing the amount of charge that finally reaches the liquid crystal pixel.

Tantalum pentoxide (Ta_2O_5) is an especially good material for MIM devices because it exhibits a self-healing process during fabrication and when heated. Due to surface diffusion, pinholes in thin films of tantalum pentoxide are found to diminish after annealing. This property is expected to ease the fabrication of high-quality products at high-volume levels and is probably what allowed MIM displays to be first demonstrated years before any other type of thin-film device addressed display was possible.

There are two types of tantalum pentoxide MIM devices currently under investigation, identified by their structure. One is called a *lateral MIM*, because the active area is grown on the side of a thin-film layer of tantalum. The other type is called *a cross-patterned MIM*; the semiconductor is grown on the surface of a thin film of tantalum. A cross-patterned MIM display is shown in Figure 7.

Although the cross-patterned MIM is very simple in construction, parasitic capacitance at the intersection of the Cr and Ta conductors through the tantalum pentoxide layer is a serious problem. As noted previously, the larger the capacitance, the more difficult it is to charge the pixel capacitor and, thus, to address each pixel. There are only two straightforward ways to reduce capacitance: increase the thickness of the dielectric layer or decrease the area of the electrodes. The thickness, however, is fairly well defined by the desired voltage characteristics needed to address the display, so the only remaining choice is to reduce the area of the electrodes. This means reducing the line widths of either the Cr or the Ta spur. Both choices increase the likelihood of problems with open circuits and reproducibility of very small relative geometries.

The lateral MIM uses a modified fabrication technique to reduce the area of the capacitor/junction electrodes, resulting in a much lower device capacitance. This is accomplished by anodizing the tapered edge of a thin film of Ta, allowing submicron geometries to be achieved without actually using submicron photolithography. The resulting lateral width of the insulator is less than a micron, so the capacitance of the crossed conductors is small, yet no photolithography below 30-micron accuracy is required (TFTs require 5-micron linewidths).

The lack of any photolithographic patterning below 30 microns is the most significant advantage of the lateral MIM. The only process that needs to be consistently controlled for device performance is the anodization of the Ta electrode to form the insulator. This is a significant

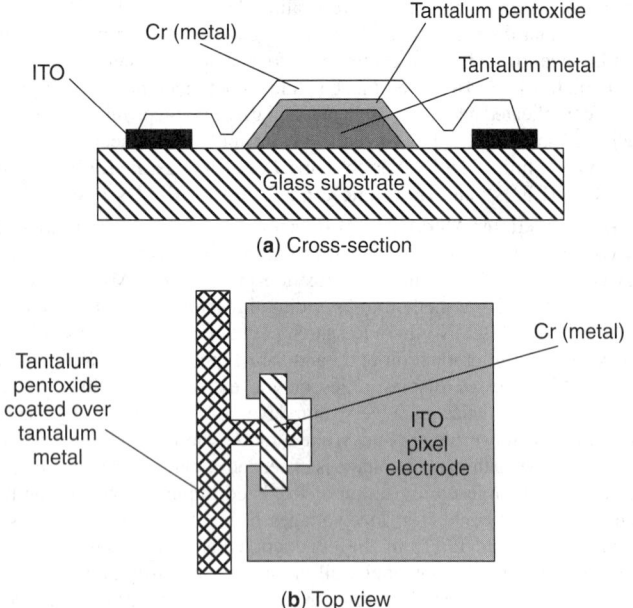

(a) Cross-section

(b) Top view

Fig. 7.   Metal-insulator-metal (MIM) device (**a**) cross section, (**b**) top view.

advantage over TFTs and even over diodes that require annealing and hydrogenation. The very high nonlinearity of the I-V characteristics and the low leakage through the insulator allow for highly multiplexed displays with performance similar to directly driven displays. The driving voltages for MIMs are necessarily high (15 to 20 volts), due to the conduction mechanism and properties of the insulator film. However, this level is still within the range of highly integrated CMOS drivers.

The University of Stuttgart, Germany, has developed a new geometry for tantalum pentoxide MIMs. This geometry has the low parasitic capacitance of lateral MIMs, can be made with only two photolithography steps, and has independent adjustment of both the gain (nonlinearity) and threshold of the device. The key to this efficient procedure is to use the ITO layer that is already present on the LCD as one of the MIM electrodes. In a later refinement, a second dielectric layer (SiN$_x$) on top of the Ta$_2$O$_5$ was used to further reduce the parasitic capacitance of the MIM without affecting the gain or threshold. The same two-step process was used, except a layer of SiN$_x$ was deposited by RF plasma chemical vapor deposition (CVD) by mixing silane gas (SiH$_4$) and ammonia (NH$_3$). The SiN$_x$ and Ta$_2$O$_5$ were etched simultaneously. This was used to make a 4.8-inch display with 96×128 pixels, a maximum contrast ratio of 35:1, and a response time of 20 milliseconds, which is fast enough for television rates. The display had eight gray levels, with excellent uniformity.

An entirely new MIM material is being developed at the Tokyo University of Agriculture and Technology, Tokyo, Japan. It uses a 20-nanometer-thick layer of Langmuir-Blodgett (LB) deposited polyimide film for the insulator. LB deposition is a liquid process that is far less expensive than vacuum deposition and can more easily cover large areas. One of the advantages of LB polyimide is its small dielectric constant (3.2), meaning the MIM parasitic capacitance is kept to a minimum. The LB polyimide MIMs are less nonlinear than other types of MIMs, but early samples had acceptable nonlinearity. A sample pixel has been made with a contrast ratio of 20:1 when driven with a frame frequency of 30 Hz and a duty ratio of 1/100.

Thin-film transistors (TFTs) devices are the most popular and widely investigated three-terminal devices for display applications; more than 30 companies around the world have made significant investments in research and production of this technology. Active matrix LCDs using TFTs provide a combination of performance, form factor, and affordability unmatched by any other display technology, and they dominate the flat-panel display market.

A schematic of a TFT is shown in Figure 8; it operates as an electronically controlled switch. By modulating the gate (G) voltage (with respect to ground), the maximum current flow from the source (S) to the drain (D) can be controlled. Larger gate voltage leads to larger maximum source-drain currents. The TFT is a very good conductor between source and drain if the gate voltage is large, and it is a very poor conductor if the

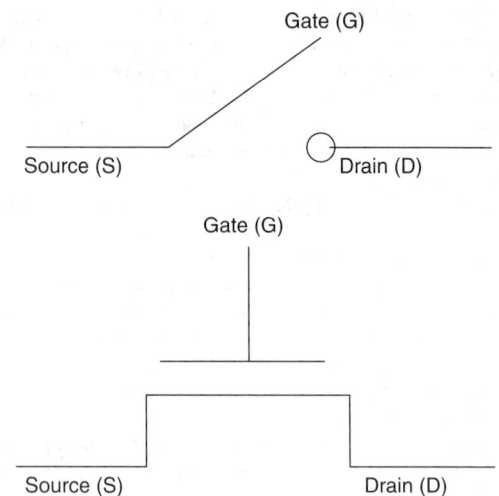

Fig. 8.   Schematic of a Thin-Film Transistor. (*Operation as a Switch*.)

gate voltage is small. No matter how large the drain voltage, no current will flow if the gate voltage is near zero. This behavior contrasts with diodes, which have a set threshold voltage.

A schematic of standard TFT layout and addressing is shown in Figure 9. Typically, the rows are connected to the gates, the columns to the source, and the ITO pixel electrode to the drain. However, assigning the source to the columns and the drain to the pixel is somewhat arbitrary; in the case of a TFT, there is symmetry between the source and the drain so that they can be swapped without any change in electrical behavior. (This symmetry does not apply to regular junction and field-effect transistors).

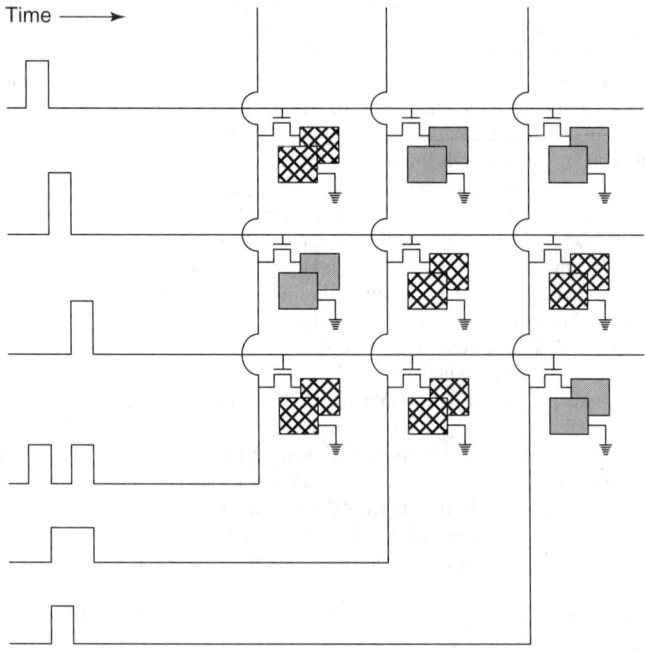

Fig. 9.   Standard TFT layout and addressing.

The upper pixel electrodes are all connected to ground. Thus, in contrast to most two-terminal displays, the bottom electrode contains both row and column lines, necessitating crossovers. The liquid crystal cell is usually the standard TN type. A row address line is selected by raising it above the gate saturation voltage; this electrically connects the source and the drain. A pixel can now be written to by raising or lowering the source voltage beyond the switching threshold. To maintain voltage symmetry, the column lines are alternately raised and lowered with successive frame cycles. Because the TFT conducts symmetrically in both directions, this addressing scheme ensures that there is no net-D.C. current. One feature of TFT addressing is that only the column (source) voltages need to be inverted. The row (gate) voltages can have a D.C. bias because they are only control voltages and are not directly attached to the pixel.

Contrast is retained between frame cycles because a nonselected pixel is electrically isolated from the driving circuitry and only loses its charge from small leakage currents. Like two-terminal displays, a small storage capacitor is sometimes added at the pixel site, parallel to the pixel, to help retain charge. Thus, the voltage at the pixel is essentially the same as if it were directly driven.

Operation of TFTs. TFT behavior closely approximates the performance of metal-oxide semiconductor field-effect transistors (MOSFETs), the main difference being the source-drain symmetry noted previously.

However, there is a limitation to the analogy with bulk silicon FET devices, due to the thin-film nature of TFTs. Bulk silicon devices use high-purity, single-crystal materials that exhibit very high mobilities. Thin films, on the other hand, are highly imperfect materials, usually having the same purity level, but lower degrees of crystallinity, especially in the case of a-Si (see "TFT Materials"). Furthermore, the mere thinness of the film acts as a defect, strongly influencing its properties. This limitation should be kept in mind for the remainder of this section.

A typical TFT cross section is shown in Fig. 10, next to a conventional metal oxide semiconductor (MOS) transistor for comparison. Conductive doped-semiconductor contacts are used to ensure ohmic (linear) voltage-current flow. If the metal electrodes were placed directly against the undoped semiconductor, the current flow would be similar to the current flow in a Schottky device, due to the poor matching of the electron energy levels between the very different materials.

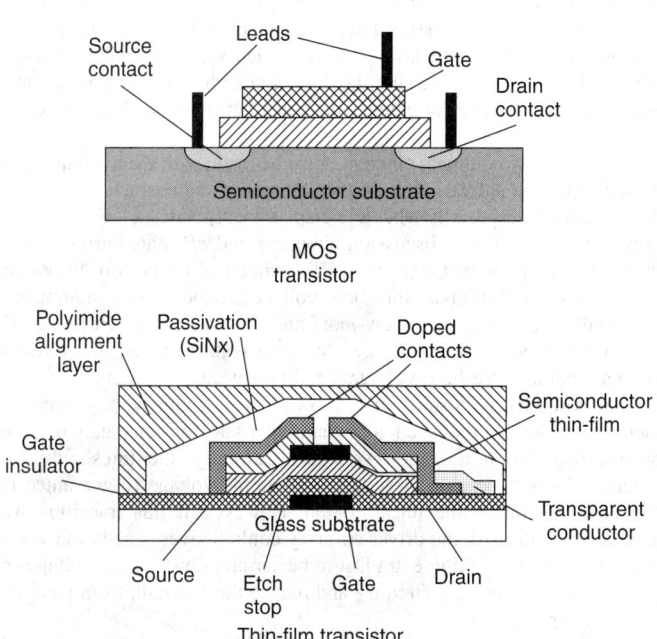

Fig. 10. Transistor cross sections.

The TFT can be thought of as a simple capacitor with the added feature of being controllable. When the gate-source voltage is zero, there is no field between the gate and the channel area. The purified semiconductor has so few free electrons that all current is effectively blocked. The semiconductor electrons are all firmly held in place. The capacitance is effectively infinite.

If, however, the gate voltage is increased such that an upward field is established between the gate and the channel, extra electrons will move from the gate to the channel, facilitating conduction. It is as if the capacitance has decreased to a finite value. The drain current is proportional to the source voltage (that is, ohmic) until the device reaches the saturation voltage and current, at which point all available conduction electrons are being used. Source voltages above saturation will not appreciably increase drain current. The greater the gate voltage, however, the more free electrons will be available and the higher the saturation current will be. The gate controls the supply of current-enabling electrons.

A small reverse-biasing of the gate pulls electrons to the gate plate, which even further reduces conduction from the low value at zero voltage. Any electrons from the semiconductor lattice with enough thermal energy

to break free will be sucked up by the gate before they reach the drain. Further reverse-biasing of the gate voltage, though, increases conductivity by creating holes. In other words, conduction switches from majority carriers (electrons) to minority carriers (holes). See also **Transistor (Invention and Development)**.

TFT Size Limitations. In theory, TFT-LCDs can be built as large as desired; in practice, though, there are limitations that become progressively more severe as the display size increases.

Finite On State Conductance. One of the most important performance criteria of TFT-LCDs is the maximum on current. This current is determined by the source-drain resistance, which must be low enough (that is, the on conductance high enough) that the pixel capacitor can be charged during the row-select time. A higher on current assures this, with the caveat that the gate voltage must also be high (15 to 20 volts) so that the TFT will remain in the linear region and will not be limited by a saturation current. Larger displays have shorter select times for each row of pixels, meaning more current must flow in a shorter amount of time.

The larger the TFT, the larger the on current. However, this simple solution is not a good option because it will greatly increase the parasitic capacitance (see below).

There are four main strategies for ensuring that on state resistance and current requirements have been met. In practice, parasitic effects discussed later will further degrade performance. In large displays these parasitic effects can become highly significant. The methods for increasing source-drain current flow are as follows:

1. *Increase source driving voltage.* A given driver chip has an upper limit to the amount of voltage and current it can practically provide. However, this solution has a limited ability to improve the on current. Even though TFTs have been made to withstand up to 500 volts, the typical maximum high-speed addressing voltage that standard CMOS chips handle is approximately 40 volts.
2. *Increase semiconductor mobility (m).* This second option is also of limited use. Mobility is primarily a function of the semiconductor used in the channel and can only be increased by employing a different semiconductor or by altering its phase. In general, higher mobility semiconductors have higher *off* leakage currents (which is undesirable), offsetting the benefit of using higher-mobility semiconductors. High-mobility semiconductors also usually require higher substrate deposition temperatures.
3. *Increase channel capacitance per unit area (C).* The only ways to increase the capacitance per unit area are to decrease the thickness of the insulating layer or to use a higher dielectric constant insulator. Channels that are too thin, in addition to being difficult to manufacture, run the risk of pinhole defects that can short out the TFT. A high dielectric strength insulator increases parasitic capacitance, which has its own undesirable effects. Again, this is a limited solution.
4. *Increase aspect ratio (width/length ratio).* This is the easiest and most straightforward way to improve the *on* current. A wider channel means there is a larger cross section in which current can flow, and a shorter channel means current has to flow a shorter distance. TFTs can be made large enough to pass sufficient current and still be small compared to the pixel size.

Table 2 illustrates width/length (W/L) ratios for various mobilities that result in proper *on* current.

TABLE 2. WIDTH/LENGTH RATIOS FOR MOBILITIES RESULTING IN PROPER ON CURRENT

Mobility (cm^2/volt-sec)	W/L Ratio
0.2	11
20	0.11
30	0.007

As the table shows, low mobilities require very wide channels relative to the length. The capabilities of the photolithographic processes and the expected yields limit the minimum feature size that can be practically fabricated. Device geometries smaller than a few microns are not economical for any type of large-area devices, including flat panel displays. In designing TFTs, the lower limit for L is typically fixed at 5 microns for practicality, and then the width is adjusted to achieve the correct W/L ratio. A TFT made with a material with a mobility of 0.2 (amorphous Si:H) and with a channel length of 5 microns would have to be 55 microns wide in order to provide the minimum current to drive the cell described previously. The length of the channel runs alongside the pixel. For exceptionally long channels, a serpentine pattern is sometimes used.

TFTs made with high-mobility materials, which could operate with W/L ratios of 0.01 or less, are still subject to the minimum device geometries imposed by fabrication and yield requirements. Using the guideline of a 5-micron-minimum feature size, the channel dimensions for a high-mobility device would be minimized at a 5×5 micron square, even though a much shorter channel width would be acceptable with this length. Such a design has problems, however, with *off* state leakage current.

Off State Leakage Current. The property that gives a TFT a high *on* state conductance also gives it a low *off* state resistance, which is undesirable because it allows the pixel charge to leak off the pixel before the next refresh period. There are two sources of leakage: the TFT and the liquid crystal material itself. The visual result of such leakage is reduced contrast.

A general rule is that the charge should not decrease by more than 10% during the interval between frame updates. The most common solution is to place a capacitor in parallel with the liquid crystal cell in order to help maintain the voltage while the display is not being addressed. However, storage capacitors have several drawbacks. First, the capacitor decreases the aperture ratio of the display by including additional inactive area. Second, the addition of another component makes fabrication more complex and increases the potential for yield losses. Large capacitors are also difficult to charge and discharge because of the large currents involved. This places an upper limit on the storage capacitor size.

Integrated drivers fabricated directly on the display substrate can help with leakage current by shortening the link from pixel to driver. High-mobility semiconductor materials are attractive because they can be used for integrated driving circuits mounted directly on the display substrate. However, geometry considerations complicate the matter. High-mobility materials tend to result in channel dimensions that are hundreds of times larger than required to drive the cell, aiding current flow. Low-mobility materials, however, require such large W/L ratios in the driving circuitry that the device dimensions can become too large to be practical. Significant advances in surface-mounted drivers, though, are offsetting some of the need for integrated drivers.

Finally, the liquid crystal material itself contributes to the leakage currents. TFTs with very high *off* state resistance may still require storage capacitors if the liquid crystal material passes a significant current. State-of-the-art liquid crystal materials have resistivities from two to three orders of magnitude higher than the minimum TFT *off* resistance, however, meaning the liquid crystal leakage current can be neglected.

Gate-Drain Parasitic Capacitance. The source-gate, drain-gate, and source-drain links can contribute unwanted (parasitic) capacitance that can bypass the TFT during nonselect periods, affecting the nonselect state of the pixel. The parasitic effect that receives the most attention is the capacitance between the gate and the drain (the TFT electrode connected to the display pixel). It occurs because of the overlap between the gate and source electrodes. This gate-drain capacitor forms a capacitive voltage divider, with the pixel receiving a portion of voltage between the gate and the drain. If the parasitic voltage exceeds the threshold voltage of the pixel, then the row voltages going to the gate could erroneously put the TFT into the on condition without the source being activated. Furthermore, the gate-drain capacitance can create a net D.C. drive component, which leads to ghosting and more rapid aging of the liquid crystal cell.

Common methods of reducing the side effects of parasitic capacitance are as follows:

1. *Improved manufacturing process.* Self-alignment procedures, in which the gate electrode is used as a photomask for the source and drain electrodes, can be accomplished by using more precise steppers that minimize the overlap area.
2. *Lower drive voltage.* Because the drain voltage is proportional to the gate driving voltage, reducing the driving voltage also reduces the voltage shift caused by the unwanted voltage divider. The lower limit for the gate driving voltage is set by the minimum on state current requirement. Greater TFT uniformity allows a lower drive voltage without affecting performance.

3. *New addressing schemes.* Techniques in which nonselected rows adjacent to selected rows are given different voltages than the more remote nonselected rows have been shown to be effective in compensating for the voltage shift from the parasitic capacitance. If the nth row is being selected, then the $(n-1)_{th}$ row and the $(n+1)_{th}$ row receive a different voltage than the other nonselect rows. In many of these new addressing schemes, the top-substrate counterelectrode is connected to a variable or nonzero voltage.
4. *Multiple TFTs.* In addition to offering redundancy against defects and lower off state leakage currents, multiple TFTs per pixel can be configured to offset the drain voltage shift. Usually these multiple TFT schemes require a modified driving scheme, like the one described previously. Multiple TFT driving schemes often need additional row or column lines to address the extra TFT terminals, increasing the general complexity of the device.

Uniformity and Other Parasitic Effects. Other parasitic effects are also important, even though they do not receive as much attention as those mentioned previously. One example is poor contacts between the semiconductor channel and the metal electrode. Ideally, the contact should be ohmic, meaning that the current flowing through the interface is linearly proportional to the voltage across it. Furthermore, the interface resistance should be as low as possible.

However, in practice the interface resistance can be high, and it can obey a nonlinear current-voltage law, such as Schottky or Poole-Frenkel conduction. Nonideal contacts, generally caused by such manufacturing imperfections as low-purity materials or contamination between deposition layers, limit the maximum on state current. TFT architectures that use only two or three mask steps have a lower probability of contamination because several thin-film layers are deposited without breaking vacuum. Other solutions include varying the doping amounts of the semiconductor-electrode interface to ensure better matching between the electronic energy levels.

TFTs, like any nonlinear element, must be made with a maximum degree of uniformity. Small variations in the voltage-current characteristics reduce display quality, especially when the display incorporates a large number of gray levels. The above discussion about *on* and *off* state currents makes the tacit assumption that all of the TFTs were equally on or off. In practice, though, small variations in threshold voltage produce above-minimum off state leakage currents and below-maximum on state charge currents. TFTs are known to vary the most when the gate voltage is near the threshold voltage, making threshold-voltage stability especially important.

Increased TFT uniformity also allows lower driving voltages, because there is no longer the need to overcompensate for uncertain threshold or linear-saturation transition voltages. Currently, the threshold is not precisely known, so gate drivers deliver ample voltage to guarantee that the TFT remains in the linear region. However, if this transition were accurately controlled, the driver circuitry could deliver exactly the voltage needed, without requiring extra just to be certain. Lower drive voltages put less strain on the driving circuitry and reduce the crosstalk from gate-drain parasitic capacitance.

The only way to improve device uniformity is through more careful fabrication, although the effects of nonuniformity can be decreased by alternate drive schemes (which will be discussed later). For video rates, uniformity in threshold voltage should be kept below 0.025 volt for small distances and below 0.05 volt for the entire display. Greater nonuniformity leads to visible flicker (at half the drive frequency) because each pixel charges and discharges slightly differently during the positive and negative bias write cycles. There is one positive side to nonuniformity, however: Unless the drive frequency is above 64 Hz, which is rare for current LCDs, the flicker component from nonuniformity will be below the visible threshold of 32 Hz.

Other Performance Issues

Grayscale. For low levels of grayscale (3 bits), the D.C. column data voltage can be adjusted by a digital-to-analog (D/A) converter, with the data voltage height proportional to the gray level. Therefore, the grayscale of the pixel is determined by the D.C. voltage at the pixel source during the address time. But, integrating a 4-bit (16-level) D/A converter into the drive circuitry for each column greatly increases driver chip cost, complexity, and size. One technique for obtaining a 16-level (4-bit) grayscale without using costly D/A converters is to use a ramp generator. When used in a

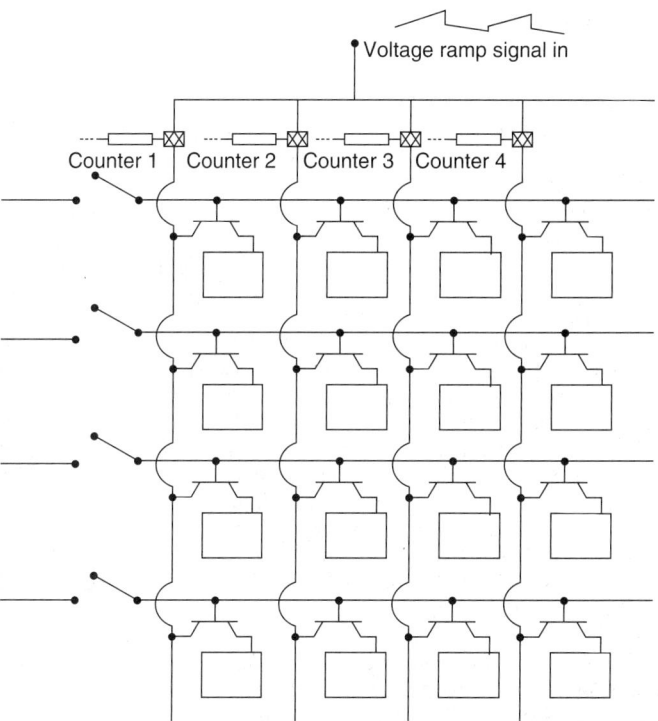

Fig. 11.   Grayscale control using ramp signal voltage.

color display, this provides up to 16 shades per color, or 4,096 different colors. The arrangement is shown in Figure 11.

Each column is connected to a common ramp waveform, which has a period that is equal to the length of time to address one line. The voltage to which the pixel charges is determined by the length of time it is connected to the ramp waveform. Because the display has been set up so the pixel voltage very closely follows the ramp voltage, the ramp cutoff time determines the gray level of the display. The ramp polarity is reversed each frame to ensure that there is no net D.C. at the pixel site.

Viewing Angle. No matter how advanced active matrix addressing schemes become, the final display quality is still limited by the viewing-angle dependence of the twisted-nematic material used in almost all active matrix pixels. As the viewing angle becomes more oblique, contrast is reduced, making it difficult to distinguish the *on* state from the *off* state. This is particularly troubling with grayscale because all the gray levels wash together. As LCDs move into the desktop monitor market, there have been serious efforts to make them more than simply detached notebook computer displays. In particular, a number of improved methods have emerged to increase the viewing angle of LCDs.

One method for improving the viewing angle and reducing the reverse-image problem at severe angles (where light pixels appear dark, and vice versa) is the *multidomain process*. This involves dividing each pixel into two parts, with one part using a high pre-tilt of the surface layer of liquid crystal molecules and the other using a low pre-tilt. Other methods use two different rub directions, but still maintain a low pre-tilt. Still others use four domains; a development by LG Electronics uses four domains combined with ultraviolet-sensitive alignment film and lighting from four directions. A process has been developed to form protrusions on the alignment layer to align the liquid crystal molecules vertically, resulting in a claimed viewing angle of 160 degrees (horizontal and vertical). The company also claimed, in 1999, to have improved transmittance by 4.5% over previous MVA displays. All of these approaches work both by counteracting the viewing-angle problems at high angles and by equalizing the viewing-angle problems over many viewing angles so that the problem is less dramatic at oblique angles. All are effective in improving the viewing angle, but at the expense of increased manufacturing costs due to additional processing and masking steps.

To date, perhaps the most effective method used to improve the viewing angle is *inplane switching* (IPS). This mode uses a structure with both electrodes on the bottom substrate in an interdigitated configuration. The top substrate (color-filter plate) has no electrodes, in contrast to the

conventional LCD. The use of interdigitated electrodes in LCDs was first demonstrated in 1970.

The structure of a Super TFT-LCD is shown in Figure 12. Unlike a conventional TFT-LCD, the device does not use transparent ITO electrodes. Instead, the source electrode is used as the pixel electrode and is formed with the same metal layer as the drain electrodes. The counterelectrode, also a metal layer, is formed at the same time as the gate electrode. There are three main advantages to this arrangement. First, the distance between the pixel and the counterelectrode can be precisely fixed; it is not determined by the cell gap (spacing between the two substrates). Second, fewer steps are involved in the fabrication process. Finally, an auxiliary capacitor is automatically formed between the pixel electrode in the source-drain layer and the counter electrode in the gate layer, so an additional capacitor is not needed.

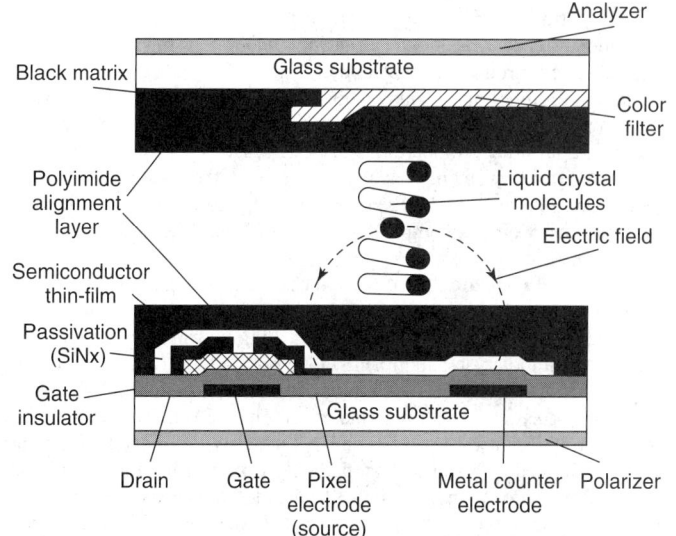

Fig. 12.   Structure of super TFT-LCD using in-plane switching.

The operation of a Super TFT-LCD is shown in Figure 13, revealing the reason for the name *in-plane switching*. The nematic liquid crystal molecules are aligned homogeneously between the substrates, with the optical axis of the molecules parallel to the polarization axis of the polarizer, but perpendicular to that of the analyzer. This is accomplished by rubbing the alignment layer on both substrates in the same direction, but at a 45-degree angle relative to the long axis of the pixel and counterelectrode stripes. In the *off* state, light entering through the polarizer passes through the cell and is blocked by the crossed analyzer, so the *off* state is black (this is the opposite of what occurs in a conventional TN-LCD).

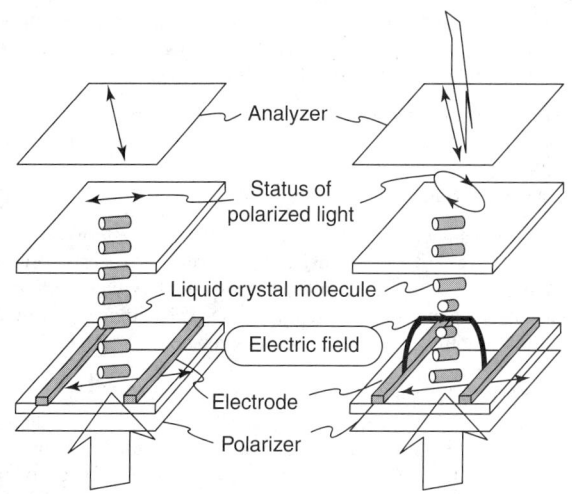

Fig. 13.   Operation of super TFT-LCD using in-plane switching.

When an electric field is applied between the pixel electrode and the counterelectrode, the field lines extend horizontally through the liquid crystal material, rather than vertically between the top and bottom electrodes. The liquid crystal molecules are rotated in-plane in the direction of the field to a maximum of 45° with respect to those firmly anchored at the surface. This sets up a twisted configuration of the molecules in the on state, so that the polarized light passing through the cell is rotated and can thus pass through the analyzer. Consequently, the *on* state in the Super TFT-LCD is white (again, this is the opposite of what occurs in a conventional TN-LCD).

The equation for light transmission through the IPS configuration differs from that for the standard TN-LCD. To obtain maximum light transmission and a high contrast ratio, the cell gap d should be adjusted such that

$$(\Delta n) \times d = (\lambda)/2$$

where l is the wavelength of the incident light and n is the optical anisotropy. Super TFT-LCDs provide a much wider viewing angle than conventional TFT-LCDs because of the greater transmission at oblique angles attained through the IPS mode.

There are three disadvantages of Super TFT-LCDs. They have a slower pixel response time, somewhat higher current consumption, and smaller aperture ratio compared to TN-LCDs. However, many design and performance improvements have been made in the past year. Hitachi's IPS display, for example, has been demonstrated to have a 25-millisecond response time, which is comparable to that of TN-LCDs and adequate for full-motion video.

Super TFT-LCDs are still quite expensive, and although they feature some impressive specifications, they have not fully penetrated the market.

Sharp takes a different approach to increasing viewing angle, combining a horizontal liquid crystal alignment with a perpendicular driving field. The *Advanced Super-V* mode offers improved performance through the use of retardation films made with new materials. The ASV LCD has the fastest response time of just 15 milliseconds (8 ms rising, 7 ms falling). Sharp has demonstrated an 18-inch-diagonal SXGA panel using this technology.

Several variations on the above techniques also exist. Hyundai has developed a variation of IPS called *fringe field switching* (FFS). The liquid crystal molecules are deliberately placed in the highly spatially varying fringe field of the electrodes. An early demonstration model of this type of display showed a transmission comparable to that of a TN-LCD, but with a far superior viewing angle. An additional type is Samsung Electronics' *patterned vertical alignment* (PVA). The development of advanced modes for wide viewing angles is continuing vigorously.

Amorphous Silicon (a-Si) Thin-Film Transistors

Amorphous silicon (a-Si) is currently the most widely used semiconductor material for AMLCD applications. However, there is growing interest in polycrystalline silicon (poly-Si), which originally required a very high temperature process over 900 °C (over 1,650 °F) but is moving toward a more manageable lower-temperature process under 500 °C (under 932 °F). There are also a number of less-common TFT materials. One reason for the popularity of a-Si is the large base of relevant knowledge resulting from intensive research performed in the 1970s on low-cost, high-efficiency a-Si solar cells. For both solar cells and TFTs, the field-effect mobility of the semiconductor material is essential in determining efficiency; in addition, the following properties are crucial:

- The ability to deposit material uniformly over a large area
- The stability of electrical properties over time and under various conditions
- Low manufacturing costs

The main advantage of a-Si is that it is deposited at low temperatures around 300 °C (572 °F), meaning that relatively inexpensive glass substrates can be used. Its low mobility makes for TFTs with very small *off* state currents, meaning that additional storage capacitors are rarely necessary, but the mobility is nonetheless high enough to provide an adequate *on* current.

The conveniently small threshold voltage (around 3 volts), combined with the high gain of the channel current, is another factor that makes a-Si a good choice for a TFT material. The result of having a low threshold is that CMOS drivers can be used to supply gate voltages to the TFT array. Typically, hydrogenated a-Si (a-Si:H) is used to reduce dangling bond defects, which are thought to increase *off* leakage current

and photosensitivity. TFTs made from a-Si:H are fabricated using common semiconductor processes, such as glow discharge decomposition of silane (SiH_4).

The main limitation of a-Si is that the low mobility makes it incompatible with an integrated driver architecture. However, this drawback is being increasingly offset by the availability of bulk-silicon driver chips, which can be surface-mounted or tape-bonded directly on the display substrate. A second limitation is that a-Si is photosensitive, increasing its conductivity to an undesirable level under visible light.

Poorly designed a-Si TFT-LCDs can produce unacceptably large leakage currents if a light shield, or some other compensating means, is not included in the manufacturing process.

The relatively low mobility of a-Si affects the channel dimensions and aperture ratio of the TFT. The mobility varies with the fabrication and post-fabrication treatment, ranging from 0.2 to 0.5 cm²/volt-second, thus requiring a large W/L ratio of the channel. Typical channels made with a-Si are 10 to 20 microns long and 100 to 140 microns wide. A typical pixel dimension for a high-resolution display is 220 × 250 microns. Thus, the semiconductor channel may span more than half of the width of a pixel. Provided that the length is minimal, the TFT will not take up an obtrusive fraction of the display area. The aperture ratio, or fill factor, typically ranges from 50% open to 75% open (obviously, the higher the aperture ratio, the better).

In general, the TFT gate electrode can be fabricated either on the substrate (*gate down* or *inverted*) or on top of the insulator layer (*normal* or *noninverted*), and the source-drain and gate electrodes can either be in the same plane (*coplanar*) or in different planes (*staggered*). Amorphous silicon TFTs are made inverted or normal and are usually staggered. Amorphous silicon offers a variety of structural possibilities because it can be deposited at low temperatures, making it suitable for use with a wide variety of materials.

The inverted-staggered structure provides good control and uniformity over TFT switching parameters, with low parasitic effects. Normal staggered fabrication is similar in terms of photolithographic steps and basic complexity. Fabrication begins with deposition and patterning of the source and drain contacts, followed by the interconnect bus lines. The rest of the TFT is fabricated in the same manner as the inverted-staggered structure.

A particular type of inverted structure is the *self-aligned* structure, in which the gate electrode is used as a photomask during subsequent fabrication steps. Self-alignment has the advantage of providing a very good alignment between the gate electrode and the source-drain electrodes, greatly reducing parasitic capacitance. Large, high-yield aspect ratios have been achieved using a self-aligned serpentine geometry. This geometry would be very difficult with standard alignment techniques because of inevitable amounts of overlap. The self-alignment can be used to fabricate TFTs in a two-photomask-step process that is both high yield and low cost. It was first developed by LETI in France in 1986 and continued by Matsushita Electric of Japan, which showed a video display with 372 × 240 pixels fabricated with this potentially economical procedure. (Because the TFT operation depends on the color-filter/black mask deposition, it really needs three photolithographic steps for proper operation. However, Matsushita does not consider this a step, because it is not electrically part of the TFT and therefore does not appreciably affect yields). A further advantage of the two-step process is that by ensuring that the black mask between the color-filter stripes covers the TFT, no additional light shield is required to guard against the a-Si's photosensitivity.

The two-step manufacturing process is far more economical than earlier methods. Each photolithographic step requires removal from the vacuum for photoresist coating, patterning, etching, and cleaning, which increases the chance of impurities becoming trapped between the thin-film layers. Because the vacuum is not broken during the deposition of the insulator and semiconducting materials in the two-step process, the interfaces between the layers are very clean and uncontaminated.

However, two-step-process displays lack some of the performance of their more complicated predecessors. They must have aluminum, tantalum, or some other metal placed over the source bus lines to lower their resistivity. Furthermore, they lack uniformity across the substrate, which can affect display quality. The threshold voltage ranges from 2 to 3 volts, a wide variation that causes problems when using large numbers of gray levels, especially on larger displays. Two-step-process displays are not yet a major factor in the market.

Photo-induced current, the property that led to the widespread investigation of a-Si for photovoltaic cells, is an undesirable quality when a-Si is used as a flat panel display control device. Functionally, a photovoltaic cell is a diode that converts light into electric current by raising the energy level of the charge carriers in the semiconductor layer. The same semiconductor layer exists in a-Si TFTs, although it usually has a lower concentration of charge carriers than do photovoltaic cells. When the a-Si TFT is exposed to light, the *off* resistance of the channel is decreased, thereby increasing undesirable leakage current. This leads to a washed-out display in bright ambient light conditions.

One remedy for the light sensitivity of a-Si is the incorporation of a light shield over the channel area of each TFT. A common shield material is a metal layer made of the same material as the gate. This type of light shield requires an additional photolithographic step, but because it is electrically passive, the chance of yield decreases is minimal. Alternatively, the black mask between the color-filter stripes can be aligned over the TFTs to act as a light shield with no additional photolithography.

There is some disagreement about whether light shields are truly necessary for a-Si TFTs. Some insist that it is a requirement to build light shields, while others claim the problem is minimal. When a-Si TFTs are used in backlit displays, particularly those with color filters that require very bright backlights, a light shield probably offers improved performance. The most cautious reports claim that the *off* resistance in light and dark conditions can differ by up to a factor of 1,000; with a light shield, this can be reduced to below a factor of 10.

There are two other approaches besides light shields. Very thin films of silicon show substantial immunity from light sensitivity. For a-Si layers less than 30 nm thick, Matsushita demonstrated a contrast ratio of greater than 100:1 in a projection television system with a very high brightness. Another method of reducing light sensitivity is to introduce atomic defects in the middle of the a-Si energy gap. The defects presumably act to stop electrons generated by incident photons in the layer before they can generate an external current.

There is a perception that the a-Si process for fabricating TFTs is nearing the lower limit in feature size. Most a-Si TFT-LCDs have a resolution of 70 to 100 dpi. This perception of diminishing returns is pushing the conversion to poly-Si (described in the next section), which saves space by allowing the TFT to be physically separated from the pixel it controls.

However, new advances in a-Si processing made in 1999 show that this material still has much to offer in terms of improved resolution. IBM announced a new LCD monitor called the Roentgen, which offers nearly 200 dpi with an a-Si process. It is only the latest improvement that began with a more modest 157 dpi (10.5-inch-diagonal SXGA) display. IBM's success is not due to any single innovation; it has come from incremental changes and a systemic improvement of the a-Si process. Also, NEC demonstrated a 9.4-inch diagonal a-Si UXGA display (1,600 × 1,280 pixels), which thus has 212 dpi. Such high resolution was achieved by fabricating the color filter on the TFT array substrate (a process called *CFonTFT*).

Polycrystalline Silicon (Poly-Si) Thin-Film Transistors

Polycrystalline silicon (also called poly-Si, or polysilicon) offers different qualities than a-Si, most of which bring both advantages and disadvantages. Poly-Si has electron mobilities of more than 100 cm^2/V-sec (much higher than a-Si, which is typically less than 1 cm^2/V-sec), resulting in increased switching speeds and reduced device size. The high mobility also leads to higher *on* current levels, reducing the liquid crystal switching time. But higher mobility also means higher *off* state leakage current. With poly-Si, the TFTs and driving circuitry can be fabricated during the same photolithographic steps, producing a self-contained unit at lower fabrication costs than if they were made separately. Connectors, driver chips, and their installation equipment are unnecessary; instead, totally integrated, onboard drivers are deposited with the TFTs. Onboard drivers offer smaller device packages; the device size of poly-Si TFTs is currently limited only by fabrication machinery, which can produce geometries as small as a few microns.

However, recent techniques for attaching drivers to a-Si devices cast doubt on the need for onboard drivers at all. Because poly-Si TFTs take up less space on the display substrate, the aperture ratio is larger, resulting in higher brightness. This is especially useful in light valves, in which a large number of pixels are packed into a small area. The high *off* state leakage current requires the use of an auxiliary capacitor, which partly offsets the aperture ratio advantage by taking up space; nonetheless, poly-Si TFTs are smaller overall than a-Si TFTs. Finally, because poly-Si TFTs are not light sensitive, no special precautions need to be taken when used in high lighting situations, such as projector displays.

Because of these properties, poly-Si is used in two major applications: video/digital camera viewfinders, which benefit from very small pixels, and projection LCDs (particularly for front projection, where poly-Si devices will dominate for the next five years).

To date, nearly all poly-Si TFT-LCDs have been fabricated using high-temperature techniques on quartz substrates, but there has been a great deal of effort applied to low-temperature manufacturing on glass, because of reduced costs.

High-Temperature Poly-Si

Poly-Si fabrication is more complex and expensive than a-Si fabrication. Most poly-Si TFT-LCDs produced to date have been made at 600 °C (1,100 °F) and higher, requiring the use of quartz substrates. High-temperature poly-Si manufacturing is similar to the manufacturing processes for silicon-on-sapphire or silicon-on-insulator integrated circuits. All poly-Si TFT fabrication geometries used to date have been noninverted coplanar. Self-aligned fabrication requires six photolithography steps. As with a-Si, self-alignment has the advantage of very little parasitic capacitance between the gate and the drain, but unlike a-Si, it requires doping of the poly-Si gate periphery to assure ohmic contact. Such doping means employing the expensive machinery used for ion implantation, which also reduces yield. An alternative fabrication procedure uses thermally grown SiO_2, which eliminates a photolithographic step and offers better alignment, but requires even higher temperatures because the oxide must be grown at 1,000 °C (1,830 °F). As is the case with a-Si, hydrogenation is shown to have several beneficial effects. Hydrogenation reduces the leaky *off* current of the material, as well as the deleterious effects of dangling bonds, which lower the response time by trapping charges along the doping interfaces.

The full benefits of poly-Si TFTs are not realized unless they are made with fully integrated drivers on the display substrate. The circuitry needed to drive an array of TFTs consists of a combination of shift registers and buffer amplifiers. These devices can be built mostly with poly-Si TFTs with channel dimensions of approximately 8 × 8 microns; one or two transistors will be required to carry larger currents (that is, dimensions up to 50 × 8 microns.) The entire amplifier and delay network can be fabricated in an area of 1,400 × 200 microns, allowing the drivers to be built in parallel to the pixel rows. The pixel rows for a high-resolution display for a video camera viewfinder can have a pitch of up to 100 microns. Driver circuits with a 200-micron pitch can be alternated from side to side in order to fit on a single display panel.

However, poly-Si integrated drivers are receiving increasing competition from driver chips mounted directly on the display substrate. These drivers can be used with a-Si TFTs, combining the advantages of onboard drivers and simple a-Si fabrication. The superiority of integrated drivers versus surface-mounted drivers promises to be one of the most debated active matrix issues in the coming years.

The effect of poly-Si's high mobility propagates through the TFT design. The high mobility has the undesirable side effect of increasing the *off* state leakage current through the TFT. Because of fewer grain boundary defects, poly-Si has a higher intrinsic conduction. Traditionally, this is compensated for by simply including a pixel storage capacitor, but newer techniques include adding a doping interface or a double-gate structure. Seiko Epson has experimented with a double-gate structure in which all current must pass through two TFTs, halving the *on* and *off* state currents, but maintaining the *on/off* current ratio. Due to the small channel dimensions of each TFT, two poly-Si TFTs are still smaller than a single a-Si device. The yield loss from having to manufacture two TFTs per pixel is less than the yield loss from including a storage capacitor, making this procedure cost effective. This technique has been applied to displays for pocket-size televisions with 240 × 240 pixels.

The higher *off* leakage current of poly-Si is somewhat offset by the higher on current, making the *on/off* current ratio comparable to a-Si devices. Therefore, it is not troublesome for TFTs to handle the additional current needed to charge and discharge the larger pixel capacitance. This, however, is not a total solution. The capacitors can be difficult to manufacture and can reduce yields by shorting out the pixels. Furthermore, the capacitors reduce the display aperture ratio, partly offsetting the advantage of smaller channels.

As noted previously, poly-Si TFTs are currently used in viewfinder and projection applications. Their prospect for penetrating other markets is extremely limited, however. The requirement for quartz substrates imposed by high-temperature processing limits display sizes to 2-inches in diameter and results in high manufacturing costs. This limitation is the driving force behind research to produce poly-Si TFTs at temperatures below $500\,°C$, which would allow for the use of standard glass types that are less expensive and are available in large-area substrates. Furthermore, a low-temperature poly-Si (LTPS) process could use fabrication equipment that could be integrated into existing a-Si manufacturing lines, saving vast sums that would otherwise go toward building new fabrication lines.

Low-Temperature Poly-Si

The most direct method for fabricating poly-Si TFTs at low temperatures is the deposition of silicon film on glass in the polycrystalline phase, typically using chemical vapor deposition (CVD). But this approach, called *as-deposited poly-Si*, often results in poor film quality, both in terms of defects and surface roughness. Research in this area has included modifying the substrate layer to eliminate surface roughness and reducing background contaminants to decrease defects, but acceptable films cannot yet be mass produced.

Given the problems associated with the as-deposited approach, most efforts have focused on phase transformation from an a-Si film (deposited on glass) to a poly-Si film. This method, called *crystallized poly-Si*, involves imparting energy, typically in the form of laser irradiation or thermal energy, to crystallize the silicon on the glass substrate. The appeal of this approach is that these films are higher quality than as-deposited LTPS and can be produced on larger areas and at lower temperatures than high-temperature poly-Si. The two types of crystallization for low-temperature poly-Si are referred to as *thermal annealing* and *laser annealing*.

Thermal annealing involves heating the substrate to recrystallize the silicon. The main challenge is, of course, to keep the temperature low. One approach is to use a silicon–germanium alloy instead of silicon for recrystallization. Another approach, developed by Intervac, is called *rapid thermal annealing* (RTA); this approach uses an arc lamp and a reflector to produce a controlled exposure of the film as it passes by in an in-line configuration. This approach requires temperatures above $600\,°C$, however, and furthermore tends to produce low mobility.

Laser annealing, which has come to be the preferred method, typically uses an excimer laser, such as XeCl (308 nm wavelength), to crystallize the silicon film on the glass substrate, heated to $300–400\,°C$ ($572–752\,°F$). The laser beam is rectangular, and it is scanned across the substrate. The long axis of the beam defines the area of recrystallized poly-Si; equipment is available with beam widths of 200 mm, which corresponds to a 12.1-inch display diagonal. Increasing the beam size incurs difficulties with laser power and beam uniformity. Films resulting from this process have shown mobilities in the range of $100–300$ cm^2/V-sec, on par with high-temperature poly-Si. However, there are some drawbacks: The laser equipment is expensive, the processing speed is slow, laser scanning must be carefully controlled, and there can be variability between laser pulses. Many companies are still working on improving the laser annealing process.

LTPS processes typically involve more mask steps than their high-temperature counterparts. Ion doping is required to suppress leakage currents at the pixel TFTs and to produce CMOS-type driver circuits. The only means of avoiding time-consuming ion implantation is to use a non-self-aligned architecture in which doped source-drain of contacts are deposited separately from the poly-Si channel such a non-self-aligned poly-Si TFT structure; in the addition of storage capacitors usually requires an extra step, as do double-layer, low-resistance bus lines. Integrated drivers can usually be made simultaneously with the TFT manufacturing process, but they may also require additional steps.

Choosing inverted (gate-down) versus normal structure for the laser-annealed LTPS TFT is a matter of efficiency versus performance. Inverted TFTs allow the laser to simultaneously recrystallize the channel and activate the doped ions, so the fabrication is faster and cheaper than for normal TFTs. However, normal orientation, with its separate crystallization and activation steps, results in higher mobility and less gate/source parasitic capacitance, which means superior performance.

Despite the great advances in LTPS processing in 1999 and early 2000, much work remains before the technology can be considered robust.

Areas of active research include identifying optimal doping formulas and temperatures, achieving the appropriate recrystallized grain size, making interfaces cleaner, and further reducing the process temperature.

With drastic price reductions for large a-Si LCDs and declines in driver chip costs, it will be hard for LTPS LCDs to compete in notebook computer and desktop monitor applications. For small and medium LCDs, it is nominally cheaper to make LTPS than a-Si because of the cost savings resulting from integrating driver functions, but low yield and throughput remain problems for poly-Si. With production improvements, poly-Si will be used increasingly in applications such as digital cameras and automobile navigation applications. TFT-LCD panels have gained acceptance in the navigation market, especially in Japan, and there are opportunities for future navigation and passenger entertainment systems.

Single-Crystal Silicon Transistors for Displays

Although single-crystal silicon has the highest mobility of all types of silicon, it is difficult to use in displays because it cannot easily be deposited on glass, which provides a poor growth surface.

An innovative approach to using single-crystal silicon transistors has been developed for a high-resolution display. Instead of fabricating standard a-Si devices, the array is built on a single-crystal silicon wafer using MOS technology, Then the array is physically removes and redeposits it on a glass substrate. The single-crystal silicon properties are preserved. Because the process is based on silicon wafer technology that uses standard semiconductor processing techniques to form TFTs on a silicon wafer, the process of building the transistors can be accomplished in any existing semiconductor fabrication facility. High-resolution devices for near-eye applications such as viewfinders and head-mounted displays have been developed.

A granular material consists of many tiny crystalline regions. Usually, these microcrystalline grains are jumbled together randomly, as is the case for poly-Si, for instance. However, in continuous-grain silicon (CGS) the silicon grains align so that the crystal axes of one grain are parallel to the crystal axes of its neighbor. The only reason CGS is not a single crystal is that the grains start growing from multiple nucleation points so that the atoms do not quite match up when the grains become large enough to touch. CGS has mobility of about 300 cm^2/volt-see for n-type and 140 cm^2/volt-see for p-type, which is three times higher than that of poly-Si and almost equal to that of single-crystal Si. The high mobility of CGS allows TFTs to be physically smaller, increasing the aperture ratio and resulting in a brighter display. CGS could also enable system-on-panel (SOP) technologies (with the interface and all other peripheral circuits incorporated into a single substrate) that allow display systems to be extremely thin. Faster driving results in a higher contrast ratio.

The successful completion of a product masks the challenges involved in using CGS to make the TFT-LCD. Sharp claims that CGS manufacturing employs "similar" processes to those involved with poly-Si. This has the advantage of possibly allowing CGS to be made on poly-Si manufacturing lines. However, considering that the time-intensive and costly nature of poly-Si manufacturing is its main limitation, it is clear that adopting CGS is no simplification of the current situation.

Further, realizing the full potential of the CGS-based display required construction of a new optical engine for the projector product. Because the device employs two lamps for high brightness, a fly eye lens system is introduced to collect and direct the light from both sources. The high heat generated from two lamps requires significant cooling to prevent degradation of the polarizers. However, standard fans introduce too much room dust into the system, resulting in an unacceptable image. Hence, the projector includes a closed cooling system with three internal fans and multiple radiator fins. The elements are arranged carefully to provide the most cooling to the blue LCD panel, since it produces the most radiative heat. The projection lens system consists of a nine-group spherical lens that is tailored to handle light with a two-lobed intensity distribution (arising from the two light sources).

Although its price means CGS is not yet competitive with present RPTVs, costs will be reduced over time.

Cadmium Selenide (CdSe) Thin-Film Transistors

Cadmium selenide is of mainly historical interest in LCD development; it has been used for TFT development since the early 1970s, when it replaced tellurium as the material of choice at Westinghouse. Research continues at several companies and universities. CdSe offers excellent performance capabilities rivaling those of poly-Si, but its disadvantages bar it from being commercially viable.

The most important feature of CdSe is its high mobility. CdSe films can be made with mobilities from 40 to 450 cm^2/V-sec (hundreds of times higher than a-Si), at temperatures below 400 °C (752 °F), making the process compatible with inexpensive soda lime glass. The high mobility allows CdSe to be used to make TFTs with channel dimensions much smaller than those of a-Si. Typical *on* current for TFTs with a W/L channel ratio of 7 is 1 microamp with 20 volts on the gate and a 10-volt source-drain voltage. The *off* current for these TFTs is approximately 0.1 nanoamps, for an *on/off* ratio of .00001. The TFT's frequency and current-carrying capability are proportional to its channel mobility, so CdSe devices can operate close to 1 MHz (compared to the order of kHz in a-Si) and carry currents closer to milliamps (rather than microamps). This allows the construction of integrated row and column drivers, and also the system-on-panel possibility noted previously for CGS.

In addition, one of the most problematic steps of silicon processes, forming doped semiconductor source-drain contacts, is not necessary with CdSe. If the CdSe is deposited, annealed, and covered with the aluminum electrodes all during the same pumpdown cycle, the CdSe-electrode contacts will be ohmic.

Despite the impressive capabilities of CdSe, three disadvantages severely limit its application. It is a very toxic material, its performance degrades quickly over time, and manufacturing has proved to be complex and costly.

Ferroelectric Liquid Crystal Display (FLCD)

Ferroelectricity is analogous to the more widely known phenomenon of *ferromagnetism* — the hysteresis, or memory effect, originally observed in iron (ferrite) magnets. The key property of hysteresis is *bistability* (two stable, distinct states). In the case of a display, bistability leads to a property called *write-and-forget*: When the driving voltage is removed from a ferroelectric liquid crystal display (FLCD), it remains in its *on* or *off* state. This is in contrast to TN- or STN-LCDs, which always revert to a unique, unpowered state. Because the liquid crystal material used is of the smectic type, the effect is referred to as the *ferroelectric-smectic effect*. Another term used is *surface stabilized ferroelectric liquid crystal (SSFLC)*, which refers to the structure of an FLCD: These displays are made such that the top and bottom substrates are spaced much more closely than the ferroelectric helix pitch, and the liquid crystal molecules are trapped in a nonhelical geometry. An applied voltage allows the material to be rapidly switched between two optically distinct configurations.

The key feature of FLCs is their fast response time, which is 10–100 microseconds, nearly 1,000 times faster than that of nematic liquid crystals. This shifts the limitation of display speed away from the liquid crystal material to the driving circuitry. For large-area displays, this speed advantage is blunted by comparable speeds achievable in plasma display panels (PDPs) and electroluminescent (EL) displays, but it is important for small displays.

The write-and-forget property of FLC is especially useful in highly multiplexed displays because the pixel remains in its state during nonselect periods; there is no problem with off state leakage. In an FLCD, the pixels respond not to the average RMS voltage, but to the most recent voltage above or below V$_{on}$ or V$_{off}$. All the driving circuitry needs to do is latch the display to its *on* or *off* state.

Researchers, encouraged by the rapid development of TN and STN-LCDs, hoped that FLCDs would follow the same course. But as FLCD technology progressed, new problems and limitations emerged that were not calculated into the original development timetables. The main trouble arises from the fact that the FLCD substrates must be very close to each other (only a few microns), and there is little tolerance for variation in the spacing. This means the glass and deposited films must be very smooth, pressure sensitivity is a concern, and cleanroom conditions are more stringent because particle contamination is so damaging. All of this adds up to a difficult fabrication process that is costly in both time and money.

Despite considerable effort, working and reliable ferroelectric displays have been much more difficult to produce than originally anticipated.

Polymer-Dispersed Liquid Crystal Display (PDLCD)

Optical scattering is the principle used in the polymer-dispersed liquid crystal display (PDLCD). The device uses nematic liquid crystals encapsulated in micron-sized polymer droplets. The droplets are suspended in an emulsified film several microns thick that is sandwiched between the substrate plates. In the unpowered *off* state, the droplets strongly disperse light, because of the lack of parallel molecular alignment both within each droplet and from droplet to droplet. The display appears cloudy or milky, rather than truly dark. Applying an electric field causes the liquid crystal within each droplet to become parallel and the droplet as a whole to align with the field, making for a transparent display. By mounting a mirror behind the cell, the mode of operation can be changed from transmissive to reflective. This system does not involve any polarizers, which are a main source of absorption on LCDs. Hence, the transmission coefficient is limited only by the small absorption and internal reflection from the glass or plastic substrate.

A major advantage of PDLCDs is their fast (submillisecond) response time. Because they have only weak anchoring bonds, they can easily realign to an electric field. Increasing the driving voltage further increases the response time. This fast response time is offset by the lack of a sharp optical threshold and a high turn-on voltage. In a typical sample, $V(B = 10\%)$ is approximately 5 volts, while $V(B = 75\%)$ is close to 12 volts. A 7-volt difference is far too high for any large, directly multiplexed display to produce visible contrast, especially when compared to STN-LCDs that accomplish their optical switching with less than a 0.05 V difference. Improved manufacturing processes, which maintain uniformity of droplet size (thus increasing nonlinear threshold behavior), may make PDLCDs suitable for small to medium multiplexed displays. It is not expected that PDLCDs will be produced in large, direct-drive multiplexed displays for some time.

Other important features of PDLCDs are their relative ease of fabrication (there is no need to seal the cell) and insensitivity to cell spacing. The cell plates simply form a container for the polymer film matrix; their spacing does not affect display performance. Thus, low-quality unpolished glass (and even plastic) can be used, creating the possibility for flexible displays. Unfortunately, there have not been as many applications for flexible displays as PDLCD developers hoped. Because spacing is not critical, very large panels can be constructed. Raychem Corporation, Menlo Park, California, used to make 1 × 3-meter (3.28 × 9.84 feet) glass panes that electronically switched between an opaque state and a transparent state, intended for use in meeting rooms as an alternative to Venetian blinds. However, production has recently ceased. Again, the number of applications has proved disappointing.

For color performance, the polymer droplets can be doped with a dichroic dye, as in *guest-host* displays. The display will modulate between a colored *off* state of strong absorption and high scattering and an on state of weak absorption and high transmittance. In general, there is a tradeoff between brightness and contrast. Dyes that produce high contrast tend to be dark in the *on* state.

In the reflective mode, an alternate technique for improving contrast without sacrificing brightness is to place a color filter between the mirror and the PDLCD cell. In the *off* state, the polymer droplets scatter enough light so that most of it does not reach the mirror, giving the cell a dark, neutral color. In the *on* state, the cell becomes transparent, allowing the bright, saturated filter color to show through. In addition to increasing the actual (measured) contrast, this type of color change greatly improves the perceived contrast. The use of fluorescent reflectors will further improve perceived contrast.

These displays offer many potential advantages such as simple manufacturing and high durability, which has sparked much interest in PCLCDs. However, such benefits have not been proven, and amidst all the research activity, there are only a few prototype PDLCDs. Research groups that have recently published papers on the subject include GEC-Marconi Ltd, Britain; Hughes Research Laboratories, California; Kent State University and Kent Display Systems, Ohio; OIS Optical Imaging Systems, Michigan; Polytronix Inc., Texas; Raychem Corp., California; and the University of Hawaii, Hawaii. See also **Colloid Systems**.

Electronically Controlled Birefringence (ECB) Liquid Crystal Displays

An electronically controlled birefringent (ECB) display is similar to an STN-LCD, except it does not incorporate a twist. There are two basic structures for an ECB display. The first uses liquid crystal material with positive dielectric anisotropy in a cell with homogeneous (parallel) alignment. The second, more popular, configuration uses liquid crystal material with negative dielectric anisotropy in cells with homeotropic (perpendicular) alignment. This geometry is also known as Vertically

Aligned Nematic (VAN). The application of an electric field causes the liquid crystal molecules to rotate perpendicular to their original resting position, effecting a change in birefringence.

When the molecules are aligned perpendicular to the cell, whether by electric field or boundary conditions, they exhibit no birefringence, because they are aligned symmetrically about an axis parallel to the normal. Birefringence is only observed as the molecules move parallel to the plates, because they lose their symmetrical alignment. Just as in STN cells, well-placed polarizers will produce contrast between the *on* state and the *off* state. Usually the polarizers are rotated 90° with respect to each other and 45° with respect to the alignment direction of the liquid crystal molecules (homogeneous director vector).

In the more common homeotropically aligned cell (VAN), it is important for the axes of rotation to be parallel to each other (that is, for the rotation to be uniform). To accomplish this, the cell boundaries are also treated with a slight homogenous rubbing so that the rotated molecules have a preferred *on* state direction. Both types of ECB usually incorporate pre-tilt to prevent degenerate rotation. ECB has a much faster response time than STN does, because the molecules' rotation is not retarded by the strong aligning forces that exist in twisted liquid crystal structures.

Unless some type of compensation is used, ECB cells have relatively poor viewing-angle dependence. Oblique angle viewing increases d and changes Δn, affecting the display contrast. Keeping the optical path length small helps minimize viewing angle dependence because of smaller absolute changes at off-angles. But if $\Delta n \times d$ is decreased too much, contrast is reduced because it is the birefringent effects that make the display work. Usually these factors are balanced in a fairly large $\Delta n \times d$, in the range of 0.5 to 1 micron. Advances in film compensation techniques are reducing these contrast and viewing-angle drawbacks, however.

Instability is also a problem with VAN devices. VAN cells are most prone to instability when they are thin cells (<5 microns) operating at a high temperature (>50 °C) and/or a low frequency (>50 Hz). Changing these parameters to reduce instability, however, negatively affects other performance criteria: Increasing cell thickness reduces turn-on and turn-off time, high temperatures cannot always be avoided, and practical limitations of circuit technology limit driving frequency. Adding dopants reduces instability, but at the cost of reducing response time. VAN ECB displays may have high quality and good color, but when left running for too long, infectious color fringes spread over the display as a result of instability. This effect is not permanent; turning the display off, and then on, will restore its original quality.

VAN displays that do not suffer from color fringes have been built and research continues.

Dynamic Scattering Mode (DSM) Liquid Crystal Display

The dynamic scattering mode (DSM) for LCDs has been commercialized but never widely adopted. It works by switching between a scattered, opaque *off* state and a transparent, clear *on* state. The addition of filters could create a colored display. DSM-LCDs were the very first LCDs developed and commercialized, occurring in the mid-1960s and early 1970s. They use liquid crystal material with negative dielectric anisotropy that is doped with a slightly conductive agent. The displays work without polarizers, but they require more than 15 volts for good operation and use more current than TN-LCDs or STN-LCDs.

Guest-Host (GH) Liquid Crystal Display

The *guest-host* (GH) display is structured like the TN-LCD, except that it uses only one polarizer and the surface alignment on both glass surfaces is in the same direction (no twist). It relies on the effects of having a dichroic dye (the guest) added to a liquid crystal material (the host) with positive dielectric anisotropy. The device is based on the fact that the dichroic dye either absorbs light or transmits light of a specific color, depending on its orientation, which is switched by the applied voltage. The effect was first reported by researchers at RCA in the late 1960s. A modification of the original design uses a chiral liquid crystal that imparts twist and eliminates the polarizer altogether. The latter types of GH displays are made in very small quantities for low information content instrument displays used in aircraft and for some clock displays.

There has been some renewed interest in these GH effects to make reflective monochrome or color LCDs. Three-layer GH displays were first developed in the late 1960s under NASA-sponsored contracts, but they were used only in the transmissive mode. A color reflective display was developed using a three-layer structure with magenta, cyan, and yellow GH cells; it showed practically no parallax and had good color purity, as well as high reflectance. More recently, technology had been developed for reflective displays (see "LCDs for Portable and Handheld Devices" for more information on reflective displays).

Cholesteric Liquid Crystal Display

Cholesteric liquid crystals (CLCs) arrange themselves in stacked planes. All molecules in a given plane are aligned, but the alignment axis of adjacent planes is rotated by a fixed amount. The result is a helical stack, with the molecules in planes like steps of a spiral staircase. This structure has two important optical properties: First, it passes light only of one handedness (a left-handed film transmits right-handed light) and in one wavelength range (a small band around the value $n p$, were n is the index of refraction and p is the helical pitch); and second, it reflects all light not transmitted (there is no absorption). Under the application of a voltage in the direction "up the staircase," the molecules all rotate 90° to align with the field, making the film transparent. In this standard configuration, a CLC film makes an excellent narrow-band polarizing filter.

CLCs can also be made into broadband polarizers, allowing the fabrication of an LCD. By adding a chiral material to the CLC, the self-aligned helix will have a varying pitch over the length of the helix and will pass light throughout the entire wavelength range implied by the pitches and index of refraction. If this range spans the visible spectrum, the CLC cell acts as a normally black LCD.

Because the CLC reflects light that is not transmitted, the display can be made very efficient by placing a mirror behind the device and its backlight. The left-handed light reflected from a left-handed CLC film travels backward, bounces off the mirror, and becomes right-handed. It can then pass through the film instead of being wasted. By employing CLC films as the polarizers and color filters that are also needed in an LCD, the entire device can be made 10 times brighter than a standard TN-LCD (that is, it passes about 50% of the backlight's brightness, compared to the usual 5%). Such a bright display is obviously valuable for today's handheld devices because of its lower power requirement and reduced battery draw.

CLC displays are very difficult to fabricate, however. The chemistry of CLCs is delicate and not yet compatible with the processing required to make TFTs. Direct-drive prototypes of CLC displays have been made.

Liquid-Crystal-on-Silicon (LCOS) Devices

Of all the types of AMLCDs, *liquid-crystal-on-silicon* (LCOS) devices most directly use semiconductor manufacturing techniques, as the active matrix array and associated driver circuits are designed and built on a silicon wafer, using complementary metaloxide semiconductor (CMOS) processing techniques. The LCOS displays are the most heavily researched types of *microdisplays*, which are devices built on silicon (or quartz) in a miniature form and designed to be viewed indirectly, either by imaging into the eye or in projection mode.

Instead of thin-film transistors on glass, LCOS displays use bulk silicon devices to control the pixels from outside the optical path. The silicon-based CMOS chip serves as both the active matrix and the reflective layer, on top of which a thin layer of liquid crystal, glass plate, and polarizer are deposited. Instead of the a-Si or poly-Si TFTs used in transmissive LCD cells, LCOS devices use the high-speed switching capability provided by single-crystal silicon. LCOS devices also have a simplified LCD structure because the device is reflective, not transmissive, like most LCDs. For example, only one polarizer is required, and the thickness of the layer of liquid crystal can be reduced, allowing for faster switching.

Two important design choices that are made when developing an LCOS device are the type of liquid crystal material and the type of control circuit to be used. The nematic and ferroelectric liquid crystal modes are the most popular types for such a display. The nematic types provide a good contrast ratio and are typically coupled with a DRAM switching array. Ferroelectric liquid crystals provide very fast switching speeds, but they do not allow for grayscale; the pixel is either black or white. Grayscale can be created temporally by taking advantage of the fast switching speed to dither the binary value within a frame period. Ferroelectric liquid crystals are typically coupled with SRAM devices, which provide the fast frame transfer rates needed to provide temporally dithered grayscales and which are also binary devices.

Liquid crystal microdisplays can also be configured to modulate unpolarized light, an approach that has the advantage of much greater light transmission by eliminating the significant absorption of polarizers. There are two ways to use unpolarized light. In one approach, light is scattered when the liquid crystal molecules are arranged in one fashion and reflected when the molecules are rearranged, a process that is controlled by the application of a voltage across the cell. Scattering displays typically use polymerdispersed liquid crystal materials. The second approach uses the principle of diffraction, in which a periodic arrangement of *on* or *off* pixels (forming a grating) causes light to constructively or destructively interfere. This interference pattern can then be filtered by a Schlieren stop to pass light when in certain grating modes; by altering the grating structure, amplitude and color may be controlled. Twisted-nematic liquid crystal material is the most common type used in diffractive systems.

Color can be achieved in LCOS devices in several ways. The most popular approach for projection applications is to use three LCOS chips and dichroic mirrors to separate red, green, and blue components. Color-filter wheels have also been used to present sequential color to a single chip. For personal viewers, red, green, and blue light sources (typically light-emitting diodes) are used to illuminate sequential frames of data at three times normal video rates.

A number of engineering issues remain with LCOS. First is the light source, which should be as close to a point source as possible because the display is so tiny. Also, improvements in the optics to handle the fine pixel pitch would increase image quality — in the rear-projection screen and the polarizing beamsplitter, in particular.

LCOS displays require a thin cell gap (1–3 μm, with a tolerance of 100–200 nm) as well as specialized filling and sealing procedures. Pixel pitches in microdisplays are as much as five times smaller than in direct-view displays. Current LCOS manufacturing is a tradeoff between circuit and assembly yield.

IBM has reported what is perhaps the most impressive device demonstrated to date. It is a CMOS device with 2,048 × 2,048 pixels and a TN-LCD layer. The device can show 65,536 colors at a frame rate of 74 Hz and has been implemented in the form of a 28-inch rear-projection monitor. One problem with the device is that it requires a rather large piece of silicon; because the pixels are 17 μm on a side, the active area is nearly 2 inches in diameter.

LCOS devices are just beginning to take hold in the display market. While there is potential for high growth in the coming years, market acceptance will be contingent on improvements in price, performance, and ergonomics of these products.

The manufacturing of LCOS microdisplays is also not yet a mature practice; manufacturing lines for common LCDs and TFTs are generally not appropriate for LCOS displays.

Reflective and Transflective Liquid Crystal Displays

The solution to making LCDs viewable in bright ambient light is the *reflective* LCD. Instead of using a backlight, this configuration relies on ambient light, incorporating a mirror-like surface behind the device to reflect the light up through the liquid crystal material.

Reflective LCDs can be made in two configurations: outer-surface reflective and inner-surface reflective. Outer-surface reflective devices are constructed similarly to transmissive devices, with the backlight replaced by a reflective surface behind the rear polarizer. While simple, this configuration has the disadvantage of poor contrast and color at a wide viewing angle, because the incident and reflected beams pass through different color filters (leading to both greater absorption of the light and color mixing). Also, there can be parallax problems associated with having the reflector behind both glass substrates. Outer-surface reflection is widely used in watch and calculator displays (and has been for more than 20 years) and may be acceptable for handheld devices, but the quality is too poor for sophisticated displays such as notebook computers.

A superior construction (although more difficult to fabricate) is the inner-surface reflective device, which incorporates the reflective layer into the bottom electrode. Because the reflection occurs much closer to the color-filter layer, the viewing-angle problems are greatly diminished.

The reflective LCD is a tremendous boon for some mobile devices because it completely eliminates the need for a backlight. However, reflective LCDs cannot, of course, be used in darkness, which is a requirement for some devices, such as cellular telephones and PDAs. Hence, a hybrid device has evolved, called the *transflective* LCD. In this case, both transmissive and reflective modes are employed; the device has a backlight, but it also makes use of ambient light through reflection.

Seiko Epson's transflective configuration, which it calls *semi-transparent*, simply replaces the reflective layer in an outer-surface reflective device with a semitransparent plate. The backlight partially shines through, and ambient light is partially reflected. A 6.5-inch thin-film diode LCD has been demonstrated. With half-VGA pixel format, the display acts like a transmissive product when the backlight is on and like a reflective product when the backlight is off. A reflective film for the display that has a 10:1 contrast ratio and a 512-color capability has been developed; power consumption is expected to be 0.12 watt.

Sharp's transflective configuration takes advantage of inner-surface reflection; each bottom electrode consists of a central transparent area surrounded by a reflective region. Sharp calls this the Advanced TFT and has two prototype displays based on this design. One is a 2-inch panel with 560 × 220 pixels, and the other is a 7-inch, wide-screen model with 1,440 × 234 pixels. Both models have a 100:1 contrast ratio in darkness with the backlight on, but only 5:1 in reflective mode.

A hybrid backlight had been developed, which collects and reflects ambient light under bright conditions and emits light under dark conditions.

In addition to changing the method of light modulation, changes can be made in other components of the LCD to increase the overall light transmission, making the display more suitable for portable and handheld applications. The division between these two approaches is less simple that it appears, however, because changing the method of light modulation often results in the need for additional or different auxiliary components.

In particular, reflecting displays require a diffuser, which spreads the reflected light for the viewer's eyes. The diffuser is a controlled reflector that distributes the intensity of the display's light over an appropriate angle so the display looks natural. The diffuser affects both the brightness and the contrast ratio of the display, and its design must be coupled to the reflection technique and type of liquid crystal used. One such diffuser is the Lumisty™ film made by Sumitomo Chemical Corp.

The two largest sources of transmission reduction are the polarizer sheets (typically, two are used) and the color-filter layer. Therefore, many manufacturers have experimented with improving or even eliminating one or both of these components. Displays have been made with various combinations of two, one, or zero polarizers and one or zero color filters, as shown in Table 3, first compiled by Matsushita.

TABLE 3. CLASSIFICATION OF REFLECTIVE/TRANSFLECTIVE COLOR LCDS

Number of Polarizers	2	1	0
Color filter	(A) "Standard" STN, TN	(B) STN, TN, and others; inner-surface reflection	(C) polymer-dispersed, guest-host
No color filter	(D) field-sequential color, electronically controlled birefringence	(E) none	(F) cholesteric, polymer-dispersed, guest-host

Group A is the "standard" configuration, with two polarizers and one color filter. Matsushita has demonstrated a color STN-LCD that uses a single polarizer instead of the usual two polarizers (Group B). This configuration takes advantage of the fact that light changes its polarization state upon reflection, so a single polarizer is all that is necessary. Using inner-surface reflection, the 7.8-inch-diagonal VGA prototype has 15% reflectance, a 14:1 contrast ratio, and a power consumption 89 mW. It is targeted for mobile business device applications. Another single-polarizer system is made by Sony using LTPS TFT technology; this prototype color display has 34% reflectance and a 19:1 contrast ratio.

Completely eliminating the polarizer sheets (Groups C and F) requires using liquid crystal materials that operate on a principle other than polarization rotation, such as the scattering mechanism of polymer-dispersed liquid crystals, guest-host effects, or electrophoresis. Scattering-type reflective displays have the advantage of simple, lowcost manufacturing, but their limited color and contrast, which have hindered development as transmissive displays, will also hinder development as reflective displays. A *double-layer guest-host* (DGH) LCD, offers a superior combination of contrast and reflectivity. Complications remain in achieving optimal color with this display (using color filters reduces the reflectivity, requiring the addition of special reflectors), so overall it is still in an early stage of development. A prototype monochrome DGH LCD had a 15:1 contrast ratio and a reflectivity of 60%. A related pursuit has been finding ways to produce a color LCD without a color-filter layer (Groups D and F). Several approaches are being investigated: field-sequential color, electronically controlled birefringence (described previously), and stacked LCD layers.

Field-sequential color is accomplished by sequentially illuminating the panel with red, green, and blue backlights at a rate of 180 Hz, or three times the frame rate. The eye integrates the three colors within the time span of a single frame of data. The backlights can be made of monochromatic fluorescent tubes, or with LEDs. One challenge has been identifying liquid crystal materials that have a sufficiently fast response rate while simultaneously allowing the generation of grayscales. Researchers at Tohoku University have developed an optically compensated bend liquid crystal material that has a response time as fast as 2 milliseconds. Another challenge has been to develop fast-reacting phosphors for fluorescent backlights. Ichiko Industries has developed red, green, and blue fluorescent lamps with 0.1 millisecond *on* and *off* times.

Stacked LCD layers use three layers of a dyed liquid crystal material (such as guest-host) in subtractive color (cyan, magenta, yellow) arrangements, which absorb half as much light as additive (red, green, blue) types do. This arrangement provides for higher throughput of the illumination, but requires great care in manufacturing to avoid a misalignment of the layers. Also, TFTs must be constructed on each of the three substrates.

There are additional benefits that arise from eliminating color filters, beyond making the LCD more suitable for handheld devices. First, color-filter fabrication is a significant fraction of the manufacturing cost of LCDs, so there is a direct cost benefit. Also, color filters require making three subpixels for each addressable pixel; eliminating this redundancy lowers manufacturing costs or allows for increased resolution at the same cost, and reduces the number of drivers required.

Overall, there is great motivation to improve the light transmission of LCDs through modifying the backlight and eliminating the color filters, which is spurring intense research and development efforts. Some companies have even launched systems-level efforts to improve many components of the LCD synergistically. An example is Casio's Hyper Amorphous Silicon TFT-LCD (HAST), which combines improvements in aperture ratio with higher-transmittance color filters and a reflective design.

Liquid Crystal Materials

Over the past 20 years, there has been an enormous effort to improve the operating parameters of liquid crystal materials. Many thousands of compounds and mixtures have been prepared to tailor the materials to particular display types or applications. Thus, the types of materials used in TN-LCDs for auto dashboards are quite different from those used in STN-LCDs for computers. Similarly, other LCD technologies (for example, AMLCDs, ferroelectric-smectic LCDs, polymer-dispersed LCDs, and electronically controlled birefringent displays) use unique materials specifically designed to optimize the performance of the display. The major classes of materials for low information content displays are the cyanobiphenyls; larger, higher information content displays use the phenylcyclohexane compounds and derivatives, as well as esters, for high-level multiplexing. The most popular type of liquid crystal material is currently the "F-system" line of products from Merck.

The explosion of new types of display applications in recent years (for notebook computers, handheld devices, cars, and specialty applications) has led to a mini-renaissance of liquid crystal development. The creation of liquid crystal materials has progressed to the point at which rational design of compounds and mixtures is routinely done. In designing materials for STN-LCDs, for example, certain structures are used to optimize the elastic constants and the viscosity in order to improve the steepness of the voltage-versus-contrast ratio curve and to increase the switching speed. Current materials are designed for use with duty ratios of 1:240 and 1:480 and response times of tens to hundreds of milliseconds, with the goal of achieving a contrast ratio greater than 20:1 over a broad viewing angle.

For TN-LCDs used in outdoor applications or in severe environments (for example, gasoline pumps, auto dashboards, aircraft cockpits, and marine instruments), a wide operating temperature range is required, in addition to other performance factors that are specific to the application.

Some materials are now available that enable LCDs to operate over a temperature range of $-40\,°C$ to $+110\,°C$. Even at $-20\,°C$, the average response time is 280 milliseconds.

Within the last few years, the development of improved materials for AMLCDs has also progressed dramatically. Materials with low threshold voltages (less than 1.5 volts) have been developed by increasing dielectric anisotrophy; these materials are needed particularly in notebook computers, which have low driving voltages to conserve power. TFT-LCDs with a broader viewing angle have been made possible by such techniques as lowering the optical anisotropy, reducing the incidence of inverse contrast, multidomain switching, in-plane switching, and retardation compensation. And, to improve the contrast ratio, materials with very high resistivity (1,012 ohm-centimeters) have been developed. The reason for this is that in an AMLCD, the RC (resistance times capacitance) time constant must be large compared to the frame time, because the charge on any one pixel must be maintained until the next addressing pulse, typically 20 milliseconds. The search for stable materials with these high resistivities led to the development of fluorinated compounds, which provide high thermal stability, low viscosity, and low birefringence, making the materials ideally suited for use in AMLCDs.

Surface Alignment Materials

An alignment layer (typically a polymer) is required for producing both TN- and STN-LCDs, in order to properly orient the liquid crystal molecules at the substrate surface and enable them to tilt slightly in a preferred direction. The process seems relatively simple to achieve, yet it is perhaps the most complex and least understood of all the processes used for LCD manufacturing.

To produce the tilt angle bias, the films are rubbed or buffed with short-nap polyester or cellulose acetate materials, either in a static or a rotating configuration. The latter may be a simple buffing machine with a paint roller attachment for small operations or it may be a more sophisticated machine with controls for buffing-wheel speed, roller pressure, and substrate travel. Buffing or rubbing of the film with materials that have higher melting points than the film is believed to produce enough localized heating to cause the long-chain polymer molecules to become oriented with their chains parallel to the buffing (or rubbing) direction. The liquid crystal molecules that come into contact with such an oriented film are then aligned in the same direction as the polymer chains. A preferred tilting of the molecules occurs as a result of the interaction of the liquid crystal molecules with the chemical structure of the film. Therefore, different liquid crystal molecules will tilt at different angles, depending on the structure of the two substances.

Early displays used poly(vinyl alcohol), polyesters, polysiloxanes, and low- to medium-molecular-weight methyl cellulose as the alignment material, but today polyimides are the materials of choice. New liquid formulations (solutions) are available in viscosities suitable for spinning, dipping, brushing, or even offset printing. The formulations for LCDs have high optical clarity and excellent adhesion to glass, silicon dioxide, and ITO. Polyimides are also used as high-temperature structural adhesives in aircraft and other applications. Care must be taken when using polyimide materials for LCDs so that only very pure starting materials with low metal ion content are used. Stability at high temperatures is the main advantage of using polyimides. The disadvantage is that expensive, potentially hazardous solvents must be used in its application.

The science of polyimide design and formation can be complex; the molecular weight and molecular structure can be altered to meet specific requirements. The most important properties of polyimides are (1) thermal stability, (2) solubility in suitable solvents, (3) resistance to solubility in the liquid crystal material, (4) adhesion to the substrate, and (5) ability to withstand the high local temperature and mechanical effects of the rubbing or buffing process. Thermal stability is important because one does not want the alignment material to decompose at the temperatures used for sealing the LCD substrates together. Also, the polyamic acid solution must be stable. Because the polyamic acid solution is to be printed on the substrates, solubility in suitable solvents is critical. Presently, N-methyl-2-pyrrolidinone and gamma-butyrolactone are the solvents commonly used in conjunction with cellosolves and carbitol additives to make the resulting solution printable.

To prevent the polyimide film from dissolving in the liquid crystal material or in the solvents used to clean the displays, various functional groups are incorporated in the diamine or dianhydride used. This is also done to improve the adhesion on the substrates and ensure the resistance to mechanical and thermal wear during the rubbing process. Because of the need to meet all of these requirements, polyimide material suppliers offer different formulations for the various types of LCDs. Thus, there are specific formulations designed for the fabrication of TN-LCDs, STN-LCDs, AMLCDs, and ferroelectric LCDs.

Color Filters for Displays

The most common way to achieve full-color displays is to extract the primary colors from a white light source and combine them as needed at each pixel. By selectively filtering and combining the three primary color wavelengths — red (700 nanometers), green (546.1 nanometers), and blue (435.8 nanometers) — full-color images can be reproduced. Displays best suited for the addition of color filters are those that switch between neutral black and white states. The materials and array production process are described in greater detail in the following subsections.

The addition of multicolor capability introduces several problems that add costs to the display module. The most obvious is that the number of pixels and driver circuits must be increased by a factor of three or four in order to maintain the same resolution as a monochrome display. This increases driver costs and complicates manufacturing by decreasing line widths or adding crossovers. A second problem is that the color filters cut the brightness of the display, requiring a more powerful backlight.

Performance standards for color filters are still elusive. The National Television Standards Committee (NTSC) has defined the primary color vertices on the standard chromaticity diagram for color television CRT phosphors. LCD manufacturers consequently strive for their displays to be as close to these vertices as possible. To provide a good representation or mixing of colors, the filter system must be definable in high resolutions with accuracy, have good transparency and chromaticity, and be resistant to light and heat, as well as the chemicals used in making the flat panel display.

The color filter is one of the most complex, expensive, and troublesome components in an LCD.

The production of the color-filter array is a technique-sensitive process that is not widely practiced. Consequently, color-filter arrays are available from only a few sources, and filter-plate prices are quite high.

The major suppliers of finished color-filter plates for LCDs are all located in Japan. One of the few non-Japanese companies known to be producing color filters for LCDs is located on Rolla, Missouri. In addition, several LCD manufacturers have onsite color-filter manufacturing facilities, which were set up by their suppliers.

The color-filter matrix manufacturing process employs many steps and consumes many intermediate materials. Besides the basic color-filter material, there is a need for photoresist, etchants, dyes, or other chemicals, depending on the process. An overcoat or planarization layer is usually put over the finished filter matrix to prevent damage to the filters, smooth the surface, increase transmission, and prevent contamination of the liquid crystal. An acrylic resin is the most commonly used overcoat material. The color-filter material that endows each subpixel with its R, G, or B color is usually a pigment dispersed in a viscous medium that is then deposited on one of the substrates. There are also dye and inorganic colorants, although they tend to have inferior resistance to heat, light, water, and chemicals; the pigment-dispersion materials became superior around 1997. The pigment is typically applied by a photolithographic process, as noted below.

The color-filter array and the black mask, a light-absorbing material that defines the colored subpixels, are usually located on the upper substrate on the inside of the liquid crystal cell. The filter fabrication process is constrained by the temperature sensitivity and the requirements for a smooth surface. The insulating properties of color-filter materials mandate that the conductive ITO layer be deposited over the filters, rather than under them (directly on the glass plate). This means that the filter materials must be capable of enduring the elevated temperature processing of the ITO deposition, usually a bit less then 200 °C. The filter material must be chemically resistant to the liquid crystal material because they are in proximity. The filter will also be illuminated by a fairly intense light source, so it must be resistant to changes due to exposure to the light, including ultraviolet light. Further, the colors must conform as closely as possible to the NTSC standards, a property known as **color purity**; state-of-the-art materials from Toppan Printing have color purity near 47%. In summary, the key requirements of color-filter materials for LCDs are optical performance (light transmittivity); dimensional precision; flatness; thermal resistance; light stability; color purity and chemical resistance.

The reproducibility and control of color-filter properties is an ongoing concern. It should be made clear that none of the color-filter manufacturers is now satisfied that the optimum process has been developed. Color-filter arrays are very expensive, due in large part to the stringent requirements for the finished product. Although many processes have been developed, only two are in widespread production: the photolithographic dye system using photosensitive gelatin and the photolithographic pigment system using pigmented photoresist.

A relatively recent lithographic/laminating process, described by T.P Brody, uses a sandwich of red, green, blue, and black layers. Pigmented, photosensitive acrylic films are successively laminated onto a plastic carrier sheet. The films are exposed through a photomask, one step each for the red, green, and blue layers. A black mask layer is also deposited. The process is said to be capable of up to 200 lines per inch and produces filters up to 20 × 24 inches. Eliminating the stepper and the resist is expected to result in a substantially lower filter manufacturing cost.

In 1999 there were many development efforts to produce color-filter plates in larger sizes than those used for notebook computers. As LCDs move into the desktop monitor market, forward-thinking color-filter-plate manufacturers are stepping up efforts to fabricate large-area plates. The newest manufacturing lines in development in 2000 will be capable of up to 42-inch-diagonal color-filter panels.

Polarizers and Film Compensators for Liquid Crystal Displays

To view the electro-optic effects of nearly all types of LCDs, a set of two polarizers is necessary (however, as noted previously, there are reflective and transflective LCDs with one or even no polarizers). These polymer films are bonded to the front and rear surfaces of the display. In advanced supertwisted-nematic types, a third optical film is added in order to neutralize or compensate for the wavelength-distorting effects of the liquid crystal cell.

Polarizers typically consist of a poly(vinyl alcohol)–iodine polarizing element sandwiched between layers of a protective film that prevents moisture from attacking the polarizing element. The protective layer typically consists of a film of cellulose triacetate or cellulose acetate butyrate. Films to be used in a transmissive LCD also have a pressure-sensitive adhesive layer, release liner, and surface protective film. For transflective or reflective LCDs, the back polarizer, when present, is combined with a reflective film.

The most prominent property of the polarizer is its light absorption; the transparency of the polarizer is one of the prime factors determining the final brightness of the LCD. Absorptive polarizers, such as the standard PVA–iodine type, convert unpolarized light to linearly polarized light by absorbing the orthogonal polarization — that is, half of the light. Further inefficiencies in the material reduce this value further; state-of-the-art polarizers transmit 44% of the incident brightness. Nitto Denko has developed a "wet drawing" process, in which the PVA is stretched (to create the dichroism) under water, in order to better align the acrylic and iodine molecules for maximum transmittance.

A newer method for increasing transparency is to use a nonabsorptive polarizer, which reflects or refracts incident light instead of absorbing it. Such polarizers have traditionally been constructed from three-dimensional optics such as beamsplitters, but these are not feasible for flat panel displays. Materials that can be made into thin-film non-absorptive polarizers are being aggressively developed as demands for increased LCD brightness and power efficiency mount. One possibility is cholesteric liquid crystal; others are stacked films with varying refractive indices.

There are several approaches to laminating the polarizing (or polarizer-retardation composite) film to the LCD glass. One method uses die-cut pieces of polarizer, while another involves the use of strips or reels of the material. Some facilities use sheets of the material to laminate an array of displays; the lamination process uses pressure, but in some cases it uses heat as well. The array is then cut into individual displays using a laser, hot knife, rotary saw, or razor. In the past, problems occurred during this process in which air bubbles or dust particles were trapped, resulting in the formation of visually unacceptable blemishes. Currently, however, the quality of the polarizer materials and the methods of application have improved to the point at which these problems are minimal.

In applications for which antireflection and antiglare properties are desired, the polarizer has a clear, hard antiglare coating, as well as an antireflection coating. Nitto Denko offers such a product, which was developed in cooperation with OCLI. Other products add a coating of ITO to the top layer to provide electromagnetic shielding.

For STN-LCDs in which optical retardation is required, a stretched polymer film is laminated to the polarizing film. While these composite films are more expensive than conventional polarizers, the cost of applying them to the LCD is less than it would be if two separate films were to be applied.

There are also films designed to broaden the viewing angle of TN TFT-LCDs, and these are sometimes integrated with polarizers to reduce the total number of films applied. While such films are not as effective at increasing viewing angles as novel LC modes, this method has the advantage of allowing improvement without modifying the LCD manufacturing process, which can be a major undertaking. Fuji Photo Film, Ltd. offers a TAC (triacetyl cellulose) film that incorporates both the polarizer and their previous product, WV film (for "wide view"), to make a thinner and more transparent optical coating for the LCD cell.

See also **Cathode-Ray Tube**; **Television (TV)**; **Computer Graphics**; and the family of articles catalogued under **Flat Panel Display Technology**.

For additional reading, refer to Flat Panel Display Technology entry.

Stanford Resources, Inc., San Jose, CA.

LIQUID CRYSTAL POLYMERS. These materials (LCPs) exhibit a highly ordered structure in the melt, solution, and solid states. A tightly packed and highly ordered morphology particularly susceptible to orientation during processing is characteristic. Commercial applications for LCP resins include chemical pumps, tower packings, coil bobbins, connectors, sockets, etc. for electronic components, and various automotive parts. LCPs have an excellent combination of chemical and flame resistance, dimensional stability, and ease of processing. Their thermal stability makes them suitable for dual ovenable cookware and where thermal resistance for both conventional and microwave oven service is important.

Compared with other polymeric materials, LCPs have very high unidirectional properties. Vectra™ (Celanese Corp.) resins are primarily aromatic polyesters based on p-hydroxybenzoic acid and hydroxynaphthoic acid monomers. Xydar™ (Celanese Corp.) injection molding resins are polyesters based on terephthalic acid, p,p'-dihydroxybiphenyl and p-hydroxybenzoic acid. Differences in monomers are primarily responsible for the differences in specific properties and end uses. The fibrous nature of the polymers imparts good impact strengths.

LIQUID CRYSTALS. Liquids that have the structural character of cybotactic liquids (see also **Cybotaxis**), but which are considerably more viscous, with viscosities extending from that of a light glue to that of a glassy solid. They also exhibit much more definite evidences of structure than the cybotactic liquids.

Liquid crystals must be geometrically highly anisotropic — usually long and narrow — and revert to an isotropic liquid through thermal action (thermotropic mesomorphism) or by the influence of a solvent (lyotropic mesomorphism). Several thousand organic compounds are now known which meet these criteria, but significant molecular features found in thermotropic liquid crystals are among the following. The molecule will be elongated and rectilinear; if "flat segments," e.g. benzene rings, are present its liquid crystallinity will be enhanced. The molecule will be rigid along its long axis and double bonds will be common in this direction. The simultaneous existence is seen in the molecule of strong dipoles and easily polarizable groups. Of lesser importance are weak dipolar groups at the extremities of the molecule.

The present day classification of thermotropic liquid crystals is threefold. Smectic liquid crystals, such as p-ethyl azoxybenzoate, have their molecules arranged in definite strata, a variety of molecular arrangements being possible within each stratification. In smectic Type A crystals, the molecules may be considered to "stand on end" with their long axes perpendicular to the plane of the layer but with their centers irregularly spaced. When the molecular centers adopt hexagonal close packing, the crystals are considered smectic Type B, and when they adopt a titled form of Type A, they are classified as smectic Type C. See Fig. 1.

In nematic liquid crystals, the molecular structures possess a high degree of long-range orientation order, but no long-range translational order. The molecules are spontaneously oriented with their long axes approximately parallel, but without the stratification seen in smectic crystals. Nematic liquid crystals like p-azoxyanisole are generally optically uniaxial, positive, and strongly birefringent, and some are composed of

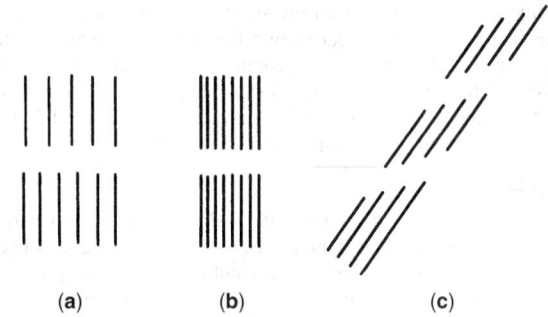

Fig. 1.   Smectic liquid crystals; types (**a**), (**b**), and (**c**).

Fig. 2.   Nematic crystals.

hundreds of molecules (cytotactic groups), the molecular centers in each group arranged in layers. See Fig. 2.

Lyotropic liquid crystals possess at least two components. One of these is water and the other is amphible (a polar head group attached to one or more long hydrocarbon chains). In the lamellar form, water molecules are sandwiched between the polar heads of adjacent layers while the hydrocarbon tails lie in a nonpolar environment. Lyotropic liquid crystals have very complex structures, but occur abundantly in nature, particularly in living systems. See Fig. 3.

Fig. 3.   Cholesteric crystals.

Polarized light is the most powerful tool for investigating liquid crystals, all of which exhibit characteristic optical properties. A smectic liquid crystal transmits light more slowly perpendicular to the layers than parallel to them. Such substances are said to be optically positive. Nematic liquid crystals are also optically positive, but their action is less definite than that of smectic liquid crystals. However, the application of a magnetic field to nematic liquid crystals lines up their molecules, changing their optical properties and even their viscosity.

Both smectic and nematic crystals split a beam of ordinary light into two polarized components whose transverse vibrations are at right angles to each other. This is the well-known phenomenon of double refraction. Cholesteric liquid crystals exhibit the phenomenon of circular dichroism. That is, they break a beam of ordinary light into two components, one with the electric vector rotating clockwise and the other with it rotating counterclockwise. The first is usually transmitted, and the second is the one to be reflected. It is this property that gives cholesteric crystals their characteristic iridescent colors when illuminated by white light.

This ability to exhibit colors is one of the most useful attributes of liquid crystals. Many cholesteric substances behave as liquid crystals only in a certain temperature range. Above it they are colorless, but as they are cooled through it they assume a succession of colors, running down the spectrum from red to violet and finally becoming colorless. However, at this final stage they still retain their molecular orientation, but it is that of smectic liquid crystals rather than cholesteric. Some cholesteric liquid crystals do not exhibit all the colors mentioned and others, which are naturally colored, simply change to another color on heating or cooling. Since the exact temperatures at which these color changes occur are invariable, these substances can be used for measuring temperatures; in fact, combinations of them cover the range from −20 to +250 °C.

Useful applications have been found for the varied effects of these crystal changes. One of the first came from the property of selectively reflecting visible light; because this is temperature-dependent, the property can be used as a temperature detector, and in gel form liquid crystals have been used for the early detection of those cancers which cause hot spots in the body. Applications of the smectic modifications arise from their ferroelectric properties; this phase can function as a fast-switching light-valve device with memory. This kind of application requires some control on the pitch of the polarized helix, which is obtained by blending together materials. Most twisted nematic field effect liquid crystal displays make use of a 90° twist between transparent conductive electrodes and crossed polarizers, as shown in Fig. 4. As randomly polarized light enters the device, only that portion which is vertically polarized may pass through the front polarizer. This, in turn, is rotated another 90° through the rear polarizer. If a reflective surface is placed behind the rear polarizer, the light will be passed back through the cell, its polarization again being rotated. By applying a voltage to the transparent electrodes, the molecules of the crystal will leave the nematic structure and align with the field, as shown in Fig. 5. Then the incoming light is no longer polarized—so extinction occurs and the area of extinction is defined very sharply by the shape of the electrode pattern, producing a dark area on a light reflective background. The reverse can also be achieved by using parallel rather than crossed polarizers. Fig. 6 shows a cross section of a typical display. A polymeric seal contains the liquid crystal material and holds the glass

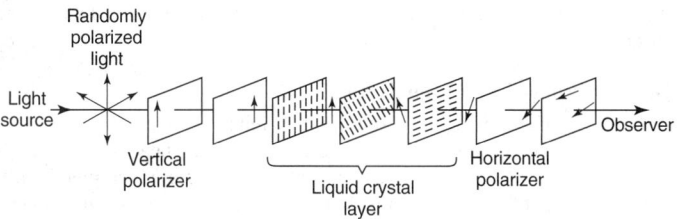

Fig. 4.   Liquid crystal display operation, unactivated.

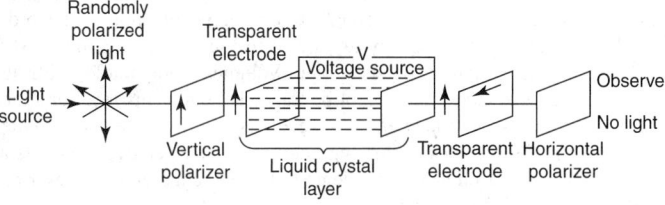

Fig. 5.   Liquid crystal display operation, activated.

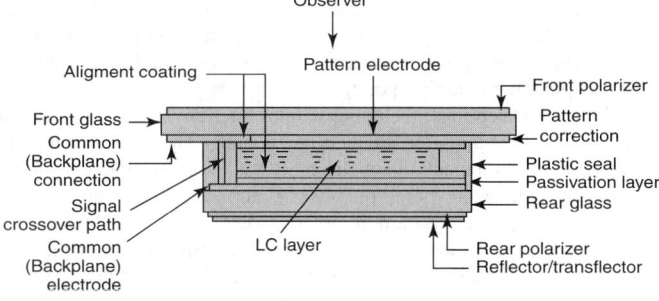

Fig. 6.   Basic liquid crystal display construction. (*Hamlin, Inc.*)

substrates together and the whole is laminated to the assembly. See also **Liquid Crystal Display Technology**.

Additional Reading

Anisimov, M.: "Critical Phenomena in Liquids and Liquid Crystals," Gordon & Breach Publishing Group, Newark, NJ, 1991.

Cladis, P.E. and P. Palffy-Muhoray: "Dynamics and Defects in Liquid Crystals," Gordon & Breach Publishing Group, Newark, NJ, 1998.

Demus, D., G.W. Gray, J.W. Goodby et al.: "Handbook of Liquid Crystals: High Molecular Weight Liquid Crystals," Vol. 3, John Wiley & Sons, Inc., New York, NY, 1998.

Demus, D., G.W. Gray, J.W. Goodby et al.: "Handbook of Liquid Crystals: Low Molecular Weight Liquid Crystals II," Vol. 2, John Wiley & Sons, Inc., New York, NY, 1998.

Kumar, S.: "Liquid Crystals: Experimental Study of Physical Properties and Phase Transitions," Cambridge University Press, New York, NY, 2000.

Langerwall, S.T.: "Ferroelectric and Antiferroelectric Liquid Crystals," John Wiley & Sons, Inc., New York, NY, 1999.

Prigogine, I. et al.: "Advances in Chemical Physics Volume 113, Advances in Liquid Crystals," John Wiley & Sons, Inc., New York, NY, 2000.

Singh, S.: "Liquid Crystals: Fundamentals," World Scientific Publishing Company, Inc., Riveredge, NJ, 2001.

Sonin, A.A.: "Freely Suspended Liquid Crystalline Films," John Wiley & Sons, Inc., New York, NY, 1999.

Yeh, P. and C. Gu: "Optics of Liquid Crystal Displays," John Wiley & Sons, Inc., New York, NY, 1999.

LIQUID-IN-GLASS THERMOMETER.   This instrument consists of a glass envelope, a responsive liquid, and an indicating scale. The envelope is in two parts fused together: a bulb completely filled with the liquid, and a capillary scale section containing the liquid in excess of that required to fill the bulb. The position of the end of the liquid capillary column or index serves, by prior calibration, to indicate the temperature of the bulb. The scale may be marked directly on the capillary tube, as in the laboratory or clinical versions, or may be on a separate member mounted alongside the capillary tube, as in the domestic and industrial forms.

Invented over three centuries ago, the liquid-in-glass thermometer reached its zenith as a temperature-measuring device in the early 1800s. It was used as a standard for the dissemination of the scale (Normal Thermometric Scale, adopted internationally) from the International Bureau of Weights and Measures to standardizing laboratories throughout the world, until the later adoption, in 1927, of the International Temperature Scale. Over the years, many practical applications were found for the liquid-in-glass thermometer in addition to its earlier use as a primary standard of temperature. Although still used widely in meteorology, medicine, and industry, and for domestic purposes, the glass thermometer has been replaced by various electrical and electronic temperature measurement methodologies for many other applications.

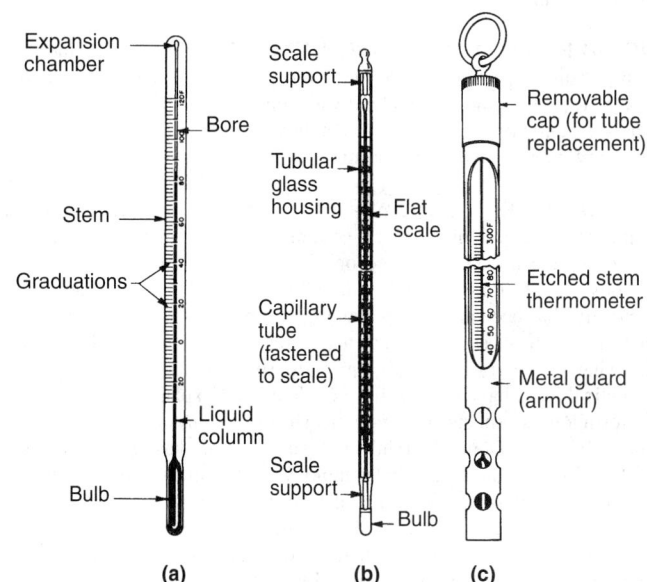

Fig. 1.   Laboratory-type liquid-in-glass thermometers: (**a**) traditional; (**b**) Einschluss; (**c**) armored.

Various liquids are used in liquid-in-glass thermometers. Mercury is the choice for higher temperatures or where accuracy is critical. Its advantages are a broad temperature span between its freezing and boiling points, a nearly linear coefficient of expansion, the relative ease of obtaining mercury in a very pure state, and its nonwetting-of-glass characteristic. For measurements below the freezing point of mercury, various organic liquids, such as toluene, other hydrocarbons, and organic phosphates, have been used. Representative versions of the glass thermometer are shown in Fig. 1. An industrial version is shown in Fig. 2.

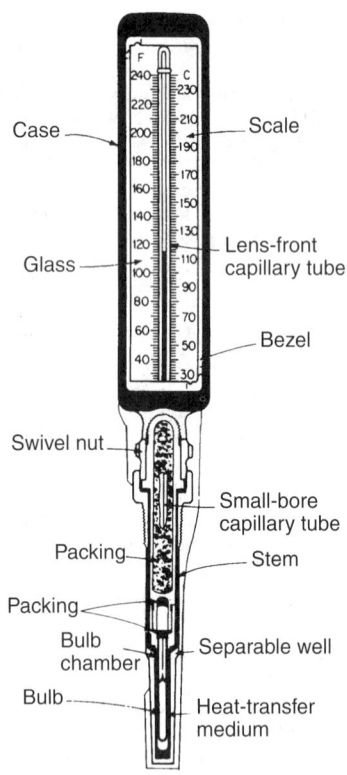

Fig. 2.   Industrial-type liquid-in-glass thermometer.

LIQUID JUNCTION.   To avoid the unknown liquid junction potential in measuring the potential of a half-cell against a reference electrode, the two half-cells are frequently connected via a salt bridge, usually a concentrated solution of potassium chloride. Since its anion and cation have almost the same velocity, a negligible diffusion potential is set up across the liquid junctions at the ends of the bridge.

LIQUID-PROPELLANT ROCKET ENGINE.   A rocket engine using a propellant or propellants in liquid form. Also called *liquid-propellant rocket*. Rocket engines of this kind vary somewhat in complexity, but they consist essentially of one or more combustion chambers together with the necessary pipes, valves, pumps, injectors, etc.

LIQUID STATE.   Because of the theoretical and practical importance to the era of electronics, which commenced nearly a half-century ago, the solid state of matter has become better known and understood than the physics of fluids (liquids and gases). Much practical engineering knowledge has been amassed pertaining to substances in the fluid state, but much research of a fundamental nature on fluids remains to be finished. Particularly, the transition of liquids to solids (and vice versa) at the theoretical level has not been fully explored and explained.

Prestigious scientists have commented on the mysteries that confront them. Russell J. Donnelly (University of Oregon) has observed, "Most flows of fluids, in nature and in technology, are turbulent. Since much of the energy expended by machines and devices that involve fluid flows is spent in overcoming the drag caused by turbulence, there is a strong practical motivation to understand the phenomenon. The study of turbulent flows, however, is one of the most formidably difficult subjects in physics and engineering. At present (1988), there is no substantial aspect of turbulent flow that can be understood fully from first principles."

Steve Granick (University of Illinois) has commented (1991), "Apart from structure, what are the dynamics of liquids in intimate contact with a solid boundary? This question has proven to be one of the most baffling aspects of liquids, in spite of long-standing interest."

Sir Samuel F. Edwards (Cavendish Laboratory, University of Cambridge) noted (1987), "Liquids are everywhere in our lives, in scientific studies and in our everyday existence. The study of their properties, in terms of the molecules of which they are made, has been the graveyard of many theories put forward by physicists and chemists. The modern student of liquids places his faith in the computer, and simulates molecular motion with notable success, but this still leaves a void where simple equations should exist, as are available for gases and solids. There is a powerful reason for the failure of analytical studies of liquids, i.e., the difficulty experienced in finding simple equations for simple liquids. We can explain the origin of the trouble and show that it does not apply to what at first might seem a much more complex system, that of polymer liquids where, instead of molecules like H_2O or C_6H_6, one has systems of molecules like $H_2(CH_2)_{10,000}$ or $H_2(CHC_6H_6)_{2,000}$ which behave like sticky jellies and yet have complex properties that can be predicted successfully."

Jacob N. Israelachvili and Patricia M. McGuiggan (University of California, Santa Barbara) observe, "The subtleties that can occur in the last few nanometers as two surfaces, particles, or solute molecules approach each other in a medium can be quite remarkable. Sometimes the forces are well described by 'continuum' or 'mean-field' theories, such as the DLVO (Derjaguin, Landau, Verwey, and Overbeck) theory, but more often they are not. Important fundamental questions remain concerning the origin of long-range attractive and repulsive hydration forces in water, the spontaneous nucleation of a bulk liquid or vapor phase between two surfaces close together, and the nature of entropic-fluctuation forces between two fluid-like interfaces. The elucidation of these interactions both at the fundamental level and when applied to specific systems (where a number of different interactions may be occurring simultaneously) present a challenge to experimentalists and theoreticians. On the purely experimental side, new techniques are constantly being introduced for extending the range and scope of surface force measurements. For example, one may anticipate that the atomic force microscope will soon provide the first direct measurements of the forces between molecules, as opposed to between surfaces."

General Properties of Liquids

A liquid is matter in a fluid state that is relatively incompressible. An *ideal liquid* offers no permanent resistance to a shear stress, but is incompressible. A liquid has a constant volume and incompletely fills any container of less than this volume. A *real liquid* is appreciably compressible, and the liquid state of a substance might be defined as the denser and less compressible phase of the two-phase fluid system that can exist in equilibrium at temperatures below the critical temperature. X-ray diffraction experiments show that, near the melting-point, the molecules of a liquid show a considerable degree of short-range order and that, in small volumes, they are arranged much as in a solid crystal. This crystalline structure persists over volumes comparable with the intermolecular distances, but cannot be traced beyond. This local or short-range order means that the average molecule is at any moment surrounded by a number of molecules occupying nearly the same relative positions as they would in the solid state. The degree of short-range order is described by the radial distribution function.

This concept of a liquid as an imperfect crystal requires that the molecules in a liquid are packed sufficiently loosely for comparatively free movement, i.e., the energy required to move a molecule from a lattice site to a vacant space is not large compared with thermal energies. Under these conditions, shear flow of the liquid resembles closely the high temperature creep of crystalline solids. A number of theories of the liquid state have this concept as their starting point.

With a few exceptions, including helium, the universal phase diagram shown in Fig. 1 applies for all pure compounds. The triple point is the single point at which all three phases (crystal, liquid, and gas) are in equilibrium. The triple point pressure is normally below atmospheric. Those substances, such as carbon dioxide, where $P_t = 3,885$ millimeters, $T_t = -56.6\,°C$, sublime without melting at atmospheric pressure. From the triple point, the melting curve defines the equilibrium between crystal and liquid, usually rising with small but positive dT/dP, and presumably always with positive dT/dP at sufficiently high P values. The line is

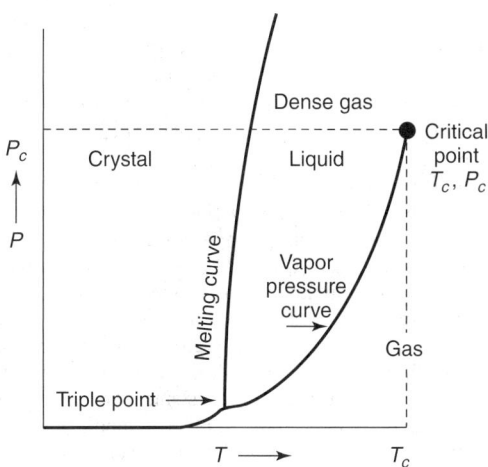

Fig. 1.   Universal phase diagram.

believed to extend infinitely without a critical point (it has been followed to $T \cong 16T_c$ for helium, and calculations indicate that hard spheres would show a gas-crystal phase change). The gas-liquid equilibrium line, the vapor pressure curve, has dT/dP always positive and greater than the melting curve. The vapor pressure curve always ends at a critical point, $P = P_c$, $T = T_c$, above which the liquid and gas phase are no longer distinguishable. Since the liquid can be continuously converted into the gas phase without discontinuous change of properties by any path in the $P - T$ diagram passing above the critical point, there is no definite boundary between liquid and gas. Two liquids of similar molecules are usually soluble in all proportions, but very low solubility is sufficiently common to permit the demonstration of as many as seven separate liquid phases in equilibrium at one temperature and pressure (mercury, gallium, phosphorus, perfluoro-kerosene, water, aniline, and heptane at 50 °C, 1 atmosphere).

Stability Limits.[1] With the exception of helium and certain apparent exceptions discussed below, Fig. 1 gives a universal phase diagram for all pure compounds. The triple point of one P and one T is the single point at which all three phases, crystal, liquid, and gas, are in equilibrium. The triple point pressure is normally below atmospheric. Those substances, e.g., CO_2, $P_t = 3885$ mm, $T_t = -56.6$ °C, for which it lies above, sublime without melting at atmospheric pressure.

From the triple point, the melting curve defines the equilibrium between crystal and liquid, usually rising with small but positive dT/dP, and presumably always with positive dT/dP at sufficiently high P values. The line is believed to extend infinitely without a critical point (it has been followed to $T \cong 16T_c$ for He, and calculations indicate that hard spheres would show a gas-crystal phase change). The gas-liquid equilibrium line, the vapor pressure curve, has dT/dP always positive and greater than the melting curve. The vapor pressure curve always ends at a critical point. $P = P_c$, $T = T_c$ above which the liquid and gas phase are no longer distinguishable. Since the liquid can be continuously converted into the gas phase without discontinuous change of properties by any path in the $P - T$ diagram passing above the critical point, there is no definite boundary between liquid and gas.

The term *liquid* is commonly reserved for $T < T_c$, and "dense gas" is used for $T > T_c$. However, certain properties, such as the ability to dissolve solids, change rather abruptly at the critical density. In many respects, the dense gas resembles the low-temperature liquid of the same density more closely than it does the dilute gas.

The slope, dT/dP, of all phase equilibrium lines obeys the thermodynamic Clapeyron equation:

$$dT/dP = \Delta V/\Delta S = T\Delta V/\Delta H \qquad (1)$$

with ΔV, ΔS, and ΔH the differences, for the two phases, of volume, entropy, and heat content or enthalpy, respectively. The quantity ΔH is

the heat absorbed in the phase change at constant P. Since always $S_{cr} < S_{liq} < S_{gas}$, and usually $V_{cr} < V_{liq} < V_{gas}$, one usually has $dT/dP > 0$; the relatively rare cases, including water, for which $V_{liq} < V_{cr}$ at low pressures leads to $dT/dP < 0$ for the melting curve near the triple point.

Figure 1 gives the P-T boundaries of the stable liquid phase. Clean liquids can readily be superheated or supercooled, and, in vessels having walls to which the liquid adheres, they can be made to support negative pressures of several tens of atmospheres. Thus the properties of the metastable liquid can be investigated outside the limits shown in the diagram.

Two apparent exceptions to the universality of the phase diagram of Fig. 1 deserve mention. First, many of the more complicated molecules decompose at temperatures below melting or boiling, and the diagram is unobservable. Secondly, some liquids, notably glycerine and SiO_2 and many multicomponent solutions, supercool so readily that crystallization is difficult to observe. In these cases, there is a continuous transition on cooling to a glass, which has the elastic properties of an isotropic solid. The structure of the glass is qualitatively that of the high-temperature liquid, lacking long-range order. Since glass and liquid are not sharply differentiated, the term *liquid* is sometimes used to include glasses, although common parlance reserves *liquid* for the state in which flow is relatively rapid.

Quantum Liquids. The one real exception to the phase diagram of Fig. 1 is that of helium, Fig. 2. Both isotopes, ^{4}He and ^{3}He, have no triple point, the liquid is stable to 0 K below about 20 atm for ^{4}He and below about 30 atm for ^{3}He. The liquids have zero entropy at 0 K in both cases. This is also the only case in which isotopic mixtures form two liquid phases at equilibrium, the isotopic solution separating below 1 K. The isotope ^{4}He has itself two phases, He I above the dotted λ-line of the diagram, and He II with remarkable properties of superfluidity, second sound, etc., below the λ-line. The phase transition along the λ-line is second order; that is, whereas S and V are continuous, heat capacity and compressibility change discontinuously across the λ-line.

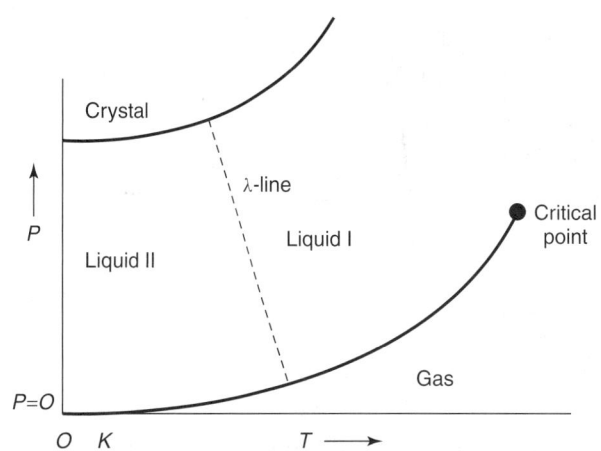

Fig. 2.   Quantum liquid exception to phase diagram of Fig. 1.

Although no completely satisfactory single theory of liquid helium has yet been formulated, one can say that most of the remarkable properties are qualitatively understood and are due to the predominance of quantum effects, including the difference in the statistics of the even and odd isotopes. Thus helium is the one example in nature of a quantum liquid, all other liquids showing only minor deviations from classical behavior.

Structure. Considerable confusion in the description of liquid structure exists, due primarily to difficulties of precise formulation of verbal concepts. The geometric arrangement of any small number (say 10 to 12) of close-lying molecules resembles the arrangement in the crystal, but the order rapidly disappears as larger groups are considered. Long-range order is lacking. The fact that numerical theories based on a lattice or cell structure have some success is evidence only that most properties depend on the configuration of near neighbors alone. Insofar as the arrangement of nearest neighbors is describable in terms of that of the crystal, the structure of the normal liquid is probably characterized best by a somewhat closer spacing than the crystal of the same molecules. The reduced density arises from a considerable number of vacancies in the lattice; the coordination

[1] The following several paragraphs by Joseph E. Mayer are part of a large article that appears in "The Encyclopedia of Physics" (Robert M. Besancon, Editor), Van Nostrand Reinhold, New York, 1984.

number, or number of nearest neighbors, is lower than in the crystal. The exception is water, in which the low coordination number, 4, of the crystal, is increased by interstitial molecules in the liquid, leading to a higher density of the liquid.

Structural descriptions of this nature usually lack the possibility of precise formulation. It is, however, possible to define for any disordered array of molecules in three-dimensional space an arrangement of contiguous cells, each containing one and only one molecule, the faces of the cells being the loci of the midpoints of neighboring molecules. The statistics of the fraction of cells with n faces and of the distances of the faces from the molecules would give the fraction of molecules having a given number of nearest neighbors and the distance distribution of these in a precisely defined manner. Neither present experimental information nor present theories lend themselves to analysis in such terms.

The only clearly defined manner of describing liquid structure in use at present involves the concept of a set of probability density functions, ρ_n, for ascending numbers, n, of molecules. The function ρ_n depends on the vector coordinates $\mathbf{r}_1, \mathbf{r}_2, \cdots \mathbf{r}_n$ of n molecules, and

$$\rho_n \mathbf{r}_1, \mathbf{r}_2, \cdots, \mathbf{r}_n, d\mathbf{r}_1, \cdots, d\mathbf{r}_n$$

is defined as being the probability that in the liquid of definite P and T, there will be, at any instant of time, one molecule at each position, $\mathbf{r}_i$, within the volume element, $d\mathbf{r}_j$. For a fluid, unlike a perfect single crystal, $\rho_i(\mathbf{r})$ is a constant independent of $\mathbf{r}$ and equal to the number density: the number, ρ, of molecules per unit volume. The first significant member of the set is then the pair density function, $\rho_2(\mathbf{r}_1, \mathbf{r}_2)$, which depends only on the distance, $\mathbf{r} = |\mathbf{r}_1 - \mathbf{r}_2|$, between the two molecules. At large distances $\rho_2(\mathbf{r} \to \infty) = \rho^2$. This function can be found experimentally from the x-ray scattering intensities of the liquid (it is the three-dimensional Fourier transform of the scattering intensity at angle θ vs $(4\pi/\lambda)/\sin(\theta/2)$). A typical plot is shown in Fig. 3. The area under the ill-defined first peak integrated over $4\pi r^2 d\mathbf{r}$ is the average number of nearest neighbors, and is of order 10 to 11 for normal liquids.

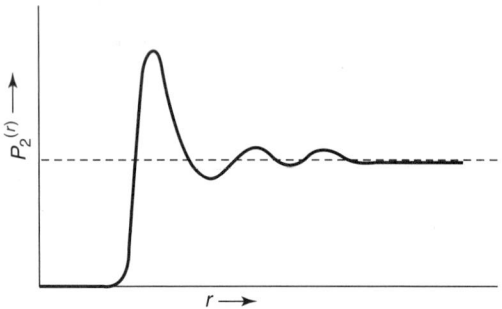

Fig. 3.   Fourier transform of scattering intensity.

The quantity of dimensions of energy,

$$W_n(\mathbf{r}_1, \cdots, \mathbf{r}_n) = -kT \ln[\rho^{-n} \rho_n(\mathbf{r}_1, \cdots, \mathbf{r}_n)]$$

can be shown to be the potential of average force of n molecules located at the positions $\mathbf{r}_1, \cdots, \mathbf{r}_n$. That is, if there are n molecules at these positions, there will be some average force, f_{xi}, along the x-coordinate of molecule i. This average is the sum of the direct force due to the other $n - 1$ plus the average of a fluctuating force due to the others, whose average position is affected by that of the n specified ones. This average force is

$$f_{xi} = -(\partial/\partial x_i) W_n(\mathbf{r}_1, \cdots, \mathbf{r}_n)$$

One frequently assumes that W_n is a sum of pair forces only,

$$W_n(\mathbf{r}_1, \cdots, \mathbf{r}_n) = \sum_{n \geq i >} \sum_{j \geq l}{}' W_2(\mathbf{r}_{ij})$$

although this assumption is known to be only approximate. With this assumption, the pair average force potential, $W_2(\mathbf{r}_{ij})$, can be computed as the solution of an integral equation, and the solutions agree quite well with the experimental curves.

The knowledge of the complete set of functions ρ_n plus that of the intermolecular forces would permit the computation of all equilibrium properties of the liquid, and indeed if the intermolecular forces are the sum of pair forces, only a knowledge of ρ_2 at all P, T values is necessary. An adequate, although numerically difficult, theory of the transport properties also exists, using the equilibrium functions, ρ_n. At present, only qualitative success is obtained in the completely *a priori* use of the equations.

Associated Liquids. The description given above is adequate only for liquids composed of spherically symmetric molecules or molecules that are nearly so. These constitute the so-called normal liquids, which obey reasonably well the law of corresponding states, for which the entropy of vaporization at the boiling point has the Trouton's rule value of approximately 21 cal/deg. For molecules containing large dipole moments, or those forming mutual hydrogen bonds, the concept of the probability density functions must be extended to include angles or other internal degrees of freedom in the coordinates. Such inclusion is conceptually easy, but incredibly complicates the already difficult numerical evaluation of any equations. However, certain qualitative statements may be made.

Liquids composed of molecules with large dipole moments are frequently referred to as associated. Although in some instances relatively stable dimer or definite polymer units of relatively fixed orientation may exist, in many cases, notably water, it is extremely doubtful if an exact knowledge of the structure would reveal any distinguishable entities of associated molecules other than that of the whole liquid. In such cases, one would, however, expect that certain mutual angular orientations between neighboring molecules will be highly preferred, whereas in the dilute gas this will not be the case. The effect of this restriction on the internal coordinates will be to decrease the entropy of the liquid markedly compared to the gas. This effect is qualitatively the same as in association, and the properties of these liquids, particularly the high entropy of vaporization, will simulate those of a liquid composed of definite associated complexes.

Traditional Views of Forces Between Surfaces in Liquids

For many years, four kinds of forces have been recognized to operate between surfaces or particles in liquids:

1. *Van Der Waals forces* — Normally, these are monotonically attractive and occur between all molecules. See also **Van Der Waals Forces**.
2. *Repulsive electrostatic (double-layer) forces* — These forces are apparent when ionizable surfaces have a net electric charge, the common case in water. See also **Electrostatics**.
3. *Structural, hydration, or solvation forces* — These forces may be attractive, repulsive, or oscillatory and depend upon the structuring or ordering of liquid molecules. (Solvation may be defined as the adsorption of a microlayer of film of water or other solvent on individual dispersed particles of a solution or dispersion.)
4. *Repulsive entropic (steric or fluctuation) forces* — As defined by Israelachvili and McGuiggan, these are "forces which arise from the thermal motions of protruding surface groups (such as polymers or lipid head groups) or from the thermal fluctuations of flexible fluid-like interfaces (or surfactant or lipid bilayers)."

Although, in a vacuum, only the Van Der Waals forces are important; in liquids, all forces may operate simultaneously. In liquids, it is extremely difficult to separate the effects of each of the aforementioned forces.

In the 1950s, the DLVO theory was based largely upon forces (1) and (2) defined above. The DLVO theory became the basis for studying the properties of colloidal and biocolloidal systems.

Electrorheological Fluids

Complications continue in the theoretical exploration of liquid behavior, but, in attempting to learn about the complexities, leads toward a more fundamental understanding of liquids may emerge. One of these complexities is a class of fluids referred to as *electrorheological*.

In a normal setting, these liquids are liquid in the conventional sense, but, when they are subjected to a strong electric field, they become solids. A common example is a mixture of cornstarch and vegetable oil. The viscous, sticky starting mixture of these two components can be converted into a hardened solid material with the application of an appropriate electric field. In recent years, through random researching, investigators have found numerous combinations of materials that qualify as electrorheological fluids, but to date no satisfactory explanation of the effect from a theoretical standpoint has been developed. The effect of the ER effect can be observed readily by microscopic examination. In a normal liquid or in an electrorheological mixture not in an electric field, examination shows particles moving in random fashion throughout a container, as may be expected. But, when a field is applied, long strands of particles appear to be "solidly" linked together, thus providing rigidity to the mix. No retentivity is involved, however, because liquid normalcy is returned immediately upon cessation of the electric field. This action intrigues a number of designers of equipment, as in the electronics and valve fields, where such materials may be used in future circuits, valves, and any number of other "on-off" devices.

Artificial Magnetic Fluids. The concept of magnetizable liquids dates back over several decades. The first breakthrough occurred in the mid-1960s, with the production of stable colloids of subdomain solid ferromagnets, which variously were called magnetic fluids, magnetizable fluids, or simply ferrofluids. These may be prepared by size reduction or precipitation. It has been a rather remarkable feat to grind bulk material down to a size of 100 micrometers. Grinding is done in the presence of a dispersing agent and a solvent. In chemical precipitation, iron(II) and iron(III) ions in aqueous solution are coprecipitated in the molar ratio of about 2:1 using ammonium hydroxide. To maintain the magnetic product in a small colloidal size range, a peptization step is included in which the particles are transferred to a heated organic phase containing the dispersing agent. The behavior of these artificial fluids offers techniques for achieving efficient heat and mass transfer, drag reduction, wetting, fluidization, sealing, damping, and other process and product potential uses.

Polymer Liquids

Sir Samuel F. Edwards of the Cavendish Laboratory, previously mentioned, observes, "Polymer liquids are liquid because the temperature is high enough to change the configuration of the molecules easily." See Fig. 4.

Scientists and engineers experienced in the production and application of polymeric liquids have learned from thousands of examples of how polymeric liquids behave, and can even classify some of them into behavioral families. But the complexity of these substances to date has eluded the achievement of precise designs and predictive behavior.

Mixing of Liquids

In the chemical process industries and in food manufacturing, the mixing of liquids is an important and frequently used operation. Over the years, the mixing operation has been poorly understood and essentially remains so. Progress in equipment design has been achieved mainly through the development of extensive empirical information rather than upon the creation of precise mathematical and theoretical relationships. J.M. Ottino (University of Massachusetts) observes, "There are many fundamental questions regarding mixing in slow three-dimensional flows, and unfortunately some of the intuition we have obtained from our study of two-dimensional flows does not necessarily carry over to flows in three dimensions." The Ottino reference listed describes studies of chaotic and nonchaotic flows in laboratory setups.

Multiphase Fluid Flow

It is quite common in industrial and cross-country liquid transport problems to encounter mixtures of liquids, vapors, and gases — that is, the presence of two phases of matter. As pointed out by D.I. Koch (Cornell University), "Research programs in this area at Cornell involve studies that fall outside of the traditional realm of chemical engineering, including blood flow in capillaries, the transport of contaminants in groundwater, the dynamics of geothermal reservoirs, enhanced oil recovery, the processing of fibrous composites, melt-spinning processes, and the growth of silicon crystals."

As previously mentioned in this article, liquid behavior presents an immense variety of puzzling problems that are difficult to comprehend and hence difficult to forecast precisely. As just one example, in the study of large drops of liquid at high flow rates, the inertia of the drops and the surrounding fluids play an important role. When such drops are propelled toward one another, they may coalesce into a single drop or may rebound like a pair of elastic balls, a phenomenon that is partially (but not fully) dependent upon the comparative velocities of the two drops.

In traditional industrial two-phase flow situations, as occur in process piping, engineers classify flow patterns, as indicated in Fig. 5. This type of classification and the development of empirical data from past experimentation and practice assist much in simplifying the problems from a practical, if not from a theoretical, standpoint. Multiphase flow behavior considerations are essential in calculating pipe diameters, pumping capacities, and energy consumption. Multiphase flows also exhibit different characteristics

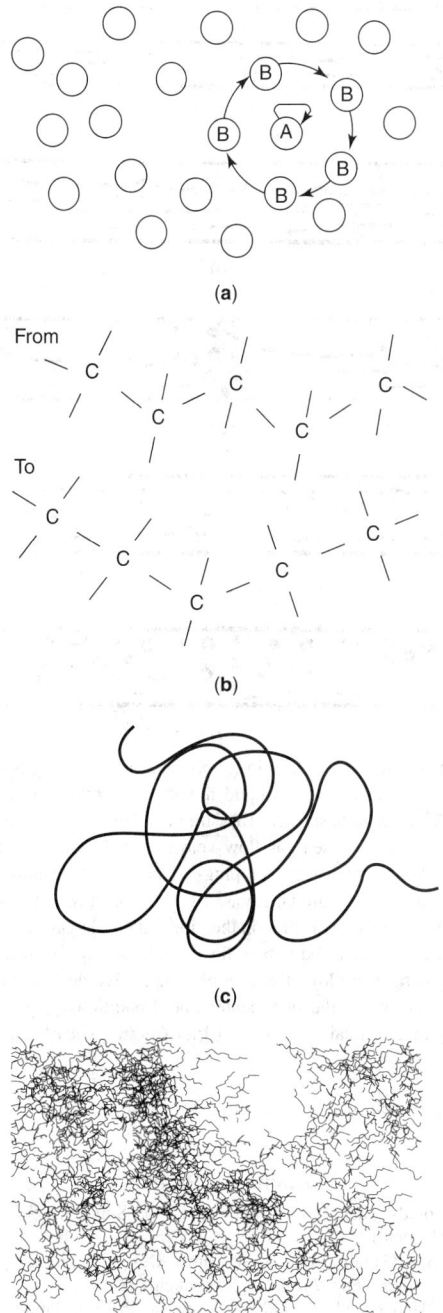

Fig. 4. Sketches by Edwards for schematically illustrating the behavior of polymer liquids. (**a**) Central molecule A moves around a (temporary) average position, occasionally escaping the barrier of the molecules B when some fluctuation of their positions permits it to do so. But, in addition to this single molecule motion, the molecules B can move around cooperatively, one of many cooperative motions that the observer may "invent." The numerous possibilities immediately derail the formulation of a simple theory. As explained by Edwards, "If the motion only involved one molecule at a time, quite a reasonable theory can be put together, but it will always be inadequate to describe the whole motion, and possibly, this always will be the case. The very apparatus of mathematical equations cannot handle this level of complexity, even though the human mind has no difficulty in seeing where the trouble is. The best current way to deal with this problem is to put all the molecules onto the computer and study their molecular dynamics." Continual changes are possible, and the long-chain molecule can be regarded as a flexible string in continual motion. A single coil may be drawn, as shown in (**c**). A computer model could be generated to show many polymers projected onto a plane, as indicated in (**d**). Edwards observes, "So the liquid looks like a seething pit of wriggling motion, a kind of living spaghetti, but with less smooth shapes than real spaghetti. How can one describe such a system? How do the molecules move, and how does the material flow?" (*Sketches after Sir Samuel F. Edwards.*)

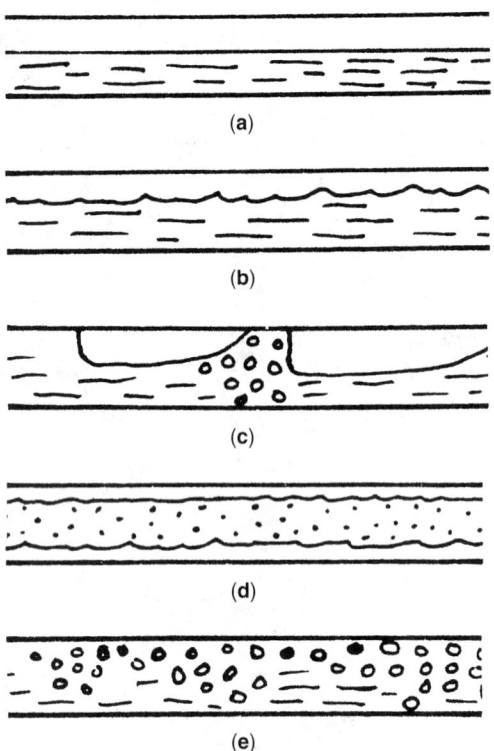

Fig. 5. Types of two-phase flow in a horizontal pipeline: (**a**) Stratified smooth flow where gas velocity is low. Liquid flows along bottom portion of pipelines with essentially a smooth surface. (**b**) Stratified flow with a wavy surface, the waviness caused by increased gas flow velocity. (**c**) Liquid bridges the pipeline cross section, thus causing slugs or plugs of liquid, which move at a velocity approximately that of the flowing gas. (**d**) Annular flow, in which the liquid essentially flows as an annular film on the pipe wall while gas flows as in a central core of the pipe. (**e**) Dispersed bubble flow usually results when liquid flow rates are high and gas rates are low. Because of comparative density differences, most bubbles are found above the pipe center line. Conditions vary somewhat when the pipeline is in a vertical orientation. (*After Cindric, Gandhi, and Williams.*)

when being pumped uphill or flowing downhill. See also **Fluid and Fluid Flow**.

Additional Reading

Amato, I.: "Liquids That Tiptoe on the Edge of Solidity," *Science News*, 342 (December 1, 1990).

Cindric, D.T., S.L. Gandhi, and R.A. Williams: "Designing Piping Systems for Two-Phase Flow," *Chem. Eng. Progress*, 51 (March 1987).

Coker, A.K.: "Understand Two-Phase Flow in Process Piping," *Chem. Eng. Progress*, 60 (November 1990).

Donnelly, R.J.: "Superfluid Turbulence," *Sci. Amer.*, 100 (November 1988).

Edwards, S.F.: "Polymer Liquids," *Review (University of Wales)*, 58 (March 1987).

Egelstaff, P.A.: "Introduction to the Liquid State," Oxford University Press, Inc., New York, NY, 1994.

Granick, S.: "Motions and Relaxations of Confined Liquids," *Science*, 1§374 (September 20, 1991).

Grimmett, G., B. Eckmann, S.S. Chern, and H. Hironaka: "Percolation," 2nd Edition, Vol. 321, Springer-Verlag, Inc., New York, NY, 1999.

Heyes, D.M.: "The Liquid State: Applications of Molecular Simulations," John Wiley & Sons, Inc., New York, NY, 1998.

Israelachvili, J.N. and P.M. McGuiggan: "Forces Between Surfaces in Liquids," *Science*, 795 (August 12, 1988).

Koch, D.L.: "Fluid Dynamics in Multiphase Systems," *Chem. Eng. Progress*, 74 (November 1989).

Langer, J.S.: "Dendrites, Viscous Fingers, and the Theory of Pattern Formation," *Science*, 1150 (March 3, 1989).

Lounasmaa, O.V. and G. Pickett: "The ^{3}He Superfluids," *Sci. Amer.*, 104 (June 1990).

Luessen, L.H., L.G. Christophorou, and E.E. Kunhardt: "The Liquid State and Its Electrical Properties," Perseus Books, Boulder, CO, 1988.

March, N.H., M.P. Tosi and R.A. Street: "Amorphous Solids and the Liquid State," Kluwer Academic Publishers, Norwell, MA, 1985.

McComb, W.D.: "The Physics of Fluid Turbulence," Oxford University Press, Inc., New York, NY, 1992.

Monastersky, R.: "Stretching Liquid to Its Physical Limit," *Science News*, 87 (August 11, 1990).

Ottino, J.M.: "The Mixing of Fluids," *Sci. Amer.*, 56 (January 1989).

Ottino, J.M. et al.: "Morphological Structures Produced by Mixing in Chaotic Flows," *Nature*, 419 (June 2, 1988).

Pool, R.: "The Fluids with a Case of Split Personality," *Science*, 1180 (March 9, 1990).

Schmidt, W.F.: "Liquid State Electronics of Insulating Liquids," CRC Press, LLC., Boca Raton, FL, 1997.

Snedden, R.: "States of Matter: Solids, Liquids and Gases," Heinemann Library, Oxford, UK, 2001.

Stixrude, L. and M.S.T. Bukowinski: "A Novel Topological Compression Mechanism in a Covalent Liquid," *Science*, 541 (October 26, 1990).

Tabor, D.: "Gases, Liquids, and Solids: And Other States of Matter," 3rd Edition, Cambridge University Press, New York, NY, 1991.

Thompson, P.A. and M.O. Robbins: "Origins of Stick-Slip Motion in Boundary Lubrication," *Science*, 792 (November 9, 1990).

LIQUIDUS CURVE. In a temperature-concentration diagram, the line connecting the temperatures at which fusion is just completed for the various compositions.

LISSAJOUS, JULES ANTOINE (1822–1880). Lissajous was a French mathematician. He is remembered for frequency patterns (Lissajous figures). The generation of Lisajous figures on a cathode-ray tube is a common method of frequency comparison.

See also **Frequency (Electric) Measurement**; and **Harmonic Motion**.

J. M. I.

LISTERIOSIS. A disease of animals including humans caused by *Listeria monocytogenes*, a thin Gram-positive bacillus having several serotypes. The organism is a soil saprophyte which is present in the intestines of many animals and birds. These reservoirs are potential sources of exposure to humans, but despite the ubiquity of *Listeria sp.*, human disease caused by the organism is uncommon. Most infections occur in the first month of life where, in early onset, the mortality may be as high as 40–60%; or beyond the age of 55, where there is usually an underlying predisposing illness.

The most common form of listeriosis is meningitis, with bacteremia occurring in 5–30% of cases, and endocarditis, osteomyelitis, and cholecystitis also sometimes evidenced.

Ampicillin or penicillin are the antibiotics of choice, with treatment being continued for at least ten days after the patient becomes afebrile.

R. C. V.

LITCHI TREE. Of the family *Sapindaceae* (soapberry family), the *Litchi chinensis* is probably best known for its fruit, which when dried is called the litchi "nut." A native of southern China, the tree has spread extensively in cultivation through many southern Asiatic countries. The tree has pinnately compound leaves, the leaflets of which are lanceolate and leathery. The small flowers are borne in panicles, and have no petals. The fruit is roughly spherical, 1 to $1\frac{1}{2}$ inches in diameter, with a hard, brittle rind. Within this rind is a fleshy, translucent pulp, the aril, the part which is eaten. When fresh, it is delectable; when dried into "nuts," it is much shrunken. The fruit contains a single seed.

LITHIFICATION. To lithify is, literally, to turn to stone. Lithification is a term commonly applied to the consolidation and hardening of sediments so as to form a sedimentary rock.

LITHIUM. Chemical element, symbol Li, at. no. 3, at. wt. 6.941, periodic table group 1, mp 180.54 °C, bp 1342 °C (at 760 torr), density 0.534 g/cm³ (20 °C). Lithium is lightest in weight of all the chemical elements that are solid at standard conditions. Elemental lithium in the solid phase has a body-centered cubic crystal structure. In comparison with other members of the alkali metal series, lithium has the smallest ionic radius, the highest ionization potential, the highest electronegativity, and the greatest heat capacity. Generally, lithium is the least reactive of the alkali metals. Lithium is a silver-white metal, harder than sodium, but softer than lead. It is tough and may be drawn into wire or rolled into sheets. The element tarnishes rapidly in air and often is preserved under naphtha. The reaction with H_2O is vigorous, producing LiOH (lithium hydroxide) and hydrogen. There are two naturally occurring isotopes, ^{6}Li and ^{7}Li. They are not radioactive. Two radioactive isotopes have been identified, ^{5}Li and ^{8}Li, both with very short half-lives, measured in fractions of seconds. Among elements occurring naturally in the earth's crust, lithium ranks 28th with an estimated average content of about 10–20 ppm. In terms of content in seawater, lithium ranks 17th with an estimated content of approximately 950 tons of lithium per cubic mile of seawater. The element was first identified by Johann August Arfvedson in 1817 in the laboratory of Berzelius. The name of the element is accredited to Berzelius.

First ionization potential 5.39 eV. Oxidation potential Li → Li$^+$ + e$^-$, 3.02 V. Other physical properties of lithium are given under **Chemical Elements**.

The main sources of lithium are pegmatites and brines. The most important pegmatite mineral is *spodumene*, LiAlSi$_2$O$_6$, which contains a theoretical content of 8.03% Li$_2$O. *Petalite*, LiAlSi$_4$O$_{10}$, contains between 4 and 4.5% Li$_2$O. *Lepidolite*, a complex mica, contains between 3 and 4% Li$_2$O. See also entries on **Lepidolite; Petalite;** and **Spodumene**. Brines contain normally a few hundred to a few thousand parts per million (ppm) of lithium. The only commercial source of spodumene in North America is located in North Carolina. Abundant resources of lithium pegmatites occur in Canada, the African continent, and unconfirmed sources in Russia and China. Significant quantities of lithium (as carbonate) are produced from the brines of Clayton Valley, Nevada. A recently discovered, lithium-rich brine deposit has been located in the Atacama desert of Chile. More detail on lithium resources is given in the next entry.

There are three major processes for extracting lithium from pegmatite ores. (1) An acid process, wherein the spodumene concentrate, after calcining at about 1095 °C, is reacted with sulfuric acid, followed by water leaching of the resulting lithium sulfate, Li$_2$SO$_4$. The sulfate is then converted to the carbonate with soda ash. (2) An alkaline process, wherein the ore is reacted with lime or limestone at high temperatures followed by water leaching of the resulting lithium hydroxide. (3) A base exchange method, whereby the ore is reacted with an alkaline chloride or sulfate at a high temperature in an aqueous phase to yield a soluble lithium salt. The sulfuric acid leaching method is the only commercial process for extraction of lithium from spodumene in practice today.

Lithium metal was first prepared by Sir Humphry Davy in 1818 by electrolyzing lithium oxide. At about that same time, Brande also isolated the metal. In 1855, R. Bunsen and A. Matthiessen prepared gram amounts by electrolyzing fused lithium chloride. Modest commercial quantities were first made in Germany during World War I when the metal was considered as a potential alloying material. Limited production did not commence in the United States until the early 1930s. Present commercial methods were pioneered by Guntz in 1893 and involve electrolyzing a low-melting mixture of LiCl and KCl. Graphite anodes and mild steel cathodes are used. Lithium is formed at the cathode and rises to the surface, from which it is skimmed periodically. Pure lithium chloride is added to the bath as required. Chlorine gas is liberated at the anodes. The process yields a lithium metal of about 99.8% purity. The metal normally is cast into ingots of different sizes, but is also available as extruded rod, ribbon, or wire. The metal also is available as "sand" — fine dispersions in the 10–30 μm range.

Lithium in Metallurgy and Alloys

In metallurgy, lithium metal is used as a deoxidizer, desulfurizer, and degasifier in the production of a number of molten metals, notably copper and copper alloys. Lithium also is an ingredient of an increasing number of alloys, particularly with aluminum. Early alloys included aluminum alloy X2020 (1% Li), which is a structural alloy with improved high-temperature strength. In another early Li alloy, about 14% Li is alloyed with magnesium in the LA 141 alloy, designed for very light-weight structural applications, notably in aerospace applications.

In late 1989, a new proprietary (Martin Marietta Corp.) family of weldable, high-strength Al-Li alloys was introduced. With a 690-MPa $(100 \times 10^3$ psi) yield strength, the material is claimed to be twice as strong as the previous leading Al-Li alloys. This alloy was developed specifically for space launch systems. The alloy is claimed to maintain a high strength under thermal conditions ranging from cryogenic to elevated temperatures. A primary use is for fuel and oxidizer tanks, where its weldability is a marked advantage. Sheet, plate, extrusion, and ingot products of the new alloy also are available.

The addition of lithium to aluminum castings has been found to be particularly advantageous. Lithium produces a lower density and higher stiffness over conventional aluminum alloys used in aerospace applications.

Lithium has one of the highest solubilities of any aluminum alloying element. About 4.2% Li can be dissolved in Al at the 602 °C (1116 °F) eutectic temperature. The hardness of Al-Li alloys improves with aging temperature. Al$_3$Li precipitates are formed, producing higher hardness. Yield strength also increases with higher aging temperature and higher Li content.

Lithium alloyed with silver has been used for fluxless brazing.

Lithium Batteries. For many years, lithium has been considered for use in batteries, particularly with the growing emphasis on the electric car. See also **Battery**.

Chemistry and Compounds

Lithium has the highest ionization potential (i.e., Li → Li$^+$ in the vapor) of the alkali metals. However, the measured value of its oxidation potential against a normal aqueous solution of its ion is 3.02 V, which does not differ from those of the other main group I metals by as much as the difference in ionization potentials. That difference, attributed to the high heat of hydration of Li$^+$, explains why lithium is a vigorous reductant in aqueous systems, but reacts slowly with H$_2$O, and not at all with dry oxygen except above 100 °C.

The single $2s$ electron in the outer shell of lithium is easily removed to form the positive ion, and stability of the remaining $1s^2$ electron pair requires too high a potential (75.62 eV) for any further ionization (by chemical means) so that lithium is exclusively monovalent in its compounds.

Because of the reactivity of lithium with water to form its hydroxide, LiOH, and hydrogen, its properties when dissolved in other solvents have been studied extensively. It does not decompose liquid NH$_3$, but does form a blue solution, which decomposes to yield its amide, LiNH$_2$, and hydrogen, when catalyzed by metallic salts. With the elements of main groups 2 to 7, lithium in liquid NH$_3$ reacts to form binary compounds, which may vary from simple halides, as with the halogens, to intermetallic phases, as with cadmium and mercury. Lithium amide in liquid NH$_3$ is regarded in the same class as a hydroxide in aqueous solution.

Many other lithium compounds not obtainable in aqueous solution can be produced from the solution of lithium in liquid ammonia. Thus the acetylide is obtained by action of acetylene.

$$C_2H_2 + LiNH_2 \longrightarrow LiC_2H + NH_3$$

$$2\ LiC_2H \longrightarrow Li_2C_2 + C_2H_2$$

The amide, as stated above, is produced by catalyzed decomposition of the liquid NH$_3$ solution, and the nitride, Li$_3$N, by heating the amide or by direct combination of the elements.

Lithium salts exhibit general high solubility and a high degree of dissociation in other nonaqueous solvents than liquid ammonia, such as liquid sulfur dioxide and acetic acid.

Like the other alkali metals lithium forms compounds with virtually all of the anions, inorganic as well as organic. The lithium salts are in many instances different in their solubility properties from the corresponding salts of the other alkali metals. Thus lithium fluoride, phosphate, and carbonate are the least soluble alkali metal fluoride, phosphate, and carbonate, the solubilities for the other alkali metals increasing with increasing ionic radius. Lithium chlorate and dichromate are, on the other hand, the most soluble alkali chlorate and dichromate, the solubilities for the other alkali metals decreasing with increasing ionic radius. These differences are partly explained, as was that in the oxidation potential, by the considerable hydration of the lithium ion, which also explains the fact that lithium salts generally crystallize as hydrates. Lithium salts, probably because of the small size of the lithium ion, do not form mixed crystals with the other alkali salts, but they do form double salts, notably the two series of lithium-sodium and lithium-potassium sulfates.

Lithium forms several organic compounds. Most of them are lithium salts or lithium acid salts of organic acids or other oxygen-connected lithium compounds. The number of lithium-carbon bonded compounds that have been reported is very small including, in addition to the carbide, methyllithium, CH$_3$Li, ethyllithium, C$_2$H$_5$Li, *n*-propyllithium, C$_3$H$_7$Li, *n*-butyllithium, C$_4$H$_9$Li, benzyllithium, C$_6$H$_5$ · CH$_2$Li, and methylenedilithium LiCH$_2$Li.

The alkyllithium compounds are usually colorless, soluble in organic solvents, and capable of distillation or sublimation. They are nonelectrolytes and are widely used in synthetic organic chemistry, since, like other lithium compounds, they resemble in their properties the corresponding magnesium compounds.

Lithium carbonate. Li_2CO_3, mp 72.6 °C, slightly soluble in H_2O. Used in glass, enamel, and ceramic formulations, in the electrowinning of aluminum, and in the manufacture of other lithium compounds. The compound also has been used in the treatment of manic-depressive psychoses.

Lithium hydride. LiH, mp 686.4 °C, reacts vigorously with H_2O. With NH_3, it forms the amide. The compound is used to produce $LiAlH_4$ and other double hydrides. Lithium hydride is an excellent light-weight source of hydrogen. One pound yields 45 cubic feet of hydrogen (one kilogram yields 2.8 cubic meters of hydrogen) at standard conditions. The compound also can serve as a light-weight shield for thermal neutrons.

Lithium hydroxide monohydrate. $LiOH \cdot H_2O$ loses water at 101 °C. LiOH melts at 450 °C. The compound is soluble in water. The compound is used in the formulation of lithium soaps used in multipurpose greases; also in the manufacture of various lithium salts; and as an additive to the electrolyte of alkaline storage batteries. LiOH also is an efficient, light-weight absorbent for carbon dioxide.

Lithium bromide. LiBr, mp 550 °C, soluble in H_2O or alcohols. The compound is very hygroscopic and forms four hydrates. Major use has been in absorption-refrigeration air-conditioning systems in which H_2O is the refrigerant—strong LiBr is used to absorb H_2O vapor.

Lithium chloride. LiCl, mp 608 °C, soluble in H_2O or alcohols. Very hygroscopic and forms four hydrates like the bromide. The compound is a component of brazing fluxes for aluminum and magnesium. It is used in dehumidification systems, as an additive to the electrolyte of dry cells for low-temperature applications; and it is used in low-freezing fire-extinguishing systems; as an ingredient of fused-salt baths to lower fusing temperature; and, as a coating, in humidity-sensing instruments.

Lithium fluoride. LiF, mp 848 °C, soluble in H_2O (slight). Used in enamel and glass formulations; as a component of welding and brazing fluxes; in the electrowinning of aluminum; and as an ingredient of molten salts.

Lithium for Thermonuclear Fusion Reactors

The possible long-range role of lithium in fusion reactor technology has caused concern over the future availability of lithium. As early as 1975, a lithium subpanel, formed under the auspices of the National Academies of Science and Engineering, evaluated the Li raw materials available in the western world. The study indicated that reserves (adjusted for mining losses) are 2.54×10^6 metric tons. This figure does not include potentially large sources contained in South American salares, geothermal brines, oil field brines, and lithium-rich clays. Tritium, required in the fusion reactor core, would be produced from lithium. Lithium will be used in the blanket surrounding the core. Assuming 100% efficiency, the Li consumption for a 1000-MW(e) fusion power plant would be approximately 200 kg/year. See also **Nuclear Power Technology**.

Lithium in Biological Systems

Although much remains to be learned, there is considerable evidence that lithium can play an active role (positive and negative) in biological systems. Possibly most widely known is the use of lithium salts, notably lithium carbonate, in the therapy for mania (a condition where the patient is mentally and physically hyperactive, associated with an elevated mood and disorganized behavior). Frequently associated with mania is the broad swinging of the patient's mood (*bipolar disorder*). Studies have shown that persons with mania have a defect in the transmissions of impulses between and along nerve cells in the brain, which depends upon the regulated movement of ions across the membranes of those cells. Lithium antagonizes synaptic transmission of catecholamines in the brain by inhibiting norepinephrine and dopamine release. This is the result of weakly increasing their re-uptake by the presynaptic neuron and by decreasing storage. Lithium also interferes with the ability of several hormones to stimulate adenylate cyclase, a property that is believed to decrease the action of catecholamines at the postsynaptic receptor sites. See also **Central and Peripheral Nervous Systems**.

Additional Reading

Birch, N.J., V.S. Gallicchio, and R.W. Becker: "Lithium: 50 Years of Psychopharmacology: New Perspectives in Biomedical and Clinical Research," Weidner Publishing Group, Riverton, NJ, 1999.

Carr, S. et al.: "Increase in Glomerular Filtration Rate in Patients with Insulin-Dependent Diabetes and Elevated Erythrocyte Sodium-Lithium Countertransport," *N. Eng. J. Med.*, 500 (February 22, 1990).

Davis, J.R.: "Metals Handbook," 2nd Edition, ASM International, Materials Park, OH, 1998.

Greenwood, N.N. and A. Earnshaw: "Chemistry of the Elements," 2nd Edition, Butterworth-Heinemann, Inc., Woburn, MA, 1997.

Julien, C. and Z. Stoinov: "Materials for Lithium-Ion Batteries," Kluwer Academic Publishers, Norwell, MA, 2000.

Krebs, R.E.: "The History and Use of Our Earth's Chemical Elements: A Reference Guide," Greenwood Publishing Group, Inc., Westport, CT, 1998.

Kubel, E.J., Jr.: "New Al-Li Alloy," *Advanced Materials 7 Processes*, 10 (October 1989).

Lewis, R.J. and N.I. Sax: "Sax'x Dangerous Properties of Industrial Materials," 10th Edition, John Wiley & Sons, Inc., New York, NY, 1999.

Lide, D.R.: "CRC Handbook of Chemistry and Physics 2000–2001," 81st Edition, CRC Press, LLC., Boca Raton, FL, 2000.

Sapse, Anne-Marie and P. von Rague Schleyer: "Lithium Chemistry: A Theoretical and Experimental Overview," John Wiley & Sons, Inc., New York, NY, 1994.

Schou, M.: "Lithium Treatment of Manic Depressive Illness: A Practical Guide," S. Karger Publishers, Inc., Farmington, CT, 1993.

Swartz, C.M.: "Serum Lithium During Treatment of Bipolar Disorder," *N. Eng. J. Med.*, 1159 (April 19, 1990).

Taketani, H.: "Properties of Al-Li Alloy 2091-T3 Sheet," *Advanced Materials and Processes*," 113 (April 1990).

Wakihara, M. and O. Yamamoto: "Lithium Ion Batteries: Fundamentals and Performance," John Wiley & Sons, Inc., New York, NY, 1998.

LITHOLOGY. Literally, the graphic study of rocks, hence a synonym for petrography, but not petrology. This term is usually restricted, however, to the purely descriptive macroscopic study of rocks, without the aid of the petrographic microscope.

LITHOMETEOR. Solid matter suspended in the atmosphere, as smoke, dust, dry haze, etc., as contrasted with hydrometeor.

LITHOPRISM. See **Prism (Optics)**.

LITHOSPHERE. The solid part of the Earth or other spatial body. Distinguished from the atmosphere and the hydrosphere. See also **Earth**.

LITTORAL. Inhabiting the shoreline of the ocean in shallow waters and in the tidal zone, which is periodically exposed to the air.

LITUUS. A transcendental plane curve, a special kind of spiral. Its equation in polar coordinates is $r^2\theta = a^2$. It begins at infinity, constantly approaches the pole but never reaches it, and has the polar axis as an asymptote. See Fig. 1.
See also **Curve (Plane)**.

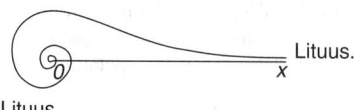

Fig. 1.   Lituus.

LIVER. The largest and one of the most complex organs in the human body, consisting of four lobes and located in the upper abdominal cavity. The liver performs multiple functions, including: (1) secretion and excretion; (2) blood-related activities, including regulation of blood volume and the formation and disposal of various blood components; (3) storage for certain vitamins and minerals; (4) metabolic functions, including fat and protein processing; and (5) detoxification. The approximate size of the adult human liver is 8–9 inches (20–22.5 centimeters) side to side, 4–5 inches (10–12.5 centimeters) front to back, and 6–7 inches (15–17.5 centimeters) top to bottom along the thickest portion. The organ weighs between 42 and 56 ounces (1.2 and 1.6 kilograms). Five ligaments attach the liver to the anterior walls of the abdomen and undersurface of the diaphragm. The lobes are usually identified as the *right* (largest); the *left* (somewhat smaller and wedge shaped; the *quadrate* (roughly square-shaped); and the *caudate* (tail-like configuration). Principal diseases and disorders with liver involvement are cirrhosis, hepatitis, and jaundice. Primary hepatic carcinomas are also seen, but less frequently. (The word *hepatic* indicates a condition of, or affecting the liver.) The liver is subject to adverse effects caused by a number of substances, including certain antimicrobials, such as isoniazid, rifampin, and pyrazinamide.

The liver secretes bile, which is discharged into the intestine: absorbs from the blood the products of carbohydrate digestion and stores them as glycogen; acts on nitrogenous wastes and returns them to the blood in the form of urea and related compounds; and destroys "worn-out" red corpuscles. The liver also produces fibrinogen and prothrombin. The bile discharged through the intestine plays an important role in the digestion of fat and carries with it some of the more complex waste products of the body. See also **Bile**. In structure, the liver is very complex. It develops as a hollow outgrowth of the embryonic gut just behind the stomach, which forks to produce the gallbladder and the liver. The connection with the gut persists as the common bile duct. In the adult, the liver cells are arranged in cords, separated by blood channels with incomplete lining known as sinusoids. Within the cords, minute bile capillaries between the cells converge to larger and larger ducts, which ultimately form the main hepatic duct. The gland receives blood from an arterial supply and also from the portal vein. The latter drains blood from the capillaries of the intestine and breaks up into sinusoids in the liver. These small passages are drained by the hepatic vein. The formation of stones in the biliary tract is described in **Gallbladder and Biliary Tract Diseases**.

A more detailed description of the physiology and biochemistry of the liver is given toward the end of this entry.

Familial Hypercholesterolemia. This disease has been treated by way of liver transplantation. It is a genetic disease caused by mutations in the gene encoding the low-density lipoprotein (LDL) receptor. This cell-surface receptor normally removes cholesterol-carrying LDL from the circulation. Persons with two mutant LDL-receptor genes produce few or no LDL receptors and, therefore, remove LDL from plasma at a reduced rate. As a result, LDL accumulates in plasma to levels up to 8 times normal. Patients almost always have severe atherosclerosis in childhood, with death from myocardial infarction often occurring before age 20 years. A single mutant LDL-receptor gene is inherited and occurs at a frequency of 1 in 500 in the general population: affected persons accumulate twice the normal level of LDL and symptomatic atherosclerosis usually occurs in the fourth and fifth decades of life. Typically, each heterozygote for familial hypercholesterolemia will transmit the mutant gene to half of the children, who then become heterozygotes. When two heterozygotes marry (estimated to be 1 in 250,000 marriages), one-fourth of the offspring will inherit a copy of the mutant LDL-receptor gene from both parents, and these offspring will be homozygotes.

Until recently, traditional approaches to the disease have not been effective. Inasmuch as the liver manufactures large numbers of LDL receptors (because the organ requires a large amount of cholesterol for secretion into the bile, for conversion to bile acids, and for the production of lipoproteins), some authorities reason that transplantation of a normal liver, with its normal receptors, should theoretically lower LDL levels profoundly in homozygotes. In early patients so treated, this has proved to be a correct assumption.

In mid-1990, Reihner and associated researchers (Karolinska Institute, Stockholm, Sweden) reported on the use of an inhibitor of cholesterol biosynthesis in the treatment of hepatic metabolism disorders. The cholesterol production rate-limiting enzyme (pravastatin) is 3-hydroxy-3-methyl glutaryl coenzyme A (HMG-CoA) reductase. In a study of ten patients over a period of three weeks, the group found that pravastatin therapy reduced total plasma cholesterol by 26 percent and LDL cholesterol by 39 percent. The report concludes: "Inhibition of hepatic HMG-CoA reductase by pravastatin results in an increased expression of hepatic LDL receptors, which explains the lowered plasma levels of LDL cholesterol."

Diseases of the Liver

Cirrhosis. In the Western Hemisphere, *chronic* diseases of the liver are comparatively infrequent—with exception of cirrhosis of the liver, which occurs frequently in Europe and the United States. In cirrhosis, the hepatic parenchyma (functioning tissue, as contrasted with connective tissue) is progressively destroyed and replaced by collagen (gelatinous substance found in connective tissue and bone). During surgery or autopsy, the organ will exhibit bands of collagen extending between the lobes and connecting portal areas. This process, if left untreated, ultimately grossly disorganizes the liver and leads to cessation of the organ's metabolic functions.

Alcoholic cirrhosis is the most common type of cirrhosis seen in the United States. Infrequently, the liver will shrink in this disease, but more frequently the organ will enlarge and may weigh two kilograms or more. Ingestion of large quantities of alcohol over a period of years[1] is the primary cause of alcoholic cirrhosis. It is no longer generally believed that poor nutrition, which often accompanies heavy alcohol consumption, is a primary cause of the disease, although it may be a secondary contributing factor to degeneration of the health of the individual.

In treatment, the logical first step for the patient is to stop drinking alcohol. Statistics indicate that the 5-year survival for patients who continue to drink is less than 50 percent. This may reach 80% in cases where the patient maintains abstinence. Treatment is directed toward maintaining good nutrition and preventing serious complications. Total fluid intake is supervised to effect an optimum fluid and electrolyte balance. Diuretics may be used, particularly where massive ascites (accumulation of serous fluid in abdominal cavity) are present. Vitamin K therapy is sometimes used. Protein intake may be restricted where hepatic encephalopathy may be suspected.

Cirrhosis increases the risk of gallstones as the result of elevated bilirubin. See also **Bile**. The risk of peptic ulcer is also increased twofold in cirrhosis. Also, in well-established cirrhosis, the *hepatorenal syndrome* may be seen. Simply defined, this is functional renal (kidney) failure. Mortality can range from 60 to 90 percent. Treatment is directed toward eliminating exogenous sources of ammonia, usually accomplished by restricting dietary protein. Means are also taken to control gastrointestinal bleeding. Where alcoholic liver disease is well established, some 10% of patients may develop *spontaneous bacterial peritonitis*. The exact mechanism underlying this condition is not fully understood. *Bleeding esophageal varices* are a serious complication of alcoholic cirrhosis.

The veins of the portal venous system transport all blood from the abdominal gastrointestinal tract, spleen, pancreas, and gallbladder, returning it to the heart by way of the liver. Portal hypertension in patients with cirrhosis causes gastrointestinal bleeding and esophageal varices. T. Poynard and a team of researchers (Franco-Italian Multicenter Study Group) conducted a study to determine the effectiveness of beta-andrenergic-antagonist drugs in the prevention of gastrointestinal bleeding in patients with cirrhosis and esophageal varices. Prior studies had not been conclusive. Generally, it had been reported that the continuous administration of beta-adrenergic-antagonistic drugs had induced a sustained decrease in portal pressure in patients with cirrhosis. Conclusion of the Poynard team findings (March 1991): "Propranolol and nadolol are effective in preventing first bleeding and reducing the mortality rate associated with gastrointestinal bleeding in patients with cirrhosis regardless of severity."

As reported by Massimo Colombo, et al. "Hepatocellular carcinoma is a highly malignant tumor with an extremely poor prognosis and an estimated incidence of about 1 million cases per year worldwide. Patients with cirrhosis of the liver have been identified as being at risk for this carcinoma." The causation of this particular type of tumor in association with cirrhosis is poorly understood. A study group at the University of Milan has concluded: "In the West, as in Asia, patients with cirrhosis of the liver are at substantial risk for hepatocellular carcinoma, with a yearly incidence rate of 3 percent. Our screening program did not appreciably increase the rate of detection of potentially curable tumors."

Effective treatment is difficult and bleeding from this source may be fatal in 70% of cases.

Primary biliary cirrhosis, relatively uncommon, is typically seen in women during the fourth to sixth decade of life. Symptoms in the early phase include a generalized itching of the skin (pruritus) and minor, continuing fatigue. Often the condition persists for a long period before a physician is consulted. In this disease, there is a significant drop in biliary secretion, causing a marked rise in serum cholesterol level. Xanthomas (small, yellow neoplastic growths) may occur about the eyes and tuberous xanthomas may be seen over the extremities. Bone pain, resulting from chronic malabsorption of fat-soluble vitamins A, D, E, and K, may be apparent. Diagnosis can be difficult because of similarities of primary biliary cirrhosis with subclinical cholangitis or other biliary tract diseases. See also **Gallbladder and Biliary Tract Diseases**.

Although the disease progresses slowly, the long-term prognosis is usually not good (10–20 years). Since there is no effective and specific therapy for the condition, treatment is directed toward alleviating symptoms.

---

[1] Ten percent of heavy drinkers (1 pint of whiskey daily for a number of years) run a high risk of developing cirrhosis.

Hemochromatosis. This is an infrequent liver disease in which inordinately large quantities of iron are deposited in the parenchymal cells of the organ. Eventually, these deposits destroy and scar the liver as in cirrhosis. Males have this disease at ten times the rate of females. Although onset may be earlier, symptoms usually develop during the fourth and fifth decades of life. Because of malfunctions in processes which govern iron absorption, the iron deposits not only in the liver, but also in the skin, pancreas, and heart muscle. The symptoms include a brown cutaneous pigmentation (caused by melanin) and grayish appearance (due to iron). The liver may be enlarged. There may be weight loss, decrease in body hair, and weakness. Other symptoms parallel those of diabetes mellitus, congestive heart failure or arrhythmias, and stiffness in the joints. Therapy at one time was directed toward ameliorating the aforementioned symptoms (related disorders), but comparatively recently it has been found that removal of a point (0.47 liter) of blood at regular intervals (*phlebotomy*), depending upon the patient's specific condition, is effective in depleting the iron stores. General improvement occurs if arthropathy or pituitary insufficiency are not present.

Wilson's Disease. Related to excessive deposits of copper, this is a liver-related disease and discussed in the entry on **Wilson's Disease**.

Hepatitis. An inflammatory and necrotic disease of liver cells. Commonly, the disease will be virus-induced, although hundreds of drugs are also known to cause hepatitis. It is difficult to determine the source. Among the known viruses that produce acute hepatitis are: (1) Hepatitis type A, once called infectious or short-incubation hepatitis virus; (2) hepatitis type B, serum or long-incubation hepatitis virus; (3) the non-A, non-B hepatitis viruses; (4) Epstein-Barr virus, which also causes infectious mononucleosis; and (5) cytomegalovirus. See also **Virus**. Hepatitis continues to appear after blood transfusions—in about 30,000 cases per year in the United States even though research has centered on preventing these occurrences. It should be pointed out that, although excellent tests are available for hepatitis B, unfortunately most post-transfusion cases (up to 90%) develop as the result of the presence of non-A, non-B viruses.

There is some evidence that type A and type B hepatitis may be decreasing in the United States. The fatality rate of type A hepatitis is low, probably not exceeding 0.2 percent. The rate is higher in type B hepatitis, ranging from 0.3 to 15 percent. It is estimated that nearly 45% of the population has an immunity to type A infections; 5–10% of the population for type B infections. Immunity for non-A, non-B infections is unknown. Type A virus is transmitted by the fecal-oral route. Sewage-contaminated shellfish are sometimes implicated. The transmission of type B virus may be percutaneous (penetration through skin), oral-oral, or venereal. With non-A, non-B viruses, the route is percutaneous. The incubation period varies with each type of virus: Type A hepatitis, 20–37 days, but in extremes may range from 15 to 49 days; type B hepatitis, 60–110 days, but in extremes, from 25 to 160 days; non-A, non-B viruses, 37–70 days, but in extremes, from 21 to 84 days.

The course of hepatitis infections also varies with the causative agent. Type A hepatitis does not progress to chronic liver disease, whereas about 10% of cases of Type B infections will lead to chronic liver disease. The risk of chronic liver disease with non-A, non-B virus infections is higher, ranging from 10–40% of cases. In situations where exposure to the virus is known, but disease has not developed, the administration of pooled gamma globulin is effective in the cases involving type A and non-A, non-B viruses, but is not effective in type B cases. Where there has been exposure to type B virus, the use of specific hepatitis B immune globulin is effective.

It is not surprising that the onset and course of acute viral hepatitis vary considerably because of the several possible causative factors. Onset may be sudden or gradual. In general, all or some of the early symptoms will include combinations of fatigue, lassitude, drowsiness, loss of appetite, nausea and, most specific to the disease, dark urine. Headache and very mild fever, in the absence of chills, may be present. There is usually mild and generalized abdominal discomfort. Movement tends to aggravate this discomfort. Itching of the skin may occur. Mild arthritis may develop, although this symptom is usually limited to type B virus infections. As the disease progresses, jaundice will be evident. See also **Jaundice**.

Particularly in older people with type B infections, recovery may be quite long—several months to a year—with recurring intermittent symptoms (relapse). A relapse is milder and of shorter duration than the initial attack. Rarely, the course of the disease will be rampant, leading to coma and even death. Fatal complications of hepatitis may include aplastic anemia,

hemolytic anemia, hypoglycemia, and polyarteritis. Some people do not recover completely from type A and non-A, non-B viruses and develop *chronic hepatitis*.

Many authorities agree that treatment seldom alters the course of acute viral hepatitis. Sensible suggestions are made to the patient—bed rest when there is excessive fatigue and serious discomfort, which may be present in the initial phase. Controlled studies have shown that vigorous physical exercise during the recovery phases does not increase the risk of relapse or chronic disease. It has been reported that a high-calorie diet (3000 + calories/day) for a few days at the appropriate time in the recovery stage may shorten the duration of the disease by a few days. Low-fat diets have not been shown to alter the course of the disease. During periods of nausea and vomiting, hospitalization may be required so that intravenous fluid and electrolyte replacement can be effected. Although alcohol has not been shown to aggravate the disease, abstinence is usually suggested in the interest of limiting any additional load on the hepatic and related systems.

In instances of severe acute viral hepatitis where the patient becomes encephalopathic, corticosteroids (not proven effective), hyperimmune gamma globulin, keto analogues of essential amino acids, and exchange transfusions, among other drugs and procedures, have been used. However, some authorities currently feel that acute encephalopathy responds little, if any, to treatment.

Chronic hepatitis may be *chronic active* or *chronic persistent*. Diagnosis is important because the therapy differs for the two conditions. Frequently, a percutaneous liver biopsy will be required. In chronic active hepatitis, the disease is variably progressive and eventually causes cirrhosis and hepatic failure. On the other hand, chronic persistent hepatitis does not progress. Persons with untreated chronic active hepatitis have a 5-year survival expectancy of less than 50%—possibly up to 90–95% where corticosteroid therapy is effective. This therapy in responsive patients brings about improvement of liver function within several months. In contrast, chronic persistent hepatitis is benign and usually patients lead a normal, active life, even though serum aminotransferase abnormalities may continue for many years.

Drug-induced hepatitis is often difficult to differentiate from the disease caused by a virus. Although uncommon, in one case in 9,000–10,000 patients the administration of halothane (see also **Anesthesia**) will produce hepatic necrosis. This occurrence is most common in overweight females or after a second exposure to the drug, and is frequently fatal. Drugs that also are hepatitis related include isoniazid, methyldopa, phenytoin, and the sulfonamides. Although such side effects are uncommon, the physician will be on guard for signs of hepatitis and liver damage in deciding on starting or continuing therapy with these drugs. There is some evidence that cysteamine or acetylcysteine may reverse the actions of these drugs if noted promptly. Poison derived from the wild mushroom *Amanita phalloides* is capable of producing overwhelming hepatic necrosis. See also **Foodborne Diseases**. The adverse effects of certain antimicrobial agents were mentioned earlier in this entry.

Reye's Syndrome. This is an often-fatal systemic disorder that follows viral infection in children. A number of cases of what appears to be Reye's syndrome have also been described in recent years in young and middle-aged adults. Present in the syndrome are encephalopathy and fatty liver. The syndrome develops suddenly, with onset of intractable vomiting occurring a few days after the viral illness. Sensorial impairment appears and soon afterward may progress to coma. Seizures also may occur. The liver is usually enlarged. Specific therapy is not available. Supportive measures include lactulose to control hyper-ammonemia; fresh frozen plasma to replenish clotting factors; mannitol or dexamethasone to lower increased intercranial pressure; and mechanical ventilation. A case fatality rate of 23% has been reported. Epidemiologic evidence strongly links Reye's syndrome with outbreaks of viral disease, especially influenza. Although the mechanism is unknown most physicians recommend that salicylates (aspirin and aspirin-containing mendicants) not be given to children with chicken pox or influenza.

Physiology and Biochemistry of the Liver

The blood returning from the intestine to the heart is shunted through a capillary system, the hepatic sinusoids, which are surrounded by epithelial cells arranged in plate forms. These plates cross each other in space at different angles, to permit the greatest possible contact between the blood and these polygonal epithelial cells. The resulting sponge-like organ located under the diaphragm and covered by the connective tissue capsule of Glisson is the largest organ of the body. Under normal circumstances, the major part of its blood, between 66 and 75%, comes from the portal vein, which drains the splanchnic capillaries, particularly those of intestine, pancreas and spleen. Approximately one quarter to one-third of the hepatic blood comes from the hepatic artery originating from the aorta at the celiac axis. Both hepatic artery and portal vein enter at the hilus of the liver and divide in a dichotomic fashion into parallel running branches. They are surrounded by ramified extensions of Glisson's capsule. The hepatic artery sends branches to the capillary plexus of the portal tracts, whereas the bulk of its blood is released into the sinusoids parenchyma, as does the portal vein, which forms by confluence of superior mesenteric, inferior mesenteric and splenic vein, and receives additional internal radicles from the portal capillary plexus. The sinusoids are blood spaces, normally without the basement membrane otherwise seen in capillaries; they are, therefore, characterized by great permeability for serum protein. Moreover, some of their lining endothelial cells, which are star-shaped and called Kupffer cells, are part of the reticuloendothelial system. The lining cells form the sinusoidal wall and leave small stomata open through which macromolecular substances pass into a tissue space between liver cell plates and sinusoids and extending between neighboring hepatocytes almost to the bile canaliculus. Tissue fluid is drained toward the lymphatics in either the central canal or, in the human, mainly the portal tract. Arterial and venous blood, mixed to a varying degree, flows toward the tributaries of the hepatic veins which combine to larger veins into which frequently small branches enter at almost right angles. The largest branches enter into the vena cava inferior behind the liver. Vascular sphincter mechanisms in various locations regulate hepatic blood flow and thus function. The portal tracts and the central canals around the tributaries of the hepatic veins cross each other in space and are throughout the liver about 0.3 mm apart. The direction of the blood flow from the portal tracts to the central canals produces the concentric arrangement of the liver cell plates characterizing the liver lobule, which conventionally is considered the structural unit of the liver.

The liver forms bile, which is released into slits between the liver cells, the bile canaliculi, which are arranged in a chicken-wire-like fashion; the wall of the canaliculi is formed by part of the hepatocellular plasma membrane. The bile canaliculi are drained by small tubes with an independent cuboidal epithelial lining, the ductules or cholangioles. Under normal circumstances hardly any are found within the lobule, the majority being in the periportal zone or in the portal tract. Under abnormal circumstances, they increase in number and are then found deep within the lobule. The ductules continue into the bile ducts located in the portal tracts, which unite in dichotomic fashion to finally form the common hepatic duct; this duct leaves the liver where hepatic artery and portal vein enter it. It combines with the cystic duct draining the gallbladder, which concentrates bile by water reabsorption to form the common duct running toward the duodenum. This entrance is controlled by the choledochoduodenal sphincter of Oddi. Bile is produced at an almost constant rate but released from the biliary system in human beings and many animals only if food appears in the duodenum. As a result of this or other mechanisms, the sphincter of Oddi relaxes and the gallbladder contracts. This leads to proper utilization of bile which, while being partly an excretory product, is a secretion essential in intestinal digestion and absorption.

In the liver, several fluid currents exist. Blood and some tissue fluid flow toward the central canal, while bile and most of the tissue fluid (at least in the human) are flowing toward the portal canal. The normal liver consists of approximately 60% hexagonal epithelial cells (hepatocytes), 30% littoral endothelial or Kupffer cells, and about 2% each of bile duct cells, connective tissue and blood vessels. The hepatocytes have three types of borders. Where they are in contact with each other, the border is straight indicating limited, if any, exchange of substance between individual cells. The border toward the tissue space is elongated by narrow extensions of the space between neighboring hepatocytes, and particularly by the formation of irregularly shaped finger-like projections in the form of microvilli. This tremendous elongation of the border of the hepatocytes and the preferential location of enzymes in this location reflects structurally the extensive exchange of substances between hepatocytes, tissue space and blood. Much shorter is the border toward the bile canaliculus, also thrown into microvilli, which are far more regular and disappear upon impairment of biliary secretion. Preferential accumulation of ATPase in the villi indicates the intensity of the metabolic processes in bile secretion.

The nucleus is normally vesicular and has conspicuous nucleoli. It varies considerably under normal and pathologic conditions, the majority being tetraploid in adult rodents. Binucleated cells increase in regeneration. The cytoplasm normally contains many and relatively large mitochondria in the matrix, of which the citric acid cycle enzymes and, in the double membrane, the electron transfer enzymes can be demonstrated. Ribosomes as ribonuclear protein are arranged around messenger RNA usually in helix form as polysomes. These polysomes as the site of protein biosynthesis may be either free in the cytoplasm or attached to the extensive endoplasmic reticulum, which thus becomes granular, and the site of secretion of protein such as serum proteins. The endoplasmic reticulum is also the site of steroid synthesis, and the smooth endoplasmic reticulum is the site of detoxification and of glucose-6-phosphatase. In addition, one notes the Golgi apparatus responsible for secretion and the perinuclear dark bodies, the lysosomes, which are the site of various hydrolytic enzymes mainly with peak activity in acid medium. They serve to segregate intracellular material after pinocytosis or for storage, secretion and separation of organelles undergoing destruction in the form of autophagic vacuoles. The soluble fraction of the cytoplasm, the hyaloplasm, corresponding to the supernatant fluid in cytochemical analysis, contains proteins and enzymes and cofactors related to carbohydrate metabolism and activation of amino acids and nucleic acids. In addition, in the normal liver, glycogen and few fat droplets are found as well as some ferritin crystals which, under abnormal circumstances, become hemosiderin deposits giving histochemical iron reaction.

The main functions of the liver cells are: (1) secretion of substances into the bloodstream of which the serum proteins particularly albumins, alpha-globulin, the proteins concerned with blood clotting, haptoglobin and transferrin, as well as some blood enzymes (e.g., esterase), serum cholesterol, and blood glucose are the most important; in contrast to all other tissues which utilize but do not form blood glucose, the liver cells are the main source of the blood glucose because of a specific phosphatase system; (2) storage of various metabolites particularly glycogen, proteins, fat and vitamins; (3) transformation of various compounds into each other, e.g., fats into carbohydrates and vice versa; (4) detoxification mainly by oxidation or conjugation, the latter mainly for better solubility and urinary excretion; (5) formation of the bile into which bile pigment is transmitted by conjugation and bile acids and cholesterol by transformation.

A variety of sinusoidal cells are seen. Some are flat endothelial cells, others are Kupffer cells with a cytoplasm of varying and irregular outlines and ameboid extensions. They contain few mitochondria but varying inclusions. They are engaged in phagocytosis of circulating exogenous and endogenous macromolecular or corpuscular elements, including bacteria, as well as of hepatocellular breakdown products, they are active in transformation of blood pigment to bile pigment. Other sinusoidal lining cells, rare under normal circumstances, form serum gamma-globulin and correspond to plasma cells. Also fibroblasts can be seen around the sinusoids.

The liver, as a whole, because of its strategic situation near the right heart and because of its sheer bulk, influences circulating blood volume, as well as electrolyte and water metabolism.

Liver Transplantations

It was just about two decades ago when a liver transplantation was covered during the prime time news. By the early 1990s, liver transplantation had become an established therapy, with such transplantations numbering in the thousands. Because of a very serious shortage of livers for donation, the waiting list numbers in the thousands. Liver transplantation ethics has become a major topic of discussion among medical and health professionals. A simple question typifies the current dilemma: "Who should receive a liver for transplantation? A young mother, whose prospects for surviving with a transplant is only 20 percent, or a 65-year-old alcoholic, whose chance of surviving is 80 percent if the patient stops drinking?" Some medical professionals have observed that, if the criteria for selecting patients become too narrow, this could have a dampening effect on liver transplantation technology.

A few experts believe that alternate non-transplantation therapies ultimately will develop, thus making liver transplantation procedures obsolete in the long run. However, for the immediate time, simple optimism for the future does not suffice.

When it was established that a human can regenerate a partly removed liver, some researchers proposed that pieces of liver tissue may be regenerated in the laboratory, thus maximizing the effectiveness of available whole organs from donors, who have been in short supply since the beginnings of liver transplant technology. Research along these lines commenced in the chemical and biochemical engineering laboratories at Massachusetts Institute of Technology. In the early phase of the project, liver cells were mounted on polymer mesh and treated with enzymes. The aim of the program is to grow the cells to about 10% of the mass of a whole liver. This tissue then would be transplanted into the patient and would, within several weeks, become fully-grown and replace the original liver. Research is continuing.

In a large series of childhood liver transplantations, nearly one-third have been performed because of metabolic disorders. Liver transplantation now is a well-accepted treatment for inherited metabolic disorders. Glycogen storage (Type IV) disease can be reversed by successfully transplanting a liver with normal amounts of branching enzyme. Inherited abnormalities of glycogen metabolism have been recognized for many years, mainly as the result of pathologists identifying gross accumulations of glycogen in tissues during postmortem examination. As pointed out by R.R. Howell (University of Miami), "The specific patterns of tissue involvement permitted recognition of a number of clinical types of glycogen storage disease long before the biochemical pathways had been identified and the specific causes of the inborn errors of metabolism understood." Some researchers had postulated that, after liver transplantation, progressive and probably fatal myopathy, cardiomyopathy, or encephalopathy would develop, since the enzyme defect is present in all the affected tissues, and that the abnormal glycogen would continue to accumulate in them. Surprising and fortuitous are findings that the predicted and ultimately fatal conditions have not occurred. A satisfactory explanation remains to be developed.

R.W. Strong and associates (Royal Children's Hospital, Brisbane, Australia) have reported that orthotopic liver transplantation is an effective therapy for end-stage liver disease and has proved to be a major advance in treating liver disease in children. The foremost obstacle faced by transplantation units worldwide has been the relative paucity of infant and child donors. The principle of transplanting a portion of the liver from an adult into a child has been accepted in many centers. Ethical issues must be considered when contemplating liver transplantation from parent to child. These issues are similar to those associated with the transplantation of renal grafts from living, related donors. Experience in many centers has shown that the risk to the donor is considered minimal. The risk to the recipient is considered no greater than that with the transplantation of a reduced-size graft from a cadaver donor. In the opinion of many specialists, living-donor liver transplantations are not justified when there are sufficient numbers of cadaver donors. In special instances, however, the procedure can be justified, as, for example, with a patient with fulminant hepatic failure when no cadaver donor is available and when the recipient has a reasonable chance of a successful outcome.

Cyclosporine is the basis of the immunosuppressive regimen in most patients undergoing orthotopic liver transplantation. The drug is a cyclic polypeptide that is produced by two species of fungi and is essentially insoluble in water. Its oral form consists of a solution of 100 mg of cyclosporine per milliliter of olive oil containing 12.5% ethanol by volume. Cyclosporine is absorbed much as fat and other fat-soluble substances are. Children, particularly infants, require much higher doses of orally administered cyclosporine than adults after liver transplantation. Bowel length appears to be the main determinant of the large difference between the adult and infant populations. These factors are reported by Whitington and associates (University of Chicago).

Additional Reading

Bacon, B.R. and A.M. Di Bisceglie: "Diseases of the Liver," Churchill Livingstone, Inc., Philadelphia, PA, 1999.

Blumberg, B.S.: "Hepatitis B and the Prevention of Cancer of the Liver: Selected Publications of Baruch S. Blumberg," 3rd Edition, World Scientific Publishing Company, Inc., Riveredge, NJ, 2000.

Blumgart, L.H. and Y. Fong: "Surgery of the Liver and Billary Tract," Vol. 1, 3rd Edition, W.B. Saunders Company, Philadelphia, PA, 2000.

Clavien, Pierre-Alain and K. Lyerly: "Malignant Liver Tumors: Current and Emerging Therapies," Blackwell Science, Inc., Malden, MA, 1999.

Colombo, M. et al.: "Hepatocellular Carcinoma in Italian Patients with Cirrhosis," N. Eng. J. Med., 675 (September 5, 1991).

Geller, S.A. and L. Petrovic: "Biopsy Interpretation of the Liver," Lippincott Williams & Wilkins, Philadelphia, PA, 2001.

Hellerstein, M.: "Influence of Pravastatin on Hepatic Metabolism of Cholesterol," N. Eng. J. Med., 128 (January 10, 1991).

Howell, R.R.: "Continuing Lessons from Glycogen Storage Diseases," N. Eng. J. Med., 55 (January 3, 1991).

Killenberg, P.G. and Pierre-Alain Clavien: "Medical Care of the Transplant Liver Patient," 2nd Edition, Blackwell Science, Inc., Malden MA, 2001.

Krawitt, E.L.: "Medical Management of Liver Disease," Marcel Dekker, Inc., New York, NY, 1999.

Krawitt, E.L., R.H. Wiesner, and M. Nishioka: "Autoimmune Liver Disease," 2nd Edition, Elsevier Science, New York, NY, 1999.

Maddrey, W.C. and M. Feldman: "Atlas of the Liver," 2nd Edition, Current Medicine, New York, NY, 2000.

Paumgartner, G. and U. Leuschner: "Immunology and Liver," Kluwer Academic Publishers, Norwell, MA, 2000.

Poynard, T. et al.: "Beta-Adrenergic-Antagonist Drugs in the Prevention of Gastrointestinal Bleeding in Patients with Cirrhosis and Esophageal Varices," N. Eng. J. Med., 1532 (May 30, 1991).

Reihner, E. et al.: "Influence of Pravastatin, A Specific Inhibitor of HMG-CoA Reductase, on Hepatic Metabolism of Cholesterol," N. Eng. J. Med., 323(4), 224 (July 26, 1990).

Sauerbruch, T., W.H. Caselmann, and U. Spengler: "Digestion: Complications of Liver Cirrhosis," S. Karger Publishers, Inc., Farmington, CT, 1999.

Sherlock, S. and J. Dooley: "Diseases of the Liver and Billary System," 11th Edition, Blackwell Science, Inc., Malden, MA, 2001.

Sorrell, M.F. and W.C. Maddrey: "Schiff's Diseases of the Liver," 8th Edition, Lippincott Williams & Wilkins, Philadelphia, PA, 1999.

Spital, A. and M. Spital: "The Ethics of Liver Transplantation from a Living Donor," N. Eng. J. Med., 549 (February 22, 1990).

Spital, A.: "The Shortage of Organs for Transplantation," N. Eng. J. Med., 1243 (October 24, 1991).

Steigmann, G.V. et al.: "Endoscopic Sclerotherapy as Compared with Endoscopic Ligation for Bleeding Esophageal Varices," N. Eng. J. Med., 1527 (June 4, 1992).

Strong, R.W. et al.: "Successful Liver Transplantation from A Living Donor to Her Son," N. Eng. J. Med., 1505 (May 24, 1990).

Suchy, F.J.: "Liver Diseases in Children," 2nd Edition, Lippincott Williams & Wilkins, Philadelphia, PA, 2000.

Veatch, R.M.: "An Alternative to Presumed Consent," N. Eng. J. Med., 1246 (October 24, 1991).

Whitington, P.F. et al.: "Small-Bowel Length and the Dose of Cyclosporine in Children After Liver Transplantation," N. Eng. J. Med., 733 (March 15, 1990).

Wright, R., G.H. Millward-Sadler, and M.J. Arthur: "Wright's Liver and Billary Disease: Pathophysiology, Diagnosis and Management," 3rd Edition, W.B. Saunders Company, Philadelphia, PA, 1999.

LIZARDS *(Reptilia, Sauria)*. Animals closely related to the snakes but having eyelids and the ventral surface of the body as well as the upper covered with small scales. They are usually elongate, with short legs and a long tail. Most lizards are small, though some species attain a length of several feet.

The classification and nomenclature of these animals is confused. The lizards are sometimes grouped with the snakes but some authorities regard them as a separate order of *Reptilia*. See also **Fossil Reptiles**.

There are many species of lizards but with the exception of the poisonous Gila monster and the edible iguanas they are of no economic importance. A few of the smaller species are eaten to a limited extent.

Table 1 gives the classification of lizards. Containing over 3000 species, the order *Squamata* (scaly reptiles) embraces about one-half of all modern reptiles. As shown by the table, there are several infraorders, families, and subfamilies, some of which are described in separate articles in this encyclopedia. Thus, please refer to **Agamids** (the agamas and chameleons); **Geckos** (geckos, snake lizards, and diabinids); **Iguanids** (iguanas, basalisks, and anoles); and **Skinks**.

This immediate article describes the remaining principal lizards, including girdled lizards, whiptails, so-called true lizards, shovel-snouted legless lizards, beaded lizards, monitors, and earless and ringed lizards, among others.

In many families of lizards, there are special "break points" in the bodies of the tail vertebrae; the tail beyond these can be discarded in various emergencies. Although the tail may be regenerated later, the new tail is usually shorter than the old and is supported not by vertebrae, but by a central rod of cartilage. The pattern of scales and coloration of the new tail frequently differ from the discarded tail.

Girdle-Tailed Lizards. Of the family *Cordylidae*, these lizards are essentially limited to sub-Saharan Africa. These lizards are well armored. The legs are reduced. The body is covered with longitudinal and transverse rows of rectangular scales, each having a bony element beneath it. The

TABLE 1. CLASSIFICATION OF LIZARDS

CLASS: Reptilia (Reptiles)
ORDER: Squamata (Scaly Reptiles)
SUBORDER: Sauria (Lizards)
INFRAORDER: Geckos (Gekkota)

FAMILY: Geckos (Gekkonidae)
Examples: House Geckos (*Hemidactylus*); Common Gecko (*Tarentola mauritanica*), Asian Tokay (*Gekko gekko*); Banded Gecko (*Coleonyx variegatus*); Banded Leaf-toed Gecko (*Hemidactylus fasciatus*); Smooth Gecko (*Thecadactylus rapicauda*); Madagascar Geckos (genus *Phelsuma*); Japanese Gecko (*Gekko japonicus*); Least Geckos (genus *Sphaerodactylus*); African Tropical Gecko (*Hemidactylus mabouia*); Turkish Gecko (*H. turcicus)*; European Leaf-fingered Gecko (*Phyllodactylus europaeus*); Sand Gecko (*Chondrodactylus angulifer*); Bibron's Gecko (*Pachydactylus bibronii*); Spotted Gecko (*Pachydactylus maculatus*); Web-footed Gecko (*Palmatogecko rangei*); Emerald Gecko (*Gekko smaragdinus*); Leaf-tailed Gecko (*Uroplatus fimbriatus*); Kuhl's Gecko (*Ptychozoon kuhli*); Panther Gecko (*Eublepharis macularius*); Padless Gecko (*Gonatodes albogularis*); Fat-tailed Gecko (*Oedura marmorata*).

FAMILY: Snake Lizards (Pygopodidae)
Examples: Western Scaly-foot (*Pygopus nigriceps*); Sharp-snouted Snake Lizard (*Lialis burtonis*); New Guinean Snake Lizard (*L. jicari*); Common Scaly-foot (*Pygopus lepidopodus*); Bailey's Scaly-foot (*Pygopus baileyi*).

FAMILY: Dibamids (Dibamidae)
Includes only one genus with three species. Some zoologists consider the dibamids as offshoots of the skinks.

FAMILY: Iguanids (Iguaninae)
SUBFAMILY: Sceloporinae
Examples: Southern Fence Lizard (*Sceloporus undulatus*); Western or Pacific Fence Lizard (*S. occidentalis*); Desert Spiny Lizard (*S. magister*); Tree Lizard (*Urosaurus ornatus*); Side-blotched Lizard or Ground Uta (*Uta stanburiana*); Banded Rock Lizard (*Petrosaurus mearnsi*); Fringe-toed Lizard (*Uma notata*); Greater Earless Lizard (*Holbrookia texana*); Zebra-tailed Lizard (*Callisaurus draconoides*); Short-horned Lizard or Pigmy Horned Lizard (*Phrynosomas douglasii*); Texas Horned Lizard (*P. cornutum*); Collared Lizard (*Crotaphytus collaris*); Leopard Lizard (*Gambelia wislizenii*).
SUBFAMILY: Tropidurinae
Examples: Spiny-tailed Iguanid (*Uracentron azureum*); Smooth-throated Lizards (genus *Liolaemus*); Crested Keeled Lizards (genus *Leiocephalus*); Narrow-tailed Lizards (genus *Stenocercus*); Madagascan Iguanid (*Oplurus sebae*); Weapon-tailed (*Hoplocercus spinosus*).
SUBFAMILY: Iguanidae
Examples: Common Iguana (*Iguana iguana*); West Indian Iguana (*I. delicatissima*); Rhinoceros Iguana (*Cyclura cornuta*); Marine Iguana (*Amblyrhynchus*); Spiny-tail Iguana (*Ctenosaura pectinata*); Desert Iguana (*Dipsosaurus dorsalis*); Chuckwalla (*Sauromalus ater*);
SUBFAMILY: Basiliscinae
Examples: Common Basilisk (*Basiliscus basiliscus*); Double-crested Basilisk (*B. plumifrons*); Banded Basilisk (*B. vittatus*); Helmeted Lizard (*Corytophanes*); Casque-headed Lizard (*Laemanctus serratus*).
SUBFAMILY: Anolinae
Examples: Long-legged Lizard (*Polychrus marmoratus*); Brazilian Tree Lizard (*Enyalius catenatus*); Patagonian Lizard (*Diplolaemus darwinii*); Sword-tailed Lizard (genus *Xiphocercus*); Cuban Water Anoles (genus *Deiroptyx*); False Chameleon (*Chamaeliolis chamaeleontides*); Carolina Anole (*Anolis carolinensis*); Knight Anole (*A. equestris*).

FAMILY: Agamids (Agamidae)
Examples: Common Agama (*Agama agama)*; Black-necked Agama (*A. atricollis*); Kirk's Rock Agama (*A. kirkii*); Bibron's Agama (*A. bibroni*); Desert Agama (*mutabilis*); Hardun (*A. stellio*); Caucasian Agama (*A. caucasica*); African Spiny-tailed Lizard (*Uromastyx acanthinurus*); Egyptian Spiny-tailed Lizard (*U. aegypticus*); Indian Spiny-tailed Lizard (*U. hardwickii*); Toad-headed Agamids (genus *Phrynocephalus*); Bearded Lizard (*Amphibolurus barbatus*); Spotted Agama (*A. maculatus*); Australian Bloodsucker (*A. muricatus*); Lake Eyre Agama (*A. maculosus*); Frilled or King's Lizard (*Chlamydosaurus kingii*); Lesuer's Water Dragon (*Physignathus lesueurii*); Oriental Water Dragon (*P. cocincinus*); Soa-soa (*Hydrosaurus amboinensis*); Philippine Water Lizard (*H. pustulatus*); Weber's Sailing Lizard (*H. weberi*); Bornean Angle-headed Lizard (*Gonocephalus liogaster*); Lyre-headed Lizard (*Lyriocephalus scutatus*); Indian Bloodsucker (*Calotes versicolor*); Bornean Bloodsucker (*C. cristatellus*); Ceylon Deaf Agamid (*Cophotis ceylanica*); Butterfly Lizard (*Leiolepis belliana*); Sita's Lizard (*Sitana ponticeriana*); Flying Dragon (*Draco volans*); Black-bearded Dragon (*D. melanopogon*); Five-lined Dragon (*D. quinquefasciatus*); Indian Dragon (*D. dussumieri*).

FAMILY: Chameleons (Chamaeleontidae)
Examples: European Chameleon (*Chamaeleo chamaeleon*); African Chameleon (*C. africanus*); Common Chameleon (*C. dilepis*); Two-lined Chameleon (*C. bitaeniatus*); Dwarf Chameleon (*C. pumilus*); Oustalet's Chameleon (*C. oustaleti*); Panther Chameleon (*C. pardalis*); Short-horned Chameleon (*C. brevicornis*); Mountain Chameleon (*C. montium*); Armored Chameleon (*Leandria permeata*).

INFRAORDER: Skinks and Allies (Scincomorpha)

FAMILY: Skinks (Scincidae)
SUBFAMILY: Tiliquinae
Examples: Giant Skink (*Corucia zebrata*); Cape Verde Skink (*Macroscincus cocteaui*); Stump-tailed Skink (*Tiliqua rugosa*); Blue-tongued Skink (*T. scincoides*); Spiny-tailed Skink (Genus *Egernia*).

SUBFAMILY: (Scincinae)
Examples: Common Skink (*Scincus scincus*); Eastern Skink (*S. mitranus*); Hemprich's Skink (*S. hemprichi*); Arabian Skink (*S. philbyi*); Persian Sand Skink (*Ophiomorus persicus*); Speckled Sand Skink (*O. punctatissimus*); Algerian Skink (*Eumeces algeriensis*); Schneider's Skink (*E. schneideri*); Five-lined Skink (*E. fasciatus*); Broad-headed Skink or Greater Five-lined Skink (*E. laticeps*); Great Plains Skink (*E. obsoletus*); Cylindrical Skinks (genus *Chalcides*).
SUBFAMILY: (Lygosominae)
Examples: East Indian Brown-sided Skink (*Mabuya multifasciata*); Keeled Indian Mabuya (*M. carinata*); Müller's Tree Skink (*Sphenomorphus muelleri*); Emerald Skink (*Dasia smaragdina*); Spotted Skink (*D. vittata*) Schmidt's Helmeted Skink (*Tribolonotus schmidti*); Florida Sand Skink (*Neoseps reynoldsi*).

FAMILY: Feylinidae

FAMILY: Anelytropsidae
Example: Mexican Blind Lizard (*Anelytropsis papillosus*).

FAMILY: Girdle-tailed Lizards (Cordylidae)
SUBFAMILY: Cordylinae
Examples: Sungazer Giant Girdled Lizard (*Cordylus giganteus*); Armadillo Lizard or Armadillo Girdled Lizard (*C. cataphractus*); Common Cape Girdled Lizard (*C. cordylus*); Blue-spotted Girdled Lizard (*C. caeruleopunctatus*); Leathery Crag Lizard (*Pseudocordylus microlepidotus*); Transvaal Red-tailed Rock Lizard (*Platysaurus guttatus*); Transvaal Snake Lizard (*Chamaesaura aenea*); Cape Snake Lizard (*C. anguina*).
SUBFAMILY: Gerrhosaurinae
Examples: Smith's Plated Rock Lizard (*Gerrhosaurus validus*); Whip Lizards (genus *Tetradactylus*); Girdled Lizards (genus *Zonosaurus*).

FAMILY: Night Lizards (Xantusiidae)
Examples: Cuban Night Lizard (*Cricosaura typica*); Granite Night Lizard (*Xantusia henshawi*).

FAMILY: Whiptails (Teiidae)
Examples: Chilean Spotted Lizard (*Callopistes maculatus*); Six-lined Racerunner (*Cnemidophorus sexlineatus*); Spotted Whiptail or Blue-bellied Racerunner (*C. sackii*).

FAMILY: True Lizards (Lacertidae)
Examples: Sand Lizard (*Lacerta agilis*); Dwarf Lizard (*L. parva*); Jewelled Lizard (*Timon lepida*); Common Lizard (*Zootoca vivipara*); Plated Lacertids (genus *Psammodromus*); Snake-eyed Lacertids (genus *Ophisops*); Fringe-toed Lacertids (genus *Acanthodactylus*).

INFRAORDER: Anguimorpha

FAMILY: Lateral Foldl Lizards (Anguidae)
SUBFAMILY: Diploglossine Lizards (Diploglossinae)
SUBFAMILY: Alligator Lizards (Gerrhonotinae)
Examples: Glass Lizards (genus *Ophisaurus*); Sheltopusik (*O. apodus*); Eastern Glass Lizard (*O. ventralis*).
SUBFAMILY: Anguine Lizards (Anguinae)
Example: Slow-worm (*Anguis fragilis*).

FAMILY: Shovel-snouted Legless Lizards (genus *Anniella*).

FAMILY: Xenosaurids (Xenosauridae)
Example: Crocodile Lizards (genus *Shinisaurus*); Chinese Crocodile Lizard (*S. crocodilurus*).

INFRAORDER: Varanomorphs

FAMILY: Aigalosauridae

FAMILY: Dolichosauridae

FAMILY: Mosasauridae

FAMILY: Bearded Lizards (Helodermatidae)
Examples: Gila Monster (*Heloderma suspectum*); Mexican Bearded Lizard (*H. horridum*).

FAMILY: Monitors (Varanidae)
Examples: Desert Monitor (*Psammosaurus griseus*); Nile Monitor (*Polydaedalus niloticus*); Cape Monitor (*Empagusia albigularis*); Yellow Monitor (*E. flavescens*); Two-banded Monitor (*Varanus salvator*); Giant Monitor (*V. giganteus*); Dwarf Monitor (*Odatria storri*).

FAMILY: Earless Monitors (Lanthanotidae)
Example: Borneo Earless Monitor (*Lanthanotus borneensis*).

INFRAORDER: Worm Lizards (Amphisbaenia)

FAMILY: Bipedidae
Example: Common Two-Legged Worm Lizard (*Bipes biporus*).

FAMILY: Ringed Lizards (Amphisbaenidae)
Examples: White-Bellied Worm Lizard (*Amphisbaena alba*); Darwin's Ringed Lizard (*A. darwini*); King's Worm Lizard (*A. kingi*); Florida Worm Lizard (*Rhineura floridana*).

FAMILY: Trogonophids (Trogonophidae)
Example: Wiegmann's Worm Lizard (*Trogonophis wiegmanni*).

Note: Some of the better known as well as other species selected at random are included in the various examples given. The examples represent only some of the species of lizards. The process of classifying the lizards continues at a relatively slow pace toward refinements in classification as well as with the nomenclature used. A few classifications remain controversial among authorities.

tongue is simple, with only a slight notch and covered with papillae. Because these animals have adapted to a dry habitat, they are not sensitive to reasonable extremes of temperature. Thus, most of these species are well adapted to terrariums. They can also adapt to a wide variety of easily available foods. Principal species include the club-tailed lizard (*Cordylus*); the yellow-brown sungazer (*C. giganteus*), which is about 38 centimeters long and lives in South Africa (Fig. 1); the armadillo lizard (*C. cataphractus*), a rather slow, heavily armored animal, the nostrils of which are conspicuously elongated; and the common cape girdled lizard, a rather flat animal about 20 centimeters long, with a somewhat spiny tail and black through dark-brown coloration, and that lives in the Cape Province of South Africa; and the blue-spotted girdled lizard (*C. caeruleopunctatus*), which is about 18 centimeters long and found in eastern Africa. The aforementioned lizards feed on insects and small animals.

Fig. 1.   Yellow-brown sungazer (*Cordylus giganteus*). Length is approximately 38 centimeters.

Of the genus *Pseudocordylus* (false club tails), the leathery crag lizard is representative. This animal (*P. microlepidotus*) is about 32 centimeters long, with a rather broad head. Dorsal scales are small and rounded. The upper side of the body is dark, usually with a dark upper side and frequently with pale bands or transverse patterns. The flanks are yellow-orange, and the belly is light. The animal inhabits the coastal mountains of Cape Province, South Africa. The habitat is characterized by shale, in deep crevices of which the lizard can hide. The diet is one of insects, other small animals, and lichens and other plants.

Of the genus *Platysaurus* (flat lizards), there are several species, including Wilhelm's red-tailed lizard and the Transvaal red-tailed lizard. The latter is the smallest of the genus and has a transparent window in the lower eyelid. In two subspecies, the males are a glowing green, with a red tail. Other species are red-brown. They live in the Soutpansberg and Drakensberg Mountains of South Africa. The females lay two elongated eggs in a crack in the rock during mid-winter. When the sun has warmed the rock surface in early morning, the *Platysauarus* emerges from an overnight shelter. Most of the daylight hours are spent under the sun. However, the mid-day heat drives the animals into shade. They eat primarily locusts and beetles. Males define their territories and display their colors to competitors by raising up. When sought, these lizards rush to narrow crevices and expand their bodies, thus making it virtually impossible to retrieve one alive.

Of the genus *Chamaesaura* (snake lizards), these animals are quite snakelike in appearance. They have pointed heads, slender bodies, and gradually tapering tails, which can be three times longer than the head and trunk together. The Cape snake lizard ranges up to 63 centimeters in length. Snake lizards prefer the grassy mountain slopes and high plateaus in southern and central Africa, where they feed mostly on grasshoppers. They move in grass with essentially the same ability as snakes. Their tiny limbs are seldom used. Generally, they are of a brown coloration.

The subfamily Gerrhosaurinae (plated or rock lizards) is distributed across Africa and is also found in Madagascar. They have well-developed bony plates under large horny scales. When well gorged with food, a fold makes it possible to expand the width of the body. This provision is also utilized when the female is carrying eggs. These animals are oviparous. The tail is long and can be autotomized. The better understood members are the African plated lizard (*G. major*), which is cylindrical in shape and grows to approximately 56 centimeters in length. The animals frequent southeastern Africa. Smith's plated lizard (*G. validus*) is the largest species in the genus. Its diet consists of insects, spiders, millipedes, or scorpions and smaller lizards. The female usually lays four eggs with leathery shells, most frequently locating them in the cracks of rock.

Night Lizards (Xantusiidae). These lizards range in length from 12 to 15 centimeters. The animals have four normally developed limbs, each with five toes. The scales are small. The ventral shields are large and rectangular. The pupil is vertical. In place of movable eyelids, there is a transparent "spectacle" something like that found in the geckos. Also, the vertebrae, skull, tongue, and eyes resemble those of the geckos. These animals feed on insects and spiders. The common night lizard also consumes vegetation. Their range extends from the southwestern United States and the offshore islands as far as Panama. One species is found in Cuba. Their hiding places are cracks in rocks, under roots, and in the bark of trees. All are viviparous, bearing from one to nine young at a time. There are four genera, but only a total of twelve species.

The Cuban night lizard (*Cricosaura typica*) was not discovered until 1863 (by German zoologist Gundlach). The large-headed or common night lizard occurs on the rocky islands of San Clemente, San Nicholas, and Santa Barbara off the coast of California. They consume both seeds and flowers. This particular animal is not strictly nocturnal.

Of the three species of *Xantusia*, the granite night lizard (*X. henshawi*) is native to California. Total length ranges up to about 17 centimeters. It is found not only in California, but also in the southwestern United States. The animal is quite flat, allowing it to creep into very small crevices, often to avoid sunshine. The desert night lizard (*X. vigilis*) prefers soft yucca plants for its diet. This species has been studied in detail on the southwestern edge of the Mojave Desert in California. They are frequently found at the foot of yuccas. In 1950, it was discovered that the formation of a placenta, rare among lizards, occurred in the Xantusiidae. After a gestation period of 90 days, from one to three young are born.

Whiptails (Teiidae). These animals vary widely in length, from 7.5 to 140 centimeters. They are sometimes referred to as the New World counterpart of the true lizards (Lacertidae). Some species, such as *Ameiva*, have well-developed limbs and scales and are formed so as to appear very much like European lizards. In the Teiidae family, there are some 45 genera and about 200 species. With the exception of northern and northeastern states, the teiids range widely throughout the United States and in Central America, the West Indies, and parts of Argentina and Chile. Their wide distribution accounts for so many species that have evolved through adaptation to numerous living and survival conditions, including arboreal, aquatic, and ground forms in rain forests, plains, deserts, seacoasts and inland regions, lowlands, and high mountains, such as the Andes, to the tropical forests along the Amazon river.

Generalizing, the teiids have lizard-like tails, ranging from cylindrical in form to a laterally flattened form found in aquatic environments. Their scales range widely—rounded, elliptical, elongated, hexagonal, smooth, or keeled. Depending upon particular species, a variety of bands may run longitudinally, transversely, or diagonally. Often, the teiid body is covered with large scales. Often, the tongue can be extended considerably and possesses a deep notch at the end. Depending upon location, the teiid ranges from herbivorous to insectivorous. Mode of reproduction has not been observed in detail for many of the species.

Racerunner (genus *Cnemidophorus*) are whip-tail lizards, the most commonly encountered teiid in the United States, and are also found in southern regions extending to northern Argentina. The term *racerunner* stems from the fact that these animals run fast, stop suddenly to scan for their enemy, and continue immediately in a series of rapid dashes, frequently in a different direction. It appears difficult for these animals to resist almost constant motion. Members of *Cnemidophorus* include the six-lined racerunner (*C. sexlineatus*), which is usually close to 8 centimeters; the checkered whiptail with a pale blue body; the seven-lined racerunner (*C. deppei*), which can achieve a length of 24 centimeters, is found from Mexico to Costa Rica, and has a light blue throat, a turquoise belly, with blue spots on the sides and a brick red lateral stripe; the spotted whiptail (*C. sackii*), which is about 25 centimeters long. Those animals with striking colorations use these features when reacting to threats from pursuers.

A striking member of the teiid family is the common tegu (*Tupinambis teguixin*), which can achieve a length of 120 to 140 centimeters. The animal is essentially black, with numerous crossbands that incorporate round yellow spots. See Fig. 2. Other animals of this type include the northern tegu or jacura (*T. nigropunctatus*), with a length of up to 120 centimeters. With their rather squat bodies, these teiids live in wooded areas, with dense undergrowth and sunny clearings and an obviously abundant food supply. Their meat is prized by local natives, who fish for them with meat-baited hooks. The yellow fat of the lizard's legs is particularly prized. Medicinal properties are ascribed by local people to several parts of the common

Fig. 2.   Common tegu (*Tupinambis teguixin*), which achieves a length of about 120 to 140 centimeters and which is sometimes referred to by native peoples in Central America and northern South America as a "chicken wolf" or "egg thief."

tegu. Locally terms ascribed to this lizard testify to its reputation among the local populace—"chicken wolf" and "egg thief," among others.

Another teiid of the genus *Crocodilurus*, the dragon lizardet (*C. lacertinus*), achieves a length of about 50 centimeters and is found in Central America and northern South America. The animal prefers a swampy habitat. The tail is somewhat flattened and features a double keel. The animal prefers fish and frogs in its diet. The animal seeks prey by hiding underwater in holes of a stream bank or under the roots of trees. Once seizing its prey, the animal returns to its hiding place to consume its catch. See Fig. 3.

Fig. 3.   Dragon lizardet (*Crocodilurus lacertinus*), achieves a length of about 50 centimeters. The animal prefers the swampy habitats of Central America and northern South America.

True Lizards (Lacertidae). These lizards are the Old World counterparts (analogues) of the Teiidae of the New World. Although these animals have not extended to Madagascar, New Guinea, and Australia, they are found elsewhere on the European and Asian continents and in most of South America as well. They range from the tiniest of Mediterranean islands to the far north of the Arctic Circle. All species of lacterids have well-developed legs and a long tail. They range in length from about 12 to 90 centimeters. The lacertids lack many of the characteristics found in other lizard families. They do not have dorsal crests or dewlaps. There are no other movable or expandable skin appendages. They have little or no ability to change their color. However, most feature an autotomous tail, part of which can be sacrificed to an enemy and later regenerated.

Many lacertids are found in dry regions. They adapt well to alternative habitats. Some species live in loose ground and elongated scales under the toes form combs at one or both sides, facilitating rapid locomotion on shifting sand. Some have a modified snout for digging. In some species there is a movable lower eyelid, often with a transparent window. Nearly all lacertids are egg layers. The species of the genus *Lacerta* are familiar to Europeans. Although subject to considerable research by noted herpetologists, the classification and relationships of species remains unclear. General characterizations include an unspecialized body structure, the band of enlarged scales around the neck, and the round or slightly compressed fingers and toes, which lack fringes or scales. There are numerous examples of the lacertids, including the sand lizard (*L. agilis*), an animal that is about 30 centimeters long; the emerald lizard (*L. viridis*), with a length up to 45 centimeters; the dwarf lizard (*L. parva*); and the largest lacertid, the jeweled or eyed lizard (*L. lepida*), which reaches a length up to 80 centimeters; among many others.

Lateral Fold Lizards (genus *Anguidae*). These animals historically are considered a rather young group that arose in the Cretaceous period. They are closely related to the monitors, which are described later. These animals generally lack limbs, but American species have four well-developed limbs. As with many other lizard families, limb degeneration in the lateral fold lizards usually can be correlated with adaption to a wide variety of native habitats. Lateral fold lizards may be shaped like a snake or like a lizard. The limbless species can achieve a length of from 50 to 100 centimeters, whereas the four-limbed species are smaller, some 20 to 40 centimeters in length. The most familiar lateral fold lizard is the slow-worm (*Anguis fragilis*), which ranges from 35 to 50 centimeters in length. This animal is not to be confused with the worm lizards (*Amphisbaenia*), to be described later.

The eastern glass lizard (genus *Ophisaurus ventralis*) is the longest lizard encountered in the United States. See Fig. 4. It can achieve a length of about 1 meter. Closely related species found in the Mississippi basin include the slender glass lizard (*O. attenautus*) and the island glass lizard (*O. compressus*), all generally confined to the southeastern United States, including Florida. Also closely related to the lateral fold lizards are the xenosaurids (*Xenosauridae*). There are two genera. One genus (*Shinisaurus*) embraces only one species, the Chinese crocodile lizard, which was first captured as recently as 1928 and first described in 1930 by the German herpetologist, Ernst Ahl. This lizard inhabits Kwangsi Province in southwestern China. The horny scales on the tail form a double row, as found in crocodiles. The teeth are fang-like. The animal inhabits areas near water. Chinese term the species the "lizard with great sleepiness." However, when threatened it will bite very quickly. When attacked, the lizard usually attempts to escape to water because of its diving and swimming agility.

Fig. 4.   Eastern glass lizard (*Ophisaurus ventralis*), the largest lizard occurring in the United States, achieves a length of 1 meter.

Beaded or Venomous Lizards (Helodermatidae). These animals are the only poisonous lizards. They are closely related to the ancestors of snakes. They are characterized by massive, un-snakelike bodies, and by the presence of four fully developed limbs. The head is broad and somewhat flattened. It is joined with a short neck and an elongated, cylindrical body, ending with a thick, rounded tail. The back is covered with large, bony scales. The legs are short and powerful. Each foot has five clawed toes. The back is covered with large, bony scales. The belly bares flat, regularly arranged scales that are not fully ossified. The lower jaw teeth have single grooves on the front and back that permit the venom to flow into the wound. The two species of the beaded lizard are (1) the gila monster (*Heloderma suspectum*), which achieves a length up to 60 centimeters; and (2) the Mexican beaded lizard (*H. horridum*), which may be up to 80 centimeters long.

The gila monster (Fig. 5) has a distinctive coloration of pink and black spots. Of two species, these animals are encountered from southern Nevada and southeastern Utah to Sonora in northwestern Mexico. Generally, these beaded snakes are active at night, but during cold weather they may appear by day. The diet consists of nestling rodents and birds and bird and reptile eggs. Beaded lizards first move very slowly and clumsily, but as the night hours progress, they become quite agile. During

Fig. 5.   Gila monster. (*A.M. Winchester.*)

long fast periods, demanded by their habitat, the animals survive on fat reserves stored up during more favorable times of the year. Fat is stored in the tail, which swells noticeably, but after a long fasting period, the tail becomes quite thin. Although seldom near water, they are good swimmers. Observations have shown that the gila monster can survive years of drought. In captivity, gila monsters can achieve an age of 20 years. They may be fed raw chicken eggs mixed with lean meat and supplemented with lime and vitamins.

Herpetologists have reported that snakebite antidote has no neutralizing action against gila monster venom. Typical localized consequences of a gila monster bite may include severe swelling of the victim's arm or part bitten and may be extremely painful for nearly 2 weeks. Evidence indicates that there is no damage to the victim's nervous system or vital organs, such as kidneys and liver. Experts insist that these animals belong only in the hands of very experienced keepers.

Monitors (Varanidae). As indicated by Fig. 6, the members of this genus have somewhat the appearance of legendary dragons. The Nile monitor has been known since antiquity. Herodotus (who died around 424 B.C.) described the desert monitor as a "terrestrial crocodile," while the ancient Egyptians, who often depicted monitors on their monuments, knew the Nile monitor so well that they never confused it with crocodiles. Monitors occur in numerous sizes. However, they all are classified in one genus (*Varanus*). They range from 30 centimeters up to 3 meters in length. The body is usually quite massive, with four powerful legs, each bearing five clawed toes. The tail is thick and may be used as a prehensile organ or as a potent weapon. Monitors are diurnal and reach their full activity level when the sun is up and their habitat has warmed. They are good runners and climbers. Some species are arboreal; others prefer proximity to water. Such species dive and swim well. Their best weapons are their sharp teeth and dagger-sharp claws, which can inflict dangerous wounds. All monitors feed on other animals, such as insects, small lizards, and the nestlings of small mammals. The larger species seek prey, such as crustaceans, fishes, frogs, birds, rats, and snakes. It is known that the giant Komodo dragon can even take small deer and wild pigs, and it has been reported that two-banded monitors have fed on human corpses that have been interred in trees. Monitors especially like eggs in their diet and are known to eat

Fig. 6.   The Nile monitor (*Varanus niloticus*) is found south of the Sahara, both in deserts and on the plains and especially along rivers. The animal feeds heavily on eggs.

eggs of their conspecifics. Most monitors, in their native habitat suffer at the hands of humans. Their meat and eggs are eaten, and the animals are often used to produce various "medications" and amulets. The fat and oil from the paired fatty organis is used by Chinese druggists, whose buying agents travel as far as Australia. The skin of larger monitors is processed into leather.

Monitors are found in the tropical and subtropical parts of Africa, the Near East, southern Asia, the Indo-Australian islands, and Australia.

Earless Monitors (Lanthanotids). In 1878, Viennese zoologist Franz Steindachner described the single animal before him as a *new* lizard from Borneo. Since it lacked external ears, Steindachner named it *Lanthonotus borneensis*, from the Greek (hidden ear). He realized that this new lizard species belonged to a unique family, which he classified as relatives of gila monsters. During the following 80 years, only six additional earless

monitors were identified. Even experienced herpetologists failed to find like specimens of the animal in Borneo. It was not until 1961 that archeologists on the island of Borneo dug out and captured alive an unusual lizard. This was the first earless lizard to be captured within a 45-year period. The local Dayak tribesmen were promised a bounty for capturing live lanthanotids. By 1960, over 60 were found. These were sent to zoo collections in Europe and the United States. Zoologists are much interested in earless monitors because they may be survivors of the animal group that gave rise to snakes. Many experts believe that snakes progressed from subterranean lizards in which the limbs and eyes regressed, the body became longer, and the number of vertebrae increased. the earless monitors possess these same characteristics.

Worm Lizards (Amphisbaenia). Questions persist concerning the accurate classification of worm lizards, which combine several characteristics of snakes and only some of the characteristics that typify lizards. Some herpetologists have even questioned whether these animals are reptiles, let alone lizards. It is notable that the characteristics are not intermediate between snakes and lizards, but rather represent evolvement over some special path. Worm lizards have been observed and their full identity determined only since the early 1800s. Identification was difficult because they are subterranean and come out of their underground locations only after sunset and return just before sunrise. Their size and coloration are such that they can be easily mistaken for earthworms.

As described by Carl Gans (Department of Biology, State University of New York at Buffalo), the typical amphisbaenian body is cylindrical, bearing a loose skin with fairly defined rings. Their color may be reddish or brownish and many incorporate a pattern of dark-brown or black spots on a lighter background. Length ranges from 8 to 80 centimeters; diameter is between 1.5 and 30 millimeters. Their shovel-like head structure is well adapted for digging. The eyes and ears are buried beneath the skin. Typically, worm lizards lay eggs, although a few species bear fully developed young. Worm lizards occur in several tropical areas of both the Old and New World.

In lizard folklore, one finds references to unusual association between ants and amphisbaenians. For example, in Brazil, some species are referred to as "ant mothers" or "ant kings," where it is alleged that worm lizards are raised and fed by ants. "Since worm lizards are often found in ant and termite colonies, the lizards probably have no trouble catching their prey. However, subterranean ant and termite colonies are not just food resources, but also serve as incubation chambers for egg-laying lizards. Most, if not all, worm lizard eggs discovered have been found in ant and termite colonies."

Some protection for the worm lizard is the enemy's difficulty in distinguishing the head from the tail. Both head and tail wag through the air in some species. Where Portuguese is spoken, the amphisbaenians may be called *cobras de dois cabeccas*, or two-headed snakes. Upon being grabbed by a carnivore, the outer tip of the tail may break off. The broken part twists about extensively and may momentarily divert the attention of the enemy. Unlike many other kinds of lizards, however, the broken tail does not regenerate. In the days of the early explorers, any worm lizard found was mistakenly regarded as poisonous.

By way of a series of underground tunnels, many worm lizards can hear the sounds of oncoming prey; their very long tongues can detect surrounding odors. In terrariums, amphisbaenians have been observed to crawl forward and backward with ease in their tunnels when they seek food. Also, their long tongue also assists in gathering insects for their diet. Their shovel-like head allows them to enter tunnels with surprising speed. The shape of their head can be a disadvantage, however, because it requires short jaws and reduces the number of teeth. Thus, in the overall, it has been suggested that species with less-effective digging mechanisms enjoy wider distribution and numbers, which are definite advantages in the long term.

The teeth of worm lizards interlock with an odd number of teeth in the upper jaw and an even number in the lower jaw. Thus, a biting action leaves a cut like that made by jagged scissors. The jaws tear or rip triangular pieces of flesh from the prey. The cheeks have strong muscles to assist the biting. Unlike the case of snakes, the amphisbaenians can rip out comparatively large pieces of flesh from small mammals.

Additional Reading

Manthey, U. and N. Schuster: "Agamid Lizards," TFH Publications, Neptune, NJ, 1997.

Roughgarden, J.: "Anolis Lizards of the Caribbean: Ecology, Evolution and Plate Tectonics," Oxford University Press, Inc., New York, NY, 1994.

Sprackland, R.: "Giant Lizards," TFH Publications, Neptune, NJ, 1992.

Vitt, L. and E. Pianka: "Lizard Ecology: Historical and Experimental Perspectives," Princeton University Press, Princeton, NJ, 1994.

LNG (Liquefied Natural Gas).   See **Natural Gas**.

LOACHES (*Osteichthyes*). Of the order *Ostariophysi*, suborder *Cypriniformes*, and family *Cobitidae*, loaches are small bottom-feeding fishes of Europe and Asia. They have a long slender body and a group of barbels near the mouth. The European species are eaten. A few fishes of similar form, but not closely related are called African loaches.

Several species are quite popular with tropical-fish fanciers. A brilliantly colored orange species from Borneo and Sumatra, the clown loach (*Botia macracanthus*) is frequently found in small aquariums, despite a habit of uprooting vegetation in the tank. Some clown loaches may reach a length of about 12 inches (30 centimeters) and live up to a quarter-century. Under normal aquarium conditions, the length is limited to a maximum of about 4 inches (10 centimeters). Other small loaches include the coolie loach (*Acanthophthalmus kuhli*), which is somewhat snake-like and found in Thailand and environs; and the spotted weatherfish or spined loach (*Cobitis taenia*) which is noted for its tolerance to wide temperature variations. A spotted species (*Cobitis biwae*) is found in Japan and is sometimes considered a food item. *Misgurnus fossilis* is the European weatherfish, known for its sensitivity to barometric pressure — hence the name, weatherfish. It is quite a large loach, sometimes attaining a length of about 20 inches (51 centimeters), but usually not exceeding 10 inches (25 centimeters).

LOAD FACTOR (Electric).   The ideal electric load, from the standpoint of equipment needed and operating routine, is one of constant magnitude and steady duration. A load of this type is shown in the upper part of Fig. 1(a). The cost to produce an elementary area of this load curve could be from one-half to three-fourths of that to produce the same unit under the more frequently realized condition shown in the lower curve (b). The problem of variable load is of vital importance to utilities whose chief concern is to put each kilowatt-hour on the transmission line at the lowest possible production cost. Interconnections between utilities and regional power generating facilities help in matching load with generation. See also **Electric Power Production and Distribution**.

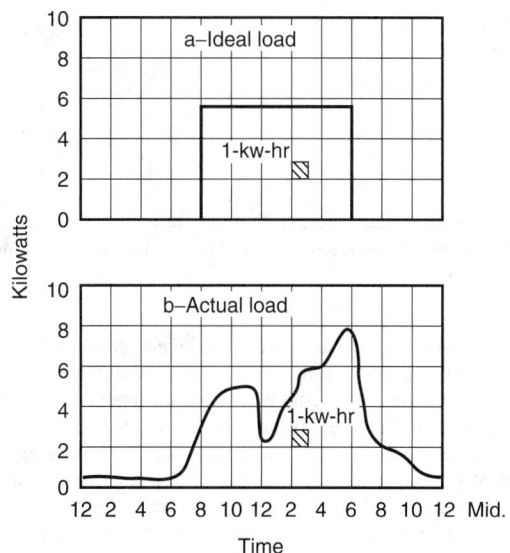

Fig. 1.   Comparison of (**a**) ideal and (**b**) actual electric power loads.

LOADING COIL.   1. An inductance inserted at regular intervals along a long transmission line or cable to increase the line's characteristic impedance and reduce its attenuation constant.

2. An inductance inserted in series with an antenna to increase its electrical length.

LOAD MATCHING.   1. Maximum power is delivered to a load when the impedance of the generator is the image impedance of the load. The adjustment of a circuit to provide this condition is called load matching.

2. In induction heating and dielectric heating usage, the process of adjustment of the load circuit impedance to produce the desired energy transfer from the power source to the load.

LOBECTOMY.   Surgical removal of a lobe of a gland or organ, such as the lung.

LOCAL APPARENT TIME.   The arc of the celestial equator, or the angle at the celestial pole, between the lower branch of the local celestial meridian and the hour circle of the apparent or true sun, measured westward from the lower branch of the local celestial meridian through 24 hours; local hour angle of the apparent or true sun, expressed in time units, plus 12 hours.

LOCAL AREA NETWORKS.   As automation in the factory increases, the need for communication between computers, controllers, and other "intelligent" machines has become critical. In the past, when factory automation was limited to the use of programmable controllers, numerical control of machines, and similar traditional approaches, communication was not a major limiting factor. Each tool was essentially self-contained and the communication requirement was mainly a user interface for controlling and updating machine operation. With the accelerated growth of automated tools and processes, communication between these entities is required to control not only their operation, but also their interrelationships. To this is added the desire to overlay environmental control, energy management, and materials requirement planning (MRP) to the factory operation. The end result is that intercomputer/controller communication has become the largest single problem to be addressed for factory automation. These interrelationships are shown in a generalized fashion in Fig. 1.

Early communication needs were served with point-to-point data links, as simply indicated in Fig. 2. Communication was relatively simple. The *star topology* for the communication system (Fig. 3) was developed so that multiple computers could communicate. The central or "master" node uses a communications port with multiple drops as shown in Fig. 4. The master is required to handle traffic from all the nodes attached, poll the other nodes for status, and, if necessary, accept data from one node to be routed to another. This heavy software burden on the master is also shared to a lesser degree among all the attached nodes. In addition, star topology requires routing a separate wire for every piece of equipment attached. This makes it difficult to wire and even more difficult to change. Also, the star topologies are inflexible regarding the number of nodes that can be attached. Either the user must invest in unused connections for further expansion, or have a system that cannot grow.

To overcome some of these disadvantages, multidrop protocols were established and standardized. Data loops, such as SDLC (Synchronous Data Link Control), were developed as well as other topologies, including buses and rings. Some of the early standards are shown in Fig. 5. The topology of these standards makes it easy to add (or subtract) nodes on the network. The wiring is also easier because a single wire is routed to all nodes. In the case of the ring and loop, the wire is also returned to the master. Inasmuch as wiring and maintenance are major costs of data communication, these topologies have virtually replaced star networks. These systems do have a common weakness, i.e., one node is the "master," with the task of determining which station may transmit at any given time. As the number of nodes increases, throughput becomes a problem because: (1) a great deal of "overhead" activity may be required to determine who may transmit, and (2) entire messages may have to be repeated, because some protocols only allow master-slave communications. That is, a slave-to-slave message must be sent first to the master and then repeated by the master to the intended slave receiver. Reliability is another problem. If the master fails, communication comes to a halt. The need for multinode networks without these kinds of problems and restraints led to the development of *Local Area Networks* (LANs), which use peer-to-peer communication. In this system, no one node is in charge and all nodes have an equal opportunity to transmit.

A LAN is a distributed communication network with the following characteristics: (1) peer-to-peer oriented (no master); (2) from 2 to 200 data devices (nodes) may be incorporated; (3) distance is limited to less than 2 km (1.2 mi); and (4) from 1 to 20 M bits per second data rates.

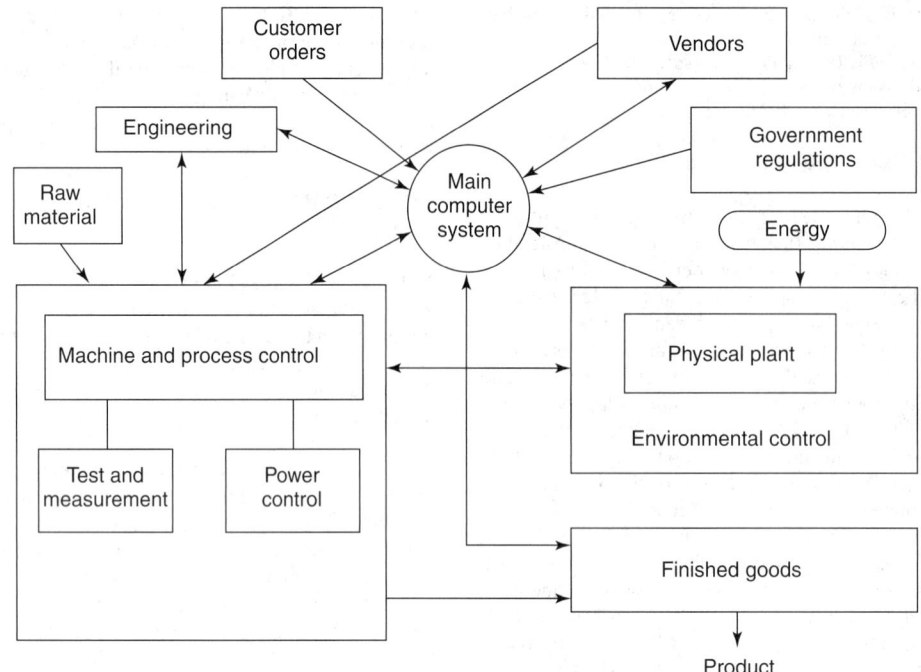

Fig. 1.   Intercommunications in a generalized automated factory situation.

Fig. 2.   Point-to-point communication.

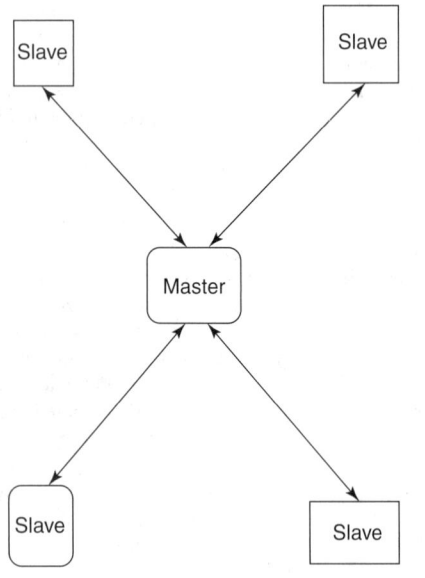

Fig. 3.   Star topology.

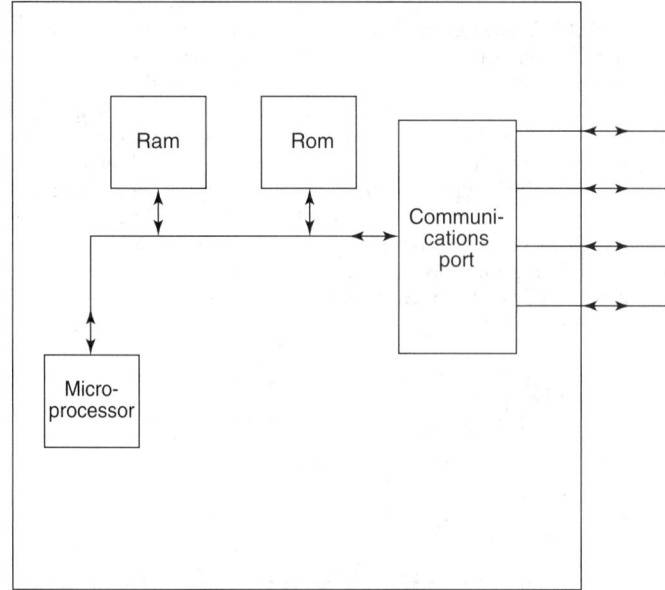

Fig. 4.   Master node for star topology. RAM = random access memory; ROM = read only memory.

In a local area network, each node is an independent computer system. Since there is no master to control traffic, each node must determine when it has the right to transmit. In a typical system, the host computer of the node is free to perform its job while the LAN protocol unit is moving data on and off the network. See Fig. 6.

As previously shown in Fig. 1, the need for LANs is driven by the proliferation of computer functions. The real motivating factor is that computers are now so cost-effective and relatively inexpensive that they are used throughout the factory floor, processing plant, and office. These computers not only must communicate with the large mainframe computers, but also with each other. The demands of materials

requirement planning, such as scheduling, inventory control, management, etc. require constant monitoring and data acquisition. Further, factory floor management requires coordination of machine operation; environmental control requires constant monitoring, among numerous other factors that go well beyond the traditional tasks of regulatory or sequencing controllers. The need to communicate status between devices (for a total integrated environment) and the sharing of costly resources, such as large-capacity disk storage, line printers, etc., have driven the requirement for LAN communication networks.

Requirements of the Industrial Environment. Although the need for LANs exists both in the office area and on the factory floor, the latter environment imposes some special restraints. The needs of the industrial environment require:

1. *Noise Tolerance.* Since a LAN will have long cables running throughout the factory, the amount of noise picked up can be large. The LAN must be capable of performing reliably in an electrically noisy area. The physical interface should be defined to provide a significant degree

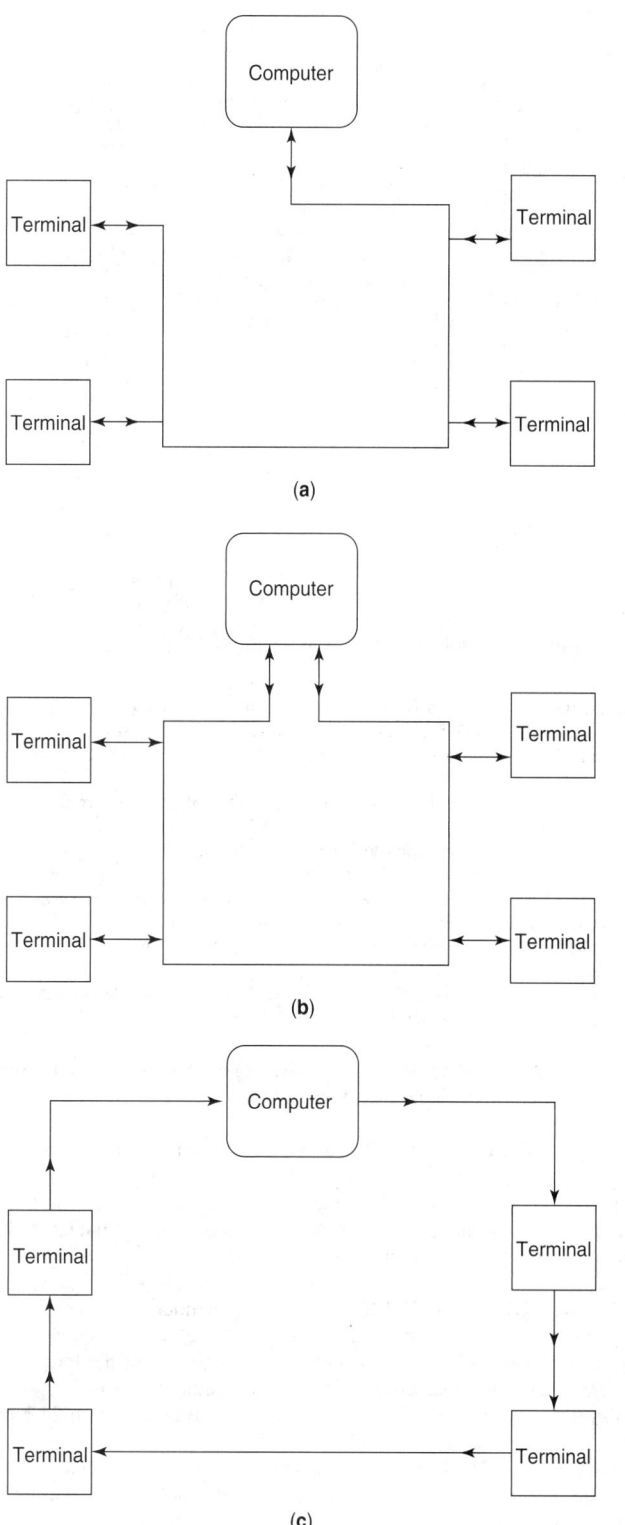

Fig. 5. Communication protocols and standards: (**a**) Data loop topology; (**b**) bus topology; (**c**) ring topology.

of noise rejection. The protocol must allow for easy recovery from data errors.

2. *Fast Response*. The LAN in an industrial situation should have a guaranteed maximum response time, i.e., the network must be able to transmit an urgent message within a specified time limit. The real-time characteristic of industrial control demands communication within a known time frame.

3. **Ability to Handle Priority Messages**. On the factory floor, both control and status data will be carried over the same network. A control message should have a higher priority and be transmitted before other messages.

Common Standards. A local area network standard that serves the harsh factory environment well can also be used for the less demanding office and administrative areas. Unless the requirements for the factory add too much cost, the factory floor standard can be the choice for the entire network. A common standard is advantageous inasmuch as system and information handling elements located in both office and factory environments must communicate with each other.

To meet current communication needs, different types of LANs are possible. Many of these already have been developed. All LANs provide the same basic service—to allow computers to pass data. However, since the major function of a network is to connect many different computers, standards are needed. The standard not only should describe how nodes are connected, but the protocol followed in transferring data as well. Ideally, the standard should be sufficiently comprehensive to permit any computer following it to pass data to any other computer that follows the same standard.

Several organizations have been working on the standards problem for a number of years, including the International Standards Organization (ISO), the Institute of Electrical and Electronics Engineers (IEEE), as well as some major users of LANs, such as General Motors Corporation, which has developed MAP (Manufacturing Automation Protocol).

Examples of Protocols

Carrier Sense Multiple Access with Collision (CSMA/CD) This is a baseband system with a bus architecture. See Fig. 5(b). Baseband is a term used to describe a system where the information being sent over the wire is not modulated, i.e., the "ones" are represented by one voltage level and the "zeros" by another voltage level. Normally, only one station transmits at any one time. All other stations hear and record the message. The receiving stations then compare the designated address of the message with their address. The one station, which matches, will pass the message to its upper layers, while the others will throw it away. Obviously, if the message is affected by noise (detected by the frame check sequence), all stations will throw the message away.

The CSMA/CD protocol requires that a station listen before it can transmit data. If the station hears another station already transmitting (*carrier sense*), the stations wanting to transmit must wait. When a station does not hear any other station transmitting (*no carrier sense*), it can start to transmit. Since more than one station can be waiting, it is possible for multiple stations to start transmitting at the same time. This causes the messages from both stations to become garbled (called a "collision"). A collision is not a freak accident, but is a normal way of operation for networks using CSMA/CD. The chances of collision are increased by the fact that signals take a finite amount of time to travel from one end of the cable to the other. If a station on one end of the cable starts transmitting, a station on the other end will "think" that no other station is transmitting during this travel-time interval and that transmission can be resumed. After a station has started transmitting, it must detect when another station is also transmitting. If this happens (*collision detection*), the station must stop transmitting. Before quitting the transmission, however, the station must make sure that every other station is aware that the last frame is in error and must be ignored. To do this, the station sends out a "jam" which is simply an invalid signal. This jam guarantees that the other colliding station also detects the collision and quits transmitting. Each station that was transmitting must then wait before trying again. To make certain that the two (or more) stations that just collided do not collide again, each station selects a random time to wait. The first station to time out will then look for silence on the cable and retransmit its message.

Token Bus. This standard has been selected for use in the previously mentioned MAP protocol. Token bus is also a bus topology, but differs in two ways from CSMA/CD: (1) The right to talk is controlled by passing a "token," and (2) the data on the bus are always carrier modulated. In the token bus system, one station is said to have an imaginary token. This station is the only one on the network that is allowed to transmit data. When this station has no more data to transmit (or it has held the token beyond the specific maximum time limit), it passes the token to another station. This token pass is accomplished by sending a special message to the next station. After this second station has used the token, it passes it to the next station, and so on. After all other stations have used the token, the original station is passed the token again. A station (example, station A) will normally receive the token from one station (B) and pass the token to the third station (C). The token ends up being passed around in a logical

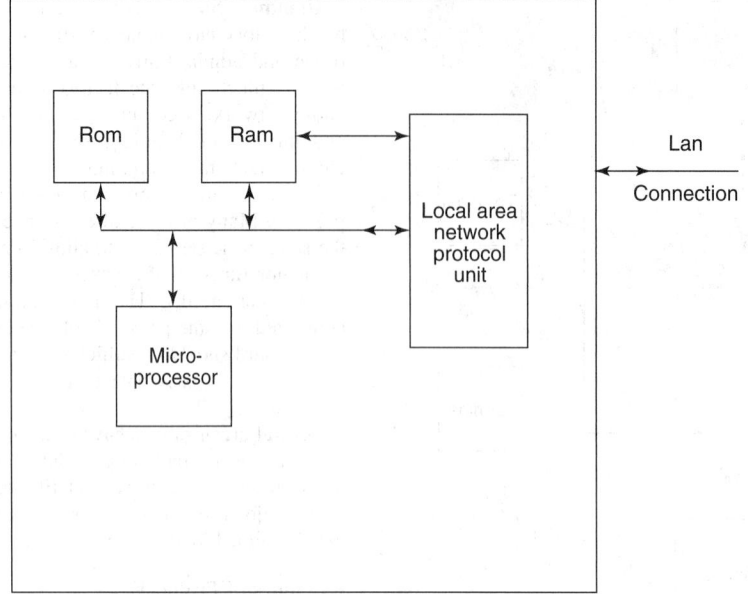

Fig. 6.   A local area network node. RAM = random access memory; ROM = read only memory.

token ring (A to C to B to A to C to B ...). The exception to this is when a station wakes up or dies. For example, if a fourth station (D) gets in the logical token ring between A and C, station A would then pass the token to D so that the token would go: (A to D to C to B to A to D ...). Only the station with the token can transmit so that every station gets its turn to talk without interfering with anyone else. Obviously, the protocol also has provisions that allow stations to enter and to leave the logical token ring.

The second difference between token bus and CSMA/CD is that with the token bus, data are always modulated before being sent. The data are not sent out as a level, but as a frequency. There are three different modulation schemes allowed—two single-channel and one broadband. Single-channel modulation permits only the token bus data on the cable. The broadband method is similar to CATV (community antenna television) and allows many different signals to exist on the same cable, including video and voice, in addition to the token bus data. The single-channel techniques are simpler, less costly, and easier to implement than broadband. Broadband, however, is of higher performance, permitting much longer distances and, very important, satisfies both present and future communication needs by allowing as many channels as needed (within the bandwidth of the cable).

Token Ring. Originally, token ring and token bus used the same protocol with different topologies. The two systems still remain rather similar. As shown by Fig. 5, any one node will only receive data from the "upstream" node and will only send data to the "downstream" node. All communication is done on a baseband point-to-point basis. This would seem to imply that one node can talk only to its downstream node. This is not the case, inasmuch as each station repeats what it hears from the upstream station to the downstream station. Since the last station is connected to the first (forming a ring), any station can send data to any other station. Precaution must be taken to prevent a short message from being retransmitted around the ring forever. This is prevented, by having the transmitting station remove its messages from the ring once they have gone around the ring one time.

The "right to talk" for the token ring scheme is also an imaginary token. The simplicity of the token ring system is that the station with the token simply sends it to the next downstream station, which either uses the token or passes it on to the next station. Space here does not permit the inclusion of numerous other pros and cons pertaining to these systems.

Additional Reading

Felt, S.: "Local Area High Speed Networks," Macmillan Publishing, New York, NY, 2000.

Goldman, J.E. and P.T. Rawles: "Local Area Networks," 2nd Edition, John Wiley & Sons, Inc., New York, NY, 2000.

Held, G.: "Internetworking LANs and WANs: Concepts, Techniques, and Methods," 2nd Edition, John Wiley & Sons, Inc., New York, NY, 1998.

King, J.P.: "Distributed Control Systems," in Process/Industrial Instruments and Controls Handbook (D.M. Considine, Editor-in-Chief), McGraw-Hill, New York, NY, 1993.

Loyer, B.A.: "Local Area Networks" in Standard Handbook of Industrial Automation (D.M. and G.D. Considine, Editors) Chapman & Hall, New York, NY, 1986.

Parnell, T.: "Building High-Speed Networks," The McGraw Hill Companies, Inc., New York, NY, 1999.

Slone, J.P.: "Local Area Networks Handbook," 6th Edition, CRC Press, LLC., Boca Raton, FL, 1999.

Stallings, W.: "Local and Metropolitan Area Networks," 6th Edition, Prentice-Hall, Inc., Upper Saddle River, NJ, 2000.

Stamper, D.A., J.C. Van Horne, and J.M. Wachowicz: "Local Area Networks," 3rd Edition, Prentice-Hall, Inc., Upper Saddle River, NJ, 2000.

Tanenbaum, A.S.: "Computer Networks," 3rd Edition, Prentice-Hall, Inc., Upper Saddle River, NJ, 1996.

Thompson, A.: "Understanding Local Area Networks: A Practical Approach," Prentice-Hall, Inc., Upper Saddle River, NJ, 1999.

LOCAL ASTRONOMICAL TIME.   Mean time reckoned from the upper branch of the local meridian.

LOCAL LUNAR TIME.   The arc of the celestial equator, or the angle at the celestial pole, between the lower branch of the local celestial meridian and the hour circle of the moon, measured westward from the lower branch of the local celestial meridian through 24 hours; local hour angle of the moon, expressed in time units, plus 12 hours.

LOCAL SIDEREAL TIME (LST).   Local hour angle of the vernal equinox, expressed in time units; the arc of the celestial equator, or the angle at the celestial pole, between the upper branch of the local celestial meridian and the hour circle of the vernal equinox, measured westward form the upper branch of the local celestial meridian through 24 hours.

LOCOMOTION.   The process of moving from place to place, a characteristic power of most animals and a lesser distinction between them and the majority of plants.

Locomotion is necessary to animals because their food is organic and in most environments does not reach the animal through external forces. Even the sessile animals, which may or may not be capable of some locomotion, often accomplish the same end by bringing food within reach through their own activities.

In the water the weight of the surrounding medium is so great that the animal may float, and the resistance offered to its body is sufficient to be utilized for propulsion. The body is so shaped that resistance is little in the direction of locomotion but great where propulsive effort is expended. Projections from the surface that beat against the water like oars or push or pull by undulating are common organs of locomotion here. They include cilia and flagella in one-celled and small multicellular forms, specialized jointed appendages of arthropods, and fins and flippers of vertebrates. Undulation of the body itself is a sufficient means of propulsion in some animals.

Some aquatic forms rest on the bottom and the terrestrial animals are forced to rest on some solid support at least intermittently because the air is too light to float them. In many of these forms the friction of contact with a solid is utilized by the development of movable supporting appendages which are shifted alternately to change the animal's position. This means of locomotion is known as walking. Other animals, notably the worms, creep through the action of muscles in the body wall. The body is progressively elongated and shortened, parts being thrust ahead and then drawn up to the maximum point of advance. In this type of locomotion they are aided by suckers, or setae in some cases, to grip the supporting surface.

Running may involve no other difference from walking than more rapid movement or it may also involve a change in the order of movement of the appendages and in their position when used, as in the various gaits of a horse. Jumping always differs in that the appendages set farthest back must be powerful enough to project the entire animal through the air. It is highly developed in such insects as the flea beetles and the grasshoppers and in the frogs and kangaroos among the vertebrates. In this class a gallop is no more than a series of leaps.

Locomotion in the terrestrial vertebrates also shows progressive change in the manner of using the appendages. The entire sole of the foot rests on the ground in man and the apes, and they are said to be plantigrade. This posture is well adapted to walking but not to running. Animals that need speed are digitigrade, resting on the toes. This position adds the length of the feet to that of the legs and permits a longer stride. It also adds the springiness incidental to the greater freedom of the ankle joint. The final expression of this position of the leg appears in the unguligrade (Ungulata) hoofed animals where only the hoof comes into contact with the ground. Man is plantigrade in walking and at rest but rises to his toes when he runs.

The locomotion of snakes is a highly specialized creeping process in which the ribs serve as the movable appendages and the grip of the body on the ground is provided by the broad scales of the ventral surface which project backward.

Climbing animals may merely run along branches, aided by sharp claws to provide a secure grip. The sloths, however, have the claws developed as great hooks, which suspend them in an inverted position. They walk as well as hang upside down. The primates show the most extreme specialization for a form of locomotion in the trees called brachiation. Their pectoral appendages are arms, adapted for grasping and suspension, and the pelvic appendages are supporting legs. They move by swinging from branch to branch or by shifting from one hold to another, as human beings climb.

Locomotion in the air is the highly specialized process of flight.

Additional Reading

Blake, R.: "Fish Locomotion," Cambridge University Press, New York, NY, 1983.
Kimura, T. et al.: "Development and Control in Primate Locomotion," S. Karger AG, Bassel, Switzerland, 1996.
Leach, D. and H. Schamhardt: "Animal Locomotion," S. Karger AG, Bassel, Switzerland, 1993.
Patla, A.: "Adaptability of Human Gait: Implications for the Control of Locomotion," Elsevier Science, New York, NY, 1991.

LOCUST (Grasshopper; Insecta, Orthoptera). The term "locust" is more properly applied to the "short-horned" grasshoppers that include the migratory locusts appearing in Africa, Asia and the plains of North America as serious crop pests. These animals have always been an important food of aboriginal peoples, and are still consumed in many parts of the world. The term grasshopper is now generally restricted to the common non-migratory forms. The name locust is also commonly but wrongly applied to the Cicada or tree cricket.

The adults measure about 2 inches (5 centimeters) in length; some are much larger. The body is thick, strong, tapering to the end of its folded wings. Large eyes are high on the head and located just above the two short antennae. A stiff, thick jacket joining the head covers and protects the back of the neck. The wings, shaped much like some aircraft wings, are held close to the body when at rest. The wings are strong, large, heavily veined and twice as wide at the center as at the ends, and about $\frac{1}{2}$ inch (12 millimeters) longer than the body.

The six legs of the locust are comprised of several segments, connected to the body; the two shorter front legs are located under the shoulders. The two legs located near the center of the body are long and powerful and used for making long hops. The thigh of the leg is oar-shaped and heavily veined. It is connected to a strong joint at the body, which aids greatly in the thrust of the kick when jumping.

The "song" of the locust is created by spurs or spines on the inner area of the hind legs which are rubbed against the wings which have a raised, spurred area on the outside surface. This gives a rasping sound.

The desert locust (Schistocera gregaria) of India swarms during the summer monsoons. In autumn, these insects migrate to Iran, Arabia, Soviet Asia, Syria, and Egypt. In early winter, they return to India and East Africa to breed.

Possibly the most impressive of all insect flights will be that of a swarm of billions of locusts. In 1889, a flight crossed the Red Sea estimated to be 2000 square miles (5180 square kilometers) in extent. Desert locusts have appeared in England, apparently flying from southern Algeria, possibly assisted by a tail wind. Swarms of locusts have been reported since ancient times. The Book of Joel describes the army that blackened out the sun—behind them a desolate wilderness and nothing escaping them.

There are about 2,000 species of locusts, of which about 20 important species are capable of causing crop desolation and accompanying famine in some areas of the world. The huge migrations are caused by hunting for food, and where locusts find food they will eat their weight daily if it is plentiful. Fortunately, the mortality rate of the insect is high when a swarm encounters stormy weather. See also **Cicada**.

LOCUST (Seventeen year; Insecta, Homoptera, Cicadidae). The 17-year locust (Magicicada septendecim, Linne), not a true locust, is named for its life cycle of 17 years. There are, however, many broods that overlap, with adults appearing in different years. There is also the 13-year locust (M. septendecim tridecim, Riley). These insects are sometimes confused with the dog-day cicada (Tibicen linnei) (Smith and Grossbeck), also commonly termed a locust, which has a 9-year cycle. The latter insect is not nearly so damaging as the other two species just mentioned. See also **Cicada**.

The 17-year and 13-year locusts (properly known as periodical cicadas) make rough punctures in twigs and small branches of apple and numerous other fruit trees. Damage does not result from the feeding of the insects, but from the puncturing of the twigs when the female deposits her eggs. In the United States, the 13-year locust ranges in a line from Virginia to southern Iowa and thence southward to the Gulf of Mexico and eastward to the Atlantic shore. The 17-year locust is found from the New England states westward to Wisconsin, southwestward into Kansas and Missouri, and south as far as Alabama and northern Georgia. It is most abundant east of the Mississippi River and greatly infests much of Wisconsin, Iowa, Tennessee, and South Carolina, and nearly all of Illinois, Indiana, Ohio, Virginia, Kentucky, West Virginia, Pennsylvania, Connecticut, and New Jersey.

The 17-year locust has the longest period of development of any insect known. As previously described, the eggs are laid in twig punctures in late spring and mid-summer. Each female deposits 400–600 eggs, with about 20 eggs per puncture. Within 6–7 weeks the eggs hatch. The young are ant-like in appearance and, when hatched, drop to the ground where they enter into cracks in the soil. They feed on sap from tree rootlets. Feeding is very slow and their presence cannot be detected by any apparent deterioration of the tree, even though there may be many thousands of these creatures at the base of an affected tree. Depending upon the species, these nymphs require from 13 to 17 years to achieve maturity. At that time, the insects are about an inch (2.5 centimeters) long and appear something like a crayfish. The insects burrow to the surface, sometimes forming mud cones or chimneys that may protrude 2–3 inches (5–8 centimeters) above ground level. Massive numbers of these nymphs emerge within a very short period, usually climbing the tree after sunset. They temporarily take hold of the bark until the adult insect removes itself from the nymphal shell. Leaving the empty skins on the tree, they take flight and are ready to mate during a period of 30–40 days. It is estimated that over 40,000 adults may emerge from a single large tree within a period of a few days.

LOCUST TREES. Several leguminous trees are commonly called locust trees. Most notable is the black locust, Robinia pseudoacacia. It is a medium-sized tree native to the Appalachian and Ozark mountains. The twigs have a pair of spines about $\frac{1}{2}$ inch (1.3 centimeters) long at the base of each leaf, although spineless varieties have been developed. The leaves are compound, 8–14 inches (20.3–35.5 centimeters) long. In the spring the tree produces its flowers, which are white or pink and very fragrant, hanging in clusters 4–8 inches (10–20.3 centimeters) long. The wood is hard and tough, and is used for fence posts, mine timbers, and rough

TABLE 1. RECORD LOCUST TREES IN THE UNITED STATES[1]

Specimen	Circumference[2]		Height		Spread		Location
	Inches	Centimeters	Feet	Meters	Feet	Meters	
Black locust (1974) (*robinia pseudoacacia*)	280	711	96	29.3	92	28	New York
Clammy locust (1996) (*Rabinia viscosa*)	19	48	35	10.7	21	6.4	North Carolina
Honeylocust (1999) (*Gleditsia triacanthos*)	226	574	100	30.5	88	26.8	Maryland
Honeylocust (1993) (*Gleditsia triacanthos*)	233	592	90	27.4	88	26.8	Pennsylvania
New Mexico locust (1997) (*Robinia neomexicana*)	90	229	71	21.6	28	8.5	Arizona
Waterlocust (1993) (*Gleditsia aquatica*)	110	279	74	22.6	73	22.3	Pennsylvania

[1]From the "National Register of Big Trees," American Forests (by permission).
[2]At 4.5 feet (1.4 meters).

construction. The tree is commonly planted for ornament and shade, or for erosion control.

The honey locust, *Gleditsia*, is also frequently known simply as locust. A variety of this genus, the Moraine locust, has become popular as a shade tree.

Record locust trees, as reported by American Forests, are listed in the Table 1.

LOESS. Loess is a buff-colored, wind-blown deposit of fine silt or marl, usually unstratified, which is often exposed in bluffs with steep to vertical faces. Loess is found in the United States in the Mississippi valley from Louisiana to Iowa, and along the course of the Missouri. The average thickness of the loess here is about 20 feet (6 meters), but may range to 50–100 feet (15–30 meters). Loess also occurs in central Europe, Mongolia and China where it is said to attain a thickness as great as 300 feet (90 meters). The loess of the United States and Europe is believed to be the finer materials first transported and deposited by the waters of the melting ice sheets of the glacial period, and later blown to considerable distances and sometimes deposited in lakes. The Asiatic loess seems to be wholly wind transported, the source of the dust being, perhaps, the great deserts of central Asia. In the latter case the accumulation of such thick deposits is attributed to the binding power of successive generations of grasses whose former existence is suggested by a network of narrow tubes.

See also **Erosion**.

LOG (Navigation). A term used with two different meanings: a speed-measuring device, and a record book. Prior to the middle of the nineteenth century, the speed of a ship relative to the surface of the water was measured by the log chip and line. "Heaving the log" was a duty performed every time the ship's bell was struck, i.e., every half-hour, and the speeds determined were entered in a book, which came to be known as the log book. Since this log book was always available to the watch officer, it became customary to enter all important incidents relating to the operations of the ship, behavior of members of the crew, conditions of the weather and sea, and, in fact, anything the watch officer might think worthy of recording. This "deck log" was turned over to the captain, who abstracted all important material and made up the ship's log. At present, practically every department of the ship, deck, engine room, ordnance, steward, etc., keeps its individual rough log, and the ship's log is made up from these under the supervision of the captain.

The oldest and perhaps even now, the most accurate and reliable method for determining the speed of a ship relative to the water is by use of the log chip and line. The log chip is a wooden quadrant, loaded along its circular edge so that it will float upright in the water. It is attached to the log line by a 3-legged bridle, the upper leg of which is attached to the apex of the log chip in such a manner that a sharp jerk will free it. Then the log chip may be easily hauled back to the ship. When the log chip is thrown overboard, it floats, nearly submerged, with the flat surface perpendicular to the motion of the ship, and the line runs out over the stern with the speed at which the ship is moving through the water. For measuring the speed at which the line runs out, a sandglass was originally used. For speeds under 4 knots, a 28-second-glass was employed, and for higher speeds, a 14-second-glass

was available. The line was divided into lengths by threading pieces of fish line through it at specified distances. In the pieces of fish line, 1, 2, 3, etc., knots were tied to indicate the number of lengths that had run out in the given time. The distance between markers was determined by the ratio between 28 seconds and the number of seconds in an hour:

$$\frac{\text{distance}}{6{,}080 \text{ ft.}} = \frac{28 \text{ sec.}}{3{,}600 \text{ sec}}; \text{hence, distance} = 47'3.5''$$

In order that the log chip may be well clear of the turbulence in the wake, a certain amount of stray line is provided, with a red marker indicating the beginning of measurement.

To "heave the log," two men are necessary, one to tend line and the other to operate the sandglass. The first operator throws the chip overboard and gives the word "tip" when the red marker passes through his fingers. When the last grain of sand runs through the glass, the timekeeper calls "check," and the line-tender grabs the line. This gives the sudden jerk necessary to free the bridle. The line-tender then notes the number of "knots" and approximate fractions thereof that have passed through his fingers and reports that number to the officer of the deck, the number of knots being equal to the speed of the ship in nautical miles per hour.

The use of the log chip and line gives the instantaneous speed of the ship, and the values obtained over a day must be averaged to obtain the total distance run. Many different types of patent logs have been devised to give both speed and distance run. They are all subject to unexpected errors and should be checked frequently by comparison with the log chip and line. On the high-speed ships of the modern navy, the old-fashioned line is impracticable. To check patent logs, the number of seconds required for a given length of line to run out is determined by means of a stop watch, and the speed is calculated.

See also **Course**; **Navigation**; and **Patent Log**.

LOGARITHM. If B is an arbitrarily chosen number greater than unity, then the logarithm L of any other number N is defined by $N = B^L$; $L = \log_B N$. The chosen number B is the base of the system of logarithms. For any base, $\log_B 1 = 0$; $\log_B B = 1$. The fundamental properties of logarithms are $\log ab = \log a + \log b$; $\log a/b = \log a - \log b$; $a^n = n \log a$, where a, b are positive numbers and n may be greater than unity or less than unity (thus a power or a root of the number, rational or irrational). Two systems of logarithms are generally used: common and natural.

The former, also called *Briggs logarithms*, uses the base 10 and is particularly useful for numerical calculations. The common logarithm of a number N could be indicated by $\log_{10} N$ but the notation $\log N$ is more usual. Since any number may be written in the form $N = 10^n \times M$, where n is an integer, positive or negative, its common logarithm is $\log N = n + \log M$. The first part of this sum is the *characteristic* of the logarithm, and it may be obtained by inspection of the given number. The second part, called the *mantissa*, is an irrational number, less than unity and usually given in decimal form. Tables of the mantissas of common logarithms are available with four, five, six, etc., significant Figures so that any required accuracy can be obtained in numerical calculations.

If a logarithm is given, the number that corresponds to it is the *antilogarithm*. The *cologarithm* is the logarithm of the reciprocal of a number. Thus $\operatorname{colog} N = \log(1/N) = -\log N$. In principle, it could simplify calculations involving quotients, for negative mantissas would not occur. This is not a serious handicap, however, and although cologarithm tables are sometimes given in handbooks, they are seldom used.

The *natural logarithm* has the irrational base $e = 2.71828\ldots$ and is also called the Napierian or hyperbolic system. Such logarithms occur as the result of differentiation, integration, etc., and they often appear in equations representing physical phenomena. Instead of the more exact symbol $\log_e N$, the abbreviated notation $\ln N$ is customary, especially in scientific work. The modulus of the common system relative to the natural system of logarithms is $\log e = 0.434294\ldots$ and, inversely, $\ln 10 = 2.302585\ldots$ is the modulus of the natural system. Conversion of one system of logarithms to another is made from the resulting relations: $\log N = 0.434294\ldots \ln N$; $\ln N = 2.302585\ldots \log N$. Although tables of natural logarithms are available, it is normally simpler to compute with common logarithms and then convert to the natural base.

A logarithmic scale is one in which the distance from the origin to any scale mark is proportional to the logarithm of the number attached to that mark.

Thus, Fig. 1 shows a (common) logarithmic scale with the numbers $1, 2, 3, \ldots, 10, \ldots$ attached to the division marks; if we take OI as the unit of length, then the distances OA, OB OC, etc., are represented by log 2, log 3, log 4, etc., so that $OI = \log 10 = 1$. In going from left to right, the scale marks will become closer and closer together.

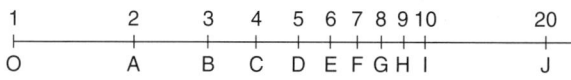

Fig. 1.   Logarithmic scale.

The logarithmic scale is applied in the slide rule and in logarithmic paper. In the latter case, if both abscissa and ordinate are marked logarithmically, the paper is called log-log paper; if the abscissa is equally spaced and only the ordinate is logarithmic, it is semilog paper. See also **Curve Fitting**; and **Logarithmic Function**.

LOGARITHMIC CHART.   A graph where one or both axis are scaled in terms of logarithms of the variables. The chart may be called a semi- or double-logarithmic chart according to whether only the ordinate or both the ordinate and abscissa are on a logarithmic scale. In general, the logarithmic method of plotting is used when relative changes are important, since equal linear displacement on a logarithmic scale indicates equal proportional changes in the variable itself.

LOGARITHMIC DECREMENT.   For a system having an oscillatory response that decays with increasing time, the logarithmic decrement is defined as the negative natural logarithm of the ratio of two consecutive excursions of the response about the bias or steady-state value. The ratio is taken as shown in Fig. 1. This quantity is used as a measure of the internal friction in an oscillating body. The logarithmic decrement equals one half the specific damping capacity.

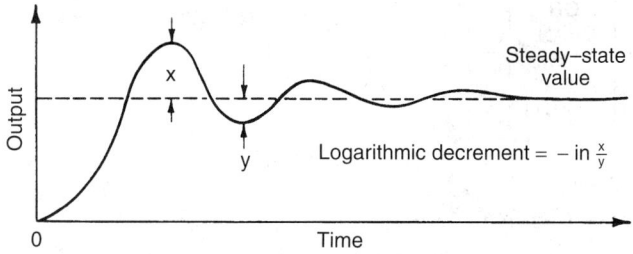

Fig. 1.   Logarithmic decrement of oscillatory response.

LOGARITHMIC FUNCTION.   Its equation in rectangular coordinates is $y = a \ln x$ and the exponential function, $x = e^{y/a}$ is its inverse. When plotted on ordinary graph paper, the ordinate increases in arithmetic progression and the abscissa in geometric progression. Semi-logarithmic paper is more convenient for such a graph (see also **Logarithm**). A plot

of $x + \log y = $ constant or $y + \log x = $ constant is then a straight line. See also **Curve Fitting**.

LOGARITHMIC SPIRAL.   A transcendental plane curve, with equation in polar coordinates, $\ln r = a\theta$. It is also known as an equiangular spiral since it cuts the radius vectors at a constant angle. See Fig. 1. As θ increases, the curve winds around the pole at ever-increasing distances; when θ is negative, the curve continually approaches, but never reaches, the pole. Its evolute is a congruent logarithmic spiral. It is said that the tombstone of James Bernoulli (1654–1705), the celebrated Swiss mathematician, is inscribed with this curve and the words "Eaden mutato resurgo." See also **Bernoulli Number and Polynomial**.

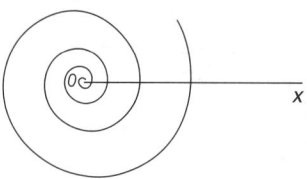

Fig. 1.   Logarithmic spiral.

LOGICAL DESIGN.   1.   The planning of a computer or data processing system prior to its detailed engineering design.

2.   The synthesizing of a network of logical elements to perform a specified function.

3.   The result of 1 and 2 above, frequently called the *logic of the system, machine*, or *network*.

LOGICAL ELEMENT.   In a computer or data processing system, the smallest building blocks which can be represented by operators in an appropriate system of symbolic logic. Typical logical elements are the AND gate and the flip-flop, which can be represented as operators in a suitable symbolic logic.

LOGICAL OPERATION (Computer System).   1. A logical or Boolean operation on N-state variables, which yields a single N-state variable; e.g., a comparison on the 3-state variables A and B, each represented by $-$, 0, or $+$, which yields: $-$ when A is less than B, 0 when A equals B, and $+$ when A is greater than B. Specifically, operations such as AND, OR, and NOT on two-state variables which occur in the algebra of logic; i.e., Boolean algebra.

2. The operations of logical shifting, masking, and other nonarithmetical operations of a computer.

See also **AND (Circuit)**; **NAND (Circuit)**; **NOR (Circuit)**; **NOT (Circuit)**; and **OR (Circuit)**.

LOGIC (Computer System).   In hardware, a term referring to the circuits that perform the arithmetic and control operations in a computer. In designing digital computers, the principles of Boolean algebra are employed. The logical elements of AND, OR, INVERTER, EXCLUSIVE OR, NOR, NAND, NOT, etc. are combined to perform a specified function. Each of the logical elements is implemented as an electronic circuit which, in turn, is connected to other circuits to achieve the desired result. The word logic is also used in computer programming to refer to the procedure or algorithm necessary to achieve a result.

LOGIC DIAGRAM (Computer System).   A drawing that indicates the interconnection of the individual logic elements in a computer. A logic diagram incorporates all of the information needed for wiring a computer. The logic diagram given in Fig. 1 shows the logic blocks and electrical interconnections. The logic block in the diagram contains the name of the function performed by it, such as AND, OR, and FLIP-FLOP; and the physical location of the circuit within the computer. Input and output signal connections for the function are given by the logic block. The logic diagram also shows the connection of a given logic block to other logic blocks. The computer supplier often provides wiring lists for the machine and a printed logic diagram. The logic diagram is used as an aid in troubleshooting the system. When utilizing large-scale integrated (LSI) circuits, each block may contain a complex logic function, such as an adder or register, rather

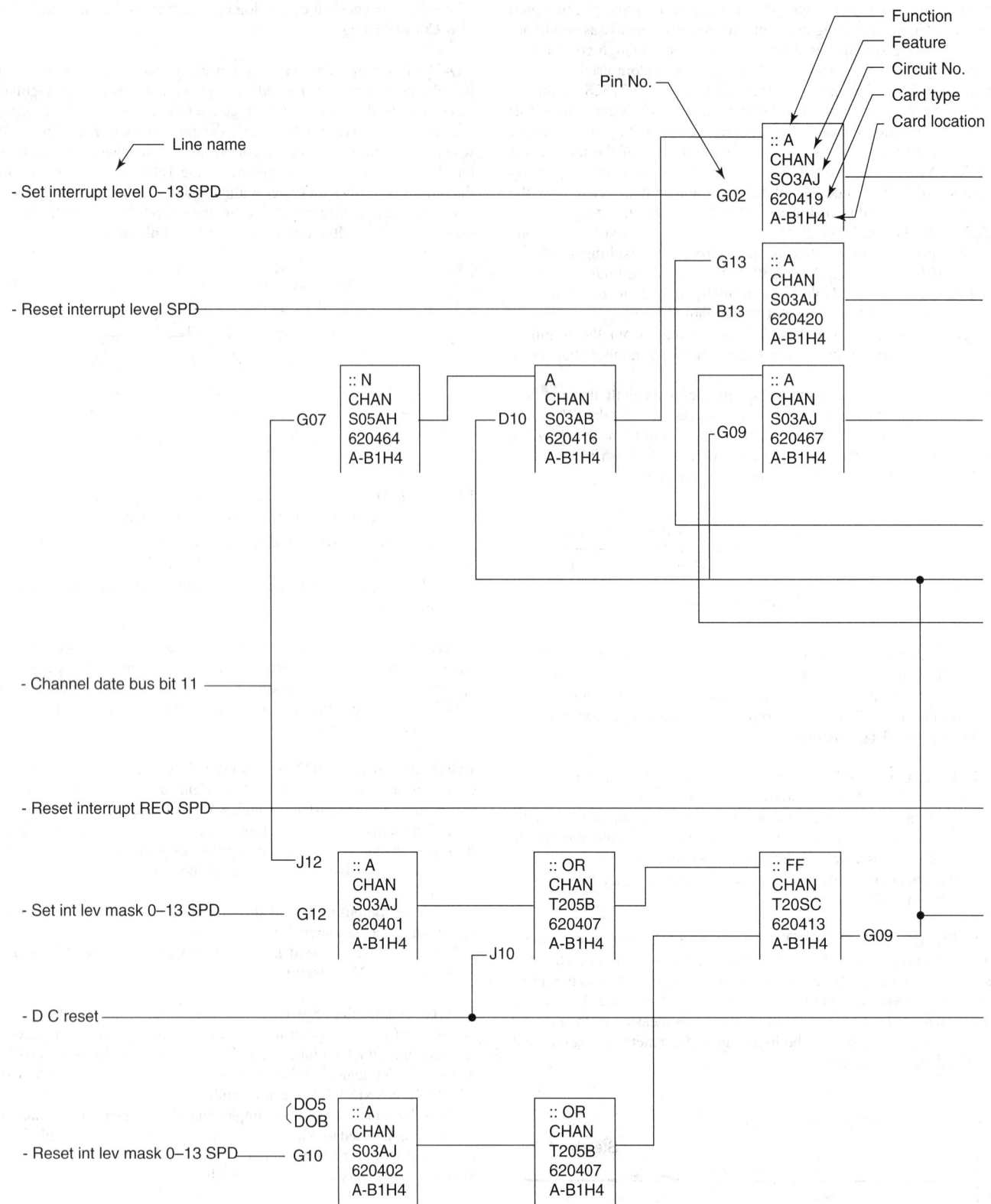

Fig. 1.   Computer logic diagram.

than single elemental logic functions. Although not needed for physical wiring or troubleshooting, these complex logic functions can, in turn, be expressed in terms of elemental logic similar to that shown in Fig. 1.

THOMAS J. HARRISON, IBM Corporation, Boca Raton, FL.

LOGISTIC CURVE. The logistic curve is a growth curve used to describe functions which continually increase, gradually at first, more rapidly in the middle growth period, and slowly again, reaching a maximum at the end of the growth. We write its equation

$$y = \frac{k'}{1 + e^{a+bt}}, \quad b < 0$$

the symmetrical logistic, where t represents time and y is the population size. A more general form is

$$y = k_1 + \frac{k_2}{1 + e^{a+bt-ct2}}, \quad c < 0$$

the asymmetrical or skew logistic. This usage of the word "logistic" has nothing to do with the military connotation or with the European meaning of "formal logic."

LOGIT.   A transformation of a variable used particularly in the analysis of dose-response relationships. If p is the probability of a certain response on dose x, the logit is defined as

$$y = \log_e\{p/(a - p)\}$$

and analysis proceeds by considering the relation between y and x.

LOG MEAN TEMPERATURE DIFFERENCE (LMTD).   An average temperature difference between the hot and cold side of a piece of equipment, e.g., a heat exchanger, for use where the temperature difference may vary along the equipment. Its form is

$$\frac{\Delta T_1 - \Delta T_2}{2.3 \log \dfrac{\Delta T_1}{\Delta T_2}}$$

where ΔT_1 is the temperature difference at one end of the equipment and ΔT_2 is the temperature difference at the other.

LOGWOOD TREE.   Of the family *Caesalpiniaceae* (senna family), the *Haematoxylon campechianum* is a small tree native to central America and has been extensively planted there and in the West Indies and South America. Rarely exceeding 25 feet (7.5 meters) in height, the tree has pinnately compounded leaves with smooth obovate leaflets, and fragrant yellow flowers in terminal racemes. The fruit is a dry two-seeded pod. The wood is very hard and yellow. Upon exposure to air, it turns red. It has a rather pleasant scent. Dye is obtained from the heartwood, which is cut into chips. In earlier years, logwood dye was extensively used. To make the dye, mordants must be added, in this case the salt of some metal, usually iron. Haematoxylon has been used in the manufacture of inks and as a histological stain in the preparation of organic tissues for microscopic examination.

LONDON DIPOLE THEORY.   A theory that accounts for the attractive forces between molecules by considering the interactions between the instantaneous dipole moments of the molecules. By considering the first order perturbation, it is shown that the interaction energy varies inversely as the sixth power of the distance between the molecules.

Weak forces of attraction (also called dispersion forces) are exerted on each other by inert atoms. London, on the basis of quantum mechanics, has shown that they are due to the perturbation of the repulsive ground state by the higher electronic states of the system consisting of the two atoms. This perturbation at large internuclear distances r, gives a potential energy decreasing as $-(1/r^6)$ toward smaller r values. At smaller distances r the strong repulsion of the zero-valent atoms sets in, so that only a very shallow minimum at a relatively large internuclear distance results. Analogous forces also add to the mutual attraction of atoms with free valences and of molecules with or without permanent dipole moments. See also **Van der Waals Forces**.

LONGITUDE.   The longitude of a point on the surface of the earth is the angular distance measured along the earth's equator from the meridian, through Greenwich, England, to the point where the local meridian of the point cuts the equator. Longitude may be expressed either in units of time (hours, minutes, and seconds) or of angle (degrees, minutes, and seconds). It is measured east or west from Greenwich, through 12 hours or 180°. For convenience in navigation, west longitude is marked plus (+) and east longitude minus (−).

LONG-RANGE ACCURACY (LORAC).   A two-dimensional radio navigation system using continuous-wave transmission to provide hyperbolic lines of position through radio frequency phase comparison techniques from four transmitters. The system is used for surveying or ship-positioning. Lorac uses the frequency band, 1.7 to 2.5 megacycles.

LONG-RANGE NAVIGATION (LORAN).   A two-dimensional pulse-synchronized radio navigation system to determine hyperbolic lines of position through pulse-time differencing from a master compared to two slave stations. Loran uses the frequency band 1.7 to 2.0 megacycles; loran

C (Cytac) uses transmission at 100 kilocycles and phase compares the continuous wave in the pulse envelopes for greater accuracy using pulse technique for resolving ambiguities.

LONG RANGE ORDER.   A system may be said to possess long range order if it is possible to assign letters A, B, C, etc., to the sites of the lattice in such a way that there is a greater probability of finding an atom of type A on an A-site, of type B on a B-site, and so on, than any other arrangement. Such order is characteristic of order−disorder transitions in binary alloys. It is measured by the parameter

$$S = \frac{r - w}{r + w}$$

where r and w are the numbers of atoms on right and wrong sites respectively.

LOOMING.   A mirage effect produced by greater-than-normal refraction in the lower atmosphere, thus permitting objects to be seen that are usually below the horizon. This occurs when the air density decreases more rapidly with height than in the normal atmosphere. If the rate of decrease of density with height is greater in the region followed by the ray from the top of the object than for the ray from the bottom of the object, the image will be stretched vertically. This stretching is often called *looming*, but is more properly termed t*owering*. The antonym of looming is sinking and that of towering is stooping.

LOON *(Aves, Gaviiformes, Gaviidae)*.   Large birds sometimes called divers which range in length from 58 to 90 centimeters (23 to 35 inches) and in weight from 1 to 6.4 kilograms (2.2 to 14.1 pounds). See also **Gaviiformes**. There are extensive webs between the three front toes, and they have a short tail with 16 to 20 tail feathers and 11 primaries. In flight the head and neck are somewhat lowered.

There is only 1 genus, *Gavia*, with four species found in the northern tundra and forest zones of the Old and New Worlds. Three of the species have a black and white checkered pattern on the wings of the breeding plumage: (1) Arctic Loon (Black-Throated Diver, *Gavia arctica*; Fig. 1); the length is 70 centimeters (28 inches) and the weight is 2−3.5 kilograms (4−8 pounds); the nape is gray. (2) The Common Loon or Great Northern Diver (*Gavia immer*). The length is 75 centimeters (30 inches) and the weight is 4 kilograms (9 pounds); the nape and beak are black. (3) The Yellow-Billed Loon or White-Billed Northern Diver (*Gavia adamsii*). The length is 87 centimeters (34 inches) and the weight is 4.5−6.4 kilograms (10−14 pounds). The beak is ivory-colored in adults. (4) The fourth species, with small white stripes on the nonbreeding plumage, is the Red-Throated Loon or Diver (*Gavia stellata*). The length is 58 centimeters (23 inches) and the weight is 1−2.4 kilograms (2−5 pounds).

Fig. 1.   Arctic loon. (*Sketch by Glenn D. Considine.*)

The loon or diver is found in Canada, the northern parts of the United States, particularly the mountain lakes of New York and Pennsylvania and the lakes of Michigan as well as in California and south to Mexico. Some species of loon are found in Europe and essentially in most parts of the world.

LOOP ANTENNA.   An antenna consisting of a conducting coil, of any convenient cross section (generally circular), which emits or receives radio energy. The principal lobe of the radiation pattern is wide and is in the direction perpendicular to the plane of the coil. Also called *loop*.

LOOP (Computer System). A sequence of instructions that may be executed repeatedly while a certain condition prevails. The productive instructions in the loop generally manipulate the operands, while bookkeeping instructions may modify the productive instructions, and keep count of the number of repetitions. A loop may contain any number of conditions for termination, such as the number of repetitions or the requirement that an operand be nonnegative. The equivalent of a loop could be achieved by the technique of straight line coding, whereby the repetition of productive and bookkeeping operations is accomplished by explicitly writing the instructions for each repetition.

See also **Program (Computer)**.

LOOP (Mathematics). A closed path. An immediate consequence of this definition is that a finite graph contains only a finite number of loops, a conclusion which is critical to the practical application of Kirchoff's law for voltage.

LOOP GAIN. In feedback terminology, the gain around the feedback loop, numerically equal to the product of the forward gain by the gain of the feedback network when the circuit configuration permits meaningful identification of these two separate transmissions. The feedback network is also called the *beta-network*.

See also **Gain (Magnitude Ratio)**.

LORENTZ FRAME. Any of the set of coordinate systems in Minkowski space for which the square of the interval between two events is $c^2 dt^2 - (dr)^2$. Any such coordinate system may be obtained from another by means of a Lorentz transformation (together perhaps, with an orthogonal transformation of the space axes). With each Lorentz frame may be associated a point observer, each of whom moves with constant velocity relative to the others.

LORENTZ, HENDRIK ANTOON (1853–1928). Lorentz was born in the Netherlands. He received a Ph.D. from Leiden in the area of electromagnetism. He became the first chair of theoretical physics at the University of Leiden in 1877.

From Lorentz stems the conception of the electron; his view that his minute, electrically charged particle plays a role during electromagnetic phenomena in ponderable matter made it possible to apply the molecular theory to the theory of electricity, and to explain the behaviour of light waves passing through moving, transparent bodies.

The so-called Lorentz transformation (1904) was based on the fact that electromagnetic forces between charges are subject to slight alterations due to their motion, resulting in a minute contraction in the size of moving bodies. It not only adequately explains the apparent absence of the relative motion of the Earth with respect to the ether, as indicated by the experiments of Michelson and Morley, but also paved the way for Einstein's special theory of relativity.

It may well be said that Lorentz was regarded by all theoretical physicists as the world's leading spirit, who completed what was left unfinished by his predecessors and prepared the ground for the fruitful reception of the new ideas based on the quantum theory.

Lorentz's most important contributions include his proposal that charged particles in matter oscillate when struck by light waves, and his applying his electron theory to explain the Zeeman effect. In 1902, he shared the Nobel Prize in Physics with Pieter Zeeman.

See also **Field Theory**; **Lorentz Transformation**; and **Relativity and Relativity Theory**.

J. M. I.

LORENTZ INVARIANCE. The equivalence principle of special relativity, which states that physical principles must be invariant under a transformation from one coordinate system to another. Since this is a Lorentz transformation, the invariance itself is often called by the name of Lorentz.

LORENTZ TRANSFORMATION. Relations connecting the space and time coordinates of an event as observed from two Lorentz frames. If S' moves relative to S with velocity v in the x-direction, x, y, z, t denote position and time coordinates of two events as measured by S, and x', y',

z', t' the corresponding quantities for S', then

$$x' = \gamma(x - vt)$$
$$y' = y, z' = z$$
$$t' = \gamma \left(t - \frac{x}{c^2} \right)$$

where $\gamma = (1 - v^2/c^2)^{-1/2}$. These relations were shown by Einstein (1905) to be a consequence of the special relativity theory. See also **Relativity and Relativity Theory**; and **Vector**.

LORISOIDS *(Mammalia).* Primitive animals of the order *Primates* and similar to lemurs. See also **Lemur**. Lorises are found in the warmer parts of southeastern Asia and the East Indian islands. They have large staring eyes and from their very slow movements have sometimes been referred to as slow lemurs. The lorises are forest animals of nocturnal habits. The slender loris is known as *Loris gracilis*; the slow loris as *Nycticebus tardigradus*.

LOSSER. A dielectric material that dissipates energy. A dissipative element placed in a circuit to prevent oscillation.

LOSS FACTOR. The rate at which heat is generated in a dielectric is proportional to its loss factor, which is equal to the product of its dielectric constant by its power factor. Both the dielectric constant and power factor are usually functions of frequency; therefore, the loss factor changes with changing frequency.

LOSS FUNCTION. In decision theory, a function of the decision and the true underlying distributions which expresses the loss incurred in taking that decision. If there are a number of possible situations, the array of losses according to situation and decision is called the *loss matrix*. It is analogous to the payoff matrix of games theory.

LOSS (Transmission). A general term used to denote a decrease in signal power in transmission from one point to another. Usually expressed in decibels.

LOUDNESS LEVEL. The loudness level of a sound is the sound-pressure level of a standard tone (usually 1,000 Hz) which sounds equally as loud as the sound under measurement. In 1933, Fletcher and Munson published their *loudness level contours* for pure tones. See Fig. 1. The curves commonly are referred to as Fletcher-Munson curves or equal-loudness contours. The numbers on the contours are the loudness levels of the sound in phons (the sound-pressure level of a 1,000 Hz tone that is equally loud). Thus, if a certain complex sound wave sounds equally as loud as a 1,000 Hz tone having a sound-pressure level of 60 db re 0.0002 dyne/cm^2, the complex wave is said to have a loudness level of 60 phons regardless of its sound-pressure level. Other sets of equal-loudness

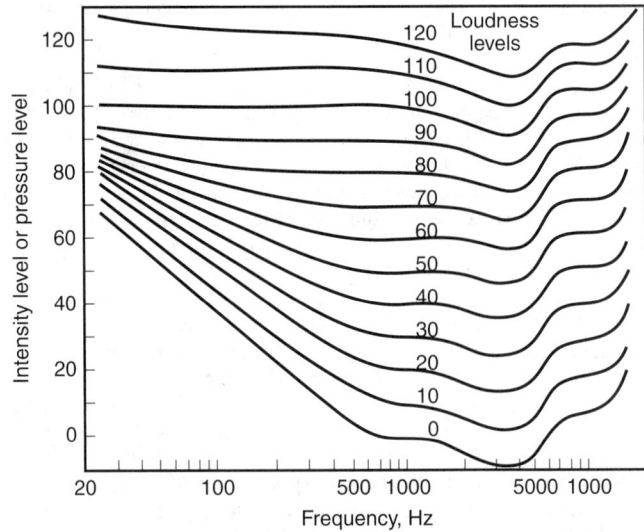

Fig. 1.   Fletcher-Munson curves indicating equal-loudness contours.

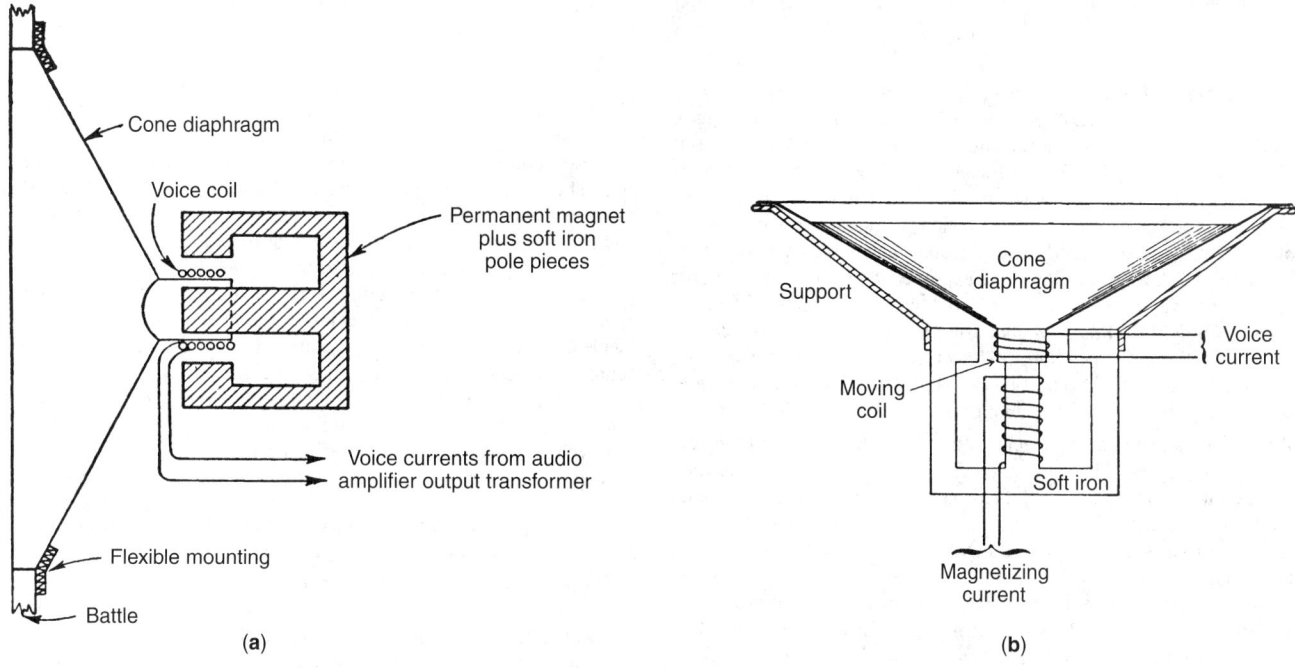

Fig. 1.   Two common types of loudspeakers: (**a**) permanent magnet type; (**b**) electromagnetic type.

contours which deviate in some respects from the Fletcher-Munson curves have been developed since by other investigators.

See also **Acoustics**.

LOUDSPEAKER.   A transduction device, usually based on the dynamic (moving-coil) principle, that converts electrical energy into mechanical energy or sound. A coil of wire located in the magnetic field of a permanent magnet is attached to a paper cone. See Fig. 1. The cone, at its outer edges, is flexibly attached to a support ring. When an electric current is passed through the coil, a force is created that acts upon the cone. Cone movement generates sound waves that are proportional to the frequency of the exciting current. Loudspeakers usually are low in efficiency, about 5%. By using horns of a gradually increasing cross sectional area, efficiencies of 30 to 50% can be achieved.

Liquid-Phase Projectors. Piezoelectric, magnetostrictive, and electro-magnetic are the principal types of transduction used in transmitters to excite acoustical waves in liquids. Designed to be resonant, these devices usually operate at their fundamental frequency. Piezoelectric liquid-phase projectors are effective from 20 kHz to above 100 MHz; while magne-tostrictive transducers will handle the range from 10 to 100 kHz. With bandwidth ratios of 5 to 20, both types have efficiencies on the order of 0.5 to 4 watts per square centimeter. Increasing the pressure of the liquid mass and providing special cooling will raise radiation intensities up to 50 watts per square centimeter. Many liquid-phase projectors are made up of arrays of individual transducers to control directivity and to increase power-handling capacity.

A large number of projectors cannot be used as receivers. The latter are designed to excite acoustical vibrations in air and are not reversible. Nonreversible projectors include modulated air-flow speakers, whistles, and sirens. In a modulated air-flow speaker, a valve controlled by an electrical signal modulates the flow of the airstream. Used for public address systems, these speakers have high efficiency and power output, but also have high distortion and poor frequency response. Whistles have high efficiency, but suffer from frequency and amplitude instability unless driven by an auxiliary device, such as the resonant cavity of an organ pipe. A siren, in which a stream of compressed air is chopped by a series of rotating blades, combines high efficiency and high intensities, with stable, easily controlled frequencies.

LOUSE *(Insecta).*   An external parasitic insect found on warm-blooded animals, both mammals and birds. Two kinds of lice occur: (1) The true or sucking lice (order *Anoplura*); and (2) the bird or biting lice (order *Mallophaga*). As the titles suggest, the sucking lice have piercing and sucking mouth parts; the bird lice have biting mouth parts. See Fig. 1.

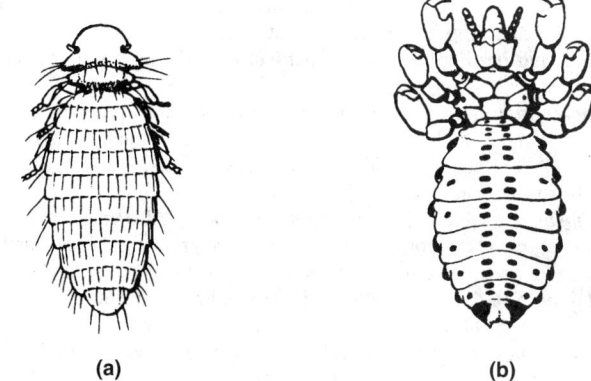

Fig. 1.   (**a**) Biting louse (*Mallophaga*); (**b**) sucking louse (*Anoplura.*) (*USDA.*)

Livestock Lice

Hog louse (Haematopinus suis or *adventicius,* Linne). Of the order *Anoplura.* This is a wingless, rather large, flat, gray-colored louse, about $\frac{1}{4}$-inch (6 millimeters) long that infests hogs. It does not affect other livestock. It is the largest of the blood-sucking lice. The head and legs are comparatively long. The insect is equipped with a hook-like member for clasping hairs of the host. A favorite habitat is in between folds of the hog's skin. The entire life cycle of the insect occurs on the host. The hog has only two commonly encountered insect parasites—the hog louse and the mange mite. Control is by dipping or spraying with Co-Ral, lindane, malathion, methoxychlor, ronnel, or toxaphene, where available and permitted. Medicated hog wallows, which incorporate a surface film of petroleum or pine oil, can be effective. The louse dies within a few days if not on a host.

Sheep and Goat Lice.   There are several species:

Bloodsucking body louse (Haematopinus orvillus, Neumann). Of the order *Anoplura.*

Bloodsucking foot louse (Linognathus pedalis, Osborn). Of the order *Anoplura.* A pale, rather slender louse, about $\frac{1}{12}$ inch (2 millimeters) long.

Sheep-biting louse (Boricola or *Trichodectes oris,* Linne). Of the order *Mallophaga.* A pale-brown insect with a red head, about $\frac{1}{20}$ inch (1 + millimeter) in length. This louse eats the wool fibers, causing fibers to become soiled and tangled. Its full life cycle is spent on the host. Treatment for this and other sheep and goat lice mentioned may include dipping or spraying lindane, methoxychlor, rotenone, or toxaphene. All of these lice can be quite damaging to the wool of the animals. The clip of mohair from an Angora goat, for example, may be reduced severely if there is an infestation.

Cattle and Horse Lice. These also affect mules and donkeys.

Long-nosed cattle louse (*Linognathus* or *Haemotopinus vituli*, Linne). Of the order *Anoplura*. This is a red louse, about $\frac{1}{8}$ inch (3 millimeters) long. The insect pierces the skin and sucks blood. In heavy infestations over long periods, animals become emaciated. Patches of skin become bare and sore. The insect seems to prefer unhealthy, poorly fed animals. Control is by spraying or dipping with lindane, malathion, methoxychlor, toxaphene, or ronnel, when available and approved. Special directions must be followed for spraying dairy cows to avoid any contamination of milk.

Horse-sucking louse (*Haemotopinus asini*, Linne or *H. macrocephalus*, Burmeister). Of the order *Anoplura*. This is a medium-size louse, about $\frac{1}{8}$ inch (3 millimeters) long. The bite is very painful. There are several generations per year. Treatment is about the same as for cattle louse.

Poultry Lice. A number of species attack poultry and wild fowl and birds, including pigeons. These lice are of the order *Mallophaga* and they do not suck blood, but rather they feed on bits of skin, scabs, feathers, and other organic debris found on the bird's body. Irritation is caused by the sharp mouth parts and sharp claws. Nibbling of the lice prevents rest and sleep, and causes loss of appetite, diarrhea, droopy wings, leading to progressive emaciation, reduction in egg production, and death of the birds when left unattended. Poultry lice are distributed worldwide. They are wingless, with six legs, flat bodies, and round heads. Treatment is by spraying or dusting with malathion or rotenone, where approved. Nests and litter also should be sprayed. Painting the roosts with 40% nicotine sulfate or 3% malathion can be effective. Some of the species include:

Chicken-head louse (*Cuclotogaster heterographus*, Nitzsch). An insect about $\frac{1}{10}$ inch (2.5 millimeters) long that severely irritates the birds around the neck and head area. They are particularly irritating to young turkeys and chicks.

Chicken-body louse (*Menacanethus stramineus*). An insect about $\frac{1}{8}$ inch (3 millimeters) long that irritates areas under the wings and about the vent. It attacks both young and old birds. Records indicate that over 35,000 of these lice were found on one chicken.

Common body louse (*Menopan gallinae*, Linne). Also called *shaft louse* or *small body louse*, about $\frac{1}{16}$ inch (1.5 millimeters) long. Lives mostly on the feathers and one of the most commonly encountered on fowl.

Fluff louse (*Goniocotes gallinae*, De Geer). Prefers operating in the fluff and under the vent; about $\frac{1}{16}$ inch (1.5 millimeters) long.

Brown chicken louse (*Goniodes dissimilis*, Denny). A medium-size louse, about $\frac{1}{10}$ inch (2.5 millimeters) long.

Large chicken louse (*Goniodes gigas*, Taschenberg). A comparatively large louse, about $\frac{5}{32}$ inch (2.5 millimeters) long.

Wing louse (*Liperus caponis*, Linne). Prefers the barbules of the wing feathers; about $\frac{1}{10}$ inch (2.5 millimeters) long.

Small pigeon louse (*Campanulotes* or *Goniocotes bidentalus*, Burmeister). Inhabits feathers of both young and old pigeons.

Slender pigeon louse (*Columbicola columbae*, Linne). Much like the small pigeon louse.

NOTE: Plant lice are not true lice, but members of the order *Homoptera* and commonly called aphids.

LOW VACUUM. The condition in a gas-filled space at pressures less than 760 torr and greater than some lower limit. It is recommended that this lower limit be chosen as 25 torr corresponding approximately to the vapor pressure of water at 25 degrees C and to 1 inch of mercury.

L-SECTION. This refers to an elementary section of a network such as a filter where the components are connected in the form of an L, i.e., one component in series with one side and the other in shunt across the two sides of the circuit. See Fig. 1.

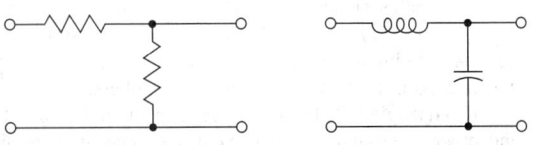

Fig. 1.   L-section.

LUBRICANT. A material used to diminish friction between the moving surfaces of machine parts; also to decrease friction between a cutting tool and the material being cut. A wide variety of materials is used for manufacturing lubricants. Animal lubricants are obtained from the fat of common animals and can be classified as hard fats (stearin) and soft fats (lard) or naturally occurring combinations. Vegetable lubricants include rape seed oil, cottonseed oil, soybean oil, castor oil, and linseed oil. They range in properties from solid to liquid. Petroleum and mineral oil lubricants, because of their greater stability, are usually preferred for machine applications. Lubricants range from light oils to very heavy solid greases. Graphite, a solid, is also used as a lubricant.

Because of increased requirements for lubricants, including higher temperature and pressure applications, greater durability, and tolerance to wide changes in ambient temperature conditions, numerous synthetic lubricants have been developed. These include synthetic hydrocarbons, carboxylic acid esters, silicones, polyethers (polyalkylene glycols), phosphate esters, silicate esters, highly fluorinated compounds, and polyaromatics (polyphenyls and polyphenyl ethers). In selecting a lubricant, the following characteristics are considered: (1) lubricity and antiwear properties; (2) fluid range; (3) viscosity index; (4) additive response of base oil; (5) oxidation stability; (6) thermal stability; (7) hydrolytic stability; (8) fire resistance; (9) compatibility with petroleum products; (10) compatibility with paints, plastics, and elastomers; and (11) cost.

Over a number of years, the early polyol esters, the formulas of some of which are shown below, appeared to be adequate for coping with the increasing rigorous properties required for increasingly difficult lubrication problems. They continue to be used, but some professionals in the field have developed a number of proprietary formulations that are claimed to be superior to the polyol esters.

Polyaromatics (polyphenyls)

$$R-O-\underset{\underset{O}{\|}}{C}-(CH_2)_n-\underset{\underset{O}{\|}}{C}-O-R$$
Diester

Polyether (polyalkylene glycols)

$$Si-(O-R)_4$$
Silicate ester

$$[-CH_2-CH_2-CH_2-]_2$$
Synthetic hydrocarbon

$$C-(CH_2-O-C-R)_4$$
Neopentyl polyol ester

More attention has been given to tribology, the scientific discipline of friction, wear, and lubrication. The underlying principles of tribology have been investigated intensively by a number of research groups, including the U.S. Naval Research Laboratory and physicists at the Georgia Institute of Technology, Atlanta, Georgia. Researchers are attempting to develop a better theoretical basis for understanding the processes that occur when two solid bodies move past each other at close to overlapping distances. Most of this new knowledge has stemmed from working models as well as from computer models. Researchers at the Georgia Institute of Technology have used a large Cray computer to predict what occurs when the tip of a thin nickel needle, for example, is pressed repeatedly onto a flat gold surface. The investigators initially employed quantum mechanics for answering such questions as adhesion, cohesion, and the making and breaking of chemical bonds. Researchers at the Naval Research Laboratory have used an atomic force microscope as a tool to check their experiments. The present goal is to model more complex systems. As pointed out by one researcher, "To make progress in the molecular engineering of lubricants, you need to know the molecular details of the process."

Additional Reading

Klaman, D.: "Lubricants and Related Products: Synthesis, Properties, Applications, International Standards," John Wiley & Sons, Inc., New York, NY, 1984.

Rudnick, L.R. and R.L. Shubkin: "Synthetic Lubricants and High-Performance Functional Fluids," 2nd Edition, Marcel Dekker, Inc., New York, NY, 1999.

Sequiera, A.: "Lubricant Base Oil and Wax Processing," Marcel Dekker, Inc., New York, NY, 1994.

Staff: Society of Automotive Engineers: "Heavy Duty Diesel Engine Lubricants," Society of Automotive Engineers, Warrendale, PA, 1996.

LUMINESCENCE. A characteristic nonthermal emission of electromagnetic radiation by a material upon some form of excitation. Some luminescent materials are called phosphors. E. Wiedemann defined the term in 1888 as "all those phenomena of light not solely conditioned by the rise in temperature."

Whereas the output from blackbody radiators consists of broad-band emissions which follow the Stefan-Boltzmann temperature relationships, luminescence emission from phosphors consists of relatively narrow bands, which do not follow the blackbody laws. Thus, light emission due solely to the temperature of a source is referred to as *incandescence*, while *luminescence*, unlike incandescence, is a function of the specific material involved. Although *fluorescence* and *phosphorescence* are sometimes used synonymously with luminescence, a more rigid definition of *fluorescence* would be luminescence having a persistence (afterglow) shorter than about 10^{-8} second, with *phosphorescence* being longer than 10^{-8} second.

The luminescence process itself involves (1) absorption of energy; (2) excitation; and (3) emission of energy, usually in the form of radiation in the visible portion of the spectrum. The *type* of luminescence is usually defined by the excitation means, i.e., *cathodo*luminescence where excitation is by cathode rays, as in a television kinescope. The most commonly encountered types of luminescence are listed in Table 1.

TABLE 1. TYPES OF LUMINESCENCE

Luminescence Type	Excitation Source	Example
Photoluminescence	Photons	$ZnS \cdot Ag$
Cathodoluminescence	Cathode Rays	$Zn_2SiO_4 \cdot Mn$
Electroluminescence	Electric Fields	$Zn \cdot (S \cdot Se) \cdot Cu$
Chemiluminescence	Chemical Reactions	Oxidation of Luminol
Bioluminescence	Biochemical Reactions	Luciferin
Triboluminescence	Mechanical Disruption	$ZnS \cdot Mn$

The luminescent material may be considered as a transformer of energy, i.e., from ultraviolet photons to photons of lower energy; from cathode rays to photons; from electric fields to photons, etc. An inorganic luminescent material, or phosphor, usually consists of a crystalline host material to which is added a trace of an impurity (activator and coactivator).

Chemiluminescence

Numerous chemical reactions produce heat, but relatively few release their energy as light. This latter phenomenon is termed *chemiluminescence*. As pointed out by researchers at the University of Wales College of Medicine, Cardiff, "Absorption of energy by an atom or molecule raises an electron to a higher energy level. This is known as an 'excited state' and is inherently unstable. When the electron drops back to its ground state the energy must either be transferred to another atom or molecule, be released as heat or be emitted as light. The decay of the electron to ground state is very fast, occurring within 1–10 nanoseconds (10^{-9}–10^{-8} second)."

In chemiluminescence, the chemical reaction raises an electron to a higher level, which then decays back to the ground state, releasing a photon of light, the energy of which is predictable by Einstein's equation. When the energy drop is large, the light is blue; if small, the light is red. Inasmuch as the electronic excitation-decay process is extremely fast, the intensity of light in chemiluminescence is determined by the kinetics of the chemical reaction.

The distinction of chemiluminescence from fluorescence results from two factors:

1. When an atom or molecule fluoresces, it remains chemically unchanged and can be immediately be reexcited once light emission has occurred.
2. In chemiluminescence, each molecule only reacts once to form an excited state, while the excited product (actual emitter) has a different chemical structure from the initial substrate.

See also **Bioluminescence.**

Light-Emitting Diode (LED)

Recombination of injection electroluminescence was first observed in 1923 by Lossew, who found that when point electrodes were placed on certain silicon carbide crystals and current passed through them, light was often emitted. Explanation of this emission has been possible only with the development of semiconductor theory. If minority charge carriers are injected into a semiconductor, i.e., electrons are injected into *p*-type material or "positive holes" into *n*-type material, they recombine spontaneously with the majority carriers existing in the material. If some of these recombinations result in the emission of radiation, electroluminescence results. Minority-carrier injection may occur not only at point contacts, but also at broad area rectifying junctions; in this case, the junction must be biased in the forward or "easy flow" direction, and the electric field in the junction is lower when the voltage is applied than in its absence. This type of emission has been observed in several materials, including SiC, diamond, Si, Ge, CdS, ZnS, ZnSe, ZnO, and some of the so-called III-V compounds, such as AlN, GaSb, GaAs, GaP, InP, and InSb. The emission of many of these materials lies in the infrared region of the spectrum. For radiation in the visible region (instead of the infrared), the energy difference between the holes and electrons (band gap of the semiconductor) must be more than 1.8 eV. Numerous materials satisfy this requirement, notably those used for cathode-ray tube phosphors, but the materials present difficulties in fabricating *p-n* junctions and thus are not candidates for light-emitting diodes.

The list of materials for LEDs includes GaP, GaAsP, GaAlAs, GaN, and SiC. The two materials of choice to date have been GaP and GaAsP. Early commercial LEDs were made from $GaAs_{0.6}P_{0.4}$ deposited epitaxially as a thin layer on a GaAs crystal substrate. With these, *p-n* junctions were made, using diffusion techniques similar to those used in making silicon diodes. The band gap is 1.92 eV. There is an emission band of red light with a peak at about 650 nm, resulting from direct recombination of electrons and holes.

GaAsP has a high index of refraction, and consequently only light emitted toward the surface (4%) is usable — the remainder is reflected back. A diode can be encapsulated in epoxy material to take on the shape and form of a lens. These diodes are particularly effective where a number are fabricated in close proximity on a single-crystal chip.

Diodes that emit light in shorter wavelengths (green, yellow, etc.) can be made by increasing the phosphorus content, but only up to about 40% because of rapid decrease in efficiency. Efficiency can be increased by incorporating nitrogen atoms into crystals. The N atoms act as isoelectronic centers, trapping electron-hole pairs in an excited state. Three types of nitrogen-doped diodes have gained some importance: $GaAs_{0.65}P_{0.35}$ (orange light); $GaAs_{0.85}P_{0.15}$ (yellow light); and GaP (green light). Zinc and oxygen doping are also used. Diodes operate more efficiently if driven with periodic pulses of high current rather than with constant current. The short response time of junction diodes to current pulses (a fraction of a microsecond) and their rectifying property (they block current flow in the reverse, nonemitting direction) combine to make the diodes a good choice for *X-Y* addressing arrangements. See also **Inorganic Light-Emitting Diode Displays**; and **Organic Light Emitting Diodes (OLEDS).**

Additional Reading

Aitken, M.J.: "An Introduction to Optical Dating: The Dating of Quaternary Sediments by the Use of Photon-Stimulated Luminescence," Oxford University Press, Inc., New York, NY, 1998.

Burgess, C. and D. Jones: "Spectrophotometry, Luminescence and Colour: Science and Compliance: Papers Presented at the 2nd Joint Meeting of the UV Spectrometry Group of the U.K. and the Council for Optical Radiation Measurements of the U.S.A., Rindge, NH, U.S.A., 20–23," Elsevier Science, New York, NY, 1995.

Ropp, R.: "Luminescence and the Solid State," Elsevier Science, New York, NY, 1991.

Schulman, S.G.: "Molecular Luminescence Spectroscopy: Methods and Applications," Vol. 3, John Wiley & Sons, Inc., New York, NY, 1993.

Stanley, P.E. and L.J. Kricka: "Bioluminescence and Chemiluminescence: Fundamental of Applied Aspects," John Wiley & Sons, Inc., New York, NY, 1996.

Vij, D.R. and N. Singh: "Luminescence and Related Properties of II-VI Semiconductors," Nova Science Publishers, Inc., Huntington, NY, 1997.

Vij, D.R.: "Luminescence of Solids," Kluwer Academic Publishers, Norwell, MA, 1998.

Ziegler, M.M. and T.O. Baldwin: "Bioluminescence and Chemiluminescence, Part C," Vol. 305, Academic Press, Inc., San Diego, CA, 2000.

LUMINOSITY FUNCTION. Because of the variable sensitivity of the human eye to radiation of different wavelengths, a standard function has been established. For the standard conditions chosen in establishing this standard luminosity function (photopic vision), the luminously effective radiant intensity in lumens of radiation of spectral energy distribution J_Λ watts/unit wavelength is given by

$$680 \int_{\lambda=0}^{\lambda=\infty} y_\lambda J_\lambda \, d\lambda$$

where y_Λ is the standard luminosity function normalized to a value of unity at 555 nanometers. The numerical values for y_λ are commonly given as a luminosity curve.

For very low levels of intensity (scotopic vision) the peak of the luminosity function curve shifts toward the violet for young eyes (507 nanometers) with an absolute value of 1,746 lumens/watt.

LUMINOUS COEFFICIENT. A coefficient that measures the integrated fraction of the radiant power that contributes to its luminous properties as evaluated by means of the standard luminosity function.

Luminous coefficient

$$= \int_{\lambda=0}^{\lambda=\infty} y_\lambda J_\lambda \, d\lambda \Big/ \int_{\lambda=0}^{\lambda=\infty} J_\lambda \, d\lambda$$

where y_λ is the standard luminosity function and J_λ is the spectral energy distribution of the radiant intensity. The luminous coefficient is unity for a narrow band of wavelengths at 555 nanometers.

LUNAR ATMOSPHERIC TIDE. An atmospheric tide due to the gravitational attraction of the moon. The only detectable components are the 12-lunar-hour or semidiurnal, as in the oceanic tides, and two others of very nearly the same period. The amplitude of this atmospheric tide is so small that it is detected only by careful statistical analysis of a long record, being about 0.06 millibar in the tropics and 0.02 millibar in the middle latitudes.

LUNAR ECLIPSE. See **Eclipse**.

LUNAR PROSPECTOR MISSION. The *Lunar Prospector*, the first dedicated lunar mission in 25 years, was a tremendous success. Following a near flawless launch on Jan 6, 1998, a four-day journey to the Moon and entry into lunar orbit, the tiny spin-stabilized spacecraft sent data back to Earth. Lunar data from the circular polar-mapping orbit started arriving January 15.

On March 5, 1998 *Prospector* scientists captured the public's imagination by announcing the discovery of a definitive signal for water ice at both of the lunar poles. At that time, a conservative analysis of the available data indicated that a significant quantity of water ice, possibly as much as 300 million metric tons, was mixed into the regolith (lunar soil) at each pole, with a greater quantity existing at the north pole. The first competitively selected Discovery class mission had conclusively demonstrated that, not only could a cost-capped, fast-development mission succeed, it could do ground-breaking science in the process.

The first operational gravity map of the Moon was announced at the same time. Since then, Lunar Prospector engineer's have taken advantage of the missions own science results and the gravity data have been used to facilitate orbit maintenance.

Mission Profile

At 9:28 p.m. (EST) on January 6, 1998, *Lunar Prospector* (LP) blasted off to the Moon aboard a Lockheed Martin solid-fuel, three-stage rocket called Athena II. It was successfully on its way to the Moon for a one-year, polar orbit, primary mission dedicated to globally mapping lunar resources, gravity, and magnetic fields, and even outgassing events. About 13 minutes after launch, the Athena II placed the Lunar Prospector payload into a "parking orbit" 115 miles above the Earth. Following a 42-minute coast in the parking orbit, *Prospector's* Trans Lunar Injection (TLI) stage successfully completed a 64-second burn, releasing the spacecraft from Earth orbit and setting it on course to the Moon, a 105-hour coast. See Fig. 1. The official mission timeline began when the spacecraft switched on 56 minutes, 30 seconds after liftoff. Shortly after turning the vehicle on, mission controllers deployed the spacecraft's three extendible masts, or booms. Finally, the spacecraft's five instruments—the gamma-ray spectrometer, alpha particle spectrometer, neutron spectrometer, magnetometer and electron reflectometer—were turned on. On Sunday, January 11, at 7:20 a.m. (EST), *Lunar Prospector* was successfully captured into lunar orbit, and a few days later began its mission to globally map the Moon. See Fig. 2.

The *Lunar Prospector* spacecraft is shaped like a drum, 4.25 feet (1.3 meters) high, with a diameter of 4.5 feet (1.4 meters). When full of fuel, Prospector weighed 650 pounds (295 kilograms). The three science masts carrying its five science instruments and isolating them from the spacecraft's electronics are each 8 feet (2.4 meters) long. The spacecraft was built by Lockheed Martin Missiles & Space, of Sunnyvale, California. See Fig. 3.

The *Lunar Prospecter* was a small spin-stabilized spacecraft in a polar orbit with a period of 118 minutes at a nominal altitude of 100 km (63 miles). Since the Moon rotates a full turn beneath the spacecraft every lunar cycle (~27.3 days) as it zips around the Moon every 2 hours, Prospector visited a polar region every hour and completely covered the

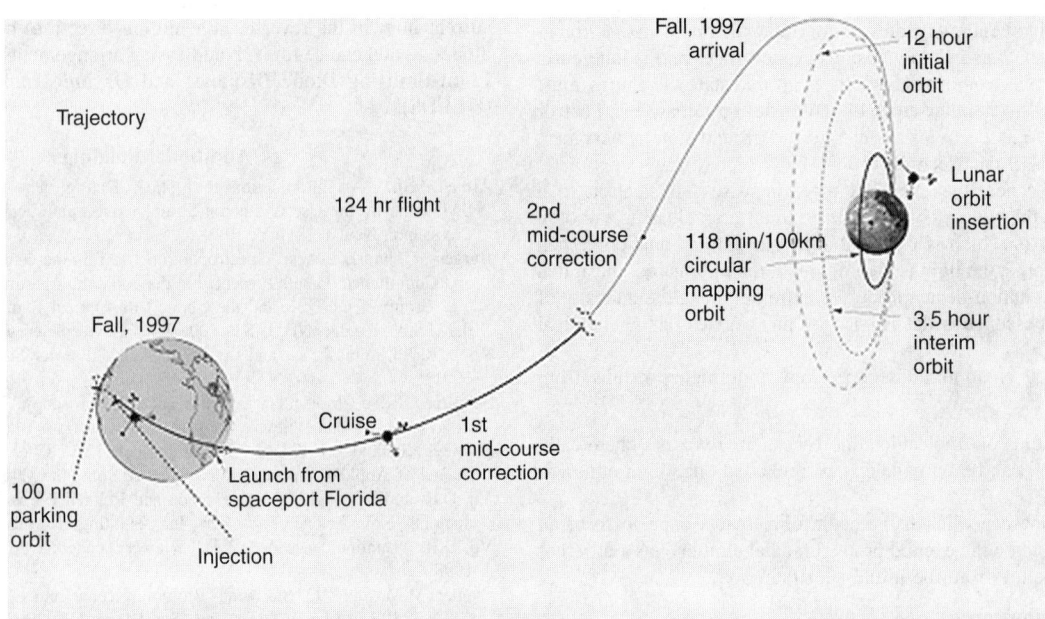

Fig. 1.   This diagram shows Prospector's flight path to the Moon. The spacecraft circles the Earth before achieving injection over Australia. There will be at least two mid-course corrections before reaching the Moon. Prospector is inserted in an initial large elliptical orbit around the Moon and goes through an interim orbit before achieving its final circular mapping orbit. This method of gradual orbit insertion assures the fuel efficiency vital to Prospector's low-cost design. (*NASA, Goddard Space Flight Center.*) *http://lunar.arc.nasa.gov/lunar/* Archives-Images Choose a Mission (Lunar Prospector) Select A Category: (Mission) Fig. 2.

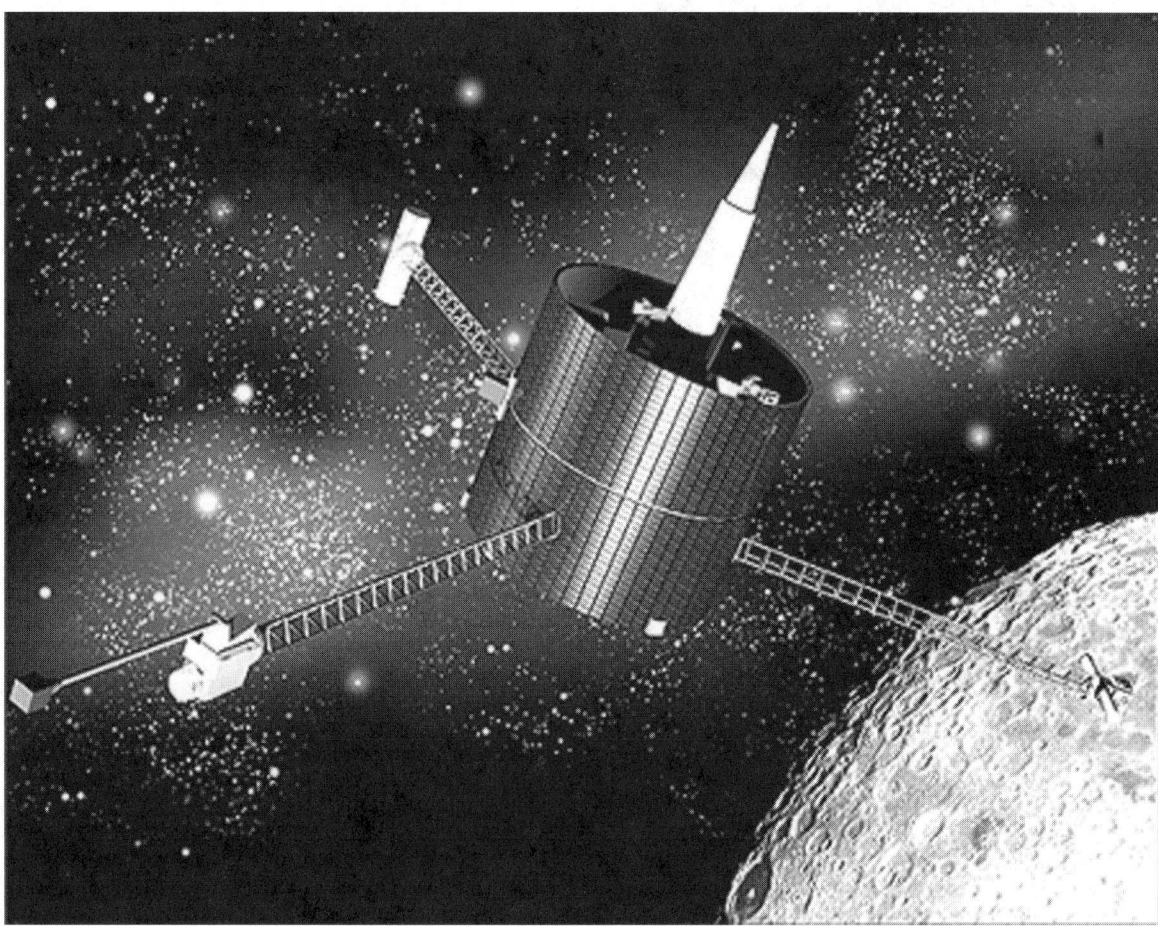

Fig. 2.   This artist's conception of Lunar Prospector shows the Spacecraft in lunar orbit with its instrument masts fully deployed. Prospector's primary mission will keep the spacecraft in a 100 km polar mapping orbit for a full year or more. This orbit will provide higher quality science data than has previously been obtained. Prospector has an Extended Mission option of a 10 km, very high resolution orbit for a brief period of time. (*NASA, Goddard Space Flight Center.*) *http://lunar.arc.nasa.gov/lunar/* Archives-Images Choose a Mission (Lunar Prospector) Select a Category (Spacecraft) (Fig. 13).

lunar surface twice a month. *Prospector's* one-year-long primary mission with an optional extended mission of a further 6 months at an even lower altitude enabled large amounts of data to collect over time. For some science instruments, a significant amount of time was required to obtain high quality usable data. Thus, *Prospector's* polar orbit and long-mission time rendered it ideal from the standpoint of globally mapping the Moon.

*(1.3 m in diameter X 1.4 m tall bus with three 2.5 meter science masts carrying its five science instruments and isolating them from the spacecraft's electronics)

Lunar Prospector Scientific Goals

As a Discovery-class mission, Prospector's scientific goals were carefully chosen to address outstanding questions of lunar science both efficiently and effectively. In the Post-Apollo era, NASA convened the Lunar Exploration Science Working Group (LExSWG) to draft a list of the most pressing, unanswered scientific riddles still facing the lunar-science community. In 1992, LExSWG produced a document, entitled "A Planetary Science Strategy for the Moon." The following lunar science objectives were listed: How did the Earth-Moon system form? How did the Moon evolve? What is the impact history of the Moon's crust? What constitutes the lunar atmosphere? What can the Moon tell us about the history of the Sun and other planets in the Solar System?

Lunar Prospector mission designers carefully selected a set of objectives and a payload of scientific instruments which would address as many of LExSWG's priorities as possible, while remaining within the tight budget confines of NASA's "faster, better, cheaper" Discovery Program.

Lunar Prospector's identified critical science objectives were:

- "Prospect" the lunar crust and atmosphere for potential resources, including minerals, water ice and certain gases,
- Map the Moon's gravitational and magnetic fields, and
- Learn more about the size and content of the Moon's core.

The six experiments (five science instruments) which address these objectives were:

- Neutron Spectrometer (NS) — Map hydrogen using neutrons having several different energy ranges and thereby infer the presence or absence of water.
- Gamma Ray Spectrometer (GRS) — Map 10 key elemental abundances, several of which offer clues to lunar formation and evolution.
- Magnetometer and Electron Reflectometer (Mag/ER) — These two experiments combine to measure lunar magnetic field strength at the surface and at the altitude of the spacecraft and thereby greatly enhance understanding of lunar magnetic anomalies.
- Doppler Gravity Experiment (DGE) — Make an operational gravity map of the Moon for use by future missions as well as LP by mapping gravity field measurements from changes in the spacecraft's orbital speed and position.
- Alpha Particle Spectrometer (APS) — Map out-gassing events by detecting Radon gas (current outgassing events) and Polonium (tracer of recent, i.e. 50 years).

The Basics of Lunar Selenology (Geology)

Formation

The current most widely held theory of how the Moon was formed is the impact theory. This theory suggests that a large body, possibly the size of Mars impacted the Earth some 4 billion years ago and the material, which was thrown off, a combination of Earth material and the impactor, coalesced into Earth's Moon. The body was originally molten. As it cooled, light elements such as calcium and aluminum and silica crystallized into feldspar and other minerals and floated upwards. Heavier minerals such as iron and magnesium-rich olivine and pyroxene formed and sank downward forming the lunar mantle. Incompatible elements such as KREEP (Phosphorous-K, Rare Earth Elements, and Potassium) were

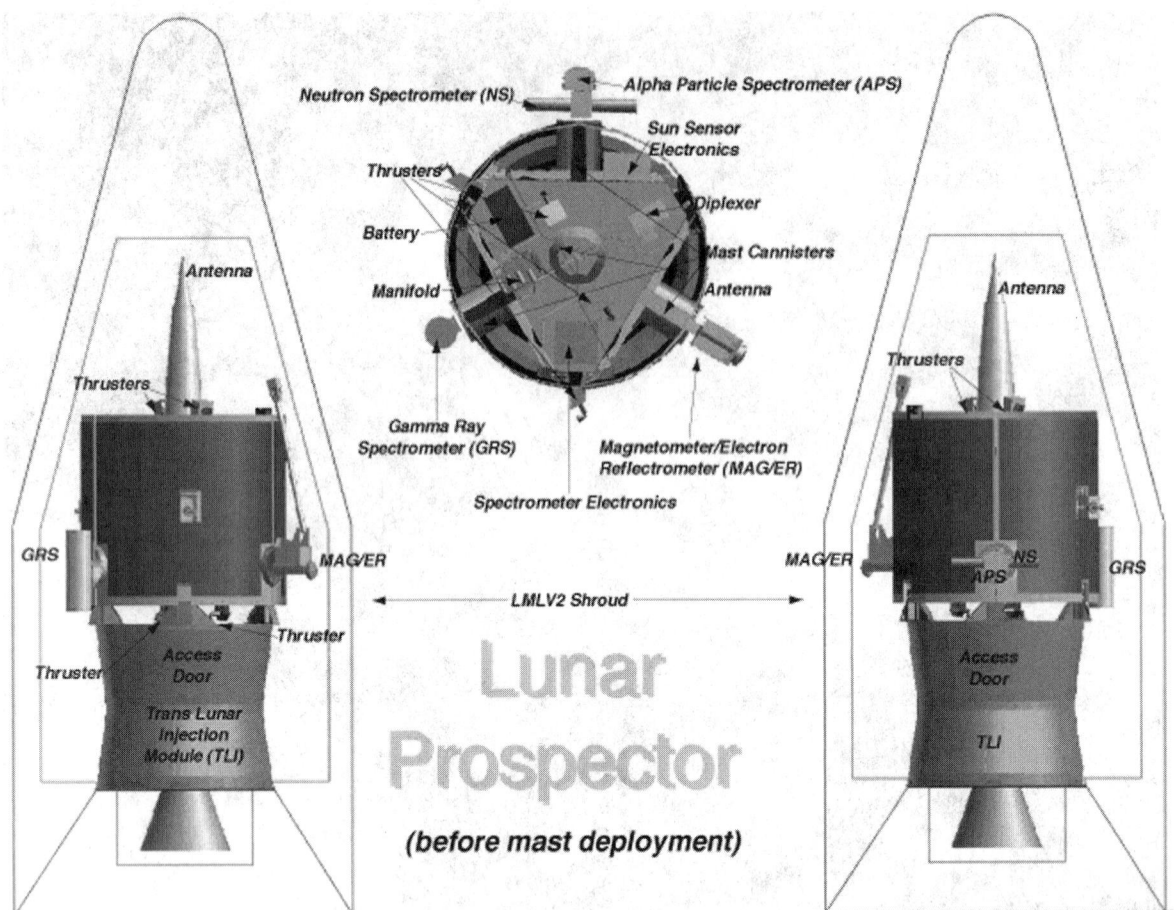

Fig. 3.   This diagram of the Lunar Prospector shows the Spacecraft with its instrument masts stowed. The three sets of science instruments: (1) Gamma Ray Spectrometer (GRS), (2) Neutron Spectrometer (NS) with Alpha Particle Spectrometer (APS), and (3)Magnetometer with Electron Reflectometer (Mag/ER) are identified. The antenna which will facilitate the Doppler gravity experiment, as well as command and control is identified. The major electronics boxes are also identified. (*NASA, Goddard Space Flight Center.*) http://lunar.arc.nasa.gov/lunar/ Archiver-Images Choose a Mission (Lunar Spacecraft) Select a Category (Spacecraft) Fig. 2.

trapped between the layers. See also **Rare-Earth Elements and Metals**. Repeated large impacts on the lunar surface itself, around 4.1 to 3.9 billion years ago, formed craters, which gradually filled with basaltic lava. These include the large dark areas that can be seen from Earth, the maria such as Mare Tranquilitatus-the Sea of Tranquillity and Mare Imbrium-the Sea of Rains. Over time, the Moon has continued to be bombarded from space by crater forming impacts. As the Moon cooled, however, lava ceased to fill the craters, and the impacts began to "garden" or break up the brittle lunar surface, creating the powdery dust, which is known as regolith or lunar "soil."

Lunar topography can be roughly divided into two main types—the maria with their distinctive dark color, and the highlands which are significantly different in composition as shown in Fig. 4.

Photographic Experiments

Ground-based astronomy provided early views of the Moon's near side, the side that perpetually faces the Earth. In addition, thirty-eight previous lunar missions have photographed the Moon, supplying scientists with a plethora of still imagery and television footage of the lunar surface. In 1959, the Soviet's *Luna 3* spacecraft was the very first to capture a full composite view of the far side of the Moon. During the 1960s and 1970s, NASA successfully flew 22 missions to the Moon, including the historic Apollo missions. The Soviets sent another 20. Throughout the 1960s and 1970s, these missions continued to amass images of the Moon and, in 1965, the American public watched for the first time, live pictures transmitted from NASA's *Ranger 9* spacecraft. Cameras aboard NASA's *Lunar Orbiter 1* spacecraft took the first pictures of Earth from the Moon. All of the Apollo missions returned lunar photographs to Earth; even the failed *Apollo 13* mission yielded a limited amount of photographic data. The U.S. Department of Defense *Clementine* spacecraft, which briefly orbited the Moon in 1994, obtained a significant set of lunar imagery.

The polar orbit was an ellipse, roughly 400 kilometers above the surface at the closest point. The payload aboard *Clementine* was a camera and a topographic mapper.

TABLE 1. APPROACHES TO LUNAR SCIENCE AND PREVIOUS MISSIONS

Experiment Type (* = aboard LP)	Previous lunar missions
Photographic Studies	U.S. Missions: Lunar Orbiter 1–5, Surveyor 1, Surveyor 3, Surveyor 5–7, Ranger 7–9, Apollo 8, Apollo 10–17, Clementine Soviet missions: Luna 3, Luna 9, Luna 12–13, Luna 16–17, Luna19–22, Zond 3, Zond 6–8
Surface Sampling/Soil Analysis	U.S Missions: Surveyor 3, Surveyor 5–7, Apollo 11–12, Apollo 14–17 Soviet Missions: Luna 16–17, Luna 20–21, Luna 24
Magnetic Studies*	U.S. Missions: Apollo 12, Apollo 14–17 Soviet Missions: Luna 10, Luna 21–22
Gravity Studies*	U.S. Missions: Apollo 12, Apollo 14–17, Clementine Soviet Missions: Luna 10–11, Luna 14, Luna 19, Luna 22
Alpha Particle Spectroscopy*	U.S. Missions: Apollo 15–16
Gamma-Ray Spectroscopy*	U.S. Missions: Apollo 15–16 Soviet Missions: Luna 10–11, Luna 22
Neutron Spectroscopy*	U.S. Missions: none, although Apollo 17 had a "neutron probe"

Lunar Prospector's Design Philosophy

Lunar Prospector was never intended to carry a camera, nor did it require a true onboard computer for two key reasons. *Prospector*, from its inception,

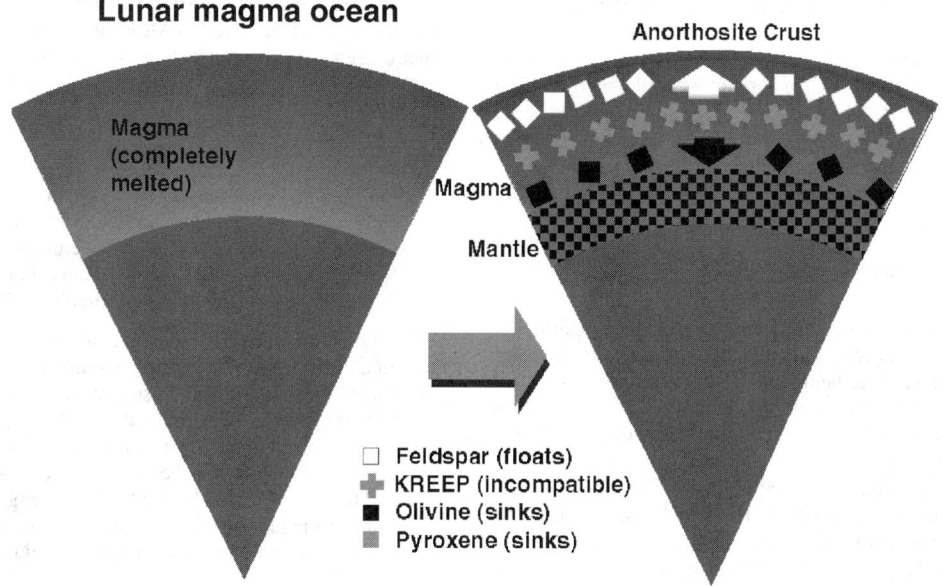

Lunar magma ocean

Magma
(completely
melted)

Anorthosite Crust

Magma

Mantle

☐ Feldspar (floats)
✚ KREEP (incompatible)
■ Olivine (sinks)
▇ Pyroxene (sinks)

Fig. 4.   A model diagram of the formation of the lunar crust, showing an incompatible KREEP layer trapped between the mantle and the crust. (*NASA, GSFC.*)

was designed to be a simple, cost-efficient spacecraft. Cameras require pointing in a given direction. The *Lunar Prospector* spacecraft was spin-stabilized, an engineering strategy that is well understood, highly successful and simple and cheap to achieve. With a camera, Prospector would have required a whole different engineering strategy, as well as a computer to control the camera. The result would have been a far more complex, heavier, costlier spacecraft that may well not have survived the Discovery selection process. Secondly, *Lunar Prospector* was a science-focused mission. That is, *Prospector* was designed to answer the highest-priority outstanding questions the planetary science community still has about the Moon. The spacecraft's five instruments were chosen according to that criterion. Previous Moon missions have, in fact, taken thousands of pictures of the Moon. We already have quite detailed knowledge of what the Moon looks like. What we did not know were details about its composition, volcanic activity, magnetic and gravitational fields and/or its resources.

Lunar Sampling/Soil Analysis Experiments

The first spacecraft to dig into the soil of an extraterrestrial body (which happened to be the Moon) was NASA's *Surveyor 3*, which landed on a geographic region of the Moon called the *Oceanus Procellarum* in April 1967. Over the next several months, each of the *Surveyor landers* (5–7) probed the lunar surface with instruments called alpha-scattering surface analyzers and soil mechanics surface samplers. Each was designed to physically probe the lunar soil and gather measurements about the abundances of the various elements that constitute lunar rocks and soil. Data gathered with these two instruments was valuable, but limited in the sense that only specific regions of the Moon were analyzed (the landing sites). Some of the Soviet missions also excavated lunar samples for further analyses on Earth.

All of the *Apollo landers* (11–17), with the exception of the failed *Apollo 13*, performed experiments with lunar surface samples. With tools such as hammers, scoops, rakes and tongs, the Apollo astronauts collected many lunar rock samples and performed a series of tests on the mechanics of the lunar soil. In addition, they returned over 800 ponds of lunar rock to Earth to enable further detailed study.

Lunar Prospector's Approach

Lunar Prospector, a polar-orbiting spacecraft, did not touch down on the lunar surface during either its primary (first year) or extended mission (up to an additional six months after that, until its fuel supply runs out). While previous lunar landing missions have returned an immense amount of physical data, the main limitation of the experiments thus far had applicability to the Moon as a whole. Lunar soil and rocks at the Moon's equator, where the majority of such experiments have been conducted tell a limited story. Global elemental abundances need to be determined. Until Prospector's confirmation of polar water ice, that valuable resource and potential sample was not seriously considered as the basis for a

sample return mission. Before a future sample return mission could be contemplated, it was necessary to have an understanding of the entire global situation in order to profitably plan such a mission. Hence, over time, scientists rely on data from both orbiters, such as LP and landers, to completely understand a planet and develop an approach to its study.

Remote Sensing Experiments

In contrast to direct geologic studies of actual lunar rocks and soil, remote sensing experiments analyze the lunar surface and atmosphere from a distance. Photography is, of course, a kind of remote sensing study, one with which we are familiar. Spectroscopy, however, provides information not discernible to the human eye. Using gamma-ray spectroscopy, *Apollo 15* and *16* were the first missions to attempt to measure the concentrations of elements that make up the lunar crust using spectroscopy. (A primitive version of the Apollo GRS was flown aboard the Soviet's *Luna 10* spacecraft in 1966, but this instrument gathered only limited data). The general findings from the *Apollo 15/16* GRS experiment indicated that there are two distinct geographic regions of the Moon — the highlands and the mare — with chemical compositions that are, in fact, markedly different (for example, iron and magnesium being relatively enriched in the mare). Those data also indicated that the concentrations of certain other elements, such as thorium and titanium, are non-uniformly distributed and that the far side of the Moon differs from the near side.

The *Apollo 15/16* GRS acquired data on the abundance of four elements — thorium, potassium, iron, and titanium — near the equatorial region of the Moon. In concert, the landing modules for those missions collected samples from the same region, permitting comparison of chemical composition data acquired via remote sensing techniques (GRS) and direct sampling (mass spectrometry of soil and rock samples). *Lunar Prospector* mission scientists, of course, would not be able to conduct similar studies comparing in situ samples with remotely acquired data, however, global coverage offers insights not achievable with single point investigations. In addition, because we do have an existing store of lunar rock, LP data is being compared to current samples and may reveal important information.

Lunar Prospector's Role

Since *Lunar Prospector* was an orbiting spacecraft, its entire science payload consists of remote sensing instruments, three spectrometers and the two magnetic field instruments. As with any form of spectroscopy, the quality of the data improves with the time of sampling (number of counts detected). Statistically, it is very important for mission scientists to continually gather spectral information over the same lunar surface regions in order to subtract-out background "noise" caused by not only the cosmic environment but also by the elements which make up the materials on the spacecraft and the instruments themselves. However, with LP, interference from the spacecraft and its electronics were kept to a minimum by the fact that the instruments were separated from the body of the vehicle by eight-foot booms.

Using spectroscopy and other remote sensing techniques, the Prospector mission can perform global-mapping studies not possible with landing craft, such as were typical of the Apollo series of missions. For the scientific community, such an approach represented the logical next step in lunar exploration. Based on the scientific findings of the *Lunar Prospector*, future lunar landers may someday return to the Moon to more thoroughly investigate specific sites, say at the 'icy' lunar poles.

Lunar Prospector's Experiments
Background

When planning the *Lunar Prospector* mission, designers chose six experiments which would most closely fulfill the priorities outlined by LExSWG, while at the same time being sufficiently economical to meet the stringent budgeting requirements of the Discovery Program. Five instruments aboard Prospector are conducting six experiments. Three spectrometers (gamma-ray, alpha-particle and neutron) were globally mapping the Moon's surface and tenuous atmosphere (consisting mainly of gas release events) to determine which minerals and gases are present, and in what abundancy. Two instruments, the magnetometer and electron reflectometer, were measuring the Moon's anomalous magnetic properties, and in doing so, began to characterize the nature and size of a possible lunar core. Finally, the vehicle's own telemetry (communication system) served as the "instrument" for *Lunar Prospector's* Doppler Gravity Experiment, which globally mapped the Moon's non-uniform gravitational fields.

Why Use Spectroscopy?

Physicists use spectroscopy for a variety of research pursuits. A rather general term, spectroscopy is nothing more than the process of visualizing a substance by splitting its light and other emissions into their constituent parts. One of the simplest spectra to consider is a rainbow: visible light is dispersed into its different energies (called wavelengths) when it passes through droplets of rainwater in the atmosphere. Each wavelength absorbs (and thus reflects) a different color, which is readily apparent to the naked eye. This is so because each wavelength has a characteristic energy level. Spectrometers are instruments, which record such dispersions of energy, called spectra.

Since it is no trivial matter to collect rocks and soil samples from other planets, break them up into their individual elements and thereby determine their chemical makeup, spectrometers play a major role in planetary exploration. Knowing what planets, asteroids and moons are made of helps scientists piece together the history and evolution of the Solar System. One type of spectroscopy, gamma-ray spectroscopy, is a technique in which scientists can measure (from orbit) the composition of surface material (down to several inches) around any celestial body possessing little or no atmosphere (such as is the case for the Moon or for Mars, for instance).

The way all of *Lunar Prospector's* three spectrometers worked is by detecting (remotely, from orbit) "signature" energies emitted by various elements in the lunar soil and atmosphere. Such energy emitted by the Moon comes from two sources: "natural" radiation and "induced" radiation. Just like on Earth, certain elements are naturally radioactive and give off radiation (energy) on a constant basis. Other elements, while not naturally radioactive, also emit energy, but in response to constant bombardment of the Moon's surface by galactic and solar cosmic radiation. This is induced radiation.

Unlike Earth, which has a thick, protective atmosphere, the Moon is recipient to most, if not all, passing solar and cosmic energy causing many interactions. The key to spectroscopy's value is that each element emits a unique level of energy. By using a spectrometer, planetary scientists can discern the chemical makeup of the surface of a planet by looking for signature "peaks" (in energy emission) of an energy output plot—an identifying barcode of sorts for the actual elements, which make up the crust and atmosphere. Because solids and gases emit vastly different levels of energy, spectrometers are usually calibrated to detect ranges of energies. Each of *Prospector's* three spectrometers "sees" a characteristic energy range, permitting the detection of energy particles such as fairly-low-energy neutrons (the neutron spectrometer's range of detection is from less than 0.3 eV [electron volts] to hundreds of keV [thousand electron volts]), high-energy gamma rays (the gamma-ray spectrometer's range of detection is ~0.3 MeV [million electron volts] to 9 MeV), higher energy neutrons (a

portion of the gamma-ray spectrometer is used to detect neutrons with energies from 0.5 MeV to 9.5 MeV) and alpha particles (the alpha particle spectrometer's range of detection is ~4.1 MeV to 6.6 MeV).

Gamma-Ray Spectroscopy

Lunar Prospector's gamma-ray spectrometer (GRS) mapped the abundances of ten elements on the Moon's surface:

thorium (Th)	silicon (Si)
potassium (K)	aluminum (Al)
uranium (U)	calcium (Ca)
iron (Fe)	magnesium (Mg)
oxygen (O)	titanium (Ti)

The GRS is especially sensitive to the heavy, radioactive element thorium and the lighter element potassium. These elements are particularly plentiful in the part of the crust that is last to solidify. Thus, mission scientists are able to determine the global distribution of KREEP (K-potassium, Rare Earth Elements, and P-phosphorous), a chemical "tracer" of sorts which helps to tell the story of the Moon's volcanic and impact history. The data produced by the GRS are helping scientists to understand the origins of the lunar landscape, and may also tell future explorers where to find useful metals like aluminum and titanium.

How Lunar Prospector's GRS Works

A gamma ray is a very energetic photon (a tiny parcel of light)—more energetic than a visible light ray or an X-ray. When gamma rays reached the orbiting *Lunar Prospector* spacecraft, they passed through a crystal of bismuth germanate (BGO crystal) in the GRS.

The various atoms inside this detector give off a flash of light when the radiation hits them. Gamma-ray photons with high energies produce brighter flashes than gamma-ray photons with low energies. The light produced by the gamma-ray is then measured by a photomultiplier tube (PMT) which converts the light signal into an amplified electronic signal. Finally, this electronic signal is sent back to Earth for scientists to analyze. The energy of a given gamma ray tells scientists exactly which kind of atom emitted it.

To fully appreciate the potential of LP's GRS, a useful comparison can be drawn with the earlier Apollo GRS experiments. In contrast to *Lunar Prospector*, which mapped the elemental composition of the entire lunar surface, the mapping performed by *Apollo 15* and *16* only covered about 20 percent of the lunar surface—specifically, the region around the Moon's mid-portion or equator. Another difference between the Apollo-era and the *Lunar Prospector* gamma-ray spectroscopy studies is the detecting crystal inside the GRS instrument itself. The *Apollo 15/16* instrument used a sodium iodide (NaI) crystal, whereas *Lunar Prospector's* crystal was composed of bismuth, germanium, and oxygen atoms (BGO crystal).

Since the combined atomic weight of bismuth, germanium and oxygen exceeds that of sodium and iodine, a BGO crystal is significantly denser than an NaI crystal. As a result, a BGO crystal is better able to stop gamma rays in their tracks and, as such, offers greater detection sensitivity—on the order of two- to eight-fold higher—than an NaI crystal. What that means is that energy spectra measured with a BGO crystal can be more cleanly separated (lines can be more easily distinguished from one another on an energy plot) than spectra measured with a NaI crystal as illustrated in Fig. 5. In addition, certain elements, which were unmappable by the Apollo GRS, such as uranium, aluminum, calcium and magnesium, were detected with Lunar Prospector's more sensitive instrument. This offered mission scientists more opportunities to distinguish subtle features of the lunar landscape. The ability to measure concentrations of aluminum and calcium, for instance, may unearth clues as to the makeup of certain types of highland rock formations. Similarly, the distribution of titanium serves as a useful "probe" for mare regions. Other geochemical clues to planetary evolution include the presence of iron stores, the ratio of iron oxide (FeO) to magnesium oxide (MgO), the ratio of potassium to uranium (which hints at remelting rates of primordial condensates), and the ratio of thorium (Th) to uranium, which serves as a marker for the relative abundance of volatile compounds.

While gamma-ray data is highly informative, relatively few gamma-rays leave the Moon's surface and escape into space. Much as a camera working in conditions of low light can compensate by increasing exposure time, gamma ray spectroscopy generally benefits from a significantly long detection period, allowing spectral lines to fill in over time. One sweep—or even a few sweeps—over the Moon's surface would not give

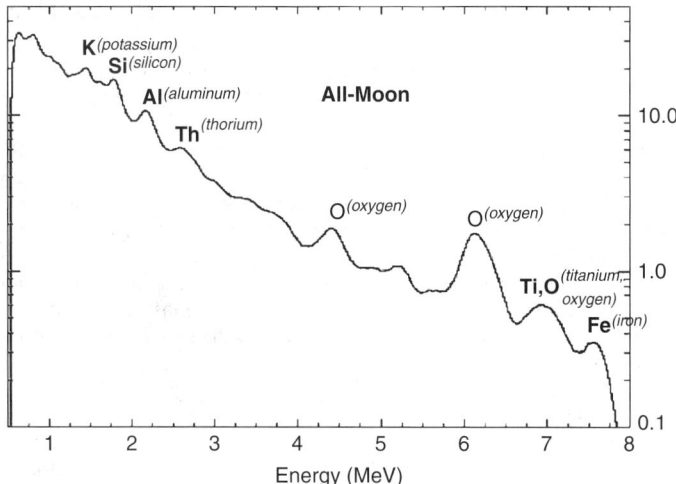

Fig. 5. Sample data plot from Lunar Prospector's gamma-ray spectrometer. (*NASA, GSFC.*)

mission scientists enough information to determine the concentration of radioactive elements. In addition, the stable (non-radioactive) elements do not emit gamma rays as readily as do naturally radioactive ones.

The GRS also contributed indirectly to the search for water on the Moon. The bismuth germanate crystal is surrounded by a shield of borated plastic (anti-coincidence shield) that detects high-energy (fast) neutrons. The main purpose of this shield is to allow correction for background signal caused by solar and galactic cosmic rays. It differentiates between gamma rays and the cosmic ray background. In addition, because it is borated, it also measures fast (high energy) and epithermal (medium-energy) neutron fluxes. Mission scientists are using this information, in concert with the lower-energy (thermal and epithermal) neutron counts detected by Lunar Prospector's neutron spectrometer, to detect water ice at the lunar poles.

Alpha-Particle Spectroscopy

The Apollo series of missions revealed that the Moon had not been perpetually cold and dead, as once believed, but rather was host to a series of dramatic volcanic eruptions in which vast seas of molten lava flooded much of the lunar surface. While the majority of such activity most likely occurred very early in the Moon's history over three billion years ago, the Moon is thought to still harbor some remnant volcanic and tectonic activity. Outgassing events, in which alpha-particle emissions of radon leak out from the lunar interior, are scientific evidence of such activity. Polonium, a natural-decay product of the heavier element radon, itself a natural-decay product of the still heavier element uranium, collects around vents and provides keys to their recent history. This is due to the fact that the entire decay chain takes over 21 years, so that Polonium is evidence of vent activity over the last half century. Radon gas and polonium were likely to be detected because of their relatively long half-lives (3.8 days for [222] radon and 138 days for [210] polonium; one half-life is the amount of time it takes for half of a given radioactive sample to decay into another substance).

Ancient volcanic vents, seismic fractures, impacts and pore openings in the lunar soil all provide paths for radon to find its way to the lunar surface. Actually, radon itself is present in very small quantities, but thought to be mixed in with other gases, such as nitrogen, carbon monoxide and carbon dioxide. Determining where and when such gas-release events take place can tell scientists just how active the Moon actually is, as well as help to identify the source(s) of the Moon's small and tenuous atmosphere.

An alpha particle is the nucleus of a helium atom: two protons and two neutrons bound together. Like gamma rays, alpha particles escape from radioactive elements as part of their natural-decay process. The alpha particles are emitted with a precise energy that serves as a fingerprint for the atom from which they came. *Lunar Prospector's* Alpha-Particle Spectrometer (APS) detected these events.

Housed inside the APS instrument are ten separate wafers of silicon. Silicon is a semiconductive material. When an alpha particle hits a silicon wafer, it creates a small track of charge. When a 25-volt bias is applied to the silicon wafer, the alpha particle's charge is funneled into an amplifier which then increases the charge. Since that pulse of charge is directly

proportional to the signature energy of the alpha particle, scientists can infer the identity of the element, which emitted the alpha particle. The APS contains ten such silicon detectors, each sandwiched between gold and aluminum disks, and arranged on five out of six sides of a cube, enabling nearly a complete field of detection.

Before Lunar Prospector

Apollo 15 and *16* scientists, using an APS instrument for the first time, found evidence for a spatially variable distribution of radon and polonium. Those studies identified a striking correlation between polonium and the edges of lunar maria, especially Mare Fecunditatus, but also at nearly all maria investigated.

Earlier astronomical studies noted that the Aristarchus crater region was the site of phenomena dubbed "transient optical events," in which regions of the lunar surface glowed and changed color for short periods of time. Some scientists believe that the light flashes may represent transient venting of volatile materials.

Factors in Analyzing APS Data

Data acquired by the APS aboard Prospector appeared in the form of counts — very similar to the way the GRS instrument works. As with the GRS data, the number of counts accumulated (and thereby the time of sampling), were the key determinants of the sensitivity of the data. The *Apollo 15* and *16* missions gathered only about 8 days' worth of data around the Moon's equator. The Lunar Prospector gathered 18 months of mapping time.

One issue mission planners had to take into account when designing alpha-particle spectroscopic experiments (and in fact any type of experiment in which a spectrometer measures cosmic radiation) was the timing of the solar cycle. Repeating every 11 years, the solar cycle is a periodic phenomenon in which the overall extent of radiation (in the form of solar particles called *protons* and *alpha particles*) produced by the Sun varies in a predictable manner. A new solar cycle began in 1997, at which time sunspot activity was at a minimum and galactic cosmic rays (GCR) were at their 11-year maximum. Since more GCR protons imply more induced planetary gamma-ray emission, it was an optimal time for performing global spectrographic mapping experiments, such as gamma-ray and neutron spectroscopy, because the inherent signal to be detected would be higher than usual. However, getting closer to solar maximum, solar activity and its associated solar energetic-particle population increases, leading to higher background radiation "noise", so mission scientists had to take into account such stray counts and subtract them from the overall data.

Neutron Spectroscopy

Lunar Prospector mission scientists devised the neutron spectroscopy experiment to search for water ice at the poles of the Moon. As the world found out on March 5, 1998, at the mission's first science data return press conference, preliminary results from the experiment were indeed positive: water ice does exist on the Moon, and there appears to be more of it at the North than at the South pole. *Lunar Prospector* had detected a significant amount of hydrogen, which is inferred to be in the form of water. This was the first direct evidence of the presence of water ice at the Moon's frigid poles. *Lunar Prospector* is also the first interplanetary mission ever to use the neutron spectroscopy technique to detect water. *Prospector's* neutron spectrometer (NS) works by detecting hydrogen, by way of subatomic particles called *neutrons*.

Neutron Science

The materials we use every day are made up of molecules, which are made up of atoms. Inside the atoms are even tinier pieces of matter called subatomic particles. Neutrons are one type of subatomic particle — they are present in every atom of every molecule in our bodies as well as in all of the synthetic and natural substances in our environment. Besides being a basic building block of matter, neutrons serve as a useful experimental tool for physicists. Materials scientists, for instance, are interested in how atoms are packed within the molecules of different materials. How the individual atoms of a given material stack up against each other in large part determines the properties (strength, plasticity, etc.) of that particular material. One way scientists study molecular structure is to bombard atoms with high-energy neutrons and then wait and see where and how fast the neutrons scatter.

The same thing happens naturally in space. When cosmic rays collide with atoms in the lunar crust, they violently dislodge neutrons and other subatomic particles. Some of the neutrons escape directly into space—essentially unchanged—as "fast" neutrons. Other neutrons shoot off into the crust, where they slam into other atoms, bouncing around like pinballs. If they only run into heavy atoms, they do not lose very much energy in the collisions, and are still traveling at close to their original speed when they finally bounce off into outer space. But if a neutron encounters a hydrogen atom—which is the same size as a neutron—it will slow greatly or even stop, much like a speeding billiard ball running into a stationary one. If the Moon's crust contains an abundance of hydrogen at a certain location—say, a crater with water ice in it—any neutron that bounces around in the crust before heading out to space will cool off (slow down) rapidly. When *Lunar Prospector* flew over such a crater, the NS detected a definitive dropoff in the number of these ("epithermal" or medium energy) neutrons.

How Lunar Prospector's NS Works

The NS has two different counters—a cadmium-wrapped canister of [3]helium and a tin-wrapped canister of [3]helium. When a neutron collides with an atom of [3]helium, a nuclear reaction takes place, producing a burst of energy. That burst of energy tells mission scientists that they have detected a neutron. Except for the outside wrapping, the two counters are nearly identical. The cadmium-wrapped counter filters out all but the epithermal neutrons, because cadmium is good at screening away the slow-moving thermal neutrons, whereas the tin-wrapped counter lets all of the neutrons through. Since the two counters are otherwise identical, counts can be subtracted, and any difference between the two must be attributable to thermal neutrons.

Lunar Prospector measured "fast" and thermal plus epithermal neutron flux with a separate instrument (the anti-coincidence shield of the GRS). The respective count rates of the different types of neutron fluxes are an indicator of hydrogen, and hence the presence of water ice, embedded within the lunar soil.

Mission scientists received data from the spacecraft every 32 seconds. Since the data contained random noise, several passes over the surface and careful statistical analysis were required to analyze the data. However, since the spacecraft passed over the poles every orbit (whereas it passed over any given region on the equator only a few times a month), the NS produced the most accurate data in the polar regions.

Magnetometer/Electron Reflectometer Studies

The magnetometer and electron reflectometer aboard the *Lunar Prospector* collected valuable data to help unravel puzzles that have intrigued scientists for more than a quarter of a century. What kind of magnetic field(s) exists on the Moon? What kind of natural resources are buried in the Moon's crust and is there a core? If so, what are its characteristics? Can we build a lunar base? How did the Moon form and evolve—what is its history?

The MAG/ER experiment aboard *Lunar Prospector* was designed with two primary goals in mind: scanning the lunar crust for signs of permanent magnetization, and searching for electrical currents flowing deep within the lunar interior—the sign of a conductive metallic lunar core. The two instruments combined to calibrate the Moon's global magnetic field strength: the magnetometer measures the field surrounding the spacecraft, and the electron reflectometer surveys the lunar surface.

How Much Do We Already Know about the Moon's Magnetic Field?

Scientists have known for years that the Earth is magnetic—there is a strong magnetic field surrounding our own planet, originating with electrical currents swirling inside the Earth's iron-rich metallic core. The boundary between the Earth's magnetosphere (a dipole field) and the influence of the Sun's charged particle activity (from the solar wind) for the most part balances out. At times, however, especially during a particularly active segment of the 11-year solar cycle when solar flares are raging, charged solar wind particles get trapped by the Earth's magnetic field and slam into the Earth's atmosphere at extremely high energies. See Fig. 6. As a result, the atmosphere glows, and a strange but beautiful phenomenon called the Aurora Borealis is formed. Such events are most prominent at the Earth's poles; hence, Alaska is an oft-cited viewing spot for such celestial fireworks.

The Moon is a different story. Based upon previous, albeit limited research, scientists have suspected that the Moon has very little, if any, magnetic field of its own. The first spacecraft to effectively measure lunar magnetic fields was *Explorer 35*, several years prior to *Apollo*. The *Apollo 15* and *16* subsatellites orbited the Moon with magnetometer instruments and concluded that the Moon possessed a vanishingly small global magnetic field; however, other experiments aboard *Apollo 12, 14, 15* and *16* with either hand-held or stationary magnetometers detected small but significant surface fields. In particular, one unresolved issue facing the lunar scientific community is the origin of swirl-like color markings (called "albedos") visible from orbit on the surface of the Moon that are thought to be due to magnetic anomalies.

Why Are We Interested in the Magnetic Field?

The presence of such non-uniformly distributed magnetic regions has led planetary scientists to conjecture about the Moon's impacts (which may have imparted some of these locally distributed magnetic properties) and the possible presence of a small, iron-laden core (roughly estimated to a maximum of about 500 km in radius). From the perspective of planetary science, better understanding the size and nature of the Moon's core can tell a lot about its evolution, history and beginnings—such as, did the Moon really arise as the result of a cataclysmic collision between the Earth and a Mars-sized body? In addition, metals discovered in the crust may be an extremely valuable resource. Now that mission scientists have found a potential source of water on the Moon, the presence (and relative abundance) of other materials may become paramount to the eventual establishment of a future possible lunar base.

How it Works

Lunar Prospector's magnetometer/electron reflectometer (MAG/ER) measurements were affected by three sources: the Earth's magnetic field (35,000 nano Teslas), the relatively weak field of the Moon (0 to 300 nano Teslas) and the very weak field carried from the Sun by the solar wind (approximately 10 nano Teslas). Mission scientists were able to compare these measurements to determine which magnetic field variations are caused by surface features (local deposits of magnetic material) or alternatively, by the Moon's core. Both instruments were copies of detectors housed aboard the *Mars Global Surveyor* spacecraft, launched in December 1996, with some modifications to adapt them for a spinning spacecraft.

The triaxial fluxgate MAG is a standard device that is also used to measure magnetic fields on Earth. "Triaxial" means that it includes sensors

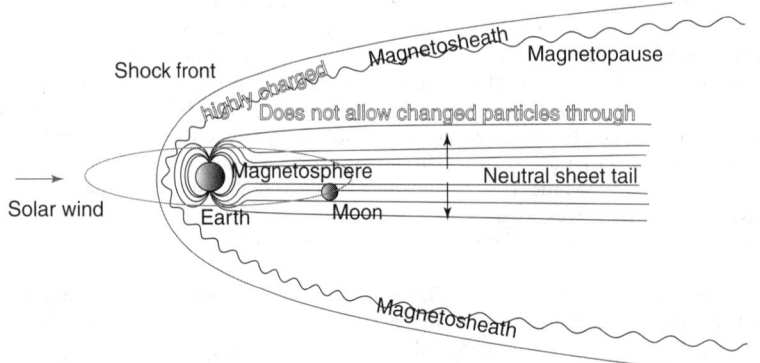

Fig. 6.   Illustration showing the Earth's magnetosheath and the shock front it creates. (*NASA, GSFC.*)

to measure the strength of the field in three different directions. This enables scientists to determine not only the maximum strength of the field but also the direction in which it points. The "fluxgate" is an electric coil through which the magnetic field passes. By measuring the variation of the current passing through the coil, the MAG determines the strength of the magnetic field. It can measure magnetic fields as weak as one-millionth the strength of the Earth's magnetic field.

Prospector's ER measured the magnetic field at the surface of the Moon. The Moon, like every other body in the Solar System, is constantly barraged by electrons from the solar wind. Unlike the Earth, the Moon was not thought to have a magnetic field strong enough to repel these tiny charged particles, and they would be expected to spiral toward the surface in giant loops, typically several miles wide. The electrons that descend in a tighter spiral make it to the surface of the Moon and are absorbed there. But if there is magnetic material on the Moon, it will reflect some of the electrons back into space. When those reflected electrons reached the *Lunar Prospector* spacecraft, the ER measured their pitch angle (the angle at which they bounce). Scientists saw an abrupt cutoff above a certain angle — because all the electrons with a larger pitch angle were absorbed by the Moon. That cutoff tells mission scientists just how intense the magnetic field is at the lunar surface.

Together, the two instruments detected local variations in the Moon's magnetic field that arise from selenological features on the lunar surface. *Lunar Prospector's* MAG instrument not only measured lunar magnetic fields, but also external fields present in space plasmas (a plasma is an electrically conductive gas comprised of both neutral and ionized particles and free electrons). As a result, only a small fraction (about 10 percent) of the MAG's data was pertinent to lunar scientists. What that means is that culling relevant data posed a significant challenge for *Prospector's* MAG/ER team. Essentially, this involved careful pruning of the data as well as paying attention to what times of the month the data was gathered, as certain times (and thereby location relative to the Earth and its own magnetic field) were better than others for measuring lunar fields.

The interaction between charged particles in the solar wind and the Earth's magnetic field (about 220 miles above the surface of the Earth) is a region scientists call the magnetosphere. Within the magnetosphere are two separate regions: the magnetosheath (which is magnetically "noisy") and the geomagnetic tail (which, by comparison, is "quiet"). It was important that mission scientists take lunar magnetic measurements while the spacecraft passed through the quieter geomagnetic tail. The data was expressed in units called "gammas" (which are the same as nano Teslas, or one-billionth of a Tesla) and were tagged with spacecraft positional data to produce computer files which demonstrate magnetic field components as a function of location relative to the Moon as well as spacecraft altitude.

As is the case with all of *Prospector's* scientific experiments, the spacecraft's polar orbiting pattern permitted complete global analysis of the Moon. So, for the first time, scientists had a handle on all of the Moon's magnetic features, not just those associated with isolated geologic regions. Such analyses also permitted scientists to correlate magnetic properties with lunar surface features — the MAG data was essential component in analyzing the data from the ER instrument. Data from the ER was plotted as electron flux (number of particles per unit area per unit time) as a function of time. The ER "probes" (from orbit, of course) the lunar surface by recording the behavior of space-born electrons after they bounce off of the lunar crust. But while the ER was capable of measuring the strength of electron paths ("bent" by tiny, locally dispersed lunar magnetic fields), this instrument is incapable of determining where on the Moon it was measuring electrons at any given time. Positional data from the MAG put the ER data into perspective, allowing mission scientists to match up electron flux strengths with the magnetic fields along which they travel. Thus, the two instruments (which were housed together on one of Prospector's eight-foot booms) provided mission scientists with a complete picture of the Moon's magnetic and electrical properties.

Gravity Studies

The gravity field of the Moon strongly influences the altitude of a spacecraft in low-circular orbit. The most dramatic example is the *Apollo 16* subsatellite. After being deployed in a near-circular orbit from the command and service module, the eccentricity increased quickly and the spacecraft impacted the lunar surface 35 days after the release strictly due to the force of the gravity field. Understanding the precise nature of a planet's gravity field is vital to all exploration and experimentation.

As presented at the March 5, 1998 science return press conference, *Lunar Prospector's* Doppler Gravity Experiment (DGE) has provided the first polar low-altitude measurement of the lunar gravity field. This provided the spacecraft with the first truly operational gravity map of the Moon and immediately improved orbit and fuel efficiency. Improved gravity information, will not only help scientists build better models of the role of impact processes on the history and evolution of the Moon, but will also help in estimating the lunar core size and metallic iron content. A more practical benefit of the new lunar gravity data provided by *Prospector's* DGE experiment is that a more precise gravity map of the Moon will inevitably aid future mission planners in planning fuel-efficient journeys to the Moon, and may even help identify potential resources.

Lunar Prospector's Doppler Gravity Experiment

The Moon has a large asymmetry due to the fact that the lunar crust is thicker on the far side than on the near side and a much "bumpier" gravitational field than the Earth, with small anomalies due to mass concentrations on the surface. The Apollo missions helped demonstrate such sizable positive gravity anomalies. Interestingly, they exist within the topographically low, large circular mare basins. This discovery was unexpected and opposite of any physical model at that time and started the development of new models of the Moon's interior. The features were called *mascons* (short for "mass concentrations"). These bumps cause an orbiting spacecraft to speed up or slow down. The DGE is, in effect, drawing a map of the bumps.

The DGE, unlike the other experiments aboard *Lunar Prospector*, required no extra instrumentation. All of the data was, collected simply by communicating with the spacecraft. As the spacecraft orbited the Moon, its speed was determined by the Doppler effect, the same effect that causes a police siren to sound higher when the police car is moving toward you and lower when it is moving away from you. The "siren," in this case, is the spacecraft's radio signal, whose frequency shifts slightly as it moves toward Earth or away from it. Relative to the near side, lunar farside gravity is poorly determined because the spacecraft is not in view from the Earth when over the lunar far side. However, some information is obtained by observing changes in the LP orbit due to the accumulated acceleration of the farside gravity as the spacecraft comes out of occultation (back into view).

By tracking the velocity of the spacecraft, mission scientists were able to infer the forces acting upon it. For over 99 percent of the duration of the mission (excepting only periods when the engines are being fired) the only force on *Lunar Prospector* was gravity. Thus, by simply circling the Moon and sending signals back to Earth, *Lunar Prospector* has mapped the Moon's global gravitational field. *Lunar Prospector* completed this gravitational map in the first two months of the mission. However, the results of the DGE, were greatly improved with data from the extended, low-altitude phase of the mission. At this low altitude of 6 miles (10 km), the precision of the gravity data was improved by a factor of over 100.

Lunar Prospector End of Mission Profile

The controlled crash of NASA's *Lunar Prospector* spacecraft into a crater near the south pole of the Moon on July 31, 1999 produced no observable signature of water, according to scientists digging through data from Earth-based observatories and spacecraft such as the *Hubble Space Telescope*.

This lack of physical evidence leaves open the question of whether ancient cometary impacts delivered ice that remains buried in permanently shadowed regions of the Moon, as suggested by the large amounts of hydrogen measured indirectly from lunar orbit by *Lunar Prospector* during its main mapping mission.

In a low-budget attempt to wring one last bit of scientific productivity from the low-cost *Lunar Prospector* mission, NASA worked with engineers and astronomers at the University of Texas to precisely crash the barrel-shaped spacecraft into a specific shadowed crater. NASA accepted the team's proposal based on successful scientific peer review of the idea and the pending end of the spacecraft's useful life, although the chances of positive detection of water were judged to be less than 10 percent.

Worldwide observations of the crash were focused primarily on using sensitive spectrometers tuned to look for the ultraviolet emission lines expected from the hydroxyl (OH) molecules that should be a by-product of any icy rock and dust kicked up by the impact of the 354-pound spacecraft.

"There are several possible explanations why we did not detect any water signature, and none of them can really be discounted at this time," said Dr. Ed Barker, assistant director of the university's McDonald Observatory at UT Austin, who coordinated the observing campaign. These explanations include:

- the spacecraft might have missed the target area;
- the spacecraft might have hit a rock or dry soil at the target site;
- water molecules may have been firmly bound in rocks as hydrated mineral as opposed to existing as free ice crystals, and the crash lacked enough energy to separate water from hydrated minerals;
- no water exists in the crater and the hydrogen detected by the Lunar Prospector spacecraft earlier is simply pure hydrogen;
- studies of the impact's physical outcome were inadequate;
- the parameters used to model the plume that resulted from the impact were inappropriate;
- the telescopes used to observe the crash, which have a very small field of view, may not have been pointed correctly;
- water and other materials may not have risen above the crater wall or otherwise were directed away from the telescopes' view.

Although the crash did not confirm the existence of water ice on the Moon, "this high-risk, potentially high-payoff experiment did produce several benefits," said Dr. David Goldstein, the aerospace engineer who led the UT Austin team. "We now have experience building a remarkably complex, coordinated observing program with astronomers across the world, we established useful upper limits on the properties of the Moon's natural atmosphere, and we tested a possible means of true 'lunar prospecting' using direct impacts." See also **Moon (Earth's)**.

Web References

Lunar Prospector: *http://lunar.arc.nasa.gov/*
Lunar Prospector Archives: *http://lunarprospector.arc.nasa.gov/*
Lunar Prospector Mission: *http://nssdc.gsfc.nasa.gov/planetary/lunarprosp.html*

LUNG CANCER.   See **Cancer and Oncology**.

LUNGFISHES (*Osteichthyes*).   Of the order *Dipneusti*, the lungfish has an air bladder opening from the pharynx which can be filled with air gulped through the mouth and serving as a lung. It is the only known species of fish that can live out of water for a period as much as four years. The dipnoids live in transient streams and swamps of Australia, South America, and South Africa. Some varieties pass the dry season in cells, which they form in the muddy bottom as the water dries up. They resemble amphibians in some details of structure and are reminiscent of creatures that existed during Devonian times. See also **Fossil Fishes**. Probably the most primitive species is the lungfish of Australia (*Neoceratodus forsteri*). Originally, the fish was found only in the Burnett and Mary rivers (northeastern Australia). However, it has been transported quite successfully to several lakes in Queensland where it thrives.

The average lungfish measures up to $3\frac{1}{2}$ feet (1 meter) in length when grown, although specimens up to 6 feet in length and weight of 100 pounds (45.4 kilograms) have been recorded. Possibly the largest recorded specimen was an African lungfish (*Protopterus aethiopicus*), measuring some 7 feet (2.1 meters) and found in Lake Victoria. See Fig. 1.

Fig. 1.   African lungfish.

It is interesting to note that similarities between the African and South American lungfishes have contributed to the hypothesis that there may have been a land connection between South America and Africa at an earlier period. See also **Earth Tectonics and Earthquakes**.

Lungfishes form mud-ball cocoons within which they encase themselves. This is a mucous cocoon, which becomes quite leathery after hardening. During estivation in their cocoons, lungfishes lose both weight and length, factors that are quickly recovered in about a month after coming out of the cocoon.

LUPUS.   See **Systemic Lupus Erythematosus**.

LUPUS (the wolf).   A southern constellation located near Libra.

LUSTER.   This term is used by mineralogists to describe the appearance of the surface of a mineral, usually a crystal face, in reflected light. The principal types of luster are: metallic, adamantine, vitreous, resinous, greasy, pearly. The degrees of luster may be defined as: splendid, shining, glistening, or dull. Schillerization is a peculiar form of submetallic luster observed in different directions in certain minerals such as schillerspar, diallage, hypersthene, etc.

LUTEIN.   Lutein is an antioxidant found in several areas of the body including the skin and eyes. In recent years, researchers have discovered that age-related eye diseases such as cataracts and macular degeneration might benefit from the antioxidant effects found in vitamins and minerals from fruits or vegetables.

Lutein belongs to a chemical class of compounds called carotenoids and is found in the central area of the retina called the macula. The macula is responsible for acute central vision, and damage to this portion of the retina severely limits a patient's ability to read, recognize faces, and perform any other task that requires straight-ahead vision. Some research suggests that lutein may protect the macula from potentially damaging forms of light.

Antioxidants act as scavengers, preventing the formation of and neutralizing the damage from free radicals-substances thought to be responsible for acceleration of the aging process and causing damage such as lowered immune system responses, heart attacks, arthritis, eye diseases, cancers, and other disorders. Free radicals are produced by the body in response to environmental pollutants such as cigarette smoke, pesticides, smog, radiation, and many drugs. Although free radicals do have certain beneficial effects, there is extensive evidence to indicate that they can damage both the structure and function of cell membranes, causing degenerative diseases and conditions.

Lutein deposited in the retina and lens of the eye is thought to protect the macula from damaging oxidation by filtering blue light. The substance apparently absorbs and dissipates dangerous ultraviolet light rays that are associated with age-related macular degeneration, the leading cause of irreversible blindness among those 65 years of age and beyond.

The body does not make its own lutein and must depend on dietary sources to get an adequate supply. Nutritionists suggest 6 milligrams per day, which would be two salad bowls of spinach, for instance. Generally, a diet rich in yellow and orange fruits and leafy green vegetables will provide an adequate supply, although most people do not get enough to reach the 6 milligram level. The foods richest in lutein are spinach, collards, kale, mustard greens, and turnip greens. Other fruits and vegetables that contain lutein are: Romaine lettuce, leeks, pears, broccoli, brussels sprouts, butternut squash, cabbage, carrots, celery, corn, cornmeal, cucumbers, green beans, honeydew, kiwi, mango, okra, oranges and orange juice, peppers, (green and orange bell), persimmons, pumpkins, red grapes, yellow squash and zucchini squash.

Cooked vegetables appear to be a better source of lutein because cooking unlocks the cell walls and releases the substance making it more available. Lutein also appears to be more easily absorbed when vegetables are served with a source of fat, such as cooking oil or butter.

Lutein has been the subject of much interest over the past few years, with the general conclusion being that a diet rich in lutein may reduce the risk of developing macular degeneration. In any event, lutein-rich foods should be a part of any balanced diet. See also **Carotenoids**; and **Macula**.

Vision Rx, Inc., Elmsford, NY.

LUTETIUM.   Chemical element symbol Lu, at. no. 71, at. wt. 174.97, fourteenth and last element in the Lanthanide Series in the periodic table, mp. 1,663 °C, bp 3.402 °C, density 9.841 g/cm^3 (20 °C). Elemental lutetium has a close-packed hexagonal crystal structure at 25 °C. The pure metallic lutetium is silver-gray in color and retains its luster at room temperature indefinitely. Although not as extensively studied as most of the other lanthanides, most of the basic properties of lutetium are known, and it behaves as a normal trivalent metal with no magnetic transitions because its $4f$ levels are completely filled. There are two natural isotopes ^{175}Lu and ^{176}Lu. The latter isotope is radioactive with a half-life of 2.2×10^{10} years. Fourteen artificial isotopes are known. The element was first identified by G. Urban in 1907 and independently by C.A. von

Welsbach in 1908. Although not investigated fully, lutetium is classified with a low acute-toxicity rating. Lutetium is the least abundant of the Lanthanide elements, estimated as present on the average of 0.5 ppm in the earth's crust. Potentially, however, it is more plentiful than mercury, cadmium, or any of the precious metals. Electronic configuration

$$1s^2 2s^2 2p^6 3s^2 2p^6 3d^{10} 4s^2 4p^6 4d^{10} 4f^{14} 5s^2 5p^6 5d^1 6s^2$$

Ionic radius Lu^{3+} 0.848 Å. Metallic radius 1.735 Å. Other important physical properties of lutetium are given under **Rare-Earth Elements and Metals**.

The source of lutetium to date has been the processing of the other heavy rare-earth metals. Because of very limited availability, little research was conducted on lutetium until the mid-1960s. Most of these studies now are concentrating on prospective uses in phosphors, semiconductor, and other electronic circuitry components. A lutetium dithalocyanine complex has received much consideration recently for application in large, thin screens for television projection.

See references listed at ends of entries on **Chemical Elements**; and **Rare-Earth Elements and Metals**.

K.A. GSCHNEIDNER, Jr. and B. EVANS
Iowa State University, Ames, IA.

LUTH *(Reptilia, Chelonia).* Also called the leathery turtle. A large marine turtle, *Dermochelys coriacea*, which reaches a length of 6 feet (1.8 meters). Its carapace is formed of bony plates connected together but not joined to the spinal column or ribs. Its flesh is not palatable.

LYASES. See **Enzyme**.

LYCOPSIDA. A group of plants (Club Mosses) that contains about 500 species, most of which are included in two genera, *Lycopodium* and *Selaginella*. The species of *Lycopodium* are trailing plants often called ground pines, or ground hemlock, as well as club mosses. Many of them are common plants of dry open places in the temperate zone. The plants have long creeping stems growing on the surface of the ground or several inches beneath it. From this prostrate stem short dichotomously branched roots extend down into the ground, and erect branches grow upward. The stems are covered with many small, pointed, dark green leaves. In the more primitive species the reproductive structures or sporangia are found in the axils of ordinary leaves. In other species the sporangia are borne in the axils of modified leaves aggregated at the tip of an erect branch, forming a slender cone or strobilus. The many spores borne within the sporangia are all alike and for this reason Lycopodium species are said to be homosporous. The spores are disseminated by the wind, and in time develop into gametophytes. The gametophytes of *Lycopodium* species are extremely small tuberous bodies, which grow slowly and reach maturity only if they are invaded by an endophytic fungus. Generally the gametophyte or prothallus develops underground. In the upper surface of the prothallus both antheridia and archegonia are found. Each antheridium contains many straight, biciliate sperms. These swim to the egg, with which one unites, forming a zygote. From this the new sporophyte develops. At first the sporophyte depends on the gametophyte for its food substances. Thus it obtains nutriment by means of a special absorbing structure called a suspensor which grows into the tissue of the gametophyte.

Lycopodium plants are widely used as material from which to make Christmas wreaths. For this purpose the entire plant is often ripped from the ground. The spores of *Lycopodium* are also gathered and sold under the name of Lycopodium powder. Formerly these spores were used in making explosive mixtures and for flashlights.

The genus *Selaginella*, containing some 400 species, is most abundant in the tropics. A few species of small plants are found in temperate regions. In the tropics there are both terrestrial and epiphytic species. The general habit of the plant is much like that of *Lycopodium*. The sporangia are formed in the axils of leaves at the tips of the branches, forming terminal strobili or cones. At the base of the sporophylls there is also a small scale, called a ligule, of unknown function. The sporangia are of two kinds, one, a megasporangium, containing four large spores, called megaspores; the other a microsporangium containing many small spores or microspores. Species of *Selaginella* are therefore heterosporous, a character which distinguishes them from *Lycopodium* species.

LYELL, SIR CHARLES (1797–1875). Lyell was born in Scotland and graduated from Oxford University. He is remembered for establishing geology as a science. He published *Principles of Geology* in the 1830's and *The Geological Evidences of the Antiquity of Man* in 1863.

J. M. I.

LYMAN-ALPHA RADIATION. The radiation emitted by hydrogen at 1216 angstrom, first observed in the solar spectrum by rocket-borne spectrographs. Lyman-alpha radiation is very important in the heating of the upper atmosphere thus affecting other atmospheric phenomena.

LYME DISEASE. Lyme disease is a multi-stage, multi-system infection caused by *Borrelia burgdorferi*, a member of the family of spirochetes, or corkscrew-shaped bacteria. See Fig. 1. *Borrelia burgdorferi* is transmitted

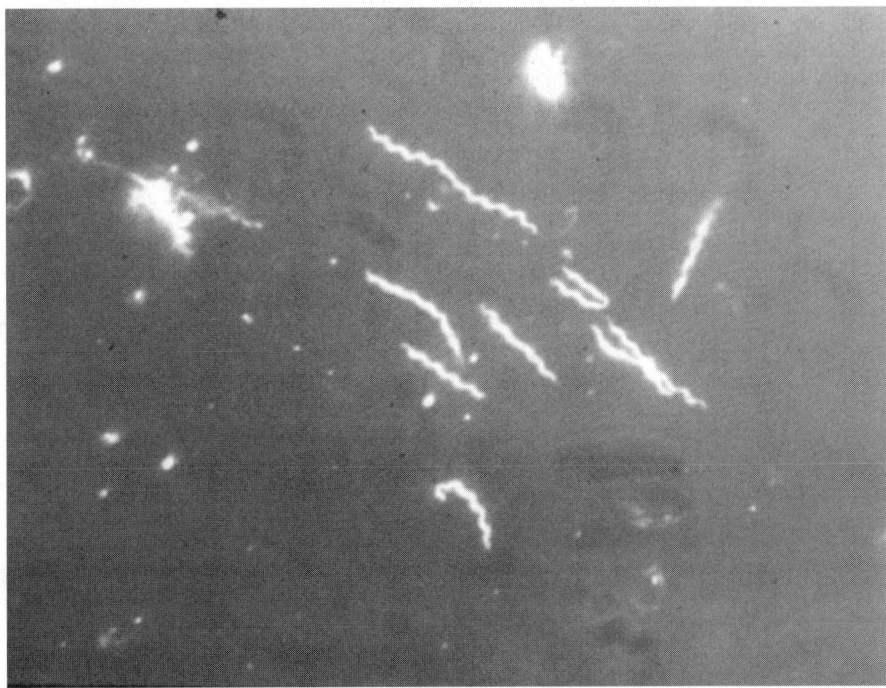

Fig. 1.   The corkscrew shaped bacteria, Borrelia burgdorferi, are the cause of Lyme disease. (Center for Disease Control, National Center for Infectious Diseases, Division of Vector-Borne Infectious Diseases, Atlanta, Georgia.)

to humans by bites from the ticks colloquially referred to as "deer ticks" in the north-eastern and north central regions of the United States. *Ixodes scapularis* (black-legged tick), *Ixodes pacificus* (western black-legged tick), *Ixodes dammini* and *Amblyomma americanus* are the primary vectors for Lyme disease. (A Vector is an organism (as an insect) that transmits a pathogen (or virus) that causes a disease).

Both of the latter ticks are known to infest white-tailed deer (*Odocoileus virginianus*) and the whitefooted mouse (*Peromysus leucopus*). Ixodes ticks are significantly smaller than common dog and cattle ticks. They are no bigger than a pinhead in their larval and nymphal stages, and are only slightly larger as adults. See Figs. 2, 3, and 4. Lyme disease was first identified in 1975 in the town of Lyme, in southeastern Connecticut, USA. There, a clustering of cases with arthritis-like symptoms were observed to have distinctive skin lesions, which led to the recognition of a new tick-borne disease.

Ixodes scapularis on a grass stem.

Fig. 3.   An adult female is shown on a grass stem. (Center for Disease Control, National Center for Infectious Diseases, Division of Vector-Borne Infectious Diseases, Atlanta, Georgia.)

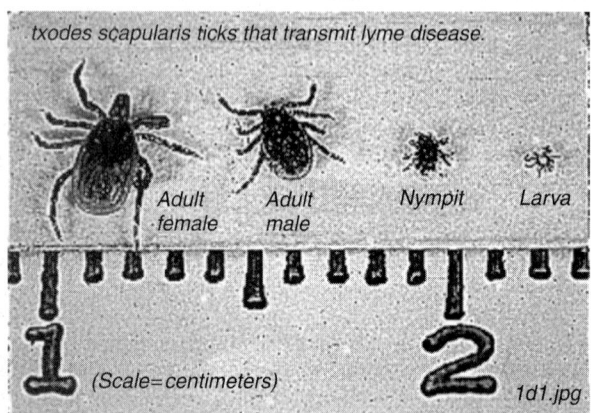

Fig. 2.   The tick vectors of Lyme disease have four life stages: egg, larva, nymph, and adult. Compare the differences in size of the three later stages on a centimeter scale. (Center for Disease Control, National Center for Infectious Diseases, Division of Vector-Borne Infectious Diseases, Atlanta, Georgia.)

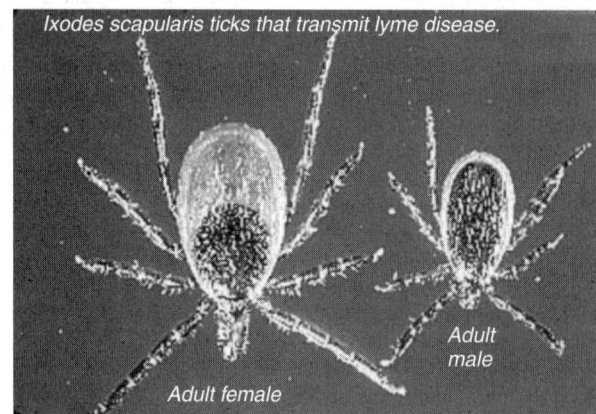

Fig. 4.   Adult female and male ticks are different in size and coloring. Note that the adult female is almost two times larger than the adult male. The female has black on red coloring, whereas the male is all black in appearance. (Center for Disease Control, National Center for Infectious Diseases, Division of Vector-Borne Infectious Diseases, Atlanta, Georgia.)

The earliest stage of the illness is commonly characterized by the appearance of a distinctive skin rash called *erythema migrans*. Approximately 3 to 20 days following a tick bite, a red *macule* or papule appears at the bite site. It then expands to a large, annular erythematous lesion (*erythema chronicum migrans*) about 6 to 16 centimeters or larger in diameter. The lesion sometimes shows central clearing, and secondary concentric rings may develop within the original ring. The lesions are not pruritic and may be multiple. Other symptoms that often occur at the onset of early Lyme disease may include fever, chills, malaise, headache, aching in the muscles (*myalgias*) and neuralgic pain in the joints (*arthralgias*). In addition, early

symptoms can include secondary skin lesions, facial nerve palsy, and *lymphocytic meningitis*. Early treatment with an appropriate antibiotic therapy is almost always successful. If left untreated, or inadequately treated, the early Lyme disease progresses to late Lyme disease, which is characterized by distinctive arthritic, neurologic, and cardiac manifestations. These

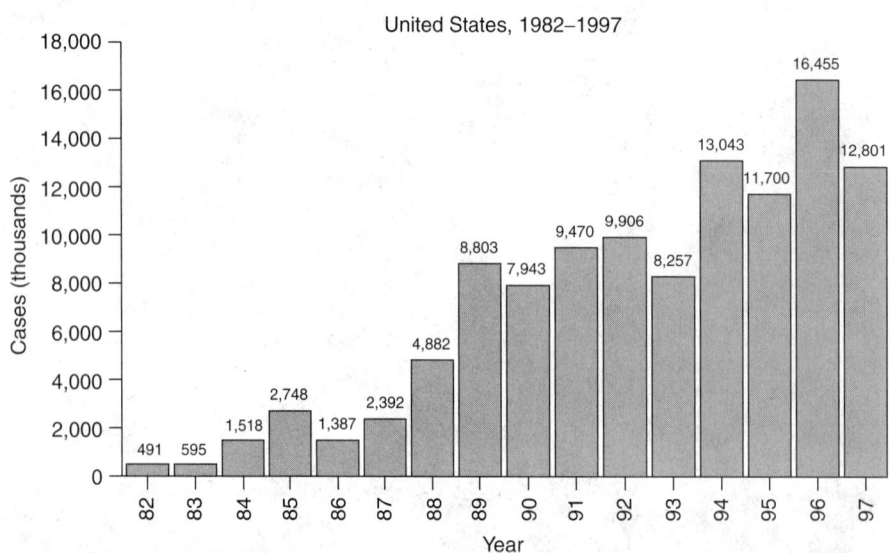

Fig. 5.   The reports of Lyme disease in the United States have increased dramatically since the first cases were recognized in southeastern Connecticut in 1975. (Center for Disease Control, National Center for Infectious Diseases, Division of Vector-Borne Infectious Diseases, Atlanta, Georgia.)

symptoms require a more intensive therapy, and permanent after-effects may occur.

The incidence of Lyme disease has been steadily increasing, making Lyme disease an important public health problem in some areas of the United States. The total of cases reported each year increased 25-fold between 1982 and 1997, with a total of more than 103,000 cases reported in that 15-year time-span. See Fig. 5. As a rapidly emerging infectious disease, Lyme disease accounts for more than 90% of all vector-borne illness reported in the United States. Cases are most concentrated in the northeastern, north central and pacific coastal regions of the United States, although 48 states and the District of Columbia, parts of Europe and Australia have reported Lyme disease. The geographic distribution of *Ixodes dammini* and *Amblyomma americanus*, the principal vectors of *Borrelia burgdorferi*, continues to spread, and with it, there has been an increase in the reported incidence of Lyme disease.

The diagnosis and confirmation of Lyme disease is difficult. Clinical manifestations can be confused with those of other illnesses and diseases, especially in late Lyme disease. Researchers continue to work to identify antigens as well as new methods that can improve upon the current diagnostic tools for Lyme disease. As the disease is more intensively studied, the recognized clinical spectrum widens, yet major factors in the origination and development of the disease remain unknown.

LYMPH. A clear fluid that circulates in the tissue spaces of vertebrates and passes into the venous system by way of a tubular lymphatic system. It is derived from the liquid plasma of the blood but is more watery and contains no red corpuscles. It serves as an intermediary between the blood itself and the tissues of the body.

Lymph is found lying free in the serous sac cavities of the body, i.e., the peritoneum, pleura, and the spaces in the brain filled with cerebrospinal fluid, which may also be classified as lymph, although it differs in composition from the fluid found in the lymph vessels.

Lymph is derived from the plasma of the blood either by filtration, diffusion, or osmosis through the capillary walls or by active secretion of endothelial cells making up capillary walls. Its composition is very similar to that of the blood plasma.

The function of lymph is to bring nourishment to the tissue cells and to return waste matter and other toxic material to the blood stream by way of the lymphatic vessels, or directly into the blood stream through the capillary walls. Lymph has been compared with the body fluids of invertebrates, commonly designated as hydrolymph and hemolymph.

The small tubules of the lymphatic system resemble capillaries and the larger ducts, called lymphatics, are similar to veins, although of a more delicate structure in relation to their size. Like veins, they have valves that aid in promoting flow through the movements of the surrounding muscles. They are irregular in diameter, forming reservoirs at some points and dilating in the amphibians, reptiles, and birds to form lymph hearts whose pulsations propel the lymph toward the heart. The smaller vessels converge like blood vessels to form larger trunks. In humans, the chief vessels are the thoracic duct, and the right lymphatic duct, which empty into the large veins at the sides of the neck. Along the course of the lymph vessels are groups of lymph nodes. They serve as filters, which localize and retard the spread of toxic and infectious elements that are being returned to the blood stream. The lymph nodes also serve as centers for formation of lymphocytes, which form one of the main divisions of blood cells.

LYMPHOCYTIC CHORIOMENINGITIS (LCM). Lymphocytic choriomeningitis, or LCM, is a rodent-borne viral infectious disease that presents as aseptic meningitis (inflammation of the membrane, or meninges, that surrounds the brain and spinal cord), encephalitis (inflammation of the brain), or meningoencephalitis (inflammation of both the brain and meninges). Its causative agent is the lymphocytic choriomeningitis virus (LCMV), a member of the family Arenaviridae, that was initially isolated in 1933. Although LCMV is most commonly recognized as causing neurological disease, as its name implies, asymptomatic infection or mild febrile illnesses are common clinical manifestations. Additionally, pregnancy-related infection has been associated with abortion, congenital hydrocephalus and chorioretinitis, and mental retardation.

LCM and milder LCMV infections have been reported in Europe, the Americas, Australia, and Japan, and may occur wherever infected rodent hosts of the virus are found. However, the disease has historically been under reported, often making it difficult to determine incidence rates or estimates of prevalence by geographic region. Several serologic studies conducted in urban areas have shown that the prevalence of LCMV infection among humans ranges from 2% to 10%.

LCMV is naturally spread by the common house mouse, *Mus musculus*. Once infected, these mice can become chronically infected by maintaining virus in their blood and/or persistently shedding virus in their urine, a common characteristic of other arenavirus infections in rodents. Chronically infected female mice usually transmit infection to their offspring, which in turn become chronically infected.

Humans become infected by inhaling infectious aerosolized particles of rodent urine, feces, or saliva, by ingesting food contaminated with virus, by contamination of mucus membranes with infected body fluids, or by directly exposing cuts or other open wounds to virus-infected blood. LCMV infection has also been documented among staff handling infected hamsters. Person-to-person transmission has not been reported, with the exception of vertical transmission from an infected mother to fetus.

The incubation period of LCMV infection is usually between 8 and 13 days. A characteristic biphasic febrile illness then follows. The initial phase, which may last as long as a week, typically begins with any or all of the following symptoms: fever, malaise, anorexia, muscle aches, headache, nausea, and vomiting. Other symptoms that appear less frequently include sore throat, cough, joint pain, chest pain, testicular pain, and parotid (salivary gland) pain. Following a few days of remission, the second phase of the disease occurs, consisting of symptoms of meningitis (for example, fever, headache, and a stiff neck) or characteristics of encephalitis (for example, drowsiness, confusion, sensory disturbances, and/or motor abnormalities, such as paralysis). LCMV has also been known to cause acute hydrocephalus, which often requires surgical shunting to relieve increased intracranial pressure. In rare instances, infection results in myelitis (inflammation of the spinal cord) and presents with symptoms such as muscle weakness, paralysis, or changes in body sensation. An association between LCMV infection and myocarditis (inflammation of the heart muscles) has been suggested.

During the first phase of the disease, the most common laboratory abnormalities are a low white blood cell count (leukopenia) and a low platelet count (thrombocytopenia). Liver enzymes in the serum may also be mildly elevated. After the onset of neurological disease during the second phase, an increase in protein levels, an increase in the number of white blood cells or a decrease in the glucose levels in the cerebrospinal fluid (CSF) is usually found.

Previous observations have shown that most patients who develop aseptic meningitis or encephalitis due to LCMV recover completely. No chronic infection has been described in humans, and after the acute phase the virus is cleared. However, as in all infections of the central nervous system, particularly encephalitis, temporary or permanent neurological damage is possible. Nerve deafness and arthritis have been reported. Infection of the human fetus during the early states of pregnancy may lead to developmental deficits that are permanent. LCM is usually not fatal. In general, mortality is less than 1%.

Aseptic meningitis, encephalitis, or meningoencephalitis requires hospitalization and supportive treatment based on severity. There is no specific drug therapy for LCM. Anti-inflammatory drugs, such as corticosteroids, may be considered under specific circumstances. Although studies have shown that ribavirin, a drug used to treat several other viral diseases, is effective against LCMV *in vitro*, there is no established evidence to support its use for treatment of LCM in humans.

Individuals of all ages who come into contact with urine, feces, saliva, or blood of the house mouse are potentially at risk for infection. Laboratory workers who themselves handle infected animals are also at risk. However, this risk can be minimized by utilizing animals from sources that regularly test for the virus, wearing proper protective laboratory gear, and following appropriate safety precautions. Owners of pet mice or hamsters may be at risk for infection if these animals originate from colonies with circulating LCMV, or if the animals become infected from other wild mice. Human fetuses are at risk of acquiring infection vertically from infected maternal blood.

Like many other rodent-borne infectious diseases, LCMV infection can be prevented by avoiding or minimizing direct physical contact with rodents or exposure to their excreta. Adequate ventilation should be provided to any heavily infested, previously unventilated enclosed room or dwelling prior to cleanup. A liquid disinfectant, such as a diluted household bleach solution, should be applied to visible rodent droppings and their immediate surroundings. Gloves should be worn when disinfecting and cleaning up rodent excreta. Rodent spring traps may be set up in households or dwellings where rodent infestations are a concern.

The geographic distributions of the rodent hosts are widespread both domestically and abroad. However, infrequent recognition and diagnosis, and therefore under reporting, of LCM, have limited scientists' ability to estimate incidence rates and prevalence of disease among humans. Understanding the epidemiology of LCM and LCMV infections will help to further delineate risk factors for infection and develop effective preventive strategies. Increasing physician awareness will improve disease recognition and reporting, which may lead to better characterization of the natural history and the underlying immunopathological mechanisms of disease, and stimulate future therapeutic research and development. See also **Arenaviruses**; and **Viral Hemorrhagic Fevers**.

Additional Reading

Galasso, G.J., T.C. Merigan, and R.J. Whitley: "Antiviral Agents and Human Viral Diseases," 4th ed., Lippincott Williams & Wilkins, Philadelphia, PA, 1997.

Jahrling, P.B. and C.J. Peters: "Lymphocytic Choriomeningitis Virus: A Neglected Pathogen of Man," *Arch. Pathol. Lab Med.* **1 16**, 486–488 (1992).

Leland, S.S. and S. Ozmat: "Clinical Virology," W.B. Saunders Company, Philadelphia, PA, 1996.

Love C.B. and P.B. Jahrling: "Viral Hemorrhagic Fever," DIANE Publishing Company, Collingdale, PA, 1996.

Pattison, J.R., J.E. Banatvala, and A.J. Zuckerman: "Principles and Practice of Clinical Virology," 4th Ed., John Wiley & Sons, Inc., New York, NY, 2000.

Peters, C.J., M. Buchmeier, P.E. Rollin, and T.G. Ksiazek: "Arenaviruses," in R.B. Belshe, ed., *Textbook of Human Virology*, 2nd ed., Mosby-Year Book, Inc., St. Louis, MO, 1991.

Peters, C.J. et al.: "Arenaviridae: Biology of Viruses," in: B.N., Fields, D.M. Knipe, and P.M. Howley, eds., *"Fields Virology,"* 3rd ed., Lippincott Williams & Wilkins, Philadelphia, PA, 1996.

Richman, D.D., R.J. Whitley, and F.G. Hayden: "Clinical Virology," Harcourt Brace & Company, San Diego, CA, 1998.

Centers for Disease Control and Prevention (CDC), Atlanta, GA.

LYMPHOGRANULOMA VENEREUM. Caused by strains of *Chlamydia trachomatis*, this is a venereal disease which features the appearance of a transitory primary genital lesion, a vesicle or a papule, followed by other stages. In the male, the primary genital lesion, which occurs within 5 to 21 days after the implicating sexual exposure, usually takes the form of a painless vesicle or papule, or as a chancriform lesion (*chancroid*). In the male, it is commonly located on the coronal sulcus of the penis; in females, on the labia or posterior vaginal wall. The situation is usually self-limiting and healing occurs within a few days. Extragenital infections occur in persons who deviate from normal sex profiles. The infection may continue in a second and a third stage, where the disease is extended to include regional lymph nodes. These nodes enlarge, produce pus, and ultimately form buboes. In a third stage, lymphatic obstruction may occur. Lymphatic obstruction may infrequently produce dilation of lymphatic channels and hypertrophy of subcutaneous connective tissue and of skin (*genital elephantiasis*).

Treatment may include oral tetracycline or sulfonamides, such as sulfisoxazole. It is a good practice to test all patients with lymphogranuloma venereum for syphilis because the two infections are frequent companions.

LYMPHOKINES. These are soluble substances (factors) produced by lymphocytes which aid in regulating a variety of immune responses. See also **Immune System and Immunology**. They are not immunoglobulins, but are synthesized by lymphocytes of undetermined structure and are classified according to the target cells they affect.

Chemotactic factors are lymphokines that attract certain cells, e.g., monocytes, neutrophils, etc., to a particular site. They are synthesized and released within 24 hours of lymphocyte stimulation.

Migration inhibition factor (MIF), the first lymphokine discovered, inhibits macrophage migration and is produced by certain sensitized lymphocytes stimulated by an exquisite antigen. Mitogens and antigen-antibody complexes can, however, nonspecifically trigger lymphocytes to produce MIF and antigen need not be present for MIF to inhibit macrophage migration.

Macrophage-activating factor (MAF) induces morphologic, metabolic, and functional changes in macrophages, which enhance the cell's ability to kill microorganisms and tumor cells.

Leucocyte-inhibitory factor (LIF) inhibits neutrophil (but not monocyte) migration in vitro.

Interleukin. Macrophages produce a monokine termed Interleukin-1 (lymphocyte activating factor) which combines with an antigen or mitogen to stimulate T lymphocytes to produce Interleukin-2 (T-cell growth factor) which in turn causes the proliferation of other T cells, such as helper, suppressor, or cytotoxic cells.

Cytotoxic and Cystostatic Factors. Lymphocytes from several species produce lymphokines, which kill or inhibit the growth of susceptible target cells. Among these lymphokines are several proliferation-inhibiting or colony-inhibiting factors and inhibitors of DNA synthesis.

Tissue Factor. Lymphocytes stimulated in vitro by an antigen or mitogen produce a procoagulant material that is biologically similar to tissue thromboplastin, but is antigenically distinct from it. The pathophysiologic importance of tissue factor is undetermined.

Lymphokines that enhance antibody production are termed helper factors and are produced by helper T lymphocytes. Lymphokines that suppress antibody production are termed suppressor factors and are produced by suppressor T lymphocytes.

Colony-stimulating activity is a lymphokine that induces differentiation of bone marrow cells in vitro into granulocytes and agranular mononuclear cells.

Osteoclast activating factor (OAF) is a lymphokine which causes bone resorption by activating osteoclasts.

Lymph node lymphocytes, from rats immunized with myelin basic protein, release immunoglobulin-binding factor that combines with antigen-antibody complexes, causing IgG-sensitized sheep erythrocytes to agglutinate.

Interferons. These are a family of glycoproteins that act against a wide range of viruses. These types are known and can be distinguished on the basis of their physicochemical characteristics, the cell types producing them, and the types of stimuli causing their production. They do not however act directly upon viruses, but render cells resistant to viral infection.

Additional Reading

Cohen, S.: "Lymphokines & the Immune Response," CRC Press, LLC., Boca Raton, FL, 1989.

Gills, S.: "Recombinant Lymphokines and Their Receptors," Marcel Dekker, Inc., New York, NY, 1987.

Goldstein, A.L.: "Thymic Hormones and Lymphokines: Basic Chemistry and Clinical Applications," Perseus Books, Boulder, CO, 1984.

ANN C. DEBALDO, Ph.D., University of South Florida, Tampa, FL.

LYMPHOMA. A lymphoma is defined as a tumor of lymphoid tissue. Some diseases are also referred to as lymphomas and comprise a group of malignant disorders that usually arise in the lymph nodes. These diseases present a broad spectrum of clinical features. Sometimes the term non-Hodgkin's lymphomas is used to distinguish other lymphomas from Hodgkin's disease because they do not have the characteristic giant cells of Hodgkin's disease. The causes of the lymphomas remain poorly understood. One suspected cause is viral infection. The epidemiology of the lymphomas may provide evidence as to the etiology of the diseases. Some lymphomas occur with much higher frequency in some geographical regions than in others. Diagnosis of a lymphoma is by biopsy. Treatment of the non-Hodgkin's disease lymphomas tends to parallel the therapy used in treating Hodgkin's disease. See also **Hodgkin's Disease**; and **Lymphosarcoma**.

LYMPHOSARCOMA. A disease, similar to leukemia and Hodgkin's disease in its symptoms. Although still generally used, the term lymphosarcoma is pathologically obsolete; it basically refers to what are now known as lymphocytic lymphomas and includes what were formerly known as lymphocytic lymphosarcomas, the so-called lymphoblastic lymphosarcoma, follicular lymphoma, and Burkitt's tumor.

It is important to realize that the distinction between chronic lymphocytic leukemia and those cases of lymphocytic lymphoma in which some of the tumor cells enter the circulating blood stream is not made easily and indeed may be impossible.

The lymphocytic lymphomas may arise in any group of lymph nodes in a tonsil, the spleen, or thymus, but lymph nodes are by far the most common site of their origin. The invasive growth of the sarcoma leads to fusion and coalescence of affected nodes and their adherence to adjacent tissues—these latter may be deeply penetrated by the tumor. Microscopically the normal structure of the affected tissue is obliterated by masses of closely packed tumor cells and in many cases the cells cannot be distinguished from normal lymphocytes. The cells of any given tumor are, however, usually uniform in type—either small or large. Occasionally a well differentiated tumor that consists predominantly of small lymphocytes may be found to include areas in which the cells are larger lymphocytes; more rarely, pleomorphic areas are seen. Whatever the type of cell, the lymphomas with mixed cytological constitution trend to be more rapidly progressive. An early indication of developing malignancy by cell differentiation is enlargement of the nucleolus. There is seldom any tendency for necrosis to occur in the lymphomas unless they have become anaplastic.

Acute lymphatic leukemia of childhood may be lymphosarcoma with the abnormal cells overflowing into the circulating blood. Whereas usually this disease is fatal if untreated, when patients receive early and proper treatment, both blood and bone marrow may revert to normal condition with full restoration of health for variable periods. In lymphosarcoma, x-ray therapy is the major and most universally effective form of treatment. Chlorambucil and nitrogen mustard have been effective chemotherapeutic agents in earlier stages of the disease; cyclophosphamide and vinblastine sulfate, in later stages. Prednisone may prove beneficial in patients with fever and hematologic disturbances no longer suitable for treatment with x-ray or other drugs.

Additional Reading

Cabanillas, F. and M. Rodriguez: "Advances in Lymphoma Research," Vol. 85, Kluwer Academic Publishers, Norwell, MA, 1997.
Crocker, J.: "Advances in Lymphoma Research," Vol. 85, Blackwell Science, Inc., Malden, MA, 1993.
Katz, R.: "Current Issues in Lymphoma," S. Karger AG, Basel, Switzerland, 1994.

R. C. V.

LYNX.   See **Cats.**

LYRA (the harp).   One of the small constellations, which contains a number of most interesting objects for viewing through a small or large telescope. Lyra is most easily distinguished by an equilateral triangle having the star Vega at one of its apexes. This star is the brightest in the northern celestial hemisphere. It lies almost in the direction in which the sun and all the planets are moving, due to solar motion. And although it is a long way from the pole of rotation of the celestial sphere at present, it will be the pole star about 12,000 years hence, due to precession. The star Epsilon Lyrae is one of the most famous multiple stars in the entire sky. It can be resolved into two components through a field glass and, on a clear night, into four components through a 6-inch (≈15-centimeter) telescope. Also to be found in this constellation are several other double stars, and the famous ring nebula, an interesting object when viewed through a 6-inch telescope. (See Fig. 1, and map accompanying entry on **Constellations.**)

LYRE BIRD (*Aves, Passeriformes*).   This bird ranges from Queensland to Victoria and is highly regarded in Australia, sometimes pictured on stamps and official seals. The male lyre bird is well known for its display of plumage and performance to attract females of the species. The "lyre frame" feathers are something like those of a bird of paradise or peacock when displayed. Of several species, there is the superb lyre bird (*Menura novaehollandiae*) and Albert's lyre bird (*M. alberti*). The tail feathers range from $1\frac{1}{2}$ to $2\frac{1}{2}$ feet (0.5 to 0.8 meter) in length and usually are white and brown. The head is small, the legs are long, the claws are strong. The male is known for its incredible mimicking of other birds. There is one pale-purple, thick-shelled egg of about the size of a chicken egg. The incubation period is 6 weeks. The young do not leave the nest for 12 weeks after hatching.

LYRIDS.   A name given to certain meteor showers that are observed about April 20 of each year. The orbit of the radiant point was definitely associated with the orbit of comet 1861 I. by Weiss. Records of showers from this radiant are found as far back as 687 B.C. A report written by the Chinese in 15 B.C. indicates that during the Lyrid shower of that year, "after the middle of the night, stars fell like rain." Several other accounts of striking showers during April are on record; in particular, we find many newspaper accounts of the Lyrid shower of 1803, which was observed over the United States from North Carolina to New Hampshire. The Richmond, Va., *Gazette* of April 23, 1803, gives a long and vivid account of the shower occurring on the morning of April 20, stating that "from one until three those starry meteors seemed to fall from every point of the heavens, in such numbers as to resemble a shower of sky rockets."

A few scattered members of this shower are observed coming from the radiant point in the constellation of Lyra every year, but there has not been any very striking display since 1803. Because there have been striking showers in the past, the assumption is that a large swarm of meteors exists at some undetermined point along the orbit, and that we may be treated to another brilliant display during some April in the future.

LYSIS (Bacteriology).   The dissolution of cells, e.g., bacteria, or red blood cells; their breakdown from structural form to a structureless fluid.

LYSIS (Physiology).   The gradual decline of the symptoms of disease, referring especially to the gradual abatement of fever.

LYSOSOMES.   Subcellular organelles believed to contain digestive enzymes capable of breaking down many of the cellular constituents. Disruption of the lysosomes and liberation of these enzymes may occur under certain conditions and can lead to lysis of the cell. See also **Cell (Biology).**

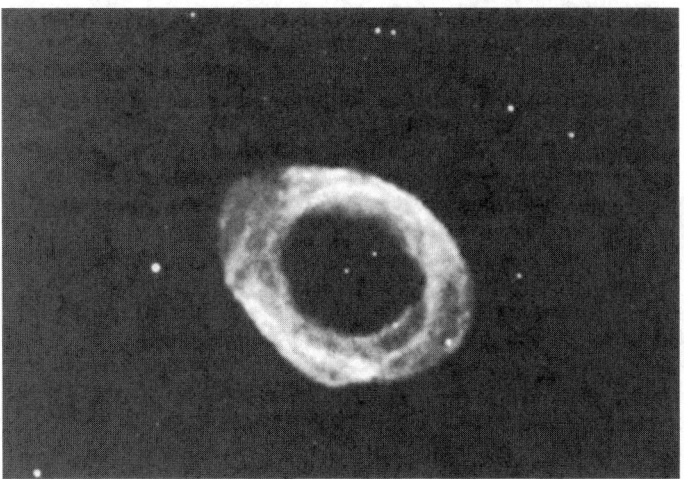

Fig. 1.   Ring nebula in Lyra. This nebula is gradually expanding, and its edges are quite red. (*Lick Observatory.*)

M

MA (or Ma).   Millions of years.

MAAR.   A volcanic explosion crater without a prominent volcanic cone.

MACADAMIA TREE.   Of the family *Proteaceae*, the macadamia tree (*Macadonia ternifolia*) is native to New South Wales and South Queensland. It was introduced into the United States (Hawaii, California, and Florida), South Africa, the Mediterranean countries, and the West Indies in the late-1900s. The tree is best known for its tasty and rich nut, now highly valued as a confection in many parts of the world. The tree must have drained rich soil, a warm climate, and considerable care, particularly in removing suckers that grow around the roots. The tree can attain a height of from 35 to 50 feet (10.5 to 15 meters) and the trunk may measure a foot or more in circumference in a fully-grown specimen. The leaf is dark green, glossy, and deciduous. The most highly regarded varieties, out of a total of over 80, in Hawaii are the Pahau, Keauhou, Kapea, Kohala, and Nuuomir. The nut of the tree was first popularized by Brisbane nurserymen in the early 1900s. Many varieties are of no significant commercial value.

MACAQUE.   See **Monkeys and Baboons**.

MACH ANGLE.   The angle a shock wave makes with the direction of motion as determined by the velocity of the object and the velocity of shock propagation.

MACH, ERNST (1838–1916).   Mach was born in Chirlitz-Turas, Moravia (now Chrlice-Turany, Czech Republic). He earned his Ph.D. in physics from the University of Vienna. Beginning in 1867, he worked as a professor in experimental physics at Charles University in Prague. He did research in projectile motion. He is best known for his theory of science called positivism. The basis of this theory is that no concept can enter into a scientific hypothesis unless it can be explained completely in terms of observations. His most important published works include *History and Root of the Principle of the Conservation of Energy* and *The Science of Mechanics*.

See also **Mach Angle**.

J. M. I.

MACHINE.   A mechanical device wherein mechanical energy is applied at one point and delivered in a more useful form at another point. The term machine applies traditionally in physics to any one of the basic devices of an elementary nature. These fall into two classes: (1) those dependent upon the vector resolution of forces, such as the inclined plane, wedge, screw, and toggle joint; and (2) those in which there is an equilibrium of torques, such as the lever, pulley, and wheel-and-axle.

MACHINE LANGUAGE.   1.   A language, occurring within a computer, ordinarily not perceptible or intelligible to persons without special equipment or training.

2.   A translation or transliteration of sense 1 into more conventional characters, but frequently still not intelligible to persons without special training.

MACHINE (Simple).   Examples of simple machines include levers, pulleys, the inclined plane, gears and gear trains, and the screw. Most of these concepts have been known almost since antiquity.

Levers.   A lever consists of a bar of nearly rigid material, either straight or bent, a fulcrum F, a weight W, and a force P. The components F, W, and P can be applied, in any position relative to each other, to the bar as shown in Fig. 1. Levers are used to move large forces (weights) by means of smaller forces, or are used to either amplify or diminish arc motion (Fig. 2). The arms A and B must be perpendicular to the lines of actions (directions) of their respective forces P and W. Then, by a balanced moment equation about F,

$$PA = WB$$

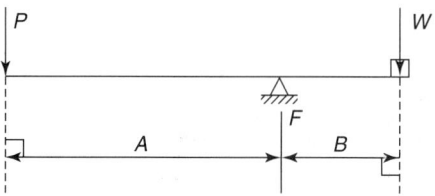

Fig. 1.   Simple first-class lever.

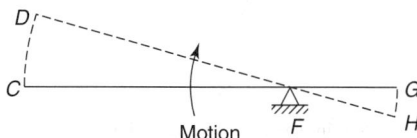

Fig. 2.   Simple lever in motion.

The mechanical advantage of a lever, which defines the ability of an available force P to overcome a resisting force W, is given by the ratio W/P or A/B and results in the expression

$$\text{Mechanical advantage} = \frac{W}{P} = \frac{A}{B}$$

The use of a lever to amplify (or decrease) motion is shown in Fig. 2. When bar CG rotates about F, point G moves through the arc GH, and C will move through the arc CD, and

$$\frac{CD}{CF} = \frac{GH}{FG}$$

Levers continue to be used in many modern kinds of machinery and are notably visible in large weighing scales, conveyors and other automated equipment, and earth-moving equipment. The crowbar is an excellent example of a simple lever-type tool.

In a *first-class lever*, the fulcrum is always between the power and the load and thus the mechanical advantage can be greater than one, one, or less than one. In a *second-class lever*, the power is always less than the load and the mechanical advantage is greater than one. In a *third-class lever*, the power is always greater than the load and the mechanical advantage is less than one. In second-class and third-class levers, the fulcrum is always to one side of the load and power—never in between the application of these two forces.

Pulleys.   In pulley systems, forces are transmitted by ropes, chains, etc. in conjunction with pulley wheels and axles. As noted from the lever examples, a wheel and axle is an adaptation of a lever rotating about its fulcrum.

Considering a theoretical frictionless system of pulleys, the force (pull) in any part of a continuous rope is constant and equal to P. Then, by establishing the number of supporting forces (ropes) and the weight W

which is being moved, $nP = W$, where n is the number of supporting ropes. In Fig. 3, two forces (ropes) support the weight W, hence $2P = W$ and $P = W/2$. In Fig. 4, four forces (ropes) support the weight W, and $4P = W$ and $P = W/4$. In Fig. 5, five forces (ropes) support W, and $5P = W$ and $P = W/5$.

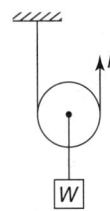

Fig. 3.   Simple rope pulley.

Fig. 4.   Pulley system in which the mechanical advantage is $W/(W/4) = 4$.

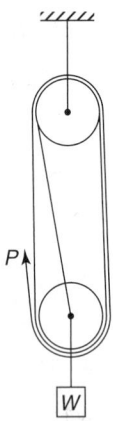

Fig. 5.   Pulley system in which the mechanical advantage is $W/(W/5) = 5$.

The mechanical advantage of a pulley system is the ratio of the weight to be moved to the applied pull in the rope, or $W/P = n$, the same as the number of supporting ropes. In Fig. 3, $W/(W/2) = 2$, or for Fig. 4, the mechanical advantage is $W/(W/4) = 4$. In Fig. 5, the mechanical advantage is $W/(W/5) = 5$.

Pulley systems can be analyzed by the use of work done by the force P and its relation to the work done by W. The displacement (Fig. 4) of P is four times that of W. Thus the work done by P is $4PS$, where $4S$ is the distance that P moves. The work done by W is WS. The mechanical advantage is $4S/S = 4$ and $4P = W$, as given above.

Differential Pulley. This device makes use of two pulleys of different radii, r_1 and r_2, attached to each other and rotating about a common axle. An endless chain connects the dual pulley to a second free pulley wheel,

as shown in Fig. 6. The chain and the corresponding teeth on the dual pulleys prevent slipping between the chain and pulleys.

From the previous analysis of pulleys, a force equal to $W/2$ acts in the chain at points A and B. A moment equation about axle C then gives

$$Pr_1 + \frac{W}{2}r_2 = \frac{W}{2}r_1$$

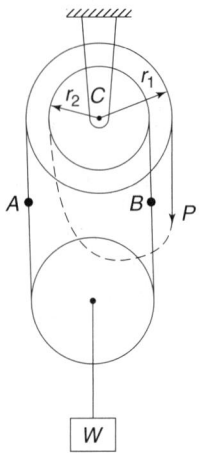

Fig. 6.   An endless chain connects a dual pulley to a second free pulley wheel.

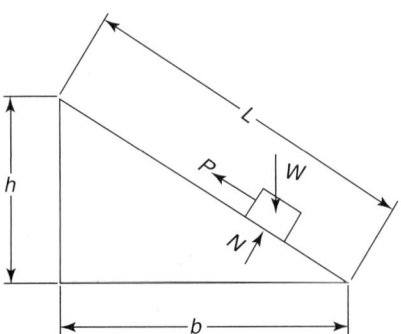

Fig. 7.   Inclined plane.

from which

$$P = \frac{W(r_1 - r_2)}{2r_1}$$

The mechanical advantage of the differential pulley is the ratio W/P.

Inclined Plane. As a simple machine, the inclined plane is presumed to be rigid and smooth. The weight W which moves along the incline is a vertically downward force partially supported (N) by the frictionless plane assumed for this example. If the weight W is to be at rest on the incline (Fig. 7), the force system must be balanced in directions normal and parallel to the inclined plane. Balancing the forces parallel to the incline,

$$P - W\frac{h}{L} \text{ or } PL + Wh$$

This is equivalent to a work equation, where P is displaced a distance L, and W is lifted a distance h. When the force P acts to the left in a horizontal direction, P moves an equivalent distance b, and then

$$PB = WH$$

The mechanical advantage is the ratio W/P or L/h, where the force is parallel to the incline. A *wedge* is equivalent in its analysis to an inclined plane. When t is the thickness of a wedge, $P/W = t/L$.

The Screw. This device is an inclined plane wrapped around a cylinder in such a way that the height h is parallel to the axis of the cylinder. If p is the height of travel in one circumference of the screw thread, r is the radius of the thread, and friction is neglected, by the work method of analysis one obtains

$$P2\pi r = W p$$

The mechanical advantage is

$$\frac{W}{P} = \frac{2\pi r}{p}$$

Gears and Gear Trains. A gear is a wheel with projections uniformly spaced around its circumference. It is usually meshed with a second similar wheel of a different diameter so that the circumferential forces and rotational speeds are different. Two meshed gear wheels (Fig. 8), the teeth of which are not shown are used as an example. From a balanced moment equation,

$$Pr_1 = Wr_1$$

and the mechanical advantage is the ratio W/P or R_1/r_1.

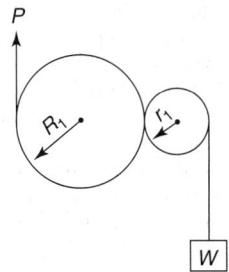

Fig. 8.   Two meshed gear wheels (teeth not shown.)

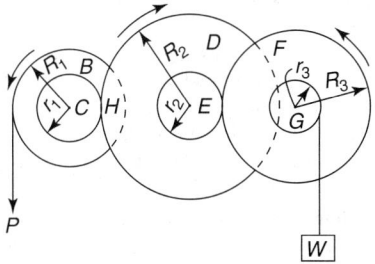

Fig. 9.   Simple gear train.

Gear trains consisting of more than two gear wheels have the same relationship. In Fig. 9, letting the wheels be represented by letters, one has

$$PR_1R_2R_3 = Wr_1r_2r_3$$

and the mechanical advantage is

$$\frac{W}{P} = \frac{R_1R_2R_3}{r_1r_2r_3}$$

By inspection, and understanding that the points of contact (such as point H) have a common speed, the directional relations can be determined. If wheels B and C are rotating counterclockwise, D and E rotate clockwise, and F and G rotate counterclockwise.

Spur gears are those which have their teeth cut parallel to parallel axes of rotation. Bevel gears are used when the axes of rotation intersect; their teeth are cut on conical surfaces with the apex at the point of intersection of the axes. Speed ratios are inversely proportional to pitch diameters, number of teeth, or number of revolutions. Referring to Fig. 9,

$$\frac{\text{rpm of } E}{\text{rpm of } F} = \frac{2R_3}{2r_2} = \frac{R_3}{r_2}$$

When a screw meshes with a cogged wheel, it is known as a *worm* and *worm wheel.* When the worm has a single continuous thread, one revolution of the worm will cause the wheel to rotate through a circumferential distance equal to the distance between two consecutive teeth. Thus, a worm must rotate 48 revolutions if a worm wheel with 48 teeth is to rotate one revolution. The same relationship is true if speeds are considered.

Efficiency. The efficiency of a machine is defined as the ratio of the output to the input. The efficiency of an inclined plane, for example, is given by the expression:

$$\text{Efficiency } (\%) = \frac{h}{h + fb}100$$

for the case where the applied pull tends to move the block up the incline (Fig. 7), and where f is the coefficient of friction between the incline and the flock. See also entry on **Efficiency**.

J.W. BRENEMAN (deceased).

MACHINE VISION (Recognition and Applications).   Since its initial serious recognition in the late 1970s, machine vision (MV) has been variously defined:

> MV is the process of extracting information from visual sensors for the purpose of enabling machines to make intelligent decisions.
> or
> MV is part of the larger technology of artificial visual perception that substitutes (partially or totally) the human visual capability by instruments that are backed up by electronic data processing (notably complex computations).

As will be noted from the foregoing, the real and practical objectives of MV tend to be nebulous and of far-reaching expectations.

MV became a "buzz word" of the 1980s — to the extent that MV, combined with robots, would bring about the most ambitious goals in terms of production automation. The fact is that, by the early 1990s, MV had lost much of its earlier luster. With relatively few exceptions, most contemporary MV systems are confined to sophisticated and demanding object-detection problems, with the majority of such detection problems currently being solved by the less exotic types of sensors previously described in this article. Because total MV systems in general still remain overly complex and costly, other nonvisual sensors have been greatly improved and applied more imaginatively to achieve many of the objectives that once appeared to lie within the province of MV. Cost, once again, has been the traditional motivating factor.

"Seeing" robots, once a major goal of MV, has proved disappointing from the standpoint of MV. To be true, in practice, robots, once programmed, do perform in many instances as though they could "see," but this has been brought about more by other instrumental detectors (tactical, for example) than by literal "viewing" of parts, pieces, machines, and entire manufacturing scenes. For example, in the late 1980s a large automotive manufacturer originally invested rather heavily in MV systems in connection with robotic operations, but after a year or so of trials, abandoned them for possible consideration at some future date.

To be sure, MV developments will continue, and systems will be installed where other simpler, lower-cost approaches do not suffice. Coupled with trends toward reducing computation costs, MV may become more competitive.

Artificial Visual Perception

Industrial MV techniques initially stemmed from the interests of the military in a technology known as pattern recognition. This may be defined as seeing, analyzing, and interpreting patterns, as of scenery, juxtaposition, dimensional magnitudes, color, and other characteristics of the visual environment.

Human visual communication (information transfer) with the outside environment depends on the interactions (predominantly absorption, reflection, and refraction) of light (visual) radiation emanating from physical objects (in point, two, or three dimensions), thus enabling the human observer to cope with the surrounding environment in a safe and efficient manner. The vision activity (human or machine) must be complemented by some form of processor (human brain or electronic counterpart) to make a final identification of what has been seen. This process, defined by a few words, is pattern recognition.

The means that are applied to perform pattern recognition is known as the pattern processor. The human brain, in processing a pattern, does an amazing job of sorting out extraneous information in the input to quickly identify what really is present (the objects of interest that are in the pattern).

Interest in pattern recognition dates back some 70 years. The more the process is studied, the more one finds how fundamental the process really is to the human brain function. The amount of information (inputs) that the human system can observe and process is tremendous, but nevertheless remains poorly understood. This explains why MV, from a technological viewpoint, is closely associated with the study of artificial intelligence.

Elements of Pattern Recognition

Any pattern recognition system contains the same three basic elements: (1) sensing, (2) processing, and (3) implementing actions based upon input data.

The primary objective of pattern recognition is to classify a given unknown pattern as belonging to one of several classes of patterns. The applications of pattern recognition are many and varied. In the case of character recognition, for example, the patterns are easily generated and recognized by humans, and the basic goal has been to improve the human-machine communication. In other situations, the patterns are difficult for humans to recognize rapidly, as, for example, the very rapid interpretation of an electrocardiogram. The wide range of applications, as well as the relationship to diverse disciplines, including machine vision, communication and control, and the area of linguistics (word and sentence reading), have broadened the interest basis in pattern recognition.

Information Content of an Image

In terms of MV as used in industrial production situations, the information content of the image falls into three basic categories:

1. *Geometry*, which in turn portrays shape, position, dimension, and a number of other associated properties, including density and texture (which can be inferred from the known geometry in most cases).
2. *Color*, which is very helpful when present. There are, of course, color-blind persons and instruments that preclude its use.
3. *Movement*, which is present in two or more images of a dynamic process.

Extracting Information from Images

As with other forms of industrial instrumentation, the MV system senses (reads may be a better term) the image, but before the data obtained from the image can become meaningful, the sensed information must be compared with some form of standard, or prelearned, pattern. In more familiar terms, the combined actions of sensing and comparing constitute measurement.

As compared with the usual type of industrial sensor, such as a thermocouple, where measured data are conveniently available in the form of a ready-made electrical signal, in MV one deals not with just one or a very few points or locations of emanating data, but with thousands-plus bits of information—because what is being observed (measured) is comprised of a multitude of points (picture pixels[1]). In an artificial vision system, image data, as may be gathered by an electronic camera, must be compared electronically with information, that is, in some fashion with data that are stored in electronic memory.

Initially the application of a general-purpose computer to MV was the only choice available. The problem immediately encountered was the fact that the computer was designed to process computational data, *not* data patterns. The data from a video camera in an MV system are pattern data and, thus, are very different from computational data as exemplified by financial balance sheets or linear regression analysis. The situation is summarized as follows:

Image data quantity 484 × 320 pixel density
Hence, 154,880 pixels per frame
At 6 bits of gray-level data per pixel, 929,280 pixel values per frame
At a 30 times a second refresh rate, 27,878,400 bits per second

Since the early days of MV, a number of innovations have contributed to the simplification of this problem, including improvements in gray-scale systems, better algorithms, the pipelined image processing engine, the geometric arithmetic parallel processor, and associative pattern processing.

Machine Vision Sensors

Factors that usually must be considered are (1) optics and lighting, (2) field of view, (3) resolution, (4) signal-to-noise ratio, (5) time and temperature stability, and (6) cost. MV sensors that have been or are currently in use include the following.

[1] A pixel is a picture element, a small region of a scene within which variations of brightness are ignored.

Line Scanners. These include solid-state arrays, flying-spot scanners, and prism, mirror, or holographically deflected laser cameras. These scanners are fast high-resolution devices, which are relatively free of geometric distortion in one dimension. In order to capture a complete two-dimensional scene, the second dimension is obtained either by motion of the object past the scanner or by mirror or prism deflectors. Mechanical motion tends to slow down data acquisition and, in some cases, produces geometric distortion.

Area-Type Scanners. These scanners were used in early systems utilizing closed-circuit television cameras with vidicon image sensors. Advantages were comparatively low cost and relatively high resolution (300 to 500 television lines). They suffer geometric distortion, temperature instability, lag (requiring several television frames for full erasure), and sensitivity to nearby magnetic fields.

Solid-State Cameras. These cameras represent a major advance in image-sensing technology. Scenes can be digitized onto an array of photosensitive cells. Charge-coupled devices (CCDs) or charge injection devices (CIDs) have been used in these cameras. The arrays form a pixel grid containing the data currently appearing on camera. The solid-state camera is available in a variety of pixel densities, notably, 128 × 128, 256 × 256, 512 × 512, 1024 × 1024, and so on.

The limiting factor of cameras for imaging systems is not necessarily the resolution or speed. Camera technology essentially has kept up with the ability of computationally based processes to handle information, especially when multiple cameras are used. Traditionally, in MV systems sensors can detect information in real time a lot faster and in larger quantities than processors can handle.

Machine Vision Image Processing

From the beginning, a major task of MV engineering has been that of reducing the amount of information actually needed, and thus lessening and simplifying the processing task. Shortcuts are frequently taken to reduce data volume. Some processing systems that have been or are in current use include the following.

Binary MV System. The representation of all combinations possible in every pixel of every scene is unrealistic. Conversion of each pixel point into a binary value reduces the pixel data, but it also reduces the accuracy of the analysis. Binary systems evaluate each pixel as black or white. A threshold-adjusting capability can allow users to select what intensity of signal is to be the black-white border.

Gray-Level System. This system can interpret each pixel's value as a specific gray tone. These systems vary in precision, with each pixel being evaluated as 16, 64, or even 256 different values.

Windowing the Scene. This system can reduce the total data system load. Since analysis of an entire scene often is unnecessary for proper recognition, the use of a window can eliminate unneeded image data.

Segmentation. A common means of data reduction, this technique divides the image data into areas of interest and then interpolates

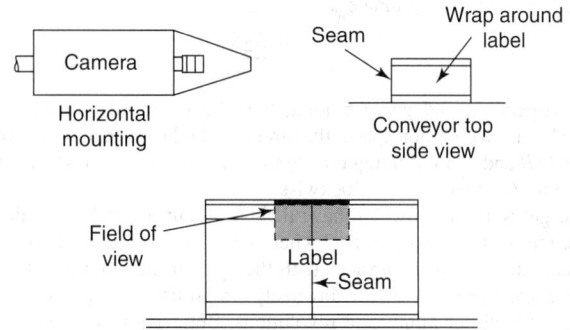

Fig. 1.   Use of a template or pattern match algorithm to check the accuracy of a label on a can of food. A label is wrapped (360°) around a tuna fish sized can (upper diagram). The camera is placed so that the field of view is located horizontally across the area where the label ends touch. The camera's field of view should be located as shown in the lower diagram. To set up the reference template, the operator adjusts the horizontal/vertical axis of the camera by viewing a display on the controller, during which time the operator also adjusts the mounted camera to obtain the view shown in Fig. 2. (*Cutler-Hammer Products, Eaton Corporation.*)

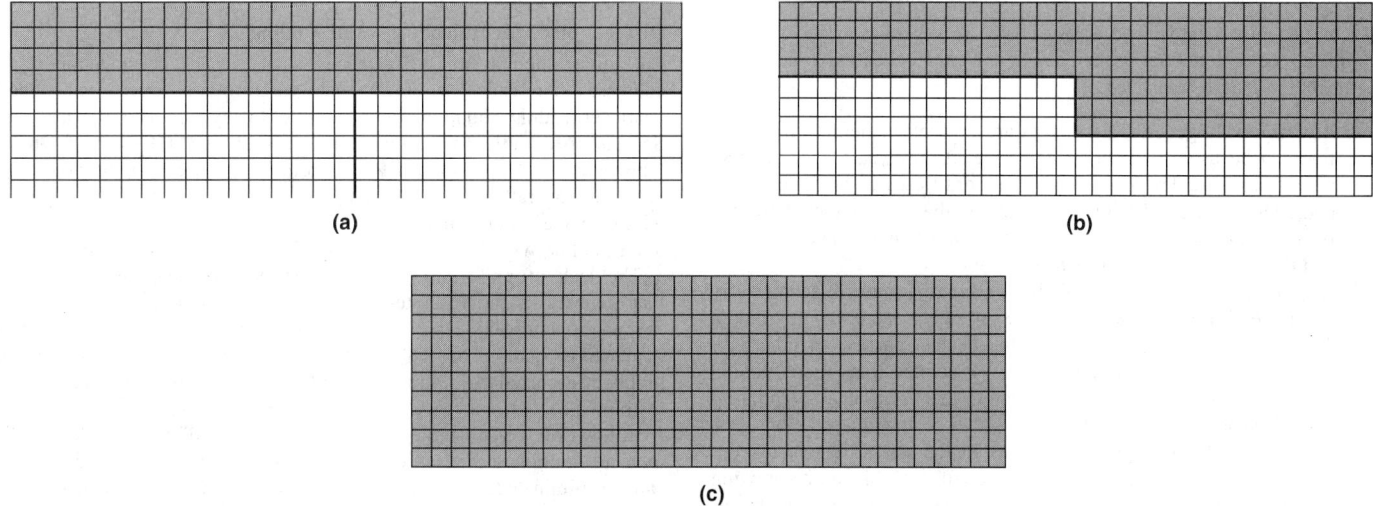

Fig. 2. Displays under three possible conditions: (**a**) Label is on can and is properly placed (can appears black; label, white); (**b**) a poorly placed label, indicating a mismatch area of 3 pixel rows; and (**c**) no label on can (screen appears black). (*Cutler-Hammer Products, Eaton Corporation.*)

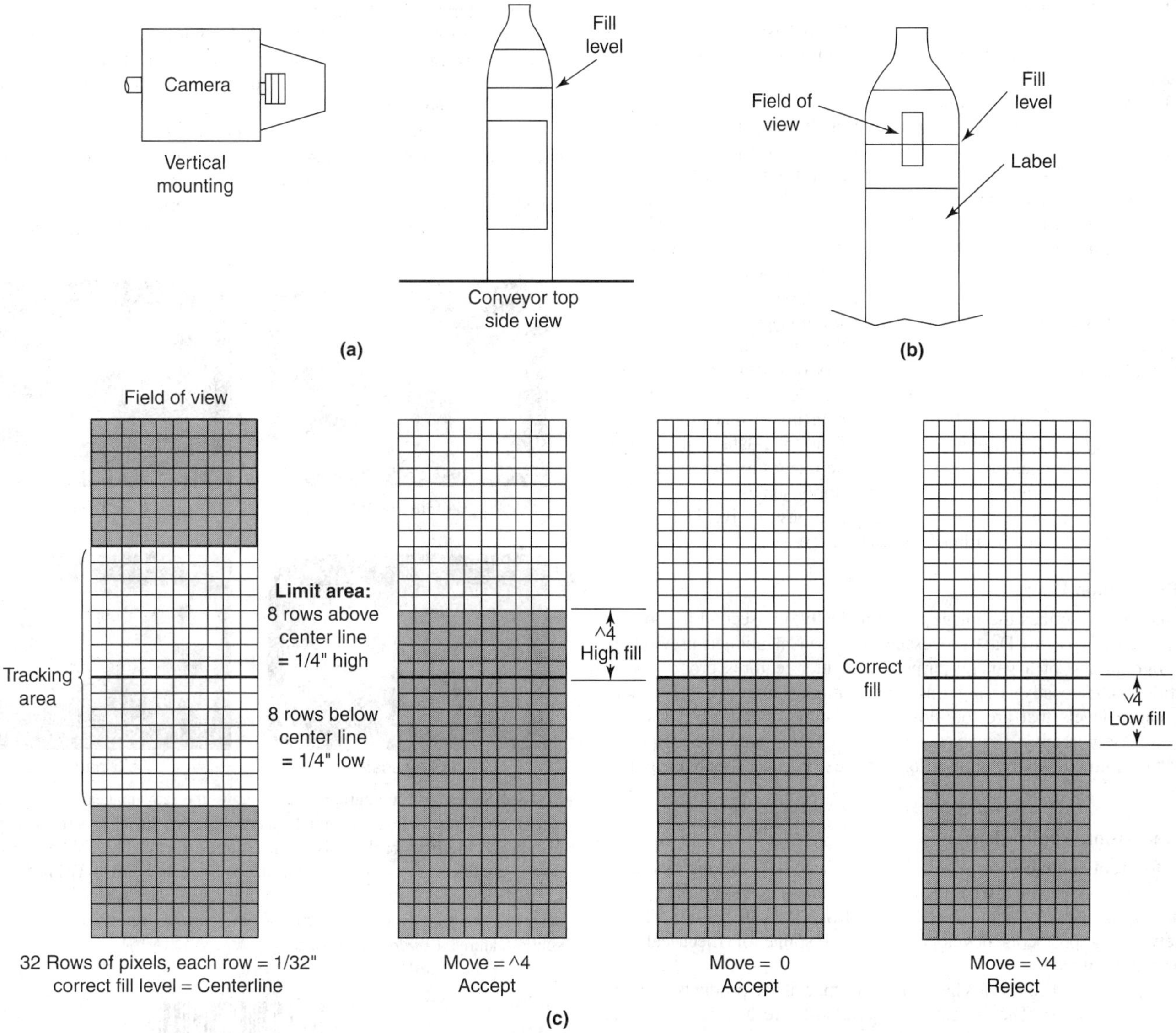

Fig. 3. Use of a Y-axis algorithm to check accuracy of filling: (**a**) arrangement of camera and conveyor juxtapositions; (**b**) location of field of view for given high and low levels; (**c**) reference images, showing high fill, correct fill, and low fill. (*Cutler-Hammer Products, Eaton Corporation.*)

surrounding pixels into those areas. There are several ways to segment an image, but all detract from accuracy. Four procedural options for segmentation may be considered:

1. *Algorithms*. In general, algorithms are a set of mathematical models that can be used to describe an image. A few years ago, the SRI algorithms were developed by the Stanford Research Institute. These consisted of about 50 different features that are extracted from a binary image, such as the size of blank areas (holes) or objects (blobs) and their *centroids* or *perimeters*.

2. *Neighborhood processing*. This is an averaging method, achieved by treating each pixel value as if it were part of a group or neighborhood. This gray-level data-reduction technique can change an otherwise complicated image into areas with well-defined lines. Each pixel's point value is calculated by considering the value of a neighboring pixel. The process effectively averages the scene into regions or areas.

3. *Convolution*. If an image is represented by the rate of light change per pixel instead of light density, the image will look like a line drawing because the greatest intensity of a pixel change is at the boundaries.

4. *String encoding*. In run-length, string, or connectivity encoding, the values of the first and last pixel positions of each scan line are compared to see if they are equal and, therefore, belong to the same region. The Tables generated by this process also consider the vertical changes in pixel state and, as may be expected, are rather short because most simple binary images contain relatively few transition borders.

5. *Associative pattern processing*. Rather than increasing the capability of processing large amounts of data in a shorter time span through the route of improving computer and data-handling techniques (hardware and software) as just described, a fundamentally different approach has three functional sections (as developed by APP (pattern processing technologies). (1) An image analyzer contains all the pixel data for the current image on the camera. (2) It converts the data into a unique statistical fingerprint, which is a unique set of response memory addresses. The response memory contains the statistical fingerprints for all trained images. (3) A matching histogram compares the fingerprints of the current image with those stored in the response memory. The theory of operation is based on a statistical phenomenon that reduces an image's data into a precise fingerprint, which is a set of values for an image that differs from other images and is a minute fraction of the total data of the image. Instead of one image being trained, all the images needed are trained and no programming is involved. One simply shows an image to the camera, allows the fingerprint memory locations to drive the response memory, and loads in a label for each image.

The Recognition Process

Once the user describes the image mathematically, the second half of the process step can occur. Data processing has been mentioned previously. The programming of a general-purpose computer requires programming expertise that is costly. Using this computational approach may entail performance levels that are too slow for real-time operation, unless the application only involves very basic procedures. Gains in processing power can be taken up quickly by application complexity with its increasing data demands.

Machine Vision Applications

Some form of actuation or control is the last step in implementing MV. Reasonably standardized methods of communication usually can be used with such common interfaces as RS-232 or IEEE-488. These are networking protocols developed by the Institute of Electrical and Electronics Engineers.

MV suppliers in recent years have turned to specialized products oriented to specific industries. The advent of generic software has been possible because of numerous application similarities.

MV applications essentially fall into four categories: (1) inspection, (2) location, (3) measurement, and (4) recognition.

MV can inspect to determine if parts are acceptable. Inspection tasks can be further categorized into one of four functions: (1) surface flaw detection, (2) presence or absence of certain features, (3) dimensional tolerance verification, and (4) shape verification. Three examples that use a gray scale are shown in Figs. 1, 2, 3, and 4.

Discrete-piece Identification — Bar Coding

The initial attempts to codify pieces, such as bank check numbers and addresses (zip codes) on mail, took the form of optical or magnetic character reading, and numerous systems of this type remain in place today. See also **Magnetic Ink Character Recognition (MICR)**; and **Optical Character Recognition (OCR)**. In principle, these systems now are a part of the overall concept of machine vision. However, more closely allied with machine vision is the machine-readable bar code, which depends only upon the line width and line location of printed bars, as shown in Fig. 5.

The lack of uniformity of parts, pieces, subassemblies, and so on, that persists in the discrete-piece manufacturing industries imposes a myriad of complexities, as just alluded to in the preceding description of machine vision. Thus MV applications, as generally defined, must be highly customized to very specific applications. This imposes a crucial cost restraint.

If, however, pieces and parts could be "packaged" in a way to simplify their recognition and exact identification, most of the complexities of MV can be eliminated. That is, of course, exactly what has been done for decades in terms of a large variety of manufactured end products, such as individual boxes of nuts and bolts, screws, cases of canned goods, boxed machine parts, containerized detergents, paints, hardware, plastic-encapsulated items, and even palletized groupings of identical merchandise.

The general concept of containerization, of course, originally was developed to simplify shipping and distribution to the end user. Later these packaging concepts became the basis for vastly improved inventory control and for automated warehousing by manufacturers, distributors, wholesalers, and, indeed, the final point of purchase.

With many millions of items produced, it became obvious that an extensive codification system was needed if piece identification were ever to be automated. Early trials with color coding proved that color-based systems were grossly inadequate. Entering from another direction (commercial and financial), the concepts of magnetic and optical coding

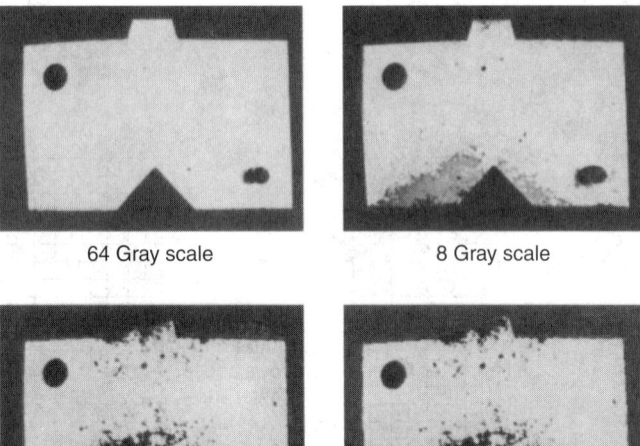

64 Gray scale   8 Gray scale

4 Gray scale   Binary

Fig. 4. Series of photographs that show the advantages of high-intensity resolution. With 64 levels of gray, for example (upper left), part boundaries are distinguishable from background. Fine features, such as the shadows cast by the post in the lower right-hand corner of the part, are discernible. Using fewer levels of gray (8,4, and 2 as shown) renders the scene more subject to the effects of shading. The slightly shaded portion of the part has deteriorated, as have other subtle features. (*Analog Devices, Inc.*)

Fig. 5. Universal bar code.

were developed and adopted rather quickly by financial institutions. However, early magnetic and optical systems required a certain closeness of the object with the detecting means, as, for example, in identifying a check or reading the address on an envelope. Specially designed alphanumeric characters have served financial and commercial needs, but it is expected that bar coding ultimately may be used.

In the late 1960s, the first supermarket checkout stand, developed by the Hughes Aircraft Company, was installed in a Los Angeles supermarket and operated successfully. The system did not get underway nationally for several years, however, because of the unwillingness of grocery manufacturers to print a bar code on each product package. By the late 1980s, however, the system had become a favorite of manufacturers and store operators. Uses also were found in automated warehousing and inventory accounting. Bar code readers became an important segment of the electronics industry.

Invention of the laser scanner has made it possible to read bar codes on boxes in a warehouse as far as 10 feet (3 meters) away. Well-designed bar code readers are considered to be very close to error free. Replacement of older style code readers with the hand-held wand has increased the acceptance and versatility of the system. Bar-coded tags for unpackaged materials, such as fabrics, have received wide acceptance. See also **Artificial Intelligence: Machine Vision**.

Additional Reading

Burke, M.: "Handbook of Machine Vision Engineering: Image Processing," Vol. 2, Chapman & Hall, New York, NY, 1997.

Cipolle, R., and A. Pentland: "Computer Vision and Human-Computer Interaction," Cambridge University Press, New York, NY, 1998.

Davies, E.R.R.: "Machine Vision: Theory, Algorithms Pacticalities," 2nd Edition, Academic Press, Inc., San Diego, CA, 1996.

Jain, R.C.: "Introduction to Machine Vision," The McGraw-Hill Companies, Inc., New York, NY, 2000.

Kanellopoulos, I., G.G. Wilkinson, and T. Moons: "Machine Vision and Adcanced Image Processing in Remote Sensing," Springer-Verlag, Inc., New York, NY, 1999.

Myler, H.R.: "Fundamentals of Machine Vision," SPIE International Society for Optical Engineering," Bellingham, WA.1999.

Parker, J.R.: "Algorithms for Image Processing and Computer Vision," John Wiley & Sons, Inc., New York, NY, 1996.

Pietikainen, M.K.: "Texture Analysis in Machine Vision," World Scientific Publishing Company, Inc., Riveredge, NJ, 2000.

Sanz, J.L.C.: "Image Technology: Advances in Image Processing, Multimedia and Machine Vision," Spriner-Verlag, Inc., New York, NY, 1996.

Sonka, M., V. Hlavac, and R. Boyle: "Image Processing, Analysis, and Machine Vision," Brooks/Cole Publishing Company, Pacific Grove, CA, 1999.

Zuech, N.: "Understanding and Applying Machine Vision," Marcel Dekker, Inc., New York, NY, 1999.

MACHINE WORD.   For a given computer, the number of information characters handled in each transfer. This number is usually fixed, but may be variable in some computers.

MACH NUMBER.   See **Aerodynamics and Aerostatics**; **Airplane**; and **Supersonic Aerodynamics**.

MACH PRINCIPLE.   The inertia of any system arises from the interaction of that system and the rest of the universe, including distant parts thereof. Postulated by applying the Mach criterion to the concept of absolute space. Applied by Einstein to the hypothesis that the metric of space-time is determined by the distribution of matter and energy.

MACKERELS (*Osteichthyes*).   Of the order *Scombridae*, mackerels have an elongated, fusiform (spindlelike) body, which is only slightly compressed. The tail typically has one to three longitudinal keels; it is long and rather thin. Scales are absent or tiny, sometimes forming a corselet on the front of the body. A wavy lateral line is present. The large head is tapered. The wide mouth opening extends at least to beneath the eyes; the jaws have large or small, sharp teeth. The wide gill opening has four gill arches. There are 31 to 61 vertebrae, and the vertebral column is well ossified.

There are 33 genera with a large number of species. All mackerels are epipelagic (high-sea) fishes. Many of them undertake extensive feeding migrations; a few species swim to coastal waters to spawn, while others are never in shallow water. The anatomy of these extremely fast swimmers reflects their great maneuvering ability. The pectoral and pelvic fins are recessed into shallow grooves. They click into place with a jerk of the fin base. The dorsal fin has a similar recess. The large majority of mackerels live in tropical and subtropical regions.

Atlantic Mackerel. This fish (*Scomberomorius scombrus*) is the best-known species. Its mature length ranges between about 10 and 20 inches (25 and 50 centimeters). This species is distributed from the Mediterranean and Black Sea along Europe's Atlantic coast to the arctic, and from there across the Atlantic Ocean to Labrador and the American East Coast to Cape Hatteras. The back is grass green, with numerous irregular dark stripes; the sides and underside have a mother-of-pearl hue with a reddish shimmer. The first and second dorsal fins are widely separated. The tail shaft lacks a median keel. The bony eye rings are well developed. There is no swim bladder. See Fig. 1.

Fig. 1.   Atlantic mackerel.

Like many other mackerels, the Atlantic mackerel occurs in dense schools just beneath the water surface, and sometimes there are so many of them together that they churn up the water at the surface. Since they lack a swim bladder, they can dive quickly to evade predators, such as sharks, tuna, and dolphins. They feed chiefly on small crustaceans, juvenile herring, sardines, and anchovies, as well as sand lances.

After the winter rest in deep waters, sexually mature mackerel seek coastal waters in April and May, where they spawn during the early summer months (March to April in the Mediterranean). They always migrate in schools. One female can lay up to 500,000 eggs, each of which has an oil bubble and floats on the water surface. The young hatch after about six days, and grow very rapidly. After 2 years, they are already over 7.8 inches (20 centimeters) long and at the end of the third year, they are about 11.8 inches (30 centimeters) long and reach sexual maturity.

Since they are highly valued because of their fine-tasting meat, Atlantic mackerels are commercially fished just about wherever they occur. Of all the numerous salt water food fishes, the Atlantic mackerel has presented marine researchers with one of the most fascinating and elusive puzzles. The curious periodic cycle of the fish, from scarcity to overabundance, has been known since the time of the American colonists in the mid-1600s. At that time, the fish had been established as an important staple commodity. Unpredictable fluctuations in the catches occurred from year to year and, as early as 1670, American settlers had enacted laws to prevent overfishing. In many parts of the world, changes in the mackerel's migration habits bring these same periodic fluctuations and thus affect the economy of fisheries and nearby ports.

Out of every million eggs one fish must survive to keep the population constant. To be considered a successful year, survival of more than four fish at the 2-inch (5-centimeter) size per million eggs is needed. In the egg and larval stages, many factors influence mortality. Adverse winds may push the waters in an unfavorable direction during the time when the fry lack sufficient motility and prevent them from reaching suitable nursery grounds. A lack of desirable zooplankton when needed reduces survival. Both conditions occurring together in one season are disastrous. Besides contending with the caprices of nature, mackerel are preyed upon by many forms of sea life.

As an example of poor survival, Atlantic mackerel mortality from the fertilized egg to a 2-inch (5-centimeter) stage is 99.9996%. Hence, a fluctuation in survival of several thousandths of 1% may make the difference between a dominant and a weak year. The mackerel catch in the Gulf of Maine may vary by a factor of 100 between a good year and a poor one.

Fresh mackerel is considered by many to be one of the choicest of food fishes. In the 1800s, mackerel were caught close inshore, dressed, and placed in tubs of salt water, which was changed frequently to keep the fish cool. The object was to catch the mackerel and get them to market before daylight, where they were sold in the cool of the morning.

Chub Mackerel. This fish (*Scomber japonicus*) is closely related to the Atlantic mackerel and occurs worldwide, but in less abundance. Chub mackerel are smaller and not as valuable commercially, except between the Yellow Sea and Sakhalin Island, where the fish has large commercial importance. It is the most fertile of the mackerel species, in that a single female can lay over 1 million eggs.

Rastrelliger mackerels are closely related to *S. japonicus*. For some Indian fisheries, the *Rastrelliger kanagurta* is practically as important as the oil sardine. Although not restricted to the west coast of India, like the sardine this species of mackerel forms a major fishery along that coast, where the fishing grounds and seasons of the two species overlap. This species moves in dense shoals into coastal waters toward the end of the southwest monsoon and is captured in shore seine (*rampani*), drift net, boat seine, and purse seine. When the catch of mackerel in the rampani is very large, the fish are impounded for up to a week by staking the net in a semicircle from the shore; the fish are then bailed out as per demand.

Spanish Mackerel. This fish (*Scomberomorius maculatus*) is actually a tuna and probably is the most elongated of tunas. These fishes often appear in huge schools in the Atlantic and Pacific Oceans. See Fig. 2. Other tunas are described in entry on **Tuna.**

See also **Fishes.**

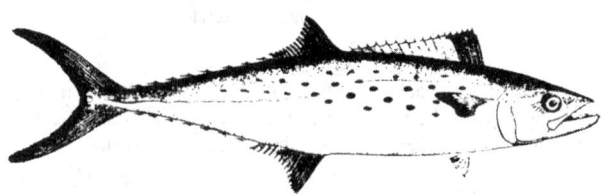

Fig. 2.   Spanish mackerel.

Additional Reading

Bond, C.E.: "Biology of Fishes," 2nd Edition, Harcourt Brace College Publishers, San Diego, CA, 1996.
Eschmeyer, W.N., C.J. Ferraris, M.D. Hoang, and D.J. Long: "Catalog of Fishes," California Academy of Sciences, San Francisco, CA, 1998.
Evans, D.H.: "The Physiology of Fishes," 2nd Edition, CRC Press, LLC., Boca Raton, FL, 1998.
Paxton, J.R. and W.N. Eschmeyer: "Encyclopedia of Fishes," 2nd Edition, Academic Press, Inc., San Diego, CA, 1998.

MACLAURIN SERIES.   A special case of the Taylor series. If $f(x)$ and all of its derivatives remain finite at $x = 0$, then $f(x)$ may be expanded to

$$f(x) = f(0) + f'(0)x + f''(0)\frac{x^2}{2!} + \cdots + f^{(n-1)}(0)\frac{x^{n-1}}{(n-1)!} + R_n$$

where R_n is the remainder after n terms. When the remainder converges towards zero as the number of terms increases, the result is the Maclaurin series for $f(x)$ at $x = 0$.

Equations for R_n are:

$$\frac{x^n}{n!}f^{(n)}(\theta x), 0 < \theta < 1$$

$$\frac{1}{(n-1)!}\int_0^x f^{(n)}(x-t)t^{n-1}dt$$

See also **Series.**

MACLEOD EQUATION.   A constant relationship between the surface tension of a liquid, its density, and the density of its vapor, of the form:

$$\frac{\gamma^{1/4}}{\rho - \rho'} = \text{constant}$$

where γ is the surface tension of the liquid, ρ is its density, and ρ' is the density of its vapor.

MACROASSEMBLER.   An assembler which permits the user to define pseudo computer instructions, which may generate multiple computer instructions when assembled. The source statements that may generate multiple computer instructions are termed *macrostatements* or *macroinstructions*. On a process-control digital computer, a macroassembler can be a most significant tool. By defining a set of macrostatements, for example, a process-control engineer can define a process-control programming language that is specifically oriented to the process.

Macrolibraries. This is comprised of a set of defined macrostatements and the associated computer instructions. A macrolibrary results for

example, when a process-control engineer defines a language for a system. The effectiveness of a macroassembler depends largely on the ease of creating, manipulating, modifying, and linking various macrolibraries.

See also **Assembler (Computer System).**

MACROMOLECULAR SCIENCE.   Macromolecules or polymers are now such a common feature of modern life that it is sometimes forgotten that the rapid development of such materials in industry has taken place in the last 60 years. In fact, the idea that such molecules exist was proposed by Staudingser, Mark, and coworkers in the 1920s and it was some time before these concepts were accepted by the scientific community.

Polymers are long-chain molecules synthesized by the linking together of a large number of identical or similar small units termed *monomers*. The large size of macromolecules leads to their many useful properties. Macromolecular science is the study of these properties in relation to their chemical and physical structure. Naturally occurring or biological macromolecules, such as proteins, carbohydrates, nucleic acid, and natural rubber, are studied by the same methods as synthetic polymers, resulting in a number of industrial and biomedical applications.

In response to the needs of industry, a number of centers for graduate study in macromolecular science have developed. The first such center was at Brooklyn Polytechnic Institute (now Polytechnic University) and large centers now exist at Case Western Reserve University in Cleveland Ohio (Fig. 1), University of Massachusetts (Amherst) and University of Akron (Ohio). The first accredited engineering undergraduate degree program in polymer science was developed in the Department of Macromolecular Science at Case Western Reserve University.

In the past, research emphasis has centered on the synthesis of new macromolecules. This is still a major emphasis, especially as applied to the production of new polymers or composite systems having unique electrical and/or mechanical properties and to the synthesis and fabrication of unique transport and barrier membranes. Such research areas relate directly to the current emphasis on high tech, high price products rather than on bulk commodity polymers.

Fig. 1.   Automated Langmuir-Blodgett deposition station used for production of ultrathin films and located in a clean room at the Polymer Microdevice Laboratory, Case Western Reserve University, Cleveland, Ohio.

At least equally important in aiding in the solution of current problems of society has been the development, primarily over the past ten years, of the fields of polymer physics and engineering. The general thrust of this development has been to gain an understanding of the physical structure and morphology of macromolecules and to develop methods to modify these parameters to produce useful properties in both polymer and composite systems. See accompanying illustration. See also **Molecular and Supermolecular Electronics**; and **Colloid System.**

JOHN BLACKWELL, Case Western Reserve University, Cleveland, OH.

MACROPHAGE.   See **Immune System and Immunology.**

MACROSCOPIC. Large enough to be visible to the naked eye or under low order of magnification.

MACULA. The macula is a small area, less than $\frac{1}{4}$ inch, in the center of the retina at the back of the eye. It is responsible for sharp, clear central vision and the ability to perceive color.

Like the film in a camera, the retina receives light rays from the front of the eye and transmits those light rays through the optic nerve to the brain where the rays are converted into images. The densely packed photoreceptor (light-sensitive) cells in the macula control all of the eye's central vision and are responsible for the ability to read, drive a car, watch television, see faces, and distinguish detail. The rest of the retina handles peripheral vision that enables your eyes to see objects off to the side while you are looking forward.

There are two types of photoreceptor cells in the cornea: rods and cones. The rods provide vision at low light levels, while the cones provide sharp vision and discrimination. Because the macula contains a high concentration of cones, straight-ahead vision is in sharp focus, particularly in bright light. Most of the rods are located in the periphery of the retina, so faint objects are more visible if you do not look directly at them. A dim star, for instance, is best seen when your eyes are not aimed directly at it.

The most common cause of functional blindness in people over the age of 60 is macular degeneration, a deterioration or breakdown of the macula. Damage to the macula results in the loss, either partial or complete, of ability to see objects clearly in the center of vision. Although not totally blind, the person has difficulty performing tasks that require "straight-on" vision, such as driving a car, reading, or watching television. Because peripheral vision is not affected, the person can adapt somewhat to the loss of central vision and continue to pursue some normal daily activities, such as walking, without assistance.

There are two types of macular degeneration. The "dry" form is usually the result of aging and thinning of the macula's layers, and the "wet" form occurs when abnormal blood vessels under the retina leak fluid and blood, causing scarring. Vision loss with dry macular degeneration occurs gradually over a number of years, and the affected person may not be aware of any problem. Dry macular degeneration is the less serious of the two forms. With the wet form of this disease, central vision capabilities can be damaged rapidly. Early detection usually results in more successful treatment. See also **Retina**.

Vision Rx, Inc., Elmsford, NY.

MADAR. See **Silk Cotton Trees**.

MADDER. See **Rubiaceae**.

MADRONE TREE. See **Heather Shrubs and Trees**.

MAFIC. Said of an igneous rock composed chiefly of one or more ferromagnesian, dark-colored minerals in its mode; also, said of those minerals.

MAGELLAN MISSION TO VENUS. The *Magellan* spacecraft, named after the sixteenth-century Portuguese-born explorer whose expedition first circumnavigated the Earth. NASA's *Magellan* spacecraft used sophisticated imaging radar to make the most highly detailed maps of Venus ever captured during its four years in orbit around Earth's sister planet from 1990 to 1994.

After concluding its radar mapping, *Magellan* made global maps of Venus's gravity field. Flight controllers also tested a new maneuvering technique called aerobraking, which uses a planet's atmosphere to slow or steer a spacecraft.

Craters shown in the radar images that *Magellan* sent to Earth tell scientists that Venus's surface appears relatively young—resurfaced about 500 million years ago by widespread volcanic eruptions.

The planet's present harsh environment has persisted at least since then, with no features detected suggesting the presence of oceans or lakes at any time in the planet's past.

Scientists also found no evidence of plate tectonics, the movements of huge crustal masses on Earth that cause earthquakes and result in the drifting of continents over time spans of hundreds of millions of years.

Magellan's mission ended with a dramatic plunge to the planet's surface, the first time an operating planetary spacecraft has ever been intentionally crashed. Contact was lost with the spacecraft October 12, 1994, at 10:02 Universal Time (3:02 a.m. Pacific Daylight Time). The purpose of the maneuver was for Magellan to gather data on Venus's atmosphere before it ceased to function during its fiery descent.

Mission Overview

Magellan was the first planetary spacecraft to be launched by a space shuttle. The shuttle *Atlantis* from Kennedy Space Center in Florida carried aloft *Magellan* on May 4, 1989. *Atlantis* took *Magellan* into low Earth orbit, where it was released from the shuttle's cargo bay. A solid-fuel motor called the Inertial Upper Stage (IUS) then fired, sending *Magellan* on a 15-month cruise looping around the Sun 1-1/2 times before it arrived at Venus on August 10, 1990. A solid-fuel motor on *Magellan* then fired, placing the spacecraft in orbit around Venus.

Magellan's initial orbit was highly elliptical, taking it as close as 294 kilometers (182 miles) from Venus and as far away as 8,543 kilometers (5,296 miles). The orbit was a polar one, meaning that the spacecraft moved from south to north or vice versa during each looping pass, flying over Venus's north and south poles. *Magellan* completed one orbit every 3 hours, 15 minutes.

During the part of its orbit closest to Venus, *Magellan's* radar mapper imaged a swath of the planet's surface approximately 17 to 28 kilometers (10 to 17 miles) wide. At the end of each orbit, the spacecraft radioed back to Earth a map of a long ribbon-like strip of the planet's surface captured on that orbit. Venus itself rotates once every 243 Earth days. As the planet rotated under the spacecraft, *Magellan* Collected strip after strip of radar image data, eventually covering the entire globe at the end of the 243-day orbital cycle.

By the end of its first such eight-month orbital cycle between September 1990 and May 1991, *Magellan* had sent to Earth detailed images of 84 per-cent of Venus's surface. The spacecraft then conducted radar mapping on two more eight- months cycles from May 1991 to September 1992. This allowed it to capture detailed maps of 98 percent of the planet's surface. The follow-on cycles also allowed scientists to look for any changes in the surface from one year to the next. In addition, because the "look angle" of the radar was slightly different from one cycle to the next, scientists could construct three-dimensional views of Venus's surface.

During *Magellan's* fourth eight-month orbital cycle at Venus from September 1992 to May 1993, the spacecraft collected data on the planet's gravity field. During this cycle, *Magellan* did not use its radar mapper but instead transmitted a constant radio signal to Earth. If it passed over an area of Venus with higher than normal gravity, the spacecraft would slightly speed up in its orbit. This would cause the frequency of *Magellan's* radio signal to change very slightly due to the Doppler effect—much like the pitch of a siren changes as an ambulance passes. Thanks to the ability of radio receivers in the NASA/JPL Deep Space Network to measure frequencies extremely accurately, scientists could build up a detailed gravity map of Venus.

At the end of *Magellan's* fourth orbital cycle in May 1993, flight controllers lowered the spacecraft's orbit using a then-untried technique called aerobraking. This maneuver sent *Magellan* dipping into Venus's atmosphere once every orbit; the atmospheric drag on the spacecraft slowed down *Magellan* and lowered its orbit. After the aerobraking was completed between May 25 and August 3, 1993, *Magellan's* orbit then took it as close as 180 kilometers (112 miles) from Venus and as far away as 541 kilometers (336 miles). *Magellan* also circled Venus more quickly, completing an orbit once every 94 minutes. This new, more circularized orbit allowed *Magellan* to collect better gravity data in the higher northern and southern latitudes near Venus's poles.

After the end of that fifth orbital cycle in April 1994, *Magellan* began a sixth and final orbital cycle, collecting more gravity data and conducting radar and radio science experiments. By the end of the mission, *Magellan* captured high-resolution gravity data for about 95 percent of the planet's surface.

In September 1994, *Magellan's* orbit was lowered once more in another test called a "windmill experiment." In this test, the spacecraft's solar panels were turned to a configuration resembling the blades of a windmill, and *Magellan's* orbit was lowered into the thin outer reaches of Venus's dense atmosphere. Flight controllers then measured the amount of torque control required to maintain *Magellan's* orientation and keep it from spinning. This experiment gave scientists data on the behavior of molecules in Venus's upper atmosphere, and lent engineers new information useful in designing spacecraft.

On October 11, 1994, *Magellan's* orbit was lowered a final time, causing the spacecraft to become caught in the atmosphere and plunge to the surface; contact was lost the following day. Although much of Magellan was believed to be vaporized, some sections probably hit the planet's surface intact.

Venus

One of the handful of planets known to the ancients, Venus is often called Earth's sister planet because of its similar size and distance from the sun. Earth is 12,756 kilometers (7,926 miles) in diameter, compared to Venus at 12,103 kilometers (7,520 miles); Earth orbits the sun at an average 149.6 million kilometers (93 million miles), compared to Venus at 108.2 kilometers (67.2 million miles). The two planets' densities are also similar — 5.52 grams per cubic centimeter for Earth, compared to 5.24 grams per cc for Venus. Because Venus is closer to the sun than Earth is, it always appears close to the sun from our point of view as either a glistening, bright evening or morning "star." See also **Venus**.

Despite the similarities, however, in other ways Venus is very much unlike Earth. Venus has a surface temperature of about 470 degrees Celsius (about 900 degrees Fahrenheit); the atmospheric pressure at the surface is 90 times greater than Earth's. Venus's atmosphere is nearly devoid of water, made up of 97 percent carbon dioxide; its upper clouds contain sulfuric acid. Venus has no moons, and no magnetic field has been detected. It rotates on its axis in a retrograde direction — that is, opposite that of Earth and most of the other planets — very slowly, once every 243 Earth days.

The first spacecraft mission to another planet was the NASA/JPL spacecraft *Mariner 2*, which executed a flyby of Venus in December 1962. Other U.S. spacecraft to visit Venus have included *Mariner 10*, which flew by Venus in 1974 on its way to Mercury in the first mission to more than a single planet; and *Pioneer Venus*, a 1978 mission that included an orbiter with an altimeter and imaging radar that functioned at lower resolution than *Magellan's*, as well as multiple probes that descended into Venus's atmosphere. The then-Soviet Union also sent a number of spacecraft to Venus, including four — *Venera 9, 10, 13* and *14* — that landed on the surface and made closeup pictures of the rocky terrain briefly before the searing heat caused them to stop functioning. Two other Soviet missions, *Venera 15* and *16*, used orbiting imaging radar similar to *Magellan's* but at a lower resolution.

The Magellan Spacecraft

Built partially with spare parts from other missions, the *Magellan* spacecraft was 4.6 meters (15.4 feet) long, topped with a 3.7-meter (12-foot) high-gain antenna. Mated to its retrorocket and fully tanked with propellants, the spacecraft weighed a total of 3,460 kilograms (7,612 pounds) at launch.

The high-gain antenna, used for both communication and radar imaging, was a spare from the NASA/JPL *Voyager* mission to the outer planets, as were *Magellan's* 10-sided main structure and a set of thrusters. The command data computer system, attitude control computer and power distribution units are spares from the *Galileo* mission to Jupiter. *Magellan's* medium-gain antenna is from the NASA/JPL *Mariner 9* project. Martin Marietta Corp. was the prime contractor for the *Magellan* spacecraft, while Hughes Aircraft Co. was the prime contractor for the radar system. See Fig. 1.

Magellan was powered by two square solar panels, each measuring 2.5 meters (8.2 feet) on a side; together they supplied 1,200 watts of power. Over the course of the mission the solar panels gradually degraded, as expected; by the end of the mission in the fall of 1994 it was necessary to manage power usage carefully to keep the spacecraft operating. See Fig. 2.

Imaging Radar

Because Venus is shrouded by a dense, opaque atmosphere, conventional optical cameras cannot be used to image its surface. Instead, Magellan's imaging radar uses bursts of microwave energy somewhat like a camera flash to illuminate the planet's surface.

Magellan's high-gain antenna sends out millions of pulses each second toward the planet; the antenna then collects the echoes returned to the spacecraft when the radar pulses bounce off Venus's surface. The radar pulses are not sent directly downward but rather at a slight angle to the side of the spacecraft, the radar is thus sometimes called "side-looking radar." In addition, special processing techniques are used on the radar data to result in higher resolution as if the radar had a larger antenna, or "aperture"; the technique is thus often called "synthetic aperture radar," or SAR.

Synthetic aperture radar was first used by NASA, on JPL's Seasat oceanographic satellite in 1978; it was later developed more extensively on the Spaceborne Imaging Radar (SIR) missions on the space shuttle in 1981, 1984 and 1994. Another imaging radar mission is also planned

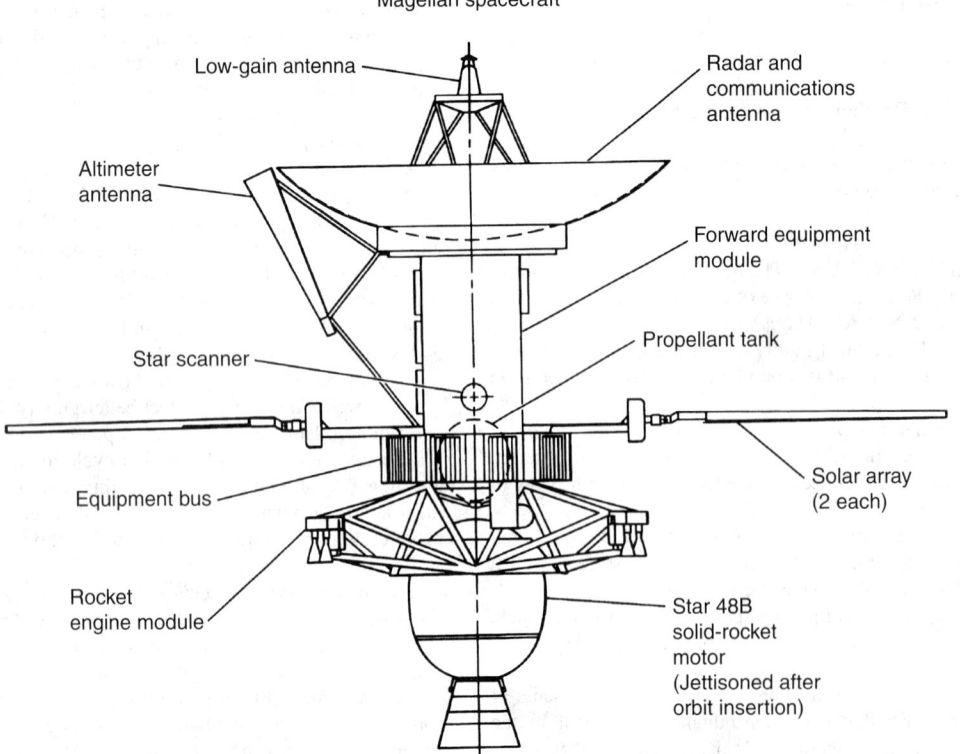

Fig. 1.   *Magellan* spacecraft (*NASA, GSFC*).

Magellan spacecraft

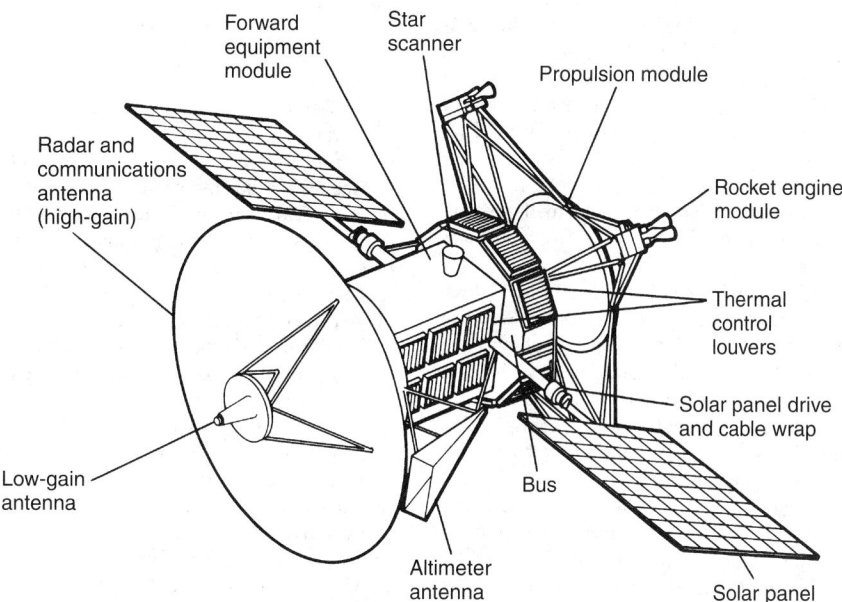

Fig. 2. *Magellan* spacecraft with solar panels extended (*NASA, GSFC*).

as part of the NASA/JPL *Cassini* mission to Saturn in 1997 to map the surface of the ringed planet's major moon Titan. See also **Cassini Mission to Saturn**; and **Saturn**.

Besides its use in imaging, *Magellan's* radar system was also used to collect altimetry data showing the elevations of various surface features. In this mode, pulses were sent directly downward and Magellan measured the time it took a radar pulse to reach Venus and return in order to determine the distance between the spacecraft and the planet.

Science Results

Magellan returned maps of Venus's surface and its gravity field in unprecedented detail that will be a resource for many years for scientists studying the planet. The mission held many surprises for scientists, and resulted in a number of theories about the planet being revised.

From the craters visible in Magellan's Venus maps, scientists believe they are looking at a relatively young planetary surface, perhaps about 500 million years old. Since Venus formed at the same time as Earth 4.6 billion years ago, some event or events 500 million years ago must have resurfaced the planet. Scientists believe that this may have been the work of massive outpourings of lava from planet-wide volcanic eruptions. Although Venus may still have active volcanoes, no visible outpourings of lava have yet been detected in comparisons of *Magellan* images between one eight-month orbital cycle and another.

Although some scientists speculate that Venus may have once been a temperate planet that fell victim to a runaway greenhouse effect creating enormously high temperatures, *Magellan's* maps show no telltale signs of past major water bodies such as shorelines or ocean basins. Also, surface features show no evidence of being eroded by water—although there is evidence of wind erosion in the form of numerous sand dunes and wind streaks.

One of the hopes that scientists had for *Magellan* was to find out if Venus, like Earth, has plate tectonics—movements of crustal masses that on Earth cause earthquakes and result in the drifting of continents over time periods of hundreds of millions of years. They in fact found no evidence of plate tectonics in the data returned by the mission. Scientists suspect that, although Venus is very similar in size to Earth, its interior is probably different in major ways. In particular, Venus seems to lack an "asthenosphere," a buffer layer within Earth between the outer part of the planet and the mantle beneath. As a result, the gravity signature of features on Venus closely reflect surface topography, whereas on Earth such a correspondence does not always occur.

Scientists are also intrigued by the distribution of volcanoes around Venus. On Earth, volcanoes occur in groups such as the so-called "Ring of Fire" around the Pacific Rim. Venus, by contrast, is peppered with hundreds of thousands to millions of volcanoes distributed more or less randomly around the planet. Scientists were also surprised to see huge channels thousands of kilometers long on Venus. These appear to be lava channels, and frequently show a fan of lava at their mouths.

Additional Reading

Arvidson, R.E. et al.: "Magellan: Initial Analysis of Venus Surface Modification," *Science*, 270 (April 12, 1991).

Head, J.W. et al.: "Venus Volcanism: Initial Analysis from Magellan Data," *Science*, 276 (April 12, 1991).

Jenkins, J., P. Steffes, J. Twicken, D. Hinson, and G.L. Tyler: "Radio Occultation Studies of the Venus Atmosphere with the Magellan Spacecraft: 2. Results from the October 1991 Experiment," *Icarus* 110, 79–94, 1994.

Pettengill, G.H. et al.: "Magellan Radar Performance and Data Products," *Science* 260 (April 12, 1991).

Saunders, R.S. etal: "The Magellan Venus Radar Mapping Mission," *J. Geophys. Res.* 95, No. B6, 8339–8355, (June 1990).

Saunders, R.S., G.H. Pettengill: "Magellan: Mission summary," *Science* 252, 247–249, (April 12, 1991).

Saunders, R.S. etal: "Magellan Mission Summary," *J. Geophys. Res.* 97, No. E8, 13067–13090, (Aug. 1992.)

Steffes, P., J. Jenkins, R. Austin, S. Asmar, D. Lyons, E. Seale, and G.L. Tyler: "Radio Occultation Studies of the Venus Atmosphere with the Magellan Spacecraft: 1. Experimental Description and Performance," *Icarus* 110, 71–78, 1994.

Web References

Magellan Fact Sheet: *http://nssdc.gsfc.nasa.gov/planetary/factsheet.html*

Magellan Mission to Venus: *http://nssdc.gsfc.nasa.gov/planetary/magellan.html*

MAGGOT *(Insecta, Diptera).* The soft-bodied larva of many species of two-winged flies. Maggots are often white, but some species are brightly colored. No organs of locomotion are present. Most maggots hatch from the egg in the midst of an abundant food supply of decaying organic matter or living plant or animal tissues and are able to move about sufficiently for their needs by wriggling the body.

The *apple maggot* (*Rhagoletis promnella*) burrows into fruit, distorts the shape, and causes it to rot and drop prematurely. The maggot is yellowish-white in color and ranges up to $\frac{3}{8}$ inch (9–10 millimeters) in length. The adult version is the black fly. The apple maggot is found in the United States from the Dakotas to New England and from the southeastern Canadian border into Arkansas, Ohio, and Georgia. However, the pest is uncommon in the southern parts of this range.

This maggot, one of the most serious insect pests of apple, either ruins the fruit entirely, or makes it unappetizing for consumption. Heavily infested fruit will be reduced to a brown, rotted mass, filled with yellowish, legless maggots. When the fruit is slightly infested, there is no external indication of maggots within. However, when the fruit ripens, burrows made by the maggots show as dark lines under the skin of the fruit. Larvae in prematurely dropped fruit can continue to live and become adult flies that will reinfest the fruit on the trees. Thus, it is important to remove and burn such fruit immediately after it has dropped. The adult flies appear in late June and early July, at which time they insert eggs under the skin of the fruit. Hibernation occurs in small puparia located just below the fruit surface. The sweet and subacid varieties of apple are the most frequently attacked varieties.

The *artichoke stem maggot* (*Straussia longipennis*) is a small, yellow-colored maggot which bores into the pith of the stems. The adults are yellow flies with two banded wings.

The *cabbage root maggot* (*Pegomya brassicae*) is a headless, legless, white maggot, about $\frac{1}{4}$ inch (6 millimeters) long, which destroys seed in the soil while also attacking the underground parts of plants which have germinated. For control, a suitable chemical insecticide, such as diazinon or chlordane, should be applied to the soil at the base of the plants when the leaves appear; the treatment should be repeated soon after transplanting or thinning. Such chemicals should not be applied to the edible parts once they are formed on the cabbage. This maggot also attacks radish.

The *onion maggot* (*Phorbia cepetorum*) is a legless, white, root-eating maggot which attains a length of about $\frac{1}{3}$ inch (8 millimeters). Distribution is in the northern United States. A dust spray containing malathion provides effective control, but this should not be applied within three days of harvesting.

The *orange maggot* (*Trypeta ludens*) is a dirty-white maggot that attains a length of about $\frac{1}{2}$ inch (12–13 millimeters). The maggot burrows into the pulp of the fruit and up to 20 maggots may be found in a single orange. The adult version is a light-yellow fly with brown markings and bands on the wings. The orange maggot is particularly serious in the Mexican citrus groves. Control is essentially by picking infested fruit and immediately destroying to prevent reinfestation.

The *mushroom maggot* (*Sciara sp.*) is a small maggot, white to yellow in color, with a black head. Treatment is mainly by prevention, keeping flies out of the mushroom growing area, fumigating regularly, and sterilizing the manure growing medium by heating to at least 150 °F (66 °C).

The *raspberry cane maggot* (*Phorbia rubivora*) is a small maggot, white, that burrows in new canes and girdles the shoot. During April and May, a fly deposits the eggs.

The *seed-corn maggot* (*Pegomya fusciceps*) attacks the germinating seeds and roots of many plants, notably bean and pea. This maggot is headless, legless, whitish in color. The best prevention is to use seed that have been commercially treated for seed-corn maggot control.

Other maggots that are quite destructive to food crops include the rice-stem maggot and the seed maggot.

MAGMA. The term for molten material. A natural, complex, liquid, high-temperature, silicate solution ancestral to all igneous rocks, both intrusive and effusive. The locus of a magma is within the lithosphere (crust) under great pressure and an impenetrable cover which helps the magma to retain its original gases and water vapor in solution. The origin of magma is not known but it is generally assumed that separate magma chambers may exist within the lithosphere.

MAGNESITE. The mineral magnesite is carbonate of magnesium, $MgCO_3$. It is a hexagonal mineral, but usually found massive. It has a rhombohedral cleavage; conchoidal fracture; brittle; hardness, 3.5–4.5; specific gravity, 3.75–4.25; luster, vitreous to dull; color, white, gray, yellow, or brown; transparent to opaque. Most magnesite is believed to have been derived from the action of carbonated waters upon rocks rich in magnesium. Magnesium-bearing waters, on the other hand, may have in some cases acted upon calcite or dolomite. Magnesite deposits are known in Greece, Austria, Norway, India, Australia, and the Republic of South Africa. In the United States, magnesite is found in California and Nevada, some of which deposits seem to be of original sedimentary character. Magnesite is in demand for the manufacture of refractories and various compounds of magnesium.

MAGNESIUM. Chemical element, symbol Mg, at. no. 12, at. wt. 24.305, periodic table group 2, mp 649 °C, bp 1,090 °C, critical temperature (calculated) 1,867 °C, density 1.74 g/cm³ (20 °C), 1.64 g/cm³ (solid at 650 °C), 1.57 g/cm³ (liquid at 650 °C). Elemental magnesium has a close-packed hexagonal crystal structure, as do the common alloys of magnesium except those that contain lithium in excess of 11%.

Magnesium is a silver-white metal, malleable and ductile when heated; unattacked by dry oxygen, by H_2O or alkalis at room temperature; when heated to about 800 °C reacts in air or steam and emits a brilliant white light of high actinic power; reactive with acids including carbonic at room temperature; reactive upon heating with nitrogen, phosphorus, arsenic, sulfur, in some cases with such vigor as to constitute a hazard.

Magnesium occurs extensively in the earth's crust, ranking 8th among the chemical elements in terrestrial abundance. An average composition of igneous rocks contains 2.09% magnesium. Of the elements present in seawater, magnesium ranks 5th with an estimated 6,125,000 tons of magnesium per cubic mile (1,323,000 metric tons per cubic kilometer) of seawater, its content exceeded only by hydrogen, oxygen, sodium, and chlorine. Magnesium is a constituent of over 150 minerals and also is found in bitterns and subterranean brines and salt beds. Only a few magnesium minerals are important commercially, notably dolomite, $CaO \cdot MgO \cdot 2CO_2$, as a source of magnesium. See also **Dolomite**. More than half of metallic magnesium produced is extracted from seawater. There are three naturally occurring isotopes, ^{24}Mg through ^{26}Mg; and three radioactive isotopes have been identified, ^{23}Mg, ^{27}Mg, and ^{28}Mg, all with comparatively short half-lives measured in seconds, minutes, or hours. The first known magnesium compound to be isolated was Epsom salt, $MgSO_4$, which Nehemiah Grew obtained in 1695 by evaporating the mineral waters at Epsom, England. In 1754, Joseph Black demonstrated that magnesia and lime were two different substances, but the exact identify of magnesia was not reported until 1808 by Sir Humphrey Davy who demonstrated that magnesia was an oxide of a heretofore unknown element. He first termed the element magnium. Metallic magnesium was first isolated by A. Bussy in 1828 when he fused magnesium chloride with potassium. Michael Faraday produced the first magnesium metal electrolytically in 1883. First ionization potential 7.64 eV; second, 14.97 eV. Oxidation potential Mg → $Mg^{2+} + 2e^-$, 2.375 V; $Mg + 2OH^- \rightarrow Mg(OH)_2 + 2e^-$, 2.67 V. Other important physical properties of magnesium are given under **Chemical Elements**.

Production

There are two principal magnesium production processes: (1) electrolytic, and (2) thermal. Electrolytic processes account for 80% of commercial production. In this process, seawater is pumped into large settling tanks where it is treated with lime. Roasted oyster shells sometimes are used if a convenient source is nearby. The lime precipitates the magnesium as the insoluble hydroxide. The hydroxide is filtered and then converted into a slurry with fresh H_2O. Subsequent treatment with HCl converts the $Mg(OH)_2$ into $MgCl_2$. The latter compound is dried and then electrolyzed in the fused state to produce molten magnesium and chlorine gas. The latter is recycled. The magnesium is cast into ingots. In the thermal or ferrosilicon process, used in some European countries, a mixture of magnesium oxide and powdered ferrosilicon (an iron-silicon alloy) is fed into a retort and heated under vacuum to about 1,200 °C. The magnesium is freed in the form of vapor and condenses into crystals at the cool end of the retort. The crystals then are remelted and cast into pigs.

Uses of Magnesium

Magnesium finds principal uses as a primary metal to which other metals are added in various alloying amounts to enhance the properties of magnesium. Magnesium is the lightest of all structural metals and consequently the metal has enjoyed much attention over the years in connection with the transportation industry, notably for applications in the aircraft, aerospace, and automotive industries. Vehicle designers constantly are aware of the additional power requirements for simply moving "dead weight" that wastes fuel and contributes to air pollution.

In addition to its use as a structural metal, magnesium is an important metallurgical chemical in the form of a deoxidizer and desulfurizer and as the constituent of numerous industrial and laboratory chemical compounds.

Magnesium Alloys. Even prior to the use of magnesium as a structural metal in the aerospace field, in 1921, Louis Chevrolet put a set of magnesium-alloy pistons in the Ford racing car that won the Indy 500 for him that year. The magnesium pistons gave racing and sports cars faster acceleration and deceleration. This application of magnesium was not intended so much as a dead-weight savings feature for the car, but rather more in terms of inertia (obviously also relative to weight). Although the magnesium pistons provided better acceleration/deceleration because of smaller inertia, the early designers encountered what is known as piston slap, which results when the piston material has a considerably higher coefficient of thermal expansion than the cylinder material does.

The use of magnesium castings for auto wheels was introduced a few years later and also serves the principal purpose of reducing inertia. With wheels, it is not just faster acceleration/deceleration that can be achieved, but also minimizing the amount of unsprung weight for a smoother and easier-to-control ride and minimizing the problem of gyroscopic action of the rapidly spinning wheels. Designers of racing cars switched from wire-spoke wheels to magnesium-alloy wheels in the early 1950s. The use of magnesium has increased not only in racing cars, but some passenger cars, both for the purpose of reducing inertia and weight. Today, magnesium is used for transmission and differential housings and a variety of other racing car parts. Serious attention continues to be given to major engine components, such as the cylinder block, head, and oil sump, all of which are candidates for reducing dead weight and increasing fuel economy.

The most extensive use of magnesium castings in automobiles commenced in 1936, with the introduction of the Volkswagen Beetle. Each Beetle used from 40 to 50 pounds (18 to 23 kg) of primary magnesium ingot plus scrap metal.

Magnesium has a density only $\frac{2}{3}$ that of aluminum, $\frac{1}{4}$ that of zinc, and about $\frac{1}{5}$ that of irons and steels. In addition to the obvious aerospace and automotive applications, other applications include hand trucks, containers, materials-handling equipment, portable electric and pneumatic tools (such as chain saws), hand tools, luggage, sporting goods, dockboards, and tooling jigs and fixtures. It has been found that lighter-weight equipment significantly reduces accidents and lost time due to injuries. On an arbitrary scale, where the power required to machine magnesium alloys is 1.0, the Figures for other metals are: aluminum alloys, 1.8; brass, 2.3; cast iron, 3.5; mild steel, 6.3; and nickel alloys, 10.0.

Some magnesium alloys are listed in Tables 1 and 2.

Magnesium in Other Metal Alloys. Magnesium is an important alloying ingredient in the production of other base metal alloys. When added during metallurgical processing, magnesium in small amounts has a marked effect on final properties of the metals:

Aluminum — Magnesium increases resistance to corrosion, facilitates heat treatment, and increases most mechanical properties. If magnesium-containing aluminum is remelted, the magnesium may be

TABLE 1. REPRESENTATIVE MAGNESIUM ALLOYS

Alloy Designation	Elements Added	Tensile Strength 1,000 psi	Brinell Hardness	Melting Point °C	Forms Available	Features
AZ31B	3% Al 1% Zn	29	49	627	Sheet, plate, extrusions, forgings.	Moderate strength, good formability, general-purpose alloy. Dent resistant, weldable.
AZ91B	9% Al 0.6% Zn	33	67	596	Die casting alloy.	Good strength and castability. Popular for portable tools, business machines, vehicles.
AZ91C	8.7% Al 0.7% Zn	40	53	596	General-purpose sand and permanent-mold casting alloy.	Good castability, pressure tightness, and weldability. Moderate strength.
HK31A	3% Th 0.7% Zn 0.7% Zr	38	57	649	Sheet and plate for aerospace uses. (200–370°C). Sand and permanent-mold castings.	Good short-time, elevated temperature characteristics. Weldable without stress relief. Low microporosity in cast form.
HM21A	0.6% Mn 2% Th	35	56	650	Sheet, plate, forgings for aerospace uses. (200–425°C)	Very stable at elevated temperatures. Good creep strength and formability. Weldable without stress relief.
HM31A	1.2% Mn (min) 3% Th	44	63	605	Extrusions for aerospace uses. (200–425°C)	Excellent elevated temperature properties. Weldable without stress relief.
QE22A	2% Pr 0.7% Zr 2.5% Ag	40	78	549	Castings for aerospace uses. (up to 260°C)	Superior tensile strength plus excellent creep and fatigue strength.
ZK60A	5.7% Zn 0.5% Zr	47	—	635	Highly stressed parts of aerospace and military uses. Used as a forging alloy.	High strength, good toughness, good spot-weldability. Limited arc-weldability.

Note: 1 psi (pounds/square inch) = 0.0069 megapascal (MPa).

Designation of Magnesium Alloys (an ASTM system now accepted by the SAE). A four-part system is used:

1. Letters indicate the two principal alloying elements: A, Aluminum; E, Rare-Earth; H, Thorium; K, Zirconium; M, Manganese; Q, Silver, S, Silicon; T, Tin; Z, Zinc. Thus HK signifies a thorium-zirconium magnesium alloy.
2. The approximate amounts (percent, wt) of the two principal alloying materials follow to the immediate right of the alloying element letters. Thus HK31 indicates approximately 3% thorium, and 1% zirconium.
3. The next two letter symbols to the right are used to distinguish two different alloys of the same chemical composition. Any letter may be used except I and O.
4. A fourth part of the designation (not indicated in this table) is separated by a dash from the foregoing parts and is used to indicate temper and other characteristics, such as F (as fabricated), 0 (annealed), H10 and H11 (slightly strain hardened), H23, H24, and H26 (strain hardened and partially annealed), T4 (solution heat treated), T5 (artificially aged only), T6 (solution heat treated and artificially aged), and T8 (solution heat treated, cold worked, and artificially aged). Thus, the complete designation may appear as: AZ91C-T6 for an aluminum-zinc-magnesium alloy containing 9% Al, 1% zinc, C indicating that this is the third alloy standardized with the same percentages of Al and Zn and T6 indicating that the alloy is solution treated and artificially aged.

TABLE 2. MAGNESIUM CASTING ALLOYS FOR AUTOMOTIVE APPLICATIONS

AM60B[2]	Die-casting alloy for uses needing toughness and ductility.		
	5.5–6.5% Al	0.25% Mn (min)	0.002% Ni (max)
	0.010% Cu max	0.22% Zn (max)	0.10% Si (max)
		0.005% Fe (max)	
AZ91D[2]	Provides an optimum combination of properties with die castability.		
	8.3–9.7% Al	0.15% Mn (min)	0.02% Ni (max)
	0.030% Cu (max)	0.35–1.0% Zn	0.10% Si (max)
		0.005% Fe (max)	
AZ91E[2]	A sand and permanent-mold casting alloy with properties and castability similar to AZ91B.		
	8.1–9.3% Al	0.17–0.5% Mn	0.0010% Ni (max)
	0.015% Cu (max)	0.40–1.0% Zn	0.20% Si (max)
		0.005% Fe (max)	
ZE41A	A sand and permanent mold casting alloy for applications to 175 °C (350 °F). Low microporosity and good pressure tightness.		
	0.0% Al	0.15% Mn (max)	0.40–1.0% Zr
	0.010% Cu (max)	0.75–1.75% Re	0.01% Ni (max)
		3.5–5.0% Zn	
ZC63	A proprietary sand and permanent-mold casting with properties similar to ZE41A, but less expensive.		
	0.0% Al	0.25–0.75% Mn	
	2.4–3.0% Cu	5.5–6.5% Zn	

lost and should be replaced by adding pure magnesium to the casting ladle or pot.

Copper — Magnesium improves tensile strength and allows age hardening. Magnesium is used mainly as a deoxidizer, notably in copper-nickel-zinc alloys and in leaded brasses and bronzes. The magnesium is added during melting.

Lead — Magnesium increases hardness, strength, and resistance to creep. Magnesium also is used as a debismuthizer in refining primary lead.

Nickel — Magnesium, in combination with carbon, forms an age-hardenable alloy. The main use of magnesium is to deoxidize and desulfurize the melts, including pure nickel, nickel-chrome, and nickel-copper alloys.

Tin — Magnesium increases hardness and tensile strength. The effect of magnesium on tin can be dramatic. However, too much magnesium will reduce corrosion resistance and ductility.

Zinc — Magnesium improves dimensional stability and reduces the intergranular corrosion of zinc die castings. Magnesium refines the grain and increases hardness and creep strength of zinc sheet. Magnesium also is used in zinc-base bearing metals and in zinc alloy metalworking dies.

Magnesium alloy extrusions have become very popular for numerous items in recent years. Extrusion is particularly attractive as a parts making method — when extruded parts and sheet can be easily joined to form an assembly, where the desired shapes are too costly to machine from castings, and where pieces cut from extrusions can replace individually cast or forged parts. Final products with outstanding performance qualities coupled with light weight include concrete hand finishing tools, tennis racquets, portable shelters for the military, snowshoes, and improved luggage, among others.

The use of magnesium composites has become popular for rotary engine parts. Rotary engines remain attractive for business aircraft, boats, industrial equipment and compressors, and well over a million rotary-engine-powered cars have been built. In a research program (NASA Lewis Research Center) rotary engine parts are made from graphite-fiber-reinforced magnesium. An AZ91C magnesium alloy is reinforced by 30% (vol) graphite fibers.

Progress has been in the early 1990s toward the development of metal-matrix composites (MMCs) that blend liquid magnesium alloys with ceramic particles, such as silicon carbide (SiC) and alumina (Al_2O_3). The method is similar to methods that have been developed for aluminum composites in that blending is accomplished by way of a high-shear process. Major differences of the new process result from increased general reactivity of magnesium and the difference in surface chemistry between the Al-SiC and Mg-SiC systems.

The particulate-reinforced MMCs are lightweight and demonstrate a significant increase in modulus and tensile strength at both ambient and elevated temperatures of the unreinforced material. The process was announced in late 1992 by Magnesium Elektron Ltd., Manchester, U.K.

Chemistry and Compounds

The behavior of magnesium is intermediate between that of beryllium and the higher alkaline earths. While it reacts readily with halogens, oxygen, and sulfur to form halides, oxide, and sulfide, it reacts with cold water only when the formation of protective oxide is prevented by amalgamation. All its compounds are divalent. Its oxide does not react with water to form the hydroxide, and it does not normally form a peroxide. Its major difference from the higher elements of the group is its much greater number of complexes. Anhydrous magnesium halides, especially, combine easily with many oxygen-functional organic compounds to form addition compounds. These reactions usually suggest covalent or dative bonding (both electrons from the oxygen) of the magnesium. Magnesium salts often form amines and amine complexes, though these are less stable than beryllium complexes. Magnesium also forms some basic salts, and many more of its salts are hydrated than are those of the higher alkaline earths. The metal reacts with alkyl and aryl halides to form the Grignard reagents, through which many organic reactions are conducted. The Grignard reagents themselves form complexes with ethers, tertiary amines, tertiary phosphines and many other type compounds. See also **Grignard Reactions**.

Important compounds of magnesium include the following:

Magnesium acetate, $Mg(C_2H_3O_2)_2 \cdot 4H_2O$, white solid, soluble, formed by reaction of magnesium carbonate and acetic acid.

Magnesium ammonium arsenate, $MgNH_4AsO_4$, white precipitate, solubility 0.0013 molar, formed by reaction of soluble magnesium salt solution and sodium arsenate in the presence of excess ammonium hydroxide, and upon igniting yields magnesium pyroarsenate, $Mg_2As_2O_7$, white solid.

Magnesium borate, $Mg_3(BO_3)_2$, or $Mg(BO_2)_2$, white precipitate, by reaction of soluble magnesium salt solution and sodium borate.

Magnesium boride, Mg_3B_2, brown solid, by reaction of boron oxide and magnesium powder ignited.

Magnesium bromide, $MgBr_2 \cdot 6H_2O$, white solid, soluble, formed by reaction of magnesium carbonate and hydrobromic acid.

Magnesium carbonate, $MgCO_3$, white solid, $K_{sp}4.0 \times 10^{-5}$, formed by reaction of soluble magnesium salt solution and sodium carbonate or bicarbonate solution. Present in carbonate minerals and rocks, magnesite (more or less pure magnesium carbonate), dolomite (magnesium-calcium carbonate mixtures), dolomitic limestone. When ignited yields magnesium oxide and CO_2; when treated with acids yields the corresponding magnesium salt and CO_2, but with carbonic acid yields soluble magnesium bicarbonate. Magnesium bicarbonate, $Mg(HCO_3)_2$, colorless solution, by reaction of magnesium carbonate and carbonic acid, yields, upon boiling, magnesium carbonate and CO_2; magnesium ammonium carbonate, $(MgCO_3 \cdot NH_4)_2CO_3 \cdot 4H_2O$, white precipitate (soluble in ammonium chloride solution) by reaction of soluble magnesium salt solution and excess ammonium carbonate.

Magnesium chloride, $MgCl_2 \cdot 6H_2O$, white solid, soluble, formed by reaction of magnesium carbonate (or hydroxide, oxide, or metal) and HCl, loses hydrogen chloride when heated, yielding magnesium oxychloride; anhydrous magnesium chloride, $MgCl_2$, white solid, soluble, formed: (1) by heating hydrated magnesium chloride crystals in a current of dry hydrogen chloride, and (2) by heating magnesium ammonium chloride, mp 712 °C. Magnesium ammonium chloride, $MgCl_2 \cdot NH_4Cl \cdot 6H_2O$, white solid, soluble, when heated yields anhydrous magnesium chloride; magnesium potassium chloride, $MgCl_2 \cdot KCl \cdot 6H_2O$, white solid, soluble, when heated fuses to anhydrous magnesium potassium chloride; magnesium oxychloride, white solid, insoluble, formed (1) by heating hydrated magnesium chloride crystals, and (2) by mixing magnesium chloride solution and magnesium oxide.

Magnesium chromate, $MgCrO_4 \cdot 7H_2O$, yellow solid, soluble, formed by reaction of magnesium carbonate and chromic acid solution.

Magnesium citrate, $Mg_3(C_6H_5O_7)_2 \cdot 4H_2O$, white solid, soluble, formed by reaction of magnesium carbonate and citric acid.

Magnesium fluoride, MgF_2, white precipitate, $K_{sp}6.5 \times 10^{-9}$, formed by reaction of soluble magnesium salt solution and sodium fluoride solution.

Magnesium hydroxide, $Mg(OH)_2$, white precipitate, $K_{sp}9.0 \times 10^{-12}$, formed by reaction of soluble magnesium salt solution and NaOH solution.

Magnesium hypophosphite, $Mg(H_2PO_2)_2 \cdot 6H_2O$, white solid, soluble, formed by reaction of magnesium carbonate and hypophosphorous acid.

Magnesium iodide, $MgI_2 \cdot 8H_2O$, white solid, soluble, formed (1) by reaction of magnesium carbonate and hydriodic acid, (2) anhydrous, by heating magnesium metal and iodine.

Magnesium lactate, $Mg(C_3H_5O_3)_2 \cdot 3H_2O$, white solid, soluble, formed by reaction of magnesium carbonate and lactic acid.

Magnesium nitrate, $Mg(NO_3)_2 \cdot 6H_2O$, white solid, soluble, formed by reaction of magnesium carbonate and HNO_3.

Magnesium nitride, Mg_3N_2, yellow solid, with moist air or water yields ammonia and magnesium hydroxide, formed by heating magnesium to a high temperature in nitrogen or NH_3 (hydrogen gas evolved).

Magnesium oleate, $Mg(C_{18}H_{33}O_2)_2$, yellow solid, insoluble, formed by reaction of soluble magnesium salt solution and sodium oleate.

Magnesium oxalate, $MgC_2O_4 \cdot 2H_2O$, white solid, insoluble, $K_{sp}8.6 \times 10^{-5}$, formed by reaction of soluble magnesium salt solution and ammonium oxalate solution.

Magnesium oxide, MgO, white solid, reacts slowly with H_2O to form magnesium hydroxide, has cubic structure, absorbs CO_2 from the air to form magnesium carbonate, is readily soluble in acids, insoluble in alkalies; formed (1) by heating magnesium carbonate to high temperature (CO_2 gas evolved), (2) by heating magnesium hydroxide, nitrate, sulfate, or oxalate, (3) by burning magnesium metal in air or oxygen.

Magnesium peroxide, MgO_2, white solid, insoluble, formed by reaction of soluble magnesium salt solution and sodium peroxide.

Magnesium ammonium phosphate, $MgNH_4PO_4$, white precipitate, $K_{sp}2.5 \times 10^{-12}$, by reaction of soluble salt solution and sodium phosphate in the presence of excess ammonium hydroxide, upon igniting yields magnesium pyrophosphate, $Mg_2P_2O_7$, white solid.

Magnesium salicylate, $Mg(C_7H_5O_3)_2 \cdot 4H_2O$, white solid, soluble, formed by reaction of magnesium carbonate and salicylic acid in H_2O.

Magnesium sulfate $MgSO_4 \cdot 7H_2O$, white solid, soluble, formed by reaction of magnesium carbonate and H_2SO_4.

Additional Reading

Avedesian, M.M. and H. Baker: "Magnesium and Magnesium Alloys," ASM International, Materials Park, OH, 1999.

Davis, J.R.: "Metals Handbook," 2nd Edition, ASM International, Materials Park, OH, 1998.

Greenwood, N.N. and A. Earnshaw: "Chemistry of the Elements," 2nd Edition, Butterworth-Heiemann, Inc., Woburn, MA, 1997.

Kainer, K.U.: "Magnesium Alloys and Their Applications," John Wiley & Sons, Inc., New York, NY, 2000.

Kaplan, H.I., J. Hryn, and B. Clow: "Magnesium Technology 2000: Proceedings of the Symposium Sponsored by the Light Metals Division of the Minerals, Metals and Materials Society (TMS) and the International Magnesium," Warrendale, PA, 2000.

Krebs, R.E.: "The History and Use of Our Earth's Chemical Elements," Greenwood Publishing Group, Inc., Westport, CT, 1998.

Lide, D.R.: "CRC Handbook of Chemistry and Physics 2000–2001," 81st Edition, CRC Press, LLC., Boca Raton, FL, 2000.

MAGNESIUM (In Biological Systems). Magnesium is an integral part of the molecule of chlorophyll, the green pigment in plants that absorbs solar energy. See also **Chlorophylls**. Magnesium deficiency is a fairly common cause of poor crop yields, especially among crops produced on sandy soils. Magnesium is a prosthetic ion in enzymes that hydrolyze and transfer phosphate groups. Hence it is essential for energy-requiring biological functions, such as membrane transport, generation and transmission of nerve impulses, contraction of muscles, and oxidative phosphorylation. See also **Phosphorylation (Oxidative)**. Magnesium is essential for the maintenance of ribosomal structure and thus protein synthesis. Magnesium may be related to the incidence of ischemic heart disease among Western populations.

The accumulation of magnesium from the soil by plants is strongly affected by the species of plant. The leguminous plants, such as clovers, beans, and peas, usually contain more magnesium than grasses, tomatoes, corn (maize), and other nonleguminous plants, regardless of the level of available magnesium in the soil where they grow.

A very high level of available potassium in the soil interferes with the uptake of magnesium by plants, and magnesium deficiency in plants is often found in soils that are very high in available potassium. High levels of available potassium may occur naturally, especially in soils of subhumid and semiarid regions; or they may be caused by heavy applications of certain commercial fertilizers or animal manure. On sandy and loamy soils, applications of magnesium fertilizers are often effective in increasing crop yields and the concentration of magnesium in the crop, but on fine-textured, clay-containing soils, especially those with substantial reserves of potassium, the application of a magnesium fertilizer may not cause higher magnesium concentration in crops. Since magnesium is not a highly toxic element in either plants or animals, precautions against its overuse are rarely necessary. When animals are fed diets primarily of grains, a proper balance among magnesium, calcium, and phosphorus should be maintained to minimize danger from urinary calculi.

The biological functions of magnesium, such as its essential role as a nutrier, its activation of enzyme systems, and its pharmacological properties, have been widely investigated. Nevertheless, some aspects of its critical physiological role remain obscure.

Distribution in System. Magnesium, primarily an intracellular ion, is distributed among all tissues. It constitutes about 0.05% of the animal body and, of this, 60% occurs in the skeleton and only 1% in extracellular fluids.

Reported serum magnesium values for most species range from 1.0 to 3.5 meg/liter, with a mean value of about 2. Between 65% and 80% of the plasma magnesium is ultrafilterable, and most of this exists as the free ion. The nonfilterable portion is reversibly bound to plasma protein. Cerebrospinal fluid contains slightly more than plasma. Interstitial fluid is similar to plasma ultrafiltrate.

The magnesium content of soft tissues varies from 0.06 to 0.13% of dry weight and remains remarkably constant regardless of the magnesium status of the animal. Normally, the intracellular concentration is more than 20 times that of the interstitial fluid, and the highest concentration occurs in the cell nucleus. Maintenance of such a large concentration gradient across the cell membrane suggests an active transport mechanism.

In late 1990's, R.R. Preston (University of Wisconsin–Madison) reported that recent reappraisals of the role of ionized magnesium in cell function suggests that many cells maintain intracellular free Mg^{2+} at low concentrations and that external agents can influence cell functions via changes in intracellular Mg^{2+} concentration. There is considerable evidence to suggest that intracellular free magnesium ions may be a key physiological regulator of cell activity.

The relatively large proportion of magnesium found in the skeleton, which amounts to about 0.6% of dry, fat-free bone, serves in part as a body reserve. It occurs largely as Mg^{2+} and $MgOH^+$ ions held by electrostatic attraction to the apatite crystal surface. During deficiency in young animals, 30% or more of bone magnesium can be mobilized for metabolic functions. Calcium ions appear to replace the magnesium that occupied the original adsorption sites.

Metabolism. The rate of absorption from the intestine exerts an important role in magnesium metabolism. Whereas in vitro studies show that magnesium absorption is positively correlated with the concentration of magnesium, it does not appear to be a purely passive process. Magnesium absorbed in excess of body needs is excreted primarily by way of the kidney. Urinary excretion is controlled primarily by a filtration-reabsorption mechanism so that magnesium appears in the urine only when glomerular filtration exceeds tubular reabsorption. Acute renal failure is accompanied by hypermagnesemia. In some species, considerable endogenous magnesium is lost by way of the feces, the amount depending upon the magnesium status of the animal and upon other dietary factors, such as the digestibility of the diet. The endogenous fecal magnesium in calves has been estimated at 3.5 milligrams/kilogram of body weight.

In contrast with the metabolism of calcium, no one endocrine gland exerts a primary regulatory function on magnesium. Thyro-parathyroidectomy in dogs causes only a temporary lowering of plasma magnesium. Adrenalectomy causes a rise, whereas hyper-aldosteronism produces a fall in the plasma level. Administrative of deoxycorticosterone or aldosterone to sheep lowers the magnesium concentration in plasma. Magnesium-deficient animals exhibit a higher metabolic rate than normal, and the toxic effect of excess thyroxine is partially overcome by increasing the dietary level of magnesium.

Function. Although magnesium activates isolated enzymes, in most cases an absolute requirement is difficult to establish because the enzymes are partially active without added magnesium. The stimulating effect is not always specific for magnesium. In some cases, manganese or calcium will also activate the system.

TABLE 1. DISTRIBUTION OF MAGNESIUM IN VARIOUS FOOD GROUPS

Group (Types of Samples Tested)	Magnesium Concentration (Milligrams/100 grams (wet))	Magnesium-Calorie Ratio (Micrograms/Kilocalorie)
Milk products (cheeses, ice cream, milk, puddings)	6.8–25.7	18–198
Meat and meat alternates (chicken, dried beef, eggs, fish, sausage)	9.8–37.6	20–353
Vegetables (cabbage, carrot, onion, turnip)	6.7–20.6	196–1000
Breads and cereals (buns, cereals, cornbread, crackers, croutons, English muffins, pasta, taco shells)	10.6–126.0	27–325
Baked desserts (cakes, cookies, doughnuts, pastries, sweet rolls)	4.6–53.2	18–307
Candies	21.8–89.9	63–225

Magnesium is particularly concerned with enzyme-catalyzed reactions involving the cleavage of phosphate esters and the transfer of phosphate groups. Magnesium ions activate phosphatases and the phosphorylation reactions involving adenosine triphosphate (ATP). Among the latter group may be mentioned glucokinase, phosphoglucokinase, phosphofructokinase, myokinase, creatine transphosphorylase, arginine transphosphorylase, and flavokinase. It has been suggested that an ATP-Mg complex is the active substrate inasmuch as ATP forms a 1:1 complex with magnesium and maximal activation occurs when the ATP: Magnesium ration is 1. Alkaline phosphatases, pyrophosphateses, and ATPase are activated by magnesium, as are enolase, certain peptidases, and pyruvic oxidase. Since magnesium is tied to ATP utilization, it follows that magnesium plays a role in important metabolic processes, including the synthesis of protein, fat, and nucleic acids, and in the trapping and utilization of energy derived from catabolism of carbohydrate and fat.

There is little change in magnesium concentration of soft tissues from deficient animals even at the point of expiration. This does not preclude the possibility that a small component of the cell, such as the nucleus or a cell particulate, is deprived of its critical level, but the dramatic drop in extracellular magnesium suggests that a function outside the cell is of greatest significance. It appears that tetany and convulsions in deficient animals result from a derangement of neuromuscular transmission. Magnesium ion possesses strong pharmacological properties, depressing both the central and peripheral nervous systems. These effects are counteracted by calcium. In the presence of normal calcium levels, a reduction of extracellular magnesium is believed to increase the release of acetylcholine and to decrease the rate of its hydrolysis. Such effects would increase the irritability of the neuromuscular system.

Magnesium generally has not been considered a major factor in bone formation and strength, but recent studies suggest closer attention be given to dietary levels of magnesium in this regard. Because of the close interrelationship with calcium, it is not surprising to see research findings of magnesium interfering with calcium entry into cells of the islets of the pancreas in studies of diabetes. The recognized presence of magnesium as part of numerous enzyme systems has led to observations of the reduction in carbohydrate metabolism associated with a deficiency and to beneficial effects in reducing blood cholesterol and lipids associated with other dietary agents when supplemental magnesium is added to the diet. The relationship to calcium also shows up in a study showing that adding magnesium to the rations of laying hens causes an increase in shell thickness, with a consequent reduction in the number of broken eggs.

Magnesium-Induced Diarrhea. It is well established that an excessive intake of magnesium causes diarrhea. This source of diarrhea is difficult to differentiate from other causes of diarrhea. Consequently, the diagnosis may be long and costly unless the physician questions a patient on possible excessive magnesium intake, which may result from large dosages of antacids or off-the-shelf food supplements.

Pathology of Magnesium Deficiency. Although there are numerous clinical symptoms, two cardinal aspects of pathology have been observed in all species of higher animals. These are hyperirritability and soft tissue calcification. While there are species differences as to the dominating syndrome, this is determined in part by the severity of the deficiency. Metastatic calcification is more likely to occur in a chronic deficiency in which the animal does not succumb at an early age. Hyperirritability,

terminating in convulsions and death, has been observed in rat, rabbit, pig, calf, chick, and duck. Magnesium deficiency in humans is characterized by muscle tremors and twitching, often accompanied by delirium and occasionally by convulsions. The guinea pig, calf, dog, and cotton rat are prone to metastatic calcification and develop grossly visible deposits in and around joints, along the muscles of the rib cage, and also in the heart, great vessels, and other critical organs. Most soft tissues show an elevated ash content and marked histopathology. Some researchers have hypothesized that long-term intakes of marginal dietary levels of magnesium may be related to the incidence of ischemic heart disease.

The first clinical symptom of magnesium deficiency is a hypomagnesemia which occurs in cattle and less frequently in sheep and is described by such names as grass tetany, grass staggers, lactination tetany, and wheat pasture poisoning. It is observed most frequently when animals are first grazed on lush grass or wheat pastures. The disease is characterized by irritability, tetany, and convulsions, and all animals have subnormal plasma magnesium. Symptoms can be relieved by administration of magnesium salts and cam be prevented by providing extra magnesium in the diet.

Hypomagnesemia, often associated with hypocalcemia, is frequently encountered in heavy users of alcohol. Alcoholism sometimes is ascribed to impaired intestinal calcium absorption.

Nutritional Requirements and Dietary Supplementation. As is true of many mineral nutrients, the requirement for magnesium is affected by other dietary constituents, by the age and species of the animal, and by the criterion of adequacy applied. An allowance for magnesium has been included in the Recommended Dietary Allowance since 1968. Calcium and magnesium have an important effect upon magnesium availability. Either of these ions in excess increases the requirements for magnesium, and their effects are additive. Since calcium is known to compete with magnesium pharmacologically, it is reasonable to believe that it also competes with magnesium for absorption sites in the intestine. It is believed that phosphate decreases magnesium absorption by formation of insoluble magnesium phosphates, and excess of calcium aggravates the effect of creating a more alkaline intestinal medium. Excess magnesium can be considered toxic, but this effect is largely due to the induction of a calcium deficiency. Magnesium deficiency in humans generally has not been fully documented except in cases of predisposing and complicating disease states.

Distribution of magnesium in food classes is summarized in Table 1.

Additional Reading

Theophanides, T.M. and J. Anastassopoulou: "Magnesium, Current Status and New Developments: Current Status and New Developments: Theoretical, Biological, and Medical Aspects," Kluwer Academic Publishers, Norwell, MA, 1997.

Tsang, R.C.: "Calcium and Magnesium Metabolism in Early Life," CRC Press, LLC., Boca Raton, FL, 1998.

Vedral, J.L.: "Dietary Reference Intakes: For Calcium, Phosphorus, Magnesium, Vitamin D, and Fluoride," National Academy Press, Washington, DC, 1997.

Web References

Health World Online: *http://www.healthy.net/asp/templates/article.asp?Page=Article&ID=541*

Magnesium Metabolism: *http://www.merck.com/pubs/mmanual/section2/chapter12/12f.htm*

Magnesium Update Information: *http://www.krispin.com/magnes.html*

The Absolute need for Magnesium: *http://members.tripod.com/~headachepa-infree/index.html*

MAGNET (Superconductivity).   See **Superconductivity**.

MAGNETIC DECLINATION.   See **Isogonic Line**.

MAGNETIC DOUBLE REFRACTION.   The splitting, into two components, of a radio wave traveling in a region of free electrons. This is due to the interaction of the Earth's magnetic field and the alternating field of the radio wave. Except for waves near the gyrofrequency, the components of the split wave, the ordinary ray and the extraordinary ray, will travel with slightly different velocities and be reflected at different heights. See **Magnetoionic Theory**.

MAGNETIC EQUATOR.   See **Aclinic Line**.

MAGNETIC FIELD.   A region of space wherein any magnetic dipole would experience a magnetic force or torque; often represented as the geometric array of the imaginary magnetic lines of force that exist in relation to magnetic poles.

MAGNETIC FIELD INTENSITY.   The magnetic force exerted on an imaginary unit magnetic pole placed at any specified point of space. It is a vector quantity. Its direction is taken as the direction toward which a north magnetic pole would tend to move under the influence of the field. If the force is measured in dynes and the unit pole is a cgs unit pole, the field intensity is given in *oersteds*. Also called *magnetic intensity, magnetic field*, or *magnetic field strength*. Prior to 1932 the *oersted* was called the *gauss*; but the latter term is now used to measure magnetic induction (within magnetic materials), whereas *oersted* is reserved for magnetic force.

By definition, one magnetic line of force per square centimeter (in air) represents the field intensity of 1 *oersted*.

MAGNETIC FLOWMETER.   See **Flow Measurement (Liquids and Gases)**.

MAGNETIC INDUCTION.   A measure of the strength of a magnetic field existing within a magnetic medium. The relation between the magnetic induction and magnetic field intensity is such that the magnetic induction within a small mass of material of magnetic permeability μ is, except for possible hysteresis effects, μ times greater than the external magnetic field intensity. Whereas magnetic field intensity is measured in *oersteds*, magnetic induction is measured in *gausses*.

MAGNETIC INK CHARACTER RECOGNITION (MICR).   Developed primarily as the common machine language for bank check handling, the early work on this system dates back to the early 1950s, when the American Bankers Association commenced writing specifications for a suitable system to be used with the then rapidly developing electronic business machines. The system is used by nearly all banking institutions.

As shown in Fig. 1, specifically designed symbols are used for all numerals and zero. A few additional symbols for special coding are also shown. To take full advantage of the system, it was necessary to modify to some degree the conventional shapes of the numbers, but generally the numbers can be easily read with the naked eye, thus serving two purposes. The waveform for each number is shown below each numeral in the diagram and it will be noted that there is a marked contrast between each waveform, thus assuring reliability of reading accuracy.

As the MICR-coded documents pass a special reading head, the symbol is converted to its relevant waveform and with further electronic translation can be introduced into a data processing system.

Other systems seriously considered by the American Bankers Association before final selection of MICR included binary or bar codes, spot code (decimal system, fluorescent ink, magnetic bar code, perforations, and notches). The principal criteria against which each system was judged included accuracy, tolerances, printing practicability, customer acceptance, verification, cost, format, and resistance to mutilation and obliteration.

For some applications, optical character recognition (OCR) is preferred over MICR. Each approach has its advantages and disadvantages and sometimes selection of the most effective approach is quite difficult. See also **Optical Character Recognition (OCR)**.

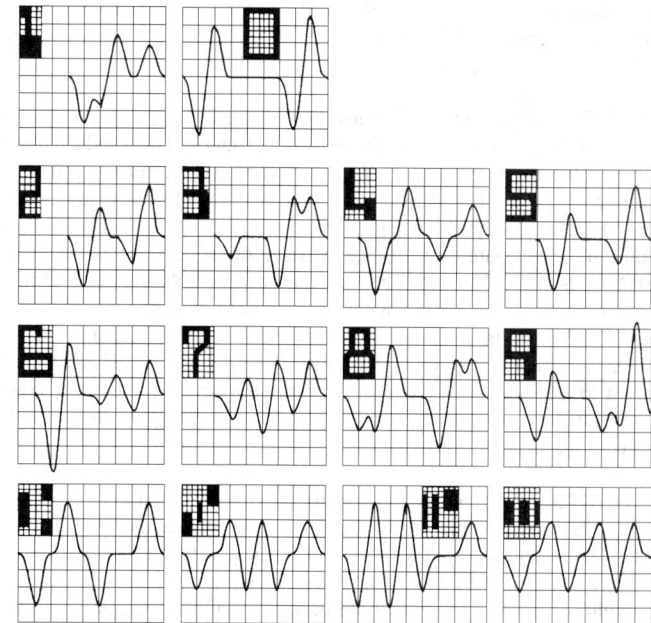

Fig. 1.   Magnetic ink numerals and associated electron waveforms. The symbols in lower row are used for special coding purposes.

MAGNETIC K-INDICES.   An approximately logarithmic measure of geomagnetic disturbance activity based on the range of the most disturbed magnetic element during each 3-hour interval of the day. The K-indices are assigned integers from 0 to 9. The K-indices averaged over the observatories of the earth are called planetary indices K_p and divided into 28 grades.

MAGNETIC LUNAR DAILY VARIATION (symbol L).   A periodic variation of the Earth's magnetic field that is in phase with the transit of the moon. This variation is essentially a tidal effect. The amplitude of this variation changes with the phase of the moon, the seasons, and the sunspot cycle.

MAGNETIC MOMENT.   1.   The quantity obtained by multiplying the distance between two magnetic poles by the average strength of the poles.

2.   A measure of the magnetic flux set up by the gyration of an electric charge in a magnetic field. The moment is negative, indicating it is diamagnetic, and equal to the energy of rotation divided by the magnetic field.

3.   (symbol m). In atomic and nuclear physics, a moment, measured in Bohr magnetons, associated with the intrinsic spin of the particle and with the orbital motion of the particle in a system. Also called *magnetic dipole moment*.

MAGNETIC MOMENT (Particle).   Use of the term magnetic moment for a nuclear or atomic particle or system of particles usually denotes the magnetic dipole moment. For a particle or system in a magnetic field, the interaction energy is the negative of the product of the field strength H by the component of the magnetic dipole moment μ_H of the particle in the direction of the field ($\mu_H H$). A magnetic moment is associated with the intrinsic spin of a particle and with the orbital motion of a particle in a system, e.g., nuclei with finite spins have finite magnetic moments between about -2 and $+6$ nuclear magnetons.

MAGNETIC NORTH.   See **North**.

MAGNETIC POLES (Earth).   See **Earth**.

MAGNETIC PRESSURE.   The energy density associated with a magnetic field. In a very real sense, there is energy stored in a magnetic field, and since energy per unit volume is equivalent to force per unit area or pressure, one may speak of the pressure exerted by a magnetic field. For plasma containment in a thermonuclear device, the magnetic pressure must be greater than the kinetic pressure of the plasma. A pressure of 1 atmosphere corresponds approximately to 5,000 *gausses*, and the pressure is proportional to the square of the field.

MAGNETIC STORM.

A worldwide disturbance of the Earth's magnetic field. Magnetic storms are frequently characterized by a sudden onset, in which the magnetic field undergoes marked changes in the course of an hour or less, followed by a very gradual return to normality, which may take several days. Magnetic storms are caused by solar disturbances, though the exact nature of the link between the solar and terrestrial disturbances is not understood. They are more frequent during years of high sunspot number. Sometimes a magnetic storm can be linked to a particular solar disturbance. In these cases, the time between solar flare and onset of the magnetic storm is about 1 or 2 days, suggesting that the disturbance is carried to the Earth by a cloud of particles thrown out by the sun. When these disturbances are observable only in the *auroral zones*, they may be termed polar magnetic storms.

MAGNETIC TAPE STORAGE (Computer).

Information is recorded on tape by magnetizing narrow (lengthwise) strips (termed tracks) in a pattern corresponding to a sequence of binary states (1s and 0s). Binary data, often corresponding to a single byte, are stored in column form across the width of the tape. A read/write head is usually associated with each row of magnetized material, so that one column can be read or written at a time as the tape traverses the head. The density of recorded data is most commonly 800, 1600, or 6250 bits per linear inch (approximately 315, 630, or 2461 bits per linear centimeter). A nine-track, 2400-foot ($\sim$732-meter) tape will carry from 20 to 156 million characters, dependent upon the density. Magnetic tape storage is most often used to provide offline archival storage of data. A drawback to tape storage is the fact that it must be read serially. Retrieval of randomly distributed information can be time consuming, compared, for example, with disk storage.

Tape material usually is polyester plastic, one side of which is coated with a suspension of ferrite or magnetic oxide particles. The common tape size is $\frac{1}{2}$ inch (12.7 millimeters) in width, $1\frac{1}{2}$ mils (0.0015 inch; 0.4 millimeter) in thickness, and 2400 feet ($\sim$732 meters) in length. Tape widths of $\frac{1}{4}$, $\frac{3}{4}$, and 1 inch (0.64, 1.91, and 2.54 centimeters); and lengths of 200, 300, 600, 1200, and 3600 feet ($\sim$61, 91, 183, 366, and 1097 meters) also are available for use in the computer field.

Recorded material can be removed from the tape by passing the tape through a strong, constant magnetic field (dc erase), or through a high-frequency alternating magnetic field (ac erase). Information is stored on the tape by magnetizing the magnetic film in one direction or the other in a pattern determined by the binary data. The information is recorded on the tape in a parallel by bit, serial by digit, format, i.e., the bits which comprise a digit are recorded across the width of the tape and the digits are recorded sequentially along the length of the tape.

The magnetic tape is stored on reels and is transferred from one reel to the other past two-gap read/write heads in the course of reading or writing data. Vacuum columns or other tension mechanisms are provided so that the tape can be moved rapidly a few inches at a time without waiting for the movement of the reels. The two-gap head allows automatic error checking of the tape data while it is being written. The first gap is used for writing, the second gap for reading.

The reading or writing of data on the tape is controlled by clock circuits within the logic of the tape unit. In writing, the frequency at which the clock is stepped is a function of the tape speed and the recording density. In reading, the oscillator that drives the read clock is also gated by the first bit being read. To compensate for the time it takes for the tape motion to get up to speed, read delay and write delay counters are designed into the logic of the tape unit.

On a tape read operation, the character is stored in a buffer register as it is read from the tape and is then transferred to the digital computer while the next character is being read. On a write operation, the next character is fetched from the computer while the previous character is being written. Since data are not written on the tape until the tape gets up to speed, the information to be written is blocked into records to minimize the number of gaps caused by the write delay.

See also **Storage (Computer)**.

THOMAS J. HARRISON, IBM Corporation, Boca Raton, FL.

MAGNETISM.

A magnet is a body possessing the property of attracting magnetic substances. The so-called permanent magnet should be used where a constant magnetic field is to be produced, inasmuch as a well-made permanent magnet loses its magnetism very slowly, and then only up to a certain value, after which it is said to be aged. Therefore, it essentially maintains a constant degree of magnetism unless subjected to strong demagnetizing effects. Permanent magnets for precision apparatus are artificially aged during manufacture.

A bar that has been magnetized is found to have poles, which are centers where magnetic attraction is strongest. If the magnet is free to turn, the pole which points northerly is aptly termed the north pole; the other, the south pole. Like poles repel; unlike poles attract. Thus, because of the magnetic properties of the earth, a magnet can serve as a compass. The poles, of course, have no physical reality, but provide a convenient concept for describing certain magnetic phenomena.

Magnetic Field. The region surrounding a magnet (or an electric current) is endowed with specific properties. The most familiar is the torque experienced by a small magnet when placed in such a region. For any point of the field, there is only one direction in which the small magnet will reach stable equilibrium. The direction in which the north pole of the magnet points when in equilibrium is termed the direction of the field. By moving a small magnet in a magnetic field, it is found that the field direction will follow curved lines of force. If the field is due to the current flowing in a conductor, the lines form completely closed curves enclosing the conductor. If the field is due to a magnet, they appear to enter the iron at the south pole and emerge at the north pole, inferring that they complete themselves through the iron as they do through a coreless, current-carrying helix. A magnetic field may be defined as a vector function field described by the magnetic induction. The term magnetic field is used interchangeably to refer to magnetic induction and magnetizing force.

The difference in the magnetic potential at two points in a magnetic field is measured by the work necessary to move a unit magnetic pole against the field from one point to the other. This difference is sometimes called a magnetomotive force, in analogy to electromotive force.

The lines of magnetic intensity around a current-carrying wire are circular, having the plane of the circle perpendicular to the axis of the wire. The direction of the lines of force is determined by the "right-hand rule." When the wire is grasped by the right hand, the fingers encircling the wire, and the thumb pointing along the wire in the direction of the current, the fingers encircle the wire in the direction of the lines of force. This rule is used to determine in which direction the north pole would lie in a helix or solenoid of wire, for, instead of having a straight wire encircled by magnetic lines, the wire itself is bent into a circular form by being wound in a solenoid or helix. The lines of force will then produce an axial magnetic field, so that one end of the solenoid is equivalent to a north pole, the other to a south pole.

In the vicinity of magnets, electric currents, or time-varying electric fields, it is found that a small current-carrying wire loop, if free to turn, will come to an equilibrium orientation. This behavior is attributed to an auxiliary field vector **B** called the magnetic induction or the magnetic flux density. The magnetic moment of a current loop or a magnetized body is a measure of the magnetizing force **H**, produced by the current or magnetized body. The magnetic moment of a plane current loop is a vector (**m**), normal to the plane of the loop and directed so that the current has a clockwise rotation around **m**. The magnitude of **m** is the product of the current and loop area. The magnetic moment of a magnetized body is the vector summation of the magnetic moments of the internal current loops and spins of the body. The magnetizing force produced at a displacement **r** from a small source of moment **m**

$$\mathbf{H} = -\nabla \frac{\mathbf{m} \cdot \mathbf{r}}{r^3}$$

where ∇ is del. the vector differential operator.

The ratio of the magnetic moment of a magnetized body to its volume, expressed by the equation $\mathbf{B} = \mu_0(\mathbf{H} + 4\pi\mathbf{M})$, where μ_0 is the permeability of free space, **B** is magnetic induction, **H** is the external magnetizing force, and **M** is the magnetization. (The symbol I is sometimes used for intensity of magnetization.) The application of an increasing magnetizing force to a ferromagnetic substance yields a resulting intrinsic induction that asymptotically approaches a constant value known as the saturation magnetization. Spontaneous magnetism is the magnetic saturation of the domains of a ferromagnetic material, even in the absence of an applied magnetizing force. Each domain is magnetized to saturation at all temperatures below the Curie point, although the material as a whole may be unmagnetized because of the differing orientations of the various domains. See also **Ferromagnetism**.

The intensity of magnetization I induced at any point in a body is proportional to the strength of the applied field H:

$$I = \kappa H \text{ (or } \kappa = I/H)$$

where κ is a constant of proportionality depending on the material of the body. It is called the magnetic susceptibility per unit volume, and may be defined qualitatively as the extent to which a material is susceptible to induced magnetism. For an isotropic body, the susceptibility is the same in all directions. However, for anisotropic crystals, the susceptibilities along the three principle magnetic axes are different, and measurements on their powder samples give the average of the three values.

Magnetic susceptibility is obviously related to magnetic permeability, and the following relationships may be derived:

$$B = \mu_H = 4\pi I + H$$

Therefore,

$$\mu = 4\pi I/H + 1 = 4\pi\kappa + 1$$

or

$$\kappa = (\mu - 1)/4\pi$$

Magnetic Flux. The magnetic flux through any closed figure, such as a circle, a rectangle, or a loop of wire, is the product of the area of the figure by the average component of magnetic induction normal to that area. Thus, if a rectangle 5 cm × 8 cm is placed in a region where there is a uniform magnetic induction of 2,500 gauss (other units defined shortly), and at an angle of 30° with the lines of induction, the magnetic flux through it is 2,500 gauss ° 40 cm^2 × sin 30° = 50,000 gauss-cm^2 or "maxwells." The magnitude of this quantity is often conventionally represented by imagining the lines of induction to be so spaced that the number of them through a given area is equal to the number of gauss-cm^2 or maxwells of flux through that area. The flux in the above example would be commonly expressed as 50,000 "lines." When a coil has several (n) turns and each turn has approximately the same flux (ϕ) through it. This product, which is called the "linkage," is expressed in "maxwell-turns" or "line-turns."

The magnetic flux or the linkage through a loop or a coil may be measured by putting into the circuit a ballistic (undamped) galvanometer and then suddenly removing the flux (or the coil). If the resistance of the whole circuit and the constant of the galvanometer are known, the flux may be calculated from the "throw" of the galvanometer.

The *gauss* (G) is the electromagnetic CGS unit of magnetic flux density. The *tesla* (T) is the **SI** unit. The *maxwell* (**Mx**) is the electromagnetic CGS unit of magnetic flux. The weber (Wb) is the SI unit.

Magnetic Circuit. The flux from a bar magnet or from a straight electromagnet issues from one end of the magnet or coil, bends around, and reenters at the other end. As mentioned before, this can be exhibited by exploring the region with a small magnet or compass needle. If there is an iron frame or ring extending from one pole of the magnet or coil around to the other, and in the case of the coil, running clear through it, the magnetic flux is not only concentrated largely in the iron, but is much greater in total amount than if the induction is entirely in the air. Even a short gap in the iron reduces the flux considerably.

The analogy of such a magnetic path to an electric circuit is easily seen. The magnetic flux corresponds to a current. The magnet or coil corresponds to a battery, and provides magnetomotive force just as a battery supplies electromotive force. The amount of flux produced by a given magnetomotive force depends upon the dimensions and material of the "magnetic circuit," e.g., the length and cross section of the iron ring followed by the flux and the permeability of the iron; just as the dimensions and material of the electric conductor determine its resistance. The attribute of the magnetic circuit (corresponding to resistance) is called its reluctance.

It must be remembered that this analogy is purely mathematical, not physical. In magnetism, there is no flow of charge, as in electricity.

For a single magnetic circuit consisting of two parts, magnetic material having a permeability μ_1, length l_1, and A_1, and an air gap of permeability μ_0, length l_2 and area A_2, Ampere's circuital law gives $\int H \cos\theta\, dl = H_1 l_1 + H_2 l_2 = NI$, N is the number of turns and I the current through the coil producing the magnetic flux Φ_m. Now as

$$H_1 l_1 = B_1 l_1/\mu_1 = \Phi_m l_1/A_1\mu_1$$

and

$$H_2 l_2 = B_2 l_2/\mu_0 = \Phi_m l_2/A_2\mu_0$$

So

$$\Phi_m[l_1/A_1\mu_1 + l_2/A_2\mu_0] = NI$$

or

$$\Phi_m = \frac{NI}{[l_1 A_1 \mu_1 + l_2/A_2\mu_0]} = \frac{F}{R}$$

where F is the magnetomotive force in ampere-turns and R, the reluctance of the magnetic circuit. This relation is a magnetic analogy of Ohm's law known as Bosanquet's law.

The *gilbert* (Gb) is the electromagnetic CGS unit of magnetomotive force. The *ampere* (*or ampere turn*) (A) is the SI unit. The *oersted* (Oe) is the electromagnetic CGS unit of magnetic field strength. The *ampere per meter* (A/m) is the SI unit.

Magnetic Materials. The electrical industry is dependent upon four basic types of materials: (1) good conductors of electricity; (2) high-resistivity conductors capable of withstanding high temperatures; (3) insulators; and (4) magnetic materials. The development of better magnetic materials has contributed to marked improvements in motors, generators, transformers, and instruments of all kinds.

The ferromagnetic elements are iron, nickel, and cobalt. Of these, iron is the only one that has important magnetic applications in pure, or commercially pure form. All three elements, and many others which themselves are nonmagnetic, are used in special alloys in which certain magnetic characteristics are developed to a high degree. It is even possible, by alloying, to develop magnetic material from certain nonmagnetic elements.

There are two distinct groups of ferromagnetic materials, those that are easily demagnetized, and those that are not. These are often designated as soft and hard magnetic materials. In many respects, iron is an excellent soft, magnetic material. It attains the highest saturation value of magnetic induction found in any material, with the exception of certain cobalt alloys. However, its maximum permeability (maximum ratio of magnetic induction to magnetizing force) is surpassed by other materials. When used as a core in a field induced by ac, the so-called "iron losses" are relatively high. Iron losses consist of energy losses related to the area of the hysteresis loop, and of eddy-current losses caused by induced current circuits within the metal. High values of residual induction (B_r on the typical hysteresis curve) and coercive force (H_c) are indications of high hysteresis loss. See also **Hysteresis**. A low value of electrical resistivity results in high eddy current losses. However, other factors such as the thickness of the sheets in laminated cores, and to a lesser degree grain size also affect eddy-current losses. By special heat treatments and by the use of specially purified iron, much higher values of permeability and lower hysteresis losses can be obtained.

With a good grade of silicon electrical steel, the hysteresis loss can be cut in half and the eddy-current loss reduced even more because of the high resistivity compared with iron. This results in much more efficient operation of alternating-current equipment.

Certain high-nickel alloys of the Permalloy type have very high permeabilities at low and moderate inductions and give low hysteresis losses. For many special applications in instruments and communications equipment, these higher cost alloys are economically justified. Power transformers, motors and generators, etc., are designed for silicon electrical steels.

In the use of any of these materials for ac applications, it is desirable to reduce the thickness of individual laminations to reduce eddy-current losses. For special applications at higher than usual power frequencies, extremely thin strips are used, approaching $\frac{1}{1000}$ of an inch (0.025 millimeter) in thickness. The most common thickness at 60-cycle power frequencies is about $\frac{14}{1000}$ inch (0.35 millimeter). In the communications field, where the use of very high frequency current greatly aggravates the problem of eddy-current losses, alloys of the Permalloy type are used. They are produced in powdered form, given a very thin insulating film, and compacted under high pressure into a suitable core shape.

In the production of magnetically soft materials, it is important that elements such as carbon and sulfur, which disturb the continuity of the base-metal crystal structure, be reduced to low residual amounts, and that the material be used in a strain-free condition. High annealing temperatures resulting in coarse grain structures are used. These conditions are, for the most part, reversed when magnetically hard materials are sought. These materials must be capable of reaching high inductions, but upon removal of the direct magnetizing force the magnetic flux remaining in the circuit should be high, therefore a hysteresis loop of large area is desired. A good overall measure of the quality of a permanent magnet is the "maximum external energy." This is the maximum value which can be obtained for the product of the coordinates of a point on the curve between B_r and H_c on the hysteresis curve.

Progress in Development of Magnetic Alloys. During the last 100 years of experience in producing and using permanent magnets, alloy magnets as of the early 1980s are the most powerful by a factor of 30. Until the late 1930s, magnets were produced from steel. Then considerable research was devoted to producing and testing various magnetic alloys. Alnico magnets became available in the mid-1940s. representative present-day Alnico magnets contain a number of other metals. For example, Alnico 5 contains 8% aluminum, 14% nickel, 24% cobalt, and 3% copper, the remainder being iron. Alnico 9 contains 7% aluminum, 15% nickel, 35% cobalt, 4% copper, 5% titanium, the remainder being iron. Work on the development or magnetic alloys containing various rare-earth elements commenced in the late 1950s. A number of magnetic alloys containing rare-earth elements and cobalt are available today. As pointed out by Chin (1986), the magnetism of the R-Co compounds is due to the interatomic exchange between the spins of the two sublattices plus the spin-orbit coupling within the rare-earth atoms. The spins of the lighter rare-earth elements, such as cerium, praseodymium, neodymium, and samarium, align parallel with the cobalt atoms, resulting in high saturation magnetization values at room temperature. Other rare-earth elements align antiparallel with the cobalt atoms and thus the values are lower. In the mid-1970s, the alloy $Sm_2(Co, Fe, Cu)_{17}$ was introduced. See also **Rare-Earth Elements and Metals**.

The importance of improved magnetic alloys is sometimes underestimated. Magnets are used widely in telecommunications and in scores of electrical and electronic instruments and appliances, and it has been estimated that the demand for permanent magnets has increased by a factor of three since 1970. Chromium-cobalt-iron alloys have increased markedly in application. It is interesting to note that these magnets have essentially about the same coercive force and maximum energy product as Alnico 5, but with only half the cobalt content.

See also **Magnetostriction**.

Molecular Magnets. Ordinary iron and alloy magnets are made of atomic constituents that are difficult to modify. By contrast, molecular magnets may be customized much more easily, and they account for the past decade of research into formulating superior molecular magnets. In mid-1991, A.I. Epstein (Ohio State University) and J.S. Miller (DuPont Central Research and Development Facility) appear to have synthesized an organic-based magnet that retains its magnetism at room temperature and up to and including an elevated temperature of 350°K. At the latter temperature, the polymeric material commences to break down chemically. M. Hoffman (Northwestern University) observes that, if this is a molecular magnet, it represents an enormous leap. No researcher in the field has come even close to 350°K.

This particular magnet, although of exceeding interest in the laboratory, is not considered for practical applications because the material deteriorates rapidly unless it is contained within an inert atmosphere. According to a paper released by J.M. Manriquez, et al., the magnet is the result of a reaction of bis (benzene) vanadium with tetrocyanoethylene (TCNE). The material is an amorphous black solid that exhibits field-dependent magnetism and hysteresis at room temperature.

Quantum Effects and Tiny Magnets. Since early computer designs of the 1950s, there has been tremendous pressure exerted to microminiaturize magnetic information systems. This marked trend continues as designers strive for increased information density. But is there a limit? Some scientists now are showing serious concern that, as magnetic devices become smaller and smaller, the point may be reached where quantum effects could cause upheavals in the magnetic storage system. It is interesting to note that, in the computers of the 1950s, an estimated 100 billion atoms were needed to store one bit of information. In the early 1990s, this figure had dropped to an estimated 1 billion atoms. Experts then predicted that, by the year 2000, the storage of one bit of information would require only 100,000 or fewer atoms. Researchers at the IBM Thomas J. Watson Research Center and other researchers in the computer field have pioneered techniques for studying magnets composed of 100,000 atoms at temperatures approaching absolute zero. Some of these researchers, who

have instrumentation that can measure a magnetic field with a million times the sensitivity of traditional instruments, have observed what they believe may be evidence of tunneling and other effects of quantum mechanics in experimental magnetic systems. One group claims that, when a tiny magnet is cooled to absolute zero, the north and south poles of this magnets can be reversed effortlessly, causing havoc to enter the magnetic storage system. Another group clings to the concept of classical mechanics and the idea that "spins" cannot occur without receiving energy — thus a so-called "energy barrier." By contrast, consistent with quantum mechanics, there is a "chance" that spins will be capable of breaking through the energy barrier. As pointed out by R. Ruthen, "This phenomenon is analogous to the ability of an electron to tunnel through an energy barrier."

Thus, a profound difference of opinion remains and will require continued research to find the truth.

An excellent description of the early investigations of electricity and magnetism is given by L. Pearce Williams (reference listed).

Additional Reading

Aharoni, A.: "Introduction to the Theory of Ferromagnetism," Oxford University Press, Inc., New York, NY, 2000.

Amato, I.: "Some Molecular Magnets Like It Hot," *Science*, 1379 (June 7, 1991).

Beeteson, J.S.: "Visualising Magnetic Fields," Academic Press, Inc., San Diego, CA, 2000.

Campbell, W.H.: "Earth Magnetism," Academic Press, Inc., San Diego, CA, 2000.

Chin, G.Y.: "Magnetic Materials," in Encyclopedia of Materials Science and Engineering, MIT Press, Cambridge, MA, 1986.

Comstock, R.L.: "Magnetic Recording," John Wiley & Sons, Inc., New York, NY, 1999.

Craik, D.J.: "Electricity, Relativity and Magnetism," John Wiley & Sons, Inc., New York, NY, 1999.

Edmonds, D.: "Electricity and Magnetism in Biological Systems," Oxford University Press, Inc., New York, NY, 2001.

Hamilton, D.P.: "A Reprieve for MIT's Magnet Lab," *Science*, 850 (August 23, 1991).

Kubler, J.: "Theory of Itinerant Electron Magnetism," Oxford University Press, Inc., New York, NY, 2000.

Majils, N.: "The Quantum Theory of Magnetism," World Scientific Publishing Company, Inc., Riveredge, NJ, 2000.

Manriquez, J.M. et al.: "A Room-Temperature Molecular/Organic-Based Magnet," *Science*, 1415 (June 7, 1991).

Miller, J.S., A.J. Epstein, and W.M. Reiff: "Molecular/Organic Ferromagnets," *Science*, 40 (April 1, 1988).

Miller, J.S., M. Drillo: "Advances in Magnetism: From Molecules to Materials," John Wiley & Sons, Inc., New York, NY, 2000.

Prinz, G.A.: "Hybrid Ferromagnetic-Semiconductor Structures," *Science*, 1092 (November 23, 1990).

Ruthen, R.: "Quantum Magnets," *Sci. Amer.*, 28 (July 1991).

Williams, L.P.: "André-Marie Ampère," *Sci. Amer.*, 90 (January 1989).

Winters, A.J., et al.: "Large-Scale Superconducting Separator for Kaolin Processing," *Chem. Eng. Progress*, 36 (January 1990).

MAGNETITE. The mineral magnetite, ferroferric oxide, Fe_3O_4, is isometric, commonly occurring in octahedrons, dodecahedrons, and massive, granular, and laminated forms. It is brittle with an uneven fracture; cleavage is not distinct, but with pressure an octahedral parting may develop; hardness, 5.5–6.5; specific gravity, 5.18; luster, metallic to dull; color, iron black, streak, black. It is opaque and strongly magnetic; when possessing polarity it is known as lodestone. Important large ore bodies are products of magmatic segregations, with titanium a prominent constituent of such deposits. Magnetite is a common mineral in the igneous rocks, especially those of the ferromagnesian varieties, and is found in many metamorphic types. It is associated with corundum in emery.

In northern Sweden are located what may be the largest magnetic deposits in the world, believed to have been formed by segregation in the magma. Magnetite is also found in Norway, in the Urals, Italy, Switzerland, Australia, and Brazil. In the United States, the Precambrian rocks of the Adirondacks contain large beds of magnetite, as well as extensive deposits of titaniferous magnetite, and the mineral is found also in New Jersey, Arkansas, and Utah. In Canada it is found in Quebec and Ontario. The lodestone or natural magnet is found in Siberia, the Harz Mountains, Germany; the Island of Elba; and at Magnet Cove Arkansas. The name magnetite is said to be derived from the district of Magnesia, near Macedonia. There is, however, a fable that it was named for a shepherd, Magnes, whose iron-bound staff and shoes with iron nails struck to the ground in which magnetite was present.

This mineral is an important ore of iron, 72% being metallic iron. Magnetite sometimes is referred to as magnetic iron ore.

See also **Ocean Resources (Mineral)**.

ELMER B. ROWLEY, Union College, Schenectady, NY.

MAGNETO. An alternating-current generator that uses one or more permanent magnets to produce its magnetic field. Also called a magneto-electric generator.

MAGNETOELECTRIC. Of or pertaining to electricity produced by or associated with magnetism. Electromagnetic pertains to magnetism produced by or associated with electricity.

MAGNETOHYDRODYNAMIC GENERATOR. The greatest concentration of research, development, and design effort in the technology of converting sources of energy into electricity (for use by the electric utility field) has been concentrated on rotating equipment. However, since about the mid-1970s, there also has been considerable research and experimentation with new kinds of batteries, fuel cells, thermionic converters, and magnetohydrodynamic generators. The possible use of magnetohydrodynamic generators in times of crisis that affect vast, interconnected electrical networks and which have resulted in wide regional blackouts in a few instances has received considerable attention.

The magnetohydrodynamic (MHD) generator is one in which a thermally ionized gas is forced at high temperature, pressure, and velocity through a duct situated in a transverse magnetic field. An induced voltage appears in the third mutually perpendicular direction (the Hall effect), and this voltage may be tapped by electrodes within the duct. If the exhaust gas from the MHD generator is used to heat steam for a conventional generator, a larger portion of the thermal spectrum will be utilized and the system efficiency may be raised from the present approximately 40% to possibly 50 or 55%. Heat for the system may come from the use of fossil fuel or nuclear reactors. An additional advantage of the MHD generator would be the absence of rugged moving parts and close tolerances in the MHD portion of the cycle.

The comparative conceptual simplicity of the MHD generator is evident from Fig. 1. The steady-state operation of this unit is governed approximately by the following one-dimensional flow equations:

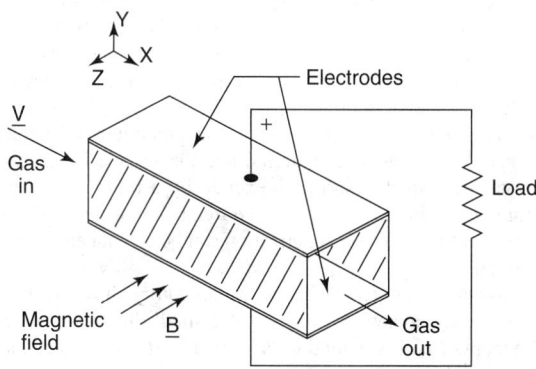

Fig. 1.   Highly simplified diagram of MHD generator.

Force/Unit Volume

$$\rho v \frac{dv}{dx} + \frac{dp}{dx} - (\underline{J} \times \underline{B})_x + F = 0$$

Power/Unit Volume

$$\rho v \left(\frac{dH}{dx} + \frac{dv}{dx} \right) - \underline{J} \cdot \underline{E} + G = 0$$

Mass Flow

$$\rho v A = \text{constant}$$

and by Ohm's law, modified to account for magnetic fields and Hall effect:

$$J + \mu (\underline{J} \times \underline{B}) = \sigma [E + (\underline{v} \times B)]$$

where ρ = gas density
  $\underline{v}$ = gas velocity factor
  ρ = gas pressure
  $\underline{J}$ = current density vector
  $\underline{B}$ = applied magnetic field
  F = wall friction force/unit volume
  H = gas enthalpy
  $\underline{E}$ = electric field vector
  G = heat loss through walls/unit volume
  A = cross sectional area of duct
  μ = electron mobility in gas
  σ = electrical conductivity in gas

The last equation is of most interest in working with the electrical load, since it greatly influences the way in which power is drawn from the unit. The axial component of $(\underline{J} \times \underline{B})$ produces an axial field, called the Hall field, which may be several thousand volts/meter. This causes currents to flow along the electrodes, creating high ohmic losses. To minimize these losses, the electrodes are segmented and separated by strips of insulating material, as shown in Fig. 2. A large MHD unit would typically have over 100 segments.

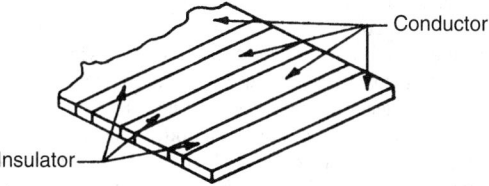

Fig. 2.   Section of segmented electrode.

For a duct with segmented electrodes, there are several ways to connect the load. Two of these connections are shown in Fig. 3. In the Faraday connection (a), the total load is divided among the electrodes; each electrode segment then behaves like a separate MHD generator. The electrical properties can be derived from the last equation by setting J_x (the axial component of current) = 0 because of the segmenting, and $J_y = -\alpha E_y$, where a represents the load admittance. The Faraday connection makes most efficient use of the thermal energy of the gas, but the requirement of multiple load connections is a drawback. This shortcoming can be overcome, however, by paralleling inverter-transformers at the ac line. The Hall connection uses the induced Hall potential to drive the load. The top and bottom of each electrode segment are joined, so that no transverse electric field exists.

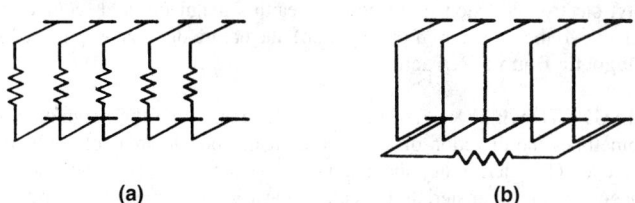

(a)   **(b)**

Fig. 3.   Load configuration for segmented electrode generator: (a) Faraday connector; (b) Hall connection.

Principal Applications to Date. Particularly in connection with the investigation of the effectiveness of MHD units to supplement or reduce required system load shedding in a situation where an island is formed with insufficient generation and with resulting declining frequency, it is believed that the use of quick-responding MHD units can significantly reduce the amount of required underfrequency load shedding, mainly in situations where the sum of the rating of the MHD units plus fast-acting system reserve is approximately equal to the load/generation imbalance at the time of separation. The benefits appear to be less pronounced, but are still significant even if this sum is only one-half of the imbalance. The MHD unit seems most effective when all of the bootstrap power output of the unit is used to build up the unit magnetization to rate level before providing power to the system. Bus voltage oscillations seem less severe for this procedure than for the approach where the MHD magnet is pre-energized, or where the unit magnetization is allowed to build up more slowly (by using only part of the bootstrap startup power for energizing purposes, with the remainder allowed to flow into the power system).

MHD units (as proposed for quick-start, short-service devices) do not seem particularly appropriate for subduing short-term (0–3 second) power system swings for two reasons. (1) In order to provide the response time required to be effective, 15 megawatts of continuous excitation would be required to obtain quick availability (20–40 milliseconds) of 200 megawatts for each unit. (2) Quite a few units would have to be scattered throughout the system to provide coverage for even a relatively small number of possible contingencies. It is obvious that the costs associated with the required continuous excitation power and the number of units needed for reasonable system coverage would not make this application of MHD units attractive.

It is generally concluded that quick-start MHD units should not be considered as an alternative to underfrequency load-shedding practices, but rather as a supplement to these practices. Underfrequency load shedding is attractive for combating rapidly dropping frequency because it is effective, has comparatively low installation costs, can be implemented independently at many scattered substations throughout the system with resulting high reliability, and can be tailored rather easily to the power system's requirements. However, if a particular power system operator did not choose to incorporate underfrequency load shedding, a sufficient number of quick-responding MHD units could be used to boost drooping frequency during the time period (up to several minutes), when the reserve capability of the other prime movers is being realized. Hence, this would potentially allow time for centrally controlled manual load shedding, if required. In this situation, the rating of MHD units would be more effective than the spinning reserve of conventional prime movers. This is because the MHD units are much faster to respond and do not suffer from such problems as the steam depletion problem inherent to many steam units after about one minute of significantly increased output.

The potential use of MHD units as infrequently-used peaking units may be considered, but other types of peaking units, such as gas turbines and diesel units, become very competitive for usages exceeding 100 hours/year.

MAGNETOHYDRODYNAMICS (MHD). The study of the interaction that exists between a magnetic field and an electrically conducting fluid. Also called *magnetoplasmadynamics, magnetogasdynamics,* or *hydromagnetics.*

MAGNETOIONIC THEORY. The theory of propagation of electromagnetic radiation through a medium containing ions in the presence of an external magnetic field. It applies to the propagation of radio waves in the ionosphere, and provides theoretical relationships among such aspects of the subject as the index of refraction, radiofrequency, free-electron density, electron collision frequency, the earth's magnetic field (components relative to the direction of propagation), the nature of polarization, etc. See **Magnetic Double Refraction**.

MAGNETOMETER. An instrument for measuring the magnitude and sometimes the direction of a magnetic field. Specific areas of application include: (1) determining the magnetic moment of a specimen and the magnetization of materials (2) calibrating electromagnets and permanent magnets; and (3) measuring the strength of magnetic fields and their components on or near the surface of the earth, as well as in space. Numerous physical principles are applied in a variety of magnetometers.

The *classical astatic magnetometer* is used to determine the magnetic moment of rod-shaped samples. The specimen can be exposed to the controllable homogeneous magnetic field produced in the center region of a solenoid that is appreciably longer than the specimen. Its magnetic moment is characterized by strength and direction of the field at a known distance from the sample. Two equivalent permanent magnet needles horizontally placed and rigidly linked by a nonmagnetic rod in a vertical position comprise the measuring system. Arranged in antiparallel alignment, the needles have a wide distance between them, as compared to length. This astatic system, unaffected by the earth's field, is suspended on a calibrated torsion wire. The axis of the test specimen is placed perpendicular to the axis of the lower needle and in its plane of rotation. It is possible to cancel the field of the magnetizing solenoid at the location of the sensing needle

by means of a magnetically opposing solenoid located in the same plane. The magnetic moment of the specimen can be derived from the angular deviation of the needle occurring from the action of the static field. With a modified astatic magnetometer, the magnetic properties of very small specimens in the form of fine wires or films have been measured. The specimen is arranged parallel to the axis of the astatic system and a short distance from the closely spaced short needles, in order that each needle will sense the pole of the neighboring sample. With this method, films as thin as 10^{-5} centimeter and weighing less than 1 milligram have been studied.

A more recent instrument for measurement of very small samples is based upon the relative periodic displacement of the dipole field of the specimen against pickup coils and subsequent amplification of the small, induced alternating current voltages.

The *vibrating sample magnetometer* makes use of a mechanical oscillator that can be either a loudspeaker or a motor. The sample, attached to the lower end of a nonmagnetic shaft, moves up and down, between and parallel to the pole faces of an electromagnet. Amplitude of the oscillation is approximately 2 millimeters, the frequency is approximately 80 Hz with loudspeaker drive. The sample stays within the homogeneous region of the magnetizing field. Two series opposing signal coils with approximately 20,000 turns each are located on each side of the sample, with axes parallel to its motion and perpendicular to the exciting field. The magnetic moment can be derived directly by comparison with the voltage excited by a standard sample of known magnetization, such as nickel.

In the *vibrating coil magnetometer*, the sample is kept in a fixed position in the homogeneous region of the magnetic field between the pole faces. The signal coil oscillates at approximately 40 Hz, its axis and velocity vector being collinear with the dipole axis of the specimen. The distance between coil and sample is large enough to allow for installation of temperature- and pressure-generating apparatus to enclose the sample. Special precautions have to be taken to eliminate the signal produced by the curvature of the magnetizing field. The sensitivity of this method is considered excellent.

The *pendulum magnetometer* is a rather simple apparatus, which utilizes the ponderomotive force that a sample experiences in an inhomogeneous magnetic field. The device is based on the concept that this force is, for small deviations from the position of maximum field strength, proportional to the displacement perpendicular to the field lines between the hemispherical pole pieces of an electromagnet. This condition creates a simple harmonic motion. The specimen is fastened to a lightweight bar whose movement is constrained only in the direction of its length and without rotation by a quinquefilar suspension. The magnetization is determined from measurements of the periods of oscillation with and without magnetic field.

In the *vibrating reed magnetometer*, the pendulum is replaced by a metallic reed of nonmagnetic material, and the sample is attached to one end. The vibration of this spring is excited by a piezoelectric transducer driven from an oscillator. The resonance frequency of the reed is observed with and without field. The magnetization can be calculated by comparison with a reference sample.

The *gaussmeter* or *fluxmeter* is a laboratory instrument used for calibrating electromagnets. It consists of a small direct-current generator. A small generator coil is wound on a nonmagnetic core, placed at the end of a 3- to 4-foot long axis, and driven by an alternating current motor with constant speed of revolution. The induced sinusoidal voltage is rectified by a commutator and read on a voltmeter calibrated in magnetic field units. The device is inherently linear.

The *earth inductor* is used for measurement of the inclination (magnetic dip) of the earth field. A coil, connected through commutator and brushes to a sensitive galvanometer, is rotated about its diameter by a hand-powered flexible shaft. When the rotation axis of the coil is brought in line with the earth field, the galvanometer will read zero. The inclination of the axis against the horizontal plane can be read with an accuracy of approximately 0.1 minute.

The *classical magnetometer of Gauss* can be used to determine the horizontal field intensity of the earth field in stationary observatories. Two measurements are required: (1) the measurement of the period of a permanent magnet, which is vertically suspended on a torsion fiber, when it oscillates in the horizontal plane about the magnetic meridian; and (2) the measurement of the deflection angle, which a magnetic needle experience attached to the same suspension, when the permanent magnet acts at a present distance in a preferred position together with the earth field upon the needle. This is an absolute method.

The *sine galvanometer* is an absolute instrument for determination of horizontal intensity. With it, the suspended detector magnet is acted upon by a field produced by a calibrated coil and the horizontal field component. Accurate measurements of the coil current and deflection angle of the needle are required.

The *flux-gate magnetometer* is well suited to airborne applications and is used for detecting magnetic anomalies and for exploring the earth's surface in search of mineral deposits. The instrument also has been used in space probes, and in underwater research.

The operation is based upon the change of permeability of a highly sensitive material in weak fields. The device incorporates two permalloy cores in parallel position. A coil is wound on each of the cores. The two windings are opposed in their polarities and connected in an impedance bridge circuit in such a manner that an alternating current voltage supplying the bridge will not produce a diagonal voltage. The bridge is balanced in the absence of an external field. See Fig. 1. If a component of the earth's field parallels the axis of the core, the bridge is unbalanced due to the opposing magnetic biases. In use, three mutually perpendicular flux gates are mounted on a platform that can be rotated by servomotors in two perpendicular planes. Activated by the diagonal voltage from one flux gate, each servomotor rotates its twin core, decreasing the unbalance and holding the cores in zero position. The combined action of two gates then brings the cores of the third gate into the direction of the field. An additional field winding encloses both cores of the third gate. A servomechanism controlled by the diagonal voltage of this gate provides a current through the field winding, in order to annul the earth field. The magnitude of this current is a measure of the strength to the earth's field. The instrument is sensitive to about 10^{-5} oersted.

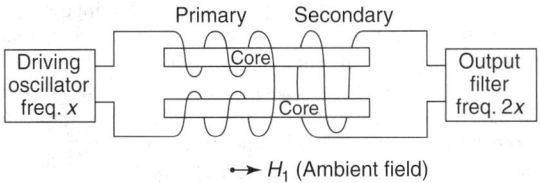

Fig. 1. Schematic of fluxgate magnetometer.

In the *Hall-effect magnetometer*, a Hall voltage occurs in a current-carrying sample of semiconducting material perpendicular to the current and perpendicular to an applied static magnetic field. The magnitude of this voltage is proportional to the field. The Hall voltage amounts to about 100 millivolts at 10,000 gauss, with a 100 milliampere current. This is primarily a laboratory device.

Nuclear magnetometers are also well adapted to airborne measurements. They utilize the magnetic properties of atomic structures. The two principal types are: (1) Proton precession and (2) alkali vapor. Both types have the advantage of being insensitive to the direction of the field and of producing absolute measurements in terms of frequency, the physical quantity most easily measured with high accuracy. See Fig. 2. The *alkali-vapor magnetometer* makes use of the fact that if circularly polarized light

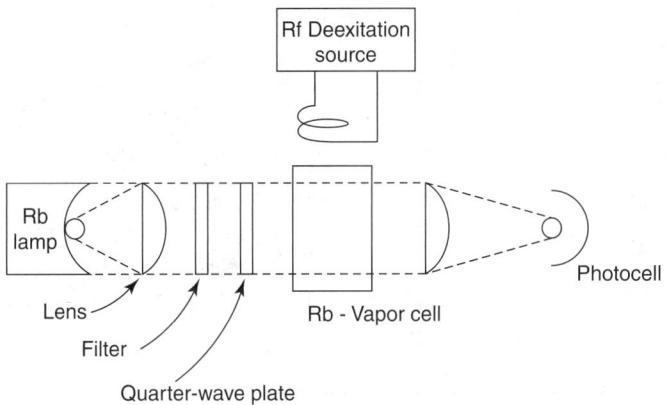

Fig. 2. Schematic of rubidium-vapor magnetometer with no signal feedback.

is shone through vapor of an alkali metal that is being excited by a varying radio-frequency magnetic field, light is absorbed at certain frequencies related to the energy levels in the atoms of a steady external magnetic field. By measuring the absorption frequencies, it is possible to determine continuously the external magnetic field.

The *metastable-helium magnetometer* was developed for spaceborne applications, which require measurement of changes in planetary magnetic fields accurate to 1 part in 5 million and a sensitivity of 0.01 orested. In operation, the system measures the absorption of infrared radiation by helium in a metastable state. Mechanically, a helium spectral-emission lamp and a helium-filled absorption cell are excited by a radio frequency source. A sensitive infrared detector then measures the amount of infrared radiation absorbed by the metastable helium. This amount is proportional to variations in planetary magnetic fields. The system also is used in antisubmarine warfare detection systems.

MAGNETO-OPTICAL ROTATION. Some substances are in themselves optically active, that is, they rotate the polarization plane of polarized light passed through them. In 1845, Faraday discovered that glass and other substances devoid of this property acquire it when placed in a strong magnetic field. This is the so-called Faraday effect. The light must traverse the substance along the lines of force. The direction of the rotation is reversed if the field is reversed, but is the same with respect to the observer whether the light is going or coming, so that a beam passing one way and reflected back has its rotation thereby doubled (which is not the case with natural activity). Some substances produce right-handed, some left-handed, rotation in light traveling in the direction of the field.

MAGNETOSPHERE. The region of the Earth's atmosphere where ionized gas plays an important part in the dynamics of the atmosphere and where the geomagnetic field, therefore, plays an important role. The magnetosphere begins, by convention, at the maximum of the *F layer* at about 350 kilometers and extends to 10 or 15 Earth radii to the boundary between the atmosphere and the interplanetary plasma. See also **Earth**.

MAGNETOSTATIC THEORY. See **Electromagnetic Phenomena**.

MAGNETOSTRICTION. When a polycrystalline nickel sample is placed in a magnetic field, it contracts along the field direction by about 30 parts per million and elongates in the transverse direction by about half that amount. There is also a small volume change. Such changes in dimension of magnetic materials with variation of magnetic field strength or direction are termed *magnetostriction*. They are measured by strain gages, optical dilatometers, capacitance variation, and x-ray analysis. Below the Curie Temperature, magnetostriction in weak fields is caused by domain rotation, becoming appreciable at fields near the knee of the $B - H$ curve. See also **Hysteresis (Magnetic)**. The saturation magnetostriction of a single crystal depends upon the direction of the (sublattice) magnetization and the direction of measurement with respect to the crystal axes. Magnetostriction coefficients vary greatly, depending upon the material, temperature, and the magnetization state. The source of magnetostriction is the dependence of magnetic energy on strain. Because the elastic energy is quadratic in strain while the magnetoelastic energy is linear in strain, the minimum free energy occurs at nonzero strain.

Magnetostriction has been put to practical use in the magnetostrictive resonator. This is essentially an iron rod maintained in longitudinal elastic vibration by a high-frequency current in a helix wound upon it, and used, through the joint operation of the Joule and Villari effects, to control the frequency of the current, somewhat after the manner of the familiar piezoelectric (crystal) resonator. It is also used in band-pass electrical wave filters.

MAGNETOSTRICTIVE DELAY LINE. In electronic computers, a device in which a wave is induced by the characteristic, possessed by nickel and certain other materials, of shortening in length when placed in a magnetic field. The wave travels at the speed of sound through the material.

MAGNETRON. See **Microwave Tubes**.

MAGNIFYING POWER. Crudely defined, the magnifying power of an optical instrument is the ratio of the apparent size of an object as seen through the instrument to the apparent size of the same object as seen without the instrument. For a telescope, the magnifying power may be defined as the ratio of the size of the retinal image obtained with the instrument to the size of the retinal image obtained without any optical aid. When a positive eyepiece is used, the magnifying power may be shown to be directly equal to the ratio, of the focal length of the object glass of the telescope, to the focal length of the eyepiece. For example, a telescope with an object glass of 10-foot (3-meter) focal length will have a magnifying power of 60 when used with a positive eyepiece of 2-inch (~13-millimeter) focal length. It should be noted that changing the eyepiece will change the magnifying power of the telescope. That is, if an eyepiece with 0.5-inch (5-centimeter) focal length were used with the above telescope, the magnifying power of a telescope with interchangeable eyepieces is meaningless unless the particular eyepiece is specified for the particular telescope.

MAGNITUDE (Stellar). See **Stellar Magnitude**.

MAGNOLIA TREES. Members of the magnolia family (*Magnoliaceae*), there are some 35 species, nearly all of which prefer the warm climates found in the southern United States and in parts of India, China, and Japan. However, the evergreen form (*Magnolia grandiflora*) ranges from New Brunswick in Canada south in the eastern United States to Florida. The tree requires rich soil and considerable water. Adult trees may attain a height of over 100 feet (30 meters). The general form is pyramidal. The bark is gray, rough, and has thin scales. Because of the wide spread of the tree, it is a favorite for shade and landscaping. The leaf is large, elliptical, and from 5 to 8 inches (12.7 to 20 centimeters) in length. The leaf is smooth, of a lustrous green color, and has a leathery feel. The flower is creamy white, very showy, fragrant, and from 6 to 8 inches (15.2 to 20 centimeters) across. Depending upon area, the trees bloom from April to June. The fruit is ovoid, 3 to 4 inches (7.6 to 10 centimeters) in length, and of a brown color when ripe. The wood is soft, light, satiny, close-grained, and of light ocher color and of no commercial value. Several record magnolias have been selected by American Forests. See Table 1. The tulip tree, with tulip-shaped, bluish-green leaves, greenish-yellow flowers, and a long, conelike fruit, is also a member of *Magnoliaceae*.

Breakthrough in Magnolia's DNA. The Clarkia shale deposit near Moscow, Idaho, differs from most rich fossil sources by the manner in which ancient fossils are preserved. Most fossils do not contain DNA because they are completely mineralized over long periods of time. The fossils at Clarkia are classified as compression fossils—that is, cold, oxygen-free sediments in a lake bottom squeezed a magnolia leaf for an estimated 20 million years. Fossils from this region contain biomolecules and subcellular structures, like those of modern plants. Researchers from the University of California at Riverside and the University of Georgia

found the leaf unchanged. D.E. Giiannasi observed, "When we first cracked open the sediment, the leaf was still dark green." Analysis of the magnolia leaf has yielded some important genetic information. Scientists now are directing their attention to other trees. This research represents an early step in the melding of molecular biology and paleontology to shed light on plant evolution.

MAGPIE (*Aves, Passeriformes*). Moderately large long-tailed birds related to the crows. The common North American species, *Pica pica*, ranges from Alaska to Arizona and eastward into Iowa. It is a black and white bird. The yellow-billed magpie, *P. nuttalli*, flies only in California. The European and Asiatic species of magpies are more brightly colored.

The magpies build very large untidy nests and are noted for their curiosity, adaptability, and noisiness.

MAHOGANY TREES. Members of the mahogany family (*Meliaceae*), these trees are found in Central America and the West Indies. A few related species are found in tropical Africa. Species of the genus *Swietenia*, are large trees with pinnately compound leaves like those of ash trees and small flowers in panicles in the leaf axils.

The trees grow in a variety of habitats, often in most inaccessible places, so that it is very difficult to get the cut logs to the market. Mahogany is usually classified according to the region from which it comes, as Cuban mahogany, Honduras mahogany, etc.

The first mahogany to appear in Europe was brought in as ballast in a ship, the heavy logs being very suitable for that purpose. In port it was necessary to remove these in order to get in more cargo. So it was offered to English woodworkers. At first it was rejected as too hard to work and of little use, being heavy and dark-colored. Gradually it found favor, until it became a highly prized wood for fine furniture making. The first mahogany used in cabinet work was Spanish mahogany, *Swietenia mahagoni*. For a long time mahogany from Santo Domingo held first rank and was eagerly sought after. It was a very hard dark wood, which could be given a very high polish, and was very durable. The supply is now nearly exhausted. Cuban mahogany is another variety which gives a dark red wood, and which finishes with a very fine glossy surface. In this, as in some other varieties, the wood is frequently marked with very small white pores of chalk-like substance. With age the wood of these varieties gradually darkens; it does not, however, lose its beautiful smooth finish. Other species of *Swietenia* are shipped from Panama, from Mexico, and from South American countries.

The heavy logs of mahogany are removed from their native forests and shipped to American or foreign markets. Especially valuable are those logs which, when cut, show a wavy grain or other irregularities. Even more valuable are blocks of mahogany that come from a large forking of the stem; from these the beautifully grained crotch mahogany is obtained. This is usually cut into thin veneers. In drying, mahogany

TABLE 1. RECORD MAGNOLIAS IN THE UNITED STATES[1]

Specimen	Circumference[2]		Height		Spread		Location
	Inches	Centimeters	Feet	Meters	Feet	Meters	
Ashe magnolia (1993) (*Magnolia ashei*)	55	140	52	15.8	37	11.3	Pennsylvania
Bigleaf magnolia (1999) (*Magnolia macrophylla*)	55	140	67	20.4	32	9.8	Georgia
Cucumbertree magnolia (1985) (*Magnolia acuminata*)	293	744	75	22.9	83	25.3	Iowa
Frasier magnolia (1998) (*Magnolia acuminata*)	118	300	121	36.9	33	10.1	Tennessee
Pyramid magnolia (1988) (*Magnolia pyramidata*)	62	157	65	19.8	32	9.8	Florida
Pyramid magnolia (1999) (*Magnolia pyramidata*)	46	117	84	25.6	30	9.1	Florida
Southern magnolia (1994) (*Magnolia grandiflora*)	268	681	98	29.9	90	27.4	Mississippi
Sweetbay magnolia (1991) (*Magnolia virginiana*)	173	439	92	28	52	15.8	Arkansas
Umbrella magnolia (1993) (*Magnolia tripetala*)	122	310	50	15.2	50	15.2	Pennsylvania

[1]From the "National Register of Big Trees," American Forests (by permission).
[2]At 4.5 feet (1.4 meters).

TABLE 1. RECORD MAHOGANY FAMILY TREES IN THE UNITED STATES[1]

Specimen	Circumference[2]		Height		Spread		Location
	Inches	Centimeters	Feet	Meters	Feet	Meters	
Tree-of-Heaven (1999) (*Ailanthus altissima*)	248	630	67	20.4	64	19.5	Tennessee
Chinaberry (1967) (*Melia azedarach*)	222	564	75	22.9	96	29.3	Hawaii
West Indies mahogany (1992) (*Swietena mahagoni*)	175	445	79	24.1	96	29.3	Florida

[1] From the "National Register of Big Trees," American Forests (by permission).
[2] At 4.5 feet (1.4 meters).

shrinks very little, and once dry, it is very durable, twisting or warping very little.

The record West Indies mahogany (*Swietenia mahagoni*) growing in the United States is located in Everglades National Park, Florida. See Table 1.

African mahogany is becoming increasingly valuable. It is obtained from large trees of the genera *Khaya* and *Entandrophragma*, both members of the same family as mahogany, and yielding woods very similar to it. Numerous other tropical woods are imported under the name of mahogany, because of a similarity in texture or color. Some of these are valuable, but many are definitely inferior substitutes.

The so-called Tree of Heaven or Tree of the Gods, commonly known as ailanthus, is also a genus of trees in the *Meliaceae* family. They have large pinnate leaves, small flowers, and winged fruit. One species, *Ailanthus glandulosa*, is a native of China and Japan. It has been introduced into the eastern United States, largely because of its excellent resistance to atmospheric pollutants. One objectionable feature is the unpleasant odor of the male flowers. One specimen singled out by American Forests is located in Hamilton Tennessee. See Table 1.

The lauan tree, which grows in the Philippines, Malaya, and Sarawak, is also of the *Meliaceae* family. Several genera of the trees grow in these areas. On the American market, the wood is known as Philippine mahogany, obtained chiefly from *G. shorea*. Dark-red Philippine mahogany, called tangil, is obtained from *S. polysperma* and most closely resembles true mahogany. It is also called Bataan mahogany. The wood from numerous varieties of the lauan tree ranges widely in color and other characteristics—so that when specifying mahogany, one must be quite exacting as to source.

The Chinatree (*Melia azedarach*) is also of the *Meliaceae* family. The tree is native to Asia, but is found in the southern parts of the United States and in Mexico. The tree generally is considered an attractive, ornamental shade tree. The top is nearly flat. The leaves are deciduous, alternate, and bi-pinnately compound. They are about 15 in. (38 cm) long, pointed, and light green. The tree grows rapidly, but has a rather short life span. The fruit is a fleshy, yellow berry, less than 1 in. (2.5 cm) in diameter. The flower is purple and fragrant and hangs in long clusters.

The tree is commonly referred to in the southern United States as the chinaberry tree, although other names include pride of India tree, umbrella tree, Cape lilac, Indian lilac, Persian lilac, China tree, and paradise or paraiso tree. In an article by M.D. Hodgins (American Forests, 22–61, May 1979), the very extensive plantings of this tree along the streets of Savannah, Georgia are described. The pale lilac blossoms of the tree perfume the air. Although still proliferating in suburban and rural areas throughout the southeastern United States, urban use was progressively discouraged because, as pointed out by H.S. Traub, "the 'smelly, gooey mess' the trees made when they dropped their berries led to their removal from the historic district of Savannah."

The record chinaberry tree (*melia azedarach*) growing in the United States is located in Kaobe, South Kona, Hawaii. See Table 1.

MAIDENHAIR TREE. The sole surviving member of an order of *Gymnosperms* which in Mesozoic times was very abundant and widely distributed. Also called the Ginkgo tree, it is of a single-species family *Ginkgoaceae* (Ginkgo family). There is uncertainty as to whether the tree still may be growing wild in any region today. Occasional reports are received indicating that the tree has been seen on the mountain slopes of China. The tree has been cultivated in the temple gardens of China and Japan for centuries. In late years, the Ginkgo has been widely planted in Europe, America, and other parts of the world. Because the tree dates back

some 200 million years, some authorities observe that the tree has outlived its natural enemies. But even in modern times, the tree seems quite resistant to human-created environments, including the fumes from buses and cars along streets where it occasionally will be seen planted in a masonry tub or opening in the sidewalk. The normal height of an adult tree is about 70 to 80 feet (21 to 24 meters). One specimen in Milan, Italy was reported to have reached 125 feet (37.5 meters). Planters like to avoid female trees because the covering over the nut (edible) radiates a most unpleasant odor' when it decomposes. Although the tree does not have needles or cones, it frequently is classified with the conifers because of its assumed ancestral relationship to the conifers.

The tree often has a tall slender pyramidal shape when young; others, and especially older specimens, are wide spreading. The branches are of two kinds; a long shoot that grows rapidly in length, composed mostly of woody tissue; and short shoots or spurs which elongate very slowly. These short shoots have a large pith, a thick cortex and very little wood. A short shoot may sometimes (especially in the case of injury to the long shoot) become a long shoot. The leaves of the ginkgo tree are somewhat variable in shape. Those on the long shoot are wedge-shaped and deeply notched, those of the short shoot broadly wedge-shaped and little or not at all notched. The veins are furcate, forked, as in the ferns. It is to the leaves that the tree owes its common name, Maidenhair tree, since their shape suggests that of the maidenhair fern. The trees are dioecious, the two types of reproductive organs being borne on different trees. The male strobilus is composed of many sporophylls, each with two sporangia. The female counterpart consists of a long slender peduncle or stalk bearing two ovules. In most cases, one of those aborts early. The pollen grains, which consist of three small disk-shaped cells and one relatively large one, are carried to the ovule by the wind. There each forms a pollen tube, which digests its way through the tissue surrounding the gametophyte. A pollen tube that has nearly reached the gametophyte contains two large sperms, each of which has a spiral coil of cilia at its anterior end. One of these sperms passes to one of the large eggs contained in the female gametophyte, and joins with it. The nucleus of the sperm unites with that of the egg, which is then said to be fertilized. The ovule containing the fertilized egg enlarges. When mature it is about an inch in diameter and green. Its outer covering is fleshy and has a curious rancid odor, which is very noticeable when this coating is crushed. Within this fleshy coat is a dry covering surrounding the gametophyte and the embryo plant. The seeds germinate readily, forming a long tap root and a short erect shoot. Young plants are rather susceptible to low temperatures, requiring some protection in the northern states.

MAILLARD REACTION.   See **Amino Acids**.

MAIZE.   See **Corn (Maize)**.

MAKO.   See **Sharks**.

MAKSUTOV-BOUWERS TELESCOPE. A system, independently devised by Maksutov and Bouwers, wherein a negative meniscus lens is used to introduce spherical aberration, causing a small chromatic effect with very little focusing effect. The primary mirror is then made spheroidal, which cancels the spherical aberration introduced by the lens. This catadioptic system is finding wide applications because of the ease with which one can compute the surface points of a spheroid. The telescope can be designed for use at different foci.

See also **Telescope (Astronomical-Optical)**.

MALACHITE.   The mineral malachite is a basic carbonate of copper corresponding to the formula $Cu_2(CO_3)(OH)_2$. It is monoclinic, crystals tending to be acicular, but usually found massive. It is a brittle mineral; hardness, 3.5–4; specific gravity, 4.05; luster, vitreous to silky or dull; color, green; streak, green; translucent to opaque. Malachite is an alternation product found associated with other copper-bearing minerals. It is a rather common mineral and is found quite widely distributed. Large quantities have been found in the Ural Mountains; it is also found in Germany, France, England, Zaire, Rhodesia, and Australia. In the United States, beautiful radiated masses of fibrous crystals have been found in Berks County, Pennsylvania, as well as in Tennessee at Ducktown, and in Arizona, Nevada, and Utah. Malachite, besides being an ore of copper, has been used for various ornamental purposes. The word malachite is derived from the Greek, meaning a *mallow*, because of its green color.

MALARIA.   Essentially a disease of the tropics, malaria is known to infect 200 to 400 million people of whom 10% will probably die of the disease in adulthood. The mortality in children is about 50% and some 2 billion people (about one-third of the world's population) are threatened by the disease. Distribution of the disease is roughly indicated in Fig. 1.

The cause of malaria was first discovered in 1880 by a French Army surgeon, Charles Laveran, who identified the malaria parasite while examining with the aid of a microscope the fresh blood of a patient infected with falciparum malaria. When Laveran made this discovery in Constantine, Algeria, the leading European medical professionals were under the spell, so to speak, of Louis Pasteur. The bold concept that malaria could be caused by the presence of millions of minute animal parasites in the blood, and not by bacteria, was difficult to accept. Thus six years were required to convince the sceptical medical profession of the validity of Laveran's discovery. By 1886, Camillo Golgi in Pavia, Italy identified two specific human malaria parasites, namely, *P. vivax* and *P. malariae*. A third species, *P. falciparum*, was identified by Ettore Marchiafava in Rome in 1889. An English physician, Patrick Manson, first suggested (1894) that mosquitoes may be the means for transmission of the disease — by drawing

out the malaria parasites from human blood — and that transmission would occur by ingestion of water contaminated by infected dead mosquitoes. Manson was prompted to this observation by prior experience in noting that mosquitoes could suck up the microscopic threadlike worms from the blood of patients infected with filariasis. Of course, Manson's theory of transmission accurately implicated the mosquito, but erred in the details of transmission. Manson desired to prove his theory, but realized that England was not a suitable location. Manson motivated a British Army surgeon, Ronald Ross, to carry on the research in India. For a few years, Ross experimented with mosquitoes at random, but it was not until 1897 that his attention was brought to the *Anopheles* mosquito. Late in that year, while making a microscopic examination of the gut of mosquitoes that had fed on a patient with malignant malaria, he noted protozoa growing only in the Anopheles. However, because of a transfer from Madras to Calcutta, Ross commenced to work with the malaria parasite of birds, which is transmitted by a *Culex* species. Ross proved that the spindle-shaped malaria organisms (sporozites) freed by rupturing of the fertilized egg, migrate from the gut of the mosquito to the salivary glands to be injected into the victim by the bite of the mosquito. To Ross goes the credit for the discovery that malaria is transmitted by the bite. In the same year (1898), a Canadian pathologist, William McCallum, also working with birds, was able to interpret and describe the fertilization process of the parasite, taking place in the gut of the mosquito.

During the period 1886–1899 a group of four Italian scientists worked in Rome on the problem of transmission of malaria in humans. In 1898, a breakthrough came with the observations of Grassi, a physician with a keen interest in zoology and mosquitoes in particular. Grassi noticed that when malaria was present there was always a large population of *Anopheles*, while in areas of large *Culex* populations there was no malaria in humans. From the Campagna Romana near Rome, Grassi collected *Anopheles* mosquitoes which his colleague, Amico Bignami, let feed on a volunteer patient of the Santo Spirito Hospital near Saint Peter's Basilica. On November 1, 1898, the patient had developed the classical symptoms of falciparum malaria. In other volunteers, they proved that only Anopheles mosquitoes transmit malaria in humans.

Even with the significance of this discovery, there remained a link missing in the cycle of the malaria parasite. Still unexplained was the time lapse between the introduction of the parasites through the bite of the

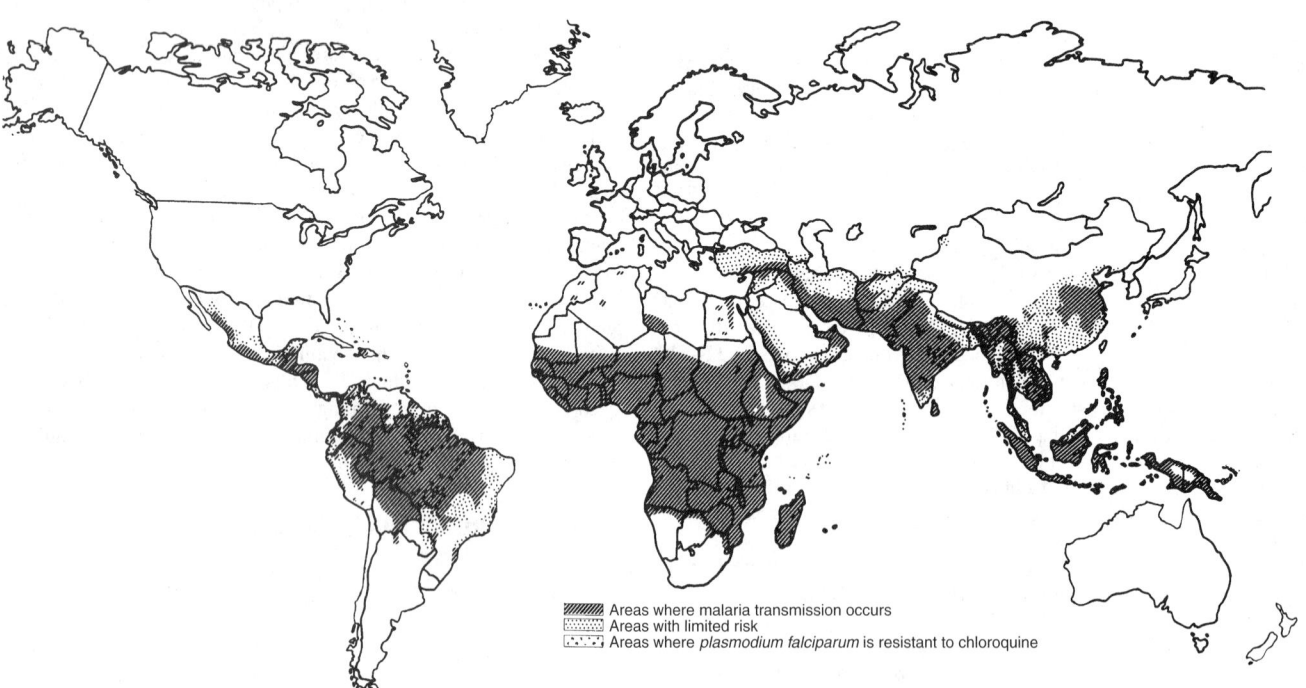

Fig. 1.   World malaria risk chart, including geographical distribution of principal vectors, geographical distribution of falciparum malaria, and areas where Plasmodium falciparum is resistant to chloroquine. Chart prepared by International Association for Medical Assistance to Travellers (IAMAT). Among the several laudable purposes of IAMAT is "To provide world wide medical assistance . . . to travelers who, while absent from their home country, may find themselves in need of medical or surgical care or of any form of medical treatment by providing these travelers with the names of centers in countries other than that to which they are native, the names of locally licensed medical practitioners who have command of the native language spoken by the traveler and who agreed to a stated and standard list of medical fees for their services." IAMAT headquarters is located in Guelph, Ontario, Canada, N1H 7L5.

mosquito and the appearance of the symptoms of malaria. It was not until 1936 that an Italian scientist. Giulio Raffaele, discovered, while working with birds, that the malaria parasites entering the host first undergo a cycle of transformation within the blood-forming cells of the liver. Later, in 1948, British researchers H.E. Shortt and P.C.C. Garnham demonstrated the liver cycle of the malaria parasite in humans. Following a period of extensive trials on monkeys, a human volunteer was bitten during three days by nearly 800 *Anopheles* infected with *P. falciparum*. On the fifth day, a surgeon removed a small piece of tissue from his liver (biopsy), which when examined under the microscope, demonstrated the growth of the parasites in the liver cells.

More than fifty different species of *Plasmodium* can cause malaria in humans, monkeys, birds, fish, and cattle. However, only four species attack humans: *P. falciparum, P. vivax, P. malariae,* and *P. ovale*.

Vivax is the mildest form of malaria and is characterized by periodic chills and fever, an enlarged spleen, anemia, severe abdominal pain and headaches, and extreme lethargy. If left untreated, the disease tends to be self-limiting within a period of 10 to 30 days, but will recur periodically. Although the fatality rate of vivax malaria is low, the disease is highly debilitating and makes the patient more vulnerable to other diseases. In addition to the aforementioned symptoms, falciparum malaria presents edema of the brain and lungs and blockage of the kidneys. Unless treated promptly, the fatality rate of falciparum malaria is high.

The life cycle of the parasite and its course in the human body proceeds in the following way. The saliva of the mosquito contains the *Plasmodium* at the lance-shaped sporozoite stage of its life cycle. Upon inoculation of the host by biting, the sporozoites quickly migrate to the liver where they divide and develop into multinucleated schizonts. Within 6 to 12 days, the schizonts disrupt and release into the blood the form known as *merozoites*. Each liver cell infected by one sporozoite releases into the blood stream from 5000 to 10,000 merozoites. These later invade the host's erythrocytes where they grow and form more schizonts which, in turn, again divide, releasing more merozoites into the blood stream to repeat the cycle. The principal symptoms of malaria are associated with the rupture of the schizonts, the periodic lysis of the blood cells with release of merozoites and toxic wastes, which cause the regular fevers and chills of malaria.

The effects of insect and parasite control chemicals in West Africa and Southeast Asia appear to be losing much of their former ability, particularly concerning *Plasmodium falciparum*, the deadliest form of malaria. For many years, despite some of their adverse effects, insecticides such as DDT, dieldrin, and HCK have been used to control the Anopheles mosquitoes. There is considerable evidence that shows that these chemicals no longer suffice as the principal control measures. This situation precipitates an even greater interest in malaria vaccine(s), which are mentioned later in this article.

As briefly outlined earlier, the quest to prevent and cure malaria and a number of similar African fevers is now in its second century of trial-and-error efforts. D.J. Wyler (Tufts University School of Medicine) places past efforts in perspective, citing a remarkable variety of substances, such as cinchona bark extracts, infusions from wormwood, quinine, and cogeners. The author describes as 40-year drug-discovery program conducted at the Walter Reed Army Institute of Research, during which time one-quarter million compounds were tested as potential antimalarial drugs. Of these, only two, mefloquine and halfantrine were licensed. The details provided by Wyler provide valuable and interesting reading for the person who is interested in tropical diseases, such as malaria.

The American Association for the Advancement of Science (AAAS) has established a Sub-Saharan Africa Program, under a cooperative agreement with the Agency for International Development, to develop and evaluate strategies to combat malaria in Africa. The AAAS is using knowledge within its affiliated societies to review sociocultural, economic, and behavioral factors; environmental and urbanization issues; health care delivery systems; and natural science applications for malaria prevention and control. Each year 80 percent of the 100 million cases of malaria worldwide occur in Africa. Included will be studies of resistance of parasites to drugs and lack of eradication efforts.

Course of the Disease. This varies with the causative protozoan. (1) *Vivax* or benign *tertian malaria* is the most common type. The incubation period ranges from 10 days to 4 weeks. Generally, paroxysms of chills and fever appear on the 14th day after the bite of an infected female anopheline mosquito. During this time the parasite has been multiplying in the liver cells of the patient. Paroxysms continue to recur every other day, as the parasite completes its 48-hour cycle of development, now in the blood. During the paroxysm, the patient first goes through a "cold stage" during which he has chilly sensations, his skin is blue, his teeth chatter and there is violent shaking. After an hour, the "hot stage" is ushered in, with a rapid rise in temperature to as high as 107 °F (41.7 °C); the skin is hot and dry and the patient complains of severe headache. The fever lasts about 2 hours, and is followed by the "sweating stage," during which there is profuse perspiration, the temperature falls to normal, the headache disappears, and although weak and drowsy, the patient feels well.

(2) *P. ovale* produces a disease very similar to tertian malaria.

(3) *Quartan malaria*, produced by *P. malariae*, has an incubation period of 18–40 days. The paroxysms occur every 72 hours, and are longer and somewhat more severe than those accompanying tertian malaria.

(4) *Falciparum, malignant tertian* or *estivo-autumnal malaria*, is the severest form of malaria and causes most of the fatalities. The paroxysms occur irregularly after a 12-day incubation period. They are severe, and accompanied by high temperatures. The so-called cerebral, algid, hemorrhagic and pernicious types of malaria represent forms of falciparum malaria with different localizations of the parasite. In the cerebral type, the onset is rapid with delirium and coma, and death may occur in several hours without return to consciousness. "Black-water fever" or hemorrhagic malaria is a type in which hemolysis or dissolution of the red cells occurs, and dark urine due to the presence of hemoglobin is an outstanding feature. In the algid form, there are vomiting, diarrhea, and subnormal temperature. Diagnosis of malaria is made by examination of blood films taken during episodes of fever, when the parasites may be seen.

Microscopy is, however, a time-consuming procedure. Immunological methods, although more rapid, cannot distinguish between past and present infections. Another procedure has now been developed. This uses a DNA probe which enables a technician in the field to process 1000 samples per day, as compared with a microscopist's 60 samples.

A.E. Greenberg (Centers for Disease Control, Atlanta, Georgia) and a group of researchers reported in July 1991 on a possible relationship between *Plasmodium falciparum* malaria and the acquired human immunodeficiency virus type 1 infection, both infections that commonly occur in Zaire. Since the cellular immune system is critical to protection against malaria, it is biologically plausible that *P. falciparum* could occur more frequently or be more severe in HIV-infected persons with profound CD4 lymphocyte depletion. Further, it was postulated that repeated malarial infection may accelerate the progression of HIV-related disease. At the end of a 13-month study involving several hundred people, the conclusion was, "In this study malaria was not more frequent or more severe in children with progressive HIV-1 infection and malaria did not appear to accelerate the rate of progression of HIV-1 disease."

Attempts to Control Malaria. A threefold approach has been taken in attempting to control malaria: the elimination of the mosquito, antimalarial drugs, and vaccines.

Mosquito Control. Attempts to destroy the breeding grounds of the mosquito, as was done in the marshes around Rome, or spraying the infected areas with an insecticide, have had limited success. Whereas area drainage has yielded good results, the use of insecticides (especially DDT) has ultimately not been so effective. After an initial period of success, *Anopheles* developed a resistance to the chemical and proceeded to reoccupy its old habitats. In recent years, the use of DDT has been banned by many countries.

Antimalarial Drugs. The use of quinine (quinine sulfate) as a prophylactic and treating drug in connection with malaria was the traditional therapy for many years. The first successful use of this drug was reported by British physicians as early as 1868. Chinchona bark, from which quinine is produced, had been recognized for its curative powers as early as the 1600s by Peruvian Indians. Quinine was first isolated in a pure form by the French chemists. Pelletier and Caventon, in 1820. During World War II, quinine was supplanted by quinacrine which, in turn, was supplanted by chloroquine and primaquine, the present drugs of choice. The role of chloroquine is that of destroying merozoites in the blood; primaquine destroys schizonts hidden in the liver. These drugs are very effective against the susceptible malaria strains that still predominate in Africa, India, and Central America.

It was discovered in 1959 that two American engineers engaged in a project in Colombia near the headwaters of the Amazon River developed a malaria that proved to be resistant to chloroquine. Other cases were reported and by 1962, there was frequent note of chloroquine-resistant malaria in Malaya, Thailand, Vietnam, and Cambodia. These occurrences led scientists at the Walter Reed Army Institute of Research to search for new antimalarial drugs. Over a period of years, they screened nearly 300,000 compounds. A key to success in identifying new antimalarials was the adaptation of human malaria strains for growth in the owl monkey, which is the only known animal in which human parasites flourish. Martin Young (Gorgas Memorial Hospital and Research Institute, Panama City) succeeded in 1966 in infecting owl monkeys with *P. vivax*. Later, Quentin Geiman (Stanford University) succeeded in infecting owl monkeys with several different strains of *P. falciparum*, some of which were resistant to chloroquine. After screening, a total of 29 drugs were submitted to clinical testing in humans. One of the most promising is mefloquine, which is structured much like other compounds found effective against malaria. Maugh reported that mefloquine has been shown to cure chloroquine-resistant falciparum malaria with only one dose and that it is almost free of side effects. Protection persists for about 30 days. It was also reported that mefloquine is effective in terminating acute attacks of vivax malaria, but must be accompanied by a schizonticidal drug, such as primaquine, for complete eradication of the disease. The wide availability of the new drugs is essentially dependent upon further testing and the economics of commercial production.

For chemoprophylaxis in areas where chloroquine-resistant *P. falciparum* malaria is present, a combination of pyrimethamine and sulfadoxine should be taken once every other week and continued for six weeks after travelers return from the malarious area. Although, as of the 1980s, this drug combination is not available in the United States, it is obtainable overseas under the tradenames Fansidar®, Falcidar®, Antemal®, and Methipox®. Areas where there is drug resistance of falciparum malaria to chloroquine include: Bangladesh, Brazil, Burma, Colombia, Ecuador, Guiana (French), Guyana, India, Indonesia, Kampuchea, Malaysia, Nepal, Pakistan, Panama, Philippines, Surinam, Thailand, Venezuela, and Vietnam. It should be mentioned that not all parts of these countries pose this risk. For more details of relevant regions, an inquiry to the International Association for Medical Assistance to Travellers, Guelph, Ontario, Canada N1H 7L5, is suggested.

Growing concern has been expressed by health officials in the United States to the possibility of introducing chloroquine-resistant *P. falciparum* malaria in connection with the thousands of refugees from southeastern Asia who have entered the country in recent years.

More recently, quinidine gluconate has been proven more effective than quinine against chloroquine-resistant *P. falciparum*, and Artemisin, an antimalarial drug extracted from the Chinese herb *Artemisia annua*, has been used against drug-resistant strains of *Plasmodium*.

Since 1991, quinidine gluconate, a class 1a anti-arrhythmic agent, has been the only parenteral antimalarial available for use in the United States [Staff: CDC]. It is indicated for the treatment of patients with life-threatening *Plasmodium falciparum* malaria [Staff: Eli Lilly Company], including those who cannot tolerate oral therapy, have high-grade parasitemia, or have complications (e.g., cerebral malaria or acute renal failure) [Miller, Greenberg, and Campbell; and Zucker and Campbell].

The limited availability of and delays in obtaining quinidine gluconate have contributed to adverse patient outcomes [Rosenthall, Peterson, and Geertsma; and Staff: CDC]. As newer anti-arrhythmics have replaced quinidine for many cardiac indications, some hospitals and other health-care facilities have dropped quinidine gluconate from their formularies and, as a result, fewer clinicians have had experience using the drug. Discussions among quinidine gluconate manufacturer Eli Lilly Company (Indianapolis, Indiana), the CDC, the U.S. Department of Defense, and the U.S. Food and Drug Administration have resulted in recommendations to improve quinidine gluconate availability for acutely ill malaria patients in U.S. health-care facilities:

Prescription Drugs for Preventing Malaria: Mefloquine/brand name **Lariam**®. The adult dosage is 250 Mg salt (one tablet) once a week. The first dose of mefloquine should be taken 1 week before arrival in the malaria risk area, and once a week, on the same day of the week, in the malaria-risk area, and once a week for 4 weeks after leaving the malaria-risk area. Mefloquine should be taken on a full stomach, for example, after dinner.

Most travelers who take mefloquine have few, if any, side effects. The most commonly reported minor side effects include nausea, dizziness, difficulty sleeping, and vivid dreams. Mefloquine has very rarely been

reported to cause serious side effects, such as seizures, hallucinations, and severe anxiety. Minor side effects usually do not require stopping the drug. Mefloquine is **NOT recommended** for travelers with a history of Epilepsy or other seizure disorders; severe psychiatric disorders; or cardiac conduction abnormalities.

Doxycycline: The adult dosage is 100 mg once a day. The first dose of doxycycline should be taken 1 or 2 days before arrival in the malaria-risk area, once a day, at the same time each day, in the malaria-risk area, and once a day for 4 weeks after leaving the malaria-risk area.

Doxycycline side effects and contraindications:

- Doxycycline may cause photosensitivity. Travelers should be advised to avoid midday sun, use a sunscreen with SPF of at least 15, wear long-sleeved shirts, long pants, and a hat.
- Patients are advised to take doxycycline on a full stomach to minimize nausea and to not lie down for 1 hour after taking the drug to prevent reflux of the drug into the esophagus.
- Doxycycline can predispose women to vaginal yeast infections. Women should be advised to bring an over the-counter vaginal yeast infection medication for use if vaginal itching or discharge develops.
- Doxycycline is contraindicated in children under the age of 8; teeth may become permanently stained.
- Doxycycline should **NOT** be used during pregnancy.

Malarone™ is a new antimalarial drug in the United States. Malarone is a combination of two drugs (atovaquone and proguanil) and is an effective alternative for travelers who cannot or choose not to take mefloquine or doxycycline.

The adult dosage is one adult tablet (250 mg atovaquone/100 mg proguanil) once a day. The first dose of Malarone should be taken 1 to 2 days before travel in the malaria-risk area, once a day in the malaria-risk area and once a day for 7 days after leaving the malaria-risk area. The dose should be taken at the same time each day with food or milk.

Although side effects are rare, abdominal pain, nausea, vomiting, and headache can occur. Malarone **should not** be taken by patients with severe renal impairment (creatinine clearance <30 ml/min); pregnant women or women breast-feeding infants weighing less than 11 kg (24 pounds) should not take Malarone to prevent malaria; and infants weighing less than 11 kg should not be given Malarone.

Chloroquine/brand name **Aralen**®: The adult dosage is 500 mg (salt) chloroquine phosphate once a week. The first dose of chloroquine should be taken 1 week before arrival in the malaria-risk area, once a week, on the same day of the week, in the malaria-risk area, and once a week for 4 weeks after leaving the malaria-risk area. Chloroquine should be taken on a full stomach, for example, after dinner, to minimize nausea.

Although side effects are rare, nausea and vomiting, headache, dizziness, blurred vision, and itching have been reported. Chloroquine may worsen the symptoms of psoriasis.

Hydroxychloroquine sulfate/brand name **Plaquenil**®: The adult dosage is 400 mg (salt) once a week. The first dose of hydroxychloroquine sulfate should be taken 1 week before arrival in the malaria-risk area, once a week, on the same day of the week, in the malaria-risk area, and once a week for 4 weeks after leaving the malaria-risk area. Hydroxychloroquine sulfate should be taken on a full stomach, for example, after dinner, to minimize nausea. Hydroxychloroquine sulfate may be better tolerated than chloroquine.

Although side effects are rare, nausea and vomiting, headache, dizziness, blurred vision, and itching have been reported. Hydroxychloroquine sulfate may worsen the symptoms of psoriasis.

Self-Treatment Medication

Travelers should be reminded that malaria can be fatal. If a traveler develops a fever or other flu-like symptoms, *and professional medical care is not available within 24 hours*, a self-treatment dose of either Fansidar® or Malarone™ is recommended. The traveler should seek professional medical care as soon as possible after self-treatment. Malaria symptoms will occur, at least seven to nine days after being bitten by an infected mosquito. Fever in the first week of travel is unlikely to be malaria; however, travelers should be advised to have any fever promptly evaluated.

Fansidar® may be used for presumptive self-treatment for travelers if: they are **NOT** allergic to sulfa drugs and their travel itinerary does **NOT** include the Amazon basin of South America, Southeast Asia, and certain countries in eastern and southern Africa (Kenya, Malawi, Mozambique,

South Africa, Tanzania, and Uganda). These countries have documented Fansidar®-resistant *Plasmodium falciparum* malaria.

Malarone™ may be used for presumptive self-treatment for travelers not taking Malarone for prophylaxis. Travelers on Malarone prophylaxis who take presumptive self-treatment should use Fansidar® if they are traveling to an area without Fansidar® resistance.

Those travelers who cannot take Fansidar® or Malarone™ for presumptive self-treatment should consult the CDC Malaria Hotline.

Note: The foregoing information on prescription drugs for preventing Malaria was furnished by the Centers for Disease Control and Prevention, Atlanta, GA.

Malaria Vaccines. Development of a malaria vaccine has been hindered mainly by a lack of a suitable source of parasites from which a vaccine could be prepared. Despite the development of a means of culturing the parasite, however, very little further progress has appeared. A large problem is the low immunogenecity of malaria parasites, which means that inducing immunity against them is difficult. In the face of the changing envelope of the trypanosome it appears unlikely that a suitable antibody against malaria will be found in the near future. Antibodies against the sporozoite sheath have been claimed to provide a lasting protection against infection by *Plasmodium*, but this claim has been strongly questioned and it is more probable that the monoclonal antibody approach will have more chance of success.

The parasite, *Plasmodium*, takes on several forms during its sojourn in the human host. At each phase, the parasite possesses a distinct protein coat. Obviously, then, an effective vaccine must induce antibodies to at least several of these, which is a very large order for one vaccine.

Additional Reading

CDC: "Morbidity and Mortality Report," Center for Disease Control, Atlanta, Georgia (Issued weekly).

Doolan, D.L.: "Malaria Methods and Protocols," Humana Press, Totowa, NJ, 2000.

Good, M.F. and A.J. Saul: "Molecular Immunological Considerations in Malaria Vaccine Development," CRC Press, LLC., Boca Raton, FL, 1994.

Greenberg, A.E., et al.: "*Plasmodium Falciparum* Malaria and Perinatally Acquired Human Immunodeficiency Virus Type 1 Infections in Kinshasa, Zaire," *N. Eng. J. Med.*, 105 (July 11, 1991).

Greene, L. and M. Danubio (Editors): "Adaptation to Malaria: The Interaction of Biology and Culture," Gordon and Breach Publishing Group, Newark, NJ, 1998.

Hoffman, S.L.: "Malaria Vaccine Development: A Multi-Immune Response Approach," ASM Press, Washington, DC, 1996.

Humar, A., S. Sharma, D. Zoutman, et al.: "Fatal Falciparum Malaria in Canadian Travelers," *Can Med Assoc J.*, **156**, 1165–1167 (1997).

Khusmith, S., et al.: "Protection Against Malaria by Vaccination with Sporozoite Surface Protein 2 Plus CS Protein," *Science*, 715 (May 5, 1991).

Kinoshita, J.: "Malaria Vaccines," *Sci. Amer.*, 34 (May 1988).

Laird, M.: "Avian Malaria in the Asian Tropical Subregion," Springer-Verlag Inc., New York, NY, 1998.

Miller, K.D., A.E. Greenberg, and C.C. Campbell: "Treatment of Severe Malaria in the United States with a Continuous Infusion of Quinidine Gluconate and Exchange Transfusion," *N. Eng. J. Med.*, **321**, 65–70, 1989.

Nagel, R.L.: "Red-Cell Cytoskeletal Abnormalities—Implications for Malaria," *N. Eng. J. Med.*, 1558 (November 29, 1990).

Poser, C.M. and G.W. Bruyn: "An Illustrated History of Malaria," Parthenon Publishing Group, New York, NY, 1999.

Rennie, J.: "Birds of a Fever: A Lethal Malaria May Have An Avian Origin," *Sci. Amer.*, 25 (July 1991).

Rosenthal, P.J., C. Peterson, F.R. Geertsma, et al.: "Availability of Intravenous Quinidine for Falciparum Malaria [Letter]," *N. Eng. J. Med.*, **335**, 138 (1996).

Schecter, J.: "Parasite Pacification," *Technology Review (MIT)*, 10 (October 1987).

Sherman, I. (Editor): "Malaria: Parasite Biology, Pathogenesis, and Protection" ASM Press, Washington, DC, 1998.

Staff: "Malaria Strategies for Africa," AAAS Sub-Saharan African Program, American Association for the Advancement of Science, Washington, DC, 1990.

Staff: "Treatment with Quinidine Gluconate of Persons with Severe Plasmodium falciparum infection: Discontinuation of Parenteral Quinine from CDC Drug Service," *MMWR*, **40** (no. RR-4: 21–23 (1991).

Staff: "Availability of Parenteral Quinidine Gluconate for Treatment of Severe or Complicated Malaria," *MMWR*, **45**, 494–495 (1996).

Staff: "Quinidine Gluconate Injection [Package Insert]," Eli Lilly Company, Indianapolis, IN, (February, 2000).

Udomsangpetch, R. et al.: "Human Monoclonal Antibodies Against *P. falciparum*," *Science*, **231**, 57 (1986).

Wahlgren, M. and P. Perlmann: "Malaria: Molecular and Clinical Aspects," Gordon & Breach Publishing Group, Newark, NJ, 2000.

Wyler, D.J.: "Bark, Weeds, and Iron Chelators—Drugs for Malaria," *N. Eng. J. Med.*, 1519 (November 19, 1992).

Zucker, J.R. and C.C. Campbell: "Malaria: Principles of Prevention and Treatment," *Infect. Dis. Clin. No. Am.*, **7**, 547–567, (1993).

Web Reference

Centers for Disease Control and Prevention, United States. *http://www.cdc.gov/default.htm*

R.C. VICKERY, M.D., D.Sc., Ph.D., Blanton/Dade City, FL.

MALEO *(Aves, Galliformes)*. A peculiar bird of Celebes and neighboring East Indian islands. The head and neck are covered with naked red skin and the crown bears a black prominence resembling a helmet. The plumage is mostly black but that of the breast and belly is salmon colored. The large eggs are buried in hot sand. See also **Galliformes**; **Megapode**; and **Mound Birds**.

MALIGNANCY. See **Cancer and Oncology**.

MALLEUS. See **Hearing and the Ear**.

MALLOPHAGA. The bird lice or biting lice, constituting a small order of insects. They have flattened bodies with many spines directed backward, short legs, and biting mouths. Wings are lacking. Most species live as external parasites on birds but a few are found on mammals.

MALLOW FAMILY. See **Malvaceae (Mallow Family)**.

MALT. Unless otherwise specified, malt usually connotes barley malt. Malt, however, can be prepared from other cereal grains. Between 75 and 80% of the malt produced in the United States goes into the manufacture of beer and associated beverages. Nearly 15% of the production goes to the manufacture of distilled alcohol products; and the remainder (about 5–6%) goes into the preparation of malt syrups, breakfast foods, malted milk concentrates, and coffee substitutes.

Barley malt is barley that has been germinated by moisture under controlled conditions and for a specified time, after which the germinated plants are carefully dried under controlled conditions. The drying or kilning operation and other operations in the total malting process are customized to the final product in which the malt will be used. The principal stages of the malting process are diagrammed in Fig. 1.

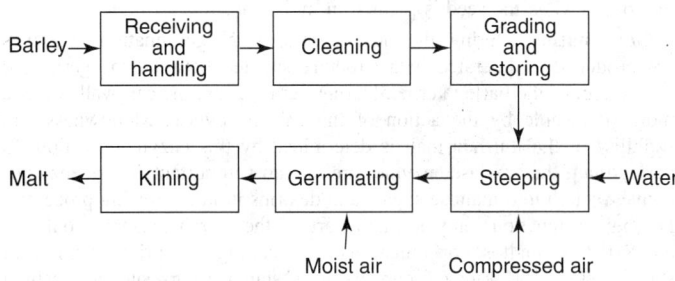

Fig. 1.   Simplified flowsheet of operations involved in preparing malt.

Upon receipt, the barley must be inspected carefully to make certain that it fully meets the minimum acceptable specifications established by the malt manufacturer. The qualities desired in grain for malting are described in the entry on Barley.

Prior to processing, the barley is stored for up to 6 weeks to permit any further ripening (after-ripening) to take place. This enhances later germination. The cleaning operation that follows removes impurities, unwanted foreign seeds, and damaged and broken kernels. Because the size of the kernels affects handling during the germinating operation, the barley is graded into 2 or 3 size ranges, each of which is malted separately.

The role served by malt in the later production of beer and distilled spirits is that of furnishing enzymes, which convert starches and other ingredients during the brewer's and distiller's mashing operations. J. de Clerck states: "Quantitatively the most significant chemical constituent of the grain is starch which constitutes some two-thirds of the dry weight of barley. Apart from a comparatively small contribution from other sugars in barley it is the starch which eventually furnishes some 85–90% of malt extracts, of which some 70% is fermented in brewing and sometimes nearly all in distilling. Starch, therefore, occupies a key position in relation to the brewing and distilling industries and its chemistry is, apart from its intrinsic biochemical interest as the final product of photosynthesis, of outstanding importance." The starch granules of barleys are first seen in the cells of the endosperm a few days after development of the seed begins as small spheres later developing into the bean-shaped and lenticular forms characteristic of mature starch. Luers observed that barley starch undergoes enzymic degradation during malting and that malt starch is different from barley starch. This was recognized as early as 1902. The loss in starch that occurs during malting has been estimated by various researchers. Luhder (1908) first established that the starch content of barley declined considerably during malting. For example, Moravian barley before malting contains 65.57% starch and after malting, 54.89%. Of the two starch constituents amylopectin and amylose, the former is more susceptible to enzymolysis during malting. Several theories have been proposed to explain this difference. The principal factors that influence the amount of starch in barley are environmental during growth and also relate to the variety planted.

Brewers' and Distillers' Malts. Barleys of large grain size are desired for preparing brewers' malt. The larger sizes usually have a greater percentage of starch and relatively little protein. The grains, however, do contain adequate amyloytic enzymes for solubilizing native starch and also some of the other ingredients of the mashing operation at the brewery.

In contrast, the primary objective of distillers' malt is to furnish enzymes in larger quantities for later use in converting grains and starch present in other substances. Barley best suited for this purpose is of a relatively high nitrogen content, a factor that relates to the ability to produce amylases. In the United States, certain barley varieties, such as *Kindred, Manchuria, Montcalm,* and *Odessa,* are specifically selected for making distillers' malt.

Steeping. Once size-graded, the barley is transferred to large steeping tanks equipped with conical or funnel-shaped bottoms. Cool, clean water is added to partially fill the tank. Air agitation assists in cleaning the barley. Any light-weight kernels present automatically rise to the surface of the water and are skimmed off. During a period of from 45 to 60 hours, the barley soaks up water (steeps). At the end of the steeping period, the moisture content of the barley will range between 45 and 48%, representing sufficient moisture to commence germination. Distillers' malt usually requires a somewhat longer period of steeping in order to bring the moisture content up to a minimum of 50%. The process that takes place during steeping is, not surprisingly, essentially the same as the changes that occur when the seed is planted in moist soil.

Germinating. During the next process step, germination, enzymes are produced or liberated in a structure situated between the germ and endosperm of the barley kernel. During germination, the cell wall is made more permeable by the action of the enzyme cytase. Mellowness and friability of the finished malt is determined by this enzymatic action. In barley malt, the amylase enzymes are the most important. These enzymes convert starch into maltose sugars and dextrins in later brewing processes. During germination, only a small part of the starch present should be converted to maltose or other sugars, recalling the prior mention of starch losses. The principal conversion of starch occurs later in mashing operations at the brewery or distillery. Other enzymes that transform proteins also are produced or activated during malting.

Over the years, three principal malting procedures have been developed. The oldest of these is the germinating floor or floor malting, still practiced in some European countries. The steeped barley, once removed from the steeping tanks, is spread in heaps about 1 foot (0.3 meter) in height. The first spreading occupies perhaps only 40% of the available floor area. The barley commences to dry out and sprout, producing small hairlike fibers (rootlets). It is necessary to aerate the early-sprouting barley (known as *green malt*), which is accomplished by turning the heap over and over with forks. The barley is spread over a larger floor area. These actions, plus the use of forced aeration, cool the green malt. This process is repeated at regular intervals over a period of 5 to 7 days. The process is confined to the cooler seasons of the year because germinating temperature should be maintained within a span between 60° and 70°F (15.6° and 21.1°C). Excessive temperature accelerates germination and generates waste, increasing starch losses and excessively large rootlets. Technically, the optimum cut-off time for germination is when the plumule or acrospire reaches the length of the kernel. An extension of the floor malting process to avoid any seasonal limitations on the process is pneumatic floor or compartment malting. Instead of placing the barley on the floor in heaps,

it is placed in box-shaped compartments, which hold the grain during the entire germinating period. Purified air is circulated through these compartments. The air is both temperature and humidity controlled. The green malt is turned by mechanical screws within the compartments. Such compartments will contain up to 5000 bushels (108 metric tons) of green malt.

In *drum malting*, widely used in modern malting plants, there are two concentric hollow metal cylinders. Grain is placed in the spaces between the cylinders, which are perforated for introduction and circulation of humidity-and-temperature-controlled air. To permit maximum movement of the grain, the equipment is only partially filled. The revolving action of the cylinders keeps the grain tumbling, constantly exposing new surfaces of the grain to the air stream. Capacity of these drums is up to 650 bushels (about 13.5 metric tons).

Drying or Kilning. These are not high-temperature or rotary kilns as one may visualize in connection with a cement or ore plant, but they are usually 2-story buildings. Drying is commenced on the upper story by spreading the green malt in layers of 2 to 3 feet (0.6 to 0.9 meter) in depth. The primary purpose of the kilning operation is to halt further germination, although the conditions of kilning also affect the final end-use properties of the malt. Hot air is drawn through the malt at a relatively low temperature for the first 24 hours to accomplish a partial drying. The malt is then transferred to the lower story, where the air temperature is increased. In the case of brewers' malt, the final kilning temperature is in the range of 160° to 180°F (71° to 82°C), and the malt is retained in the kiln until the moisture has reached a content of about 4%. The entire process requires from 48 to 72 hours. In the case of distillers' malt, the final kilning temperature is lower—in the range of 120° to 140°F (49° to 60°C) and the final moisture content is from 5 to 7%. The lower temperature preserves higher enzymatic activities. The higher temperature for brewers' malt introduces a more intense malt flavor and aroma, usually desired for the brewing process. In the case of porter, bock beer, and stout, the kilning temperature is higher and a caramel malt is produced. This imparts a dark color and distinctive flavor during the brewing of these malted beverages.

After kilning, the malt is cleaned to remove rootlets and any broken kernels that may remain in the batch.

Malting Process Innovations. In recent years, several steps to modernize and improve the operations just described have been made. Sophisticated control and conveying systems have assisted in more exacting control of processing conditions and reducing hand labor costs. There is a growing trend toward continuous operations. Some consideration has been given to the use of gibberellic acid as part of the steepwater or in the form of a spray during the germinating process. The objective of these steps would be that of increasing yield. It also has been found that potassium bromate in the steepwater will depress respiration and rootlet formation.

A nutritional profile is given in Table 1.

TABLE 1. NUTRITIONAL PROFILE OF MALT (100-gram samples)

	Dry Malt	Dried Malt Extract
Water	18.7 g	11.5 g
Food energy	374	374 cal
Protein	1.3 g	1.7 g
Fat	1.8 g	trace g
Carbohydrate	78.8 g	91.1 g
Calcium	—	50 mg
Phosphorus	—	299 mg
Iron	4.0	9.0 mg
Sodium	—	83 mg
Potassium	—	234 mg
Vitamin A	—	—
Thiamine	0.50	0.36 mg
Riboflavin	0.32	0.47 mg
Niacin	9.4	10.1 mg
Vitamin C (ascorbic acid)	—	—

Source: U.S. Dept. of Agriculture, Washington, DC.

Malt from Other Grains. Sometimes wheat malt is used as a flour supplement for bakery products. The malt provides a source of a-amylase, which degrades starch to sugars, the fermentation of which causes rising of the dough. Of course, barley malt is also used in some bakery product flours. Wheat malt imparts a characteristic flavor to beer. Most consumers find this undesirable. However, *Weissbier*, made with wheat malt, is quite popular in certain regions of Germany. In producing wheat malt, kilning temperatures are more moderate.

Additional Reading

Hough, J.S.: "The Biotechnology of Malting and Brewing," Cambridge University Press, New York, NY, 1992.
Hough, J.S., D.E. Briggs, R. Stevens, and T.W. Young: "Malting and Brewing Science," Vol. 2, Chapman & Hall, New York, NY, 1999.
Hough, J.S. and T.W. Young: "Malting and Brewing Science," Vol. 1, Kluwer Academic Publishers, Norwell, MA, 1999.
Staff, Briggs Corporation: "Malts and Malting," Aspen Publishers, Inc., Gaithersburg, MD, 2000.

MALTOSE. See **Carbohydrates**.

MALUS COSINE-SQUARED LAW. A law applying to the intensity of polarized light as affected by the polarizing apparatus. If a beam of plane-polarized light is passed through a Nicol prism, for example, the intensity (flux density) of the emergent beam falls off, as the prism is rotated, from a maximum value when the transmission plane of the prism coincides with the plane of vibration of the light, to zero when it is at right angles to that direction. The intensity varies as the square of the cosine of the angle through which the prism has been thus rotated. The same law applies to the effect of a glass reflector, reflecting always at the polarizing angle, as the plane of reflection is rotated around the stationary, polarized incident beam.

MALVACEAE (Mallow Family). The plants of this family include herbs, shrubs and trees (the latter tropical), and are rich in mucilaginous substance. The leaves are alternate, and in most cases palmately lobed and veined, with small deciduous stipules. The flowers are regular and perfect, often large and showy, and variously borne. They have five (or

Fig. 1.   Okra plant (*Hibiscus esulentus*) showing pods ready to harvest. (USDA photo.)

rarely, fewer) more or less united sepals, five petals, and numerous stamens, which characterized the family by having their filaments joined to form a tube which surrounds the styles.

The most important member of this family is the cotton plant, whose fibers outrank in commercial importance all others. See also **Cotton**. Okra or Gumbo, *Hibiscus esculentus*, a native of tropical Africa, is another member of importance. See Fig. 1. It is a coarse annual plant with large veiny leaves and showy axillary flowers. The slender 5-ribbed pods are used in soups, or when young, are cooked and used in salads. Okra has also been used as a source of fibers for paper manufacture. Another member of some importance is *Althaea officinalis*, the Marsh Mallow. The underground rootstock of this plant is not only rich in mucilage, but is also used medicinally in ground form, the bark being removed before grinding.

Many members of the family are grown as ornamental plants, among them being the Hollyhock, *Althaea rosea*, species of *Malva*, the true Mallows, and the Rose of Sharon, *Hibiscus syriacus*, which becomes a large bush or even a small tree, with showy pink or white flowers.

MAMMALIA. The Latin word *mamma* (meaning breast) provides the key to that large class of animals designated *Mammalia*. There is an exceedingly great variety of characteristics among mammals. Their size ranges from the tiny shrew of some 2 inches (5 centimeters) in length, to the whales, which may attain 100 feet (30 meters) in length and a weight of 130 tons (117 metric tons) or more. The principal point of commonality is that all are vertebrates, the females of which possess mammary glands, milk-secreting glands for feeding the young. Within the framework of this definition, humans are mammals. There are 19 orders and some 5,000 species of mammals, as shown in Table 1. Several examples of each order and reference to specific entries in this volume on mammals are given in the Table 1.

TABLE 1. MAMMALS (Alphabetical Listing of Orders)

Order and Examples	Major Entries in this Encyclopedia
ARTIODACTYLA	Artiodactyla
(Even-toed Hoofed Animals)	
Antelopines	Antelope
(Antelopes and gazelles)	
Antilocaprines	Pronghorn Antelope
Bovines	Bovines
(Oxen, cattle, buffalo, bison, and duikers)	
Camelines	Camels and Llamas
Caprines	Goats and Sheep
Cervines	Deer
(Deer, muntjacs, moose, and reindeer)	
Giraffines	Giraffe and Okapi
Hippopotamines	Hippopotamus
Suines	Suines
(Pigs and peccaries)	
Tragulines	Tragulines
(Chevrotains)	
CARNIVORA	Carnivora
(Flesh-eating Mammals)	
Felines	Cats
(Lions, tigers, leopards, jaguars, ocelots, domestic cats, lynxes, servals, jaguarondis, and cheetahs)	
Viverrines	Viverrines
(Civets, hemigales, and mongooses)	
Hyaenines	Hyena
Procyonines	Raccoons
Canines	Canines
(Wolves, jackals, foxes, and dogs)	
Ursines	Bears
Mustelines	Mustelines
(Weasels, badgers, skunks, and otters)	
Ailuridae	Pandas
CETACEA	Whales, Dolphins, and Porpoises
(Hairless, fish-like Water Mammals)	
CHIROPTERA	Bats
(Flying Mammals)	
DERMOPTERA	Dermoptera

(continued)

TABLE 1. (Continued)

Order and Examples	Major Entries in this Encyclopedia
(Gliding Mammals)	
(Flying Lemur or Kobego)	
EDENTATA	Edentata
(Anteaters, sloths, and armadillos)	
HYRACOIDEA	Hyraxes
INSECTIVORA	Moles and Shrews
(Insect-eating Mammals)	
LOGOMORPHA	Rabbits and Hares
(Leaping Mammals)	
MARSUPIALIA	Marsupialia
(Pouched Mammals)	
(Opossums, bandicoots, phasogales, phalangers, koalas, wombats, and kangaroos)	
MONOTREMATA	Monotremata
(Egg-laying Mammals)	
(Duckbills and spiny anteaters)	
PERISSODACTYLA	Perissodactyla
(Odd-toed Hoofed Mammals)	
Equines	Horses, Asses, and Zebras
Tapirines	Tapir
Rhinocerotines	Rhinoceros
PHOLIDOTA	Pholidota
(Scaly Mammals)	
(Pangolins)	
PINNIPEDIA	Sea-Lions and Seals
(Fin-footed Mammals)	
PRIMATES	Primates
(Top Mammals)	
Tupaioids	Moles and Shrews
(Tree-shrews)	
Lorisoids	Lorisoids
(Lorises and bush-babies)	
Lemuroids	Lemur
Tarsioids	Tarsioids
(Tarsiers)	
Hapaloids	Marmoset
Ceboids	Monkeys and Baboons
(Half-monkeys and hand-tailed monkeys)	
Simioids	Monkeys and Baboons
(Colobine monkeys, long-tailed monkeys, dog-faced monkeys, the black ape, baboons, the gelada, and drills)	
Anthropoids	Anthropoids
(Lesser apes, gibbons, Greater apes, gorillas, chimpanzees, and orangutans)	
PROBOSCIDEA	Elephant
RODENTIA	Rodentia
(Gnawing Mammals)	
Sciuromorphs	Rodentia
Squirrels	Squirrels and Other Sciuromorphs
Beavers	Beaver
Myomorphs	Rodentia
(Mice, rats, and jerboas)	
Hystricomorphs	Rodentia
Porcupines	Rodentia
SIRENIA	Sea-Cows
(Manatees and dugongs)	
TUBULIDENTATA	Aard-Vark

Because of the variety among mammals, generalizations are difficult and can be misleading, but with these factors in mind, the following observations may be made. (1) The young of most species develop in the body of the mother and all are nourished by milk secreted by special glands known as the mammary glands. (2) There are typically two sets of teeth, the so-called milk teeth or temporary teeth which the young retain for a varying length of time—followed by permanent teeth set in sockets in the jawbones. The teeth are of four kinds, i.e., incisors, canines, premolars, and molars. (3) Mammals are one of the more dominant forms of animal life of the phylum *Chordata*, this dominance greatly aided by superior intelligence, locomotion, and well-developed sensory organs in most cases. (4) Most mammals have a covering of skin, out of which grows hair. (5) Although predominantly creatures of the land, both above and below the surface, species of mammals are found in the seas, and a few species can take to the air. (6) Mammals are found practically everywhere on earth—if not in their native habitat, often in areas of extreme climates, where their great adaptability after migration often assures survival. Humans create artificial shelters and, in essence, often create their own environment through the application of energy for heating and cooling, as well as applying technology for raising and processing a food supply. They can exist for long periods of time in the most severe of climates and, of course, in recent years have proved their ability to survive in the extremely hostile environment of interplanetary space.

Much knowledge pertaining to the adaptation of mammals has been learned from the study of fossils. See also **Paleontology.**

Additional Reading

Balog, J.: "A Personal Vision of Vanishing Wildlife," *Nat'l. Geographic*, 84 (April 1990).

Dixson, A.F.: "The Natural History of the Gorilla," Columbia University Press, New York, NY, 1981.

Dutcher, K. "The Secret Life of America's Ghost Cat," *Nat'l. Geographic*, 38 (July 1992).

Gould, E., G. McKay: "Encyclopedia of Mammals," 2nd Edition, Academic Press, Inc., San Diego, CA, 1998.

Grzimek, B.: "Grzimek's Animal Life Encyclopedia," Vols. 10–13, Van Nostrand Reinhold, New York, NY, 1975. (A Classic Reference.)

Heinz-Georg, K., E.M. Lang: "Handbook of Zoo Medicine," John Wiley & Sons, Inc., New York, NY, 1982.

Linden, E., F. Lanting: "Bonobos, Chimpanzees With a Difference," *Nat'l. Geographic*, 46 (March 1992).

Linden, E., M. Nichols: "A Curious Kinship: Apes and Humans," *Nat'l. Geographic*, 2 (March 1992).

Macdonald, D.: "Encyclopedia of Mammals," Barnes & Noble Books, New York, NY, 1999.

Maple, T.L., M.P. Hoff: "Gorilla Behavior," John Wiley & Sons, Inc., New York, NY 1981.

Napier, J.R., P.H. Napier: "The Natural History of the Primates," MIT Press, Cambridge MA, 1994.

Preuschoft, H., et al.: "The Lesser Apes," Edinburgh University Press, Edinburgh, U.K., 1984.

Schaller, G.B., et al.: "The Giant Pandas of Wolong," University of Chicago Press, Chicago, Illinois, 1985.

Staff: "Encyclopedia of Mammals," Marshall Cavendish, Inc., Tarrytown, NY, 1997.

Walther, F.R., et al.: "Gazelles and Their Relatives," Noyes, Park Ridge, New Jersey, 1983.

Zhi, Lu: "Newborn Panda in the Wild," *Nat'l. Geographic*, 60 (February 1993).

MAMMARY GLAND. A large gland, which secretes milk for the nourishment of the young of mammals.

MAMMOTH. See **Elephant.**

MANAKIN *(Aves, Passeriformes).* Brightly colored, small wren-size bird of Central and tropical South America. There are several species. The males are known for their singing and dancing to exhibit their beautiful plumage to the usually drab, green-colored females. Much like the bowerbird, the male clears away an area for his demonstrations. Most species are polygamous. The female builds the nest, incubates and hatches the eggs, and rears the young. An open-type nest is constructed in the fork of a tree. There are usually two pastel eggs with dark markings. Incubation period is from 19 to 21 days. Manakins feed mostly on insects.

MANDIBLE. See **Fishes.**

MANGANESE. Chemical element, symbol Mn, at. no. 25, at. wt. 54.9380, periodic table group 7, mp $1,244 \pm 3\,°C$, bp $1,962\,°C$, density 7.3 g/cm^3 (solid), 7.21 (single crystal) $(20\,°C)$. Manganese has a cubic (complex) crystal structure.

Manganese is a silver-white metal, not notably hard (becomes hard on alloying with carbon), brittle, capable of taking a brilliant polish but readily oxidized upon heating, reacts with water upon boiling, soluble in dilute acids. Discovered by Scheele in 1774.

In terms of abundance, manganese is present in igneous rocks to an average extent of 0.10% (weight). In terms of cosmic abundance, in the estimate by Harold C. Urey (1952), using a base figure of 10,000 for silicon, the figure for manganese is 75. Manganese is estimated as 34th among the elements in its content in seawater, an estimated 9.5 tons per cubic mile of seawater. There are eight isotopes of manganese, ^{50}Mn through ^{57}Mn, all radioactive with exception of ^{55}Mn. Half-lives range from a fraction of a second for ^{50}Mn to approximately 140 years for ^{53}Mn. Electronic configuration $1s^2 2s^2 2p^6 3s^2 3p^6 3d^5 4s^2$. Ionic radius Mn^{2+} 0.83 Å. Metallic radius 1.365 Å. First ionization potential 7.32 eV; second, 15.7 eV. Oxidation potentials Mn $\rightarrow$ Mn^{++} + 2e$^-$, 1.18 V; Mn^{2+} + 2H$_2$O $\rightarrow$ MnO$_2$ + 4H$^+$ + 2e$^-$, -1.28 V; Mn^{2+} $\rightarrow$ Mn^{3+} + e$^-$, 1.51 V; Mn^{2+} + 4H$_2$O $\rightarrow$ MnO$_4^-$ + 8H$^+$ + 5e$^-$, -1.52 V; MnO$_2$ + 2H$_2$O $\rightarrow$ MnO$_4^-$ + 4H$^+$ + 3e$^-$, -0.168 V; Mn(OH)$_2$ + OH$^-$ $\rightarrow$ Mn(OH)$_3$ + e$^-$, 0.40 V; MnO$_4^=$ $\rightarrow$ MnO$_4^-$ + e$^-$, -0.54 V; MnO$_2$ + 4OH$^-$ $\rightarrow$ MnO$_4^-$ + 2H$_2$O + 3e$^-$, -0.58 V. Other important physical properties are given under **Chemical Elements.**

Occurrence. The most common manganese ore is pyrolusite, MnO$_2$. Other commercial ores include braunite, Mn$_2$O$_3$; hausmannite, Mn$_3$O$_4$; and rhodochrosite, MnCO$_3$. Although not of industrial value, manganese also exists in nature as the silicate, sulfate, sulfite, and tungstate. See also **Pyrolusite**; and **Rhodochrosite.**

Manganese Nodules. These are rocks composed largely of ferromanganese oxides formed by precipitation at the bottom of lakes and the oceans. They range in size from micrometers to meters. Their morphology is highly variable. They contain up to 55% manganese, 35% iron, and 2% nickel, cobalt, and copper. Manganese nodules were first discovered in the open ocean by Thompson, Murray, and Renard during the *Challenger* expedition (1873–1876). Buchanan reported the occurrence of nodules in the Firth of Clyde, a shallow-water area, and by the end of the century at least five additional occurrences of manganese nodules in shallow marine environments had been discovered. Early workers chemically analyzed about a score of manganese nodules and hypothesized about their mechanism of growth. Two principal concepts emerged: (1) they grow by the slow precipitation of manganese from seawater; and (2) they are formed by the rapid precipitation of manganese released in submarine volcanism.

Until the 1950s, little additional work was done except for some early measurements of manganese nodule growth rates. During recent years, however, there has been a strong revival of interest in manganese nodules, stimulated both by the expansion of oceanographic facilities and the realization of the economic importance of the nodules as ores. It has been found that in large areas of the ocean floor, manganese nodules may be absent. In other areas, they may cover nearly 100% of the area. In all of the Pacific Ocean, nodules have been estimated to cover approximately 10% of the ocean floor. The estimated coverage in the Indian and Atlantic Oceans is less. The local variability in manganese nodule concentration is large. Two ocean bottom photographs only a few meters apart may show very different nodule concentrations. In some locations, the weight concentration of nodules ranges up to 5 g/cm^2.

Manganese nodules are composed of cryptocrystalline minerals. They are known to consist of three major manganese phases: (1) δ MnO$_2$ (birnessite); (2) 10-Å manganite; and (3) 7-Å manganite. The first is the most highly oxidized form, and has a chemical composition of about Mn$_{1.9}$. Barnes (1967) examined the depth dependence of the mineralogy in nodules taken from the Pacific. His data indicate that above 3,500 m in depth, the only important manganese phase is δ MnO$_2$, but, below the 3,500 m depth, both 10-Å manganite and 7-Å manganite coexist with the δ MnO$_2$. The observed phase changes may be pressure induced.

During recent years, the growth rates of manganese nodules have been determined by various methods. Results all indicate that the nodules measured grow at a rate of a few millimeters per million years. This does not exclude the possibility that nodules in certain areas evolve more rapidly, but it appears that most deep-sea nodules grow slowly. There is some belief that the nodules are primarily the result of bacterial fixation of manganese. Other investigators believe that the nodules are formed by inorganic precipitation of metals supersaturated in sea water. There is some experimental and theoretically tenable evidence to support both concepts.

Research gathered during the DeepSea Drilling Project (DSDP) and the International Decade of Ocean Exploration (IDOE) programs is described in the entries on **Ocean**; and **Ocean Resources (Mineral).**

Hypotheses continue to evolve concerning the manner in which manganese nodules develop. In the early 1980s, in an effort to sweep away some of the mysteries concerning the Mn nodules, a consortium of American researchers participated in the MANOP (Manganese Nodule Project) program. The program involved the creation of mathematical models, not a simple task because it has been estimated that nodule attrition is about one atomic layer of the Mn-O structure per year. As reported by J. Dymond and colleagues (Oregon State University), two different processes operate within sediments, the particular process depending on whether any oxygen remains below the sediment surface. The amount of oxygen present, of course, is a function of the biological productivity of the overlying surface waters. Where winds and currents mix the sea in the needed manner, microscopic plants and animals do well and part of their inorganic skeletons and probably about one percent of their organic tissues will sink to the bottom — along with clay washed from the land. Regardless of how deep the sea floor is, bacteria and animals dwelling on and in the sediment will oxidize organic matter. However, a small percentage of their organic matter does reach the bottom. If the falling organic matter is light, not all of the sediment oxygen will be consumed. It is proposed that toxic chemical alterations (oxic diagenesis) of the sediment can then supply metals to nodules. It is further reasoned that oxic sediments must be altered chemically to produce nodules because under such oxidizing conditions, Mn and Fe are tied up as insoluble oxides, which cannot move and thus cannot be incorporated in nodules. Several possible diagenetic reactions, such as those involving volcanic ash and skeletal opal, have been suggested. As reported by Kerr (1984), the Oregon State University team concluded that the nodule composition most typical of growth under oxic diagenesis is that of the nodule bottom most rich in trace metals from a siliceous sediment in the tropical North Pacific. In terms of the overall picture, some scientists have observed that manganese nodules should not be there, but are! Obviously Mn nodule research will require many more years in supplying answers. See also **Ocean Resources (Mineral).**

Todorokites may be defined as calcium-bearing manganese oxides, which are found in terrestrial Mn ore deposits, in weathering products of Mn-bearing rocks, and in some Mn nodules. In some cases, todorokites are the principal constituents of Mn nodules. Knowledge collected concerning todorokites has contributed and will continue to contribute to a better understanding of Mn nodule formation in ocean waters. See also **Todorokites.**

Processing. Manganese metal can be obtained from oxide ores by reduction with carbon, aluminum, magnesium, or sodium in an electric furnace. The main form in which manganese is used is *ferromanganese*. This material contains approximately 80% manganese and 20% iron. Ferromanganese is generally made in a blast furnace or an electric-arc furnace. Usually a mixture of ores is used, proportioned to yield the final desired specifications of the alloy. To reduce slag volume, low-silica ores are preferred. It is also desirable to maintain a low phosphorus content in the alloy. The charge to the electric furnace process for making ferromanganese is the manganese ore, coke, and limestone. The loss of metal to the slag is determined by the silica present. Usually about 75% of the metal is recovered. Where high-purity manganese is produced, the ore is first roasted to MnO, then leached with H$_2$SO$_4$ to form the sulfate. The solution is then neutralized to precipitate iron and aluminum. Other impurities are removed as the sulfides. Electrolysis of the resulting solution yields a 99.94% pure manganese metal.

The high-purity (electrolytic) manganese is used as a deoxidizing agent and sometimes as a constituent of nonferrous metals where it improves strength, ductility, and hot-rolling properties. Because of their very high temperature thermal coefficient of expansion, manganese-base alloys with 72% manganese (balance is copper and nickel) are used in bimetals for switching applications. Manganese (60–80% Mn) and copper alloys find application because of their vibration-damping properties.

The standard ferromanganese (7% carbon; 74–78% Mn) is used both to produce a manganese alloy steel and as a deoxidant. As early as 1856, Robert Mushed used *spiegeleisen* (10–23% Mn; 4–5% C) in alloys. Where the carbon content of steel is critical, low-carbon ferromanganese is added. Silicomanganese is used as a blocking agent to stop the reaction of carbon and oxygen in steel. Developed in 1888, Hadfield steel contains about 13% manganese. It finds use where a very hard material is needed and it has the interesting property of increasing its hardness when subjected to repeated impacts. In the 200 series of stainless steels, manganese is replacing nickel in order to achieve more economical austenitic materials.

Manganese Inorganics. A number of chemical processes have been developed to upgrade Mn ore which produce an intermediate Mn compound. These intermediates usually are free of most siliceous matter. Although these processes were designed to convert the compound to an oxide for use in metallurgical applications, the purity of the compounds often renders them suitable for commercial use.

The ammonium carbonate process (developed by Manganese Chemicals Corp.) is the first such upgrading process that has reached commercial application. The high-grade manganese carbonate produced is sold to the chemical industry. The process involves reducing the ore to MnO by roasting with gases rich in CO as the initial step. The calcine is then ground and leached in an aqueous solution containing 18 moles of NH_3 and 3 moles of CO_2. The resulting product is decomposed to yield $MnCO_3$ and NH_3.

The manganese nitrate process is the second upgrading process that has reached commercial application. The high-purity manganese oxides produced are sold to the chemical and ferrite industries. The process involves the reaction between NO_2 and manganese ore to form manganese nitrate solution. The resulting aqueous solution is then thermally decomposed to produce MnO_2 and nitrogen oxides. The nitrogen oxides are recycled to the leaching step, while the MnO_2 is recovered and processed by reduction to Mn_2O_3, Mn_3O_4, or MnO. Processes of lesser importance include the chloride and sulfur oxide processes and bacterial leaching.

Chloride Process. In 1985, investigators at the Argonne National Laboratory reported on the success of a two-step process that extracts cobalt and manganese from low- and medium-grade ores that are mined mainly for other metals. Inasmuch as cobalt and manganese are strategically critical minerals for the United States and several other industrial nations, a viable process for secondary sources of Co and Mn is attractive. A molten salt is used to dissolve more than 90% of the Co and Mn found in common nickel- and copper-bearing ores. The salt mixture contains the chlorides of sodium, potassium, and magnesium (mp 750 °C; 1382 °F). Approximately one part (wt) of the ore requires four parts of the chloride mixture. The latter is recyclable. The desired metals are subsequently separated electrolytically.

Chemistry of Manganese. Manganese has a $3d^5 4s^2$ electron configuration, and compounds in all oxidation states from 0 to 7+ are known, although those of 1+ and 5+ are uncommon. The reducing power of the manganese atom (Mn → Mn^{2+}, 1.18 V) is less than that of magnesium, although the first and second ionization potentials are closely similar, due to the higher heat of sublimation of manganese. However, manganese is oxidized by the halogens, H+ or even H_2O to the dipositive state.

Like so many other metals manganese forms compounds with nitrogen, carbon and even oxygen that exhibit unusual valences, or are even of nonstoichiometric character. With nitrogen manganese combines with unusual valence of 5+ to form Mn_3N_5; with carbon it forms Mn_3C, while with free oxygen it forms first MnO, then Mn_3O_4, and finally Mn_2O_3. An exception to this rule is the MnO_2 produced by thermodecomposition of concentrated manganese nitrate solutions where the oxygen-to-manganese ratio is 1.99+.

Manganese (0) compounds are exemplified by the carbonyls discussed below.

Manganese(I) is found chiefly in the few complex ions, such as the hexacyanomanganate(I) ion $[Mn(CN)_6]^{5-}$, produced by vigorous reduction (e.g., by aluminum in alkaline solution) of the corresponding manganese(II) ion $[Mn(CN)_6]^{4-}$, or in isocyanide complexes (formed by reduction of the diiodide with alkyl isocyanides) $[Mn(RNC)_6]^+$ where R is an alkyl radical.

Manganese(II) (manganous) compounds are obtained, as stated above, by action of water, halogens (except fluorine) or acids upon the metal, or by reduction of more highly oxidized compounds in acid solution. Many salts of Mn^{2+} are known, including all four of the common halides, the nitrate, the sulfate, the sulfite, various phosphates, the arsenate, and many salts of organic acids, e.g., the acetate, butyrate, citrate, lactate, oleate, and tartrate. The manganese(II) compounds are in general relatively resistant to oxidation, due to the stability of the half-filled $3d$ subshell. However, the oxide, MnO, and the hydroxide, $Mn(OH)_2$, are rather easily oxidized by air.

This stability of the Mn(II) state is also reflected in the relatively strong oxidation potential of Mn^{3+} (manganic) ion (the value for Mn^{3+}/Mn^{2+} being −1.51 V), and the readiness with which Mn(III) compounds disproportionate. Manganese(III) fluoride, produced by the action of fluorine on lower compounds, reacts with H_2O to produce the difluoride, hydrogen

fluoride, and MnO_2. In general, however, the manganic compounds such as dimanganese trisulfate and manganese triacetate, decompose in H_2O to divalent manganese ions and Mn_2O_3, forming the MnO_2 only if the pH is definitely below 7. The phosphate, $MnPO_4$, is easily formed by action of nitric acid on manganese(II) phosphate in concentrated phosphoric acid. The fluoro salt K_3MnF_6 is formed by reduction of potassium permanganate, $KMnO_4$, in 40% hydrofluoric acid by an excess of diethyl ether or manganese(II) salt. Manganese(III) also forms a variety of complexes with chelating agents, e.g., oxalate, glycine, acetylacetone, and the like. Like other tripositive transition metal ions it forms alums. The cyanide $K_3Mn(CN)_6$ is stable. All Mn(III) compounds undergo hydrolysis except the complexes.

In addition to the dioxide and the manganites, formed by fusion of MnO_2 with alkali, manganese(IV) forms a number of complexes, such as K_2MnF_6 by reduction of potassium permanganate in 40% hydrofluoric acid with a limited amount of diethyl ether or manganese(II) salts; and Cs_2MnCl_6, by action of cold concentrated HCl containing cesium chloride on MnO_2. Complex iodates are known, e.g., $M_2^I[Mn(IO_3)_6]$, as are cyanides, formed by the action of potassium cyanide on potassium permanganate and said to be $K_4Mn(CN)_8$ (cf. $K_4Mo(CN)_8$ and $K_4Re(CN)_8$).

Manganese rarely occurs with an oxidation number of 5+. In addition to the nitride, there is another compound of interest, in that it can be formed in solution. It is an oxyanion of pentavalent manganese, MnO_4^{3-}, which occurs in the compound, $Na_3MnO_4 \cdot 7H_2O$, formed by reduction of the manganate in strongly alkaline formate or sulfite solutions or by heating MnO_2 in alkali hydroxide at very high temperature. Upon neutralization, it disproportionates to the manganate (and MnO_2).

The manganates, containing MnO_4^{2-}, and produced by alkaline oxidation of MnO_2, are the principal known compounds of hexavalent manganese. They are unstable in neutral or acidic solution, undergoing disproportionation to permanganate (MnO_4^-) and MnO_2. In basic solution, the reaction is reversible. The equilibrium is displaced toward the MnO_4^- by the action of strong oxidants.

The permanganates are strong oxidizing agents, and are usually reduced down to Mn^{2+} under acidic conditions, but to MnO_2, manganate, MnO_4^{2-}, or even hypomanganate, MnO_4^{3-}, under progressively more alkaline conditions. Permanganic acid, $HMnO_4$, and its anhydride, Mn_2O_7, can be obtained at lower temperatures, but they are unstable, decomposing above 0 °C. Permanganyl fluoride, MnO_3F, formed by the action of liquid hydrogen fluoride on potassium permanganate, decomposes at about 0 °C. In strongly acidic media, such as 100% H_2SO_4, manganese (VII) appears to exist as permanganyl ion, MnO_3^+. The sigma bond hybridization in MnO_4^-, MnO_4^{2-} and MnO_4^{3-} is best represented as $d^3 s$.

The only compound of manganese with carbon monoxide alone is the decacarbonyl dimanganese, $(CO)_5MnMn(CO)_5$, but several hydrogen-containing carbonyls, such as $HMn(CO)_5$, halogen-containing carbonyls, such as $Mn(CO)_5Br$, alkyl carbonyls, such as $C_2H_5Mn(CO)_5$ and oxygen-function organometallic compounds, such as

$$[CH_3C(=O)O]_3Mn$$

are known. With the exception of the dicyclopentadienyl compounds, $C_5H_5MnC_5H_5$ manganese does not combine with unsubstituted hydrocarbons or their radicals.

It is interesting to note that some bacteria found near manganese ore plants have the ability to dissolve manganese oxides in solutions of pH 5–6 by the slow addition of H_2SO_4. The only requirement other than the organisms is a nutrient solution. The extraction of manganese as a sulfate is on the order of 71.7–99.9%, depending on the ore. The action of the bacteria is not fully understood.

See also **Manganese (In Biological Systems)**.

Additional Reading

Davis, J.R.: "Metals Handbook," 2nd Edition, ASM International, Materials Park, OH, 1998.

Greenwood, N.N. and A. Earnshaw: "Chemistry of the Elements," 2nd Edition, Butterworth-Heinemann, Inc., Woburn, MA, 1997.

Kerr, R.A.: "Manganese Nodules Grow by Rain from Above," *Science*, **223**, 576–577 (1984).

Klimas-Tavantzis, D.: "Manganese in Health and Disease," CRC Press, LLC., Boca Raton, FL, 1994.

Krebs, R.E.: "The History and Use of Our Earth's Chemical Elements," Greenwood Publishing Group, Inc., Westport, CT, 1998.

Lewis, R. and N. Sax: "Sax's Dangerous Properties of Industrial Materials," 10th Edition, John Wiley & Sons, Inc., New York, NY, 1999.

Lide, D.R.: "CRC Handbook of Chemistry and Physics 2000–2001," 81st Edition, CRC Press, Boca Raton, FL, 2000.

Varentsov, I.: "Manganese Ores of Supergene Zone: Geochemistry of Formation," Kluwer Academic Publishers, Norwell, MA, 1996.

The major portion of this entry was furnished by
J.Y. WELSH and D.F. DE CRAENE, Chemtals Corporation,
Baltimore, MD.

MANGANESE (In Biological Systems). Manganese is required by both plants and animals. Although its deficiency is normally a problem in small areas of fields, it has caused economic losses in the production of cereal small grains on some alkaline soils. In acid soils, manganese is more soluble and plants may be damaged by excessive uptake of the element. Reduced crop yields due to manganese toxicity on acid soils are probably responsible for greater economic losses in a number of regions of the world than are reduced crop yields as the result of manganese deficiency. Measurement of the total manganese concentration in any soil is of little value for predicting possible manganese deficiency or toxicity. The amounts of soluble manganese are more directly related to the level of manganese in plants, but soluble manganese in the soil may fluctuate over short periods because of flooding or drying of the soil or the addition of fresh organic matter. The concentration of manganese in food and in feed plants varies widely; it is more dependent upon the acidity or alkalinity of the soil than on the amount of manganese used in fertilizers.

Although established as an essential trace element, manganese is less well understood than many of the other trace elements. The evidence for its essentiality rests extensively on the consequences of limiting or curtailing the supply of the element of various organisms. Manganese deficiency has induced in most organisms studied a diminished life expectancy. The element is associated with reproductive processes. Additional manifestations depend upon the kind of organism under observation, its age, the degree and duration of manganese deficiency, as well as the coexistence of still another deficiency. For example, in plants, a striking manifestation is chlorosis in which the leaves become pale or yellow, while the veins of the leaves remain green. Manganese plays a significant role in photosynthesis by plants, but it also participates in the regulation of several other enzymic processes.

In poultry, manganese deficiency causes a different clinical picture when it affects the egg than when it affects the hatched bird. In the case of the egg, the embryo become swollen and deformed, and their skeletons become defective and fragile ("chondrodystrophy"). Adult birds develop perosis (slipped tendon) which is an enlargement and malfunction of the tibial metatarsal joint, followed by slipping of the Achilles tendon from the condyles. The bone deformities seen in poultry also can be induced in mammals.

Manganese deficiency also results in the birth of "crooked calves" that are born with enlarged joints, stiffness, and twisted legs. See also **Bovines**.

If the deficiency is induced prior to birth, there is a high intrauterine mortality and whatever young are born alive tend to suffer from an inability to coordinate their muscles (ataxia). These young also have convulsions, delayed growth, and defective bone formation. Adequate manganese intake after birth will correct many of these anomalies, but not the ataxia. If the deficiency is imposed on adult female mammals, ataxia develops infrequently. Instead there appear anemia, defective bone formation, infertility, a tendency to miscarry, and a tendency to absorb the embryo which die within the uterus. The sickly offspring are jeopardized after birth by a disinterest on the part of the manganese-deficient mothers. These avoid nursing their young even when they produce adequate milk. In males, in addition to poor growth, bone deformities and anemia, impotence and infertility develop. Adult animals also develop defects in metabolizing body fat, which are reflected in abnormal amounts and abnormal distribution of body fat. This liptropic effect of manganese extends also to the metabolism of cholesterol and, in this particular role, it can be antagonized by vanadium. The bone deformities are ascribed primarily to poor synthesis of the mucopolysaccharides that make up the matrix of the bones. The infertility is a consequence of death of the testicle's germ cells.

On the other hand, manganese in excessive amounts can cause manganese toxicity. In the past, this disease has mainly affected miners who work either in manganese mines or in ore crushing mills. The manganese ore enters the body by inhalation of the dust. Among the many miners exposed throughout the world, some develop brain symptoms. Involvement of the brain manifests itself first in mental aberrations. Later, neurological changes occur in the form of trembling, rigidity, salivation, mask-like face, and a general appearance of a person afflicted with Parkinson's disease. Chronic manganese poisoning has occurred in epidemics. The condition is incurable, but not necessarily life-limiting.

It is believed that high manganese diets in cattle will decrease fiber digestion. Manganese interferes with iron metabolism by antagonizing the enzyme system that oxidizes or reduces iron at the absorption site, thus affecting iron availability. It is also suggested that manganese may cause a condition in ruminant cattle (hypomagnesia). Most of the foregoing observations can be explained on the basis of manganese activating various cellular enzymes. Much importance has been given to the particular enzyme systems responsible for oxidative phosphorylation. These systems determine the generation and utilization of energy from foodstuffs by the cells. Additionally, manganese appears to activate many other enzymes (arginase, enolase, peroxidases). It also appears to participate in the structure of the nucleic acids responsible for the manufacture of enzymes and other proteins. Manganese probably plays a number of unique roles. No other metal replacement for the element in biological functions has been identified to date.

Manganese occurs in the liver of the animal body. Even though the amount of manganese present in mammalian tissues is very small, its concentration seems to be accurately controlled by elaborate mechanisms. These mechanisms function primarily by promoting the excretion of excesses of the element from the body rather than by regulating the amounts of manganese the body absorbs. The mechanisms are located in the liver and on the mucosa of the gut. In cases of manganese toxicity, it is assumed that these mechanisms become saturated.

Cereals and pulses (peas and beans) are the major sources of manganese in human diets, and diets containing these foods can be expected to provide adequate manganese. Dietary supplements for manganese (feeds and foodstuffs) include: manganese chloride, manganese gluconate, manganese glycophosphate, manganese hypophosphate, and manganese sulfate.

Additional Reading

Adriano, D.C.C.: "Biogeochemistry of Trace Metals," Lewis Publishers, Boca Raton, FL, 1992.

Klimas-Tavantzis, D.: "Manganese in Health and Disease," CRC Press, LLC., Boca Raton, FL, 1994.

Staff: "Handbook of Inorganic and Organometallic Chemistry," Gmelin Institute Series, Springer-Verlag Inc., New York, NY, 1997.

Staff: "Dietary Reference Intakes: Vitamin A, Vitamin K, Arsenic, Boron, Choromium, Copper, Iodine, Iron, Manganese, Molybdenum, Nickel, Silicon, Vanadiu," National Academy Press, Washington, DC, 2001.

Underwood, E.J. and W. Mertz: "Trace Elements in Human and Animal Nutrition," 5th Edition, Academic Press, Inc., San Diego, CA, 1990.

MANGANITE. The mineral manganite is a hydrous oxide of manganese corresponding to the formula $MnO(OH)$. It occurs in prismatic monoclinic crystals, sometimes in massive columnar forms, granular, concretionary, and stalactitic. It is a brittle mineral, with perfect prismatic cleavage; hardness, 4; specific gravity, 4.33; luster, submetallic; color, steel gray to iron black; streak, red-brown to almost black; opaque. Manganite is of secondary origin and it may itself alter to pyrolusite. It is usually associated with other manganese minerals. It is found in the Harz Mountains, Germany; Sweden; Cornwall and Cumberland, England; and in the United States in Michigan. It is an ore of manganese. See also **Pyrolusite**.

MANGO TREE. See **Cashew and Sumac Trees**.

MANGROVE TREE. Of the family *Rhizophoraceae*, the mangrove is a moderate-sized tree which grows on low, often submerged, coastal lands. It is found, for instance, in all tropical American coasts. The leaves of the plant are opposite, entire, dark green, and rather tough. The flowers are borne in small clusters and are perfect, with four sepals, four pale yellow linear petals, four to twelve stamens and single two-celled inferior ovary. Only one ovule develops. The seed usually germinates while the fruit is still attached to the tree. A long thick hypocotyl grows from the fruit, and attains a length of 5–10 inches (12.7 to 25 centimeters) and a diameter of less than 3/4 inch (1.9 centimeters). Eventually an abscission layer develops, so that the fruit, in which the young seedling is well advanced in germination, falls to the soft muddy ground, in which the new tree will grow. In a favorable location the hypocotyl puts out many roots, which anchor the young plant; then the epicotyl quickly grows. It is characteristic of the mangrove that from the stem and branches there

TABLE 1. RECORD MANGROVE TREES IN THE UNITED STATES[1]

Specimen	Circumference[2]		Height		Spread		Location
	Inches	Centimeters	Feet	Meters	Feet	Meters	
Black mangrove (1996) (*Avicennia germinans*)	101	257	43	13.1	57	17.4	Florida
Red mangrove (1995) (*Rhizophora mangle*)	47	119	58	17.7	42	12.8	Florida

[1]From the "National Register of Big Trees," American Forests (by permission).
[2]At 4.5 feet (1.4 meters).

grow out arching prop roots which soon form an intricate mass in which is deposited silt and all sorts of debris floating in the water. Because of this the mangrove causes a gradual building up of the land around it, until eventually the black slimy mud in which it grows gives place to a low coastal land which gradually becomes usable by man.

In addition to its land-forming function, the mangrove has other uses. The wood is dark red or reddish-brown, fine-grained, and hard; it is used in charcoal making. The bark contains tannin and so is employed in tanning hides. From the young shoots, a reddish dye may be obtained, which, however, is of little value.

Several other species of similar habitat and growth pattern are also called mangroves.

Two record mangrove trees selected by American Forests are described in the Table 1.

MANHATTAN PROJECT (The).   More than 55 years ago, work at Los Alamos and elsewhere in the world set in motion developments in military and civil applications of nuclear science and technology. Over the years these ongoing developments have shaped history. The resulting "Nuclear Age" has had a significant impact on many aspects of society — nationally and internationally.

The Manhattan Project of the "Second World War" represents the most remarkable congregation of scientific minds in human history. New scientific ground was broken which helped to produce numerous additional discoveries. Modern computer theory largely grew from bomb-related research with the first huge mainframe computers being used mainly for bomb design.

On August 2nd 1939, just before the beginning of World War II, Leo Szilard persuaded Albert Einstein to write to then President Franklin D. Roosevelt. Einstein and several other scientists told Roosevelt of efforts in Nazi Germany to purify U-235 with which might in turn be used to build an atomic bomb. It was shortly thereafter that the United States Government began the serious undertaking known only then as the Manhattan Project. Simply put, the Manhattan Project was committed to expedient research and production that would produce a viable atomic bomb.

General Leslie R. Groves, Deputy Chief of Construction of the U.S. Army Corps of Engineers, was appointed to direct this top-secret project. Groves established three large engineering and production centers at remote U.S. sites: the Clinton Engineer Works at Oak Ridge, Tenn.; the Hanford Engineer Works in eastern Washington State; and Project Y, a code-named site 100 miles north of Albuquerque at Los Alamos, N.M. All three sites still exist and contribute to America's nuclear arsenal.

The most complicated issue to be addressed was the production of ample amounts of "enriched" uranium to sustain a chain reaction. At the time, Uranium-235 was very hard to extract. In fact, the ratio of conversion from Uranium ore to Uranium metal is 500:1. An additional drawback is that the 1 part of Uranium that is finally refined from the ore consists of over 99% Uranium-238, which is practically useless for an atomic bomb. To make it even more difficult, U-235 and U-238 are precisely similar in their chemical makeup. This proved to be as much of a challenge as separating a solution of sucrose from a solution of glucose. No ordinary chemical extraction could separate the two isotopes. Only mechanical methods could effectively separate U-235 from U-238. Several scientists at Columbia University managed to solve this dilemma.

A massive enrichment laboratory/plant was constructed at Oak Ridge, Tennessee. H.C. Urey, along with his associates and colleagues at Columbia University, devised a system that worked on the principle of gaseous diffusion. Following this process, Ernest O. Lawrence (inventor of the Cyclotron) at the University of California in Berkeley implemented a process involving magnetic separation of the two isotopes.

Following the first two processes, a gas centrifuge was used to further separate the lighter U-235 from the heavier non-fissionable U-238 by their mass. Once all of these procedures had been completed, all that needed to be done was to put to the test the entire concept behind atomic fission.

Over the course of six years, ranging from 1939 to 1945, more than 2 billion dollars were spent on the Manhattan Project. The formulas for refining Uranium and putting together a working bomb were created and seen to their logical ends by some of the greatest minds of our time. Among these people who unleashed the power of the atomic bomb was J. Robert Oppenheimer.

Oppenheimer was the major force behind the Manhattan Project. He literally ran the show and saw to it that all of the great minds working on this project made their brainstorms work. He oversaw the entire project from its conception to its completion.

Finally the day came when all at Los Alamos would find out whether or not *The Gadget* (code-named as such during its development) was either going to be the colossal dud of the century or perhaps end the war. It all came down to fateful morning of midsummer, 1945. The first nuclear explosion in history took place in New Mexico, at the Alamogordo Test Range, on the Jornada del Muerto (Journey of Death) desert, in the test named Trinity. Trinity, was the conclusion of the Manhattan Project to build the bomb in a frantic race with Adolf Hitler's scientists.

This test was intended to prove the radical new implosion weapon design that had been developed at Los Alamos during the previous year. This design, embodied in the test device called *Gadget*, involved a new technology that could not be adequately evaluated without a full scale test. The gun-type uranium bomb, in contrast, was certain to be effective and did not merit testing. In addition, since no nuclear explosion had ever occurred on Earth, it seemed advisable that at least one should be set off with careful monitoring to test whether all of the theoretical predictions held.

At 5:29:45 (Mountain War Time) on July 16th, 1945, in a white blaze that stretched from the basin of the Jemez Mountains in northern New Mexico to the still-dark skies, *The Gadget* ushered in the Atomic Age. The light of the explosion then turned orange as the atomic fireball began shooting upwards at 360 feet per second, reddening and pulsing as it cooled. The characteristic mushroom cloud of radioactive vapor materialized at 30,000 feet. Beneath the cloud, all that remained of the soil at the blast site were fragments of jade green radioactive glass. All of this caused by the heat of the reaction.

The brilliant light from the detonation pierced the early morning skies with such intensity, that residents from a faraway neighboring community would swear that the sun came up twice that day. Even more astonishing is that a blind girl saw the flash 120 miles away.

Upon witnessing the explosion, reactions among the people who created it were mixed. Isidor Rabi felt that the equilibrium in nature had been upset — as if humankind had become a threat to the world it inhabited. J. Robert Oppenheimer, though ecstatic about the success of the project, quoted a remembered fragment from Bhagavad Gita. "I am become Death," he said, "the destroyer of worlds."

Several participants, shortly after viewing the results, signed petitions against loosing the monster they had created, but their protests fell on deaf ears. As it later turned out, the Jornada del Muerto of New Mexico was not the last site on planet Earth to experience an atomic explosion.

Detonation

As many know, atomic bombs have been used only twice in warfare. The first and foremost blast site of the atomic bomb is Hiroshima. A Uranium bomb (which weighed in at over 4 1/2 tons) nicknamed "Little Boy" was dropped on Hiroshima August 6th, 1945. The Aioi Bridge, one of 81 bridges connecting the seven-branched delta of the Ota River, was the aiming point of the bomb. Ground Zero was set at 1,980 feet. At 0815 hours, the bomb was dropped from the Enola Gay. It missed by only 800 feet. At 0816 hours, in the flash of an instant, 66,000 people were killed and 69,000 people were injured by a 10 kiloton atomic explosion.

The point of total vaporization from the blast measured one half of a mile in diameter. Total destruction ranged at one mile in diameter. Severe blast damage carried as far as two miles in diameter. At two and a half miles, everything flammable in the area burned. The remaining area of the blast zone was riddled with serious blazes that stretched out to the final edge at a little over three miles in diameter.

On August 9th 1945, Nagasaki fell to the same treatment as Hiroshima. Only this time, a Plutonium bomb nicknamed "Fat Man" was dropped on the city. Even though the "Fat Man" missed by over a mile and a half, it still leveled nearly half the city. Nagasaki's population dropped in one split-second from 422,000 to 383,000. 39,000 were killed, over 25,000 were injured. That blast was less than 10 kilotons as well. Estimates from physicists who have studied each atomic explosion state that the bombs that were used had utilized only 1/10th of 1 percent of their respective explosive capabilities.

While the mere explosion from an atomic bomb is deadly enough, its destructive ability doesn't stop there. Atomic fallout creates another hazard as well. The rain that follows any atomic detonation is laden with radioactive particles. Many survivors of the Hiroshima and Nagasaki blasts succumbed to radiation poisoning due to this occurrence.

The atomic detonation also has the hidden lethal surprise of affecting the future generations of those who live through it. Leukemia is among the greatest of afflictions that are passed on to the offspring of survivors.

While the main purpose behind the atomic bomb is obvious, there are many by-products that have been brought into consideration in the use of all weapons atomic. With one small atomic bomb, a massive area's communications, travel and machinery will grind to a dead halt due to the EMP (Electro-Magnetic Pulse) that is radiated from a high-altitude atomic detonation. These high-level detonations are hardly lethal, yet they deliver a serious enough EMP to scramble any and all things electronic ranging from copper wires all the way up to a computer's CPU within a 50 mile radius.

At one time, during the early days of The Atomic Age, it was a popular notion that one day atomic bombs would one day be used in mining operations and perhaps aid in the construction of another Panama Canal. Needless to say, it never came about. Instead, the military applications of atomic destruction increased. Atomic tests off of the Bikini Atoll and several other sites were common up until the Nuclear Test Ban Treaty was introduced. Photos of nuclear test sites here in the United States can be obtained through the Freedom of Information Act.

See also **Nuclear Fission**; and **Oppenheimer, J. Robert (1904–1967)**.

Additional Reading

Gonzales, D.: "The Manhattan Project and the Atomic Bomb in American History," Enslow Publishers, Inc., Berkeley Heights, NJ, 2000.

Groueff, S.: "Manhattan Project: The Untold Story of the Making of the Atomic Bomb," iUniverse.com, Inc., New York, NY, 2000.

Sparks, R.C., B.G. Storms: "Twilight Time: A Soldier's Role in the Manhattan Project in Los Alamos," Los Alamos Historical Society, Los Alamos, NM, 2000.

Web References

Hanford Site: *http://www.hanford.gov/doe/culres/historic/index.htm*

History of the Plutonium Production Facilities at the Hanford Site Historic District, 1943–1990, Manhattan Project: 1943-1946, Cold War Era: 1947–1990: *http://www.hanford.gov/docs/rl-97-1047/index.htm*

Los Alamos National Laboratory: *http://lib-www.lanl.gov/infores/history/history.htm*

Nuclear History Site: *http://geocities.com/RainForest/Andes/6180/*

The US National Atomic Museum's virtual tour of the Manhattan Project: *http://www.atomicmuseum.com/tour/manhattanproject.cfm*

MANIC-DEPRESSIVE (Bipolar) ILLNESS.

This is a major mental illness, estimated as of the 1990s to affect 5 persons per 100,000 population, a number that is considerably less than for depression without mania. Researchers at Washington University have established six criteria for the diagnosis of manic-depressive illness. (1) Hyperactivity and aggressiveness. This applies to some or all personal activities—motor, social, and sexual. (2) Incessant talking—with no toleration for interruptions, resulting in a "pushy" behavior that can lead to much frustration in persons attempting to communicate with the manic-depressive patient. (3) Rapid transfer and alteration of thought patterns, resulting in racing from one idea to the next, often precipitated by mention of a key word. (4) An air of grandeur—promotion of schemes requiring large sums of money, with little basis in fact, dramatized by excessive use of credit cards, telephone calls, etc. (5) Reduced need to sleep. (6) Overreaction to numerous stimuli—the patient, easily distracted by unimportant events. In using the foregoing criteria, the examiner must make certain that there have been no preexisting psychiatric disturbances as may result from alcoholism or schizophrenia. The etiology of this illness appears to be connected with genetic factors as well as environment. Frequently, familial connections are apparent, but no genetic connection has been revealed thus far.

In 1987, a group of researchers announced that a statistical analysis of members of the Amish community, using a genetic technique (linkage analysis), indicated that a gene located at the top of the short arm of chromosome 11 seemed to be responsible for the heritability of certain forms of manic-depressive illness. This series of experiments later was challenged and refuted. The Amish community had been selected because of its meticulous family records, dating back to the early 18th century when 30 Amish couples increased in numbers to about 15,000 persons today.

Role of Lithium. In 1949, a researcher, J.F.J. Cade (Australian psychiatrist), accidentally discovered that lithium had a calming effect on laboratory animals. This led to the investigation of lithium as a possible drug for use in treating manic-depressive illness. Subsequent research indicated that lithium antagonizes synaptic transmission of cathecholamines (specifically by inhibiting norepinephrine and dopamine release). See also **Central and Peripheral Nervous Systems**. Lithium carbonate is the drug of choice and is widely used. Because of several side effects of lithium, considerable clinical evaluation of the patient is required prior to its administration. Inasmuch as many cases of mania require immediate or very early hospitalization, various antipsychotic drugs may be used as an interim measure. Because manic-depressive patients are also subject to periods of depression (thus a biopolar situation), it is fortunate for many patients that lithium also is an effective prophylaxis for depression. In less responsive patients, more dramatic therapy, such as electroconvulsive therapy, may be required. Complications of prolonged lithium therapy include benign diffuse nontoxic goiter, usually treatable with L-thyroxine. The physician must monitor all factors which would increase sodium loss (increases lithium retention). Lithium has been shown to cause serious renal damage when administered over long periods. See also **Schizophrenia**.

The bidirectional therapeutic effect of lithium carbonate has puzzled investigators for several years. In 1972, D.S. Janowski (Vanderbilt University) proposed that adrenergic-cholinergic unbalances are the root cause of bipolar illness. A group of researchers (Ben Gurion University of the Negev, Israel) suggested in 1988 that lithium blocks the activation of G proteins by neurotransmitters binding to both adrenergic and cholinergic receptors. It had been established previously that lithium cripples a messenger system linked to cholinergic receptors. In studies of rats, the Israeli team reported that lithium also is effective in modulating adrenergic receptors. This, at least, assists in explaining why lithium functions bidirectionally. But other researchers, who are aware of the wide distribution of G proteins, wonder why lithium's effectiveness is limited to the central nervous system. This part of the puzzle, however, has not been solved because researchers in the United Kingdom observe, "At therapeutic doses, the side effects of lithium are negligible." W.R. Sherman (Washington University) has observed, "Studies of lithium and manic depression, like all studies of drug action in psychiatry, are disarmed by their reliance on subjective information. In spite of the fact that thousands of Americans lead normal lives because of lithium therapy, it's almost impossible to get real evidence of what's going on!"

Additional Reading

Gabbard, G.O.: "Treatments of Psychiatric Disorders," 3rd Edition, American Psychiatric Press, Inc., Washington, DC, 2001.

Goodwin, F.K. and K.R. Jamison: "Manic-Depression Illness," Oxford University Press, Inc., New York, NY, 1990.

Halleck, S.L.: "Evaluation of the Psychiatric Patient: A Primer," Kluwer Academic Publishers, Norwell, MA, 1991.

Hawton, K. and P. Cowen:"Dilemmas and Difficulties in the Management of Psychiatric Patients," Oxford University Press, Inc., New York, NY, 1991.

Lam, D.H., P. Hayward, S.H. Jones, and J.A. Bright: "Cognitive Therapy for Biopolar Disorder," John Wiley & Sons, Inc., New York, NY, 1999.

Mirin, S.M., J.T. Gossett, and M.C. Grob: "Psychiatric Treatment: Advances in Outcome Research," American Psychiatric Press, Inc., Washington, DC, 1991.

Mondimore, F.M.: "Biopolar Disorder: A Guide for Patients and Families," Johns Hopkins University Press, Baltimore, MD, 2000.

Pearlman, T.: "The Threatened Medical Identity of Psychiatry: The Winds of Change," Charles C. Thomas, Springfield, IL, 1992.

Reid, W.H., G.U. Balis, and B.J. Sutton: "The Treatment of Psychiatric Disorders," 3rd Edition, Brunner/Mazel, New York, NY, 1997.

Staff: "Practice Guideline for the Treatment of Patients with Bipolar Disorders," American Psychiatric Press, Inc., Washington, DC, 2000.

Swartz, C.M.: "Serum Lithium During Treatment of Bipolar Disorders," *N. Eng. J. Med.*, 1159 (April 19, 1990).

Walden, J. and H. Grunze: "Bioplar Illnesses: New Ways of Treatment," S. Karger Publishers, Inc., Farmington, CT, 2000.

Weller, M. and M. Eysenck: "The Scientific Basis of Psychiatry," 2nd Edition, W.B. Saunders, Philadelphia, PA, 1992.

Yonkers, K. and B.B. Little: "Management of Psychiatric Disorders during Pregnancy," Oxford University Press, Inc., New York, NY, 2001.

Web Reference

American Psychiatric Association: *http://www.psych.org/*

MANOMETER.   A relatively simple instrument that provides direct measurement of positive pressure, vacuum, and differential pressure. The manometer is also used indirectly for the measurement of flow by sensing the output of a pressure-differential producing device, such as a venturi or orifice plate. The manometer operates on the fundamental principle of displacing a liquid column by the unknown pressure force to be measured. Two types of manometer are illustrated in Fig. 1. See also **Barometer**.

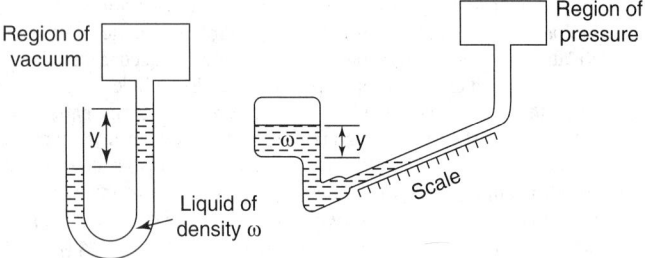

Fig. 1.   Types of manometer: U-tube manometer at left; inclined-tube manometer at right.

Fig. 1.   Praying mantis (mantis religiosa). Length ranges up to 7.5 cm.

MANTIS *(Insecta, Orthoptera).*   A large insect of predatory habits. The body is moderately broad and bears four wings. The first segment of the thorax is long and rather slender, adding to the reach of the powerful raptorial front legs. The head is prominent and has large eyes.

From the fancied suppliant air of these voracious insects as they await their prey with the forelegs uplifted they are called praying mantises, and their owlish expression has given them the rarer name of soothsayers. The common species is *Mantis religiosa.*

The mantis is found principally in America and southern Europe. There are about 800 species. The common species (*Mantis religiosa*) measures about 2 inches in length, but in South America the species is larger and sometimes these insects will attack small birds, frogs, and lizards. Sometimes, a mantis will devour its young and the female may eat the male.

The mantis is a great help in destroying flies, grasshoppers, caterpillars, and other insects. They are harmless to human beings.

See Fig. 1.

MANTIS FLY *(Insecta, Neuroptera).*   Small predacious insects superficially like the mantis in form. They make up the family *Mantispidae* and are also called mantispas.

MANTISSA.   See **Logarithm**.

MANTLE (Earth).   See **Earth**.

MANUFACTURING MESSAGE SPECIFICATION (MMS).   MMS (Manufacturing Message Specification) is an internationally standardized application layer messaging service for exchanging real-time data and supervisory control information between networked intelligent electronic devices (IED) and/or computer applications. MMS is used in manufacturing automation for communications between programmable logic controllers (PLC), robots, computer numerical controls (CNC), and computers. In the electric utility industry, MMS is used for communications between utility control centers, substations, and intelligent electronic devices (IED) such as remote terminal units (RTU), protection relays, and meters.

History of MMS

Work on MMS was originally begun during the Manufacturing Automation Protocol (MAP) effort sponsored by the General Motors Corporation in the 1980s. The MAP effort was undertaken to develop a standardized automation protocol for use across a range of industrial applications in both the discrete[1], continuous[2], and batch[3] processing industries. A broad group of engineers from the programmable logic control (PLC), computer numerical control (CNC), robotic, automotive, chemical, oil processing, and communications industries came together under the auspices of the International Organization for Standardization's (ISO) technical committee 184 (TC184) to develop an application layer protocol suitable to satisfy this broad range of interests. The result was the MMS standard published as ISO9506 part 1 (Services) and part 2 (Protocol) in December of 1988.

In 1990 the Electric Power Research Institute (EPRI) began an effort to develop a standard for electric utilities to use for building a modern real-time communications architecture called the Utility Communications

[1] Discrete manufacturing processes are characterized as producing products that are measured in discrete units (e.g. automotive parts and assembly, appliances, toys, and tools).

[2] Continuous manufacturing processes are characterized by producing products that are measured in bulk units (e.g. oil refining, power generation, fertilize, and chemicals).

[3] Batch manufacturing processes are characterized by producing discrete units of continuously produced products (e.g. food processing).

Architecture (UCA™). One of the objectives of the UCA effort was to leverage existing technology and adapt it to the needs of electric utilities. The UCA effort selected the use of Ethernet, fiber optics, TCP/IP-Internet networking, and MMS as an application layer protocol for real-time communications. In addition to the use of MMS at the application level, UCA also defines:

- A set of profiles for running MMS over networks (TCP/IP and OSI) and serial links for spread spectrum (SS) or multiple address system (MAS) radios.
- Object model and service mappings to MMS for facilitating data exchange between utilities to support a deregulated electric energy market called the Intercontrol Center Communications Protocol (ICCP).
- A Common Application Service Model (CASM) for substation automation that specifies how to use MMS to perform electric utility industry specific functions such as select before operate (SBO) and report by exception (called *reporting*).
- Defined device and object models for electric utility substation devices called the Generic Object Models for Substation and Feeder Equipment (GOMSFE).

The International Electro-Technical Commission (IEC) TC57 working group 6 (WG6) published the results of the ICCP effort as IEC60870-6 Telecontrol Application Service Element Number 2 (TASE.2) in 1996. The UCA CASM and GOMSFE specifications were then used as the basis for the international standard for substation communications published by IEC TC57 WG10, WG11, and WG12 in 2001.

In 1998 Gas Research Institute (GRI) began working on a gas industry version of UCA that builds upon the concepts of GOMSFE to build gas industry specific device and object models. In 1999, the American Water Works Association (AWWA) developed a version of UCA for water utility applications. More recently, the United States Postal Service (USPS) has utilized MMS as the basis for its specifications for communicating with mail processing equipment.

The messaging services provided by MMS are generic enough to be appropriate for a variety of devices, applications, and industries. Applications as diverse as material handling, fault annunciation, energy management, electrical power distribution control, inventory control, and deep space antenna positioning in industries as varied as automotive, aerospace, petrochemical, electric/gas utility, office machinery, and space exploration have put MMS to useful work.

As of this writing, most new applications of MMS are in electric utility and postal equipment applications. The manufacturing networking/communications industry has since fragmented into a large number of mostly incompatible fieldbus technologies for connecting controls to I/O systems and numerous TCP/IP based application protocols for communications between automation controllers and business information systems. These systems tend to be either very simplistic, making them unsuitable for the more complex interactions required by a typical MMS application, or are specific to a very narrow application niche such as motion control and distributed I/O. MMS remains the only internationally standardized application level protocol that has a proven track record of being effective across a broad range of industries.

The VMD Model

The primary goal of MMS was to specify a standard communications mechanism for devices and computer applications that would achieve a high level of interoperability[4]. In order to achieve this goal, it would be necessary for MMS to define much more than just the format of the messages to be exchanged. A common message format, or protocol, is only one aspect of achieving interoperability. In addition to protocol, the MMS standard also provides definitions for:

- **Objects**. MMS defines a set of common objects (e.g., variables) and defines the network visible attributes of those objects (e.g., name, value, type).

™ UCA is trademark of EPRI.

[4] Interoperability is the ability of two or more networked applications to exchange useful information between them without the user of the applications having to create the communications environment. Most communication protocols can provide some level of interoperability. However, many of them are too specific to work outside of a very narrow appliction niche. Others are not specific enough and offer too many choices for developers. This results in incompatible systems.

- **Services**. MMS defines a set of communications *services* (e.g., read, write) for accessing and managing these objects in a networked environment.
- **Behavior**. MMS defines the network visible behavior that a device should exhibit when processing these services.

This definition of objects, services, and behavior comprises a comprehensive definition of how devices and applications communicate in which MMS calls the *Virtual Manufacturing Device* (VMD) model. The VMD model is the key feature of MMS. The VMD model specifies how MMS devices, also called servers, behave as viewed from an external MMS client application point of view. The VMD model only specifies the network visible aspects of communications. The internal detail of how a real device implements the VMD model (i.e., the programming language, operating system, CPU type, input/output (I/O) systems) are not specified by MMS. By focusing only on the network visible aspects of a device, the VMD model is specific enough to provide a high level of interoperability. At the same time, the VMD model is still general enough to allow innovation in application/device implementation and making MMS suitable for applications across a range of industries and device types. See Fig. 1.

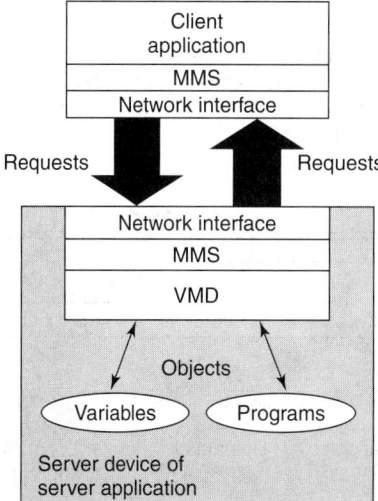

Fig. 1. The VMD model provides a consistent and well defined view for client applications of the objects contained in the VMD. Clients use MMS services to access and manipulate those objects. MMS requires that all servers behave according to the VMD model.

Client/Server Relationship. MMS is a client/server based protocol A *server* is a device or application that contains data and executes commands. A *client* is a networked application or device that accesses and manipulates data or issues command requests to a server (Fig. 1). While MMS defines the services for both clients and servers, the VMD model defines the network visible behavior of servers only. See Fig. 2. Many MMS applications and devices can provide both MMS client and server functions. The VMD model would only define the behavior of the server functions of those applications. Any MMS application or device that provides MMS server functions must follow the VMD model for all the network visible aspects of the server application or device. MMS clients are only required to conform to rules governing message format or construction and sequencing of messages (the protocol).

Real and Virtual Devices and Objects

There is a distinction between a real device and a real object (e.g., a PLC with a part counter) and the *virtual* device and objects (e.g. VMD, domain, variable, etc.) defined by the VMD model. Real devices and objects have peculiarities (a.k.a. product features) associated with them that are unique to each brand of device or application. Virtual devices and objects conform to the VMD model and are independent of brand, language, operating system, etc. Each developer of a MMS server device or MMS server application is responsible for "hiding" the details of their real devices and objects, by providing an *executive function*. The executive function translates the real devices and objects into the virtual ones defined by

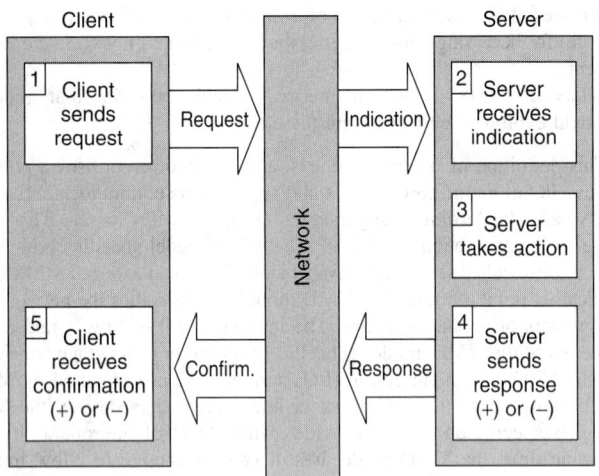

Fig. 2.   Client/Server Interactions. MMS clients and servers interact with each other by sending/receiving request, indication, response, and confirmation service primitives over the network. The figure depicts the interactions between a client and server for a MMS confirmed service where (1) the client sends a request, (2) the server receives an indication, (3) the server performs the desired action, (4) the server sends a positive (+) response if the action was successful or a negative (−) response if there was an error, and (5) the client receives the confirmation (+) or (−). An Unconfirmed Service is send by the server and has only the request and indication service.

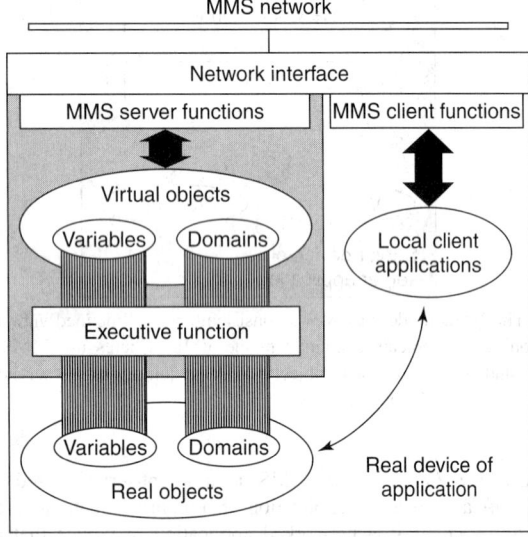

Fig. 3.   Real and Virtual Objects. The executive function provides a translation, or "mapping" between the MMS defined virtual objects and the real objects used by the real device. Applications local to the VMD, and the objects contained in them, are only accessible to a remote MMS client application if the executive function provides the mapping function for those objects and applications. Client applications local to the VMD may access and manipulate the real objects without using MMS.

the VMD model when communicating with MMS client applications and devices. See Fig. 3.

Because MMS clients always interact with the virtual device and objects defined by the VMD model, the client applications are isolated from the specifics of the real devices and objects. A properly designed MMS client application can communicate with many different brands and types of devices in the exact same manner. This is because the details of the real devices and objects are hidden from the MMS client by the executive function in each VMD. This virtual approach to describing server behavior does not constrain the development of innovative devices and product features and improvements. The MMS VMD model places constraints only on the network visible aspects of the virtual devices and objects, not the real ones.

MMS Device and Object Modeling

The implementor of the executive function (the application or device developer) must decide how to "model" the real objects as virtual objects. The manner in which these objects are modeled is critical to achieving interoperability between clients and servers among many different developers. Inappropriate or incorrect modeling can lead to an implementation that is difficult to use or difficult to interoperate with. One of the key benefit of using UCA and ICCP is the additional object modeling definitions these standards provide for electric utility specific applications.

Objects

MMS defines objects that are found in many typical devices and applications requiring real-time communications. For each object MMS defines a set of properties or *attributes* that describe various aspects of the object such as its name, status/state, value/contents, etc. The objects defined by MMS are[5]:

- **VMD**. The device itself is an object.
- **Domain**. A resource (e.g. memory) represented by a block of untyped data.
- **Program Invocation**. A runnable program consisting of one or more domains.
- **Variable**. An named (or unnamed) element of typed data.
- **Type**. A description of the format of a variable's data.
- **Named Variable List**. A list of variables that is named as a list.
- **EventCondition, EventEnrollment, Event Action**. These are objects that are related to the control, processing and notification of events.
- **Semaphore**. An object used to control access to a shared resource.
- **Journal**. An object used to keep a time-based record of variables and events.
- **Operator Station**. A display and keyboard for use by an operator.
- **File**. Data stored in files on a file server.

MMS Services

The MMS client uses MMS services to access and manipulate the objects and their attributes. Each class of object has a unique set of services available to the client. In general, these services support the following actions on the objects[5]:

- **Create Object**. Many MMS objects can be created clients using MMS services. Examples of object creation services are: CreateNamedVariableList, InitiateDownloadSequence (for domains), CreateProgramInvocation, etc.
- **Delete Object**. Objects that are created by clients can also be deleted by clients. Some examples are DeleteNamedVariableList, DeleteDomain, DeleteProgramInvocation, etc.
- **Get Attributes**. The MMS client can determine the attributes of a given MMS object are by using MMS services. Examples of MMS services used to obtain object attributes include GetVariableAccessAttributes (retrieves the definition of a variable, Read (retrieves the value of a variable), GetDomainAttributes, etc.
- **Change Attributes**. The MMS client can also modify the attributes of an MMS object by using MMS services. The MMS services that can be used to change an object's attributes include Write (change the value of a variable), Start/Stop (changes the state of program invocations), WriteJournal, etc.

ICCP-TASE.2 Objects

The ICCP (IEC60870-6 TASE.2) standard defines the following additional objects that are useful for inter-utility data exchange:

- **Bilateral Table**. An object that represents an agreement between a local and remote node regarding the data to be exchanged and how the data exchanges will be controlled. The use of the bilateral table allows both sides of a data exchange to carefully control the information to be exchanged.
- **Data Set**. An object, represented by an MMS Named Variable List that is used to collect data into logical groups.
- **Transfer Set**. The object, represented by one or more data sets, that is sent in a transfer report that the server sends to the client on a periodic

---

[5] More detailed descriptions of MMS objects, attributes, and services is available on the Internet at: *http://www.sisconet.com/techinfo.htm*

basis or when the data changes (called "report by exception") using the MMS unconfirmed service of InformationReport.

- **Account**. A special type of transfer set that contains information, stored in a row/column format, related to energy exchange schedules and device outages of high-voltage electrical transmission systems.

UCA Objects

The UCA (IEC61850) standard defines additional objects and services for SCADA, distribution and substation automation applications:

- **Select Before Operate**. The SBO object is used as an interlock to coordinate control commands issued by multiple UCA clients to the same point. The SBO object is mapped to a structured MMS variable.
- **Log Object**. The UCA log is an object that is used to store a time based record of events and variables as a sequence of events (SOE) log. UCA clients can then use the log to retrieve the data for archiving. The UCA log is mapped to a MMS journal object.
- **UCA Report**. The UCA report allows UCA clients to control how unsolicited report by exception (RBE) data is sent to them by an IED. The UCA report is mapped to a structured MMS variable using MMS named variable lists for the actual reports.
- **GOMSFE Objects**. UCA defines numerous object models for many common IED functions used in the electric utility. GOMSFE object models are mapped to MMS domains and structured MMS variables with standardized names.
- **Generic Object Oriented Substation Event**. A GOOSE object is broadcast over a local area network (LAN) by protection relays to exchange protection signals as a replacement for individually hard-wired signals. See Fig. 4.

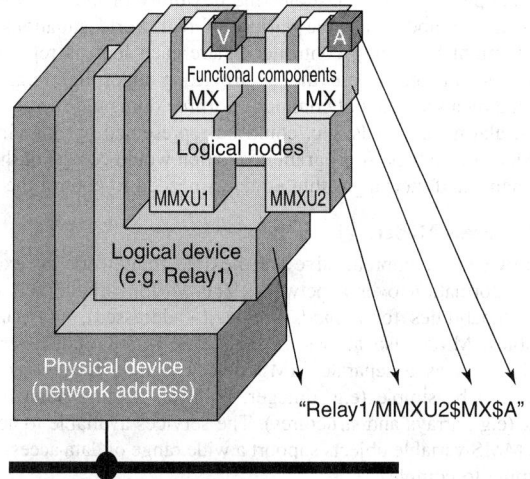

Fig. 4. A GOMSFE name preserves the hierarchy of the date. Logical Devices, modeled as MMS domains, are broken down into functional nodes for measurement functions (MMXU), basic protection relay functions (PBRO), and numerous other device specific models. Each logical node is further broken down into functional components such as measurements (MX), status (ST), descriptions (DC), etc. Each functional component consists of one or more groupings of variables such as amps (A), volts (V), watts (W), vars., (VA), etc. A MMS client specify the entire name to access a specific element or can specify just the logical node designation (e.g. MMXU2), to get all date associated with that logical node.

Object Attributes and Scope

Associated with each object are a set of attributes that describe that object. MMS objects have a name attribute and other attributes that vary from object to object. Variables have attributes such as, name, value, type. Other objects, program invocations for instance, have attributes like name and current state.

Subordinate objects exist only within the scope of another object. For instance, all other objects are subordinate to, or contained within, the VMD itself. Some objects, such as the *operator station* object, may be subordinate only to the VMD. Some objects may be contained within other objects, such as variables contained within a *domain*. This attribute of an object is called its *scope*. The object's scope also reflects the lifetime of an object. An object's scope may be defined to be:

- **VMD-Specific**. The object has meaning and exists across the entire VMD (is subordinate to the VMD). The object exists as long as the VMD exists.
- **Domain-Specific**. The object is defined to be subordinate to a particular domain. The object will exist only as long as the domain exists.
- **Application-Association-Specific**. Also referred to as *AA-Specific*. The object is defined by the client over a specific application association and can only be used by that specific client. The object exists as long as the association between the client and server exists on the network.

The name of a MMS object must also reflect the scope of the object. For instance, the object name for a domain-specific variable must specify the name of the variable within that domain and the name of the domain. Names of a given scope must be unique. For instance, the name of a variable specific to a given domain must be unique for all domain specific variables in that domain. When an object like a domain is deleted, all the objects subordinate to that domain are also deleted.

VMD Object

The VMD itself can be viewed as an object to which all other MMS objects are subordinate (variables, domains, etc., are contained within the VMD). Because the VMD itself is an object, it has attributes associated with it such as status, capabilities, and the list of subordinate objects. The MMS client has several services available to interact with the VMD object. These services include Status, UnsolicitedStatus, Identify, GetNameList, and others.

Self Describing Devices

One of the unique capabilities of MMS devices is their ability to support self-description. This means that the device can completely describe all the objects contained within it without the client having to be configured in advance. Self description is accomplished by a MMS client using the GetNameList service to first obtain a list of the named objects that are defined in the device such as domains, variables, journals, etc. Then the client can issue "get object attribute" requests to the device to obtain detailed information about the individual objects in the device. This allows the client to automatically retrieve from the device directly all the information it needs to know about the objects in the device. This can greatly simplify client application configuration by eliminating many of the manual configuration steps required by devices that do not support MMS.

The VMD Execution Model

The VMD has a flexible execution model that provides a definition of how the execution of programs by the MMS server can be controlled. Central to this execution model are the definitions of the domain and program invocation objects.

Domains

The MMS domain is a named MMS object that is a representation of some resource within the real device. This resource can be anything that is appropriately represented as a contiguous block of untyped data (referred to as load data). In many typical applications, domains are used to represent areas of memory in a device. For instance, a PLC's ladder program memory is typically represented as a domain. Some applications allow blocks of variable data to be represented as both domains and variables. MMS provides no definition for, and imposes no constraints on, the content of a domain. To do so would be equivalent to defining a "real" object (i.e., the ladder program). The content of the domain is left to the implementor of the VMD.

MMS provides a set of services that allow domains to be uploaded from the device or downloaded to the device. The MMS domain services do not provide access to subordinate objects individually within the domain. The domain can only be uploaded or downloaded as a single contiguous block of untyped data.

UCA Usage of Domains. Within UCA a domain is treated as a representation of a logical device. This allows a single VMD to represent multiple devices each containing their own domain-specific objects, particularly variable access objects. This is important for UCA because of the extremely large number of devices in a typical utility. For instance, a single substation could have many thousands of customers connected to it and therefore many thousands of meters. To require client applications to directly address each device using a unique network would require that the network infrastructure support many millions of addresses to accommodate all the potential devices. Although this is possible to do using current networking technology, it would require higher levels of performance on the network, thereby increasing its cost. The utility industry has addressed this concern historically by using Remote Terminal Units (RTU) as data concentrators and gateways. Modeling data in devices as domain-specific objects allows UCA to easily support a data concentrator architecture while still conforming to the VMD model. By remaining within the bounds of the VMD model, generic non-UCA MMS clients are still able to interoperate with a UCA device without any prior knowledge of the UCA specific object models (GOMSFE) and application service models (CASM).

Program Invocations

It is through the manipulation of *program invocations* that a MMS client controls the execution of programs in a VMD. A program invocation is an execution thread which consists of a collection of one or more domains. Simple devices with simple execution structures may only support a single program invocation containing only one domain. More sophisticated devices and applications may support multiple program invocations containing several domains each.

As an example, consider how the MMS execution model could be applied to a personal computer (PC). When the PC powers up, it downloads a domain called the operating system into memory. When you type the name of the program you want to run and hit the <return> key, the computer downloads another domain (the executable program) from a file and then creates and runs a program invocation consisting of the program and the operating system. The program by itself cannot be executed until it is bound to the operating system. MMS services are provided that allow the client to create, delete, get the attributes, and control the execution of program invocations. See Fig. 5.

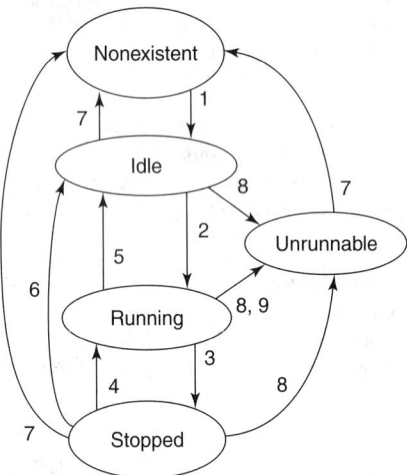

Fig. 5.   MMS Clients use MMS services to cause state transitions in the program invocation as follows: 1. CreateProgramInvocation, 2. Start, 3. Stop or program stop, 4. Resume, 5. End of program and reusable = true, 6. Reset, 7. DeleteProgramInv., 8. Kill or error condition, and 9. End of program and reusable = false.

Example Batch Controller. As an example of how the MMS execution model can be applied to a typical device, let us look at a VMD model for a simple batch controller. See Fig. 6. The example in the figure depicts how the VMD model could be applied to define a set of objects (e.g., domains, program invocations, variables) appropriate for a batch controller. This model will provide clients an appropriate method of controlling the batch process using MMS services. In order to startup and control these two processes, a MMS client using this controller would perform the following actions:

1.   Initiate and complete a domain download sequence for each domain: recipe and data domains A and B, I/O domains A and B, and the control program domain.

2.   Create program invocation A consisting of I/O domain A, the control program domain, and recipe and data domain A.

3.   Start program invocation A. Create program invocation B consisting of I/O domain B, the control program domain, and recipe and data domain B.

4.   Start program invocation B.

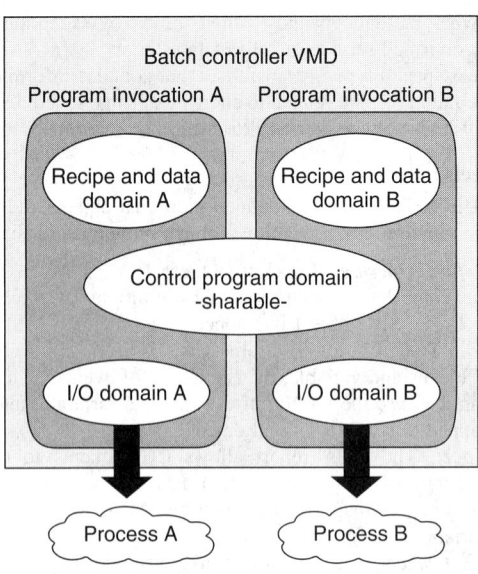

Fig. 6.   Our example batch controller allows us to control two identical batch oriented processes. An I/O domain, specific to each process ties the data in the corresponding recipe/data domain to the process. The control program domain in sharable and contains the control algorithm. The MMS client can create a separate program invocation to control each process. This allows the client to control the recipe and perform supervisory control (start, stop, etc.) of both processes independently from each other.

The example above demonstrates the flexibility of the VMD execution model to accommodate a wide variety of real world situations. Further examples might be a loop controller where each loop is represented by a single domain and where the control loop algorithm (e.g., PID) is represented by a separate but common *sharable* domain. Process variables, setpoints, alarm thresholds, etc. could be represented by domain-specific (control loop) variables. A program invocation would consist of the control loop domains and their algorithm domains needed to control the process.

Variable Access Model

MMS provides a comprehensive and flexible framework for exchanging variable information over a network. The MMS variable access model includes capabilities for *named, unnamed* (addressed), and named lists of variables. MMS also allows the *type description* of the variables to be manipulated as a separate MMS object (named type object). MMS variables can be simple (e.g., integer, boolean, floating point, string) or complex (e.g., arrays and structures). The services available to access and manage MMS variable objects support a wide range of data access methods from simple to complex.

MMS Variables

A real variable is an element of typed data that is contained within a VMD. A MMS variable is a virtual object that represents a mechanism for MMS clients to access the real variable. The distinction between the real variable (which contains the value) and the virtual variable is that the virtual variable represents the access path to the variable, not the underlying real variable itself.

The named variable object describes the access to the real variable by using a MMS object name. MMS clients need only know the name of the object in order to access it. Remember that the name of a MMS variable must also specify the *scope* of the variable (the scope can be VMD, domain, including the domain name, or association specific).

MMS also defines a *named variable list* object that provides an access mechanism for grouping both named and unnamed variable objects into a single object for easier access. A named variable list is accessed by a MMS client by specifying the name (which also specifies its scope) of the named variable list. When the VMD receives a Read service request from a client, it reads all the individual objects in the list and returns their value within the individual elements of the named variable list.

Because the named variable list object contains independent subordinate objects, a positive confirmation to a Read request for a named variable list may indicate only partial success. The success/failure status of each individual element in the confirmation must be examined by the client to ensure that all of the underlying variable objects were accessed without error. In addition to its name and the list of underlying named and unnamed variable objects, named variable list objects also have a MMS deletable attribute that indicates whether or not the named variable list can be deleted via a DeleteNamedVariableList service request.

Variable Type

Simple types are the most basic types and cannot be broken down into a smaller unit via MMS. The other type forms (arrays and structures) are constructed types that can eventually be broken down into simple types. Simple type descriptions generally consist of the *class* and *size* of the type. The size parameter is usually defined in terms of the number of bits or bytes that a variable of that type would comprise in memory. See Fig. 7.

The various classes of simple types defined by MMS consists of:

- *Boolean*. Variables of this type can only have the values of *true* (value = non-zero) or *false* (value = zero). There is no size parameter for Boolean types and are generally represented by a single byte or *octet*.
- *Bit String*. A Bit String is a sequence of bits. The size of a Bit String indicates the number of bits in the Bit String. The most significant bit of the most significant byte in the string is defined as Bit0 in MMS terminology.
- *Boolean Array*. A Boolean Array is also a sequence of bits where each bit represents a *true* or *false*. It differs from an array of Booleans in that each element in a Boolean Array is represented by a single bit while each element in an array of Booleans is represented by a single byte. The size parameter specifies the number of Booleans (number of bits) in the Boolean Array.
- *Integer*. MMS Integer's are signed integers. The size parameter specifies the number of bits of the integer in 2's complement form.
- *Unsigned*. The Unsigned type is identical to the Integer type except that it is not allowed to take on a negative value. Because the most significant bit of an Integer is essentially a sign bit, an Unsigned with a size of 16 bits can only represent 15-bits of values or values of 0 through 32,767.
- *Floating Point*. The MMS definition for floating point can accommodate any number of bits for the format and exponent width including the IEEE 754 single and double precision floating point formats commonly in use today.
- *Octet String*. An Octet String is a sequence of bytes (*octet* in ISO terminology) with no constraint on the value of the individual bytes. The size of an Octet String is the number of bytes in the string.
- *Visible String*. The Visible String type only allows each byte to contain a printable character. The character set is defined by ISO 10646 and is compatible with the common ASCII character set. The size of the Visible String is the number of characters in the string.
- *Generalized Time*. This is a representation of time specified by ISO 8824. It provides millisecond resolution of date and time.
- *Binary Time*. This is a time format that represents time as either (1) time of day by a value which is equal to the number of milliseconds from midnight or (2) date and time by a value that is equal to the time of day and the number of days since January 1, 1984.
- *BCD*. Binary Coded Decimal format is where four-bits are used to hold a binary value of a single digit of zero to ten. The size parameter is the number of decimal digits that the BCD value can represent.
- *Object Identifier*. This is a special class of object defined by ISO 8824 that is used to define ISO registered network objects.

An *array* type defines a variable that consists of a sequence of multiple identical (in format not value) elements. Each element in an array can also be an array or even a structured or simple variable. MMS allows for arbitrarily complex nesting of arrays and structures.

A *structured* type defines a variable that consists of a sequence of multiple, but not necessarily identical, elements. Each individual element in a structure can be of a simple type, an array, or another structure. MMS allows for arbitrarily complex nesting of structures and arrays. A structured variable consisting of individual simple elements requires a nesting level of 1. A structured variable consisting of one or more arrays of structures containing simple variables requires a nesting level of 3.

Variable Access Features

The Read, Write, and InformationReport services provide several features for accessing Variables. The use of these service features, as described

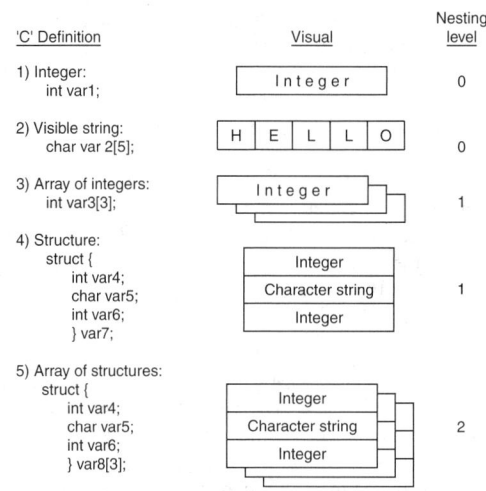

Fig. 7. The level of complexity that a VMD can support is defined by its nesting level. A simple variable has a nesting level of zero. An array of simple variables (e.g., an array of Integers) requires a nesting level of one. An array of structured variables, each consisting of simple variables, requires a nesting level of two and so forth.

below, by MMS clients can provide enhanced performance and very flexible access to MMS Variables.

- *List of variables* is a function that allows a list of named variable, unnamed variable, and named variable list objects to be accessed in a single MMS Read, Write, or InformationReport service. Care must be taken by the client to ensure that the resultant MMS service request message does not exceed the maximum message size (maximum segment size) supported by the VMD. This option also requires that a client examine the entire response for success/failure for each individual element in the list of variables.
- *Access specification in result* is an option for the Read service that allows a MMS client to request that the variable's access specification be returned in the Read response. The access specification would consist of the same information that would be returned by a GetVariableAccess-Attributes service request.
- *Alternate Access* allows a MMS client to 1) partially access only specified elements contained in a larger arrayed and/or structured variable, and 2) rearrange the ordering of the elements contained in structured variables.

Event Management Model

In a real sense, an event or an alarm is easy to define. Most people have an intuitive feel for what can comprise an event within their own area of expertise. For instance, in a process control application, it is common for a control system to generate an alarm when the process variable (e.g., temperature) exceeds a certain preset limit called the high alarm threshold.

The MMS event management model provides a framework for accessing and managing the network communication aspects of these kinds of events. This is accomplished by defining three named objects that represent 1) the state of an event (*event condition*), 2) who to notify about the occurrence of an event (*event enrollment*), and 3) the action that the VMD should take upon the occurrence of an event (*event action*).

For many applications, the communication of alarms can be implemented by using MMS services other than the event management services. For instance, a system can notify a MMS client about the fact that a process variable has exceeded some preset limit by sending the process variable's value to a MMS client using the InformationReport service. The InformationReport service is how the typical ICCP-TASE.2 and UCA applications communicate event information. Other schemes using other MMS services are also possible. When the application is more complex and requires a more rigorous definition of the event environment the MMS event management model can be used.

Event Condition Object

A MMS *event condition* object is a named object that represents the current state of some real condition within the VMD. It is important to note that MMS does not define the VMD action (or programming) that causes a change in state of the event condition. In the process control example given above, an event condition might reflect an *idle* state when the process variable was not exceeding the value of the high alarm threshold and an *active* state when the process variable did exceed the limit. MMS does not explicitly define the mapping between the high alarm limit and the state of the event condition. Even if the high alarm limit is represented by a MMS variable, MMS does not define the necessary configuration or programming needed to create the mapping between the high alarm limit and the state of the event condition. From the MMS point of view, the change in state of the event condition is caused by some autonomous action on the part of the VMD that is not defined by MMS.

Event Actions

An *event action* is a named MMS object that represents the action that the VMD will take when the state of an *event condition* changes. An event action is optional. When omitted, the VMD would execute its event notification procedures without processing an event action. An event action, when used, is always defined as a confirmed MMS service request. The event action is linked to an event condition when an *event enrollment* is defined. For example, an event action might be a MMS Read request. If this event action is attached to an event condition (by being referenced in an event enrollment), when the event condition changes state and is enabled, the VMD would execute this Read service request just as if it had been received from a client. Except that the Read response (either positive or negative) is included in the EventNotification service request that is sent to the MMS client enrolled for that event. A confirmed service request must be used (i.e., Start, Stop, Read). Unconfirmed services (e.g., InformationReport, UnsolicitedStatus, and EventNotification) and services that must be used in conjunction with other services (e.g., domain upload-download sequences), cannot be used as event actions.

Event Enrollments

The *event enrollment* is a named MMS object that ties all the elements of the MMS event management model together. The event enrollment represents a request on the part of a MMS client to be notified about changes in the state of an *event condition*. When an event enrollment is defined, references are made to an event condition, an *event action* (optionally), and the MMS client to which EventNotification should be sent. See Fig. 8.

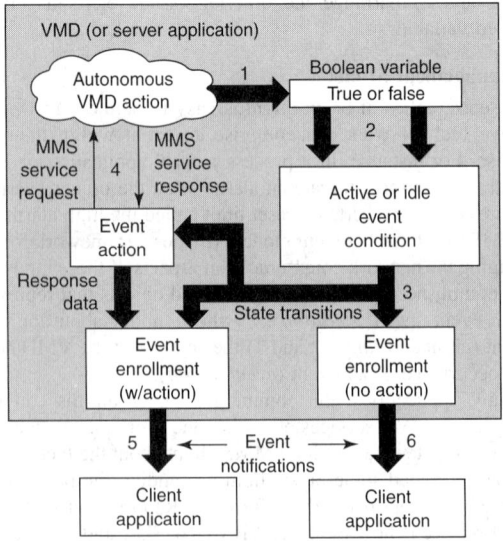

Fig. 8.   A Monitored event condition has a Boolean variable associated with it that the VMD sets (1) via some form of local autonomous action. The VMD periodically evaluates this variable (2) to determine the state of the event condition. MMS clients "enroll" to be notified of event condition state transitions (3) by defining an event enrollment. If an event action is defined for the event enrollment, the VMD obtains the response (4) to the event action (a MMS service request) and inserts the response data into the event notification (5) that is sent to the client. Event enrollments without an event action have their event notifications sent to the client (6) without any event action response data.

Semaphore Management Model

In many real-time systems there is a need for a mechanism by which an application can control access to a system resource. An example might be a workspace that is physically accessible to several robots. Some means to control which robot (or robots) can access the workspace is needed. MMS *semaphores* are named objects that can be used to control access to other resources and objects within the VMD. For instance, a VMD that controls access to a setpoint (a variable) for a control loop could use semaphores to only allow one client at a time to be able to change the setpoint (e.g., with the MMS Write service). The MMS semaphore model defines two kinds of semaphores. *Token* semaphores are used to represent a specific resource within the control of the VMD. *Pool* semaphores consist of one or more named tokens each representing a set of similar but distinct resources under the control of the VMD.

Because semaphores are used solely for the purpose of coordinating activities between multiple MMS clients, the scope of a semaphore cannot be *AA-specific* where the object exists only an association between a single VMD and a single MMS client.

Token Semaphores

A token semaphore is a named MMS object that can be a representation of some resource, within the control of the VMD, to which access must be controlled. A token semaphore is modeled as a collection of tokens that MMS clients take and relinquish control of using MMS services. When a client *owns* a token, the client may access the resource that the token represents. This allows both multiple or exclusive ownership of the semaphore. An example of a token semaphore might be where two users want to change a setpoint for the same control loop at the same time. In order to coordinate their access to the setpoint a token semaphore, containing only one token, can be used to represent the control loop. When a user "owns" the token, they can change the setpoint. The other user would have to wait until ownership is relinquished. See Fig. 9.

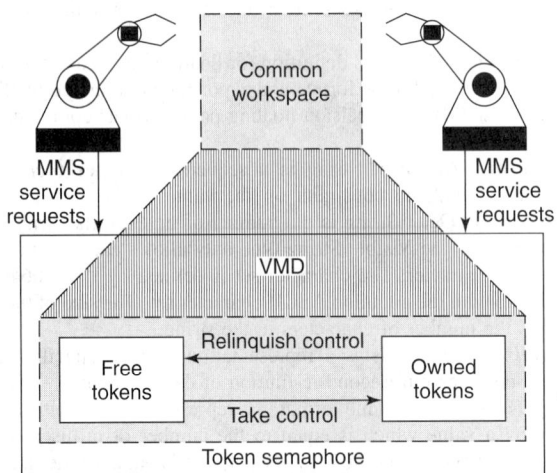

Fig. 9.   A token semaphore is modeled as a collection of free tokens and owned tokens. When a robot (a MMS client in this example) wants to access the common workspace, it will issue a TakeControl request to the VMD controlling the workspace. If there is a free token available, the VMD will grant control by moving a token from the free state to the owned state and then responding positively to the TakeControl request. If the other robot had already owned the token, then the VMD would respond negatively to the TakeControl. The token representing the common workspace would remain under the control of the robot until control was released with a RelinquishControl request or upon a control timeout by the VMD. The total number of tokens availables how many simultaneous owners can exist for the same semaphore.

A token semaphore can also be used for the sole purpose of coordinating the activities of two MMS clients without representing any real resource. This kind of "virtual" token semaphore looks and behaves the same except that they can be created and deleted by MMS clients using the DefineSemaphore service.

Pool Semaphores

A *pool semaphore* is similar to a token semaphore except that the individual tokens are identifiable and have a name associated with them. These named tokens can optionally be specified by the MMS client when issuing TakeControl requests. The pool semaphore itself is a MMS object. The named tokens contained in the pool semaphore are not MMS objects. They are representations of a real resource in much the same way an unnamed variable object is. Pool semaphore objects are used when it is desired to represent a set of similar resources where clients that need control of such a resource may or may not care which specific resource they desire control over. For instance, the individual vehicles in an automated guided vehicle (AGV) system can be represented by a pool semaphore. MMS Clients at individual work centers may desire to control an AGV to deliver new material but may not care which specific AGV is used. The AGV system VMD would decide which specific AGV, represented by a single named token, would be assigned to a given MMS client. The name of the pool semaphore is independent of the names of the named tokens (see Fig. 10). Pool semaphores can only be used to represent some real resource within the VMD. Therefore, pool semaphores cannot be created or deleted using MMS service requests and cannot be *AA-specific* in scope.

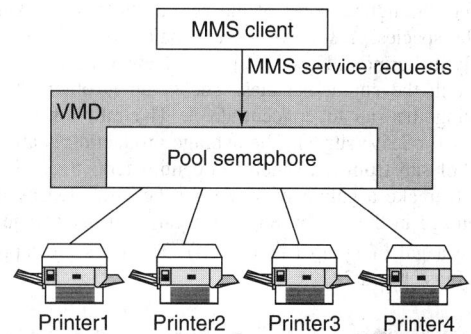

Fig. 10. A pool semaphore can be useful to control access to similar but distinguishable resources, such as a printer pool. In the example, each printer is represented as a separate named token. A MMS client can request control of a specific printer by issuing a TakeControl request specifying a named token. Alternately, if the client does not care which specific printer it is granted control of, it can issue the TakeControl request without specifying a named token.

Semaphore Entry

When a MMS client issues a TakeControl request for a given semaphore, the VMD creates an entry in an internal queue that is maintained for each semaphore. Each entry in this queue is called a *semaphore entry*. The attributes of a semaphore entry are visible to MMS clients and provide information about the internal semaphore processing queue in the VMD. The semaphore entry is not a MMS object. It only exists from the receipt of the TakeControl indication by the VMD until the control of the semaphore is relinquished or if the VMD responds negatively to the TakeControl request. See Fig. 11.

Other MMS Objects

Journal Objects. A MMS journal represents a log file that contains a collection of records (called a *journal entry*) that are organized by time stamps. Journals are used to store time based records of tagged variable data, user generated comments (called *annotation*), or a combination of events and tagged variable data. Journal entries contain a time stamp that indicates when the data in the entry was produced, not when the journal entry was made. This allows MMS journals to be used for applications where a sample of a manufactured product is taken at one time, analyzed in a laboratory off-line, and then placed into the journal at a later time. In this case, the journal entry time stamp would indicate when the sample was taken.

MMS clients read the journal entries by specifying the name of the journal (which can be *VMD-specific* or *AA-specific* only) and either 1) the date/time range of entries that the client wishes to read or 2) by referring to the *entry ID* of a particular entry. The entry ID is a unique binary identifier assigned by the VMD to the journal entry when it is placed into the journal. Each entry in a journal can be one of the following types:

- *Annotation*. This type of entry contains a textual comment. This is typically used to enter a comment regarding some event or condition that had occurred in the system.

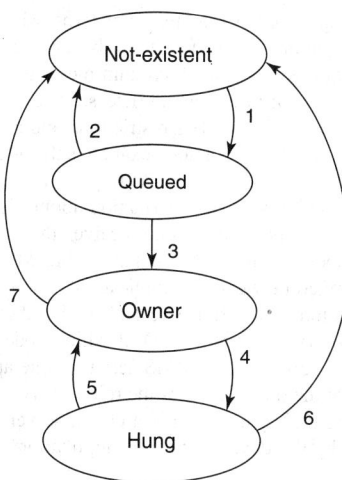

Fig. 11. A semaphore entry is created each time a client attempts to take control of a semaphore. The semaphore entry reflects the state of the relationship between the client and the semaphore. State transitions can be caused by local action by the VMD or by the following events: 1. TakeControl request received, 2. Timeout, Cancel request, or abort; 3. Semaphore available (token free), 4. Application association aborted, 5. TakeControl with preempt, 6. Control timeout, or 7. RelinquishControl request received or timeout.

- *Data*. This type of entry would contain a list of variable tags and the data associated with those tags at the time indicated by the time stamp. Each variable tag is a 32-character name that does not necessarily refer to a MMS variable, although it might.
- *Event-Data*. This type of entry contains both variable tag data and event data. Each entry of this type would include the same list of variable tags and associated data as described above, along with a single event condition name and the state of that event condition at the time indicated by the time stamp.

Operator Station Object

The *operator station* is an object that represents a means of communicating with the operator of the VMD via a keyboard and display. An operator station is modeled as character based input and output devices that may be attached to the VMD for the purpose of communicating with an operator local to the VMD. Because the operator station is a representation of a physical feature of the VMD, it exists beyond the scope of any domain or application association. Therefore, MMS clients access the operator station by name without scope. There can be multiple operator stations for a given VMD.

Files

MMS also provides a set of simple file transfer services for devices that have a local file store but do not support a full set of file services via some other means. For instance, an electric meter may use the file services for transferring oscillography (waveform) files to an interested MMS client. The MMS file services support file transfer only, not file access. Although these file services are defined in an annex within the MMS standard, they are widely supported by many commercial MMS implementations.

Context Management

MMS provides services for managing the context of communications between two MMS nodes on a network. These services are used to establish and terminate application *associations* and for handling protocol errors between two MMS nodes. The terms *association* and *connection* are sometimes used interchangeably although there is a distinction from a network technology point of view. The node that initiates the association with another node is referred to as the *calling* node. The responding node is referred to as the *called* node.

In a MMS environment, two MMS applications establish an application association between themselves using the MMS Initiate service. This process of establishing an application association consists of an exchange of some parameters and a negotiation of other parameters. The exchanged parameters include information about restrictions that pertain to each node that are determined solely by that node (e.g. which MMS services are supported). The negotiated parameters are items where the called node either accepts the parameter proposed by the calling node or adjusts it downward as it requires (e.g., the maximum message size).

The calling application issues an Initiate service request that contains information about the calling node's restrictions and a proposed set of the negotiated parameters. The called node examines the negotiated parameters and adjusts them as necessary to meet its requirements. It then returns the results of this negotiation and the information about it's restrictions in the Initiate response. Once the calling node receives the Initiate confirmation, the application association is established and other MMS service requests can then be exchanged between the applications.

Once an application association is established, either node can assume the role of *client* or *server* independent of which node was the calling or called node. For any given set of MMS services, one application assumes the client role while the other assumes the role of server or VMD. Whether or not a particular MMS application is a client, server (VMD), or both is determined solely by the developer of the application.

Associations vs. Connections

Although many people may refer to network connections and application associations interchangeably, there is a distinct difference. A connection is an attribute of the underlying network layers that represents a virtual circuit between two nodes. For instance, telephone networks require that two parties establish a connection between themselves (by dialing and answering) before they can communicate. An application association is an agreement between two networked applications governing their communications. It is analogous to the two telephone parties agreeing to use a particular language and to not speak about religion or politics over the telephone. Application associations exist independent of any underlying network connections (or lack thereof).

In a connection oriented environment (e.g. TCP/IP) the MMS Initiate service is used to signal to the lower layers that a connection must be established. The Initiate service request is carried by the network through the layers as each layer goes through its connection establishment procedure until the Initiate indication is received by the called node. The connection does not exist until after all the layers in both nodes have completed their connection establishment procedures and the calling node has received the Initiate confirmation. Because of this, the association and the connection are created concurrently in a connection oriented environment.

In a connectionless environment (e.g. 3-layer), it is not strictly necessary to send the Initiate request before two nodes can actually communicate. In an environment where the Initiate service request is not used before other service requests are issued by a MMS client to a VMD, each application must have prior knowledge of the other application's exchanged and negotiated parameters via some local means (e.g., a configuration file). This foreknowledge of the other MMS application's restrictions is the application association from a MMS perspective. Whether an Initiate service request is used or not, application associations between two MMS applications must exist before communications can take place. In some connectionless environments such as the UCA 3-layer for serial link communications, MMS nodes still use the Initiate service to do the application association negotiation before communicating data or control information.

Summary

MMS provides a very flexible real-time messaging architecture for a wide range of applications. The concept of the VMD model of MMS is used in many lower level networking systems such as Foundation Fieldbus and Profibus where a more limited subset of MMS, called the Fieldbus Messaging System (FMS), is used. MMS has been used for years as a messaging system for Ethernet networks in automotive manufacturing, pulp-paper, aerospace, and other large and complex material handling systems. Recent technological innovations have allowed MMS to be applied into smaller and more resource limited devices like small PLCs, RTUs, meters, breakers, and relays. With the backing of EPRI, GRI, U.S. Postal Service, and a large number of utility users, the use of MMS is gaining momentum in these industries as a considerable number of equipment suppliers now support MMS and UCA. MMS is an effective bridge between the plant floor and the management information systems (MIS) as well as between the process control systems of the power plant and the distribution systems of the utility. With the additional refinements

of GOMSFE and CASM, the level of interoperability between dissimilar equipment that is being achieved is welcome relief to an industry long plagued with numerous incompatible proprietary communication methods.

Ralph E. Mackiewicz, SISCO, Inc., Sterling Heights, MI.

MANY-BODY FORCE. An interaction between two particles that becomes modified when a third particle is present, e.g., the forces between polarizable molecules. A large part of the experimental data of physics is concerned with natural objects that may be looked upon as being made up from smaller bodies. Thus, results the so-called many-body problem; there are several approaches to its solution. This highly theoretical topic is beyond the scope of this volume, but reference is suggested to the "Many-Body Problem," in *The Encyclopedia of Physics* (R.M. Besancon, editor), 3rd Edition, Van Nostrand Reinhold, New York, 1985.

MAPLE TREES. Of the family *Aceraceae* (maple family), the maple tree is native to the United States and Canada. There are 13 or more species of these trees in the United States. The sugar maple is found in large numbers throughout most of the central parts of the United States. However, the species, *A. saccharum*, the major source of maple syrup is found mainly in Vermont, New Hampshire, New York, and Ohio, and to a lesser extent in the southern states. The syrup is obtained from maple sap by boiling the sap to concentrate it. The ratio of sap to sugar is about 40 to 1 (or 2.5% sugar). The average production is about 20 quarts (18.9 liters) of sap from a 15-inch (38 centimeters) tree. The amount of sap needed to make a gallon of syrup can be estimated by dividing the sugar percentage in the sap by 86. About $34\frac{1}{2}$ gal of average-quality sap will produce a gal of syrup ($\frac{86}{2.2} = 34.4$). The trees are tapped in early spring. See Figs. 1 and 2.

Fig. 1. Excellent specimen of sugar maple growing in Pisgah National Forest, North Carolina. (U.S. Forest Service photo.)

As reported by Behlen, it is believed that only in North America (and on a single sugar-producing island in northern Japan) do farmers

Fig. 2.   Tapping of sugar maples. (USDA photo.)

collect sap and make maple syrup. In addition to the sugar maple, the black maple (A. nigrum) also is an outstanding syrup producer. The silver maple (A. saccharinum), red maple (A. rubrum), bigleaf maple (A. macrophyllum), and the boxelder (A. negundo) also produce a sweep sap that can be processed into syrup. Sometimes, collectively, the sugar producing trees are referred to as *sugarbushes*. Production of maple syrup in the United States is well in excess of 1.2 million gal/year.

Tapholes ($\frac{7}{16}$ in.; 1.1 cm) are drilled into the wood $2\frac{1}{2}$ to 3 in. (6 to $7\frac{1}{2}$ cm) in depth. Properly made, these holes do not harm the tree. It is general practice not to tap a tree under 10 in. ($25\frac{1}{2}$ cm) in diameter. A tree with a 10 to 17 in. ($25\frac{1}{2}$ to 43 cm) diameter can handle a single tap; between 17 and 24 in. (43 and 61 cm) diameter, a tree can handle two taps. Larger trees can handle up to four producing taps. Taps should not be placed closer than 6 in. (15 cm) to unhealed tap holes. From two to three years are required for a hole to heal completely. As reported by the U.S. Forest Service Experiment Station in Vermont,

some 200-year-old maples in that state have been tapped every year for the past 150 years and still show no ill effects. The processing of the sap and many practical hints for syrup production are given in "Maple Sugaring" by D. Behlen (American Forests, 12–15, March 1980). The article also contains a helpful list of other references. The characteristics and production of maple syrup are also described in the Foods and Food Production Encyclopedia (D.M. and G.D. Considine, Eds.), Van Nostrand Reinhold, New York, 1982.

The maple when fully grown is quite a large tree, spreading, pendulous, and graceful. The branches are long, trim, and narrow. The leaf is about 3 to 4 inches (7.6 to 10 centimeters) across, 3- or 5-lobed, and relatively dense. Distinctive of shape, the leaf is beautifully portrayed on the flag of Canada. The leaf may be simple or compound, depending upon the species. The fruit is green, winged, and full centered. The wood is close-grained, hard, and light in color. It weighs nearly 40 pounds per cubic foot (640 kilograms per cubic meter) and is strong. The wood is used for numerous objects that require a durable and high-quality material. The species A. macrophyllum is native to the western United States and is found in abundance in the Columbia River area. Several million board feet are shipped annually. The trees are quite large. The leaf is very long, from 8 to 12 inches (20.3 to 30.4 centimeters) wide, with stalks about 10 to 12 inches (25 to 30.4 centimeters) long. Along the Pacific coast, moisture from winter rains and summer fogs usually assures full, healthy growth.

The box elder, a fast-growing tree that drops its leaves early, is also a member of the family Aceraceae.

In recent years, tree scientists have become increasingly concerned with a condition that is sometimes referred to as *maple decline*, an insidious disease, which seems to prefer street and landscape plantings. Dr. Robert Norgren, a plant pathologist with the Wisconsin Department of Agriculture who compiles statistics on diseased trees has observed, "Maple decline will surpass Dutch elm disease, in both complexity and severity, for many of these [North-Central and Northeastern] states." Diseased trees show a decay of the trunk near the base, frequently below the soil line. Inspection has shown that the affected areas are frequently located on the root collar and sometimes on the large roots extending from the trunk. The decay is sometimes characterized by bark discoloration and looseness. Underneath healthy-appearing bark will be found infected wood and cambium that is dark brown or red in contrast with the normal healthy white. Laboratory

TABLE 1. RECORD MAPLE TREES IN THE UNITED STATES[1]

Specimen	Circumference[2]		Height		Spread		Location
	Inches	Centimeters	Feet	Meters	Feet	Meters	
Bigleaf maple (1995) (Acer macrophyllum)	419	1064	101	30.8	90	27.4	Oregon
Black maple (1987) (Acer nigrum)	198	503	118	36	127	38.7	Michigan
Canyon maple (1998) (Acer grandidentatum)	139	353	75	22.9	50	15.2	Arizona
Chalk maple (1999) (Acer leucoderme)	33	84	54	16.5	50	15.2	Georgia
Florida maple (1997) (Acer barbatum)	140	356	75	22.9	52	15.8	Georgia
Mountain maple (1982) (Acer spicatum)	33	84	58	17.7	31	9.4	Michigan
Norway maple (1994) (Acer platanoides)	180	457	120	36.6	66	20.1	New York
Red maple (1997) (Acer rubrum)	276	701	141	43	88	26.8	Tennessee
Rocky Mountain maple (1996) (Acer glabrum)	107	272	67	20.4	55	16.8	Washington
Silver maple (1996) (Acer saccharinum)	293	744	115	35.1	110	33.5	Wisconsin
Striped maple (1984) (Acer pensylvanicum)	50	127	77	23.5	28	8.5	New York
Striped maple (1987) (Acer pensylvanicum)	44	112	77	23.5	31	9.4	Tennessee
Sugar maple (1996) (Acer saccharum)	274	696	65	19.8	54	16.5	Maryland
Vine maple (1992) (Acer circinatum)	67	170	46	14	35	10.7	Oregon

[1] From the "National Register of Big Trees," American Forests (by permission).
[2] At 4.5 feet (1.4 meters).

analyses of infected trunk samples have shown two fungi to be involved, *Phytophthora* and *fusarium*. As pointed out by R. Dries (American Forests, 49–63, May 1982), for trees that show no decline symptoms, treatment may include carefully removing the soil from the root-collar area, leaving this area exposed, or filling it with a compost of hardwood bark. During dry periods, the tree should be watered. Further details are given by Dries in his article, "The Battle Against Maple Decline."

Record maple trees in the United States are listed in Table 1.

MAP-MATCHING GUIDANCE.

1. The guidance of a rocket or aero-dynamic vehicle by means of a radarscope film previously obtained by a reconnaissance flight over the terrain of the route, and used to direct the vehicle by aligning itself with radar echoes received during flight from the terrain below.

2. Guidance by stellar map matching.

MAPPING (Mathematics).

A mapping f from a set D to a set R is a correspondence which associates with each point x in the set D a unique point $f(x)$ in R called the *image* of x. The set D is said to be the *domain* of f, while R is said to be the range of f. The mapping f is said to be *onto* or subjective if every point y in R is the image of some point x in D, and *one-to-one* or injective if for x_1 and x_2 in D with $x_1 \neq x_2$ it follows that $f(x_1) \neq f(x_2)$. See also **Isomorphism**.

MAP PROJECTIONS.

The numerous methods for representing the surface of the earth on a plane. The term originated from the geometric-projection methods of drawing lines from some point out through the surface of the earth to a developable surface. The surfaces most commonly used in the geometric projections were the plane, the cylinder, and the cone. In most cases, the distortions introduced by the strictly geometric projections are intolerable for modern purposes, and the term map projection has been expanded to include all sorts of purely mathematical methods for representing the surface of the earth on a plane. The pattern of lines on the map or chart representing meridians of longitude and parallels of latitude form what is known as the graticule of the projection.

Additional Reading

Pearson, F.: "Map Projection: Theory and Applications," CRC Press, LLC., Boca Raton, FL, 1999.
Synder, J.P.: "Flattening the Earth: Two Thousand Years of Map Projections," University of Chicago Press, Chicago, IL, 1997.
Yang, O.H. and J.P. Snyder: "Map Projection Transformation: Principles and Applications," Taylor & Francis, Inc., Philadelphia, PA, 1999.

MARANGONI EFFECT.   See **Foam**.

MARBURG-EBOLA VIRUS DISEASE.

First discovered in Marburg, West Germany, this highly fatal disease is caused by the Marburg and Ebola viruses carried by vervet monkeys. The disease was recognized in 1967 among German animal handlers and laboratory personnel, as well as by hospital workers who treated victims of the disease. Cases also appeared among similar workers in Yugoslavia. Investigation indicated that the disease was spread among persons who were in contact with monkeys shipped from Uganda in 1967. In 1976, 500 cases of the disease were reported in Zaire and the Sudan, particularly in persons who lived in the Ebola River valley. Mortality with the Marburg agent is estimated to be 30%; with the Ebola virus, 85%.

The onset of the disease is abrupt with severe headache, sore throat, high fever and myalgia, followed by rapid prostration, and dehydration from diarrhea and vomiting. A generalized rash appears about the fifth day and lasts for two to three days. Severe bleeding occurs from the gastrointestinal tract and lungs. Death usually eventuates between the 7th and 16th days. Autopsy shows necrotic cellular lesions in the liver and other organs, including the kidneys.

Although monkey organs were the source of the first infection, primates are not thought to be the natural reservoirs since no infection developed in people who had handled whole monkeys. Experimental infection of several species of primates with body fluids from patients with the disease produced a uniformly fatal infection and the course of the illness was similar to that in humans. No natural reservoir of the virus has yet been discovered.

Care is primarily supportive, although antiviral agents may have a place in therapy. Barrier nursing and infection-control precautions are required; it is most important to handle blood and other body fluids and contaminated needles with great care.

Travel into endemic areas during periods of disease activity should be minimized and quarantine of vervet monkeys coming from such areas is mandatory.

R. C. V.

MARBURG HEMORRHAGIC FEVER.

Marburg hemorrhagic fever is a rare, severe type of hemorrhagic fever which affects both humans and nonhuman primates. Caused by a genetically unique zoonotic (that is, animal-borne) RNA virus of the filovirus family, its recognition led to the creation of this virus family. The four species of Ebola virus are the only other known members of the filovirus family.

The Marburg virus was first recognized in 1967, when outbreaks of hemorrhagic fever occurred simultaneously in laboratories in Marburg and Frankfurt, Germany and in Belgrade, Yugoslavia (now Serbia). A total of 37 people became ill; they included laboratory workers as well as several medical personnel and family members who had cared for them. The first people infected had been exposed to African green monkeys or their tissues. In Marburg, the monkeys had been imported for research and to prepare polio vaccine.

Recorded cases of the disease are rare, and have appeared in only a few locations. Although the 1967 outbreak occurred in Europe, the disease agent had arrived with imported monkeys from Uganda. No other case was recorded until 1975, when a traveler, most likely exposed in Zimbabwe, became ill in Johannesburg, South Africa and passed the virus to his travelling companion and a nurse. Two other cases were seen in 1980. One in Western Kenya not far from the Ugandan source of the monkeys implicated in the 1967 outbreak. This patient's attending physician in Nairobi became the second case. Another human Marburg infection was recognized in 1987 when a young man who had traveled extensively in Kenya, including western Kenya, became ill and later died.

The Marburg virus is indigenous to Africa. While the geographic area to which it is native is unknown, this area appears to include at least parts of Uganda and Western Kenya, and perhaps Zimbabwe. As with the Ebola virus, the actual animal host for the Marburg virus also remains a mystery. Both of the men infected in 1980 in western Kenya had traveled extensively, including making a visit to a cave, in that region. The cave was investigated by placing sentinels animals inside to see if they would become infected, and by taking samples from numerous animals and arthropods trapped during the investigation. The investigation yielded no virus. The sentinel animals remained healthy and no virus isolations from the samples obtained have been reported.

Just how the animal host first transmits Marburg virus to humans is unknown. However, as with some other viruses which cause viral hemorrhagic fever, humans who become ill with Marburg hemorrhagic fever may spread the virus to other people. This may happen in several ways. Persons handling infected monkeys, who come into direct contact with them or their fluids or cell cultures, have become infected. Spread of the virus between humans has occurred in a setting of close contact, often in a hospital. Droplets of body fluids, or direct contact with persons, equipment, or other objects contaminated with infectious blood or tissues are all highly suspect as sources of disease.

After an incubation period of 5–10 days, the onset of the disease is sudden and is marked by fever, chills, headache, and myalgia. Around the fifth day after the onset of symptoms, a maculopapular rash, most prominent on the trunk (chest, back, stomach), may occur. Nausea, vomiting, chest pain, a sore throat, abdominal pain, and diarrhea then may appear. Symptoms become increasingly severe and may include jaundice, inflammation of the pancreas, severe weight loss, delirium, shock, liver failure, massive hemorrhaging, and multiorgan dysfunction.

Because many of the signs and symptoms of Marburg hemorrhagic fever are similar to those of other infectious diseases, such as malaria or typhoid fever, diagnosis of the disease can be difficult, especially if only a single case is involved.

Antigen-capture enzyme-linked immunosorbent assay (ELISA) testing, IgM-capture ELISA, polymerase chain reaction (PCR), and virus isolation can be used to confirm a case of Marburg hemorrhagic fever within a few days of the onset of symptoms. The IgG-capture ELISA is appropriate for testing persons later in the course of disease or after recovery. The disease is readily diagnosed by immunohistochemistry, virus isolation, or PCR of blood or tissue specimens from deceased patients.

Recovery from the Marburg hemorrhagic fever may be prolonged and accompanied by orchititis, recurrent hepatitis, transverse myelitis or uvetis. Other possible complications include inflammation of the testis, spinal cord, eye, parotid gland, or by prolonged hepatitis. The case fatality rate for the Marburg hemorrhagic fever is between 23–25%.

A specific treatment for this disease is unknown. However, supportive hospital therapy should be utilized. This includes balancing the patient's fluids and electrolytes, maintaining their oxygen status and blood pressure, replacing lost blood and clotting factors and treating them for any complicating infections.

Sometimes, treatment also used has been transfusion of fresh-frozen plasma and other preparations to replace the blood proteins important in clotting. One controversial treatment is the use of heparin (which blocks clotting) to prevent the consumption of clotting factors. Some researchers believe the consumption of clotting factors is part of the disease process.

People who have close contact with a human or nonhuman primate infected with the virus are at risk. Such persons include laboratory or quarantine facility workers who handle nonhuman primates that have been associated with the disease. In addition, hospital staff and family members who care for patients with the disease are at risk if they do not use proper barrier nursin-Q techniques.

Due to our limited knowledge of the disease, preventive measures against transmission from the original animal host have not yet been established. Measures for prevention of secondary transmission are similar to those used for other hemorrhagic fevers. If a patient is either suspected or confirmed to have Marburg hemorrhagic fever, barrier nursing techniques should be used to prevent direct physical contact with the patient. These precautions include wearing of protective gowns, gloves, and masks; placing the infected individual in strict isolation; and sterilization or proper disposal of needles, equipment, and patient excretions.

In conjunction with the World Health Organization, the CDC has developed practical, hospital-based guidelines, titled *Infection Control for Viral Haemorrhagic Fevers in the African Health Care Setting*. The manual can help health-care facilities recognize cases and prevent further hospital-based disease transmission using locally available materials and few financial resources.

Marburg hemorrhagic fever is a very rare human disease. However, when it does occur, it has the potential to spread to other people, especially health care staff and family members who care for the patient. Therefore, increasing awareness among health-care providers of clinical symptoms in patients that suggest Marburg hemorrhagic fever is critical. Better awareness can help lead to taking precautions against the spread of virus infection to family members or health-care providers. Improving the use of diagnostic tools is another priority. With modern means of transportation that give access even to remote areas, it is possible to obtain rapid testing of samples in disease control centers equipped with Biosafety Level 4 laboratories in order to confirm or rule out Marburg virus infection.

A fuller understanding of the Marburg hemorrhagic fever will not be possible until the ecology and identity of the virus reservoir are established. In addition, the impact of the disease will remain unknown until the actual incidence of the disease and its endemic areas are determined. See also **Ebola Hemorrhagic Fever**; **Filoviruses**; and **Viral Hemorrhagic Fevers**.

Additional Reading

Fields, M.B.N., D.M. Knipe, P.M. Howley, and R.M. Chanock: "Fields Virology," 3rd Edition, Lippincott Williams & Wilkins, Philadelphia, PA, 1996.
Galasso, G.J., T.C. Merigan, and R.J. Whitley: "Antiviral Agents and Human Viral Diseases," 4th Edition, Lippincott Williams & Wilkins, Philadelphia, PA, 1997.
Gear, J.H.S.: "Handbook of Viral and Rickettsial Hemorrhagic Fever," CRC Press, LLC., Boca Raton, FL, 1988.
Love, C.B. and P.B. Jahrling: "Viral Hemorrhagic Fever," DIANE Publishing Company, Collingdale, PA, 1996.
Pattison, J.R., J.E. Banatvala, and A.J. Zuckerman: "Principles and Practice of Clinical Virology," 4th Edition, John Wiley & Sons, Inc., New York, NY, 2000.
Richman, D.D., R.J. Whitley, and F.G. Hayden: "Clinical Virology," Harcourt Brace & Company, San Diego, CA, 1998.
Voyles, B.A.: "The Biology of Viruses," Mosby-Year Book, Inc., St Louis, MO, 1993.

Centers for Disease Control and Prevention (CDC), Atlanta, GA.

MARCASITE. The mineral marcasite, sometimes called white iron pyrites, is, like ordinary pyrites, disulfide of iron corresponding to the same formula, FeS_2. Marcasite, however, crystallizes in the orthorhombic system often yielding serrate, spear-shaped twins, hence the name "cock's comb pyrites." It is a brittle mineral; hardness, 6–6.5; specific gravity, 4.92; luster, metallic; color, light bronze-yellow; streak, greenish-black; opaque. Marcasite alters very easily and may disintegrate with the formation of sulfuric acid and iron sulfate. Fossils replaced by marcasite are therefore often destroyed after being placed in collections. Marcasite is found in numerous places in Europe, notably in Czechoslovakia, France, and England; in Mexico; and in the United States in the lead districts of Illinois, Wisconsin, and Missouri. The name marcasite is believed to be of Arabic origin and formerly was applied to common pyrite.

MARCHING PROBLEM. A differential equation with initial conditions solved numerically by computing the values of the dependent variable recursively for systematically increasing values of the independent variable. For example, the wave equation is solved at each time-step before advancing to the next time-step. Hyperbolic equations may be formulated as marching problems. See also **Jury Problem**.

MARCONI, GUGLIELMO (1874–1935). Marconi was born in Bologna, Italy. He was educated privately at Bologna, Florence and Leghorn. Even as a boy he took a keen interest in physical and electrical science and studied the works of Maxwell, Hertz, Righi, Lodge and others. He began his scientific experiments in 1895, by setting up an electrical laboratory in the attic of his father's country estate, where he succeeded in sending wireless signals over a distance of one and a half miles, thus becoming the inventor of the first practical system of wireless telegraphy.

In 1896 Marconi took his apparatus to England where he was introduced to Mr. (later Sir) William Preece, Engineer-in-Chief of the Post Office, and later that year was granted the world's first patent for a system of wireless telegraphy. He demonstrated his system successfully in London, on Salisbury Plain and across the Bristol Channel, and in July 1897 formed The Wireless Telegraph & Signal Company Limited (in 1900 re-named Marconi's Wireless Telegraph Company Limited). In 1899 he established wireless communication between France and England across the English Channel. He erected permanent wireless stations at The Needles, Isle of Wight, at Bournemouth and later at the Haven Hotel, Poole, Dorset.

In 1900 he took out his famous patent No. 7777 for "tuned or syntonic telegraphy." In December 1901 determined to prove that wireless waves were not affected by the curvature of the Earth, he used his system for transmitting the first wireless signals across the Atlantic between Poldhu, Cornwall, and St. John's, Newfoundland, a distance of 2100 miles.

Between 1902 and 1912 he patented several new inventions. In 1902 the "magnetic detector" which then became the standard wireless receiver for many years. In 1905 he patented his "horizontal directional aerial" and in 1912 a "timed spark" system for generating continuous waves.

He received the Nobel Prize in Physics in 1909, which he shared with Professor Carl Ferdinand Braun. He is remembered best for his development of radio wave communication.

See also **Electronics**.

J. M. I.

MARE. (Plural: maria.) 1. One of several dark, low-lying, level, relatively smooth, plains-like areas of considerable extent on the surface of the earth's moon, having fewer large craters than the highlands, and composed of mafic or ultramafic volcanic rocks, e.g., Mare Imbrium (a circular mare) and Mare Tranquillitatis (a mare with an irregular outline). Lunar maria are completely waterless.

2. By extension, a dark area of the surface of Mars or other planets and satellites whose origin is not known. See also **Mars**; and **Moon (Earth's)**.

MARGARINE. See **Vegetable Oils (Edible)**.

MARGAY. See **Cats**.

MARICULTURE. See **Aquaculture**.

MARIJUANA. A drug derived from *Cannabis indica*, a variety of common hemp. The much more potent hashish is also derived from this plant. The active substance is located in the glandular hairs of the leaves and stems. From the pistillate flowers and fruits is obtained a resinous substance, which is smoked under the name of marijuana, hashish, or bhang, depending upon its concentration and particular mode of preparation. Synthetic cannabis, or synhexyl, is a pyrahexyl with an action more severe than the natural material.

Marijuana has been used as an agent for achieving euphoria since ancient times; it was described in a Chinese medical compendium traditionally considered to date from 2737 B.C. Its use spread from China to India and then to N Africa and reached Europe at least as early as A.D. 500.

The first direct reference to a cannabis product as a psychoactive agent dates from 2737 BC, in the writings of the Chinese emperor Shen Nung. The focus was on its powers as a medication for rheumatism, gout, malaria, and oddly enough, absent-mindedness. Mention was made of the intoxicating properties, but the medicinal value was considered more important. In India though it was clearly used recreationally. It was the Muslims who introduced hashish, whose popularity spread quickly throughout 12th century Persia (Iran) and North Africa.

In 1545 the Spanish brought marijuana to the New World. The English introduced it in Jamestown in 1611 where it became a major commercial crop alongside tobacco and was grown as a source of fiber.

By 1890, hemp had been replaced by cotton as a major cash crop in southern states. Some patent medicines during this era contained marijuana, but it was a small percentage compared to the number containing opium or cocaine. It was in the 1920's that marijuana began to catch on. Some historians say its emergence was brought about by Prohibition. Its recreational use was restricted to jazz musicians and people in show business. "Reefer songs" became the rage of the jazz world. Marijuana clubs, called tea pads, sprang up in every major city. These marijuana establishments were tolerated by the authorities because marijuana was not illegal and patrons showed no evidence of making a nuisance of themselves or disturbing the community. Marijuana was not considered a social threat.

Marijuana was listed in the United States Pharmacopoeia from 1850 until 1942 and was prescribed for various conditions including labor pains, nausea, and rheumatism. Its use as an intoxicant was also commonplace from the 1850s to the 1930s. A campaign conducted in the 1930s by the U.S. Federal Bureau of Narcotics (now the Bureau of Narcotics and Dangerous Drugs) sought to portray marijuana as a powerful, addicting substance that would lead users into narcotics addiction. It is still considered a "gateway" drug by some authorities. In the 1950s it was an accessory of the beat generation; in the 1960s it was used by college students and "hippies" and became a symbol of rebellion against authority.

The Controlled Substances Act of 1970 classified marijuana along with heroin and LSD as a Schedule I drug, i.e., having the relatively highest abuse potential and no accepted medical use. Most marijuana at that time came from Mexico, but in 1975 the Mexican government agreed to eradicate the crop by spraying it with the herbicide paraquat, raising fears of toxic side effects. Colombia then became the main supplier. The "zero tolerance" climate of the Reagan and Bush administrations resulted in passage of strict laws and mandatory sentences for possession of marijuana and in heightened vigilance against smuggling at the southern borders. The "war on drugs" thus brought with it a shift from reliance on imported supplies to domestic cultivation (particularly in Hawaii and California). Beginning in 1982 the Drug Enforcement Administration turned increased attention to marijuana farms in the United States, and there was a shift to the indoor growing of plants specially developed for small size and high yield. After over a decade of decreasing use, marijuana smoking began an upward trend once more in the early 1990s, especially among teenagers.

The use of marijuana is one of numerous problem interfaces between science and society. Solutions are for society to devise. Science assists society in constructing enlightened public policy.

Many marijuana users are unaware of several fundamental factors concerning the drug. This is particularly true of adolescents. For example: (1) the smoking of marijuana produces over 2000 separately identifiable chemicals. Many of these compounds remain in fat stores for several weeks, commonly with unknown consequences. The half-life in humans of a single marijuana cigarette containing 2% concentration of THC (delta-9-tetrahydrocannabinol) ranges from 3 to 7 days, meaning that several weeks are required for the THC to exit the body completely. (2) On occasion, marijuana is adulterated with various substances. One of these compounds is PCP (phencyclidine, an established dangerous drug). (3) Other adulterants may include insect sprays (some with known carcinogenic properties); and dried shredded cow manure. The latter may contain salmonella, a major cause of food poisoning. Each of the fundamental ingredients of marijuana, as well as adulterants, has its own intrinsic toxicity.

Marijuana, when smoked or otherwise ingested, is reported by users to produce a dreamlike state of relaxation, a sense of contentment, improved social interaction, loss of inhibitions, and feelings of heightened self-awareness. To achieve such effects, marijuana acts upon the brain, including the initiation of chemical and electrophysiologic changes.

As pointed out by Schwartz, there are also the adverse effects of acute panic, flashback phenomena, and acute toxic psychosis, such as excitement, disorientation, confusion, delusions, depersonalization, delirium, and visual hallucinations, which can occur unpredictably in some individuals. Acute panic reactions may be accompanied by abdominal discomfort, headache, anxiety, depression, fear of dying, restlessness, uncontrollable hostility, anxiety, and paranoia. Also, accurately intoxicated individuals may have impaired reflexes, decreased attention, altered depth perception, and reduced long-term memory.

The drug causes acute changes in the heart and circulation which are characteristic of stress. Acute exposure to marijuana smoke causes bronchodilation. In chronic heavy smoking of the drug, there is inflammation and preneoplastic change in the airways, not unlike those produced by heavy tobacco smoking. Thus, there is a strong possibility that prolonged heavy smoking of the drug will lead to cancer of the respiratory tract and to serious impairment of lung function.

Professionals who treat marijuana users generally focus on the disturbed interpersonal relationships of the abuser. For example, adolescents may use marijuana as an act of defiance or to conceal and cover up particular difficulties in their lives. Fortunately, many chronic users ultimately outgrow their habit, often when they enter into family formation and make an increasing number of friendships with drug-free persons.

Additional Reading

Abel, E.L.: "Marijuana: The First Twelve Thousand Years," Perseus Publishing, Boulder, CO, 2000.

Grinspoon, L. and J.B. Bakalar: "Marijuana: The Forbidden Medicine," Yale University Press, New Haven, CT, 1997.

Iversen, L.L.: "The Science of Marijuana," Oxford University Press, Inc., New York, NY, 2000.

Mack, A. and J. Joy: "Marijuana as Medicine? The Science beyond the Controversy," National Academy Press, Washington, DC, 2000.

Marx, J.: "Marijuana Receptor," Science, 624 (August 10, 1990).

Musto, D.F.: "Opium, Cocaine, and Marijuana in American History," Sci. Amer., 40 (July 1991).

Nahas, G.G., S. Agurell, K.M. Sutin, and D.J. Harvey: "Marijuana and Medicine," Humana Press, Totowa, NJ, 1999.

Schwartz, R.H.: "Marijuana: A Crude Drug with a Spectrum of Underappreciated Toxicity," Pediatrics, 73, 455 (1984).

Schwenk, C.R. and S.L. Rhodes: "Marijuana and the WorkPlace," Greenwood Publishing Group, Inc., Westport, CT, 1999.

Smith, A. and E. Tanner: "Highlights: An Illustrated History of Cannabis," Ten Speed Press, Berkeley, CA, 1999.

Somdahl, G.L.: "Marijuana Drug Dangers," Enslow Publishers, Inc., Berkeley Heights, NJ, 1999.

Staff: "Facts About Marijuana and Smoking," American Lung Association, New York, NY (revised periodically).

Stanley, D.: "Marijuana and Your Lungs," Rosen Publishing Group, Inc., New York, NY, 2000.

Watson, J., J.E. Joy, and J.A. Benson: "Medical Use of Marijuana: Assessment of the Science Base," National Academy Press, Washington, DC, 1999.

MARINE BIOLOGY.   See **Ocean**.

MARINE CLIMATE.   See **Climate**.

MARKHOR.   See **Goats and Sheep**.

MARKOV, ANDREI ANDREYVICH (1856–1922).   Markov was a renowned Russian mathematician working in number theory and probability theory. His name is best remembered for the concept of Markov chain, a series of events in which the probability of a given event depends only on the immediately previous event.

See also **Markov Process**.

J. M. I.

MARKOV CHAIN.   This expression is used in two different senses, both relating to a Markov process. In one sense, a process (x_i) is called a chain if the time parameter is discontinuous. In the other it is called a chain if the values of x are discontinuous. The former appears preferable.

MARKOV PROCESS.   A stochastic process such that the conditional probability distribution for the state at any future instant, given the present state, is unaffected by any additional knowledge of the past history of the system. See also **Autoregression**.

MARLIN.   See **Billfishes**.

MARMOSET *(Mammalia, Primates).*   Small monkeys of Central and South America. They have only thirty-two teeth, four less than the other American monkeys, and the thumb is not opposable to the fingers. All digits but the great toe bear claws instead of nails. They constitute the family of Hapaloids.

Most of these monkeys are called marmosets but the common Brazilian species, *Hapale jacchus,* is also known as the ouistiti and the group of long-tusked marmosets, *Mystax,* are called tamarins. One species, *Midas aedipus,* of the Isthmus of Panama, bears the French name pinché.

MARMOT.   See **Squirrels and Other Sciuromorphs**.

MARS.   The fourth planet from the Sun, Mars is the first planet in the solar system beyond Earth. Compared with Earth, Mars is a small planet. With a value of 1.00 for Earth, the mass of Mars is 0.107; the density, 0.719; volume, 0.149. The equatorial diameter of Mars is 4,226 miles (6,772 kilometers) as compared with that of earth of 7,960 miles (12,757 kilometers). Although Mars has been observed since early times, reliable and detailed data did not become available until commencement of the *Mariner* exploratory program in the 1960's. With the *Viking* programs of the last half of the 1970s, some important *Mariner* data had to be revised.

The orbit of Mars is noticeably eccentric (0.093). The distance from Mars to the sun varies from a minimum of 129 million miles (208 million kilometers) at perihelion to a maximum of 155.3 million miles (249 million kilometers) at aphelion. Earth-bound observations of Mars are best when the planet is within a few months of opposition (when the Earth lies between the planet and the sun). During the remainder of the present century, Mars will be closest to earth on February 11, 1995, March 20, 1997, and May 1, 1999.

Mars has two satellites, or moons, Phobos and Deimos, both discovered in 1877 by Hall. Spacecraft have shown these bodies to be cratered, rocky, and chunky, and in recent years there has been serious speculation that these may not be moons in the usual sense, but rather captive asteroids. See also **Asteroid**. Phobos is quite small, with dimensions of approximately 12.4 × 17.4 miles (20 × 28 kilometers) and Deimos even smaller, 6.2 × 9.9 miles (10 × 16 kilometers).

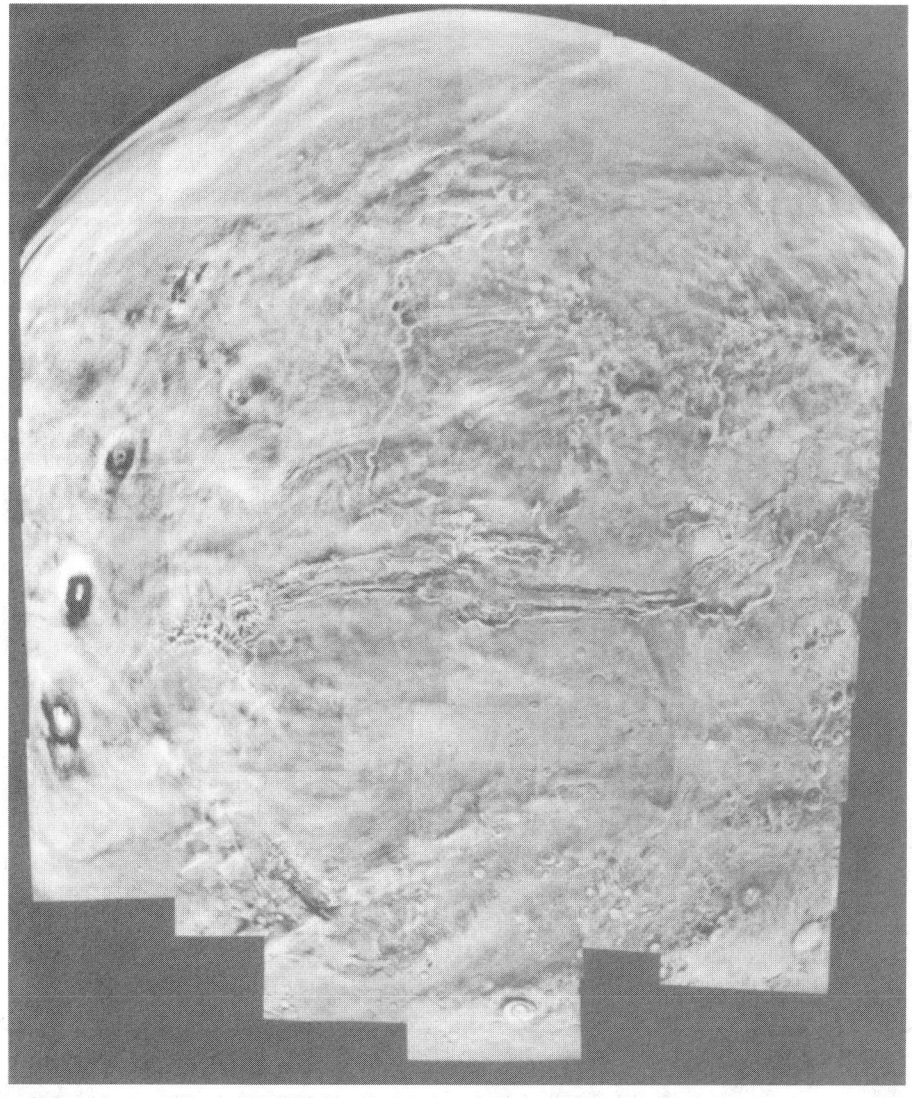

Fig. 1.   Mosaic of 102 photos of Mars taken on February 22, 1980 by *Viking Orbiter 1.* Several prominent Martian features and at least two unusual weather phenomena are visible. Valles Marineris (Mariner Valley), as long as the North American continent from coast to coast, stretches across the center. Three huge volcanoes of the Tharsis Ridge are visible at the left: Arsia Mons, Pavonis Mons, and Ascraeus Mons, proceeding from south to north. A sharp line, either a weather front or an atmospheric shock wave, curves north and east from Arsia Mons. This is the first time a feature like this had been seen on a planet. Four tiny clouds can be seen in the southernmost frame, just north of a large crater named Lowell. While the clouds are too close together to be resolved, even under high magnification, their shadows can be separated easily. The largest cloud is nearly 32 kilometers (20 miles) long. Measurements show the elevation of the clouds at nearly 28 kilometers (91,000 feet). Such distinct cloud-shadow patterns apparently are quite rare on Mars. (*NASA; Jet Propulsion Laboratory, Pasadena, California.*)

Missions to Mars. The twin *Viking* missions to Mars, each with its own lander, represented a very sophisticated and successful venture. Among some scientists there remains perplexity regarding some of the main features of the planet, notably numerous channels and rifts at one time called "canals" by Earth-bound observers several decades ago. Knowledge of how these features look (including full-color) and their dimensions have been greatly enhanced, but the mysteries of their origins remains unknown. Earth-bound estimates and *Mariner's* measurements of Mars comparatively thin atmosphere were confirmed, a factor which detracted from the possibility of organisms living on the planet. The polar ice cap once thought to be frozen carbon dioxide has been found to be ice with possibly some frozen carbon dioxide with it. Biological experiments designed to detect living organism proved negative, but the apparently oxidizing characteristic of Martian soil has introduced new puzzles.

An interesting view of Mars taken by *Viking Orbiter 1* showing the huge Mariner Valley (Valles Marineris) is given in Fig. 1. A close-up of this extremely impressive Martian feature is given in Fig. 2.

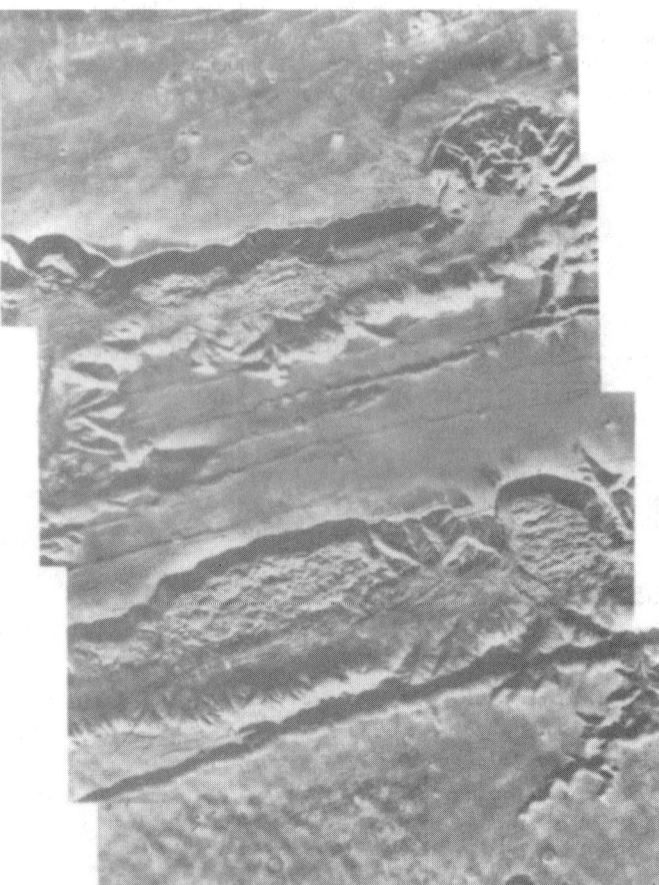

Fig. 2.   Mosaic of the surface of Mars showing the west end of the Valles Marineris (all of which is shown in Fig. 1) from a range of about 4300 kilometers (2700 miles). These two canyons, running east-west across the picture, are each about 60 kilometers (37 miles) wide and more than 1 kilometer (0.6 mile) deep. Some scientists suggest that the canyons were originally formed by downfaulting of the crust along parallel faults. Other faults and collapsed depressions with the same trend are seen between the two canyons. After they were formed, it is suggested that the canyons were modified by erosion that formed great slumps on the walls and also cut side valleys to the main canyons. A few comparatively recent impact craters will be noted, particularly at the bottom right of the view. (*NASA; Jet Propulsion Laboratory, Pasadena, California.*)

The *Viking* missions are discussed in greater detail in a later section of this entry. Many missions preceded the *Viking* missions to Mars, and several have followed. The list below presents the chronology of missions to Mars:

Mars 1960A — USSR Mars Probe was launched on October 10, 1960, however, it failed to reach Earth orbit.

Mars 1960B — USSR Mars Probe. Launched on October 14, 1960. It also failed to reach Earth Orbit.

Mars 1962A — USSR Mars Flyby. Launched on October 24, 1962, this spacecraft failed to leave Earth orbit after the final rocket stage exploded.

Mars 1 — USSR Mars Flyby. Launched on November 1, 1962. The spacecraft weighed 1,969 pounds (893 kilograms). This mission was not successful due to communications failure.

Mars 1962B — USSR Mars lander. Launched on November 4, 1962. This spacecraft also failed to leave Earth orbit.

Mariner 3 — Launched on November 5, 1964 at a weight of 572 pounds (260 kilograms) by the USA, the solar panels did not open, preventing a successful flyby. *Mariner 3* remains in solar orbit.

Mariner 4 — Launched on November 28, 1964 at a weight of 572 pounds (260 kilograms) by the USA, *Mariner 4* reached Mars on July 14, 1965. It passed within 5,952 miles (9,920 kilometers) and returned data confirming that the atmosphere was composed of carbon dioxide, and identifying a small magnetic field. *Mariner 4* obtained 22 close-up photos of the surface of Mars clearly showing surface features, notably craters. *Mariner 4* remains in solar orbit. See Fig. 3.

Zond 2 — USSR Mars Flyby launched on November 30, 1964, which was unsuccessful. Contact with the spacecraft was lost and its fate is unknown.

Mariner 6 — USA Mars Flyby launched at a weight of 910 pounds (413 kilograms), the spacecraft reached Mars on July 31, 1969. It passed within 2,062 miles (3,437 kilometers) of the surface of the planet. *Mariner 6* remains in solar orbit.

Mariner 7 — USA Mars Flyby launched at a weight of 910 pounds (413 kilograms), the spacecraft reached Mars on August 5, 1969. It passed within 2,131 miles (3,551 kilometers) of the surface of Mars at the south pole region. Both *Mariner 6* and *Mariner 7* obtained data related to the atmosphere and surface composition. Over 200 photos were obtained during these two missions. *Mariner 7* remains in solar orbit. See Fig. 4.

Mariner 8 — USA Mars Flyby launched May 8, 1971, this mission was unsuccessful as it failed to reach Earth's orbit.

Kosmos419 — Launched by the USSR May 10, 1971, this mission was unsuccessful as it failed to reach Earth's orbit.

Mars 2 — This spacecraft was a USSR Mars Orbiter/Soft Lander launched May 19, 1971 that weighed 10,230 pounds (4,650 kilograms). It failed in its landing mission as the *Mars 2 Lander*, which was released from the Orbiter on November 27, 1971, crash-landed on the surface of the planet. It is known that the breaking rockets failed, but no data was returned. The *Mars 2 Orbiter* returned data until 1972.

Mars 3 — This spacecraft was another USSR Mars Orbiter/Soft Lander that weighed 10,215 pounds (4,643 kilograms). It reached Mars on December 2, 1971, and successfully released the lander to the surface of the planet. It was the first successful landing on the surface of Mars, but the *Mars 3* failed to record and transmit more than 20 seconds of data to the orbiter. The *Mars 3 orbiter* collected data related to the surface temperature and atmospheric conditions until August 1972.

Mariner 9 — Launched by the USA May 30, 1971, the spacecraft weighed 1,116 pounds (506 kilograms). *Mariner 9* was the first US spacecraft to enter orbit around a body other than the Earth's moon, and it entered this orbit on November 24, 1971. Among the data obtained were the first high-resolution images of the Martian moons, Phobos and Deimos, and surface data detailing river and channel-like features. *Mariner 9* remains in Martian orbit. See Fig. 5.

Mars 4 — Another of the USSR Mars Orbiter/Soft Lander vehicles, this mission was not wholly successful. Although it arrived at Mars in February 1974, it failed to enter orbit due to failure of the breaking rockets. A flyby at a distance of 1,320 miles (2,200 kilometers) returned limited data.

Mars 5 — A USSR Mars Orbiter/Soft Lander vehicle, the spacecraft weighed 10,230 pounds (4,650 kilograms) and entered Martian orbit in February 1974. Data obtained during this mission set the stage for the *Mars 6* and *Mars 7* missions.

Mars 6 — This USSR Mars Orbiter/Soft Lander vehicle, which weighed 10,230 pounds (4,650 kilograms) entered Martian orbit on March 12, 1974 and launched its lander. The lander successfully transmitted atmospheric data during its descent, but failed prior to landing.

Fig. 3. *Mariner 4* was the fourth in a series of spacecraft used for planetary exploration in a flyby mode. It was designed to conduct close-up scientific observations of the planet Mars and to transmit these observations to Earth. (*Courtesy of the Jet Propulsion Laboratory and NASA's National Space Science Data Center.*)

Fig. 4. *Mariner 6* and *7* were designed to fly over the equator and southern hemisphere of the planet Mars. They were solar powered and capable of continuous telemetry transmission. Each spacecraft weighed 910 pounds (413 kilograms) and measured 11 feet (3.35 meters) from the scan platform to the top of the low-gain antenna. The width across the solar panels was 19 feet (5.8 meters). The eight-sided body of the spacecraft carried seven electronic compartments. A small rocket engine, used for trajectory corrections, protruded through one of the sides. The planetary experiments aboard the spacecraft were two television cameras, an infrared radiometer, and infrared spectrometer and as ultraviolet spectrometer. (*Courtesy NASA.*)

Mars 7 — Another USSR Mars Orbiter/Soft Lander vehicle that weighed 10,230 pounds (4,650 kilograms), failed both to enter Martian orbit and to set the lander vehicle on the Martian surface. The *Mars 7 orbiter* and lander remain in solar orbit.

Viking 1 — Designed after the *Mariner* spacecraft, the USA Mars Lander/Orbiter was launched on August 20, 1975 weighing 7,478 pounds (3,399 kilograms). The orbiter weighed 1,980 pounds (900 kilograms) and the lander weighed 1,320 pounds (600 kilograms). *Viking 1* entered Martian orbit June 19, 1976, and its lander successfully set on the surface one day later on July 20, 1976 on the western slopes of Chryse Planitia. The lander and orbiter obtained data related to the weather on Mars, the Martian terrain, and microorganisms on the planet. The *Viking 1 orbiter* ran out of altitude control propellant August 7, 1980 and was deactivated. The *Viking 1 lander* was

accidentally shut down and neither communication nor activation was ever regained. See Fig. 6.

Viking 2 — Also designed after the *Mariner* spacecraft, the USA Mars Lander/Orbiter was launched on September 9, 1975 weighing 7,478 pounds (3,399 kilograms). See Fig. 7. The orbiter weighed 1,980 pounds (900 kilograms) and the lander weighed 1,320 pounds (600 kilograms). Viking 2 entered Martian orbit on July 24, 1976 and its lander set down at Utopia Planitia on August 7, 1976. See Fig. 8. While both landers had experiments to search for and identify microorganisms on Mars, the results of the experiments are still subject to debate. Both landers together obtained over 52,000 images while mapping the planet's surface. The *Viking 2 orbiter* ran out of altitude control propellant July 25, 1978 and was deactivated. Because the *Viking 2 lander* used the *Viking 1 orbiter* for communications,

Fig. 5.   The *Mariner 9* spacecraft was built on octagonal magnesium frame 18 inches (45.7 centimeters) deep and 54.5 inches (138.4 centimeters) across a diagonal. Four solar panel each 85 × 35 inches (215 × 90 centimeters), extended out from the top of the frame. Each set of two solar panels spanned 23 feet (6.89 meters) from tip to tip. Also mounted on the top of the frame were two propulsion tanks, the maneuver engine, a 5-foot (1.44 meters) long low gain antenna mast and a parabolic high gain antenna. A scan platform was mounted on the bottom of the frame, on which were attached the mutually bore-sighted science instruments (wide-and narrow-angle TV cameras, infrared radiometer, ultraviolet spectrometer, and infrared interferometer spectrometer). The overall height of the spacecraft was 7.5 feet (2.28 meters). (*Courtesy of NASA's National Space Science Data Center.*)

Fig. 6.   This image shows a model of one of the *Viking* spacecraft, which were made of two parts: an orbiter and a lander. The orbiter's initial job was to survey the planet for a suitable landing site. Later the orbiter's instruments studied the planet and its atmosphere, while the orbiter acted as a radio relay station for transmitting lander data. (*Courtesy NASA/JPL.*)

it had to be shut down the same time the *Viking 1 orbiter* was deactivated on August 7, 1980.

Phobos 1 — USSR Mars Orbiter/Lander weighing 11,000 pounds (5,000 kilograms) that was launched on July 7, 1988. The spacecraft was lost on the way to Mars due to a command error on September 2, 1988. See Fig. 9.

Phobos 2 — USSR Phobos Flyby/Lander, which weighed 11,000 pounds (5,000 kilograms), was launched on July 12, 1988. The spacecraft entered Martian orbit January 30, 1989, but failed at a distance of 480 miles (800 kilometers) from the Martian moon Phobos.

Mars Observer — USA Mars Orbiter was launched September 25, 1992 but failed to enter Martian orbit. Communication with the Mars Observer was lost August 21, 1993.

Mars Global Surveyor — USA Mars Orbiter was launched November 7, 1996 to complete the mission of the Mars Observer. See Fig. 10.

Mars 96 — Russia Orbiter and Lander which was launched November 16, 1996, consisted of an orbiter, two landers and two soil penetrators. The fourth stage of the rocket that launched the Mars 96 spacecraft ignited prematurely as the vehicle entered orbit. The spacecraft crashed into the ocean and sank between the coast of Chile and Easter Islands.

Mars Pathfinder — USA Lander and Surface Rover launched on December 4, 1996. The lander weighed 581 pounds (264 kilograms) and the rover vehicle weighed only 23 pounds (10.5 kilograms). Mars Pathfinder reached Mars July 4, 1997 and impacted the surface at a velocity of approximately 40 miles per hour (18 meters per second).

Fig. 7.   Launch of the *Viking 2* spacecraft from Cape Canaveral, Florida. (*Courtesy NASA/JPL.*)

Fig. 8.   Captured in this rendering is a *Viking* lander just before it touched down on the Martian surface. The parachute and upper aeroshell can be seen in the upper left corner of the image. At this stage of the descent, the lander's terminal descent propulsion system (three retro-engines) had slowed the craft down so that velocity at landing was about 7 miles per hour (2 meters per second). Seconds after the lander reached the surface it began transmitting images back to the orbiter for relay to Earth. (*Courtesy NASA/JPL.*)

It bounced into the air about 50 feet (15 meters), bounced another 15 times, and rolled to a stop approximately 2.5 minutes after impact about one-half mile (about 1 kilometer) from the site of initial impact. The landing site, in the Ares Valley region at 19.33 °N, 33.55 °W, was named the Sagan Memorial Station. The rover, a six-wheeled vehicle named Sojourner, hit the Martian surface July 6. See Figs. 11 and 12. This highly successful mission returned 2.6 billion bits of information including over 16,000 images from the lander, 550 images

Fig. 9.   This artist's concept depicts the *Phobos 1 & 2* spacecraft destined for Mars. They were the next-generation in the Venera-type planetary missions, succeeding those last used during the Vega 1 and 2 missions to comet P/Halley. (*Courtesy of NASA's National Space Science Data Center.*)

from Sojourner, 15 chemical analyses of rocks, and extensive data on climatic conditions. See also **Pathfinder Mission to Mars**.

Planet B — Japan Mars Orbiter launched August 1998 by Japan's Institute of Space and Astronautical Science (ISAS) will be the first Japanese spacecraft to reach another planet.

Mars Surveyor '98 Orbiter — Scheduled for launch December 1998, this spacecraft will study the planet from polar orbit for one to two years using a variety of highly sophisticated instruments. See Fig. 13.

Mars Surveyor '98 Lander — The companion vehicle to the Mars Surveyor '98 Orbiter is scheduled to be launched January 1999. It will study the Martian environment at the south pole region specifically climate, atmospheric conditions, and soil. It will be equipped with advanced meteorological equipment and a robotic arm for digging into the soil. See Fig. 14.

Mars Surveyor 2001 — USA Mars Probe scheduled to be launched in 2001 as part of NASA's ten-year program to launch probes to Mars as favorable launch opportunities arise. See Figs. 15, 16 and 17.

NASA Unveils Its 21st Century Mars Campaigns

The seven-month retooling of its Mars campaign was prompted by the back-to-back loss last year of two spacecraft at the Red Planet. Subsequent investigative reports, including one authored by retired Lockheed Martin executive Tom Young, found bad management, a lack of training and an inadequate system of checks and balances, as well as too-tight budgets, doomed the Mars Climate Orbiter and Polar Lander missions, a $300 million-plus loss.

NASA will halt ambitious plans to send a lander/orbiter pair to Mars every 26 months, when the Earth and the Red Planet are closely aligned. Instead, it will now stagger the pace dispatching just one of each at the roughly two-year intervals.

The revised program also looks out beyond returning a sample of Martian soil to Earth for study. That goal has been pushed back to 2011 or later.

This program will represent a long-term strategy. It won't just end with Mars sample return like the previous one did. Officials, said the new program, allows for NASA to respond to any new discoveries on Mars, like the evidence that suggests water may have flowed on the planet's surface in the recent past, as well as accommodate the prospect of any of the planned missions failing.

What's missing from the equation are humans. NASA has already scrapped plans to launch in 2001, a package of experiments that would have laid some of the groundwork for future human missions to Mars, including experiments to produce oxygen from the Martian atmosphere and to assess the threat of its dust and radiation environments. Now, similar experiments might not make to Mars until 2007 at the earliest.

The agency plans on six major Mars missions for this decade alone, spending as much as $450 million a year on its near-term efforts. The missions include:

2001 – The Mars Odyssey Orbiter, for high-resolution mapping and imaging.

2003 – Two water-sniffing Mars Exploration Rovers, for detailed field geological work.

2005 – A Mars Reconnaissance Orbiter: an orbiter modeled on NASA's successful Mars Global Surveyor, but capable of imaging objects as small as 30 centimeters (a foot) in diameter. Jim Garvin, NASA's Mars exploration program scientist, likened it to putting a microscope in orbit.

2007 – A "smart" surface lander equipped with a hazard avoidance system, precision landing capability and designed to deliver a rover laden with up to 270 kilograms (600 pounds) of scientific instruments; also in 2007, a "Scout" mission, which could entail a small Beagle 2-type lander, a balloon or an airplane. Both balloon and airplane Mars missions have been submitted as proposals in the current round of Discovery-class NASA projects.

The Mars Pathfinder landing in 1997 was within a 100×300-kilometer (60×200-mile) landing ellipse. "Where we want to be by 2007 is down to something that's 1 kilometer by 3 kilometers (0.62 by 2 miles) — a reduction by a factor of 100," Lavery said. The eventual goal is to land spacecraft on the equivalent of a Martian dime — within a tight ellipse just a few hundreds of yards (meters) across, he said.

2007 – NASA could also kick off an international collaborative effort in 2007, teaming up with the Italian space agency on a telecommunications orbiter for Mars or with the French on a network of small landers.

2009 – NASA could team up again with the Italians on a follow-on to the European Space Agency's 2003 Mars Express mission. The probe would carry ground-penetrating radar to prospect for water on the Red Planet.

2011 – As early as 2011, but perhaps slipping to 2014, NASA could start a long-term project to return multiple samples of Martian soil and rock to

Fig. 10.   Captured in this rendering the *Mars Global Surveyor (MGS)* is designed to orbit Mars over a two year period and collect data on the surface morphology, topography, composition, gravity, atmospheric dynamics, and magnetic field. This data will be used to investigate the surface processes, geology, distribution of material, internal properties, evolution of the magnetic field, and the weather and climate of Mars. (*Courtesy of NSSDC.*)

Earth. The effort, which could cost as much as $2 billion a pop, had been on tap for 2005 under the previous plan.

The Martian Atmosphere

As pointed out by Anders and Owen (1977), the thinness of the Martian atmosphere has been one of the great disappointments of the space age. At one time, Lowell, Dollfus, and others had suggested a surface pressure near 80 millibars (1 standard atmosphere on earth = 1013 millibars). Even with an atmosphere only about one-tenth the value of earth's, it was envisioned that Mars possibly could sustain some forms of life. *Viking 1* established a figure of 7.65 millibars. Substantiation of this relatively low pressure, indicating that Mars has only about 3% of the volatiles found on earth, revised scientific thinking in terms of the development of Mars. Anders and Owen suggested five processes, in combination, which may have been responsible for the thin martian atmosphere: (1) a small initial endowment of volatiles; (2) incomplete outgassing from the interior; (3) recondensation or trapping in surface regions; (4) catastrophic loss of an early atmosphere; and (5) gradual escape of the lighter constituents. Although these investigators do not describe a detailed model for how the Martian atmosphere passed from an early, dense state to its present condition, they offer a schematic scenario:

The basic notion is that the atmosphere gradually decreased in density as a result of the deposition of carbon dioxide in the form of carbonates and the escape of nitrogen from the upper atmosphere. While the latter process was critical for the ultimate nitrogen abundance and isotope ratio, it should have played a small role in determining the total atmospheric pressure, since carbon dioxide was probably the most abundant gas. The depositional process (which may have included formation of nitrates or nitrites) was most active during the time when liquid water was most abundant—the cutting of the sinuous channels was thus a premonition of the end of the dense atmosphere. The apparent absence of an active Martian biota has prevented the recycling of volatiles through biological processes. Moreover, there is evidence that carbonates may form even under the present arid conditions on Mars.

Anders and Owen proceed in their interesting paper to make comparisons between the large and the small planets, with earth and Mars as paradigms.

Water and Water Vapor. An infrared spectrometer operating at the 1.38 micrometer region, mounted on the scan platform, was used to detect water vapor in the Martian atmosphere.[1] This scanning device was used

[1] Information extracted from official NASA Langley Research Center report.

Fig. 11.   Artist view of *Pathfinder* on Mars. (*Courtesy of NASA's National Space Science Data Center.*)

to measure the latitudinal variations and diurnal variations. By operating over a complete martian year, the instrument was able to measure the seasonal changes. The southern hemisphere, which was at the onset of winter, was found to have very little water vapor (0 to 0.3 precipitable micrometer). In contrast, the northern hemisphere showed a significant amount of water (up to 75 micrometers at 70–80 °N), a range of almost three orders of magnitude. The north polar region showed a slight drop in water vapor abundance. A strong diurnal repetitive cycle in certain regions, peaking out in the local mid-afternoon, was also found. Negative correlation existed between the elevation and the water vapor abundance, as would be expected. On the basis of the abundance of water vapor in

the polar region, a lower limit can be put on the atmospheric temperature, namely 205 K ($-68\,°C$; $-90\,°F$). This value indicates that the permanent polar cap consists of water ice, a factor that was confirmed by the infrared thermal mapper (IRTM) also part of the *Viking* instrumentation package.

Observations of the latitude dependence of water vapor made from the *Viking 2 Orbiter* showed peak abundances in the latitude band 70–80 °N in the northern midsummer season. Total column abundances in the polar regions appeared to require near-surface atmospheric temperatures in excess of 200 K ($-73\,°C$: $-99\,°F$) and are incompatible with the survival of a frozen carbon dioxide cap at martian pressures. The remnant or residual on the north polar cap and the outlying patches of ice at lower latitudes are thus believed to be predominantly water ice whose thickness can be established within widely spaced limits, between 1 meter and 1 kilometer ($\sim$3 and 3280 feet). Broadband thermal and reflectance observations of the Martian north polar region in late summer yielded temperatures for the residual polar cap near 205 K ($-68\,°C$; $-90\,°F$). Residual cap and several outlying smaller deposits appeared to be water ice with included dirt. No evidence was found for a permanent carbon dioxide polar cap.

The first evidence of the direct visible exchange of water between the martian surface and atmosphere was obtained by the *Viking 1 Orbiter* on July 24, 1976, as shown by Fig. 18.

Ground Ice on Mars. As reported by Squyres (NASA Ames Research Center) and Carr (U.S. Geological Survey), many martian landforms suggest the former presence of ground ice or water, including fretted and chaotic terrain, valley systems, outflow channels, and, with less certainty, various types of patterned ground and rampart craters. If sufficient ice is present now, the regolith should undergo quasi-viscous flow due to creep deformation of the ice. Accordingly, to determine where ice may be present, the researchers examined approximately 24,000 *Viking Orbiter* images taken with 5000 km of the surface and mapped the distribution of three types of features—lobate debris aprons, concentric crater fill, and terrain softening—that may indicate creep of near-surface materials.[2] In summary, the researchers observe that the origin of ice in the martian

---

[2] *Lobate debris aprons*—accumulations of erosion debris at the base of steep escarpments. *Concentric crater fill*—develops where debris aprons are confined within impact craters, and inward flow of material produces a pattern of concentric ridges. *Terrain softening*—a distinctive style of landform degradation apparent in high-resolution images. Definitions as given by Squyres and Carr.

Fig. 12.   The rover *Sojourner* is a six-wheeled vehicle launched with the Mars Pathfinder mission. It is controlled by an Earth-based operator, who uses images obtained by both the rover and lander systems. Note that the time delay is about 10 minutes, requiring some autonomous control by the rover. The primary objectives were scheduled for the first seven sols (1 sol = 1 martian day = $\sim$24.7) (*Courtesy of NASA's National Space Science Data Center.*)

Fig. 13. The *Mars Climate Orbiter* spacecraft was launched from the Cape Canaveral Air Force Station (CCAFS) Space Launch Complex 17 (SLC-17) on December 11, 1998. The Mars Orbit Insertion (MOI) propulsive maneuver will occur in September 1999 and will place the orbiter into a highly elliptical, near polar orbit around Mars. Peripapse will be lowered approximately 110 kilometers (68 miles) altitude to initiate the aerobraking maneuvers. Successive passes of the orbiter through the upper atmosphere of Mars will slow the vehicle and lower the apoapse of the orbit to 450 kilometers (280 miles) over the course of the 2 month aerobraking phase. The orbit then will circularized using the orbiter's onboard propulsion resulting in the design 400 kilometers (249 miles) altitude, circular, polar science mapping orbit. Science operation of the PMIRR and MARCI instruments will be conducted over the course of the one Martian year (687 Earth day) mapping mission. The orbiter will continue operations in a relay only mode following the science mission in support of any future U.S. or international Mars surface missions ending December 1, 2004. (*Courtesy of NASA/JPL.*)

Fig. 14. The *Mars Polar Lander* spacecraft was launched from Cape Canaveral Air Force Station (CCAFS) Space Launch Complex 17 (SLC-17) during a 25 day launch period beginning on January 3, 1999. Dimensions of the spacecraft 1.06 meters (3.5 feet) tall by 3.6 meters (12 feet) wide. The total spacecraft mass is 576 kilograms (1,270 pounds). The Lander weighs 290 kilograms (639 pounds). Propellant 64 kilograms (141 pounds). Cruise Stage 82 kilograms (181 pounds). Aeroshell and Heat Shield 140 kilograms (309 pounds). Mars Landing is scheduled on Dec. 3, 1999 with the end of the primary mission on March 1, 2000. The science payload includes Deep Space 2 Microprobes, Mars Descent Imager (MARDI), Light Detection and Ranging (LIDAR), New Millennium Microprobes, Mars Volatiles and Climate Surveyor (MVACS), Stereo Surface Imager (SSI), Robotic Arm and Camera, Meteorological Package (MET), and Thermal and Evolved Gas Analyzer (TEGA). (*Courtesy of NASA/JPL.*)

regolith is unclear at this time. The many lines of evidence implying that ice was common in the cratered uplands early in Martian history suggest that the ice was emplaced during an early period of intense outgassing. An alternative scenario would be the continuous outgassing throughout the planet's history at a rate substantially lower than the low-latitude depletion rates in order to keep the low latitudes ice-free. In either case, intense early meteoritic brecciation was probably largely responsible for the apparent capability of the deep regolith to hold large amounts of water.

Juvenile Water from Volcanism. In studying *Viking* data. Greeley (Arizona State University) reports that volcanism played a dominant role in the evolution of the Martian surface and environment. It is estimated that volcanism has occurred from at least the close of the period of heavy impact cratering (~3.9 billion years ago) to the age of the youngest rocks visible on the planet. Materials that are considered to be of volcanic origin cover more than half the surface of Mars. Perhaps the assumption can be made that, as with the Earth, juvenile water[3] was released on Mars in

association with the eruption of these volcanic materials. By determining the volumes and ages of volcanic units and inferring the volatile content for the magmas, the amounts and timing of associated water release can be estimated. Tentative conclusions indicate the amount of juvenile water release on Mars would equal a layer some 46 meters deep over the entire planet. Most of this water was released in the first 2 billion years of Martian history. There are several uncertainties in estimates made to date, including lack of knowledge of volatile contents for magmas; even terrestrial values, as used, have large uncertainties, and extrapolation to martian values is difficult. Uncertainties that stem from estimates of volcanic unit volumes can be reduced through more detailed mapping and determination of flow thicknesses, which will include additional new data, obtained from the *Mars Observer* in the 1990s.

Carbon Dioxide. At both *Viking* landing sites, it was found that the temperature was appreciably above the carbon dioxide condensation boundary, thus precluding the occurrence of carbon dioxide hazes in northern summer at latitudes at least 50°N. Thus, the ground level mists seen in these latitudes would appear to be condensed water vapor. Neutral mass spectrometers carried on the aeroshells of both *Viking* spacecraft indicated that carbon dioxide is the major constituent of the martian atmosphere over the height range of 120 to 200 kilometers (74 to 124 miles). Densities for carbon dioxide measured by the upper

---

[3] In terms of Earth, *juvenile water* refers to water which has been derived from the crystal rocks or from the interior of the Earth, and at the time of its appearance in the circulating water of the hydrosphere represents an accretion to the available water supply. Juvenile water is, therefore, water which has not previously been a part of the hydrosphere. Although many investigators have tried to device means for identifying juvenile water, no satisfactory method has been found. It is difficult

to be certain that a particular water, such as that charged by hot springs, geysers, furaroles, etc., is of juvenile origin. Magmatic water is derived from a magma or included in a magma (rock melt).

Fig. 15. The *Mars Surveyor 2001 Orbiter* was scheduled for launch on April 18, 2001. It will arrive at Mars on Oct. 27, 2001, if launched on schedule. The 2000 Orbiter will be the first to use the atmosphere of Mars to slow down and directly capture a spacecraft into orbit in one step, using a technique called aerocapture. It will then reach a circular mapping orbit within about 1 week after arrival. The Orbiter will carry 2 main science instruments, the Thermal Emission Imaging System (THEMIS) and the Gamma Ray Spectrometer (GRS). THEMIS will map the mineralogy and morphology of the Martian surface using a high resolution camera and a thermal infrared imaging spectrometer. The GRS will achieve global mapping of the elemental composition of the surface and the abundance of hydrogen in the shallow subsurface. The gamma ray spectrometer was inherited from the lost Mars Observer mission. The 2001 Orbiter will also support communication with the Lander and Rover scheduled to arrive on Jan. 22, 2002. (*Courtesy of NASA/JPL, 2001 Artwork by Corby J. Waste.*)

atmospheric mass spectrometers on both *Viking* spacecraft were analyzed to yield height profiles for the temperature of the Martian atmosphere between the aforementioned range. The upper atmosphere of Mars was found to be surprisingly cold with an average temperature of less than 200 K (−73 °C; −99 °F). The atmosphere contains detectable concentrations of nitrogen, argon, carbon dioxide, molecular oxygen, atomic oxygen, and nitric oxide. The upper atmosphere exhibits a complex and variable thermal structure and is well mixed to heights in excess of 120 kilometers (74 miles).

Carbon Dioxide and Heat Balance at Polar Caps. As reported by Paige and Ingersoll (California Institute of Technology), the Infrared Thermal Mappers (IRTM) aboard the two *Viking* orbiters obtained solar reflectance and IR emission measurements of the martian north and south polar regions during an entire martian year. The observations were used to determine annual radiation budgets and to infer annual carbon dioxide frost budgets. The results provide further confirmation of the presence of permanent CO_2 frost deposits near the south pole and show that the stability of these deposits can be explained by their high reflectivities. In the north, the observed absence of solid CO_2 during summer was primarily the result of enhanced CO_2 sublimation rates due to lower frost reflectivities during spring. The results suggest that the present asymmetric behavior of CO_2 frost at the Martian poles is caused by preferential contamination of the north seasonal polar cap by atmospheric dust. The investigators emphasize that the *Viking* results have made it clear that the annual heat balance at the Martian poles is not purely a local phenomenon, but may be strongly influenced by the complex, global scale geologic and atmospheric processes that bring dust to the polar regions each year. The forthcoming *Mars Observer* mission, which will follow a polar orbit about the planet, will furnish much needed additional data.

Nitrogen. Results from the neutral mass spectrometer carried on the aeroshell of *Viking 1* spacecraft showed evidence for NO in the upper atmosphere of Mars and indicated that the isotopic composition of carbon and oxygen is similar to that of earth. Mars is enriched in ^{15}N relative to earth by about 75%, a consequence of escape that implies an initial abundance of nitrogen equivalent to a partial pressure of at least 2 millibars. The initial abundance of oxygen present either as carbon dioxide or water must be equivalent to an exchangeable atmospheric pressure of at least 2 bars in order to inhibit escape-related enrichment of [18]O. McElroy, Yung, and Nier (1976) constructed models for the past history of nitrogen on Mars based upon *Viking* measurements showing that the atmosphere is enriched in [15]N. The enrichment is attributed to selective escape, with fast atoms formed in the exosphere by electron impact dissociation of N_2 and by dissociative recombination of N_2^+. The initial partial pressure of N_2 should have been at least as large as several millibars and could have been as large as 30 millibars if surface processes were to represent an important sink for atmosphere HNO_2 and HNO_3.

Krypton and Xenon. These gases were discovered in the Martian atmosphere with the mass spectrometer on *Viking Lander 2*. Krypton is more abundant than xenon. The relative abundances of the krypton isotopes appear normal, but the ratio of xenon-129 to xenon-132 is enhanced on Mars relative to the earth value for this ratio. The mass spectrometer on *Viking Lander 1* had previously reported the detection of [36]Ar and the establishment of upper limits on Ne, Kr, and Xe in the atmosphere. The upper limit of krypton was close the value that would be predicted if the [36]Ar/Kr ratio on Mars were identical to that on earth. As pointed out by Owen et al. (1976), the Earth's atmosphere is deficient in xenon compared with the primordial gas in meteorites, and this is exactly the situation found on Mars. The xenon deficiency on earth has been attributed to the preferential adsorption of xenon in shales and other sedimentary material after it was outgassed. One is thus led to the tentative conclusion that similar processes have been active on Mars, perhaps in association with the epochs of fluvial erosion that have left their imprint on the planet's surface. Owen et al. suggest as an alternative or supplementary suggestion that some of the xenon could be absorbed in the regolith.

Weather

The atmosphere of Mars appears to favor stability, much unlike earth in that regard. The annual temperature range for the Martian surface at the *Viking* landing sites was computed on the basis of thermal parameters derived from observations made with infrared thermal mappers. Viking Lander 1 site showed small annual variations in temperature, whereas VL-2 site showed larger annual changes. (Locations of the sites are described in the latter portion of this article.) At both sites, daily temperature ranges at the top of the soil were 183 to 268 K (−90 to −5 °C; −130 to +23 °F). Diurnal variations decreased with depth in an exponential manner. The maximum temperature of soil sampled beneath rocks at the VL-2 site was computed to be 230 K (−43 °C; −45 °F). Daily patterns of temperature, wind, and pressure were highly repetitive at both sites during the early summer period. Wind was found to have a vector mean of 0.7 meter/second from the southeast with diurnal amplitude of 3 meters/second. Pressure exhibited both diurnal and semidiurnal oscillations, although of substantially smaller amplitude than those of VL-1. It should be mentioned that Mars does not have an ozone layer in its atmosphere as a shield against ultraviolet radiation and the absence of it, of course, has some effect on its climate.

It will be obvious to the reader that a satisfactory model of Martian weather cannot be formulated with so many unanswered questions as thus far indicated in this article. One popular concept, based upon incomplete data, suggests that the Earth and Mars commenced largely with similar initial conditions, but that over the course of some 4 billion years, the two planets have evolved differently. Both of these premises seem reasonable and logical. The images from *Viking* certainly prove that the two planets are distinctly different, yet when compared with other planets in the solar system, Earth and Mars present more similarities. Fundamental differences (not a direct function of the passage of time) between the two planets include: (1) *size* (Mars is only a little more than half the size of Earth); (2) the much greater distance from the sun of Mars than of Earth (hence less radiation received); and (3) the orbit of Mars is more eccentric, or elliptical, than that of Earth. These factors all have a fundamental bearing on a planet's atmosphere and weather system.

Fig. 16.   The *Mars Surveyor 2001 Lander* was scheduled for launch on April 10, 2001. It will land on Mars on Jan. 22, 2002, if launched on schedule. The 2001 Lander will carry an imager to take pictures of the surrounding terrain during its' rocket-assisted descent to the surface. The descent imaging camera will provide images of the landing site for geologic analyses, and will aid planning for initial operations and traverses by the rover. The 2001 Lander will also be a platform for instruments and technology experiments designed to provide key insights to decisions regarding successful and cost-effective human missions to Mars. Hardware on the Lander will be used for an in-situ demonstration test of rocket propellant production using gases in the Martian atmosphere. Other equipment will characterize the Martian soil proper and surface radiation environment. (*Courtesy of NASA/JPL, 2001 Artwork by Corby J. Waste.*)

Fig. 17.   As part of the 2001 mission redesign, the plans for the 2001 Rover have changed significantly. Current plans are to send the Marie Curie rover to Mars on the 2001 lander. This rover is very similar to the Pathfinder Sojourner Rover, and in fact is the same rover that was used within the Pathfinder "sandbox" test bed pictured above. (*Courtesy of NASA/JPL.*)

It has been generally established that a planet's size determines the strength of the internal heat sources, coming mainly from radioactive decay processes and gravitational energy released during accretion, that drive tectonic and volcanic activity. Even though, as a smaller planet. Mars sustained volcanic activity (as evident from *Viking* images), it was not imbued with the volcanic potential of Earth, which continues apace. Plate tectonics on Earth permits frequent access to internal heat sources, whereas *Viking* images indicate no recent evidence of plate tectonics; the entire crust appears to be a single plate. (Not full agreement among observers regarding this point.)

The greater distance of Mars from the sun obviously is a factor that must be built into any model of Mars. However, it may not be a major determining factor. With essentially general agreement that liquid water once existed on Mars and with suspicions that it may exist as ice in the regolith today (still very speculative), it follows that Mars received significant radiation from the sun in its early, formative periods, prior to losing its primitive atmosphere, to maintain water in the liquid state. (It has been theorized that once volcanic activity essentially abated on the planet, the atmosphere CO_2 level and hence greenhouse effect declined, causing cooling to the point where liquid water could no longer exist.) Kahn (Washington University) suggests that the surface pressure on the planet is so low today because CO_2 continued to be removed from the atmosphere and stored as carbonates by transitory pockets of liquid water. As interpreted by Haberle (NASA Ames Research Center), such pockets could have existed long after the global mean temperature had dropped below the freezing point; specifically, they could form as long as the surface pressure was sufficiently high to limit evaporation. Through the action of transitory water pockets the pressure was gradually reduced to its present value of 6.1 millibars. Haberle further observes that the small size of Mars probably had at least as much influence on its climate as has its distance from the sun. Moreover, the size of Mars has determined the fate not only of water and CO_2, but also of nitrogen, which is relatively scarce in the planet's atmosphere. The lower level of volcanic activity on Mars meant that less nitrogen was outgassed than on Earth. The smaller gravitational pull of Mars also made it easier for nitrogen to escape. (Although nitrogen does not have enough thermal energy to escape, it can acquire the necessary energy by dissociative recombination.)

Because of the eccentricity of Mars' orbit, the seasons on the planet are of unequal duration and intensity. (The Martian year is 687 Earth days long.) Perihelion (closest approach to the sun) occurs late in the southern hemisphere, making it 52 Earth days shorter than in the north. When at perihelion, Mars receives some 40% more solar radiation than when at aphelion. In terms of the Earth, this difference is only 3%. This asymmetry of seasons markedly affects the weather on Mars as it is known today,

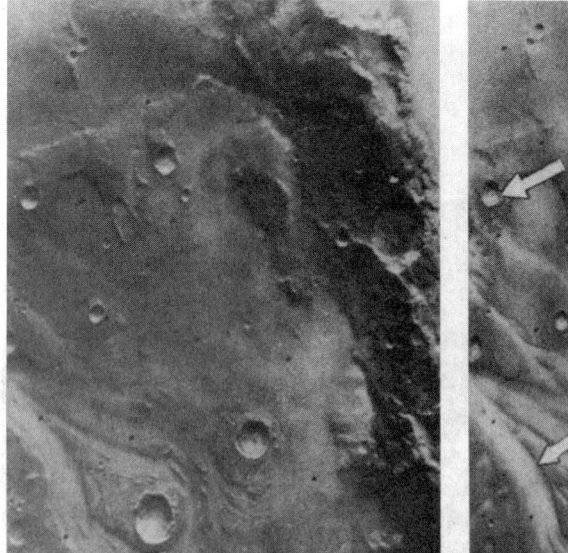

Fig. 18.   Two pictures taken a half-hour apart by *Viking 1 Orbiter* shows the development on Mars of early morning fog in low spots, such as crater and channel bottoms (see arrows on view at right). The scene at left was photographed shortly after martian dawn on July 24, 1976 from 12,400 kilometers (7700 miles) and, at right, 30 minutes later from 9800 kilometers (6100 miles). Slight warming of the sub-zero surface by the rising sun evidently drove off a small amount of water vapor which recondensed in the colder air just above the surface. Brightness measurements of the resulting fog patches indicated that a film of water about one micrometer thick had condensed. These fog patches were the first direct, visible evidence as to where the exchange of water between the martian surface and atmosphere may occur. (*NASA; Jet Propulsion Laboratory, Pasadena, California.*)

Fig. 19.   Operation of the surface sampler in obtaining martian soil was closely monitored by one of the Lander cameras because of the precision required in trenching a small area (8 × 10 inches; 20 × 25 centimeters) surrounded by rocks. The exposure of thin crust appeared in unique contrast with surrounding materials and became a prime target for organic analysis in spite of potential hazards. The large rock in the foreground is only 8 inches (20 centimeters) high. At left, the sampler scoop has touched the surface, missing the rock at upper left by a comfortable 6 inches (15 centimeters), and the backhoe had penetrated the surface about 0.5 inch (13 millimeters). The scoop was then pulled back to sample the desired point and (second view) the backhoe furrowed the surface, pulling a piece of thin crust toward the spacecraft. The initial touchdown and retraction sequence was used to avoid a collision between a rock (in the shadow of the arm) and a plate joining the arm and scoop. The rock was cleared by 2 to 3 inches (5 to 7.5 centimeters). The third picture was taken 8 minutes after the scoop touched the surface and shows that the collector head has acquired a quantity of soil. With the surface sampler withdrawn (right), the foot-long (0.3-meter) trench is seen between the rocks. The trench is 3 inches (7.5 centimeters) wide and about 1.5 to 2 inches (3.8 to 5 centimeters) deep. The scoop reached to within 3 inches (7.5 centimeters) of the rock at the far end of the trench. Penetration appears to have left a cavernous opening roofed by the crust and only about one inch (2.5 centimeters) of undisturbed crust separates the deformed surface and the rock. (*NASA; Jet Propulsion Laboratory, Pasadena, California.*)

particularly influencing the cycles of CO_2, of water, and of dust (an important factor in the planet's current climate.) These cycle have been plotted by several researchers and are contained in the Haberle reference listed.

Obvious major differences in the weather of the two planets cast doubts on attempting to compare Earth models with those of Mars. The thin atmosphere of Mars today eliminates any consideration of a greenhouse effect. The absence of oceans eliminates the Earth's familiar hydrologic cycle. As pointed out later, these major differences in the way the martian atmosphere and "weather" system has evolved helps to explain some of the very unusual geologic features found in the Viking images.

Soil and Rocks

Nergal, Ares, and Mars were legendary names for a pinpoint of reddish light in the night sky, observed to move relative to the star field even in ancient times. Because of its color, Mars was an important part of the mythology of early civilizations, serving as an abode for gods of fire, war and terror in the minds of many populaces through the centuries. The ancients would not have been disappointed in the coloring (reddish-brown) of most of the Martian soil and rocks.

Steps taken in the operation of the surface sampler on the *Viking* landers is shown in Fig. 19. As pointed out by the x-ray analysis team,[4] elemental analyses of fines in the martian regolith at the two widely separated landing sites were remarkably similar. At both sites, the uppermost regolith was found to contain abundant silicon and iron, with significant

[4] Scientists with Martin-Marietta Aerospace Corp., NASA Langley Research Center, Pomona, College, the University of New Mexico, and the U.S. Geological Survey.

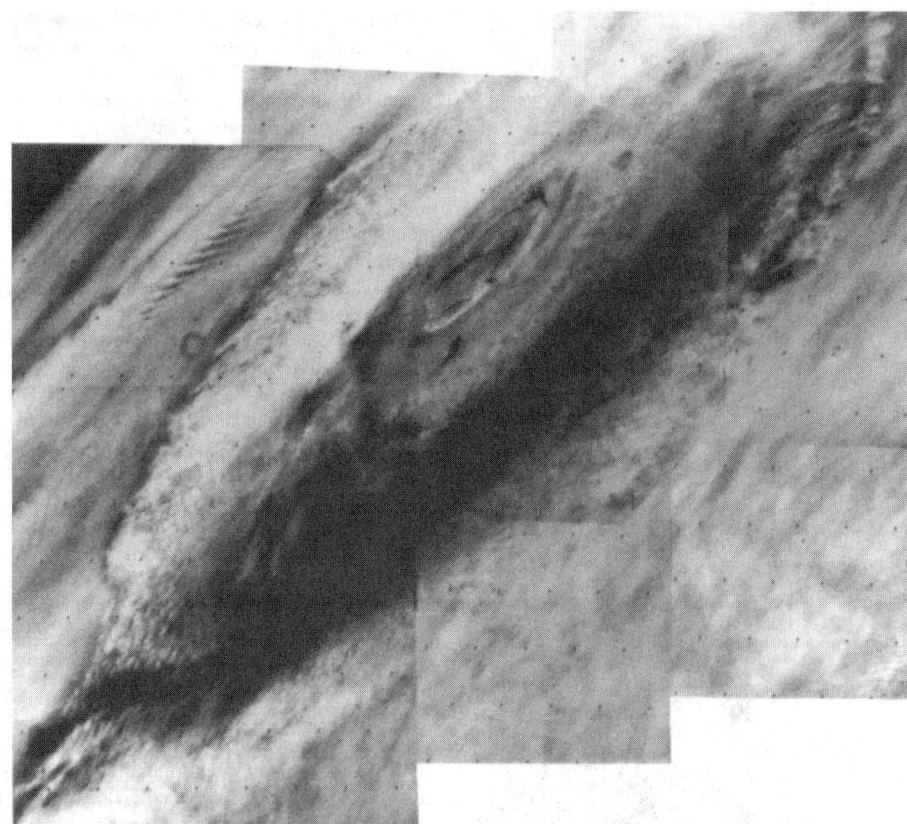

Fig. 20.   The great Martian volcano Olympus Mons was photographed by the *Viking 1 Orbiter* on July 31, 1976 from a distance of 8000 kilometers (5000 miles). The 24-kilometer-high (15-miles) mountain is seen in mid-morning, wreathed in clouds that extend up the flanks to an altitude of about 19 kilometers (12 miles). The multi-ringed caldera (volcanic crater), some 80 kilometers (50 miles) across, pushes up into the stratosphere and appears cloud-free at this time. The cloud cover is most intense on the far western side of the mountain. A well-defined wave cloud train extends several hundred miles beyond the mountain (upper left). The planet's limb can be seen at the upper left-hand corner of the view. It also shows extensive stratified hazes. The clouds are thought to be composed mainly of water ice condensed from the atmosphere as it cools while moving up the slopes of the volcano. In the martian afternoon, the clouds develop sufficiently to be seen from earth and it is known that they are a seasonal phenomenon largely limited to spring and summer in the northern hemisphere of the planet. Olympus Mons is about 600 kilometers (375 miles) across at the base and would extend from San Francisco to Los Angeles. (*NASA; Jet Propulsion Laboratory, Pasadena, California.*)

concentrations of magnesium, aluminum, sulfur, calcium, and titanium. The sulfur concentration is one to two orders of magnitude higher, and potassium (<0.25% by weight) is at least five times lower than the average for the earth's crust. The trace elements strontium, yttrium, and possibly zirconium, were detected at concentrations near or below 100 parts per million. Pebble-sized fragments at VL-1 site were found to contain more sulfur than the bulk fines, and were thought to be pieces of a sulfate-cemented duricrust. It is interesting to note that no phosphorus was found on Mars.

Each *Viking* lander carried an energy-dispersive X-ray fluorescence spectrometer for elemental analysis of samples of the Martian surface. This composition is best interpreted as representing the weathering products of mafic igneous rocks. A mineralogic model, derived from computer mixing studies and laboratory analog preparations, has suggested that the martian fines could be an intimate mixture of about 80% iron-rich clay; about 10% magnesium sulfate (kieserite perhaps), about 5% carbonate (calcite?) and about 5% iron oxides (hematite, magnetite, maghemite, goethite?). The mafic nature of the fines, which appear to be distributed globally, and their probable source rocks seem to preclude large-scale planetary differentiation of an earthly nature. The samples were characterized by abundant red-colored fine material and scattered blocks of generally angular rocks. More diversity was found at VL-1 than at VL-2.

Terrain

The Martian terrain in the vicinity of the two *Viking* lander sites is well illustrated in latter part of this article. Of course, these sites represented only a portion of the planet. There are ten or more volcanoes prominent on Mars, with scientific estimates of their age ranging from 100 million years to 1 billion years. See Figs. 20 and 21.

Evidence of erosion, including dry channels resembling riverbeds and tributaries, has led many analysts to conclude that Mars may have had a warmer, more water-rich climate in the past. Photographic evidence from

spacecraft indicates that the once-reported "canals" are mostly illusory and that the dark patchy markings once suspected to be vegetation along these canals, varying with the seasons, are in reality simply deposits of wind-blown dust that may be altered from time to time. That alterations can and do occur on Mars is shown by Fig. 22. With the benefit of a few years to assimilate observational data from *Mariner* and the *Viking* missions, the findings of Pieri et al. (1980) are exceptionally interesting. As explained by Pieri, *Mariner 9* orbital reconnaissance mission discovered ubiquitous valley networks in heavily cratered terrain. Branching valley networks throughout the heavily cratered terrain of Mars exhibit no compelling evidence for formation by rainfall-fed erosion (one of the popular hypotheses). Rather, the networks are diffuse and inefficient, with irregular tributary junction angles and large, undissected intervalley areas. The deeply entrenched canyons, with blunt amphitheater terminations, cliff-bench wall topography, lack of evidence of interior erosion of flow, and clear structural control, suggest headward extension by basal sapping.[5] The size-frequency distributions of impact craters in these valleys and in the heavily cratered terrain that surrounds them are statistically indistinguishable, suggesting that valley formation has not occurred on Mars for billions of years.

Pieri observes that the branching and coalescent character of the channels provoked immediate comparison to terrestrial riverine networks produced by fluvial[6] erosion, a process driven primarily by rainfall. Liquid water, however, cannot now exist at the surface of Mars for more than a few minutes owing to a very low atmospheric pressure and very cold

---

[5] The undercutting, or breaking away of rock fragments, along the headwall of a cirque (semicircular, amphitheater-like, or armchair-shaped hollow of nonglacial origin), due to frost action at the bottom of the bergschrund (a deep and often wide gap or crevasse).

[6] Of or pertaining to a river or rivers — produced by the action of streams.

Fig. 21.   Fine detail in the interior of a martian crater can be seen in this photo taken by *Viking 1* of an area near the *Viking 2* landing site. The crater (on the left margin of the view) is about 40 kilometers (25 miles) in diameter and shows many features found in lunar craters. The central portion is crossed by numerous cracks. Similar features are seen at the huge lunar impact basin, Orientale. Their origin is unknown, but it has been suggested that the cracks were formed either by consolidation of lava that filled the crater after it formed, or by fallback from the impact process. Alternatively, the cracks may have formed long after the impact event by uplift of the crater floor. Between the cracked terrain and the crater rim is a region of chaotic debris. Beyond the rim there is no evidence of an ejecta blanket (rock material which is blasted from the crater by the shock of the impacting meteorite). The ejecta blanket is presumably overcovered by later deposits. (*NASA; Jet Propulsion Laboratory, Pasadena, California.*)

temperatures during most of the year at most latitudes. It has been suggested that these features, if formed by processes similar to those that operate in the formation of earthly river systems, are relics of a more clement epoch. Thus, a major problem in the study of Mars is whether the valley networks could have evolved under current surface conditions, or whether a major shift in Martian climate occurred.

There are certain unifying characteristics of Martian valleys. (Valleys are distinguished from channels by the absence in the former of direct evidence of fluid erosion often found in the latter.) There is no clear evidence (streamlined obstacles, interior channels) of direct fluid erosion in any Martian valley. It is possible that such features are too small to be observed by present instrumentation, although *Viking Orbiter* images as small as 100 meters (328 feet) can be resolved. Walls of the valleys are typically rugged and clifflike, with some debris accumulation and talus, and the floors are generally flat. Mantling by materials of eolian and volcanic origin is common. Some valleys display cliff-bench interior topography, similar in character and scale to features in the Grand Canyon of the Colorado River. The most striking morphological characteristics, however, are the presence of steep-walled, cuspate terminations at the heads of the smallest tributary valleys. These steep-walled, amphitheater terminations suggest headward extension (sapping) by basal undermining and wall collapse, as in the predominant mode of headward extension for many earthly canyons. A variety of Martian terrain is shown in Figs. 23, 24, 25, and 26.

Martian valley networks lack the dendritic pattern so common to terrestrial streams. The Martian valley patterns show remarkable parallelism and lack of tributary competition for undissected intervalley terrain, and thus appear diffuse compared to terrestrial systems. Viewed from spacecraft, terrestrial drainage systems have a fine, filigreed texture, whereas Martian systems appear coarse.

Pieri has suggested that the valleys were formed on Mars during an ancient epoch by erosional processes involving not rainfall, but the movement of groundwater and its participation as a liquid or a solid in the undermining of less competent strata, causing progressive headward collapse. These processes, combined with modification by impact and eolian (wind) processes, have produced the degraded valleys seen on Mars today.

Even a brief description of the physical features of Mars is not complete without mention of the so-called "Spokane Flood" concept. As summarized by Baker (1978), in a series of papers published between 1923 and 1932, J.H. Bretz described an enormous plexus of proglacial stream channels that eroded into the loess and basalt of the Columbia Plateau in eastern Washington state. Bretz argued that this region (which he termed the Channeled Scabland) was the product of a cataclysmic flood, which he called the Spokane flood. Considering the nature and vehemence of the opposition to his hypothesis, which at one time was considered highly imaginative, its eventual scientific verification constitutes a fascinating episode in the history of modern science. The discovery of possible catastrophic flood channels on Mars has given new relevance to Bretz's insights. The connection between Bretz's proposal and parts of the Martian surface is well developed by Baker.

Volcanism. As readily apparent from *Viking* images, volcanism on Mars was widespread. According to Lucchitta (U.S. Geological Survey), volcanism has formed enormous shields, large composite cones, lobate lava flows, and possibly small cones and pseudocraters. Flood basalts similar to those filling lunar maria may have resurfaced ridged highland plateaus. Large deposits of pyroclastic material may also exist, although their presence is controversial. Dark patches are common on the planet as they are on Earth's moon, where they have been interpreted as pyroclastic materials. The association of dark patches with pyroclastic volcanism on Mars has been largely overlooked, because most dark patches are inside craters and were obviously accumulated by wind; the possibility was neglected that some Martian dark patches, like lunar ones, may reflect pyroclastic vents. Lucchitta describes dark patches in Valles Marineris that may be such vents and may reflect young mafic volcanism. The evidence for past volcanism on Mars is commonly accepted, but none has been documented in the Valles Marineris equatorial rift system. A recent survey of the troughs in this valley revealed dark patches that are interpreted to be volcanic vents. The configuration and association of these patches with tectonic structures suggest that they are of internal origin; their albedo and color ratios indicate a mafic composition; and their stratigraphic position, crispness of morphologic detail, and low albedo imply that they are young, perhaps even recent. If this volcanism is indeed as young as it seems, Mars has been an active planet throughout most of its history.

Case for an Early "Wet" Mars. Certain features of the Martian surface, as observed by the Viking missions, have continued to intrigue and confuse analysts of the data returned from Mars. Included are ancient valleys, channels, and what appear from images to be tributary systems. What appear to be numerous extinct volcanoes and meteoritic craters over which more recent geologic features have been superposed are found in the images. For many of these features, the presence of water in relatively large quantities on the planet during its earlier phases offers the most tempting solution. Many scenarios have been developed. For example, some scientists at the Jet Propulsion Laboratory (Pasadena, California) suggested, at a symposium of the Lunar and Planetary Institute (see reference listed), that lakes, or a shallow sea or ocean, may have encompassed as much as 10 to 15% of the Martian surface and of a generous portion of the northern plains from 2 to 3 billion years ago. Timothy Parker (JPL) estimates that the water would have been about 100 meters deep (or less), making the sea's volume equivalent to a layer of about 10 meters deep covering the globe. This volume of water in a hypothetical sea would equal all the water that some geochemists have allowed for the entire planet. But, all do not agree. Michael Carr (U.S. Geological Survey) reported that the latest estimates of the amount of water hidden beneath the surface may be several times greater. Some investigators believe that the surface of Mars, something comparable to Earth's moon, is made up of rubble and porous soil well capable of

Fig. 22.   Changes observed by the *Viking Landers* over a period of time included water-ice snow seen by VL-2 during the winter at Utopia Planitia, and a thin dust layer deposited at both sites during the dust storms of 1977. As shown here, a change occurred by Chryse Planitia over a 4-day period in September 1978. Top photo is the "before" and bottom photo is the "after" view. Change (A) appears as a small circle-like formation on the side of a drift in the lee, or downward, side of "Whale Rock." This is believed to have been a small-scale landslide of an unstable dust layer which had accumulated behind the rock. Interpretation of this feature would be difficult without an earlier change (B) near "Big Joe," a slump. The new slump is observed approximately 25–35 meters (82–115 feet) from the lander craft and just under 1 meter (3.3 feet) across. This slumping was probably initiated by the daily heating and cooling of the surface by solar radiation. More importantly, it is now believed that, based upon the repeated occurrence of such slumping features, a dust layer which overlies the surface may, in fact, be redistributed fairly regularly during periods of high wind activity. (*NASA; Jet Propulsion Laboratory, Pasadena, California.*)

storing ice, water, or brine. Stephen Clifford (Lunar and Planetary Science Institute) estimates that this megaregolith has the capacity to hold water equivalent to a global layer 200 to 500 meters deep. The main remaining question, of course, is how much of that capacity is actually filled? Certainly, the unanswered question of the amount of water that was and still may be trapped on the planet is central to preparing a satisfactory model of the planet.

Age Determination. Mars has been mapped extensively by *Mariner 9* and later by the *Viking* missions. One major goal in planetary science is to determine the chronology of development of the surfaces of the terrestrial planets, particularly Mars. As indicated in an excellent paper by Neukum and Wise (1976), cratering links to lunar time suggest that Mars died long ago. Fortunately, for the purpose of age determination from photographs, Mars is impact-cratered. Differences in impact crater frequencies at different sites reflect differences in age. Two attempts have been made to determine absolute age for Mars from its measured crater frequencies, based on extrapolations from the cratering chronology of the lunar surface (Hartman, 1973; Soderblom et al., 1974). Unfortunately, a straightforward comparison of martian and lunar crater frequencies does not necessarily yield true ages; relative impact rates and the time dependence of the martian cratering rate are not known; and it is not certain whether the same meteoroid population bombarded both planets.

At *Mariner 9* resolution, the impact crater production size-frequency distribution of Mars is generally similar to that of the moon for crater diameters in the range 0.8 to 50 kilometers (0.5 to 31 miles), and it appears to have been relatively stable through time. The lunar and Martian crater curves can be brought into near coincidence by a diameter shift appropriate to reasonable impact velocity differences between bodies hitting Mars and the moon. This indicates that a common population of bodies impacted both planets and suggests the same or a very similar time dependence of impact flux. Constraints on relative lunar and Martian fluxes can be obtained by comparing crater frequency data for the lunar and martian highlands and for Mars' satellite Phobos.

These cratering constraints, as pointed out by Neukum and Wise, provide the basis for a tentative Martian time scale derived from lunar data. Previous time scales have painted a picture of a disorderly planetary evolution of Mars, punctuated by a strange pulse of Tharsis Ridge tectonic and volcanic activity late in geological history. The new scale suggests a much more orderly evolution with Mars, like the moon, winding down most of its major planetary tectonic and volcanic disturbances in the first 1.5 billion years of its history. By 2.5 billion years ago the volcanic-tectonic era on Mars had ended.

Other Physical Characteristics of Mars

Doppler radio-tracking data have provided detailed measurements for a Martian gravity map extending from 30°S to 65°N latitude and through 360° of longitude. The feature resolution is approximately 500 kilometers (310 miles), revealing a huge anomaly associated with Olympus Mons, a mascon in Insidis Panitia, and other anomalies correlated with volcanic structures. Olympus Mons has been modeled as a disk of 600-kilometer (372-mile) surface area, having a mass of 9.7×10^{21} grams. The very

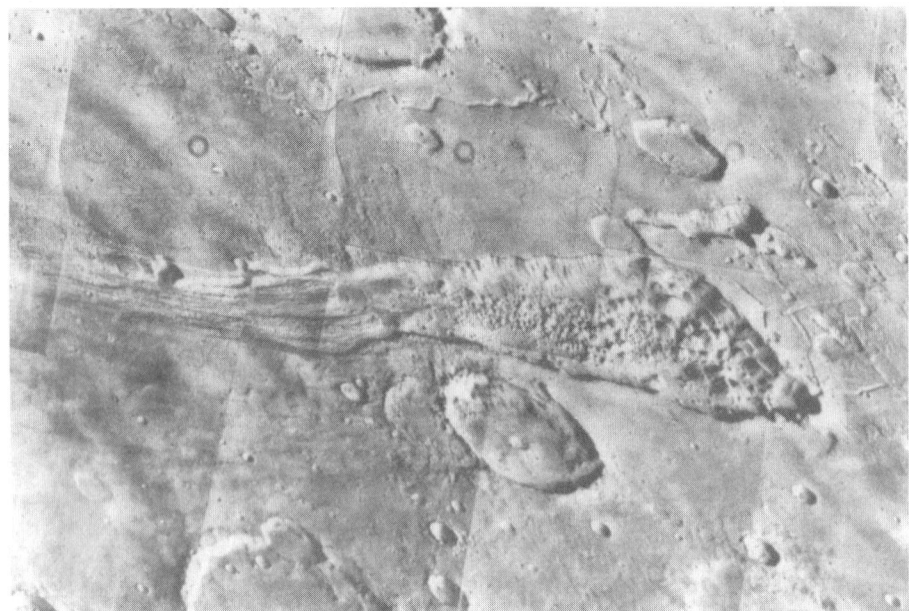

Fig. 23.   The sinous rille (a relatively long, narrow, trench- or cracklike valley) at the top of this mosaic of 8 photos is believed by some scientists as indicative of flooding of the high plateau in the vicinity of an alternative landing site (known as Capri) for *Viking Lander 2*. In the foreground is a valley that may have been caused by downfaulting of the martian crust. The hummocks (rounded or conical knolls or mounds of comparatively small elevation) on the valley floor look like chaotic terrain. Some scientists have suggested that the subsidence may be partially caused by melting of the subsurface ice. The large areas of the collapsed terrain show the regional extent of this phenomenon. These views were taken by *Viking 1* on July 3, 1976 from a range of 2300 kilometers (1400 miles) and cover an area of about 300 × 300 kilometers (180 × 180 miles). South is toward the top as seen from the spacecraft. (*NASA; Jet Propulsion Laboratory, Pasadena, California.*)

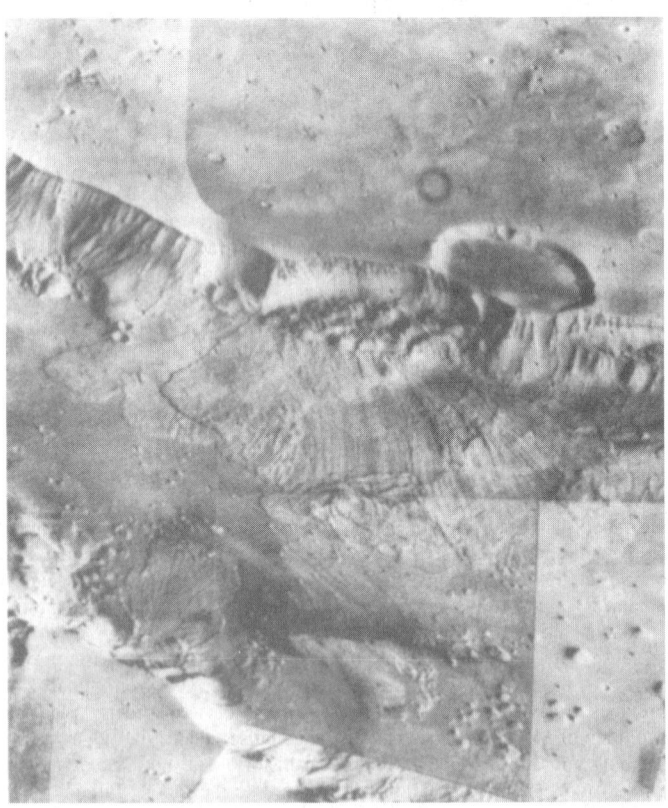

Fig. 24.   View taken by *Viking 1* on July 3, 1976 from a range of 2000 kilometers (1240 miles), looking southward across Valles Marineris. This huge equatorial canyon is about 2 kilometers (1.2 mile) deep. The area shown is 70 kilometers (43 miles) by 150 kilometers (94 miles). Aprons of debris on the canyon floor indicate how the canyon may have enlarged itself. The walls appear to collapse at intervals to form huge landslides that flow down and across the canyon floor. Linear striations on the landslide surface show the direction of flow. On the canyon's far wall in this view, one landslide appears to have ridden over a previous one. Streaks on the canyon floor, aligned parallel to the length of the canyon, probably are evidence of wind action. Layers in the canyon wall indicate that the walls are made up of alternate layers of lava and ash or wind-blown deposits. (*NASA; Jet Propulsion Laboratory, Pasadena, California.*)

large Olympus Mons anomaly should have a very significant impact on geophysical modeling of the planet. Similarly, the Elysium anomaly and the Insidis mascon should place constraints on the internal structure. Gravity in the southern hemisphere remains poorly resolved.

A three-axis short-period seismometer was delivered to the surface of Mars by *Viking Lander 2* on September 3, 1976. Noise background correlated well with wind gusts. Data returned to earth indicated that Mars is a very quiet body.

The amounts of magnetic particles held on the reference test chart and backhoe magnets on *Viking Landers 1* and *2* were comparable, indicating the presence of an estimated 3–7% (weight) of relatively pure, strongly magnetic particles in the soil. It is argued that the results indicate the presence, now or originally, of magnetite, which may be titaniferous.

Dust Devils on Mars. Several scientists, after studying Viking data, have reported the existence of dust devils (columnar, cone-shaped, and funnel-shaped clouds rising 1 to 6 km above the surface) on Mars. Dust devils result from atmospheric conditions that occur close to the ground and are, therefore, sensitive to surface topography. Dust devils on Mars may be responsible for the initiation of large dust storms on the planet and for increasing the general atmospheric dust content.

Dust devils, as observed by Thomas and Gierasch (Cornell University), have meteorological as well as geological significance. Fluid motions in an atmospheric boundary layer can be driven either by stresses due to the mean wind (forced convection) or by buoyancy due to heating of the gas adjacent to the surface (free convection). Dust devils are an example of the latter. On Earth, large-scale eolian transport is generally due to forced convection. The investigators report that moderate to high winds characterize forced convection, and on Mars, where the atmospheric density is only about 1% of that on Earth, it is estimated that winds must exceed about 25 to 40 meters sec^{-1} to initiate soil movement.

One of the major geologic processes on Mars is the entrainment and transportation of dust by winds. Observations on the genesis and development of local and global dust storms on Mars are sparse.

Tornadolike Tracks on Mars. Some images from the *Viking Orbiter* reveal well defined, dark filamentary lineations in numerous locations on the Martian surface. On Earth, tornadic-intensity vortices commonly leave distinctive tracks whose appearance is similar to that of the Martian lineations. A high-resolution imaging system, as proposed for the *Mars Observer* mission, could resolve these ground tracks. The filamentary lineations, as reported by Grant and Schultz (Brown University) are from 2 km to at least 75 km long and less than 1 km wide. Most are straight to

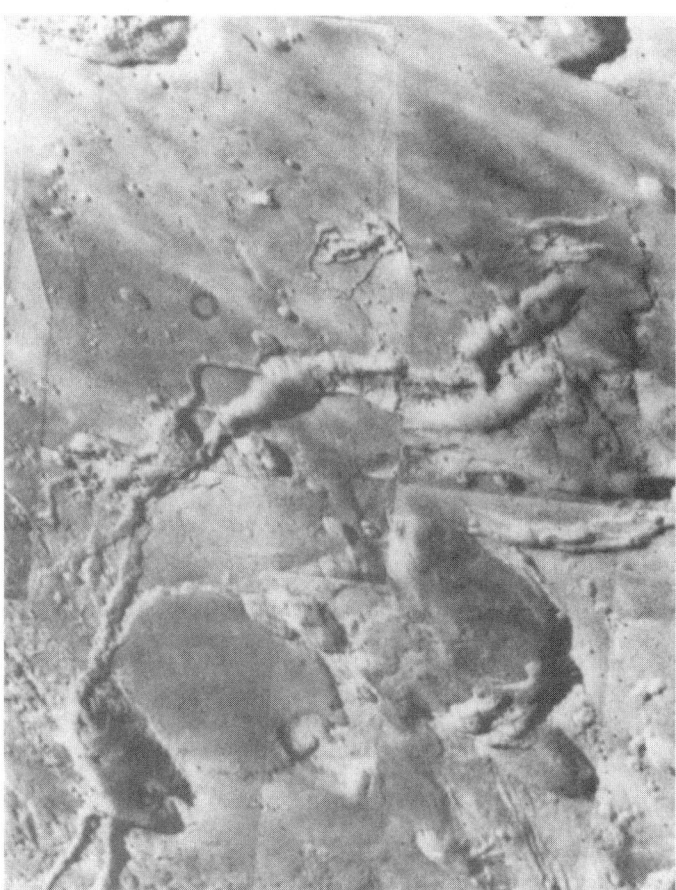

Fig. 25.   A view obtained by *Viking 1* on July 8, 1976 showing what appear to be fault zones in the martian crust in an area two degrees south of the equator. The fault valleys are widened by mass wasting and collapse. Mass wasting is the downslope movement of rocks due to gravity (possibly hastened by seismic shaking if present). (*NASA; Jet Propulsion Laboratory, Pasadena, California.*)

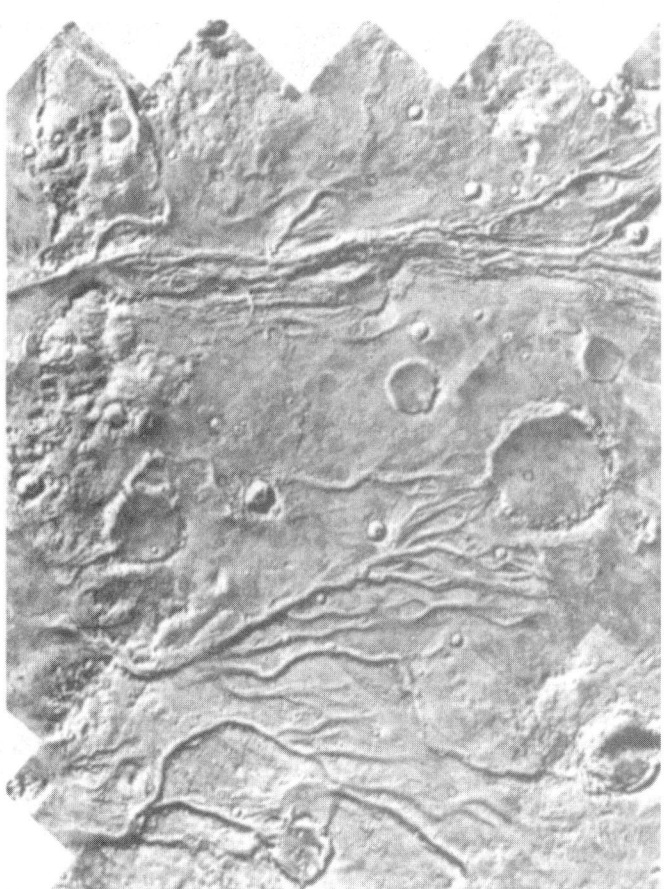

Fig. 26.   Mosaic of martian surface made by *Viking Orbiter 1* over a period between August 4 and 9, 1976. The area is centered at 17 °N, 55 °W, to the west of the Viking 1 Lander site in Chryse Planitia. Just to the west of this area are the plains of Lunae Planum. The terrain shown in this view slopes from west to east with a drop of about 3 kilometers (1.9 mile). The channels are a continuation of those to the west of the VL-1 landing site and, to some scientists, suggest a massive flood of waters from Lunae Planum, across this intervening cratered terrain, and into the general region of the VL-1 landing site. In several cases, it will be noted that channels cut through craters; in others, the craters are clearly later than the assumed flood and are superimposed in the channels. (*NASA; Jet Propulsion Laboratory, Pasadena, California.*)

curvilinear, and some have obvious nontopographically initiated gaps in their path. The visible occurrence of the lineations appears to be seasonal. In the southern hemisphere, they were visible (from *Mariner 9*) only from midsummer into early fall. After formation, they were rapidly modified and were no longer visible by midfall. In the northern hemisphere, lineations appear from early to midsummer. By late summer, these lineations also become smeared and faint.

Natural Laser Phenomenon Noted on Mars.   Based upon observations made with the Goddard infrared heterodyne spectrometer during the period of January to April 1980, when the planet was near opposition, astronomers M.J. Mumma and colleagues (NASA-Goddard Space Flight Center) and D. Zipoy (University of Maryland) noted natural gain amplification in the mesosphere of Mars, probably representing the first definite identification of a natural infrared laser. Natural microwave amplifiers (masers) are abundant in interstellar clouds and some circumstellar shells, primarily among the rotational level populations of certain molecules, such as OH, SiO, and H_2O, but no optical lasers in nature had previously been observed. As reported by Mumma et al., many examples of natural nonthermal optical emission have been found, such as the infrared and ultraviolet auroras or the day glows of Earth, Jupiter, Mars, and Venus. Details are reported in the reference listed.

Pole Wandering and Crustal Shifts on Mars

Careful study of *Viking* images has revealed a number of features of the planet that are very difficult to explain. For example, regions at the planet's equator seem once to have been near a pole. As observed by Schultz (see reference), in certain areas of the surface, erosion appears to have occurred at a very low rate (perhaps less than a millimeter in a million years). But, in other areas, at the same latitude, there are regions that have been heavily stripped and etched by the wind. Also, very old networks of narrow valleys, once cut into the surface by flowing water, suggest a warm climate, although such networks are seen within 10 degrees of the southern polar

ice cap. While many details remain to be worked out, Schultz suggests that one hypothesis may explain all or most of the contradictions: the orientation in space of the Martian crust has not always remained the same throughout geologic time, but rather, it has shifted with respect to the planet's axis of spin. This would require that the spin axis, which intersects the planet's surface at the north and south poles, would appear to have wandered over the planet's crust. This would indicate that certain regions of the crust, presently far from the poles, may have been at some time in the past within the polar regions. In introducing a detailed paper on this topic, Schultz observes that if Mars had undergone polar wandering, then martian geology may have to be viewed in the context both of a dynamically changing planet like the Earth and of a stable, rigid body like the Earth's moon. In this sense, the Martian equivalent of plate tectonics might simply be the movement of the entire lithosphere, the solid outer portion of the planet, as one plate.

Martian Satellites

Mars has two satellites, Phobos, the inner and larger of the two moons, and Deimos. These satellites were visited during the *Viking* missions.

Phobos.   This satellite revolves around Mars in an orbit of about 9330 kilometers (5800 miles) from the center of Mars (some 5950 kilometers; 3700 miles above the planet's surface). The diameter was stated in the introductory section of this entry. Its orbital period is 7 hours, 40 minutes. Because its orbital period is in the same direction as, but is less than, that of Mars, it rises in the west and sets in the east as seen from Mars. Phobos is heavily cratered and dark in color,

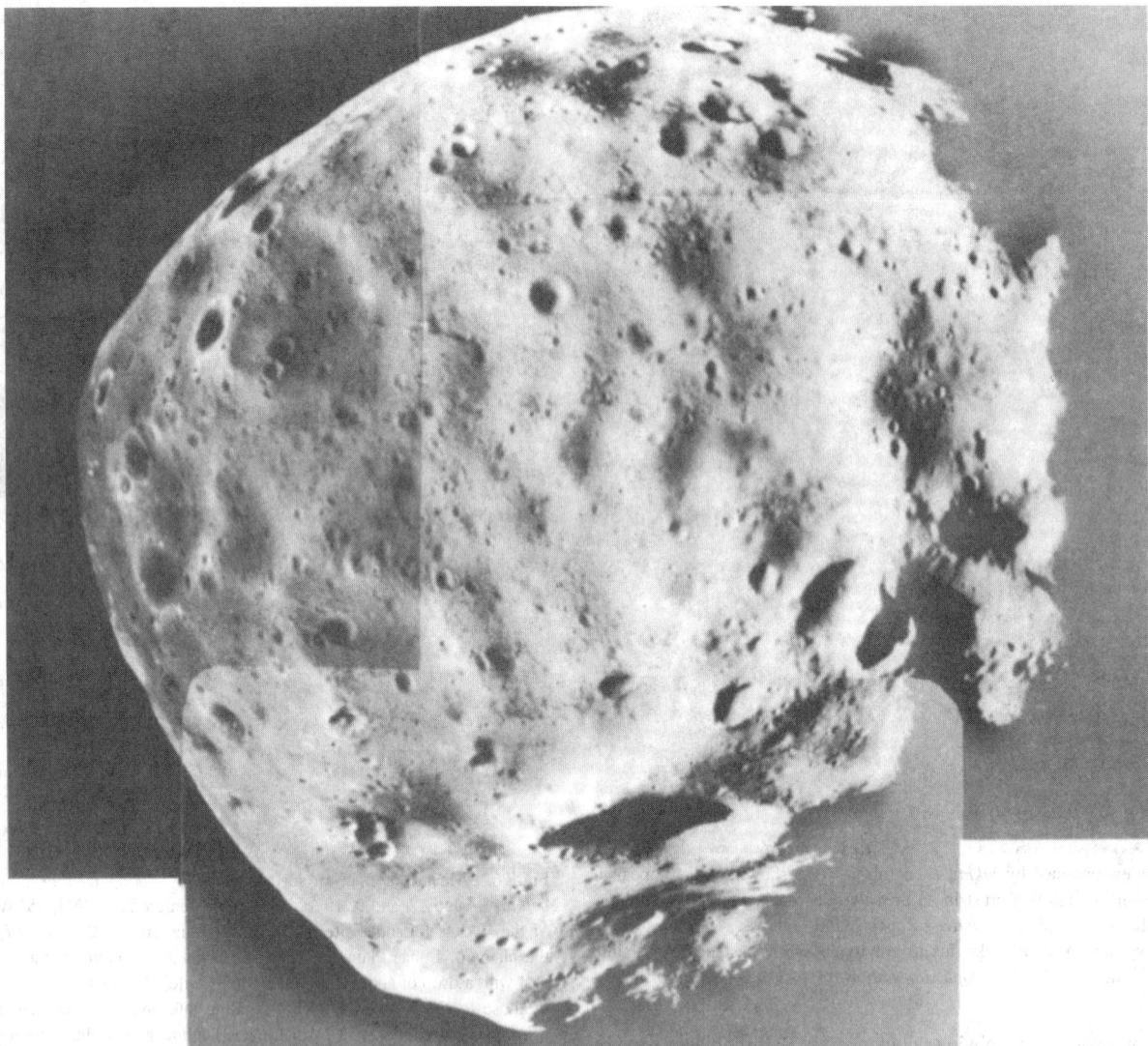

Fig. 27.   View of Phobos taken by *Viking Orbiter 1* from a distance of 480 kilometers (300 miles). This mosaic of 3 pictures was made in February 1977. As seen here, Phobos is about 75% illuminated and is about 21 kilometers (13 miles) across and 19 kilometers (11.8 miles) from top to bottom. North is at top. The south pole is within the large crater (Hall) with a diameter of 5 kilometers (3.1 miles) and will be noted at bottom center where the pictures overlap. Some features as small 20 meters (65 feet) across can be seen. Remarkable features include striations, crater chains, a linear ridge, and small positive features which appear to be resting on the surface. A long linear ridge is seen starting near the south pole and extending to the upper right. A very sharp wall at the intersection of two craters (about 1 kilometer; 0.6 mile across) is seen along this ridge at right. A series of craters runs horizontally in the picture which is parallel to the orbit plane of Phobos. These crater chains are commonly associated with secondary cratering by ejecta from larger impacts. A surprising discovery has been made of what apparently resembles hummocks or small positive features. These features, primarily seen near the terminator (right), are about 50 meters (165 feet) in size and may be surface debris from previous impacts. (*NASA; Jet Propulsion Laboratory, Pasadena, California.*)

of a material resembling carbonaceous chondrite meteorites. A system of grooves, possibly marking fractures, is associated with the largest crater, Stickner, which is about 10 kilometers (6.2 miles) across.

Viking Orbiter 1 flew within 480 kilometers (300 miles) of Phobos to obtain the view given in Fig. 27. A view much closer to the satellite is given in Fig. 28. A considerably later view, made in 1978, is given in Fig. 29.

Deimos. This satellite revolves around Mars in an orbit about 23,000 kilometers (14,260 miles) from the center of Mars. Five close flybys, within 1000 kilometers (620 miles) of Deimos, were made in October 1977. The closest encounter was on October 5, 1977 when the spacecraft passed within 50 kilometers (30 miles) of the moon's surface. Images indicated that the surface of Diemos differs considerably from that of Phobos. Deimos has many craters, but appears smoother than Phobos. See Fig. 30. With reference to the peculiar blocks observed on Deimos, which are visible in the illustration, Duxbury and Veverka (1978) suggest: "If the bright patches and blocks represent ejecta, then it is puzzling why apparently so much of it was retained by such a small satellite and why the process seems to be so much more efficient on Deimos than on Phobos. It is conceivable that the very close proximity of Phobos to Mars makes it

easier for impact ejecta to escape from the inner satellite, but the mechanics of such a preferential process remain to be worked out."

Illustrations of Phobos and Diemos are also given in the entry on **Asteroid.**

Additional Post-Viking Mission Studies and Hypotheses

Further studies of the Viking information and observations made from Earth in recent years have posed interesting new questions pertaining to Mars.

Radar Images of Mars. In late 1991, D.O. Muhleman and a team of researchers (California Institute of Technology) conducted aperture synthesis mapping of Mars by using the Very Large Array (VLA) in New Mexico as the imaging instrument to detect continuous wave signals transmitted at 9.5 GHz (3.5 cm) from the Jet Propulsion Laboratory (JPL) 70-meter antenna in Goldstone, California. See also **Antenna (Communications).** Summary of the project: "The surface of Mars was illuminated with continuous wave radiation. The reflected energy was mapped in individual 12-minute snapshots with the VLA in its largest configuration; fringe spacings as small as 67 km were obtained. The images reveal near-surface features, including a region in the Tharsis volcano area,

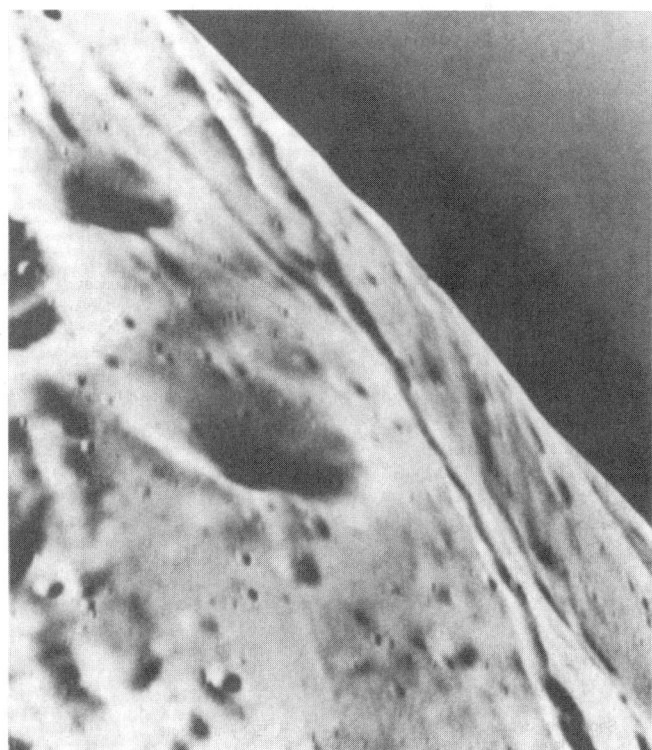

Fig. 28. *Viking Orbiter 1* took this close-up picture of Phobos from a range of 120 kilometers (75 miles) on February 20, 1977. This is the closest range at which a spacecraft has photographed the tiny satellite. At that range, Phobos is too large to be captured in a single frame. This picture covers an area 3 × 3.5 kilometers (1.86 × 2.17 miles). A single picture element is about 3 meters (7.5 feet) across. However, the high relative speed of Orbiter 1 and Phobos caused some image smear so that the smallest surface feature identifiable is between 10 and 15 meters (32 and 49 feet). The picture shows a region in the northern hemisphere of Phobos that has striations and is heavily cratered. The striations, which appear to be grooves rather than crater chains, are about 100 to 200 meters (328 to 656 feet) wide and tens of kilometers long. Craters range in size from 10 meters (32 feet) to 1.2 kilometers (0.75 mile) in diameter. The surface of Phobos appears similar to the highland regions of the earth's moon, which also is heavily cratered and an ancient terrain. The dark region above the limb of Phobos is an artifact of processing and does not indicate an atmosphere. (*NASA; Jet Propulsion Laboratory, Pasadena, California.*)

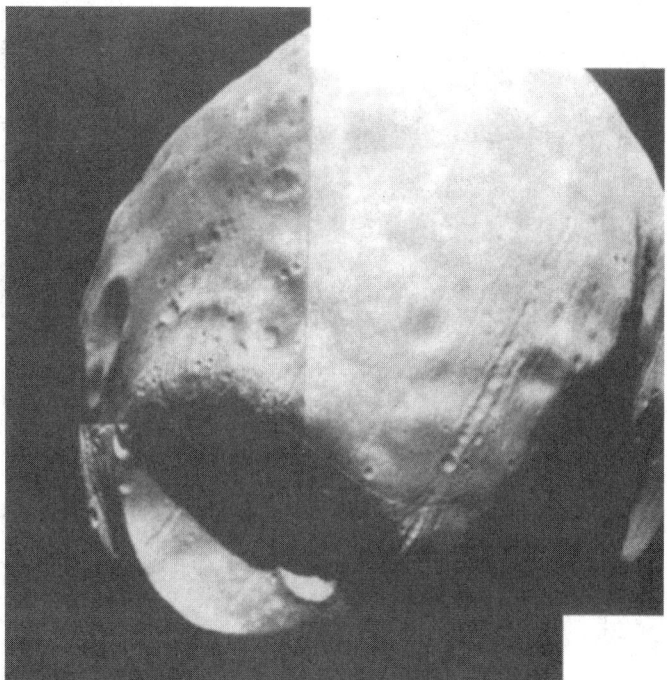

Fig. 29. This view of Phobos was made by *Viking Orbiter 1* on October 19, 1978 at a range of 612 kilometers (379 miles) during the spacecraft's 854th revolution of Mars. This view was made just before Phobos entered the shadow of the planet. The photomosaic shows the front side of Phobos which always faces Mars—from about 10° below the equator with north at the top. Stickney, the largest crater on Phobos (10 kilometers; 6.2 miles across) is at the left near the morning terminator. Linear grooves coming from and passing through Stickney appear to be fractures in the surface caused by the impact which formed the crater. Two earlier new encounters with Phobos brought Viking Orbiter 1 within close range of the satellite, but had not provided scientists with good opportunities to observe Stickney as well. This view provides new high-resolution coverage of the front side of Phobos (approximately 19 × 22 kilometers; 11.8 × 13.6 miles as seen here) as well as the highest resolution yet achieved of the western wall of Stickney. Kepler Ridge is casting a shadow in the southern hemisphere which partially covers the large crater (Hall) at the bottom. (*NASA; Jet Propulsion Laboratory, Pasadena, California.*)

over 2000 km in east-west extent, that displayed no echo to the very low level of the radar system noise. This feature (called *Stealth*) is interpreted as a deposit of dust or ash with a density less than about 0.5 grams/cubic centimeter and free rocks larger than 1 cm across. The deposit is envisioned to be several meters thick and may be much deeper. The strongest reflecting geological feature was the south polar ice cap, which was reduced in size to the residual south polar ice cap at the season of observation. The cap image is interpreted as arising from nearly pure carbon dioxide or water ice with a small amount of Martian dust (less than 2 percent by volume) and a depth greater than 2 to 5 meters. Only one anomalous reflecting feature was identified outside of the Tharsis region, although the Elysium region was poorly sampled in this experiment and the north pole was not visible from Earth." More detail is given in reference listed.

Radar Detection of Phobos. During the exceptionally close approach of Mars to Earth in the autumn of 1988, the Goldstone 70-meter antenna was used as a radar telescope to observe Phobos. A total of 117 transmit/receive cycles were completed. Radar echoes from the Martian satellite provided information about the object's surface properties at scales near the 3.5-cm observing wavelength. In summary, "Phobos's surface apparently resembles those of many (if not most) large, C-class asteroids in terms of bulk density, small-scale roughness, and large-scale topographic character, but differs from the surfaces of the moon and at least some small, Earth-approaching objects. Additional 3.5-cm and 13-cm radar observations of asteroids, comets, and the martian satellites can clarify these relations."

Simulating the Surface of Phobos. Researchers (University of Arizona, Lunar and Planetary Laboratory) in recent years have been attempting to simulate how certain distinct features of Phobos may have been formed and as imaged by the Viking orbiters. Lines (rows) of comparatively small craters resemble a beaded chain unlike other features found in the planetary system thus far. An initial hypothesis described the features as being formed out of secondary ejecta—that is, debris resulting in a crater-causing impact. In attempting to duplicate the unusual feature in the laboratory, the researchers have developed an apparatus consisting of a pair of narrow, rigid glass plates and a variety of materials, including expanded vermiculite, silica sand, and small glass spheres. Intense interest in unusual surface conditions is not new in connection with planetary space explorations. Ponder, for example, the variety of scientific opinions that were expressed prior to touchdown of *Surveyor*, the first unmanned spacecraft to land on the Moon. From their work to date, the researchers have suggested that Phobos may be covered with a regolith some 300 meters thick!

A Peopled Mission to Mars. Although, in early 1990s, it was delayed for want of funding and other political considerations (due in part to some lack of interest on the part of the public) an *Apollo*-type mission to Mars is in advanced planning phase. Conservative scientists have suggested, however, that another *Viking*-type venture may be the safest and most sensible step prior to putting human lives at risk. Probably the most intense interest in a fully blown venture stemmed from Russian scientists, who have been preparing for a "Soviet" Phobos mission. A major concern has been and continues to be that of prolonged crew interest and mental and physical reactions to a sojourn in space that would require a currently calculated minimum of 15 months from launch to return on Earth. This has

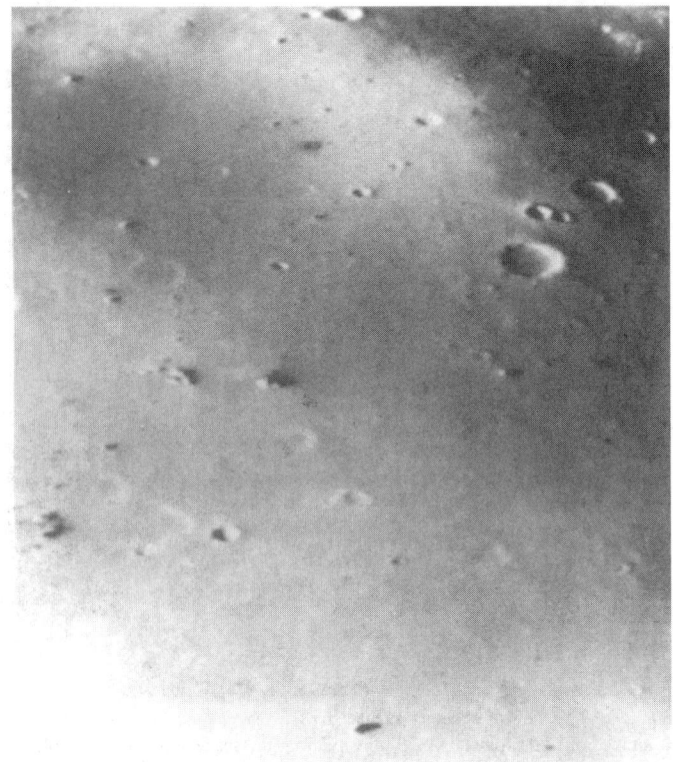

Fig. 30. View of the Martian moon Deimos taken on October 15, 1977 when *Viking Orbiter* 2 passed within 50 kilometers (30 miles) of the satellite. The picture covers an area 1.2 × 1.5 kilometers (0.74 × 0.93 mile) and shows features as small as 3 meters (10 feet). Deimos is saturated with craters, but a layer of dust appears to cover craters smaller than 50 meters (165 feet) in diameter, making Deimos look smoother than the other Martian moon, Phobos. Boulders as large as houses (10 to 30 meters; 33 to 100 feet) across are strewn about the face. It is suggested that these objects may be blocks ejected from nearby craters. The spacecraft would have been clearly visible to an observer standing on the surface of Deimos. (*NASA; Jet Propulsion Laboratory, Pasadena, California.*)

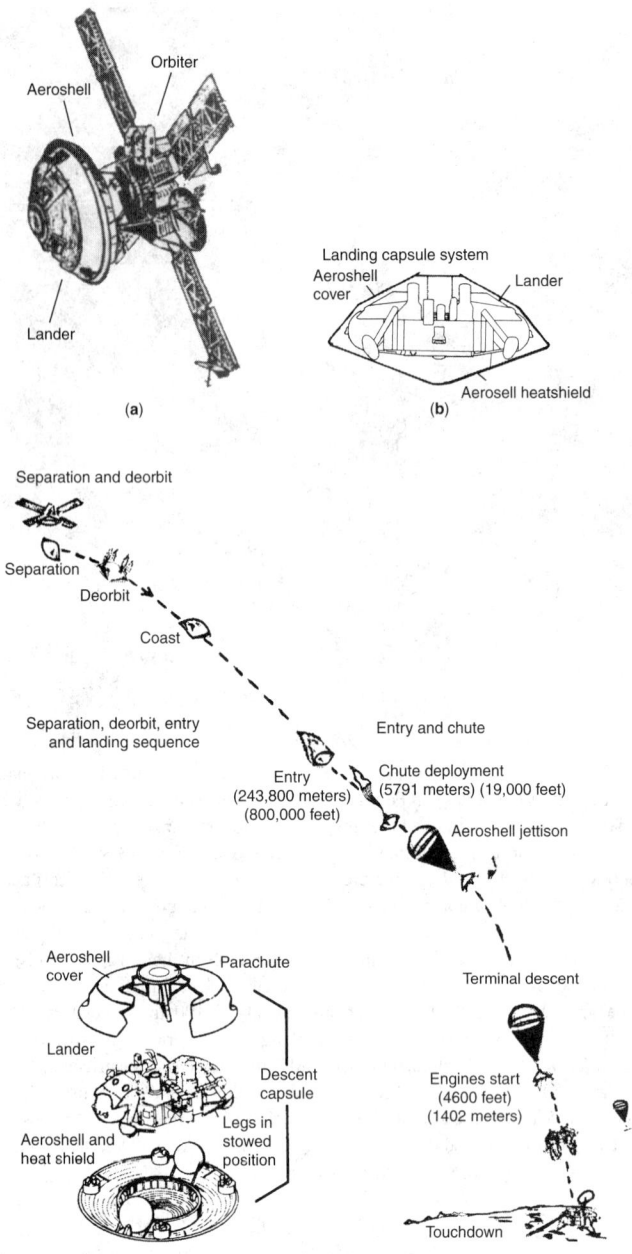

Fig. 31. Principal subsystems of the *Viking* spacecraft: (**a**) Orbiter and Lander linked together as they travel through space; (**b**) landing capsule system; (**c**) aeroshell cover, parachute, and descent capsule; and (**d**) separation, deorbit, entry, and landing sequence. (*NASA; Jet Propulsion Laboratory, Pasadena, California.*)

been referred to as the "Sprint" mission. For example, if it were assumed that the mission would leave Earth on 19 November 2004, it would reach at Mars on 30 July 2005 and depart Mars on 20 August 2005, returning to Earth on 2 February 2006. Time of stay on Mars would be less than 1 month and fuel costs would be at a maximum. A much longer (31 months) mission could be much more fuel efficient, and the stay on Mars could be considerably longer.

The Viking Missions to Mars

Two identical spacecraft, *Viking 1* and *Viking 2*, were launched in 1975 to explore Mars. In actuality, there were four spacecraft in all—a *Viking 1* orbiter and lander and a *Viking 2* orbiter and lander. Each orbiter and lander traveled together as one unit to rendezvous with Mars. The principal subsystems of the Viking spacecraft and separation, deorbit, entry, and landing sequences are shown in Fig. 31. The principal *Viking* events occurred as follows:

	Viking 1	*Viking 2*
Date of launch	August 20, 1975	September 10, 1975
Placed in elliptical orbit around Mars	June 6, 1976	August 7, 1976
Touchdown of Lander	July 20, 1976	September 3, 1976

Viking 1 traveled nearly 676 million kilometers (420 million miles) and *Viking 2* nearly 713 million kilometers (443 million miles) in their heliocentric Mars intercept trajectories prior to their respective insertions into elliptical orbits around the planet. See Figs. 32 and 33. Timing of the Viking missions was planned to achieve the trajectory situation shown. The Mars orbit insertion maneuvers for the Vikings require significant engine burns—in the case of *Viking 1*, for example, 38 minutes, consuming 2330 pounds (1057 kilograms) of propellant. Once in the Martian vicinity, radio signals required 22 minutes in either direction between earth and Mars,

thus a total of 44 minutes to execute a command and receive confirmation of that command. The general plan successfully followed in the *Viking* program involved orbiting the spacecraft in their respective orbits around Mars for several days, not only to commence imaging of the planet and certain scientific experiments, but to reconfirm earlier decisions concerning the best landing sites to finally elect for the two landers. Then, much as the *Surveyor* soft lunar landing craft had been placed on the earth's moon several years before, the landers were released from the aeroshells on the orbiting spacecraft, using parachute deployment at an elevation above the Martian surface of about 19,400 feet (5,913 meters) and the firing of terminal-descent engines at about 4,600 feet (1,402 meters) above the planet's surface. Both landers touched down successfully.

Later, adjustments were made to the *Viking* orbiters to provide better imaging of additional areas of the planet. Transmissions from the orbiters and the landers extending over an extensive period and information from one of the spacecraft was still being received during the early 1980s.

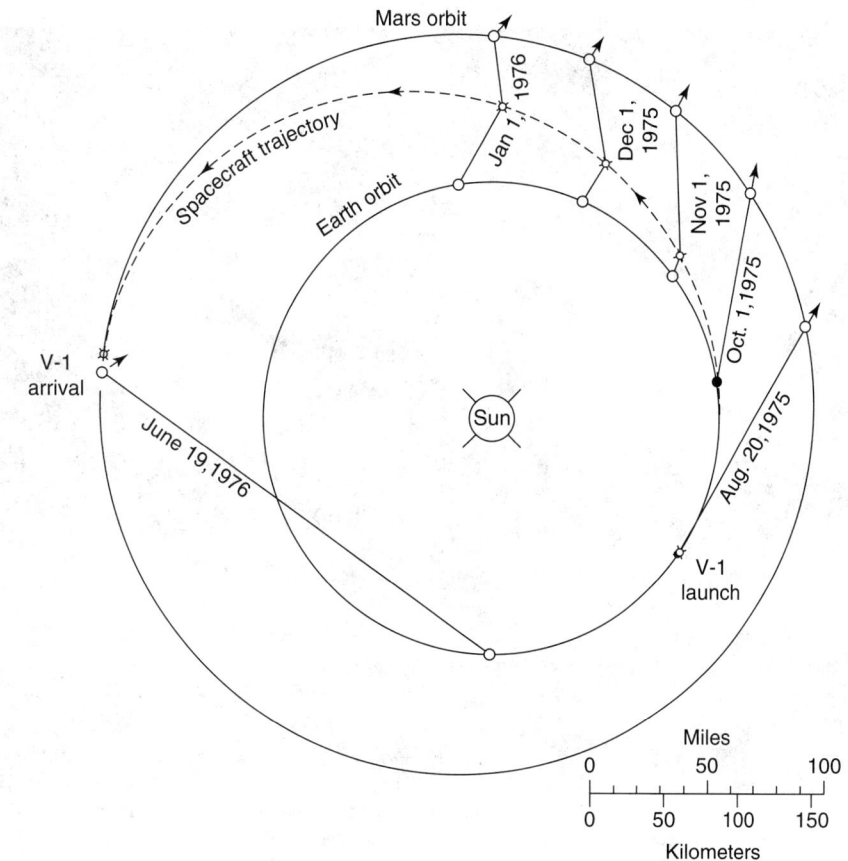

Fig. 32. Trajectory followed by *Viking 1*. Dates given show relationship of earth, spacecraft, Mars, and the sun at specific times.

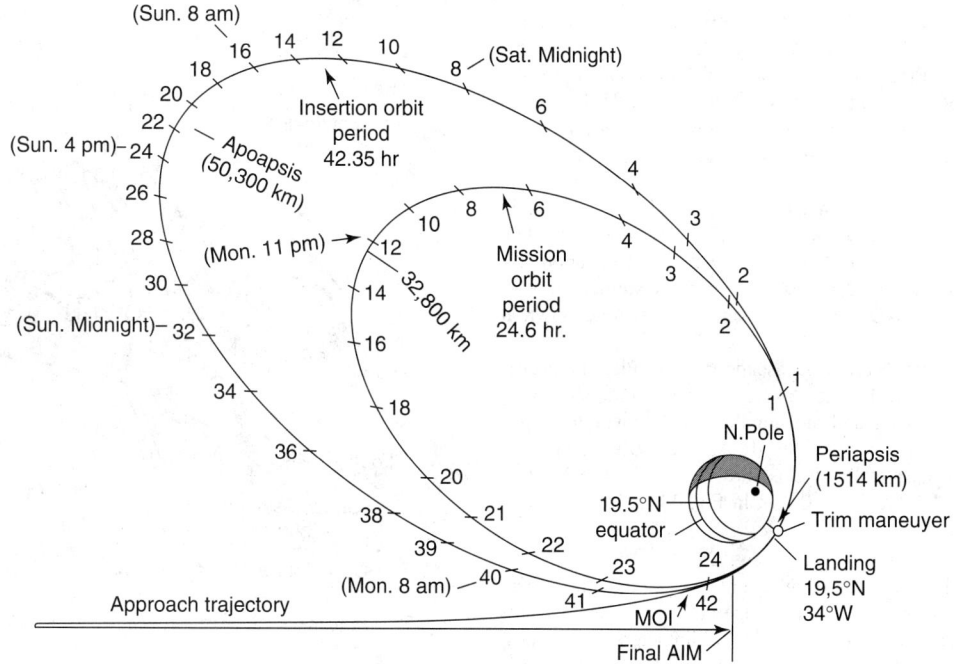

Fig. 33. Orbit geometry for the insertion and mission orbits. *Viking 1* completed only one revolution on the insertion orbit before the trim maneuver placed it on the mission orbit. The tick marks indicate spacecraft flight hours with periapsis as the zero point. (Periapsis = the orbital point nearest the focus of attraction; apoapsis = the farthest point.) Additional information at selected points along the insertion orbit indicate where the spacecraft was that day relative to Earth Pacific Daylight Time. A complete revolution of Mars on the mission orbit required 24.6 hours, the length of a martian day. The oribt was synchronized with the landing site in that the spacecraft passed over the site once each day near periapsis, allowing maximum resolution orbital photography of that region for site certification and surface-data (after landing) correlation.

Fig. 34.   One of the *Viking* Landers (test model) in a simulated Martian setting. This spacecraft was set up in the auditorium at NASA's Jet Propulsion Laboratory to thoroughly familiarize the many scientists on the project with the detailed operation of the spacecraft's mechanical sampling system and other scientific experiments aboard. (*NASA; Jet Propulsion Laboratory, Pasadena, California.*)

The *Viking* Landers

To assist in familiarizing many scientists at the control center in the detailed operation of the complex *Viking* landers, a test lander was installed in the auditorium at NASA's Jet Propulsion Laboratory in Pasadena, California. See Fig. 34. A diagram of one of the Landers is given in Fig. 35. Locations on Mars of the final landing sites are shown in Fig. 36. A view of the landing site for *Viking Lander 1*, taken from the orbiting *Viking 1*, prior to the landing is shown in Fig. 37.

The first photograph ever taken on the surface of Mars is shown in Fig. 38. This picture was taken just minutes after *Viking Lander 1* touched down successfully at Chryse Planitia. The center of the image is about 1.4 meters (5 feet) from camera No. 2 of the spacecraft. A similar view of the martian surface taken by *Viking Lander 2* shortly after touchdown at Utopia Planitia is shown in Fig. 39.

The first photograph of the Martian landscape (Chryse Planitia site) is shown in Fig. 40. In real color, this view is predominantly reddish brown. A diagram of the *Viking 1 Lander* and showing field of view with reference to equipment components is given in Fig. 41. Another striking view of the Martian landscape of Chryse Planitia is shown Fig. 42. The Martian landscape at the Utopia Planitia site is shown in Fig. 43.

Viking Scientific Experiments

Thirteen scientific investigations yielded information about the atmosphere and surface of Mars. Two orbiters and landers operating for several months photographed the surface extensively from 1500 kilometers (930 miles) and directly on the surface. Measurements were made of the atmospheric composition, the surface, elemental abundance, the atmospheric water vapor, temperature of the surface, and meteorological conditions: direct tests were made for organic material and living organisms.[7]

Inorganic Chemistry. An x-ray fluorescence spectrometer was used to determine the elemental composition of samples at each lander site. Both sites yielded analyses of the fine-particle materials that are strikingly similar. Silica and iron in large amounts and magnesium, aluminum, calcium, and sulfur in significant amounts. More detail pertaining to the inorganic aspects of the martian surface and rock materials are given in

---

[7] Information extracted from official NASA Langley Research Center report.

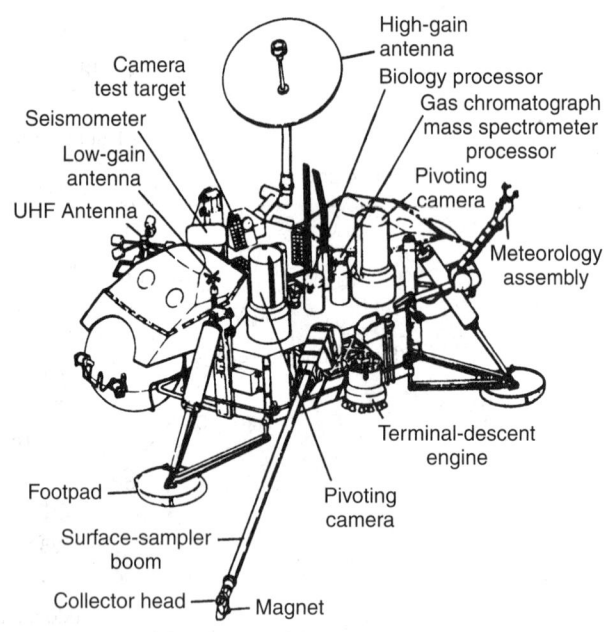

Fig. 35.   Principal features of the *Viking* Lander. (*Martin Marietta Corporation.*)

the entry on Mars. The trenching and sampling equipment on the Lander spacecraft are shown in the foreground of Fig. 20. Plan views of the sampling apparatus and procedures followed the *Viking* Landers are given in Figs. 44 and 45.

Molecular Analysis. Two samples from each lander site were analyzed for organic material with successive use of volatilization, pyrolysis, and detection by gas chromatography-mass spectrometry (GCMS). The sensitivity of the method is of the order of parts per billion. No organic compounds were detected at that level. The instrument was not designed to detect life—neither the quality or sensitivity permitted detecting biomass directly. The absence of organics in the sample was somewhat surprising considering the likelihood of carbonaceous chondrites reaching the Martian

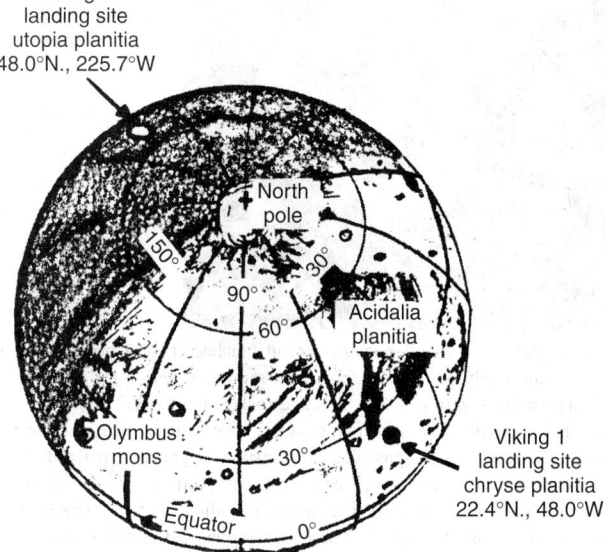

Viking 2
landing site
utopia planitia
48.0°N., 225.7°W

North pole

Acidalia planitia

Olympus mons

Equator

Viking 1
landing site
chryse planitia
22.4°N., 48.0°W

Valles marineris

Fig. 36.   Locations of the two *Viking* landing sites on Mars.

abundances, it is believed unlikely that a history of Mars outgassing will emerge.

Biology Experiments.   Three experiments were conducted to test directly for life on Mars. The tests revealed a surprisingly chemically active surface — very likely oxidizing — but no evidence concerning the existence of life on the planet. The biological experiments were conducted with fully programmed and automated miniature laboratory equipment installed in each *Viking* lander. In the pyrolytic release (PR) experiment, Martian soil was placed in a chamber, after which carbon dioxide and carbon monoxide were added. These compounds were traceable because of the addition of radioactive carbon-14. The soil was incubated beneath a lamp that simulated Martian sunlight, but with no ultraviolet radiation present, as is the actual case on the planet today. If microorganisms were present, they would take up the radioactive gases. The chamber was heated to pyrolyze (decompose) any microbes present in the organic gases. The gases then were forced into an organic vapor trap, allowing other gases to pass to a radiation detector for "first count." With additional heating, the "organic vapors" were released and, if radioactive, they would indicate that living organisms were present. Results were negative.

In the labeled release (LR) experiment radioactive nutrient was added to a soil sample. Microorganisms present would digest the nutrient and release radioactive carbon dioxide. The soil is permitted to "incubate" for a period of a week or more, with further additions of nutrient. Results were negative.

In the gas exchange (GEX) experiment, scientists were looking for changes that Martian microbes might cause in gas levels over a long period. Soil was placed in a chamber, which was sealed to prevent gas leakage. Just sufficient nutrient flows admixed with water vapor would awaken spores or seeds, changing the gas level in the experiment. The results were negative. Considerable production of oxygen was noted from the GEX experiment, more or less explained as unusual Martian exotic chemistry.

surface, or the possibility of de novo synthesis. Explanations involving dilution in the regolith and destruction by ultraviolet light or oxidation are all plausible. The GCMS was also used to measure the Martian atmosphere. It was ideally suited to measure isotopic ratios. Based upon measurements of nitrogen, argon, xenon, and krypton and their isotopic

Fig. 37.   View from *Viking* 1 Orbiter showing two candidate landing sites. White circle indicates prime site in Chryse where *Viking Lander 1* touched down a few days later. In this group of five adjoining photos taken from about 32,000 kilometers (20,000 miles) through a violet filter, Chryse is shown lying at the mouth of the channels, which proceed southward on the planet. An alternative site lies on a plateau adjacent one of the canyons in the lower (foreground) part of the picture. A bit of the planet's limb can be seen in the upper right-hand corner. Near the lower right-hand corner is a white cloud, believed to be ice crystals. From a comparison with pictures taken three minutes apart, the cloud was found to be moving about 97 kilometers (60 miles) per hour toward the upper left of view. Overall, the picture spans about 40° in longitude and 35° in latitude. The prominent feature in the lower left frame is Grangis Chasma, an arm of the great equatorial rift. North in these views is toward upper-right-hand corner. (*NASA; Jet Propulsion Laboratory, Pasadena, California.*)

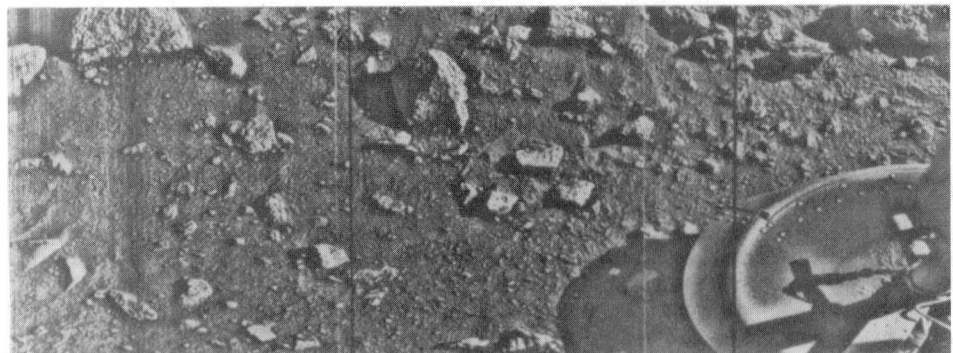

Fig. 38.   First photograph ever taken on surface of Mars, obtained by *Viking 1 Lander* just a few minutes after its successful touchdown on Chryse Planitia. The center of the image is about 1.4 meters (5 feet) from camera #2 of the spacecraft. Both rocks and finely granulated material—sand and dust—are observed. Many of the small foreground rocks are flat with angular facets. Several larger rocks exhibit irregular surfaces with pits and the large rock at the top left shows intersecting linear cracks. Extending from that rock toward the camera is a vertical linear dark band, which may be due to a one-minute partial obscuration of the landscape due to clouds or dust intervening between the sun and the surface. Associated with several of the rocks are apparent signs of wind transport of granular material. The large rock in the center is about 10 centimeters (4 inches) across and shows three rough facets. To its lower right is a rock near a smooth portion of the Martian surface, probably composed of very fine-grained material. It is possible that the rock was moved during the *Viking 1* descent maneuvers, revealing the finer-grained basement substratum; or that the fine-grained material had accumulated adjacent to the rock. There are numerous other furrows and depressions and places with fine-grained material elsewhere in the view. At right is a portion of footpad #2. Small quantities of fine-grained sand and dust are seen at the center of the footpad near the strut and were deposited at landing. The shadow to the left of the foodpad clearly exhibits detail, due to scattering of light either from the Martain atmosphere or from the spacecraft, observable because the Martian sky scatters light into shadowed areas. (*NASA; Jet Propulsion Laboratory, Pasadena, California.*)

Fig. 39.   First photograph of Martian surface taken by *Viking Lander* 2 at the Utopia Planitia site. The scene reveals a wide variety of rocks littering a surface of fine-grained deposits. Boulders in the 10–20 centimeter (4–8 inch) size range—somewhat vesicular (holes) and some apparently fluted by wind—are common. Many of the pebbles have tubular or platy shapes, suggesting that they may be derived from layered strata. The fluted boulder just above the Lander's footpad displays a dust-covered or scraped surface, suggesting it was overturned or altered by the foot at touchdown. Brightness variations at the beginning of the picture scan (left edge) probably are due to dust settling after landing. A substantial amount of fine-grained material kicked up by the descent engines has accumulated in the concave interior of the footpad. Center of the image is about 1.4 meters (5 feet) from the camera. Field of view extends 70° from left to right and 20° from top to bottom. This second landing location is in the northern latitudes about 7500 kilometers (4600 miles) northeast of the Viking 1 Lander site, where touchdown occurred 45 days earlier. (*NASA; Jet Propulsion Laboratory, Pasadena, California.*)

No plausible ties with living organisms were established. There continues to be some speculation that the chemical oxidative qualities of Martian soil may support microorganisms of a sort which earth-bound scientists have not described even on a sound theoretical basis.

Meteorology. A meteorological weather station to measure changes in pressure, temperature, wind speed and direction operated well on both landers. Generally, the weather at both sites was repetitive, with a daily temperature variation of between 190 and 240 K, the peak usually occurring in mid-afternoon Martian time. The pressure at each site was in the range of 7 to 8 millibars. Daily pressure variations were about 0.3 millibar.

Seismology. The seismometer on Viking Lander 1 failed to be uncaged, but the Viking 2 functioned normally. Little or no quake activity was detected.

Atmospheric Water Detector. An infrared spectrometer operating at the 1.38 micrometer region was mounted on the scan platform of each

lander. The device was used to measure the latitudinal variations and diurnal variations and, by operating over a complete Martian year, it was able to indicate seasonal changes. More moisture was found in the northern hemisphere than in the southern portion of the planet. These measurements helped to confirm that the permanent polar cap of Mars consists of water ice.

Infrared Thermal Mapper. An infrared radiometer measured thermal emission of the surface and atmosphere. It was found that the atmospheric temperature above 20 kilometers (about 12 miles) varies from 165 K (near dawn) to 185 K at about 2:15 in the afternoon Martian time. This variation is believed to be initiated at the lower levels and radiatively propagated by dust in the atmosphere. The temperature of the surface was found to be highly variable.

Physical and Magnetic Properties. Cameras were used to determine certain physical and magnetic properties of the soil. Pictures of stroke gages, sample digging, footpad movement, areas underneath the lander,

Fig. 40.   First panoramic view of Martian surface taking by *Viking 1 Lander*. The out-of-focus spacecraft component toward left center is the housing for the Viking sample arm, which is not yet deployed. Parallel lines faintly seen in the sky are an artifact and are not real features. However, the change of brightness from horizon toward zenith and toward the right (west) is accurately reflected in this picture, which was taken in the late Martian afternoon. At the horizon to the left is a plateaulike prominence much brighter than the foreground material between the rocks. The horizon features are approximately three kilometers (1.8 miles) away. At left is a collection of fine-grained material reminiscent of sand dunes. The dark sinous markings in the left foreground are of unknown origin. Some unidentified shapes can be perceived on the hilly eminence at the horizon toward left center of view. The horizontal cloud stratum can be made out halfway from the horizon to the top of the picture.

At the center is seen the low-grain antenna for receipt of commands from earth. The projections on or near the horizon may represent the rims of distant impact craters. In the right foreground are color charts for Lander camera calibration, a mirror for the Viking magnetic properties experiment and part of a grid on the top of the Lander body. At upper right is the high-gain dish antenna for direct communication between the landed spacecraft and earth. Toward the right edge is an array of smooth, fine-grained material which shows some hint of ripple structure and may be the beginning of a large dune field off to the right of the view, which joins with dunes seen at the top left in this 300° panoramic view. Some of the rocks appear to be undercut on one side and partially buried by drifting sand on the other (*NASA; Jet Propulsion Laboratory, Pasadena, California.*)

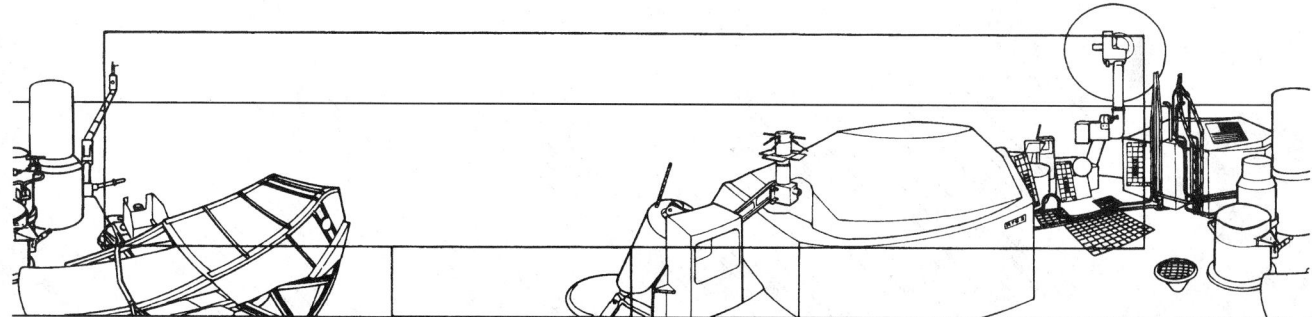

Fig. 41.   This diagram illustrates a full 360° image from camera #2 on the *Viking Lander* spacecraft. The outlined image areas represent the Viking 1 Lander's first two pictures (Figs. 10 and 12). This type of diagram helped analysts at the mission control center to identify photo orientation in terms of parts of the Lander spacecraft components when they appeared in various views. (*NASA; Jet Propulsion Laboratory, Pasadena, California.*)

Fig. 42.   Martian landscape as viewed by *Viking 1 Lander*, showing a dune field with features remarkably similar to many seen in the deserts on earth. The early morning lighting (7:30 A.M. local Mars time) reveals subtle details and shading. The picture covers 100°, looking northeast at left and southeast at right. Viking scientists observed that this area is reminiscent of regions in Mexico, California, and Arizona (Kelso, Death Valley, Yuma). The sharp dune crests indicate the most recent wind storms capable of moving sand over the dunes in the general direction from upper left to lower right. Small deposits downwind of rocks also indicate this wind direction. Large boulder at left is about 8 meters (25 feet) from the spacecraft and measures about 1 × 3 meters (3 × 10 feet). The meteorology boom, which supports the spacecraft's miniature weather station, cuts through the center of the picture. The sun rose two hours earlier and is about 30° above the horizon near the center of the view. In real color, the landscape is predominantly reddish brown. (*NASA; Jet Propulsion Laboratory, Pasadena, California.*)

and rock movement were used to determine the bulk density, particle size, angle of internal friction, cohesion, adhesion, and penetration resistance of the Martian soil. Small permanent magnets on the sampler collected material indicating that the surface contains a few percent of magnetic material, very likely magnetite.

See also **Odyssey Mission to Mars**; **Pathfinder Mission to Mars**; and **Viking Mission to Mars**.

Classic References to Mars and Missions to Mars

The *Viking* and other missions to Mars by the United States and the former U.S.S.R. were quite thoroughly documented and represent a wealth of resource information. Information collected between 1970 and 1990, which appeared in the 7th Edition of this encyclopedia, is re-listed here for the serious scholar of Mars and its exploration.

Fig. 43. High-resolution photo of the Martian surface taken by Viking Lander 2 at the Utopia Planitia landing site. View was made on May 18, 1979 and relayed to earth by Orbiter 1 on June 7. The "rolling hill" Marscape is an artifact of the Lander's 8-degree tilt; the horizon is generally flat. There is some relief of the flat terrain, however, at lower scales locally and at greater scales on the horizon. In stereo, a few gullies and depressions take on considerable depth and dimension. A thin coating of water ice on the rocks and soil is visible. The time the frost appeared corresponded almost exactly with the build-up of frost one Martian year (23 earth months) earlier when a similar view had been taken. The frost remained on the surface for about one hundred days. Some scientists believe dust particles in the atmosphere may pick up bits of solid water. But carbon dioxide present in the Martian atmosphere also may freeze and adhere to particles, becoming sufficiently heavy to settle. Warmed by the sun, the surface evaporates the carbon dioxide and returns it to the atmosphere, leaving behind water and dust. The ice seen in this picture, like that which formed during the earlier martian year, is extremely thin, perhaps no more than 1/1000 inch (0.03 millimeter) thick. (*NASA; Jet Propulsion Laboratory, Pasadena, California.*)

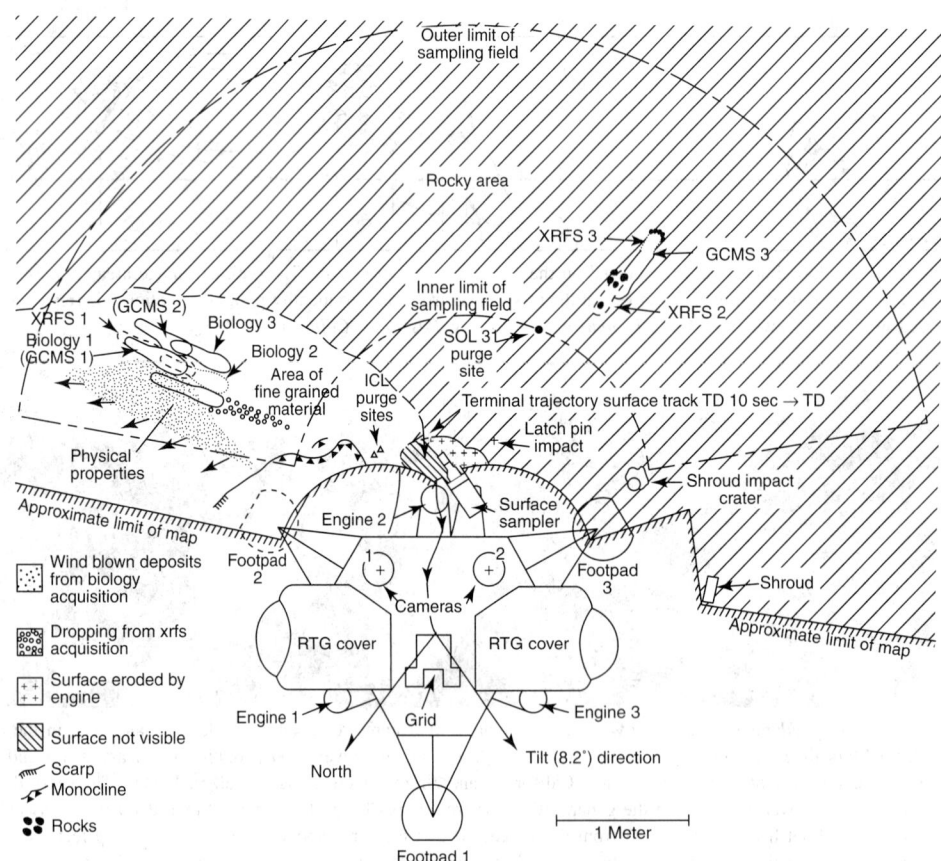

Fig. 44. Plan view of *Viking Lander* 1, showing the spacecraft and its orientation, location of sample sites, locations of selected rocks for analytical experiments. 1 Sol = 1 complete Martian day and night; XRFS = x-ray fluorescence spectrometer, GCMS = gas chromatograph-mass spectrometer, ICL = initial rock to be investigated. (*NASA; Jet Propulsion Laboratory, Pasadena, California.*)

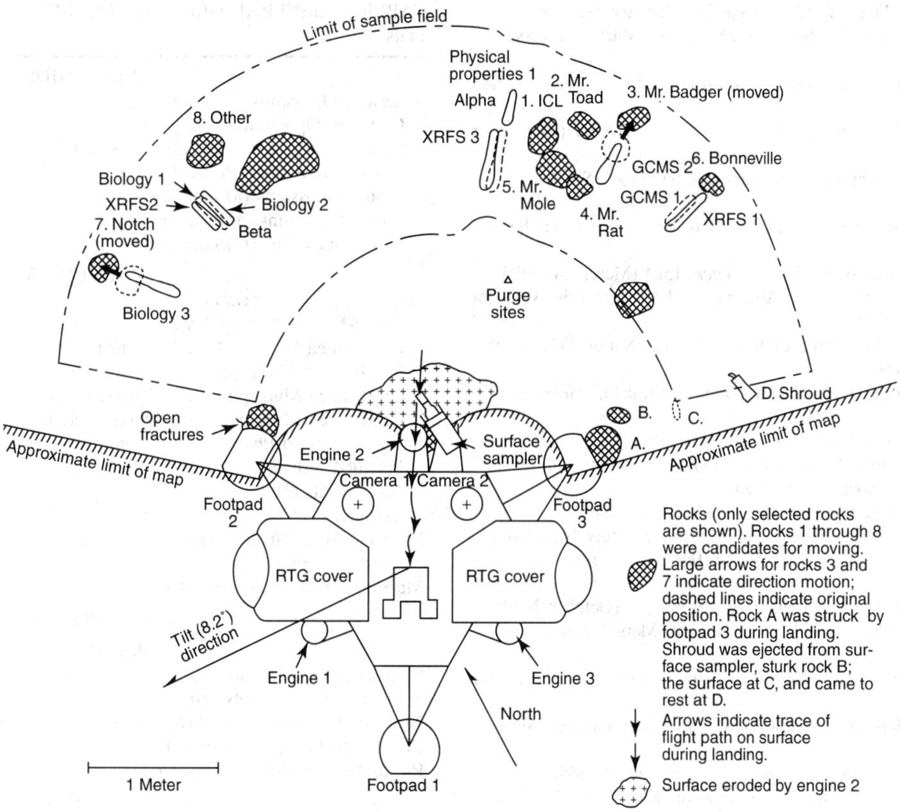

Fig. 45.   Plan view of *Viking Lander* 2, similar to that of Fig. 29. (*NASA Jet Propulsion Laboratory, Pasadena, California.*)

Additional Reading (1976–1987)

Anders, E. and T. Owen: "Mars and Earth: Origin and Abundance of Volatiles," *Science*, **198**, 453–465 (1977).

Arvidson, R.E. et al.: "Three Mars Years: *Viking Lander 1* Imaging Observations," **222**, 463–478 (1983).

Baker, Victor R.: "The Spokane Flood Controversy and the Martian Outflow Channels," *Science*, **202**, 1249–1257 (1978).

Baker, V.R.: "The Channels of Mars," Univ. of Texas Press, Austin, Texas, 1982.

Bogard, D.D. and P. Johnson: "Martian Gases in an Antarctic Meteorite?" *Science*, **221**, 651–654 (1983).

Carr, M.H.: "The Surface of Mars," Yale University Press, New Haven, CT, 1981.

Carr, M.H., R.S. Saunders, R.G. Strom, and D.E. Wilhelms: "The Geology of the Terrestrial Planets," Spec. Pubn. SP-469, National Aeronautics and Space Administration, Washington, DC, 1985.

Duxbury, T.C. and J. Veverka: "Deimos Encounter by Viking: Preliminary Imaging Results," *Science*, **201**, 812–814 (1978).

Ezell, E.C. and L.N. Ezell: "On Mars," National Aeronautics and Space Administration, Washington, DC, 1984.

Greeley, R.: "Release of Juvenile Water on Mars," *Science*, **236**, 1363–1364 (1987).

Haberle, R.M.: "The Climate of Mars," *Sci. Amer.*, 54–62 (May 1986).

Hartman, W.K.: *J. Geophys. Res.*, **78**, 4096 (1973).

LPI: "Mars: Evolution of Its Climate and Atmosphere," Proceedings of Symposium, Washington, DC, Lunar and Planetary Institute, Houston, TX (July 17–19, 1986).

Lucchitta, B.K.: "Recent Mafic Volcanism on Mars," *Science*, **235**, 565–567 (1987).

McElroy, M.B. et al.: "Composition and Structure of the Martian Upper Atmosphere: Analysis of Results from Viking," *Science*, **194**, 1295–1298 (1976).

Mumma, M.J. et al.: "Discovery of Natural Gain Amplification in the 10-Micrometer Carbon Dioxide Laser Bands on Mars: A Natural Laser," *Science*, **212**, 45–49 (1981).

Mutch, R.A. et al.: "The Geology of Mars," Princeton Univ. Press, Princeton, NJ, 1977.

Neukum, G. and D.U. Wise: "Mars: A Standard Crater Curve and Possible New Time Scale," *Science*, **194**, 1381–1387 (1976).

NEWS: "Launch of *Mars Observer* — 1992," *Science*, **235**, 743 (1987).

Owen, T. et al.: "The Atmosphere of Mars: Detection of Krypton and Xenon," *Science*, **194**, 1293–1295 (1976).

Paige, D.A. and A.P. Ingersoll: "Annual Heat Balance of Martian Polar Caps: *Viking* Observations," *Science*, **228**, 1160–1168 (1985).

Pieri, D.C.: "Martian Valleys: Morphology, Distribution, Age, and Origin," *Science*, **210**, 895–897 (1980).

Soffen, G.A. and C.W. Snyder: "The First *Viking* Mission to Mars," (contains 13 papers relating to *Viking 1 Lander* on Mars), *Science*, **193**, 759–815 (1976).

Soffen, G.A.: "Scientific Results of the Viking Missions," (contains 20 papers relating to Viking missions to Mars), *Science*, **194**, 1274–1353 (1976).

Soffen, G.A.: "Status of the *Viking* Missions," (contains 15 papers relating to the Viking mission to Mars), *Science*, **194**, 57–105 (1976).

Squyres, S.W.: "The History of Water on Mars," *Ann. Review of Earth and Planetary Sciences*, **12**, 83–106 (1984).

Squyres, S.W. and M.H. Carr: "Geomorphic Evidence for the Distribution of Ground Ice on Mars," *Science*, 231–252 (1986).

Thomas, P. and P.J. Gierasch: "Dust Devils on Mars," *Science*, **230**, 175–177 (1985).

Additional Reading (1988–2000)

Beardsley, T.M.: "U.S.–Soviet Collaboration in Space Science Is Improving," *Sci. Amer.*, 21 (August 1988).

Beardsley, T.: "Slow Boat to Mars," *Sci. Amer.*, 14 (April 1990).

Bergreen, L.: "Voyage to Mars: NASA's Search for Life Beyond Earth," The Putnam Publishing Group, New York, NY, 2000.

Collins, M.: "Mission to Mars," *Nat'l. Geographic*, 732 (November 1988).

Cornell, J.: "Red Weather (Mars)," *Technology Review (MIT)*, 19 (February/March 1990).

Eberhart, J.: "Soviet Findings from Phobos and Mars," *Science News*, 286 (October 28, 1989).

Eberhart, J.: "Phobos: Moonlet of the Pits," *Science News*, 301 (November 4, 1989).

Eberhart, J.: "Powerful Appeal of Mars' Missing Field," *Science News*, 150 (March 10, 1990).

Eberhart, J.: "Episodic Oceans: Mars," *Science News*, 283 (May 5, 1990).

Eberhart, J.: "The Sandy Face of Mars," *Science News*, 268 (October 27, 1990).

Eberhart, J.: "Mars: Let It Snow, let it snow…," *Science News*, 286 (November 3, 1990).

Flam, F.: "Swarms of Mini-Robots Set to Take on Mars Terrain," *Science*, 1621 (September 18, 1992).

Greeley, R. and B.D. Schneid: "Magma Generation on Mars: Amounts, Rates, and Comparisons with Earth, Moon, and Venus," *Science*, 996 (November 15, 1991).

Hamilton, D.P.: "NASA to Explore Three Possible Mars Missions," *Science*, 863 (February 22, 1991).

Keating, G.M.M.: "Exploration of Venus and Mars Atmospheres," Elsevier Science, New York, NY, 1995.

Kerr, R.A.: "Soviet Failure at Mars a Reminder of Risks," *Science*, 26 (April 7, 1989).

Kerr, R.A.: "Planetary Science Funds Cut," *Science*, 282 (January 19, 1990).

Kieffer, H.H., C. Snyder, B.M. Jakosky, and M.S. Matthews: "Mars," University of Arizona Press, Phoenix, AZ, 1997.

Kiernan, V.: "Reactor Project Hitches onto Moon-Mars Effort," *Science*, 1482 (June 22, 1990).

Kiernan, V.: "Sailing to Mars," *Technology Review (MIT)*, 20 (November/December 1990).

McKay, C.P. and R.H. Haynes: "Should We Implant Life on Mars?" *Sci. Amer.*, 144 (December 1990).

Muhleman, D.O. et al.: "Radar Images of Mars," *Science*, 1508 (September 27, 1991).

Ostro, S.J. et al.: "Radar Detection of Phobos," *Science*, 1584 (March 24, 1989).

Owens, T. et al.: "Deuterium on Mars: The Abundance of HDO and the Value of D/H," *Science*, 1767 (June 24, 1988).

Raeburn, P.: "Mars: Uncovering the Secrets of the Red Planet," National Geographic Society, Washington, DC, 2000.

Rubincam, D.P.: "Mars: Change in Axial Tilt Due to Climate," *Science*, 720 (May 11, 1990).

Schwartz, B.D.: "Muddy Evidence," *Sci. Amer.*, 28 (June 1989).

Sheehan, W.: "The Planet Mars: A History of Observation and Discovery," University of Arizona Press, Phoenix, AZ, 1996.

Staff: "Mars Mission," *Technology Review (MIT)*, 19 (May/June 1989).

Staff: "Mars Magnetism: A Moot Question?" *Science News*, 31 (July 14, 1990).

Staff: "Martian Atmosphere Eyed by Hubble Space Telescope," *Hughes News*, 5 (May 17, 1991).

Staff: "Instruments Help Unlock Mars' Secrets," *Hughes News*, 1 (October 2, 1992).

Touma, J. and J. Wisdom: "The Chaotic Obliquity of Mars," *Science*, 1294 (February 26, 1993).

Waldrop, M.M.: "Jet Propulsion Lab Looks to Life After Voyager," *Science*, 1037 (September 8, 1989).

Waldrop, M.M.: "Phobos at Mars: A Dramatic View — and Then Failure," *Science*, 1042 (September 8, 1989).

Waldrop, M.M.: "Asking for the Moon," *Science*, 637 (February 9, 1990).

Walker, M.: "Evolution of Hydrothermal Ecosystems on Earth and Mars," John Wiley & Sons, Inc., New York, NY, 1996.

Walters, M.: "The Search for Life on Mars," Perseus Publishing, Boulder, CO, 1999.

Web References

Mars Exploration: *http://Mars.jpl.nasa.gov/*

National Aeronautics and Space Administration: *http://sse.jpl.nasa.gov/missions/mars_missions/mgs.html*

MARSH GAS. See **Natural Gas**.

MARSUPIALIA. Pouched mammals, including the kangaroos, opossums, koala, and many less familiar forms. The order is characterized by the presence of a marsupium or pouch on the abdomen of the female in which the young complete their development. They are born in a very early stage. With few exceptions, such as the opossums of the Americas, marsupials occur in the Australian region. The general organization of Marsupialia is given in Table 1.

Didelphids. Opossums of various species make up this family. They are of moderate size, comparable to a large domestic cat or small dog. They are of a gray color with deep fur. The tail is long and scaly; the nose is sharp. See Fig. 1. There usually are from 6 to 16 young, usually more than the number of teats provided for in the mother's pouch. Hence, some of the young frequently die of starvation. The weaker and smaller young (approximately $\frac{1}{2}$ inch; 1.3 centimeters in length) are carefully placed into the pouch by the mother immediately after birth. As they grow larger and stronger, the mother will carry them on her back. The young use their prehensile tail very early for hanging on to the mother's tail or fur coat.

Fig. 1.   Opossum. (*A.M. Winchester.*)

The common opossum, *Didelphys virginiana*, ranges from the Great Lakes to the Gulf and westward into Oklahoma. Other species range from

TABLE 1. GENERAL ORGANIZATION OF MARSUPIALIA Pouched Mammals

DIDELPHIDS

American Opossums (*Didelphidae*)
  Common Opossums (*Didelphis*)
  Four-eyed Opossums (*Metacheirus....*)
  Woolly Opossums (*Philander....*)
  Mouse-Opossums (*Marmosa*)
  Shrew-Opossums (*Monodelphis....*)
  Water Opossums (*Chironectes*)

DASYURIDS

Phascogales (*Phascogalinae*)
  Broad-footed Phascogales (*Antechinus*)
  Flat-headed Phascogales (*Planigale*)
  Brush-tailed Phascogales (*Phascogale*)
  Crest-tailed Marsupial Mice (*Dasycercus*)
  Crest-tailed Marsupial Rats (*Dasyuroides*)
  Narrow-footed Phascogales (*Sminthopsis*)
  Pouched Jerboas (*Antechinomys*)
  Native Cats (*Dasyurus*)
  Tasmanian Devil (*Sarcophilus*)
Pouched Wolf (*Thylacininae*)
Numbats (*Myrmecobiinae*)
Marsupial Moles (*Notoryctidae*)

CAENOLESTIDS

PERAMELIDS

Long-nosed Bandicoots (*Perameles*)
Short-nosed Bandicoots (*Thylacis*)
New Guinea Bandicoots (*Echymipera*)
Rabbit-Bandicoots (*Marcrotis*)
Pig-footed Bandicoots (*Choeropus*)

PHALANGERIDS

Honey-Suckers (*Tarsipedinae*)
Phalangers (*Phalangerinae*)
  Dormouse-Phalangers (*Cercoertus*)
  Striped Phalangers (*Dactylopsila*)
  Feather-tailed Phalangers (*Distoechurus*)
  Gliding Feather-tails (*Acrobates*)
  Leadbeater's Phalanger (*Gymnobelideus*)
  Flying-Phalangers (*Petaurus*)
  Ring-tailed Phalangers (*Pseudocheirus....*)
  Great Gliders (*Schoinobates*)
  Brush-tailed Phalangers (*Trichosurus*)
  Scaly-tailed Phalanger (*Wyulda*)
  Cuscuses (*Phalanger*)
The Koala (*Phascolarctidae*)
Wombats (*Wombatidae*)
Musk Rat-Kangaroos (*Hypsiprymnodontinae*)
Rat-Kangaroos (*Potoroinae*)
  Long-nosed Rat-Kangaroos (*Potorous*)
  Short-nosed Rat-Kangaroos (*Bettongia*)
  Desert Rat-Kangaroos (*Caloprymnus*)
  Rufous Rat-Kangaroos (*Aepyprymnus*)
Kangaroos (*Macropodinae*)
  Tree-Kangaroos (*Dendrolagus*)
  Hare-Wallabies (*Lagorchestes*)
  Rock-Wallabies (*Petrogale* and *Peradorcus*)
  Nail-tailed Wallabies (*Onychogale*)
  Pademelons (*Setonyx* and *Thylogale*)
  True Wallabies (*Wallabia*)
  Wallaroos (*Osphranter*)
  True Kangaroos (*Macropus*)

Mexico to southern Brazil. With exception of the water opossum, they are arboreal animals which eat insects, small birds, and fruit. The water opossum or yapock, *Chironectes*, of South America is a swimming animal and lives on fishes and other aquatic life, thus in these habits resembling the behavior of the mink and otter. The hind toes are webbed.

Dasyurids. The phascogales generally are rodent-like creatures. The Tasmanian Devil, *Sarcophilus ursinus*, resembles the badger in its stout build and large head. It is nocturnal in habits, hiding during the day in a burrow or in natural crevices. The animal eats all kinds of living creatures, even killing forms much larger than itself. The Tasmanian Wolf, *Thylacinus*, is a pouched animal closely resembling the wolves and of similar habits. Like the wolves of the northern hemisphere, this species has been killed in large numbers for its attacks on domestic animals and is now restricted to the wilder mountainous parts of Tasmania. The pouched mole is a rare mammal of the Australian deserts. The species, *Notoryctes typhlops*, is known for its highly specialized burrowing. Like the true moles, the pouched mole has enormous claws and rudimentary eyes.

Peramelids. Various species of the bandicoot comprise this family. The bandicoot is a medium-to-small burrowing animal of the Australian region, stoutly built and quadrupedal.

Phalangerids. Although there are exceptions, the phalangers are comprised generally of small-to-medium-size animals with thick woolly fur. Some resemble mice or squirrels superficially and all are arboreal. With exception of the koala, they have prehensile tails. In addition to the true phalangers, the group includes the cuscuses, the flying phalangers, and long-snouted phalangers. The koala, shown in Fig. 2, is a curious pouched marsupial of eastern Australia. Because of its resemblance to a teddy bear, it is sometimes called the native bear. The animal is robust and achieves a length of about 2 feet (0.6 meters). An arboreal animal, the koala feeds on eucalyptus leaves, preferring those leaves with the greatest oil content. The animal can digest from 2 to 3 pounds (0.9 to 1.4 kilograms) of this foliage per day. The koala matures at 4 years of age and lives for about 20 years. Breeding occurs every other year. The gestation period is 35 days. After a short time in the pouch, the mother carries the young on her back. Highly regarded by Australians, the animal is protected by law.

The wombat, also a phalangerid, is a stoutly built pouched animal, broad of body and with short thick legs. The wombat has a pair of chisel-like incisor teeth in each jaw, like the rodents, and displays similar feeding habits. The wombat lives in burrows or in rock crevices. There are several species.

Possibly best known of all marsupials is the kangaroo. The typical kangaroo, *Macropus*, has large hind legs and a strong tail. The animal sits upright and moves by spring leaps, not touching the front feet to the ground. The largest kangaroos reach a height of more than 6 feet (1.8 meters) and a weight of 200 pounds (91 kilograms). See Fig. 3. Kangaroos usually travel in groups. They are easy to frighten and tend to panic under tension. However, they are known to bite rather severely and to box with their forefeet. The feeding habits are much like deer, foraging on grasses and tender shoots of plants. Some species can cause considerable crop damage.

Moderate and small-size members of the kangaroo family, which resemble the true kangaroo in form, are called wallabies. See Fig. 4. The distinction between kangaroos and wallabies is not rigidly scientific, the large species being called kangaroos and the smaller species wallabies, with a transition in the larger wallabies which are also known as brush kangaroos. Wallabies, like their large relatives, have powerful hind legs and small forelegs, and are bipedal in locomotion. They vary from the hare wallaby, *Lagorchestes*, less than 2 feet (0.6 meter) long, to the rednecked wallaby whose body is $3\frac{1}{2}$ feet (1 meter) long, exclusive of tail. The spur-tailed wallabies, *Onychogale*, are peculiar in having the tail tipped with a horny spur.

Fig. 2.   Female koala with baby.

Fig. 4.   Wallaby.

Fig. 3.   Kangaroo.

Additional Reading

Aitkin, L.: "Hearing-the Brain and Auditory Communication in Marsupials," Springer-Verlag, Inc., New York, NY, 1998.
Austad, W.N.: "The Adaptable Opossum," *Sci. Amer.*, 98–105 (February 1988).
Dawson, T.J.: "Kangaroos: The Biology of the Largest Marsupials," Cornell University Press, Ithaca, NY, 1995.
Degabriele, R.: "The Physiology of the Koala," *Sci. Amer.*, **243**, 1, 110–117 (July 1980).
Grzimek, B.: "Grzimek's Animal Life Encyclopedia," Vol. 10, Mammals 1, Van Nostrand Reinhold, New York, NY, 1975.

Lee, A.K. and A. Cockburn: "Evolutionary Ecology of Marsupials," Cambridge University Press, New York, NY, 1985.

Sharman, G.: "*National Geographic*," **155**, 2, 192, 209 (1979).

Szalay, F.S.: "Evolutionary History of the Marsupials and an Analysis of Osteological Characters," Cambridge University Press, New York, NY, 1994.

MARTEN. See **Mustelines**.

MARTIN (*Aves, Passeriformes*). Any of several species of swallows, represented on all continents except Australia. They have the short beak and wide mouth of the swallows and either a forked or a square tail. They nest about houses, on cliffs, and in burrows and hollow trees. North America has one species, the purple martin, *Progne subis*. The male is glossy blue-black and the female somewhat duller. These birds commonly nest in the cornices of buildings or in bird houses where they live in colonies. See also **Swallow**.

MARTINGALE. Originally, a process known to gamblers, under which the loser at a fair game doubled his stakes for the next, and so on at each loss. The paradox is that in the long run he appeared certain to win sooner or later and at that point would have a net gain. More recently, the term has been given a precise significance in the theory of stochastic processes. A stochastic process $\{x_t\}$ is called a martingale if $E\{|x_t|\}$ is finite for all t, and

$$E\{x_{t_{n+1}}|x_{t_1}, \ldots, x_{t_n}\} = x_{t_n}$$

with probability unity for all $n \geq 1$ and $t_1 < \cdots t_{n+1}$.

MASCON. A large-scale, high-density lunar (or planetary, as Martian) mass concentration below a ringed mare. See also **Mare**.

MASER. An acronym for microwave amplification by stimulated emission of radiation. The device is identical in theory of operation to the laser except that it operates at frequencies in the microwave region of the electromagnetic spectrum, rather than in the light range. See also **Laser**.

Consider a stream of atoms in equilibrium for a given energy transition, that is, some are emitting radiation at the frequency of the transition and others are absorbing it. This process can be represented by the equation

$$h\nu_{21} = E_2 - E_1$$

where E_2 represents the energy of an atom in an excited state, E_1 represents its energy in a lower state, such as the ground state, h is Planck's constant, ν_{21} represents the frequency of a photon which would excite the atom from the state E_1 to state E_2, or conversely, the frequency of a photon that would be emitted in the reverse transition. Now consider a large number of atoms in thermal equilibrium in a closed box. The radiation inside the box will be black body radiation; in other words, the number of photons of given energy will be uniquely determined by the temperature. In addition, the temperature will fix the proportion of atoms in the excited state, which will be given by the Boltzmann distribution

$$\frac{N(E_2)}{N(E_1)} = \exp\left(\frac{-(E_2 - E_1)}{kT}\right)$$

where $N(E_2)$ and $N(E_2)$ are the populations of the states E_1, E_2, respectively. These two assertions, governing the number of excited atoms and the number of photons, respectively, are both fundamental thermodynamic principles and should fit together into a consistent picture. To bring about the energy balance necessary for equilibrium, an extra term is introduced, corresponding to a second process of radiation emission. The second term is quite different from the first in that it represents a process whose rate (or probability) is proportional to the intensity of radiation of frequency ν falling on the atom. The radiation field stimulates the excited atom to emit, and we can contrast this process of stimulated emission with the more familiar process of spontaneous (random) emission. The latter is more familiar because it is overwhelmingly dominant so long as $h\nu/kT$ is large, and this is so in the optical and infrared regions of the spectrum, where most of our accumulated experience lies. On the other hand, if $h\nu/kT$ is small (as it is in the microwave region), stimulated emission is the more important; however, stimulated emission in the visible region may be significant under certain conditions.

In terms of photons, stimulated emission appears as the interaction of the photon with an excited atom, which leads to the emission by the atom of a second photon, identical with the primary. In other words the photon has multiplied.

To accomplish amplification it is only necessary to produce more photons by stimulated emission than are lost by absorption. The details of the balance at equilibrium show that the probability of an atom in the ground state absorbing a photon, and of one in the excited state emitting one by stimulation, are equal. Hence, if amplification is to be possible, the number of atoms in the excited state must be greater than the number in the ground state. A look at the Boltzmann equation shows that this cannot be attained at any (positive) value of temperature, which means that it cannot be attained in thermal equilibrium. The problem then is essentially one of disturbing equilibrium so as to bring about the population inversion required.

Although having a number of limitations, the ammonia maser can be used to describe the principles involved. What is involved is picking out the excited atoms from the unexcited ones, and segregating them. This implausible procedure is in certain cases possible, the most notable involving the ammonia molecule, NH_3. The structure of this molecule is pyramidal, with the nitrogen atom at the vertex and the three hydrogen atoms forming the base. It is possible for the molecule to execute vibrations in which the nitrogen atom vibrates back and forth through the plane of the hydrogen atoms. The energy difference between this excited state and the ground state (no vibration) corresponds to a wave-length in the microwave region, i.e., about 3 centimeters. Because of its lack of symmetry, the molecule in the ground state will have a dipole moment, but the average dipole moment of the molecule executing the vibrations is described as zero. If ammonia molecules are formed into a beam by allowing the gas to stream out of a collimating tube into a vacuum, and this beam is passed through a nonuniform electric field, the separation can be effected. The ground state molecules will be deflected by the nonuniform electric field, and lost to the beam; the excited molecules, however, by virtue of their zero dipole moment, experience no deflecting force, and are introduced into a resonant cavity resonating at the appropriate frequency. The cavity will then contain a preponderance of excited molecules, provided the lifetime of the excited state is long enough (as it will be at microwave frequencies). See Fig. 1.

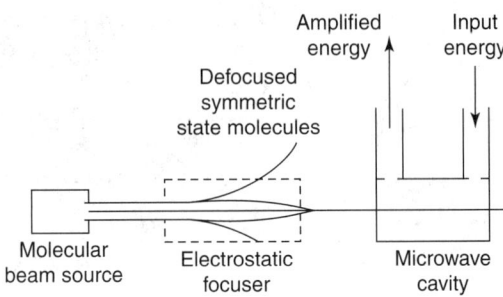

Fig. 1. Schematic representation of an ammonia beam maser. (*After Gordon, Zeiger, Townes.*)

If now an input signal of the resonant frequency is fed into the cavity, it can bring about spontaneous emission, and amplification will occur. The shortcoming of this arrangement is that the ammonia resonance cannot be tuned and it represents an impractically narrow bandwidth. On the other hand, like any other amplifier, it can be made to oscillate, so that it can be used as a very stable frequency standard (ammonia clock).

The three-level maser provides a more versatile arrangement for achieving inversion. Three energy levels are associated with the same atomic or molecular system. See Fig. 2. A strong microwave signal of frequency ν_{31} corresponding to $E_3 - E_1$ raises some of the atoms to E_3. A limit is reached when the populations of E_3 and E_1 are the same, since then the radiation absorbed is just balanced by stimulated emission, a state of affairs known as "saturation." Not all the atoms in state E_3 will return directly to E_1 — some will return via E_2. By the choice of states of suitable lifetimes ($\tau_2 > \tau_3$) it is possible to arrange that the number of atoms in E_2 exceeds that in E_1 because the "pump" frequency maintains a "head" of atoms in E_3. Thus E_2 and E_1 are inverted and amplification is possible.

Maser amplifiers are used where the requirement for a very low noise amplifier outweighs the technological problems of cooling to low temperatures. They have been used in passive and active radiostronomical

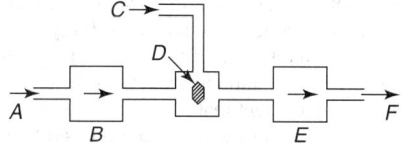

Fig. 2. Three-level maser amplifier, where A is signal input; B is the first isolator; C is the power input for pumping; D is the crystal; E is the second isolator, usually operated at low temperature as a noise reflector; and F is the signal output.

work, in satellite communications, and as preamplifiers for microwave spectrometry. The ammonia and the atomic hydrogen masers have been studied as frequency standards and have been used in accurate tests of special relativity.

Additional Reading

Arecchi, F. et al.: "Instabilities and Chaos in Quantum Optics II," Perseus Publishing, Cambridge, MA, 1988.

Bertolotti, M.: "Masers and Lasers: An Historical Approach," Institute of Physics (IoP), London, UK, 1988.

Elitzur, M.: "Astronomical Masers," Kluwer Academic Publishers, Norwell, MI, 1992.

MASK (Computer System). A pattern of digits used to control the retention or elimination of portions of another pattern of digits. Also, the use of such a pattern. For example, an 8-bit mask having a single i bit in the ith position, when added with another 8-bit pattern, can be used to determine if the ith bit is a 1 or a 0; that is, the ith bit in the pattern will be retained and all other bits will be 0s.

As another example, there are situations where it is desirable to delay the recognition of a process interrupt by the digital computer. A mask instruction permits the recognition of specific interrupts to be inhibited until it is convenient to service them.

MASS. The physical measure of the principal inertial property of a body, i.e., its resistance to change of motion. At speeds small compared with the speed of light, the mass of a body is independent of its speed. Under these circumstances, the masses m_1 and m_2 of two bodies may be compared by allowing the two bodies to interact. Then

$$m_1/m_2 = |a_2|/|a_1|$$

where $|a_1|$ and $|a_2|$ are the magnitudes of the respective accelerations of the two bodies as a result of the interaction. This permits the measurement of the mass of any particle with respect to a standard particle (for example, the standard kilogram). At higher speeds, the mass of a body depends on its speed relative to the observer according to the relation:

$$m = m_0/\sqrt{1 - v^2/c^2}$$

where m_0 is the mass of the body as found by an observer at rest with respect to the body, v is the speed of the body relative to the observer who finds its mass to be m, and c is the speed of light in empty space as prescribed by the theory of relativity.

When relativistic mechanics is appropriate, e.g., when speeds comparable to the speed of light are involved, mass may be converted into energy and vice versa, hence the energy of the system must be converted into mass through the Einstein equation

$$E = mc^2$$

where c is the speed of light in empty space, before the conservation law may be applied. See also **Gravitation**; and **Inertia**.

MASS (Center of). The center of mass is that point in a collection of mass-particles which moves as if the total mass of the collection were concentrated there and the resultant of all the external forces were acting there. The position vector of such a point is given by

$$\mathbf{r} = \frac{\sum_{i=1}^{n} m_i \mathbf{r}_i}{\sum_{i=1}^{n} m_i}$$

where m_i is the mass of ith discrete particle, $\mathbf{r}_i$ is the position vector of ith discrete particle.

MASS DEFECT. The difference Δ between the atomic number A and the atomic mass M of a nuclide, $\Delta = A - M$. The negative of the mass defect, $-\Delta$, is known as the mass excess.

MASS-ENERGY EQUIVALENCE. The formula $E = mc^2$, which equates a quantity of mass m to a quantity of energy E. The conversion factor c_2 is the square of the speed of light. The relationship was developed from the relativity theory (special), but has been experimentally confirmed. See also **Relativity and Relativity Theory**.

MASS EXTINCTIONS. Estimates of the number of species of plants and animals that have lived on Earth during the past 600 million years (my) range from hundreds of millions to several billion, yet fewer than two million are known to live on Earth today. It is apparent that most species that have lived are extinct. Paleontologists have calculated extinction rates for most of the groups of geologically important organisms that have lived during the past 600 my, and these data suggest some predictability of rates of extinction and evolution. However, the normal or "background" extinction rates of Earth's inhabitants are punctuated by a few intervals when extinction rates were much greater. These were times of drastic reduction of biologic diversity and resulted in biologic catastrophes that are called *mass extinctions*. The occurrence of mass extinctions is well known, and the major divisions of geologic time for the past 600 my are largely based on the intervals of time between mass extinctions. Thus, the Paleozoic Era ended approximately 245 my ago at a mass extinction event that may have affected between 77 and 96% of all marine invertebrate species while significantly reducing Earth's terrestrial fauna. The younger Mesozoic Era ended approximately 65 my ago when approximately 60 to 75% of marine species became extinct; this mass extinction is thought to have included dinosaurs and many other reptiles. There were other major extinction events during the past 600 my, the most important of which are noted on the Fig. 1. In addition to the events noted, different paleontologists believe that there may have been 1 to 3 additional times of major mass extinction and perhaps 8 or more times of minor mass extinction. The events noted at 245 my and 65 my are the most severe and dramatic of all mass extinctions. Most of the data available are based on marine organisms, but include information on terrestrial vertebrates as well. Although the data are small, it appears that plants may have been more resistant to mass extinctions than animals.

Documentation of extinction is produced from specialized actuarial Tables that are calculated to demonstrate background extinction levels for genera and families of organisms. These calculations are convincing and suggest that extinction of any group of organisms may be the expected part of the evolutionary process. The explanations for the extinction of a large number of groups during a very brief interval of geologic time are much more complicated. Background extinction is the result of assorted kinds of ecologic interactions and biologic changes, and the extinction of an entire family comprised of multiple species of organisms may or may not be due to some shared deficiency. In contrast, mass extinctions may be the result of a single event that, together with feedback, amplifies the catastrophe and cuts across ecologic boundaries in the elimination of a diversity of unrelated organisms.

One of the most profound hypotheses of the past 10 years is that the Cretaceous mass extinction (an event that occurred approximately 65 my ago) may have resulted from the impact of an asteroid of approximately 10 km in diameter. It is proposed that the direct damage caused by such an impact fractured Earth's crust (and perhaps the upper mantle), damaging an area of several hundred or thousand km. Heat generated and released from the impact would affect a large part of Earth, perhaps burning large vegetated regions. If the initial impact occurred in the ocean, this would result in the generation of tsunamis that could destroy life living in low-lying coastal regions. Release of crust/mantle chemicals following impact may have poisoned surface waters, affecting additional forms of marine life. It is theorized that debris from the impact would result in the formation of a gigantic dust cloud consisting of pulverized crustal particles ejected from the impact and that the cloud probably would be so dense that it could effectively block solar radiation and plunge Earth into darkness. Calculations on the period of darkness resulting from the dust cloud range from a few days to a few years. Just how long would depend on poorly understood factors of atmospheric circulation 65 my ago. Much smaller volcanic eruptions during historic times have produced atmospheric dust that has affected solar radiation and altered normal Earth weather patterns, sometimes for a period of several years. Reduction or even elimination of photosynthesis could occur if the duration of such a cloud was prolonged. Any reduction of photosynthesis, in turn, would affect the food chain both in the oceans and on land. The demise of autotrophic organisms would have a chain reaction on heterotrophic organisms at several different levels. Organisms that survived the immediate effects of heat and/or tsunamis would be affected by such a massive disruption of the food chain. An eventual cleansing of the atmosphere by an acid rain of Armageddon proportions might be the final straw for survivors of the earlier sequence of events. It is proposed that Earth's climate 65 my ago might even have been changed by such a chain of events, shifting from what is known to have been a mild climate to a worldwide cooling that is also known to have occurred.

But other than the general Earth cooling known to have occurred at this time, what is the evidence to support such a scenario? There appears to be general agreement that between 60 and 75% of marine species living during the Cretaceous Period became extinct during a very short period of time at the close of the Cretaceous, generally dated as 65 my ago. Terrestrial organisms, including the dinosaurs, flying reptiles, and a few other types, became extinct at approximately the same time. Most important for this theory is the 1980 discovery of the chemical element iridium (Ir) in abundances well above that of normal Earth crustal levels concentrated in the sediment that is believed to have accumulated during the time of the 65 my mass extinction event. Additional investigations showed similar non-Earth abundances of Ir worldwide, all occurring in rocks dated at 65 my. Such concentrations of Ir are most common in asteroids, but not in Earth's crust. Associated with the anomalously high Ir at many sites, geologists found a fractured sediment that is most easily explained as a result of tremendous pressures that could result from an asteroid impact. Additionally, soot accumulations interpreted to be the evidence of widespread burning on Earth have been found at the 65 my old layer at many localities.

Of considerable interest during the past few years has been the search for the actual impact crater that should have resulted from the theorized 65-my-old asteroid collision with Earth. After years of tabulating information on the various impact craters that are still preserved on Earth, a group of researchers believe that the site of the asteroid impact 65 my ago is located in the Gulf of Mexico, adjacent to the north coast of the Yucatan peninsula centered at Chicxulub. This 180-km wide submarine fossil impact scar is difficult to study, but, fortuitously, it had been drilled by oil companies so that a great deal of the geologic framework for the structure is known.

Evidently the Chicxulub crater is the right size and age to accommodate almost everything that has been proposed by the proponents of an asteroid impact as the cause of the mass extinction 65 my ago. The evidence from this crater structure includes the right geologic age for the impacted sediment (Late Cretaceous), believable age melt rocks within and surrounding the crater, and the remarkable similarity in the age of the crater melt rocks and ejecta that have been found great distances away. In fact, all of the stratigraphic, sedimentologic, petrographic, geochemical, and geophysical data that are studied at the Chicxulub crater are consistent with the idea that this is the site of an asteroid impact 65 my ago and may be the "smoking gun" that proves the asteroid impact/mass extinction theory.

However, it is extremely important to note that the imprecision of dating events occurring millions of years ago casts a degree of uncertainty on the precise contemporaneity of all the data considered to be critical for the theory. If higher resolution of geologic time is accomplished, it is possible that the 65-my-old event (or other mass extinction events that are considered contemporaneous) may turn out to have occurred over a longer time interval than could be explained by the single impact theory. Proponents of the theory are aware of this and suggest that multiple impacts over a short period of geologic time (perhaps up to 1 million years) could also explain the data that are accumulating on mass extinction.

At almost the same time that the impact evidence was accumulating for the Late Cretaceous extinction, it was suggested that perhaps all or most of the major mass extinctions during the geologic past may have occurred as the result of similar asteroid impacts. See Fig. 1. These were produced by periodic perturbations of our solar systems by cyclic encounters with an asteroid swarm or by some planet whose orbit has been obscured by

the sun. Periods of between 26 and 30 my, were calculated for the known times of mass extinctions (at least for the last 100 my), and this periodicity is assumed to be that of the source of the perturbations. In fact, the reality of periodicity is still debated, and the arguments for periodicity of asteroid impacts are tempered by the fact that a worldwide Ir anomaly as definitive as that found in the sediment of the Late Cretaceous (65 my) extinction has not been found at any of the other mass extinction levels of the Paleozoic or Mesozoic.

Even though most of the publicity for mass extinctions has featured the Late Cretaceous event (most likely because of the extinction of dinosaurs at that time), this event was not as significant as the mass extinction that marks the close of the Paleozoic Era, 245 my ago. This event, some 180 my earlier than the Cretaceous event, is considered to be the most profound of all of Earth's mass extinctions. It is reported to have resulted in the extinction of up to 96% of all marine species, and many terrestrial species were affected as well. There is no Ir anomaly for this event as there is for that of the Late Cretaceous, and some evidence supports the idea of multiple causes for the mass extinction at this time. A great deal of evidence points to widespread volcanic eruptions 245 my ago. The volcanic theory is supported by recent dating of Siberian basaltic flows and volcanic tuff that are within the theoretical time range of the Late Permian (245 my) extinction event. According to this theory of mass extinction, widespread volcanism may have injected large amounts of sulfur dioxide into the atmosphere, resulting in a decrease in solar radiation and consequent global cooling that affected marine and terrestrial life. This extinction event of 245 my ago is not as well studied, nor are the data as well controlled as that of the mass extinction of 65 my ago. In addition to the known volcanic activity of this time, other specialists point out that there may have been catastrophic ecosystem devastation produced by reorganization of Earth's continental shelves during this time that also was a time of major plate tectonic activity.

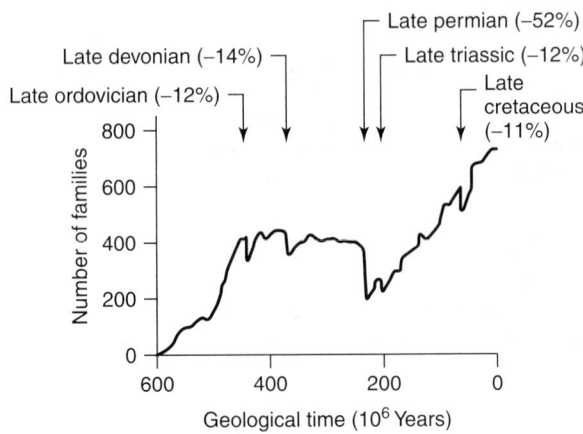

Fig. 1.   Chart indicating five of the six to eight major mass extinction events, with the relative magnitudes of these events shown in terms of percentage of families of marine invertebrates and vertebrates affected. (After Raup and Sepkoski, 1982.)

An even older major mass extinction is that of the Late Devonian (~367 my). Recent reports of glass spherules, interpreted to represent material similar to the microtektites (known from the 65-my-old extinction event), and the tentative identification of a crater in the Baltic Shield of approximately the same age, suggest that an asteroid impact may be responsible for this mass extinction. This is perhaps the closest match to date for the Late Cretaceous event.

The interest generated by these kinds of studies has stimulated a great amount of research for each of the major and minor mass extinction events. Paleontologists and geologists continue to study possible causes, and, although the reality of multiple mass extinctions of Earth's organisms at intervals during the past 600 my is fairly well established, certainly there is no consensus on the actual mechanism of extinction. The asteroid impact proposed to explain the Late Cretaceous event is attractive, but has a few problems that bother some specialists. The volcanic theory for the Late Permian event is supported by a variety of data, but even these can be interpreted differently. The precise timing of the extinction of different groups of organisms is difficult to establish, and the older the extinction event, the more difficult it is to unequivocally establish synchroneity of extinction of different species. Understanding and organizing all of the components involved in mass extinction is analogous to the creation of a picture from pieces of a jigsaw puzzle. Because some of the pieces of the mass extinction puzzle have been lost during the millions of years since the occurrence of the event, we may never be able to complete the whole picture. But we are beginning to understand the picture of mass extinction to a much greater extent than was possible even a few years ago.

Additional Reading

Alvarez, L.W., W. Alvarez, F. Asaro, and H.V. Michel: "Extraterrestrial Cause for the Cretaceous-Tertiary Extinction," *Science*, **208**, 1095–1105 (1980).

Bernhard, T.: "Extinction," University of Chicago Press, Chicago, IL, 1996.

Campbell, I.H., G.K. Czamamske, V.A. Fedorenko, R.I. Hill, and V. Stepanov: "Synchronism of the Siberian Traps and the Permian-Triassic Boundary," *Science*, **258**, 1760–1763 (1992).

Claeys, P., J-G Casier, and S.V. Margolis: "Microtektites and Mass Extinctions: Evidence for a Late Devonian Asteroid Impact," *Science*, **257**, 1102–1104 (1992).

Clark, D.L., C.Y. Wang, C.J. Orth and J.S. Gilmore: "Conodont Survival and Low Iridium Abundances Across the Permian-Triassic Boundary in South China," *Science*, **223**, 984–989 (1986).

Courtillot, V.: "Evolutionary Catastrophes: The Science of Mass Extinction," Cambridge University Press, New York, NY, 1999.

Eldredge, N.: "The Miners Canary: Unraveling the Mysteries of Extinction," Princeton University Press, Princeton, NJ, 1994.

Friedman, G. et al. (Editors): "Extinction Events in Earth History: Proceedings of the Project 216: Global Biological Events in Earth History," Springer-Verlag New York, Inc., New York, NY, 1990.

Kitchell, J.A. and D. Pena: "Periodicity of Extinctions in the Geologic Past: Deterministic Versus Stochastic Explanations," *Science*, **226**, 689–692 (1984).

Ramino, M.R. and R.B. Stothers: "Geological Rhythms and Cometary Impacts," *Science*, **226**, 1427–1431 (1984).

Raup, D.M. and J.J. Sepkoski, Jr.: "Mass Extinctions in the Marine Fossil Record," *Science*, **215**, 1501–1503 (1982).

Swisher, C.C.III, J.M. Grajales-Nishimura, A. Montanari, S.V. Margolis, P. Claeys, W. Alvarez, P. Renne, E. Cedillo-Pardo, F.J-M.R. Maurrasse, G.H. Curtis, J. Smit, and M.O. McWilliams: "Coeval^{40}Ar/^{39}Ar Ages of 65.0 Million Years Ago from Chicxulub Crater Melt Rock and Cretaceous-Tertiary Boundary Tektites," *Science*, **257**, 954–958 (1992).

DAVID L. CLARK, Ph.D., University of Wisconsin–Madison.

MASS-LUMINOSITY RELATION. Observationally, it is found that a plot of absolute bolometric magnitude against mass for giant and main sequence stars reveals a definite correlation. The observational diagram is shown in Fig. 1. Using only visual binaries, Eddington observed this trend and then deduced the relation theoretically. One may conveniently use the approximation

$$\log L = 3.3 \log M$$

where L and M are in solar units.

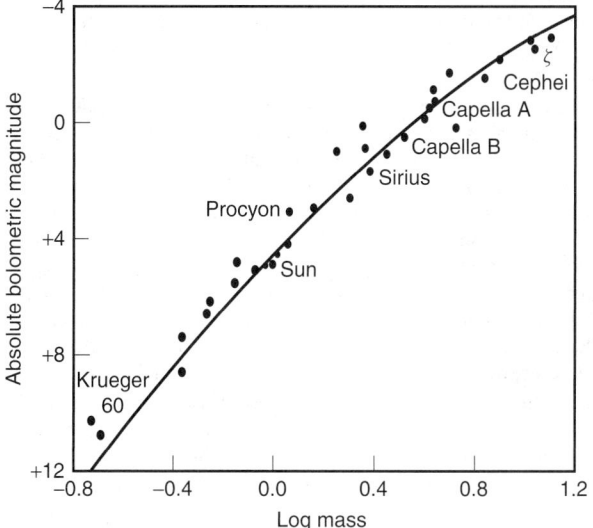

Fig. 1.   Mass-luminosity curve. (*Eddington.*)

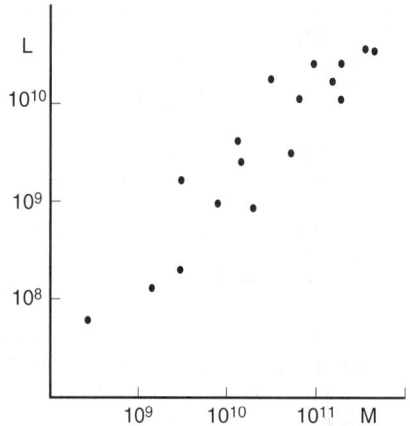

Fig. 2.   Mass-luminosity relation for selected galaxies.

A similar diagram can be constructed for galaxies, and is shown in Fig. 2. Again, a clear relationship is exhibited, but there is no theoretical explanation for this fact.

MASS NUMBER. The total number of nucleons in the nucleus of an atomic species is its mass number, which then is numerically equal to the sum of the atomic number and the neutron number of the species. See also **Chemical Elements**.

MASS SPECTROMETRY. A general quantitative and qualitative analyzer for most components in all types of samples — gas, liquid, or solid, but with some volatility limitations. A complete analysis is obtained from nanogram samples in a few seconds or minutes. The range is from parts per billion to 100% purity. Accuracy is ±1%; the specificity is good. Conventional methodology is first described after which recent trends are presented.

Operating Principle. With reference to Fig. 1, the sample to be studied is introduced into an evacuated area (ion source), where it is ionized, accelerated by an electrostatic field, and separated according to mass. The various masses are collected and measured.

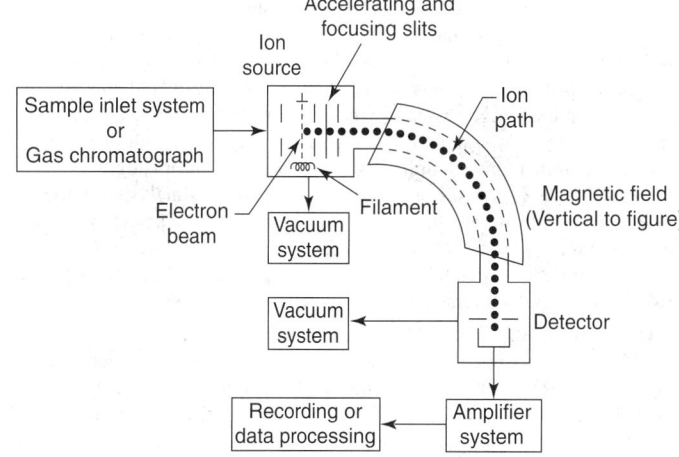

Fig. 1.   Basic elements of conventional mass spectrometer system.

Production of Ions. Several methods are used: (1) by bombardment with electrons from a heated filament; (2) by application of a strong electrostatic field (field ionization, field desorption); (3) by reaction with an ionized reagent gas (chemical ionization); (4) by direct emission of ions from a solid sample that is deposited on a heated filament (surface ionization); (5) by vaporization from a crucible and subsequent electron bombardment (e.g., Knudsen cell for high-temperature studies of solids; and (6) by radio-frequency spark bombardment of sample for parts-per-billion (ppb) elemental analysis of solids as encountered in metallurgical, semiconductor, ceramics, and geological studies. Ions also are produced by photoionization and laser ionization.

Fragmentation. Ionization usually is accompanied by partial fragmentation of the molecule. The fragmentation pattern is constant for a specific molecule and operating conditions. Fragmentation complicates computation, but permits distinguishing between isomers and gives molecular-structure information.

Ion Separation. After acceleration, the ions are focused and separated according to mass. The most common separating means is a magnetic field, which causes the ions to follow curved paths of radii proportional to their masses. Many different geometrics are used. The masses may be scanned by varying the accelerating voltage or magnetic field. Other separation means include: (1) combinations of electrostatic fields (double focusing cycloidal focusing); (2) crossed alternating electrical fields (quadrupole mass filter); and (3) use of a filled-free drift tube combined with pulsed ion source and gated detection (time-of-flight). Still additional means are omegatron, radio frequency, and cyclotron resonance.

Detection. Commonly used detection means are: (1) *electrical* — ion beams are successively scanned across a collector where they pick up electrons. The resulting current is amplified with an electron multiplier and/or electrometer and recorded or computer-processed. (2) *photographic plate* — ions strike a photographic plate, activating the emulsion and thus giving a line for each mass, after development. Line intensity is proportional to ion abundance. All masses are recorded simultaneously.

Sample Introduction. A variety of inlet systems is available for gases, liquids, and solids: (1) *heated batch inlet.* The sample is expanded into a volume at about 50 micrometers pressure and bled into the ion source through a molecular or viscous leak. This method will handle gases, liquids, and solids with vapor pressures above approximately 1 torr at 350 °C. (2) *Direct introduction system.* The sample is inserted directly into the ion source through a vacuum lock on a heatable probe. Liquids and solids with vapor pressures above approximately 10^{-7} torr at 350 °C can be handled. (3) *Direct insertion by venting*, or through a vacuum lock (spark or surface ionization). This method is well suited for involatile samples. (4) *Gas chromatograph interface.* A continuous inlet which permits mass spectrometric analysis of the separated components as they emerge from the gas chromatograph.

The combination of gas chromatograph and mass spectrometer provides a separating and identifying capability not achievable by other means, particularly for very small samples.

Data Processing. The high data output rate of many mass spectrometer systems requires data processing to fully utilize the capability of the instrument. The present trend is toward systems using small dedicated computers with digital tape, core, or disk memory, and printer, plotter, and cathode-ray tube output. Typical outputs available are: (1) mass and abundance printouts; (2) mass versus abundance plots; (3) elemental composition printouts (high-resolution mass spectrometer); (4) "total ionization chromatogram" plot (summed ion plot, similar to gas chromatogram); (5) mass chromatogram (similar to total ionization chromatogram, but for a selected mass or masses only); and (6) quantitative analysis printout.

Outputs are available from raw data, data with background or other spectra subtracted, and normalized data. A number of other options of data selection, manipulation, and output are usually available from the instrument manufacturers.

Uses and Applications. Common uses for mass spectrometers include: (1) compound identification; (2) elemental formula determination; (3) molecular-structure determination; (4) quantitative mixture analysis; (5) ppb solids elemental analysis; (6) isotope ratio determination; (7) leak detection (helium tracer); (8) residual-gas analysis in vacuum systems; and (9) age dating.

Mass spectrometers are widely used for studies and determinations in: organic chemistry; petroleum and biological laboratories; nuclear investigations; geochemistry and cosmochemistry; metallurgical, semiconductor, and ceramics investigations; pollution control; space programs; agricultural and pesticide research and manufacture; flavors and fragrances chemistry; and as a basic research tool in studies of ion-molecule reactions, high-temperature chemistry kinetics, free radicals, and thin films. Broad classes of instruments are summarized in Table 1.

Advancements in Mass Spectrometers

The great strides made by mass spectrometry since its inception several years ago are aptly put forth by Delgass and Cooks (see reference), who describe the status of mass spectrometry as of the late 1980s. The

TABLE 1. TYPES AND CHARACTERISTICS OF MASS SPECTROMETERS

General Class	Type	Major Uses
Leak detector	Magnetic analyzer (helium only)	Detects small leaks, using helium gas tracer.
Residual-gas	Magnetic or quadrupole analyzer	Analysis of gases in vacuum systems
Low-resolution	Magnetic, cycloidal, or quadrupole analyzer	Analysis of gases or light liquids
Medium-resolution	Magnetic analyzer	Identification, molecular-structure studies, mixture analysis, isotope ratios.
High-resolution	Magnetic and electric (mass and energy) focusing	Molecular-structure determinations, ppb solids analysis, isotope ratios of solids, high temperature chemistry

applications of mass spectrometry have penetrated into physical chemistry (bond dissociation energies, ion enthalpies, proton affinities); organic chemistry (structure studies, organic ion structure and fragmentation); biology (drug metabolism, stable isotope tracer work, modifications in biopolymers); the earth sciences (chronology/dating of geological and life extinction events); and environmental science (trace organic analysis).

Mass spectrometry is undergoing a rapid development that shows little indication of abating either in the areas of instrument refinement or of extended applications. Mass spectrometry is expected to play a major role in the revitalized science of materials and surface phenomena.

Tandem Mass Spectrometry. Coupling mass spectrometers in series has many advantages for the analysis of specific organic compounds in complex mixtures. Sensitivity to picograms of targeted compounds can be achieved with high specificity and almost instantaneous response. As reported by McLafferty (see reference), the targeted compound is selectively ionized, and its characteristic ions are separated from most others of the mixture in the *first* mass spectrometer. The selected primary ions then are decomposed by collision and, from the resulting products, the *final* (or second) mass analyzer selects secondary ions characteristic of the targeted compound. Tandem mass spectrometry (MS-MS) can achieve specificities and sensitivities equivalent to those of methods such as radioimmunoassay and gas chromatography/mass spectrometry, while performing analyses in much shorter times. Just a few of the materials and systems successfully studied by tandem mass spectrometry include: polynuclear aromatics, DNA pyrolysis, steroid mixtures, ion structures, stereoisomers, pyrolysis of bacteria, alkaloids in plants, penicillins, polychlorodibenzodioxins, petrochemicals, drug metabolites, enkephalins, peptide mixtures, diesel exhaust, odors in the air, concealed drugs, parathion in lettuce, and ion plasmas.

Tandem mass spectrometry has been particularly effective in molecular structure determinations. To increase the number and absolute abundance of peaks in the secondary mass spectrum, it is necessary to add energy to the separated primary ions. Collisionally activated dissociation (CAD) is frequently used.

Fourier Transform Mass Spectrometry (FTMS). This technique enables chemists to use mass spectrometry in expanding and new ways. The technique enables the measurement of high molecular masses and, by application of ion manipulation (MS-MS), molecules can be degraded into more manageable pieces for which accurate mass measurement is feasible. As observed by Gross and Rempel, the unique applications of FTMS result from its ability to store ions. Ions in the trap can be reacted in very specific ways (chemical ionization) or activated by using lasers to give photodissociation spectra. Instead of operation at 1 torr of pressure as in the case of conventional mass spectrometers, FTMS chemical ionization must be conducted at 10^{-6} torr of reagent gas and 10^{-8} torr of sample.

Coupling Lasers with Mass Spectrometers. The analysis of inorganic atomic species can be facilitated by using lasers with mass spectrometers. A tunable dye laser, by itself or in combination with a pump laser, ionizes atoms by resonant excitation processes. The ions are then analyzed in the mass spectrometer. This combination of techniques has much potential for overcoming traditional limits of sensitivity and selectivity and will lead to increasing applications in analytical chemistry. As observed by

Fassett, et al. (see reference), a large potential exists for the application of *multiphoton resonance ionization mass spectrometry*, ranging from basic spectroscopic studies of atoms and molecules to the detection of solar neutrinos and quarks. Discovery of the optogalvanic effect (Green, et al.) and earlier work (Hurst, et al.) led to the development of *laser-enhanced ionization* (*LEI*). It has been determined that elements suitable for resonance ionization by the one-photon-resonant, two-photon ionization scheme (wavelengths between 260 and 355 nm) include: Na, Ca, Ba, Cr, Mn, Sc, Cr, Mn, Ru, Rh, Pd, Pt, Au, Ga, Ge, Sn, Bi, Eu, Gd, Tb, Ho, Tm, and Yb. Elements for which resonance ionization has been demonstrated in the laboratory include: Li, Mg, Sr, Y, Ti, Zr, Hf, V, Ta, Mo, W, Re, Fe, Os, Co, Ni, Al, In, Ge, Sn, Pb, Dy, Er, Lu, Th, and U. Some elements also have been ionized by the two-photon-resonant, three-photon ionization scheme, including: Be, C, and I.

It is reported by Fassett, et al. that the resonance ionization process has an inherently high elemental selectivity. Mass spectrometric detection provides an increased selectivity that is a practical necessity for analytical problems in which nonspecific background ionization must be characterized and differentiated. Resonance ionization is ideally suited to mass spectrometry; ionization is well defined in both time and space, and only a small excess of translational kinetic energy is added to the atom by the process.

Nuclear Accelerators and Mass Spectrometers. An interesting area is that of using nuclear accelerators as high-sensitivity mass spectrometers. As recently as a decade ago, the Grenoble cyclotron was used as a mass spectrometer to measure ratios of $^{10}Be/^9Be$ of $10^{-8}, 10^{-9}$, and 10^{-10} in standardized beryllium oxide samples. Measurements of this type also can be used to determine cosmogenic ^{10}Be profiles in various geophysical reservoirs, such as sea sediments and polar ice. See also **Particles (Subatomic)**.

Mass Spectrometers as Process Analyzers. The use of mass spectrometers in process analysis has not been widespread because of its perceived complexity and high cost. Technological advancements over the past decade or two have reduced costs and simplified mass spectrometry to the point where it is suitable for a number of process analytical applications in place of infrared absorption or gas chromatography (GC) techniques. The *quadrupole mass spectrometer* is well accepted in such applications because of low cost, good reliability, and ease of computer-controlled interfaces.

Simply, a quadrupole mass spectrometer can be divided into three parts: (1) the ionizer, (2) the mass filter, and (3) the detector, all of which are contained in a vacuum chamber maintained at a low pressure. When a gaseous sample is introduced into the system's ionizer, it is bombarded with a stream of electrons, producing positively charged parent ions (ions with the same molecular weight as the neutral molecule), and fragment ions. The ionizer has a series of lenses that serve to collimate the cloud of sample molecules toward the mass filter.

A quadrupole mass filter is a set of four rods disposed parallel and symmetrically with one another; opposite rods are electrically connected. An rf and dc voltage of equal potential, but opposite charge, is applied to each set of rods (Fig. 2). By varying the absolute potential applied to the rods, it is possible to stop all ions except those of a given m/e (mass-to-charge) ratio.

Finally, the ions flowing down the quadrupole strike the Faraday plate detector. In some cases, the signal is amplified further by an electron multiplier. Thus, there is obtained a spectrum of signal intensity versus m/e value. Each molecule has a unique fragmentation pattern so that a spectrum can be used as a fingerprint for compound identification. In addition, it is possible to quantitate the amount of a particular compound by comparing sample signal intensity with the intensity produced by a known amount of the compound.

Many applications of the quadrupole mass spectrometer use a gas or liquid chromatograph to introduce the sample into the ionizer. When the spectrometer is used in this manner, it is most common to scan a wide mass range (50–1000 amu) at rates on the order of 1000 amu/second for compound identification. For process analyses, it is most common to introduce the sample directly into the ionizer and scan a shorter mass range. For both applications, computer systems are needed to collect the enormous amounts of data produced.

The computer is also used to control the potential applied to the quadrupole in order to scan the mass range of interest. Alternatively, the computer can command the appropriate potential required to focus

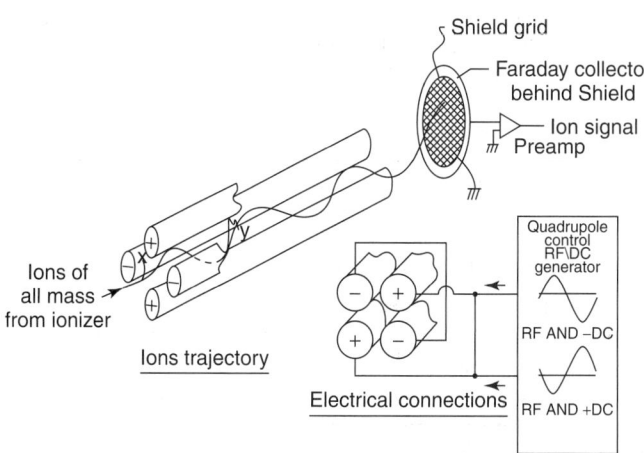

Fig. 2.   Schematic diagram of quadrupole mass spectrometer of type used for process analysis. (*Extranuclear Laboratories.*)

a particular mass, a technique referred to as Selected Ion Monitoring (SIM). Process analysis is essentially continuous SIM to achieve online quantitation of the components in the process stream.

Once the stream components to be analyzed have been identified, the mass spectrum of each component must be compared to the spectra of all other components to find unique mass peaks. Not all constituents of the mixture will have unique peaks; some may have intense peaks that might be useful to measure that compound except that this mass peak also exists in another compound. Considering a stream containing N_2, CO, H_2O), and CO_2, Table 2 shows the mass peaks in each component's spectrum and their relative intensities using the largest peak from each compound as 100% in each case. Inspection of the table shows that m/e 18 and m/e 44 are unique peaks for H_2O and CO_2, respectively. However, the parent molecular ions of both N_2 and CO appear at m/e 28. Thus, some other method of measuring these components must be applied.

TABLE 2. MASS PEAKS IN EACH COMPONENT'S SPECTRUM AND THEIR RELATIVE INTENSITIES. (Largest Peak from each compound is used as 100% in each case)

Mass	H_2O	CO	N_2	CO_2
1	6.00			
12		4.00		9.00
14			5.00	
15			0.03	
16	4.00	1.00		10.00
17	28.00			
18	100.00			
28		100.00	100.00	11.00
29		1.00	0.73	0.09
30			0.02	
44				100.00
45				1.00
46				0.04

One might choose to measure CO at m/e 12 even though CO_2 has a fragment at this mass. Since the CO_2 intensity can be determined by measuring the peak at m/e 44, the CO_2 contribution to mass 12 is known and can be subtracted. The resultant at m/e 12 is then a measurement of the CO present in the mixture.

Two possibilities exist for the N_2 measurement; measure m/e 28 and subtract the CO contribution since CO intensity is known from the step above, or measure the intensity at m/e 14 and subtract the contribution from CO.

In practice, the spectral information used for calibration is obtained by actually introducing the compound of interest into the mass spectrometer and measuring the spectrum. This is usually done by using a known binary mixture of the particular compound in nitrogen or argon. The ratio of each mass peak compared to some standard ion, for example, m/e 28 of nitrogen, is measured. The fragmentation of the peaks of interest are then programmed into a fragmentation matrix. For each compound to be measured, only one mass peak is programmed.

For the analysis of N_2, CO, H_2O, and CO_2, the fragmentation matrix becomes:

	12	18	28	44
CO	4.00*	0.00	100.00	0.00
H_2O	0.00	100.00*	0.00	0.00
N_2	0.00	0.00	100.00*	0.00
CO_2	9.00	0.00	11.00	100.00*

The peaks to be measured for each component are indicated by an asterisk. The subtraction of interfering mass peaks is accomplished by inverting the matrix to yield:

	12	18	28	44
CO	25	0	0	−2.25
H_2O	0	1	0	0
N_2	−25	0	1	2.14
CO_2	0	0	0	1

In the actual analysis, the amplitudes of peaks at masses 12, 18, 28 and 44 are measured and multiplied by the inverted matrix. For example:

	12	18	28	44	
CO	25	0	0	−2.25	I(12)
H_2O	0	1	0	0	I(18)
N_2	−25	0	1	2.14	I(29)
CO_2	0	0	0	1	I(44)

Therefore, CO intensity is determined by:

$$25[I(12)] + 0[I(18)] + 0[I(28)] - 2.25[I(44)]$$

H_2O, N_2, and CO_2 are similarly calculated. Intensities may be converted to percent concentration by dividing by the component's sensitivity and normalizing to 100%. The formulae for these calculations are therefore:

$$INT(J) = IBKG(J) - BKG(J)$$

$$CONC(J) = \text{Sum over J } [INVER(I,J) * INT(J)]/SENS(J)$$

$$\%CONC(J) = CONC(J) * 100/\text{Sum over J } [CONC(J)]$$

A modern mass spectrometric process analyzer should be designed for rapid response time, wide dynamic range, accuracy, and ease of computer operation. Because the instrument is a nonspecific detector, it is useful in detecting almost any gaseous compound that produces ions in the mass spectrometer's range (typically 200 amu). The instrument is adaptable to process analysis because it is easily and rapidly directed to monitor any mass ion peak within its range. One modern instrument is controlled by a computer system, which commands masses, gathers and reduces data, and presents useful information almost instantly to the operator. The instrument enables the operator to design analysis, calibration, and sampling sequence methods to optimize conditions for a particular application, An analysis method is designed by specifying the molecular weight and the mass to be monitored for each of the components to be analyzed. The data system then searches a library (electronically stored) for the spectra and sensitivity of the compounds of interest. The fragmentation matrix used for the analysis is thereby constructed. Changing the parameters for any or all compounds can be easily accomplished by the operator. Complex streams may require more than one calibration gas to properly calibrate all stream components.

The analyzer has three modes of operation: (1) *Sweep Mode* — the system can be commanded through the CRT keyboard to sweep the quadrupole over a specified range with the data displayed on the oscilloscope. This mode is useful for tuning or to determine the masses present in a particular stream. (2) *Manual Mode* — during real-time operation, the operator can display and change parameters in order to fine-tune a particular analysis. Data can be presented as intensity (raw signal voltages), percent concentration, or taken as a ratio to a specific component. Optionally, data can be printed or electronically stored. (3) *Automatic Analysis*?

Additional Reading

Adams, F., R. Gibbels, and R. Van Grieken: "Inorganic Mass Spectrometry," 2nd Edition, John Wiley & Sons, Inc., New York, NY, 2001.

Barker, J.: "Mass Spectrometry," 2nd Edition, John Wiley & Sons, Inc., New York, NY, 1997.

Chait, B.: "The Application of Matrix-Associated Laser Desorption Time-of-Flight Mass Spectrometry to the Analysis of Proteins," Pittcon, Pittsburgh, PA, (March 1992).

Conzemius, R.J.: "Prospects for Radically New TOF-MS Instrumentation; Making Single-Ion Concepts Provide Higher Performance," Pittcon, Pittsburgh, PA, (March 1992).

Cotter, R.J.: "Designing Time-of-Flight Instruments to Solve Biological Problems," Pittcon, Pittsburgh, PA, (March 1992).

Dass, C.: "Principles and Practice of Biological Mass Spectrometry," John Wiley & Sons, Inc., New York, NY, 2000.

Delgass, W.N. and R.G. Cooks: "Focal Points in Mass Spectrometry," *Science*, **235**, 545–552 (1987).

Enke, C.G.: "Time-of-Flight Mass Spectrometry with Time-Array Detection," Pittcon, Pittsburgh, PA, (March 1992).

Fassett, J.D. et al.: "Laser Resonance Ionization Mass Spectrometry," *Science*, **230**, 262–267 (1985).

Fenn, J.B. et al.: "Electrospray Ionization for Mass Spectrometry of Large Biomolecules," *Science*, 64 (October 6, 1989).

Goeringer, D.E., G.L. Glish, and S.A. McLuckey: "Fixed-Wavelength Laser Ionization/Tandem Mass Spectrometry for Mixture Analysis in the Quadrupole Ion Trap," Pittcon, Pittsburgh, PA, (March 1992).

Grant, E.R. and R.G. Cooks: "Mass Spectrometry and Its Use in Tandem with Laser Spectroscopy," *Science*, 61 (October 5, 1990).

Green, R.B. et al.: *J. Am. Chem. Soc.*, **98**, 8517 (1976).

Hoffmann, E., De, V. Stroobant, and J. Charette: "Mass Spectrometry: Principles and Applications," John Wiley & Sons, Inc., New York, NY, 1996.

Hurst, G.S., J.H. Nayfeh, and J.P. Young: *Phys. Rev. A*, **15**, 2283 (1977).

McLafferty, F.W.: "Studies of Unusual Simple Molecules by Neutralization-Reionization Mass Spectrometry," *Science*, 925 (February 23, 1990).

Siuzdak, G.: "Mass Spectrometry for Biotechnology," Academic Press, Inc., San Diego, CA, 1996.

Watson, J.T.: "Introduction to Mass Spectrometry," Lippincott-Raven Publishers, Philadelphia, PA, 1997.

MASTICATORY SUBSTANCES. The property of chewiness is one of several components that comprise so-called mouth feel experienced by the consumer of a food product. Whereas chewiness may be highly undesirable in a cut of roast beef, this property is the predominant rewarding factor of some food products, notably certain kinds of novelties, such as chewing gum. Biting and deformation resistance can be created or improved by the use of a number of essentially rubber-like substances. These are commonly termed *masticatory substances*. Chewiness is the main advantage contributed by such substances and thus other ingredients, such as sweeteners, flavorings, colorants, etc. are admixed with them to result in an overall attractive product for particular consumers.

Masticatory substances are of (1) vegetable origin, or (2) the products of organic synthesis. In the case of anhydrous lanolin, the source is fat from the wool of sheep.

Vegetable Substances. Masticatory substances derived from vegetables are gums from various plants and trees of the families *Apocynaceae* (dogbane family), *Euphorbiacea* (spurge family), *Moraceae* (mulberry family), and *Sapotaceae* (sapodilla family). Most of the naturally derived substances have unfamiliar names, usually known well only by persons in the trade. For example, from the *Apocynaceae* family are obtained jelutong, leche caspi or sorva, pendare, and perillo. From the *Euphorbiaceae* family, there are candelilla wax, chilte, and natural rubber (latex solids). From the *Moraceae* family, there are leche de vaca, Niger gutta, and tunu or tuno. From the *Sapotaceae* family, there are chicle, chiquibul, crown gum, gutta hang kang, gutta katiau, massaranduba balata, nispero, rosidinha, and Venezuelan chicle. Other naturally derived substances include lanolin, petroleum wax, rice bran wax, and natural terpene resin.

Synthetic Substances. During the last several decades, naturally derived masticatory substances have been displaced to a considerable degree by synthetic materials — for reasons of availability, economics, and, frequently, better control over purity. These developments essentially paralleled the development of the synthetic rubbers for industrial uses. Some of the synthetic substances now used and listed in the "Food Chemicals Codex," published by the National Academy of Sciences (Washington, DC) include: butadiene-styrene 75/25 rubber; butadiene-styrene 50/50 rubber; glycerol ester of partially dimerized rosin; glycerol ester of partially hydrogenated wood rosin; glycerol ester of tall oil rosin; isobutylene-isoprene copolymer (butyl rubber); methyl ester of rosin (partially hydrogenated); paraffin (synthetic, by Fischer-Tropsch process);

pentaerythritol ester of partially hydrogenated wood rosin; polyethylene; polyisobutylene, polyvinyl acetate; and terpene resin (synthetic).

MASTITIS. Acute infection of the breast. The breast becomes swollen, red, and tender. Fever and malaise generally accompany the infection. Lactation is halted and antibiotics are administered. Further relief may be provided by a breast support and cooling with ice bags. In some cases, formation of a breast abscess will require incision and drainage.

Staphylococcal mastitis is most frequently seen in nursing mothers in the early postpartum period, but may occur in neonates or women who are not lactating, or have not been recently pregnant. Group B streptococci, once considered exclusively as animal pathogens causing bovine mastitis, have been clearly established as important human pathogens.

Bovine Mastitis. This is an inflammation of the udder and is the most costly disease to dairypeople.

Most of the economic loss from the disease results: (1) when cows produce less milk, because of permanent udder damage; (2) when cows are culled, because they do not respond to therapy or do not produce enough milk; (3) when milk must be discarded, from cows showing signs of inflammation and from those undergoing antibiotic treatment; and (4) when treatment with antibiotics or other drugs is expensive. Market milk is examined regularly to assure more uniform compliance with regulations prohibiting the sale of milk from diseased cows. When milk is found to contain an excessive number of leucocytes, the producer is warned to take corrective action. If the excessive cell count continues, the producer will be shut off from the market. It is estimated that, from all effects, the disease costs the American dairy industry over $1/2 billion per year.

Over 20 different organisms cause mastitis. All of these can be transmitted from cow to cow. Two types of bacteria most frequently found in any herd are *Streptococcus agalactiae* and *Staphylococcus aureus*. The former bacteria live only in the cow's udder, but *S. aureus* can be harbored in a variety of places in the cow's environment.

With this disease, the cow's udder becomes inflamed, a condition that can be acute or chronic. Acute mastitis is easy to recognize and is dangerous. This condition is usually fatal. In the acute form, the entire body of the cow reacts to the infection. Effects are depression, rough coat, dull eyes, loss of appetite, constipation, fever, and, eventually, death.

In the chronic form, the affected quarter is hot, very hard, and tender. The hardness of the udder is caused by an influx of white blood cells and fluids from the blood to fight the bacterial infection. If the action of the bacteria goes unchecked, some of the normally functioning tissue will be destroyed. It will then be replaced by scar tissue, which can be felt as hard lumps or knots. Eventually, the entire gland atrophies and milk production may stop completely.

MASU SALMON. See **Salmon**.

MATAMATA. See **Turtles**.

MATHEMATICAL PHYSICS (Equations of). The name is sometimes given to a set of partial differential equations of second order, of which the following are the most commonly met with:

(1) the Laplace equation,

$$\nabla^2 \phi = 0$$

and its inhomogeneous analogue, the Poisson equation;
(2) the equation of wave motion,

$$c^2 \nabla^2 \phi = \partial^2 \phi / \partial t^2;$$

(3) the diffusion equation, which also applies to thermal conduction,

$$a^2 \nabla^2 \phi = \partial \phi / \partial t;$$

(4) the equation of telegraphy,

$$a \partial^2 \phi / \partial t^2 + b \partial \phi / \partial t = \partial^2 \phi / \partial x^2.$$

The parameters are observable quantities and t is the time. In modern theoretical physics, the differential equations of quantum mechanics, particularly the Schrödinger wave equation, must be included.

Many special functions (described elsewhere) owe their importance to the fact that they are useful in constructing solutions of these equations. In fact, the name "equations of mathematical physics" is sometimes also

given to the linear ordinary differential equations, which arise when the above partial differential equations are solved by the method of separation of variables. These ordinary equations are all specializations arising from confluence from the generalized Lamé equation.

See also **Bessel Function**; **Lamé Equation (Generalized)**; **Laplace Equation**; and **Poisson Equation**.

MATHEMATICS. Several score entries pertaining to mathematics and related topics appear in this encyclopedia. Consult the alphabetical index. Particularly, check the following key words and phrases: angles, triangles, squares, and polygons; arithmetic and algebra; charts, graphs, tree maps; coordinates; curves; sections and solids; determinants and matrices; differential equations; equations; functions; groups; integrals and integration; interpolation and approximation; series; space and topology; statistics, distribution, and probability; tensors; and vectors.

Mathematical Symbols. The symbolism of mathematics dates back at least three centuries. To a large degree, there is standardization of these symbols. However, one may find optional symbolic devices, each equally acceptable. Simply because of limitations imposed by the English and Greek alphabets, there is some duplication of symbol usage between mathematics and other scientific fields. An abridged list of some of the more commonly used mathematical symbols is given in the Abridged List of Mathematical Symbols.

MATHEMATICS (State of the Art Reviews). During the summer of 1985, a group of state-of-the-art reviews was initiated by the National Research Council (NRC) at the request of the National Science Foundation (NSF). The purpose of these reviews was to assess and monitor world trends, relative strengths, and competitiveness of the United States in rapidly evolving areas of science and technology. Particular emphasis was placed on developments that influence the rate at which these scientific fields evolve. *Applied Mathematics* was one of three such studies undertaken.[1] The study on mathematics was conducted by the Panel on Mathematical Science under the auspices of the Board of Mathematical Sciences of the NRC's Commission on Physical Sciences, Mathematics, and Resources. In its final report the panel described major trends in modern mathematics, illustrating these trends with a few examples. The report, in full, was published in 1986. The study was funded by U.S. Government resources. Chairman of the Panel on Mathematical Sciences for the report was Phillip A. Griffiths, Duke University.[2]

Current Summary. In this 9th Edition of the encyclopedia, an attempt is made to update the 1985 findings.

The executive summary of the 1985 report stresses the following trends in the mathematical sciences:

1. Mathematical sciences research is strong worldwide; the United States is maintaining the leading role.
2. Mathematics is unifying internally.
3. Applications of mathematics in both traditional and new areas are flourishing and involving more central areas of mathematics.
4. Mathematics is the driving force behind new areas of computational science and, in turn, is profoundly influenced by high-speed computing.

There are several corollaries of the trends, which are especially notable in view of a declining number of Ph.D. degrees in the mathematical sciences earned in the United States. The expanding number and sophistication of tools needed for successful research as areas of mathematics become intertwined now require protracted study often beyond the Ph.D. degree. This corollary development is especially critical for those working at high levels of applications of mathematics. The difficulty in reaching mathematical research frontiers with the requisite deep, broad range of knowledge, without the opportunity for extended study, may partially

---

[1] The other two studies: Cell Biology; Materials Science.

[2] Other panel members: Hyman Bass (Columbia University); Peter Bickel and Alexandre Chorin (University of California, Berkeley); Richard Dudley (Massachusetts Institute of Technology); Wendell Fleming (Brown University); Ronald L. Graham (AT&T Bell Laboratories); David Kazhdan (Harvard University); Cathleen S. Morawetz (New York University); Richard Schoen (University of California, Los Angeles); and Michael E. Taylor (State University of New York at Stony Brook).

explain the decreasing Ph.D. production. Continued concern should be expressed over the decline in the number of Ph.D.s in the mathematical sciences in the U.S., and in view of the critical role of mathematics, this trend must be reversed. The competition for students among the various fields of science is keen. Thus it is critical that mathematics be able to provide sufficient inducements to candidates to maintain its vitality. That the U.S. maintains mathematical preeminence is due in part to the commitment and investment made in the past to an outstanding group of researchers. Unfortunately, this commitment has been considerably weakened over the past decade, as was documented in the 1984 NRC report, *Renewing U.S. Mathematics: Critical Resource for the Future*, released by the National Academy in 1984.

Unity of Mathematics

The unification that is taking place within mathematics is obvious to people in the field and will be apparent in the examples (vignettes) given in this article, particularly with reference to D-modules, computational complexity, the Yang-Mills equations, and operator algebras. Unification occurs when there is a confluence of seemingly independent phenomena, motivating cooperative study of the significant underlying patterns. One symptom of this accelerating unification is the increasing difficulty that agencies are having in assigning proposals to their discipline programs, which now substantially overlap. The trend burdens young investigators with a need to pursue increasingly broad training, as is noted in several of the examples given later. In mathematics, it is becoming critical to lengthen the training period substantially.

An example of this confluence of areas is the Korteweg-de Vries (KdV) equation $u_t = uu_x = u_{xxx} = 0$ [where u is an unknown function $u(x, t)$ in one space dimension and time], which arises both as the simplest nonlinear dispersive equation in shallow-water wave theory and as the equation of isospectral evolution of the potential in the Schrödinger equation $\psi_{xx} = (k^2 - u)\psi = 0$ of quantum mechanics. The intensive study of the KdV equation during the past quarter century has affected many major areas of mathematics. For example, recently a young Japanese mathematician, using a development in the study of KdV equations initiated a decade before by the Moscow school, solved a major problem in algebraic geometry that was first discussed more than a century ago by Riemann, a German.

Mathematics and Other Sciences

The symbiotic relationship between mathematics and its areas of applications is ever deepening as more areas of science and engineering become almost indistinguishable from subareas of mathematics, and this relationship is producing exciting and intriguing new mathematics. Cross-disciplinary collaboration between mathematicians and professionals in other fields is accelerating and deserves encouragement. An important number of interactions between mathematics and science and engineering are not exactly interdisciplinary, but might be more accurately described as resonance phenomena in that advances in one field spur development in another. Examples of this important trend are described in the examples on Yang-Mills equations and operator algebras described later in this article. The broadening and deepening of these applications, as noted in the example on nonlinear hyperbolic conservation laws, create pressure for mathematicians to pursue significant postdoctoral study.

See Table 1 for a list of mathematical symbols.

New Opportunities for Mathematics

There is a growing trend for mathematics to be incorporated into the language, not only in the physical sciences and technology, but also into other fields, such as the social sciences. Mathematical models (descriptions of real-world events that use mathematical language) form the basis of econometrics and health policy analysis. The survival analysis example, described later, demonstrates this. It examines statistical and mathematical methods used to provide a realistic analysis for problems in medical research, reliability theory, actuarial computations, and demographic studies. Mathematical analyses contribute substantially to decisions about economic and health policies, which in turn have enormous financial and social consequences.

TABLE 1. ABRIDGED LIST OF MATHEMATICAL SYMBOLS

Symbol	Definition
$\times$ or $\cdot$	Multiplied by; or times
$\div$ or;	Divided by
$+$	Plus; added to; positive
$-$	Minus; subtracted from; negative
$\pm$	Plus or minus; positive or negative
$\mp$	Minus or plus; negative or positive
$=$ or ::	Equals; is equivalent to
$\equiv$	Identical with
$\cong$	Is approximately equal to; approximates
$\neq$	Unequal; does not equal; not equivalent to
$>$	Greater than
$\gg$	Much less than
$<$	Less than
$\ll$	Much greater than
$\geq$	Greater than or equal to
$\leq$	Less than or equal to
$)$	Therefore
$\angle$	Angle
Δ	Increment or decrement
Δx	Change in x
Δf	Change in value of f
$\perp$	Perpendicular to
$\parallel$	Parallel to
$a \in A$	a is an element of set A
$\{\}$	Notation of a set
$\{3\}$	The set of which 3 is the only element
$\{2, 7, 15, 36\}$	The set whose elements are 2,7,15,36
$X = \{x \mid x \text{ is a real number}\}$	The set of all real numbers
$X \times Y$	The set of all ordered pairs (x, y) of real numbers
$\emptyset$	The empty set
$\cup$	The universal set
$A \subseteq B$	Set inclusion
$A \subset B$	Proper set inclusion
$A \cap B$	Intersection of A and B
$A \cup B$	Union of A and B
A'	Complement of A
$|a|$	Absolute value
$a + bi$	Complex number in conventional notation
(a, b)	Complex number in ordered pair notation
$r(\cos \theta + i \sin \theta) = r \operatorname{cis} \theta$	Complex number in polar form
$a \equiv b, \mod m$	a congruent to b modulo m
$1.\overline{142857}$	Repeating decimal
$n!$	factorial
f	Function
$f(x)$	Value of f at x
$f : X \to Y; f : (x, y)$	Notations for a function
f^{-1}	Inverse of f
$e^x = \exp x$	Exponential function
$\log x$	Logarithmic function
$\sin x$	Restricted sine function
$\operatorname{arc} \sin x = \sin^{-1} x$	Inverse sine function
(r, θ)	Polar coordinates
$\lim_{x \to \infty} f(x)$	Limit of the function f as x approaches ∞
$\lim_{x \to \alpha} f(x)$	Limit of the function f as x approaches a
$\lim_{x \to \infty} S_n$	Limit of a sequence as n approaches infinity
p, q, r	Propositions
p_x, q_x, r_x	Open sentences
P, Q, R	Truth sets of p_x, q_x, r_x
$\forall x$	For all x
$\exists x$	For some x
$p \wedge q$	Conjunction of p and q
$\sim p$	Negation of p
$p \to q$	The implication, "If p then q"
$p \leftrightarrow q$	p is equivalent to q
$[\ ]$	Greatest integer function
$g \circ f\, g(f(x))$	Composite of g and f
$\sum$	Summation
$A_a^b = \lim_{n \to \infty} \sum f(x_i)\Delta x$	Area under $y = f(x)$ from $x = a$ to $x = b$
$\int_a^b f(x)\,dx$	Definite integral
$D_x f = f'(x) = df/dx$	Derivative of f
$P(A)$	Probability of the event A
$P(A|B)$	Probability of A, given B
$P_{n,r}$	Number of permutations of n things, taken r at a time

TABLE 1. (*Continued*)

Symbol	Definition
$P_{n,n}^{r_1, r_2 \cdots}$	Number of permutations of n things, n at a time, of which $r_1 r_2 \ldots$ are alike
$C_{n,r}$	Number of combinations of n things, taken r at a time
$\binom{n}{r}$	Binomial coefficient
$\begin{pmatrix} a & b \\ c & d \end{pmatrix}$	2×2 matrix
$\circ$	Group operation
$\oplus$	Sum of vectors
$\begin{pmatrix} 1 & 2 & 3 \\ 3 & 1 & 2 \end{pmatrix}$	Permutation
$P(x)$	Polynomial
(x, y)	Rectangular coordinates of a point in the plane
$\sqrt{x}$	Square root of x
x^n	x raise to a power of n. n is referred to as the exponent

There is no longer any question as to whether mathematical analysis will substantially influence discussions of public policy, but only whether it will be used appropriately and effectively. It is essential that those making the decisions understand and influence the assumptions used to form the mathematical model and that mathematicians comprehend the applications sufficiently well that they address and solve the *correct* problem. In fields where mathematical models are not subject to experimental verification—such as those dealing with the most drastic consequences—it is especially essential that the mathematics be critically scrutinized.

The Role of Computation

Computational mathematics is an integral part of the mathematics discussed in the examples given later about complexity and nonlinear hyperbolic equations, and there are two important observations worthy of emphasis: (1) Computational methods pervade almost all aspects of science, and mathematics is the foundation of these methods. Today's complex problems, involving computational solutions, range from the design of computer architecture itself through mathematical modeling of physical, chemical, biological, and engineering processes. Mathematics, the intellectual basis of computational science, has been and will continue to be the key to the dynamic revolution being created by the computer in science and engineering. (2) Computational results provide the insight for the development of mathematical theory. For example, the behavior of the solutions to the KdV equation previously mentioned was first discovered numerically. The mathematical theory, in turn, provides a deeper understanding of the models, revealing phenomena that enable people to analyze and test previous computational results and conceive of new computations that will facilitate further theory.

Core Mathematics now consist of three basic operations—*computation*, *abstraction*, and *generalization*. Raw information leading to mathematical discovery comes from concrete examples, and it is increasingly the role of computation to provide such examples, although they are frequently formulated mathematically. Abstraction is the process of distilling the essential features from such examples. Number and space are, respectively, abstractions of the process of counting and of our experience in the physical world, and the mathematical idea of functions similarly abstracts human ideas of measurement and motion. In these contexts, ideas from one manifestation of the abstraction are often relevant in solving problems in seemingly dissimilar situations. Generalization uncovers hidden analogies between abstract patterns of mathematics and frequently extends the range of applicability of such patterns.

Thus, in the development of mathematics, periods of computation often alternate with periods of theorizing. During the former, new raw data are generated and horizons expanded. Eventually, there is a plethora of information in need of an intellectual framework on the basis of which masses of material can be comprehended simultaneously. As reunification occurs, seemingly disparate examples are often revealed as different aspects of the same phenomenon.

Computation, abstraction, and generalization need each other to be meaningful. Recent mathematics has focused more on concrete problems than on abstractions as it revels in the computing power suddenly available to it, and evidence of a new unification is now appearing. At the same time,

our computing power has matured to the point where it can be an enormous asset in the investigation of the still more complex mathematical examples that will inevitably be suggested by pending research.

Organization of Mathematics

More and more, the mathematical community is now organizing itself by areas of interest instead of fields of traditional study. A variety of mathematicians traditionally educated in the disparate fields of topology, algebra, differential geometry, algebraic geometry, complex variables, and partial differential equations now must cooperate in solving some of the newer, exciting equations. In numerous organizations, research is accomplished by teams frequently gathered around major pieces of scientific equipment or laboratories. Each team consists of at least one senior scientist and many, sometimes a great many, junior investigators working on related problems. Mathematicians likewise are often informally grouped around a common research interest. Such mathematical groups, however, are usually geographically separate, and the analogue of access to a common major piece of scientific equipment is their ability to gather together for sustained periods of collaboration. (Possibly this type of process in broadening the student mathematicians can be targeted in educational institutions.)

Language of Mathematics

At least three centuries ago, mathematics developed a language of its own, which has become thoroughly distinctive and international. Just as it takes years for students from one country to become fluent in the spoken language of another, any aspiring student, discounting his/her nationality, spends years studying the *common language* of mathematics. As a result, each can open the other's mathematics books and recognize the topic under discussion (even if ignorant of the other's verbal language).

Language Problem for the Nonmathematician. The complex and dynamic language of mathematics that has developed over the past brings great satisfaction in terms of international understanding for those who are fluent in it, but the professional language does present substantial obstacles to those who have not devoted years to its study. The history of mathematics is one of progressively less translatability for nonmathematicians. In contrast, over a long period, there has been an everbroadening use of mathematics in what might be termed nontraditional mathematical areas. (What may be the solution to this problem?)

A Sampler of Interesting New Topics in Mathematical Science

Traditionally, core mathematics has been broadly divided into analysis, algebra, and geometry/topology, although all three subfields include extensive applied mathematics components. (Geometry and topology are not synonymous, but each term is often used for areas in their broad confluence.) However, as the interplay of analysis, algebra, and geometry/topology becomes more complicated, even the division of core mathematics into subfields (not to mention its distinction from applied mathematics) seems artificial. Indeed, the essential unity of mathematics is vivid as one reviews some important recently discovered relationships among these traditional subfields.

The Concept of Chaos—A Fresh Perspective

Since the early formative years of science, the precepts of classical mechanics were entrenched firmly in the pursuit of dynamic systems and guided by the unwavering notion that the behavior of complex systems could be predicted accurately, provided that one had enough information and intelligence. The concept (or theory) of chaos has challenged this historic approach. The ground rules are changing!

The "sufficient information" doctrine first was challenged at the atomic level by quantum mechanics in the 1920s; then, during the 1980s, prior tenets received another blow with the emergence of chaos theory. This theory holds that, for microscopic or macroscopic systems, tiny variations in initial conditions sometimes may create unexpected, radically different outcomes, seemingly making it impossible to predict fully the behavior of some systems. Perhaps most startling of all, such behavior can arise in relatively simple systems governed by a few uncomplicated equations. Thus, relatively simple or highly complex systems can exhibit chaos.

During the course of the first score of years of its existence, chaos theory has generated wide interest in academia, and, although relatively few practical applications have emerged, research is intensifying. As pointed out by a number of researchers in the field, the science of system dynamics is entering a new era, one that is comparable to the time frame when quantum mechanics was "fleshing out." A period of intellectual ferment transpired years before the transistor made its debut. Presently, chaos is being searched for its rightful scientific underpinnings.

Numerous researchers in recent years have stressed the ubiquity of chaos, ranging from fluid dynamics to electric power networks to physiology—ad infinitum. Before describing, within the next few pages, the searches underway for the practical application of chaos theory, a variety of general observations made by some of the leaders who are pursuing this strange blend of mathematics and physics may serve as an introduction.

John Dorning, University of Virginia: "For decades, engineers, scientists, and mathematicians alike, for the most part, when confronted with nonlinearity looked the other way or looked and shrugged their shoulders, or worse yet, looked and saw nothing at all beyond which their intellectual tunnel vision allowed."

Jong Kim, Electric Power Research Institute: "With chaos, we're on the brink of a new classical dynamics, and people thought that classical physics was dead."

John Stringer, Electric Power Research Institute: "It's called the curse of dimensionality—the amount of data you need to understand a system rises exponentially with the system's dimensionality, that is, the number of independent variables or degrees of freedom needed to describe it. Some of the projects involving what we thought would be simple questions have turned out to be very difficult. And, of course, there's the problem of noise. In many cases, it may be very hard to get data sets that are sufficiently tidy for understanding chaos. On the other hand, chaos theory can help us learn the limits of predictability for very complex systems, such as the weather, and may even give us new tools for controlling these systems."

Chaos in Electric Power Generation and Distribution. John Douglas (Electric Power Research Institute, EPRI) observes, "The implications of chaos theory for electric power equipment and networks are both disturbing and exciting. On the one hand, an unsuspected potential for instability may lurk among the operating conditions of systems thought to be well understood."

Sudden voltage collapses on power grids, for example, may indicate the presence of underlying chaotic dynamics. On the other hand, understanding chaos may provide unprecedented control over some of the most complex and elusive natural processes, such as combustion, corrosion, and superconductivity.

Along these lines, EPRI (Palo Alto, California) established a workshop consisting of specialists selected from a variety of disciplines, ranging from physics and engineering to physiology and computer science, in an effort to bring chaos theory to bear on practical problems of the power industry.

Jong Kim and John Stringer, both leaders of this workshop, astutely observe, "Is chaotic behavior impossible to understand? Not necessarily, according to current theory, which describes an underlying order in seemingly random phenomena. Using the tenets of deterministic chaos, EPRI is doing exploratory research on several dynamic processes of importance to the utility industry—searching for points of departure from linear behavior and for the reasons why predictable dynamics become chaotic. For power delivery systems, this research may help define the difference between a stable network and a system failure. For combustion processes, on the other hand, chaotic behavior may actually be encouraged in order to optimize the turbulent mixing of fuel and air that leads to higher combustion efficiency. Convection and metal passivation also may have chaotic aspects. Understanding chaotic behavior in such processes—learning how to control it and, if desirable, reverse it—could

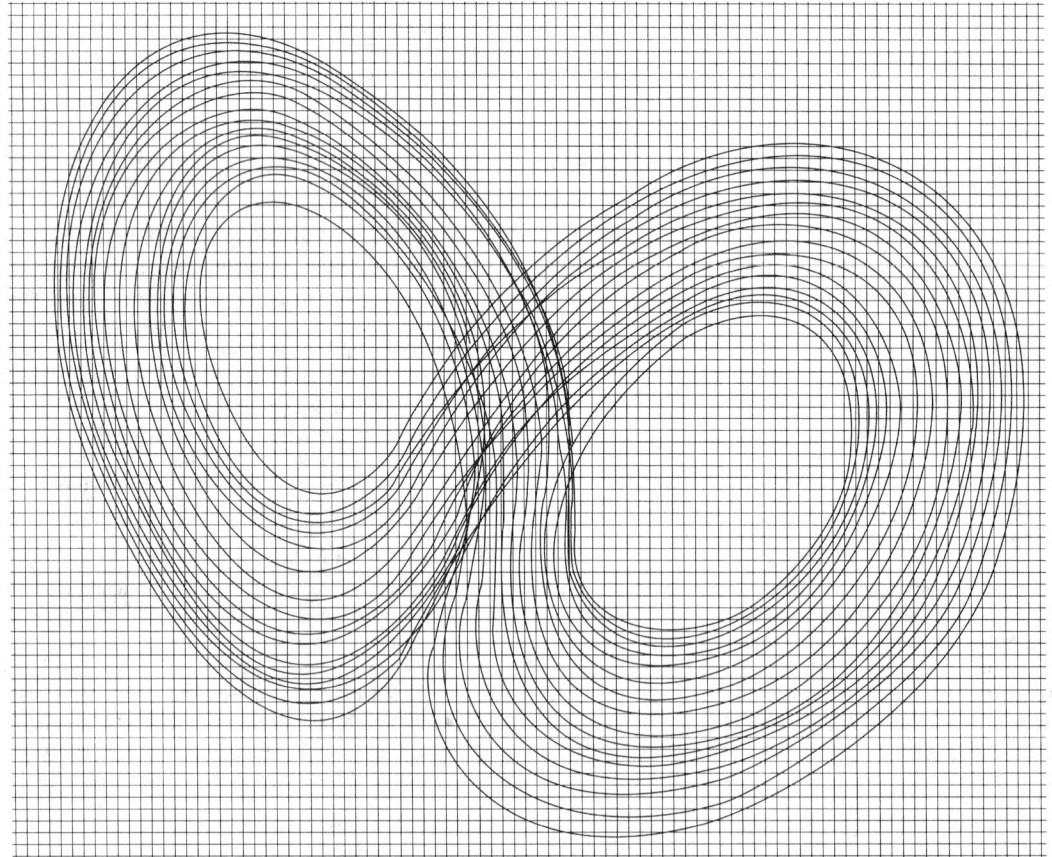

Fig. 1.   Reasonable facsimile of a computer-generated butterfly pattern. This is a representation of the Edward Lorenz butterfly pattern (1961) that displays an "orderly disorder." The pattern exhibits deterministic chaos. The pattern was created by Lorenz as he attempted to develop a simple computer model of weather based on convection currents in the atmosphere. In an elegantly created computer graphic of the pattern, it would be noted that the lines in the shape never quite touch each other, thus imputing the pattern as an infinitely complex microstructure.

When, in the investigation of a chaotic phenomenon, a recognizable pattern of data emerges, the pattern may be indicative of the types of equations that generate the pattern's structure. Such patterns frequently are referred to in the literature of chaos as *strange attractors*.

A powerful tool of chaos theory, known as the Takens embedding theorem, may create an identifiable pattern simply by analysis of how any one of the key variables evolves.

lead researchers to a better grasp of complex natural phenomena and to very practical fixes as well."

The EPRI researchers stress that the application process probably will be long and complex. The problem is how to distinguish "deterministic chaos" from stochastic, or totally random, behavior. John Stringer emphasizes, "Chaos has an underlying order, a pattern that's not periodic, but isn't completely random either. In any real system, however, some stochastic processes are also likely to be present as noise. It's like looking for a fuzzy pattern through a fog!" Indeed, what eventually led to the current revolution in the science of dynamic systems was the slow realization that chaotic behavior is ubiquitous. Today, chaotic behavior is found in the fluctuations in predator-prey populations, fibrillation of the heart, the dripping of a faucet, and trends in the price of cotton; these all demonstrate telltale signs of chaotic behavior. They are amenable to the same mathematical formalities.

In commenting on the EPRI programs targeted for study, researchers observe that the equations that describe potentially chaotic systems often have at least one thing in common: *nonlinearity*. The proportionality and easy-to-solve aspects of linear equations are *not* present. In addition to lacking proportionality, nonlinear equations differ qualitatively as conditions change.

Currently, EPRI is exploring four specific chaos-focused research projects: (1) convective flows, which have potential application in nuclear safety and thermal power plant operating efficiency; (2) fluidized beds, which are an attractive model for understanding and potentially improving control over chaotic flow processes, keeping in mind that chaos can be a desirable condition because it leads to good mixing; (3) power grids—learning how to avoid any combination of conditions that could lead to chaos; and (4) the kinetics of metal passivation, in an effort to reveal how chaotic processes influence corrosion.

Convection (The Problem of Nonlinearity). As early as 1899, the French mathematician Jules-Henri Poincaré first recognized the possibility for chaotic behavior in dynamic systems. However, it was not until 1961 that the meteorologist Edward Lorenz observed the phenomenon when he was attempting to construct a simple computer model of weather on the basis of convection currents in the earth's atmosphere. Lorenz was puzzled by the sensitivity of the model to what appeared to be insignificant differences in starting conditions. Even after Lorenz developed a very abstract version of convection using only three variables and three equations, he still encountered unexpected complex behavior. Because the variables changed in a very complex manner, Lorenz found it impossible to forecast their values with any degree of certainty over long periods of time. As Lorenz mapped their long-term trends, he found that the variables produced a three-dimensional pattern, as shown in Fig. 1. Because in three dimensions it appeared something like the outstretched wings of a butterfly, this type of pattern is now dubbed the "butterfly" and is frequently mentioned in the current literature on chaos.

The development of chaos science to where it is today is exquisitely summarized by John Douglas: "It took more than a decade and a half for this phenomenological pattern (Lorenz butterfly)—globally organized but locally unpredictable—to gain enough recognition to be named and it took even longer for investigations of chaos to earn scientific respectability."

As pointed out by Douglas, "Convection starts out as a smooth flow that speeds up as the temperature difference between the top and bottom of a fluid increases. But, beyond a certain point, instabilities begin to appear and at great enough temperature differences, the flow becomes turbulent—that is, *chaotic*." See Figs. 2 and 3.

The EPRI program on chaos in convective systems, now underway at the University of Virginia, is targeting on the development of a generic model of nonlinear flow in critical electric power devices, such as transformers, heat exchangers, and boiling water (nuclear) reactors. The three original Lorenz equations, previously mentioned, are being used as a starting point. Researchers are attempting to determine what conditions lead to chaotic flow and what effect this transition has on the heat-removal efficiency of the devices in question. Researchers also will investigate the feasibility of reversing this process—that is, driving the system back from chaotic to periodic flow.

Jong Kim, project leader, comments, "We need to understand what conditions can result in chaotic behavior in major power systems. It's not always to be avoided, of course. Chaotic flow is an aid when you want more fluid mixing, for example. But, we do need to learn how to control chaos, including how to reverse it. Such research will be particularly important

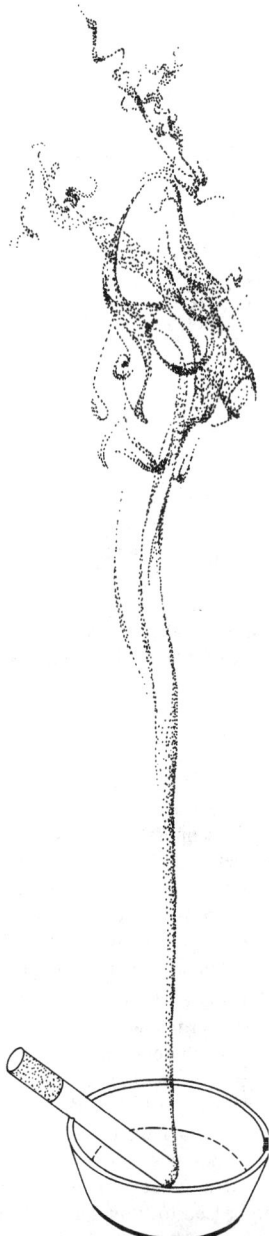

Fig. 2. For a considerable height, the smoke from a lighted cigarette at rest initially rises in typical laminar flow. But then the flow of smoke takes on a rather violent, turbulent flow behavior. The fine structure of the turbulent flow pattern currently is beyond exact prediction. This is an example of chaos as related to fluids.

for ensuring the stability of the next generation of so-called passively safe reactors, which rely on natural convection to provide emergency cooling capability."

Chaos in Fluidized-Bed Combustion (FBC). This program has at least two important targets: (1) instabilities in low-nitrogen oxide (NOx) burners, and (2) the formation of air pollutants at various stages in conventional boilers. The EPRI plan notes that an FBC unit makes an ideal model for studying chaos, since coal and limestone particles are lifted by upward-rushing gases to form a suspended bed of material that acts like a turbulent fluid.

The researchers, in particular, are seeking characteristic patterns (strange attractors) that represent the signature of unique kinds of chaotic behavior. For the FBC project, once a strange attractor has been found, it may be possible to recognize its shape and perhaps identify the types of equations that generate its structure. In addition, a powerful principle of chaos theory (the Takens embedding theorem) implies that, at least theoretically, the form of a strange attractor should be identifiable simply by examining how any of the key variables evolves over time. This is regarded as pertinent

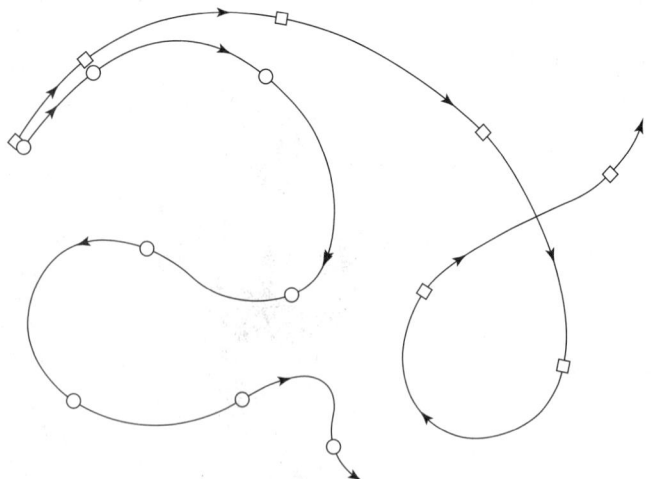

Fig. 3.   A tenet of chaotic phenomena is the tendency for divergence. The dynamics of a chaotic system in most cases are extremely sensitive to starting (initial) conditions. For example, two adjacent wind-driven particles may commence their journey in an essentially parallel manner, but diverge radically after a short distance. As shown by diagram, the divergence is not slight, but is a radical departure. Chaos theory holds that the positions of the two particles, as related to each other, cannot be predicted. On a much larger scale, convection currents of a weather system behave similarly.

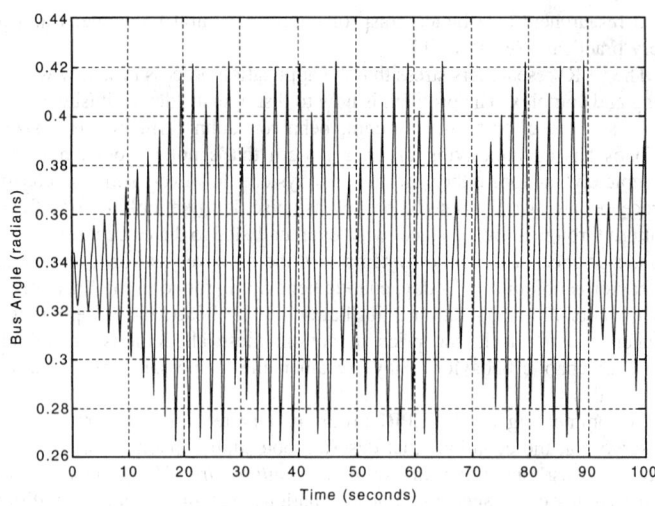

Fig. 4.   Power system stability. Shown here is the relative generator bus angle on a simulated system, driven to an unstable, chaotic state by increasing the reactive power by only about 8%, again demonstrating the characteristic of chaos in which small divergences result in what appear to be disproportionately large changes. (*Electric Power Research Institute.*)

to the FBC project because the simultaneous measurement of multiple variables, such as pressure, particle velocity, et al., in a particular small region of interest is extremely difficult, if not impractical, to attempt.

Also, there is the ever-present problem of noise in a system. The question is posed: "How can one detect a recognizable chaotic pattern when it may be obscured by the random meanderings of stochastic events?"

Investigators at Oak Ridge National Laboratory believe that they have found a way to separate major dynamic effects from noise, producing a series of somewhat messy, but nevertheless recognizable, strange attractors.

Chaos in Power Grids. As emphasized by John Douglas (EPRI), "The very idea that chaos may occur in electric power grids is about as welcome to utility planners as a heart flutter—and for many of the same reasons. If confirmed, it would at least prompt a fundamental rethinking of the analytical methods used to ensure network stability. At worst, it could mean that power systems harbor an unappreciated potential for voltage oscillations and collapse—the network equivalent of a heart attack."

EPRI research reports that a study commenced in 1989 focuses on the chain of events that might lead to fully chaotic behavior. Simple network models indicate that the onset of chaos is preceded by system bifurcations, in which operating characteristics suddenly can oscillate between two sets of conditions. Problems multiply if one or both of the new operating conditions lead to instability. Also, successive bifurcations (dividing into two or more branches) can so disrupt a system that eventually the number of possible operating states becomes infinite—that is, chaos ensues. See Fig. 4.

A common type of bifurcation observed in utility power systems is *period doubling*, in which line frequency jumps between 60 and 30 Hz. Mark Lauby (EPRI power system engineering manager) observes, "It's like plucking the D string on a violin and watching the G string start to vibrate in response. As long as the violin is constructed to withstand both kinds of vibration, there's no problem. But some power systems might be hypersensitive to the new frequency and collapse after a bifurcation—which would be like having the violin fall apart in your hands. With chaos there are an infinite number of possible frequencies, and the system could collapse from experiencing any of these outside its design limits. What we hope to do is to learn how to eliminate conditions that could set up network bifurcations and thus block the path to chaos."

Research along these lines is being conducted at Cornell University and the University of California at Berkeley. Investigators are using highly simplified power system models, from which they have identified various kinds of bifurcations as well as chaos, among the myriad operating conditions that affect the behavior of variables in the "state space" of the system. Although the researchers consider these observations rather tentative, the problems observed could conceivably result from modeling

inadequacies, but nevertheless they raise important warning "flags." The studies do establish the presumption that chaotic behavior will exist in most power system models; however, it is not yet clear if chaos occurs for parameters in regions sufficiently near (ordinary, conventional) operating regimes to affect the stability region of utility power systems to a significant extent.

Utility engineers long have recognized that power systems have nonlinearities. Now that chaos and bifurcations exist in these systems, ways must be developed to reliably identify their presence. Then remedial measures for controlling systems and steering them away from these conditions must be tackled. Analysis tools must be developed to enable the design of power systems that will be free of the threats of chaotic behavior.

Exploring Chaos at the Microscopic Level. It is well-established knowledge that the corrosion of metals exposed to reactive gases and liquids can be prevented by improving *metal passivation* (that is, by reducing the chemical activity of a metal surface as its electric potential increases). This is extremely important, not only to the power industry, but to other industrial applications. It is estimated, for example, that 4% of the gross national product of the industrialized nations is lost to corrosion.

Researchers have found that the sets of equations used by various models to describe metal passivation are all nonlinear and thus pose an inherent possibility for chaotic phenomena. Work underway through EPRI sponsorship at Batelle Memorial Institute and Ohio University has led to the development of a new model of metal passivation. The model predicts that chaos will occur during the passivation process and identifies several different bifurcation routes to full chaotic behavior. To date, the physical implications of the model remain unclear because of the surprisingly complex ways in which chaos appears and disappears again in response to small changes in key parameters. This type of research, of course, requires extremely precise control of electrochemical conditions at the metal surface.

Chaos and Fractal Geometry. The concepts of chaos and fractal geometry have been researched essentially during the same time frame. The interrelationship of the two disciplines has been observed by numerous scientists during the past few years. Fractional dimensionality contributes an additional complexity in one's understanding of chaos. See also article on **Fractal Geometry**.

This connection is being explored by numerous researchers, including the EPRI group.

As pointed out by John Douglas, "The idea of multiple and fractional dimensionality is not as unfamiliar as it might seem." In addition to the traditional three geometric dimensions used to describe the size or location of an object in space, other dimensions usually are required to describe the behavior of a dynamic system.

For example, in considering the several independent variables that determine the flight of a baseball, to the three aforementioned (x, y, z) data values, one must add the three directions of linear velocity along with three spin coordinates. Thus, as pointed out by Douglas, there is a total of nine dimensions, if each variable is assigned its own axis for graphing purposes. In analyzing the flight of a *curve ball*, Douglas observes that "dimensions" researchers attempt to explain that real-world phenomena are not simply the dimensions of physical length, width, and height, but rather the parameters that drive the dynamics of a system, such as pressure, velocity, and temperature.

D-Modules

Algebraic geometry has been one of the most lively areas of research in algebra during recent decades. It is the study of geometric objects that are the loci of points satisfying polynomial equations in two or more variables, such as the familiar conics from classical geometry. Meanwhile, algebraic topology has become a leading area of geometry. Considerably more general geometric objects than loci of polynomial equations are studied in algebraic topology; in the 1950s and 1960s this was an especially active field and was discussed on a number of occasions in the popular scientific literature.

Any geometric object has a group of symmetries. For example, a cube is invariant under a finite set of rotations. Similarly, a sphere is symmetric under (the infinite set of) all rotations and reflections around its center. A continuous symmetry group, such as the latter example, is called a Lie group. Lie groups can also be viewed as certain groups of matrices with their usual matrix multiplication. This multiplication is not commutative; that is, in general $XY \neq YX$. The set of derivatives along curves through the identity matrix of a particular Lie group can be viewed as an additive set of matrices and is called a Lie algebra. The theory of Lie groups and algebras, one of the great achievements of modern mathematics, originated from questions in differential equations, which has always been central in analysis. But "group" and (of course) "algebra" are quintessentially algebraic notions. Thus Lie theory, based on geometric symmetries and increasingly useful for physicists, incorporates aspects of algebra, analysis, and geometry.

Another place where these three fields meet is in *D*-modules, which have been developed recently in Japan. A module (with or without the *D*) is an algebraic structure consisting of a group such as a vector space whose elements can be multiplied by another set of mathematical objects such as matrices. *D*-modules are modules whose vectors can be multiplied by partial differential operators with analytic coefficients. One motivation for their study was to focus attention on the equations themselves rather than solutions to differential equations. They were investigated using methods of algebraic geometry invented in France in the early 1960s.

The theory of *D*-modules is also related to the Riemann-Hilbert problem, posed in Germany around the turn of the century. Suppose we have a linear, homogeneous system of n first-order ordinary differential equations for n functions $f_1, \ldots f_n$ of one complex variable z. It might be written $f' = Af$, where the coefficient A is an n by n matrix whose elements are analytic functions of z. In general the system will have n independent solutions $(f_{11}, \ldots, f_{1n}), (f_{21}, \ldots, f_{2n}), \ldots, (f_{n1}, \ldots, f_{nn})$ such that each solution $(g_i, \ldots, g_n)$ can be written uniquely as a linear combination $g_i = c_1 f_{1i} + c_2 f_{2i} + \cdots + c_n f_{ni}$ for some constants $c_1, \ldots, c_n$. But there may be singularities, points at which one or more of the elements in $A(z)$ tend to infinity or are otherwise irregular. Not only may solutions be undefined at the singularity, but when we follow a curve around the singularity, one solution may be transformed into a different one. Going around the given singularity (and no others) once in a clockwise direction has the effect of transforming a solution by multiplication by an invertible matrix, depending only on the system and not the path traversed. The Riemann-Hilbert problem, a question in analysis, was to show a converse for the theorem just presented — that for any admissible map from a set of singularities into invertible matrices, there is a system of n differential equations, as described, such that encircling each singularity changes the solution by the corresponding matrix. The solution, now several decades old, has been extensively generalized.

Mathematicians working in this country and France were investigating ostensibly a quite different area — an area at the intersection of algebraic geometry and algebraic topology that focuses on the concept of duality in geometry. In 1977 researchers in the United States conjectured the existence of a relation between this geometric theory and a fundamental problem from Lie groups. Their conjecture was then proved independently by two pairs of mathematicians, one in France and the other in the Soviet Union. Their proof surprised the mathematical world by using

the generalized Riemann-Hilbert results, thus again linking algebra with geometry through differential equations.

Papers about *D*-modules are impossible to classify into the three traditional fields of analysis, algebra, and geometry/topology, a problem for the editors of *Mathematical Reviews*. It also suggests how very tentative any division of mathematics into subfields may be. See also **Lie Group**.

Computational Complexity — Algorithms

An algorithm is a procedure for solving a given class of problems with a specified set of mathematical tools. In classical geometry the tools were straight edge and compass, and the ancient Greeks provided a simple algorithm for trisecting any segment using only these. Their extensive attempts to trisect a general angle were proved futile in the nineteenth century, when it was shown that there can be no such algorithm.

If the tools in algebra are addition, multiplication, division, and taking radicals (square roots, cube roots, etc.), then there is a familiar algorithm for solving any quadratic equation, $ax^2 + bx + c = 0$. However, a question posed by Renaissance Italians was answered when it was proved by Abel, a Norwegian in the early nineteenth century, that there can be no such algorithm for solving equations of degree five or more. The tenth problem posed by Hilbert at the 1900 International Congress of Mathematicians asked whether there is an algorithm for deciding if a general polynomial equation, $f(x_1, \ldots, x_n) = 0$, with integer coefficients has a solution in the positive integers. In the 1960s a Soviet mathematician, strongly incorporating the work of two Americans, proved that there is no such algorithm.

Such decision questions were formerly addressed primarily by logicians, but computers have given new urgency to algorithmic questions. Computers after all operate with only a few primitive tools, and the programs that instruct them are essentially algorithms. For most problems that computers are asked to solve the existence of some algorithm is usually evident; instead, the problem is to find algorithms that are efficient and reliable. The mathematical analysis of these practical considerations has spawned the field now called "complexity theory."

This field measures the complexity of an algorithm by the maximum number $f(n)$ of basic steps that the algorithm needs to solve those cases of the problem requiring n digits in their statement. The theoretically tractable problems, called type P, are those for which there is an algorithm with $f(n)$ growing no faster than some polynomial in n. There are adjustments to this assertion, most notably in the well-publicized linear programming problem discussed below, where the simplex algorithm, though exponentially long in the worst cases, is remarkably efficient in the vast majority of its application.

Each generation of computer scientists has quickly encountered pressing problems that overwhelm available computational resources. Unable to obtain exact answers, they resort to simulation, approximation, and sampling, perhaps by using Monte Carlo techniques. Such approaches are not always appropriate, as when a massive computation is used to make a momentous yes/no decision. In the development of antiballistic missile software during the early 1960s scientists tried to cope with such situations by the simultaneous use of many processors in a multiprocessor environment. It was found that this could result in unpredictable and often serious deterioration in the performance of the system as a whole. This discovery generated a serious study of such anomalies. The resulting work in the mid-1960s provided rigorous bounds for these deleterious effects. Performance guarantees of this type, usually called "worst-case bounds," remain a major focus of algorithm analysis.

The fundamental class of *NP*-complete problems was defined in 1971 independently by a Canadian and by a former Russian citizen who now resides in the United States. This class includes thousands of basic computational problems arising in computer science, mathematics, physics, biology, economics, business, and the social sciences. The *NP*-complete problems are of equivalent complexity in the sense that if one of them submits to an algorithm of polynomial complexity, then so can all the others. Proving the widespread conviction that no such algorithm exists (that $P \neq NP$) is considered the most fundamental of the open problems in theoretical computer science. It is known that any algorithm for solving an *NP*-complete problem can be modeled by an appropriate circuit. Until recently the only established lower bounds on circuits for *NP*-complete problems were linear. However, recently a Taiwanese mathematician now living in the United States made a dramatic breakthrough by establishing the conjectured exponential lower bound, but under the assumption that the number of "levels of the circuits is bounded". Since then a Swedish graduate student in the United States simplified and strengthened these arguments. Almost simultaneously, two Soviet mathematicians independently established an exponential lower bound assuming that the circuit is "monotone," a related but distinct development. This work has been extended by an Israeli doing postdoctoral work in the United States, and many complexity specialists now believe that the tools are finally becoming available to prove the corresponding results for unrestricted circuits, thereby finally settling this fundamental problem.

As the difficulty of *NP*-complete problems was realized, attention shifted to other approaches. These included approximation algorithms, average-case instead of worst-case performance analysis, and randomized algorithms that give a confident guess rather than a firm answer. Timely examples of the latter are the twin problems of deciding if a large integer n is prime or, if it is not, of factoring n. One such algorithm randomly selects an integer k less than n, performs a simple test, and announces either that n is definitely composite or that the problem is still undecided. About three quarters of the possible k's will establish that a composite n is not prime. Thus, performing the test with 100 independent k's that do not prove n to be composite justifies the conclusion that n is prime with a mere one in 4^{100} chance of error.

The study of primes has long been central to number theory, a field that has been pursued for its own splendid beauty but traditionally was considered to have few pragmatic consequences. Recent innovations in cryptography have completely reversed the latter perception. The security of important cryptosystems depends crucially on the belief that the problem of finding the prime factors of a random number with a thousand decimal digits or more is, and will be for decades, a computationally infeasible problem. However, faith in this belief is beginning to erode because of some recent unexpected advances in primarily testing and factoring. Essentially overnight, a Dutch mathematician produced the currently best factoring algorithm by using the theory of elliptic curves from algebraic geometry, a field mentioned in the previous vignette. It is not yet clear whether this will lead to even more effective algorithms that will undermine the security of these cryptosystems.

This vignette concludes with a discussion of linear programming (LP), a subject that occurs widely in discrete optimization. For example, a linear profit function, possibly depending on a large number of variables, is to be maximized in a region defined by linear constraints. This question is geometrically equivalent to finding the highest point in some high-dimensional convex solid or polyhedron in n-dimensional space, where n can be quite large. The importance of these problems, which have many applications in such varied areas as airline scheduling, meteorology, portfolio management, and telephone traffic routing, was first observed more than 40 years ago by Dantzig in the United States and Kantorovitch in Russia. Dantzig developed the very effective "simplex algorithm" for solving LP problems, which examines first one vertex and then another, moving along the outside edges of the high-dimensional polyhedron in such a way as to improve the function that is to be optimized at each step, until the optimal vertex is reached. Even though these n-dimensional polyhedrons, which typically arise in practical problems, can have exponentially many vertices, over 40 years of experience with the simplex algorithm indicates that only rarely are more than $4n$ or $5n$ vertices tested before the optimum point is attained, despite the fact that pathological examples can be constructed that do indeed require that all vertices be tested.

After the concept of *NP*-completeness was introduced in 1971, researchers struggled without success to find either a polynomial-time algorithm for LP or a proof that it was *NP*-complete. In 1979 the polynomial "ellipsoid algorithm" was produced in the Soviet Union, but the bound on this method grows as n^6, rendering it impractical for large problems, e.g., those having hundreds of thousands of variables. Subsequently, however, the ellipsoid algorithm has had a significant impact in the theory of combinatorial optimization.

In 1984 a young Indian mathematician in the United States made a striking breakthrough when he discovered an iterative method that plunges through the interior of the polyhedron, transforming it nonlinearly at each step so as to stay as far as possible from the boundaries of the changing solid. The number of steps required by this method is on the order of $n^{3.5}$, a significant improvement over the earlier n^6, and a variety of applications have shown that when implemented cleverly, this algorithm seems to

perform significantly faster than the simplex algorithm. See also **Linear Programming**.

Very recent efforts in understanding this new method indicate that an n-dimensional polyhedron can be equipped with a coordinate system that transforms its interior into a certain quasi-hyperbolic geometry so that the trajectories to the optimal vertex form geodesics in this space. Obviously, this area is extremely active worldwide, and much more work is needed before a full understanding can be achieved. Nevertheless, it is already apparent that complexity theory addresses many practical problems and that sophisticated core mathematics, including algebraic geometry, number theory, and geometry, is vital to complexity theory.

Nonlinear Hyperbolic Conservation Laws

The development of both theory and numerical methods for solving nonlinear hyperbolic conservation laws (NLHCLs) has been an exciting area of recent mathematical research. There are many applications of NLHCLs because they describe many important physical systems, including some in aerodynamics, meteorology, water waves, plasma physics, and combustion. In gas dynamics these are the laws of conservation of mass, momentum, and energy.

The major technical obstacle to both solving and analyzing these systems is the fact that their solutions are not smooth, that is, they do not have derivatives of all orders. Many standard approximation procedures require smoothness, and, furthermore, solutions tend to be unique only if certain physical constraints (called entropy conditions) are satisfied. Since there are other important equations in hydrodynamics, relativity, and optics that do not have smooth solutions, NLHCLs are providing a testing ground for innovations with potentially wide applicability.

The first systematic computational attempts to solve NLHCLs, motivated by problems of jet propulsion and in the Manhattan Project, were made in the United States during World War II. John von Neumann and others provided a clear formulation of the problem and introduced several crucial ideas, in particular artificial viscosity (a justifiable technique for smoothing the problem) and a first analysis of stability. These ideas were greatly extended after the war and gave rise to an elegant theory of weak solutions that treated the lack of smoothness without smoothing. In particular, the Lax-Wendroff scheme and its many variants yielded acceptable solutions for many practical problems. However, many other problems remained unsolved, and the theory was incomplete.

In the 1950s the relevance of the Riemann problem, well known to chemists and engineers working with the Riemann shock tube, became fully recognized. The Riemann problem contains the pathology of the general NLHCL problem but is in a more tractable form. An American mathematician gave a mathematical analysis of the Riemann problem, and then a Russian mathematician incorporated the American's analysis into a numerical construction that described the misbehavior of the solutions in a natural way. These developments led to a proof, developed in the United States, that solutions exist for one-dimensional NLHCLs subject to some technical restrictions. This result gave rise to practical algorithms that in turn generated sharper existence results.

The Russian's construction has been generalized in several ways, and dramatic progress has arisen in the last three years from a combination of these ideas. In particular, reliable and efficient solutions of the equations of gas dynamics in any number of space dimensions are now available, and they reveal and explain intricate physical phenomena that previously had been only dimly comprehended. Examples include the disclosure of unexpected complexity in flows involving interacting discontinuities, the discovery of transition criteria from regular to Mach reflection for waves impinging on a surface, and the revelation of instability mechanisms for supersonic jets. Both theory and experiments in NLHCL have been aided by elaborate computations from experiments using new laser technologies.

Recently, much computational activity has been directed toward solving systems of NLHCLs that are structurally more complex than those of gas dynamics, in particular those that arise in combustion theory or in the analysis of flow of porous media—a subject of great relevancy for oil recovery. New methodologies have appeared involving front tracking, mesh refinement, and piecewise parabolic approximations. Some of these are related to higher-order versions of the aforementioned Russian construction.

Many important practical problems remain open. For example, there is as yet no reliable numerical method for solving the equations of combustion theory in more than one space dimension except in the low-Mach-number limit. The newer numerical methods are so complex that computer science questions regarding their implementation, similar to those discussed in the previous vignette, have become crucial. Also, perturbations of NLHCLs, for example by boundary layers, are beginning to be considered.

Why numerical methods fail and why they sometimes succeed so spectacularly are questions that have been studied successfully in recent years, especially through the reexamination of the precise role and possible forms of artificial viscosity. New theoretical tools for understanding practical algorithms and new ideas, such as the notion of variation diminishing schemes, are producing a slow confluence of algorithms rooted in disparate a priori notions of what is important for practical calculations.

New techniques of functional analysis, in particular the compensated compactness method that originated in France, have given new impetus to existence theory, removing some of the earlier limitations. The compensated compactness analysis is significant to the broader context of homogenization and order/disorder phenomena, two other areas in which the French have been involved. In addition, partial existence and stability results have appeared for nonsmooth solutions in more than one space dimension for both convex and nonconvex systems.

The strong interaction between mathematics and practical applications in NLHCL is clear from this account. All the major advances in computation have been anchored in theoretical developments. Indeed, most of the more innovative practical algorithms are due to mathematicians, and more mathematicians have been involved than is apparent. An explosion of knowledge could at this time be safely forecast if there were more high-caliber people active in the field.

However, there are too few people with a combined understanding of the abstruse physical, mathematical, computer science, and related aspects of NLHCLs. Such an understanding requires a broad education in several fields that is not easily available in the United States. One explanation for the strong role played by Israelis in this field may be that in Israel many mathematicians are exposed to engineering problems and to programming during a lengthy military service. In other countries, such as Japan, China, Russia, and in much of Europe, most students complete algebra by the seventh grade and soon begin calculus, leaving time during secondary school for those with mathematical and technical talent to pursue advanced topics. Given that changes in precollegiate education will take a long time to evolve, the most immediate solutions in the United States would seem to be an extension of the graduate student years through supporting young investigators with adequate postdoctoral fellowships and providing enrichment for talented undergraduate students.

Yang-Mills Equations

In 1954, Yang in the United States and Mills in England constructed a nonlinear version of Maxwell's equations that incorporated a non-Abelian group, typically SU(2), the group of two by two unitary complex matrices with a determinant one. [SU(2) is a three-dimensional Lie group.] This was first conceived as a classical theory transplanted to Lorentz space-time; but when it is used in quantum theory, it is convenient to use Euclidean space-time. The theory has been incorporated into nearly every model of particle physics since the construction in 1975 of instanton solutions and the more recent construction of multi-instanton solutions in four-dimensional space. Shortly thereafter, the Penrose twistor theory was shown to transform an apparently very messy nonlinear system of partial differential equations into an elegant problem in algebraic geometry. Then the equations themselves were noticed to be natural geometric objects.

The Yang-Mills equations depict the curvatures (fields) of connections (potentials) as a principal bundle over a Riemannian manifold. By now, Yang-Mills theory is prominent in pure mathematics. It has already affected subjects as diverse as differential geometry, algebraic geometry, the topology of four-dimensional manifolds, the calculus of variations, nonlinear partial differential equations, index theory (or anomalies), and even the representation of infinite-dimensional groups, and it remains fertile research ground.

The extended impact of Yang-Mills theory does not yet involve the complete equations but concentrates on their role as nonlinear extensions of Laplace's equations, which are well known to be fundamental to earlier mathematics, physics, and engineering. In two variables Laplace's equation is related to the Cauchy-Riemann equations, part of the foundation of complex analysis. One form of the Yang-Mills equations, called the self-dual Yang-Mills equations (SDYM), is a four-variable analogue of the Cauchy-Riemann equations. The instantons are solutions of these equations, and they appear to have properties nearly as basic as solutions of the Cauchy-Riemann equations. The SDYM equations are in turn important in algebra, geometry/topology, and analysis, respectively.

At first glance, the SDYM equations seem absolutely intractable for writing explicit solutions. Even for SU(2), a small essentially non-Abelian Lie group, there are nine first-order equations with twelve unknown dependent variables as well as the four independent variables on the space. There are three extra degrees of freedom, due to gauge symmetries, so it is not surprising that insight from algebraic geometry had to precede progress in topology. The Penrose twistor methods were used to transform these equations in four-dimensional space to a problem concerning holomorphic bundles on a six-dimensional manifold. These methods from algebraic geometry are surprisingly general and can lead in many directions, for example, toward Kac-Moody Lie algebras and models with loop groups.

Learning about the structure of the space of instantons over four-dimensional manifolds has generated profound insight into the topological structure of general four-dimensional manifolds, demonstrating the fundamental value of an equation specific to a low dimension like four. This was helpful to physicists, who have since developed the fundamental intuition of instantons as solitons, the wavelike solutions of KdV and Sine-Gordon that superimpose nonlinearly.

It is interesting to note that mathematicians had been able to deal with the structure of manifolds in dimensions two, three, five, and greater. The Cauchy-Riemann equations are used in two dimensions, geometric methods are employed in three, and mathematicians find five and more dimensions amenable to standard methods of algebraic topology. The gap of the fourth dimension appears to be filled by SDYM theory.

In analysis, one of the key properties of Yang-Mills theory is its conformal invariance. Some of the basic equations about instantons can be formulated in the context of the calculus of variations. The conformal invariance of the theory implies that the variational problem does not satisfy the conditions that are required in order to use the method of direct steepest descent, so helpful in three-dimensional work. The attempt to understand the failure of the steepest descent method for the Yang-Mills problem has led to the development of new variational techniques, which are useful on a variety of problems. It is interesting, and possibly significant, that the three fundamental scale-invariant geometric problems coincide with three basic models of quantum fields: the Yamabe problem (phi-four theory), harmonic maps (sigma models), and the Yang-Mills equations.

In any case, Yang-Mills theory is a beautiful example of the intense bonds between current theoretical physics and all subfields of core mathematics. Yang-Mills theory is a young discipline that will undoubtedly attract many more mathematicians in the near future. Its results to date are primarily due to the efforts of English, American, and Russian researchers. Although Americans cannot claim to dominate the field, they have certainly contributed significantly to its development. The necessity of extremely broad training, mentioned in the final paragraph of the preceding vignette, also applies to successful research in this field. Increasingly, postdoctoral or protracted study is necessary in order to become a successful researcher in many areas of mathematics.

Operator Algebra

The area of operator algebras is currently very active and provides another excellent example of unexpected interactions between areas of core mathematics and the natural sciences. Quantum physics originated the Heisenberg Uncertainty Principle, which forces consideration of quantities P and Q satisfying $PQ - QP = h/2\pi i$. It motivated a search for appropriate mathematical systems containing infinitely many such noncommuting variables. In the 1920s, M.H. Stone and J. von Neumann demonstrated that such systems require a general theory of algebras of operators on Hilbert spaces, generalizations of finite-dimensional vector spaces. These were studied extensively in the 1930s by F. Murray and von Neumann.

The mathematics of operator algebras is characterized by a profound blend of noncommutative algebra and infinite-dimensional analysis. Although operator algebras exhibit a very rich structure, no serious work following that of Murray and von Neumann appeared until after World War II.

The simplest example of an operator algebra is the entire set of n by n dimensional matrices for a specific n. A general operator algebra is a subset of the bounded linear transformations on a (generally infinite-dimensional) Hilbert space that is closed under addition, multiplication, an adjoint operation, and a suitable limiting process. If the system is closed only for the strongest limiting process, it is called a C^*-algebra. If it also contains the most general limits, it is said to be a von Neumann algebra. The full algebras of all linear transformations on Hilbert spaces of arbitrary dimensions are the building blocks of the simplest operator algebras. However, consideration of proper subalgebras of the universal operator algebra over some Hilbert space reveals far more complex situations. This complexity can be slightly relieved and the study reduced to three basic types of von Neumann algebras simply called Types I, II, and III.

During the early postwar years, both American and French mathematicians made substantial progress in operator algebras, including some important applications to the theory of infinite-dimensional group representations. The French school became relatively inactive by the early 1960s and reemerged in the mid-1970s when a young mathematician, subsequently a Fields Medalist, began working in the area. During the late 1950s and early 1960s, some Japanese mathematicians entered the field. Simultaneously, numerous theoretical physicists became involved and provided valuable insights derived from their physical applications. It was shown that there exist infinitely many distinct Type II and Type III factors. The latter, especially, remained mysterious because they did not possess the special functionals called traces that were so fundamental to studying the structure of the other two types. When it was discovered that the physically interesting algebras are of Type III and certain classes were explicitly parameterized by physicists, parts of the mystery began to unravel.

Major international conferences began having a crucial influence on operator algebra theory in the mid-1960s. In the late 1960s, a conference held in the United States removed a key obstacle to analyzing Type III factors by displaying results proved independently by a Japanese mathematician and some Dutch and German physicists. At a subsequent conference, the French Fields Medalist, then a student, was motivated to work on the subject. He opened entirely new vistas by investigations that combined algebra and analysis in new ways and led to the classification of the algebraic structure of Type III factors. In the 1970s, researchers turned toward the geometric aspects of the subject. A new extension theory of C^*-algebras generated a successful synthesis of geometry and algebra, resulting in a unified view of features from these two subjects. At about the same time, the fundamental Atiyah-Singer Index Theorem was extended from locally trivial families to the more general foliations by using operator algebras. Additionally, a Soviet mathematician solved specific cases of a long-standing problem in differential topology concerning smooth deformations by developing powerful new techniques in C^*-algebras.

Recently, an American mathematician, born in New Zealand and educated in Switzerland, has found a totally unexpected connection between three apparently diverse fields—knot theory, the classification of subfactors of Type II factors, and the theory of Hecke algebras. This activity has occurred between 1984 and 1986, a catalyst being the fortuitous meeting of the Mathematical Science Research Institute at Berkeley in 1985, sponsored by the National Science Foundation, where experts in operator algebras were meeting concurrently with those in low-dimensional topology, the field that contains knot theory. The excitement of the connection between von Neumann algebras and knot theory may be overshadowed by investigations of its utility to biologists in describing large-scale structures of DNA. A knot is a closed curve in three-space, and a link is a (possibly interlocking) system of knots. Although they can be surprisingly complicated, knots and links can be adequately represented by a projection onto the plane. Two relatively simple examples are shown in Fig. 5.

The knot 6_2

The link 6_2^3
(also called the Borromean rings)

Fig. 5.   Examples of relatively simple knots. (*National Research Council.*)

Although people probably have always used knots, the first known attempt to list and classify them mathematically (as opposed to mechanically) resulted from an erroneous hypothesis of Lord Kelvin in the late

nineteenth century that atoms were knotted vortices in the ether. His vain hope of deriving the periodic table by classifying knots stimulated scientific advances entirely different from his vision—in the characteristic but unpredictable manner that pure research, stimulated by attractive ideas, often yields unanticipated harvests quite different from their original intent.

Knot theory considers two knots or links to be the same if one may be deformed without crossing strings until it is identical to the other. It is exceedingly difficult (and obviously fundamental) to decide when two given links are the same. Trial and error methods are rarely satisfactory. One approach to classifying two mathematical entities in some class is to assign each entity in the class some mathematical label (called an invariant) that coincides on the two entities considered to be the same. Thus a search for appropriate invariants for links began. In the 1920s polynomial invariants were developed by studying the topology of the space remaining when the link is removed from ordinary three-space. The corresponding polynomial for the knot 62 in Fig. 5 is $1 - 3x + 3x^2 - 3x^3 + x^4$.

Another approach to knots first studied in the 1920s was a mathematical structure called the braid group. Braids can be spliced together thus forming a group. However, splicing the top of a braid to its bottom forms a knot or link, and indeed all knots and links may be so formed (in possibly many ways). See Fig. 6.

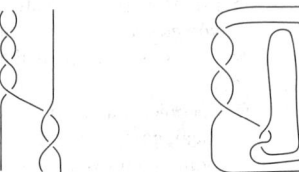

Fig. 6.   Examples of knots of the braid group. (*National Research Council.*)

Braids and their invariants have rewarded investigators with considerable valuable insight over the years. However, their connection with operator algebras remained unnoticed until the proof of a deep result about subfactors of Type II factors required an analysis of one representation of the braid group. Coincidentally, almost the same representation, arising from a special case of the Hecke algebras, had been discovered by mathematical physicists in the 1960s as they partially solved the Potts model of statistical mechanics. The connection with operator algebras generated a trace function for the braids that could be used to recover numerical information. The trace of a braid then provided a new (Laurent) polynomial invariant for the associated knot or link. It was more sensitive than the earlier polynomial in that it separated knots that were previously indistinguishable. It could even distinguish knots from their mirror images. Computers can be used to compute this polynomial for many links. For the knot 6_2, it is $x^{-1} - 1 + 2x - 2x^3 - 2x^4 + x^5$ and, for the link 6_2^3, it is $-x^{-3} + 3x^{-2} - 2x^{-1} + 4 - 2x + 3x^2 - x^3$.

Both operator algebra and topology already have produced substantial generalizations of this new invariant that have been used to solve many venerable problems of knot theory. Similar work is expected to shed new light soon on the Potts model, von Neumann algebras, knot theory, statistical mechanics, quantum physics, and possibly even basic structures of life via the DNA application. In any case, these developments emphasize again the eternity of a good mathematical result, the harmony of all mathematics, the unpredictable relationships between fields, and the many bridges across the humanly created gap between core mathematics and basic science.

At present, only a few institutions in the United States offer training in operator algebras. In this area, as with other areas of mathematics, the United States has become increasingly dependent on hiring people from other countries. In addition, some related areas with potential for interaction with engineering have been underemphasized in the United States. This is an ideal time to put a greater effort into providing the correct conditions for new young researchers and strong leadership in the United States in the subject.

Survival Analysis

Since problems in collecting, analyzing, and interpreting data are universal, it is not surprising that people in many countries (including England, France, Scandinavia, Russia, and the United States) have made significant contributions both to statistics and also to probability, the branch of core mathematics that has until recently provided the major theoretical support for statistics. Expansions in technology have both motivated and enabled the most striking progress in statistics during the past decade. Advances in instrumentation and communication have generated enormously complex sets of data, and the growth of computing power permits the collection and management of such data. The United States has been the unquestioned world leader in statistical computing, mainly because the availability and sophistication of its equipment has been unmatched elsewhere.

However, other countries have made significant contributions; a notable example from England is David Cox's proportional hazards model for life history data. Life history analysis is a body of statistical techniques useful in medical research, reliability theory, actuarial computations, and demographic studies. John Graunt initiated this field in 1662 with his invention of life Tables for analyzing English mortality data, but recent clinical studies are far more complex. Typically, these begin with patients who have an unpleasant disease (possibly at different stages) being assigned randomly to two or more different treatments; they are then followed until they die or disappear or the study ends. Observers record variables, called covariates, that might affect the survival of the patients, including some that do not change, such as sex and age at diagnosis, and some that do, such as blood pressure and glucose level. Statisticians use these observations to study the relationships between the covariates and the survival time of patients after contracting the disease, and especially the comparative merits of the treatments. One goal may be to predict the effect of different treatment on life expectancy of future sufferers of the disease, another may be to determine which factors prolong the patient's normal functioning as long as possible. Subtleties such as the role of the individuals who are still alive at the end of the study and those who withdraw or disappear complicate the analysis.

One model for the life history of a patient (or piece of equipment) includes a vector $Z(t)$ of covariates that vary (possibly randomly) with time, from zero, when the study beings, to the time T when the study is terminated, and a finite-valued process $Y(t)$ designating the state of the dependent variable. These states might be "alive-and-functioning," "alive-not-functioning," "dead," or "lost." The treatments appear as coordinates of $Z(t)$. Analysis of these models concentrates on the intensity of changes; these intensities correspond to a stochastic process $J(t, y_0, y)$, where $j(t, y_0, y)dt$ represents the probability that Y changes from y_0 to y between time t and $t + dt$ assuming a past of $Z(t)$. The life histories of the n patients are regarded as a set of independent observations that can be used to estimate J.

Until the early 1970s, research concentrated on the extremes of either relatively simple situations, with few assumptions, or on vastly more complicated parametric models, with heavy assumptions on j. An example of the former is the setting for the product limit estimator for the probability of surviving beyond time t, when no covariants are measured and only questions of life expectancy are of interest. The assumptions of the parametric models were found to be too unrealistic by experts in biostatistics.

In 1973 Cox proposed his "proportional hazards model" for survival-time data, basing his work in part on a model developed during the 1960s at the National Cancer Institute. In this model

$$J(t, \text{Alive}, \text{Dead}) = \exp[\theta^T Z(t)]\lambda(t)$$

where $\lambda \neq 0$ is a (nonrandom) function and θ is a vector of unknown parameters. $J(t, \text{Alive}, \text{Dead})$ can be an essentially arbitrary function of $Z(s), 0 \leq s \leq t$.

The attractive features of the Cox model are easily seen if we specialize to fixed covariates. The model permits an arbitrary lifetime distribution for a control population and postulates a linear approximation for the log of the ratio of intensities corresponding to two different values of Z. The effects of the covariates can then be measured in terms of the components of θ.

The inferential procedures proposed by Cox were quickly applied to heart transplant and other data by statisticians and biomedical scientists in the United States. Evidence of these applications appears in numerous citations of the model, mostly in the medical literature. Although the model seemed to give sensible answers in applications, its highly nonlinear nature and the complex probabilistic structure of the data prevented rigorous theoretical analysis for some time.

The theory for the case of time-independent covariates was independently developed in the United States and Denmark in the mid- to late

1970s, but these methods could not handle the much more difficult case of time-varying covariates. In 1975, the thesis of a Norwegian student studying in the United States with a United States statistician who emigrated from France showed how to attack the analytic problems in this area. He applied a multivariate counting process framework to both survival analysis models and more general life history models. Most significantly, he exhibited applications to these models of the deep results in continuous time martingales and stochastic integrals, which were introduced by researchers in Japan, the United States, and France. Then others, primarily in The Netherlands, Norway, Denmark, England, and the United States, used these techniques, both to analyze the heuristic suggestions of Cox, and to devise and investigate new inferential procedures in more complicated life history models. Despite their flexibility, the Cox models are still burdened by the questionable assumption that the ratio of intensities for two individuals can be modeled parametrically. It is not clear that the counting process techniques will prove adequate for the analysis of the newer, more flexible, semiparametric models that have been proposed, but they should provide a good starting point.

The analysis of life history data is a rapidly expanding field widely used in a variety of disciplines, including biomedical science, demography, and sociology, fields that reciprocate by continually presenting statisticians with data for which previous methods are inadequate. Although the interaction of theory and application and the international nature of this work are hardly new features of statistics, they have become more prominent recently.

The United States and England are the primary world centers of statistical activity. An important pattern in the field, which is illustrated here, is that foreign scientists and students come to study, lecture, and meet in the United States, and subsequently many of the best remain as citizens or permanent residents.

Note: Some portions of this article also appeared in the publicly funded report, "Mathematical Sciences: A Unifying and Dynamic Resource," prepared by the National Research Council, Washington, DC.

Additional Reading

Papers and Reports, Board of Mathematical Sciences, Washington, DC.
"A Challenge of Numbers: People in the Mathematical Sciences," 1990.
"Actions for Renewing U.S. Mathematical Sciences Departments," 1990.
"Applications of the Mathematical Sciences to Materials Science," 1991.
"Calculus for a New Century: A Pump, Not a Filter," 1987.
"Chairing the Mathematical Sciences Department of the 1990s," 1990.
"Combining Information: Statistical Issues and Opportunities for Research," 1992.
"Discrimination Analysis and Clustering," 1988.
"Educating Mathematical Scientists: Doctoral Study and the Postdoctoral Experience in the United States," 1992.
"Everybody Counts: A Report to the Nation on the Future of Mathematics Education," 1989.
"Mathematical Foundations of High-Performance Computing and Communications," 1991.
"Mathematical Opportunities in Nonlinear Optics," 1992.
"Mathematical Sciences: A Unifying and Dynamic Resource," 1986.
"Mathematical Sciences: Some Research Trends," 1988.
"Mathematical Sciences, Technology, and Economic Competitiveness," 1991.
"Moving Beyond Myths: Revitalizing Undergraduate Mathematics," 1991.
"Number Theory, Proceedings," 1990.
"Probability and Algorithms," 1992.
"Renewing U.S. Mathematics: A Plan for the 1990s," 1990.
"Report of the Advisory Panel to the Mathematical and Information Sciences Directorate, Air Force Office of Scientific Research," 1987.
"Selected Opportunities for Mathematical Sciences Research Related to the Navy Mission: An Update," 1990.
"Select Opportunities for Mathematical Sciences Research Related to the Navy Mission," 1987.
"Spatial Statistics and Digital Image Analysis," 1991.
"Statistical Models and Analysis in Auditing," 1988.
"Statistics — A Guide to Assessing Societal Risk, Proceedings," 1991.
"Symposium on Statistics in Science, Industry, and Public Policy, Proceedings," 1989.
"The Future of Statistical Software," 1991.
"The Impact of Mathematics: Nonlinear Mathematics, Chaos, and Fractals in Science, Proceedings," 1990.

Papers Pertaining to Chaos

Bennett, C.H.: "Quantum Cryptography: Uncertainty in the Service of Privacy," *Science*, 752 (August 7, 1992).

Campbell, A.: "That Uncertain Something," *Case Western Reserve University Magazine*, 18 (November 1992).

Cipra, B.: "Cross-Disciplinary Artists Know Good Math When They See It," *Science*, 748 (August 7, 1992).

Cipra, B.: "Putting the Pedal to the Metal in a Controlled Chaotic Laser," *Science*, 1309 (November 20, 1992).

Douglas, J.: "Seeking Order in Chaos," *Electric Power Research Institute Journal*, 5 (June 1992).

Fischer, P. and W.R. Smith: "Chaos, Fractals, and Dynamics," Marcel Dekker, New York, NY, 1985.

Flam, F.: "The Quest for a Theory of Everything Hits Some Snags," *Science*, 1718 (June 12, 1992).

Garfinkel, A. et al.: "Controlling Cardiac Chaos," *Science*, 1230 (August 28, 1992).

Gleick, J.: "New Images of Chaos That Are Stirring a Science Revolution," *Smithsonian*, 122 (December 1987).

Goldberger, A.L., D.R. Rigney, and B.J. West: "Chaos and Fractals in Human Physiology," *Sci. Amer.*, 42 (February 1990).

Gutzwiller, M.C.: "Quantum Chaos," *Sci. Amer.*, 78 (January 1992).

Horgan, J.: "Nonlinear Thinking," *Sci. Amer.*, 26 (June 1989).

Kerr, R.A.: "Does Chaos Permeate the Solar System?" *Science*, 144 (April 14, 1989).

Krasner, S., Editor: "The Ubiquity of Chaos," American Association for the Advancement of Science, Waldorf, MD, 1990.

Lundqvist, S., N.H. March, and M.P. Tosi, Editors: "Order and Chaos in Nonlinear Physical Systems," Plenum, New York, NY, 1988.

McCauley, J.L.: "Chaos, Dynamics, and Fractals: An Algorithmic Approach to Deterministic Chaos," Cambridge University Press, New York, NY, 1994.

Pool, R.: "Chaos Theory: How Big an Advance?" *Science*, 26 (July 7, 1989).

Pool, R.: "Putting Chaos to Work," *Science*, 626 (November 2, 1990).

Staff: "Nonlinear Mathematics, Chaos, and Fractals: Symposia Proceedings," Board of Mathematical Sciences, Washington, DC (April 28, 1988).

Stewart, I.: "Does God Play Dice? — The Mathematics of Chaos," Science News Books, Washington, DC, 1989.

Sussman, G.J., and J. Wisdom: "Chaotic Evolution of the Solar System," *Science*, 56 (July 2, 1992).

Papers Pertaining to Strings and Knots

Cipra, B.A.: "To Have and Have Knot: When are Knots Alike?" *Science*, 1291 (September 9, 1988).

Cipra, B.A.: "From Real Numbers to Strings of Zeros," *Science*, 1140 (March 3, 1989).

Cipra, B.: "Knotty Problems — and Real-World Solutions," *Science*, 403 (January 24, 1992).

Horgan, J.: "The Pied Piper of Superstrings," *Sci. Amer.*, 42 (November 1991).

Jones, V.F.R.: "Knot Theory and Statistical Mechanics," *Sci. Amer.*, 98 (November 1990).

Papers Pertaining to Factoring

Cipra, B.A.: "Mathematicians Reach Factoring Milestone," *Science*, 374 (October 21, 1988).

Cipra, B.A.: "PCs Factor a Most Wanted Number," *Science*, 1634 (December 23, 1988).

Cipra, B.A.: "Math Team Vaults Over Prime Record," *Science*, 815 (August 25, 1989).

Cipra, B.A.: "Big Number Breakdown," *Science*, 1608 (June 29, 1990).

Fackelmann, K.A.: "Fermat-Number Factors," *Sci. News*, 389 (June 23, 1990).

Ruthern, R.: "Factoring Googols," *Sci. Amer.*, 22 (December 1988).

Papers Pertaining to Geometry

Banavar, J.R., A. Maritan, and A. Stella: "Geometry, Topology, and Universality of Random Surfaces," *Science*, 825 (May 10, 1991).

Berry, M.: "The Geometric Phase," *Sci. Amer.*, 46 (December 1988).

Borwein, J.M., and P.B. Borwein: "Ramanujan and Pi," *Sci. Amer.*, 112 (February 1988).

Cipra, B.A.: "Getting a Grip on Elliptic Curves," *Science*, 30 (January 6, 1989).

Cipra, B.A.: "The Circle Can Be Squared!" *Science*, 528 (May 5, 1989).

Cipra, B.A.: "Another Piece of 3.14159," *Science*, 1260 (June 16, 1989).

Cipra, B.A.: "Packing Your n-Dimensional Marbles," *Science*, 1035 (March 2, 1990).

Cipra, B.A.: "A Speedier Way to Decompose Polygons," *Science*, 270 (July 19, 1991).

Corcoran, E.: "Geometry Decrees a New Dome," *Sci. Amer.*, 102 (September 1989).

Dewdney, A.K.: "The Theory of Rigidity," *Sci. Amer.*, 126 (May 1991).

Hey, A.J.G. et al.: "Topological Solutions in Gauge Theory and Their Computer Graphic Representation," *Science*, 1163 (May 27, 1988).

Horgan, J.: "Pythagoras's Bells," *Sci. Amer.*, 25 (July 1990).

Peterson, I.: "The Color of Geometry," *Sci. News*, 408 (December 23/30, 1989).

Peterson, I.: "Curves for a Tighter Fit," *Science News*, 316 (May 19, 1990).

Strauss, S.: "Impossible Matter," *Technology Review (MIT)*, 19 (January 1991).

Wallich, P.: "Simple Geometry Brings Supercomputers to Their Knees," *Sci. Amer.*, 126 (December 1990).

Papers Pertaining to Computer Mathematics, Algorithms, Networks

Bauschlicher, C.W., Jr. and S.R. Langhoff: "Quantum Mechanical Calculations to Chemical Accuracy," *Science*, 394 (October 18, 1991).

Bern, M.W. and R.L. Fraham: "The Shortest-Network Problem," *Sci. Amer.*, 84 (January 1989).

Cipra, B.A.: "Do Mathematicians Still Do Math?" *Science*, 760 (May 19, 1989).

Cipra, B.A.: "Combinatorial Arguments," *Science*, 595 (August 11, 1989).

Cipra, B.A.: "Algebra: A Hotbed of Radicalism," *Science*, 1190 (September 15, 1989).

Cipra, B.A.: "In Math, Less is More — Up to a Point," *Science*, 1081 (November 23, 1990).

Cipra, B.A.: "The Breaking of a Mathematical Case," *Science*, 165 (January 11, 1991).

Dewdney, A.K.: "Old and New Three-Dimensional Mazes," *Sci. Amer.*, 136 (September 1988).

Hau, Feng-hsiuung et al.: "A Grandmaster Chess Machine" *Sci. Amer.*, 44 (October 1990).

Peterson, I.: "Natural Selection for Computers," *Science*, 346 (November 25, 1989).

Peterson, I.: "Little Fermat: Scheme for Multiplication Leads to A Unique Computer," *Sci. News*, 222 (October 6, 1990).

Pool, R.: "Computing in Science," *Science*, 44 (April 3, 1992).

Rowe, D.E. and J. McCleary: "The History of Modern Mathematics," Academic Press, San Diego, CA, 1994.

Singh, J. and M. McBride: "Successfully Model Complex Hazard Scenarios," *Chem. Eng. Progress*, 71 (October 1990).

Wallich, P.: "Computers Are Changing the Spirit of Mathematics," *Sci. Amer.*, 24 (March 1989).

Yam, P.: "Math Exorcise: A New Algorithm Lifts the Curse of Dimensionality," *Sci. Amer.*, 29 (July 1991).

MATRIX (Adjacency). Let G be a nonoriented graph possessing e elements $\varepsilon_1, \varepsilon_2, \ldots, \varepsilon_e$ and v vertices $\beta_1, \beta_2, \ldots, \beta_v$. The adjacency matrix **A** of G is a square matrix of order v in which $a_{ij} = 1$ if β_i and β_j are adjacent (see also **Vertex**) and zero otherwise; a_{ij} denotes the entry in A located in the ith row and jth column of **A**. Note that $a_{ji} = 0(j = 1, 2, \ldots, v)$, i.e., the diagonal of **A** is composed solely of zeros.

Let **B** be the degree matrix of G. The *Matrix Tree Theorem* states that all the primary cofactors of the matrix $\mathbf{M} = \mathbf{B} - \mathbf{A}$ of a connected nonoriented graph G are equal to the number of trees in G. A graph having the associated matrices **A, B, M** below is shown in Fig. 1.

$$\mathbf{A} = \begin{bmatrix} 0 & 1 & 1 & 1 \\ 1 & 0 & 1 & 1 \\ 1 & 1 & 0 & 1 \\ 1 & 1 & 1 & 0 \end{bmatrix}$$

$$\mathbf{B} = \begin{bmatrix} 3 & 0 & 0 & 0 \\ 0 & 3 & 0 & 0 \\ 0 & 0 & 3 & 0 \\ 0 & 0 & 0 & 3 \end{bmatrix}$$

$$\mathbf{M} = \begin{bmatrix} 3 & -1 & -1 & -1 \\ -1 & 3 & -1 & -1 \\ -1 & -1 & 3 & -1 \\ -1 & -1 & -1 & -3 \end{bmatrix}$$

Observe that the (1,1) cofactor of **M** is precisely equal to 16, the number of trees in G. See also **Digraph**; and **Tree (Mathematics)**.

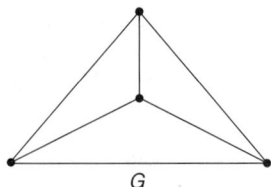

Fig. 1. Concept of adjacency matrix.

MATRIX (Mathematics). Consider a set of elements, finite in number, which may be arranged in rows and columns. If A_{ij}, B_{ij} are the elements in ith row and jth column of two such arrays and if these arrays combine to form a product with elements $C_{ij} = \Sigma A_{ik}B_{kj}$, then they are called matrices. A convenient symbolism is $\mathbf{A}, \mathbf{B}, \mathbf{C} = \mathbf{AB}$.

A matrix containing m rows and n columns is of order $(m \times n)$ and it has mn elements. If $m \neq n$, the matrix is rectangular; if $m = n$, it is square;

if either m or n is unity, the matrix is called a vector, for its n elements can be regarded as vector components in an n-dimensional space. A row vector, of order $(1 \times n)$, will be denoted by $[\mathbf{x}]$; a column vector, of order $(n \times 1)$ by $\{\mathbf{x}\}$.

Matrix elements may be real, complex, or pure imaginary quantities of a very general nature. The concepts may be extended to matrices of infinite order.

If two matrices are equal, $\mathbf{A} = \mathbf{B}$, and $A_{jk} = B_{ij}$. The sum or difference of two matrices is defined by $\mathbf{A} \pm \mathbf{B}$, and $A_{ij} = B_{ij}$. The sum or difference of two matrices is defined by $\mathbf{A} \pm \mathbf{B} = \mathbf{C}$, $A_{ij} \pm B_{ij} = C_{ij}$. Also $\mathbf{A}$, where c is a constant, has elements cA_{ij}.

Two matrices $\mathbf{A}$ and $\mathbf{B}$ can form a product $\mathbf{AB}$ only if the number of columns in $\mathbf{A}$ equals the number of rows in $\mathbf{B}$. Thus, if $\mathbf{A}$ is of order $(n \times h)$ and $\mathbf{B}$ of order $(h \times m)$; the product $\mathbf{C} = \mathbf{AB}$ is of order $(n \times m)$ and $\mathbf{A}, \mathbf{B}$ are conformable. The following cases can also occur: $\mathbf{Ax} = \mathbf{y}$; $[\mathbf{s}]\mathbf{A} = [\mathbf{y}]$; $[\mathbf{x}]\mathbf{y} = $ a scalar; $\mathbf{x}[\mathbf{y}] = \mathbf{B}$, where $\mathbf{B}$ is square with rows or columns proportional to each other. Matrix multiplication is not necessarily commutative, but the distributive and associative laws hold. If $\mathbf{AB} \neq \mathbf{BA}$, the quantity $(\mathbf{AB} - \mathbf{BA})$ is called the commutator. A special combination called the direct product can be defined. It is of interest in group theory.

Matrices with special properties are named as follows: (a) Null, indicated by $\mathbf{0}$, with zero for all its elements. In this case, $\mathbf{A} + \mathbf{0} = \mathbf{A}$; $\mathbf{A0} = \mathbf{0A} = \mathbf{0}$. (b) Unit, $\mathbf{E}$ (for German, *Einheit*). Its elements are δ_{ij}, the Kronecker delta, and $\mathbf{EA} = \mathbf{AE} = \mathbf{A}$. (c) Diagonal, all elements are zero except those with equal subscripts; thus, if $\mathbf{D}$ is diagonal, $D_{ij} = D_i\delta_{ij}$ or in another symbolism $\mathbf{D} = \text{diag}(D_1, D_2, \ldots, D_n)$. (d) Triangular, all elements above or below the main diagonal are zero ($T_{ij} = 0$, $i > j$ or $i < j$). (e) Transposed, given $\mathbf{A}$ with elements A_{ij}, its transposed matrix is $\tilde{\mathbf{A}}$ or $\mathbf{A}'$ with elements A_{ji}. If $\mathbf{ABC}\cdots\mathbf{X} = \mathbf{Y}$, then $\tilde{\mathbf{Y}} = \tilde{\mathbf{X}}\cdots\tilde{\mathbf{C}}\tilde{\mathbf{B}}\tilde{\mathbf{A}}$. (f) Singular, the determinant of $\mathbf{A}$ vanishes. All rectangular matrices are singular since their determinant, by definition, does not exist. (g) Adjoint, find the cofactor A_{ij} of A_{ij} in $|A|$ and transpose to get $\mathbf{A}$ with elements A_{ji}. It follows from the properties of determinants that $\mathbf{A}\hat{\mathbf{A}} = \hat{\mathbf{A}}\mathbf{A} = |A|\mathbf{E}$. (h) Reciprocal or inverse, $\mathbf{A}^{-1} = \hat{\mathbf{A}}/|A|$, as defined from (g). Only a square matrix, thus a nonsingular one, has a reciprocal. If $\mathbf{ABC}\cdots\mathbf{X} = \mathbf{Y}$, then $\mathbf{Y}^{-1} = \mathbf{X}^{-1}\cdots\mathbf{C}^{-1}\mathbf{B}^{-1}\mathbf{A}^{-1}$, as for (e). (i) Complex conjugate, only if $\mathbf{A}$ has complex elements. Using an asterisk for this operation, the complex conjugate to $\mathbf{A}$ is $\mathbf{A}^*$ with elements A_{ij}^*. Unlike some previous cases, $\mathbf{A}^*\mathbf{B}^*\mathbf{C}^*\cdots\mathbf{X}^* = \mathbf{Y}^*$. (j) Associate (also called adjoint or Hermitian conjugate) $\mathbf{A}\dagger = (\mathbf{A}^*) = (\tilde{\mathbf{A}})^*$. In this case, $\mathbf{Y}\dagger = \mathbf{Y}\dagger\cdots\mathbf{C}\dagger\mathbf{B}\dagger\mathbf{A}\dagger$.

Certain other special matrices lead to names and relations as shown in Table 1.

TABLE 1. SPECIAL MATRICES

Name	Relation	Matrix Elements
Symmetric	$\mathbf{A} = \tilde{\mathbf{A}}$	$A_{ij} = A_{ji}$
Skew symmetric	$\mathbf{A} = -\tilde{\mathbf{A}}$	$A_{ii} = 0$; $A_{ij} = -A_{ji}$
Orthogonal	$\mathbf{A} = \tilde{\mathbf{A}}^{-1}$	$\sum A_{ij}A_{kj} = \sum A_{ji}A_{jk} = \delta_{ik}$
Real	$\mathbf{A} = \mathbf{A}^*$	$A_{ij} = A_{ij}^*$
Pure imaginary	$\mathbf{A} = -\mathbf{A}^*$	$A_{ij} = iB_{ij}$; B_{ij} real
Hermitian	$\mathbf{A} = \mathbf{A}\dagger$	$A_{ij} = A_{ji}^*$
Skew Hermitian	$\mathbf{A} = -\mathbf{A}\dagger$	$A_{ii} = 0$; $A_{ij} = -A_{ji}^*$
Unitary	$\mathbf{A} = (\mathbf{A}\dagger)^{-1}$	$\sum A_{ij}A_{kj}^* = \sum A_{ji}A_{jk}^* = \delta_{ik}$

MATRIX (Triangular). A matrix in which all elements are zero above the diagonal (a lower triangular matrix), or else below the diagonal (an upper triangular matrix). It is a unit upper or lower triangular matrix when also every diagonal element = 1. The product of two triangular matrices of the same type is again of that type, i.e., (unit) upper or lower; and the reciprocal of a triangular matrix (when it exists) is also of the same type. A triangular matrix is properly triangular if the diagonal is null.

MATRIX INVERSION. The determination for a given matrix $\mathbf{A}$ of its inverse matrix $\mathbf{A}^{-1}$ such that the product $\mathbf{AA}^{-1} = \mathbf{A}^{-1}\mathbf{A} = \mathbf{E}$, the unit matrix. The unit matrix is also symbolized by $\mathbf{I}$. All its elements are zero except those with equal subscripts, which are unity. Inversion is defined only for square matrices.

MATTER-ANTIMATTER SYMMETRY. See **Quantum Mechanics**.

MAXIMUM AND MINIMUM (Mathematics). If is often of interest to study the behavior of a plane curve, described by the equation $y = f(x)$, as the independent variable x is continually increased. The possibilities are: (1) the curve continuously rises or falls; (2) the curve reaches a maximum and then falls from it or reaches a minimum and then rises; (3) an abrupt break in the curve appears at a certain pair of values (x_0, y_0); (4) some peculiar change in the slope or direction of the curve occurs. This article is concerned with the conditions for a maximum or a minimum; for cases (1) and (3), see also **Continuous Function**; for case (4), see also **Singular Point of a Curve**.

If the curve $y = f(x)$ is plotted, it is easy to see that the curve rises or increases if its tangent with the X-axis is acute or if the slope, dy/dx, is a positive number. Conversely, if the curve decreases or falls, the tangent or dy/dx is negative. In order for a maximum to occur, the derivative must change its sign and at the maximum point (x_0, y_0), $dy/dx = 0$. Similar considerations show that a minimum also occurs at $f'(x) = 0$ but the sign of the derivative must there change from minus to plus.

A more careful investigation of the situation shows that the derivative can also vanish when neither a maximum nor a minimum occurs, for the curve could momentarily become parallel to the X-axis and then continue to rise or fall without passing through either a maximum or a minimum. One is thus led to consider the second derivative and one then finds that a rule, generally applicable, is: $f(\ )$ is a maximum at $x = x_0$ if $dy/dx = 0$; $dy/dx < 0$ for $x > x_0$; $dy/dx > 0$ for $x < x_0$; $d^2y/dx^2 < 0$ for $x = x_0$. The corresponding rule for a minimum is $dy/dx = 0$ at $x = x_0$; $dy/dx < 0$ for $x > x_0$; $dy/dx > 0$ for $x > x_0$; $d^2y/dx^2 > 0$ for $x = x_0$.

A third possibility arises when both dy/dx and d^2y/dx^2 vanish at $x = x_0$. Before consideration of this case, let us refer to the concavity or convexity of a curve. These terms are to be interpreted in their usual non-technical way with respect to the X-axis but technically *concavity* means that the curve lies above its tangent and *convexity* means that the tangent is above the curve. Hence, if d^2y/dx^2 is positive (negative) at $x = x_0$, the curve has passed from *concavity* to *convexity*, or the reverse. Such a situation is called a point of inflection.

These arguments can be extended to the case of several independent variables and, in fact, both a maximum and a minimum, called a saddle point or a minimax, can occur simultaneously. A more refined treatment of such problems is the subject matter of the calculus of variations. A neutral term to designate either a maximum or a minimum is *extremal*.

See also **Variations (Calculus of)**.

MAXWELL EQUATIONS. See **Electromagnetic Phenomena**.

MAXWELL, JAMES CLERK (1831–1879). A Scottish physicist, Maxwell worked in electromagnetism and the kinetic theory of gases. Maxwell studied molecules of gases in rapid motion. He treated them statistically and is remembered for the Maxwell-Boltzmann kinetic theory of gases. Maxwell's most important achievement was showing that a few simple mathematical equations could express the behavior of electric and magnetic fields and their interrelated nature. His work is known as Maxwell's equations and is a great achievement for 19th century physics. His important publications include, *Theory of Heat, Treatise on Electricity and Magnetism*, and *Matter and Motion*.

See also **Boltzmann's Distribution Law**; **Bridge Circuits (Electrical/Electronic)**; **Color**; **Electromagnetic Phenomena**; **Electron Theory**; **Equilibrium**; **Gravitation**; **Kinetic Theory**; **Photography and Imagery**; **Quantum Mechanics**; and **Statistical Mechanics**.

J. M. I.

MAYER, MARIA GOEPPERT (1906–1972). Mayer was born in Germany and immigrated in 1930 to the United States. Her early work is in molecular spectra and is important for laser physics. She also proposed, with Teller, a theory for the origin of the elements. In 1948, she developed the nuclear shell model. She won the Nobel Prize in Physics in 1963.

J. M. I.

MAYFLY *(Insecta, Ephemeroptera)*. Also known as shad flies, salmonflies, and lake flies. The adults are sluggish insects with slender filaments at the caudal end of the body and large, triangular front wings. The hind wings are much smaller, in some species rudimentary. The immature insect is aquatic and in most species feeds on decaying vegetable matter. It may live for several years, while the adult stage lasts only a few days.

Mayflies of one species emerge as adults in large numbers within a short period and are sometimes very abundant near favorable bodies of water. They fly at twilight and can sometimes be seen in large gray clouds at a distance of more than a mile over the islands of Lake Erie, where they are especially abundant. In the cities bordering the lake they are attracted to lights and their dead bodies are sometimes swept up in bushels after a heavy flight. Under such conditions they are a nuisance but not a serious pest. Their value as food for fishes more than offsets what little harm they do. See Fig. 1.

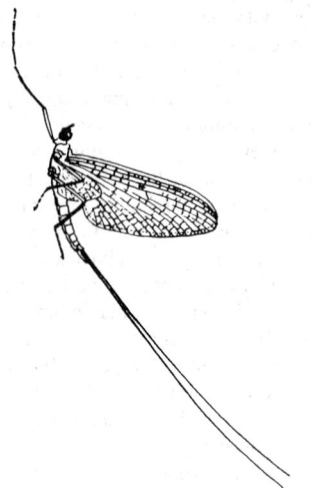

Fig. 1.   Mayfly. (*USDA.*)

MCLEOD GAGE.   A liquid-level vacuum gage in which a known volume of gas, at the pressure to be measured, is compressed by the movement of a liquid column to a much smaller known volume, at which the resulting higher pressure is measured.

M-DISPLAY.   In radar, a display in which target distance is determined by moving an adjustable blip along the baseline until it coincides with the horizontal position of the target signal deflections. The control which moves the blip is calibrated in distance. Also called *M-scan, M-scope, or M-indicator.*

MEADOW-BROWN (*Insecta, Lepidoptera*).   Butterflies of dull gray-brown color marked with eye-like spots, in some species set in a yellow patch on the fore wings. Family *Satyridae.*

MEADOWLARK (*Aves, Passeriformes*).   A common North American bird (Aves) more closely related to the blackbirds and orioles than to the true larks. The eastern meadowlark, *Sturnella magna*, is less attractive than the western, *S. neglecta*, which has a brief but glorious song. Both have the characteristic yellow breast with a black chevron at the throat.

MEALYBUG (*Insecta, Homoptera*).   A small, very damaging insect on citrus and in glasshouse operations. The insects, of several species as indicated below, are distributed worldwide.

Citrophilus mealybug (*Pseudococcus gahani* (Green)).
Citrus mealybug (*Pseudococcus citri* (Risso)).
Coconut mealybug (*Pseudococcus nipae* (Maskell)).
Grape mealybug (*Pseudococcus maritimus* (Enrohorn)).
Long-tailed mealybug (*Pseudococcus adonidum* (Linne)).
Mexican mealybug (*Phenacoccus gossypii* (Townsend and Cockerell)).

In the United States, they are notably destructive of citrus crops in California. All stages of development of the insect can be found throughout the year.

The life cycle of all species is about the same. The appearance differs slightly. The citrus mealybug has a dense white powder over its back. The long-tailed mealybug is ovoviviparous. Adult females of the other species, ranging from $\frac{1}{8}$ to $\frac{1}{3}$ inch (3 to 8 millimeters) in length, deposit from 300 to 600 eggs. The females secrete a waxy, cottony mass into which the eggs are placed. From 1 to 3 weeks are required for hatching. The mealybug

young feed on the sap and juices from fruit and leaves. Their movement is slow and limited. It requires from 1 to 4 months to complete their growth. The female passes through a pupa stage before becoming an adult. The females are wingless. Usually there are from 2 to 4 generations per year.

Biological methods have been quite successful in controlling mealybugs. In the California citrus orchards, great numbers (millions) of the coccinellid beetle (*Cryptolaemus montrouzieri*), have been very effective in controlling the mealybug. Several hymenopterous parasites have been imported and are effective: *Coccophagus gurneyi* and *Tetracnemus pretiosus* against the citrophilus mealybug; *Leptomastidea abnormis* against the citrus mealybug; *Anaphopus sydneyensis* and *T. peregrinus* against the long-tailed mealybug.

MEAN DEVIATION.   The mean deviation of n sample values x about a point a is defined as $\Sigma |x - a|/n$. Used as a measure of dispersion, a is usually chosen to be the sample mean, or the sample median about which the mean deviation is a minimum.

MEAN LIFE.   For any unstable system that decays in accordance with the laws of probability, such as either excited states of atoms or radioactive nuclides, the number N of individual units of the system in existence at any time t is $N = N_0 e^{-\lambda t}$, in which N_0 is the number of units existing at time $t = 0$ and λ is the decay constant. Then the number of units that have a lifetime between t and $t + dt$ is $N\lambda dt = N_0 \lambda e^{-\lambda t} dt$ and the total lifetime of all N_0 nuclides is

$$L = \int_0^\infty tN\lambda \, dt = \int_0^\infty tN_0 \lambda e^{-\lambda t} dt = N_0/\lambda$$

The average lifetime, or mean life τ, of a single nuclide is $\tau = 1/\lambda$. Another commonly used unit of lifetime is the half life, $t_{1/2}$ which is less than the mean life by a factor $\ln 2 = 0.693$, such that $\tau = t_{1/2}/0.693$. The mean life is used as a measure of the lifetime of many other systems besides radioactive nuclides, such as metastable states, mesons, hyperons, the recombination of carriers of opposite sign in semiconductors, and the rate of absorption of thermal neutrons in matter. See also **Decay Constant**; and **Radioactivity**.

MEAN SEA LEVEL.   The average height of the sea surface, based upon hourly observation of the tide height on the open coast or in adjacent waters that have free access to the sea. These observations are to have been made over a "considerable" period of time. In the United States, mean sea level (MSL) is defined as the average height of the surface of the sea for all stages of the tide over a 19-year period. See also **Altimetry**.

MEAN SEA LEVEL PRESSURE.   See **Altimetry**.

MEAN SIDEREAL TIME.   See **Time**.

MEAN SOLAR DAY.   The duration of one rotation of the Earth on its axis, with respect to the mean sun. The length of the mean solar days is 24 hours of mean solar time or 24 hours 3 minutes 56.555 seconds of mean sidereal time. A mean solar day beginning at midnight is called a civil day; and one beginning at noon, 12 hours later, is called an astronomical day.

MEAN SOLAR SECOND.   Prior to 1960 the fundamental unit of time, equal to 1/86,400 of the mean solar day. Now replaced by the *ephemeris second*. In radar, a display in which target distance is determined by moving an adjustable blip along the baseline.

MEAN SQUARE ERROR.   If a statistic t estimates a population parameter θ, the mean square error of t is defined as $E(t - \theta)^2$, when E denotes expectation. It is equal to the sum of the variance of t plus the square of the bias.

MEAN VALUE THEOREMS.   The first law of the mean for integrals is

$$\int_a^b f(x) \, dx = (b - a)f(z)$$

where $a \le z \le b$ and $f(x)$ is a continuous function.

The second law of the mean is

$$\int_a^b f(x)\phi(x) \, dx = \phi(a) \int_a^z f(x) \, dx$$

with z and $f(x)$ restricted as before; $\phi(x)$ is also continuous and a positive monotonic decreasing function in the interval (a, b). Another form of the second law is

$$\int_a^b f(x)\phi(x)\,dx = \phi(a)\int_a^z f(x)\,dx + \phi(b)\int_z^b f(x)\,dx$$

where $\phi(x)$ is not necessarily always positive.

There are similar formulas for the case where $\phi(x)$ is an increasing function. The two forms of the second theorem are known as the forms of *Bonnet* and of *DuBois-Reymond*, respectively.

A mean value theorem also exists for a derivative. Let $f(x)$ be a function which has a finite derivative at all points of the interval (a, b). Then there exists a value of z between a and b such that $f(b) - f(a) = f'(z)(b - a)$. The theorem may be interpreted geometrically, for it states that the tangent to a smooth curve is parallel to an intermediate point on a chord of the curve. The procedure can be generalized to give the extended mean value theorem

$$f(b) = f(a) + (b - a)f'(a) + \frac{(b-a)^2}{2!}f''(a) + \cdots$$
$$+ \frac{(b-a)^{n-1}}{(n-1)!}f^{(n-1)}(a) + \frac{(b-a)^n}{n!}f^{(n)}(a)$$

See also **Taylor Series**.

MEASLES. A viral infection that remains a major cause of childhood morbidity and mortality in developing countries. A World Health Organization estimate (1990) indicates that more than 2 million children die of measles each year. Measles also is an important cause of blindness, diarrhea, and malnutrition in many children who survive the acute illness. Vaccines are effective in reducing the incidence of measles in many countries. Infants in areas endemic for measles, however, lose maternal antibody before they are 9 months old and thus remain at high risk for measles and account for 20 to 30% of all patients with measles in some large urban areas.

In the advanced and industrialized nations, the incidence of measles has decreased markedly over the last few decades, mainly because of the increase in the number of vaccinations and, in particular, because of the increased effectiveness of vaccines. For example, with the introduction of live measles immunization, the number of cases of measles in the United States declined from 450,000 in 1964 to 22,400 in 1967. With this decline, the serious complications of the disease and resulting fatalities from encephalitis also dropped in 1979. The decline continued throughout the 1980s, and professionals set a target to eliminate indigenous measles from the United States. As of the early 1990s, measles remained a notifiable disease in the United States, with approximately 2100 cases reported in 1992. In contrast, the incidence of measles in many parts of the world remains one of major concern.

As reported by Georges Peter (Rhode Island Hospital and Brown University School of Medicine), "The marked decline in the incidence of vaccine-preventable diseases in the United States has correlated with rates of immunization of approximately 95% or more in school-age children. These rates can be attributed in part to the enactment and enforcement of school immunization laws in each state. Among children in the first two years of life, however, rates of immunization in some areas are substantially below the national goal of 90 percent for completion of the recommended immunizations by the second birthday. The gap is especially prominent in some intercity populations." In a study by Cutts, et al. (see reference), "The principal cause of the measles epidemic of 1989 through 1990 was failure to vaccinate children at the recommended age." The study notes that this deficiency also caused concern over possible outbreaks of other vaccine-preventable diseases in the United States. In 1990, there was an outbreak of 25 cases of congenital rubella infection in Southern California. In 57% of the women who delivered infants with the congenital rubella syndrome, one or more missed opportunities for rubella-susceptibility testing and vaccination were identified. See also **Immune System and Immunology**.

The etiologic agent is a single-stranded RNA paramyxovirus of which only one strain is known; symptomatic variations in the disease course are related to local environmental factors rather than the causative agent. The virus is spread by droplet infection and gains access through the respiratory tract and conjunctivae. It replicates mainly in the pharynx and regional lymph nodes. After 2 or 3 days of infection, primary viremia (virus in blood) develops. In the fifth to sixth day, a secondary viremia occurs. This produces a rash, fever, and conjunctivitis, with coryza (acute head cold) present in the tenth to fourteenth day after incubation. The incubation period runs from 7 to 14 days. Characteristically, the rash begins at the hairline and spreads down over face, neck, trunk, and eventually over the entire body. The lesions start as tiny flat red spots; they enlarge and spread to become confluent in many areas. At the height of the disease, the temperature may be as high as $105\,°F$ $(40.6\,°C)$. The patient suffers from itching and burning of the skin. Marked sensitivity of the eyes to light, and cough, are present at this stage. Koplik's spots appear in the oral mucosa as whitish-blue centers in red erythematous backgrounds. After the eruption reaches its peak, it begins to fade, usually in the order of its appearance and, during this period, the patient improves dramatically.

Therapy is mainly supportive. Although usually benign, measles may develop some untoward complications, otitis media, giant cell interstitial pneumonia, and encephalitis being the most common. These should be treated as separate disease entities, with penicillin or other antibiotics as necessary.

In July 1990, G.D. Hussey (University of Cape Town, South Africa) and M. Klein (Red Cross War Memorial Children's Hospital, Rondebosch, South Africa) reported on a study of 189 children in an attempt to determine the possible effectiveness of vitamin A therapy in the treatment of measles. This therapy first was proposed in the mid-1980s. Conclusions of the report: "Treatment with vitamin A reduces morbidity and mortality in measles, and all children with severe measles should be given vitamin A supplements, whether or not they are thought to have a nutritional deficiency." Other than vaccination for prevention, this is one of the few therapies recommended other than careful supportive care.

Additional Reading

Cliff, A.D., M. Smallman-Raynor, and P. Haggett: "Measles: A History," Blackwell Science, Inc., Malden, MA, 1993.
Cutts, F.T. et al.: "Monitoring Progress Toward U.S. Preschool Immunization Goals," *J. Amer. Med. Assn.*, 1952 (May 1992).
Diaz-Ortega, J. et al.: "Immunization of Six-Month Old Infants with Different Doses of Edmonston-Zagreb and Schwarz Measles Vaccines," *N. Eng. J. Med.*, 580 (March 1, 1990).
Eobbins, A. and P. Freeman: "Obstacles to Developing Vaccines for the Third World," *Sci. Amer.*, 126 (November 1988).
Hussey, G.D. and M. Klein: "A Randomized Controlled Trial of Vitamin A in Children with Severe Measles," *N. Eng. J. Med.*, 160 (July 19, 1990).
Kurstak, E.: "Measles and Poliomyelitis: Vaccines, Immunization, and Control," Springer-Verlag, Inc., New York, NY, 1993.
Oehen, S., H. Hengartner, and R.M. Zinkernagel: "Vaccinations for Disease," *Science*, 195 (January 11, 1991).
Peter, G.: "Childhood Immunizations," *N. Eng. J. Med.*, 1794 (December 17, 1992).

Web Reference

Center for Disease Control and Prevention: *http://www.cdc.gov/health/diseases.htm*

MEASURE OF LOCATION. A quantity calculated from a frequency distribution intended to indicate the position of the distribution on the scale of measurement. The commonest measure of location is the arithmetic mean.

MEASURING WORM *(Insecta, Lepidoptera).* A caterpillar that loops the body by drawing the hind legs up close to the front legs as it crawls. The movement is associated with the lack of most of the legs near the middle of the body. Caterpillars of the family *Geometridae* and a few species of *Noctuidae* are of this type. The adult moths of the *Geometridae* family are relatively large, having a wingspan of one inch (2.5 centimeters) on the average, but some species exceeding 2 inches (5 centimeters). Their bodies are slender and they have a rather delicate appearance, something like a butterfly. However, as with other moths, their wings are spread when they are at rest. Closely related are the spring and fall cankerworms, the currant spanworm, and the snow-white linden moth.

MECHANICAL EQUIVALENT OF HEAT. The conservation of heat per se is observed only for systems not involving the performance of mechanical or electrical work. Count Rumford (ca. 1800) was the first to establish this fact in his well-known cannon-boring experiments carried out in the arsenal of the Duchy of Bavaria in Munich. He observed that when his drills became dull, heat was produced in great quantities limited only by the amount of work done against friction. He concluded that the large-scale mechanical energy used in overcoming friction could only be converted into the motions of the ultimate particles of matter, a motion not directly observable, but detected by our senses as heat. His results were confirmed and extended by the late work of Joule and Helmholtz, in particular, and also provided a more reliable value for the so-called *mechanical equivalent of heat.*

This is taken as the amount of mechanical (or electrical) energy which when converted into heat is equivalent to exactly 1 calorie. The value for this important constant is 4.185 joules per 15° calorie. Here the joule is the work performed when power is expended at the rate of 1 watt for 1 second. Thus, a 100-watt lamp bulb converts 100 joules of electrical energy to thermal energy each second; this amounts to 100/4.185, or about 24 calories.

The foregoing experiments emphasized that heat is merely another form of the universal quantity *energy.* Its transformation always occurs at the rate of 4.185 joules per calorie whether heat goes into external work or work is dissipated through friction into heat.

MECHANICS.   The science that deals with the effects of forces upon bodies at rest or in motion. The laws and phenomena of gases and liquids and solid bodies have a part in this subject, and it is one of the basic studies of engineering, physics and astronomy. It is customary to subdivide mechanics into the study of fluids and the study of particles or bodies of solid materials. It is to the latter field that the term mechanics is frequently restricted. For convenience it is subdivided into statics and dynamics. Dynamics is usually further subdivided into kinematics and kinetics. Statics deals with bodies at rest, in equilibrium under the action of forces or of moments; kinematics deals with abstract motion and kinetics treats of the effect of forces or of moments upon the motions of material bodies.

Fluid mechanics is that branch of mechanics that deals with those fundamental laws that apply to all fluids (liquids or gases) at rest or in motion.

See also **Dynamics**; **Kinematics**; **Kinetics**; and **Rotation**.

MECOPTERA.   Insects with four narrow wings with numerous veins. The head is prolonged downward in a beak bearing biting mouth parts at the tip. The relatively few species inhabit moist woods and are not commonly known.

This order includes two chief forms, the scorpion flies and a group of slender long-legged insects usually known by their generic name, *Bittacus.* In the former the tip of the abdomen is modified so that it resembles that of a scorpion slightly. *Bittacus* has the general appearance of the crane flies but for its four wings, and is chiefly remarkable for the grasping joints at the tips of the legs.

MEDIAN.   See **Quantile**.

MEDIATOR.   See **Immune System and Immunology**.

MEDITERRANEAN WATER.   A study of the oceanic water masses contained in the Mediterranean Sea is complicated by the fact that, due to the numerous peninsulas and the high rate of evaporation, four distinct bodies of water can be identified by temperature and salinity measurements. These are the Algiers-Provencal and Tyrrhenian on the West and the Ionian and Levantine on the East. Moreover, in each area all three types, surface water, intermediate and bottom water, can be identified. The oxygen content is also a valuable differential property, since the salinity is generally consistently over 36%. In fact, as the Mediterranean water leaves the Straits of Gibraltar its salinity averages over 37%, and its temperature over 18 °C (64.4 °F). It forms currents, which, because of these properties, can be traced far into the South and North Atlantic Oceans.

MEDLAR TREES.   See **Rose Family**.

MEDULLA OBLONGATA.   See **Central and Peripheral Nervous Systems**.

MEDUSA (or Hydromedusa).   A form of coelenterate in which the body is shortened on its principal axis and broadened, sometimes greatly, in contrast with the hydroid or polyp. It varies from bell-shaped to a thin disk, scarcely convex above and only slightly concave below. The upper or aboral surface is called the ex-umbrella and the lower surface, on which the mouth opens, the subumbrella. The latter may be partly closed by a

membrane extending inward from the margin. This structure is called the velum. The digestive cavity consists of a central chamber, the stomach, and radiating canals, which extend toward the margin. These canals may be simple or branching and few or many. The margin of the dish bears tentacles and sensory organs.

In the class *Hydrozoa*, medusae are the sexual individuals of many species, alternating in the life cycle with asexual polyps; but in the *Scyphozoa* or jellyfish proper, the medusa alone is well developed.

MEDUSOID.   The medusa of certain coelenterates of the class *Hydrozoa*. A hydromedusa. Medusoids differ from the free-swimming jellyfishes to which the term medusa is applied in the usual presence of the velum and in the simpler digestive cavity. The term is also applied in some cases to the young medusae budded from the larval polyp stage of the jellyfishes.

MEERKAT.   See **Viverrines**.

MEGAPODE *(Aves, Galliformes)*.   Dull-colored birds of the Pacific islands, from the Philippines to Australia. They have strong legs and feet and resemble turkeys slightly. The eggs are deposited in mounds of decaying vegetation which generate the heat necessary for incubation. The family includes the brush turkeys of the Australian region and the maleo in addition to the true megapodes. The brush turkey also is called the zebra bird, shell bird, and warbling grass parakeet. See also **Galliformes**; and **Mound Birds**.

MEISSNER EFFECT.   When a superconductor is cooled in a magnetic field the lines of induction are pushed out at the transition, as if it exhibited perfect diamagnetism, an effect essentially distinct from the zero resistivity of the metal, which must be considered as a separate phenomenon.

MEL.   A unit of acoustic pitch. By definition, a simple tone of frequency 1000 cycles per second, 40 decibels above a listener's threshold, produces a pitch of 1000 mels. The pitch of any sound that is judged by the listener to be n times that of a 1-mel tone is n mels.

MELALEUCA TREE.   In an effort to afforest the Everglades at the turn of the century, the melaleuca tree was introduced from Australia in 1906. The effort was stepped up in the 1930s when melaleuca seeds were scattered over the region by aircraft. In recent years, the effort has backfired because the melaleuca has damaged and continues to threaten the diversity of life that once characterized the Everglades. The melaleuca is a large, bushy tree that can reach 80 feet (24+ meters) in height and have a diameter of 40 in. (102 cm). The tree is well adapted to damp areas. Like the casuarina (see also **Casuarina Tree**), the melaleuca is extremely rugged and very difficult to eradicate. One biologist has observed, "Cut one down and you can wind up instead with three or four — every cut piece sprouts back. Bulldoze one, and it will grow back from the roots. Disturb one in any way and its seed pods open up." The melaleuca is equipped with mechanisms that make it a natural marvel. During flooding, it produces new roots up the stems as the water level rises; seedlings produce new leaves and continue to grow underwater. during fires, the pods burst from the heat and disperse their seeds, while a thick, punky bark acts as excellent insulation. During high winds, seeds are released and broken branches quickly sprout and take root. Frost seems to be the tree's only principal adversary.

The melaleuca is not without some benefits. For example, the trees provide abundant pollen and nectar over much of the year and it is estimated that beekeepers maintain some 200,000 bee colonies on the melaleucas. Some authorities believe that melaleuca wood may have potential as a hardwood raw material. The heartwood is resistant to decay and termites. The wood can be seasoned and finished to rival the attractiveness of cherry, black walnut, or mahogany.

The record melaleuca tree, melaleuca quinquenervia, in American Forests is located in new Ft. Denaud, Florida. Also referred to as the cajeput tree, this record holder is 62 feet (18.9 meters) high, has a spread of 28 feet (8.58 meters), and a circumference of 231 rule (554.5 centimeter).

MELAMINE.   ($N\equiv C-NH_2$)$_3$, formula weight 126.12, white solid, mp 355 °C, sp gr 1.56. The compound may be considered the trimer of cyanamide, or as the triamide of cyanuric acid. Melamine resembles an amide more than an amine. Liebig first prepared melamine in 1834. In early production methods, melamine was prepared from calcium cyanamide through conversion to the cyanodiamide and then to the trimer, melamine. The compound now is synthesized from urea. The production of melamine exceeded 450 metric tons annually in the early 1970s and has been growing at a rate of about 5% annually. Most of the melamine made is condensed with formaldehyde or other aldehydes to form resins. These resins possess particularly outstanding resistance to heat, water, and many chemicals. The electrical properties and surface hardness also are rated high. The consumption of melamine for these resins is: (1) protective and decorative laminates, 45%; (2) molding compounds, 30%; (3) textile resins, 9%; (4) coatings, 7%; (5) paper-treating resins and various adhesives, 9%. Typically, in a modern synthesis process, (1) urea is thermally decomposed into a gas mixture of cyanic acid and NH_3: $H_2N \cdot CO \cdot NH_2 \rightarrow HCNO + NH_3$; (2) cyanic gas is thermally decomposed into a melamine-CO_2 vapor:

$$6 \ HCNO \longrightarrow (N\equiv C-NH_2)_3 + 3 \ CO_2.$$

Step 1 is endothermic; step 2 is exothermic; the overall reaction is endothermic. Because of the large quantities of CO_2 and HN_3 generated, the process often is undertaken in connection with urea manufacture, which permits the off-gases to be recycled usefully. The melamine synthesis may be carried out at low or medium pressures with the assistance of a catalyst; or at higher pressures without a catalyst.

MELANIN.   See Dermatitis and Dermatosis.

MELIOIDOSIS.   The unusual causative agent of this disease, *Pseudomonas mallei*, is a free-living organism widely distributed in stagnant ponds, streams, rice paddies and soils in endemic areas of the tropics. Most infections have emanated from southeast Asia, but indigenously acquired disease has been reported from India, Korea, the Philippines, Australia, Panama, Turkey, and the United States. The organism is a small, pleomorphic, Gram-negative, strictly anaerobic bacillus, which shows a prominent bipolar pattern on staining.

Human infection usually results from contact of the broken skin with infected water, or by inhalation of contaminated dust. Five patterns of the disease have been observed. The *subclinical* pattern is generally asymptomatic and manifest only by elevated antibody levels. The *acute localized infection* takes the form of a localized pustule at the site of the skin injury and may be self-limited, develop into lymphangitis and regional adenopathy, or progress to acute septicemia. *Pulmonary infection* is the most common clinical presentation, ranging from a mild pneumonitis to a fulminant necrotizing pneumonia. *Acute septicemia* may derive from any manifestation of the disease and is the most mortal of the variations. Metastatic abscesses may be seen in the liver, spleen, lungs, lymph nodes, bone, and brain. Hepatosplenomegaly is felt and rales and friction rubs are heard. *Chronic suppurative infection* characterizes the fifth form, in which chronic cavitary lesions in the upper lungs may develop within months or years after the primary infection. Chronic abscesses can also develop and a protracted intermittently febrile wasting illness may result.

Melioidosis should be considered in any patient who has ever resided in an endemic area and who has an acute or chronic illness fitting one of the aforementioned patterns. Chloramphenicol, tetracycline, trimethoprim-sulfamethoxazole and kanamycin are the preferred therapeutic agents.

MEMBRANE (Semipermeable).   See Semipermeable Membrane.

MEMBRANE SEPARATIONS TECHNOLOGY.   The separation of materials (solids from liquids; liquids from other liquids; solids of one size from solids of another size; gases from liquids; etc.) is one of the fundamental unit operations needed by the chemical, food, and related fluid processing industries. Such operations include adsorption, absorption, distillation, evaporation, extraction, filtration, ion exchange, settling, and preparative chromatography, among others. In recent years, membrane technology has made important inroads into several of these more traditional unit operations, and on adsorption, distillation, and filtration in particular. Either membrane technology has replaced traditional separation operations or is used in connection with them—with resulting improvement of the separation (in terms of product quality) and in operating efficiency as well. The membranes used are synthetic polymers, the pores of which are made in a number of interesting ways. Membrane separations are not a single technology, but rather they differ in the methodology

used and in the degree of separation that can be effected. Membrane-using subtechnologies include electrodialysis, electrolysis, microfiltration, ultrafiltration, and reverse osmosis.

The general design of the equipment in which membranes are used is commonly (1) *perpendicular flow*, i.e., where the flow of unprocessed material approaches the "filtering" medium in an effrontal manner, passing through the medium, the processed material exiting the medium on the opposite side, and the material removed remaining on the surface of the medium. This is the common figuration that applies to traditional filters, such as cartridge, bag, diatomaceous-earth precoated, backwashable sand, and backwashable mixed-media filters. Some of these configurations allow regeneration by way of backflushing; some do not—the used media must be discarded. See also **Filtration**. (2) In *crossflow*, the influent unprocessed stream is separated into two effluent streams, known as the *permeate* and *concentrate*, respectively. The permeate is that fraction which has passed through a semipermeable membrane; the concentrate is that stream which has been enriched with the solutes or suspended solids, i.e., those materials which have not passed through the membrane. This design permits the membrane medium to operate continuously in a self-cleaning mode, with solutes and solids swept away by the concentrate stream which is running parallel to the membrane (hence the term "crossflow"). As with conventional filtration, sometimes the trapped material (filter cake) is the principal desired end product; in other cases, the desired product is clarified effluent. However, in contrast with conventional filtering, membrane, methodologies not only separate solids (or gases) from liquids (or gases), but proper selection of the membrane will allow separation of solids (particles) by size range. Thus, the permeate (what passes through the membrane) will contain much smaller particle sizes (in the molecular range) than will the concentrate. Again, depending upon the objective of the process, the permeate or the concentrate will be the principal product of interest. In some instances, both products may be of vital interest. See Fig. 1.

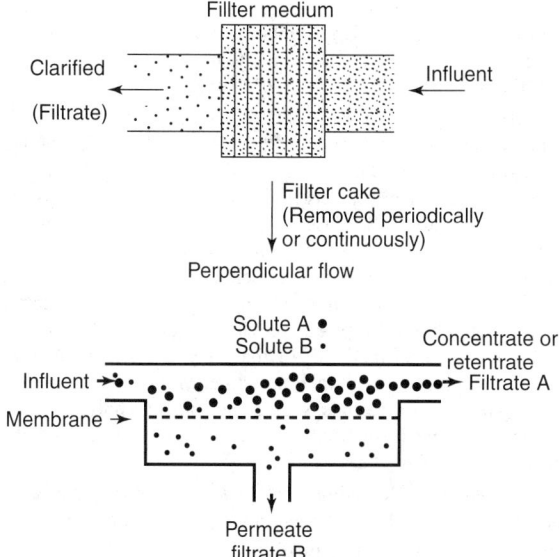

Fig. 1.   Perpendicular flow contrasted with crossflow. Perpendicular flow is shown here in the familiar terms of conventional filtering, but the principle is the same if a membrane is used. There is one influent and a single effluent. In crossflow, there are two effluents, a concentrate (or retentate) and a permeate, usually both liquids of different concentrations in terms of particle size. However, membranes are also used for separating acid gases and hydrocarbons.

Membranes may be *isotropic*, i.e., their pore structure and material are the same throughout the membrane; or they may be *anisotropic*, i.e., they have a dense skin layer on top which defines the degree of separation effected, with a spongy support layer underneath. The dimensions of the pores range widely as detailed later in this article. Membranes are made by a number of processes, including solvent casting and mechanical stretching to form pores in an otherwise impervious film. Irradiation, followed by acid etching, has been used to create pores. The cellulose acetate membrane (Fig. 2) is made by casting thin sheets of polymer dissolved in a water-miscible solvent on a flat plate, usually glass. Shortly after casting, the cast

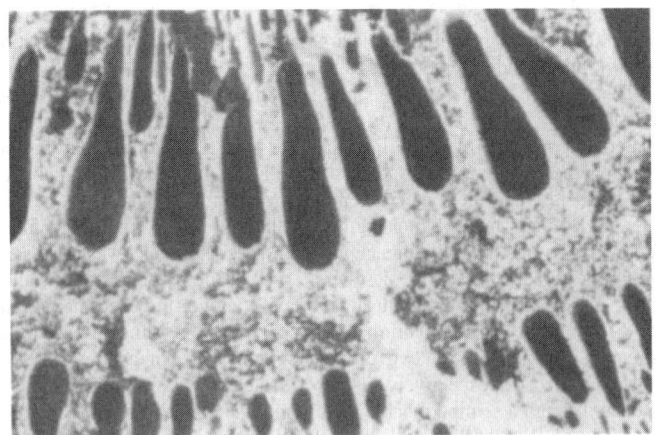

Fig. 2.   Facsimile of photomicrograph of cellulose acetate membrane, prepared by solvent casting process, clearly showing pore structure.

solution is immersed in water. The water diffuses into the solution and causes the polymer to coagulate at a rate that is a complex function of the polymer and solvent properties. These membranes are porous throughout, possessing a thin, relatively dense skin near one surface.

The range of small particles, molecules, and ions dealt with by membrane separation technology, particularly as encountered in the biochemical field are as follows.

Microfiltration. This process effects separations in the 0.02–2.0 micron range and historically has been run in the perpendicular flow mode, requiring disposal of the membrane medium as a result of binding by the retained material. Crossflow technology is increasing, as it proves practical. Microfiltration membranes are of an isotropic and homogeneous morphology, i.e., the pore structure is consistent throughout. There is some movement, however, toward the use of "skinned" anisotropic membranes. Microfiltration membranes are available in a wide variety of polymers, including some that are quite chemically inert. They also are available as tubular, hollow fiber, or capillary fiber elements.

Ultrafiltration. This process effects separations in the 0.002 to 0.2 micron range—more specifically described as the 500–300,000 molecular-weight cutoff range, requiring pore sizes of from 15 to 1000 angstroms. For practical reasons, ultrafiltration almost always requires a crossflow configuration. Because of the size and the gelatinous nature of many of the solutions and particles handled by this process and that the membrane retains, an ultrafiltration membrane would have a very short life if used in the perpendicular flow mode. Nearly all membranes used in ultrafiltration are anisotropic, as previously defined. Membranes usually are of a homogeneous material, in that they consist of the same polymer or copolymer throughout their structure. Membranes must be made of tough, relatively inert materials, such as polysulfone polymer, or cellulose acetate polymer, the latter used more extensively in the earlier days of ultrafiltration.

Reverse Osmosis. Sometimes called *hyperfiltration*, this is the most technically complex class of membrane technology. Reverse osmosis effects separations both at the micromolecular and ionic size ranges. Pore sizes range from 5 to 15 angstroms. These membranes can effect separation of solutions down to a molecular weight of 150 and sometimes lower. Reverse osmosis membranes are anisotropic. Cellulose acetate has long been the favorite, especially for many industrial and medical applications. However, homogeneous polyamide-type membranes are finding a large share of the market and are increasingly favored for seawater desalting purposes.

Applications of Membrane Separations Technology. The food, biochemical, and petrochemical industries are the largest users. In the food processing industry, there are three main categories of use — processing, waste treatment, and pure water make-up. Processing applications include milk concentration and fractionation, numerous fermentation operations, and the production of colorants, among many others. Waste handling operations include corn processing byproducts, soybean protein reclamation, and handling meat processing oils and fish processing proteins and oils.

Membrane separations technology is also widely used in the pharmaceutical industry in connection with recovering antibiotic products. The list of references given at the end of this article is rather long. There the reader will find a wealth of information on applications, including the growing recognition by the petroleum and petrochemical industries of the viability of membrane technology as a replacement or companion for the more traditional separation processes.

Membranes Configurations for Use in Separation Equipment. For insertion into separation equipment, frequently with large throughputs, membranes are available in several configurations: (1) the tubular form ($\frac{1}{2}$ in; 12.5 mm) is common; (2) hollow-fiber; (3) plate-and-frame; and (4) spiral wound. The Paulson reference covers membrane formats in considerable detail.

See also **Desalination**.

Additional Reading

Abelson, P.H.: "Synthetic Membranes," *Science*, 1421 (June 23, 1989).
Beaudry, E.G. and K.A. Lampi: "Membrane Technology for Direct-Osmosis Concentration of Fruit Juices," *Food Technology*, 121 (June 1990).
Bedzyk, M.J. et al.: "Diffuse-Double Layer at a Membrane-Aqueous Interface Measured with X-ray Standing Waves," *Science*, 52 (April 6, 1990).
Carroll, L.E.: "New Process Concentrates Juices, Preserving'Fresh Notes," ' *Food Technology*, 148 (October 1989).
Dziezak, J.D.: "Membrane Separation Technology Offers Processors Unlimited Potential," *Food Technology*, 107 (September 1990).
Friedman, R.: "Seawater to Drink," *Technology Review (MIT)*, 14 (August/September 1989).
Hsieh, H.P.: "Inorganic Membranes for Separation and Reaction," Elsevier Science, New York, NY, 1996.
Kosenoglu, S.S., J.T. Lawhon, and E.W. Lusas: "Use of Membranes in Citrus Juice Processing," *Food Technology*, 90 (December 1990).
Kosenoglu, S.S., J.T. Lawhon, and E.W. Lucas: "Vegetable Juices Produced with Membrane Technology," *Food Technology*, 124 (January 1991).
Matsurra, T.: "Synthetic Membranes and Membrane Separation Processes," CRC Press, LLC., Boca Raton, FL, 1994.
Noble, R.D., C.A. Koval, and J.J. Pellegrino: "Facilitated Transport Membrane Systems," *Chem. Eng. Progress*, 58 (March 1989).
Noble, R.D. and S.A. Stern: "Membrane Separations Technology: Principles and Applications," Elsevier Science, New York, NY, 1995.
Paulson, D.J., R.L. Wilson, and D.D. Spatz: "Crossflow Membrane Technology and Its Applications," *Food Technology*, 77–87 (December 1984).
Rousseau, R.W.: "Handbook of Separation Process Technology," John Wiley & Sons, New York, NY, 1987.
Singh, R.: "Surface Properties in Membrane Filtration," *Chem. Eng. Progress*, 59 (June 1989).
Spillman, R.W.: "Economics of Gas Separation Membranes," *Chem. Eng. Progress*, 41 (January 1989).

MEMBRANE TECHNOLOGY.   See **Desalination**; and **Ocean Resources (Energy)**.

MEMORY, ELECTRONIC — CHRONOLOGY TO 1990.   The need to remember information — to store and later to retrieve data — is one of the principal chores of electronic information processing equipment (computers, etc.). The time interval during which information must be retained (stored) ranges from extremely short time spans to very long periods of time. The time interval permitted to store and retrieve information also varies. Costs of memory systems also vary. Consequently, over a number of years, several information storage formats have emerged. In addition to memory system capacity and speed, protection of memory from physical destruction has been of major concern. Also, in numerous cases, a memory must be safeguarded against unauthorized access.

Considering a digital data processing system (Fig. 1), the basic functional elements consist of the central processing unit (CPU), main memory, input/output interface (I/O), and the associated peripherals for the input and output of information. Some, but not always all of these functional elements may be used in a given system. In the case of some industrial control applications, the I/O simply may be a display panel with switches and light-emitting diode (LED) display. *All* of the elements in the diagram use memory, including the peripherals. The characteristics and type of memory used varies greatly for each element.

The relative features and physical residences of those memory types used in each of the identified functional elements are shown in Fig. 2, commonly referred to as the *memory hierarchy*. Note that the memory characteristics used in each system element differ considerably.

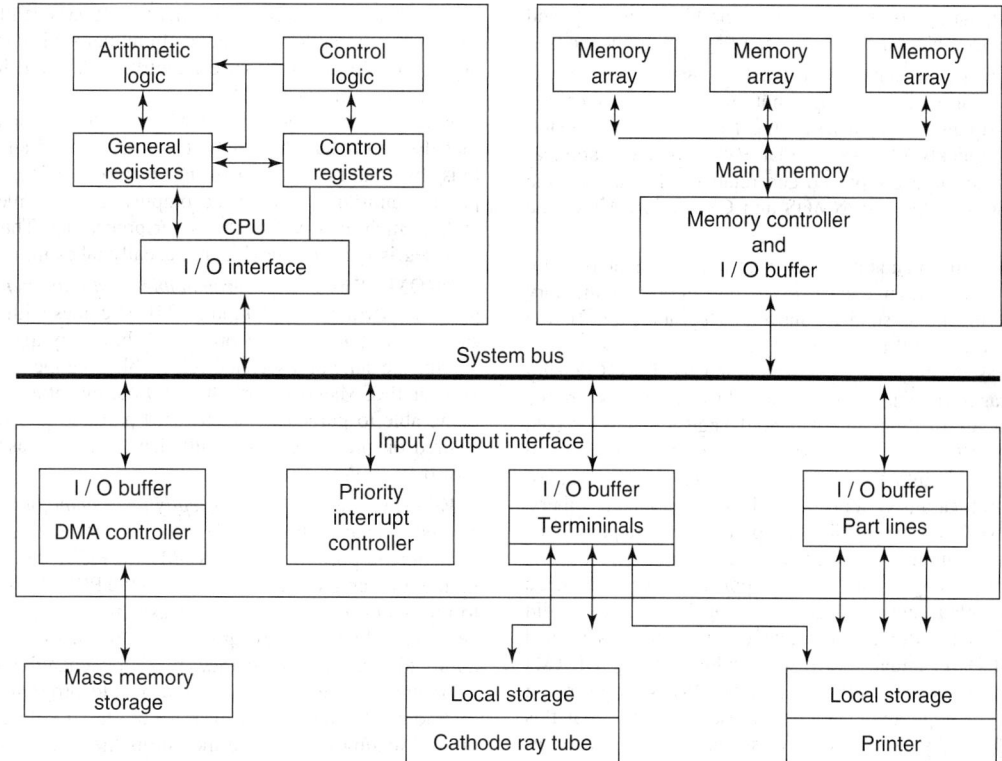

Fig. 1.   All elements shown in this diagram require electronic memory.

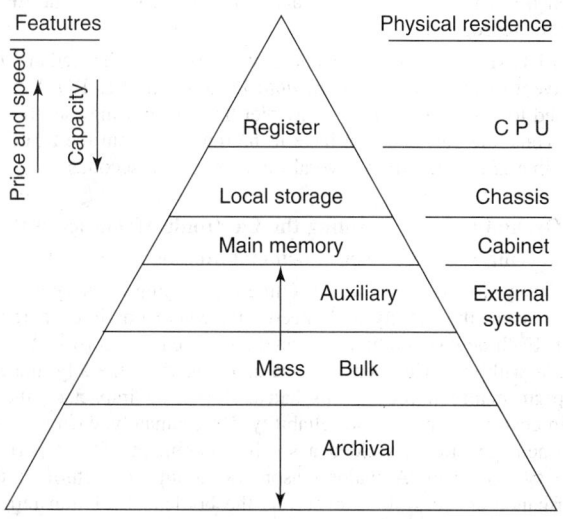

Fig. 2.   Memory system hierarchy.

Register Memory. For example, the register memory is used for the *temporary storage* of data being modified by the CPU. Thus, to insure fast execution of operations, the register storage should be as fast as the logic used to implement the CPU. The size of storage needed is usually small because only a few bits of data can be utilized on each execution cycle of the CPU.

Local Storage. This is similar in usage to that of the registers, here the emphasis is on the *temporary storage* of small amounts of data being moved between elements, such as the peripherals noted in Fig. 1. Generally, the performance needed for these applications does not require the high speed of register storage. However, the memory size will be two to three orders of magnitude larger than that of the register storage memory.

Main Memory. As the name implies, this is the main working data storage device of the system. This random access read/write memory acts as a buffer between the CPU and the bulk or mass storage and is used to hold current resident programs, subroutines, and data. Hence, this memory is usually quite large (from thousands to millions of words). Since the

current instructions and data of the programs being executed are stored in this memory, high speed is also essential. Thus, in most systems, the size and performance of the main memory will determine the overall performance of the system.

Mass Memory. This memory is used to store data and programs when they are not being used. In general, this data storage medium is the least expensive; it is also of very high density and slow. These low-cost memories employ *sequential address accessing*, which results in variable access times. The access time is variable because the time needed to get from one address to the next depends on the location of the last address versus the location of the next address. With the exception of Mass Memory, other memory technologies use the faster *random access* method, which permits all addresses to be selected in the same time regardless of location. As noted in Fig. 2, there are three categories of mass memory: (1) auxiliary, (2) bulk, and (3) archival. Both bulk and archival memories are used for permanent data storage, whereas an auxiliary memory is referred to as *semipermanent*.

Historical Perspective

Prior to the mid-1970s, magnetic storage media predominated. Functions and access times of electronic memories of that period are given in Fig. 3. The most dramatic changes in memory technology occurred in connection with mainframe applications. The dominant main memory technology had switched to semiconductors by 1980. The rapid changeover to semiconductor memory started with the introduction of the 1024 bit *dynamic* random access memory (DRAM) in 1970. Dynamic storage is accomplished by storing charge on a capacitor and, therefore it is necessary to periodically recharge to maintain data. Reasons for quick acceptance of the DRAM for main memory applications were the increase in density and performance with lower cost. Since 1970, the density of these memories has increased manyfold.

Bipolar Technology. Bipolar static RAM (random access memory) technology historically has been used for register storage. Static RAM storage is accomplished through cross-coupled gates. Thus data as such does not require recharging to retain the data. Although the bipolar technology has gone through dramatic changes in density, speed, and power, the MOS (metal-oxide semiconductor) technology is strongly competitive. Through clever design techniques and process changes, such as high-performance *N*-channel metal oxide semiconductors (NMOS) and complementary MOS (CMOS), MOS static RAMs (SRAMs) approached the performance of bipolar units. Hence, with their higher density and lower power, they replaced bipolar RAMs in many system applications.

Local Storage. In a similar manner, the local storage memory applications went the route of register storage, but much faster. Since the performance of local storage is much less critical than that for register, static MOS memories quickly replaced bipolar RAM for these applications. Due to the high density, low power, and relatively fast access, the local storage applications widely use NMOS and CMOS SRAMs (static RAMs).

Mass Storage. This still remains the domain of magnetic material, but there have been numerous dynamic changes in this important memory segment. Considering the three subdivisions noted earlier (Fig. 2), the progress of auxiliary storage will be examined first.

Auxiliary storage has the highest performance and the lowest density of the three mass storage segments. The primary objective of this storage is to have a large storage medium that can hold large amounts of data that are needed sufficiently often that it becomes necessary to have the data on line. Data that fall into this category are large lookup Tables, data files, and backup operating systems with diagnostics, among others. Since auxiliary memory is always on line, its performance requirements differ from those of bulk and archival memory (Fig. 3). In the 1960s and early 1970s, memory technologies used for this application were magnetic bubbles and charge coupled devices (CCDs). Inasmuch as the CCD did not retain the density and lower cost advantages over DRAMs as had been expected, the CCDs eventually gave way to high-density DRAMs for this application. Thus, by the 1980s, the popular DRAMs were being used in many auxiliary storage applications. The use of DRAMs for this application developed rapidly. The system is sometimes referred to as semiconductor DISK.

Bubble memory can be used for auxiliary storage, but it does not have performance comparable to DRAM. Magnetic bubble memory does have one valuable advantage over DRAM in mass storage applications, namely, *nonvolatility*. When power is lost, bubble memory retains data, whereas DRAMs lose data and must be reloaded on powerup. Although bubble is magnetic, it is not mechanical in nature like magnetic tape or disk. These properties (nonvolatile and nonmechanical) make the bubble memory ideal for mass storage in harsh environments, such as encountered in military, industrial control, and portable systems.

For several decades, the primary storage device for bulk and archival storage have been magnetic—tape or disk. Fixed- and movable-head disks are used for bulk applications, whereas both movable-head disks and tape are used for offline archival storage.

Read-Only Memories (ROMs)

The memories discussed thus far have been read/write memories. There are other memory types. Their performance, density, and cost characteristics follow the memory hierarchy (Fig. 2). Read-only (ROM) memories utilize both the MOS and bipolar process technologies. These memories are used to hold code data that will *not change* with time. They also can be used for virtually all of the functional building blocks (Fig. 1). The ROM is used in a variety of applications, including the microcodes in a CPU (instruct CPU what to do), code converters, and frequently used Tables and constants, among many other functions. ROMs can be used as a substitute for combinational logic elements. In numerous instances, because of its bit density, the ROM can replace a controller design using many logic devices with just one or two ROMs.

A major advantage of a ROM over other semiconductor memories is that the ROM has the smallest storage cell of any memory device and thus the smallest die area of any memory of identical bit capacity. The programming of a ROM is accomplished with a metal mask interconnect and, as such, is a part of the wafer processing. Thus, its major drawback is that it is a custom device that usually takes many weeks to produce.

PROM. This is a *programmable read-only memory* that can be programmed by the user. A bipolar PROM comes with all storage cells in the same logic state. All the storage cells have tiny metal links so the user can change the state of a selected cell by blowing the link with a current pulse. Bipolar PROMs cost more than ROMs, but often the added flexibility of being able to purchase one standard part and generating the code pattern desired on the same day is sufficient in many cases to offset the higher cost over ROM.

EPROM. This is an *electrically programmable read-only memory* that first became available in 1971. The impact on the electronics industry was of major proportion. The EPROM is a ROM that can be reprogrammed, giving it much greater flexibility than the PROM just described. The ability to reprogram means that the mistakes made in program development can be corrected without having to discard parts. Another key advantage is the ability to perform field upgrades by simply changing the code. The EPROM is programmed electrically as one would program a PROM except that instead of blowing a metal fuse, an electrical voltage pulse traps electrons onto a floating gate, which then turns the selected cell transistor on. The unprogrammed cells leave the transistor floating. Before the part can be reprogrammed, the die must be exposed to ultraviolet light (UV) for a set period of time to allow the trapped charge to leak off. This returns all cells to their original unprogrammed state. After this procedure, the EPROM is ready for reprogramming.

EEPROM. The EPROM technology was further enhanced in 1980 with the advent of the *electrically erasable* PROM. The EEPROM overcomes the need to expose the die to ultraviolet light for erasing the programmed cells. Thus, the part does not have to be removed from the board and the erase time is reduced from several minutes to milliseconds.

The Groundwork for Creating the Electronic Memories of the 1990s
Transferring Data Between Functional Modules

For many years, the trend was to refine the existing memory technologies by increasing the density and speed and, where possible, to reduce the power. Each new generation was essentially a carbon copy of the existing one, but with enhanced performance and density. Generally, the memory cost performance improvements had a significant impact on the overall system cost performance and reliability. Unfortunately, each new memory component generation yielded a smaller increment of cost performance at the system level. A major reason for diminished return of the cost performance at the system level was the bottleneck of transferring data between the functional modules of the computer (Fig. 1). As indicated in

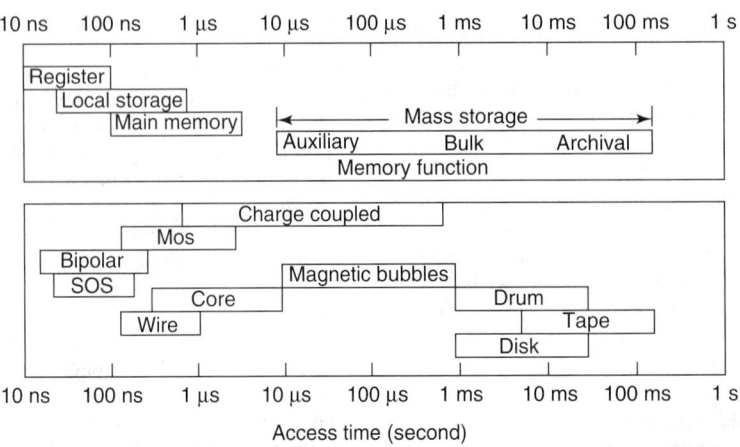

Fig. 3.   Historical perspective (mid-1970s) of functions and access times of electronic memories.

that figure, the modules communicate with each other over a single data bus. With a single bus system, only two modules can use the bus at any one time. Due to long lines and heavy loading, the bus data bandwidth of most systems is limited to the range of 2 to 5 MHz. Widening the data bus is one obvious way to improve overall system performance.

An extension of adding intelligence to the modules is to build a multi-processing system with multiple buses that can have both distributed and common shared memory so that many instructions can be processed simultaneously. A key problem (limitation) of these system schemes is the development of software that will take advantage of all the processing power. Among other system concepts used to enhance performance are pipeline architecture, cache, and interleaving.

Enhancing Main Memory Performance. As previously mentioned, the main memory performance has the largest impact on the overall performance of a computer system. Thus, it is logical that one would expect system designers to look at ways to modify this memory technology. An evolutionary process essentially commencing in the mid-1970s.

The initial DRAMs had only one addressing mode, but the number increased markedly. The first special addressing mode to appear was *page mode* on the 4 K DRAM. This dramatically reduced the access and cycle times of the DRAM (Fig. 4). This simple feature reduced the time required to transfer large blocks of contiguous data between disk storage and main memory. Other peripherals, such as graphic terminals, benefited from this feature by providing a greater bandwidth needed for screen refresh.

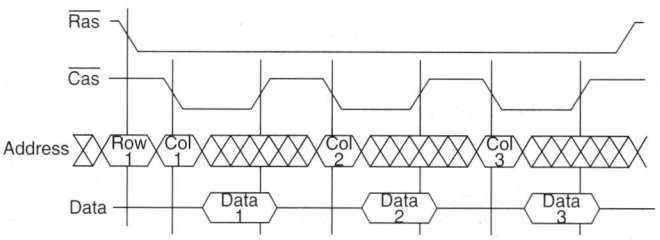

Fig. 4.   Early page addressing mode. (Motorola Inc.)

Since the appearance of page mode, a variety of other popular addressing modes and cycles (bearing some resemblance to page mode) were added. These include *ripple mode, static column, nibble, CAS* (column-address strobe) before *RAS* (row-address strobe), and *hidden refresh*, among others. Although the foregoing features improve system performance, the limit to what can be accomplished is determined by the bus, which is a major bottleneck to system improvement.

Bus Limitations and Improvements. An obvious way to overcome the bus problem is to have more than one bus per system. This approach has

been slow in acceptance, however, because of cost for all but the very high end high performance computers. Until relatively recently, the bit density of semiconductor memory was not sufficient to allow for special features that would require a lot of silicon to implement (a cost restraint). With low-bit-density memory, even small systems will require many memory components for main memory, i.e., into the hundreds for many systems. With this situation, the support electronics required to interface to the bus and control the operation of the memories can be amortized over many memory components and thus make this a feasibly cost-effective approach.

However, with the fourfold increase in density with each new generation of semiconductor memory, more memory becomes available per system with fewer components required. For example, with 64 K, one needs 32 memory components for a fourth of a megabyte, while at the 256 K bit density, only eight memory components are needed. It follows that with 1 M bit density, only two memory components are required, etc. With the high bit density and low package count, alternate solutions at the silicon level become economically feasible, which only a few years ago would have been impossible.

The logical solution to an economical multibus system lies in the ability to make a cost-effective multiport semiconductor memory. The ability to implement such a memory is no longer visionary, but is a reality as dual-port memories with higher and higher densities become available. This is exemplified by Fig. 5. In this case, the new memory components are actually two memories in one. One is a 256 K DRAM organized as 64 K × 4; the other is a 256 × 4 random access serial port. The DRAM port has performance characteristics similar to standard DRAMs. The serial port is very high performance, with cycle times in the 25 ns range. Although these two memories can operate totally independent of each other, with a special cycle, the DRAM port can be connected to the serial port and 1024 bits of data can be transferred between the two ports in one DRAM cycle time.

A representative system block diagram using this type of memory is shown in Fig. 6. To take full advantage of the high-speed data bus, it is advantageous to also employ a cache memory. To accomplish this requires a dual-port static RAM. Since both ports of this memory operate independently, the real advantage to the system architecture is that it is possible to transfer data between the disk and the main memory serial port or the serial port and cache at the same time the CPU is executing a program from data out of the DRAM port. The performance cost tradeoffs of this system are far superior to that of the conventional single-bus system. It is interesting to note that the dual port DRAM was not designed specifically for this application, but rather for the high-speed data storage required for high-resolution graphics.

Memory with On-Chip Control. Also with densities of 1 M bit and beyond, it is possible to design memories with on-chip support logic so that the memory requires little if any support logic to function in the system. Examples of this are *intelligent* DRAMs, which include the refresh function on chip. As the chip density increases, it is more economically

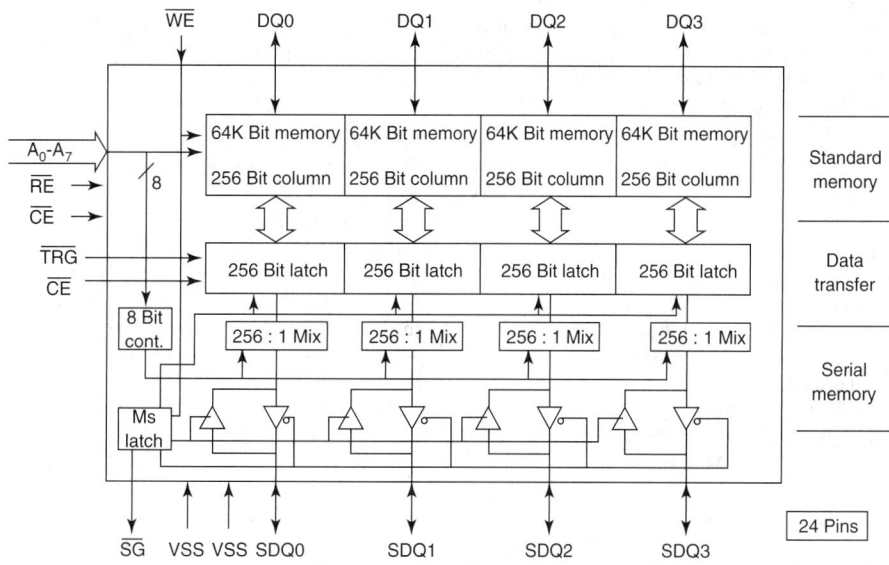

Fig. 5.   Early dual-port system block diagram. (Motorola Inc.)

feasible to add the control logic that does the clock timing for the memory chip. Thus, the memory will require no support electronics whatever and will interface directly to the CPU bus. With this level of integration, the word width of the memory component is not ×1 or ×4, but rather $\times \frac{8}{9}$ or $\times \frac{16}{18}$. Thus, a complete memory system with a half to one megabyte can be accomplished with a half-dozen chips or less.

Similar changes at the other end of the semiconductor spectrum emphasize high performance. In both the bipolar and high-speed HCMOS and GaAs SRAMs, dramatic changes are occurring. In addition to such architectural changes as dual port, the support electronics, which includes the address latching, internal timing generation, such as the write pulse, can accommodate memories with access and cycle times below 10 ns. Other features include comparator logic so that the SRAM can function as a content-addressable RAM.

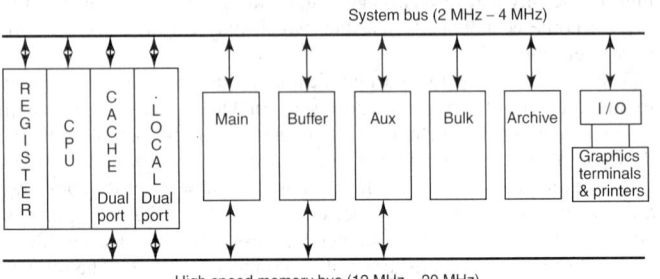

Fig. 6. Early multiport dynamic RAM (64 K × 4). (Motorola Inc.)

Chipmaking technology is also changing. In order to meet the higher density, with ever increasing speed and manageable power, designers are turning to GaAs (gallium arsenide) and BiCMOS. With GaAs, it is possible to achieve sub-nanosecond speeds. This places even greater emphasis on having on-chip latching and control features in order to take advantage of these speeds at the system level.

Gate-Array with SRAM. Another relatively recent development is the use of small on-chip SRAM storage with large gate arrays and a form of wafer scale integration that will make very large storage arrays possible in one package. Some of the latest very large gate arrays (1 K to 20 K range) have an on-chip SRAM array. Since these gate arrays are used in the design of very high-performance custom CPUs and controllers that require the need for temporary storage of processed data, having the SRAM storage on-chip eliminates the costly time required to store the data off-chip as is traditional.

Cluster of Memory Chips. Some semiconductor makers are investigating methods of using a cluster of memory chips that test good at wafer probe and then interconnect them at assembly into one package. As an example, take the entire wafer and segment it into eight adjacent 64 K DRAM chips and test each before scribing. Next, interconnect the good chips into a variety of organizations. Examples of possible memory organizations are: For two good chips, you would have only 128 K × 1; 3 good chips would yield 192 × 1; four good chips could be 64 K × 4, 256 K × 1, or 128 K × 2, and so on, up to all 8 chips testing good. The scheme is probably marginal at the 64 K bit density, but at 256 K and above, it would be much more attractive. As the process technologies converge, this scheme may be accomplished more readily with an EEPROM interface controller that interconnects the good die through programming.

Editor's Note: This summary of the chronology of electronic memory development during the 1970s–1980s illuminates the remarkable engineering capabilities that led to the contemporary memories of the early 1990s. The present electronic memories are described in connection with numerous entries in this encyclopedia relating to electronic equipment and computer technology. Consult alphabetical index.

R. BRUNNER, Semiconductor Products Sector, Motorola Inc., Phoenix, AZ.

MEMORY LOSS. See **Amnesia**.

MENDEL, GREGOR JOHANN (1822–1884). Mendel, an Austrian monk and naturalist and botanist, is often referred to as the father of genetics. By studying about 28,000 pea plants in his garden peas he developed the basic principles of inheritance. Although, the importance of Mendel's work was not recognized in his lifetime, it is from his research that much of modern day genetic terminology comes including recessive traits.

See also **Genetics and Gene Science (Classical)**.

J. M. I.

MENDELEVIUM. Chemical element, symbol Md, at. no. 101, at. wt. 256 (mass number of known isotope), radioactive metal of the Actinide series, also one of the Transuranium elements. The element was produced synthetically and first identified by A. Ghiorso, B.G. Harvey, G.R. Choppin, S.G. Thompson, and G.T. Seaborg at the University of California at Berkeley in 1955. ^{256}Md was produced by the bombardment of ^{253}Es on gold foil with 48-MeV alpha particles in the 60-inch cyclotron at Berkeley. By ion exchange treatment of the dissolved gold foil, only one or two atoms of ^{256}Md were obtained, which decayed (half-life 1.3 hours) by K-electron capture to ^{256}Fm, which underwent its characteristic spontaneous fission.

Probable electronic configuration $1s^2 2s^2 2p^6 3s^2 3p^6 3d^{10} 4s^2 4p^6 4d^{10} 4f^{14} 5s^2 5p^6 5d^{10} 5f^{13} 6s^2 6p^6 7s^2$. Ionic radius $Md^{3+} 0.96$ Å.

Another isotope, ^{255}Md, is also formed during the bombardment of ^{253}Es by alpha particles. It also decays by electron capture, and has a half-life of 30 minutes.

Regarding the first identification, scientists considered it notable in that only in the order of 1 to 3 atoms per experiment were produced, thus making Md the first element to be discovered on an atom-at-a-time basis. The techniques developed in the search for Md served as a prototype for the discovery of subsequent elements in the Transuranium series.

See also **Chemical Elements**.

Additional Reading

Fuger, J. and L.R. Morss: "Transuranium Elements: A Half Century," American Chemical Society, Washington, DC, 1992.

Ghiorso, A., B.G. Harvey, G.R. Choppin, S.G. Thompson, and G.T. Seaborg: "New Element Mendelevium, Atomic Number 101," *Phys. Rev.*, **98**, 5, 1518–1519 (1955). (A classic reference.)

Hulet, E.K. et al.: "Mendelevium: Divalency and Other Chemical Properties," *Science*, **158**, 486–488 (1967).

Lide, D.R.: "CRC Handbook of Chemistry and Physics 2000–2001," 81st Edition, CRC Press, LLC., Boca Raton, FL, 2000.

Seaborg, G.T.: "Transuranium Elements," Dowden, Hutchinson & Ross, Stroudsburg, Pennsylvania, 1978. (A classic reference.)

Seaborg, G.T. and W.D. Loveland: "The Elements beyond Uranium," John Wiley & Sons, Inc., New York, NY, 1990.

MENDELEYEV, DIMITRI (1834–1907). Mendeleyev was a Russian chemist who recognized the regular variation in the chemical and physical properties of the elements and classified the 63 elements known at the time into groups. He devised the periodic table of the elements by placing the elements in order of increasing atomic weight.

See also **Periodic Table of the Elements**.

J. M. I.

MENHADEN. Of the genus *Brevoortia*, the menhaden is one of the most important commercial fishes in North America. Of the four species of menhaden occurring along the coasts of the United States, two contribute nearly all of the commercial catch. The Atlantic coast catch consists principally of the *Atlantic menhaden* (*B. tyrannus*), which ranges from Nova Scotia to northern Florida. The *Gulf menhaden* (*B. patronus*), which ranges from the west coast of Florida to Mexico, contributes most of the catch from that area. Two other species, the *yellowfin menhaden* (*B. smithi*), which occurs mainly along the east and west coasts of Florida; and the *finescale menhaden* (*B. gunteri*), which occurs from the western Gulf of Mexico to Florida, are of relatively minor importance. Other species are known.

Menhadens are small, oily, herring-like fishes closely related to shad, alewife, herring, and sardine. The largest authenticated specimen was about 19 inches (48 centimeters) long, but most of those caught are less than 12 inches (30.5 centimeters) in length and weigh somewhat less than 1 pound (0.45 kilogram). The average size of Gulf menhaden at any particular age is considerably less than that of the Atlantic species.

Menhaden are prolific spawners. The number of eggs produced by a single female varies, depending on the age and size of the fish. Individual estimates of the number of eggs spawned by Atlantic menhaden range from 38,000 for a medium-size fish to hundreds of thousands for a larger fish. Adult menhaden feed on microscopic plants and animals rather than on other fish. As the fish swims through the water, the planktonic particles are effectively strained from the water by sieve-like gills.

The abundance of menhaden fluctuates greatly. Although these fluctuations in abundance in certain local areas most likely are controlled to a large degree by such environmental factors as water temperature and food availability, variations in abundance of the entire resource appear to be largely caused by changing survival of individual yearly spawnings (year classes). For example, records show that the highly successful 1958 spawning of Atlantic menhaden yielded about seven times as many fish to the fishery as did the 1964 spawning. Since the late 1950s, the catch of menhaden has been decreasing.

In the United States, over the years, the menhaden catch has accounted for a large percentage of the marine oils and fish solubles produced. The fish is caught for its reduction products. Processing plants are usually located quite close to the fishing areas.

See also **Fishes**.

MENINGITIS. The meninges are the covering membranes of the brain and spinal chord. Inflammation of these membranes, especially of the *piamater* and the *arachnoid mater*, is called *meningitis*. See also **Central and Peripheral Nervous Systems**. Meningitis takes a number of forms, including: (1) *aseptic meningitis*, (2) *bacterial meningitis*, (3) *fungal* or *cryptococcal meningitis*, and (4) a number of less common forms, such as *Mollaret's meningitis*.

Aseptic Meningitis. The virus or viruses that cause this disease are positively identified in only about one-quarter of cases. However, increasing emphasis is being placed on more thorough diagnosis and clinical testing for the treatment of specific viral infections.

It is generally believed that enteroviruses cause the majority of cases, but numerous other viruses are seen in this disease. Without identifying the specific virus, aseptic meningitis is relatively simple to diagnose particularly during a spring or summer enterovirus epidemic and where the patient has a rash, a symptom of enterovirus infection. Positive determination of viral etiology is a relatively complex procedure and requires, among other procedures, the inoculation of suckling mice with pharyngeal washings, a stool specimen, and cerebrospinal fluid from the patient.

Symptomatic of aseptic meningitis are fever, headache, and rigidity or stiffness of the neck. Further confirmation can be obtained from examination of the cerebrospinal fluid. Upon completion of the initial diagnosis, treatment and support must be commenced because a full identification of the causative virus requires several days. Statistics indicate that in about three-fourths of the cases of enteroviral meningitis, the virus has been spread by the fecal-oral route and that coxsackieviruses A and B and echoviruses are implicated. See also **Coxsackie Virus**; and **Virus**. The next most frequent causative agents are mumps and varicella zoster viruses. Because of the presence of parotitis (inflammation of salivary glands), the diagnosis is simplified. A less frequent agent, herpes simplex type 2 virus, is seen particularly in women, often with an accompanying primary herpes simplex genital infection. An infrequent agent is lymphocytic choriomeningitis virus, an arenavirus. This is found most often in young adults.

Mollaret's meningitis is a relatively uncommon disease characterized by recurring episodes (self-limited) of aseptic meningitis. Episodes persist for 2 to 5 days and remit spontaneously. The patient may be free of symptoms for weeks or even months. As reported by L.J. Yamamoto (University of Colorado) and a research team, measures have been taken to establish a microbiologic cause, but thus far this has been elusive. However, in at least one case, herpes simplex virus (HSV) has been found in cerebrospinal fluid.

Bacterial Meningitis. Three microorganisms account for over three-fourths of the cases of bacterial meningitis. These are influenza bacillus, meningococcus, and pneumococcus. Meningitis is most common among infants less than one year old. Epidemics of meningococcal meningitis are not uncommon. Areas where many people live in close proximity, as in crowded urban neighborhoods or in military barracks, are particularly productive of this disease. The meningococci reside in the nose and throat

(present in about 5% of population) and spread of the agent via the respiratory route is believed to be the principal pathway of transmission from human to human. The highest incidence of the disease occurs during the winter and spring. The majority of cases among adolescents and young adults is caused by meningococcus, whereas in adults, pneumococcus accounts for most cases. Influenza bacillus rarely causes meningitis in adults except in cases of head injury.

In the United States, approximately 2000 cases were reported (mandatorily) in 1992 to the Centers for Disease Control in Atlanta, Georgia.

Symptoms of bacterial meningitis include headache, fever, and stiff neck. Infection of the respiratory tract and/or middle ear may also be present. With progress of the disease, more dramatic features, including stupor, coma, and seizures, may be presented. Symptoms vary in degree among patients. Although clinical features may be quite indicative of the disease to the physician, positive diagnosis requires examination of the cerebrospinal fluid. At one time, sulfonamides were extensively used in the treatment of meningococcal meningitis, but the organisms have become resistant to these drugs and hence penicillin is the usual drug of choice.

Much less frequent, but of growing concern is *listerial meningitis*, caused by a thin gram-positive bacillus found extensively in soil and in the intestinal tract of many birds and animals. Infection of humans with the bacillus is called *listerosis* and the most common form (60–70% of cases) of listerosis is listerial meningitis. The symptoms of this disease are sometimes less pronounced than with other types of bacterial meningitis, frequently with only fever and headache presented. The usual neck stiffness is absent. Onset of the disease is frequently subacute, extending over a period of several weeks. Antibiotic therapy (ampicillin or penicillin G) is the usual course of treatment followed. Alternative drugs are tetracycline or erythromycin.

Meningitis frequently develops as a consequence of gonococcemia. See also **Gonorrhea and Gonococcemia**.

Cryptococcal meningitis is a common *opportunistic infection* found in patients in the United States who have the human immunodeficiency virus (HIV). This occurs in 5–10% percent of patients with the acquired immunodeficiency syndrome (AIDS). A few years ago, it was believed that a full cure of cryptococcal disease was highly improbable and that most AIDS patients who had completed primary therapy for cryptococcal meningitis remained at risk for relapse. In 1992, W.G. Powderly (Washington University School of Medicine) and a team from other hospitals and medical departments reported on a controlled trial of fluconazole or amphotericin B in preventing relapse in such cases. Conclusions of the study: "Fluconazole taken by mouth is superior to weekly intravenous therapy with amphotericin B to prevent relapse in patients with AIDS-associated cryptococcus meningitis after primary treatment with amphotericin B."

Additional Reading

Berkow, R. and M.H. Beers: "The Merck Manual," 17th Edition, Merck & Company, Inc., Whitehouse Station, NJ, 1999.

Cartwright, K.: "Meningococcal Disease," John Wiley & Sons, Inc., New York, NY, 1995.

Davies, P.A. and P.T. Rudd: "Neonatal Meningitis," Cambridge University Press, New York, NY, 1995.

Lange, J.M.A. and P. Reiss: "Suppressive Therapy for Cryptococcal Meningitis," *N. Eng. J. Med.*, 565 (August 20, 1992).

Powderly, W.G. et al.: "A Controlled Trial of Fluconazole or Amphotericin B to Prevent Relapse of Cryptococcal Meningitis in Patients with the Acquired Immunodeficiency Syndrome," *N. Eng. J. Med.*, 793 (March 19, 1992).

Roos, K.L.: "Meningitis: 100 Maxims in Neurology," Vol. 4, Oxford University Press, Inc., New York, NY, 1996.

SchFonfeld, H. and H. Helwig: "Bacterial Meningitis," S. Karger Publishers, Inc., Farmington, CT, 1992.

Staff: "Morbidity and Mortality Weekly Report," Massachusetts Medical Society, Waltham, Massachusetts (issued weekly).

Taylor, H.G. et al.: "The Sequelae of Haemophilus Influenzae Meningitis in School-Age Children," *N. Eng. J. Med.*, 1657 (December 13, 1990).

Tunkel, A.R.: "Bacterial Meningitis," Lippincott Williams & Wilkins, Philadelphia, PA, 2001.

Yamamoto, L.J. et al.: "Herpes Simplex Virus Type 1 DNA in Cerebrospinal Fluid of a Patient with Mollaret's Meningitis," *N. Eng. J. Med.*, 1082 (October 10, 1991).

Web Reference

Center for Disease Control and Prevention: *http://www.cdc.gov/health/diseases.htm*

MENISCUS. The curved surface of a liquid, particularly noticeable in vessels of tubes of small diameter and due to the surface tension of the liquid. If the liquid wets the containing vessel, the meniscus is concave; otherwise it is convex. The meniscus of mercury in glass is convex.

MENOPAUSE.   See **Gonads**.

MENTAL DISORDERS.   See **Brain Disorders**.

MERCAPTANS.   Hydrogen sulfide yields two classes of organic compounds: (1) hydrosulfides, and (2) sulfides. The hydrosulfides are termed mercaptans, a name derived from *mercurium captans*, because of their ability to react with mercuric oxide to form crystalline compounds. Mercaptans also are termed thioalcohols and sulfur-alcohols. A more general term *thiols* also is used. This term not only embraces mercaptans, but also covers thioethers, sulfhydrates, and thiophenols.

Ethyl mercaptan C_2H_5SH, one of the better known mercaptans, is an odorous liquid, mp $-121\,°C$, bp $36-37\,°C$, sp gr 0.839. The compound is very slightly soluble in H_2O; soluble in alcohol and ether. It is prepared by distilling ethyl potassium sulfate with potassium hydrogen sulfide. Additional mercaptans can be prepared in a similar manner with the corresponding proper ingredients. All mercaptans have unpleasant garlic-type odors; when oxidized with HNO_3 they yield sulfonic acids.

Some formulations for styrene-butadiene rubber (GR-S) contain dodecyl mercaptan which plays the role of a chain-transfer agent used to control the molecular weight of the final synthetic product.

MERCATOR PROJECTION.   The mercator method of projecting the surface of the earth on a plane was invented by Gerard Mercator, who lived in Flanders during the latter part of the seventeenth century. At the present time, 90% of the chart work of the deep-sea navigator is done on the mercator chart. To realize the general outline of the method, say that we have a terrestrial globe and wish to peel off the surface. One method of procedure is to make cuts through the surface along meridians of longitude and remove the sectors of the surface thus obtained. If the material of the surface construction has sufficient elasticity, these segments, with a little stretching, can be placed on a plane. The segments will be tangent along the equator, but will separate as the poles are approached, thus forming a discontinuous map. To make the map continuous requires stretching the segments along parallels of latitude, the stretching increasing as higher latitudes are reached. This east-west stretching will introduce distortion in the shape of objects, e.g., a circular object will be distorted into an ellipse. To maintain the shape of objects, the same amount of stretching must be introduced in the north-south direction as is necessary in the east-west direction, in any latitude, to make the segments tangent. This will result, finally, in objects retaining their true shape, but also, in considerable alteration of the relative sizes of surface features in different latitudes.

On the completed mercator chart, meridians of longitude and parallels of latitude appear as perpendicular straight lines. The meridians are equally spaced, but the distance between successive parallels becomes greater as we proceed away from the equator. This distance becomes so great that mercator projection is not used beyond 70° latitude. The actual computation of the distances of the successive parallels from the equator, taking into account the ellipticity of the earth, is a complicated mathematical task.

MERCATOR SAILING.   If the distance run by a ship does not exceed 300 nautical miles, the various problems of dead reckoning and the sailings, may be solved without sensible error by methods that assume the surface of the earth to be a plane. The distances run by modern steamships and aircraft during a single day frequently exceed this limit, and the true shape of the earth's surface must be taken into consideration. The method of solving the various problems connected with course and distance in such cases is usually one that depends for its theory upon the mercator projection, and is known as mercator sailing.

In Fig. 1, we have the representation of the general problem, drawn on a section of a mercator chart. The vessel is proceeding from A in latitude ϕ_1 and longitude λ_1, to B in latitude ϕ_2 and longitude λ_2. The course is extended to cross the equator, EE', at the point E. The angle C is approximately the rhumb-line course between the two points. M_1 and M_2 represent the stretched distances of the parallels ϕ_1 and ϕ_2 from the equator, the values being taken from tables of meridional parts. D_1 and D_2 represent the difference of longitude of the points A and B from the point of equator crossing of the extended course.

Now call $M_2 - M_1 = m$ and $D_2 - D_1 = \delta\lambda$ (the longitude difference between A and B), and we have at once, from the figure, $\delta\lambda = m \tan C$. The

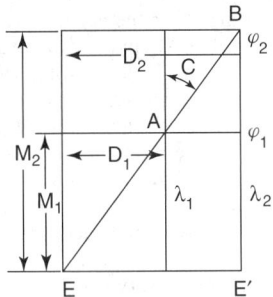

Fig. 1.   Mercator sailing.

value of C thus computed, taking into account the mercator stretching, is frequently referred to as the mercator course between A and B. The distance along AB is a stretched distance and, if calculated from the above figure, or obtained graphically, would be greater than the distance actually traveled between A and B. To avoid this discrepancy, the distance is computed by the methods of plane sailing; the plane triangle is solved using the mercator course as computed above for the vertex angle and the difference of latitude as one leg. The so-called mercator distance is obtained as $d = (\phi_1 - \phi_2)$ secant C, the value of $(\phi_2 - \phi_1)$, the difference of latitude being expressed in minutes of arc, which is practically identical with nautical miles.

See also **Course**; **Navigation**; and **Rhumb Line**.

MERCURY.   Chemical element, symbol Hg, at. no. 80, at. wt. 200.59, periodic table group 12, mp $-38.84\,°C$, bp $356.58\,°C$, density 13.546 g/cm^3 (liquid), 14.193 g/cm^3 (solid). Solid mercury has a rhombohedral crystal structure. The element, sometimes referred to as quicksilver, is a silver-white liquid metal at standard conditions. There are seven stable isotopes of mercury, ^{196}Hg, ^{198}Hg through ^{202}Hg, and ^{204}Hg, and seven radioactive isotopes, ^{192}Hg through ^{195}Hg, ^{197}Hg, ^{203}Hg, and ^{205}Hg. With exception of ^{194}Hg (half-life of approximately 130 days) and ^{203}Hg (half-life of approximately 46 days), the half-lives of the radioactive isotopes are short, measured in terms of minutes or hours.

First ionization potential 10.434 eV; second, 18.65 eV; third, 34.3 eV. Oxidation potentials $Hg \rightarrow \frac{1}{2}Hg_2^{2+} + e^-$, -0.7986 V; $Hg \rightarrow Hg^{2+} + 2e^-$, -0.852 V; $Hg_2^{2+} \rightarrow 2Hg^{2+} + 2e^-$, -0.905 V. $Hg + 2OH^- \rightarrow HgO + 2H_2O + 2e^-$, -0.098 V; $2Hg + 2OH^- \rightarrow Hg_2O + H_2O + 2e^-$, -0.123 V. Other important physical properties of mercury are given under **Chemical Elements**.

Mercury forms alloys, called amalgams, with most metals, but not with iron or platinum; does not wet glass but forms a convex surface when in a glass container; is slightly volatile at ordinary temperatures and a health hazard due to its poisonous effect; slowly tarnishes in moist air; upon heating in air or oxygen, somewhat below its boiling temperature of 357 °C, forms mercuric oxide slowly, as in the classical experiment by Lavoisier on the composition of air; may be purified by distillation and condensation (health hazard); unattacked by dilute HCl or H_2SO_4, but dissolved by dilute or concentrated HNO_3 with the formation of mercurous and mercuric nitrates, respectively, and by hot concentrated H_2SO_4 with the formation of both mercurous and mercuric sulfates; unattacked by alkalis. Discovery ancient.

When cooled to sufficiently low temperatures, mercury becomes *superconducting*, virtually conducting electricity with no resistance. See also **Superconductivity**.

Mercury was mined as early as 500 B.C. and currently ranks tenth in worldwide production of nonferrous metals. The unusual combination of physical properties possessed by the element give it an importance exceeding its production rating. The chief source of mercury is cinnabar, HgS, the red sulfide that contains 86.2% mercury. See also **Cinnabar**. Although the mineral occurs widely throughout the world, relatively few deposits are of commercial importance, notably those in China, Italy, Mexico, the Philippines, Peru, Spain, the former Soviet Union, the former Yugoslavia, and the United States. World reserves of mercury currently are estimated at 4 million flasks, of which sources in the United States account for about 300,000 flasks. A flask contains 76 pounds (34.5 kilograms) of the liquid metal.

The ore is concentrated to about 25–50% mercury by flotation. Beneficiation of mercury ores is not commonly practiced. The concentrate is roasted: $HgS + O_2 \rightarrow Hg + SO_2$. The process is essentially one of distillation because the freed mercury quickly volatilizes, after which it is condensed with a resulting purity of 95% (furnace plants) to 98% (retort plants). The mercury is further refined, through filtering, oxidation, acid leaching of the impurities, or by distillation, to yield prime or virgin mercury with an average purity of 99.9%. This purity is satisfactory for all but the most exacting requirements. Some special mercury chemicals may require that the virgin mercury be triple distilled. Significant quantities of secondary mercury are recovered from waste products, such as dental amalgams, sludges, used batteries, used instruments, and other mercury-bearing materials. Secondary recovery accounts for close to 20% of the total domestic production of mercury in the United States.

Toxicity: Mercury and its compounds, with few exceptions, are highly poisonous to living organisms. Particularly in connection with finely divided mercury metal, extreme care must be taken to avoid inhalation of and contact with the element. The chances of poisoning are increased because awareness of the presence of the metal is reduced when it is finely divided. The fine gray mercury powder is easily generated when liquid mercury is rubbed against or agitated with grease, chalk, sugar, ether, and numerous other substances.

Mercury can cause acute renal failure and nephrotic syndrome. Chronic exposure to or ingestion of mercury may lead to polyneuropathy. Confirmation of poisoning is sometimes made by analysis for the presence of mercury in hair, fingernails, serum, and urine. Removal of the metal may be hastened by the oral administration several times a day for a limited period of D-penicillamine. As pointed out by Beary (1979), with reference to the relatively high mercury content of whale meat, when humans consume contaminated whale meat, the lipid-soluble methylmercury is concentrated in the cells of the nervous system and very slowly eliminated from the body, even when all intake is stopped.

One of the early examples of mercury poisoning may have involved the so-called "dark year" (1693) in the life of Sir Isaac Newton, when it was reported that Newton broke away from friends and associates, accusing them of plotting against him, when he slept very little, and when he reported conversations that actually did not occur. Initial explanations of Newton's erratic behavior included his failure to obtain certain appointive government positions, overwork, and the traumatic fire that destroyed a number of his valuable manuscripts. This evidence was not convincing, however, to investigators P.E. Spargo and C.A. Pounds. Through contacts with Newton's descendants, this team obtained four hairs taken from the head of the master and subjected them to laboratory tests. In the very interesting account by Broad (1981), the investigators concluded that Newton's madness was "due principally to poisoning by the metals which he used so frequently and with such cavalier disregard for his own safety." Locks of hair were located at Trinity College, Cambridge. One hair was found to contain 197 ppm Hg (even quite high in terms of modern-day Hg poison cases). Broad concludes his article, "Perhaps poisoning by mercury was not only the cause of Newton's brief lunacy, but was also the pivotal event that nudged the superstitious genius away from his researches in the laboratory to the seemingly less dangerous ways of the world." There are several doubters of the Spargo-Pounds hypothesis and the controversy is likely to persist.

Uses of Mercury. Because of the poisonous nature of Hg, wherever possible the industrial uses of the element have been reduced and research to find suitable substitutes for Hg for many applications has accelerated during recent years. Traditionally, mercury's largest usage has been in connection with electrical apparatus (switching, etc.), the electrolytic preparation of chlorine and caustic soda (mercury cells), in antifouling and mildew-proofing agents and paints, in industrial and clinical thermometers, in pharmaceutical preparations, in agricultural herbicides and pesticides, in dental preparations, in the preparation of amalgams, and in use of Hg as a catalyst for certain chemical reactions.

The consumption of mercury by the chlor-alkali industry may range from 15% to as high as 35% of the total in any given year, depending upon new construction. Large amounts of mercury are required for the start-up of mercury cell operations, whereas replacement requirements are quite low. Several areas of mercury use are declining gradually, particularly in the pharmaceutical field where sulfa drugs, iodine, and various antiseptics and disinfectants have made inroads on mercury chemicals. Mercury compounds, used for many years in the treatment of syphilis, for example, largely have been displaced by antibiotics and other treatments. Because of the fundamentally toxic nature of mercury and its compounds, the agricultural uses of mercury formulations are being deemphasized, with constant research for substitute materials. In the dental field, a number of metal powders, porcelain, and plastic materials have displaced mercury amalgams in many dental applications. In the explosives field, several compounds, such as lead azide, diazodinitrophenol, and other organic initiators, are serving the same function as mercury fulminate. The use of mercury as a heat-transfer medium in boilers was essentially abandoned a number of years ago. On the other hand, mercury-base catalysts are increasing in application and, to date, suitable substitutes for mercury in the antifouling and mildew-proofing formulations area have not been found. The essential properties of mercury that will be difficult to replace are its high specific gravity, fluidity at room temperatures, and excellent electrical conductivity.

In addition to some diminishment of mercury usage for various products because of increasing awareness of its toxicity potential, conservation-minded technologists also have pointed to the relatively limited world resources of the metal. Considerable ingenuity has been used to replace mercury. For example, diaphragm cells can be used in caustic-chlorine production, organic biocides are replacing mercury-containing compounds, gold recovery is accomplished by the cyanide process rather than by the amalgamation process, and plastic paints and copper oxide paints can be used in place of mercuric biocidal paints. The use of the diaphragm cell for chlorine-alkali production possibly is the most dramatic substitution in terms of mercury conservation. Diaphragm cells require no mercury, whereas the traditional mercury cells once accounted for as much as 35% of the mercury use in some years.

Chemistry and Compounds. The apparent anomaly between mercury and the lighter elements of transition group 2, in that mercury regularly forms both univalent and divalent compounds, while zinc and cadmium do so very rarely, is partly understood from the observation that mercury(I) salts ionize even in the gaseous state to Hg_2^{2+}, rather than Hg^+. Evidence for this double ion is provided by its Raman spectral line, by the lineal $Cl-Hg-Hg-Cl$ units in crystals of mercury(I) chloride, and by the emf of mercury(I) nitrate concentration cells. The anomaly is further removed by the observation that cadmium also forms a (much less stable) diatomic ion Cd_2^{2+}, e.g., in $Cd_2(AlCl_4)_2$.

Oxides. Heating of mercury in air yields the (divalent) oxide HgO, which at higher temperatures decomposes into its elements. Mercury(II) oxide is also precipitated from solutions of mercury(II) salts by alkaline solutions. Alkalies precipitate a yellow form, while alkali carbonates give a red one. The yellow is apparently a finely divided form of the red, since they are crystallographically identical, but differ slightly in certain chemical and physical properties, including solubility. Mercury(II) oxide exhibits solubility in solutions of alkali salts, which is attributed to formation of complex ions such as $[Hg(OH)_2NO_3]^-$ and $[Hg(OH)_2SO_4]^{2-}$.

Halogen Compounds. All eight compounds of univalent and divalent mercury with the single halogens are known, as well as several compounds of mercury(II) with two halogens, such as HgBrI and HgClI. The mercury(I) halides are insoluble in water, with the exception of the fluoride which, like mercury(II) fluoride, is hydrolyzed by water. Like the zinc and cadmium halides, mercury(II) halides behave anomalously in aqueous solution, and for similar reasons, i.e., the presence of complex ions and un-ionized molecules. In the case of mercury(II) halides, with their more covalent character than zinc or cadmium halides, the ionization is somewhat less, and the concentration of Hg^{2+} relatively low. Thus, in aqueous solution, mercury(II) chloride, $HgCl_2$, is present largely as un-ionized molecules, but also ionizes $HgCl^+$ and Cl^-, and only secondarily and to a slight extent to give Hg^{2+}. In the presence of added Cl^-, an $HgCl_2$ solution is a complex system involving equilibria between $HgCl_2$, $HgCl^+$, Cl^-, Hg^{2+}, and the complex ions $HgCl_3^-$ and $HgCl_4^{2-}$. Similarly, the hydrolysis of $HgCl_2$, though slight, involves several equilibria whose relative importance varies with the concentration of the solution. In more concentrated solutions the hydrolysis of $HgCl_2$ to HgOHCl, Cl^- and H^+ is prominent, while in more dilute solutions the most important equilibria involve the ionization $HgCl_2$ to $HgCl^+$ and Cl^-, and the hydrolysis of $HgCl_2$ to $[HgOHCl_2]_2^{2-}$ and H^+, and that of $HgCl^+$ to HgOHCl or $[HgOHCl]_2$ and H^+. Finally, oxyhalides, such as $HgBr_2 \cdot 3HgO$, $HgCl_2 \cdot 2HgO$, $HgCl_2 \cdot 3HgO$ and $HgCl_2 \cdot 4HgO$ are also obtainable, usually by action of alkali hydroxides upon mercury halides. Mercury(II) iodide, like the oxide, is polymorphic. It has three forms, yellow, red, and white, the second being the most stable up to 127°C, where it undergoes a definite transition to the yellow. The colorless HgI_4^{2-} is very stable, especially to alkalies, and is used in Nessler's reagent.

Salts. Mercury forms many salts, both of mercury(I) and mercury(II). In general, action of oxidizing acids upon the metal yields the latter, while the former requires either a limited amount of the oxidant or indirect methods. Mercury(I) salts are made by treating a solution of a soluble mercury(II) salt with metallic mercury. Thus, heating mercury with H_2SO_4 or HNO_3 yields mercury(II) sulfate or nitrate, $HgSO_4$ or $Hg(NO_3)_2$, respectively, crystallizing as hydrates, while mercuric(I) nitrate results from the use of cold acid in limited amount, and mercuric(I) sulfate is produced by the last method as well as from mercury(I) nitrate and sulfuric acid. The other salts of both univalent and divalent mercury include the acetates, antimonates, arsenates, bromates, carbonates, chlorates, chromates, fluorosilicates, iodates, oxalates, perchlorates, periodates, phosphates, tartrates, thiocyanates, tungstates, uranates, and vanadates. Also known in both valences are the arsenides (from arsine and the mercury solutions), azides (from hydrozoic acid and the mercury solutions), nitrides and phosphides. Only mercury(II) selenide and telluride exist. Hg_2S_2 has been reported to be obtained as a black powder, but is believed to be a mixture of Hg and HgS. The latter exists in two forms, the black form that is usually precipitated by hydrogen sulfide, and the red, cinnabar, precipitated by H_2S from a solution of mercury(II) acetate and ammonium thiocyanate. The black changes to the red in liquid H_2S, and the red to the black on heating to $386\,^\circ C$. Cinnabar is the thermodynamically stable form at room temperature.

Compounds with Nitrogen and Sulfur. The reactions of mercury(I) compounds and NH_3 are complex, and published results vary. Recent (x-ray) studies show that this reaction, modified by the presence of ammonium chloride, NH_4Cl, yields three aminobasic compounds containing divalent mercury: $Hg_2NCl \cdot H_2O$, $HgNH_2Cl$, and $Hg(NH_3)_2Cl_2$. The first of these is the chloride of Millon's base, $Hg_2NOH \cdot 2H_2O$, which is produced by warming HgO with aqueous ammonia.

Mercury is the least active of the elements of its group as an electron acceptor from oxygen; however, with sulfur it is more active, the mercury halides forming dialkyl sulfide addition products, $R_2S \cdot 2HgX_2$, and HgS dissolves in alkali sulfides forming $[HgS_2]^{2-}$ or $Hg(SH)_4^{2-}$. Like zinc and cadmium, mercury forms a series of ammines, which with the mercury halides are principally the diammines, $[Hg(NH_3)_2]X_2$, where X is a covalently bonded halogen atom, and with more ionic mercury compounds, e.g., the nitrate and sulfate, especially in the presence of high concentrations of ammonium salts, the tetrammines, e.g., $[Hg(NH_3)_4](NO_3)_2$. These complexes also form with amines and diamines, e.g., ethylenediamine, which contributes three molecules per Hg^{2+} ion. Besides forming univalent and divalent cyanides, mercury forms complex cyanide ions, $[Hg(CN)_3]^-$ and $[Hg(CN)_4]^{2-}$, as well as additional compounds with the mercury(II) halides of the structure HgCNX. Mercury forms insoluble thiocyanates and complex ions, e.g., $Hg(SCN)_3^-$, $Hg(SCN)_4^{2-}$. Mercury cyanates and fulminates are insoluble. Mercury(II) mercaptides, $Hg(SR)_2$, decompose on heating to give HgS and R_2S. Mercury(II) hydrogen sulfite, $Hg(HSO_3)_2$, is actually mercuridisulfonic acid, $Hg(SO_3H)_2$, with Hg–S bonds. Like the halides, compounds such as $Hg[C(NO_2)_3]_2$. $Hg[C(CN)_3]_2$, $Hg(NO_2)_2$ (dinitromercury), $Hg(CF_3)_2$ are noteworthy for their lack of ionic dissociation.

Organometallic Compounds. Mercury also forms a large number of organometallic compounds of the type HgR_2, where R may be not only an alkyl radical, but an alkoxy radical, an acyl radical, a halogenated alkyl radical, an alkylthio radical, an aryl radical or a perfluoroalkyl radical. In addition, mercury also forms numerous organometallic compounds of structure RHgX, where X is a halogen atom, and R one of the foregoing organic radicals.

The organic compounds containing mercury that are used as disinfectants, germicides, and antiseptics are known as mercurials. Among these are Merthiolate, Mercurochrome, and Metaphen. Merthiolate is the sodium salt of ethylmercurithiosalicylic acid, $C_2H_5HgS \cdot C_6H_4 \cdot COONa$. It contains 49.5% of mercury. It is a crystalline, cream-colored powder which is very soluble in water, for about 1 gram dissolves in 1 milliliter of water. It is much less soluble in alcohol, 1 gram in 8 milliliters. It is insoluble in organic solvents like benzene and ether. It is used as an antiseptic for tissues in concentrations of the order of 1:1,000 to 1:30,000. It is commonly used as an antiseptic in biologics.

Mercurochrome, also known by the name of *Merbromin* and by many other trade names, is the disodium salt of 2,7-dibromo-4-hydroxymercurifluorescein, $C_{20}H_8Br_2HgNa_2O_6$. It forms green, iridescent scales or granules, which are freely soluble in water, yielding a bright red solution with dilute solutions having a yellow-green fluorescence. It is generally used in a 2% aqueous solution as a mild antiseptic. It is nearly insoluble in alcohol, and is insoluble in organic solvents like acetone and ether. Most other common mercurials and iodine solution are considered to be better antiseptics.

Additional Reading

Agocs, M.M. et al.: "Mercury Exposure from Interior Latex Paint," *N. Eng. J. Med.*, 1096 (October 18, 1990).

Broad, W.J.: "Sir Isaac Newton: Mad as a Hatter," *Science*, **213**, 1341–1344 (1981).

Clarkson, T.W.: "Mercury—An Element of Mystery," *N. Eng. J. Med.*, 1137 (October 18, 1990).

Fackelmann, K.A.: "Painting a Perilous Picture of Mercury," *Science News*, 244 (October 20, 1990).

Greenwood, N.N. and A. Earnshaw: "Chemistry of the Elements Revised and Updated," 2nd Edition, Butterworth-Heinemann, Inc., Woburn, MA, 1997.

Krebs, R.E.: "The History and Use of Our Earth's Chemical Elements: A Reference Guide," Greenwood Publishing Group, Inc., Westport, CT, 1998.

Lide, D.R.: "CRC Handbook of Chemistry and Physics 2000-2001." 81st Edition, CRC Press, LLC., Boca Raton, FL, 2000.

Raloff, J.: "Mercurial Risks from Acid's Reign," *Science News*, 154 (March 8, 1991).

Stone, R.: "Mercury's Metabolic Fingerprint," *Science*, 29 (April 3, 1992).

Stwertka, A. and E. Stwertka: "A Guide to the Elements," Oxford University Press, Inc, New York, NY, 1998.

MERCURY (Planet). The first of the planets in order of distance from the Sun. At maximum elongation, Mercury is not more than 28 degrees away from the Sun as seen from earth. Thus, the planet does not rise above the horizon for more than 2 hours before sunrise, or remain above the horizon for more than 2 hours after sunset. Consequently, observations of Mercury from earth are less than ideal. This behavior of Mercury also led early observers to believe that Mercury was actually two planets—Apollo observed in the early morning and Mercury as seen in the early evening.

The mean distance of Mercury from the Sun is about 58 million kilometers (40 million miles). The planet's orbit is inclined some $7°$ to the plane of the ecliptic and has a large eccentricity of 0.206. Because of this eccentricity, the distance of Mercury from the sun varies from a minimum of 46 million kilometers (28.5 million miles) at perihelion to a maximum of 70 million kilometers (43.4 million miles) at aphelion. The orbital velocity of Mercury varies accordingly, ranging from 39 kilometers (24.2 miles) per second at aphelion to 57 kilometers (35.2 miles) per second, the fastest of all of the planets.

Orbits computed for Mercury on the basis of Newtonian mechanics, and allowing for all perturbations, revealed a progressive eastward advance of the longitude of perihelion of 531 seconds per century. Later observations of the actual advance of perihelion of 574 seconds per century were made. The discrepancy of 43 seconds led to a search for an intra-mercurial planet. However, the application of the theory of relativity, or relativistic mechanics, to the problem showed that the added advance of perihelion is due to a relativistic mass effect, and was one of the early, classical proofs of Einstein's theory of relativity.

The albedo of Mercury is quite low, on the order of 0.056. This is much like the albedo of the moon, which led to the conclusion for many years that the two surfaces are similar and, indeed, recent observations also indicate this to some degree. The brightness of Mercury is on the order of -0.4 apparent visual magnitude, and its color is somewhat redder than that of the Sun. Like Venus and the moon, Mercury shows changes of phase.

Exploration of Mercury. The absence of special features, such as satellites and rings, coupled with the limiting viewing time of Mercury from Earth, have contributed to a better understanding and interest in Mercury as compared with the outer planets. The most dramatic and thorough exploration occurred in 1974 by way of the *Mariner 10* spacecraft. Between the time of that encounter and the early 1990s, additional findings of Mercury have been made from Earth-based instrumentation, including spectroscopy and radar imaging. In terms of long-range studies, Mercury is not high on the list of exploratory priorities. However, prior to the breakup of the U.S.S.R., Soviet scientists had proposed in 1989 that a Mercury "lander" be planned for launching sometime during the period, 1998 to 2000. Of course, as of mid-1993, the extent to which Russia may participate in space exploration programs is vague. At the Conference on Solar System Exploration held at California Institute of Technology in August 1989, Valery Barsukov, a leading Soviet space scientist, pointed out that Mercury holds interest for exploration on several points. Mercury is the closest planet to the sun and thus presents great extremes of daytime heat and nighttime cold. Further, Mercury is the only one of the inner planets whose surface has not been probed by a landing craft.

Mariner 10 **Mission.** The principal scientific objective of the *Mariner 10* mission was the exploratory study of Mercury. Thus, the instruments selected were chosen to serve these objectives. However, *Mariner 10* encountered with Venus on February 5, 1974 because of the use of Venus' gravity field as a "third stage" so to speak to achieve the encounter with Mercury. Some very useful information was obtained concerning Venus. See also **Venus.**

The first pass (closest encounter) of *Mariner 10* with Mercury occurred on March 29, 1974, after a flight of 146 days, but photographing commenced on March 23 from a distance of 3.36 million miles (5.4 million kilometers). Periodic photographic operations continued until April 3 when the probe was 2.17 million miles (3.5 million kilometers) beyond the planet. Over 2,000 television frames were transmitted from twin, high-resolution cameras. Because the study of the interaction between Mercury and the solar wind was given a high scientific priority, a darkside passage was selected for the first flyby. Nevertheless, excellent computer-enhanced pictures were taken at about 145,000 miles (234,000 kilometers) away from the planet, about 6 hours before the closest approach.

Mariner 10's second encounter, after a long voyage around the sun, with Mercury occurred on September 21, 1974. Much additional information and many photographs of surface features of the planet were gained on the second flyby. See Figures 1, 2, and 3.

Landforms. The principal landforms on Mercury are basins, scarps, ridges, and plains that resemble analogous forms on the moon. Where the plains are absent, overlapping craters and basins form rugged terrain. The plains materials have many of the characteristics of the lunar maria and have been cratered to approximately the same degree. This twofold division of the surface morphology of Mercury is strikingly similar to that of the moon. Prominent structural features of the planet include irregular scarps which are up to 1 kilometer high (0.6 mile), extend for hundreds of kilometers, and cut across large craters and intercrater areas. Similar features are absent on the moon. No features suggestive of either

Fig. 3. Mosaic of 18 pictures, selected from over 2,300 photographs of Mercury taken by *Mariner 10*, shows the largest structural feature on the planet discovered to date. In the left half of this photomosaic taken on March 20, 1974, is a ring basin 1300 kilometers (800 miles) in diameter and bounded by mountains which rise as high as 2 kilometers (1.2 miles). The structure is centered at 190°W. longitude and 30°N latitude and is intensely disrupted with many fractures and ridges. The feature is similar in size and appearance to the Imbrium Basin on Earth's moon and most likely originated by impact of a body at least tens of kilometers in diameter. The huge basin has provisionally been named *Caloris*, or hot basin, for its position near one of the subsolar points when the planet is nearest the sun. (*NASA; Jet Propulsion Laboratory, Pasadena, California.*)

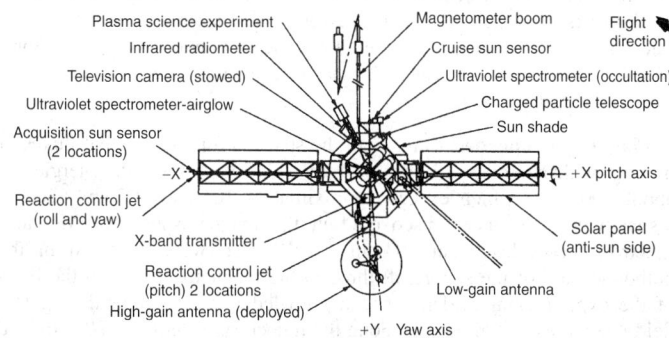

Fig. 1. *Mariner 10* spacecraft looking sunward along the Z axis (normal to plane of diagram at the intersection of the X and Y axes). (*Jet Propulsion Laboratory.*)

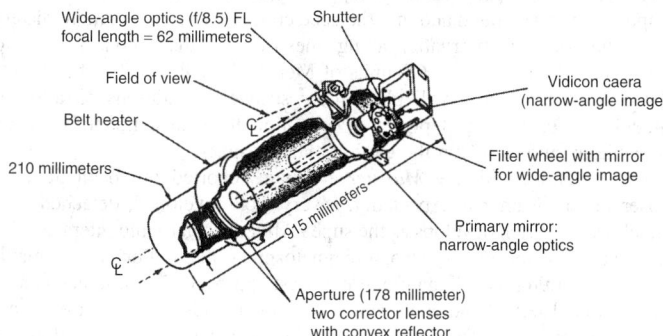

Fig. 2. Schematic diagram of television camera used on *Mariner 10*. (*Jet Propulsion Laboratory.*)

earthlike plate tectonics or large-scale tensional faulting in the crust have been recognized. Mariner spacecraft observations included: (1) extensive flooding by rock materials at least grossly similar to those of the lunar maria has occurred on Mercury; (2) the heavily cratered surfaces on Mercury record the final periods of heavy impact bombardment at Mercury; (3) in the half of the planet observed by Mariner 10, Mercury, like the moon, seems to exhibit a hemispherically nonuniform distribution of flooded basins; and (4) mare-like surfaces now have been formed on the moon, Mars, and Mercury which show a surprising similarity in accumulated impacts, although only those of the moon have been dated radiometrically.

Volcanism on Mercury? In addition to *Mariner* data, an estimated one-third of the planet (mainly the equatorial region) has been mapped from radar data in recent years. Many of these areas were not visible by Mariner. Initially, based mainly on *Mariner* data, some of the surface features of Mercury definitely appeared to some astronomers as being of volcanic origin. However, during the interim, counterarguments have developed, and it is highly likely that a final answer must await the next encounter by a spacecraft with the planet. One argument that has been proposed is that the plains surrounding Caloris (See Fig. 3) were the result of ejecta from an impact—that is, the plains were formed from debris flows.

During the course of diagnosing Mercury's past history and current appearance, frequent comparisons of the planet with earth's moon have been made. More recently, some scientists point out that this analogy may have been overworked.

Atmosphere of Mercury. It appears that there is virtually no atmosphere on Mercury. Findings of the *Mariner* program showed that the atmosphere of the planet, like that of earth's moon, is maintained in an extremely tenuous minimum state by weak solar wind accretion and radioactive decay processes, and depleted by strong removal mechanisms. Unlike the moon, Mercury has a high daytime surface temperature in the range of 100–325 °C (212–617 °F) that promotes the production of water vapor, which may be the dominant constituent derived from solar wind protons.

In 1985, Andrew Potter (NASA Johnson Space Center) and Thomas Morgan (Southwestern University) reported that sodium had been observed in the atmosphere of Mercury. The spectrum of Mercury at the Fraunhofer sodium D lines shows strong emission features that are attributed to resonant scattering of sunlight from sodium vapor in the atmosphere of the planet. The total column abundance of Na was estimated to be 8.1×10^{11} atoms per square centimeter, which corresponds to a surface density at the subsolar point of about 1.5×10^5 atoms per cubic centimeter. The most abundant atmospheric species found by the *Mariner 10* mission to Mercury was helium, with a surface density of 4.5×10^3 atoms per cubic centimeter. It now appears that Na vapor is a major constituent of Mercury's atmosphere. The atomic weight of sodium is sufficiently high so that the thermal escape of Na from Mercury should be negligible. However, it has been suggested that the solar wind can sweep away ions produced by photoionization of Mercury's atmosphere. This may be a major loss process for atmospheric gases that otherwise would not escape and may account for the manner in which the planet originally lost most of its atmosphere. It has been further suggested that Mercury's atmosphere may resemble a cometary coma rather than an Earthlike planetary atmosphere. More information is needed before the sources and sinks of Na in Mercury's atmosphere are understood. By measuring the variation of Na emission with phase, time, and distance of the planet from the sun, it may be possible to determine what processes control it.

In May 1990, A.E. Potter and T.H. Morgan (NASA), using a Bowen image slicer and a charge-coupled device (CCD), probed the spectra of Mercury. Their report concluded, "Monochromatic images of Mercury at the sodium D_2 emission line showed excess emission in localized regions at high northern and southern latitudes and day-to-day global variations in the distribution of sodium emission. These phenomena support the suggestion that magnetospheric effects could be the cause. Sputtering of surface minerals could produce sodium vapor in polar regions during magnetic substorms, when magnetospheric ions directly collide with the surface. Another important process may be the transport of sodium ions along magnetic field lines toward polar regions, where they impact directly on the surface of Mercury and are neutralized to regenerate neutral sodium atoms. Day-to-day variations in planetary sodium distributions could result from changing solar activity, which can alter the magnetosphere in time scales of a few hours. Observations of the sodium exosphere may provide a tool for remote monitoring of the magnetosphere of Mercury."

During the period between June 1986 and January 1988, A.L. Sprague, R.W.H. Kozlowski, and D.M. Hunten (University of Arizona), using a 1.5-meter Cassegrain reflecting telescope and spectrographic instrumentation, observed enhanced abundances of neutral potassium (K) in the atmosphere of Mercury above the longitude range containing the Caloris Basin. Results of large data sets showed typical K column abundances of $\sim 5.4 \times 10^8$ K atoms/cm². The scientists observe that this enhancement is consistent with an increased source of K from the well-fractured crust and regolith associated with this large impact basin. The phenomenon is localized because, at most solar angles, thermal alkali atoms cannot move more than a few hundred kilometers from their source before being lost to ionization by solar ultraviolet radiation.

Thermal Properties. The infrared radiometer on *Mariner 10* measured the thermal emission from Mercury with a spatial resolution element as small as 40 kilometers (25 miles) in a broad wavelength band centered

at 45 micrometers. The minimum brightness temperature (near local midnight) in these near-equatorial scans was 100 K. Along the track observed, the temperature declined steadily from local sunset to near midnight, behaving as would be expected for a homogeneous, porous material with a thermal inertia of 0.0017 cal^{-2}sec$^{-1/2}$ degrees K^{-1}, a value only slightly larger than that of the moon. From near midnight to dawn, however, the temperature fluctuated over a range of about 10 degrees Kelvin, implying the presence of regions having thermal inertia as high as 0.003 cal^{-2}sec$^{-1/2}$ degrees K^{-1}.

Polar Ice Caps? M.A. Slade, B.J. Butler, and D.O. Muhleman (Jet Propulsion Laboratory) conducted the first unambiguous full-disk radar mapping of Mercury at 3.5-centimeter wavelength with the Goldstone 70-meter antenna transmitting and 26 antennas of the Very Large Array receiving. This study is reported to provide evidence for the presence of polar ice. The radar experiments were designed to image that half of the planet that was not photographed by *Mariner 10*. The investigators reported, "The orbital geometry allowed viewing beyond the north pole of Mercury, a highly reflective region clearly visible on the north pole during the experiments. The team initially was surprised to find that ice could exist on Mercury (planet nearest the sun). The radar data, however, were consistent with reflections from ice. Later, the existence of the bright north polar feature was confirmed at Arecibo, and a compact feature near the south pole also was discovered. See Fig. 4. The plausibility of ice at the poles, not apparent in conventional telescopic images, is supported by the fact that radar can penetrate beneath the surface, where ice can exist because it is well protected from the baking hot surface temperatures. The concept of ice at the poles of Mercury still is somewhat debatable. For example, why haven't the ices evaporated over billions of years?"

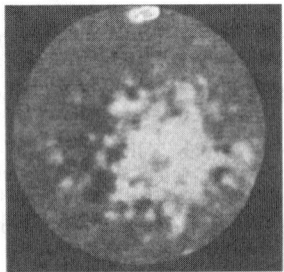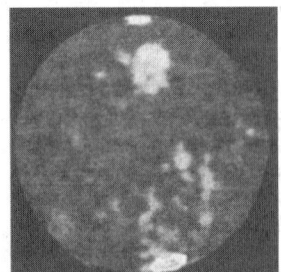

Fig. 4.   Radar images of Mercury made in 1992. Suspected ice cap is shown at Mercury's north pole in both images. In a view taken at a somewhat later date (right), a bright image also is evident at the south pole of the planet. Some scientists propose that this region also may be made up of ice.

Magnetic Properties. Data from the spacecraft indicated that fluxes of protons with energies of about 550 keV and electrons with energies of about 300 keV, which exceed approximately 104 and 105 cm^{-2} sec^{-1}, respectively, have been discovered in the magnetosphere of Mercury. Electron fluxes less than 103 cm^{-2} sec^{-1} also were observed in the outbound pass of the spacecraft through the magnetosheath. On the basis of the experimental evidence and a knowledge of the general magnetic field intensities and directions along the trajectory of *Mariner 10* provided by the magnetic field observations, it is shown that the radiation events observed in the magnetosphere and magnetosheath are transient and are not interpretable in terms of stable trapped particle populations. Furthermore, experimental evidence strongly supports the view that the particles are impulsively accelerated and that the acceleration source is not more distant from the point of observation along lines of force than about 8×10^3 to 16×10^3 kilometers (3 to 6.5 units of Mercury's radius). The phenomena discovered at Mercury will place more stringent conditions on allowed models for electron and proton acceleration than have heretofore been possible in studies within the earth's magnetosphere.

One scientist with the *Mariner* program reported on magnetic field observations. Rather unexpectedly, a very well-developed, detached bow shock wave, which develops as the super-Alfvénic solar wind interacts with the planet, was observed. Also, a magnetosphere-like region, with a maximum field strength of 98 gammas at closest approach (704 kilometers altitude), was observed and contained within boundaries similar to the terrestrial magnetosphere. The obstacle deflecting the solar wind flow is global in size, but the origin of the enhanced magnetic field was not uniquely established. The field may be intrinsic to the planet and distorted by interaction with the solar wind. It may also be associated with a complex induction process whereby the planetary interior-atmosphere-ionosphere interacts with the solar wind flow to generate the observed field by a dynamo action.

A magnetic field about 1% as strong as the earth's, together with the planet's high mean density of 5.4 grams per cubic centimeter, suggests that Mercury may have an iron core similar to earth's core, in which the magnetism is generated.

The concept of an intrinsic magnetic field and, in fact, the nature of the core of Mercury, essentially remain an enigma. A group of scientists, including J.O. Burns, at the University of New Mexico have used radio emissions to probe Mercury's subsurface. These investigators have observed, "There does not appear to be any excess heat arising from the core of Mercury. The temperature map can be explained by reradiation of solar energy." J.O. Burns (New Mexico State University), in combining the data of the mapping studies with Mercury's slow rotation, observes, "This indicates that Mercury does not have a large, hot molten core that many believe is needed to produce the strong magnetic field via the dynamo model."

Classic References to Mariner 10 Mission (1974) to Mercury

Broadfoot, A.L. et al.: "Mercury's Atmosphere from Mariner 10: Preliminary Results," *Science*, **185** (4146), 166–169 (1974).

Chase, S.C. et al.: "Preliminary Infrared Radiometry of the Night Side of Mercury from Mariner 10," *Science*, **185** (4146), 142–145 (1974).

Dunne, J.A.: "Mariner 10 Mercury Encounter," *Science*, **185** (4146), 141–142 (1974).

Howard, H.T. et al.: "Mercury, Results on Mass, Radius, Ionosphere, and Atmosphere from Mariner 10 Dual-Frequency Radio Signals." *Science*, **185** (4146), 179–180.

Metz, W.D.: "Mercury: More Surprises in the Second Assessment," *Science*, **185** (4146), 132 (1974).

Murray, B.C. et al.: "Venus: Atmospheric Motion and Structure from Mariner 10 Pictures," *Science*, **183** (4131), 1307–1315 (1974). (Describes Mariner 10 cameras and photographic techniques.)

Murray, B.C. et al.: "Mariner 10 Pictures of Mercury: First Results," *Science*, **184** (4135), 459–461 (1974).

Murray, B.C.: "Mercury's Surface: Preliminary Description and Interpretation from Mariner 10 Pictures," *Science*, **185** (4146), 169–179 (1974).

Ness, N.F. et al.: "Magnetic Field Observations near Mercury: Preliminary Results from Mariner 10," *Science*, **185** (4146), 151–159 (1974).

Ogilvie, K.W. and J.D. Scudder, et al.: "Observations at Mercury Encounter by the Plasma Science Experiment on Mariner 10," *Science*, **185** (4146), 145–151 (1974).

Simpson, J.A. et al.: "Electrons and Protons Accelerated in Mercury's Magnetic Field," *Science*, **185** (4146), 160–169 (1974).

Additional Reading

Eberhart, J.: "Cold Message from Mercury's 'Hot Poles'," *Science News*, 375 (June 16, 1990).

Holden, C.: "An Ice Cap on the Hottest Planet," *Science*, 935 (November 15, 1991).

Potter, A.E. and T.H. Morgan: "Evidence for Magnetospheric Effects on the Sodium Atmosphere of Mercury," *Science*, 835 (May 18, 1990).

Slade, M.A., B.J. Butler, and D.O. Muhleman: "Mercury Radar Imagine: Evidence for Polar Ice," *Science*, 635 (October 23, 1992).

Sprague, A.L. et al.: "Caloris Basin: An Enhanced Source for Potassium in Mercury's Atmosphere," *Science*, 1140 (September 7, 1990).

Sprague, A.L., R.W.H. Kozlowski, and D.M. surHunter: "Detecting Potassium on Mercury," *Science*, 974 (May 17, 1991).

Staff: Time-Life Books, "The Near Planets," Time-Life, Inc., New York, NY, 1992.

Strom, R.: "Mercury: The Elusive Planet," Smithsonian Institution Press, Washington, DC, 1987.

Vilas, F. et al. (Editors): "Mercury," University of Arizona Press, Phoenix, AZ, 1997.

Waldrop, M.M.: "A New Soviet Plan for Exploring the Planets," *Science*, 211 (October 13, 1989).

Web References

http://sse.jpl.nasa.gov/features/planets/mercury/mercury.html (National Aeronautics and Space Administration, United States).

http://sse.jpl.nasa.gov/missions/merc_missns/m10.html (National Aeronautics and Space Administration, United States)

MERGANSER. See **Waterfowl**.

MERIDIAN. On the celestial sphere, a great circle that passes through the poles of rotation and, hence, is perpendicular to the celestial equator. The *local meridian* is the great circle passing through both the poles of rotation and, also, through the zenith; it is both a vertical circle and an

hour circle. However, the local meridian differs from the ordinary circles on the sphere in that it apparently remains fixed as the celestial sphere apparently rotates once each day. An object on the local meridian is said to be in culmination. If it is on that section of the meridian between the horizon and above the pole, it is said to be in upper culmination; if below the horizon, or below the pole and still above the horizon, it is said to be in lower culmination. The spherical coordinates both of hour angle and astronomical azimuth are measured from the local meridian in the direction of apparent rotation of the celestial sphere.

A *terrestrial meridian* is a curve cut on the surface of the earth by a plane containing the earth's axis of rotation. The local meridian of a point on the surface of the earth is the meridian passing through that point. The plane of the local terrestrial meridian coincides with the plane of the local celestial meridian. Terrestrial longitude is measured from the local meridian of Greenwich, England.

See also **Bearing (Navigation)**; and **Celestial Sphere and Astronomical Triangle**.

MERIDIAN CIRCLE. A telescope, adjusted so that the collimation plane of the instrument is in the plane of the local meridian, and which may be rotated about a horizontal axis. The instrument is usually fitted with a circle, which is accurately graduated in degrees, minutes, and seconds, and is perpendicular to the axis of rotation, hence, in the plane of the meridian. If the instrument does not carry the circle in the meridian, it is known as a transit circle.

At the principal focus of the telescope is placed a reticle with an odd number of vertical wires, the middle one of which is in the collimation plane. One horizontal wire through the optic axis of the telescope is usually present.

The instrument is used to determine the equatorial coordinates of the stars, when the local sidereal time and terrestrial latitude are known; or, conversely, to determine accurate local sidereal time by observation of stars of known right ascension. The local sidereal time of the instant of passage of a star across the middle wire of the reticle must be the right ascension of the star. The declination is obtained from the readings of the graduated circles when the star passes through the field of view along the horizontal wire. The instrument is also used for the accurate determination of terrestrial longitude by determining the local sidereal time and knowing the corresponding Greenwich time. See also **Telescope (Astronomical-Optical)**.

MEROMORPHIC FUNCTION. A function of a complex variable for which every point in the finite plane is either a regular point or a pole. For example,

$$f(z) = \frac{P(z)}{Q(z)}$$

where P and Q are polynomials, without common factor, is a meromorphic function.

MESA. A flat-topped, steep-sided, table-like mountain capped with a formation or stratum, which is relatively horizontal and resistant to erosion. When such a topographic feature is less than 1 square mile (2.6 square kilometers) in area it is usually called a butte.

MESODESMIC STRUCTURE. A type of ionic crystal in which one of the cation-anion bonds is equal in strength to all the bonds from the cation to the other anions. The silicates are important members of this class.

MESONS. The *mesons* are subatomic particles of the hadron family. See also **Hadrons**. Fermionic hadrons are called *baryons*; the others are called *mesons*. The meson family consists of eight members which fall into a triplet of *pions*, a singlet *eta*, a doublet of *kaons*, and a doublet of *antikaons*. (They are all pseudoscalar (spin zero and odd parity) and exhibit strong interactions.) The charged particles are coupled to the photon, but even the neutral members can participate in electromagnetic interaction by virtue of the large probability for virtual dissociation into charged particles. They participate in a variety of weak interactions including the nuclear beta decay interaction.

It is found that the kaons, the hyperons (baryons other than the neutron and proton) and their antiparticles, collectively known as *strange particles*, can decay by weak interactions not involving leptons or photons, with a lifetime which is large compared to the natural periods appropriate to strong interactions. On the other hand, these particles are produced copiously in high-energy nuclear collisions. These two circumstances can be understood in terms of the existence of another additive quantum number (*hypercharge*) which is conserved in strong and electromagnetic interactions, but violated in weak interactions.

The meson-baryon system exhibits further regularities as far as strong interactions are concerned. The neutron and the proton have very nearly the same mass and similar nuclear interactions although their electromagnetic properties are quite different. The three *pions* have different electric charges, but again they have approximately equal masses and similar nuclear interactions. This kind of multiplet structure is evident for other strongly interacting particles; the *kaons* form a doublet, the *sigma hyperons* form a triplet, the *xi hyperons* form a doublet, and the *lambda hyperon* remains a singlet. See also **Particles (Subatomic)**.

The pion (*pi meson*) was first recognized in 1947 in photographic films made by C.F. Powell, P.S. Occhialini, and their collaborators of cloud chamber tracks made by cosmic rays high in the Andes. The masses of these pions were greater than those of the previously discovered muons, corresponding more closely to those predicted by K. Yukawa in his theory of the nuclear structure of the atom. The positive or negative pion has a mass 273 times that of the electron, and a charge equal in magnitude to that of the electron. Both positively charged and negatively charged pions are found in cosmic rays. Neutral pions may also be present in cosmic rays, but are produced in much greater abundance by high-energy particle accelerators and are therefore more easily detected in these laboratories. The first artificially produced pions were made in 1948 by the impact of 380 MeV alpha particles, from the Berkeley synchrocyclotron, on a target of carbon or certain metals. These were charged pions. Others were produced later by beams of protons and deuterons. The first evidence of neutral pions were the gamma-rays produced in 1950 by the impact of 175 MeV protons upon similar targets (carbon, beryllium, etc.). These gamma-rays had a minimum energy of about 140 MeV, which would be expected from the decay of a (neutral) pion into two gamma-rays. The same method is used today to produce beams of charged pions. The life of the neutral pion is so much shorter (about 10^{-16} seconds against 2.6×10^{-8} seconds) that beams cannot be produced. Unless it is captured by an atom, or reacts with another particle, a charged pion decays into a muon of the same sign and a neutrino or antineutrino. Like the other mesons the pion is a boson. It has zero spin.

Evidence of the *kaon* was found in 1944 by L. Laprince-Ringuet and M. Lhéritier in a cloud chamber photograph of a cosmic ray event. It was found again in 1947 by Rochester and Butler as a V-shaped track in a cloud chamber, the particle forming the other side of the V being probably a pion. For that reason it was first called a charged V-particle, which has been superseded by kaon.

MESOSPHERE.   1.   The atmospheric shell, in which temperature generally decreases with heights, extending from the stratopause at about 50 to 55 kilometers to the mesopause at about 80 to 85 kilometers.

2.   The atmospheric shell between the top of the ionosphere (the top of this region has never been clearly defined) and the bottom of the exosphere See **Atmosphere (Earth)**.

MESOZOIC.   A major subdivision of the geologic time-scale. The era of "Middle" life, or the age of reptiles. Subdivided from the base up, into the following periods: Triassic, Jurassic, Comanchean, Cretaceous. The era was characterized by: the rise of dinosaurs (Triassic); rise of birds and flying reptiles (Jurassic); rise of flowering plants (Comanchean); great development of ammonites which became extinct at the end of the Cretaceous; culmination and extinction of most reptiles, and rise of the archaic mammals between the Cretaceous and the Tertiary. The Mesozoic era began 200 million years ago and lasted 140 million years.

MESQUITE TREE.   Commonly found in the southwestern United States, Mexico, other parts of Central America, the West Indies, and parts of South America, the mesquite (*Prosopis L.*) is a very hardy, small-to-medium-size tree. Examples of mesquites growing in the United States, as reported by American Forests are shown in Table 1.

Some beholders of the tree consider it scraggly and unkempt; to others, the dropping habit of its branches is attractive. The tree does not offer full shade. Its general appearance is very suggestive of its location in a desert and drought-stricken locale. The leaves are small, compound pinnately, and drought and heat resistant. Flowers are small and greenish white and occur in 2-to-5 in. (5–13 cm) spikes. Seeds, which develop in large 6 to 12 in. (15–30 cm) flat pods, are dropped to the ground between June and August. Succulent and containing sugar, the seed pods are welcomed by livestock. In Mexico, the pods are used to form a type of tamale (*mesquitamales*), which are sun dried and can be preserved as food for long periods.

The trunk of the mesquite generally is short, often twisted, and features limbs that are gnarled at the base. Thorns (11/2 in; 4 cm) long are found among the leaves and along the branches, but rarely on the trunk. The silhouette of the mesquite varies immensely—from that of a weeping willow to unlimited grotesque configurations. In a tree that is well developed and balanced above ground, the root system will be large, but not abnormally so. But where there is limited above ground development, the root system will be huge. Some taproots will reach out for water down to 60 feet (18 meters) below ground level. It has been observed by some foresters in the arid American Southwest, that the mesquite is one of the region's most interesting trees because of its heat and drought resistance resulting from its huge root system.

Along with its other survivability characteristics, the mesquite is relatively free from insects, pests, and fungus diseases. Its natural enemy is mistletoe, which hangs in heavy masses from the branches and causes deformation and atrophy of the branches, as well as asymmetrical burls on the trunk and major limbs. The tree is notably resistant to heartwood decay. Formation of a natural resin (closely related to acacia gum) covers over pruning cuts and other wounds, contributing to the longevity of the tree amidst very adverse conditions. Very important, as a legume, the mesquite can fix nitrogen in the soil. Further descriptions of the properties and lore of the mesquite tree can be found in "The Indomitable Mesquite," by J.H. Haller, American Forests, 20–24, August 1980.

MESSENGER RNA.   See **Cell (Biology)**; and **Human Genome Project**.

METACHROSIS.   The change of colors in animals by the expansion or contraction of pigment cells.

TABLE 1. RECORD MESQUITE TREES IN THE UNITED STATES[1]

Specimen	Circumference[2]		Height		Spread		Location
	Inches	Centimeters	Feet	Meters	Feet	Meters	
Honey mesquite (typ.) (1997) *(Prosopis glandulosa var. glandulosa)*	172	437	57	17.4	87	26.5	Texas
Screwbean mesquite (1983) *(Prosopis pubescens)*	39	99	30	9.1	36	11	Texas
Screwbean mesquite (1983) *(Prosopis pubescens)*	35	89	28	8.5	40	12.2	Texas
Velvet mesquite (1993) *(Prosopis velutina)*	196	498	46	14	60	18.3	Arizona
Western honey mesquite (1997) *(Prosopis glandulosa var. torreyana)*	76	193	47	14.3	39	11.9	California

[1]From the "National Register of Big Trees," American Forests (by permission).
[2]At 4.5 feet (1.4 meters).

METALLOBIOMOLECULES. Natural products, the biologically active forms of which contain one or more metallic elements. Metallobiomolecules may be transport and storage proteins, such as cytochromes (Fe), ferritin (Fe), transferrin (Fe), cruloplasmin (Cu), myoglobin (Fe), or hemoglobin (Fe); or they may be enzymes, such as carboxypeptidases (Zn), aminopeptidases (Mg, Mn), phosphatases (Mg, Zn, Cu), hydroxylases (Fe, Cu, Mo), or isomerases and synthetases, such as coenzymes (Co). Metallobiomolecules also may be nonproteins, such as siderophores (Fe) or chlorophyll (Mg). Ibers and Holm provide an excellent review of metallobiomolecules in *Science*, **209**, 223–235 (1980).

METALLOGRAPHY. Study of the structure and properties of metals and alloys, principally by microscopic and x-ray diffraction methods. The term is also used in a broader sense to include the processing of metals by mechanical and heat treatments and the fabrication and testing of finished products. In this usage it is synonymous with physical metallurgy.

METALLOID. A chemical element that may exhibit physical and chemical properties both of a metal and a nonmetal sometimes is referred to as a metalloid. Antimony, arsenic, and tellurium are examples. Less frequently, metalloid refers to elements, such as carbon, silicon, phosphorus, and sulfur, which are added in small amounts in the manufacture of iron and steel.

METALLOPROTEINS. Proteins, especially in solution, readily participate in a greater variety of chemical reactions than any other class of compounds of biological interest. This reactivity is a function primarily of the many polar side chains containing −OH, −COOH, −NH$_2$, −SH, and other groups, all of which can, to varying extents, interact with metal ions. Proteins can bind metals, some of them very tightly. However, relatively specific and nonspecific binding should be differentiated.

A negatively charged protein molecule exerting a nonspecific electrostatic attraction on metal ions would not qualify as metalloprotein. The term metalloprotein is restricted to compounds in which under natural conditions a metal ion is relatively specifically and strongly bound to a protein molecule in such a way that the compound can be isolated and shown to contain a stoichiometric amount of metal.

A variety of metal ions are found in biologically important metalloproteins. Metalloproteins occur in a wide range of biological systems. The function of the metalloproteins, as indicated in Table 1, varies widely from one compound to another.

TABLE 1. REPRESENTATIVE METALLOPROTEINS AND THEIR FUNCTIONS

Name	Metal	Source	Function
Hemocuprein	Cu	Erythrocytes	Unknown
Ceruloplasmin	Cu	Serum	Oxidase (probable)
Hepatocuprein	Cu	Liver	Unknown
Polyphenol oxidase	Cu	Mushroom	Enzyme
Hemocyanin	Cu	Mollusks	Respiratory pigment
Tyrosinase	Cu	Mushroom	Enzyme
Metallothionein	Cd + Zn	Kidney	Na Reabsorption (probable)
Xanthine oxidase	Mo	Liver	Enzyme
Carbonic anhydrase	Zn	Erythrocytes	Enzyme
Alcohol dehydrogenase	Zn	Yeast	Enzyme
Ferritin	Fe	Spleen	Fe storage
Transferrin	Fe	Plasma	Fe transport
Conalbumin	Fe	Eggs	Fe storage
Ferredoxin	Fe	Bacteria	Electron transport
DPNH-cytochrome *c* reductase	Fe	Heart muscle	Electron transport
Hemovanadin	V	Tunicates	Respiratory pigment

The chemical properties of the metal in these compounds may be greatly affected by bonding to a protein ligand. The bound metal can play one of many roles. Thus, in an enzyme, the metal ion may permit the formation of a ternary complex between protein, metal and substrate or coenzyme. An instance of this role is provided by the enzyme enolase, which is unable to catalyze the equilibrium between 2-phosphoglycerate and 2-phosphopyruvate in the absence of Mg ions. In other enzymes, the metal may actually participate in electron transport by cyclic oxidation and reduction. Such is probably the case with the Cu in polyphenol oxidase. The metal may serve primarily for the maintenance of a specific spatial folding of the polypeptide chains in the protein molecule.

The strength of metal-protein bonds in metalloproteins may vary from relatively loose association to very tight binding. When the metal ion is able to dissociate with some ease from the protein, it is usually a single ligand responsible for the metal binding. Such ligand groups are mainly found in the amino acid side chains of the protein molecule (e.g., −NH$_2$ or −OH groups). The interaction of metal and ligand may exhibit strong pH dependence because of competition between metal and hydrogen ions. Of the single ligand groups, by far the strongest is the −SH group in the amino acid cysteine. Even stronger metal bonding to protein may be observed when a divalent or trivalent metal forms chelate complexes with the protein. Chelation is often indicated not only because of the strength of the bond, but also because of the specificity of the reacting site on the protein molecule for one particular metal. Such a specificity may reflect the coordination requirements of the various metals. The preferred electron donor in the formation of protein-metal coordination compounds is N, such as that of the imidazole nucleus of histidine, but S and O may also participate in this process. If the protein contains carboxyl or phosphoryl groups, strong ionic bonds between metal and protein may be formed. A completely different type of protein-metal interaction is illustrated by the Fe-containing protein ferritin. Basically, this compound consists of a coat of protein (apoferritin) surrounding a micelle of hydrated iron hydroxide. The metal can be readily and reversibly removed from the apoprotein.

See also **Chelates and Chelation**.

METALLOTHIONEINS. These are low-molecular weight, cysteine-rich proteins that bind metal ions. As reported by Furey, et al.(*Science*, **231**, 704, 1986), metallothioneins and their genes have several potential kinds of physiological activity, including: (1) the genes are induced by metal ions and glucocorticoid hormones; (2) transcription is modulated during embryonic development; (3) the genes may be involved in control of cell differentiation and proliferation; (4) the proteins may function to activate Zn requiring apo-enzymes and regulate cellular metabolism; and (5) the proteins may act as free-radical scavengers. It appears that metallothioneins are synthesized in response to ultraviolet radiation. Cadmium, in addition to zinc, is found in metallothioneins.

METALLURGY. The science and technology of metals and alloys. Process metallurgy is concerned with the extraction of metals from their ores and with the refining of metals; physical metallurgy, with the physical and mechanical properties of metals as affected by composition, processing, and environmental conditions; and mechanical metallurgy, with the response of metals to applied forces. (*From Glossary of Metallurgical Terms and Engineering Tables, American Society for Metals, with permission.*)

Scores of entries in this encyclopedia deal directly or indirectly with various aspects of metallurgy. Check the alphabetical index. In addition to each of the individual metals (aluminum, cadmium, iron, vanadium, etc.) described in separate articles in this encyclopedia, also check the following key words and phrases: alloys; amalgam; annealing; brazing; brittle fracture; calorizing (and other metal treating processes); corrosion; creep; metallography; phase diagram; powder metallurgy; temper; welding; wire drawing; wrought iron; etc.

METAL-MARK (*Insecta, Lepidoptera*). Butterflies of small or moderate size, in many species marked with metallic spots and dashes. They constitute the family *Riodinidae* (*Erycinidae*). Relatively few species occur in temperature climates but in the tropics they are numerous and varied.

METALS (The). In terms of classification, several of the chemical elements are referred to as metals, principally because of the metallic qualities they exhibit. The new group designating system is used here. There are several subclassifications:

Group 11: In order of increasing atomic number, these are copper, gold, and silver. Sometimes, these metals also are referred to as noble metals, principally because they sometimes occur in nature in elemental form. Gold and silver also are frequently referred to as "coinage" metals. The elements of this group are characterized by the presence of one electron in an outer shell. Although copper and gold also have other valences, all of the elements in this group have a 1+ valence in common.

Group 12: In order of increasing atomic number, these are zinc, cadmium, and mercury. The elements of this group are characterized by the presence of two electrons in an outer shell. Although mercury also has a valence of 1+, all of the elements in this group have a 2+ valence in common.

Group 4: In order of increasing atomic number, these are titanium, zirconium, and hafnium. The elements of this group are characterized by the presence of two electrons in an outer shell. Although titanium and zirconium also have other valences, all of the elements in this group have a 4+ valence in common.

Group 5: In order of increasing atomic number, these are vanadium, niobium (sometimes called columbium), and tantalum. Vanadium and tantalum have two electrons in an outer shell; niobium has one electron in its outer shell. Although niobium and vanadium also have other valences, all of the elements in this group have a 5+ valence in common.

Group 6: In order of increasing atomic number, these are chromium, molybdenum, and tungsten. Chromium and molybdenum have one electron in their outer shells; tungsten has two electrons in its outer shell. Although chromium and molybdenum also have other valences, all of the elements in this group have a 6+ valence in common.

Group 7: In order of increasing atomic number, these are manganese, technetium, and rhenium. Manganese and rhenium have two electrons in their outer shells; technetium has one electron in its outer shell. Although manganese and rhenium also have other valences, all of the elements in this group have a 7+ valence in common.

Groups 8, 9, 10: In order of increasing atomic number, these are iron, cobalt, nickel, ruthenium, rhodium, palladium, osmium, iridium, and platinum. Ruthenium, rhodium, and platinum have one electron in their outer shells; iron, osmium, cobalt, and nickel have two electrons in their outer shells; iridium has 17 outer electrons and palladium 18 outer electrons. Although all of these elements fall into one group, they appear in the classification in three subgroupings (hence the sometimes-used term *triads*): (1) iron, cobalt, and nickel each have valences of 2+ and 3+; (2) ruthenium, rhodium, and palladium each have valences of 4+, in addition to other valences; (3) osmium, iridium, and palladium each have valences of 4+, in addition to other valences.

In terms of the periodic classification, all elements here designated as the metals fall between highly alkaline elements (alkali metals and alkaline earths) at the left end of the table and the acidic elements, ending with the halogens at the right end of the table. Thus, the term "transition elements" sometimes is used to describe these in-between elements. Actually, the term transition can be applied to the differences between any series of elements within the overall classification, or between individual elements within a group—because of the gradual alteration in chemical behavior that takes place between groups and between elements.

METAMERE.   A division of the animal body occurring as one of a series along the principal axis. Segments of this type are well developed in the earthworms. As a general rule, each segment of such a body contains similar internal organs but a certain amount of specialization of the segments is evident both internally and externally in different parts of the body.

METAMORPHIC ROCKS.   Metamorphism of rock material, whatever its original nature and origin, may produce such profound changes that the resulting mass has distinctive characters which warrant a new classification. The chief agents of metamorphism are pressure and heat, and circulating liquids and gases, and usually a long time interval is required for their operation. Metamorphic changes affect mineral composition, texture, and structure of rocks. Minerals that compose most rocks are definite chemical combinations, which commonly are stable only under definite conditions. If these minerals are subjected to radically new conditions, they will tend to change slowly into new chemical combinations, stable under the new conditions. For example, bituminous coal that has been subjected to long-continued pressure and increased temperature in a belt of deformation becomes changed into anthracite coal through loss of volatile constituents and concentration of fixed carbon.

METAMORPHOSIS.   Development of the individual after birth or hatching, involving marked change in form as well as growth and differentiation.

Metamorphosis usually accompanies change of habitat or of habits. In some forms, however, it is merely development through a succession of forms, which probably represent ancestral stages in the evolution of the species. The first type is illustrated by many insects and by amphibians. Immature dragon flies are aquatic although the adults are flying insects, and frogs undergo a transition from the aquatic tadpole to the air-breathing, if not entirely *terrestrial*, adult. Change of habits is illustrated by the transformation of free-swimming young of many aquatic invertebrates into sessile adults, and by the development of adult butterflies and moths with sucking mouths from caterpillars which eat solid food. The crustaceans afford many examples of transformation through several immature forms without conspicuous change of habits or of habitat.

The immature stages are wholly or partly designated by the term larva. In the complex metamorphosis of insects, however, only the first stage is called the larva and even it sometimes bears a different name. The distinction depends on the nature of the metamorphosis.

Some insects hatch from the egg with the general form of the adult and the attainment of the adult stage is marked chiefly by the completion of the wings. This type of metamorphosis is said to be gradual and in its early stages the insect is called a nymph. The orders that develop in this way are grouped as the Paurometabola. A few orders are aquatic in early life and are then called naiads. They transform directly into the terrestrial adult and are known as the Hemimetabola, or insects with incomplete metamorphosis. The Holometabola, with complete metamorphosis, pass through a larval stage, then enter an inactive stage known as the pupa, and finally become the conspicuously different adult. A few beetles undergo hypermetamorphosis with a sequence of different larval forms preceding pupation.

METASTABLE NUCLEI.   Nuclei in excited nuclear states that have measurable lifetimes (exceeding 10^{-10}–10^{-9} second).

METASTABLE STATE.   Three common uses of this term denote: (1) A peculiar state of pseudo-equilibrium, in which the system has acquired energy beyond that for its most stable state, yet has not been rendered unstable. Thus, by using great care, water at 760 millimeters pressure may be heated several degrees above its normal boiling point, say to $105\,°C$, yet not boil. In this condition it has received heat energy beyond that normally required for liquid-vapor equilibrium, energy which it might be expected to release by spontaneously exploding into steam; and only a slight disturbance will precipitate that change, but the disturbance must come from some external source.

(2) The term has been used in atomic physics for various excited states, but its most general usage today is for an excited state from which all possible quantum transitions to lower states are forbidden transitions by the appropriate selection rules.

(3) In nuclear physics, the term is used to denote the states in which metastable nuclei are found.

METASTABLE SYSTEM.   A system that has a measurable lifetime in an energy state that is not its lowest.

METASTASIS.   See **Cancer and Oncology**.

METEOROIDS AND METEORITES.   A *meteroid* is one of countless small, solid particles existing in interplanetary space, which, when encountering the Earth's atmosphere, burn up. During the short instant of ablation, a brilliant track of light often can be observed, commonly called "falling star," "shooting star," or "fireball." When ablation is accompanied by an explosion, the meteoroid may be called a bolide. Sometimes a loud detonation following the explosion may be heard. Considering the large number of meteoroids entering the Earth's atmosphere, the *bolides* are rare. The brilliance of a meteoroid trail normally does not exceed that of zero stellar magnitude, i.e., no brighter than the planet Jupiter. Most debris from meteoroids ultimately falls to the Earth as dust. Sometimes meteoroids enter the Earth's atmosphere in swarms at given times each year when the Earth is passing an area of space where there is a high population of meteoroids. An example is the Leonid Meteor Show, which occurs in November each year. See also **Leonids**.

A *meteorite* is that portion of a meteoroid that survives passage through the Earth's atmosphere and lands on or penetrates the surface of the Earth. Many thousands of meteorites have been identified over several decades. These represent but a small portion of the total number estimated to have struck the Earth. Very little research has been conducted to determine at what average rate meteorites may strike the Earth.[1] Meteorites may strike the earth as isolated instances, or they may fall in swarms. Evidence of several very large swarms has been recorded. The Tungusta fall in Russia in 1908 leveled trees within a radius of nearly 48 kilometers (30 miles). In 1947, also in Russia, the Sikhote-Aline fall caused heavy damage. In the Holbrook, Arizona area, a meteorite shower (1912) produced more than 20,000 stones, many of which were recovered. Large meteorites can produce immense craters, as exemplified by the large Barringer meteorite crater in Arizona. See Fig. 1.

Information returned from planet-orbiting satellites has shown extensive cratering on Mercury, Mars, and several other planets and their moons. Telescopic observation of craters on the Earth's moon were first made a few centuries ago. The comparative absence of large craters on earth is most likely due to the earth's atmosphere, which causes meteoroids to "burn up" before they can impact the earth's crust, but other factors also contribute. Among these factors is the location of the earth within the solar system, which may be unfavorable insofar as the potential for impacts is concerned. The geological and tectonic history of the earth may have contributed to covering over or otherwise obliterating the existence of craters caused during the earlier periods of the developing earth. See also **Astrobleme**.

Fig. 1. Barringer meteorite crater located near Canyon Diablo in northern Arizona. It is a classic example of a huge meteor crater, approximately three-fourths of a mile (1.2 km) in diameter and more than 600 feet (183 m) deep. (*Meteor Crater Society.*)

Meteorites found on earth range in size from that of a pea (and probably smaller, considering the difficulty in noting them) up to tons. The Hoba West meteorite, discovered in South Africa, weighs 70 tons (63 metric tons) and measures about 3×3 meters (10×10 feet). The ahnighito meteorite, one of the largest, was found in Greenland and brought to the American Museum of Natural History (New York) by Admiral Peary many years ago. A large meteorite weighing about 1050 kilograms (2300 pounds) fell on Kansas in 1948. In addition the large meteorite crater in northern Arizona, previously mentioned, the Ries Basin in Germany has a notable meteorite crater, originally measuring about 27 kilometers (17 miles) in diameter, and estimated to have been formed between 15 and 20 million years ago.

[1] In 1984, Halliday, Blackwell, and Griffin (Herzberg Institute of Astrophysics, National Research Council of Canada) conducted a study involving a network of 60 cameras in western Canada to determine the influx rate of meteorites. By way of calculations too complex to describe here, the researchers derived the following figures: 19,000 impacts per year of meteorites having a mass of about 0.1 kg; 4100 impacts (1 kg); and 830 impacts (10 kg). See reference listed.

There is no really characteristic shape to meteorites except that many look like stones and could be casually observed without comment. Many fragment during passage through the atmosphere, producing a variety of shapes. Some meteorites exhibit intense sculpturing caused by the intense heat when the object passes through the earth's atmosphere.

Many larger and much older craters have been located in recent years. In particular, the Chubb Crater in northern Canada is an example of an old, filled-in, meteor impact crater.

The 2-ton (1.8-metric-ton) Allende meteorite, which fell in northern Mexico in 1969, has been of particular fascination to scientists. From these studies, several concepts pertaining to the origin of the solar system have evolved. Central to most of these studies of the Allende and other meteorites is the determination of isotopic ratios of various elements, such as oxygen-16 to oxygen-18; of magnesium-26 to magnesium-25 to magnesium-24.

It has been estimated that meteorites impact the earth's surface with a velocity ranging from about 11 kilometers (7 miles) per second up to a theoretical velocity (maximum if object arrives from within the solar system) of nearly 72 kilometers (45 miles) per second. It is believed that the average velocity, however, is about 16 kilometers (10 miles) per second.

When Antarctic exploration revealed the presence of numerous meteorites, this broadened the database for scientists who are pursuing this special discipline. A connection with oceanographic exploration may yield additional rich evidence of meteor impacts and craters. Such studies, of course, also interlock with the scientific exploration of past mass extinctions on Earth.

In 1990, researchers with the Geological Survey of Canada serendipitously located an exceptionally large crater buried beneath the waters of Lake Huron at a point that straddles the Canadian–United States border. The researchers captured an excellent magnetic image of what may be largest meteorite crater on record. It has been estimated that a hole the size of the "Can-Am" would have required a projectile about 3 miles (4.5 kilometers) in diameter. In observing this discovery, J. Melosh (University of Arizona) commented, "We would expect 100 craters this size or greater during the last 500 million years. We know of about six." Melosh further observed that impact data to date have been too scattered to prove any correlation between the collisions and species extinctions. In summary, Melosh observed, "With more craters like Can-Am to study, we can test those correlations."

Over a period of several years, some scientists have expected to find a "killer" crater near Cuba. A number of geologists visited the suspected southwestern corner of Cuba and reported at the annual meeting of the Geological Society of America that investigators may best conduct their research elsewhere. However, some scientists remain unconvinced that a massive crater does not exist in the Caribbean region.

Classes of Meteorites

There are three broad classes of meteorites that strike the Earth:[2]

1. *Iron meteorites* (sometimes called siderites) account for about 6% of the meteorites found. These objects are essentially a nickel-iron alloy containing from 4% to 20% Ni (rarely greater), together with several accessory minerals, of which troilite (FeS), schreiberside (Fe, Ni, Co)$_3$P, and graphites are important. Iron meteorites tend to weather more rapidly than other types once they impact the earth. The cross section of an iron meteorite exhibits a thin layer of vitreous material, caused by fusing of the substance during entry, which may be up to about 2 millimeters ($\frac{1}{16}$ inch) in thickness. Iron meteorites, when viewed sectionally and particularly after they have been ground flat, polished, and etched, will show intersecting bands which are known as Widmanstatten lines. The Widmanstatten structure has been defined as a triangular pattern of iron meteorites composed of parallel bands or plates of kamacite bordered by taenite and intersecting one another in two, three, or four directions. The kamacite bands, arranged parallel to the octahedral planes in the host taenite, are produced by exsolution from an originally homogeneous taenite crystal. As the bands become finer (thinner), the nickel content increases. The structure is named after A.B. Widmanstatten, the Austrian mineralogist who discovered the structure in 1808.

[2] Some of the definition material is from the "Glossary of Geology," American Geological Institute, Washington, DC (by permission).

2. *Stony-iron meteorites* (sometimes called siderolites) account for 1% to 2% of the meteorites found. These objects contain at least 25% and approximately equal amounts (by weight) of nickel-iron and heavy basic silicates, such as pyroxene and olivine. Geological and geochemical analysis indicates that these meteorites formed inside parent bodies resembling asteroids, typically some hundreds of kilometers in diameter, about 4.5 billion years ago. Some scientists consider them to be fragments of these bodies broken apart by collisions among themselves during the formation of the solar system. Further evidence indicates that they are related to the asteroids observed today. Studies show that their spectra resemble those of certain asteroids, especially the Apollos, whose orbits cross that of the earth and theoretically could at some time intersect the Earth. Orbits observed for two or three meteorites resemble Apollo-type orbits and reach into or across the asteroid belt, suggesting that some of these bodies may have somehow been deflected out of the belt, eventually falling to Earth. See also **Asteroid**.

3. *Stony meteorites* constitute over 90% of the meteorites known to strike Earth. These objects consist largely or entirely of silicate minerals, chiefly olivine, pyroxene, and plagioclase. Stony meteorites are sometimes called *chondrites*. See also Chondrite. Most of the meteorite material has been modified from primitive chemical forms by geological processes, such as heating, exposure to pressure and radiation; some have been entirely melted and resolidified. Approximately 6% of the stony meteorites are *carbonaceous chondrites*. They are characterized by the presence of hydrated, clay-type silicate minerals (usually fine-grained serpentine or chlorite); by considerable amounts and a great variety of organic compounds (hydrocarbons, fatty and aromatic acids, porphyrins, etc.) believed to be of extraterrestrial origin; by an absence or almost total absence of free nickel-iron; and by abnormally high contents of inert gases, particularly xenon. Much of the organic matter is a black insoluble complex of compounds of high molecular weight; the water content (usually water of hydration)may be up to 20% (weight). Carbonaceous chondrites are grouped in three types, each characterized by the amount of organic material they contain and by other compositional features:

Type I	contain the greatest amount of water and organic matter (3% of 5% combined carbon; 24% to 30% ignition loss)
Type II	are of an intermediate composition (12% to 24% ignition loss)
Type III	contain high-temperature minerals and some metallic components (2% to 12% ignition loss)

Possibly such objects may represent surviving material from the matter that formed in the ancient solar system and aggregated into planets.

Nearly all meteorites contain small amounts of *troilite* (FeS). It is a variety of pyrrhotite with almost no ferrous-iron deficiency.

Diamond in Meteorites. S.S. Russell (Oxford University) and C.T. Pillinger reported in 1992 that the presence of diamond in meteorites has been known for at least a century. Scientists have theorized that the mechanism for diamond formation in meteorites probably does not compare with the manner in which diamond was formed on Earth, but that meteoritic diamond probably is associated with *shock processing* — that is, upon impact. Analysis of material from the iron meteorite crater in Canyon Diablo, Arizona, tends to confirm this hypothesis.

In contrast, however, diamond polymorphs found in an Antarctic iron meteorite shows no evidence of terrestrial shock. This, then, would suggest that the meteoritic diamond had been present prior to impact. An alternative mechanism, chemical vapor deposition (CVD), also has been suggested. In studying a particular meteorite (Abbe, an enstatite chondrite), Russell and co-researchers observe, "Because the Abee diamonds have typical solar system isotopic compositions for carbon, nitrogen, and xenon, they are presumably nebular in origin rather than presolar. Their discovery in an unshocked meteorite eliminates the possibility of origins normally invoked to account for diamonds in ureilites and iron meteorites and suggests a low-pressure synthesis. The diamond crystals are "100 nanometers in size, of an unusual lath shape, and represent "100 ppm of Abee by mass."

Analysis of Meteoritic Materials. As pointed out in 1989 by R. Zenobi (Stanford University) and a research team, "Meteorites are highly hetero-genmous mineralogically and chemically, even on a submillimeter scale. Techniques for elemental analysis are well established for the investigation of microgram amounts of solids and the investigation of surfaces with micrometer-scale resolution. In contrast, analysis of organic molecules has been impossible even in a few milligrams of carbonaceous chondrites, the meteorite group with the highest carbon content." Reasons stated for

this difficulty in analysis include: (1) very low concentrations of organic materials (ppm range) and (2) the large number of different organic compounds found in meteorites. Usually, multiple extraction and purification must precede analysis. The variety of organic compounds is exemplified by naphthalene, 2-methylnapthalene, 2,3-dimethylnaphthalene, phenanthrene, anthracene, 2-methylanthracene, 9-methylantracene, 2-tertbutylacene, fluorene, fluoranthene, m poyrene, perylene, and coronene.

In analyzing the Allende meteorite, the aforementioned team, the technique used was that of two-step laser desorption spectrometry. The analysis of each meteorite, of course, presents special problems, and thus numerous meteorite analytical techniques have been devised and customized for the problem.

SNC Meteorites. In addition to the three main classes of meteorites just described, a relatively new group (very small in number at present) is attracting considerable scientific attention, mainly because they are considered of lunar or Martian origin. Currently attention is being given to about eight-odd rocks that have been collected around the world. Three of these rocks are known to be meteorites and they had been named Shergotty, Nakhla, and Chassigny. Thus, the acronym is SNC.

These meteorites are youthful geologically speaking (1.3 billion years old), whereas 4.0 billion years is considered a minimum age for other meteorites. Thus far, the age determination tends to disqualify them as coming from an asteroid, although not all experts agree. The SNC meteorites resemble certain basaltic rocks found on Earth, also tending to disqualify asteroids as a source. Further, their features, including their oxygen isotope composition, indicate that indeed they are meteorites and not Earth rocks.

In 1992, H.R. Karlsson (NASA) and a team of researchers reported that recent evidence suggests that the planet Mars once had a water-rich atmosphere and flowing water on its surface. The paper observes, "The SNC meteorites, purportedly of Martian origin, contain 0.04 to 0.4 percent water by weight. Oxygen isotopic analysis can be used to determine whether this water is extraterrestrial or terrestrial. Such analysis reveals that a portion of the water is extraterrestrial and furthermore was not in oxygen isotopic equilibrium with the host rock. Lack of equilibrium between water and host rock implies that the lithosphere and hydrosphere of the SNC parent body formed two distinct oxygen isotopic reservoirs. If Mars was the parent body, the maintenance of two distinct reservoirs may result from the absence of plate tectonics on the planet." See also **Mars**.

Antarctic Meteorites. A meteorite, now called Allan Hills 81005 (only 31 grams and about 3 cm across), was found in the Antarctic region. From studies of lunar rocks returned by the *Apollo* mission, experts felt from the start that the meteorite was of lunar origin even though the concept of lunar meteorites had not been popular for a number of years. When the meteorite's oxygen isotope composition was checked, it was found to be like that of moon rocks and unlike most meteorites. But, upon further analyses and hypothesizing, lunar origin was questioned. One problem was that of timing. The age of the meteorite is about 2 billion years more recent than when the moon was volcanically active. Further, there were suspicions about certain characteristics which indicated that the object had to come from a body of sufficient size to permit the kind of melting and crystallization processes that produced Earth's basalt lavas.

J.T. Wasson (University of California, Los Angeles) reported in 1990, "Eighty-five percent of the iron meteorites collected *outside* Antarctica are assigned to 13 compositionally and structurally defined groups; the remaining 15 percent are ungrouped. Of the 31 iron meteorites recovered from Antarctica, 39 percent are ungrouped. This major difference in the two sets is almost certainly not a stochastic variation, a latitudinal effect, or an effect associated with differences in terrestrial ages." Wasson suggests that, during impacts on asteroids, smaller fragments tend to be ejected into space at higher velocities than larger fragments, and thus, on the average, small meteroids undergo more changes in orbital velocity than the larger ones. As a result, the set of asteroids that contributes small meteoroids to Earth-crossing orbits is larger than the set that contributes larger meteroids. It is suggested that most small iron meteorites may escape from the asteroid belt as a result of impact-induced changes in velocity that reduce their perihelia to values less than the aphelion of Mars.

A.J.T. Jull (University of Arizona) and a team of researchers reported in 1988 that weathering products have been observed on many Antarctic meteorites, but their consequences for cosmochemical analysis have remained poorly understood. The time of formation has not been quantitatively evaluated. The research team observed, "Weathering products are typically in the form of 'rust' or hydrated iron oxides and hydroxides; however, sulfate and carbonates have also been observed. Most workers have presumed that weathering proceeds slowly in the cold, dry Antarctic environment. Most Antarctic meteorites have been exposed to terrestrial conditions for 10^4 to 10^5 years and small increments of weathering can accumulate into measurable effects. Many specimens, however, are remarkably well preserved."

The Jull team reported, in particular, on studies of the Antarctic meteorite LEW 85320 (H5 chondrite). The research paper concludes that "Results from carbon-14 dating suggest that, although the meteorite has been in Antarctica for at least 3.2×10^4 to 3.3×10^4 years, nesquehonite[3] formed after A.D. 1950."

Meteoric Communications. It has been estimated that tiny meteors enter Earth's atmosphere, producing an ionized residue at the rate of over a billion events per day. The practical use of this phenomenon as a communications medium for bounding very high frequency radio signals has been under investigation during the past 15–20 years. A number of firms are commercializing the method, which much has appeal for trucking firms. The system competes directly with satellite communications for such purposes. The Conservation Service of the U.S. Department of Agriculture has used meteoric communications for monitoring snowpack and other weather data collected by instruments in remote areas. The system also has been used in connection with military communications.

Ionospheric Interference. In late 1989, P. Kaufmann (Universidade de Sao Paulo) reported on the findings of a study of the effects that the large June 1975 meteoroid storm had on communications worldwide. A brief summary of the report: "The June 1975 meteoroid storm detected on the moon by the Apollo seismometers was the largest ever observed. Reexamination of radio data taken at that time showed that the storm also produced pronounced disturbances on Earth, which were recorded as unique phase anomalies on very low frequency (VLF) radio propagation paths in the low terrestrial ionosphere. Persistent effects were observed for the major storm period (20–30 June 1975), including reductions in the diurnal phase variation, advances in the nighttime and daytime phase levels, and reductions in the sunset phase delay rate. Large nighttime phase advances, lasting a few hours, were detected on some days at all VLF transmission, and for the shorter propagation path, they were comparable to solar Lyman alpha daytime ionization. Ion production rates attributable to the meteor storm were estimated to be about 0.6 to 3.0 ions per centimeter cubed per second at the E and D regions, respectively. The storm was a sporadic one with a radiant (that is, the point of apparent origin in the sky) located in the Southern Hemisphere, with a right ascension 1 to 2 hours larger than the sun's right ascension."

K. Brecher (Boston University) and J. Crouchley (University of Queensland, Australia) corroborated in the study.

Additional Reading

Benoit, P.H. and D.W.G. Sears: "The Breakup of a Meteorite Parent Body and the Delivery of Meteorites to Earth," *Science*, 1685 (March 27, 1992).

Eugster, O.: "History of Meteorites from the Moon Collected in Antarctica," *Science*, 1197 (September 15, 1989).

Flam, F.: "Seeing Stars in a Handful of Dust," *Science*, 580 (July 26, 1991).

Halliday, I. et al.: "The Frequency of Meteorite Falls on the Earth," *Science*, **223**, 1405–1407 (1984).

Hodge, P.: "Meteorite Craters and Impact Structure," Cambridge University Press, New York, NY, 1994.

Hoon, J. et al.: "Application of Two-Step Laser Mass Spectrometry to Cosmogeo-chemistry: Direct Analysis of Meteorites," *Science*, 1523 (March 25, 1988).

Horan, M.F. et al.: "Rhenium-Osmium Isotope Constraints on the Age of Iron Meteorites," *Science*, 1118 (February 28, 1992).

Hutchinson, R. et al.: "Catalogue of Meteorites," Cambridge University Press, New York, NY, 1999.

Jull, A.J. et al.: "Rapid Growth of Magnesium-Carbonate Weathering Products in a Stony Meteorite from Antarctica," *Science*, 417 (October 21, 1988).

Jurewicz, A.J.G., D.W. Mittlefehldt, and J.H. Jones: "Partial Melting of the Allende (CV3) Meteorite: Implications for Origins of Basaltic Meteorites," *Science*, 695 (May 3, 1991).

Karlsson, H.R. et al.: "Water in SNC Meteorites: Evidence for a Martian Hydro-sphere," *Science*, 1409 (March 13, 1992).

Kaufmann, P. et al.: "Effects of the Large June 1975 Meteoroid Storm on Earth's Ionosphere," *Science*, 787 (November 11, 1989).

---

[3] A hydrogen magnesium carbonate that occurs as a weathering product on the surface of Antarctic meteorite LEW 85320.

Lowe, D.R. et al.: "Geological and Geochemical Record of 3400-Million-Year-Old Terrestrial Meteorite Impacts," *Science*, 959 (September 1, 1989).

Margolis, S.V., P. Claeys, and F.T. Kyte: "Microtektites, Microkrystites, and Spinels from a Late Pliocene Asteroid Impact in the Southern Ocean," *Science*, 1594 (March 29, 1991).

Mark, K.: "Meteorite Craters," University of Arizona Press, Phoenix, AZ, 1995.

McSween, H.Y. Jr.: "Meteorites and Their Parent Planets," Cambridge University Press, New York, NY, 1987.

Melosh, H.J.: "Impact Cratering: A Geologic Process," Oxford University Press, Inc., New York, NY, 1996.

Russell, S.S., J.W. Arden, and C.T. Pillinger: "Evidence for Multiple Sources of Diamond from Primitive Chondrites," *Science*, 1188 (November 22, 1991).

Russell, S.S. et al.: "A New Type of Meteoritic Diamond in the Enstatite Chondrite Abee," *Science*, 206 (April 10, 1992).

Stix, G.: "Meteoric Messages," *Sci. Amer.*, 167 (September 1990).

Stolzenburg, W.: "Impact Crater May Lie Beneath Lake Huron," *Science News*, 133 (September 1, 1990).

Walker, R.J. and J.W. Morgan: "Rhenium-Osmium Isotope Systematics of Carbona-ceous Chondrites," *Science*, 519 (January 27, 1989).

Wasson, J.T.: "Ungrouped Iron Meteorites in Antarctica: Origin of Anomalously High Abundance," *Science*, 900 (August 24, 1990).

Weaver, K.F.: "Meteorites," *National Geographic*, 390 (September 1986).

Zenobi, R. et al.: "Spatially Resolved Organic Analysis of the Allende Meteorite," *Science*, 1026 (November 24, 1989).

Web Reference

Ames research Center, National Aeronautics and Space Agency, United States. *http://www.arc.nasa.gov/index.html*

METEOROLOGY. The study dealing with the phenomena of the atmosphere. This includes not only the physics, chemistry, and dynamics of the atmosphere, but is extended to include many of the direct effects of the atmosphere upon the earth's surface, the oceans, and life in general. The goals often ascribed to meteorology are the complete understanding, accurate prediction, and artificial control of atmospheric phenomena. In a more restricted sense, meteorology is the science of weather, being particularly concerned with the physics of the elements that make the weather. A distinction can be drawn between meteorology, in this sense, and climatology, the latter being primarily concerned with average, not actual, weather conditions.

Meteorology may be subdivided, according to the methods of approach and the applications to human activities, into a large number of specialized sciences. For example, the aeronautical, agricultural and industrial branches of meteorology are concerned with the application of meteorological information and techniques within their respective fields. *Hydrometeorology* is the branch directly concerned with hydrologic problems, particularly flood control, hydroelectric power, irrigation and water resources. *Radio meteorology* (also known as *radioelectric meteorology*) is concerned with the propagation of radio energy through the atmosphere and with the use of radio equipment in meteorological studies; while radar meteorology embraces all meteorological matters connected with radar.

There are specializations in terms of atmospheric regions to which study is devoted. *Aerology* is the study of the free atmosphere throughout its vertical extent, as distinguished from studies confined to the layer of the atmosphere adjacent to the earth's surface. *Aeronomy* is a recently introduced term denoting, basically, the physics of the upper atmosphere. It is concerned with upper-atmospheric composition (i.e., nature of constituents, density, temperature, etc.) and chemical reactions. Studies of atmospheric phenomena are designated from the largest-scale aspects to the smallest, respectively, as *macrometeorology, mesometeorology,* and *micrometeorology.*

Still other specializations include: *synoptic meteorolgy,* the study and analysis of weather information obtained from synoptic reports, charts, and weather observations; *applied meteorology,* the application of current weather data, analyses, and/or forecasts to specific practical problems; *dynamic meteorology,* the study of atmospheric motions as solutions of the fundamental equations of hydrodynamics or other systems of equations appropriate to special situations, as in the statistical theory of turbulence; and *physical meteorology* (also called *atmospheric physics*), dealing with optical, electrical, acoustical, and thermodynamic phenomena of the atmosphere, its chemical composition, the laws of radiation, and the explanation of clouds and precipitation. Mathematical theory of the motions of the atmosphere and the forces responsible falls in the field of dynamic meteorology. Atmospheric thermodynamics lies so near the borderline of physical and dynamical meteorology that it is treated as often in one as in the other branch.

There are numerous entries in this encyclopedia relating directly to meteorology. Check the alphabetical index. Particularly check such topical terms as: air mass; altimetry; atmosphere (Earth); atmosphere-ocean interface; aurora; barometer; climate; clouds; fog; fronts and storms; gust front; jet streams; polar front theory; precipitation and hydrometeors; psychrometric chart; sky cover; visibility; weather; winds; and wind shear. See also **Bjerknes, Vilhelm Frimann Koren (1862–1951)**.

PETER E. KRAGHT, Certified Consulting Meteorologist, Mabank, TX.

METEOR SHOWER.   A term applied to indicate a number of meteoroids coming from the same general part of the sky known as a radiant point. It has been theorized that these large concentrations of falling meteoroids are the remains of disintegrated comets. According to this theory, the earth's orbit may intersect the orbits of any one of several comets, and, due to the attraction of earth's gravity, the meteoroid particles would be attracted and fall into the atmosphere.

Multitudes of these meteoroid particles revolve together around the sun; if clustered together, they are known as "meteor swarms"; if more evenly distributed throughout the orbit, they are called "meteor streams." The orbits of meteor swarms or streams and the orbit of the earth may intersect; and it is then that a meteor shower may be observed from earth, the intensity and duration of which depends upon the distribution of meteoroids within the intersecting portion of the orbit. Some meteor showers occur annually (or biannually), and some periodically but at longer intervals. The showers are commonly named after the constellations or stars in which their radiant points occur. Thus, the most conspicuous and dependable annual display, with trails visible for 2 or 3 weeks, is known as the Perseids (after the constellation Perseus). The Orionids (after the constellation Orion) and the Geminids (after the constellation Gemini) also occur annually. The Draconids, on the other hand, occur only every 13 years, and the Leonids at 33-year intervals. Because meteor showers so often occur in the orbits of comets, they are commonly named after their associated comet as well as after a star or constellation.

Certain meteor showers observed in the daytime (and sometimes only then) have been observed in the course of systematic surveys of meteor activity by means of radar echoes.

See also **Leonids**; and **Perseids**.

METHANE. CH_4, formula weight 16.04, colorless, odorless (when pure) gas, mp $-182.6\,°C$, bp $-161.4\,°C$, sp gr 0.415 (at $-164\,°C$). Sometimes referred to as *marsh gas* or *fire damp*, methane is practically insoluble in H_2O, and moderately soluble in alcohol or ether. The gas burns when ignited in air with a pale, faintly luminous flame, forming an explosive mixture with air between gas concentrations of 5% and 13%. Methane is the principal constituent of natural gas, averaging 75% by weight. Natural gas from the Pennsylvania fields is almost 99% methane, but some gas from Kentucky fields contains as little as 23% methane. Pipeline gas from several fields typically will contain about 78% methane, 13% ethane, 6% propane, 1.7% butane, and 0.6% pentane. The remaining fraction consists of gases higher in the alkyl series. While generally not referred to as such, methane can be classified as a major fuel. The heating value of pure methane is 995Btu/ft^3 (8856 Calories per cubic meter).

Methane, as the major constituent of natural gas, is an extremely important raw material for numerous synthetic products. For most processes, it is not required to isolate and purify the methane, but the natural gas as received may be used. The high percentage of CH_4 in various feedstocks makes possible the formation of synthesis gas: $CH_4 + H_2O \rightarrow CO + 3H_2$. The percentages of CO and H_2 in synthesis gas vary depending on the end product to be made. Synthesis gas is used widely in the manufacture of NH_3, oxo-chemicals, and methyl alcohol. See also **Synthesis Gas**.

In addition to the preparation of synthesis gas, which is used so widely in various organic syntheses, methane is reacted with NH_3 in the presence of a platinum catalyst at a temperature of about $1,250\,°C$ to form hydrogen cyanide: $CH_4 + NH_3 \rightarrow HCN + 3H_2$. Methane also is used in the production of olefins on a large scale. In a controlled-oxidation process, methane is used as a raw material in the production of acetylene.

Most artificial gases, such as producer gas, coal gas, water gas, manufactured gas, and town gas contain a high content of methane. In addition to its use as a basic chemical and fuel, methane is of notable interest because of its role as the anchor compound of the *alkanes* (paraffin or aliphatic hydrocarbons). All of these compounds may be considered derivatives of methane.

Carbon monoxide and hydrogen react to form CH_4 in the presence of a nickel catalyst. Methane also is formed by reaction of magnesium methyl iodide in anhydrous ether (Grignard's reagent) with substances containing the hydroxyl group. Methyl iodide (bromide, chloride) is preferably made by reaction of methyl alcohol and phosphorus iodide (bromide, chloride).

Additional Reading

Clever, H. et al.: "Methane," Elsevier Science, New York, NY, 1987.
Lee, S.: "Methane and Its Derivatives," Marcel Dekker, Inc., New York, NY, 1996.
Mastalerz, M. et al.: "Coalbed Methane: Scientific, Environmental, and Economic Evaluation," Kluwer Academic Publishers, Norwell, MI, 1999.

METHANOGENS.   Cells that resemble bacteria in a superficial way, but that have unique genetic and metabolic characteristics. Methanogens are anaerobic, methane-producing microorganisms that occur in a wide variety of places—the gastrointestinal tract of animals, including humans, in the sediments of natural waters, in sewage treatment plant vessels and piping, and in natural hot springs. As proposed by Woese (University of Illinois) and Fox (University of Houston), the methanogens probably make up a third line of descent of cells in addition to the prokaryotes (bacteria and blue-green algae cells which do not have a well-defined nucleus) and eukaryotes (more complex cells with a nucleus). See also **Cell (Biology)**. These researchers also have suggested that there may be still other kinds of cells that do not meet the criteria set down for prokaryotes and eukaryotes.

Methanogens are distinguished from bacteria on at least three counts:

(1) The cell walls do not contain muramic acid, the characteristic constituent of the peptidoglycans that form bacterial cell walls. (2) Their metabolism differs markedly from bacteria. A number of coenzymes apparently unique to methanogens have been identified. Some of these enzymes are involved in methyl transfer reactions, including the formation of methane. One of the coenzymes is possibly the smallest coenzyme yet to be discovered. The methanogens also differ in the manner in which carbon dioxide is fixed into cellular carbon. However, the pathway has not been clearly identified. (3) The RNA sequences of methanogens differ from those of other organisms. These observations have indicated to Woese and Fox that although the methanogens share a common ancestor with prokaryotes and eukaryotes, an independent line of descent branched off at possibly about the same time the other cell types diverged.

Although not fully understood, the methanogens place new challenges to the evolutionary biologists for further explanation in terms of the development of early life on earth and may be very valuable toward understanding life on extraterrestrial bodies as these may be explored over future years.

Barker (University of California at Berkeley) and Huntgate (University of California at Davis) as early as the mid-1950s noted that methanogens differ radically from bacteria.

Methanogens take part in the terminal stages of organic matter degradation and survive on carbon dioxide and hydrogen yielded by anaerobic bacteria and converting them to methane.

METHANOL.   See **Methyl Alcohol**.

METHOD OF COMPONENTS.   A trigonometrical procedure for adding forces in which the components of each force along a chosen set of orthogonal coordinates axes generally symbolized by x, y, and z are determined. The components along the x, y, and z axes are then added up separately to give the components of the resultant force. The magnitude and direction of the resultant force can then be determined from its components. These procedures can be used to add any vector quantities.

METHYL ALCOHOL. CH_3OH, formula weight 32.04, colorless, mobile liquid with mild characteristic odor, mp $-97.6\,°C$, bp $64.6\,°C$, sp gr 0.792. Also known as *methanol*, the compound is miscible in all proportions with H_2O, ethyl alcohol, or ether. When ignited, methyl alcohol burns in air with a pale blue, transparent flame, producing H_2O and CO_2. The vapor forms an explosive mixture with air. The upper explosive limit (% by volume in air) is 36.5 and the lower limit is 6.0.

Methyl alcohol possesses distinct narcotic properties. It is also a slight irritant to the mucous membranes. The principal toxic effect is exerted on the nervous system, particularly the optic nerves and possibly the retinae. The effect upon the eyes has been attributed to optic neuritis, which subsides, but is followed by atrophy of the optic nerve. Once absorbed, methyl alcohol is only very slowly eliminated. Coma resulting from massive exposures may last as long as 2 to 4 days. In the body, the products formed by its oxidation are formaldehyde and formic acid, both of which are toxic.

Chemical Properties. Methyl alcohol is a versatile material, reacting (1) with sodium metal, forming sodium methylate, sodium methoxide CH_3ONa plus hydrogen gas, (2) with phosphorus chloride, bromide, iodide, forming methyl chloride, bromide, iodide, respectively, (3) with H_2SO_4 concentrated, forming dimethyl ether $(CH_3)_2O$, (4) with organic acids, warmed in the presence of H_2SO_4, forming esters, e.g., methyl acetate CH_3COOCH_3, methyl salicylate $C_6H_4(OH) \cdot COOCH_3$, possessing characteristic odors, (5) with magnesium methyl iodide in anhydrous ether (Grignard's solution), forming methane as in the case of primary alcohols, (6) with calcium chloride, forming a solid addition compound $4CH_3OH \cdot CaCl_2$, which is decomposed by H_2O, (7) with oxygen, in the presence of heated smooth copper or silver forming formaldehyde. The density of pure methyl alcohol is 0.792 at 20 °C compared with H_2O at 4 °C (the corresponding figure for ethyl alcohol is 0.789), and the percentage of methyl alcohol present in a methyl alcohol-water solution may be determined from the density of the sample.

A common test for methyl alcohol is by its oxidation in air with a hot copper wire to form formaldehyde.

At one time, most methyl alcohol was obtained by the destruction distillation of hardwoods (hence the name *wood alcohol*) at about 350 °C, along with a yield of acetic acid and small percentages of acetone in the water condensate. Interest in returning to wood as a source has revived because of fossil fuel shortages.

Production of Methyl Alcohol.[1] Synthetic methanol is one of the major raw materials of the organic chemical industry. Methanol has economic stability and a steady growth rate owing to the low costs of production and diversity of applications. Nearly all the methanol producers also make formaldehyde, which is the main end use (more than 50%) of methanol. The other main end uses are dimethyl terephthalate, methacrylates, methylamines (for resins, herbicides, and fungicides), methyl halides (for silicones, tetramethyl lead, butyl rubbers, paint removers, photographic

films, aerosol propellants, (diminishing use), and degreasing compounds), acetic acid, and solvents.

An important process for production of synthetic protein uses methanol as feedstock. The use of methanol as a fuel, either as pure methanol, as a mixture (approximately 15%) with gasoline, or as a feedstock for synthetic gasoline is envisaged for possible large-scale application; as well as use in gas turbines for electricity generation. See also **Gasahol**.

There are three principal commercial grades of methanol (as defined in U.S. Federal Specification O-M-232f: June 5, 1975): *Grade A*, synthetic, 99.85% by weight (solvent use); *Grade AA*, synthetic, 99.85% by weight (hydrogen and carbon dioxide generation use); and *Grade C*, wood alcohol (denaturing use).

The most recent advances in methanol synthesis are the low- and intermediate-pressure processes of the type shown in Fig. 1. The synthesis step of this process[2] relies upon a copper-based catalyst, which gives good yields of methanol at pressures of 50 and 100 atmospheres. These pressures are substantially below those of the 250–350 atmospheres required by earlier processes. The high catalyst activity allows the synthesis reaction to take place at a relatively low temperature of 250–270 °C. As a result, methanation is avoided, and byproduct formation is lower, giving increased process efficiency.

The development of this low-pressure technology has caused a major reassessment of the economics of methanol production. The energy required to compress the synthesis gas from its production pressure to the synthesis unit is reduced by a factor between 2 and 3. The lower synthesis pressure allows the exclusive use of centrifugal compressors in plants with capacities as low as 15 million gallons (0.57 million hectoliters) per year. Small producers find attractive the savings in investment, operating, and maintenance costs made possible by low-pressure operation. Plants range in capacity from 15 million gallons (0.57 million hectoliters) to 250 million gallons (9.46 million hectoliters) per year.

Synthesis gas is prepared by the steam reforming or partial oxidation of a liquid or gaseous hydrocarbon feedstock, or by direct combination of carbon dioxide with purified hydrogen-rich gases. Economic considerations usually favor the steam-reforming route for a naphtha or natural gas feedstock. In this instance, desulfurized feedstock is preheated, mixed

[1] Remainder of this entry prepared by J.R. Masson, Process Engineering Consultant, Davy McKee (Oil & Chemicals) Ltd., London, England.

[2] Developed by Imperial Chemical Industries, Ltd.

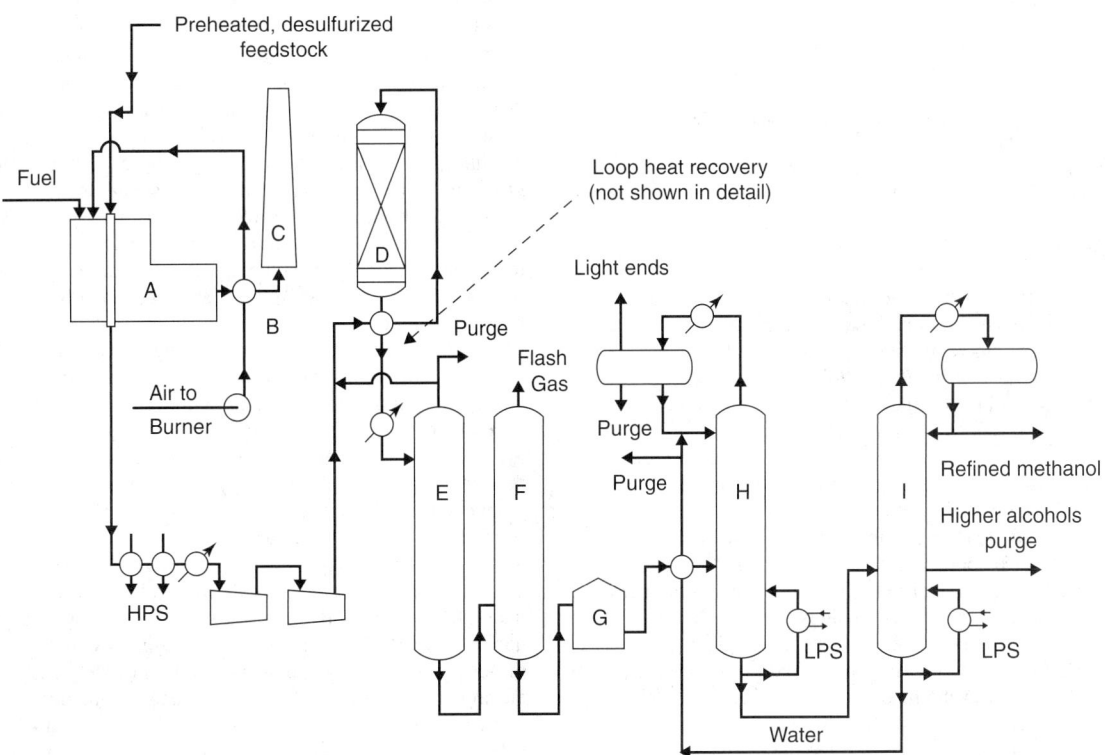

Fig. 1. Low-pressure methanol production. (A) Burner and superheater; (B) air preheater; (C) stack; (D) methanol converter; (E) separator; (F) flash vessel; (G) crude storage; (H) topping column; (I) refining column. HPS = high-pressure steam; LPS = low-pressure steam. (*Imperial Chemical Industries, Ltd.*)

with superheated steam, and reacted over a conventional catalyst (normally nickel-based) in multitubular reformer. The reformer usually is operated at between 15 and 30 atmospheres and at a tube outlet temperature of 840–900 °C. The reforming conditions are chosen to give the most economic overall production costs. Methane slip (amount of unconverted methane) usually is greater than for conventional high-pressure synthesis processes, since the cost of compressing the additional methane is less significant with the low-pressure process. With a naphtha feedstock, an almost exact stoichiometric ratio of carbon oxides to hydrogen in the synthesis gas is achieved, but when natural gas is the feedstock, there is an inherent deficiency of carbon. Established practice for many years has been to add carbon dioxide from an external source in preparing a stoichiometric synthesis gas. Development of the low-pressure process has shown that this addition of carbon dioxide is not required and that, depending upon the cost of carbon dioxide production, the production of methanol from natural-gas feedstock alone is economic.

After heat recovery and cooling, the synthesis gas is compressed to the required synthesis pressure and passed into the synthesis loop at the suction of a circulator. The circulator, which boosts the pressure of the circulating gases to make up the total loop pressure drop, also is a centrifugal machine. Feed-gas preheating is carried out by heat exchange with the hot gases leaving the converter. Heat recovery is incorporated into the loop to recover the heat of reaction of methanol synthesis.

Synthesis takes place in a hot-wall converter over the low-pressure methanol-synthesis catalyst at 250–270 °C. Temperature control of the converter is effected by injecting cold gas at appropriate levels in the catalyst bed, using specially developed distributors that provide excellent gas mixing while allowing free passage of the catalyst for easy charging and discharging. After leaving the converter and passing through the feed-gas preheater the converted gases are cooled, and crude methanol is condensed and separated from the uncondensed gases, which are recycled with makeup synthesis gas to the converter. A continuous gas purge is taken from the synthesis loop in order to remove an accumulation of inert gases. This purge is recycled to the synthesis-gas preparation section as reformer fuel. The crude methanol is reduced in pressure before passing forward to the methanol-purification section, where methanol of the required purity is produced by conventional distillation methods.

Economics in fuel gas consumption are achieved by use of recovered heat in reboiling in the distillation columns. In addition, distillation schemes involving three or four columns have also been developed with reduced reboil heat requirements.

Additional Reading

Chang, C.: "Hydrocarbons from Methanol," Marcel Dekker, Inc., New York, NY, 1983.
Cheng, W. and H. Kung (Editors): "Methanol Production and Use," Marcel Dekker, Inc., New York, NY, 1994.
Murrell, J. and H. Dalton: "Methane and Methanol Utilizers," Kluwer Academic Publishers, Norwell, MI, 1992.

METRIC SPACE. A generalization of distance and related concepts to abstract sets.

A metric space is a set X such that to each pair of points x and y, there is associated a nonnegative real number $\rho(x, y)$, called the *distance* from x to y. This distance satisfies the conditions: (1) $\rho(x, x) = 0$, (2) $\rho(x, y) = 0$ implies $x = y$, (3) $\rho(x, y) = \rho(y, x)$, and (4) $\rho(x, y) + \rho(x, z) \geq \rho(x, z)$. The latter is the familiar triangular inequality.

The function $\rho(x, y)$ is said to be a metric for X. Examples of metric spaces are the unit circle in the plane and three-dimensional space with the usual distances. Less familiar examples are Hilbert space and any set X with the distance function $\rho(x, y) = 1$ for $x \neq y$ and $\rho(x, x) = 0$.

A metric can be used to define a topology for X called the *metric topology*. The set $S(x, \varepsilon)$ of points in X at distance less than ε from the point x is said to be the *open ball* of radius ε centered at x. A subset U of X is then said to be *open* in the metric topology if for each x in U there is an $\varepsilon > 0$ so that $S(x, \varepsilon)$ is contained in U. A topological space (X, τ) is said to be *metrizable* if a metric can be defined on X in such a way that open sets in τ are open in the metric topology and, conversely, most familiar topological spaces are metrizable.

A sequence of points, $x_1, x_2, x_3, \ldots$ in the metric space X is said to *converge* to the point x in X if the sequence of real numbers $\rho(x, x_1), \rho(x, x_2), \rho(x, x_3), \ldots$ converges to 0. A sequence converges in this sense if, and only if, it converges in the metric topology.

A sequence of points $x_1, x_2, x_3, \ldots$ in X is said to be a *Cauchy sequence* if for $\varepsilon > 0$ there exists an integer N so that $n, m \geq N$ implies $\rho(x_n, y_m) < \varepsilon$. A sequence that converges to a point in X can be shown to be a Cauchy sequence. A metric space is said to be *complete* if every Cauchy sequence converges to some point in X. The real number line is an example of a complete metric space, while the set of points between 0 and 1 with the usual distance is not complete.

See also **Topology**.

METROLOGY. The science of dimensional measurement; sometimes includes the science of weighing.

METRONIC CYCLE. A period of 19 years, after which the various phases of the moon fall on approximately the same days of the year as in the previous cycle. The Metonic cycle is the basis for the golden numbers used to determine the data of Easter. Four such cycles form a Callippic cycle.

MICA. The mica group of minerals includes several closely related species, having a highly perfect basal cleavage; all are monoclinic with a tendency toward pseudohexagonal crystals, and are closely similar in chemical composition. The highly perfect cleavage, the most prominent characteristic of the group, is explained on a basis of x-ray studies of these minerals, which seems to show a sheet-like arrangement of the atomic structure and a hexagonal grouping of atoms which apparently explains the pseudohexagonal crystals above mentioned. The word mica is believed to have been derived from the Latin *micare*, meaning to shine, in reference to the brilliant appearance of this mineral, especially when in small scales.

MICELLE. See **Colloid System**.

MICRODISPLAYS. Microdisplays are very small displays that are viewed through the use of optics. Although no formal definition exists for the size of a microdisplay, most in the industry would agree on a diagonal measurement of one inch or less. The two most competitive features of microdisplay technology are the ability to display large numbers of pixels and to do so in a lightweight package that occupies a small volume. Traditional display technologies tend to get larger, heavier, and more expensive as image size and pixel count increase. The fact that microdisplays can be highly compact without sacrificing image quality gives them an advantage over existing products. In addition, these features of microdisplays enable a number of new products, such as lightweight headsets, that could not be served by existing technologies.

High-density information displays currently have from 100,000 to more than 1 million pixels. Clearly, a one-inch display with such pixel counts must be coupled to an optical system to create reasonable viewing conditions. There are two broad methods used to implement microdisplays: *projection* and *near-eye*. Projection systems are designed to magnify a small, real image onto a screen for viewing by one or more users. In contrast, near-eye applications use an optical system to project a virtual image intended to be viewed by a single person (hence, they are also called personal viewers). These systems are best suited for compact handheld or head-mounted devices.

Within the category of head-mounted displays (HMDs), there are a number of divisions. The display may be monocular (presented to only one eye) or binocular (presented to both eyes). A binocular display can be stereo or biocular, where the first consists of a single image that the two eyes naturally see from different perspectives (as in the real world), and the second presents separate, slightly different images to each eye. Both monocular and binocular HMDs can be immersive (blocking the viewer from seeing anything but the display) or transparent (allowing the viewer to see both the displayed image and the real world).

Most microdisplays are fabricated on silicon (or quartz) substrates rather than glass, as in the production of direct-view flat panel displays, a quality that has both technical and business consequences. By building the display directly on semiconductor substrates, designers can integrate electronic components — such as row and column drivers, digital and video interface, and control circuits — directly alongside the display on the same substrate. This can lead to higher performance with lower manufacturing costs. On the business side, this has led to production arrangements that are a departure from those typically seen in the display industry, more closely resembling those of the semiconductor industry. Rather than require new investments in production facilities that use increasingly large area glass substrates, microdisplays can be fabricated on existing semiconductor equipment, leading to the possibility of flexible arrangements between display developers and semiconductor firms.

There are four types of microdisplays: *transmissive*, which modulate an external light source as it is transmitted through the device; *reflective*, which modulate an external light source by varying the properties of a reflecting surface; *emissive*, which produce light internally; and *scanning*, which write images directly onto the retina.

Transmissive Microdisplays

Direct-view displays based on the light-modulating properties of liquid crystals are ubiquitous. Over the past decade, new materials and manufacturing techniques have evolved to allow for very small (10–20 micron) pixel sizes, resulting in the development of transmissive microdisplays that function much as their direct-view counterparts. In this section, detailed descriptions are given only for twisted-nematic (TN) displays, but many of the same principles apply to other types, such as ferroelectric.

Amorphous silicon (a-Si) is currently the most widely used semiconductor channel material for direct-view active matrix LCD applications. It is especially convenient because it is deposited at a low temperature around 300 °C (572 °F), meaning that relatively inexpensive glass substrates can be used. In addition, its low mobility (although still high enough to provide adequate *on* current) makes for TFTs with very small *off* state currents, meaning supplementary storage capacitors are rarely necessary. However, this low mobility makes a-Si incompatible with integrated driver architecture, which is required for the compactness of LCD microdisplays. Integrated drivers are made possible by using the higher-mobility poly-crystalline silicon (poly-Si, or polysilicon) as the channel material.

All transmissive LCD microdisplays, with the exception of those from Sharp and Kopin Corporation *http://www.mdreport.com/summaries/october99.html* (see the following sections on continuous-grain silicon and single-crystal silicon), use poly-Si TFTs. As noted previously, the systems called microdisplays are near one inch in diagonal size and are viewed indirectly through external optics, disqualifying very small LCDs that are viewed directly (such as in some viewfinders).

Poly-Si microdisplays are used in two major applications: video/digital camera viewfinders, which benefit from very small pixels, and projection LCDs (particularly for front projection, where poly-Si devices will dominate for the next five years). To date, nearly all poly-Si TFT-LCDs, of all sizes, have been fabricated using high-temperature techniques on quartz substrates, but there has been a great deal of effort applied to low-temperature manufacturing on glass, because of reduced costs.

Reflective Microdisplays

Liquid Crystal on Silicon. Of all the types of AMLCDs, liquid-crystal-on-silicon (LCOS) devices most directly use semiconductor manufacturing techniques, as the active matrix array and associated driver circuits are designed and built on a silicon wafer, using complementary metal oxide semiconductor (CMOS) processing techniques. The LCOS displays are the most heavily researched microdisplays.

Instead of thin-film transistors on glass, LCOS displays use bulk silicon devices to control the pixels from outside the optical path. See Fig. 1. The silicon-based CMOS chip serves as both the active matrix and the reflective layer, on top of which a thin layer of liquid crystal, glass plate, and polarizer are deposited. Instead of the a-Si or poly-Si TFTs used in transmissive LCD cells, LCOS devices use the high-speed switching capability provided by single-crystal silicon. LCOS devices also have a simplified LCD structure because the device is reflective, not transmissive, like most LCDs. For example, only one polarizer is required, and the thickness of the layer of liquid crystal can be reduced, allowing for faster switching.

Two important design choices are made when developing an LCOS device: the type of liquid crystal material and the type of control circuit to be used. The nematic and ferroelectric liquid crystal modes are the most popular types for such a display. The nematic types provide a good contrast ratio and are typically coupled with a DRAM switching array. Ferroelectric liquid crystals provide very fast switching speeds, but they do not allow for grayscale; the pixel is either black or white. Grayscale can be created temporally by taking advantage of the fast switching speed to dither the binary value within a frame period. Ferroelectric liquid crystals are typically coupled with SRAM devices, which provide the fast frame

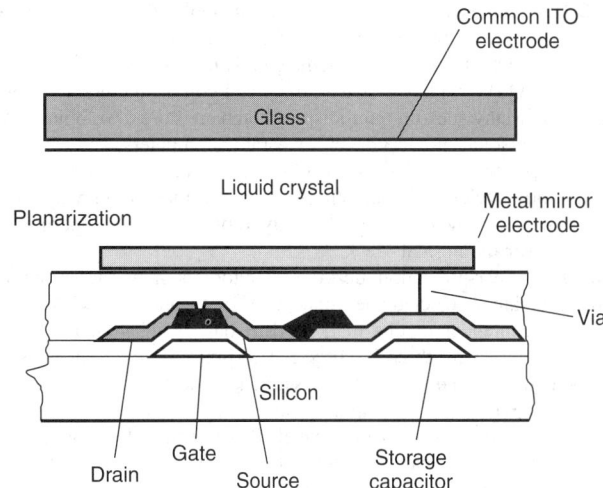

Fig. 1.   Structure of a LCOS Microdisplay (*Courtesy of CMD*).

transfer rates needed to provide temporally dithered grayscales and which are also binary devices.

Liquid crystal microdisplays can also be configured to modulate unpolarized light, an approach that has the advantage of much greater light transmission by eliminating the significant absorption of polarizers. There are two ways to use unpolarized light. In one approach, light is scattered when the liquid crystal molecules are arranged in one fashion and reflected when the molecules are rearranged, a process that is controlled by the application of a voltage across the cell. Scattering displays typically use polymer-dispersed liquid crystal materials. The second approach uses the principle of diffraction, in which a periodic arrangement of *on* or *off* pixels (forming a grating) causes light to constructively or destructively interfere. This interference pattern can then be filtered by a Schlieren stop to pass light when in certain grating modes; by altering the grating structure, amplitude and color may be controlled. Twisted-nematic liquid crystal material is the most common type used in diffractive systems.

Color can be achieved in LCOS devices in several ways. The most popular approach for projection applications is to use three LCOS chips and dichroic mirrors to separate red, green, and blue components. Color-filter wheels have also been used to present sequential color to a single chip. For personal viewers, red, green, and blue light sources (typically light-emitting diodes) are used to illuminate sequential frames of data at three times normal video rates.

A number of engineering issues remain with LCOS. First is the light source, which should be as close to a point source as possible because the display is so tiny. Also, improvements in the optics to handle the fine pixel pitch would increase image quality, particularly in the rear-projection screen and the polarizing beamsplitter. The manufacturing of LCOS microdisplays is also not yet a mature practice; manufacturing lines for common LCDs and TFTs are generally not appropriate for LCOS displays. These reflective displays require a thin cell gap (1–3 microns), with a tolerance of 100–200 nm, as well as specialized filling and sealing procedures. Pixel pitches in microdisplays are as much as five times smaller than in direct-view displays, adding to the engineering challenge. Current yields are far below acceptable levels. Applications for LCOS devices are grouped into four main types: Front projectors; viewfinders (camera and camcorder) and viewers integrated into other handheld devices (cellular telephones and PDAs); head-mounted displays; and rear-projection monitors and televisions.

Most companies use TN liquid crystals, although two use ferroelectric liquid crystals (FLCs).

The first devices reached the market in the form of front projectors for business applications. The LCOS front projectors developed so far are not very reflective and thus require bright illumination to function. These illumination requirements restrict LCOS projectors from competing in the fast-growing market segment of ultralight projectors, due to the need for a hefty lamp and optics and a specialized cooling system.

LCOS displays for viewfinders (in camcorders or digital cameras, for instance) have the prospect of reaching the market in the near term.

LCOS projectors are the first successful application of this type of microdisplay, providing credibility for further development.

The HMD (which could be implemented with a variety of display types, including LCOS) is widely touted as an up-and-coming method for viewing information. Many start-up microdisplay companies are developing HMDs for games, personal movie viewing, wearable computers, and screens for cellular telephones and PDAs. Some of these systems require total attention (such as movie viewers), but others strive for integration with normal vision, either by showing the display only to one eye or by superimposing the image over the normal view.

Interest in rear-projection desktop monitors is now increasing after several companies rejected the concept in 1999, and the LCOS display could be a contender in this market. IBM has demonstrated a 28-inch-diagonal monitor with three $2,048 \times 2,048$-pixel TN microdisplays. It is yet to be demonstrated that LCOS-based rear projectors can compete with direct-view CRT or LCD in the desktop monitor market or with the rear-projection CRT or plasma display panel (PDP) in the television market.

These activities indicate that there is currently great interest in (and resources devoted to) microdisplays on the supply side. So far, performance has been adequate but not stellar, and the price of products is still quite high. Both of these factors must improve in order to generate significant demand.

LCOS devices are just beginning to take hold in the display market. Although there is potential for high growth in the coming years, market acceptance will be contingent on improvements in price, performance, and ergonomics of these products.

Microelectromechanical systems (MEMS). Devices use semiconductor fabrication equipment and processes to build miniature systems that have both mechanical and electrical components. They rely on physical movement of a display element (pixel or line) to modulate the amplitude or phase of an external light source. The most common type of MEMS device used in the display industry is the Digital Micromirror Device™ (DMD), but two other types the Grating Light Valve (GLV) and the Actuated Mirror Array (AMA) are also relevant.

Digital Micromirror Device (DMD). The DMD is a monolithic, micromechanical spatial light modulator integrating discrete, tilting mirror elements with CMOS addressing circuits at each pixel location. The device was invented by Larry J. Hornbeck of Texas Instruments in 1987. Devices with 800×600 pixels and $1,024 \times 768$ pixels were available in 1998, and systems with $1,280 \times 1,024$ pixels were introduced in 1999. The technology is capable of providing projected images from all currently operational or proposed high-definition standards; demonstration systems have been built with $1,920 \times 1,080$ pixels. Hence, the DMD is a potential technology for HDTV projection systems. The devices can be made in high volume at relatively low cost in a specialized semiconductor-manufacturing environment. This DMD is sold by Texas Instruments to numerous projection-system vendors who use it as the core of their systems. The subsystem driving the products built around the DMD chip is called a Digital Light Processing™ (DLP) engine.

The DLP circuit board accepts an incoming signal from a computer and, using an array of digital-signal processors, converts the signal to digital sequences. This digital system, controlled by software and algorithms, converts text, graphics, and video into a digital signal that yields tones and color values that rival the image quality of photographic film. No analog signals are used in the control circuit.

After processing, the digital signal moves to the heart of the DLP system, the DMD. The DMD functions like an array of light switches. It consists of an array of 17 micron $\times$ 17 micron mirrors, fabricated on hinges on top of a static RAM (SRAM) chip. The DMD decodes the incoming signal, providing each mirror with its own unique instruction. Then each mirror is responsible for creating a single pixel in the projected image and receives a new instruction approximately every millisecond.

Depending on the state of the SRAM cell, the mirror is electrostatically attracted to either one or the other address electrodes by a combination of bias and address voltages. Thus, the mirror rotates until its tip touches a landing electrode, which is held at the same potential as the mirror. A "1" in the memory cell causes the mirror to rotate $+10$, while a "0" in the memory cell allows the mirror to rotate -10. Figure 2 shows an array of micromirrors in three positions: flat, on, and off. The mirrors tilt into the *on* or *off* states when 5 volts are applied to one of the two offset electrodes while the other is grounded. Although the DMD can be operated in the analog mode, the mirrors are biased in such a way that only the digital

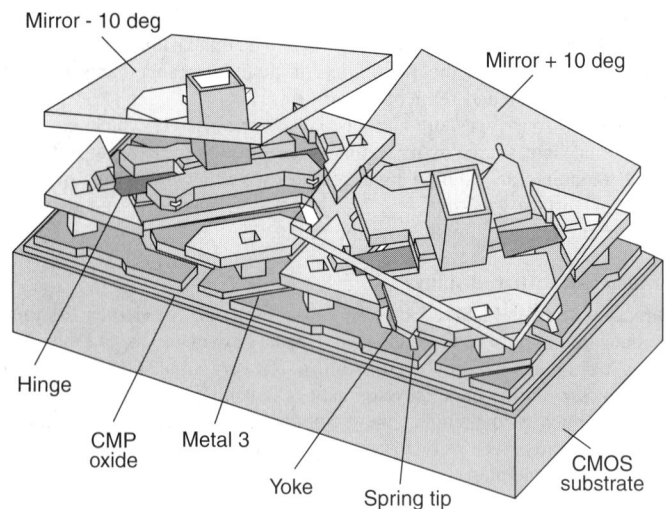

Fig. 2.   DMD Pixel Structure. (*Courtesy of Texas Instruments.*)

landing sites of $+10$ or -10 are possible. The digital mode of operation permits operation within the low-voltage CMOS framework and ensures uniform deflection angles.

Because the process is completely digital, DLP projection systems based on one-, two-, or three-DMD configurations can serve as the digital display interface for not only presentations but also computers, televisions, or any future application where digital light can be harnessed and effectively used. In the three-chip system, one chip is used for each of the primary colors (red, green, and blue), similar to other projection systems. But, because of the DMD's high speed and light throughput (optical efficiency of the DMD pixels have been measured to be 61% by Texas Instruments, and no polarizers are needed), it is also practical to create a single-chip system, which is less costly and is self-converged. Diagrams of the three system configurations are shown in Fig. 3.

In a one-chip system, color is added to the image by filtering light through a color system. Typically, a white light source is focused onto a color wheel through the use of condensing optics. The color wheel is a red, green, and blue (RGB) filter system that spins at 60 Hz to give 180 color fields per second. In this configuration, DLP operates in a sequential color mode. As white light hits the red filter, the red light is transmitted and the blue and green lights are absorbed. The same holds true for the blue and green filters: The blue filter transmits blue and absorbs red and green, while the green filter transmits green and absorbs red and blue. Thus, when a color wheel is used, two-thirds of the light is blocked at any given time.

The light that passes through the color wheel is then imaged onto the surface of the DMD. The integrator rod shown in Figure 3 serves to match the beam cross section to the shape of the DMD, and the total internal reflection (TIR) prism simplifies the optical layout; both are optional in low-end versions. The input signal is broken down into RGB data, and data are sequentially written to the DMD's SRAM. If a mirror is controlled such that it reflects red light 50% of the time and blue light for the other 50% the resulting projected pixel will be purple. Color shades are created by varying the amount of time the mirror is rotated toward or away from the incident light. As the wheel spins, sequential red, green, and blue lights hit the DMD. The color wheel and video signal are in sequence so that when red light is incident on the DMD, the mirrors tilt according to where and how much red information is intended for display. The same is done for the green and blue lights and the video signals. The human visual system integrates the red, green, and blue information and sees a full-color image. Using a projection lens, the image formed on the surface of the DMD can be projected onto a large screen.

Because a National Television System Committee (NTSC) television field is 16.7 milliseconds (1/60 of a second), each of the primary colors must be displayed in about 5.6 milliseconds. Given that the DMD has less than a 15 ms switching time, an 8-bit-per-color grayscale (256 shades) is possible with a single DMD system. This gives 256 shades for each of the primary colors, or $2,5633$ (16.7 million) possible colors that can be generated.

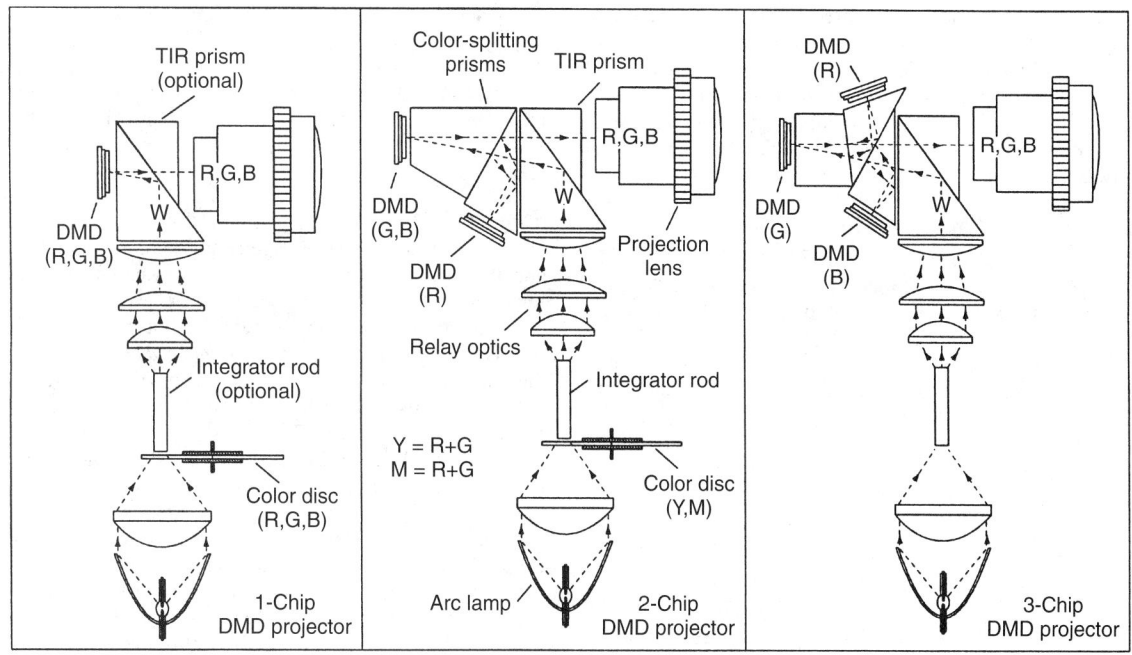

Fig. 3.   Three DMD Systems. (*Courtesy of Texas Instruments.*)

The two-chip DMD system uses a color-filter wheel that passes two of the secondary colors, magenta and yellow. The magenta segment of the color wheel allows both red and blue to pass through, while the yellow segment passes red and green. The result is that red light is passing through the filter system at all times. Blue and green alternate with the rotation of the magenta-yellow color wheel and are each essentially on half of the time.

By biasing light transmission toward the red, the two-chip subsystem compensates for the red-light deficiency in long-lifetime metal halide lamps. Compared to a one-chip system, the two-chip system yields roughly three times the amount of red-light output. Because the color wheel is made up of only two filters instead of three, the blue and green light output is increased by roughly 50%.

Once through the color wheel, the light is directed to a system consisting of a TIR and dichroic prisms. At this point, the constant red light is split off and directed to a DMD dedicated to handling red light and red component-video signals. The sequential blue and green light is directed to another DMD that is driven by blue and green component-video signals and is configured to handle the sequential colors.

The three-chip system uses one chip for each of the primary colors: red, green, and blue. Light is separated solely by dichroic prisms. No color-filter wheel is required, so light output is maximized. In addition to the higher brightness achieved by the three-chip system, the image quality—a function of not only sophisticated signal processing but also higher-bit color depth—is also increased. Because light is directed to each DMD for the whole field, 10-bit grayscale per color is possible, allowing three-chip systems to reproduce 1,024 shades of gray compared with the one-chip system's 8-bit color and 256 shades of gray.

The grating light valve (GLV), invented at Stanford University and being developed by Silicon Light Machines (Sunnyvale, California), is a silicon-based integrated circuit with a micromachine reflection phase grating. The structure consists of a series of suspended silicon-nitride beams, referred to as ribbons, which are coated with a reflective aluminum layer. The ribbons are approximately 3 microns wide, 100 microns long, and 100 nanometers thick, and they are arranged in parallel. Each ribbon is supported at both ends and is suspended above an air gap of approximately 650 nanometers. When a potential difference is applied between the aluminum coating on a ribbon and a conductive layer below the air gap, the ribbon is pulled down by electrostatic force. This effect is very fast; the switching time can be as short as 20 nanoseconds.

A typical pixel consists of a set of six ribbons. If no voltages are applied, the set of ribbons functions as a mirror; when a pixel is addressed, alternate ribbons are deflected, forming a grating surface, and incoming light is diffracted. All that is required is the application of a voltage to the selected pixel; no transistors are required to switch and hold a pixel. The hysteresis effect inherent in the device also allows the deflection to be maintained at a voltage level in between the turn-on and turn-off voltages. The other three ribbons remain in a fixed position, and need only a fixed bias voltage. The amount of diffraction from a pixel is related to the depth of the grating at that pixel, which is proportional to the applied voltage. A Schlieren optics system is used to allow only the diffracted light to pass.

The distinctive feature of a GLV projection system is that it consists of linear arrays of pixels, which are scanned perpendicular to the array to form an image rather than a typical two-dimensional pixel array. The fast switching time of the GLV allows for the horizontal dimension to be accomplished through scanning. A projection system using a white-light laser source, separated into red, green, and blue components has been demonstrated. The primary colors illuminate three GLV arrays with 1,080 pixels and are scanned with a galvanometric mirror to produce an image with $1,920 \times 1,080$ pixels. The company reported that this system had a contrast ratio of 200:1, showed 16.7 million colors, and had a maximum refresh rate of 96 Hz. The pixel elements can be fabricated in as few as 7 mask steps, promising low tooling and manufacturing costs relative to more complex MEMS devices.

The actuated mirror array (AMA) was developed by Aura Systems, El Segundo, California, in the early 1990s, and licensed to Daewoo Electronics, which has been developing the concept further under the name thin-film micromirror array-actuated (TMA). The device is similar to the DMD in concept and system design, although the TMA requires a Schlieren stop to operate properly, whereas the DMD's larger rotation angles allow the display to be coupled directly to the projection lens. The key difference is that the TMA uses piezoelectric materials, rather than electrostatic forces, to tilt the mirror/pixels. Piezoelectric materials constrict or expand in response to an applied voltage; each mirror sits on a pair of piezoelectric posts, and when one post is constricted and the other is expanded, the mirror tilts toward the constricted post.

A linear response to an applied voltage was reported, up to a tilt of 4 at 10 volts, and a frequency response up to 30 kHz, or approximately 33 ms switching time, allowing for a gamut of 16.7 million colors using three chips. A device with 640×640 pixels, each 97×97 microns was also reported, subsequently announced a $1,032 \times 774$–pixel device with 50-micron square pixels was announced. Due to the large chip size (approximately 3 inches in diagonal), three-chip demonstration systems have achieved 5,000 ANSI lumens.

Emissive Microdisplays

Emissive microdisplays produce their own light, obviating external light sources or reliance on reflection. All emissive microdisplays are based on technologies that were first developed for direct-view displays and then applied to chip-based systems later. There are three main types of emissive displays: thin-film electroluminescent (TFEL) displays; vacuum fluorescent displays (VFDs), which are called Vacuum Fluorescent on Silicon (VFOS) when they are in microdisplay form; and organic light-emitting diode (OLED) displays. The microdisplay version of the field emission display (FED) was briefly pursued, but is no longer being developed.

Thin-Film Electroluminescent Microdisplays

Active matrix thin-film electroluminescent (AM-TFEL) displays are miniature AC-TFEL structures constructed on an integrated circuit backplane. By using active matrix addressing, AM-TFEL displays can provide higher information content than direct-view TFEL displays; microdisplays with up to $1,280 \times 1,024$ pixels have been demonstrated, using a 0.7-inch-diagonal substrate and a resolution of 2,000 lines per inch. The high voltages (80 volts AC) used in such devices require the use of silicon-on-insulator substrates that isolate the high voltage signals from nonselected pixels and logic. Color AMEL devices have been produced using filtered white phosphor and field-sequential liquid crystal shutters.

In 1996, Planar announced the first AM-TFEL head-mounted display, which has since evolved into a line of microdisplays called MicroBrite™. These devices are designed for such applications as hands-free maintenance computers, military targeting and communications systems, portable personal communications and computing, medical imaging and virtual reality for commercial applications, and consumer entertainment. The 2000 model has an active area of 0.61 inch $\times$ 0.45 inch, is 1.7 mm thick, and weighs less than 3 grams. The monochrome (amber) image has a VGA pixel format (640×480 pixels) with 32 gray levels and is sold as a developer's kit with interface electronics for digital LCD VGA, analog VGA, or video. Integrators such as Liteye and Kaiser Electro-Optics offer products based on Planar's technology. A color version was expected to be available in late 2000. See also **Electroluminescent Displays**.

Vacuum Fluorescent on Silicon (VFOS) Microdisplay

The VFOS display is a blend of semiconductor technology and conventional VFD technology. The display is a phosphor dot matrix on top of a 5.4 mm $\times$ 6 mm semiconductor chip that integrates memory function and display driver circuits. The basic structure of this display device is similar to the conventional VFD except for the structure of the anode. In a typical VFD, wiring, insulation, and phosphor layers are fabricated directly on the glass plate and no semiconductor devices are arranged internally. With VFOS, phosphor matrices are fabricated on a semiconductor chip and arranged on a glass plate. The chip and the outer leads are connected by wire bonding. These types of graphic displays contain semiconductor driver ICs with embedded memory.

Display Research Laboratories (Los Altos, California) has pioneered VFOS. Claimed benefits of VFOS displays over conventional VFDs include lower noise, longer life, higher brightness, and more stable operation under varied environmental conditions. See also **Vacuum Fluorescent Displays**.

Organic Light-Emitting Microdisplays

A new direction in OLED development is to build the device on a silicon backplane for use as a microdisplay. This technology is being pursued by eMagin Corporation, which has licensed Kodak's small-molecule patents. The company's active matrix OLED display has 12-micron pixels in a $1,280 \times 1,024$ (SXGA) pixel format on a 0.77-inch-diagonal chip. It is capable of showing full-motion video. The prototype shown in May 2000 was monochrome, but there are plans to add color filters (which will reduce the pixel format). The device is intended for consumer products such as cellular telephones and movie viewers and also for use as a virtual computer monitor.

Because OLEDs emit light perpendicularly from an areal source (that is, the emission is Lambertian), they are very well suited for near-eye applications. No special optics are needed to make the light resemble the light that normally strikes the eyes. The flip side of this property is that OLED microdisplays are poorly suited for projection applications, because they are not bright enough.

Many challenges remain in commercializing OLED microdisplays. As with other microdisplays, the manufacturing process is not yet fully automated, but OLEDs have the additional burden of the lack of large-scale manufacturing even for the direct-view version of the technology.

Equipment has not been standardized or refined for large-area OLED displays as it has been for LCDs, VFDs, and TFEL panels. Also, the organic light-emitting materials have limited lifetimes and are sensitive to environmental factors like moisture, heat, and oxygen. Also, there are interface and driver issues; standard chips and connectors are not yet established for this technology. See also **Organic Light-Emitting Diodes (OLEDs)**.

Scanning Microdisplays

Microdisplays for projection applications produce a *real* image on a screen, and those discussed so far for near-eye applications produce a *virtual* image that appears to float before the viewer's eyes. The virtual image is a magnification of a tiny real image. But there is an alternative method for near-eye applications: *scanning microdisplays*. These devices "paint" the image directly on the retina, so in some sense, there is no image.

Scanning microdisplays work by sweeping a light beam to create an image, much as a CRT sweeps an electron beam. However, unlike electrons, which can be directed with electromagnetic fields, photons must be directed with mirrors, because they have no charge. By coordinating the sweep and some form of light modulation, the image is produced.

Only four components are needed: a controllable light source, a scanner, drive electronics to control the scanning, and viewing optics. The light source should be compact and low power; a laser diode, laser, or LED is a standard choice (note that the intensity of the light used cannot harm the eye). The brightness is controlled by one of several types of optical attenuator. In the case of color displays, three light sources are used (red, green, and blue). The light mix is controlled separately from the brightness control by an acousto-optic or electro-optic modulator.

There are two kinds of devices that sweep light beams: line scanners and pixel scanners. Line scanners are as wide as the image they produce, sweeping the light beam only from top to bottom. These simple scanners are widely used in fax machines and document scanners. Although common line scanners are too slow for display applications, faster ones for displays are being developed.

Pixel scanners, which are analogous to CRTs and are the only kinds used as microdisplays so far, control pixels one by one, or in groups of a few, using drive electronics that change the mixture of red, green, and blue light appropriately. The beam sweeps in a two-dimensional raster pattern to create the image. The horizontal sweep must be much faster than the vertical sweep and is typically accomplished with a magnetically controlled mechanical resonant scanner or a MEMS device. A galvanometer or a MEMS device can perform the slower vertical sweep. See also **Cathode-Ray Tube**.

A mechanical resonant scanner consists of a reflective plane surface that can tilt like a seesaw under the influence of magnets on either side. Varying the fields controls the position continuously within a range that defines the sweep angle. Operating frequencies between 15 and 20 kHz can be achieved. Since this is a uniaxial scanner, it must be combined with another scanner (usually a galvanometer) to produce the two-dimensional raster motion.

Using a MEMS device allows for a much more sophisticated implementation of the same concept. In one arrangement, a silicon micromirror is etched away from the wafer surface, suspended by thin flexures on all four sides. About one axis (the fast, horizontal sweep), the mirror vibrates in a torsional oscillation mode, while in the other direction (the slow, vertical sweep), the flexures are moved capacitively. The scanner is biaxial and also operates between 15 and 20 kHz. (Note that the other type of micromirror device commonly used in display technology flips between two distinct states, rather than sweeping continuously).

As in all display systems, image quality in a scanning microdisplay results from a complex trade-off of many variables. The two factors with the greatest influence on the system's image properties are the size of the micromirrors and their deflection angle. The product of these two parameters (at a given wavelength) is directly proportional to the resolution of the image. Clearly, there is a trade-off: the bigger the mirror, the less deflection is necessary. There are also manufacturing trade-offs that limit the design space. Large mirrors carry the penalty of large inertia and hence slow response. Increasing the stiffness of the flexure mechanism (the lever that moves the mirror) to speed the mirror's response results in higher stress during mirror motion, which reduces the acceptable deflection angle (which was small anyway, because the mirror is large). Large mirrors are also less likely to be adequately flat and in general take up more space in the microdisplay system, which should be as compact as possible.

Small mirrors are fast and have larger deflection angles, but they also have limitations. If the deflection angle gets too large, there are problems with excessive stress in the mirror, which can lead to reduced lifetime. Changing the shape of the flexure to a long, thin profile helps mitigate stress, but can result in coupling to undesirable modes of the mirror. The ultimate effect of all these problems is a blurry, shifting, or flickering image.

Many other factors play a role in the MEMS system design also. The optimization of mirror size and deflection angle must be tailored to each wavelength, so color displays must have three different micromirror setups. Other important variables include device materials, mirror thickness, and flexure dimensions. Together with the mirror properties discussed above, these parameters interact to determine the overall cost and performance of the microdisplay.

To finally produce the image, the scanned beam must be directed into the viewer's eye using optics. The beam size and sweep angle must be defined to correctly enter the pupil while preserving the diffraction-limited nature of the original beam. There are a number of optical arrangements that can be employed, each with trade-offs in size/weight, complexity, field of view, and distortion effects.

Microvision produces scanning microdisplays based on both magnetically controlled scanners and its proprietary MEMS technology, which uses silicon micromirrors less than one millimeter on a side. The company has demonstrated horizontal scanners that operate at up to 19 kHz. Each half cycle of oscillation sweeps one line. If the scan is performed *bidirectionally*, meaning that the first line is scanned left to right, the second is scanned right to left, and so on, then the effective horizontal sweep frequency is 38 kHz. Painting a complete SVGA image (600 lines) thus requires 15.8 ms, for a frame rate of 63 Hz. The scanners can be improved to operate above 30 kHz, allowing for at least SXGA format (1,024 lines).

The compact version of Microvision's display, intended for consumer use as a wearable computer or gaming unit, is a monochrome, monocular VGA (640×400 pixels) HMD weighing about 1.5 pounds. The company is also engaged in designing and building high-performance binocular (and necessarily biocular) versions for surgery or military applications. These systems are more substantial helmet-type units and will incorporate full color and SVGA (800×600) or higher pixel formats. Handheld units for the wireless market are planned; studies show that retinal scanning displays indeed work when simply held near the eye. These products use either LEDs or lasers as the light source.

MicroOptical Corporation, *http://www.microopticalcorp.com/* an integrator that makes glasses-mounted display systems based on liquid crystal microdisplay technology, is also developing MEMS scanning technology through a grant from DARPA. It will use lasers as the light source. The company claims that MEMS display systems can outperform LCD systems in terms of brightness, power efficiency, and reduced cabling.

Research work on scanning displays is also going on at numerous universities. The founders of Microvision originated at the Human Interface Technology Lab (HIT) at the University of Washington, Seattle, *http://www.hitl.washington.edu/* where topics ranging from interface software design to human factors in wearing HMDs are investigated. There are also studies and design work proceeding at the University of California, San Diego, and Delft University, the Netherlands.

The experience of using a scanning microdisplay is different from viewing the virtual image that is seen in most HMDs. Everyone who has used a direct-view computer monitor (or looked at photograph) knows that even the most accurate representation of three-dimensional space on a flat surface cannot be mistaken for the real world. This experience has to do with a multitude of visual cues and physiological phenomena that people use to judge distance. Flat, finite-sized surfaces cannot reproduce all the cues people need to perceive an image as three dimensional. Even though a scanning microdisplay writes directly on the retina (as the real world does), it too cannot provide the full experience of reality; however, sophisticated optical processing of the scanning beam can satisfy more of the brain's criteria for three-dimensionality than a flat virtual image. Thus, scanning microdisplays can potentially produce more realistic stereoscopic images than any other type of microdisplay.

One key feature of scanning microdisplays is that the brightness and contrast are adjustable over a wide range by changing the intensity of the light source (a feature that is also available on CRTs but not reflective displays, other emissive displays, or LCDs). Hence, a person with low vision could use a scanning display to read or to watch television. A monochrome VGA display that is modified for patients with low-vision

by has been produced. Although it is still being refined, this application is the subject of continued research at medical centers.

Although much research and development has focused on producing scanning microdisplays for playing games, performing field work, or for military use, a significant portion is devoted to the medical use of these displays for patients with low vision.

In summary, scanning microdisplays, like other types of microdisplays, are finally finding applications as interest in high-resolution near-eye displays is increasing. As with most new technologies, price is an issue; scanning microdisplays are very expensive and face the usual problem of needing volume production to achieve lower prices but needing lower prices to achieve high demand.

However, the crucial concern is whether consumers will like using near-eye displays. Exploring this issue means looking at the ergonomic factors of near-eye microdisplays, an area in which there have been few significant studies. The companies that succeed in developing the near-eye microdisplay market will be those that attend to human factors first. See also **Television (TV)**; **Optical Character Recognition (OCR)**; and **Semiconductor**.

For additional reading, refer to Flat Panel Display Technology entry.

Stanford Resources, Inc., San Jose, CA.

MICROELECTRONICS. A constantly evolving technology, microelectronics is concerned with the design and manufacture of semiconductor electronic circuitry–containing components measured in microscopic dimensions. Also referred to as *chips, microchips,* and *integrated circuits,* microelectronic circuits usually are subsystems of larger systems, such as computers, telecommunications switches and transmission systems, guidance systems for aircraft and missiles, and satellites. The tiny chips also are found in commonplace items, such as personal computers, automobiles, household appliances, digital wristwatches, calculators, toys, and video and audio entertainment systems.

Brief Chronology. The birth of microelectronics can be traced to the invention of the *transistor* in 1947 at Bell Laboratories in Murray Hill, New Jersey. With a name coined from its ability to *transfer* electric signals across a *resistor*, the transistor initially replaced the hot and bulky vacuum tube. The scope of transistor applications later expanded dramatically. In 1958, the integrated circuit (IC) was invented. In addition to combining numerous electrical components into one device, integrated circuits or microchips offered several advantages over discrete diodes and transistors, including smaller size, greater reliability, improved performance, lower cost, and reduced power requirements. See also **Diode**; and **Transistor (Invention and Development)**.

As of the mid-1990s, electronic systems were composed of many different microelectronic components formed from various types of semiconductor materials. Each component is selected because of a specific ability, such as amplifying, switching, or blocking electrical current flow. Based on its ability, an electrical component is classified as either *passive* (resistors, inductors, and capacitors) or *active* (diodes and transistors). As the word implies, semiconductor materials are not good or bad conductors of electrical current, but fall somewhere in between.

The two most commonly used semiconductor materials are silicon and gallium arsenide. Silicon, which is the second most abundant element in the earth's crust and the primary ingredient of sand, has become to the electronics revolution what steel was to the Industrial Revolution. Gallium arsenide, which is more expensive to manufacture than silicon, is useful for a variety of high-frequency, low-power applications. GaAs is faster than Si, can convert electronic signals to laser light and back again, and is radiation hardened, meaning it is less susceptible to radiation, a trait that is useful in satellite applications.

Function of a Chip. A chip may function as a *memory storage* device or a *processor*. There are several types of memory, with the two most common being random access memory (RAM) and read-only memory (ROM). RAM is the basic memory storage unit in a computer. It stores information temporarily in binary form and can be changed by the user. ROM, on the other hand, cannot be changed. It permanently stores binary information that is used repeatedly. See Fig. 1.

The chip may also be a processor or microcomputer, which in reality is a *computer-on-a-chip*. The application of the integrated circuit (IC) may be continuous, as in a digital wristwatch, or occasional, as in a stereophonic audio amplifier. The area of a chip may be as large as a thumbnail or as small as a grain of salt. The chip may be purely electronic, or it may be an

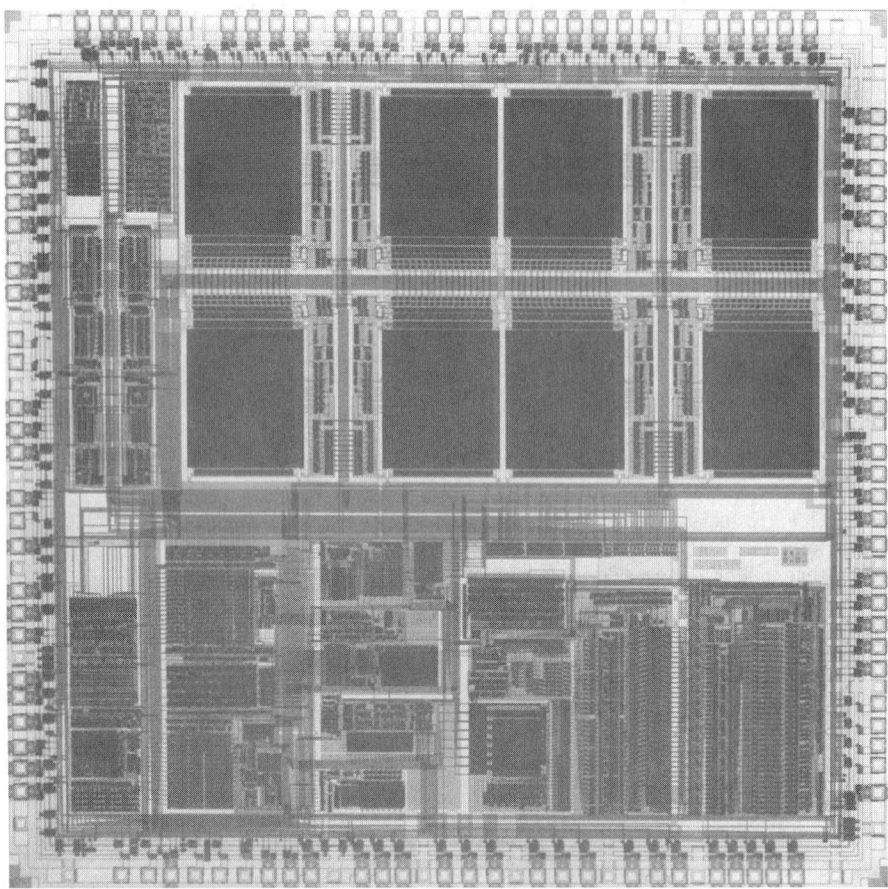

Fig. 1. AT&T's Digital Signal Processor (DSP) 1610, one of the world's fastest and most easily customized chips, operates at speeds up to 40 million instructions per second. Specifically designed to process speech for transmission and reception in digital cellular telephone networks, the DSP 1610's modular architecture gives designers the ability to customize the chip for different applications with relative ease and speed, without sacrificing performance. Actual size of the DSP 1610 is slightly less than a square centimeter. (*AT&T Bell Laboratories.*)

electro-optical IC that combines electronic circuitry with photonics (i.e., which involves processing information with light pulses).

Manufacture of a Chip. To be useful, a microchip must be mounted in a "carrier" or "package" (i.e., a housing that protects the delicate chip while connecting its circuit inputs, outputs, power supply, and ground to conductive pins that are pressure-fitted or soldered to other conductors). See Fig. 2. With the gradual shift from analog to digital technology, which represents information as a superfast stream of digital on-off pulses coded in binary notation, a single IC may now contain a million or more logic gates (on and off switches composed of combinations of transistors, diodes, and their connecting circuits).

The replacement of discrete diodes and transistors with integrated circuits based on more advanced microelectronics technology has significantly lowered the costs of sophisticated electronic functions. A logic gate that cost several dollars in 1950 now costs a fraction of a penny, and the cost is still declining. In addition, reliability of electronic circuits has improved by a factor of about 100,000 from the changeover of vacuum tubes to discrete transistors to integrated circuits.

The development of microelectronic devices is made possible by computer-based systems that design a circuit and its implementation in physical structures and simulate its operation. This process, which uses software on personal computers or workstations, is called computer-aided design (CAD). Essentially, it allows the chip designer to "construct" a chip step by step, function by function, using powerful software to simulate the operation and speed of an IC. Some of the power of CAD comes from the use of libraries of previously designed and tested circuit modules. The explosive growth in chip complexity over the years has been largely fueled by increasingly powerful software that not only facilitates chip design, but vastly shortens the interval needed to bring a new chip to the market.

The manufacturing process for integrated circuits typically consists of more than a hundred steps, during which dozens of copies of an integrated circuit are formed on a single wafer of silicon. Generally, each copy of the circuit is constructed by the creation of 8 to 20 or more patterned

layers on the wafer, also known as a substrate. This layering process, which uses a printing technique called lithography, creates electrically active structures on the semiconductor wafer surface, along with the circuit patterns interconnecting them.

Fabrication of microelectronic circuits relies on complex machinery housed in "clean rooms." The air in these rooms is constantly filtered to remove particles and gases that might contaminate the circuits during the manufacturing process. See Fig. 3 on p. 2338. Clean rooms are rated according to the maximum number and size of particles allowed per cubic foot of air; a Class 100 room has only 100 particles per cubic foot that are 0.5 micrometers in size and is regarded as cleaner than a hospital operating room. (A micrometer is a millionth of a meter. Human hairs average 75 micrometers in width.) New chip fabrication systems that are now being built for ever-more complex integrated circuits require Class 1 clean rooms with particle sizes of 0.2 micrometers to prevent contamination that leads to defective chips. Such rooms and the fabrication lines they contain are extremely expensive to construct, with costs reaching $500 million in the early 1990s. As a result, only the largest chip makers will be able to afford an investment of that magnitude.

The fabrication sequence for a microelectronic circuit starts with molten purified polycrystalline silicon that is used to form a single crystal ingot. The next step involves slicing the silicon ingot into thin wafers, up to 8 inches (20 cm) across. These are polished to a mirror-like luster.

Fabrication of a chip is accomplished through a series of repeated steps involving patterning and processing. In the first step, called thermal oxidation, the wafers are heated and exposed to ultra-pure oxygen in diffusion furnaces under carefully controlled conditions. This forms a silicon dioxide film of uniform thickness on the surface of the wafer. A photoresist or light-sensitive film is applied to the wafer. Next a pattern mask containing a distinctive design is used to expose a portion of the silicon to light while covering other areas. A mask resembles a photographic negative on a glass sheet, usually about 6 × 6 inches (15 × 15 cm). A wafer is mounted into a step-and-repeat machine that

Fig. 2.   Nestled in the heart of 44 conductor leads is a tiny AT&T UNITE® semiconductor chip, only $\frac{1}{4}$-inch (0.6 cm) square. This chip is a key component in the digital telephone, personal computers, workstations, and other terminals used with the new digital telecommunications network (the Integrated Services Digital Netork [ISDN] now being implemented throughout the world. The UNITE® chip formats (organizes) digitized voice and data signals so that they can share the same four-wire line for ISDN operations. The network will simplify the world's public communications systems by eliminating the need for separate networks to handle voice, data, and images. (*AT&T Bell Laboratories. UNITE is a registered trademark of AT&T.*)

holds a mask above a special reducing lens. Because of a reduction ratio as large as 10 to 1, the machine exposes the mask to only a small portion of the wafer and then moves to the next position and exposes it again. Each time, the mask pattern is duplicated on the surface of the wafer. This process is repeated many times until that pattern has filled the wafer's surface area, thus creating multiple copies of the integrated circuit being fabricated.

The next step is called etching. It involves dissolving a portion of the silicon dioxide insulating layer on the surface of the silicon wafer and washing away the unexposed resist material. This leaves the desired circuit features on the wafer. Another step, called doping, exposes the surface of the wafer to impurities that alter the electrical characteristics of the silicon. The thermal oxidation, masking, etching, and doping steps are then successively repeated for each new pattern, or layer. A given chip may have 20 or more layers precisely registered with each other for component and conductor locations.

After all the layers of circuits have been constructed, a process called metallization imprints metal wiring patterns on the wafer so that the individual devices are connected as needed. Finally, each chip on the wafer is tested electrically, and a diamond saw cuts the wafer into single chips.

The chips that perform correctly are then assembled into a package that provides the contact leads for the chip. A wire bonding machine attaches wires — each a fraction of the width of a human hair — to the leads of the package. Encapsulated with a plastic coating for protection, the chip is tested again prior to delivery to the customer.

Variety of ICs Produced. The semiconductor industry currently manufactures a wide variety of ICs that can be used in a virtually limitless array of products. The most common ICs include:

Microprocessor/central processing unit of a computer on a single chip. A chip with processing capability, but without extensive random access memory (ROM). A new architectural concept for microprocessors uses reduced instruction set computing (RISC), which speeds the processing of data by limiting the number of instructions the processor has to execute. The fastest RISC microprocessor processes information in chunks of 64 bits, each at a speed of 200 megahertz.

Digital signal processors (DSPs) — These devices represent the fastest-growing segment of the semiconductor industry. DSPs are specialized chips that perform high-speed addition and multiplication, digital filtering, data decoding, and formatting. One of the DSPs developed

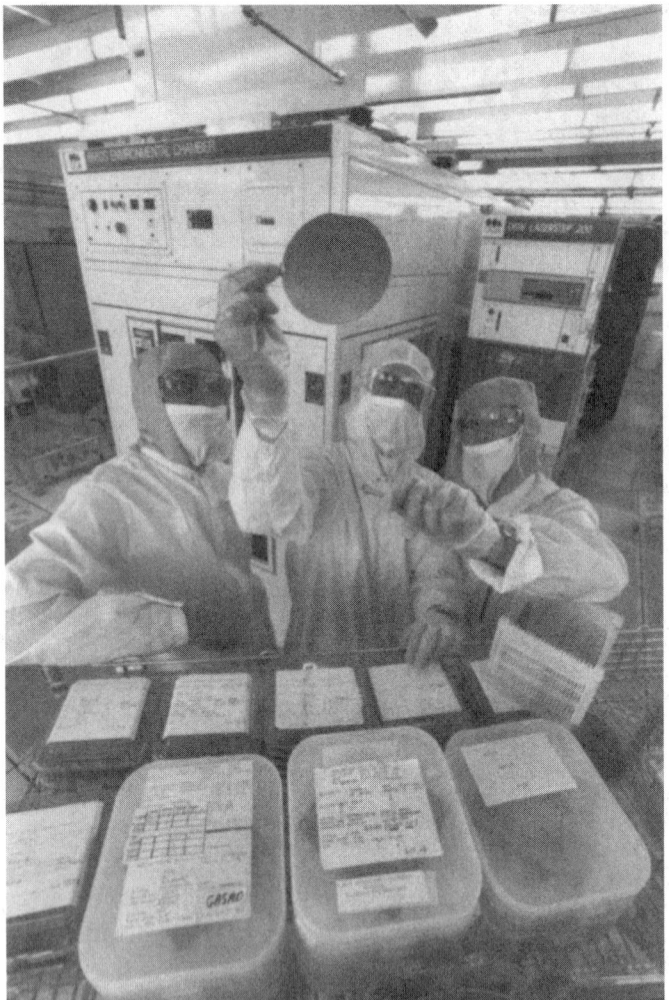

Fig. 3.   More complex and powerful integrated circuits are possible using a photolithographic resist that relies on deep ultraviolet radiation in the wavelength range of 200 to 300 nanometers. The resist, a thin film of which is coated onto the surface of a semiconductor wafer, can be used to produce highly complex chips with features as small as 0.3 micrometer, or roughly $\frac{1}{400}$ the diameter of a human hair. Three AT&T Bell Laboratories researchers in Murray Hill, New Jersey, inspect a wafer made using the advanced resist. (*AT&T Bell Laboratories.*)

by AT&T Microelectronics performs about 40 million instructions a second, with up to five basic operations within each instruction. This is substantially faster than a typical desktop computer. The power of DSPs to do digital computations on analog information — both sounds and images — ideally suits them for consumer electronic products, cellular telephones, and laptop computer modems. In addition, they also can be used in large products like high-definition televisions.

Telecommunications integrated circuits — A broad line of specialty chips used for managing signal traffic in telecommunications equipment and networks. Telecommunications ICs usually perform dedicated functions, such as converting voice signals into digital format for faster transmission and then, at the receiving end, back into voice signals again.

Hybrid integrated circuits and multichip modules — Essentially a different way to package ICs, hybrids and multichip modules are complete electronic circuits in which numerous, separately manufactured items are arrayed on a suitable passive substrate, such as ceramic or polymer, and interconnected by metallization patterns, very fine wires, or both. Hybrids and multichip modules afford faster communications between chips and faster system performance compared to printed circuit boards.

Application specific integrated circuits (ASICs) — Semicustom chips designed to be tailored by individual customers to meet specific requirements, thus saving time and cost. One type of ASIC, called standard cell, is constructed from basic circuit modules that can be

selected and wired together by the designer. Another type, called gate array, consists of identical circuit elements constructed on the wafer and later wired together according to customer requirements. An alternative approach, called field-programmable gate arrays, allows customers to program their own connections among circuit elements.

Memory chips — In addition to read-only memory, memory chips include two principal types: dynamic random access memory (DRAM) and static random access memory (SRAM). DRAMs store information in the form of binary digits, represented by the presence or absence of an electrical charge in capacitors. The stored charge must be refreshed periodically by circuitry to counteract the leakage of the capacitors. SRAMs hold binary information in a digital logic circuit, whose charge does not have to be periodically refreshed. Like DRAMs, SRAMs lose memory when power is turned off. See Fig. 4.

One problem facing researchers is that ever-increasing circuit density and performance depend on ever-shrinking lines, or features, that define circuit elements. These now measure less than 1 micrometer. Soon the lines won't be visible with optical microscopes because they will be smaller than the wavelength of light. As a result, light-based lithography, or photolithography, likely will give way to new forms of lithography based on new energy sources. Among the options being examined at Bell Laboratories and elsewhere are the use of X-rays and electron beams to carry on the lithography process. In the meantime, scientists have continued to extend the life of silicon technology by shrinking the size of features using photolithography. As of early 1992, the microelectronics industry was manufacturing the most advanced chips, with feature sizes of 0.5 micrometers. Researchers estimate that, by about the year 2010, chips may have line sizes of 0.05 micrometers, well below the 0.125-micrometer level that is now under research.

Developers double the number of components on a microchip every few years, which means two or three new generations of chips were produced during the 1990s. That trend will have matured by the year 2010, with chips containing up to a few billion components each for highly packed circuits, such as memory. For typical custom logic circuits, the range is in the 10^8 to 10^9 region. The actual number of components per chip produced may be constrained by economics to be somewhat less than the physical limit. With advanced semiconductor technology, complete electronic systems will be put on a single chip, even systems as complex as today's supercomputers. Ideally, researchers look for ways to lengthen the lifespan of the existing technology, an approach consistent with good economics and good engineering. Besides the physical processing to make the chips, chip designers need additional automated tools to help them design and test advanced chips. High-level, modular design capabilities will continue to allow developers to focus on what they want the chip to do, rather than worrying about the details of the chip's 50 million or more transistors, an impossible task for humans.

Future of Microelectronics. The familiar "bulk-effect" solid-state devices may mature in the early part of the next century. At that point, the smallest functional bulk-effect transistor operating at room temperature will measure about 400 atoms by 400 atoms. The next frontier may be based on single-electron devices and ballistic transistors, whose behavior is described by the laws of quantum physics as applied to single particles.

Semiconductors will play a central role in the future of the U.S. electronics industry — chips provide the fundamental building blocks in hundreds of products. According to the National Advisory Committee on Semiconductors (NACS), in 1990 semiconductors annually represented a $21 billion industry in the United States and a $63 billion world market. U.S. market share, which has declined steadily in the past 25 years, was 37%. Semiconductors are critical to the $384 billion U.S. electronics industry and the $751 billion world market for electronics. According to NACS, by the year 2000, world semiconductor sales are expected to rise to $200 billion, and the world electronics industry to $2 trillion. Thus, electronics, already the largest employer in the United States, holds the promise for strong growth in the decade ahead. And a key linkage is the dependence of all electronic systems on semiconductors.

In the "Information Age," the amount of information transmitted in communications and stored and processed in computers continues to expand at a phenomenal rate. To cope with this growth, faster, more powerful microchips are essential for the advancement of information products during the 1990s and beyond. The increasing computing power of ICs also promises to deliver a wide range of easy-to-use consumer and

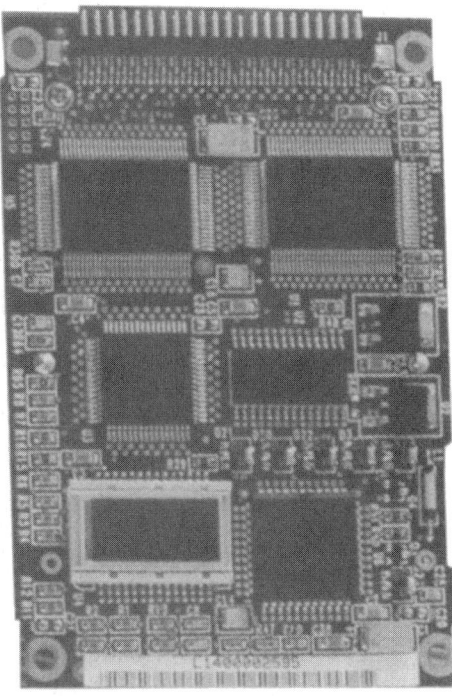

Fig. 4. At some future date when chips will be able to store data in large volumes and at a sufficiently low cost to compete with rotating-disk storage systems, a flash memory may be used. This type of memory has no moving parts and, due to its ruggedness, is already able to compete with electromechanical drive systems in products that are used in demanding environments. (*AT&T Bell Microelectronics.*)

Fig. 5. As chips become smaller and incorporate more functions, fewer of them are required to accomplish a task, such as sending and receiving digital information over a telephone line. Such is the job of a modem, and the few chips here are able to do that task in a package the size of a credit card and only 5 millimeters thick. Just a few years ago, modems were the size of a hardcover novel. (*AT&T Microelectronics.*)

hand-held information products that may, for example, respond to speech and handwritten input. See Fig. 5.

Advanced chips are needed for national and global high-speed networking of voice, data, images, and video for business and commerce, academia, research, and entertainment. New applications will become possible, such as intelligent systems for toll booths and highways that will read chips built into vehicles. By shortly after the turn of the century, it is likely that people will be able to talk directly with their computers as an adjunct to using keyboards. This will be possible, thanks to digital signal processors that can help understand human speech and generate a synthesized voice.

Video, speech recognition and synthesis, and character-recognition capabilities will be relatively easy just by adding a few chips to a product such as a laptop computer. See Fig. 6.

Fig. 6. The AVP1000 three-chip video codec makes it cost-effective for manufacturers to bring advanced video telephony and multimedia features, such as full-motion video and video-conferencing, to desktop personal computers and workstations. (Codec refers to "compression/decompression," a technique required to process image and sound information in a manageable digital form.) The chip set shown here compresses image information with sufficient rapidity to enable video to be sent through digital communication lines. The set, which replaces printed circuit boards (PCBs), also decompresses stored or transmitted video information to recreate a motion display. (*AT&T Microelectronics.*)

Products using such capabilities will range from individual electronic student tutors to high-performance military systems. Speech recognition and speech synthesis will mature in a variety of service capabilities based on the ability of intelligent machines to talk and listen much as people do. For example, by the year 2010 we should have the ability to transfer unique personal speech characteristics across languages, to make possible customized speech translators. Speech in one language would be automatically translated into a second language, which might then be synthesized with the voice characteristics of the original speaker. And this translation would be done in real time.

During its lifetime, microelectronics has been one of the most spectacular forces in the history of science and engineering. It has been the most powerful driver of progress in the communications field and also was at the forefront of breakthroughs in many other industries, including aerospace, computers, and consumer electronics. Microelectronics promises to continue as a major force well through its maturity during the coming decade, reaffirming that "smaller is better."

Additional Reading

Jaeger, R.: "KC's Problems and Solutions for Microelectronic Circuits," Addison Wesley Longman, Inc., Reading, MA, 1988.
Sedra, A. and K. Smith: "Microelectronic Circuits," Oxford University Press, Inc., New York, NY, 1997.
Smith, K.: "KC's Problems and Solutions for Microelectronic Circuits," Oxford University Press, Inc., New York, NY, 1997.

THOMAS K. LANDERS, AT&T Bell Laboratories, Short Hills, NJ.

MICROFILM AND MICROFORM.   Prior to the advent of convenient computer systems for storing long-term information, microfilm (photographic reduction on 16- or 35 mm film) was widely used. Microform, microfiche, and microcard were variations of this technique. For some applications, some users still use these techniques.

MICROGRAVITY AND MATERIALS PROCESSING.   The development of space transportation systems during the past several years has drawn attention to materials processing in a reduced gravity environment. Actually, exploratory work in this area has been proceeding since the early *Apollo* missions to the moon. For example, materials processing experiments were carried out during the *Skylab* program and on the *Apollo-Soyuz* test program fight. More recently, tests have been conducted on some of the *Space Shuttle* projects.

Effects of Reduced Gravity. The reduced gravity of space offers a unique environment for materials processing. It is untrue, of course, to claim that a state of zero gravity is achieved as, for example, on space flights as currently experienced. Rather, the gravity is greatly reduced — to about 10^{-6} g. Hence, the term *micro* rather than *zero* gravity is more appropriate to use here. It is interesting to note that spacecraft in low earth orbit may experience changes in the gravitational field, referred to as *g-jitter*, which are caused by such factors as maneuvering the craft while in orbit, atmospheric drag, and even movement of the crew within the spacecraft. The *g-jitter* phenomenon can cause "spikes" in the gravitational field, ranging from 10^{-2} g to 10^{-1} g. These may have an adverse effect upon on-board experiments and processes.

Manned or unmanned orbital spaceflight is not the only possibility for reduced-gravity studies of materials processing. Drop tubes and towers, aircraft, and sounding rockets also offer opportunities for varying levels and time periods of reduced gravity. Time spent in low gravity and payload size are the limiting factors for earth-based facilities. Drop tube experiments provide 2.5 to 4.5 seconds at 10^{-8} g to 10^{-9} g for free falling droplets, whereas a drop tower allows an entire experimental package, weighing 100 kg, to be tested. By flying in parabolic flight paths, aircraft ranging in size from a KC-135 to a single-seat F-104 can provide 10^{-1} g to 10^{-2} g for 15 to 60 seconds for payloads ranging from 10 to 35 kg. Sounding rockets provide up to 300 seconds at 10^{-5} g. When compared with the duration time provided by an orbiting space platform, the earthbound methods are quite brief.

One effect of the reduction of gravitational forces experienced in space is the virtual elimination of *buoyancy-driven convection*. Convection within a fluid medium arises when the medium is subjected to a nonvertical thermal gradient. Temperature differences create density gradients as gases or liquids expand upon heating. In the presence of a gravitational field, the less dense volume of the medium is displaced by a denser, cooler volume, resulting in the circulation of gases or liquids, commonly known as convective flow.

A more unstable form of convection occurs when a denser fluid lies above a less dense fluid, corresponding to a situation where the medium is heated from below. If viscous forces outweigh the buoyant forces within the medium, this unstable condition can be maintained. If the buoyancy is greater than the viscosity, however, the volume element rises too quickly, resulting in spontaneous flow, which takes the form of cells or vortex rolls.

Concentration gradients caused by chemical reaction within a fluid medium may also cause convection. In this case, density gradients occur when a particular chemical component is consumed or produced by reaction.

Convective flow is mathematically characterized by the Grashof number *Gr*, which represents the ratio of buoyant to viscous forces. The Grashof number is given by the expression

$$Gr = \frac{gl^3 \beta \Delta T}{\mu^2}$$

where g is the gravitational acceleration; β the coefficient of expansion; ΔT the temperature gradient; and μ the kinematic viscosity. In an earth-based experiment, convection can be reduced somewhat within practical limits by altering the geometry of a given system or minimizing temperature gradients. Conducting a similar experiment in low earth orbit can reduce the value of *Gr* by six orders of magnitude.

Because convection and diffusion occur simultaneously in an earth-based process, corresponding studies in a microgravity environment can help identify the effects of these two phenomena on a given process such as crystal growth.

The reduced gravity of space can also be used to process materials in a container-free environment. This feature of microgravity processing is particularly advantageous when a material of high purity is desired, or for achieving a high degree of supercooling in a sample. Containerless processing is especially useful for obtaining glasses and alloys.

The presence of a gravitational field causes substances of differing densities to separate out. In microgravity, however, this gravity-induced separation is eliminated, thereby producing a more uniform mixture. This effect is useful in the processing of alloys and organic polymers.

Representative Microgravity Experiments

Biological Materials. The degree of purity of biological materials severely limits their usefulness. Electrophoresis is a commonly used method of separation and purification of substances such as cells, enzymes, and proteins. This technique relies upon the fact that surface charge distribution, and thus mobility in an electric field, vary from one material to another. The degree of separation, product yield, and purity are limited by convection which is caused by concentration gradients within the process medium.

A continuous flow electrophoresis (CFE) process has been used to effect separation of biological materials in microgravity. The absence of convection permits continuous processing of relatively large volumes of material, higher yields, finer separation, and higher product purity than are possible on earth. Erythropoietin, which is produced by the kidneys and controls red blood cell production in the body, has been produced by CFE on board the *Shuttle*. The first CFE experiment performed in 1982 in *Spacelab* yielded 463 times more material than comparable earth-based processes. Separation rates were boosted in later flights to yield 700 times more material having a fourfold increase in purity over products obtained on earth.

Polymers. Research efforts in the area of organic polymer growth in space seek to take advantage of the absence of phase separation due to density differences. On earth, density differences in nonhomogeneous mixtures of organic liquids produce buoyancy-driven convection and cause immiscible liquids to separate. These phenomena affect the growth of organic polymers, causing flaws in the final product. In the absence of phase separation and convection, more uniform mixtures can be produced. Secondary effects such as surface tension can also be utilized to obtain more perfect polymers and organic compounds.

One type of experiment performed in space was the diffusive mixing of organic solutions (DMOS) study conducted by the 3M Corporation. The DMOS experiments mixed different types of organic solutions to yield crystalline material. The purpose of the study was to determine the effect of microgravity upon the ordering of organic molecules upon crystallization. The crystals grown in the experiment were not only significantly larger than similar crystals grown on earth, but possessed much better optical and electrical properties as well.

The dominance of surface tension, due to the lack of convection in microgravity, has been used to produce perfectly round spheres of polystyrene-latex. The spheres are grown in space by the coalescence of an emulsion. Under conditions where surface tension controls the process, droplets do not readily break up, thereby allowing large spheres to coalesce. As a result, large, perfect spheres, having a diameter of up to 30 mm, can be produced in space. In comparison, a maximum diameter of 5 mm can be produced on earth. These spheres are offered commercially for reference and calibration applications. As such, they are the first commercial products to be made in space.

Physical Metallurgy. One obstacle to the processing of alloys on earth is that components of a given mixture are often immiscible. As a result, density differences cause the components to separate as the bulk melt cools. By eliminating this gravity-driven separation, the manufacture of alloys can benefit from a microgravity environment.

Several alloy systems have been studied in space. In general, these experiments have yielded promising results, showing that finer, more homogeneous mixtures of components can be obtained in microgravity. In space, reduced convection in the melt apparently reduces microsegregation and heat transfer. This allows materials possessing highly directional physical properties, such as magnetic coercivity and microstructure, to be produced.

Containerless processing is also of interest in physical metallurgy, as it provides opportunities to study thermophysical properties of high temperature metals and alloys, avoid sample contamination due to contact with container walls, and observe the solidification of materials that have been rapidly cooled from the melt.

Containerless processing is accomplished on earth under the influence of gravity by using electrostatic, acoustic, or electromagnetic energy to levitate a sample. Sample size is limited, however, by power requirements for levitation. The application of the forces necessary to levitate a substance also induces a certain amount of mixing and heating. Gravity-driven convection is also present in this situation and can cause unwanted mixing of liquid samples.

To date, containerless processing in near zero gravity has been limited to drop tube experiments, which provide a few seconds of low gravity for small drops of material. Alloys studied in this fashion have been undercooled as much as 500 °C, which corresponds to a cooling rate of greater than 10^6 K/sec. Samples obtained in these experiments have exhibited metastable or peritectic phases which are extremely difficult to obtain under normal conditions.

Some levitation of samples will still be necessary to carry out containerless processing in space, although the magnitude of the forces necessary to do so will be small relative to those required in earth-based work. Levitation would be necessary only to avoid contact between sample and container, so larger samples could be used in space-based experiments.

Other metallurgy experiments scheduled for space will examine the role of macrosegregation in the processing of metals, the feasibility of using directional solidification in the processing of different classes of alloys, and the manufacture of alloys which cannot be produced on earth due to density induced separation.

Glasses. The manufacture of glasses also benefits from containerless processing in space. As with alloys, the purity of glasses can be affected by contact and subsequent reaction with container walls. In the case of glass processing, however, contact between sample and reactor wall also causes crystallization, and hence loss of the amorphous glassy state. In addition, less viscous glasses require high cooling rates in order to prevent crystallization. By avoiding contact-induced nucleation during cooling, containerless processing may be used to obtain larger, high purity samples of such glasses.

As mentioned previously, some levitation is required in microgravity to maintain sample positioning. Several designs, some of which are capable of processing temperatures up to 1600 °C, have been developed for use in space experiments.

Glass processing experiments that have already been flown on the *Shuttle* have been concerned with melt homogenization, bubble behavior in molten glass spheres, preparation of glass microballoons, and comparing properties of space-produced glasses with those manufactured on earth. Results to date indicate that glasses having different microstructures than those of glasses processed on earth can be obtained. Galliacalcia, sodium-borate, and lead-silica glasses have been selected for the above experiments. The list of glasses to be studied will be expanded to include materials that are particularly difficult to produce on earth. Among these materials are heavy cation (Zr, Hf, Th) glasses that tend to react with containers, and silica glasses that must be processed at high temperatures.

Crystal Growth. Single crystals of both organic and inorganic substances can be grown from either the vapor or liquid phase, using several different experimental techniques. Gravity-driven convection affects the motion of these fluid media, greatly affecting the mixing and transport of individual chemical components. Experiments aimed at examining the effects of eliminating buoyancy-driven convection upon different crystal growth techniques have been performed on *Skylab*, the *Apollo-Soyuz* mission, and several *Space Shuttle* flights.

Growth of crystals from the vapor relies upon the presence of a temperature gradient. Concentration gradients of vapor species and subsequent migration from a source region to a seed crystal, substrate, or deposition region are caused by this temperature difference. In the case of physical vapor transport (PVT), solid source material vaporizes at one temperature. The gaseous vaporization products migrate, usually through an atmosphere of inert gas, to another temperature where solid material condenses. In space, the PVT method has been used to grow highly ordered organic thin films onto silicon wafer substrate and large single crystals of germanium selenide.

Another PVT experiment examined the growth of HgI_2 onto a seed crystal. This substance has potential for use as a radiation detector. Due to its high density, however, the HgI_2 crystal structure readily deforms during earth-based processing. Large crystals of HgI_2 have been grown by the PVT method aboard *Spacelab 3*. Growth times in space were considerably less than those normally required on earth. Performance of the space-grown crystals as radiation sensors is matched by only the very best crystals obtained on earth.

Unlike PVT, chemical vapor transport (CVT) utilizes a highly reactive gaseous substance—such as a halogen or metal halide—to transport source material to a region of the reaction container where single crystals condense from the vapor. In earth-based CVT studies, under conditions where buoyancy-driven convection drives the overall transport process, crystal size is generally small. The morphology of crystals grown under these conditions is often poor; surfaces are marked by large numbers of defects and irregular growth steps. In contrast, space-grown crystals grown by chemical vapor transport are much larger, have smoother growth steps, and fewer defects. Chemical homogeneity within these crystals is also considerably better than in similar crystals grown on earth.

The presence of convection also affects crystal growth from the melt. Single crystals of Te-doped InSb were grown from the melt on *Skylab*. The crystals obtained in space were free of striations caused by convection-driven growth rate fluctuations that are normally seen on earth. Future space experiments will examine the growth of electronic materials such as GaAs from a solution subjected to an electric current.

ROBERT P. SANTANDREA, Ph.D., Los Alamos National Laboratory, Los Alamos, NM.

MICROMANOMETER. A manometer capable of measuring very small pressure changes or differences.

MICROMETEORITE. A very small meteorite or meteoritic particle with a diameter in general less than a millimeter. See also **Moon (Earth's)**; and **Poynting-Robertson Effect**.

MICROMETER. The micrometer represents a general principle of physical measurements, used on various instruments such as comparators; spherometers, compensators, interferometers, etc. It is essentially a screw of accurately known, uniform pitch (commonly 1 millimeter or 0.5 millimeter), provided with a large head whose periphery is divided into equal parts, forming a scale. Turning the screw through a given number of these parts causes the shaft to travel through a distance which is a proportionate fraction of the pitch. For example, if the pitch is 0.5 millimeter and the head is divided into fiftieths, each scale division corresponds to a travel of 0.01 millimeter.

Micrometer also is a unit of length. See also **Units and Standards**.

MICROPHONE. An electroacoustic transducer that responds to sound waves and delivers essentially equivalent electric waves. Making this conversion as faithfully as possible is one of the most important criteria in microphone design. The transducer should be linear for minimum waveform distortion and should have the greatest possible dynamic range. The transducer should have minimal effect on the dynamic phenomenon and whatever effect it does have should be determined by calibration or computation so that data can be appropriately corrected.

Microphones are of two basic types: (1) *pressure-sensing*, used to observe the pressure waveform of sound; and (2) *velocity-sensing*, used to observe the velocity waveform. For measurement purposes, the pressure-sensing microphone usually is preferred because of its superior frequency response and dynamic range.

Pressure-sensing microphones used for measurement purposes operate on the capacitive, piezoelectric, or moving-coil principle. In a pressure microphone, the electric output substantially corresponds to the instantaneous sound pressure of the impressed sound wave. When stability and a broad flat frequency response are required for accurate measurement, a condenser (capacitive) microphone is used.

Condenser microphone. In a capacitive transducer, mechanical-to-electrical conversion is effected by mechanically induced changes in electrical capacitance. The capacitance between two conducting plates, separated by an insulator, is inversely proportional to the distance between the plates, provided that the plates are large enough relative to their separation so that edge effects can be ignored.

When the metal diaphragm of a condenser microphone is displaced by acoustic pressures, the diaphragm produces a capacitance change between the diaphragm and a rear electrode. Air is the insulator between the conductors. The components of a simplified condenser microphone are shown in Fig. 1. The small hole in the rear cavity behind the diaphragm is to equalize static pressure across the diaphragm. The most common method of sensing capacitance change in this type of microphone is to apply, between the diaphragm and the rear electrode, a large dc voltage. This is called the polarizing voltage and should come from a high-resistance source. The voltage decreases as the distance between the conductors decreases and vice versa. This voltage change can be amplified and observed on a display instrument as the electric analog of the acoustic pressure waveform.

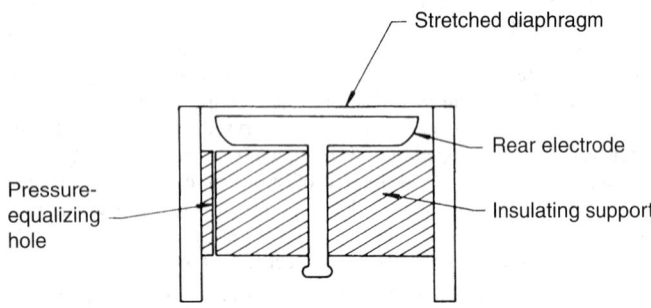

Fig. 1.   Components of a simplified condenser microphone.

The sensitivity of the transducer is the ratio of its electric output to its mechanical input. Thus, the sensitivity is the slope of the linear range of the input-output characteristic. Condenser microphone sensitivities fall in the range from -50 dB to -100 dB re 1 volt per dyne/cm^2. The flat response range, wherein sensitivity is relatively constant, generally is from less than 100 cycles/sec (Hz) to several thousand Hz. While the polarizing voltage technique is simple and practical, it requires very high electrical resistance for adequate low-frequency response and to ensure that the charge does not change as the diaphragm moves.

The resonant circuit technique also has been used with capacitive transducers. This technique does not require high-resistance circuitry and responds to static pressures, but the response is less linear than the polarizing voltage technique and requires more complex electronic circuits.

Piezoelectric Microphones. These can be substituted when it is not convenient to use the peripheral electrical circuitry essential to the capacitive type—and when some loss of accuracy can be tolerated; or when the sound pressure is high enough to overload the condenser microphone. A piezoelectric transducer normally used is an accelerometer. Voltage sensitivities of piezoelectric accelerometers typically range from 1 to 100 mV/gram. The piezoelectric accelerometer is used widely because it is simple, rugged, and requires no outside power supply.

Carbon-button Microphones. These employ a packet of carbon granules, which the sound field acts upon through a linkage to vary the resistance of the packet. With a source of direct current, this device produces a dc voltage proportional to the pressure.

Moving-coil Microphone. This device operates on the same principle as an electric motor/generator and the loudspeaker. That is, when an

electrical conductor is moved through a magnetic field, a voltage is induced in the conductor that is proportional to (1) the strength of the magnetic field; (2) the length of exposed conductor; and (3) the velocity of motion. Loudspeakers sometimes are used in reverse as receivers of sound, but their size and uneven frequency response limits their utility. See also **Loudspeaker**. Microphones designed on the moving-coil principle are used widely on sound stages and in recording studios. Moving-coil (dynamic) microphones work well with low-impedance electrical circuitry, but since their electrical output is proportional to the velocity of coil motion, the devices are unresponsive to very low frequency or quasi-static pressure changes.

Ribbon Velocity Microphone. Also termed the pressure gradient microphone, this device employs a metal ribbon suspended between the poles of a magnet. The pressure gradient in the acoustic field, acting on both sides of the ribbon, induces a voltage across the ribbon that is proportional to the sound pressure.

Electrostatic Microphone. This is a capacitor type microphone wherein variations in sound pressure, acting on a stretched metal diaphragm, effect variations in capacity. This microphone normally is used in a dc circuit with a voltage source and a resistor to develop a voltage across the resistor; or in an electronic-tube circuit, to modulate the frequency of an oscillator. The output is proportional to pressure in either application.

Directivity of Microphones. This is the relative sensitivity of the microphone to the angle of arrival of a plane wave perpendicular to a specified axis. Most commonly, microphones are omnidirectional; bipolar; or cardioid. See Fig. 2. Omnidirectional microphones usually are of the pressure type. Velocity microphones have bipolar response, showing high sensitivity to the front and rear, but little to the sides. Cardioid microphones are unidirectional, highly indifferent to sounds from outside the cone of prime sensitivity.

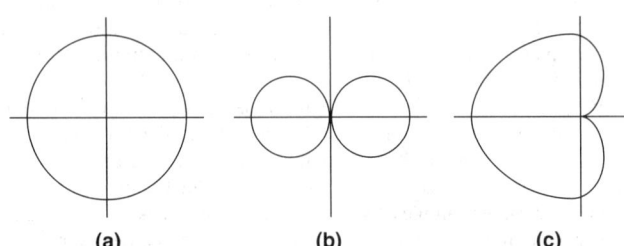

Fig. 2.   Directivity of microphone: (**a**) omnidirectional; (**b**) bipolar; (**c**) cardioid.

See also **Acoustics** for description of electret and other recent developments in microphone design.

MICROPROCESSOR (Chronology). The microprocessor, in terms of concept and reduction to practical applications in thousands of end products, will be recorded in the history of the electronics industry as a truly major achievement. The microprocessor also has had a profound effect upon the design of nearly all forms of electronic equipment, including *smart* or *intelligent* sensors, transducers, and controllers.

A microprocessing unit (MPU) is a computer central processing unit (CPU) built as a single semiconductor chip. (In some early designs, an MPU consisted of a small number of chips.) The difference between a microprocessor and a microcomputer has progressively grown fuzzy. Traditionally, the microcomputer is somewhat more complete than a MPU. This is because the microcomputer contains not only the CPU logic, but also memory for storing programs and I/O (input/output) data interfaces for exchanging data with peripheral devices, and timing circuits to control the flow of data within the computer, and the Advanced Micro Devices *Am29000*, among others. Several giant firms entered the microprocessor manufacturing arena, producing a highly competitive market during the early 1990s. The microprocessor has continued to advance in capacity and performance.

A microcomputer is frequently a single-board computer containing several to many MPUs. Marked enhancement of the MPU in the relatively recent past has further muddied the distinction.

History records that the first commercial microprocessor (the *4004*) was introduced by Intel in 1971. Attention to standardized architecture lowered the cost of the chip design and development, permitting amortization over many more units. Intel introduced the *4040* and *8008* shortly thereafter, and the *8080* in 1974. By 1976, several manufacturers were supplying 8-bit (basic word length) microprocessors in large volumes, significantly reducing unit cost. In 1980, more powerful 16-bit MPUs were developed, examples of which were Texas Instruments' *9900*; Intel's *8086*, Zilog's *Z8000*, and Motorola's *MC 68000*. One of the first 32-bit MPUs was introduced in 1984, the Motorola *MC 68020*, with its emphasis on higher execution speed, larger address space (4 billion locations) and more instructions relative to the 16-bit *MC 68000* chip. The *MC 68020* used the equivalent of about 200,000 transistors on an integrated circuit chip measuring about 9.5×8.9 mm. In 1986 and early 1987, a number of other manufacturers announced the availability of 32-bit MPUs, including the Fairchild *Clipper*, the Motorola *68030*. See Fig. 1.

See also **Microelectronics**.

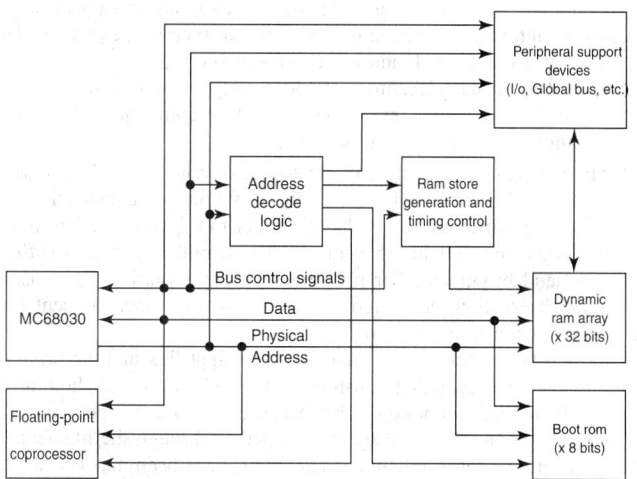

Fig. 1. Circuit organized around a microprocessor—in this case, the Motorola 68030 introduced in 1986 (32-bit). Block diagram illustrates a representative example, one that is useful in numerous cost-sensitive, performance-oriented applications that may range from real-time process control to the execution of multitasking virtual-memory operating systems. The system may consist of some address decode logic, a RAM control section, a RAM array, a group of standard input/output devices and interface modules, and optional elements, such as a floating-point coprocessor and boot ROM.

The address code logic monitors the physical address outputs of the 68030. During a bus cycle, it generates a select signal for the specific device being accessed by the 68030. The logic unit could also provide timing control for simple devices with fixed access times, as in the case of the boot ROM. The on-chip cache memory of the 68030 and paged memory-management unit do much of the work that, in another system, would be handled by the decode logic. The peripheral section shown is the same type found in prior existing systems. It may contain serial and parallel I/O devices, DMA controllers, an interface to a global or extension bus, and other functions. The RAM control section is the standard control logic required to maintain and access the RAM array as modified to support the 68030's burst-mode bus operations, this being a feature of the microprocessor's efficient on-chip bus-interface unit. The burst mode takes advantage of dynamic RAMs with a nibble mode or equivalent to shorten access time to the array and to burst-fill the 68030's two on-chip caches with fresh data or instructions. Modification to the RAM control increases the cost slightly as well as complexity, but yields a significant increase in overall performance. Delivery of the 68030 commenced in quantity in 1987. (*High-End Microprocessor Products Group, Motorola, Inc.*)

MICROPROGRAM. Microprogramming is a technique of using a special set of instructions for an automatic computer. Elementary logical and control commands are used to simulate a higher-level instruction set for a digital computer. The basic machine-oriented instruction set in many computers is comprised of commands, such as add, divide, subtract, and multiply. These are executed directly by the hardware. The hardware actually implements each function as a combination of elementary logical functions, such as AND, OR, and EXCLUSIVE OR. The manner of exact implementation usually is not of concern to the programmer. Compared with the elementary logical functions, the add, subtract, multiply, and divide commands are a higher-level language set in the same sense that a macrostatement is a higher-level instruction set when compared with the machine-oriented language-instruction set. See also **Macroassembler**.

In a microprogrammed computer, the executable (micro) instructions which may be used by the (micro) programmer, are comprised of a logical function, such as AND and OR and some elementary control functions, such as shift and branch. The (micro) programmer then defines microprograms, which implement an instruction set analogous to the machine-oriented language-instruction set in terms of these microinstructions. By use of this derived instruction set, the systems/application programmer writes programs for the solution of a problem. When the program is executed, each derived instruction is executed by transferring control to a microprogram. The microprogram is then executed in terms of the microinstructions of the computer. After execution of the microprogram, control is returned to the program written in the derived instruction set. The microprogram typically is stored in a read-only storage (ROS) and thus is permanent and cannot be changed without physical replacement.

An advantage of microprogramming is increased flexibility. This can be realized by adapting the derived, machine-oriented instruction set to a particular application. The technique enables the programmer to implement a function, such as square root, directly without subroutines or macrostatements. Thus significant programming and execution efficiency is realized if the square-root function is commonly required. Also, the instruction set of a character-oriented computer can be implemented on a word-oriented machine where adequate microprogramming is provided.

See also **Program (Computer)**.

THOMAS J. HARRISON, IBM Corporation, Boca Raton, FL.

MICROSCOPE (Traditional-Optical). An instrument capable of enlarging the angle under which a small object appears to the unaided eye at the distance of distinct vision (250 millimeters). The result is a magnified image of the object on the retina of the human eye. This image reveals details that are invisible without the magnifying power of the microscope. See also **Electron Microscope**; and **Scanning Tunneling Microscope**.

Optical Microscope

A microscope usually refers to a compound microscope. The compound microscope forms the image by a two-stage process and consists of an objective and an ocular. A simple microscope uses only objectives to form an image, e.g., a loop for visual use, or a camera set-up for photomicrography.

The total magnification of a compound microscope is

$$M = \frac{250}{f_m} = M_{\text{objective}} \times M_{\text{ocular}} = \frac{t}{f_{\text{objective}}} \times \frac{250}{f_{\text{ocular}}}$$

where f_m = focal length of the microscope, millimeters
$M_{\text{objective}}$ = magnification of the objective
M_{ocular} = magnification of the ocular (eyepiece)
t = optical tube length, millimeters
$f_{\text{objective}}$ = focal length of the objective, millimeters
f_{ocular} = focal length of the ocular, millimeters

Any magnification changer, e.g., a system with continuously variable focal length (zoom system) between the objective and eyepiece increases the total magnification by the tube factor of this system. With objectives ranging in power from about $1\times$ to $100\times$ (and even higher) and eyepieces ranging in power from about $4\times$ to $20\times$ and higher, it is possible to cover a magnification range from approximately $5\times$ to $2,000\times$. The total magnification M should be in a range given by $500 \times N.A. \leq M \leq 1,000 \times N.A.$ to obtain best results. $N.A.$ is the numerical aperture of the objective. Since the largest $N.A.$ of an objective is approximately 1.4, the upper limit of M is in the neighborhood of $1,400\times$. As M exceeds this figure, one enters the range of "empty magnification" without gaining additional resolution (or resolving power). The latter factor is expressed for incoherent light by

$$d = \frac{\lambda}{2 \times N.A.}$$

where d = smallest resolvable distance or detail,
$N.A.$ = $n \sin \alpha$ (n = efractive index of medium between object and objective; α = half-angle of cone of light entering the objective).

This formula indicates several ways to improve the resolving power, either by reducing the wavelength of light, or by increasing the *N.A.* (immersion objective), or both at the same time.

Besides sufficient resolution as one prerequisite for a satisfactory image, the object must be imaged with an optimum of contrast. To enhance the contrast, the specimen is either stained (amplitude object, contrast generated by absorption of light), or, if this is not possible (e.g., a living specimen, a phase object with negligible absorption, contrast near zero), the contrast is optically created. Common methods, for example, are phase contrast and interference contrast which utilize the interference phenomenon of light. Another method is darkfield illumination (illumination of the specimen by inclined pencils of light, thus permitting only light scattered by the object to enter the objective). The object appears bright on a dark background. Oblique illumination was often used before these methods became available.

Types of Microscopes. These instruments can be divided into two large groups: (1) microscopes for transmitted light, and (2) microscopes for incident (reflected) light. Within these groups, microscopes are commonly distinguished either according to their application (e.g., polarizing microscope, fluorescence microscope, and so on), or to their illumination system (e.g., brightfield microscope or darkfield microscope), or even by a combination of the foregoing factors (e.g., darkfield-fluorescence microscope).

Function of Microscope. The cross section of a microscope for transmitted light is shown in Fig. 1. The light path at the left explains the formation of the image of the specimen. The path at the right is that of the rays for formation of the images of the light sources and the pupils of condenser and objective. In reality, both paths have to be superimposed to obtain the true conditions of the optical system and its participation in image formation.

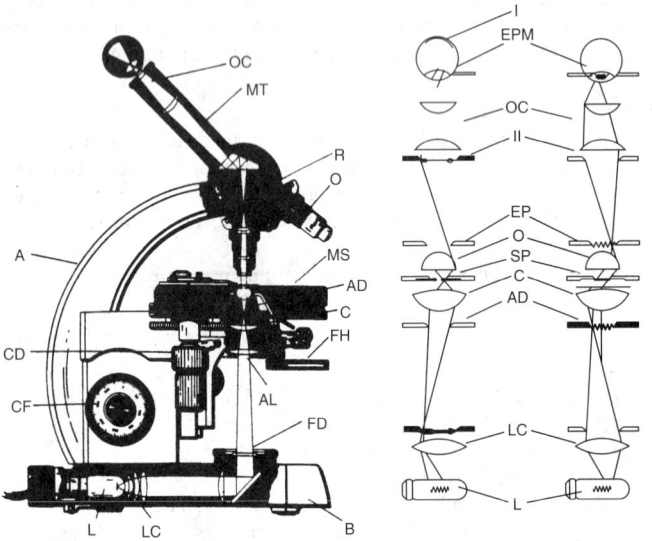

Legend:

A	Arm of microscope	FD	Field diaphragm (iris)
CD	Coaxial stage drive	AL	Auxiliary lens
CF	Coaxial coarse and fine motion knobs	FH	Swing-out filter holder, diaphragm (iris)
L	Tungsten light source	B	Base of microscope
LC	Lamp condenser	I	Image of object
OC	Ocular	EPM	Exit pupil of microscope
MT	Inclined monocular tube	II	Intermediate image plane
R	Revolving nose piece	EP	Exit pupil of objective
O	Objectives	SP	Specimen
MS	Microscope stage with built-in mechanical stage	SP	Specimen
AD	Aperture	C	Abbe condenser

Fig. 1. Microscope for transmitted light showing two main light paths. (*Carl Zeiss.*)

The light source is normally built into the base of the microscope. Koehler illumination is commonly used. This system utilizes the uniformly illuminated area of the lamp condenser and the field diaphragm as a light source. This diaphragm is imaged by the condenser into the specimen plane and by the objective into the intermediate image plane. The filament of the bulb is imaged by the lamp condenser into the plane of the condenser diaphragm. Thus, one illuminates only the field of the specimen imaged by

the objective and protects the specimen from excessive heat and reduces stray light. Reduced stray light results in improved contrast. An advantage of the Koehler illumination system is its capability of fully illuminating the exit pupil of the objective. Only then is its maximum resolving power available. Optimum contrast occurs with the aperture diaphragm closed to approximately $\frac{2}{3}$ to $\frac{3}{4}$ of its entire diameter. The microscopist, therefore, has to make a compromise between required resolution and contrast.

The critical illumination system images the filament of the light source into the specimen, thus creating there a higher intensity. However, it is very difficult to homogeneously illuminate the specimen (photomicrography). Critical illumination, therefore, is seldom used.

The light traverses the condenser, which illuminates a small field of the specimen. The objective forms an inverted and reversed magnified image of the object in the intermediate image plane of the microscope. The ocular further magnifies this image by the magnification factor of the eyepiece.

The human eye can be replaced by a photographic camera with or without an automatic exposure device. Several microscopes are available, with built-in cameras with automatic exposure control.

Microscopes usually feature a wide variety of interchangeable condensers, objectives, tubes, eyepieces, and other components to permit a rapid change from one application to the next.

Light Sources. The common light source of conventional microscopes for transmitted or reflected light is the low-voltage tungsten lamp. Its electric rating varies from 6 volts, 15 watts to 12 volts, 100 watts. For visual work in brightfield, 15 watt or 30 watt bulbs provide a sufficient light flux and brightness. For photomicrography, polarizing microscopy, darkfield observation, and interference microscopes, incandescent lamps of a higher wattage are preferred.

Tungsten lamps permit a variation of their light flux and the brightness of the image by a regulating transformer. In black-and-white photomicrography, this method is not applicable because the color temperature of the lamp is a function of the voltage and amperage through the filament. The color temperature must remain constant after it has been matched with the color temperature of the film by inserting various color-balancing filters in the beam. If this is not done, the color rendition of the film will deteriorate. The color temperature of tungsten lamps ranges from 2,850 K to 2,400 K. Special types reach 3,600 K.

Tungsten lamps are sensitive to overload. If the voltage increases by 10%, the illuminance will rise by 45%, but the life of the bulb will decrease by about 70%. In the other direction, if the voltage is 10% less than the rated voltage, the illuminance is reduced by 35%, but the life of the bulb is increased by 450%. The average life of a tungsten microscope bulb at rated power is approximately 100 hours.

The quartz iodide lamp is a special tungsten lamp, which produces a very high illuminance.

Projection microscopes require "white" light sources (color temperature, 5,000–6,000 K) and of an even higher illuminance. Discharge lamps (xenon lamps and metal halide lamps, but rarely mercury lamps) are used. These lamps require special electric power supplies and high voltage for ignition. Because of the large amount of heat generated by these lamps, they are installed in special lamp housings. The light flux of these lamps cannot be regulated by electrical means, such as a transformer. The life of the lamps ranges from 200 to 2,000 hours, depending upon the type of discharge lamp used and the power at which they are operated.

Xenon lamps also are used as continuous ultraviolet radiators in ultraviolet microscopy. The metal halide lamps emit a nearly white light and are used mainly for projection and polarizing microscopy. Mercury lamps are commonly used in fluorescence microscopy. They emit a blue visible radiation at the short wavelength end of the visible spectrum as well as the long wavelength ultraviolet radiation required to excite the fluorescence in the visible spectrum. Carbon arc lamps seldom are used as light sources.

Condensers. The condenser concentrates light on the specimen. Its numerical aperture must be variable so that it may be adjusted to that of the objective in order to obtain optimum resolution and contrast. The condenser images the field diaphragm into the specimen plane. Each method of illumination requires its own condenser. Condensers in modern microscopes are, in most cases, multipurpose units, e.g., one condenser for brightfield, phase contrast, and darkfield; or for brightfield, phase contrast, and interference contrast. Classes of condensers include: (1) condensers for transmitted light, (2) condensers for reflected light, and (3) special condensers for either transmitted or reflected light (for phase contrast,

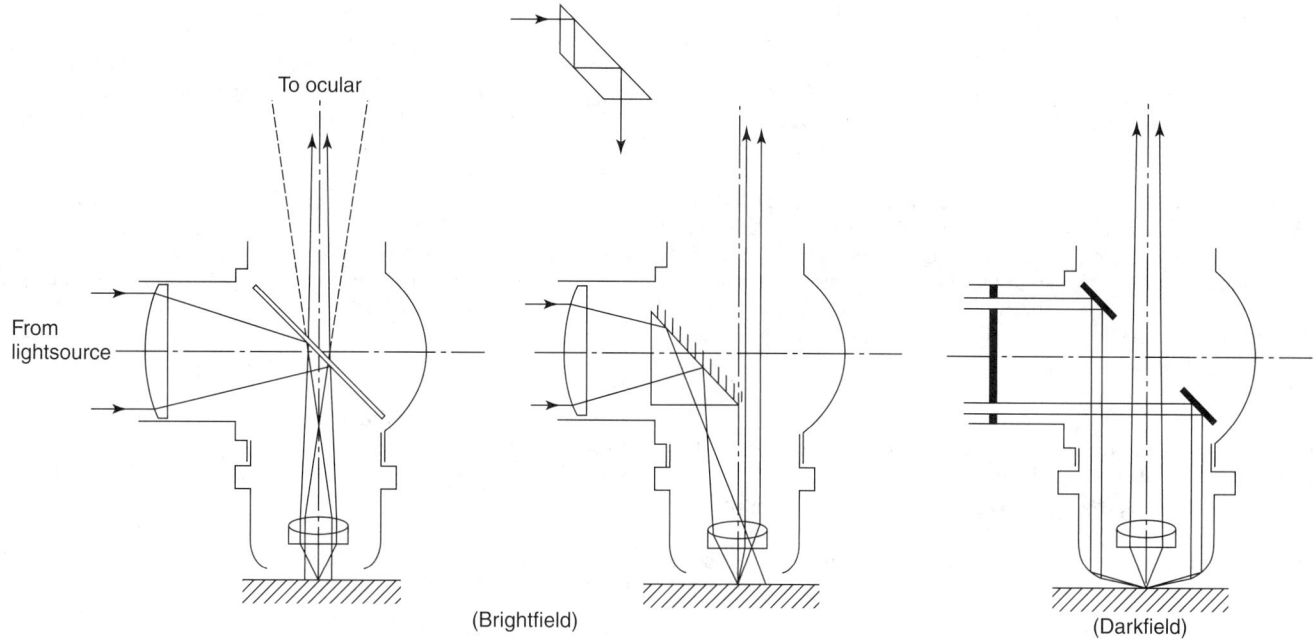

Fig. 2.   Illuminators for incident light.

interference contrast, phase contrast-fluorescence, ultraviolet microscopy, interference microscopy, pancratic condensers).

The simplest condenser for transmitted light brightfield is the Abbe condenser, consisting of one lens for numerical apertures up to 0.6 and of three lenses for numerical apertures up to 1.3. Since neither configuration is chromatically nor spherically corrected, they should be supplemented by highly corrected achromatic-aplanatic condensers (*N.A.* up to 1.4) for high quality microscopy, e.g., color photography with apochromatic objectives.

Incident light illuminates opaque specimens. The reflected fraction of the incident light forms the image according to the same principle as previously described. The objective (either for brightfield or for brightfield and darkfield) is its own condenser, as shown in Fig. 2. The reflector either is a thin plane parallel glass plate mounted at 45°, or a prism, the latter used mainly for polarizing microscopy.

Objectives. The microscope objective is the most important component of the optical system of the microscope inasmuch as it essentially determines the image quality. Typical objectives for one-stage imagery in photomicrography are *Luminars, Micro Tessars, Macro objectives, Summars, Milars, Photars*, and so on.

The majority of microscopes use the two-stage magnification previously described. The objectives are divided into different categories according to their optical design and type of correction.

Achromats. These are the simplest microscope objectives, corrected for equal back focal length of one blue and one red wavelength falling in the green region of the visual spectrum. Spherical aberration is fully corrected for one wavelength in the green region. They are ample for all routine work, although they show some field curvature. Planachromats, Flatfield Achromats, Plano Objectives form a "flat" or plane intermediate image, important for photomicrography.

Fluorite Objectives. When compared with achromats, the fluorite objectives incorporate one or two fluorite lenses to obtain an improved chromatic correction (reduced chromatic difference of magnification). The result is a higher *N.A.* than that of an achromat of the same magnification. Fluorite objectives form an image with a contrast superior to the achromats and their image quality is very close to that of apochromats. They are, therefore, often called *semiapochromats*. Synthetic fluorspar is used mainly for fluorite objectives. Fluorite objectives still show a slight field of curvature. They are rarely available as flatfield objectives.

Apochromats. These are the best and most expensive objectives. The back focal length is identical for three specific wavelengths in the blue, green, and red regions of the spectrum. Further, the apochromats are aplanatic for at least two wavelengths. A consequence of the excellent correction is the high *N.A.* obtainable with these objectives as compared

with fluorites of the same magnification. Apochromats also are available as flatfield objectives.

Objectives for Reflected Light. This group of objectives is usually provided as special flatfield achromats. They are corrected for use with specimens without cover glass. Any objective with *N.A.* > 0.3 forms a poor image due to increased spherical aberration when it is used for an uncovered specimen, but has been corrected for use with standard cover glasses of 0.17 millimeter (Europe) to 0.18 millimeter (United States) thickness.

Special Objectives. These systems are for use in the ultraviolet and consist of either quartz lens systems corrected for a wide wavelength (Carl Zeiss *Ultrafluars*, 250–600 nanometers), or a single wavelength (Vickers *Monochromats*); or they are mirror objectives (Leitz, Bausch and Lomb, and so on). To this group also belong objectives for interference microscopes, and special immersion objectives (water, methylene iodide, glycerine, and so on).

The described objectives also are available as phase contrast objectives.

Any of these objectives is corrected not only for use with or without cover glass, but also for a certain mechanical tube length (distance between shoulder of objective and shoulder of eyepiece—upper end of straight microscope with no optics in between objective and eyepiece). The commonly applied distance is 160 millimeters or infinity (American Optical Company) for transmitted light. For reflected light, the mechanical tube length varies between 180 millimeters and 250 millimeters and infinity, depending upon the manufacturer of the instrument. The objectives usually are designated by their type (except achromats), magnification, numerical aperture, tube length, and cover glass correction.

Oculars. The ocular at the upper end of the microscope tube enlarges the immediate image formed by the objective. Together with an attachment camera, it can also form the final image on a film plane. All oculars are derived either from the Huygens or Ramsden type.

Due to the low numerical aperture of oculars, only astigmatism, field curvature, distortion and the chromatic difference of magnification need be corrected. The commonly used type of oculars is the compensating eyepiece, whose chromatic difference of magnification is equal but opposite to that of the objective. For use with flatfield objectives, differently corrected oculars are used to fully utilize the performance of the flatfield objectives. Many eyepieces have a high eyepoint to permit the microscopist to wear his corrective glasses.

Special eyepieces are available for measurements (filar eyepieces, micrometer eyepieces, interference eyepieces, image-splitting eyepieces, and so on) and for teaching purposes (pointer eyepieces and demonstration eyepieces). The magnifications range from 5× to 25×, of which 8×, 10×, 12.5×, and 15× eyepieces are mainly used.

Photographic Equipment. Photography, both black-and-white and color, through a microscope plays an ever increasing role in modern microscopy. Attachment cameras are either manual, or semiautomatic, or fully automatic devices for all common film formats. Standard sizes are 35 millimeters, $2\frac{1}{4}$ inches $\times 4\frac{1}{4}$ inches, 4 inches $\times$ 5 inches. In addition in Europe sizes include 56×72 millimeters, 6.5×9 centimeters, and 9×12 centimeters. In a modern 35-millimeter attachment camera, a photomultiplier tube measures the light flux, an exposure device keeps the shutter open for the exact exposure time, depending upon the brightness of the image and the speed of the film emulsion. After the shutter is closed, an electric motor advances the film to the next frame. Filter factors, coverage of the field, and other factors can be compensated. In some cases, the automatic camera is built into the microscope stand. Cutaway view of a photomicroscope is shown in Fig. 3.

Special Microscopes

The basics of the transmitted light microscope just described apply, with few exceptions, to most special microscopes.

Metal Microscopes for Incident Light. The classic instrument of this type utilizes the principles of Le Chatelier. It is an inverted microscope with the objective underneath the specimen, which is placed with its polished surface facing down on the microscope stage. The size of these instruments varies from simple routine laboratory microscopes to the big bench-type (Reichert, E. Leitz, American Optical Company, Vickers, Unitron, and so on) to the console type (Bausch and Lomb) research instruments. Also, *upright* microscopes with incident illumination are used as metallographs (Zeiss), a term often used for research camera microscopes.

Plankton and tissue culture microscopes form the second group of inverted microscopes.

Phase contrast microscopes (after Zernicke) and *interference contrast microscopes* (after Nomarski, Françon, and so on). These utilize the interference of light to yield an image of high contrast for nonabsorbing phase objects. These objects alter only the phase of light according to the differences in optical thickness of the specimen. Since the human eye and the photographic emulsion are insensitive to phase variations, the phase contrast and the interference contrast system convert these phase variations into visible contrast.

The Phase Contrast Microscope. This microscope requires both a special phase condenser and phase objectives. Interference contrast is accomplished by means of polarized light, a Wollaston prism in the condenser, and a second prism in the exit pupil of the objective, or in a plane conjugate to it. The prisms are matched to certain regular objectives.

Interference Microscopes. These are used for quantitative microscopy in transmitted or reflected light to measure the roughness or flatness of

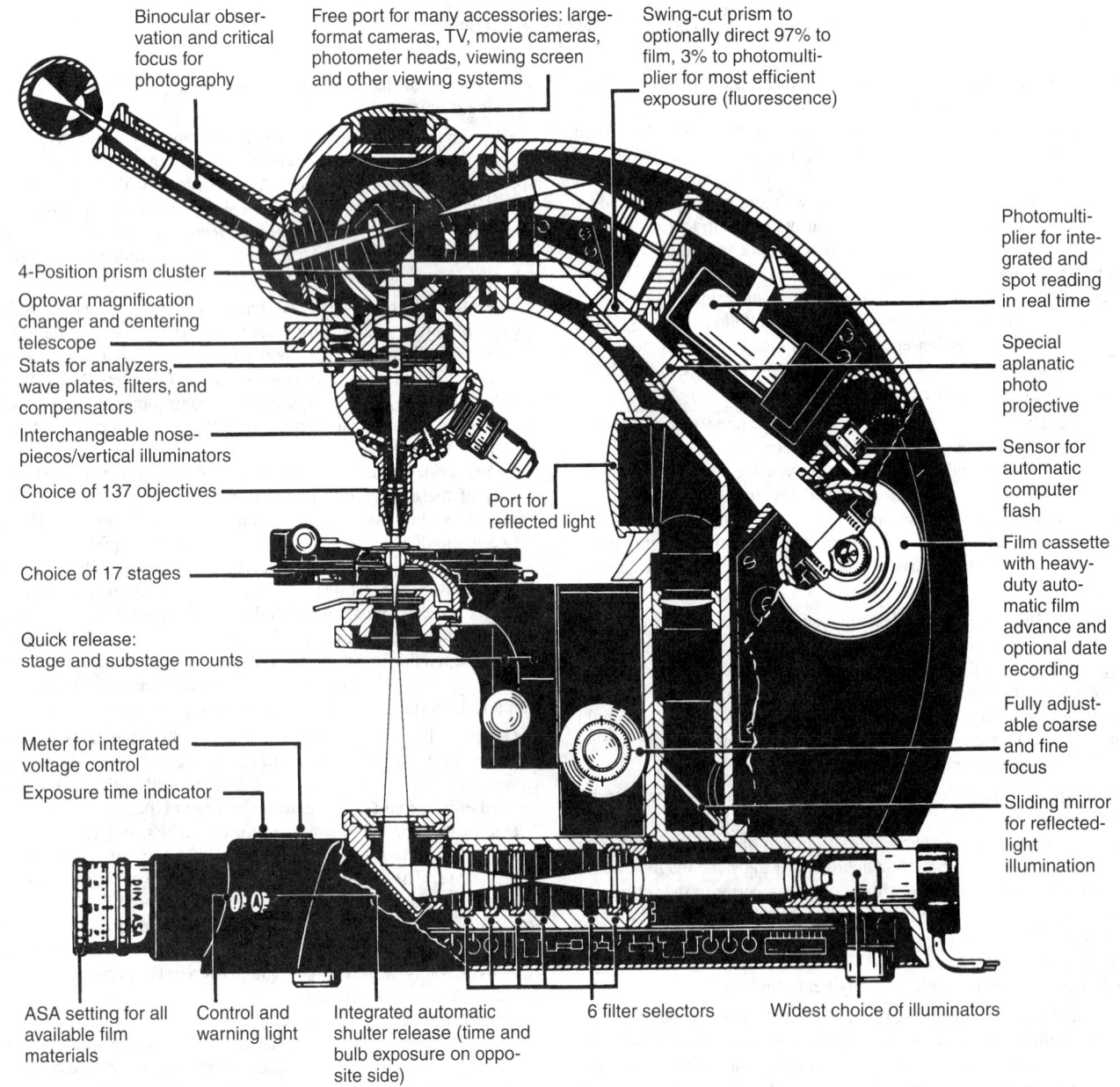

Fig. 3.   Cutaway view of Zeiss Photomicroscope III.

surfaces, thickness of coating, refractive index and dispersion of liquids or solid materials, and the dry mass of living or nonliving biological material—to name only the most important applications. The interference microscopes are either double beam (one beam acts as a reference beam; the second, traversing the object, as the measuring beam), or multiple beam instruments (interference by multiple reflection of rays). Due to the great variety of interference microscopes, further details are beyond the scope of this review.

Polarizing Microscopes. These are instruments for qualitative or quantitative work in either transmitted or reflected light. The majority of polarizing microscopes are used for quantitative studies of birefringent substances (e.g., in mineralogy, geology, metallurgy, fiber research, medicine, zoology) in linearly polarized light.

The substage polarizer converts the unpolarized light of the light source into linearly polarized light. Between objective and eyepiece, there is a second *polarizer*, called the analyzer. Both are in calibrated mounts rotatable to 180° or 360°, and they can be swung out of the path of light. All optical elements between the polarizers must be strain free. The objectives are individually centerable. The tubes permit the use of an Amici-Bertrand lens to observe interference figures in the exit pupil of the objectives (conoscopic path of rays), because the conoscopic image (e.g., together with a gypsum or quartz plate red first order) renders information on the type of crystal (examined e.g., uniaxial or biaxial, positive or negative). Most of the investigations require the polarizers crossed (directions of oscillation of polarizer and analyzer oriented at 90° to each other). The crosshair of the oculars is congruent with these directions. The stage, rotatable by 360°, is of high precision.

Compensators (Senarmont, Berek, Ehringhaus, Brace-Koehler). These measure the path difference between the ordinary and extraordinary beams introduced by the specimen. This difference value is the basis for computation of the characteristic features or constants of a crystal.

Fluorescence Microscopes. These are routine and research instruments for transmitted and reflected light and used to study specimens showing either primary (natural) or secondary (induced by fluorochromes) fluorescence.

In both cases, fluorescence must be excited by light sources emitting wavelengths below approximately 420 nanometers (as from a mercury lamp). The specimen absorbs this radiation and emits a fluorescence radiation of longer wavelengths in the visible range. Barrier filters between objective and eyepiece prevent light of the light source from hitting the observer's eye and allow only the fluorescence light to pass.

High aperture brightfield, or darkfield condensers illuminate the specimen, which is imaged through objectives having a numerical aperture as high as possible because the fluorescence is normally relatively weak. The image brightness increases with the square of the *N.A.*

Fluorescence microscopy is a very sensitive method. Fluorochromes diluted to 1 ppm can be detected without difficulty.

Ultraviolet Microscopes. These are equipped with quartz optics and work in the wavelength range from approximately 240 nanometers through the visible range, unless objectives corrected for a fixed wavelength are used. Unstained material on quartz slides showing no absorption in the visible often absorb considerably in the ultraviolet. Strong xenon light sources provide the ultraviolet light and ultraviolet-image converters present a visible image of the otherwise invisible specimen to the observer.

Additional Reading

Bradbury, S.: "An Introduction to the Optical Microscope," Oxford University Press, New York, NY, 1999.
Marmasse, C.: "Microscopes and Their Uses," Gordon & Breach Publishing Group, Newark, NJ, 1980.
Wilson, C.: "The Invisible World: Early Modern Philosophy and the Invention of the Microscope," Princeton University Press, Princeton, NJ, 1997.

W.G. HAVEMANN, Carl Zeiss, Inc., New York, NY.

MICROWAVE RADIATION. That portion of the electromagnetic spectrum that lies adjacent to the far-infrared region, is commonly identified by the term microwave. The range of microwaves extends from a wavelength of approximately 1 millimeter and frequency of 300,000 MHz to wavelength of 10 centimeters and frequency of 3,000 MHz. The early interest in microwaves was associated with the development of radar during the World War II years. The advent of the magnetron, an electric generator of high-power microwaves, made possible wartime radar at approximately

3,000 MHz and led to the utilization of waveguides for the efficient transmission of microwaves from the generator to the transmitting antenna and from the receiving antenna to the detector. See also **Microwave Tubes**.

The propagation of radio signals in space between transmitting and receiving antennas can be described in terms of ground waves, sky waves, and space waves. At microwave frequencies, ground waves attenuate completely within a few feet (meters) of travel; sky waves are influenced by the ionosphere and can penetrate into outer space, and space waves behave like light waves as they travel through the atmosphere immediately above the surface of the earth. Microwave space waves travel in a direct line of sight, can be reflected from smooth conducting surfaces, and can be focussed by reflectors or lenses. Their behavior is similar to light waves and they follow many of the rules of optics. See also **Waveguide**.

If a space wave is radiated from a point antenna, the radiated energy spreads out like an ever expanding sphere, and the energy of the wave front decreases inversely with the square of the distance from the antenna. The power that can be extracted from a wave front by a similar point antenna varies inversely with the square of the frequency. Thus, a point antenna receives power that is inversely proportional to both the square of the distance from the source and the square of the frequency. The ratio of the power received to the total power radiated is known as *path attenuation*.

When the receiving antenna is a parabola-shaped dish, the power extracted from the wave front is greatly increased. The ratio of the power received by such an antenna to the power received by a theoretical point antenna is defined as antenna gain. The gain of a parabolic antenna increases with the antenna area and the operating frequency. Thus, for a given microwave path with fixed-size antennas, the path attenuation increases with frequency, the antenna gain increases with frequency, and the over-all result is that one tends to offset the other.

Because microwave transmission follows essentially a straight line, reflectors are often used to redirect a beam over or around an obstruction. The simplest and most common reflector system consists of a parabolic antenna mounted at ground level that focuses a beam on a reflector mounted at the top of a tower. This reflector inclined at 45° redirects the beam horizontally to a distant site where a similar "periscope" reflector system may be used to reflect the beam down to another ground level. If two sites are separated by a hill, it may be necessary to use a large, flat-surface reflector referred to as a "billboard" reflector. In a typical system, a "billboard" reflector might be located at a turn in a valley, effectively bending the beam to follow the valley. Many arrangements are possible which, in effect, resemble huge mirror systems.

Microwaves are ideally suited for communication systems where a broad frequency bandwidth of the order of several megacycles is required for the rapid transmission of signals that contain a large amount of information, such as television. See also **Radar**; and **Telephony (Telecommunications)**.

MICROWAVE REGION.   That portion of the electromagnetic spectrum lying between the far infrared and the conventional radiofrequency spectrum. The microwave region is commonly regarded as extending from 300,000 MHz to 1,000 MHz (1 millimeter to 30 centimeters in wavelength). See also **Electromagnetic Phenomena**.

MICROWAVE TUBES.   To avoid the difficulties resulting from the short transit time of electrons as well as the effects of interelectrode capacitances, microwave tubes have been designed to make use of velocity-modulated beams of electrons. Traveling-wave tubes, magnetrons, and klystrons are special-purpose tubes designed to operate in the microwave region.

A *traveling-wave tube* is a broad-band microwave tube which depends for its characteristics upon the interaction between the field of a wave propagated along a waveguide and a beam of electrons traveling with the wave. In this tube, the electrons in the beam travel with velocities slightly greater than that of the wave, and on the average are slowed down by the field of the wave. The loss in kinetic energy of the electrons appears as increased energy conveyed by the field of the wave. The traveling-wave tube may, therefore, be used as an amplifier or as an oscillator. See Fig. 1.

A *magnetron* is a vacuum tube, which functions under the joint action of an externally applied magnetic field and the electric field between its anode and cathode. In one form, it consists of a cylindrical cathode and a coaxial anode structure. The anode may be a single cylinder or be split lengthwise for all or part of its length. These tubes were originally designed for use in ultra high-frequency oscillator circuits where conventional vacuum tubes could not operate effectively. The electric field is created by applying a

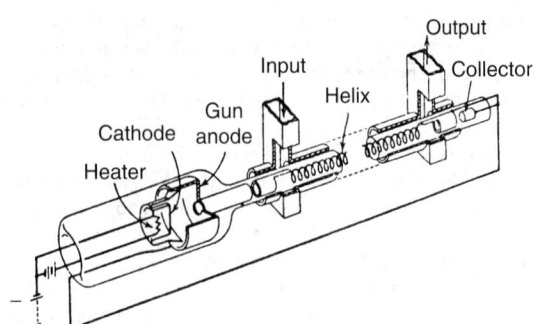

Fig. 1.   Schematic diagram of early traveling-wave amplifier demonstrates basic principles of device.

high, direct-current potential between the filament and anode structure, while the magnetic field is applied longitudinally by external permanent or electromagnets. When the tube is properly connected to a resonant line it can be made to operate as an oscillator for certain values of the applied fields. A magnetron of suitable dimensions can be made to generate frequencies measured in thousands of MHz (wavelengths of a few centimeters).

A *klystron* is an electron tube of the velocity-modulated type used in ultra-high frequency circuits. At these very high frequencies (hundreds or thousands of MHz), conventional vacuum tubes proved useless because of lead and electrode inductance and capacitance, and transit-time effects. The klystron was a solution to that problem. The tube may consist of a cathode, grid, and perforated anode somewhat like the electron gun of a cathode ray tube, followed by two cavity resonators separated by a calculated distance, and finally a collector. Except for the collector, all electrodes have grid-like surfaces, so that electrons can pass on through them. The beam of random-velocity electrons passing through the grid is accelerated by the positive potential applied to the first resonant cavity structure, causing this structure to serve as an anode. These electrons pass through the grids into the cavity (buncher). The standing waves in the cavity act on the electrons and cause them to change speed so that they arrive at the second cavity (catcher) in bunches, having passed out of the first into a field-free space, and then into the second, through the grids in the sides of the cavities. Here the energy of the electrons is absorbed by the field and contributes to the useful output and normally supplies also the driving energy for the buncher. The electrons then pass on to the collector and return to the cathode. By proper adjustment of the voltages and spacings of the cavities, the circuit may be made to oscillate or amplify as desired. The circuit of a modulated klystron oscillator is shown in Fig. 2.

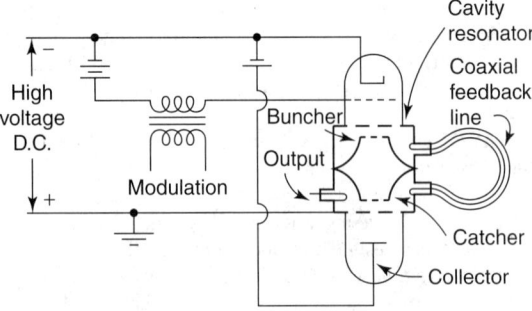

Fig. 2.   Modulated klystron oscillator.

MIDAS.   A two-object trajectory measuring system whereby two complete Cotar antenna systems and two sets of receivers at each station, with the multiplexing done after phase comparison, are utilized in tracking more than one object at a time.

MIDGE (*Insecta, Diptera*).   Of the family *Chironomidae*, several species of very small insects (from $\frac{1}{10}$ to $\frac{1}{5}$ inch; 2.5 to 5 millimeters long) that resemble in build and habit the much larger mosquitoes and crane flies. Generally, they are scavengers, but a few species are of serious economic interest in food production.

Clover seed midge (Dasyneura leguminicola, Lintner*).*   This insect, when present in relatively large numbers, can devastate a clover seed crop, but is not damaging if the clover is produced for hay. The maggots infest the clover seed. Red clover is the principal target of attack, while alsike, mammoth, crimson, white, and sweet clover are relatively immune to attack. A related species, *D. gentneri* (Pritchard), however, does attack Ladino and alsike clover. Distribution of the insect is throughout the United States and in the southern part of Canada.

Pear midge (Contarinia pyrivora, Riley*).*   This insect causes blotched, deformed, and prematurely dropping fruit. The maggots are white to-orange in color and only about $\frac{1}{7}$ inch (3 to 4 millimeters) in length. On occasion over 100 maggots per fruit may be found, with almost complete consumption of the interior of the fruit.

Wheat midge (Sitodiplosis mosellana, Gehin*).*   The maggots are pink-to-reddish in color and only about $\frac{1}{12}$ inch (2 millimeters) long. They are found among the bracts and feed on the kernels of the wheat. Cultural methods usually are adequate protection. These include crop rotation, fall plowing to bury and destroy the larvae, and destruction of all debris from infested fields. The control chemical etrimfos is also effective.

The punkie (Culicoides spp.*).*   Also called sand fly or no-see-um, the punkie is a small biting midge, found in abundance at certain times along streams in the eastern mountains of the United States and at some parts of the seashore. Although not injurious to food crops, the insects can be extremely annoying to persons who work in the fields. See Fig. 1.

Fig. 1.   A punkie or "no-see-um."

Sorghum midge (Centarinia sorghicola).   This insect affects sorghum over a widespread area of the Central and South American countries, the southern United States, southern Europe, notably Italy, and Egypt and the Sudan. In some areas, sorghum production is unprofitable because of damage caused by the insect. This is a small, orange-colored fly, which deposits its eggs in the spikelets of sorghum, as well as in Johnson grass and related plants. Egg depositing is done at the time of flowering. When the white larvae hatch within a few days, they feed on the young seed of the plant. The insect pupates after 9 to 12 days, after which the adult midge emerges to commence a new generation. Two or more generations may be produced per month and thus control is extremely difficult. The insect overwinters in the larval stage in a cocoon. Helpful control measures are early planting and overplanting to allow for a certain amount of this damage, as well as the usual good practice of destroying plant residues and debris. Grass species such as Johnson grass and related weeds that act as hosts for the midge also must be controlled.

MILANKOVITCH, MILUTIN (1879–1958).   The Serbian astrophysicist Milutin Milankovitch is best known for developing one of the most significant theories relating Earth motions and long-term climate change. Born in the rural village of Dalj, Serbia, Milankovitch attended the Vienna Institute of Technology and graduated in 1904, with a doctorate in technical sciences. After a brief stint as the chief engineer for a construction company, he accepted a faculty position in applied mathematics at the University of Belgrade in 1909, a position he held for the remainder of his life.

Milankovitch dedicated his career to developing mathematical theory of climate based on the seasonal and latitudinal variations of solar radiation received by the Earth. Now known as the Milankovitch Theory, it states that as the Earth travels through space around the sun, cyclical variations in three elements of Earth-sun geometry combine to produce variations in the amount of solar energy that reaches Earth: (1) Variations in the Earth's orbital eccentricity—the shape of the orbit around the sun: (2) Changes in obliquity—changes in the angle that Earth's axis makes with the plane of Earth's orbit: and (3) Precession—the change in the direction of the Earth's axis of rotation, i.e., the axis of rotation behaves like the spin axis of a top that is winding down; hence it traces a circle on the celestial sphere over a period of time. Together, the periods of these orbital motions have become known as Milankovitch cycles.

Orbital Variations

Changes in orbital eccentricity affect the Earth-sun distance. Currently, a difference of only 3% (5 million kilometers) exists between closest approach (perihelion), which occurs on or about January 3, and furthest departure (aphelion), which occurs on or about July 4. This difference in distance amounts to about a 6% increase in incoming solar radiation (insolation) from July to January. The shape of the Earth's orbit changes from being elliptical (high eccentricity) to being nearly circular (low eccentricity) in a cycle that takes between 90,000 and 100,000 years. When the orbit is highly elliptical, the amount of insolation received at perihelion would be on the order of 20 to 30% greater than at aphelion, resulting in a substantially different climate from what is experienced today. See Figs 1 and 2.

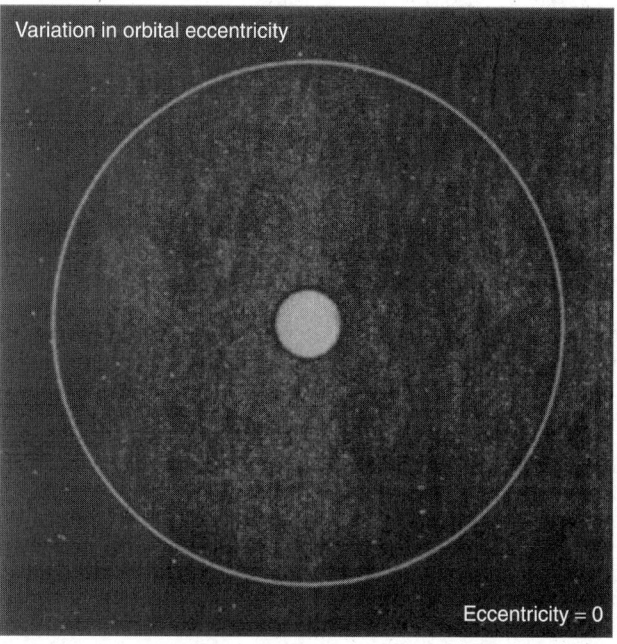

Fig. 1. The eccentricity of the Earth's orbit changes slowly over time, from nearly zero to 0.07. As the orbit gets more eccentric (oval) the difference between the distance from the Sun to the Earth at perihelion (closest approach) and aphelion (furthest away) becomes greater and greater. Note that the Sun is not at the center of the Earth's orbital ellipse, rather it is at one of focal points. (Image by Robert Simmon, *NASA GSFC*.)

Obliquity Change in Axial Tilt

As the axial tilt increases, the seasonal contrast increases so that winters are colder and summers are warmer in both hemispheres. Today, the Earth's axis is tilted 23.5° from the plane of its orbit around the sun. But this tilt changes. During a cycle that averages about 40,000 years, the tilt of the axis varies between 22.1 and 24.5°. Because this tilt changes, the seasons as we know them can become exaggerated. More tilt means more severe seasons—warmer summers and colder winters; less tilt means less severe seasons—cooler summers and milder winters. It is the cool summers that are thought to allow snow and ice to last from year-to-year in high latitudes, eventually building up into massive ice sheets. There are positive feedbacks in the climate system as well, because an Earth covered with more snow reflects more of the sun's energy into space, causing additional cooling. See Fig. 3.

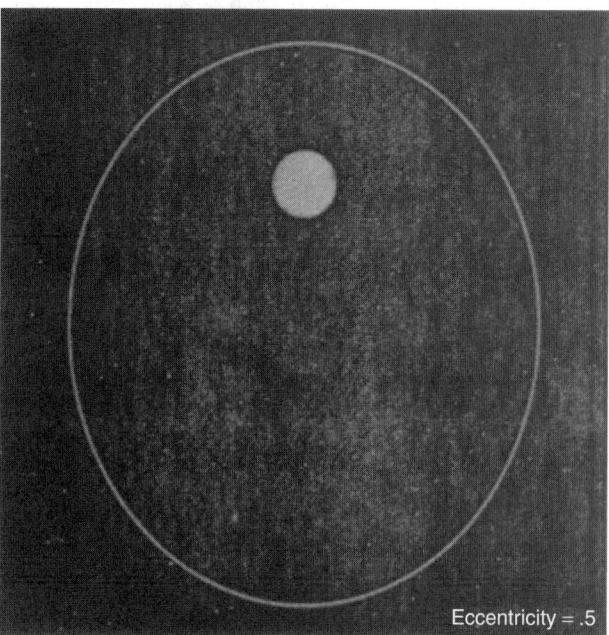

Fig. 2. The eccentricty of the orbit shown in this image is a highly exaggerated 0.5. Even the maximum eccentricity of the Earth's orbit (0.07) it would be impossible to show at the resolution of a web page. Even so, at the current eccentricity of. 017, the Earth is 5 million kilometers closer to Sun at perihelion than at aphelion. (Image by Robert Simmon, *NASA GSFC*.)

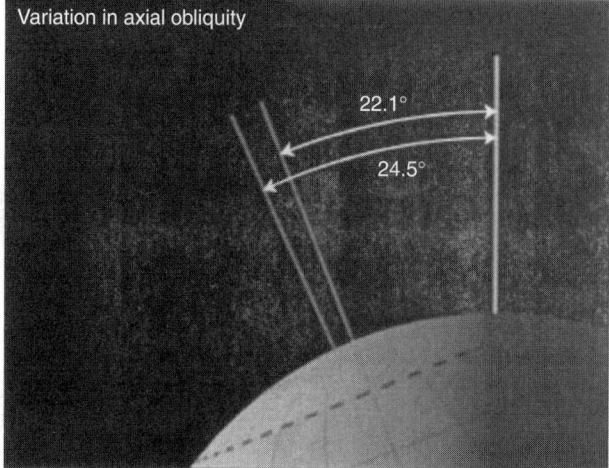

Fig. 3. The change in the tilt of the Earth's axis (obliquity) effects the magnitude of seasonal change. At higher tilts the seasons are more extreme, and at lower tilts they are milder. The current axial tilt is 23.5°. (Image by Robert Simmon, *NASA GSFC*.)

Precession

The change in orientation of the Earth's rotational axis—alters the orientation of the Earth with respect to perihelion and aphelion. If a hemisphere is pointed towards the sun at perihelion, that hemisphere will be pointing away at aphelion, and the difference in seasons will be more extreme. This seasonal effect is reversed for the opposite hemisphere. Currently, northern summer occurs near aphelion. See Fig. 4.

Using these three orbital variations, Milankovitch was able to formulate a comprehensive mathematical model that calculated latitudinal differences in insolation and the corresponding surface temperature for 600,000 years prior to the year 1800. He then attempted to correlate these changes with the growth and retreat of the Ice Ages. To do this, Milankovitch assumed that radiation changes in some latitudes and seasons are more important to ice sheet growth and decay than those in others. Then, at the suggestion of German Climatologist Vladimir Koppen, he chose summer insolation at 65° North as the most important latitude and season to model, reasoning that great ice sheets grew near this latitude and that cooler summers might

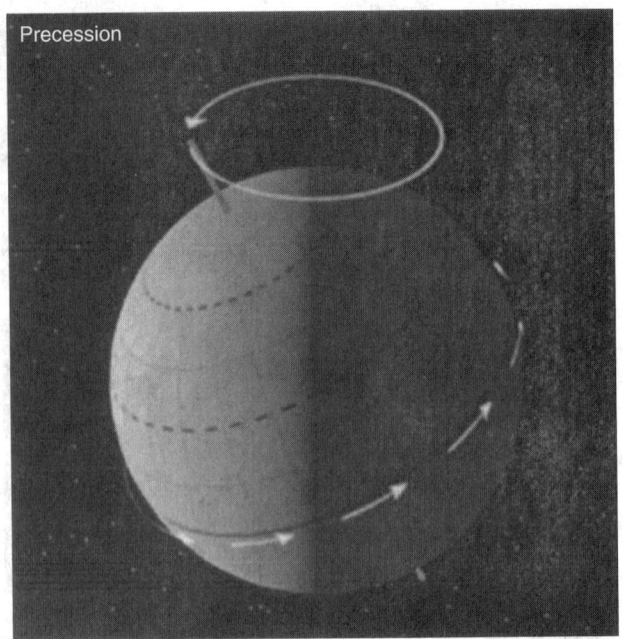
Precession

Fig. 4.   (Image by Robert Simmon, *NASA GSFC*).

reduce summer snowmelt, leading to a positive annual snow budget and ice sheet growth.

But, for about 50 years, Milankovitch's theory was largely ignored. Then, in 1976, a study published in the journal *Science* examined deep-sea sediment cores and found that Milankovitch's theory did in fact correspond to periods of climate change (Hays et al.). Specifically, the authors were able to extract the record of temperature change going back 450,000 years and found that major variations in climate were closely associated with changes in the geometry (eccentricity, obliquity, and precession) of Earth's orbit. Indeed, ice ages had occurred when the Earth was going through different stages of orbital variation.

Since this study, the National Research Council of the U.S. National Academy of Sciences has embraced the Milankovitch Cycle model.

...orbital variations remain the most thoroughly examined mechanism of climatic change on time scales of tens of thousands of years and are by far the clearest case of a direct effect of changing insolation on the lower atmosphere of Earth (National Research Council, 1982).

See also **Climate**.

Additional Reading

Hays, J.D., J. Imbrie, and N.J. Shackleton: "Variations in the Earth's Orbit: Pacemaker of the Ice Ages," *Science* **194**(4270), 1121–1132 (1976).

Hays, J.D.: "Encyclopedia of Weather and Climate," S.H. Schneider, Editor, Oxford University Press, New York, NY, 1996.

Lutgens, F.K. and E.J. Tarbuck: "The Atmosphere," Prentice-Hall, Inc., Upper Saddle River N.J., 1998.

Staff: National Research Council, "Solar Variability, Weather and Climate," National Academy Press, Washington, DC, 1982.

Web References

About.com: Milankovitch Cycles—Changes in Earth-sun Interaction: *http://geography.about.com/science/geography/library/weekly/aa121498.htm?once=true&*
Alaska Science Forum: The Earth's Changing Orbit: *http://www.gi.alaska.edu/ScienceForum/ASF8/825.html*

MILDEW. See **Fungus**.

MILKFISH (*Osteichthyes*). Of the order *Isospondyli*, family *Chanidae*, the milkfish occurs in the tropical waters of America and Africa. It can be described as a silvery-appearing fish which attains a length of about 5 feet (1.5 meters). The milkfish is considered a valuable food fish, notably in the Philippines. The fish is known for its ability to withstand very hot water, as may be found in shallow-water areas.

MILKWEEDS (*Asclepiadaceae*). A family of some 325 genera, with over 1,700 species of shrubs, woody vines, and perennial herbs. All contain a milky juice from which rubber may be made. The family is particularly abundant in the tropics, especially in Africa. The forms are extreme: many are lianas of great length; others have leaves modified into pitcher-like forms; some are epiphytes; while many, especially in Africa, are very much like Cacti in appearance. Indeed; many species are sold under the name of cactus. The uses made of various milkweeds are many and varied. The young shoots of many species are eaten as greens; other species yield dyes. The juices of several are violent poisons, as *Gonolobus*, from which an arrow-poison is obtained, and *Cynanchum*, which is used to stupefy fish. Many are grown for their weird shape, or because of their beauty, as the Waxplant, *Hoya carnosa*.

MILLER INDICES. In mineral crystallography the identity of a crystal face consists of a series of whole numbers which are the products of their parameters relating to that face by their inversion, and where required the clearing of fractional values. A parameter is the relative intercept of a crystallographic axis on a given crystal face.

Assuming parameter values on a given crystal face to be $1a$, $1b$, $\frac{1}{2}c$ would on inversion yield 1, 1, $\frac{2}{1}$; parameters $1a$, $1b$, $2c$ would on inversion yield 1, 1, $\frac{1}{2}$; and parameters of $3a$, $2b$, $6c$ would on inversion yield $\frac{1}{3}$, $\frac{1}{2}$, $\frac{1}{6}$. Clearing the fractions in each instance would yield Miller indices of (112), (221) and (231) respectively.

The three Miller indices for a crystal face in all systems except the hexagonal, which requires four indices, are always given in the same order as their crystallographic axes, a, b, c, respectively; a^1, a^2, a^3, in the isometric system; a^1, a^2, c, in the tetragonal system; and a^1, a^2, a^3, c, in the hexagonal.

If the parameter intercepts for a given face are unknown, general indices (hkl) may be used if that face intercepts all three axes; four in the hexagonal, with general indices ($hkil$). If a crystal face cuts two axes and parallels the third, general indices would be identified as ($h0l$), ($0kl$), or ($hk0$) as applicable to that face; in the hexagonal system as ($h0hl$), etc.

See also **Crystal**; and **Mineralogy**.

MILLERITE. The mineral millerite is nickel sulfide, NiS, whose slender hexagonal interwoven crystals so suggestive of hairs has led to the application of the name "capillary pyrites." It occurs also as radiated masses and coatings. It is brittle; hardness, 3–3.5; specific gravity, 5.48–5.52; luster, metallic; color, brass-yellow, often with an iridescent tarnish. Millerite is found in association with other nickel-bearing minerals and other sulfides. European localities are Bohemia, Westphalia, Wales, etc.; and in the United States at Antwerp, New York; with pyrrhotite in Lancaster County, Pennsylvania; at St. Louis, Missouri; Keokuk, Iowa; and Milwaukee, Wisconsin. In Canada millerite occurs in Oxford, Quebec, and in the famous Sudbury District, Ontario. It is used as an ore of nickel. Millerite was named for the English mineralogist, W.H. Miller.

MILLET (*Gramineae*). Cereal and forage grasses of several different genera are included in the term *millet*. All have a fibrous root system, ample foliage, and rather small grains. They are grown extensively in the occidental countries as forage crops and in some of the oriental countries for human food as well. Of the world's cereal crops, millet ranks sixth in terms of quantity produced.

Millets are members of *Gramineae* (grass family). Compared with other major cereal crops, the millets produce better yields under highly adverse conditions, such as those brought on by inadequate fertility of soils, intensity of heat, and drought. Millets are grown in numerous regions of the world where severe environmental conditions preclude the successful planting of other cereal crops. Parts of Russia, China, India, and Africa are examples. From the standpoint of consumption, the millets tend to be less desirable than other cereals, but there often is no other choice, giving rise to the term "poor man's cereal" for the millets. To a considerable extent, over a long period of years, the millets have become less important in those areas where genetic research and plant-breeding experiments have improved the performance of maize (corn) and wheat when planted in areas previously considered marginal. On the other hand, research of the millets, considered inadequate for many years, is increasing and is pointing toward new areas of interest in millets.

Finger millet. (*Eleusine coracana*). Also known as African millet, birdsfood millet, coracana millet, nagli, and ragi, this plant is grown in Africa (Ethiopia, Somalia, the Sudan) and in southern Asia (notably India's states of Madras and Mysore) for human consumption. The plant is hardy and tolerates poor soils. Unlike millets in general, finger millet is tolerant of moisture. In India it is frequently found near rice-growing areas. The plant does not do well, however, in frequent heavy rains. Moist mountainous slopes (the foothills of the Himalayas, for example) are particularly favorable to the plant.

Ditch millet. (*Paspalum scrobiculatum*). Also known as koda millet (India), this is a forage grass principally grown in India and New Zealand. Several grasses (genus Paspalum) are found in the southeastern United States, such as Bahia grass and Dallis grass.

Browntop millet. (*Panicum ramosum*). This plant was introduced into the United States from India in 1915. There are relatively limited plantings in the southeastern states (Alabama, Florida, Georgia) for hay and pasture. As a dividend, the seed supports wild birds (quail, doves, etc.). The plant is self-seeding, a disadvantage for use in rotation with other crops. At one time, the plant was known as German hay grass. The plant is not to be confused with *browntop panicum (Panicum culatum)*, which is a native of Central and South America.

Foxtail millet. (*Setaria italica*). Also called German millet, Italian millet, and Hungarian grass, foxtail millet is widely grown in China, Japan, India, and Manchuria. In the United States, it is planted in limited quantities in the Great Plains states as a forage crop. In some areas, it is sown as an emergency crop when all other plants have failed. In China, this millet is next to rice and wheat as a cereal for human consumption. In some regions, such as Armenia and Turkey, this millet is used as a substitute for proso millet in areas of very poor soils and during extended hot, dry periods. Foxtail millet is frequently mixed with proso millet in such areas. Foxtail millet is an annual grass and is slender and erect, reaching a height of from 1 to 6 feet (0.3 to 1.8 meters).

Japanese millet. (*Enchinochloa crusgalli* var. *frumentacea*). Also called billion dollar grass (United States), Japanese barnyard millet, and Sawan millet (India), this plant is often present (as a weed) where other millets are cultivated. It is purposely planted as a catch crop in parts of Australia (New South Wales and Victoria) for hay and pasture. Some limited plantings have been made in the United States (New York, Pennsylvania, Pacific Northwest) as a green feed.

Most Japanese millet grown in India is consumed by low-income populations. It is often boiled with milk and sugar. In Japan and eastern India, the millet is mixed with rice and used for making beer. It also serves well as a forage grass. As many as eight forage crops may be produced in a single year.

Pearl millet. (*Pennisetum glaucum* or *P. typhoideum*). This is an annual, warm-weather grass that has been cultivated in Africa and Asia since antiquity. The plant is also known as cattail millet, penicillaria, and Mands forage plant (in the United States); as bullrush millet or Dukha (in Africa); as candle millet or dark millet (in Europe); and as baira, cumbo, or sajja (in India).

Pearl millet was probably domesticated somewhere in the dry savanna fringing the southern Sahara desert between western Sudan and Senegal-Mauritania. The plant is extremely drought-resistant and can be grown at the limits of agriculture near the deserts of Africa and India, where millions of people depend upon it for survival. It is one of the most nutritious cereals, containing good quantities of phosphorus, minerals, and vitamins A and E.

In Africa, pearl millet serves as a substitute for sorghum on sandy soils and in dry regions. It is estimated that about 20.5 million hectares (50.6 million acres) are planted in India, and in the Sudan, at the edge of the Sahara, 1.2 million hectares (3 million acres) are planted, making millet second only to sorghum as the major cereal crop. Pearl millet is well adapted to the extensive dry periods found in the western Ghats and the plains of Rajputana in India. Limited quantities of pearl millet are grown in the sandy soils of the eastern coastal plain of the United States. Research tests have shown that pearl millet produces more beef per area planted than any other grazing crop used in that area.

Pearl millet has coarse, pithy stems that grow from 6 to 12 feet (1.8 to 3.6 meters) in height. See Fig. 1. The stems are about 1 inch (2.5 centimeters) in thickness. Blades are up to 3 feet (0.9 meters) in length and up to 3 inches (2.5 centimeters) in width. The spike ranges from 8 to 18 inches (20 to 46 centimeters) in length and about 0.5 inches (12 millimeters) in thickness. The heads, as illustrated, appear as beads glued to a round stick. The seeds of pearl millet are somewhat larger than those of other millets, ranging from 3 to 4 millimeters long and about 2.25 millimeters wide. They are of a gray-yellow color and of an obovoid shape. The plant is cross-pollinated.

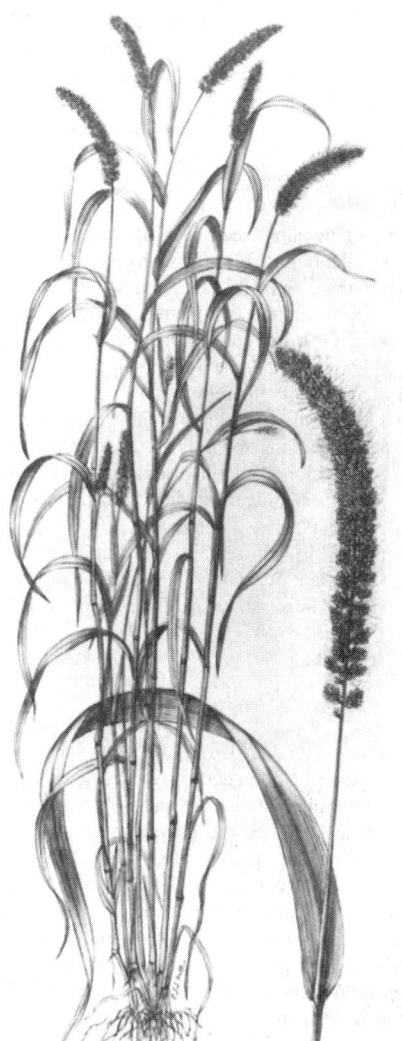

Fig. 1.   Pearl millet (*Pennisetum glaucum*). (*USDA diagram.*)

The *Starr* variety was developed in Georgia and made available in 1950. A later hybrid, the Gahi (Georgia Hybrid No. 1) became available. This yields about 50% more than the *Starr* variety.

In Africa and Asia, pearl millet is ground into meal, from which gruel, porridge, and baked products are prepared.

Proso millet. (*Panicum milliaceum*). Also called broom-corn millet hog millet, and Hershey millet (in the United States); and common millet (in Europe). There are three main subspecies of the plant: (1) *P. miliaceum effusion* (characterized by broad panicles that spread out in all directions; (2) *P. miliaceum contractum* (one-sided panicles that have a limited spread); and (3) *P. miliaceum compactum* (compact, thick, erect panicles). Most varieties of proso millet grown in the Middle Eastern countries are derived from southeastern Russian varieties.

Additional Reading

Buerkert, B. et al. (Editors): "Wind Erosion in Niger: Implications and Control Measures in a Millet-Based Farming System," Kluwer Academic Publishers, New York, NY, 1996.

Khairwal, I. and C. Ram: "Pearl Millet, Seed Production and Technology," South Asia Books, Colombia, MO, 1990.

Leslie, J. and R. Frederikson (Editors): "Disease Analysis through Genetics and Biotechnology: Interdisciplinary Bridges to Improved Sorghum and Millet Crops," Iowa State University Press, Ames, IA, 1995.

Web Reference

United States Department of Agriculture website. *http://www.usda.gov/*

MILLIKAN, ROBERT (1863–1953). Millikan was an American experimental physicist who earned his doctorate at Columbia University (1895) for research on the polarization of light emitted by incandescent surfaces using for this purpose molten gold and silver at the U.S. Mint.

Millikan made numerous momentous discoveries, chiefly in the fields of electricity, optics, and molecular physics. During an oil-drop experiment, he devised a way of measuring the size of an electric charge and confirmed that electric charge comes in identical chunks, or quanta. His earliest major success was the accurate determination of the charge carried by an electron, using the elegant "falling-drop method"; he also proved that this quantity was a constant for all electrons (1910), thus demonstrating the atomic structure of electricity. From 1912–1915, Millikan verified experimentally Einstein's all-important photoelectric equation, and made the first direct photoelectric determination of Planck's constant h. In addition his studies of the Brownian movements in gases put an end to all opposition to the atomic and kinetic theories of matter. During 1920–1923, Millikan occupied himself with work concerning the hot-spark spectroscopy of the elements (which explored the region of the spectrum between the ultraviolet and X-radiation), thereby extending the ultraviolet spectrum downwards far beyond the then known limit. The discovery of his law of motion of a particle falling towards the earth after entering the earth's atmosphere, together with his other investigations on electrical phenomena, ultimately led him to his significant studies of cosmic radiation (particularly with ionization chambers).

Throughout his life Millikan remained a prolific author, making numerous contributions to scientific journals. He held honorary doctor's degrees from some twenty-five universities. Millikan received the Nobel Prize for physics in 1923.

See also **Electron Theory**.

J. M. I.

MILLIPEDE *(Diplopoda).* Also known as the thousand-legged worm, there are several species. Millipedes are damaging to some plants in much the same way as certain wireworms and white grubs. Vegetable crops attacked include bean, beet, carrot, cauliflower, corn (maize), cucumber, lettuce, muskmelon, parsnip, pea, potato, radish, squash, and tomato. The pest is not an insect, but a member of the class *Diplopoda*. Millipedes are widely distributed. They appear to be more of a nuisance in New York, the Great Lakes area, especially Ohio, and the Pacific coastal states. The species *Julus impressus* is damaging to vegetable crops; the species *Orthomorpha gracilis* (Koch) is injurious to plants raised in glasshouses.

In addition to tender plant parts, millipedes consume all kinds of decayed vegetable materials, manure, decayed leaves, and seeds. Millipedes are egg-laying and the generations develop slowly, producing about one complete generation per year. Control measures are similar to those used for sowbug and wireworms.

MILT. See **Fishes**.

MIMETITE. The mineral mimetite is a chloro-arsenate of lead corresponding to the formula $Pb_5(AsO_4)_3Cl$. It is monoclinic (pseudohexagonal); brittle; hardness, 3.5; specific gravity, 7.0–7.25; luster, resinous; color, usually yellow to brown but may be colorless or white; translucent. Mimetite is a rather rare secondary mineral occurring in altered lead deposits. Found in Bohemia; Saxony; Cornwall and Cumberland, England; South West Africa; Mexico; and in the United States, in Pennsylvania and Utah. The name mimetite is derived from the Greek word meaning imitator, because of the similarity of mimetite and pyromorphite. See also **Pyromorphite**.

MIMICRY. The resemblance of an animal to some other living thing or inanimate object. Mimicry may involve both color and form, hence it is closely related to coloration. It is supposed to benefit the mimic either by concealing it from its enemies, by causing them to mistake it for something undesirable, or by enabling it to approach its prey without giving alarm.

MIMOSA TREE. See **Acacia Trees**.

MINDANAO TRENCH. See **Ocean**.

MINERALOGY. The science of mineralogy is concerned with the formation, occurrence, properties, composition, and classification of minerals. Various definitions of a mineral have been proposed. Possibly, the most acceptable may be, "a naturally occurring inorganic substance, usually crystalline, possessing a relatively definite chemical composition and physical characteristics." It should be pointed out that some naturally formed organic substances, particularly of an economic resource nature, are sometimes classified as minerals.

Although in its broadest application, mineralogy is as ancient as human civilization, mineralogy is a modern science, the mineralogist taking full advantage of all modern tools and instruments for exploration, analysis, testing, and study of minerals. Several major scientific advances in the materials field have stemmed from the study of minerals as will be pointed out shortly.

Presumably in the early ages man used minerals as weapons. Through the passage of time and attainment of knowledge regarding certain mineral characteristics man learned, notably initially by accident but later by design, that the content of minerals provided essential materials for his expanding needs. Very early, the natural form and beauty of certain minerals became objects for personal adornment. Later it was found that both the form and the innate beauty of minerals were enhanced by cutting and polishing them. Although the science of mineralogy touches the life of every person, a fundamental understanding of minerals is not common.

Modern mineralogy is the product of research and discovery by many persons. Robert Hooke (1665) foretold the atomic theory by constructing models of alum crystals out of leaden musket balls. Nicolaus Steno (1669) discovered the constancy of interfacial angles between corresponding faces of quartz crystals from many localities. This was later formalized by Rome de l'Isle (1764) under his *law of constancy of crystal interfacial angles*. In 1784, René Just Haüy proposed the theory of *integral molecules* by stacking calcite rhombs to show that structural units could produce exact external facial planes of various forms of calcite crystals. Haüy is known as the "father" of geometrical crystallography. A great advance in mineral studies was made in 1828 when Nicol invented the nicol prism for investigating the behavior of polarized light in crystallized minerals. In 1912, Max von Laue, a student of Roentgen, theorized that the wavelengths of x-rays and atomic spacing of crystals may be of the same magnitude. Laue found that the diffracted rays, when passed through a crystal, substantiated his theory. This discovery opened up an entirely new field of mineral research, i.e., *crystal chemistry*.

Origin of Minerals

Minerals are products of formation and deposition from Earth's natural open solution systems, as opposed to substances formed as products of closely controlled laboratory systems, and thus may vary in many instances from an exact chemical content and formula. Minerals crystallizing from such open solution systems utilize effectively the ions required to form the mineral, but may incorporate within that structure other nonessential ions foreign to it. The incorporation of those foreign ions may, under favorable conditions of chemical affinity, replace certain specific elements named in the given formula. Or there may be defects in the space lattice of the mineral, e.g., vacancies in the crystalline structure comparable to a classroom with X number of seats orderly arranged where, under ideal conditions, each seat would be occupied by a student. Owing to some extraneous circumstance one or more students may be absent, yet the total seating capacity and orderly arrangement would remain constant. Deviations of the first type may be a product of included impurities in the structure; of the second type, by occupancy of a foreign compatible element in the vacant space(s) of the host crystal structure. See "**Isomorphism and Diadochy**" later in this article.

In other minerals, the formula may vary within restricted limits, such as (Zn, Fe) S for sphalerite, where ferrous iron substitutes for zinc within the sphalerite structure. Two factors prevail in such chemical substitution within a given mineral structure: (1) reasonably comparable ionic radii (approximately, but not strictly limited to ($\pm 15\%$); and (2) maintenance of electrical neutrality of the compound. In the event the substituting ions have a different valence or charge, electrical neutrality may be obtained by an accompanying substitution elsewhere within the crystal structure. Basically, a mineral is a homogeneous inorganic solid with an ordered atomic arrangement, which places it in the category of a crystalline material and possessing a definite, though not a fixed chemical formula.

Natural systems strive toward a state of equilibrium when all component units attain their lowest energy level. The respective energy level of the elements within any mineral is dependent upon the physical environment, principally the temperature, pressure, and chemical substances present, at the time and place of its formation. Any later change in its environment may cause a change in the mineral's composition and form. Whatever the primary or intermediate environmental conditions may have been, the mineral as observed, represents its present equilibrium energy state or crystal structure.

Matter exists in three states — gaseous, liquid, and solid.* In the gaseous state, the elements move freely about within their environment, their only contact being haphazard collisions. Elements in the liquid state are in closer contact with each other, but still retain freedom of mobility. The solid state is characterized by the chemical elements combining under atomic bonds of various types and strengths into a structured system. For any element within the structure, there are equivalent elements in a definite three-dimensional crystalline pattern. Under restricted circumstances, two or more minerals may form in contact with one another in such a manner as to preclude their complete development as single crystals. Those minerals which develop fully and which exhibit well-formed polyhedral (many sides) facial planes are referred to as being *euhedral*. Those minerals that exhibit no facial planes are referred to as *anhedral*. Imperfectly formed crystals are referred to as *subhedral*. Crystalline minerals, where the crystallinity can be determined only by aid of a microscope, are said to be *microcrystalline*. Where the materials are so finely divided as to be discernible only by x-ray analysis they are referred to as *cryptocrystalline*. The rate of crystallization plays an important part in the resultant crystalline character of minerals.

Crystalline solids are bonded together by electrical forces, which originate in the constituent atoms. The position of the respective unit particles within the crystal structure are determined by geometric factors, along with considerations of electrical neutrality and lattice energy. One type of bond is produced by the lending or borrowing of electrons in an attempt to complete the outer electron shell of the atom. A stable compound is formed when the outer shell is complete and the mutual attraction is termed an *ionic bond*. For example, sodium with one electron in its outer electron shell lends that electron to a neutral chlorine atom with seven electrons in its outer electron shell. The result is a stable electron shell for each ion (charged atom) which then join by electrostatic attraction in a crystal structure to form sodium chloride, NaCl. Minerals so bonded are of moderate hardness, readily soluble, and poor conductors of electricity and heat.

The bond produced by the sharing of electrons by which a stable configuration is achieved is termed a *covalent bond*. For example, chlorine with seven electrons in its outer shell needs one additional electron to achieve stability. If its nearest neighbor is another chlorine atom, the two atoms combine in such a way that the one electron serves in the outer shell of each atom, thus achieving a stable configuration. Diamond is another example of this type. Carbon with four electrons in its outer shell mutually shares the four electrons with four adjoining carbon atoms, thus achieving a stable configuration by completing the shell with eight electrons. This type of electron-sharing or covalent bonding is the strongest of all chemical bonds. Minerals so bonded are generally insoluble, possess a high degree of stability, and are nonconducting.

Metallic bonds consist of a structure of positive ions through which free electrons can drift. The structure of metals may be envisioned as a mesh of electrons that surround the atomic nuclei and bind them together. This electron mobility produces minerals of generally a low hardness and of high electrical and thermal conductivity.

The *Van der Waals* bond is a very weak attraction between atoms, usually between essentially neutral atoms or groups of atoms. These neutral and essentially uncharged structure units are held together within the crystal lattice by virtue of small residual charges on their surfaces. This type of bonding is rare in minerals. Minerals so bonded often display ready cleavage and low hardness. For example, the foliated scales of graphite are linked weakly together by Van der Waals bonds.

Every mineral is a product of the redistribution or recombination of its component chemical elements to form a stable substance. The process is known as *crystallization*. The process may involve *precipitation* of chemical elements from aqueous solutions at the earth's surface; or from siliceous melts (magmas) from the earth's interior. In either situation, the process is dependent upon the degree of concentration of the constituent chemical elements present and the temperature/pressure conditions. Precipitation from vapor also is possible. An example is the hot vapor, rich in sulfur dioxide, which is emitted from vents associated with volcanoes. Upon becoming exposed to the cooler atmosphere, crystal

sulfur is deposited around those vents. Snow crystals are another example of precipitation from vapor.

Crystal Structure

Associated chemical units become systematically arranged in the crystal structure, which is constructed from a single motif that develops repetitively. The resulting three-dimensional array is called the *space lattice* of the crystal. The lattice or framework is defined by three directions and by the distance along those directions where the motif repeats itself. Because the units within the structure adhere to a strict arrangement, the external facial planes of a crystal represent the limiting surfaces of that growth and are an external expression of its internal atomic order. Crystals are formed, therefore, where constituent atoms or ions are free to combine in constant chemical proportions and are an expression of the environmental conditions that promote their formation.

Periodic repetitions of a space lattice cell in three dimensions from the original cell will completely partition space without overlapping or omissions. It is possible to develop a limited number of such three-dimensional patterns. Bravais, in 1848, demonstrated geometrically that there were but fourteen types of space lattice cells possible, and that these fourteen types could be subdivided into six groups called *systems*. Each system may be distinguished by symmetry features, which can be related to four symmetry elements:

1. *Symmetry with respect to a point*. If through a central point in a geometric figure, lines are drawn from a point on one side of the figure to a similar point equidistant on the other side, the figure is symmetrical to a point.

2. *Symmetry with respect to a plane*. A geometrical figure is symmetrical with respect to a plane when for each edge, solid angle, or face on one side, there is a corresponding edge, solid angle, or face on the other side of that plane. One side is, in fact, a mirror image of the opposite side. That plane is called a *plane of symmetry*.

3. *Symmetry with respect to a line*. If during a complete revolution of 360° about a given axis, a geometrical figure repeats itself in appearance two or more times, it is said to be symmetrical with respect to a line, or to an *axis of symmetry*. Possible axes of symmetry are twofold, threefold, fourfold, and sixfold. A rotation of onefold or 360° is equivalent to no rotation at all.

4. When a crystal is rotated about an axis and inverted about the central point and at that point repeats itself, it is said to have an *axis of rotary inversion*. It is a twofold axis of rotary inversion if the geometrical figure is rotated 180° and then inverted. Additionally, there are threefold, fourfold, and sixfold axes of rotary inversion possible.

Within each of the six crystal systems, there are specific *crystal classes*. Each class displays distinctive symmetry elements. There are 32 possible classes distributed among the six crystal systems. One of the crystal classes within each system possesses all of the symmetry elements that are characteristics of its space lattice cell. These are called the *holohedral* class of that system. Other classes within each system possess somewhat fewer symmetry elements and are called *merohedral* classes.

It is significant to note that in most geometrical situations regarding crystal study, the distinguishing characteristic is symmetry, not geometry. This is especially true in the case of a mineral such as pyrite (FeS_2) which assumes a cubic shape, but its symmetry, controlled by the three sets of opposing striations on its crystal faces, identify it as belonging to a much lower order symmetry than that of a true cube.

The six crystal systems are identified by hypothetical lines of reference known as *crystallographic axes*, and their angular relationship to each other. Crystal orientation and axial order are given as: (1) front to back; (2) right side to left side; and (3) top side to bottom side. The front, right, and top side axial ends represent the positive ends of those respective axes. The back, left, and bottom side ends are designated as the negative ends. The relative intensity of crystal growth along those axes gives to each crystal face a distinctive identifying character. This character is evidenced by the physical similarity of equivalent facial planes on the crystal and their relationship to the crystallographic axes of that form. The six systems, their crystallographic axes, and interaxial angular relationship are defined and illustrated in the Fig. 1.

Crystallographic identification of facial planes on a crystal become possible through assignment of numerical values to a face which represents its relationship to the crystal axes. *Parameters*, or relative intercepts, are obtained by plotting coordinates of crystal faces with respect to their

* Traditional concepts. Details on more advanced concepts on the states of matter are given elsewhere in this volume.

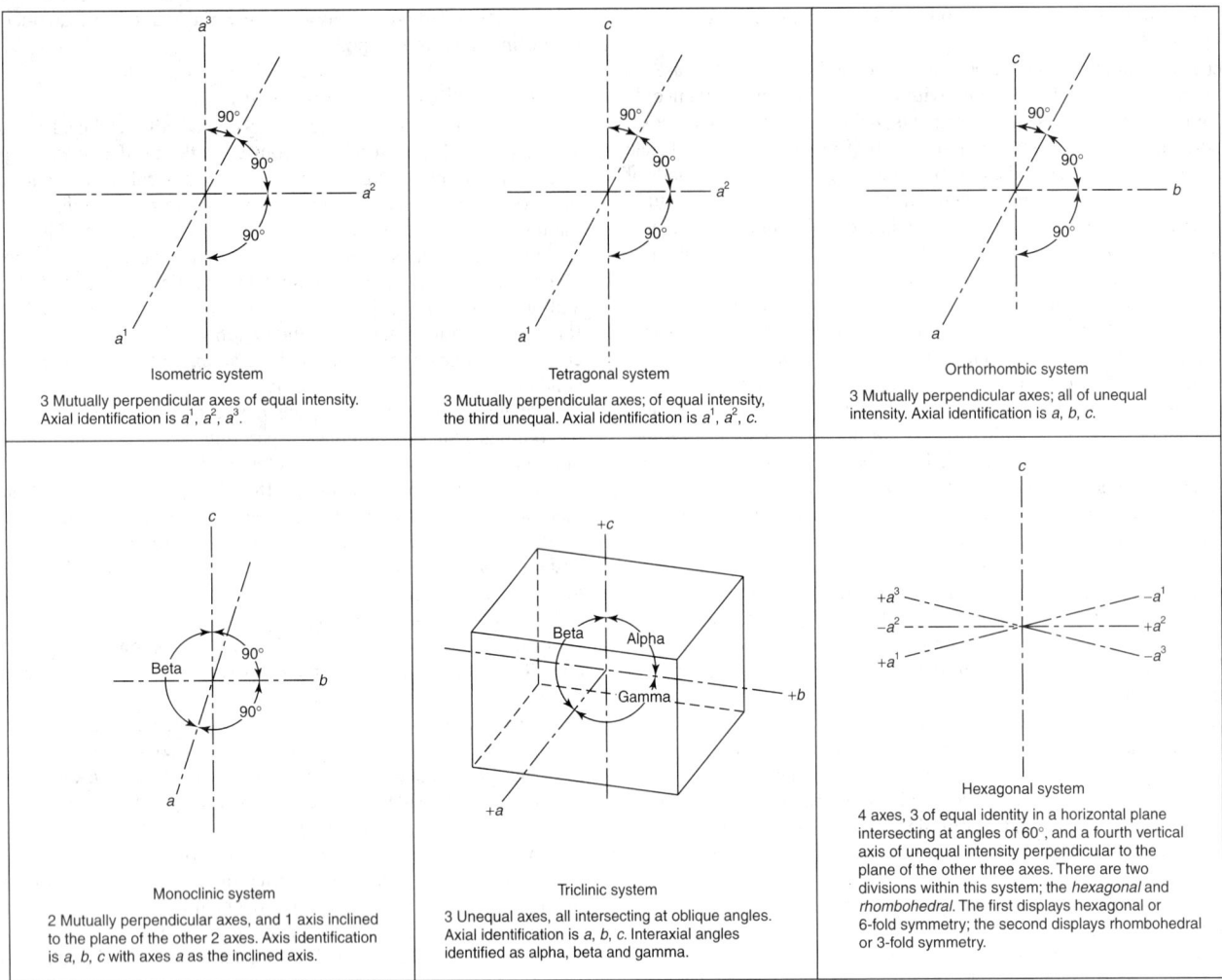

Fig. 1.   Crystallographic axes and interaxial angular relationship of six crystal systems.

crystallographic axes. The actual distances of axial intercepts of the crystal face are determined and expressed as a unit of measurement. The product of these values is known as *Miller indices*.

A Miller indices face on the front face of a cube would be (100), signifying that face intersects the a^1 axis at 1 unit length from the center of the crystal, and is parallel to axes a^2 and a^3, or intersects those axes at infinity. In this system, zero (0) is the numerical substitute for infinity. A (111) Miller indices face identifies that facial plane as intersecting each of the three crystallographic axes of that form at 1 unit length from the crystal center.

The Miller Indices identify the orientation of a face in relationship to its axes of reference regardless of its size and position on the crystal. Haüy first proposed this basic law of crystallography—"crystal faces make simple rational intercepts on suitable crystal axes." Inasmuch as the intercepts are simple, it follows that the Miller indices should likewise be simple whole numbers.

See **Crystal**; and **Miller Indices**.

Crystal Forms

Form is used here to designate the general outward appearance of a crystal, specifically to a group of crystal faces that bear identical relationships to the crystal's symmetry elements. It is essentially a geometric form with equivalent facial planes in their relationship to the crystal lattice symmetry elements. In this regard, an octahedron in the isometric system would be identified as a "closed form," inasmuch as its eight equivalent faces totally enclose the crystal space; its form identification would be {111}, enclosed in braces. Even though the crystal faces may be equivalent they may vary widely in size and distance from the crystal center, owing to irregular or distorted development during their formation. Braces shown around the Miller indices, e.g., {111} signify form identification, as opposed to Miller indices shown in parentheses, e.g. (111), which identify a specific crystal

face only. In crystals possessing different lengths of their crystallographic axes general form {*hkl*} would be used. The hexagonal system with four axes requires a four-unit form identification.

In the isometric system, crystals are recognized by their geometrical forms, e.g., cube, octahedron, trapezohedron, pyriteohedron, and so on. In all other systems, crystal faces are given specific names, which refer to their relationship with their respective crystallographic axes. The most common facial forms in these systems are *pedion, pinacoid, dome, prism*, and *pyramid*. Crystal forms bear specific relationships to symmetry axes or planes, and not to the general shape of the crystal.

Physical Mineralogy

This aspect of mineralogy is concerned with several observable physical characteristics, such as color, hardness, lustre, fracture, cleavage, magnetic properties, radioactivity, fluorescence, and specific gravity.

Color. The color of a mineral is a product of selective absorption of certain wave lengths of visible white light by atoms within the mineral structure. A mineral color may be indicative of a species, but more often is of descriptive value only. There are two broad classification categories of color in minerals—*allochromatic* and *idiochromatic*. Allochromatic minerals are those which occur with variable colors, such as quartz, SiO_2; corundum, Al_2O_3; and calcite, $CaCO_3$. Color in such minerals may be a product of included foreign elements, rate of crystallization, or a defective lattice structure. Idiochromatic minerals are those in which the color is a characteristic constant, such as the green of malachite, $Cu_2CO_3(OH)_2$; the blue of azurite, $Cu_3(CO_3)_2(OH)_2$; the black of magnetite, Fe_3O_4; and the red of cinnabar, HgS. The actual color of most minerals can be obtained by rubbing the mineral on an unglazed porcelain plate and observing the powder (streak) of that mineral left on the plate. This test is essential in certain instances inasmuch as many minerals display a physical color foreign to their actual color, e.g., hematite may appear black or blue, but its streak color is always red.

Hardness. The hardness of a mineral is its resistance to scratching. Testing for hardness is based upon the premise that all minerals possess such resistance to a lesser or greater degree. The Mohs scale of hardness has been universally adopted to test this physical property. In the list below, the ascending numeric order of the mineral named will scratch all of those of lower order.

1 Talc
2 Gypsum
3 Calcite
4 Fluorite
5 Apatite
6 Orthoclase
7 Quartz
8 Topaz
9 Corundum
10 Diamond

In a relative sense, minerals from 2 to 2.5 hardness can be scratched by a fingernail; 4 by a penny, and 5 to 6 by a knife blade or piece of glass.

This scale is strictly relative and nonlinear because there is a much wider differential, for example, between corundum and diamond than between topaz and corundum. Hardness also varies according to crystallographic direction in certain minerals. This test should always be made on a freshly broken area because certain minerals are subject to surface alteration. A hardness test on an altered area can be misleading.

The hardness (Mohs scale) of several minerals is given in Table 1.

TABLE 1. HARDNESS OF REPRESENTATIVE MINERALS

(Mohs Scale)

Agate	6–7	Galena	2.5
Alabaster	1.7	Garnet	6.5–7
Andalusite	6.5–7.5	Graphite	0.5– 1
Aragonite	3.5	Kaolinite	2.0–2.5
Asbestos	5.0	Magnetite	6
Barite	3–3.5	Marble	3– 4
Beryl	7.8	Mica	2.8
Corundum	9	Opal	5.5–6.5
Diatomaceous earth	1–1.5	Pumice	6
Dolomite	3.5–4	Pyrite	6–6.5
Emery	7–9	Serpentine	3– 4
Flint	7	Tourmaline	7–7.5

Lustre. The lustre of a mineral is a product of both light reflection and refraction. Reflection is the governing factor in translucent to opaque minerals; refraction in transparent minerals. There are two broad categories in describing lustre—*metallic* and *nonmetallic*. Minerals of nonmetallic lustre are further subdivided into categories, such as vitreous (glass); adamantine (diamond); resinous (sphalerite); silky (asbestos); waxy (chalcedony and opal); greasy (some quartz and diamonds); and pearly (talc or mica).

Fracture. The product of irregular breaking of a mineral is termed fracture. Categories include conchoidal or conch-like (quartz); uneven (serpentine); and hackly (copper). Fracture in a mineral is unrelated to its crystal structure.

Cleavage. The cleavage of a mineral is the product of regular breaking of the mineral along specific surfaces related to the mineral's internal structure. The cleavage plane is always parallel to a possible crystal face and, therefore, is a reflection of the atomic structure of the mineral. Three planes of cleavage are present in the cube and rhombohedron. Four cleavage planes produce an octahedron, six planes a dodecahedron, both of which fall within the isometric system.

Magnetic Properties. A few minerals possess the property of being attracted by a magnet and they are known as *ferromagnetic* minerals. The more common ferromagnetic minerals are magnetite, Fe_3O_4, and pyrrhotite, $Fe_{2-x}S$. Those minerals which are natural magnets are known as *lodestones*. Most minerals are affected to some extent in a magnetic field. The minerals that are repelled are known as *diamagnetic*; those that are weakly attracted are known as *paramagnetic*.

Radioactivity. Basically, this involves the spontaneous disintegration of uranium and thorium minerals. Both of these mineral species disintegrate at a steady rate that is completely unaffected by external chemical or mechanical conditions. The end-product of this disintegration is lead within the mineral structure. Geophysicists are able to determine the geologic age of a specific uranium or thorium mineral and its host environment by measuring the amount of lead present and computing the known time required to produce this amount. With the impetus of atomic energy, many previously unknown uranium minerals were uncovered and identified.

Fluorescence. This is the property of certain minerals to absorb ultraviolet radiation and convert that energy into visible light. The rays energize certain elements within the mineral, causing the excitation of electrons in the orbital shell. If the activating source is removed and visible light continues for a period of time following, the phenomenon is known as *phosphorescence*. The visible light continues until the electrons return to their normal orbital electron shell.

Specific Gravity. The density or specific gravity of a substance is the amount by which that substance decreases in apparent weight when first weighed in air, followed by weighing in water. The density is obtained by dividing the weight of the substance in air by the loss of its weight in water.

Isomorphism and Diadochy

Isomorphism is the substitution of an atom, ion, or radical for another within a mineral structure. The degree of substitution is controlled largely by two factors: (1) temperature, and (2) ionic size. Only ions of similar size will readily substitute for one another.

Under certain conditions, isomorphism may be either partial or complete. The spinel, garnet, and amphibole mineral groups represent a complete isomorphous series. Partial isomorphism is exemplified by chemical components displaying extensive ionic substitution at high temperature levels, as evidenced by the nearly complete random mixing of potassium and sodium in microcline $KAlSi_3O_8$ and albite $NaAlSi_3O_8$ feldspars. Reducing temperature causes segregation of their component content according to their respective radii, which then group together, forming two separate mineral phases. This phenomenon is known as *exsolution* and is shown by the mixed feldspar known as perthite, common in igneous and high-grade metamorphic rocks. Other isomorphous examples include calcite and siderite; magnesite and siderite.

Diadochy is a term applied to the substitution of foreign atoms/ions within the same space lattice of a crystal atomic structure. Diadochic substitution refers only to specific crystal structures, as substitution elements may be diadochic in one structure and not in another. For example, excess Al^{3+} (0.57kX) may substitute diadochically for Fe^{3+} (0.64kX) in the epidote structure, thereby producing clinozoisite; when within the dolomite structure, Fe^{2+} (0.83kX) substitutes for Mg^{2+} (0.78kX), ankerite is the product; marmatite (ferroan sphalerite) results when Fe^{2+} (0.83kX) replaces Zn^{2+} (0.83kX) within the sphalerite structure. Diadochy involves the actual substitution/replacement of a given compatible element for another within a crystal structure, which, as in isomorphism, may be either partial or complete. In like manner, high temperature contributes materially to such substitution.

Polymorphism

Polymorphism is the capability of a substance to exist in more than one crystal form. The basic controlling factors appear to be temperature-pressure conditions at the time of formation, which controls the type of atomic packing within the structure. Examples of polymorphism are given in Table 2.

TABLE 2. REPRESENTATIVE POLYMORPHIC FORMS

Chemical Composition	Mineral	Crystal System	Specific Gravity
Carbon	Diamond	Isometric	3.5
Carbon	Graphite	Hexagonal	2.2
Al_2SiO_5	Sillimanite	Orthorhombic	3.2
Al_2SiO_5	Kyanite	Triclinic	3.7
FeS_2	Pyrite	Isometric	5.0
FeS_2	Marcasite	Orthorhombic	4.9

Pseudomorphism

Pseudomorphism is the change of the original chemical composition of a substance into some other equally definite compound by the action of natural agencies. Pseudomorphism exists when the external crystalline form of a mineral is inconsistent with its internal chemical composition and atomic structural arrangement. It is always a secondary process. The altered substance is known as a *pseudomorph*.

Types of pseudomorphic alteration include:

Substitution. This is a process whereby silica replaces wood fiber to form silicified (petrified) wood. Quartz replacing fluorite is another example. In the latter instance, the original fluorine in the fluorite is removed by silica-rich solutions that first remove the fluorine and then substitute silica in its place.

Incrustation. This is a process whereby one mineral forms a crust over another mineral, e.g., prehnite over anhydrite crystals. Later solutions may remove the anhydrite, but the space occupied by the anhydrite remains as a cast surrounded by the prehnite.

Alteration results from the addition of new material or partial removal of original material from a mineral, e.g., anglesite after galena; or gypsum after anhydrite.

Paramorphism results when the internal crystal structure of a mineral is changed to a polymorphous form, yet retaining the external crystal form of the original mineral, e.g., aragonite after calcite.

Twinning

Twinning in crystals results from the intergrowth of two or more individuals in such a way as to yield parallelism in the case of certain parts of the different individuals and, at the same time, other parts of the different individuals are in reverse positions in respect to each other. For example, an octahedral crystal of magnetite is twinned when one-half of the crystal is rotated 180° parallel to an octahedral facial plane. This type of twinning is known as *spinel twinning*, owing to its common occurrence in the spinel group of minerals.

Mineral aggregates are often grouped together to form compound crystal structures. Such groupings may be a product of irregular and accidental growth which do not conform to basic twinning laws. Two or more intergrown crystals should not arbitrarily be labeled as twins unless their twin relationship can be established.

Rock Types, Associated Minerals and Their Uses

Minerals are the basic building blocks of all rock types found in the earth's crust. Rocks are classified into three broad categories: *igneous, sedimentary*, and *metamorphic*.

Igneous rocks have their origin deep within the earth's interior from molten magmas of siliceous (silica-rich) melts. The rocks resulting from deep-seated solidification are known as *plutonic*; while those that have been extruded on the surface as lava flows are termed *extrusives*. Plutonic rocks are products of slow cooling and possess well-formed crystalline structure, e.g., granite. Extrusive rocks are products of fast cooling and occur as glassy or microcrystalline masses, e.g., obsidian and basalt.

Minerals occurring in igneous rocks possess crystalline character, but their rate of precipitation from the parent melt prevented their development as euhedral crystals and thus they occur as granular (anhedral) aggregates within these rocks. The more prominent component minerals include quartz, feldspar, and feldspathoid family members, micas, pyroxenes, amphiboles, and olivine. Zircon, magnetite, ilmenite, hematite, apatite, pyrite, and garnet are commonly associated in these rock types.

Pegmatites represent a residual phase of igneous depositions, characterized by extremely coarse crystalline material, that results from the presence of associated volatiles, e.g., water vapor, carbon dioxide, sulfur dioxide, and others, which decrease the viscosity and facilitate crystallization. Quartz, feldspar, and mica are the more common minerals found in this environment, but such bodies are also hosts for many rare minerals and several types of gem stones, e.g., beryl, tourmaline, and topaz.

Sedimentary rocks are products of deposition from either the mechanical or chemical breakdown of all preexisting rock types. Precipitation from aqueous bodies rich in soluble salts produces economically valuable beds of halite and gypsum. Weathering of iron-bearing rocks has produced extensive deposits of hematite. Limestone, chalk, and diatomite are products of biochemical precipitation. Minerals commonly associated with sedimentary rocks include calcite, galena, sphalerite, pyrite, marcasite, fluorite, barite, celestite, and quartz.

Metamorphic rocks are those that have undergone a reconstitution or redistribution of the chemical elements contained in the original formations to new mineral species. The process of change involves attaining a state of equilibrium of the constituent elements with the newly imposed environment. Chlorite, biotite, garnet, staurolite, andalusite, kyanite, and sillimanite, with ubiquitous quartz, are a few of the more common minerals associated with metamorphic rocks.

Minerals of Primary Industrial and Economic Importance

Economically valuable mineral occurrences are the products of particular types of worldwide geological formations that are, or have been, hosts to such minerals. In each occurrence, the minerals found represent the products of geological processes, which include:

1. The character and concentration of the mineral components within the source magmas from which the minerals were formed.

2. Secondary deposition from either percolating solutions or gaseous emanations from intrusive formations, causing chemical reactions within the intruded formations.

3. Precipitation from chemically supersaturated solutions.

4. Alluvial deposits resulting from the erosion weathering of the original host rocks.

Minerals that are a product of the first type of formation include the basic native metals, gold, silver, and copper.

Gold is primarily a product of deposition from ascending hydrothermal solutions associated genetically with siliceous-rich igneous rocks. Pyrite and other sulfide minerals are common associates within which the gold is often physically admixed. Surficial weathering of such deposits removes the sulfides, leaving free gold as a residual deposit. Erosion of these deposits results in alluvial deposits of placer gold, both as flakes and nuggets. Other characteristics and the uses for gold are described under **Gold**.

Silver occurs both as native ore and in combination with various silver sulfide minerals. Native silver is predominantly a product of primary deposition from hydrothermal solutions. Minor occurrences are products of oxidation of silver sulfide minerals with which the native ore is secondarily associated. See also **Silver**.

Native copper commonly occurs in the oxidized zones of copper deposits in association with cuprite, malachite, and azurite. The native copper deposits on the Michigan Keeweenaw Peninsula represent an exceptional occurrence. The copper occurs there as veins within igneous trap rocks interbedded with conglomerates. See also **Copper**.

Similar and valuable minerals include cinnabar (mercury); antimony (for type metal and battery plates); galena (lead and silver); argentite, pyrargyrite, and proustite (silver); sphalerite (zinc); chalcocite, chalcopyrite, bornite, malachite, azurite, and cuprite (copper); nickeline and pentlandite (nickel); bauxite (aluminum); magnetite, hematite, and goethite (iron).

Diamonds were found originally as loose crystals in geologically ancient alluvial stream beds. Later, their host formations were found to be a basic igneous rock (kimberlite) in the Republic of South Africa. Diamonds are the products of extremely high-temperature, high-pressure environment and are composed of pure carbon. See also **Diamond**.

Trap rocks (basalts) are products of volcanic action, either as extensive lava flows, or as intrusive dikes in preexisting rocks. Secondary mineralization within such rocks from circulating waters produces interesting suites of zeolitic minerals, such as analcime, heulandite, natrolite, stilbite, mesolite, and others.

These minerals possess the ability to exchange ions contained in the mineral structure for those in solutions. This facility promotes the use of zeolitic minerals (or their synthetic counterparts) as water softeners. Water rich in calcium (hard water), when passed in solution through a tank containing zeolites, loses the calcium ions by absorption in the zeolite structure, with substitution of calcium ions by sodium ions. A reverse process may be initiated with the sodium ions replacing calcium ions in the structure, thereby reconstituting the original zeolite composition. See also **Zeolite Group**.

Halite, gypsum, and anhydrite are products of precipitation from large bodies of supersaturated salt water. The salt and gypsum of commerce are derived from such deposits. See also **Gypsum**; and **Sodium Chloride**.

Landlocked inland seas and lakes become enriched with various soluble elements from waters draining into those basins. Sylvite and carnallite are valuable for their potassium content. They represent the final evaporation products of landlocked bodies of supersaturated sea water. Two famous localities are Stassfurt, Germany and near Carlsbad, New Mexico. Minerals formed from the evaporation of boron-rich waters include borax, kernite, colemanite, and ulexite. See also separate alphabetical entries for these minerals. The only known locality for kernite is the Mohave Desert in California, where a deposit of great extent exists — with potential reserves of millions of tons beneath the desert floor.

Pegmatites are valuable mineral sources. These formations represent the residual phase of igneous crystallization from magmas rich in siliceous content. As crystallization of their component elements proceeds, these magmas become increasingly enriched with volatile substances (mineralizers), such as water vapor, carbon dioxide, chlorine, fluorine, phosphorus, and others. The volatiles reduce the viscosity of the residual magmas and facilitate crystallization, as previously mentioned. When the residual liquids are injected into cooler rocks, they crystallize from their peripheral borders inward. The great mobility of the constituents enhances the growth of large mineral crystals — a characteristic feature of pegmatite bodies. Beryl and spodumene crystals from pegmatites attain sizes in terms of feet and tons. Feldspar, quartz, and mica crystals of comparable character are not uncommon.

There are two genetic pegmatite types — *simple* and *complex*. Simple pegmatites are recognized by their coarse texture and normal granite components, e.g., quartz, feldspar, and mica. Pegmatites produce the feldspar of commerce and mica for industrial and commercial uses. See also **Feldspar**; and **Mica**. Complex pegmatites are characterized by the presence of rare elements, in addition to the normal feldspar, quartz, and mica. Such bodies are also hosts for many semiprecious gem stones, such as amethyst, rose quartz, topaz, tourmaline, beryl, and chrysoberyl. Many rare-earth minerals obtained from complex pegmatite minerals include columbite/tantalite (columbium or niobium; tantalum), lepidolite, triphylite, spodumene, ambylgonite (lithium), zircon (zirconium), and monazite (thorium oxide).

A most unusual pegmatite occurs near Ivigtut, Greenland. This consists of a cryolite with subordinate siderite, chalcopyrite, galena, and sphalerite. Cryolite is a fluoride of sodium and aluminum. For many years, cryolite was mined from this single occurrence for use as a flux in the electrolytic recovery of aluminum from bauxite, the major ore source of aluminum. Synthetic sodium aluminum fluoride essentially has replaced the need for natural cryolite. See also **Aluminum**; **Bauxite**; and **Cryolite**.

Quartz and tourmaline crystals once were commercially important for their piezoelectric properties as radio oscillation wafers and other electronic and instrumental uses. Synthesized quartz crystals have largely replaced the need for natural quartz for such applications.

Nuclear fission reactors are supplied with materials from uranium-bearing minerals of primary origin, e.g., uraninite/pitchblende, and other uranium-bearing minerals of secondary origin, e.g., carnotite, tyuyamunite, torbernite, and autunite.

Minerals of economic importance within sedimentary formations include, but are not limited to fluorite, barite, phosphorite, and oolitic hematite. Fluorite is utilized as a flux in steelmaking and when of high quality as lenses and prisms in the optical industry. Barite is an essential mineral used in gas- and oil-well drilling. Phosphorite, a product of chemical precipitation from seawater, when treated with sulfuric acid, produces superphosphate fertilizer. Oolitic hematite deposits of extensive size are important sources of iron ore.

Garnet is a common mineral component in metamorphic rocks. A major occurrence of this type is at the summit of Gore Mountain, North River, in Warren County, New York. The garnet is a composite of almandine and pyrope with a hardness exceeding that of most world garnets. Gore Mountain garnet retains sharp cutting edges even when crushed to sub-micron size, making it an outstanding abrasive. It is used extensively as an abrasive (garnet paper) and as a glass-polishing agent in the optical industry.

Titaniferous iron ores, represented by the mineral ilmenite, occur within crystalline metamorphic environments. These ores are the major source of titanium. See also **Titanium**.

Major sources of industrial (non-fuel) minerals are shown in Table 3.

Ocean Sources of Minerals

Mineral requirements for future world needs has focused increased attention to potential ocean resources. Major attention has been directed to petroleum resources. Associated with the petroleum are salt domes or bedded salt deposits, often with anhydrite and sulfur. Their potential is dependent upon development of economically feasible recovery methods.

TABLE 3. MAJOR SOURCES OF INDUSTRIAL (Nonfuel) MINERALS

ALUMINUM
Major source is bauxite (gibbsite and boehmite). Deposits are found worldwide, except in Antarctica. Major producers: Australia: Caribbean countries; Venezuela; Brazil; Indonesia.

CHROMIUM
Nearly all chromium ores are found in the Republic of South Africa and Zimbabwe.

COBALT
Zaire, Zambia, and Russia.

COPPER
Chile, United States, Russia, Canada, Zambia, and Zaire.

GOLD
Republic of South Africa: Russia, Brazil, and the United States (not extensive).

IRON
Russia, the United States, Brazil, Australia, and China.

LEAD
United States, Russia, Australia, and Canada

MANGANESE
United States, Japan, and Western Europe. (Major reserves are in South Africa and Russia)

NIOBIUM (Columbium)
Brazil and Canada

NICKEL
Russia, Canada, Australia, and Indonesia.

PLATINUM-GROUP METALS
Russia and Republic of South Africa.

SILVER
United States. (Silver is mined in more than 53 countries.)

TANTALUM
Thailand, Australia, and Brazil.

TITANIUM
Russia, Japan, United States

ZINC
Canada, Russia, Australia, Peru, and United States.

Source: U.S. BUREAU OF MINES.

Not enough is known at this time about the origin of sulfur to satisfactorily predict precise occurrences. The Frasch process is presently being utilized in certain offshore deposits of this type. The economics of sulfur also may be affected by availability of large quantities of the element from Claus recovery units used in connection with the desulfurization of flue gases.

Valuable deposits of detrital sands and lime muds occur on the continental shelves of many world areas. These can be recovered by dredging operations. Diamonds are presently being recovered by means of vacuum suction tubes from detrital subsea sands adjacent to the Orange River section off the south-western African coast. Dependent upon the nearshore geology, it is known that iron, copper, and coal deposits extend into the subsea areas. In several world areas (Scotland and Japan for coal; Finland and Canada for iron ore; the English coast for tin and copper) deposits have been mined from underground entrances from the adjacent land areas. Sphene and zircon, plus other heavy minerals, have been noted in Texas offshore sediments.

Phosphorite, a major source of phosphorus, is known to occur both as nodular masses and crusts on rocks in subsea areas. Although enormous amounts of phosphorite are accessible in relatively shallow water, marine phosphorites have not been economically competitive with terrestrial supplies.

Metallic sulfides of copper, zinc, and iron have been found in central oceanic rocks and muds under conditions that indicate their deposition from hydrothermal solutions. Such solutions, rich in carbon dioxide, leached metallic elements from both basic rock masses and sedimentary formations with which they came in contact. When such solutions ascended, with concomitant cooling, the minerals were precipitated in the overlying sediments.

Manganese and iron oxides occur as nodular masses in many subsea world areas. They presently are of more interest for their copper, nickel, and cobalt content than for the manganese. Most extensive occurrences are at great ocean depths, as much as 3,500 to 4,500 meters. Fullest exploitation of these deposits and the metallic sulfides will require not only additional technical knowledge for their initial recovery, but also for their refinement to a marketable form. Again, the economics depends upon future demands for the metals as the continental deposits become depleted. Beyond these considerations, the persistent problem of ownership of oceanic resources must be solved as a condition of large-scale recovery. See also **Manganese**.

Lunar Rocks

Geological specimens collected on the *Apollo* lunar missions are indicative of an anhydrous igneous origin. There are three major rock types: (1) a potassium-rich basalt; (2) anorthosite; and (3) an iron, titanium-rich basalt. The first two types are prevalent in the highland areas; the latter in the maria terrain. They occur as crystalline vescicular masses, breccias, and regolithic mantle dust. The absence of an atmosphere and weathering processes on the moon has left the rocks and their component minerals unchanged through eons of time since their formation. Secondary mineralization, therefore, is generally absent and the rocks exhibit a rather limited mineralogy.

Lunar rocks differ in their chemical content rather than type of rock from their terrestrial counterparts. They consistently contain more titanium and chromium, less sodium, and most are richer in iron content. Lunar plagioclase, the major mineral component of anorthosite, is almost always the calcium-rich anorthite, $CaAl_2Si_2O_8$, indicating extreme magmatic differentiation. Lunar basalts are olivine-rich and have been found to be from 3 to 10 times richer in ilmenite (titaniferous iron) as compared with terrestrial basalts.

Clinopyroxene materials, which are common in terrestrial basalts, are well represented in lunar rocks. They include diopside, hedenbergite, johannsenite, aegerine-augite, spodumene, jadeite, augite, pigeonite, omphacite, and fassaite. The more prominent lunar mineral species noted include ilmenite, with rutile intergrowths in certain subfloor maria basalts, cristobalite/tridymite, and pyroxferroite, a mineral closely related in both structure and composition to terrestrial pyroxmangite. Accessory minor minerals include troilite, chromite, ulvospinel, apatite-whitlockite, potash feldspar, quartz, hafnium-rich baddeleyite, and perovskite. Two newly classified species have been recorded—armalcolite, an iron-magnesium titanate, named after *Apollo* astronauts *Arm* strong, *Al* drin, and *Col* lins; and zirkelite, an oxide of calcium, iron, thorium, uranium, with titanium, niobium, and zirconium. Euhedral iron crystals in a pyroxene-rich vug of recrystallized breccia were recovered on the *Apollo 15* mission.

Tiny translucent-to-opaque glassy spherules are prominent in the lunar regolith, within which ilmenite (as thin plates) with minor olivine are present.

Lunar mineralogy is generally analogous to that of terrestrial basalts, the major difference being the lack of oxygen during crystallization which has resulted in the presence of free iron and the exotic minerals, such as troilite, pyroxferroite, armalcolite, and zirkelite. The bulk mineralogy, however, is quite similar to terrestrial rocks with pyroxene, plagioclase, ilmenite, and olivine as the dominant minerals.

Phase equilibrium research of mineral solids has revealed vital information regarding their molecular structure. Application of knowledge gained from this research has extended into the fields of metallurgy, glass and ceramics, and a more adequate interpretation of mineral geology. See also **Moon (Earth's)**.

Mars Surface Geology/Mineralogy

Mariner 9 fly-by of Mars revealed a surface terrain of massive blocks of tumbled character cut by ridges and graben-type troughs. Huge volcanic peaks dominate a pock-marked landscape. Extensive channels characteristic of concentrated erosive powers of torrential floods were also evident, as were braided stream systems emanating from what were resolved to be plateau-type elevations.

Viking spacecraft equipped with a hoe-type scoop and spectrometer analyzed the surface soil to be of a character suggestive of an igneous mafic rock origin, rich in magnesium and iron. The *Viking Landers'* spacecraft analysis of the surface soil chemical composition by x-ray fluorescence revealed a low SiO_2 concentration ($\sim$45%) with iron as Fe_2O_3 near 20%. Further analysis revealed that the regolith mantle soil consisted essentially of iron-rich clay mineral with iron hydroxide, and minerals of sulfate and carbonate content with approximately 1% water by weight. Magnets attached to *Viking's* scoop attracted magnetic material aggregates. It is quite probable that the magnetic material represents a component part of that regolith and possibly the soil is enriched by both magnetite (Fe_3O_4, color black) and maghemite (γ-Fe_3O_4, color yellowish-brown). The yellowish-brown surface color may be the product of thin coating of

hydrated iron oxides, with nontronite/montmorillionite as the host soil. See also **Mars**; and **Pathfinder Mission to Mars**.

Much remains to be resolved before final definitive answers can be given in this area of planetary investigation and evaluation.

Classification of Minerals

Minerals are classified in groups, according to their chemical composition, based upon the dominant anion or anionic group. The system works well in various ways. Generally, the dominant anion or anionic group brackets minerals of corresponding characteristics which tend to occur in quite similar environments. The dominant chemical subdivisions are:

Elements. Minerals composed of uncombined chemical elements, e.g., Au, gold; Ag, silver; Cu, copper; although minor impurities may be present within the structure.
Sulfides. Minerals composed of compounds of metals with sulfur.
Sulfosalts. Minerals composed of compounds of semimetals with sulfur.
Halides. Minerals composed of compounds of metals with fluorine, chlorine, bromine, and iodine.
Oxides and Hydroxides. Minerals composed of compounds of the metallic elements with oxygen.
Carbonates. Minerals composed of compounds of a metal with the carbonate radical CO_3.
Borates. Minerals composed of compounds of a metal with the borate radical, BO_3.
Nitrates. Minerals composed of compounds of a metal with the nitrate radical, NO_3.
Sulfates. Minerals containing the sulfate radical, SO_4.
Chromates. Minerals containing the chromate radical, CrO_4.
Molybdates. Minerals containing the molybdate radical, MoO_4.
Tungstates. Minerals containing the tungstate radical, WO_4.
Phosphates. Minerals containing the phosphate radical, PO_4.
Arsenates. Minerals containing the arsenate radical, AsO_4.
Vanadates. Minerals containing the vanadate radical, VO_4.
Silicates. This mineral classification encompasses the largest group of mineral species and includes most of the important rock-forming minerals, such as the feldspars, feldspathoids, pyroxenes, amphiboles, micas, olivine, and quartz. Silicon is the basic chemical element, as the name implies. The small silicon cation combines with four oxygens to form an SiO_4 tetrahedral structure. The SiO_4 formula leaves a net negative charge, which requires additional combinations with other tetrahedra or anions to effect a neutral balance. The type and degree of such tetrahedral combinations control the final structural character and act as a convenient classification of the silicate mineral family.

Subclassification of silicates are:

Nesosilicates, with each tetrahedron existing within the structure as isolated SiO_4 units.
Sorosilicates involve the pairing of SiO_4 tetrahedra. The shared-oxygen anion represents the link between these tetrahedra.
Cyclosilicates involve two oxygens from each SiO_4 tetrahedron combining with oxygen in adjacent tetrahedral units to form *ring* structures.
Inosilicates are the product of oxygen sharing between adjacent tetrahedra to form *single* or *double* chains. In the single-chain structure, two oxygens from each tetrahedra combine with adjacent tetrahedra. In the double chain, half of the tetrahedra share three oxygens, while the other half share only two.
Phyllosilicates involve the sharing of three oxygens in each tetrahedron with adjacent tetrahedrons to form *sheet* structures. Minerals in this classification are usually flaky in character and relatively soft.
Tectosilicates involve the sharing of all four oxygens in each tetrahedral unit with adjacent tetahedrons to form a three-dimensional *framework* of SiO_4 units linked together. The product is a strongly bonded structure with a silicon-oxygen ratio of 1:2. The greater portion of the earth's crust is composed of minerals found within this classification.

Most minerals exhibit a variation in their chemical composition, with the exception of the elements (see preceding list). The substitution of one ion for another is common, since minerals crystallize in solutions of complex composition.

A full listing of all minerals described separately in this encyclopedia is given in Table 4.

TABLE 4. MINERALS DESCRIBED IN THIS ENCYCLOPEDIA

ARSENATES	OXIDES AND	SILICATES (Inosilicates)	SILICATES (Sorosilicates)	Calaverite
Annabergite	HYDROXIDES	Acmite-Aegerine	Allanite	Chalcopyrite
Erythrite	Alabandite	Actionolite	Clinozoisite	Chalcopyrite
Mimetite	Alexandrite	Aegerine	Epidote	Cinnabar
Scorodite	Anatase	Amphibole	Hemimorphite	Cobaltite
	Bauxite	Anthrophyllite	Lawsonite	Covellite
BORATES	Brookite	Augite	Prehnite	Galena
Boracite	Brucite	Babingtonite	Vesuvianite	Gersdorffite
Borax	Cassiterite	Bustamite	Zoisite	Greenockite
Colemanite	Cat's-Eye	Crocidolite		Hessite
Inyoite	Chromite	Cummingtonite	SILICATES (Tectosilicates)	Krennerite
Kernite	Chrysoberyl	Diallage	Agate	Mercasite
Ulexite	Columbite	Diopside	Amethyst	Millierite
	Corundum	Enstatite	Analcime	Molybdenite
CARBONATES	Cuprite	Glaucophane	Bloodstone	Nickeline
Aragonite	Diaspore	Hornblende	Cairngorm Stone	Orpiment
Azurite	Emery	Hypersthene	Cancrinite	Pentlandite
Barytocalcite	Fergusonite	Jade	Carnelian	Pyrite
Bastnasite	Franklinite	Jadeite	Chabazite	Pyrrhotite
Calcite	Gahnite-Zinc Spinel	Pyroxene	Chalcedony	Realgar
Cerussite	Geikielite	Rhodonite	Citrine	Skutterudite
Chalk	Goethite	Riebeckite	Danburite	Sperrylite
Dolomite	Hematite	Serandite	Desmine	Sphalerite (Biende)
Magnesite	Ilmenite	Spodumene	Feldspar	Stannite (Mineral)
Malachite	Limonite	Tremolite	Flint	Stibnite
Phosgenite	Magnetite	Uralite	Harmotome	Sylvanite
Rhodochrosite	Manganite	Wollastonite	Heulandite	Tetradymite
Siderite	Perovskite		Hyalite	Wurtzite
Smithsonite	Pitchblende	SILICATES (Nesosilicates)	Jasper	
Strontianite	Psilomelane	Andalusite	Lazurite	SULFOSALTS
Travertine	Pyrolusite	Chrondrodite	Leucite	Boulangerite
Witherite	Pyrophanite	Datolite	Natrolite	Bournonite
	Rutile	Dumortierite	Nepheline	Enargite
CHROMATES	Spinel	Fayalite	Opal	Geocronite
Crocoite	Tantalite	Forsterite	Perthite	Jamesonite
ELEMENTS	Tenorite	Garnet	Petalite	Polybasite
Amalgam	Thorianite	Kyanite	Phillipsite	Poustite
Antimony	Uraninite	Olivine	Pollucite	Pyrargyrite
Bismuth	Wad	Phenacite	Quartz	Stephanite
Copper	Zincite	Silimanite	Scolecite	Tetrahedrite
Diamond		Sphene	Sodalite	
Electrum	PHOSPHATES	Staurolite	Stilbite	TUNGSTATES
Gold	Amblygonite	Thorite	Tridymite	Scheelite
Graphite	Apatite	Topaz	Wernerite	Wolframite
Mercury	Autunite	Willemite	Zeolite Group	
Platinum	Lazulite	Zircon		VANADATES
Plumbago	Monazite		SULFATES	Carnotite
Quicksilver	Pyromorphite	SILICATES (Phyllosilicates)	Alabaster	Tyuyamunite
Silver	Torbernite		Alunite	Vanadinite
Sulfur	Triphylite	Apophyllite	Anglesite	
	Turquois	Asbestos	Anhydrite	OTHER MINERALOGICAL
HALIDES	Vivianite	Biotite	Antlerite	TERMS
Atacamite	Wavelite	Chlorite	Barite	Abrasion pH
Carnallite		Chloritoid	Brochantite	Carbonado
Chlorargyrite	SILICATES (Cyclosilicates)	Chrysotile	Celestite	Diamond
Cryolite	Axinite	Garnierite	Chalcanthite	Diatomite
Fluorite	Beryl	Glauconite	Epsomite	Clay
Halite	Chrysocolla	Kaolinite	Glauberite	Fuller's Earth
Sylvite	Cordierite	Lepidolite	Gypsum	Gangue
	Dioptase	Mica	Jarosite	Gem Stones
MOLYBDATES	Dravite	Muscovite	Polyhalite	Kimberlite
Wulfenite	Elbaite	Phlogopite		Peridotite
	Emerald	Pyrophyllite	SULFIDES	Tripolite
NITRATES	Euclase	Sepiolite	Argentite	Vitrophyre
Niter	Iolite	Serpentine	Arsenopyrite	
Soda-Niter	Liddicoatite	Talc	Bismuthinite	
	Tourmaline		Bornite	

Additional Reading

Arem, J.E.: "Color Encyclopedia of Gemstones,"3rd Edition, Chapman & Hall, New York, NY, 1994.

Boyd, F.R. and J.J. Gurney: "Diamonds and the African Lithosphere," *Science*, **232**, 472–477 (1986).

Boyle, R.W.: "Gold: History and Genesis of Deposits," Chapman & Hall, New York, NY, 1990.

Brierley, C.L.: "Microbiological Mining," *Sci. Amer.*, 44–53 (August 1982).

Brown, W.L., Editor: "Feldspars and Feldspathoids," Reidel, Boston, 1984.

Campbell, A.N. et al.: "Recognition of a Hidden Mineral Deposit by an Artificial Intelligence Program," *Science*, **217**, 927–929 (1982).

Carmichael, R.S. and S. Robert: "Practical Handbook of Physical Properties of Rocks and Minerals," CRC Press, LLC., Boca Raton, FL, 1990.

Cornelis Klein, C. and C.S. Hurlbut: "Manual of Mineralogy," 21st Edition, John Wiley & Sons, Inc., New York, NY, 1998.

Crowson, P.: "Minerals Handbook: 1996–1997," Groves Dictionaries, Inc., New York, NY, 1996.

Derry, D.R.: "A Concise World Atlas of Geology and Mineral Deposits," John Wiley & Sons, Inc., New York, NY, 1981.

Dietrich, R.V., B.J. Skinner, and R. Vincent: Cambridge University Press "Gems, Granites, and Gravels: Knowing and Using Rocks and Minerals," Cambridge University Press, New York, NY, 1990.

Glusker, J.P.: "Structural Crystallography in Chemistry and Biology," John Wiley & Sons, Inc., New York, NY, 1982.

Goeller, H.E. and A. Zucker: "Infinite Resources: The Ultimate Strategy," *Science*, **223**, 456–462 (1984).

Golden, D.C., C.C. Chen, and J.B. Dixon: "Synthesis of Todorokite," *Science*, **231**, 717–719 (1986).

Hein, J.R.: "Siliceous Sedimentary Rock-Hosted Ores and Petroleum," John Wiley & Sons, Inc., New York, NY, 1987.

Holland, H.D. and M. Schidlowski: "Mineral Deposits and the Evolution of the Biosphere," Springer-Verlag, Inc., New York, NY, 1982.

Kahle, A.B. and A.F.H. Goetz: "Mineralogic Information from a New Airborne Thermal Infrared Multispectral Scanner," *Science*, **222**, 24–27 (1983).

Kelly, E.G. and D.J. Spottiswood: "Introduction to Mineral Processing," John Wiley & Sons, Inc., New York, NY, 1982.

Lide, D.R.: "CRC Handbook of Chemistry and Physics 2000-2001," 81st Edition, CRC Press, LLC., Boca Raton, FL, 2000.

Meyer, C.: "Ore Metals Through Geologic History," *Science*, **227**, 1421–1428 (1985).

Nancollas, G.H.: "Biological Mineralization and Demineralization," Springer-Verlag, Inc., New York, NY, 1982.

Nesse, W.D.D.: "Introduction to Mineralogy," Oxford University Press, Inc., New York, NY, 1999.

Newton, R.C., A. Navrotsky, and B.J. Wood: "Thermodynamics of Minerals and Melts," Springer-Verlag, Inc., New York, NY, 1981.

O'Reilly, W.: "Rock and Mineral Magnetism," Chapman and Hall, New York, NY, 1984.

Ozima, M. and S. Zashu: Primitive Helium in Diamonds," *Science*, **219**, 1067–1068 (1983).

Park, C.F.: "The Geology of Ore Deposits," 4th Edition, W.H. Freeman Company, New York, NY, 1998.

Parker, S.P.: "McGraw-Hill Dictionary of Geology and Mineralogy," The McGraw-Hill Companies, Inc., New York, NY, 1997.

Robinson, E.S.: "Basic Physical Geology," 3rd Edition, John Wiley & Sons, Inc., New York, NY, 1991.

Rona, P.A.: "Mineral Deposits from Sea-Floor Hot Springs," *Sci. Amer.*, 84–92 (January 1986).

Sawkins, F.J.: "Metal Deposits in Relation to Plate Tectonics," 2nd Edition, Springer-Verlag, Inc., New York, NY, 1989.

Sohn, H.Y. et al.: "Processing of Energy and Metallic Minerals," American Inst. of Chemical Engineers, New York, NY, 1982.

Swanson, E.A., D.F. Strong, and J.G. Thurlow: "The Buchans Orebodies," Geological Association of Canada, Toronto, Ontario, 1981.

Touloukian, Y.S. et al.: "Physical Properties of Rocks and Minerals, Vol. 2, The McGraw-Hill Companies, Inc., New York, NY, 1989.

Wills, B.A.: "Mineral Processing Technology: An Introduction to the Practical Aspects of Ore Treatment and Mineral Recovery," Butterworth-Heinemann, Inc., Woburn, MA, 1997.

Web Reference

Mineralogical Society of America: *http://www.minsocam.org/*

ELMER B. ROWLEY, F.M.S.A., Union College, Schenectady, NY.

MINIMAX METHOD OF ESTIMATION.   Suppose we wish to estimate a population parameter ϕ from a sample, and that we can specify the amount of loss we shall make if we adopt any value ϕ_i when the true value is ϕ_j. Denoting the sample values collectively by x, suppose also that we have some rule R that tells us which value of ϕ to adopt for any particular sample x. We can then calculate the expected loss or risk as a function of ϕ. For any given rule R we can find the maximum value of the risk, and the minimax estimate is provided by the particular rule R_0 that minimizes this maximum risk. (For application in **Game Theory**, see that entry.)

MINNOWS.   See **Viviparous Topminnows**.

MINOR (Of a Matrix).   If a matrix **B** is formed from a matrix **A** by striking out certain rows and columns, then **B** is called a minor of **A**. The matrix **C** formed by the deleted rows and columns is called the complementary minor of **B** in **A**. If **A**, **B**, **C** are all square, then $(-1)^k$ **C** is called the algebraic complement of **B**, where k is the sum of the indices of the rows and columns of **C**.

MINT FAMILY (*Labiatae*).   The greater number of the 3,000 species of this family are herbaceous plants, widely distributed in the temperate regions. The family also includes a few small trees and shrubs, found mostly in the American tropics. The characters distinguishing these plants are so outstanding as to make identification easy. Commonly the stem is 4-angled, appearing square in cross section, with the leaves opposite, simple, and without stipules. The flowers are borne in racemes or more frequently in dense axillary cymes. The flowers are irregular, with the calyx composed of five sepals united to form a tube, and the corolla composed of fused petals which form a 2-lipped tube, the upper lip of three, the lower of two lobes. The stamens, either two or four in number, are inserted on the corolla tube. The pistil is composed of a bicarpellate ovary, each of whose carpels becomes constricted early in development to form a 4-parted ovary, from which a 2-lobed style arises. The fruit is commonly composed of four achenes or nutlets. The flowers of this family are mainly cross-pollinated by insects: in some species the corolla tube is short enough to allow bees to obtain the nectar located in a disk at the base of the ovary. Pushing into the corolla tube to reach this nectar causes the pollen in the anthers to be shaken onto the insect's back, where it will be carried to the stigmas of another flower. Other species having longer corolla tubes are cross-pollinated by butterflies.

Many members of this family have volatile oils located in epidermal glands on the leaves, giving to the plants characteristic odors. Because of these volatile oils many of them are useful to man, some as condiments, some as perfumes, and some as drugs. Food products are rare in this family, the genus *Stachys* having species which form tubers, eaten in some European countries. *Salvia* species are widely grown because of the showy scarlet flowers, often accompanied by brightly colored bracts. The leaves of the garden sage, *Salvia officinalis*, are used as flavoring for poultry dressing.

The leaves of *Ocimum basilicum* (basil), *Thymus vulgaris* (thyme), and *Origanum vulgare* (majoram), as well as *Salvia*, are used for flavoring. *Rosamarinus officinalis* gives rosemary oil; *Lavandula vera* and other species, oil of lavender; and *Pogostemon patchouly*, oil of Patchouli — all these oils being used in making perfumes.

Certain species of Mints, commonly called nettles, are troublesome weeds, often with strong rank odors.

MIRA (Omicron Ceti).   A star that has the distinction of being the first variable star initially announced as such. In 1596, the Dutch astronomer Fabricius noticed a star of about the third magnitude that had not previously been recorded. The star faded within a few weeks, but was again seen and recorded by Bayer, in 1603. In 1638, Holwarda, another Dutch astronomer, again observed the star, found it to disappear and to then return to visibility about 11 months later. At maximum brightness, the star is easily visible to the naked eye, having a magnitude of about 3.5; but at minimum, it can be seen only with a telescope of aperture greater than 1 inch (2.54 centimeters), because its magnitude is only about 9. In the near infrared, its magnitude changes very little.

Many determinations of the period of variability have been made, the time from maximum to minimum averaging about 330 days, with variations between times of successive maxima amounting to as long as one month. The diameter of the star is of the order of magnitude of 4.2×10^8 kilometers or large enough to contain the sun and all the members of the solar system in their orbits to beyond the planet Mars. Coupled with the variation in brightness is a variation in diameter amounting to about 5.1×10^7 kilometers, together with a temperature variation of roughly 500 K (900 °F).

The spectrum displays fascinating changes in the near infrared region, due primarily to the absorption bands of VO and TiO. See Fig. 1.

Fig. 1.   Spectrum of Mira. (*University of Michigan Observatory.*)

MIRAGE.   See **Atmospheric Optical Phenomena**.

MIRROR NUCLIDES.   Pairs of nuclides, having their numbers of protons and neutrons so related that each member of the pair would be transformed into the other by exchanging all neutrons for protons and vice versa.

MIRRORS AND LENSES.   This article is chiefly concerned with spherical or plane reflecting and refracting surfaces. A spherical mirror may be treated as a 1-base zone of a spherical surface, the axis of which is the straight line passing through the center of curvature and the pole of the zone. When the diameter of the mirror is small compared with its radius of curvature, and when the rays make only small angles with the axis, so that spherical aberration may be neglected, such a mirror produces fairly sharp images which are easily calculated from the laws of reflection. Taking the pole O as origin and the mirror axis as X-axis (Fig. 1), and representing the radius OC by r, the image of a point $P_1(x_1, y_1)$ in the plane XY is the point P_2 whose coordinates, for either a concave or a convex mirror, are

$$\left. \begin{array}{l} x_2 = \dfrac{rx_1}{2x_1 - r} \\[2mm] y_2 = \dfrac{-ry_1}{2x_1 - r} \end{array} \right\}$$

r is $+$ or $-$ according as the mirror is convex or concave. If x_2 turns out positive, it means that the image is virtual.

In Fig. 1, r, x_1 and x_2 are all *negative*, as is y_2, so that the image is real and inverted. If the incident rays are parallel to the axis, the focus of the reflected rays, called the focal point of the mirror, is on the axis at $x = r/2$. For a plane mirror, $r = \infty$ and $x_2 = -x_1$, $y_2 = y_1$.

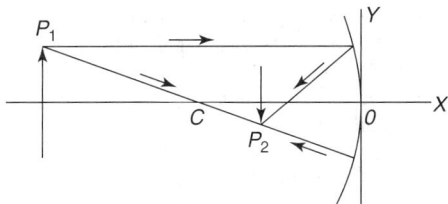

Fig. 1.   Formation of real image by concave mirror.

Spherical lenses have various combinations of convex, concave, or plane surfaces. There is always a point on the axis, called the optical center, such that if a ray in traversing the lens is in line with this point, the entering and emerging parts of the ray are parallel. For a very thin lens this point may be considered at the center of the lens and is a suitable origin. If the radii of curvature r_1, r_2 are large compared with the diameter of the lens, and if the refractive index of the lens is n, the equations giving the image of the point x_1, y_1 made by a thin lens are

$$\left. \begin{array}{l} x_2 = \dfrac{r_1 r_2 x_1}{r_1 r_2 + (n-1)(r_2 - r_1)x_1} \\[3mm] y_2 = \dfrac{r_1 r_2 y_1}{r_1 r_2 + (n-1)(r_2 - r_1)x_1} \end{array} \right\}$$

r_1 and r_2 refer to the left and right surfaces, respectively, which is the order in which the light encounters them. P_1 and P_2 are called conjugate points of the system. The focal length is obtained by letting $x_1 + \infty$ and calculating x_2 from (2), which gives

$$f = \frac{r_1 r_2}{(n-1)(r_2 - r_1)}$$

This enables us to write (2) in simpler form:

$$\left. \begin{array}{l} x_2 = \dfrac{f x_1}{f + x_1} \\[2mm] y_2 = \dfrac{f y_1}{f + x_1} \end{array} \right\}$$

If x_2 turns out negative, the image is virtual; if y_2 is negative, it is inverted. For lenses of appreciable thickness or of strong curvature, the calculations are not so simple, and in general there are two unequal focal lengths, depending upon which way the rays pass through the lens.

The reciprocal of the focal length of a lens, called its "focal power," is a measure of the converging or diverging effect of the lens. It is commonly expressed in diopters or reciprocal meters; thus if the focal length is 50 centimeters or $\frac{1}{2}$ meter, the focal power is 2 diopters.

MISCIBILITY.   The ability of two or more substances to mix, and to form a single, homogeneous phase.

MISSILE PROPELLANTS.   See Rocket Propellants.

MISSILRY.   The art or science of designing, developing, building, launching, directing, and sometimes guiding a rocket missile; any phase or aspect of this art or science. This term is sometimes spelled missilery, but is then pronounced as a three-syllable word.

MISSISSIPPIAN PERIOD.   A geologic period in the Paleozoic Era. Type locality, Mississippi Valley. The period began about 280 million years ago and lasted for about 25 million years. The term Mississippian, first proposed by H.S. Williams in 1891, is roughly equivalent to the more general term, Lower Carboniferous. In Britain, the formations of this system are grouped under the terms Culm and Mountain Limestones, which, as in the United States, immediately succeed the Devonian and are followed by the upper Carboniferous, or Pennsylvania System (U.S.), and Coal Measures (Britain). The formations of this system are chiefly sandstones and shales in the Appalachian Geosyncline, representing delta and estuarine deposits of considerable thickness which pass Westward into thinner facies of marine shales and limestones. In the Rocky Mountain region occurs a great thickness of marine Mississippian called, locally, the Madison Limestone. The marine life of the Mississippian is chiefly characterized by echinoderms and foraminifera. Petroleum occurs in the Mississippian formations of Southeastern Ohio, West Virginia, Southeastern Pennsylvania and Eastern Kentucky.

MITE (*Arachnida, Acarina*).   Minute animals related to spiders, ticks, and scorpions. Control chemicals used against these creatures are called *acaricides*. As is evident from the following descriptions, there are many species and varieties of the mites, with a remarkable degree of specialization exhibited. Also, many of the mites have shown outstanding capacity to develop genetic resistance to various control chemicals after long periods of exposure. Since the mites are so small, they are difficult to identify, but this is sometimes necessary to determine what chemicals may or may not be effective against a given infestation. Various species of mites are severely damaging to fruit crops, both deciduous and citrus, as well as certain field crops, such as alfalfa and clover. Further, the mites are serious pests to humans and domestic animals.

Citrus Mites

Among the most damaging mites on citrus are the red spider mite, the citrus rust mite, and the citrus bud mite.

Red Spider Mite (family *Tetranychidae*, many species). This is one of the most damaging pests on citrus and several other fruit crops. These insects are found in all stages all year long in the warmer citrus growing regions. Bright red eggs are deposited by females on the underside of leaves or in silken webs spun by the mite and located near twigs and fruit. The egg laying goes on for 2 weeks, the female laying two or three eggs per day. The total life cycle is from 3 to 5 weeks and is temperature-dependent. The larva has 6 legs and the protonymph and deutonymph have 8 legs. The adult mite ranges in color from purple to red and appears to be of a velvety texture. The six-spotted mite larvae are yellowish-green.

Citrus Rust Mite (Phyliocoptrula oleivora, Ashmead*); Citrus Bud Mite (Aceria sheldon*, Ewing). The citrus rust mite is the second (to purple scale insect) most severe economic pest on citrus in the Gulf States. The pest sucks sap from the leaves and fruit skins, with resulting russeting of oranges and silvering of lemons. This mite attacks orange, grapefruit, lemon, and lime. Sulfur is particularly effective in controlling the rust mite.

Citrus Bud Mite. This mite found in California is a more recent species. It attacks buds and blossoms, causing poorly shaped fruit. Oil emulsions are effective against bud mite.

Deciduous Fruit Mites

Several species of mite are injurious to deciduous fruit trees. They tend to specialize, but their life cycles are similar.

Brown Mite (Bryobia arborea, Morgan and Anderson). Particularly damaging to apple during dry seasons. Habilitates the foliage and sucks sap from buds and leaves, causing foliage to turn yellow. In some cases, the twigs of a tree may have so many tiny spherical red eggs attached that they will have a red aura about them. Until the late 1950s, this mite was confused with the *clover mite* (*Bryobia praetiosa*, Koch). The brown mite is found in Canada and in the northern and southwestern parts of the United States. Target trees of the mite include almond, apple, cherry, peach, pear, plum, prune, and walnut — as well as the raspberry plant.

European Red Mite (Panonychus ulmi, Koch*), or Paratetranychus pilosus).* A very significant pest of deciduous fruits in North America. The mite was first reported in the United States in 1911 and, although it occurs throughout the United States, it is most common in regions that lie north of 37 °N latitude (about San Francisco in the West; Saint Louis in the Midwest; and Richmond, Virginia in the East). Light invasions cause speckling of leaves. Heavy infestations cause paling and discoloration of foliage, causing leaves to drop. Fruit bud formation is difficult and fruit may be deformed, smaller than normal, and of poor color. All deciduous fruit trees can be affected; the mite is most severe against apple, pear, plum, and prune.

Pacific Mite (Tetranychus pacificus, McGregor*).* This species is often a severe economic pest of deciduous fruit, causing extensive webbing and discoloration of foliage. Fruit drops prematurely and is poorly colored. A thousand or more mites may be found on a single leaf. Fruits attacked include almond, apple, blackberry, grape, pear, plum, prune, and walnut. The mite also damages alfalfa, bean, clover, and cotton. The mite, about $\frac{1}{60}$ inch (0.4 millimeter) in length, occurs along the Pacific coast of North America, south to California from British Columbia. Adult females (summer) are of a green color and have brownish-black spots. Overwintering is by adult females (bright orange) that habitate in leaves, trash, or under pieces of bark.

Vegetable Plant Mites

Many species of spider mite cause extensive damage of certain vegetable crops. See Fig. 1. They are of the family *Tetranychidae.* Particularly active during hot, dry seasons, these mites cause damage to bean and a number of other vegetables. They tend to specialize. When infested, the leaves of a plant appear pale and progressively drop off. Prior to dropping, the leaves will appear yellow and red-brown in splotches of varying size. The undersides of the leaves take on a whitish powdery appearance. These whitened areas are made up of the wrinkled and empty skins of very tiny eggs, which are suspended by hardly visible silken threads. On the silk and on the leaf will be found many very small, eight-legged mites, ranging in color from white to green to red, and which are only about $\frac{1}{60}$ inch (0.4 millimeter) long. These mites pierce the leaves and consume the sap.

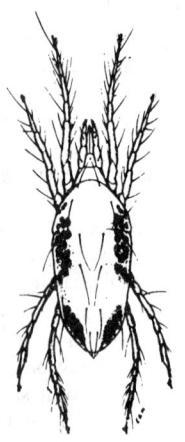

Fig. 1.   Adult female of the two-spotted spider mite. (*USDA.*)

The mites overwinter as females, usually of an orange color. Because there are so many species and the mites are so small, it requires an expert to identify a given species with certainty.

Control measures against the spider mite include mechanical removal of the intricate and continuously made webbing, spraying the plant with cool, clear water at frequent intervals, tasks that can be done more conveniently under glasshouse conditions and for plants that can withstand a wet soil. Control chemicals can be effective.

Processed Food Product Mites

These mites tend to specialize in certain substances as their sources of food.

Flour Mite (Acarus sino, Linne, or *Tyroglyphus farinae,* De Geer*).* Also known as the *grain mite,* this mite can be extremely troublesome to workers who handle flour, meal, and grain in various stages of processing. The itching resulting from bites is sometimes called "grocer's itch." Infestation in flour mills and grain-processing plants not only is a hazard to workers, but also severely damages product, sometimes requiring reprocessing or disposal.

Cheese Mite (T. catellanii, Hirst*).* Damaging to cheese in storage and often associated with the *cheese skipper,* description of which follows.

Cheese Skipper (Piophila casei, Linne*).* This insect infests and seriously damages or destroys cheese as well as smoked and cured meats. Distribution of the mite is believed to be worldwide. The adult skipper is a two-winged fly, about $\frac{1}{6}$ inch (4 millimeters) in length. Each female lays as many as 500 eggs, often in groupings of 40 to 50.

Ham Mite (family Tyroglypuhus). This mite is very similar to the cheese mite except that it attacks smoked and cured meats. Curing temperature below a range of 30–36 °F (−1.1 to +2.2 °C) is required to prevent the presence of this mite in storage rooms. The mite is very adaptable in terms of its life cycle, and can inhabit a space for a long time without feeding. However, they do require a minimum relative humidity of 11%. Control is by fumigation with sulfur dioxide. Allethrin is also effective.

Mushroom Mite (Tyroglyphus lintneri, Osborn*).* Of an off-white coloration, this mite is about $\frac{1}{32}$ inch (1.5 millimeters) long. When present in large numbers, the mites produce a brownish powder which has an unpleasant, musty odor that is easily imparted to adjacent products. The mite is responsible for eating holes in the caps and stems of mushrooms and also for eating the spawn. Handlers also can be bothered with grocer's itch when handling infested products.

Animal Mange and Human Mites

Several species of mite produce serious problems for animals and humans. As with other mites, they tend to specialize.

Itch or Mange Mite (Sarcoptes scabiei, De Geer*).* This mite attacks humans, horses, hogs, and cattle. See Fig. 2. In animals, they cause *mange.* Several strains of this mite exist, each adapted best to a given host. The damage is caused by the female mite burrowing into the skin of its host and depositing eggs in the tunnels. A century or so ago, when the cause of the resulting itching was not known and treatments were not available, a number of generations of the mites could be produced in a given host, including a human being. This gave rise to the term "seven year itch." These mites are very small. The female is about $\frac{1}{60}$ inch (1.5 millimeters) long and the male is only half that length. The tunnels made by the female mite are about $\frac{1}{50}$ inch (0.2 millimeter) in diameter and about 1 inch (2.5 centimeters) long. In this tunnel, up to 24 eggs may be laid. The mite goes through an egg stage, a larva stage (6-legged creature), two 8-legged nymphal stages. Once the eggs are laid, the females prepare new tunnels and mating occurs within the tunnels. The main indication of the disease is extreme itching. Sometimes the tunnels can be seen just below the surface of the skin of the host. Common locations include the back of the knee, genitalia, and in between fingers and toes. Pimples and pustules form; they are subject to infections. The infested host can become quite ill as a result of the intense itching, concern, and stress, and the venom introduced by the presence of large numbers of mites.

The infestation can be transmitted to another host because the creatures tend to crawl outside at night and a person occupying the same bed with an infested person has an excellent opportunity of also becoming infested. Until control methods take effect, isolation of the infested host is highly desirable. One control measure, the United States Army NBIN formulation consists of: Benzyl benzoate, 68% (weight); DDT (or suitable substitute), 6%; Benzocaine (ethyl para-aminobenzoate), 12%; and sorbitan monooleate poloxyalkalene derivative (*Tween 80*), 14%. With exception of the head, about 2.5 ounces (75 cubic centimeters) of the diluted emulsion should be applied to the body. The compound should not be washed off for 24 hours. A second application may be required. Commercial preparations are also available.

In hogs, the mite can be diagnosed by first noting if hogs are scratching extensively and if their hair is standing erect. If there are no gray hog lice present, it is most likely that the hogs are infested with the mites. If so, inspection of the neck and back will show inflammation. The skin should be scraped to permit the mites to escape from the burrows. A close examination will show the 8-legged mites moving about.

The same mite attacks horses, hogs, mules, cattle, foxes, rabbits, squirrels, sheep and other common mammals. It is important to note that

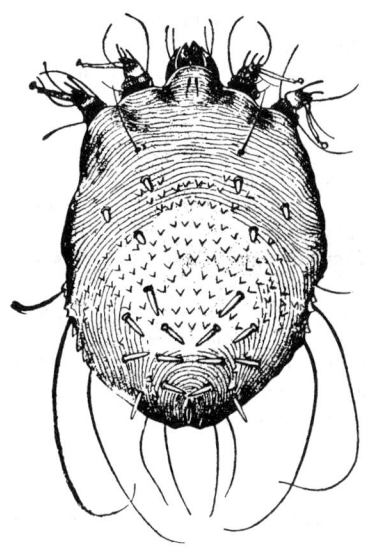

Fig. 2.   Female sarcoptic mange mite. (*USDA.*)

the resulting condition, mange, is highly contagious and transferable from one host to the next, regardless of species of host.

Poultry Mites

Poultry Mite or Chicken Mite (Dermanyssus gallinae, De Geer*).* This mite feeds on the blood of poultry during the night and usually leaves the fowl during the day. Since the presence of the mite cannot be detected by inspection during the day, the cause of an unhealthy flock may not be immediately traced to this mite. A telltale is the excrement of the mites, which appears like salt and pepper dusted about. Closer inspection will indicate the presence of many very tiny gray-and-brown mites, which appear as specks. The mite devitalizes its host, causing the fowl to be inactive, droopy, and listless. Some will cease laying eggs, and chicks and sitting hens may ultimately succumb to this persistent attack. Persons working in poultry houses also can be irritated by the mites. Control is essentially prevention—keeping the houses extremely clean. Where mites are suspected (or as prevention), the area can be sprayed with malathion. The birds can be sprayed with dilute formulations. Isolation of birds when moved from one house to the next is also suggested.

Northern Fowl Mite or Feather Mite (Ornithonyssus or Liponyssus sylviarum, Canestrini and Fanzago*).* This mite causes a drop in egg production and, if an infestation is severe, many fowl may perish. Eggs are laid in the fluff feathers of the bird.

MIXER. 1. In a transmission, recording, or reproducing system, a device having two or more inputs, usually adjustable, and a common output, which operates to combine linearly, in a desired proportion, the separate input signals to produce an output signal. The term is sometimes applied to the operator of the above device.

2. In a superheterodyne receiver, the first detector (or transducer, heterodyne conversion).

MIZAR (*ζ Ursae Majoris*). An interesting star in the Big Dipper, and probably the first double star ever observed. See also **Double Star**. Mizar forms with the fourth magnitude star Alcor a naked-eye double, and Mizar, itself, has a close companion that is telescopically visible. It was first observed as a double star by Riccioli, in 1650. Tradition says that observation of the pair Mizar-Alcor was considered a test of good eyesight among the American Indians. However, if this was a difficult pair for the Indians to separate, their eyesight could not have compared very favorably with that of modern times, because this is an easy double for most people.

As well as being the first visual double star to be discovered. Mizar also has the distinction of being the first spectroscopic binary discovered. In 1889, E.C. Pickering discovered that the spectral lines of this star were alternately double and single, a phenomenon that can be adequately explained only by the star being a close binary. In 1908, both the fainter companion of Mizar and the more distant, bright companion Alcor were found to be spectroscopic binaries.

MODEL (Scientific). Fundamentally, a model may be defined as a representation of some or all of the properties of a device, system, or object. The more properties represented by the model, the more complex the model becomes and thus one constantly faces tradeoffs in constructing a satisfactory model. There are three basic classes of models: (1) *mathematical models*, wherein the representation is comprised of procedures (algorithms), mathematical equations, and so on; (2) *physical models*, such as models of rivers and dam systems, airfoils and ship contours for use in wind tunnels and similar apparatus, and construction projects of extreme three-dimensional complexity (see Fig. 1); and (3) *logical models*. Simulation is closely associated with modeling, in that electrical circuits may be set up to correspond with mechanical, thermal, and fluid systems. See also alphabetical index.

Fig. 1.   Physical model ($\frac{1}{32}$ of full scale) of a large crude refining unit. (*Foster Wheeler Energy Corporation.*)

MODERATOR. A substance used to slow down neutrons by means of collisions. Moderators play an important part in the design and operation of nuclear reactors. Moderators *thermalize* neutrons to an energy of about 0.025 eV. See also **Nuclear Power Technology.**

MODULATION. The process, or the result of the process, whereby some characteristic of one wave is varied in accordance with some characteristic of another wave. Usually one of these waves is considered to be a carrier wave while the other is a modulating signal. The various types of modulation, such as amplitude, frequency, phase, pulse width, pulse time, and so on are designated in accordance with the parameter of the carrier which is being varied.

Amplitude modulation (AM) is easily accomplished and widely used. Inspection of Fig. 1 shows that the voltage of the amplitude modulated wave may be expressed by the following equation

$$v = V_c(1 + M \sin \omega_m t) \sin \omega_c t$$

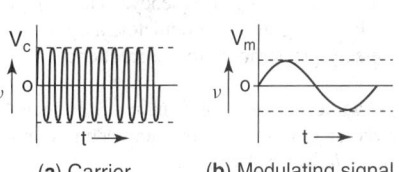
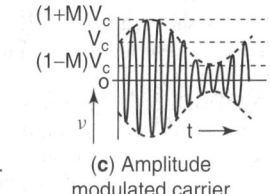

(**a**) Carrier.   (**b**) Modulating signal.   (**c**) Amplitude modulated carrier

Fig. 1.   Amplitude modulation.

where ω_c and ω_m are the radian frequencies of the carrier and modulating signals, respectively. The modulation index M may have values from zero to one. When the trigonometric identity $\sin a \sin b = \frac{1}{2}\cos(a-b) - \frac{1}{2}\cos(a+b)$ is used in the equation above, this equation becomes

$$v = V_c \sin \omega_c t + \frac{MV_c}{2}\cos(\omega_c - \omega_m)t - \frac{MV_c}{2}\cos(\omega_c + \omega_m)t$$

This equation shows that new frequencies, called side frequencies or side bands, are generated by the amplitude modulation process. These new frequencies are the sum and difference of the carrier and modulating frequencies.

Amplitude modulation is accomplished by mixing the carrier and modulating signals in a nonlinear device such as a vacuum tube or transistor amplifier operated in a nonlinear region of its characteristics. The nonlinear characteristic produces the new side-band frequencies. Frequency converters or translators and AM detectors are basically modulators. The various types of pulse modulation are actually special types of amplitude modulation.

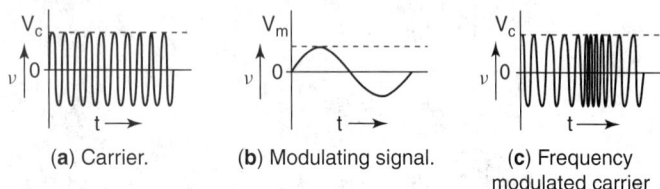

(a) Carrier.   (b) Modulating signal.   (c) Frequency modulated carrier

Fig. 2.   Frequency modulation.

Frequency modulation (FM) is illustrated by Fig. 2. The frequency variation, or deviation, is proportional to the amplitude of the modulating signal. The voltage equation for a frequency modulated wave follows.

$$v = V_c \sin (\omega_c t + M_f \sin \omega_m t)$$

The modulation index M_f is the ratio of maximum carrier frequency deviation to the modulating frequency. This ratio is known as the deviation ratio and may vary from zero to values of the order of 1,000. FM requires a broader transmission bandwidth than AM but may have superior noise and interference rejection capabilities. A large value of modulation index provides excellent interference rejection capability but requires a comparatively large bandwidth. The approximate bandwidth requirement for a frequency modulated wave may be obtained from the following relationship

$$\text{Bandwidth} = 2(\text{Modulating frequency})(M_f + 1)$$

The noise and interference characteristics of FM transmission are normally considered satisfactory when the modulation index or deviation ratio is five or greater.

Phase modulation is accomplished when the relative phase of the carrier is varied in accordance with the amplitude of the modulating signal. Since frequency is the time rate of change of phase, frequency modulation occurs when the phase-modulating technique is used and vice versa. In fact, the equation given for a frequency-modulated wave is equally applicable for a phase-modulated wave. However, the phase-modulating technique results in a deviation ratio, or modulation index, which is independent of the modulating frequency, while the frequency modulating technique results in a deviation ratio which is inversely proportional to the modulating frequency, assuming invariant modulating voltage amplitude in each case.

The phase-modulating technique can be used to produce frequency-modulated waves, providing the amplitude of the modulating voltage is inversely proportional to the modulating frequency. This inverse relationship can be obtained by including, in the modulator, a circuit that has a voltage transfer ratio inversely proportional to the frequency.

MODULUS. 1. The absolute value of a complex number. It may be interpreted as the length of a vector representing the number in complex space. Thus, the modulus of $(a+ib)$ is $(a^2+b^2)^{1/2}$.

2. The modulus of common logarithms is $\log e = 0.434294\ldots$, the factor which converts a natural logarithm to a common logarithm. Similarly, the modulus of natural logarithms is $\ln 10 = 2.302585\ldots$

3. A parameter which occurs in integrals or elliptic functions.

4. A formula, coefficient, or constant that expresses a measure of a property, force, or quality, such as of elasticity, efficiency, density, or strength.

MODULUS OF ELASTICITY.   The ratio of the unit stress to the unit deformation of a structural material is a constant, as long as the unit stress is below the proportional limit, and is called the modulus of elasticity. The shearing modulus of elasticity is frequently called the modulus of rigidity. See also **Elastic Constants and Moduli**; and **Elasticity**.

MODULUS OF RUPTURE.   The modulus of rupture in bending of a material is found by testing a transversely loaded beam of constant cross section to failure, and substituting the maximum bending moment, the moment of inertia of the cross section, and the distance from the neutral axis to the extreme fiber in the flexure formula:

$$S_M = \frac{Mc}{I}$$

The torsional modulus of rupture is obtained by testing a shaft of constant, circular cross section to failure and then substituting the maximum torque, polar moment of inertia of the cross section and the radius in the torsion formula:

$$S_s = \frac{Tc}{J}$$

The bending or torsional modulus of rupture may be used to predict the maximum bending or torsional moment which a member can resist.

MOHAIR.   This very resilient hair is obtained from the Angora goat. The staple length ranges from 5 to 8 inches (12.5 to 20 centimeters), but Turkish fibers go up to 10 inches (25.5 centimeters). Mohair provides a characteristic crisp, resilient, and slightly scratchy hand to fabrics even when used in very low percentages with other fibers. See also **Fibers**; and **Goats and Sheep**.

MOHOROVICIC, ANDRIJA (1857–1936).   Mohorovici was born in Volosko, Istria, Austrian Empire. He earned his degree in mathematics and physics at the University of Prague. He studied seismic waves and is best remembered for his discovery of the Mohorovicic Discontinuity, which led to the understanding that the Earth has a thin and brittle crust.
    See also **Earth**.

J. M. I.

MOIRE PATTERN.   If one draws a regular pattern, such as vertical lines of a given width and spacing, on a transparent sheet and then overlays this sheet onto another sheet that is ruled with lines, but of somewhat differing line widths and spacings—and then moves the two ruled surfaces horizontally with relation to each other, a shimmering effect will be noted. This is because of differences in the reinforcement of the lines and spaces one over the other. Walker (1978) suggests an interesting experiment with a comb and mirror. As one views the handheld comb against the mirror image of the comb (at just the right distances from the mirror), various periodic patterns will be noted along the length of the comb. This occurs because the teeth of the comb are "in step" in some locations and "out of step" in other locations. By changing the distance of the comb from the mirror, keeping the comb parallel to the mirror, different Moire patterns will be observed. Possibly this principle was first put to use in connection with various novel devices for Victorian entertainment. A present very useful application of the principle is in connection with dimension measurement and guidance systems for automated machine tools. A crisp scientific definition is "a pattern resulting from interference beats between two sets of periodic structures in an image."

Additional Reading

Cassin, C.: "Visual Illusions in Motion: With Three Different Moire Screens," Dover Publications, Mineola, NY, 1997.
Grafton, C. Belanger: "Optical Designs in Motion: With Moire Overlays," Dover Publications, Mineola, NY, 1990.
"Moire Effects, the Kaleidoscope and Other Victorian Diversions," by Jearl Walker, *Sci. Amer.*, **239**, 6, 182–186 (December 1978).
Patorski, K.: "Handbook of the Moire Fringe Technique," Elsevier Science, New York, NY, 1993.

MOLAL CONCENTRATION.   A one molal solution contains one mole of a particular substance (the solute) in 1,000 grams of solvent. Thus, a 0.5 molal solution of potassium chloride in water contains $0.5 \times$ (gram-molecular weight of KCl = 74.555), or 37.278 grams of the salt in 1,000 grams of H_2O. See also **Molar Concentration**; and **Normal Concentration**.

MOLAR CONCENTRATION.   A one molar solution contains one mole of a particular substance (the solute) in 1,000 milliliters of solution. Thus, a 0.5 molar solution of potassium chloride in water will be prepared by placing $0.5 \times$ (gram-molecular weight of KCl = 74.555), or 37.278 grams of the salt in a vessel and then adding H_2O, while thoroughly mixing to assure complete solution of the salt, until a total volume of 1,000 milliliters of solution is obtained. Molar is abbreviated M. Thus, the solution in the foregoing example would be $0.5M$ KCl. Molar solutions sometimes are referred to as *formal* solutions, not to be confused with normal solutions. See also **Molal Concentration**; and **Normal Concentration**.

MOLAR HEAT.   The product of the gram-molecular weight of a compound and its specific heat. The result is the heat capacity per gram-molecular weight.

MOLD.   See **Fungus**; and **Yeasts and Molds**.

MOLE (Stoichiometry).   Sometimes spelled *mol*, a mole is a quantity of a substance, expressed in specified mass units, that is equal to the molecular weight of the substance. For example, a *gram-mole* or *gram-molecular mass* of hydrogen H_2 will have a mass of $2 \times$ (atomic weight of hydrogen), or 2.016 grams. A gram-mole of carbon dioxide CO_2 will have a mass of $1 \times$ (atomic weight of carbon) plus $2 \times$ (atomic weight of oxygen), or 12.011 plus 31.998 = 44.009 grams. A pound-mole of ammonia gas NH_3 will have a mass of $1 \times$ (atomic weight of nitrogen) plus $3 \times$ (atomic weight of hydrogen), or 17.031 pounds. See also **Avogadro Constant**.

MOLE CRICKET *(Insecta, Orthoptera)*.   Burrowing crickets whose large forelegs give them a superficial resemblance to moles.

MOLECULAR AND SUPERMOLECULAR ELECTRONICS.   Professor Gareth Roberts FRS, Director of Research, Thorn EMI plc, and Professor of Engineering Science at the University of Oxford.

The microelectronics and optoelectronics industries will continue to grow vigorously well into the 21st Century. Until now, they have relied largely on inorganic materials such as silicon and lithium niobate in single crystal form. However, as the perceived limitations inherent in these materials begin to restrict the realization of more complex system designs, more attention is being focussed on the *organic* solid state. The richness of the variety of organic molecular materials offers enormous potential compared with the relative paucity of structures achievable with inorganic compounds, even when due allowance is made for the recent exciting developments in inorganic quantum well semiconductors (Kelly and Weisbach 1986).

The ability to enlist the assistance of synthetic organic chemists to produce organic materials with tailored properties has, of course, already been used to advantage in several applications. The best known is that of liquid crystals and their use in displays and digital thermometers. New phenomena and types of molecule are still being discovered and seem likely to lead to successful large area displays for high definition television and to high density information stores. Other examples are piezoelectric polymers as very sensitive hydrophones for submarine detection, photoconducting polymers for electrocopying, and photochromic molecules for reversible high density optical storage and signal processing. Biosensors and chemical sensors for converting specific biochemical or chemical solute or gas interactions into electrical signals for use in industrial or medical diagnostics can also be mentioned. All are examples of 'Molecular Electronics,' that is, they are fields in which organic molecular materials perform an *active* function in the processing of information and its transmission and storage. This definition does not embrace their use in possible roles such as insulation, adhesion or encapsulation. Thus, molecular electronics is interpreted broadly and is not limited to phenomena concerning the movement of electrons only. Electromagnetic radiation, polarization phenomena, and various forms of electromechanical and electrochemical energy transfer are also included in the definition. A common feature of all the examples cited and of the area in general is that progress is achieved *via* 'molecular engineering' that is, using the ability to manipulate the architecture of a material to optimize a specific physical parameter.

An alternative definition exists for molecular electronics; this is formulated in terms of switching on a molecular scale and is aimed more at the long term problem of fabricating molecular electronic devices suitable for assembly into a computer (Carter 1986). It is interesting to note that only a modest diminution in the size of electronic circuit components is required before the scale of individual molecules is reached; in fact, many existing circuit elements could already be accommodated within the area occupied by a leukaemia virus. An illustration of the rapid evolution of silicon based microelectronics may be gained by studying Figure 1. If this systematic reduction in feature size suggested by the good log-linear graph is sustained, then the extrapolated line indicates device geometries with nanometer dimensions in approximately thirty years' time! The requirements of reliability and testing of complex structures suggest a system approach rather than the traditional one, which uses the properties of individual circuit elements. it appears likely that sequential designs, because of their vulnerability, will be abandoned in favor of supermolecular arrays acting as concurrent processor networks. For this reason, and to differentiate it from 'molecular electronics,' signal transport and control in nanometer scale assemblies is referred to as 'supermolecular electronics'.

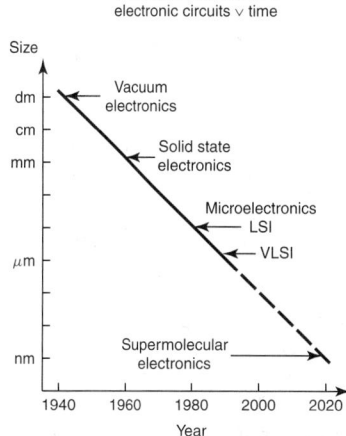

Fig. 1.   Linear feature size of commercial electronic circuits versus time; the bottom arrow speculatively points to an era where switches on a molecular scale will have application in computer systems.

Animals can solve with little apparent effort the tasks for which advanced supermolecular information processors are required; accordingly, some enthusiasts have speculated that the thirty-year time scale could be foreshortened using biological molecules. However, such thoughts are misplaced, for nature constructs organic materials for purposes other than those required for logic and memory functions. What can be learnt from studying biological systems are the scientific principles of organization and assembly; eventually, it may be possible to apply these concepts to construct synthetic supermolecular arrays. These will require nonlinear interactions between neighbors so that configurations of 'on' and 'off' elements propagate with time. However, it will be a difficult task to control the correlations between the parallel processing elements whether they be based on electronic, vibrational or magnetic effects. A great deal of fundamental research will be required to identify molecular systems with the necessary degree of cooperativity and to develop the routine analytical techniques designed for investigations in three dimensions with molecular scales of resolution.

Three-dimensional integration is extremely difficult using silicon technology while chemically nonspecific methods such as molecular beam epitaxy are relatively crude. A self-assembly technique is a far more attractive alternative if regular, three-dimensional, ordered structures are required. This involves the construction of unique assemblies whose architectures depend on the shapes and charge distributions of the units from which they are built, as distinct from the methods used to assemble them. There is a considerable degree of self assembly associated with organic monomolecular films deposited using the Langmuir trough technique (Roberts 1985). About half a century after the report of their discovery, intense interest is now being displayed in these so called Langmuir-Blodgett films. Their precise thickness, coupled with the degree of control over their molecular architecture has now firmly established a role for such layers in thin film technology. It seems likely that an understanding of their physical properties and utilization in molecular based devices will assist generally in the transition from molecular to supermolecular electronics. The likely pattern for this evolutionary process is given in Figure 2. The plan is speculative and envisages three main stages before the advent of applied, complex supermolecular systems. In the near term it predicts that current research and development will result in organic materials ousting inorganic materials in existing applications, e.g. optoelectronics. Hybrid technologies, comprising a novel device partly based on conventional solid state materials and partly on organic compounds, could possibly follow in the mid-term, say 5 or 8 years. Thereafter, at some stage dictated by the emergence of reliable, stable supermolecular assemblies, a true watershed will occur. When this occurs, materials and process technologies of conventional solid state devices will be superceded by radically new types of devices. This era will be equivalent to that witnessed about forty years ago when inorganic semiconductor materials were developed. Just as then, novel effects should be discovered and these in turn will lead to the fabrication of novel devices that can be integrated in novel systems.

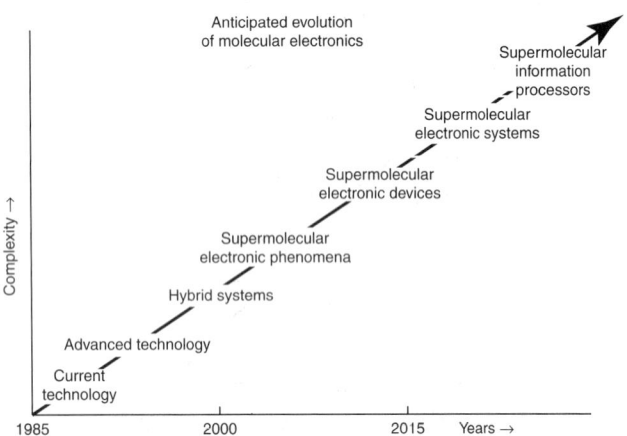

Fig. 2. The anticipated evolution of molecular electronics to supermolecular electronics.

Molecular and supermolecular electronics are broad and interesting subjects. Moreover, they require a multidisciplinary approach where collaboration between biologists, chemists, computer scientists, electronic engineers and physicists is of paramount importance. To illustrate these features, this article concentrates on Langmuir-Blodgett films. Equal emphasis is placed on their importance in basic science and on their potential applications, especially in the area of electronics.

Historical Review of Langmuir-Blodgett Films

According to Tabor (1980) the earliest written record of observations of the spread of oil on water is in cuneiform on clay tablets, dating from Hammurabi's period (18th century B.C.) in Babylonia. The earliest technical application of organic monolayer films is believed to be the Japanese printing art called subminagashi, involving a suspension of submicron carbon particles and protein molecules spread on the surface of water. The distinctive patterns so formed can be transferred by lowering a sheet of paper onto the water surface. There are also many references dating from the classical times of Plutarch. Aristotle and Pliny describing the ability of oil spread on water to dampen surface waves and ripples. It was this property which attracted Benjamin Franklin, the versatile American statesman, to the subject. During his frequent visits to Europe in the 18th century A.D. to negotiate the sovereignty of his country with the French and the British he carried out his famous 'teaspoonful of oil' experiment. Often quoted and picturesque, Franklin's (1774) account to the Royal Society included the following phrases: "At length, being at Clapham where there is, on the common, a large pond, which I observed to be one day very

rough with wind, I fetched out a cruet of oil, and dropped a little of it on the water. I saw it spread itself with surprising swiftness upon the surface ... I then went to the windward side, where (the waves) began to form and there the oil, though not more than a teaspoonful, produced an instant calm over a space of several yards square, which spread amazingly, and extended itself gradually till it reached the lee side, making all that quarter of the pond, perhaps half an acre, as smooth as a looking glass. After this, I contrived to take with me, whenever I went into the country, a little oil in the upper hollow joint of my bamboo cane, with which I might repeat the experiment as opportunity should offer; and I found it constantly to succeed."

Franklin must have been too preoccupied with political affairs to place his observations on a quantitative basis. Had he done so he might well have calculated that a volume of one teaspoonful (approximately 2 ml) spread over an area of nearly half an acre (200 m^2) leads to a surface coating approximately 1 nm thick. However, it is Lord Rayleigh (1890) who is given the distinction of first suspecting that the maximum extension of an oil film on water represents a layer one molecule in thickness. For a direct measurement on molecular sizes he was indebted to Pockels (1891) whose simple apparatus later became the model for what is now called a Langmuir trough; using very simple equipment he calculated the precise thickness of a monomolecular layer of castor oil on water to be 1 nm. This significant observation was not fully followed up until the pioneering work by Langmuir on the adsorption of gases or solutes by solids. In order to test the general applicability of his hypothesis about the involvement of short range forces, he turned his attention to liquids and he essentially repeated (Langmuir 1917; 1920) the earlier measurements of Pockels and Rayleigh, and extended them to include the transfer of molecules from a water surface to a solid support. The first detailed description of sequential monolayer transfer was given by Blodgett (1935), his collaborator at the General Electric Company laboratories. These built up monolayer assemblies are now called Langmuir-Blodgett (LB) films, while the floating monolayer is referred to as a Langmuir film. The extensive list of publications resulting from the pioneering experiments carried out by these two investigators during the period 1934 to 1952 has been compiled by Gaines (1983).

The Chemistry and Preparation of Langmuir Films

Most of the experiments reported by Langmuir and Blodgett were on a well-defined series of fatty acids and their salts. Figure 3(a) shows stearic acid, a molecule in which sixteen CH_2 groups form a long hydrophobic chain; the other end of the molecule terminates in a hydrophilic carboxylic acid group. When dissolved in a suitable solvent and spread on the surface of water, molecules may be compressed with the aid of a barrier. Figure 4 shows a plot of the surface pressure (differential surface tension) versus area occupied per molecule for stearic acid. The monolayer undergoes a number of phase transformations during compression; the well-defined sequence can be viewed as the two-dimensional analogue of the classical transitions observed with pressure-volume isotherms. However, it should be emphasized that some materials, while forming acceptable quality LB films, do not display the well-defined break points shown in Fig. 4.

Generally speaking, the approach to the synthesis of suitable molecules for examination with a Langmuir trough has been an ad hoc one and has relied on the modification of known materials. For example, the alkyl group of fatty acids may be replaced by chains containing one or more double bonds. The ω-tricosenoic acid (Barraud 1983) molecule shown in Figure 3(b), which is similar to stearic acid but contains a terminal double bond, displays all the essential film-forming qualities including solubility in convenient organic solvents, stability at the surface of water, shear resistance, stability against collapse, and suitable orientation features. It is relatively straightforward to attach long aliphatic chains to a molecule and spread a monolayer. However, this may well dilute the desirable properties of the basic molecule; moreover, for stability reasons, the presence of long side groups will severely restrict their practical applicability. It has therefore been recognised that the scope of the Langmuir trough technique would be considerably enhanced if interesting materials containing only short, stable, side groups could be formed into LB films. A good example is provided by the anthracene derivative (Vincett et al. 1979) shown in Figure 3(c); multilayers of excellent quality can be obtained, even though the alkyl group contains only four aliphatic carbons and the hydrophobic group is attached to the ring structure via only two methylene groups. Extremely robust monolayer assemblies can be constructed using dye molecules such as the porphyrins and phthalocyanines. In general their

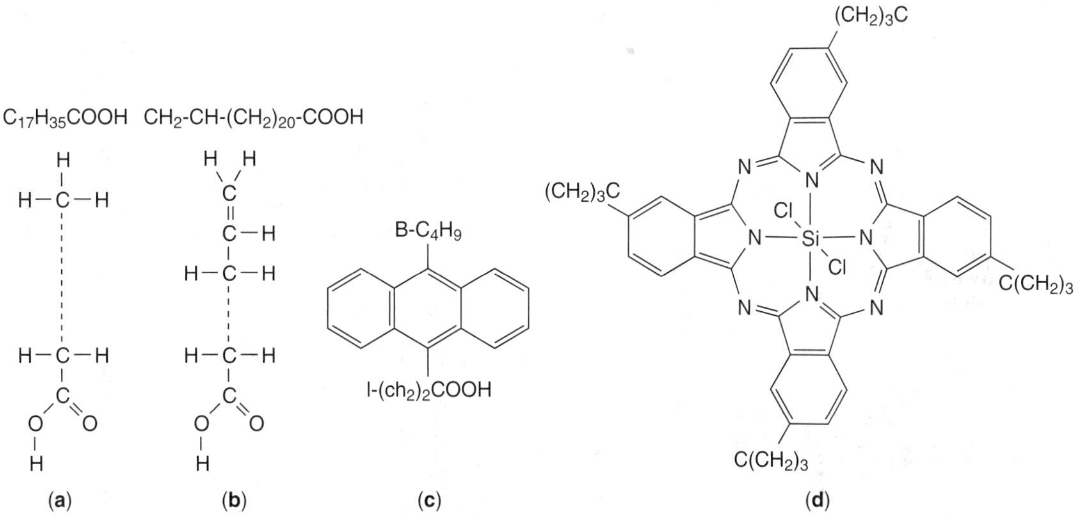

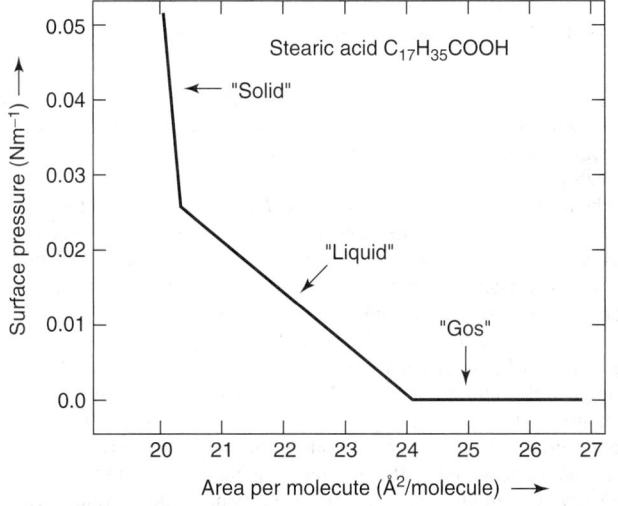

C₁₇H₃₅COOH   CH₂-CH-(CH₂)₂₀-COOH

Fig. 3.   A selection of molecules used to form LB films: (**a**) fatty acid; (**b**) ω-tricosenoic acid; (**c**) 9-butyl-10-anthrylpropionic acid; (**d**) tetra-4-tert-butyl-10-phthalocyaninato silicon dichloride.

quality is relatively imperfect compared with those of the classic film forming materials but their significant advantages lie in their thermal and mechanical stabilities. An example of a substituted phthalocyanine molecule (Hue et al. 1986) that can be deposited in monolayer form is shown in Figure 3(d).

Fig. 4.   Surface pressure versus area characteristic for stearic acid.

The molecules shown in Figure 3 represent only a few of the materials that have been studied in LB film form. Nonetheless, a great deal more needs to be done to tap the vast wealth of opportunities available with organic systems. There will inevitably be short-term opportunistic attempts aimed at discovering molecules for specific devices. However, there is a more pressing need for a systematic approach that will yield rules governing structure-property correlations, so as to enable scientists confidently to predict the molecular architecture of monolayer assemblies. See also **Macromolecular Science**.

Langmuir-Blodgett Film Deposition

An LB film is formed by transferring a floating monolayer onto a solid substrate. The quality of the Langmuir film and the surface pressure at which 'dipping' occurs is established using the type of isotherm shown in Figure 4. The subphase is normally ultra-pure water, because it is readily available and it has an exceptionally high value of surface tension. The composition of the subphase, including its purity, pH, and ionic strength, can have a profound influence on factors such as the solubility of the monolayer and segregation effects resulting in molecular aggregates or domains.

Using conventional LB film technology, the substrate is raised and lowered vertically through a compact floating monolayer; the surface pressure at which this occurs is normally just above the 'knee' in the steeply rising sector of the isotherm indicating low compressibility in the monolayer. At this stage, if conditions are carefully controlled and appropriate molecules are used, one monolayer is transferred during each excursion through the subphase surface. The most common deposition mode (Y-type) is illustrated in Figure 5(a), where the molecules can be seen to stack in a head to head and tail to tail configuration. The floating molecules on a water surface are shown at the top left in this diagram. With a hydrophobic substrate (for example, a group III-V compound semiconductor), no pick-up occurs during the first immersion and the first monolayer is therefore deposited during the first withdrawal as shown in Figure 5(b). The surface is now hydrophobic and deposition does now occur during the next immersion into the water. Thus, one monolayer coverage is obtained on each traversal of the liquid surface. With a hydrophobic surface such as freshly etched silicon, pick up also occurs during the first insertion. Sometimes, the common deposition mode illustrated in Figure 5 is not followed and one of the other two possible configurations, X and Z-type, is observed, where transfer occurs only during immersion or withdrawal, respectively. The surface quality and chemical composition of the substrate is bound to control the nature of the deposited layer. When adhesion is poor, some researchers have resorted to the less satisfactory method of placing the substrate flat on the liquid surface, a technique first used by Langmuir and Schaefer.

Many modifications of the very early film balances have been described by Gaines (1983). However, the upsurge in interest in LB films has led to greater attention being placed on trough design and control systems to meet the stringent requirements of scientists and engineers. Therefore, modern instruments are relatively sophisticated and, for device-related work, need to be situated on anti-vibration Tables in clean environments. Although it is possible to automate most features, the primary benefit at the present time lies in efficient data collection and the ease with which data can be manipulated. For example, phase transitions are more apparent when the differential of the pressure-area isotherm is plotted. No difficulties are envisaged in scaling up the Langmuir trough or in the design of continuous fabrication arrangements. When an important practical application is discovered there will be a need to produce a specially designed trough capable, for example, of coating a moving belt or multiple wafers of silicon.

A recent development in trough design is worthy of special attention as it could have important commercial significance. It has arisen because of the need to produce non-centrosymmetric structures that display interesting non-linear physical effects. The conventional Y-type films are symmetrical in character and experience has shown that X and Y type layers, although non-symmetrical, are usually imperfect. Therefore, an alternative approach to producing noncentrosymmetric structures is to use alternate layers of two different materials where the contributions of adjacent molecules do not cancel. See Fig. 6. The additions of a fixed beam and a revolving center

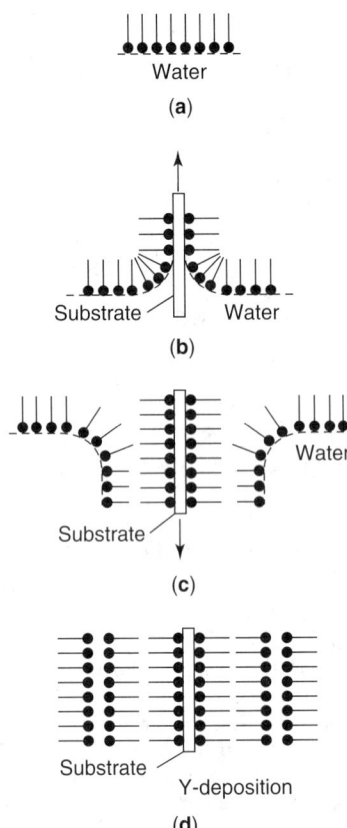

Fig. 5. Langmuir-Blodgett film deposition (Y-type) on a hydrophilic substrate: (a) monolayer on the surface of water; (b) first layer on withdrawal; (c) second layer (second insertion); (d) substrate with three layers (after second removal).

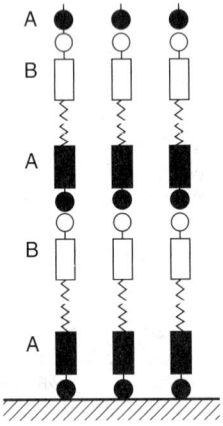

Fig. 6. Alternating organic multilayer structure which enables a Y-type LB film to be produced of non-centrosymmetric character (molecular lego!).

section to an automated constant perimeter barrier Langmuir trough enables the formation of an alternating Y-type structure of two different molecules spread in the two distinct areas of the subphase. The structural qualities of the LB films prepared in this way can be of high quality (Holcroft et al. 1985); another advantage of the rotating substrate arrangement, which is conducive to fast dipping, is that the meniscus, unlike that in the vertical dipping method, is always in the same direction.

Many different experimental techniques indicate that carefully prepared films of appropriate molecules do indeed possess a high degree of structural order. The reader is referred to the proceedings of the two international conferences on LB films for literature references describing the vast range of characterization experiments that have been employed (Roberts and Pitt 1983; Gaines 1985). These include ellipsometry, electron spin resonance, infrared dichroism, photoacoustic spectroscopy, secondary ion mass spectrometry, surface potential, polarized X-ray and electron diffraction, neutron reflection and diffraction. Most of the electrical data

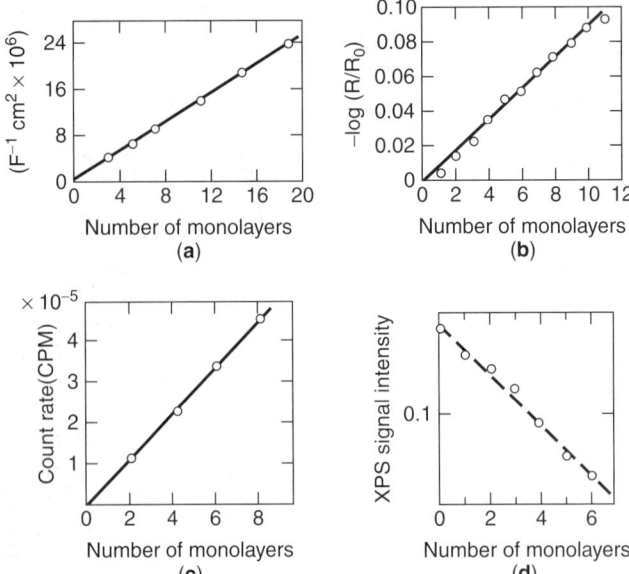

Fig. 7. These diagrams are designed to emphasize the reproducibility of various physical parameters in monolayer assemblies of different thicknesses: (a) reciprocal capacitance per unit area versus number of monolayers of cadmium arachidate on an aluminum substrate (Roberts et al., 1978); (b) absorption intensity versus number of monolayers for the symmetric carboxylate stretching mode of cadmium arachidate at $1432\,cm^{-1}$ (Allara and Swalen, 1982); (c) count rate of ^{14}C rays versus number of layers of barium stearate labelled with ^{14}C (Mori et al., 1980); (d) X-ray photoelectron signal intensity versus number of layers of cadmium dimethyl arachidate on silver (Brundle et al., 1979).

for LB films are suspect in that they have been obtained for films deposited onto metals that are invariably coated with semi-insulating native oxides. A comprehensive account of these studies was presented in a review by Vincett and Roberts (1980).

The four separate diagrams in Figure 7 all describe results for fatty acids or their derivatives and are designed to emphasize the reproducibility of various physical parameters from one monolayer to the next. Figure 7(a) shows the capacitance (C) as a function of film thickness for cadmium arachidate deposited onto aluminium. The linear dependence of $C^{-'}$ versus the number of monolayers demonstrates clearly the repeatability of the dielectric thickness of each monolayer. In Figure 7(b) it is a band in the infrared reflection spectrum of the same material that has been used to demonstrate the uniformity of successive monolayers. The reason for the scatter around the origin is not understood but it probably reflects in this case that the structure of the first few monolayers is affected by the metal underlay. Figure 7(c) is based on experiments using barium stearate as the absorber for L shell Auger electrons. By labelling the molecules in these overlays with ^{14}C and examining their autoradiographs it is possible to confirm the uniformity of the deposition process by plotting the count of ^{14}C rays versus the number of monolayers. The final diagram in the set, Figure 7(d), illustrates another powerful tool for investigating organic coatings on metals. In this case different thicknesses of cadmium dimethyl arachidate have been used to attenuate the substrate X-ray photoemission signal.

During the next few years many techniques-oriented scientists will be attracted to work on LB films because they provide interesting novel structures whose molecular architecture can be systematically controlled. The quality of the floating monolayer is also important and needs to be characterized as does the interface between the first deposited monolayer and the substrate. Fluorescence microscopy and Brillouin and Fourier transform infrared spectroscopies are currently being used to address these problems.

Applications of Langmuir-Blodgett Films

Following the pioneering work of their famous employees, the General Electric Company introduced several simple applications of LB films including step-thickness gauges, anti-reflection coatings and soft X-ray gratings. Since that time, stimulated no doubt by the availability of well engineered troughs and a wider range of suitable materials, researchers have suggested other applications for monolayer and multilayer films. A selection of areas where LB films may find practical use is given in Table 1. Further details are given in the review by Roberts (1985). However, it should be remembered that one of the principal virtues of LB films is their usefulness in fundamental research. Therefore, before discussing more applied areas we shall mention a few areas of science that can benefit from investigations of model systems based on monomolecular assemblies.

TABLE 1. PROMISING APPLIED RESEARCH AREAS

Topic	Molecular Electronics Applications
Model Systems in Fundamental Research	Spectroscopy of Complex Monolayers: spectral sensitization, fluorescence quenching, energy transfer between excited states. Model membranes to mimic photosynthetic systems. Modification of solid surface properties. Examination of lipids, proteins and membrane phenomena; organic semiconductors.
Applied Chemistry	Surface chemistry and behavior of surfactants: catalysis; filtration/reverse osmosis membranes; adhesion; surface lubrication, e.g. magnetic tape; encapsulation.
Electron Beam Microlithography	Good sensitivity and contrast, acceptable plasma etching resistance; less scattering of electrons and therefore better resolution; negative and positive resists possible.
Integrated Optics and Storage Optics	Film thickness plus refractive index of film and hence guided wave velocity can be controlled with great precision; acceptable attenuation loss. Possible uses in conventional optics and optical data storage: photochromic and ablative systems. Optical sensors e.g. based on coated fibres.
Nonlinear Physics	Control of molecular architecture to produce asymmetric structures with high non-linear coefficients, e.g. in electro-optics, pyroelectric detectors, or acoustoelectric devices.
Dilute Radioactive Sources	Radioactive nuclide incorporated in conventional LB film; used to measure the ranges of low energy electrons.
Electronic Displays	Large area capability of LB films is an advantage; the monolayers can either be the active electroluminescent layer or used to enhance efficiency of an inorganic diode; passive application to align liquid crystal displays. Deposition of liquid-crystal type molecules also possible.
Photovoltaic Cells	Used as a tunnelling layer in an MIS solar cell or as an active layer in p-n junction diode, perhaps involving an inorganic/organic junction.
Two Dimensional Magnetic Arrays	Magnetic atoms e.g. Mn, periodically spaced in LB film; possible applications include magnetic control of superconducting junctions and bubble and magneto-optical devices.
Field Effect Devices	Accumulation, depletion and inversion regions possible with a variety of semiconductors; can therefore form the basis of several devices e.g. CCD, bistable switch, gas detector or pyro/piezo FET, is suitable LB films are used.
Biological Membranes	Attractive supporting membranes for commerical exploitation of biological material, e.g. immobilization of membrane bound enzymes in solid state sensors; ISFET type structures.
Supermolecular Structures	Speculative work aimed at superconductors, organic metals, 3D memory storage, molecular switches.

Model System in Basic Research

(a) **Energy transfer in complex monolayers:** The Langmuir trough technique provides a method of constructing simple artificial systems of co-operating molecules on a substrate. The pioneer in this field has been Kuhn (1983); the elegance of his and his colleagues' work is evident in their reviews of the subject. These describe the use of LB films to investigate intermolecular interactions and various photophysical and photochemical processes. Their supermolecular structures have mainly involved long chain fatty acids as matrices for appropriate synthetic dyes and have been ingeniously designed to clarify the different interactions that can

occur between various molecules via photon, electron and proton transfer. An example of their research, designed to investigate the Förster type of energy transfer from a sensitising molecule, S, to acceptor molecule, A, is given in Figures 8 and 9. If S is a compound that absorbs in the ultraviolet part of the spectrum and fluoresces in the blue, while A absorbs in the blue and fluoresces in the yellow, then interesting effects are observed when the system is irradiated with ultraviolet light. If there is a sufficient distance between S and A, as in Figure 8(a), the fluorescence of S appears since A does not absorb UV radiation. However, below a certain threshold distance, as in Figure 8(b), the excitation energy of S is transferred to A and the yellow fluorescence of A is expected. Similar experiments based on fluorescence quenching indicate that the rate constant of the electron transfer decreases exponentially with increasing barrier thickness separating a donor chromophore and an electron acceptor. In the example shown in Figure 9, N,N'-dioctadecylthiacyanine has been used in conjunction with a viologen acceptor layer to observe the steady fluorescence intensities of the cyanine dye monolayer in the absence (I_0) and in the presence of the acceptor layer (I). The quantity, $[I_0/I] - 1$ is proportional to the rate constant of the electron transfer; its linear dependence with d, the distance between the chromophores, is evidence of electron tunnelling. In a similar series of experiments it has been possible to investigate the energy transfer mechanism responsible for spectral sensitization.

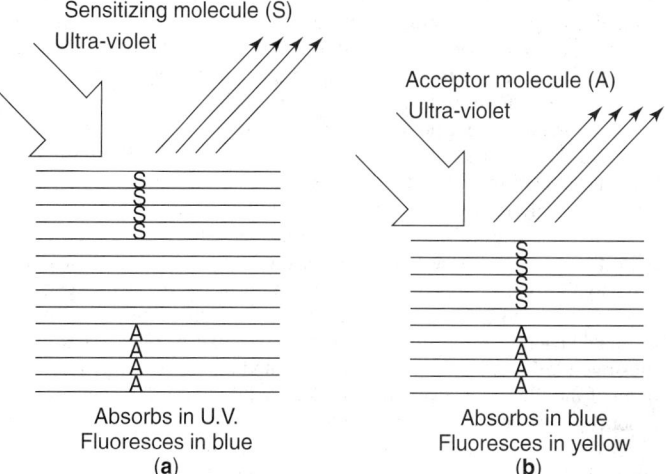

Fig. 8. Schematic diagram showing basis of experiments designed to investigate energy transfer from a sensitizing molecule (S) to an acceptor molecule (A). The number of monolayers separating the two species governs the spectral response of the fluorescence spectrum. In (**b**) the separation distance is sufficiently small for the excitation energy of S to be transferred to A. In (**a**) the acceptor molecules does not absorb the ultraviolet radiation.

(b) **Biological membranes:** The physical structure and chemical nature of classical LB films gives them a close resemblance to naturally occurring biological membranes. For example, because the two ends of a lipid molecule have incompatible solubilities, they spontaneously organize themselves in the form of a bilayer, or essentially a two layer LB film. Scientists have suggested that they might provide a suitable model of the lipid membrane for probing the cooperative interactions between its constituents. However, caution must be exercised in assessing the biological relevance of this type of work and associated studies aimed at incorporating ionophores into phospholipid layers. Much of this research is targetted at novel integrated solid state devices incorporating biological molecules such as enzymes.

In some cases, LB films are useful to facilitate physical studies of biological molecules e.g. to measure the ionic permeability of reconstituted membranes. Supermolecular structures have also been designed to mimic the primary process in photosynthesis and for achieving an efficient photoinduced charge separation by appropriate modelling of potential profiles. Chlorophyll has been studied in this context. Some of the results may have relevance to solar photochemical conversion devices.

(c) **Metal-ion incorporation:** The addition of divalent ions into the liquid subphase in a Langmuir trough can increase both the shear resistance

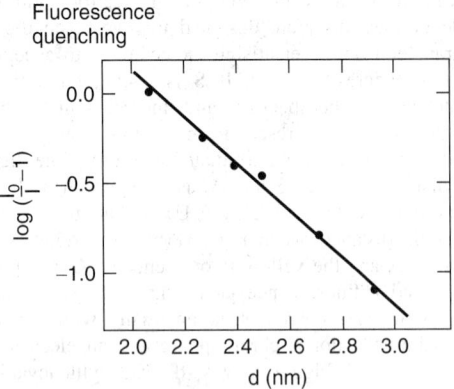

Fig. 9.   The fluorescence intensity I_0 of a donor dye is reduced to a value I in the presence of an acceptor dye. The logarithm of $[(I_0/I) - I]$ is shown as a function of d, the spacing between the donor and acceptor planes. (Mobius, 1981.)

and cohesion of the monolayer. For this reason it is more common to find reports of studies on fatty acid salts than their acids. By adjusting the pH of the subphase it is possible to assemble multilayers containing metal ions separated by the width of an integral number of monolayers (for Y-type deposition, two monolayers). The ability to do this has been capitalized upon in several fundamental investigations, three of which will be mentioned here. The first of these relates to two-dimensional magnetic monolayers involving iron or manganese ions. Using electron spin resonance Pomerantz (1980) has demonstrated that at temperatures near 2 K, the resonance field and line-shapes were affected, thus signifying the rapid development of a large internal magnetic field. His results have been interpreted in terms of a predominantly anti-ferromagnetic state, but with a weak ferromagnetic component. Further experiments are required to clarify magnetic ordering in two-dimensional space.

In the area of surface science, X-ray photoemission is now used extensively to study organic materials on surfaces. In such experiments it is important to establish electron mean free path lengths as a function of kinetic energy. An example, of this type of investigation in LB films is illustrated in Figure 7(d); generally it is found that the mean free paths for ordered multilayers are significantly longer than those for conventionally produced polymers (Clark et al. 1981).

A third example showing the usefulness of using the Langmuir trough technique to provide matrices containing regularly spaced metal ions lies in radioactivity. Mori et al. (1980) used radioactive stearate monolayers in which some of the hydrogen atoms had been replaced by nuclides such as ^{51}Cr, ^{54}Mn, ^{55}Fe, ^{57}Co, ^{65}Zn and ^{109}Cd, to produce dilute and standard radioactive sources. By labelling the molecules with ^{14}C and examining autoradiographs he was able to confirm the uniformity of the deposition process as shown in Figure 7(c). Using conventional monolayers with well-controlled dimensions as overlays they were also able to demonstrate that Auger electrons from the L shell with an energy of approximately 0.5 keV are almost completely absorbed by fifteen monolayers of barium stearate. Experiments of this kind are of importance in fields such as medical physics and upper atmosphere science.

Promising Applied Research Areas

Langmuir-Blodgett films may have value in many applied areas of traditional interest to the industrial chemist such as adhesion, encapsulation and catalysis. The permeability characteristics of monolayer assemblies may also find application as synthetic membranes for ultra fine filtration, gas separation and reverse osmosis. For example, Albrecht et al. (1985) have proved the efficiency of polymeric diacetylene monolayers on semi-permeable supports in reducing the flow of CH_4. One interesting possibility lies in using LB monolayers as lubricants in magnetic tape technology. Unpublished reports have indicated that frictional coefficients can be reduced markedly when the tape is coated with a few monolayers. In applications such as those listed above, difficulties may well be encountered with the mechanical stability of the films. To date relatively little research has been carried out in this area.

For commercial applications, the LB films will need to play an essential, integral role. That is, one must capitalize on their special features such as the degree of control over their molecular architecture, their thinness

and the selective way in which they might react with their environment. Some of the potential areas of interest are listed in the table. The long term interest as far as the applied physicist is concerned, lies in the possible uses of supermolecular assemblies for memory storage, molecular switching, and superconducting devices. However, at the present time it is in potential improvement areas where monomolecular films show most promise and where the prospects of commercial exploitation seem reasonable in the medium term. A few of these areas are described below; most of the illustrations are based on work carried out in the author's research laboratory:

(a) Nonlinear physics: There is evidence that many organic molecules possess very high non-linear coefficients and therefore, if LB films with the required architecture can be formed, these could form the basis of novel devices. In order to avoid the symmetry inherent with conventional Y-type deposition, X and Z type films have been studied; some have displayed a permanent polarization with a strong component in a direction perpendicular to the substrate. However, as has already been mentioned, films produced in this way with their dipoles supposedly aligned in a common direction, are invariably of poorer quality than Y-type layers. a possible method of improving the structure is to use electric or magnetic fields to help align the molecules but efforts to orient films on the subphase and substrate have met with only limited success. The problem can be overcome by using organic superlattices based on alternating layers of two different materials (See Fig. 6). A good example is given in Figure 10 which shows a superlattice comprising acid and amine molecules whose dipole moments are in opposite senses but when deposited in Y-type LB film form are aligned in the same direction. Two areas of particular interest that would capitalize on this feature of organic superlattices are pyroelectricity and optoelectronics. Each of these will now be considered in turn.

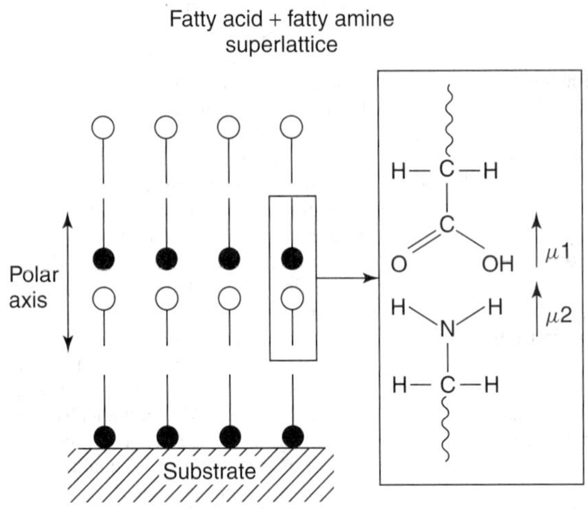

Fatty acid + fatty amine
superlattice

Fig. 10.   The left-hand diagram shows an organic superlattice with a unique polar axis. The two types of molecule involved could be a fatty acid and a fatty amine. The insert is designed to show that these two materials have dipole moments in opposite senses with respect to the hydrophobic chain. Thus, the Y-type film has a resultant dipole moment.

Pyroelectric devices respond to a rate of change of temperature rather than to changes of temperature as in other, types of thermal detector. This gives them inherent advantages but their full potential has yet to be realized. For applications where both high speed and sensitivity are required, conventional materials have been unsuccessful. The desirable pyroelectric properties of inorganic materials appear to vanish for thicknesses less than 10 μm, and pyroelectric organic single crystals, such as triglycine sulfate, which possess a high pyroelectric coefficient, cannot be produced in thin film form. Thus the future development of relatively cheap thermal imaging systems with reasonable performance and an optimum thickness of approximately 0.5 μm, requires a materials breakthrough. A normal detector consists of a capacitor whose dielectric is an oriented pyroelectric material; using this type of device Christie et al. (1986a, b) have observed encouraging results using the alternate layer structure shown in Figure 10. The pyroelectric coefficient, p, has been determined using both dynamic and static detection techniques.

For the simple fatty acid/fatty amine superlattice $p \simeq 1$ C cm^{-2} K^{-1} but more recent results involving a system where proton transfer occurs from the acid to the amine have yielded higher values, comparable with those observed for triglycine sulfate. The exploitation of this work will depend not only on the properties of the LB films but also on our ability to deposit the layers onto reticulated structures with low thermal mass.

Although many of the potential optical applications of LB films are in transmission optics employing the linear response properties of molecules, it is in the area of non-linear optics where the most exciting applications are perceived. Highly efficient nonlinear optic materials permit functions such as those illustrated in Figure 11 to be performed in a totally optical manner without the need for electron-photon conversion processes. Second harmonic generation and parametric amplification can be obtained using inorganic single crystal materials such as lithium niobate, but recently, organic crystals such as 3-methyl-4-nitroaniline have been shown to possess exceptionally large second order electrooptic coefficients (Zyss 1982). However, a thin film geometry is preferred; then it will be possible to integrate nonlinear interactions, linear filtering and transmission functions into one precision monolithic structure. Therefore, researchers are currently substituting such molecules with appropriate side groups to enable LB film deposition to occur. To date, second harmonic generation has been reported in multilayers of nitrooctadecylazobenzene and mercocyanine and hemicyanine dyes. Nonlinear coefficients comparable to those of inorganic materials have been achieved but there are many other considerations involved (e.g. phase matching, suitable spectral response and refractive index, good optical damage threshold, low scattering coefficients and mechanical stability) before practical objectives can be accomplished. The most interesting observation to date is that of second harmonic generation in LB organic superlattices of the two molecules displayed in Figure 12. Neal et al. (1986) have shown that the nonlinear response of a hemicyanine/nitrostilbene layer is greater than that expected from the simple addition of contributions arising from the individual (separated) monolayers. The coefficient for second harmonic generation of the alternate layer structure is approximately five times the average value of the same parameter measured for hemicyanine and nitrostilbene. This superadditive effect is best explained in terms of improved film structure with adjacent molecules influencing each other to orient more vertically with respect to the substrate.

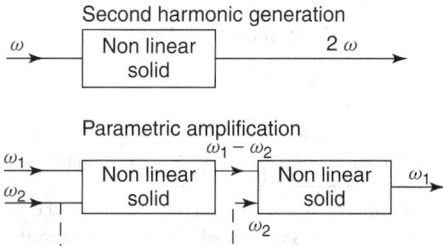

Fig. 11. Schematic diagram illustrating the ability of nonlinear materials to double the frequency of incident laser radiation and also function as a parametric amplifier when two beams of different frequency are involved. Most of the non-centrosymmetric solids currently in use are inorganic, but they may be replaced by organic single crystal thin films.

(b) Enhanced device processing: In integrated circuit technology the quest for faster speeds and larger memories has led to a gradual refinement of microlithographic methods for producing smaller and more closely spaced circuit elements. Sub-micron resolution is now required and this has necessitated a move away from conventional photolithography to techniques involving electron or ion beams and x-rays. The main disadvantage of electron beam systems lies with their scattering characteristics; this enforces the requirement that the resist materials be pinhole free and less than 1 μm thick. Conventional spin-coated polymers display unacceptably large pin-hole densities and variations of thickness; however, LB films have already demonstrated their capability in this regard. There are good examples of both positive and negative resists but the best material reported to date is the ω-tricosenoic acid molecule shown in Figure 3(b). In purified form this material has adequate sensitivity, may be deposited at a rate of 0.5 cm s^{-1} to produce uniform coatings in the range 30–90 nm, and

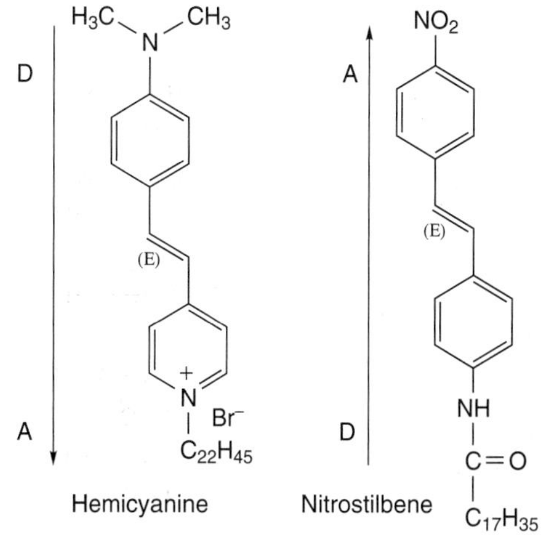

Fig. 12. These two molecules, one a hemicyanine and the other a nitrostilbene dye, can be used to form an organic superlattice displaying a high coefficient for second harmonic generation.

has been shown capable of a line resolution of 60 nm. The main property which requires some improvements is the etch resistance in plasma processing (Barraud et al. 1983).

Interconnects become increasingly important as the size of electronic circuits reduces. It has been suggested that protein layers might be useful in this regard. The method involves first depositing a synthetic protein and patterning it using a conventional resist. The NH$_2$ groups in the exposed protein layer can then be used to adsorb silver ions from a silver nitrate solution; this physical development stage converts the ions to metallic silver. An alternative approach is to fabricate supermolecular assemblies that conduct. The most successful attempt to date uses the molecule shown in Figure 13. Barraud et al. (1985) have succeeded in producing close packed multilayers of this TCNQ derivative, where the polar planes are separated by insulating lamellar regions. The resistivity of this N-stearylpyridinium$^+$1TCNQ$^-$ film is approximately 1Ωcm. Of course, there are many other attractions in producing conducting LB films, including their uses.

There are several niches where LB films could have a useful role in semiconductor technology. Probably the most important is the ability of an oriented monolayer to change the effective barrier height at a semiconductor surface. Researchers at the University of Durham first demonstrated the effect on cadmium telluride and showed how it could be used to improve the efficiency of a photovoltaic diode (Dharmadasa et al. 1980). Similar effects have now been confirmed on a variety of materials including ZnSe, InP, GaP and GaAs and related III-V semiconductor alloys. The control afforded by the Langmuir-Blodgett technique permits the degree of band bending to be adjusted to suit the particular application, e.g. the increase in efficiency of an electroluminescent diode. In most applications it has been necessary to use robust monolayers of phthalocyanine which can withstand the large current densities involved. Figures 14 and 15 show data for gallium phosphide (Batey et al. 1983; Petty et al. 1985); for convenience the error bars have been removed. The LB thickness required to optimize the electroluminescence efficiency is found to be approximately 21 nm; this value is determined by the ability of minority carriers to cross the semi-insulating phthalocyanine film. Similar results have recently been achieved using zinc selenide layers grown using MOCVD. Blue electroluminescence is observed provided the organic film is present.

There are many methods of producing a surface layer on a semiconductor. However, experience has shown that when an energetic process such as evaporation, sputtering, or growth from a plasma, is used to deposit a thin film onto a semiconductor, a surface damaged layer is produced which invariably dominates the electrical characteristics of the junction so formed. However, the Langmuir trough technique, being a low temperature deposition process, provides a means of circumventing this particular difficulty. On the other hand, it does mean that how the substrate is prepared before dipping is of considerable importance in determining the quality of the interface produced. That is, the nascent 'oxide' layer formed during the etching procedure remains relatively undisturbed and this can play a vital role even after it has been coated with an LB film. For this reason it

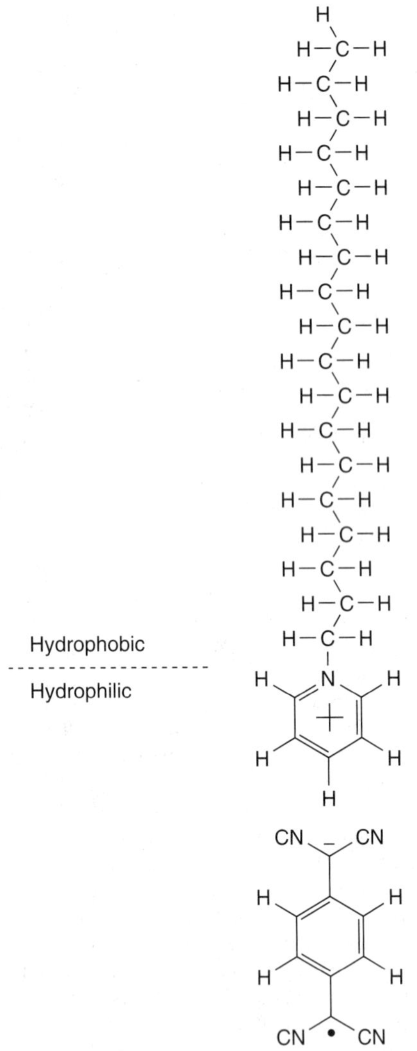

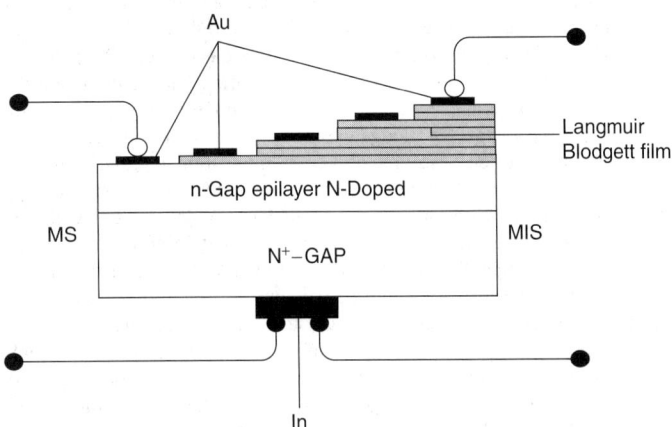

Fig. 13.   A substituent, pyridinium tetracyanoquinodimethane (TCNQ) molecular system.

Fig. 14.   Schematic diagram (not to scale) of a Gold-LB film — Gallium Phosphide device structure.

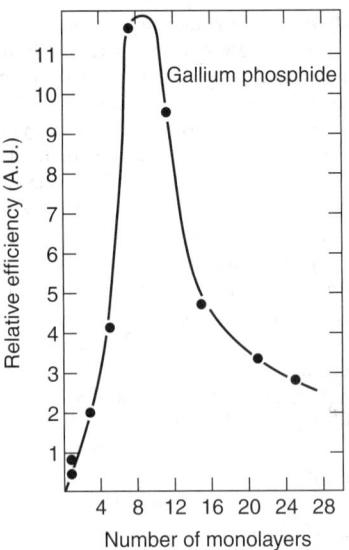

Fig. 15.   The electroluminescent efficiency versus number of monolayers or substituted phthalocyanine for the device shown in Fig. 14.

(MIS) diode. Conventionally these devices are made using inorganic materials; silicon holds a pre-eminent position mainly because of the insulating qualities of its native oxide. The first transistor incorporating LB monolayers as the insulator was reported several years ago (Roberts et al. 1978); using the type of three terminal device shown in Figure 16, on indium phosphide and cadmium stearate, they showed that the channel conductivity between the source and the drain could be modulated by the action of a gate electrode.

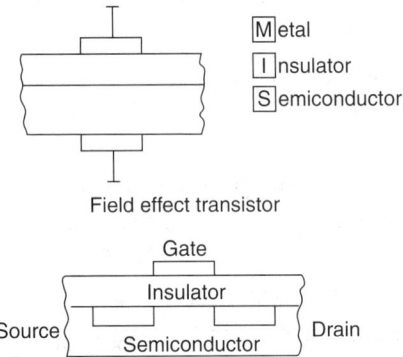

Fig. 16.   Schematic diagrams (not to scale) showing, in the upper diagram a metal-insulator-semiconductor structure which forms an integral part of the field effect transistor shown in the lower diagram.

Subsequently, other semiconductors have been used and results have confirmed the ease with which a range of single crystal surfaces can be accumulated, depleted or inverted with an applied voltage. It is recognised that in all cases, the LB film is deposited on top of a nascent 'oxide' layer and that the insulation is provided essentially by a double dielectric structure. In the case of silicon, this can be used to advantage, in that a closely packed LB film layer can seal its surface from the atmosphere and thereby greatly retard the development of interface states. The organic film is also efficient in increasing the dielectric strength of a leaky silicon oxide film.

The fact that organic compounds normally respond more positively than inorganic materials to external stimuli such as pressure or radiation, provides a means of making sensitive transducers. Moreover, by controlling the architecture of the LB film, the interactions can be designed to be of a lock-key type, thus enhancing the selectivity of the device. Following the non-linear physics work described earlier, it is now possible to envisage pyro- or piezo- FETs based on insulating LB films with an inbuilt polarization, or field effect devices incorporating biological membranes. Another advantage of using ultra-thin organic films is their fast response and recovery times because so little material is present. In the

is important first to carry out a systematic study of the surface chemistry of the semiconductor substrates.

(c) Sensors: The good insulating properties of LB films suggest their possible use in field effect devices, not so much to compete with existing semiconductor technology but more to capitalize on the advantages of being able to incorporate an organic layer within a semiconductor structure. Figure 16 shows schematic diagrams of both a field effect transistor (FET) and the 'heart' of this device, which is the metal-insulating-semiconductor

example shown in Figure 17, an eight monolayer LB film of a substituted phthalocyanine has been exposed to volume parts per million of nitrous oxide. It may be seen that the saturation current of the device scales linearly with gas concentration. Even though operation was at room temperature, the recovery times were shorter than those for evaporated film devices used at higher temperatures.

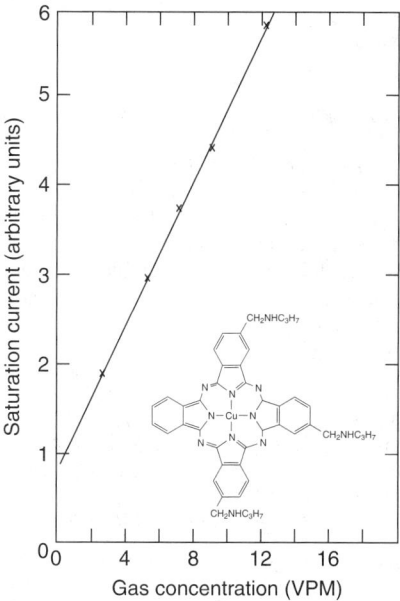

Fig. 17. Saturation current versus NO_2 gas concentration for a device incorporating eight monolayers of the asymmetrically substituted phthalocyanine molecule shown in the insert (Baker et al., 1983.)

It is not necessary to confine the discussion of microelectronic LB film based sensors to MIS or FET structures. For example, the switching voltage of a bistable switch or the characteristics of a gate-controlled diode could be very sensitive to an ambient. Moreover, optical and acoustic devices frequently show interesting threshold or resonance effects which could form the basis of a sensor. One area receiving particularly strong attention is that of surface plasmon resonance (SPR). The principle of this optical detection method is illustrated in Figure 18. A surface plasmon is a surface charge density wave at a metal surface. If the metal is sandwiched between two materials of different dielectric constant, then resonance can occur; this is observed as a very sharp minimum of the light reflectance when the angle of incidence is varied. The resonance angle is ultrasensitive to variations in the refractive index of the medium adjacent to the metal film. For example, the small change in an organic material due to gas absorption can easily be monitored even for concentrations in the part per billion range. In a practical situation, one normally selects an angle of incidence approximately half way down the reflectance minimum curve when no special gas is present; the change in intensity of the reflected light is then monitored at a constant angle.

There is widespread interest in the potential of LB films as biosensors as many believe that the incorporation of biological molecules such as enzymes will lead to novel devices. Some are exploring the deposition of biologically active molecules onto the gate electrodes or oxides of field effect transistors but optical sensors, probably based on fiber optics, are the most favored technique. In all cases the aim is to couple the specificity of interaction of chemicals or biochemicals with proteins or enzymes e.g., the change in their molecular conformation, with the sensitivity and signal transduction properties of the device. It is recognized that stability and lifetime may be problem areas and, for this reason, cross linked polymers are being explored as the hosts for the active species.

Some of the most convenient types of sensor are based on acoustoelectric devices; these can either be conventional bulk piezoelectric oscillators normally made of quartz, or surface acoustic wave (SAW) devices. The resulting change in the quiescent resonant frequency of a quartz oscillator coated with LB films of different thickness provides a very simple and elegant way of monitoring the reproducibility of monolayer deposition. Figure 19 illustrates how the quartz oscillator functions as a microgravimetric sensor; from the change in frequency it is possible to determine the density of the thin films. There is a further change when the organic films are exposed to minute concentrations of gas. The results presented in Figure 20 are for ω-tricosenoic acid in the presence of acetic acid, a species used for the detection of heroin. Greater effects can be obtained if the organic film is specially sensitised. Another way of increasing the sensitivity is to use acoustic surface waves. In such devices, input and output interdigitated electrodes are formed on a piezoelectric substrate usually made of quartz or lithium niobate. These perform the conversion between electric and acoustic energy. The single crystal surface region between the transducers serves as a propagation path for acoustic waves and thus forms a delay line which has application in electronic signal processing. Figure 20 shows a dual line configuration specially designed for sensing purposes. Basically, the device comprises two identical SAW (surface acoustic wave) oscillators positioned alongside each other. The hatched regions are earth shields to minimize reflections and cross talk between the two oscillators. The selective coating is placed in the propagation path of one of the oscillators thus affecting the delay time; the relative shift in frequency between the two oscillators is measured using a mixer circuit to obtain the difference frequency and then passing the resulting signal through a low pass filter. The change in frequency (Δf) between the two channels is then directly attributable to the sensor layer and other extraneous effects such as those due to temperature changes, are eliminated or very much reduced. Other geometries, for example, a surface acoustic wave resonator, are also possible. The device shown in Figure 21 has been constructed by Roberts et al. (1987) using quartz and interdigitated electrodes 24 μm wide separated by gaps of 25 μm; the operating frequency of the device is 98.4 MHz. Results have been obtained for both insulating (ω-tricosenoic acid) and conducting (pyridinium TCNQ) LB films as the sensor layer. In the case of the ω TA, only a mass loading effect is observed but with the TCNQ films, electric field effects are also apparent; these are associated with interactions between the surface acoustic wave and mobile charge carriers in the LB film. The combination of a device capable of measuring mass changes as low as femtograms per square centimeter, and the ability to detect minute changes in the electrical characteristics of a monolayer, augers well for sensors based on surface acoustic waves and monolayers.

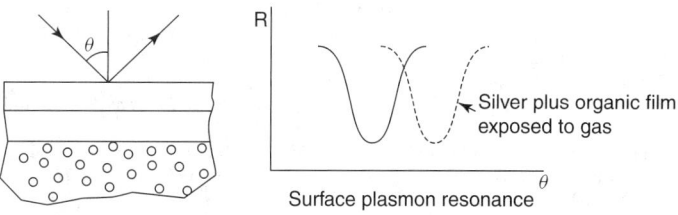

Fig. 18. Schematic diagrams illustrating the basis of the surface plasmon resonance technique. (Left) a beam of radiation striking the back surface of a glass prism coated with a metal film (usually evaporated silver). The reflected intensity and angle of reflection are extremely sensitive to variations in the dielectric on the metal surface. (Right) The shift in the R/θ plot when the organic film is exposed to a gas.

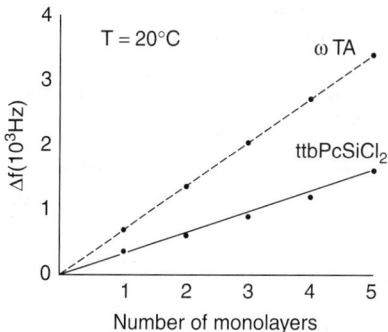

Fig. 19. The change in resonance frequency of 18 MHz piezoelectric quartz crystals coated with LB films of different thickness. The molecular structures of the organic films are shown in Figs. 3(b) and 3(d).

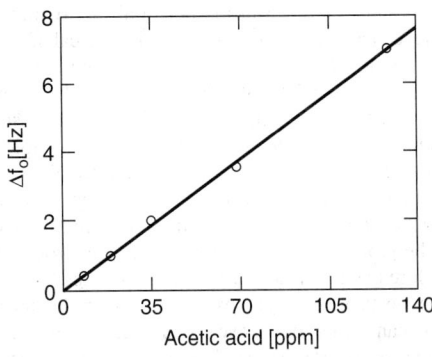

Fig. 20. Response characteristics of ω-tricosenoic acid coated quartz crystal oscillator at 22 °C exposed to acetic acid. A greater change in the resonance frequency can be obtained if the film is specially sensitized to detect acetic acid.

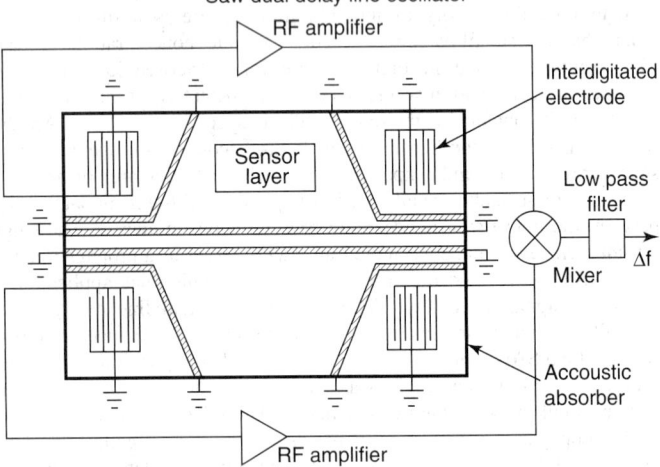

Fig. 21. A surface acoustic wave dual-delay line oscillator. The sensitive layer is placed in the propagation path of one of the two SAW devices. The difference in frequency (Δf) between the two channels provides a direct result of the mass loading and electric field effects associated with the sensor layer.

Conclusions

We have discussed several research areas where there appears to be a tangible benefit in using monomolecular assemblies rather than organic or inorganic thin films deposited by other means. There are many ways of producing organic films and the onus will be on the Langmuir-Blodgett enthusiasts to demonstrate the special advantages to be gained using their technique for a particular application. There is every likelihood that this will occur but it will require the combined efforts of high caliber teams with knowledge of physics, chemistry, electronics and biology. The ability of the synthetic chemist to manipulate the molecular architecture of a material to optimize a specific physical parameter or figure of merit will be vital. However, the role of the physicist or engineer in identifying the targets and guiding the main thrust of the research program will also be essential. Industry is already well able to organize interdisciplinary activities. The relatively inexpensive equipment requirements associated with the Langmuir trough technique coupled with the elegance of the fundamental science and the interesting applied prospects for LB films provides an excellent opportunity for the academic community to similarly break down the traditional barriers between disciplines. At the present time it would appear that non-linear optics, electron beam microlithography, magnetic tape lubrication, sensors, and optical storage, are the areas most likely in the short term to benefit from LB film technology.

The above comments referring to the need for multidisciplinary activity apply equally well to other fields covered by the definition of Molecular and Supermolecular Electronics. The goal of a supermolecular information processor based on organic assemblies is a very ambitious one. Nevertheless, pursuing this target should serve to enumerate technologies and identify areas of basic scientific research that would otherwise remain dormant and unexplored. Far simpler but novel devices based on organic molecular materials should appear as a result of this research effort. It seems

likely that Langmuir-Blodgett films will play a key role in helping scientists identify basic physical phenomena in supermolecular assemblies and at the same time enable engineers to become more familiar with devices incorporating organic films.

Author's Acknowledgments

The author is indebted to his many colleagues and former colleagues at the Universities of Durham and Oxford for their important contributions to the research described in this article. Particular thanks are due to Dr. M.C. Petty and Mr. B. Holcroft.

Editor's Acknowledgments

This article, originally published in the *University of Wales Science and Technology* Review, is reprinted here with approval of J.H. Purnell, Honorary Editor. This prestigious journal, issued quarterly, contains dissertations on various topics of science and technology, written by professionals who have refreshingly new viewpoints. The *Review*, which commenced publication in March 1987, is a welcome contribution to the scientific community throughout the English-speaking world. Publication Office: MBN2 Cardiff, Marketing Department, 17th Floor, Pearl House, Greyfriars Road, Cardiff, Wales CF1 3XX, United Kingdom.

Additional Reading

Albrecht O., A. Laschewsky, and H. Ringsdorf: *J. Membrane Sci.*, **22**, 186 (1985).
Allara D. and J.D. Swalen: *J. Phys. Chem.* **86**, 2700 (1982).
Baker S., G.G. Roberts, and M.C. Petty: *Proc. IEE Pt. 1*, **130**, 260 (1983).
Barraud, A.: *Thin Solid Films*, **99**, 317 (1983).
Barraud A., P. Lesieur, A. Ruaudel-Teixier, and M. Vandevyver: *Thin Solid Films*, **134**, 195 (1985).
Batey J., G.G. Roberts, and M.C. Petty: *Thin Solid Films*, **99**, 283 (1983).
Blodgett, K.B.: *J. Amer. Chem, Soc.*, **57**, 1007 (1935).
Brundle C.R., H. Hopster, and J.D. Swalen: *J. Chem. Phys.*, **70**, 5190 (1979).
Carter, F.L.: *Superlattices and Microstructures*, **2**, 113 (1986).
Christie P., G.G. Roberts, and M.C. Petty: *Appl. Phys. Letts.*, **48**, 1101 (1986a).
Christie P., C.A. Jones, M.C. Petty, and G.G. Roberts: *J. Phys.*, **D19**, L167 (1986b).
Clark D.T., Y.C.T. Fok, and G.G. Roberts: *J. Electron. Spectroscopy*, **22**, 17 (1981).
Dharmadasa I.M., G.G. Roberts, and M.C. Petty: *Electronics Letts.*, **16**, 201 (1980).
Franklin, B.: *Phil. Trans. Roy. Soc.*, **64**, 445 (1974).
Gaines, G.L.: *Thin Solid Films*, **99**, ix (1983).
Gaines, G.L.: *Insoluble Monolayers at Liquid-Gas Interfaces*, Interscience, New York, NY, 1966.
Gaines, G.L.: *Thin Solid Films*, **132, 133, 134** (1985).
Holcroft B., M.C. Petty, G.G. Roberts, and G.J. Russell: *Thin Solid Films*, **134**, 83 (1985).
Hua Y.L., G.G. Roberts, M.M. Ahmed, M.C. Petty, M. Hanack, and M. Rein: *Phil. Mag.*, **B53**, 105 (1986).
Kelly M.J. and Weisbach, C.: *The Physics and Fabrication of Microstructures and Microdevices*, Springer-Verlag, Berlin (1986).
Kuhn, H.: *Thin Solid Films*, **99**, 1 (1983).
Langmuir, I.: *J. Amer. Chem. Soc.*, **39**, 1848 (1917).
Langmuir, I.: *Trans. Faraday Soc.*, **15**, 62 (1920).
Mobius, D.: *Accts. Chem. Res.*, **14**, 63 (1981).
Mori C., H. Noguchi, M. Mizuno, and T. Watanabe: *Jap. J. Appl. Phys.*, **19**, 725 (1980).
Neal D., M.C. Petty, G.G. Roberts, M.M. Ahmed, W.J. Feast, I.R. Girling, N.A. Cade, P.V. Kolinsky, and I.R. Peterson: *Proc. Int. Symp. on the Applications of Ferroelectric Materials*, Philadelphia (1986).
Petty M.C., J. Batey, and G.G. Roberts: *IEE Proc. Pt. 1*, **132**, 133 (1985).
Pockels, A.: *Nature*, **43**, 437 (1891).
Pomerantz, M.: Phase Transitions in Surface Films, Plenum Press, New York, NY, 1980, p. 317.
Rayleigh, Lord: *Proc. Roy. Soc.*, **47**, 364 (1890).
Roberts, G.G., K.P. Pande, and W.A. Barlow: *Proc IEE Pt. 1*, **2**, 169 (1978).
Roberts, G.G. and C.W. Pitt: *Thin Solid Films*, **99** (1983).
Roberts, G.G.: *Advances in Physics*, **34**, 475 (1985).
Roberts, G.G., B. Holcroft, J. Ross, and A. Barraud: *British Polymer Journal*, to be published (1987).
Tabor, D.: *J. Colloid and Interface Science*, **75**, 240 (1980).
Vincett P.S., W.A. Barlow, F.T. Boyle, J.A. Finney, and G.G. Roberts: *Thin Solid Films*, **60**, 265 (1979).
Vincent, P.S. and G.G. Roberts: *Thin Solid Films*, **68**, 135 (1980).
Zyss, J.: *J. Non. Cryst. Solids*, **47**, 211, (1982).

GARETH ROBERTS, FRS, Thorn EMI plc and University of Oxford.

MOLECULAR BEAM. A unidirectional stream of neutral molecules passing through a vacuum, generally with thermal velocity. Such a beam may be produced by emergence from a pinhole in a chamber containing low pressure gas or vapor, and it may be defined by a system of slits. By passing the beam through known electric or magnetic fields, quantities such as nuclear magnetic moments can be determined.

MOLECULAR BIOLOGY. A self-defining term — the study of biological substances and phenomena at the molecular level.

For many years, biologists, aided by numerous laboratory instruments and methodologies (X-ray crystallography, electron microscopy, chromatography, electrophoresis, etc.) constructed a vast databank of biochemistry. This has been and continues to be of inestimable value to those professionals who deal with health, medicine, the industrial use of biochemicals, among other areas that require such information. There remained, however, a dissatisfaction concerning the absence of knowledge pertaining to the manner in which biochemical processes are directed and take place at the molecular level. See also **Industrial Biotechnology**.

The first break occurred early in the 1950s when very illuminating findings were made concerning the complex biomolecule, DNA. The research of DNA, in essence, constituted the beginnings of *molecular biology*. A chronology of these early years is given in the article on **Genetics and Gene Science**. This pioneering work altered the thrust of biochemical research, as previously typified by traditional biochemistry, to a concentration and emphasis on studying biological events at the molecular level.

A second break occurred in the early 1970s when it was discovered that a strand of the DNA molecule can be cut by restriction enzymes and that the sticky ends can be reassembled by a new technology known as *genetic recombination* or *recombinant DNA*. This is frequently the starting point for projects engaged in biological research at the molecular level. It is no surprise that molecular biology has become a highly diverse, wide-spectrum, multifaceted discipline. The many hundreds of research projects in this area underway today obviously cannot be delineated here. The highlights of just a few research targets are given here to provide a representative view of projects underway and inroads made to date.

Cell Membrane Research. Cellular membranes serve as the interface between the cell and the organism of which it is a part. It has been established that the movement of cells, as directed during growth and development, must involve, in some way, the plasma membranes. It also has been observed that plasma membranes play some role in cancerous growth, where cell multiplication and migration proceed at an uncontrolled rate. Membranes are composed of lipids that interact with each other in a watery medium to form a closed and flexible compartment. This formation process occurs at a relatively rapid rate. Bretscher reports that an area of membrane equivalent to the area of the entire surface of the cell requires less than an hour to form. A good start has been made by a number of investigators toward understanding the detailed molecular structure of membranes. In retrospect, it is interesting to note that recent findings pertaining to the basic frame of membranes was qualitatively proposed as early as 1925 by Gorter and Grendel (University of Leiden). This framework is composed of a double layer of lipid molecules. One end of the molecule is hydrophilic, that is, soluble in water; the other end is hydrophobic, a hydrocarbon that is oily and insoluble in water. Most commonly, membrane lipids are phospholipids. See also **Lipids**.

Cytoplasm Research. The existence of fibers in cytoplasm was first observed microscopically about a century ago. It has been learned during the last few years as the result of biochemical and immunological studies, that a distinct set of proteins characterizes each filament system. Proteins in the cytoplasm make up a highly structured, yet changeable matrix. The matrix determines the cell shape, its division and motion, as well as the transport of vesicles and organelles. Research in molecular genetics is targeted to reveal the function of cytoskeletal proteins. Weber and Osborn report that current knowledge of the cell matrix is contributing to diagnosis and research in human pathology. For example, typing of intermediate filaments by way of immunofluorescence microscopy will distinguish the major tumor groups. Keratins are found in carcinomas. What is missing is detailed knowledge at the molecular level of why and how the protein molecules in the cytoplasm function.

Immune System Research. This is one of the most active research areas as of the late 1980s. It has been observed that proteins responsible for recognizing foreign invaders are the most diverse of all known proteins. Tonegawa observes that these proteins are encoded by hundreds of scattered gene fragments and these can be combined in millions, probably

billions of ways. This gives some idea of how complex research in this area really is. It is one of the most difficult of tasks for molecular biologists, but is proceeding apace. See also **Immune System and Immunology**.

Cellular Intercommunications Research. It is well known that the majority of higher organisms utilize hormones or systems of neurons for intercellular communication. See also **Central and Peripheral Nervous Systems**; and **Hormones**. Snyder observes that chemical messengers mediate long-range communication via hormones; short-range communication via nerve cells. The two systems differ in directness, but some messenger molecules are common to both. Hormonal molecules are usually peptides and steroids. Communication between cells or groups of cells is mandatory for the survival of all multicellular organisms. Snyder also observed that as investigators come to know the properties and functions of highly specialized messenger molecules, it will be possible to develop better therapeutic agents for a number of diverse conditions, including hormonal abnormalities, heart disease, and mental illness.

Cellular Intracommunications Research. The communication within a cell is being investigated at the molecular level. The mechanism by which a cell receives external signals by way of receptors has been reasonably well understood for a number of years. But what comprises the cell's internal mechanism for reacting to received signals by way of the receptors? Berridge refers to receptors as "molecular antennas." Obviously, the barrier to the flow of information from receptor to mechanisms within the cell is the plasma membrane, previously mentioned. How are external signals acted upon internally within the cell to cause it to secrete, to contract, to metabolize, or to grow, among several other biological reactions? What are the "second messengers" within the cell? How do they function? A second messenger, cyclic adenosine monophosphate (cyclic AMP) has been identified. Other second messengers include Ca^+ ions, inositol triphosphate (IP_3), and diacylglycerol (DG). Berridge also reported that some evidence has been gained to the effect that the aforementioned substances are cannibalized from the plasma membrane itself. Obviously, the cell must include genes required for the synthesis of the proteins used in the internal pathways. Aberration of these gene functions could lead to abnormalities of cellular growth and hence to uncontrolled growth and structural transformations typical of cancer. Cells related in some way to tumor growth revealed thus far are called *oncogenes*. Berridge and other researchers are attempting to identify additional second messengers that may be present and how they function in internal communication.

Biological Development and Growth Research. For many years, the processes which take place within an organism to cause it to grow (the early embryonic phase), leading to maturity (adulthood) and later to initiate the aging phase have been among the least understood of biological phenomena. As aptly observed by Gehring: How is the basic architecture of an embryo laid down? How does the linear information of DNA generate a three-dimensional organism? One of the most recent and intriguing discoveries is that of the so-called *homeobox*, which is found as a short stretch of DNA. When the gene containing the homeobox is translated into a protein, the homeobox yields a string of amino acids that are believed to bind to the DNA double helix. By binding to the DNA of particular genes, the protein may be able to turn them on or off. The homeobox was first noted in studies of *Drosophila*; subsequent studies have revealed a homeobox within a range of organisms extending from worms to humans. It is well established, of course, that animals develop in numerous ways. Researchers now question whether or not the molecular mechanisms underlying development may be more universal than previously suspected. Again, it appears that the answers to still another of the biological secrets may lie in better understanding the process at the molecular level.

Adaptive and Evolutionary Processes. Wilson observes that in the past most biologists working in this area have concentrated their attention at the level of the whole organism. Mutation frequently has been suggested as the cause of adaptive changes. Prior to studies at the molecular level, however, it was *not* known that mutations may accumulate at steady rates over time in the genes of all lineages. Wilson, a proponent of the molecular evolution concept, reports on two assumptions: (1) Heritable differences among organisms result from differences in their DNAs, and (2) molecular evolutionists must not only measure differences in DNA, but also explain the origin of the differences and their relation to organismal differences. Wilson also points out that two critical elements of molecular evolution are (1) *point mutations*, specifically those occurring in the genes coding, and (2) *regulator mutations*. A point mutation is defined as a single replacement of a DNA base; a regulatory mutation is any change in a gene or within the vicinity of a gene that determines whether the gene is active or inactive. Research on point mutations led to the concept of a *molecular clock* and in the discovery of still another kind of genetic change, which Wilson calls a *neutral mutation*. This mutation is neither advantageous nor disadvantageous for an organism. Wilson also refers to the "pressure" to evolve, noting that this stems from geologic forces as well as from the brains of mammals and birds. It is obvious that the study of molecular evolution occupies a special position in contemporary biology.

See also **Biological Timing and Rhythmicity**; and **Evolution**.

Additional Reading

Ausubel, F. et al.: "Short Protocols in Molecular Biology: A Compendium of Methods from Current Protocols in Molecular Biology," John Wiley & Sons, Inc., New York, NY, 1999.
Berridge, M.J.: "The Molecular Basis of Communication within the Cell," *Sci. Amer.*, 142–148 (October 1985).
Lackie, J. and J. Dow: "The Dictionary of Cell and Molecular Biology, 3rd Edition, Academic Press, Inc., San Diego, CA, 1999.
Stahl, W. and A. Hershey: "We Can Sleep Later: Alfred D. Hershey and the Origins of Molecular Biology," Cold Spring Harbor Laboratory Press, Cold Spring Harbor, NY, 2000.
Synder, S.H.: "The Molecular Basis of Communication Between Cells," *Sci. Amer.*, 132–137 (October 1985).
Tonegawa, S.: "The Molecules of the Immune System," *Sci. Amer.*, 122–128 (October 1985).
Weber, K. and M. Osborn: "The Molecules of the Cell Matrix," *Sci. Amer.*, 110–118 (October 1985).
Wilson, A.C.: "The Molecular Basis of Evolution," *Sci. Amer.*, 164–170 (October 1985).

MOLECULAR DISTILLATION.   A special form of distillation conducted at pressures of 1–7 micrometers in the laboratory and 3–30 micrometers in industrial applications. Compared with conventional laboratory vacuum distillations carried out between 1 and 10 millimeters of mercury pressure, this is a very high vacuum. One micrometer equals 0.001 millimeter of mercury pressure. The other feature of the molecular still is that the condenser is located within a distance less than the mean free path of the evaporating molecules from the evaporator portion of the apparatus. Thus, although a molecule may return to the distilland many hundred times before reaching the exit of a conventional vacuum still, 50% of the molecules in a properly functioning molecular still will reach the exit on their first try. Thus, efficiency is remarkably high.

Because of the absence of convection due to ebullition and because high viscosities and high molecular weights may impede diffusion within the distilland, the surface of the distilland in a molecular still may not always represent the total liquid. Therefore, efficient molecular distillation requires the mechanical renewal of the surface film. This is achieved by vigorous agitation, as in the *stirred-pot still*; by employing a *falling film*; or by using centrifugal force, as in the *centrifugal* molecular still. Commercial installations of falling-film stills achieve throughputs of many tens of liters per hour, whereas the centrifugal still is capable of several hundred liters per hour. Centrifugal stills usually are arranged in groups of from three to seven. This permits fractionation by multiple redistillation. Among the uses of molecular distillation are: the separation of mono and diglycerides for bread and paraffin wax for milk cartons; the distillation of plasticizers, fatty acid dimers, and synthetics; the distillation of vitamin A esters and intermediates; and the stripping of α, β, γ, and δ-tocopherols and sitosterols from vegetable oils.

See also **Distillation**.

MOLECULAR SIEVE.   Materials with special porous and absorbing properties that chemically lock materials into their pores, as used in purification and separation processes, are sometimes called molecular sieves. The sodium aluminum silicates (synthetic zeolites) are examples of materials of this type.

MOLECULE.   In the traditional sense, a molecule is the smallest particle of a chemical substance capable of independent existence with retention of all its chemical properties. Molecules comprise one or more atoms which need not be of the same kind. Only the rare, or noble gases form single-atom or monatomic molecules. All other elements form bi-, tri-, quadri-, etc. atomic molecules, e.g., hydrogen, H_2; ozone, O_3; phosphorus, P_4; and sulfur, S_8; or hydrogen chloride, HCl; sodium sulfide, Na_2S, aluminum chloride, $AlCl_3$, carbon tetrachloride, CCl_4, and so on.

Structurally, a more specific definition would be that a molecule is a local assembly of atomic nuclei and electrons in a state of dynamic stability. The cohesive forces are electrostatic, but, in addition, relatively small electromagnetic interactions may occur between the spin and orbital motions of the electrons, especially in the neighborhood of heavy nuclei. The internuclear separations are of the order of $1-2 \times 10^{-10}$ meter, and the energies required to dissociate a stable molecule into smaller fragments fall into the 1–5 eV range. The simplest diatomic species is the hydrogen molecule, H_2^+, with two nuclei and one electron. At the other extreme, the protein ribonuclease contains 1876 nuclei and 7396 electrons per molecule.

Another form of molecule is known, however, and this is formed by atomic nuclei alone. Although these have not yet been found in nature, they may well play a role in stellar evolution. Under special conditions in high-energy interactions, they can be momentarily held together by effective bonds. Whether these bonds are the result of exchange or sharing of valence protons and neutrons is as yet moot. In these *nuclear molecules* a somewhat unstable balance is attained between long-range electrostatic repulsion of positively charged nuclei and the much stronger short-range nuclear force which determines the motions of protons and neutrons. Nuclear molecules are significant entities because they live much longer ($\sim 10^{-21}$ second) than the time usually taken for nuclei to collide ($\sim 10^{-23}$ second).

The molecular, or kinetic, theory of matter makes four assumptions: (1) that the molecules of which matter is composed are constantly in motion; (2) that their energy is increased by the addition of heat; (3) that they undergo elastic collision with each other and with the walls of a containing vessel; and (4) that they exert forces upon each other. As first developed by Heisenberg, Schrödinger, and Dirac, reduction of these theoretical assumptions to mathematical bases is somewhat inadequate and can relate only to interaction between hypothetical electron clouds. Except for the simplest of systems, the Schrödinger equation cannot be solved exactly. On the other hand, a more manageable understanding of atom interactions in molecules is afforded by the Valence Shell Electron Pair Repulsion Theory, independently enunciated by Nyholm and Gillespie. This theory proposes that both bonding and nonbonding pairs of outer atomic shell electrons in a molecule repel each other and establish themselves as far apart as possible.

Historically, molecules were regarded as being formed by the association of individual atoms. This led to the concept of *valency*, i.e., the number of individual chemical bonds or linkages with which a particular atom can attach itself to other atoms. When the electronic theory of the atom was developed, these bonds were interpreted in terms of the behavior of the valence, or outer shell, electrons of the combining atoms. Each atom with a partly-filled valence shell attempts to acquire a completed octet of outer electrons, either by electron transfer, as in (a) shown below, to give an electrovalent bond, resulting from coulombic attraction between the oppositely charged ions; or, as in (b) and (c) to give a covalent bond. The concept of (a) was proposed by Kossel in 1916; that of (b) and (c) by Lewis, also in 1916.

$$Na^+ \; [:\!\ddot{C}\!\ddot{l}\!:]^- \qquad :\!\ddot{C}\!\ddot{l}\!:\!\ddot{C}\!\ddot{l}\!: \qquad R:\!\overset{\overset{R}{\displaystyle|}}{\underset{\underset{R}{\displaystyle|}}{N}}\!:\!\ddot{O}\!: \qquad R = CH_3$$

    (a)   (b)   (c)

In (b), each chlorine atom donates one electron to form a *homopolar bond*, which is written Cl—Cl where the bar denotes on this theory one single bond, or shared electron pair. In (c), the nitrogen-oxygen bond is formed by two electrons donated by only the nitrogen atom, giving a *semipolar*, or *coordinate-covalent bond*, which is written $R_3N \rightarrow O$, and which is electrically polarized. Double or triple bonds result from the sharing of 4 to 6 electrons between adjacent atoms. More information on these bonding theories is given in the entries on **Chemical Composition; Chemical Elements**; and **Compound (Chemical)**.

However; difficulties arise in describing the structures of many molecules in this manner. For example, in benzene, C_6H_6, a typical aromatic compound, the carbon nuclei form a plane regular hexagon, but the electrons can only be conventionally written as forming alternate single and double bonds between them. Furthermore, an electron cannot be identified as coming specifically from any one of these bonds upon ionization. Such difficulties disappear in the quantum-mechanical theory of a polyatomic molecule, whose electronic wave function can be constructed from nonlocalized electron orbitals extending over all of the nuclei. The concept of valency is not basic to this theory, but is simply a convenient approximation by which the electron density distribution is partitioned in different regions in the molecule.

Molecular compounds consist of two or more stable species held together by weak forces. In *clathrates*, a gaseous substance, such as SO_2, HCl, CO_2, or a rare gas is held in the crystal lattice of a solid, such as beta-quinol, by Van der Waals-London dispersion forces. The gas hydrates, e.g., $Cl_2 \cdot 6H_2O$, contain halogen molecules similarly trapped in ice-like structures. The hydrogen bond, with energy ~ 0.25 eV, is responsible not only for the high degree of molecular association in liquids, such as water, but also for such molecules as the formic acid dimer, which contains two hydrogen bonds indicated by dashed lines.

Molecular complexes vary greatly in their stability; in donor-acceptor complexes, electronic charge is transferred from the donor (e.g., NH_3) to the acceptor (e.g., BF_3), as in a semipolar bond. The $BF_3 \cdot NH_3$ complex has a binding energy with respect to dissociation into NH_3 and BF_3 of 1.8 eV. The bond here is relatively strong; the electron transfer can occur between the components in their electronic ground states. On the other hand, in weaker complexes, such as $C_6H_6 \cdot I_2$, with binding energy of about 0.06 eV, there is only a fractional transfer of charge from benzene to iodine. The actual ionic charge-transfer state lies at much higher energy than the ground state of the complex.

Complete pairing of all electrons present in a molecule and absence of any bonding orbitals was long taken to be a stable, unreactive state exemplified by the inert, or rare gases. In 1962, however, Bartlett unequivocally synthesized $XePtF_6$, and this was rapidly followed by the synthesis of other rare gas compounds whose existence was not predicted by classical valency theories. Compounds such as XeF_2, XeF_4, XeF_6, and $XeOF_4$ are quite stable, the average Xe-F bond energy in the square planar XeF_4 being 1.4 eV.

A molecule is characterized by (1) a stoichiometric formula; (2) the spatial distribution of the nuclei in their mean equilibrium or "rest" positions; and (3) the dynamical state.

The ratio $a: b: c: \ldots$ in a formula $A_aB_bC_c$, where a, b, c, $\ldots$ are the numbers of atoms of elements A, B, C, $\ldots$ that it contains is found by chemical analysis for these elements. The absolute values of a, b, c, $\ldots$ are then fixed by determination of the molecular weight. This principle is further described under the entry on **Compound (Chemical)**.

The spatial distribution of the nuclei in their mean equilibrium positions, at an elementary level, is described in geometrical language. For example, in carbon tetrachloride, CCl_4, the four chlorine nuclei are disposed at the corners of a regular tetrahedron, and the carbon nucleus is at the center. In the $[CoCl_4]^{2-}$ ion, the arrangement of the chlorine nuclei about the central metal nucleus is also tetrahedral, whereas in $[PdCl_2]^{2-}$, it is planar. For example, the pyramidal ammonia molecule NH_3 has a threefold rotation axis C_3 through the nitrogen nucleus and three reflection planes σ_v intersecting at the axis, and belongs to the $C_{3v}(3m)$ point group. Tetrahedral molecules CX_4 belong to the $T_d(\bar{4}3m)$ point group. Linear diatomic and polyatomic molecules belong to either of the continuous point groups $D_{\infty h}$ or $C_{\infty v}$ according to whether a center of symmetry is present or not.

The symmetry classification does not define the geometry of a molecule completely. The values of certain bond lengths or angles must also be described. In carbon tetrachloride, it is sufficient to give the C—Cl distance (1.77×10^{-10} meters), since classification under the T_d point group implies that all four of these bonds have equal length and the angle between them is $109°28''$. In ammonia, both the N—H distance (1.015×10^{-10} meters) and the angle HNH ($107°$) must be specified. In general, the lower the molecular symmetry, the greater is the number of such independent parameters required to characterize the geometry. Information about the symmetry and internal dimensions of a molecule is obtained experimentally by spectroscopy, electron diffraction, neutron diffraction, and x-ray diffraction.

The dynamic state is defined by the values of certain observables associated with orbital and spin motions of the electrons and with vibration and rotation of the nuclei, and also by symmetry properties of the corresponding stationary-state wave functions. Except when heavy nuclei

are present, the total electron spin angular momentum of a molecule is separately conserved with magnitude Sh, and molecular states are classified as singlet, doublet, triplet ... according to the value of the multiplicity ($2S + 1$). This is shown by a prefix superscript to the term symbol, as in atoms.

The Born-Oppenheimer approximation permits the molecular Hamiltonian H to be separated into a component H_e that depends only on the coordinates of the electrons relative to the nuclei, plus a component depending upon the nuclear coordinates. This in turn can be written as a sum $H_v + H_r$ of terms for vibrational and rotational motion of the nuclei, translation being ignored. The eigenfunctions Ψ of H may correspondingly be factorized as the product $\Psi_e\Psi_v\Psi_r$ of eigenfunctions of these three operators, and the eigenvalues of E decomposed as the sum $E_e + E_v + E_r$. In general, $E_e > E_v > E_r$.

Molecular Spectra. The spectra of substances in the molecular state, like atomic spectra, are made up of lines, although more complex. The transitions in a molecule which release the most energy (largest quanta) are due to electron changes, as in atoms, and the results of these changes are observed as lines in the ultraviolet region. But there are other ways in which a molecule can release or absorb energy. Thus, the component atoms oscillate with reference to each other within the molecule, and this motion apparently is "quantizied," i.e., changes abruptly from one state to another of different energy. But these "vibrational" energy changes are much less than the electronic, so that the resulting quanta and spectrum lines are of much lower frequency, and appear in the extreme red or near-infrared. Again, the molecule rotates, and the quantization of its rotational energy results in the emission of quanta of still lower frequency, appearing as lines in the far-infrared.

Molecular and laser spectroscopic approaches are making possible a deeper resolution of the dynamics of atomic and molecular motions and the potential energy surfaces governing energy transference within a molecule or group of molecules as chemical bonds are made and broken.

Integration of evolutionary biology with molecular biology, epitomized by the later work of Szent-Györgyi, is, in effect, translating morphogenesis into molecular language. Along metabolic pathways, intermediate molecules usually appear in such an order that thermodynamic stabilities increase progressively from starting materials to final products. Such stabilities are also associated with the influence of solvent water on chemical systems. The qualitative effects of this have been known for many years, but the quantitative data have so far been scarce. The solvent molecules tend to be reorganized in the neighborhood of the interacting groups, but not all molecules are attracted enough to overcome the self-cohesive properties of water.

Free energies of two or more moderately polar groups are often approximately additive and departures from this suggest special interactions between parts of the solute molecules and the surrounding solvent.

Polymerization. This term is used to designate a reaction in which a complex molecule of high molecular weight (or macromolecule) is formed from a number of simpler molecules. Thus the monomer formaldehyde, CHOH, can form the trimer trioxane $(CHOH)_3$, or the long-chain polymer paraformaldehyde, $HO(CHOH)_nH$, where $n = 8-100$. But the combining molecules may be of the same or different sorts. An *additional polymerization* is one in which like or unlike molecules combine without the elimination of any atoms or molecules. A *simple polymerization* involves only one species of molecule. *Copolymerization* is an addition polymerization in which two or more distinct molecular species are involved, each one of which is capable of polymerizing by itself. The high polymer formed contains each molecular constituent or an essential portion, as a distinct unit in the structure of the polymer. *Heteropolymerization* is an addition polymerization in which two or more molecular species are involved, one of which species will not polymerize by itself. It does, however, form distinct units in the high polymer.

A condensation polymerization is one in which the molecules undergoing polymerization react with the elimination of simple molecules like water, ammonia, and the like.

Polymerizations are also characterized by the state in which they are carried out such as: *gaseous polymerization* or those carried out in the vapor or gaseous phase; *mass polymerizations* in the liquid state; *solution polymerizations* carried out by first dissolving the material to be polymerized in an adequate solvent; *emulsion polymerizations* carried out in which one of the components is in an emulsion as in the case of rubber polymerizations; and *bulk polymerizations* in which the polymerization takes place without the use of a solvent or other medium. The wide variety of methods by which the process of polymerization can be used in the manufacture of plastics is exemplified by the production of polystyrene from styrene (vinylbenzene, $CH_2 \cdot CH \cdot C_6H_5$). One factor that conditions all these processes is the highly exothermic (high heat production) nature of this polymerization. This fact, together with the low heat conductivity of polystyrene, determines certain characteristics of an industrial process. This is particularly the case because the extent of the polymerization of styrene, like that of many other plastics, depends on the temperature. When the process is conducted at higher temperatures, the resulting polymer has a low molecular weight and low physical strength properties (i.e., it is weak and brittle). Extremely high molecular weight polymers, although mechanically tough, are more difficult to fabricate. Since the higher polymerization temperatures also result in faster polymerization rates, economic considerations dictate a compromise between faster production and better physical properties. Practical experience indicates a temperature range from 60 to 150 °C.

The polymerization of styrene illustrates the application of several of the methods defined above.

In *batch mass polymerization* the reaction vessel is loaded with styrene monomer and heated to a temperature sufficient to initiate polymerization within a reasonable time. As polymerization proceeds, the temperature within the vessel rises, thus increasing the rate.

In *continuous mass polymerization* the reaction vessel contains monomer and polymer. As more polymer is formed it is drawn off the bottom of the reaction vessel while monomer is added to the top of the vessel. The temperature of the reaction is controlled by cooling coils within the vessel.

In *solution polymerization* the styrene monomer is diluted with a solvent. The solvent acts as a diluent, which decreases the rate of polymerization and also serves as a heat transfer medium for removing the excess heat developed by the reaction.

In *suspension polymerization* water is used as a diluent and as a heat transfer aid. Suspending agents such as starch and methylcellulose are used to keep the styrene monomer particles in suspension. The more efficient heat transfer of this process also allows for a narrower molecular weight distribution.

In *emulsion polymerization* the styrene monomer is emulsified with water by the addition of certain emulsifying agents. This results in very small particles and rapid polymerization rates. The heat of polymerization is dissipated by the water ingredient.

See also **Fibers**; and **Macromolecular Science**.

Additional Reading

Barondes, S.: "Molecules and Mental Illness," Scientific American Library, New York, NY, 1999.

Eisberg, R. and R. Resnik: "Quantum Physics of Atoms, Molecules, Solids, Nuclei, and Particles, 2nd Edition," John Wiley & Sons, Inc., New York, NY, 1990.

Parr, R. and Y. Weitao: "Density-Functional Theory of Atoms and Molecules," Oxford University Press, New York, NY, 1994.

MOLE FRACTION. As applied to a system, the *mole fraction* (sometimes spelled *mol fraction*) of a given substance in the system is found by dividing the number of moles of that substance by the total number of moles in the system. For a mixture of ideal gases, the mole fraction is equal numerically to the volume fraction. The *volume fraction* of a component in a mixture is found by dividing the volume of that component at the total pressure and at the temperature of the mixture by the volume of the mixture at the same pressure and temperature.

In a binary solution consisting of components X, and Y, the mole fractions, F_X and F_Y, respectively, are:

$$\text{Mole fraction of } X = F_X \frac{f_X}{f_X + f_Y}$$

$$\text{Mole fraction of } Y = F_Y \frac{f_Y}{f_X + f_Y}$$

where f = number of moles of specific component present.

It is apparent, of course, that the mole fraction of X plus the mole fraction of Y must equal unity, or, if expressed as a percentage (mole percent), must equal 100. In the instance of three or more components, the denominators of the prior expressions must reflect the additional moles present.

In considering a solution containing 50 grams of methyl alcohol (m.w., 32) in 1,000 grams of H_2O (m.w., 18), the mole percent of each component will be:

Moles of CH_3OH = 50/32 = 1.562

Moles of H_2O = 1,000/18 = <u>55.556</u>

Total Moles   57.118

Mole percent CH_3OH = 1.562/57.118 × 100 = 2.735%

Mole percent H_2O = 55.556/57.118 × 100 = 97.265%

The same solution, expressed in terms of weight percentage, is:

Weight percent CH_3OH = 50/1,050 = 4.762%

Weight percent H_2O = 1,000/1,050 = 95.238%

In nuclear chemistry, the term *mole fraction* may be used to indicate the number of atoms of a given isotope in an isotopic mixture, as a fraction of the total number of atoms of that element in the mixture.

MOLES AND SHREWS. Important members of the order of mammals known as *Insectivora*. Generally, members of this order are small animals, mostly nocturnal in habits. They live on insects and other small invertebrates and, in a few cases, on vegetation. The general organization of the *Insectivora* is given in Table 1.

Tenrecids. The tenrec is an animal of Madagascar. The species resembles the other members of the order in its compact body, short legs, and long sharp muzzle. Some specimens attain a length of 16 inches

TABLE 1. GENERAL ORGANIZATION OF TENRECS, MOLES, AND SHREWS

INSECTIVORA	
TENRECIDS	SORICIDS
Tenrecs (*Tenrecidae*)	Shrews (*Soricidae*)
Common Tenrec (*Tenrec*)	Common Shrews (*Sorex ...*)
Striped Tenrecs (*Hemicentetes*)	Short-tailed Shrews (*Blarina ...*)
Hedgehog-Tenrecs (*Setifer*)	Common Water-Shrews (*Neomys*)
Rice-Tenrecs (*Oryzorictes*)	Musk-Shrews (*Crocidura*)
Long-tailed Tenrecs (*Microgale ...*)	Forest-Shrews (*Sylvisorex ...*)
Water-Tenrecs (*Limnogale* and *Geogale*)	Fat-tailed Shrews (*Suncus*)
Giant Water-Shrew (*Potamogalidae*)	Mole-Shrews (*Anourosorex*)
Solenodons (*Solenodontidae*)	Asiatic Water-Shrews (*Chimarrogale*)
Golden Moles (*Chrysochloridae*)	Web-footed Shrews (*Nectogale*)
	Girder-backed Shrew (*Scutisorex*)
ERINACIDS	MOLES
Hedgehogs (*Erinaceinae*)	Asiatic Shrew-Moles (*Uropsilus* and *Nasillus*)
Gymnures (*Echinosoricinae*)	Desmans (*Desmana* and *Galemys*)
	Eurasian Moles (*Talpa ...*)
MACROSCELIDS	Pacific Moles (*Scapanus ...*)
	Star-nosed Mole (*Condylura*)

(41 centimeters) and are clothed with a mixture of spines, bristles, and hair. Tenrecs are nocturnal burrowing animals. The geogale is a waterten-rec, a small animal of Madagascar resembling a mouse. The animal has webbed feet and a long tail. The geogale lives in and along streams and rivers. The solenodon is also a tenrecid. It is an animal of about the size of a rabbit, but with a long slender nose and a long naked tail. The claws are strong, those of the forefeet being much larger than those of the hind feet. There are two species, both of which occur in the West Indies.

Erinacids. The principal member of this family is the hedgehog. This mammal is small and compact, about 10 inches (25 centimeters) in length, with short legs and tail and a sharp nose. The entire upper surface is covered with sharp spines which protect the animal's body when it is curled up. However, a fox will roll the animal into a pool of water. To swim, the hedgehog must uncurl and thus lose much of its protection from the projecting spines. Then, the hedgehog is relatively easy for the fox to capture. Numerous species of hedgehogs live in Africa, Europe, and India; others are found in Asia north of the Himalayas. They are nocturnal animals. The European hedgehog, *Erinaceus europaeus*, is also called the urchin.

Soricids. This family of *Insectivora* is made up of several species of shrew. Shrews are small animals closely resembling mice in general appearance. They are common in Europe, Asia, Africa, and North America. Many species burrow. Some live near the water and have developed aquatic habits. A species found in Africa is characterized by the great development of the hind legs and is known as the jumping shrew. See Fig. 1. They also have very prolonged snouts, characteristic of many insect-eating mammals. The tree shrews or tupaias of the Oriental region constitute still another family of shrews. They have arboreal habits and are somewhat squirrel-like in appearance. All of these animals subsist on a diet of worms, insects, and plant shoots.

Fig. 1.   African jumping shrew. (*Painting by Charles R. Knight, American Museum of Natural History.*)

Moles. Possibly the best known of the *Insectivora* is the mole. See Fig. 2. The mole is a burrowing animal of small size, and is highly specialized for life underground. The legs are short, but the front pair of legs is powerful, with broad feet and strong claws. The eyes are rudimentary and there are no external ears. The term *mole* is used in connection with mammals other than the true moles. In the order *Marsupialia*, a single rare Australian animal adapted for subterranean life is called the pouched mole or marsupial mole. The term also appears in mole voles, mole rats, Cape mole rats — all burrowing animals of the Old World belonging to the order *Rodentia*. These animals are much less extensively adapted (specialized) than the true moles. See also **Rodentia**.

It should be pointed out that one of the best known of insect-eating mammals is the anteater. However, the anteater is a member of the family *Edentata*. See also **Edentata**. The spiny anteater is a member of Monotremeta. See also **Monotremeta**.

MOLE VOLUME.   A mole of gas will occupy a definite volume under definite conditions regardless of the nature of the gas. This definite volume

Fig. 2.   A mole. The hairless nose is sensitive to touch and is used as a guide. (*John H. Gerard from National Audubon Society.*)

is called the *mole volume*. Under a pressure of 760 torr and at a temperature of 0 °C, a gram-mole of gas will occupy 22.41 liters. This situation also applies to a mixture of gases. A pound-mole of gas will occupy 359 cubic feet at a pressure of 760 torr and at a temperature of 32 °F (0 °C).

Because the volume of one mole of gas at any specific pressure and temperature contains the same number of molecules even though there may be several different gases in the mixture, the percent by volume of any given gas is equal to the percent pressure exerted by that gas and is also equal to the mole percent of that gas. Mole percent equals volume percent equals pressure percent.

MOLLIER CHART.   A thermodynamic diagram for a homogeneous system possessing two independent properties, in which enthalpy is the ordinate and entropy is the abscissa. Mollier charts are used widely in engine calculations, particularly those in which the working fluid is steam.

MOLLISOLS.   See Soil.

MOLLUSCA.   A major division of the animal kingdom containing the snails, oysters, clams, mussels, squids, octopus, nautilus and related forms. Mollusks are the most highly developed of the unsegmented invertebrates and are both diverse in form and numerous in species.

The phylum is characterized by the following structures. (1) The body is unsegmented. (2) A well-developed head is found in most species. (3) The body bears a ventral muscular protuberance, the foot. (4) A fold extends in most species from the dorsal wall, enclosing a cavity associated with respiration. The fold is the mantle and the cavity, the mantle cavity. (5) In many species the mantle secretes a shell. (6) The circulatory system consists of tubular vessels and open spaces, with a heart made up of a ventricle and two auricles.

Some mollusks are important as food. Clams, oysters, and scallops are the most familiar of the edible species but others are eaten. Pearls and mother-of-pearl are also molluscan products.

The phylum is divided into the following classes:

Class *Amphineura*. Without a distinct head. Shell absent or composed of a series of plates. Chitons.
Class *Gasteropoda*. With a distinct head. Shell absent, conical, or spiral. Snails and related forms.
Class *Scaphopoda*. Head indistinct. Shell cylindrical.
Class *Lamellibranchiata* (*Pelecypoda*). Head indistinct. Shell of two lateral parts (bivalve). Clams, mussels, oysters, etc.
Class *Cephalopoda*. Head distinct, with long tentacles. Squids, octopus, nautilus, etc.

Fig. 1.   Mollusk shells. (*A.M. Winchester.*)

Several mollusks are described under separate alphabetical entries in this encyclopedia. See illustration of mollusk shells in Fig. 1.

MOLLUSCUM CONTAGIOSUM.
An orthopoxvirus infection characterized by multiple, painless, umbilicated nodules (2–5 mm diam) which can appear anywhere on the body except the palms and soles. The nodules, which are most commonly found in anogenital regions, rupture easily and may spread by sexual routes. Spread may also occur by close family contact under conditions of poor hygiene. Lesions may clear rapidly or persist for up to 18 months. When near the eye, lesions may be complicated by chronic conjunctivitis or superficial keratitis. Lesions resolve spontaneously without scarring and no antiviral therapy or vaccine is available.

MOLLUSKS.
The mollusks are greatly diverse in structure and represented by many curious forms, from the highly active squids to the slow and sluggish snails. Yet all of these creatures have one or more basic morphological and embryological features that unite them into the phylum *Mollusca*. The mollusks are the second largest animal phylum, with thousands of species. Only a relatively few of these species are of importance as food substances—either as food for human consumption, or as bait for catching marine fishes, or for use as fertilizer. About 40% of the total world catch of mollusks is cephalopods, mainly squids, which are pelagic. The catches of the benthonic octopuses are less important. The gastropods only represent 1% (abalone and some small snails). The bivalves are by far the most important resource of mollusks, with 59% of the total. In fact, most of the landings result from farming (oysters, mussels, clams), but the natural types are also important, for instance, scallops. The nomenclature and taxonomy of the mollusks tends to be complex, this accentuated by numerous changes in systematics over the years. Table 1 is offered to provide some assistance in deciphering the nomenclature.

Bivalves
From a commercial standpoint, the bivalves make up the largest economic class of mollusks. The bivalves are one of the most peculiar groups of animals. There is no recognizable head section and, therefore, it is difficult for a non-expert to differentiate between the front and back of the animal. Although most bivalves are able to move freely, one only rarely sees them changing their position or locality. The bivalve's body is surrounded by the laterally enlarged mantle folds and the two shell plates it secretes. An uncalcified ligament holds the two shell valves together without influencing their movement against each other. Along the hinge line, the two shell halves usually possess interlocking teeth. Muscular action is responsible for the tight and firm closure of the bivalve shell. A diagram is given in the entry on **Clam**.

Clams. The external appearance of the soft-shell clam is an irregular ellipsoid with the two valves or shells of nearly equal size. The right shell is usually slightly larger than the left. When closed, the two valves may touch along the ventral edge but gape widely in front, allowing passage of the foot, and in the rear, allowing extrusion of the siphons. The outer surface of the shell is covered with rough striations. The annular marks or "rings" result from the change in rate of growth during the year. Since cold winter temperatures restrict clam growth, the rings can indicate the age of the clam. The interior of the shell is relatively smooth, but has certain areas scarred at the points where major muscles are attached. Along the dorsal edge of each shell are the prominent hinge teeth; one valve carries a large, projecting, spoon-shaped structure, and the other valve has an inverted bowl. The bowl lies directly over the spoon. Between the two structures is a tough, rubber-like ligament that acts to keep the shells apart by opposing the closing action of the adductor muscles, much as a rubber eraser squeezed in a door hinge would force the door open.

Unlike the higher animals, the clam has blood circulating in an open system. The two-chambered heart pumps blood through a system of arteries that eventually empty into large, open lacunae. The blood is returned to the heart by a loose network of veins. Considerably more detail on the physiology of the clam can be found in the Hanks reference.

Life for a soft-shell clams begins as sperm and egg join in the coastal waters. The fertilized egg develops into a swimming trochophore larva in about 12 hours in cold New England water, probably sooner in warmer, southern waters. The larva moves by hair-like projections (*cilia*) arranged in distinct bands around the body. The larva has a mouth and a minute shell gland, which gives rise, within the following 24 to 36 hours, to the two calcified valves that envelop and protect the clam body throughout its life. After further physiological changes, the clam reaches a setting stage when metamorphosis is possible and the final adaptation to sedentary bottom existence can begin. This period is extremely critical in the life cycle for, once removed from transportation by water, the clam must establish itself within a relatively small area and its future success is influenced by the type of bottom material on which it settles.

Upon setting, the clam immediately attaches to sand grains, plants, or other materials, by a tough, horny thread (*byssus*). At the onset of cold weather, the young clam burrows and then becomes less active through the winter. Spring tides tend to move many clams from their winter burrows.

The soft-shell clam is found in bays, coves, estuaries, and other protected areas. It lives in a wide variety of sediments, ideally a mixture of mud and sand where it is less likely to be exposed to predation and climate. Clams live deeply buried in the bottom sediments, and the greatly extensible siphons may reach lengths of more than four times the length of the shell, drawing water and food from just above the bottom. Microscopic plants, animals, and other organic food substances are filtered from the water, possibly at a rate of about 45 liters (12 gallons) per day. Most soft-shell clam populations in New England are found in intertidal flats, exposed at low tide, whereas most clams in the Chesapeake Bay are in subtidal regions and are seldom exposed. In productive areas, clams occur in vast but erratically distributed beds.

Sexual maturity may be reached at one year of age (at a length of about 12 to 19 millimeters; 0.5 to 0.75-inch) in northern waters. The sexes are separate and can be determined only by microscopic examination. Each clam can produce millions of eggs or billions of sperm. Sex cells are exuded from the gonads through the siphon in successive puffs. Spawning is temperature-related and occurs from June to mid-August in northern areas. Clams north of Cape Cod have only one reproductive cycle per year, but south of Cape Cod, spawning may occur twice each year.

Southern clams may reach a shell length of 5 centimeters (2 inches), an acceptable commercial size, within 1.5 to 2 years. Maine clams require from 5 to 6 years to achieve this size. The green crab (*Carcinus maenas*) is the most serious predator (other than humans). Mass mortalities of some

TABLE 1. ABRIDGED CLASSIFICATION OF EDIBLE MOLLUSKS

Common Name	Genus And Species	Family	Order	Subclass	Class
CLAMS Soft shell clams	*Mya arenaria* *Mya truncata*	*Myidae*	*Myoida*	*Heterodonta*	*Bivalvia* (The bivalves)
Surf clams	*Spisula solidissima*	*Spisula*	*Veneroida*		
American quahog	*Mercenaria mercenaria*	*Veneridae* (Venus clams)			
Razor clam	*Solen marginatus* *Solen pellucidas*	*Solenidae* (Razor shells)			
Ocean quahog	*Arctica islandica*	*Arcticidae*			
Common edible cockle Devon cockle	*Cerostoderma edule* *Cerostoderma aculeatum*	*Cardiidae* (Cockles)			
MUSSELS Mediterranean mussel Blue mussel California mussel Asian mussel Philippine mussel Indian mussel New Zealand mussel Brazilian mussel	*Mytilus galloprovincialis* *Mytilus edulis* *Mytilus californianus* *Mytilus crassitesta* *Mytilus smaragdinus* *Mytilus viridis* *Mytilus canaliculus* *Mytilus perna*	*Mytilidae* (Marine mussels)	*Mytiloida*	*Anisomyaria*	
SCALLOPS American scallop Pilgrim's scallop Giant scallop	*Pecten magellanius* *Pecten jacobaeus* *Pecten maximus*	*Pectinadae* (Scallops)	*Pteroidea*		
OYSTERS Common European oyster Japanese oyster Portuguese oyster American oyster Chilean oyster	*Ostrea edulis* *Ostrea lurida* *Crassostrea gigas* *Crassostrea angulata* *Crassostrea virginica* *Crassostrea commenalis* *Crassostrea chilensis*	*Ostreidae* (True oysters)			
GASTROPODS Abalone	*Haliotis refuscens* *Haliotis gigantea* *Haliotis tuberculata*	*Haliotidae* (Abalones)	*Archaeogastropoda* (Primitive univalves)	*Prosobranchia* (Front-gilled snails)	*Gastropoda* (The univalves)
Giant conch Queen conch Crown conch	*Syrinx goliath* *Syrinx gigas* *Melongena melongena*	*Melongenidae*	*Caenogastropoda* (Modern univalves)		
Louisiana conch	*Thais haemastoma*	*Muricidae*			
Escargot snail	*Helix pomatia*	*Helicidae*	*Stylommatophora* (Land snails)		
CEPHALOPODS Common squid No. American common squid No. European squid Opalescent squid	*Loligo vulgaris* *Loligo pealei* *Loligo forbesi* *Loligo opalescens*	*Loliginidae*	*Decabrachia*	*Dibranchiata*	*Cephalopoda* (Squids and Octopods)
Giant squid	*Architeuthis dux* *Architeuthis harvey* *Architeutis princeps*	*Architeuthidae* (Giant squids)	*Teuthoidei*		
Short-finned squid	*Illex illecebrosus*	*Ommatostrephidae*			
Common octopus	*Octopus vulgaris*	*Octopodidae* (Octopuses)	*Octobrachia*		
Common cuttlefish	*Sepia officinalis*	*Sepiidae* (The cuttlefishes)	*Decabrachia*		

soft-shell clam populations have been reported from time to time, but causes are poorly understood.

Coastlines in the Northern Hemisphere on both sides of the Atlantic and Pacific Oceans support large numbers of soft-shell clams. The fishery on the Atlantic coast of North America is of the greatest importance, partly because of the heavy demands for these products in the nearby northeastern United States. The soft-shell clam (*Mya arenaria*) is found from the coast of Labrador to the region of Cape Hatteras, North Carolina. On the European coast, it has been recorded from northern Norway to the Bay of Biscay. In the late 1870s, young clams were accidentally introduced into San Francisco Bay, with shipments of seed oysters from the Atlantic coast. Since that time, the clams have spread from Monterey, California to Alaska. The soft-shell clam is also found along the western Pacific coast from the Kamchatka Peninsula to the southern regions of the Japanese islands. The southern range limits are well defined, but the northern limits are less marked because of the presence of a closely related species, *Mya truncata*, a northern form of which some populations exhibit similar external shell morphology.

Surf Clam. The Atlantic surf clam (*Spisula solidissima* Dillwyn) is estimated to have lived on the earth for many millions of years. The oldest known fossils are from deposits of the upper Miocene age in North Carolina. Surf clams are the largest bivalve mollusks in the region; some have a maximum length of over 20.3 centimeters (8 inches). Their shells are a familiar sight along ocean beaches after storms. Many names are in common use for the surf clam, such as *bar clam* (Canada); *hen clam* (Maine); *sea clam* (Massachusetts); and *beach clam* or *skimmer clam* (middle-Atlantic states). The surf clam industry has gained considerable importance since the 1950s.

The surf clam is found from the southern Gulf of Saint Lawrence to the northern Gulf of Mexico. In the northern part of its range, it is commonly abundant in the turbulent waters off outer beaches, just beyond the breaker zone, and is occasionally found to depths of 126 meters (420 feet) on Georges Bank. Surf clams have been taken at depths of 60 meters (200 feet) on the middle Atlantic continental shelf, but commercially useful concentrations are in depth of 11 to 55 meters (36 to 180 feet). South of Cape Hatteras, the clams are small and inhabit shallow water.

American Quahog. This hard-shell clam (*Mercenaria mercenaria*) should not be confused with the ocean quahog described later. The American quahog is an important commercial species along the northeastern Atlantic coast of the United States. Clams of the family *Veneriidae* are sometimes collectively referred to as the Venus clams. Many years ago, the American quahog was introduced to northwestern French waters from the eastern coast of North America. The range of the American quahog is from Cape Cod southward, although it is sometimes found as far north as the Maine coast (Casco Bay) and the Gulf of Saint Lawrence at Shediac. Young and tender quahogs are sometimes called *little-neck clams*.

Ocean Quahog. The natives of Iceland have eaten this hard-shell clam for at least two centuries, where it is called *ku-skioezl* and *krokfishur*. This use of ocean quahogs by Icelanders is the basis for its scientific name, *Arctica islandica* Linne. In North America, the clam is sometimes called *black quahog* and *mahogany clam*. The *Arcticidae* were abundant in earlier geological times; fossils of several genera and over 100 species have been found, largely in the North Atlantic region.

Mussels. Mussels (*Mytilidae*) enjoy a worldwide distribution and are extremely adaptable. Their meat is tasty and contains a high percentage of vitamins, protein, minerals, and other important nutritive substances. Mussels are popular throughout Europe, and are a particular favorite in France and the Netherlands. Certain countries cultivate edible mussels. In the Mediterranean region, the Mediterranean mussel (*Mytilus galloprovincialis*) is raised. Natural occurrence of these mussels in this tidal region greatly facilitates their cultivation. The related bearded mussel (*Modiolus barbatus*) is also a valued food item in the Mediterranean countries. A principal predator of the mussel is the starfish.

Mussel culture is required when natural beds are too small to support an expanding fishery. It is a two-part process: (1) The quality of natural stocks is improved by moving them to good growing areas; and (2) when supplied of seed mussels run short, new supplies sometimes can be obtained by bringing about the settlement of planktonic mussel spat, by various methods still being developed. As planktonic spat is quite abundant in European waters, the vast size of this resource, plus the fact that mussels live on phytoplankton, gives mussel culture an outstanding biological potential. See also **Aquaculture.**

The distribution of *Mytilus edulis* and *M. galloprovincialis* ranges from the Pacific and Atlantic coasts of North America to most of Europe and the African Mediterranean coast. *Mytilus perna* (or perna perna) is found in Brazil and Venezuela, while on the Pacific coasts of the Americas, there are *Mytilus californianus* in the north, and *Choromytilus choros* and *Aulocomya* species in Chile. Along Asiatic coasts and Japan is found *M. crassitesta*. *M. smaragdinus* is found in the Philippines as well as in India, where *M. viridis* and the Brown Mussel also occur. *Mytilus viridis* also has been reported in Malaysia, and *M. canaliculus* in New Zealand.

In several of these locations, mussel fisheries exist, and the prospects of further development of them are worthy of examination. Potential mussel culture in Asia and South America appears particularly promising and important because of local protein shortages.

Data on the growth rates of mussels is not highly definitive. In the Philippines, *M. smaragdinus* is harvested at lengths of 5 to 10 centimeters (2 to 4 inches). In India, specimens of *M. viridis* of about 15 to 20 centimeters (6 to 8 inches) have been reported. Reports from Venezuela indicate that *Mytilus perna* will grow from 10 to 20 millimeters (0.4 to 0.8 inch) to 10 to 15 centimeters (4 to 6 inches) over a 5-month period, yielding 50% of flesh by weight. By comparison, *M. edulis* requires a year to grow from 10 to 75 millimeters (0.4 to 3 inches) and well over 3 years to reach the larger size in Wales. These figures suggest that mussel culture in Asia and South America has advantages.

Freshwater mussels are eaten but they are not an important source of food. They are taken in large numbers from the larger rivers of the United States for their shells, which are used in making buttons, and for the pearls they contain. As many as 50,000 tons of shells have been marketed in a single year, chiefly from the Mississippi River.

Scallops. Although the beautifully colored and sculptured shells of some scallops are much admired for their aesthetic qualities, scallops are important because of the economic value of the muscle, which is used for food. Populations of the largest American scallop (*Placopecten magellanius* Gmelin) occur from Labrador to New Jersey. Individual animals with an estimated age of 18 to 20 years and a diameter of 22.5 centimeters (9 inches) by 21 centimeters (8.25 inches) wide have been recorded in Maine. Some authorities believe that the greatest known sea scallop grounds are found between the 20- and 50-fathom curves on Georges Bank. Preceded by harvesting by American Indians, the commercial scallop fishery of Maine dates back to 1880. Scallop populations occur discretely in major estuaries and embayments from the Piscataqua River, separating Maine and New Hampshire, to the Saint Croix River, which forms the international boundary between Maine and New Brunswick, Canada. Vertical distribution in Maine waters ranges from mean low water in some areas to depths of over 100 meters in others. Concentrations of commercial importance are generally limited to the area from Penobscot Bay eastward. Sometimes, scuba divers are able to gather scallops in commercial quantities from rocky bottoms that are impossible to drag with conventional scallop gear.

Production of scallops tends to fluctuate widely from year to year. The principal lead for indication of why abundance fluctuations have taken place is the record of seawater temperatures taken at Boothbay Harbor, Maine since 1905. Since the offspring of any year's spawning—August to October—becomes of major importance to the fishery six years later, it appears from a study of temperature and production records that an association exists between seawater temperature six years earlier and highs and lows of scallop landings. Research indicates that an optimum temperature of 7.8–8.1 °C (46–46.5 °F) is required for peak landings and that temperature on either side of this narrow range seriously affects production. Some authorities, however, do not agree with this conclusion because a generally inverse relationship between Maine and New England (Georges Bank) landings suggest that seawater temperature is not an overriding influence.

In Maine inshore waters, sea scallops grow rapidly. At 6 months of age, a scallop may be only 2 millimeters (about $\frac{1}{16}$ inch) long. Within 4 years, it will have reached a size of 7.4 centimeters (2.9 inches); and by 10 years, 13 centimeters (5.1 inches).

Possibly the best-known scallop species worldwide is the pilgrim's scallop (*Pecten jacobaeus*), which reaches a size of 5 to 15 centimeters (1.9 to 5.7 inches). In decades past, many crusaders returning from the Mediterranean carried these shells on their hats or garments—thus the name for the scallop.

The giant scallop (*Pecten maximum*) is well known in the Mediterranean region. This scallop is also found in the North Sea.

In general, the scallop is mobile. A series of prominent, highly developed, deep-blue eyes are situated among the tentacles along the mantle margin. The location and degree of development of these visual organs can vary with species and within a species. The pilgrim's scallop detects the difference between light and dark as well as motion. The animal thus can perceive approaching starfishes or octopuses; in addition, it can recognize its enemies by their scent. When endangered, the scallop escapes by its thrusting manner of swimming. However, the visual acuity of these bivalves is not sufficient to help them orient toward a source of light during swimming.

Oysters. Oysters (*Ostreidae*) are found worldwide in a broad band between latitudes 64 °N and 44 °S. They are distributed from the intertidal zone down to approximately 39 meters (130 feet). Commercial edible oysters belong to two genera (*Ostrea* and *Crasostrea*). The more important species are listed in the table. Pearl oysters are species of different families found only in warmer oceans, while the edible oysters include a European species (*Ostrea edulis*), a species of the Atlantic coast of North America (*O. virginica*), and the Pacific coast oyster (*O. lurida*). The common oyster of commerce was transplanted successfully to the Pacific coast many years ago. All are marine.

The soft body of an oyster is covered by two shells or valves. The right or top valve is flat, and the left or bottom valve is heavier and cupped. An oyster attaches and usually rests on its left valve. The two shells are joined together at the hinge by an elastic material known as the ligament. The shape of the shell is highly variable in *Crassostrea* and subcircular in *Ostrea*. More than 95% of the shell is calcium carbonate. Growth of the shell depends directly upon water temperature. Along the east coast of North America, the growing season is longer in the south than in the north. In Canada and New England, the season of shell growth is from 4 to 5 months; in the Chesapeake Bay, 6 to 7 months; and in the warm waters of Florida, oysters grow nearly all year. Oysters in northern waters require about five years to reach market size (10 centimeters; 4 inches). In Florida, this size is attained in some areas within 18 months.

Oyster meats are said to contain about every element found in seawater. The amount of meat yielded varies with geographic location and season. Meat quality drops during the spawning season. After spawning, oysters build up glycogen and are in their best condition at the height of this buildup. These seasonal changes are sometimes measured by calculation of percentage solids (dry weight of meat multiplied by 100 divided by weight of meat). In the spring, glycogen is converted into sex chemicals.

In American waters, the two principal predators are the Atlantic oyster drill (*Urosalpinx cinera*), and the thick-lipped drill (*Eupleura candata*). These are carnivorous gastropods. Certain species of snails are also detrimental to oysters. The type of predator is largely determined by water salinity. Numerous diseases affect oysters, as well as parasites.

For a successful industry, seed oysters must be readily available. Methods of collection vary with region. The Japanese catch seed oysters on shells suspended from rafts or on shells draped over wooden racks. In France, large quantities of seed are caught on lime-coated tiles, which are laid in rows in the intertidal zone. Australian seed oysters are caught on sticks. In the United States, most oysters are caught on shells scattered over the bottom. There is also some production of seed by suspending strings and bags of shells from rafts.

The development of off-bottom culture in Japan during the 1930s caused oyster production to increase at a rapid rate in that country. Presently, most areas in Japan, especially in protected bays, are being utilized. There is considerable potential for greater oyster production in South Atlantic waters.

Cockle. Of the family *Cardiidae*, the cockles are worldwide in distribution and contain some 200 species. The common edible cockle (*Cardium edule*) attains a length of 3.5 to 5 centimeters (1.4 to 1.9 inches). This is one of the most common and best-known bivalves in Europe. Shells of this form have been found in superabundance swept up on sandy beaches from Iceland to western Africa. The animals are characterized by whitish-yellow shells with regularly spaced ribs. There is an exhalent and inhalent siphon. The long foot is bent and tapers to a point. It enables the bivalve to move in a peculiar way. The animal extends the foot as far out of the shell as possible, up to nearly 5 centimeters (2 inches) and gropes for resistance from any object. When the bent portion of the foot is suddenly jerked into a straight position, which results in a push, the bivalve flings itself a distance of over 50 centimeters (20 inches). Some cockles possess eyes. The common cockle occurs along the coasts of the British Isles and is also found in the Baltic Sea. It is well regarded as food in various regions. A number of nonedible American varieties occur in the Cape Cod and Long Island waters.

Univalves

The univalves or *Gastropoda* are snails. Of this class of mollusks, probably the two most important species for food are the abalone, a marine species that inhabits the rocky shores of all continents and among many of the

islands in the Pacific, Atlantic, and Indian Oceans; and the edible snail found in many parts of the world.

Abalone. All gastropods undergo a twisting, or torsion, early in their development, wherein the visceral mass rotates 180° counterclockwise and the gills, mantle cavity, and anus, which originally faced to the rear, come to lie behind the head. Abalone development typically proceeds from a short (3 to 6 days) pelagic free-swimming stage to a benthonic creeping phase and finally to a more sedentary and retiring adult existence.

The greatest concentrations of abalone populations, both in numbers of species and inhabitants, are off the coasts of Australia, Japan, and western North America. The largest species (*Haliotis rufescens*) occurs on the California coast; mature individuals average between 17.8 and 22.9 centimeters (7 and 9 inches). Some may exceed 28 centimeters (11 inches) in diameter. The next largest species (*H. gigantea*) occurs near Japan and attains a length of about 25.5 centimeters (10 inches). Other large haliotids live off south Australia, New Zealand, and South America.

Regional names for abalone vary widely — *abalone* in the United States; *Oreille de Mer* in France; *Ormer* or *Venus ear* in England; *mutton fish* in Australia; *paua* in New Zealand; *aulone* in Mexico; *awabi* in Japan.

Peeled abalones are sliced into steaks approximately 1.3 centimeters (0.5 inch) thick, usually with a slicing machine. The steaks have to be pounded to break down their connective fibers. Premium prices are received for large, white-meat steaks. Occasionally abalones will have grayish colored meat. They taste just as good, but bring a lower price.

Conch. The term *conch* is of Indo-European, origin and refers to the shell of mollusks. Presently, the term is predominantly used in reference to certain large prosobranch gastropids, particularly in the families *Strombidae*, *Cassidae*, and *Galeondidae*. The most important species from an economic standpoint belong to *Strombidae*.

Distributed throughout the world in tropical and subtropical waters, this group is represented by nearly 40 species in the Indo-Pacific region. The most extensive fishery, however, exists in the Caribbean area, where some seven species of *Strombus* thrive. The largest stromb, the rare and coveted *S. goliath*, lives off the coast of Brazil and attains a length of over 33 centimeters (13 inches). It is sometimes available in native markets near Recife. The queen conch (*S. gigas*), also known as pink-lipped conch, is the most important commercial species in the family and is used as an important source of protein by natives of Haiti and the Bahamas. The ornamental values of the conch shell are also highly regarded and exploited. Pearls also are formed from the mantle of the species.

The usual habitat of adult *S. gigas* is in beds of turtle grass on sandy bottoms. The animals have been found in depths to 60 meters (200 feet), but normally occur from the low-tide line to 30 meters (100 feet) because the plants in which they are found are limited to these depths. Almost exclusively herbivorous, *S. gigas* prefers a diet of algae and will move about to find better grazing areas — nearly 300 meters (1000 feet) in a relatively short period. *S. Gigas* reaches maturity in about 2 years. Some individuals live up to 10 or 25 years of age.

In Florida and along the Atlantic seaboard, *Busycon carica* and *B. canaliculatum* of the family *Galeodidae* are often referred to as conchs.

Edible Snail. Commonly referred to as escargot, the *Helix pomatia* inhabits vineyards and all regions that are not too moist, particularly brushy. With the onset of winter, the edible snail burrows into loose soil to a depth of nearly 30 centimeters (1 foot). The shell aperture becomes covered by an epiphragm, and thus the animal survives the cold season. *Helix pomatia* becomes active again with the warmth of spring, recuperates, and above all equalizes its loss of water. Crosscopulation occurs during the moist days of May or June. Edible snails produce sperm throughout the entire warm season, but eggs for only a limited time.

Even in ancient times, the *Helix* snail was a favorite food item, and the edible snail also played a role in folk medicine.

Cephalopods

The *Cephalopoda* represent the most complex of the classes of phylum *Mollusca*. One authority has stated, "From the standpoint of complexity of structure and behavior, cephalopods stand at the apex of invertebrate evolutionary development." The cephalopods include squid, octopus, and cuttlefish, all important either as direct food substances for human consumption, as bait, or as fertilizers for growing crops. Japan leads the world in the fishing of squid, and the great majority of the catch is a single species. Canada, especially Newfoundland and Labrador, is next in importance to the Orient. A single decapod species dominates, i.e., *Illex illecebros Illecebrosus* Leseur. Spain exceeds Canada in total catch of all cephalopods. See also **Cephalopoda**.

Squid. The short-finned squid (*Illex illecebrosus*) is fished commercially in Newfoundland, the entire catch being taken close to shore, particularly in Conception, Trinity, and Bonavista bays. In recent years, the squid have been used for bait in the local line fishery for cod, and exported to Portuguese, Norwegian, and Faroese ships longlining in the northwestern Atlantic for cod and sharks. Large quantities of dried squid were once exported to China for the human diet. Most of the world's commercial catch of squid is taken in Japan.

Octopuses. The octopod fishery is more specialized. Fishing is still largely a hand operation. In general, octopods are docile and timid in manner and behavior, as contrasted with the decapods (squid) which are aggressive and extremely active.

Additional Reading

Boon, D.D.: "Coloration in Bivalves," *J. Food Sci.*, **42**, 4, 1008–1015 (1977).

Boss, K.J.: "Conchs," in "The Encyclopedia of Marine Resources," (F.E. Firth, editor) Van Nostrand Reinhold, New York, NY, 1969.

Davies, G.: "Mussels as a World Food Resource," in "The Encyclopedia of Marine Resources," (F.E. Firth, editor), Van Nostrand Reinhold, New York, NY, 1969.

Ebert, E.E.: "Abalone," in "The Encyclopedia of Marine Resources," (F.E. Firth, editor), Van Nostrand Reinhold, New York, NY, 1969.

Edmunds, W.J. and D.A. Lillard: "Sensory Comparison of Aroma Precursors in Marine and Terrestrial Animals," *J. Food Sci.*, **42**, 3, 843–844 (1977).

Grzimek, B.: "Grzimek's Animal Life Encyclopedia," Vol. 3, "Mollusks and Echinoderms," Van Nostrand Reinhold, New York, NY, 1974.

Hanks, R.W.: "Clams," in "The Encyclopedia of Marine Resources," (F.E. Firth, editor), Van Nostrand Reinhold, New York, NY, 1969.

Sidwell, V.D. et al.: "Composition of Edible Portion of Raw (Fresh or Frozen) Crustaceans, Finfish, and Mollusks," *Marine Fisheries Review*, **36**, 3, 21 (1974).

Staff: "Atlas of the Living Resources of the Seas," Food and Agriculture Organization (United Nations), Rome, 1976.

Staff: "Yearbook of Fishery Statistics," Food and Agriculture Organization (United Nations), Rome (Issued annually).

MOLYBDENITE. The mineral molybdenite is sulfide of molybdenum, MoS_2. Its hexagonal crystals are usually tabular to short prismatic, but if in massive form it may be foliated or granular. Has a perfect basal cleavage; is sectile; hardness, 1–1.5; specific gravity, 4.52–5.06; luster, metallic; color, very slightly bluish, lead gray; streak, greenish-gray; opaque. Molybdenite is one of the few minerals soft enough to give a distinctly greasy feel. Molybdenite is found as a contact mineral with cassiterite and wolframite, in granite pegmatites and sometimes in granites, syenites, or gneisses. It is found associated with tin ore in Saxony and Bohemia; in Norway; England; Australia; and in the United States in Colorado, Washington County and Oxford County, Maine, in New Hampshire, Connecticut, Pennsylvania, and Washington. Its name, derived from the Greek meaning lead, was formerly applied to minerals containing lead, to graphite and to molybdenite as well. Later the term was restricted to the latter mineral. It is an ore of molybdenum.

MOLYBDENUM (In Biological Systems). Molybdenum is required in very low amounts by both plants and animals. Nutrient imbalances involving molybdenum and copper have caused serious problems in cattle and sheep production.

Molybdenum deficiencies are found in plants grown on certain acid soils, and sometimes the deficiency can be corrected by adding either small quantities of manganese compounds or larger quantities of limestone to the soil. The limestone makes the soil more alkaline and increases the availability of the native molybdenum in the soil. In certain parts of the world (including the eastern United States), small amounts of molybdenum fertilizer are used regularly for producing some vegetables, notably cauliflower. In Australia, large areas have been changed from near-desert conditions to productive agriculture through the application of molybdenized superphosphate.

In alkaline soils, molybdenum is more available to plants. Forage crops growing on some alkaline soils (as in the western United States) may take up high concentrations of molybdenum. The element is not toxic to the plants. They grow normally and may produce excellent yields. But cattle and sheep that eat these forages may suffer from molybdenum toxicity. It is now well established that what appears to be molybdenum toxicity is actually a copper deficiency that is induced by the molybdenum. Thus, the symptoms of molybdenum toxicity are the same as those of copper

deficiency and include fading of the hair and diarrhea. The condition may be prevented by supplementing the animal diet with extra copper, or by injecting copper compounds into the animal body, usually by an experienced veterinarian. Cattle are more susceptible to molybdenum-induced copper deficiency than other types of livestock. Horses and pigs are rather tolerant of high levels of dietary molybdenum.

High levels of molybdenum are generally considered to be 20 parts per million (ppm) or more in dry forage. Some symptoms of interference with copper metabolism in cattle may be evident when the forage contains as little as 5 ppm molybdenum if the forage is also low in copper. The effects of high-molybdenum forage in interfering with copper metabolism in animals are generally more severe if the animal diet is also high in sulfates.

In terms of humans, some research in New Zealand and the United Kingdom indicates that diets containing moderately high levels of molybdenum help to prevent dental decay. The high-molybdenum soils in the United States are seldom used for production of food crops and thus the effects of molybdenum toxicity from food substances are not well known.

Restriction of the molybdenum intake by young rats in a synthetic purified casein diet results in a decreased level of tissue, particularly small intestinal, xanthine oxidase. The enzyme levels are restored to normal by the inclusion of sodium molybdate and other molybdate compounds. Sodium tungstate is a competitive inhibitor of molybdate, and dietary intakes of tungstate greatly reduce the molybdenum and xanthine oxidase concentrations in tissues.

Legumes, cereal grains, and some green leafy vegetables are good sources of molybdenum, whereas fruits, berries, and most root or stem vegetables are poor sources. Vertebrate tissues are generally low in molybdenum, with concentrations in liver and kidney being higher than in other organs and cells. Excess molybdenum intake by cattle causes the disease known as "teart," characterized by severe diarrhea and degradation of general health.

For many years, it has been established that molybdenum is a catalyst for biological nitrogen fixation, that is, reducing molecular nitrogen to ammonia by a nitrogenase which exists in soil- and water-dwelling microorganisms. This was considered an exclusive process. However, in 1985, some British investigators (University of Sussex) confirmed that a second nitrogen-fixation scheme, one not involving molybdenum, also exists. Their experiments were conducted with *Azotobacter vinelandii*. As reported by Marx, the genes that code for the enzymes of the alternative system are being sought. Recent work has shown that the alternative system is activated in soils where there is a molybdenum deficiency. The detailed role of Mo in the principal pathway still is not well understood. Whether or not further research may lead to better growth of some plants in molybdenum-deficient soils remains unanswered.

Additional Reading

Allaway, W.H.: "The Effect of Soils and Fertilizers on Human and Animal Nutrition," Cornell University Agricultural Experiment Station, Agriculture Information Bulletin 378, U.S. Department of Agriculture, Washington, DC, 1975.

Braithwaite, E.R. and J. Haber: "Molybdenum: An Outline of Its Chemistry and Uses," Elsevier Science, New York, NY, 1994.

Considine, D.M. and G.D. Considine: "Foods and Food Production Encyclopedia," 633–634, 674, 1307–1308, Appendix 2, Table 1, Van Nostrand Reinhold, New York, NY, 1982.

Kirchgessner, M. (editor): "Trace Element Metabolism in Man and Animals," Institut für Ernahrungsphysiologie, Technische Universität München, Freising-Weihenstephan, Germany, 1978.

Lewis, R.J. and N.I. Sax: "Sax's Dangerous Properties of Industrial Materials," 10th Edition, John Wiley & Sons, Inc., New York, NY, 1999.

Marx, J.L.: "Fixing Nitrogen without Molybdenum," *Science*, **229**, 956–957 (1985).

Mertz, W.: "The Essential Trace Elements," *Science*, **213**, 1332–1338 (1981).

Underwood, E.J.: "Trace Elements in Human and Animal Nutrition," 5th Edition, Academic Press, Inc., San Diego, CA, 1990.

MOMENT. In physics and engineering, the term moment denotes the product of a quantity and a distance to some significant point connected with that quantity. The principal moments are moments of forces, moments of lines, moments of areas, and moments of masses. Two types of moments are statical moment and the moment of inertia. Unless specifically stated to be otherwise, the word moment would be taken to mean statical moment. A physical picture of moment may be obtained by considering the moment of a force (called torque). It is the magnitude of the force multiplied by the moment arm which is perpendicular dropped from the moment center to the line of action of the force. This moment is the turning effect on a body against which the force is applied. The moment of an area is the magnitude of the area multiplied by the perpendicular distance from the centroid of the area to the axis of moments. Similarly, the moment of a solid is its weight multiplied by the distance from its center of mass to the axis of moments. See also **Statics**.

MOMENTUM.   The momentum of a body is the product of its mass and linear velocity, while the moment of momentum (or angular momentum) of a body is the product of its moment of inertia and angular velocity. Thus linear momentum is MV, and angular momentum is $I\omega$. Both linear and angular momentum are vector quantities.

$$M = mass$$

$$I = moment\ of\ inertia$$

$$\omega = angular\ velocity$$

Because of the relation of momentum to force as set forth in the second of Newton's laws, this is a fundamental concept of dynamics.

A body tends to continue unchanged in momentum unless acted upon by external forces such as applied working forces, resistance, friction or air drag. The force F acting on a body for t seconds alters its momentum by Ft. The product Ft is known as the impulse, with a free body equal to change of momentum. In rotation, the corresponding quantity is the moment of impulse, that is, the product of the time by the applied torque or force moment.

The Law of Conservation of Momentum forms one of the basic cornerstones in physics and engineering. Its application is all-pervading, from the motion of the stars to the encounter and scattering of molecules, atoms, and electrons. The Law applies to either one particle or to a system of particles. See also **Conservation Laws and Symmetry**.

MONAZITE.   The mineral monazite is essentially a phosphate of the rare-earth metal cerium. (Ce, La, Nd, Th)PO_4; but other rare-earth metals are usually present. So constant is the presence of thorium that monazite is the chief source of thorium dioxide. It is monoclinic, but found ordinarily as translucent yellow to brown grains with a resinous luster, often as sand. Its hardness is 5.0–5.5; specific gravity, 4.6–5.4. Monazite is found in granites, pegmatites and similar rocks, but rarely in any concentration. The commercial deposits are residual sands. The Ilmen Mountains in the U.S.S.R., Norway, India, Madagascar, the Republic of South Africa, and Brazil are well known for their monazite deposits. In the United States monazite is known in Connecticut, New York, Virginia, North Carolina and Idaho. Monazite derives its name from the Greek word meaning solitary, in reference to the relative rarity of this mineral.

MONGOOSE.   See **Viverrines**.

MONKEYS AND BABOONS.   As shown by the organization of the monkeys and baboons in Table 1, there are two large families, the *Ceboids* and the *Simioids*. These mammals, in terms of position among the primates, rate just below the anthropoids, the latter including the lesser apes, gibbon, greater apes, gorillas, chimpanzees, and orangutans. The *ceboids* are also sometimes called the New World monkeys because they are found only in tropical America. The *simioids*, on the other hand, are sometimes called Old World monkeys because they are found in Africa and Asia and occur nowhere in the natural state in the New World. See Fig. 1. There is another group of small monkeys of Central and South America that constitute the family of *Hapaloids*, commonly known as marmosets. See also **Marmoset**.

General characteristics of monkeys, realizing that there are exceptions, include: (1) one pair of pectorally placed mammary glands, (2) frequent dilation of the cheeks into pouches for storing food; (3) the presence of two incisors on either side of each jaw, and well-developed, large canine teeth, the teeth being well adapted to crushing vegetables and fruits; (4) limbs that are of nearly equal length; (5) when present, the thumb (pollex) of the forelimbs is opposable to the other fingers; and (6) the great toe is opposable to other digits of the foot, permitting the feet, in essence, to perform as hands.

The *ceboids* generally have a broad partition in the nose; they possess an extra premolar on each side of the jaw; most species have a long, prehensile tail, providing much assistance in climbing and agile movement in tree tops. The face is free of hair; there are no callosities on the buttocks, the latter being a marked characteristic of the *simioids*. The thumb usually is not opposing, but resembles the other fingers. However, the big toe is more readily opposed to the other digits of the foot. The *ceboids* are essentially vegetarians and arboreal.

Probably no other mammal excels the spider monkey (*Ateles*) in arboreal life. This monkey was named because of the great length of its limbs in comparison with the rest of its body. The body is about 12 inches (30.5 centimeters) long and the tail is twice that length. The tail is used in assisting the legs when swinging from tree to tree. The fur usually is black, but is gray or brown in some species. Spider monkeys sometimes are used for pets and are considered intelligent. The *coaita* has a red face and is found in the Lower Amazon.

Capuchin monkeys are any of several species of South American monkeys of the genus *Cebus*, which are characterized by their moderately long prehensile tail. They differ from the other species with prehensile tails in having this organ fully covered with hair and not bare on the lower surface near the tip. This animal is also called the *sapajous*.

The Saki is a New World monkey of the genus *Pithecia*. Most of the species bear the name of the group, as the white-headed saki and the whiskered saki, but native names have been adopted for some. The hairy saki is called the parauacu and the black saki is also known as the cuxio. No members of this genus have prehensile tails.

The *simioids* are rated next to the anthropoid apes on the zoological scale. The nasal apertures are close together, the nasal septum being very narrow. With exception of the Colobine monkeys, all simioids have opposing thumbs and big toes. They have four incisors, two canines, four premolars, and six molars in each jaw. The canines are large and strong. In contrast to the ceboids, the simioids do not have a prehensile tail and,

TABLE 1. GENERAL ORGANIZATION OF MONKEYS AND BABOONS

Ceboids	Simioids
Half-Monkey (*Pithecinae*)	Colobine Monkeys (*Colobinae*)
Dourocoulis (*Aotes*)	Guerezas (*Colobus*)
Sakiwinkis (*Pithecia*)	Langurs (*Presbytis,...*)
Bearded Sakis (*Chiropotes*)	Snub-nosed Monkeys (*Rhinopithecus*)
Uacaris (*Cacajao*)	Proboscis Monkey (*Nasalis*)
Hand-tailed Monkeys (*Cebinae*)	Long-tailed Monkeys (*Cercopithecinae*)
Squirrel-Monkeys (*Saimiri*)	Guenons (*Cercopithecus*)
Capuchin Monkeys (*Cebus*)	Allen's Swamp Monkey (*Allenopithecus*)
Wooly Monkeys (*Lagothrix*)	Military Monkeys (*Erythrocebus*)
Wooly Spider Monkey (*Brachyteles*)	Mangabeys (*Cercocebus*)
Spider Monkeys (*Ateles*)	Dog-faced Monkeys (*Cynopithecinae*)
Howler Monkeys (*Alouatta*)	Macaques (*Macaca*)
	The Black Ape (*Cynopithecus*)
NOTE:	Baboons (*Papio*)
1. Ceboids are found in South America.	The Gelada (*Theropithecus*)
2. Simioids are found in Africa and Asia, south of the	Drills (*Mancrillus*)
desert areas and the Himalayas in the east. One tribe	
of Barbary Apes, the only wild monkey in Europe	
(*Macaques*) lives on Gibraltar.	

Fig. 1.   Golden monkey (*Cercopithecus mitis kandri*), a rather rare subspecies of monkey that lives in the high-altitude forests near the Virunga volcanoes of east central Africa. The habitat is near that of the mountain gorilla. There are some 100 species of Old World monkeys.

in fact, the tail may be quite rudimentary or essentially absent. They have cheek pouches. In some species, there may be no hair on the buttocks, but instead the presence of naked, hardened areas called callosities. In some species, the callosities are brilliantly colored. The simioids are considered to have superior intelligence and exhibit greater variety in food and living habits than the ceboids. The majority of pet and performing monkeys are simioids. Most species breed and otherwise do well in captivity, but are susceptible to respiratory diseases in northern climates unless suitable shelter precautions are taken.

The langurs represent a group of Old World monkeys characterized by slender build and extremely long tails. The legs are longer than the arms. They eat principally the leaves and young shoots of trees. Langurs live only in Asia and in some of the East Indian islands. Among the species of langurs that bear distinctive names are the hanuman, lutong, douc, negro monkey, leaf monkeys, bear monkey, white monkey, and purple-faced monkey. The latter species is found in Ceylon and is sometimes called the wanderoo.

Macaques are monkeys of several Asiatic and one African species. They are related to the mangabeys, but are stouter and have a slightly longer muzzle. The tail ranges from long to rudimentary. These monkeys make up the genus *Macacus* (Rhesus-like monkeys). Among the included species are the bonnet monkey, lion-tailed monkey, pig-tailed monkey, and magot. The Indian macaque is known as the rhesus monkey. A colony of macaques (also called Barbary apes) has inhabited the Rock of Gibraltar for many years. The crab-eating macaque (*kra*), *Macacus cynomolgus*, is found in the Oriental region. The mangabey is an African monkey of a small group of species, also called the white-eyelid monkeys. They are slender animals with a fairly long muzzle and long tail. The proboscis monkey, *Nasalis larvatus*, is a moderately large monkey of Borneo, characterized by the long, fleshy, and somewhat drooping nose. This organ is largest in adult males, but is relatively small and upturned in the young. The nostrils open near each other on the lower surface. The species is closely related to the langurs. The thumbless monkeys constitute a group of African monkeys of the genus *Colobus*. They are named from the reduction of the thumb, which is either entirely absent or reduced to a small projection, with or without vestigial nail. Some of the included species are called colobs, one is a guereza, and one is the king monkey. See also **Marburg-Ebola Virus Disease**.

Baboons. The true baboons are characterized by a large head with the elongate, dog-like muzzles having nostrils at the tip. They walk on all fours. These animals constitute the genus *Pipio*, but the term baboon is also applied to the gelada baboon, *Theropithecus gelada*, of southern Ethiopia, also a dog-like species although the nostrils are some distance from the tip of the snout. Baboons sometimes are called dog-faced monkeys. Among the baboons, several species are known by special names, including the mandrill, *P. spinx*; the drill, *P. leucophaeus*; and the chacma, *P. porcarius*. All are found in the Ethiopian region. Baboons can be ferocious. Individually, they are able fighters and their defenses are augmented by their habit of living in bands. A group of males is reported to be a good match for some of the larger predators.

Additional Reading

Estes, R.: "The Behavior Guide to African Mammals," The University of California Press, Berkeley, CA, 1991.
Smuts, B.: "Sex and Friendship in Baboons," Harvard University Press, Cambridge, MA, 1999.
Strum, S.: "Almost Human: A Journey into the World of Baboons," W.W. Norton and Company, Inc., New York, NY, 1990.

Web Reference

http://www.primates.com/welcome.htm (Photo gallery of monkeys and other primates).

MONKFISHES.   See **Anglerfishes**.

MONOCEROS.   An equatorial constellation that lies between Canis Minor and Canis Major.

MONOCHROMATIC.   1. Having one color, strictly one frequency or wavelength of optical radiation. Actually, no finite amount of radiation will ever be strictly monochromatic. It will, at best, contain a narrow band of frequencies.

2. By analogy, a beam of particles, such as beta-particles or neutrons, is said to be monochromatic if all the particles have the same, or nearly the same, energies.

MONOCHROMATOR.   An instrument used to supply a beam of light having some desired, narrow range of wavelengths. Although sometimes used in photochemistry experiments as a single instrument, it is usually part of a spectrophotometer. In that case, it has the following component parts: (1) an entrance slit producing a well-defined beam of heterochromatic radiation; (2) a prism or diffraction grating dispersing the incident radiation into a continuous spectrum; (3) some device to rotate the prism or grating so that the desired wavelengths of exit radiation are obtained; (4) an exit slit producing a narrow band of wavelengths. A monochromator becomes a spectrophotometer if it is preceded by a source of continuous radiation and followed by a sample holder, a detector, an amplifier, and a device for measuring the amplified output signal.

MONOCOQUE.   A type of construction, as of a rocket body, in which all or most of the stresses are carried by the skin. A monocoque may incorporate formers but not longitudinal members such as stringers.

MONOECIOUS PLANTS.   The flowers of many plants are unisexual; that is, they contain only stamens or pistils, but not both. When these two kinds of flowers are borne on the same plant, the plant is said to be monoecious. Familiar monoecious plants are oaks, corn, squash, begonia, and castor beans. The corresponding zoological term is hermaphrodite.

MONOMER.   A single molecule, or a substance consisting of single molecules. The term monomer is used in differentiation of dimer, trimer, etc., terms designating polymerized or associated molecules, or substances composed of them, in which each free particle is composed of two, three, etc., molecules.

MONOMOLECULAR LAYER.   The early work of Rayleigh, Langmuir, Hardy and others has shown that it is possible to deposit on solid or liquid surfaces films which are one molecule thick. Any such layer is called a monomolecular layer, *unilayer* or *monolayer*. See also **Molecular and Supermolecular Electronics**.

MONOTREMETA. A class of egg-laying mammals, containing the duck-billed platypus and the spiny anteaters of the Australian region. The duckbill, *Ornithorhynchus anatinus* (translated — "here is a creature with a bill like a duck"), also called platypus and duckmole, is about 18 inches (46 centimeters) long and has close fur something like that of the moles. The coat is glossy black hair with a waterproof undercoat of fur. The broad and flattened tail is covered with coarse hair, but is essentially bare underneath as in a beaver. The animal is aquatic. The muzzle is broad, flat, and naked, resembling the beak of a duck, but not horny. It is formed as part of the jaw bone and is very sensitive to touch. The animal also has a keen sense of smell. The legs are short, the web on the feet is loose, the claws are long and sharp and slightly curved. See Fig. 1. Duckbills are found in the streams of southern and eastern Australia. The animal nests in burrows in the banks and females deposit two small eggs at a time. The young, born naked, and blind, are nourished with milk secreted from glands on the mother's abdomen. There are no teats. The animal has pouches; like those of a squirrel, for collecting food. Spurs on the hind feet can inflict a poisonous venom, useful in capturing small animals, but not injurious to humans. The duckbill's voice sounds something like an angry puppy's growl. The diet consists of insects, crustaceans, and worms. The duckbill is considered shy by nature and, upon hearing the slightest noise, will disappear under water.

Fig. 1.   Duckbill or platypus.

The spiny anteater of Australia ranges from a few inches to 20 inches (51 centimeters) in length when fully grown, depending upon species. The body is covered with hair and spines. The animal has a slender snout, short legs, and strong claws. See Fig. 2. The spiny anteater is a burrowing animal. See also **Edentata**.

Fig. 2.   Australian echidna or spiny anteater.

MONSOON.   See Winds and Air Movement.

MONTE CARLO METHOD. A method for resolving problems in mathematics, statistics or operations research by the use of random sampling. For example, in the theory of integration it may be impossible to derive an explicit numerical value of a particular integral

$$I = \int_a^b f(x)\,dx$$

by classical mathematics. If the domain from a to b is sampled at random, that is to say n values x_i are chosen from the uniform distribution from a to b and $f(x_i)$ evaluated at each point, then the mean of the values of $f(x_i)$ multiplied by half the interval from x_{i-1} to x_{i+1} will tend to I as n increases.

In practice it is more usual to divide the range a to b into equal intervals and evaluate by quadrature, but if the integral is multiple a systematic lattice of points may involve too much computations and a smaller sample of random points may be easier to handle.

The method, however, has more obvious advantages in statistical situations where the integral or summatory processes cannot be explicitly written down. For example, many sampling distributions, say of a statistic t, are not derivable in tractable form; it is then possible to draw a random sample from the parent population, calculate t for each and hence to derive an empirical estimate of the distribution of t by repeating the operation a large number of times.

Analogous procedures apply to the study of a system in which the determinantal equations or inequalities can be written down but explicit solutions cannot be derived. The behavior of the system can be stimulated by feeding in values of the variables, and repeating the operation over different sets of values so as to explore the system under a variety of conditions. Such sets of values may be chosen systematically or by a random mechanism.

Usage varies, but it seems preferable to confine the term "Monte Carlo" to those cases where probabilistic sampling is employed.

Sɪʀ Mᴀᴜʀɪᴄᴇ Kᴇɴᴅᴀʟʟ, International Statistics Institute, London.

MOON (Earth's). The earth's natural satellite. The maximum distance from earth that may be reached by the moon is placed at 384,321 kilometers (238,857 miles); the minimum distance at 356,334 kilometers (221,463 miles). Allowing for the radius of the earth and moon, the least surface-to-surface distance between the two bodies is about 348,200 kilometers ($\sim$216,370 miles).

The tide-raising power of the moon upon the seas of the earth is $2\frac{1}{5}$ times the tidal influence of the sun. The force of gravity at the moon's surface is approximately $\frac{1}{6}$th that of the earth.

The diameter of the moon is 3476.6 ± 2.2 kilometers (2160 ± 1.4 miles). Other statistics of the moon are given in the entry on **Planets and the Solar System**. Data from the *Apollo* program indicate that the moon was formed 4.6 billion years ago and was strongly heated during its first few hundred million years, causing melting and formation of igneous rocks in its outer layers. Much of the outer crust, still preserved in the cratered uplands, formed from a low-density type of rock known as anorthosite or anorthositic gabbro. Large impact basins and other depressions are believed to have been flooded about 3.8 to 3.2 billion years ago by basaltic lavas generated beneath the surface, forming the dark mare plains that cover roughly 18% of the moon. Few geological processes appear to have occurred since that time, as there is virtually no atmosphere or water to permit erosive processes. The modest seismic activity of the moon tends to confirm the essential absence of geologic processes as found on earth. The rate of meteorite impact on the moon is low today (comparable to earth), but heavy cratering indicates that the rate was very high in the first half-billion years of lunar history. See Fig. 1.

Orbit and Phases of the Moon. The moon, in orbiting around the earth, exhibits a strikingly interesting phase change. Since the moon shines by reflected light from the sun (and a very small amount of earth light), the phases are easily explained, as aided by Fig. 2. The moon is called *new moon* (position 1) when it is positioned between the earth and the sun, at which time the light of the sun does not reach the side of the moon that faces the earth. At full moon (position 3), the side of the moon facing the earth is completely flooded with sunlight. This is the position where the moon has half-completed a revolution about the earth. The various other phases are caused by this revolution about the earth, with a complete revolution occurring approximately once every $29\frac{1}{2}$ days. At position 2, essentially one-half of the surface is illuminated, although at that position, the phase is termed the *first quarter* (one-fourth of a revolution completed). Between positions 1 and 2, the lighted portion of the moon takes on a crescent shape and this is known as the *crescent phase*. Between positions 2 and 3, more than half the moon surface is illuminated and the term *gibbous phase* is used. At position 3, the moon is fully illuminated (*full moon*). Between positions 3 and 4, over half of the surface is illuminated and the term gibbous phase also applies here. At position 4, half of the surface

Fig. 1. The near-full moon as photographed shortly after trans-earth injection (TEI) by the Fairchild Metric Camera mounted in the Scientific Instrument Module (SIM) bay of the Apollo 16 Service Module. This view is looking generally westerly toward the large, circular Mare Crisium (Sea of Crises) on the horizon. Immediately east of Mare Crisium is Mare Marginis (Border Sea): and the more circular mare area south of Mare Marginis is Mare Smythii (Smyth's Sea). Most of the lunar area in this photo is on the far side of the moon. The most conspicuous crater, the smooth-floored one northeast of Mare Marginis, is Lomonosov. Neper is the large crater between Mare Marginis and Mare Smythii. The larger crater, Tsiolkovsky, is barely visible on the horizon at the southeastern edge of the Moon. (*National Aeronautics and Space Administration.*)

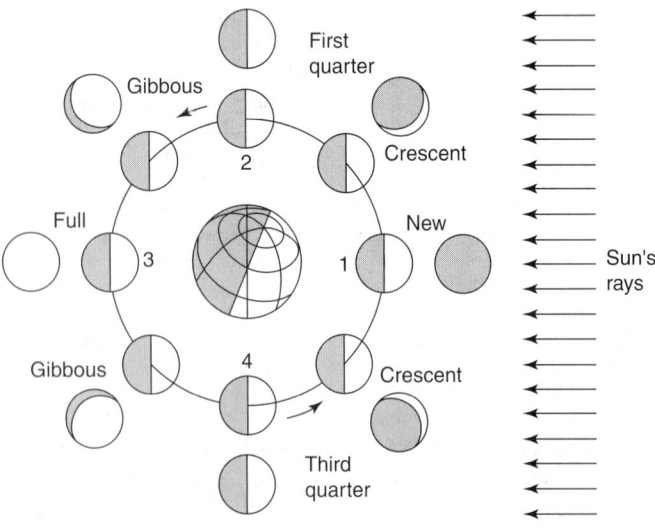

Fig. 2.   Phases of the moon.

surface will be noted to be very faintly lighted. This is due to reflection of sunlight from the earth back on to the moon and is called *earth-shine.*

The moon rotates slowly about its axis, which is tipped about 6.5° from the perpendicular to the plane of its orbit. This results in the remarkable circumstance that the period of rotation matches the moon's sidereal period of revolution and is in the same direction. Consequently, the moon always turns the same face toward the earth. The *sidereal month* is defined as the time required by the moon to complete one revolution about the earth with reference to the stars and is 27 days, 7 hours, 43 minutes, 11.5 seconds. The interval between new moon and new moon or full moon and full moon is called the *synodic month* and is 29 days, 12 hours, 44 minutes, and 2.8 seconds long.

The intensity of moonlight is quite weak even though the light of the full moon makes it possible for a person to move about on foot quite well on a clear night. The light from the full moon has an intensity of about 1/465,000th of sunlight on a clear day. It has been estimated that were the entire sky packed with full moons, limb to limb, they would give only about 1/6th the light of the sun. The intensity falls off rapidly from the full moon, either in the earlier or later phases, bearing a nonlinear relationship with the area of the moon illuminated. Illumination from either the first or third quarters is only about 1/10th as intense as that from the full moon. See also **Eclipse.**

Total lunar eclipses, which occur when the moon passes completely into the Earth's shadow, occur on the average of about once a year. Kepler suggested in 1604 that the moon is visible during the eclipse because sunlight is refracted (and to a much lesser extent, scattered) into the shadow by the Earth's atmosphere. Rays of sunlight passing between 5 and 25 km above the Earth's surface are most effective at illuminating the eclipsed moon. As pointed out by Keen (see reference listed), the brightness of the moon will be affected by the refractive properties of this layer, the presence of absorbing media such as clouds, ozone, and aerosols in and above the layer, and, of course, the position of the moon within the shadow. Large differences in brightness between eclipses have been noted for centuries. The *Anglo-Saxon Chronicle* described a markedly dark eclipse in 1110. Kepler reported that the moon essentially disappeared during an eclipse in 1588, which he attributed to mists and smoke in the Earth's atmosphere. It was not until 1883, however, that the relative brightness of the eclipsed moon was related to volcanic aerosols. During that year, the eruption of Krakatoa occurred. Researcher Keen and others in going back into historical records have been able to relate a number of "dark" eclipses with major eruptions. In the Keen report, details of these calculations are given. Observed brightnesses of 21 lunar eclipses during the 1960–1982 period were compared with theoretical calculations based on refraction by an aerosol-free atmosphere. Results indicate the global aerosol loading from the 1982 eruption of El Chicón is similar in magnitude to that from the 1963 Agung eruption.

Because of the tipping of the moon's axis, it is possible to see a bit past the one pole for half of the month and, conversely, to see past the other pole during the other half. The maximum that can be seen is about 6.5° of latitude on the moon. This is termed *libration in latitude.*

The gravitational attraction of the earth and sun cause the moon's orbit to change. The greatest influence is exerted by the sun and the second greatest by the earth's tidal bulge.

Laser altimeter data obtained from the *Apollo 15* and *16* missions provided two elevation cross sections of the moon separated by 35° of latitude. Prior to this technique, lunar topography was determined from earth by photography and radar techniques. Shadows were measured and stereoscopic effects were accounted for to determine surface elevations. Errors were large over distances of hundreds of kilometers. Results were very poor near the limbs. Numerous independent measurements of lunar elevations were made, including the elevations of Ranger impact point, Surveyor landing sites, *Apollo* survey sites, and *Saturn* booster impact points, in addition to elevations determined from velocity-height data of the *Lunar Orbiter,* and from earth-based radar. These results were obtained from front-side observations at random points with the exception of ten far-side landmark points obtained with the *Apollo* sextant. *Apollo 16* data put a 2-kilometer bulge toward the earth. However, combined data from this and other prior data are best fit by a sphere of radius 1,737 kilometers. The offset of the center of gravity from the optical center is about 2 kilometers toward the earth and 1 kilometer eastward.

is illuminated and this phase is termed the *third quarter* (three-quarters of a revolution completed). This phase is also commonly called the *last quarter.* Between position 4 and position 1, the crescent phase is resumed until a full return to position 1 occurs. If the plane of the moon's orbit were not inclined to the plane of the ecliptic (average value, 5°9'), there would be an eclipse of the sun at each new moon, and an eclipse of the moon at each full moon.

Also, because of the inclination of the moon's orbit, there is seasonal variation in the path in the sky followed by the moon. During wintertime in the Northern hemisphere, the moon will be higher in the sky than during the summertime. Soon after the new moon, a thin crescent of the moon will be brightly visible and, on a clear night, the remainder of the moon's

Lunar Exploration

Probably the first major and highly successful step toward direct exploration of the moon was the soft lunar landing of the unmanned NASA spacecraft *Surveyor 1* on June 2, 1966. Prior to the findings of the first *Surveyor* spacecraft lander, scientists were uncertain about the moon's surface and its ability to support a landing craft. The impressive touchdown of *Surveyor* demonstrated that the lunar crust was strong and hard and that future spacecraft would not settle into deep layers of dust. Subsequent *Surveyor* missions collected important information preparatory to the launching of manned spacecraft. The next major aspect of the program was that of a manned lunar orbiter. During a two-day period (December 24–25, 1968), two U.S. astronauts made ten orbits around the moon and were the first persons to see the back side of the moon. On July 20, 1969, two U.S. astronauts (*Apollo 11* mission) set foot on the moon (Sea of Tranquility region). Five more manned missions, thus making a total of six, followed: November 18, 1969 (Ocean of Storms region); February 5, 1971 (Fra Mauro region); July 31–August 2, 1971 (Hadley Apennine region); April 20–23, 1972 (Descartes region): and December 11–14, 1972 (Taurus-Littrow region). Although the Soviet soft lunar lander *Luna 9* reached the moon on February 3, 1966, a few months prior to the landing of *Surveyer 1*, Russia did not follow through with a scientific program of lunar exploration comparable to the *Apollo* program. Costs of the U.S. lunar programs were approximately $25 billion. More than 2000 samples of lunar rocks and soil, weighing in the aggregate almost 400 kilograms (882 pounds), were returned to earth. These lunar materials were placed in storage chambers at the Manned Spacecraft Center near Houston, Texas under precisely controlled conditions. Samples for research by various investigators throughout the world were released. Elaborate care is taken not to lose track of the rocks' original orientations and shapes. The working assumption of persons in charge of the lunar samples is that they are essentially irreplaceable and that even more intelligent use of them will be made as the related sciences generally progress, but only if they are kept in a pristine state.

Lunar history turned out to be a much more subtle affair than most knowledgeable people had expected. As observed by Gillette (1972), the notion that the moon was a dead body made up of the primordial material from which the terrestrial planets coalesced now seems as quaint as the Ptolemaic idea that it was all shining crystal. It appears likely that if any bits of primordial crust do remain, they will have to be tracked down laboriously in the samples of soil, a task comparable to finding the proverbial needle in a haystack.

This process of image and rock analysis continues at various laboratories throughout the world, partly because these tasks are very demanding and partly because no new exploratory ventures to the moon are currently scheduled.

On numerous occasions researchers in the field have suggested that further data are needed from lunar exploration, in an effort to study the origin of the solar system, to build giant observatories on the far side of the satellite, and to commence further expeditions to Mars and the asteroids from the moon. More ambitious researchers are insisting that we also exploit lunar resources (minerals, etc.), as well as process certain materials under low-gravity conditions. A principal contention is using the moon as a base instead of building an entirely new space station. It has been suggested that using the moon as a platform for space astronomy would have at least three advantages. (1) Radio telescopes on the far side of the moon would be shielded from terrestrial radio emissions, greatly improving observational sensitivity and accuracy. (2) The moon provides a solid, seismically stable, high-vacuum platform for interferometric arrays. (For example, a lunar optical array might resolve astronomical details about a million times finer than those seen from Earth.) (3) The moon lies beyond the Earth's radiation belts. This lowers the background radiation environment for investigating cosmic rays, the solar wind plasma, cosmic neutrinos, and gravitational radiation, among other phenomena.

Origin of the Moon

Just as one of the major scientific incentives to explore Mars was to learn if any life existed on that planet, a major driving force for the earlier *Apollo* missions to the moon was that of attempting to determine the origin of the moon. Although the knowledge of the moon was greatly enhanced by the *Apollo* missions, the question of lunar origin was answered only very partially. While the moon is a specific situation, other planets have satellites, and it is strongly suspected that other "solar systems" exist and

also most likely will be found to have satellites. Thus, the origin of the moon is best not approached as an isolated happening, but still in some unexplained way should relate to other like happenings in the solar system. See also **Cosmology**. It is interesting to note that while there are similarities there also are inconsistencies with reference to the satellites of the other planets in the solar system. For example, Venus and Mercury have no moons; Mars has two large boulders as its only moons; the satellites of Earth and Pluto are of sufficient size to be considered sister planets; the satellites of Uranus revolve in a plane perpendicular to all others; Neptune's major satellite orbits in "reverse." The latest "giant impact" hypothesis of the moon's origin accommodates, to some degree, this variety.

Few areas of astronomy have nurtured so much controversy over such a long period of time as has been that dealing with the moon's origin. For many years, numerous investigators have researched this topic.[1]

Potpourri of Hypotheses. During the period ranging from the 1800s to the mid-1900s, three so-called classical hypotheses were discussed. During this period, there was marked vacillation among scientists regarding their support of this or that hypothesis. One or more of these classical hypotheses reigned until the more recent (circa 1975) concept that is currently called the *giant impact* was formulated. The three classical hypotheses were:

1. *Binary Planet or "Sister" Hypothesis.* Scenario: The moon was formed by condensation from a cloud of material surrounding the Earth. A corollary (proposed by Gilbert); the moon was formed from a ring of small solid particles, the first stage of which produced the lunar craters.

2. *Rotational Fission or "Daughter" Hypothesis.* This concept was based on the classically known dynamical instability of incompressible, inviscid bodies in uniform rotation. Scenario: Originally, Earth and moon were one. The outer layers of the proto-Earth were spinning too rapidly to achieve dynamic stability. Because of tidal dissipation, angular momentum was transferred from the Earth's rotation to the orbital motion of the moon. The moon commenced to recede from the Earth as the moon's angular speed (as seem from Earth) diminished. (G.H. Darwin, who conceived the hypothesis, "traced" the process back some 50 million years, at about which time he suggested the moon was only about 6000 miles (9654 km) from the surface of the Earth. At that time, the period of revolution of the moon and the Earth's period of rotation would have been about 5 hours.) The initiation of separation of the moon from the Earth was caused by solar influences, suggesting that the tides raised on the proto-Earth had a peaking time each 5-hour day. Any given place where this peaking occurred could resonate with the free oscillations, thus causing distortions sufficient to disrupt the body. In 1882, Osmond proposed a supplemental concept, suggesting that the scar left by the moon's initial separation did not fully "heal"—hence the Earth's ocean basins are the holes in the Earth left behind. This spawned the popular concept that the moon was "taken" out of the Pacific Ocean basin.

   This hypothesis was reasonably well received for a number of years, but tended to be replaced by the "capture" hypothesis proposed by Thomas See in the early 1900s. The fission concept was revived for a few years in the 1930s and, even today, probably lurks in the study halls. There are three convincing arguments against the fission hypothesis: (1) Modern calculations indicate that the spinning rate of the proto-Earth would have to be 2 hours or less in order to exceed the limit for dynamic stability. Currently, the achievement of such a high rate is considered improbable. (2) The amount of angular momentum needed for the fission instability would have to be approximately

---

[1] Several well known scientists devoted much time and effort to the problem of the moon's origin and these included, in chronological order: Edouard Roche (1873); George Howard Darwin, son of Charles Darwin (1878); Osmond Fisher (1882); Grove Karl Gilbert (1890s); Thomas See (1909); H. Jeffreys (1930); Harold Urey (1951); J.A. O'Keefe (1963); T.C. Chamberlin (1963); J.F. Simpson (1964); D.U. Wise (1969); A.E. Ringwood (1970); A.G.W. Cameron (1973); Donald Davis (1975); W. Hartman (1975); A.P. Boss (1986); and S.J. Peale (1986); among many others. Dates given are the periods of early formulation or active support for a given hypothesis. Listings of major papers given by these and other scientists on the subject are listed at the ends of the articles by Brush (1982) and Boss (1986) cited for **Additional Reading** at the end of this encyclopedia article.

four times that of the present Earth-moon system. Boss explains this critical problem in his paper (reference listed). (3) Because fission must occur through a dynamic instability, a fission origin of the moon would be apparently impossible unless the proto-Earth was nearly inviscid.

3. *Capture or "Wife" Hypothesis.* This concept was first strongly advocated by Thomas See in the early 1900s and long supported by Harold Urey. Scenario: The moon was initially formed in the outer reaches of the solar system (in the vicinity of the orbit of Neptune). As this proto-moon moved through space it lost energy and, in some way, approached Earth. (This concept removed some of the objections to the other hypotheses, in that the capture theory would not necessarily conflict with certain inconsistencies, as previously mentioned, in connection with the satellites of other planets.) Upon reaching the vicinity of Earth, the proto-moon, by a combination of gravitational and other forces, reached a balance and formed its own orbit around the Earth. In a 1949 paper, Urey suggested that the moon represents more nearly the composition of the original dust cloud relative to the nonvolatile elements than does the Earth. Urey also argued that the fact that the moon does not have a shape corresponding to isostatic equilibrium indicates that it was frozen a long time ago. Urey prepared the way for the argument that exploration of the moon could provide information about the early history of the Earth. In lunar rock analysis, the moon's oxygen isotopic composition tends to support the simultaneous formation of the Earth and moon from the same portion of the solar nebular, perhaps while in orbit around each other. As found by researchers, the ratios of all three oxygen isotopes in lunar rocks and in terrestrial rocks are essentially identical. They do differ, however, from those ratios found in meteorites. But, the correlation tends to stop here because numerous other chemical differences between lunar and terrestrial materials exist.

Giant Impact Hypothesis. Proposed by A.G.W. Cameron (Harvard University) and researched by several other scientists who used computer modeling and simulation techniques over the past decade (Benz and Slattery of the Los Alamos Laboratory; Boss, Minzum, Peale, and Wetherill of the Carnegie Institution of Washington; and others), this relatively new hypothesis avoids many of the "hangups" that characterized prior hypotheses. As described in the Boss paper referenced, the scenario is about as follows: The origin of the moon is considered within the theory of formation of the terrestrial planets by accumulation of planetesimals. That theory predicts the occurrence of giant impacts, suggesting that the moon formed after a roughly Mars-size body impacted on the proto-Earth. The impact blasted portions of the proto-Earth and the impacting body into geocentric orbit, forming a prelunar disk from which the moon later accreted. Although other mechanisms for formation of the moon appear to be dynamically implausible, fundamental questions must be answered before a giant impact origin can be considered both possible and probable. The general theory of terrestrial planet formation provides the framework in which to consider lunar formation. The theory could change in the future, in which case certain revisions may have to be made in the giant impact hypothesis. A key element in the giant impact hypothesis of lunar origin is the mass distribution of the smaller bodies, called *planetesimals*, from which the terrestrial planets formed. Because of their physical and orbital similarities, Mercury, Venus, Earth-moon, and Mars each may have been formed by the same fundamental process. Much of the information from research on terrestrial planet formation compiled to date is based on the assumption that formation took place by accumulation of dust grains rather than by the gravitational collapse of gas in the solar nebula and subsequent removal of a gaseous envelope. See also **Cosmology.**

Boss estimates that in order to deposit the angular momentum of the Earth-moon system, a Mars-size body would have had to strike Earth nearly tangentially with a relative velocity of about 10 km sec^{-1}. This is consistent with the relative velocities of giant impacts now envisioned to have occurred in the late phases of accumulation.

Also, as observed by others, in the computer simulation, the collision shatters the impactor, and the rocky debris of the mantle subsequently become separated from the iron core. The gravitational fields of the three bodies — Earth, impactor core, and impactor mantle — through interacting forces keep the mantle debris in orbit. Within a very short period (hours), the gravitation of the debris itself draws it together into a roughly spherical moon. Thus, according to the giant impact scenario, the moon consists mainly of material from the impactor's mantle minus the volatile elements,

which are vaporized by heat from the impact. The impactor core is sacrificed and it crashes back into the Earth. Because it is heavy, it soon settles to the Earth's own iron core. The estimated date for the primeval cataclysm is about 4.6 billion years ago.

Lunar Topography

Craters. Very numerous on the moon are circular depressed structures (craters) with raised rims believed to be caused mostly by meteorite impacts, although some may be of volcanic origin. At one time, the largest craters were believed to be some 100 to 200 kilometers (62 to 124 miles) in diameter, but improved observations by earth-based instruments over the last few decades and the findings of the lunar orbiter and landing programs indicated that large mare-filled basins up to 1000 kilometers (620 miles) in diameter are also impact craters. Small craters are exceedingly numerous on the moon. The smallest are microscopic pits that dot the surfaces of exposed rocks. Ejected material blasted out of craters form bright rays (or radial streaks of bright material) and secondary impact craters (formed by ejected debris hitting the surface) approach the size of the parent primary crater in some cases.

Copernicus is a bright-rayed impact crater 95 km (60 miles) in diameter. It was recognized as stratigraphically important and features associated with it were used to define the most recent major time period of lunar history. Copernicus contains several prominent central peaks that rise about 800 meters above the floor of the crater; the largest is 12×5 km (3×7.5 miles) at the base. Images of these peaks taken from lunar orbit showed them to be massive and blocky. They were initially mapped as deep-seated bedrock. Subsequently, as observed by Pieters (Brown University), studies of crater dynamics indicated that the material of the central peaks in craters the size of Copernicus had likely been uplifted from an original depth of about 10 km ($6\frac{1}{4}$ miles) by dynamic rebound of local material in the terminal stages of the impact event.

The system for naming craters is generally accredited to Riccioli, (an Italian astronomer, who in 1651 proposed that the craters be named after famous scientists and other persons of prominence). Some craters on the moon so named include Copernicus, Archimedes, Newton, Plato, etc. Because the moon can be observed in considerable detail by earth-based instruments, detailed maps of crater locations as well as of other lunar features have existed for decades and were very helpful in the very early phases of planning the lunar exploratory missions.

Lunar Maria. Early astronomers thought that the lunar lowlands, which are dark, flat plains, were liquid bodies and thus the word sea (Latin = *mare*) was adopted to identify these features. The lunar maria are broad depressions filled with dark volcanic basalts, rich in iron and titanium. The maria were sampled at five sites during the *Apollo* program. The basalt ages fall into a surprisingly narrow time range, estimated between 3.15 and 3.85×10^9 years. It is reasoned that the mare basalts may have been generated by radioactive heating and partial melting in an iron-rich, plagioclase-poor region in the interior of the moon. Thus, they were not a product of the primary differentiation that gave rise to the lunar crust. That the basins were filled by a succession of flows is indicated by numerous overlapping flow fronts. Intervals between them indicate periods of cooling and solidification. Small dark steps or ledges are sometimes visible at the base of the highlands where the mare basalts lap against highland masses. This is particularly apparent on the western near sides. These have been called a form of "bath-tub ring," remains of the highest level attained by the lava as it was emplaced. The Planning Team report indicates that the most recent flows emanated from sources at the margins of the maria; they followed definite, but subdued axial channels, which may be collapsed lava tubes. Some of the flow units are as much as 350 kilometers in length. It is believed that older flows may have extended up to 1,200 kilometers north across the center of the basin (Mare Serenitatis). Prominent wrinkle ridges are found in the southern Imbrium region. These are believed to be compressional structures superimposed before complete solidification of the lavas.

It appears that the thicknesses of basalt are largest in the centers of maria and taper toward the margins, a condition possibly attributable to load-induced subsidence. In many small areas on the lunar nearside, usually at the margins of maria, a very dark material appears to have blanketed craters and other structures. Because of their darkness in comparison with the rest of the moon, they are apparent to the untrained eye when captured on a lunar photograph of good quality. Earlier it was conceived that the dark material comprised a layer of volcanic ash because of its mantling effect on other structures — as well as the fact that not many craters appear to postdate the dark material. This was considered as evidence of relatively recent volcanism on the moon.

Partly because the southeast margin of Mare Serenitatis includes a prominent deposit of dark mantling material, the *Apollo 17* landing site was selected at Taurus-Littrow. It turned out that the valley floor on which the *Challenger* landed is fully blanketed with this dark substance. However, no prominent component of ash-like material was found in these soil samples. It was learned that the blackness of the Taurus-Littrow soils derives from a high titanium and iron content in the underlying bedrock. The latter is a mare-type basalt similar chemically to that collected by *Apollo 11* in Mare Tranquillitatis. Apparently, the iron and titanium cause the bedrock and associated soil derived from it by comminution to have a high content of the black mineral ilmenite ($FeTiO_3$). They deeply color glasses generated by impacts on the local soil. Since the local basalts crystallized 3.8×10^9 years ago, the dark mantling material at the one place where it was sampled is attributed to early volcanic activity, i.e., in terms of eruptions that may have occurred at the beginning of the epoch of basin-filling volcanism, rather than later volcanism as has been anticipated.

The most unusual material discovered by the *Apollo 17* astronauts was an "orange soil," which constituted a 25-cm layer on the rim of Shorty Crater (110 m in diam. × 10 m deep). This was one of the few colorful spots observed on the moon. The orange color is due to a relatively high abundance of trivalent titanium. The soil also contains an excess of fission xenon isotopes attributable to neutron-induced fission of ^{235}U. It is estimated that these volcanic glasses were formed 3.63×10^9 years ago. Examination indicates that the glasses resided on the lunar surface for about 38 million years before they were deeply buried. The glass spherules were reexcavated by the impact that formed Shorty Crater some 17 million years ago and remained undisturbed until they were collected. As reported by Eugster and colleagues (Physikalisches Institut, University of Bern), the glass droplets from the bottom of the core were probably produced by lava fountaining a few tens of millions of years after the end of the lava flooding of Mare Serenitatis. They were then exposed to cosmic rays for about 38 million years.

Lunar Highlands. Many rocks obtained from the lunar highlands (See Fig. 3.) are feldspathic breccias.[2] Six or more categories of highland breccia rocks have been identified. Differences in these categories include: (1) variation in the composition of the fragments or clasts which they contain; (2) distribution of sizes; (3) the content of glass; and (4) the degree of thermal sintering or recrystallization that occurred since aggregation. However, some truly igneous rocks were also returned from the highlands. Such rocks may have resulted from melting in the interior of the moon, or from impact melting.

Lunar Crust and Rocks. It is generally believed that the moon's crust, estimated thickness in tens of kilometers, was produced by magmatic formation. This would have taken place when extensive melting occurred in the outermost layers of the moon (hundreds of kilometers deep). To date, the source of this heat has not been adequately explained. Concepts proposed include: (1) accretional heating during rapid formation of the moon; (2) electrical heating that would have resulted from an intense solar wind; or (3) tidal heating at the time the moon was in close proximity of the earth.

On a chemical basis, three broad classes of terra rocks have been recognized, as listed in Table 1. In elemental plots, these categories do not display well-separated clusters. Rather, they tend to grade into one another. No simple correlation of chemical class with petrographic texture has been formulated.

Considerations of phase equilibria and trace element partitioning make it most likely that the moon first separated a plagioclase-rich crust, presumably by upward flotation of plagioclase as it crystallized from an early surface magma system of monumental proportions, and that subsequently the KREEP-rich and KREEP-poor noritic rock types were formed by partial melting of a feldspar-rich parent material (presumably the roots of the anorthsitic crust) and erupted onto the surface as lavas. See Table 1.

---

[2] Feldspar is a group of minerals of the general formula, $Mal(Al, Si)_3O_8$, where M = K, Na, Ca, etc.: breccia is a coarse-grained elastic rock composed of particles greater than 2 millimeters in diameter, angular, and broken rock fragments that are cemented together in a fine-grained matrix.

Fig. 3. A near vertical view of the Hadley-Apennine area, as photographed by the Fairchild metric mapping camera mounted in the Scientific Instrument Module (SIM) bay of the Apollo 17 Service Module in lunar orbit. Hadley-Rille meanders through the center of the picture. The Apennine Mountains dominate the picture. The Lunar Module touchdown point is on the right (east) side of the "chicken beak" in the lower center of the photograph. The craters Aristillus (top) and Autolycus are near the horizon at upper left. The small, distinct crater next to the rille is Hadley C. (*National Aeronautics and Space Administration.*)

TABLE 1. THREE CHEMICAL CLASSES OF TERRA ROCKS — SELECTED OXIDE AND ELEMENTAL ABUNDANCE (parts per million unless otherwise indicated)

Constituent	Anorthositic Rocks	KREEP-Poor Norites and Troctolites	Kreep-Rich Norites
Al_2O_3	>25%	20–25%	15–20%
FeO	0–5%	4–9%	8–10%
MgO	2–8%	8–16%	7–13%
P_2O_5	0–0.06%	0.1–0.3%	0.3–2%
K_2O	0.01–0.2%	0.05–0.1%	0.2–2.0%
Uranium	<0.4	0.4–1.0	2–6
Lanthanum	0.1–4.5	10–30	40–80
Europium	0.6–1.2	1–2	2–3
Europium (anomaly)	Positive	Negative	Negative
Hafnium	<0.01–1.5	4–10	10–30

Source: Lunar Sample Analysis Planning Team.
Note: 1. The use of the terrestrial rock names, anorthosite, norite, and troctolite refer to chemical similarities with no implication of an origin in deep-seated igneous bodies as in the case of the terrestrial counterparts.
2. Anorthositic rocks contain more than 65% plagioclase. Norites and troctolites contain roughly equal amounts of plagioclase and a mafic mineral. The latter is mainly orthopyroxene in the case of norites; olivine in the case of troctolites.
3. KREEP refers to potassium, rare-earth elements, and phosphorus.

Radiometric dating of terra rock samples has yielded interesting data. Ages derived from Rb-Sr internal isochrons and the ^{40}Ar-^{39}Ar technique consistently fall in the range of 3.85 to 4.05×10^9 years. The rocks are neither as old nor as scattered in age as was expected. Because it has been believed that the moon is of the same age as the rest of the solar system (about 4.6×10^9 years), it was expected that terra rocks, most of which suffered a series of shock brecciations and reheatings, would show a spectrum of apparent ages reaching back to that time. But, because of the consistency of the terra rocks in the 3.9 to 4.0×10^9 years age range, it is now believed that some major cataclysm affected the moon at that time, or perhaps had affected it continuously until about 3.9×10^9 years ago. Such events would have reset the clocks of the rocks throughout that part of the crust which to date has been accessible to exploration. It is visualized that the cataclysm may have been a highly intense and violent epoch of meteroid bombardment, possibly ending in the giant impacts that caused excavation of the Imbrium and Orientale basins. It is possible that the Orientale basin may have been the source of much of the highland material gathered at the *Apollo 16* and other landing sites. See also **Mineralogy**.

Previously, it had been apparent that debris from Orientale, the youngest of the major ringed basins, may have projected over the entire surface of the moon. However, it was felt that little more than a veneer of Orientale material would have been deposited at distances as far away from the basin as the Descartes landing site of *Apollo 16* (3,000 kilometers). As the Planning Team report indicates, the debris-moving capability of giant impacts has not been calibrated, and it is possible that such a thickness of Orientale ejecta was deposited at Descartes as to preclude substantial dilution by indigenous materials up to the present day. Depositions elsewhere on this scale would account for the remarkable similarity of terra rock types recovered from one site to the next.

Lunar Atmosphere

The *Apollo 17* station included a mass spectrometer for measuring the density and chemical composition of the lunar atmosphere. During the early observational period, ^{4}He, ^{40}Ar, and possibly Ne were detected. The daytime and nighttime ^{4}He concentrations at the lunar surface were found to be 3×10^3 and 6×10^4 atoms per cubic centimeter, respectively. This factor of 20 in the variation between daytime and nighttime concentrations corresponds with expected behavior of a noncondensable gas in the lunar atmosphere. However, the ^{40}Ar concentration exhibited unexpected behavior. This concentration dropped during the night to below 100 atoms per cubic centimeter, rising orders of magnitude a few hours before the sunrise terminator reached the measurement site. It has been proposed that ^{40}Ar is absorbed at the lunar surface during the night and released again at sunrise.

Fortunately, a record of past and present lunar atmospheres is preserved in the lunar fines and breccias. Solar ultraviolet radiation and charge exchange with the solar wind cause ionization of atmospheric atoms. Fine-grained lunar surface materials trap some of these accelerated ions. It has been recognized since *Apollo 11* that lunar soil particles contain a component of ^{40}Ar the abundance of which is propisitional to the surface area of the particles and it is believed that this is retrapped gas from the lunar atmosphere. The amount of ^{40}Ar present in the lunar atmosphere is reflected by the ^{40}Ar/^{36}Ar ratio of trapped gas. Observation of these ratios indicate a variation that suggests that perhaps ^{40}Ar may have been more abundant in the ancient lunar atmosphere than prevails today.

Structure and Seismic Activity

Other than the fact that the moon is considerably less dense than other terrestrial planetary bodies (mean density of about 3.4 times that of water), very little was known pertaining to the interior of the moon prior to the *Apollo* explorations. Laser altimeter measurements from the *Apollo* orbiting command module indicated the lunar center of mass is displaced by about 2 kilometers (1.2 mile) earthward and 1 kilometer (0.6 mile) east of the center. Possibly this may be explained by the mean thickness of the low-density lunar crust being about a factor of 2 greater on the lunar farside than on the nearside. A crustal heterogeneity of this magnitude possibly also can account for the differences in the principal moments of inertia of the moon. These had been known before the *Apollo* missions from observations of the lunar librations. Previously, the moment differences had been attributed to a distorted, nonequilibrium form of the moon.

A network of four seismic stations deployed on the moon during the *Apollo* missions had recorded more than 1000 moonquakes between 1969 and 1976. The moonquakes are very small, with estimated magnitudes of less than about 2. The annual release of seismic energy is estimated at about 10^{15} ergs, or about ten orders of magnitude less than that of the earth. Except for a small number of shallow events, moonquakes occur deep inside the moon at depths of between 600 and 1000 kilometers (965 and 1610 miles). Moonquakes appear to occur repeatedly at distinct foci and events from each focus produce matching waveforms. During the aforementioned time span, nearly seventy such repeating foci were identified. The events at each focus appear to occur at 27-day intervals and there is evidence of a 206-day periodicity. These periodicities clearly relate the moonquake occurrences to the tidal forces acting on the moon. Further details are reported by Toksöz, Goins, and Cheng (1977).

The *Apollo* seismometer network has detected compressional seismic waves from meteoroid impacts and internal moonquakes on all sides of the moon. However, shear waves have been received only from the nearside events. Attenuation of shear waves by the interior of the moon indicate that it is probably hot and weak, possibly with a small fraction of the rock melted. The relatively rigid layer which transmits shear waves, the lunar lithosphere, is about 1,000 kilometers (621 miles) in thickness.

Magnetic Properties

An unexpected finding of the analysis of returned rocks, both crystalline and breccia, was their possession of a stable remanent magnetization. No dipole magnetic field in excess of 5 gammas (1 gamma $= 10^{-5}$ gauss) had been detected by lunar orbiting satellites. In contrast, magnetometers landed during the *Apollo* missions indicated local magnetic field of up to 300 gammas. The *Apollo 15* and *16* subsatellites also had indicated complex magnetic anomalies up to 1 gamma. Since magnetization appears widespread on the lunar surface, it is estimated that a field of from 1,000 to 10,000 gammas was required when the rocks were formed some 3.2 to 4.1×10^9 years ago. Investigators do not fully agree on the mechanism that may have provided such an early magnetic field. It is believed that the rocks may have been heated above their Curie temperatures, either by volcanism or by cataclysm at the aforementioned time period. This would have taken place long after any transient magnetic field that may have been associated with the origin of the solar system would have passed. One proposal is that the moon had a small, fluid, metallic core and that this performed as a self-exciting dynamo at that time. Another proposal is that a primeval solar system field may have magnetized the cool interior of the moon while the surface layers were still molten and that later the field of the lunar interior magnetized the outer layers as they cooled, and that finally radioactive heating eliminated the interior magnetization.

Investigators Hood and Coleman (Univ. of California, Los Angeles) and Wilhelms (U.S. Geological Survey) reported in 1979 of additional studies of previously unmapped *Apollo 16* subsatellite magnetometer data collected at low altitudes over the lunar near side. The scientists found that medium-amplitude magnetic anomalies exist over the Fra Mauro and Cayley Formations (primary and secondary basin ejecta emplaced 3.8 to 4.0 billion years ago) but are nearly absent over the maria and over the craters Copernicus, Kepler, and Reiner and their encircling ejecta mantles. In their summary of this study, these investigators suggest that a single hypothesis that is consistent with the subsatellite data as well as with studies of returned samples and surface magnetic fields is that basin and crater ejecta deposits are major sources of the orbital anomalies. The level of magnetization is weak peripheral to the impact region inside the crater, but it rises beyond the rim and may rise sharply in the case of some ejected materials transported ballistically to large distances. Also, as suggested by Strangway et al. (1973), the ejected materials are probably also the materials most strongly shocked and heated by the impact event, a relative increase in the volume fraction of free iron grains capable of retaining a strong and stable magnetic remanence is reasonable to expect.

Meteorites from the Moon?

Because meteorites tend to weather less in the atmosphere of the Antarctic, an effort has been made since the 1970s to locate meteorites in that region. Code number ALHA 81005, a meteorite discovered in the Allan Hills region in January 1982, was initially considered as being of lunar origin. The appearance and chemical composition checked well with lunar rocks returned by the Apollo mission. It was initially hypothesized that the meteorite is a tiny piece of breccia from the lunar regolith. But further study created many questions and a large body of scientists now proposes that the meteorite is of Martian origin. See also **Meteroids and Meteorites**; and **Lunar Prospector Mission**.

Additional Reading

Belton, M.J.S. et al.: "Lunar Impact Basins and Crustal Heterogeneity," *Science*, 570 (January 31, 1992).

Boss, A.P.: "The Origin of the Moon," *Science*, 341 (January 24, 1986).

Brush, S.G.: "Nickel for Your Thoughts: Urey and the Origin of the Moon," *Science*, **217**, 891–898 (1982).

Burke, B.F.: "Astrophysics from the Moon," *Science*, 1365 (December 7, 1990).

Burns, J.O. et al.: "Observations on the Moon," *Sci. Amer.*, 42 (March 1990).

Eberhart, J.: "Going for the Whole Moon: The Lunar Observer Would Study What Other Spacecraft Missed," *Sci. News*, 264 (April 28, 1990).

Eugster, O.: "History of Meteorites from the Moon Collected in Antarctica," *Science*, 1197 (September 15, 1989).

Garber, M.: "New View of Universe to Come from Moon," *Hughes News*, 5 (May 3, 1991).

Gillette, R.: "The Aftermath of Apollo: Science on the Shelf?" *Science*, **178**, 1265–1268 (1972).

Gribbin, J.R. and S. Goodwin: "Empire of the Sun: Planets and Moons of the Solar System," New York University Press, New York, NY, 1998.

Hartmann, W.K.: "Moons and Planets," 4th Edition, International Thomson Publishing, New York, NY, 1999.

Hood, L.L., P.J. Coleman, Jr., and D.W. Wilhelms: "The Moon: Sources of the Crustal Magnetic Anomalies," *Science*, **204**, 53–57 (1979).

Horgan, J.: "Blame It on the Moon," *Sci. Amer.*, 18 (February 1989).

Keen, R.A.: "Volcanic Aerosols and Lunar Eclipses," *Science*, **222**, 1011–1013 (1983).

Kerr, R.A.: "Making the Moon, Remaking Earth," *Science*, 1433 (March 17, 1989).

Kerr, R.A.: "Treasuring the Moon for 20 Years," *Science*, 1552 (March 24, 1989).

Kerr, R.A.: "A Rare View of the Moon," *Science*, 1651 (December 21, 1990).

Kerr, R.A.: "Did the Moon Suffer a Cataclysmic Bombardment?" *Science*, 1634 (June 19, 1992).

Marvin, U.B., J.W. Carey, and M.M. Lindstrom: "Cordierite-Spinal Troctolite, a New Magnesium-Rich Lithology from the Lunar Highlands," *Science*, 925 (February 17, 1989).

Morrison, D.C.: "An Unsung Legacy of the First Lunar Landing," *Science*, 447 (October 27, 1989).

Ottwell, G.: "The Understanding of Eclipses," Furman University, Greenville, SC, 1991.

Porter, A.E. and T.H. Morgan: "Discovery of Sodium and Potassium Vapor in the Atmosphere of the Moon," *Science*, 675 (August 5, 1988).

Strangway, D. et al.: "Magnetism and Magnetic Materials," (C. Graham, Jr., and editors), pp. 1178–1196, American Institute of Physics, New York, NY, 1973.

Toksoz, M.N., N.R. Goins, and C.H. Cheng: *Science*, **196**, 979–981 (1977).

Waldrop, M.M.: "Asking for the Moon," *Science*, 637 (February 9, 1990).

Wilhelms, D.E.: "The Geologic History of the Moon," U.S. Geological Survey, Denver, CO, 1988.

Web References

Chronology of Lunar and Planetary Exploration: *http://nssdc.gsfc.nasa.gov/planetary/chrono.html*

Exploring the Moon: *http://www.space.edu/moon/*

MOOSE.   See **Deer**.

MORELS.   See **Ascomycetes**.

MORERA THEOREM.   The converse of the Cauchy integral theorem. If $f(z)$ is a continuous function of the complex variable z, defined in finite simply connected domain,

$$\int_C f(z)\,dz = 0$$

for any closed contour on D, then $f(z)$ is an analytic function of z in D.

MORMON CRICKET *(Insecta, Orthoptera)*.   A large wingless long-horned grasshopper of the western United States. It varies in color from pale green or yellow to black. It eats vegetation of all kinds and is a cannibal and scavenger, even eating dead animals. When abundant it is a serious crop pest and one that is difficult to combat.

MORMYRIDS *(Osteichthyes)*.   Of the order *Isospondyli*, family *Mormyridae*, there are well over 100 species of these fishes. They are characterized by medium-long to very long snouts. They are bottom feeders, foraging worms out of the mud, as well as various other small creatures and insect larvae. There is a wide range of size, from 6 inches (15 centimeters) in length to 5 or more feet (1.5 + meters). The larger mormyrids are food fish, usually in the dried form. The very long, extended snout of the *Gnathonemus numerous* (elephant-nose mormyrid) is a particularly striking feature, nothing else quite like it in the world of fishes. A much less pronounced but still prominent snout is exhibited by the related species *Mormyrops deliciosus*. Only in recent years has it been found that most or all species of mormyrids possess an electrogenic capability, with voltages discharged in the microvolt range. It is assumed that this feature is useful in locating food sources. These electrogenic abilities in the *Gymnarchus niloticus* have been recognized over a longer period of time. This fish can attain a length of 5 feet (1.5 meters) and is well considered as an edible fish. Electrical discharges in this species are by way of the tail. Biologists are interested in some apparent parallels exhibited by the African mormyrids and the Southern American gymnotids. The latter group includes the electric eel. In terms of classification, they are not related, but do display somewhat similar behavior patterns.

MORPHINE.   About 10% of the weight of opium is morphine which was the first of the vegetable alkaloids to be isolated in 1805 by Sertürner. Since the source of the natural alkaloids is opium, all narcotics whose actions resemble those of morphine are sometimes referred to as opiates. Semisynthetic agents are usually made by altering the morphine molecule, and include such agents as diacetylmorphine (heroin), ethylmorphine (*Dionin*), dihydromorphinone (*Dilaudid*), and methyldihydromorphinone (metopon). Synthetic narcotics include agents with a wide variety of chemical structures. Some of the important synthetic agents are meperidine (piperidine type), levorphanol (morphinian type), methadone (aliphatic type), phenaxocine (benzmorphan type), and their derivatives. The structures of the various narcotics are given in Fig. 1.

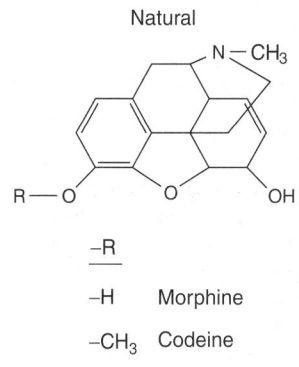

Fig. 1.   Chemical structures of various narcotics.

Since morphine is responsible for the major actions of opium and the actions of all narcotics are qualitatively similar, morphine can be used as a model for discussing narcotic agents. The most prominent effects of morphine in the human body are on the central nervous system and the gastroenteric tract. The principal central action of morphine is the relief of pain, and this occurs in at least three ways: (1) morphine reduces central perception of pain probably at the thalamic level; (2) it alters the reactions to pain probably at the level of the cerebral cortex; and (3) it elevates the pain threshold by inducing sedation or sleep. In the medulla, morphine depresses the respiratory, cough and vasomotor centers and indirectly stimulates the vomiting center. The nuclei of the occulomotor (III) and vagus (X) nerves are stimulated by sufficient doses of morphine causing myosis (constriction of the pupils), bradycardia (slowing of the heart rate), and increased gastroenteric tone. The overall effect of morphine on the gastroenteric tract is spasmogenic and constipative. Morphine causes the constipative action by several means, including increased segmental movement of the large bowels, spastic tonus of the sphincters, decreased defecation reflex, and increased reabsorption of water in the large intestines to cause drying of feces.

The metabolic effects of morphine are not marked and are clinically unimportant. The metabolic rate may be decreased slightly due to the lowered activity and tone of the skeletal muscles resulting from the central depression. A rise in blood sugar may be observed after the injection of morphine. The hyperglycemia is due to glycogenolysis in the liver resulting from the release of epinephrine from the adrenal medulla. The lowering of urine production noted after the administration of the drug is due mainly to the release of antidiuretic hormone from the posterior pituitary gland.

Morphine is detoxified or biotransformed mainly in the liver by conjugation with glucuronic acid. Morphine is conjugated by a series of reactions involving the formation of uridine diphosphoglucose (UDP-glucose), the oxidation of carbon-6 of glucose to form uridine diphosphoglucuronic acid (UDP-glucuronic acid) and the transfer of glucuronic acid to morphine to form the morphine glucuronide. This reaction is diagrammed in Fig. 2. The following enzymes catalyze the sequential reactions; reaction (1), UDP-glucose pyrophosphorylase; reaction (2), UDP-glucose dehydrogenase; reaction (3), glucuronyl transferase; reaction (4) nucleoside diphosphokinase.

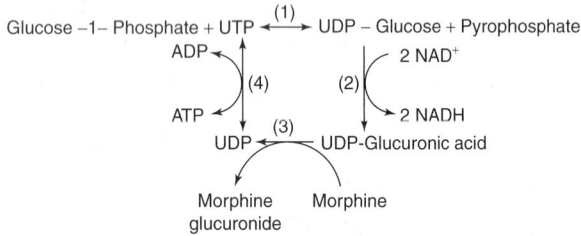

Fig. 2. Formation of morphine glucuronide. NAD^+ = nicotinamide adenine dinucleotide; NADH = reduced NAD^+; ATP = adenosine triphosphate; ADP = adenosine diphosphate; UTP = uridine triphosphate; UDP = uridine diphosphate.

The most serious drawback in the use of morphine and other narcotic analgesics is their addictive potentiality. The characteristics of drug addiction include psychological need or habituation, tolerance and physical dependence. Habituation consists of an emotional and psychic dependence, and in addiction, the habituation becomes an overpowering desire to take the drug. Tolerance is a phenomenon whereby the dosage of the drug must be continually increased to maintain equivalent pharmacological effects. Physical dependence develops when the tissues of the body become so adapted to the effects of the drug that the cells of the tissues cannot function normally without the drug in the environment. This is the most vicious characteristic of drug addiction.

The mechanisms underlying the development of tolerance are not fully understood. Biochemically, it may be attractive to explain tolerance by decreased absorption, altered distribution, increased biotransformation, and/or increased excretion of the drug. However, these processes have been shown to be unrelated to the development of tolerance. Thus, cellular adaptation offers the greatest likelihood for clarifying the phenomenon. Evidence for cellular adaptation is the finding that the respiration of chemically stimulated cortical slices of brain from normal rats is markedly depressed by morphine, whereas the respiration of those from rats chronically dosed with morphine is unaffected.

Heroin is diacetylmorphine (diamorphine hydrochloride) and is prepared by the action of acetic anhydride on morphine, possessing four times the analgesic affect of morphine, but having a considerably less depressant effect. Addiction is common, the drug being taken in the form of snuff, or by injection.

Nalorphine, the allyl ($-CH_2-CH=CH_2$) derivative of morphine (N-allylnormorphine) is remarkable in that it is antagonizing to almost all the effects of narcotics. The antagonizing action is specific for the narcotic analgesics. For instance, nalorphine will antagonize the respiratory depression due to morphine or other narcotics, but not that caused by other depressants, such as hypnotics or anesthetics. This property of nalorphine makes it a particularly useful antidote in cases of acute morphine poisoning. The agent can also precipitate acute withdrawal symptoms if administered to persons addicted to narcotics. The agent has become a useful biochemical tool for studying the mechanism of action of narcotics and tolerance. Since the chemical structures between morphine and nalorphine are so similar, it has been suggested that nalorphine acts by competing with morphine for the receptor site. The antagonistic effect of nalorphine cannot be explained by a simple competitive inhibition if equal affinity for the receptor site with the agonist and antagonist is assumed, because small doses of nalorphine antagonize the effects of much higher doses of the narcotic. Nalorphine also antagonizes the effects of synthetic narcotics of varying chemical structures, such as methadone and meperidine.

Morphine, $C_{17}H_{19}NO_3 \cdot H_2O$, is a white powder melting at 253 °C and is derived from opium which is the dried juice obtained from unripe capsules of the poppy plant (*Papaver somniferum*), variously cultivated in the Near East and Far East. The opium poppy is an annual. When the petals drop from the white flowers, the capsules are cut. The juice exudes and hardens, forming a brownish mass which is crude opium. It contains a total of about 20 narcotics, including morphine. See also **Alkaloids**; and **Analgesics**.

Natural morphine-like substances generated within the human brain are described in the entry on **Central and Peripheral Nervous Systems**.

Additional Reading

Courtwright, D.: "Dark Paradise: Opiate Addiction in America before 1940," Harvard University Press, Cambridge, MA, 1982.
Gianino, J. et al.: "Intrathecal Drug Therapy for Spasticity and Pain: Practical Patient Management," Springer-Verlag New York, Inc., New York, NY, 1995.

Web Reference

U.S. Food and Drug Administration Center for Drug Evaluation and Research. *http://www.fda.gov/cder*

MOSAIC VISION. A theoretical interpretation of the action of compound eyes of arthropods. Each visual unit (ommatidium) of such an eye forms an image of part of the object toward which it is directed and the total image seen by the animal is supposed to be composed of many such partial images formed by the many units in the eye.

MOSQUITO (*Insecta, Diptera*). A small two-winged fly with slender body, long legs, and narrow wings bearing scales along the veins. The larvae are aquatic. Male mosquitoes feed on plant juices and only the females suck blood, but in some species neither sex sucks blood.

Mosquitoes are well known as a nuisance and in warmer climates they are also dangerous because some species transmit disease. Malarial parasites are carried by mosquitoes of the genus *Anopheles* and yellow fever by species of *Aedes*. In regions where these diseases occur the destruction of mosquitoes by draining swampy areas where they may breed, and by applying oil to water that cannot be drained, is important.

The protection of patients from the attack of mosquitoes so that the disease cannot be carried to others is accomplished by adequate screening, a measure that is valuable both for comfort and safety in all dwellings.

Occurrence of the mosquito is quite widespread. They are found in the Americas, Europe, Asia, and Africa. They are abundant in Alaska and the Arctic regions. They also have been found in the Himalayas amid rock and snow at altitudes of 20,000 feet (6100 meters). Generally, mosquitoes are found wherever warm-blooded animals are to be found. For raising their young, mosquitoes prefer lakes, ponds, stagnant pools and puddles.

Dimensions vary with species. On the average, the mosquito is approximately $\frac{3}{4}$ inch (1.9 centimeters) long and from wing tip to wing tip about 1 to $1\frac{1}{2}$ inches (2.5 to 4.1 centimeters).

Some species of mosquito cannot produce fertile eggs without a prior meal of blood. Many types of mosquitoes have definite preferences in type of animal to feed on and will reluctantly feed upon another. Some mosquitoes excrete a sticky liquid and construct rafts on water by gluing their eggs together. The eggs are pointed. By constructing the raft with the pointed ends down, the raft will right itself automatically if overturned. The top part of the raft is rendered waterproof by the sticky exudate.

The female mosquito may lay as many as 40 to 200 eggs on water or near it. Only a very small amount of water is needed for the eggs to hatch. However, without water the eggs cannot survive.

Some mosquitoes may produce a generation every three weeks; while other species may produce only one generation per year.

Mosquitoes cannot breathe under water. The mosquito's breathing tube (sometimes called a snorkle) is kept dry by an oily secretion on it that repels water. If a thin film of petroleum is poured over the water, the petroleum enters the snorkle causing the mosquito larvae to smother. For many years, this has been an effective procedure in controlling mosquito population.

Some important species of mosquitoes include: *Anopheleini maculipennis*, common in Europe and the most numerous mosquito in the tropics; the *Aedes aegypti*, which transmits yellow fever in the Americas, Africa, and all tropical regions; the *A. albopictus*, which transmits dengue fever; the *A. obturbans*, which is found in the Pacific Islands and the coast of Asia; and the *Sabethes* and *Haemagogus*, which transmits jungle and yellow fever in jungle areas. See Fig. 1.

Fig. 1.   Yellow-fever mosquito, Stegomyic fasciata. (*F.B. Howard, USDA.*)

MÖSSBAUER EFFECT. The phenomenon of recoilless resonance fluorescence of gamma rays from nuclei bound in solids. It was first discovered in 1958 by R.L. Mössbauer. The extreme sharpness of the recoilless gamma transitions and the relative ease and accuracy in observing small energy differences make the effect an important tool in nuclear physics, solid-state physics, and chemistry.

If a gamma ray is emitted by an atomic nucleus, the system to which the emitting atom belongs must recoil, in order to conserve momentum, in a direction opposite to that in which the gamma rays is emitted. Similarly, if an atomic nucleus absorbs a gamma ray, the system must continue to move, following absorption, in such a way that momentum is conserved. If the recoiling system is a single atom, such as in a gas and shown schematically in Fig. 1(a), the emitting atom carries away enough energy from the transition for the observed energy $E_o - R$ of the emitted gamma ray to be measurably less than the energy E_o of the nuclear transition that caused the gamma ray to be emitted, also indicated in (a) of the diagram. Furthermore, a gamma ray that is absorbed by a single atom must transfer a measurable kinetic energy to that atom, as well as the energy of the nuclear transition. On the other hand, if the emitting or absorbing nucleus belongs to an atom that is bound into a crystalline structure, such that the structure as a whole can recoil, and as indicated schematically in Fig. 1(b), the kinetic energy that must be given to the crystalline system to conserve momentum is greatly reduced, compared to the energy that must be given to a single atom, because of the much larger mass of the system. The recoil energy is then so small that the gamma ray carries away essentially the

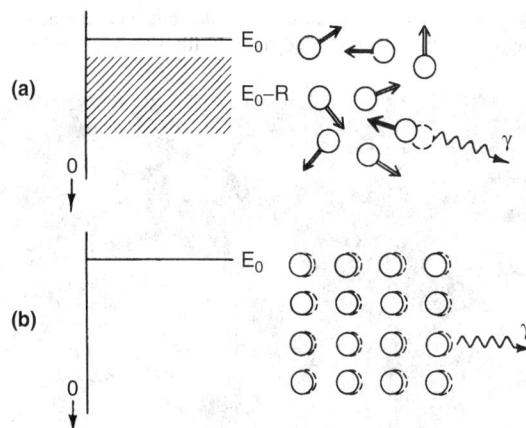

Fig. 1.   (a) Emission of a gamma-ray by a single atom moving randomly in a gas transfers appreciable energy to the emitting atom in the form of recoil kinetic energy, and reduces the energy of the gamma ray from the transition energy E_0 to some lower energy $E_0 - R$. (b) Emission of a gamma ray by an atom bound into a crystalline structure may sometimes cause recoil of the whole crystal, in which case, the loss of energy in the form if kinetic energy of the crystal is negligibly small and the gamma ray appear to be emitted with an energy E_0.

full energy E_o of the transition in the case of emission, transferring such a small fraction of its energy to the absorbing system that emission and absorption appears to be recoil-free.

This process is observed, of course, in the analogous case of the resonance radiation in atomic transitions, in which case, the photons have energies in the range of light, commonly visible light. However, the protons of gamma radiation are so much more energetic that their energy loss by recoil of the nucleus emitting them is great enough, in the case of free atoms, for the resonance effect not to occur.

Mössbauer discovered, however, that in the case of atoms which are not free, but bound in a solid, the effect can often be observed. It is easily demonstrated when the normal, free-atom recoil energy is comparable to the energy of the quantized lattice vibrations. Under these conditions, zero-phonon processes are possible in which the entire energy of the nuclear transition goes into the gamma ray and the recoil momentum is taken by the solid as a whole. The resulting gamma rays then have the proper energy to be resonantly absorbed or scattered in an analogous zero-phonon process.

The Mössbauer effect is useful in determining nuclear level widths and Doppler effects. Another application is based upon the measurement of nuclear hyperfine structure, a measurement which is possible when the line-width of the gamma ray is smaller than the hyperfine interaction (that due to the coupling of the nuclear moments with external fields). In this application, the Mössbauer effect is almost unique because one obtains the splitting of both the ground and the first excited nuclear states. This effect, in turn, makes possible the determination of nuclear moments of the excited states, which can be important tests of nuclear models. Another important feature is the so-called isomer or chemical shift (terms used interchangeably) which measures the simple electrostatic interaction of the nucleus with its own s-electron and has given information about the difference in the nuclear radii of the ground and excited states.

MOSSBAUER, RUDOLF LUDWIG (1929). Mossbauer was a German physicist who discovered and had named after him the Mossbauer effect. He received the Nobel Prize in Physics in 1961.

See also **Mössbauer Effect**.

J. M. I.

MOTH (*Insecta, Lepidoptera*). Insects with the four wings at least partly scaly, the mouth parts formed for sucking, and the antennae rarely clubbed near the tips. No one character serves to distinguish all moths from the butterflies and skippers. Most moths are nocturnal but many are diurnal or crepuscular. Most of them have the antennae slender and tapering or broadened by setae or by processes from the segments, forming a comb-like (pectinate) structure, but a few have an expansion just before the tip like that found in most skippers. Most of the caterpillars of moths pupate in a cocoon or a subterranean cell or in the tissues of plants but a few form a brightly colored naked pupa like those of the butterflies. The term does not apply to a principal division of the order but includes many families making up one entire suborder and most of the second. See Fig. 1.

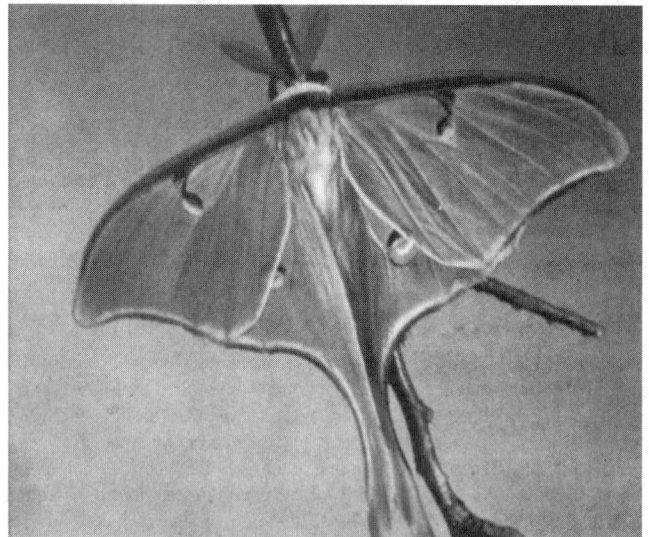

Fig. 1.   The luna moth. (*A.M. Winchester.*)

There are at least 140,000 species of moth. The size varies considerably—the large Atlas moth of India measures about 10 to 12 inches (22.5 to 30 centimeters) (wing tip to wing tip) and the South American giant moth may be even larger. Many species have wing-tip spreads of only about $\frac{1}{2}$ inch (12 millimeters). Some of the numerous species of moth have been investigated quite intensively. Nerve impulses have been measured and habits studied in detail. Some moths go into a wing-beating ritual before taking to flight, warming up so to speak, much as a pilot may rev up an aircraft before taking off.

The moth larvae are soft and quite helpless when they emerge from their cocoon. However, the tiny parachute-like wings unfold and become useful wings within less than an hour. The moth achieves adult size, is firm, dry, and in full color within a very short period. Moth hormonal systems have been studied as well as habits. Some moths mimic other moths or even other insects, and may change coloration for purposes of camouflage almost instantly.

As in some other insects, the taste receptors of the moth are on the forelegs. The moth tastes prospective food by walking about on it. Each moth species has particular preferences in food and will follow the eating habits of its parents. In searching for a particular food, a moth may scan an entire garden within a short period without eating. The moth has a keen sense of smell and it is believed able to detect its mate over a distance of $\frac{1}{2}$ mile (0.9 kilometer). Although of the same order, moths differ basically from butterflies in the following major respects: When at rest, moths spread their wings, whereas the butterfly presses its wings close together over its head; the body of the moth is shorter and stouter than that of the butterfly.

Several species of moth are particularly damaging to food crops, this damage occurring mainly as the result of the larvae. Female moths commence the destructive activity when they find locations in which to place their eggs. Often, a thousand or more eggs will be deposited. The developing larvae are voracious eaters.

Some of the economic moths described in this book include: **Brown-Tail Moth**; **Bud Moth**; **Clear-Winged Moth**; **Codling Moth**; **Grain-Storage Insects**; **Gypsy Moth**; **Tussock Moth**; and **Wax Moth**. For food-destructive moths, see also **Dried-Fruit Insects**.

Additional Reading

Mitchell, R. et al.: "Butterflies and Moths: A Guide to the More Common American Species," Western Publishing Company, Inc., North Platte, NE, 1987.

Tuskes, P. et al.: "The Wild Silk Moths of North America: A Natural History of the Saturniidae of the United States and Canada," Cornell University Press, Ithaca, NY, 1995.

Young, M.: "The Natural History of Moths," Morgan Kaufmann Publishers, San Mateo, CA, 1996.

MOTION (Gauss Principle of Least Constraint).   1. The motion of connected points is such that, for the elementary motion actually taken, the sum of the products of the mass of each particle into the square of the distance of its deviation from the position it would have reached if free, is a minimum.

2. The motion of a system of material points interconnected in any way and submitted to any influences, agrees at each instant as closely as possible with the motion the points would have if they were free. The actual motion takes place so that the constraints on the system are the least possible. For the measurement of the constraint, during any infinitesimal element of time, take the sum of the products of the mass of each point and the square of its deviation from the position the point would have occupied at the end of the element of time, if it had been free.

MOTMOT.   See **Kingfishers and Other Coraciiformes**.

MOTOR (Electric).   In an electric motor, electrical energy is converted into mechanical energy (torque in a rotary system). (Linear motors are described in the article on **Servomotors**). By way of a series of experiments conducted by Michael Faraday in the 1820s and culminating with his famous experiment of October 17, 1831, when he discovered the principle of electromagnetic induction, he had in essence invented the electric generator, among other important machines including the motor, dynamo, and transformer. Later, in explaining his experiments, Faraday developed the lines-of-force theory, although he did not refer to it by that name. He wrote in his diary in 1846, "All I can say is, that I do not perceive in any part of space, whether vacant or filled with matter, anything but forces and the lines in which they are exerted." In retrospect, going far beyond the immediate realm with which he was concerned, Faraday's vision was truly remarkable. Physicists prior to Faraday's time traditionally explained all physical processes by the laws of Newtonian motion and forces of mutual interaction working upon the particle. For the first time in physics, Faraday concentrated his thoughts on lines of force operating in space as of paramount concern, not the electric or magnetic particle. Thus, Faraday, unknowingly, was a forerunner of the modern relativistic era of physics. A few sketches taken from Faraday's diary are shown in Fig. 1.

From its modest start in 1831, the electric motor was years in development. The basic principles were understood, but creating practical operating machines required the ideas of many investigators. The commutator was developed by Henry, Pixii, and Wheatstone (1841); the drum armature by Siemens, Pacinotti, and von Alterneck (1867); ring armatures by Gramme (1870); disk armatures by Desroziers (1885); revolving magnetic fields and ac theory by Ferraris (1885); polyphase motors by Tesla (1888); the squirrel-cage rotor by Bradley (1889); and ac commutator motors by Eickmeyer, Thomson, and Atkinson (1889), among other important aspects of motor and generator design. During the decades that followed, refinement of electric machines (motors, etc.) continued apace. (Electricity on a large scale did not become available to major cities in the eastern United States until 1880s and 1890s, and somewhat later as the so-called electrical age moved westward.) Broad areas of motor improvements have included marked reductions in size and weight, greater efficiency, overload protection, and safety features. At present, much research is directed toward further improving efficiency by way of improved magnetic materials, conductors, thermal insulation, and even the refinement of relatively new concepts, such as linear and planetary motors. See also **Superconductivity**.

This article concentrates on the fundamentals of electric motors and on the major designs that have been traditionally available for a wide variety of uses. The somewhat specialized fields of servomotors and variable-speed motors used in industrial processes, robotics, etc. are discussed in the article on **Servomotors**. The principal classifications of electric motors are listed below and are diagrammed schematically in Fig. 2.

Electric motors are built in a range, varying from outputs of $\frac{1}{100}$ horsepower up to over 1,000 horsepower. A 50-horsepower motor is considered a large one, and the majority of electric motors used are in the range between $\frac{1}{4}$ and 10 horsepower. Standard motor sizes, above the flea power and fractional sizes, are $\frac{1}{4}$, $\frac{1}{3}$, $\frac{1}{2}$, $\frac{3}{4}$, 1, $1\frac{1}{2}$, 2, 3, 5, $7\frac{1}{2}$, 10, 15, 20, 25, 30, 40, and 50 horsepower. Sixty-cycle synchronous speeds are 3,600, 1,800, 1,200, 900, 720, 600, 514, and 450 rotations per minute. Full-load induction motor speeds are 2–5% less than these. The efficiency of the electric motor ranges from 75–95%. It is higher in large motors than in small. Induction motors are more efficient the higher the rated speed, but do motor efficiency is little affected by speed. Efficiency is often secondary to reliability; nevertheless, it is a factor to be considered, particularly if

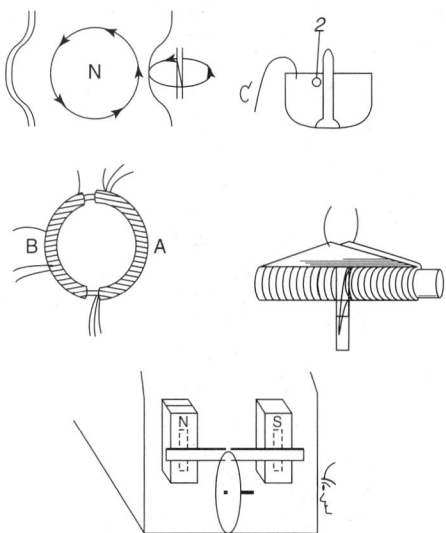

Fig. 1. Series of experiments depicted by Faraday in his *Diary*. (*Top Left*) Copper wire was bent into the form of a carpenter's brace, after which one end of the wire was pressed into a cork that floated on a pool of mercury. The other end of the wire was connected to a battery to complete a wire-mercury circuit. A bar magnet was placed into the curved part of the wire. Faraday noted that when current passed through the circuit, the curved wire moved around until it contacted the fixed magnet. This was the first demonstration of *electromagnetic rotation*. (*Top Right*) Faraday redesigned the experiment, using a straight wire, again with one end impressed into a cork floating in the mercury container. Passage of current through the wire this time made the wire revolve continuously around the magnet and a reversal of current caused the wire to revolve in the opposite direction. This demonstrated the principle of the *electric motor*. (*Center Left*) In an effort to determine if a magnet would turn around a fixed conductor, Faraday wound two coils of wire (A and B) on an iron ring. Then he connected coil A to a battery and found that an intermittent current would flow in coil B. This demonstrated the principle of *electrical induction*. (*Center Right*) In another experiment, Faraday plunged a magnet in and out of a hollow cylinder and coil connected to a galvanometer. This showed that current can be induced as the result of relative motion of a conductor and a magnetic field. This demonstrated the principle of the *electric generator*. (*Bottom*) In another experiment, Faraday rotated a copper disk between the poles of a compound magnet, thus inducing a continuous current. This demonstrated the principle of the *dynamo*. (*Faraday's Diary, circa 1831.*)

the drive is heavy and the motor is well loaded over a considerable part of the time.

Direct-current motors are much less frequently employed than alternating current, because of the preponderance of ac over dc systems. However, speed control and starting torque are so excellent with dc that it is frequently used in ac territory where these characteristics are important. Dc power for motors is commonly obtained from the ac supply through motor-generators, mercury-arc tubes, or grid-controlled rectifiers, with voltages ranging from 110 to 600. The extra expense of the converter installation lays some handicap upon the employment of dc motors, and a number of methods have been devised to vary the speed of ac types, but, in the main, the latter are constant speed.

The losses sustained by a motor in converting electrical to mechanical power arise chiefly through the electrical and magnetic characteristics, but also, in some degree, through bearing friction and windage. The losses are, then, the resistance losses occasioned by current flowing through the conductors of the armature, the field, and the controller, the core losses of hysteresis and eddy currents, friction and windage. The cores of all motors must be built up of laminations insulated from one another by lacquer or enamel, otherwise the core loss becomes excessive.

The relation between input, output, and efficiency is expressed by the following equations, wherein

P = horse power output
η = efficiency
I = line current
V = lin voltage
$p.f.$ = power factor

For all dc motors,

$$P = \frac{\eta I V}{746}$$

For single-phase ac motors,

$$P = \frac{\eta I V\, p.f.}{746}$$

For three-phase ac motors,

$$P = \frac{\sqrt{3}\ \eta I V\, p.f.}{746}$$

Direct-Current Shunt Motor. With reference to Fig. 2, the shunt motor has a wound armature, the ends of the windings of which are brought to a commutator, upon which rest brushes. The incoming leads are connected to these brushes so that line voltage is impressed across the windings of the armature. The stationary field coils are connected across the brushes in shunt arrangement so that they receive a constant voltage. In the illustration, only one coil is shown, but any practical machine would be multipolar. When the motor is running, the coils of the armature cut the lines of force of the magnetic field, and so generate an internal voltage known as the counter-electromotive force. The sum of this counterelectromotive force and the resistance drop through the armature must equal the impressed voltage. Consequently, the current taken is much larger when the motor is revolving slowly than when it is up to speed. This also explains why weakening the shunt-field current (and thereby the magnetic field) causes the armature to increase its speed, since the weakened field causes less counter-voltage, hence allows more current to flow in the armature. Then more torque is produced, which increases the speed until the back voltage allows just the right current to flow to carry the existing load. The torque of the motor is produced by the magnetic reaction existing between the magnetism of the stationary field and the electromagnetic field surrounding the armature conductor.

Direct-Current Series Motor. Unlike the shunt motor, whose field current is practically constant at all speeds, the series motor produces a field, which is maximum during starting and decreases as the motor comes up to speed. For this reason, the series motor has a powerful starting torque, and is used for hoists, traction motors, and the like. The shunt motor is essentially a constant-speed type; the series, a variable speed type. A motor having better speed regulation and starting torque can be obtained by a compound winding having both shunt and series fields. However, the simplicity of the shunt-field motor, coupled with the possibility of effecting a reasonable variation in speed by a variable resistance in the field circuit, has caused it to be widely used. It has been found, though, that any considerable weakening of the field is accompanied by sparking at the commutator, due to the demagnetizing armature reaction. Small poles, located between the shunt-field poles, and wound with series coils, will compensate for the distortion of the field flux, and such motors are known as interpole motors.

Alternating-Current Motors. The foregoing classification of motors indicates a primary division into synchronous and induction type alternating-current motors. Of these, the induction motor is the most important for many applications. However, the strictly constant-speed feature of the synchronous motor, dictates its selection for certain kinds of applications.

Synchronous Motors. The synchronous motor is practically an alternator operated inverted. It has a polyphase stator winding which carries the main line current. The field is wound on the rotor and is excited by dc brought to it by brushes resting on slip rings. The synchronous motor is stable only when operating at a synchronous speed corresponding to the frequency of the system, and if it is loaded to where it lags ever so slightly behind this synchronous speed, it quickly "falls out of step" and comes to rest. The disadvantages of the synchronous motor are principally two: (1) its constant speed, and (2) it requires dc excitation. Modern construction of polyphase synchronous motors results in good starting torque. The single-phase synchronous motor has no starting torque, but with three-phase the motor may be made self-starting if copper bars similar to the rotor of a squirrel-cage motor are embedded in the rotating field and connected to end rings. To start the motor, the dc field is open-circuited and the stator windings connected to the line. The motor will then come up to speed, operating as a squirrel-cage induction motor, after which the field current may be applied, upon which the rotor will lock itself into step with the frequency of the system. A synchronous motor is generally used only

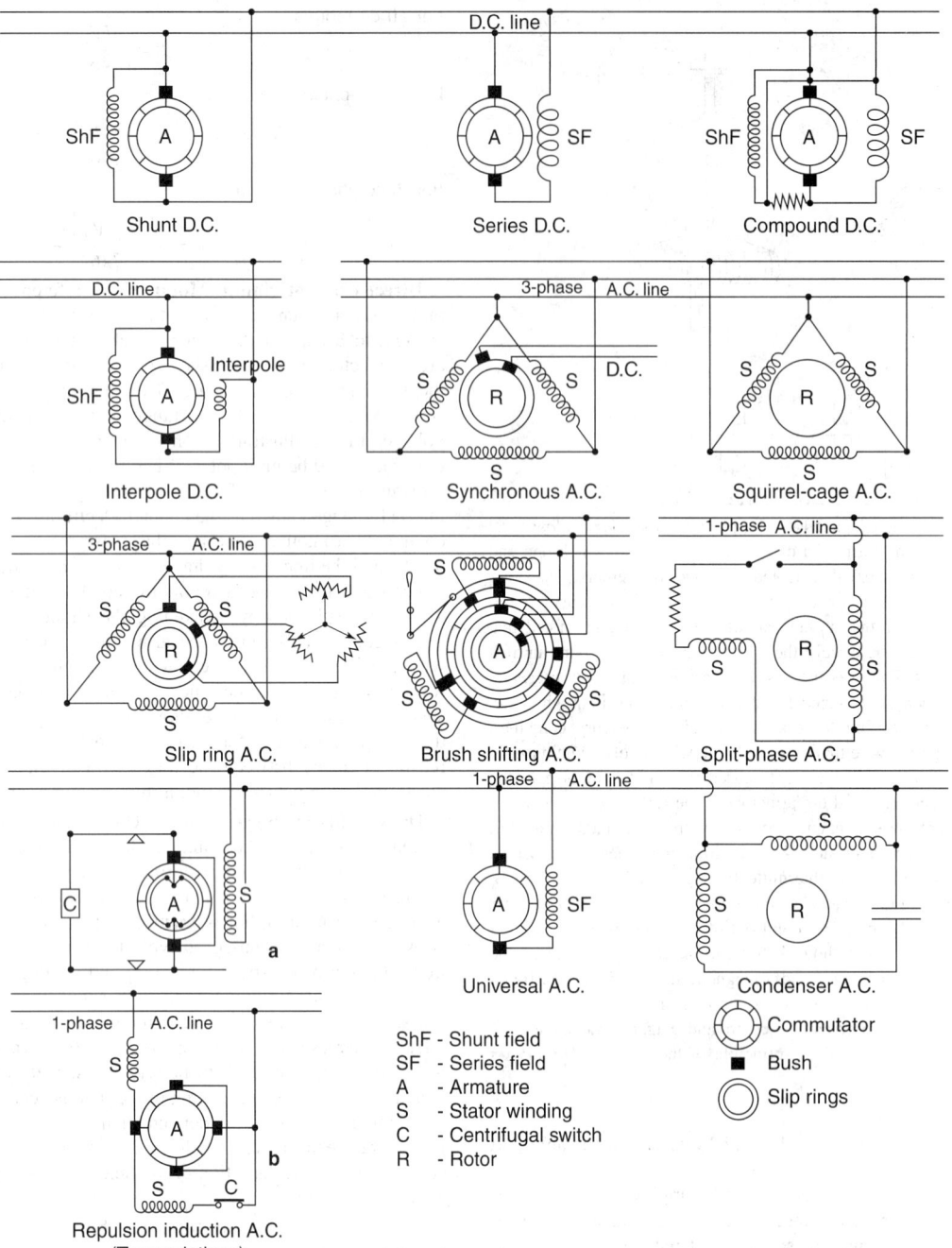

Fig. 2.   Fundamental circuits of some major types of electric motors.

in large sizes where it has the advantage of providing some power factor correction, since one of the characteristics of this motor is that a leading current will be drawn if the dc field is overexcited, and this can be adjusted to neutralize the lagging current drawn by induction type motors.

Induction Motors. The three-phase squirrel-cage motor is one of the simplest and most reliable electric motors made. It has a powerful starting torque, and good efficiency, and would probably replace all other types were it not for the following reasons: It is essentially a constant speed motor, it draws a lagging current, and it is not built single-phase. The stationary windings are connected either in Delta or Y, as may suit the individual design, and are so arranged as to produce a rotating field in the space occupied by the rotor. The rotor is a shaft upon which is built up a laminated steel core carrying embedded in its surface copper or aluminum bars which are parallel to the shaft. The inductive action of the field on this "cage" (if the core were removed, the bars would resemble the familiar squirrel exercising cage) sets up in the latter induced currents whose magnetic field reacts against the rotating field set up by the stator winding, producing a torque. If the rotor were turning in synchronism with the

rotating field there would be no induction and no rotor currents. Therefore it is seen that the rotor cannot possibly operate it at full synchronous speed, even though idling. The difference in speeds is expressed by slip, i.e., difference in speeds divided by synchronous speed, and a certain amount of slip is necessary to secure inductive action. As mentioned before, this varies from 2–5% of synchronous speed. A squirrel-cage motor with rotor blocked acts like a transformer with short-circuited secondary, thus explaining the high starting torque.

Polyphase Motors. There are applications where a polyphase motor is required, having some degree of speed control, and which may be brought up to speed more slowly than is customary with the squirrel-cage motor. For this service the more expensive wound-rotor and brush-shifting types may be used on three-phase circuits. The wound-rotor principle is employed chiefly on large motors. The wound rotor has polar windings in the rotor, the ends of which are joined either in Y or Delta, and brought to three slip rings. The currents induced in the rotor are brought out through these slip rings to an external three-phase resistance, which may be varied at will from zero to maximum. The operation is much like that of a squirrel-cage motor, except that, for starting, the rotor current is decreased by inserting the maximum of resistance in the external circuit. This is gradually decreased as the motor comes up to speed, until all of the resistance is short circuited and the motor is operating inductively with a normal slip. Given constant torque, this motor may be varied in speed by varying the external resistance, but it is somewhat less efficient than the brush-shifting type, because of the energy consumed in the resistance. The brush-shifting motor is used where considerable speed variation is desired at good efficiency, as, for instance, when driving fans or pumps of large size. The brush-shifting motor has the primary winding on the rotating armature, similar to dc practice. This winding is connected to the three-phase line through slip rings. Another winding, called an adjusting winding, is also placed on the rotor; in fact, in the same slots, but is connected with a commutator, which is made fairly wide. The three-phase stator secondary windings are brought out individually to six brushes, which bear on the commutator, and are connected as shown in the diagram. Each set of three brushes is joined by a yoke, so that they may be moved simultaneously, and each pair is placed on opposite ends of the commutator. When these yokes are moved with respect to one another, they cause to be included a certain number of commutator segments in each secondary coil. When each pair of brushes is on a common commutator bar, the motor runs as a straight induction motor at slip frequency. By moving the brushes apart by rotating a yoke, the voltage induced in the commutator coil is added to that in the secondary, and the motor speeds up. These voltages may be subtracted by moving the brush in the opposite direction, resulting in slowing down the motor. Since the forces needed to move the yoke are very small, it may be readily operated by the light pressures produced in an automatic control system.

Single-Phase Motors. It is possible for a single-phase motor to operate inductively like the squirrel-cage motor, provided that it can be brought up to speed, but a single-phase squirrel-cage motor has no starting torque, so that numerous ways have been developed to overcome this limitation. In the split-phase motor, an inductance and resistance are used to displace the voltage at the mid-point so as to get an arrangement resembling a two-phase impressed voltage. Of course the starting torque obtainable is inferior to that of a polyphase motor, but is sufficient to start a motor attached to a drive requiring low starting torque. A fan illustrates this service. For heavy starting duty the starting torque of a single-phase motor is created by repulsion, which shifts over to induction as the motor comes up to speed. Several systems have been invented, two of which are illustrated. At *a*, in the lower left part of the diagram, the armature windings are brought out to a commutator, upon which rest two brushes connected externally by a low resistance conductor. A stator winding is connected across the line. The short-circuited armature has induced in it the large current necessary to secure starting torque. As the motor comes up beyond a certain speed, a centrifugally operated switch lifts the brushes from the commutator and applies to it a ring that short-circuits all the segments. When this is done the motor operates as a straight induction motor. This principle, known as repulsion-induction—i.e., repulsion starting and induction running—is employed in most small motors which are to produce large starting torques on single-phase supply. At *b* is another repulsion-induction principle, less complicated mechanically. Here, the switch operates during starting, and is closed for induction operation.

Universal Motor. This is a motor of the series type, which may be operated on either direct- or alternating- current. It is usually used in small sizes only, there being a compensating coil to prevent armature sparking and to improve the power factor.

Capacitor Motor. This is a split-phase motor, having the phase displaced by capacitance rather than inductance. It is superior to the former in starting torque, efficiency, and power factor, but more expensive and slightly more bulky. The starting torque of the modern capacitor motor compares favorably with the repulsion-start motors, and, since the cost is less, the capacitor motor is replacing the repulsion type in many applications. In some of these motors the capacitor is disconnected by a centrifugal switch after starting; in others, it is left in the circuit to improve the operating characteristics and the power factor of the motor.

See also **Stepper Motors.**

Additional Reading

Gottlieb, I.: "Electric Motors and Control Techniques," The McGraw-Hill Companies, Inc., New York, NY, 1994.
Hughes, A.: "Electric Motors and Drives: Fundamentals, Types and Applications," Butterworth-Heinemann, Inc., Woburn, MA, 1993.
Kaiser, J.: "Electric Power: Motors, Controls, Generators, Transformers," Goodheart-Willcox Publisher, Tinley Park, IL, 1998.

MOTOR FUELS.   See **Petroleum**.

MOULT.   1.   Ecdysis.

2.   The shedding of old feathers by birds preparatory to the development of new. Most birds moult at least once a year, beginning just after the breeding season. Most flying birds shed the large flight feathers in pairs but a few shed them all together and temporarily lose the power of flight. The rest of the plumage is also shed and renewed little by little. Moulting is accompanied in many species by the seasonal changes in plumage which make some species so different in summer and winter. See also **Birds**; and **Poultry**.

3.   The shedding of the outer layer of the skin by reptiles.

MOUND BIRDS *(Aves, Megapodiidae)*.   This family of birds belongs to the suborder *Galli*. They are dark-colored and differ from all other birds in their particular method of incubating their eggs. They are between the size of domestic fowl and turkey. There are 7 genera with 12 species in the southwest of the Old World, with New Guinea as the focus. They are divisible into two tribes according to size.

The mound birds proper *(Megapodiini)* are small dark birds with short tails, often insular: They include (1) the Scrub Fowl *(Megapodius)* with the species Micronesian Scrub Fowl *(Megapodius laperouse)*, found on the Marianas, the Niuafoo Scrub Fowl *(Megapodius pritchardii)* found on Niuafoo (central Polynesia), and the Australian Scrub Fowl *(Megapodius freycinet)* in many subspecies from the Nicobar islands to northern Australia and Polynesia; (2) Maleo *(Macrocephalon maleo)* is found on Celebes; and (3) Wallace's Eulipoa *(Eulipoa wallacei)* is found on the Moluccas.

The large mound builders *(Alecturini)* are larger and are mostly tied to definite habitats. Their distribution is limited: (1) Latham's Brush Turkey *(Alectura lathami)* is found in eastern Australia; (2) Talegallas *(Talegalla)*, with three species on New Guinea; (3) the Combed Talegallas *(Aepypodius)*, with two species on New Guinea; and (4) the Mallee Fowl *(Leipoa ocellata)* inhabits dry areas in central, southern Australia. See Fig. 1.

Fig. 1.   Mallee fowl, *Leipoa ocellata*. *(Sketch by Glenn D. Considine.)*

The ancient Egyptians first built ovens to incubate fowl eggs artificially and we do this today in electrically heated incubators. However, the mound birds, also called incubator birds, "discovered" this method much earlier. Many species lay their eggs near hot volcanic springs or even hotter lava. Others use the heat generated by rotting leaves and vegetation. Still others go to the seashore and lay their eggs in the sand where they can be warmed by the sun.

Mound builders use various means to incubate their eggs. The maleo *(Macrocephalon maleo)* and Wallace's eulipoa *(Eulipoa wallacei)* use only the sun's heat. They emerge from the dark forest and dig into their areas along the shore which are in sunshine. In the jungle fowl *(Megapodius)* the method of egg disposal varies greatly. The Australian scrub fowl *(Megapodius freycinet)* in particular, often selects places heated by volcanic action, such as New Britain and the Solomons. There it digs burrows, which may be up to a meter deep (3.3 feet), into warm soil. In other places these birds build large brood heaps, which have a diameter of 12 meters (39 feet) and a height of 5 meters (16 feet), of sandy soil and leaves. These are the largest structures built by birds. The incubation heat is provided partially by the sun and partially by fermentation of the leaves. In those places directly exposed to the sun, the heaps consist almost entirely of soil. In dense jungle they consist, however, almost entirely of leaves. Each egg is laid in the passage, which has been dug out at a place where there is a suitable incubation temperature. After egg-laying, the birds take no further steps to control the temperature. See also **Maleo**; and **Megapode**.

MOUNTAIN. Mountains may be classified in three chief groups: (1) mountains of accumulation (volcanoes); (2) mountains formed by crustal movements; (3) residual mountains (erosional remnants). Structural mountains are those whose form and relief have not, as yet, been particularly modified by erosion. The ridges are still anticlinal and the valleys synclinal. Later, in the cycle of erosion, the synclines may become mountains and the anticlines, valleys. If the folded mountainous region is ultimately reduced to a peneplain, and the peneplain is then lifted without further folding, a new cycle of erosion operating on the same, but base-leveled, structure will develop the type of topography now seen in the Appalachian mountains. See also **Antecedent Stream**; and **Earth**.

MOUNTAIN BUILDING.   See **Earth**.

MOUNTAIN LION.   See **Cats**.

MOURNING CLOAK (*Insecta, Lepidoptera*).   A large butterfly, *Nymphalis* (*Euvanessa*) *iopa*, of Europe and North America. The wings are very deep maroon above, bordered with yellowish-white, and are slightly angular. The species hibernates in the adult stage and is often in flight on the first warm days of spring. See also **Butterflies**.

MOUSEBIRD (*Aves, Coliiformes*).   Peculiar bird of central Africa, also known as colies, and the only bird in its family. The mousebird is gray with dark tintings and, in a few species, spots of bright colors. The bird is small with a finch-like beak, short weak wings, and long tail. The feet are exceptionally strong. The mousebird only infrequently takes to the air, habituating the lower branches of trees and displaying mouse-like behavior—hence the name. They roost in very large groups, with their heads hanging downward. The diet is principally fruit. Cup-like nests are located in low shrubs and trees. See also **Coliiformes**.

MOVING AVERAGE.   A term used in time-series analysis. Given a series of values $u_1, u_2, \ldots, u_n$, the values

$$v_j = \sum_{k=0}^{m} u_{j+k} w_k$$

where w_k are a set of weights, are defined for $j = 1, 2, \ldots, n - m$. The average of a group of $m + 1$ consecutive terms moves, so to speak, along the series. The moving average is principally used to smooth the series, or to determine a trend in it.

MUCILAGES.   See **Gums and Mucilages**.

MUCORMYCOSIS.   Also called invasive phycomycosis or zygomycosis, this is a rare invasive disease caused by mucor-like fungi of the class Phycomycetes, particularly the genera *Rhizopus* and *Mucor*, which are common airborne molds associated with decaying vegetable matter. They may invade the human body via lungs, sinuses, gastrointestinal tract, or through damaged skin, producing the disease of worldwide distribution. The characteristic trait of the infection is extensive localized necrosis followed by blood vessel invasion, possibly leading to infarction through thrombus formation.

In the compromised host, infection may lead to paranasal sinus destruction (often seen in diabetics with ketoacidosis), necrotic lung or skin lesions, and disseminated disease.

Basic symptoms of infection are fever and unilateral face pain with facial swelling, nasal obstruction, and proptosis. Palatal ulceration may occur, and the organisms may invade the brain.

Localized infections have been seen when surgical or burn wounds receive contaminated dressing packs.

Once infection has spread beyond the original site of invasion, phycomycosis is almost invariably fatal despite treatment. The most heroic measures are wide surgical debridement and amphotericidin B in maximum dosage.

MUCUS.   A clear, slimy secretion secreted by animals where surfaces must be lubricated or moistened. It is produced at the surface of the body of fishes, and in some amphibians. Terrestrial vertebrates secrete it in the linings of the respiratory and digestive systems in abundance. A mucous membrane is a membrane composed of mucus-secreting epithelial cells. Mucous membranes line those canals, cavities, and tracts that communicate with the external air, such as the nose and throat, respiratory tract, generative and urinary passages, and the digestive system.

MUD DAUBER (*Insecta, Hymenoptera*).   Any of several species of wasps that make their nests of mud.

The mud dauber has four wings, two antennae, two forelegs, with four additional legs attached to the hind part of the thorax. The insect is about $1\frac{1}{2}$ inches (4 cm) in length. The insects are referred to as "threadwaisted" because of the thin connection of the thorax and stomach. The mud dauber is a black-blue satin color and is not harmful. Small balls of mud are carried to the insect's nest which is often artistically designed. Separate cells are constructed. Each cell contains an adequate number of paralyzed spiders in which the mud dauber has deposited an egg. This is sufficient food for the larvae until its wings have developed.

MUELLER BRIDGE.   See **Bridge Circuits (Electrical/Electronic)**.

MULBERRY FAMILY.   Widely scattered in all but the coldest regions of the world, members of the mulberry family (*Moraceae*) include over 900 species of trees, shrubs, and herbs. A number are of economic importance. The many species of *Moraceae* contain a milky juice. The flowers are borne in axillary spikes or heads. Many members have dioecious flowers, i.e., the staminate and pistillate flowers are borne on different plants; while others are monoecious. The staminate flower has a variously three- to six-parted calyx, no petals, and one to four stamens with filiform filaments. The pistillate flower has a calyx of three to five, essentially united sepals and a single one- to two-celled superior ovary. The fruit varies greatly in different members of the family.

True Mulberries. Trees and shrubs of the genus *Morus* are widely distributed plants of temperate regions. The leaves are alternate. On a single tree one may observe interesting variations of the leaves; on one shoot they may all be entire, while on a nearby shoot they are variously and irregularly lobed or divided. The flowers develop early in the growing season. The plants are either monoecious or dioecious. The staminate flowers are borne in long catkins and soon fall from the tree. The calyx is divided into four lobes and there are four stamens inserted at its base. The pistillate inflorescence is a short dense catkin. The flowers have a four-lobed calyx and a single one-celled ovary. After pollination the calyx lobes become greatly swollen and fleshy, the individual fruits pressing together tightly to form a multiple fruit. These fruits may be white or pink in the White Mulberry, red in the Red Mulberry, and black in the Black Mulberry. The white mulberry, *Morus alba*, is grown in the Orient largely to supply food for silk worms. The roots of the tree yield a yellow dye, and the wood is used for various purposes. The black mulberry, *Morus nigra*, is grown largely for the fruits, which are greedily eaten by birds, including domestic poultry. The wood is also valuable. Red mulberry, *Morus rubra*, furnishes wood used in making shoe lasts, and for other purposes. Numerous species are planted as decorative trees.

Banyan Tree. The banyan (*Ficus benghalensis*) is found in tropical areas over most of the world. In the United States, the tree prefers parts of Florida and California along the warm coastal areas. The tree is evergreen with leaves from 4 to 8 inches (10 to 20 centimeters) long, alternate and ovate. The fruit occurs in pairs, is red and about $\frac{1}{2}$ inch (1.3 centimeters) in diameter. Banyan trees are considered sacred in parts of India. These trees have a habit of sending down from their spreading branches, aerial roots which enter the ground and increase in size until they resemble trunks. A single tree may have scores and, in some cases, hundreds of such "trunks," thus spreading the base of the tree over an unusually large area.

Breadfruit Tree. Native of the East Indian and Pacific islands, the breadfruit tree (*Artocarpus altilis*), is widely planted in tropical regions elsewhere. The tree not only is attractive and excellent for shade, but is also an important source of staple food, the fruit being used for making poi. When ripe, the fruit, weighing up to 30 pounds (13.6 kilograms), may be used in puddings and other dessert foods. The large ovoid fruits usually have a rough surface, but may be smooth in some varieties. Each fruit is composed of many achenes, each surrounded by a fleshy perianth and growing on a fleshy receptacle. Usually yellowish-brown when ripe, the fruit is roasted or boiled before being eaten. Some improved varieties of breadfruit are seedless. Whereas the fruit of the mulberry tree contains sugar, the breadfruit is rich in starch. The large leaves, from 1 to 2 feet (0.3 to 0.6 meter) in length, are deeply cut into pinnate lobes. Condensation dripping from the deeply-lobed leaves aids in maintaining the soil under the tree moist during dry periods. There are over 40 species of the breadfruit.

Fig Trees. The familiar fig is the fruit of a shrub or small tree (Ficus carica) which is probably native to southwestern Asia. Figs have been cultivated since earliest recorded times, being widely used by the Hebrews, greatly improved by the Greeks, and highly valued by the Romans. The fig tree is now grown in cultivation in nearly all tropical countries and many subtropical regions; in the United States it is grown to some extent in the southwestern states, notably in California.

The plants have alternate leaves, which are rather thick and rough surfaced above, but soft and hairy beneath. The leaves are deeply lobed in the cultivated varieties. The minute flowers are borne on the inside of the hollow receptacle, which develops into a pear-shaped body with a minute opening at its apex. The fruit developing from this is a synconium, composed of many small fruits inserted in the inner wall of a hollow fleshy receptacle. The narrow passage into this is partly closed by numerous small bracts.

Four kinds of flowers occur in fig trees: (1) Staminate flowers, each having four pollen-bearing stamens, occur in the wild "caprifig." A few cultivated forms have staminate flowers. Pollination is brought about by using pollen from caprifigs, and is called caprification. (2) Pistillate flowers, each with a single pistil which if pollinated produces a seed; these flowers are short-stalked. (3) The third flower type is the gall-flower, so-called because a small wasp, Blastophaga grossorum, lays its eggs in them. The developing larvae cause the ovaries to become swollen galls, incapable of developing seeds. This type of flower occurs only in caprifigs, in the basal portion of the synconium. (4) Lastly, in varieties of cultivated figs there are found sterile flowers, which will neither produce seeds nor become galls; these are called mule flowers. Caprifigs contain the first three types of flowers. If pollen is needed to insure fruit development, caprifigs must be planted, since they alone have pollen-bearing flowers. So, among Smyrna fig trees, the fruits of which fail to develop unless pollinated caprifigs must be planted.

In Mediterranean countries, where figs are grown in abundance, three crops are produced each year. The first fruits, known as profichi, are formed in the spring. In the pistillate flowers of these the female wasp lays her eggs, so that galls are formed in the profichi. When the young wasps emerge, these profichi are gathered and hung among Smyrna figs. Wasps escaping have to crawl past the staminate flowers near the aperture of the synconium and so are dusted with pollen. The wasp then enters and pollinates a flower of the second crop, thus insuring fruit development. This second crop is known as mammoni. The third crop, the mammae, remain on the trees. It is in these that the wasp passes the winter.

Fig fruits are gathered and sometimes eaten fresh, but more frequently dried in the sun, then pressed together and shipped. Smyrna figs are often enlarged, by pulling during drying. The fruits produce a mild laxative effect, and so are sometimes prescribed in cases of chronic constipation. From them a wine is sometimes made, and alcohol produced. The wood of the fig is occasionally used in cabinet work.

Propagation is usually by stem cuttings, but sometimes by budding or grafting.

Strangler Fig (Ficus aurea). Closely related to the wild banyan, a single specimen has been known to spread its mass of foliage on stiltlike roots over several acres. The plant bears fruit, but the figs are not edible. The tree has a rough, leathery bark. It is also sometimes called rubber tree; this name arises from the white, sticky sap that exudes when the bark is bared. The leaves alternate on the branches with red fruit; the leaves are shiny and oval in shape. Propagation of the tree is unusual. To turn flowers into seed, an insect must be of minute size to penetrate a tiny hole in bottom of each fig. Since the strangler fig is a killer of other trees, Florida, for example, is fortunate in that it has no insects that will carry out the difficult pollination. Each of the many varieties of F. aurea requires its own kind of pollinating insects. The strangler fig proliferates in locations such as south Florida, because the seeds are dropped by birds, or are carried inland by ocean currents from nearby islands, or by fierce gales that come out of the Caribbean. A seed that falls on bare ground or porous limestone (plentiful in Florida) will germinate into a tree. The strangler fig is an epiphyte (can subsist on air only), but if seed sprouts in a host tree or plant, the early roots of the F. aurea work their way into the bark of the host, thus slowly but steadily depriving the host of light and moisture. As one observer (J. Fix) notes, "Woe betide the Florida tree that gets intimate with F. aurea." Trees frequently affected include the Sabal palm and the Florida oak. The strangler fig is extremely difficult to kill—because of its rampant root system that may reach hundreds of feet in all directions from the tree. This root system also damages water lines and residential drain fields.

Common Hop. The perennial climbing plant, Humulus lupulus, is another member of the mulberry family of economic importance. The plant has an extensive underground stem, or rhizome, from which rise the annual climbing stems, which twine in a clockwise direction around any supporting object. The hollow stem is ridged, with downward pointing hairs along each of the ridges. The opposite leaves are large and palmately veined. Hops are usually dioecious plants. The staminate inflorescence is a loose panicle; the pistillate, spike-like, with conspicuous bract-like structures subtending each branch. The staminate flowers have a five-parted calyx and five stamens; there is no corolla. The pistillate flower is a single ovary, partially surrounded by a small bract and having two long hairy

TABLE 1. RECORD TREES OF THE MULBERRY FAMILY IN THE UNITED STATES[1]

Specimen	Circumference[2]		Height		Spread		Location
	Inches	Centimeters	Feet	Meters	Feet	Meters	
MULBERRY							
Black mulberry (1999)	252	640	78	23.8	76	23.2	Maryland
(Morus nigra)							
Paper mulberry (1991)	157	399	75	22.9	55	16.8	Florida
(Boussonetia papyrifera)							
Red mulberry (1999)	301	765	52	15.8	52	15.8	Tennessee
(Morus rubra)							
Texas mulberry (1997)	36	91	29	8.8	31	9.4	Arizona
(Morus microphylla)							
White mulberry (1992)	292	742	59	18	73	22.3	Missouri
(Morus alba)							
FIG							
Florida strangler fig (1993)	360	914	63	19.2	72	21.9	Florida
(Ficus aureo)							
Shortleaf fig (1993)	248	630	41	12.5	57	17.4	Florida
(Ficus citrifolia)							

[1]From the "National Register of Big Trees," American Forests (by permission). [2]At 4.5 feet (1.4 meters).

stigmas; around the ovary is a cup-like perianth. Hops are wind-pollinated. After fertilization the bracts enlarge greatly. On their outer surface, and also on the surface of the perianth and the subtending bract, yellow grains develop. These grains are called hop-meal and are multicellular cup-shaped bodies developing from single epidermal cells. The cells of these bodies secrete a yellow substance which fills the cup-shaped hollow and which contains the substances which makes hops valuable—an essential oil, resins, tannin, and a bitter substance, probably alkaloidal in nature. The resins are bitter and germicidal.

The principal use of hops is in the brewing of beer. Hops are prepared for the brewing process by drying and bleaching.

Some record trees of the mulberry family growing in the United States are listed in Table 1.

MULE.   See **Horses, Asses, and Zebras**.

MULLETS (*Osteichthyes*). Of the order *Mullus*, there are several species of mullet, distributed mainly in the Mediterranean and Black Seas and the Atlantic Ocean. From there, they move up to the Norwegian coast. The upper jaw has no teeth, but teeth are present on the vomer and the gums.

The *red mullet* (*Mullus ba rbatus*) has a very steep forehead and prefers muddy or sandy ground. The *striped mullet* (*Mullus surmuletus*), with a less steep forehead, is usually found above sandy ground. Both species attain sexual maturity in 2 to 3 years. Spawning generally occurs during summer, off the coast. The eggs float in the water because of their oil bubbles. In the fall, the young move into greater depths. Mullet were famous as food in ancient Rome, despite their small size, bringing a high price. They were brought into the banquet halls alive.

There is another *striped mullet* (*Mugil cephalus*); a much larger and important commercial fish. This species achieves a length up to nearly 3 feet (90 centimeters) and a weight of 31.7 pounds (7 kilograms). See Fig. 1. The color is ash gray with a dark blue shimmer. The sides of the body have 9 to 10 lighter longitudinal stripes. Distribution extends across all warmer seas, including the Mediterranean. Striped mullet are often found in lagoons or river mouths. Their meat is tasty and often, young striped mullet are kept in salt-water or brackish-water ponds. They are fed until autumn, when they reach a marketable size and are shipped off to be sold. There are several other species of mullets (suborder *Mugiloidei*; family *Mugilidae*). They are coastal fishes and have a high adaptability to salt water, brackish water, or fresh water. These lively and schooling fishes are generally found in tidal zones with a rich plant supply above soft ground. Their diet consists of plankton, snails, mussels, and other small organisms associated with algal colonies. They also feed on detritus and very small organisms found on the floor. This is the basis for the generic name *Mugil* (sucker).

See also **Fishes**.

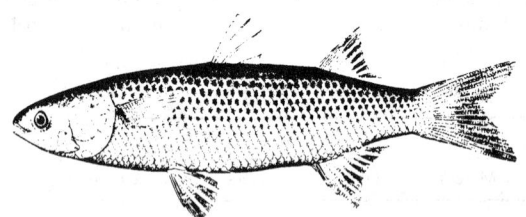

Fig. 1.   Striped mullet.

MULTIPLE INTEGRAL.   The definition of an indefinite integral can be extended to cover the case of a function of several variables and then more than one integration will be required to evaluate the integral. The subject can be considered in various ways but one simple approach is that of partial integration as the inverse to partial differentiation. Thus, given the double integral

$$u = \iint f(x, y)\, dx\, dy$$

one wishes to determine u so that it will satisfy the partial differential equation $u_{xy} = f(x, y)$. The first integration is performed with respect to x, for example, holding y constant, and the second with respect to y, although the order of integrating does not matter. Constants of integration added

to the result complete the work, although these are not really constants but arbitrary functions of the variables. Further generalization to triple, quadruple, etc., integrals offers no further difficulty in principle. Often, for ease in printing, a single integral sign is used for multiple integrals.

The definite multiple integral is commonly of more importance and it may be interpreted geometrically. For example, a function of three variables $f(x, y, z) = 0$ can be considered as a surface. Double integrals, with the appropriate limits, can then be formulated to give the volume of a solid bounded by two or more surfaces, the area of the surface itself, and the moment of inertia of a plane area. Similarly, a triple integral may be used to obtain the volume of a solid or of a closed surface. When given these geometrical interpretations double and triple integrals are often called surface and volume integrals. See also **Area**; **Vector Integral**; and **Volume (Geometry)**.

MULTIPLE INTERFEROMETER DETERMINATION OF TRAJECTORIES (MIDOT).   A trajectory measurement system with multiple-object-tracking capability utilizing two or more short-baseline stations and a data output consisting of a series of amplitude nulls that represent direction cosines at given times in the flight.

MULTIPLE MYELOMA.   See **Bone**.

MULTIPLE OBJECT PHASE TRACKING AND RANGING (MOPTAR).   A short-baseline continuous-wave phase comparison, trajectory measuring system, similar to the *Cotar* which consists of a crossed-baseline angle-measuring-equipment (AME) system and a distance-measuring-equipment (DME) system, wherein time sequencing of the ground station and transponders is used to track multiple targets.

MULTIPLE SCLEROSIS.   A *demyelinating* disease of the central nervous system presently of unknown cause, the onset of which seldom occurs before 15 or after 40 years of age. *Myelin* is the white, fatty substance that forms a sheath about certain nerve fibers. *Sclerosis* indicates the hardening (induration) of the sheath substance, forming zones of demyelination (*plaques*) that range in size and location. Multiple indicates *multiple sites*. Myelinated nerve fibers are minute in diameter, but may be very long. The myelin sheath conducts nervous impulses at a rate which enables muscles to make precise and delicate movements. It is the alteration or destruction of the sheath that interferes with these impulses and thus creates the features of multiple sclerosis (MS). The axon is not damaged. See also **Central and Peripheral Nervous Systems**.

Early symptoms of MS include clumsiness, slowness, stiffness, and weakness (42% of cases). There may be visual disturbances (hazy, misty, blurred vision), sometimes with pain associated with the eyeball and frequently a loss of central vision of one eye (34% of cases). Some patients describe tingling sensations, numbness, and a band-like tightness (18% of cases). There may be nausea, vomiting, and general light-headed feeling (7% of cases). In a minority of MS patients, there may be incontinence, loss of bladder sensation, and diminished or no sexual function (4% of cases). Following a general course, the disease can be highly variable among patients, particularly as to time and severity and mixture of symptoms.

Much has been learned about the nature and occurrence of MS, but the exact cause or causes remain elusive. Certain paths of research are converging on a probable profile of causes and the beginnings of an understanding of the pathophysiology of the disease. As of this time, the etiology can be summarized by the following superficial definition—MS is an exceptionally complex disease, resulting from poorly understood and intricate interaction of genetics, environment, geography, viruses, and the immune system.

Related Factors. For reasons not yet understood, there is a coarse relationship between climate and the incidence of MS. For example, in temperate latitudes, the disease is much more prevalent (50–100 cases per 100,000 population) than in the tropical latitudes, where the incidence is relatively low (5–10 cases per 100,000 population). This immediately suggests that the causative factors must be more prevalent in the temperate than in the tropical zones. But to date the relationship has not led researchers to the cause. Persons who emigrate from tropical to temperate zones apparently bring with them their low risk for the disease. This particularly applies to emigrants over 15 years of age. Multiple sclerosis also occurs more often among urban dwellers *not* from areas of poverty;

this is in marked contrast to the pattern of so many other diseases, particularly of an infectious nature. These epidemiologic aspects of MS have tended to give support to a viral etiology. Also, a coarse relationship of a familial nature exists. It has been estimated that the risk of disease is about 15 times greater among those persons with a family history of the disease than for the population as a whole. The genetic connection, however, has not been worked out in a predictable, clear-cut manner.

Although a number of parallels have been observed among MS patients, such as an elevated level of IgG in the spinal fluid (80–90% of cases), the presence of antibodies to myelin and oligodendroglia, and alterations in lymphocyte distribution, among other factors, diagnosis of MS essentially must be made on the basis of clinical observations.

Although there is no specific treatment for MS, attention can be directed toward partially or temporarily alleviating some of its features. It has been generally believed that corticosteroid therapy is helpful, as in hastening the recovery of optic neuritis. General counsel to patients usually includes the desirability of avoiding excessive fatigue, emotional stress, and significant temperature changes. Pregnancy has not been found to carry a higher risk among MS patients, but usually is not advised because of the long-term prognosis for the mother. Elective surgical procedures are seldom indicated. A number of drugs have been used for controlling spasticity or painful muscle spasms, including diazepam, dantrolene, and baclofen. The physician will pay particular attention to monitoring bladder dysfunction.

Role of Viruses. There was reasonable consensus in the 1980s that probably several viruses play some causative role in MS. Until the mid-1970s, the evidence that a virus may be a causative factor was essentially epidemiological. Data patterns indicated that the disease may arise from a viral infection early in life. Research continues in an effort to prove the virus connection conclusively. Positive proof requires demonstrations such as the production of MS in a laboratory animal after injection with a suspect virus; or isolating a virus from a human with active MS. During the 1970s and continuing into the 1980s, some progress was made toward identifying several viruses in active MS patients. Some investigators found traces of measles virus at different sites in MS patients. Measles virus as a possible causative agent was suggested as early as 1962 when the virus was found in the blood and cerebrospinal fluid of an MS patient. Other investigators subsequently identified abnormal concentrations of antibodies to measles in the blood of many MS patients. More recently, such antibodies were found in the brain tissues of MS patients after death. But these findings are not considered clear-cut evidence of a singular connection with measles virus because abnormal concentrations of antibodies to other viruses, such as rubella, vaccinia, and herpes complex, among others, also have been found. There are also differences in the results of various investigators. Some investigators have found a substance called *blocking factor* in the blood of MS patients. This substance prevents leukocytes from destroying cells infected with measles virus. The substance is not found in normal individuals. Still other investigators have implicated a virus known as *MS-associated agent* (MSAA), which was first studied in 1972. Parainfluenza virus (called 6/94 virus) has been considered suspect by researchers in Germany and Japan. However, when the virus is injected into mice, while a chronic neurological disease is produced, it is quite unlike MS. The apparent persistence of evidence implicating several rather than one virus has bolstered the hypothesis that MS may be the result of several viral infections and that possibly one or two viruses, such as measles or 6/94 virus, may precipitate the disease. Findings also tend to support the hypothesis that MS may be caused by an inborn defect in the immune system of MS patients that permits viruses to proliferate in the central nervous system.

In 1986, a scientist at the University of Chicago observed that there is a serious derangement of immune function in multiple sclerosis. It appears to be an immune attack, but the mechanism is not understood. The immune attack may be against myelin basic protein, a component of the fatty myelin sheath. Other observers continue to associate an inherited predisposition in some persons toward the disease.

In 1985, workers at the National Cancer Institute detected in some MS patients traces of what may be a new viral relative of the human leukemia virus (HTLV-I—human T-cell lymphotropic virus-I). However, there is no evidence that the virus is a cause of MS. The high antibody concentrations found in MS patients may simply indicate their autoimmunity. However, there are precedents for suggesting that the HTLV viruses may produce neurological diseases. HTLV-III not only infects and destroys helper T cells, thereby causing the severe immune deficiency of AIDS, but also

can be found in the brain. Attempts are being made to isolate the HTLV-I related viruses. Should the cloning or virus isolation be successful, this will provide researchers with a specific probe for future studies.

Additional Reading

Cook, S.D.: "Handbook of Multiple Sclerosis," 3rd Edition, Marcel Dekker, Inc., New York, NY, 2001.
Donald W., D.W. Paty, and G.C. Ebers: "Multiple Sclerosis," Vol. 51, Oxford University Press, Inc., New York, NY, 1997.
Feinstein, A.: "The Clinical Neuropsychiatry of Multiple Sclerosis," Cambridge University Press, New York, NY, 1999.
Goodkin, D.E. and R.A. Rudick: "Multiple Sclerosis: Advances in Clinical Trial Design, Treatment and Future Perspectives," Springer-Verlag, Inc., New York, NY, 1998.
Hawkins, C.P. and J.S. Wolinsky: "Principles of Treatment in Multiple Sclerosis," Butterworth-Heinemann, Inc., Woburn, MA, 2000.
Staff: "Multiple Sclerosis: Current Status and Strategies for the Future," National Academy Press, Washington, DC, 2001.

Web Reference

Multiple Sclerosis Foundation: *http://www.msaa.com/*

MULTIPLET. A complex energy level in an atom or molecule, which gives rise to a corresponding series of spectral lines. In an atom, energy levels with a given resultant electronic angular momentum $\mathbf{L}$ and nonzero resultant electron spin $\mathbf{S}$ split into a number of fine structure components with quantum numbers

$$\mathbf{J} = (\mathbf{L} + \mathbf{S}), (\mathbf{L} + \mathbf{S} - 1), \ldots, (\mathbf{L} - \mathbf{S})$$

The resulting multiplet has $2\mathbf{S} + 1$ components for $\mathbf{L} > \mathbf{S}$, and $2\mathbf{L} + 1$ components for $\mathbf{L} < \mathbf{S}$.

The term multiplet is applied also to the narrowly spaced groups of lines corresponding to transitions between the multiplet components of the same or of two different atomic energy levels. A line multiplet may have more components than either energy state involved (*compound multiplet*). See also **Atomic Spectra**.

MULTIPLICATION. An operation which is the inverse of division, used in arithmetic, algebra, and other branches of mathematics. The result of multiplying two or more numbers is a product. Sometimes, but not always, multiplication obeys the commutative and associative laws; in combination with addition, also the distributive law. The commutative law often fails for matrix or group multiplication.

Multiplication can also be considered as successive addition.

MULTIPLICITY. 1. The number $2S + 1$, representing the number of ways of vectorially coupling the orbital angular momentum vector L with the spin angular momentum vector S of an atom. This value represents the number of relatively closely spaced energy levels or terms in an atom which result from the coupling process. The value of the multiplicity is added as a left superscript to the term symbol, as ^3P (triplet P), ^4D (quartet D), etc. The multiplicity of molecules is analogous to that of atoms, and is expressed also by the number $2S + 1$. The multiplicity of an atomic or nuclear level is $2J + 1$, where J is the total angular momentum quantum number of the level.

2. The term is used in biology to indicate that in some cases of irradiation studies a number of targets must be inactivated before any effect is observed: for example, a small colony of bacteria must have hits on all members to prevent growth of the colony. However, the phrase *extrapolation number* has been suggested because factors that cannot affect the number of targets in an irradiated material can sometimes affect the multiplicity: for example, toxic agents released from one cell which may affect another. In either case, one obtains a dose-response curve where the initial low dose gives rise to nearly no observable effect followed by a region of exponential decline of activity with increasing dose. When one extrapolates this latter portion of the dose-response curve to zero dose, the point at which it intersects the biological activity coordinate is referred to as the extrapolation number of multiplicity.

MULTIPOSITION CONTROLLER. An automatic controller that has two or more discrete values of output. Curves indicating this type of control action are given under **Control Action.** In this action, the final controlling element is moved to one of two or more fixed positions, each corresponding to a definite range of values of the controlled variable. The controller can produce a three-position action if it is used with a final controlling element that takes a third position when the variable value is within the differential gap. Inasmuch as multiposition control is capable only of a limited number of corrections, this control action seldom produces an exact correction for any load condition and thus produces continuously cyclic control.

Two Position Controller. This is a special case of multiposition control wherein the controller has two discrete values of output. Curves indicating this type of control action are given under **Control Action**.

On-Off Controller. This, in turn, is a special case of two position control wherein the controller has two discrete values of output, i.e., either *fully on*, or *fully off*.

In the two position mode of control, the final controlling element is moved relatively quickly from one of two fixed positions to the other at a single value of the controlled variable. In the case of an actual controller, as in the instance of an electric temperature controller, when the temperature is at or above the setpoint value, the contact is closed and the valve closes; when the temperature is below the setpoint, the contact is opened and the valve opens. A two position controller cannot make an exact correction. The correction must be greater or less than exact. Thus, no stable balanced condition of input to output energy can be achieved. The controlled variable must cycle.

MULTIPROGRAMMING (Computer System). The essentials of a multiprogramming system in connection with digital computer operation are: (1) several programs are resident in main storage simultaneously, and (2) the central processing unit (CPU) is time shared among the programs. This makes for better utilization of a computer system. Where only one program may reside in the main storage at any given time, inefficient use of CPU time results when a program requests data from an input/output device. The operation is delayed until the requested information is received. In some applications, such delays can constitute a large portion of the program-execution time. In the multiprogramming approach, other programs resident in storage may use the CPU while a preceding program is awaiting new information. Multiprogramming practically eliminates CPU lost time due to input/output delays. Multiprogramming is particularly useful in process control or interactive applications which involve large amounts of data input and output.

Input/output delay-time control is a basic method for controlling the interplay between multiple programs. Where multiprogramming is controlled in this manner, the various programs resident in the storage are normally structured in a hierarchy. When a given program in the hierarchy initiates an input/output operation, that program is suspended until such time as the input/output operation is completed. A lower-priority program is permitted to execute during the delay time.

In a time-slice multiprogramming system, each program resident in the storage is given a certain fixed interval of the CPU time. Multiprogramming systems for applications where much more computation is done than input/output operation usually use the time-slice approach.

Multiprogramming systems allow multiple functions to be controlled simultaneously by a single process-control digital computer. A multiprogramming system allows a portion of the storage to be dedicated to each type of function required in the control of the process and thus eliminates interference between the various types of functions. In addition, it provides the means whereby asynchronous external interrupts can be effectively serviced on a timely basis.

See also **Program (Computer)**; **Time Sharing**; and **Time Slicing**.

THOMAS J. HARRISON, IBM Corporation, Boca Raton, FL.

MULTIVARIATE ANALYSIS. The analysis of data which are multivariate in the sense that each member bears the values of *p* variates. The principal techniques of multivariate analysis, beyond those admitting of straightforward generalism, e.g., regression correlation and the variance analysis, are cluster analysis, component analysis, factor analysis, and discriminatory analysis.

MUMPS. Also known as *epidemic parotitis*, mumps is a contagious disease characterized by swelling of the salivary glands located below the ear and below the angle of the jaw (parotid, submaxillary, and sublingual glands). The parotid glands, located just below and in front of the ears, are the ones principally affected and this swelling is often the first recognizable symptom of the disease. A sudden rise in temperature, to between 40 and 40.5 °C (104 and 105 °F), with or without vomiting and headache may also be a first symptom. The swelling of mumps is firm, and typically obliterates the angle of the jawbone, giving the face a pale, shiny, and somewhat bloated appearance. Enlargement may extend along the neck; the degree of swelling varies with the severity of the attack. Characteristically, swelling appears first on one side and then the other. The interval between enlargement of the opposite sides may be up to 12 days or more; in some cases, the second side never swells. The swelling in each side generally lasts from a week to 10 days; usually the swelling reaches its peak on the third day and gradually subsides thereafter. The early stages of the disease are marked by high fever, headache, pain in the back, reddened taste buds, and loss of appetite. There may be an excess of saliva, or the mouth and throat may be abnormally dry. The initial high temperature gradually subsides, but the patient usually has a mild fever so long as there is any swelling.

Mumps is primarily a disease of children and young adults. It occurs much less commonly later in life. The disease usually occurs in children between the ages of 5 and 15; children between the ages of 7 and 9 seem particularly susceptible. Infants appear to be immune for the first 8 to 10 months of life, and those under 2 years of age are considered just slightly susceptible. Although there can be serious complications of mumps in adults, particularly males, the disease and its complications are rarely fatal.

The causative agent, paramyxovirus, invades the salivary glands. The virus is transmitted mainly through direct contact with an infected person, although about 40% of exposed persons may not have apparent infection, but can infect others. The virus is present in the saliva and in the secretions of the nose. It may be present in secretions for up to 7 days before symptoms develop and for 9 days after swelling subsides. The incubation period of the disease is usually 13 to 21 days.

Firm diagnosis of mumps can be gained by detecting a rise in complement-fixing antibodies. Early in the disease, there is a rise in antibodies to S antigens and later an elevation of antibodies to V antigens. A mumps skin test antigen is not considered reliable. Management of the disease is essentially supportive.

The mumps virus is one of the most widely spread of all disease-causative agents and the disease is encountered throughout the world with exception of isolated locations, such as sparsely populated islands and essentially untraveled jungle communities. However, persons in those isolated areas, not having any immunity, are attacked in large numbers.

In from 5 to 20% of mumps-infected males, orchitis will occur. One-third of the males infected may have unilateral testicular atrophy following the disease. To relieve the pain of orchitis, an anesthetic block of the spermatic cord may be used. Ovarian involvement may occur in about 5% of adult women with mumps. Other organs, such as the pancreas, prostate, seminal vesicles, breasts, thyroid and thymus are less commonly involved. Central nervous system involvement is found in males at a rate 5 times that of women. One of the most serious complications is meningoencephalitis, which can be fatal. Involvement of the pancreas as a complication of mumps has suggested a tie between mumps and diabetes in some patients.

Mumps Vaccine. A live mumps vaccine has been available for several years. In many developed countries, such as the United States, the vaccine is administered almost routinely to children at about 15 months of age — along with vaccine for measles and rubella. The vaccine can be administered at any age. However, the vaccine will not prevent disease if a person has been exposed because the contact will have been shedding virus for several days before parotitis is evident. Unless contraindicated (for example, in persons who are immunosuppressed, immunodeficient, or pregnant), mumps vaccine is considered effective in almost 95% of persons vaccinated even though the degree of immunization induced is only about 20% as great as that resulting from natural infection. The vaccine has been instrumental in markedly reducing the occurrence of mumps as a general disease. Although mumps vaccine has been shown to reach the placenta, there is no evidence to date that the fetus will become infected.

Mumps virus is enveloped, containing single-stranded RNA. It averages 140 nanometers (0.14 micrometer) in diameter. Mumps hyperimmune globulin is sometimes administered to persons who have been exposed to the virus. However, there is little evidence that this is an effective preventive measure.

Web Reference

Center for Disease Control and Prevention: *http://www.cdc.gov/health/diseases.htm*

ANN C. DEBALDO, Ph.D., University of South Florida, Tampa, FL.

MUON. The *muon* (μ^-) is an elementary particle of the lepton family. Properties include: Spin, $\frac{1}{2}$; mass (MeV), 105.66; lifetime, 2.20×10^{-6} second. The antiparticle is the positive muon (μ^+). The muon neutrino (ν) has spin, $\frac{1}{2}$; 0 mass; and is stable. The muon family appears to be simply a duplicate of the electron family except for a change in the unit of mass. See also **Particles (Subatomic)**.

The positive muon was discovered in cloud chamber photographs made by C.D. Anderson and S.H. Neddermyer on Pike's Peak in 1935, and the negative muon almost simultaneously in cloud chamber photographs made by J.C. Street and E.C. Stevenson. These particles have long been called mu-mesons, but since they are fermions (spin $\frac{1}{2}$) while all other mesons are bosons, the name *muon* is preferred, as is their classification with the leptons because of their small rest mass, which is about 206 m_e, where m_e is the mass of the electron. Another reason is their inability to interact with other particles through the nuclear forces.

Their charges are equal in magnitude to that of the electron. They are produced by the decay of pions (*pi mesons*) and (to a limited extent) by the decay of kaons and hyperons. Positive-negative muon pairs also can be generated by the action on matter of gamma-rays of energy greater than the rest masses of the particles, i.e., exceeding 211 MeV. Their lives are short, about 2.2×10^{-6} seconds in the free state, and the negative muon usually decays into an electron, a neutrino, and an antineutrino, while the positive muon usually gives a positron, as well as a neutrino and antineutrino. As explained in the entry on neutrino, there are two types of neutrinos and antineutrinos (ν_e or $\bar{\nu}_e$) like that produced in the decay of radionuclides, and a muon-associated neutrino or antineutrino (ν_μ or $\bar{\nu}_\mu$) so that these reactions would be written

$$\mu^- \rightarrow e^- + \nu_\mu + \bar{\nu}_e$$
$$\mu^+ \rightarrow e^+ = \bar{\nu}_e + \nu_e$$

Muons can easily penetrate many meters of iron and can sometimes cause problems in particle physics research. For example, the upsilon experiment at Fermilab in 1977, conducted by L.M. Lederman and others, required building a simple magnetic system that would remeasure each muon's energy after it emerged from the main detector. See also **Upsilon Particle**; and **Particles (Subatomic)**.

MUONIUM. The atom consisting of a positive muon and an electron. Thus, muonium may be regarded as a light isotope of hydrogen in which the positive muon replaces the proton. When a beam of positive muons is stopped in a gas (argon under such pressures as 50 atmospheres has been used in much of this research), muonium is formed directly in its ground state by the capture of an electron by a positive muon. The reaction is important because of its bearing upon the nature of the muon-electron interaction and the muon itself. The study has included measurement of the hyperfine structure interval in the ground state of muonium, and measurement of muon polarization as a function of time and impurity concentration. By adding such gases as oxygen (O_2) and nitric oxide (NO) as impurities to the argon, data on spin exchange of electron and muon is obtained, while with impurities such as nitrogen dioxide (NO_2) and ethylene (C_2H_2), evidence of such reactions as $NO_2 + M \rightarrow NO + OM$ and $C_2H_4 + M \rightarrow C_2H_4M$ is obtained.

MURINE. Referring to rodents, notably mice and rats.

MUSA ANTENNA. A multiple-unit steerable antenna consisting of a number of stationary antennas, the composite major lobe of which can be aimed electrically.

MUSCLE. An organ formed of a bundle of contractile fibers attached to parts of the body which are moved in relation to each other when it shortens. Among the invertebrates, the fibers of a muscle are loosely associated, but in the vertebrates they are bound together and enveloped by special tissues.

The typical vertebrate muscle is surrounded by a connective tissue sheath, called the external perimysium, which continues into it a series of septa (septum), called the internal perimysium. Between the septa lie bundles of muscle fibers, each surrounded by a delicate continuation of the connective tissue closely joined to the surface of the fiber. Blood vessels and nerves supplying the muscle course through the perimysium.

Muscular tissue is made up of long slender cells specialized for contraction. It is usually mesodermal in origin. There are three kinds of muscular tissue: smooth, cardiac, and striated or skeletal. Among the invertebrates all muscle may be smooth, as in many worms, or striated, as in most arthropods. All three kinds are found in vertebrates and cardiac muscle is characteristic of the vertebrate heart. Muscle cells may contain a few scattered myofibrils, or a large number, occupying most of the cytoplasm. The myofibrils consist fundamentally of protein chains which are responsible for the contractile properties of muscles. See also **Contractility and Contractile Proteins**.

The smooth muscle tissue of vertebrates consists of slender tapering cells with a nucleus placed centrally. These cells are involuntary in action and are found in the walls of the alimentary tract excretory system, blood vessels, and other organs.

Striated muscle is so named because the many fibrils in its cytoplasm are made up of altering zones of different refractive quality which cause the fiber to show light and dark transverse bands. The fiber is long and cylindrical and contains many nuclei located at the periphery. It is surrounded by a delicate membrane, the sarcolemma. Most striated muscle is the foundation of voluntary movements and from its extensive association with the skeleton it receives the name skeletal muscle.

Cardiac muscle, like skeletal, is striated, although the striations are much finer. It differs in the general branching of its fibers and in its centrally placed nuclei. Cardiac muscle is rhythmically contractile, initiating the heartbeat of vertebrates. The nerves that innervate the vertebrate heart serve to modify but not to initiate the contraction of the heart muscle.

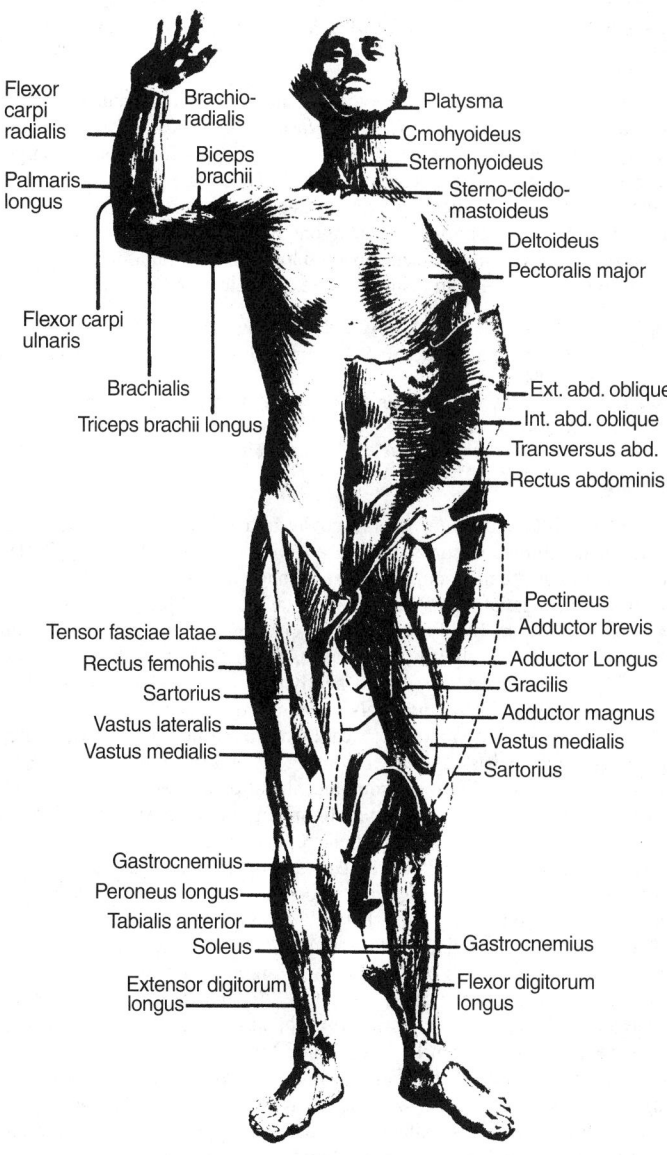

Fig. 1.   Ventral musculature of human.

In the simplest animals having special muscular tissues, the worms, muscle cells pass across the loose tissues within the body and form layers in the body wall. Their disposal is often in a circular layer, with fibers running around the body, and a longitudinal layer with fibers parallel to the main axis. The contractions of these layers lengthen and shorten the body and so carry on the creeping movements characteristic of worms. Special groups of muscle fibers also govern such special structures as the setae of earthworms. These muscles have an attachment to the body wall called the origin, and an insertion in the tissue surrounding the part to be moved.

In animals having a rigid skeleton, an arrangement of muscle fibers, like that of the body wall of worms, persists in hollow organs like the alimentary tract. In organs like the heart and urinary bladder, a less regular arrangement provides for the uniform contraction of all parts of the wall. Locomotion and similar movements, however, are due to the action of muscles on skeletal supports that serve as systems of levers. Usually, the shortening of a single muscle regulates the movement of one part in relation to another; it is said to have its insertion in the part moved and its origin at the other point of attachment. When a muscle is used for slowing a movement, it is stretched; and when it is used for equilibrium, as in standing, the length does not change. These activities are all included under the term contraction for historical reasons.

The nature of movements varies greatly. Extension of jointed append-ages, flexion, retraction, rotation, and other movements are carried out by opposed systems of muscles. Any movement is positive, due to the contraction of a specified muscle, and the return of the part to its former position results from the contraction of an opposed muscle, sometimes aided by the action of gravity. The anatomy of the muscular system and the relations of specified muscles have been worked out in great detail in the human body, in other vertebrates, and in some of the arthropods. See Fig. 1.

Important disorders affecting muscle and the neuromuscular junction include alcoholic myopathy, dermatomyositis, drug- and endocrine-induced myopathies, glycogen storage disorder, malignant hyperthermia (a compli-cation of anesthesia), mitochondrial and lipid storage myopathies, mus-cular dystrophy, myasthenia gravis, myastenic (Eaton-Lambert) syndrome, mycoblobinuria, and polymyositis, among others. See also **Muscular Dys-trophy**; and **Myopathy**. Electromyography is an important diagnostic tool for differentiating muscle disorders. See also **Electromyography**.

Additional Reading

Burke, E.: "Optimal Muscle Recovery," Avery Publishing Group, Inc., Garden City Park, NY, 1999.
Kendall, F. et al.: "Muscles: Testing and Function," Williams and Wilkins, Philadel-phia, PA, 1994.
Matthews, G.: "Cellular Physiology of Nerve and Muscle," Blackwell Science, Inc., Malden, MA, 1997.

MUSCOVITE. The mineral muscovite is a hydrated silicate of potas-sium and aluminum corresponding to formula $KAl_2(AlSi_3)O_{10}(OH)_2$, crystallizing in the monoclinic system although frequently hexagonal found in pseudohexagonal forms. Usually tabular in habit, the most prominent characteristic is the highly perfect basal cleavage yielding remarkably thin laminae, which are often highly elastic. Hardness, 2.5–3; specific grav-ity, 2.77–2.88; luster, vitreous to pearly; color, colorless through grays, browns, greens, yellows, and rarely violet or red; transparent to translucent.

Muscovite is the commonest mica, being found in granites, pegmatites, gneisses, and schists, and as a contact metamorphic mineral, or as a secondary mineral resulting from the alteration of topaz, feldspar, and kyanite. In pegmatites it is often found in immense sheets which are commercially valuable. A complete list of occurrences of muscovite would be impossible. In the United States, excellent specimens are found in the pegmatites of New England, where they are associated with rarer minerals like tourmaline, and beryl. Pennsylvania, Maryland, Virginia, North Carolina, Georgia, South Dakota, and New Mexico also furnish large and fine examples of this mineral. Foreign sources include Canada, Brazil, Norway, Sweden, the former U.S.S.R, and India. A single crystal weighting 85 tons, 10 feet (3 meters) in diameter, 15 feet (4.6 meters) in length was obtained from Inikurite Mine, Nellore, India.

The name muscovite comes from Muscovy-glass, a name formerly much used for this mineral because of its use by Russians for windows. It is in much demand for the manufacture of insulating and fireproofing materials, and to some extent as a lubricant. Muscovite sometimes is called *potash mica*.

MUSCULAR DYSTROPHY. One of a group of muscle disorders (myopathy). Dystrophy distinguishes this disease from other myopathies because of the heritable, progressive nature of the condition. In general terms, muscular dystrophy (MD) is a progressive weakness and wast-ing of muscles. There is no known effective treatment. Because of its genetic origin, genetic counseling prior to marriage and family formation is absolutely essential among thoughtful couples of childbearing age. Bio-chemically, the complex changes that occur during the progressive course of the disease are rather poorly understood even though much research has been devoted to such studies. As of the late 1980s, a number of stud-ies are being directed toward membrane structure and function. Dystrophy is usually accompanied (probably preceded) by a number of abnormali-ties—changes in erythrocytes (red blood cells), alterations in membrane surface ultrastructure, as well as changes of phospholipids, membrane-associated enzymes, and the presence of higher concentrations of calcium (Ca^{2+}) in dystrophic muscle as compared to normal muscle. An alteration of membrane fluidity occurs, but the cause is not explained. Strongly sus-pected are the roles of proteases (enzymes associated with protein) and antigen-antibody complexes. Much more is understood concerning protein synthesis than protein catabolism (destruction-decomposition). See also **Contractility and Contractile Proteins**; and **Protein**. Investigators (State University of New York), working with chickens, showed that certain pro-tease inhibitors (leupeptin and pepstatin) at least partially inhibit muscle degeneration, with success shown by these substances in delaying muscle atrophy. It also has been shown in studies at a number of laboratories that the symptoms of dystrophy can be partially alleviated by means of peni-cillamine, methysergide, and Dilantin. However, in the absence of detailed cell studies, the mechanism by which these drugs were effective is difficult to postulate.

Muscular dystrophy can be subcategorized into:

(1) **Duchenne's (Pseudohypertrophic) Dystrophy**. The sex and genetic linkage of this disorder is relatively well understood. Males between 2 and 5 years of age may develop hip-girdle weakness. This is characterized by a waddling gait and certain muscles, particularly of the calf, will become enlarged. The Duchenne form of MD is the most common type in children and occurs once in every 3000 to 4000 births. It is estimated that one-third of these cases arise from new mutation. By comparison, another X-linked disease, hemophilia, occurs only once in every 10,000 births. Although MD has been considered a disease of males, there are rare case where females also have the disease. The disease was diagnosed in 1977 (Lindenbaum, Oxford University) and it was determined that the child's X chromosome was broken at position Xp21. Later, a number of other girls were found to have similar broken X chromosomal genes. Other scientists who have made traditional linkage analyses trace to Xp21 as the site of the gene. (The general area of location constitutes about 10 million base pairs and represents about 20% of the short arm of the X chromosome.)

Traditionally, genetic researchers know in advance what protein products to seek. In the case of the MD gene, they are seeking the gene in the absence of an understanding of just how the gene functions. For several years, there has been evidence that indicates its approximate location. At least two approaches have been taken. Kunkel (Children's Hospital, Boston) and colleagues have been analyzing in great detail the region of the X chromosome in youths who have deletions of the MD gene. Worton and colleagues (Hospital for Sick Children, Toronto, Ontario) have been searching for the gene in children with translocations that interrupt the MD gene. It is beyond the scope of this encyclopedia to report on the fine details of these and other researches. Probes have now been developed which now can detect carriers with 98% accuracy, an invaluable achievement for assisting families with at least one child with the disease who want to know who carries the gene and who does not.

Currently, the prognosis is discouraging—confinement to a wheelchair during the teens, with fatalities often occurring prior to age 25. Not uncommon in association with this disorder are weakness of the heart muscles and a moderate mental retardation. Confirmation of the condition is found in the high serum creatine phosphokinase (CPK) levels of these individuals. A high percentage (up to 70%) of the female carriers of the disorder will also have high CPK levels.

(2) **Facioscapulohumeral Dystrophy**. This is a disorder that is not usually evident until adolescence. Because of the slow course of the disease, patients may live for many years. The disorder is characterized by a wasting of facial, pectoral, and shoulder-girdle muscles, making it difficult for the patient to smile, close the eyes, whistle, etc. In most cases,

the lower limbs also will be involved. The heart muscle is not usually involved and CPK serum levels are not usually elevated.

(3) **Limb-Girdle Dystrophy**. This disorder involves the shoulder and/or hip girdle and varies in intensities among families. The onset usually occurs between the age of 20 and 50 years of age. There is no known effective treatment. The physician will be careful to differentiate this disorder from chronic polymyositis because it has been found that corticosteroids are sometimes helpful for the latter condition.

(4) **Ocular Dystrophy**. This is an uncommon disorder that is associated with disturbances of the retina, peripheral nerves, and central nervous system. Progressive external ophthalmoplegia may be the only feature of the disease. Like other dystrophies, there is considerable variation among families.

(5) **Oculopharyngeal Dystrophy**. This is also uncommon and usually commences during middle life. Symptoms include seeing, hearing, and swallowing difficulties.

(6) **Myotonic Dystrophy**. This is also heritable, usually becoming apparent during the teen years. Principal features are a wasting of facial muscles (usually producing facial features that are characteristic of the disorder), with accompanying difficulties in swallowing and cardiac arrhythmias. There are several other features which may develop as the result of neuromuscular degradation, including involvement of the muscles that serve the extremities, cataracts, premature balding, and sometimes various endocrine abnormalities, including diabetes mellitus and gonadal atrophy.

Additional Reading

Burnett, G. and S. Rioux: "Muscular Dystrophy," Silver Burdett Press, Englewood Cliffs, NJ, 1995.
Emery, A.: "Muscular Dystrophy: The Facts," Oxford University Press, Inc., New York, NY, 2000.
Siegel, I.: "Muscular Dystrophy in Children: A Guide for Families," Demos Medical Publishing, New York, NY, 1999.

Web References

Muscular Dystrophy Association: *http://www.mdausa.org/disease/als.html*
United States Centers for Disease Control and Prevention *http://www.cdc.gov/search.htm*

MUSHROOM. See **Agarics**; **Basidiomycetes**; and **Fungus**.

MUSKEG. A bog, usually a sphagnum bog, frequently with tussocks of deep accumulations of organic material, growing in wet, poorly drained, boreal regions, often areas of permafrost. Tamarack and black spruce are commonly associated with muskeg areas.

MUSKELLUNGE. See **Pike**.

MUSKMELON. See **Cucurbitaceae**.

MUSK OX. See **Goats and Sheep**.

MUSKRAT. See **Rodentia**.

MUSSEL. See **Mollusks**.

MUSTARD. See **Flavorings**.

MUSTARD FAMILY. See **Brassica**.

MUSTELINES (*Mammalia, Carnivora*). As shown by the organization of the *Mustelines* in Table 1, there are four main groupings—the weasels, the badgers, the skunks, and the otters.

The *weasel* is a small animal, slim and short-legged, related to the minks, ferrets, and martens. Several species of weasels occur in Eurasia and a dozen species in North America have been identified. The common or long-tailed weasel, *Mustela noveboracensis*, ranges from Illinois to the Carolinas and northward into Canada. The animal is brown above and yellowish below in the summer, becoming white in winter, with a black-tipped tail when in the northern part of its range. The winter phase also has been called ermine. The name long-tailed weasel applies also to another species, *M. longicauda*, found only on the plains from Kansas northward. The short-tailed weasel, *M. cicognanii*, resembles the common species, but

TABLE 1. GENERAL ORGANIZATION OF THE WEASELS, BADGERS, SKUNKS, AND OTTERS

Mustelines	
Weasels (*Mustelinae*)	Skunks (*Mephitinae*)
True Weasels (*Mustela, . . .*)	Hog-nosed Skunks (*Conepatus*)
Polecats (*Putorius*)	Striped Skunks (*Mephitis*)
Minks (*Lutreola*)	Spotted Skunks (*Spilogale*)
Martens (*Martes, . . .*)	
Tayras (*Tayra*)	Otters (*Lutrinae*)
Grisons (*Grison, . . .*)	
Striped Weasels (*Poecilogale* and *Poecilictis*)	Common Otters (*Lutra*)
	Simung (*Lutrogale*)
Zorilles (*Zorilla*)	Clawless Otters (*Amblonyx*)
	Small-clawed Otters (*Aonyx* and *Paraonyx*)
Badgers (*Melinae, . . .*)	Saro (*Pteroneura*)
	Sea-Otter (*Enhydra*)
Wolverines (*Gulo*)	
Ratels (*Mellivora*)	
Eurasian Badgers (*Meles*)	
Sand-Badgers (*Arctonyx*)	
Teledu (*Mydaus*)	
American Badgers (*Taxidea*)	
Tree-Badgers (*Helictis*)	

Fig. 1. Weasel. (*A.M. Winchester.*)

is white below in summer. It ranges from the northern states to Alaska. The common weasel is shown in Fig. 1.

The *stoat*, *M. erminea*, is a slender, short-legged animal related to the weasels and sometimes called the greater weasel. It has been regarded by some zoologists as common to both hemispheres, living in northern latitudes, but authorities now consider the common North American weasel to be distinct, although a closely related species. Except in the extreme southern parts of their range, both of these animals turn white in winter.

The *polecat* is a long-bodied animal with short legs and is closely related to the weasels and minks. The polecat is known for its disagreeable odor, a characteristic responsible for the occasional use of this name for a skunk. The European polecat, also known as the foumart or foul marten and as the fitchet, fitcher, or fitcheu, is the source of the fur known on the market as fitch. Three other species occur in Europe and Asia. One, the black-footed polecat, commonly called the black-footed ferret, is found in the plains region of North America. The Cape polecat of South Africa is more closely related to the skunks than to the true polecats. The polecat feeds on small animals, birds, poultry, snakes, lizards, and frogs. The animal is considered bloodthirsty and hunts at night.

The *mink* is a slender semiaquatic animal with short legs, partially webbed toes, and a short, moderately bushy tail. It is related to the polecats.

One species, *Mustela sibiricus*, occurs in Siberia; the *M. lutreola* is found in Eastern Europe; and the *M. vison* is found throughout North America. The fur varies from light to very dark brown and is thick and soft, mixed with long glossy hairs.

The *marten* is a slender animal with short legs and a moderately long bushy tail. Related to the weasels, the marten is larger. Several species live in the Northern Hemisphere. Some of the most valuable furs marketed are from martens, and all species bear fur of fine quality. The sable of northern Asia, *Martex zibellina*, and the marten of northern North America, *M. americana*, produce the most valuable fur, although that of the fisher, *M. pennanti*, is also good. The fisher formerly ranged over most of North America, but is now found only in the wilder northern areas. It is also called the pekan and the American martens, as well as the American sable.

The *tayra* is a South American animal of the weasel family. It is among the larger species, comparing with the otter in size, and has the characteristic long body and short legs of the weasels, with a long and rather heavily furred tail. The animals range from white to black in color. The grison is a small South American animal with long slender body and tail, short legs, and related to the weasels.

The *wolverine* is classified with the badgers (*Melinae*) and is a stoutly built short-legged animal, *Gulo luscus*, of northern North America, ranging from the mountain forests of Colorado and Pennsylvania to the Arctic region. The animal is over 3 feet (0.9 meter) long, with large feet, a short bushy tail, and shaggy fur. It is also known as the glutton. In French Canada, it is termed the carcajou. The wolverine is noted as a vicious killer and as a despoiler of camps. It not only damages foods that it cannot eat, but steals objects that it cannot use and so becomes a general nuisance to campers and trappers in the northern woods. Since it climbs, gnaws, and digs with great facility, protection of supplies is difficult. Wolverines eat dead animals as well as prey that they kill, and often they rob trap lines. A second species of wolverine, *G. luteus*, of slightly smaller size and different proportions, has been found as far south as the Sierra Nevadas in California.

The *ratel* is a short-legged animal with a broad depressed body and head, resembling the badger in form. The ratels (*Mellivora indica*) live in India and *M. capensis* in south and west Africa. They are nocturnal burrowing animals and eat other animals and insects. From their love of honey, they have been called the honey badger.

The *badger* is a stoutly built and short-legged burrowing animal. There are several species found throughout the Northern Hemisphere. The term is extended to the ferret-badgers, which are related to the skunks and to the ratels of India and Africa. Badgers are courageous and able fighters when attacked and are sometimes hunted for sport, but they are peaceful animals unless molested. The fur is moderately valuable and the long hairs, especially of the European species, *Meles taxus*, are used in making brushes. Badger-hair shaving brushes widely used for many years are an example. The badger measures about 28 inches (71 centimeters) long. It has a semiflat body, broad and flat tail. The fur is long, gray to red-brown. The feet and ears are black. The diet is mainly rodents. The ferret-badger occurs in several species in the forests of the Oriental region. They are intermediate between the ferret and the more stocky badger.

Skunks are animals of the Americas and widely known for their foul odor. The species are black and white, with a long bushy tail, under which are located the two glands that produce the defensive secretion. The skunks of North America have been divided into 15 species and a number of varieties. These species are grouped as the striped or common skunks and the small spotted skunks, belonging respectively to the two genera, *Mephitis* and *Spilogale*. Both genera are widely distributed. The fur, especially of the larger striped skunks, is of commercial value.

The *otter* is an elongate animal with short legs, broad head, and webbed toes. The otter is an excellent swimmer, sometimes appearing to catch fish simply for the pleasure of the pursuit. Otters are found on all continents except Australia. Different species vary from 2 to 3 feet (0.6 to 0.9 meter) in length, exclusive of the long tail, but there is relatively little difference in their general appearance. *Lutra vulgaris* is the common otter of Europe and the closely related *L. canadensis* is found in North America. Otter fur is thick, soft, and glossy, and among the valuable commercial furs. The fur of the sea otter, *Enhydris* (*Latax*) *lutris*, one of the largest species, is regarded as highly valuable. The species inhabits the northern Pacific, in both Asiatic and American waters, where it feeds on marine invertebrates. Once abundant, indiscriminate trapping has greatly reduced the population. Sea otters are known to travel under water for a half-mile (0.8 kilometers)

without surfacing. They can travel on land faster than man can run. For a den, they prefer a dirt cave or a place dug out of the ground or old logs. They are reported to be very playful, sliding down mud or snow banks until the bank becomes smooth. It is also reported that they throw stones in water, diving in after them before they sink. The sea otter is readily trained, learning fast, exhibiting great curiosity, and displaying a friendly personality along with unusual intelligence.

MUTATION.   A change in a gene which results in an altered effect of its expression. The great majority of mutations have effects that are harmful to the organism expressing them, but a very few may make the organism better adapted to its environment. Through selection, these favorable mutations are established in a population. The capacity to mutate is an important property of the genetic material since it is the origin of variation in the gene pool. See also **Evolution**; and **Genetics and Gene Science (Classical)**.

Mutations can be of several types. When the chromosome structure is involved, it is referred to as a chromosomal mutation or aberration. When the locus is restricted, the result is called a gene mutation. This type of mutation can be the result of a change in a single nucleotide by substitution of one purine or pyrimidine for another (transition), or of a purine by a pyrimidine or vice versa (transversion). Insertions and deletions of one or more nucleotides can also occur.

Although a gene consists of hundreds and even thousands of nucleotide bases, a change in just one of these bases can cause a change in the effect of the entire gene. One base alteration would cause a difference in one codon of the genetic code and a different amino acid might be inserted in the polypeptide chain formed. This could change the entire protein greatly. An example to illustrate is found in human hemoglobin. Normal hemoglobin consists of two pairs of polypeptide chains. Each chain contains about 150 amino acids. A change in one right of the DNA ladder causes the substitution of one amino acid, valine, for glutamic acid in the pair of beta chains. This small change is sufficient to alter the entire hemoglobin molecule and cause it to form in chains with other molecules when the oxygen tension is reduced. A severe anemia, sickle-cell anemia, results when this reaction takes place in the body. The red blood cells form into sickles and do not carry oxygen properly. Another change in the DNA ladder causes lysine to be substituted at this point on the polypeptide chain, resulting in hemoglobin C, which is also abnormal.

Spontaneous mutations are caused by unknown and omnipresent mutagenic agents (mutagens). Induced mutations are produced at will by subjecting the genetic material to a variety of known mutagens. The comparatively high mutation rates so obtained are superimposed on the spontaneous mutation rates which, as a rule, are quite low. Study of the action of these known mutagens on DNA (deoxyribonucleic acid—see also **Cell (Biology)**) furnishes information as to the chemical nature of gene action in general, and mutation in particular. Some typical mutagens and their possible effects on DNA are listed in Table 1.

ANN C. DEBALDO, Ph.D., University of South Florida, Tampa, FL.

MYALGIA.   Aches and pains in the muscles.

MYASTHENIA GRAVIS.   A disease characterized by excessive fatigability of the muscles after mild or moderate exertion. The muscles about the head and neck are the most seriously involved. In some instances, toward the end of a meal, the jaw muscles may be so tired that further chewing is impossible. After a few minutes of rest, there is sufficient strength to start again, but the fatigue returns rapidly. Difficulty in swallowing or keeping the eyes open may be noted. The disease affects about 5 persons per 100,000 population. In women, the onset usually is during the third decade, whereas for men, onset usually is in the seventh decade. The incidence is about 20% higher among females. The illness takes many forms, ranging from a mild and restricted disturbance of ocular muscles to that of a rapidly developing, generalized, and sometimes fatal, result. About 10% of myasthenia gravis cases are ultimately fatal. For persons who survive the first 3 years of the disease, the disease may stabilize, with a reasonable possibility for some recovery from symptoms.

The disease is currently poorly understood, but as of the early 1980s some scientists believe that monoclonal antibody techniques may ultimately lead to a cure. In the early stages, these techniques will lead to a better understanding of the mechanism of the disease.

TABLE 1. TYPICAL MUTAGENS AND THEIR POSSIBLE MUTAGENIC EFFECTS ON DNA

Mutagen	Possible Mutagenic Effect on DNA
X-rays	Largely unknown — probably caused by free radicals produced in aqueous solution
Ultraviolet radiation	Largely unknown — dimerization of thymine has been suggested
Auto-oxidizing agents, such as Fe^{2+}	Largely unknown — probably caused by free radicals produced in aqueous solution
Heat	Depurination or ionization of guanine at N-1, followed by a change in base pairing.
Alkylating and esterifying agents:	Alkylation of N-7 of guanine, possibly followed by depurination or ionization of guanine, and a change in base pairing
Mustards Dimethyl- and diethyl sulfate Beta-propiolactone Ethylmethane sulfonate	
Nitrous-acid	Deamination of adenine, guanine, and cytosine, followed by a change in base pairing
Hydroxylamine	Attachment to the C−C double bond of cytosine, followed by a change in base pairing
Base analogues:	Incorporation into DNA in place of thymine, followed by a change in base pairing
5-Bromouracil 5-Iodouracil 5-Chlorouracil	
Deuterium	Replacement of hydrogen, causing an unknown disturbance
Ionizing radiation	See also **Cancer Research**.

Confirmation of the clinical diagnosis can be made with the edrophonium (Tensilon) test. Tensilon is a cholinesterase inhibitor with a brief (5 to 10 minute) duration of action. A marked improvement in power, which occurs in 30 to 60 seconds and lasts a few minutes, will confirm the diagnosis of myasthenia gravis in a majority of cases. Electromyography also may be an aid in diagnosing the disorder. Routine chest x-ray may yield helpful information as well as evaluation of thyroid function. The incidence of thyroid disorders is about 13% among myasthenic patients. Thymectomy has been used in a number of instances where myasthenia persists.

MYELOMA.   Essentially a tumor composed of cells normally found in the bone marrow (plasma cells). The stimulus for malignant conversion of the plasma cells in humans is not known, but the original mutation and much of the subsequent cell multiplication probably occurs at an early stage in the B cell maturation sequence. The cells grow relatively slowly with a doubling time of about four months. Electron microscopic studies of malignant plasma cells show megaloblastosis, that is, cytoplasmic maturity and nuclear immaturity.

Although myeloma may be classified, on the basis of the secreted protein, into a number of different immunochemical categories, there is little clinical difference between these varying groups. Myeloma resembles other malignancies in having its peak incidence in the seventh decade. The frequency is about five new cases per 100,000 of population, with commonality between sexes.

In about 10% of cases, the diagnosis is made by chance during blood chemistry analysis. Some 60% of patients present with skeletal pain, particularly affecting the back and ribs with deterioration of general health, fatigue, malaise, and anorexia. About 50% of myeloma patients also show some evidence of renal impairment and bruising and epistaxis may be seen. A peripheral neuropathy may occur and about 10% of patients present with paraplegia.

Soft tissue plasmacytomas may be found but they are of little prognostic significance. Occasionally patients present with a solitary erosive bone deposit of myeloma.

In myeloma, there is an increased susceptibility to infection, associated with depression of normal serum immunoglobulin levels and a defective antibody response. In some patients, the underlying diagnosis may be made first at the time of an infection.

The diagnostic triad of myeloma includes the presence of osteolytic lesions in bones (well resolved by x-ray), atypical plasma cells in the bone marrow, and demonstration of an M protein band in the blood serum.

In treatment, efforts should be made to avoid hospitalization. A graded exercise program should be encouraged for increasing muscular tone and reducing osteoporosis. Solitary plasmacytomas may be treated by radiotherapy and a proportion of apparent cures is achieved. Where chemotherapy is indicated, melphalan and cyclophosphamide are the principal drugs used under careful monitoring. About 16% of patients fail to respond. The general management of patients is extremely important. Much can be done to relieve discomfort and pain with appropriate analgesics, but sooner or later the patient's condition escapes from control and death is ultimately due to renal disease, infection, hyperviscosity, or hemorrhage.

MYNA *(Aves, Passeriformes)*.   There are at least 110 species of this Old World family of birds, which are found in India and in many parts of Southeast Asia. They are related to the starlings. Most species have glossy, showy plumage, often with orange-yellow wattles. The face is partly naked. Most of the body is a purplish- or sooty-black. The myna feeds mostly on fruit, insects, and grain. Large groups numbering in the thousands may destroy crops and otherwise become a nuisance to communities, as do their relatives, the starlings. In some species of myna, the wings have odd notchings, which assist in making various mechanical noises characteristic of these birds. The Indian species (*Gracula acridotheres*) is quite adept at mimicking human speech and is frequently kept as a pet. Most species appear to like people. The species, *Artamus maximus*, sometimes called the Papuan wood swallow, is found in New Guinea. These are among the few species of birds that exhibit communal nesting and it is not uncommon to witness three adult birds sitting on a nest at one time. These birds are sooty black with white breasts. There remains much knowledge to be gained of some of the primitive species of myna which occur on many of the isolated Pacific islands of Melanesia and Micronesia.

MYOPATHY.   Disease and degeneration of muscle and the neuromuscular junction make up a number of specific disorders, the majority of which are, at least in part, of genetic origin. Probably the best known of these is muscular dystrophy, which in itself has several subcategorizations. See also **Muscular Dystrophy**. *Myopathy* is a general term, taking in all disorders of muscle. *Dystrophy* refers to a specific kind of heritable, progressive myopathy.

Sometimes muscular weakness may be seen in association with various metabolic disorders, also usually found to be of genetic origin. It is sometimes difficult to associate the relationship — that is, which of the factors may be causative, which may be results, and which may simply occur together. There are inflammatory muscle diseases, such as *polymyositis* and *dermatomyositis*. *Myasthenia gravis* appears to be related to the function of the immune system. Some drugs, including a number of antibiotics, may contribute to muscle weakness, particularly in individuals who are prone to muscular disorders.

McArdle's disease is probably the better known of the glycogen storage diseases. In this disease, there is a deficiency of the enzyme *myophosphorylase*. This is a heritable disorder that usually appears during the late teen years. Commencing with pronounced muscle cramping and stiffness, which is accentuated by exercise, this disorder normally develops slowly into progressive muscular atrophy with consequent weakness of a permanent nature. *Pompe's disease* is due to a deficiency of the enzyme, α-1,4-glucosidase (*acid maltase*) and is usually fatal in infants. The disease usually involves the brain, liver, and heart in children. In an adult, there may be a chronic, indolent myopathy without involvement of other organs. In *Forbe's disease*, there is a deficiency of the enzyme amylo-1,6-glucosidase (*debrancher enzyme*) whereas in *Tarui's disease*, the deficiency is of the enzyme *phosphofructokinase*. These are uncommon diseases with definite genetic connections.

There are several disorders of energy metabolism in muscle, the causes of which are not fully understood. These include *mitochondrial* and *lipid storage* myopathies.

In certain individuals with a myopathic disorder, there appears to be a genetically determined susceptibility to *malignant hyperthermia*, an extremely rare and often fatal complication of anesthesia, particularly when halothane or succinylcholine are administered. Onset is sudden during anesthesia. A temperature as high as 110.3 °F (43.5 °C) may develop, frequently accompanied by acidosis, cardiac arrhythmias, circulatory collapse, coma, muscular rigidity, and frequently death. Muscular rigidity, although not the only major factor in the disorder, does interfere with procedures to assist the patient's ventilation. Prompt treatment measures, including immediate cessation of anesthesia, forced cooling of the body, oxygen therapy, and intravenous infusion with bicarbonate have been life saving in a number of cases.

Myopathy is sometimes associated with heavy consumption of alcohol over protracted periods. Certain drugs, such as colchicine, corticosteroids, some antimalarials, including chloroquine, and fluorinated compounds, such as triamcinolone, have been implicated in inducing myopathy.

Poorly understood are inflammatory muscle diseases. Most research has been concentrated on finding viral and immunologic mechanisms. In *polymyositis*, symptoms include muscle weakness as manifested by difficulties in rising from a chair or toilet, stair climbing, etc. Difficulty of swallowing may be present because of muscle involvement in the pharynx. Malaise and low-grade fever may be present. The disease is rarely fatal. *Dermatomyositis* is another inflammatory muscle disease which is accompanied by a skin rash, usually in the area of knuckles, elbows, and knees. The disease varies in intensity. There is a relatively high association with carcinoma in persons over 50 years of age (estimated at 25% of cases). Although the malignancy can appear in any organ, the lung is the most common site.

Myasthenia gravis is relatively uncommon, affecting an estimated 5 persons out of 100,000 population. See also **Myasthenia Gravis**.

Web References

Inflammatory and Immune Myopathies. *http://www.neuro.wustl.edu/neuromuscular/antibody/infmyop.htm*

Mitochondrial Myopathies United States National Institute of Neurological Disorders and Stroke. *http://www.ninds.nih.gov/health_and_medical/disorders/mitochon_doc.htm*

United States Centers for Disease Control and Prevention. *http://www.cdc.gov/search.htm*

MYOPIA (Nearsightedness).   Myopia is the ability of the eye to clearly see objects that are up close, but not far away. A nearsighted person can usually see well to read, but has trouble with distance vision, such as that used in driving. This is a common refractive eye condition created when the eyeball is more elongated than normal from front to back, or the cornea is too steep or dome-shaped. Myopia is an inherited condition that affects about one person in five.

The usual treatment for myopia is prescription eyeglasses with concave (inwardly curved) lenses or contact lenses that counteract the distortion created by corneas that are too outwardly curved in shape. A concave lens moves the image of a distant object backward onto the retina, thereby bringing it into proper focus.

Refractive eye surgery, which flattens the cornea, has also become a popular option for the correction of nearsightedness in recent years. The most popular of those procedures is Laser In-Situ Keratomileusis (LASIK), which uses an Excimer laser to reshape the cornea, often eliminating the need for corrective lenses. Another option that is becoming popular is Intrastromal Corneal Ring Segments (ICRSs), tiny plastic arcs that are implanted in the peripheral area of the cornea, causing the center of the cornea to flatten. These segments can be removed and/or replaced if needed. Refractive eye surgery is usually not recommended for people under 18 years of age. See also **Intrastromal Corneal Ring Segments (ICRS)**; **Laser In-Situ Keratomileusis (LASIK)**; **Refractive Eye Surgery**; **Retina**; and **Vision and the Eye**.

Vision Rx, Inc., Elmsford, NY.

MYRISTIC ACID.   Also called tetradecanoic acid, formula $CH_3(CH_2)_{12}COOH$. At room temperature, it is an oily, white crystalline solid. Soluble in alcohol and ether; insoluble in water. Specific gravity 0.8739 (80 °C); mp 54.4 °C; bp 326.2 °C. Combustible. The acid is derived by the fractional distillation of coconut oil. Myristic acid is used in soaps; cosmetics; in the synthesis of esters for flavorings and perfumes; and as a component of food-grade additives. Myristic acid is a constituent of several vegetable oils. See also **Vegetable Oils (Edible)**.

MYRMECOPHILE.   An animal that makes its home in the nests of ants. Among the insects that have become adapted to this mode of life are crickets, beetles, and larval flies. In most cases the myrmecophiles seem to feed at the expense of the ants, which may in turn eat secretions produced by their guests.

N

NADIR. The point exactly opposite the zenith on the celestial sphere is termed the nadir. See also **Celestial Sphere and Astronomical Triangle**.

NAIAD. An aquatic insect larva.

NAND (Circuit). A computer logical decision element which has the characteristic that the output F is 0 if, and only if, all the outputs are 1's. Conversely, if any of the input signals A or B or C of the three-input NAND element shown in Fig. 1 is not a 1, the output F is a binary 1. Although the NAND function can be achieved by inverting the output of an AND circuit, the specific NAND circuit requires fewer circuit elements. A two-input transistor NAND circuit is shown in Fig. 2. The output F is negative only when both transistors are cut off. This occurs when both inputs are positive. The number of inputs, or fan-in, is a function of the components and circuit design. See also **AND (Circuit)**. NAND is a contraction of NOT AND.

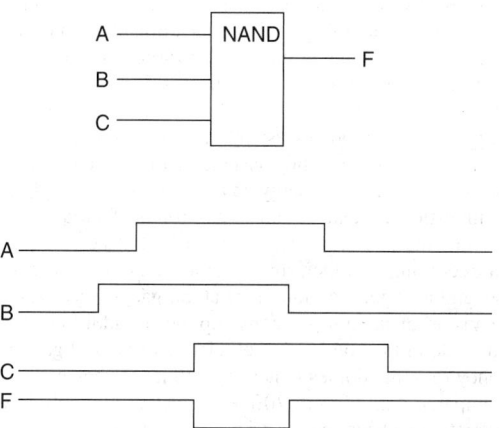

Fig. 1. Schematic of a NAND circuit.

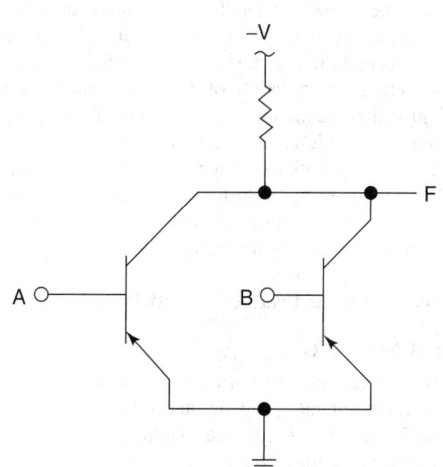

Fig. 2. Transistor-type NAND circuit.

THOMAS J. HARRISON, IBM Corporation, Boca Raton, FL.

NANNOPLANKTON. The portion of the floating and drifting aquatic animals (plankton), including minute species, which pass through ordinary nets and must be secured by centrifuging.

NANO (abbr n). A prefix meaning multiplied by 10^{-9}.

NANOSECOND (abbr nsec). 10E-9 second. Also called *milli-microsecond*.

NANSEN BOTTLE. A bottle used for collecting samples of seawater at any desired depth. After the bottle has been lowered to the desired depth, a weight is sent sliding down the wire to which the bottle is attached. This releases a catch when it strikes the bottle, which in turn closes the valves and traps the water inside. Special thermometers may be used to record the temperature at the desired depth.

NAPHTHALENE. See **Coal Tar and Derivatives**; and **Organic Chemistry**.

NAPIER'S RULES. See **Spherical Trigonometry**.

NARWHAL. See **Whales, Dolphins, and Porpoises**.

NASOLACRIMAL. The lacrimal gland located under the upper eyelid is what makes it possible for the eyes to produce tears, a necessity for keeping eyes moist and a functioning part of our emotions. Because the eyes produce fresh tears continuously, an efficient drainage system is necessary to drain the used tears from the eyes. This system is called the *nasolacrimal system*.

The nasolacrimal system starts with tiny openings on the brim of the upper and lower eyelids near the inner edge of the nose. These lead to the nasolacrimal tear ducts next to the bridge of the nose. The tears move from these ducts into the nasal cavity where they are either swallowed or discharged through the nose.

Occasionally, lacrimal or nasolacrimal functions become impaired, causing irritation and infection, and, sometimes, signifying other vision problems. One of the most common nasolacrimal problems is congenital nasolacrimal duct obstruction, where an infant's blocked tear duct becomes infected, causing matter to collect in the corner of the eyes and between the eyelids. In adults, excessive tearing and drainage problems can signify other areas of concern. It's important to distinguish between the two, because drainage and tearing are two separate issues that must be evaluated by an eye care professional.

The most common symptoms of lacrimal and nasolacrimal impairments are excess tearing (tears may run down the face) and mucous discharge. Eye infections, eyelashes, exposure to the wind, yawning, glaucoma, certain drugs, eyestrain, or even dry eyes can contribute to excessive tearing.

Treatment depends on the cause of the excessive tearing. Sometimes, simply removing an irritant or other environmental conditions contributes to a decrease in tear production. Other times, small plugs can be placed in the opening of the tear duct to decrease the amount of tears produced.

A blocked tear duct forces tears to build up in the eye and run down the cheeks. And because of the lack of drainage, leftover tears can remain within the eye and become infected. Injury, birth defects, consistent nasal infections, narrowing of the nasolacrimal system associated with age, and other infections can lead to improper tear drainage.

An ophthalmologist often instructs the patient to massage the lacrimal sac area several times a day to help unblock the duct. Congenital nasolacrimal duct obstruction almost always resolves itself as children grow, usually between 6 months and 1 year.

Probing and irrigating the nasolacrimal system may be sufficient to relieve the blockage. In some cases, an eye care professional will recommend surgery to open a blocked tear duct. See also **Tears**.

Vision Rx, Inc., Elmsford, NY.

NATIONAL BUREAU OF STANDARDS (U.S.).   The NBS was established by the U.S. Congress in 1901. The name of the institution was changed **just a few years ago** to the National Institute of Standards and Technology (NIST). See also **NIST**.

NATIONAL CENTER FOR ATMOSPHERIC RESEARCH (NCAR). This organization was founded in 1960 by a consortium of 14 universities with doctoral programs in atmospheric sciences. Its staff numbers about 750 people. The headquarters is in Boulder, Colorado. By the early 1990s, the consortium had grown to more than 50 members in the United States and Canada. NCAR's chief source of funding is the U.S. National Science Foundation.

NCAR's research consists of four basic areas: (1) atmospheric chemistry; (2) climate and its links with other environmental systems; (3) solar and solar-terrestrial physics; and (4) microscale and mesoscale meteorology (the study of phenomena as small as the formation of ice crystals in clouds and as large as major thunderstorm systems). Additionally, one group of researchers looks at how societies and individuals respond to changes in their environment. All of the foregoing activities involve field and laboratory studies, theoretical work, and computer modeling.

Supercomputing power and observing facilities are among the services provided by NCAR because they are too costly and specialized for individual universities to acquire. NCAR also adapts research technology for practical applications, particularly in the area of aviation safety. In 1990, NCAR acquired a CRAYY-MPS/864 supercomputer, which, at the time, was the most powerful computing system in existence. With such computer power, researchers are able to construct models, or mock worlds, allowing them to explore speculative scenarios of such events as the fates of chemicals in the atmosphere, the formation of clouds, the global circulation of winds, and the movements of ocean eddies. These models are important for exploring phenomena that cannot be observed or studied firsthand.

Much of the information for NCAR's projects is gained from advanced observing systems. Included in such systems are advanced radar technology based on the Doppler effect and a fleet of research aircraft, which carry a variety of sophisticated instruments. Projects directed toward early practical application include improvement of aircraft flight safety and the efficient use of U.S. airspace. Surface wind shifts, microbursts, tornadoes, hail, lightning, and heavy rain pose hazards and compromise the efficiency of the aviation industry. So-called "Nowcasts" are designed to enable predictions within a few minutes to an hour, as contrasted with longer-term hazardous weather forecasts. A program known as Terminal Doppler Weather Radar detects and warns of wind shifts near the airport terminal, approach, and departure zones, and runways.

NATIONAL RESEARCH COUNCIL (NRC).   The Council was established by the National Academy of Sciences in the United States in 1916 to associate the broad community of science and technology with the Academy's purpose of furthering knowledge and of advising the federal government. The Council operates in accordance with general policies determined by the Academy under the authority of its congressional charter of 1863, which establishes the Academy as a private, nonprofit, self-governing membership corporation. The Council has become the principal operating agency of both the National Academy of Sciences and the National Academy of Engineering in the conduct of their services to the government, the public, and the scientific and engineering communities. It is administered jointly by both academies and the Institute of Medicine. The National Academy of Engineering and the Institute of Medicine were established in 1964 and 1970, respectively, under the charter of the National Academy of Sciences.

NATROLITE.   The mineral natrolite, one of the zeolites, is a sodium aluminum silicate corresponding to the formula $Na_2Al_2Si_3O_{10} \cdot 2H_2O$. It is orthorhombic, crystallizing in slender prisms of nearly square cross-section which are terminated by relatively flat pyramids. There are also fibrous to compact varieties. Natrolite is a brittle mineral; hardness, 5–5.5; specific gravity, 2.2; luster, vitreous; color, red, yellow, white, or colorless; transparent to opaque. Natrolite is found with other zeolites in fissures and cavities in basaltic and related rocks. Czechoslovakia, France, Italy, Norway, Scotland, Ireland, Iceland, Greenland, and South Africa contain well-known localities for natrolite. In the United States it is found in the Triassic traps of New Jersey; also from Oregon, Washington, Montana, Colorado, and as exceptional crystals from San Benito County, California. Superb crystals occur at Mt. St. Hilaire, Quebec, Canada, and from an asbestos mine in Quebec, crystals up to 3 feet (0.9 meters) long and 4 inches (10 centimeters) in diameter have been found. The name natrolite refers to its soda content.

NATURAL COORDINATES.   An orthogonal, or mutually perpendicular, system of curvilinear coordinates for the description of fluid motion, consisting of an axis T tangent to the instantaneous velocity vector and an axis N normal to this velocity vector to the left in the horizontal plane, to which a vertically directed axis Z may be added for the description of three-dimensional flow.

NATURAL FREQUENCY (Mathematics).   A term broadly applied to any system whose transfer function approximates a second-order differential equation of the form $s^2 + 2\zeta\omega_n s + \omega_n^2$, where ω_n is the natural frequency, ζ is the damping ratio, and s is the Laplace transform operator. In control and feedback systems, for example, continuous oscillation or hunting may occur. The frequency of this oscillation is the natural frequency.

NATURAL GAS.   A major source of energy for industrial, commercial, and domestic needs, natural gas is consumed by numerous countries worldwide. Because natural gas is comparatively easy to transport over long distances, usage is not confined to regions that produce it. In addition to energy, natural gas also is a critically important source of industrial chemicals, including numerous hydrocarbon-based organics that find ultimate usage in plastics, films, fibers, solvents, and coatings.

In terms of interest as a fossil fuel for generating electrical power and other energy-conversion processes, natural gas is gaining favor. Compared with coal, natural gas frequently is termed the "clean-burning" fuel. As compared with coal as an energy raw material, the "add-on" costs for treating combustion effluents to satisfy environmental requirements are less for natural gas than for coal, and, consequently, the lower cost benefits of coal are eroding. Further, improvements in natural gas production technology and a brighter outlook for natural gas reserves are contributing to the expansion of natural gas consumption. In addition, large advances have been made in the combustion efficiency of natural gas. For example, the efficiency of some domestic heating appliances has increased to about 95% as compared with 60% or 70% efficiency a relatively few years ago. The cogeneration of heat and electricity in industrial utilities is tending to favor natural gas as the raw fuel. In terms of local natural gas distribution, utilities are finding the use of polyethylene pipe an important cost-saving factor.

In a summary of the Gas Research Institute (GRI)[1], the following observation is made: "The United States, the gas industry, and the gas consumer have entered a dynamic new decade filled with change, challenge, and opportunity. These include the reemergence of the environmental movement, the continuing deregulation of the U.S. natural gas market, the expansion of global trade in the former Soviet Bloc, the emergence of more competitive international and national energy markets, and the rapid expansion of technology options. Three strategic needs are likely to dominate the 1990s for the U.S. gas industry and the gas consumer: (1) ensuring gas deliverability while controlling costs to the consumer; (2) responding to increased concern for the environment; and (3) satisfying a demand for higher quality of energy service."

See also **Electric Power Production and Distribution**.

Composition of Natural Gas

The composition of natural gas varies with the source, but essentially is made up of methane, ethane, propane, and other paraffinic hydrocarbons, along with small amounts of hydrogen sulfide, carbon dioxide, nitrogen, and, in some deposits, helium. Natural gas is found underground at various depths and pressures, as well as in solution with crude-oil deposits. Principal gas deposits are found in the United States, Canada, the former

[1] "1993–1997 Research & Development Plan," Gas Research Institute, Chicago, Illinois.

Soviet Bloc, and the Middle East. The analysis of a gas sample taken from the Panhandle natural gas field in Texas is given in Table 1. Because numerous parts of the earth do not have natural gas at all, or where supply is less than demand, much natural gas is transported, notably by pipeline in the gaseous or liquid phase and across the seas in specially-designed LNG (liquefied natural gas) carriers.

TABLE 1. ANALYSIS OF NATURAL GAS FROM NATURAL GAS FIELD IN TEXAS PANHANDLE

Component	Mole Percent
Methane	76.2
Ethane	6.4
Propane	3.8
Normal butane	1.3
Isobutane	0.8
Normal pentane	0.3
Isopentane	0.3
Cyclopentane	0.1
Hexane plus other hydrocarbons	0.35
Nitrogen	9.8
Oxygen	Trace
Argon	Trace
Hydrogen	0.0
Hydrogen sulfide	0.0
Carbon dioxide	0.2
Helium	0.45

Note: Heating value of various natural gases averages between 975 and 1180 Btu/cubic foot (8678-10,502 Calories/cubic meter) at 60 °F (15.6 °C) and 30 inches (76.2 centimeters) mercury pressure.

Origin and Geology of Natural Gas

The most commonly accepted theory concerning the formation of natural gas is the organic theory. Methane is a product of decaying vegetable matter and in areas of stagnant water is found as *marsh gas* or *swamp gas*. It is theorized that over millions of years, the remains of plants and animals were washed down into lakes, the accumulations covered with layers of mud and stone. The latter became stone while the organic matter decayed through the action of heat and pressure and perhaps from effects of bacteria and radioactivity, forming various hydrocarbons. The hydrocarbons were held in tiny spaces between the particles of sand and porous rock and formed natural gas and petroleum. Often the natural gas so formed made its way through the rock to the surface and escaped. In some areas, however, the layers of sand and porous rock were covered by impermeable rock to form huge reservoirs of natural gas at various levels of pressure. See also **Petroleum**.

The organic theory as usually presented is rather general and vague in many respects. As observed by Ourisson (Université Louis Pasteur, Strasbourg) and colleagues, natural gas, as well as coal and petroleum, are fossil fuels, but fossils of what? Fossil fuels form only if the organic matter is buried before it can become completely oxidized to carbon dioxide by microorganisms. According to the microbial origin concept, as the carbon compounds sink deeper into the Earth under accumulating sediments, they are subjected to high temperatures and undergo chemical reaction, during which oxygen and most other elements are eliminated. This yields a mixture composed in the case of gas and petroleum almost entirely of hydrocarbons (carbon in the case of coal). Since the beginnings of photosynthesis on Earth, it is estimated that 10 quadrillion (1016) tons of carbonaceous material has been stored in sediments. Most of this material is stored in very dilute form and only under exceptional geologic conditions, is it concentrated to become a viable fuel source. In a twenty-year study, which might be called molecular paleontology, Ourisson and coworkers have been studying the detailed genesis of fossil fuels. Thus far, chemical analysis of the most varied organic sediments reveals a surprising commonality — all appear to derive much of their organic matter from once unknown microbial lipids. This topic is presented in more detail in entry on **Petroleum**.

A few scientists, notably Thomas Gold (Cornell University), have proposed that, in contrast with the organic sediments theory, the prime source of natural gas is primordial, abiotic methane rising from deep within the Earth's mantle. This is sometimes referred to as the "deep-earth gas" hypothesis. In this view, methane flows up around the edges of the shield and is responsible for the oil and gas fields in the North Sea and the southern Baltic. Admittedly, this reservoir of natural gas is presently out of the reach of any foreseeable drilling technology. In some areas, it is suggested that the granite crust may have been fractured and subsequently became porous to the extent that methane may have risen into the crust and have become trapped at accessible depths. Other scientists point out that no evidence supports the concept that a large amount of methane was incorporated in the Earth when it was formed. Available geochemical evidence suggests that the early atmosphere, produced by outgassing of the planetary interior, could not have been rich in hydrogen. Further, if the Earth did at one time contain primordial methane or other hydrocarbons, most of that volatile material would have long since escaped by way of volcanism and diffusion. It is also suggested that the analysis of volcanic basalts shows that the rock in the upper mantle is highly oxidizing, in which case any methane present would have been converted to carbon dioxide.

Some experimental drilling programs underway in Sweden, including a well some 5000 meters in granite bedrock, may shed further light on Gold's hypothesis.

In searching for new fields during the 1960s and 1970s, drillers seeking gas and oil in traditional suspect source reservoirs, would find gas and/or oil in only about 9 of 100 wells drilled. Usually, only 2 or 3 of these wells produced sufficient gas and/or oil to be of commercial value. Whereas the average depth of gas well drilled during this period ranged between 5000 and 6000 feet (1524–1829 meters), some drillers are now aiming at the 30,000-foot (9144-meter) depth. For many years, the deepest gas well in Texas was 28,600 feet (8717 meters). In the early days of offshore drilling, operations were conducted in water only 20–25 feet (6–7.5 meters) deep. There are now many platforms in waters that are deeper by a factor of 20–30 times.

In terms of the conventional or traditional sources, natural gas is found in areas close to exposed or buried mountain ranges. Major deposits of natural gas are found in inclined strata where the rock formations dip away from the crest of a buried hill or the ridge of a buried mountain. Some of the common types of formations in which natural gas is found are shown in Fig. 1. Although natural gas and crude oil are frequently found together, the largest natural gas reserves (about 70% of the estimated reserves) are in deposits neither in contact with, nor dissolved in, oil.

Forecasting Natural Gas Reserves

Since the 1920s, natural gas reserves have been based upon the amount of gas that most likely will be found with oil. Although numerous experts in the field have claimed over the years that this methodology overlooks large amounts of "unassociated" gas, the professionals have been slow to revise their procedures. The concept that much natural gas occurs quite apart from oil is now taking hold. The U.S. Geological Survey, which has proclaimed a gas resource base at about 400 trillion cubic feet, undertook a major study to be completed in the late 1990s. Survey techniques have undergone several dramatic improvements, including the use of computer techniques that will project three-dimensional survey information, as contrasted with past reliance on two-dimensional mapping. A number of experts now believe that the reserves are well over 1,200 trillion cubic feet (34 trillion cubic meters). One independent gas producer has estimated the figure at about 1,500 trillion cubic feet (42.5 trillion cubic meters).

Traditional estimates of oil-associated natural gas reserves historically have rated the former Soviet Bloc as holding nearly 40% of the reserves, Iran about 14%, and the United States nearly 6%, with additional fairly high reserves in Qatar, Algeria, and Saudi Arabia. At one time, North America was attributed to have a 60-year supply, but with revisions in estimating procedures, coupled with increased efficiency in natural gas production and gas combustion efficiency, the future is now believed to be in terms of at least a few centuries. Interest continues, however, in upgrading low-Btu natural gases and developing so-called *substitute natural gas*.

The principal gas fields within the continental United States are indicated in Fig. 2.

Ultimate Recovery of Natural Gas Reserves. It has been traditional for many years to categorize ultimate recovery of gas reserves by reservoir lithology which involves the compilation of such data by three types of reservoirs: (1) *Sandstone reservoir* — consisting of sedimentary rock composed predominantly of quartz grains or other noncarbonate mineral or rock detritus. Included in this reservoir type are unconsolidated sand, sandstone, siltstone, graywacke, arkose and granite wash, conglomerate, and breccia. (2) *Carbonate reservoir* — composed of sedimentary rock made

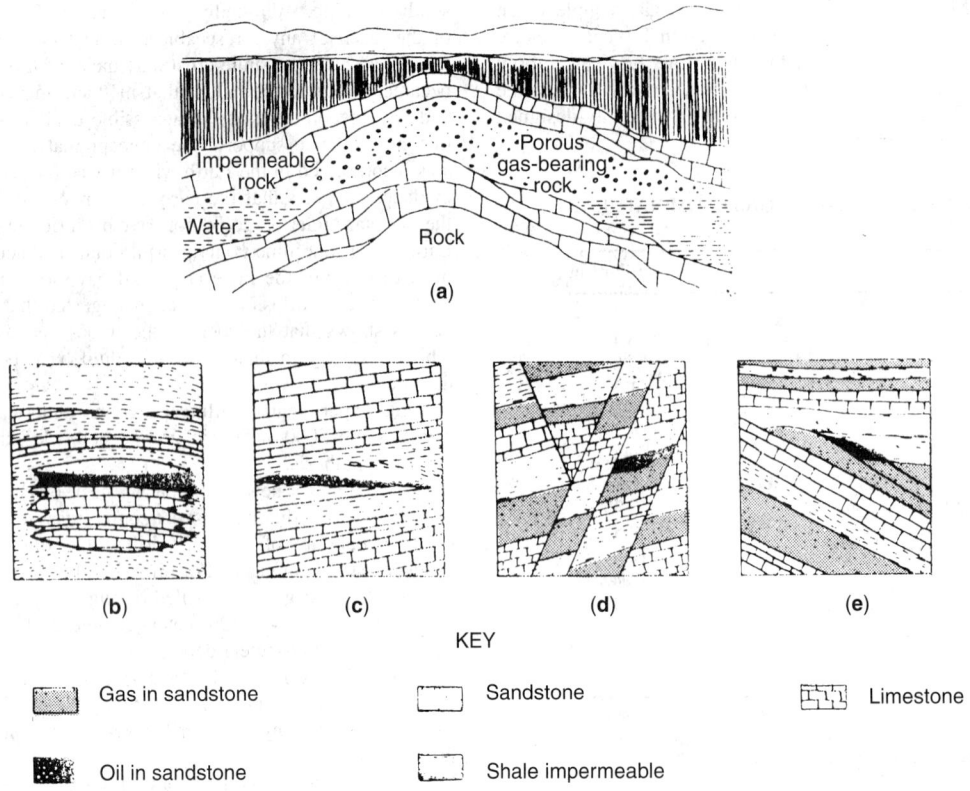

KEY

Gas in sandstone

Sandstone

Limestone

Oil in sandstone

Shale impermeable

Fig. 1.   Types of natural gas reservoirs and entrapments: (**a**) anticlinal trap; (**b**) coral reef trap; (**c**) stratigraphic trap; (**d**) fault trap; and (**e**) unconformity.

Fig. 2.   Location of principal natural gas fields in the lower 48 of the United States. Alaska, not shown, ranks third in terms of holding estimated reserves. (*Batelle Memorial Institute.*)

up predominately of calcite (limestone) and/or dolomite. (3) *Other reservoirs* — including igneous and metamorphic rocks and some sedimentary rocks, such as fractured shale.

Estimated ultimate recovery is also reported by type of entrapment, of which there are two major types: (1) *Structural trap* — an entrapment in which migration of hydrocarbons in the reservoir rock has terminated primarily because of closure induced by structural deformation, such as folding or faulting. Within this category should also be included entrapments attributed to hydrodynamic forces. (2) *Stratigraphic trap* — an entrapment in which migration of hydrocarbons has terminated because of the pinchout of reservoir rock due either to truncation or to nondeposition or to a facies change in the form of diminished permeability of reservoir rock. Also included in this category are entrapments in which a pinchout of facies change provides part of the barrier to migration of hydrocarbons, with structural elements providing the remaining closure for the entrapment. In these cases, it is recognized that the dominant cause of the accumulation is the lenticularity of truncation of the reservoir rock.

Some estimates indicate that about 65.8% of ultimately recoverable natural gas in the United States will be found in structural traps; the remaining 34.2% in stratigraphic traps.

The estimated ultimate recovery is also reported by the geologic age of the reservoir. It is recognized that problems may arise where the geologic age of a reservoir cannot be determined specifically, such as Permo-Pennsylvanian and Cambro-Ordovician, or where production from reservoirs of different geologic age is combined.

Natural gas liquids occur in either the gaseous phase or in solution with crude oil in the reservoir. They are recovered at the surface as liquids by separation from produced natural gas, by such processes as condensation and absorption in field separators, gasoline plants, and other surface facilities. In this processing, valuable by-products are recovered, such as light oils, natural gasoline, and other petroleum gases such as ethane, propane, and butane. Natural gasoline is blended with gasoline from petroleum refineries to improve starting properties, especially desirable in cold weather. Ethane is a major petrochemical raw material. Propane and butane are made available as LPG (liquefied petroleum gas). Processing of natural gas also removes unwanted material, such as nitrogen, sulfur compounds, carbon dioxide, and water vapor. Some gas fields produce helium, which is extracted cryogenically.

Unconventional Sources of Natural Gas. These include: (1) tight sandstones, (2) Devonian shales, (3) geopressured zones, (4) deep basins, (5) gas associated with coal seams, and (6) gas in the form of methane hydrates.

1. *Tight Sandstones.* In the United States, tight sandstones of the western basins range from the northern tier states to the Mexican border. Some tight gas sands also occur in the eastern United States. To date, resource development has occurred only in the limited areas characterized by thick, fairly uniform, blanket-type formations which, when hydraulically fractured, provide sufficient gas production rates to merit commercial exploitation. In these areas, as pointed out by Sharer (Gas Research Institute, GRI), state-of-the-art technologies can be used because only a limited knowledge of the formation characteristics is required for economic production. However, a majority of the resource base is associated with lower permeability and more complex blanket and lenticular sand formations, for which current technology is not adequate. GRI is concentrating research in these areas.

2. *Devonian Shales.* The large eastern Devonian gas shales resource base underlies approximately 174,000 square miles (453,000 km^2) of the eastern U.S. Estimates of recoverable gas range from 2 to 15% of the gas in place. Natural gas has been produced from these shales for decades. Well production rates are relatively low, but after the first few years of production it does not usually decline rapidly with time. A major constraint to present-day exploitation has been the extraordinary inability to predict with confidence the gas production rates that may be obtained in wells drilled outside the traditional production areas. Presently, the GRI is studying the systematics of historically successful fields, including the Appalachian, Illinois, and Michigan Basins.

3. *Geopressured Zones.* A test well in a geopressure zone was drilled some years ago in Tigre Lagoon in the coastal marshes of southern Louisiana. Known as Edna Delcambre #1, this well produced at a rate of up to 10,000 barrels of water per day from a sandstone aquifer some 12,600 feet (3840 meters) below the surface. Pressure at that depth is nearly 11,000 pounds per square inch (748 atmospheres) and the temperature is 116 °C. Quite an elaborate manifold system is required to collect the gas. The water is disposed by forcing it by its own pressure into another well bore, which penetrates to a depth of 2500 feet (762 meters). Scientists associated with this project had expected about 20 cubic feet/barrel (42 gallon); about 0.6 cubic meter/barrel (159 liters). In actuality, reports indicate that the yield of gas was about 2.5 times that amount.

As explained by specialists in geopressure technology, at great depths (in terms of present technology), the solubility of natural gas in water may be as much as 1000 cubic feet/barrel (28.3 cubic meters/159 liters) at depths of 30,000 feet (9144 meters), whereas that solubility will be reduced by a factor of ten at a depth of 20,000 feet (6096 meters). Under the right combination of geologic and hydrologic conditions, this gas-laden water will move toward the surface, during which process some of the gas will be released from the water in the form of very small bubbles. Ultimately, this gas collects beneath a geologic trap, where conventional free-gas reservoirs are formed. Some authorities now believe that the very deep aquifers are much more extensive than the free-gas reservoirs. It is this gas-saturated water that some scientists believe will be a great source of future natural gas.

The GRI has been investigating the coproduction of gas and water for a number of years. Natural gas from watered-out reservoirs, geopressured aquifers, and high-water-saturated gas-bearing reservoir strata are prime targets. This natural gas is trapped by water such that special production techniques must be used to move the water and remobilize the gas. Although some gas is also dissolved in the water, it is of less significance than the free gas trapped as dispersed bubbles or in pockets or stringers of various sizes in the reservoir rock matrix.

4. *Deep Basins.* These are found at depths between 15,000 and 30,000 feet (4572–9144 meters) and are estimated to contain significant quantities of gas, but generally await the development of advanced production technology and economic incentive.

5. *Gas Associated with Coal Seams.* Methane, the principal constituent of natural gas, is generated during the geologic process of coal formation. A significant portion of this gas is trapped by impermeable strata, and it is present within the fractures and micropores of the coal. (The presence of methane is an ever-present hazard in coal mining.) Major variations in resource estimates are due to uncertainties in the gas content and size of the deeper, not minable coal deposits in the western states that form the major portion of the resource base. Seeking such gas may involve depths as great as 6000 feet (1829 meters) underground. Except for reasons of safety, little effort has been made to recover any of this resource due to high recovery costs, potential uncertainties in production, and deficiencies in state-of-the-art equipment, particularly for the deeper coals. The GRI is concentrating its research efforts on unminable coal because of its large potential as a resource base. While the gas resource associated with mining amounts to about 10% of the energy value of the coal, producers rarely apply new gas recovery technology except where safety is a requirement. Targets of the GRI program are deep coal seams, multiple seams interbedded with shales and sandstones, and deep multiple beds that are too thin to mine.

6. *Methane Hydrates.* Within a certain range of pressures and temperatures, methane and water form hydrates. Described as icelike substances, these hydrates are believed to occur in very substantial quantities, particularly beneath permafrost and in deep-ocean bottoms. Although slush has occurred in gas pipelines under certain conditions for many years, the existence of hydrates in nature was not made known until the mid-1960s. Geologists and hydrologists had previously assumed that gas of this type would have dissipated during earlier geologic ages. This is another area of natural gas resource research awaiting economic incentives.

Exploratory Methods. The principal exploratory methods used are: (1) *Airborne magnetometers*, which seek out anomalies in the magnetic field. Experienced geologists relate these irregularities to the probability of gas reservoirs below the surface. See also **Magnetometer**. (2) *Satellite imagery*, from which surface structures and patterns can be related to previous pattern recognition studies made of surfaces below which gas reservoirs exist. (3) *Gravitometers* are used to detect subtle variations in gravitational pull inasmuch as this is less for a gas reservoir than for continuous dense rock formations. (4) *Seismic methods*, which constitute the most widely-used of exploration methods. See also **Earth Tectonics and Earthquakes**. (5) *Data logging methods*, wherein an instrument is lowered into the borehole and which telemeters back to the surface readings of sonic absorption in an effort to determine the nature and thickness of rock formations. Data loggers operate on the basis of several physical phenomena. (6) *Fossil inspection.* The careful examination of microfossils can assist in fixing the age of rocks that are being penetrated. The condition of the fossils also can be related to probable temperatures to which they have been exposed over geologic periods and these, in turn, can be advantageous in locating possible gas deposits. Usually a combination of two or more exploratory techniques is used.

Liquefied Natural Gas (LNG)

The liquefaction of natural gas for storage and transportation and regasification for final distribution dates back several decades. A few major accidents in the handling of LNG thwarted the progress of the field for a while, but in the early 1970s, LNG was again considered in a major way because of energy-short nations. One of the more serious LNG accidents occurred in Cleveland, Ohio on October 20, 1944, when a storage tank developed a leak with spillage and subsequent fires in the surrounding neighborhood in which 135 persons lost their lives. While liquefaction offers marked storage space savings and convenience, the predominant advantage occurs in connection with both pipeline and ship transportation. Energy-short nations, such as Japan, and some of the European nations, have turned in recent years to the concept of shipping LNG by ship. For example, a large LNG plant at Lumut, Brunei, Borneo went onstream in mid-1974 essentially to furnish LNG to Japan.

Oil- and gas-rich nations, who at one time flared to the atmosphere much of the natural gas that accompanied the production of crude oil, have turned toward conservation—either through reinjection of much of the natural gas underground or through constructing LNG production facilities for shipment of the product overseas. Concurrent with such planning was reevaluation by a number of nations of their own valuable resources and a growing reluctance toward exporting inordinate quantities of gas and oil strictly for money. As of the early 1990s, the shipment of LNG overseas competes with other ways and means for alleviating energy shortages, including coal conversion and gasification, nuclear energy, solar energy, etc.

Three types of liquefaction processes may be used for production of LNG. The standard cascade process, which uses three refrigerants—methane, ethylene, and propane—all circulating in closed cycles, is shown in Fig. 3. There is a separate compressor for each of these refrigerants. The methane and propane are available from the feed gas (natural gas). The ethylene must be furnished separately. Ethane may be used in place of ethylene at a subatmospheric suction pressure. The cascade process has the highest rank in terms of thermal efficiency. As a possible improvement over the cascade process, the mixed refrigerant process was developed in the early 1960s. A single-pressure mixed refrigerant cascade (MRC) system is shown in Fig. 4. In one plant using this process, a hydrocarbon-plus-nitrogen mixture of relatively wide boiling range (N2 through C5) is used as the refrigerant. All of these components can be recovered from natural gas in separate apparatus. In still another system, shown in Fig. 5 a propane and mixture-refrigerant cycle is used. In this process, the cooling load is divided horizontally at about −34.4 °C into an upper portion absorbed by propane and a lower portion absorbed by the

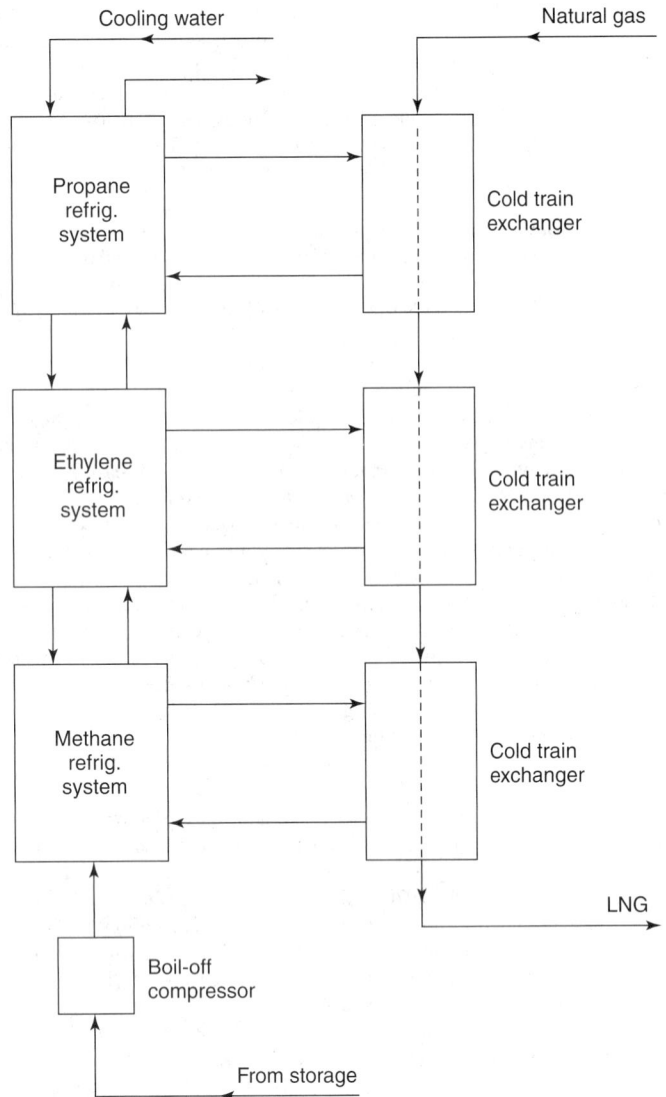

Fig. 3.   Conventional or standard cascade system for producing liquefied natural gas (LNG).

mixed refrigerant. In essence, the system is a dual refrigerant cascade in which the lower boiling fluid is a mixture refrigerant. The cascade combination with propane makes it possible to reduce the boiling range of the mixture refrigerant substantially, which improves the thermodynamic efficiency over that of the straight MRC process.

Cryogenic Upgrading of Low-Btu Natural Gases

Worldwide, there are substantial reserves of natural gas in which the reservoir formation hydrocarbons are contaminated with nonburning components. The presence of components, such as helium, nitrogen, or carbon dioxide, reduces the heating value of the gas mixture. This can result in the gas being unsuitable for existing transmission and distribution systems. Such contaminated mixtures are termed low-Btu gases if their heating values fall below the minimum standards, regulations, or contract heating value requirements.

Cryogenic processing can be used to upgrade some of these low-Btu gases so as to produce an acceptable high-Btu product. Cryogenic upgrading is a physical process in which subambient temperatures are employed to bring about a separation between the hydrocarbons and nonhydrocarbons in the mixture. The reduction of temperature occurring during cryogenic processing produces a two phase (gas-liquid) mixture. The relative volatilities between the components in the mixture result in selective mass transfer between the two phases. One phase becomes enriched with hydrocarbons and then has a heating value higher than the original gas. The second phase becomes denuded of hydrocarbons and has a heating value below that of the original gas mixture. Frequently, the mass transfer operation requires several theoretical stages in order to achieve

the desired product heating value and high hydrocarbon recoveries. While cryogenic upgrading can be applied to gas mixtures containing carbon dioxide or hydrogen sulfide, it has so far only been applied commercially to those hydrocarbon mixtures contaminated with nitrogen and helium.

One of the main considerations in the design and operation of cryogenic upgrading plants is to identify and remove any component from the gas that could adversely affect the operation of the cold sections of the plant. Such components are carbon dioxide, water vapor, and heavy hydrocarbons that have high solidification temperatures and low solubilities. In general, if these components are allowed to remain in the gas to the cryogenic unit, they will form solids during the cooling process that will be deposited on the heat exchanger surfaces. This will lead to fall-off in performance and, possibly, to blockages and plant shutdown.

Particular attention should be paid to identifying any high freezing point components in the low-Btu gas. There exists a range of absorption and adsorption processes to pretreat the low-Btu gas to remove these undesirable components.

A simplified flowsheet of cryogenic upgrading plant is given in Fig. 6. The plant, consisting of two identical trains, is capable of processing 260 million standard cubic feet (7.3 million cubic meters) per day of low-Btu gas (580 Btu/standard cubic foot) (5162 Calories/cubic meter) and upgrading the gas into 143 million standard cubic feet (4 million cubic meters) of high-Btu gas (980 Btu/standard cubic foot; 8722 Calories/cubic meter). The plant stream parameters are indicated in Table 2.

The low-Btu gas is available at 800 psig (54 atmospheres) and is mainly a nitrogen-methane mixture. In addition to a small quantity of helium, the gas also contains small quantities of carbon dioxide, water vapor, and heavy hydrocarbons. The carbon dioxide is removed by washing with monoethanolamine (MEA), the water is taken out on molecular sieve, and the heavy hydrocarbons by adsorption on activated carbon. The gas is then cooled in aluminum plate-fin exchangers against the returning high-Btu product gas and vent gas. The gas is then expanded to 380 psig (26 atmospheres) and a vapor-liquid mixture passes into the H.P. (high pressure) fractionator. The purpose of this fractionator is to bring about an initial separation of the nitrogen-methane and to produce a liquid reflux for the L.P. (low pressure) fractionator.

A nitrogen-enriched vapor flows up the H.P. fractionator, while methane is returned to the sump of this column by a nitrogen reflux stream produced in the tubes of the overhead condenser. The refrigeration required to produce this nitrogen reflux is provided by evaporating some of the liquid methane, from the L.P. column, in the shell of the overhead condenser. Two liquid streams are taken from the H.P. fractionator, and these become the feed and reflux for the L.P. fractionator. The L.P. feed is an enriched methane stream taken from the base of the H.P. fractionator. The L.P. reflux is a high purity nitrogen liquid taken off the H.P. fractionator just below the condenser. The upgrading is completed in the L.P. fractionator. The feed stream is stripped to produce a high-Btu liquid containing 4% nitrogen and having a heating value of 980 Btu/standard cubic foot (8722 Calories per cubic meter). The liquid is pumped from the column sump, evaporated, and superheated against the incoming low-Btu gas. The gas from the top of the L.P. column is mainly nitrogen and is also heated to ambient temperature against the incoming low-Btu gas. By using this arrangement of two distillation columns, the separation of nitrogen and methane can be achieved using only the pressure energy available in the low-Btu gas.

Transportation of Natural Gas

The mode selected for gas transportation depends mainly on: (a) the distance over which the gas must be moved; (b) the geographical and geological characteristics of the terrain (considering both overland and overseas [underseas]) across which the gas must be moved; (c) environmental factors directly associated with the gas transportation mode; (d) the physical characteristics of the gas to be transported, notably, the phase — whether gaseous or liquid; and (e) the construction and projected operating costs of the transportation system, based upon trading off the advantages and limitations over which some flexibility of selection may be present. Aside from economic factors, a system can be engineered to transport either the gaseous or liquid phase, thus giving rise to considerable flexibility in certain situations.

Overland Pipelines.   Detailed maps of gas pipelines in the United States and other parts of the world can be found in several references, particularly among the periodicals serving the pipeline industry. Notable among these references is the international petroleum encyclopedia and atlas issued

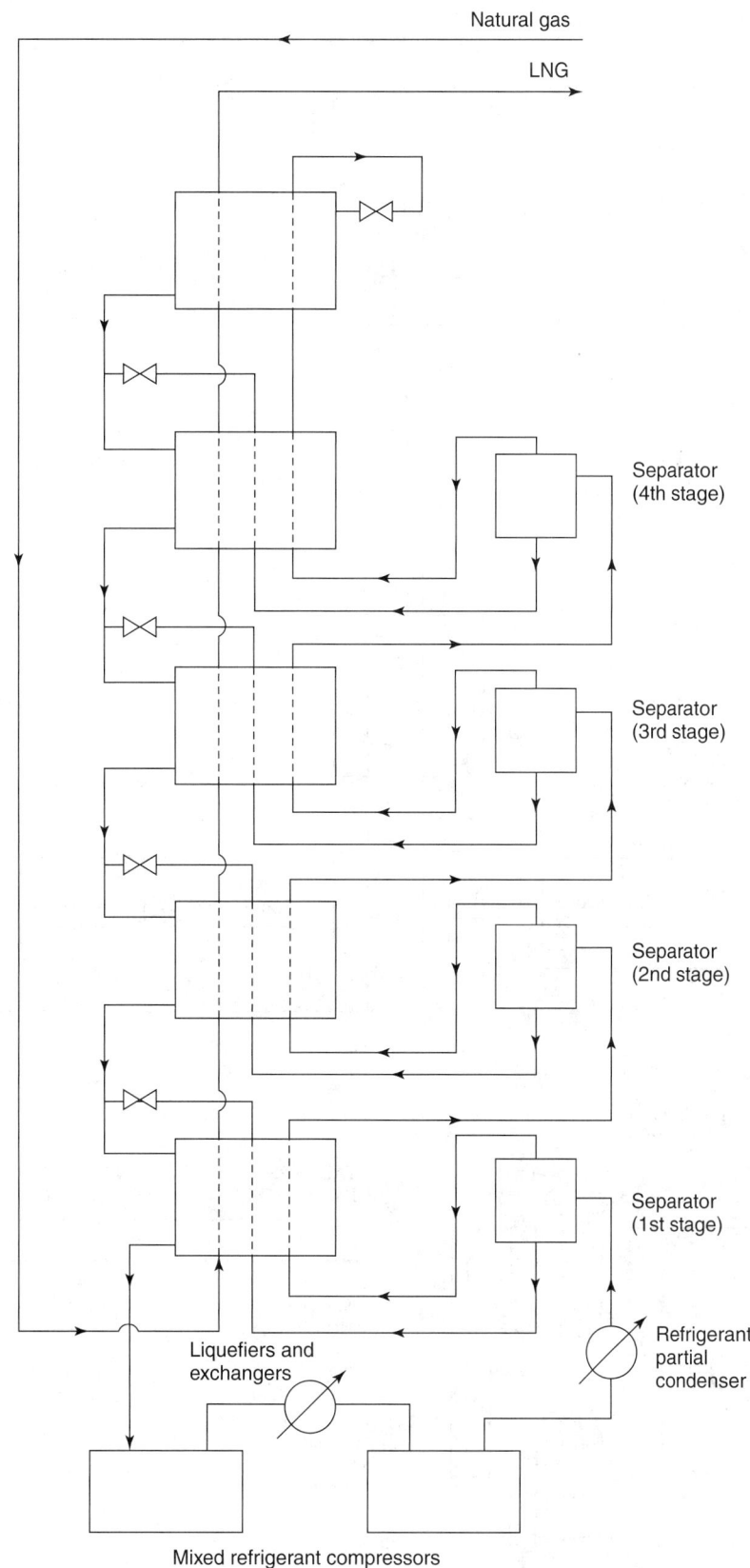

Natural gas

LNG

Separator
(4th stage)

Separator
(3rd stage)

Separator
(2nd stage)

Separator
(1st stage)

Liquefiers and
exchangers

Refrigerant
partial
condenser

Mixed refrigerant compressors

Fig. 4.   Single-pressure, mixed refrigerant cascade system for producing LNG.

periodically by Petroleum Publishing Co., Tulsa, Oklahoma. Numerous trade associations serving the pipeline industry are also excellent sources on pipeline statistics. There are so many pipelines that presentation of this type of information is beyond the scope of this encyclopedia.

Historically, Texas, Louisiana, Oklahoma, and New Mexico have been large producers of natural gas, as well as some significant fields in the West Virginia-Ohio-Pennsylvania area. New developments in Alaska are

and will continue to influence the gas transportation and distribution pattern.

Much of the installed gas pipeline ranges from 14 to 30 inches (36 to 76 centimeters) in diameter, the most common ranging from 20 to 35 inches (51 to 89 centimeters) but there is a strong trend toward larger-diameter lines, from 42 inches (107 centimeters) upward. Line pipe is made from high-strength plates, 3/38 inch to 1 inch (1 to 2.5 centimeters)

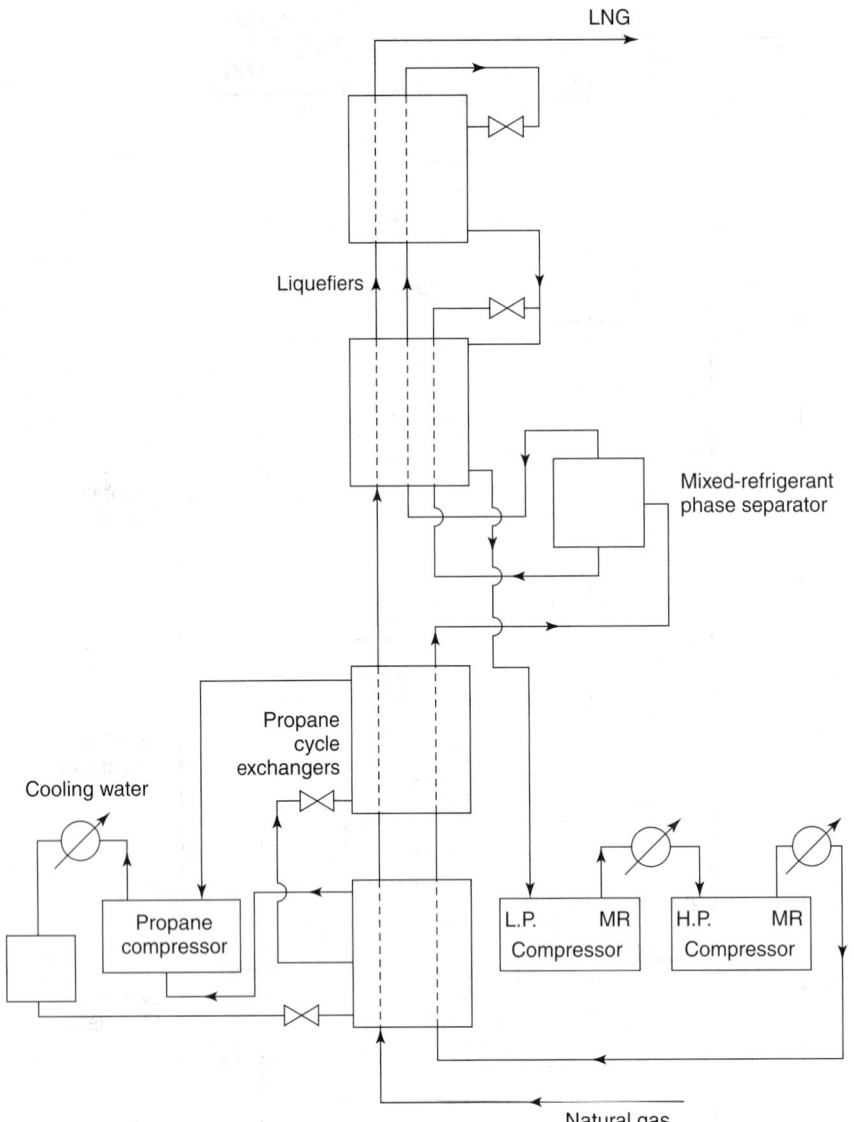

Fig. 5.   Propane-mixed-refrigerant liquefaction system for producing LNG. L.P. = low pressure; H.P. = high pressure; MR = mixed refrigerant.

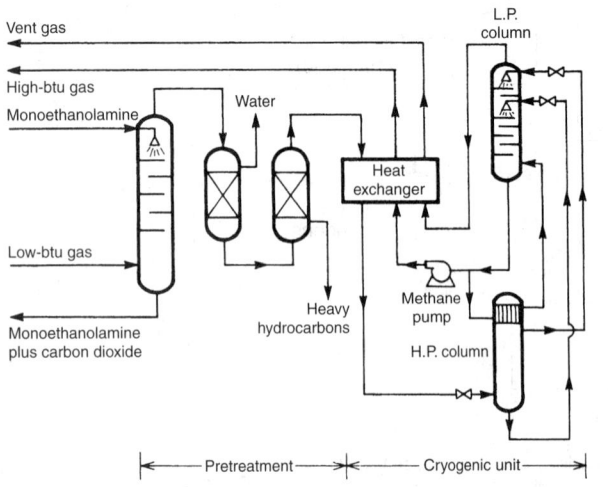

Fig. 6.   Plant for nitrogen removal from natural gas using cryogenic upgrading. (*Petrocarbon Developments, Ltd.*)

TABLE 2. CRYOGENIC UPGRADING OF NATURAL GAS—STREAM PARAMETERS Composition—Mol.%

	Low-Btu Gas	High-Btu Gas	Vent Gas	Helium
Helium	0.40	—	0.09	100.00
Nitrogen	42.75	4.00	98.95	—
Methane	56.02	95.09	0.96	—
Ethane +	0.53	0.91	—	—
CO_2	0.30	—	—	—
Flow (million standard cubic feet/day)	246	143	100	0.43
Flow (million cubic meters/day)	7	4	2.8	0.012
Heating value (Btu/standard cubic foot)	580	980	—	—
Heating value (Calories/cubic meter)	5162	8722	—	—

in thickness. Sections of pipe are usually 40 feet (12 meters) long, minimum, ranging up to 60 or 80 feet (18 or 24 meters). Lengths of pipe arrive at the scene most often by truck and are strung out by special pipe carriers along the right-of-way so that the construction crews will find them near the place where they are to be installed. Helicopter delivery of pipe is sometimes used where it is impossible for trucks to do the job. The total weight of steel going into a long-distance pipeline is impressive. For example, a pipe with a wall thickness of 1/2 inch (13 millimeters) and a diameter of 30 inches (76 centimeters) will weigh more than 400 tons (360 metric tons) per mile.

In building very long pipelines, the pipeline company usually employs several construction contractors. The total length of line is divided into a number of sections with separate equipment and crews. Usually, each crew works on not more than 100 miles (161 kilometers). By partitioning the construction task, the entire operation can be speeded up, particularly important in areas where freezing temperatures or rain and mud may interfere with the work.

The numerous machines needed to dig the trench, weld the sections of pipe, apply protective coating to prevent corrosion, lower the pipe into the trench, and cover the trench with earth are known collectively as a *main line spread*. The trench is usually 3 or more feet (1 meter or more) in depth, sufficiently deep to prevent damage by plowing and earth-moving equipment. Depending on the size of the pipe, the trench will range from 2 to 4 feet (0.6 to 1.2 meters) or more in width.

Teams of welders join the pipe sections into a continuous tube. The most modern welding techniques involve automatic welding machines. X-ray equipment is used to inspect welds. When several sections of pipe have been welded together, the continuous tube is lowered gently into the trench by *sideboom tractors*. These machines have cranes or derricks slanted over to one side so that they can pick up the pipe and lower it several feet away from the tractor itself. Pipe purchased from steel mills may come with a coating and wrapping already applied. The thick coating may be of coal tar or asphaltic material, which is then covered with heavy paper or fiberglass. This protective coat-and-wrap is needed to prevent rusting. If bare pipe is used, there are special machines that coat and wrap right on the job just before the pipe is lowered into the trench.

A special piece of equipment, known as the *holiday detector*, is a hoop of metal placed around the pipe after it is coated and wrapped. A small electrical current flows through the hoop. If there is a "holiday," i.e., a spot where there is no coating, the detector alerts the operator. This is brought to the attention of a special crew that coats and wraps bare spots in the pipe.

Since pipelines do not follow an absolutely straight line, bending machines are used to curve the pipe in the vertical, horizontal, or both directions. When a pipeline must cross a river, the contractor will dig or dredge a deep trench in the river bed. The pipe is then surrounded by heavy weights and encased in concrete so that it will not be carried away by the current. If there is a suitable bridge across the river, the pipeline may be hung from the underside of the steel girders of the bridge. In some cases, a special bridge is constructed to carry the pipe across the stream. In crossing a highway or railroad, the pipeline must be put through a tunnel under the structure. A giant auger will be used to bore under the road to accommodate a section of somewhat larger-diameter pipe, forming the tunnel through which the main pipeline passes.

Gas pressures in long-distance pipelines may range from 500 to 5,000 pounds per square inch (34 to 340 atmospheres) with 1,000 psi (68 atmospheres) being quite common. Pressure is boosted to make up for frictional losses by use of compressor stations located every 50 to 100 miles (80 to 161 kilometers) along the pipeline. In terms of lineal velocity, natural gas may travel at a rate of about 15 miles (24 kilometers) per hour; thus, about three days are required to move a molecule of gas over a distance of 1,000 miles (1609 kilometers).

All along the pipeline, there are valves and regulators that may be opened or shut to control the internal pressure, or to cut off the flow entirely if an unexpected break in the line is caused by a flood, earthquake, or other disaster. The valves and regulators can be operated by microwave radio long before any crew could reach them. Stations for reducing the pressure, located near points of consumption, frequently are called *city gates*. These stations measure the amount of gas leaving the main pipeline at this point as well as reducing the pressure.

Marine Pipelines for Gas

With some alterations, the techniques that apply to construction and laying of marine pipelines for gas also apply to fluids, such as oil. Marine pipelines can be underwater in a river, marsh, or ocean, but the predominant industry effort in recent years is the construction of pipelines in the open ocean at increasingly deeper levels. The trend toward deepwater pipelining and construction in harsher environments naturally follows the expansion of the search for offshore gas and oil. This search began in earnest after World War II and is expanding at an ever-increasing rate; even if slowed to some extent by some environmental concerns in the United States, the rate is rapid in other parts of the world. Worldwide energy needs have caused

oil companies to move into areas that only a few years ago would have been too expensive to develop on a practical basis. Lines are now being laid in water depths of several hundred feet (meters) and cover distances of 200 miles (320 kilometers) or more from field to shore. These longer lines are major trunk lines bringing gas and oil to land terminals. Other lines are necessary out at the field to connect platforms to each other; or possibly to connect platforms to sea berths.

The sizing of the pipeline, the design of the pumping and compression systems needed to move the products, the design of the automation systems, and many of the corrosion control procedures are the same regardless of whether the pipeline is on land or at sea. The two major areas of design difference between land and marine pipelines are (1) the stresses incurred in getting the pipeline to the sea bottom, and (2) the necessity of keeping the line stable and in place while it is exposed to forces induced by current and wave.

The stability problem is theoretically simple but is complicated somewhat by the uncertainty of precise values for some of the coefficients used in the calculations. Basically, it is a matter of providing enough weight in the pipe and pipe coating system to provide a net downward force when balanced against the buoyance and the lift force caused by the seawater moving by the pipe. This net downward force, in conjunction with the coefficient or friction for the particular pipe-soil combination under examination, can then mobilize a horizontal resisting force. This should be somewhat larger than the drag force exerted on the pipe by the water motion in order to give the desired safety factor.

Different safety factors or horizontal water velocities may be utilized depending on the operating conditions that will be encountered during the life of the pipeline. For example, many pipelines will be buried beneath the sea bottom at some time interval, ranging from a few weeks to a year or two, after their construction. The exposure of this line to maximum horizontal water velocities caused by storm current and waves is obviously much less than that of a line that will remain on the surface of the sea bottom. It is also obviously necessary to consider whether the line will contain gas, oil, or other substance at the time the design loads may occur.

It is important to carefully consider the foregoing points in the design of the weight coating since the ability of a contractor to safely construct the line relates very closely to the negative buoyancy of the pipe and coating.

The most common method of marine pipeline construction utilizes a floating vessel on which the pipe is assembled in a horizontal position. As additional joints or sections of pipe are added to the already-completed segment, the barge is moved forward, actually moving out from under the completed pipeline. This is sometimes called the "stovepipe" method, named after the manner the pipe sections are added, one after another. This pipeline extends off the stern of the vessel and spans down to the sea bottom. It is supported part of the way down by a construction aid called a "pontoon" or "stinger." This is basically a slender structure pinned to the vessel on one end and with built-in buoyancy that can be controlled so that it floats at the proper angle to the water surface to provide support to the pipeline.

In shallow water, the pipeline is then allowed to span from the end of the pontoon to the sea bottom as a simple beam. As water depths increase, it becomes necessary to add tension to the pipe on the barge. This, of course, changes the analytical problem from one of a simple beam to one of a beam under tension. This analysis must take into account the weight of the pipe, the wall thickness, the type of steel in the pipe, the tension on the pipe, the support of the pontoon, the geometrical configuration of the tension on the pipe, the geometrical configuration of the pipe-pontoon-barge system, and the pipe end condition at the sea bottom.

There are three basic configurations of pipelay vessels in common usage: (1) the barge-type hull; (2) the ship-shape hull; and (3) the semi-submersible vessel. The barge-type hull (Fig. 7) is the most common because of its economy and simplicity, its ability to provide the space and stability for heavy lifts and deck cargo, including pipe, and its shallow draft, permitting work close inshore. The primary disadvantage is its relative sensitivity to sea conditions. In particular, roll and heave motions will shut down pipelay operations in 6-foot to 14-foot (1.8- to 4.2-meter) waves, depending on wave direction and period.

Overseas Shipping of LNG. A key feature of most LNG carriers in operation is the insulation system, which maintains the cargo at $-162\,^\circ$C. In one type of ship, the cargo is carried in five tanks constructed of a thin welded membrane of special steel. Each tank is separated from the inner hull by insulating material. The small fraction of the cargo that boils off

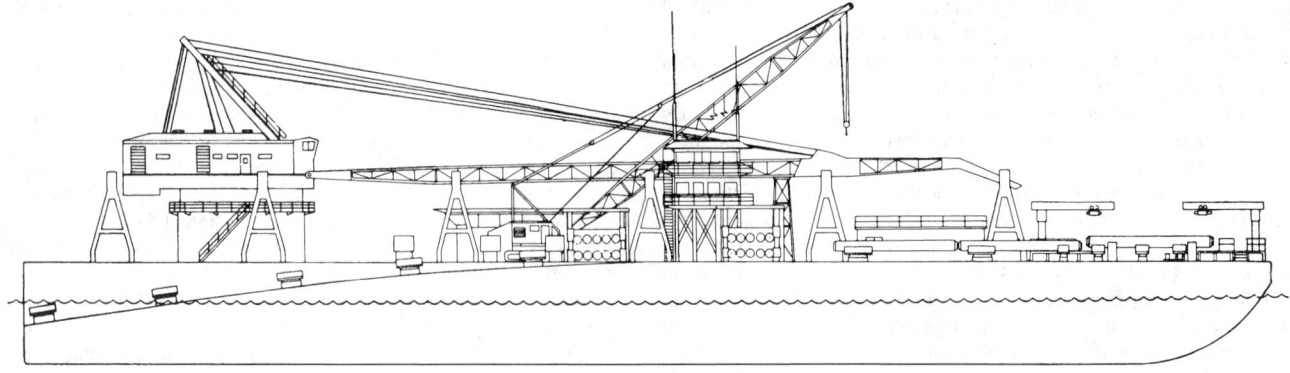

Fig. 7.   Pipelay vessel with barge-type hull.

because of heat leakage is used as boiler fuel for the propulsion of the ship. On a loaded voyage, this may provide about 90% of the fuel needed. The ships are ballasted for return voyage with seawater carried in separate wing tanks. Some LNG is left in the cargo tanks to ensure a nonexplosive gaseous atmosphere and to keep the tanks cool for the next voyage. Again, boil-off gas provides part of the propulsion fuel. One configuration of an LNG ship-loading system is shown in Fig. 8.

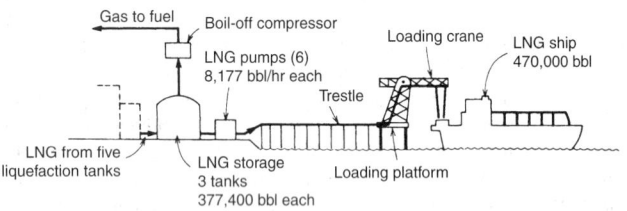

Fig. 8.   LNG ship-loading system. Trestle is 2.6 miles (4.3 km) long.

Safety in handling LNG in ships and at loading and unloading terminals of large scale has been a matter of constant concern. The observation has been made that the LNG gas carried would bury a football field under 125 feet (38 meters) of liquefied gas, or, after conversion to the gaseous phase, 600 football fields to the same depth. One factor that has not been routinely considered in the past is a phenomenon called a *flameless vapor explosion*. It is well known that, if water, for example, could be heated without nucleation occurring on the sides of the vessel, the water temperature could be raised well above the boiling point of 100 °C. If this could be done, and with the continued application of heat, the liquid would suddenly explode in its transition from the liquid to the vapor phase. Although not probable, it is possible that conditions favoring flameless vapor explosion could occur if liquefied natural gas were permitted to escape over a water surface. One scientist has observed that an explosion of this kind is possible when a liquid is 4 to 6% (no more, no less) above its normal temperature of vaporization. It is further observed that an explosion of this nature would not occur when LNG first spreads across a volume of water, but with time and the warming of the LNG such a hazard could occur. While an explosion of this type is not comparable to that from a chemical reaction, the explosion could greatly disperse the LNG over a greater area, thus spreading the zone of risk.

Underground Storage

The largest additional supply of natural gas for peak demands comes from underground storage reservoirs located, for example, close to the northern cities, as compared with the producing wells which may be located in the southwestern area of the country. Some of the storage pools are operated by pipeline companies, but most of the gas in underground storage is owned by the local gas companies that serve metropolitan areas.

The underground reservoirs are filled with gas from the pipelines during the summer months, when all of the fuel that the lines can deliver is not consumed. This method allows the producing wells and the pipelines to operate at fairly steady rates at all times of the year. Also, it is established that a gas field will produce more gas over a longer period if the gas is withdrawn at a steady rate.

Four states — Michigan, Pennsylvania, Illinois, and Ohio — have half of the total underground gas storage pools in the United States, with a total capacity of over 5 trillion cubic feet (142 billion cubic meters). In a typical year, about one-fourth of this volume will be used during cold waves to furnish the additional gas needed to supply homes and apartments.

The most common type of underground reservoir now storing gas is a previously producing gas or oil field. The supplies remaining in these pools are too small, and at too low a pressure, to justify continued production. But, the reservoir rock can hold gas pumped down through the same wells that once took gas out of the ground.

About 90% of the storage pools being used once produced gas or oil. In Pennsylvania there are over 60 such pools close to the large industries and centers of population. There are over 30 such pools in West Virginia, Michigan, Ohio, Kansas, Indiana, New York, and Kentucky, as well as smaller numbers in 13 other states. The gas to be stored is pumped into the old wells by compressors similar to those used to move gas in pipelines. The gas is stored under about the same pressure as originally existed in the field. In developing a gas storage reservoir, a company obtains a lease from the landowners in much the same manner that gas producers do.

The gas industry has been developing underground storage reservoirs for more than 60 years. The first known experiment in storing gas underground was conducted in 1915 in Welland County, Ontario, Canada by the National Fuel Gas Company. The success of this effort prompted the Iroquois Gas Corporation, a subsidiary of National Fuel, to develop, in 1916, the Zoar field south of Buffalo, New York. It was the first storage operation in the United States and is the oldest continuously used reservoir.

During the past 60 years, over 80 companies have invested several billions of dollars in underground storage facilities.

Another kind of underground storage reservoir is called an aquifer. An aquifer is an underground rock structure holding large quantities of water. The underground rock is porous and permeable. The pore spaces are filled with water, and impermeable rock covers the porous rock. Wells are drilled into such formations, and gas is forced into the pores under pressure. As the gas pressure increases, the gas pushes the water farther down into the porous rock, making room for the gas.

There are over 40 aquifers in the United States, located in Illinois, Indiana, Iowa, Kentucky, Minnesota, Missouri, Utah, and Washington. Three unusual reservoirs have been developed: an abandoned coal mine in Colorado; and salt domes in Michigan and Mississippi.

History of Natural Gas as an Energy Resource

It is reported that, perhaps 2,000 years ago, the Chinese piped natural gas from shallow wells through bamboo poles, for burning under large pans to evaporate seawater for salt. The first commercial use of natural gas in the western world was for lighting the streets of Genoa, Italy, circa 1802. The first evidence of natural gas deposits in the United States is found in reports of "burning springs" in various parts of New York, Pennsylvania, Ohio, and West Virginia. As early as 1626, French missionaries visiting the Indians in northwestern New York recorded that they could ignite gases rising from shallow waters. Many early reports were given of the presence of natural gas along the shores of Lake Erie and in the streams flowing into it. There also are references to "burning springs" in the Ohio River valley and along the Pacific shores of California. It is reported that General George Washington was fascinated by a "burning spring" in the Kanawha Valley, near Charleston, West Virginia, in 1775. Early settlers who drilled wells for water often reported the presence of traces of natural gas. The generally accepted birthplace of the natural gas industry in the United States is Fredonia, New York. Fredonia is located on Canadaway Creek, which empties into Lake Erie in the northwest corner of New York State. William A. Hart is reported to have dug a well in 1825 and obtained sufficient natural gas to light two stores, two shops, and a grist mill. Hollow logs were used for piping. Sufficient gas would accumulate in the well riser during the day to supply the gas lights at dusk. Hart was also instrumental in building the first natural gas lighthouse in 1829 along Lake Erie. The lighthouse, consisting of 13 gas lamps and reflectors in two tiers, served until 1859. In 1858, the first natural gas company in the United States was formed, The Fredonia Gas Light Company.

The consumption of natural gas gradually increased prior to World Wars I and II as more and more small pipelines brought communities within reach of natural gas fields accompanied by the retirement of previous manufactured or town gas facilities (the early forerunners of the substitute natural gas).

Natural Gas–powered Vehicles. The concept of natural gas–powered vehicles has become a *limited* reality in terms of the millions of gasoline- and diesel-fluid-power vehicles. In 1993, Mack Trucks (Allentown, Pennsylvania) and the Gas Research Institute have teamed to research and develop a natural-gas version of the Mack E7™ heavy-duty engine. A prototype of the design was scheduled for testing on a refuse vehicle in the Boston area. If successful, the engine also could be applied to a variety of heavy-duty vehicles, including long-haul tractor/trailers, construction equipment, and road maintenance trucks. The development was propelled by the needs of truck fleet owners who may be required by legislation to operate alternatively fueled vehicles. The engine will be required to meet applicable U.S. Environmental Protection Agency and California Air Resources Board emissions standards while maintaining the performance and reliability of its diesel-fueled counterpart. A 6-cylinder, 12-liter engine will be developed. The vehicle's onboard gas storage will hold the energy equivalent of about 45 gallons (170 liters) of diesel fuel.

Substitute Natural Gas. The oil crisis of the 1970s spawned a number of attempts to create synthetic natural gas. Some of these processes reached pilot and demonstration plant stages and beyond in their development. For example, substitute natural gas (SNG) from sewage wastes has enjoyed impressive success. The anaerobic digestion of a solid waste and water or sewage sludge slurry will produce a methane-rich gas.

Additional Reading

Abelson, P.H.: "The Gas Research Institute," *Science*, 1715 (December 11, 1992).
Bethke, C.M. et al.: "Supercomputer Analysis of Sedimentary Basins," *Science*, 261 (January 15, 1988).
Burnett, W.M. and S.D. Ban: "Changing Prospects for Natural Gas in the United States," *Science*, 305 (April 21, 1989).
Castaneda, C.J.: "A History of the Natural Gas Industry," Macmillan Library Reference, New York, NY, 1999.
Caton, J. (Editor): "Alternative Fuels and Natural Gas, Volume 3," American Society of Mechanical Engineers, New York, NY, 1995.
Considine, D.M.: "Energy Technology Handbook," The McGraw-Hill, Companies, Inc., New York, NY, 1977.
Fischetti, M.: "There's Gas in Them Thar Hills!" *Technology Review (MIT)*, 17 (January 1993).
Fulkerson, W., R.R. Judkins, and M.K. Sanghvi: "Energy from Fossil Fuels," *Sci. Amer.*, 136 (September 1990).
Holtberg, P.: "1993 Policy Implications of the GRI Baseline Projection of U.S. Energy Supply and Demand to 2010," Gas Research Institute (Washington Operations), Washington, DC, 1993.
Jensen, B.A.: "Improve Control of Cryogenic Gas Plants," *Hydrocarbon Processing*, 109 (May 1991).
Lyons, W.C.: "Standard Handbook of Petroleum and Natural Gas Engineering," Vol. 1, Butterworth-Heinemann, Inc., Woburn, MA, 2001.
Lyons, W.C.: "Standard Handbook of Petroleum and Natural Gas Engineering," Vol. 2, Butterworth-Heinemann, Inc., Woburn, MA, 2001.
McCabe, K.A., S.J., Rassenti, and V.L. Smith: "Natural Gas Pipeline Networks," *Science*, 534 (October 25, 1991).
Melvin, A.: "Natural Gas: Basic Science and Technology," Adam Hilger (London), Taylor & Francis (Philadelphia), 1988.
Ourisson, G., P. Albrecht, and M. Rohmer: "The Hopanoids: Paleochemistry and Biochemistry," *Pure and Applied Chemistry*, **51**(4), 709–729 (April 1979).
Ourisson, G., P. Albrecht, and M. Rohmer: "Predictive Microbial Biochemistry: From Molecular Fossils in Procaryotic Membranes," *Trends in Biochemical Sciences*, **7**, 236–238 (1982).
Ourisson, G., P. Albrecht, and M. Rohmer: "The Microbial Origin of Fossil Fuels," *Sci. Amer.*, 44–51 (August 1984).
Sharer, J.C. and P. O'Shea: "Gas Research Institute's Research Program on Unconventional Natural Gas," *Chem. Eng. Progress*, (February 1986).
Sweetser, R.: "The Fundamentals of Natural Gas Cooling," Prentice-Hall, Inc., Upper Saddle River, NJ, 1997.
Willett, R. (Editor): "1996 Natural Gas Yearbook," John Wiley & Sons, Inc., New York, NY, 1995.
Woods, T.J.: "The Long-Term Trends in U.S. Gas Supply and Prices: 1992 Edition of the GRI Baseline Projection of U.S. Energy Supply and Demand to 2010," Gas Research Institute, Chicago, Illinois, December 1991.

Web References

Gas Research Institute: *http://www.gri.org/*
The American Petroleum Institute: *http://www.api.org/*

NATURAL RUBBER.   See **Rubber (Natural)**.

NAUTICAL MILE. The fundamental unit of distance used in navigation, once defined, for purposes of convenience, as 6,080 feet (1,853 meters). Rigorously, the nautical mile was defined as the length of 1 minute of arc on a great circle drawn on the surface of a sphere with the same area as the earth; thus as 6,080.27 feet (1,853.27 meters).

But, owing to the fact that the earth is an oblate spheroid and flattened at the poles, the length of 1 minute of arc measured along a meridian varies in different latitudes. It is shortest at the poles and longest at the equator, having an average length of 6,076.82 feet (1,852.21 meters). To resolve the confusion caused by this variation of the nautical mile with latitude, various countries established standard figures. At last an international agreement was reached whereby the nautical mile was defined, and is presently accepted, as 1,852 meters. For navigation purposes, a good approximation for the number of nautical miles is the number of minutes of arc along a great-circle route.

The nautical mile is frequently confused with the *geographical mile*, which is defined as the length of 1 minute of arc on the earth's equator and has a length of 6,087.15 feet (1,855.36 meters). See also **Units and Standards**.

NAUTICAL TWILIGHT. That period when the upper limb of the sun is below the visible horizon and the center of the sun is not more than 12 degrees below the celestial horizon.

NAVIER-STOKES EQUATIONS. The equations of motion of a Newtonian fluid that are applicable to the motion of simple liquids and nondissociating gases. In tensor form, they are

$$\frac{du_i}{dt} = \frac{\partial u_i}{\partial t} + u_j \frac{\partial u_i}{\partial u_j} = -\frac{\partial p}{\rho \partial x_i} + \mu \frac{\partial^2 u_i}{\partial x_j^2} + \frac{\mu}{3} \frac{\partial}{\partial x_i} \left(\frac{\partial u_i}{\partial x_i} \right)$$

where du_i/dt is the time rate of change of the velocity of a fluid with velocity vector u_i, u_j is the velocity vector perpendicular to u_j, ρ is fluid density, p is the hydrostatic pressure, and μ is viscosity. See also **Fluid and Fluid Flow**.

NAVIGATION. Since development of the early navigation satellites (INMARSAT, et al.) just a few decades ago, the approach to the science and equipment of navigation has undergone a major anatomical change that is geared to ultrasimplification for the end user. The operator no longer requires sophisticated support instrumentation or an understanding of and appreciation for the geometry and mathematics that have typified pre-satellite navigation systems. Refinements of navigation satellite systems, as exemplified by the Global Positioning System (GPS), first applied by the military during the Persian Gulf War (1990–1991), are occurring apace and ultimately to a significant extent will obsolete many of the early and contemporary navigation systems. Because of the installed investment in contemporary systems, possible security restrictions that may be imposed on GPS during wartime, and the ability of some users throughout the world operating on limited budgets, the time of phaseout for contemporary non-satellite–based systems is unknown. Further, there is much interest in the lore of navigation methodologies. This article incorporates, as in prior editions, a chronology of navigation system developments. The article commences, however, with a condensed overview of the GPS system.

Space-Age Triangulation

In the Global Positioning System, satellites with precisely synchronized atomic clocks continuously broadcast their time and location to suitable earthbound computerized receivers. In a fully self-contained manner, the receiver calculates and displays its Earth position. Data received from at least three satellites is required to yield the *latitude-* and *longitude* of the receiver. Data from a fourth satellite is required to yield *altitude*. The receiver is synchronized to the satellites' clocks. Distance from the receiver to the satellite is calculated by measuring the time interval required for the signal to travel from satellite to receiver. When three satellites are in "view" simultaneously, a computer in the receiver quickly works outs the precise position by triangulation because only one point on a plane can yield a particular combination of distances from the three satellites. See Fig. 1. Receivers are small and may be handheld.

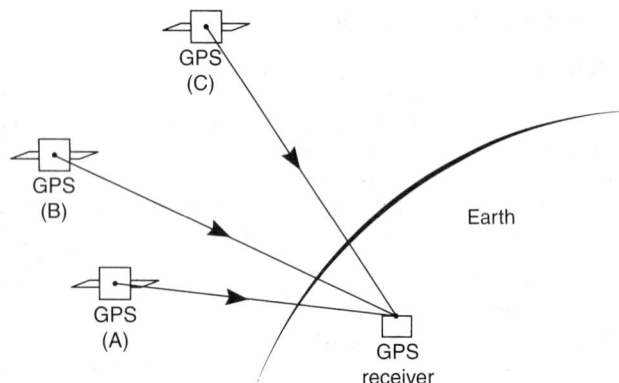

Fig. 1.   Global Positioning Satellites (A, B, C), each equipped with a precisely synchronized atomic clock, continuously broadcast their time and location to earthbound receivers. Computers in the receiver calculate distances of receiver from each satellite, this information derived from the time required for transmission of signal from satellite to receiver. By employing the methodology of *triangulation*, the receiver determines the only *acceptable location* that will geometrically satisfy the signal time differences. Only three GPS readings are required to determine *latitude* and *longitude*. Additional data from a fourth GPS will yield *altitude* information. The GPS thus combines advanced satellite technology with simple geometry. See **Satellites (Communications and Navigation)**.

Quite a large number of satellites in orbit are needed to ensure that at least three satellites are in sight at any particular Earthpoint and four in view if altitude information is required as well. As of 1994, the U.S. military intended to have 24 satellites in place, including three spares. Each satellite orbits Earth once every 12 hours at an altitude of 10,900 miles (17,538 km). The ultimate total cost of the GPS system is estimated at $10 billion upward.

Because of the advantages of the GPS to air, sea, and ground operations during the U.S. military venture Desert Storm, equipment was released for use in the field at least 2 years prior to an established schedule. The GPS markedly increased the accuracy of bombing runs, accurately guided sea-launched missiles to targets, steered ships and personnel safely through mine fields, and advised tank and land-based commanders with reliable information on their exact location even in featureless desert scenarios.

Had the enemy been aware of more detail receiver design and had developed suitable receivers, they could have taken advantage of the GPS. Subsequent analyses of the various encounters did not reveal, however, any enemy use of the system. But this possibility in terms of future military actions has prompted the U.S. military to offer GPS information on two levels by way of an arrangement known as *differential GPS*. For most military uses of GPS, the maximum available accuracy is needed. Reports on early design of the system indicated that accuracy could be expressed in terms of a few millimeters rather than meters. An early satellite-based system had provided scientists in various fields, notably in connection with movement of Earth's crust, with a *fine signal* in the millimeter range.

Although subject to future alteration, the military developed the concept of *selective availability* — that is, (1) a fine and very accurate position reading, and (2) a course location fix, the accuracy of which would be in the range of 100 meters. In this system, a slight distortion of the satellite signal is introduced but still permits a finer, undistorted signal for specially designed receivers for military applications. This approach basically denies location information of the best achievable accuracy to scientists and for

a host of demanding applications of the system for commercial use. One scientist has observed, "The military has come up with a system that is useful to mankind. Now they don't want mankind to have full use of it." This scientist is engaged in studies for predicting seismic risks.

For example, commercial uses that require information of high accuracy are found in the monitoring of railway cars on tracks that are quite close to each other and for navigating ocean-going vessels through rock-infested straits where, at a minimum, an accuracy of a few meters is required.

Some scientists have developed electronic means to circumvent the signal distortion problem by using schemes that are identified as *differential GPS*. In one version of this technique, GPS information from numerous satellites and using several receivers will cancel out minor errors. One scientist has observed, "We've learned to live with selective availability, but the solution is costly, by perhaps an order of magnitude increase in the GPS readings required. This in turn is reflected by increases in the costs of data storage, transmission, and analysis."

Although GPS is a major breakthrough for navigation and position-related analyses, the system, like communications satellites, can be plagued with satellite instability problems, faulty antenna pointing, and failure of solar panels for power, among other factors.

The early work required to develop navigation satellites, dating back to the 1970s and ultimately leading to the GPS, is described in considerable detail in the Seventh Edition of this encyclopedia.

Classification of Navigation Sensors

The term *navigation* stems from the Latin word *navigare*, -meaning "to conduct a ship." Traditionally, navigation has been treated under three main classifications: (1) the *sailings* — to find the direction or directions in which to head a ship in order that it may proceed on the most practical course from one place to another (see also **Sailings (The)**), (2) *dead reckoning* — to find the position of a ship in any particular instant, provided that its position at some prior time is given and that, since leaving this position, the headings, distances run on each heading, and the effects of environment and motions of the medium supporting the ship are known (see also **Dead Reckoning**), and (3) *celestial navigation* — to find the position of a ship at any instant by means of observations, not necessarily visual, of one or more objects on the surface of the earth, or on the celestial sphere. Although still a helpful classification, the foregoing obviously is dominated by consideration of ships at sea, not taking into full account air and space navigation. Also, within the last few decades, navigational instrumentation has been put onto a much more sophisticated footing so that a classification of the type shown in Fig. 1 is much more useful.

Further, there are two broad categories of navigation: (1) *absolute navigation*, wherein knowledge of present position is known in relation to an overall earth-coordinate system (latitude and longitude, for example); and (2) *relative navigation*, wherein position is known relative to some special local coordinate or grid system. The accuracy attainable in a relative navigation situation usually is significantly better than that which can be obtained on an absolute basis.

Navigation Sensors

The primary navigation sensors and their functions are listed in Table 1. The more commonly used sensors for both marine and aircraft navigation are included. Frequently, a navigation system will be made up of various combinations of these sensors.

As will be noted from Fig. 2, there are self-contained navigation systems (self-processed information) and externally-controlled systems wherein a data link is required between the vehicle (aircraft, ships, etc.) being navigated and some reference or data point. An abridged list of examples of these two general classes of systems is given in Table 2.

Many factors are involved in selecting the most appropriate navigation system, including the usual parameters of accuracy, cost, size, weight, power requirements, reliability, operational simplicity/complexity, degree of automaticity desired, ruggedness in environmental extremes — factors which have given rise to numerous standards of equipment. Other very important factors for many situations include, for example, worldwide availability, all-weather operability, and continuous versus periodic availability. Certain radio navigation systems, for example, may not embrace transmitters that cover the entire world. All-weather, night-and-day operation rules out, for example, such systems as optical star trackers (as sole sensors); clouds below an aircraft, for example, rule out laser doppler systems. In some navigational satellite systems, there may not always be a

TABLE 1. NAVIGATION SENSORS BY PRIMARY FUNCTION

Meters

Velocity Meters	Altitude or Depth Meters
Pilot log	Pressure altimeter
Electromagnetic log	Pressure-depth meter
RPM tachometer	Radar altimeter
Airspeed (pilot) meter plus wind	Sonar fathometer
Doppler radar	Inertial systems
Laser radar	Ground or other vehicle
Doppler sonar	Trackers and data link
Inertial systems	

Time Meters
Chronometer
Radio-synchronous signals
Time standards (crystal oscillators)

Heading Reference

Magnetic compass	Radiometric celestial tracker
Gyro compass	Inertial systems
Optical celestial tracker	

Position-fix Devices

Sextant	Navigation satellite	Radio transponders
Radio aids	Landmarks	Ground or "mother" ship
LORAN	Optical sighting	Trackers and data link
OMEGA	Radar sighting	Gravity anomaly, map matching
SHORAN	Seamarks	Magnetic anomaly, map matching
RAYDIST	Sonar bottom fixing	
DECCA	Optical celestial tracker	
VOR/DME TACAN	Radiometric celestial tracker	

satellite within range, and a fix may have to await radio view when one comes into position. Thus, it is rare when a single navigation sensor can satisfy all desired requirements. Even magnetic compasses and normal-mode gyrocompasses are essentially unusable in the polar regions of the Earth.

TABLE 2. SELF-CONTAINED AND EXTERNALLY CONTROLLED SYSTEMS

SELF-CONTAINED SYSTEMS
— Completely self-contained
— Self-processed information; no radiation required.
 All measurements internal or in local vicinity of vehicle, such as inertial navigation.
— Self-processed information
— Vehicle is passive; natural earth or natural sky references.
 Natural can include constructed landmarks (buildings, etc.) not purposely placed as navigation aids.
 Optical star trackers.

SELF-CONTROLLED SYSTEMS
— Self-processed information; vehicle passive; earth passive or artificial
— Purposeful navigation landmarks.
 Navigational buoys.
— Self-processed information; vehicle passive; active earth or active sky

LORAN
— Self-processed information; vehicle active; earth passive or artificial; or artificial celestial
 Ranging to lead ship.
— Self-processed information; vehicle active; earth active
 Radar transponders.

EXTERNALLY CONTROLLED SYSTEMS
— Ground or other vehicle processed information and data link; vehicle passive; external tracker passive
 Theodolite optical tracker.
 Skin tracking radar.
— Ground or other vehicle processed information and data link; vehicle active; external tracker passive
 Optical tracking of light on vehicle.
— Ground or other vehicle processed information and data link; vehicle active; external tracker active
 Beacon tracking radar.
Foregoing are examples only; not all-inclusive.

Celestial Navigation

In this system, navigation is achieved by means of observing celestial objects. In the year 1837, Captain Thomas Sumner discovered what has since been known as the Sumner Line, and modern celestial navigation may be said to date from that discovery. The methods for determining terrestrial latitude and longitude from sextant observations of altitude of celestial objects are briefly described in **Sumner Line**. During intervening years, much research has been carried out on the theory of celestial line of

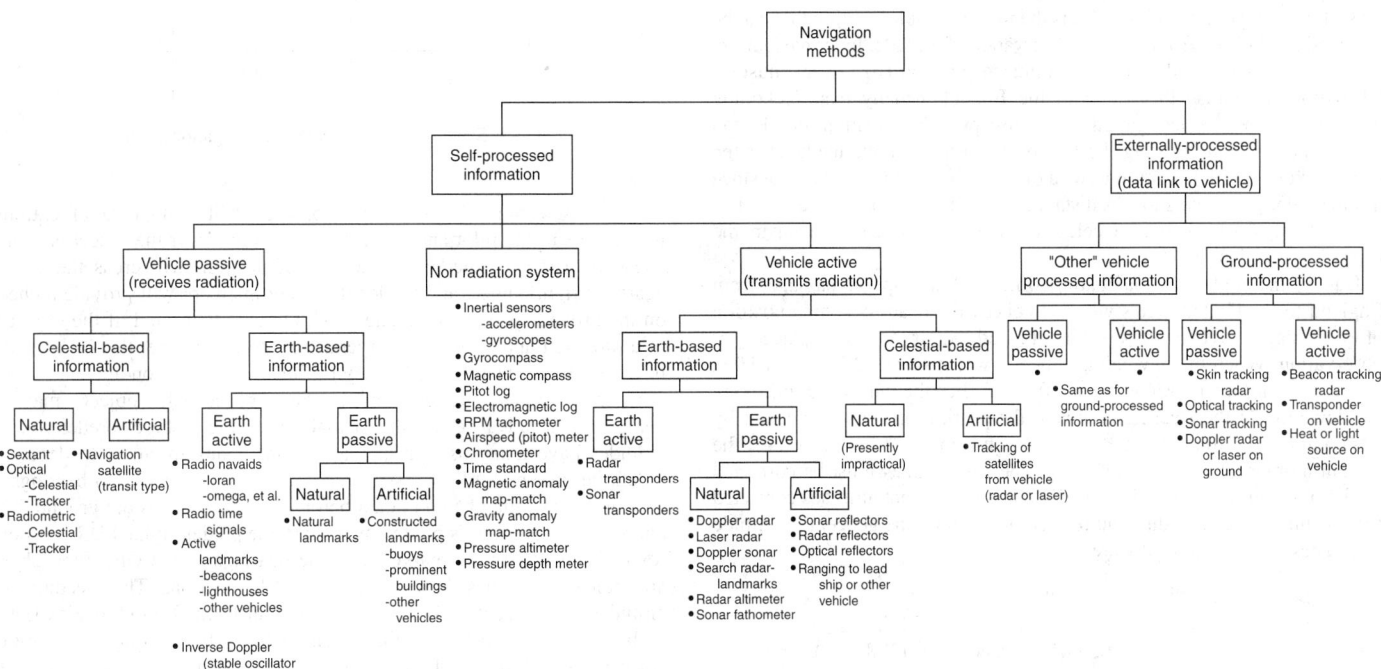

Fig. 2.   Classification of sensors used in navigation systems.

position, and numerous methods for calculating the data necessary to plot that line have been developed.

It is in order to briefly describe what is meant by a celestial line of position. At any instant, any celestial object is directly at the zenith for some particular spot on the surface of the earth. This point in years past was known as the subsolar, sublunar, or substellar point, depending upon whether the object observed was the sun, the moon, or a planet or star. The term *Ground Position* (GP) is now used no matter what celestial object is used. The terrestrial coordinates of GP may be expressed in terms of celestial coordinates of the object, the latitude being equal to the declination of the object, and the longitude equal to its Greenwich hour angle. Both of these quantities are tabulated for the sun, moon, planets, and navigator's stars in various readily available almanacs. The tabulations are given in terms of Greenwich Civil Time, and this time is used by navigators for recording the times of sextant observations for altitude.

In practice, the line of position is drawn on a Mercator chart, a plotting sheet, or a small-area plotting sheet by employing the geometric proposition that a radius of a circle is always perpendicular to an arc. At some particular Greenwich Civil Time (GCT), the altitude (h_s) of a celestial object is obtained with the sextant or bubble octant. Then, using the dead-reckoning position, or some position close to it, which leads to simplified computations, the values of the altitude (h_c) and the bearing (Z_n) that the object would have at the assumed position and GCT of observation are computed or taken from suitable tables. The dead-reckoning, or assumed, position is now set down on the plotting sheet and a line drawn through it in the direction Z_n. This line is a section of the radius of a circle drawn about the GP, which, in reality, is usually off the plotting sheet. Next, the difference between the computed and observed altitudes ($h_c - h_t$) is taken; this is called the "intercept." If the intercept is zero, the ship must be on a line of position perpendicular to the bearing line and passing through the plotted position. If the intercept is plus (+), the line of position must pass through a point ($h_c - h_t$) minutes of arc, or nautical miles, away from the GP along the bearing line; and if the intercept is minus (−), the line must be between the plotted position and the GP. In either case, the line of position must be perpendicular to the bearing line.

In spite of the fact that the line of position is actually a circle with radius ($90° - h_t$) miles, the line may be drawn as straight in practically all cases. If we assume that the altitude of the object is 80°, the value of ($90° - h_t$) is 10° or 600 miles (965.6 kilometers). In this case, a straight line 60 miles long (96.56 kilometers), perpendicular to the radius, will differ from the actual circle by less than a mile at its extremities. Accordingly, the assumption of a straight line will not lead to appreciable errors, if the altitude is less than 80° and the drawn line is less than 60 miles (96.56 kilometers) long.

Simple statistical analysis shows that the point on the line of position closest to the dead-reckoning point is the most probable position that can be obtained for the ship from a single observation of altitude. In air navigation, this position is referred to as the estimated position (EP). Care must be taken not to confuse this EP with the EP obtained by dead-reckoning navigation in marine navigation. This most probable position, or EP, can be obtained without plotting the line by any of the above methods if the dead-reckoning (DR) position instead of the assumed position is used, since the intercept gives the shortest distance from the DR position to the line.

An example of the use of celestial navigation at sea is given in the following practical case:

During the night, the navigating officer of a ship on passage from England to the United States wishes to check the dead-reckoning position of the ship. The two stars Alpheratz and Altair are well placed for observation. When the navigating officer's watch reads 23 h 40 m 10.0 s, the sextant altitude of Alpheratz is 50° 34′.3. For the purpose of checking the deviation of the steering compass, the bearing of the star is taken by this compass and found to be 121°. At watch reading 23 h 46 m 15.4 s, the sextant altitude of Altair is 50° 20′.7. The watch times must be corrected to obtain Greenwich Civil Time (GCT), and the sextant altitudes corrected for instrumental errors, dip, and refraction to obtain the geocentric altitude (h_t). These corrected results are:

Star	GCT	h_t
Alpheratz	02h 39m 34.0s	52° 27′.4
Altair	02h 45m 39.4s	50° 13′.8

Using the average of the watch times, the dead-reckoning position of the ship is found as latitude 43° 24′.6N and longitude 48° 27′.4W. Since the

ship is proceeding at only 16 knots, and since the DR position is probably somewhat in error, this position is used for computing the altitudes and bearings that the stars should have at the GCT observation. These values are found to be:

Star	Bearing	h_c
Alpheratz	098°.4	52° 32′.5
Altair	216°.2	50° 07′.4

The intercepts ($h_c - h_t$) are found: for Alpheratz +5′.1 and Altair −6′.4. These yield two "most probable" positions of the ship, one 5.1 miles from the DR position in the direction 278°.4 (i.e., away from the GP of Alpheratz), and the other 6.4 miles in the direction 216°.2 (i.e., toward the GP of Altair). To determine the fix, the DR position is plotted, the bearing lines to the two GP's drawn through it, the intercepts measured off in the proper direction, and the two lines of positions drawn through the intercepts perpendicular to the lines to the GP points. The point of intersection of the lines of position is the fix at 2345. The actual plotting, on small area plotting sheet, is shown in Fig. 3. From the figure, the fix is found to be in latitude 43° 20′.7N and longitude 48° 35′.2W. To determine the compass deviation, the difference between the observed compass bearing of Alpheratz and the computed bearing is found to be 121° − 098° = 23°. Since the variation in this region is 26°W, the deviation must be 3°E.

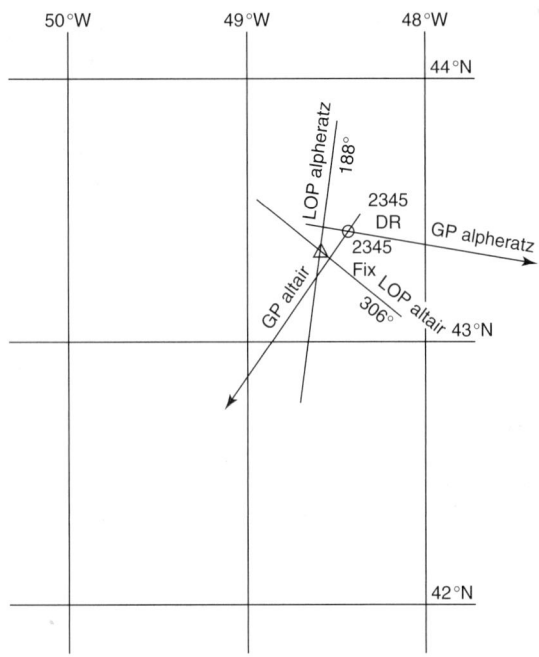

Fig. 3.   Scale drawing for celestial navigation example.

Proper selection of stars to be observed will yield data of extreme importance to the pilot and navigator. For example, if the object is nearly ahead or astern of the plane, the line of position will cross the course nearly at right angles, and the length of the intercept will provide a check on the ground speed being made good. On the other hand, if the object is in a direction approximately perpendicular to the course, the value of the intercept will indicate the accuracy of the wind correction angle.

In many cases, altitudes and bearings of celestial objects may be computed in advance of the actual observing. These predetermined altitudes have many uses in air navigation. If an aircraft is to depart at a definite time and follow a specified course, the altitudes and bearings at indicated times may be computed before the plane leaves the ground. The course to be followed is plotted on a chart, the predetermined DR positions for indicated times are marked, and the bearings of the GP of the object are indicated by lines drawn through the DR positions. The precomputed altitudes are geocentric, but they may be transformed into those expected to be read to the octant at the specified times by applying the various corrections with reversed signs. The navigator measures the altitude at an indicated time, obtains the difference between his value and that predicted, and lays off this distance along the drawn bearing line either toward or away from the GP. In this way, an EP is determined within a few seconds after the observation is completed, and no computing is required during the flight. The pilot is notified to alter heading and air speed to bring the plane back to schedule. If, due to unforeseen conditions, the plane gets so far off scheduled position that the intercepts are more than 150 miles, the predetermined altitudes must be abandoned and regular celestial navigation adopted.

Under some conditions, it may be necessary for a plane to make an accurate landfall (e.g., locate a small island, life raft, etc.) under conditions where celestial navigation must be relied upon. In such cases, the use of precomputed altitudes gives great assistance. First, an estimated time of arrival (ETA) is obtained. Then, using the latitude and longitude of the landfall, a series of altitudes and bearings of a celestial object is computed. The interval of time between computed values depends somewhat on the rapidity with which the values are changing, but is usually about 10 minutes. The series begins at least half an hour before ETA and extends beyond that value. Two curves are then drawn on graph paper, showing altitude and bearing as a function of time. The plotted altitudes are those expected with the octant, i.e., with corrections applied to computed geocentric values. If possible, an object that is approximately ahead or astern of the plane should be selected. About half an hour before the predicted ETA, the pilot alters heading 10° or 15° to the right or left of that predicted for the true course, so that there will be no question as to which side of the landfall he is approaching, and the navigator begins taking altitudes of the object. The navigator plots his observed values, as a function of time of observation, on the same graph as that showing the predetermined values, and obtains a curve of observed values. At the instant that the observed curve intersects the curve obtained from precomputation, the plane must be on a line of position running through the landfall. The bearing of the celestial object at this instant is read off the plotted bearing curve, the line is drawn at right angles to this bearing through the destination, and the pilot is instructed to alter heading to run down the line.

Radio Navigation

The use of radio aids in navigation for checking the dead-reckoning position of a ship is known as radio navigation. Radio direction-finders were used very early in the development of radio technology to avoid the difficulties of celestial navigation from a ship or aircraft and for emergencies in bad weather. The simplest system uses a directional antenna to locate the direction to several radio stations. A simple triangulation then locates the ship or aircraft with respect to the location of the stations, usually well known and in the map of reference with which the pilot is familiar. The method is complicated by the aircraft velocity, but not by accelerations, weather, or the availability of tables. The pilot can tune in a station near his destination and simply follow the signal to it. Two deficiencies are present: (1) the location of the stations; and (2) the errors inherent in a directional antenna. These problems led to improved radio systems.

The use of radio bearings as lines of positions is best explained by an example from ship navigation. A ship is proceeding on heading 330° at 12 knots. Three radio beacons, A, B, and C, are located in the following positions:

Station	A	B	C
Latitude	29° 30′N	30° 00′N	28° 40′N
Longitude	83° 20′W	81° 40′W	81° 52′W

At 0812 the dead-reckoning position of the ship is L = 28° 32′N & Lo = 82° 42′W and at that time, radio bearings, corrected for deviation of the radio compass, are A = 000°, B = 063°, and C = 117°. These must be changed to true bearings by adding to each the heading of the ship, obtaining: A = 330°, B = 033°, and C = 087°. Since these are great-circle bearing, they must be reduced to rhumb-line bearings by applying the correction factors for A − 0.05, B + 0°.2, and C + 0°.1. Then, working either on a mercator chart, mercator plotting sheet, or small-area plotting sheet, the corrected rhumb-line bearings are plotted, and the fix determined as the center of the triangle of intersection of the three lines of position. The position of the fix is L = 28° 37′N & Lo = 82° 44′W, and, since the sides of the triangle are less than 3 miles, we can assume that the fix is probably correct to within 1 mile. The complete solution is illustrated in Fig. 4 which is drawn on a small-area plotting sheet and labeled in accordance with standard procedure.

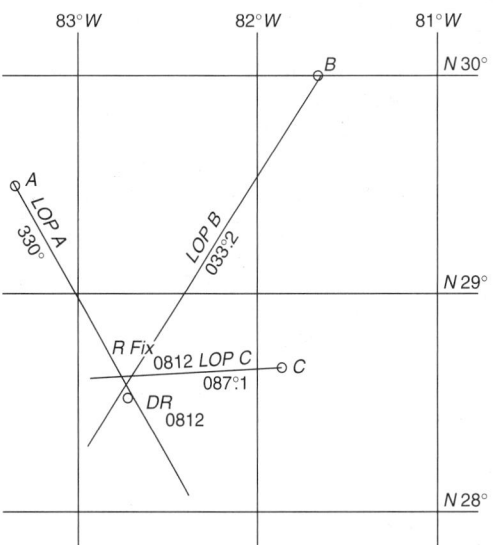

Fig. 4.   Radio navigation scale diagram.

Three corrections must be applied to a radio bearing before it can be used as a line of position on a mercator chart or small-area plotting sheet. Loop antennae and radio direction-finders have deviation corrections, due to the magnetic field of the ship. These must be determined in advance for different headings of the ship and applied to radio bearings as obtained. Then the radio bearing, which is relative, must be changed to true bearing by adding the true heading of the ship. Finally, since radio follows great circles, the bearing must be converted from great-circle to mercator, or rhumb-line, bearing. Figure 5 shows two points, X and Y, plotted on a mercator chart, with the rhumb line, XMY, and the great circle, XGY, connecting the two points. The great circle will always be convex toward the nearest pole, and we have drawn the figure for the Northern Hemisphere with true north indicated both at X and Y. The lines gX and g′Y are tangents to the great circle at X and Y, respectively, and are the directions in which the signal from Y will arrive at X, and that from X will arrive at Y. Let us consider X to be the receiving station. Then the angle R (NXg) represents the great-circle bearing of Y, and the angle B (NXM) the rhumb-line bearing. In this case, it is noted that a correction must be added to the great-circle bearing to obtain the rhumb line. Reversing stations and considering Y the receiver, we see that at this point the correction must be subtracted to obtain rhumb line from great circle.

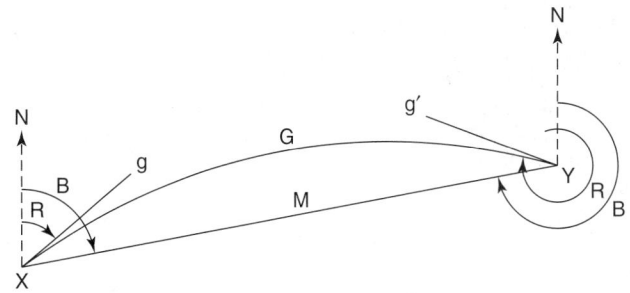

Fig. 5.   Plotting of X and Y on a mercator chart.

In case a navigator is working on a Lambert chart, as is frequently the case in air navigation, the great-circle bearing is close enough to a Lambert line to be plotted without correction other than for deviation and heading.

A radio range is a system of radio signals designed for the purpose of guiding a ship or plane along a designated track toward or away from a specified location. Relatively low frequency (200–400 kHz) has been used. Although there are various modifications, the track is indicated by the intersection of two field patterns from the range antenna system. The usual antenna arrangement is two pairs of cross antennae set 90° in space from one another. This gives two figure-eight field patterns as shown in Fig. 6(a). The patterns overlap in narrow wedge-shaped regions, which

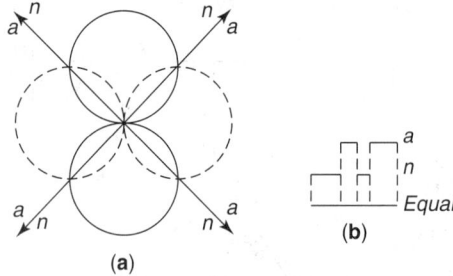

Fig. 6.   Radio ranging.

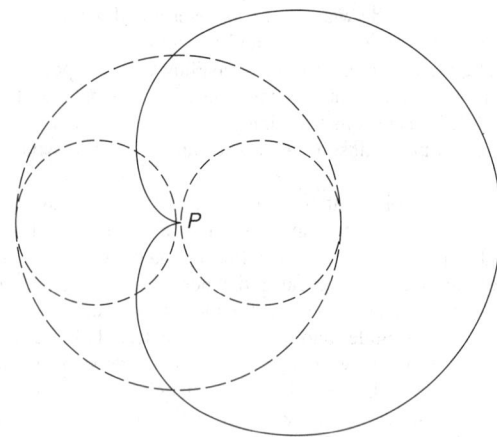

Fig. 7.   Radio compass antenna patterns.

have their apices at the transmitting station. These overlapping sectors are known as the range, or "the beam." Both an aural and visible system may be used for keeping a ship "on the beam."

In the aural system the carriers from the two antennae are so modulated with some audio note, say 1,000 Hz. The code signal (letter *a*, dot dash) is transmitted from one antenna, and the code signal (letter *n*, dash dot) from the other, so timed that on the center of the overlapping region the two signals blend together in a continuous note. This is indicated in part (a) of Fig. 6 and the time sequence of the code characters is shown in part (b). To remain directly "on the beam," all that is necessary is for the navigator to head his plane so that the continuous note is heard in his receiver. If he drifts to the right or left of the center, he hears either the *a* or *n* signal superimposed on the steady note. This also provides a method for planes proceeding in opposite directions to remain on the proper side of the airway. The range signals are interrupted at frequent intervals to give the name of the station and, when necessary, weather reports and information to aviators or navigators.

In the visual method, the carriers from the two antennae are modulated with different frequencies, say 65 and 85 Hz. The antennae are then excited alternately so that the plane receiver gets first one signal and then the other very rapidly. The demodulated output of the receiver is fed to a tuned reed instrument, so that, if the two signals are received with equal strengths, indicating on course, both reeds will vibrate with equal amplitude, while off-course flight will cause a greater vibration of one or the other, depending upon which side of the course the plane is flying. The exact angular relation of the courses laid down by the range station may be altered in several ways. Feeding the two antennae with different strength signals, feeding in different phases, utilization of additional antenna elements, etc., all serve to alter the field pattern so the lines of equal strength can be varied in direction. Where it is desired to rotate the courses after their angular relation has been fixed, a double goniometer may be used to feed the antennae. This gives a continuous 360° control of the direction of the beams.

The term *radio compass* has been used loosely over the years. When the loop antenna was first applied to the determination of radio bearings, the term *radio-compass station* was applied to shore installations that would forward, upon request, the bearing of a ship from the station. Next, the term was applied to a group of shore installations, each equipped with a loop antenna, from which the navigating officer of a ship within range could obtain the latitude and longitude of his ship. After the loop antenna and receiving sets had been developed to a state where they could be carried by the ships themselves, the term radio compass was applied to the loop. As new and improved radio equipment became available, the term radio compass was successively applied to any radio device that could be used to determine bearing. A glance through any textbook on navigation, particularly those dealing with air navigation, will yield at least two, and sometimes as many as five, different instruments for radio compass.

Radio compass is also applied to a direction-finding instrument which has a dial and looks in many respects like an ordinary compass and which is used for heading the ship in much the same manner as a compass. Two radio antennae are used with the instrument — a loop and a nondirectional antenna. The volume controls for signal intensity are so adjusted that the signal strength from the loop and the nondirectional antenna are the same when the station is in the plane of the loop. The two antennae are then fed into a single receiver, and the signal intensity is illustrated in Fig. 7. The resultant signal intensity from any station is shown on the instrument panel.

With reference to the figure, the length of a line from *P* to any one of the three curves is proportional to the strength of the signal from a station in the direction toward which the line is drawn. The dotted (figure eight) curve represents the relative intensities in various directions for the loop alone; the dashed (circle) curve represents the intensity for the nondirectional antenna alone; and the full curve that for the combined loop and nondirectional antenna, i.e., for the radio compass.

Hyperbolic Navigation. Hyperbolic navigation is a general method for determining a line of position by measuring the difference in the distance from the navigator of two stations of known position. The difference in distance is found by measuring the difference in time between arrival of signals transmitted from the two stations. A great variety of signaling methods is theoretically possible. Some of these systems are described shortly. Since electromagnetic waves travel with a speed of about ~186,000 miles (299,274 kilometers)/second, the difference between arrival times of the signals will be very small. The unit of time used in these systems is the microsecond (0.000001 second); a difference between arrivals of one of these units indicates a distance difference of 0.186 miles (0.299 kilometers) from the two transmitters.

Points of constant difference of time between arrival of the two signals will fall on spherical hyperbolas, with the transmitters at the foci. For navigational purposes, one need only consider the lines of intersection of these surfaces with the surface of the earth. The total number of distinguishable lines in any system is equal to the time required for the signal to travel from the master to the slave and back again, divided by the smallest time interval that can be measured by the receiving equipment.

In some systems, two or more slave stations may be used with a single master. The cycle of transmission always begins at the master station, and the signal travels out in all directions. The arrival of the master signal at the slave "triggers off" the slave. Operators continuously monitor both stations, each monitoring the signals for the other, to detect the slightest variations of frequency carrier waves, intervals between pulses, and characteristics of the signals. At the slave station, adjustable delay circuits are available, so that, once the cycle has been started, the slave station may transmit simultaneously with the master or be delayed by any desired amount.

As shown in Fig. 8, certain fundamental lines, representing integral multiples of distance or time difference, are superimposed on regular

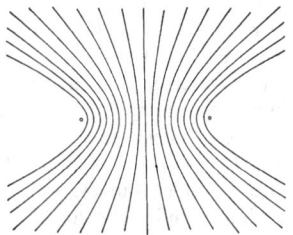

Fig. 8.   Fundamental curves for hyperbolic navigation.

navigational charts. Tables are published that contain the data for determining the fundamental lines. By graphical interpolation on the chart, or by mathematical interpolation from tables and stored data, the navigator can determine a hyperbolic line of position, using the observed difference in time of arrival of the signals, and the particular stations being used. The accuracy of the line varies from about 200 yards (183 meters) up to about 2 miles (3.2 kilometers), depending upon the distance of the observer from the base line between stations, and the type of equipment in use.

A fix as determined by a hyperbolic navigation system employing one master station, M, and two slaves, S_1 and S_2 (as in GEE navigation) is shown in Fig. 9. The diagram also indicates the value of hyperbolic navigation for "homing" on point A. The navigator obtains a fix at P and then sets his indicating equipment so that the pips from M and S_1 are in coincidence on the display instrument. Then the navigator heads the plane or ship so that, when these pips remain in coincidence, the ship will be following the hyperbola PA. By taking observations on the MS_2 pair at intervals, the navigator can determine the rate at which the objective is being approached.

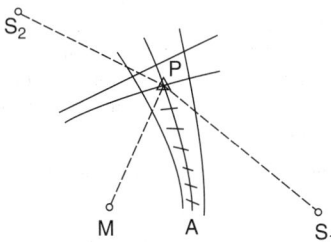

Fig. 9. Fix (hyperbolic navigation).

DECCA Navigation. This is a system of hyperbolic navigation that employs low-frequency continuous-wave radiation.

The master and any slave station radiate continuous waves, whose frequencies are related by a simple fraction, say one-fourth. The radiations from the two will be in phase when the distances from the stations differ by even multiples of a specified unit. This unit is a function of the wavelengths radiated.

In practice, a master and two slave stations are used. The receiving set reduces all three to a common frequency, and phase meters indicate the relative phase of each slave to the master. The accuracy of the setting of the phase meter is such that differences in distance may be determined with an accuracy of the order of magnitude of 100 feet (30.5 meters), and this is independent of distance from the base line. However, there is complete ambiguity of position, since there is no positive method of determining the number of complete phase changes between the observer and either station.

In spite of the ambiguity mentioned above, the system is of great accuracy and value when used in proceeding to some specified objective. A line of position, involving the master and one of the slave stations, is selected, which passes through the desired objective. The pilot must get the ship onto this line and then set his phase meter. If the pilot then proceeds so that the phase meter setting remains constant, the craft must be following the hyperbola directly to the objective. The hyperbolic lines from the master and the other slave will intersect the hyperbolic track along which the ship is proceeding. The pilot computes the number of complete phase changes that are to be expected, between the point of departure and the objective, along the line that the craft is following. When this number, plus any remaining fractional phase change, has been completed, the pilot must be directly over the objective.

LORAN Navigation. This is a long-distance radio-navigation system for aircraft and ships, utilizing synchronized pulses transmitted simultaneously by widely spaced transmitting stations. Hyperbolic lines of position are determined by measuring the difference in the time of arrival of these pulses. The intersection of two of these lines of position, obtained from either three or four stations, gives a position fix. Standard LORAN operates on frequencies between 1,800 and 2,000 kHz. LORAN C is a widely used version of LORAN that uses pulse signals for more precise time-delay measurement and operates at a frequency of 100 kHz. The range is

2,000 nautical miles (3706 kilometers). LORAN D is a tactical LORAN system that uses the coordinate converter of low-frequency LORAN C.

In standard LORAN, the time systems of the master and slave stations are such that the signal from the master always reaches a ship during the first half of the recurrence cycle, and that from the slave during the second half. This is accomplished by including a delay circuit in the slave timing system that delays the retransmission of the signal received from the master until half the recurrence period has elapsed.

Standard receiving equipment has been designed for ships and planes in which both the receiving and timing units are present, with selector switches permitting the operator to set on the frequency and recurrence rate assigned to any LORAN pair he wishes to use. Differential amplifiers, synchronized by the timing circuit to the recurrence rate of the station, act on both the master and slave signals to deliver them at equal strength to the indicator unit.

The slow sweep of the oscilloscope ("viewing scope") appears as two parallel lines, one covering the first half of the recurrence cycle and the other second half. Hence the signals received from the master appear on one line and from the slave on the adjacent parallel line (see Fig. 10(c)). An adjustment is provided to allow for correction of slight variations in the crystal control of the timing unit and, when this is properly set, the desired signals remain stationary on the scope. The signals from other stations, which may be within range and operating on the same frequency, will drift along the line since their recurrence interval will be different from that of the pair being used.

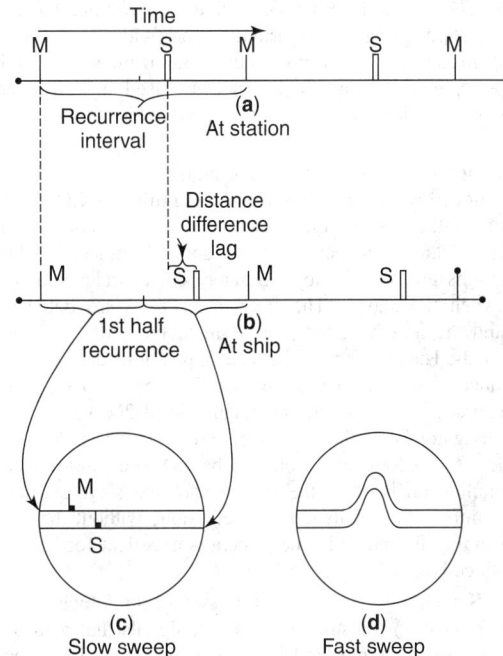

Fig. 10. Characteristics of LORAN signals.

When the signals are properly "set up" on the scope, a set of time markers is thrown on the screen, and a determination is made of the time interval between reception of signals from the master and slave. This will include the delay interval at the slave station, but, since this is standard for each recurrence rate, it may be allowed for. A delay circuit is now introduced, and the signals brought into approximate coincidence. With this adjustment made, a fast sweep spreads out the signals so that close coincidence may be established (see Fig. 10(d)), and the time interval measured to within one microsecond.

Using this measured difference in time of arrival of the two signals, the navigator then uses either tables or LORAN charts to obtain one line of position. The selector switches are then set to the characteristics of another LORAN pair, a second line of position determined, and a fix obtained (see Fig. 11). The accuracy of the determined fix is of the same order of magnitude as that obtained from good celestial navigation and, of course, is independent of the state of visibility.

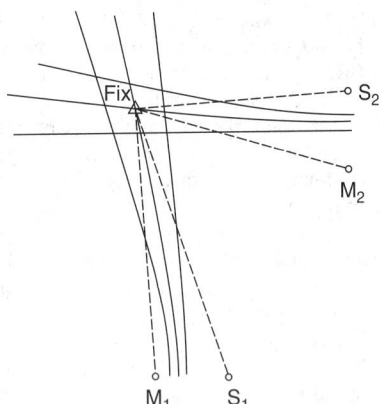

Fig. 11.   Obtaining a fix with LORAN.

LODAR. This is a direction finder with which the direction of arrival of LORAN signals is determined free of night effect by observing the separately distinguishable ground and skywave LORAN signals on a cathode ray oscilloscope and positioning a loop antenna so as to obtain a null indication of the component selected to be most suitable.

OMEGA Navigation. This is a long-range system that can provide worldwide coverage with only eight stations. The operating frequency is 10 kHz, and the estimated accuracy is 5 nautical miles (9.3 kilometers) with typical receivers. The system was developed for totally submerged submarines. The operating principle is similar to *delrac*, a British radio-navigation system designed to provide worldwide coverage by using 21 pairs of master-slave stations, with a 3,000-mile (4827 kilometers) range for each pair of stations. Frequencies used in *delrac* are in the band from 10 to 14 kHz. DECCA indicating equipment can be used with *delrac*.

GEE Navigation. This is a vhf (very high frequency) radio navigation system developed in Great Britain and is similar to LORAN. For the transmission of the signals, one master and two or more slave stations are used. The distance between stations is about 75 miles (121 kilometers) and the stations are located approximately on a circle, with the master station between the slaves. The frequencies used are between 20 and 88 MHz, and the length of the pulses are of the order of magnitude of 6 microseconds. The accuracy of the lines of position varies with the square of the distance of the ship from the base lines between stations. On this line, the accuracy is of the order of magnitude of 200 yards (183 meters) when the navigator is on the base line, and about one mile at a distance of 400 miles (644 kilometers) from the base. Since time differences can be read simultaneously from the master and two slave stations, the fix can be determined by simultaneous observations without the necessity of using the running fix method. The system is excellent for "homing" on a particular objective.

TACAN System. An air-navigation system in which a single uhf (ultra-high frequency) transmitter sends out signals that actuate airborne equipment to provide range and bearing indications with respect to the transmitter location when interrogated by a transmitter in the aircraft. Each TACAN station broadcasts a location-identifying Morse-code signal at regular intervals. Also termed tactical air navigation.

SHORAN System. A precision short-range position-fixing system using a pulse transmitter and receiver in an aircraft or other vehicle and two transponder beacons at fixed points. A receiver in the aircraft measures the round-trip times of the signals and converts these into distances to the fixed ground stations. Ordinary triangulation on a map then gives position.

OMNI System. This is a radio system that includes the directional information by modulating its radio signal with a simple dot-and-dash code, one code for a position to the left of the beam and another for a position to the right of the beam. The stations are located on airways and at the approaches to airways and to airports. With the system, the pilot selects his OMNI way point or destination on the radio and listens for the code that tells when the aircraft is to the left or right of the path. The system has been highly refined with onboard computers and displays. The system also is supplemented with distance-measuring equipment (DME). With the latter, the pilot interrogates the station with a transmitter and receives a distance indication from the station on a receiver. The measurement is made by determining the travel time of the radio wave. DME complicated the airborne equipment considerably but opened the way for easy, continuous navigation by using only a single set of equipment. Use of two stations can provide coverage for all locations within their range and thus free the pilot from flying the designated lines toward the OMNI stations. Very-high-frequency omnirange operates in the band from 112 to 118 MHz.

Doppler Navigation System. This is a navigation system for aircraft which makes use of the doppler effect as a means for determining drift and ground speed. In one configuration, there are four beams of pulsed microwave energy, which are beamed toward the ground (along the corners of an imaginary pyramid). The peak of the pyramid is at the aircraft. The echoes from the front-pointing beams experience an upward doppler shift, whereas the echoes from the rearward beams experience a downward doppler shift. Any drift is determined by doppler shift of echoes from beams on either side of the aircraft. The doppler shifts are compared in a computerized system, enabling all necessary navigation information under adverse weather conditions, various altitudes, and with need for reference to ground stations.

In a navigation satellite system, the satellite transmits accurate time signals and position data to a receiver on board a ship or aircraft. A central ground tracking station transmits correction signals to the satellite many times each day to sustain high accuracy of the system. The objective of such satellites, from which radio doppler shift measurements can be made under all weather conditions, is to provide the position of a ship or aircraft anywhere on earth with an accuracy of about 0.5 nautical mile (0.93 kilometer) or better.

NOTE: Numerous other articles in this encyclopedia relate directly or indirectly in navigation. Check alphabetical index.

Additional Reading

Baker, D.J.: "Toward a Global Ocean Observing System," *Oceanus*, 76 (Spring 1991).

Beardsley, T.: "Messages from on High," *Sci. Amer.*, 112 (July 1988).

Bjerklie, D.: "The Electronic Transformation of Maps," *Technology Review (MIT)*, 54 (April 1989).

Carron, M.J. and K.A. Countryman: "Developing Oceanographic Products to Support Navy Operations," *Oceanus*, 67 (Winter 1990/91).

Frye, D.W., W.B. Owens, and J.R. Vales: "Ocean Data Telemetry," *Oceanus*, 46 (Spring 1991).

Garver, J.G., Jr.: "A Love Affair with Maps," *Nat'l. Geographic*, 130 (November 1990).

Grewal, M.S., A.P. Andrews, and L.R. Weill:"Global Positioning Systems, Inertial Navigation, and Integration," John Wiley & Sons, Inc., New York, NY, 2001.

Grosvenor, G.M.: "New Atlas Explores a Changing World," *Nat'l. Geographic*, 126 (November 1990).

Kiernan, V.: "Guidance from Above in the Gulf War," *Science*, 1012 (March 1, 1992).

Koehr, J.E.: "The United States Navy's Role in Navigation and Charting," *Oceanus*, 82 (Winter 1990/91).

Lawrence, A.: "Modern Inertial Technology: Navigation, Guidance, and Control," Springer-Verlag, Inc., New York, NY, 1998.

Marden, L.: "Tracking Columbus Across the Atlantic," *Nat'l. Geographic*, 572 (November 1986).

McVey, V.: "The Sierra Club Wayfinding Book," Little, Brown and Co., Boston, MA, 1989.

Monastersky, R.: "Satellite Secrecy Doesn't Sink Scientists," *Science News*, 358 (December 8, 1990).

Richardson, P.L. and R.A. Goldsmith: "The Columbus Landfall: Voyage Track Corrected for Winds and Currents," *Oceanus*, 2 (Fall 1987).

Tetley, L. and D.M. Calcutt: "Electronic Navigation Systems," 3rd Edition, Butterworth-Heinemann, Inc., Woburn, MA, 2001.

Turbank, G.: "A New Way to Find Your Way," *Amer. Forests*, 10 (August 1989).

Waldrop, M.M.: "Flying the Electric Skies," *Science*, 1532 (June 30, 1989).

Wolper, J.S.: "Understanding Mathematics for Aircraft Navigation," McGraw-Hill Professional Book Group," New York, NY, 2001.

Wood, D.: "The Power of Maps," *Sci. Amer.*, 88 (May 1993).

NAVIGATORS' STARS.   A list of 55 stars has been designated as the "navigators' stars." The list was selected to cover the entire celestial sphere in such a manner that a navigator, no matter at what season or in what part of the earth he may be operating, will have two or three navigators' stars available for observation. The names and positions of these stars are listed in the Air Almanac, published by the U.S. Naval Observatory, and in a number of other publications. See also **Celestial Sphere and Astronomical Triangle**; and **Navigation**.

NAZCA PLATE.   See **Ocean**.

N-DISPLAY.   In radar, a display similar to the K-display in which the target appears as a pair of vertical deflections or blips from the horizontal time base. Direction is indicated by the relative amplitude of the vertical deflections; target distance is determined by moving an adjustable signal along the baseline until it coincides with the horizontal position of the vertical deflections. The horizontal control is calibrated in distance. Also called *N-scan, N-scope*, or *N-indicator*.

NEARSIGHTEDNESS.   See **Vision and the Eye**; and **Myopia**.

NEBULA.   The term nebula (*stella nebulosa*) was originally used by astronomers to describe any luminous spot that remained fixed relative to the stars. Before the application of the telescope, probably the only objects to which the term applied where those that are now referred to as star clusters, although reference was made in the tenth century to the great spiral in Andromeda. Following the application of the telescope, many more nebulous objects were discovered. Originally, these were grouped into three classes: the diffuse nebulae (Fig. 1); the planetary nebulae; and the spiral nebulae. Further research indicated that the spiral nebulae were very different from the other two. Application of the large telescopes proved that, in reality, they are groups of stars. Thus, for this class, the term nebula was dropped and the term galaxy or spiral galaxy used.

Fig. 1.   The Great Nebula in Orion (NGC 1976), with NGC 1977 below. (*Lick Observatory.*)

Some nebulae shine only by reflecting the light of stars contained in them ("reflection nebulae," which are dust in the environment of cooler, young stars). Others contain extremely hot stars whose radiation knocks electrons into high energy levels in the nebula gas, exciting gas atoms and allowing them to radiate as the electrons return to their ground states, producing soft glows also called HI regions. Some nebulae are stellar birthplaces; inside them, fresh stars are being produced out of the nebular gas and dust. Other nebulae are debris of explosions marking disruptions of unstable stars, supernova remnants like the Crab Nebula.

With the tendency for a mixture of gas and dust to collect in clouds and condensations, obscuring clouds of interstellar material, known as dark nebulae, may collect. When one of these clouds is in the vicinity of a bright star, the intense radiation from the star will illuminate the cloud, and a bright diffuse nebula is observed. Studies of the spectra of these objects have shown that the light is made up of both reflected starlight and of radiation from the interstellar material. The character of the reflected light is similar to that from the nearby stars. The radiation from the nebulous material itself, however, is quite different in character from starlight. This nebular spectrum is the bright-line type, and is probably produced by the absorption of radiation from the star by the gas atoms, and then reradiation of this energy in frequencies characteristic of the gas atoms and their states of excitation and ionization. This hypothesis has been considered sound because the character of the nebular spectrum depends upon the spectral type of the nearby stars. If the star is hot (B-type), the nebular spectrum is rich with bright lines, but in the vicinity of a relatively cool star (A-type or later), the nebular spectrum is almost entirely that of reflected starlight.

When the spectra of the bright nebulae were first studied, two bright lines were observed that could not be identified with any known terrestrial or solar element. At one time, it was believed that the lines were due to some material that existed only in the nebulae and the name of an element Nebulium was coined. These lines turned out to be due to forbidden transitions of doubly ionized oxygen, [OIII], which, at that time, had not been produced in the laboratory because of the very low density required to prevent collisions from de-exciting the atoms.

In addition to the diffuse nebulae, there are the planetary nebulae (see photograph accompanying entry on **Lyra**), so-called because they have quite definite shapes and look more or less like planets when observed through a telescope. The general appearance of these objects, on detailed photographs, is that of shells of gaseous material. They are generally elliptical in form, and may appear to be made up of several elliptical shells having their axes at various angles to each other or as helices. Frequently, a very blue, and hence, a very hot star is observed at the center of the shell, which is a hot white dwarf, the core of the star whose envelope the nebula represents. A number of these stars prove to be very short (less than one hour) period binaries. The spectra of the planetaries is, in general, the same as that of the diffuse nebulae found in the vicinity of B-type stars. In some cases, where the central star can be studied, variations in its light have been found, and the nebular radiation is found to vary with that of the star. It is evident that the planetary nebulae are actually stars with very extended and attenuated atmospheres. Careful studies of the spectral lines from the planetaries indicate that the shell may be expanding. Expanding shells of gas have been observed around some novae, but the rate of expansion is far greater than that for the planetaries.

Since the advent of radio astronomy, a significant number of different kinds of molecules have been found in the interstellar medium, mostly in dark clouds of dust where light from stars cannot penetrate. However, the first interstellar molecule to be discovered was the radical CH, found accidentally in a star spectrum in 1937 with conventional earth-based optical instrumentation. The radicals CN and CH^+ were found with the Mount Wilson Observatory 100-inch (254-centimeter) optical telescope in 1939. The discovery of other interstellar molecules had to await the development of radio telescopes. The OH radical was the first to be added to the list by radio techniques in 1963. Since that time, and particularly since the early 1970s, a number of new interstellar molecules have been added to the list. These include NH_3, H_2O, H_2CO (formaldehyde), CO, HCN, HC_3N (cyanoacetylene) CH_3OH (methanol), CH_2O_2 (formic acid), CS, CH_3CN (methyl cyanide), SiO, HNC, CH_3C_2H (methyl acetylene), NH_2CHO (formamide), CH_3CHO (acetaldehyde), HNCO (isocyanic acid), H_2CS (thioformaldehyde), H_2S, H_2CNH (methylene amine), and SO (sulfur monoxide), the latter found in 1973.

At one time, some investigators felt that the most important process in interstellar molecule formation was the combination of neutral molecules by radiative association, that is, reaction with photon emission. However, because the chemical activation energy required for neutral molecule reactions acts as a barrier to formation of more complex molecules, it was agreed that the reaction rates would be too slow for interstellar molecule formation. In contrast, the majority of reactions between ions and neutral molecules do not have such a barrier. Heavier ions can be built from lighter ones. Exemplary reactions include: $H_2^+ + H_2 \rightarrow H_3^+ + H$; or $H_3^+ + CO \rightarrow HCO^+ + H_2$. Ions can recombine with electrons to form neutral molecules, thus completing the reaction pathway. Electron density in dense clouds is estimated at 10^{-7} cm^{-3}. Commencing with the foregoing reactions, the building of complex molecules in dense clouds may require the assumption that the source of ionization is the flux of cosmic rays at energies of 100 MeV and upwards. The reactions also require solid surfaces, like dust grains, on which to occur. Densities greater than 10^4 cm^{-3} also appear necessary. In addition to a chain of reactions commencing with hydrogen, other possible chains could commence from ionized helium, carbon, and oxygen. However, some investigators feel that the hydrogen reactions are probably basic and that the ion-molecule scheme probably would not take place unless HCO^+ is present in the dense cloud.

Theories of interstellar molecule formation include the theory of formation on dust grains and the ion-molecule theory. Estimates indicate that many of the simple diatomic molecules may be formed by collisions in space; the more complex ones, such as H_2CO_2, CH_3CN, and NH_3CO, may be formed on grains of interstellar dust in clouds where the concentration of hydrogen molecules (H_2) is $10^6/cm^3$ or higher. See also list of entries given in **Astronomy**. See also **Hubble Space Telescope (HST)**.

STEVEN N. SHORE, New Mexico Institute of Mining and Technology, Socorro, NM.

NECROSIS. The local death of cells results in changes in the tissue known as necrosis. These consist of disintegration of the cellular structure with destruction of the nucleus and coagulation or liquefaction of the cytoplasm. The causes of necrosis include interference with the blood supply of a tissue physical injury, and deleterious actions by bacteria or their toxins.

NÉEL TEMPERATURE. The transition temperature for an antiferromagnetic material. Maximal values of magnetic susceptibility, specific heat, and thermal expansion coefficient occur at the Néel temperature. See also **Antiferromagnetism**.

NEGATRON. A term sometimes applied to the normally occurring negatively charged electron when it must be distinguished from a positron. In many parts of the world the name *negaton* is used instead of negatron. The word negatron is used in this encyclopedia wherever distinction is made between positively and negatively charged electrons.

NEGRO BUG *(Insecta, Hemiptera)*. Small shining black bugs with a smooth convex upper surface. They resemble beetles superficially. Most of the abdomen is covered by a greatly enlarged sclerite of the thorax, which also conceals most of the wings.

NEIGHBORHOOD OF A POINT. The interior of some bounded geometric figure (such as a square or circle in the plane) which contains the point. A neighborhood or a point on a line, plane, or surface is usually taken as the set of points within a stated distance of the point (e.g., an open interval on the line or the interior of a circle in the plane, with the point as center). One speaks of a property as holding *in the neighborhood of a point* if there exists a neighborhood of the point in which the property holds, or of a numerical quantity (e.g., curvature) depending on the nature of a curve or surface in the neighborhood of a point if the value of the quantity can be determined from knowledge of the portion of the curve or surface in an arbitrarily small neighborhood of the point. See also **Point**.

NEKTON. The portion of a population made up of animals capable of directive locomotion through a fluid medium. Usually applied only to aquatic animals, including the fishes, although flying creatures constitute a similar part of the terrestrial fauna and may be called an aerial nekton. See also **Ocean**.

NEMATIC LIQUID CRYSTALS. See **Liquid Crystals**.

NEMATODES. Of the phylum *Nemata* or *Nematoda*, these are roundworms or threadworms. They are abundant in fresh and salt water and in the soil; many are internal parasites of animals and plants. Some are parasitic in humans and the domestic animals and are important in relation to human welfare. The body of these worms lacks a spiny proboscis and is marked by slender longitudinal lines along the sides. These lateral lines follow the excretory tubes. There is a wide range of variations in the life cycle. To place *Nemata* in their proper perspective, it is in order to mention the other important phyla of worms: Flatworms (*Platyhelminthes*);

ribbon worms (*Nemertea*); spiny-headed worms (*Acanthocephala*); hairworms (*Nematomorpha*); and segmented worms (*Annelida*). Reference to the entry on **Intestinal Nematodes** is suggested.

Nematode Damage to Food Crops

Nematodes are very important economic pests on food crops. Very few crops are immune to attacks of these creatures, which inhabit the soil about the roots of plants. Nematode populations number into the millions and billions, in field crops, orchard operations, greenhouse (glasshouse) facilities, and truck and home gardens. Actually, nematodes have been one of the last of the major crop pests to be well understood, and aggressive research in the field only dates back some 50 to 60 years. Research progress was impeded mainly by difficulties in isolating and preparing the nematodes for detailed examination. Some authorities place nematology just about one-half century behind entomology, but much progress has been made during the past decade or two and thus the technological gap is narrowing. Among the challenges facing the nematologist today are: (1) development of nematode-resistant varieties of crop plants; (2) cooperation with agricultural engineers in development of more effective means for applying nematicides; (3) education of farmers and large food producers on cultural methods for controlling nematodes, including fallowing and rotation of crops; (4) development of improved systemic nematocides as well as synthetic plant diffusates which stimulate early emergence of nematodes from dormancy; and (5) a continuing program of identifying yet undiscovered species. Continuing work on the classification and nomenclature of nematology also is important.

Although the importance of nematodes to food growth economics was relatively late in being appreciated, a knowledge of the existence of nematodes dates back to ancient times. It is believed that the Guinea worms mentioned in the Old Testament (*Numbers* 21:6–9) as "fiery serpents" were nematodes. Parasitic nematodes were alluded to by Aristotle as early as about 350 B.C. Free-living nematodes were observed in vinegar, and referred to as vinegar eels, as early as the mid-15th century. In the late 1700s, Linnaeus, Scopoli, Steinbuch, and Needham showed a causal relationship between the nematode *Anguina tritici* and the disease known as "cockles" of wheat and other cereal plants. In the late 1800s and early 1900s, Julius Kühn and associates in Germany intensively researched the sugar beet nematode and some authorities credit Kühn with the first use of soil fumigation. He used carbon disulfide on infested sugar beet fields. The life cycle of this pest (*Heterodera schachtii*) was ascertained, providing knowledge upon which effective cultural practices could be established.

Galls on the roots of cucumbers were noted by Berkeley in England as early as 1855. It is believed that the term root-knot nematode was first used in 1879 by Cornu. In 1887, root galls on coffee were described by Goeldi. Bastian, who wrote a monograph on the *Anguillulidae* in 1886, is considered the father of nematology. One of the first full texts on the subject, "Nematodes That Are Important for Agriculture," was authored by the Russian I.N. Filipjev in 1934. Outstanding early work was done by N.A. Cobb at the U.S. Department of Agriculture in the early 1900s. An excellent summary of the history of nematology from its beginnings to the 1960s is given in *Principles of Nematology*, by Gerald Thorne, McGraw-Hill, New York, 1961. Additional and more recent references are listed at the end of this article.

Nature of Nematodes

These economic pests are found essentially wherever soil is found—from deserts and tropical areas to cold, high-altitude mountainous terrains. For example, in 1929 Thorne found several hundred specimens in soil samplings taken from Colorado mountain soils at levels of over 14,000 feet. There appears to be an almost infinite variety of nematodes, the greatest variations probably occurring in marine waters near shallow coastal beaches.

The typical nematode may be described as a slender, quite active animal that ranges from 0.2 to 10 millimeters in length, although the majority are less than 2 millimeters long. The body is usually cylindrical in shape, although several other forms are known, including pear- and lemon-shaped forms. The nematode body is covered with cuticle, a tough, flexible layer of material. In some species, the coating is marked or texturized, which helps the nematologist greatly in identification. However, many are not so conveniently marked, thus requiring detailed microscopic examination to yield identity. On the average, a nematode will undergo four moults in developing from egg to adult. During these stages, the body increases in diameter and length.

Control chemicals for use against nematodes are various fumigants, such as carbon bisulfide, chloropicrin, D-D, EDB, formaldehyde, Fumazone, hydrogen cyanide, methyl bromide, and Nemagon. These and other chemical materials either are banned or are subject to rigid control in some countries.

Nematodes are both *endo-* (inhabit and consume internal organs of host) and *ectiparasitic* (live on surface of host). They can be classified roughly in terms of the portions of the host plant they prefer for habitation. Some of these include:

Bud and Leaf Nematodes (genus *Aphelenchoides*). These nematodes exhibit both endo- and ectoparasitism, a factor determined by Franklin in 1950. The endoparasites are found in leaves; the ectoparasites are found in plant crowns, leaf axils, or inflorescences, where these parts are protected by other folding tissues of the host plant. Nematodes of this type were discovered by Ormerod in England in 1889.

Bulb and Stem Nematodes (genus *Ditylenchus dispaci* (Kühn) Filipjev). There are over 30 species. The symptoms of their presence were first observed by Schwertz in 1855 on clover, oats, and rye, but the nematode causative agent was not revealed until further studies by Kühn in 1857 and others at later dates.

Burrowing Nematodes (genus *Radopholus* (Thorne)). These nematodes were named and classified in 1949. They are endoparasitic, attacking plant roots. They probably are found in tropical and subtropical regions, but have been found in cooler regions as well. *Radopholus similis* (Cobb, 1893) causes banana plant disease (*Musa sapientum*), first noted in Fiji (1890). Also noted on diseased coffee roots in Java (1898) and on diseased sugarcane roots in Hawaii (1907).

Cyst-Forming Nematodes (genus *Heterodera* (Schmidt, 1871)). These include the first of the important nematodes to be associated with plant diseases, namely, the sugar beet nematode (*Heterodera schachtii*) first observed by Schmidt. The female cuticle of this pest transforms into a light-to-dark brown, cyst-like sac. This sac protects the eggs. These cysts are oval or spheroidal in shape and range from 0.4 to 0.8 millimeter in length. Inasmuch as the cyst material does not decompose readily, there are often great accumulations of these bodies in the soil. While older cysts will not have eggs, those of more recent years may contain as many as 600 eggs.

Root-Knot Nematodes (genus *Meloidogyne* (Goeldi)). At one time these were considered to be a single species. Five or more species were established by 1949. These pests are among the most economically important of the nematodes and include the coffee root-knot nematode *Meloidogyne exigua* (Goeldi 1892); the Japanese root-knot nematode, *M. japonica* (Treub, 1885); the northern root-knot nematode *M. hapla* (Chitwood, 1949a); the peanut (groundnut) root-knot nematode *M. arenaria* (Neal, 1889); the Thames root-knot nematode *M. arenaria* thamesi (Chitwood, 1952); the southern root-knot nematode *M. incognita* (Chitwood, 1949a); and the cotton root-knot nematode *M. incognita var. acrita* (Chitwood, 1949a), among others. The root-knot nematodes are often associated with various fungus diseases. Damage is caused by formation of galls on plant roots, causing stunting and wilting and, frequently, expiration of the plant if not controlled.

Root-Lesion Nematodes (genus *Pratylenchinae* (Thorne, 1949)). These have been described since the late 1920s. Because of openings in plant roots caused by the pests, bacteria and fungi may enter and thus these nematodes are often associated with serious diseases from these causes. The root-lesion nematodes cause openings or lesions rather than knots or galls as in the case of the genus *Meloidogyne*. There are about 10 major species and they infest a variety of very important crops, such as coffee, citrus, pineapple, potato, rice, and sugar beet, among many others.

Nematode Damage

Important diseases caused by nematodes affect many crops. Nematodes are found almost universally in the soil and, at one time or another, they contribute to the damage, minor or major, to nearly all plants. There are however, numerous situations where nematode damage can be severe and even catastrophic to some crops if effective control measures are not taken. Only three examples are given here. Much more detail will be found in the Considine (1981) reference listed.

Citrus. A number of nematode species damage various citrus crops. Possibly the most serious situation occurs on citrus in Florida and is a condition known as the *spreading decline of citrus*. See Figs. 1 and 2. Caused by a burrowing nematode, the damage has ranged into the many millions of dollars. Experiments in treating diseased trees, however, point to a factor, still unknown, that also is active in causing spreading decline disease. In some experiments, destruction of the nematodes did not fully prevent spreading of the disease. Symptoms of the disease include stunted trees with subnormal foliage, small fruit, and retarded terminal growth. Wilting is excessive during dry, hot periods. There is also a reduction of young feeder roots. The term *spreading* derives from the fact that the nematodes spread out or migrate from one tree to the next. A study made by Suit and Ford in 1950 indicates that the advance is at the rate of 1.6 trees per year. The nematodes have been known to migrate under highways and railroad rights of way. Rather aggressive methods have had to be used in attempts to eradicate the disease in Florida, including systematic and frequent soil inspections and the destruction by burning of infested trees

(a)

(b)

Fig. 1.   Young Navel orange trees on Troyer citrange rootstock 2 years after planting: (**a**) trees planted in soil in which citrus nematodes had been killed by preplant soil fumigation; (**b**) tree infected with the citrus nematode soon after planting in nonfumigated, nematode-infested soil in same field as tree at left. (*Agricultural Extension Service, University of California.*)

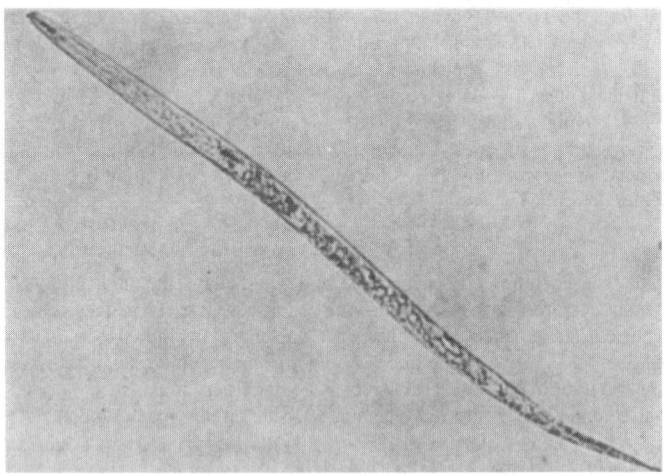

Fig. 2. Citrus nematode larva. Actual length is $\frac{1}{70}$ inch (0.4 mm). Stylet of feeding mechanism is at head (lower right-hand corner of view). (*Agricultural Extension Service, University of California.*)

and planted areas; and the extensive use of very strong chemicals to an average depth of 12 inches (0.3 meter). The persistence of the nematodes is exemplified by the finding of live nematodes within a depth of 12 feet (4 meters) of a very heavily treated area. However, inasmuch as most of the nematodes are found within the top 5 feet (1.5 meters) of the soil, diffusion of the treatment to a depth of 6 to 8 feet (2 to 2.5 meters) is usually effective.

Some authorities have observed a close parallel between the spreading decline of citrus disease in Florida and the yellows disease of pepper on Bangka.

The nematode species *Hemicycliophora arenaria* has been found on rough lemon root-stock in California. This pest causes an enlargement of terminal and lateral root-tips, which appear as small knobs. Damage may be the result of secretion of enzymes by the nematodes.

Potato. Nematodes of the genus *Ditylenchus* (Filipjev, 1934) injure potato by producing a progressive dry rot of the tuber. While this damage is proceeding, there is no evidence to be seen from observation of stems and foliage. There may be several strains that cause this condition. Known since 1888, *Ditylenchus destructor* has been a major cause of injury to potato in Europe for a long time. For many years, United States officials intercepted numerous shipments containing the strain.

Presence of the pest can be determined only by cutting into the tuber. In some instances, much of the tuber can be infested. Crop rotation provides no effective control. Fumigation is the main control, and of the fumigants, ethylene dibromide is perhaps most effective. With sufficient fumigant applied, virtual eradication can be accomplished. Resistant varieties have not been successful. Where potatoes are used for silage, the fermentation of the silage kills the pests.

The golden nematode of potatoes, *Heterodera rostochiensis* (Wollenweber, 1923), is very damaging to potato. This nematode was discovered by Kühn in 1881 when he was doing research on the sugar beet nematode. Somewhat later reports were made in Germany and Scotland. The presence of the nematode in much of Europe was confirmed and, in the British Isles, the pest was called the potato root eelworm. During the interim, the nematode has been reported by Israel and, in 1941, the pest was first found in the United States on Long Island, New York. The nematode forms pear-shaped cysts on the potato, thus differing from the lemon-shaped cysts of the sugar beet nematode. This difference removed any doubt that the two pests are different. Since this was a new find in nematology and one affecting a huge market crop, research on the pest was intensive and many countries instituted crop regulatory and quarantine measures.

Rice. The rice stem nematode (*Ditylenchus angustus* (Filipjev, 1936)) is the cause of ufra disease in rice and was observed for the first time by Butler in 1913. This pest and resulting disease poses the greatest threat to success of the rice crop in India. It is found most commonly in an area north of the Bay of Bengal and east of the Ganges River. Even though many hundreds of varieties of rice are grown in these areas, all appear to be susceptible to this pest. Massive infestations in India have been known since 1916. The pest climbs the stems and interferes with the growing

process of the plant, after which it consumes leaves and stems — in essence devastating the plant.

Additional Reading

Anderson, R.C.: "Nematode Parasites of Vertebrates: Their Development and Transmission," 2nd Edition, Oxford University Press, Inc., New York, NY, 2000.

Atkinson, H.J.: "The Physiology of Nematodes," 2nd Edition, Columbia University Press, New York, NY, 1977.

Considine, D.M., Editor: "Foods and Food Production Encyclopedia," Van Nostrand Reinhold, New York, NY, 1982.

Croll, N.A.: "The Organization of Nematodes," Academic Press, Inc., Boca Raton, FL, 1976.

Dropkin, V.H.: "Introduction to Plant Nematology," 2nd Edition, John Wiley & Sons, Inc., New York, NY, 1980.

Hollis, J.P.: "Action of Plant-Parasitic Nematodes on Their Hosts," *Nematologica*, **9**, 475–494 (1963).

Kontaxis, D.G.: "Nematicides Improve Sugar Beet Yields," *California Agriculture*, **31**(4), 10–11 (1977).

Lee, D.L.: "Biology of Nematodes," Gordon & Breach Publishing Group, Newark, NJ, 2001.

Radewald, J.D. et al.: "Citrus Nematode Disease and Its Control," Agricultural Extension Service, Univ. of California, Berkeley, CA, Bul. AXT-211 (revised periodically).

Rhode, R.A.: "The Nature of Resistance in Plants to Nematodes," *Phytopathology*, **55**, 1159–1162 (1965).

Stone, A.R. et al., Editors: "Concepts in Nematode Systematics," Academic Press, Inc., Boca Raton, FL, 1983.

Thorne, Gerald: "Principles of Nematology," The McGraw-Hill Companies, Inc., New York, NY, 1961.

Zuckerman, B.M. et al.: "Plant Parasitic Nematodes: Morphology, Anatomy, Taxonomy, and Ecology," Academic Press, Inc., Boca Raton, FL, 1981.

NEMERTEA. Marine worms with a flattened body, often long and ribbon-like in form. They are unsegmented and have no body cavity, hence they are sometimes included with the flatworms as a class of the phylum *Platyhelminthes*. More often they are made a separate phylum because the alimentary tract is a tube opening with an anterior mouth and a posterior anus. Like the free-living flatworms, they have ciliated (cilia) integument. They are also provided with an eversible proboscis enclosed in a dorsal tubular cavity, the rhynchocoel, and associated with but not derived from the alimentary tract.

These worms live among seaweed or at the bottom of the ocean and prey on living animals or eat dead ones. They are not economically important. A few freshwater species and a few parasitic forms are known.

NEODYMIUM. Chemical element symbol Nd, at. no. 60, at. wt. 144.24, third in the Lanthanide Series in the periodic table, mp 1,016°C, bp 3,068°C, density 7.004 g/cm^3 (20°C). Elemental neodymium has a close-packed hexagonal crystal structure at 25°C. The pure metallic neodymium is silver-gray in color, the luster becoming dull upon exposure to moist air at room temperatures. When pure, the metal is soft and malleable and may be worked with ordinary equipment. Because the metal is pyrophoric, it must be stored in an inert atmosphere or vacuum. There are seven natural isotopes, ^{142}Nd through ^{146}Nd, ^{148}Nd, and ^{150}Nd. ^{144}Nd is mildly radioactive with a half-life of $10^{10} - 10^{15}$ years. Seven artificial isotopes have been produced. Of the light (or cerium-group) rare-earth metals, neodymium is the third most plentiful and ranks 60th in abundance of elements in the earth's crust, exceeding tantalum, mercury, bismuth, and the precious metals, excepting silver. The element was first identified by C.A. von Welsbach in 1885. Electronic configuration $1s^2 2s^2 2p^6 3s^2 3p^6 3d^{10} 4s^2 4p^6 4d^{10} 4f^3 5s^2 5p^6 5d^1 6s^2$. Ionic radius Nd^{3+} 0.995 Å. Metallic radius 1.821 Å. First ionization potential 5.49 eV; second 10.72 eV.

Other important physical properties of neodymium are given under **Rare-Earth Elements and Metals**.

Primary sources of the element are bastnasite and monazite, which contain from 15 to 25% neodymium. Plant capacity involving liquid-liquid or solid-liquid organic ion-exchange processes for recovering the element is in excess of 200,000 pounds (90,720 kilograms) Nd$_2$O$_3$ annually. Metallic neodymium is obtained by electrolysis of fused anhydrous NdCl$_3$ or by the electrolytic reduction of the oxide in molten NdF$_3$.

Use of elemental neodymium as a colorant for glass was one of the early applications. The color ranges from pure violet to purple and finds use in sunglasses, protective glasses for industry, art objects of glass, tableware, and decorative fiber optics. Use of neodymium in amounts of

3–5% by weight imparts dichroic properties to glass. Neodymium-doped single-crystal yttrium-aluminum oxide garnets (Nd:YAG) have been used in lasers. Research has shown the Nd ion to exhibit laser characteristics in a wide range of compounds and glasses. A formulation of 75% neodymium and 25% praseodymium, frequently called didymium, is used as a metallurgical additive. Within the last several years, it has been found that the use of Nd_2O_3 in barium titanate capacitors increases the dielectric strength of these electronic components over a wider temperature range. Neodymium also has been used as an ingredient of phosphate-type phosphors. Investigations continue into further electronic and optical uses of the element and its compounds.

EDITOR'S NOTE: Extensive research during the early 1980s led to the development of a new and powerful magnet material with the probable composition, $R_2Fe_{14}B$ (where R = a light rare earth). The rare earth predominantly used thus far is neodymium. The recent neodymium-iron-boron material exhibits extremely powerful magnetic qualities as compared with traditional magnet materials. More detail is given under **Rare-Earth Elements and Metals**. Also see **Magnetism**.

Scientists (California Institute of Technology) reported that the isotopic composition of Drake Passage (Antarctica) seawater had been determined. The Antarctic Circumpolar Current, which controls interocean mixing, flows through the Drake Passage. The ratio, $^{143}Nd/^{144}Nd$, was found to be uniform with depth at two experiment stations—with an intermediate value between those of the Atlantic and Pacific Oceans. Further, Piepgras and Wasserburg determined that the Antarctic Circumpolar Current is made up of approximately 70% Atlantic water. It was further reported that cold bottom water from a site in the south-central Pacific has the Nd isotopic signature of the water in Drake Passage. The investigators used a box model to emulate the exchange of water between the Southern Ocean and ocean basis to the north with the isotopic results. An upper limit of about 33 million cubic meters/second was calculated for the rate of exchange between the Pacific and the Southern Ocean. Further determinations of samarium and neodymium were made and found to increase approximately linearly with depth. In essence, the findings suggest that Nd may be a valuable tracer in oceanography and possibly useful in paleo-oceanographic studies. See also Ocean; and Polar Research.

Additional Reading

Anderson, D.L.: "Composition of the Earth," *Science*, 367 (January 20, 1989).

Cherfas, J.: "Proton Microbeam Probes the Elements," *Science*, 11500 (September 28, 1990).

DePaolo, D.: "Neodymium Isotope Geochemistry," Springer-Verlag New York, Inc., New York, NY, 1988.

Greenwood, N.N. and A. Earnshaw: "Chemistry of the Elements," 2nd Edition, Butterworth-Heinemann, Inc., Woburn, MA, 1997.

Letokhov, V.S.: "Detecting Individual Atoms and Molecules with Lasers," *Sci. Amer.*, 54 (September 1988).

Lewis, R.J., Sr.: "Hawley's Condensed Chemical Dictionary," 13th Edition, John Wiley & Sons, Inc., New York, NY, 1999.

Lide, D. (Editor): "CRC Handbook of Chemistry and Physics 2000-2001: 81st Edition - A Ready- Book of Chemical Reference and Physical Data," CRC Press, LLC., Boca Raton, FL, 2000.

Lugmair, G.W. et al.: "Samarium-146 in the Early Solar System: Evidence from Neodymium in the Allende Meteorite," *Science*, **222**, 1015–1017 (1983).

Piepgras, D.J. and G.J. Wasserburg: "Isotopic Composition of Neodymium in Waters from the Drake Passage," *Science*, **217**, 207–214 (1982).

Robinson, A.L.: "Powerful New Magnet Material Found," *Science*, **223**, 920–922 (1984).

Staff: "ASM Handbook—Properties and Selection: Nonferrous Alloys and Pure Metals," ASM International, Materials Park, OH, 1990.

White, R.M.: "Opportunities in Magnetic Materials," *Science*, **229**, 11–15 (1985).

K.A. GSCHNEIDNER, Jr., and B. EVANS, Iowa State University, Ames, IA.

NEON. Chemical element, symbol Ne, at. no. 10, at. wt. 20.183, periodic table group 18, mp $-248.68\,°C$, bp $-246.01\,°C$, density 1.204 g/cm^3 (liquid). Specific gravity compared with air is 0.674. Solid neon has a face-centered cubic crystal structure. At standard conditions, neon is a colorless, odorless gas and does not form stable compounds with any other element. Due to its low valence forces, neon does not form diatomic molecules, except in discharge tubes. It does form compounds under highly favorable conditions, as excitation in discharge tubes, or pressure in the presence of a powerful dipole. However, the compound-forming capabilities of neon, under any circumstances, appear to be far less than those of argon or

krypton. No known hydrates have been identified, even at pressures up to 260 atmospheres. First ionization potential, 21.599 eV.

Neon occurs in the atmosphere to the extent of approximately 0.00182%. In terms of abundance, neon does not appear on lists of elements in the earth's crust because it does not exist in stable compounds. However, because of its limited solubility in H_2O, neon is found in seawater to the extent of approximately 1.5 tons per cubic mile (324 kilograms per cubic kilometer). Commercial neon is derived from air by liquefaction and fractional distillation. For most applications, the gas need not be in a highly pure form, but may be supplied along with small quantities of the other rare gases, such as argon and krypton. The gas finds principal applications in various electronic devices and lamps, but the most familiar application is the neon tubes used mainly in signs. The use of neon signs for identification and advertising signs reached the Iron Curtain countries at a date much later than in the Western countries. Neon emits the familiar orange light. Neon also has been used in certain lasers.

In the 1983 Luberoff reference, the author observes that neon signs, once considered vulgar symbols of a consumer society, are fast becoming icons of a bygone era. However, in recent years a group of preservationists, people who formerly decried the impact of neon advertising, now often defend it. Luberoff points out how a blue-lettered sign (5878 neon-filled glass tubes) became an integral part of the Boston skyline. A study of Boston's signs and lights by the Boston Redevelopment Authority showed that this sign (Citgo) was the only commercial sign that the public thought should remain.

There are three natural isotopes, ^{20}Ne through ^{22}Ne, and four radioactive isotopes, ^{18}Ne, ^{19}Ne, ^{23}Ne, and ^{24}Ne, all with half-lives of less than 5 minutes. Ramsay and Travers first found the element when investigating the properties of liquid air in 1898. The element is easily identified spectroscopically. Neon emits characteristic red and green lines in its spectrum.

Neon in Meteorites. As pointed out by Lewis and Anders, the noble gases are unique among the elements found in meteorites. They are highly volatile and unreactive and they did not condense in even the most primitive meteorites and thus are present at only a minute fraction of their proportion in the sun, ranging from about 10^{-5} for xenon to 10^{-9} for neon and helium. However, very small quantities of these gases are tightly bound in the meteorite and are freed when the host mineral begins to melt or decompose at high temperatures.

Scientists have found three types of neon in meteorites: (1) Primordial or planetary neon (called neon *A*); (2) solar neon (neon *B*), which consists of solar-wind neon ions implanted in meteorites that happen to have been at the surface of their parent body; and (3) cosmogenic neon (neon *S*), formed when cosmic rays passing through the meteorite spall, or shatter, atomic nuclei in their path. Each type has different proportions of the three isotopes of neon. Although the procedure is too detailed for inclusion here, Lewis and Anders explain how, through the use of stepped heating of meteorite materials, the types of neon can be measured. Their ratios to each other provide clues as to what type of star may have been the source of a given meteorite.

Additional Reading

Anderson, D.L.: "Composition of the Earth," *Science*, 367 (January 20, 1989).

Cherfas, J.: "Proton Microbeam Probes the Elements," *Science*, 11500 (September 28, 1990).

Greenwood, N.N. and A. Earnshaw: "Chemistry of the Elements," 2nd Edition, Butterworth-Heinemann, Inc., Woburn, MA, 1997.

Letokhov, V.S.: "Detecting Individual Atoms and Molecules with Lasers," *Sci. Amer.*, 54 (September 1988).

Lewis, R.S. and E. Anders: "Interstellar Matter in Meteorites," *Sci. American*, **249**(2), 66–77 (1983).

Lewis, R.J., Sr.: "Hawley's Condensed Chemical Dictionary," 13th Edition, John Wiley & Sons, Inc., New York, NY, 1999.

Lide, D.R.: "CRC Handbook of Chemistry and Physics 2000-2001), 81st Edition, CRC Press, LLC., Boca Raton, FL, 2000.

Luberoff, D.: "*But Is It Art? (Neon Signs), Technology Review (MIT)*, **86**(5), 76–77 (July 1983).

NEONATAL. Related to or affecting the newborn human child, particularly during the first month after birth.

NEOPLASM. Any new or abnormal overgrowth of cellular tissue. A neoplasm is a cellular tumor and may be either benign or malignant.

NEOPRENE.   See **Elastomers**.

NEPHELINE.   Nepheline, of hexagonal crystallization, is a sodium-potassium aluminum silicate $(Na, K)(AlSiO_4)$. It is found in silica-poor geological environments, where there had been insufficient silica to form feldspar. Nepheline rocks are characterized by the absence of quartz within them. They constitute a mineral family group known as the *feldspathoids*. Crystals are extremely rare; usually occurs massive to compact. Luster, is greasy in the massive varieties; vitreous in crystals. Color grades from yellowish to colorless in crystals; gray, green and reddish in massive material. It ranges from transparent to translucent. Hardness is of 5.5–6, specific gravity of 2.55–2.65.

Immense masses of nepheline-rich rocks occur on the Kola Peninsula, the former U.S.S.R., in Norway and in the Republic of South Africa; also in the Bancroft, Ontario, Canada region. Smaller deposits are found in Maine and Arkansas in the United States. Fine crystals are found in lavas on Mt. Vesuvius, Italy.

Nepheline is used extensively in the manufacture of glass.

ELMER B. ROWLEY, Union College, Schenectady, NY.

NEPHELOMETRY.   Sir John Tyndall noted that particles that are invisible when directly in the path of a strong light become discernible when viewed from the side. Now known as the Tyndall effect, the phenomenon derives from reflection of part of the incident light by the particles. The reflected light is directly proportional to the number of particles in suspension. An instrument for measuring the intensity of reflected light so produced in a nephelometer and may be used for the quantitative determination of small amounts of diverse materials that have the ability to reflect light when in liquid suspension. Examples include the measurement of traces of silver wherein the chloride ion is added to a solution of material containing silver to produce insoluble silver chloride in suspension form. Small amounts of calcium in titanium alloys may be determined by measuring suspensions of the stearate formed in a suitable medium. Nephelometry also finds application in the measurement of bacterial growth rates; for the analysis of cholesterol, glycogen, and enzymes; for controlling the clarity of beverages, water, and wastewater; for solution control in tanning operations; and for any measurement situation where an unknown composition may be transformed into, or related to, a form of suspension.

Nephelometric methods are similar to fluorometric methods in that both involve measurement of scattered light. However, the scattering is inelastic in nephelometry and elastic in fluorometry. Thus, the scattered light measured in fluorometry is of a longer wavelength than the incident light, and both incident and scattered light are of the same wavelength in a nephelometric determination. In fact, the two functions sometimes are combined into one instrument, which may be termed a nefluoro-photometer. When the instrument operates as a nephelometer, it utilizes two Tyndall windows, located opposite each other in a cylindrical sample cell and with their common axis perpendicular to the path of the entering light. The concentration of suspended particles is determined by summing the photocurrents of the two cells. When used as a fluorometer, the instrument measures light emitted by a sample that is excited by incident radiation in the appropriate spectral band. Further, the same instrument can be set up for use as a photometer to measure light transmitted by the sample. Three light sources may be used—an incandescent source for colorimetric or nephelometric applications; a mercury-arc source for fluorometry; and a sodium-arc source with principal emission at 320 and 590 nanometers, when a sharp peak at either of these wavelengths is required, as in the instance of vitamin A determinations.

See also **Analysis (Chemical)**; **Fluorometers**; **Photometers**; and **Turbidimetry**.

NEPHOMETER.   See **Precipitation and Hydrometeors**.

NEPHOSCOPE.   An instrument for determining the direction of cloud motion. There are two basic designs of nephoscope: the direct-vision nephoscope and the mirror nephoscope. Also called *nepheloscope*.

NEPHRITIS.   See **Kidney and Urinary Tract**.

NEPHRON.   See **Kidney and Urinary Tract**.

NEPTUNE (Planet).   Eighth planet from the sun (~3 billion miles; 4.5 billion km), Neptune is about four times the size (diameter) of Earth, with a mass slightly greater than 17 times that of Earth. Its mean density is less than that of Earth. This low density, coupled with the high value of the planet's albedo (0.52), is indicative of the planet's thick atmospheric layer. Neptune has seven confirmed satellites, the best understood of which is Triton.

Neptune is invisible to the naked eye, having a stellar magnitude of only about 7.7. The planet can be observed with a telescope having an aperture greater than 2.5 cm (1 in), but it can be distinguished with such a small instrument only by its change in position against the starry background from night to night.

The planet was discovered in 1846 and will not complete one trip around the sun from its position at that time until the year 2010, a total of 164 years later. Viewed by a large telescope, Neptune appears as a small, circular, somewhat greenish disk without distinctive markings. Thus, for many years, it was difficult to estimate the planet's rotation period. However, in 1928, Moore and Menzel (Lick Observatory) found, from spectroscopic observations employing the Doppler principle, that the planet rotates in the same directional sense as most other members of the solar system and with a rotation period of about 16 hours.

Voyager 2 **Encounter with Neptune**

The *Voyager 2* spacecraft, initially conceived to visit Jupiter (July 9, 1981) and Saturn (August 26, 1981), was launched from Cape Canaveral, Florida on August 20, 1977, and performed so well that its mission was extended to include later encounters with Uranus (January 24, 1986) and Neptune (August 25, 1987). The dates given are for closest approaches.

Upgrading of System Instrumentation.   With the pending investigation of Neptune in mind, after completion of the encounter with Uranus, project managers took advantage of the available time (approximately 3 years) to improve ground system instrumentation as much as possible to upgrade the data return for the Neptune encounter. These efforts included:[1]

- *Attitude control* was altered to reduce angular rates by approximately 25% below those experienced at Uranus, in an attempt to improve compensation for impulses from tape recorder starts and stops, to permit an additional scan platform rate useful for motion compensation near the closest approaches to Neptune and Triton, and to provide for "nodding" image motion compensation (NIMC). This permitted acquisition of motion-compensated images without disrupting the communication link with Earth or utilizing limited tape recorder resources. Instrument control for the imaging system also was changed to permit exposure durations between 15 and 96 seconds and real-time images with exposure durations in multiples of 48 seconds—capabilities that were used extensively during the Neptune encounter, where light levels were only 40% of those at Uranus.

- *Signal strength* is notably less from Neptune than from Uranus because of the great transmission distance between the spacecraft and Earth-based tracking systems. Each of the three 64-meter–diameter tracking antennas of the National Aeronautics and Space Administration (NASA) Deep Space Network (DSN) were enlarged to 70-meter diameter and improved in shape. A high-efficiency, 34-meter tracking station was added to the Madrid DSN complex. Extensive arraying of antennas was used to further increase the effective collector area. The 64-meter Parkes Radio Telescope in Australia again was made available to enhance the capability of the Canberra DSN tracking antennas. Similarly, Voyager signals collected by the National Radio Astronomy Observatory's Very Large Array (VLA) near Socorro, New Mexico, were combined with signals collected at the Goldstone, California DSN complex. The 27 25-meter antennas of the VLA provided a collecting area equivalent to 2 70-meter antennas. During closest-approach operations on August 25, 1989, the Japanese Institute of Astronautical Science utilized its 64-meter tracking antenna at Usuda, Japan, to augment Voyager science data collection. The spacecraft and all of the ground systems worked flawlessly during the Neptune encounter, testifying to the high level of expertise and teamwork within the Voyager project and its supporting organizations.

[1] As reported by E.C. Stone (California Institute of Technology) and E.D. Miner (Jet Propulsion Laboratory).

- *Trajectory design* was engineered to maximize the information returned during the Neptune-Triton encounters. The team had maximum freedom in this regard because no further missions for Voyager 2 were contemplated. Three primary objectives were sought: (1) a close approach to Triton, including both sun and Earth occultations as viewed from the spacecraft; (2) a close polar passage of Neptune, including both sun and Earth occultations; and (3) timing of the closest approach so that both of the Neptune-Triton occultations occurred at relatively high elevation angles over the Canberra DSN complex.

***Voyager 2* Instrument Systems and Management.** The investigatory disciplines and their management for the Neptune encounter are outlined in Table 1.

TABLE 1. *VOYAGER 2* INVESTIGATIONS AND MANAGERS

Investigation	Principal investigator and affiliation
Imaging (ISS)	B.A. Smith, University of Arizona, Tucson, AZ
Photopolarimetry (PPS)	A.L. Lane, Jet Propulsion Laboratory, California Institute of Technology, Pasadena, CA
Infrared spectroscopy (IRIS)	B.J. Conrath, Goddard Space Flight Center, Greenbelt, MD
Ultraviolet spectroscopy (UVS)	A.L. Broadfoot, University of Arizona, Tucson, AZ
Radio science (RSS)	G.L. Tyler, Stanford University, Stanford, CA
Magnetometry (MAG)	N.F. Ness, Bartol Research Institute, University of Delaware, Newark, DE
Plasma (PLS)	J.W. Belcher, Massachusetts Institute of Technology, Cambridge, MA
Low-energy charged particles (LECP)	S.M. Krimigis, Applied Physics Laboratory, The Johns Hopkins University, Laurel, MD
Cosmic rays (CRS)	E.C. Stone, California Institute of Technology, Pasadena, CA
Plasma waves (PWS)	D.A. Gurnett, University of Iowa, Iowa City, IA
Planetary radio astronomy (PRA)	J.W. Warwick, Radiophysics, Inc., Boulder, CO

Chronology of the Neptune Encounter. Activities of the *Voyager 2* encounter commenced on June 5, 1989, when the spacecraft was 117×10^6 km from the center of the planet. The closest approach (center of Neptune) was a distance of 29,240 km and occurred on August 25, 1989. Radio signals from this position required a transit time to Earth of 4 hours, 6 minutes. The closest approach to Triton occurred on September 10, 1989, when the spacecraft was 39,800 km from the center of the satellite. This encounter period extended to October 2, 1989.

Design of the Neptune science sequences relied mainly on prior telescopic observations from Earth and on Voyager findings at Uranus, although some early *Voyager* data were used to make revisions to later observations. See Fig. 1. Provision was made for late retargeting of imaging frames to newly discovered satellites and rings. Timing of close Neptune and Triton observations also was adjusted at the last possible instant to take advantage of the most recent estimates of geometric event times made by the navigation team.

An initial, comprehensive report of the *Voyager 2* encounter was released in December 1989. Detailed information is available from the Jet Propulsion Laboratory, Pasadena, California. Some of these reports were included in *Science* magazine, issue of December 15, 1989. Analysis of these data, including the formulation of explanations and postulations pertaining to the overall Neptunian system, will be forthcoming for several years into the future. A number of post-encounter papers issued during the early 1990s are listed at the end of this article.

The spacecraft's plutonium power sources may hold out until approximately 2015. Somewhat before that time, *Voyager 2* is expected to encounter the heliopause (very edge of the solar system where the solar wind collides with interstellar media). See also **Voyager Missions to Jupiter and Saturn.**

Atmosphere of Neptune

Images of Neptune were obtained by the narrow-angle camera of Voyager 2 and indicated large-scale cloud features that persist for several

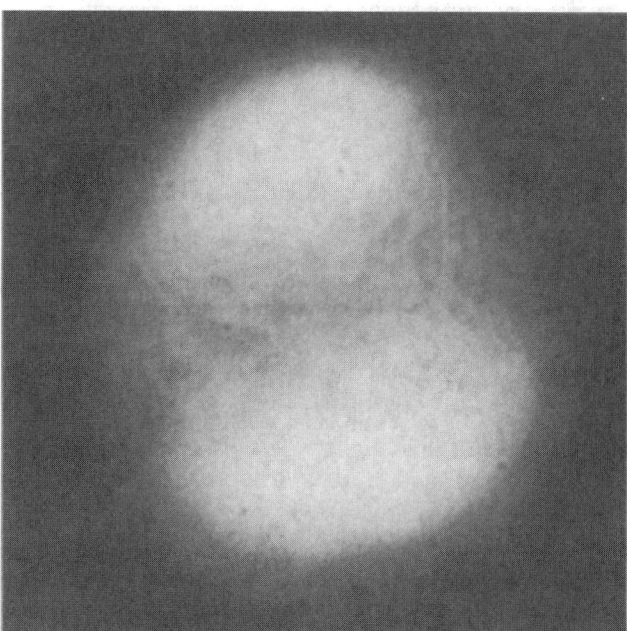

Fig. 1.   Features of Neptune in which camera incorporating a charge-coupled device (CCD) was used with a 1.54-meter Catalina telescope. (*University of Arizona*, 1979.)

months or longer.[2] The periods of rotation of these features about the planetary axis range from 15.8 to 18.4 hours. The atmosphere equatorward of $-53°$ rotates with periods longer than the 16.05-hour period deduced from *Voyager's* planetary radio astronomy experiment. This is presumably the planet's internal rotation period. The wind speeds computed with respect to this radio period range from 20 meters per second eastward to 325 meters per second westward. Thus, it was found that the cloud-top wind speeds are approximately the same for all the planets ranging from Venus to Neptune, even though the solar energy inputs to the atmospheres vary by a factor of 1000.

Neptune has an effective temperature of about 59.3 K. Derivation of Neptune's Bond albedo continues to require a more thorough study of *Voyager 2* instrument data. Neptune, however, appears to emit about 2.7 times as much energy as it absorbs from the sun. This greater contribution of internal heat may be the cause of the greater activity in the Neptunian atmosphere relative to that of Uranus. The horizontal temperature structures of the two atmospheres are quite similar, with the poles and equator at very nearly equal temperatures, while mid-latitudes are several degrees cooler. Temperature in the extreme upper atmosphere is nearly 750 K, but, because of Neptune's larger mass, colder atmosphere, and greater ring distances, the effects of gas drag on ring material are less at Neptune than at Uranus.

A number of prominent cloud features are apparent in images of Neptune's atmosphere, including an Earth-size "Great Dark Spot" (GDS) which occurs near $-10°$ latitude. The GDS is located and is of approximately the same size as Jupiter's Great Red Spot. See Figs. 2 and 3. The GDS rolls in a counterclockwise direction, with a 16-day period. Another, but smaller, dark spot with a bright central core is located at $-55°$ latitude. Bands of lower reflectivity extend from $+6°$ to $+25°$ latitude and from $-45°$ to $-70°$ latitude. The GDS is flanked by cirruslike cloud features. Other similar features occupy relatively narrow latitude ranges near latitudes of $-27°$ and $-71°$. These are believed to be optically thick upward extensions of the methane (CH_4) cloud deck. The details of these features change with time scales that are smaller as compared with the 16-hour rotation period. One of these features, referred to as the "*Scooter*," is bright and found near the $-42°$ latitude. It is deeper within the atmosphere than the aforementioned cirrus clouds and is postulated to be an upward extension of the deeper cloud deck. Velocities measured with respect to the internal rotation of Neptune indicate wind speeds ranging from about $+20$ m s^{-1} (prograde) at $-54°$ latitude to -325 m s^{-1} (retrograde) at $-22°$. The GDS

---

[2] As reported by H.B. Hammel (Jet Propulsion Laboratory) and associated team members (see papers listed).

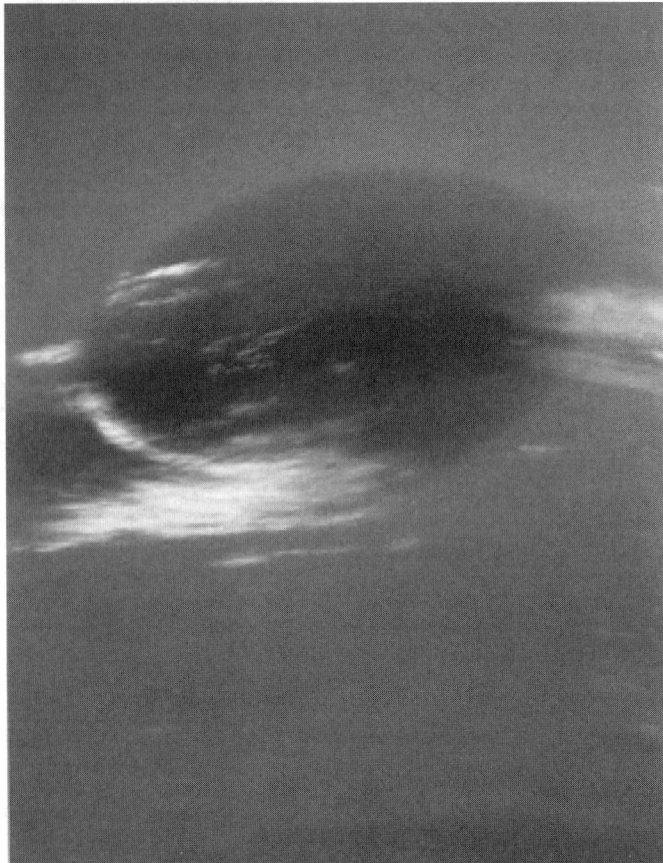

Fig. 2. This photograph shows the last face-on view of the "Great Dark Spot" that *Voyager* will make with the narrow-angle camera. The image was shuttered 45 hours before closest approach at a distance of 2.8 million kilometers (1.7 million miles). The smallest structures that can be seen are of an order of 50 kilometers (31 miles). The image shows feathery white clouds that overlie the boundary of the dark and light blue regions. The pinwheel (spiral) structure of both the dark boundary and the white cirrus suggest a storm system rotating counterclockwise. Periodic small- scale patterns in the white cloud, possibly waves, are short-lived and do not persist from one Neptunian rotation to the next. This color composite was made from the clear and green filters of the narrow-angle camera. (*National Aeronautics and Space Administration, Jet Propulsion Laboratory, Pasadena, California.*)

Fig. 3. This picture of Neptune was produced from the last whole planet images taken through the green and orange filters on the *Voyager 2* narrow angle camera. The images were taken at a range of 4.4 million miles from the planet, 4 days and 20 hours before closest approach. The picture shows the Great Dark Spot and its companion bright smudge; on the west limb the fast moving bright feature called Scooter and the little dark spot are visible. These clouds were seen to persist for as long as Voyager's cameras could resolve them. North of these, a bright cloud band similar to the south polar streak may be seen. (*Jet Propulsion Laboratory, Pasadena, California.*)

resides in a region with strong wind shears and is believed to lie at a lower level than most of the brighter cloud features, virtually independent of the higher-altitude winds. The cirruslike clouds are found at altitudes of 50 to 100 km above the lower cloud layer. Optically thin layers of haze, believed to be produced photochemically from CH_4, are found at still higher altitudes.

As with the other giant planets, hydrogen (H_2) predominates the Neptunian atmosphere. Although subject to further calculations from measurements taken, the mole fraction of atmospheric helium $[He]/[H_2]$ is estimated to be less than 0.25. Methane is more abundant in Neptune's upper atmosphere than in the atmosphere of Uranus. The absorption of red light by CH_4 gives Neptune its characteristic blue color. Deeper in the atmosphere, acetylene was detected. The signature of an optically thin cloud deck of methane ice was observed in radio occultation data. Strong absorption of radio waves may indicate the presence of small amounts of ammonia (NH_3).

Some further analyses of wind speed data are in order. Some scientists believe that the winds on Neptune are faster than found on any other planet, a surprising situation because of the small amount of energy received from the sun or from the interior of the planet. Past conceptual models of the general circulation of the giant planets do not readily provide a simple explanation as to why the highest winds appear to occur on Neptune, with its assumed low-energy sources. Neptune receives an estimated $\frac{1}{900}$ of the Earth's input of solar energy, but may have wind speeds of nearly 600 meters per second. How the near-supersonic winds can be maintained has puzzled some investigators. Scientists at the Space

Science and Engineering Center of the University of Wisconsin–Madison have offered the following hypothesis: "Based on principles of angular momentum and energy conservation in conjunction with deep convection, leads to a regime of uniform angular momentum at low latitudes. In this model, the rapid retrograde winds observed are a manifestation of deep convection, and the high efficiency of the planet's heat engine is intrinsic from the room allowed at low latitudes for reversible processes, the high temperature at which heat is added to the atmosphere, and the low temperatures at which heat is extracted." (See Suomi/Limaye/ Johnson reference listed.)

A scientist (California Institute of Technology) observes, "Neptune's supersonic winds are not a certainty. The altitudes of the different cloud features, for example, are difficult to confirm, making it hard to compute the clouds' speeds. Moreover, it is difficult to tell from the photos whether the movements represent actual fluid motion of atmospheric masses or merely a wave moving through the atmosphere." (See Eberhart reference listed.)

Rings of Neptune

Earth-based stellar occultation measurements made during the early and middle 1980s alerted investigators to the probable existence of rings or, at least, partial rings (ring arcs) at a number of radial distances from the center of the planet. *Voyager 2* imaging confirmed the presence of a system of at least six rings of prograde, equatorial, and circular rings. The outermost ring occurs at a distance of 62,900 km from the center of the planet. It is described as being composed of three bright, dusty areas. Data on Neptune's rings are given in Table 2.

Narrow rings are believed to be confined by the actions of relatively nearby satellites (shepherds). These rings may serve to prevent material from spiraling inward toward the planet. No ring shepherds have been noted thus far in the Neptunian system. The *Voyager 2* instrumentation, however, was limited to observing satellites of a diameter of 12 km or greater. Thus, tiny satellites may have escaped attention. Thus, it is not known whether or not additional shepherding satellites exist. Such material, if azimuthally unrestrained within the ring, should spread

TABLE 2. PROPERTIES OF NEPTUNE'S RINGS

Feature	Distance (10³ km)	Distance (R_N)	Width (km)	Optical Depth	Comments
1-bar atmosphere	24.76	1.000			Equatorial radius of Neptune
	38.	1.5		<0.0001	Inner extent of 1989N3R?
1989N3R	41.9	1.69	1700*	0.0001	High dust content
	49.	2.0		<0.0001	Outer extent of 1989N3R?
1989N2R	53.2	2.15	†	0.01	High dust content
1989N4R (inner)	53.2	2.15		0.0001	Inner edge of "plateau"
1989N4R (outer)	59.	2.4		0.0001	Outer edge of "plateau"
1989N1R	62.9	2.54	15	0.01–0	Contains three bright dusty arcs

*Tabulated width of 1989N3R is full width at half maximum. †1989N2R is narrow and unresolved in Voyager images.
Source: California Institute of Technology and Jet Propulsion Laboratory.

relatively uniformly around the ring within a time span of a few years. Additional encounters with largely enhanced resolution may be required at some future date to fully explain the dynamics of the planet's ring system.

Particles within the rings appear to be smaller than those found in the rings of Uranus. The dust content of one ring (1989N3R) is nearly double that of the other rings and thus compares better with that of the rings of Saturn and Uranus.

From analysis of *Voyager 2* data, C. Porco (Department of Planetary Sciences, University of Arizona) proposes an interesting explanation for Neptune's ring arcs, "A radial distortion with an amplitude of approximately 30 km is traveling through the ring arcs, a perturbation attributable to the nearby satellite Galatea. Moreover, the arcs appear to be azimuthally confined by a resonant interaction with the same satellite, yielding a maximum spread in ring particle semimajor aces of 0.5 km and spread in forced eccentricities large enough to explain the arcs' 15-km radial widths." Additional ring arcs were discovered during the course of

the study and provide further support to this model. (See Poroco reference listed.)

TABLE 3. PROPERTIES OF NEPTUNE'S SATELLITES

Satellite Name	Distance (10³ km)	Distance (R_N)	Period (Hours)	Diameter (km)	Resolution (km per Line Pair)	Normal Albedo
1989N6	48.0	1.94	7.1	54 ± 16	47.2	0.06?
1989N5	50.0	2.02	7.5	80 ± 16	34.8	0.06?
1989N3	52.5	2.12	8.0	180 ± 20	36.8	0.06?
1989N4	62.0	2.50	10.3	150 ± 30	33.8	0.054
1989N2	73.6	2.97	13.3	190 ± 20	8.2	0.056
1989N1	117.6	4.75	26.9	400 ± 20	2.6	0.060
Triton	354.8	14.33	141.0	2705 ± 6	0.8	0.6–0.9
Nereid	5513.4	222.65	8643.1	340 ± 50	86.6	0.14

Source: California Institute of Technology and Jet Propulsion Laboratory.

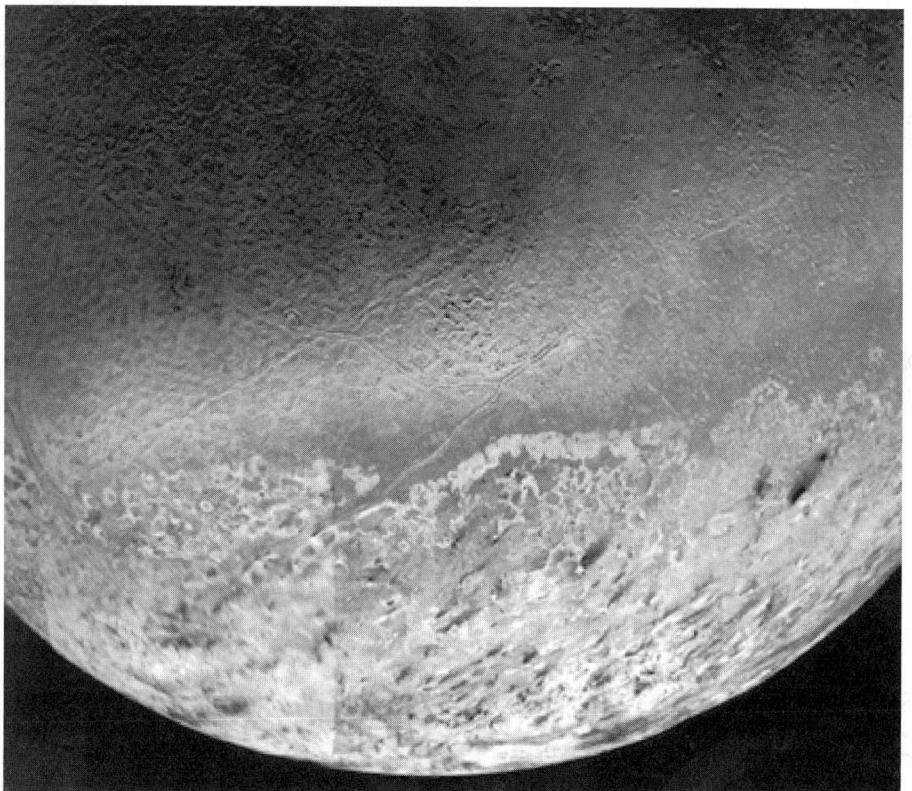

Fig. 4. *Voyager 2* obtained this high-resolution color image of Neptune's large satellite Triton during its close flyby on Aug. 25, 1989. Approximately a dozen individual images were combined to produce this comprehensive view of the Neptune-facing hemisphere of Triton. Fine detail is provided by high-resolution, clear-filter images, with color information added from lower-resolution frames. The large south polar cap at the bottom of the image is highly reflective and slightly pink in color; it may consist of a slowly evaporating layer of nitrogen ice deposited during the previous winter. From the ragged edge of the polar cap northward the satellite's face is generally darker and redder in color. This coloring may be produced by the action of ultraviolet light and magnetospheric radiation upon methane in the atmosphere and surface. Running across this darker region, approximately parallel to the edge of the polar cap, is a and of brighter white material that is almost bluish in color. The underlying topography in this bright band is similar, however to that in the darker, redder regions surrounding it. (*Jet Propulsion Laboratory, Pasadena, California.*)

Triton and Other Satellites[3]

During its approach to Neptune, *Voyager 2* images revealed six new satellites. All satellites orbit the planet in prograde, circular orbits of low inclination. Characteristics of the satellites are summarized in Table 3. Five of the six satellites orbit within 1° of Neptune's equatorial plane. The 1989N6 satellite has an inclination of nearly 6°. Data from Earth-based observations and *Voyager 2* (Figs. 4–10). Fig. 11 shows Triton's inclination to be 157°. Nereid's inclination is 29°. The respective orbital eccentricities of Triton and Nereid are 0.00 and 0.75. Nereid's distance from the center of the planet ranges from 1.39×10^6 to 9.64×10^6 km. Nereid's highly elliptical orbit makes it theoretically unlikely that its rotation and orbital periods are equal. However, *Voyager 2* detected no rotational brightness variations in excess of 10%.

Fig. 5. View of about 300 mi (483 km) across Triton's surface. (*National Aeronautics and Space Administration, Jet Propulsion Laboratory, Pasadena, California.*)

Triton. Much scientific interest has been directed toward this, by far the largest of Neptune's satellites and the existence of which has been known by earthbound observations for several years. Of all known natural bodies in the solar system, Triton has the lowest surface temperature (38 ± 4 K). Triton's atmosphere is predominantly nitrogen (N_2), with the presence of CH_4 in the lower atmosphere. The surface pressure, as measured by the radio science instrumentation aboard *Voyager 2*, is 16 ± 3 microbars. It is believed that a thermal inversion may exist in the lower 5 km of the atmosphere, and thus a tropopause altitude of 25 to 50 km is inferred. It is uncertain whether the clouds and haze layers observed in this region of the atmosphere result from simple condensation or from surface eruptions. The temperature at altitudes above 400 km is 95 ± 5 K. Atmospheric nitrogen is transported from the illuminated polar regions to the unilluminated polar regions.

The surface of Triton is that of a geologically young body and is devoid of heavily cratered terrain. The polar regions (south of latitude $-15°$) are covered with seasonal ice, believed to be N_2. Spring in Triton's southern region extends for several years—for example, in the present time frame, from about 1960 to the year 2000. Seasonal ice presents a slightly reddish tint believed to result from organic compounds photochemically produced by interaction between methane and nitrogen. Energetic particle bombardment also may assist in producing these reactions. The equatorial regions of Triton at most latitudes contain a thin layer of nitrogen frost. The layer appears as a bright, slightly blue coloration, a layer that does not obscure underlying topography. Observations indicate that northward of the equator there is a variety of terrains. Some scientists have referred to this topology as reminiscent of the "skin of a cantaloupe."

This terrain dominates the western (trailing) hemisphere of Triton and is believed to consist of a dense concentration of pits (dimples) that are

[3] Principal information source: Initial report of Jet Propulsion Laboratory, Pasadena, California.

Fig. 6. This photo-mosaic of Neptune's major satellite Triton was constructed from high-resolution images obtained by the *Voyager 2* spacecraft during its close flyby August 25, 1989. Images taken through Voyager's orange, violet, and ultraviolet filters are displayed as red, green, and blue respectively. The large, pinkish, highly reflective south polar cap is in the lower part of the picture; it may consist of a slowly evaporating layer of nitrogen ice deposited during the previous winter. North of the polar cap's ragged edge, the surface is darker and redder, perhaps because of the action of ultraviolet and magnetospheric radiation upon atmospheric and surface methane. A bluish band, running just north of the edge of the cap, may be freshly deposited nitrogen frost. The western (left) part of the region is dotted with small dimples with raised rims and shallow central depressions; long fractures, which have allowed icy material to ooze up, cross each other and extend into the polar region. Eastward, in the bluish region, an area of smooth plains and low hills, shows dense cratering, extensive resurfacing, and two areas resembling lunar maria, which may have been formed by large scale volcanic flooding. Triton's volcanic and tectonic activities involved icy materials such as frozen methane, nitrogen and water. The dark streaks prominent in the south polar region probably result from active venting, discovered in Voyager 2 images. (*U.S. Geological Survey, Flagstaff, Arizona. Jet Propulsion Laboratory, Pasadena, California.*)

crisscrossed by ridges. Few impact craters are discernible in these areas. The terrain of the eastern (leading) hemisphere of Triton is made up of a series of much smoother units, including caldera-like structures of water ice. Frozen lakes are surrounded by successive terraces, possibly due to a series of flooding actions.

Within the polar regions of Triton are numerous wind streaks with albedos that are 10–20% lower than the polar ices. The streaks, which overlie deeper ice deposits, appear to be young, possibly less than 1000 years of age. Two active geyserlike plumes were discovered near the subsolar latitude ($-55°$). As determined by stereoscopic viewing, these plumes rise to an altitude of about 8 km. Above a plume, dense clouds form and serve as a source for a westward wind-driven trail of material more than 100 km long. It has been suggested that the plumes may result from the explosive release of N_2 gas, which carries ice-entrained dark material in the exit nozzle to high altitudes.

In a report by L.A. Soderblom (U.S. Geological Survey, Flagstaff, Arizona) and a team of investigators, they explicitly describe the plume

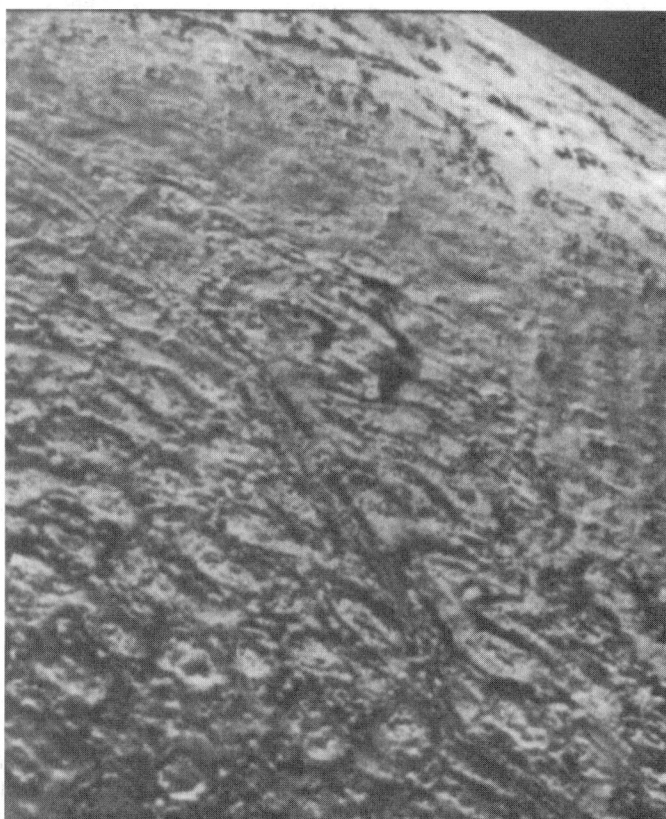

Fig. 7. Triton from 25,000 mi (40,225 km). Depressions may be caused by melting and collapsing of icy surface. (*National Aeronautics and Space Administration, Jet Propulsion Laboratory, Pasadena, California.*)

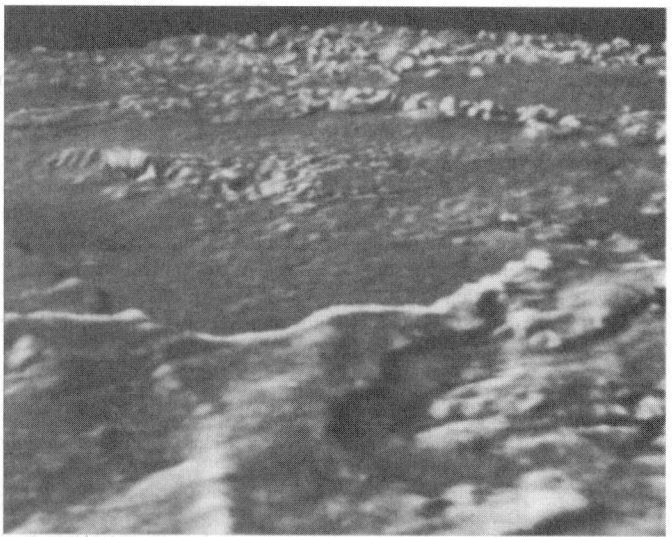

Fig. 8. Computer-generated perspective view of one of Triton's calderalike depressions. (*National Aeronautics and Space Administration, Jet Propulsion Laboratory, Pasadena, California.*)

phenomenon: "The radii of the rising columns appear to be in the range of several tens of meters to a kilometer. One model for the mechanism to drive the plumes involves heating of nitrogen ice in a subsurface greenhouse environment; nitrogen gas pressurized by the solar heating explosively vents to the surface carrying clouds of ice and dark particles into the atmosphere. A temperature increase of less than 4 kelvins above the ambient surface value of 38 ± 3 K is more than adequate to drive the plumes to an 8-km altitude. The mass flux in the trailing clouds is estimated to consist of up to 10 kg of fine dark particles per second, or twice as much nitrogen ice and perhaps several hundred or more kilograms of nitrogen gas per second. Each eruption may last a year or

Fig. 9. Satellite 1989N1, discovered by *Voyager 2*. (*National Aeronautics and Space Administration, Jet Propulsion Laboratory, Pasadena, California.*)

more, during which on the order of a tenth of a cubic kilometer of ice is sublimed."

In another approach to the plume phenomenon, A.P. Ingersol and K.A. Tryka (Division of Geological and Planetary Sciences, California Institute of Technology) observe, "Their structure suggests that the plumes are an atmospheric rather than a surface phenomenon. The closest terrestrial analogs may be dust devils, which are atmospheric vortices originating in the unstable layer close to the ground. Since Triton has such a low surface pressure, extremely unstable layers could develop during the day. Patches of unfrosted ground near the subsolar point could act as sites for dust devil formation because they heat up relative to the temperature of 48 K or higher, as observed by the *Voyager* radio science team. Assuming that velocity scales as the square root of temperature difference times the height of the mixed layers, a velocity of 20 m per second is derived from the strongest dust devils on Triton. Winds of this speed could raise particles provided they are a factor of 10^3 to 10^4 less cohesive than those on Earth."

Impact craters are rare on Triton—that is, those that are observable at the 3 to 1.8 km resolution acquired during the mapping sequence of *Voyager 2*. The highest-resolution images obtained show various degrees of smear, making analysis difficult. Thus, it is difficult to compare Neptunian cratering with other observed planets. The largest and uncontested impact crater viewed by *Voyager 2* on Triton is only about 27 km in diameter. Several large quasi-circular features exist, but are believed to be of internal origin. Fresh impact craters on Triton have morphologies similar to those on other icy satellites seen at comparable resolutions. These features include simple sharp-rimmed and bowl-shaped interiors and a few craters with flat floors and central peaks. Based mainly on the similarity of size distribution on Triton and Miranda (satellite of Uranus) and the relatively young surface of Triton, comets are believed to be the primary source of cratering. On the other hand, the peculiar size distribution of sharp craters on the "cantaloupe" terrain and other evidence suggests that they are of volcanic origin.

Neptune's Magnetosphere

Eight days prior to *Voyager 2's* closest approach to Neptune, a distance of about 470 Neptune radii (R_N), the first indication of a Neptune magnetic field was obtained from radio emissions. Subsequently, the spacecraft crossed a well-defined, detached bow shock at 34.9 R_N. The inbound magnetopause was not as well defined because *Voyager* entered a highly tilted magnetic field at very high magnetic latitude, permitting the first observation of a "pole-on" magnetosphere in which the solar wind is incident on the magnetic polar region rather than the equatorial.

It has been determined that, as the magnetic field rotates with the planet each 16.11 hours, satellites and ring particles sweep through large ranges of magnetic latitude. The incident solar wind deforms the magnetic field, resulting in a well-developed magnetic tail behind the planet. However, as the planet rotates, the magnetosphere configuration changes from pole-on with a cylindrically shaped magnetotail plasma sheet to a more normal

Fig. 10. These two 591-second exposures of the rings of Neptune were taken with the clear filter by the *Voyager 2* wide-angle camera on Aug. 26, 1989 from a distance of 280,000 kilometers (175,000 miles). The two main rings are clearly visible and appear complete over the region imaged. The time between exposures was one hour and 27 minutes. [During this period the bright ring arcs in the outer bright ring were not visible in either picture (they were unfortunately on the opposite side of the planet for each exposure).] Also visible in this image is the inner faint ring at about 42,000 kilometers (25,000 miles) from the center of Neptune, and the faint band which extends smoothly from the 53,000 kilometer (33,000 miles) ring to roughly halfway between the two bright rings. Both of these newly discovered rings are broad and much fainter than the two narrow rings. These long exposure images were taken while the rings were back-lighted by the sun at a phase angle of 135 degrees. This viewing geometry enhances the visibility of dust and allows fainter, dusty parts of the ring to be seen. The bright glare in the center is due to over-exposure of the crescent of Neptune. The two gaps in the upper part of the outer ring in the image on the left are due to blemish removal in the computer processing. Numerous bright stars are evident in the background. Both bright rings have material throughout their entire orbit, and are therefore continuous. (*Jet Propulsion Laboratory, Pasadena, California.*)

Fig. 11. Neptune and Triton 3 days after flyby. Triton is smaller crescent and is closer to viewer. (*National Aeronautics and Space Administration, Jet Propulsion Laboratory, Pasadena, California.*)

planar plasma sheet. Because of this unique geometry and the timing of the flyby, the spacecraft did not cross the plasma sheet.

No evidence exists for Neptunian electrostatic discharges of the kind observed on Saturn and Uranus by planetary radio astronomy (PRS). Many typical plasma waves were detected by the plasma wave instrumentation during the encounter. These included electron plasma oscillations in the solar wind upstream of the bow shock, and chorus, hiss, electron cyclotron waves, and upper hybrid resonance waves in the inner magnetosphere. There was no indication of lightning-generated whistlers.

See also **Voyager Missions to Jupiter and Saturn**; and **Hubble Space Telescope (HST)**.

Additional Reading

Belcher, J.W. et al.: "Plasma Observations Near Neptune: Initial Results from Voyager 2," *Science*, 1478 (December 15, 1989).

Broadfoot, A.I.: "Ultraviolet Spectrometer Observations of Neptune and Triton," *Science*, 1459 (December 15, 1989).

Brown, R.H. et al.: "Energy Sources for Triton's Geyser-Like Plumes," *Science*, 431 (October 19, 1990).

Brown, R.H. et al.: "Triton's Global Heat Budget," *Science*, 1465 (March 22, 1991).

Conrath, B. et al.: "Infrared Observations of the Neptunian System," *Science*, 1454 (December 15, 1989).

Cruickshank, D.P. et al.: "Triton: Do We See to the Surface?" *Science*, 283 (July 21, 1989).

Eberhart, J.: "Neptune Marvels Emerge from Data Deluge," *Science News*, 391 (December 16, 1989).

Goldreich, P. et al.: "Neptune's Story," *Science*, 500 (August 4, 1989).

Gore, R.: "Neptune — Voyager's Last Picture Show," *Nat'l. Geographic*, 34 (August 1990).

Gurnett, D.A. et al.: "First Plasma Wave Observations at Neptune," *Science*, 1494 (December 15, 1989).

Hammel, H.B.: "Neptune Cloud Structure at Visible Wavelengths," *Science*, 1165 (June 9, 1989).

Hammel, H.B. et al.: "Neptune's Wind Speeds Obtained by Tracking Clouds in Voyager Images," *Science*, 1307 (September 22, 1989).

Hansen, C.J. et al.: "Surface and Airborne Evidence for Plumes and Winds on Triton," *Science*, 421 (October 19, 1990).

Helfenstein, P. et al.: "Large Quasi-Circular Features Beneath Frost on Triton," *Science*, 824 (February 14, 1992).

Hillier, J. et al.: "Voyager Disk-Integrated Photometry of Triton," *Science*, 419 (October 19, 1990).

Hubbard, W.B. et al.: "Interior Structure of Neptune: Comparison with Uranus," *Science*, 648 (August 9, 1991).

Hunt, A.E. and P. Moore: "Atlas of Neptune," Cambridge University Press, New York, NY, 1994.

Ingersoll, A.P. and K.A. Tryka: "Triton's Plumes: The Dust Devil Hypothesis," *Science*, 435 (October 19, 1990).

Kerr, R.A.: "Triton Steals Voyager's Last Show," *Science*, 928 (September 1, 1989).

Kerr, R.A.: "The Neptune System in Voyager's Afterglow," *Science*, 1450 (September 29, 1989).

Kerr, R.A.: "Neptune's Triton Spews a Plume," *Science*, 213 (October 13, 1989).

Kerr, R.A.: "A Passion for the Little Things Among the Planets," *Science*, 998 (November 24, 1989).

Kerr, R.A.: "A Geologically Young Triton After All?" *Science*, 1563 (December 22, 1989).

Kerr, R.A.: "Geysers or Dust Devils on Triton?" *Science*, 377 (October 19, 1990).

Kinoshita, J.: "Neptune," *Sci. Amer.*, 82 (November 1989).

Kirk, R.L., R.H. Brown, and L.A. Soderblom: "Subsurface Energy Storage and Transport for Solar-Powered Geysers on Triton," *Science*, 424 (October 19, 1990).

Krimigis, S.M. et al.: "Hot Plasma and Energetic Particles in Neptune's Atmosphere," *Science*, 1483 (December 15, 1989).

Lane, A.L. et al.: "Photometry from Voyager 2: Initial Results from the Neptunian Atmosphere, Satellites, and Rings," *Science*, 1450 (December 15, 1989).

Lunine, J.I.: "Voyager at Triton," *Science*, 386 (October 19, 1990).

Lyons, J.R., Y.L. Yung, and M. Allen: "Solar Control of the Upper Atmosphere of Triton," *Science*, 204 (April 10, 1992).

McElheny, V.: "Neptune's Magnetic Looks," *Technology Review (MIT)*, 12 (January 1990).

Nelson, R.M. et al.: "Temperature and Thermal Emissivity of the Surface of Neptune's Satellite Triton," *Science*, 429 (October 19, 1990).

Ness, N.F. et al.: "Magnetic Fields at Neptune," *Science*, 1473 (December 15, 1989).

Pollack, J.B., J.M. Schwartz, and K. Rages: "Scattering in Triton's Atmosphere: Implications for the Seasonal Volatile Cycle," *Science*, 440 (October 19, 1990).

Polvani, L.M. et al.: "Simple Dynamical Models of Neptune's Great Dark Spot," *Science*, 1393 (September 21, 1990).

Poroco, C.C.: "An Explanation for Neptune's Ring Arcs," *Science*, 995 (August 30, 1991).

Smith, B.A. et al.: "Voyager 2 at Neptune: Imaging Science Results," *Science*, 1433 (December 15, 1989).

Smith, B.A.: "Voyage of the Century," *Nat'l. Geographic*, 34 (August 1990).

Soderblom, L.A. et al.: "Triton's Geyser-Like Plumes: Discovery and Basic Characteristics," *Science*, 410 (October 19, 1990).

Stone, E.C. et al.: "Energetic Charged Particles in the Magnetosphere of Neptune," *Science*, 1489 (December 15, 1989).

Stone, E.C. and E.D. Miner: "The Voyager 2 Encounter with the Neptunian System," *Science*, 1417 (December 15, 1989).

Strom, R.G., S.K. Croft, and J.M. Boyce: "The Impact Cratering Record on Triton," *Science*, 437 (October 19, 1990).

Stromovsky, L.A.: "Latitudinal and Longitudinal Oscillations of Cloud Features on Neptune," *Science*, 684 (November 1, 1991).

Suomi, V.E., S.S. Limaye, and D.R. Johnson: "High Winds of Neptune: A Possible Mechanism," *Science*, 929 (February 22, 1991).

Thompson, W.R. and C. Sagan: "Color and Chemistry on Triton," *Science*, 415 (October 19, 1990).

Tyler, G.L. et al.: "Voyager Radio Science Observations of Neptune and Triton," *Science*, 1466 (December 15, 1989).

Warwick, J.W. et al.: "Voyager Planetary Radio Astronomy at Neptune," *Science*, 1498 (December 15, 1989).

Yelle, R.V.: "The Effect of Surface Roughness on Triton's Volatile Distribution," *Science*, 1553 (March 20, 1992).

Web References

Jet Propulsion Laboratory: *http://www.jpl.nasa.gov/solar_system/solar_system_index.html#*

NASA: *http://www.nasa.gov/projects.html*

NEPTUNIAN SERIES. See **Radioactivity**.

NEPTUNIUM. Chemical element, symbol Np, at. no. 93, at. wt. 237.0482 (predominant isotope), radioactive metal of the Actinide series, also one of the Transuranium elements. Neptunium was the first of the Transuranium elements to be discovered and was first produced by McMillan and Abelson (1940) at the University of California at Berkeley. This was accomplished by bombarding uranium with neutrons. Neptunium is

produced as a by-product from nuclear reactors. ^{237}Np is the most stable isotope, with a half-life of 2.20×10^6 years. The only other very long-lived isotope is that of mass number 236, with a half-life of 5×10^3 years.

^{237}Np is parent of the neptunium $(2n + 1)$ alpha decay series. Other isotopes include those of mass numbers 229–235 and 238–241; metastable forms of ^{236}Np, ^{240}Np and two of ^{237}Np are known. Electronic configuration $1s^2 2s^2 2p^6 3s^2 3p^6 3d^{10} 4s^2 4p^6 4d^{10} 4f^{14} 5s^2 5p 5d^{10} 5f^5 6s^2 6p^6 - 6d^1 7s^2$. Ionic radii Np^{4+} 0.88 Å; Np^{3+} 1.02 Å (Zachariasen). Oxidation potential $Np \rightarrow Np^{3+} + 3e^-$, 1.85 V; $Np^{3+} \rightarrow Np^{4+} + e^-$, −0.155 V; $Np^{4+} + 2H_2O \rightarrow NpO_{2+} + 4H^+ + e^-$, −0.739 V; $NpO_2^+ \rightarrow NpO_2^{2+} + e^-$, −1.137 V. See also **Chemical Elements**.

Neptunium has the oxidation states (VI), (V), (IV), and (III) with a general shift in stability toward the lower oxidation states as compared to uranium. The compounds which are formed are very similar to the corresponding compounds of uranium.

The ionic species corresponding to the oxidation states vary with the acidity of the solution; in acid solution of moderate strength the species are Np^{3+}, Np^{4+}, NpO_2^+, and NpO_2^{2+} as in the case of uranium and plutonium. The potential scheme in $1 - M$ HCl is as follows:

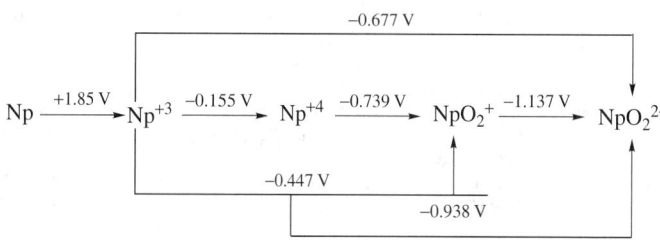

It will be seen that the metal is highly electropositive, in common with the other actinide elements. The $Np^{3+} \rightarrow Np^{4+}$ couple is reversible and this oxidation can be accomplished by the oxygen of the air. The (IV) state is stable, not oxidized by air, and only slowly oxidized to NpO_2^+ by nitric acid. The $Np^{4+} \rightarrow NpO_2^+$ couple is not readily reversible, whereas the $NpO_2^+ \rightarrow NpO_2^{2+}$ couple is reversible; this is reasonable on the basis that the former involves making or breaking the neptunium-oxygen bonds, whereas the latter does not. The oxidation of NpO_2^+ to NpO_2^{2+} requires moderately strong oxidizing agents. Neptunium differs from uranium and plutonium in that its potential relations are such as to render NpO_2^+ moderately stable with respect to disproportionating, even in solutions containing moderate concentrations of hydrogen ion.

The potentials are altered extensively by change in the hydrogen ion concentration and by the presence of any of a number of anions capable of forming complex ions.

Neptunium ions in aqueous solution possess characteristic colors: pale purple for Np^{3+}, pale yellow-green for Np^{4+} green-blue for Np^{5+}, while NpO_2^{2+} varies from colorless to pink or yellow-green depending on the acid present.

The precipitation reactions of Np^{3+} are similar to those of the tripositive rare earths, those of Np^{4+}, to the other tetrapositive actinides and to Ce^{4+}, and those of NpO_2^{2+} to the corresponding ions of uranium and plutonium. All of the simple salts of NpO^{2+} appear to be soluble.

The neptunium oxide system exhibits complexity similar to that found in the uranium oxide system. Thus, the important oxide is NpO_2 and there exists a range of compositions, depending upon conditions, up to Np_3O_8.

As a metal, Np has a relatively low melting point ($\sim 640 °C$), is very dense (20.45 g/cm^3), and is ductile. The alpha form reacts with hydrogen, carbon, oxygen, sulfur, the halogens, and phosphorus to yield a number of binary compounds.

The important halides of neptunium are the trifluoride, NpF_3, purple or black and hexagonal, the hexafluoride, NpF_6, brown and orthorhombic, the trichloride, $NpCl_3$, white and hexagonal, the tetrachloride, $NpCl_4$, red-brown and tetragonal, and the tribromide, $NpBr_3$, α-form green and hexagonal, β-form green and orthorhombic.

In research at the Institute of Radiochemistry, Karlsruhe, West Germany, during the early 1970s, investigators prepared alloys of neptunium with iridium, palladium, platinum, and rhodium. These alloys were prepared by hydrogen reduction of the neptunium oxide in the presence of finely divided noble metals. The reaction is called a *coupled reaction* because the reduction of the metal oxide can be done only in the presence of noble metals. The hydrogen must be extremely pure, with an oxygen content of less than 10^{-25} torr.

Industrial utilization of neptunium has been very limited. The isotope ^{237}Np has been used as a component in neutron detection instruments. Neptunium is present in significant quantities in spent nuclear reactor fuel and poses a threat to the environment. A group of scientists at the U.S. Geological Survey (Denver, Colorado) has studied the chemical speciation of neptunium (and americium) in ground waters associated with rock types that have been proposed as possible hosts for nuclear waste repositories. See Cleveland reference.

Additional Reading

Cleveland, J.M., K.L. Nash, and T.F. Rees: "Neptunium and Americium Speciation in Selected Basalt, Granite, Shale, and Tuff Ground Waters," *Science*, **221**, 271–273 (1983).

Fuger, J. and L.R. Morss: "Transuranium Elements: A Half Century," American Chemical Society, Washington, DC, 1992.

Keller, C. and B. Erdmann: "Preparation and Properties of Transuranium Element–Noble Metal Alloy Phases," Proc. 1972 Moscow Symp. Chem. Transuranium Elements, 1976.

Krot, N.N. and A.D. Gel'man: "Preparation of Neptunium and Plutonium in the Heptavalent State," *Dokl. Chem.* **177**, 1–3, 987–989 (1967).

Lewis, R.J., Sr.: "Hawley's Condensed Chemical Dictionary," 13th Edition, John Wiley & Sons, Inc., New York, NY, 1999.

Lide, D.: "CRC Handbook of Chemistry and Physics 2000-2001," 81st Edition, CRC Press, LLC., Boca Raton, FL, 2000.

Magnusson, L.B. and T.J. LaChapelle: "The First Isolation of Element 93 in Pure Compounds and a Determination of the Half-life of 93Np237," *Amer. Chem. Soc. J.*, **70**, 3534–3538 (1948).

Marks, T.J.: "Actinide Organometallic Chemistry," *Science,* **217**, 989–997 (1982).

McMillan, E. and P.H. Abelson: "Radioactive Element 93," *Phys. Rev.*, **57**, 1185–1186 (1940).

Seaborg, G.T.: "The Chemical and Radioactive Properties of the Heavy Elements," *Chem. Engng. News*, **23**, 2190–2193 (1945).

Seaborg, G.T. and W.D. Loveland: "The Elements beyond Uranium," John Wiley & Sons, Inc., New York, NY, 1990.

Thayer, J.S. and F.E. Brinckman: "Environmental Chemistry of the Heavy Elements: Hydrido and Organo Compounds," John Wiley & Sons, Inc., New York, NY, 1995.

NERNST EFFECT.   If heat is flowing through a strip of metal and the strip is placed in a magnetic field perpendicular to its plane, a difference of electric potential develops between the opposite edges. This phenomenon, discovered by Nernst in 1886, is analogous to the Hall effect, but with a longitudinal flow of heat replacing the longitudinal electric current. See also **Hall Effect**.

NERNST HEAT THEOREM.   For a homogeneous system, the rate of change of the free energy with temperature, as well as the rate of change of heat content with temperature, approaches zero as the temperature approaches absolute zero.

NERNST-THOMPSON RULE.   A solvent of high dielectric constant favors dissociation by reducing the electrostatic attraction between positive and negative ions, and conversely a solvent of low dielectric constant has small dissociating influence on an electrolyte.

NERNST, HERMANN WALTHER (1864–1941).   Nernst was a German Chemist and Physicist. In 1894 he received invitations to the Physics Chairs in Munich and in Berlin, as well as to the Physical Chemistry Chair in Göttingen. He accepted this latter invitation. At Göttingen Nernst founded the Institute for Physical Chemistry and Electrochemistry and became its Director. In 1905 he was appointed Professor of Chemistry, later of Physics, in the University of Berlin, becoming Director of the newly-founded "Physikalisch-Chemisches Institut" in 1924. He remained in this position until his retirement in 1933.

He made major contributions to electrochemistry, thermodynamics, and photochemistry. Nernst's early studies in electrochemistry were inspired by Arrhenius' dissociation theory which first recognized the importance of ions in solution. His heat theorem, known as the Third Law of Thermodynamics, was developed in 1906. In 1918 his studies of photochemistry led him to his atom chain reaction theory. In later years, he occupied himself with astrophysical theories, a field in which the heat theorem had important applications.

He is remembered best for the Nernst Effect, named after him. For his work in thermochemistry Nernst won the Nobel Prize in Chemistry in 1920.

See also **Nernst Effect**; **Nernst Heat Theorem**; **Nernst-Thompson Rule**; and **Thermodynamics**.

J. M. I.

NERVOUS SYSTEM (Fishes).   See **Fishes**.

NERVOUS SYSTEM AND THE BRAIN.   See **Central and Peripheral Nervous Systems**.

NET.   A set of intervals such that every point of a closed linear interval [*a*, *b*] is contained in at least one interval of the set, each interval being called a *mesh* of the net.

NETTLE HAIR.   A form of hair-like scale, found on some caterpillars, which causes an irritation of human skin resembling that of nettles. Caterpillars of the io and the brown-tail moths have larger poisonous spines of similar properties. The irritation caused by some species is very mild but others are much more severe.

NETWORK.   In elementary terms, a network consists of three or more points that are connected physically by some medium for the purpose of transferring information, energy, or materials from one point to another. Networks may occur naturally, as in the case of the brain and nervous system of living creatures, or networks may be contrived to serve human needs. *Material-transfer* networks are exemplified by gas, oil, and water pipelines. An *energy-transferring* network is typified by the distribution over wires and cables of centrally generated electricity. An *information-transfer* network relies on various segments of the electromagnetic spectrum to convey intelligence from transmitters to two or more receivers. A simple connection between two points usually is referred to as a *link* rather than a network. As of the early 1990s, the word *network* most frequently refers to a *communication network*, the subject addressed by this article.

The points on a network may be referred to as stations or nodes at which information is received/transmitted instrumentally or humanly for storage, action, or sometimes retransmittal. Ultimately, unless wasted, the information will become part of the human need for that information. Information transmitted and received may be of analog or digital form, but the latter predominates by far. Thus, the term *digital network* is very common.

Networking equipment is costly, and the total cost of some nationwide and global networks is tremendous. Networks must be planned for expansion because of the need for more and more information exchange. The concept of information networks as they are known today essentially commenced with telegraphy and telephony. For decades, these were person-to-person communication. See also **Telephony (Telecommunications)**.

Chronology of Computer Networks. The concept of a computer network in the United States and worldwide stemmed from research work at the Rand Corporation in the early 1960s. Late in the 1960s, a network sponsored by the U.S. Defense Advanced Research Projects Agency (DARPA) and known as ARPANET was established. According to D. Van Honweling (University of Michigan), "The notion when ARPANET was established was that it was primarily to share computing resources. As things turned out, that wasn't the way it got used. It got used by human beings who wanted to work with other human beings." In the mid-1970s, a variety of networks joined ARPANET to offer connectivity. These subsidiary networks had the specific objective of linking government and academic institutions together. Concurrently, NASA and the U.S. Department of Energy established their own networks.

Taking note of the lack of a national framework for networking as of the early 1980s, the National Science Foundation (NSF) established the NSFNET designed around the existing six NSF-supported supercomputer centers. The concept was that of including not only the supercomputer centers, but also to link together the several other important networks that had been developed during the mid-1980s. Operation of the network was assigned to private contractors when this "backbone" operated at 56 kilobits per second. A rate of 1.5 million bits per second had been achieved by early 1990, with goals as of 1993 set in excess of 50 million bits per second.

A *packet* is defined as bits that contain address and some fraction of the particular message being sent. On this standard, over 100 million packets per month had been sent in 1988. By 1990, this figure had increased to

2.5 billion packets. By February 1990, 10% of all information ever sent on the network had been sent during 1 month. A survey of NSFNET use indicates: (1) networked mail, 30%, (2) file exchange, 29%, (3) interactive applications, 20%, (4) name lookup, 15%, and (5) other services, 6%.

Of growing value to users of computer networks is the transmission of *motion graphics*. Most scientists agree that moving pictures are more eloquent than words. One scientist has observed, "Most people's comprehension works around visualization. Networks are making that comprehension tremendously more transferable."

Several years ago, the concept of a "Global Village" was envisioned, wherein communications, including graphics—essentially as an extension to traditional radio and television—would permit persons living almost anywhere to participate globally in science, politics, the arts, and other information exchange. Scientists who work on a single project, but at different locations, use networks to exchange concepts and ideas, problems and solutions, thus conserving on the expenses of establishing single quarters or of traveling a lot. But, from the shadows, the vision of computer hackers is seldom lost. Means to prevent system tampering must continue to be high on the list of designing future networked systems. One scientist has observed, "We really need to develop a set of cultural and behavioral protocols for using the networks!" See **Internet (The History)**.

Data Transmission Needs Are Virtually Unlimited. Considering the information worthy of recording that has been generated in centuries past and the macro amounts of data generated each second of every day in present times that are important to some segment of civilized society, the word *overwhelming* seems inadequate. Thousands of needs to retrieve data can be cited, of which just one case—that is, the needs of the National Aeronautics and Space Administration (NASA)—are exemplary. Considering just the time frame that embraced NASA data gathering from spacecraft (*Pioneer* to the *Ultraviolet Explorer*), 6 trillion bytes (1 byte = 8 bits or binary units) were collected. Spacecraft launched during the last few years have been estimated to more than quadruple that amount. The *Hubble Space Telescope* will generate several trillion bytes each year. Introduction of the *Earth Observing System* (EOS) is estimated to generate a trillion bytes of information every few days. (The battery of sensors planned for EOS have been so designed.) Because of data storage and retrieval costs, these are a prerequisite cost accounting factor in budgeting for total mission coverage and, in fact, for selecting only those missions for which adequate data handling support will be available the years that the mission may be productive.

By way of comparison, it has been estimated that the U.S. Library of Congress, with over 19 million books, represents the storage of 6 trillion bites of information.

The data storage and retrieval components of the human genome project also are exemplary of how failure to adequately budget for electronic storage and networking of monstrous quantities of information can threaten ambitious research projects per se. An accounting of the *Bionet* project (National Institutes of Health) is given in the Wallich reference listed.

Examples of network databases are those of the Scientific & Technical Information Network, including such topics as: (1) BEILSTEIN, a comprehensive database in organic chemistry offering a wide range of information for over 1,700,000 substances reported over the past 150 years; (2) JANAF, a joint U.S. Army-Navy-Air Force file produced by the National Institute for Standards and Technology (NIST), containing the chemical thermodynamic properties for over 1000 substances; (3) NBSFLUIDS, which is an NIST file that contains calculation and data generation software useful for creating tables of temperature-dependent thermodynamic and thermophysical properties for 12 cryogenic fluids; and (4) NBSTHERMO, an NIST file of ambient temperatures and chemical thermodynamic properties of 8000 inorganic and numerous two-carbon organic substances.

Expanding the Points in a Network. Even though most of the time communications and transportation networks usually are initially designed to accommodate future as well as present needs, in recent years the satiation of network users is seldom accommodated. The frequent requirement to expand networks today is regarded by many network engineers as virtually inevitable. What initial steps can be taken to assure a minimum cost for network expansion? As reported by B. Cipra (see reference listed), F. Hwang (AT&T Bell Laboratories) and Ding Zhu Du (Princeton University) have proposed the method indicated in Fig. 1. This materializes a conjecture with which mathematicians have been attempting to prove for

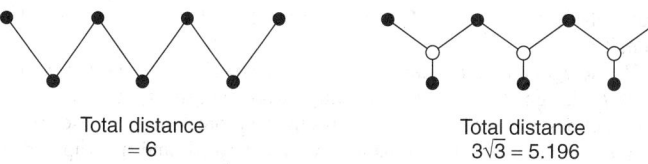

Total distance = 6   Total distance $3\sqrt{3} = 5.196$

Fig. 1. Designing the "shortest" network. As shown here, a network can be shortened by adding additional points. The example shows a savings of about 13%. Whether or not savings of over 13% can be achieved has not been settled among mathematicians. As indicated by this figure, the greatest savings result when the points are at the vertices of an equilateral triangle. Using the center of the triangle as a hub reduces the length of the network by $\frac{\sqrt{3}}{2}$ ($\sim$13%).

many years. In formulating their approach, the engineers have borrowed from the "minimax" problem of game theory. See also **Game Theory**.

The "shortest network" problem has been analyzed in scholarly detail by M. Bern (Xerox Palo Alto Research Center) and R. Graham (AT&T Bell Laboratories). In their paper (see reference listed), the researchers conclude: "Although knowledge about algorithms (including Melzak's) has progressed greatly in recent years, the shortest-network problem remains tantalizingly difficult. The problem can be stated in simple terms, and yet solutions defy analysis. A tiny variation in the geometry of a problem may appear to be insignificant, and yet it can radically alter the shortest network for the problem. The shortest-network problem will continue to frustrate and fascinate us for years to come."

Classification of Network Uses. To ask the question, "Where are communication networks used?" is tantamount to asking, "Where is information used?" An appreciation for the phenomenal needs and uses of networking may be piqued by considering the following:

- Statistical information for business and government. One of the early pioneering users of networks was the travel industry for determining space availability and making reservations. Computerized banking and trading also are early examples of early network use.
- Educational institutions, where most universities today have extensive private networks and the ability to tap into national and global networks. This is expanding rapidly to include public and vocational schools.
- Scientific research, notably for efforts conducted by teams of researchers who represent several disciplines and who do not have their offices and laboratories at a centralized location.
- Transportation and shipping, where goods movements must be closely time related. See also **Navigation**.
- Medical practice, particularly pertaining to crises that must tap distant sources of special expertise and in order to expedite the location and ultimate receipt of organs and fluids for transplantation. Numerous hospitals operate their own networks, which can readily be tapped into national and global networks.
- Earth and space research, as previously mentioned.
- Entertainment program retrieval.
- Manufacturing production, described later in this article.
- Distribution systems, to which much attention has been devoted by economists recently.
- And this list could go on and on.

Distribution and Networking. Information networking has and will continue to impact upon the very fundamental life-styles of business, government, scientific, and other professional people. Quite recently, particular attention has been given to the manner in which networking ultimately may affect materials distribution, selling, and trading. Depending upon a nation's form of government, prices may be fixed or tightly regulated by a central organization or allowed to float in accordance with supply and demand. Even in free-trade economies, some commodity prices (notably of utilities, such as the gas and electricity supply) are regulated by national or state and provincial authorities. As observed by K. McCabe, S. Rassenti, and V.L. Smith (University of Arizona), "Domestically, we have witnessed in the last decade (1980s) uncommon political and economic forces that have resulted in increased reliance on markets to discipline prices, output, and the entry and exit of firms in industries traditionally regulated by state and federal agencies. This has been part of a worldwide move toward privatization in the socialist and command economies of Great Britain, New Zealand, Eastern Europe, and the (former) Soviet Union. In the United

States the extent of deregulation has been less than complete in all of them."

The deregulation movement motivated an experimental study of *auction markets* designed for interdependent network industries, such as natural gas pipelines or electric power systems by the aforementioned researchers. Through the use of an information-transfer network and a computerized dispatch center, decentralized agents would submit bids to buy commodities and offers to sell transportation and commodities. Computer algorithms would determine prices and allocations that maximize the gains from exchange in the system relative to the submitted bids and offers. Details are described in the McCabe reference listed.

Networked Communications for Manufacturing

Although information-exchange networks for manufacturing and processing facilities frequently include wide-area networks (WANs), which exchange information between plants at different locations and with corporate headquarters, the principal need for information exchange occurs at the factory floor level through the use of *local-area networks* (LANs), which provide the coordination and govern the actions of specific machines and processes. See also article on **Local Area Networks**.

As described in the article **Control System**, industrial manufacturing is based largely on the measurement and control of numerous variables (temperature, pressure, dimension, count, material flow, to mention a few). In a typical manufacturing scene, hundreds to thousands of feedback control loops may be present. Information coordination and balance are achieved by way of local area networks.

Because of the multitude of connections required and the great variety of sensors used, severe standards for component design are imposed. Development of these standards is described in the article on Control System Architecture.

Other Networks

The word *network* is used to designate systems other than information-exchange networks. For example, *electric network* stems back to Kirchhoff (1850) and Maxwell (1870).

An electric network may be described as a combination of elements, either as (1) a combination of interconnected devices, such as inductors, capacitors, resistors, and generators, or (2) the abstraction of interconnected branches having the properties of inductance, resistance, capacitance, etc.

An *active network* is one whose output waves are dependent upon sources of power, apart from that supplied by any of the actuating waves, which power is controlled by one or more of these waves.

An *all-pass network* is designed to introduce phase shift or delay without introducing appreciable attenuation at any frequency.

A *differentiating network* is one whose output is the time derivative of its input waveform. Such a network preceding a frequency modulator makes the combination a phase modulator; or, following a phase detector, it makes the combination a frequency detector. Its ratio of output amplitude to input amplitude is proportional to frequency, and its output phase leads its input by 90 degrees.

An *equivalent network* is one in which, under certain conditions of use, it may replace another network. The networks need not be of the same form. For example, one may be electrical, the other mechanical. If one network can replace another network in any system whatsoever without altering in any way the operation of that portion of the system external to the networks, the networks are said to be of *general equivalence*.

A *linear-passive network* is a network such that (1) if currents of any waveform are fed to the terminals of the network, the total energy delivered to the network is non-negative; (2) no voltages appear between any pair of terminals before a current is fed to the network.

Neural networks are described in the articles **Central and Peripheral Nervous Systems**; and **Artificial Intelligence: Neural Networks**.

Additional Reading

Bern, M.B. and R.L. Graham: "The Shortest-Network Problem," *Sci. Amer.*, 84 (January 1989).

Cerf, V.G.: "Networks," *Sci. Amer.*, 72 (September 1991).

Cipra, B.: "In Math, Less is More — Up to a Point," *Science*, 1081 (November 23, 1990).

Dertuozos, M.L.: "Building the Information Marketplace," *Technology Review(MIT)*, 29 (January 1991).

Haavind, R.: "The Smart Tool for Information Overload — Hypertext," *Technology Review (MIT)*, 42 (November–December 1990).

Horgan, J.: "Jukeboxes for Scientists," *Sci. Amer.*, 24 (July 1989).

Kauffels, J.: "Practical LANs Analyzed," Halstead Press, New York, NY, 1989.

Kay, A.C.: "Computers, Networks and Education," *Sci. Amer.*, 138 (September 1991).

King, J.P.: "Distributed Control Systems," 3.5 in Process/Industrial Instruments and Controls Handbook, 4th Edition (D.M. Considine, Editor), The McGraw-Hill, Companies, Inc., New York, NY, 1993.

Malone, T.W. and J.F. Rockart: "Computers, Networks and the Corporation," *Sci. Amer.*, 128 (September 1991).

McCabe, K.A., J. Rassenti, and V.I. Smith: "Smart Computer-Assisted Markets," *Science*, 534 (October 25, 1991).

Naugle, M.G.: "The Illustrated Network Book: A Graphic Guide to Understanding Computer Networks," Thomson Publishing Company, New York, NY, 1997.

Negroponte, N.P.: "Products and Services for Computer Networks," *Sci. Amer.*, 106 (September 1991).

Palca, J.: "Getting Together Bit by Bit," *Science*, 160 (April 13, 1990).

Peterson, I.: "The Electronic Grapevine," *Science News*, 90 (August 11, 1990).

Steen, M.V. and H. Sips: "Computer and Network Organization: An Introduction." Prentice-Hall, Inc., Upper Saddle River, NJ, 1995.

Tesler, L.G.: "Networked Computing in the 1990s." *Sci. Amer.*, 86 (September 1991).

Wallich, P.: "Who's Minding the Store," *Sci. Amer.*, 20 (October 1989).

White, G.W., U.W. Pooch, and E.A. Risch: "Computer System and Network Security," DIANE Publishing Company, Collingdale, PA, 1999.

NETWORK SYNTHESIS. Network synthesis is that branch of the theory of electric networks that deals with the systematic determination of the structure and element values of an electric network possessing preassigned characteristics.

NETWORK (Telecommunications).   See **Telephony (Telecommunications)**.

NEURALGIA.   Pain occurring along the course of a cranial or peripheral nerve or posterior root. Neuralgia is difficult to differentiate sharply from neuritis. The primary distinction is that neuritis best describes the acute and chronic inflammation of the peripheral nerves, while neuralgia is due to acute and chronic inflammation of the pathway stations (ganglia) lying along the course of nervous pathways. The pain in neuralgia is usually of a sharp, shooting, intermittent type and accompanied by increased sensitivity of the skin supplied by the nerve. Many varieties of neuralgia are differentiated according to the body part affected. Common forms include: (1) trigeminal neuralgia, a very severe form marked by agonizing pain over branches of the trigeminal nerve in the face; (2) intercostal neuralgia, which can be mistaken for pleurisy because the intercostal nerves, after leaving the spinal cord, run around each side of the chest between the ribs; (3) Morton's neuralgia — pain in the joint of the third and fourth toe caused by pinching of a nerve in this region; and (4) sciatic neuralgia.

Trigeminal neuralgia also is known as tic douloureux and involves one or more branches of the fifth cranial (trigeminal) nerve. A closely associated neuralgia involves the ninth cranial nerve, often causing severely sharp pains in the throat, spreading to the ear on the same side of the face. Palliative relief in some cases can be obtained from injections of alcohol into the ganglion of the nerve. In more severe cases, surgical procedures, as cutting the sensory root of the nerve, may be required to provide permanent relief. Drugs such as *Dilantin*® Sodium slow the rate of peripheral nerve conduction and have been used in the treatment of trigeminal neuralgia. The successful use of injections of vitamin B_{12} also has been reported. Intercostal neuralgia is sometimes accompanied by a skin eruption in herpes zoster or shingles.

NEURITIS.   An inflammation of a nerve, either chronic or acute. Mononeuritis or localized neuritis is the term used when one nerve is involved. When several or many nerves are involved the term multiple neuritis or polyneuritis is used. Localized neuritis develops from injury, infection, chronic intoxication, or metal poisoning. Neuritis, either localized or multiple, complicates many of the infectious diseases, such as typhoid, diphtheria, tuberculosis, smallpox, etc. Other causes include pressure on a nerve by a tumor or calcium deposits in osteo-arthritis, etc.

Multiple or polyneuritis is inflammation and degeneration affecting the peripheral nerves, usually in their distal portions. The commonest cause is dietary deficiency, primarily of vitamin B_1 (thiamine), which is usually associated with chronic alcoholism; other causes are various toxins, either chemical agents such as arsenical drugs, lead, nitrobenzol, coal-tar products, or toxins produced by bacteria.

The clinical picture is similar no matter what the cause: there is a gradual onset of numbness and tingling of the hands and feet, weakness of the limbs, altered sensation to touch, and pain, loss of deep reflexes, and eventually paralysis and atrophy of the extremities. Treatment is directed first toward removal of the cause.

NEUROFIBROMATOSIS. A condition characterized by multiple tumors in the skin or along the course of peripheral nerves. These tumors occasionally become malignant. Also called von Recklinghausen's disease.

NEUROLEPTIC MALIGNANT SYNDROME. First described and named by Delay and Deniker in 1968, neuroleptic malignant syndrome (NMS) was thought to be a variant of drug fever in which hyperpyrexia and autonomic and other neurological abnormalities developed during phenothiazine therapy. The principal characteristics of NMS are hyperthermia, hypertonicity of skeletal muscles, and fluctuating consciousness, along with instability of the autonomic nervous system. As described by Guzé and Baxter (see reference), common autonomic dysfunctions include pallor, diaphoresis, blood-pressure instability, tachycardia, and cardiac dysrhythmias. The muscular hypertonia consists of a generalized "lead-pipe" increase in tone, which increased muscle tone may result in decreased chest-wall compliance and, as a consequence, breathing problems severe enough to require respiratory support. Disturbances of consciousness may range from agitation or alert mutism to stupor and coma. Estimates of mortality range from 20 to 30%, this depending upon the type of neuroleptic drugs responsible and upon duration of exposure to offending drug. Deaths usually occur between 3 and 30 days after presentation of symptoms. Death may result from respiratory failure, cardiovascular collapse, renal failure, arrhythmias, and thrombo-embolism. Respiratory failure may result from aspiration pneumonia or tachypneic hypoventilation. A primary factor in the causation of NMS may be a decrease in the availability of dopamine in the brain.

Neuroleptic drugs, also referred to as antipsychotic agents or major tranquilizers have been commonly used in the treatment of psychotic illnesses and also used in general medicine as antiemetics; in connection with dissociative anesthesia; and for the treatment of diseases, such as Tourette's syndrome. Although neuroleptic agents have been known for side-effects, NMS, frequently fatal, has not been well recognized.

In 1985, the seriousness of the side effects of neuroleptic or antipsychotic drugs, such as phenothiazines, butyrophenones, and thioxanthenes, among others, prompted the American Psychiatric Association to issue a warning to psychiatrists and physicians, advising them to carefully weigh the benefits of the drugs against the potential for developing *tardive dyskinesia* (or NMS).

Additional Reading

Caroff, S.N.: "The Neuroleptic Malignant Syndrome," *J. Clin. Psychiatry*, **41**, 79–83 (1980).

Culliton, B.J.: "Antipsychotic Drugs (and Tardive Dyskinesia)," *Science*, **229**, 1248 (1985).

Diamond, J.M. and A.B. Santos: "Unusual Complications of Antipsychotic Drugs," *Am. Fam. Physician*, **26**, 153–157 (1982).

Guzé, B.H. and L.R. Baxter, Jr.: "Neuroleptic Malignant Syndrome," *N. Eng. J. Med.*, **313**(3), 163–165 (July 18, 1985).

Hopkins, P.M. and F.R. Ellis: "Hyperthermic and Hypermetabolic Disorders: Exertional Heat Stroke, Malignant Hyperthermia and Related Syndromes," Cambridge University Press, New York, NY, 1996.

Tollefson, G.I.: "The Neuroleptic Syndrome and Central Dopamine Metabolites," *J. Clin. Psychopharmacology*, **4**, 150–153 (1984).

Yassa, R., D.V. Jeste, and N.P.V. Nair: "Neuroleptic-Induced Movement Disorders," Cambridge University Press, New York, NY, 1996.

NEURON. See **Central and Peripheral Nervous Systems**.

NEUROPTERA. Insects of varied form and habits, including the dobson fly, the golden eyes or lacewings, the alder flies, and the ant lions. The order is characterized by the four membranous wings, usually with a large number of branching veins that are united in some species to form a network. The mouth is formed for biting but in some larvae the mandibles and maxillae fit together to form a piercing and sucking organ. See Fig. 1. The insects are predacious.

Neuroptera are of minor economic importance as food for fishes. The larvae of lacewings prey on aphids, hence they are of some assistance in holding these pests in check.

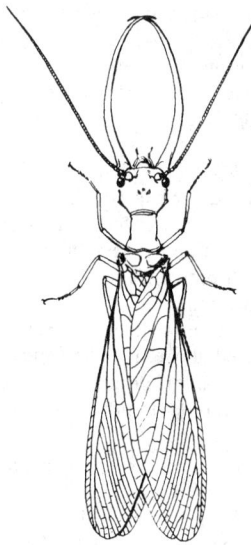

Fig. 1. Dobson fly, valued as a food for fishes. (*USDA*.)

The order contains about 1700 species. The main families of Neuroptera include:

Chrysopidae	Lacewings, aphid-lions, golden-eyed flies
Hemerobiidae	Brown lacewings
Mantispidae	Mantid flies
Myrmeleonidae	Ant lions or doodle bugs
Rhaphidiidae	Snake flies
Sialidae	Dobson flies, alder flies, fish flies

NEUROSYPHILIS. See **Syphilis**.

NEUROTENSION. See **Central and Peripheral Nervous Systems**.

NEUTRAL. 1. Having no electric charge, or no net electric charge. (Thus, an atom, in which the total negative charge of the electrons is equal to the positive charge of the nucleus which they surround, is neutral.)

2. According to the ionization hypothesis, a concentration of hydrogen ions equal to 1×10^{-7}. (The figure varies a little according to the temperature and the method of determining the degree of ionization of water.) Hydrogen-ion concentrations greater than this figure confer acid properties; lower concentrations occur in alkaline systems.

NEUTRALIZATION. See **Acids and Bases**.

NEUTRINO. A neutral particle of very small (presumed zero) rest mass and of spin quantum number $\frac{1}{2}$. This particle was initially postulated to account for the continuous energy distribution of beta particles and to conserve angular momentum in the beta-decay process. Experimental evidence indicates that, for the linear momentum to be conserved in the beta process, there must be a contribution from a departing neutrino. Presumably, a neutrino (or antineutrino) is emitted in every beta transition. The energy of a neutrino emitted in a beta disintegration is assumed equal to the difference between the energy of the particular beta particle and the energy corresponding to the upper limit of the continuous spectrum for that beta transition. The neutrino has also been postulated as one of the particles in pion (π) decay and as two of the particles in muon (μ) decay. These processes, however, lead to two types of neutrinos: an electron-associated neutrino ν_e and a muon-associated neutrino ν_μ. For example, $\pi^+ \rightarrow \mu^+ \rightarrow \bar{\nu}_\mu + e^+ + \nu_e$ or $\pi^+ \rightarrow e^+ + \nu_e$, whereas neutron decay obeys only the process $n \rightarrow p^+ + e^- + \bar{\nu}_e$, where the bar over ν indicates an anti-particle. The difference between neutrinos was established at Brookhaven in 1962 when it was shown that a beam of neutrinos from the process $\pi^+ \rightarrow \mu^+ + \nu_\mu$ gave rise to the process $\nu_\mu + n \rightarrow p + \mu^-$ but not to $\nu_\mu + n \rightarrow p + e^-$. There is also a neutrino associated with the tau particle. Because of its properties, the neutrino has negligible interactions with matter and has proved difficult to detect. It was first positively identified experimentally in 1956 by Reines and Cowan, Jr. See also **Particles (Subatomic)**.

The term antineutrino usually denotes an antiparticle whose emission is postulated to accompany radioactive decay by negatron emission, such as, for example, in neutron decay into a proton p^+, negatron e^- and antineutrino $\bar{\nu}_e$, expressed by the equation $n \rightarrow p^+ + e^- + \bar{\nu}_e$. Capture of a neutrino by the neutron, $\nu_e + n \rightarrow p^+ + e^-$ would be an equally good description of the process. Positron emission is accompanied by a neutrino, as in the decay $^{64}Cu \rightarrow {}^{64}Ni + e^+ + \nu_e$. Orbital electron capture also involves a neutrino, as for example, $e^- + {}^{64}Cu \rightarrow {}^{64}Ni + \nu_e$. Since there is no possibility of charge differentiation between the antineutrino and the neutrino, differentiation between these two particles can be made only on the basis of such properties as the sign of the ratio of magnetic moment to angular momentum.

In the past the terms neutrino and antineutrino were sometimes used in reverse sense to that stated above, i.e., the neutrino is said to accompany negatron emission and the antineutrino, positron emission. The preferred usage has been accepted in order to provide conservation of leptons in the conservation laws.

Neutrino Astronomy. For several years, there has been a marked trend in astronomy to expand the use of the electromagnetic spectrum beyond the visual range in investigating the universe. It now appears that, in addition to infrared, ultraviolet, gamma ray, x-ray, and radio astronomy, among others, increasing attention will be given to neutrino astronomy, i.e., the detection and measurement of neutrinos emanating from such celestial bodies as supernovas and x-ray double stars. Traditionally, neutrinos have been created in large accelerators and used for investigating the characteristics of other elementary particles. It is now reasoned that neutrinos entering the earth's atmosphere from celestial distances may furnish new, heretofore unavailable information.

Search for Neutrinos

Certain mysteries continue to surround the neutrino, emphasizing how far physicists still may be from a full comprehension of the complete array of subatomic particles. Numerous detectors in different locations have been constructed for detecting neutrinos, particularly as they emanate from the sun.

The first recorded attempt to construct a neutrino telescope was undertaken by Davis (Brookhaven National Laboratory) in the 1960s. Davis' principal objective was detection of low-energy neutrinos emitted by the sun. These neutrinos are generated deep within the sun as the result of thermonuclear reactions. It has been estimated that nearly 10% of the energy released by the transmutation of hydrogen in the sun is carried away by neutrinos, which have energies ranging from $\frac{1}{2}$ million eV to about 14 million eV. The solar flux of neutrinos is tremendous. It is estimated that 10^{14} solar neutrinos pass through the human body every second. Davis' detector consisted of a large tank containing 610 tons of tetrachloroethylene, C_2Cl_4. Of the chlorine atoms present, 25% are of the isotope chlorine 37. When this atom captures a neutrino, it is transformed to an atom of argon 37.

An early detector was established in the shaft of the Homestake Mine (South Dakota) in 1968. The data gathered did not correspond with the scientists' expectation. The neutron flux measured was less than one-third that predicted. This discrepancy challenged researchers to question perhaps the "established" concepts of particle physics and indeed the manner in which the sun functions.

Some years later, a Japanese-built detector (*Kamiokande II*), designed to detect the more energetic solar-emitted neutrinos (~5 mil electron volts), came up about 50% short of the expected counts, thus reconfirming a shortage of solar neutrons.

Neither of the aforementioned detectors was designed to sense comparatively low-energy (proton-proton) neutrinos, which result from the fusion of two protons.

Subsequently, two additional detectors were built with the objective of seeking lower-energy neutrinos. One of the detectors was located in the Caucasus of the former Soviet Bloc and was named the Soviet-American Gallium Experiment (SAGE). The detector used a gallium metal detector believed to be sensitive to neutrinos, with energy as low as 0.23 mil electron volts. Another similar detector (*Gallox*) was constructed in Italy. At both sites, difficulties with calibration were expected and did occur. Some tenuous observations later were made to suggest a correlation of neutrinos detected with the occurrence of sun spots and with solar acoustic oscillations.

Based upon the foregoing experiences, some researchers observed that the same reluctance to interact with matter is responsible for the neutrino's

long range and ability to resist detection. Thus, it was reasoned that an apparatus for detecting neutrinos should be massive and shielded from the interference of other particles and radiation. As a solution to these problems, some researchers proposed a deep underwater muon and neutrino detector (acronym DUMAND).

Fiber optic data cable will stretch from the shore for 30 km to a connector box some 4800 meters below the ocean surface. Strings of nine separate cables will rise vertically about 280 meters above the ocean floor. Each cable, held up by a float, will contain 24 detectors. The apparatus, referred to as a neutrino telescope, will have to pick up at least ten muon events per year from any given $1°$ patch of sky for the DUMAND scientists to be confident that they have a significant neutrino source, not just a few background pulses from non-neutrino cosmic rays.

Construction of DUMAND, with Japanese and German collaborators, got underway off the west coast of Kona (Hawaii) in 1990 and was expected to be operational within a few years. V. Stenger (University of Hawaii) observes, "The idea of using the ocean floor as a detector actually goes back to the 1960s and people started taking it seriously back in the mid 1970s."

A number of scientists are hopeful that the neutrino telescope will ultimately yield much additional information on neutron stars, supernovas, quasars, etc., which are presumed to be large emitters of neutrinos. Some astronomers estimate that SS 433 puts out 1000 times more energy per second than the brightest stellar object known in the galaxy. Why should such a powerful object have been discovered so recently? The fact is that SS 433 is a comparatively weak source of photons. The accreting matter that gives SS 433 its great power also serves to screen its bright central region from view. Much remains to be learned concerning SS 433 and perhaps neutrino astronomy may supply some answers at a future date.

Laboratory Experimentation on Neutrinos

During the 1960s, L.M. Lederman, M. Schwartz, and J. Steinberger conducted the well-known two-neutrino experiment, which established a relationship between particles, muon and muon neutrinos, electron and electron neutrino. This later evolved into the standard model of particle physics. The Nobel prize in physics was shared by these researchers in 1988.

Massive Neutrino Proposed. J. Simpson (University of Guelph, Ontario) in the late 1980s presented evidence of the existence of a *heavy neutrino* having a mass of 17,000 electron volts (keV, the units of energy that are interchangeable with mass). A renowned neutrino physicist for several years prior, in 1985 Simpson conducted a series of experiments in his laboratory. The objective of his initial experiment was that of measuring the energy of electrons emitted from heavy hydrogen (tritium) in the radioactive process of beta decay. In this process, the energy of the emitted electrons should appear as a spectrum (smooth curve) from zero to a maximum endpoint. However, Simpson noted "occasional" aberrations in the plotted data. This so-called "kink" corresponded with an energy of 17 KeV of the normal plot. This indicated the probable presence of an unknown massive force. In addition to repeated experiments with tritium, Simpson also used Sulfur-35. Simpson then suggested that the aberration, when it occurred, could be attributed to a heavy 17 KeV neutrino.

Simpson's carefully prepared data, when published, attracted wide attention and drew a wide spectrum of professional reactions. Meanwhile, experiments by other laboratories have confirmed Simpson's results. Researchers at the Lawrence Berkeley Laboratory, using Carbon-14, have reported evidence for a 17.2 KeV neutrino in 1.4% of their experiments. Researchers at the Ruder Boskovic Institute in Zagreb (formerly Czechoslovakia), using Iron-55 and Germanium-71, also have found the evidence appearing in 1.5% of their experiments. Thus, confidence in Simpson's claim appeared to be gaining momentum in the early 1990s.

As pointed out by M. Turner (University of Chicago), a massive neutrino would "violate every theoretical prejudice we have in particle physics, astrophysics, and cosmology." J. Bahcall (Institute of Advanced Study at Princeton) observes, "It's a surprise. If it's true, then it's pointing us in a different direction than previous physics suggested."

See also **Particles (Subatomic)**.

Additional Reading

Bahcall, J.N.: "Neutrino Astrophysics," Cambridge University Press, New York, NY, 1990.

Breuker, H. et al.: "Tracking and Imaging Elementary Particles," *Sci. Amer.*, 58 (August 1991).

Brown, L.M., M. Dresden, and L. Hoddeson: "Pions to Quarks: Particle Physics in the 1950s," Cambridge University Press, New York, NY, 1989.

Cence, R.J. et al.: "Neutrino 81," University of Hawaii, Honolulu, HI, 1981.

Dar, A.: "Astrophysics and Cosmology Closing in on Neutrino Masses," *Science*, 1529 (December 14, 1990).

Dehmelt, H.: "Experiments on the Structure of an Individual Elementary Particle," *Science*, 539 (February 2, 1990).

Florini, E.: "Neutrino Physics and Astrophysics," Plenum, New York, NY, 1982.

Gutbrod, H. and H. Stöcker: "The Nuclear Equation of State," *Sci. Amer.*, 58 (November 1991).

Horgan, J.: "Three Americans Honored for 1960's Neutrino Experiment," *Sci. Amer.*, 31 (December 1988).

Kim, C.W. and A. Pevsn: "Neutrinos in Physics and Astrophysics," Gordon & Breach Publishing Group, Newark, NJ, 1993.

Lederman, L.M.: "Observations in Particle Physics from Two Neutrinos to the Standard Model," *Science*, 664 (May 12, 1989).

Mohapatra, R.N. and P.B. Pal: "Massive Neutrinos in Physics and Astrophysics,"World Scientific Publishing Company, Inc., River Edge, NJ, 1997.

Powell, C.S.: "Looking for Nothing: The Taciturn Neutrino Keeps Physicists Guessing," *Sci. Amer.*, 22 (April 1991).

Schwartz, M.: "The First High-Energy Neutrino Experiment," *Science*, 1445 (March 17, 1989).

Selvin, P.: "Is There a Massive Neutrino?" *Science*, 1426 (March 22, 1991).

Stenger, V.J. and J.G. Learned: "High Energy Neutrino Astrophysics: Proceedings of the Workshop," World Scientific Publishing Company, Inc., Riveredge, NJ, 1992.

Traweek, S.: "Beamtimes and Lifetimes. The World of High Energy Physics," Harvard University Press, Cambridge, MA, 1992.

Waldrop, M.M.: "A Nobel Prize for the Two-Neutrino Experiment," *Science*, 669 (November 4, 1988).

Waldrop, M.M.: "Solar Neutrino Deficit Confirmed?" *Science*, 1607 (June 29, 1990).

Waldrop, M.M.: "Astrophysics in the Abyss," *Science*, 208 (October 12, 1990).

NEUTRON. The discovery of the neutron by Chadwick in 1932 represented a great step forward in the investigation of nuclei of atoms. Chadwick found that a radiation emitted when α-rays from polonium reacted with beryllium could project protons from a thin sheet of paraffin wax. Although the radiation itself produced no observable ionization when passing through a gas, the protons released from the paraffin were detected in an ionization chamber. Inability to produce ionization was interpreted as a lack of electric charge. From measurements of the ionization from the protons, Chadwick deduced that the so-called beryllium radiation must consist of neutral particles with a mass very nearly equal to that of the proton. He announced the discovery of the neutron, a previously unknown particle. It has been confirmed that the neutron has no charge and a mass of 1.088665 atomic mass units. Thus, it is heavier than the proton by 0.00139 mass unit. The introduction of the neutron into nuclear structure produced a sharp change in previously held concepts. Lacking knowledge of the neutron, masses of atomic nuclei had been attributed solely to protons. The number of protons required on this basis for most nuclei greatly exceeded the known charge number. In an attempt to solve this dilemma, a number of electrons were assigned to each nucleus to adjust the charge number to the proper value. This compromise created an even greater problem, that of accommodating so many electrons in the small space occupied by a nucleus. Bringing the neutron into the picture meant that a nucleus contains only enough protons to equal the charge number, with the rest of the mass contributed by neutrons. No additional electrons were required.

Decay. The neutron in the free state undergoes radioactive decay. Elaborate experiments by Robson were required to identify the products of the decay and to measure the half-life of the neutron. He showed that the neutron emits a β-particle and becomes a proton. The half-life was found to be 12.8 minutes. In stable nuclei, neutrons are stable. In radioactive nuclei, decaying by β-emission, the neutrons decay with a half-life characteristic of the nuclei of which they are a part. See also **Radioactivity**.

Detection. Because it is a neutral particle the neutron is detected by means of a secondary charged particle which it releases in passing through matter or by means of the radioactivity which the neutron can induce in stable elements. Protons may be projected by collisions with neutrons in hydrogenous material and the ionization from the protons can be measured in an ionization chamber, as in the original experiment with neutrons. Secondary charged particles may be the direct result of nuclear disintegration produced by neutrons, as in the case of the reaction $^{10}B + ^{1}n \rightarrow ^{7}Li + \alpha$. Commonly, the radioactivity induced in a stable element by neutron capture serves to detect neutrons, and this technique is known as the *activated foil method.* Also, fission may be utilized for detection of neutrons by placing fissionable material inside an ionization chamber and observing the ionization generated by the fission fragments.

Energies. The kinetic energy of neutrons has an important bearing on their behavior when interacting with nuclei. These kinetic energies may range from near zero to as much as 50 MeV. It is therefore natural to classify neutrons in terms of energy according to their properties in each range of energy. For example, energies from zero to about 1000 eV are usually called *slow neutrons.* Because they are more readily captured by nuclei than faster neutrons, slow neutrons are responsible for a large number of nuclear transformations. When slow neutrons have velocities in equilibrium with the velocities of thermal agitation of the molecules of the medium in which they are situated, they are called *thermal neutrons.* The distribution of these velocities approaches the Maxwell distribution

$$dn\ (v) = Av^2 e^{-(Mv^2/2kT)}\, dv$$

where v is the neutron velocity, M its mass, k is Boltzmann's constant, and T is the absolute temperature. In the slow neutron range of energies, various atomic nuclei show strong absorption (capture) of neutrons at fairly well-defined energies. Neutrons having energies corresponding to those of the absorption bands are called *resonance neutrons.* Frequently, neutrons with energies greater than 1000 eV and less than 0.5 MeV are termed *intermediate neutrons.* In more general terms, all neutrons with energies greater than 0.5 MeV are called *fast neutrons.* The practical upper limit of neutron energy is set by the devices thus far developed for accelerating charged particles to extremely high energies.

Magnetic Moment and Spin. Alvarez and Bloch succeeded in measuring the moment of the magnetic dipole associated with the known spin of $\frac{1}{2}$ possessed by the neutron. More refined measurements by Cohen, Corngold, and Ramsey of the magnetic moment μn yielded a value of

$$\mu n = 1.913148 \text{ nuclear magnetons}$$

Interactions with Nuclei. Neutrons may be scattered or captured by heavy nuclei. Scattering may be elastic, resulting only in the change of direction of the neutrons, or inelastic, in which the neutron loses part of its energy to the scattering nucleus. Collisions with light nuclei, in the absence of capture, result in communicating considerable fractions of the neutron energy to the target nucleus. A neutron colliding head-on with a proton will give practically all its kinetic energy to the proton. As the mass of the target nucleus increases, the transfer of energy decreases, in accordance with the laws of conservation of energy and momentum. The loss of energy by mechanical impact is utilized in slowing down fast neutrons, a process known as *moderation.* Slow neutrons are most useful, for example, in the production of radioelements from stable elements by neutron capture. A good moderator should have low mass and a small capture cross section. The rate r of capture of neutrons from a neutron flux F (neutrons $\text{cm}^{-2}\ \text{sec}^{-1}$) incident on a layer of matter having N nuclei per square centimeter is given by

$$r = F\sigma N$$

where σ is the complete probability of capture. Replacing r by dN/dt and writing the flux as nv, where n is the number and v is the velocity of the neutrons, we have

$$dN/dt = nv\sigma N$$

which integrated gives

$$N = N_0 e^{-nv\grave{e}t}$$

where N is the number of unchanged nuclei in the target area at time t and N_0 is the number at time $t = 0$. The cross section σ is so named because it has the dimensions of an area. The unit for the cross section is the *barn*, equal to $10^{24}\ \text{cm}^2$. When, as is often the case, σ is proportional to $1/v$, the advantage of slow neutrons in capture interactions becomes apparent. When the value of s departs sharply from that predicted by the $1/v$ law, it usually increases over a narrow range of energies, and we have what is called a *resonance.* Slow neutron cross sections are customarily quoted for thermal neutrons at 20 °C, corresponding to a value of v of 2200 m/sec. Under these conditions, representative thermal neutron capture cross sections are: boron, 759 barns; cobalt, 38 barns; cadmium, 2450 barns; gadolinium, 46,000 barns; gold, 99.8 barns; helium, 0; lead, 0.170 barn; and oxygen <0.0002 barn.

Additional interactions of neutrons with nuclei include the release of charged particles by neutron-induced nuclear disintegration. Commonly known reactions are n-p, n-d, and n-α. In these cases, the incident neutrons may contribute part of their kinetic energy to the target nucleus to effect the disintegration. Hence, more than mere neutron capture is involved. Then, there is usually a lower threshold for the neutron energy below which the reaction fails to occur. Another important reaction involving neutrons is fission, which may occur under different conditions for either slow or fast neutrons with appropriate fissionable material.

Sources of Neutrons. Any nuclear reaction in which neutrons are released might serve as a source of neutrons. In the initial experiments with neutrons, an α-n reaction was used. Because of the charge on the α-particle, it must have a high kinetic energy to penetrate a nucleus. Thus, polonium α-particles could release neutrons from beryllium. Such a natural source produces relatively few neutrons. The yield of neutrons from charged particle reactions can be increased manyfold by the use of particle accelerators. Here large numbers of charged particles of high energy can be used in the bombardment of the target to release numerous neutrons. Frequently deuterons or protons are used for the bombardment. A far more prolific source is the nuclear reactor. Fission of uranium is usually the source of the neutrons in this case. A nuclear reactor as usually constructed generates neutrons of different energies in various parts of its structure. Neutrons of suitable energy for a given experiment may be brought outside the reactor through channels into appropriate sections of the reactor. See also **Nuclear Power Technology**.

Traditionally, the neutron is regarded as a particle which is a component of nuclei and which exists only briefly in the free state. For many purposes, this view is sufficient. However, it became obvious some years ago from various experiments, for example, in very high energy accelerators, that the neutron must have a complex structure. This view was reinforced by the nature of the decay of the neutron. A ς particle is ejected from the neutron on decay, but it is quite certain that the electron did not exist within the neutron prior to the decay. Rearrangements of an internal structure of the neutron must provide the energy for the formation and ejection of the ς particle. An early theory would have the neutron consist of a proton and a π^- meson bound together so that they oscillate between a completely bound state and a more loosely bound state. This concept might explain the feeble interaction that has also been observed between electrons and neutrons at very short range.

Fermi Age Model. This is a model for the study of the slowing down of neutrons by elastic collisions. It is assumed that the slowing down takes place by a very large number of very small energy changes. Phenomena due to the finite size of the individual losses are ignored. In this model, the word age is somewhat of a misnomer, since its units are those of area rather than time. The name arises because the variable τ, the Fermi age, appears in the Fermi age equation in the same way that time appears in the standard heat-diffusion equation. The equation for the Fermi age in a unit volume of nuclear reactor is $\tau = D\phi/q$, in which D is the diffusion coefficient for fast neutrons, $\phi = nv$ is the fast neutron fluence, and q is the number of neutrons thermalized per second per cubic centimeter. For this purpose fast neutrons are inclusively all neutrons with energies between those acquired at fission and that energy at which they are thermalized.

Ultracold Neutrons. As pointed out by King (Massachusetts Institute of Technology), there is probably as much to be learned between the lowest energy yet reached and zero energy as there is between the highest energy attained and infinite energy. Ultracold neutrons may provide an avenue to very-low-energy research. It should be recalled that a neutron with an energy of 10^{-7} electron volt is at the low end of the energy scale. A neutron has the energy that would be imparted to an electron by a potential difference of one ten-millionth of a volt (0.1 microvolt). As pointed out by Golub et al. in 1979, this is the amount of energy of a particle in a gas whose temperature is one millidegree K. Unlike high-energy particles, ultracold neutrons move at a rate measured in a few meters per second. Golub et al. have proposed that inasmuch as ultracold neutrons cannot penetrate a solid surface, they can be confined in a metal bottle and by storing over long periods, it may be possible to measure the fundamental properties of the neutron.

Outside the nucleus, the neutron is an unstable particle. Free neutrons are rare in nature. By the process of beta decay, the neutron breaks down into a proton, an electron, and a neutrino (a massive particle). For probing atomic and molecular structures, thermal neutrons have traditionally been used. As explained by Golub et al., ultracold neutrons can be employed in a similar way, but their low energy and long wavelength adapt them to the examination of materials on a somewhat larger scale. At the

Technical University (Munich), Steyeri and associates have developed a neutron spectrometer. In conventional neutron spectrometers, particles are analyzed by a magnet that bends their trajectories. In an ultracold-neutron spectrometer, the earth's gravitational field is used. In the device, the neutrons enter the spectrometer in a horizontal movement and are accelerated as they fall a fixed distance to a specimen. Those neutrons that rebound from a target are collected by an exit slit of the instrument. An exchange of energy with the specimen is reflected by the maximum height to which the neutrons rebound.

During the past decade, since their detection, much research has been directed toward methods of extracting, storing, and manipulating ultra-cold neutrons. It now appears that the next period will be one of investigating the neutron per se and possibly of using this new knowledge for the study of other systems of particles.

See **Particles (Subatomic)** for more recent views. See also **Proton**.

Glossary of Neutron Terms

Neutrons are designated according to their energies, including the following:

Thermal neutrons, or neutrons in thermal equilibrium with the substance in which they exist; most commonly, neutrons of kinetic energy about 0.025 eV, which is about $\frac{2}{3}$ of the mean kinetic energy of a molecule at $15\,°C$.

Epithermal neutrons, or neutrons having energies just above those of thermal neutrons; the epithermal neutrons energy range is between a few hundredths eV and about 100 eV.

Slow neutrons (a less definite classification), which may mean either neutrons having energies up to about 100 eV, or thermal neutrons.

Intermediate neutrons, which are neutrons having energies in a range that extends roughly from 100 to 100,000 eV. This range is above that of epithermal neutrons and below that of fast neutrons.

Fast neutrons, which are neutrons with energies exceeding 10^5 eV, although sometimes a lower limit is given.

Resonance neutrons may be either of the following: (1) for a specified nuclide or element, neutrons that have energies in the region where the cross section of the nuclide or element is particularly large because of the occurrence of a resonance. For example, cadmium resonance neutrons have energies between 0.05 and about 0.3 eV. (2) Neutrons having kinetic energies in the region of values for which prominent resonances are encountered in many nuclides; loosely, epithermal neutrons.

Prompt neutrons are those neutrons released coincident with a fission process.

Delayed neutrons are those neutrons released subsequently in a fission process, or, more generally, neutrons emitted by excited nuclei formed in any radioactive process (beta disintegration, in all cases so far known). The neutron emission itself is prompt, so that the observed half life is that of the preceding beta emitter. The situation is similar to that involving gamma-ray emission, which is a competing process. Delayed neutron emission is possible only if the excitation energy of the product nucleus exceeds the neutron binding energy for that nucleus. The chemistry of the delayed neutron emitter is that of the beta activity; thus ^{87}Br, ^{137}I, and ^{17}N are delayed neutron precursors, although the neutron emission actually takes place from excited nuclei of the products ^{87}Kr, ^{137}Xe, and ^{17}O.

Neutron cycle is the average life history of a neutron in a nuclear reactor. The gain in the number of neutrons in a reactor during any individual neutron cycle is given by $n(k-1)$, where n is the number of neutrons in the reactor of the beginning of the cycle and k is the multiplication factor.

Neutron excess is the difference between the number of neutrons and the number of protons in an atomic nucleus. This is found by subtracting the atomic number of that nuclide from the neutron number; or by subtracting twice the atomic number from the mass number.

Neutron flux density is the number of neutrons that enter a sphere of unit cross-sectional area per unit of time. This quantity is sometimes defined in terms of a unidirectional beam of neutrons incident perpendicularly upon a unit area, but this definition is less general. It is also sometimes called neutron current density.

Neutron number is the number of neutrons in a nucleus. Its symbol is N. The neutron number for a given nuclide is equal to the difference between the mass number and the atomic number for that nuclide.

Additional Reading

Breuker, H. et al.: "Tracking and Imaging Elementary Particles," *Sci. Amer.*, 58 (August 1991).

Brown, A.: "The Neutron and the Bomb: A Biography of Sir James Chadwick," Oxford University Press, Inc., 1997.

Gutbrod, H. and H. Stöcker: "The Nuclear Equation of State," *Sci. Amer.*, 58 (November 1991).

Hamilton, J.H.: "Fission and Properties of Neutron-Rich Nuclei," World Scientific Publishing Company, Inc., River Edge, NJ, 1998.

Harding, A.K.: "Physics in Strong Magnetic Fields Near Neutron Stars," *Science*, 1033 (March 1, 1991).

Ruthen, R.: "Out of Its Field: How the Neutron Responds to A Field That Exerts No Force," *Sci. Amer.*, 26 (October 1989).

Schopper, H.: "Low Energy Neutrons and Their Interaction with Nuclei and Matter: Low Energy Neutron Physics," Springer-Verlag, Inc., New York, NY, 2000.

Shepherd, G.M.: "Foundations of the Neutron Doctrine," Oxford University Press, Inc., New York, NY, 1998.

Soloway, A.H., R.F. Barth, and D.E. Carpenter: "Advances in Neutron Capture Therapy," Kluwer Academic Publishers, Norwell, MA, 1993.

Traweak, S.: "Beamtimes and Lifetimes: The World of High Energy Physics," Harvard University Press, Cambridge, MA, 1992.

NEUTRON ACTIVATION ANALYSIS. This is a method of elemental analysis based upon the quantitative detection of radioactive species produced in samples via nuclear reactions resulting from neutron bombardment of samples. The neutron-induced reactions are of two main types: (1) those induced by very slow (thermal) neutrons, having energies of about 0.025 eV; and (2) those induced by fast neutrons having energies in the range of MeV. The method is used in two different forms. The purely instrumental form is fast and nondestructive and is based upon the quantitative detection of induced gamma-ray emitters by means of multichannel gamma-ray spectrometry. The amount of the element present is usually computed from the photopeak (total absorption peak) height or area of its gamma ray, or one of its principal gamma rays, compared with that of the standard. Where interferences from other induced activities are serious and cannot be removed by decay, spectrum subtraction, or computer solution, one must turn to the radiochemical separation method. Here the activated sample is put into solution and equilibrated chemically with measured amounts (typically 10 milligrams) of added carrier of each of the elements of interest, before chemical separations are carried out. The element to be detected needs then to be recovered in chemically and radiochemically pure form, but it need not be quantitatively recovered, since the carrier recovery is measured and the counting data are then normalized to 100% recovery. This form is slower, but it applies to pure beta emitters, as well as to gamma emitters, and it does eliminate interfering activities.

NEUTRON INTERFEROMETER.   See **Gravitation**.

NEUTRON STARS.   One of the end points in stellar evolution, in which a star of 1.4 to approximately 3 times the mass of the sun is supported by internal pressure generated by compressed "degenerate" neutrons.

Normal stars result from a balance of forces: the force of gravity pulling inward and a force generated by the energy of the nuclear reactions inside pushing outward. When the star exhausts its core hydrogen, it leaves the "main sequence" of normal stars and contracts under the force of gravity. The increased fusion that follows provides energy for the outer layers to swell, and the star becomes a "red giant." Later, after the helium is exhausted, stars still contain more than about 4 solar masses expand again to become "red supergiants."

When the supergiants have built all their nuclear fuel into iron, they explode as *supernovae* (q.v.). Those that have more than about 1.44 solar masses left after this stage are not stopped by the force of electron degeneracy that supports *white dwarfs* (q.v.). 1.44 solar masses is a theoretically derived value known as the "Chandrasekhar limit," after S. Chandrasekhar of the University of Chicago. Those stars that have more than 1.44 but less than about 3 solar masses (with the value of this upper limit being less accurately known than the value of the lower limit) will contract until the neutrons in the star resist being pushed together any further. This generates a pressure known as "neutron degeneracy pressure." The resulting object is a *neutron star*.

This situation was predicted by J. Robert Oppenheimer and colleagues in the 1930s. It comes from the application of quantum mechanical rules, including the Pauli exclusion principle and the Heisenberg uncertainty principle, to the neutrons. A neutron star contains approximately 2 solar masses in a volume only 20 km or so across, giving a density of billions of tons per cubic centimeter.

Neutron stars may have solid, crystalline crusts about 100 m thick. The neutron stars' atmospheres may be only a few centimeters thick. The concentration of a normal stellar magnetic field by gravitational collapse would correspond to a neutron star magnetic field of billions of gauss.

Though no neutron stars were known until 1968, neutron stars are now detected in various ways. Most of the known neutron stars are pulsars (q.v.), of which over 400 are now known. In these cases, we detect a beam of radio radiation, possibly emitted along the magnetic axis of the star, as it sweeps across space. Periods of rotation of the neutron stars in pulsars range from 1.4 ms to about 3 s. A very few of these pulsars have been detected in the optical or x-ray part of the spectrum. The best-studied example is the pulsar in the Crab Nebula, the remnant of the supernova explosion of 1054. This object is detected across the entire spectrum. In addition to its radio pulsations, its pulsations have been recorded in the x-ray region with the Einstein Observatory spacecraft and in the optical region with telescopes on earth. Another unusual pulsar, in a binary system, has been used as a test of the general theory of relativity. See also **Cosmology**.

Other neutron stars appear in binary x-ray sources. Hundreds were mapped by NASA's High Energy Astronomy Observatory 1 in the late 1970s, and many were then studied individually by the Einstein Observatory, High Energy Astronomy Observatory 2. Perhaps the best studied had been discovered by an early x-ray spacecraft, UHURU. This source, Hercules X-1, shows rapid pulsations every 1.24 seconds, but the pulses turn on and off with a 1.7-day period. This longer period is presumably caused by eclipses by the primary object in the system, the variable star HZ Herculis, while the shorter period is presumably caused by rotation of the neutron star. A third period of about 35 days also exists in Hercules X-1.

The most unusual astronomical object now known is SS433, whose spectral lines vary with an 164-day period over a Doppler range corresponding to velocities of about 25% of the speed of light. SS433 is in our galaxy, and such a velocity in our galaxy is unprecedented. A 13-day periodicity has also been discovered in SS433, indicating that it is a binary system of that period. In leading models, an accretion disk of material has formed around the neutron star, and jets of gas are emitted above and below the plane of the disk, which is inclined to us. As the disk precesses, we alternately see each of the beams approaching and receding with the longer period of the system. It is the material in the beams that has the high velocities.

Discovery of neutron stars has spurred work in stellar evolution of binary systems. Each object in a binary system has a Roche lobe around it, within which the gravitational force of the object is strong enough to resist the tidal forces from outside objects. The Roche lobes define a figure-of-8 around the two members of a binary system. In some binaries, one or both objects swell to fill their Roche lobes, at which time material can flow from one object to the next. The exchange of mass can severely alter the evolution of the objects, and the relative masses of the objects can interchange. When one of the objects is a compact object like a neutron star, mass from the companion flowing through the neck of the figure-of-8 and falling on the neutron star can lead to a surface nuclear explosion that is a powerful emitter of x-rays.

Additional Reading

Buccheri, R., J. Van Paradijs, and M.A. Alpar: "The Many Faces of Neutron Stars," Kluwer Academic Publishers, Norwell, MA, 1998.

Kawaler, S.D., G. Srinivasan, I.D. Novikov, G. Meynet, and D. Schaerer: "Stellar Remnants," Springer-Verlag, Inc., New York, NY, 2000.

Meszaros, P.: "High-Energy Radiation from Magnetized Neutron Stars," University of Chicago Press, Chicago, IL, 1992.

Rothschild, R.E. and R.E. Lingenfelter: "High Velocity Neutron Stars and Gamma-Ray Bursts," Springer-Verlag, Inc., New York, NY, 1996.

JAY M. PASACHOFF, Williams College, Williamstown, MA.

NEUTROSPHERE.   The atmospheric shell from the Earth's surface upward in which the atmospheric constituents are for the most part not ionized, i.e., it is electrically neutral. The region of transition between the neutrosphere and the ionosphere is somewhere between 70 and 90 kilometers, depending on latitude and season. See also **Earth**.

NEWCASTLE DISEASE.   Also sometimes called *fowl pest, pseudo-fowl pest, Ranikhet disease,* and *avian pneumoencephalitis,* the disease is of viral origin. The disease was first noted in the Dutch East Indies. A series of outbreaks of the disease occurred in England near Newcastle-on-Tyne, from which the name of the disease was derived.

In his excellent reference, R.F. Gordon devotes several pages to the history and characteristics of Newcastle disease and summarizes the situation as of the mid- to late 1970s as follows: "By 1966, a fairly stable situation seemed to have developed in which fully virulent virus had become endemic in the tropics, milder disease in North America and western Europe and an intermediate form in Iran and Arab countries to the West. In 1968, an upsurge of the disease was reported in Iraq. Reports of disease, difficult to control then followed from Lebanon, Israel, Greece and in 1970 from England and Holland and in 1971 to other countries in western Europe. Workers have given the strain the designation Essex '70. In 1971 the United States reported cases of fully virulent Newcastle disease occurring along its southern border and designated such strains as velogenic viscotropic Newcastle disease (VVND). A comparison of the two types of isolate has been made. The Essex '70 type of virus has been observed to be highly pneumotropic and to spread rapidly while the VVND type of isolate seems on average, more lethal, more likely to give rise to visceral lesions and to spread more erratically. Although no distinction can accurately be made by laboratory tests, there is considerable epidemiological evidence to suggest their mode of spread is significantly different. By 1974, the world use of vaccine had increased greatly and in countries such as the UK where slaughter on infected farms was practiced until 1962, there had been heavy expenditure on control. Currently losses from the disease are not high in countries where extensive vaccination is practiced although virulent forms of the virus are now much more widespread than ever before."

NEWT *(Amphibia, Urodela)*.   Air-breathing salamanders that are at least partially aquatic in habits. The many species are found chiefly in the Northern hemisphere.

NEWTON-COTES FORMULA.   A method of numerical integration. Assume that the integral

$$\int_a^b f(x)\, dx$$

may be approximated by

$$\int_a^b \phi(x)\, dx = A_0 y_0 + A_1 y_1 + \cdots + A_n y_n$$

where the quantities A_i are independent of y_i. By proper choice of these quantities, which may be found by the method of undetermined coefficients, the numerical result may be made very close to the true value of the integral. Special cases of the formula are the trapezoidal rule, the Simpson rule, and the Weddle rule.

NEWTONIAN FLUID.   See **Colloid System; Fluid**.

NEWTONIAN LIQUIDS.   See **Viscosity**.

NEWTONIAN TELESCOPE.   A reflecting-type telescope having a 45° mirror located just inside the focus, so that the primary image is observed through a hole in the side of the tube. See also **Telescope (Astronomical-Optical)**.

NEWTON RINGS.   An interference phenomenon, easily observed by laying a slightly convex lens upon a flat glass plate. When the lens and plate are arranged so that monochromatic light is reflected at a suitable angle to the observer's eye, the point of contact is seen to be surrounded by a series of concentric, alternately bright and dark rings, which become closer together with increasing radius. The rings are due to the interference of light at the film of air between the glass surfaces, which film increases in thickness with increasing distance from the contact point. If the radius of curvature of the convex surface is R, and if, counting the central contact-spot as the zero ring, we number the rings in order, both bright and dark, from the center out, the radius of the Nth ring in monochromatic light of wavelength λ is approximately

$$a = \sqrt{\frac{NR\lambda}{2}}$$

With white light, the bright rings become colored spectra, the overlapping of which at larger values of N causes the system to become indistinct and disappear.

NEWTON'S FORMULA FOR INTERPOLATION.   Let a difference table be given with numerical values of $y_0, y_1, y_2, \ldots$; equally spaced values of the argument, $x_0, x_1, x_2, \ldots$; $h = (x_n - x_0)/n$ and the finite differences $\Delta^n y_k$. Then a value of y for $x = x_k + hu$, not contained in the table, may be found by Newton's formula for forward interpolation

$$y = y_k + u\Delta y_k + \binom{u}{2}\Delta^2 y_k + \cdots + \binom{u}{2}\Delta^n y_k$$

As its name implies, this equation is used for calculation near the beginning of a difference table. Near the end of such a table, Newton's formula for backward interpolation is appropriate

$$y = y_k - v\Delta y_{k-1} + \binom{v}{2}\Delta^2 y_{k-2} - \cdots \binom{v}{k}\Delta^k y_0$$

where $x = x_k - hv$. These equations are also known as the Gregory-Newton formula.

See also **Interpolation**.

NEWTON, SIR ISAAC (1642–1727).   Sir Isaac Newton was one of the greatest scientific thinkers of all ages. Newton's Laws of Dynamics, set forth as three "laws" of force and motion changed the way people viewed the world.

Newton was born in Woolsthrope, Lincolnshire, England. His birth was premature and he was not expected to live. In his early elementary school years, he seemed like an average student. It was not until he was in Trinity College at Cambridge University that Newton began to display his genius for mathematics. It was perhaps, however, the outbreak of the bubonic plague that gave Newton the time to do his greatest thinking. When Cambridge closed due to the plague, Newton went home and during 1665–1666 he was at the height of his creative power. While "thinking" he discovered the calculus, the law of gravity, his three laws of motion, and various properties of optics including measurable, mathematical patterns in the phenomenon of color. During this time, he also invented a reflecting telescope.

In 1684, astronomer Edmond Halley encouraged Newton to put his research on gravity into a book. Newton wrote *Principia*, a work that is now considered one of the greatest of all scientific books.

He was elected a Fellow of the Royal Society of London in 1671, and in 1703, Newton was elected president of the Royal Society of London. He was annually re-elected for the rest of his life. In 1704, his major work, *Opticks* was published and also works on the quadrature of curves and the classification of the cubic curves. In 1705, he was knighted in Cambridge by Queen Anne and was the first scientist with such an honor.

From about 1714 on, Newton became a highly esteemed philosopher and scientist throughout Europe. Today Newton's name is like an historic marker in man's history. Almost all modern physical science and technological advances since the seventeenth century can be traced back to the great thinking of Sir Isaac Newton. See also **Acceleration (Due to Gravity)**; **Calculus**; **Gravitation**; **Newton-Cotes Formula**; **Newton Rings**; **Newton's Formula for Interpolation**; **Newton's Laws of Dynamics**; **Relativity and Relativity Theory**; **Rheology**; and **Telescope (Astronomical-Optical)**.

J. M. I.

NEWTON'S LAWS OF DYNAMICS.   The classical or Newtonian dynamics rests upon certain propositions first enunciated in systematic form by Sir Isaac Newton, which he set forth as three "Laws" of force and motion.

The first Law states, in effect, that bodies of matter do not alter their motions in any way except as the result of forces applied to them. A body at rest remains at rest, or if in motion it continues to move in the same direction with the same speed, unless a force is impressed upon it. It is quite conceivable that Newton's interpretation of "force" was the primitive concept we all have, based on muscular effort, and that he regarded this statement as the expression of a natural law connecting force with motion. On the other hand, he may have recognized in this first Law, as we now do, an objective definition of force, namely, that which is capable of altering bodily motions in the face of an opposition called inertia whose nature is even now not fully understood.

The second Law is made up of two distinct parts. (1) When different forces are allowed to act upon free bodies, the rates at which the momentum changes are proportional to the forces applied. (2) The direction of the change in momentum caused by a force is that of the line of action of the force. Part 1 may now be regarded, from the more rigorous viewpoint of dimensional analysis, as a definition of the standard measure of force. Two forces are judged equal if they produce change of momentum at equal rates; one force is twice as great as another, if it changes the momentum at twice the rate; etc. Moreover, two bodies have equal mass if equal forces produce change of momentum at equal rates; and one body has twice the mass of another if an equal force changes its momentum at half the rate. This is the inertial concept of mass mentioned above. Part 2 emphasizes the vector character of force, and points out that no single force, acting alone, can cause a change of motion in any direction save that of its own line of action. If the effect is apparently in some other direction, as when a string operates over a pulley, the force is always combined with one or more auxiliary forces, the resultant of all of them being in the direction of the observed change of motion.

The third Law asserts the equality of "action and reaction." In the case of forces acting on bodies at rest, the principle is easily illustrated. When a steel truss rests on a pier and presses downward upon it with a force of 100,000 units, the pier exerts an upward thrust or "reaction" against the truss, also of 100,000 units, and this thrust, tending to bend the truss upward, is a most important factor in computing the stresses in the truss members. The Law also applies to forces acting upon bodies free to yield and to receive acceleration; a fact not explicitly stated by Newton, and discussed more fully elsewhere as d'Alembert's principle.

These propositions constituted the unquestioned foundation of dynamics until about the beginning of the twentieth century. So far as practical operations with bodies of ordinary size are concerned, they still answer every purpose. It is only when we consider motions with velocities comparable to that of light, or attempt to analyze the mechanics of bodies of the atomic and electronic order of magnitude, that the Newtonian dynamics breaks down and must be replaced by a system founded upon the postulates of relativity and the concepts of the quantum theory.

NEYMAN-PEARSON THEORY. In statistical inference, the theory of tests of hypotheses developed by J. Neyman and E.S. Pearson. It is based on the delimitation of two types of error, the rejection of a true hypothesis and the acceptance of a false hypothesis. Errors of the first kind are controlled at assigned probability levels. Errors of the second kind cannot simultaneously be so controlled, but can be explored for different values of parameters alternative to the true one. Tests that have a smaller probability of the second kind of error are said to be more powerful and part of the theory is devoted to seeking for most powerful tests. The probability of rejection of a false value of a parameter, graphed as ordinate against the value of the parameter as abscissa, is the Power Function of the test.

NIACIN. Sometimes referred to as nicotinic acid or nicotinamide and earlier called the P-P factor, antipellagra factor, anti-blacktongue factor, and vitamin B$_4$, niacin is available in several forms (niacin, niacinamide, niacinamide ascorbate, etc.) for use as a nutrient and dietary supplement. Niacin is frequently identified with the B complex vitamin grouping. Early in the research on niacin, a nutritional niacin deficiency was identified as the cause of pellagra in humans, blacktongue in dogs, and certain forms of dermatosis in humans. Niacin deficiency is also associated with perosis in chickens as well as poor feathering of the birds.

Varying in degree in relationship to the length and severity of diet deficiency of niacin, pellagra is clinically manifested by skin, nervous system, and mental conditions. The disease occurs most frequently among the economically deprived and particularly in areas where the diet may be high in maize (corn) intake. The disease was first described by Gaspar Casal in 1735 and was common in many areas, including Europe, Egypt, Central America, and the southern portion of the United States for many years. The largest outbreak occurred in the United States during the period 1905–1915 and resulted in a high mortality. The medical awareness and understanding of vitamins and dietary deficiencies, coupled with the availability of dietary supplements in staple foods, have resulted in a great lessening in the occurrence of pellagra. Where the disease is found, niacin is a specific for the treatment of acute pellagra.

Those afflicted are accustomed to a diet low in protein and made up largely of carbohydrates. Predisposing causes are found idiosyncrasies, chronic alcoholism, and diseases that interfere with the assimilation of a proper diet.

Huber first synthesized nicotinic acid in 1867. In 1914, Funk isolated nicotinic acid from rice polishings. Goldberger, in 1915, demonstrated that pellagra is a nutritional deficiency. In 1917, Chittenden and Underhill demonstrated that canine blacktongue is similar to pellagra. In 1935, Warburg and Christian showed that niacinamide is essential in hydrogen transport as diphosphopyridine nucleotide (DPN). In the following year, Euler et al. isolated DPN and determined its structure. In 1937, Elvhehjem et al. cured blacktongue by administration of niacinamide derived from liver. In the same year, Fouts et al. cured pellagra with niacinamide. In 1947, Handley and Bond established conversion of tryptophan to niacin by animal tissues.

In the physiological system, niacin and related substances maintain nicotinamide adenine dinucleotide (NAD) and nicotinamide adenine dinucleotide phosphate (NADP). Niacin also acts as a hydrogen and electron transfer agent in carbohydrate metabolism; and furnishes coenzymes for dehydrogenase systems. A niacin coenzyme participates in lipid catabolism, oxidative deamination, and photosynthesis.

Nicotinic acid can be converted to nicotinamide in the animal body and, in this form, is found as a component of two oxidation-reduction coenzymes, NAD and NADP, as previously mentioned. Structurally, these are:

Nicotinic acid   Nicotinamide

Nicotinamide adenine dinucleotide (NAD) R* = H

Nicotinamide adenine dinucleotide phosphate (NADP) R* = P—OH

The nicotinamide portion of the coenzyme transfers hydrogens by alternating between an oxidized quaternary nitrogen and a reduced tertiary nitrogen as shown by:

$$NAD \xrightleftharpoons[-2H]{+2H} NADH + H^+$$

NAD
(oxidized)

NADH + H$^+$
(reduced)

Enzymes that contain NAD or NADP are usually called dehydrogenases. They participate in many biochemical reactions of lipid, carbohydrate, and protein metabolism. An example of an NAD-requiring system is lactic dehydrogenase which catalyzes the conversion of lactic acid to pyruvic acid. Numerous NAD-dependent enzyme systems are known.

Lactic acid   Pyruvic acid

Distribution and Sources. In plants, niacin production sites occur in leaves, germinating seeds, and shoots. In humans, niacin is not available from intestinal bacteria, but some conversion is made from tryptophan which occurs in tissues.

High niacin content (10–100 milligrams/100 grams) Chicken (white meat), groundnut (peanut), halibut, heart (calf), kidney (beef, pork), liver (beef, calf, chicken, pork, sheep), meat extracts, rabbit (white meat), swordfish, tuna, turkey (white meat), yeast

Medium niacin content (1–10 milligrams/100 grams) Almond (dry), asparagus, avocado, barley, bean (kidney, lima, snap, wax), beef, broccoli, cashew, cheeses (camembert, roquefort, Swiss), chestnut, chicken (dark meat), clam, date (dry), duck, fig (dry), fishes (except those listed under "High"), kale, lamb, lentil (dry), maize (corn), molasses, mushroom, oats, oyster, parsley, pea, potato, prune (dry), rice (brown), rye, soybean (dry), shrimp, walnut, wheat, wheat germ

Low niacin content (0.1–1.0 milligram/100 grams) Apple, apricot, banana, beet, beet greens, berries (black-, blue-, cran-, rasp-, straw-), Brussels sprouts, cabbage, carrot, cauliflower, celery, cherry, chicory, coconut, cucumber, currant, dandelion greens, eggs, eggplant, endive, fig, grape, kohlrabi, lettuce, lemon, melons, milk, onion, parsnip, peach, pear, pecan, pepper, pineapple, plum, pumpkin, radish, raisin (dry), rhubarb, spinach, sweet potato, tangerine, tomato, turnip, watercress

Precursors in the biosynthesis of niacin: In animals and bacteria, tryptophan; and in plants, glycerol and succinic acid. Intermediates in the synthesis include kynurenine, hydroxyanthranilic acid, and quinolinic acid. In animals, the niacin storage sites are liver, heart, and muscle. Niacin supplements are prepared commercially by: (1) Hydrolysis of 3-cyanopyridine; or (2) oxidation of nicotine, quinoline, or collidine.

Bioavailability of Niacin. Factors which cause a decrease in niacin availability include: (1) Cooking losses; (2) bound form in corn (maize), greens, and seeds is only partially available; (3) presence of oral antibiotics; (4) diseases which may cause decreased absorption; (5) decrease in tryptophan conversion as in a vitamin B_6 deficiency. Factors that increase availability include: (1) alkali treatment of cereals; (2) storage in liver and possibly in muscle and kidney tissue; and (3) increased intestinal synthesis.

Antagonists of niacin include pyridine-3-sulfonic acid (in bacteria); 3-acetylpyridine, 6-aminonicotinamide, and 5-thiazole carboxamide. Synergists include vitamins B_1, B_2, B_6, B_{12}, and D, pantothenic acid, folic acid, and somatotrophin (growth hormone).

In humans, overdosage of niacin causes a limited toxicity (1 to 4 grams/kilogram) with individual variations in sensitivity.

Additional Reading

FAO United Nations: "Requirements of Vitamin A, Thiamine, Riboflavin and Niacin: Report of a Joint FAO-Who Expert Group", FAO United Nations, Geneva, Switzerland, 1967. *http://www.fda.gov/ (United States Food and Drug Association).*

Institute of Medicine, Food Nutrition Board Staff: "Dietary Reference Intakes: Thiamin, Riboflavin, Niacin, Vitamin B6, Folate, Vitamin B12, Pantothenic Acid, Biotin, and Choline," National Academy Press, Washington, DC, 1998.

Web Reference

United States Food and Drug Association. *http://www.fda.gov/*

NIACINAMIDE. See **Pyridine and Derivatives**.

NICHE. See **Ecology**.

NICKEL. Chemical element, symbol Ni, at. no. 28, at. wt. 58.69, periodic table group 10, mp 1453 °C, bp 2732 °C, density 8.9 g/cm^3 (solid, 20 °C), 9.04 g/cm^3 (single crystal). Elemental nickel has a face-centered cubic crystal structure. Nickel is a silver-white metal, harder than iron, capable of taking a brilliant polish, malleable and ductile, magnetic below approximately 360 °C. When compact, nickel is not oxidized on exposure to air at ordinary temperatures. The metal is soluble in HNO_3 (dilute), but becomes passive in concentrated HNO_3. The metal does not react with alkalis. Finely divided nickel dissolves 17 × its own volume of hydrogen at standard conditions. There are five naturally occurring stable isotopes, ^{58}Ni, ^{60}Ni through ^{62}Ni, and ^{64}Ni. Six radioactive isotopes have been identified, ^{56}Ni, ^{57}Ni, ^{59}Ni, ^{63}Ni, ^{65}Ni, and ^{66}Ni. ^{59}Ni has a half-life of 8×10^4 years, and ^{63}Ni has a half-life of 80 years. The half-lives of the remaining radioactive isotopes are relatively short, expressed in hours and days. The element ranks 21st among the elements in terms of abundance in the earth's crust, the estimated average content of igneous rocks being about 0.02%. In terms of cosmic abundance, nickel ranks 28th among the elements. Nickel ranks 40th in terms of concentration in seawater, the estimated content being about 2.5 tons of nickel per cubic mile (540 kilograms per cubic kilometer) of seawater. Awareness of nickel probably dates back to antiquity, but the element was not firmly identified until 1751 when Axel Fredric Cronstedt isolated the metal from the sulfide ore NiAsS.

First ionization potential 7.33 eV, second, 18.13 eV. Oxidation potentials $Ni \rightarrow Ni^{2+} + 2e^-$, 0.230 V; $Ni^{2+} + 2H_2O \rightarrow NiO_2 + 4H^+ + 2e^-$, -1.75 V; $Ni + 2OH^- \rightarrow Ni(OH)_2 + 2e^-$, 0.66 V; $Ni(OH)_2 + 2OH^- \rightarrow NiO_2 + 2H_2O + 2e^-$, -0.49 V.

Other physical properties of nickel are given under **Chemical Elements**.

In the early 1800s, the principal sources of nickel were in Germany and Scandinavia. Very large deposits of lateritic (oxide or silicate) nickel ore were discovered in New Caledonia in 1865. The sulfide ore deposits were discovered in Sudbury, Ontario in 1883 and, since 1905, have been the major source of the element. The most common ore is pentlandite, $(FeNi)_9S_8$, which contains about 34% nickel. Pentlandite usually occurs with pyrrhotite, an iron-sulfide ore, and chalcopyrite, $CuFeS_2$. See also **Chalcopyrite**; **Pentlandite**; and **Pyrrhotite**. The greatest known reserves of nickel are in Canada and Russia, although significant reserves also occur in Australia, Finland, the Republic of South Africa, and Zimbabwe.

Principal producers and/or exporters of nickel include, in diminishing order, Canada, Russia, the United Kingdom, Norway, and Indonesia. Main consumers are the United States, Japan, the United Kingdom, Norway, Germany, Canada, and France.

After beneficiation of the raw ore to form a sulfide concentrate, the latter is roasted to achieve partial oxidation of iron and partial removal of sulfur. The roasted material then is smelted with a flux to eliminate the rock content. At this point, part of the iron goes into the slag. The remaining material is a copper-bearing nickel-iron matte, made up mainly of the sulfides of these metals. The matte is then treated in a Bessemer converter to achieve further removal of iron and sulfur. After controlled cooling, which assists separation, the Bessemer product is finely ground and subjected to magnetic separation and differential flotation. The separated product is an impure nickel sulfide. The sulfide then is sintered to nickel oxide. This product may be marketed for some applications, but the majority of the oxide is cast into anodes for refining into nickel metal by one of two major processes.

In (1) the electrolytic process, a nickel of 99.9% purity is produced, along with slimes which may contain gold, silver, platinum, palladium, rhodium, iridium, ruthenium, and cobalt, which are subject to further refining and recovery. In (2) the Mond process, the nickel oxide is combined with carbon monoxide to form nickel carbonyl gas, $Ni(CO)_4$. The impurities, including cobalt, are left as a solid residue. Upon further heating of the gas to about 180 °C, the nickel carbonyl is decomposed, the freed nickel condensing on nickel shot and the carbon monoxide recycled. The Mond process also makes a nickel of 99.9% purity.

Uses. The three main commercial forms of primary nickel are: (1) electrolytic sheets, (2) pellets resulting from the decomposition of nickel carbonyl, and (3) ferronickel. Traditionally, pellets are favored in Europe, whereas electrolytic nickel is favored in North America. Additional forms of commercial nickel are powder, ingots, shot, and briquettes. Ferronickel, containing 24–48% nickel with the remainder iron, is used mainly in the production of stainless steel. More than half of the nickel produced is used in stainless steels and high-nickel alloys. Additional uses include nickel plating, iron and steel castings, coinage, and copper and brass products.

The main consumer of nickel is austenitic stainless steel, which contains from 3.5 to 22% nickel and 16 to 26% chromium. In these steels, nickel stabilizes the austenite and enhances the ductility of the steel. Nickel, along with chromium, contributes to corrosion resistance. Up to amounts of about 9%, nickel adds strength, hardness, and toughness to many alloy steels. Alloys in the 9% nickel range remain stable at low temperatures and are capable of handling liquefied gases. The lower-nickel steels (0.5 to 0.7%) are ductile, strong, and tough, and find use for many automobile parts, in power machinery, and construction equipment. There are hundreds of nickel-containing alloys, running the gamut from hardenable silver alloy (0.02% Ni) up to malleable nickel (99% Ni).

Wrought Nickel and High-Nickel Alloys. Some of the major nickel alloys, along with wrought nickel, are described in Table 1.

Commercially pure wrought nickel in the form of sheets, wire, and tubing has many uses because of its corrosion resistance. These uses include utensils, food-processing equipment, marine hardware, coinage, and chemical equipment. Electroplated nickel also is used as a protective coating on steel. *Nimonic* alloys, not shown on the table, are based on an 80% Ni, 20% Cr composition. They are high-strength, heat-resistance metals that are age-hardened to increase strength at elevated temperatures—with a useful range of 700–825 °C. *Monel* metal (several types) is a high-strength corrosion-resistant alloy available in many wrought and cast forms for use in processing equipment, marine construction, and household appliances. *K Monel* can be heat treated by precipitation hardening to about 2 × the strength of annealed *Monel*. *Hastelloy*-type alloys are well known for their excellent resistance to HCl, H_2SO_4, and other acids. The *Incoloy*-type alloys (35% Ni approximately) are heat-resistant alloys used mainly as castings for furnace parts. The lower-nickel/higher-chromium alloys generally are classified as stainless steels. See also **Iron Metals, Alloys, and Steels**.

Although not of high-tonnage production, several nickel metals serve important uses, such as:

Permalloy, 78.5% Ni, 21.5% Fe; *Hipernik*, 50% Ni, 50% Fe; and *Perminvar*, 45% Ni, 30% Fe, 25% Co—are representative of a group of high-nickel magnetic alloys.

Constantan, 45% Ni, 55% Cu, has high electrical resistivity and a very low temperature coefficient of resistivity. It is extensively used with copper as a thermocouple element.

Nichrome, 80% Ni, 20% Cr (several types with variations of these percentages and additions of other elements, such as silicon in small amounts), is used as resistance wire for heating elements.

Calorite, 65% Ni, 8% Mn, 12% Cr, 15% Fe, also is used in electric heating elements.

Alumel, 94% Ni, 2.5% Mn, 0.5% Fe plus small amounts of other elements, is used in thermocouples

Chromel, 35–60% Ni, 16–19% Cr, generally with the balance Fe, also is used as resistance wire and for thermocouples.

Invar, 36% Ni, 64% Fe, has a very low temperature coefficient of expansion and is used for measuring tapes, instruments, and bimetallic thermostats.

Elinvar, 34% Ni, 57% Fe, 4% Cr, 2% W, has a very low temperature coefficient of elasticity which makes it useful for springs in watches and precision instruments.

There are hundreds of special nickel-bearing alloys of proprietary formulations and tradenames.

Alloy with Memory. In seeking a way to reduce the brittleness of titanium, U.S. Navy researchers serendipitously discovered a nickel-titanium alloy having an amazing memory. Previously cooled clamps made of the alloy (*nitinol*) are flexible and can be placed easily in position. When

TABLE 1. WROUGHT NICKEL AND REPRESENTATIVE NICKEL ALLOYS

	Melting Range °C	Poisson's Ratio
Wrought nickel 99% Ni, 0.25% Cu, 0.15% C	1,435– 1,445	0.31
Duranickel 301 93.9% Ni, 0.05% Cu, 0.15% C, 0.15% Fe, 0.5% Ti, 4.5% Al	1,400– 1,440	0.31
Monel 400 66.0% Ni, 31.5% Cu, 0.12% C, 1.35% Fe	1,300– 1,350	0.32
Hastelloy B 63.5% Ni, 0.05% C, 5.0% Fe, 2.5% Co, 1.0% Cr, 28.0% Mo, 0.3% V	1,320– 1,460	—
Hastelloy F 45.5% Ni, 0.05% C, 20.5% Fe, 2.5% Co, 22.0% Cr, 6.5% Mo, 1% W, 2% (Nb + Ta)	1,290– 1,295	0.305
Inconel 600 72% Ni, 0.5% Cu, 0.15% C, 8.0% Fe, 15.5% Cr	1,370– 1,425	0.29
Incoloy 800 32.5% Ni, 0.75% Cu, 0.10% C, 45.6% Fe, 21.0% Cr	1,355– 1,390	0.30
Illium G 56.0% Ni, 6.5% Cu, 22.5% Cr, 6.5% Mo	1,255– 1,340	0.29

Note: Recently introduced new or improved nickel alloys include:

Inconel alloy 625—Low-cycle fatigue resistance has been increased from 70–80,000 to 110–120,000 psi at 10 cycles. This has been achieved through grain size control and improved product cleanliness. Major applications are bellows and expansion joints.

Inconel alloy 725—An age-hardenable alloy for deep sour gas well service, combining height strength with the attributes of Inconel alloy 625, such as pitting resistance and stress corrosion cracking resistance to salt, hydrogen sulfide, and sulfur at temperatures up to about 230 °C (450 °F) and to sulfide stress corrosion cracking.

Inconel alloy 622—Modified composition and special thermal mechanical processing give this alloy superior thermal stability and resistance to intergranular attack and localized corrosion. The alloy is particularly suited to acidified halide environments, especially those containing oxidizing acids.

Inconel alloy 925—An age-hardenable nickel-iron-chromium alloy providing high strength up to 540 °C (1000 °F). Developed for use in gas production applications, such as tubular products, tool joints, and equipment for surface and downhole hardware in gas industry.

Inco alloy 25–6MO—Used for its corrosion resistance in many environments, this is an austenitic nickel-iron-chromium alloy with a substantial (6%) addition of molybdenum. Especially useful for resisting pitting and crevice corrosion in media containing chlorides or other halides. Applications include equipment for handling sulfuric and phosphoric acids, offshore platforms and other marine equipment, and for bleaching circuits in pulp and paper plants.

warmed to a given temperature, the alloy hardware then exerts tremendous pressure. Use of conventional clamps for holding bundles of wires or cables in a ship or aircraft structure requires special tools. For this and other applications in industry and medicine, nitinol has been in demand. The alloy, however, is not easy to produce because only minor variations in composition can affect the "snap back" temperature by several degrees of temperature.

Nickel Powders. The use of nickel in powdered form has increased markedly during the last few years. As shown by Fig. 1, nickel powders are available in several types and are used in a variety of products.

Nickel in Nanometer Materials. Coating a metal with an ultrathin layer of another metal creates properties not found separately in either of the materials. Considerable recent research has been directed toward improving the mechanical properties of bimetallic laminates, sometimes called *composition modulated films*, which have interlayers only a few nanometers thick. Attractive properties also have been found for similar systems, called *nanometer materials*. Nickel has been used in combination with copper, ruthenium, and other metals for producing these new materials.

Production of High-Performance Nickel Alloys. In the production of high-performance alloys, the critical first step of alloying requires sophisticated equipment, stringent controls, and expertise. Several production methods are used.

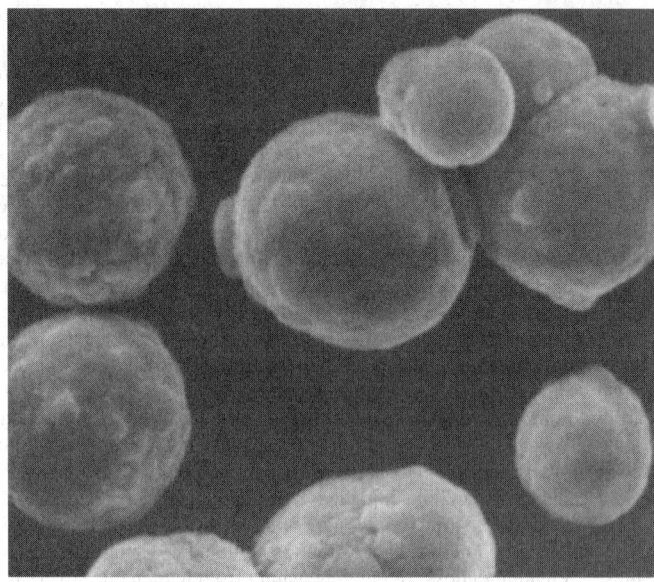

(c)

Fig. 1.   (*Continued*)

Air melting in electric-arc or induction furnaces is used for many alloys, sometimes for final alloying, with further refining by argon-oxygen decarburization. Melting in air can result in impurities in some alloys, a problem eliminated by vacuum induction melting, used to produce ingots for direct rolling or for remelting. Remelting is accomplished by two methods, both with precise, computerized control. *Electroslag* remelting uses electrical resistance heating to remelt an ingot (electrode) under molten slag containing fluxes that remove impurities. *Vacuum arc* melting refines the structure of cast electrodes in a contaminant-free chamber. Remelting yields alloys of the highest level of refinement.

Nickel Chemistry and Compounds

With its $3d^8 4s^2$ electron configuration, nickel forms Ni^{2+} ions. Having a nearly complete $3d$ subshell, nickel does not yield a $3d$ electron as readily as iron and cobalt, and trivalent and tetravalent forms are known only in the hydrated oxides, Ni_2O_3 and NiO_2, and a few complexes.

Nickel(II) oxide, NiO, produced by heating the carbonate, is thermally stable. Higher oxides of nickel, including Ni_2O_3 and NiO_2, are known only as hydrates, being prepared by vigorous oxidation of NiO in alkaline solution.

Nickel(II) sulfide, precipitated from Ni^{2+} solutions by ammonium sulfide, may show quite a little departure from stoichiometric composition. Like iron(II) and cobalt(II) FeS and CoS, it has in crystal form an electrical conductivity and other properties similar to a metal or alloy. There is no conclusive evidence that Ni_2S_3 can be prepared, but NiS_2 is known and believed to be, like FeS_2, a compound of Ni^{2+} and the S_2^{2-} ion.

All four dihalides of nickel with the common halogens are known: NiF_2, formed by reaction of hydrofluoric acid or nickel(II) chloride or by thermal decomposition of $[Ni(NH_3)_6][BF_4]_2$, is greenish yellow, while the other three dihalides, formed directly from the elements, are green for the chloride, yellow for the bromide, and black for the iodide. In general, anhydrous Ni^{2+} salts are yellow and the ion $Ni(H_2O)_6^{2+}$ in aqueous solution is green.

Other elements with which nickel forms binary compounds, especially at higher temperature, are boron, carbon, nitrogen, silicon, and phosphorus. Like NiO, these compounds may depart slightly or even considerably from daltonide composition, frequently being interstitial compounds, and with higher elements of transition groups 5 and 6, merging into the interstitial compound-solid solution picture which nickel exhibits with the other transition metals.

Divalent nickel forms two main types of complexes. The first consists of complexes of the spin-free ("ionic" or outer orbital) octahedral type (see also **Ligand** for their discussion) in which the ligands are principally H_2O, NH_3, and various amines such as ethylenediamine and its derivates, e.g., $Ni(H_2O)_6^{2+}$, $Ni(NH_3)_6^{2+}$, $Ni(en)_6^{2+}$. These complexes usually have colors toward the high-frequency side of the spectrum, i.e., violet, blue,

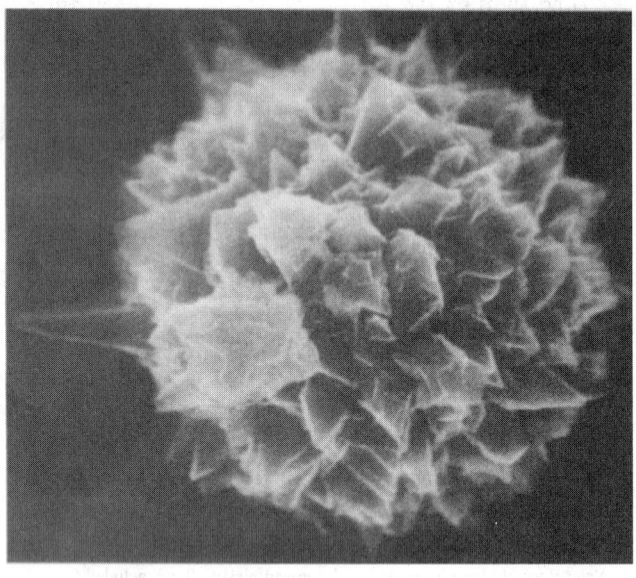

(a)

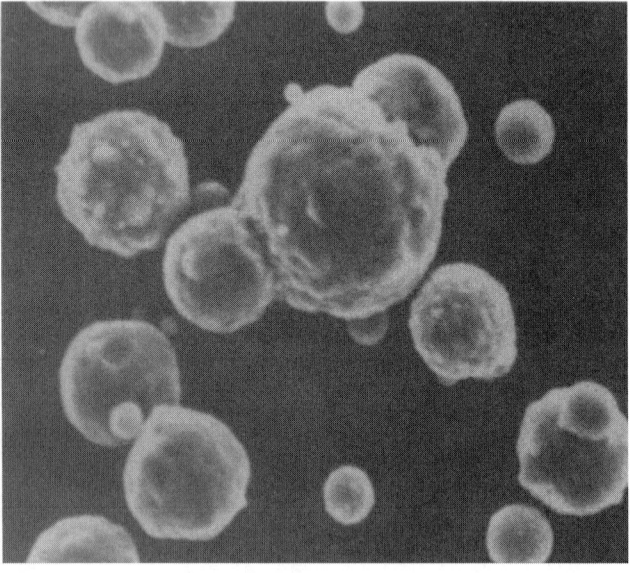

(b)

Fig. 1.   Types of available nickel powders. (**a**) With a surface area of 0.4 m²/g. this is a spiked nickel powder of single particles 3–7 microns in diameter, with a bulk density of 2 g/cc. The powder is used in both metal and chemical systems for powder metal parts, getters, magnets, electronic strip, flake, and organo-nickel compounds and nickel salts and soaps. (**b**) High-density nickel powder consisting of 8–12 micron semismooth particles, offering mixability with both metallic and nonmetallic powder systems. Applications include welding rods, nickel aluminide, nickel-columbium additives, abradable seals, powder metal parts, carbide binders, and conductive plastics. (**c**) Spherically shaped, high- purity nickel powder, with a surface area of 0.15 m²/g and a Fisher particle size of 8–9 microns. Applications include friction materials, plasma spraying, metal injection molding, welding electrodes, magnets, cemented carbides, and powder metal nickel steels. (*Source: INCO Specialty Powder Products, Saddle Brook, New Jersey.*)

NICKEL (In Biological Systems)   2455

and green. The other class consists of tetracovalent square complexes with ligands such as CN^-, the dioximes and their derivatives, and other chelates, which usually have colors on the low frequency side of the spectrum, i.e., red, orange, and yellow. The structure of the nickel-dimethylglyoxime complex is

$$
\begin{array}{c}
H_3C \qquad\qquad CH_3 \\
C = C \\
N - O \qquad O - N \\
\vdots \qquad Ni \qquad \vdots \\
H \qquad\qquad H \\
O - N \qquad N - O \\
C \qquad C \\
H_3C \qquad\qquad CH_3
\end{array}
$$

This compound is of interest not only in analysis, but because by limited oxidation with the halogens it yields a unipositive ion containing trivalent nickel and also because the hydrogen bonds formed to the oxygen atoms are among the shortest known. Similarly, the tetracyanide complex of nickel, $Ni(CN)_4^{2-}$, may be reduced by sodium amalgam to give an ion of composition $Ni(CN)_4^{3-}$, or $(NC)_3Ni\text{---}Ni(CN)_3^{4-}$ containing Ni(I). This latter ion forms a potassium salt of nickel(I) of the formula $K_4Ni_2(CN)_6$ which is reduced in liquid NH_3 by metallic potassium to give the compound $K_4Ni(CN)_4$ in which the nickel has an effective valence of zero. Of course, this zero valence also exists in the carbonyls of nickel (and other elements) which, however, are covalent. $Ni(CO)_4$ is prepared by reaction of carbon monoxide with freshly reduced nickel, which occurs at ordinary temperatures and pressures. As with the carbonyls of other metals, the CO groups may be directly or indirectly, partially or completely, replaced by other groups. Derivatives of trivalent phosphorus form many such compounds of general formula $Ni(CO)_{4-x}(PR_3)_x$, where R may be one or more of such groups as F, Cl, Br, I, alkyl, aryl, alkoxy, aryloxy, etc.

See also **Nickel (In Biological Systems)**.

Additional Reading

Carter, G.F. and D.E. Paul: "Materials Science and Engineering," ASM International, Materials Park, OH, 1991.

Greenwood, N.N. and A. Earnshaw: "Chemistry of the Elements," 2nd Edition, Butterworth-Heinemann, Inc., Woburn, MA, 1997.

Hanson, A.: "The Metal That Remembers," *Technology Review(MIT)*, 26 (May–June 1991).

Houston, J. and P. Feibelman: "Ultrathin Metal Coatings Yield Unique Properties," *Advanced Materials & Processes,* 31 (March 1991).

Lancaster, J.R., Jr.: "The Bioinorganic Chemistry of Nickel," John Wiley & Sons, Inc., New York, NY, 1988.

Lide, D.R.: "CRC Handbook of Chemistry and Physics 2000–2001," 81st Edition, CRC Press, LLC., Boca Raton, FL, 2000.

Sax, N.R. and R.J. Lewis, Sr.: "Sax'x Dangerous Properties of Industrial Materials," 10th Edition, John Wiley & Sons, Inc., New York, NY, 1999.

Staff: "A Quick Reference Guide to Nickel and High-Nickel Alloys," *Advanced Materials and Processes,* 54 (October 1991).

Staff: "ASM Handbook — Properties and Selection: Nonferrous Alloys and Pure Metals," ASM International, Materials Park, OH, 1991.

Staff: "Metals Forecast," *Advanced Materials & Processes,* 17 (January 1991); 24 (January 1992); 18 (January 1993).

NICKEL (In Biological Systems). Despite its many pharmacological and in vitro actions, convincing evidence showing that nickel is an essential element for some animal species did not appear until the early 1960s. There has been considerable further and more convincing research during the 1970s and early 1980s.

Like most trace elements, nickel can activate various enzymes in vitro, but no enzyme has been shown to require nickel, specifically, to be activated. However, urease has been shown to be a nickel metalloenzyme and has been found to contain 6 to 8 atoms of nickel per mole of enzyme (Fishbein et al., 1976). RNA (ribonucleic acid) preparations from diverse sources consistently contain nickel in concentrations many times higher than those found in native materials from which the RNA is isolated (Wacker-Vallee, 1959; Sunderman, 1965). Nickel may serve to stabilize the ordered structure of RNA. Nickel may have a role in maintaining ribosomal structure (Tal, 1968, 1969). These studies and other information have led to the suggestion that nickel may play a role in nucleic acid and/or protein metabolism.

Nickel also may act to stimulate or inhibit the release of various hormones (Nielsen, 1971, 1972; Dormer et al., 1973; Clay, 1975; Horak-Sunderman, 1975). Nickel has been found to inhibit insulin release from the pancreas (Dormer et al., 1973; Clay, 1975), and stimulates glucagon secretion (Horak-Sunderman, 1975).

Nickel as an essential element in ruminant nutrition has not been proved conclusive as of the early 1980s. However, with nonruminants, some evidence indicates that certain species fed low-nickel diets have a greater infant mortality rate and a general degradation of the reproductive process (Nielsen, 1975; Anke et al., 1973).

Zinc and nickel appear to behave similarly at certain sites in the biological system. Both elements are capable of activating certain enzymes; for example, arginase is an enzyme which can be activated by either element (Parisi-Vallee, 1969). Stimulation of enzyme activity is at a site at which trace element substitutions or interactions may occur. However, some sacrifice of activity usually results when normally occurring metal is replaced by a trace metal. Nucleic acids as well as the ribosomes are likely sites of interaction between nickel and zinc. Both metals are consistently found in high concentrations firmly bound to RNA. It has been suggested that they function in maintaining the structure of RNA, thus preventing conformational changes. Nickel appears to be as effective as zinc at equal concentrations in this respect. Nickel and zinc are also found in ribosomal ash and studies have indicated that both can contribute to ribosomal conformation. The white blood cell is another possible site at which nickel and zinc may interact. Leukocytes are high in zinc and total leukocyte counts as well as differential white cell counts change drastically during a zinc deficiency. The interrelationship between nickel and zinc has been studied in vitro primarily in swine and rats. Their relationship has been studied largely from a substitution standpoint. Nickel appears to substitute for zinc to a certain extent in both species.

Similarly, the relationship between nickel and copper has been under study. One of the major functions of copper is in hemoglobin formation. Hemoglobin and hematocrit values decline rapidly during a copper deficiency. Copper is currently believed to exert its effect on hemoglobin metabolism through ceruloplasmin. Early work also indicated that nickel might be involved in hematopoiesis. Investigators in 1974 found a decreased concentration of copper in the lung and spleen of rats receiving 5 parts per million of nickel in drinking water. High levels of dietary nickel in rats and mice have been reported to decrease the activity of cytochrome oxidase, a copper-containing enzyme.

As pointed out by Eskew, Welch, and Cary in 1983, in contrast with the situation in animals, for which four new essential trace elements were identified in recent years, no new generally essential micronutrient for higher plants has been found since the mid-1950s. When it was found that urease is a nickel-metalloenzyme, this suggested that Ni may play a role in higher plants. Nickel has evidenced a stimulation of growth when urea is the sole source of nitrogen, but has slight or no effect when other nitrogen enrichment sources are used. The aforementioned investigators claim that Ni is essential for nitrogen metabolism in soybeans (*Glycine max* (L.) Merr.), either when nitrogen is furnished as NO_3^- and NH_4^+ or when plants depend upon nitrogen fixation. In experiments, soybean plants deprived of Ni accumulated toxic concentrations of urea (2.5%) in necrotic lesions on their leaflet tips. This occurred regardless of whether the plants were furnished with inorganic N or were dependent on N fixation. Nickel deprivation resulted in delayed nodulation and in a reduction of early growth. The addition of Ni (1 microgram/liter) to the nutrient media prevented urea accumulation, necrosis, and growth reductions. Extrapolating these findings, it is suggested that Ni may be essential for other higher plants.

Toxicity. Nickel contact dermatitis can occur among wearers of nickel-containing jewelry, more common among females than males. This is particularly true of nickel sulfate present in some jewelry. Localization of sites unexpectedly involves the ear lobes, neck, fingers, and wrists. Nickel is a major offender in connection with AECD (allergic eczematous contact dermatitis).

As mentioned earlier, nickel carbonyl is a volatile intermediate in the Mond process for nickel refining. This compound also is used for vapor plating of nickel in the semiconductor industry, and as a catalyst in the chemical and petrochemical industries. The toxicity of the compound has been known for many years. Exposure of laboratory animals to the compound has induced a number of ocular anomalies, including anophthalmia and microphthalmia, and has been shown to be a carcinogenic for rats.

Additional Reading

Anke, M. et al.: "Low Rations for Growthe and Reproduction in Pigs," in "Trace Element Metabolism in Animals," (W.G. Hockstra et al., editors), University Park Press, Baltimore, MD, 1073.

Clay, J.J.: "Nickel Chloride-induced Metabolic Changes in the Rat and Guinea Pig," *Toxicol. Appl. Pharmacol.*, **31**, 55 (1975).

Considine, D.M. and G.D. Considine, Eds.: "Foods and Food Production Encyclopedia," in "Van Nostrand Reinhold," New York, NY, 1982.

Dormer, R.L. et al.: "The Effect on Nickel on Secretory Systems," *Biochem.J.*, **140**, 135 (1973).

Eskew, D.L., R.M. Welch, and E.E. Cary: "Nickel: An Essential Micronument for Legumes and Possibly All Higher Plants, *Science*, **222**, 621–623 (1983).

Fishbein, W.N. et al.: "The First Natural Nickel Metalloenzyme: Urease, *Fed. Proc.*, **35**, 1680 (1976).

Horak, E. and F.W. Sunderman, Jr.: "Effects on Ni(II) upon Plasma Glucagon and Glucose in Rats, "*Toxicol. Appl. Pharmacol.*, **33**, 388 (1975).

Neilsen, F.H.: "Studies on the Essentiality of Nickel," in "Newer Trace Elements in Nutrition," (W. Mertz and W.E. Cornatzer, editors), Marcel Dekker, New York, NY, 1971.

Parisi, A.F. and B.L. Vallee: "Zinc Metalloenzymes: Characteristics and Significance in Biology and Medicine," *Amer. J. Clin. Nutr.*, **22**, 1222 (1969).

Spears, J.W and E.E. Hatfield: "Role of Nickel in Animal Nutrition, "*Feedstuffs*," 24–28 (June 13, 1977).

Sunderman, F.W., Jr.: "Measurements of Nickel in Biological Materials by Atomic Absorption Spectrometry," *Amer. J. Clin. Path.*, **44**, 182 (1965).

Sunderman, F.W., Jr. et al.: "Eye Malformations in Rats:Induction by Prenatal Exposure To Nickel Carbonyl," *Science*, **203**, 550–552 (1979).

Tal, M.: "On the Role of Zn^{2+} and Ni^{2+} in Ribosome Structure, *Biochem. Biophys. Acta*, **169**, 564 (1968).

Wacker, E.E.C. and B.L. Vallee: "Nucleic Acids and Metals. I. Chromium, Manganese, Nickel, Iron and Other Metals in Ribonucleic Acid from Diverse Biological Sources," *J. Biol. Chem.*, **234**, 3257 (1959).

NICKELINE. A nickel arsenide mineral, NiAs, crystallizes in the hexagonal system but is usually found massive. Color, light copper; hardness, 5.0–5.5; specific gravity, 7.784; luster, metallic; opaque. Found in several European localities and in the Province of Ontario, Canada; in the United States at Franklin, New Jersey, and Silver Cliff, Colorado. It is an ore of nickel.

NICOTINAMIDE. See **Pyridine and Derivatives**; and **Vitamins**.

NICOTINE. See **Alkaloids**.

NICOTINIC ACID. See **Pyridine and Derivatives**; and **Vitamins**.

NIGHT BLINDNESS. See **Vision and the Eye**.

NIGHTINGALE (*Aves, Passeriformes*). A warbler, *Luscinia megarhyncha*, of western Europe, noted for its song. Farther east two other species, the eastern, *L. pheilomella*, and Persian, *L. hafizi*, nightingales, are found.

The nightingale is a trim bird, about 6 to 7 inches (15 to 18 centimeters) in length. The coloring is brown with lighter brown underneath. The female does the nesting and brooding, although the male helps with feeding the young. Incubation period is from 13 to 14 days. The eggs are blue or off-white with some markings.

NIGHTJARS AND NIGHTHAWKS (*Aves, Caprimulgiforme*). These birds make up the majority of the *Caprimulgiformes*, an order of nocturnal birds with very wide mouths. Nightjars have mottled plumage and a short beak. They are insect-eaters, flying chiefly at night. All continents have some of the numerous species with the exception of Australia. In North America, the nighthawk and whippoorwill are the most widely known representatives of the group, with the poorwill, chuck-will's widow (*Antrostomus carolinensis*), and the Merrill parauque (*Nyctidromus albicollus*) as less widely distributed species. Nightjars have been known since the time of Aristotle and are mentioned in the Bible. The poorwill (*Phalaenoplilus nuttallii*) is about 8 to 10 inches (20 to 25 centimeters) long and makes no formal nest. Eggs are laid in grasslands. This species uses the tactic of displaying a "broken" wing to distract the attention of

predators. The legs are short and weak. The species is known for squatting lengthwise on limbs of trees and also for going into hibernation. During such periods, respiration is barely detectable and the temperature drops from a normal figure of about 100 °F to 66 °F (38 °C to 18.9 °C).

The whippoorwill, is a nocturnal bird (*Caprimulgus (Antrostomus) vociferus*), with a short beak and wide mouth, adapted for taking insects in flight. Its call has been likened to the words used in its name. Often the three syllables are repeated over and over scores of times without cessation.

The common nighthawk (*Chordeiles minor (virginianus)*) is widely distributed in North America and winters far into South America. A second species, the Texan nighthawk (*C. acutipennis*) enters the southwestern United States. Nighthawks are characterized by very short beaks and by wide mouths. They are well adapted for catching insects in flight. They commence their flights late in the day. Their flight is easy and powerful and their long dives, terminating in a peculiar hollow boom, are a memorable exhibition.

Goatsuckers were given their odd name in the mistaken belief during Aristotle's time that the birds took milk from domesticated goats. They, like the other nightjars, are characterized by weak legs, a short weak beak, a very wide mouth, and crepuscular habits. The *Uropsalis lyra* is a small bird about 4 inches in length (10 centimeters), but with a lyre-type tail some 27 inches (69 centimeters) in length. This species inhabits the environs of Colombia in South America. The habits are nocturnal. The call is penetrating. Another species is the parauque, a large bird of Mexico and Texas which resembles the poorwill. Also related to the goatsuckers is the frog mouth (*Podargus*) found in the Oriental and Australian regions. See also *Caprimulgiformes*.

NIMBOSTRATUS. See **Clouds and Cloud Formation**.

NIMBUS. See **Clouds and Cloud Formation**.

NIOBIUM. Chemical element, symbol Nb, at. no. 41, at. wt. 92.906, periodic table group 5, mp 2,458–2,468 °C, bp 4,742 °C, density 8.6 g/cm^3 (20 °C). Elemental niobium has a body-centered cubic crystal structure. The metal has a slightly bluish tinge, is ductile and malleable, and when polished resembles platinum. The metal burns upon being heated in air. There is one natural isotope ^{93}Nb. Seven radioactive isotopes have been identified ^{90}Nb through ^{92}Nb and ^{94}Nb through ^{97}Nb, with a wide range of half-lives. ^{94}Nb has the longest half-life (2×10^4 years). The element was first identified by C. Hatchett in 1801 and was originally called columbium which name persisted for many years. The name still appears widely in the literature, particularly in connection with alloys bearing the element, such as columbium steels.

First ionization potential 6.77 eV; second 13.895 eV; third 24.2 eV. Oxidation potential $Nb \rightarrow Nb^{3+} + 3e^-$, ca. 1.1 V; $2Nb + 5H_2O \rightarrow Nb_2O_5 + 10H^+ + 10e^-$, 0.62 V.

Other important physical properties of niobium are given in the Table 1 and under **Chemical Elements**.

Niobium occurs, usually with tantalum, in columbite $Fe(NbO_3)_2$, (80% Nb_2O_5), pyrochlore (50% Nb_2O_5), samarskite (50% Nb_2O_5), chiefly found in western Australia, and South Dakota. Recovered along with tantalum by fusion with potassium bisulfate, and obtained in the residue after subsequent extraction with H_2O. Niobium and tantalum are separated by fractional crystallization of the potassium fluorides, niobium concentrating in the mother liquid and tantalum in the crystals.

The principal uses for the element are in alloys. Niobium also has gained prominence in research as a superconducting material. At the temperatures of liquid helium, niobium becomes a superconductor and, in the form of a fine wire, has been incorporated in a superconducting cell. The element has both size and cost advantages over electronic materials. The alloy Nb$_3$Sn becomes superconducting at a somewhat higher temperature. Niobium-titanium and niobium-zirconium alloys also have potential as superconductors.

Alloys. Niobium is used in steel, notably stainless steels, to stabilize the carbon present (as carbide) and for preparing niobium carbide, used for dies and cutting tools. Ferroniobium is a strong carbide-forming material and, when added to 18-8 stainless steel, stabilizes areas that may be heat-affected during welding and thus cause subsequent intergranular corrosion. Niobium steels are used for rotors in gas turbines where temperatures up to 700 °C must be withstood. Niobium-base alloys find application in fast reactors. Superalloys for very demanding use, as in

TABLE 1. REPRESENTATIVE PROPERTIES OF REFRACTORY METALS

Property	Tungsten	Tantalum	Molybdenum	Niobium
Density, g/cm^3	19.3	16.6	10.2	8.7
Melting point, °C	3,390–3,420	2,996	2,617	2,458–2,468
Boiling point, °C	5,660	5,325–5,525	4,612	4,742
Linear coefficient of expansion per °C	4.3×10^{-6}	6.5×10^{-6}	4.9×10^{-6}	7.2×10^{-6}
Thermal conductivity, 20°C (cal/cm^2/cm/°C/s)	0.40	0.13	0.35	0.13
Specific heat, 20°C (cal/g/°C)	0.032	0.036	0.061	0.065
Working temperature, °C	1,700	ambient	1,600	ambient
Electrical conductivity, % IACS	31	13	30	12
Nuclear cross section (thermal neutrons, Barns/atom)	19.2	21.3	2.4	1.1
Tensile strength, 1000 psi				
20°C	100–500	100–150	120–200	75–150
500°C	175–200	35–45	35–65	35
1000°C	50–75	15–20	20–30	13–17
Young's Modulus of Elasticity, psi				
20°C	59×10^6	27×10^6	46×10^6	14×10^6
500°C	55×10^6	25×10^6	41×10^6	7×10^6
1000°C	50×106	22×106	39×10^6	—
Poisson's Ratio	0.284	0.35	0.32	0.38
Corrosion resistance, 100°C				
Dilute HNO$_3$		N	R	N
Dilute H$_2$SO$_4$		N	S	VS
Concentrated H$_2$SO$_4$		N	S	R
Dilute HCl	See	N	S	—
Concentrated HCl	Tungsten	N	SL	SL
Concentrated Hydrofluoric acid		R	SL	R
Phosphoric acid, 85%		N	SL	VS
Concentrated NaOH		R	N	R

N = no appreciable corrosion.
VS = <0.0005 inch (0.013 millimeter) per year.
SL = 0.0005 − 0.005 inch (0.013–0.13 millimeter) per year.
S = 0.005 − .01 inch (0.13 0.25 millimeter) per year.
R = >0.01 inch (0.25 millimeter) per year.

military applications contain niobium with cobalt and zirconium. When alloyed with titanium, molybdenum, and tungsten, the elevated-temperature hardness of niobium in enhanced, whereas when alloyed with vanadium and zirconium, the strength of niobium up to temperatures of 500°C is increased. Metallurgically, niobium is attractive because of its density, good workability, retention of tensile strength at high temperatures, and its high melting point. In the temperature range 920–1,200°C, niobium has been found superior to most other metals on a strength-to-weight basis for aerospace applications. In multicomponent alloys, zirconium and hafnium when added with niobium add effectively to strength, even more so than molybdenum or tungsten, but there is some sacrifice in ductility.

In metallurgy, niobium is classified as a refractory metal, along with tungsten, tantalum, and molybdenum. A comparison of the four metals is given in the accompanying table.

Niobium in Tool Steels. In the matrix method of tool-steel development, the composition of the heat-treated matrix determines the steel's initial composition. Carbide volume-fraction requirements then are calculated, based upon historical data, and the carbon content is adjusted accordingly. This approach has been used to design new steels in which niobium is substituted for all or part of the vanadium present as carbides in the heat-treated material. Niobium provides dispersion hardening and grain refinement, and forms carbides that are as hard as vanadium, tungsten, and molybdenum carbides.

Chemistry and Compounds. Elemental niobium is insoluble in HCl or HNO$_3$, but soluble in hydrofluoric acid or a mixture of hydrofluoric and HNO$_3$.

As might be expected from its $4d^4 5s^1$ electron configuration, niobium forms pentavalent compounds. However, the stability of its compounds of lower valence is greater than that of the corresponding tantalum compounds, in keeping with the group 5 position of niobium and tantalum. Nevertheless the similarity of the properties of the compounds of the two metals is so great that special methods are required for their separation, such as solvent extraction of the pentachlorides or chromatographic removal of adsorbed TaF$_5$ with an ethylmethyl ketone-water system. In addition, divalent and tetravalent compounds are known, and an interstitial, nonstoichiometric hydride.

Niobium forms a divalent oxide, NbO, insoluble in water, but readily soluble in acids or NH$_4$OH. It also gives by direct combination of the metal on heating with oxygen, the pentoxide, Nb$_2$O$_5$, which can be reduced by hydrogen at high temperature to NbO$_2$, and on heating with magnesium to Nb$_2$O$_3$.

Niobium(III) halides are known, notably the chloride, NiCl$_3$, which is of particular interest because its solution has been shown to contain Nb^{3+} ions (in equilibrium with NbCl$_6{}^{3-}$ complex ions).

Tetravalent niobium is believed to occur in the form of NbOCl$_4{}^{2-}$ ions in a solution obtained, with color change, by reduction of HCl solution of NbCl$_5$, and by inference in similarly reduced solutions of the other pentahalides. Tetravalent niobium also is found in the dioxide (see above) and the carbide, NbC.

Four pentahalides of niobium, NbF$_5$, NbCl$_5$, NbBr$_5$, and NbI$_5$ have been prepared by heating the pentoxide with carbon in a current of the halogen. They are hydrolyzed in H$_2$O, and even in concentrated aqueous solution of the respective halogen acids; the Nb^{5+} ion is apparently not present, but rather complex ions such as [NbOCl$_4$]$^-$ or [NbOCl$_5$]$^{2-}$. The products of partial hydrolysis of the pentahalides are oxyhalides, such as NbOF$_3$, NbOCl$_3$, and NbOBr$_3$. They are designated in the older literature as columbyl or columboxy compounds. The more stable oxyhalogen compounds of niobium are complexes, such as NbOF$_3 \cdot 3$NaF, NbOF$_3 \cdot$ ZnO $\cdot 6$H$_2$O, and NbOF$_3 \cdot 2$KF $\cdot$ H$_2$O.

Further complexes of Nb(V) are formed with oxygen-function compounds, such as o-dihydroxybenzene and acetylacetone.

The so-called niobic acid is the hydrated pentoxide, Nb$_2$O$_5 \cdot x$H$_2$O, insoluble in H$_2$O.

The metaniobates of the alkali metals, MNbO$_3$, the orthoniobates M$_3$NbO$_4$ and the pyroniobates, M$_4$Nb$_2$O$_7$, where M is an alkali metal, can be prepared by various alkali carbonate or hydroxide fusion processes.

Niobium forms a nitride, NbN, and a carbide, NbC.

Niobium forms a diamino compound, (NH$_2$)$_2$NbCl$_3$, and an ammine complex, NbCl$_5 \cdot 9$NH$_3$. It forms two cyclopentadienyl compounds, (C$_5$H$_5$)$_2$NbBr$_3$ and (C$_5$H$_5$)Nb(OH)Br$_2$. Its other organometallic compounds are essentially oxygen-functional ones, such as Nb(OCH$_3$)$_5$, Nb(OC$_2$H$_5$)$_5$, Nb(O)(OC$_5$H$_{11}$)$_3$, and Nb(OC$_5$H$_{11}$)$_5$. These compounds are named as substituted niobanes (thus, the last is pentabutoxy niobane) or as alkyl niobate esters.

Additional Reading

Carter, G.F. and D.E. Paul: "Materials Science and Engineering," ASM International, Materials Park, OH, 1991.

Gupta, C.K. and A.K. Suri: "Extractive Metallurgy of Niobium," CRC Press, LLC., Boca Raton, FL, 1994.

Lide, D.R.: "CRC Handbook of Chemistry and Physics 2000–2001," 81st Edition, CRC Press, LLC., Boca Raton, FL, 2000.

Staff: "ASM Handbook—Properties and Selection: Nonferrous Alloys and Pure Metals," ASM International, Materials Park, OH, 1990.

Staff: "Tool-Steel Developers Take Note of Niobium," *Advanced Materials 7 Processes*, 15 (June 1991).

Titran, R.H.: "Niobium and Its Alloys," *Advanced Materials & Processes*, 34 (November 1992).

NIST. The National Institute of Standards and Technology, the headquarters of which is located in Gaithersburg, Maryland, 20899. NIST replaces the former National Bureau of Standards (NBS), which was established by the U.S. Congress in 1901, with the objectives of: (1) serving as the basis for the nation's physical measurement system; (2) providing scientific and technological services for industry and government; (3) establishing a technical basis for equity in trade, and (4) providing technical services to promote public safety.

As of 2000, NIST is comprised of several divisions and departments, including:

- Technology services
- Manufacturing technology centers
- Standards services
- Standards Code and information
- Standards management
- Weights and measures
- Laboratory accreditation
- Measurement services
- Standard reference materials
- Physical measurement services
- Research and technology applications
- Technology development and small business
- National technology workshop
- Information services
- Research resources development
- Research information services

NITER. This potassium nitrate mineral KNO_3 of orthorhombic crystallization usually occurs as thin crusts, or as silky acicular crystals. It has a hardness of 2, and specific gravity of 2.09–2.14, is of white color, translucent with vitreous luster. It occurs as a surface efflorescence, or in soils rich in organic material in arid regions. World occurrences include Spain, Italy, Egypt, Arabia, India, Russia, the western United States, the Republic of South Africa, and Bolivia, South America. Large quantities were recovered from limestone caves in Tennessee, Kentucky, Alabama and Ohio during the Civil War for use in the manufacture of gunpowder. It is used as a source of nitrogen compounds, for explosives and fertilizers.

NITRATION. The process of adding nitrogen to a carbon compound, generally to create a nitro-derivative (adding a —NO_2 group) is termed nitration. An example is the formulation of nitrobenzene from benzene: $C_6H_6 + HNO_3 \rightarrow C_6H_5NO + H_2O$. In most instances, the —NO_2 group replaces a hydrogen atom. More than one hydrogen atom may be replaced, but each succeeding hydrogen represents a more difficult substitution. The nitrogen-bearing reactant may be: (1) strong HNO_3; (2) mixed HNO_3 and H_2SO_4; (3) a nitrate plus H_2SO_4; (4) nitrogen pentoxide N_2O_5; or (5) a nitrate plus acetic acid. Both straight-chain and ring-type carbon compounds can be nitrated. The alkanes yield nitroparaffins.

Various rules of addition govern the position of the entering nitro group, depending upon the conditions. For example, in the nonsubstituted benzene series, the nitro group can enter in the ortho, meta or para position, but the presence of some other group usually fixes the position of the entering nitro group. For example, it enters meta to a nitro, sulfonic, or carbonyl group, and ortho and para to a chloro, bromo, or hydroxy group. (These statements apply to the principal product formed, since in most substituted benzene reactions, a limited quantity of all ring positions are entered.) Various other rules govern other conditions in other aromatic series.

One of the great uses of nitration is to break into a pure hydrocarbon, which is usually more difficult to do by other means. The nitro group may then be changed and another group may take its place. Typical examples are nitration of ethane to form nitroethane, and of benzene to form nitrobenzene, which is easily changed to aniline.

An important economic consideration in any nitration process is the recovery of the spent acid. Since the nitration reaction forms H_2O, the reagents gradually become diluted to a point where they will not react any more. The water may be taken up during the reaction by removing it with oleum or acetic anhydride, a practice that still leaves large amounts of the reagents at the end of the process.

The HNO_3 is usually concentrated by distilling it from H_2SO_4 solution, which retains the H_2O. After the HNO_3 has been driven off, the temperature is raised and the H_2O is driven off the H_2SO_4, thereby concentrating the latter.

As an example of nitration, let us consider the preparation of nitrobenzene. Mixed acid consisting of strong H_2SO_4 plus HNO_3 is slowly added to benzene in a closed iron vessel provided with stirrer and reflux condenser. The acid must be added to the benzene. If it were done the other way, the benzene which was added first would be quickly nitrated all the way to a trinitrobenzene. The temperature is maintained from 45–55 °C. After the nitration is finished, the nitro compound is separated from the acid by decantation, since the nitrobenzene is lighter and does not mix with the acid. The nitrobenzene is washed with water and with dilute caustic or sodium carbonate solution and then again with water to give a neutral product. To obtain dinitrobenzene the reaction would be run with stronger acid and at a temperature of about 100 °C.

Since the reaction used concentrated H_2SO_4, ordinary iron vessels can be used, but the neutralization process must be carried out in leadlined tanks. Good agitation and adequate cooling facilities are necessary to avoid any local overheating and the formation of higher nitrated compounds.

NITRIC ACID. This important industrial chemical has been known for at least 1000 years. The acid was known to alchemists as *aqua fortis* (strong water) or *aqua valens* (powerful water). Nitric acid was of particular interest to the early experimenters because of its ability to dissolve a number of metals, including copper and silver. Early chemists were also fascinated by the fact that addition of sal ammoniac (ammonium chloride) gave *aqua regia* (royal water) which dissolves gold as well as silver.

Nitric acid is a colorless liquid, sp. gr. 1.503 (25 °C), freezing point −41.6 °C, and boiling point 86 °C. The 100% acid is not entirely stable and must be prepared from its azeotrope (constant-boiling mixture) by distillation with concentrated sulfuric acid. Reagent grade HNO_3 is a water solution containing about 68% HNO_3 (weight). This strength corresponds to the constant-boiling mixture of the acid with water, which is 68.4% HNO_3 (weight) at atmospheric pressure and boils at 121.9 °C. Nitric acid is completely miscible with water. It forms two solid hydrates, $HNO_3 \cdot H_2O$ and $HNO_3 \cdot 2H_2O$, with corresponding melting points of approximately −38 and −18.5 °C. Nitric acid is a strong acid and a powerful oxidizer. In dilute solutions, it is almost completely ionized to H^+ and NO_3^- ions and behaves like a strong acid.

With organic compounds, HNO_3 may act as a nitrating agent, as an oxidizing agent, or simply as an acid. The classic example of nitration is its reaction with benzene or toluene in the presence of concentrated H_2SO_4 to form nitrobenzene or nitrotoluene (TNT). An example of oxidation properties is in the oxidation of cyclohexanol by HNO_3 to produce adipic acid, an intermediate of nylon. Behaving like an acid, it forms nitroglycerin by esterification of glycerol in the presence of concentrated sulfuric acid.

An interesting property of HNO_3 is its ability to passivate some metals, such as iron and aluminum. This property is of significant industrial importance, since modern processes for producing the acid depend on it. Modern suitability formulated stainless steel alloys are usefully resistant to nitric acid through a wide range of conditions. The acid's passivity or the metal's resistance to attack is attributed to the formation of a protective oxide layer on the surface of the metal.

Nitric acid is a high tonnage industrial chemical. Much of the production is used in the manufacture of agricultural fertilizers, largely in the form of ammonium nitrate, NH_4NO_3. See also **Fertilizer.** About 15% of the nitric acid produced is used in explosives (nitrates and nitro compounds), and about 10% is consumed by the chemical industry. As the red fuming acid or as nitrogen tetroxide, HNO_3 is used extensively as the oxidizer in propellants for space rockets and missiles.

Production of Nitric Acid. Three commercial methods have been developed for nitric acid production: (1) the reaction between sulfuric acid and sodium nitrate, (2) the thermal combination of oxygen and nitrogen in air, and (3) the catalytic oxidation of ammonia and absorption of the gaseous products in waters. There are numerous variations of these fundamentals processes. The principal process used today is based on the catalytic oxidation of ammonia and absorption of the gaseous products in water. This process was developed by Ostwald (Germany) and based on earlier work of Kuhlmann (France). In the Ostwald process, HNO_3 is produced in a 3-stage operation: (1) Ammonia is oxidized to nitric oxide, (2) the nitric oxide is further oxidized to nitrogen dioxide, and (3) the gases are absorbed in water to yield HNO_3 according to

$$4\, NH_3 + 5\, O_2 \longrightarrow 4\, NO + 6\, H_2O$$

$$2\, NO + O_2 \longrightarrow 2\, NO_2$$

$$3\, NO_2 + H_2O \longrightarrow 2\, HNO_3 + NO$$

The nitric oxide formed in the last equation returns to the gas phase, is reoxidized to nitrogen dioxide, and reabsorbed. These reactions are highly exothermic. In actuality, numerous complex reactions occur in addition to the main reactions just outlined.

In a manufacturing plant, air is preheated, mixed with superheated ammonia vapor, and reacted catalytically over a gauze composed of 90% platinum and 10% rhodium at a temperature of $800-960\,°C$ and operating pressures between 1 and 8.2 atmospheres. The reaction produces nitrogen dioxide, NO_2 and nitric oxide, NO. The latter is oxidized to NO_2 in the reaction train. The NO_2 actually exists in equilibrium with its dimer, N_2O_4. This equilibrium mixture, sometimes referred to as nitrogen peroxide, is absorbed in water in a cooled absorber tower to form HNO_3 at a strength of $55-60\%$ HNO_3.

NITRIDING. Surface hardening of alloy steels by heating the metals to a temperature of $490-650\,°C$ in an atmosphere of partially dissociated NH_3 (ammonia). As in cyaniding, hardening results from the formation of nitrides of iron and of certain alloying elements that may be present in the steel. Much longer heating time is required than in carburizing practice, and while the depth of penetration is generally less, the maximum hardness at the surface is higher, $900-1,100$ D.P.H. (Vickers Brinell) compared to $800-900$ D.P.H. for an average carburized case. Nitriding also differs from carburizing in that the parts are fully heat-treated to develop the required core properties before the nitriding treatment. Because of the comparatively low temperature of the process, distortion and dimensional changes are at a minimum. Nitrided steels have good corrosion-resistance when used for valves, pump parts, shafting, and bearing surfaces operating in steam, crude oil, gasolines, and gaseous products of combustion. The fatigue strength is also improved by nitriding.

Other typical applications are piston pins, crankshafts, cylinder liners, timing gears, gauges, and ball and roller bearing parts.

NITRILE RUBBER. See **Elastomers**.

NITRILES. See **Amines**.

NITRO- AND NITROSO-COMPOUNDS. Nitro-compounds contain the nitro-group ($-NO_2$) attached directly to a carbon atom; nitroso-compounds contain the nitroso-group ($-NO$) similarly attached. A very important member of this group is nitrobenzene, which upon reduction yields a variety of products, important in the synthesis of drugs and dyes. See Table 1.

TABLE 1. REPRESENTATIVE NITRO- AND NITROSO COMPOUNDS

Compound	Formula	Melting Point, °C	Boiling Point, °C
REPRESENTATIVE NITRO COMPOUNDS			
Nitrobenzene	$C_6H_5 \cdot NO_2$	6	211
1,3-Dinitrobenzene	$C_6H_4(NO_2)_2$ (1,3)	90	302
2-Nitrotoluene	$CH_3C_6H_4(NO_2)$ (2)	−11	222
2,4-Dinitrotoluene	$CH_3C_6H_3(NO_2)_2$ (2,4)	70	300
Trinitrotoluene (TNT)	$CH_3C_6H_2(NO_2)_3$ (2,4,6)	81	240 expl.
3-Nitrophenol	$HOC_6H_4 \cdot NO_2$ (3)	96	194 (70 torr)
2,4,6-Trinitrophenol (picric acid)	$HOC_6H_2(NO_2)_3$ (2,4,6)	122 expl.	>300
4-Nitrobenzaldehyde	$C_6H_4(COH)(NO_2)$ (1,4)	58	164 (23 torr)
4-Nitrobenzoic acid	$C_6H_4(COOH)(NO_2)$ (1,4)	240	subl.
4-Nitrobenzyl alcohol	$C_6H_4(CH_2OH)(NO_2)$ (1,4)	93	185 (12 torr)
2-Nitronaphthalene	$C_{10}H_7(NO_2)$ (2)	79	165 (15 torr)
1-Nitroanthraquinone	$C_6H_4(CO)_2C_6H_3(NO_2)$ (1)	230	subl.
2-Nitropropane	$(CH_3)_2CHNO_2$	−93	120
Nitroethyl alcohol	$CH_2OHCH_2NO_2$	<−80	194
Nitrobromoform (bromopicrin)	NO_2CBr_3	10	expl.
Nitrochloroform (chloropicrin)	NO_2CCl_3	−64	112
Nitrofurane	$C_4H_3O \cdot NO_2$	28	
Nitrourea	$OC \big\langle {}^{NH_2}_{NHNO_2}$	155 dec.	
Nitroguanidine	$HNC \big\langle {}^{NH_2}_{NHNO_2}$	246	
1,3-Nitroaniline	$C_6H_4(NO_2)(NH_2)$ (1,3)	114	>285
REPRESENTATIVE NITROSO COMPOUNDS			
Nitrosobenzene	C_6H_4NO	68	58 (18 torr)
4-Nitrosophenol (4-quinoneoxime)	$C_6H_4(OH)(NO)$ (1,4)	125	144 dec.
4-Nitrosonaphthol-1 (4-naphthaquinoneoxime)	$C_{10}H_6(OH)(NO)$ (1,4) or $C_{10}H_6(O)(NOH)$ (1,4)	193	
2-Nitrosonaphthol-1	$C_{10}H_6(OH)(NO)$ (1,2)	163 dec.	
N-Nitrosomethylaniline	$C_6H_5N \big\langle {}^{CH_3}_{NO}$	13	128 (20 torr)
4-Nitrosophenylaniline	$C_6H_5NH \cdot C_6H_4NO$	145	
1-Nitrosonaphthylamine-2	$C_{10}H_6(NH_2)(NO)$ (2,1)	151	
Diphenylnitrosamine	$(C_6H_5)_2N \cdot NO$	66	

dec., decomposes; expl., explodes; sub., sublimes

Alkylnitro-Compounds:

Primary	Secondary	Tertiary
$CH_3CH_2 \cdot NO_2$	$(CH_3)_2CH \cdot NO_2$	$(CH_3)_3C \cdot NO_2$
Nitroethane	Nitrodimethylmethane (2−nitropropane)	Nitrotrimethylmethane

Isomeric Nitrites:

$CH_3CH_2 \cdot ONO$	$(CH_3)_2CH \cdot ONO$	$(CH_3)_3 \cdot ONO$
Ethyl nitrite	Isopropyl nitrite	1,1−dimethylethyl nitrite

Alkylnitroso-Compounds:

$(CH_3)_3C \cdot NO$
Nitrosotrimethylmethane

Nitrates:

$CH_3CH_2 \cdot ONO_2$	$(CH_3)_2CH \cdot ONO_2$	$(CH_3)_3C \cdot ONO_2$
Ethyl nitrate	Isopropyl nitrate	1,1−dimethylethyl nitrate

Nitrosamine:

$(C_2H_5)_2N{:}NO$
Diethylnitrosamine

Benzenoid Nitro- and Nitroso-Compounds:

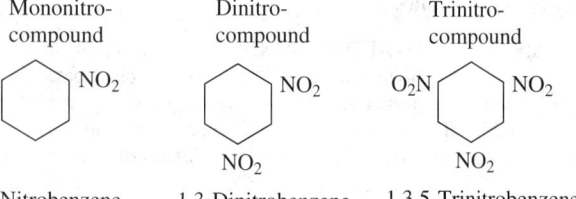

Mononitro-compound	Dinitro-compound	Trinitro-compound
Nitrobenzene	1,3-Dinitrobenzene	1,3,5-Trinitrobenzene

Nitroso-compounds

Nitrosobenzene   Diphenylnitrosamine

Under the proper conditions of concentration of HNO_3 and of temperature, benzene forms mainly nitrobenzene, nitrobenzene forms mainly 1,3-dinitrobenzene, and 1,3-dinitrobenzene, mainly 1,3,5-trinitrobenzene.

When nitrobenzene is treated (1) with zinc and calcium chloride or ammonium chloride solution, beta-phenylhydroxylamine, C_6H_5NHOH, is formed, and from this by treatment with chromic acid or ferric chloride nitrosobenzene is formed, (2) with tin or iron and HCl, aniline, $C_6H_5NH_2$, is formed and from this by treatment with nitrous acid followed by treatment with stannous chloride plus HCl phenylhydrazine, $C_6H_5NH \cdot NH_2$, is formed.

Mono- or poly-substituted nitro-compounds are changed in whole or in part to the corresponding amino-compounds by proper choice of reducing agent and temperature, e.g., in acid medium 1,3-dinitrobenzene yields 1,3-phenylenediamine, $C_6H_4(NH_2)_2(1, 3)$, and with ammonium sulfide yields 3-nitroaniline $(1)H_2NC_6H_4NO_2(3)$. When diphenylnitrosamine is reduced, 1,1-diphenylhydrazine, $(C_6H_5)_2N \cdot NH_2$, is formed.

See also **Nitration**.

NITROCELLULOSE. See **Cellulose**.

NITROGEN.
Chemical element, symbol N, at. no. 7, at. wt. 14.0067, periodic table group 15, mp $-209.86\,°C$, bp $-195.8\,°C$, critical temperature $-147.1\,°C$, critical pressure 33.5 atmospheres, density 1.14 g/cm^3 (solid), 1.25057 g/L (0 °C, 760 torr), 0.9675 (air = 1.0000). Solid nitrogen has a hexagonal crystal structure. Nitrogen at standard conditions is a colorless, odorless, tasteless gas. The gas is slightly soluble in H_2O (2.35 parts nitrogen in 100 parts H_2O at 0 °C), the solubility decreasing with increasing temperature (1.55 parts nitrogen in 100 parts H_2O at 20 °C). Nitrogen is slightly soluble in alcohol and is essentially insoluble in most other known liquids. There are two naturally occurring isotopes, ^{14}N and ^{15}N,

with ^{14}N by far the most abundant (99.635%). Four radioactive isotopes have been identified, ^{12}N, ^{13}N, ^{16}N, and ^{17}N, all with extremely short half-lives measured in seconds or minutes. In terms of abundance in igneous rocks in the Earth's crust, nitrogen does not appear among the first 37 most abundant elements. In terms of abundance in seawater, nitrogen ranks 16th, with an estimated 2,300 tons of nitrogen per cubic mile of seawater. In terms of cosmic abundance, nitrogen ranks 7th. For comparison, assigning a value of 10,000 to silicon, the figure for nitrogen is 160,000 and that for hydrogen, estimated the most abundant, a figure of 3.5×10^8. Of dry air in the earth's atmosphere, disregarding pollutants, 78.09% is nitrogen by volume and 75.54% by weight. In the atmosphere, the nitrogen is mixed with oxygen, argon, the rare gases, CO_2, and H_2O vapor. Nitrogen was first identified as an element by Daniel Rutherford in 1772. Lavoisier further confirmed Rutherford's findings in 1776. Like oxygen, nitrogen is essential to practically all forms of life, making some of the compounds of this element extremely important as foods and fertilizers. Nitrogen serves the important function of diluent in the earth's atmosphere, controlling natural burning and respiration rates that otherwise would proceed much faster with higher concentrations of oxygen. Nitrogen is an important ingredient of numerous inorganic and organic compounds, including alkaloids, amides, amines, cyanides, cyanogens, diazo compounds, hydrazines, imides, nitrates, nitrides, nitrites, nitriles, oximes, purines, pyridines, and ureas. In terms of high-tonnage production, the nitrogen compound NH_3 (ammonia) ranks first with worldwide production exceeding 50 million tons annually.

First ionization potential 14.84 eV; second, 29.47 eV; third, 47.17 eV; fourth, 73.5 eV; fifth, 97.4 eV. Oxidation potentials $H_2N_2O_2 + 2H_2O \rightarrow 2HNO_2 + 4H^+ + 4e^-$, -0.80 V; $N_2O_4 + 2H_2O \rightarrow 2NO_3^- + 4H^+ + 2e^-$, -0.81 V; $HNO_2 + H_2O \rightarrow NO_3^- + 3H^+ + 2e^-$, -0.94 V; $NO + 2H_2O \rightarrow NO_3^- + 4H^+ + 3e^-$, -0.96 V; $NO + H_2O \rightarrow HNO_2 + H^+ + e^-$, -0.99 V; $2NO + 2H_2O \rightarrow N_2O_4 + 4H^+ + 4e^-$, -1.03 V; $2HNO_2 \rightarrow N_2O_4 + 2H^+ + 2e^-$, -1.07 V. $N_2O + 3H_2O \rightarrow 2HNO_2 + 4H^+ + 4e^-$, -1.29 V; $N_2O + H_2O \rightarrow 2NO + 2H^+ + 2e^-$, -1.59 V; $N_2 + H_2O \rightarrow N_2O + 2H^+ + 2e^-$, -1.77 V; $N_2O_4 + 4OH^- \rightarrow 2NO_3^- + 2H_2O + 2e^-$, 0.85 V; $NO + 2OH^- \rightarrow NO_2^- + H_2O + e^-$, 0.46 V; $N_2O_2^{2-} + 4OH^- \rightarrow 2NO_2^- + 2H_2O + 4e^-$, 0.18 V; $NO_2^- + 2OH^- \rightarrow NO_3^- + H_2O + 2e^-$, -0.01 V; $N_2O_2^{2-} \rightarrow 2NO + 2e^-$, -0.10 V; $N_2O + 6OH^- \rightarrow 2NO_2^- + 3H_2O + 4e^-$, -0.15 V; $N_2O + 2OH^- \rightarrow 2NO + H_2O + 2e^-$, -0.76 V. $2NO_2^- \rightarrow N_2O_4 + 2e^-$, -0.88 V.

Other physical properties of nitrogen are given under **Chemical Elements**.

Industrial Nitrogen

Like many of the elements, the compounds of nitrogen by far exceed the use of elemental nitrogen (discounting its important role as diluent in the atmosphere). Industrially, nitrogen gas is produced as a by-product in the liquefaction of air to produce pure oxygen. For some applications, nitrogen provides an excellent inert atmosphere for electric furnace operations and for the gaseous insulation of transformers. An inert atmosphere is required where air must be excluded. Nitrogen is one of the three main gases used for such atmospheres, the other two being carbon monoxide and hydrogen. In providing an inert atmosphere, nitrogen reduces the velocities of reactions, lowers the partial pressure and reduces the flammability of any active gases that may be present. Since commercial nitrogen usually contains traces of oxygen, H_2O vapor, and CO_2, sufficient to cause some oxidation at high temperatures, methane may be added to make the gas fully inert.

Nitrogen gas also is required for nitriding certain alloy steels, but pure gas is not required. The nitrogen is provided by dissociating ammonia at the process temperatures ranging from 475–650 °C. Metals treated in this manner are hardened by the formation of nitrides on their surface (casehardening). In cyaniding, iron-base alloys simultaneously absorb carbon and nitrogen by heating the metals in a cyanide salt. Again, the nitrogen is not required in initial gaseous form. See also **Nitriding**. Several powder metallurgy techniques also utilize dissociated NH_3 atmospheres.

Environmental Aspects of Nitrogen

The oxides of nitrogen are among the most critical of air pollutants — both in their effects and in their abatement. These aspects of nitrogen are discussed under **Pollution (Air)**.

Chemistry and Compounds

Most of the high-tonnage nitrogen-bearing compounds are described elsewhere in this volume. See also **Ammonia**; **Ammonium Chloride**; **Ammonium Hydroxide**; **Ammonium Nitrate**; **Ammonium Phosphates**; **Ammonium Sulfate**; and **Fertilizer**.

In the laboratory, nitrogen, mixed with argon, neon, krypton, and xenon, is obtained from the air by passing it over heated copper to remove the oxygen, or pure by fractional distillation of liquid air whereby the nitrogen distills off before the oxygen. Pure nitrogen may also be obtained by heating such compounds as ammonium nitrite and ammonium dichromate, and collecting the gas. Mixed with carbon monoxide in producer gas, nitrogen may be utilized without separation by first making methyl alcohol from carbon monoxide and hydrogen and then using hydrogen and nitrogen for ammonia. When nitrogen at low pressure is subjected to the silent electric discharge, activated nitrogen is produced. Activated nitrogen displays a golden yellow afterglow upon cessation of the current, increased by cooling and decreased by heating. This form of nitrogen is very active with phosphorus, with alkali metals (forming azides), with the vapor of zinc, mercury, cadmium, arsenic (forming nitrides), with many metallic chlorides (forming a green fluorescence), and with hydrocarbons (forming hydrocyanic acid and cyanides). The transformation of nitrogen to activated nitrogen is partial, and its return to ordinary nitrogen takes place rapidly, in about one minute.

The metal amides and imides are important in the nitrogen system. The amides of the active metals are produced by (1) reaction of the metal with NH_3, (2) reaction of the metal hydride with NH_3, (3) reaction of the metal nitride with ammonia, (4) reaction with another amide, as $KNH_2 + NaI \rightarrow NaNH_2 + KI$ (in liquid NH_3). This last method is generally useful for the preparation of the heavy metal amides and imides from halides and binary halogenoids of the heavy metals. Cadmium amide, $Cd(NH_2)_2$ and lead imide, PbNH, for example, are readily prepared in this way. In some cases neither the amide nor the imide is stable, and the reaction proceeds to the nitride.

$$3\ HgBr_2 + 6\ KNH_2 \xrightarrow[\text{liq.}]{NH_3} Hg_3N_2 + 6\ KBr + 4\ NH_3$$

The metal amides and imides are very reactive with oxygen, and are often unstable or even explosive. Some nitrides (e.g., of silver, gold, and mercury) are explosive, but others are stable. The latter may be obtained, (1) by reaction with the metal with nitrogen or ammonia at higher temperatures, e.g., aluminum nitride and magnesium nitride, AlN and Mg_3N_2, (2) by deamination of the metal amide or azide on heating, e.g., Ba_3N_2. The great thermal stability of certain nitrides, e.g., those of boron, silicon and phosphorus, BN, Si_3N_4 and P_3N_5, is attributed to polymerization. Many of the transition metal nitrides are interstitial compounds and are hard and metal-like in their properties.

In the nitrogen system, hydrazine is analogous to hydrogen peroxide in the oxygen system, its structure being

It is readily oxidized, even undergoing auto-oxidation under many conditions, and it is a powerful reducing agent. Like hydrogen peroxide it readily disproportionates (e.g., with a platinum catalyst), giving nitrogen and NH_3. Its reactivity (and other properties) makes it, and its derivative, unsymmetrical dimethylhydrazine, important rocket fuels. It forms addition compounds with many substances, including a monohydrate with H_2O. Hydrazine ($pK_{B1} = 6.04$, $pK_{B2} = 14.88$) forms hydrazinium(1+) compounds, containing the $N_2H_5^+$ ion, analogous to ammonium, and hydrazinium(2+) compounds containing the $N_2H_6^{2+}$ ion.

Hydroxylamine is related in its structure both to hydrazine (see formula above) and to hydrogen peroxide. Its bond lengths

The chemical properties of hydroxylamine also suggest a compound intermediate between hydrazine and hydrogen peroxide. Its bond lengths

are, N–O, 1.46 Å, N–H, 1.01 Å, O–H, 0.96 Å, and its angles are H–O–N, 103°, H–N–O, 105°, and H–N–H, 107°. It is a base ($pK_B = 9.02$), forming salts containing the hydroxylammonium ion $HONH_3^+$.

Hydrazoic acid, HN_3, $pK_A = 4.72$, and most of its covalent compounds (including its heavy metal salts) are explosive. It is formed (1) in 90% yield by reaction of sodium amide with nitrous oxide, (2) by reaction of hydrazinium ion with nitrous acid, (3) by oxidation of hydrazinium salts, (4) by reaction of hydrazinium hydrate with nitrogen trichloride (in benzene solution). Hydrazoic acid forms metal azides with the corresponding hydroxides and carbonates. It reacts with HCl to give ammonium chloride and nitrogen, with H_2SO_4 to form hydrazinium acid sulfate, with benzene to form aniline, and it enters into a number of oxidation-reduction reactions.

The azides, except those of mercury(I), Hg(I), thallium(I), Tl(I), copper, Cu, silver, Ag, and lead, Pb, are readily prepared from hydrazoic acid and the oxide or carbonate of the metal, or by metathesis of the metal sulfate with barium azide. They are all thermally unstable, giving nitrogen and free metal or occasionally nitride. The azide ion appears to resonate between four structures:

These structures are in accord with a spacing of 1.15 Å and electronic charges of -0.83, 0.66, and 0.83 on the three nitrogen atoms.

N(I) compounds. Hydration of nitrogen(I) oxide, N_2O to hyponitrous acid, $H_2N_2O_2$, is not possible. However, the latter decomposes (in three steps) to yield the former, which is thus its anhydride. Spectroscopic studies indicate a linear structure for N_2O, resonating between

However, heat capacity measurements give a higher entropy at low temperatures than spectroscopic studies do, which is explained by a partial randomness of the structure at low temperatures.

Hyponitrous acid ($pK_{A1} = 7.05$, $pK_{A2} = 11.0$) and its salts are obtained by: (1) reduction of sodium nitrite with (a) sodium amalgam, (b) by electrolysis, (c) by stannous or ferrous salts; (2) by reduction of alkyl nitrates; (3) by reduction of hydroxylamine by noble metal oxides; and (4) by reduction of sodium hydroxylamine monosulfonate in alkaline solution.

Explosive salts such as NaNO can be prepared by the reaction of NO and liquid ammonia solutions of alkali metals. The unstable free acid, HNO, is thought to be an intermediate in many redox reactions of nitrogen compounds.

Nitramide, NO_2NH_2, a weak acid ($pK_A = 6.59$), is relatively more stable than its isomer hyponitrous acid.

N(II) Compounds. Nitrogen(II) oxide is formed in many reductions of nitrous acid, but is best prepared pure by reduction with ferrous ions, Fe^{2+}, or iodide ions, I^-. It undergoes many types of addition reactions, but its very slight tendency to dimerize and its low reactivity under ordinary conditions suggest that its odd electron lies in an antibonding orbital of very low energy; and the molecular orbital formulation is

$$NO[KK(z\sigma)^2(y\sigma^*)^2(x\sigma)^2(w\pi)^4(v\pi^*)]$$

The nitrosyl compounds can be readily classified on the basis of three modes of reaction of the NO molecule in accordance with the above formulation.

1. It can lose (or partly lose) the odd electron to form an ion of the formula $:N{\equiv}O:^+$. This formula gives rise to ONF, ONCl and ONBr by direct reaction of NO and the halogen. These are covalent compounds. Such salts as $NOBF_4$, $NOPF_6$, $NOAuF_4$, $NOSO_3F$, and $NOHSO_4$, on the other hand, are ionic. These may be considered the salts of nitrous acid acting as a base, $ONOH \rightleftharpoons NO^+ + OH^-$, $pK_B = 18.2$.

2. It can gain an electron to form a negative ion of the formula

$$N \equiv \overset{..}{\overset{..}{O}}:^-$$

Thus dry NO reacts with sodium in liquid ammonia to form sodium nitrosyl, NaNO (empirical formula).

3. It can share a pair of electrons to form a coordinate link, as it does in coordination compounds. In most of these, it appears to coordinate as the positive ion, by transfer of an electron to an acceptor metal, which is thereby reduced by 1 unit in oxidation state. This causes, in some cases, the need for placing a negative charge on the metal. To avoid this, Pauling assumed the presence of four bonding electrons, involving structures of the type

$$M = \overset{+}{N} = \overset{..}{O}:$$

Nitrogen(III) Compounds. Nitrogen(II) oxide, NO, readily enters into equilibrium with NO_2 to form N_2O_3, nitrogen sesquioxide. The latter is unstable even at room temperature and consists of an equilibrium mixture of the three compounds. Its structure appears to be $O=N-NO_2$. If an equimolar mixture of NO and NO_2 is cooled and condensed, a blue liquid, bp $3.5\,°C$, largely N_2O_3, is obtained. The latter readily combines with H_2O to form nitrous acid, HNO_2 ($pK_A = 3.29$). Nitrous acid is unstable, forming the equilibrium mixture, $3HNO_2 \rightleftharpoons NO_3^- + 2NO + H_3O^+$, which in concentrated solution or on warming is largely displaced to the right ($K = 39.6$ at $30\,°C$). Moreover, the NO undergoes further reactions, so that the actual system is complex. One of these reactions is: $NO + OH \rightarrow NO^+ + OH^- \rightleftharpoons NO \cdot OH \rightleftharpoons HNO_2$.

The existence of NO^+ and NO^- helps to explain the kinetics of nitrous acid as an oxidizing agent. It oxidizes I^-, Sn^{2+}, Fe^{2+}, Ti^{3+}, $S_2O_3^{2-}$, SO_2, and H_2S. It reacts with NH_3, urea, sulfonates and some other nitrogen compounds to produce nitrogen. With aromatic amines in the cold, it gives diazo compounds, while with secondary amines it gives nitroso compounds. Nitrous acid also functions as a reducing agent, as in the reactions with permanganate and hydrogen peroxide, in which nitrate ion is formed.

The nitrites vary widely in solubility, those of the alkalies and alkaline earths being very soluble, while those of the heavy metals are only slightly so. Moreover, the latter are relatively unstable, some decomposing at room temperature. The nitrites, like nitrous acid, function either as oxidizing or reducing agents. X-ray and spectroscopic studies give a triangular structure for the nitrite ion, with the N—O bond length 1.13 Å and the O—N—O angle $120-130°$. Values of 1.23 Å and $116°$ have also been reported. Complex ions containing the NO_2 group may be either nitrito complexes (e.g., $Co(NH_3)ONO^{2+}$) or nitro complexes (e.g., $Co(NH_3)NO_2^{2+}$). The former of these two examples readily isomerizes to the latter.

Nitrosyl fluoride, NOF, and nitrosyl chloride, NOCl, are quite stable, but the bromide decomposes at room temperature. They are prepared by direct union of NO and the halogen, among other methods. Three trihalides, NF_3, NCl_3, and NI_3, are known. The first is a colorless stable gas; NCl_3 is a yellow liquid and NI_3, a brown solid; both are explosive. The contrast in stability is attributed to the large amount of ionic resonance energy of the N—F bond, which gives NF_3 a negative heat of formation.

Nitrogen(IV) Compounds. Nitrogen dioxide, NO_2, readily associates to form the tetroxide, N_2O_4, so that at ordinary temperatures and pressures both forms are present in equilibrium. Since nitrogen dioxide has an unpaired electron, it is paramagnetic and colored (red). N_2O_4 is diamagnetic and colorless. As with NO, the odd electron is in an antibonding orbital but of higher energy so that NO_2 is more reactive and more readily undergoes dimerization. The N—O bond length is 1.20 Å and the angle is $132°$ (electron diffraction). The structure of N_2O_4 is, on the basis of spectral and entropy considerations,

This formula is at variance with Pauling's stability argument, but is supported by Ingold's evidence (Nature, **159**, 743, 1947). Longuet-Higgins has proposed the structures

Nitrogen dioxide molecules react with NO to form N_2O_3, in an equilibrium mixture. The equilibrium mixture of NO_2 and N_2O_4 also reacts with water in a series of reactions

$$2\,NO_2 + H_2O \rightleftharpoons H^+ + NO_3^- + HNO_2$$

$$3\,HNO_2 \rightleftharpoons H^+ + NO_3^- + 2\,NO + H_2O$$

In warm solution, at high acidity, the second reaction is very rapid. In basic solutions the simple disproportionation $N_2O_4 + 2OH^- \rightarrow NO_2^- + H_2O$ takes place.

Nitrogen(V) Compounds. Nitrogen(V) oxide, N_2O_5, the anhydride of nitric acid, is a white solid subliming at $32.4\,°C$ and 760 mm. It hydrates readily to HNO_3, is a strong oxidizing agent, and decomposes at $20\,°C$ slowly into NO_2 and O_2. Its structure in the gas state consists of the molecules

However, x-ray, Raman, and infrared spectra show the crystalline solid to consist of NO_2^+ and NO_3^- ions.

Pure nitric acid, HNO_3, is a colorless liquid boiling with decomposition at $86\,°C$ and 760 torr. Upon continued heating it decomposes into NO_2, O_2 and H_2O. It is a fairly strong acid ($K_A = 22$), showing dissociation in concentrated solutions, and the presence of nitryl cation, NO_2^+ (nitronium ion). Solutions of HNO_3 in H_2SO_4 owe many of their properties to ions such as NO_2^+ and NO^+, as well, of course, as to HSO_4^- and oxonium ions.

The properties of HNO_3 are in accordance with resonance between the three electronic structures:

in which the last formula contributes a relatively small proportion to the overall structure. The two N—O bond lengths are 1.22 Å, and N—O—H bond lengths 1.41 Å and 0.96 Å. The N—O—H angle is $90°$ and the O—N—O angle $130°$.

The reactions of nitric acid are of three types: (a) acid-base reactions which are typical of a strong acid; (b) oxidation reactions, such as those with metals and organic materials, the latter often involving carbonization; (c) substitution reactions such as the replacement of —H by —NO_2 in aromatic hydrocarbons, to form nitro compounds, or of hydroxyl hydrogen by —NO_2 to produce esters of HNO_3.

These esters of nitric acid form one of the two groups of nitrates, the covalent group, which are also exemplified by nitryl hypofluorite and hypochlorite ($FONO_2$ and $ClONO_2$), often called fluorine and chlorine nitrate. Most nitrates, however, are ionic, e.g., salts of HNO_3. All metal nitrate are soluble in H_2O. Anhydrous metal nitrates, such as $Cu(NO_3)_2$, $Ti(NO_3)_4$, $VO(NO_3)_3$, $CrO_2(NO_3)_2$, $Si(NO_3)_4$, can be made by the action of liquid N_2O_4 on the metal (e.g., Cu) or of $ClONO_2$ on the corresponding chloride (e.g., the other examples given above).

The nitrate ion is considered to resonate between three equivalent structures of the form:

Two nitryl halides, NO_2F and NO_2Cl, are known, as well as nitryl salts, such as NO_2AsF_6, NO_2SbF_6, $(NO_2)_2SiF_6$, NO_2ClO_4, etc. Nitrogen also forms higher oxides, such as NO_3, and possibly NO_4, under action of the electric discharge.

Nitrate Losses from Disturbed Forest Ecosystems

Nutrient losses occur following a forest harvest or other disturbance, whether natural or anthropogenic. Studies have shown a variety of patterns of such losses. Vitousek et al. (1979) report on a systematic examination of nitrogen cycling in disturbed forest ecosystems and show that at least 8 processes, operating in 3 stages in the nitrogen cycle, can delay or prevent solution losses of nitrate from disturbed forests. The study involved 19 forest sites in the United States, including Pack Forest, Findley Lake, and Cascade Head in the northwest; Tesuque Watersheds in the southwest; Lake Monroe in southern Indiana; Coweeta in southwestern North Carolina; and Harvard Forest, Mount Mossilauke, and Cape Cod in the northeastern United States.

The 3 stages and 8 operative processes identified are:

Stage 1. Processes preventing or delaying ammonium accumulation.
(a) Nitrogen immobilization
(b) Ammonium fixation
(c) Ammonia volatilization
(d) Plant nitrogen uptake

Stage 2. Processes preventing or delaying nitrate accumulation.
(e) Lag in nitrification
(f) Denitrification to: $-N_2$, N_2O, or NO_x, $-NH_4$

Stage 3. Processes preventing or delaying nitrate mobility.
(g) Lack of water
(h) Nitrate sorption
(f) Denitrification at depth

The researchers stress that the net effect of all of these processes, except uptake by regrowing vegetation, is insufficient to prevent or delay losses from relatively fertile sites and thus such sites have the potential for very high nitrate losses following disturbance.

Nitrogen Fixation

A positive balance of usable nitrogen on earth depends upon nitrogen fixation which is the process by which atmospheric nitrogen, N_2, is converted either by biological or chemical means to a form of nitrogen, such as ammonia, NH_3, that can be used by plants and other biological agents. Insofar as the total amount of N_2 fixed, the biological processes for converting from N_2 to NH_3 are the most significant. In biological nitrogen fixation, microorganisms, either free-living or in symbiosis with plants (mainly in root nodules), reduce N_2 to NH_3 at atmospheric pressure and within the temperature range of 20–37 °C. This natural process is to be contrasted with industrial chemical conversion processes, which may require up to 300 atmospheres of pressure and a reaction temperature range of 200–300 °C.

Biological Nitrogen Fixation. The occurrence and importance to soil fertility of biological nitrogen fixation have been known since the early 1800s. The first major finding did not occur until 1960, however, when it was shown that cell-free extracts of the anaerobic bacterium *Clostridium pasteurianum* could be made to fix nitrogen if molecular oxygen, O_2, were rigorously excluded—and also if pyruvic acid, a source of energy and electrons, was supplied. This finding demonstrated that studies no longer were restricted to whole cells, as previously indicated, but that it should be possible to isolate and chemically identify the components of the nitrogen-fixing system.

The first demonstrable product of cell-free N_2 fixation is NH_3, as had been strongly suggested by previous whole-cell studies. Since the reduction of N_2 to $2NH_3$ requires six electrons and since most electron transfer systems known in biochemical pathways involve either a one-or a two-electron transfer, it could be expected that either six one-electron or three two-electron transfer steps would be involved in nitrogen fixation. This would also suggest the existence of nitrogen compounds of valence states (reduction states) intermediate between N_2 and NH_3. However, no such intermediates have been found even in systems using cell-free extracts.

Because of failure to detect intermediates, attention was focused on the mechanism in extracts of *Clostridium pasteurianum* through which electrons were transferred from pyruvic acid to the nitrogen-fixing system. These investigations led to the discovery and isolation of the new electron carrier ferredoxin (Fd) which functioned by accepting electrons released during pyruvate oxidation by enzymes present in the clostridial extracts. The electrons from reduced Fd were transferred to a variety of different acceptors as directed by the cell. For example, some of the electrons from reduced Fd were transferred to hydrogenase, an enzyme which combined the electrons with protons (H^+) to produce molecular hydrogen, H_2, a major by-product of this anaerobe. Other electrons from reduced Fd were transferred via a flavoprotein carrier to nicotinamide adenine dinucleotide phosphate ($NADP^+$) to yield NADPH, a reduced electron carrier shown to be important in the metabolism of all biological agents. It was also found that electrons from Fd were required for nitrogen fixation when pyruvate was present as supporting substrate.

A major finding was that H_2, through hydrogenase, would act as an electron source for reducing ferredoxin. Thus, in these extracts, H_2 could be used to reduce $NADP^+$ to NADPH and NO_2^- to NH_3, and Fd was necessary as an intermediary electron carrier. Since Fd is required for pyruvate-supported N_2 fixation, it may be expected that H_2 would support nitrogen fixation, since reduced Fd is readily produced from H_2 in these extracts. Molecular H_2 alone, however, did no support N_2 fixation. This suggested either that a component other than reduced Fd was required, or that H_2, although capable of reducing Fd, was inhibitory to N_2 fixation as prior whole-cell studies had indicated. If an additional component were required, it appeared that it was produced from pyruvic acid, since pyruvic acid supported active N_2 fixation.

Several unsuccessful attempts were made to obtain N_2 fixation in extracts to which H_2, N_2, and one of the other products of pyruvate metabolism, ATP, were added. Active N_2 fixation did occur, however, when another product of pyruvate metabolism, acetyl phosphate, was added in addition to H_2 and N_2. When compounds such as ADP were removed from cell extracts by dialysis, no N_2 fixation occurred unless ADP was added together with phosphate, H_2, and N_2. Acetyl phosphate then was acting as a source of ATP. The reason ATP did not work directly was that a continuous supply of ATP was required, and a high concentration of ATP, if added directly to a cell-free extract, was highly inhibitory to N_2 fixation. In whole cells that are fixing N_2, a continuous supply of ATP is made available during sugar metabolism.

Genetic Manipulation. High on the list of many researcher's agendas for projects using the practical application of recombinant DNA research has been the possible development of a living organism that will produce ammonia—in an effort to lessen dependence upon costly and highly energy-consuming synthetic ammonia fertilizers. However, at symposia held on this topic, these achievements are considered by most researchers as quite long-range. There are fundamental problems difficult to overcome, including: (1) the possibility that increasing biological nitrogen fixation, for which the plant furnishes the energy, can cause a net decrease in crop yields by depriving the plant of nitrogen for the production of certain critical growth elements; and (2) the very rapid-acting inactivation by oxygen of nitrogen-fixation mechanisms. Cloning techniques may be a path toward introducing nitrogen-fixation genes into certain bacteria. One objective is that of developing new forms of bacteria that will enter into symbiotic relationships with crop plants, such as corn (maize) and wheat, that do not possess their own nitrogen fixation symbionts.

In addition to recombinant DNA and molecular cloning techniques, some scientists have combined their research with more conventional genetic techniques. An *E. coli* plasmid capable of carrying nitrogen-fixation genes of *K. pneumoniae* has been developed. Some researchers also believe that nitrogen-fixation genes may be introduced directly into plant cells to result in a plant that requires no nitrogen fertilizer.

In research activities such as these, much knowledge has been gained concerning the energy needs for biological nitrogen fixation. More energy is used than originally contemplated; for example, 20 moles of adenosine triphosphate (ATP) are required to fix one mole of nitrogen. This contributes largely to the first problem mentioned earlier, namely, the great amount of energy required for the plant to fix its own nitrogen, possibly leading to yield reduction.

The well-known nitrogen fixation by rhizobia depends on photosynthesis by the plant. See also **Leguminosae**. Although the method is essentially impractical, photosynthesis can be increased by blanketing the plant with an atmosphere enriched in carbon dioxide. When this is done in the laboratory, increased legume yields are reported. Some investigators postulate that this effect is the result of a reduction photorespiration, a rather wasteful process in which carbon dioxide gained through photosynthesis is diverted into a series of less productive pathways in the plant. Investigators have also found that 30% of the energy used by the nitrogenase of most rhizobial species goes to producing hydrogen rather than ammonia. Research has also shown that the organisms that perform the nitrogen fixation function in plants are indeed quite diverse in themselves. Thus, new combinations of plants and organisms may increase efficiency in some cases.

As pointed out by Evans-Barber (1977), nitrogen is fixed by a variety of microorganisms in addition to those associated with legumes. Some of these include bacteria located in soils, in decaying wood, and on the surfaces of plant roots. They also include free-living blue-green algae, with fungi, ferns, mosses, liverworts, and higher plants (Hardy-Havelka, 1975). Reviews of numerous nitrogen-fixing organisms are given by Silvester (1976), Dalton (1974), Bond (1974), and Stewart (1974).

Role of Molybdenum in Nitrogen Fixation. Traditionally, molybdenum has been considered a key to the reduction of molecular nitrogen to ammonia by soil- and water-dwelling microorganisms. The metal is believed to be a part of the catalytically active site of nitrogenase, which is the enzyme that accomplishes the reduction. As early as 1980, researchers (North Carolina State University) suggested that the bacterium *Azobacter vinelandii* may have an alternative system for nitrogen fixation, a mechanism that may not require molybdenum. The proposal was regarded with some skepticism until the findings by researchers (Agriculture and Food Research Council Unit of Nitrogen Fixation, University of Sussex, England) in 1985 that confirmed the fact that *A. vinelandii* does have a second fixation system. Mutants of *A. vinelandii* were studied. Genes coding for the nitrogenase proteins were specifically deleted or inactivated. It was found that deletion of all three nitrogenase structural genes did not interfere with the fixation process. However, the process was effective only when molybdenum was not present. It was also found that the "wild" type of bacterium must have Mo in order to reduce nitrogen. Thus, it appears that the alternative mechanism is activated only when the system is subjected to molybdenum starvation. It has been suggested that the alternative system represents an adaptation to molybdenum-poor soils. The extension of these findings to other nitrogen-fixing microorganisms remains to be accomplished. Some further details pertaining to the Sussex investigation are given by Marx (1985).

Madigan (1979) and associates found that photosynthetic purple bacteria can grow with dinitrogen gas as the only source of nitrogen under anaerobic conditions, with light as the energy source. They also found that *Rhodopseudomonas capsula* can fix nitrogen in darkness with alternative energy conversion systems.

See also **Fertilizer**.

Additional Reading

Bond, G.: in "The Biology of Nitrogen Fixation," (A. Quispel, editor), North-Holland, Amsterdam, 1974.

Cheung, H. and J.H. Royal: "Efficiently Produce Ultra-High-Purity Nitrogen On-Site," *Chem. Eng. Progress,* 64 (October 1991).

Dalton, H.: *Crit. Rev. Microbiol,* **3**, 183 (1974).

Evans, H.J. and L.E. Barber: "Biological Nitrogen Fixation for Food and Fiber Production," *Science,* **197**, 332–339 (1977).

Graham, P.H. and S.C. Harris: "Biological Nitrogen Fixation," Unipub, New York, NY, 1984.

Greenwood, N.N. and A. Earnshaw: "Chemistry of the Elements," 2nd Edition, Butterworth-Heinemann, Inc., Woburn, MA, 1997.

Hardy, R.W.F. and M.D. Havelka: *Science,* **188**, 633 (1973).

Knowles, R. and T.H. Blackburn: "Nitrogen Isotope Techniques," Academic Press, Inc., San Diego, CA, 1997.

Legocki, A., H. Bothe, and A. Puhler: "Biological Fixation of Nitrogen for Ecology and Sustainable Agriculture," Springer-Verlag, Inc., New York, NY, 1997.

Lide, D.R.: "CRC Handbook of Chemistry and Physics 2000-2001," CRC Press, LLC., Boca Raton, FL, 2000.

Madigan, M.T., J.D. Wall, and H. Gest: "Dinitrogen Fixation by Photosynthetic Microorganisms," *Science,* **204**, 1429–1430 (1979).

Marx, J.L.: "Fixing Nitrogen without Molybdenum," *Science,* **229**, 956–957 (1985).

Metz, C.B.: "Biology of Fertilization, Academic Press, Inc., San Diego, CA, 1985.

Meyers, R.A.: "Handbook of Chemicals Production," The McGraw-Hill Companies, Inc., New York, NY, 1986.

Postgate, J.R.: "Nitrogen Fixation," Cambridge University Press, New York, NY, 1998.

Silvester, W.B.: in Proceedings of the 1st International Symposium on Nitrogen Fixation, (W.E. Newton and C.J. Nyman, editors), Washington State University Press, Pullman, WA, 1976.

Staff: "Nitrogen Plant Opens in South Carolina," *Chem. Eng. Progress,* 10 (February 1990).

Staff: "Can Catalytic Combustion in Jet Engines Zap NOx?" *Chem. Eng. Progress,* 19 (March 1992).

Staff: "Advanced Catalyst Zaps Nitrogen," *Chem. Eng. Progress,* 21 (July 1992).

Stevenson, F.J. and M.A. Cole: "Cycles of Soils: Carbon, Nitrogen, Phosphorus, Sulfur, Micronutrients," 2nd Edition, John Wiley & Sons, Inc., New York, NY, 1999.

Stewart, W.D.P.: in "The Biology of Nitrogen Fixation," (A. Quespel, editor), North-Holland, Amsterdam, 1974.

Vitousek, P.M. et al.: "Nitrae Losses from Disturbed Ecosystems," *Science,* **204**, 469–474 (1979).

NITROGEN (Fertilizer).   See **Fertilizer**.

NITROGEN GROUP (The).   The elements of group 15 of the periodic classification sometimes are referred to as the Nitrogen Group. In order of increasing atomic number, they are nitrogen, phosphorus, arsenic, antimony, and bismuth. The elements of this group are characterized by the presence of five electrons in an outer shell. The similarities of chemical behavior among the elements of this group are less striking than hold for some of the other groups, e.g., the close parallels of the alkali metals or alkaline earths. Although all of the elements of this group have valences in addition to 5+, all do have the 5+ valence in common. Unlike the alkali metals or alkaline earths, for example, the elements of the nitrogen group are not so similar chemically that they comprise a separate group in classical qualitative chemical analysis separations. Three of the five, however, antimony, arsenic, and bismuth are members of the second group in terms of qualitative chemical analysis.

NITROGLYCERIN.   See **Explosive**.

NMR SPECTROSCOPE.   See **Nuclear Magnetic Resonance**.

NOBEL, ALFRED BERNHARD (1833–1896).   Nobel was a Swedish industrialist and European munitions maker. He was born in Stockholm in 1833. Most of his early education was from tutors and what he learned as he traveled during his teenage years through much of North America and Europe. He learned to speak several languages fluently and studied mechanical engineering.

After his travels, he joined his father's business and worked developing mines, torpedoes, explosives, and other war materials for the Russian czar. The Nobel factory was financially successful until the end of the Crimean War when it fell into bankruptcy. Alfred then went into business with his brother manufacturing drilling tools for oilfields.

In the 1860's Alfred revolutionized the explosives industry by developing the Nobel detonator, a new fuse for nitroglycerin followed by a safe way to handle nitroglycerin by using an organic packing material to reduce its volatility and producing dynamite. In the late 1880's, Nobel produced ballistite, a smokeless powder.

Nobel's inventions brought him a monetary fortune as he controlled most of the explosive manufacturing factories throughout the world. He was, however, disillusioned that his inventions had mostly applications for war. Nobel personally desired world peace. Throughout his life, he had acted in a humanitarian manner. In his will, Nobel directed that the great majority of his estate be used for the purpose of giving yearly prizes to persons whose personal efforts made outstanding contributions to the advancement of physics, chemistry, medicine and physiology, literature, and world peace. The first Nobel Prizes were awarded in 1901 to Wilhelm K. Roentgen (discovery of X-rays in 1895), J.H. van't Hoff (chemical thermodynamics and osmotic pressure), and E.A. von Behring (diphtheria antitoxin). See also **Nobel Prizes**.

Web Reference

Nobel Foundation: *http://www.nobel.se/nobel/nobel-foundation/index.html*

J. M. I.

NOBELIUM.   Chemical element, symbol No, at. no. 102, at. wt. 254 (mass number of ^{254}No), radioactive metal of the Actinide series, also one of the Transuranium elements. Nobelium has valences of 2^+ and 3^+. In 1957, a group of American, English, and Swedish scientists bombarded a target of several curium isotopes (largely ^{244}Cm) with a beam of ^{13}C ions from the cyclotron at the Nobel Institute for Physics. They obtained a few alpha particles of 8.5 MeV energy and half-life of 10 minutes. This was considered to indicate the presence of element 102 with a probable mass number of 251 or 253. At that time, the element was named nobelium with assignment of the symbol, No. Further experiments at the University of California, however, failed to confirm this discovery. In April 1958, Ghiorso, Sikkeland, Walton, and Seaborg, working with the heavy ion linear accelerator (HILAC) at Berkeley, showed the isotope 102^{254} to be a product of the bombardment of ^{246}Cm with ^{12}C ions. Confirming experiments at Berkeley in 1966 showed the existence of ^{254}No with a 55-second half-life; ^{252}No with a 2.3 second half-life; and ^{257}No with a 23-second half-life. Four other isotopes are now recognized, including ^{255}No with a half-life of 3 minutes.

In 1973, scientists at Oak Ridge National Laboratory and Lawrence Berkeley Laboratory, produced a relatively long-lived isotope of nobelium through the bombardment of ^{248}Cm with ^{18}O ions. A total half-title of 58 ± 5 minutes was computed from the combined data of both laboratories. See also **Chemical Elements**.

Additional Reading

Note: The following classic references as listed in prior editions of this encyclopedia are preserved here.

Ditmer, P.F. et al.: "Identification of the Atomic Number of Nobelium by an X-ray Technique," *Phys. Rev. Lett.*, **26**, 17, 1037–1040 (1971).

Fields, P.R. et al.: "Production of the New Element 102," *Phys. Rev.*, **107**, 5, 1460–1462 (1957).

Flerov, G.N. et al.: "Experiments to Produce Element 102," *Sov. Phys. Dokl.*, **3**, 3, 546–548 (1958).

Ghiorso, A., T. Sikkeland, J.R. Walton, and G.T. Seaborg: "Attempts to Confirm the Existence of the 10-minute Isotope of 102," *Phys. Rev. Lett.*, **1**, 1, 17–18 (1958).

Ghiorso, A., T. Sikkeland, J.R. Walton, and G.T. Seaborg: "Element No. 102," *Phys. Rev. Lett.*, **1**, 18–20 (1958).

Hammond, C.R.: "The Elements," in "Handbook of Chemistry and Physics," 67th Edition, CRC Press, Boca Raton, Florida, 1986–1987.

Maly, J., T. Sikkeland, R. Silva, and A. Ghiorso: "Nobelium: Tracer Chemistry of the Divalent and Trivalent Ions," *Science*, **160**, 1114–1115 (1968).

Marks, T.J.: "Actinide Organometallic Chemistry," *Science*, **217**, 989–997 (1982).

Mikheev, V.L. et al.: "Synthesis of Isotopes of Element 102 with Mass Numbers 254, 253, and 252," *Sov. At. Energy*, **22**, 93–100 (1967).

Seaborg, G.T. (editor): "Transuranium Elements," Dowden, Hutchinson & Ross, Stroudsburg, PA, 1978.

Silva, R.J. et al.: "The New Nuclide Nobelium-259," *Nucl. Phys.*, **A216**, 97–108 (1973).

NOBEL PRIZES. In 1895, the will of Alfred Bernhard Nobel, a successful Swedish industrialist and European munitions maker, directed that the great majority of his estate be invested for the purpose of yielding annual prize money to be awarded to persons who, as the result of their personal efforts, made outstanding contributions to the advancement of science, literature, and peace. Initially, the awards were confined to five domains—Physics, Chemistry, Physiology or Medicine, Literature (of an idealistic tendency), and Peace (to promote the fraternity of nations and the abolition or diminution of standing armies and the formation and increase of peace congresses). In later years, the governors of the fund added Mathematics as a qualifying discipline and, in 1968, a Nobel Prize for the Economic Sciences was established. Although a prize for each of the foregoing categories is awarded each year, it is not mandatory that an award be made for each category every year, a factor determined by the governors of the fund. The will became effective when Nobel died on December 10, 1886, but the first prizes were not awarded until 1901 because of the need for legal clarification of Nobel's wishes as demanded by family members who also were mentioned in the will.

In studying the will, attorneys recognized the possible requirement for splitting a given award among two or even three individuals, but with a maximum of three persons per award. Nobel also mentioned certain criteria for selection of award recipients in each field. In connection with the prize for Physics, Nobel mentioned "discovery" or "invention," whereas the words "discovery" or "improvement" were stipulated concerning the prize for Chemistry. In terms of the prize for Physiology and Medicine, the key word was "discovery." These were guiding factors for application by the governors of the fund, especially during the early years.

If adjusted for inflation over the years, the original fund of 27,716,243 Swedish kroners ($7,427,953) was quite significant, especially when allowing for capital gains through investments realized over subsequent years. Money available for the annual prizes is essentially determined by the annual capital generated by the fund, but with the stipulation that at least 10% of that gain be reinvested each year. It is interesting to note that the honor associated with the prize is never split—that is, a Nobel Laureate is so designated even though a prize may be shared by two or three persons. The original will specified that the Royal Academy of Science

(Sweden) select winners in Physics and Chemistry, that the Karolinska Institute of Medicine (Stockholm) make the selections in Physiology and Medicine, and that the Swedish Academy (of Letters) select winners in Literature. A committee of the Norwegian Parliament was specified to select winners of the Peace prize, noting "no consideration whatever be paid to the nationality of the candidates."

Over the years, with the addition of new prize categories, the governors of the Nobel Foundation necessarily have amended certain procedures. The Foundation is governed by a five-person board of control made up of one appointment by the King of Sweden and one each as appointed by the aforementioned organizations (now referred to as Nobel Institutes). Members of the selection board are appointed for a period of $4\frac{1}{2}$ years. The science awards are presented on December 10 (anniversary of Nobel's death) of each year at the Stockholm Concert Hall, with personal felicitation of the King of Sweden. The Peace prize is presented in a formal ceremony in Oslo.

The first Nobel prizes were awarded in 1901 to Wilhelm K. Roentgen (discovery of X-rays in 1895), J.H. van't Hoff (chemical thermodynamics and osmotic pressure), and E.A. von Behring (diphtheria antitoxin). Listings of scores of Nobel prizes awarded over a century of progress can be found in a number of references, such as "The Information Please Almanac," 45th Edition, 701–709, Houghton Mifflin Company, Boston, Massachusetts. Specific 1992 winners are described in "U.S. Researchers Gather a Bumper Crop of Laurels," Science, 542 (October 23, 1992). An excellent review of the first half-century of the Nobel Foundation is given in an article by George W. Gray," The Nobel Prizes," *Sci Amer.*, 1 (September 1949).

Although a chronological listing of Nobel prizes in the sciences provides a good source of tracing the progress of science over the years, not all major discoveries and inventions have been so honored. There are several outstanding scientists who have not been included. Traditionally, Nobel prizes are given for achievements that date back a few to several years rather than for discoveries of the immediate past.

Web Reference

The Nobel Foundation: *http://www.nobel.se/nobel/nobel-foundation/index.html*

NOBLE FIR. See **Fir Trees**.

NOBLE GASES. See **Inert Gases (The)**.

NOCTILUCA (*Protozoa, Mastigophora*). A genus of protozoans with one phosphorescent species. This minute form is sometimes so abundant in the ocean that the water appears luminous at night.

NODAL LINES. On a vibrating diaphragm, lines along which no vibration takes place. If the diaphragm is circular they consist of two kinds, concentric nodal circles and nodal diameters.

NODAL POINT. Of all the rays that pass through a lens from an off-axis object point to its corresponding image point, there will always be one for which the direction of the ray in the image space is the same as that in the object space. See Fig. 1. The two points at which these segments, if projected, intersect the axis are called the nodal points, and the transverse planes through them are called the nodal planes. Only if n and n'', the refractive indices in the object and image spaces, are identical are the nodal planes also the principal planes. C is the optical center of the lens.

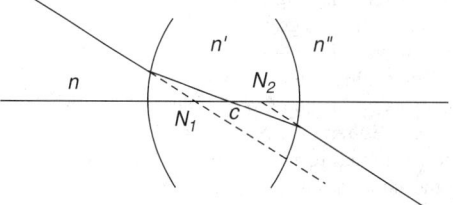

Fig. 1.   Nodal points of lens.

NODE. The point, line, or surface in a standing wave system where some characteristic of the wave field has essentially zero amplitude. A node is also a terminal of any branch of a network, or a terminal common to two or more branches of a network. The terms junction point, branch point, and vertex are synonymous.

NODES (Line of).   See Line of Nodes.

NOISE.

Any undesired sound. By extension, noise is any unwanted disturbance within a useful frequency band, such as undesired electric waves in any transmission channel or device. Such disturbances, when produced by other services, are called interference. Noise is also accidental or random fluctuation in electric circuits due to motion of the current carriers. From this concept of noise, the term is used as an adjective to denote unwanted fluctuations in quantities that are desired to remain constant, or to vary in a specified manner. For example, *noise voltage* is a term applied to spontaneous fluctuations in voltage in a component, device or system. The root-mean-square value of these fluctuations then gives a quantitative measure of the noise voltage. With this figure, such other measures as noise power and noise temperature are defined with the necessary stipulation of standard conditions for the acoustic, electric or electroacoustic system.

Background noise is (1) noise due to audible disturbances of periodic and/or random occurrence; (2) in receivers, the noise in absence of signal modulation on the carrier; (3) in recording and reproducing, the total system noise independent of whether or not a signal is present. The signal should not be included as part of the noise.

Thermal noise or Johnson noise is the noise produced by thermal agitation of charges in a conductor. The available thermal noise power produced in a resistance is independent of the resistance value, and is proportional to the absolute temperature and the frequency bandwidth over which the noise is measured, as indicated by the formula

$$N_t = 1.37 \times 10^{-23} T \Delta F$$

in which N_t is the available thermal noise power, T is the temperature of the resistance in degrees Kelvin, and Δf is the bandwidth in cycles per second.

Shot noise is the fluctuation in the current of charge carriers passing through a surface at statistically independent times. It has a uniform spectral density W_i given by

$$W_i = \frac{eI_o}{2\pi}$$

Random noise, exemplified by thermal noise and shot noise, has a uniform energy versus frequency distribution. *White noise* (or *Gaussian noise*) is a random noise having a constant energy per unit bandwidth that is independent of the central frequency of the band.

Over a number of years, society has been plague with ever-increasing noise from traffic, machinery, overamplified and discordant music, rocket and jet engine blasts, and sonic booms. Excessive and constant exposure of the ear to noise can and does cause impairment or loss of hearing. In addition other physical and frequently psychological changes can also occur as the result of noise disturbing sleep, impairing efficiency, and otherwise causing emotional disturbances. During wartime, military personnel often receive partial or total hearing loss after exposure to blasts and gunfire. Nearly 60,000 veterans of World War II, for example, have received compensation for ear damage considered to be a service-connected disability. Occupational deafness (earlier referred to as boiler maker's deafness) is a hearing loss resulting from prolonged exposure to industrial noise. Such damage is permanent, and may be partial or total. Overamplified music, particularly favored by teenagers, also is a cause of hearing impairment although the problem has not been subject to the same degree of analysis as in the case of industrial noises. See also **Hearing and the Ear**; and **Supersonic Aerodynamics**.

One of the frustrating results of noise is the masking effect it produces in reducing the intelligibility of speech. For example, if the speaker and listener are separated by 5 feet (1.5 meters), the levels of noise that will barely permit reliable word intelligibility are 50 decibels (dB) for normal conversation; 57 dB for raised speech; 63 dB for very loud speech; and 69 dB for shouting. As shown in Table 1, these levels are approached or exceeded in several day-to-day industrial and commercial activities. See also **Acoustics**.

TABLE 1. NOISE LEVELS FOR VARIOUS SOURCES AND LOCATIONS

Description of Noise	Noise Level (dB)
Threshold of hearing	0
Rustle of leaves in gentle breeze	10
Quiet whisper (distance of 5 feet)	10
Average whisper (distance of 4 feet)	20
House in country (average situation)	30
House in city (average situation)	40
Apartment (average situation)	40
Hotel	42
Theater (between performances)	42
Small retail establishment	52
Commercial garage	55
Medium-size office	58
Residential street	58
Restaurant	60+
Medium-size retail establishment	62
Factory or warehouse office	63
Large retail establishment	63
Ordinary conversation (distance of 3 feet)	65
Large office	65
Traffic on busy street	68
Factory (light-to-medium work)	78
Riveter (distance of 35 feet)	97
Hammer blows on steel plate (distance of 2 feet)	114
Threshold of pain	130

Additional Reading

Beranek, L. and I.L. Ver: "Noise and Vibration Control Engineering: Principles and Applications," John Wiley & Sons, Inc., New York, NY, 1992.

Fahy, F.J. and J.G. Walker: "Fundamentals of Noise and Vibration," Routledge, New York, NY, 1998.

Kryter, K.D.: "The Effects of Noise on Man," 2nd Edition, Academic Press, Inc., San Diego, CA, 1985.

Lord, H. and H.A. Evensen: "Noise Control for Engineers," Krieger Publishing Company, Melbourne, FL, 1987.

Norton, M.P.: "Fundamentals of Noise and Vibration Analysis for Engineers," Cambridge University Press, New York, NY, 1990.

Ott, H.W.: "Noise Reduction Techniques in Electronic Systems," 2nd Edition, John Wiley & Sons, Inc., New York, NY, 1991.

Sankar, B.V.: "Vibration and Noise Control," American Society of Mechanical Engineers, New York, NY, 1998.

NOISE GENERATOR.

Electronic noise, by definition, is an unwanted disturbance and its reduction in communications circuits is a constant aim of the electronics engineer. When supplied by a properly controlled generator, however, noise becomes a very useful test signal. Random noise contains no periodic components and its future value is completely unpredictable. It is described by its amplitude distribution and its spectrum. In its most common form, it has a gaussian distribution of amplitudes. Two spectral varieties are useful: (1) "white noise" has a uniform spectral level over a specified frequency range, i.e., equal energy in each hertz of frequency; (2) "pink noise" has equal energy in each octave and, therefore, the energy in each hertz decreases at a rate of 3 decibels per octave. The random-noise signal, embracing a wide range of frequencies and having a randomly varying instantaneous amplitude, closely approximates the signals normally encountered in many electronic circuits and, particularly, in busy communication systems.

Specific applications for electronic noise generators include, as a broad-band signal source: (1) intermodulation and cross-talk tests, (2) simulation of telephone line noise, (3) measurements on servo amplifiers, (4) noise interference tests on radar, (5) determining meter response characteristics, (6) setting transmission levels in communication circuits, (7) frequency response measurements; and, as a signal source, for (8) the measurement of reverberation, (9) sound attenuation of ducts, walls, panels, or floors, (10) acoustical properties of materials, (11) room acoustics, and (12) classroom or laboratory demonstrations. With a suitable power amplifier, such devices (13) can be used to drive a loudspeaker to produce high level acoustic noise for fatigue testing of structures and components, and (14) can drive a vibration shaker for structural tests of components and assemblies.

Two noise sources for the audio-frequency range are in common use. One is the semiconductor diode, operated at low current in the reverse breakdown mode. In the instrument illustrated in Fig. 1, noise from a semiconductor diode is amplified, filtered to establish the spectral shape, and further amplified to produce an output level of 3 volts. Internal circuits provide clipping of the peaks of the noise, if desired, or change the spectrum from white to pink noise as required.

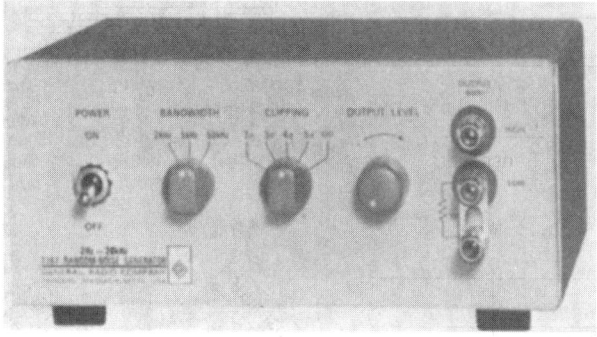

Fig. 1.   Random-noise generator. (*GenRad, Inc.*)

The other noise source is the pseudo-random noise generator, constructed of digital circuit elements connected as a shift register with feedback connections. When the feedback and initialization are such as to produce the maximum length sequence when the shift register is clocked, the rectangular wave output can be taken from the shift register and filtered with a low-pass filter to produce a waveform with good approximation to Gaussian or normal amplitude distribution. Such noise is called pseudo-random noise, because the pattern is precisely repeated each time the shift register sequence is repeated.

The most common noise source for the radio-frequency range is the temperature-limited thermionic diode. Its shot noise, although very low in level, is used because of its excellent spectral flatness and true Gaussian amplitude distribution.

JAMES J. FARAN, Jr., Lincoln, MA.

NOISE (Statistics).   Disturbance terms, analogous to noise in the engineering sense, which appear in time-series. They are usually regarded as random elements superposed on the systematic components of the series and have a tendency to obscure its nature.

NOMENCLATURE (Organic Chemistry).   See **Organic Chemistry**.

NONDESTRUCTIVE TESTING (NDT).   The examination of materials and objects for the purpose of detecting defects without in any way harming the test object. NDT contrasts vividly with destructive testing methods, which chemically consume or physically damage the test object, rendering it unfit for use. Whereas destructive testing must be confined to statistical sampling procedures, NDT enables 100% on-line inspection if desired. The trend in recent years has been in this direction, with emphasis on automating and increasing the speed of NDT operations. Another significant trend has been that of *testing work in progress*, as contrasted with earlier procedures which concentrated on testing raw materials and final products. In this way, very helpful information for step-by-step quality control can be provided. There remain, of course, numerous examples of where statistical destructive testing is needed—for example, in determining the ultimate compressive and tensile strengths of materials and parts or checking the corrosion resistance of materials. In recent years, it has proven possible to combine the results of NDT with computerized simulation in some instances to predict failure of test objects under certain conditions. For obvious economic reasons, NDT is preferred by manufacturers over destructive testing this accounts for high acceptance and many advancements which have occurred in NDT methods. For research and development applications, nondestructive methods are sometimes referred to as NDE (nondestructive evaluation).

Traditionally, NDT has been associated with metals and materials of construction for finding potentially unsafe conditions, such as cracks, voids, holes, inclusions, and other inconsistencies, as may be found in metal sheets, plates, bars, tubes, castings, forgings, and weldments. Such defects may arise from faulty manufacturing, or from later use, as the result of corrosion, abrasion, vibration, mishandling, and inattention to required maintenance procedures. In recent years, the applications for NDT have broadened to include all manner of materials—films, coatings, polymers, composites, and ceramics as encountered in a wide variety of industries, including numerous uses in the electronics manufacturing industry. Also, NDT is widely used for on-site inspection of large and heavy equipment, which cannot be detached for testing, after installation, but where periodic checks are required. Examples include the inspection of weldments in pipelines, aircraft engines and structural components, military equipment, bridge structures, etc.

During the last half of the 1980s and well into the 1990s, NTD enjoyed the benefits of measurement and computer technologies that contributed immensely to the speed, accuracy, and reliability of NTD, even though instrument costs have risen markedly as a result. However, the ability to make more measurements within shorter periods of time probably has not increased the unit costs proportionately. A number of measurement techniques entirely new to the NDT field have been added to increase the variety of choices.

Radiographic Methods

Radiographic (X- and gamma-ray technology) method using film was one of the earliest NTD schemes used. Although early systems are undergoing modernization, this comparatively simple method still enjoys acceptance for certain applications. This basic technique has taken on a number of new formats.

Film Images.   Images made by the traditional film technique are shown in Fig. 1. To reduce costs and meet environmental restrictions, a dry-silver system was introduced in 1991. The system produces a silver-based image without the use of wet chemistry, using photothermographic technology. The image is developed on exposed film by thermal energy rather than by the traditional method of immersing film in a liquid developer and fixer. Three elements required for dry processing are a specially coated film, fluorescent exposing screen, and a thermal processor. The film has a translucent polyester base similar to that of conventional film. Its ultrafine grain produces detailed images of archival quality.

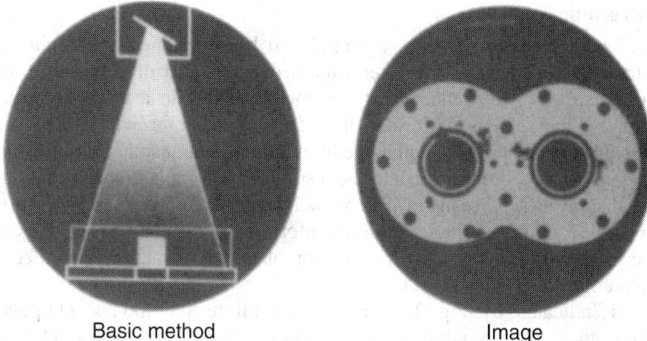

Basic method   Image

Fig. 1.   X-rays or gamma rays are used to create a shadow image of light and dark that reveals any flaws or inclusions in a test part.

Traditional radiographic methods use two-dimensional film to record the attenuation of X-rays passing through a three-dimensional object. The result is a shadowgraph in which all object features are superposed. To improve the totality of information obtained, backscattering methods were introduced several years ago.

Principal applications for radiography include the inspection of castings, electrical assemblies, weldments, small, thin, and complex wrought products, some nonmetallics, solid propellant rocket motors, cans or containers, composites, and nuclear reactor fuel rods, among many others.

Chronology of Radiographic Methods.   In 1985, researchers at John Hopkins University described a flash X-ray system that uses increased-power X-ray sources to generate very intense short pulses. High-gain X-ray intensifier detectors are used. Exposure times as short as 30 ns are possible and thus microstructural changes due to explosions, heat pulses, and shock waves can be detected. An indirect and direct method are used. In the indirect method, the X-ray diffraction image is converted into a visible light image by a fluorescent screen. The researchers have found that for the indirect method, a multiple-stage image-intensifier system coupled to an external fluorescent screen is the most sensitive and almost instantaneous system. Multiple stages of amplification allow individual X-ray photons to be detected. In more advanced systems, there is inclusion of a microchannel plate where electrons strike the output phosphor and are converted into a strong, visible image.

In the direct method, an X-ray-sensitive vidicon TV camera directly converts the X-ray image into an electronic charge pattern on a photoconductive target, which is read out by a scanning electron beam and displayed visually on a TV monitor.

In addition to testing uses per se, flash-X-ray techniques have been used to study the orientation of single crystals, to study lattice rotation accompanying plastic deformation, to measure the grain boundary migration during recrystallization annealing, and to determine the physical state of exploding materials.

In another technique known as X-ray transmission asymmetric crystal topography, changes in defect structure during polymerization of single crystals have been studied.

Digital Radiography. In this technique, the traditional film is replaced by a linear array of detectors and the X-ray beam is collimated into a fan beam. The object is moved perpendicularly to the detector array, and the attenuated radiation is sampled digitally by the detectors. Data are processed by stored information in the computer's memory to yield a two-dimensional image of the part being inspected.

X-Ray Computed Tomography (CT). In computed tomography, penetrating radiation from many angles is used to reconstruct cross-sectional images of an object. The advantages of CT are exemplified by the inspection of aircraft/aerospace castings for internal defects. Advantages of CT include increased reliability, elimination of unnecessary rejects, and wider use of castings instead of forgings and parts machined from wrought stock. CT has been found to have greater sensitivity (dependent on part size and geometry) than conventional film. CT can spatially define flaw distribution. Aerospace test engineers claim that castings can be measured with an accuracy of better than 0.05 mm (0.002 in), but is adversely affected by the amount of image noise and the edge-detection method used. Computed tomography systems are costly. The general principles of CT are described in article on **X-Ray Scan and Other Medical Imagery**.

Ultrasonic Methods

Typically, ultrasonic images are produced by mechanically scanning an ultrasonic transducer in a raster pattern over an area of a structure and then displaying the reflected or transmitted energy in a suitable format. Usually, the scan is performed in a tank of water or with some form of squirter nozzle. The liquid medium serves to transmit the ultrasonic energy from the transducer into the test material. Conventionally, the data are displayed as C-scans (a plan view image where a color scale is used to display signal amplitude or depth information) or as B-scans (image of a cross section at one particular location of interest, typically with a color indicating signal amplitude).

As indicated by Fig. 2, there are several testing modes: (1) pulse-echo mode, (2) through-transmission mode, (3) reflector-plate mode, and (4) angle-beam mode.

Sonic (<0.1 MHz) and ultrasonic (0.1 to 25 MHz) radiation have been used for many years in NDT. In a simple testing scheme, sonic or ultrasonic vibrations are generated and sent by way of a pulse beam through the part to be tested. The beam travels unimpeded through large parts, may be angled for testing sheet stock, and can impact materials immersed in a liquid. Any flaw reflects vibrations back to the instrument, which indicates the location and size of the discontinuity on a CRT (cathode-ray tube). Access is required to only one side of the material being tested. Although energy can be lost from the ultrasonic beam due to geometrical effects, these can be controlled to increase the sensitivity of attenuation measurements. Hence, microstructural alterations, such as microcracks, foreign particles, precipitates, grain boundaries, interphase boundaries, and dislocation defects, can be detected. Research has shown that attenuation measurements have detected microstructural change during fatigue testing, therefore giving early warning to fatigue-induced failure, as well as measuring oxygen content in titanium welds.

Acoustic methods are applicable to numerous kinds of materials. The method can be used, for example, to reveal fiber/matrix bond strength in polymer-matrix composites. In one method (Wan-li Wu, National Institute

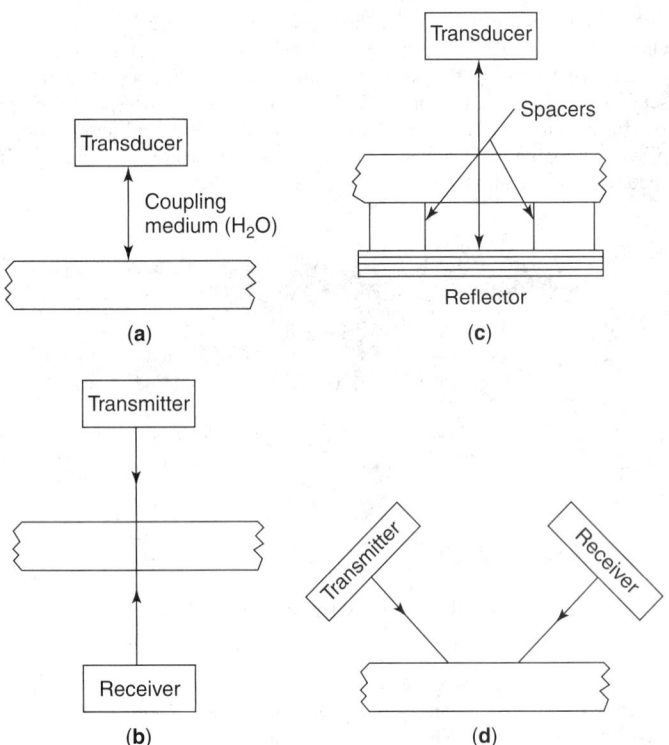

Fig. 2. Ultrasonic NDT methods: (**a**) pulse-echo; (**b**) through-transmission; (**c**) reflector-plate (double-through transmission); and (**d**) angle-beam.

of Standards and Technology), a continuous wave argon-ion laser is used to heat a very small area of the composite. The resulting thermal expansion between fiber and resin produces a measurable change due to debinding. Conventional methods of evaluating bond strength are time consuming and tedious. Instead of measuring the thermal stress, the laser power level at which debonding occurs is used as the index of debonding stress. Although sonic scanning techniques can be used to detect voids and cracks at interfaces in polymer-matrix composites, they do not measure the strength of interfacial bonds.

As pointed out by D. Sturges (General Electric Aircraft), "Modern ceramic materials offer many attractive physical and mechanical properties for use in a rapidly growing variety of industrial applications. The critical nature of many applications, however, imposes technical challenges in manufacturing and inspection. One nondestructive evaluation (NDE) technique of major relevance to inspecting ceramics is ultrasonic microscopy (also termed acoustic microscopy), that is, the use of tightly-focused, high-frequency sound beams to form images of the point-to-point reaction of a material to periodic stress waves. This technique offers high sensitivity for the detection of small defects, and often is a complementary technique to X-ray inspection."

Computer-assisted ultrasonic microscopy (CAUM) has been of particular significance in the testing of new materials developed for more fuel-efficient engines, wherein one objective is that of maximizing the high thermal efficiency of gas turbine engines by way of incorporating high-temperature ceramic components and exhaust-heat recovery. The object of NDE is that of assuring that ceramic components are free of both surface and internal flows that limit component life. Surface flaws can be generated during production by machining and normal handling.

Penetrant Method. This method does not depend upon radiation interactions with the test object and is essentially noninstrumental. A special penetrant substance is applied freely on the test object and allowed to work into tight cracks. See Fig. 3. The penetrant is removed from all surface areas and the piece is sprayed with a developer. The developer dries to an even white coating, while the penetrant bleeds up from any flaws through the developer, forming bright-red or fluorescent indications on the white surface. The size of the defect is indicated by the richness of color, speed of bleed-out, and dimensions observed.

Because of environmental concerns, a new generation of biodegradable penetrants having sensitivity levels ranging from 1 to 4 has been developed.

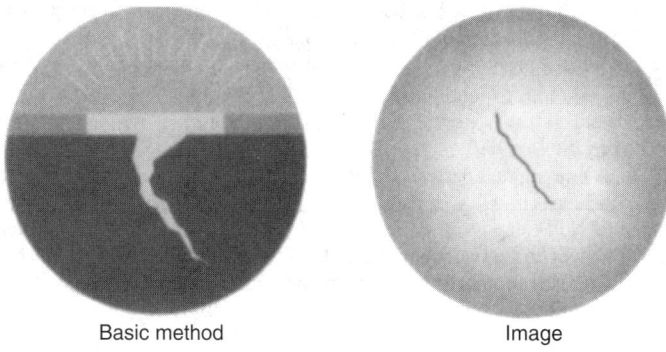

Fig. 3. Penetrant method for detecting flaws.

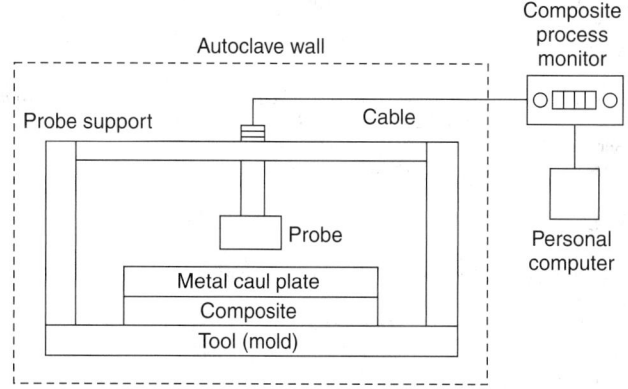

Fig. 5. Use of eddy-current testing for monitoring a composite cure. Typically, eddy-current testing uses an electronic instrument having a small probe on the end of a flexible electrical cable. The probe is placed either against or close to the target. The target must be electrically conductive to allow the generation of eddy currents. The time-variant nature of the probe's magnetic field causes electric currents to flow in the target material. Higher field frequencies or a more electrically conductive material increase the depth of the eddy-current penetration into the material. The concentration of eddy currents near the surface of the material is referred to as the "skin effect." Eddy currents generate their own magnetic field that opposes the probe's magnetic field. Detection circuits in the instrument sense the impedance changes in terms of phase/amplitude changes in probe-coil voltage. (*Suggested by Bar-Cohen and Nguyen.*)

The new penetrants are water washable and, in most instances, can be directly discharged into sewers. They are free of petroleum-based solutions.

Magnetic-Particle Method. This method makes use of iron powder to reveal the leakage magnetic field created at a flaw or break when any part is magnetized. The familiar horseshoe magnet best illustrates this principle. (1) If a horseshoe magnet is bent into a circle, the field between the ends attracts and holds magnetic iron powder. (2) If a magnet is made completely closed, the field will be contained entirely within the ring and no iron powder will be attracted. (3) However, if the round magnet is cracked, poles are created at the break, and iron powder is instantly attracted to the cracked area to pinpoint the defect. See Fig. 4.

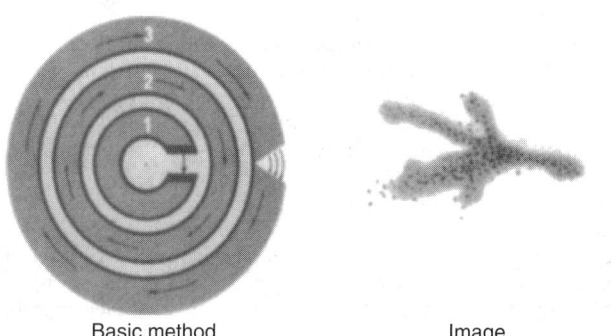

Fig. 4. Magnetic particle method for detecting flaws.

Eddy-Current Methods. This is one of the earliest NDT methods and is still used. Basically, this method reveals any differences in electrical impedance between parts to be tested and a reference sample. Parts to be examined are passed through a coil or explored with a probe, and a trace appears on a CRT. Since magnetic and electrical characteristics are closely related to metallurgical quantities, a trace position or pattern or a meter reading clearly shows variations in metal hardness and composition, as well as defects. Both ferrous and nonferrous parts can be tested, and various coils, probes, and detector tips are available.

Aside from more sophisticated electronics, a major contribution to improve eddy-current instrumentation has come from the development of the eddy current resonance digitizing (ECRD) method. With this method, eddy-current instrumentation can separate nonferrous alloys based upon characteristics other than simply their conductivity. See Fig. 5.

NDT Outlook

Improvements in current, established technologies and the introduction of new ways to test materials, nondestructively are expected to continue apac. One promising method is *positron annihilation*. The positron is the antiparticle of the electron; thus a positron/electron pair is unstable and will annihilate. In this process, two gamma rays at approximately 180° to one another are emitted from the center of the mass of the pair. A very slight departure from 180° is directly proportional to the transverse component of the momentum of the pair. The momenta of the electrons involved in such collisions can be calculated from the geometry and intensity of the gamma rays. The dynamics of the electron/positron system underlie the use of the technique for the study of defects in materials.

Additional Reading

Adams, T.E. and A.C. Wey: "Nondestructive Sectioning: Alternative to Physical Sectioning," *Advanced Materials & Processes*, 54 (February 1992).

Akuezue, H.C. and S.K. Verma: "Positron Annihilation: NDE at the Atomic Level," *Advanced Materials & Processes*, 26 (March 1992).

Altshuler, T.L.: "Atomic-Scale Materials Characterization," *Advanced Materials & Processes*, 18 (September 1991).

Bar-Dohen, K.H. Nguyen, and R. Botsco: "Eddy Currents Monitor Composites Cure," *Advanced Materials & Processes*, 41 (April 1991).

Bindell, J.B.: "Elements of Scanning Electron Microscopy," *Advanced Materials & Processes*, 20 (March 1993).

Blitz, J.: "Electrical & Magnetic Methods of Nondestructive Testing," Institute of Physics Publishing, London, UK, 1991.

Bray, D.E. and D. McBride: "Nondestructive Testing Techniques," John Wiley & Sons, Inc., New York, NY, 1992.

Carter, G.F. and D.E. Paul: "Materials Science and Engineering," ASM International, Materials Park, OH, 1991.

Cartz, L.: "Nondestructive Testing: Radiography, Ultrasonics, Liquid Penetrant, Magnetic Particle, Eddy Current," ASM International, Materials Park, OH, 1995.

Cormia, R.D.: "Problem-Solving Surface Analysis Techniques," *Advanced Materials & Processes*, 16 (December 1992).

Dulski, T.R.: "Residual-Element Analysis: Measuring the Minuscule," *Advanced Materials & Processes*, 20 (February 1992).

Engl, H.W. and W. Rundell: "Inverse Problems in Medical Imaging and Nondestructive Testing," Springer-Verlag, Inc., New York, NY, 1997.

Evans, N.J.: "Impedance Spectroscopy Reveals Materials Characteristics," *Advanced Materials & Processes*, 41 (November 1991).

Hauk, V. and H. Behnken: "Structural and Residual Stress Analysis by Nondestructive Methods: Evaluation, Application, Assessment," Elsevier Science, New York, NY, 1997.

Hellier, C.J.: "Handbook of Nondestructive Evaluation," McGraw-Hill Professional Book Group, New York, NY, 2000.

Malhotra, V. and N. Carino: "CRC Handbook on Nondestructive Testing of Concrete," CRC Press, LLC, Boca Raton, FL, 1990.

McGonnagle, W.: "International Advances in Nondestructive Testing, Vol. 16," Gordon & Breach Publishing Group, Newark, NJ, 1991.

Michaels, T.E. and B.D. Davidson: "Ultrasonic Inspection Detects Hidden Damage in Composites," *Advanced Materials & Processes*, 34 (March 1993).

Prask, H.J.: "Neutron Probes Tackle Industrial Problems," *Advanced Materials & Processes*, 26 (September 1991).

Staff: "Testing for Materials Selection," *Advanced Materials & Processes*, 5 (June 1990).

Staff: "Computed Tomography Details Casting Defects," *Advanced Materials & Processes*, 54 (November 1990).

Staff: "Nondestructive Examination," *Advanced Materials & Processes*, 63 (January 1992).

Staff: "Nondestructive Testing," American Society for Testing & Materials, West Conshohocken, PA, 1999.

Sturges, D.: "Sounding Out Ceramic Quality," *Advanced Materials & Processes*, 35 (April 1991).

Webb, S.C.: "PCs Help Optimize Materials Testing," *Advanced Materials & Processes*, 21 (November 1991).

Wu, Wen-li: "Acoustic Emissions Reveal Fiber/Matrix Bond Strength," *Advanced Materials and Processes*, 39 (August 1991).

Xavier Maldague, P.V.: "Theory and Practice of Infrared Technology for Nondestructive Testing," John Wiley & Sons, Inc., New York, NY, 2001.

NONEUCLIDEAN GEOMETRY.   See **Geometry**.

NONNEWTONIAN LIQUIDS.   See **Viscosity**.

NONNUTRITIVE SWEETENERS.   See **Sweeteners**.

NONSINUSOIDAL WAVE.   Any periodic wave that is not a pure sine wave. Such waves may, however, be analyzed into numerous sine components, utilizing Fourier series, and often circuits may be analyzed by considering these one at a time.

NOR (Circuit).   A computer logical decision element that provides a binary 1 output if all the input signals are a binary 0. This is the overall NOT of the logical OR operation. Output F is positive only when both transistors are cut off. This occurs when both inputs A and B are negative. See Fig. 1.

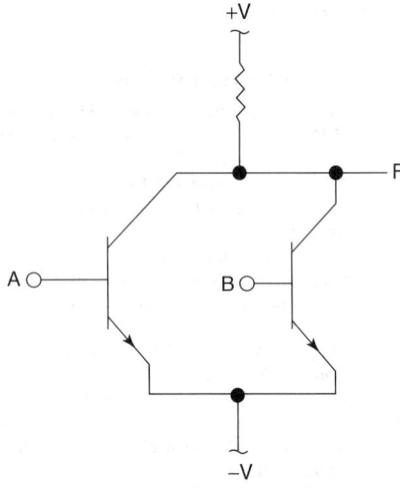

Fig. 1.   Transistor-type NOR circuit.

NORFOLK ISLAND PINE.   See **Araucarias**.

NORMAL CONCENTRATION.   A one normal solution (often abbreviated 1N) contains one gram-equivalent weight of a particular substance dissolved in 1 liter of *solution*. The equivalent weight of a substance may be defined as that weight of the substance that will involve, in a chemical reaction, one atomic weight of hydrogen, or that weight of any other element or portion of a substance, which, in turn, would involve in reaction one atomic weight of hydrogen.

As an example, the chlorine atom of potassium chloride (KCl) also is found in hydrochloric acid (HCl) in combination with one hydrogen atom. Thus, the gram-equivalent weight of KCl is 74.555, which is the same as its gram-molecular weight. A one normal solution of KCl will contain 74.555 grams of the salt per liter of solution.

For a particular solution, the molar and normal concentration are the same only when the gram-molecular and gram-equivalent weights are the same. Sulfuric acid H_2SO_4 represents a case where these values are not the same. This acid contains two active hydrogen ions and, therefore, its gram-equivalent weight is one-half of its gram-molecular weight. Phosphoric acid H_3PO_4 contains three active hydrogen ions. Consequently, the gram-equivalent weight for this acid is one-third of the gram-molecular weight. Calcium hydroxide $Ca(OH)_2$ contains two active hydroxyl ions, each being equivalent to a hydrogen ion. Therefore, the gram-equivalent weight of $Ca(OH)_2$ is one-half of its gram-molecular weight.

NORMAL EQUIVALENT DEVIATE.   The normal equivalent deviate of a proportion p is the deviate in a normal distribution with unit variance that exceeds a proportion p of the total frequency. Thus N.E.D. $(p) = x$ where

$$p = (2\pi)^{-1/2} \int_{-\infty}^{x} \exp(-\tfrac{1}{2}t^2) \, dt$$

For ease in computation, x is often replaced by y, the probit of p, where $y = x + 5$.

NORMAL (Gaussian) DISTRIBUTION.   One of the standard distributions of statistical theory. The mathematical form may be written

$$f(x) = \frac{1}{\sigma(2\pi)} \exp\left\{ -\frac{1}{2\sigma^2}(x - \mu)^2 \right\}$$

This has mean μ and variance σ^2 and the form of the frequency curve is shown in Fig. 1.

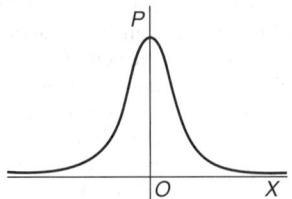

Fig. 1.   Typical Gaussian distribution curve.

The distribution was considered by Laplace and Gauss and is known by various names, that of "normal" being conferred by Karl Pearson, although he admitted that the practical occurrence of data of exactly normal type was somewhat abnormal. However, a number of distributions occurring in practice are approximately of the normal form, such as that of heights of men or women, or errors of measurement in reported observations of a magnitude.

The distribution owes its theoretical importance to a number of features, notably that under the **Central Limit Theorem** (q.v.) a great many of the statistics in common use tend to have a normal distribution in large samples, so that asymptotically the standard error can be used to set probability limits to estimates of parent parameters even in data from non-normal populations.

The distribution is a simple transfer of the "error curve." See also **Error**.

The distribution can be generalized to the case of p variables, being then of the form

$$f^\infty \exp\left\{ -\frac{1}{2} \sum_{j,k=1}^{p} \alpha_{jk}(x_j - \mu_j)(x_k - \mu_k) \right\}, \quad -\infty \leq x_j \leq \alpha$$

That is to say, the quantity in the exponential is a general quadratic form, limited only by the fact that the form must be positive definite in order to ensure the convergence to unity of the total frequency.

SIR MAURICE KENDALL, International Statistical Institute, London.

NORMAL (Geometry).   Perpendicular. At right angles to a given line or plane.

NORMALIZE (Mathematics).   1. To change in scale so that the sum of squares, or the integral of the squares of the transformed quantity is unit. (See also **Orthogonal Function**.)

2. To transform a random variable so that the resulting random variable has a normal distribution.

3. In computer operations, to adjust the exponent and coefficient of a floating-point result, so that the coefficient is in the prescribed normal range. Also called standardize.

NORMALIZING.   See **Iron Metals, Alloys, and Steels**.

NORMAL (Principal, to a Curve at a Point *P*).   The normal to the curve at P which lies in the osculating plane at P. A unit vector in the direction of the principal normal is called the *unit (principal) normal*. It is usually taken as directed from P to the concave side of the curve.

NORTH.   For many centuries, north has been the fundamental direction used by navigators and surveyors. During this long period, a loose usage of the term has become prevalent. It seems desirable to set down certain standard meanings accepted by the majority of modern navigators and astronomers.

True north (unless a qualifying adjective is used with north, true north is to be assumed) is the direction along the geographical meridian of the observer, in the plane of the observer's horizon, toward the North Pole of rotation of the earth. When the observer is facing the setting sun, the North Pole is to his right. *Compass north* is the direction of the plane of the horizon toward which the north-seeking end of the compass points. Unless otherwise stated, compass north refers to north defined by the magnetic compass; if another type of compass is used, it should be clearly indicated, e.g., gyrocompass north, etc. *Magnetic north* is the direction in the plane of the observer's horizon toward the north magnetic pole of terrestrial magnetism. For methods of conversion from any one of these three "norths" to any other, see **Compass (Navigation)**. See also **Navigation**.

NORTH AMERICAN HIGH.   See **Atmosphere (Earth)**.

NORTH ATLANTIC FLOUNDER.   See **Flatfishes**.

NORTH ATLANTIC WATERS.   In studies of oceanography, numerous bodies of ocean water are designated. The principal bodies in the North Atlantic are:

North Atlantic Central Water. A shallow oceanic water mass extending roughly from the southern parts of Greenland and Iceland to a region described by a line drawn from the northern end of South America to Africa. Temperature range is 8–19 °C (46.4–66.2 °F), salinity ranges from 35.1 to 36.7%.

North Atlantic Deep and Bottom Water. This dense ocean current arises in the Atlantic Ocean near the southeastern tip of Greenland where it meets the warmer water of the Gulf Stream below which it sinks to a depth of from 7,000 to 13,000 feet (2,100 to 3,900 meters) as it creeps southward. This water has been traced as far south as 60 degrees where the colder and heavier Antarctic water forces it to the surface.

North Atlantic Intermediate Water. An oceanic water mass lying at depths between the North Atlantic Deep and bottom water and the North Atlantic central water. Temperature range is from 2.5 to 4.0 °C (36.5 to 39.2 °F), salinity range is from 34.7 to 34.9%. The area of this water is more limited than those of the water masses above and below it. It has a high oxygen content.

NORTHERN LIGHTS.   See **Aurora and Airglow**.

NORTHERN PIKE.   See **Pike**.

NORTH PACIFIC WATERS.   In studies of oceanography, numerous bodies of ocean water are designated. The principal bodies in the North Pacific are:

North Pacific Central Water. Due to the great size of the Pacific Ocean, it contains more well-developed oceanic water masses than the Atlantic Ocean. Thus, there are both eastern and western north Pacific central surface water masses. They are relatively shallow, extending from the subarctic Pacific water on the north to the Pacific equatorial water on the south, and covering the width of the ocean except for a transition zone on the eastern side. Their temperature ranges are 8 to 18 °C (46.4 to 64.4 °F) and their salinity ranges from 33.8 to 34.9%, the lower values of each existing at lower depths (400 to 700 meters) (1,320 to 2,310 feet) where they meet the north Pacific intermediate water.

North Pacific Deep Water. Measurements at great depths show the existence of an oceanic water mass of practically constant salinity and low temperature below 2,500 to 3,000 meters (8,250 to 9,900 feet) in the south Pacific Ocean, and still deeper in the north Pacific Ocean. Since the Bering Straits are so narrow and shallow, this mass cannot be produced by currents from the Arctic Ocean, as in the north Atlantic deep and bottom water. Since the oxygen content of the Pacific deep water is less in the north Pacific than the south Pacific Ocean, this water mass is believed to be supplied from the antarctic bottom water or the Atlantic deep and bottom water.

North Pacific Intermediate Water. An oceanic water mass lying below the north Pacific central and equatorial waters at depths ranging from 600 to 800 meters (1,980 to 2,640 feet) in the north and to about 200 meters (660 feet) and 1,000 meters (3,300 feet) (two layers) in the south. It is characterized by low salinity and especially low oxygen content.

NORWALK VIRUS.   An epidemic of winter vomiting disease in 1968 involved many residents of Norwalk, Ohio. At that time, the causative agent was unknown. However, the illness resembled a vomiting and diarrhea syndrome observed in some regions as early as 1929. The syndrome occurred during the winter and affected mainly teenagers and adults. Similar episodes among close-living groups of people were reported from boarding schools in England during the 1950s. A number of enteroviruses and rotaviruses were identified in earlier years, such as the coxsackieviruses and echoviruses, among others, but not all epidemics had been satisfactorily explained. The Norwalk agent did not exactly fit any of the prior patterns.

Field investigators from the Center for Disease Control (Atlanta, Georgia) were despatched to Norwalk to collect stool specimens and early and convalescent blood. These were frozen for future examination. The National Institute of Health (United States) established a program involving volunteers and electron microscopic examinations, directed at identifying the Norwalk agent. A specific virus was identified. It resembles paroviruses, measures about 27 nanometers (0.027 micrometer) and has since been officially called the Norwalk agent. Three serologically different forms have been identified. One of these is now called the Hawaiian agent. More recently, additional varying agents have been identified in England. A widespread outbreak of Norwalk virus gastroenteritis in Australia was attributed to contaminated shellfish.

Persons infected with the Norwalk agent display (after 12 hours incubation) fever, vomiting, diarrhea, cramps, malaise, and leukopenia over a period of 2 to 3 days. Biopsy of the intestine during the acute phase of the disease has demonstrated that the villi of jejunum become flattened and infiltrated with mononuclear cells. Absorption of xylose and fat is reduced during and after the illness. Treatment is supportive with rehydration therapy for any major loss of fluids. Experience with volunteers indicates that natural immunity as the result of infection may persist only for a matter of weeks or months in many cases. Several volunteers who were rechallenged after 27 and 42 months developed symptoms of the disease. Because of wide variation in the results of the volunteer program, some authorities believe that there may be genetic differences in susceptibility.

Additional Reading

Blacklow, N.R., R. Dolin, and D.S. Fedson: "Acute Infectious Nonbacterial Gastroenteritis: Etiology and Pathogenesis," *Ann. Intern. Med.*, **76**, 993 (1972).

Melnick, J.L.: "Enteroviruses. Viral Infections of Humans: Epidemiology and Control," (A.S. Evans, editor), Plenum Medical Book Co., New York, NY, 1976.

Web Reference

Center for Disease Control and Prevention (CDC): *http://www.cdc.gov/health/diseases.htm*

NORWAY SPRUCE.   See **Spruce Trees**.

NORWEGIAN TOPKNOT.   See **Flatfishes**.

NOSE FLY (*Insecta, Diptera*).   Also known as the *sheep bot*, this species, *Oestrus ovis* (Linne), acts against sheep, goat, and wild deer and is found throughout North America; it is particularly abundant in Idaho, Montana, New Mexico, and Texas, and generally in the areas lying between these states. The nose fly strikes at the nose of the animal and deposits eggs in the nostrils. The maggots lodged in the nostrils and head sinuses create an inflamed condition and accompanying catarrhal discharge. The insect is known, on occasion, to attack humans.

Where the insect is present, the nostrils of the animal should be treated with pine tar. An early application is required in mid-April in most areas. Specially treated salt logs are obtainable which permit the animals to smear their noses with pine tar in an "automatic" fashion. In times of abundant presence of the flies, some producers herd the animals into darkened retreats where the flies are not active.

NOSE (Odor Receptor).   See **Flavorings**; and **Olfactory System**.

NOT (Circuit).   Also known as an inverter circuit, this is a circuit which provides a logical NOT of the input signal. If the input signal is a binary 1, the output is a binary 0. If the input signal is in the 0 state, the output is in the 1 state. Referring to Fig. 1, if A is positive, the output F is at 0 V inasmuch as the transistor is biased into conduction. If A is at 0 V, the output is at +V because the transistor is cut off. Expressed in Boolean algebra, $F = A'$, where the prime denote the NOT function.

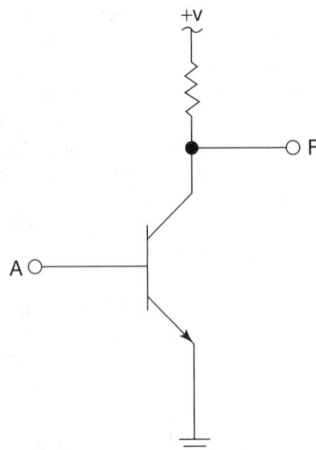

Fig. 1.   Inverter or NOT circuit.

See also **Diode Transistor Logic**.

NOTOCHORD.   A longitudinal stiffening rod found in all embryonic chordates and in the adults of some of the lower members of this phylum (*Chordata*). It lies between the central nervous system and the alimentary tract (digestive system) and is the axis around which the spinal column develops. As the bony structure forms, the notochord is almost crowded out of existence. In the human body small remnants of it form the nuclei pulposi in the intervertebral disks.

NOVA AND SUPERNOVA.   Traditionally, a nova (an early term meaning "new star") has been described as a star that suddenly displays an increasing brilliance and then over a period of time grows fainter. Because of the tremendous distances between the earth and such objects, it is interesting that while some of these events of a cataclysmic nature can be seen by the naked eye from earth as though occurring now, they actually extend far back in time. Most novae that do not attain naked-eye brilliancy are discovered and studied instrumentally.

Supernova 1987A.   On the night of 23/24 February 1987, Ian Shelton, working at the University of Toronto Las Campanas Station in northern Chile, discovered the brightest supernova seen since 1604. A separate article on this historic discovery will be found in this encyclopedia: **Supernova 1987A**.

Brightness of Supernova.   There is no known procedure by which an estimate of the total number of novae appearing each year can be predicted. It has been estimated that ten or more novae reach a brightness of the ninth stellar magnitude or greater each year, based upon historical records. Between 1900 and 1935, only five novae reached conspicuous brightness. More recent bright novae included V1500 Cyg (Nova 1975 Cyg) and the recurrent nova RS Ophiuci.

In Fig. 1, the light curves of three bright novae of the present century are represented. The ordinate scale of brightness is expressed in stellar magnitude. Since magnitude 6 is the limit of naked-eye visibility, the length of time that each was visible to the naked eye may be determined from the time scale given at the bottom. These curves are characteristic of most novae, with the very rapid rise to maximum and then the relatively slow and irregular decline. Examination of photographic records indicates that novae are not actually new stars at all, but rather are faint stars that suddenly increase in intensity. An increase of ten magnitudes is by no means uncommon, and this represents an increase of light intensity amounting to 10,000-fold. The total emitted energy in one outburst is about 10^{45} ergs.

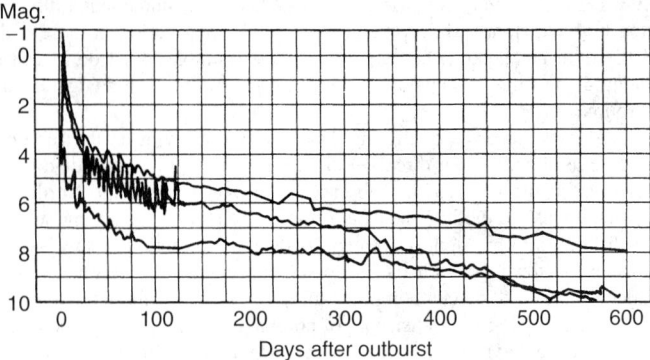

Fig. 1.   Light curves of Nova Aquilae, 1918; Nova Persei, 1901; and Nova Geminorum, 1912. They are designated in order of decreasing height. (*Harvard College Observatory Annals.*)

Coupled with the increase in light intensity of a nova is a correspondingly remarkable change in spectral characteristics. Although the spectral changes in different novae vary to a considerable extent, there are certain stages of development that are more or less characteristic of them all. During the period of rise, in the few cases where increasing novae have been detected in time for observation, the stars is of the hot, blue, A-type, with the absorption lines displaced very strongly to the violet. As the star starts to decline, the color changes from white to yellow, and bright lines, particularly of hydrogen and ionized iron, appear. The bright lines then broaden out to bands of irregular structure, which soon completely mask the continuous spectrum of the star. A few days later, dark lines again make their appearance, and these are displaced far to the violet. Soon, bright lines again appear, frequently of the type characteristic of the gaseous nebulae, except that they are broad. Often multiple components, corresponding to separate shells or blobs, are observed. As the brightness of the star further decreases, it eventually settles down to a peculiar O-type spectrum, with bright lines superimposed on a continuous and dark-line absorption spectrum.

Recent infrared observations have revealed that considerable amounts of dust can also be present in the nova ejecta. Radio observations argue for the presence of extensive circumstellar shells pre-existing in the nova environment. Ultraviolet observations show that the period immediately following the nova ejection event is dominated by a strong stellar wind emanating from the still-hot compact star.

Novae have been observed telescopically in some of the exterior galaxies such as the great spiral in Andromeda. Since the distances of some of these objects are at least very approximately known, it is possible to get an approximation to the absolute magnitudes at maxima of the novae observed in them. For these extragalactic novae, we find absolute magnitudes of the order of −4, a value that compares favorably with those determined for the few cases where the distance of a galactic nova is known.

Supernovae fall into two broad categories, mainly on the basis of their light curves. These are:

Type I.   These supernovae are around absolute photographic magnitude −18.6 maximum, or more than 200 million times as luminous as the sun. If one of them were placed at the distance of 10 parsecs from Earth, at which distance the sun would be barely visible, the nova would appear 14 times as bright as the full moon does to Earth. The spectra show extremely broad emission bands. The light variations show a rapid rise to maximum, followed at first by a rapid, and later by a slower, decline, usually with a characteristic time scale of 50 to 70 days. See Figs. 2 and 3.

Type II.   The members of this group reach a maximum luminosity equal to about 20 million suns. After maximum, they fade more slowly at first than do the members of the other group, followed by a more rapid decline after about 100 days. These supernovae (SN) show greater diversity than Type I. See Fig. 4.

Type V.   Two extragalactic examples of this class are known. They are more massive than Type II and appear to be associated with very massive (of order 100 $M_{\odot}$) progenitors.

Historical records indicate that supernovae were observed in AD 185, 1006, 1181, 1572, 1604, and possibly in AD 396. The radio and x-ray source, Cas A, is the remnant of a supernova that may have occurred in the 1670s, although the exact date is not certain. The first extragalactic

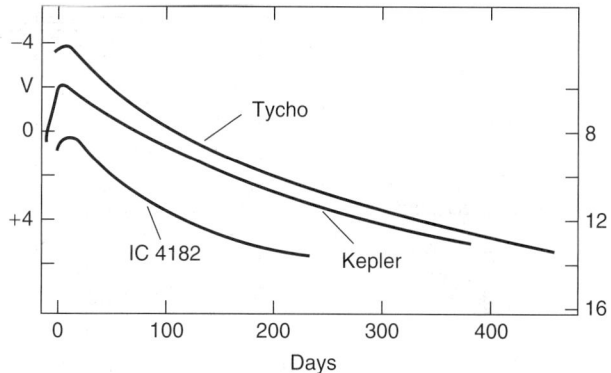

Fig. 2.   Light curves of Kepler's and of Tycho's supernovae, as reconstructed by Baade (1945). Scale on right is for the supernova IC 4182. The light curves show that the supernovae of 1572 and 1604 were of Type I. (*After van den Bergh.*)

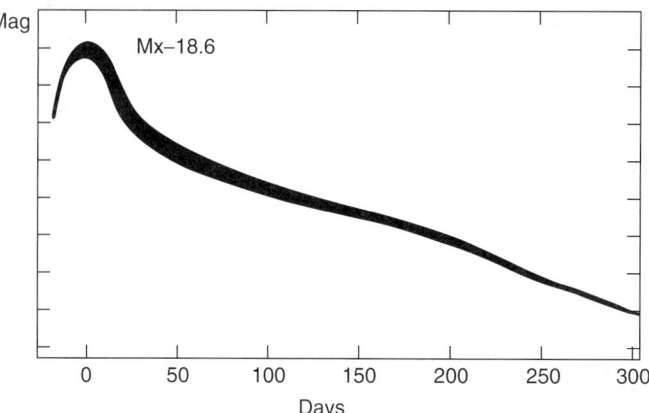

Fig. 3.   Composite blue light curve obtained by fitting observations of 38 Type I supernovae. One-magnitude intervals are marked on the ordinates. (*After van den Bergh.*)

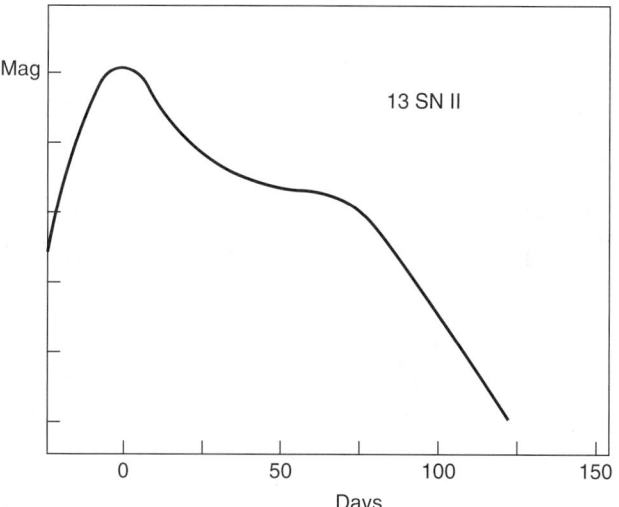

Fig. 4.   Composite light curve obtained from 13 Type II supernovae. (*After van den Bergh.*)

supernova observed was S and, in M31, which occurred in 1885. Since that time, due to the efforts of groups of many observatories, several hundred extragalactic supernovae have been recorded in a wide class of galaxies. Most galactic supernovae are identified on the basis of their radio and optical remnants, which are caused by the rapid expansion of the ejected matter of the exploded star into the interstellar medium. This shell compresses the magnetic field and generates high-energy electrons which, due to trapping, are swept up with the expanding blast wave. These electrons radiate at radio frequencies, while the shock heats the interstellar

gas to temperatures often in excess of a million degrees K, causing the gas to radiate X-rays.

The Crab and Vela supernova remnants have pulsars associated with them, while W 50, a weak optical nebula, has been associated with the peculiar x-ray and optical source SS 433. Few other supernova remnants can, however, be unambiguously linked to pulsars at this writing.

Observations indicate that all novae are members of short-period, low-mass binary systems, in which one of the stars in a white dwarf on which mass lost by the companion (usually a red dwarf or subgiant) is accreting. The slow accumulation of mass eventually causes fusion to begin on the surface of the white dwarf, resulting in the violent ejection of the outer envelope of this hydrogen-rich material. Depending upon the mass of the stars in question, their orbital period and the rate of mass transfer between them, a wide spectrum of explosive behavior can be derived, which appears to cover most of the novae observed thus far. The two types of SN appear to arise from different mechanisms. SNI may be due to the collapse of a white dwarf star, induced by the accretion of matter from a low mass companion in a binary system by a white dwarf near its maximum stable mass. SNII are likely due to the explosion of unstable, massive, recently formed red supergiants. It is suggested that some x-ray pulsars (Cen X-3, HZ Her, for example) represent supernova explosions by members of a binary system which did not disrupt the binary. These are, however, rare exceptions compared with the observed number of supernova remnants.

Additional Reading

Asimov, I.: "The Exploding Suns: The Secrets of the Supernovas," St. Martin's Press, Inc., New York, NY, 1985.

Bethe, H.A. and G. Brown: "How a Supernova Explodes," *Sci. Amer.*, 60–68 (May 1985).

Boss, A.P.: "Collapse and Formation of Stars," *Sci. Amer.*, 40–45 (January 1985).

Chandrasekhar, S.: "On Stars, Their Evolution and Their Stability," *Science*, 226, 497–505 (1984).

Harris, M.J.: "Short-Lived s-Process Gamma-Ray Lines in Type II Supernovae," *Science*, 60 (April 1, 1988).

Imshennik, V. et al.: "Astrophysics and Space Physics Reviews: Supernova 1987a, Vol. 8," Gordon & Breach Publishing Group, Newark, NJ, 1989.

Kirshner, R.F.: "Supernova—Death of a Star," *National Geographic*, 618 (May 1988).

Mann, A.K.: "Shadow of a Star: The Neutrino Story of Supernova 1987a," W.H. Freeman Company, New York, NY, 1997.

Marschall, L.: "The Supernova Story," Princeton University Press, Princeton, NJ, 1994.

McCray, R. and Z. Wang: "Supernovae and Supernova Remnants: IAU Colloquium 145," Cambridge University Press, New York, NY, 1994.

News: "Automated Spotting of Supernovae," *Sci. Amer.*, 65 (August 1986).

Peterson, I.: "A Supernova Story in Clay," *Science News*, 397 (June 23, 1990).

Rees, M.J. and R.J. Stoneham: "Supernovae," Reidel, Boston, MA, 1982.

Seward, F.D. et al.: "Young Supernova Remnants," *Sci. Amer.*, 88–96 (August 1985).

Sparke, L.S. and J.S. Gallagher: "Galaxies in the Universe: An Introduction," Cambridge University Press, New York, NY, 2000.

Spergel, D.N. et al.: "A Simple Model for Neutrino Cooling of the Large Magellanic Cloud Supernova," *Science*, **237**, 1471–1473 (1987).

Staff: "Supernova Yields Cosmic Yardstick," *Science News*, 59 (January 26, 1991).

Thompson, G.D. and J.T. Bryan, Jr.: "Supernova Search Charts and Handbook," Cambridge University Press, New York, NY, 1989.

Waldrop, M.M.: "Supernova 1987A: A Mysterious Stranger," *Science*, **237**, 25–26 (1987).

Waldrop, M.M.: "Sighting of a Supernova," *Science*, **235**, 1143 (1987).

Waldrop, M.M.: "The Supernova 1987A," *Science*, **235**, 1322–1323 (1987).

Waldrop, M.M.: "Feeding the Monster in the Middle," *Science*, 478 (January 27, 1989).

Waldrop, M.M.: "And Now for a Real Crab Nebula," *Science*, 1140 (March 3, 1989).

Weiss, P.L.: "Seeing Supernovas in Galactic Chimneys," *Science News*, 133 (September 1, 1990).

Wheeler, J.G. and R.P. Harkness: "Helium-rich Supernovas," *Sci. Amer.*, 50–58 (November 1987).

Williams, R.E.: "The Shells of Novas," *Sci. Amer.*, 120–131 (April 1981).

Wilson, O.C. et al.: "The Activity Cycle of Stars," *Sci. Amer.*, 104–121 (February 1981).

Web Reference

United States National Aeronautics and Space Administration, Space Science homepage. *http://universe.gsfc.nasa.gov/*

STEVEN N. SHORE, Indiana University, South Bend, South Bend, IN.

NOVOLATILE. Of a computer or computer component. The ability to retain information in the absence of power as *nonvolatile memory*, or *nonvolatile storage*.

NUCLEAR FISSION. A type of nuclear reaction in which the compound nucleus splits into two nearly equal parts, rather than ejecting one or a few small nuclear particles, as in most nuclear reactions. Our knowledge of nuclear fission dates back to the mid-1930s when Fermi and his coworkers showed that the number of distinctly different radioactive nuclides that could be induced by neutron bombardment of uranium far exceeded the number expected, unless some previously unknown pattern of isomerism could be found. Furthermore, the radiochemical properties of many of these radio-elements different quite markedly from expectations. For example, both Hahn and Strassman in Germany and Curie and Savitch in France found that certain unknown activities, thought to be radioactive radium, always followed the chemically separated barium fraction rather than the radium fraction. Hahn and Strassman found several other similar examples and were able to show that uranium, when bombarded by neutrons, undergoes what then appeared to be a very unusual nuclear reaction in that the products are radio-elements with about half the atomic number of uranium. These findings were interpreted by Meitner and Frisch as the division of an excited nucleus into nuclei of medium mass, a process that was given the name *nuclear fission*.

The first such process to be extensively studied was fission induced in ^{235}U by thermal neutrons (neutrons with energies of about 0.03 eV). This reaction, symbolically represented by the equation

$$^{235}U + n \longrightarrow\ ^{236}U \longrightarrow \text{fission},$$

produces an unstable system which achieves stability by splitting into two large fragments, not by ejecting one or a few small particles.

An individual fission does not produce a unique pair of fragments, but in a large number of such processes, the mass distribution of the fragments can be predicted with reasonable certainty, leading to predictable fission yields. A fission yield, usually expressed as a percentage, describes that fraction of nuclear fission processes that give rise to a specified nuclide or group of isobars. The yields of single nuclides are known as independent yields and those of a set of isobars as mass yields or chain yields. Since two fragments are produced by each fission, the total of all fission yields for a given fission process is 200%. The fission yield curve is different for each mode of induced fission, the most commonly known one being that for thermal neutron induced fission of ^{235}U, shown in Fig. 1. The chemical characteristics of the two fragments vary within limits, so that many elements are formed. Analysis of the fission products shows that most of them are in two mass groups, a "light" group consisting of elements having mass numbers between 85 and 104, and a "heavy" group consisting of elements having mass numbers between 130 and 149. Fragment mass numbers that have been detected range from around A = 70 to around A = 160. The determination of independent yields is made more difficult by the fact that many of the products are highly radioactive and undergo extensive secondary changes, sometimes in extremely short times, a very small fraction of a second.

A most significant aspect of nuclear fission is its great release of energy. The source of this energy is the loss of mass between the initial and final products of the reaction. The total mass of all atoms and nuclear particles produced in a single fission process in less than the original mass of the ^{235}U atom and the neutron that combined with the ^{235}U to induce fission. During fission of ^{235}U, the total energy released because of loss of mass is about 200 MeV. In practical units, the fissioning of 1 gram of ^{235}U yields 24,000 kilowatt-hours of energy.

Another important feature of fission is the presence of neutrons among the reaction products, slightly more than two for each fission of a ^{235}U atom. These neutrons are not an immediate consequence of fission, but are boiled off the original fission products, their release being possible because of the very large amount of available energy. If all these neutrons were captured by other ^{235}U nuclides, the number of available neutrons would multiply by factors of two for every generation of fission processes, a very rapid increase. However, some neutrons escape from the region containing the ^{235}U and others are absorbed in nonfission capture processes. The minimum conditions for a self-sustaining chain reaction is that at least one neutron from each nucleus undergoing fission must cause fission of another nucleus, a multiplication factor of one or greater. Maintenance of a chain reaction is essential to the proper functioning of both nuclear weapons and nuclear reactors.

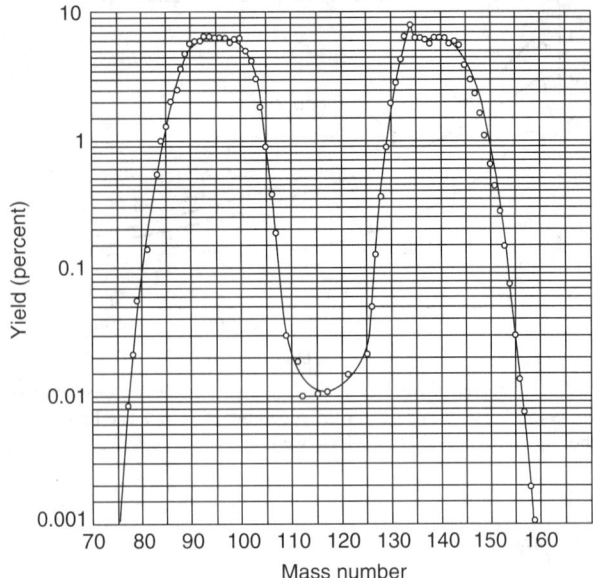

Fig. 1.   Mass fission yield curve for $^{235}U + n$ (thermal).

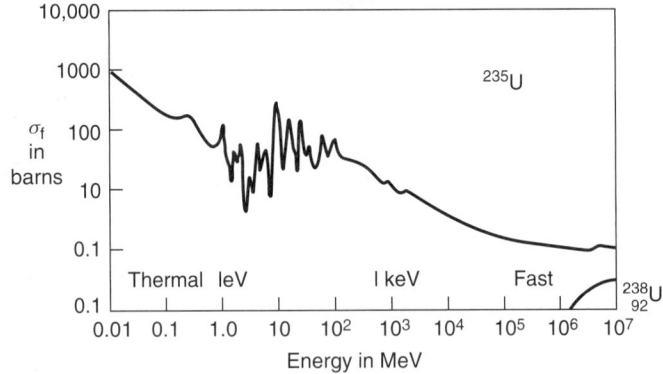

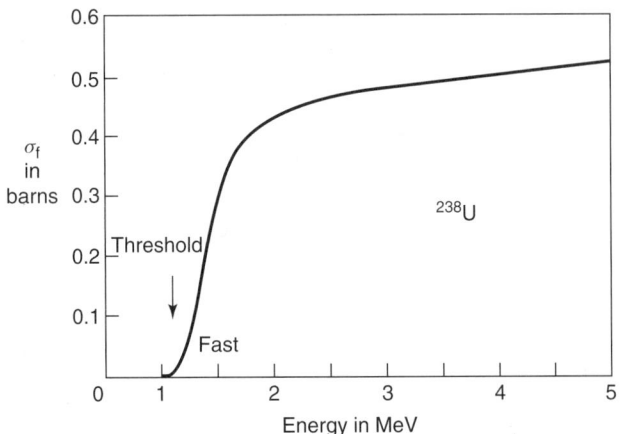

Fig. 2.   Fission cross section as function of energy for ^{235}U and ^{238}U.

The probability that fission can occur (generally called the cross section for fission) varies widely among different nuclides. Only a few nuclides, such as ^{235}U, have a high probability of undergoing fission when they capture a neutron. In other nuclides, the probability of fission is generally much smaller. As an example, the cross section as a function of incident neutron energy is shown in Fig. 2 for fission of ^{235}U and of ^{238}U. Although fission can be induced in ^{238}U, such a process is possible only if the incident neutron has an energy greater than 1 MeV, whereas neutrons of any energy can induce fission in ^{235}U. The characteristic double hump yield curve of Fig. 1 (asymmetric fission) is common only for low neutron excitation energy and targets consisting of highly fissile elements. For

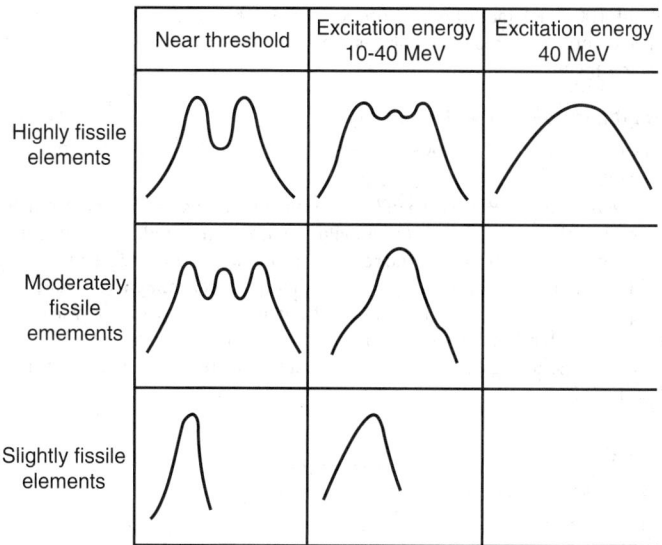

	Near threshold	Excitation energy 10-40 MeV	Excitation energy 40 MeV
Highly fissile elements			
Moderately fissile emements			
Slightly fissile elements			

Fig. 3.   Mass fission yield curves as function of excitation energy and degree of fission probability.

either higher excitation energies or less fissile elements, such as actinium or radium, symmetric fission becomes much more important, creating a triple humped fission-yield curve, shown in Fig. 3. Slightly fissile elements, such as lead and bismuth, or very high excitation energies further emphasize the symmetric mode of fission, also illustrated in Fig. 3. Nuclear fission may be induced by particles other than neutrons, such as alpha particles and photons. In some nuclides, it also occurs spontaneously, although the probability of such occurrence is so low that it has almost no effect on the radioactive decay characteristic of the nuclide.

Nuclear fission has generally been explained theoretically in terms of the liquid-drop model of the nucleus. In this model, the incident neutron combines with the target nucleus to form a compound nucleus at a high excitation energy. A small part of this excitation energy can be attributed to the kinetic energy of the incident neutron, but most of it usually comes from the binding energy of the incident neutron. This added energy initiates oscillations in the drop, which then sometimes assumes an elongated shape, similar to B in Fig. 4. If oscillations become sufficiently violent that a form similar to D is reached, fissioning (form E) becomes inevitable, since the positive charge at the two ends of the dumbbell-shaped nucleus then produces an electrostatic repulsive force greater than the attractive nuclear force holding the neck of the dumbbell together. The reason for asymmetric fission is not clearly understood. The liquid drop model predicts symmetric fission. Most people believe that asymmetric fission results because of the effects of the closed shells of the nucleus. See also **Nuclear Power**

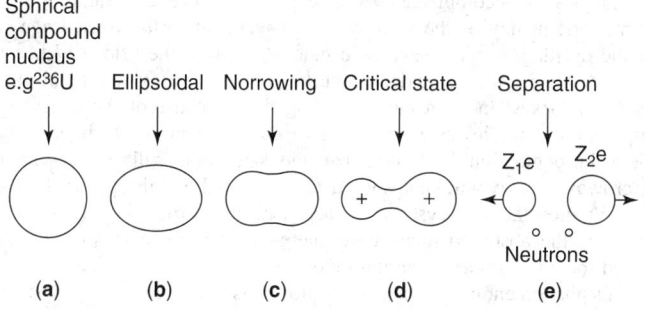

Fig. 4.   Fission mechanism according to liquid-drop model of the nucleus.

Technology; and **Nuclear Structure**.

C. SHARP COOK, The University of Texas at El Paso.

NUCLEAR FORCES.   Strong, short-range, attractive forces that interact between the individual nucleons of an atomic nucleus. Unfortunately, despite several decades of research, a clear and unambiguous description cannot be given for the forces that hold individual protons and neutrons

together in an atomic nucleus. Unlike the electrostatic force that holds electrons in an atom, no equation can be written that completely describes the nature of the force that holds an atomic nucleus together, or the nature of its associated potential energy. A description of the detailed structure of a nucleus cannot, therefore, be derived directly from calculations based on knowledge of nuclear forces. Instead, detailed knowledge of the structure of atomic nuclei has been derived from nuclear models. These models have been constructed by using results from other fields of physical science which display the same or similar characteristics as those observed in nuclear reactions and in radioactive decay. From such analogies, construction of a partial description of nuclear structure and of the nature of nuclear forces has been possible.

Because of the unknown characteristic of nuclear forces, many different suppositions have been made, using available experimental evidence, regarding the nature of the potential energy V of a nuclear particle as a function of its position in the field of a nucleus, or of another nuclear particle. To a first approximation, the nuclear potential is assumed to be spherically symmetric, such as V is a function only of the distance r from the center of the field, thus being the same in all directions, and is representable by a curve as in Fig. 1 curves (a) to (f).

A *potential well* is the name given to a region in which a minimum in the potential is formed; it results from attractive forces. A *potential barrier* is the name given to a region in which there is a maximum in the potential; it results from repulsive forces, either alone or in combination with attractive forces. Some central potentials commonly used as approximations to nuclear potentials are illustrated in the curves. Curve (a) shows a square well potential, which has a constant negative value $-V_0$ for $r \leq r_0$ and zero value for $r \geq r_0$. When this curve represents the potential between two nucleons, r_0 is called the *range of nuclear forces*; when it represents the potential of a nucleus, as this nucleus interacts with

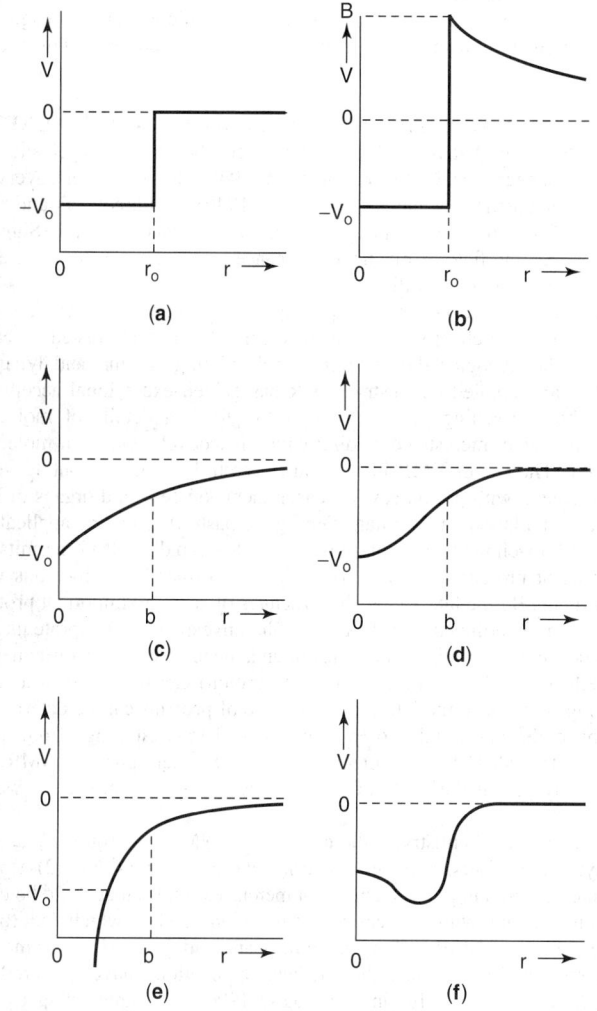

Fig. 1.   Potential energy of a nuclear particle versus distance from the center of the field.

an individual nucleon, r_0 is called the *nuclear radius*. Curve (b) shows a square well potential for $r \leq r_0$ with a Coulomb potential resulting from repulsive electrostatic forces, for $r > r_0$. The resulting barrier is called a *Coulomb barrier*, and the maximum energy b is called the barrier height. Such a potential approximates that of a positively charged particle in the field of a nucleus, and is often used in the theory of alpha particle distintegration and nuclear reactions. Curve (c) shows an exponential well, $V = V_0 e^{-r/b}$; curve (d) shows a Gaussian well, $V = V_0 e^{-r/b2}$. Curve (e) shows a Yukawa potential, $V = -(V_0/r)e^{-r/b}$ used in the meson theory of nuclear forces for the interaction between two nucleons; and (f) shows a wine-bottle potential, characterized by a low central elevation. If a high central elevation is present, the resulting barrier is called a *central barrier*, or *repulsive core*.

Although the predominant part of the nuclear potential is the part described above that is derived from the central force produced by the average effects of all other nucleons in the system on the individual nucleon under observation, evidence exists that nonsymmetric tensor and spin-orbit coupling terms must be included in the description of the nuclear potential. These are derived from a tensor force resulting from a coupling between individual pairs of nucleons and from the coupling between spin and orbital angular moments of the individual nucleus, as described by the shell model of the nucleus.

A considerable amount of evidence indicates that nuclear forces are charge-independent, i.e., the neutron-neutron, neutron-proton, and proton-proton forces are identical. The meson theory of nuclear forces, originated by Yukawa, postulates the atomic nucleus being held together by an exchange force in which particles, now called mesons, are exchanged between individual nucleons within the nucleus.

C. Sharp Cook, The University of Texas at El Paso.

NUCLEAR MAGNETIC MOMENT.

An electrically charged particle of finite size that possesses angular momentum, acts like a small magnet and thus possesses a magnetic moment. For atomic nuclei, the magnitude of this moment is within a range of values between zero and a few nuclear magnetons.

NUCLEAR MAGNETIC RESONANCE (NMR) AND MAGNETIC RESONANCE IMAGING (MRI).

Since the discovery of electron spin resonance (ESR) by Zavoiski in 1945 and the codiscovery of nuclear magnetic resonance (NMR) in 1946 by Purcell, Pound, and Torrey (Harvard University) and Bloch, Hansen, and Packard (Stanford University), the field of magnetic resonance has reached a high degree of sophistication and versatility.

Soon after its discovery, the principle was applied by physicists to a variety of research programs. Chemists also became interested in NMR because they recognized its potential for elucidating structure and dynamics in pure and applied chemistry. NMR has gained exceptional acceptance over the intervening years as a tool for yielding details of molecular structure, such interests coinciding with other developments in molecular biology. The use of NMR in medical research dates back about 40 years, commencing with proton NMR measurements on cells and organs of both humans and laboratory animals. During the past 20 years, the applications of NMR to clinical medicine have expanded rapidly. NMR permits the analysis of protein structures, for which the former X-ray methods were inadequate. By the late 1980s, the structures of some 50 important proteins had been determined. NMR allows the investigation of proteins and peptides and other macromolecules in an aqueous medium. Consequently, the effects of pH (hydrogen ion concentration) can be observed and the binding of water at the interior and surface of proteins can be determined.

The molecular weights of proteins studied thus far range from about 5000 to 15,000. The use of three-dimensional Fourier transform NMR may ultimately permit the study of macromolecules up to a molecular weight of 40,000.

In analytical chemistry in the early 1990s, NMR was routinely used to study: (1) polymers, polymer networks, and copolymers, and (2) asphalt, bitumens, tars, and pitches, among numerous examples that could be cited.

Commencing with clinical medicine (circa 1981), which led to the development of NMR and magnetic resonance imaging (MRI) for medical research and diagnostics, the medical applications have captured the "headlines," particularly since the early 1990s. It is interesting to note that in 1981 there were only two MRI devices in the United States. By 1987, the installations had reached about 600 in the United States and

about 100 MRI facilities in Europe. As of 1994, these Figures expanded into several hundred installations, representing a tremendous investment by the health care field.

Fundamentals of the Technology

Discovery of the principles of ESR and NMR led to the development of several subtechnologies.

Electron Spin Resonance (ESR). Fundamentally, ESR spectrometers consist of three main parts: (1) a magnet with corresponding power supply to provide a steady dc magnetic field at about 3300 G; (2) a microwave bridge capable of producing an oscillating electromagnetic field at a frequency of about 0.5 GHz (*X* band), which is coupled via waveguide to a high-*Q* microwave reflection cavity; and (3) the associated signal detection including dc field modulation, amplification, and display systems. See Fig. 1.

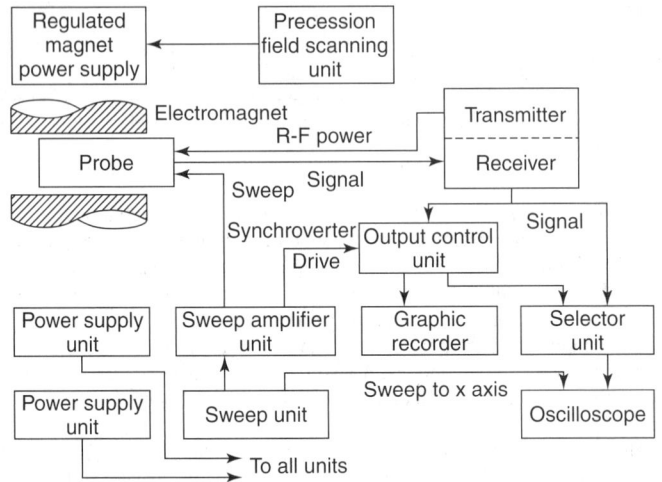

Fig. 1.   Schematic of ESR spectrometer.

Observation of ESR from a particular sample is contingent upon the presence of a macroscopic spin magnetic moment $\overline{\mu}$; i.e., the sample under investigation must contain some minimum number of unpaired electron spins. Upon insertion into the cavity, the sample is subjected to the dc magnetic field H_0, and the unpaired electrons align themselves both parallel (high energy) and antiparallel (low energy) to H_0. The ratio of the spin populations of the high to low energy states is given by the Botzman distribution, $e^{-hv/kT}$, where hv is the energy difference of the two states. Because of the torque on $\overline{\mu}$ produced by H_0, the spin magnetic moment will precess about H_0, at the Larmor frequency $\omega = \gamma_e H_0$, where γ_e is the gyromagnetic ratio of the electron.

Analysis is accomplished by sweeping the magnetic field until the Larmor frequency of the spin system is identical to the fixed frequency of the oscillating microwave field emanating from the bridge. When the two frequencies are coincident, a net microwave energy will be absorbed by the spin system from the oscillating field because of the excess spin population in the lower energy state. If energy continues to be absorbed, the spin populations will equalize and saturation will occur—no net microwave energy will be absorbed and the signal will disappear. To avoid this situation the spin system interacts with its surroundings (lattice) and transfers the absorbed microwave energy to the lattice in some interval called the spin-lattice relaxation time.

This phenomenon, known as relaxation, acts to restore the spin system to its original Botzman distribution of populations. At equilibrium, microwave energy is being absorbed by the spin system and then transferred to the lattice. This process is monitored via microwave rectification of the reflected cavity signal by a diode detector, preamplification, and lock-in phase detection at the dc field modulation frequency. The signal which appears on the recorder may be a single line from an unpaired electron or a group of lines (hyperfine structure). The latter are caused by neighboring nuclei with nonzero nuclear spin (hyperfine structure) and by surrounding electric field gradients (quadrupole interactions).

In this manner the unpaired electron spins may be used as a probe for analysis of their immediate microscopic surroundings. Typical instrument sensitivity is such that approximately 10^{-11} mole of a paramagnetic species can be detected, but generally concentrations of 10^{-4} mole give optimum ESR spectra.

Nuclear Magnetic Resonance (NMR). This technique is essentially based on the same principle as ESR, but NMR is capable of detecting nuclei (MHz) instead of electrons (GHz). (Lack of a standardized nomenclature has resulted in numerous modifiers in connection with magnetic resonance instrumentation—electron, proton, nuclear, etc., plus application-related terms, such as silicon-29, oxygen-17, ^{13}C, ^{31}P NMR, etc.)

In nuclear magnetic resonance spectroscopy, a nucleus possessing a magnetic moment when placed in a homogeneous magnetic field will precess about the field axis at a rate which is dependent upon the strength of the field. See Fig. 2. If these nuclei are then brought into contact with an oscillating electromagnetic field of the same frequency as the Larmor precessional frequency of the nuclei, energy will be absorbed by the nuclei spin system from the oscillating ratio frequency field. As stated above, the net energy absorption is proportional to the Boltzman distribution of spin populations and the effective nuclear relaxation times.

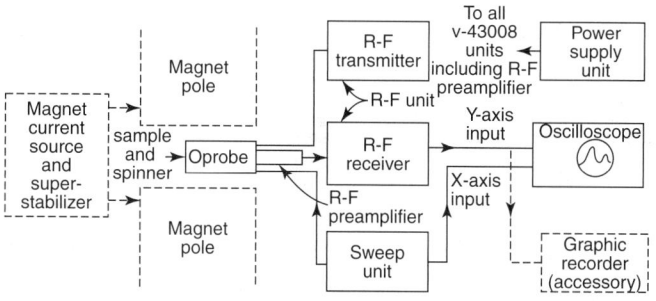

Fig. 2.   Schematic of NMR spectrometer.

Different nuclei will have different precessional frequencies and therefore at a particular field will absorb energy at certain characteristic radio frequencies. Also, nuclei of the same nuclear species (such as hydrogen) will absorb energy at slightly different frequencies, dependent upon their molecular environment. The latter observation makes possible an entire subfield of NMR spectroscopy, termed high-resolution NMR, which is an appropriate method for determining chemical structure and identifying and measuring similar nuclei in two or more different compounds (mixtures).

The major components of a magnetic resonance spectrometer are: (1) a magnet capable of producing a very strong homogeneous field which may be continuously varied over a very small range; (2) a low-power radio frequency oscillator which supplies rf power to a small transmitter coil surrounding the sample; (3) a small receiver coil which also surrounds the sample (but is orthogonal with respect to the transmitter), and feels (4) a sensitive radio receiver (tuned to the same frequency as the transmitter) capable of amplifying any signal which might be induced in the receiver; and (5) a recorder of oscilloscope which can display the resulting spectra.

Various decoupling techniques are used to simplify complex NMR spectra. For example, in a simple two-spin system (homonuclear) giving rise to four NMR lines it is possible to saturate a nucleus at a particular rf frequency and collapse the remaining doublet to a single NMR line. This occurs because the second nucleus sees an averaged interaction from the first rather than two distinct interactions from spins in the high and low energy states. Noise decoupling is predominantly used to decouple different nuclei—for example, hydrogen nuclei C^{13} spectra (heteronuclear). Here the entire hydrogen spectrum is saturated over a range of frequencies (noise) leaving behind the much simplified C^{13} spectrum.

In internuclear double resonance one nuclear line at a given frequency is observed while a second frequency is swept through the remainder of the NMR spectrum. All nuclei which are in some way coupled to the line being observed will enhance or de-enhance the latter as the second frequency passes through the resonance condition. Also, the Fourier transform technique has nicely complimented NMR. Any complex waveform can be converted to a spectrum of frequencies by Fourier transformation.

In NMR, the waveform is a superposition of a set of nuclear precession frequencies with amplitudes decaying due to relaxation and field inhomogeneity. The transformation may be carried out by analog means (spectrum analyzer) or on a small dedicated laboratory computer. The latter appears

to be the most convenient solution. Since the free induction decay signal decays with time, whereas the instrumental noise remains constant, the noise content is higher in the tail of the transient signal, and it is possible to improve the overall signal-to-noise ratio by weighting the transient signal with an exponentially decaying function of time. The shorter the time constant of this exponential, the greater the improvement in sensitivity, but this increases the linewidths of the transformed spectrum. The reversal of this procedure can enhance resolution at the cost of sensitivity.

High-Resolution Nuclear Magnetic Resonance of Solids. Important developments in solid-sample NMR techniques of the early 1980s have made NMR of significant interest as a tool for characterizing solid samples—as it has been in the past for the study of liquids. As observed by Maciel, the development of line-narrowing techniques, such as magic-angle spinning (MAS) and high-power decoupling, has led to powerful high-resolution NMR for studying solids. In favorable cases (for example, where high abundances of protons are present) cross polarization (CP) provides a means of circumventing the time hurdle caused by inefficient spin-lattice relaxation in many solids. Combining the CP and MAS approaches for carbon-13 with proton decoupling has become a popular and routine experiment for organic solids. For many nuclides with spin quantum number $1 > \frac{1}{2}$, the central nuclear magnetic resonance transition can be used in high-resolution experiments that involve rapid sample spinning. A continuing stream of other advances in NMR technology bodes well for the characterization of solids by a wide range of nuclides. The complexities of this topic unfortunately are beyond the scope of this encyclopedia.

It is interesting to note that several of the concepts for improving NMR technology, as listed by Levy and Craik, in 1988, already have been partially or fully achieved: (1) two-dimensional Fourier transform (FT NMR); (2) high-resolution NMR in solids; (3) new types of pulse sequences; (4) chemically induced dynamic nuclear polarization; (5) multiple quantum NMR; and (6) NMR imaging (MRI).

Two-Dimensional NMR. Bax and Lerner report on how two-dimensional Fourier transform pulse NMR (2-D FT NMR) has extended the range of applications of NMR spectroscopy into the area of large, complex molecules, such as DNA and proteins. Great spectral simplification can be obtained by spreading the conventional one-dimensional NMR spectrum in two independent frequency dimensions, thus removing spectral overlap, facilitating spectral assignment, and providing additional information. Conformational information related to interproton distances is available from resonance intensities in certain types of two-dimensional experiments. Two-dimensional NMR spectroscopy also has been applied to the study of ^{13}C and ^{15}N to provide connectivity information and greatly improving the sensitivity of these determinations. A traditional NMR spectrum of a sugar and a 2D spectrum are contrasted in Fig. 3.

Correlative spectroscopy (COSY) is an approach that involves correlating groups believed to be coupled to each other and to prove that this coupling does exist. *Spin-echo correlation spectroscopy* (SECSY) is a variation of the COSY technique. Sometimes called *J-resolved spectroscopy*, it is a technique that allows a separation of the chemical shift of a nucleus from the coupling to other nuclei. This simplifies the spectrum and permits one to assign each resonance to a specific nucleus. Sometimes referred to as *nuclear Overhauser effect* (NOESY), this is a technique that makes it possible to determine distances between nonadjacent residues in a peptide chain. Another application, *multiple quantum transitions*, takes advantage of the fact that molecules in a sample are forced to absorb or emit several quanta of energy at one time. For example, the technique can be used to determine which carbon atoms in a molecule are connected to other specific atoms in the molecule.

Magnetic Resonance Imaging (MRI). In 1973, Lauterbur added a new aspect to NMR basics, namely that of image formation based upon NMR principles. This led to NMR imaging, now commonly referred to as magnetic resonance imaging (MRI).

The source of the imaging photons is not a Roentgen tube, nor an unstable isotope, nor an electron storage ring, but rather it is a radio transmitter. The commonly referred *radio photon* radiation allows photons to pass into the tissues under study, where they are absorbed by nuclear particles rather than by electrons. The nucleons are momentarily excited by the process and then return to their resting state by emitting photons of the same or nearly the same energy that they had absorbed. Because the nuclei of the different chemical elements absorb radio photons of different frequencies, it is possible to use the method to detect the presence of a single element in a sample. When protons are irradiated by radio

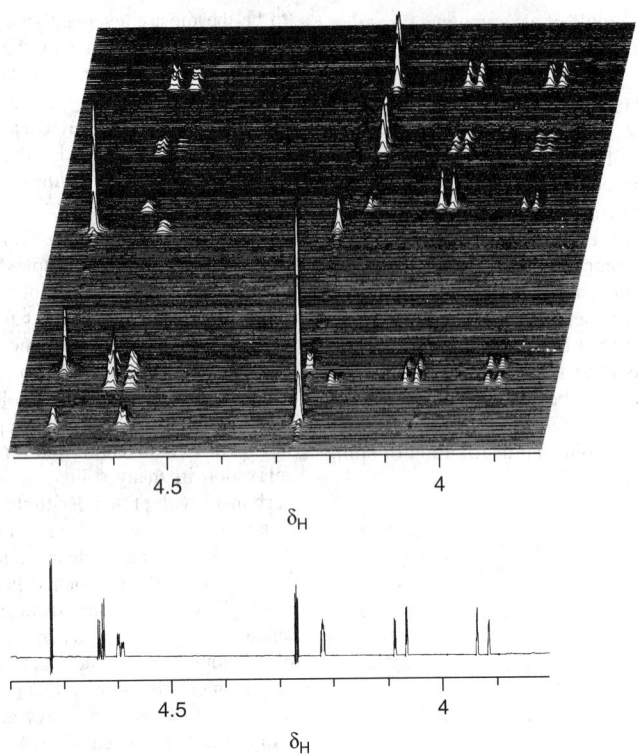

δ_H

δ_H

Fig. 3.   (*Top*) A two-dimensional nuclear magnetic resonance spectrum of a sugar; (*Bottom*) conventional NMR spectrum of same sugar. The two-dimensional spectrum also can be plotted as a contour map with intensities denoted by color. (*JEOL Inc.*)

photons of a frequency that precisely matches their own precessional frequency, resonance occurs. The resonant frequency (Larmor frequency) is determined by the natural rotational velocity (gyromagnetic ratio) characteristic of each species of atomic nucleus and by the strength of the applied static magnetic field, as expressed by: Larmor frequency = Magnetic field × Gyromagnetic ratio. See Fig. 4.

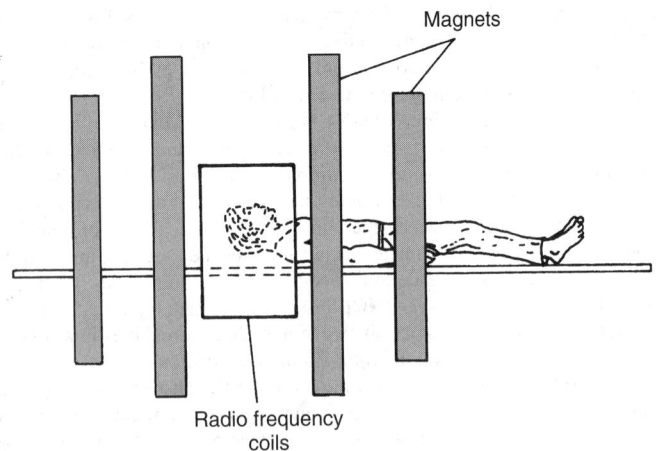

Fig. 4.   General arrangement of MRI equipment in a medical setting.

Magnetic resonance imaging is well suited for the imaging of soft tissues. It has been found particularly effective in (1) studying skeletal musculature, notably in the male and female pelvic regions, (2) delineating tumor development and assisting in planning surgery or therapy, and (3) precisely locating tumors, particularly in the brain where, with time, the evolution of hematomas can be studied. For clinical work, MRI installations may be used in two staff shifts per day, holidays, and weekends.

The magnetic fields used are usually 1.5 to 2 teslas (15,000 to 20,000 gauss). Barriers to using higher magnetic fields are magnet cost, additional expense for building the site, and notation by some persons

(volunteers) to discomfort when over 4 teslas are used, which also causes the patient to move.

In 1986, Fossel and McDonough (Beth Israel Hospital, Boston) proposed that proton NMR spectroscopy of human plasma possibly could be a "potentially valuable approach to the detection of cancer and the monitoring of therapy." Subsequent investigations through 1990 failed to develop a convincing correlation.

Traditional MRI of the human body relies mainly on the detection of the most abundant type of nuclei, the hydrogens in water, and, to some extent, fat.

As pointed out by C. Moonen (National Institutes of Health), "For discrimination of healthy and diseased tissues, adequate contrast is essential. Such contrast depends not only on differences in water concentration, but also on the NMR relaxations, which, in turn, are related to local mobilities and interactions."

Magnetic resonance imaging, in addition to providing detailed information about the macroscopic structure and anatomy, also permits the noninvasive spatial evaluation of various biophysical and biochemical processes in living systems. These include the motion of water in processes, such as vascular flow, capillary flow, diffusion, and exchange. Further, the concentrations of various metabolites can be determined for the assessment of regional regulation of metabolism. In the scholarly Moonen paper, examples are given of flow imaging, diffusion imaging, and the imaging of tissue perfusion, of exchange, and of metabolites. These aspects of MRI imaging sometimes are referred to as *functional MRI*.

Imaging of the central nervous system by MRI technology is of great interest. Advantages of MRI include its sensitivity to soft tissue, the contrast between gray and white matter, the paucity of signals from the skull, and the availability of coronal, sagittal, and transverse sections. The procedure has been used to detect posterior fossa tumors, such as acoustic neuromas, and pituitary and parasellar tumors, orbital tumors, multiple sclerosis, and a number of other lesions in the craniovertebral junction, and of the cord and spine. MRI technology, according to a number of authorities, has large potential for use in *noninvasive* cardiological studies, particularly of blood flow imaging. Contrast agents are not required because nuclei in rapidly flowing blood move out of the volume of interest during the interval required for application of the rf pulse and of gradient magnetic fields. The parameters that influence MRI signals at various flow rates are being studied.

In addition to hydrogen atoms, MRI can create, for example, ^{31}P images which are excellent indicators of energy metabolism. ^{23}Na images reflect extracellular and intracellular fluid fluxes.

Echo-Planar Imaging. A major problem with MRI in the past has been the long data acquisition times (up to several minutes). Consequently, MR images are subject to so-called *motional artifacts*, caused by physiological motions (heartbeat, blood flow, bowel peristalsis, breathing) as well as by voluntary movements in severely ill and uncooperative patients, including children. Echo-planar imaging (EPI) permits faster scan times, thus effectively reducing imaging to a fraction of a second as compared with minutes. Although a technical description of how EPI is implemented is too complex for coverage here, echo-planar imaging uses only one nuclear spin excitation per image.

EPI has broadened the use of MRI to include the evaluation of cardiac function in real time, mapping of organ blood pool and perfusion, functional imaging of the central nervous system, depiction of blood and cerebrospinal fluid flow dynamics, and motion picture imaging of the mobile fetus in utero. EPI also has the practical advantages of increasing patient throughput at a lower cost per MRI examination. With these advantages, it is expected that EPI will become an established tool for early diagnosis of some common and potentially treatable diseases, such as ischemic heart disease and stroke.

Comparison with Ultrasonography. The health care community is aware of the high costs of MRI and continues to seek other techniques that may be effective at lower cost. One such comparison was made in connection with prostate cancer. The approach to treatment varies and depends on the extent of cancer at the time of diagnosis.

In a specific comparative study over a period of 15 months, 230 patients were evaluated with identical imaging techniques. It was found that MRI correctly staged 77% of cases of advanced disease and 57% of cases of localized disease. The corresponding Figures for ultrasonography were 66% and 46%. MRI identified only 60% of all malignant tumors measuring more than 5 mm on pathological analysis, while ultrasonography identified only 59%. The study concluded, "The MRI and ultrasonography equipment is not highly accurate in staging early prostate cancer, mainly because neither technique has the ability to identify microscopic spread of disease."

Superconducting Quantum Interference Device (SQUID). These devices are sensitive detectors of magnetic fields. Low-temperature superconductors have been used in the past to sense weak magnetic signals from the brain for medical diagnosis. Such devices, however, required cooling with liquid helium and thus have resulted in costly, unwieldy apparatus. Research is now underway toward using higher-temperature superconductors, such as $YBa_2Cu_3O_7$. Although the higher-temperature semiconductor still requires refrigeration, this can be accomplished with liquid nitrogen, thus resulting in a less costly, simpler, and more portable detecting device.

Additional Reading

Abelson, P.H.: "New Horizons in Medicine," *Science*, 1109 (November 25, 1988).

Ancreasen, N.C.: "Brain Imaging: Applications in Psychiatry," *Science*, 1381 (March 18, 1988).

Bain, L.: "MRI — Safety Issues Stimulate Concern," *Science*, 1245 (May 31, 1991).

Bax, A. and L. Lerner: "Two-Dimensional Nuclear Magnetic Resonance Spectroscopy, "*Science*, **232**, 960–967 (1986).

Edelman, R. et al.: "Clinical Magnetic Resonance Imaging, 2nd Edition" W.B. Saunders, Philadelphia, PA, 1996.

Hari, R. and O.V. Lounasmaa: "Recording and Interpreting Cerebral Magnetic Fields," *Science*, 432 (April 28, 1989).

Kirkwood, J.R.: "Essentials of Neuroimaging, 2nd Edition," Churchill-Livingstone, Inc., New York, NY, 1995.

Krestel, E. (Editor): "Imaging Systems for Medical Diagnosis: Fundamentals and Technical Solutions — X-Ray Diagnostics — Computed Tomography — Nuclear Medical Diagnostics — Magnetic Resonance Imaging — Ultrasound Technology," John Wiley & Sons, Inc., New York, NY, 1990.

Levy, G.C. and D.J. Craik: "Developments in Nuclear Magnetic Resonance Spectroscopy, "*Science*, **214**, 291–299 (1981).

Magin, R. et al. (Editors): "Biological Effects and Safety Aspects of Nuclear Magnetic Resonance Imaging and Spectroscopy, Vol. 649," New York Academy of Sciences, New York, NY, 1992.

Moonen, C.T. et al.: "Functional Magnetic Resonance Imaging in Medicine and Physiology," *Science*, 53 (October 6, 1990).

Pool, R.: "Putting SQUIDs to Work," *Science*, 862 (August 24, 1990).

Randal, J.: "NMR: The Best Thing Since X-Rays," *Technology Review (MIT)*, 59 (January 1988).

Rifkin, M.D. et al.: "Comparison of Magnetic Resonance Imaging and Ultrasonography in Staging Early Prostate Cancer," *New Eng. J. Med.*, 623 (September 6, 1990).

Shulman, R.: "NMR — Another Cancer-Test Disappointment," *New Eng. J. Med.*, 1002 (April 5, 1990).

Sochurek, H. and P. Miller: "Medicine's New Vision," *National Geographic*, 2 (January 1987).

Stehling, M.K., R. Turner, and P. Mansfield: "Echo-Planar Imaging in a Fraction of a Second," *Science*, 43 (October 4, 1991).

Sutton, D. and J.W.R. Young: "A Short Textbook of Clinical Imaging," Springer-Verlag, Inc., New York, NY, 1991.

Tamraz, J. and Y. Comair: "Atlas of Regional Anatomy of the Brain Using MRI: With Functional Correlations," Springer-Verlag New York, Inc., New York, NY, 2000.

Wilson, M. and F. Ruzicka (Editor): "Modern Imaging of the Liver: Applications of Computerized Tomography Ultrasound, Nuclear Medicine and Magnetic Resonance Imaging," Marcel Dekker, Inc., New York, NY, 1989.

NUCLEAR MAGNETIC RESONANCE SPECTROSCOPY. See **Magnetic Resonance Spectroscopy**.

NUCLEAR MAGNETON. A unit of magnetic moment used in atomic and nuclear physics, defined as

$$\mu_N = eh/4\pi M = 5.050 \times 10^{-27} \text{ ampere-meters}^2$$

in which e is the electronic charge ($1.602 \times 10^{-19}c$), h is Planck's constant, and M is the rest mass of the proton. This unit is derived for nuclear particles by analogy with the Bohr magneton, which is applicable to the magnetic moment associated with electron orbital angular momentum in an atom. No real nuclear particle, however, has a magnetic moment of exactly one nuclear magneton. The magnetic moment of the proton is +2.793 nuclear magnetons, and of the neutron −1.913 nuclear magnetons.

NUCLEAR POTENTIAL. The potential energy V of a nuclear particle as a function of its position in the field of a nucleus or of another nuclear particle. A central potential is one that is spherically symmetric, so that V is a function only of the distance r of the particle from the center of force. A noncentral potential, on the other hand, is one that is not spherically symmetrical, or one that depends upon the relative directions of the angular momenta associated with the particle and the center of force, as well as upon the distance r. A negative potential corresponds to an attractive force, while a positive potential corresponds to a repulsive force.

Although the expression can certainly be applied to the problem of nuclear forces, the usual meaning of a nuclear potential refers to the interaction of a nucleon (neutron or proton) with a complex nucleus. Although the potential energy of a single nucleon inside a nucleus is clearly a rapidly varying function of position and time (since it represents the interaction with a large number of closely packed, fast-moving particles), one may nevertheless speak of the average potential energy, and one may regard this as a smoothly varying function. For a neutron, the nuclear potential is essentially negative inside the nucleus, rising rapidly to zero outside the nuclear radius R. For a proton, the long-range electrostatic repulsion must, of course, be added. Owing to the Pauli exclusion principle, and to the exchange nature of nuclear forces, however, such a potential cannot in general be regarded simply as a function of position, $V = V(r)$; it depends in addition upon the momentum of the particle, which in quantum mechanics does not commute with the position. Hence, the potential must be regarded as a nondiagonal matrix operator $V = \langle r|V|r'\rangle$ in configuration space, or a similar operator in momentum space.

Although the concept of a nuclear potential in this latter sense cannot be defined in a precise way, it has nevertheless been useful, both qualitatively and quantitatively, in the investigations of nuclear structure and nuclear reactions. It has been of particular usefulness in the optical model of nuclear reactions.

NUCLEAR POWER TECHNOLOGY. After a terse review of the physics and chemistry of nuclear fission and the nature of these reactions, some of the design features and operating parameters of the four families of reactors currently installed are described. These units, some of which are approaching their retirement are, of course, the starting basis for the design of next-generation reactors, which appeared prior to the end of the 20th century. Design improvements that constitute these forthcoming systems are described in terms of their status as of the mid-1990s. The topic of radioactive waste handling is summarized. Based upon available information, nuclear power technology in the United States as well as in Canada, France, Japan, the U.K., and other leading industrialized nations is covered.

The first nuclear power plant was installed at Shippingport, Pennsylvania in 1957 and, after serving as a test facility for several years, was dismantled because of "old age." The plant had a capacity of 60,000 kilowatts. Indeed, the plant was extremely small by comparison with hundreds of units installed today in the United States and other major countries of the world.

NATURE OF NUCLEAR FISSION REACTIONS

The energy of a nuclear fission reaction can be computed from the change in mass between reactants and products according to Einstein's law:

$$\Delta E = \Delta m c^2$$

where E is the energy in ergs, m is mass in grams, and c is the velocity of light in centimeters per second. For example, the mass difference in this equation is $\Delta m = 0.2058$ amu (atomic mass units). Therefore, $\Delta E = 931$ MeV/amu $\times 0.2058$ amu $= 191.6$ MeV. The average amount of energy released in the various fission reactions is about 200 MeV (million electron volts). This energy is distributed in the fission process as:

	MeV
Kinetic energy of fission fragments	165
Radioactive-decay energy	23
Kinetic energy of neutrons	5
Prompt gamma-ray energy	7

The energy of a chemical reaction, approximately 3–4 eV, is dramatically lower than that of a nuclear reaction. Hence, the fission of ^{235}U yields 2.5 million times as much energy as the combustion of the same weight of carbon.

The importance of fission in energy (power) production lies in two facts: (1) an exceedingly large amount of energy is released in the fission reaction; and (2) the production of excess neutrons permits a chain reaction. These two circumstances make it possible to design nuclear reactors in which self-sustaining reactions occur with the continuous release of energy. Although described later, it may be pointed out here that nuclear fission is not the only energy-releasing nuclear reaction. The fusion of light nuclides, like hydrogen, into heavier elements is also an energy-producing process.

The heat generated in nuclear power plants is transferred to a working fluid and from this point on the nuclear power plant and the conventional fossil-fueled power plant are essentially similar.

Fission Reaction. In nuclear fission, the nucleus of a heavy atom is split into two or more fragments. The reaction is initiated by the absorption of a neutron. A typical reaction is

$$^{235}_{92}\text{U} + ^{1}_{0}n \longrightarrow ^{137}_{56}\text{Ba} + ^{97}_{36}\text{Kr} + 2^{1}_{0}n + \Delta E$$

In this reaction, a ^{235}U atom absorbs one neutron, becomes unstable, and subsequently fissions into two fission fragments plus two neutrons. This is just one of the many ways in which ^{235}U might fission. The number of neutrons produced in a fission reaction is usually 2 or 3. The excess neutrons produced by the fission reaction provide the means of self-sustaining the chain reaction. Nuclides including ^{233}U, ^{235}U, and ^{239}Pu, which are fissionable by neutrons of all energies, are termed fissile nuclides.

Nuclear Fuels. There are two broad categories of nuclear fuels: (1) the fissile nuclides previously mentioned; and (2) the fissionable nuclides, ^{232}Th (thorium) and ^{238}U. Thermal reactors use fissile nuclides as fuel, while fast reactors are designed to burn fissionable materials. In fast reactors, only a small portion of the ^{232}Th and ^{238}U are fissioned directly. A larger portion of these materials is converted into ^{235}U and ^{239}Pu, respectively, through neutron absorption. Thus, this type of reactor not only consumes fuel, but also produces (breeds) new fuel material. Hence, the term *breeder reactor* is used for reactors designed to take advantage of this phenomenon. Breeding is possible in thermal reactors also, but to a lesser extent. The fuel material in a fast reactor must contain a significant amount (about 10%) of one of the fissile materials. The remainder of the fuel must have a high mass number in order to avoid slowing down the neutrons. The natural reserves of fissionable materials are more than 100 times greater than the reserves of fissile materials. Consequently, from the viewpoint of utilization of available energy resources, fast reactors are of great importance. Breeder reactors are described later.

Moderators. The most important slow-down mechanism is elastic scattering on elements of low mass number. Materials like light and heavy water, beryllium oxide, and graphite are used to slow down, or *thermalize* the neutrons to an energy of about 0.025 eV. As neutrons collide with the nuclei of these atoms, their kinetic energy and speed are gradually reduced until thermal equilibrium is achieved with the reactor structure. The fewer such collisions before deceleration is complete, the less chance of ^{238}U atoms absorbing neutrons.

Critical Mass. Thermal neutrons, which move like atoms in a low-pressure gas, diffuse throughout the reactor. They must be absorbed by a nucleus of the reactor structure, in which case they merely make that nucleus radioactive. Or, they may strike a fissionable atom of ^{235}U, causing fission and, in turn, releasing more neutrons to maintain the reaction. Should the number of neutrons absorbed by the moderator and ^{238}U be greater than about 1.5 excess neutrons emitted from each fission, the chain reaction will not be maintained. Therefore, the reactor core must be designed so that the mass of fuel will be just sufficient to ensure one neutron from each fission causing fission in another atom. A mass and configuration of fissionable material in which this occurs is termed the *critical mass* — or a reactor in which this condition is achieved is said to have "gone critical."

To measure a chain reaction, a multiplication factor k is used to indicate the ratio of neutrons in one generation to those in the preceding generation. Thus, in a constant chain reaction where the total number of neutrons neither increases nor decreases, the heat output is constant and $k = 1$. Should k rise above unity, the rate of fission, and hence the rate of heat productivity, steadily rises. This is so even if k is held constant at its new value. Here lies one *major difference between nuclear reactors and conventional steam generators*. In the latter, heat output is proportional to firing rate. If the firing rate is increased, the steam output is increased; but it remains constant at its new level. In a nuclear reactor, an increase in k results in continuously rising heat output. Only by returning the rate of neutron production to its original ratio can heat output be maintained at its new level.

Reactivity Control. Absorption of excess neutrons, above those needed to maintain a constant reactivity level, provides close control over the degree of reactivity. This is accomplished by inserting materials having a high neutron-capture rate into the core. Control rods of special alloy metals are moved into and out of the cores as required. To start the reactor from shutdown (black start), control rods are partially withdrawn until k becomes greater than one. Neutron flux and heat output grow until the desired level is reached. At this point, control-rod movement is quickly reversed to keep k at unity. The reactor is shut down by inserting the rods to their full extent. In this position, the rods absorb more than 1.5 excess neutrons per fission and the chain reaction quickly stops. Heat production continues for a time, but is usually dissipated by an auxiliary cooling system.

It is interesting to contrast a nuclear power reactor and a nuclear bomb. The designer of a nuclear fission bomb seeks to release as much fission energy as possible within the shortest possible time (milliseconds). Thus, a bomb is designed to favor *prompt neutrons*. By contrast, the normal operating mode of a nuclear power reactor is one in which prompt neutrons alone cannot sustain a chain reaction, but prompt neutrons together with delayed neutrons can. Only the delayed neutrons are controllable. A power reactor is designed to release fission energy *slowly and smoothly* and in just the right amounts to convert water into steam. Whereas the "fuel" in a bomb is used up essentially in an instant, in a power reactor the energy release is spread over months and years. It has been agreed by physicists for many years that it is physically impossible for a power reactor to explode in the manner of an atomic bomb.

TYPES AND MAJOR CHARACTERISTICS OF NUCLEAR POWER REACTORS

In order, the following types of nuclear fission reactors are described in this section: (1) light water reactors, (a) pressurized water reactors, (b) boiling-water reactors; (2) high-temperature gas-cooled reactors; (3) heavy water reactors; and (4) fast breeder reactors. Military reactors are not described.

Contemporary Light-Water Reactors (LWRs)[1]

These reactors are of two principal designs: (1) *pressurized water reactors* (PWR), and (2) *boiling-water reactors* (BWR). In a PWR, heat generated in the nuclear core is removed by water (reactor coolant) circulating at high pressure through the primary circuit. The water in the primary circuit both cools and moderates the reactor. Heat is transferred from the primary to the secondary system in a heat exchanger, or boiler, thereby generating steam in the secondary system. The BWR differs from the PWR primarily in that boiling takes place in the reactor itself. Comparable steam temperatures are possible at pressures of about 1000 pounds per square inch (6.9 mPa) as contrasted with 2000 psi (13.8 mPa) for pressurized reactors.

Contemporary Boiling Water Reactor (BWR)

Aside from its heat source, the boiling water reactor (BWR) generation cycle is substantially similar to that found in fossil-fueled power plants. One of the first BWRs was the Vallecitos BWR, a 1000 psi (6.9 mPa) reactor which powered a 5 MW electric generator and provided power to the Pacific Gas & Electric Company grid through 1963. Power output capabilities have increased many times during the intervening years as shown by tabular summaries given later in this entry.

The direct-cycle boiling water reactor nuclear system (Fig. 1) is a steam generating system consisting of a nuclear core and an internal structure assembled within a pressure vessel, auxiliary systems to accommodate the operational and safeguard requirements of the nuclear reactor, and necessary controls and instrumentation. Water is circulated through the reactor core producing saturated steam which is separated from the recirculation water, dried in the top of the vessel, and directed to the steam turbine-generator. The turbine employs a conventional regenerative cycle with condenser de-aeration and condensate demineralization. The direct-cycle system is used because of its inherently simple design, contributing to reliability and availability.

The steam from a BWR is, of course, radioactive. The radioactivity is primarily ^{16}N, a very short-lived nitrogen isotope (7 seconds half-life) so that the radioactivity of the steam system exists only during power generation. Extensive generating experience has demonstrated that shutdown maintenance on a BWR turbine, condensate, and feedwater components can be performed essentially as a fossil-fuel plant.

The reactor core, the source of nuclear heat, consists of fuel assemblies and control rods contained within the reactor vessel and cooled by the recirculating water system. A 1,220-MWe BWR/6 core consists of 732 fuel assemblies and 177 control rods, forming a core array 16 feet (4.8 meters) in diameter and 14 feet (4.2 meters) high. The power level is maintained

---

[1] In nuclear power technology, ordinary water, in contrast with *heavy water*, is termed *light water*.

---

or adjusted by positioning control rods up and down within the core. The BWR core power level is further adjustable by changing the recirculation flow rate without changing control rod position, a feature that contributes to excellent load-following capability.

The BWR is the only light water reactor system that employs bottom-entry control rods. From the very first BWRs, bottom-entry control rods have been used because reactivity and moderator density is highest in the lower part of the core. They provide optimum power shaping characteristics for the type of core where moderator density is varied as a function of power level. Bottom-entry and bottom-mounted control rod drives also allow refueling without removal of rods and drives, and allow drive testing with an open vessel prior to initial fuel loading, or at each refueling operation. The hydraulic system, using reactor system pressure, provides rod insertion forces that are greater than gravity or mechanical systems.

The BWR requires substantially lower primary coolant flow through the core than pressurized water reactors. The core flow of a BWR is the sum of the feedwater flow and the recirculation flow, which is typical of any boiler. Unique to the BWR is the application of jet pumps inside the reactor vessel. See Fig. 2. The jet pumps deliver their driving force from the external recirculation pumps and generate about two-thirds of the recirculation flow within the reactor vessel. The jet pumps also contribute to the inherent safety of the BWR design under loss-of-coolant emergency conditions because they continue to provide internal circulation with one or both external recirculation loops out of service. The BWR can deliver about one-third power through this natural jet pump circulation mode, a vital capability in effecting a "black start" (a fully fresh start-up of a reactor) of the plant without external power.

The BWR operates at constant pressure and maintains constant steam pressure similar to most fossil-fueled boilers. The BWR primary system operates at pressure about one-half that of a pressurized water reactor primary system, while producing steam of equal pressure and quality.

The integration of the turbine pressure regulator and control system with the reactor water recirculation flow control system permits automated changes in steam flow to accommodate varying load demands on the turbine. Power changes of up to 25% can be accomplished automatically by recirculation flow control alone, at rate of 15% per minute increasing and 60% per minute decreasing. This provides a load-following capability that can track rapid changes in power demand.

Nuclear Boiler Assembly. This assembly consists of the equipment and instrumentation necessary to produce, contain, and control the steam required by the turbine-generator. The principal components of the nuclear boiler are: (1) reactor vessel and internals—reactor pressure vessel, jet pumps for reactor water circulation, steam separators and dryers, and core

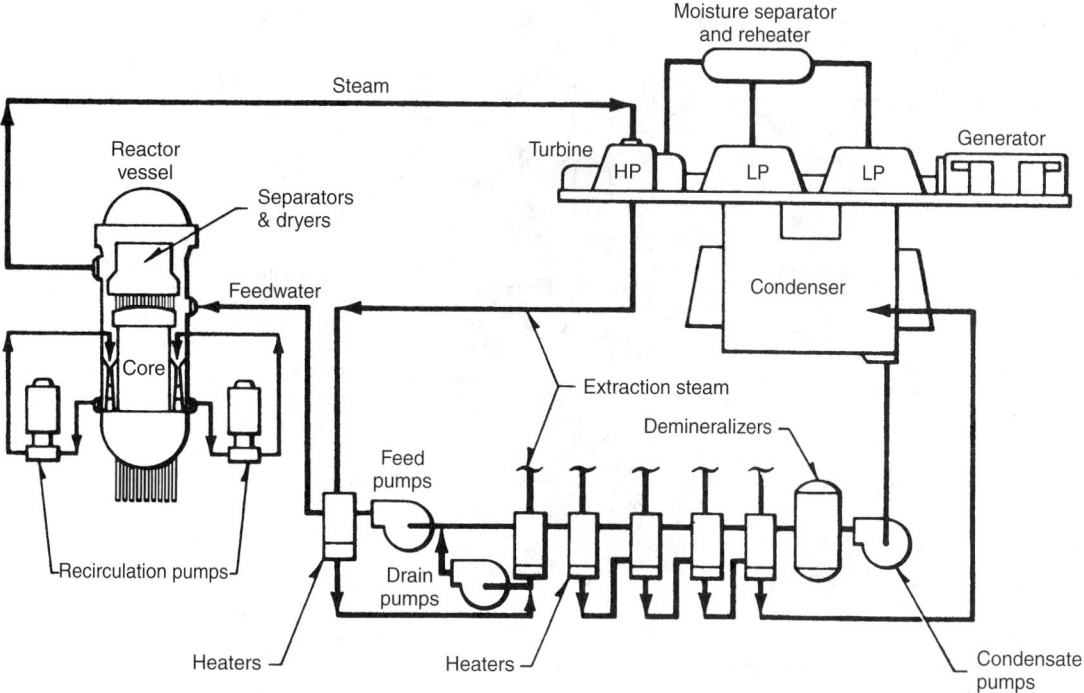

Fig. 1.   Contemporary direct-cycle reactor system used in boiling water reactor. (*General Electric.*)

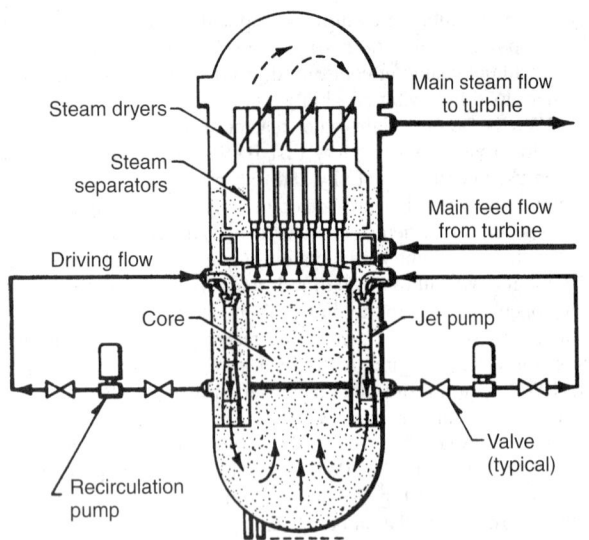

Fig. 2.   Steam and recirculation water flow paths of contemporary boiling water reactor. (*General Electric*.)

support structure; (2) reactor water recirculation system — pumps, valves, and piping used in providing and controlling core flow; (3) main steam lines — main steam safety and relief valves, piping, and pipe supports from reactor pressure vessel up to and including the isolation valves outside of the primary containment barrier; (4) control rod drive system — control rods, control rod drive mechanisms and hydraulic system for insertion and withdrawal of the control rods; and (5) nuclear fuel and in-core instrumentation.

Reactor Assembly. This assembly (Fig. 3) consists of the reactor vessel, its internal components of the core, shroud, top guide assembly,

core plate assembly, steam separator and dryer assemblies and jet pumps. Also included in the reactor assembly are the control rods, control rod drive housings and the control rod drives.

Each fuel assembly that makes up the core rests on an orificed fuel support mounted on top of the control rod guide tubes. Each guide tube, with its fuel support piece, bears the weight of four assemblies and is supported by a control rod drive penetration nozzle in the bottom head of the reactor vessel. The core plate provides lateral guidance at the top of each control rod guide tube. The top guide provides lateral support for the top of each fuel assembly.

Control rods occupy alternate spaces between fuel assemblies and may be withdrawn into the guide tubes below the core during plant operation. The rods are coupled to control rod drives mounted within housings welded to the bottom head of the reactor vessel. The bottom-entry drives do not interfere with refueling operations. A flanged joint is provided at the bottom of each housing for ease of removal and maintenance of the rod drive assembly.

Except for the Zircaloy in the reactor core, the reactor internals are stainless steel or other common corrosion-resistant alloys. The reactor vessel is a pressure vessel with a single full-diameter removable head. The base material of the vessel is low alloy steel, which is clad on the interior, except for nozzles with stainless steel weld overlay to provide the necessary resistance to corrosion.

The shroud is a cylindrical, stainless steel structure that surrounds the core and provides a barrier to separate the upward flow through the core from the downward flow in the annulus. Two ring spargers, one for low pressure core sprays and the other for high pressure core spray are mounted inside the core shroud in the space between the top of the core and steam separator base. The core spray ring spargers are provided with spray nozzles for the injection of cooling water under emergency conditions. A nozzle for the emergency injection of neutron absorber (sodium pentaborate) solution is mounted below the core in the region of the recirculation inlet plenum.

The steam separator assembly consists of a domed base, on top of which is welded an array of standpipes with a 3-stage separator located at the top

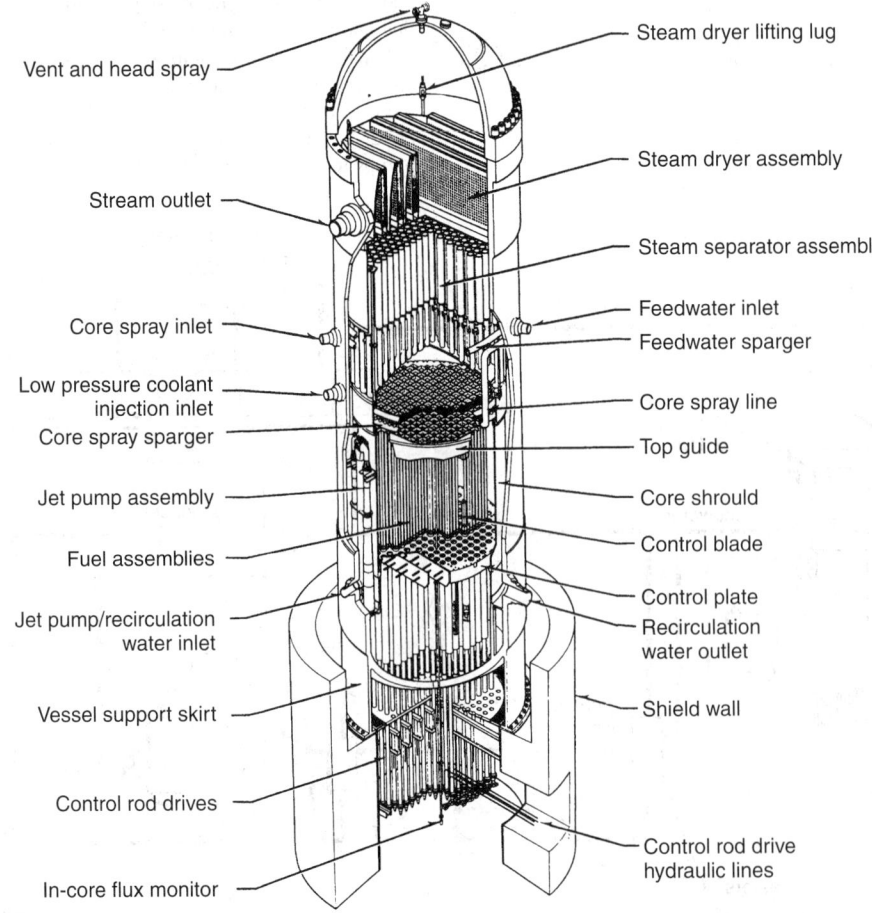

Fig. 3.   Reactor assembly of contemporary boiling water reactor. (*General Electric*.)

of each standpipe. The steam separator assembly rests on the top flange of the core shroud and forms the cover of the core discharge plenum region. In each separator, the steam-water mixture rising through the standpipe impinges on vanes which give the mixture a spin to establish a vortex wherein the centrifugal forces separate the water from the steam in each of three stages. Steam leaves the separator at the top and passes into the wet steam plenum below the dryer. The separated water exits from the lower end of each stage of the separator and enters the pool that surrounds the standpipes to join the downcomer annulus flow.

The steam dryer assembly is mounted in the reactor vessel above the separator assembly and forms the top and sides of the wet steam plenum. Vertical guides on the inside of the vessel provide alignment for the dryer assembly during installation. The dryer assembly is supported by pads extending inward from the vessel wall and is held down in position during operation by the vessel head. These vanes are attached to a top- and bottom-supporting member forming a rigid, integral unit. Moisture is removed and carried by a system of troughs and drains to the pool surrounding the separators and then into the recirculation downcomer annulus between the core shroud and reactor vessel wall.

Control Rod Drive System. Positive core reactivity control is maintained by the use of movable control rods interspersed throughout the core. These control rods thus control the overall reactor power level and provide the principal means of quickly and safely shutting down the reactor. The rods are vertically moved by hydraulically actuated, locking piston type drive mechanisms. The drive mechanisms perform both a positioning and latching function, and a scram function with the latter overriding any other signal (scram signifies prompt shutdown).

Core Configuration. The reactor core of the BWR is arranged as an upright cylinder containing a large number of fuel assemblies and located within the reactor vessel. The coolant flows upward through the core. The plan of a typical core arrangement of a large BWR is shown in Fig. 4. The lattice configuration is shown in Fig. 5.

Fuel Rod. A fuel rod consists of uranium dioxide (UO_2) pellets and a Zircaloy-2 cladding tube. The UO_2 pellets are manufactured by compacting and sintering UO_2 powder into cylindrical pellets and grinding to size. The immersion density of the pellets is approximately 95% of theoretical UO_2 density. A fuel rod is made by stacking pellets into a cladding tube that is evacuated, back-filled with helium to atmospheric pressure, and sealed by welding Zircaloy end plugs in each end of the tube. The pellets are stacked to an active height of 148 inches (376 centimeters) with the top 12 inches (30.5 centimeters) of tube available as a fission gas plenum. A plenum spring is provided in the plenum space to exert a downward force on the pellets, the spring keeping the pellets in place during the pre-irradiation handling of the fuel bundle.

Fuel Bundle. Each fuel bundle contains 63 fuel rods, which are spaced and supported in a square (8×8) array by a lower and upper tie plate. Three types of rods are used in a fuel bundle: (1) tie rods; (2) a water rod; and (3) standard fuel rods. The third and sixth fuel rods along each outer edge of a bundle are tie rods. The eight tie rods in each bundle have threaded-end plugs, which screw into the lower tie plate casting. A stainless steel hexagonal nut and locking tab is installed on the upper end plug to hold the assembly together. The water rod not only serves as a spacer support rod, but also provides a source of moderator material near the center of the fuel bundle. This flattens the neutron flux across the bundle, and leads to lower local peaking factors and better utilization of uranium in the interior rods of the fuel assembly.

The initial core will contain fuel assemblies having a common average enrichment ranging from approximately 1.6% (weight) of ^{235}U to 2.2%, depending upon initial cycle requirements. Each assembly will contain different enrichment rods. Selected rods in each assembly will, in addition, be blended with gadolinium burnable poison. The reload fuel will also contain four different enrichment rods with an average enrichment in the range of 2.4 to 2.8%. Different ^{235}U enrichments are used in fuel assemblies to reduce the local power peaking. Low enrichment rods are used in the corner rods and in the rods nearer the water gaps; higher enrichment uranium is used in the central part of the fuel bundle.

Fuel Channel. A fuel channel encloses the fuel bundle. The combination of a fuel bundle and a fuel channel is called a fuel assembly. See Fig. 6. The channel is a square-shaped tube fabricated from Zircaloy 4. The outer dimensions are 5.518 inches (14 centimeters) by 5.518 inches (14 centimeters) by 166.9 inches (424 centimeters) long. The reusable channel makes a sliding seal fit on the lower tie plate surface. It is attached to the upper tie plate by the channel fastener assembly, consisting of a spring and a guide, and a cap screw secured by a lock washer. The fuel channels direct the core coolant flow through each fuel bundle and also serve to guide the control rods.

Neutron Sources. Several antimony-beryllium start-up sources are located within the core. They are positioned vertically in the reactor by "fit up" in a slot (or pin) in the upper grid and a hole in the lower core support plate. The active portion of each source consists of a beryllium sleeve enclosing two antimony-gamma sources. The resulting neutron emission strength is sufficient to provide indication on the source range neutron detectors for all reactivity conditions equivalent to the condition of all rods inserted prior to initial operation. The active source material is entirely enclosed in a stainless steel cladding with an outside diameter of approximately 0.7 inch (1.8 centimeter). The source is cooled by natural circulation of the core leakage flow in the annulus between the beryllium sleeve and the antimony-gamma sources.

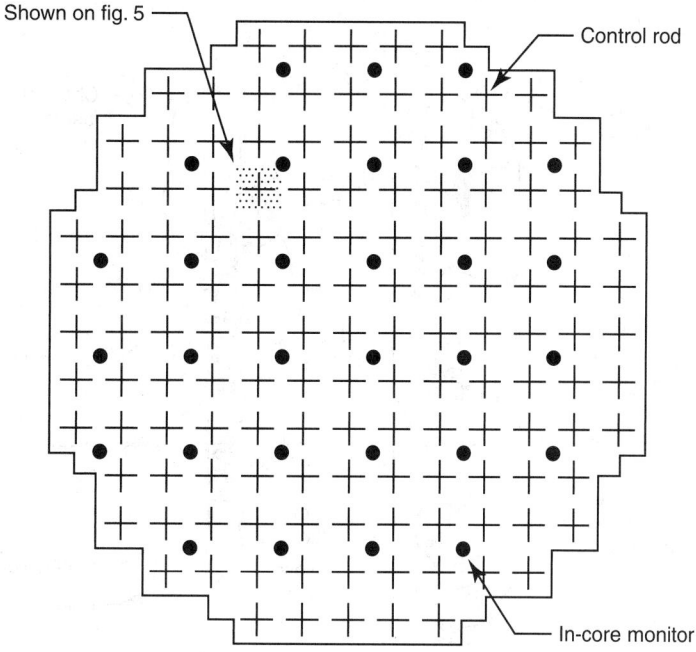

Fig. 4. Typical core arrangement in contemporary boiling water reactor. (*General Electric.*)

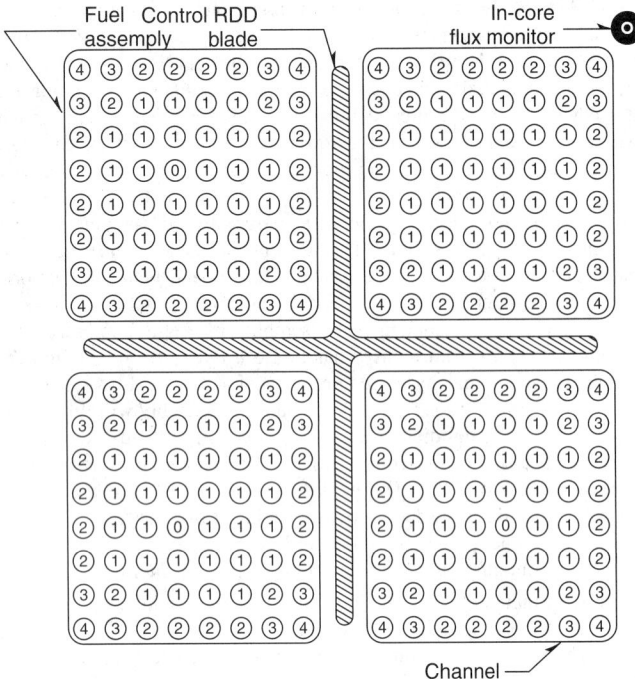

Fig. 5. Core lattice arrangement in contemporary boiling water reactor. (*General Electric.*)

Core Design Margins. The reactor core is designed to operate at rated power with sufficient design margin to accommodate changes in reactor operations and reactor transients without damage to the core. In order to accomplish this objective, the core is designed, under the most limiting operating conditions and at 100% rated power, to meet the following bases. (1) The maximum linear heat generation rate, in any part of the core, is always less than 13.4 kW/foot (43.97 kW/meter). (2) The minimum ratio between critical heat flux and fuel operating heat flux, in any part of the core, is always greater than 1.9.

Power Distribution. The design power distribution is divided for convenience into several components: (1) relative assembly power; (2) local; and (3) axial. The relative assembly power peaking factor is the maximum fuel assembly average power divided by the reactor core average assembly power. The local power peaking factor is the maximum fuel rod average heat flux in an assembly divided by the assembly average fuel rod heat influx. The axial power peaking factor is the maximum heat flux of a fuel rod divided by the average heat flux in that rod. Peaking factors vary throughout an operating cycle, even at steady-state full-power operation, since they are affected by withdrawal of control rods to compensate for fuel burnup. The design peaking factors represent the values of the most limiting power distribution that will exist in the core throughout its life.

Because of the presence of steam voids in the upper part of the core, there is a natural characteristic for a BWR to have the axial power peak in the lower part of the core. During the early part of an operating cycle, bottom-entry control rods permit a partial reduction of this axial peaking by locating a larger fraction of the control rods in the lower part of the core. At the end of an operating cycle, the higher accumulated exposure and greater depletion of the fuel in the lower part of the core reduces the axial peaking. The operating procedure is to locate control rods so that the reactor operates with approximately the same axial power shape throughout an operating cycle.

Reactivity Control. The movable boron-carbide control rods are sufficient to provide reactivity control from the cold shutdown condition to the full-load condition. Supplementary reactivity control in the form of solid burnable poison is used only to provide reactivity compensation for fuel burnup or depletion effects. The movable control rod system is capable of bringing the reactor to the subcritical when the reactor is an ambient temperature (cold), zero power, zero xenon, and with the strongest control rod fully withdrawn from the core. In order to provide greater assurance that this condition can be met in the operating reactor, the core is designed to obtain a reactivity of less than 0.99, or a 1% margin on the "stuck rod" condition. See Fig. 7.

Reactor Auxiliary Systems include: (1) a reactor water cleanup system for maintaining high reactor water quality by removing fission products, corrosion products, and other soluble and insoluble impurities; (2) a fuel and containment pool cooling and cleanup system—a system which accommodates the beta and gamma radiation heating from the fission products that remain in the spent fuel, as well as drywell heat transferred to the upper containment pool; (3) a closed cooling water system for reactor service consisting of a separate, force circulation loop; (4) emergency equipment cooling system; (5) standby liquid control system; (6) reactor core isolation cooling system; (7) emergency core cooling system; (8) high-pressure core spray system; and (9) residual heat removal system.

Contemporary Pressurized Water Reactor (PWR)

In a typical pressurized water reactor (PWR), heat generated in the nuclear core is removed by water (reactor coolant) circulating at high pressure through the primary circuit. The water in the primary circuit cools and moderates the reactor. The heat is transferred from the primary to the secondary system in a heat exchanger, or boiler, thereby generating steam in the secondary system. The steam produced in the steam generator, a tube-and-shell heat exchanger, is at a lower pressure and temperature

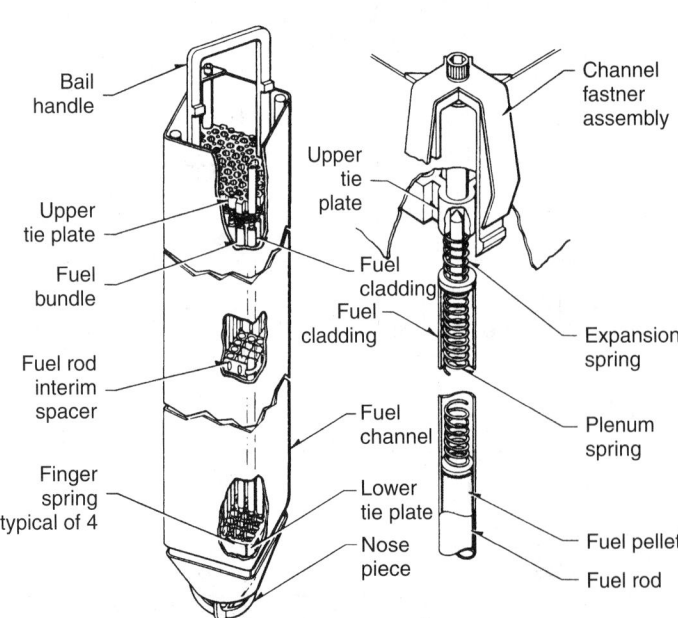

Fig. 6. Fuel assembly of contemporary boiling water reactor. (*General Electric.*)

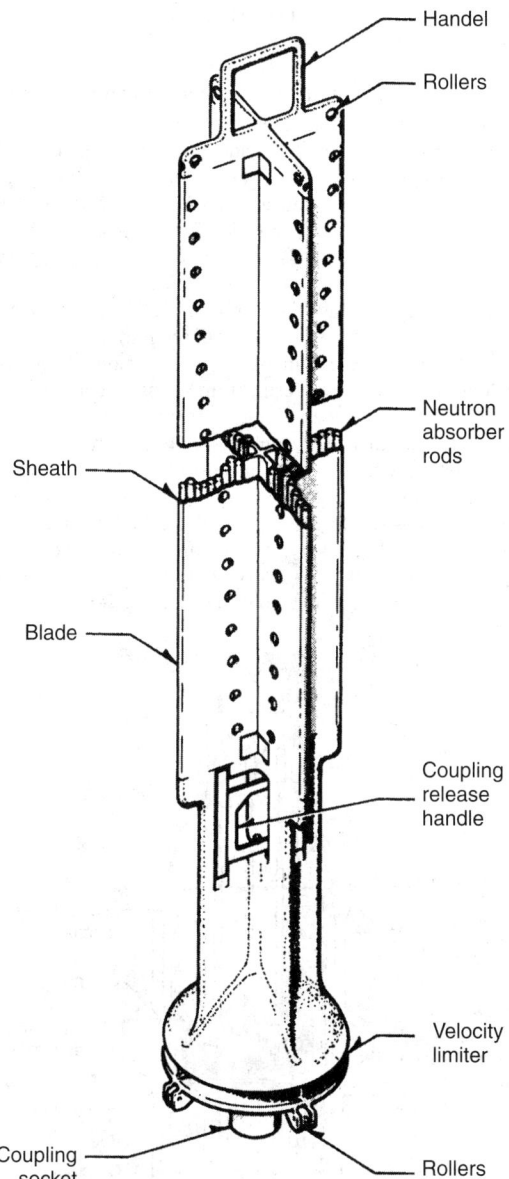

Fig. 7.   Control rod used in contemporary boiling water reactor. The cruciform control rods contain 76 stainless steel tubes (19 tubes in each wing of the cross). These tubes are filled with boron carbide powder compacted to approximately 65% of theoretical density. The tubes are seal-welded with end plugs on either end. The individual tubes act as pressure vessels to contain the helium gas released by the boron-neutron capture reaction. The control rods have an active length of 144 inches (365.8 centimeters) of boron carbide, a span of 9.75 inches (24.8 centimeters), and an overall length of 173.75 inches (441.3 centimeters). The control rods can be positioned at 6-inch (15-centimeter) steps and have a nominal withdrawal and insertion speed of 3 inches (7.5 centimeters) per second. Control rods are cooled by the core leakage (bypass) flow. In addition to satisfying initial control effectiveness requirements, it is expected that the control rods will have an average lifetime of approximately 15 full-power years. (*General Electric.*)

than the primary coolant. Therefore, the secondary portion of the cycle is similar to that of the moderate-pressure fossil-fueled plant. In contrast, in boiling-water or direct-cycle systems, steam is generated in the core and is delivered directly to the steam turbine.

The similarities of basic pressurized water reactor design from one manufacturer to the next are more striking than the differences. Therefore, the description of one particular configuration (Combustion Engineering, Inc.) can suffice to convey the general operating principles. The major components of a PWR are: (1) the reactor vessel which contains the oxide fuel core, core intervals, control element assemblies, and in-core instruments; (2) the electrically-heated pressurizer; (3) the electric-motor-driven primary coolant pumps; and (4) the U-tube type steam generators.

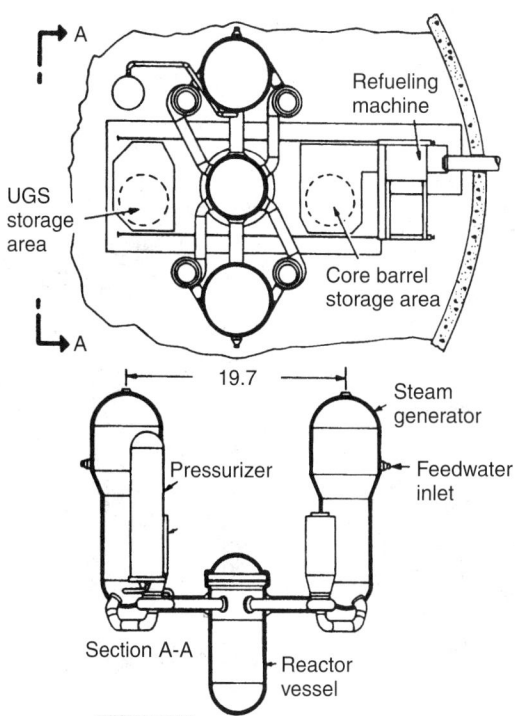

Fig. 8.   Nuclear steam supply system for contemporary pressurized water reactor. (*Combustion Engineering.*)

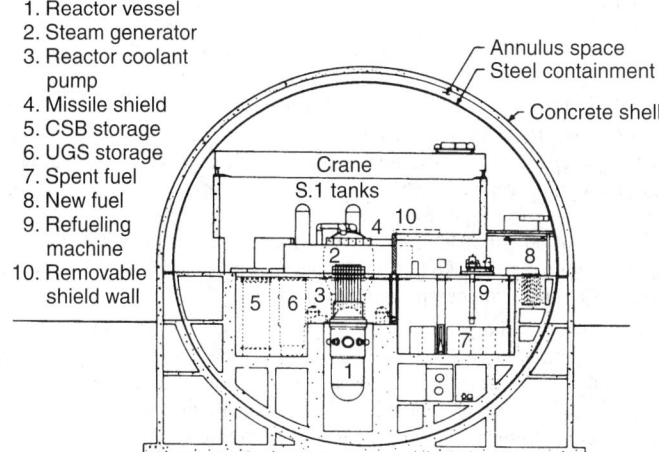

1. Reactor vessel
2. Steam generator
3. Reactor coolant pump
4. Missile shield
5. CSB storage
6. UGS storage
7. Spent fuel
8. New fuel
9. Refueling machine
10. Removable shield wall

Fig. 9.   Spherical containment for contemporary pressurized water reactor. (*Combustion Engineering.*)

See Fig. 8. The primary coolant system layout can be fitted into a variety of containment types and concepts. A prestressed cylindrical containment is common. Figure 9 shows the arrangement in a spherical containment. This type of building lends itself to separation of safeguards equipment, steam lines, and emergency power supplies.

Steam Generators. The basic geometry is shown in Fig. 10. With the nuclear steam supply system operating at 3,817 MW, two steam generators produce a total of 17.18×10^6 pounds (7.89×10^6 kilograms) of steam per hour at 1,070 psia (72.8 atmospheres). The steam generators are constructed, using carbon steel pressure-containing members and Inconel-600 tubes. The tube-sheet is clad with weld deposit for maximum strength; tongue and groove construction of the divider plate places no stress on the tube-sheet cladding. Fusion welding of the end of each tube to the tube-sheet primary cladding provides an effective seal for leakage control, and "expanding" (explosively expanding) the tubes in the full length of the tube-sheet eliminates corrosion-prone crevices. An economizer section on the units improves heat transfer by preheating the incoming feedwater, using the low (primary side) temperature heat transfer area of the U-tubes. Multiple feed nozzles allow the economizer flow distribution to be optimized for each power level.

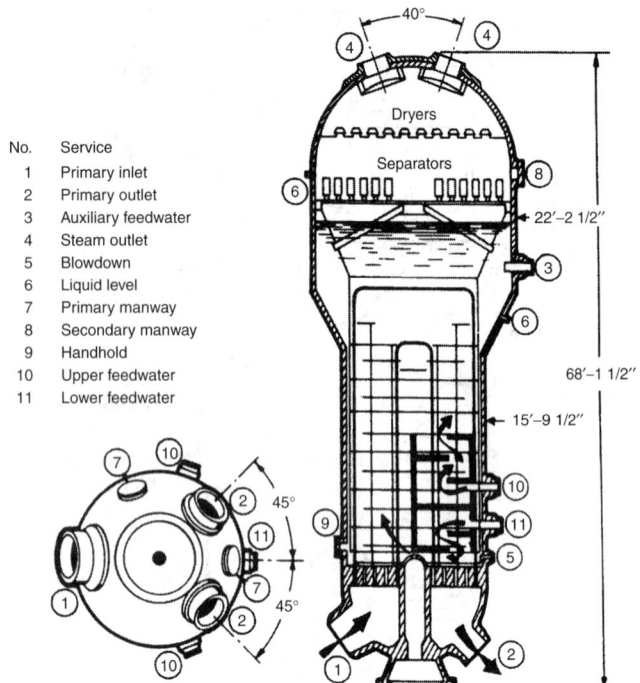

No.	Service
1	Primary inlet
2	Primary outlet
3	Auxiliary feedwater
4	Steam outlet
5	Blowdown
6	Liquid level
7	Primary manway
8	Secondary manway
9	Handhold
10	Upper feedwater
11	Lower feedwater

Fig. 10.   Steam generator for contemporary pressurized water reactor. (*Combustion Engineering.*)

Reactor Coolant Pumps. As indicated by Fig. 8, four reactor coolant pumps are used, two for each steam generator. The pumps are vertical, single-bottom-suction, horizontal-discharge, motor-driven centrifugal units. The pump impeller is keyed and locked to its shaft. A complex system of seals is used to prevent any leakage. The motors are designed to start and accelerate to speed under full load with a drop to 80% of normal rated voltage at the motor terminals. Each motor is provided with an anti-reverse rotation device. Each reactor coolant pump is provided with four vertical support columns, four horizontal support columns, and one vertical snubber. The structural columns provide support for the pumps during normal operation, earthquake conditions, and any hypothetical loss-of-coolant accident in either the pump suction or discharge line.

Pressurizer. The pressurizer is a cylindrical pressure vessel, vertically mounted and bottom supported. Energy to the water is supplied by replaceable direct-immersion electric heaters, which are inserted from the bottom head of the pressurizer. Nozzles are provided for spray, surge, relief, and instrumentation connections. The pressurizer maintains reactor coolant system operating pressure and, in conjunction with the chemical and volume control system, compensates for changes in reactor coolant volume during load changes, heat-up, and cool-down. During full-power operation, the pressurizer is about $\frac{1}{3}$ full of saturated steam.

Reactor Vessel. This vessel is designed to contain the fuel bundles, the control element assembles, and the internal structures necessary for support of the core. The reactor is a stainless clad, thick-walled, carbon steel pressure vessel comprised of a cylinder with two hemispherical heads. The lower head is integrally welded to the vessel shell and contains in-core instrumentation nozzles. The upper closure head, containing the control element drive mechanism nozzles, is attached to the vessel by means of a bolted flange, thus permitting the head to be removed to provide access to the reactor internals. The head flange is drilled to match the vessel flange stud bolt locations.

The vessel flange is a forged ring with a machined ledge on the inside surface to support the core support barrel. The flange is drilled and tapped to receive the closure studs and is machined to provide a mating surface for the reactor vessel closure seals. Sealing is accomplished by using two silver-plated, NiCrFe-alloy, self-energizing O-rings. The space between the two rings is monitored to detect any inner-ring coolant leakage. The inlet and outlet nozzles are located radially on a common plane below the vessel flange. Extra thickness in the vessel course provides the reinforcement required for the nozzles. Snubbers built into the lower portion of the vessel shell limit the amplitude of any displacement of the core support barrel.

Core stops are also built into the reactor vessel to limit the downward displacement of the core support barrel.

Cladding for the reactor vessel is a continuous integral surface of corrosion-resistant material, having $\frac{7}{32}$-inch (0.56 centimeter) nominal thickness, and a $\frac{1}{8}$-inch minimum thickness. The reactor vessel is supported by four vertical columns located under the vessel inlet nozzles. These columns are designed to flex in the direction of horizontal thermal expansion and thus allow unrestrained heat-up and cool-down. The columns also act as a hold-down device for the vessel. The supports are designed to accept normal loads and seismic and pipe rupture accident loads.

The reactor arrangement is shown in Fig. 11. The barrel-calandria guide structure is a rugged (3-inches thick, barrel section) unit that can withstand and protect all control element fingers from the combined effects of seismic and blowdown loads that may result from a loss-of-cooling accident. The calandria structure fits over the control element guide tubes of the fuel assemblies, aligning all fuel assemblies, and laterally restraining the top ends of the fuel assemblies. With the upper guide structure in place, a continuous guide tube for each control finger is formed, extending from the top of the tube-sheet to the bottom of the fuel assembly. Because of this feature, which isolates every control finger from the coolant crossflow, flexibility is obtained in the number of control fingers that can be attached to one control assembly, i.e., one control element assembly can serve more than one fuel assembly.

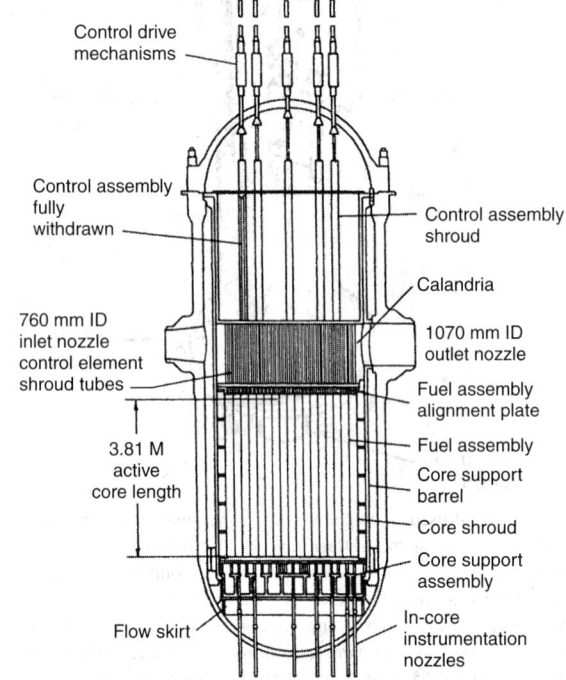

Fig. 11.   Reactor arrangement in contemporary pressurized water reactor.

Severe emergency core cooling system criteria require that the builders of water reactors increase the linear feet of fuel in the reactor core for the same power in order to reduce LOCA (loss-of-cooling accident) fuel temperatures. In the unit described here, an assembly with a 16×16 fuel rod array of smaller diameter rods is used in the same assembly envelope that was occupied by a 14×14 assembly in earlier designs. This results in a maximum linear heat rate decrease in the assembly of about 25%.

As shown in Fig. 12, the active core is made up of 241 fuel assemblies, all of which are mechanically identical. As indicated by Fig. 13, each fuel assembly contains 236 Zircaloy-clad, UO_2 fuel rods retained in a structure consisting of Zircaloy spacer grids welded at about 15-inch (38.1-centimeter) intervals to five Zircaloy control element assembly guide tubes which, in turn, are mechanically fastened at each end to stainless steel end fittings. The overall length of the fuel assembly is about 177 inches (450 centimeters) and the cross section is about 8 inches (20.3 centimeters) by 8 inches (20.3 centimeters). Each fuel assembly weighs about 1,450 pounds (657.7 kilograms). With reference to Fig. 13,

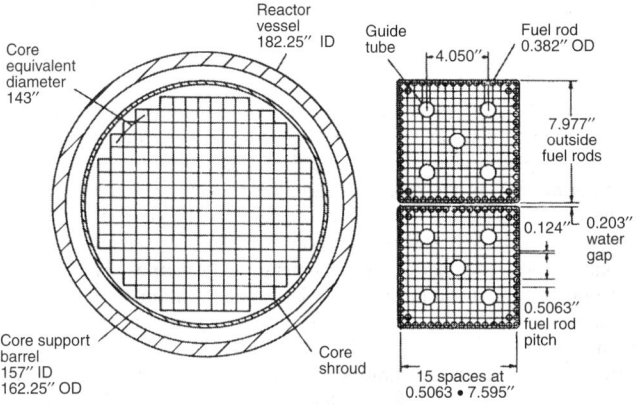

Fig. 12. Reactor core cross section of contemporary pressurized water reactor with 241 fuel assemblies. (*Combustion Engineering.*)

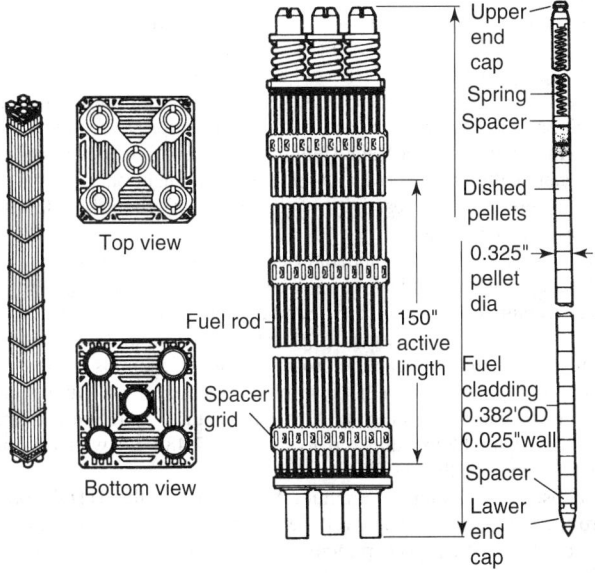

Fig. 13. Fuel assembly used in contemporary pressurized water reactor.

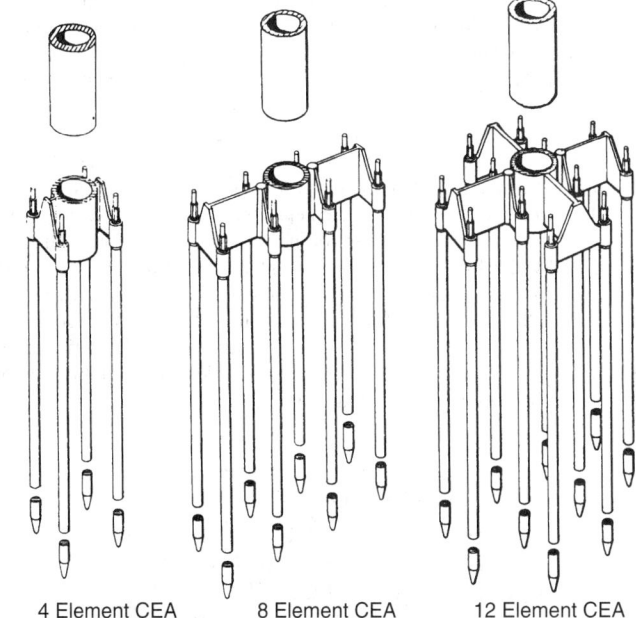

Fig. 14. Four 8- and 12-element control element assemblies. (*Combustion Engineering.*)

fuel rods, consisting of uranium dioxide (UO_2) pellets of low enrichment canned in thin-walled Zircaloy-4 tubing, are designed to achieve average burnups of about 33,000 MWD/MTU (thermal megawatt days/metric tons of uranium) and peak burnups of about 50,000 MWD/MTU. The design factors limiting burnup of the fuel are the effects on the clad of volumetric changes of the fuel pellet and fission gas release.

As indicated in Fig. 13, the fuel rod consists essentially of 0.325-inch (0.82-centimeter) diameter, 0.390-inch long UO_2 pellets canned in a 0.382-inch (0.97-centimeter) outside diameter Zircaloy-4 tube. The high density fuel pellets are dished at both ends to allow for axial differential thermal expansion and fuel volumetric growth with burnup.

The control element assemblies consist of an assembly of 4, 8, or 12 fingers approximately 0.8-inch (2-centimeter) outside diameter and arranged as shown in Fig. 14. The use of cruciform control rods, as in boiling water and early pressurized water reactors, necessitates large water gaps between the fuel assemblies to ensure that the control rods will scram (prompt shutdown) satisfactorily. These gaps cause peaking of the power in fuel rods adjacent to the water channel compared to fuel rods some distance from the channel.

A five-hole assembly design was evolved from consideration of the lower peaking effect of smaller (removal of one fuel rod) water holes versus the mechanical advantages of larger (four fuel rods removed) water holes. The larger water holes allowed the use of rugged, 0.9-inch (2.3-centimeter) outside diameter by 0.035-inch (0.89-centimeter) thick Zircaloy guide tubes for the fuel assembly structure. These water holes are distributed relatively uniformly in the reactor core when placed in the 16×16 fuel rod lattice. The particular arrangement of water holes was selected in consideration of the water gap between fuel assemblies; the effect of the central water hole

in the fuel assembly is balanced by the water gap between fuel assemblies. The mechanical simplicity and ruggedness outweigh the advantage of obtaining a small decrease in local peaking by using very small fingers. The slightly higher peaking associated with the design can be compensated, to a large extent, by varying the enrichment of the fuel in the rods adjacent to the control channel and/or by using water displacers in local hot spots.

The control element assembly, shown in Fig. 15, consists of 0.8-inch (2-centimeter) outside diameter Inconel tubes containing boron carbide pellets as the neutron absorbing material. A gas plenum is provided in order to limit the maximum stress due to generation of internal gas pressure. The individual control fingers are attached mechanically and locked to the various spider assemblies. This allows for simplifications in manufacture, shipping, and assembly of the control element assembly. Because all

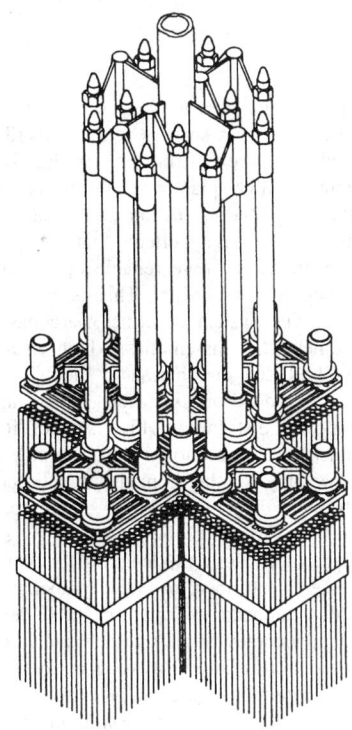

Fig. 15. Control element assembly and fuel for contemporary pressurized water reactor.

fingers are removable and replaceable, servicing and disposal problems are decreased. It is intended that the spider assembly and its extension shaft be reused whenever possible.

Design of the upper guide structure permits flexibility in the number of control fingers that can be attached to one control assembly. The standard pattern of control assemblies is shown in Fig. 16. In the standard design, power changes at close to full power, shaping of the radial power distribution, and control of the axial power distribution are best handled by the low worth 4-finger control element assembly entering a single fuel assembly. Shutdown reactivity control in the peripheral region of the core is handled by the 8-finger control element assembly and in the central region of the core it is handled by the 12-finger control element assembly. The need for the two types of shutdown control element assemblies is to obtain "stuck rod worths" in the high reactivity fuel on the periphery of the core which are about equal to the control element assembly control worth in the lower reactivity central zone of the core.

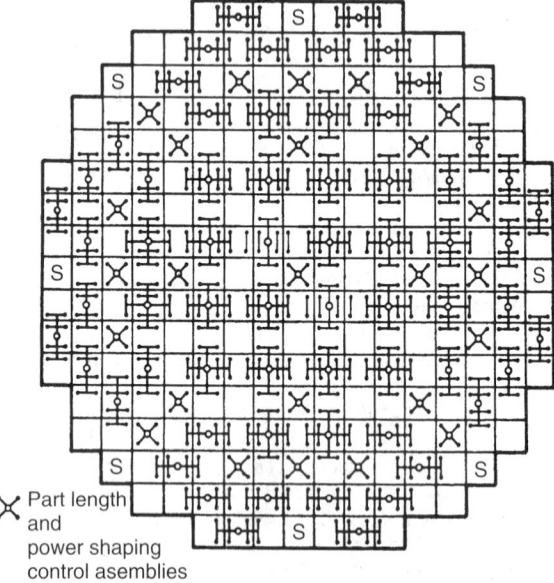

Fig. 16. Standard pattern of control assemblies in contemporary pressurized water reactor core. The pattern provides more-than-sufficient control for self-generated plutonium recycle. For complete open-market plutonium recycle, 4-element control assemblies are added in positions marked S.

Instrumentation. The large size of present water reactors and the nuclear effects which can occur, such as xenon redistribution, stuck rods, and reactivity anomalies require that emphasis be placed on instrumentation and control systems if high plant availability is to be maintained, while providing the necessary protection due to abnormal occurrences. Because there are many reactivity effects that can produce changes in the reactor power distribution, more reliable operation can be obtained by on-line monitoring of the reactor. This is best achieved with in-core instrumentation. The system described here has provision for up to 61 in-core instrument (ICI) assemblies, which enter from the bottom of the reactor vessel. Radial distribution of the ICI is such that every type of fuel assembly, rodded and unrodded, is instrumented, assuming symmetrical core power distribution. Five sets of four symmetrically located ICI assemblies are included to monitor core power tilts. Also, every instrumented fuel assembly is either immediately adjacent to or diagonally adjacent to an instrumented fuel assembly to obtain good radial coverage of the core. Each of the ICI assemblies contains five self-powered fixed detectors distributed axially along the length of the core, a thermocouple at the end of the assembly to monitor outlet temperature, and a dry-well instrument tube which can accommodate a movable detector. This allows for high measurement accuracy of the on-line fixed detector.

The continuous monitoring and processing of the data from over 300 fixed detectors by the core monitoring computer provides the operator with information on core power distribution, maximum linear heat rate in the fuel, departure from nucleate boiling ratio, and fuel exposure. These data can then be used to obtain improved maneuvering of core power using the

relative low worth 4-finger control element assemblies for power changes and power distribution control.

High-Temperature Gas-cooled Reactor (HTGR)

Although there have been comparatively few gas-cooled reactors installed for generating commercial nuclear electric power, the concept has a number of operating advantages over light-water reactors and could play an important role in the reactor designs for the next century.

The high-temperature gas-cooled reactor (HTGR) is a thermal reactor that produces desired steam conditions. Helium is used as the coolant. Graphite, with its superior high-temperature properties, is used as the moderator and structural material. The fuel is a mixture of enriched uranium and thorium in the form of carbide particles clad with ceramic coatings.

The high-temperature conditions and high thermal efficiency (approximately 39%) of the (HTGR) result in high performance. The amount of cooling water required to carry away the waste heat is significantly less than in a light-water reactor (LWR). The use of thorium in the fuel cycle decreases fuel cost, improves the conservation of fuel, and adds the large deposits of thorium to available fuel reserves. The HTGR has significant environmental advantages, including: (1) lower thermal discharge because of its high efficiency; (2) low release of radioactive waste because of the high-integrity fuel and the inert coolant; and (3) low consumption of raw materials because of high efficiency and use of thorium in the fuel cycle.

High operating temperatures at moderate pressures are achieved through the use of helium as the coolant. Helium is attractive as a coolant because it: (1) is chemically inert; (2) absorbs essentially no neutrons; and (3) makes no contributions to the reactivity of the system. Carbon dioxide also has been used as a coolant.

Graphite is used as the moderator and core structural material because of (1) excellent mechanical strength at high temperatures; (2) very low neutron-capture cross section; (3) good thermal conductivity; and (4) high specific heat. Graphite has a long history of use in thermal reactors. Because of low neutron-capture cross section, no neutrons are lost within the core through absorption in metallic fuel cladding or structural supports. Graphite also is well suited to high-temperature operations, increasing in strength with temperature up to a point (2,482 °C) well beyond the operating range of the HTGR.

The use of the thorium-uranium fuel cycle in the HTGR provides improved core performance over the plutonium/uranium low-enrichment cycle used in LWRs. The principal reason for this is that fissile ^{233}U produced from neutrons captured in thorium during reactor operation is neutronically a better fuel than ^{239}Pu, produced from ^{238}U in the low-enrichment cycle. The excellent neutronic characteristics of the graphite-moderated thorium/uranium cycle leads directly to high conversion ratios and low fuel inventories. Reduced ^{235}U inventories and make-up requirements spell reduced sensitivity to uranium prices.

Early Development of the HTGR

Work on the gas-cooled reactor has been underway essentially since the dawning of the nuclear power industry. The earliest developments were in Britain and France, at which time carbon dioxide gas was used as the coolant. In 1965, Britain opted for an advanced gas-cooled reactor (AGR). In 1969, France swung away from the HTGR (because of high construction costs) and targeted to the employment of more LWRs as well as commencing a concerted effort to develop a fast breeder type reactor. West Germany has been active in the development and testing of HTGRs since the early 1960s, but only recently (late 1980s) have the Germans indicated serious efforts toward commercialization of the HTGR. In the United States, a HTGR was installed at Peach Bottom, Pennsylvania, commencing commercial operation in 1974. As of the late 1980s, only one other HTGR was installed in the United States, the Fort St. Vrain plant near Denver, Colorado.

A simplified flow diagram of this station, which generates 842 MW (thermal) to achieve a net output of 330 MW (electrical), is given in Fig. 17. The helium coolant, at a pressure of about 700 psi (47.6 atmospheres), flows downward through the reactor core, where it is heated to 777 °C. The coolant flow can be trimmed by the use of orifice valves located at the top of the core that are integral with the control rod drive mechanisms. From the reactor core, the coolant flows through the steam generators. After passing through the steam generators, the helium is returned to the core at a temperature of about 404 °C by four steam-turbine-driven helium circulators. Two identical loops are used, each including

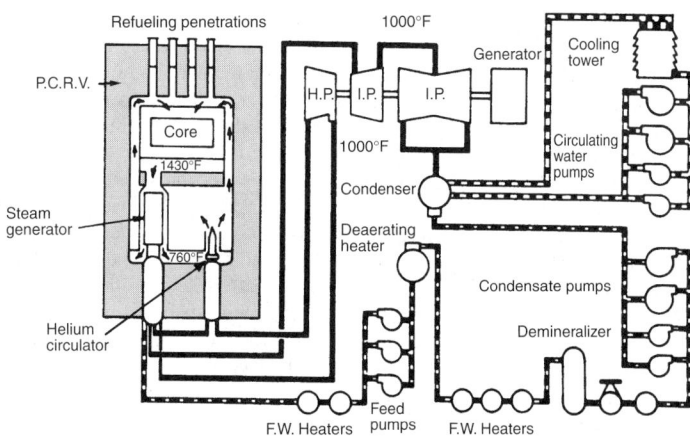

Fig. 17.   Simplified flow diagram of Fort St. Vrain Nuclear Generating Station. (*GA Technologies.*)

a six-module steam generator and two helium circulators. Each loop contributes half of the total output of the nuclear steam supply system, which produces steam at 2,400 psig (163.3 atmospheres) and 538 °C with single reheat to 538 °C. The helium circulators are driven by the exhaust steam from the high-pressure turbine. This steam is then reheated and returned to the intermediate-pressure turbine. The circulators are also equipped with a Pelton water wheel drive so that they may be driven using the boiler feed pumps for emergency conditions.

The general reactor arrangement is shown in Fig. 18. The prestressed concrete reactor vessel (PCRV) is 31 feet in internal diameter with a 75-foot (23 meters) internal height. The upper and lower heads are nominally 15 feet (4.5 meters) thick, and the walls have a nominal thickness of 9 feet (2.7 meters). Thus, the PCRV provides the dual function of containing the coolant at operating pressure and also providing radiological shielding.

Reactor Core. The HTGR fuel element is a graphite block, hexagonal in cross section and having a grid of longitudinal fuel holes and coolant channels. The fuel element blocks are stacked in columns of eight blocks each and grouped into fuel regions consisting of a central column surrounded by six columns. Each region rests on a large core support block which, in turn, rests on graphite posts standing on the liner of the central cavity. Hexagonal graphite reflector elements are located above, below, and around the active core. These elements are surrounded by permanent side-reflector blocks to give the entire assembly a circular configuration. The fuel holes contain a rod consisting of ceramic-coated fuel particles in a graphite matrix. The coatings, applied by pyrolitic techniques, are multilayered to ensure a high degree of fission-product confinement. A porous interlayer, or buffer zone, accommodates the expansion of the irradiated fuel and provides storage space for gaseous fission products. The outer layer acts as a fission-product retention barrier and provides structural strength. In effect, the particle coating functions as a miniature spherical pressure vessel.

To achieve a fuel management scheme with the lowest fuel cycle cost consistent with the current thermal and material performance limits, the following parameters are selected: (1) a fuel cycle incorporating uranium/thorium; (2) a fuel lifetime of four years; (3) an average power density of 8.4 W/cm^3; and (4) a refueling frequency of once a year.

The reactor is controlled by two control rods located in each refueling region. All control rod pairs have scram (quick shutdown) capability and are driven by gravity. A backup reserve shutdown system is included. This consists of boronated graphite pellets that can be introduced from hoppers located in each refueling penetration into the core via the cylindrical channels in the central fuel element of each refueling region.

Safeguard Systems. The design of HTGR incorporates many inherent safety features and a number of engineered safeguards. The inherent safety characteristics include negative power and temperature coefficients, assured by the thorium content of the fuel. In addition, the high heat capacity of the large mass of graphite ensures that any core temperature transient resulting from reactivity insertions or interruptions in cooling will be slow and readily controllable. This important safety feature eliminates the need for an emergency core cooling system. Only a residual heat removal system is required for the long-term decay heat, and control of the

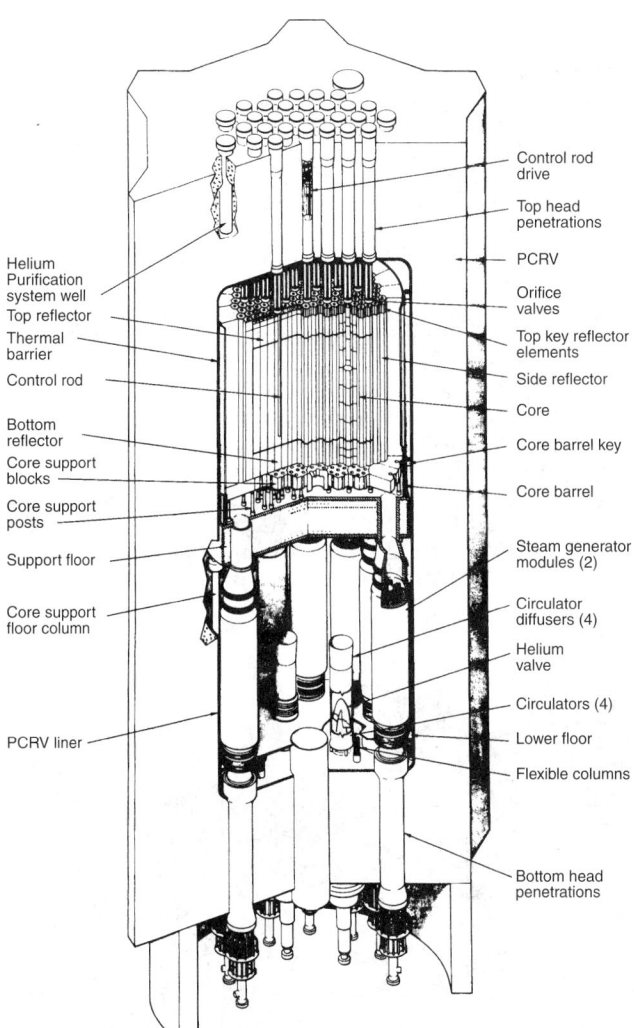

Fig. 18.   General reactor arrangement of Fort St. Vrain high-temperature gas-cooled reactor. (*GA Technologies.*)

HTGR is inherently easier than in reactors in which the coolant functions as the moderator. The uranium/thorium fuel contained in the ceramic-coated particles is not susceptible to sudden release of the stored-up fission products as a result of melting. Since the entire primary coolant system is contained within the PCRV, external piping, which might be subject to sudden rupture, is eliminated. Structural strength and integrity of the PCRV is enhanced by the redundant reinforcing steel and prestressed wire tendons. At the maximum credible pressure, the prestressing elements are not stressed above levels experienced during their initial tensioning. As a result, sudden loss of coolant due to prestress failure is not credible.

Second-Generation HTGRs

In addition to upgrading the HTGR at the Fort St. Vrain nuclear power station, efforts have been underway for several years to make both larger and smaller gas-cooled high-temperature reactors. Smaller, modular units could provide the flexibility needed by the public utilities as they plan their expansions for projected increases in electricity requirements. Inherent safety, already a feature of the HTGR, would be enhanced because of the smaller size and low power density of modular units. For example, it is estimated that the power density of a modular gas-cooled reactor would be only 3 kW/liter, as compared with 6 kW/liter for a large reactor and 100 kW/liter for a conventional pressurized-water reactor (PWR) as previously described.

In the new designs, if coolant were lost, the nuclear chain reaction would be terminated by the reactor's negative temperature coefficient after a modest temperature rise. Core diameter of the modular units would be limited so that decay heat could be conducted and radiated to the environment without overheating the fuel to the point where fission products might escape. Thus, inherent safety would be realized without operator or mechanical device intervention.

Large Commercial HTGRs[2]

Following construction of the Fort St. Vrain facility, the HTGR was marketed commercially in direct competition with large pressurized water reactors (PWRs) and boiling water reactors (BWRs). Between 1971 and 1975, ten such reactors were ordered by U.S. utilities. The designs of the commercial HTGR were similar to Fort St. Vrain in that they used the graphite based core structure, helium coolant, prestressed concrete reactor vessel (PCRV) and superheated steam cycle. However, the designs differed in that power outputs were significantly larger and the reactor system was rearranged to accommodate the larger-size components.

The large HTGRs had power ratings of 2000 and 3000 MWt which corresponded to net electrical outputs of 770 and 1160 MWe, respectively. An example of the rearranged reactor system is shown in Fig. 19. A multi-cavity PCRV was used to enclose the reactor system instead of the single-cavity PCRV used in Fort St. Vrain. This was a major advancement in PCRV technology and necessitated the development of a circumferential wire-wrap prestressing system instead of circumferential tendons, although the longitudinal tendons were retained. The sizes of the multi-cavity PCRV were approximately 100 feet (30 m) high by 120 feet (36 m) in diameter for the 3000 MWt plant and 100 feet (30 m) high by 105 feet (32 m) in diameter for the 2000 MWt plant.

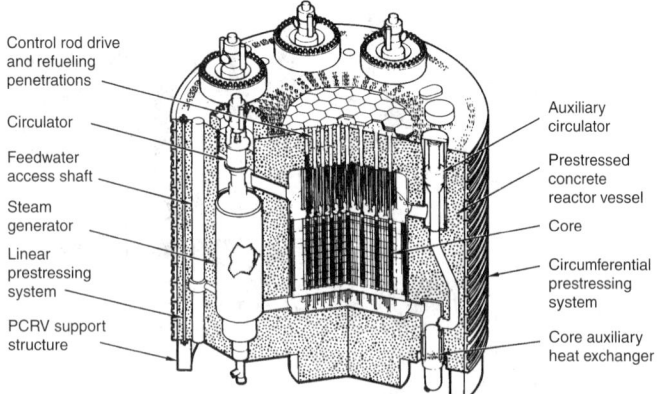

Fig. 19.   Integrated HTGR nuclear steam system (1170 to 3360 MW thermal). (*GA Technologies.*)

The graphite reactor core was located in the central cavity of the PCRV. The steam generators and steam-driven main helium circulators were located in vertical cavities arranged around the periphery of the core. The 2000 MWt system had four steam generator-circulator side cavities while the 3000 MWt unit had six such cavities. The hot primary coolant helium (1366 °F; 741 °C) exiting from the bottom of the core collected in the lower

[2] This portion of article on HTGRs contributed by R.A. Dean, Sr. Vice President, GA Technologies Inc., San Diego, California.

core plenum from which it was distributed to the steam generators through the lower cross-ducts. The circulators, located above the steam generators, returned the cool helium (710 °F; 377 °C) through the upper cross-ducts to the upper core plenum. The helium in the upper plenum then flowed down through the core where it was heated.

Another feature of the large HTGRs was the core auxiliary cooling system which provided an independent means of core afterheat removal in the event that the main coolant loops (i.e., steam generators and steam-driven circulators) were shut down. The core auxiliary cooling system consisted of two redundant cooling loops, each capable of removing 100% of the afterheat, for the 2000 MWt plant and three redundant cooling loops, each capable of removing 50% of the decay heat, for the 3000 MWt plant. Each loop contained a motor-driven circulator and water-cooled heat exchanger and circulated flow through the core just at the main loops. Shutoff valves were located in both auxiliary and main loops to assure that helium would not bypass the core through the one system while the other system was in operation.

All ten large commercial HTGRs were ordered during the early 1970s as an indirect consequence of the energy crisis brought about by the oil embargo; all were cancelled by 1976. The combination of the recession plus new emphasis on conservation brought about a rapid reduction in electric energy demand which, in turn, resulted in cancellation of over 100 nuclear power plant orders, including the large HTGRs.

Small Modular HTGRs

In the early 1980s, the major influence on new designs was the renewed emphasis on safety brought about by the accident at the Three Mile Island nuclear plant. The experience from licensing and operation of nuclear plants during the 1970s indicated a need to reduce design complexity and develop passive approaches to reactor safety rather than rely on complex emergency safety systems. HTGR designers in the United States and Europe determined that a substantial reduction in plant size could enable the HTGR to be entirely inherently safe by virtue of the high temperature structural integrity of the graphite core and ceramic coated fuel. This means that a small HTGR would not require any active safety equipment or any action by the operator in order to prevent release of radioactivity for any accident condition.

A major concern with reducing plant size was the economic impact from reversing the economy of scale. However, it was learned that economy-of-scale effects could be offset by several beneficial factors that apply to smaller nuclear plants. This includes the shift of major portions of the work from the site to the factory; the learning effects appreciated by replication of a larger number of smaller units in a factory environment and the elimination or simplification of many components/systems no longer required for smaller plants.

These considerations led to the reconfiguration of the HTGR plant into a system of one or more downsized 350 MWt modular reactors. The physical arrangement of a single reactor module, designed for installation in a below-grade silo, is shown in Fig. 20. The primary components are contained within two vertically oriented metal pressure vessels connected by coaxial cross-duct. Thus, the field-erected PCRV, which was used on previous large HTGRs, was eliminated in favor of shop-fabricated metal pressure vessels. The use of metallic pressure vessels also facilitates installation in underground silos, which enhances the safety of the plant.

The reactor vessel, which is approximately 72 feet (22 m) high by 22.5 feet (6.9 m) in diameter, contains the graphite core, reflector, and shutdown heat removal system (non-safety). The other vessel contains a single helical coiled steam generator and motor-driven circulator with magnetic bearing instead of water-lubricated bearings as in previous concepts. The size of both vessels is within allowable limits for barge, rail, and overland transportation.

During normal operation, the main circulator transports hot helium at 1266 °F (686 °C) from the bottom of the core to the steam generator which, in turn, produces superheated steam at 1005 °F (541 °C) and 2500 psia. The cold helium at 496 °F (258 °C) is returned to the top of the reactor core. During normal shutdown and refueling, the non-safety auxiliary shutdown heat removal system removes core afterheat if the main heat transport system is not operational.

A principal feature of the modular HTGR is its capacity for safely rejecting core afterheat in a completely passive manner (i.e., without the need for any active core cooling systems) such that any release of fission products from the fuel is prevented during severe accident conditions.

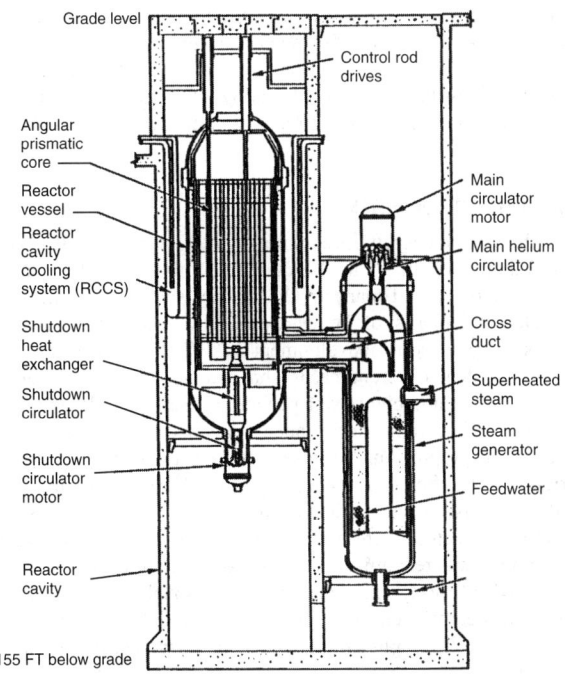

Fig. 20. Elevation of 350 MW modular HTGR. (*GA Technologies.*)

This feature is a result of both the reactor system configuration and the high temperature capability of the fuel. In the event of a loss of forced circulation cooling of the core via either the main circulator/steam generator or auxiliary shutdown circulator/heat exchanger, core afterheat will continue to be safely removed by direct conduction through the core and reflector to the reactor vessel wall. The heat is then dissipated from the reactor vessel surface by radiation and natural convection to cooling panels surrounding the interior surface of the reactor cavity. See Fig. 20, previously mentioned. These panels are part of the Reactor Cavity Cooling System (RCCS) which consists of natural convection air ducts that ultimately transport the core afterheat directly to the atmosphere.

In order to achieve this passive core cooling capability in a reactor with a power level and power density that are economically attractive, the annular core arrangement shown in Fig. 21 was adopted. The active core consists of an annular region of hexagonal graphite fuel blocks containing standard HTGR fuel. Unfueled graphite reflector blocks make up the region inside the active core annulus and the region surrounding the outside of the annulus. This arrangement results in a higher radial heat conductance for fuel at the innermost radius than for a solid cylindrical active core. Thus, the annular core arrangement permits operation at a higher power for a given volume than a solid active core.

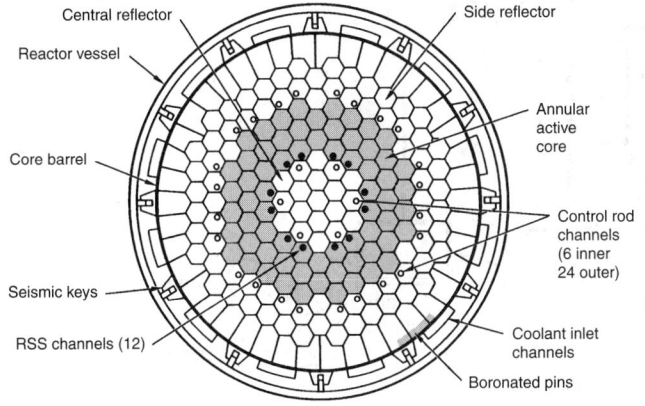

Fig. 21. Reactor core cross section of 350 MW modular HTGR. (*GA Technologies.*)

The modular HTGR uses the same form of fuel as the large HTGR and Fort St. Vrain installation except for the important difference that the

^{235}U enrichment was reduced to about 20% from the previous value of 93%. The fuel is in the form of coated fuel and fertile particles which are bonded into graphite rods and inserted into the hexagonal graphite fuel blocks. See Fig. 22. The fuel and fertile particles consist of uranium oxycarbide and thorium oxide kernels (about 350 micrometers in diameter), respectively, first coated with a porous graphite buffer, followed by three successive layers of pyrolytic carbon, silicon carbide, and pyrolytic carbon. The outer diameter of the coated particles is about 800 micrometers for the uranium particles and slightly larger for the thorium particles. The coatings essentially form a high-temperature refractory-based pressure vessel around each fuel/fertile kernel for the purpose of retaining fission products. Extensive operation and test data on these particles confirm that essentially no failure of the refractory coating occurs if the fuel is maintained below 3272 °F (1800 °C). As previously mentioned, the reactor design parameters were selected such that passive core afterheat removal will prevent this temperature from being reached during any credible accident condition. Thus, the modular HTGR is inherently, passively safe.

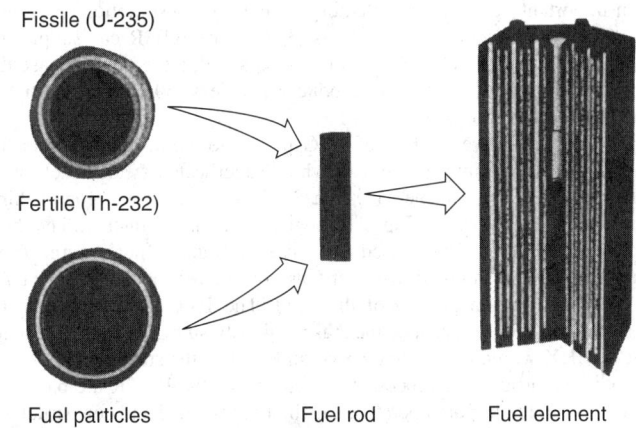

Fig. 22. Fuel components of HTGR. (*GA Technologies.*)

The reference plant arrangement features four 350 MWt HTGR modules supplying steam to two turbine generators that produce a net electrical output of 558 MWe at a new plant efficiency of 39.9%. Each reactor module is housed in an independent, vertical, cylindrical, concrete confinement, which is fully embedded in the earth. The four reactor modules share common systems for fuel handling, helium processing, and other essential services. A common control room is used to operate all four reactors and the turbine plant. Operation of the entire complex is completely automated. Human operator actions are not required for control during power production or to assure safe shutdown during hazardous conditions.

Potential for HTGRs

The HTGR's use of ceramic-coated fuel and graphite moderator enables operation of the reactor core at much higher temperatures than are required for electric power production via a steam Rankine cycle. Core outlet helium temperatures in excess of 1800 °F (982 °C) are achievable without impacting the integrity of the HTGR fuel or core structures. This very high temperature capability opens up the possibility for more efficient methods of power production or direct use of high-temperature thermal energy for process heat applications. See also **Cogeneration (Electricity and Thermal Energy)**.

An attractive electric power producing concept for the 21st century is the HTGR gas-turbine (HTGR-GT) which has the potential for thermal efficiencies over 50% by taking advantage of the high HTGR core outlet temperature. Although several variations in system configuration are possible, the most straightforward HTGR-GT concept is the direct Brayton cycle, illustrated in Fig. 23. This cycle is closed and the helium primary coolant is also the working fluid for the power conversion system. The entire heat source and power conversion system of an HTGR-GT, which is capable of a net electric output of 170 MWe, can be enclosed within the two pressure vessels and cross-duct arrangement similar to the modular steam cycle HTGR previously shown in Fig. 20. The recuperator and precooler would occupy the same space as the steam generator and the turbo-compressor would replace the circulator.

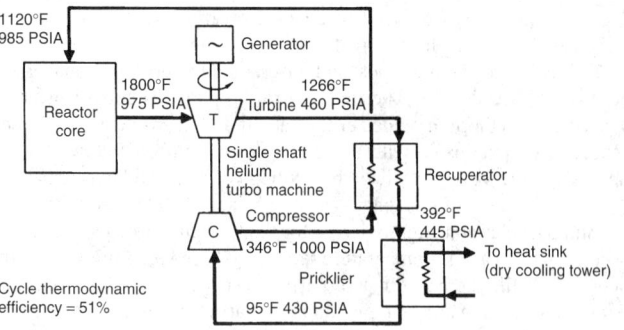

Fig. 23. HTGR gas turbine system with exceptional cycle thermodynamic efficiency. (*GA Technologies.*)

Perhaps the most significant potential use of the HTGR's high temperature capability is the production of synthetic fuels from coal. The HTGR is an important option for supplanting the current consumption of oil and natural gas with synthetic natural gas (SNG). The HTGR can supply the necessary energy for this endothermic process and, therefore, increase the recoverable energy in the SNG product by at least 60% over traditional coal combustion processes.

The most effective method of SNG production with an HTGR is the steam-carbon reforming process in which superheated steam reacts with pulverized coal to form methane-rich SNG. A system for accomplishing this process is shown in Fig. 24. In this system, an intermediate heat exchanger (IHX) has been used to isolate the nuclear heat source from the process steam, thus allowing the use of conventional equipment for the SNG production portion of the plant. The IHX and reactor can be configured in the same arrangement as previously shown in Fig. 23, except that the IHX would occupy the space allocated to the steam generator.

Both gas turbine and process heat versions of the HTGR are based on the demonstrated high-temperature capability of the fuel and core structure. However, some development in the metallic components, such as the turbine, hot ducts and intermediate heat exchanger is necessary. Present

commercial alloys would have limited lifetime under service conditions at 1650 °F (899 °C) and above. However, currently envisioned advancements in ceramics and carbon-carbon composites indicate that high-temperature nonmetallic substitutes for metallic alloys will soon be available. These materials advances are the key to making future application of the HTGR a reality.

Heavy Water Reactor (HWR)

During the atomic energy developments in the World War II years and for a period thereafter, the United States, the United Kingdom, and Canada cooperated closely and many of the nuclear scientists of these countries appreciated the merits of heavy water as a moderator. Each of these countries pursued some development of HWRs for commercial power generation, but at different paces and dedication. Only Canada took to the HWR for commercial power generation. See Figs. 25 and 26.

One of the first high-priority nuclear applications of the United States was for naval propulsion. Because of a very tight minimal physical size criterion, LWRs offered advantages over the HWR. The United Kingdom placed emphasis on the production of plutonium for weapons programs. Gas graphite reactors were a reasonable early choice. When commercial nuclear power was recognized as a needed source of energy, because of the accumulated operating experience it was reasonable to adapt the reactors which had already been developed in the United States and the United Kingdom for military purposes. Long-term savings at that time was not a major criterion.

In the postwar years, hydroelectric power amply met a large portion of Canada's power needs and its abundance made nuclear power quite noncompetitive. Canadian utility operators were used to capital-intensive plants combined with low operating costs. In analyzing the prospects for nuclear power in Canada, utility planners and engineers placed a significant value on low fueling costs, and thus neutron economy was paramount. Therefore, when commercial nuclear power studies commenced in Canada in the mid-1950s, the choice was the HWR. This choice was bolstered by experience with, and knowledge about, heavy water production plants gained when Canadian scientists were trading experience from the heavy water-moderated NRX research reactor when the United States was

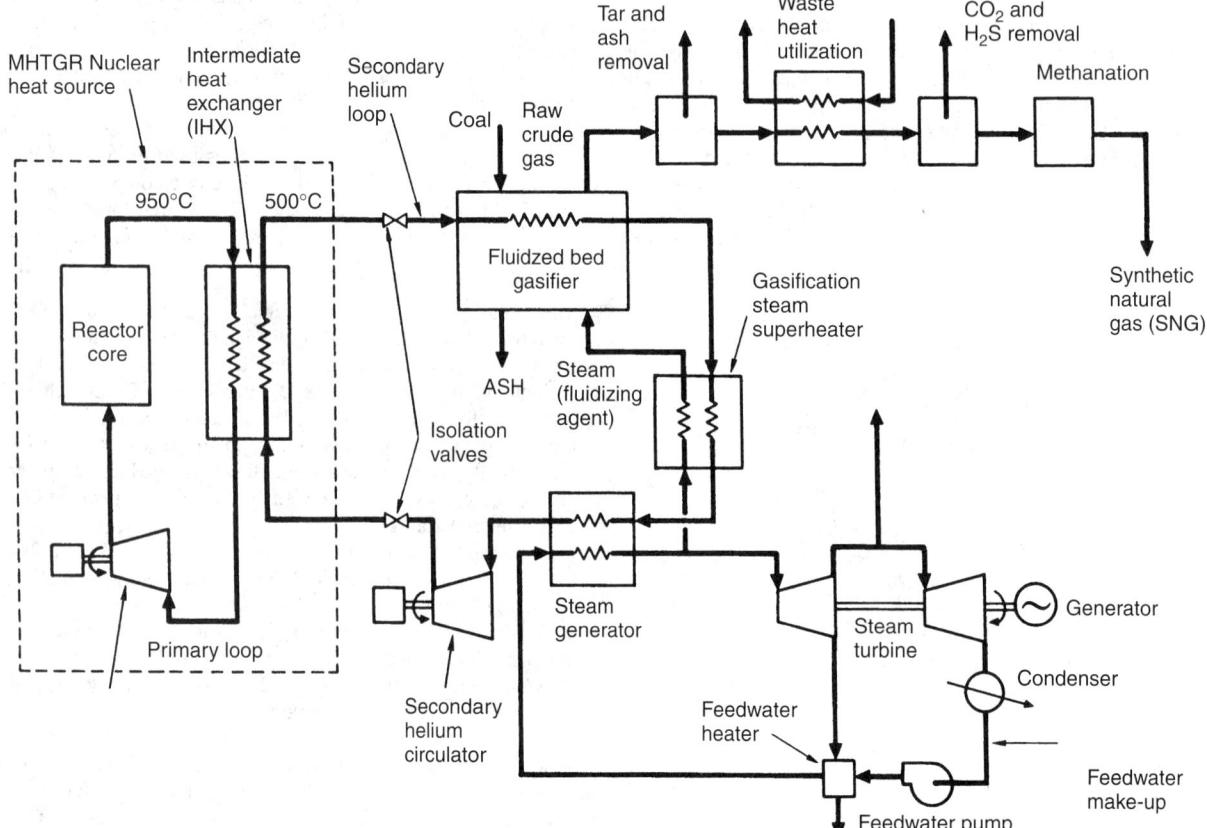

Fig. 24.   Advanced process heat HTGR with intermediate loop for producing synthetic natural gas (SNG) by steam gasification of coal. (*GA Technologies.*)

Fig. 25.   Series of towers comprising part of the heavy water production plant at Ontario Hydro's Bruce nuclear power complex near Tiverton on the shores of Lake Huron. Heavy water is a clear, colorless liquid that looks and tastes like ordinary water. It occurs naturally in ordinary water in the proportion of approximately one part heavy water to 7000 parts of ordinary water. While ordinary water is a combination of hydrogen and oxygen (H_2O), heavy water (D_2O) is made of up of deuterium — a form, or isotope, of hydrogen — and oxygen. Deuterium is heavier than hydrogen in that it has an extra neutron in its atomic nucleus, so heavy water weighs about 10% more than ordinary water. It also has different freezing and boiling points. It is the extra neutron that makes heavy water more suitable than ordinary water for use in CANDU nuclear reactors as both a moderator and a heat transport medium. (*Ontario Hydro, Toronto, Ontario, Canada.*)

developing the Savannah River production facility, which was dismantled in the early 1990s.

Principal advantages of heavy-water reactors are: (1) more efficient absorption of the energy released in the reactor, (2) greater fuel "burn-up" and, therefore, fuel economy, and (3) refueling can take place while the reactor is in service.

The first Canadian nuclear power demonstration (NDP) reactor was of 20-MWe capacity and was configured similarly to a light water reactor. Because of limited facilities for making large pressure vessels, a modular pressure-tube design of the configuration shown in Fig. 24 was investigated. Zircaloy-2 had become available at that time for fabrication of the pressure tubes. Hence the NPD was constructed using Zircaloy as cladding material and uranium dioxide as fuel. The NPD reactor has been in operation since 1962. The CANDU (Canada Deuterium Uranium) power reactors, including the NPD, number over twelve facilities. See Figs. 27, 28, and 29.

CANDU power reactors are characterized by the combination of heavy water as moderator and pressure tubes to contain the fuel and coolant. Their excellent neutron economy provides the simplicity and low costs of once-through natural uranium cycling. Future benefits include the prospect of a near-breeder thorium fuel cycle to provide security of fuel supply without the need to develop a new reactor, such as the fast breeder. The CANDU system is appropriate for countries of intermediate economic and industrial capacity, such as Canada. Producing heavy water is fundamentally simpler than enriching uranium and commercial heavy water plants have been built in smaller sizes than would be possible for uranium enrichment plants.

Although Canada has rather generous supplies and reserves of uranium, there is increasing pressure on Canada to export uranium, a pressure that will probably intensify further if the introduction of fast breeder reactors in other countries is delayed. The current simplest possible fuel cycle for the CANDUs, which is not dependent upon fuel reprocessing, will probably be retained in Canada so long as uranium remains plentiful and comparatively economical. However, for future planning, research to date has indicated that a "self-sufficient thorium cycle" may be practicable in the CANDUs with minimal modification. It has been observed that, at equilibrium, the thorium cycle would require no further uranium. Only small quantities of thorium, which is more abundant than uranium, would be required. Also of interest for the future is *electronuclear breeding*, i.e., the use of electric power to convert fertile to fissile material for neutron economy.

Fast Breeder Reactors

The fast breeder reactor derives its name from its ability to breed, that is, to create more fissionable material than it consumes. This ability stems from the fact that neutrons travel faster than they do in a thermal reactor. The breeding process depends, in part, upon the neutrons maintaining a high speed, or high energy. If their speed or energy is allowed to degrade as occurs in thermal reactors, the number of neutrons produced per absorption in uranium or plutonium decreases. Furthermore, at lower velocities, neutrons tend to be captured in various structural materials of the reactor, and this further reduces the breeding potential. It is important, therefore, in fast reactors to keep the velocity of the neutrons high. Water, which is used as a coolant in some thermal reactors, tends to slow the neutrons

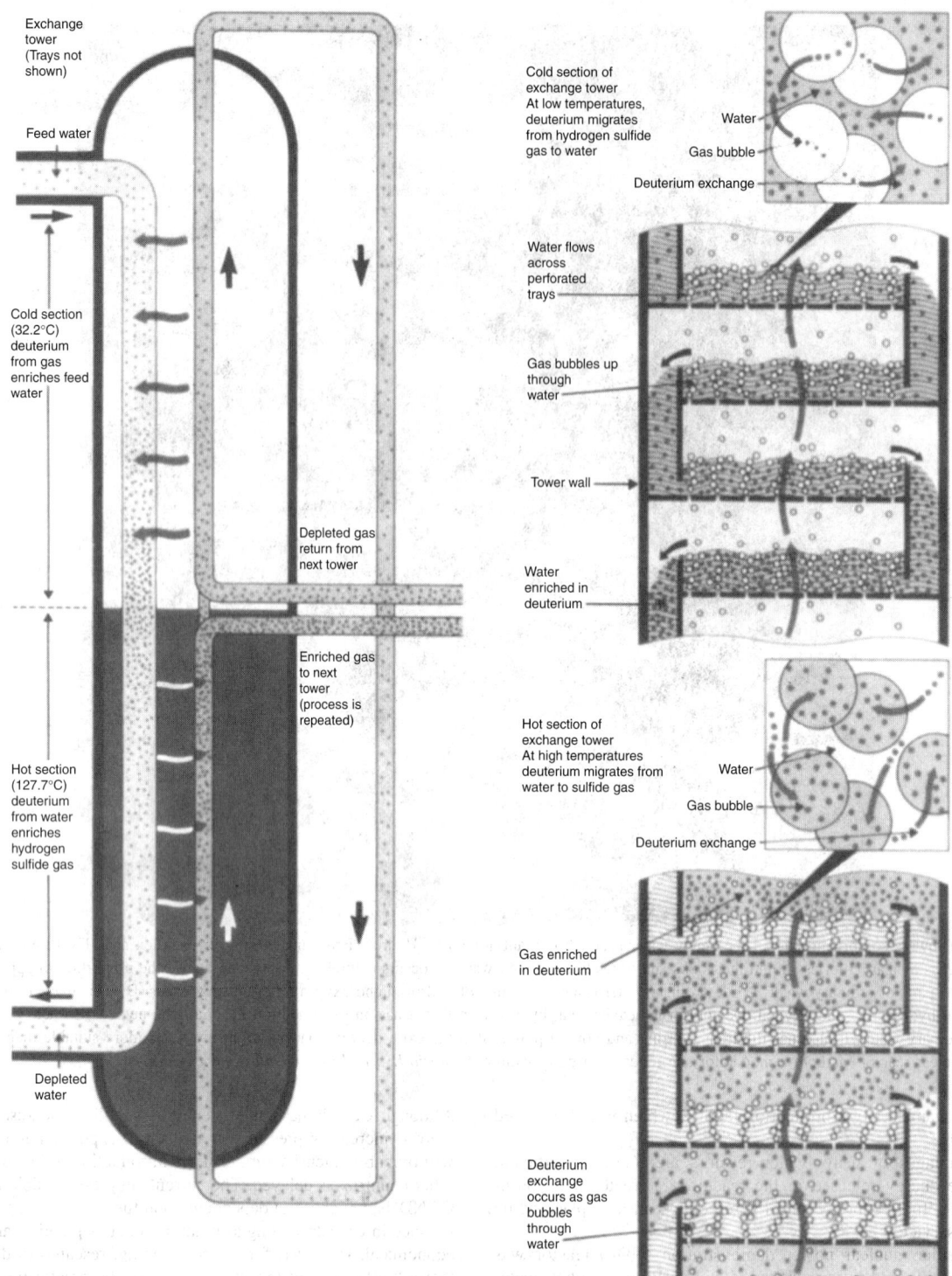

Fig. 26. The production of heavy water is based upon the behavior of deuterium in a mixture of water and hydrogen sulfide. When liquid H_2O and gaseous H_2S are thoroughly mixed, the deuterium atoms exchange freely between the gas and the liquid. At high temperatures, the deuterium atoms tend to migrate toward the gas, while they concentrate in the liquid at lower temperatures. In the first and second stages of production, the towers of a heavy water plant are operated with the top section cold and the lower section hot. Hydrogen sulfide gas is circulated from bottom to top and water is circulated from top to bottom through the tower. In the cold section, the deuterium atoms move toward the water and are carried downward, while in the hot section, they move toward the gas and are carried upward. The result is that both gas and liquid are enriched in deuterium at the middle of the tower. A series of perforated trays are used to promote mixing between the gas and water in the towers. A portion of the H_2S gas, enriched in deuterium, is removed from the tower at the juncture of the hot and cold sections and is fed to a similar tower for the second stage of enrichment.

The first stage of the process enriches the gas from 0.015% deuterium to 0.07%. A second stage further enriches it to about 0.35%. Again, the enriched gas is fed forward to a third stage. The product from this third stage, now in the range of 10 to 30% heavy water, is sent to a distillation unit for finishing to 99.75% purity "reactor-grade" heavy water. Because the production of heavy water uses a toxic gas, H_2S, safety is a top priority at heavy water plants. H_2S is a colorless gas, slightly heavier than air. To safely expel H_2S from the system, it is directed to a flare tower where it is burned off. Initially, each of Ontario Hydro's reactors requires about 800 megagrams, or one year's production from a heavy water plant. After that, less than 1% of the heavy water is lost and has to be replenished each year. (*Ontario Hydro, Toronto, Ontario, Canada.*)

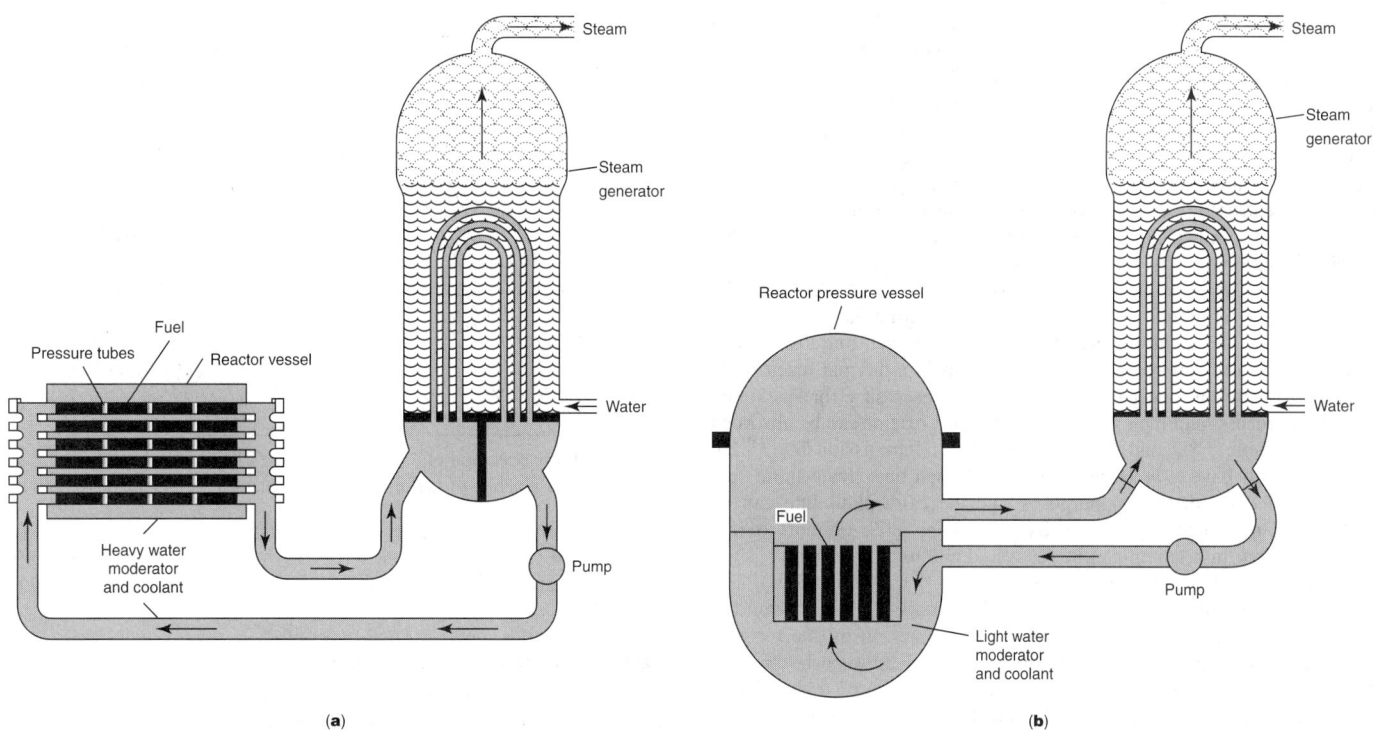

Fig. 27.   Comparison of heavy water reactor (**a**) with light water reactor (**b**). (*After Robertson, Atomic Energy of Canada Limited.*)

Fig. 28.   Pickering (Ontario) "A" generating station's initial performance was outstanding and the station was hailed as a major Canadian technological achievement at the time of its commissioning in February 1971. It reached full power in 3 months, well ahead of schedule. The final Unit 4 went on line in May 1973. At full output (2,160,000 kW), Pickering "A" generates enough power to supply more than 1.5 million homes. In 1974, construction was begun on a twin station, Pickering "B," also shown in this view. The "B" station has the same capacity as its forerunner and all four units became operational in 1986. (*Ontario Hydro, Toronto, Ontario, Canada.*)

Fig. 29.   Darlington, Ontario, nuclear generating station, one of the largest energy projects undertaken in North America. The plant is shown in late stages of construction in 1986. It is located on the shores of Lake Ontario, 5 km southwest of Bowmanville in the Town of Newcastle. Currently nuclear power plants provide one-third of Ontario's electricity needs (the other two-thirds comes from hydro installations). Selected because of its proximity to the residential, industrial, and commercial energy markets of Ontario, the 1200-acre (485-hectare) site has good transportation access, an abundant supply of cooling water from Lake Ontario, relative isolation, and excellent bedrock for station foundations. When the four-unit station is completed it will provide 3,524,000 KW of electricity, enough to serve a city of 3 million people. Electricity from the four units (each 881,000 KW net) will be fed into the Ontario Hydro electricity grid system through 500,000 V transmission lines already crossing the site. The Darlington generating station was approved by the provincial government in July 1977. Site preparation began in 1978 and by 1981, the first concrete was poured for the station's foundations. Construction activity peaked in 1986–1987 with approximately 6800 workers on site. (*Ontario Hydro, Toronto, Ontario, Canada.*)

down and thus prevent efficient breeding. Therefore, it is necessary to use a coolant that does not slow the neutrons or capture them as they travel through the coolant. Liquid sodium and gaseous helium under pressure are the two principal coolants used to date.

Fuel Cycle Considerations. Approximately 99.3% of uranium as it is found in nature is the isotope ^{238}U and 0.7% is ^{235}U. Uranium-235 is a fissile isotope, that is, if it is struck by a neutron it will split; this fission yields on the average approximately two neutrons and 200 MeV of energy. This amount of energy corresponds to approximately 78 million Btu for every gram of uranium which fissions (3.5×10^{10} Btu/pound)

(1.95 Calories/kilogram). Most reactors today are largely dependent upon ^{235}U for their energy. However, some of the neutrons released in fission of ^{235}U also are absorbed in nonfissionable ^{238}U. As the ^{238}U absorbs a neutron, it is transformed into fissionable ^{239}Pu (plutonium). Thus, while the reactor is sustaining the fission process and thereby creating energy, it is also generating fresh fuel, which can later be used to create more energy. Unfortunately, this is an inefficient process in present thermal reactors where the neutron velocity is established by the temperature, or thermal energy, so only limited amounts of additional energy are made available by transformation of ^{238}U into ^{239}Pu.

The fast breeder reactor makes possible the recovery of most of the available energy in uranium. This occurs because during fission in the fast breeder nearly three neutrons are released for every neutron absorbed as compared with only approximately two neutrons in a thermal reactor. On the average, between one and two neutrons are necessary for sustaining the fission process, and the extra neutron in a fast reactor can be absorbed in nonfissionable ^{238}U and thereby transformed into fissionable ^{239}Pu. Reactors which have a breeding ratio greater than one create more fuel than they need for their own purposes, and the extra plutonium can be used to fuel new breeder reactors. By this means, 80% or more of the available energy in uranium can be recovered and used in reactors.

In a typical fast breeder, most of the fuel is ^{238}U (90 to 93%). The remainder of the fuel is in the form of fissile isotopes, which sustain the fission process. The majority of these fissile isotopes are in the form of ^{239}Pu and ^{241}Pu, although a small portion of ^{235}U can also be present. Normally, the fissile isotopes are located in a central "core" region that is surrounded by the fertile isotopes in the "blanket" region. This is illustrated in Fig. 30.

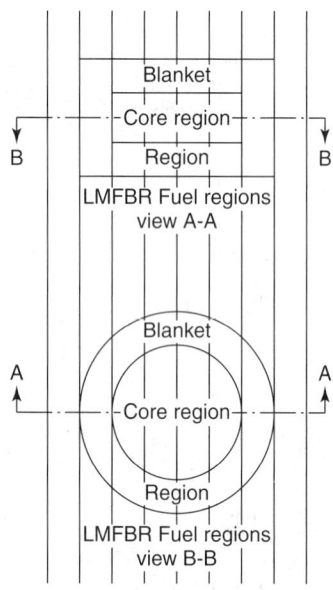

Fig. 30.   Liquid-metal fast breeder reactor core and blanket arrangement.

When the fuel is initially loaded into the reactor, the core region will typically contain from 10 to 15% fissile isotopes with the remainder being ^{238}U. Essentially all of the blanket will be ^{238}U. As energy is extracted from the fissile isotopes, they become depleted (the initial plutonium is gradually used up). However, in a breeder reactor, new plutonium will be formed in the core and blanket regions faster than it is consumed. Additionally, undesirable fission products are formed which must ultimately be removed. This process is schematically illustrated in Fig. 31. The "before" chart represents the new fuel condition and the "after" chart corresponds to the situation when the fuel is removed for reprocessing. Typically, the fuel removed for reprocessing will contain from 1 to 3% new plutonium. It is in this manner that the fast breeder can recover from 80 to 90% of the available energy in uranium resources. Most present reactors require some enrichment of the ^{235}U isotope used to fuel them. This enrichment process requires a plant, which, in turn, uses large amounts of electrical energy. Because the fast breeder converts the fertile isotope ^{238}U into the fissile isotope ^{239}Pu, no enrichment plant

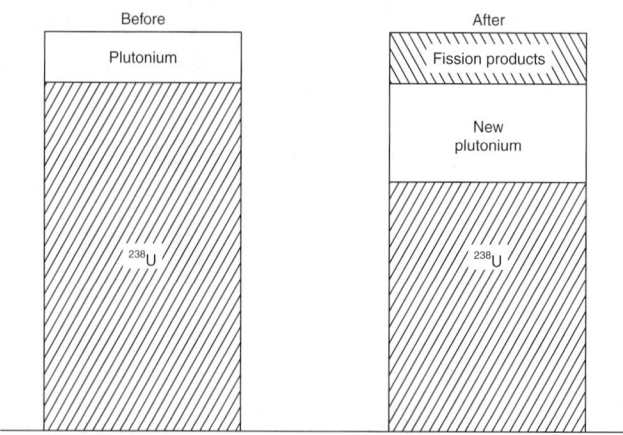

Fig. 31.   Basic operation of the breeder reactor. The illustration does not include geometrical disposition of fuel in the core and blanket system. (*General Electric.*)

is necessary. The fast breeder serves as its own enrichment plant. The need for electricity for supplemental uses in the fuel cycle process is thus reduced.

Fast Breeder Reactors in Perspective

Of the several fundamental ways to use nuclear fission reactions to generate electric power, the fast breeder reactor (notably the liquid-metal-cooled fast breeder reactor, LMFBR) probably has the most checkered history. The fast breeder reactor received its early impetus when there was serious concern over what appeared to be a limited supply of uranium and consequent increasing prices of uranium fuels. With the fast breeder concept offering up to a 100-fold increase in the utilization efficiency of uranium, it appeared to be the logical replacement (second generation) for light-water reactors. It would present a technical solution in time for the expected tight supplies of uranium. The United States funded LMFBR research quite generously until a serious reduction in 1982. Then, it was determined that the shortage of uranium fuels no longer posed a serious threat in the short-term, thus establishing the general consensus in the U.S. that the breeder, if needed at all, could be delayed until well into the next century. The interest of the French and Japanese in breeders also stemmed from early concerns with uranium shortages and prices, but was much more serious because these countries are extremely uranium poor and, further, these countries have fewer energy options. For example, the cost of coal ranges from 1.6 to 2.5 times the cost of coal in the United States. Continued progress in fast breeder development, particularly in France, also has been accelerated by a very heavy past investment in the technology coupled with a desire to be fully self-contained as regards the generation of electric power.

Particularly, in the United States, because of its several energy fuel options and its current "rethinking of nuclear power," the principal nuclear power research targets no longer include uranium supplies. Rather, the targets are the lowering of capital costs for existing light-water reactor technology, shortening the plant construction and licensing lead times, achieving higher plant availability (essentially eliminating long power outages), and increasing plant safety. From this "rethinking" process, the U.S. Congress suspended funding for the Clinch River Breeder Reactor.

It is interesting to note that, as of the early 1990s, most authorities were not quite willing to "forget" the fast breeder altogether. The differences in views essentially reside in timing. The fast breeder, even though costly to build today, does offer several temptations. For example, experience gained from the Experimental Breeder Reactor (Idaho Falls, ID) which commenced operation in 1964 is reported to operate better now than when it was first built, showing no evidence of corrosion. It is suggested that if such experience were extrapolated the useful life of an LMFBR could be between 100 and 150 years! Weinberg (1986 reference listed) recalls, "that one of Newcomen's original steam engines, built in the mid-18th century, continued to operate until 1918." It does appear that the economy of the LMFBR is somewhat analogous to large hydroelectric projects that require very large investments of capital, but coupled with low operating costs, provide really cheap electric power once they are fully amortized. However, one usually must wait a generation before this situation occurs. Weinberg also observes, "Because the breeder requires little, if any, mining of uranium, its environmental impact is much smaller, at least at the front end of the fuel cycle, than is the impact of the LWR. The roughly 300,000 tons of depleted uranium stored outside the diffusion plants, if used in breeders, could fuel our (U.S.) entire electric system for centuries!"

Davis (1984 reference listed) observes that there is sufficient know-how today to build and operate fast breeder reactors with confidence of their safety and reliability, as exemplified with large units in France. Davis further suggests that the principal problems remaining in fast breeder technology include: (1) a reduction in overall costs to make the fast breeder competitive with coal and the light-water reactors — an estimated reduction factor of 1.5 to 2; (2) assurance of adequate safeguard on the plutonium fuel cycle; (3) more demonstrations of commercial-scale operations, such as the engineering scale-up of important system components — pumps and steam generators; (4) implementing a large, overall system which must include parallel facilities for fuel processing and refabrication; and (5) setting in place the reprocessing of light-water reactor fuel to provide the "start up" plutonium for the fast breeders.

Some technical successes with the fast breeder have occurred in France: (1) demonstrations of a positive breeding gain of 0.15 ± 0.04 in a complete breeder fuel cycle, and (2) demonstration at several laboratories of uranium and plutonium oxide fuel elements that can sustain more than 10% burnup of the original mixture of $^{238}U/^{239}Pu$ before the fuel has to be reconstituted, representing a tenfold improvement (Weinberg, 1986). It has also been observed that, compared with a number of so-called alternative fuels, the new Super-Phoenix (France) plant can produce electricity at costs that are markedly less than electrical energy from solar-powered photovoltaics, for which funding remains significant. Another proposed inexhaustible power source, nuclear fusion, still remains in an early stage of development. See **Fusion Power**.

LMFBR Design Principles. There are many design differences among the reactor designs, including: (1) primary coolant system arrangement; (2) refueling mechanism design and arrangement; (3) steam generator type and arrangement; (4) core support method; (5) structural material choices; and (6) safety features. Perhaps the most noticeable difference is that of the primary system arrangement. This difference is schematically illustrated in Figs. 32 and 33. The system of Fig. 32 corresponds to a "loop" or "piped" arrangement where the reactor, pumps, and intermediate heat exchangers are located separate from each other and piping carries the sodium from one point to the other. The "pool" or "tank" arrangement of Fig. 33 includes the reactor, intermediate heat exchangers and pumps in one large pool of sodium which is contained in a separate tank. Each concept has advantages and disadvantages. The pool concept is somewhat easier to design for certain hypothetical accident situations. The loop concept is easier to construct and to maintain.

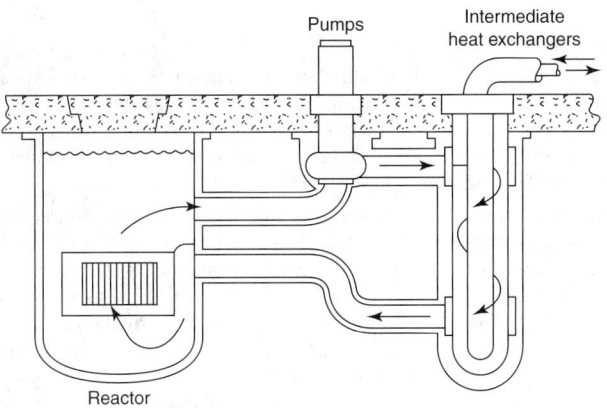

Fig. 32.   Loop arrangement in the liquid-metal fast breeder reactor. (*General Electric.*)

The flow circuit for an LMFBR where two sodium circuits are included is shown schematically in Fig. 34. The reactor is cooled by the primary sodium, which becomes radioactive as it picks up heat in passing through the core or fueled region. In this particular arrangement, the sodium is heated to $560\,°C$ and flows through pipes (schematically shown as a single line in the figure) to the intermediate heat exchangers. In the heat exchangers, the primary sodium transfers heat to the nonradioactive sodium. After being cooled to $393\,°C$ in the heat exchangers, the primary

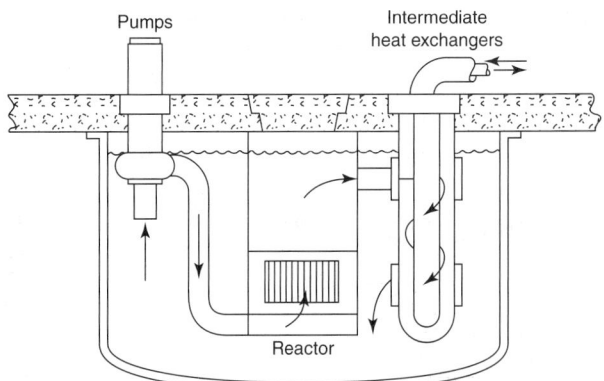

Fig. 33.   Pool arrangement in the liquid-metal fast breeder reactor. (*General Electric.*)

sodium is pumped back into the reactor where it again repeats the circuit. The nonradioactive secondary sodium is circulated from the intermediate heat exchangers through steam generators where the heat from the sodium is transferred to water, which becomes superheated steam for use in the turbine. The cooled secondary sodium is pumped back through the intermediate heat exchangers where the process is repeated. Steam from the steam generators is used to turn the rotor of the turbine generator to generate electricity. In the arrangement shown, 1,200 MW of electricity are generated at a net overall efficiency of 39%. This relatively high efficiency is possible because of the excellent thermal characteristics of sodium.

Nuclear Power Innovations for 1995 and Beyond

Increasing concerns over the impact of fossil fuel-burning electric power generating plants on the environment, the accelerating demands for electric power, and a growing dependence on foreign oil supplies brought about a resurgence of interest in nuclear power reactor research during the mid-1980s. Advanced programs were established to create new plant designs that would reflect past experience to achieve an extremely high degree of operating safety, competitive construction, and operating costs, including the streamlining of the licensing process and consumer confidence in nuclear power technology. Thus, during the past decade, much private and government-sponsored research has been directed toward nuclear reactor research in three areas, namely, light water reactors (LWRs), high-temperature gas-cooled reactors (HTGRs), and liquid metal-cooled reactors (LMRs).

The innovative program commenced with an impressive nuclear power base — some 107 nuclear plants with full-power licenses operating in the United States and producing 18% of the nation's electricity — worldwide, 414 plants in 26 countries generate 298,000 megawatts of electricity (MWe), accounting for 16% of the world's generating capacity.

A survey of reactor developments of the three aforementioned types reveals a number of common generic technical features. These include passive stability, simplification, ruggedness, ease of operation, and modularity.

Overall goals for the innovative program can be summarized as follows:

1. Assured safety with features that minimize the negative consequences of human error, especially a reduction in the chance of occurrence of severe core damage by at least a factor of 10 less than former, contemporary designs.
2. Significantly simpler designs, with increased safety and performance margins in key operational parameters.
3. High reliability throughout a lifetime on the order of 60 years and an increase in plant availability to 85% or greater than the contemporary average of less than 70%.
4. Reduction in capital, operating, maintenance, and fuel costs to meet the economic competition with coal-burning generators. A reduction in construction time to the range of 3 to 5 years as compared with more than 10 years, which has been the experience with some of the later contemporary reactors.
5. A modular design that is standardized at a highquality level and thus predictably licensable.

Passive Stability. Passive design characteristics ensure core stability by eliminating the potential for a runaway chain reaction. In the innovative

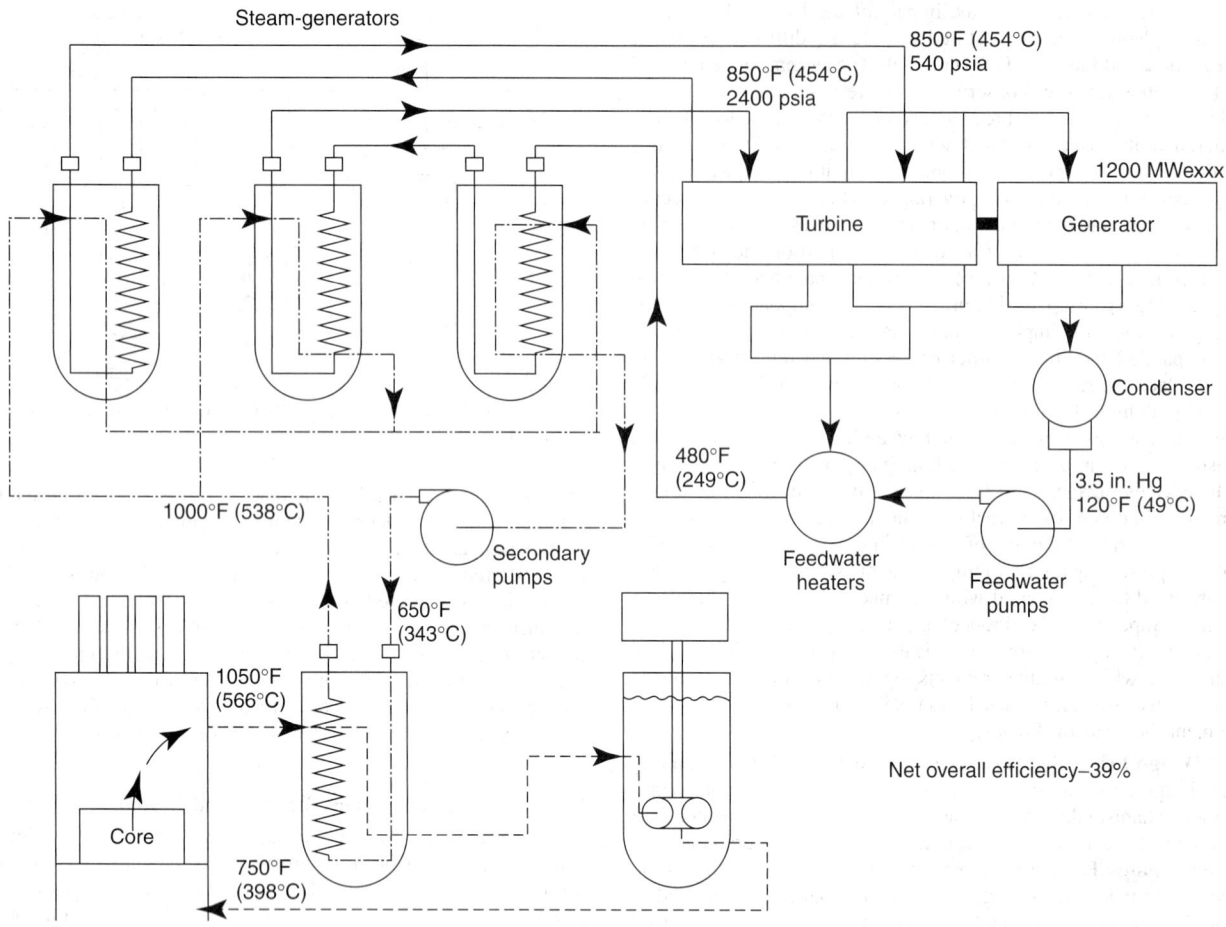

Fig. 34.   Liquid-metal fast breeder reactor flow circuit. (*General Electric.*)

program, this has been a hallmark from the outset of the program. Passive characteristics are internal governors—that is, physical laws ensure that the reaction rate decreases instantaneously as the temperature of the coolant or fuel or the power of the reactor increases, without the need for external control devices.

Ruggedness. In some past designs, long-term reliability has been impaired by attempts to achieve the highest in efficiency and economic performance. In response to this negative experience, the margin in certain key performance parameters is being increased in order to lessen the burden on the equipment. By reducing power densities and coolant temperatures, higher reliability will be achieved over a longer lifetime. Past field experience has identified more effective methods of coolant chemistry control and materials selection, factors that will contribute to the long-lived reliability of the components of future systems. Greater emphasis is being given to the selection of proven, high-quality materials and components and on improved methods and quality control over assembly and construction.

Ease of Operation. Thorough investigation of the Three Mile Island incident some years ago showed that a lack of attention had been given to the human factor. The innovation program is addressing this problem in several ways. The computer and telecommunication revolution has made it more practical to use improved technology and human engineering methodologies to revamp the control room and the reactor instrumentation system. These improvements will make the plant easier to operate and provide the operator with a greatly increased amount and quality of information on plant conditions. Graphic displays, diagnostic aids, and expert systems are being developed for such advanced control rooms.

The other design goals complement the new technology to make the operator's task even easier. The passive safety features substantially extend the response time required of the operators in an emergency condition. The margin being built into the systems provides broader normal operating regimes and longer response times for operator action. Greater emphasis is being placed on simplification of operating procedures.

Modularity. Economic competitiveness requires that the construction time be shortened dramatically. Modular construction techniques are a

key contributor to achieving this goal and are a proven approach to cost control in major construction projects. Modularization provides for a larger percentage of factory construction, rather than field construction. New innovative concepts will rely heavily on modularization and will be centered around lower unit power outputs, factory assembly, and transportation of modules to the plant site. The overall plant size target is 600 MWe net electrical output.

The AP600 Advanced Plant

As of 1994, the AP600 (Westinghouse) plant was in the most advanced stage of the innovative program. The first unit may be put on line just prior to the expiration of the last century. This design satisfies all of the previously mentioned goals of the innovative program. A sectional view of the AP600 is shown in Fig. 35. The site plan is shown in Fig. 36.

AP600 Passive Safety System Details. These features reduce operator responsibilities and add an extra margin of safety over contemporary PWR designs. See Fig. 37 on p. 2501. Large volumes of water stored in the containment eliminate the need for operator action to assure make-up water, either for small leaks that may occur during normal operation or for a major loss of coolant accident (LOCA). A passive plant is a system that assures public safety even if the operators fail to act.

The passive residual heat removal heat exchangers remove core decay heat if steam generator heat removal is not available. Passive residual heat removal heat exchangers in the in-containment refueling water storage tanks are connected to the reactor coolant system (RCS) piping, forming a full-pressure, closed, natural circulation cooling loop.

Two core make-up tanks provide borated make-up water whenever the normal make-up system is unavailable. The tanks are located above the reactor coolant system loop piping and kept at system pressure by steam lines from the pressurizer. These tanks function at any system pressure, using only gravity as a motive force. If the reactor protection system detects a need for make-up water, core make-up tanks discharge and isolation valves open automatically, allowing the tanks to drain into the reactor vessel.

Two accumulators provide the high make-up flows initially required by a large LOCA. These tanks contain 1700 cubic feet (about 48 cubic meters) of borated water pressurized with 300 cubic feet (about 8.5 cubic meters) of nitrogen at 700 psi (4.8×10^6 Pa). The accumulators are isolated from the RCS by check valves. Each accumulator is paired with a core make-up tank, the pair sharing an injection line to the reactor vessel downcomer. This ensures that at least one accumulator/core make-up tank pair would be available following an injection line LOCA.

The in-containment refueling water storage tank provides 500,000 gallons (about 1900 cubic meters) of water with a gravity head above the core. This water inventory is sufficient to flood the containment above the level of the reactor core and provide decay heat removal by natural circulation.

The automatic depressurization system depressurizes the RCS if core make-up tank level is low. Depressurization allows gravity injection from the in-containment refueling water storage tank, which is at atmospheric

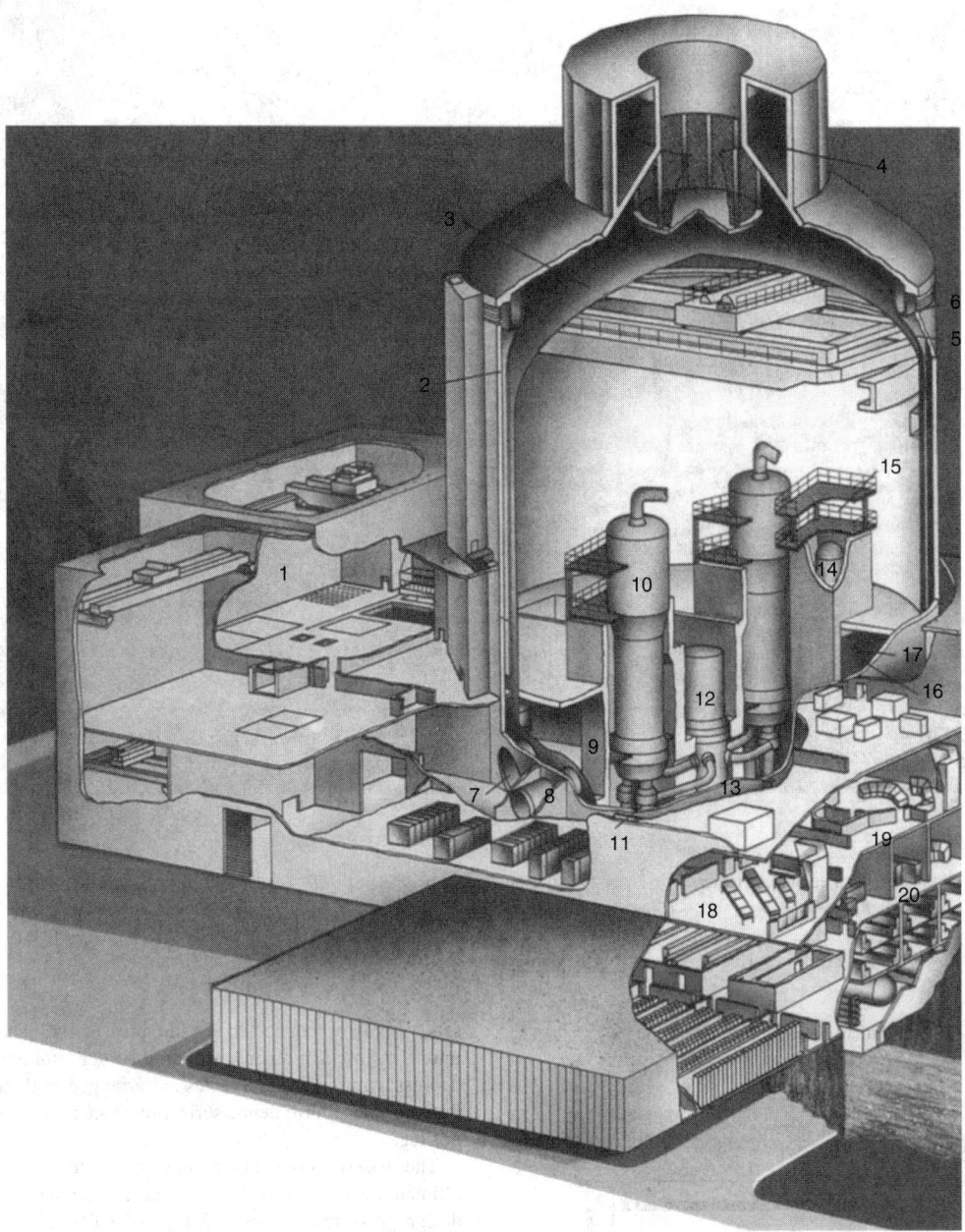

Fig. 35.   Sectional view of the 600 MWe pressurized water reactor (PWR) expected to be operational by 1995. Considered the PWR of the future, the plant is designed for a minimum useful life span of 60 years and features numerous economic and safety features, including passive systems for ultimate protection. (*Joint project of Westinghouse, the Electric Power Research Institute, and the U.S. Department of Energy.*)

Legend:

1. Fuel Handling Area
2. Concrete Shield Building
3. Steel Containment
4. Passive Containment Cooling Water Tank
5. Passive Containment Cooling Air Baffles
6. Passive Containment Cooling Air Inlets
7. Equipment Hatches (2)
8. Personnel Hatches (2)
9. Core Make-up Tanks (2)
10. Steam Generators (2)
11. Reactor Coolant Pumps (4)
12. Integrated Head Package
13. Reactor Vessel
14. Pressurizer
15. Depressurization Valve Module Location
16. Passive Residual Heat Removal Heat Exchangers
17. Refueling Water Storage Tank
18. Technical Support Center
19. Main Control Room
20. Integrated Protection Cabinets

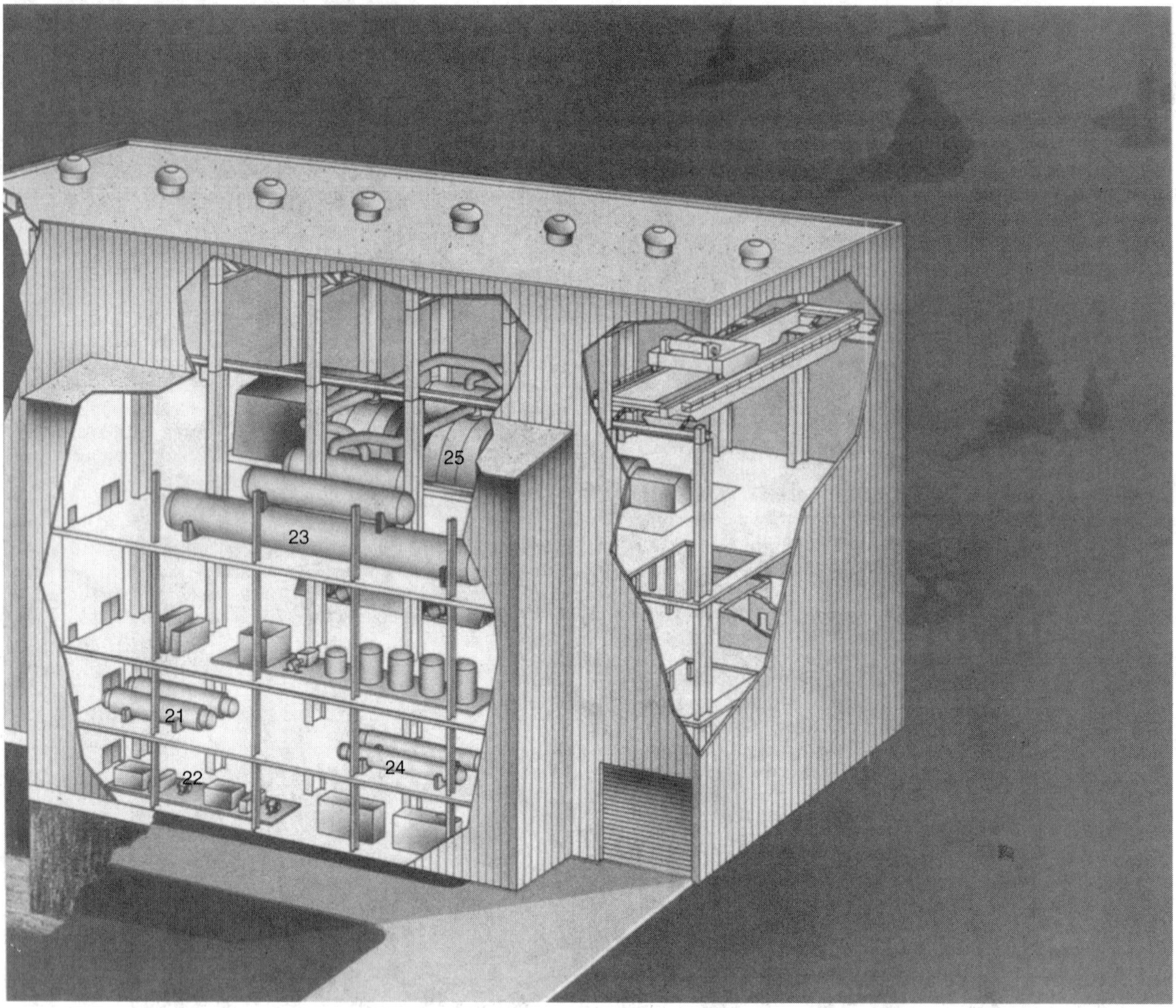

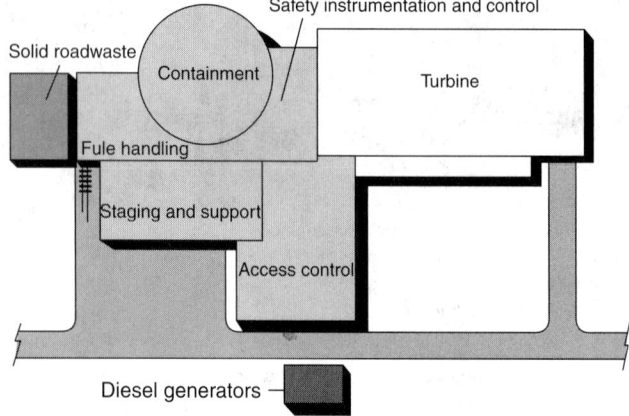

Fig. 35.   (*Continued*)

21. High Pressure Feedwater Heaters	25. Turbine/Generator	29. Feedwater Line
22. Feedwater Pumps	26. Spargers (2)	30. Passive Containment Cooling Air Flow
23. Deaerator	27. Accumulators (2)	
24. Low Pressure Feedwater Heaters	28. Main Steam Line	

Fig. 36.   Site plan for the advanced pressurized water reactor (PWR). (*Westinghouse Electric Corporation, Energy Systems.*)

pressure. To ensure that the automatic depressurization system works when needed, while minimizing the consequences of spurious valve operation, the system provides phased depressurization with two redundant sets of valves connected to the pressurizer. The discharge is sparged into the in-containment refueling water storage tank. The automatic depressurization system valves are arranged in three stages to reduce peak flow rates. A fourth depressurization stage is provided directly on the RCS hot leg.

The passive containment cooling system provides the safety grade ultimate heat sink that prevents the containment shell from exceeding its design pressure of 45 psig [3.1×10^5 Pa (g)]. The system uses natural air circulation between the steel containment shell and the concrete shield building. During postulated accidents, air cooling is enhanced by draining water onto the steel containment shell. The water is provided by gravity from a 350,000-gallon (about 1300 cubic meters) annular tank in the roof of the shield building. This tank has sufficient water to provide three days of cooling.

RADIOACTIVE WASTES

Decisions pertaining to the location of radioactive waste sites mainly derive from political and sociological sources. A condensed overview of the technical aspects is given here.

The radioactive wastes associated with nuclear reactors fall into two categories: (1) *commercial wastes*—the result of operating nuclear-powered electric generating facilities; and (2) *military wastes*—the result of reactor

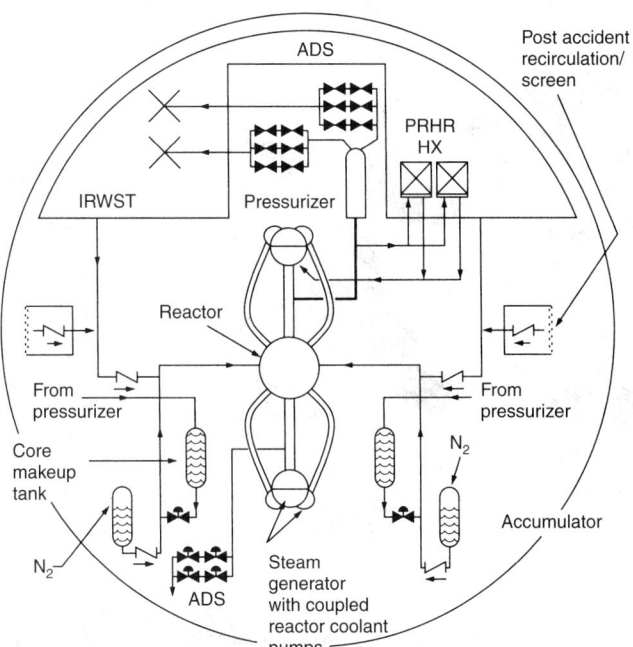

Fig. 37. Schematic representation of the in-containment passive safety injection system (PSIS). IRWST = in-containment refueling water storage tank. PRHR-HX = passive residual heat removal heat exchanger. ADS = automatic depressurization system (four stages). (*Westinghouse.*)

operations associated with weapons manufacture. Because the fuel in plutonium production reactors, as required by weapons, is irradiated less than the fuel in commercial power reactors, the military wastes contain fewer fission products and thus are not as active radiologically or thermally. They are nevertheless hazardous and require careful disposal.

Nuclear power plants use fuel rods with a life span of about three years. Each year, roughly one-third of spent fuel rods are removed and stored in cooling basins, either at the reactor site or elsewhere. Typical modern nuclear power plants discharge about 30 tons of the spent fuel per reactor per year. Comparatively little of the radioactive wastes, as is currently reliably known worldwide, has been processed for return to the fuel cycle. Actually, fuel reprocessing causes a net increase in the volume of radioactive wastes, but, as in the case of military wastes, they are less hazardous in the long term. Nevertheless, the wastes from reprocessing also must be disposed of with great care.

Spent fuel from a reactor contains unused uranium as well as plutonium-239 which has been created by bombardment of neutrons during the fission process. Mixed with these useful materials are other highly radioactive and hazardous fission products, such as cesium-137 and strontium-90. Since reprocessed fuels contain plutonium, well suited for making nuclear weapons, concern has been expressed over the possible capture of some of this material by agents or terrorists operating on behalf of unfriendly governments that do not have a nuclear weapons capability.

Categories of Wastes by Content

In addition to the two source categories previously mentioned, radioactive wastes are classified in accordance with their content:

High-level wastes contain 99.9% of the nonvolatile fission products, 0.5% of the uranium and plutonium, and all the actinides formed by transmutation of the uranium and plutonium in the reactors. Among the actinides are neptunium and americium. High-level wastes are either the aqueous wastes resulting from reprocessing; or the spent-fuel rods to be disposed of in the absence of reprocessing.

Cladding wastes are comprised of solid fragments of Zircaloy and stainless steel cladding (tube in which the fuel is placed) and other structural elements of the fuel assemblies remaining after the final cores have been dissolved.

Low-level transuranic wastes are solid or solidified materials which contain plutonium or other long-lived alpha-particle emitters in known or suspected concentrations higher than 10 nanoCuries per gram and external radiation levels after packaging sufficiently low to allow direct handling.

Intermediate level transuranic wastes are solids or solidified materials that contain long-lived alpha-particle emitters at concentrations greater than 10 nanoCuries per gram and which have, after packaging, typical surface dose rates between 10 and 1000 mrems/hour due to fission product contamination.

Nontransuranic low-level wastes are diverse materials which are contaminated with low levels of beta- and gamma-emitting isotopes, but which contain less than 10 nanoCuries of long-lived alpha activity per gram.

Permanent Disposal Methodologies

Because of many doubts among various authorities pertaining to the permanent geologic depositories, as previously mentioned, considerable effort continues as regards semipermanent and permanent types of depositories. Many methodologies involve the so-called "sequence of barriers" approach. The first barrier is the form in which radioactive materials are embedded — vitrification, calcination, etc. The requirements for the first barrier are that it not be corrosive and possess excellent thermal stability and mechanical integrity. Wastes generate much heat during their initial decade of confinement. This affects decisions as regards the wasteform and the second barrier, the frequently mentioned canister which encapsulates the wasteform. The principal function served by the canister is protection of the material during the collection and transportation (to geologic site) phases. The canisters also should provide excellent protection of their contents for a minimum of 50 years, just in case it is desired to retrieve the wastes at some future date. Canisters must resist corrosive chemicals, they must withstand extremely high radiation fluxes caused by fission-product decay and the heat generated by the decaying wastes. It is interesting to note that an unprotected stainless steel canister will not resist structural deterioration arising from salt brines for that long a period. Provision must be made for cooling the canisters, either by air or water. The canisters should be designed to permit maximum heat transfer and, currently, the cylinder and annulus configurations are preferred. A third barrier would be the geologic site itself, obviously impervious to water penetration and in a seismically stable location. To fulfill all the foregoing requirements (and more), consideration is being given to phasing the waste storage procedure, possibly storing the canisters for water cooling during the first few years, after which air cooling would suffice.

The predominant preference appears to be the underground depository option. Along these lines, it is interesting to review the rock formations in the United States, as shown in Fig. 38.

Commenting briefly on other proposed methodologies, as shown in Fig. 39: (a) In *solution-mined cavities*, it is proposed that chemical solutions would be used to mine cavities in appropriate media, such as rock salt; (b) in the *drilled-hole matrix*. A series of large-diameter holes would be drilled into the geologic media to depths up to 2 kilometers to form a grid of holes. The solid wastes would be packed into these holes, then sealed. (c) In the *rock-melting concept*, liquid wastes (no solidification) are poured into a subterranean cavity, which would be created by an underground explosion. (d) In the *hydrofracture concept*, liquid radioactive wastes are converted into a type of grout (cement or cementlike materials used). This grout is pumped under high pressure into shale as deep as 1 kilometer. The pressure of the operation causes the underlying shale to fracture and the wastes fill up the cracks so formed. This procedure has been used for years in the petroleum field. See also **Petroleum.** (e) In the *polar ice concept*, the wastes would melt through the ice (although this approach would require considerable new technology); or the wastes would be placed on the surface of the ice or anchored within the ice. Advantages include long distances from populations and excellent thermal cooling. Disadvantages include extensive transport and poor retrievability. This method is not high on the list of choices mainly because of too many unknown factors that will require considerable research and experimentation. (f) *Oceanic disposition*, in addition to the polar ice cap concept, are subduction zones and other deep sea trenches and rapid sedimentation areas. A "Seabed Disposal Program Annual Report" states, "Placing high-level wastes on the seafloor, i.e., in the water column, effectively puts the waste contained directly into the biosphere. Since it is difficult to conceive of a practical man-made wasteform/container system that would survive without releasing radionuclides for hundreds of thousands of years in a marine environment, one must assume the

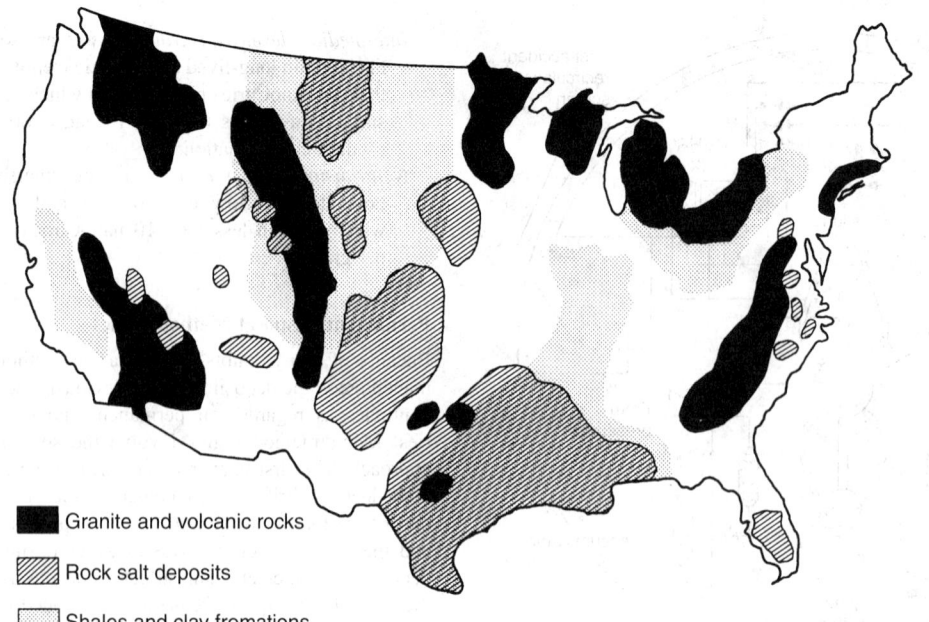

■ Granite and volcanic rocks

▨ Rock salt deposits

▢ Shales and clay fromations

Fig. 38.   Some authorities prefer rock salt deposits for the permanent disposal of nuclear wastes on the assumption that the heat generated by radioactive decay would fuse salt and wastes into an impermeable mass. Other experts question the integrity of salt formations. Increasing attention has turned to hard media, such as the granitic and basaltic rocks, and to shale and clay formation, with the hope that the extensive occurrence of such formations would minimize the need to transport wastes over long distances.

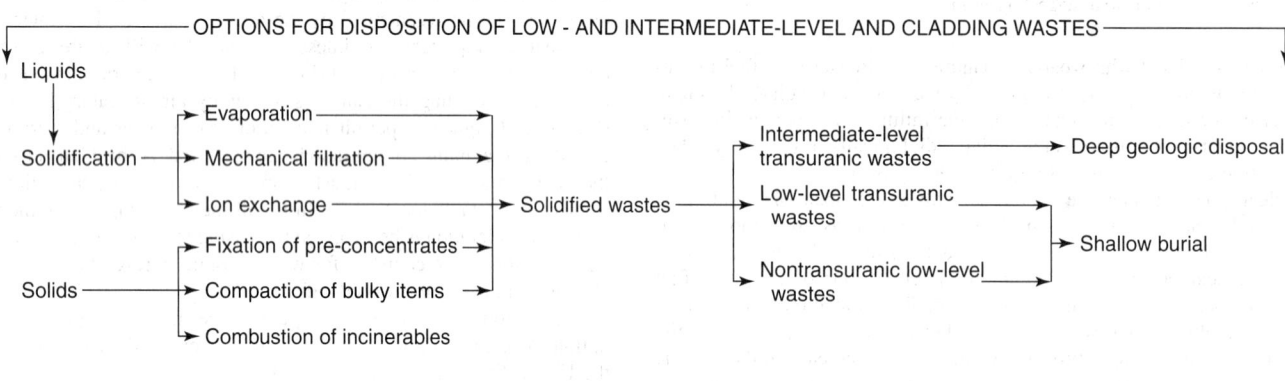

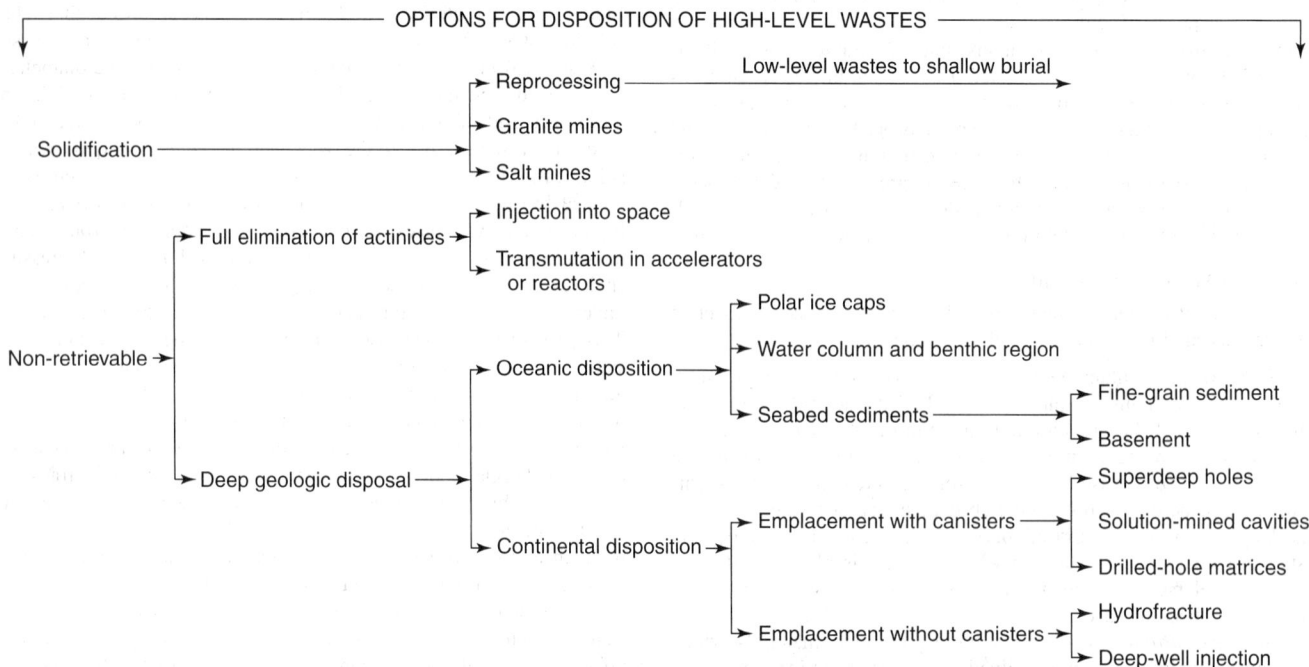

Fig. 39.   Panorama of options for consideration in the disposition of radioactive wastes. (*Top portion of chart*: Options for disposition of low- and intermediate-level and cladding wastes. *Bottom portion of chart*: Options for disposition of high-level wastes.)

radioactive material would eventually enter the ecosystem." The subseabed sediments in the central North Pacific are loosely packed, fine-grained, deepsea "red clays" and have not been fully dismissed from continuing investigation.

Additional Reading

Apostolakis, G.: "The Concept of Probability in Safety Assessments of Technological Systems," *Science*, 1359 (December 7, 1990).

Bodansky, D.: "Nuclear Energy: Principles, Practices, and Prospects," American Institute of Physics, College Park, MD, 1996.

Bothwell, R.: "Nucleus. The History of Atomic Energy Limited," Univ. of Toronto, Toronto, Canada, 1988.

Bruschi, H. and T. Andersen: "Turning the Key," *Nuclear Engineering International*, 1 (November 1991).

Cobb, C.E., Jr. and K. Kasmauski: "Living with Radiation," *National Geographic*, 403 (April 1989).

Cottrell, A.: "Safe Energy," *Review (University of Wales)*, 38 (Autumn 1987).

Davis, W.K.: "Problems and Prospects for Nuclear Power," *Chem. Eng. Progress*, **80**(6), 11–16 (June 1984).

Golay, M.W. and N.E. Todreas: "Advanced Light-Water Reactors," *Sci. Amer.*, 82 (April 1990).

Golay, M.W.: "Longer Life for Nuclear Plants," *Technology Review (MIT.)*, 25 (May/June 1990).

Goldschmidt, B.: "Atomic Rivals," Rutgers University Press, New Brunswick, NJ, 1990.

Green, S.J.: "Solving Chemical and Mechanical Problems of PWR Steam Generators," *Chem. Eng. Progress*, 31 (July 1987).

Hansen, K. et al.: "Making Nuclear Power Work: Lessons from Around the World," *Technology Rev. (MIT)*, 30 (February 1989).

Hodgson, P.E.: "Nuclear Power, Energy and the Environment," World Scientific Publishing Company, Inc., River Edge, NJ, 2000.

Jerome, F.: "Yo-Yo Journalism and Nuclear Power," *Technology Review (MIT)*, 73 (April 1989).

Kairi, S.P.: "Outage Risk Management," *EPRI J.*, 34 (April/May 1992).

Kurdsunoaeglu, B., A. Perlmutter, and S.L. Mintz: "The Challenges to Nuclear Power in the Twenty-First Century," Kluwer Academic Publishers, Norwell, MA, 2000.

Lester, R.K.: "Rethinking Nuclear Power," *Sci. Amer.*, 31 (March 1986).

Marshall, E.: "Counting on New Nukes," *Science*, 1024 (March 2, 1990).

Miller, P. and R.H. Ressmeyer: "A Comeback for Nuclear Power? Our Electric Future," *National Geographic*, 60 (August 1991).

Roberts, L.: "British Radiation Study Throws Experts into Tizzy," *Science*, 24 (April 6, 1990).

Shoup, R.L.: "International Waste Management Symposium," *Nuclear Safety*, **18**(4) (1977).

Skerret, P.J.: "Will the Public Say Yes to Nukes?" *Technology Review MIT*, 8 (April 1991).

Slovic, P., J.H., Flynn and M. Layman: "Perceived Risk, Trust, and the Politics of Nuclear Waste," *Science*, 1603 (December 13, 1991).

Spinard, B.I.: "U.S. Nuclear Power in the Next Twenty Years," *Science*, 707 (December 12, 1988).

Staff: "AP600—A Cost Competitive Power Source," *Energy Digest*, 4 (Pittsburgh, Pennsylvania (Fall 1991).

Staff: International Atomic Energy Agency, "Choosing the Nuclear Power Option: Factors to Be Considered," Bernan Associates, Lanham, MD, 1998.

Staff: University Press Cambridge, "Resistance to New Technology: Nuclear Power, Information Technology and Biotechnology," Cambridge University Press, New York, NY, 1997.

Staff: IAEA, "Nuclear Power Reactors in the World, April 1997," Bernan Associates, Lanham, MD, 1997.

Suzuki, T.: "Japan's Nuclear Dilemma," *Technology Review (MIT)*, 41 (October 1991).

Taylor, J.J.: "Improved and Safer Nuclear Power," *Science*, 318 (April 21, 1989).

Weinberg, A.M. and I. Spiewak: "Inherently Safe Reactors and a Second Nuclear Era," *Science*, **224**, 1398–1402 (1984).

Web References

History of Nuclear Power Plant Safety: *http://users.owt.com/smsrpm/nksafe/*
International Atomic Energy Agency: *http://www.iaea.org/worldatom/*
NRC Short History: *http://www.nrc.gov/SECY/smj/shorthis.htm*
Nuclear History Site: *http://geocities.com/RainForest/Andes/6180/*
U.S. Nuclear Regulatory Commission: *http://www.nrc.gov/*

NUCLEAR SPIN. The intrinsic angular momentum of the atomic nucleus due to rotation about its own axis. It is usually designated I and has the magnitude, $\sqrt{I(I+1)}h/2\pi \approx I(h/2\pi)$, where I is the nuclear spin quantum number which has different (integral or half-integral) values (including zero) for different nuclei. In spectroscopy, the nuclear

spin is of importance for the explanation of the hyperfine structure, and of the intensity alternation in band spectra.

NUCLEAR STRUCTURE. The nucleus of an atom of atomic number Z and mass number A contains Z protons and $A-Z$ neutrons, bound together under the influence of shortrange nuclear forces much as molecules are bound together in a drop of liquid. The strength of binding may be determined by subtracting the actual mass of the atom from the mass of its constituent particles considered as free particles. The binding energy E_B is then related to this mass defect ΔM by Einstein's relation, $E_B = \Delta Mc^2$, where c is the velocity of light. The precise value of E_B depends upon the nucleus concerned, and upon how many neutrons and protons it contains but it is of the order of 8 MeV per nucleon in most nuclei.

The binding energy determines whether the nucleus is stable or unstable. Among the lighter nuclei, the ones which are stable are those in which the number of protons is approximately equal to the number of neutrons, so that $A \approx 2Z$. In heavier stable nuclei, there is an excess of neutrons over protons owing to the repulsive electrostatic forces between the protons. Thus, the most stable oxygen nucleus is ^{16}O, containing 8 protons and 8 neutrons, while the most nearly stable uranium nucleus is ^{238}U, containing 92 protons and 146 neutrons. Nuclei containing a disproportionate number of neutrons tend to be unstable and decay radioactively by emission of electrons whereby neutrons are converted into protons; those containing an excess of protons similarly tend to decay by emission of positrons or by capture of orbital electrons.

It has been shown that the nucleus is approximately spherical in shape and of volume proportional approximately to its mass. It is, however, capable of executing oscillations about the spherical form, and in certain circumstances may even acquire a permanent deformation. The heaviest nuclei are unstable under deformation, as a result of which they undergo spontaneous fission. These properties may be described qualitatively by regarding the nucleus as an electrically charged drop of liquid possessing volume energy and surface tension.

Although the nucleus is normally found in its lowest energy state, it may be produced as the result of a nuclear reaction, or through radioactivity in a number of excited states whose detailed properties may differ quite markedly from the lowest state. If formed in an excited state, it will decay, normally by the emission of electromagnetic radiation (gamma rays) to the lowest state, or by the emission of particles to another nucleus.

NUCLEAR TRANSMUTATION. The transformation of one nuclide into another, which differs from it in nuclear charge, mass, or stability, i.e., in a nuclear reaction or a process of radioactivity. Such changes occur in natural radioactive processes, but the general need for a systematic notation for expressing them came only with the great number of transmutations discovered after particle accelerators provided high-energy ion beams capable of penetrating the Coulomb barrier of all stable atomic nuclei.

Two representative transmutation equations are

$$^{27}Al + n \longrightarrow {}^{27}Mg + {}^1H$$

which shows the transmutation of aluminum atoms of mass number 27, by bombardment with neutrons, to magnesium atoms of mass number 27, with the emission of a proton, and

$$^{9}Be + \gamma \longrightarrow {}^8Be + n$$

which shows the transmutation of beryllium atoms of mass number 9, under gamma-ray bombardment, to beryllium atoms of mass number 8, with the emission of a neutron.

The two reactions above may also be expressed in condensed from as

$$^{27}Al(n, p)^{27}Mg$$

and

$$^{9}Be(\gamma, n)^{8}Be.$$

NUCLEATE BOILING. See **Boiler (Steam Generator)**; and **Boiling**.

NUCLEIC ACIDS AND NUCLEOPROTEINS. Nucleic acids are compounds in which phosphoric acid is combined with carbohydrates and with bases derived from purine and pyrimidine. Nucleoproteins are conjugated proteins consisting of a protein moiety and a nucleic acid. Originally, nucleoproteins were thought to occur only in the nuclei of cells, but it was later established that they are far more widely distributed, being found in cells of all types, animal and plant. They are found in the chromosomes, in the genes, in viruses, and bacteriophages.

The protein portion of the nucleoproteins is basic in nature and being complex in structure may form several types of linkage, depending upon the type of nucleic acid. In gastric digestion or hydrolysis with weak acid, nucleoproteins yield protein and nuclein. The latter in pancreatic digestion or hydrolysis with weak alkali yields additional protein and nucleic acid.

Upon additional hydrolysis, nucleic acids yield four characteristic constituent groups: (1) heterocyclic nitrogenous bases of the purine type; (2) heterocyclic nitrogenous bases of the pyrimidine type; (3) a carbohydrate, either ribose or deoxiribose; and (4) phosphoric acid.

Hydrolysis of nucleic acid with enzymes belonging to the group of nucleases gives nucleotides, nucleosides, and the constituent groups already mentioned. Thus, polynucleotidase catalyzes the hydrolysis of nucleic acid to give nucleotides which consist of purine or pyrimidine, ribose or deoxyribose, and phosphoric acid. Nucleotidase catalyzes the hydrolysis of nucleotides to nucleosides (which consist of a purine or a pyrimidine and ribose or deoxyribose) and phosphoric acid. Nucleosidase catalyzes the acid hydrolysis of nucleosides to the respective base or carbohydrate. Since the nucleic acids are composed of nucleotides, they may be considered polymers of nucleotides and thus be called polynucleotides.

There are two groups of nucleic acids differentiated by the carbohydrate present. Those which contain D-ribose,

$$CHO \cdot CHOH \cdot CHOH \cdot CHOH \cdot CH_2OH,$$

are generally known as ribonucleic acids, usually termed RNA. Those which contain D-2-deoxyribose,

$$CHO \cdot CH_2 \cdot CHOH \cdot CHOH \cdot CH_2OH,$$

are usually known as *deoxyribonucleic acids*, generally termed DNA. Sometimes the latter are also termed deoxyribonucleic acid or DRNA.

Deoxyribonucleic acid is found not only in the nucleus of normal cells but is also present in mitochondria of both plants and animals as well as in plant chloroplasts. DNA occurs in circular, double standard structures. Formerly, this type of nucleic acid was thought to be a constituent only of animal cells and was known as thymus nucleic acid. Ribonucleic acid is found principally in the cytoplasm, although small amounts are found the nucleus, nucleoli, and the chromosomes. RNA at one time was known as yeast nucleic acid and was thought to be the characteristic nucleic acid of plants.

Deoxyribonucleic acid is a constituent of the chromosomes of the cell nucleus. Since the number of chromosomes of a cell and its daughter cells are equal, the quantity of DNA in the normal cells of any given species or type should be and is remarkably constant. This quantity is not changed by starvation or other action or form of stress. The quantity in a normal diploid nucleus is twice that of a normal haploid nucleus and in polyploid cells, such as cultivated wheats, the quantity of DNA is some multiple of the quantity in the haploid cell. This quantity is of the order of 6×10^{-9} milligrams per nucleus in the case of mammals. Birds and fishes have lesser amounts.

At one time, the chemical synthesis of DNA and the resultant ability to construct totally synthetic genes appeared to be impractical. However, under the influence of recombinant DNA technology, the field of synthetic DNA has matured. Two chemical methods — diester and triester — and one enzymatic method — polynucleotide phosphorylase — are generally used for the synthesis of oligodeoxyribonucleotide. Once made the resultant gene can be cloned (see also **Clone**) and then used to produce useful peptide products. Products such as insulin and somatostatin have been made in *Escherichia coli* following insertion of the chemically synthesized DNA into the bacterial gene.

The quantity of RNA varies in different tissues. It also varies in amount in the cytoplasm and nucleus of the same cells. The amount of RNA is affected by the nutritional state of the cells, the type of tissue, and the metabolic action in which they participate. Thus, there is more ribonucleic acid in growing embryonic tissue, in tumor tissue, in pancreas, salivary glands, and other gland with secretory functions. All of these indications led to the concept that RNA controls the rate of the production of protein by cells, while DNA controls the transmission of hereditary characteristics from one generation to the next.

At one time, it was believed that nucleic acids were tetranucleotides, but it has been established that they are much larger molecules. Ribonucleic acids have been found to have molecular weights as high as 300,000 (implying approximately 1,000 nucleotides per molecule), while deoxyribonucleic acids have molecular weights of the order of 1 or 2 million. It has been shown that in the living organism, RNA and DNA are long double chains, with each chain spiraling around the other, also a characteristic of some proteins.

Each of the two spiral chains that constitute the DNA molecule is composed of alternating sugar (deoxyribose or ribose) and phosphate groups, and attached to each sugar is one of the four bases, usually adenine, guanine, thymine, and cytosine. (Adenine and guanine are pyrimidines, while thymine and cytosine are purines). Linkage between the two chains is effected by hydrogen bonds formed between the adenine groups of each chain and the thymine groups of the other, and between the guanine groups of each chain and the cytosine groups of the other. Variations occur in the sequence in which these four bases appear in each chain. These variations are the means of transmission of genetic "information," e.g., a basic mechanism of heredity.

Delineation of DNA as the bearer of genetic information stems from the findings of Chargaff et al. that the base composition of DNA is related to the species of origin. Indeed Chargaff's chromatographic studies enabled the establishment of the following conclusions:

1. The base composition of DNA varies from one species to another.
2. DNA specimens isolated from different tissues of the same species have the same base composition.
3. The base composition of DNA in a given species does not change with age, nutritional status, or changes in environment.
4. In nearly all DNAs examined, the number of adenine residues is always equal to the number of thymine residues, i.e., A = T, and the number of guanine residues is always equal to the number of cytosine residues, G = C. As a corollary, it is clear that the sum of purine residues equals the sum of pyrimidine residues, i.e., A + G = C + T.
5. The DNAs extracted from closely related species have similar base composition, whereas those of widely different species are likely to have widely different base composition.

See also **Adenosine**; **Adenosine Phosphates**; **Amino Acids**; **Bacteriophage**; **Cell (Biology)**; **Genetics and Gene Science (Classical)**; **Heredity**; and **Protein**.

ANN C. DeBALDO, Ph.D., College of Public Health, University of South Florida, Tampa, FL.

NUCLEI (Cloud Formation).   See **Atmosphere-Ocean Interface**.

NUCLEI (Sublimation).   See **Precipitation and Hydrometeors**.

NUCLEONICS.   The applications of nuclear science in physics, chemistry, biology, and other sciences, including military science, and in industry, and the techniques associated with these applications.

NUCLEONS.   Two nuclear particles, the proton and the neutron, and their antiparticles are known as *nucleons*. The rest mass of the proton is 1.0076 amu; that of the neutron, 1.0089 amu. The antiproton bears the same relation to the proton that the electron does to the positron, i.e., its charge is equal and opposite and its mass is the same, the charge being equal in magnitude to the electronic charge. Protons and antiprotons also annihilate each other when they collide, the reaction of a single pair producing positive and negative pions or kaons. If the proton and antiproton do not collide, but experience a "near miss," then an exchange of charge can occur, resulting in the formation of a neutron-anti-neutron pair. The four nucleon particles are fermions and have a spin angular momentum quantum number of $\frac{1}{2}$. See also **Neutron**; **Particles (Subatomic)**; and **Proton**.

NUCLEOPHILE.   An ion or molecule that donates a pair of electrons to an atomic nucleus to form a covalent bond. The nucleus that accepts the electrons is called an *electrophile*. This occurs, for example, in the formation of acids and bases according to the Lewis concept, as well as in covalent bonding in organic compounds.

NUCLEOPHILIC REACTION.   A reaction in which a nucleophilic reagent attacks an electrophilic compound. The reagent is taken to be the inorganic substance (in the case of reactions of inorganic and organic substances) or the simpler of two reacting organic compounds. The electron pair for the bond formed is furnished by the nucleophilic reagent.

NUCLEUS. 1. The nucleus of an atom is the positively charged core, with which is associated practically the entire mass of the atom, but only a minute part of its volume.

2. The nucleus of a molecule is a group of atoms connected by valence bonds so that the atoms and their bonds form a ring or closed structure, which persists as a unit through a series of chemical changes.

3. A group of cell bodies in the central nervous system of vertebrates. Examples of such groups are the red nucleus in the midbrain, through which impulses are routed for the control of subconscious muscular movements, and Deiter's nucleus, lying at the junction of the medulla with the hindbrain. Through this center impulses pass for muscular action involved in the maintenance of equilibrium.

4. A somewhat spherical or oblong body in most living cells. This nucleus contains the chromosomes, which, in turn, bear the genes of heredity. The nucleus also contains a nucleolus, or sometimes two or more nucleoli and a basic ground substance, the nucleoplasm. A nuclear membrane surrounds it on the outside, but this membrane is very porous, allowing materials to pass through rather freely.

5. Condensation nuclei take part in phase changes, as in the formation of clouds; in seeding concentrated solutions to bring about crystallization, precipitation, etc.

NUCLIDES. See **Chemical Elements**.

NULL. In direction-finding systems wherein the output amplitude is a function of the direction of arrival of the signal, the minimum output amplitude (ideally zero). The null is frequently employed as a means of determining bearing. The term *minimum* is often used to indicate an imperfect null.

NULL VECTOR. A vector $\mathbf{A}_\mu$ of zero length ($\mathbf{A}_\mu \mathbf{A}_\mu = 0$). In special relativity theory the displacement between two events on the path of a photon is a null vector.

NUMBER THEORY. The theory of numbers is concerned primarily with the properties of the integers, or whole numbers, $0, \pm 1, \pm 2, \ldots$. The integers form a ring, i.e., when one or more of the operations addition, subtraction, and multiplication are applied to any two integers, the result is always an integer.

Numbers can be classified in many ways. For example, odd and even numbers, prime numbers, square numbers, and perfect numbers. These classes of numbers were mentioned by Euclid (circa 300 B.C.). The Chinese knew in 500 B.C. that $2^p - 2$ is divisible by p, for prime numbers p. Euclid proved that there exist infinitely many primes. The proof is by contradiction. Assume p is the largest prime. Let M be the product $2 \cdot 3 \cdot 5 \cdots p$ of all primes up to p. Then $M + 1$ is either a prime number itself, or is divisible only by primes greater than p.

While most results of number theory are easy to understand, it has taken brilliant mathematicians many years to construct proofs of many and some of the most interesting conjectures remain unproved.

Number theory has been extended to the study of the specific properties of other classes of numbers, such as rational, algebraic, and transcendental numbers. Discussed here are elementary number theory, algebraic number theory, analytic number theory, and Diophantine approximations.

Elementary Number Theory. The main concern of this branch of number theory is divisibility. If a and b are integers, and $a = bc$ for some c, then b is a factor or divisor of a, written symbolically as $b|a$. Thus, $1|a$ and $a|a$ for every integer a. A prime number $p > 1$ is an integer that has only 1 and p as divisors. Other integers greater than 1 are called composite.

A concept known as the fundamental theorem of arithmetic which was proved by Euclid states that every integer n greater than 1 can be expressed as a product of primes in only one way when the order of prime factors is disregarded.

A perfect number is an integer that is equal to the sum of its proper divisors less than itself. The five least are: 6; 28; 496; 8,128; 33,550,336. Euclid showed that $(2^n - 1)2^{n-1}$ is a perfect number if and only if 2^{n-1} is a prime number. Numbers of the form $2^p - 1$ are known as Mersenne numbers. All even perfect numbers are of Euclid's type. No odd perfect numbers have been found.

Let m denote a positive integer. Every integer a can be represented in the form $a = qm + r$, where q is an integer and r is an integer which has one of the m values $0, 1, \ldots, m - 1$. Two integers a and b are congruent modulo m if the value of r is the same for a and b when expressed in the above form. Symbolically, $a \equiv b (\text{mod } m)$. If $a \equiv b (\text{mod } m)$, then $m|(a - b)$. The congruence relationship is an equivalence, i.e., it is reflexive, symmetric, and transitive. The relation defines equivalence classes of numbers congruent to each other, which are called residue classes. There are m residue classes modulo m.

The symbol (a, b) is frequently used to denote the greatest common divisor of integers a and b, assumed at least one of which is nonzero. If $(a, b) = 1$, a and b are coprime.

If $a \equiv b (\text{mod } m)$, $(a, m) = (b, m)$. If $(a, m) = 1$, the residue class of a modulo m is said to be prime to m. Those residue classes which are prime to m form a reduced system of residue classes mod m. Euler introduced the function $\Phi(m) = m\Pi_{p/m}(1 - 1/p)$ to denote the number of reduced residue classes modulo m. Some of the properties of Φ are:

$$\Phi(mn) = \Phi(m)\Phi(n) \text{ if } (m, n) = 1$$

and

$$\Phi(p^r) = p^r(1 - 1/p)$$

where p is any prime and r is a positive integer.

Two important congruences are Wilson's Theorem

$$(p - 1)! \equiv -1 (\text{mod } p)$$

and Fermat's theorem

$$a^{p-1} \equiv 1 (\text{mod } p), \text{ where } p \text{ is a prime.}$$

Euler generalized the latter to

$$a^{\Phi(m)} \equiv 1 (\text{mod } m), \text{ where } (a, m) = 1$$

The linear congruence $ax \equiv b (\text{mod } m)$ is solvable for x if and only if $d|b$, where $d = (a, m)$. There are exactly d solutions (incongruent to each other).

If $m = p$ is a prime number then $ax \equiv 1 (\text{mod } p)$ has exactly one solution for each a which is coprime with p. This solution is the reciprocal of a modulo p which shows that residue classes modulo p form a finite field.

The congruence $a_0 x^n + a_1 x^{n-1} + \cdots + a_n \equiv 0 (\text{mod } m)$ is of degree n, if m does not divide a_0. If $m = p$, the number of solutions cannot exceed n but a solution may not exist.

If $(a, m) = 1$, the least positive e such that $a^e \equiv F (\text{mod } m)$ must divide $\Phi(m)$. If $e = \Phi(m)$, a is called a primitive root of m and the powers of $a, a^2, \ldots, a^{e-1}$ form a reduced set of residues modulo m. A modulus m has primitive roots only when $m = 2, 4, p^k, 2p^k$, where p is an odd prime. The period of the periodic decimal into which the reduced fraction a/m can be expanded has the length $\Phi(m)$ or a divisor thereof. It has the length $\Phi(m)$ only when the base (10 for decimal expansion) is a primitive root modulo m.

Fibonacci numbers were discovered by Leonardo of Pisa in 1202 A.D. The first two Fibonacci numbers are 1; subsequent Fibonacci numbers are obtained by adding the previous two Fibonacci numbers. The first seven are 1, 1, 2, 3, 5, 8, and 13. The sequence grows exponentially with the nth number in the sequence being approximately equal to $(1/\sqrt{5})((1 + \sqrt{5})/2)^n$.

A theorem known to the Chinese in ancient times and known as the Chinese remainder theorem can be stated as:

If $(m_i, m_j) = 1$ for $1 \leq i < j \leq n$, then the system of linear congruences

$$x \equiv a_1 (\text{mod } m_1)$$

$$x \equiv a_2 (\text{mod } m_2)$$

$$\cdots$$

$$x \equiv a_n (\text{mod } m_n)$$

has a unique solution modulo m, where $m = \Pi_{i=1}^n m_i$.

Gödel showed how to use this theorem as a coding trick, in which an arbitrary finite sequence of numbers can be encoded as a single number.

In 1970, Yuri Matyasevich completed the proof that no procedure can be devised to determine the existence of a solution to a Diophantine equation. This proof involves the Chinese remainder theorem, Fibonacci numbers, and Diophantine equations. In 1900, David Hilbert posed 23 major problems for mathematicians. Matyasevich's results complete the solution of Hilbert's tenth problem.

If $x^2 \equiv a \pmod{m}$ is solvable, a is called a quadratic residue modulo m, otherwise a quadratic nonresidue. Let $m = p$ be an odd prime. The Legendre symbol (a/p) is defined to be $+1$, if a is a quadratic residue modulo p and as -1 if a is a quadratic nonresidue. It is assumed that $a \not\equiv 0 \pmod{p}$.

The Legendre symbol is a character of the multiplicative group of residue classes modulo p:

$$\left(\frac{a}{p}\right) \cdot \left(\frac{b}{p}\right) = \left(\frac{ab}{p}\right)$$

If q is another odd prime

$$\left(\frac{p}{q}\right)\left(\frac{q}{p}\right) = (-1)^{(p-1)/2 \cdot (q-1)/2}$$

Which is known as the reciprocity law of quadratic residues. Also,

$$\left(\frac{-1}{p}\right) = (-1)^{(p-1)/2}$$

Jacobi generalized the Legendre symbol by defining

$$\left(\frac{a}{p_1 p_2 \cdots p_n}\right) = \left(\frac{a}{p_1}\right)\left(\frac{a}{p_2}\right)\cdots\left(\frac{a}{p_n}\right), \text{ and } \left(\frac{a}{-k}\right) = (ak),$$

if a is positive.

The reciprocity law still holds for any odd numbers p and q not both negative.

An expression, $f(x, y) = ax^2 + bxy + cy^2$, is a binary quadratic form. A number m is represented by the form f if there exists a pair of integers u and v such that $f(u, v) = m$. The representation is called primitive if $(u, v) = 1$. The determinant of the form f is defined as $d = ac - (\frac{1}{4})b^2$. The discriminant $D = -4d$.

The forms, $f = ax^2 + bxy + cy^2$ and $F + AX^2 + BXY + CY^2$, are equivalent if there exists a transformation $x = \alpha x + \beta y$, $Y = \gamma x + \delta y$ with $\alpha\delta - \beta\gamma = \pm 1$, α, β, δ all integers so that $F(X, Y) = f(x, y)$. The determinants of equivalent forms are equal.

Since $af = (ax + \frac{1}{2}by)^2 + dy^2$, if $d > 0$, $a \neq 0$ then af is positive for any x and y not both zero. Therefore, f represents numbers of only one sign, that of a. If $d < 0$, f represents both positive and negative numbers.

For binary forms with a positive determinant, the number of classes of integral, positive-definite, binary quadratic forms with a given determinant is finite.

An automorph of f is a unimodular transformation carrying f into itself. For example, $x = -x_1$, $y = -y_1$.

In the above, a, b, c may denote any real numbers. Now consider f integral; i.e., a, b, c, integers. The greatest common divisor (g.c.d.) of a, b, c is called the divisor of f. If the g.c.d. is 1, f is called primitive.

All the automorphs of the primitive form $ax^2 + bxy + cy^2$ of discriminant D are given by $x = \frac{1}{2}(t - bu)X - cuY$, $y = auX + \frac{1}{2}(t + bu)Y$ where t, u range over all the integral solutions of the so-called Pell's equation, $t^2 - Du^2 = 4$.

It can be shown that if D is a positive nonsquare integer, all integral solutions of Pell's equation are given by $\frac{1}{2}(t + u\sqrt{D}) = \pm[\frac{1}{2}(T \pm U\sqrt{D})]^k$, $k = 0, 1, 2, \ldots$, and T and U are the least solution in positive integers.

Fermat stated and Euler proved that every prime that is congruent to 1 mod 4 is the sum of two squares. For such a prime number, -1 is a quadratic residue. Lagrange found that every natural number is the sum of at most 4 squares.

Algebraic Number Theory. The notions of primality, integers, divisibility, etc., which originated in arithmetic, have been extended to systems other than ordinary integers. The ordinary integers are referred to as rational integers to distinguish them from integers in fields other than the real number field. Gauss defined a complex number as $a = a_0 + a_1 i$, where a_0 and a_1 are rational and $i^2 = -1$. The conjugate of a denoted by a^* is given by $a^* = a_0 - a_1 i$.

The product aa^{**} is rational and is called the norm of a denoted by $N(a)$ which is equal to $a_0^2 + a_1^2$. If $a = bc$, then $a^* = b^* c^*$ and $aa^{**} = (bb^*) \cdot (cc^*)$ which shows that the norm of b is a factor of the norm of a. As with ordinary integers $b|a$ if $a = bc$ in complex integers. A complex integer is a number of the form $a = a_0 + a_1 i$ where a_0 and a_1 are rational integers.

The only complex numbers that divide all complex integers are those of norm 1, ± 1 and $\pm i$. If $b|a$ the associates ub of b, where u is any unit,

are also factors of a. If the complex integer p has no factors other than associates and units, p is called a prime.

The complex number $p = p_0 + p_1 i$ is a complex prime if $N(p)$ is a rational prime such that $N(p) \equiv 1 \pmod{4}$, $p = \pm(1 \pm i)$, or if p is the associate of a rational prime which is congruent to 3 modulo 4. One can prove that $p|ab$ only if $p|a$ or $p|b$ and thus deduce the fundamental theorem of arithmetic for complex integers. Virtually all theorems in the theory of ordinary numbers have analogues for complex numbers.

An algebraic number field $R(\theta)$ of degree n is generated by the root θ of an algebraic equation $a_0 x^u + a_1 x^{u-1} + \cdots a_n = 0$ where all a_i are rational and $a_0 \neq 0$, is an algebraic integer if all a_i are rational integers and $a_0 = 1$. The algebraic integers in an algebraic number field form an integral domain. Consider numbers of the form $a = a_0 + a_1\theta$, where $\theta^2 = -5$. The conjugate a^* and the norm of a are defined in a manner analogous to the complex numbers. The number 21 can be expressed as $21 = (1 + 2\sqrt{-5})(1 - 2\sqrt{-5})$ or as $21 = (4 + \sqrt{-5})(4 - \sqrt{-5})$, where the factors on the right hand side of each equation are prime in this system. This example demonstrates that the fundamental theorem of arithmetic (uniqueness of factorization) does not always hold true for such algebraic number fields.

Uniqueness of factorization can be restored by the introduction of ideals. An ideal a in the algebraic number field $K = R(\theta)$ is a set of integers of K such that: (1) if α and β are in a, then $(\alpha + \beta)$ is in a; and (2) if α is in a and ξ is any algebraic integer in K than $\alpha\xi$ is in a. A number α is replaced by its principal ideal (α) which consists of all numbers $\alpha\xi$, where ξ runs through all integers of K.

Ideals, which were invented for use in number theory, are now used extensively in higher algebra. Other generalizations of numbers occur in the theory of linear algebras. The algebra of quaternions is an example. See also **Quaternion**.

Analytic Number Theory. This involves the use of the methods of calculus and function theory to study the properties of the integers. The most famous problem is to determine the number $\pi(x)$ of prime numbers up to X. Gauss and Legendre suggested asymptotic formulas for $\pi(x)$. The simplest formula is $\pi(x) \sim (x/\log x)$. The symbology $f(x) \sim g(x)$ indicates that $f(x)$ is asymptotic to $g(x)$. Two functions, $f(x)$ and $g(x)$, are said to be asymptotic when the ratio $f(x)/g(x)$ tends to one as a limit as x tends to infinity.

This formula was not proved until 1896. The methods involved the use of the Riemann-Zeta Function $\zeta(s)$ defined by the series $\zeta(x) = \Sigma^\infty 1/n^s$. Where s is a complex variable $s = \sigma + it$. The series is convergent only for $\sigma > 1$ but $\zeta(s)$ can be defined for $\sigma \leq 1$ by using analytic continuation and is found to be a function with only a simple pole of residue 1 at $s = 1$. By using the theorem of unique prime number factorization, it can be shown that $\zeta(x) = \Pi_p 1/(1 - p^{-s})$, where p runs over all prime numbers, $\zeta(s)$ has no zeros for $\sigma = 1$. Riemann knew that $\zeta(s) = 0$ for s equal any negative even integer and that all the remaining zeros satisfied $0 < \sigma < 1$. Riemann conjectured that all the nontrivial zeros of $\zeta(s)$ lie on the line $\sigma = \frac{1}{2}$. This conjecture was contained in a memoir of 1859 and remains unproved.

P.G.L. Dirichlet proved that there exist infinitely many prime numbers which satisfy $p \equiv a \pmod{m}$ where a and m are given coprime numbers. Dirichlet defined a more general set of functions $L(s, x) = \Sigma_{n=1}^\infty X(n)/n^s$ which are known as Dirichlet L-functions. Where $s = \sigma + it$ as before and $X(n)$ is a character of the multiplicative group of residue classes modulo m and coprime to m. Therefore, X satisfies the equations $X(a_1) = X(a_2)$ for $a_1 \equiv a_2 \pmod{m}$ and $X(a)X(b) = X(ab)$. $X(n)$ is zero if n and m contain a common factor greater than 1. $X(n)$ can have values ± 1, 0, or roots of unity.

The conjecture that there are no zeros of the L-functions with real part $\sigma > \frac{1}{2}$ is known as the Generalized Riemann Hypothesis.

Additive Number Theory. Srinivasa Ramanujan was a protegé of G.H. Hardy. Hardy once visited Ramanujan in a hospital in England. Hardy remarked that the taxi in which he came had the rather uninteresting number 1729 for its license number. Ramanujan immediately remarked that 1729 was interesting because it was the smallest positive integer that could be expressed as the sum of two cubes in two different ways, namely $1729 = 12^3 + 1^3 = 10^3 + 9^3$. This is an example of an additive number theory result. Partitions and polygonal numbers, which occur in additive number theory, are discussed below.

A partition of n is a decomposition of the number n into additive parts. Repetitions in the additive parts are allowed and the order is irrelevant.

Let $p(n)$ denote the number of partitions of n. Euler gave

$$1 + \sum_{n=1}^{\infty} p(n)x^n = \prod_{k=1}^{\infty} (1 - x^k)^{-1}$$

as a generating function for $p(n)$. Generating functions can be constructed for partitions that are subject to various restrictions. Euler observed that if $E(n)$ is the number of partitions of n into an even number of unequal parts and $U(n)$ is the number of partitions into an odd number of unequal parts, then $E(n) - U(n) = 0$ unless n is of the form $\left(\frac{1}{2}\right)m(3m \pm 1)$, when $E(n) - U(n) = (-1)^m$.

An asymptotic form for $p(n)$ was developed by Hardy and Ramanujan, the first term of which is

$$p(n) \sim \frac{1}{4n\sqrt{3}} \exp\left(\pi \frac{2n}{\sqrt{3}}\right).$$

Rademacher gave an exact expression for $p(n)$ as the sum of a convergent infinite series. Ramanujan observed that $p(n)$ possesses various congruence properties such as

$$p(5n + 4) \equiv 0 \pmod 5$$

and

$$p(7n + 5) \equiv 0 \pmod 7.$$

A polygonal number of order m is given by $x + \frac{1}{2}(m - 2)(x^2 - x)$, where $x = 0, 1, 2, \ldots$. The value of m is equal to the number of sides of a polygon; i.e., $m = 3$ gives triangular numbers, $m = 4$ gives square numbers, etc. Fermat stated that every positive integer is the sum of m polygonal numbers of order m. This theorem was proved by Legendre for $m = 3$, Lagrange for $m = 4$, and by Cauchy for the remaining values of m.

Polygonal numbers of order 4 are squares of integers. Fermat's theorem says that every positive integer can be expressed as the sum of one, two, three, or four nonzero squares. Waring stated an extension of this theorem to higher powers. Every positive integer can be expressed as the sum of at most $g(k)$ of kth powers. Another function $G(k)$ which gives the least number of kth powers required to represent all but a finite number of positive integers is of even greater interest. Clearly, $G(k) \leq g(k)$. Lagrange's result indicates that $G(2) = g(2) = 4$. Other results include $g(3) = 9$, $G(3) \leq 7$, $19 \leq g(4) \leq 35$, $G(4) = 16$, $37 \leq g(5) \leq 54$, and $G(5) \leq 23$. Vinogradow proved that $G(k) \leq k(3 \log k + 10)$.

Vinogradow also proved that every sufficiently large odd number is the sum of three odd primes.

Diophantine Approximations. Every number processed by a computer or written in decimal form with a finite number of digits is a rational number. Therefore, there is some practical interest in the errors introduced by approximating irrational numbers by rational fractions. By using continued fractions, it can be shown that if θ is any irrational number, then there are infinitely many fractions p/q such that $|\theta - p/q| < 1/q^2$. A continued fraction is an expression of the form

$$a_1 + \cfrac{b_1}{a_2 + \cfrac{b_2}{a_3 + \cdots}}$$

which is abbreviated $a_1 + (b_1/a_2) + (b_2/a_3) + \cdots$. If all b_i's are equal to 1, the continued fraction is called a simple continued fraction. The simple continued fraction

$$a_1 + \cfrac{1}{a_2 + \cfrac{1}{a_3 + \cdots}}$$

is sometimes abbreviated as $[a_1, a_2, a_3, \ldots]$. The finite simple continued fractions

$$c_1 = [a_1] = a_1$$

$$c_2 = [a_1, a_2] = a_1 + \frac{1}{a_2}$$

$$c_3 = [a_1, a_2, a_3] = a_1 + \cfrac{1}{a_2 + \cfrac{1}{a_3}}$$

are called the convergents of the simple continued fraction $[a_1, a_2, \ldots a_k]$.

In general, it can be shown that $c_1 < c_3 < c_5 \cdots < x < \cdots c_3 > c_4 > c_2$, where x is any irrational number. If x is rational the last convergent is equal to x (in this case the continued fraction is finite).

A periodic continued fraction is a continued fraction of the form $[a_1, a_2, \ldots, a_n, \overline{b_1, b_2, \ldots b_m}]$ where n is a nonnegative integer and m is a positive integer. The period of the continued fraction is the sequence of repeating terms $b_1, b_2, \ldots b_m$. The length of the period is m.

A quadratic irrational is an irrational number that is a solution of $ax^2 + bx + c = 0$ where a, b, and c are integers. Every periodic continued fraction represents a quadratic irrational and every quadratic irrational can be expressed as a periodic continued fraction. It can be shown that if k is not a square then $\sqrt{k}$ can be expressed as

$$\sqrt{k} = [a_1, \overline{a_2, a_3, a_4, \ldots, a_4, a_3, a_2, a_1}].$$

Liouville proved that if ξ is any real algebraic number of degree n and k is any constant, then there exist at most a finite number of rational fractions p/q such that $|p/q - \xi| < (k/q^{n+1})$. This result can be used to construct transcendental numbers. Choose a number λ with a sufficiently rapid sequence of rational approximations such that there are an infinite number of rational fractions p/q that satisfy $|p/q - \lambda| < (k/q^{n+1})$.

DONALD R. HODGE, The BDM Corporation, Vienna, VA.

NUMERICAL CONTROL. Since the Industrial Revolution (circa 1760), there has been a continuing search for more effective manufacturing methods. As part of this unrelenting process, numerically controlled (NC) machines were developed in the early 1960s and have been and continue to be widely used. These developments occurred before the widespread use of electronic computers, and, of course, the personal computer was unknown at that time. The simplest NC machines of the 1960s were *open-loop* — that is, the two or three axes of a machine tool were directed to manipulate the machining of a part in two or three dimensions by controlling the axes of the machine. There was no feedback to the controller. See Fig. 1. However, *closed-loop* systems with feedback to the controller were introduced shortly thereafter. See Fig. 2. Numerically coded coordinate data were fed to the early machines by means of punched paper (later mylar) tape. The tapes proved to be cumbersome to prepare and very difficult to store and retrieve over long periods of time. As a consequence of these inconveniences, but limited to rather large installations, data were transmitted to the controller by means of digitized electric signals. Unfortunately, these modernized early systems were limited to sharing mainframe computers because smaller computers with the required capacity were still unavailable. Later, less expensive computers with the required data-handling capacity were introduced. These led to the tying of NC systems with CAD/CAM data centers. In what is now termed *distributed numerical control* (DNC), one computer can feed part programs to *multiple* machine tools. The advent of mini- and microcomputers, plus the ability to address multiple serial ports on a real-time basis, gained acceptance of this technology by manufacturers. Contemporary NC systems continue to cut down job lot sizes, reduce setup time, and trim machining time and the skill level of the direct labor force. It should be emphasized further that NC systems are now tied together at the factory floor level by *local area networks* and that these, in turn, can receive and feed information to plant-wide and corporate-wide networks.

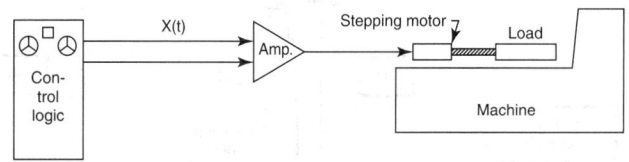

Fig. 1.   Open-loop numerical control system.

NC Components. In the early days of applying NC, machines were not designed with NC in mind and lacked such characteristics as stiffness, accuracy obtainable by a highly skilled worker, and often the required production speed. Today, many machines are designed specifically with NC

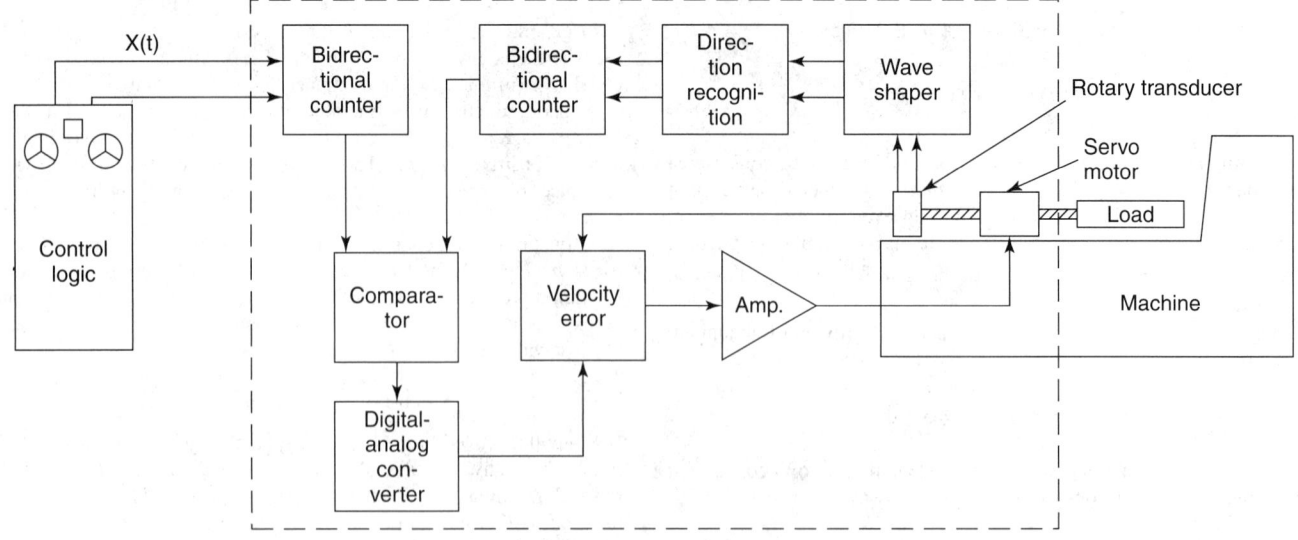

Fig. 2.   Closed-loop numerical control system.

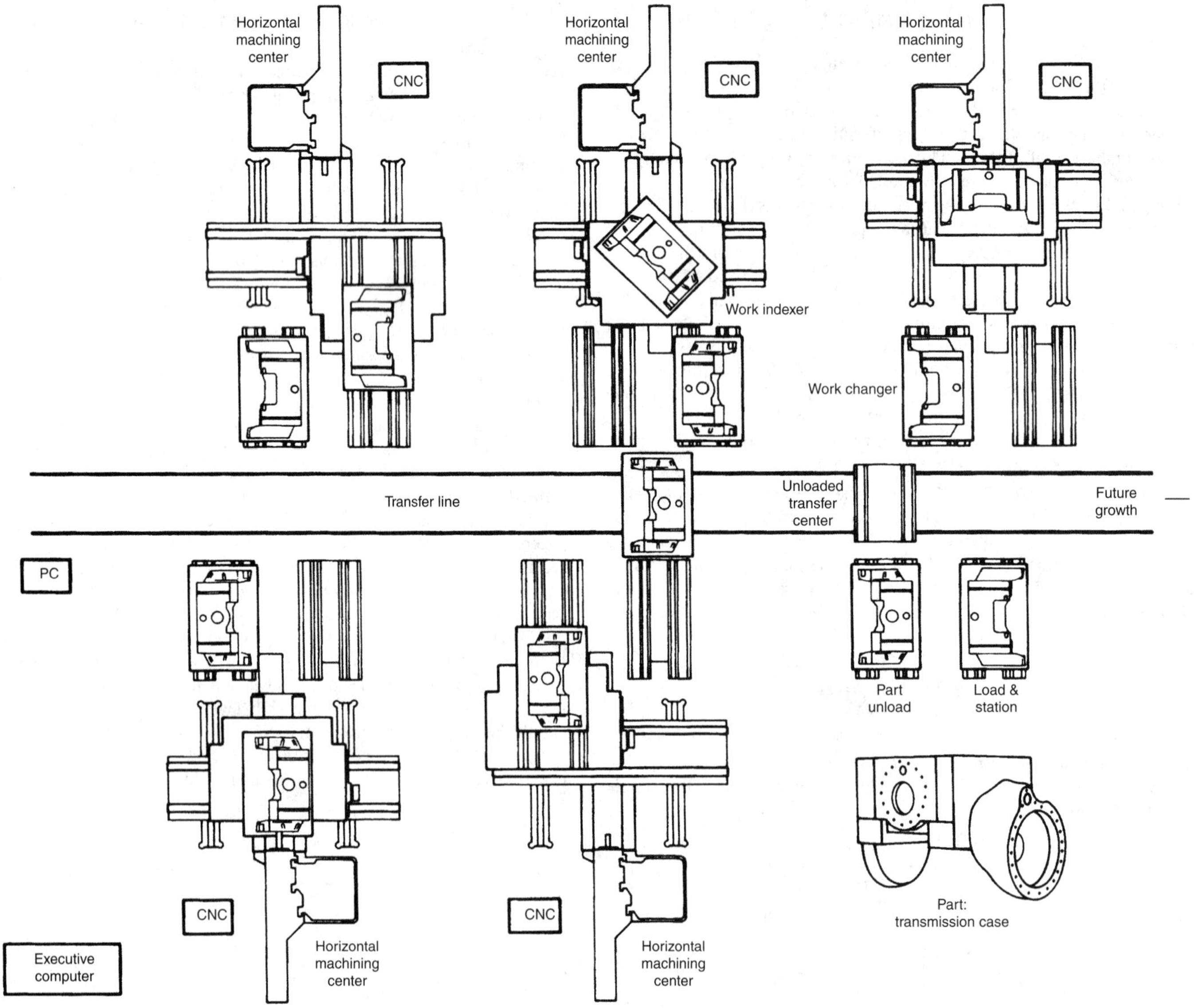

Horizontal machining center

CNC

Horizontal machining center

CNC

Work indexer

Horizontal machining center

CNC

Work changer

Transfer line

Unloaded transfer center

Future growth

PC

Part unload

Load & station

CNC

CNC

Horizontal machining center

Horizontal machining center

Executive computer

Part: transmission case

Fig. 3.   Some manufacturing systems are basically machining centers equipped with work handling options and joined by transfer equipment. Each machining center in itself is a minisystem that processes parts through a single station rather than through sequential stations. PC = programmable controller; CNC = computerized numerical control. (*Giddings & Lewis.*)

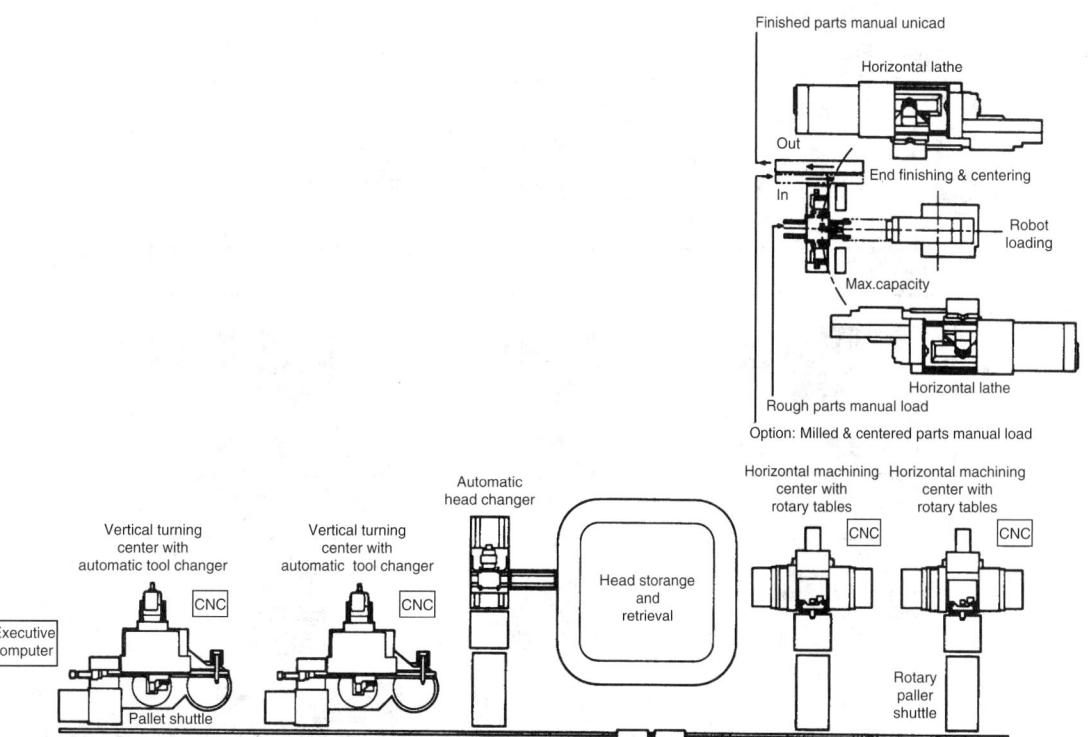

Fig. 4.   Mid-volume manufacturing is frequently best served with general-purpose equipment and controls arranged as production modules. Thus, familiar and proven parts and components can be programmed by a master computer that has been preprogrammed for any number of parts. (*Giddings & Lewis.*)

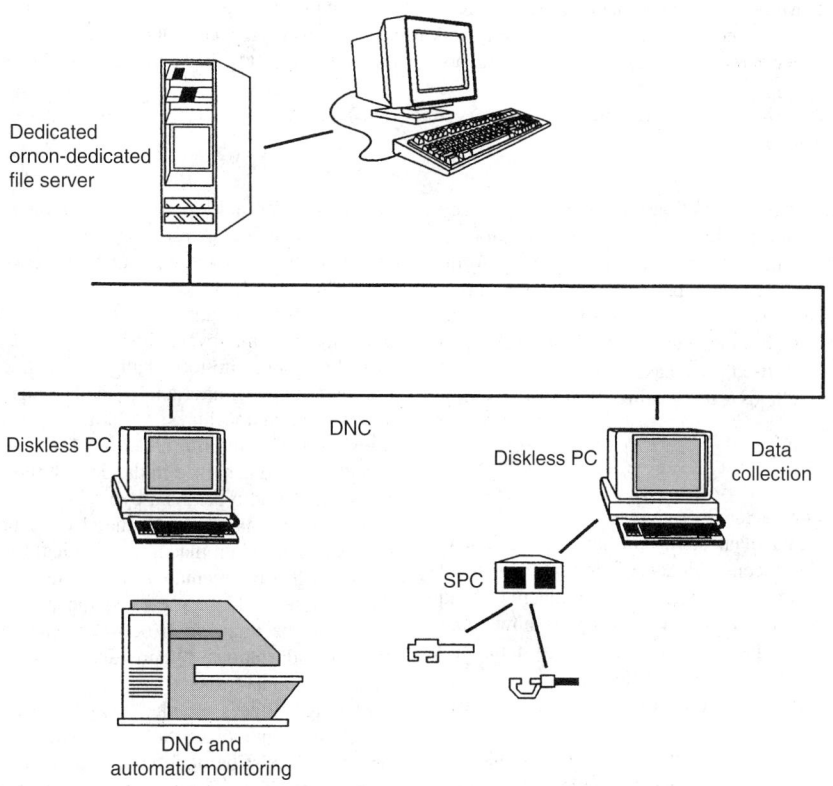

Fig. 5.   Entry-level network for direct numerical control (DNC).

Basic system: (1) Entry-level standard network, (2) dedicated or nondedicated file server, (3) based on Intel 286 microprocessor, (4) supports up to five active users, (5) controls application information over network, (6) applications can be integrated in seamless environment through factory network control system.

Features: (1) High performance — files are transferred at 10 Mbs, (2) easy-to-use menu-driven utilities put the network supervisor in control, (3) security — allows supervisor to restrict network access, (4) single-source database for complete control of all files, (5) cost-effective network uses diskless PCs, (6) virtually unlimited expansion capabilities.

Applications: Statistical process control (SPC), (2) direct numerical control (DNC), (3) data collection, (4) view graphics and documentation, (5) automatic monitoring. (*CAD/CAM Integration, Inc.*)

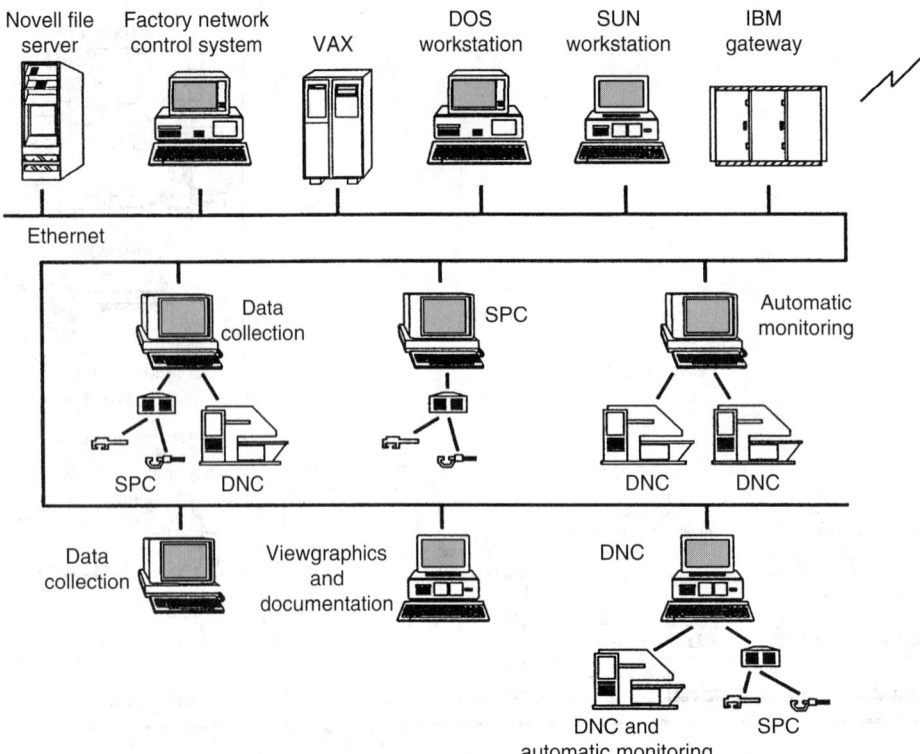

Fig. 6.   Advanced NC network.

Basic system: (1) Controls application information of network, (2) applications can be integrated in seamless environment through factory network control system, (3) designed for medium to large business, (4) dedicated 386-based file server technology, (5) supports up to 100 active users on network, (6) high-performance network operating system.

Features: (1) Provides high functionality while maintaining high performance levels, (2) multiuser, multi-tasking architecture allows user to perform many operations simultaneously, (3) enhanced features for security, system reliability, and network management, (4) cell configurations decrease exposure to manufacturing down-time, (5) single-source database for complete control of files, (6) cost-effective network—uses diskless PCs, (7) simple-to-use menu-driven utilities make learning and adding applications easy.

Applications: (1) Statistical process control (SPC), (2) direct numerical control (DNC), (3) data collection, (4) view graphics and documentation, (5) automatic monitoring. (*CAD/CAM Integration, Inc.*)

as an objective. Much attention is paid to position measurement accuracy and by devices that have electronic outputs. Two types of positioning devices have emerged. They are available in both linear and rotary form. Linear devices possess very good accuracy because they are a direct measure of axis position, but they also are somewhat difficult to mount and protect. Rotary types are more compact, but may be more subject to wear and accuracy deterioration because they measure position by way of lead screws or rack-and-pinion methods with gearing. See also **Position and Displacement Measurement**.

Designing for NC Production

Although NC has been a major factor in achieving flexible factory automation, this does not mean that plant engineers have carte blanche in terms of designing parts without consideration of currently installed manufacturing methods. Indeed, for the high levels of productivity and quality needed for a business to survive, this knowledge is now more important than ever. While cells and flexible manufacturing systems add to the flexibility of a plant, they also have considerations that the designer needs to be aware of in order to make optimum use of these high-priced installations.

The size and shape of the part are among the most obvious and important criteria. Most manufacturers define their machine capabilities in terms of "cube" of workspace that can be accommodated. This does not mean that the overall part size must fit into this cube, but rather that machining operations must be limited to it. Workpiece weight is another consideration for table-type machines. Where floor-type machines are available, these problems are less critical. Cells and flexible manufacturing systems introduce the additional consideration of the material handling equipment. Size and weight are particularly important where parts are to be moved around automatically, and these limitations must also be known by the designer.

The number and type of cutting tools required to machine a workpiece must also be in the forefront of the designer's mind. The capacity of automatic tool-change equipment ranges from about ten to one-hundred tools. Significant increases in manufacturing efficiency can be achieved if these do not require frequent change. In particular, tool-change time and thus machine-cycle time is reduced and errors caused by incorrect tools being put into tool-changer sockets can be eliminated. These factors are particularly important in cells and flexible manufacturing systems where several machines operating on several parts or families of parts are simultaneously in production. The major saving in tool quantity is still probably to be achieved in the area of tapped holes that often require three tools per thread type. As a byproduct of saving tool magazine space, tool-change time is also reduced, thereby cutting down on total machining time.

Another problem made more critical by the advent of multiple machine cells and flexible manufacturing systems is that of holding the workpieces for machining. Not only must the part be held while being cut, but it also must be transported between machines and the system or cell load and unload stations. Where designers can help is by keeping required machining operations to as few sides of the part as possible and by providing surfaces or other features to locate and clamp on quickly and effectively. In most systems, all load and unload as well as refixturing operations must be done at special stations and not on the machine tools themselves. Thus, each refixturing requires the part leaving a machine, traveling to a special station, and then traveling back again to continue being machined. This involves significant time as well as making the material transporter unavailable for other operations and can have an important detrimental effect on the efficiency of the system as a whole.

The designer must also know the rotary axis capability of the manufacturing plant. Many rotary Tables can only position in 15-degree steps, and any subdivisions beyond this angle can result in severe difficulties for the shop floor. The cost increment between Tables with indexing as opposed to full positioning capability to any angle can be very large, and thus designs that call for this requirement in order to be manufactured may be very expensive.

The designer must also keep in mind the part programmer and the capabilities available to simplify the task of producing effective NC tapes. Features such as origin shift, subroutines, probing, etc., previously mentioned are targeted at making programming easier in certain commonly found manufacturing situations. By designing the part to take advantage of these features, the designer can enhance the success of the manufacturing process.

Numerical control can be an effective production tool, not only for the very large manufacturing complexes, but for smaller shops as well. A panorama of contemporary systems is shown by Figures 3 through 6.

NUSSELT NUMBER.   The nondimensional parameter, defined as

$$N_u = \frac{Q}{\Delta T} \frac{d}{k}$$

where Q is the heat loss or heat transfer from a solid body, ΔT is the difference of temperature between the body and its surroundings, d is the scale size of the body, k is the thermal conductivity of the surrounding fluid. The Nusselt number is useful in the reduction of measurements of free and forced convective loss of heat either from the same body in different conditions or from different bodies of geometrically similar shapes. See also **Heat Transfer**.

NUTATION.   In the case of a spinning object (e.g., a top or gyroscope, a particle, or an astronomical body), the inclination of the axis to the vertical will vary periodically between certain limiting angles. This motion is called nutation. It is a variation in precession. In astronomy, nutation is caused by the attracting force of the sun and the moon tending to pull the equatorial bulge of the earth into the plane of the ecliptic. The amount of this force is changing throughout the year as the declinations of the sun and the moon change. For example, twice during each year, both the sun and the moon are on the equator, and at those times, their precessional forces are zero. The principal nutation is due to the periodic change in the plane of the moon's orbit, and has a period of about 19 years. Most of the nutation effects are periodic in character.

NUTHATCH *(Aves, Passeriformes)*.   Small climbing birds which cling in any position to the bark of trees as they search for food. They eat both insects and seeds. The most common North American species is the white-breasted nuthatch, *Sitta carolinensis*, found throughout the United States east of the Rockies. The red-breasted nuthatch, *S. canadensis*, is more common in Canada and the mountains but migrates in winter as far as the southern states. The pigmy nuthatch, *S. pygmaea*, is a western species. Nuthatches are found on all other continents except South America, although in Africa they are confined to the north.

NUTMEG TREE.   Of the family *Myristicaceae*, the nutmeg tree grows to about 60 feet (18 meters) in height and has pointed lanceolate leaves. Nutmegs are the seeds of this tree which is native to the Molucca Islands. The trees are dioecious, pistillate and staminate flowers being borne on separate trees. Since the trees are frequently grown in cultivation, especially in favorable localities, it is necessary to plant some of both sexes to ensure cross-pollination and seed formation. The flowers are pale yellow, the fruit a dark orange-colored berry containing a single large brown seed. Surrounding the seed is a branched deep-red aril, which upon drying

becomes pale brown. The seed is the nutmeg of commerce; the aril is mace. Both the seed and the aril contain an aromatic oil, but only the poor-quality seeds are used for extraction of oil. The tree begins to bear at about 6 years of age and may produce for nearly a century. The tree produces best when planted at an altitude of from 700 to 1,800 feet (213 to 549 meters) in a tropical climate. The average tree will produce about 20 pounds (9 kilograms) of nutmegs per year. An acre produces about 1,200 pounds (544 kilograms) of green nutmegs annually. From this quantity, about 150 pounds (68 kilograms) of mace are produced; and 720 pounds (327 kilograms) of dried nutmegs. See also **Flavorings**.

NUTRIENTS (Soil).   See **Fertilizer**.

NUX VOMICA TREE.   Of the family *Loganiaceae*, the nux vomica tree is a small-to-moderate size tree found in India, Sri Lanka, Burma, Thailand, and Australia. The trunk is often crooked, short and thick. The leaf is smooth with three to five veins and ovate. The flower is light green, small and of a tubular shape. The fruit is hard with a gelatinous pulp. The tangerine-size fruit contains from one to five disk-shaped seeds. The seeds contain powerful alkaloids, including strychnine and brucine.

NYMPH.   If the wing pads of an insect are developed on the outside of its body, the insect is said to have a *simple* or *gradual metamorphosis*. The insect during this growing stage is called a *nymph*. This situation is contrasted with the case where the wing pads are developed internally during the growing stage. In this case, the growing stage is called a *larva*, and the insect is said to have a *complete* or *complex metamorphosis*. Nymphs generally appear something like miniature adults and have the same general life style. In most cases of larva, the size, shape, locomotion, and eating habits of the larva contrast sharply with the adult form. See also **Larva**.

NYQUIST FREQUENCY.   For data defined at equal time-intervals t, the frequency of a sine or cosine term with a period double the interval t. Frequencies higher than this will not be directly detectable by spectral analysis.

NYQUIST, HARRY (1889–1976).   Nyquist, born in Sweden, was an American engineer, mathematician and scientist. His work on frequency response has been used in analysis and design of electronic amplifiers and servomechanisms.

See also **Frequency Response**; **Nyquist Frequency**; **Nyquist Rate (Signaling)**; and **Stability (System)**.

J. M. I.

NYQUIST RATE (Signaling).   In transmission, if the essential frequency range is limited to B cycles per second, $2B$ is the maximum number of code elements per second that can be unambiguously resolved, assuming the peak interference is less than half a quantum step. This rate is generally referred to as signaling at the Nyquist rate, and $\frac{1}{2}B$ is called the Nyquist interval.

NYSTAGMUS.   See **Vision and the Eye**.

NYTRIL.   See **Fibers**.

O

OAK TREES.   Of the *Fagaceae* family (beech family), the oaks are trees and shrubs (*Quercus*) essentially of the north temperate region. All of the more northern species are deciduous plants. Many of those in the southern part of the range have evergreen leaves and often are called live oaks. In Asia and the Pacific coast of North America, oaks are found in regions approaching tropical conditions. The oaks make up the greater part of the beech family.

The oak sends down a root system as deep as 15 feet (4.5 meters) into the ground. The over 300 species of oaks have simple alternate leaves. The flowers are of two kinds, borne on the same tree. The pistillate flowers are borne singly and are surrounded by an involucre of many scales beyond which the stigmas protrude. The staminate flowers are borne in long slender pendant catkins. Pollination is by wind. The fruit is an acorn, a nut of characteristically cylindrical shape, capped by the small persistent style-base, and seated in the scaly involucre, which forms a cup partially or almost wholly surrounding the nut.

Many of the oaks are valuable trees, yielding woods that have a variety of uses. In early times, before the day of the sawmill, oaks were much used in the construction of buildings. Often the oaks used for this purpose were split into thin planks, a method of preparation that served well to bring out the attractive grain of the wood. This grain is due partly to the numerous large vessels which are formed periodically every spring and appear as very evident dark lines or streaks in the wood, and partly to the large vascular rays which appear as irregular flakes, especially when the wood is split in a radial plane. In modern construction, oak is often used as paneling or flooring. To obtain the best grain, the wood is quarter-sawed, that is, cut in such a way that the flat surfaces shall be as nearly radial as is possible. Because of its beauty and also its durability, oak wood is also much used in furniture making. In America the principal species used for wood is white oak, *Quercus alba*. In Europe several species are used, among them the English oak, *Quercus robur*. Often these European oaks are trees of remarkable size, and are preserved because of their rugged beauty. The wood is very strong and durable, and finds considerable use in ship construction. Formerly much more was used for this purpose.

Some engineering properties of oak wood are given in Table 1.

Accidents sometimes cause the formation of oak wood of special properties and value. The trunks of fallen trees may lie buried for long periods of time in bogs or elsewhere. Sometimes, when removed, these logs are found to be perfectly sound and to have developed a rich dark brown or nearly black color, which makes them especially sought after for furniture making. Such oak is known as bog oak. Living trees frequently develop large irregular growths, known as burls, in which a very irregular much-contorted grain is found. The custom of cutting back the top of the tree, causing the development of numerous adventitious buds, a practice known as pollarding, causes a similar irregular grain. These burls are used for making veneers. Another species of oak, *Quercus Suber*, yields cork.

Oaks are valuable sources of tannin. In many species the bark is the source of the tannin, but in *Quercus aegilops*, a native of Eastern Europe and Asia, a tannin known as valonia is obtained from the cup and the young acorns.

African oak, a strong, heavy wood, comes from African trees of other genera than *Quercus*. This wood is rarely used, due to the difficulty of removing the heavy wood from its native forest.

Oak wilt, a serious disease caused by the fungus *Endoconidiophora fagacearum*, has become widespread in the Central and Eastern States. Great damage to the oak forests there is threatened unless some means of controlling the disease can be found.

Periodically, American Forests revises its National Register of Big Trees. The numerous kinds of oaks holding records in the United States (1986) are given in Table 2. The record holding blue oak is shown in Fig. 1 on p. 2517; the swamp white oak in Fig. 2 on p. 2517.

White Oak. Considered by many foresters as the outstanding tree of the many oaks, the white oak (*Quercus alba*) receives its name from its light-colored bark. The white oak occurs widely throughout the eastern United States, ranging from central Maine to northern Florida, and west from southern Quebec through southern Ontario and southern Michigan, through Wisconsin, southeastern Minnesota, much of Iowa, eastern Kansas, eastern Oklahoma to eastern Texas, excluding a narrow belt along the Gulf of Mexico. The white oak prefers well-drained soil. The largest white oaks are found in the valleys of the western slopes of the Allegheny Mountains as well as the bottom lands of the lower Ohio Basin. Some white oaks live for several centuries, a few are known to date back over 800 years.

Black Oak. This is another common and large oak which is found in the eastern United States, ranging from southern Maine and northern Vermont westward through southern Ontario, southern Michigan to southeastern Minnesota, and south to northern Florida and eastern Texas, eastern Oklahoma, eastern Kansas, southeastern Nebraska and Iowa. The black (*Quercus velutina*) oak prefers rich, well-drained, gravelly soils. However, it is not usually found in great numbers in areas with exceptionally rich soils because of the black oak's intolerance for shade caused by competitive trees that thrive in such soils. The ash and tuliptree are frequently found associated with the black oak.

Pin Oak. The natural range of the pin oak (*Quercus palustris*) extends from southwestern New England to northern North Carolina, and from Ohio to Kentucky and western Tennessee. Its distribution also includes southeastern Iowa, eastern Kansas, northeastern Oklahoma and northern Arkansas. The pin oak is not considered an important source of lumber, but rather it is preferred for urban and ornamental plantings. If encountered in normal logging operations, however, it may be marketed as red oak.

Northern Red Oak. The range of this tree (*Quercus rubra*) extends from Maine south and westward, including New England, New York, Pennsylvania, Maryland, Delaware, Virginia and West Virginia, Ohio, Michigan, Indiana, Wisconsin, all but extreme western Minnesota, all of Iowa except the northwest corner, all of Missouri, Illinois, Kentucky, Tennessee, all but the southern portion of Alabama, the northwestern portions of Georgia and South Carolina, and much of North Carolina except the coastal and southeastern portions of that state. Timber from the tree is widely used for general construction, flooring, furniture, railroad ties, posts, and poles, and interior finish. It is estimated that West Virginia, North Carolina, and Tennessee probably have the largest stands.

Bur or Mossycup Oak. This is generally considered second to the white oak in terms of size and grandeur. The tree (*Quercus macrocarpa*) is predominant in the midwestern United States and southern Quebec, Ontario, Manitoba, and parts of Nova Scotia in Canada. Although the bur oak is found in New England, its principal range includes New York, Michigan Ohio, Indiana, Illinois, much of Kentucky and a small part of Tennessee, as well as North and South Dakota, Minnesota, Wisconsin, Iowa, and Missouri. The tree also is found in eastern Nebraska, eastern Kansas, much of Oklahoma except the extreme western portion of that state, and in a north-south corridor of Texas, essentially in the central portion of that state. The bur oak grows slowly and some authorities do not consider it mature prior to an age of 200 to 300 years. The tree is popular for urban plantings and has demonstrated an unusual ability to withstand the smoke and pollution of cities.

NOTE: A concise and convenient reference that details the major oaks (and over a score of other principal trees found in the United States) is "Knowing Your Trees," edited by G.H. Collingwood, Warren

TABLE 1. SELECTED ENGINEERING PARAMETERS OF OAK WOODS

Parameter	Red Oak		White Oak	
	Green	Oven-Dried	Green	Oven-Dried
Moisture content, %	80	12	70	12
Specific Gravity	—	0.66	—	0.70
Modulus of rupture, psi	8500	14,400	8100	13,900
MPa	58.7	99.4	55.9	95.9
Modulus of elasticity, psi	1360	1810	1200	1620
MPa	9.4	12.5	8.3	11.2
Crushing strength (compression parallel to grain), psi	3520	6920	3520	7040
MPa	24.3	47.7	24.3	48.6
Shear, psi	1220	1830	1270	1890
MPa	8.4	12.6	8.8	13.0
Compressive strength (compression perpendicular to grain), psi	800	1260	850	1410
MPa	5.5	8.7	5.9	9.7
Tensil strength (perpendicular to grain), psi	740	760	820	770
MPa	5.1	5.2	5.7	5.3
End hardness, pounds	1050	1490	1110	1300
kilograms	476	676	503	590
Side hardness, pounds	1030	1300	1070	1330
kilograms	467	590	485	603

Source: U.S. Forest Products Laboratory.

TABLE 2. RECORD OAK TREES IN THE UNITED STATES[1]

Specimen	Circumference[2]		Height		Spread		Location
	Inches	Centimeters	Feet	Meters	Feet	Meters	
Ajo oak (1998) (*Quercus turbinella* var. *ajoensis*)	82	208	32	9.8	40	12.2	Arizona
Arizona white oak (1999) (*Quercus arizonica*)	133	338	56	17.1	47	14.3	Arizona
Arkansas oak (1996) (*Quercus arkansana*)	140	356	88	26.8	100	30.5	Mississippi
Bear oak (1992) (*Quercus ilicifolia*)	34	86	41	12.5	30	9.1	West Virginia
Bigelow oak (1999) (*Quercus durandii* var. *breviloba*)	111	282	54	16.5	38	11.6	Texas
Black oak (1999) (*Quercus velutina*)	322	818	86	26.2	105	32	Connecticut
Blackjack oak (1999) (*Quercus marilandica*)	144	366	94	28.7	65	19.8	Georgia
Blue oak (1974) (*Quercus douglasii*)	243	617	94	28.7	48	14.6	California
Bluejack oak (1992) (*Quercus incana*)	119	302	54	16.5	54	16.5	Florida
Bur oak (1995) (*Quercus macrocarpa*)	322	818	96	29.3	103	31.4	Kentucky
California black oak (1972) (*Quercus kelloggii*)	338	859	124	37.8	115	35.1	Oregon
Canyon live oak (1998) (*Quercus chrysolepis*)	422	1072	95	29	126	38.4	California
Chapman oak (1989) (*Quercus chapmanii*)	81	206	45	13.7	50	15.2	Florida
Cherrybark oak (1991) (*Quercus falcata* var. *pagodaefolia*)	324	823	124	37.8	136	41.5	Virginia
Cherrybark oak (1993) (*Quercus falcata* var. *pagodifolia*)	342	869	110	33.5	108	32.9	Virginia
Chestnut oak (1997) (*Quercus prinus*)	222	564	144	43.9	70	21.3	Tennessee
Chinkapin oak (1995) (*Quercus muehlenbergii*)	258	655	110	33.5	92	28	Kentucky
Chisos oak (1982) (*Quercus graciliformis*)	65	165	66	20.1	36	11	Texas

TABLE 2. (*Continued*)

Specimen	Circumference[2]		Height		Spread		Location
	Inches	Centimeters	Feet	Meters	Feet	Meters	
Coast live oak (1999) (*Quercus agrifolia*)	338	859	58	17.7	75	22.9	California
Darlington oak (1992) (*Quercus hemisphaerica*)	234	594	96	29.3	95	29	Georgia
Delta post oak (1988) (*Quercus stellata* var. *paludosa*)	118	300	108	32.9	56	17.1	Texas
Dunn oak (1995) (*Quercus dunnii*)	83	211	37	11.3	36	11	Arizona
Dunn oak (1999) (*Quercus dunnii*)	85	216	35	10.7	40	12.2	California
Durand oak (typ.) (1997) (*Quercus durandii* var. *durandii*)	189	480	95	29	106	32.3	Georgia
Emory oak (1998) (*Quercus emaryi*)	192	488	54	16.5	86	26.2	Arizona
Emory oak (1993) (*Quercus emaryi*)	186	472	56	17.1	92	28	Arizona
Engelmann oak (1968) (*Quercus engelmanni*)	129	328	78	23.8	100	30.5	California
English oak (1997) (*Quercus robur*)	187	475	88	26.8	91	27.7	Ohio
English oak (1993) (*Quercus robur*)	178	452	102	31.1	89	27.1	Washington
Gambel oak (1981) (*Quercus gambelii*)	216	549	47	14.3	85	25.9	New Mexico
Georgia oak (1999) (*Quercus georgiana*)	73	185	75	22.9	63	19.2	Georgia
Graves oak (1982) (*Quercus gravesii*)	154	391	42	12.8	40	12.2	Texas
Graves oak (1976) (*Quercus gravesii*)	145	368	51	15.5	41	12.5	Texas
Gray oak (1993) (*Quercus grisea*)	216	549	45	13.7	73	22.3	New Mexico
Havard oak (1986) (*Quercus havardii*)	40	102	30	9.1	23	7	Texas
Interior live oak (1982) (*Quercus wislizeni*)	268	681	90	27.4	69	21	California
Lacey oak (1989) (*Quercus glaucoides*)	107	272	58	17.7	96	29.3	Texas
Laurel oak (1993) (*Quercus laurifolia*)	267	678	93	28.3	122	37.2	Alabama
Live oak (typ.) (1976) (*Quercus virginiana* var. *virginiana*)	439	1115	55	16.8	132	40.2	Louisiana
Mexican blue oak (1999) (*Quercus oblongifolia*)	120	305	65	19.8	69	21	New Mexico
Myrtle oak (1986) (*Quercus myrtifolia*)	69	175	36	11	35	10.7	Florida
Netleaf oak (1998) (*Quercus rugosa*)	88	224	47	14.3	36	11	Arizona
Northern pin oak (1999) (*Quercus ellipsaidalis*)	184	467	128	39	92	28	Ohio
Northern red oak (1999) (*Quercus rubra*)	294	747	98	29.9	97	29.6	Massachusetts
Northern red oak (1997) (*Quercus rubra*)	257	653	134	40.8	81	24.7	North Carolina
Nuttall oak (1991) (*Quercus nuttallii*)	280	711	118	36	85	25.9	Louisiana
Oglethorpe oak (1999) (*Quercus oglethorpensis*)	117	297	69	21	69	21	Georgia
Overcup oak (1987) (*Quercus lyrata*)	258	655	156	47.5	120	36.6	North Carolina

(*continued*)

TABLE 2. (*Continued*)

Specimen	Circumference[2]		Height		Spread		Location
	Inches	Centimeters	Feet	Meters	Feet	Meters	
Pin oak (1991) (*Quercus palustris*)	240	610	110	33.5	112	34.1	Tennessee
Post oak (typ.) (1987) (*Quercus stellata* var. *stellata*)	236	599	85	25.9	88	26.8	Virginia
Post oak (typ.) (1996) (*Quercus stellata* var. *stellata*)	237	602	84	25.6	88	26.8	Georgia
Sand live oak (1995) (*Quercus virginiana* var. *geminata*)	189	480	81	24.7	106	32.3	Florida
Sand live oak (1995) (*Quercus virginiana* var. *geminata*)	181	460	94	28.7	100	30.5	Florida
Sand post oak (1995) (*Quercus stellata* var. *margaretta*)	157	399	87	26.5	92	28	Florida
Scarlet oak (1995) (*Quercus caccinea*)	248	630	120	36.6	93	28.3	Kentucky
Shingle oak (1997) (*Quercus imbricaria*)	208	528	105	32	62	18.9	Ohio
Shumard oak (typ.) (1994) (*Quercus shumardii* var. *shumardii*)	249	632	190	57.9	88	26.8	Tennessee
Silverlear oak (1994) (*Quercus hypoleucoides*)	123	312	69	21	52	15.8	Arizona
Southern red (typ.) (1999) (*Quercus falcata* var, *falcata*)	312	792	150	45.7	156	47.5	Georgia
Swamp chestnut oak (1998) (*Quercus michauxii*)	276	701	105	32	216	65.8	Tennessee
Swamp chestnut oak (1989) (*Quercus michauxii*)	197	500	200	61	148	45.1	Alabama
Swamp white oak (1992) (*Quercus bicolor*)	228	579	120	36.6	92	28	Maryland
Texas oak (1999) (*Quercus shumardii* var. *texana1*)	108	274	60	18.3	59	18	Texas
Texas live oak (1999) (*Quercus virginiana* var. *fusiformis*)	295	749	42	12.8	98	29.9	Texas
Toumey oak (1994) (*Quercus toumeyi*)	68	173	27	8.2	33	10.1	Arizona
Turbinella (typ.) (1993) (*Quercus turbinella* var. *turbinella*)	160	406	43	13.1	49	14.9	Nevada
Turkey oak (1994) (*Quercus laevis*)	127	323	72	21.9	75	22.9	Florida
Valley oak (1984) (*Quercus lobata*)	348	884	163	49.7	99	30.2	California
Vasey oak (1982) (*Quercus pungens* var. *vaseyana*)	45	114	48	14.6	40	12.2	Texas
Vasey oak (1996) (*Quercus pungens* var. *vaseyana*)	61	155	39	11.9	32	9.8	Texas
Water oak (1996) (*Quercus nigra*)	278	706	120	36.6	111	33.8	Louisiana
White oak (1996) (*Quercus alba*)	382	970	96	29.3	119	36.3	Maryland
Willow oak (1986) (*Quercus phellos*)	318	808	73	22.3	132	40.2	Mississippi

[1]From the "National Register of Big Trees," American Forests (by permission).
[2]At 4.5 feet (1.4 meters).

Fig. 1.   Record blue oak tree. See Table 2. (*E. Logel.*)

Fig. 2.   Record swamp white oak tree. See Table 2. (*D.J. Preston.*)

D. Brush and Devereux Butcher, published by American Forests, Washington, DC, (revised periodically).

Cork Oaks. The cork oak (*Quercus suber*) is found in Europe and Africa. Cork cells are found in the outer bark of most woody-stemmed plants, but in amounts too small and with too-brittle walls to be of any commercial value. But in the *Quercus Suber*, the cork cells become a very large part of the tissue of the bark, and have been used for centuries. The cork oak tree is a medium-size tree seldom much over 50 feet (15 meters) in height, growing particularly in those countries bordering the Mediterranean Sea. The evergreen leaves are small, ranging from $1\frac{1}{2}$ to 3 inches (3.8 to 7.6 centimeters) long, and about an inch wide, with

slightly toothed margins. The bark of the tree soon becomes rough and deeply furrowed, but is of little value except as ground cork or as a source of tannin. When the tree is about 20 years old, this first-formed bark is removed, care being taken not to injure the phloem and cambium layers. Within 10 years a new cork layer has formed. This layer is the first of many layers which are removed once every 10 years or so throughout the life of the tree. Removal is generally done in the early summer at a time when hot dry winds will not cause injury to the unprotected phloem and cambium.

After removal, the cork is air-dried for a time, then boiled to soften it and to remove some of the tannin. The outer part of the bark is scraped off, and the rest pressed out flat and dried. It is then ready to ship.

The physical properties of cork account for its many uses. It is very light and buoyant, more than 50% of its volume being air, and hence is used in the manufacture of floats, life-preservers, and so forth. The living protoplasm of the cork cells dries up early in their development, leaving hollow cells, each containing a small mass of air which expands after compression. Therefore, cork is very resilient, and is frequently used as a core on which to wind yarn or string in the manufacture of baseballs. In the early stages of their formation, the walls of cork cells are cellulose, but this is soon impregnated by a waterproof and nonabsorbent lipoid substance, suberin. Therefore cork is used in making handles for fishing rods, shoe-soles, and cork stoppers. Since the hollow cork cells are poor conductors both of heat and sound, cork is much used as insulating material. For this use cork is ground up and then pressed into sheets with various binding materials, giving much larger sheets than can be obtained from the tree. Ground cork is also an important constituent of linoleum, gaskets, and other products.

Cork is traversed by lenticels, loose masses of porous tissue, which appear as dark spots or holes in stoppers. Usually in making stoppers the bark is cut so that these will be transverse in the stopper. In making stoppers, the forms are first punched out as cylinders, and then trimmed down by machine to the required tapering shape. New packaging materials and techniques have displaced cork from a number of formerly exclusive applications for it.

English Oak. This is a comparatively small oak, the height ranging from 20 to 40 feet (6 to 12 meters). The spread is about 20 feet (6 meters). The tree is cone-shaped with a life expectancy of about 50 years. It is easily cultivated and is often used as a windbreak or as an ornamental tree between sidewalks and curbs in large cities. Careful grooming produces beautiful specimens. The bark is light-gray.

OARFISH *(Regalecus glesne).* The longest bony (rather than cartilaginous) fish known to exist in the ocean. The fish may reach a length of more than 17 meters (56 feet) and weigh up to 300 kilograms (~660 pounds). The fish, seldom observed, is called the "king of the palace under the sea" by the Japanese and a sea serpent by others. The oarfish prefers warm, temperate waters worldwide and lives at depths ranging from 20 to 200 meters (66 to 660 feet). The lifespan of the fish is unknown. The fish has a long red dorsal fin that rises to a manelike crest atop its head.

The oarfish is described as very elusive and secretive, with few sightings reported over the centuries. Morton Brunnich, a Dutch naturalist, first described the fish in the scientific literature. In 1771, Brunnich found a specimen washed up on a beach in Norway. Fewer than 25 sightings subsequently have been reported. Another was found on the shore off Orkney (Scotland) in 1808. In 1906, possibly the closest encounter with a live oarfish occurred off the shore of the Island of Sumbawa (Indonesia). Although the crew of a nearby ship was not successful in enticing the fish, a crew member described it, "A long and very beautiful fish came to the surface at the ship's bow. Baited rigs were thrown to it, but it took no notice of them. With its vivid red crest and dorsal fin, scarlet streamers on its sides, and blue of its head and intense shine of silver on its body, it was probably the most beautiful creature I've ever seen."

The last sighting of an oarfish (well-preserved remains) was found on the beach at Malibu, California, in 1963. The specimen is displayed at the Los Angeles County Museum of Natural History. More detail can be found in article by Cheryl Lyn Dybaa (*Oceanus*, 98, Spring 1993).

OATS *(Avena sativa; Gramineae).* Oats are annual cereal grasses native in temperate regions of the Old World. The several species of oat plants are characterized by their closed leaf-sheath, a wide-branched panicle, and by a special type of inflorescence. In the florets of wild oat plants the lemma has the midrib prolonged as a prominent awn, the basal portion of which is spirally twisted. In many cultivated forms this awn has been eliminated. The grain of oats is not easily separated from the surrounding husk, composed of the lemma and palea, which are neither palatable nor digestible. See also **Grasses**.

Oats are principally adapted to growing in a climate having cool summers and abundant moisture. However, the plants are very hardy and tolerant of adverse conditions.

Oats are an important livestock feed, with approximately 95% of the United States crop used for this purpose. Rolled oats and oatmeal are important breakfast cereals for human consumption in some regions of the world. Use of oats in bakery products is quite limited because oat flour contains no gluten. However, oat flour can be mixed in relatively small portions with other flours and thus impart a distinct flavor. Also, because of a natural antioxidant contained in oat flour, the substance finds use for preserving certain foods that contain fats, such as peanut (groundnut) butter, oleomargarine, and lard. Oat flour also has been used for dusting potato chips and salted nuts, as well as for a coating on papers used to contain fatty products, such as bacon and coffee.

Oat hulls are the primary source of furfuraldehyde, which is used industrially as a solvent and chemical intermediate. Over the years, limited amounts of oatmeal have been used in soaps.

Botany. The several species of oat plants are characterized by their closed leaf-sheath, a wide-branched panicle, and by a special type of inflorescence. See Fig. 1. In the florets of wild oat plants, the lemma has the midrib prolonged as a prominent awn, the basal portion of which is spirally twisted. In many cultivated forms, this awn has been eliminated, mainly in the interest of processing for livestock feed. The grain of oats is not easily separated from the surrounding husk, composed of the lemma and palea, neither of which is palatable or digestible.

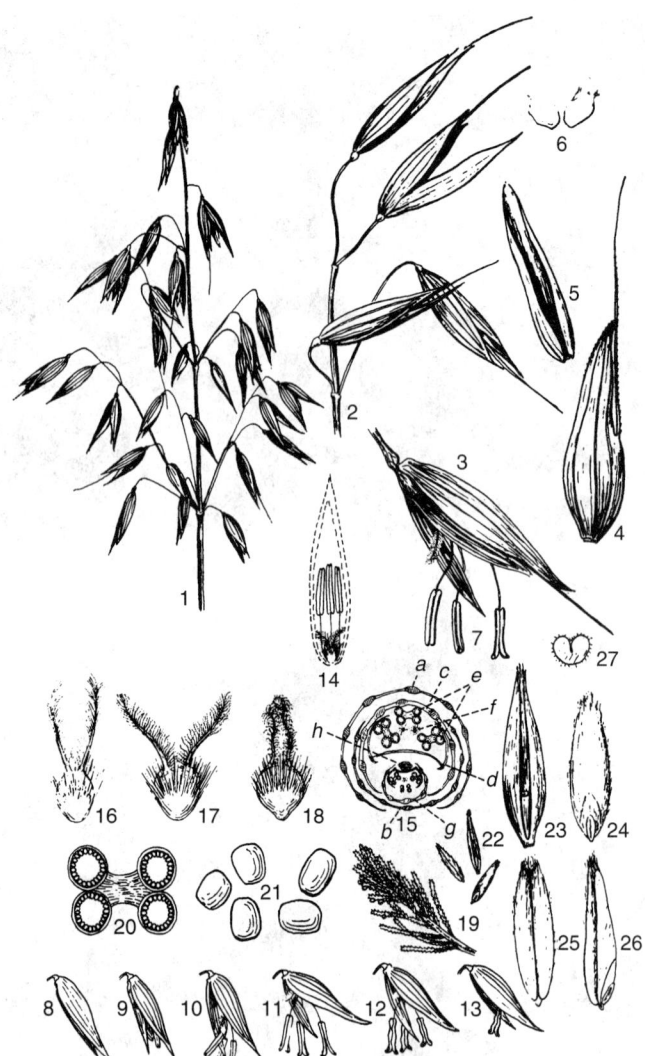

Fig. 2.   Oat inflorescence and details of components. (1) Panicle of Avena sativa. (2) Distal or top part of panicle, bearing four spikelets. (3) Lateral view of spikelet (one-flowered) in anthesis, showing separated lemma and pales with one branch of plumose stigma and three stamens (item 7) protruding. (4) Lateral view of lemma, showing dorsal attachment of awn. (5) Ventral view of palea. (6) Lodicules. (7) Stamens. (8 to 13) Lateral views of a floret before, during, and after anthesis. (14) Diagrammatic longitudinal dorsal-ventral section of floret, showing lemma (X) palea, androecium, and gynoecium. (15) Diagrammatic cross section of spikelets of three-flowered spikelet before anthesis: (a) lower or outer glume; (b) upper or inner glume; (c) lemma or flowering glume; (d) palea; (e) anthers; (f) stigma; (g) lemma or secondary floret with enclosed pales, stamens, and stigma; (h) rudimentary tertiary floret. (16 to 18) Pistil before, during, and after anthesis. (19) Apical portion of stigma, greatly enlarged, showing adhering pollen grains. (20) Cross section of anther showing four lobes. (21) Pollen grains (enormously enlarged). (22) Floret, ventral, and dorsal view of kernel and caryopses (smaller than natural size). (23) Oat kernel. (24 to 26) Caryopsis or groat (enlarged). (27) Cross section of caryopsis. (*USDA diagram from original sketch by Boettcher and Hughes.*)

Fig. 1.   Oat panicles: (left) unilateral, "side," or "horse-mane" oat with 7 whorls or branches; (right) equilateral, or spreading, or "tree" panicle with 5 whorls or branches. (*USDA photo.*)

The principal parts of the common oat (*A. sativa*) are shown in Fig. 2. Oats are naturally self-pollinated.

As shown by Fig. 3, the oat kernel or caryopsis is spindle-shaped and furrowed on one side. The color is a light buff and the hairs are fine and silky. The kernel usually is from $\frac{5}{16}$- to $\frac{7}{16}$-inch (7.5 to 10.5 millimeters) long and from $\frac{1}{16}$- to $\frac{1}{8}$-inch (1.5 to 3 millimeters) wide. Of the total grain, the kernel makes up between 65 and 75% of the weight. The two main parts of the kernel are the endosperm and embryo. The term *groat* is frequently used to designate the oat caryopsis.

Species and Varieties. The manner of designating and identifying the various species and subspecies of oats is technically complex and mainly of importance to plant breeders.

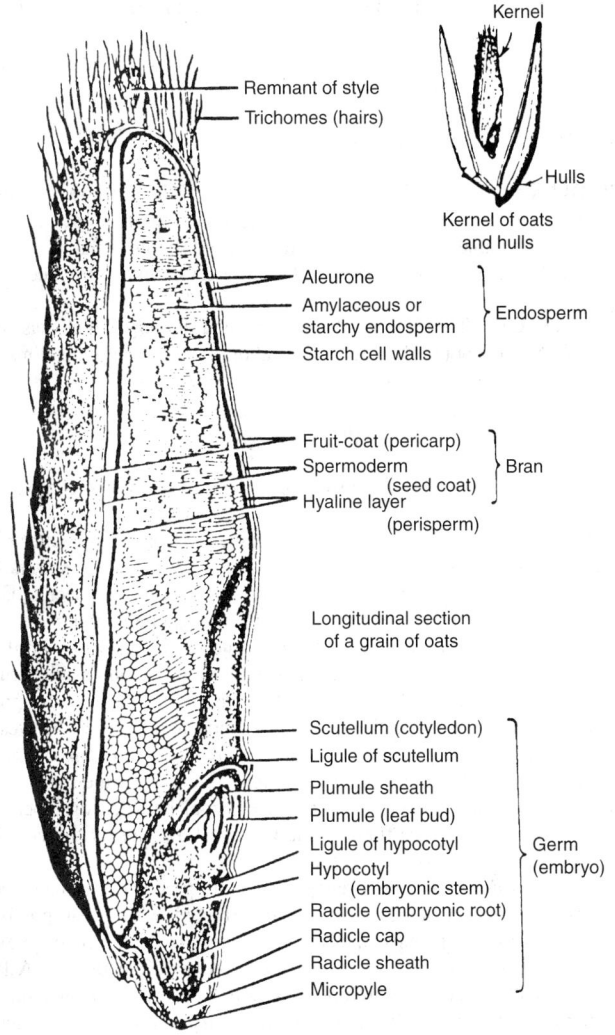

Kernel

Remnant of style
Trichomes (hairs)

Hulls

Kernel of oats
and hulls

Aleurone
Amylaceous or
starchy endosperm
Starch cell walls
} Endosperm

Fruit-coat (pericarp)
Spermoderm
(seed coat)
Hyaline layer
(perisperm)
} Bran

Longitudinal section
of a grain of oats

Scutellum (cotyledon)
Ligule of scutellum
Plumule sheath
Plumule (leaf bud)
Ligule of hypocotyl
Hypocotyl
(embryonic stem)
Radicle (embryonic root)
Radicle cap
Radicle sheath
Micropyle
} Germ
(embryo)

Fig. 3.   Parts of the oat caryopsis. (*The Quaker Oats Company.*)

Characteristics that are desirable in oat varieties include short straw, strong straw that resists lodging, high test weight, high yield, and resistance to crown rust, stem rust, yellow dwarf disease, and smut.

In planting oats, tillage of soil that is too wet causes severe compaction and reduced yields. When oats follow maize (corn) in rotation, the stalks should be chopped and either diced or plowed. One tandem dicing just ahead of planting provides an excellent seedbed when oats follow soybeans or other beans. Oats usually are seeded at the rate of 2 to 3 bushels per acre (71 to 107 kilograms per hectare) and at a depth of 1 to 2 inches (2.5 to 5 centimeters) to achieve 20 to 30 plants per square foot (215 to 323 plants per square meter).

Oats may be seeded alone strictly as an oats crop, or they may be used as a companion crop when seeding legumes and/or grasses. The seeding of the two crops together represents a compromise between obtaining the highest yield of oats and a good stand of the legumes or grasses. A vigorous, late-maturing oat crop may result in a poor, weak stand of forage as the two crops compete for water and soil nutrients.

Oats have the capacity to respond to fertilizer, especially nitrogen. Phosphorus and potassium are needed to balance the nitrogen. The cool, moist soils that are common early in the spring are not conducive to bacterial action that will release nitrogen, or to the high availability of phosphorus.

When oats and a companion crop of legumes and grasses are seeded, the fertility program must be adequate to feed both crops. Greatest response will be obtained when the fertilizer is drilled with the seed oats and the legume seed is placed in a band directly over the fertilizer band. One hundred bushels (1440 kilograms) of oats with 2 tons (1814 kilograms) of straw will require 90 pounds (40.5 kilograms) of nitrogen; 60 pounds (27 kilograms) of P_2O_5; and 115 pounds (51.8 kilograms) of K_2O. Various soils will release these nutrients at different rates due to the organic matter

content variations, rotation program used, soil temperature, and the parent material of the soil. A typical fertilizer application will be from 40 to 60 pounds per acre (45 to 67 kilograms per hectare) each of nitrogen, phosphate, and potash. Oats are a short-season crop. Every growing day is important, and a crop that is deficient in any nutrient even when only 1 to 4 inches (2.5 to 6 centimeters) tall will probably not recover from any significant span of inadequate nutrition.

Additional Reading

Dendy, D.A. and B.J. Dobraszczyk: "Cereals and Cereal Products: Chemistry and Technology," Aspen Publishers, Inc. Gaithersburg, MD, 2000.

Forsberg, R.A.: "Breeding Oat Cultivars Suitable for Production in Developing Countries: 1996 Report," DIANE Publishing Company, Collilngdale, PA, 1998.

Marshall, H.G. and M.E. Sorrells: "Oat Science and Technology," American Society of Agronomy, Madison, WI, 1992.

Seibold, R.: "Cereal Grass: Nature's Greatest Health Gift," Keaets Publishing, Inc., Chicago, IL, 1994.

Simons, M.: "Crown Rust of Oats and Grasses," The American Phytopathological Society, St. Paul, MN, 1970.

Webster, F.H.: "Oats Chemistry and Technology," American Association of Cereal Chemists, St. Paul, MN, 1986.

Wood, P.J.: "Oat Bran," American Association of Cereal Chemists, St. Paul, MN, 1993.

Web References

American Association of Cereal Chemists: *http://www.scisoc.org/aacc/*

The American Phytopathological Society: *http://www.apsnet.org/*

OBESITY.   The accumulation and storage of body fat in excess of that required by a statistical normal individual of same age, frame size, sex, and height. When caloric intake exceeds expenditure, excess fat accumulates—either existing fat cells are swelled with additional fat, or new fat cells are formed, or both actions may occur together. Scientists are not certain. There is the general opinion among specialists that the number of fat cells within the human body increases in the fetus and newborn and later in childhood and adolescence, while during other periods, particularly in later life, the number of cells does not increase, but rather the amount of fat in each cell may increase. However, in persons who are over 50% overweight, there is an increase in both cell numbers and fat crowding. Researchers have estimated that an average adult body contains between 3 and 5×10^{10} fat cells. It has further been estimated that each cell contains about 0.5 microgram of triglycerides. Thus, the fat content ranges between 26 and 44 pounds (12 and 20 kilograms). Obesity in individuals ranges from being just a little bit fat (plump, paunchy, etc.) to an extreme overweight condition where the person's life is threatened and where living is uncomfortable and clumsy. Treatment of the severely obese includes total starvation over a period in a hospital environment. This approach has proved successful, but many specialists now prefer to use a so-called protein-sparing diet in such cases. See also **Starvation**. In more recent years, an intestinal bypass may be considered. A majority of physicians consider this a solution of last resort because of the morbidity and mortality presently associated with the procedure. Consequently, the approach is generally reserved for situations where serious complications, such as congestive heart failure, debilitating arthritis, anoxia, among others, are present as well as the failure of prior weight reduction procedures.

The mental and emotional aspects of obesity are emphasized by the fact that only 10% or fewer of the individuals who commence a dieting regimen and who achieve early success will maintain their reduced weight beyond one, two, or three years, depending upon the individual. Thus, calorie control through dieting and the possible use of other procedures, such as jaw wiring, acupuncture, transcendental meditation, yoga, hypnosis, and gonadotropin injections, are, unfortunately, of a transient nature. Such procedures require consistent enforcement of sensible eating habits, essentially over a lifetime, once the initial objectives are achieved. Those persons who do ultimately succeed in controlling or eliminating obesity have, through a combination of physical and mental conditioning, managed to alter their fundamental eating and exercise habits.

There is no shortage of references on special diets for losing weight or counseling to participate in more vigorous exercise. Dieting, other than the simplistic, common sense approach of voluntarily reducing caloric intake, should be supervised by a physician who is well acquainted with the health of the individual. Some weight-reduction diets can seriously affect the health of some individuals. Most of the millions of words written to describe the benefits of special diets essentially are wasted because of the transient nature of responding to them. The benefits of a sensible exercise program for a majority of individuals are undeniable and exercise, of course, can contribute to a weight-loss program. In the interest of realism, however, it should be pointed out that from the standpoint of calorie burning, exercise is consistently exaggerated. For example, 10 to 15 minutes of jogging will only consume 200–300 calories. Nevertheless, regular exercise contributes to the successful maintenance of weight loss.

For many years, drug therapy for obesity has been a topic of controversy and remains so. In tests using placebos, it has been demonstrated that amphetamines, including their analogues (fenfluramine, etc.) can result in increased weight loss over a period of several months. These drugs appear to be mainly effective where emotional stress is a prime component of obesity. But such drugs are contraindicated for some individuals with other problems and thus require the judgment of a physician in their use.

The fact that obesity has been experienced by the human race since antiquity attests to the stubborn nature of the problem. The connection between obesity and inheritance is obvious, but not understood. Obviously, much additional research is required to understand the chemistry of obesity as well as of its psychiatric and heritable nature. Quite recently, some scientists have been taking bold and new approaches to developing this understanding. For example, some scientists have been investigating the systemic changes that occur during certain periods in the lives of various mammals — these initial studies indicating that probably through hormonal control, there are preprogrammed periods of fasting. These are not confined simply to hibernation, but to periods when anorexia (loss of appetite or distaste for food) functions at times for survival of individuals and preservation of species. Bull seals, for example, go without feeding for many weeks while minding their duties of defending territory and harem. Preprogrammed anorexia appears to be part of such processes as migrating and molting, i.e., not eating even when food is immediately available. Researchers are also looking to anorexia nervosa, principally a disorder of teenagers, who literally starve themselves because of a fear of fatness. This is a complex disease and still poorly understood. See also **Anorexia**; and **Diet**.

OBJECTIVE PRISM.   In the ordinary laboratory spectrograph a collimator is necessary in order that the light from the narrow slit may be sent through the dispersing agent in a parallel beam. At the principal focus of the camera lens the images of the slit in the different radiations are commonly known as the spectral lines.

The stars are at such tremendous distances that they subtend infinitesimal angles and the light from them reaches the earth in parallel beams. Hence the slit and collimating lens of the laboratory spectrograph are unnecessary and the dispersing agent, usually a prism, may be placed directly before the objective lens of the telescope or astrographic camera. With such an instrument, commonly known as an objective prism, instead of single-point images of the stars, there appear on the photographic plate a series of dots, each of them in some particular radiation from the stars, or, in other words, the spectra of the stars. In order that the spectra may have appreciable width or that the spectral lines may have length instead of appearing as mere dots, the refracting edge of the prism is set parallel to the spherical coordinate of right ascension and the telescope driven a bit too fast or too slow, slightly "trailing" the images.

OBJECT (Real).   In geometrical optics, an object (or image) is termed *real* if from each point of it, light diverges toward the optical system. The first object (O) in Fig. 1 is real. An example of a *virtual* image is I_1, which would have been formed by lens L_1 had not lens L_2 been interposed. I_1 acts as a virtual object for lens L_2. See also **Mirrors and Lenses**. The real

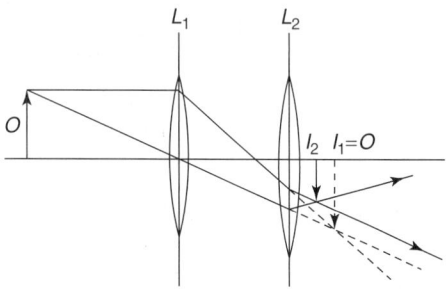

Fig. 1.   Demonstration of real and virtual objects.

or virtual point of intersection of a pencil of rays incident upon an optical system is called the *object point*.

OBLATE SPHEROID.   See **Spheroid**.

OBLIQUE ANGLE.   See **Angle**.

OBSIDIAN VOLCANIC GLASS.   Highly acidic lavas (those containing a preponderance of silica), when chilled very rapidly and congealed, without appreciable crystallization, into a rigid liquid solution. Such a solid is called a glass to distinguish it from a crystalline substance.

OBSTETRICS.   The branch of medicine and surgery which has to do with the management of pregnancy, labor and the complications and disorders arising from them.

OBTUSE ANGLE.   See **Angle**.

OCCIPITAL LOBE.   See **Central and Peripheral Nervous Systems**.

OCCULTATION.   When a large body passes in front of a smaller body, an occultation occurs. Thus, when the moon passes between a star and the earth, the light of the star is cut off, and the star is said to be occulted. The term eclipse, which is technically correct for this phenomenon, is reserved for circumstances involving the sun, earth, and moon, and for such things as eclipsing binaries, satellites of Jupiter, etc. As the moon passes between the earth and a star in its revolution about the earth, the star disappears behind the eastern edge of the moon and reappears again at the western edge. The disappearance and reappearance are practically instantaneous, proving both that the moon has no sensible atmosphere and that the star appears sensibly as a point of light. The interval between disappearance and reappearance depends fundamentally upon how closely the center of the moon appears to pass across the star.

Since the disappearance and reappearance are practically instantaneous, the times of the phenomena can be determined with great precision. Observations of the instants of occultation of stars by the moon may be used to determine the position of the moon extremely accurately. A large number of such occultations are observed both by professional and amateur astronomers, and the results are used to verify and correct the theory regarding complicated motions of the moon. The differences in the time of occultation of a given star, as observed by widely separated observers, may be used to determine both the distance of the moon and the difference in terrestrial longitude of the two observers. This method of determination of difference in longitude was the most accurate available before the development of modern methods for distribution of Greenwich time.

If the diffraction pattern differs from that of a point source, it is possible to obtain stellar diameters by observing occultations photoelectrically with high time resolution equipment. This same technique was used at radio frequencies to measure the sizes of the two components of the radio source 3C273.

OCEAN.   The surface area of the earth is approximately 510.1×10^6 square kilometers (196.9×10^6 square miles). Of this total area, the oceans and adjacent seas, forming a series of interconnected saline bodies, cover 375.55×10^6 square kilometers (145 million square miles), or over 70% of the earth. The oceans and seas are not evenly distributed. Approximately 81% of the surface of the Southern Hemisphere is covered by ocean waters; about 61% of the Northern Hemisphere is covered. Because the oceans are continuous and interconnected, the names assigned are somewhat arbitrary or depend upon natural land boundaries, such as the continental land masses or various island chains.

Traditionally, there are five oceans: the Atlantic, the Pacific, the Indian, the Arctic, and the Antarctic. The Antarctic Ocean, lacking any precise natural boundaries, is sometimes considered an extension of the Atlantic, the Pacific, and the Indian oceans. The adjacent bodies of salt water and various subdivisions of the oceans are generally known as seas, but local usage may also sanction such terms as gulfs, bays, channels, and straits, terms that are sometimes used interchangeably. The names *epicontinental, epeiric,* and *mediterranean* seas are less frequently used. An epicontinental or epeiric sea is a sea on the continental shelf or within a continent. A mediterranean sea is a type of epicontinental sea that is deep and that connects with an ocean by a narrow opening.

Among the large and important seas or designated regions of the principal oceans are the South China Sea, the Caribbean Sea, the Mediterranean Sea, the Bering Sea, the Gulf of Mexico, the Sea of Okhostk, the Sea of Japan, Hudson Bay, the East China Sea, the Andaman Sea, the Black Sea, the Red Sea, the North Sea, the Baltic Sea, the Yellow Sea, the Gulf of California, and the Persian Gulf. It should be noted that some landlocked bodies of water, usually called lakes, are historically called seas—thus, the Caspian Sea, the Dead Sea, and the Sea of Galilee, among others.

From a geochemical standpoint, the oceans and seas may be classified and named in still another way, i.e., in terms of ocean water masses. Determinations of the physical and chemical properties of water samples from the oceans have shown that they may be divided into regions in which the properties of samples are relatively constant. These regions are characterized not only by their geographical locations, but also by their depths. Thus, there is not only a water mass described as *arctic surface water*, but also *arctic deep water*, *antarctic surface water*, *antarctic intermediate water*, *antarctic bottom water*, and *antarctic circumpolar water*, among others.

The properties most widely measured to characterize the water masses are: density, expressed in terms of the unit sigma; temperature, expressed in degrees Celsius; and salinity. In actual practice, the last is not determined directly, but is computed from various measurements, such as electrical conductivity, refractive index, or some other property whose relation to salinity is well established. Chlorinity is especially useful for this purpose, since the ratio of chloride content to the other elements in seawater is virtually constant despite the variation in total salinity from one water mass to another. Since in the same water mass the salinity varies with the temperature, the most widely used function for characterizing a given water mass is its *T-S* (temperature-salinity) graph. For that reason, temperature and salinity ranges are generally given in the various entries in this encyclopedia pertaining to the major oceanic water masses. These entries include: **Antarctic Waters** (bottom, circumpolar, intermediate, and surface); **Arctic Waters** (deep and surface): **Indian Ocean Water; Mediterranean Water**; **North Atlantic Waters** (deep and bottom, central, and intermediate); **North Pacific Waters** (central and intermediate); **Pacific Equatorial Water; South Pacific Central Water; South Pacific Current**; and **Subarctic Pacific Water**.

The sizes and depths of the three major oceans are;

Pacific Ocean: 166.24 × 106 square kilometers (64,186,300 square miles). Average depth: 4188 meters (13,740 feet).
Atlantic Ocean: 86.56 × 106 square kilometers (33,420,000 square miles). Average depth: 3735 meters (12,254 feet).
Indian Ocean: 73.43 × 106 square kilometers (28,350,500 square miles). Average depth: 3872 meters (12,704 feet).

Later in this entry other physical and chemical characteristics of the oceans and seas are described in more detail.

Ocean Volume and Depth

The volume of the oceans and their seas is nearly 1.5 × 109 cubic kilometers (350 million cubic miles). The average depth is approximately 4 kilometers (2.5 miles). An average depth figure is useful for some calculations, but it should be emphasized that the depths of the oceans vary over a wide range. The oceans are not the deepest in their middle portions, as sometimes believed, but rather the greatest depths occur in trenches found along the continental margins. As indicated in Table 1, the greatest known depth is the Mariana Trench of the Pacific Ocean, which is over 11,000 meters (6.9 miles) below sea level. The continental shelves are about 200 meters (656 feet) below sea level.

Sea Level

Sea level changes occur principally as the result of glaciers forming or melting. In an ice age, the sea level drops; in an age of ice melting, the sea level rises. At present, the earth is in an interglacial period. A schematic sea level curve for the eastern United States is shown in Fig. 1. The curve indicates a stabilization of sea level (at a low point) some 15,000 B.P. (before present), followed by a relatively rapid rise in level (in terms of

TABLE 1. PRINCIPAL DEEP TRENCHES OF THE OCEAN

Name	Location		Depth	
			Meters	Feet
PACIFIC OCEAN				
Mariana Trench	11°21'N	142°12'E	11,033	36,200
Tonga Trench	23°15.3'S	174°44.7'W	10,882	35,702
Kuril Trench	44°15.2'N	150°34.2'E	10,542	34,587
Philippine Trench	10°24'N	126°40'E	10,539	34,578
Izu Trench	30°32'N	142°31'E	10,374	34,033
Kermadec Trench	31°52.8'S	177°20.6'W	10,047	32,964
Bonin Trench	24°30'N	143°24'E	9,156	30,041
New Britain Trench	06°34'S	153°55'E	9,140	29,988
Yap Trench	08°33'N	138°02'E	8,527	27,976
Japan Trench	36°08'N	142°43'E	8,412	27,600
Palau Trench	07°40'N	135°04'E	8,138	26,700
Aleutian Trench	50°53'N	176°23'E	8,100	26,574
Peru-Chile Trench	23°18'S	71°41'W	8,064	26,454
New Hebrides Trench	20°36'S	168°37'E	7,570	24,837
Ryukyu Trench	25°15'N	128°32'E	7,507	24,629
Mid-America Trench	14°02'N	93°39'W	6,669	21,880
ATLANTIC OCEAN				
Puerto Rico Trench	19°35'N	66°17'W	8,648	28,374
South Sandwich Trench	55°14'S	26°29'W	8,252	27,072
Romanche Gap	00°16'S	18°35'W	7,864	25,800
Cayman Trench	19°12'N	80°00'W	7,535	24,720
Brazil Basin	09°10'S	23°02'W	6,119	20,076
INDIAN OCEAN				
Java Trench	10°15'S	109° (approx) E	7,725	25,344
Ob Trench	(not accurately fixed)		6,874	22,553
Vema Trench	(not accurately fixed)		6,402	21,004
Agulhas Basin	(not accurately fixed)		6,195	20,325
Diamantina Trench	35°00'S	105°35'E	6,062	19,889
ARCTIC OCEAN				
Eurasia Basin	82°23'N	19°31'E	5,450	17,880
MEDITERRANEAN SEA				
Ionian Basin	36°32'N	21°06'E	5,150	16,896

Source: U.S. Mapping Agency, Hydrographic/Topographic Center

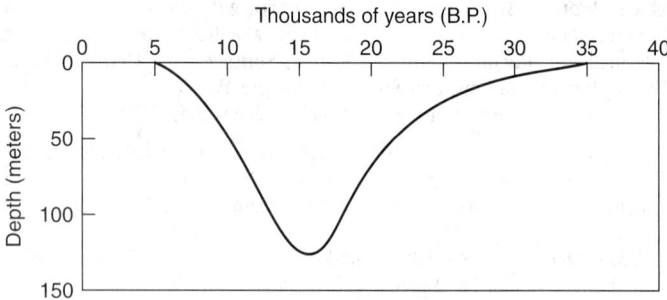

Fig. 1.   Sea-level curve for last 40,000 years. (*After Milliman and Emery.*)

geological time), which tapered off about 5000 years ago. At the latter time, it is estimated that the sea level was about 5 meters (15 feet) below its present level. On the average, since that time, the sea level has risen about one millimeter per year and thus is gradually encroaching on the continents. Such worldwide changes over long periods of time are termed *eustatic*.

There are also noneustatic sea level changes as the result of tectonism, which is prevalent along the Pacific Coast. These changes are also caused as the result of glacial rebound (local adjustments to prior glaciation), as well as seasonal changes that result from freshwater inflow and heating and cooling cycles in the ocean (*steric effects*). Records of tide gages, which reflect both eustatic and noneustatic changes, along the New England coast since about 1940 show a submergence of the coast at a rate of about 3 millimeters per year, whereas along the Gulf coast, the rate is about 15 millimeters (continental submergence) per year. Although erosion by wind and wave actions associated with hurricanes and winter storms, coupled with ecological disturbances, are principally responsible for beach and coastal area deterioration, the slow changes in sea level also participate in these processes. Coastlines are discussed later in this entry.

Much of the research on sea level changes was conducted during the International Decade of Ocean Exploration during the 1970s, particularly as part of the CLIMAP (*Climate: Long-Range Investigation, Mapping, and Prediction*) project. As part of that project, attempts were made to develop the sea level profile as it changed throughout geologic history. Researchers constructed a model of Earth's surface as per a typical August and February during the peak of the last ice age (some 18,000 years ago). Mapping delineation included: (1) extent and thickness of land- and marine-based ice sheets, (2) vegetation patterns, (3) global sea level, (4) seasonal extremes of sea ice, and (5) sea-surface temperatures.

Four types of fossils (coccoliths, foraminifera, radiolaria, and diatoms), sedimentation rates, and oxygen isotope data were analyzed and compared with conditions of the present time. Fortunately, extensive data were already available from the deep-sea core library maintained by the Lamont-Doherty Geological Observatory of Columbia University. With these data, it was possible to reconstruct many of the ocean's characteristics at the time of the last ice age. Samples documented all major temperature gradients, surficial water masses, and circulation patterns of glacial world oceans. By determining ice volumes, global sea level changes, which were caused by transfer of water from oceans to ice caps, could be calculated. Maximum and minimum ice sheet models recorded sea level changes of 150 to 100 meters. These models were further refined by using oxygen isotope data and yielded an overall value of a drop in sea level of 150 meters. The sea-surface temperatures were based upon quantitative counts of microfossils.

Scientists observed that the area covered by permanent ice was substantially different during the glacial maximum as compared with the situation today. It is believed that, in the Northern Hemisphere, the huge land-based ice sheets attained a thickness of about 3 kilometers and the extent of pack ice and marine-based ice sheets significantly increased. As regards the Southern Hemisphere, the most striking contrast was the greater extent of sea ice. Thus, it is postulated that, combined with a sea level lower by 150 meters, these developments caused significant changes in the surface characteristics of the earth. It has been reasoned that, on land, grasslands, steppes, sandy outwash plains, and deserts spread at the expense of forests; whereas the extent of snow-covered land increased significantly. The reconstruction of land vegetation was based upon pollen distribution and types. The aforementioned changes, coupled with an

increase in glaciation, are assumed to have caused an increase in surface albedo over modern values.

In more recent studies, Gornitz, et al. (Goddard Space Flight Center), have analyzed large amounts of data derived from tide-gauge stations throughout the world. These data indicate that the mean sea level rose by about 12 centimeters over the past century. The sea level change has a high correlation with the trend of global surface air temperature. A large part of the sea level rise, according to these investigators, can be accounted for in terms of the thermal expansion of the upper layers of the ocean. The data also indicate weak indirect evidence for a net melting of the continental ice sheets.

The researchers suggest that global warming due to increasing atmospheric carbon dioxide could melt the marine West Antarctic ice sheet, raising the global sea level some 5 to 6 meters. They point out that a sea level rise of as little as 15 centimeters could double the probability of damaging storm surges on the coast of Britain, causing substantial beach erosion and the intrusion of seawater into low-lying areas that are now freshwater regions. Data from over 700 stations were obtained from the Institute for Oceanographic Science, Birkenhead, England. Individual station records were reduced to a common reference point by fitting a least-squares regression line to sea level as a function of time and by defining the zero point to be the value of the regression curve for 1940. The annual mean sea level curves for stations within a geographical region were then averaged to yield a mean sea level curve for each region. The researchers also attempted to remove the long-term (usually 6000 years) sea level trends from the station data in order to obtain short-term sea level fluctuations, which are perhaps more appropriate for correlation with global climate variations in the past century. The cause of the long-term trend is uncertain. It has been argued that as much as 90% of it is residual isostatic uplift of continents due to the removal of the Wisconsin ice sheets. However, the long-term trend may contain a eustatic component, for example, due to a change in volume of the Antarctic or Greenland ice sheets.

As observed by Gornitz, et al., the estimates for long-term sea level change are based on ^{14}C dating of measured positions of shoreline indicators in the geologic records, for example, mollusks, corals, and brackish-water peats. Further adjustments in the data were made and are described in the Gornitz et al. reference (listed).

In another study, M.F. Meier (U.S. Geological Survey, Tacoma, Washington) investigated the contribution of small glaciers to global sea level. Meier observed long-term changes in glacier volume and hydrometeorological mass balance models, which yield information on the transfer of water from glaciers, excluding those in Greenland and Antarctica, to the oceans. The average observed volume change for the period 1900–1961 was scaled to a global average by the use of the seasonal amplitude of the mass balance. The data are used to calibrate models for estimating the changing contribution of glaciers to sea level for the period 1884 to 1975. Although the error band is large, these glaciers appear to account for a third to half of observed rise in sea level, approximately that fraction not yet explained by thermal expansion of the ocean. More details are given in the Meier reference (listed).

Although too detailed to describe here, it should be noted that in the mid-1970s, geologists (Peter Vail, et al., Exxon Production Corp.) announced a new approach to estimating and forecasting sea level. In deep drilling activities, mainly in connection with petroleum exploration, numerous unconformities (missing sections of sediment laid down during a given geologic time period and later eroded away) have been found with considerable consistency worldwide. The same unconformities are found on such widely dispersed, tectonically unrelated margins that it would appear only a global sea level change could link them, due mainly to the waxing and waning of major ice sheets. A part of the complexity is that while sea level changes occur, so do the levels of the continental margins, which have been estimated to be sinking at a rate of 1 to 2 centimeters per thousand years. Thus, it is concluded that even if global sea level remains unchanged, the sea still will creep up the shelf as it sinks—and even if global sea level drops fast enough to overtake subsidence and cause the sea's edge to retreat down the margin, sediments washing off the continent can stop short of the sea and fill in space created by subsidence above the waterline. For a further introduction to this topic, reference to Kerr is suggested. Also, see the Watts references.

Seawater Temperature

The mean temperature of the oceans is about 3.9 °C (~39 °F). This figure is useful in certain calculations, but obviously specific waters vary widely as affected by geographic location, season of the year, and depth. In torrid zones, the temperature of the surface waters may be between 24 and 27 °C (75 and 80 °F). Temperature decreases markedly toward the polar regions, where the temperature is about −2 °C (28 °F), except where ocean currents of higher or lower temperature exert thermal influence. With exception of these currents, it was once believed that at any given depth the ocean waters within any given region were quite uniform.

In connection with the IDOE program, previously mentioned, the NORPAX project (North Pacific Experiment) science team found the existence of enormous pools of surface water abnormally warmer or cooler than the mean by 1 to 1.5 °C. These pools were found to be up to 300 meters (984 feet) deep, some 1500 kilometers (932 miles) in lateral extent, and with time periods of some 2.5 years in duration.

It has long been established that, when compared with the atmosphere, the ocean has a great capacity to store heat and energy. The ocean also has large inertia that creates a potential for delayed feedback on climatic time scales. Thus, the time duration of the anomalous pools of water was not surprising. By using computer models and numerical techniques, the investigators were able to forecast both the shapes and intensities of anomaly patterns — up to a season in advance. These proved useful in shedding light on the upper-ocean processes as well as for predicting long-term ocean effects on the marine atmosphere. Essentially, in a continuation of the 1970s NORPAX project, scientists have more recently improved their computer modeling techniques. However, the interface between ocean and atmosphere is severely complex, particularly as the interface affects inland weather. This problem is explored in some detail in the article on **Atmosphere-Ocean Interface**.

Seawater temperature also varies with the ocean's currents, which are discussed later in this article.

Because solar energy is the most important source of heat for the oceans, there exists (except in polar regions) a very shallow layer of relatively warm water at the surface. Under this is a boundary layer called the *thermocline* in which the temperature drops quite rapidly with increasing depth until the large mass of water at its lower boundary is encountered, below which the temperature falls only very slightly with increasing depth. At the polar regions, the temperature of the water at or near the surface does not show significant differences with the waters at a greater depth. The effect of depth on seawater temperatures varies with locale. For example, in the Caribbean area, within a short distance of land, surface water temperatures may be in the 27–29 °C (~81–84 °F) range, yet the temperature at the 600–900-meter (2000–3000-foot) level may be as low as 4–7 °C (~39–45 °F). Use of seawater temperature differentials is the basis of one means for capturing solar energy for useful work. See also **Solar Energy**.

Sea-Surface Temperature (SST). As described by Cane (see reference), El Niño conditions occur when there are very marked alterations in SST. The El Niño of 1982–1983 has been very well documented in the literature and is described in some detail in this encyclopedia in the article on **Atmosphere-Ocean Interface**. Briefly, in July 1982, conditions in the eastern equatorial Pacific seemed normal. By October, however, the SST was nearly 5 °C above normal and the sea level at the Galápagos Islands had risen by 22 centimeters. The anomalies at depth were even greater. A huge influx of warm water had increased the heat content of the upper ocean at a rate that exceeded the climatological surface heat flux by a factor of more than 3, and the thickness of the warm layer was now greater than all previously observed values. Temperatures at the South American coast were near normal, but within a month they too would rise sharply. It became obvious that what had been labeled as a "warm event" had become a major El Niño. The severe nature and rather unusual developmental behavior of the 1982–1983 El Niño provided a great incentive for oceanographers and meteorologists to probe the extremely complex relationship between the oceans and climate.

Density

The density of seawater is determined by temperature and by the concentration of salts (salinity). The density increases as the temperature falls to the freezing point and also with increasing salinity. Accordingly, because the denser water tends to sink, and temperature is the most important determinant, the waters with the greatest density are found in the cold polar and arctic seas. While density of pure water at 4 °C (39 °F) is equal to 1, the density of seawater ranges over somewhat higher values, which vary with proximity to shores, rivers, etc., as well as with geographic location and depth. Representative average values are 1.026–1.028. However, seawater, like fresh water, has its maximum density about 4 °C above its freezing point, and as it becomes colder than 3–4 °C, it becomes less dense, and rises. This fact explains why water cannot freeze at the bottom of the sea.

The high density and fluidity of seawater make it difficult to lower people and equipment below the ocean surface. The immense hydrostatic pressures at great depths will crush all but the strongest of vessels. Materials selection is difficult because very high stresses can result when equipment is constructed of materials having differing compressibilities. Thus, ocean technology, like so many areas of technology is confronted with corrosion, strength, and other material-related problems.

The combination of density and low viscosity provides excellent conditions for surface travel on the oceans. There is a high lift-drag ratio for ships, making it possible to move heavy loads with reasonable speed and relatively little power. However, this same combination of characteristics contributes to major hazards of shipping, notably the creation of large wind-waves. Obviously, if the water were more viscous, as in the case of a heavy oil or syrup, the wind would not have sufficient force to build up high waves. Conversely, if the water were considerably less viscous, the force and hence hazard present in waves would not be present. It is also of interest to observe that the high surface tension of seawater contributes to its not sticking readily to surfaces, allowing removal (except salt residues) by evaporation within short periods.

Seawater Composition

Because the composition of seawater is affected mainly by the addition of dissolved salts brought to it by the great rivers, the composition differs from one region to the next. On the average, the salt content of seawater is about 3.5% (weight). Although there are many compounds and chemical elements in seawater, the principal dissolved salts are:

	% (Weight) of all Dissolved Salts
Sodium chloride (NaCl)	77.76
Magnesium chloride ($MgCl_2$)	10.88
Magnesium sulfate ($MgSO_4$)	4.74
Calcium sulfate ($CaSO_4$)	3.60
Potassium sulfate (K_2SO_4)	2.46
Magnesium bromide ($MgBr_2$)	0.22
Calcium carbonate ($CaCO_3$)	0.34

Aside from mineral values, exploitation of which may increase in the future, salinity is probably the least desirable characteristic of seawater. The water obviously cannot be consumed in any quantity by most land forms without adverse effects. Although saline waters (mainly brackish) can be used with limited success agriculturally in modern drip irrigation systems, generally they are not acceptable. The electrolytic chemical activity resulting from ionized salts, together with frequent presence of dissolved oxygen, exerts very adverse corrosive attacks on metals and other substances that come in contact with seawater. Salt water is a relatively good conductor of electricity and thus the ocean can be penetrated only slightly by radio waves. In contrast, seawater is an excellent conductor of sound waves of low frequency. For example, an explosion that may be detected only a distance of a half-mile or so on land can be heard a thousand or more miles under water. See also **Sonar**. The U.S. Navy has had a program under consideration for several years that would bring long-wave radio communication with submarines on a worldwide basis. See also **Antenna (Communications)**.

In separate articles on the many *chemical elements*, the content of each element in seawater is given. See also the table on chemical abundance in the summary article on the **Chemical Elements**.

Chlorinity is a measure of the chloride content, by mass, of seawater (grams per kilogram of seawater, or per cubic mile). Initially, chlorinity was defined as the weight of chlorine in grams per kilogram of seawater after the bromides and iodides had been replaced by chlorides. To make the definition independent of atomic weights, chlorinity is now defined as 0.3285233 times the weight of silver equivalent to all the halides present.

Pressure/Depth of Ocean

At the average depth of the oceans (4000 meters; 13,124 feet), the pressure is about 388 atmospheres. At the Mariana Trench, previously mentioned, the pressure is on the order of 1070 atmospheres. Pressure increases at the rate of one atmosphere per each 10.3 meters (33.9 feet) of depth, not considering corrections for temperature and varying composition.

Ocean Currents

Currents in the oceans represent the movement (flow) of water, usually in large amounts, as the result of tides, differences in water densities, the stress of various internal and external pressures, and of the wind. Forces of less importance include frictional and Coriolis effects. The movement may be either horizontal or vertical and includes the currents which are a part of the general pattern (*climatic circulation*) as well as the more transitory movements (*synoptic*).

Significant new information has been collected on ocean currents during the past decade, notably by programs in connection with the IDOE Program previously described. As with other areas of physical oceanography, data from the IDOE venture continues to be analyzed and scientists have not squared completely in their framing of new concepts as the result of new data. For this reason, in this edition of this encyclopedia, traditional concepts based upon many years of study and thought prior to IDOE are included. A major observation of the MODE (Mid-Ocean Dynamics Study) and the POLYMODE (Polygon Mid-Ocean Dynamics Experiment) projects, for example, was the discovery that ocean currents, such as the Gulf Stream and the Circumpolar Current, are not relatively quiescent streams, as previously assumed, but rather they spawn numerous rings and eddies.

Although ocean currents move very slowly when compared with the winds or the rivers, they are capable of moving large masses of water; indeed, it is estimated that the Gulf Stream opposite Chesapeake Bay moves between 70 and 90 million cubic meters of water per second. Our earliest knowledge of such current arose from ocean navigation in which a ship sighted on a fixed point, A, and was headed in the direction of that point. In the presence of a current, the ship will be displaced toward A' and from the difference between A and A' it is possible to determine the direction and average surface velocity of the current. Current measurement techniques improved over the years, including platforms, floating markers, and radioactive tracers. In studies of many years ago, almost exclusive dependence was placed on the drift bottle, a bottle that is partially loaded with sand (so that it is almost completely submerged to minimize the effects of the surface winds) and set free at a given time with a card to the finder enclosed requesting that it be returned, showing place and date where found. In earlier years, the Ekman current meter was used. This is a meter in which a propeller turns at a speed proportional to the water velocity and a tail fin keeps the propeller pointed into the current. See also **Ekman Spiral (Oceanography)**. The meter can be lowered to the desired depth on a steel wire and by various ingenious mechanisms the number of turns of the propeller in a given time can be determined as can the direction of the current, hence the speed of the current. Others, of the drift type, such as the Swallow floats, utilize neutral density floats that can be set to float at a predetermined depth from which they emit electronically sounds that can be detected by appropriate devices at a distance up to several miles. In every instance, however, the accuracy and utility of these and other devices depends on the accuracy with which the position of the meter or ship can be determined. If either of these is determined within as little as 300 feet (91 meters) by use of Loran or Decca, its velocity readings are accurate within 1–2%. It is also possible to use ships and buoys anchored in very deep water. Ocean currents and circulation are also studied by satellite altimetry. Major ocean currents are the *surface currents* and *density currents*.

Surface Currents. These are currents that move along the surface and are restricted to mainly the upper 600 feet (183 meters) or so of the sea and to a maximum depth of perhaps 1000 feet (302 meters). These currents vary somewhat with latitude, being shallower in the lower latitudes. These surface currents arise mainly from wind stress directly or indirectly and various internal pressure forces. Due to friction, the wind does not merely skim the surface of the sea, but introduces a wind effect, or stress, that carries the surface water along with it. This layer, in turn, sets up a turbulence in the deeper levels. However, due to the Coriolis effect, the surface water is deflected to the right of the wind direction in the Northern Hemisphere due to the Earth's rotation, and each lower layer is dragged still further to the right until the direction of the drift is opposite to that at the surface. The velocity of the drift, however, decreases rapidly downward

and is only $\frac{1}{23}$ that of the surface velocity at the point at which the drift is in opposite direction to that of the wind. In addition, of course, the average current of water will be at an even greater angle to the wind than the surface current. The Coriolis effect is slight near the equator and increases as the distance from that line increases.

Density Currents. Also called *subsurface currents*, these currents consist of those in which the water begins to flow downward when it is more dense than the water next to it. These may arise from several causes; evaporation, from polar waters, and from turbidity. An illustration of the first is the Mediterranean Sea in which the hot, dry climate evaporates far more water from the sea than it receives from rainfall or from rivers feeding into it. Accordingly, the Mediterranean has a greater concentration of salts than the Atlantic Ocean with which it is connected by the Strait of Gibraltar, the floor of which is only 900 feet (274 meters) below sea level. As a result, the much less dense Atlantic water flows in at the surface and the much more dense water from the Mediterranean moves into the Atlantic under the inflow.

Polar Creep. Polar waters are also responsible for density currents. In such cases—known collectively as *polar creep*, cold water from the polar regions gradually sinks down and moves toward the Equator and beyond until it encounters still colder water and is gradually forced to the surface. Due to the great depth at which such currents are found, it is difficult to estimate their rate of travel, but it is believed to be no more than a mile (1.6 kilometers) per day—taking no less than 20 years to pass the Equator. It is believed by oceanographers that these currents are of great importance to the animal life in the seas; first, by carrying dissolved oxygen to the great depths at which such life is found; second, by *upwelling*—that is, by carrying minerals from the bottom of the sea to the top where they are used for growth by the microscopic plant and animal life (plankton) which, in turn, provide sustenance for much larger animals of the sea, including the fish.

Upwelling is encountered along the western coasts of continents, such as the coast of southern California. The displaced surface water is transported away from the coast by the action of winds parallel to it or by diverging currents. Upwelling also may occur in the open ocean, where cyclonic circulation is relatively permanent or where southern trade winds cross the equator. In 1991, J.R. Lutjeharms and associate researchers (University of Cape Town) reported on the observation of extreme upwelling filaments in the southeast Atlantic Ocean. The researchers note, "Oceanic upwelling regimes play a principal role in the ecology of the eastern boundary areas of most ocean basins, in particular that of the northeast Pacific, the southeast Pacific, and southeast Atlantic oceans. It has been suggested that in many instances biological productivity of upwelling areas is concentrated at the fronts that separate cold upwelled water from the adjacent ocean surface water. Frontal behavior may therefore be an important element in the potential primary productivity of upwelling areas as a whole. Moreover, by raising deeper water to the sea surface where it is warmed by insolation and atmosphere-to-ocean heat transfer, upwelling regimes modify the temperature and salinity of substantial volumes of water. The full areal extent of upwelling regimes, as delineated by their upwelling fronts, is therefore a factor in global water mass modification and thus by implication in climate."

Surveillance by satellite has been an effective tool for gathering data on upwelling.

So-called *ocean cataracts* were noticed initially by oceanographers in the 1960s. Inasmuch as the ocean is in thermal equilibrium, it follows that heat flowing upward in the process must equal the heat flowing downward. As observed by J. Whitehead (Woods Hole Oceanographic Institute), "Because convection is an extremely efficient mechanism for transferring heat, the downward convective channels do not have to be very large in cross-sectional area to balance the heat transferred by the oceanwide warming of the deep water. The narrow currents of sinking cold water are in fact the precursors to the ocean cataracts."

These water-in-water flows can be enormous. The largest waterfalls on land pale by comparison with the relatively few ocean cataracts that have been located. In studying ocean cataracts, oceanographers use radioactive isotopes in addition to temperature measurements. Tritium, left over from nuclear bomb tests, has been used. Eight ocean cataracts have been studied—Denmark Strait and the Iceland-Faroe Passage in the north Atlantic; Discovery Gap and Strait of Gibraltar in the middle Atlantic; Ceara Abyssal Plain and Romansch Fracture in the mid-south Atlantic; South Shetland Islands and Filchner Shelf off the tip of South America and the coast of Antarctica. Exemplary of the tremendous flows is the rate of about 5 million cubic meters per second through the Denmark Strait cataract. It should be pointed out that the Shetland Islands cataract is powered primarily by salinity differences rather than by thermal differences.

Turbidity Currents. The third type of density current is the *turbidity current*—a current made more dense than the surrounding water by the addition of mud or silt. Their origin is the subject of much dispute, they have not ever been actually observed in the ocean but they are believed to be responsible for the deposits of sand and shells at great distances out on the continental slope. The regional currents include the Agulhas current. Antarctic bottom water, the Antarctic circumpolar current, Antarctic immediate water, the Antilles current, the Caribbean current, the Cromwell current, the East Greenland current, the equatorial countercurrent, the Florida current, the Guiana current, the Gulf Stream, the Gulf Stream countercurrent, the Humboldt current, the Irminger current, the Labrador current. North Atlantic deep water, the North Equatorial current, the South Atlantic current, the South Equatorial current, the South Pacific current, and the West-Wind drift. Many of these are described in separate alphabetical entries in this encyclopedia.

Continuing Research on Ocean Currents. Since the IDOE Program, considerable scientific research has been directed, during the past decade, to better understanding the nature of ocean currents. For example, the instantaneous California Current, as investigated by Mooers and Robinson (see reference), is seen to consist of intense meandering current filaments (jets) intermingled with synoptic-mesoscale eddies. These quasi-geostrophic jets entrain cold, upwelled coastal waters and rapidly advect them far offshore. This behavior accounts for the elongated, cool surface features that are seen extending across the California Current region in satellite infrared imagery. The associated advective mechanism should provide significant cross-shore transports of heat, nutrients, biota, and pollutants. The California Current is the major eastern boundary current of the North Pacific. Its flow regime is important for fisheries and climate-related processes; for oil and gas recovery operations and waste disposal; for biological, chemical and geological investigations; and for physical oceanographic studies. The source of these eddies and their role in the local internal dynamics of the California Current have yet to be determined, but are under serious study.

In a pioneering research effort, Weller and colleagues (Woods Hole Oceanographic Institution) have made measurements from the research platform, FLIP, which provide some of the first direct observations of three-dimensional flow within the surface mixed layer of the ocean. Relatively narrow regions of downwelling flow were found within the mixed layer, in coincidence with bands of convergent surface flow. At mid-depth in the mixed layer, the downwelling flow had magnitudes of up to 0.2 meter per second and was accompanied by a downwind, horizontal jet of comparable magnitude. There is some evidence that these motions transport heat and phytoplankton within the mixed layer. The researchers observed that during the day incoming solar radiation heats the upper ocean, but the amount of radiation that penetrates the ocean decays exponentially. As a result, a shallow, warm layer may tend to form at the surface during midday. At night the surface of the ocean loses heat to the atmosphere and fluid at the surface may be cooler than at the interior. This three-level flow may play an important role in the biology of certain oceanic life forms.

Since the beginning of the Global Change Research Program (GCRP), the World Ocean Circulation Experiment (WOCE) project has been considered one of the most important efforts to be undertaken by GCRP. See entry on **Global Change**.

Ocean-Continent Boundaries and Margins

About a century ago, Suess (German geologist) proposed that at one time in the development of the earth's surface there was no South Atlantic Ocean, but rather Africa and South America comprised one very large continent, which Suess chose to call *Gondwanaland*. Suess based his proposals largely upon observing the excellent fit (jigsaw puzzle analogy) between the eastern coast of South America and the west coast of Africa, rather vividly apparent from a cursory examination of a map of the world, particularly if constructed with polar coordinates. Then, in 1912, another German scientist, Wegener, extended the concept of the South America–Africa "connection" and proposed the prior existence of a supercontinent, which he proposed to call *Pangea*. It is interesting to note that, during the interim, researchers have demonstrated a number of similarities in rock, fossil, and land forms that provide a rather striking correspondence between eastern South America and western Africa. These include certain similarities of land forms, such as mountains, deserts, cliffs, and flat-topped peaks. See diagram in entry on **Earth Tectonics and Earthquakes**.

As stressed by McCoy/Rabinowitz, "The South Atlantic has intrigued scientists for decades—from Suess and Wegener to Bullard to many researchers today. This archetypal example of the development of an ocean basin by continental drift is particularly well displayed by Bullard's interpretation, where the small irregularities in refitting continental margins seem remarkably slight considering the millions of years of erosion and deposition along these margins. This, however, emphasizes an important point; such continental margins preserve ancient events with little modification. Technically, they define an ocean/continent boundary. In addition to the remarkable mesh of these margins is the geological link of older rocks, fossils, structural alignments, and even some ancient physiographic features that are found on the two opposing land masses, which are today separated by a vast ocean."

Although various postulates had been developed, a concerted program to obtain a better understanding of the sea floor did not commence until the mid-1940s. Important to the early efforts were maps developed by Heezen and Tharp which indicate a worldwide ridge and rift network on the ocean floor. This proved to be a submarine chain of mountains some 65,000 kilometers long and extending into all oceans. These mountains follow a global zone of shallow earthquakes as well as heat flows.

The greater depths of the oceans are found in trenches at the margins of the oceans. Excepting the Indonesia, Antilles, and Scotia deeps, the major trenches are found in the Pacific Ocean floor. However, as pointed out by *Uyeda*, in a broad sense, even these three exceptions may be regarded as part of the Pacific margin because they comprise the belts of active volcanoes, sometimes called the Ring of Fire around the Pacific. These are also belts of high seismic activity. Margins with trenches are Pacific-type or active margins.

Continental Displacement. Sometimes less appropriately called *continental drift*, continental displacement is a general term that can be used in connection with many aspects of the theory originally propounded at length by Wegener. In a pioneering way, Wegener postulated the displacement of large plates of continental (sialic) crust, moving freely across a substratum of oceanic (simatic) crust, but the mechanisms involved were so implausible to most geologists at that time that the concept was generally discredited for many decades. In more recent years, however, new evidence has been found and more acceptable alternative mechanisms have been proposed—so that the original theory has gained a greater degree of credence: (1) the continents have remained relatively fixed as the earth has expanded, leaving progressively wider gaps of oceanic areas between; (2) the continents have moved away from each other by sea-floor spreading along a median ridge or rift, producing new oceanic areas between the continents; or (3) the masses propelled away from the ridges consist of thick plates, composed of both continental and oceanic crust, which have moved in various directions independently of each other. Some contemporary geologists and oceanologists suggest that a true explanation of world tectonics will combine some or all of the foregoing explanations, to which will be added concepts yet to be created.

In deference to perspective, it is in order to review very briefly the content of the aforementioned concepts. The *expanding earth concept* suggests that the diameter of the earth has grown progressively larger through time, perhaps by a third or more during recorded geologic time, as a result of changes in atomic and molecular structure in the core and lower mantle, without change in actual mass. The theory has been linked with continental displacement and sea-floor spreading, although these have also been otherwise explained. Arguments and evidence for an expanding Earth were eloquently presented by Holmes. But, as pointed out by Burke, "For a variety of reasons, most earth scientists found and still find the idea of substantial expansion of the earth hard to accept (for one thing, the amount of work required against gravity is mind-boggling)."

In another concept, a *world rift system* is proposed as a major tectonic element of the earth consisting of mid-oceanic ridges and their associated trenches, such as those found along the Mid-Atlantic Ridge. The rift system is believed to be the locus of tensional splitting and upwelling of magma that has resulted in sea-floor spreading.

Another more recent and well-accepted concept is that of *plate tectonics*. The essence of this concept is that global tectonics is based on an Earth model characterized by a relatively small number (10 to 25) of large, broad, thick *plates* (composed of areas of both continental and oceanic crust and mantle), each of which "floats" on some viscous underlayer in the mantle and moves more or less independently of the others and grinds against them like ice floes in a river, with much of the dynamic activity concentrated along the periphery of the plates, which are propelled from the rear by *sea-floor spreading*. It has been suggested that the continents form a part of the plates and move with them, much as logs frozen in the ice floes. In 1965, Wilson, reporting in *Nature*, suggested that the rigid plates of the lithosphere moved along tectonically active boundaries of *three kinds*. These observations lead to an increased validity of the general continental displacement or drift theory as applied to an interpretation of the Atlantic margins. As observed by Burke, "These margins formed at divergent plate boundaries and, as the ocean grew and the divergent boundary continued to spread, the two continental fragments were carried symmetrically away from each other. At the same time, their edges were draped in young sediments. The contrasting character of Pacific coasts became understandable once it was recognized that they are places where the lithosphere is returning to the mantle and that most Pacific margins mark boundaries across which plates converge."

The third type of margin, known as a transfer margin, from present knowledge is believed to occur less frequently. In this type of margin, plates slide past each other.

The classification of margins into three well definable categories has provided for a convenient framework for investigating and reporting on ocean-continent boundaries. In the recent literature, margins thus are frequently described as: (1) *Atlantic-type* margins; (2) *Pacific-type* or active margins; and (3) *transfer* margins.

An understanding of continental margins is no longer solely a matter of scientific interest, but is economically very important today because exploration for oil and gas sometimes occurs along these margins and associated shelves. It should be pointed out that these exploratory activities have significantly contributed to an understanding of the margins and of ocean formation. See also **Petroleum**.

A margin is made up of three depth zones, proceeding in the following order from the continent toward the ocean basin: (1) *Continental shelf*, generally ranging from 30 to 300 + kilometers in width and usually with a gentle slope to a depth of less than 200 meters. The seaward end of the shelf is called the *shelf break*. (2) *Continental slope*, a narrow, plunging face, extending from the shelf break to the sea floor to a depth of some 3000–4000 meters. (3) At a point where the sea-floor gradient drops below 1 in 40, the *continental rise* extends further to a depth of some 4000–6000 meters in a much more gentle slope that may be from 80 to 500 + kilometers long. See also entries on **Continental Rise; Continental Shelf**; and **Continental Slope**. See also Fig. 2.

In 1953, an International Commission defined the continental shelf, shelf edge, and continental borderland as: "The zone around the continent extending from the low-water line to the depth at which there is a marked increase of slope to greater depth. Where this increase occurs the term shelf edge is appropriate. Conventionally the edge is taken at 100 fathoms (200 meters), but instances are known where the increase of slope occurs at more than 200 or less than 65 fathoms. When the zone below the low-water line is highly irregular and includes depths well in excess of those typical of continental shelves, the term continental borderland is appropriate."

Continental shelves comprise about 18% of the earth's total land area. Depending upon the advance or retreat of glaciers, the shelves are

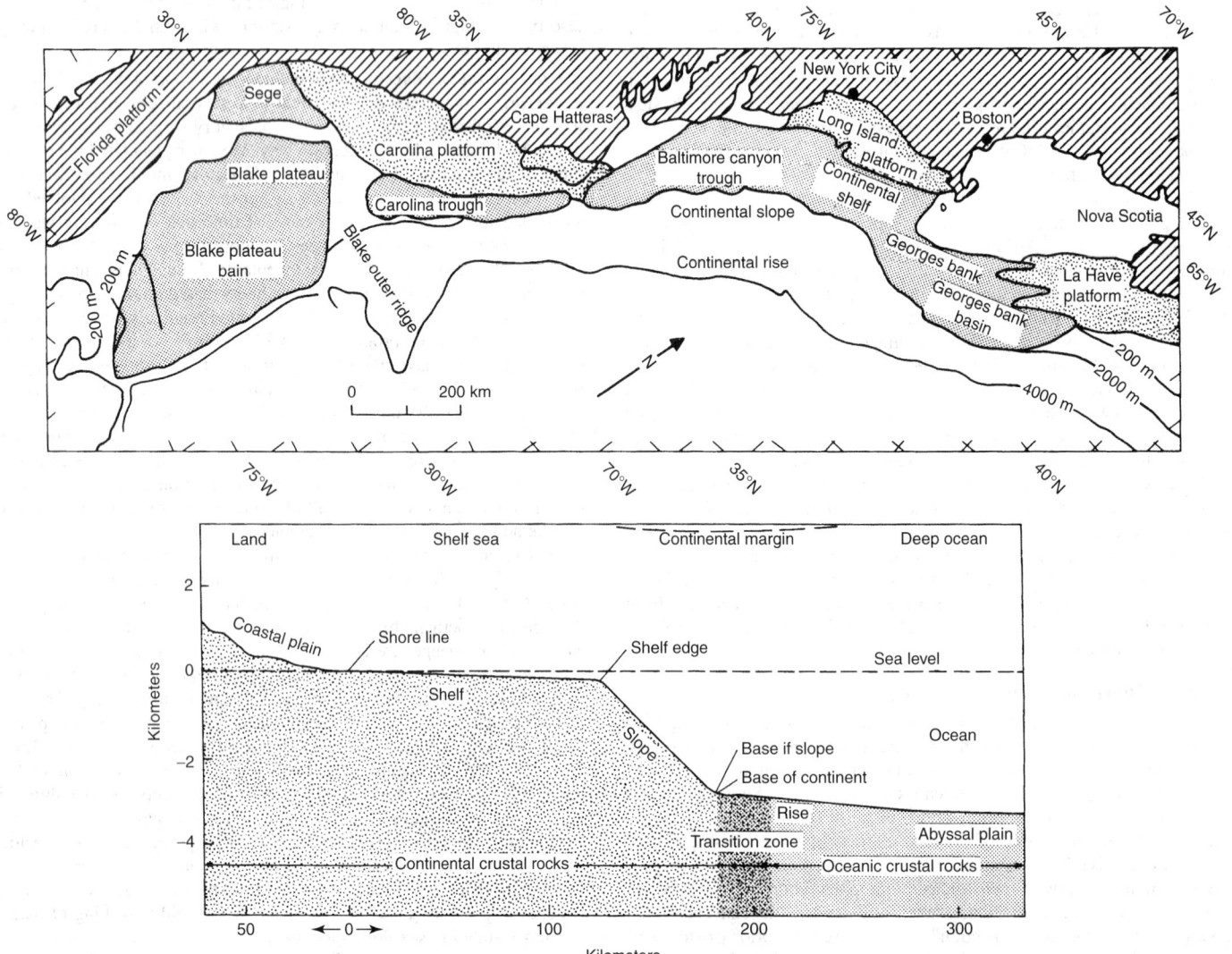

Fig. 2.   (Top) Continental margin along the eastern coast of the United States. Major sediment-filled basins and platforms are shown. *SEGE* = Southeast Georgia Embayment. (*After Schlee* et al.) (Bottom) Geomorphic features of the continental margin. (*National Petroleum Council.*)

alternately exposed and drowned. The shelves underlie about 7.5% of the total oceans. In total area, the shelves comprise about 25.9 million square kilometers.

Exploration is far from complete in terms of determining the specific parameters of the numerous continental shelves. Data are most detailed in terms of the shelves around Alaska, the southern California coast and Baha California, the northern shore of the Gulf of Mexico and the coast around Florida and northward to Nova Scotia, the coasts around Ireland and the British Isles and the northern coasts of France and Germany, and parts of the coasts of the Sea of Japan, the Yellow Sea, and the East China Sea. Continental shelf data are particularly scant pertaining to Greenland, the north central and north eastern coast of Russia (Kara Sea, Laptev Sea, etc.), most of the Canadian Arctic, Labrador, Newfoundland, Iceland, most of the coastline of South America (excepting Venezuela and Argentina), and most of Africa, Australia, and New Zealand. The gathering of continental shelf information on hand has been assisted importantly as the result of off-shore oil exploration activities and continuing studies which followed the IDOE projects.

The most accessible rocks, of course, provide information only on the most recent history of a shelf. Particular attention has to be given to properly identifying whether a rock sample is actually from the underlying bedrock of the shelf, or possibly from an adjacent projecting hill. There is also the possibility that rocks may have been laid down by ancient glaciers or streams and thus not native to the shelf under study. Unfortunately, from the standpoint of geological study, many of the drillings into the shelves have penetrated salt domes and folds rather than normal structural features. Consequently most information pertaining to the shelves has come from geophysical methods, notably seismic reflection and refraction and measurements of geomagnetism and gravity.

Investigations to date indicate two principal types of shelves: (1) Underlying sedimentary strata; and (2) underlying igneous and metamorphic rocks. Many of the continental shelves comprise the top surface of very long, essentially pyramidically-shaped sedimentary strata. Their position against the continents is maintained by long and narrow fault blocks. Along the perimeter of the Pacific Ocean, noted for its tectonic activity, there are also deep trenches, which parallel the base of the continental slope. It is estimated that a geologic dam extends for thousands of kilometers along the coastline of the western United States. Rocks found on the Farallon Islands (granitic) off San Francisco are estimated at about 100 million years of age, but it was not until about 25 million years ago that they were pushed up to form the dam of which they are a part. Several such fault dams also are estimated to have risen during the last 500 million years in the Yellow Sea. A similar dam existed along the eastern coast of the United States. With passage of time the trench was filled with sediments. These spilled over the former dam, resulting in the continental slope. The slope is maintained by the angle of slope of the sediments and, as might be expected, some instability has been evidenced by landslides and features of erosion as detected by seismic probing.

Prior to investigations of the last few decades, it was believed that sediments deposited on the shelves gradually became finer as the edge of the shelf was approached—from gravel and sand at the shore to very fine sand and ultimately silt and clay to form the so-called mud line at the edge. Samplings have not indicated this to be true. The current conclusion is that the size of sediment grains is not related to distance from shore. The former concept only holds for wave-controlled areas between the shore and perhaps depths up to 20 meters.

It is estimated that a majority of the sediment found on the continental shelves was deposited during the last 15,000 years, that is, after the last lowering of the sea level by glaciers. The finding of fossilized plant and animal life in the sediments attests to the time when the shelf areas were dry.

Atlantic-Type Margins. It is estimated that the Atlantic Ocean started to open about 200 million years ago (Triassic Period), commencing with the separation of South America from Africa and, later, the separation of North America from Europe. As pointed out by Ross, "The breakup probably started with a broad upwelling of the crust, followed by thinning, and then gradual splitting or rifting apart. The two new continents then continued spreading apart with new ocean crust forming by volcanic activity in the resulting central rift. In the early stages of the split, the new sea was long and narrow, probably with isolated depressions. Its connection with the ocean was restricted, and evaporitic conditions often developed, which led to extensive salt deposits."

The accumulation of sediment in the Atlantic-type margins is far greater than that which occurred on Pacific margins. This is as may be expected because the majority of sediment-carrying rivers (Amazon, Congo, Mississippi, Niger, Rhine, etc.) drain into the Atlantic Ocean. It is with these wedges of sediment that oil and gas are associated.

Most of the sediment introduced into the Atlantic Ocean comes from just ten rivers. A number of new concepts have been added to the original concept of displacement (drift) in an effort to further refine a model of the Atlantic-type margin. One of these refinements deals with *subsidence*, which apparently occurs when the margins cool as they age. Marginal subsidence of a mature nature, relatively speaking, is found along the eastern United States. This subsidence has been estimated at about 3 kilometers, occurring over a 150-million year period. In contrast, higher ground is found along the margins of younger oceans, such as the Red Sea, associated with which are the hills found in Egypt and the Sudan.

Many contemporary geologists and oceanologists suggest that possibly the Red Sea is a relatively new body of seawater slowly expanding to form a larger ocean. Research has indicated that the Red Sea is opening to the northeast at a rate of one to a few centimeters per year, causing a slow compression of the Persian Gulf against the Asian continent. See Fig. 3. In this process, it is postulated that the Persian Gulf region is being thrusted (subducted), along with the related coastal area of Iran, under the adjacent continental region. Ross (1979) projects that it will require some 25 million years for the Persian Gulf to be completely subducted, at which time Saudi Arabia and Iran will collide.

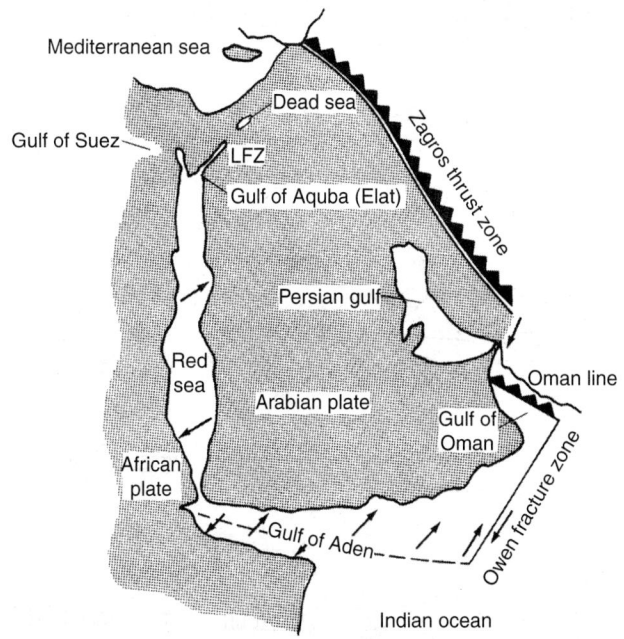

Fig. 3.   Features of the Arabian Plate showing relationship with African Plate. The spreading of the sea floor is represented by full arrows (Red Sea and Gulf of Aden). Translation is indicated by half-arrows. Subduction is indicated by wedged zone. *LFZ* = Levant fracture zone. (*After Ross.*)

It is believed by some scientists that continent-to-continent collision already is taking place along parts of the Gulf of Oman. Ross summarizes this by stating, "Translation movements, where plate boundaries slide by parallel to each other, occur in the Gulf of Aqaba (Elat), the Dead Sea, and the Jordan Rift Valley up to northern Syria. The entire length of this feature is often called the Levant Fracture Zone. A similar type of motion also occurs in the Indian Ocean along the Owen Fracture Zone and the Oman Line, which offsets the subduction along the Zagros Thrust Zone and Gulf of Oman."

Although the application of plate tectonics to a description of the Red Sea as being in an early stage of sea-floor spreading in current times is convincing to some scientists, not all investigators accept this explanation of the origin of the Red Sea. Questions include: To what extent is the current width of the Red Sea ascribable to sea-floor spreading? To what degree is this attributable to thinning and stretching of the continental crust that took place prior to and during the early spreading? These questions are explored by Francheteau and LePichon (reference listed).

An important geologic aspect of the Red Sea is the hot brine areas which occur along the bottom of the central rift. The economic value of these sediments (underlying the brines) has been emphasized by many investigators from a number of countries.

The sediments which form the thick wedges at Atlantic margins undergo changes with the passage of time and thus are good indicators for determining the age of existing oceans. See Fig. 4. For example, it is estimated that the Red Sea commenced to take form as recently as 25 million years ago, whereas it is estimated that the South Atlantic Ocean is some 130 million years old, and the Central Atlantic Ocean some 160 million years old. Some scientists believe that the type of sediments being collected today in the East African Rift typify the phenomenon of early sediment accumulation. It has been further postulated that after initial sediment accumulation, formation of limestone structures occurs. Evidence indicates that widespread limestone accumulation in the Central and South Atlantic occurred over a period of some 20 million years, and this was followed by accumulation of marine sands and muds. Of course, as an ocean matures, the distribution of sediment along the margins becomes less uniform. Generally, past alterations in sea level have been attributed to cyclic melting and freezing of the polar ice caps. In more recent years, many geologists and oceanologists now believe that another major cause of sea level changes has been the changes in the rate of sedimentation, i.e., the rate at which the ocean floors are made.

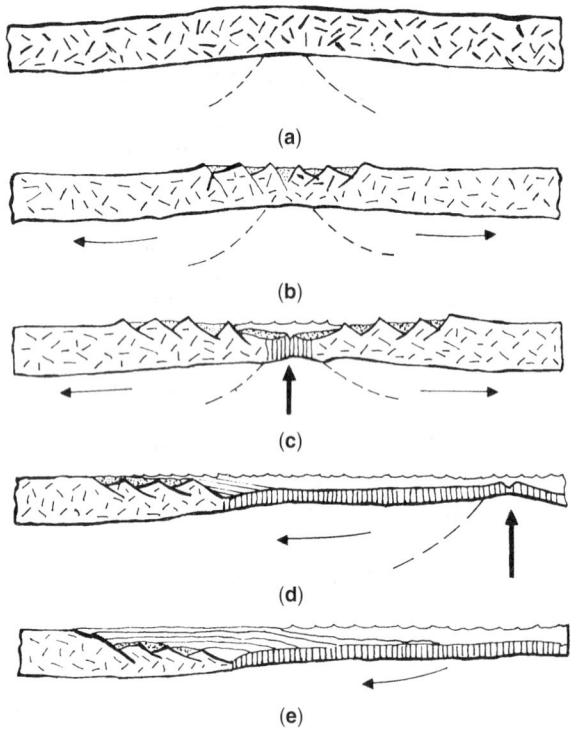

Fig. 4.   Formation and evolution of a passive continental margin: (**a**) doming, (**b**) rifting, (**c**) youthful drifting, (**d**) intermediate drifting, and (**e**) mature drifting. (*After Oceanus.*)

It is postulated that because the young ocean floor is hot, it requires a larger volume—thus displacing water and causing a rise in sea level. Relative to the sea floor, the sea level was perhaps twice as high as the present sea level during the Cretaceous Period (136–65 million years ago). It has been postulated that this striking rise in sea level was the result of a combination of large new accumulations of sediment coupled with the greater volume required for the hot, young materials forming new oceanic crust. At much later times, the crust cooled and subsided, adding another factor to the complex series of events that affects sea level.

Ocean Floor and Crust — Active Margins

At one time, geologists believed that the earth's crust was a reasonably stable layer forming an envelope around the fluid mantle and core. It was envisioned that crustal blocks floated on a plastic mantle. This concept was not supported by evidence, particularly of the kind assembled during the

past few decades. Rather, it is now envisioned that at least three concepts are at work, i.e., sea-floor spreading, continental drift, and plate tectonics. Crustal plates apparently move apart at a rate of 1 to 10 centimeters annually. As they spread apart, there is an upwelling of basaltic material that causes a ridge. Upon cooling, the molten rock adheres to the crust that is moving away on each side of the fissure. When the movement is slow (about 3 centimeters per year), a rather steep sloping of the sea floor on either side of the ridge occurs. Steep escarpments are produced because the sinking dominates the process informing the slope. On the other hand, when movement of the plates is at a faster rate, the horizontal spreading is rapid in contrast with the sinking of the crust. The steepness of the slope is determined by the balance between these two actions.

The Atlantic Ocean is known for slow spreading. The Pacific Ocean is known for fast spreading. Thus, in a continuous process running over millions of years, there is a slow but steady spreading and renewal of the sea floor, the spreading ranging from 1 to 10 centimeters per year. It has been estimated that the crust sinks about 9 centimeters per year for the first ten million years after it forms, dropping to somewhat over 3 centimeters per 1000 years for the next 30 million years, and to about 2 centimeters per 1000 years after that. However, not all crust sinks, a notable example being that of the southern Mid-Atlantic Ridge where it is estimated that the sea floor level has not changed for the last 20 million years. Regions where spreading occurs in multiple directions are termed spreading centers. These centers also move.

When a crustal plate grows, its leading edge is destroyed at an equal rate. In some instances, the edge may slide under the oncoming edge of another plate and return to the asthenosphere. Such action causes a deep trench. The Mariana Trench and Tonga Trench in the Pacific are examples. If plate movement is relatively slow (5 to 6 centimeters per year), compressional forces created upon encountering another plate may be absorbed within the plate, with resulting mountain building by buckling up, as may have occurred in the formation of the Himalayas.

Ocean floor sediments are of two principal types: (1) *Terrigenous deposits* of terrestrial origin; and (2) *pelagic deposits*, consisting of matter produced within the sea itself. Close to the edges of the great land masses a variety of sands, silts, clays, and marls will be found on the continental shelves and epicontinental seas and, to a lesser extent, on the continental slopes. This terrigenous matter is deposited by rivers or secured by the effects of erosion on the coastlines and subsequently deposited by the action of currents. Such accumulations may total many hundreds of feet in depth. Other deposits have been formed by icebergs carrying materials that are released and settle to the ocean floor as the berg melts. In certain deep parts of the ocean floor there are deposits of red clay, believed to be of terrestrial origin. These are of extremely fine texture and are carried in suspension in the ocean currents until they finally settle to the ocean floor. The maximum accumulation of this red clay is believed to be at the rate of $\frac{1}{25}$ of an inch (1 millimeter) per 1000 years. See Fig. 5.

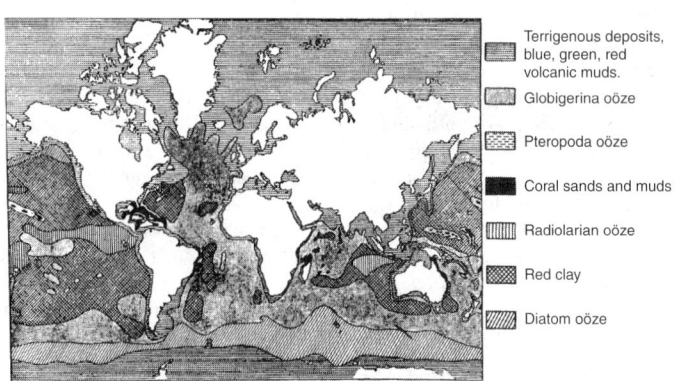

Fig. 5.   Distribution of marine-continental and oceanic deposits. (*After Scott and Collet.*)

The deep sea oozes of pelagic origin are of two kinds. The *calcareous* ooze consists of the shells of the single-celled *Globigerina*, of a small mollusk, the pteropod, or of certain plants. These calcium carbonate shells are found only in depths less than approximately 14,500 feet (4420 meters); at greater depths, the available carbon dixoide is able to dissolve such shells. The second type of *pelagic* ooze is that of *siliceous* origin, consisting of the silicon skeletons produced by organisms known as radiolaria and diatoms. Because it is not subject to dissolution, this siliceous ooze is found at very great depths. The rough average of these sediments over the entire ocean floor is estimated to be about 2,000 feet (610 meters).

For many years, the mysteries of the Mediterranean Sea have been probed by oceanographers who have found the Mediterranean an excellent laboratory for geological and geophysical studies and for physical oceanography. Scientists continue to probe the geological history of the two complex deep-sea basins that comprise the Mediterranean. As some researchers have observed, "If rocks could tell tales, the Rock of Gibraltar would be a master storyteller."

In the late 1880s, a number of European geologists suggested a number of theories pertaining to the tectonic origin of the Mediterranean, but any substantiation had to await the development of modern deep-sea drilling technology and oceanographic instrumentation. For a number of years, some researchers proposed that the Mediterranean was a huge dry desert, situated some 3000 meters below present sea level. It was likened to a "death valley" scenario. Proof of this concept has been lacking, and the theory no longer has strong support among the professionals. In 1990, D. Stanley (National Museum of Natural History) remarked, "I believe the seafloor remained almost continually covered by very saline waters, perhaps one hundred to several hundred meters deep."

Almost an entire issue of the prestigious *Oceanus* magazine (Spring 1990) is devoted to the various scientific aspects of this interesting sea. It is also noteworthy to report that, after more than a decade of research and regulations, the effects of the Mediterranean Action Plan indicate some progress in the struggle to overcome severe pollution.

Subduction. As early as the late 1920s, Holmes (University of Edinburgh) and others suspected that the tectonic features at active margins were the result of the workings of some common process. It was further suggested that the process may be a down-thrusting mantle convection current. These concepts were later refined in the light of an emerging theory of plate tectonics and it was proposed that the leading edge of a rigid oceanic plate may be the specific material being thrust downward into the mantle. Subduction, a concept originally proposed by Alpine geologists, may be defined as the process of one crustal block descending beneath another, by folding or faulting or both. A subduction zone may be defined as an elongate region along which a crustal block descends relative to another crustal block, e.g., the descent of the Pacific plate beneath the Andean plate along the Andean trench.

Because evidence indicates that mid-oceanic ridges are continually being generated, it is logical to assume that they are being consumed elsewhere in order to keep the surface area constant.* Thus, the subduction or consumption of oceanic plates may be the logical consequence of sea-floor spreading. As observed by Uyeda, active margins, with deep trenches and seismicity, were the most obvious localities. Uyeda asks, "Can the process of subduction really explain the tectonic features at active margins?" In studying a number of active regions, including the Benioff-Wadati zone underlying Japan, the seismicity of an area generally fits the concept of subduction. As of the early 1990s, the subduction theory does not offer a full explanation, particularly concerning the *thermal regime* of the subduction zones. High heat flow in the back-arc basins, the low velocity and high attenuation anomalies of the mantle wedge, and the active arc volcanism all require further explanation. Perhaps the answers lie with the formulation of two or more modes of subduction. Two such modes are described by Uyeda.

Mid-Ocean Ridges Research. Investigations accelerated markedly during the 1980s and early 1990s toward understanding the mid-ocean ridges. These ridges wrap around Earth for over 70,000 kilometers (44,500 mi)

* It is of historical interest to introduce the theory of the *contracting earth*, widely believed in the 19th and the first part of the 20th centuries—to the effect that orogenic and other structures of the earth were produced by compression of the crust during its gradual contraction on the surface of a cooling, but originally molten globe (a familiar textbook illustration of the time was a dried apple). The theory has since been discredited, as the evidence shows that the earth is not cooling and contracting in the manner then believed.

It should also be mentioned that the expanding earth concept is described earlier in this entry.

and have been likened to the "seam of a baseball." These ridges are considered to be the most volcanically active mountain chain in the entire solar system.

Contemporary researchers tend to support the concept of a hierarchy in the segmentation of mid-ocean ridges. As described by K. Macdonald (University of California, Santa Monica), "*First-order* segments are generally hundreds of kilometers long, persist for millions to tens of millions of years, and are bound by relatively rigid, plate-transform faults, called *first-order discontinuities*. A first-order segment usually is divided into several *second-order* segments." The latter are shorter, are not as long lived, and are bounded by non-rigid, second-order segments of smaller magnitude, third- and fourth-order discontinuities. It has been established that these discontinuities can migrate along the length of a ridge.

It is interesting to note that the chain of active volcanoes that comprise the mid-ocean ridges expel, during an average year, ten times more lava than that which flowed from Mt. St. Helens in 1980.

Oceanographers in the past generally have directed their research of the mid-ocean ridges on a segment-by-segment basis. In their studies, the theory of plate tectonics served as a foundation. The principal plates and ridges are indicated on a map contained in the entry on **Earth Tectonics and Earthquakes**. With improvements in drilling techniques and exploratory instrumentation, scientists now can concentrate on the macrostructure of the mid-ocean ridge system. Pertaining to the *systematics* that have emerged, Macdonald comments, "Is the architecture of the global mid-range ridge system really so orderly, or is this concept of a 'segmentation hierarchy' merely a human construct?"

As of 1992, D. Blackman and T. Stroh (InterRidge, an international project) summarize their key goals as including: (1) characterizing the global ridge structure; (2) understanding crustal accretion and upper-mantle dynamics; (3) charting the variability over time of volcanic and hydrothermal systems; (4) mapping biological colonization and evolution at ridge crests; (5) determining the properties of multiphase materials at ridge crests; and (6) developing technology for ridge-crest experimentation.

Examples of new technological advances include chemical sensors that detect minute changes in trace elements and compounds (such as hydrogen sulfide, methane, iron, manganese, and oxygen), geodetic instruments to measure uplift and tilt of volcano flanks, broadband ocean-bottom seismometers, and deep-water temperature and chemical profiling systems. Systems that can deploy and manipulate these sensitive instruments will also be required and may take the form of remotely operated seafloor vehicles or manned submersibles.

P. Lonsdale and C. Small (Scripps Institution of Oceanography) reported in 1992 of studies on seafloor spreading, a process that creates new material to fill in gaps between Earth's separating crustal plates, which results in broad elevations with spreading centers along their crests. Examples of how mid-ocean ridges and ocean basins are created are given. See Fig. 6.

In 1992, W. Bryan (Woods Hole Oceanographic Institution) reported on studies of the evolution of deep-sea volcanology. It is interesting to note that the first volcanic rocks from a mid-ocean ridge were accidentally sampled during cable-laying operations in the North Atlantic in 1874. As reported by Bryan, "Throughout the first half of the 20th century the seafloor was widely assumed to be basaltic, but evidence for this assumption was still sketchy and indirect. A 'basaltic' and therefore 'volcanic' seafloor was consistent with the arguments based on isostasy and bathymetry that remain valid today. The continents must stand high, because they are composed of relatively thick, light granitic rock that literally floats higher on the underlying mantle than does the thinner, heavier rock comprising the oceanic crust. Also, petrologists generally assumed that basalts of volcanic islands such as Hawaii or Iceland were representative of the rocks to be found on the deep seafloor. Although there are often striking differences between continental volcanic rocks and the deeper crustal rocks on which they have erupted, the shaky logic of this analogy as applied to the seafloor does not ever appear to have been challenged."

When the reality of seafloor spreading and plate tectonics became generally accepted in the mid-1960s, mid-ocean ridge volcanoes proved to be the answer needed to explain the process for creating new seafloor. Deep-sea drilling programs proved invaluable toward reaching conclusions drawn today. One of the greatest challenges remaining is that of long-term observation of several of the numerous volcanoes located along the Mid-Atlantic Ridge. Also, models will be required to prove if hot spots are the locations of upwelling "mantle plumes" that carry fresh, hot, and previously

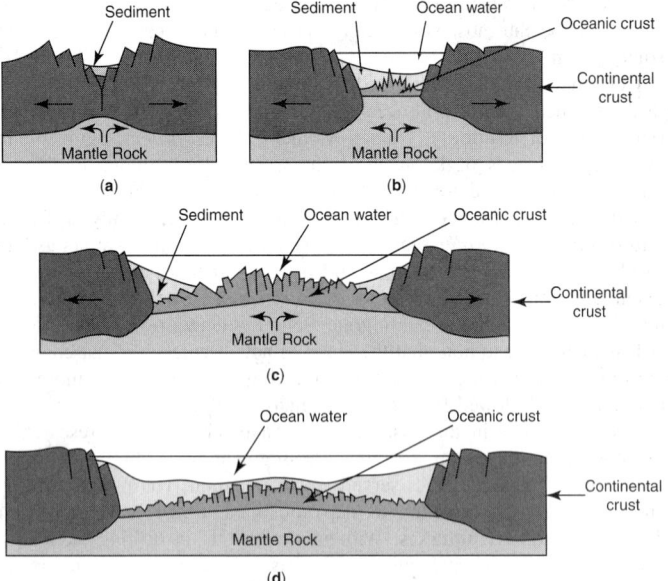

Fig. 6. Development and demise of a mid-ocean ridge: (**a**) Rift valley — the deep central cleft with a mountainous floor in the crest of a mid-ocean ridge. The valley results from plate separations, as fast-spreading ridges and upwelling magma fill the rift and smooth the topography, while at slow-spreading ridges the upwelling magma does not fill the rift, but adheres to the trailing edge of the spreading plates. Example of a rift valley is the East African Rift. (**b**) Continued crustal separation produces a gap that is partly filled by sediment washed off the continents and partly by melting of the mantle to produce oceanic crust. Example is the Gulf of California. (**c**) As the gap between separating continents continues, the gap increases and oceanic crust formation by seafloor spreading at the crest of a rifted mid-ocean ridge develops. Example is North Atlantic Ridge. (**d**) When continental separation ceases, seafloor spreading stops and the mid-ocean ridge subsides as it cools and gradually becomes covered with sediment. (*After Lonsdale and Small.*)

undepleted mantle from a deep, previously untapped source. The principal hot spots on the Mid-Atlantic Ridge now recognized include the Azores and Tristan de Cunha in the south Atlantic.

J. Karson (Duke University) is investigating the tectonics of slow-spreading ridges. Karson observes, "There is a growing awareness that fast- and slow-spreading ridges function in very different ways. The sputtering magma supply of slow-spreading ridges results in substantial periods of plate separation that involve stretching and faulting of relatively cool oceanic lithosphere with little or no magmatism. The fault patterns of the median valley appear to mimic those of continental rifts; however, at least locally, very highly stretched and thinned masses of crust and upper mantle occur. The median-valley geology and fault structure documented by near-bottom studies predicts a very heterogeneous geological structure in slow-speed crust. This result is yet to be clearly defined or reconciled with the geophysical expression of the crust away from spreading centers. Future studies of the geometry and kinematics of faulting on slow-spreading ridges will determine the nature of faulting over much larger areas than have been studied to date, and will help contribute to the overall understanding of how the lithosphere is pulled apart to form rifts in both the continents and the seafloor."

E. Bergman (U.S. Geological Survey) reported in 1992 that "Systematic studies of mid-ocean ridge earthquakes have produced many insights concerning the tectonics of accreting plate boundaries. The largest transform earthquake in the last three decades on the northern Mid-Atlantic Ridge was a magnitude 7 event on the Vena Transform in 1962." Several approaches are being investigated for deploying high-dynamic range, broadband seismometers in deep ocean basins for the Ocean Seismic Network. These types of instruments would greatly enhance the ability to monitor mid-ocean ridge seismicity. In terms of future studies, Bergman observes, "Transform faults are an end member of a spectrum of geologic features associated with offsets of mid-ocean ridge spreading segments. Little is known about the seismicity associated with very small offsets. From a seismological point of view, it is natural to define a transform as a strike-slip focal mechanism. This definition may not be consistent with one

based on morphology. The issue has yet to be investigated. Obstacles to such a study include obtaining sufficiently accurate epicenters to unequivocally place earthquakes on small ridge offsets and the lack of a reliable means to determine focal mechanisms for earthquakes with magnitudes less than about 5."

Permanent seafloor geophysical observatories are on the horizon.

It is interesting to note that *seaquakes* were reported as early as the 19th century, but not in a systematic way. In the late 1950s, seismology was investigated during the 1956–1957 International Geophysical Year (IGY) project. The importance of mid-ocean ridge seismicity soared in the late 1960s, when it provided compelling evidence for the plate-tectonic hypothesis.

Ophiolites. The concept of crust formation at mid-oceanic ridges has been largely postulated on the basis of measurements of magnetic and seismic patterns and upon the examination of sediments recovered by exploration drilling vessels — all well-established methods, but techniques which are undergoing continuous improvement. In more recent years, the study of ophiolites has made a major contribution to the understanding of crust formation.

An *ophiolite* may be described as a *group* of mafic and ultramafic igneous rocks ranging from spilite and basalt to gabbro and peridotite, including rocks rich in serpentine, chlorite, epidote, and albite derived from them by later metamorphism, whose origin is associated with an early phase of the development of a geosyncline. The term was originated by Steinman in 1905. Ophiolites are found on continents and for many years were not directly associated with oceanic crust. The existence of ancient oceans over present continental material was mainly suspected as the result of finding the fossilized remains of marine plants (diatoms) and marine animals (shells, etc.). For example, in the middle of the 15th century, the Swiss naturalist Hemerli found some fossils in Alpine rocks, well removed from the present oceans and thousands of meters above sea level. Considerably later, marine fossils from the Triassic period (200–230 million years ago) were found in the high peaks of the Himalayas. An association of mountains with the remains of marine activity became established and, during more recent periods, much additional evidence of the mountain-ocean relationship has been gathered.

An association of ophiolites with ocean crust in recent years has been established as the result of oceanographic technology, including seismic probing, drilling, dredging, and observations from submersibles. As pointed out by O'Connell, the resemblance between ophiolites and what is known about ocean crust is so strong that ophiolites are thought to be huge slabs of oceanic crust that broke off subducting ocean crust and were pushed onto continents and island arcs. O'Connell further observed that, as fragments of ocean crust, ophiolites hold intriguing scientific information about the location of ancient oceans and the formation of ocean crust, and they also can be the site of important mineral deposits. In some ophiolite complexes, such as at Hare Bay in northwestern Newfoundland, no economically recoverable minerals have been found, but in others, such as Troodos on Cyprus, there are extensive deposits. Sulfide minerals may occur as large deposits in the basalts of the ophiolites. It is interesting to note that there is archeological evidence to the effect that possibly the first smelting of sulfide ores and copper production occurred on Cyprus as early as 4000 B.C.

Considerable thought and research has gone into describing how ophiolites may be formed and become oceanic crust. See Fig. 7. Ultramafics frequently undergo a process known as serpentinization. This may be defined as the process or state of hydrothermal alteration (metasomatism) by which magnesium-rich silicate minerals (olivine, pyroxenes, amphiboles, periodotites, and other basic rocks) are converted into or replaced by serpentine minerals, forming serpentinite. This process tends to obliterate much of the original texture of the rocks. Serpentinite is a dark green color.

Deep-Sea Hot Springs and Cold Seeps

One of the most interesting findings, of which there were many, made during the IDOE program was the discovery of deep-sea hot springs and cold seeps with their associated oases of life. These phenomena have been under vigorous study ever since. Formation of oceanic crust on the Mid-Atlantic Ridge southwest of the Azores at depths of 2700 meters were observed for the first time by the IDOE French-American Mid-Ocean Undersea Study (FAMOUS) team in 1974. Along this line, the American and African crustal plates are separating at a rate of 2 to 3 centimeters per year.

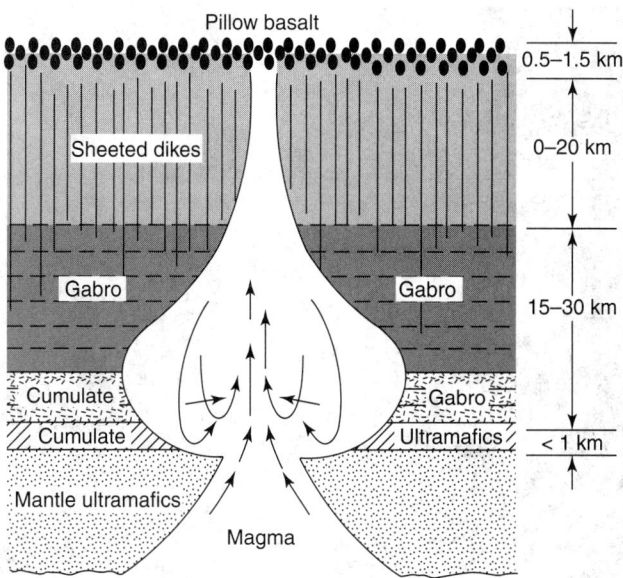

Fig. 7.   Schematic diagram of process that leads to production of cumulate rocks (ophiolites) that in some ways resembles the formation of sedimentary rocks. Magma protrudes through an opening in the mantle ultramafics (rocks of the mantle that have been partially melted, deformed, and recrystallized) and rises to form a "magma chamber," visualized as having the form of a thin-necked, upright vase. It is envisioned that convection results as the magma at the top and adjacent to the edges of the chamber cool and thus migrate downward, while the new, warmer material rises. It is theorized that this may be the manner in which oceanic crust (or some of it) is formed. (*After O'Connell, 1979; and Peterson et al., 1974.*)

Fig. 8.   Lava formation known as collapsed blister pillow is caused by tectonic activity on rift valley floor of Mid-Atlantic Ridge. (*Woods Hole Oceanographic Institution.*)

Fig. 9.   Bulbous pillow lavas with cracked crust. Note sponge growing on pillow in left center of view. To the right is elongate or "Cousteau" pillow. Submersible shown is the Alvin. (*Woods Hole Oceanographic Institution.*)

Fig. 10.   Lava formation found on Mid-Atlantic Range. (*Woods Hole Oceanographic Institution.*)

Investigations showed that the ridge crest has a fault-bounded central valley about one kilometer deep and 2 to 4 kilometers wide. Volcanic rock was found extending across the width of the floor, but was most prevalent along a line of central valley hills. Mid-Atlantic Ridge lava formations are shown in Figs. 8, 9, and 10. As observed by Davin and Gross (1980), systematic compositional variation in the lavas across the valley floor apparently reflects a zoning or evolution in the underlying magma chamber. However, it is believed that several lava flows with discrete geochemical characteristics may result from mantle-derived magma moving into the chamber. This compositional zoning is one of the most important discoveries made by the scientists of the FAMOUS project and thus far has not been documented at any other location. Two fracture zones were studied extensively through photography, dredging, and submersible dives and

little evidence of recent faulting was observed, although microearthquakes are frequent along the faults. The FAMOUS project demonstrated that detailed geological mapping of rough, deep-sea terrain can be done by submersibles, and it assisted an understanding of the principal components in the seafloor spreading process, namely, a narrow rifted valley with an axial volcanic ridge fed by an underlying magma chamber.

The Nazca Plate off the west coast of South America also was studied during IDOE for its tectonic plate cycle. This includes the generation of new crust along the East Pacific Rise and processes at the zone of continental plate collision where oceanic plate is partly subducted along the Peru-Chile Trench and assimilated beneath South America. See Fig. 11. Researchers found submarine hot springs on the seafloor, around which new mineral deposits were forming. As observed by Davin and Gross, "These features were first observed at the Galápagos Rift in 1977, and then more extensively in 1978 and 1979 in a larger-scale cooperative program among French, Mexican, and American scientists at the Rivera Fracture Zone off the west coast of Mexico near the mouth of the Gulf of California. Seawater circulates through newly formed oceanic crust, removing heat and reacting chemically with the rocks. Recently formed volcanic rock (erupted at temperatures of about 1200 °C) causes very hot waters to be discharged at temperatures of about 350 °C through vents on the ocean floor. These vents have been observed on the East Pacific Rise project off Mexico. Vents occur on fresh basalt in clusters and narrow bands about 250 meters across and several kilometers long. Individual vents form irregular chimneys nearly 10 meters high and about 4 meters across. The discharges, 1 to 2 meters per second, form dark, smokelike plumes in the overlying waters. The scientists who first saw the vents in 1979 called them "black smokers." See Figs. 12 and 13.

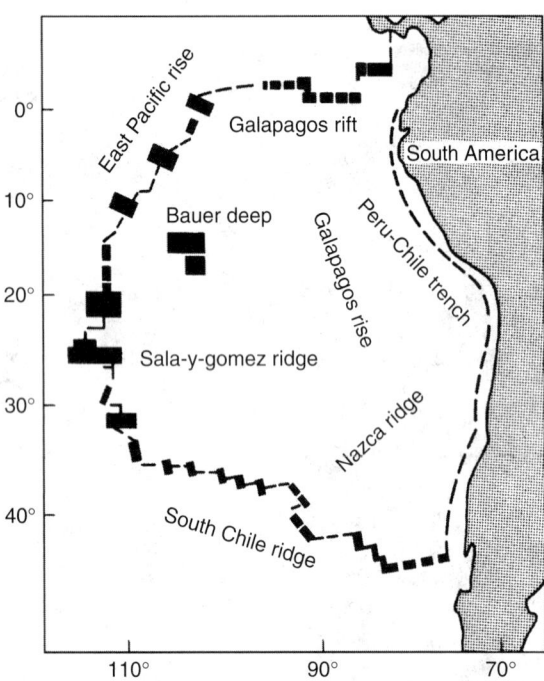

Fig. 11.   Nazca Plate off west coast of South America. New crust is being formed along East Pacific Rise. (*After Davin and Gross, 1980.*)

Made of silica and metal sulfides (copper, nickel, cadmium), the freshly precipitated metal sulfides are carried upward as plumes. At 300 °C, seawater density is only 0.7 grams/cc and therefore a buoyant plume is formed. Ultimately, the sulfide particles settle and enrich the ridge crust sediment. It is interesting to note that filter-feeding organisms, which live off bacteria that grow in hydrogen sulfide, are found at these hot springs.

In an effort to explain metalliferous deposits on the Nazca Plate, a so-called geo-still concept was developed. This is well explained by Davin and Gross: "A major portion of the sediments on the moving plate descends into the subduction zone. As the plate reaches mantle depths, materials are heated and ore-forming solutions move into the overlying rocks while refractive materials remain in the mantle. Molten rock (magma) rises in large batholiths to within a few kilometers of the surface. As the magma cools, copper is concentrated in deposits near the top of the formation. Erosion subsequently exposes these deposits for exploitation."

Based upon IDOE data and additional explorations, a highly schematic representation of the "black smoker" geochemical process is shown in Fig. 14. As explained by Edmond (Massachusetts Institute of Technology), black smokers form by the precipitation of $CaSO_4$ (anhydrite) and Fe, Zn,

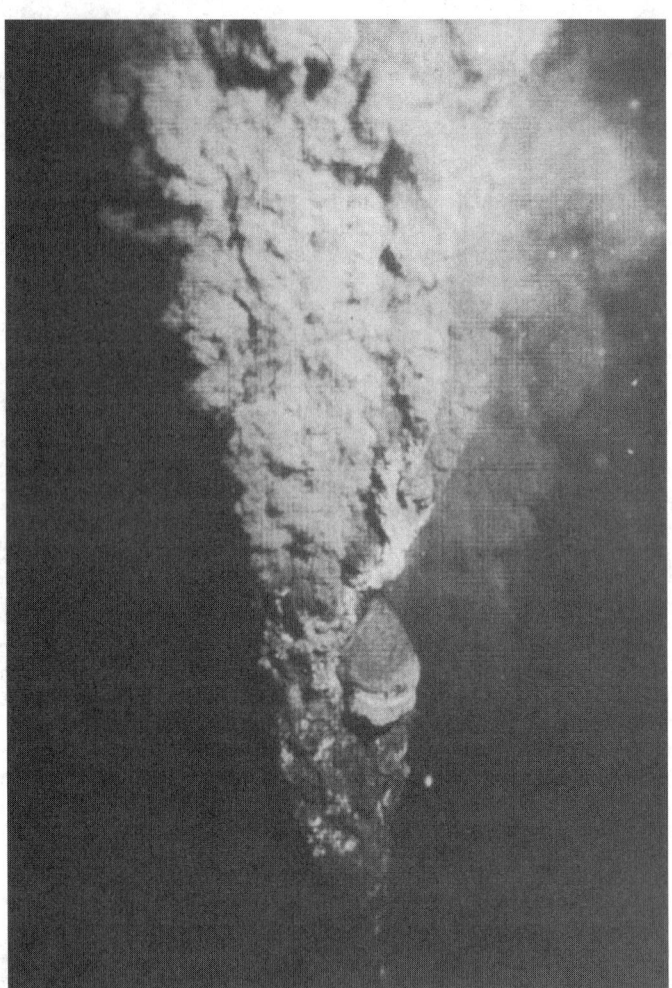

Fig. 12.   Black smoker on East Pacific Rise (21 °N). (*Photo by Dr. R.D. Ballard, Woods Hole Oceanographic Institution.*)

and Cu sulfides. The hot solutions exiting the smoker are buoyant relative to the surrounding cold water. Thus, they rise and disperse into the water column. Edmond also observes that generally void spaces in rocks are filled with water. When molten material from the mantle intrudes into the crust, this water is raised to high temperatures. If there is sufficient permeability, the water will convect to the seafloor, where it forms *hot springs*. If this condition occurs on land (for example, the geysers and hot springs at Yellowstone Park, U.S.A.), much of the hot water recirculates back into the rock, thus making it much more difficult to understand the chemistry of such springs. The situation is much simpler at mid-ocean spreading centers. Seawater enters the highly permeable crust through tectonically induced faults and through contraction cracks caused by rapid cooling. The heated seawater exits through the undersea vents, rises through the water column above the vent orifices, and dissipates. The comparatively simple chemistry of the two reactants—basalt and seawater—facilitate experimentation in the laboratory and the construction of models.

In addition to the earlier studies at the Galápagos Spreading Center, previously mentioned, hydrothermal activity at the ocean bottom in other regions has been observed and studied. These include hot vents and hydrocarbon seeps found in the Sea of Cortez (Gulf of California). In this area, a whole system of spreading centers at which new oceanic crust is formed by cooling of molten rock has been found. As described by Lonsdale (University of California, San Diego), where the system comes ashore (beneath the Colorado River delta in the Salton Trough), there are high-temperature geothermal fields (for possible later exploitation for electricity generation). See also **Geothermal Energy**. Other spreading centers have been found in the Guaymas Basin in the central gulf. Marine polymetallic sulfide deposits have also been found on the Juan de Fuca Ridge and also are presumed to exist on the Gorda Ridge nearby (in the U.S. Exclusive Economic Zone). Broadus and Bowen (Woods Hole Oceanographic Institution) are exploring the feasibility and economics of mining such sulfide deposits.

Fig. 13.   Black smokers on East Pacific Rise. (*Photo by Dudley Foster, Woods Hole Oceanographic Institution.*)

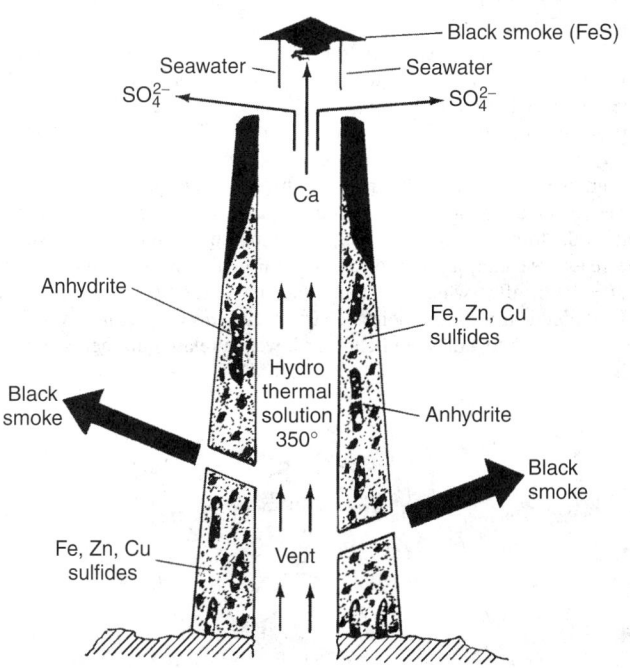

Fig. 14.   Highly schematic diagram of a black smoker found in certain hydrothermal areas of the sea floor. These smokers are formed by the precipitation of anhydrite (CaSO₄) and iron, zinc, and copper sulfides. The hot solutions exiting the smoker are buoyant relative to the surrounding cold seawater and consequently rise and disperse into the water column. (*After Oceanus.*)

When exploring the processes of formation and subsequent erosion of the passive continental margin off the west coast of Florida, a group of researchers serendipitously found exotic communities with abundant organisms at a depth of 3266 meters in the abyssal Gulf of Mexico. The Atlantic gulf community was found to contain the same types of organisms that characterize the Pacific vent communities — white bacterial mats; large, dense beds of mussels; numerous small gastropods; the shells of live mussels; thick patches of 1-meter long tube worms; red-fleshed vesicomyid clams; galatheid crabs; and eel-like zoracid fishes. The Pacific vent communities are found immediately adjacent to hydrothermal vents associated with mid-ocean ridge crest magma sources. As pointed out by the researchers, until the Florida discovery, these were the only known

large deep-sea biological communities that receive their primary energy from chemical sources rather than from solar radiation via photosynthesis. It is proposed that the chemical energy is released by bacterially mediated oxidation of reduced inorganic compounds, such as hydrogen sulfide (H₂S) coming from the hot vents. This is a process of *chemosynthesis* and is described in some detail by Jannasch (Woods Hole Oceanographic Institution).

In 1994, research of hydrothermal vent systems is continuing. As pointed out by M. Tivey (Woods Holes Oceanographic Institution), "Hydrothermal systems transfer large amounts of heat and mass from Earth's interior to the oceans. Fluids exiting the chimneys are metal-rich, hot, and acidic, and vent at velocities on the order of meters per second. A striking feature of black smoker chimneys is how remarkably thin their walls are. They range in thickness [5 in (12.7 cm) to $\frac{1}{4}$ in (6.3 mm)]. Across this thin layer is a temperature difference of 300 °C or more."

Principal questions for which answers are being sought include: (1) Where is all the fluid coming from? (2) How does it circulate? (3) How does it get so hot? (4) How do the fluids become metal-rich? (5) Where do the particulates come from? (6) How do chimneys form?

As posed by researcher Tivey, more advanced questions would include: (7) What is the extent of hydrothermal venting at mid-ocean spreading centers and back-arc basins? (see also **Volcano**) (8) What is the significance of variation in fluid composition, temperature, flow rate, and composition of solid precipitates among hydrothermal sites? (9) How long are vent sites active? (10) Does fluid composition change with time and, if so, on what time scale? (11) What proportion of minerals is deposited at the vent site versus dispersed into the water columns as black smoke?

As part of an ongoing National Science Foundation–funded project to study the genetics and dispersal mechanisms of organisms inhabiting vent environments, several biological expeditions to the East Pacific and the Gulf of Mexico were staged in 1991. One exploration was that of the *Rose Garden* hydrothermal vent site (Galápagos Rift). Previously, it had been visited in 1979, 1985, and 1988. No major changes were noted from prior visitations. Other known vents also were visited. Numerous species of life were found and cataloged. The research vessel used was DSV *Alvin*. An informative chart listing the various vent and seep regions and their known resident fauna is available from researcher R.A. Lutz (Woods Hole Oceanographic Institution, Salem, Massachusetts, 1970).

The research party also investigated seeps at several locations. A seep is a place of contact between deep-sea sediments and limestone walls where hypersaline waters seep onto the seafloor and feed sulfide-dependent biological communities.

In 1992, D. Toomey (University of Oregon) reported on progress being made in the tomographic imaging of spreading centers. The researcher observes, "In recent years, working models of oceanic spreading centers have evolved from two-dimensional, steady-state idealizations to more realistic three-dimensional, time-dependent systems. The new dimension added to the working models is the pervasive along-strike variability of mid-ocean ridge processes, notably in the production of melt beneath the spreading center. Current hypotheses suggest that ascending melt within the mantle is focused into magmatic centers separated on the order of tens to a hundred kilometers. Each magmatic center supplies the greater portion of melt and heat to a single ridge segment. Within an individual ridge segment, processes such as faulting, hydrothermal circulation, and magmatic accretion vary systematically as a function of distance from the magmatic center. The hypothetical structural unit, consisting of a local maximum of magmatism bounded by along-axis minima, became known as a spreading-center segment or cell. This simple model of cellular segmentation provides an improved, but controversial, working hypothesis for mid-ocean ridge studies."

Tomographic studies of seismic velocity structure beneath local segments of the East Pacific Rise, the Mid-Atlantic Ridge, and the Icelandic rift represent a new and powerful approach to the seismological study of divergent plate boundaries.

Wave Action in the Ocean

Wind blowing across the surface of water exerts a force on the surface of the water in the direction of the wind. This interface phenomenon between the ocean and the atmosphere (see also **Atmosphere-Ocean Interface** for description from meteorological viewpoint) is far from understood in any degree of detail. The rotation of the earth is a fundamental contributing force. Fluid friction is also an important factor and poorly understood. There is a certain depth below which both current and frictional forces associated with it become very small. In the layer above that level, friction is important. This upper layer is called the Ekman layer. See also **Winds and Air Movement**. Using l to denote the length of a wave, C, the speed of the waveform and T, the period (the time it takes to move one wavelength) for waves in general is

$$C = \frac{1}{T} \tag{1}$$

which is obvious if we substitute units in some system, such as the length of one wave in feet, the period in seconds, and the speed in feet per second.

The three principal types of ocean waves are due to (1) the forces that produce the tides, (2) the wind, and (3) earthquakes (seismic waves or tsunami).

The waves produced by the tidal forces are progressive waves because their length (one-half the circumference of the earth) is so great in comparison with the depth of the oceans. For such waves the wave velocity, C, varies with the square root of the depth, being given by the expression

$$C = \sqrt{gd} \tag{2}$$

where d is depth and g is the acceleration of gravity 32.16 feet (9.8 meters) per second. Substituting this value, we find that the wave velocity at a depth of 500 feet (152 meters) would be 127 feet (38.7 meters) per second, while at 5,000 feet (1524 meters) it would increase to 399 feet (122 meters) per second. Note that this speed is that of the waveform; that of the water is far smaller, being only about 2.3 feet (0.7 meter) per second (at crests and troughs, where it is greatest) for a wave having an amplitude of 10 feet (3 meters).

The assumption of a channel of infinite length, on which Equation (2) for a progressive wave was based, is only strictly true in the oceans surrounding Antarctica. It fails completely in tidal basins and estuaries, where the types of waves due to the tides are stationary waves rather than progressive waves. Here we are concerned with stationary waves. The periods of these stationary waves are determined by the dimensions of the basin. Assuming that these are constant, which would obviously not be true of any actual basin, a simple relationship can be derived for the period of the stationary wave:

$$T = \frac{2l_B}{\sqrt{gd_B}} \tag{3}$$

where l_B is the (constant) length of the basin and d_B its (constant) depth. (The width is not significant for such an ideal basin.) When the period of the stationary wave so computed is the same, or very close to, that of the tides, then a condition of resonance is obtained which produces the abnormally high tides that occur in certain basins. (The tidal range in the

Bay of Fundy may exceed 50 feet (15.2 meters) in spring at times when the moon is at perigee, i.e., the point in its orbit closest to the earth.)

Another tidal phenomenon, which differs from the foregoing in that it involves a traveling wave, is the *tidal* bore, a relatively massive wave that moves up a river. The height of such a wave depends upon the slope of the river bed (as well as the depth and the current). Favorable combinations of these factors, resulting in a rapid change of surface elevation, produce a large bore or a series of small ones, especially during periods of abnormal high tides.

The common waves of the ocean, as well as the greater ones occurring during storms, are produced by the wind. For such waves, a general equation is

$$C = \sqrt{\frac{gl}{2\pi}\left(\tanh\frac{2\pi d}{l}\right)} \tag{4}$$

where C is the velocity of the waveform as before, l is the wavelength, tanh means the hyperbolic tangent, and d is the depth. This equation clarifies the difference between waves in deep water and those in shallow water. If d is large relative to l so that $2\pi d/l$ is large, then the tanh term is close to 1 (since tanh x approaches 1 rapidly as x increases, being 1 to four places of decimals when x is 6.5). In that case, Equation (4) simplifies to

$$C = \sqrt{\frac{gl}{2\pi}} \tag{5}$$

which is the equation for all wind-produced waves when in deep water. When the ratio d/l is very small, as it is for the waves produced by the tidal forces, then the value of tanh x approaches x closely, and tanh $2\pi d/l$ approaches $2\pi d/l$, so that cancellation in Equation (4) gives $C = \sqrt{gd}$, as given in Equation (2) for those semidiurnal waves of the tides.

The velocity of waves in shallow water, however, cannot be represented by either Equation (2) or Equation (5) since for such values of tanh $2\pi d/l$ the value of the tanh term must be found and used in Equation (4). Of the three properties of water waves, period (T), wave velocity (C), and wavelength (l), the period is the least changed by movement into shallow water, but the wavelength and amplitude are the quantities that change. These changes are not uniformly related; as a wave moves from deep water into a depth such that $d/l = 0.5$, the amplitude (height) begins slowly to decrease, reaching a value of about 90% of the original amplitude when $d/l = 0.06$, after which the amplitude increases until the wave breaks. The theoretical breaking point occurs when $l/d = 7$, but few waves attain such heights, most of them being far lower in relation to their lengths. See Fig. 15.

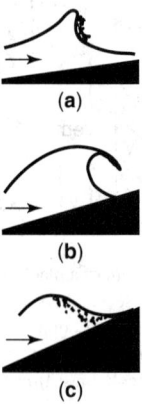

Fig. 15.   The principal types of breaking waves, as determined by beach slope and wave steepness: (**a**) Spilling or dissipative wave; (**b**) plunging wave; and (**c**) surging or reflective wave. (*After Aubrey, 1981.*)

There are a number of theories of wave generation by the wind. They usually include the effect of wind turbulence, with the resulting pressure fluctuations on the surface. Eddies occur in the wind, and under favorable conditions, enter into resonance with the motion of the water induced by the pressure fluctuations. Thus the wave tends to grow as it moves with the wind. As it becomes larger, of course, other factors complicate the picture. In fact, analysis of the waves usually is a complicated operation, although the methods of harmonic analysis have been widely applied.

The wave analyzers that have been developed analyze the wave trace into the periods that compose it. The figure shows the considerable number and range of periods in the wave pattern of a typical storm sea. The energy in the wave system is the important consideration for many purposes; this is represented by the peak period, that is, the period of maximum energy. Even this, however, is an oversimplification in that it does not take into consideration the interaction of the components and other variables. This fact explains the difficulties in wave forecasting from meteorological data.

The various formulas developed for this purpose must take into account the speeds of the surface wind and the gradient wind, their duration and the distance of the observer from the region of wave generation, since the wind-produced waves are attenuated with distance. These are the major factors; others include the motion of the storm area, the effects of tidal currents, the temperatures of air and sea, and the effects of any shallow depths between storm and observer.

In recent years wave prediction has become important for another reason, i.e., for dealing with seismic waves of tsunamis, which are waves produced by earthquakes. Incorrectly called "tidal waves," tsunamis have no relation to the tide. Since such waves, because of their great size, often cause great damage in coastal areas, the prediction of their time of arrival, and their height, is of great importance. Their height depends upon the depth of water, for essentially the same reasons as those given for wind-generated waves. Thus waves only a foot or more high in the open sea may break on the shore at heights of 30 feet (9 meters) or more. This great amplification is due to their length and speed. Their length is great, being on the order of tens of miles, or even over 100 miles (161 kilometers), and thus as explained earlier in this entry, their speed depends on the depth of the sea over which they travel. As stated there, their speed over a depth of 5,000 feet (1524 meters) would be 399 feet per (122 meters) second, or about 270 miles (434 kilometers) per hour. Since the average depth of the Pacific Ocean is much greater than 5,000 feet (1524 meters) there is less than a day after the earthquake occurs to prepare for the tsunami at distances as great as 8–10,000 miles (12,872–16,090 kilometers).

Coastlines and Coastal Waters

The shape and characteristics assumed by a stretch of oceanic coastline reflect a myriad of geological factors and continuous chemical and biochemical processes. A major geological factor controlling shoreline features is plate tectonics, which influence the width and bathymetric detail of the continental shelf as well as the local rise and fall of sea level. Principal factors of an immediate nature that contribute to coastline and beach modification include winds, waves, tides, storms (notably hurricanes), and human influences on the ecology of coastal waters. Coastlines sometimes are referred to as Atlantic-type or Pacific-type, as shown by Fig. 16.

In terms of geological time, a rising sea level results in beaches migrating landward and hence contracting. Changes of river drainage patterns, faulting, slumping, and biological changes (particularly as regards coral and mangrove beaches) are very short term geologically.

In many areas of the world, there is concern with island safety, where the coastline is a predominant factor. Pilkey and Neal (see reference)

established a set of guidelines for evaluating island safety, showing high, moderate, and low hazard potential.

Factors constituting a high hazard include: (1) an island elevation of less than 5 feet (1.5 meters); (2) the absence of dunes; (3) an island of narrow width; (4) an island located near an existing inlet and where there is no salt marsh backup; (5) a known shoreline erosion rate in excess of 3 feet (1.5 meters) per year; (6) a location where, in the past, dunes have been destroyed and new inlets created; (7) a history of overwash; (8) little or very little vegetation; (9) poor footing materials, such as compactible layers of peat and clay; (10) poor drainage during the rainy season; (11) an inadequate or contaminated water supply; (12) unsuitability for sanitary installation because of improper sediment and soil that may intersect the water table; (13) the presence of many shells in the soil; and (14) the presence of finger canals.

Coastal waters (sometimes called *coastal oceans*) may be defined as the region extending from the beaches out across the *continental shelf*, slope, and rise. As pointed out by K. Brink (Woods Hole Oceanographic Institute), "Using this definition, the offshore edge would often be near that of the purely politically defined Exclusive Economic Zone or EEZ. There is, however, a true scientific cohesiveness to the essentially geological definition above. For example, current patterns over the continental shelf and slope tend to be distinctly different from those of the open ocean, and the consequent shelf physical processes make this region the most biologically productive area of the world's ocean .... The coastal ocean absorbs most of the impact that land-based activities have on the ocean, including river outflow and wind transport of particles and chemicals into the sea. These effects make the coastal ocean important scientifically and economically."

Although the year 1980 was dedicated in the United States as the Year of the Coast in an effort to refocus attention of scientists and the lay public on the 128,772 km (80,000 mi) of the nation's coastline, an impressive degree of interest in this topic was not exhibited until 1988, as explained in the following pages.

Degradation of Coastlines and Coastal Waters

Progressive deterioration of many of the world's shorelines and beaches has been known by the scientific community and by the educated and concerned politicians and lay public for many years. But as so often is the case, some form of crisis must occur before corrective measures are taken. This crisis, which received wide attention by the news media and politicians, occurred in 1988 when several of the ocean beaches in the northeastern United States (New Jersey and New York) had to be closed to recreation. The basic cause of this crisis was water pollution of the most revulsive and frightening nature, such as the appearance of medical wastes (including syringes and other hospital wastes), not to mention ordinary garbage. Polluting wastes, of course, are *only one* of several problems that are affecting coastal waters. Some solutions to coastal problems reside in science and technology, but, to a much greater extent, answers must be provided by people and their governments. There are no technical cures of a breakthrough nature that can be called upon as a "quick fix" for coastal water problems. Rather, the role of science is to understand the infrastructure of coastal technology and, with increasingly accurate

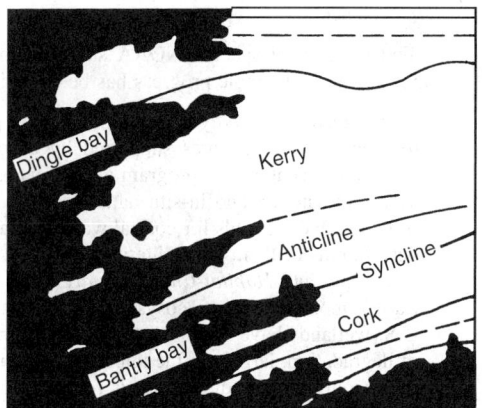
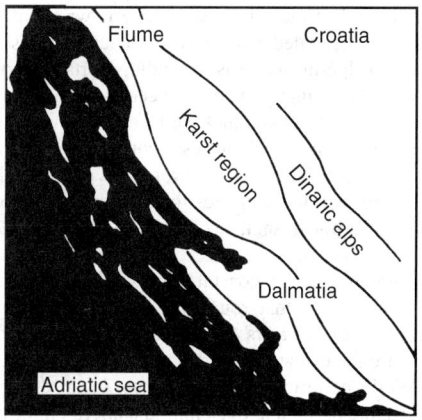

Fig. 16.   Types of coastlines. (Left) Atlantic-type in which geologic structure trends at a high angle to the coast; (Right) Pacific-type, where structure trends parallel to the coast. Suess came up with this type of classification over 100 years ago. (*After Holmes, 1965.*)

information on hand, to enhance the wisdom of political regulators and of the great numbers of people who dwell in the coastal and nearby regions, or who frequent the coastal areas as tourists and recreation seekers.

Scope of Coastal Water Problems. Worldwide, there are comparatively few countries that do not share some coastline with the oceans. Total global coastlines are estimated to extend about 440,000 km (273,400 mi). An impressive portion of the total coastline length has been affected adversely over scores of years, mainly as the result of human activities. Devastation, of course, is more obvious in some areas than in others. Degradation, for example, is quite apparent along the eastern and western United States, including the Gulf of Mexico to the south. The southern shorelines of the European countries, particularly lying along the Mediterranean, Adriatic, and Aegean seas; the islands of the Caribbean sea; the Arabian sea, including the Persian Gulf; the Baltic sea; and the North sea have been seriously affected. Some of the coastline problems are essentially universal; others are specific to certain lengths of the coastline.

This article is concerned mainly with the ocean coastlines, but degradation of freshwater coastlines, notably the Great Lakes of North America, the shores of massive rivers worldwide, and, for example, the virtual disappearance of the mineral-laden Aral Sea, which typifies the environmental neglect of the former Soviet bloc, are part of the total coastline problem.

Classes of Coastal Water Problems. The degradation of coastal waters is exemplary of what can occur when there are severely conflicting interests at work that can lead to strong pro and con positions in terms of solving problems.

Population Pressures. Demographics reveal considerable evidence for supporting population pressure as one of the root causes of present coastline problems. A poorly appreciated statistic is that 45% of the population of the United States lives in coastal counties, such as Nantucket, Massachusetts, although coastal regions constitute only 10% of the land area in the country (excluding Alaska). It is estimated that if current trends continue, by the year 2015, this coastal population will have grown by 15 million, to 127 million persons.

In recent years, Italy has experienced a tremendous immigration of people from other countries (18,000 Albanian refugees in 1991, for example), prompting future Italian restrictions. It is interesting to note that a large percentage of immigrants to countries on the ocean coasts tend to remain near the coast. People pressure exacerbates coastal water problems because of increased production of local wastes and the quest for housing along or near the shorelines. (It is interesting to note that population pressure is one of the root causes of nearly all environmental concerns facing the world today.)

Economic Pressures. Within the last several decades, the impact of the tourist and recreation industry on coastline land and coastal waters has been tremendous. There are some additional, poorly appreciated statistics that are relevant. It is estimated that 5% of the world's gross national product is expended on tourism, making that industry one of the largest of global enterprises and thus a group of economic interests that has considerable clout with politicians and regulators. Further, the construction industry, which builds hotels and associated recreational facilities (restaurants, nightclubs, casinos, etc.) has a large economic stake in coastal locations.

It is interesting to note that the major concern with coastal waters of a comparatively few years ago was beach and property protection in connection with recreational facilities. Primary concern was that of building sea walls and restoring decimated beaches. Commercial interest in coastline areas also embraces shipping interests, including ports, lagoons for pleasure boating and fishing, and simply the overall economic pressures generated by so many large cities and towns that are located immediately adjacent to a coastline. The population of such places is constantly growing. Boston's polluted harbor, considered one of the worst environmentally disturbed bodies of water in the world, dramatizes the effects of pollution on coastal waters. Hong Kong harbor is another example. See also **Water Pollution**.

Another economic pressure exerted on coastal waters management is the fisheries industry. It is estimated that roughly 75% of the world's total fish catch comes from coastal waters worldwide. The outlets for waterborne inland pollution are the estuaries and deltas of rivers and direct dumping into the ocean. While pollution may be greatest in an estuary, ocean currents mix and transport pollutants over long distances along the coastlines. These carry significant concentrations of agricultural chemicals and fertilizers along the coastline for significant distances. The dwindling fish catch in many areas is attributed to: (1) poisoning of numerous

fish and crustaceans with chemicals that kill or adversely affect the reproductive processes of these creatures; and (2) nutrient overenrichment, which nurtures undesirable plants and other organisms, thus creating a condition known as *hypoxia* (depletion of oxygen). Thus, the fisheries industry has a large stake in what occurs biologically along the world's coastlines. Particularly, this impact has adversely affected fishing along the eastern seaboard of the Atlantic ocean in recent years.

For obvious reasons, polluted beaches also affect coastline tourism and vacationing.

Positive Pressures of Coastal Science. Scientists have recognized comparatively recently that the problems of coastal waters involve a networking of actions and reactions, both scientific and geographic and that, in addition to understanding the purely physical forces to which the oceans are subject (that is, currents, tides, seiches, tropical storms and hurricanes, the geometry and tectonics of the ocean shelves), equal if not greater attention must be given to biological and biochemical forces that persist in coastal waters. By providing greater understanding of these ecological and biological factors, coastal scientists can assist effectively in developing long-term protocols for guiding future coastline development.

Coastal Ocean Processes Program (CoOP)

It is along the foregoing lines that the CoOP program was established in early 1990 and defined as a program—*to obtain a new level of quantitative understanding of the processes that dominate the transports, transformations, and fates of biologically, geologically, and chemically important matter on the continental margins.*

Specific objectives include:

- Coastal air-sea fluxes and couplings, such as how carbon dioxide finds its way from the ocean to the atmosphere or vice versa.
- Fluxes of matter through the seabed, such as sediment deposition or the release of chemicals from the bottom.
- Land-derived effects, such as the fate of river-borne nutrients.
- Chemical and biological transformations within the water column, such as how plants grow in response to a chemical change.

National Oceanic and Atmospheric Administration (NOAA) Program

The New Jersey and New York beach crises of 1988 may, in historic perspective, be regarded as a fortuitous omen of the future and a call for immediate action. In that year, the beaches were cluttered with hospital wastes, debris, and garbage; dead and dying dolphins were washed ashore; numerous waters were contaminated by toxics and nutrients — all affecting safe usage of the beaches for recreation and damaging fisheries.

Spurred by national authorities, the existing National Oceanic and Atmospheric Administration created the *Coastal Ocean Program* (COP) in 1989 to focus not only the NOAA but also the academic community on studying intensely the longstanding and rapidly emerging problems of the coastal waters of the United States. The general mandate given to the NOAA was:

1. Develop a coastal forecast system in cooperation with the National Weather Service;
2. Protect the environmental quality of the coastlines; and
3. Coordinate all U.S. federal science efforts as they pertain to coastal science.

Four of several specific NOAA subprograms are described here, and since 1989 considerable progress has been made in all areas:

- *CoastWatch* is a survey and communication program. It is a system of regional data-access sites supported by a central data processing and distribution center. The program allows managers and researchers rapid access to satellite and in-situ data. Recently developed remote sensors provide effective tools for coastal water researchers long before they see descriptions of them in the literature.
- *Watershed and Habitat-Change Analysis.* Coastal wetlands are among Earth's most productive ecosystems. It is estimated that nearly half of the U.S. wetlands have been lost through draining, filling, and other forms of degradation, largely in the interest of land recovery and development.

  Not only are breeding grounds for waterfowl and other wildlife lost, but wetlands also support commercial species of crustaceans, such as shrimp. Loss of wetlands also contributes to the pollution of coastal waters because wetlands assist in protecting the coasts from considerable water runoff. Part of the NOAA program is that of developing a protocol for use by federal and state agencies as well as for academic researchers in mapping coastal habitats and adjacent uplands. Satellite mapping, already in place by the NOAA, is an important element of this program.

This program was applied first to the Chesapeake Bay watershed to analyze changes in land cover and habitats for the world's largest estuary. Completed in 1992, this study shows the effects of population increases in the Washington, DC, and Baltimore, Maryland areas and reveals a 1% loss in wetland habitats over the time frame 1984–1989.

An important point is that the study differentiated between pollution from rivers and tributaries and atmosphere-derived pollutants.

- *Nutrient Overenrichment Studies.* Agricultural and suburban runoff, industrial waste, and raw sewage contain nutrients that enter the coastal waters in excessive amounts and can kill some organisms of value and cause some undesirable species to proliferate. A condition referred to as hypoxia (depletion of oxygen) occurs commonly, especially in the Gulf of Mexico. To address this problem, the NOAA has created the Nutrient Enhanced Coastal Ocean Productivity (NECOP) program. The Mississippi River basin, which drains one-third of the continental United States, carries huge quantities of nutrient-rich water to the Gulf.

- *Coastal Fisheries Ecosystem (CFE) Program.* This study focuses on the ecological processes that affect commercial fish populations. As pointed out by L. Wenzel (Sea Grant Fellow) and D. Scavia (NOAA), "CFE grows out of a new direction in fisheries science, an integrated approach to understanding fisheries within the context of their ecosystems. The need for such an approach is clear. In the Bering Sea pollock fishery, for example, the largest single-species fishery in the world, the Russian and the U.S. 200-mile exclusive economic zones enclose a high-seas 'doughnut hole' open to foreign fleets. Since the mid-1980s, this 'hole,' the central basin of the Bering Sea, has been heavily fished. The Northwest Fisheries Management Council needs to know how this fishery affects stock sizes in the U.S. exclusive economic zone. Recruitment (the number of adult fish added to the population each year) varies dramatically in pollock populations, creating enormous uncertainty in stock assessments. To improve these predictions, COP-funded scientists conduct genetic analyses of pollock stocks, to see how much stocks from foreign and U.S. waters mix, and they mount field studies to determine what physical oceanographic and ecological factors affect pollock survival in the critical egg and larval periods. Once these factors have been identified, monitoring programs can be established to improve the scientific basis for pollock management."

In summary, future ventures of the NOAA's COP program will be directed to:

- Developing a coastal forecast system in cooperation with the National Weather Service.
- Protecting environmental quality.
- Coordinating U.S. federal science efforts as they pertain to coastal ocean science.

Additional Reading

Aksenov, V.V. and A.B. Karasev: "Satellite Oceanography," *Oceanus*, 69 (Summer 1991).

Aubrey, D.G.: "Perspectives from a Shrinking Globe," *Oceanus*, 9 (Spring 1993).

Baker, D.J.: "Toward a Global Ocean Observing System," *Oceanus*, 76 (Spring 1991).

Baker, E.T.: "Megaplumes," *Oceanus*, 84 (Winter 1991/1992).

Bartholomew, C. and C. Mullen: "The Ocean Versus Deep-Water Salvage," *Oceanus*, 73 (Winter 1990/1991).

Bergman, E.A.: "Mid-Ocean Rise Seismicity," *Oceanus*, 60 (Winter 1991/1992).

Bischof, J.: "Ice Drift, Ocean Circulation and Climate Change," Springer-Verlag, Inc., New York, NY, 2000.

Bjerklie, D.: "Getting Heated Up Over Climate Research," *Technology Review (MIT)*, 10 (November/December 1991).

Blackman, D. and T. Stroh: "RIDGE and InterRidge," *Oceanus*, 21 (Winter 1991/1992).

Bonatti, E.: "Not So Hot 'Hot Spots' in the Oceanic Mantle," *Science*, 107 (October 5, 1990).

Bowen, M.F.: "Jason's Mediterranean Adventure," *Oceanus*, 61 (Spring 1990).

Bradley, R.S.: "Principles of Ocean Physics," 2nd Edition, Morgan Kaufmann Publishers, Orlando, FL, 2000.

Brekhovskikh, L.M. and V.G. Neiman: "The History of Soviet (Russian) Oceanology," *Oceanus*, 20 (Summer 1991).

Brink, K.H.: "The Coastal Ocean Processes (CoOP) Effort," *Oceanus*, 47 (Spring 1993).

Britton, J.C. and B. Morton: "Shore Ecology of the Gulf of Mexico," University of Texas Press, Austin, TX, 1989.

Broadus, J.M. and R.V. Vartanov: "The Oceans and Environmental Safety," *Oceanus*, 14 (Summer 1991).

Brown, N.: "The History of Salinometers and CTD (Conductivity, Temperature, Depth) Sensor Systems," *Oceanus*, 61 (Spring 1991).

Bryan, W.B.: "Exploring Pacific Seafloor Ashore: Magadan Province, USSR (Russia)," *Oceanus*, 48 (Summer 1991).

Bryan, W.B.: "From Pillow Lava to Sheet Flow: Evolution of Deep-Sea Volcanology," *Oceanus*, 42 (Winter 1991/1992).

Bulloch, D.K and G. Reiger: "The Wasted Ocean," Lyons & Burford Publishers, Inc., New York, NY, 1989.

Burke, K.: "The Edges of the Ocean," *Oceanus*, **22**, 3, 2–9 (1979).

Cane, M.A.: Oceanographic Events During El Nino, *Science*, **222**, 1189–1195 (1983).

Davin, E.M. and M.G. Gross: "Assessing the Seabed," *Oceanus*, **23**, 1, 20–32 (1980).

Deacon, M.B., Rice, T. and C. Summerhayes: "Understanding the Oceans: Marine Science in the Wake of HMS Challenger," Taylor & Francis, Inc., Philadelphia, PA, 2001.

Duedall, I.W. and M.A. Champ: "Artificial Reefs: Emerging Science and Technology," *Oceanus*, 94 (Spring 1991).

Edmond, J.M. and K. Von Damm: "Hot Springs on the Ocean Floor," *Sci. Amer.*, 78–93 (April 1983).

Edmond, J.M.: "The Geochemistry of Ridge Crest Hot Springs. "*Oceanus*, **27**(3), 15–19 (1984).

Francheteau, J.: "The Oceanic Crust," *Sci. Amer.*, 114–129 (September 1983).

Fredj, G., et al.: "Mediterranean Biology," *Oceanus*, 43 (Spring 1990).

Frisbee, K.S.: "Deep Water Over Complex Tectonics," *Oceanus*, 56 (Spring 1990).

Garrett, C. and L.R.M. Maas: "Tides and Their Effects," *Oceanus*, 27 (Spring 1993).

Given, D.: "Underwater Technology in the USSR (Russia)," *Oceanus*, 67 (Spring 1991).

Goldberg, E.D.: "Competitors for Coastal Open Space," *Oceanus*, 12 (Spring 1993).

Golden, F.: "A Quarter-Century Under the Sea (*Alvin*)," *Oceanus*, 2 (Winter 1988/1989).

Gornitz, V., Lebedeff, S., and J. Hansen: "Global Sea Lever Trend in the Past Century," *Science*, **215**, 1611–1614 (1982).

Hartwig, E.O.: "Trends in Ocean Science," *Oceanus*, 96–100 (Winter 1990/1991).

Hass, P.M. and J. Zuckman: "The Mediterranean Is Cleaner," *Oceanus*, 38 (Spring 1990).

Holmes, A.: "Principles of Physical Geology," 2nd edition, Chapman & Hall, New York, NY, 1992.

Ivanov, Y.A.: "Physical Oceanography: A Review of Recent Soviet (Russian) Research," *Oceanus*, 81 (Summer 1991).

Jensen, J.J. and J. Hovermale: "Numerical Air/Sea Environmental Protection," *Oceanus*, 40 (Winter 1990/1991).

Julian, M. and P.R. Ryan: "Introduction: The Med," *Oceanus*, 4 (Spring 1990).

Karson, J.A.: "Tectonics of Slow-Spreading Ridges," *Oceanus*, 51 (Winter 1991/1992).

Kerr, R.A.: "Vail's Sea-Level Curves Aren't Going Away," *Science*, **226**, 677 (1984).

Lacombe, H.: "Water, Salt, Heat, and Wind in the Mediterranean," *Oceanus*, 26 (Spring 1990).

LePichon, X. and J. Francheteau: "A Plate-Tectonic Analysis of the Red Sea-Gulf of Aden Area," *Tectonophysics*, **46**, 369–406 (1978).

Lindau, R.: "Climate Atlas of the Atlantic Ocean: Derived from the Comprehensive Ocean Atmosphere Data Set (Coads)," Springer-Verlag, Inc., New York, NY, 2001.

Liu, J.: "The Segmented Mid-Atlantic Range," *Oceanus*, 11 (Winter 1991/1992).

Lonsdale, P. and C. Small: "Ridges and Rises: A Global View," *Oceanus*, 26 (Winter 1991/1992).

Lutjeharms, J.R.E., Shillington, F.A., and C. M. Duncombe Ray: "Observations of Extreme Upwelling Filaments in the Southeast Atlantic Ocean," *Science*, 774 (August 16, 1991).

Lutz, R.A.: "The Biology of Deep-Sea Vents and Seeps," *Oceanus*, 75 (Winter 1991/1992).

Macdonald, K.C. and P.J. Fox: "The Mid-Ocean Range," *Sci. Amer.*, 72 (June 1990).

Macdonald, K.C., Scheirer, D.S., and S.M. Carbotte: "Mid-Ocean Ridges: Discontinuities, Segments, and Giant Cracks," *Science*, 986 (August 30, 1991).

Macdonald, K.C.: "Introduction: Mid-Ocean Ridges, The Quest for Order," *Oceanus*, 9 (Winter 1991/1992).

McCoy, F.W. and P.D. Rabinowitz: "The Evolution of the South Atlantic," Oceanus **22**, 3, (1979).

Meier, M.F.: "Contributions of Small Glaciers to Global Sea Level," *Science*, **226**, 1418 (1984).

Milliman, J.D.: "Sea Levels: Past, Present, and Future," *Oceanus*, 40 (Summer 1989).

Monastersky, R.: "Predictions Drop for Future Sea-Level Rise," *Science News*, 397 (December 18, 1989).

Mooers, C.N.K. and A.R. Robinson: "Turbulent Jets and Eddies in the California Current and Inferred Cross-Shored Transports," *Science*, **223**, 51–53 (1984).

O'Connell, S.: "Ophiolites," *Oceanus,* **22**(3), 33 (1979).

Ostenso, N.A., Metalnikov, A.P., and B.I. Imerekov: "A History of USSR (Russia)–US Cooperation in Ocean Research," *Oceanus,* 87 (Summer 1991).

Peltier, W.R. and A.M. Tushinham: "Global Sea Level Rise and the Greenhouse Effect," *Science,* 806 (May 19, 1989).

Pilkey, O.H. and W.J. Neal: "Barrier Island Hazard Mapping," *Oceanus,* 38 (Winter 1980–1981).

Potter, T.D. and B. R. Colema: "Handbook of Weather, Climate, and Oceans," The McGraw-Hill Companies, Inc., New York, NY, 2001.

Prager, E. and S. Earle: "The Oceans," The McGraw-Hill Companies, Inc., New York, NY, 2001.

Ross, D.A.: "The Red Sea: A New Ocean," *Oceanus,* **22**(3), 33–39 (1979).

Roughgarden, J., Gaines, S., and H. Possingham: "Recruitment Dynamics in Complex Life Cycles," *Science,* 1460 (1988).

Ryan, P.R.: "A Challenge to Alvin from the USSR (Russia)," *Oceanus,* 67 (Winter 1988/1989).

Schouten, H. and J. Whitehead: "Ridge Segmentation: A Possible Mechanism," *Oceanus,* 19 (Winter 1991/1992).

Siedler, G., J. Gould, and J. Church: "Ocean Circulation and Climate," Academic Press, Inc., San Diego, CA, 2001.

Smith, Walker O., Jr., "Polar Oceanography," Academic Press, Inc., San Diego, CA, 1990.

Spindel, R.C. and P.F. Worcester: "Ocean Acoustic Tomography," *Sci. Amer.,* 94 (October 1990).

Stanley, D.J.: "In Search of the Origins of the Mediterranean," *Oceanus,* 16 (Spring 1990).

Suess, E. and J. Thiede: "Coastal Upwelling," Plenum Publishing Corporation, New York, NY, 1983.

Tippie, V.K. and J.H. Cawley: "Modernizing NOAA's Ocean Services," *Oceanus,* 84 (Summer 1991).

Tivey, M.K.: "Hydrothermal Vent Systems," *Oceanus,* 68 (Winter 1991/1992).

Toomey, D.R.: "Tomographic Imaging of Spreading Centers," *Oceanus,* 92 (Winter 1991/1992).

Uyeda, S.: "Subduction Zones," *Oceanus,* **22**(3), 52–62 (1979).

Van Dover, C.L.: "Diving the Soviet (Russian) Mir Submersibles," *Oceanus,* 8 (Summer 1991).

Vartanov, R.V.: "Dynamics of Ocean Ecosystems: A National Program in Soviet (Russian) Biooceanology," *Oceanus,* 66 (Summer 1991).

Vine, A.C.: "The Birth of Alvin," *Oceanus,* 10 (Winter 1988/1989).

Watts, A.B. and J. Thorne: *Nature (London),* **311**, 365 (1984).

Weller, R.A., et al.: "Three-Dimensional Flow in the Upper Ocean," *Science,* 227, 1552 (1985).

Wenzel, L. and D. Scavia: "NOAA's Coastal Ocean Program: Science for Solutions," *Oceanus,* 85 (Spring 1993).

Whitehead, J.A.: "Giant Ocean Cataracts," *Sci. Amer.,* 50 (February 1989).

Wilson, J.T.: "A New Class of Faults and Their Bearing on Continental Drift," *Nature,* 207, 343–347 (1965).

Wright, L.: "Morphodynamics of Inner Continental Shelves," CRC Pres, LLC, Boca Raton, FL, 1995.

Zilanov, V.K.: "Living Marine Resources," *Oceanus,* 29 (Summer 1991).

Web References

Scripps Institution of Oceanography: *http://www.sio.ucsd.edu/*

Scripps Research Oceanography: *http://www.sio.ucsd.edu/research/oceanography.html*

The National Oceanic & Atmospheric Administration (NOAA): *http://www.noaa.gov/*

Woods Hole Department of Applied Ocean Physics and Engineering: *http://www.whoi.edu/science/AOPE/dept/*

Woods Hole Department of Biology: *http://www.whoi.edu/science/B/dept/*

Woods Hole Department of Geology & Geophysics: *http://www.whoi.edu/science/GG/dept/*

Woods Hole Department of Marine Chemistry and Geochemistry: *http://www.whoi.edu/science/MCG/dept/*

Woods Hole Department of Physical Oceanography: *http://www.whoi.edu/science/PO/dept/*

Woods Hole Oceanographic Institution (WHOI): *http://www.whoi.edu/science/science.html*

OCEAN-ATMOSPHERE INTERFACE. See **Atmosphere-Ocean Interface**.

OCEAN (Hydrology). See **Hydrology**.

OCEANICITY. See **Climate**.

OCEAN PERCH. See **Redfish**.

OCEAN RESEARCH VESSELS. Principal types of oceanographic research vessels are surface craft and manned or unmanned submersibles.

Some craft are quite specialized, as in the case of a drilling vessel. In recent years, scientific satellites also have contributed to oceanographic research.

Chronology of Ocean Research Vessels

The era of the early mapping of the seas and the use of vessels for other kinds of scientific studies is somewhat obscure. It is known that, as early as 1838, the U.S. Navy designed two "exploring vessels" for use in the U.S. Exploring Expedition of 1838–1842. In 1838, the Coast Survey schooner *Nautilus* was built specifically for surveying, as was the steamer *Blake* in 1874. Alexander Agassiz carried out some oceanographic studies with the *Blake.* In 1879, the U.S. Fish Commission ordered a coal-burning steamer, the *Fish Hawk,* for fisheries research. The vessel was complete with a floating fish hatchery. The *Alexander Agassiz,* designed and built in San Diego in 1907 and first assigned to the West Coast marine station, was a sailing vessel equipped with two gasoline engines. The ship was used in an area of the Pacific Ocean from Point Conception to the Mexican border and seaward to about 120 miles (193 kilometers). The region included depths to 1100 fathoms. (1 fathom = 6 feet = ~1.8 meters) The *Agassiz* was equipped with dredges, trawls, closing nets, current meters, and other oceanographic instruments of that era. The ship served the Scripps Institution of Oceanography (La Jolla, California), the successor of the West Coast marine station, for about ten years. The *Atlantis,* probably the first American ship especially designed for oceanographic research in the modern sense, entered operations in 1931 as the research vessel for the newly formed Woods Hole Oceanographic Institution (Woods Hole, Massachusetts), originally funded by the Rockefeller Foundation. At a length of 142 feet (43 meters) the *Atlantis* was quite large for a sailing ketch, but a fine size for a research vessel, striking a good balance between seaworthiness, range, personnel requirements, and research capability. The *Atlantis* was a pioneering ship and her scientific accomplishments were impressive. Work with the *Atlantis* included studies of the Gulf Stream and its meanders, investigations of the geophysical properties of the sea floor; studies of the ocean's sound properties, studies of deep midwater fauna, and investigations of submarine canyons. This vessel was replaced by the *Atlantis II* in 1963. Other earlier vessels included the R.V. *E.W. Scripps* and the *R.V. Vema.* Earlier vessels charged principally with making ocean depth surveys included the *H. M.S. Challenger,* which operated over the period 1872–1876, and the U.S. Navy surveying ship *Nero,* which made a survey for a proposed telegraphic cable between Honolulu and Manila by way of Guam and Yokohama. The greatest depths accurately measured during that survey were in the range of 5100 and 5270 fathoms.

Submersible research craft applied much of the technology developed for military submarines. Interest in submarines dates back to the early 1600s. Van Dribel invented a submarine rowboat in 1624; Le Son built the *Rotterdam* in 1652; Bushnell built a submersible boat (the *Turtle*) in 1776; Fulton built the *Nautilus* for the French government in 1800; and the partially submersible torpedo boat *David* was constructed by the Confederacy during the War Between the States, and probably sparked the potential for submersible craft in military engagements. The military submarine did not gain significant prominence until World War I. Charles Beebe pioneered the use of a bathysphere (nonnavigable) in 1934. Lowered by a steel cable from a surface ship, Beebe could attain depths of a little over 923 meters (3028 feet). The practical use of navigable submersible vessels for research work dates back only a few decades.

Contemporary Ocean Research Vessels

Robertson Dinsmore (Woods Hole Oceanographic Institution) classifies ocean research surface craft as follows:

1. Large Ships — Length, 200 feet (61 meters) or longer. These craft make expeditions of long duration and carry 20 to 25 scientists with a like number of crew. Capable of cruising 250 to 280 days at sea, or up to a year on extended voyages. These ships play a major role in research.

2. Intermediate-Size Ships — Length, 150 to 200 feet (37 to 61 meters) for carrying 12 to 16 scientists and a crew of about 12 persons. Because of their lower operating costs, they are used for most of the shorter oceanographic cruises — 2 to 3 weeks' duration. They are limited by the sea state they can operate in and have limited laboratory and storage space.

3. Small Vessels — Length, 80 to 150 feet (24 to 37 meters), generally considered coastal vessels. They carry from 9 to 12 scientists on short exploratory cruises of 1 to 2 weeks' duration. They are used mainly

for small projects close to shore. There are also a number of vessels of less than 80 feet (24 meters) in length, usually located near the main oceanographic research centers. Some are also located in the Great Lakes.

R.V. Atlantis II. This vessel was named for the Woods Hole Oceanographic Institution's first research vessel, previously mentioned. Under a grant from the National Science Foundation (NSF), the vessel was built in 1961–1962 and commissioned in 1963. The vessel has worked in all disciplines and has traveled worldwide. In recent years, the ship has been engaged in intensive geological and geophysical studies in the Atlantic. The vessel participated extensively in some of the IDOE (International Decade of Ocean Exploration) programs. See also Ocean. In 1979, the ship underwent a major mid-life refit and was converted from steam to diesel power to reduce operating costs and increase her range and selection of ports. Programs for improving the effectiveness of the vessel continue as funding permits.

A diagram of the *Atlantis II* is shown in Fig. 1. There are four laboratories aboard the ship. These encompass 404 square meters (4350 square feet) and are fitted with outlets for fresh water, seawater, oxygen, and other gases. There is a controlled-temperature aquarium aboard. The ship was designed for efficiency and versatility. She carries a general purpose shipboard computer for data analysis at sea and a precision graphic recorder to electronically record depth measurements. Gyro compass repeaters, speed indicators, and winch line pull indicators are located around the ship for the use of scientists. Hydrophones are attached to the underwater hull to receive sound, and large electric patch panels are installed in all major laboratories for rapid communication and recording of information. The ship has a fully equipped machine shop and an explosives magazine for safe storage of small depth charges used in seismic refraction profiling. A bulbous underwater observation chamber in the bow is equipped with six viewing ports. The ship is air conditioned and is equipped with underwater lights for night work, a large stern ramp, large uncluttered deck space, and an enclosed crow's nest for daytime lookout.

R.V. Knorr and Other Surface Vessels. The *Knorr* participated in a major way in the Titanic Expeditions and numerous other research projects over the past several years. The *Knorr* is 74.6 meters (245 feet) long, with a displacement of 2075 long tons, a range of 16,900 kilometers (10,000 miles) at 11 knots, and accommodations for a crew of 25 and scientific party of 24. The ship was built in 1969. There is also the *Oceanus*, which is 54 meters (177 feet) long, with a displacement of 962 long tons, a range of 12,067 kilometers (7,500 miles) at 14 knots, and accommodations for a crew of 12 and scientific party of 12. The ship was built in 1975. The *Knorr* is owned by the U.S. Navy; the *Oceanus* by the NSF.

The University-National Oceanographic Laboratory System (UNOLS). The UNOLS was established in 1971 with the responsibility of coordinating and scheduling oceanographic research ships and to provide opportunities to scientists who normally do not have access to research vessels. During the IDOE, previously mentioned, this arrangement proved extremely worthwhile. Generally, the membership in UNOLS is defined as "those academic institutions that operate significant federally funded oceanographic facilities." See Table 1. The membership also includes some 40 smaller laboratories that hold associate memberships and participate in the use of seagoing science facilities.

Surface vessels of a nature similar to *Atlantis II, Knorr*, and *Oceanus* are used by other institutions and countries for oceanographic research, but their number is quite limited in the light of research required. Much of the information gathered for oceanographic research is obtained through the use of these special research vessels which are equipped with a large variety of instruments for measuring depth, temperature, salinity, magnetism, variations in gravitational force, seismic probes, and for taking and analyzing samples. Excellent underwater photographic equipment is usually aboard, as well as excellent navigational aids. Numerous improvements have been made in recent years. An example is the sound energy used in connection with acoustic and seismic equipment. At one time dynamite was the major source of sound energy. More recent energy sources include the electric spark, compressed air and propane gas. Hydrophones that trail behind the vessel receive reflected energy from the bottom. These signals are duly processed and tape recorded, thus providing a continuous seismic-reflection profile. Information from depths of several kilometers below the sea floor can be obtained. This makes possible the construction of geological cross sections. Together with samples from dredging or drilling, much information has been gained pertaining to the structure of the continental shelves. The larger research vessels include their own drilling equipment.

Internationally Operated Ocean Research Surface Vessels. Those countries in addition to the United States that own and operate ten or more surface vessels for ocean research include: Russia[1], 194 vessels; Japan, 94; United Kingdom, 39; France, 27; Canada, 25; former West Germany, 15; Brazil, 12; Sweden, 11; and Argentina, Australia, and Italy, 10 each.

Special mention should be made of the French vessel *Le Suroit*, operated by the Institute Français de Recherche sur l'Exploitation des Mers (Toulon, France), which played an important role in the discovery of the *Titanic*.

The efficacy and sophistication of modern underwater research was widely acclaimed by the scientific community and by the public worldwide when the discovery of the resting site of the *Titanic* was announced in early September 1985. Culminating years of research and planning by an American and French team of researchers, the moment of discovery

[1] As of 1994, the status of oceanographic research in Russia and other former Soviet bloc countries is uncertain.

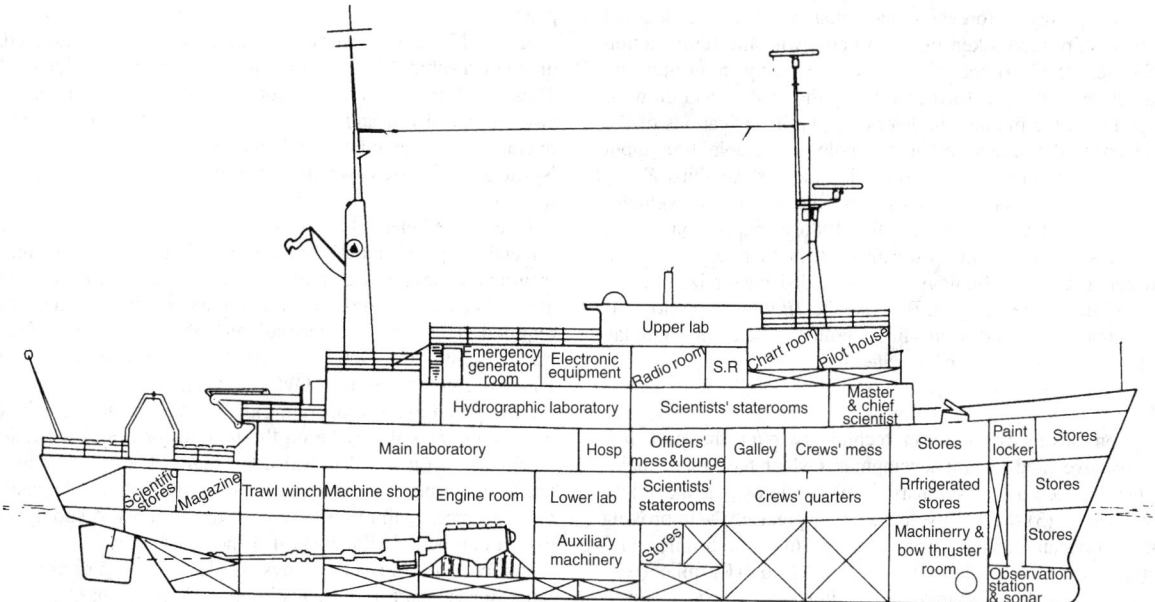

Fig. 1.   Inboard profile of the *R.V. Atlantis II*, considered a large ocean research surface vessel. Ships of this type are periodically updated as technology advances and as funding permits. (*Woods Hole Oceanographic Institution.*)

TABLE 1. UNOLS FLEET OF SURFACE VESSELS FOR OCEAN RESEARCH

Ship's Name	Length Ft.	M	Built/ Converted	Crew/ Scientists	Owner	Operating Laboratory
LARGE SHIPS						
Melville	245	75	1970	22/26	U.S. Navy	Scripps
Knorr	245	75	1969	24/25	U.S. Navy	W.H.O.I.
Atlantis II	210	64	1963	24/25	W.H.O.I.*	W.H.O.I.
T.G. Thompson	208	63	1965	22/19	U.S. Navy	U. Washington
T. Washington	208	63	1965	19/23	U.S. Navy	Scripps
Conrad	208	63	1963	25/18	U.S. Navy (Columbia U.)	Lamont-Doherty
INTERMEDIATE SHIPS						
Oceanus	177	54	1975	12/12	N.S.F.	W.H.O.I.
Wecoma	177	54	1975	12/16	N.S.F.	Oregon State U.
Endeavor	177	54	1976	12/16	N.S.F.	U. Rhode Island
Gyre	174	53	1973	11/18	U.S. Navy	Texas A. M.
Columbus Iselin	170	52	1972	12/13	U. Miami*	U. Miami
New Horizon	170	52	1978	12/13	S.I.O.	Scripps
Fred H. Moore	165	50	1967/1978	9/20	U. Texas	U. Texas
Kana Keoki	156	48	1967	12/16	U. Hawaii	U. Hawaii
SMALL SHIPS						
Cape Florida	135	41	1981	9/10	N.S.F.	U. Miami
Cape Hatteras	135	41	1981	9/12	N.S.F.	Duke U
Alpha Helix	133	40.5	1965	12/12	N.S.F.	U. Alaska
Ida Green	130	40	1965/1972	7/12	U. Texas	U. Texas
Cape Henlopen	122	37	1975	6/12	U. Delaware	U. Delaware
Velero IV	110	34	1948	11/12	U.S.C.	U. Southern California
Ridgely Warfield	106	32	1967	8/10	J.H.U.*	Johns Hopkins U.
E.B. Scripps	95	29	1965	5/8	S.I.O.	Scripps
Cayuse	80	24	1968	7/8	N.S.F.	Moss Landing Marine Lab
Longhorn	80	24	1971	5/10	U. Texas	U. Texas
Laurentian	80	24	1974	5/8	U. Michigan	U. Michigan

*Funded by National Science Foundation.
UNOLS = University-National Oceanographic Laboratory System.
W.H.O.I = Woods Hole Oceanographic Institution.
S.I.O. = Scripps Institution of Oceanography.
N.S.F. = National Science Foundation.
O.S.U. = Oregon State University.

was given as very early on the morning of 1 September. The location: approximately 360 miles (580 km) off Grand Banks, Newfoundland at a depth of 13,000 feet (396 m) of water. On 9 September 1985, Robert D. Ballard (Woods Hole Oceanographic Institution) and leader of the American team observed that "the *Titanic* lies on a gently sloping alpine-like countryside overlooking a small canyon below. [Her] bow faces north and the ship sits upright on the bottom, [two of her] mighty stacks still pointing upward. There is no light at this great depth. It is quiet and peaceful, a fitting place for the remains of this greatest of [peacetime] sea tragedies to rest. May it forever remain that way." It was learned later that the great ship had broken into two parts, with the stern section lying some 1830 feet (558 m) behind the bow and facing in an opposite direction. In addition to the great skills used by the leaders and crew of this special expedition, the finding and later close-up investigations of the ship are a testament to the oceanographic technology available throughout the 1980s and notably the research ships used—the surface ships *Knorr* (United States) and the *Le Suroit* (France) and the underwater vehicles *Argo* (U.S.), the SAR (France), and for the 1986 re-expedition to the *Titanic*, the robotic submersible, the *Jason Jr.* (*J.J.*). Other very important elements of oceanographic technology used included customized sonar, photography, and imagery techniques. Because the *Titanic* expedition has been described in almost limitless detail in both the scientific and lay literature, the fascinating story is not repeated here.

Submersibles

In commenting on advances in ocean technology over the past few decades, scientists have cited several developments which have made major contributions: (1) the deep-diving submarine (submersibles); (2) the ability to drill in deep water; (3) the ability to navigate precisely—improving knowledge of mid-ocean position from a range of 1 to 5 miles (1.6 to 8 kilometers) down to a position fix within 0.1 to 0.01 mile (0.16 to 0.016 kilometer) if within 500 miles (805 kilometers) of land and to 10 feet (3 meters) if within 10 miles (16 kilometers) of land; (4) the availability of television and side-looking sonar for inspection of the sea

floor; and (5) among the most important of advances, the vastly improved communications links between remotely located instruments and receiving stations and the efficacy of communication between submersibles and the mother research vessel.

Improved television tubes can amplify light by a factor of 30,000, thus eliminating the need for artificial lighting and the consequent problems of backscatter. Also included with the foregoing advances would be greatly improved materials of construction, such as better steels, high-strength aluminum alloys designed for marine uses, titanium, glass, fiberglass, and plastics.

As of 1994, it is estimated that well over one hundred manned and unmanned submersible craft are in use as oceanographic exploratory tools. These craft are used not only for fundamental oceanographic research, but also by petroleum and gas producers in exploratory operations, notably operating off the continental shelves of the United States and in the North Sea. The use of submersibles in connection with military operations remains obscure.

In early submersibles, personnel and instrumentation were housed in a metallic sphere about 2 meters (6.6 feet) in diameter that incorporated from one to several viewports. The inside of the sphere was maintained at atmospheric conditions and thus no special suits or decompression were required. High-pressure technology for constructing submersibles dates back at least to the 1960s. In 1960, the *Trieste* went to a depth of more than 10 kilometers (6.2 miles). Over the years, the earlier technology has been refined. In an average operating dive, a vessel may remain submerged for 6 to 10 hours and may be on the bottom for 3 to 7 hours in waters some 3 kilometers (nearly 2 miles) deep. Rate of ascent is about 2 kilometers (1.25 miles) per hour. Movement over the bottom, depending upon local circumstances, will range between 1 and 2 knots. A submersible may range over a number of kilometers if it moves steadily.

For certain kinds of surveys, in deep-sea exploration as in exploring outer space, there is no substitute for human observers. However, there are many measurements where submersible instrument packages perform well and frequently at lower cost. A number of advanced surface instruments have been adapted to submersible usage. Examples include the narrow-beam multichannel echo sounding system (SONARRAY) developed by the U.S. Navy and the surface side-scan system (GLORIA) developed in the United Kingdom. Some instruments for deep-sea research can be dropped and can be commanded remotely to become operational. Among these are seismometers and current meters. Submersible instrument packages also may be deep-towed. The Scripps Deep-Tow, for example, is equipped with multiple sensors and accurate navigation. In contrast with manned submersibles, which sometimes yield only photographic or subjective information, submersible instruments commonly will yield graphic and quantitative data. Small transponders may be attached to nearly any instrument package to provide information on its position.

In exploration of the Galápagos Rift in 1979, scientists of the Woods Hole Oceanographic Institution employed a system known as *Angus* for the first time. An acronym for acoustically navigated underwater system, *Angus* is an unmanned 2-ton sled that is towed across the bottom terrain. It can take some 3000 colored pictures of the bottom in a 15–16-hour-period. Previously it had been towed at about 3.6 meters (11.8 feet) from the bottom, giving pictures of a relatively small area. With an advanced lighting system, *Angus* can provide pictures covering one-half acre (0.2 hectare or 1765 square meters) in one frame.

DSV Alvin. Probably the world's most active manned submersible is the *Alvin*, operated by the Woods Hole Oceanographic Institution, which has made well over 1700 dives of exploration to depths up to 13,120 feet (~4000 meters). The craft undergoes regular updating to incorporate the latest technological developments. The diagram given in Fig. 2 indicates the general layout of this submersible. The term *propeller* and *thruster* with reference to a submersible essentially are synonymous. The *Alvin* was a key vehicle in the exploration of the *Titanic* remains.

Funds for the construction of *Alvin* were provided by the Office of Naval Research. The Bureau of Ships of the U.S. Navy assisted in the preparation of performance specifications for its design and construction, and the Applied Sciences Division of Litton Industries designed and built the original vehicle. *Alvin's* original 2.1-meter (7-foot) diameter pressure sphere of high-strength steel 3.39 centimeters (1.33 inches) in thickness for operational depth to 1829 meters (6000 feet) was replaced in 1973 with a 4.90-centimeter (1.93-inch) thick titanium alloy hull provided by the Naval Ship Systems Command. The titanium hull doubled the vehicle's capability without increasing the weight. The pressure sphere accommodates a pilot and two scientific observers as well as instrumentation and life support equipment for endurance up to 72 hours.

In 1978, *Alvin's* 7-meter (23-foot) aluminum frame was replaced with a 7.6-meter (25-foot) titanium frame and an optional second arm installed. The new frame allows for increased instrumentation and can accommodate a fourth battery pack for additional endurance. The submersible was designed to require minimum assistance from large vessels. A front view is given in Fig. 3.

Navigation equipment and other instruments for *Alvin* include gyro compass and gyro repeater, magnetic compass; nose-mounted horizontal-scanning sonar system; indicators for depth, speed, list, trim and variable ballast; echo sounder, battery voltmeters, ammeters and ground detector; and five viewports. Electric power is provided by banks of lead-acid batteries, 60- and 30-volt dc systems, 40.5 kWh total. Communications equipment includes closed circuit television, sonar telephone (voice or code), and marine band (VHF) radio telephone.

In the event of accident or malfunction, occupants of *Alvin* can be returned safely to the surface. Each of the batteries can be dropped to reduce the vehicle's weight, and the two mechanical arms can be dropped should they become hopelessly entangled. As a last resort, the pressure sphere itself, carrying the personnel, can be mechanically disconnected from the rest of the vehicle. The sphere is buoyant and will float to the surface. A closed circuit rebreather with a 6-hour capacity is provided for each occupant in case a fire in the sphere produces noxious fumes. Chemical fire extinguishers are carried.

Comparison of *Alvin* with submersibles capable of operating at exceptional depths can be noted from Fig. 4.

Projects in which *Alvin* is continually engaged include studies of deep-sea geology, geophysics, and biology, as well as establishment of and periodic visits to deep ocean stations. For several years the vessel has played a primary role in investigations of plate tectonic spreading centers. Seventeen dives in 1974 to the rift valley of the Mid-Atlantic Ridge gave scientists their first close-up view of the tectonic landscape now known to be common to spreading centers—as part of project FAMOUS (French-American Mid-Ocean Undersea Study). In 1976, *Alvin* continued these studies with dives to 3,658 meters (12,000 feet) in the Cayman Trough. In 1977, *Alvin* made the first of a continuous series of dives to warm water vents on the Galápagos Rift, where diving scientists discovered clusters of unusual animal life.

Robotic Undersea Vehicles

Although frequently called underwater *robots*, contemporary unmanned submersibles are not true robots, but are designated more accurately as *teleoperators*—that is, they are remotely operated vehicles under the control of human operators. Teleoperators are becoming more self-sufficient and automatic, capable of making local decisions, and are approaching the status of true robots. However, some form of master programming will be required by even the most advanced designs. At some early date in the future, it is envisioned that preprogrammed robots will

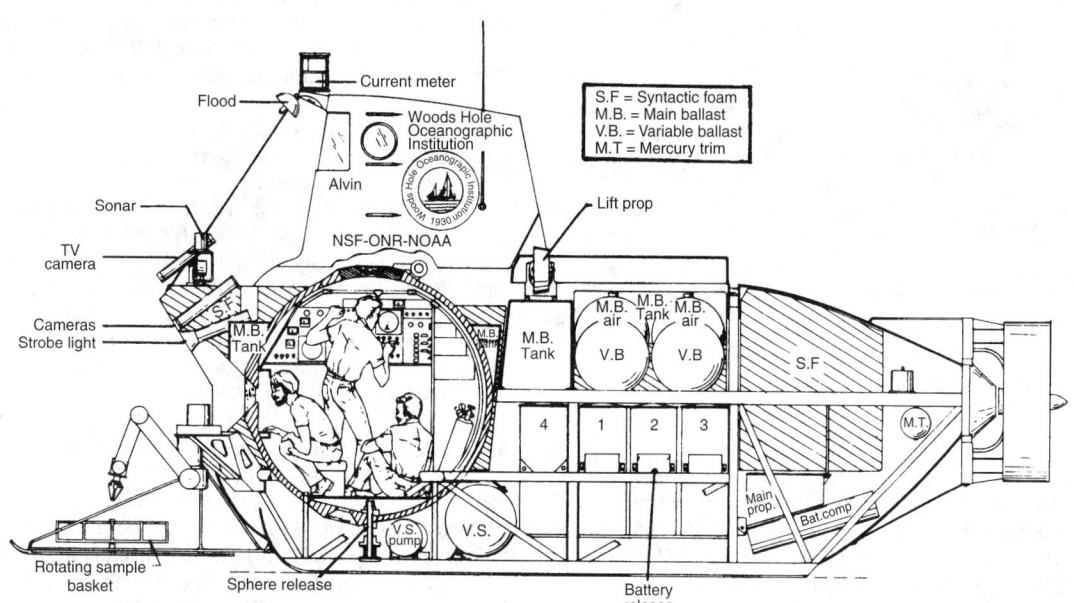

Fig. 2.   Inboard profile of the *R.V. Alvin*, one of the most active of manned submersibles. Design details are periodically updated. The vehicle is designed to carry a pilot and two scientific observers and, in recent years (for and since the *Titanic* Expedition), is linked to a robotic submersible by a 61-meter (200-foot) long tether. The submersible's images are transmitted to *Alvin*, thus enabling *Alvin's* occupants to maneuver it precisely in confined spaces. (*Woods Hole Oceanographic Institution.*)

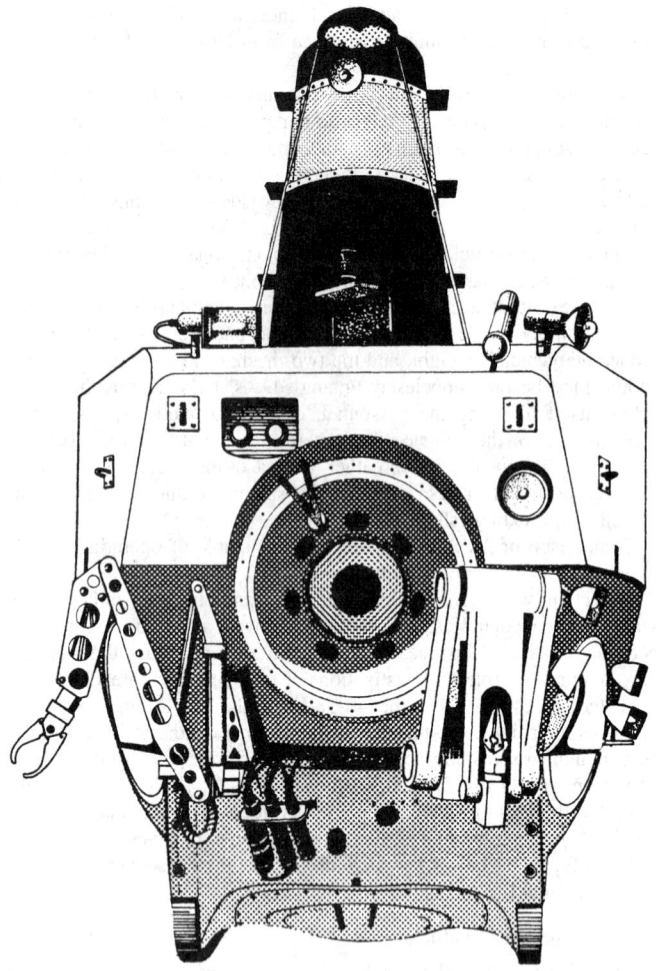

Fig. 3.   Front view of *Alvin*, showing manipulating arms, cameras, lights, and numerous other accessories. (*Woods Hole Oceanographic Institution.*)

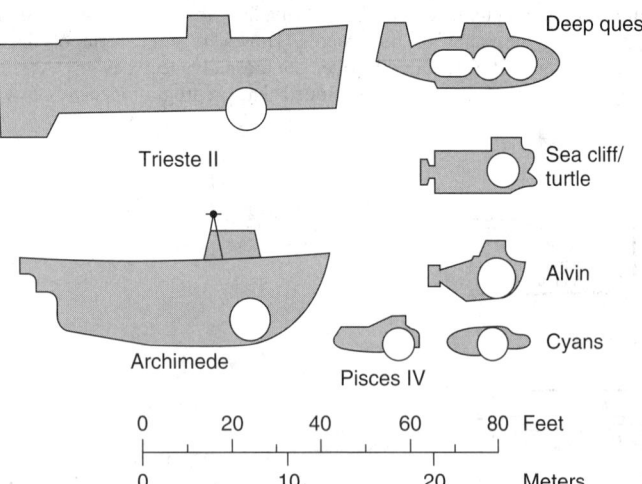

Fig. 4.   Profiles of submersibles capable of reaching exceptional ocean depths. (*After Hertzler and Grassie.*)

be able to survey large stretches of the seafloor for long periods without assistance from surface mother vessels. In such cases, the power supply will have to be self-contained within the submersible, and information collected by the craft will be transmitted or stored until the robot is retrieved. Such devices will be designed to take greater risks in some cases without endangering the well-being of any human operators.

In 1988, a Finnish firm constructed two identical submersibles capable of accommodating three people for the Russian oceanographic research program. The two craft, named *Mir 1* and *Mir 2*, were carried by a mother research vessel, *Akademik Mstislav Keldysh*, a ship that could

support 130 people and incorporate 18 laboratories. The blimp-shaped submersibles were designed to dive to 6000 meters (19,686 feet) and thus can explore the seafloor at greater depths than the *Alvin* previously described. In their first few years of use, the *Mirs* explored parts of the seafloor of the Atlantic and Pacific oceans, and plans call for their exploration of the essentially unknown seafloor of the Indian ocean. The *Mir* submersibles also completed a survey of the *Titanic* resting site in 1991.

Teleoperated craft for underwater studies and in use for well over a decade stemmed from the development of teleoperators for the handling of radioactive and other dangerous materials.

Jason Jr. This is a small robotic (unmanned) type of tiny submersible that is designed to operate in conjunction with a manned submersible, such as *Alvin*. *J.J.* is a mere 28 inches (71 cm) in length and, on land, weighs 250 pounds (113 kg). The hull of the small craft is made of syntactic foam (incorporating billions of microscopic air-filled glass spheres bonded by epoxy). The result is that it is essentially weightless in water. *J.J.* is controlled by a console held by a pilot onboard the *Alvin*. Buttons in the handgrip activate the robot's vertical thrusters and control the tether to *Alvin*. A photo trigger activates the still camera aboard *J.J.* and a companion button adjusts the tilt of the video camera on *J.J.* A video monitor in *Alvin* displays *J.J.'s* field of vision. Switches in *Alvin* remotely control the motors and lights on *J.J.* and a joystick controls horizontal movements of the robot. A larger version of *J.J.* is now being designed. It will be considerably more sophisticated than *J.J.* and will include robot arms for retrieving samples. The versatility and maneuverability (in and out of very tight spaces) were proved during the meticulous examination of parts of the *Titanic* wreckage.

As observed by D. Yoerger (Woods Hole Oceanographic Institution), "One view of teleoperated submersibles is that they are replacements for manned submersibles, with advantages in terms of endurance, safety, and cost. While these are strong points for remotely operated vehicles (ROVs), *Jason* was not intended strictly to replace manned submersibles. It was designed to perform a variety of functions, many of which complement rather than duplicate the strong points of a manned vehicle. For example, manned submersibles are rarely used for sonar survey, due in part to problems of storing or processing data in small spaces with limited manpower. *Jason's* high-bandwidth telemetry system and precise control capabilities make it a very effective sonar platform."

ROVs are capable of a variety of interesting tasks. For example, within the last few years, *Jason* has participated in several projects of an archeological nature. One of these ventures was the exploration of two War of 1812 vessels, the *Hamilton* and *Scourage*, which lay on the bottom of Lake Ontario at a depth of 90 meters (255 feet). The craft was able to move over the decks of these historic vessels for electronic still camera surveys. In an expedition conducted in 1989, *Jason's* manipulator excavated the remains of an ancient Roman shipwreck at a depth of 760 meters (2500 feet) in the Mediterranean Sea. The computer-controlled compliance of *Jason's* arm enabled recovery of delicate ceramic jars, dishes, and oil lamps without damage. An earlier ROV (*Jason Jr. (J.J.)*) received worldwide acclaim when it was deployed in the second expedition to the *Titanic* in 1986.

ROV *Jason* carries a wide variety of sensors and manipulators for operation at depths to about 7000 meters (20,000 feet). ROV *Jason* can maintain heading and depth, follow precise track lines, and move automatically under the pilot's supervision. The manipulator arm automatically reacts to contact force to grasp objects firmly yet without damage. Although the human pilot is always in charge, the control task is apportioned between the pilot and the automatic systems in the craft.

Although ROVs have participated in a number of newsworthy events, they are designed primarily for gathering scientific knowledge. Currently there is much interest in *untethered* vehicles that can move without the restriction of cable. As of the early 1990s, a number of such craft are in the design or construction and testing phase.

Operational as of 1991, the French oceanographic agency developed the *Epulard*. The craft operates in a telerobotic mode, with its activity supervised by shipboard operators through an acoustic communications link. A similar craft, *EAVE*, is under development at the University of New Hampshire. The craft will have completely autonomous capabilities for search and survey tasks.

The Woods Hole Oceanographic Institute is building a robotic vehicle called the *ABE* (Autonomous Benthic Explorer). The *ABE* is designed to perform surveys in deep-sea hydrothermal vent areas independently of a surface vessel. These vents, located at several thousand meters, are extremely dynamic. As pointed out by D. Yoerger, "Flows of hot, chemical-laden water and volcanic activity occur unpredictably. Traditional submersible or ROV expeditions cannot remain on station long enough to observe many of these changes. *ABE* will remain on the bottom for extended periods of time, mostly in a low-power "sleep" mode, eventually as long as one year. Periodically *ABE* will repeat video surveys and measure oceanographic parameters such as water temperature, salinity, and optical properties. *ABE* will complement other vent-studying techniques. While *ABE's* observations will not be as high in quality as *Jason's* or *Alvin's*, *ABE* is designed to make observations when these traditional vehicles cannot be on station. Likewise, through its mobility *ABE* will complement fixed instrumentation.

Initially the vehicle will execute preplanned tasks. With experience, it is believed that more instrumentation and measurement quality and flexibility can be added in future redesigns. For example, *ABE* may be programmed to react to data from other seafloor sensors, such as an ocean bottom seismometer, and could report on seismic activity by way of an acoustic link.

In the United Kingdom, autonomous vehicles are being planned. One will be used for geological surveys, and the other to survey the water column. Japan has plans underway to construct a diesel-powered, full-ocean-depth autonomous submersible to survey the Mid-Ocean Ridge.

Deep-Ocean Drilling

As may be gleaned from the prior article on Ocean, knowledge of ocean processes and phenomena has increased many times over the past few decades and appears to be proceeding at an accelerating rate into the mid-1990s. The combined achievements of the numerous subsciences that comprise ocean knowledge are coalescing to form a unification of past principles. There is a major unknown, however—that being how little scientists know about the composition and structure of the two-thirds of Earth's crust that underlies the oceans. As observed by H.J.B. Dick (Woods Hole Oceanographic Institution), "While it has often been said that the surface of Mars is better known than the seafloor of Earth, the situation is far worse for the oceanic crust." This is explained, of course, by the severe difficulties in retrieving evidence. Conventional geological techniques simply have not worked to solve this problem. Among these techniques, one would include the use of surface mapping with geophysical sensing at depth to interpolate Earth's deep structure. Researcher Dick further points out, "It is increasingly apparent that the only way to obtain direct and precise knowledge of the composition and structure of the oceanic crust is to drill into it."

Deep-sea drilling dates back to the 1960s and the Mohole project. The project, almost constantly plagued by budget deficiencies, produced a few successes, but insufficient data to prove old theories or propose new concepts. The project, however, can be credited for sharpening the awareness, on an international scale, of the need for information and for the establishment of the Deep Sea Drilling Project, later followed by the Ocean Drilling Program (ODP). This program represents an international partnership of scientists and governments to explore Earth's origin and evolution beneath the seafloor.

The drill ship *JOIDES Resolution* operates a continuous series of cruises for the purpose of drilling and retrieving long cylindrical samples of sediment and rock samples or *cores*. Additional information is obtained from measurements made in the drilled holes. ODP is funded by the U.S. National Science Foundation and contributions from 18 other countries. Ten leading oceanographic institutions provide planning and management. These include Texas A&M University, the Lamont-Doherty Geological Observatory, and the Scripps Institution of Oceanography (University of California, San Diego). During the time frame (1985 to 1992), 77,500 meters (254,275 feet) of cores from 683 holes at 279 sites have been recovered. This represents nearly 50 cruises to sites in the Atlantic, the Pacific, the northern and southern polar seas, and the Indian Ocean. These cores are archived at three marine institutions in the United States, from which thousands of core samples have been passed to researchers in more than 38 countries throughout the world.

As described by V. Cullen (Woods Hole Oceanographic Institution), "Many lengths of 9.5-meter (31-foot) drill pipe are attached together to lower a large drill bit to the seafloor. This takes about 12 hours in 5,500 meters (18,045 feet) of water. Core barrels are then lowered through

the drill pipe to receive and contain the core material. When a length of about 9.5 meters (31 feet) has been drilled, the core barrel is raised to the ship, where technicians recover the long cylinder of sediment or rock, cut it into 1.4 meter (5-foot) sections, and begin documenting and describing its origin, appearance, and contents."

The *JOIDES Resolution* is equipped with seven different laboratories, representing expertise in geochemistry, geophysics, paleontology, petrology, paleomagnetics, and sedimentology. After photographing and preparing meticulously written descriptions, the cores are packed and stored under refrigeration. The ship is operated by a crew of 68. Approximately 50 scientists and technicians are aboard.

The principal areas of interest of ODP research are:

- Tectonic evolution of passive and active margins.
- Origin and evolution of oceanic crust.
- Origin and evolution of marine sedimentary sequences.
- Paleo-oceanography.

Drilling cruises are planned well in advance. Sites are selected to fill out data from given areas, and new sites are selected on the probability of data returned and to avoid any areas where hydrocarbons may be found. Such latter accumulations would be dangerous to both ship and surrounding environment.

Rather than to encounter some particular information that may be a key to supporting given hypotheses or theories of Earth's crust, a number of scientists feel that, with so much data to be analyzed, the overall puzzle will unwind methodically but slowly. Still greater depths will have to be probed. As commented by H.J.B. Dick, "The composition, internal stratigraphy, and rock history of the lower oceanic crust remains one of the fundamental unanswered questions of Earth science."

For many years, the so-called "layer-cake" or "infinite onion model" was accepted as describing the processes that occur between the mantle and the surface of the oceans. The model is shown in Fig. 5 in a highly schematic fashion.

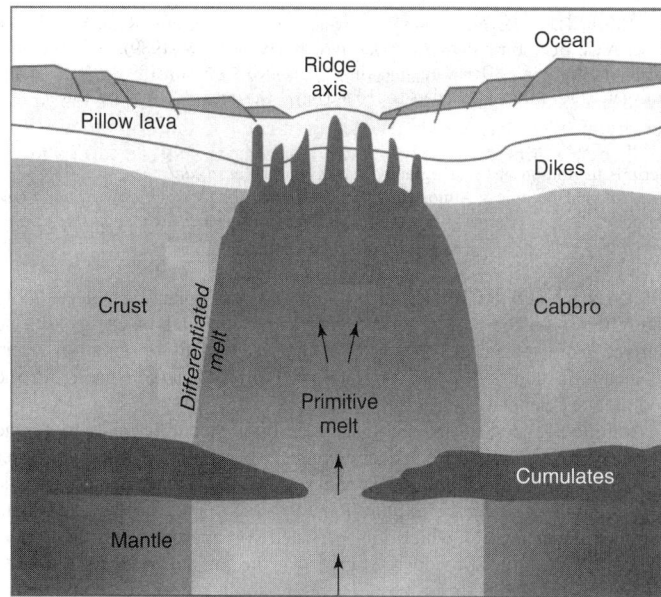

Fig. 5.   Highly schematic sketch of "layer cake" or "infinite onion" model. This model has been accepted for several years, but now is in question. The model as shown portrays a large steady-state magma chamber that underlies the ridge system. The chamber is disrupted only by very large fracture zones. Here oceanic crust grows continuously to eventually make up the lower two-thirds of oceanic crust. (*After Oberlander.*)

After drilling in 1987, a tectonically exposed section of the lower oceanic crust on the southwestern Indian Ridge, some of the plans for future drilling operations were changed. It was found that drilling lower-oceanic crust was much easier and accompanied by nearly total rock recovery, and required no new technology. It was also found that the compositional diversity of the rocks closely resembled those found in fossil magma chambers exposed on land, but that the processes for rock formation differed markedly from the dynamic and physical processes that appear on land. In 1989, leaders of the ODP proposed "that while drilling a single deep hole through the oceanic crust should be a long-term goal, a major program of drilling directly into the lower oceanic crust and mantle should be done over the next decade." Offset drilling, as used by sedimentologists, is the strategy proposed.

See also **Earth Tectonics and Earthquakes.**

Additional Reading

Ballard, R.D.: "A Long Last Look at TITANIC," *National Geographic*, 698–727 (December 1986).

Cullen, V.: "Ocean Drilling Program," *Oceanus*, 31 (Winter 1992/1993).

Detrick, R.S. and J.C. Mutter: "New Seismic Images of the Ocean Crust," *Oceanus*, 54 (Winter 1992/1993).

Dick, H.J.B.: "A New Mandate for Deep-Ocean Drilling," *Oceanus*, 26 (Winter 1992/1993).

Dinsmore, R.: "The University Fleet," *Oceanus*, 5–14 (Spring 1982).

Foster, D.: "DSV Alvin: Some Dangers and Many Delights," *Oceanus*, 17 (Winter 1988/1989).

Given, D.: "Underwater Technology in the USSR (Russia)," *Oceanus*, 67 (Spring 1991).

Golden, G.: "A Quarter-Century Under the Sea," *Oceanus*, 2 (Winter 1988/1989).

Hanson, L.C. and S.A. Earle: "Submersibles for Scientists," *Oceanus*, 31 (Fall 1987).

Herbst, D.: "In the Wake of a Modern Jason," *Oceanus*, 84 (Summer 1989).

Honjo, S.: "From the Gobi to the Bottom of the North Pacific," *Oceanus*, 45 (Winter 1992/1993).

Hotta, H.: "Deep Sea Research Around the Japanese Islands," *Oceanus*, 32 (Spring 1987).

Kaharl, V.A.: "Water Baby: The Story of Alvin, " Oxford University Press, Inc., New York, NY, 1990.

Kirn, T.F.: "Tuna Sub," *Technology Review (MIT)*, 10 (October 1991).

Odani, Y.: "Kalro, A Unique Research Vessel," *Oceanus*, 35 (Spring 1987).

Ryan, W.B.F.: "An Introduction — Down to the Sea in a Ship," *Oceanus*, 10 (Winter 1991/1993).

Smith, D.K.: "Illuminating the Seafloor," *Oceanus*, 74 (Winter 1992/1993).

Staff: *"Japanese Deepsea Research,"* 28 (January 3, 1992).

Tagawa, S.: "Deep Submersible Project," *Oceanus*, 29 (Spring 1987).

Van Dover, C.L.: "Diving the Soviet Mir Submersibles," *Oceanus*, 8 (Summer 1991).

Vine, A.C.: "The Birth of Alvin," *Oceanus*, 10 (Winter 1988/1989).

Williams, A.J., 3rd: "Ocean Engineering," *Oceanus,* 4 (Spring 1991).

Yoerger, D.R.: "Robotic Undersea Technology," *Oceanus,* 32 (January 1881).

Web References

Scripps Institution of Oceanography: *http://www.sio.ucsd.edu/*

The National Oceanic & Atmospheric Administration (NOAA): *http://www.noaa.gov/*

Woods Hole Marine Operations: *http://www.marine.whoi.edu/ships/ships_vehicles.htm*

OCEAN RESOURCES (Energy). In one way or another, it is highly likely that over the next several decades large amounts of energy will be derived from the oceans and seas of the world. This observation refers to means in addition to the production of oil and natural gas from the sedimentary deposits beneath the oceans.

Admittedly, during the past decade, both scientific and economic interests have dwindled considerably for two reasons: (1) much greater optimism regarding natural gas supplies, particularly under the continents, hence slowing undersea exploration; and (2) a glut in the petroleum supplies of the world which has persisted for approximately a decade. However, all authorities do not agree that the pursuit of oceanic energy should be permitted to atrophy. There are some schools of thought that forecast a negative turnaround in our energy supply fairly early in the 21st century, at which time aggressive interest in oceanic energy will reemerge.

Numerous sources of energy from the oceans have been proposed and investigated over a period of many decades. These include: (1) *Ocean thermal energy conversion* (OTEC), wherein advantage is taken of thermal differences in layers of seawater, a technology in which sizeable investments have been made during the past decade or so and a concept that dates back to the 1880s. (2) *Ocean wave energy conversion* — in actuality, a manifestation of wind power, because waves on the ocean surface are generated by atmospheric winds acting over large areas and long periods of time. This concept also dates back a number of years, but has received only marginal scientific and engineering attention as compared with OTEC technology. Much remains to be learned pertaining to the basic science of ocean wave action; much of the attention to date has

been focused on the interaction of waves with coastlines, where coastline conservation rather than waves as a source of energy has been the major concern. (3) *Ocean current energy conversion*, also a concept dating back many years. It was in 1835 when Gaspard Gustave de Coriolis published a paper in which he analyzed the distortions of fluid motions resulting from Earth's rotation — the Coriolis effect. See also **Coriolis Effect**. This effect combines with thermal and climatic forces as components of a mighty global heat engine to create current streams and gyres, as associated with the Gulf Stream and other major ocean currents. The wind plays an important, but not exclusive role in the creation of gyres.*. Much has been learned concerning ocean currents during the past decade (as, for example, from the IDOE programs described in entry on **Ocean**). From a practical application standpoint, energy from ocean currents is in a very early stage of investigation. (4) *Tidal energy* has been exploited on comparatively small scale for several centuries, dating back to the 11th Century when tidal mills for grinding grain were used in Gaul, Andalusia, and England. It has furnished regionally significant electrical energy for northern France since 1966, when the world's largest tidal generating station installed in the maritime estuary of the Rance River (Saint Malo-Dinard, Brittany) was dedicated. Tidal energy is unlikely to be a truly major factor in the world's future energy supplies simply because, without the expenditure of tremendous funds, there are relatively few economically viable locations to exploit. However, as explained in a separate entry on **Tidal Energy**, the science and technology required for energy from the source have been well established over a number of years. (5) *Energy from saline gradients* is a relatively new concept, even though the forces of osmosis have been understood for many years. The phenomenon of osmosis was first observed by Abbé Nollet in 1748. An osmotic pressure difference equal to a theoretical waterfall of 240 meters (787 feet) exists at the mouth of rivers where fresh water enters the sea. Assuming that energy extraction technology can be developed, a large potential would exist at the Amazon River (Brazil), the La Plata-Parana River (Argentina), the Congo, Yangtze, Ganges, and Mississippi rivers, as well as the Great Salt Lake (United States) and the Dead Sea (Israel/Jordan).

Further ocean-related energy sources include (6) *submarine geothermal springs* and *geothermal aquifers*; (7) *oceanic biomass* (it is estimated that the oceans of the world fix 10^{10} tons or more of carbon per year into organic material, mainly through photosynthesis by microscopic plants, the phytoplankton — with the possible future conversion of biomass to liquid fuels); (8) *resources for nuclear power processes*, including uranium, thorium, lithium, and deuterium (9) *direct conversion of ocean winds* by surface-located wind machines and wind farms (see also **Wind Power**); and (10) *ice from icebergs*, not only as a source of fresh water, but also used as a heat sink for nuclear, fossil, and even OTEC-type power generation. When two phases of a substance are present in a nearly isothermal environment, rather high temperature gradients and potential power can be achieved with small energy inputs, by a change in phase of one form.

Ocean Energy Thermal Conversion (OTEC)

It was d'Arsonval who in 1881 drew the attention of the scientific community to the enormous amounts of energy that may be available from thermal differences in layers of seawater. Not until 1930, however, did the French physicist Georges Claude operate the first crude ocean thermal difference power plant at the edge of Mantanzas Bay, Cuba. Besieged with fabrication, logistics, and weather problems, Claude finally succeeded in laying a 1.6-meter (diameter) by 1.75 kilometer (length) cold water pipe from an on-shore plant out into the bay, reaching a depth of about 700 meters. From a seawater temperature difference of $14\,^\circ\text{C}$, Claude's turbine generated 22 kilowatts (electric) of power. The plant operated on an open Rankine cycle system, i.e., steam generated from the seawater was used directly as the working fluid for the turbine. Whereas a conventional steam-turbine system operates with high pressures and large temperature differences, the Claude installation required a very low pressure (vacuum) to evaporate. The steam had pressures of approximately 0.02 atmosphere in the condensers, where it was condensed by mixing with cold water falling as rain.

---

* A gyre is a great, closed, circular motion of water in each of the major ocean basins, centered on a subtropical high-pressure region; its movement is generated by convective flow of warm surface water poleward, by the deflective effect of the Earth's rotation, and by the effects of prevailing winds. The water within each gyre turns clockwise in the Northern Hemisphere; counterclockwise in the Southern Hemisphere. See also **Coriolis Effect**

At these pressures, the specific volume (cubic meters/kilogram or cubic feet/ pound) of the steam was very high—so that the diameter of the turbine required for a given power level was some 30 to 50 times larger than that required for a conventional plant. Thus, with only a 1-meter (diameter) turbine available, Claude had deliberately oversized the cold water pipe by a factor of ten in cross-sectional area to avoid too much warming of the cold water as it was drawn up to the condenser. As a result, Claude put more power into the vacuum pump than he obtained from the turbine. Nevertheless, the principle of conversion of low-quality heat to power had been demonstrated. A diagram of Claude's design is given in Fig. 1.

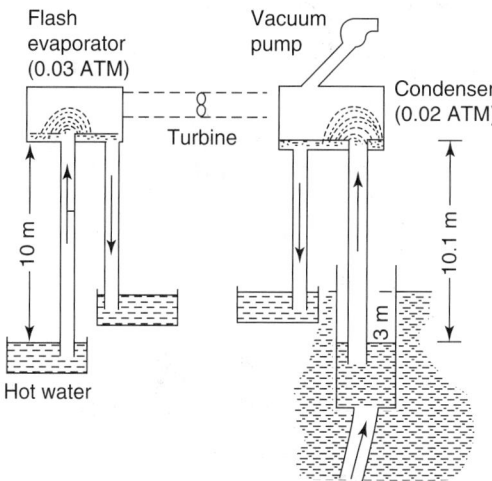

Fig. 1.   Open Rankine cycle system used by Claude in 1930 experiment. Flash-evaporated seawater drove a turbine and then was recondensed by cold water falling in the condenser. Details are available in "Power from Tropical Seas," by G. Claude, *Mech. Eng.*, 52, 1039–1044 (December 1930).

A principal key to the scientific and economic viability of an OTEC system is the temperature difference (ΔT) between the cold (low-layer) seawater and the warm surface seawater. The greater this difference, the greater the chances are for having an efficient system at lowest design and construction costs. The maps of Fig. 2 indicate the values of ΔT averaged over an annual period that exist between 40°S and 40°N. It will be noted that the most favorable regions (from the standpoint of ΔT) include a wide swath of the Pacific Ocean eastward of China, Indonesia, and northern Australia. There is also a large favorable region in the central Pacific Ocean south and west of Mexico. Another favorable region extends eastward from South America to west central Africa. Of course, there are numerous other factors that enter into consideration of a location. It will be recalled that Claude operated from a shore-located plant with a cold water pipe extending out into the ocean. An unfavorable factor for a shore-based facility is the relative shallowness of waters immediately off the coastlines in most places. Another factor that determines location selection, among other considerations, is the method selected for utilization of the power generated. There are two principal options in this regard: (1) transmission of the power by submarine electric cable to the shore; and (2) manufacture energy-intensive products aboard the OTEC platform at sea. Most likely candidate products would be hydrogen (see also **Hydrogen (Fuel)**), ammonia, or chlorine. More complex operations, such as electrowinning of magnesium and aluminum and fertilizer production, have been suggested. Precise station keeping would be mandatory for the electricity transmission scheme, whereas the product option would permit periodic location changes. However, even with the product option, mid-ocean locations, which would involve long distances, are not attractive from the standpoint of transportation economics, not only of the products, but of crews and maintenance material.

Open-Cycle System. A modern version of Claude's open-cycle system is shown in Fig. 3. The state of open-cycle system technology is somewhat behind that of the closed system approach, at least in the United States. Considerably larger and more complex equipment is required for the open-cycle system. Very large turbines, comparable to wind turbines, and de-aerators for removing dissolved gases from seawater are required. More recent open-cycle system studies have been pointing toward cost-effective solutions of the turbine and degasification problems.

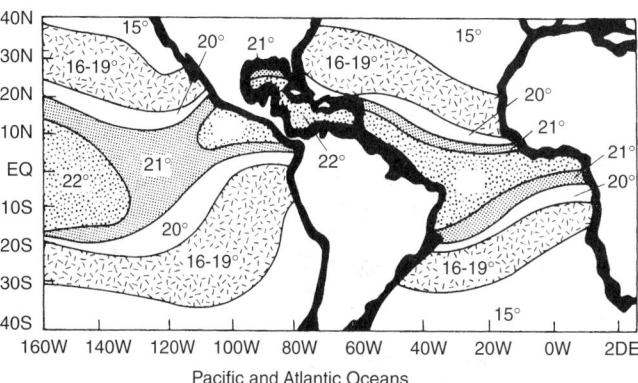

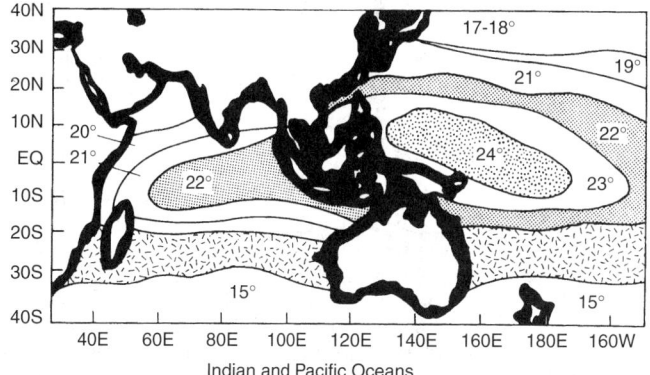

Fig. 2.   Average annual temperature difference (ΔT) between ocean surface water temperature and temperature of water at a depth of 1000 meters (3281 feet). ΔT is in degrees Celsius. Solid black areas are waters of less than 1000 meters in depth. (Suggested by R. Cohen, 1980.)

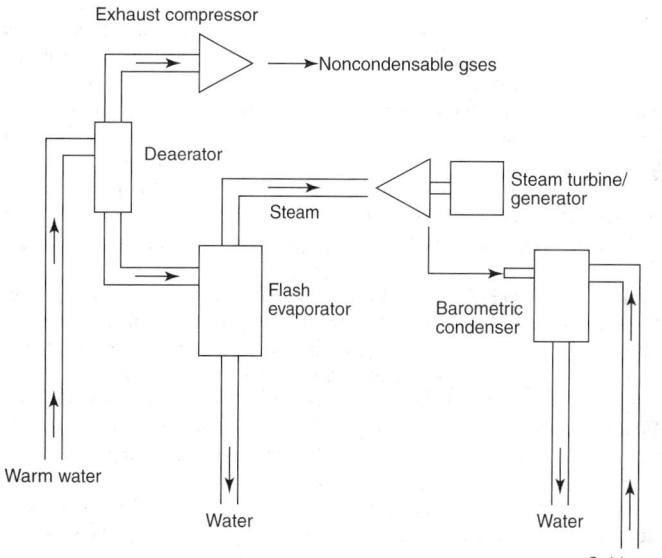

Fig. 3.   Open-cycle OTEC system. In an open system, the ocean surface temperature (warm water) is adequate to flash evaporate seawater. The escaping steam causes a turbine wheel (connected to an electric generator) to turn. Spent vapor is then cooled in a condenser by cold water pumped from depth. Unlike the closed-cycle system, the condensate need not be returned to the evaporator.

Closed-Cycle System. As shown by Fig. 4, the closed-cycle system requires a refrigeration-type working fluid, such as ammonia or propane, which is evaporated and recondensed continuously in a closed loop to drive a turbine. Because of the small ΔT, currently envisioned OTEC plants must by their very nature be of very low efficiency—in the neighborhood of 2.5%. Under normal circumstances (requiring some costs for fuel), such a plant would be considered hopelessly inefficient for serious consideration. But since this inefficiency is reflected only against

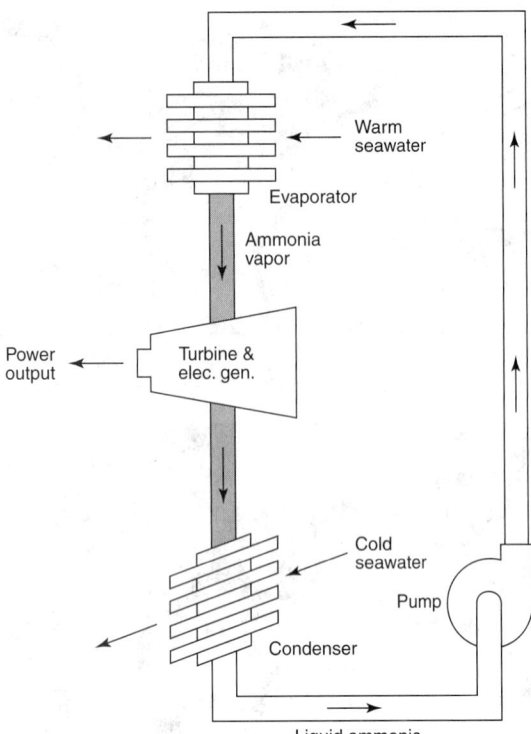

Fig. 4.   A closed-cycle OTEC system. Cold fluid, such as ammonia, is heated and vaporized in an evaporator. Vapor at high pressure moves blades of a turbine which drives a generator, as it is expanding toward the cooling system (condenser) in which the vapor is condensed back into liquid. The ammonia (fluid chosen for system shown) is vaporized by the narrow temperature differential between the warm and cold seawater.

initial costs of heat exchangers and working fluid, the approach holds promise in an energy-conscious world. Most investigators to date have found ammonia to be the best working fluid from the point of view of heat transfer—about twice as effective as propane, the next best candidate. The technology for handling ammonia has been established for many years. In these plants, usable power can be generated when there is a ΔT of as little as 14.7 Celsius degrees (27 Fahrenheit degrees), but locations providing a ΔT of a minimum of 20 Celsius degrees (36 Fahrenheit degrees) are the most desirable. So-called grazing OTEC platforms, readily movable from one location to another, could feasibly take advantage of seasonal changes in ΔT.

The state of OTEC technology is probably pretty well summarized by Isaacs and Schmitt (1980): "Problems with ocean thermal energy conversion include the design and stability of the intake pipe; biofouling of pipes, heat-exchange surfaces, and structure; construction, control, and maintenance of heat exchangers of unprecedented dimensions and required efficiency; corrosion; power transmission; and environmental effects. The best approach to these problems will probably involve testing modules of the final assembly both ashore and afloat where conditions are appropriate."

One estimate of the thermal-gradient flux available on an annual basis over the global surface is 1028 ergs. See the figure showing the spectrum of various energy quantities in the entry on **Energy**.

Ocean Wave Energy Conversion

The surfaces of the oceans and seas are rarely smooth. The forces of wind power exerted over vast expanses of the sea tend to be integrated in the form of waves that reflect wind energy from long distances. It has been estimated that wave-energy fluxes in the open sea or against the coasts may vary from a few watts to a megawatt per meter. The total wave power incident upon the coastlines of the world has been estimated by various researchers to be $2-3 \times 10^{12}$ watts. Wave energy impinging upon the coastlines varies considerably from one location to the next, as well as with seasons.

Wave energy is distributed in a thin layer of the ocean (less than 100 meters; 328 feet in depth). Wave-energy fluxes are greatest in summer and in the zones of the westerlies and trade winds. As pointed out by Newman (1980), the energy per unit of horizontal area is proportional

to the wave period and to the square of the wave height. This energy is carried along at a reduced speed, known as the group velocity, which in deep water is half the velocity of the individual wave crests. The product of the energy per unit area and the group velocity is the rate of energy flux per unit width of wave front.

One of the earliest approaches to retrieving energy from wave action is the most obvious, i.e., the conversion of vertical displacement into a useful form of energy. A simplistic device is illustrated in Fig. 5. Vertical displacement has been the basis of the use of wave energy in whistling and lighted navigational buoys for many years. It has been reported that a moderate-scale power plant is being constructed in the Orient on the basis of a sophisticated version of this principle.

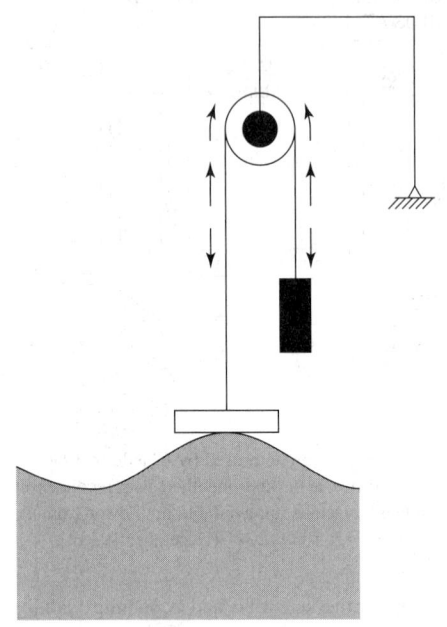

Fig. 5.   Schematic diagram of simplistic vertical-type wave-energy pick-up or absorber.

Wave motors brought out from time to time have depended for their operation largely on the lifting power of the waves. One installed at Atlantic City, New Jersey as early as 1911 consisted of six 4-foot (1.2-meter) cylindrical floats 4 feet (1.2 meters) high. These, each weighing about 3100 pounds (1406 kilograms), were lifted 2 feet (0.6 meter) by the waves about 11 times per minute. They drove a horizontal shaft by means of chains and ratchets, developing only 12 horsepower (12.2 metric horsepower). Steadiness was obtained through the use of heavy flywheels.

A wave motor employing a hydraulic ram for raising a portion of the water to a high level was proposed by Smith in 1927 (*Mech. Eng.*, September 1927, page 995). The waves were envisioned as entering a scoop, which would be connected to the ram by a long drive pipe. It was envisioned that the apparatus would be automatically adjusted for vertical level with tide changes.

A wave-operated pump was developed at the Fountain for Ocean Research (San Diego, California) in 1976. This design is based on the inertial interaction of two integrated systems—a buoy connected to a long vertical pipe flooded with water. As explained by the researchers (Isaacs, Castel, and Wick, 1976), the water column responds to such low frequencies that it cannot follow the sea surface when uncoupled from the pipe and float. Controlled by a simple check valve, the water in the pipe is forced to accelerate upward by the motion of the buoy, and it continues to flow upward as the buoy and pipe drop. The hydraulic pressure in a reservoir then increases and is limited only by the maximum vertical acceleration and the pipe length. For a 100-meter pipe and a maximum estimated acceleration of 0.5 g, the pressure will increase to about 5 atmospheres. Before this pressure is reached, the water is allowed to escape continuously under pressure and to interact with a turbine or Pelton wheel, and thus power is generated. Tests of this system at sea have indicated a 25% efficiency. Advantages of the system as claimed by the inventors include simplicity, relative invulnerability, response to a broad band of wave frequencies, and multiplication of wave pressure.

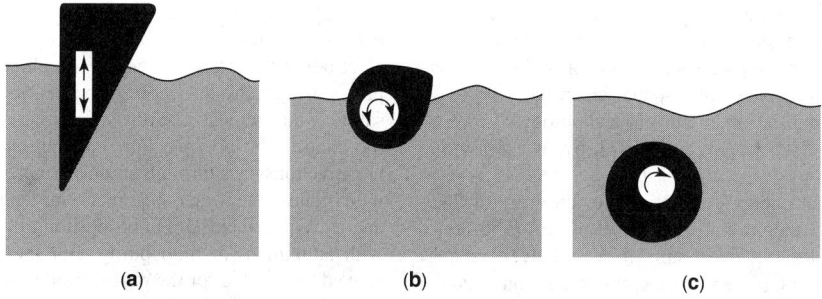

Fig. 6. Types of wavemakers, also considered as wave-energy absorbers: (a) unidirectional wedge-shaped device; (b) the Salter cam; and (c) submerged cylinder for generating unidirectional waves through orbital motion. (*After Newman.*)

In studying energy pick-up or absorbing devices that would interface in some way with energy-carrying waves, some researchers have turned to studying wave-making devices which have been used to simulate waves, required for various studies as, for example, coastline erosion research. Three such devices are shown in Fig. 6. Commenting briefly on these devices, Newman (1980) observes that if the wedge device (a) is driven in an oscillatory manner (in the direction parallel to the back side), the fluid disturbance will be confined essentially to the front. If the apex is sufficiently deep, the resulting waves will be trapped on the front side of the wedge, radiating away from this side in one direction.

The Salter cam (b), also known as Salter's Duck, is a British development. As pointed out by Isaacs and Schmitt (1980), Salter's rotor can utilize wave energy with considerable efficiency, both passing and reflecting very little of the incident energy. This wave form is utilized in the bell buoy. Early work on this design, conducted by S.H. Salter (University of Edinburgh), using a single cam rotating about a fixed axis in a narrow tank, demonstrated absorption efficiencies of 80 to 90%. More recently, a moving axis of rotation has been used to simulate the performance with a slack mooring. This research is being undertaken in a sophisticated three-dimensional wave tank at the Edinburgh facility.

As explained by Newman (1980), a submerged device presents certain advantages with respect to environmental impact and survival in storm conditions. A simple circular cylinder, as shown in Fig. 6(c), will generate unidirectional waves if it is given an orbital motion of circular form about its axis. Experiments to determine the feasibility of this scheme are under study at the University of Bristol (United Kingdom). Pairs of taut moorings with winch systems are used to impart the desired orbital motion to the cylinder.

Much of past research has been concentrated on two-dimensional types of absorbers as just described. There is also increasing interest in three-dimensional point absorbers. Based upon experience in the design of wave-power sources for buoys and lighthouses, it is well established that such devices respond equally to waves from all directions. In certain situations, wave-energy transducers sensitive to multidirectional waves are attractive. Particular attention in recent years has been given to small-scale point absorbers. An analogy can be made between a wave-energy point absorber and a simple radio antenna, the wire diameter of which is not important in terms of the power received or transmitted. A device of this type, developed by Yoshio Masuda, has been used for over a decade as a power source for navigation buoys and lighthouses, where a capacity of 70–120 watts is needed. Currently, these devices are operational in about 400 installations. Considerable current research on point devices is also underway at the Technical University of Trondheim, Norway and Chalmers University in Sweden. A British firm (Wavepower, Limited) has developed the Cockerell raft. In this configuration, a series of rafts are connected with hinges, which, in turn, are equipped with power take-off mechanisms. An advantage of this approach is that it obviates the need for rigid foundations and taut mooring systems.

Pneumatic designs have also been constructed as wave-energy transducers. Air bells have been designed by the National Engineering Laboratory in the United Kingdom. These use an oscillatory air column to extract power instead of strictly mechanical kinds of transducers. A Japanese ship (*Kaimei*) has been equipped with air chambers arranged longitudinally in the interior of the ship. This development ties back to the previously mentioned buoy systems researched by Masuda. Testing of this equipment in the Sea of Japan has received international sponsorship, including participation by Canada, the United Kingdom, and the United States.

Wirt and Morrow (Lockheed Corporation) were granted a patent in 1979 for an ocean wave energy device they call "Dam-Atoll." See Fig. 7. As reported in *Oceanus*, **22**, 4, 43 (1980), waves enter an opening at the top of the unit, just at the ocean's surface. A set of guide vanes at the opening causes the entering water to spiral into a whirlpool, held inside a 60-foot (18-meter) deep central core. The swirling column of water in the central core turns a turbine wheel, which is the only moving part of the system. The turbine is envisioned as having an output of 1 to 6 megawatts (electric). The inventors suggest that such devices could be anchored off the windy beaches of the world, where there is about 40 megawatts of power available per kilometer (64.4 megawatts/mile) of beach. This is estimated as sufficient to furnish the domestic power requirements for about 40,000 people. It has been envisioned that it would be possible to anchor 500 to 1000 such units along the coast of the Pacific Northwest, for example, that would furnish an electric generating capacity comparable to that of the Hoover Dam. In addition to electricity generation, other uses for the device suggested include a means for cleaning up and recovering oil spills, protecting beaches from wave erosion, forming calm harbors in the open sea, and desalinating seawater through the reverse osmosis process.

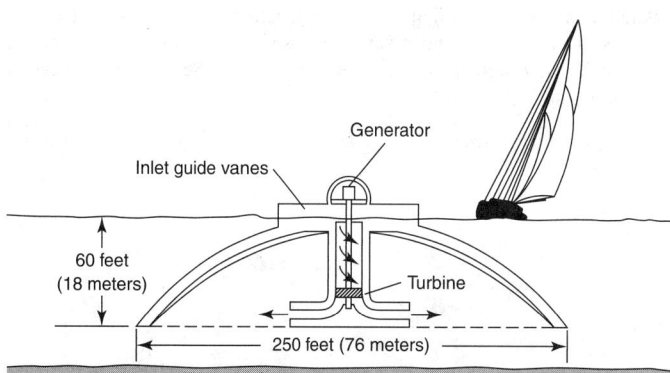

Fig. 7. Ocean wave energy device (*Dam Atoll*) for converting wind-powered wave energy into electricity. Arrows in the central cylinder indicate the whirlpool action of the water in the central cylinder, which acts like a giant flywheel to keep the turbine spinning even though wave action may be intermittent.

The design of systems to capture wave power has intrigued scientists for many years, and many other schemes and concepts have been proposed, well beyond the scope of this encyclopedia. In particular, reference is suggested to the proceedings listed (University of Kent; Gothenburg Symposium; and University of Delaware), as well as the Newman (1980), Ross (1979), and Isaacs and Schmitt (1980) reference listed.

Ocean Current Energy Conversion

The ocean currents or gyres are a major storage medium for solar energy. The major ocean current energy flux has been estimated at 1025 ergs (annually) as compared with the ocean-wave energy flux of 1027 ergs (annually). Even though the energy densities are relatively low, a number of proposals to exploit this energy source have been made over the years. A central theme of several proposals has been the application of the venturi-tube principle, wherein currents are directed to flow into throatlike configurations, thus increasing their velocity. As mentioned by Isaacs and

Schmitt (1980), if the potential for using currents of flowing water has been so attractive, one would expect the technology to be in practice along the great rivers. Instead, the practice in the past has been one of building dams to achieve higher current velocities over small spatial intervals.

A major step toward advancing the ocean current technology of recent years is the Coriolis concept first proposed in 1973 by W.J. Mouton (Tulane University).

Lissaman (1980) observes that energy calculations for an array of 242 large turbines (each rated at 83 megawatts) about 170 meters (558 feet) in diameter and moored in the Gulf Stream, in a relatively compact array of about 30 kilometers (18.6 miles) cross-stream dimension and 60 kilometers (37 miles) streamwise extent, could produce about 10,000 megawatts of electricity, a significant portion of Florida's electrical needs. This is an energy equivalent of about 130 million barrels of oil per year. Cost effectiveness studies thus far have been encouraging. Late in 1978, work funded by the U.S. Department of Energy involved the design, analysis, and water tests of a 1-meter (3.2-foot) diameter model at the David Taylor Model Basin in the U.S. Naval Ship Research and Development Center, Bethesda, Maryland. Additional studies are scheduled, with the sea test of a full-scale prototype scheduled for completion in 1984. For more details, see Lissaman (1979, 1980), Richardson et al. (1969), and Stommel (1965) references.

Energy from Saline Gradients

The phenomenon of osmotic pressure is explained in the entry on **Osmotic Pressure**. The diagram of Fig. 8 portrays the osmotic pressure difference between fresh water and salt water, this amounting to the energy equivalent of a waterfall some 240 meters (787 feet) in height, which theoretically occurs at the mouth of every river that exits fresh water into the oceans. Since one foot of water is equivalent to 0.0295 atmosphere, this pressure difference also can be expressed as 23.2 atmospheres. Although scientists agree that the energy, as the result of salient gradients, is in place, there is no consensus at this early phase of a young technology as to how this energy can be extracted and converted to a practical, usable form. Certain concepts are immediately suggested as the result of past related experiences, particularly in connection with various processes, such as electrodialysis and reverse osmosis, which have become part of desalination and permeable membrane technology. See also **Desalination**; and **Dialysis**.

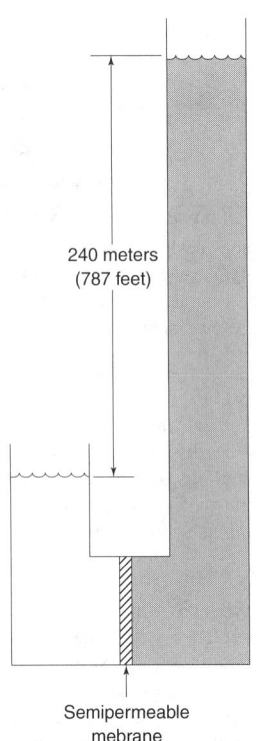

240 meters
(787 feet)

Semipermeable
mebrane

Fig. 8. Magnitude of osmotic pressure difference between fresh water and seawater.

In 1976, Leitz (U.S. Bureau of Reclamation) described a "dialytic battery" in this regard. This concept is essentially the reverse of the electrodialysis process used for desalination. Briefly, in the desalination procedure, electrodes impress an electric field across a cell through which saline water flows. Cations migrate to the cathode; anions migrate to the anode. The positive ions pass through cation-permeable membranes; negative ions pass through anion-permeable membranes. Between alternate membranes, the water becomes enriched with or depleted of salts. The energy requirements are in proportion to saline content, but increase rapidly with attainment of high purity. For the latter reason, electrodialysis has proved best suited for the purification of brackish waters rather than normal seawater.

As explained by Salisbury (1978), in the dialytic battery, fresh water is poured into half of the cells; salt water into the other half. As they mix, they generate an electric current. The cells are adjacent and separated by selective membranes. Half allow only positive ions to pass through, while the other half allow only passage of negative ions. The membranes are arranged so that the positive ions flow in one direction and negative ions travel in the other, thus generating an electric current. In a system of this type, it is estimated that the membrane would be the most costly component of the basic system. However, the costs of arranging a steady supply of fresh water, and an equal volumetric supply of seawater, coupled with the means to exit the mixed water in such a way as not to pollute the fresh water supply pose construction and engineering problems.

McCormick (U.S. Naval Academy) has suggested a "pressure retarded osmosis" process. Again, this is essentially the reverse of another desalination process, i.e., reverse osmosis. Briefly, in the desalination procedure, the saline water is pressurized above its osmotic pressure — then flows over a semipermeable membrane. The membrane is permeable to water, but not to dissolved solids. The fresh water is transported through the membranes. Energy requirements increase less rapidly with increasing saline content than with the previously described electrodialysis process. Tailoring of membranes to different saline concentrations is technologically feasible. High-pressure equipment is required for this process.

In the proposal of McCormick, brine would be pumped at about 120 atmospheres of pressure into a large chamber. A large pipe made of a semipermeable membrane (to permit passage of water, but no dissolved solids) would rest inside the chamber. The membrane pipe would be filled with fresh water, which would flow through the pipe in an attempt to dilute the brine. It is suggested that the diluted solution would then flow through a turbine from an exit port in the chamber, achieving a force twice that of the initial hydraulic pressure. Israeli scientists have applied this method, using brine from the Dead Sea. With this brine, it was found necessary to reduce the applied pressure from 120 atmospheres to below 100 atmospheres and to use a diluted brine in order to alleviate deterioration and compaction of the semipermeable membrane. However, even with these alterations, the process appears viable from an economic standpoint.

With membrane processes, it has been estimated that 70,000 square meters (753,200 square feet) of membrane are required to generate approximately 10,000 horsepower (10,139 metric horsepower). With current technology, obviously this suggests a large and thus land-based installation. However, with technological advancements in membrane materials, the system may at some future date be applicable as a source of power for ships, in which case, a ship would fuel up with salt rather than petroleum or coal prior to leaving port.

It is obvious that a salinity-gradient power plant would not have to be located on a coastline next to the sea. Rather, as proposed by Intertechnology Corporation, after mixing in a pressure retarded osmosis facility and generating power, diluted brine could be pumped to a solar still, where the fresh water and brine would be separated and thus, as renewed raw materials, would be recycled through the osmotic generator. Scientists recently have made some interesting observations as regards the salinity-gradient concept. Wick and Isaacs (Scripps Institution of Oceanography) have suggested that possibly more energy could be produced from the salt in the salt domes along the Gulf Coast than from the oil and natural gas which are associated with them. Although these salt domes already have produced on the average between 55,000 and 260,000,000 barrels of oil, it has been suggested that only in the highest yielding of these domes is the energy potential of the oil greater than that of the salt. Along these lines, some scientists have been disturbed by the suggestion that these brines be pumped out and dumped into the sea so as to prepare a place to store parts of the nation's strategic oil reserve. However, because salinity-gradient technology is only in a very early phase, it is not likely that such appeals will be listened to seriously.

Wick and Isaacs (1978) have reported estimates that, in the case of the Thompson salt dome (Ft. Bend, Texas) the oil energy (in MW-years) is 3.1 times that of the salt energy (in MW-years), but that in many other domes, the reverse situation holds. For example, it is estimated that the East Tyler dome (Smith, Texas) has a salt energy potential that is over 16,500 times that of its oil potential; that the Bethel dome (Anderson, Texas), the salt energy potential is over 1600 times that of its oil energy potential.

Reverse vapor compression as a process to utilize salient-gradient energy also has been proposed. In this concept, the vapor pressure difference between high-salinity and low-salinity waters is exploited. Membranes are not required. As in the cases of the other two major approaches just described, the reverse vapor compression has its reverse analog in the vapor compression distillation process used for desalination. It has been shown that, in principle, most desalination schemes that are reversible are capable of producing power when directed to *salination*. In vapor compression distillation, the vapor from boiling brine is compressed mechanically, thus increasing the vapor temperature and pressure. The compressed vapor is then fed back into the evaporator to distill more seawater feed. The multiple-effect approach can be applied to the concept. The process has been attractive for small-size units. As pointed out by Olsson, Wick, and Isaacs (1979), due to the lower vapor pressure of salt water, water vapor will rapidly transfer from fresh water to salt water in an evacuated chamber. If a turbine is interposed in the vapor flow between the two solutions, power can be extracted. Claude experimented with this concept as early as 1930, using vapor pressure differences resulting from the temperature gradient between ocean surface water and deep water. For obvious reasons, technological development of reverse vapor compression can benefit from any progress made in open-cycle OTEC power plants, previously described, and vice versa.

Submarine Geothermal Springs and Geothermal Aquifers

During the past decade, the discovery of a number of submarine geothermal springs and geothermal aquifers has been reported, some of these in connection with the IDOE projects of the 1970s. Discoveries have been made in the eastern Mediterranean Sea and in the Pacific Ocean off Hawaii, among other locations. These areas are not easy to find because their thermal effects usually do not penetrate upper layers of the ocean, but rather they are trapped below the thermocline. As may be expected, the most concentrated geothermal heat is associated with areas of tectonic activity, such as volcanic islands. Some of these sources are capable of releasing in excess of 106 MW-years of energy in a single eruption. It has been reported that a suspected underwater explosive event destroyed all sea-level culture in the eastern Mediterranean (1500 B.C.). Possible exploitation of submarine geothermal energy sources is in a very early stage of research.

Oceanic Biomass

Use of dried sea plants as a source of energy by simply burning them dates back many centuries in coastal areas where seaweeds were more available than wood. Modern planners are now directing their attention to the utilization of wet biomass, using such processes as anaerobic digestion to produce low-grade gas (50–60% methane). The basic chemical building blocks are also available in these materials to produce alcohol and other organic fuels. Seaweed farms as a source of energy have been envisioned. Seaweed as a source of foods and food additives is described briefly in the entry on **Seaweeds**. The giant kelp *Macrocystis pyrifera* is also a potential energy source. In the mid-1970s, an Ocean Food and Energy Farm (OFEF) program was commenced, jointly funded by the Energy Research and Development Administration, the American Gas Association, and the U.S. Navy. An open-ocean farm to cover some 100,000 acres (40,469 hectares), some 12.5 miles (21 kilometers) on a side and located about 100 miles (160 kilometers) off southern California was proposed. The farm substrate would be maintained at a depth of approximately 100 feet (30 meters). The area would be made up of flexible triangular modules, about 1000 feet (304 meters) on a side and covering about 10 acres (4 hectares). The modules would be held in position by diesel-powered propulsors. Through the use of wave-powered pumps, nutrient-rich water would be upwelled from a depth of about 300 feet (91 meters). Kelp plants would be attached to the substrate, with one plant per each 363 square feet (33.7 square

meters). It is estimated that about 4 years would be required for a crop to mature, after which the standing crop would be harvested by ship at the rate of six times per year. It is estimated that the yield would approximate 15 dry tons per acre (33.3 metric tons per hectare) per year. About 53% of this would be organic biomass.

Prior to proceeding with such a large commercial operation, a research program was commenced. The first test farm was made operative in late 1978, but the initial plantings did not survive. A second test is currently underway. As pointed out by Rhyther (1980), open-ocean energy farming of seaweeds must be regarded as a long-term prospect that cannot be expected to be realized in a time frame of less than tens of years. Much remains to be learned about the basic biology of the plants, particularly their nutrition and growth, and factors that control their organic productivity.

As reported in the entry on **Ocean**, explorations of the Galápagos Rift in 1979 excited the scientific community concerning the concept of chemosynthesis as the primary source of organic nutrients for life found in the vicinity of submarine thermal vents. It has been hypothesized that this particular food chain begins with the production of bacterial biomass, which then leads to the massive but highly localized animal communities found in the vent areas. This has opened up an entirely new concept of the chemosynthetic production of biomass. Instead of using light for the reduction of carbon dioxide, some bacteria use the energy liberated by the oxidation of certain electron sources with free oxygen. Thus, as observed by Jannasch (1980), these bacteria produce organic matter chemosynthetically rather than photosynthetically. For energy, they use hydrogen sulfide and other sulfur compounds.

Additional Reading

Blevins, R.W., J.T. Stadter, and R.D. Weiss: "At-Sea Test of Large Diameter Steel Cold Water Pipe," *IEEE Symp. Ocean Engng.*, New York, NY (February 3, 1980).

Chang, P.Y. and R.A. Barr: "A Frequency-Domain Approach to Analyzing the Dynamics of OTEC Plant Platforms and Cold Water Pipes," *IEEE Symp. Ocean Eng.*, New York, NY (February 3, 1980).

Charlier, R.H. and J.R. Justus: "Ocean Energies: Resources for the Future," Elsevier Science, New York, NY, 1992.

Cohen, R.: "Energy from Ocean Thermal Gradients," *Oceanus,* **22**, 4, 12–22 (1980).

Considine, D.M.: "Tidal Energy," in "Energy Technology Handbook," (D. M. Considine, editor), The McGraw-Hill Companies, Inc., New York, NY, 1977.

d'Arsonval, A.: "Utilisation des forces naturelles: Avenir de l'electricité," *La Revue Scientifique*, 370–372 (1881).

Duff, G.F.D.: "Tidal Power in the Bay of Fundy," *Technology Review (MIT),* **81**, 2, 34–42 (1978).

Dugger, G.L.: "Ocean Thermal Energy Conversion," in "Energy Technology Handbook," (D.M. Considine, editor), The McGraw-Hill Companies, Inc., New York, NY, 1977.

Dugger, G. (editor): "Proceedings, Sixth OTEC Conference (June 19–22, 1979)," U.S. Govt. Printing Office, Washington, DC, 1979.

Guenther, D.A., Jones, D., and W. Chiou: "Power Extraction from Deep Ocean Waves," *IEEE Symp. Ocean Engng.*, New York, NY (February 3, 1980).

Hartline, B.K.: "Tapping Sun-Warmed Ocean Water for Power," *Science,* **209**, 794–796 (1980).

Hunt, J.: "Petroleum Geochemistry and Geology," W.H. Freeman Company, San Francisco, CA, 1995.

Isaacs, J.D. and W.R. Schmitt: "Ocean Energy: Forms and Prospects," *Science,* **207**, 265–273 (1980).

Jannasch, H.W. and C.O. Wirsen: "Chemosynthetic Primary Production at East Pacific Sea Floor Spreading Centers," *BioScience,* **29**, 592–598 (1979).

Kash, D., et al.: "Energy under the Ocean: A Technology Assessment," University of Oklahoma Press, Norman, OK, 1973.

Lissaman, P.B.S.: "Energy Available from Arrays of Ocean Turbine Systems Moored in the Florida Current," Rept. AUR 7038, AeroVironment, Inc., 1977.

Lissaman, P.B.S.: "The Coriolis Program," *Oceanus,* **22**, 4, 23–28 (1980).

McCormick, M. and Y. Kim (Editors): "Utilization of Ocean Waves — Wave to Energy Conversion," American Society of Civil Engineers, Reston, VA, 1987.

Mei, C.C.: "The Applied Dynamics of Ocean Surface Waves," World Scientific Publishing, Inc., Riveredge, NJ, 1989.

Metz, W.D.: "Ocean Thermal Energy," *Science,* **198**, 178–180 (1977).

Newman, J.N.: "Power from Ocean Waves," *Oceanus,* **22**, 4, 38–45 (1980).

Olsson, M., Wick, G.L., and J.D. Isaacs: "Salinity Gradient Power: Utilizing Vapor Pressure Differences," *Science,* **206**, 452–454 (1979).

Pauling, J.R.: "An Equivalent Linear Representation of the Forces Exerted on the OTEC Cold Water Pipe by Combined Effects of Waves and Currents," *IEEE Symp. Ocean Engng.*, New York, NY (February 3, 1980).

Pierre, A.: "Tides in the Competition between Energy Sources," *Rev. Fr. Energ.* (September–October 1966).

Rau, G.H. and J.I. Hedges: "Carbon-13 Depletion in a Hydrothermal Vent Mussel: Suggestion of a Chemosynthetic Food Source," *Science,* **203**, 648–649 (1979).

Richardson, W.S., Scmitz, W.J., and P.P. Niiler: "The Velocity Structure of the Florida Current from the Straits of Florida to Cape Fear," *Deep, Sea Res.*, **16**, 225–234 (1969).

Ross, D.: "Energy from the Waves," Pergamon, New York, NY, 1979.

Ryan, P.R.: "Harnessing Power from Tides," *Oceanus*, **22**, 24–65 (1980).

Ryther, J.H.: "Fuels from Marine Biomass," *Oceanus,* **22**, 45–63 (1980).

Salisbury, D.F.: "Plugging into Salt," *Technology Review(MIT)*, **80**, 7, 8–9 (1978).

Scotti, R. and T. McGuinness: "Design and Analysis of the OTEC Cold Water Pipe," *IEEE Symp. Ocean Engng.*, New York, NY (February 3, 1980).

Staff: "Offshore Wind Systems and a New Wave Energy Device," *Oceanus,* **22**, 4, 46–47 (1980).

Stambaugh, K. and W. Jawish: "Structural Analysis and Design of a Cold Water Pipe for a 40 MWe Span OTEC Platform," *IEEE Symp. Ocean Engng.*, New York, NY (February 3, 1980).

Stommel, H.: "The Gulf Stream — A Physical and Dynamical Description," Univ. California Press, Berkeley, California, 1965.

Whitmore, W.F.: "OTEC: Electricity from the Ocean," *Technol. Rev. (MIT)*, **81**, 1, 58–63 (1978).

Wick, G.L.: "Power from Salinity-Gradients," *Energy*, **3**, 95–100 (1978).

Wick, G.S.: "Salt Power," *Oceanus,* **22**, 4, 29–37 (1980).

Web References

Scripps Institution of Oceanography: *http://www.sio.ucsd.edu/*

The National Oceanic & Atmospheric Administration (NOAA): *http://www.noaa. gov/*

Woods Hole Oceanographic Institution (WHOI): *http://www.whoi.edu/science/ science.html*

Woods Hole Department of Applied Ocean Physics and Engineering: *http://www. whoi.edu/science/AOPE/dept/*

Woods Hole Department of Geology & Geophysics: *http://www.whoi.edu/science/ GG/dept/*

OCEAN RESOURCES (Living).

OCEAN RESOURCES (Living). The oceans and seas contribute in a major way to the sustenance and wealth of Earth's human inhabitants as well as to many other life forms. Several of these resources are described in separate entries in this encyclopedia. See also **Crustaceans (Edible)**; **Fishes**; **Fish Meals, Oils, and Protein Concentrates**; **Mollusks**; **Seaweeds**; and **Turtles**. Numerous specific fishes are described separately. See list of entries in entry on **Fishes**.

Ocean Life Forms and Balances. Marine biology and ecology are concerned with the study of the plants and the animals in the sea and of their relationships with one another and with the environments. Such studies pertain to the manner in which these organisms are adapted to the various chemical and physical properties of the seawater, including various pollutants, to the motions and currents of the sea, to the availability of light at various depths, and to the solid surfaces that constitute the sea floor.

Marine life is classed in three groups: *benthos, nekton,* and *plankton.* The first are the plants and animals living on the sea bottom such as the permanently fixed or immobile forms (sponges and corals), the various creeping forms (crabs, snails), and others that burrow. Barnacles, the larger seaweeds, and sea squirts are also members of this group. *Nekton* are swimming animals that can move freely and that are capable of migration from one place to another. *Plankton* are the floating and drifting small animals (*zooplankton*) and plants (*phytoplankton*) capable of only limited locomotion, if at all.

Animal life of one kind or another exists at all depths of the oceans and in great abundance; indeed, nearly half of all classes of animals are marine. Plant life, however, is much less abundant and consists chiefly of the phytoplankton and the large seaweeds and algae. Nevertheless, the animal life is almost wholly dependent on the phytoplankton for its existence and these minute forms, in turn, have the same requirements for growth as other green plants. The most important of such requirements is sunlight without which photosynthesis cannot take place. Accordingly, these minute forms, which constitute the primary source of food in the oceans, are restricted to a rather shallow layer of water (generally no more than 450 feet (137 meters) deep under even the most favorable condition) called the *photic zone,* the zone in which there is sufficient light for plant growth. In addition, such growth depends on the availability of certain minerals that may be in short supply (thus limiting further growth), notably the phosphates and nitrates. Other salts and carbon dioxide are present in such quantities that they do not normally constitute any impediment to such growth. The shallow zone to which the plankton are restricted as well as the various limitations on several necessary nutrients has important consequences for animal life throughout the oceans.

A thin uppermost layer, of which the photic zone is a part, contains most of the animal life within the oceans. Called the *epipelagic zone,* it is no more than 700 feet (213 meters) in depth. Beneath it is the twilight zone (*mesopelagic zone*) in which there is sufficient light for animals to use, but less than required for plant photosynthesis. Below this is the *bathypelagic zone* into which small amounts of light penetrate; this is succeeded, finally, by the abyssal pelagic zone.

Within the photic zone, the phytoplankton would soon consume several of the available salts were these not replenished from the abundant supplies at lower depths. In the temperature oceans, the warmer waters rising to the surface as the colder waters sink during winter bring an ample replenishment of such nutrients and the phytoplankton grows rapidly in spring, consuming much of the replenished supply of nutrients. Growth slows during the summer and finally ceases in the absence of one or more necessary salts until the cycle is repeated the following winter and spring. In tropical regions, where light penetrates the oceans to much greater depths and no winter mixing occurs of surface waters with deeper waters richer in nutrients, the phytoplankton is much more sparsely distributed and to a much greater depth than occurs elsewhere. The balance of the oceans and, by far the greater part, is abundant in such salts since there is not sufficient light to permit utilization of such salts by plant life.

At the greater depths there is no food production, and animal life in such zones is dependent on the remains of surface organisms falling from above or brought down, together with the oxygen required for animal life, by the mixing of waters of different densities. Thus limited, animal life is also much more sparse than in the surface layer.

The phytoplankton are consumed chiefly by their animal counterparts, the zooplankton — the copepod Crustaceans and the somewhat larger euphausid crustaceans — as well as the larvae of most of the larger forms that occupy the seas. These in turn, constitute the food source for still larger animals and, in general, each form tends to feed on smaller organisms and is itself the food supply for somewhat larger organisms, the whole resting on the existence in great amounts of the simple plant forms the phytoplankton.

The central fact in the ecology of the oceans is that their plant life consists essentially of the phytoplankton, which are minute in size and can live only in those top levels of the ocean to which sufficient sunlight can penetrate to permit photosynthesis. This depth varies roughly from 350 feet (107 meters) in the clearer waters of the oceans to as little as 50 feet (15 meters) in the more turbid coastal waters.

With the exception of such negligible amounts of plant material as may be carried into the sea by rivers or as provided by algae, the phytoplankton is the sole source of the plant food upon which all the animals in the sea depend, since marine animals, like those on land, cannot synthesize living matter from the chemical materials in seawater. Of course, the supply of the food for the animals of the sea is not limited to the upper levels where the phytoplankton can live; both this plant food and food derived from upper-level animals can reach the lower levels by the sinking of the bodies of dead organisms. However, all the animals of the sea do not live directly on phytoplankton; many of them are carnivores, whose food is entirely animal. An important intermediary is the zooplankton, which preys upon the phytoplankton and in turn forms the food supply of larger organisms. The zooplankton is a general name for a large number of groups of organisms, which include small jellyfish, arrow worms and crustacea; in the latter group the genus *Calanus* is particularly important, forming as it does the principal food of the herring, which in turn is eaten by larger animals.

One of the most striking facts about marine ecology is the stratification of organisms by depth. While the fish are broadly classified into pelagic fish, which spend much of their time near the surface, and the demersal fish, which tend to stay near the bottom, this classification has been developed chiefly in fishing waters, and there is considerable evidence that in the parts of the ocean of greatest depth there are more than two zones in which the characteristic fish and other animals remain.

The general term for the organisms living on the bottom of the sea is the Benthos. Some of them live entirely on fine particles suspended in the water, others on the materials that lie on or in sea beds, while many others are carnivorous. Thus, the first of these groups includes the animals that form colonies (polyzoa), such as the corals, as well as many of the mollusca including mussels, oysters, scallops and calms. Some of these animals bury themselves in the sediment, and being provided with siphons, draw in the seawater. Many of the second class of animals, i.e., those which feed on materials in the sea bottom, also burrow into it. They include various species of worms, as well as hearturchins, brittle stars and other animals.

The carnivorous group includes many invertebrates, some of which feed on animals in the seawater, as the starfish feed on mollusks, and the bristle worms feed on many of the smaller animals. There are also carnivores, like the sea mouse, which burrow into the sea bed to eat the worms. In many parts of the sea the invertebrate predators consume so much of the available animal food that the food supply remaining for the fish seriously limits their population. This is especially true on the sea bottom, because so many important commercial fish are demersal, including cod, haddock, hake and plaice. In fact, the pelagic fish, i.e., those which obtain their food near the surface, such as the mackerel, herring and tuna, especially the first two of these, are not limited by the food consumption of the predators because they feed on plankton directly. These fish are, however, limited by themselves being prey, especially during their earlier stages of growth.

Plankton. Phytoplankton require nutrients for successful growth similar to green plants on land. Some of these are almost always present in excess of requirements, but others, e.g., phosphorus and nitrogen, are greatly depleted during the periods of active algal growth, which commonly occur in spring and autumn in northern open coastal waters. The plant population then declines and remains minimal until the nutrient supply is restored from deeper levels by vertical mixing and by decomposition of vegetation and animal remains. Light is a limiting factor to aquatic plant growth in high latitudes at certain seasons. Temperature exerts a selective influence on both plants and animals, but does not in itself appear to limit the total quantity of plankton developed.

Phytoplankton is for the most part composed of a variety of unicellular yellow-green algae dominated by diatoms, armored dinoflagellates, naked flagellates, and coccolithophorides,* which together according to some authorities account for more than half of the organic production of the earth. See Fig 1. The relative abundance of component groups varies regionally and seasonally. Diversity of species is greatest in tropical waters and least in boreo-arctic regions. Diatoms, distinguishable by their siliceous shells, occur singly and in chains, have one or more chromatophores, and during flowering season, or "blooms," flourish in great abundance in coastal waters. The dinoflagellates are more mobile, possess flagella, and like diatoms dominate at times in the phytoplankton during bloom periods, particularly in warm neritic (near surface and/or coast) waters. At these times luminous species often cause brilliant bioluminescence. Some, possessing chlorophyll, are truly phytoplankton, but others are either parasitic, or feed like animals and are classed as zooplankton. See also **Diatom**; and **Dinoflagellata**. Little is understood about the naked flagellates because of their minute size and the difficulty in preserving them. However, they are now considered the principal food of most planktonic filter feeders and may play an equally important, if not more

Fig. 1.   Diatoms (*magnification 1000×: Bausch & Lomb*).

* Coccolithophore is any of numerous, minute, mostly marine, planktonic biflagellate protists having brown pigment-bearing cells that at some phase of their life cycles are encased in a sheath of coccoliths to form a complex calcareous shell.

important, role than diatoms, particularly in neritic waters. In the open ocean, the naked flagellates tend to be augmented, if not replaced, by small calcareous coccolithophorides.

Other pelagic yellow-brown algal types at times present include the brownish colonial flagellate *Phaeocystis*, which forms gelatinous clusters, particularly in northern coastal waters, and the globular green *Halosphaera*, widespread over the open ocean. A blue-green alga *Trichodesium* appears at times as yellowish-brown patches on the surface in warm climates.

The principal research tool in plankton biology is the towed net. It has provided most of the basic information pertaining to the distribution, diversity, productivity, and behavior of zooplankton. During the 1970s, an alternative method of study was developed—direct underwater observation and collection of animals by divers. Studies have shown that much essential information about the lives of planktonic animals cannot be derived simply by collecting the animals with a net. As pointed out by Harbison and Madin (1979), this new perspective is essential to understanding the place of pelagic animals in the open ocean—one of the largest and least known of all environments on Earth.

Coccolithophore. Like any other type of phytoplankton, coccolithophores are one-celled marine plants that live in large numbers throughout the upper layers of the ocean. Unlike any other plant in the ocean, coccolithophores surround themselves with a microscopic plating made of limestone (calcite). These scales, known as *coccoliths*, are shaped like hubcaps and are only three one-thousandths of a millimeter in diameter.

What coccoliths lack in size they make up in volume. At any one time a single coccolithophore is attached to or surrounded by at least 30 scales. Additional coccoliths are dumped into the water when the coccolithophores multiply asexually, die or simply make too many scales. In areas with trillions of coccolithophores, the waters will turn an opaque turquoise from the dense cloud of coccoliths. Scientists estimate that the organisms dump more than 1.5 million tons (1.4 billion kilograms) of calcite a year, making them the leading calcite producers in the ocean.

Most phytoplankton need both sunlight and nutrients from deep in the ocean. The ideal place for them is on the surface of the ocean in an area where plenty of cooler, nutrient-carrying water is upwelling from below. In contrast, the coccolithophores prefer to live on the surface in still, nutrient-poor water in mild temperatures.

Coccolithophores do not compete well with other phytoplankton. Yet unlike their cousins, coccolithophores do not need a constant influx of fresh food to live. They often thrive in areas where their competitors are starving. Typically, once they are in a region, they dominate and become more than 90 percent of the phytoplankton in the area.

Coccolithophores live mostly in subpolar regions. Some other places where blooms occur regularly are the northern coast of Australia and the waters surrounding Iceland. In the past two years, large blooms of coccolithophores have covered areas of the Bering Sea. This surprises many scientists since the Bering Sea is normally a nutrient-rich body of water.

Coccolithophores are not normally harmful to other marine life in the ocean. The nutrient-poor conditions that allow the coccolithophores to exist will often kill off much of the larger phytoplankton. Many of the smaller fish and zooplankton that eat normal phytoplankton also feast on the coccolithophores. In nutrient-poor areas where other phytoplankton are scarce, the coccolithophores are a welcome source of nutrition.

In the long term, the plants seem to be good for the environment. Coccolithophores make their coccoliths out of one part carbon, one part calcium and three parts oxygen ($CaCO_3$). So each time a molecule of coccolith is made, one less carbon atom is allowed to roam freely in the world to form greenhouse gases and contribute to global warming. Three hundred twenty pounds of carbon go into every ton of coccoliths produced. All of this material sinks harmlessly to the bottom of the ocean to form sediment.

The coccolithophores' short-term effect on the environment is somewhat more complex. This effect again has to do with the formation of their coccoliths and the chemical reaction involved in the process. The chemical reaction that makes the coccolith also generates a carbon dioxide molecule, a potent greenhouse gas, from the oxygen and carbon already in the ocean. While much of the gas is sucked back in by the coccoliths (all plants take in carbon dioxide for food) some of it escapes into the atmosphere and immediately becomes part of the greenhouse gas problem. Scientists are concerned in the short term that greenhouse gases will cause the upper layers of the ocean to become more temperate and stagnant. This would increase the number of coccoliths in the world, which would produce more greenhouse gas.

The coccolithophores also affect the global climate in the short term by increasing the oceans' albedo. Albedo is the fraction of sunlight an object reflects–higher albedo values indicate more reflected light. Coccolithophore blooms reflect nearly all the visible light that hits them. Since most of this light is being reflected, less of it is being absorbed by the ocean and stored as heat.

Ctenophora. The comb jellies or sea walnuts, constituting a small phylum of marine pelagic animals related to the coelenterates, are sometimes included in this phylum. The phylum includes the peculiar Venus' girdle, a transparent, ribbonlike animal whose longitudinal axis is across the middle of the slender body.

The ctenophores differ from the coelenterates in the following structures: (1) The alimentary tract opens to the exterior at the end of the body opposite to the mouth. (2) The body bears rows of ciliated plates, the combs, which give the animals one of their common names. These are organs of locomotion. (3) Colloblasts or adhesive cells take the place of nematocysts.

The classification is as follows:

Class *Tentaculata*. A pair of long tentacles.
   Order *Cydippida*. Body spherical to cylindrical. Tentacles long.
   Order *Lobata*. Tentacles replaced in adult by fringe of short tentacles around mouth.
   Order *Cestida*. Body ribbon-like. Venus' girdle.
Class *Nuda*. Tentacles absent. Body conical to ovoid.
   Order *Beroida*. Wide mouth and pharynx; numerous side branches to meridional gastrovascular canal.

Ctenophores are among the most beautiful and voracious of the marine plankton. They consume enormous quantities of copepods, fish larvae, and other planktonic forms and thus are an important part of the ocean food chains. Ctenophores are highly dependent upon their ciliary paddles, called comb plates. As pointed out by Tamm (1980), each comb plate, or ctene, consists of a transverse band of thousands of long cilia, up to 2 millimeters in length, which beat together as a unit. Ctenophores are the largest known animals that use cilia for locomotion, and the longest cilia are found in the comb plates of ctenophores. Thus, ctenophores offer an excellent medium for studying cilia and flagella, as reported in considerable detail by Tamm (1980).

Mass Extinctions in the Ocean. Considerable attention has been given during the past few years to evidence that indicates the occurrence of mass extinctions of certain forms of life in the ocean. It appears that during brief intervals over the past 700 million to one billion years, a number of marine animals and plants died out. Various hypotheses have been offered to explain such mass extinctions, including loss of some species due to the cooling of the sea. Such extinctions not only have affected lower sea life forms, such as single-celled algae and plankton, but also large swimming reptiles and whales. Probably the best known of these extinctions is that which occurred about 65 million years ago at the end of the Cretaceous period. This event apparently eliminated most marine species at about the same time it is estimated that dinosaurs etc. became extinct on land. In a mass extinction, some species can recover, or new but related species may evolve. In other instances, evidence indicates the complete elimination of certain species. Changes in sea level also are implicated as causes of mass extinctions. The ratio of various chemical isotopes in skeletons, oozes, and other geologic sediments and materials in the ocean provide clues as to the timing and effects of mass extinctions. In addition to isotopic carbon dating, which has been well established for many years, other isotope ratios are now widely used, including, for example, the isotopic composition of neodymium in ocean waters and remnants. For further detail, see the separate article of **Mass Extinctions**.

International Decade of Ocean Exploration

The decade of the 1970s represented an accelerated program of research into nearly all aspects of ocean science. See also **Ocean**. Programs directed specifically to ocean biology included an intensive study of the Coastal Upwelling Ecosystems Analysis (CUEA), the initial goal of which was to predict phytoplankton distribution and growth in upwelling ecosystems on the basis of mesoscale observations of the critical forcing processes, mainly wind and circulation, but also including biological processes of grazing, predation, and nutrient regeneration of zooplankton, fish, and benthos.

On the basis of the published work, it is possible to cite a series of advances resulting from the CUEA project. This research tested the hypothesis that upwelling results from the tight coupling of a set of physical and biological processes, that this coupling is understandable, and that it can therefore be the basis for long-term management and use of the biological resources of upwelling ecosystems. The research also established the relationship between local winds and productivity and, less precisely, the relationships between very large-scale variations and productivity and nature of the ecosystem. Costlow and Barber (1980) suggest that given any coastal upwelling regime, with knowledge of its shelf width and latitude, it is possible to predict how variations in the local winds will affect primary productivity. The researchers give as an example the northwest African region at the latitude of Cape Verde or Daker, where increased storm frequency will enhance primary productivity inasmuch as in those regions there are frequent periods of nutrient depletion in the surface coastal waters.

Another IDOE biology program was the Controlled Ecosystem Pollution Experiment (CEPEX), which had three objectives: (1) to determine the effects of various pollutants on the microbial, phytoplankton, and zooplankton components of a large, field-based experimental ecosystem; (2) to evaluate changes in nutrient uptake kinetics related to pollutant stress; and (3) to identify the chemical variations that may occur in experimental ecosystems subjected to pollutant stress over specific periods of time. The effort involved scientists from nine American institutions as well as Canadian and British scientists.

Additional Reading

Attrill, M. (Editor): "Rehabilitated Estuarine Ecosystems: The Thames Estuary, Its Environment and Ecology," Chapman and Hall, New York, NY, 1998.

Betzer, P.R., et al.: "The Oceanic Carbonate System: A Reassessment of Biogenic Controls," *Science*, **226**, 1074–1077 (1984).

Capone, D.G. and E.J. Carpenter: Nitrogen Fixation in the Marine Environment," *Science*, **217**, 1140–1142 (1982).

Corliss, B.H., et al.: "The Eocene/Oligocene Boundary Event in the Deep Sea," *Science*, **226**, 806–810 (1984).

Costlow, J.D. and R. Barber: "IDOE Biology Programs," *Oceanus*, **23**, 1, 52–61 (1980).

Dawes, C.J.: "Marine Botany," John Wiley and Sons, Inc., New York, NY, 1998.

Druffel, E.M.: "Banded Corals: Changes in Oceanic Carbon-14 During the Little Ice Age," *Science*, **218**, 13–19 (1982).

Fenical, W.: "Natural Products Chemistry in the Marine Environment," *Science*, **215**, 923–928 (1982).

Grassle, J.F.: "Hydrothermal Vent Animals: Distribution and Biology," *Science*, **229**, 713–717 (1985).

Harbison, G.R. and L.P.MNadin: "Diving—A New View fo Plankton Biology," *Oceanus*, **22**, **2**, 18–27 (1979).

Haymon, R.M., Koski, R.A., and C. Sinclair: Fossils of Hydrothermal Vent Worms from Cretaceous Sulfide Ores of the Samali Ophiolite, Oman," *Science*, **223**, 1407–1409 (1984).

Hsü, K.J., et al.: "Mass Mortality and Its Environmental and Evolutionary Consequences," *Science*, **216**, 249–256 (1982).

Jannasch, H.W. and M.J. Mottl: "Geomicrobiology of Deep-Sea Hydrothermal Vents," *Science*, **229**, 717–725 (1985).

Kerr, R.A.: "The Ocean's Deserts are Blooming," *Science*, **232**, 1345 (1986).

Koehl, M.A.R.: "The Interaction of Moving Water and Sessile Organisms," *Sci. Amer.* **247**(6), 124–134 (1982).

Levinton, J.: "Marine Biology: Function, Biodiversity, Ecology," Oxford University Press, Inc., New York, NY, 1995.

Lewin, R.: "Life Thrives Under Breaking Ocean Waves," *Science*, **235**, 1465–1466 (1987).

Littler, M.M., et al.: "Deepest Known Plant Life Discovered on an Uncharted Seamount," *Science*, **227**, 57–59 (1985).

Martinez, L., Silver, M.W., King, J.M., and A.L. Alldredge: "Nitrogen Fixation by Floating Diatom Mats: A Source of New Nitrogen to Oligotrophic Ocean Waters," *Science*, **221**, 152–154 (1983).

Piepgras, D.J. and G.J. Wasserburg: "Isotopic Composition of Neodymium in Waters from the Drake Passage," *Science*, **217**, 207–214 (1982).

Rau, G.H.: "Hydrothermal Vent Clam and Tube Worm 13C/12C: Further Evidence of Nonphotosynthetic Food Sources," *Science*, **213**, 338–339 (1981).

Raup, D.M. and J.J. Sepkoski, Jr.: "Mass Extinction in the Marine Fossil Record," *Science*, **215**, 1501–1503 (1982).

Stanley, S.M.: "Mass Extinctions in the Ocean," *Sci. Amer.*, **250**(6), 64–72 (1984).

Takahashi, K., Hurd, D.C., and S. Honjo: "Phaeodarian Skeletons: Their Role in Silica Transport to the Deep Sea," *Science*, **222**, 616–618 (1983).

Tamm, S.: "Cilia and Ctenophores," *Oceanus*, **23**, **2**, 50–59 (1980).

Thiede, J.: "Reworked Neritic Fossils in Upper Mesozoic and Cenozoic Central Pacific Deep-Sea Sediments Monitor Sea-Level Changes," *Science*, **211**, 1422–1424 (1981).

Valiela, I.: "Marine Ecological Processes,"Springer-Verlag New York, Inc., New York, NY, 1995.

Vernick, E.I., et al.: "Climate and Chlorophyll — Long-Term Trends in the Central Pacific Ocean," *Science*, **238**, 70–73 (1987).

Other References

Alexander, L., et al.: "Large Marine Ecosystems: Patterns, Processes, and Yields," AAAS Books, Waldorf, MD, 1992.

Beauchamp, B., et al.: "Cretaceous Cold-Seep Communities and Methane-Derived Carbonates in the Canadian Arctic," *Science*, 53 (April 7, 1989).

Blaxter, J.H.S. and A.J. Southwardz: "Advances in Marine Biology," Vol. 30, Academic Press, Inc., San Diego, CA, 1995.

Duedall, I.W. and M.A. Champ: "Artificial Reefs: Emerging Science and Technology," *Oceanus*, 94 (Spring 1991).

Grassle, J.F.: "A Plethora of Unexpected Life," *Oceanus*, 41 (Winter 1988/1989).

Grigg, R.W.: "Paleoceanography of Coral Reefs in the Hawaiian-Emperor Chain," *Science*, 1737 (June 24, 1988).

Holden, C.: "Ocean-Slick Yardstick," *Science*, 1484 (September 27, 1991).

Holden, C.: "Picture-Perfect Plankton," *Science*, 681 (February 7, 1992).

Horgan, J.: "A Grave Tale — Do Whale Remains Help Life Spread on the Deepest Ocean Floor?" *Sci. Amer.*, 18 (January 1990).

Kasteleijn, H.W.: "Marine Biological Research in the Galapagos: Past, Present, and Future," *Oceanus*, 33 (Summer 1987).

Kerr, R.A.: "An About-Face Found in the Ancient Ocean," *Science*, 1359 (September 20, 1991).

Steele, J.H.: "The Message from the Oceans," *Oceanus*, 4 (Summer 1989).

Van Dover, C.L.: "Do 'Eyeless' Shrimp See the Light of Glowing Deep-Sea Vents?" *Oceanus*, 47 (Winter 1988/1989).

Vinogradov, M.E.: "Dynamics of Ocean Ecosystems: A National Program in Russian Biooceanology," *Oceanus*, 66 (Summer 1991).

Walbran, P.D., et al.: "Evidence from Sediments of Long-Term Acanthaster planci Predation on Corals of the Great Barrier Reef," *Science*, 847 (August 25, 1989).

Ward, F. and J. Greenberg: "Florida's Coral Reefs are Imperiled," *Nat'l. Geographic*, 114 (July 1990).

Zilanov, V.K.: "Living Marine Resources," *Oceanus*, 29 (Summer 1991).

Web References

Scripps Institution of Oceanography: *http://www.sio.ucsd.edu/*

Scripps Research Oceanography: *http://www.sio.ucsd.edu/research/oceanography.html*

The National Oceanic & Atmospheric Administration (NOAA): *http://www.noaa.gov/*

Woods Hole Oceanographic Institution (WHOI): *http://www.whoi.edu/science/science.html*

Woods Hole Department of Biology: *http://www.whoi.edu/science/B/dept/*

Woods Hole Department of Marine Chemistry and Geochemistry: *http://www.whoi.edu/science/MCG/dept/*

OCEAN RESOURCES (Mineral).   Since antiquity, the oceans and seas have been a major source of salt (sodium chloride) and continue to be so. Today, solar sea salt is produced in about 60 countries. The People's Republic of China, Australia, Mexico, India, Brazil, the Bahamas, Spain, and France are among the leading producers. At present, about 38% of the sodium chloride produced is evaporated from seawater. The value is estimated at over $400 million per year. Solar salt is very important to many countries, such as Japan, where there are few or no salt deposits. For several decades, seawater has been a significant source for bromine (production commenced by DuPont in 1931); iodine (from kelp, once very important, but no longer an economic source); magnesium (production started by Dow in 1941); potassium; sulfur; and several other elements and their compounds. Today, over 13% of the requirements for bromine come from seawater, as do over 70% of magnesium metal and 33% of magnesium compounds required by industry. Sulfur, associated with the cap rock of salt domes, has been produced from two salt domes just off the coast of Louisiana for many years. For a few decades, the continental shelves under the oceans have been producing large volumes of natural gas and petroleum. And also, for a few decades, the oceans have provided fresh water for many regions through various desalination processes. There are over 500 desalination plants in operation or under construction, with plants in arid and semiarid locations, such as the Middle East, but also in some highly urbanized regions and cities where fresh water is in short supply, such as in Italy and the Netherlands.

Many of the ore deposits found on the continents are the result of ancient oceans. Tin is found in offshore deposits, such as in Indonesia, in Cornwall (Saint Ives Bay), and Phuket Island off the west coast of the Malay Peninsula.

Within the past 20–30 years, much interest has been shown in manganese nodules on the seafloor in various locations. More recently, the discovery of hot brines in the Red Sea, "black smokers" on the East Pacific Rise and suspected in many other locations, and ophiolites has excited the scientific community and attracted industrialists because these phenomena are associated with metals, such as cadmium, copper, nickel, and zinc. These findings have largely resulted from the funding provided for geological and oceanographic research as part of the Deep Sea Drilling Project (DSDP) and the International Decade of Ocean Exploration (IDOE), projects which have been in place since the late 1960s.

More details pertaining to most of these ocean raw materials will be found in a number of specific entries in this encyclopedia. See also **Bromine**; **Chemical Elements**; **Desalination**; **Magnesium**; **Manganese**; **Natural Gas**; **Ocean**; **Ocean Research Vessels**; **Petroleum**; **Sodium Chloride**; and **Sulfur**.

The resource potential of the oceans awaits further technological development. It is interesting to note that the famous German chemist, Fritz Haber, spent more than eight years after World War I in attempts to recover gold from seawater in order to pay the German war debt. The results were disappointing, but large quantities of gold are in very large quantities of seawater. Currently, there is considerable interest in attempts to recover uranium from seawater, particularly by nations with no assured supply. Should fusion power come to fruition, after a few years, the ocean may be looked to as a source of lithium. Beach sands also have received considerable attention in recent years as sources of metals and other materials. Marine beaches may contain gold, silver, platinum, and diamonds in addition to magnetite, cassiterite, chromite, columbite, ilmenite, rutile, scheelite, zircon, monazite, and wolframite. Heavy-mineral beach sands are usually commercially worked for the titanium content of the rutile and the ilmenite. The same sands may also be processed to recover thorium from monazite and zircon for use in foundry sands. Currently, marine beaches are mined for heavy mineral production in Australia, Brazil, India, Madagascar, Mozambique, Sierra Leone, South Africa, and Sri Lanka, among other countries.

Diamonds are found in the seafloor sediments on the coast of the Kalahari Desert in southwest Africa.* The origin of the diamonds is obscure, but it is generally believed that basaltic and kimberlite pipes exist on the ocean floor as on the nearby land. There is a relative abundance of gemstones in the marine deposits and a few large stones have been recovered. Dredging began in 1961, using suction dredges capable of operating in waters to depths of 50 meters. Because of rough seas on this exposed coast, a number of barges were lost and the operation was concluded. However, in recent years a subsidiary of DeBeers is using a dredge protected by a seawall, thus permitting mining offshore about 120 meters at depths of 90 meters.

Calcium carbonate often precipitates from tropical or subtropical waters when the water becomes supersaturated due to enrichment of the carbonate content by intense biological photosynthesis and by solar heating of carbon dioxide-rich cooler waters. The aragonite precipitates as single needles in the shallow waters at a rate of about one millimeter of wet sediment per year. Continuing deposition leads to cementation and the formation of successive concentric sheaths known as oöids. The most extensively studied oölithic aragonite deposit is that distributed over the 250,000-square-kilometer (96,525-square-mile) Great Bahama Bank on the continental shelf near islands of the Bahamas. Most of the areas are less than 5 meters deep and are composed of quite pure calcium carbonate containing higher levels of strontium and uranium than are found in limestones of biological origin. Similar deposits occur in the Gulf of Batabanó (Cuba) and in the Mediterranean Sea off Egypt and Tunisia, as well as on the Trucial coast of the Persian Gulf.

Iron is a common constituent of marine sediments. Magnetite is found in beach sands and iron is common in glauconitic marine silicates. Iron oxides and sulfides occur where anaerobic conditions and elevated temperatures are found, as in the hot, salty brines found near rifts. Iron is a major constituent of the ferromanganese nodules.

Magnetite-rich iron sands have been dredged from the ocean floor just off Kagoshima Bay (Japan) in water averaging from 15 to 40 meters in depth. Iron sand concentrates were produced in Japan as recently as 1976,

* Acknowledgement of assistance obtained from W.F. McIlhenny, The Dow Chemical Company, Freeport, Texas in preparation of several of the following paragraphs is hereby made.

although a major marine iron sand operation in Kyushu ceased operation in 1966.

Marine sand and gravel for fill and for aggregate have been produced on all coasts of the United States, particularly from San Francisco and San Pedro Bays in California and from Long Island Sound. Marine sand and gravel are found in significant quantities in the United Kingdom.

Phosphorites (marine apatites) are dense, light-brown-to-black concretions, ranging in size from sands to nodules and irregular masses. Phosphorites have been found off Argentina, Chile, Japan, Mexico, Peru, South Africa, and Spain, and several islands in the Indian Ocean. Some also have been found off the west coast of North America and on the eastern North American continental shelf. These deposits occur where water upwelling transports phosphorus and where the rate of sedimentation is slow. The nodules are usually found as a monolayer on the surface. The mineralogy of the marine phosphorites is similar to western U.S. land deposits, which were almost certainly marine in origin. Phosphorites are quite constant in composition, containing 45–47% calcium oxide and 29–30% phosphorus trioxide. Seawater is generally saturated with tricalcium phosphate so that, under the oxidative conditions normally present, the phosphates precipitate in colloidal form and accrete to existing surfaces, rather than forming a phosphorite suspension. Although most of the phosphorite is believed to have formed during the Miocene epoch, it is believed that precipitation is currently taking place. The largest known seafloor phosphorite deposit is off the coast of California from Point Reyes to the Gulf of California along the inner edge of the continental shelf. Additional deposits have been found on the edges of the Blake Plateau east of Florida. A recovery project was commenced in 1962–1963, but failed to materialize.

Glauconite or green sands (a hydrated silicate with potassium, iron, and aluminum as cations) is widely distributed on the ocean floor in both ancient and more recent marine sediments. Glauconite is often found with phosphorite and occurs on the tops of banks, submerged hillcrests, and on slopes in water from 50 to 2000 meters in depth. Glauconites are known off the coasts of Africa, Australia, China, Japan, Portugal, South America, the United Kingdom (Scotland), the United States (California and the Atlantic shelf), and New Zealand. A 130-square-kilometer (50-square-mile) deposit has been identified on the Santa Monica shelf off California.

Submarine Hydrothermal Deposits

Discovery of the East Pacific Rise hot springs has created extensive interest and plans are underway to commence a four-year, multi-institutional project to explore the East Pacific Rise for additional areas of hot spring activity and ore deposition. The major objective of the program will be to examine the nature of hydrothermal processes along the mid-ocean ridge system from the slow-spreading to the very fast-spreading segments, such as at 10–30° South. The project will involve the use of surface ships, deeply towed instrument packages, new high-precision multibeam echo sounding for making highly accurate topographic maps of the seafloor, and ultimately manned submersibles, such as *Alvin*. See also **Ocean Research Vessels**.

The knowledge of submarine hydrothermal deposits was advanced by a large measure in 1979 when the hot springs on the East Pacific Rise at 21° north were discovered. Unlike the warm springs discovered on the Galápagos Spreading Center a few years ago, the springs on the East Pacific Rise are hot, with water venting at temperatures as high as 350 °C and at velocities of several meters per second. These formations are precipitating large quantities of sulfide ore and minerals rich in copper, zinc, and iron. See also **Ocean**. The precipitates form chimneys around the individual vents that spout black or white smoke composed of precipitated crystals of sulfides and other minerals. The discovery is the most exciting and significant in this field since the discovery of the Red Sea hot brines and metal deposits (Mottl, 1980).

By *hydrothermal* is meant hot water. When deposits are formed by chemical precipitation from hot solutions, they are termed hydrothermal. Hydrothermal deposits on land represent a very important class of economically retrievable ore deposits and provide a significant percentage of various metals, such as copper, zinc, lead, silver, gold, tin, molybdenum, among others. Mottl (1980) suggests that five factors are involved in forming hydrothermal ore deposits: (1) a source of the ore metals; (2) a source of water that dissolves and later precipitates the metals, concentrating them during the total cycle; (3) a source of heat; (4) a pathway between the site where metals are dissolved and precipitated and the site where they are finally deposited which is permeable and permits solution flow; and (5) the ultimate collection or deposition site.

For preservation, it is also important that ores be deposited in places where they will not be eroded away, as by weathering. Because so many factors are involved, there is a wide variety of hydrothermal ore deposits.

In terms of submarine hydrothermal ore deposits, the source of heat is the thermal energy associated with the formation of new oceanic lithosphere along the mid-ocean ridge system, where the seafloor is spreading apart and basaltic magma wells up. Because of tensional forces present, the newly formed crust becomes fractured, allowing seawater to percolate down through the fractures. During this percolation, the seawater is heated by contact with hot rock and commences to react with the rock, leaching metals that may be present. Because of the lesser density of the seawater (due to temperature), it rises and ascends to the seafloor and exists at submarine hot springs. Because there are several factors involved, there is, as on land, a wide variety of submarine hydrothermal ore deposits. Much remains to be understood and to confirm some of the early postulates, as given above, pertaining to the actual formation of submarine deposits.

For example, the concentrations of ore metals in most natural waters are quite low, particularly so in "normal" seawater. Measuring these low concentrations has been a problem of marine chemistry for many years. It is interesting to note that when artificial seawater, made up from pure reagent chemicals, is exposed to metallic elements, the ultimate solutions produced will contain from 100 to 1000 times the concentrations of these metals as compared with natural seawater.

The first submarine hot springs discovered along a mid-ocean ridge, those at the Galápagos Spreading Center, were emitting water at only 20 °C, but the chemistry of this water indicated that it had reacted with basalt at 350–400 °C. Then came the discovery of the 350° springs on the East Pacific Rise. Currently, the chemistry of this water is being studied at the Massachusetts Institute of Technology. To date, no submarine hydrothermal deposit has been sufficiently studied that all components contributing to its formation are known. Nevertheless, data at hand as of the early 1980s suggest some intriguing relationships among known deposits along mid-ocean ridges and point out the importance of special situations in producing and preserving large deposits.

Offshore Oil and Gas Resources

Although oil and gas exploration and production activities which occur offshore involve an extension of continents (the continental margins), they are nevertheless considered more in the general terms of oceanological rather than continental resources (the latter generally considered land or above-sea-level resources). Various aspects of offshore oil and gas production are described briefly in the entries on **Natural Gas**; and **Petroleum**.

As we go into the next century, because of an apparent glut of petroleum on world markets and because of much greater optimism pertaining to the ultimate natural gas reserves in the world, the emphasis on the exploration for petroleum and natural gas in underwater locations has markedly diminished.

Manganese Nodules

Deep-sea nodules, comprised mainly of manganese and iron oxides, have been found in abundance over large areas of the deep ocean floor that have been examined to date. In some locations, the nodules have been found to contain generous proportions of nickel, copper, cobalt, molybdenum, and vanadium, as well as manganese and iron. From the standpoint of potential commercial exploitation, a deposit is not considered promising unless the nickel-copper content is about 1.8% (weight) or greater. On this basis, one of the most promising areas is the Clarion and Clipperton fracture zone, an immense area some 4400 kilometers (2730 miles) long and 900 kilometers (560 miles) wide at its widest point. This zone is located southeast of Hawaii and southwest of Baja California. See Fig. 1.

An investigation of the origin and distribution of manganese nodules and the processes by which they selectively concentrate copper, nickel, and other metals was one of the first major projects under the sea beds Assessment project of the IDOE program. At a workshop attended by over a hundred scientists from various countries, the most likely locations for the exploration and study of manganese nodules were selected. The north central Pacific was identified as the zone where the nodules have the highest metal content. A team of American scientific investigators proposed that a comprehensive field and laboratory program be initiated to relate the high metal content to the local geological conditions. Data gathering was concerned along a transect that both academic and industrial

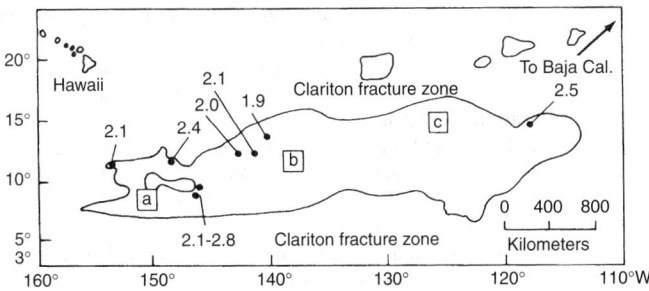

Fig. 1. Regions where manganese nodules containing more than 1.8% nickel-copper occur in the northeastern equatorial Pacific Ocean. Numbers indicate average percent of nickel-copper in one-degree squares. Areas a, b, and c indicate locations of activity carried out as part of Deep Ocean Mining Environmental Studies Program. (*After McKelvey, U.S. Geological Survey.*)

scientists agreed could serve as a potential mining site. In addition to dredge sampling and piston cores, bathymetric measurements, sidescan sonar, and high-resolution television pictures were obtained. See Fig. 2. The results provided a broad-scale picture of the conditions under which nodules form, but the mechanisms for concentrating specific metals are still not well defined.

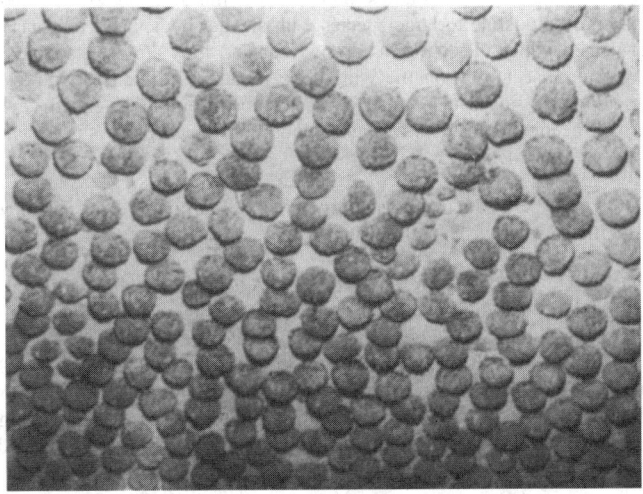

Fig. 2. Type of manganese nodules found in north central Pacific Ocean zone. (*Woods Hole Oceanographic Institution.*)

Although the nodules vary widely in their composition over the world oceans, metals are concentrated in three distinct types. One type comprises the nickel-copper-rich nodules of the Clarion-Clipperton variety, which is mainly formed in the equatorial regions. Another type, high in cobalt (1% or more) and low in nickel and copper, appears to be most commonly formed on sea mounts. The third type is high in manganese (35% or more), but low in other metals; it is known mainly on the eastern side of the Pacific Basin. As of the early 1980s, the most economically attractive were the cobalt-rich nodules.

The nodules form in a layered structure around a nucleus, which may be almost any material on the ocean floor. Most deep-sea nodules tend to be spherical or oblate in form. Nodules may occur up to 25 centimeters in diameter, but they average about 5 centimeters. The deposits may occur as slabs or agglomerates, or as incrustations on rocks or as pavement in some areas. The nodules are disorderly crystalline materials with layers of MnO_2 (mixed Mn^{2+}-Mn^{4+} oxides) alternating with $Mn(OH)_2$ and $Fe(OH)_3$. Excess iron appears as a mixture of goethite ($Fe_2O_3 \cdot H_2O$) and lepidocrocite.

The nodules are formed by the oxidation and precipitation of iron and manganese. The oxidation of Mn^{2+} is catalyzed by a reaction surface to a tetravalent state that absorbs additional Fe^{2+} or Mn^{2+} which, in turn, becomes oxidized. A surface is required and the initial deposition may be of iron oxide, possibly from volcanic or geothermal sources. Proper conditions of pH, redox potential, and metal ion concentration are found in deep ocean waters. The rate of accumulation appears to be very slow. The growth also may be discontinuous, and is estimated at a faster rater rate near the continental margins.

Precious Coral

Only a few species of coral have a combination of beauty, hardness, and luster, such the black coral species of Hawaii (*Antipathes dichotoma* and *Antipathes grandis*). These are highly valued by the jewelry trade. There are also a few red, pink, gold, and bamboo varieties that are in demand. Black coral also occurs in the Gulf of California and in the Pacific Ocean off Baja California, plus a few scattered locations in the Pacific Ocean east of Australia and north of New Zealand. Traditional sources of red and pink corals have been the Mediterranean Sea, various locations in the western Pacific Ocean, ranging from the Philippines, the Ryuku and Bonin islands and south of Japan. There is also a string of precious red and pink coral beds northwest of Hawaii and off the Cape Verde islands in the Atlantic Ocean off west Africa. A submersible vessel, the *Star II*, operated by Maui Divers of Hawaii, Ltd., is used to harvest pink coral (*Corallium secundum*) from the Makapuu bed. State regulations permit the collection of only 4400 pounds (1996 kilograms) within a 2-year period.

Additional Reading

Amsbaugh, J.K. and J.L. Van der Voort: "The Ocean Mining Industry: A Benefit for Every Risk?" *Oceanus,* 22–27 (Fall 1982).

Andreae, M.O. and H. Raemdonck: "Dimethyl Sulfide in the Surface Ocean and the Marine Atmosphere: A Global View," *Science,* **221**, 744–747 (1983).

Burroughs, T.: "Ocean Mining," *Technology Review (MIT),* **87**(3), 54–60 (1984).

Clark, J.P.: "The Rebuttal: The Nodules are Not Essential," *Oceanus,* 18–21 (Fall 1982).

Cooke, R.: "Metals in the Sea," *Technology Review (MIT),* **87**(3), 61–65 (1984).

Cronan, D.S.: "Underwater Minerals," Academic Press, Inc., San Diego, CA, 1980.

Curtis, C.: "The Environmental Aspects of Deep Ocean Mining," *Oceanus,* 31–36 (Fall 1982).

Dordrecht, Y.: "Transfer of Technology for Deep Sea-Bed Mining: The 1982 Law of the Sea Convention and Beyond," Kluwer Academic Publishers, Norwell, MI, 1995.

Fitzgerald, W.F., Gill, G.A., and J.P. Kim: "An Equatorial Pacific Ocean Source of Atmospheric Mercury," *Science,* **224**, 597–599 (1984).

Heath, G.R.: "Manganese Nodules: Unanswered Questions," *Oceanus,* 37–41 (Fall 1982).

Johnson, K.S.: "In Situ Measurements of Chemical Distributions in A Deep-Sea Hydrothermal Vent Field," *Science,* **231**, 1139–1141 (1986).

Knecht, R.W.: "Introduction: Deep Ocean Mining," *Oceanus,* 3–11 (Fall 1982).

Koski, R.A., et al.: "Metal Sulfide Deposits on the Juan de Fuca Ridge," *Oceanus,* 42–46 (Fall 1982).

MacLeish, W.H.: "The Struggle for Georges Bank," Atlantic Monthly Press, New York, NY, 1985.

Manheim, F.T.: "Marine Cobalt Resources," *Science,* **232**, 600–608 (1986).

Moore, J.G. and G.W. Moore: "Deposit from a Giant Wave on the Island of Lanai, Hawaii," *Science,* **226**, 1312–1315 (1984).

Mortlock, R.A. and P.N. Froelich: "Hydrothermal Germanium Over the Southern East Pacific Rise," *Science,* **231**, 43–45 (1986).

Mottl, M., Holland, H.D., and R.F. Corr: "Chemical Exchange During Hydrothermal Alteration of Basalt by Seawater," *Geochim. Cosmochim. Acta,* **43**, 869–884 (1980).

Mottl, M.J.: "Submarine Hydrothermal Ore Deposits," *Oceanus,* **23**, 2, 18–27 (1980).

Pendley, W.P.: "The Argument: The U.S. Will Need Seabed Minerals," *Oceanus,* 12–17 (Fall 1982).

Post, A.: "Deepsea Mining and the Law of the Sea," Kluwer Academic Publishers, Norwell, MI, 1983.

Riggs, S.R.: "Paleoceanographic Model of Neogene Phosphorite Deposition, U.S. Atlantic Continental Margin," *Science,* **223**, 123–131 (1984).

Rona, P.A.: "Mineral Deposits from Sea-Floor Hot Springs," *Sci. Amer.,* 84–92 (January 1986).

Siegel, M.C. and S. Turner: "Crystalline Todorokite Associated with Biogenic Debris in Manganese Nodules," *Science,* **219**, 172–174 (1983).

Vogt, P. and B. Tucholke: "The Western North Atlantic Region," Geological Society of America, Inc., Boulder, CO, 1986.

Web References

Scripps Institution of Oceanography: *http://www.sio.ucsd.edu/*

Scripps Research Oceanography: *http://www.sio.ucsd.edu/research/oceanography.html*

The National Oceanic & Atmospheric Administration (NOAA): *http://www.noaa.gov/*

Woods Hole Oceanographic Institution (WHOI): *http://www.whoi.edu/science/science.html*

Woods Hole Department of Marine Chemistry and Geochemistry: *http://www.whoi.edu/science/MCG/dept/*

Woods Hole Department of Geology & Geophysics: *http://www.whoi.edu/science/GG/dept/*

OCEAN (Tides).   See **Tides**.

OCEAN WAVES.   See **Atmosphere-Ocean Interface**; and **Ocean Resources (Energy)**.

OCELOT.   See **Cats**.

OCTAHEDRITE.   See **Anatase**.

OCTAL.   A number system used on binary digital computers. The system is based on a radix of 8. One octal digit corresponds to three binary digits, thus allowing easy conversion from one system to the next. The correspondence between numbers in decimal, binary, and octal notation is given in Table 1.

TABLE 1. COMPARISON OF DECIMAL, OCTAL, AND BINARY NOTATION

Decimal	Octal	Binary
0	0	0
1	1	1
2	2	10
3	3	11
4	4	100
5	5	101
6	6	110
7	7	111
8	10	1000
9	11	1001
10	12	1010
11	13	1011
12	14	1100
13	15	1101
14	16	1110
15	17	1111
16	20	10000

OCTANE NUMBER.   See **Petroleum**.

OCTAVE.   The interval between two sounds having a basic frequency ratio of two. By extension, the octave is the interval between any two frequencies having the ratio 2:1. The interval, in octaves, between any two frequencies is the logarithm to the base two (3.322 times the logarithm to the base 10 of the frequency ratio).

OCTOPUS.   See **Mollusks**.

OCUBA WAX.   See **Bayberry Shrubs and Trees**.

OCULAR HYPERTENSION.   Ocular hypertension occurs when the intraocular pressure in the eyes is above the normal range, but it has not yet affected vision or damaged the structure of the eyes. Normal eye pressure usually ranges between 10 and 21 mm of mercury. Pressure consistently above 21 indicates ocular hypertension. The condition can develop into glaucoma, a serious disease that causes damage to the optic nerve.

Most at risk for developing ocular hypertension are African Americans and those with a family history of the condition. It is also more common in those who are nearsighted, have high blood pressure, or are diabetic. Because ocular hypertension has no outward symptoms, people over the age of 40 and those in a high-risk category for glaucoma should have their pressure checked every year. A pressure check is a painless procedure that is part of any comprehensive eye examination.

In simple terms, ocular hypertension is caused by an excessive buildup of fluid inside the eye. This fluid, or aqueous humor, nourishes the cornea, iris, and lens, and maintains intraocular pressure. A typical eye produces about 4 cc of fluid a day, which circulates and then drains out of the eye. If the drainage system becomes clogged or if too much fluid is produced, pressure inside the eye can build up. The reasons for this are not fully understood.

There are normally no symptoms of ocular hypertension, which is one of the reasons why regular eye examinations are so important. Although ocular hypertension in itself is not sight threatening, if pressure within the eye builds to the point where it damages the optic nerve (glaucoma), eyesight can be permanently damaged.

An instrument called a tonometer is used to check eye pressure. There are two types of tonometers. One is called an applanation tonometer and uses an instrument that looks somewhat like a pen. After numbing eye drops are administered, the instrument is applied gently to the front surface of the eye and provides a pressure reading. The other type is a noncontact tonometer, which directs a warm puff of air toward the eye without touching it.

Neither ocular hypertension nor glaucoma can be prevented or cured, and ocular hypertension does not usually require treatment unless it progresses to glaucoma. Some doctors may, however, treat the condition with eye drops or other medicines as a precautionary measure. After you are diagnosed with ocular hypertension, your eye health must be monitored closely. See also **Glaucoma**; and **Tonometer**.

Vision Rx, Inc., Elmsford, NY.

ODD-EVEN RULE OF NUCLEAR STABILITY.   The rule, based on the number of stable nuclides, that nuclides with even numbers of both protons and neutrons are most stable; those with an even number of protons and an odd number of neutrons, or vice versa, are somewhat less stable; and those with odd numbers of both protons and neutrons are least stable.

ODONATA.   The dragonflies and damsel flies, an order of insects containing moderate to large species with slender bodies and four narrow net-veined wings. Their flight is powerful and they are well adapted to take their insect prey on the wing. The mouth is formed for biting. In the immature stages these insects are aquatic. Commonly called devil's darning needles. The order contains about 2700 species and is represented in all parts of the world.

ODOR DETECTION (Physiology of).   See **Flavors and Essences**; **Olfactory System**; and **Pollination**.

OFF-AXIS PARABOLIC MIRROR.   A mirror in the shape of a paraboloid of revolution will reflect a parallel beam to a single focus within the incident beam, if the beam is parallel to the axis of the paraboloid. In order to reflect a beam to a point off of the original beam, a part only of a paraboloidal mirror is used. Such a part is therefore known as an off-axis parabolic mirror. It is commonly manufactured by cutting it from the larger total paraboloid, because of the difficulty of making small, paraboloidal sections.

OFFLINE.   In terms of instruments and computers, offline describes situations where such devices are not directly in the dynamic control or data system loop, but rather they are auxiliary entities. Information collected by offline equipment may be used to adjust dynamic conditions within a loop, but not by the usual instantaneous feedback method. Rather, offline equipment is used in a supervisory manner for making periodic adjustments.

OFFSET.   See **Proportional Control**.

OFFSHORE DRILLING.   See **Petroleum**.

OGIVE.   A body of revolution formed by rotating a circular arc about an axis that intersects the arc; the shape of this body; also, a nose of a projectile or the like so shaped.

OHM, GEORGE SIMON (1789-1854).   Ohm was a German physicist whose early interest in mathematics was inspired by his father who educated him at home. He was a mathematics and physics teacher at the Jesuit Gymnasium of Cologne when he began his own experimental work after he learned of Oersted's discovery of electromagnetism in 1920. He wanted to teach at a university so he began to work towards the publication of his experimental results. He is remembered for finding and stating in 1827, the relationship between voltage, resistance, and current in an electric circuit referred to as Ohm's law. Ohm's law appears in his famous book *Die galvanische Kette, mathematisch bearbeitet* in which related his theory of electricity. His work was not immediately accepted in Germany. Recognition finally came in the early 1840s from British Scientists. The Royal Society of London awarded him its Copley Medal in 1841. In 1845, he became a full member of the Bavarian Academy. In 1852, Ohm achieved the position of chair of physics at the University of Munich. The unit in which resistances are measured has been named the ohm after him.

See also **Ohm's Law**.

J. M. I.

OHMIC CONTACT. A contact between two materials, possessing the property that the potential difference across it is proportional to the current passing through.

OHMMETER. See **Electrical Instruments**.

OHM'S LAW. This very familiar law of electric conduction, stated by George Simon Ohm in 1827, is expressible in various forms, of which the following is typical: The steady electric current in a metallic circuit is proportional to the constant total electromotive force operating in the circuit: $I = KE$. The constant K, known as the "conductance" of the circuit, is the reciprocal of the resistance R; so that the equation may be written in the more usual form.

$$I = \frac{E}{R}$$

OIL. See **Petroleum**.

OIL SANDS. See **Tar Sands**.

OILS (Fixed). See **Fixed Oils**.

OIL SHALE. This term refers to a carbonaceous rock that can produce oil when heated to pyrolysis temperatures of 427 to 538 °C (800–1,000 °F). The oil cannot be extracted with organic solvents at room and moderate temperatures with known technology. The oil precursor in the rock is a high-molecular-weight organic polymer, *kerogen*. An elemental analysis of kerogen derived from the upper zones of Colorado and Utah oil shales is: Carbon, 80.5% (weight); hydrogen, 10.3%; nitrogen, 2.4%; sulfur, 1.0%; and oxygen 5.8%. Usually, the host rock is comprised of dolomite, calcite, quartz, and clays.

Status of Oil Shale Technology. As with most other alternative sources of energy and notably hydrocarbons, the so-called oil glut of the 1980s essentially brought to a halt the further development of oil shale technology. With the fading of federal support in connection with the demise of the Synthetic Fuels Corporation (SFC) in the mid-1980s and the abandonment by most petroleum companies of their earlier interest in alternative energy sources, the shale oil program is essentially a "dead issue" as of the late 1980s. It is estimated that few researchers are now engaged in advancing the technology. One of the last major investments made in oil shale retorting was sponsored by Union Oil Company in connection with its plant located near Parachute Creek, Colorado.

Resources

Within a radius of about 200 kilometers (124 miles) of a center point where the borders of Colorado, Utah, and Wyoming intersect, an area comprised of four great basins—the Green River, Piceance, Uinta, and Washakie Basins as shown on the map in Fig. 1—some authorities estimate there is the equivalent of 2 trillion barrels of oil to be found—more than 50 times the total reserves for petroleum in the United States as presently known. These shales can be expected to yield from 25 to 30 gallons of oil per ton of shale. Of the explored basins, it is estimated that about 83% of these shales are in Colorado; 8.8% in Utah; and 8.2% in Wyoming. These deposits are generally referred to as occurring in the Green River formation. These oil shale resources occur in beds at least 30 meters (98 feet) in thickness. A comparatively small percentage may be mined by surface techniques. For the deeper veins, essentially underground mining technology or in-situ recovery technology will be required. Geologists consider that these deposits date from the Eocene era.

There are also Eastern oil shales, which represent petroleum locked in older shales dating back to the Devonian and Mississippian eras. It is estimated that these shales contain some 2 trillion barrels of oil and underlie an area of over a million square kilometers (400,000 square miles) in Michigan, western Pennsylvania, eastern Ohio, southwestern Indiana, eastern and southern Illinois, most of Kentucky and Tennessee, much of Oklahoma, and northern Alabama. They are sometimes broken down into what are known as the Eastern and Midwestern oil shale deposits. In

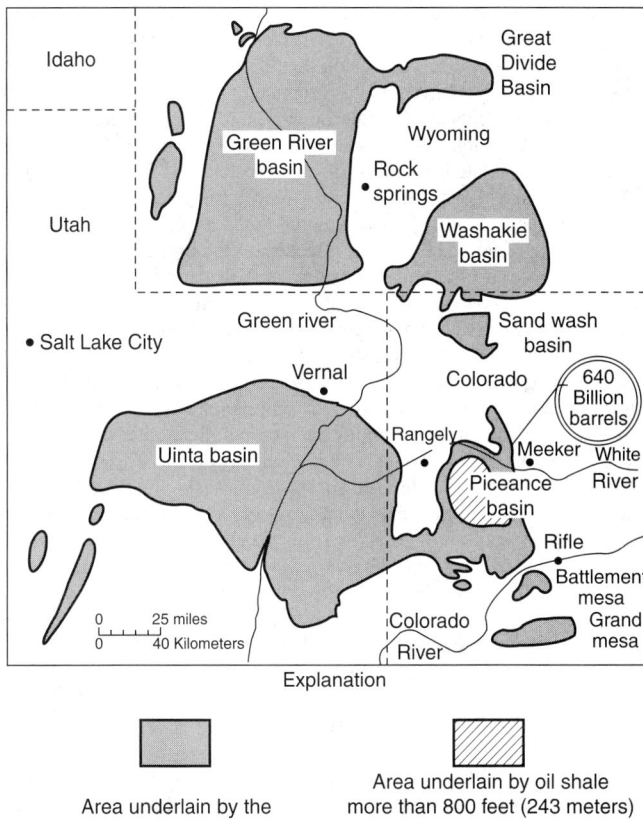

Fig. 1. Distribution of oil shale in the Green River formation. (*U.S. Geological Survey.*)

Michigan, the shale is approximately 60 meters (197 feet) thick and is in a basin at depths ranging from about 0.8 kilometer (0.5 mile) to outcroppings in three of the northern counties. In general, the Eastern oil shales become thicker and deeper toward the east.

Processing Oil Shale

Work on the production of petroleum-like materials from oil shale in the Green River formation dates back many years. One of the first major efforts was that undertaken by the U.S. Bureau of Mines in 1944. This program involved two 40-ton capacity retorts that operated between 1947 and 1951, with a production of some 920 runs and a total consumption of 37,500 tons of raw shale. This was a batch process and much was learned from this experience.

In a status report (1983), one of the largest (12,800 tons/day) and the last of numerous oil shale retorting processes (Union Oil Company) is described in considerable detail by Duir, Griswold, and Christolini in the February 1983 issue of *Chemical Engineering Progress* (pp. 45–50). This paper includes several diagrams and provides an excellent starting point for the reader who may be interested in this topic.

OKAPI. See **Giraffe and Okapi**.

OKRA. See **Malvaceae**.

OLEFIN FIBERS. See **Fibers**.

OLEIC ACID. $CH_3(CH_2)_7CH:CH(CH_2)_7 \cdot COOH$, formula weight 282.45, colorless liquid, mp 14 °C, bp 286 °C, sp gr 0.854. Sometimes referred to as red, oil, elaine oil, or octadecenoic acid, this compound is insoluble in H_2O, but miscible with alcohol or ether in all proportions. Oleic acid solidifies into colorless needle crystals.

Oleic acid differs from stearic acid chemically by possessing 33 instead of 35 hydrogen atoms in the radical $C_{17}H_{33} \cdot COOH$ (oleic acid), $C_{17}H_{35}COOH$ (stearic acid). It is possible to convert oleic acid and oleate esters into stearic acid and stearate esters by treatment with hydrogen gas in the presence of finely divided nickel as a catalyzer at 250 °C under pressure as in the hydrogenation of oils and fats. Either by careful oxidation, or by addition of ozone and splitting, oleic acid yields products of 9 carbon atoms, thus leading to the conclusion that the double bond is in the center of the carbon chain. Oleic acid adds bromine or iodine in definite amounts to confirm the conclusion that one double bond is contained. Nitric acid converts oleic acid into elaidic acid, $C_{17}H_{33}COOH$, mp 51 °C (oleic and elaidic acids are related, cis- and trans-, as maleic and fumaric acids).

Oleic acid may be obtained from glycerol trioleate, present in many liquid vegetable and animal nondrying oils, such as olive, cottonseed, lard, by hydrolysis. The crude oleic acid after separation of the water solution of glycerol is cooled to fractionally crystallize the stearic and palmitic acids, which are then separated by filtration, and fractional distillation under diminished pressure. Oleic acid reacts with lead oxide to form lead oleate, which is soluble in ether, whereas lead stearate or palmitate is insoluble. From lead oleate oleic acid may be obtained by treatment with H_2S (lead sulfide, insoluble solid, formed). With sodium oleate, a soap is formed. Most soaps are mixtures of sodium stearate, palmitate, and oleate.

Representative esters of oleic acid are: methyl oleate, $C_{17}H_{33}COOCH_3$, bp 190 °C at 10 millimeters pressure; ethyl oleate, $C_{17}H_{33}COOC_2H_5$, bp 205 °C at 10 millimeters pressure; glyceryl trioleate (triolein), $C_3H_5(COOC_{17}H_{33})_3$, bp 240 °C at 18 millimeters pressure.

Oleic acid is used in the preparation of metallic oleates, such as aluminum oleate for thickening lubricating oils, for water-proofing materials, and for varnish dryers. The glyceryl ester of oleic acid is one of the constituents of many vegetable and animal oils and fats.

See also **Vegetable Oils (Edible)**.

OLEORESINS.   See Resins (Natural).

OLFACTORY SYSTEM.   In humans, this system provides the sense of smell and also contributes to taste. Most sensations described as taste or flavor actually are aromas. When chewing or swallowing, odor-bearing molecules progress through the back of the mouth and into the olfactory cells, which are linked by nerve cells to the brain. Obviously, smell and taste are extremely important to the marketers of foods, beverages, cosmetics, deodorants, soaps, cleaners, and a variety of household products. See also **Flavors and Essences**.

As pointed out by A.R. Newman (see reference listed), most odorants are a select group of small, hydrophobic molecules with masses up to 300 Da, which is typical of volatile molecules. Some non-volatiles, such as oils, can be detected as aerosols. Complex structure-function relationships have been developed to classify odorants. Generally, many odor-producing molecules include one or two polar functional groups, often containing an oxygen or sulfur atom. Classification of odors is made difficult by the fact that they can be described only in subjective terms, such as *minty, fishy, pungent*, and so on. Moreover, the vast majority of natural odors are complex mixtures of odorants in which subtle variations in the relative ratios may, for instance, distinguish the bouquet of one wine from another.

As observed by A.R. Hirsch (Smell and Taste Treatment and Research Foundation, Chicago, Illinois), "Ability to smell and taste varies widely among individuals and even the same individual varies in sensitivity under different conditions... In general, women have greater olfactory sensitivity than men, and their sensitivity is keenest at ovulation... Young people are more sensitive than old; nonsmokers are more sensitive than smokers; hungry people are more sensitive than those who are satiated; heavy eaters especially are much more sensitive to smell and taste impressions when they are hungry."

A phenomenon sometimes referred to as "smell blindness" occurs in nearly half of the human population, particularly in the lack of sensitivity (*anosmia*) to the odor of androsterone, a volatile steroid found in human perspiration. Nearly every individual has a few deficiencies in sensing the full spectrum of odors. Aside from the odor of combustible materials and certain animal odors, such as skunks, few odors announce imminent dangers, and hence they are not regarded as serious handicaps.

Researchers at the Monell Chemical Senses Center (Philadelphia, Pennsylvania) have discovered that anosmias can be reversed. This is accomplished by continued repetitious exposure to certain "blind spots." Investigators suspect that repeated exposure to odorous molecules, such as androsterone, stimulates certain olfactory receptors to multiply. Neurons of the olfactory system have two distinguishing factors—an ability to detect specific molecules and the ability to reproduce themselves. Thus, the correction of an anosmia may be the result of a multiplication of the neurons, or more molecular receptors may be increased on each neuron.

It is important to note that these are the only neurons in the body known to regenerate and repopulate themselves. One researcher has observed, "If we could learn what causes the turnover in olfactory neurons, we might learn how to stimulate other nerves to regenerate. Some researchers suggest a possible connection with the histocomplex, which is important to immune response. Thus, olfactory-directed research may provide further understanding of the human immune system.

Newman suggests further that three properties define olfactory response: (1) *threshold value* (minimum amount detected), (2) *intensity* (response), and (3) *type of odor* (physical and/or chemical properties). See reference listed.

In 1986, the staff of *National Geographic* magazine conducted a survey of the "Human Sense of Smell" by sending out thousands of forms containing six sealed patches representing different aromas and including a questionnaire for respondents to complete. The study was designed by the Monell Chemical Senses Center. The survey questionnaire stated, "Of all our senses, smell is the least understood. You have probably seen all the colors you will ever see and heard all the tones you will ever hear. But smell (and flavor, which is mostly smell) seems to have no such limits, and how do you measure it?"

As pointed out by the Editor of *National Geographic*, "Smell and emotion are so entwined with experience that each of us may perceive the same odor with far different feelings. Depending on one's early exposure to horses, the aroma of a stable might delight one person, frighten another, and sadden a third."

The results of the survey were published in the October 1987 issue of the magazine. The highlights included the findings pertaining to six fundamental odors:

- Androsterone (*sweat*)—difficult to identify by many people; women are more sensitive to this odor than men are.
- Isoamyl acetate (*banana*)—readily detectable, but difficult to identify; discerned equally by men and women.
- Galaxolide (*musk*)—difficult to detect and identify; recognized by women more easily than by men.
- Eugenol (*cloves*)—easily detected and identified by both sexes.
- Mercaptans (*gas*)—easy to detect, but less obnoxious to older than younger people.
- Rose—easy to detect and identify by women and only slightly less so by men.

G. Lynch and R. Granger (University of California, Irvine) have combined their respective skills in neurobiology and computer science in constructing a computer model based on the olfactory receptors of rats. Basically, individual odors are simulated on bar codes. This has led the researchers to note that the neuronal cells closest to the entrance of the nose make a rough "sort" of incoming molecules, and other cells located further within the nose sort out the more subtle differences.

P. Bartlett (University of Warwick, U.K.) demonstrated an "electronic nose" at the Pittsburgh Conference (analytical chemistry) in 1991. The device is comprised of an array of twelve stannic oxide (SnO_2) sensors for discriminating the aromas of foods, beverages, and perfumes. Although having possible use in product manufacture, the device is strictly an electronic analogue and does not duplicate mammalian processes.

Olfactory Systems in Lower Animals. Sense organs of many lower animals that live in water are more logically interpreted as organs of taste or as more primitive chemoreceptors, since materials must reach them in solution. Some aquatic insects and fishes, however, have olfactory organs enclosed in cavities which open to the exterior. Whether reached by water or not, they are classed as olfactory organs from their resemblance to such organs in related terrestrial forms.

The olfactory organs of insects are most abundant on the antennae. They consist of blunt processes or flat plates, associated with sensory nerve endings. In some cases they are grouped in the lining of depressions. As many as 39,000 have been reported on a single antenna and five or six thousand are frequently present.

In the vertebrates the olfactory cells lie in the epithelium lining a pair of olfactory pits which form in the embryo as depressions at the anterior end of the head. These cells are connected with the brain by the fibers of the olfactory nerves, the most anterior pair of cranial nerves of known function. In the air-breathing vertebrates the olfactory epithelium becomes part of the lining of the nasal passages.

Additional Reading

Amato, I.: "Evolving an Electronic Schnozz," *Science,* 1431 (March 22, 1991).

Billmeyer, B.A. and G. Wyman: "Computerized Sensory Evaluation System," *Food Technology,* 100 (July 1991).

Brennan, P., Kaba, H., and E.B. Keverne: "Olfactory Recognition: A Simple Memory System," *Science,* 1223 (November 30, 1990).

Gardner, J., et al.: "NATO ASI Series," Springer-Verlag, Berlin, 1990.

Gibbons, R.: "The Intimate Sense of Smell," *National Geographic,* 324 (September 1986).

Gilbert, A.N. and C.J. Wysocki: "The Smell Survey Results," *National Geographic,* 514 (October 1987).

Hirsch, A.R.: "Smell and Taste (Foods)," *Food Technology,* 96 (September 1990).

Margolis, F.L. and T.V. Getchell: "Molecular Neurobiology of the Olfactory System: Molecular, Membranous and Cytological Studies," Kluwer Academic Publishers, Norwell, MA, 1988.

Marshall, E.: "Don't Underestimate the Nose," *Science,* 1021 (March 1, 1991).

Newman, A.R.: "Electronic Noses," *Analytical Chemistry,* 585A (May 15, 1991).

Ross, P.E.: "Smelling Better: Smell-Blindness 'Cure' May Point to Olfactory Mechanism," *Sci. Amer.,* 32 (March 1990).

Shulman, S.: "Banana Neurons," *Technology Review (MIT),* 18 (August/September 1988).

Stone, H., McDermott, B.J., and J.L. Sidel: "The Importance of Sensory Analysis for the Evaluation of Quality," *Food Technology,* 88 (June 1991).

Swanson, L.W.W., Hokfelt, T., and A. Bjorklund: "Integrated Systems of the CNS: Cerebellum, Basal Ganglia, Olfactory System," Vol. 12, Elsevier Science, New York, NY, 1996.

OLIGOCENE. A geologic period of the Tertiary, of the Lower Cenozoic era of the geologic time-scale. The term was proposed by Beyrich in 1854. Type locality near Paris, France. Maximum thickness of strata in Italy. This period began approximately 36,000,000 years ago and lasted for about 16,000,000 years. In the United States marine sediments overlap the Cretaceous and earlier Tertiary sediments of the Atlantic border of South Carolina and the Gulf of Mexico. Marine sediments also occur on the Pacific Coast. Terrestrial sediments are well developed in the easterly Great Plains and Oregon (John Day Basin). The "Bad Lands" of South Dakota, eastern Wyoming, and North Dakota (Black Hills) are important collecting localities for the fossil mammals of this period. In the Paris Basin occur fresh and brackish water deposits, which contain numerous fossil vertebrates, invertebrates, and plants (see also **Paleobotany**). The Oligocene formations are also well developed in Germany, and in the Alps. The marine invertebrates and fishes of the Oligocene are similar to those in the Eocene. Among the mammals the true carnivores have replaced the creodonts. The principal types of mammals are the Archaetherium (giant pig). Poebrotherium (ancestor of the camels), Mesohippus (early horse), Hyracodon (cursorial rhinoceros), and Hoplophoneus (progenitor of the saber-toothed cats). For mineral resources of this period see also **Tertiary**.

OLIGOCLASE. See **Feldspar**.

OLIGURIA. See **Kidney and Urinary Tract**.

OLINGOS. See **Raccoons**.

OLIVE TREE. Of the family *Oleaceae*, the olive is a small tree (*Olea europaea*) indigenous to the eastern Mediterranean region. It has lanceolate evergreen leaves, small inconspicuous flowers, and a purplish drupe, the flesh of which is very bitter in the natural state. Cultivation of the tree has continued through many centuries, gradually spreading not only to all Mediterranean countries, but abroad to suitable regions both in the Old and the New Worlds. In the United States, olive growing is largely restricted to California. The wood of the tree is used to a limited extent.

The olive fruit contains from 60 to 85% oleic acid, 7 to 14% palmitic acid, and 4 to 12% linoleic acid. It also contains stearic and myristic acids. Its iodine value is 85%.

The principal product is the oil, which is expressed from the flesh of the fruit. To obtain this oil the fruit is picked when ripe and usually allowed to dry a bit to remove some of the water contained in the flesh. Pressing freshly picked fruit yields a much higher grade of oil. Pressing is often done in a rather primitive machine. The fruit is first crushed and then firmly pressed. The best grade of oil is known as virgin oil.

Olive oil is widely used as a food or in the preparation of food, owing to its characteristic color, odor, and flavor. Cheaper grades are used in soap-making. Sardine packers require large quantities for packing their product.

OLIVINE. The mineral olivine is a silicate of magnesium and ferrous iron corresponding to the formula $(Mg,Fe)_2(SiO_4)$. Olivine is the group name for the isomorphous series forsterite, Mg_2SiO_4, and fayalite, Fe_2SiO_4. The ratio of magnesium to iron varies considerably but the more common olivines are richer in Mg than in Fe. Olivine crystallizes in the orthorhombic system, usually in flattened prismatic forms, also granular and massive. It has a conchoidal fracture and is rather brittle; hardness, 6.5–7; specific gravity, 3.22–4.39; luster, vitreous; color, olive to gray-green; may be yellowish-brown from the oxidation of the iron. It is transparent to translucent. Olivine occurs both in igneous rocks as a primary mineral as well as in certain rocks of metamorphic origin. It has also been discovered in meteorites.

Olivine crystallizes from magmas that are rich in magnesia and low in silica and which form such rocks as gabbros, norites, peridotites and basalts. The metamorphism of impure dolomites or other sediments in which the magnesia content is high and silica low seems to produce olivine.

Transparent olivines of good color are sometimes used as a gem, often called *peridot*, the French word for olivine; it is also called chrysolite from the Greek meaning gold, and stone. Olivine occurs in the lavas of Vesuvius and Monte Somma and in the Eifel district of Germany. Gem material comes from St. John's Island in the Red Sea, Upper Burma, and from Minas Geraes, Brazil. In the United States olivine localities are Orange County, Vermont; Webster and Jackson counties, North Carolina. Arizona and New Mexico have also furnished some gem material.

See also **Peridotite**.

ONAGER. See **Horses, Asses, and Zebras**.

ONCOLOGY. See **Cancer and Oncology**.

ONION. Of the family *Amaryllidaceae* (amaryllis family), the common dry onion of commerce (*Allium cepa*) is related to a great number of other species of the genus and of similar odor and taste. Closely related species are chive, garlic, leek, shallot, and Welsh onion. The great majority of the over 500 species of *Allium* are wild plants of no commercial significance, but that possess the characteristic odor and taste. The onion is consumed raw and is cooked by almost all means known. The onion is in countless thousands of cooking recipes, used by peoples of various cultures throughout the world.

Some authorities regard the general region of northwestern India, Afghanistan, western Tien Shan, and the former Soviet bloc republics of Tajik and Uzbek, as the most likely area of origin of *Allium cepa*. But the vegetable also may be native to Turkey, the Near East, and the Mediterranean region. The Welsh or Japanese onion (*Allium fistulosum*), widely grown in the Orient, is regarded as being native to central and western China. The onion is mentioned in the Old Testament, Numbers 11:5, "We remember the fish, which we did eat in Egypt freely; the cucumbers, and the melons, and the leeks, and the onions, and the garlic." A number of species of onion were described by Theophrastus as early as 322 B.C. Pliny offered directions for cultivation of the onion in 210 A.D. Chaucer mentioned the onion in 1340 and European botanists in the 1500s regarded the onion as one of "the commonest vegetables."

In the United States, there are two crops of onions per year. Spring crops in Arizona, California, and Texas account for about 21% of total annual production. The summer crop, accounting for the bulk of onion production, is spread over numerous states. The total dry onion production in the United States approaches 2 million tons. There are numerous varieties of commercially grown onions. See Fig. 1.

The two major insect pests of onion are the onion maggot and the onion thrip. Onions also are attacked by the bulb and stem nematode, causing a condition known as onion bloat. Soft rot is the most prevalent causes of onion loss in storage. It is a bacterial infection. Black mold may be destructive during storage and transit of onions grown in Texas, California, Arizona, and New Mexico. To control this condition, bulbs should be protected from moisture in the field during and after pulling and during transit. During storage, onions require adequate vertical air channels between bags to permit good ventilation and refrigeration.

ONLINE. In terms of instruments and computers, online describes situations where such devices participate directly in the dynamic control or data system loop. Information collected by online equipment is immediately processed and fed into a closed-loop system with no delay or human intervention.

Fig. 1.   Various shapes of onions: (1) flattened globe; (2) globe; (3) high globe; (4) spindle; (5) Spanish; (6) flat; (7) thick-flat; (8) Granex; (9) top. (*By permission from "Onions" by Dr. Henry A. Jones, Desert Seed Company, El Centro, California*).

ONYCHOPHORA.   Small, soft-bodied, creeping animals, slightly like caterpillars in appearance. They are of limited distribution in warm countries and have no common name. The name of one genus, *Peripatus*, is sometimes applied indiscriminately to all members of the group.

Onychophora are regarded as the most primitive of the terrestrial arthropods. Their structure suggests the ancestral form of the insects.

ONYX.   See **Agate**.

ONYX MARBLE.   See **Travertine**.

OÖLITE.   The term is from the Greek meaning egg and stone. Oölites are well-rounded sand-like particles, originally formed of calcite but sometimes subsequently altered to either dolomite, or entirely silicified. The structure is typically concentric about a nucleus, and often with radial lines. Oölites are relatively common constituents of limestones, often forming distinct beds. Oölites are now forming on the shores of Great Salt Lake, but no authentic cause is known of marine oölites being formed at the present time. Coarse-grained oölites, in which the particles are about the size of peas, are called pisolites, from the Greek, meaning pea and stone.

OOLOGY.   The study of eggs and egg-producing processes in various egg-bearing animals. A good reference on this topic is "Egg Incubation in Birds and Reptiles," by D. Charles Deeming and Mark W.J.K. Ferguson (Editors), Cambridge University Press, New York (1992).

OOZE.   Ooze is a general term used to designate the mud found on the ocean bottoms at abyssal depths and composed largely of the calcareous and silicified shells of minute surface living marine organisms, called plankton.

OPACITY.   Imperviousness to radiation, especially to light; the property of stopping the passage of light rays numerically expressed as the reciprocal of the transmittance. Density (photographic) or optical density is given by

$$d = \log O = \log 1/T$$

where d is density, O is opacity, and T is transmittance. This usage should now be discarded in favor of the term absorbance.

OPAH (*Osteichthyes*).   Of the order *Allotriognathi*, family *Lampridae*, the opah (*Lampris regius*) is the only member of this family. See Fig. 1. It may be described as a laterally compressed fish with an oval shape. The fish is noted for its spectacular coloration and patterning with a blue-to-gray upper surface, rose red undersurface, body covered with white spots, jaws, and fins vermilion, and the eyes set within a gold-colored area. The opah is a large fish, measuring up to 6 feet (1.8 meters) in length and weighing up to 600 pounds (272 kilograms). Because the flavor is excellent, the opah would make an excellent commercial item were it abundant. It is well distributed throughout the seas. The depth at which the opah may be most abundant has yet to be determined. During a number of years, fewer than 50 specimens have been taken from the waters of southern California and the Pacific northwest to Alaska, where it is known to range.

Fig. 1.   Opah (*Lampris regius.*)

OPAL.   The mineral opal, long classified as an amorphous mineral gel, has been found by X-ray analysis to consist of a microcrystalline aggregate of crystallites of cristobalite. On this basis, opal may be considered as a variety of cristobalite bearing the same relationship to that mineral as chalcedony does to quartz. Opal is hydrous silica, $SiO_2 \cdot nH_2O$, with variable water content. It never occurs in crystal form; usually as irregular veins or masses, or as pseudomorphous replacements after wood or fossilized material such as bones and shells. Opaline silica occurs in many forms: geyserite from geyser deposits, siliceous sinter (fiorite) form siliceous waters of hot springs, and diatomite (diatomaceous earth) from siliceous shells of diatoms and comparable microscopic species. It has a conchoidal fracture; hardness 5.5-6.5; specific gravity 2.1-2.3; luster, vitreous or greasy to dull; color very variable, colorless, white, milky-blue, gray, red, yellow, green, brown, and black. Often a beautiful play of colors may be observed in the gem varieties. The color play in opals is attributed to three different mechanisms: finely divided pigmentation of foreign material; light interference by open-spaced grid of cristobalite crystallization; and reflected light. It may well be that two or all three causes may contribute to the color effect in any given opal specimen. Before a more complete understanding of opal color is established these phases seem to be of prime significance.

Besides the gem varieties, which show the delicate play of colors, there are other kinds of common opal, such as: the milk opal, a milky bluish to greenish kind; resin opal, which is honey-yellow with a resinous luster; wood opal, resulting from the replacement of the organic matter of wood by opal, and hyalite, a colorless glass-clear opal sometimes called Muller's Glass. Opal is deposited at relatively low temperatures and may occur in the fissures of almost any type of rock. Hungary, Australia, Honduras, Mexico and in the United States Nevada and Idaho, have been sources of gem opals. Hyalite comes from Czechoslovakia, Mexico, Japan and British Columbia. Other common varieties of opal are widespread in their occurrence. The word opal is derived from the Latin *opallus*.

OPAQUE PLASMA.   A plasma through which an electromagnetic wave cannot propagate and is either absorbed or reflected. In general, a plasma is opaque for frequencies below the plasma frequency. The fact that a plasma is opaque over a certain frequency range will change the radiation properties within that frequency range. Any radiation emitted within the volume of the plasma is quickly absorbed. In this opaque region, therefore, the plasma can only radiate from its surface.

OPEN-CHANNEL FLOWMETERS.   See **Flow Measurement (Liquids and Gases)**.

OPEN DELTA.   This is a three-phase transformer connection using two single-phase transformers connected to form a V or open delta across the three lines. Such a connection has about 58% of the capacity of full delta using transformers of the same rating. It is often used for temporary work anticipating a later completion of the delta or for emergency service when one transformer of the complete delta requires servicing.

OPEN-LOOP CONTROL.   See **Feedback Control**.

OPEN UNIVERSE.   See **Cosmology**.

OPERAND.   An entity to which an operation is applied. The operand, for example, may be a portion of a computer instruction, or it may be identified by the address part of the instruction.

OPERATING CHARACTERISTIC.   In quality control and decision theory generally, a measure of the probabilities of accepting a false hypothesis for varying values of the parameter specified by that hypothesis. For example, if a batch of items is to be accepted or rejected on the basis of the proportion of defective items in a sample from the batch, the OC curve would graph the proportion of defectives as abscissa against the probability of acceptance as ordinate. A good acceptance rule would then have a curve falling rapidly to zero as the proportion of defectives increased. Considered upside down (i.e., graphing the probability of *rejection* as ordinate) the OC curve is equivalent to the graph of the Power Function. See also **Neyman-Pearson Theory**.

OPERATING SYSTEM (Computer).   An integrated collection of service routines for supervising the sequencing of programs by a computer and may provide debugging, input/output control, accounting, compilation, storage assignment, data management, and related services. Essentially synonymous with monitor system and executive system. See also **Program (Computer)**.

OPERATIONAL AMPLIFIER.   See **Amplifier**.

OPERATOR (Mathematics).   The symbolic direction to perform an operation such as addition, multiplication, differentiation, extraction of roots, etc., or some combination of these operations. A linear operator is the most common case. It obeys the distributive law, $\mathbf{A}[f(x) + g(x)] = \mathbf{A}f(x) + \mathbf{A}g(x)$ and $\mathbf{A} \cdot cf(x) = c\mathbf{A}f(x)$, where c is any constant. If the order of applying operators to a function is immaterial, the operators are commutative. Suppose $\mathbf{A} = a +$ and $\mathbf{B} = b+$, with a, b constant, are applied to a function of x, then $\mathbf{AB}f(x) = a + b + f(x) = \mathbf{BA}f(x)$ and $\mathbf{A}$, $\mathbf{B}$ commute. However, if $\mathbf{P} = \partial/\partial x$ and $\mathbf{Q} = x$, then $\mathbf{P}$ and $\mathbf{Q}$ are not commutative for $\mathbf{PQ}f(x) = f(x) + \mathbf{QP}f(x)$.

If $\mathbf{A}$ and $\mathbf{B}$ are noncommutative, their commutator is $(\mathbf{A}, \mathbf{B}) = \mathbf{AB} - \mathbf{BA}$. According to quantum theory, if the commutator vanishes for two operators that represent dynamical variables, then the measurement of one of these variables does not interfere with that of the other.

A differential operator involves one or more differentiations. Examples are $D = d/dx$, $D^2 = d^2/dx^2$, $D^{(n)} = d^n/dx^n$; $F(u) = f(x)u'' + g(x)u'' + (x)u$. In the more general case of an nth-order operator

$$L(u) = \sum_{i=0}^{n} [f_i(x)u^{(r_i)}]^{(S_i)}$$

the adjoint operator is

$$L(u) = \sum_{i=0}^{n} (-1)^{r_i + S_i} [f_i(x)u^{(S_i)}]^{(r_i)}$$

If $L(u) = \bar{L}(u)$, the operators are self-adjoint. Any second-order differential operator can be made self-adjoint with an integrating factor $\exp \int ((g - f')/f)dx$.

See also **Del**; and **Laplacian**.

OPERCULUM.   See **Fishes**.

OPHITIC TEXTURE.   A term proposed by Michel-Levy, in 1877, for the characteristic texture of dolerites, in which the pyroxene crystals are penetrated by laths of plagioclase feldspar. This type of texture differs from poikilitic in that in the latter type of texture the pyroxene crystals entirely enclose a number of laths of plagioclase.

OPHIUROIDEA.   The brittle stars, a class of echinoderms resembling starfishes with a well-marked disk and slender arms. The class is distinguished chiefly by this sharp demarcation of disk and arms, which accompanies the restriction of visceral organs to the disk. In addition the tube feet are without suckers and the madreporite lies on the oral surface.

OPHTHALMOLOGY.   That branch of medical science that deals with the structure, functions, and diseases of the eye and of the visual system.

OPOSSUM.   See **Marsupialia**.

OPPENHEIMER, J. ROBERT (1904–1967).   Oppenheimer was an American scientist whose areas of achievement include, invention, physics, and technology. His interest in science, is believed to have been sparked by a German grandfather who gave him a mineral collection. He wrote letters to famous geologists and was invited at age twelve to give a lecture at the New York Mineralogical Society. Oppenheimer went to Harvard University and excelled at theoretical physics. After graduation he studied at the famous Cavendish Laboratory at Cambridge, England under Ernest Rutherford. In 1929, he took positions at Berkeley and Cal Tech in the United States. He was always regarded as an exceptional teacher and excellent theoretician. He devoted early research to the study of subatomic particles such as electrons and positrons and published 16 papers on quantum physics. His work lead to many later finds including neutron stars.

Oppenheimer's name is almost synonymous with the atomic bomb. In 1941, after learning the Germans had split the atom, President Roosevelt established the Manhattan Project. In June 1942, Oppenheimer was appointed its scientific director. Oppenheimer recruited and coordinated the effort of hundreds of scientists at a research station at Los Alamos, New Mexico for the Manhattan Project. The scientists were working to produce the first atomic bomb. At 5:30 A.M., Monday, July 16, 1945, Oppenheimer witnessed the first explosion of an atomic bomb in the New Mexico dessert.

After the bombing of Hiroshima and Nagasaki, Oppenheimer received much publicity including Time magazine referring to him as "The Father of the Atomic Bomb". On January 12, 1946, he was given the Presidential Medal of Merit. Americans were full of gratitude for his work.

After the war, Oppenheimer was always concerned about atomic weapon usage. In 1947, he chaired the U.S. Atomic Energy Commission. He wrote government reports about atomic energy and gave more than 200 speeches to the public. In 1949, Oppenheimer became the director of the Institute for Advanced Study in Princeton, New Jersey. In 1953, however, at the height of U.S. anti-communist feelings, Oppenheimer lost his security clearance because of his associations with friends who were sympathetic to communism. With the loss of his security clearance, Oppenheimer lost his influence on America's science policy. However, in 1963 Oppenheimer was presented with the Enrico Fermi award (the highest award a physicist can receive) by President Lyndon B. Johnson.

Oppenheimer is considered one of the greatest scientists of the twentieth century.

See also **Manhattan Project (The)**; and **Neutron Stars**.

J. M. I.

OPTICAL AIR MASS (symbol m).   A measure of the length of the path through the atmosphere to seal level traversed by light rays from a celestial body, expressed as a multiple of the path length for a light source at the zenith. Originally called, simply, *air mass*. Also called *airpath*.

OPTICAL ANOMALY.   The behavior of certain organic compounds, such as those whose molecules contain conjugated double bonds, in which the observed values of the molar refraction are not in accord with the values calculated from the known equivalents.

OPTICAL ANTIPODES.   Two compounds composed of the same atoms and atomic linkages, which differ in their structural formulas only in that one is the mirror image of the other. The term is commonly applied to substances containing an asymmetric atom, or bond, in which the plane of polarized light is rotated to the right by one of the optical antipodes, and to the left by the other. See also **Amino Acids**.

OPTICAL CENTER (of a Lens).   A point so located on the axis of a lens that any ray, which in its passage through the lens passes through this point, has its incident and emergent parts parallel. See also **Mirrors and Lenses**.

OPTICAL CHARACTER RECOGNITION (OCR).   A system for automatically identifying handwritten or printed characters by one of several types of photoelectric devices for the purpose of providing electronic identification input to data processing systems. OCR systems are applicable where there are voluminous amounts of printed input data, as encountered, for example, by banks, insurance companies, retail credit firms, brokerage houses, warehouse accounting, mail and postal systems, etc. In addition to the several types of OCR systems, there are other character identification approaches, notably magnetic ink character recognition systems. See also **Magnetic Ink Character Recognition (MICR)**. Each approach has its advantages and disadvantages and sometimes selection of the most effective approach is quite difficult.

As with MICR, to take advantage of full electronic differentiation in character reading, some modification of the general appearance and shape of letters and numbers is required. The numerals as modified for MICR (and illustrated in that alphabetical entry in this volume) also are useful in OCR systems with the exception, of course, that magnetic ink is not required. In this approach (sometimes termed the one-dimensional approach), the signal obtained is the amount of material that is sensed through a slit—a single function of time corresponding in duration to the time required for the character to pass by the slit. One-dimensional optical approaches are limited, however, to small numbers of accurately printed characters. Two-dimensional systems, although considerably more complex, make much more sensing data available.

A number of sensing approaches have been conceived, and some have been quite successful. In the optical masking approach, an image of the character is projected on a set of masks. The total amount of light passing through the masks is collected by a photodetector. It is necessary to relate the mask designs to the expected character shapes in such a way that the mask that permits the largest signal will be the mask that identifies with a given character. Obviously, in such a system the characters being measured must be reasonably uniform both in terms of size and optical characteristics.

In spot scanning, one small character segment is covered at a time. In one spot scanner, there is a rotating disk between the light source and the character, with a pattern of slots that breaks the character into distinguishable elements. The device may be limited to from 300 to 500 characters per second and is designed to one highly stylized type font. In one type of electronically generated spot scanner, a vidicon tube similar to that used in telecasting picks up the character image. The surface of the tube is scanned by an electron beam that breaks the character down into digital components. In another method (flying-spot scanner), a cathode-ray tube generates a beam of light that moves across the character in a scan pattern. A lens system projects the reflected light to photomultipliers that translate black-and-white areas into electrical signals. Speeds up to 2,000 or more characters per second are obtainable. The systems are costly and require high printing quality of specific types of fonts.

In another system (retinal sensing), a two-dimensional matrix of photosensors is used. These sense an entire moving character rather than a segment of it. Early configurations of this system had a character resolution approaching that of the human eye and could read up to 2,400 characters per second. Essentially, the device is a mosaic-image sensor, or a mosaic of photocells onto which each character to be read by the system is focused. The photocells are physically constrained and thus their dimensions with respect to each other can be held constant to avoid character distortion. The mosaic, like the scanner, is one character width wide and three character heights high. Behind each photocell in the retina, there is a silicon chip that is sensitive to black, white, and shades of gray between. The photocells are interconnected so that each single cell "sees" not only its own portion of the character being read, but also the portions covered by other cells around it. This arrangement enables the device to judge relationships, dismiss smudges as not being part of the characters, and accept even the light portion of the character because the area next to the character is even lighter. The recognition logic establishes relative values between each cell and those surrounding it.

In recent years, particularly because of the urgent needs for automation in postal and package handling and manufacturing situations, OCR systems have been undergoing constant change and improvement. See also **Pattern Recognition**.

OPTICAL EMISSION SPECTROCHEMICAL ANALYSIS.   In this analytical technique, an optical device is used to analyze radiation from electrically excited sample atoms. The analyzing device provides monochromatic images whose intensities are measured and related to the concentration of the elements within the sample that produces the specific radiation measured. The technique is precise and rapid, and adaptable to solid, powder, or liquid samples.

More than seventy elements may be detected by standard procedures. Atomic gases, such as O, N, H, He, Ar, Ne, Kr, Xe, and Rn and the halogens are excluded. Nonmetallic substances, such as C, S, and Se, require vacuum path spectrometers for optimum detection and measurement. Analytical ranges may extend from fractional parts per million to about 40% concentration. Computer-controlled photoelectric optical emission spectrometers will output printed percent concentrations for 30 to 50 elements per sample in just a few minutes. This form of analytical instrumentation is used widely in production and quality control, as well as for research studies.

A schematic diagram of an optical emission spectrometer is given in Fig. 1. Various means are used to introduce the sample, whether solid or liquid, into an excitation stand where energy is imparted to it by some form of excitation source. The atoms composing the sample are excited and therefore emit their characteristic radiations, which are then separated by a grating in the spectrometer into line spectra. The light of selected element lines is isolated by slits and focused on phototubes. The sensitivity is adjusted by attenuating the high voltage from the high voltage supply. The intensity of a spectrum line can be correlated with the concentration of the element producing it. It is therefore necessary to measure intensities with very high precision.

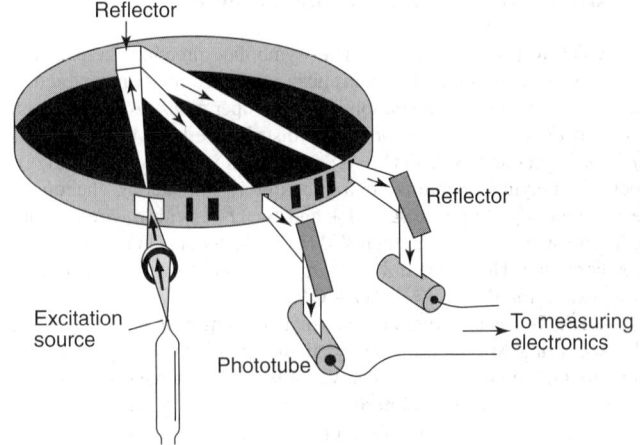

Fig. 1.   Operating principle of optical emission spectrometer.

Sample atoms may be excited by absorbing specific energies from an electric discharge. These atoms, raised to higher-than-usual energy levels, are unstable, and revert to their stable states by emitting the absorbed radiation according to the relation:

$$E_2 - E_1 = hv = hc/l \tag{1}$$

where E = energy, eV

h = Planck's constant (6.624×10^{-27} erg-second)

v = frequency, Hz

$\lambda = c/v$ = wavelength (in Å = 10^{-8} centimeter)

c = velocity of light (3×10^{10} centimeters/second).

Atomic transitions may be expressed in terms of wavelength, and qualitative analysis may be performed by wavelength determination and identification.

The commonly used dispersive device is the diffraction grating, which produces spectra by light interference according to the relation:

$$N\lambda = d(\sin \alpha \pm \sin B) \tag{2}$$

where N = an integer

λ = wavelength

d = grating constant (width of single groove)

α = angle of incident light

B = angle of diffracted light.

For constant a and the same sin B, integer values of N produce spectra of $\frac{1}{2}\lambda$, $\frac{1}{3}\lambda$, etc., called *spectral orders*.

A grating ruled on a spherical surface combines the properties of the diffraction grating with the focusing ability of the optical surface. Such a device, with radius of curvature R, focuses spectra as images of the entrance (primary) slit on the circumference of a circle of diameter R, when the entrance slit is also located on the circumference of the circle.

The usual measure of how well a grating separates individual wavelengths is given by the reciprocal linear dispersion, in angstroms per millimeter, as follows:

$$\text{Å/mm} = \frac{d \cos B}{N f} \tag{3}$$

Thus, dispersion is governed by the fineness of the grating ruling d and the focal length f of the focusing element.

The concentration C of an irradiating element is related to the intensity I of the emitted spectral line, according to the relationship:

$$I = kC^n \tag{4}$$

where k and n depend on the excitation conditions employed. Accuracy and precision are improved by use of an internal standard reference line of another element of constant concentration. The relationship becomes:

$$\frac{I_x}{I_r} = k_1 C_x^{n_1} \tag{5}$$

where I_x and I_r are the intensities of spectral lines emitted by elements x and r; C_x is the concentration of element x; and k_1 and n_1 are constants depending on the line pair and on the excitation conditions. The relative intensities of lines having different excitation energies depend on the temperature of the spark discharge column.

The source unit must vaporize and excite a portion of the sample, which is generally used as one of the electrodes between which the electric discharge takes place. No single excitation source is ideally suited for all applications of emission spectrochemistry. Trace impurities in metals, alloying constituents in high concentrations, biological substances, ceramics, slags, oils, nonconductors, refractories — all may require different excitation techniques and sample preparation procedures. Table 1 summarizes the important characteristics of the commonly used spectrochemical source units.

Photographic radiation detection may be used, but film emulsion response is not linear. Film calibrations are required to relate measured densities with the intensities producing these densities before *intensity* versus *concentration* working curves can be formulated. Although quite general in application at one time, the photographic technique is slower than photoelectric radiation detection wherein each beam whose intensity is to be measured is directed onto a photomultiplier detector through a suitably sized exit (secondary) slit. The output of the detectors is transmitted to the measuring console, where it is translated into the readout format of the system.

Many dramatic changes in the development of readout electronics have occurred over the last 15 to 20 years. Modern systems use integrated circuit and digital computing devices. The engineer is no longer required to design complex circuitry to perform the basic tasks of control. Software now becomes the tool by which timing, sequencing, and logic control are accomplished.

Generally, spectrometer systems fall into two major functional categories — system control and data handling. The digital controller with its controlling and computing capabilities is ideal for handling both tasks with a minimum of effort required by the circuit design engineer.

Particularly during the last decades, several new analytic techniques have been developed that, when appropriate, have a tendency to displace former traditional methodologies.

OPTICAL FIBER SYSTEMS. Optical fibers are hair-thin structures (usually cylindrical in shape) capable of transmitting light signals with extremely low signal loss and at very high digital pulse rates. Fibers are available in a variety of sizes and material compositions and with a wide range of optical performance. Although commercially available fibers are solid structures, they function as "light pipes" that guide rays of light and are therefore sometimes called *lightguides*. When used to connect a light source to a light receiver (photodetector) to form a communication system, the fiber carries *photons* instead of the *electrons* used in traditional metal-conductor communication links. Although a laser light source containing a narrow range of optical wavelengths is preferred for carrying signals the farthest and fastest, light-emitting diodes (LEDs) are also used, especially for short distance communication links within buildings. Thus, the three key elements of a lightwave communication system are the light source, the optical fiber, and the photodetector.

Fiber-Optic Systems in Perspective

As early as 1841, D. Colladon demonstrated light guiding by a jet of water, and in the following year J. Babinet showed the phenomenon in a

TABLE 1. SPECTROCHEMICAL EXCITATION SOURCE UNITS

Type	Voltage	Current, A	Characteristics
Dc arc	220	3–30	Most sensitive, least reproducible, quantitative analysis; trace element quantitative analysis.
Ac arc	2500 5000	5 2–5	Good sensitivity, more reproducible, best use for self-electrode metal analysis.
High-voltage spark interrupted auxiliary gap	15–40,000	3–20 RF	Least sensitive, most precise, ±1% or better. Excites higher energy lines. Parameter selection allows variations between arc-like and spark-like in spectral excitation.
Multisource	1000	Peak discharge currents from 5 to in excess of 500 A at time constants ranging from 8 to less than 1 ms.	Sensitive and precise. Parameter selection allows wide variety of controlled unidirectional and oscillatory charges variable from arc-like to spark-like in spectral excitation.

bent glass rod. However, these experiments did not receive wide publicity until John Tyndall duplicated and popularized the same effect in 1854. In 1880, shortly after the telephone was invented, Alexander Graham Bell proposed telecommunications using lightwaves. See Fig. 1. Patented as the "Photophone," the concept depended on the free propagation of light through the atmosphere. Of course, a century ago Bell had no powerful steady light source such as the laser, and even on very clear days atmospheric disturbances severely limited the practical distance over which undisturbed light could travel.

Fig. 1.   Old woodcuts showing photophone patented by Alexander Graham Bell in 1880, representing the first attempt to utilize light for the transmission of sound. (*Top*) Sunlight was reflected and focused by a lens onto a mechanism that was vibrated by sound waves (speech), thus modulating the intensity of the exiting light beam. (*Bottom*) At the receiving end of the system, variations of intensity of the light changed the resistance of a selenium photocell, thus controlling an electric current input to the receiving telephone.

These experiments were followed by the development of light pipes to illuminate homes (W. Wheeler), bent glass rods to illuminate body cavities for dentistry and surgery (Roth and Reuss), and surgical lamps (D. Smith).

As early as 1910, the possibility of guiding electromagnetic waves by internal reflection within long cylinders of dielectric material was investigated on a theoretical basis (D. Hondros and P. Debye). The first quantitative experimental investigations took place in 1920. Although some interest was shown in conducting light by glass rods, a serious rekindling of interest in lightwave communication had to await the first laboratory demonstration of a laser in 1960 (T.H. Maiman, Hughes Aircraft Company). Earnest efforts to devise shielded waveguide structures were made shortly thereafter. Because the glasses available in the early 1960s possessed prohibitively large absorption and scattering losses, some early experimental lightguides consisted of gas-filled underground conduits (some 20 centimeters (7.87 inches) in diameter) that incorporated lenses at various intervals to refocus the light and change its direction when required.

These systems were unsatisfactory on several counts: bulk, expense, and the extreme sensitivity to temperature and alignment of the components.

In the mid-1960s, Charles K. Kao of STL Laboratories, determined that the fundamental limit on glass transparency was less than 20 decibels (dB) per kilometer (km), which is low enough to make glass fibers practical for communications. (A loss of 20 dB/km means that the optical power after 1 km is 1% of the amount at the beginning.) This spurred investigators to explore methods for making purer glass. In 1970, fibers with attenuation less than 20 dB/km were made and demonstrated in the laboratory (Maurer, Keck and Schultz, Corning), and this was reduced to 4 dB/km by 1972.

Combining these fibers with emerging semiconductor lasers with lifetimes reaching 1000 hours (Bell Labs, 1973) and photodetectors, enabled lightwave communications to become practical in the mid-1970s. The first nonexperimental fiber-optic link became operational in Dorset, England (1975) to service a police station. In early 1976, Bell Labs started tests at 45 million bits per second on multimode fibers installed in Norcross, Georgia. Fibers continued to improve as their attenuation decreased to 0.47 dB/km (M. Horiguchi).

In 1980, video signals were carried by optical fibers $2\frac{1}{2}$ miles (4 kilometers) for the Winter Olympic Games in Lake Placid, New York. The first long-haul intercity installations (AT&T, Washington-New York; New York-Boston) were made in 1983. After that, the capacity of fiber optic transmission systems increased exponentially. Despite this progress, the fundamental limits predicted by the physics of photonics materials, devices, and systems have not yet been approached (Kogelnik). The challenge of future research and development continues to be a fuller exploitation of the ultimate capacity of optical fibers.

Fig. 2 shows a schematic history of the development of optical communication systems.

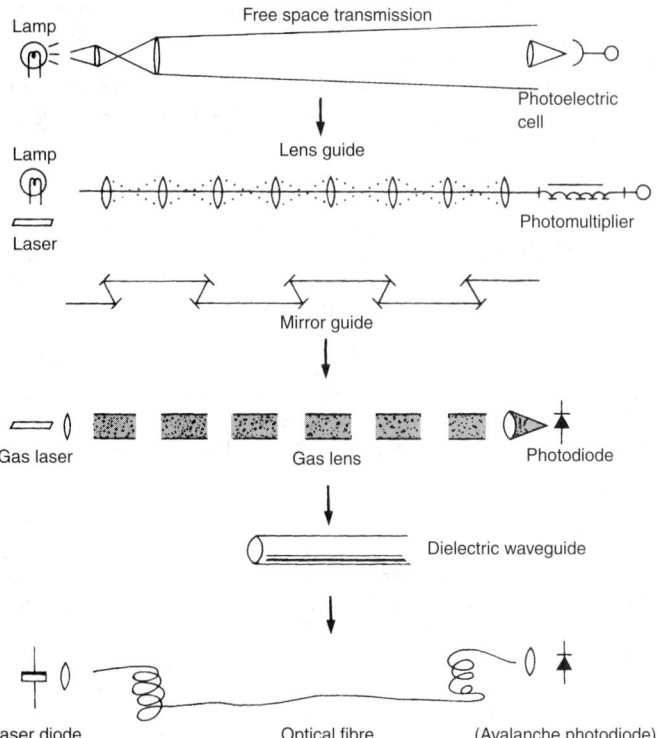

Fig. 2.   Schematic representation of the development of optical communication systems. (*Adapted from Suematsu and Iga.*)

Seldom does a new technology have as many practical advantages as found in optical fiber systems and become available in only a decade of concentrated research and development. As solid state physics and semiconductors created the electronics industry, fiber-optic systems have revolutionized the telecommunications industry.

Fiber optic systems are more economical than their alternatives—copper wire, radio relay, and satellite. The regeneration of signals sent on copper cables is necessitated at several mile intervals, whereas the distance on optical fibers can be over a thousand miles by using optical amplifiers approximately every 50 miles.

The first fiber optic systems used light in the 850 to 875 nanometer (nm) wavelength region (the so-called "first window"), and were followed with systems operating near 1300 nm (the "second window") where fiber loss is smaller, and then near 1550 nm (the "third window") where fiber loss is even smaller. In the year 2000, fibers were carrying digital signals at 10 billion bits per second on one or more (as many as 40) wavelengths in the 1550 nm window.

Because optical fibers are nonconducting, fiber optic systems provide excellent electrical isolation and immunity from electrical interference. Signal losses are much lower in fibers (as low as 0.20 dB/km) compared to other guided transmission media, such as twisted copper pairs, coaxial cable, and metallic waveguides. In addition, the bandwidth or information carrying capacity of fibers is far greater. When one or more optical fibers are packaged into cables, the cables are smaller and more flexible than their metallic counterparts.

In 1984, British Telecom laid the first submarine fiber optic cable to carry regular traffic, to the Isle of Wight. In 1988, service began on the first transatlantic fiber-optic cable (TAT-8) from Tuckerton, New Jersey to France and England. Similar undersea communication links between Japan and Guam and other Pacific locations began in 1989. With the development of optical fiber technology proceeding at such a rapid rate, decisions as to whether or not to delay undersea cables for further advancements sometimes are difficult to make.

Earlier "fantastic" claims that 24,000 simultaneous telephone calls on a single pair of fibers at rates of up to 1.7 billion bits per second could be made, that full-length motion pictures such as *Gone with the Wind* could be fed to a home memory unit in one second, or that major symphonies, such as Beethoven's Fifth, could be transmitted in less than 1/50 of a second have already been surpassed. For example, here is listing of some "hero" experiments illustrating an optical fiber's capability as of the year 2000.

- Using 160 wavelengths of light with each wavelength carrying 40 billion bits per second of information, NEC researchers transmitted 6.4 trillion bits of information per second over a special fiber 186 kilometers long.
- Using a single wavelength of light on an experimental TrueWave® fiber, Bell Labs scientists transmitted 320 billion bits of information per second over a distance of 200 kilometers.
- Bell Labs demonstrated 3.28 trillion bits per second over a 300 km of experimental TrueWave fiber. This experiment used 82 wavelengths of light with each operating at 40 billion bits per second.

Types of Optical Fibers

Optical telecommunications fibers fall into two main categories: *multimode fiber*, and *single-mode* (also called *monomode*) *fiber*. Multimode fibers receive their name because they can propagate many (hundreds) of light modes, which can be thought of as paths taken by the light as it travels in the fiber. Single-mode fibers, on the other hand, propagate only one mode. Multimode and single-mode fibers can be further divided into subset categories. For example, multimode fibers can be either *step-index* or *graded-index*, and single-mode fibers can be *dispersion-unshifted*, *dispersion-shifted*, or *nonzero-dispersion shifted*. Each of these, in turn, can be further differentiated. For example, two types of graded-index multimode fibers have either 50 micron or 62.5 micron core diameters. Two types of dispersion-unshifted fiber have either depressed clad or matched clad refractive index profiles. Or they may have a typical high loss at the "water peak" wavelength near 1385 nm or low loss at this wavelength, such as with Lucent's AllWave™ fiber.

Single-mode fibers have many advantages over multimode fibers. See Fig. 3. Compared to the 50 or 62.5 micron core diameter of typical graded-index multimode fibers, some single-mode fibers have a core diameter near 8 micrometers. Single-mode fibers also have a core-cladding refractive index difference of a few tenths of a percent compared to 1 or 2% for multimode fibers. Their smaller core diameter and refractive index difference allow single-mode fibers to propagate light in only one clearly defined path (or mode). Because this reduces multipath effects that spread the arrival times of a short input pulse, single-mode fibers have high bandwidth, meaning that they can carry signals at higher bit rates than multimode fibers.

Various types of optical fibers are used for specific applications. For example, multimode fibers are used primarily in enterprise systems: buildings, offices, campuses. Special single-mode transmission fibers exist for submarine applications, and for metropolitan and long-haul terrestrial applications. And in addition to these transmission fibers, there are various

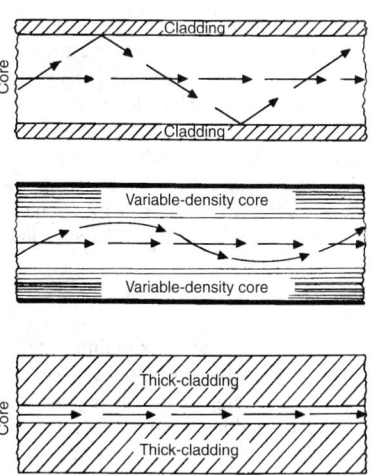

Fig. 3. Fundamental classes of optical fibers: (**a**) Step-index fiber made up of glasses of two different densities, the core and the cladding. Lightwaves travel in a zig-zag fashion down the core, bouncing off one side and then the other side of the core-cladding interface; (**b**) Graded-index fiber where the glass in the core varies in density—hence the light travels in a smooth, curving path, causing less distortion of transmitted information; and (**c**) single- or monomode fiber where the core is very small relative to the wavelength of the transmitted light causing the light to move down the fiber in a straight line—hence resulting in very low distortion. In modern designs, the distortion has been reduced to almost the theoretical low limit.

"specialty" fibers for performing dispersion compensation (dispersion compensating fiber), optical amplification (erbium-doped fiber), and other special functions.

Characteristics Affecting Optical Fiber Performance

Phase Index of Refraction. Phase index of refraction is the ratio of the phase velocity of light in a vacuum to its velocity in another medium, such as glass. The value of this parameter depends on wavelength, and the composition, temperature and pressure of the medium. The higher the refractive index of a material, the lower the phase velocity of light in the material, and the more a light ray is bent as it enters the material from air.

Numerical Aperture (*NA*). Numerical aperture describes an angle just outside the a fiber's end face that determines the largest angle that a light ray can have to the fiber axis and still be captured and propagate within the fiber. The formula from Snell's law governing the numerical aperture number of a fiber is

$$NA = \sqrt{n_1^2 - n_2^2}$$

where n_1 and n_2 are the phase refractive indices of the fiber's core and cladding, respectively.

Most optical fibers have numerical apertures between 0.15 and 0.4, and these correspond to light acceptance half-angles of about 8 and 23 degrees. Typically, fibers having high *NAs* exhibit greater loss and lower bandwidth.

Light Loss or Attenuation through a Fiber. The amount of light at the output of a fiber is smaller than at the input. This attenuation is expressed in decibels per kilometer (dB/km), which is a relative power unit according to the formula

$$\alpha(dB) = -10 \log \frac{P_o}{P_i}$$

where P_o/P_i is the ratio of optical power at the output to the optical power launched into the fiber at the input. This ratio is smaller than 1, and the logarithm of a number smaller than 1 is negative. Consequently, the minus sign in the equation makes the attenuation value a positive number. A comparison of the ratio of output power to input power in percent to the quantity in dB is as follows:

80% transmission = a loss of ~1 dB

50% transmission = a loss of 3 dB

10% transmission = a loss of ~10 dB

1% transmission = a loss of ~20 dB

Bandwidth. Bandwidth is a measure of information-carrying capacity. An optical fiber's bandwidth can be expressed either in the time domain as pulse dispersion in nanoseconds per kilometer (ns/km), or in the frequency domain as frequency passband in megaHertz-kilometers (MHz-km). Light pulses spread or broaden as they pass through a fiber depending on the material used and its design. A fiber's bandwidth limits the rate at which optical pulses can be transmitted and decoded without error at the terminal end of the optical fiber. In general, optical fibers with small core diameter and low numerical aperture have higher bandwidth and lower loss.

Glass Processing

Depending on a fiber's application, a number of glass compositions may be used. However, for low-loss applications, the options become increasingly limited. Multicomponent glasses containing a number of oxides are not suited to making very low-loss fibers. Multicomponent glasses are prepared by essentially standard optical melting procedures, but with special attention given to details for increasing transmission and controlling defects from later fiber drawing steps. In contrast, low-loss fibers are usually made from pure fused silica doped with minor constituents. These require special manufacturing techniques.

Several methods have been used for manufacturing low-loss optical fibers, including the rod-in-tube method, the double-crucible method, and the more recent and widely used chemical vapor deposition (CVD) methods. CVD methods can be divided into two main categories: inside processes and outside processes. With the inside processes, such as modified chemical vapor deposition (MCVD) and plasma-activated chemical vapor deposition (PCVD), the chemical reactions that form the glass occur inside a glass starting tube (sometimes called a "substrate" tube). This starting tube serves as a containment vessel during deposition and eventually becomes part of the fiber. With outside processes, such as outside vapor deposition (OVD) and vapor axial deposition (VAD), vapor deposition occurs on the outside surface of a starting rod, and the reaction that forms the glass occurs near a burner.

With all CVD methods, silica and other glass-forming oxides and dopants are deposited at high temperatures on an object. In the inside processes, the object is collapsed and becomes part of the fiber, whereas in the outside processes, the object is removed and the resultant *soot blank* is dried and sintered. This produces a thick cylindrical *preform*. A long, thin fiber can then be drawn from this preform at high temperature, or the preform can first be made larger by either depositing more glass on its outer surface or by placing the preform in a glass overclad tube.

There are numerous variations on these processes and these are considered proprietary by most manufacturers. See Fig. 4.

Improvements in Lasers

Just a few years ago, the so-called C^3 laser (cleaved-coupled-cavity) appeared, wherein the alignment of two conventional semiconductor lasers yields a beam of exceptional purity that enables communication systems to send signals at rates as great as billions of bits, or binary digits per second. Just as recently as the late 1980s, commercial lightwave systems were limited to somewhat less than 2 million bits per second, but nevertheless a rate that permits the transmission of 24,000 simultaneous telephone calls on a single pair of fibers.

The cleaved-coupled-cavity laser developed by W.T. Tsang and colleagues (AT&T Bell Laboratories) was designed especially for use in fiber optics telecommunication systems and probably as much as the optical fibers themselves will contribute to the gross claims, as mentioned earlier in the article, made for faster, higher-capacity optical links.

Traditionally, semiconductor lasers have been used in optical communication systems. They are about the size of a grain of salt and produce pulses of light from pulses of electric current. Their electrical requirements are minimal (generally a few mA at 1 or 2 V). These lasers can generate infrared (IR) light where optical fibers are most nearly transparent. Unlike a gas laser, for example, the semiconductor laser is mechanically stable and reliable.

For comparison, a simple semiconductor laser is illustrated and briefly described in Figs. 5 and 6. A buried-heterostructure laser is shown in Fig. 7. The alignment of two lasers to form the C^3 laser is shown in Fig. 8. The turnability of a C^3 laser is illustrated in Fig. 9.

As observed in Tsang, the C^3 laser configuration offers at least three advantages in optical communication systems: (1) The exceedingly

Fig. 4. Experimental optical fiber designs are tested by creating glass preforms, heating them in a special furnace located above the research fiber drawing tower (as shown) and drawing the fiber from the preform onto a collection drum shown in the foreground. The fiber is then used in tests conducted with light generators (lasers and LEDs) and photodetectors. The single-mode fiber is rated for high-capacity transmission, while lower-capacity multimode fibers are well matched to various economical sources and detectors. (*AT&T Bell Laboratories.*)

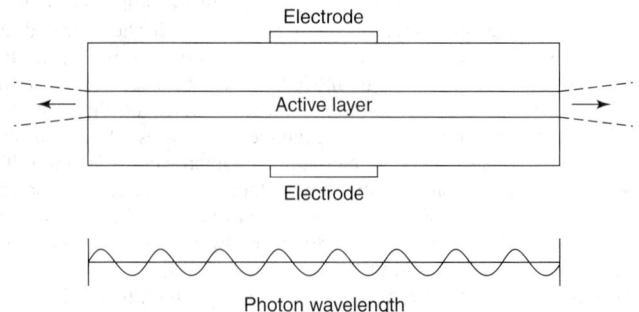

Fig. 5. A simple semiconductor laser. It is a *p-n* junction in a semiconductor crystal the end faces of which are flat and perfectly parallel. Thus, the faces form a pair of semireflecting mirrors that bounce photons back and forth through the *active layer* of the crystal. Current injection causes photons to arise by spontaneous emission. Those photons traversing the semiconductor cause an avalanche of *stimulated* emission. Reflections at the mirrors are self-reinforcing provided that the wavelength of the photon fits evenly into the *length* of the laser. See also Fig. 6.

monochromatic output of the laser eliminates the problem of chromatic dispersion in optical fibers. This facilitates the transmission of digital information in a single-mode fiber at a wavelength of 1.55 micrometers, that is, the wavelength at which a silica fiber is most nearly transparent to electromagnetic radiation. In a test, using the C^3 laser, scientists at AT&T

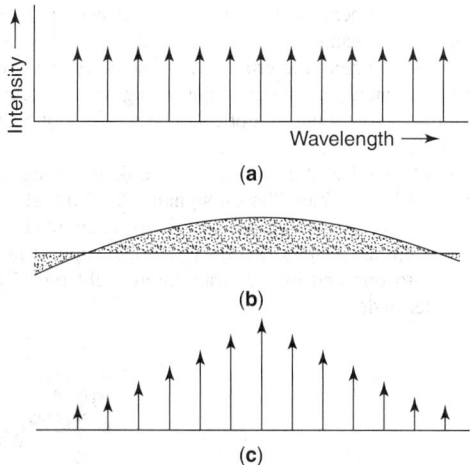

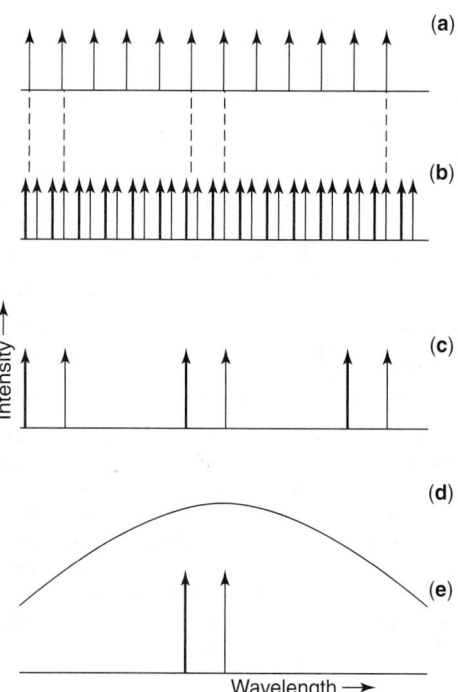

Fig. 6. Simple semiconductor laser energy diagram. This schematic diagram illustrates why the output beam of the laser jumps randomly among several wavelengths. Fundamentally, the laser resonates at the *infinite number* of wavelengths that fit evenly into the length of the laser as indicated in (**a**). On the other hand, the *p-n* junction produces photons in only a narrow range of wavelengths—the *gain profile*, as indicated in (**b**). Therefore, the beam emitted by the laser includes only the resonant wavelengths positioned within the profile, as shown in (**c**). (*After Tsang.*)

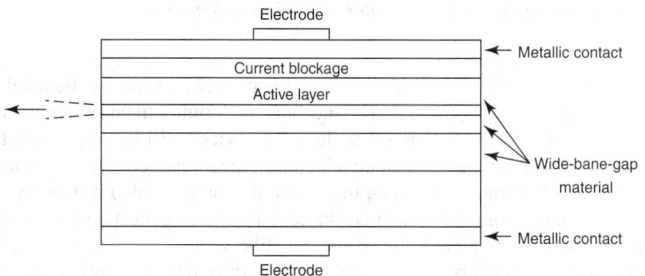

Fig. 9. The energy diagram of the C^3 laser illustrates its tunability by current injection. As explained by Tsang, the current to one of the half-lasers elevates it above its *lasing* threshold. Thus, its resonant modes are fixed as indicated in (**a**). The other (second) half-laser is maintained below its lasing threshold. Thus, it still has resonant modes, as shown by the thin black lines in (**b**). But, some of the resonant modes [thin black lines in (**c**)] match the modes of the light-emitting half-laser. The match at the peak of the laser's gain profile (**d**) determines the wavelength of the C^3 laser's output (**e**). Now the current applied to the second half-laser is changed. The change alters its resonant modes [heavy black lines in (**b**)]. Thus, a new set of matching resonances is established [heavy black lines in (**c**)] and with it a new output wavelength [heavy black line in (**e**)]. (*After Tsang.*)

Fig. 7. In an effort to improve the performance of the basic semiconductor laser, researchers developed the buried-heterostructure laser. In this configuration, the *p-n* junction is reduced to a "tube" that runs the length of the semiconductor crystal. This tube is surrounded by layers of semiconductor whose wide band gap raises the electrical barrier confining charge carriers within the tube. The wide-band-gap material also confines the photons produced at the junction. The laser beam spreads because of diffraction occurring where the beam emerges from the face of the device.

multiplexing can make it possible to carry several independent messages on the fiber. (3) As proposed by Tsang, at rates on the order of a billion switchings per second, it is possible to shunt the output wavelength of a C^3 laser among as many as 15 modes spaced about 2 nm apart. Thus, the single-wavelength transmission of data, with high-power and low-power pulses representing the binary digits 1 and 0, respectively, yields to multiple-wavelength transmission.

In the future, it is expected that the C^3 laser may become part of optical logic circuitry. This will be feasible because of the laser's tunability and from the electrical isolation of the two half-lasers. See Fig. 10.

One problem of the past results from the fact that much energy is wasted at the beginning of an optical link. H.M. Presby (Bell Laboratories) asserts in connection with the loss of about half the laser light, "It seemed like an awful waste, like throwing away half a tank of gas." Through redesigning the tips of optical fibers into carefully sculpted microlenses, a surprising amount of light was captured. Optical problems could be solved by designing an aspheric, hyperbolic shape. In a process that is akin to shaping a piece of metal in a lathe, the researchers rotate silicon fibers in and around a carbon dioxide laser beam.

Researchers also have been investigating microlasers. These lasers range from 1 to 5 microns in diameter and are "carved" from a multilayered semiconductor substrate. It is estimated that a million such lasers occupy but a square centimeter on a chip. See Fig. 11.

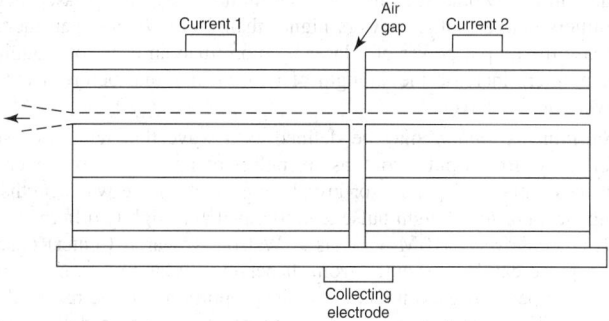

Fig. 8. The C^3 laser consisting of two aligned lasers. The half-lasers have different lengths and thus their resonant wavelengths are differently spaced. Only a few of them match. The mismatches are suppressed. Among the matches, only one is near the peak gain. Thus, the C^3 laser beam is made up almost exclusively of that wavelength. Tests have shown that the probability of the beam "jumping" to another wavelength is less than 1 in 10 billion beam samplings.

Light-Emitting Diode for Telecommunications

In telecommunications, a light-emitting diode (LED) operates similarly to a laser in that it sends pulses of light signals representing speech or data through an optical fiber. Lasers are powerful and are preferred for sending light signals over great distances, such as between cities. Lasers require a measurable amount of power and require temperature regulation. An LED system developed by GTE Laboratories (LOC-LED System) operates on less than one-tenth the power required for a typical laser transmitter and requires no temperature control or means to adjust light signals, thus greatly

Bell Laboratories transmitted digital information in an optical fiber more than 120 kilometers long at a rate of one gigabit (10^9 bits) per second without reamplification along the path. The frequency of error was less than two bits in 10^{10}. (2) By coupling several C^3 lasers, each tuned to a different wavelength, to a single optical fiber, wavelength-division

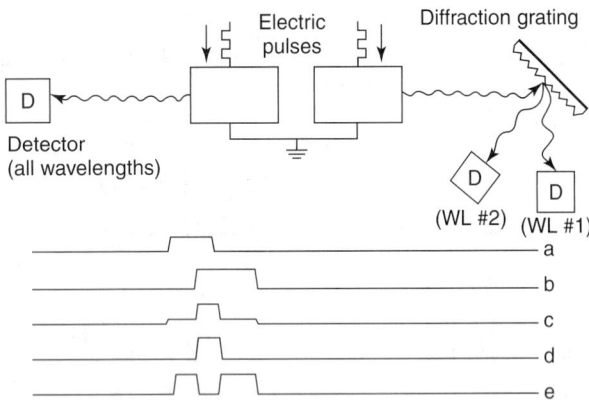

Fig. 10.   Use of the C³ laser in an optical logic circuit. The laser's tunability and the electrical isolation of the two half-lasers make this application possible. Visualize the application of independent trains of electric pulses applied to the half-lasers, as indicated in (**a**) and (**b**). Simultaneous pulses cause the emission of light at wavelength #1. A pulse to one of the half-lasers causes the emission of light at wavelength #2. Detection of light at both wavelengths is equivalent to the logic operation OR (**c**). Detection of wavelength #1 is equivalent to the logic operation AND (**d**). Detection of wavelength #2 is equivalent to the logic operation EXCLUSIVE OR (**e**). (*After Tsang.*)

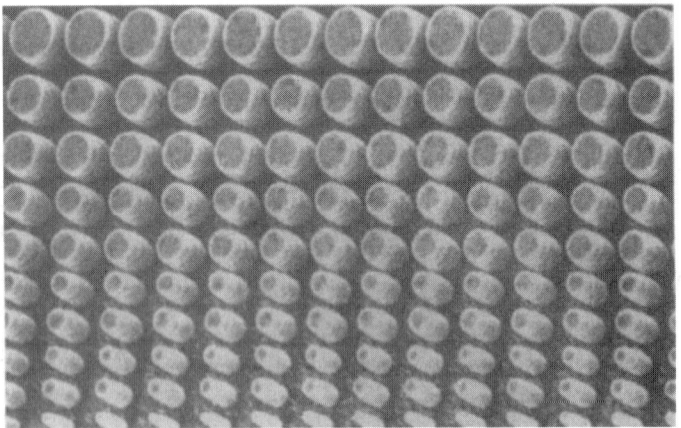

Fig. 11.   An array of microscopic lasers sculpted from a multi-layered semiconductor substrate. The lasers range from 1 to 5 microns in diameter, and a million of them occupy a square centimeter on a chip. (*AT&T Bell Laboratories.*)

simplifying the circuitry. The LOC-LED system generates sufficient optical power to fulfill most needs of fiber optic systems over a range of about 6 to 8 miles (9.7–12.9 km), which is a typical span for a local service loop. The system has been tested at temperatures from −4 to +185 °F (−20 to +85 °C). No prior commercially available LED device generates as much power while using so little energy as the LOC-LED.

Fiber optics in telecommunications is discussed further in the article on **Telephony (Telecommunications)**.

Erbium-doped Fibers for Amplification

The use of repeaters to amplify signals dates back to the earliest days of telegraphy and telecommunications. Although optical fibers offered numerous improvements over copper wire communications, the need to use repeaters was not overcome. Since the serious introduction of optical fibers in the mid-1980s, long-range research has been directed toward maintaining signals over long distances and minimizing the requirement for repeater stations. As aptly pointed out by E. Corcoran, "Repeaters have become as endemic — and as constraining — on the telecommunications freeway as tollbooths on a turnpike." Lightwaves, of course, are capable of many frequencies, but electronic repeaters can handle only one frequency at a time and thus decrease data throughput through a network. Elias Snitzer (Rutgers University) found that erbium-doped glass fibers could be used to provide optical amplification and thereby increase the number of light frequencies that can be amplified. When pumped with a light source, more

erbium ions in these fibers are forced to a higher energy level (known as *population inversion*) than remain at lower levels. As a transmission signal carrying information enters the erbium fiber, some of the excited erbium ions give up their energy to the information signal — thereby amplifying them. The amplification is purely optical. The information signal does not first need to be converted to an electrical signal.

Researchers (AT&T Bell Laboratories) have demonstrated that erbium-doped devices (Fig. 12) can "boost signals traveling at any bit rate, transmission networks can be upgraded by simply changing the transmitters (and receivers). Virtually any video data or voice signal can be dumpled 'like marbles' into one end of the 'transparent light pipes' and roll out intact at the other end."

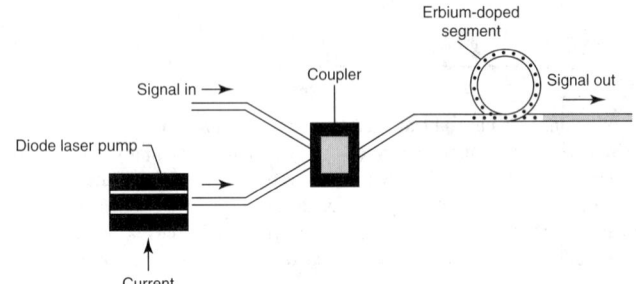

Fig. 12.   Erbium-doped light amplifier. Light from a diode laser pump excites erbium ions in segment of an optical fiber. These ions then emit light and thus boost a passing optical signal. (*After AT&T Bell Laboratories.*)

Earlier research on the erbium-doped principle had been conducted by University of Southampton, U.K., and Japanese communications engineers. Currently, scientists envision that the new devices will be applied first to cable television and transoceanic telecommunications network. As pointed out by E. Corcoran, "Fiber amplifiers could enable a video broadcaster to lay one cable from the transmission center to a neighborhood, then split up the signal, amplify it and route the fibers directly into the homes." The Japanese already are at work on an all-optical network under the Pacific Ocean to be operable by 1996. For local communications loops, L.J. Andrews (GTE) observes, "Cost is the critical issue." John Mellis (BTD, Ipswich, U.K.) says, "The challenges are all engineering now ... Because optical amplifiers are being seriously considered in transoceanic submarine systems, they will certainly happen, if not in this decade then the next." Another professional in the field observes, "There's absolutely no doubt that this is the thing to work on. It's changed how we think about fiber-optic systems."

There are three different locations in which optical amplifiers can be located in a network. See Fig. 13. Used immediately after a laser, booster amplifiers deliver output powers higher than 100 mW, and can therefore increase the output power of a laser by more than an order of magnitude. Preamplifiers increase the strength of an optical signal before it enters a conventional receiver.

Solitons. A *soliton* may be defined as a wave that retains its shape indefinitely. Bright-pulse solitons are pulses of light that travel over long distances without dispersing or broadening. Each of the wavelengths that comprise an *ordinary* light pulse tend to travel at a slightly different speed. D.R. Grischkowsky (IBM Thomas J. Watson Research Center) observes that a pulse can be prevented from dispersing by taking advantage of an optical property of glass fibers — that is, proportional to the received light intensity. This output current is AC coupled to the load to deliver the RF signal.

Optical Fibers in Aircraft

Traditionally, communications and control in modern aircraft have been accomplished by as much as 240 km (in a typical wide-body jet) of electric wires, adding significantly to the weight of the craft. Even with insulation and shielding, such wire connections constantly are subject to electromagnetic interference (EMI), not to mention increased vulnerability to lightning during storm conditions and other sources of electromagnetic radiation of particular note to military aircraft. The potential of an "electronic blizzard" over a war zone long has been recognized as a hazard

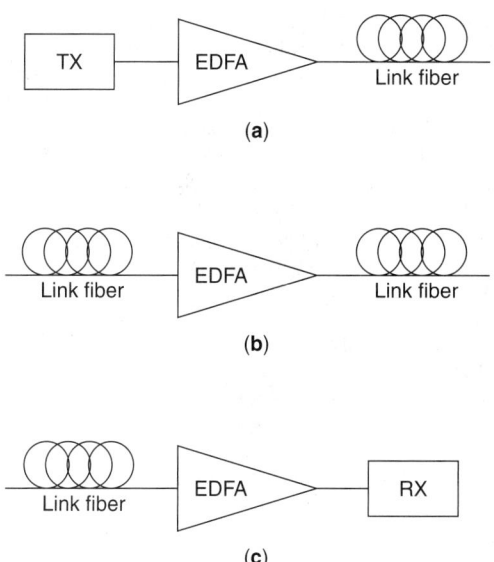

Fig. 13. Three types of optical amplifier as distinguished by their location in a network: (**a**) Postamplifier (booster), (**b**) in-line amplifier, and (**c**) optical preamplifier. (*From J. Augé.*)

by military planners. Numerous crashes of the Sikorsky Aircraft Black Hawk helicopter have been attributed to EMI. R.J. Baumbick (National Aeronautics and Space Administration, Cleveland, Ohio) notes, "Wires are becoming the dominant antennae in aircraft." By replacing copper lines with optical fibers, the weight of cabling required can be reduced by an estimated 50%. See Fig. 14.

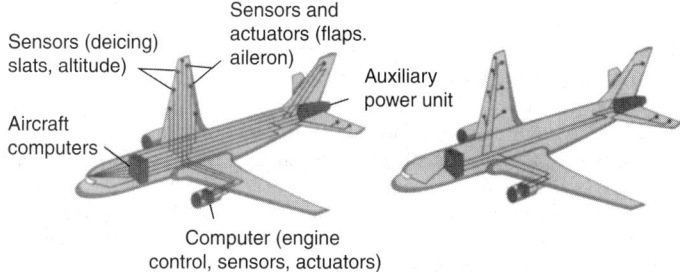

Fig. 14. Contrast of (left) hard-wire aircraft communications and controls and (right) use of fiber optics. (*After United Technologies Research Center.*)

For the use of light systems in aircraft, researchers recognize the need for developing improved light sources for transmitters. Lasers do not serve well because of the high temperatures encountered in supersonic flight and of intense heat from aircraft engines. Powerful light-emitting diodes are required. A difficult problem is that of using optical components in place of traditional electric control of hydraulic actuators (control of flaps, rudders, and other flight-control surfaces). Sensors that heat a small amount of hydraulic fluid and thus build up pressure that can be amplified by the actuator's hydraulic system are being developed and tested.

Fiber Optics in Biomedicine

Sensors incorporating glass or plastic optical fibers have demonstrated several advantages over electrosensors for biomedical applications. These sensors involve no electrical connections and hence are safe from that standpoint; the leads are quite small and flexible: they can be incorporated in catheters for multiple sensing; where required, they can be implanted for relatively long periods. The fibers are considerably less than 1 millimeter in diameter. Where designed for simplicity, they often can be considered disposable.

As reported by Peterson (National Institutes of Health) and Vurek (Sorenson Research Corp.), there are three principal types of *in vivo* fiber-optic sensing configurations—*photometric* (or bare-ended fiber), *physical parameter sensors* in which a transducer at the end of a fiber alters the

light signal in accordance with the values of the parameter measured, and *chemical probes*. In the latter, a suitable reversible reagent fixed at the end of a fiber provides spectrophotometric or fluorometric analysis. The earliest use of fiber optic sensors was in connection with reflectometry, spectrophotometry, and fluorometry where no transducer was used. In oximetry, the hemoglobin content of blood is measured spectrophotometrically. Where dyes are injected into the blood, fiber optic sensors can be used for measuring blood flow, cardiac output, and perfusion. When a microtransducer is attached at the end of a fiber optic conductor, temperature and pressure measurements can be made. A notable application is the use of a temperature sensor in connection with the hyperthermal treatment of cancer. Accuracy of $\pm 0.1\,^\circ$C can be obtained. Sometimes such devices are used in multiple locations. For example, a layer of liquid crystals at the end of optical fibers will produce changes in light scattering due to temperature change. Fiber optic sensors have been designed for monitoring intracranial and intracardiac pressure. Chemical sensors have been developed to measure pH, PO_2, PCO_2, and glucose, among other chemical variables. The details of these biomedical applications for fiber optic sensors are well developed in the Peterson-Vurek paper (reference listed).

Fiber Optic Communications in Botany

It has been known for centuries that the position of a plant relative to its source of light will affect the manner in which the plant grows and develops, including its shape. Thus, there arose expressions such as "a plant seeks or reaches for the light." Horticulturists and farmers carefully lay out their greenhouses and fields so that maximum advantage can be taken of available sunlight or, in some cases, artificial light. Only in recent years, however, has it been suspected that plants may possess means for further utilizing the radiation which they receive. Researchers Mandoli (Stanford University) and Briggs (Carnegie Institution) have observed that, in addition to depending upon light as a source of energy (photosynthesis), light is also used as a means of communication. For example, the tissues of plant seedlings can guide light through distances measured in centimeters (a distance of 4.5 centimeters has been demonstrated in the laboratory). The exact method of light transmission within plants has not been fully established, but researchers suspect that natural fibers are involved—so called "light pipes."

In the initiation and control of several physiological processes in plants, lightwave communications may serve in a way comparable to the nervous system of animals. Some of the factors influenced by light receptors include the time for a seed to germinate, the angle a shoot should take to counter gravitational forces, the rate with which a leaf should develop, and the time when a plant should bloom. Laboratory findings show that the amount of energy needed for light signaling is several orders of magnitude less than the light used for photosynthesis.

Light-sensitive detectors are pigment cells, of which the molecules making up a substance known as *phytochrome* are among the most important. These photoreceptors are sensitive to various parts of the light spectrum and thus play distinctive roles in managing different parts of a plant's physiology. In addition to the spectral distribution of light as received by the photosensitive cells, other important factors include the amount of light received, the direction from which the light is received, and the duration of the light signal.

This relatively recent area of botany and biology admittedly is in an early stage of development. Also see the article on **Etiolation**.

Fiber Optics in the Private Network

Private Networks, new and existing, must be well planned and carefully structured. Their network cabling solution may require a balance of both copper and fiber to cost effectively meet today's needs and support the high-bandwidth applications of the future, such as multimedia and full motion digital video-conferencing. Lucent Technologies, SYSTIMAX® SCS product families, supports these new and existing networks.

Rapidly evolving applications and technologies will drastically increase the speed and volume of traffic on LAN/WAN networks. Ensuring that your structured cabling solution is designed to accommodate the higher transmission rates associated with these evolving bandwidth intensive applications will be critical. Some examples include:

- Multimedia workstation
- Networked Scientific Modeling

- Imaging, Radiography, Computer Aided Design/Computer Aided Manufacturing (CAD/CAM)
- Asynchronous Transfer Mode (ATM)
- High-Definition Television (HDTV)
- Array processor workstations
- Mass memory database transfer
- Videophone
- Photonic (lightwave) switches and processors.

Local Area Networks (LANs). A LAN is a data communications system that enables users to access common data processing (PCs, minicomputers, and mainframe computers) and peripheral equipment (printers and fax machines). LANs, are created by using workstations with adapter cards and connecting them to file servers (where the operating system/software resides) and printers. Gateways are used to connect LANs to other LANs or operating systems like large mainframes where there is a need to share departmental or corporate computing systems. A LAN can be as simple as a few workstations working off a file server or as complex as putting hundreds of workstations on a network that runs between floors of a building or between a number of buildings in a campus environment. LANs, which were originally designed so that users could share and access a few expensive printers or controllers, have expanded into essential telecommunications networks. Today, LANs are used for file and printer sharing, electronic mail, shared databases, point-of-sale, and order entry systems.

LANs deployed on different floors or buildings are typically connected with multimode fiber. However, newer high-speed LAN topologies like full motion video do utilize single-mode fiber in some long-distance route applications where the excellent transmission characteristics of single-mode fiber are required.

LAN Topologies. SYSTIMAX SCS offers cabling architecture options for Fiber-to-the-Desktop installations: the traditional **Hierarchical Star architecture** and the new **Single Point Administration architecture**.

The traditional **Hierarchical Star architecture** is designed for maximum flexibility. Cross-connect facilities are provided in both the telecommunications closets and the main equipment room. The riser backbone cables can be sized with low counts which allow only distributed active equipment, or for greatest flexibility, with high counts which permit both distributed and centralized active equipment. The horizontal cross-connect facility helps ensure the greatest life span for the system by allowing the active equipment to be located closer to the work areas. The short horizontal runs can support applications at higher speeds than the longer combined horizontal/riser runs used with centralized active equipment. Also, this architecture is standards-compliant with both riser design approaches.

Single Point Administration architecture is designed for simplicity and cost-effectiveness for centralized equipment. This approach provides direct connections from all work areas (offices) to the cross-connect in the main equipment room, forming a single point of administration optimized for centralized active equipment. The single point of administration provides the simplest circuit management possible by eliminating the need to cross-connect circuits in multiple places. It also provides the ability to connect users in different areas of a building directly to the same LAN segment, reducing traffic on bottleneck-prone bridges and routers. Single point administration also provides three cost benefits. First, it eliminates the need for horizontal cross-connects, saving passive hardware costs. Second, it consolidates active equipment, reducing the number of idle ports in the system, thereby saving active equipment costs. Finally, this type of architecture eases network administration and maintenance, reducing technical support staff effort.

The LAN topology is the physical layout of a LAN—how the controllers, workstations (primarily PCs), and other equipment are connected by the cable. The three basic LAN topologies are: Bus; Star; and Ring.

The bus topology has all the workstations on the network attached (via the information outlet) to a single cable that carries the signal in both directions through the network. The bus network, can be expanded by adding several segments together with bridges, routers, and repeaters. The Institute of Electrical and Electronics Engineers Inc. (IEEE) Standard 802.3 is an example of a bus topology. In a star topology, all of the nodes or workstations are connected with unshielded twisted pair (UTP) or fiber optic cable to a centrally located common controller or concentrator. The central control point permits centralized network administration, management, and troubleshooting. StarLAN is an example of a star topology, as is IEEE 802.3. In a ring topology, the network forms a ring.

A token carries data through the network. Workstations are connected as in a star topology to a central administration point, such as a multistation access unit (MAU). Token Ring (IEEE 802.5) is an example of a ring topology.

The LANs, which were introduced back in the 1980s, were based on copper cable, with very few fiber applications being used. Most of the more recently introduced LANs (FDDI, 10BASE-F, and DFON) primarily use fiber optics but offer interfaces (bridges and routers) with the older copper-based networks. The more recent trend is in the use of "smart hubs" or concentrators, which allow both bus and ring (Ethernet and Token Ring) topologies to be mixed in the same electronics within the communications closet. In addition, the backbone network, which links the hubs/ concentrators together, could be on a higher speed LAN topology such as FDDI running at 100 Mbps.

The advent of new lower cost optoelectronics for LAN applications has spurred a growing interest in using fiber all the way to the workstation. These total fiber networks offer easy migration to even higher speed applications like ATM. See also **Local Area Networks**.

Fiber Optics in Industrial Instrumentation and Networks

The combination of light-transmitting optical cable and miniature silicon sensors has resulted in the development of a new measurement technology for various industrial processes. Three variables may be measured with this technique—temperature, pressure, and refractive index. The new systems are immune to electromagnetic and radio-frequency interference, they provide more accuracy in electrically noisy environments, and their miniature size improves response and causes minimal process disturbances.

The fiber optic sensors utilize an extrinsic Fabry-Perot interferometer to spectrally modulate light in proportion to pressure, temperature, or refractive index variations. Because they are based on spectral modulation instead of amplitude modulation, they are not affected by such common problems as fiber bending, connector losses, and aging.

Pressure Sensing. As shown by Fig. 15, a cavity resonator is constructed using a float-bottom pocket-etched into a refractory glass substrate. The pocket is 2–3 wavelengths deep and covered with a diaphragm. The two reflecting surfaces (bottom of pocket and the diaphragm) form an interferometer. The cavity is evacuated, thus permitting the diaphragm to deflect. This deflection is a function of absolute pressure. In effect, the path length changes. A parallel can be made with the action of a soap bubble. As the bubble changes size and hence film thickness, interference effects occur. These reinforce reflected light at certain wavelengths, enhancing transmission of light through the film at other wavelengths. Color (frequency) changes occur as a result.

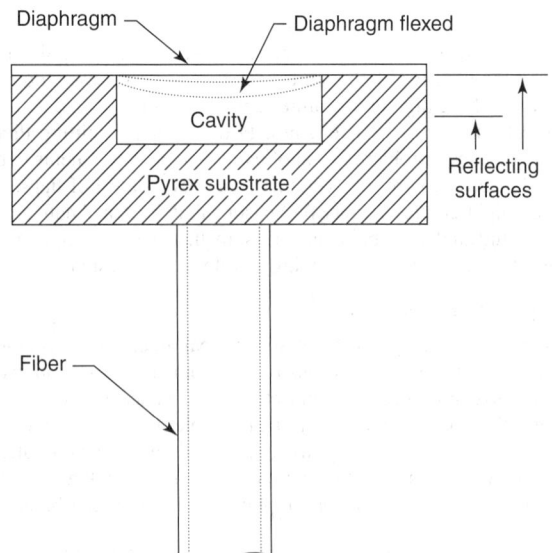

Fig. 15. Fiber-optic pressure sensor. As light passes into a cavity resonator formed by a glass substrate and a flexible diaphragm, it is reflected from both surfaces, forming an interference pattern that changes as the diaphragm flexes with pressure changes. (*Yazbak, Foxboro, Massachusetts.*)

The cavity resonator (sensor) is 0.015 in (0.4 mm) in all three dimensions. The resonator is joined to a multimode fiber by a glass capillary 0.032 in (0.8 mm) in diameter. The light source's spectrum is centered at 850 nm. As the effective cavity depth changes as a function of the measured variable, the cavity's reflectance spectrum shifts. This spectral shift modulates (skews) the incoming light spectrum. A dichroic filter and photodiodes are then used to discern differences in the returning light's spectrum strength within two wavebands.

Temperature Sensing. As shown in Fig. 16, a layer of silicon (whose refractive index changes with temperature) is placed in the optical path in place of the evacuated cavity, as previously described. The second reflector (glass) is rigid. The effective path length thus changes with temperature.

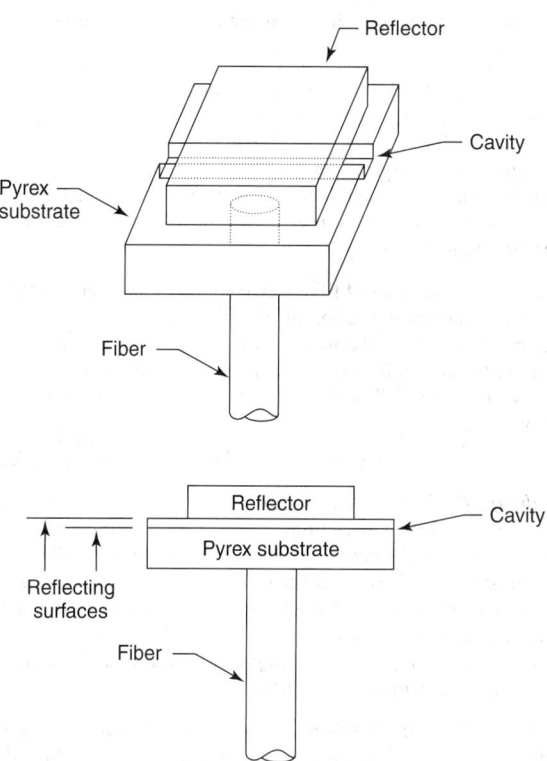

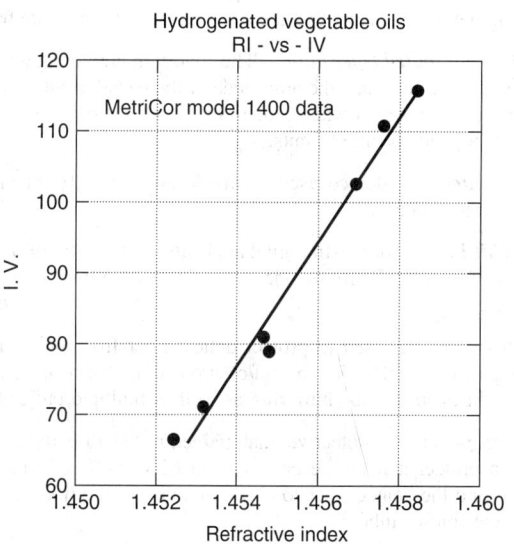

Fig. 17. Fiber-optic refractive index sensor. Fluid of interest is drawn by capillary action into a duct through a glass substrate. The effective path length varies in proportion with the refractive index. (*Yazbak, Foxboro, Massachusetts.*)

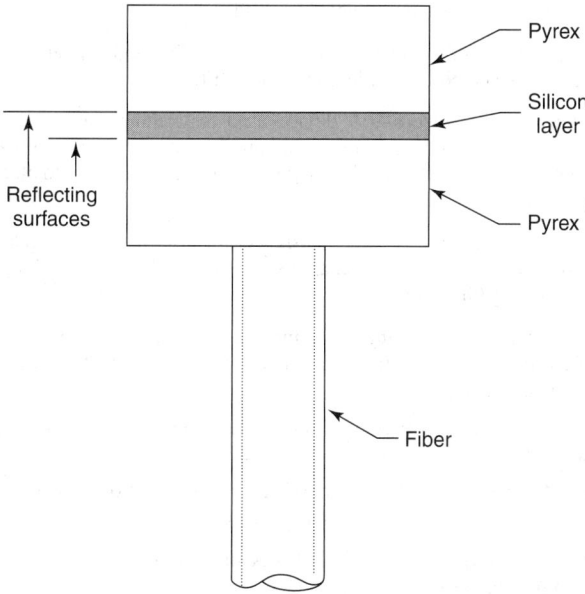

Fig. 16. Fiber-optic temperature sensor. A thin layer of silicon placed in the optical path exhibits a large change in refractive index with temperature, changing the effective path length. (*Yazbak, Foxboro, Massachusetts.*)

Refractive Index. This property is the ratio of the velocity of light in a vacuum to the velocity of light in a transparent material. The property is used extensively in the processing industries for measuring component concentrations or for tracking changes in molecular makeup, such as materials that are reacting. The hydrogenation of food oils is an example of such a process.

The extent of hydrogenation (the degree of saturation of a double bond in the ester chain of an edible oil) may be indicated by conventional laboratory titrations of the iodine number. This is a time-consuming, grab-sampling method. Continuous measurement and control via refractive index measurement is far more efficient. As shown by Fig. 17, the fluid of interest is drawn by capillary action into a duct through a glass substrate, and the effective path length varies in proportion to the refractive index of the fluid. Excellent correlation of refractive index with the iodine number of hydrogenated oil is shown in Fig. 18.

Note: The foregoing information on fiber optic sensors was furnished by Gene Yazbak, Foxboro, MA.

Fig. 18. Correlation of refractive index with iodine number of hydrogenated oil. (*Yazbak, Foxboro, Massachusetts.*)

Optical Fiber Terminology

An abridged glossary of terms used to describe optical fiber products and processes would include:

Absorption — A physical mechanism in fibers that attenuates light by converting it into heat-thereby raising the fiber's temperature. In practice the temperature increase is slight and difficult to measure. Absorption arises from tails of the ultraviolet and infrared absorption bands, from impurities such as the OH−ion, and from defects in the glass structure.

Adapter — A mechanical media termination device designed to align and join fiber optic connectors. Often referred to as a coupling, bulkhead, or interconnect sleeve.

Adapter Efficiency — The efficiency of optical power transfer between two components.

Adapter Loss — The power loss suffered when coupling light from one optical device to another.

Aramid Yarn — Strength elements that provide tensile strength and provide support and additional protection of the fiber bundles. Kevlar is a particular brand of aramid yarn.

Armor — Additional protective element beneath outer jacket to provide protection against severe outdoor environments. Usually made of plastic-coated steel, it may be corrugated for flexibility.

Attenuation — The decrease in magnitude of power, of a signal in transmission between points. A term used for expressing the total loss of an optical system, normally measured in decibels (dB) at a specific wavelength.

Attenuation Coefficient — The rate of optical power loss with respect to distance along the fiber, usually measured in decibels per kilometer (dB/km) at specific wavelength. The lower the number, the better the fiber's attenuation. Typical multimode wavelengths are 850 and 1300 manometers (nm); single-mode wavelengths are 1310 and 1550 nm. Note: When specifying attenuation, it is important to note whether the value is average or nominal.

Avalanche Photodiode (**APD**) — A photodiode designed to take advantage of avalanche multiplication of photocurrent. As the reverse-bias voltage approaches the breakdown voltage, hole-electron pairs created by absorbed photons acquire sufficient energy to create additional hole electron pairs when they collide with ions; thus a multiplication or signal gain is achieved.

Axial Ray — A light ray that travels along the axis of an optical fiber.

Backbone Cabling — The portion of premises telecommunications cabling that provides connections between telecommunications closets, equipment rooms, and entrance facilities. The backbone cabling consists of the transmission media (optical fiber cable), main and intermediate cross-connects, and terminations for the horizontal cross-connect, equipment rooms, and entrance facilities. The backbone cabling can further be classified as interbuilding backbone (cabling between buildings), or intrabuilding backbone (cabling within a building).

Bandwidth-Distance Product — The information-carrying capacity of a transmission medium is normally referred to in units of MHz-km. This is called the bandwidth-distance product or more commonly bandwidth. The amount of information that can be transmitted over any medium changes according to distance. The relationship is not linear, however. A 500 MHz-km fiber does not translate to 250 MHz for a 2 kilometer length or 1000 MHz for a 0.5 kilometer length. It is important, therefore, when comparing media, to ensure that the same units of distance are being used.

Bandwidth Limited Operation — The condition prevailing when the system bandwidth, rather than the amplitude of the signal, limits performance. The condition is reached when modal dispersion distorts the shape of the waveform beyond specified limits.

Beamsplitter — A device used to divide an optical beam into two or more separate beams.

BER (*Bit Error Rate*) — In digital applications, the ratio of bits received in error to bits sent. BERs of one errored bit per billion (I × IO-9) sent are typical.

Buffer — Material used to protect optical fiber from physical damage, providing mechanical isolation and/or protection. Fabrication techniques include tight or loose tube buffering as well as multiple buffer layers.

Buffering — (1) A protective material extruded directly on the fiber coating to protect it from the environment (tight buffered); (2) extruding a tube around the coated fiber to allow isolation of the fiber from stresses in the cable (buffer tubes).

Buffer Tubes — Extruded cylindrical tubes covering optical fibers(s) used for protection and isolation.

Bundle — Many individual fibers contained within a single jacket or buffer tube. Also a group of buffered fibers distinguished in some fashion from another group in the same cable core.

Cable — An assembly of optical fibers and other material providing mechanical and environmental protection.

Cable Assembly — Optical fiber cable that has connectors installed on one or both ends. General use of these cable assemblies includes the interconnection of optical fiber cable systems and opto-electronic equipment. If connectors are attached to only one end of a cable, it is known as a pigtail. If connectors are attached to both ends, it is known as a jumper or patch cord.

Cable Bend Radius — Cable bend radius during installation implies that the cable is experiencing a tensile load. Free bend infers a smaller allowable bend radius since it is at a condition of no load.

Central Member — The center component of a cable. It serves as an antibuckling element to resist temperature induced stresses. Sometimes serves as a strength element. The central member material is either steel, fiberglass, or glass-reinforced plastic.

Centralized Cabling — A cabling topology used with centralized electronics connecting the optical horizontal cabling with intra-building backbone cabling passively in the telecommunications closet.

Chromatic Dispersion — Spreading of a light pulse caused by the difference in refractive indices at different wavelengths.

Cladding — The dielectric material surrounding the core of an optical fiber.

Coating — A material put on a fiber during the drawing process to protect if from the environment and handling.

Composite Cable — A cable containing both fiber and copper media per article 770 of the National Electric Code (NEC).

Connecting Hardware — A device used to terminate an optical fiber cable with connectors and adapters that provide an administration point for cross-connecting between cabling segments or interconnecting to electronic equipment.

Connector Panel — A panel designed for use with patch panels; it contains either 6, 8, or 12 adapters pre-installed for use when field-connectorizing fibers.

Connector Panel Module — A module designed for use with patch panels, it contains either 6 or 12 connectorized fibers that are spliced to backbone cable fibers.

Core — The central region of an optical fiber through which light is transmitted.

Core Eccentricity — A measure of the displacement of the center of the core relative to the cladding center.

Core Ellipticity (*non-circularity*) — A measure of the departure of the core from roundness.

Critical Angle — The smallest angle from the fiber axis at which a ray may be totally reflected at the core/cladding interface.

Cutoff Wavelength — The shortest wavelength at which only the fundamental mode of an optical waveguide is capable of propagation.

Data Rate — The maximum number of bits of information, which can be transmitted per second, as in a data transmission link. Typically expressed as megabits per second (Mbps).

Dielectric — Nonmetallic and, therefore, nonconductive. Glass fibers are considered dielectric. A dielectric cable contains no metallic components.

Entrance Facility — An entrance to a building for both public and private network service cables including the entrance point at the building wall and continuing to the entrance room or space.

Fan-Out — Multifiber cable constructed in the tight buffered design. Designed for ease of connectorization and rugged applications for intra- or interbuilding requirements.

Ferrule — A mechanical fixture, generally a rigid tube, used to protect and align a fiber in a connector. Generally associated with fiber optic connectors.

Fiber — Any filament or fiber, made of dielectric materials, that guides light.

Fiber Bend Radius — Radius a fiber can bend before the risk of breakage or increase in attenuation.

Fiber Distributed Data Interface (**FDDI**) — A standard for a I 00 Mbit/s fiber optic area network

Fiber Optic Cable — An optical fiber, multiple fiber, or fiber bundle which includes a cable jacket and strength members, fabricated to meet optical, mechanical, and environmental specifications.

Fiber Optic Link — Any optical fiber transmission channel designed to connect two end terminals or to be connected in series with other channels.

Fiber Optics — The branch of optical technology concerned with the transmission of radiant power through fibers made of transparent materials such as glass, fused silica, or plastic.

Field-Effect Transistor (FET) *Photodetector* — A photodetector employing photogeneration of carriers in the channel region of an FET structure to provide photodetection with current gain.

Fresnel Reflection — The reflection of a portion of the light incident between two homogeneous media having different refractive indices. Fresnel reflection occurs at the air/glass interfaces at entrance and exit ends of an optical fiber.

Fresnel Reflection Losses — Reflection losses that are incurred at the input and output of optical fibers due to the differences in refraction index between the core glass and immersion medium.

Fundamental Mode — The lowest order mode that will travel in a waveguide.

Fusing — The actual operation of joining fibers together by fusion or by melting.

Fusion Splice — A permanent joint produced by the application of localized heat sufficient to fuse or melt the ends of the optical fiber, forming a continuous single fiber.

Graded-Index — Fiber design in which the refractive index of the core is lower toward **the** outside of the fiber core and increases toward the center of the core, thus, it bends the rays inward and allows them to travel faster in the lower index of refraction region. This type of fiber provides higher bandwidth capabilities for multimode fiber transmissions.

Graded Index Fiber — An optical fiber with a variable refractive index that is a function of the radial distance from the fiber axis.

Horizontal Cabling — That portion of the telecommunications cabling that provides connectivity between the horizontal cross-connect and the work-area telecommunications outlet. The horizontal cabling consists of transmission media, the outlet, the terminations of the horizontal cables, and horizontal cross-connect.

Horizontal Cross-Connect (HC) — A cross-connect of horizontal cabling to other cabling, e.g., horizontal, backbone, equipment.

Hybrid Cable — A fiber optic cable containing two or more different types of fiber, such as 62.5 μm multimode and single-mode.

Index Matching Fluid — A fluid with an index of refraction close to that of glass that reduces reflections caused by refractive-index differences.

Index Matching Material — A material, often a liquid or cement whose refractive index is nearly equal to the core index. Used to reduce Fresnel reflections from a fiber end face.

Insertion Loss — The attenuation caused by the insertion of an optical component; in other words, a connector or coupler in an optical transmission system.

Interbuilding Backbone — The portion of the backbone cabling between buildings. (See Backbone Cabling.)

Intermediate Cross-Connect (IC) — A secondary crossconnect in the backbone cabling used to mechanically terminate and administer backbone cabling between the main cross-connect and horizontal cross-connect.

Intrabuilding Backbone — The portion of the backbone cabling within a building. (See Backbone Cabling.)

Irradiance — Power density at a surface through which radiation passes at the radiating surface of a light source or at the cross section of an optical waveguide. The normal unit is Watts per centimeters squared, or W/cmu2d.

Jumper — Optical fiber cable that has connectors installed on both ends. (See Cable Assembly.)

Laser Diode (LD) — Light Amplification by Stimulated Emission of Radiation. An electro-optic device that produces coherent light with a narrow range of wavelengths, typically centered around 780 nm, 1320 nm, or 1550 nm. Lasers with wavelengths centered around 780 nm are commonly referred to as CD Lasers.

Lasing Threshold — The lowest excitation level at which a laser's output is dominated by stimulated emission rather than spontaneous emission.

Launching Fiber — A fiber used in conjunction with a source to excite the modes of another. fiber in a particular way. Launching fibers are most often used in test systems to improve the precision of measurements.

Leaky Modes — In the boundary region between the guided modes of an optical waveguide and the lightwaves which are not capable of propagation, there are so-called leaky modes which are not guided but are capable of limited propagation with increased attenuation. Leaky modes are a possible source of errors in the measurement of fiber loss, but their effect can be reduced by mode strippers.

Light — In the laser and optical communication fields, the portion of the electromagnetic spectrum that can be handled by the basic optical techniques used for the visible spectrum extending from the near ultraviolet region of approximately 0.3 micron, through the visible region and into the mid infrared region of about 30 microns.

Lightwaves — Electromagnetic waves in the region of optical frequencies. The term "light" was originally restricted to radiation visible to the human eye, with wavelengths between 400 and 700 manometers (nm). However, it has become customary to refer to radiation in the spectral regions adjacent to visible light (in the near infrared from 700 to about 2000 nm) as "light" to emphasize the physical and technical characteristics they have in common with visible light.

LXE — Fiber Optic Express Entry

MDPE — Abbreviation used to denote medium density polyethylene. A type of plastic material used to make cable jacketing.

Macrobending — Macroscopic axial deviations of a fiber from a straight line, in contrast to microbending.

Main Cross-Connect (MC) — The centralized portion of the backbone cabling used to mechanically terminate and administer the backbone cabling, providing connectivity between equipment rooms, entrance facilities, horizontal cross-connects, and intermediate cross-connects.

Material Dispersion — The dispersion associated with a non-monochromatic light source due to the wavelength dependence of the refractive index of a material or of the light velocity in this material.

Mechanical Splicing — Joining two fibers together by permanent or temporary mechanical means (vs. fusion splicing or connectors) to enable a continuous signal. The CamSplice is a good example of a mechanical splice.

Megahertz (MHz) — A Unit of frequency that is equal to one million cycles per second.

Microbending — Curvatures of the fiber which involve axial displacements of a few micrometers and spatial wavelengths of a few millimeters. Microbends cause loss of light and consequently increase the attenuation of the fiber.

Micrometer (gm) — One millionth of a meter; 10-6 meter. Typically used to express the geometric dimension of fibers, for example, 62.5 μm.

Mini Bundle Cable — Loose tube cable in which the buffer tube contains two or more fibers, typically 6 or 12 fibers.

Modal Dispersion — Pulse spreading due to multiple light rays traveling different distances and speeds through an optical fiber.

Modal Noise — Disturbance in multimode fibers fed by laser diodes. It occurs when the fibers contain elements with mode-dependent attenuation, such as imperfect splices, and is more severe the better the coherence of the laser light.

Mode — A term used to describe an independent light path through a fiber, as in multimode or single-mode.

Mode Field Diameter — The diameter of the one mode of light propagating in a single-mode fiber. The mode field diameter replaces core diameter as the practical parameter in single-mode fiber.

Mode Mixing — The numerous modes of a multimode fiber differ in their propagation velocities. As long as they propagate independently of each other, the fiber bandwidth varies inversely with the fiber length due to multimode distortion. As a result of inhomogeneities of the fiber geometry and of the index profile: a gradual energy exchange occurs between modes with differing velocities. Due to this mode mixing, the bandwidth of long multimode fibers is greater than the value obtained by linear extrapolation from measurements on short fibers.

Modes — Discrete optical waves that can propagate in optical waveguides. They are eigenvalue solutions to the differential equations which characterize the waveguide. In a single-mode fiber, only one mode, the fundamental mode, can propagate. There are several hundred modes in a multimode fiber which differ in field pattern and propagation velocity. The upper limit to the number of modes is determined by the core diameter and the numerical aperture of the waveguide.

Mode Scrambler — A device composed of one or more optical fibers in which strong mode coupling occurs. Frequently used to provide a mode distribution that is independent of source characteristics.

Modified Chemical Vapor Deposition (MCVD) *Technique* — A process in which deposits are produced by heterogeneous gas/solid and gas/liquid chemical reactions at the surface of a substrate. The MCVD method is often used in fabricating optical waveguide preforms by causing gaseous material to react and deposit glass oxides. Typical starting chemicals include volatile compounds of silicon, germanium, phosphorus, and boron, which form corresponding oxides after heating with oxygen or other gases. Depending on its type, the preform may be processed further in preparation for pulling into an optical fiber.

Modulation — Coding of information onto the carrier frequency. This includes amplitude, frequency, or phase modulation techniques.

Monochromatic — Consisting of a single wavelength. In practice, radiation is never perfectly monochromatic but, at best, displays a narrow band of wavelengths.

Multimode Distortion — The signal distortion in an optical waveguide resulting from the superposition of modes with differing delays.

Multimode Fiber — An optical waveguide in which light travels in multiple modes. Typical core/cladding size (measured in micrometers) is 62.5/125.

Multiplex — Combining two or more signals into a single bit stream that can be individually recovered.

Multi-User Outlet — A telecommunications outlet used to serve more that one work area, typically in open-systems furniture applications.

Nanometer (nm) — A unit of measurement equal to one billionth of a meter; 10-9 meters. Typically used to express the savelength of light, for example, 1300 nm.

Near Field Radiation Pattern — Distribution of the irradiance over an emitting surface; in other words, over the cross section of an optical waveguide.

Numerical Aperture — A measure of the range of angles of incident light transmitted through a fiber. Depends on the differences in index of refraction between the core and the cladding. (The number that expresses the light gathering ability of a fiber. Related to acceptance angle.)

Optical Time Domain Reflectometer (OTDR) — A method for characterizing a fiber wherein an optical pulse is transmitted through the fiber and the resulting backscatter and reflections to the input are measured as a function of time. Useful in estimating attenuation coefficient as a function of distance and identifying defects and other localized losses.

Optical Waveguide — Dielectric waveguide with a core consisting of optically transparent material of low attenuation (usually silica glass) and with cladding consisting of optically transparent material of lower refractive index than that of the core. It is used for the transmission of signals with lightwaves and is frequently referred to as fiber. In

addition, there are planar dielectric waveguide structures in some optical components, such as laser diodes, which are also referred to as optical waveguides.

Optoelectronic — Pertaining to a device that responds to optical power, emits or modifies optical radiation, or utilizes optical radiation for its internal operation. Any device that functions as an electrical-to-optical or optical-to-electrical transducer.

PE — Abbreviation used to denote polyethylene. A type of plastic material used for outside plant cable jackets.

PVC — Abbreviation used to denote polyvinyl chloride. A type of plastic material used for cable jacketing. Typically used in flame-retardant cables.

PVDF — Abbreviation used to denote polyvinyl difluoride. A type of material used for cable jacketing. Often used in plenum-rated cables.

Photocurrent — The current that flows through a photosensitive device, such as a photodiode, as the result of exposure to radiant power.

Photodiode — A diode designed to produce photocurrent by absorbing light. Photodiodes are used for the detection of optical power and for the conversion of optical power into electrical power.

Pigtail — A short length of optical fiber for coupling optical components. It is usually permanently fixed to the components.

PIN Diode — A semiconductor device used to convert optical signals to electrical signals in a receiver.

PIN-FET Receiver — Optical receiver with a PIN photodiode and low noise amplifier with a high impedance input, whose first stage incorporates a Field-Effect Transistor (FET).

PIN Photodiode — A diode with a large intrinsic region sandwiched between p-doped and n-doped semiconducting regions. Photons in this region create electron hole pairs that are separated by an electric field, thus generating an electric current in the load circuit.

Preform — A glass structure from which an optical fiber waveguide may be drawn.

Prefusing — Fusing with a low current to clean the fiber end. Precedes fusion splicing.

Primary Coating — The plastic coating applied directly to the cladding surface of the fiber during manufacture to preserve the integrity of the surface.

Rayleigh Scattering — Scattering by refractive index fluctuations (inhomogeneities in material density or composition) that are small with respect to wavelength.

Receiver — A detector and electronic circuitry to change optical signals into electrical signals.

Receiver Sensitivity — The optical power required by a receiver for low error signal transmission. In the case of digital signal transmission, the mean optical power is usually quoted in Watts or dBm (decibels referred to I milliwatt).

Reflection — The abrupt change in direction of a light beam at an interface between two dissimilar media so that the light beam returns into the media from which it originated.

Refraction — The bending of a beam of light at an interface between two dissimilar media or in a medium whose refractive index is a continuous function of position (graded index medium).

Refractive Index — The ratio of the velocity of light in vacuum to that in an optically dense medium.

Repeater — In a lightwave system, an optoelectronic device or module that receives an optical signal, converts it to electrical form, amplifies or reconstructs it, and retransmits it in optical form.

Riser — Pathways for indoor cables that pass between floors. It is normally a vertical shaft or space. Also a firecode rating for indoor cable.

Scattering — A property of glass that causes light to deflect from the fiber and contributes to optical attenuation.

Single-Mode Fiber — Optical fiber with a small core diameter (typically 9 µm) in which only a single-mode, the fundamental mode, is capable of propagation. This type of fiber is particularly suitable for wideband transmission over large distances, since its bandwidth is limited only by chromatic dispersion.

Spontaneous Emission — This occurs when there are too many electrons in the conduction band of a semiconductor. These electrons drop spontaneously into vacant locations in the valence band, a photon being emitted for each electron. The emitted light is incoherent.

Step Index Fiber — A fiber having a uniform refractive index within the core and a sharp decrease in refractive index at the core/cladding interface.

Stimulated Emission — This occurs when photons in a semiconductor stimulate available excess charge carriers to the emission of photons. The emitted light is identical in wavelength and phase with the incident coherent light.

Telecommunications Closet (TC) — An enclosed space for housing telecommunications equipment, cable terminations, and cross-connects. The closet is the recognized cross-connect between the backbone and horizontal cabling.

Threshold Current — The driving current above which the amplification of the lightwave in a laser diode becomes greater than the optical losses, so that stimulated emission commences. The threshold current is strongly temperature dependent.

Tight-Buffered Cable — Type of cable construction whereby each glass fiber is tightly buffered by a protective thermoplastic coating to a diameter of 900 micrometers. Increased buffering provides ease of handling and connectorization.

Total Internal Reflection — The total reflection that occurs when light strikes an interface at angles of incidence greater than the critical angle.

Transmitter — A driver and a source used to change electrical signals into optical signals.

Wavelength Division Multiplexing (WDM) — Simultaneous transmission of several signals in an optical waveguide at differing wavelengths.

Zero-Dispersion Wavelength — Wavelength at which the chromatic dispersion of an optical fiber is zero. Occurs when waveguide dispersion cancels out material dispersion.

Additional Reading

Adrian, P.: "Technical Advances in Fiber-Optic Sensors: Theory and Applications," *Sensors* 23 (September 1991).

Agrawal, G.: "Fiber-Optic Communication Systems," John Wiley & Sons, Inc., New York, NY, 1997.

Amato, I.: "The Natural Roots of Fiber Optics," *Science News* 414 (December 23–30, 1989).

Augé, J., et al.: "Progress in Optical Amplification," *Microwave J.* 62 (June 1993).

Baumbick, R.J. and J. Alexander: "Fiber Optics Sense Process Variables," *Control Eng.* **27**, 3, 75–77 (1980).

Bobb, L.C. and P.M. Shankar: "Tapered Optical Fiber Components and Sensors," *Microwave J.* 219 (May 1992).

Corcoran, E.: "Light Talk: U.S. and Japanese Compete to Put Optical Fibers in the Home," *Sci. Amer.* 74 (October 1989).

Corcoran, E.: "Light Traffic: Optical Amplifiers Promise to Unclog Lightwave Communication," *Sci. Amer.* 106 (March 1991).

Corcoran, E.: "Avoiding the Potholes on Optical Highways," *Sci. Amer.* 143 (April 1992).

Desurvire, E.: "Lightwave Communications: The Fifth Generation," *Sci. Amer.* 114 (January 1992).

Dutton, H.: "Understanding Optical Communications," Prentice-Hall, Inc., Upper Saddle River, NJ, 1999.

Furse, C. and R. Haupt: "Down to the Wire," *IEEE Spectrum* (February 2001).

Gabel, D.: "Fiber Optics on the Rise," *Electronic Buyers' News* 36 (January 28, 1991).

Grimes, G.: "Microwave Fiber-Optic Delay Lines: Coming of Age in 1992," *Microwave J.* 61 (August 1992).

Hamilton, K.J.: "Fiber Optic Sensors Grow Into Networks," *InTech* 20 (February 1991).

Hecht, J.: "City of Light," Oxford University Press, New York, NY, 1999.

Henkel, S.: "Single Optical Fiber Does It All for Smart Transmitters," *Sensors* 8 (January 1992).

Holden, C.: "Plugging Into the Pacific Ocean," *Science* 599 (August 11, 1989).

Horgan, J.: "Dark Solutions: Physicists Generate Durable Pulses of Darkness," *Sci. Amer.* 24 (May 1988).

Ito, T., K. Fukuchi, K. Sekiya, D. Ogasahara, R. Ohhira and T. Ono: "6.4 Tb/s (160 × 40 Gb/s) WDM Transmission Experiment with 0.8 bits/Hz Spectral Efficiency," European Conference on Optical Communication," post-deadline paper, September 2000.

Jones, W.B. Jr.: "Introduction to Optical Fiber Communication Systems," Oxford University Press, Inc., New York, NY, 1995.

Kazovsky, L., et al.: "Optical Fiber Communication Systems," Artech House, Inc., Norwood, MA, 1996.

Kogelnik, H.: "High-Speed Lightwave Transmission in Optical Fibers," *Science* **228**, 1043–1048 (1985).

Ledwith, A.: "Glasses for Fibre Optic Communications," *Review (University of Wales)* 15 (Spring 1988).

Mandoli, D.F. and W.R. Briggs: "Fiber Optics in Plants," *Sci. Amer.* 90–98 (August 1984).

McHugh, P.: "Fiber Optics Extend the Reach of Photoelectric Sensors," *Instruments & Control Systems* 57 (August 1989).

Nicholson, P.J.: "An Introduction to Fiber Optics," *Microwave J.* 26 (June 1991).

Nicholson, P.J.: "An Overview of the Synchronous Optical Network," *Microwave J.* 24 (December 1991).

Nielsen, T.N., et al.: "3.28 Tb/s (82 × 40 Gb/s) Transmission Over 3 × 100 km of Nonzero-dispersion Fiber Using Dual C- and L-band Hybrid Raman/Erbium-doped Inline Amplifiers," Optical Fiber Communication Conference, post-deadline paper 29, March 2000.

Papannareddy, R.: "Introduction to Lightwave Communication Systems," Artech House, Inc., Norwood, MA, 1997.

Peterson, J.I. and G.G. Vurek: "Fiber-Optic Sensors for Biomedical Applications," *Science* **224** 123–127 (1984).

Pratsinis, S.E. and S.V.R. Mastrangelo: "Material Synthesis In Aerosol Reactors (Optical Fiber Manufacture)," *Chem. Eng. Progress* 65 (May 1989).

Raybon, G., et al.: "320 Gbit/s Single-channel Pseudo-linear Transmission over 200 km of Non-zero Dispersion Fiber," Optical Fiber Communication Conference, post-deadline paper 29, March 2000.

Refi, J., "Fiber Optic Cable — a LightGuide," abc TeleTraining, Inc., Geneva, IL, 1991.

Stix, G.: "Light Flight: Optical Fibers May Be the Nerves of New Aircraft," *Sci. Amer.* 120 (May 1991).

Suematsu, Y. and K.I. Iga: "Introduction to Optical Fiber Communications," John Wiley & Sons, Inc., New York, NY, 1982.

Tsang, W.T., N.A. Olsson and R.A. Logan: "High-Speed Direct Single-Frequency Modulation with Large Tuning Rate and Frequency Excursion in Cleaved-Coupled-Cavity Semiconductor Lasers," *Applied Physics Letters* **42**(8), 650–652 (April 15, 1983).

Woracek, D.: "Fiber Optic Sensors Endure Microwaves," *InTech* 24 (February 1991).

Yazbak, G.: "Fiberoptic Sensors Solve Measurement Problems," *Food Technology* 76 (July 1991).

Web References

CoreTek Inc: *http://www.coretekinc.com/*
General Cable: *http://www.generalcable.com/*
Lucent Technologies: *http://www.lucent.com/ofs/*
Nanoptics, Inc: *http://www.nanoptics.com/*

Lucent Technologies, Optical Fiber Solutions, Norcross, GA.

OPTICAL GLASS. Glass to be useful for lenses, prisms and other optical parts through which light passes, as distinguished for mirrors, must be completely homogeneous. This includes freedom from bubbles, striae, seeds, strains, etc. In order to reduce aberrations, the optical designer needs many different kinds of glass. A few typical types are described in Table 1. The v-number is the reciprocal of the dispersive power of the glass.

TABLE 1. VARIOUS GLASSES

	Type	n_D	v-Number
Borosilicate	Crown	1.5170	64.5
Barium	Crown	1.5411	59.5
Spectacle	Crown	1.5230	58.4
Light	Flint	1.5880	53.4
Ordinary	Flint	1.6170	38.5
Dense	Flint	1.6660	32.4
Extra dense	Flint	1.7200	29.3

OPTICAL IMAGES (Graphical Construction). The image of a object point may be located to first order accuracy by drawing any two of three easily located lines.

Given a lens L, its optical axis $x - y$, its foci F_1 and F_2 and an object point O (Fig. 1.):

1. Draw a line OA parallel to $x - y$ and then the line AF_2.
2. Draw a line OF_1, extend it to the lens at B, and then extend it from B parallel to the optical axis.
3. Draw the line OC, and continue it without deviation through the lens.

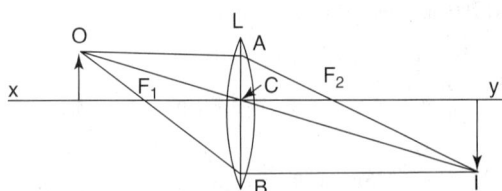

Fig. 1.   Principal pathways of simple lens system.

The three lines should meet at the point I, the image of O. If, instead of converging to a point, the three lines are diverging, trace each of them back, and they should meet at a point to the left of the lens indicating a virtual image. See also **Geometrical Optics**. If the three lines are parallel the image is at infinity. This same method of construction may be applied to any lens or curved mirror.

These are the three easily located lines. If it becomes desirable to trace some other ray (tracing a single ray through more than one lens) the following construction holds. (Fig. 2.)

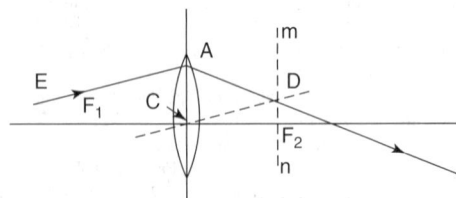

Fig. 2.   Method for tracing specific rays through lens system.

EA is the ray to be traced through the lens. Draw mm through F_2 perpendicular to the optical axis. Draw CD parallel to EA. The ray will follow the path AD after leaving the lens.

OPTICAL ISOMERISM.   See Isomerism.

OPTICAL LEVER.   A common device for amplifying and measuring small rotations. The object rotated carries a small mirror, which, reflecting a beam of light, deflects it through twice the angle of rotation to be measured. Light from a lamp is thus reflected as a bright spot moving along a scale, or the image of a fixed scale is viewed in the mirror by means of a reading telescope. The most common applications are to galvanometers, electrometers, etc., (see also **Galvanometer**; and **Electrometer**) using a torsion suspension; but the principle is also often adapted to devices such as that used for measuring Young's modulus, and in situations where a micrometer might otherwise be used.

OPTICALLY EFFECTIVE ATMOSPHERE.   That portion of the atmosphere lying below the altitude from which scattered light at twilight still reaches the observer with sufficient intensity to be discerned. Also called *effective atmosphere*. The top of this region lies between 50 and 60 kilometers (31 to 37 miles).

OPTICAL MODE.   A type of thermal vibration of a crystal lattice whose frequency is nearly independent of wave number. The optical modes may be thought of as internal vibrations of the molecules or unit cells of the lattice, loosely coupled from cell to cell. In ionic crystals, this leads to strong absorption in the infrared because of the fluctuating dipole moment as the ions of opposite sign move relative to one another. The optical modes contribute to the specific heat.

OPTICAL PATH.   In a medium of refractive index n, the product of the geometrical distance d and the refractive index. When there are several segments $d_1, d_2 \ldots$ of the light path in substances having different indices $n_1, n_2, \ldots$, the optical path is found from the relationship:

$$\text{optical path} = n_1 d_1 + n_2 d_2 + \cdots = \sum_i n_i d_i$$

and in a medium in which n varies continuously:

$$\text{optical path} = \int n \, ds$$

where ds is an element of length along the path. According to the Fermat principle, the optical path connecting two points has an extreme value.

OPTICAL PUMPING.   The process of "pumping" atoms from one hyperfine quantum state to another by a process of resonant fluorescent scattering of light. The original purpose was to facilitate the detection and measurement of the radio-frequency fine and hyperfine structure, which otherwise is observable by magnetic resonance and atomic beam methods, yet which cannot be directly observed by optical spectroscopy, primarily because of the Doppler width of spectral lines. Optical pumping also embraces any experimental work in which fine structure, radio-frequency spectroscopic measurements, polarization of electrons or nuclei, atomic cross sections, and oscillator strengths are measured or produced by means of polarized, filtered, or modulated light, and perhaps detected by the light as well. The distinction between measurements made in this way and those of conventional spectroscopy arises from the fact that the wavelength of the light is not used to measure energy splittings.

Several interesting light modulation effects may be observed in pumping experiments. Light beat and modulation effects occur when an atom scatters light while it simultaneously is undergoing hyperfine transitions. If a sample is oriented in the Z direction, and a radio-frequency field excites a Zeeman resonance, the atoms will precess coherently. An additional polarized light beam passing through the sample will be modulated at the Larmor frequency. As an inverse effect, if the pumping light itself is modulated at the appropriate radio frequencies, transitions will be induced.

Atomic gyroscopes may be constructed by using the optically pumped angular momentum and the light to detect precession. An important application also has been that of producing population inversion required in the operation of masers and lasers, particularly potassium and rubidium vapor, and chromium in ruby.

See also **Laser**.

OPTICAL PYROMETER.   A device for measuring the temperature of an incandescent radiating body by comparing its brightness for a selected wavelength interval within the visible spectrum with that of a standard source; a monochromatic radiation pyrometer. Temperatures measured by optical pyrometers are known as brightness temperatures and except for black bodies are less than the true temperatures.

OPTICAL TURBULENCE.   Irregular and fluctuating gradients of optical refractive index in the atmosphere. Optical turbulence is caused mainly by mixing of air of different temperatures, and particularly by thermal gradients which are sufficient to reverse the normal decrease in density with altitude, so that convection occurs.

OPTIC NEURITIS.   An inflammation of the optic nerve, which is the bundle of nerve fibers that starts at the back of the eye and carries light impulses from the retina to the brain. The retina receives light signals like the film in a camera, and the optic nerve transmits these signals to the brain where they are processed and turned into images, giving us the ability to see. If some or all of these nerve fibers become inflamed, the optic nerve becomes swollen and the fibers do not work properly, causing blurry vision. Depending on the number of inflamed nerve fibers, vision can range from near normal to extremely poor.

Optic neuritis is a relatively rare condition that can affect both adults and children. In adults, the condition usually affects one eye; in children it often affects both eyes at the same time. The most common age for developing the condition is in the 30s, but it can affect people of all ages.

Optic neuritis is usually associated with other medical problems, particularly viral infections and multiple sclerosis (MS). The disorder is often the first symptom of MS, and about 40% of people who experience optic neuritis eventually develop multiple sclerosis. The condition can also develop from an abuse of tobacco or alcohol, or from exposure to toxic substances such as lead or wood alcohol. Often the actual cause of optic neuritis cannot be determined.

Symptoms of optic neuritis usually come on suddenly and may include blurry or dim vision (as though the lights are being turned down) and a fading of colors. There can also be pain in the eye socket, impaired depth perception, and blind spots in the field of vision. Loss of vision usually occurs over a period of 2 to 5 days and generally improves within 4 to 12 weeks.

The condition can be difficult to diagnose because the symptoms are similar to those caused by several other eye problems. One of the first steps in diagnosing optic neuritis is an examination of the optic nerve with an instrument called an ophthalmoscope. The optic nerve enters the back of the eye and is visible to the doctor as a small disc. Any swelling of this inside part of the nerve can be detected, but if the swelling occurs behind the eye, it is not visible and other diagnostic procedures are necessary. These procedures may include tests such as an ultrasound, CT scans, or visual brain wave recordings. Other standard tests used in diagnosing optic neuritis are color vision, peripheral vision, and the reaction of the pupil to light.

Unfortunately, there is no consistently reliable treatment for optic neuritis. Steroids such as cortisone and prednisone are sometimes prescribed, but their effectiveness in treating the problem is questionable. Most people who develop the condition, however, recover normal vision without treatment. Recovery may take from a few weeks to a few months, depending on the severity of the condition.

Vision Rx, Inc., Elmsford, NY.

OPTICS. Originally that branch of physical science which treats of the phenomena of light and of vision. Today, because of the constantly increasing importance of ultraviolet and infrared radiation, optics has come to include all phenomena associated in any way with electromagnetic waves with wavelengths greater than x-rays and shorter than microwaves. Numerical limits of this wavelength region are not definitely defined.

Since the advent of devices such as the electron microscope and the cathode ray tube, in which beams of particles are focused to form images, the study of the behavior of such instruments is also called optics, usually with an appropriate modifier.

Scores of articles in this encyclopedia are devoted to specific aspects of the optical- and light-related sciences. Check the alphabetical index for subjects related to lenses and lens systems, light, mirrors and reflectors, optical fiber technology, optical instruments and materials, photography, prisms, and vision and the eye. See also **Geometrical Optics**; **Optical Fiber Systems**; **Photography and Imagery**; **Telephony**; and **Thin Films**.

OPTICS (Fiber). See **Optical Fiber Systems**.

OPTIMUM MAGNIFICATION. The maximum value of the numerical aperture of a dry lens is 1.0. Oil-immersion objectives with numerical apertures up to 1.65 have been constructed. The minimum distance between points which are just resolved is thus 2.7×10^{-5} centimeters for a dry lens, and 1.6×10^{-5} centimeters for an oil-immersion lens, using oblique illumination and a wavelength of 5,500 Å. The maximum useful magnification is thus about 800 for dry objectives, and 1,200 for oil-immersion objectives.

OPTOMETRY. A branch of optics dealing with the optical performance of the individual eye, and with measurements upon it. See also **Vision and the Eye**.

ORAL CAVITY. The cavity usually called the mouth. It is formed in the vertebrates of an embryonic depression, the stomodaeum, which forms in the ectoderm of the under side of the head and unites with the embryonic gut just behind its anterior end. The depression is deepened by the growth of processes from the body wall at the level of the pharynx, which forms the upper and lower jaws. Later the olfactory pits break through to join it and in this stage, which persists in the amphibians, the cavity is common to the respiratory and digestive systems. In a more advanced stage shelf-like partitions grow out from the lateral walls and join to form the palate, which divides the cavity into respiratory and oral portions, as in humans. Also called the buccal cavity.

ORANGE TREE. See **Citrus Trees**.

ORANGUTAN. See **Anthropoids**.

ORBIS. See **Antelope**.

ORBIT (Astronomy). The path that a celestial object follows in its motions through space, relative to some selected point, is known as the orbit of the object. The solution of the two-body problem indicates that, in the case of two objects moving under the influence of their mutual gravitational attractions, the relative orbit of one to the other will be a conic section. The character of the conic will depend upon initial conditions. In the case of members of the solar system, because of the very large mass, relatively, of the sun in comparison with any of the other members, the orbits of the members may be conveniently represented as ellipses with the sun at one focus. In the case of the satellites of the various members, the orbits of the satellites may be represented as ellipses with the primary object at one focus. Any deviations from the two-body problem may be treated as departures or perturbations from the simple conic.

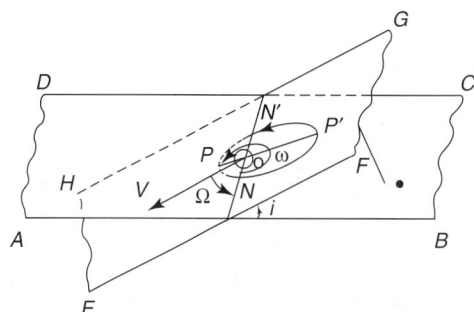

Fig. 1.   Diagram of orbit of member of solar system.

To define completely the orbit of an object at any particular instant, and to permit the determination of the position of the object in this orbit at any future time, six quantities must be calculated, known as the elements of the orbit. In Fig. 1, we have a diagram of a planetary orbit about the sun. The plane $ABCD$ represents the plane of the ecliptic, and the plane $EFGH$ represents the plane containing the orbit. These two planes must intersect in a line, NN', which must pass through the sun, since both the earth (the plane of the ecliptic being the plane of the earth's orbit) and the planet are revolving about the sun. The elliptical orbit of the planet is represented by $PNP'N'$, with the direction of motion of the planet as indicated, PP' representing the major axis of the ellipse, and P the point where the object is closest to the sun (the perihelion point). The line SV represents, in the plane of the ecliptic, the direction to the vernal equinox. We will assume that we are looking down on the plane of the ecliptic from the direction of its north pole. The points N and N', where the planet is on the ecliptic, are known as the nodes; the point N, where the planet is passing from south to north of the ecliptic, is the ascending node; and the point N' is the descending node. To define the orbit plane, the orbital element, i, the inclination, gives the angle between the orbit plane and the plane of the ecliptic; while the element, Ω (the angle VSN), gives the celestial longitude of the ascending node. The size and shape of the orbit in its plane are given by the orbital elements a, semimajor axis (usually expressed in astronomical units) and e, the eccentricity of the conic. To locate the position of the conic in the orbit plane, we use either ω (the angle NSP measured in the direction of motion of the planet), or, more commonly, $\overline{\omega}$, the longitude of perihelion, which is the sum of the two angles, Ω and ω. It should be noted that Ω is measured in the plane of the ecliptic, whereas Ω is measured in the orbit plane, so that is not strictly a longitude. To locate the position of the object in its orbit, the element T, which is some epoch or date when the object is at perihelion, and the rate at which the object is moving in the orbit are necessary. Occasionally, we find the rate of motion of the object given as an orbital element, but it is strictly not an independent element, since it may be derived from a. Listing the six elements as defined above, we have i, Ω, a, e, ω, and T. Unfortunately, several different systems of symbolic notation have been used by different authors, and great care must be exercised in interpreting any symbolic description of an orbit.

In describing the orbits of satellites, the plane frequently used is the plane of the planet's orbit about the sun, and the inclination, i, of the satellite orbit is referred to it. Also, instead of locating the position of the orbit in the plane by perihelion, the point where the satellite is closest to the planet is used. In the case of satellites of Jupiter, the point is known as perijove. In the case of binary star orbits, i gives the inclination of the orbit plane to the plane perpendicular to the line of sight, and the point where the smaller star is closest to the primary is called periastion, for obvious reasons.

The problem of orbit computation is far too complex to be included in a work of this character. In general, three observations of an object, giving the spherical coordinates of the object and the times of observation, are sufficient to compute a set of preliminary elements. From these preliminary elements, subsequent positions of the object may be computed and compared with observed positions. From the differences between observed and computed positions, the preliminary elements are corrected until a satisfactory representation of all observations is obtained. Before a definitive orbit can be obtained (i.e., an orbit that will give accurate positions for a long period of time), the perturbations due to other objects must be computed.

See also **Kepler's Laws of Planetary Motion**; and **Perturbation**.

ORCHARD GRASS.   See **Grasses**.

ORCHID (*Orchidaceae*).   The Orchids, which form the second largest family of plants, are the highest development of the Monocotyledons, just as the Composite Family marks the highest point of evolution reached by the Dicotyledons. But between the two families a most striking contrast exists. The Composites are most beautifully formed to enjoy a more abundant life; the numerous small flowers are massed in compact heads, so pollination is almost certain of accomplishment. The reduction of the fruit to a single ovule and the presence of various barbs or scales or bristles, called the pappus, greatly favor the probability of successful continuance of the species. In the 9,000–10,000 species of Orchids, the individual flowers are usually very conspicuous objects often of bizarre form and rare beauty. But they depend entirely on insects for pollination and often exhibit elaborate modifications to insure successful insect pollination. The seeds are minute and borne in tremendous numbers, few of which germinate and grow to maturity. As a result, orchid plants are relatively rare, a fact which has contributed not a little to the zeal with which collectors have sought these plants.

Originally, orchids were known only from the reports brought back by travelers from the tropics, who spoke of the brilliant colors, the curious forms and the delightful fragrance, and also of the mystery and folklore that often attached to orchids. Later, botanists gained a wider knowledge of the family from dried specimens, in which color, fragrance, and to a considerable extent, form, were lacking. In time, however, living plants were obtained by collectors and carried to western lands. There they were grown usually without much success, since it was assumed that they could grow only in a very hot humid atmosphere. Only the most tolerant species stood this, and they only partially. Gradually, however, better understanding of the plants' requirements was obtained, and successful culture followed, so that during the first third of the nineteenth century orchids became popular. Necessarily, they are expensive to collect, for they are not easy to transport, and they can be grown only in glass houses under fairly uniform conditions of temperature and humidity.

Orchids are primarily plants of the tropical rain forests, where they grow in greatest abundance. In these forests, they are found mostly as epiphytes, plants growing on other plants. Such plants grow attached to the branches of large trees, or massed in a crotch of the tree, or even attached to the trunk. Often they occur high up on the topmost branches of lofty trees, where they are inaccessible except when the supporting tree falls, bringing all its attached plants to earth. Other species of orchids, including nearly all those found in extratropical regions, are terrestrial. All orchids, wherever they grow, are herbaceous perennials. A few species are saprophytes, lacking chlorophyll entirely, and obtaining their food by absorption from the soil of complex organic substances. See also **Epiphytes**.

In terrestrial orchids the roots are rather coarse and sparsely branching. In many species one of these roots becomes greatly swollen with food, as the growing season advances, forming a tuberous body which will be used to promote rapid growth in the following season. In epiphytic orchids

two kinds of roots are found. One of these grows tightly appressed to the supporting plant, is not affected by gravity but is negatively phototropic, growing into crevices in the supporting plant. The other roots are the aerial roots, coarse branching objects that hang down, often in conspicuous masses, from the base of the plant. The epidermis and the outer portion of the cortex, which is called the velamen, are composed of dead cells with perforated walls that readily absorb any water which may come to them and retain it tenaciously. The inner tissues of the cortex are green and capable of carrying on photosynthesis. In some orchids slender absorbing branches grow out from the base of the other roots and penetrate the mass of debris that frequently collects at the base of the plant.

The stems of terrestrial orchids are erect and leafy, and terminated by the inflorescence. In many epiphytic forms the leaves are dropped at the end of the growing season, the bare stem remaining. The internodes of such stems are often conspicuously swollen, forming pseudobulbs in which are stored water and food reserves. In other epiphytic species the leaves are fleshy and serve for storage.

The most characteristic and in many species the most conspicuous part of the plant is the flower. In all orchids the flowers are irregular but formed on a very uniform pattern. In all there is a six-parted perianth composed of two groups of three members each. One of the inner three differs from all the others and is designated the lip or labellum. This becomes variously fringed in some orchids, broadly expanded in others, and even saccate (slipper-like), as in the Lady's Slipper. It is commonly much more brilliantly colored than the other five parts and gives to the flower its showiness. It serves as a landing place for insects, and is often a definite factor in bringing about pollination. Often various outgrowths such as spurs add further complexity to the flower. The essential organs of the flower, the stamens and pistil, are united into a single body, the column, which is a characteristic feature of the orchid flower. Its structure varies somewhat in the different groups of the family. In orchids there are only one or in some species, two, anthers present. The greater number of species have a single anther, which has two lobes, each filled with a mass of pollen grains held together by fine elastic threads. Such a mass is called a pollinium. The threads all unit to form a slender stalk which ends in a sticky mass resulting from the breaking down of certain cells in the rostellum. The latter is a special organ, which represents one of the stigmas of the flower.

An insect comes to the orchid flower seeking the nectar it contains. In getting this nectar the insect comes in contact with the sticky mass of cells which become firmly cemented to some part of the body, often the eyes, or the antennae. When the insect leaves the flower, it drags the pollinium out. The latter may be in a position so that when the insect enters the next flower, the pollen comes directly in contact with the stigma and pollination is accomplished. In many cases, however, a striking change occurs. The stalk of the pollinium, due to changes in its water content, bends through an angle of 90°, and so brings the pollen mass into a position which will insure its reaching the surface of the stigma of the next flower visited by the insect. There are many, often complex, variations in this process of pollination in orchids, all on this general method. In the second group of orchids the pollen is not combined into masses.

The ovary of the orchid flower is interior; that is, all the floral organs are borne at the apex of the ovary, which contains an immense number of ovules attached to its walls. After fertilization, the ovules develop into very minute, light seeds, which are easily blown about by air currents.

Additional Reading

Allikas, G. and N. Nash: "Orchids," Advantage Publications, San Diego, CA, 2000.

Pridgeon, A. (Editor): "The Illustrated Encyclopedia of Orchids," Timber Press, Inc., Portland, OR, 1992.

White, J.: "Taylor's Guide to Orchids," Houghton Mifflin Company, New York, NY, 1996.

OR (Circuit).   A computer logical decision element, which has the characteristics of providing a binary 1 output if any of the input signals are in a binary 1 state. This is expressed in terms of Boolean algebra by $F = A + B$. A diode and a transistor representation of this circuit is shown in Fig. 1. In the diode-type OR circuit, if either (or both) input signal A or B is positive, the respective diode is forward-biased and the output F is positive. The number of allowable input signals to the diode Or gate is a function of the back resistance of the diodes. The input transistors of the transistor-type OR circuit are forced into higher conductivity when the

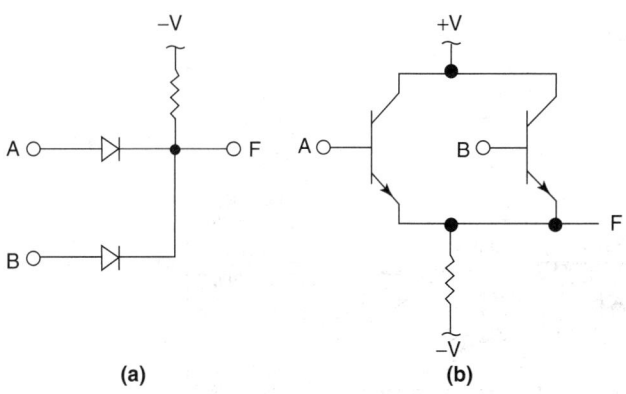

Fig. 1. OR circuits. (**a**) Diode or gate-type; (**b**) transistor-type.

respective input signal becomes positive. Thus, the output signal becomes positive when either or both inputs are positive.

See also **Exclusive OR Circuit**.

THOMAS J. HARRISON, IBM Corporation, Boca Raton, FL.

ORDER (Of). If a function $f(x)$ becomes $|x^k f(x)| < K$ as x approaches 0 or ∞, where K is a positive number independent of x and not zero, then $f(x)$ is said to be of order $1/x^k$. In symbols, $f(x) = \mathbf{O}(x^{-k})$, where this notation is called the *ordor-symbol*. The limiting process 0 or ∞ is not always stated explicitly but inferred from the context. In the special case where $\lim x^k f(x) = 0$, one writes $f(x) = \mathbf{o}(x^{-k})$. If $f(x)$ is bounded, one writes $f(x) = \mathbf{O}(1)$. In any case, $\mathbf{O}(x^k) \times \mathbf{O}(x^n) = \mathbf{O}(x^{k+n})$.

ORDER OF MAGNITUDE. For two functions u and v to be of the same order of magnitude near t_o means there are positive numbers ε, A, and B such that

$$A < \left| \frac{u(t)}{v(t)} \right| < B \text{ if } 0 < |t - t_0| < \varepsilon.$$

If this is true for $v(t)$ replaced by $[v(t)]^r$, then u is of the rth order with respect to v. If $\lim_{t \to t_0} u(t)/v(t) = 0$, u is of lower order than v and one writes $u = \mathbf{o}(v)$; if $\lim_{t \to t_0} |u(t)/v(t)| = \infty$, u is of higher order than v. Also, one says that u is of the order of v on a set S (usually a deleted neighborhood of some t_0) and writes $u = \mathbf{O}(v)$ if there is a positive number B such that $|u(t)/v(t)| < B$ if $t \varepsilon S$. The statements $u = \mathbf{o}(1)$ and $u \to 0$ are equivalent; $u = \mathbf{O}(1)$ on S is equivalent to "u is bounded on S." If $\lim_{t \to t_0} u(t)/v(t) = 1$, then u and v are asymptotically equal at t_0. These concepts also are used when $t_0 = \pm\infty$ with appropriate changes. For example, if $\{u_n\}$ and $\{v_n\}$ are sequences and there are numbers B and N such that

$$\left| \frac{u_n}{v_n} \right| < B \text{ if } n > N,$$

then u_n is of the order of v_n and $u_n = \mathbf{O}(v_n)$.

ORDINARY RAY. That magnetoionic wave component deviating the least, in most of its propagation characteristics, relative to those expected for a wave in the absence of the Earth's magnetic field. More exactly, if at fixed electron density, the direction of the Earth's magnetic field were rotated until its direction is transverse to the direction of phase propagation, the wave component whose propagation is then independent of the magnitude of the Earth's magnetic field. Also called *ordinary-wave component*. See **Magnetic Double Refraction**; and **Magnetoionic Theory**.

ORDOVICIAN. A period of the Paleozoic Era. Type locality, border of Wales and Shropshire, England.

The formations of this system were first studied and described by Charles Lapworth in 1879. The Ordovician period began 440 million years ago and lasted for 60 million years. In 1874 J.D. Dana had already proposed a post-Cambrian system, the Canadian. This system has only been recently recognized in Britain (Northwest Highlands of Scotland) where it is still considered as upper Cambrian by some British geologists (1937). In 1911, E.O. Ulrich proposed a new system (period) called the Ozarkian, to include some of the upper Cambrian and some of the lower Canadian. There is still considerable doubt as to the systemic importance of the Ozarkian,

especially as it appears to be absent in Great Britain. Ordovician formations are well exposed in North America due to the fact that a large part of the continent was submerged during this period. The marine sediments are of two principal types, limestones and shales, the former contain "shelly fossils" and the latter, principally graptolites. Thus the American Ordovician covers an enormous area and has an endless variety of local developments, for during the vast lapse of time included in the period, the shallow epeiric seas were continually shifting their positions, outlines and depths. The United States Ordovician contains no igneous rocks with the exception of volcanic ashes, called bentonites. The formations of this system are well exposed in Portugal, Switzerland, Bohemia, Austria, Hungary, Ireland, eastern Asia, China, Manchuria, Siberia, the Himalayas, Burma, Morocco, Australia, New Zealand, northern Argentina, Bolivia, and eastern Peru. The maximum thickness of Ordovician strata, 40,000 feet, occurs in Australia. All classes of marine invertebrates occur as fossils, a number of which classes are now extinct. The principal types are graptolites, corals, echinoderms, bryozoans, pelecypods, nautiloids, trilolites, and ostracods. (See also **Invertebrate Paleontology**.) The strata of eastern North America are an important source of petroleum. Because of the wide surficial distribution of the Ordovician limestones they are much quarried for foundation structures, road metal, flux for the reduction of iron ores, and especially for the lime used in mortar, whitewash, fertilizers and Portland cement. In Vermont and Tennessee occur valuable deposits of marble. Lead and zinc ores in rocks of Ordovician age are mined in Iowa, southern Wisconsin and northern Illinois. Phosphates of the same age are mined in central Tennessee. In eastern North America the Ordovician closed with a period of mountain building named the Taconic Revolution by J.D. Dana, in 1895.

ORE BENEFICIATION. See **Iron**.

ORE BLOCK. A section of an orebody, usually rectangular, that is used for estimates of tonnage and quality of the orebody.

OREBODY. A continuous, well-defined mass of material of sufficient ore content to make extraction economically feasible.

OREGANO. A savory herb which has become quite popular in the past several years. It was introduced into the United States and England from the Mediterranean area. Probably the original herb is a member of the Mint family (see also **Mint Family**), *Origanum vulgare*, and most of the oregano imported from Europe comes from this plant. Mexican oregano, of indistinguishable flavor and aroma, however, comes from species of *Lippia* of the family *Verbenaceae*. Thus the name oregano is a general term applying to a particular herb flavor, rather than to a particular species of plant.

ORES. Mineral aggregates in which the valuable metalliferous minerals are sufficiently abundant to make the aggregates worth mining. Types and origins of ore deposits are illustrated in Fig. 1. See also **Copper**; and **Iron**.

ORGAN. A multicellular structure made up of various tissues for the performance of some complex function. The stomach, for example, contains in its walls tissues of all of the five principal divisions. Its function is the digestion of food and this end is gained by the cooperative exercise of the simpler functions of all of the tissues composing it.

ORGANIC CHEMISTRY. The term *organic*, which means *pertaining to plant or animal organisms*, was introduced into chemical terminology as a convenient classification of substances derived from plant or animal sources. Early it was believed that organic compounds could arise only through the operation of a vital force inherent in the living cell. However, Wöhler discovered in 1828 that the organic compound urea, identified in urine by Rouelle in 1773, could be produced by heating the inorganic salt ammonium cyanate:

$$\text{NH}_4{}^+\text{OCN}^- \longrightarrow \text{H}_2\text{NCONH}_2$$
$$\text{Ammonium cyanate} \qquad\qquad \text{Urea}$$

Subsequently, the association of organic compounds only with living organisms was discontinued.

The term *organic* has persisted, but the modern definition of *organic chemistry* has changed to mean the *chemistry of carbon compounds*.

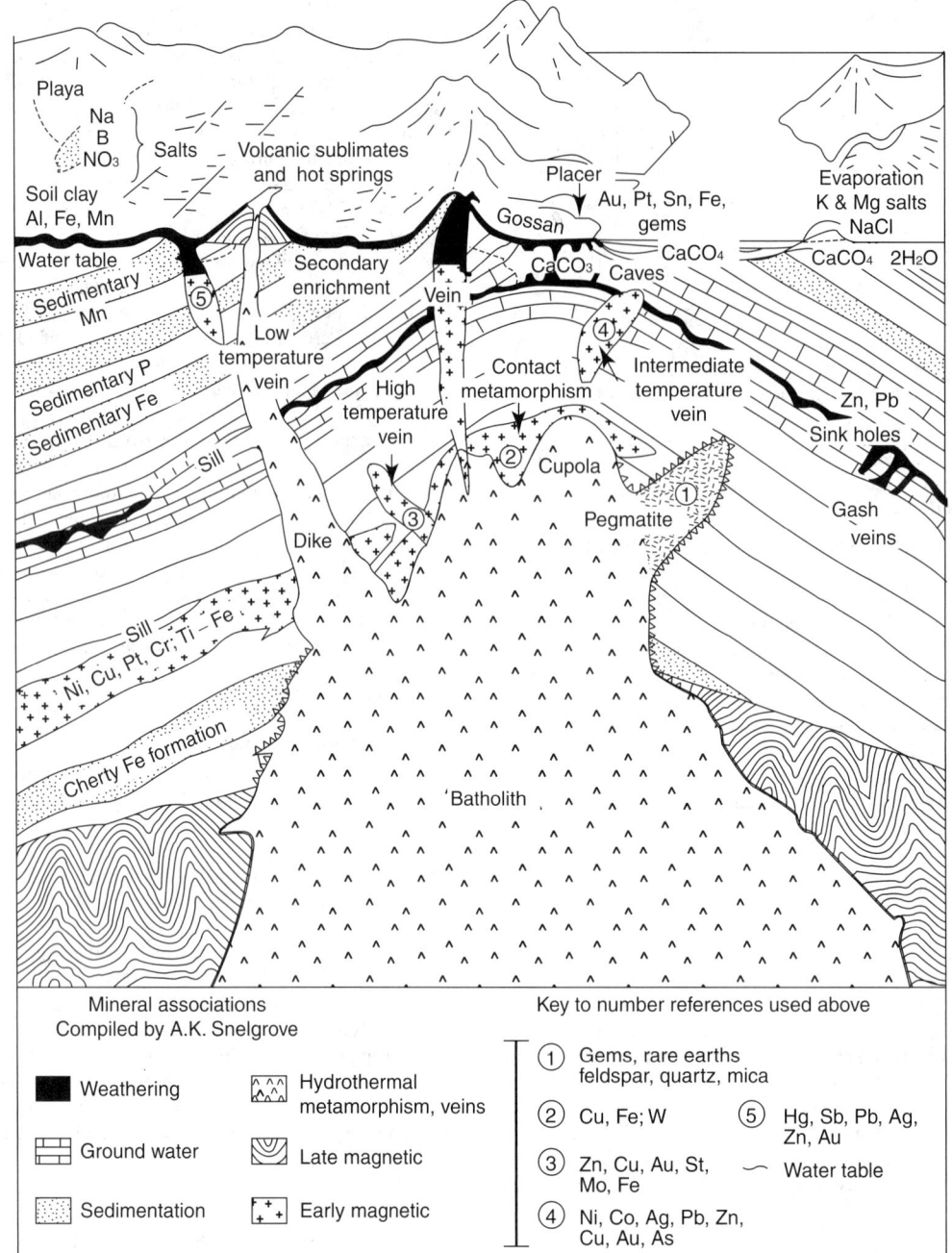

Fig. 1.   Diagrammatic illustration of the origin of ore deposits. (*Field.*)

Sometimes a few carbon compounds are excluded from this category, such as carbon dioxide, CO_2; metal carbonates, e.g., Na_2CO_3; carbonyls, e.g., $Ni(CO)_4$; cyanides, e.g., KCN; carbides, e.g., CaC_2; and a few others, but this exclusion is somewhat arbitrary. The designation *organic* is still pertinent because the chemistry of carbon compounds is more important to everyday life than that if any other element.

The uniqueness of carbon stems from its ability to form strong carbon-carbon bonds that remain strong when the carbon atoms are simultaneously bonded to other elements. Whereas both the carbon-hydrogen and carbon-fluorine compounds CH_3CH_3 and CF_3CF_3 are highly stable and relatively unreactive, the corresponding compounds in which the carbon atoms are replaced by boron, silicon, phosphorus, and others either are thermodynamically unstable or highly reactive.

Theoretically, an infinite number of different carbon compounds can exist. Carbon atoms alone or in combination with other atoms, such as oxygen, nitrogen, etc., can join to form linear, branched, and cyclic chains of nearly any length. One, two, or three bonds may be shared between two carbon atoms. In most stable organic compounds, the total number of bonds to each carbon atom is four.

Classification of Organic Compounds

A subject with the wide scope that characterizes organic chemistry requires a logical approach to its organization so that knowledge can be gathered and applied manageably. Molecular structure has become the key method for classifying this subject. Scientists use two-dimensional diagrams or three-dimensional models to depict molecular structures. Although these analogies are sometimes crude representations of actual molecules, they are useful for communicating information about the molecules. Structural diagrams characteristic of some basic types of organic molecules are shown in Table 1. A more detailed discussion on naming organic compounds is given later.

The compounds shown in Table 1 contain only carbon and hydrogen and are called hydrocarbons. Most organic compounds that contain other kinds of atoms in addition to carbon and hydrogen are considered as formally derived from hydrocarbons in which a hydrogen atom has been replaced by another atom or collection of atoms. However, these derivatives usually are not formed directly from the hydrocarbons. The other atoms usually are referred to as functional groups. A group of compounds having the same functional group is referred to as a

TABLE 1. EXAMPLES OF STRUCTURAL DIAGRAMS OF HYDROCARBONS

Type of Compound	Formula	Name
Linear alkane	H–C–C–C–C–C–C–H (each C with H's) or $CH_3CH_2CH_2CH_2CH_2CH_3$ or $CH_3(CH_2)_4CH_3$ or $n\text{-}C_6H_{14}$ or (line structure)	*n*-Hexane
Branched alkane	CH_3 / $CH_3CH_2CH_2CH_2CH_3$ or (line structure)	2-Methylpentane
Monocyclic alkane	H_2C, CH_2, H_2C, CH_2, CH_2, CH_2 (ring) or (hexagon)	Cyclohexane
Bicyclic alkane	(structure)	Bicyclo[2.2.1] heptane or Norbornane
Polycyclic alkane	(structure)	Pentacyclo $[5.3.0.0.^{2,5} - 0^{3,9} \cdot 0^{4.8}]$decane or 1,3-Bishomocubane
Alkene	$CH_3C{=}CCH_2CH_3$ (with H's) or (line structure)	*trans-* 2-Pentene
Alkyne	$CH_3C{\equiv}CCH_2CH_3$ or (line structure)	2-Pentyne
Aromatic	(benzene structure with C's and H's) or (hexagon with circle)	Benzene
Polymer	$X(CH_2)_n Y$ ($n \ge 1000$, X and Y vary according to how polymer was prepared)	Polymethylene or Polyethylene

TABLE 2. EXAMPLES OF SOME MAJOR FAMILIES OF ORGANIC COMPOUNDS

Family	General Structure	Family	General Structure
Alcohol, phenol	ROH	Isocyanate	$RN{=}C{=}O$
Ether	ROR′	Thiol	RSH
Aldehyde	$\overset{O}{\overset{\|}{RCH}}$	Sulfide	RSR′
Ketone	$\overset{O}{\overset{\|}{RCR'}}$	Sulfoxide	$\overset{O}{\overset{\|}{RSR'}}$
Carboxylic acid	$\overset{O}{\overset{\|}{RCOH}}$	Sulfone	$\overset{O}{\overset{\|}{\underset{\underset{O}{\|}}{RSR'}}}$
Ester	$\overset{O}{\overset{\|}{RCOR'}}$	Sulfonic acid	$\overset{O}{\overset{\|}{\underset{\underset{O}{\|}}{RSOH}}}$
Amine	RNH_2	Chloride	RCl
Amide	$\overset{O}{\overset{\|}{RCNH_2}}$	Bromide	RBr
Nitrile	$RC{\equiv}N$	Iodide	RI
Isonitrile	$R\overset{+}{N}{\equiv}\overset{-}{C}$	Organolithium	RLi
Nitro compound	$R\overset{+}{N}\big({\overset{O}{\diagdown}}\big)_{O^-}$	Heterocycle	(R Y ring)
Nitroso compound	$RN{=}O$		(Y is an atom other than carbon such as N, O, Si, P, S, etc.: Y is bonded to R at two or more positions.) Examples:
Imine	$\overset{NH}{\overset{\|}{RCR'}}$		
Azo compound	$RN{=}NR'$		(pyridine structure), pyridine;
Diazo compound	$RR'C{=}\overset{+}{N}{=}\overset{-}{N}$		
Diazonium salt	$RN{\equiv}N^+X^-$ (anion)		(thiophene structure), thiophene.

Although these classifications are important for education and documentation in organic chemistry, research usually is rather specialized.

It is often oriented in a practical way, not to the structure of compounds or to the kinds of atoms they contain, but to the manner in which the compounds are used. A partial list of such uses includes plastics, pharmaceuticals, insecticides, fungicides, herbicides, paints, petroleum, fuels, dyes, photography, and adhesives. See Fig. 1.

An important area of organic chemistry is that which deals with life and living substances. Organized under the title biochemistry, this is a subfield of organic chemistry, since most of the compounds involved contain carbon. Numerous advances have been made recently in biochemistry, and of all the areas of organic chemistry it probably will produce the greatest progress in the next decade. Noteworthy advances can be expected in the areas of biochemistry relating to medicine and human health. Some of the categories along which the study of biochemistry is organized are proteins, peptides, amino acids, nucleoproteins, enzymes, nucleotides, carbohydrates, lipids, steroids, carotenoids, porphyrins, nucleic acids, vitamins, and hormones. These topics are described in detail elsewhere in this volume.

Theoretical organic chemistry is another field that has progressed rapidly in recent years. Chemists have derived molecular orbital symmetry rules

family. Table 2 shows a number of such families. The R and R′ in the formulas of Table 2 represent any hydrocarbon in which a hydrogen atom has been removed from the position to which the functional group is attached.

Molecules with common features also are grouped into more specialized families. The compounds in one such family, carbohydrates, contain only carbon, hydrogen, and oxygen. The hydrogen and oxygen atoms are in the same ratio as in water (2H:1O), hence the suffix *-hydrate*. Other specialized families include terpenes, alkaloids, steroids, lipids, proteins, enzymes, vitamins, and organometallic compounds.

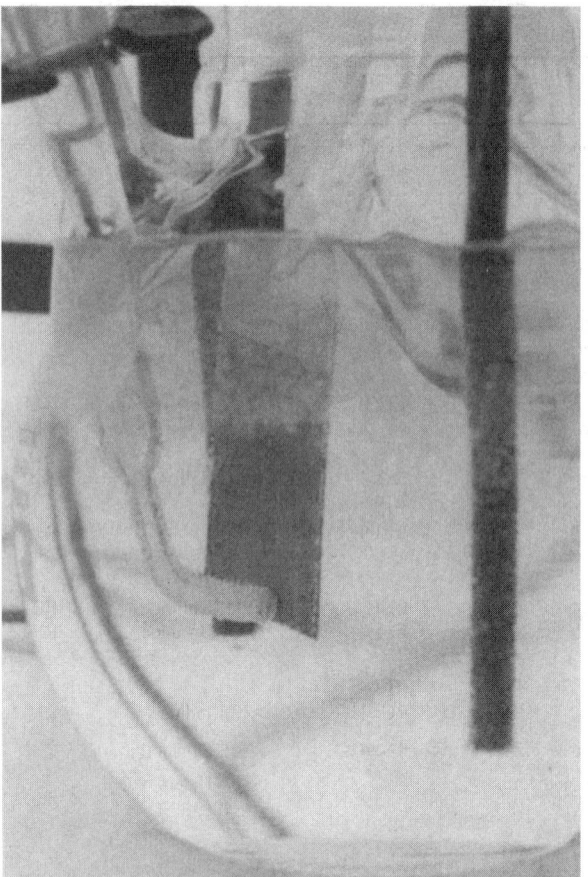

Fig. 1.   Apparatus for testing corrosion of metal coated with a plastic. (*The Dow Chemical Company.*)

that allow understanding and predicting the stereochemistry and relative rates of organic reactions in electronic ground and excited states. In a ground state molecule, all electrons are in their lowest energy levels, whereas, in an excited state molecule, at least one electron is in a higher energy level. For example, the Woodward-Hoffman orbital symmetry rules for concerted reactions predict that ground state (thermal) cycloaddition reactions involving $4n + 2$ (where n is an integer) π-electrons, and excited state (photochemical) cycloaddition reactions that involve $4n\pi$-electrons, may occur via a concerted process (Scheme 1).

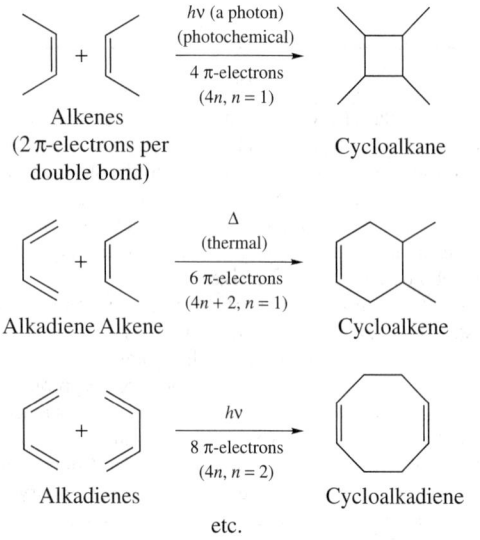

Alkenes
(2π-electrons per double bond)

$h\nu$ (a photon)
(photochemical)
4π-electrons
($4n, n = 1$)

Cycloalkane

Alkadiene   Alkene

Δ
(thermal)
6π-electrons
($4n + 2, n = 1$)

Cycloalkene

Alkadienes

$h\nu$
8π-electrons
($4n, n = 2$)

Cycloalkadiene

etc.

Scheme 1

A concerted process or reaction occurs without the involvement of an intermediate. The stereochemistry of the reactants is retained in the product,

and the reaction is usually more facile than a comparable nonconcerted reaction. The other combinations of ground state, excited state, $4n + 2$, $4n$ reactions cannot be concerted reactions.

Another area that has received increased attention is environmental organic chemistry. Reactions that organic compounds undergo when they are released to the environment are becoming as significant as the reactions by which the compounds are prepared or the reactions that take place in the use of the compounds. Some environmentally important types of reactions are hydrolysis, oxidation, sunlight-initiated photochemical decomposition, and biodegradation by microbes.

Only a limited discussion of the large field of organic chemistry can be given in the space allotted for this article. Consult alphabetical index for further information on specific topics. The reader is also directed to the references at the end of this article for examples of more detailed treatments of organic chemistry. Refs. Schmerling; Richey; and Morrison et al.represent texts that treat organic chemistry at elementary, intermediate, and advanced levels, respectively.

Nomenclature of Organic Compounds

With the foregoing review in mind, the reader will appreciate the difficulties associated with naming organic molecules. Originally, chemical names were indicative of the sources of compounds. For example, *catechol* was the name given to a compound isolated from the natural product *gum catechu*. Chemists have coined other nonsystematic names such as *cubane* or *basketene* to pictorially describe molecules. These nonsystematic names are called common or trivial names. The names *phenol, acetic acid*, and *styrene* are also nonsystematic but are widely understood in chemistry.

As the number of known organic molecules increased, a systematic approach to nomenclature was required. To minimize confusion in communicating chemical information, a name should be consistent with other systems in use and should clearly define the structure of a molecule. Specialists in organic chemistry have developed nomenclatures that are logical for their disciplines, thus devising systems for naming alcohols, antibiotics, carboxylic acids, etc.

Systematic nomenclature on a worldwide scale began in 1892 when a committee of the International Chemical Congress established a set of standards known as the Geneva Rules for naming organic compounds. The International Union of Pure and Applied Chemistry (IUPAC) was formed in 1919 and further developed this nomenclature system. In 1886 in the United States, the American Chemical Society (ACS) established a Committee on Nomenclature. The ACS and IUPAC have developed parallel rules for naming organic compounds.

Alternative rules within the latter system may allow assigning more than one unambiguous name to a compound. The controlled alphabetic listing in the *Chemical Substances Index* of Chemical Abstracts Service (CAS), a part of the ACS, requires that each compound have a unique name. This convention ensures that all information for a single compound such as $H_2N-CH_2-CH_2-OH$, which can be unambiguously named 2-aminoethanol, 2-aminoethyl alcohol, 2-hydroxyethylamine, etc., will appear in the index under one name: ethanol, 2-amino-. Because universal nomenclature systems are complex and sometimes inconvenient, some chemists have retained the older methods of naming molecules. These older names are used in the parts of this section that do not deal specifically with nomenclature.

The following paragraphs introduce some basic areas of organic chemical nomenclature. Further details can be found in *Chemical Substances Index Names* and *Nomenclature of Organe Chemistry* listed and end of this entry.

In addition to trivial names, some of the other categories of organic compound names are:

Generic name: one that indicates a class of compounds; e.g., *alkanes, esters.*
Parent name: a base from which other names are derived; e.g., *ethanol,* from *ethane; butanoic acid,* from *butane.*
Systematic name: a name composed of syllables defining the structure of a compound; e.g., *chlorobenzene, 2-methylhexane.*
Substituitive name: one describing replacement of hydrogen by a group or element; e.g., 1-*methylnaphthalene,* 2-*chloropropane.*
Replacement name: a name describing compounds that have carbon replaced by a hetero atom; also called "a" nomenclature; e.g., 2-*azaphenanthrene.*

Subtractive name: one that indicates removal of specified atoms; e.g., in the aliphatic series names ending in *-ene* or *-yne*, such as *ethene* or *ethyne*, and names involving *anhydro-, dehydro-, deoxy-, nor-,* etc.

Additive name: one that signifies addition between molecules and/or atoms without replacement of atoms; e.g., *styrene oxide.*

Conjunctive name: a combination of two names, one of which represents a cyclic structure and the other an acyclic chain, with one hydrogen atom removed from each; e.g., *benzenemethanol.*

Fusion name: a combination that results from linking with an "o" two names of cyclic systems fused by two or more common atoms; e.g., *benzofuran.*

Multiplicative name: nomenclature describing the symmetrical repetition of radicals about a central unit; e.g., *2,2′-oxybis* (*ethanol*).

Hantzsch-Widman name: a name devised by Hantzsch and Widman for describing heterocyclic systems, in which the prefix denotes a hetero atom(s) and the suffix denotes the ring size and degree of saturation; e.g., *oxirene, aziridine.*

Von Baeyer name: a name that describes alicyclic bridged systems; e.g., *bicyclo* [2.2.1]*heptane.*

The procedure for naming a compound involves some or all of the following steps, depending on the structure of the molecule under study: (1) the type of nomenclature to be used (conjunctive, multiplicative, etc.) is chosen; (2) the parent structure is named; (3) the prefixes, suffixes, and names of functional and substituent groups that were not included in (2) are attached; (4) the numbering is completed.

Hydrocarbons

Acyclic Hydrocarbons. A knowledge of the structural features of hydrocarbon skeletons is basic to the understanding of organic chemical nomenclature. The generic name of saturated acyclic hydrocarbons, branched or unbranched, is *alkane*. The term *saturated* is applied to hydrocarbons containing no double or triple bonds.

The simplest saturated acyclic hydrocarbon is called methane. Names of the higher, straight-chain (*normal*) homologs of this series contain the termination "-ane," as shown in Table 3. The structures of the first four members of the series in Table 3 are: CH_4, CH_3-CH_3, $CH_3-CH_2-CH_3$, and $CH_3-CH_2-CH_3$. The structures of subsequent members of this series are formed by inserting additional CH_2 units.

TABLE 3. EXAMPLES OF SATURATED ACYCLIC HYDROCARBONS

Molecular Formula	Name	Molecular Formula	Name
CH_4	methane	$C_{18}H_{38}$	octadecane
C_2H_6	ethane	$C_{19}H_{40}$	nonadecane
C_3H_8	propane	$C_{20}H_{42}$	icosane
C_4H_{10}	butane	$C_{21}H_{44}$	henicosane
C_5H_{12}	pentane	$C_{22}H_{46}$	docosane
C_6H_{14}	hexane	$C_{23}H_{48}$	tricosane
C_7H_{16}	heptane	$C_{30}H_{62}$	triacontane
C_8H_{18}	octane	$C_{31}H_{64}$	hentriacontane
C_9H_{20}	nonane	$C_{32}H_{66}$	dotriacontane
$C_{10}H_{22}$	decane	$C_{40}H_{82}$	tetracontane
$C_{11}H_{24}$	undecane	$C_{50}H_{102}$	pentacontane
$C_{12}H_{26}$	dodecane	$C_{100}H_{202}$	hectane
$C_{13}H_{28}$	tridecane	$C_{101}H_{204}$	henhectane
$C_{14}H_{30}$	tetradecane	$C_{102}H_{206}$	dohectane
$C_{15}H_{32}$	pentadecane	$C_{110}H_{222}$	decahectane
$C_{16}H_{34}$	hexadecane	$C_{120}H_{242}$	icosahectane
$C_{17}H_{36}$	heptadecane	$C_{132}H_{266}$	dotriacontahectane
		$C_{200}H_{402}$	dictane

Univalent groups derived from the preceding acyclic hydrocarbons by removal of one hydrogen atom from a terminal carbon atom are named by replacing the ending "-ane" with "-yl."

EXAMPLES: ethyl CH_3-CH_2-
            butyl $CH_3-CH_2-CH_2-CH_2-$

A saturated branched acyclic hydrocarbon is named by numbering the longest chain from one end to the other, and the positions of the side chains are indicated by the lowest possible numbers. The numbers precede the group, and are separated from them by a hyphen.

EXAMPLES:

$$\overset{1}{CH_3}-\overset{2}{CH}-\overset{3}{CH_2}-\overset{4}{CH_2}-\overset{5}{CH_3}$$
$$|$$
$$CH_3$$

2-methylpentane (not 4-methylpentane)

$$\overset{6}{CH_3}-\overset{5}{CH}-\overset{4}{CH_2}-\overset{3}{CH}-\overset{2}{CH}-\overset{1}{CH_3}$$
$$|\qquad\qquad|\quad\ |$$
$$CH_3\qquad CH_3\ CH_3$$

2,3,5-trimethylhexane (not 2,4,5-trimethylhexane)

If two groups are attached to the same carbon atom, the number is repeated.

EXAMPLES:

$$CH_3$$
$$|$$
$$\overset{1}{CH_3}-\overset{2}{CH_2}-\overset{3}{C}-\overset{4}{CH_2}-\overset{5}{CH_3}$$
$$|$$
$$CH_3$$

3,3-dimethylpentane (not 3-dimethylpentane)

For some purposes, such as alphabetical listing of basic skeleton names, inverted word order is used.

EXAMPLES: pentane, 2-methyl-
             benzene, chloro-

The two propyl groups are distinguished by calling them *normal*-propyl or *n*-propyl, $CH_3-CH_2-CH_2-$, and isopropyl or i-propyl, $[(CH_3)_2-CH-]$, in common usage. The latter is called 1-methylethyl in systematic nomenclature. The butyl groups are named as follows:

Structure	*Systematic Name*	*Trivial name*
$CH_3-CH_2-CH_2-CH_2-$	butyl	*n*-butyl
$CH_3-CH-CH_2-$ (with CH_3 branch)	2-methylpropyl	*i*-butyl
CH_3-CH_2-CH- (with CH_3 branch)	1-methylpropyl	*s*-butyl (*s = secondary*)
CH_3-C- (with two CH_3 branches)	1,1-dimethylethyl	*t*-butyl (*t = tertiary*)

Hydrocarbons that contain one or more double bonds are called "unsaturated," and are named by replacing the ending "-ane" of the corresponding saturated hydrocarbon with the ending "-ene," "-adiene," or "-atriene," etc. The generic names of unsaturated hydrocarbons are *alkene, alkadiene, alkatriene,* etc. The double bonds receive the lowest possible numbers.

In the following examples, the names listed in the *Chemical Substances Index* of the CAS system are given first. Other names are given in parentheses.

EXAMPLES: 2-butene
           (2-butylene) $\overset{1}{CH_3}-\overset{2}{CH}=\overset{3}{CH}-\overset{4}{CH_3}$

           1,4-hexadiene $\overset{1}{CH_2}=\overset{2}{CH}-\overset{3}{CH_2}-\overset{4}{CH}=\overset{5}{CH}-\overset{6}{CH_3}$

           ethene (ethylene) $CH_2=CH_2$

           2-methyl-1,3-
           butadiene $\overset{1}{CH_2}=\overset{2}{C}-\overset{3}{CH}=\overset{4}{CH_2}$
           (isoprene) $|$
                                  CH_3

Hydrocarbons containing one or more triple bonds are named by replacing the ending "-ane" of the corresponding saturated hydrocarbon with the ending "-yne," "-adiyne," "-atriyne" etc. The triple bonds receive the lowest possible numbers. Double bonds take precedence over triple bonds when there is a choice in numbering.

EXAMPLES: 1-butyne

$$\overset{1}{C}H \equiv \overset{2}{C} - \overset{3}{C}H_2 - \overset{4}{C}H_3$$

1-hexane-3,5-diyne

$$\overset{1}{C}H_2 = \overset{2}{C}H - \overset{3}{C} \equiv \overset{4}{C} - \overset{5}{C} \equiv \overset{6}{C}H$$

Unsaturated branched acyclic hydrocarbons are numbered in the same manner as alkanes. The longest chain is chosen as the parent. If the alkene or alkyne contains two or more chains of equal length, the chain containing the maximum number of double bonds is chosen as the parent.

Univalent or multivalent groups derived from alkenes and alkynes are named as follows:

EXAMPLES: ethenyl (vinyl)

$$\overset{2}{C}H_2 = \overset{1}{C}H -$$

ethynyl

$$\overset{2}{C}H \equiv \overset{1}{C} -$$

2-propynyl

$$\overset{3}{C}H \equiv \overset{2}{C} - \overset{1}{C}H_2 -$$

methylidyne $CH\equiv$
ethylidene $CH_3-CH=$
ethylidyne $CH_3-C\equiv$

Alicyclic Hydrocarbons. Saturated monocyclic hydrocarbons, or "cycloalkanes," are named by attaching the prefix "cyclo" to the name of the acyclic unbranched alkane.

EXAMPLES: cyclopropane

cyclohexane

Univalent groups derived from unsubstituted cycloalkanes are named *cyclopropyl, cyclohexyl*, etc., in a manner analogous to that used for naming acyclic alkanes. The carbon atom with the free valence is numbered as 1.

EXAMPLES: cyclopropyl

cyclohexyl

Unsaturated monocyclic hydrocarbons are named by substituting "-ene," "-adiene," "-atriene," "-yne," "-adiyne," etc., in the name of the corresponding cycloalkane. Double and triple bonds are given numbers as low as possible.

EXAMPLES: cyclohexane

1,3-cyclohexadiene

1-cyclodecen-4-yne

Aromatic Hydrocarbons. Aromatic hydrocarbons generally are considered those which have the characteristic chemical properties of benzene. Many such compounds are known more commonly by their trivial names than by their systematic names.

EXAMPLES: benzene

methylbenzene (toluene)

1,2-dimethylbenzene (*o*-xylene)

ethenylbenzene (styrene)

(1-methylethyl)benzene (cumene)

The terms *ortho, meta,* and *para* (abbreviated *o, m,* and *p*) refer to the location of substituents on the benzene ring and are equivalent to 1,2-, 1,3-, and 1,4-substitution in systematic nomenclature, respectively. The lowest numbers possible are given to substituents.

Fused aromatic systems are named by prefixing the largest parent trivial names with combining forms such as benz(o)- and naphth(o)-. Hydrocarbons that contain five or more fused benzene rings in a linear arrangement are named from a numerical prefix followed by "-acene."

EXAMPLES: naphthalene

anthracene

hexacene

Fusion prefixes are used to designate to which side of the parent hydrocarbon a substituent ring is attached.

EXAMPLES:

benzene

+

anthracene

=

benz[a]anthracene

Hydrogenation products of complex aromatic ring systems that are not treated as alicyclic hydrocarbons are named by prefixing "dihydro," "tetrahydro," etc., to the parent name. The lowest locants are used. "Perhydro" is used in trivial nomenclature to indicate a fully hydrogenated compound.

EXAMPLES: 1,2-dihydronaphthacene

docosahydropentacene (perhydropentacene)

Multiple unsubstituted assemblies of benzene rings are named by using the appropriate prefix with the radical name "phenyl."

EXAMPLES: 1,1'-biphenyl

1,1':4',1''-terphenyl
(*p*-terphenyl)

Indicated hydrogen in aromatic systems is assigned to angular or non-angular positions when needed to accommodate structural features in systematic nomenclature. The lowest locants are used.

EXAMPLE: 1*H*-indene (not 3*H*-)

Bridged Hydrocarbon Ring Systems. These compounds are named by prefixing the parent ring with "bicyclo," "tricyclo," etc., or in some complex cases by using prefixes that denote the nature of the bridges. The three numbers in brackets denote the number of carbon atoms in each of the three bridges in descending order.

EXAMPLES:

bicyclo[3.1.0]hexane

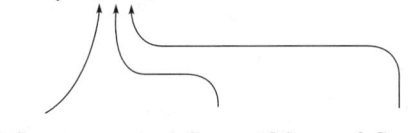

| 3 C-atoms (C2,3,4) between bridgeheads C1 and C5 | 1 C-atom (C6) between bridgeheads C1 and C5 | 0 C-atoms between bridgeheads C1 and C5 |

1,2,3,4-tetrahydro-1,4-methanonaphthalene

Spiro Hydrocarbon Ring Systems. Spiro systems contain pairs of rings or ring systems that have only one atom (a "spiro atom") in common. The name of the simplest monospiro system is formed by prefixing the acyclic hydrocarbon name with "spiro" and numerals separated by periods. The numerals are given in ascending order to define the number of atoms in each ring linked to the spiro atom. Numbering begins at the atom next to the spiro atom in the smallest ring for monospiro systems.

EXAMPLES: spiro[3,4]octane

dispiro[5.1.7.2]heptadecane

Carboxylic Acids and Their Anhydrides

Acids are named according to the Geneva ("-oic") or "-carboxylic" system. They are regarded as derived from parent hydrocarbons having the same number of carbon atoms so that CH_3 is replaced by COOH. The carbon atom of the carboxyl group is assigned number one in aliphatic monocarboxylic acids. In an alternative numbering system, the Greek letters *alpha, beta,* etc., are assigned to the second, third, etc., carbon atoms, respectively, leading away from the —COOH group. When chain branching is present, the longest chain containing the carboxylic acid group at one end is chosen for naming the molecule. Unsaturated aliphatic acids are named so the longest chain includes the maximum number of unsaturated linkages. Double bonds are given preference over triple bonds. Trivial names are retained for some common molecules.

EXAMPLES:

formic acid	HCOOH
acetic acid	$\overset{2}{C}H_3\overset{1}{C}OOH$
hexanoic acid (caproic acid)	$CH_3(CH_2)_4COOH$
octadecanoic acid (stearic acid)	$CH_3(CH_2)_{16}COOH$
2-propenoic acid (acrylic acid)	$\overset{3}{C}H_2=\overset{2}{C}H\overset{1}{C}OOH$

2-methylbutanoic acid

$$\overset{4}{C}H_3\overset{3}{C}H_2\overset{2}{C}H\overset{1}{C}OOH$$
$$|$$
$$CH_3$$

cyclobutanecarboxylic acid   ☐—COOH

benzoic acid

ethanedioic acid (oxalic acid)	HOOC—COOH
butanedioic acid (succinic acid)	$HOOC-CH_2CH_2-COOH$

1,2-benzenedicarboxylic acid (phthalic acid)

Conjunctive nomenclature may be used for naming cyclic acids. It is applied to any ring system attached by a single bond to one or more acyclic hydrocarbon chains, each of which bears only one principal functional group.

EXAMPLE: cyclohexaneacetic acid   —CH$_2$COOH

Acid anhydride names are formed from systematic, Geneva, trivial, conjunctive, or other type of acid names.

EXAMPLES: propanoic acid anhydride (propionic anhydride)

CH_3CH_2CO
$\quad\quad\quad\searrow$
$\quad\quad\quad\quad O$
$\quad\quad\quad\nearrow$
CH_3CH_2CO

benzoic acid anhydride (benzoic anhydride)

See also **Carboxylic Acids.**

Alcohols

Monohydric alcohols are named by adding "-ol" to a molecular skeleton name. Carbon chains, unsaturation, etc., are numbered in a manner analogous to that used for carboxylic acids. (See preceding section.)

EQUATIONS: methanol (methyl alcohol) CH_3OH
cyclohexanol (cyclohexyl alcohol)   —OH

2-propen-1-ol (allyl alcohol) $\overset{3}{C}H_2=\overset{2}{C}H\overset{1}{C}H_2OH$

See also **Alcohols.**

The simplest aromatic hydroxy compound is called phenol:

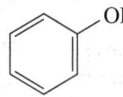

Esters

Simple esters are named on the basis of their alcohol and acid functions. Carbon chains, unsaturation, etc., are numbered in a manner analogous to that used for carboxylic acids.

EXAMPLES:

ethyl acetate $\qquad$ $CH_3COOC_2H_5$
(inverted name: acetic acid, ethyl ester)

propyl 2-butenoate

$$\overset{4}{C}H_3\overset{3}{C}H=\overset{2}{C}H\overset{1}{C}OOCH_2CH_2CH_3$$

methyl benzoate $\qquad$ ⟨benzene ring⟩—$COOCH_3$

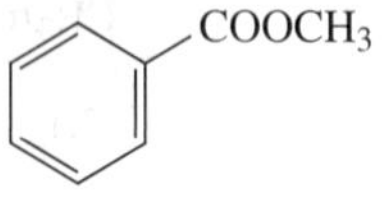

See also **Esters**.

Ethers

Alkoxy compounds are commonly called ethers. In current CAS nomenclature they are named as derivatives of functional parent compounds, hydrocarbons, etc., by use of "oxy" radicals.

EXAMPLES:

1,1′-oxybis(ethane) $\qquad$ $C_2H_5OC_2H_5$
(ethyl ether or diethyl ether)

(hexyloxy)cyclopropane (cyclopropyl hexyl ether) ⟨triangle⟩—$O(CH_2)_5CH_3$

methoxybenzene (anisole) ⟨benzene ring⟩—OCH_3

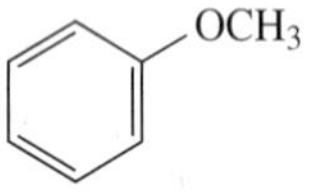

See also **Ethers**.

Aldehydes are named from the corresponding acids by use of "-carboxaldehyde" and "-al" suffixes.

EXAMPLES:

acetaldehyde $\qquad$ CH_3CHO

2-butenal

$$\overset{4}{C}H_3\overset{3}{C}H=\overset{2}{C}H\overset{1}{C}HO$$

3-methylbenzaldehyde

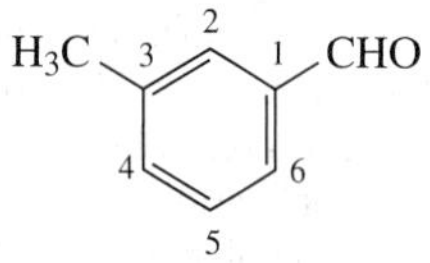

$$H_3C\overset{3}{-}\overset{2}{\underset{4\ \ 5\ \ 6}{}}\overset{1}{-}CHO$$

See also **Aldehydes**.

Ketones

Ketones are named by use of the characteristic suffix "-one."

EXAMPLES:

2-propane (acetone)

$$\overset{1}{C}H_3\overset{2}{C}O\overset{3}{C}H_3$$

1-hexen-1-one (butylketene)

$$CH_3(CH_2)_3\overset{2}{C}H=\overset{1}{C}O$$

4-cyclopentyl-2-butanone

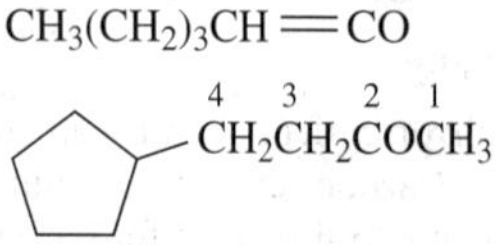

⟨cyclopentane ring⟩—$\overset{4}{C}H_2\overset{3}{C}H_2\overset{2}{C}O\overset{1}{C}H_3$

diphenylmenthanone (benzophenone)

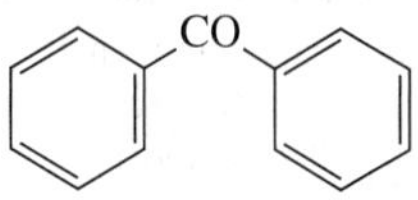

⟨benzene ring⟩—CO—⟨benzene ring⟩

See also **Ketones**.

Peroxides

Simple peroxides are named as follows:

EXAMPLES:

ethyl methyl peroxide $\qquad$ $C_2H_5OOCH_3$

benzoyl peroxide ⟨benzene ring⟩—CO—OO—OC—⟨benzene ring⟩

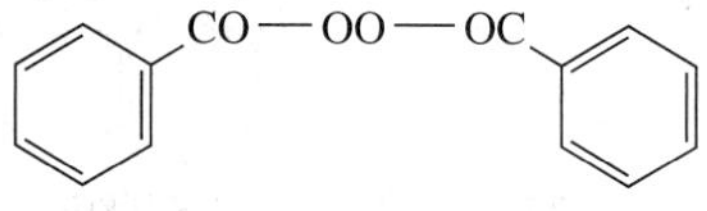

Halogenated Compounds

Hydrocarbons, esters, etc., which have one or more hydrogen atoms replaced by a halogen atom are named so that the substituents have the lowest possible numbers. When multiple functions are present, they are numbered according to precedences established by IUPAC or CAS nomenclature rules. Both trivial and systematic nomenclatures are used.

EXAMPLES:

chloromethane (methyl chloride) $\qquad$ CH_3Cl

2-iodobutane (sec- butyl iodide)

$$\overset{1}{C}H_3\overset{2}{C}H\overset{3}{C}H_2\overset{4}{C}H_3$$
$$\quad\ |$$
$$\quad\ I$$

1-bromo-2-fluorobenzene ⟨benzene ring with Br at position 1, F at position 2, numbered 1–6⟩

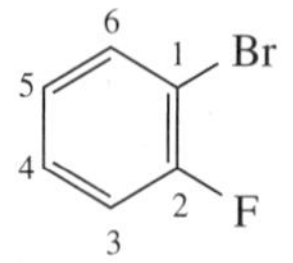

3-chloropentanoic acid (β-chlorovaleric acid)

$$\overset{5}{C}H_3\overset{4}{C}H_2\overset{3}{C}H\overset{2}{C}H_2\overset{1}{C}OOH$$
$$\qquad\quad |$$
$$\qquad\quad Cl$$

chloroethane (vinyl chloride) $\qquad$ $CH_2=CHCl$

1,1-dichloroethene (vinylidene chloride) $\qquad$ $CH_2=CCl_2$

tetrabromomethane (carbon tetrabromide) $\qquad$ CBr_4

Acid halides are named as follows:

EXAMPLES:

acetyl chloride $\qquad$ CH_3COCl

6-heptenoyl chloride

$$\overset{7}{C}H_2=\overset{6}{C}H(CH_2)_4\overset{1}{C}OCl$$

cyclohexanecarbonyl bromide ⟨cyclohexane ring, numbered 1–6⟩—$COBr$

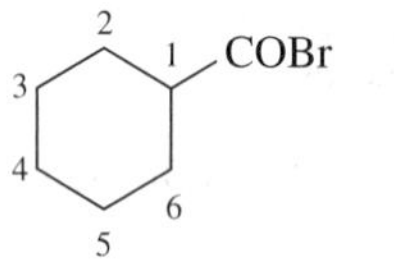

benzoyl fluoride ⟨benzene ring, numbered 1–6⟩—COF

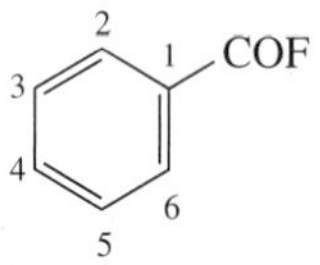

See also **Chlorinated Organics**.

Nonheterocyclic Nitrogen Compounds

Amines are named by adding the suffix "-amine" either to the name of the hydrocarbon or to the hydrocarbon radical. A second system names all amines as derivatives of primary amines.

EXAMPLES:

methanamine (methylamine) $\qquad$ CH_3NH_2

N,N-dipropyl-1-propanamine (tripropylamine) $\qquad$ $(CH_3CH_2CH_2)_3N$

benzenamine (aniline) ⟨benzene ring⟩—NH_2

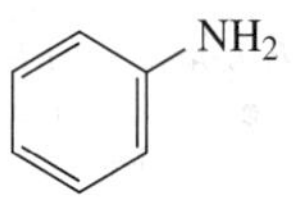

N-ethylcyclohexanamine (N-ethylcyclohexylamine) ⟨cyclohexane ring⟩—NHC_2H_5

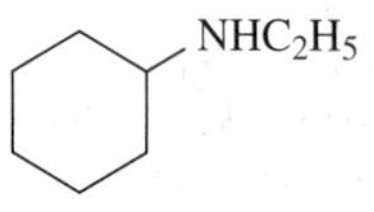

Imines are named from the hydrocarbon by the addition of the suffix "-imine."

EXAMPLES:

ethanimine

$$\overset{2}{C}H_3\overset{1}{C}H=NH$$

2,4-cyclopentadien-1-imine

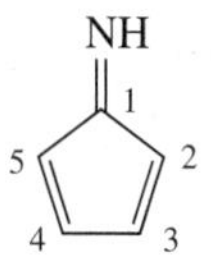

⟨cyclopentadiene ring with =NH at position 1, numbered 1–5⟩

See also **Amines**.

Names of amides are based on the corresponding acids. Thus, "-oic acid" becomes "-amide," and "-carboxylic acid" becomes "-carboxamide."

EXAMPLES:

acetamide

$$\overset{2}{C}H_3\overset{1}{C}ONH_2$$

benzamide

cyclohexanecarboxamide

N-methylpentadecanamide

$$\overset{15}{C}H_3(CH_2)_{13}\overset{1}{C}ON\overset{N}{H}CH_3$$

N,N-dimethyl-2,4-pentadienamide

$$\overset{5}{C}H_2=\overset{4}{C}H\overset{3}{C}H=\overset{2}{C}H\overset{1}{C}O\overset{N}{N}(CH_3)_2$$

See also **Amides**.

Nitro and nitroso compounds are named so that the substituents have the lowest possible numbers.

EXAMPLES: 2-nitrobutane

$$\overset{1}{C}H_3\overset{2}{C}H\overset{3}{C}H_2\overset{4}{C}H_3$$
$$|$$
$$NO_2$$

4-nitrosobenzoic acid

See also **Nitro- and Nitroso-Compounds**

Nitrile names are formed from common names of carboxylic acids.

EXAMPLES: acetonitrile CH_3CN

benzoinitrile (phenyl cyanide)

3-butenenitrile

$$\overset{4}{C}H_2=\overset{3}{C}H\overset{2}{C}H_2\overset{1}{C}N$$

ethenetetracarbonitrile (tetracyanoethylene)

In the presence of more senior functional groups, the nitrile function is expressed by the prefix "cyano."

EXAMPLES: 4-cyanobenzamide

3-cyanobutanoic acid

$$\overset{4}{C}H_3\overset{3}{C}H\overset{2}{C}H_2\overset{1}{C}OOH$$
$$|$$
$$CN$$

Nonheterocyclic Sulfur Compounds

Sulfur compounds are named similarly to oxygen compounds.

EXAMPLES: (methylthio)benzene (methyl phenyl sulfide)

1,1'-sulfinylbis(benzene) (diphenyl sulfoxide)

2-(propylsulfonyl)naphthalene (2-naphthyl propyl sulfone)

2-butanethiol (2-mercaptobutane)

$$\overset{1}{C}H_3\overset{2}{C}H\overset{3}{C}H_2\overset{4}{C}H_3$$
$$|$$
$$SH$$

2-propanethione (thioacetone)

$$\overset{S}{\underset{||}{}}$$
$$CH_3CCH_3$$

N-methylbenzenesulfonamide SO_2-NHCH_3

benzenesulfonic acid SO_3H

See also **Sulfonic Acids**.

Heterocyclic Compounds

Some of the common heterocyclic nitrogen, oxygen, and sulfur compounds and their numbering systems are shown:

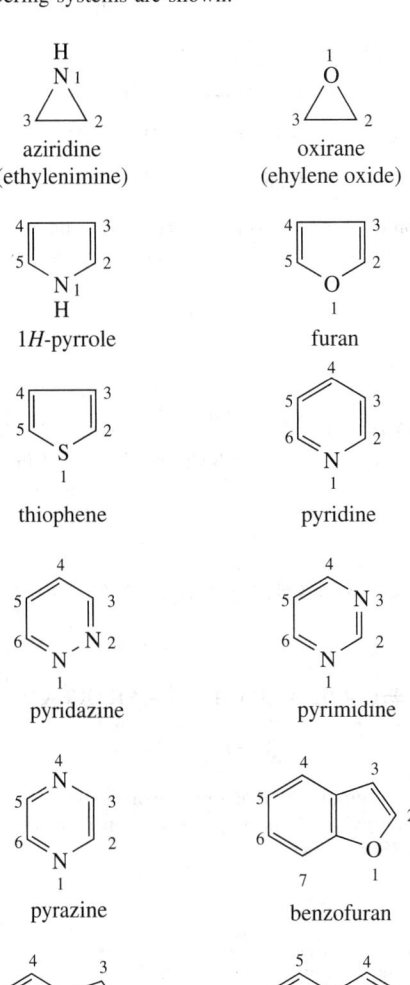

aziridine (ethylenimine)

oxirane (ehylene oxide)

1H-pyrrole

furan

thiophene

pyridine

pyridazine

pyrimidine

pyrazine

benzofuran

1H-indole

quinoline

morpholine

benzoxazole

Organic Reaction Mechanisms and Processes

Many reactions occur in which one organic compound is converted into another. The molecular details of the intermediate steps by which compounds are converted into new products are called reaction mechanisms. The four broad classes of reaction mechanisms are: cationic, anionic, free radical, and multicenter processes in which neither charged species nor odd electron species is involved. Examples of each type will be given, but many variations can exist within each type. Also, varying degrees of sophistication exist in our knowledge of the exact reaction pathways that organic compounds follow. The examples discussed show only the major steps involved.

Cationic and anionic mechanisms involve species that have either positive or negative charges, respectively (heterolytic reactions). An example of a reaction that proceeds via a cationic mechanism is the hydrolysis of *t*-butyl bromide (Scheme 2).

SCHEME 2

$$(CH_3)_3CBr + H_2O \longrightarrow (CH_3)_3COH$$
t-Butyl bromide   Water   *t*-Butyl alcohol

$$+ (CH_3)_2 C=CH_2 + HBr$$
Isobutylene   Hydrogen bromide

Mechanism:

$$(CH_3)_3CBr \longrightarrow (CH_3)_3C^+ + Br^-$$

$$(CH_3)_3C^+ + H_2O \longrightarrow (CH_3)_3COH + H^+$$

$$(CH_3)_3C^+ \longrightarrow (CH_3)_2C=CH_2 + H^+$$

The addition of methanol to methyl acrylate in the presence of sodium methoxide is an example of a reaction that proceeds by an anionic mechanism (Scheme 3).

SCHEME 3

Mechanism:
Free radical (homolytic) reactions involve species with an unpaired electron. The ultraviolet-light-initiated reaction of methane with chlorine is an example (Scheme 4).

SCHEME 4

$$CH_4 + Cl_2 \xrightarrow{h\nu} CH_3Cl + HCl$$
Methane   Chlorine   Methyl chloride   Hydrogen chloride

$$Cl_2 \xrightarrow{h\nu} 2Cl\cdot$$

$$Cl\cdot + CH_4 \longrightarrow CH_3\cdot + HCl$$

$$CH_3\cdot + Cl_2 \longrightarrow CH_3Cl + Cl\cdot$$

A number of reactions do not seem to belong to any of the above mechanistic types. Such processes are referred to as multicenter reactions. The Diels-Alder cycloaddition reaction of 1,3-butadiene with maleic anhydride is an example (Scheme 5). No charged or odd electron intermediates seemingly are involved in this reaction.

SCHEME 5

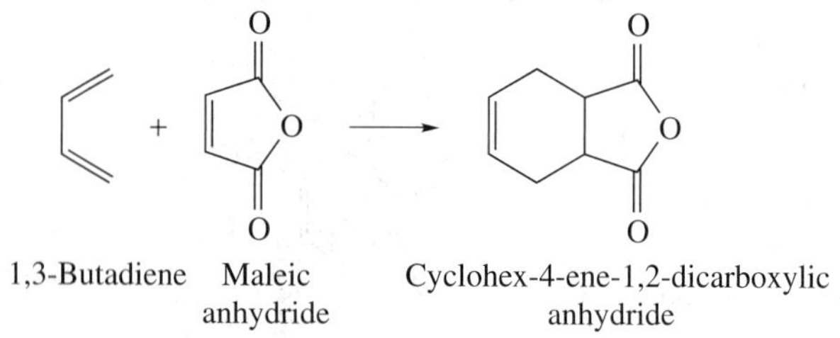

1,3-Butadiene   Maleic anhydride   Cyclohex-4-ene-1,2-dicarboxylic anhydride

The reactions shown in Schemes 2–5, with the exception of the photodissociation of Cl_2 to Cl atoms (Scheme 4), occur in molecules that are in electronic ground states. Reactions also can occur in molecules existing in excited electronic states. Commonly, these excited states are produced by irradiating the reactants with ultraviolet or visible light, hence the term *photochemistry*. When a molecule is in an excited state, its reactions are often different from those it normally exhibits in its ground state. An organic compound often can exist in more than one excited state, as shown in Scheme 6.

SCHEME 6

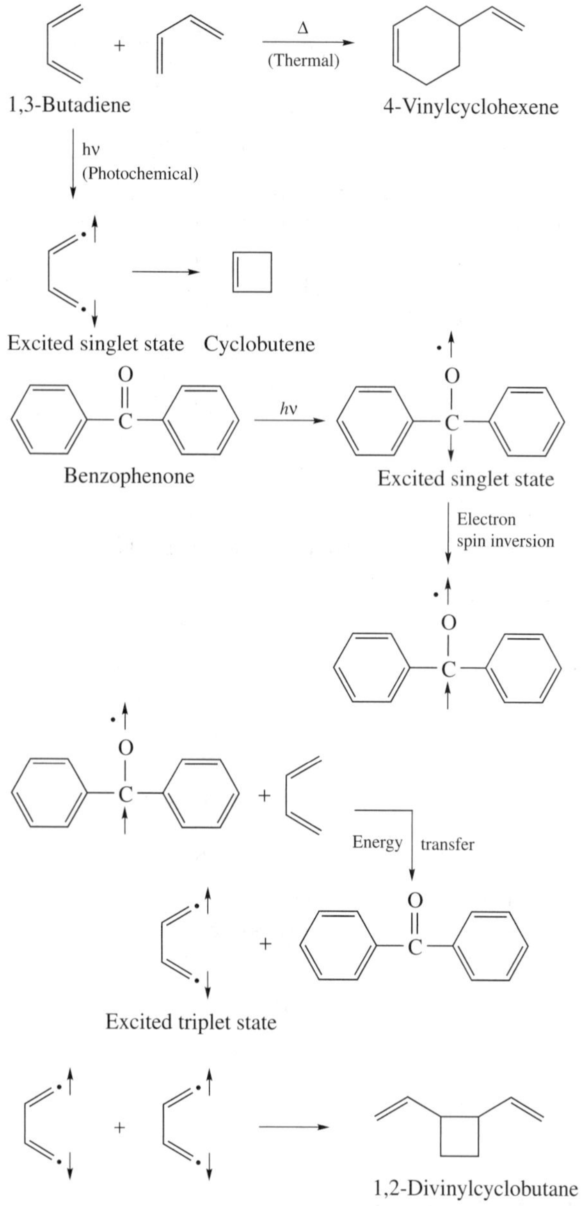

1,3-Butadiene   4-Vinylcyclohexene

Benzophenone   Excited singlet state

Excited triplet state

1,2-Divinylcyclobutane

Excited state molecules usually differ from their ground state counterparts by having dissimilar electronic and geometric configurations, and shorter lifetimes, e.g., 10^{-2} to 10^{-10} second. See Fig. 2.

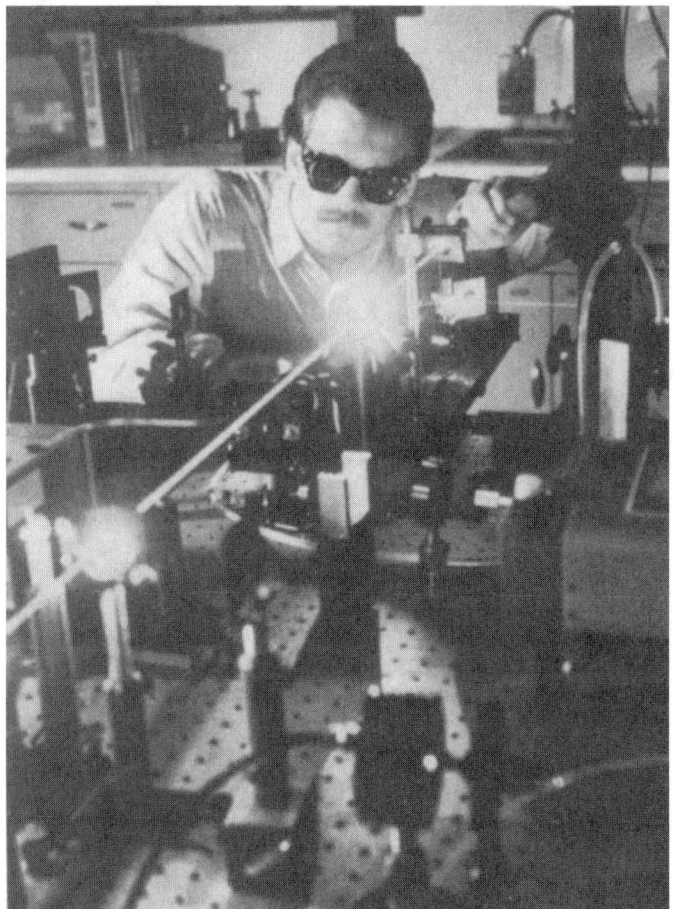

Fig. 2. Excited-state photochemical reaction being performed with a laser light source. (*The Dow Chemical Company*.)

A few additional examples of organic reactions are shown in Scheme 7. Many novel and complex compounds can be prepared by these and similar reactions.

SCHEME 7. Examples of Organic Reactions.

Free Radical Addition

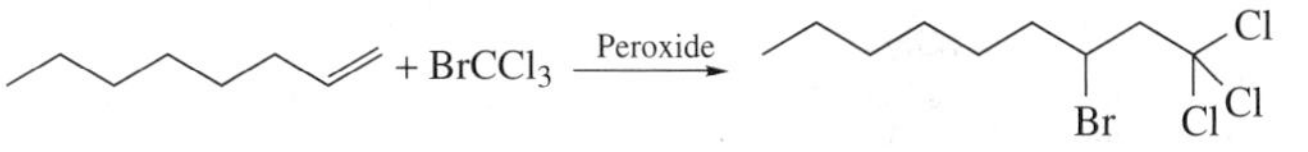
+ $BrCCl_3$
Peroxide
Cl
Br Cl Cl

1-Octene Bromotrichloromethane 3-Bromo-1,1,1-trichlorononane

Hydrogenation

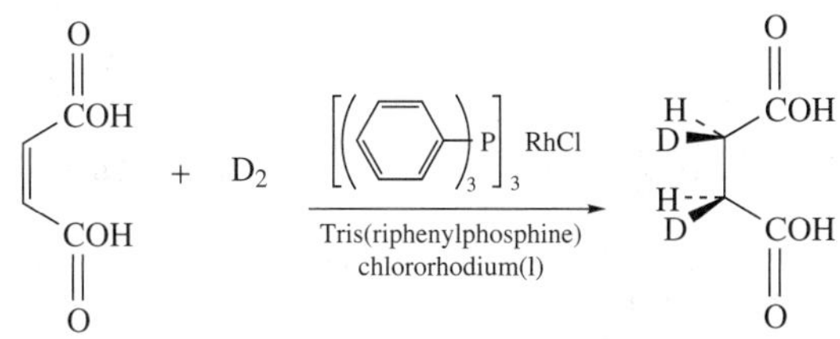
COH
COH
+ D_2
RhCl
Tris(riphenylphosphine) chlororhodium(l)
H D COH
H D COH

Maleic acid Deuterium *meso*-2,3-D_2-Butanedioic acid

Carbene Addition

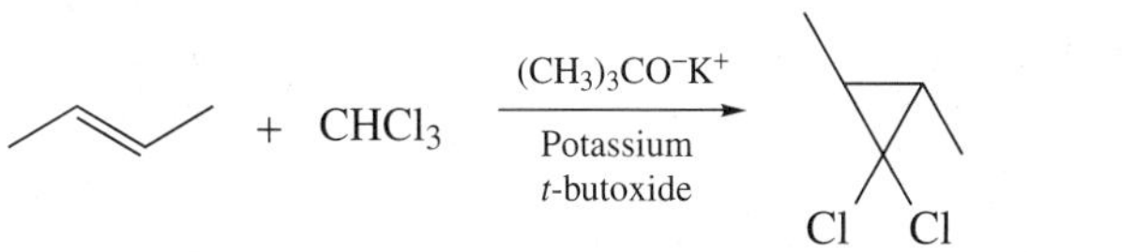
+ $CHCl_3$
$(CH_3)_3CO^-K^+$
Potassium
t-butoxide
Cl Cl

trans-2-Butene Chloroform *trans*-3, 3-Dichloro-1, 2-dimethylcyclopropane

Oxidative Cleavage

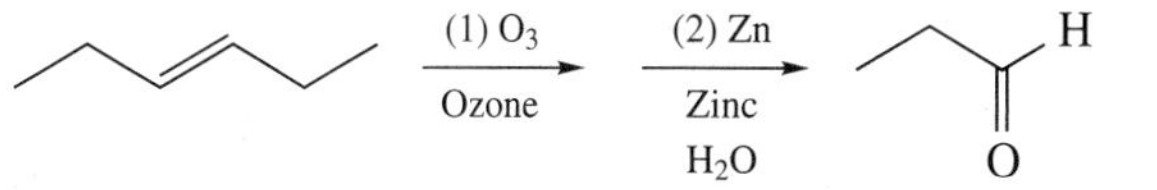
(1) O_3
Ozone
(2) Zn
Zinc
H_2O
H
O

trans-3-Hexene Propionaldehyde

Hydroboration-Oxidation

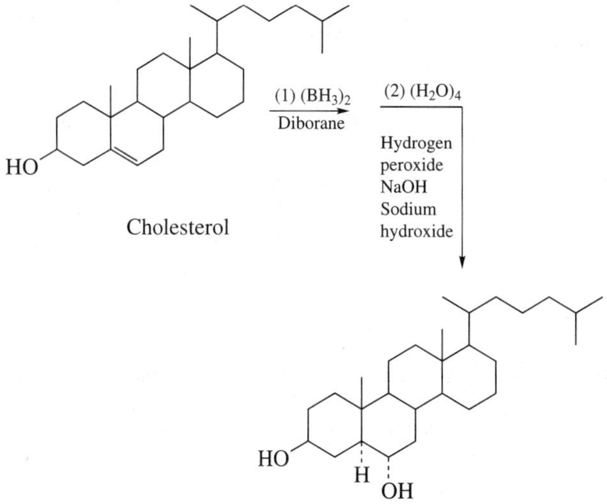
HO
Cholesterol
(1) $(BH_3)_2$
Diborane
(2) $(H_2O)_4$
Hydrogen peroxide
NaOH
Sodium hydroxide
HO
H
OH

Cholestane-3β, 6α-diol

Oxidation

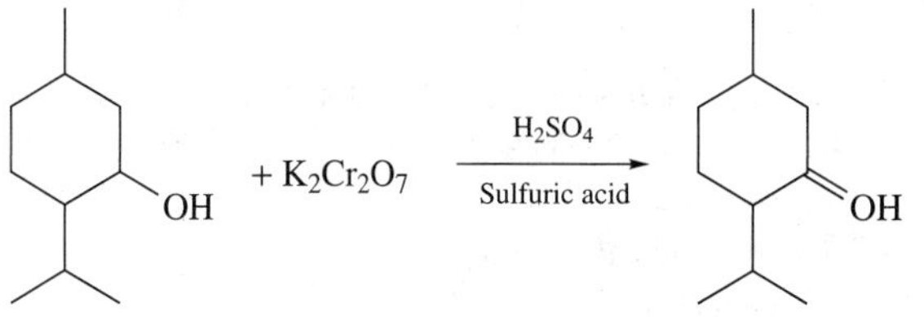
OH
+ $K_2Cr_2O_7$
H_2SO_4
Sulfuric acid
OH

Menthol Potassium dichromate Menthone

Reduction

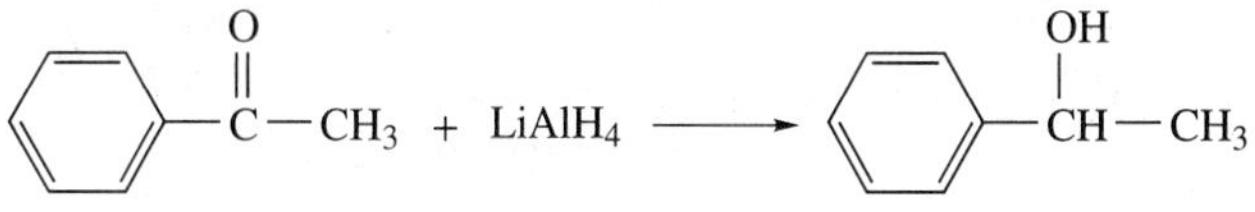
O
C — CH_3 + $LiAlH_4$
OH
CH — CH_3

Acetophenone Lithium aluminum hydride 1-Phenylethanol

Chlorination-Amination

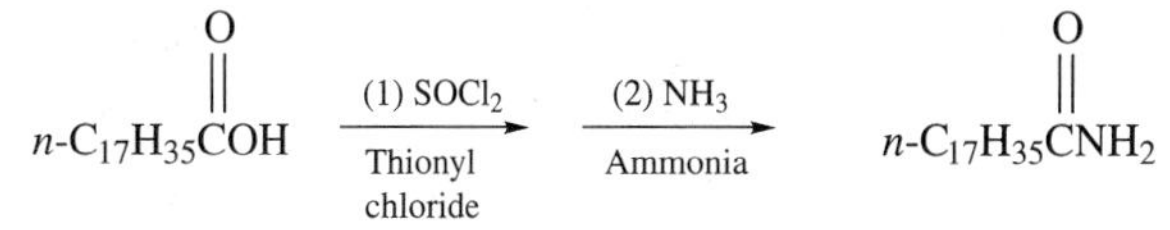
O
n-$C_{17}H_{35}COH$
(1) $SOCl_2$
Thionyl chloride
(2) NH_3
Ammonia
O
n-$C_{17}H_{35}CNH_2$

Stearic acid Stearamide

Substitution

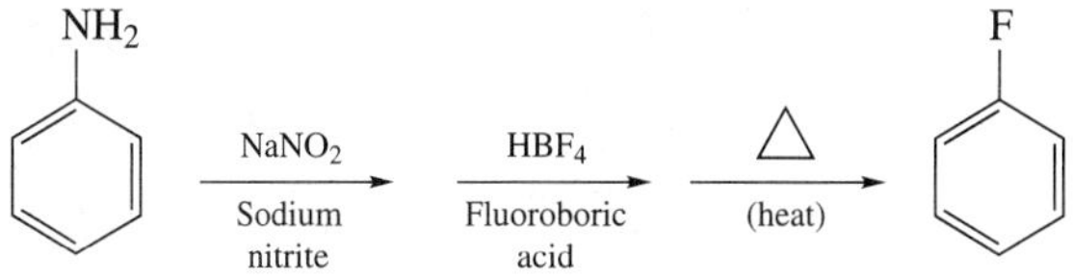
NH_2
$NaNO_2$
Sodium nitrite
HBF_4
Fluoroboric acid
△
(heat)
F

HCl
Hydrochloric acid

Aniline Fluorobenzene

Benzyne Cycloaddition

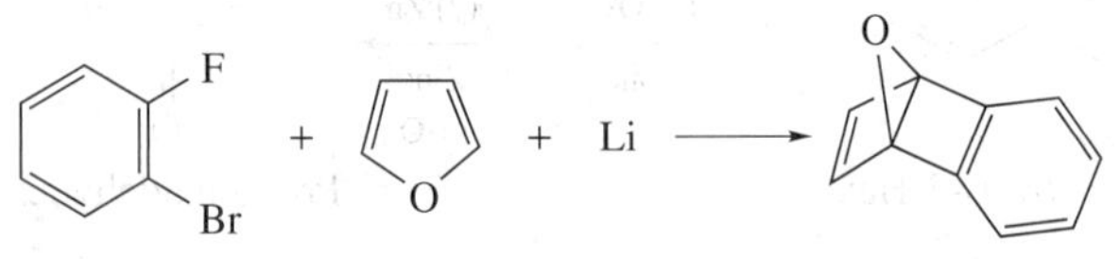

o-Bromofluorobenzene   Furan   Lithium

1,4-Dihydronaphthalene-1,
4-endoxide

Photoisomerization

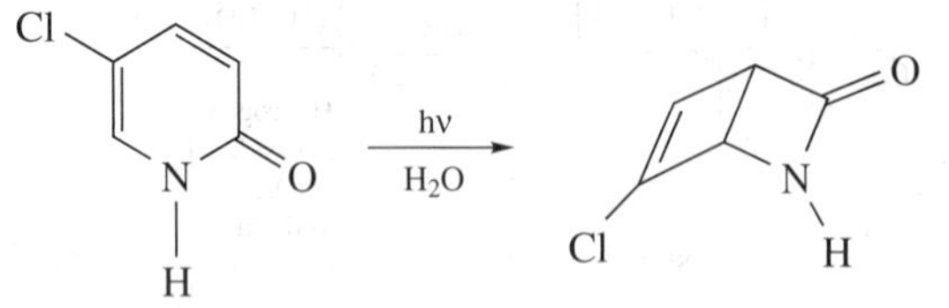

$\xrightarrow[H_2O]{h\nu}$

5-Chloro-2-pyridinone

6-Chloro-*cis*-2-azabicyclo
[2.2.0] hex-5-en-3-one

Many organic reactions are referred to by the inventor's or discoverer's name. A few name reactions are shown in Scheme 8.

SCHEME 8. Examples of Organic Name Reactions.

Friedel-Crafts Alkylation

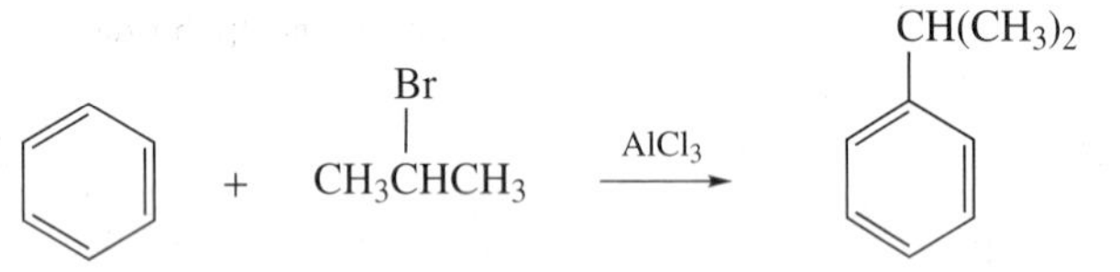

Benzene   Isopropyl   Aluminum   Isopropylbenzene
            bromide   chloride

See also **Friedel-Crafts Reaction**.

Grignard Reaction (Fig. 3).

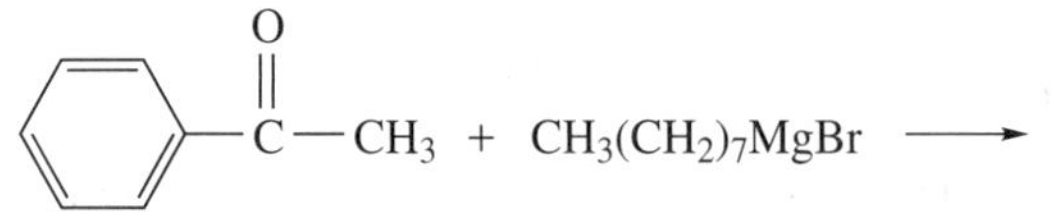

Acetophenone   *n*-Octylmagnesium
                bromide

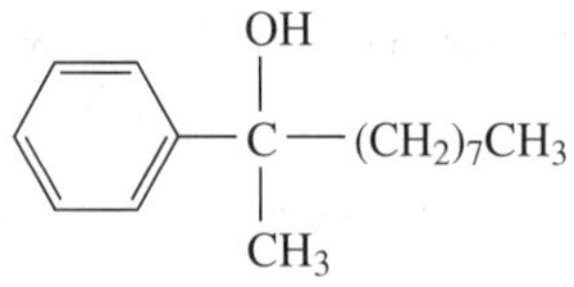

2-Phenyl-2-decanol

See also **Grignard Reactions**.

Baeyer-Villiger Oxidation

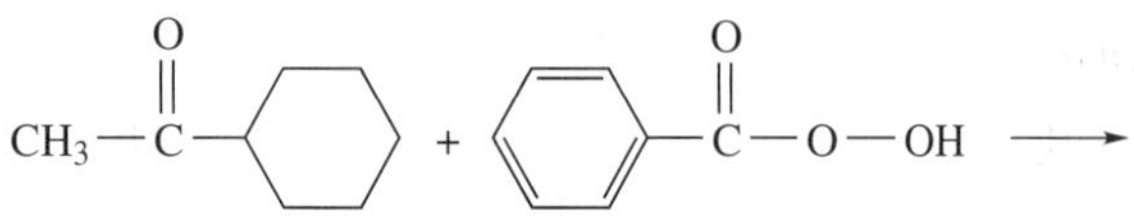

Methyl cyclohexyl ketone   Perbenzoic acid

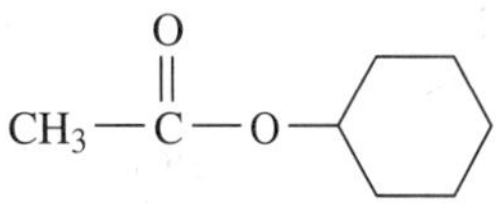

Cyclohexyl acetate

Fig. 3.   Typical laboratory apparatus showing preparation of 2-phenyl-2-decanol by a Grignard reaction. (*The Dow Chemical Company*.)

Meerwein-Ponndorf-Verley Reduction

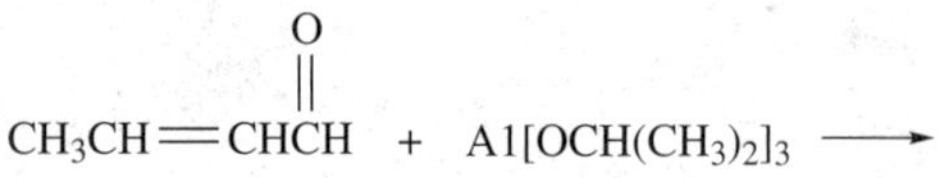

$CH_3CH=CHCH$ + $Al[OCH(CH_3)_2]_3$ $\longrightarrow$

Crotonaldehyde   Aluminum isopropoxide

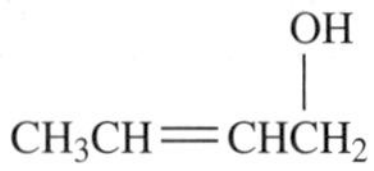

$CH_3CH=CHCH_2$

But-2-en-1-ol

Skraup Quinoline Synthesis

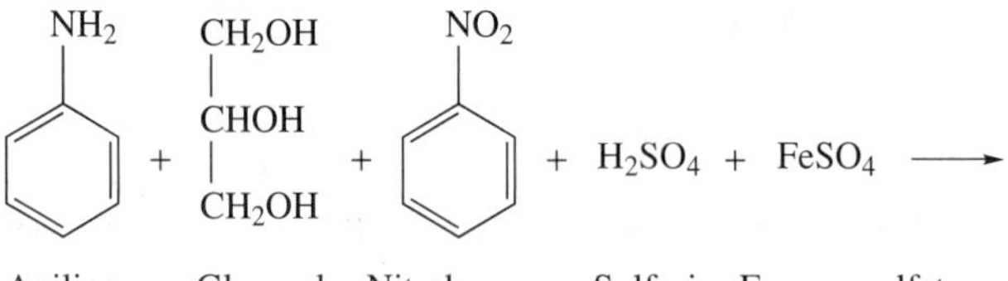

Aniline   Glycerol   Nitrobenzene   Sulfuric   Ferrous sulfate
                                       acid

+ H_2SO_4 + $FeSO_4$ $\longrightarrow$

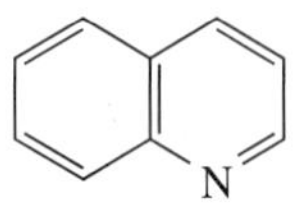

Quinoline

A very active area of organic chemistry is the synthesis of complex natural products. In these syntheses, numerous reactions, of which those in Schemes 7 and 8 are examples, are often employed serially to convert a starting compound into a final product that occurs in nature.

These syntheses are useful because they serve to verify structure that has been assigned, provide an alternate source of the compound if a larger supply is needed, or provide a route to derivatives or analogs of the natural material. The derivatives may possess enhanced properties, such as biological activity, that are not found in the natural product.

Some important industrial organic processes are shown in Scheme 9. Although most of these reactions involve mixtures of isomers or homologs as reactants and products, for simplicity only the major components are shown.

SCHEME 9. Examples of Industrial Organic Processes.

Alkylation

$(CH_3)_3CH + CH_2\!=\!CHCH_3 \xrightarrow[\text{catalyst}]{HF} CH_3CH\!=\!CHCH_2CH_3$

Isobutane Propylene 2,3-Dimethylpentane

See also **Alkylation**.

Isomerization

$CH_3(CH_2)_2CH_3 \xrightarrow[\text{catalyst}]{AlCl_3} CH_3CHCH_3$ (with CH_3 branch)

n-Butane *i*-Butane

Cracking

$CH_3(CH_2)_{14}CH_3 \xrightarrow[\text{catalyst}]{\text{Silica alumina}} CH_3(CH_2)_4CH\!=\!CH_2$

n-Hexadecane 1-Heptene

$+ CH_3CH_2CH\!=\!CH_2 + CH_3CH_2CHCH_3$ (with CH_3)

1-Butene Isopentane

See also **Cracking Process**.

Oxidation

$CH_2\!=\!CH_2 + O_2 \xrightarrow[\text{catalyst}]{Ag} H_2C\!-\!CH_2$ (epoxide O)

Ethylene Oxygen Ethylene oxide

Chlorination

Benzene $+ Cl_2 \xrightarrow[\text{catalyst}]{FeCl_3}$ Chlorobenzene

See Fig. 4; also separate entry on **Chlorinated Organics**.

Hypochlorination

$CH_3CH\!=\!CH_2 + HOCl \longrightarrow CH_3CHCH_2Cl$ (OH)

Propylene Hypochlorous acid 1-Chloro-2-hydroxypropane

Dehydrochlorination

$CH_3CF_2Cl \xrightarrow{\text{Pyrolysis}} CH_2\!=\!CF_2$
1-chloro-1,1-difluoroethane Vinylidene flouride

$+ HCl$
Hydrogen chloride

Bromination

$CH_2\!=\!CH_2 + Br_2 \longrightarrow CH_2BrCH_2Br$
Ethylene Bromine Ethylene dibromide

Fig. 4. Typical industrial reactor for the preparation of chlorobenzene by the chlorination of benzene. (*The Dow Chemical Company*.)

Hydrogenation

$CH_2OC(CH_2)_7CH\!=\!CHCH_2CH\!=\!CH(CH_2)_4CH_3$
$CHOC(CH_2)_7CH\!=\!CHCH_2CH\!=\!CH(CH_2)_4CH_3$
$CH_2OC(CH_2)_7CH\!=\!CHCH_2CH\!=\!CH(CH_2)_4CH_3$

Trilinolein

Fermentation

Sucrose $\xrightarrow[\text{(a microorganism)}]{\textit{Aspergillus niger}}$ $HOC\!-\!CH_2CCH_2\!-\!COH$

Citric acid

Coupling

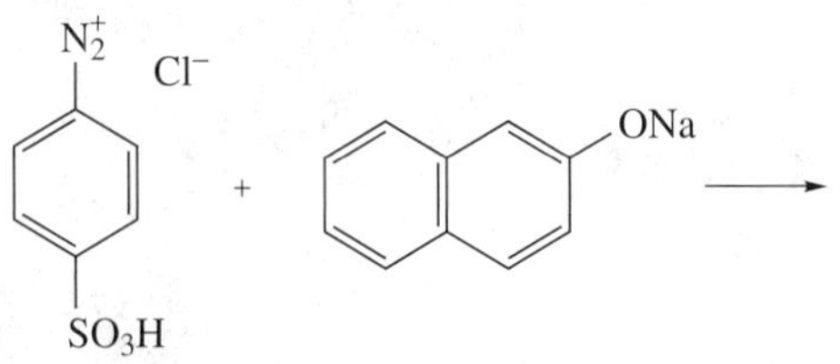

p-Sulfobenzenediazonium chloride — Sodium 2-naphtholate

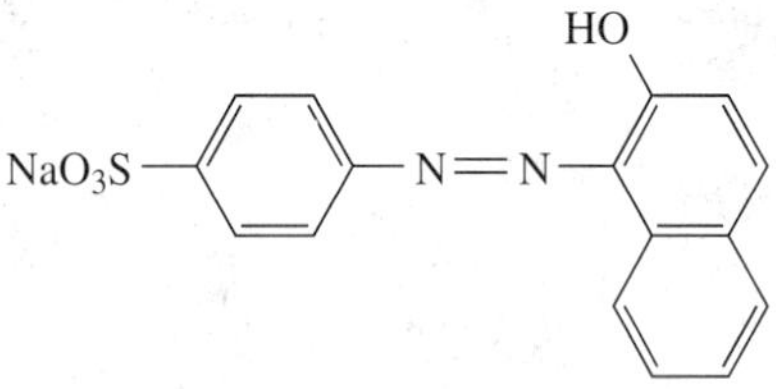

Orange II (a dye)

Sulfonation

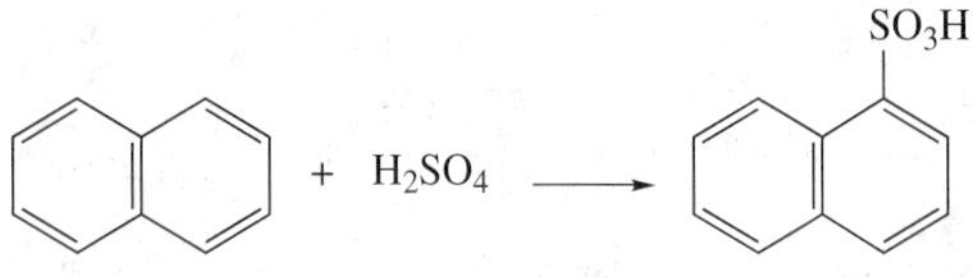

Naphthalene — Sulfuric acid — α-Naphthalenesulfonic acid

Hydrolysis

$$\underset{\text{Allyl chloride}}{CH_2{=}CHCH_2Cl} + \underset{\text{Sodium hydroxide}}{NaOH} \longrightarrow \underset{\text{Allyl alcohol}}{CH_2{=}CHCH_2OH}$$

Ammonolysis

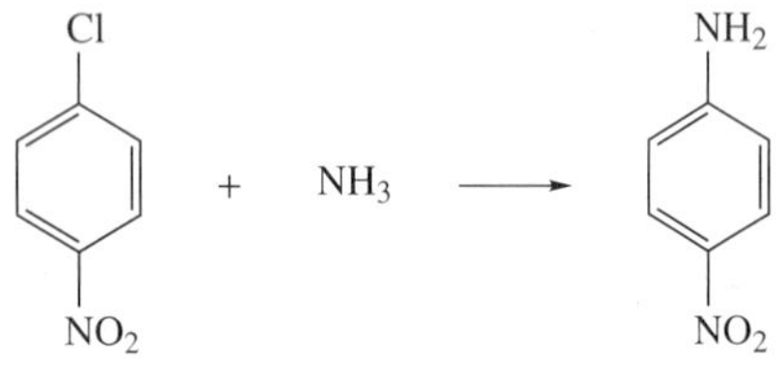

p-Chloronitrobenzene — Ammonia — *p*-Nitroaniline

See also **Amination**.

Hydration

$$\underset{\text{Ethylene}}{CH_2{=}CH_2} + \underset{\text{Sulfuric acid}}{H_2SO_4} + \underset{\text{Water}}{H_2O} \longrightarrow \underset{\text{Ethanol}}{CH_3CH_2OH}$$

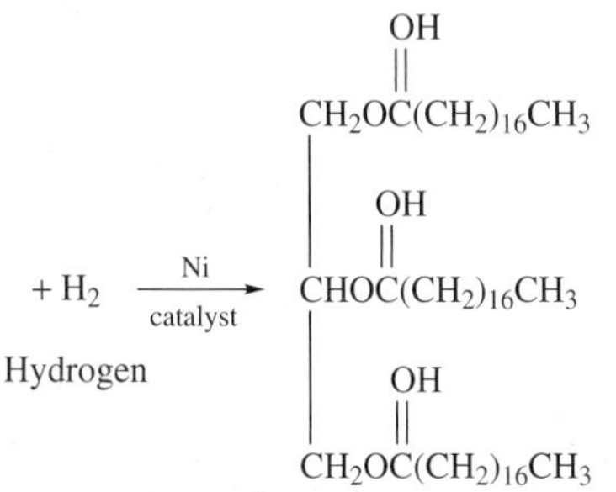

Tristearin

See Fig. 5 and also the separate entry on **Fermentation**.

Fig. 5. Laboratory fermentation process equipment. (*The Dow Chemical Company.*)

Hydrogenolysis

$$\underset{\text{Methyl laurate}}{CH_3(CH_2)_{10}\overset{\overset{\displaystyle O}{\|}}{C}OCH_3} + \underset{\text{Hydrogen}}{H_2} \xrightarrow[\text{catalyst}]{\text{Copper-chromic oxide}} \underset{\text{Lauryl alcohol}}{CH_3(CH_2)_{11}OH}$$

Dehydrogenation

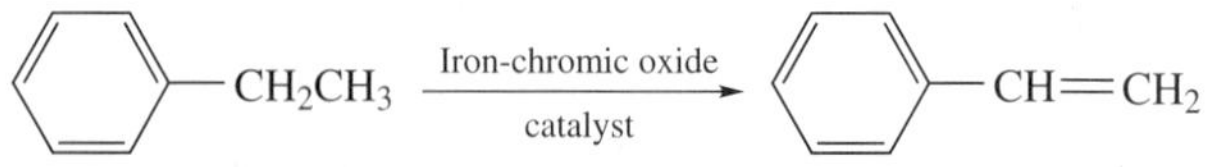

Ethylbenzene — Styrene

Nitration

Toluene — Nitric acid — Sulfuric acid — 2,4,6-trinitrotoluene (TNT)

See also **Nitration**.

Hydroformylation

$$CH_3CH = CH_2 + CO + H_2 \longrightarrow CH_3(CH_2)_2\overset{\displaystyle O}{\overset{\|}{C}}H$$

Propylene   Carbon   Hydrogen   *n*-Butyraldehyde
             monoxide

Esterification

Phthalic anhydride   Octanol   Dioctyl phthalate

Vinyl Polymerization

$$CH_2{=}CHCl \xrightarrow{\quad K_2S_2O_8 \quad} (CH_2CHCl)_{\sim 1,000-10,000}$$

Vinyl chloride   initiator   Polyvinyl chloride

Condensation Polymerization

$$HO\overset{\displaystyle O}{\overset{\|}{C}}(CH_2)_4\overset{\displaystyle O}{\overset{\|}{C}}OH + H_2N(CH_2)_6NH_2$$

Adipic acid   Hexamethylenediamine

$\downarrow \Delta \text{ (heat)}$

$$HO{\left[CH(CH_2)_4\overset{\displaystyle O}{\overset{\|}{C}}NH(CH_2)_6NH \right]}_{\sim 50-90} H$$

Poly(hexamethyleneadipamide)
[Nylon 66]

Additional Reading

Chemical Substance Index Names: Section IV from the Chemical Abstracts Index Guide, American Chemical Society, Columbus, Ohio, 1985 — and references cited therein.

Graham Solomons, G.: "Fundamentals of Organic Chemistry," John Wiley and Sons, Inc., New York, NY, 1999.

Hellwinkel, D.: "Systematic Nomenclature in Organic Chemistry: A Directory to Comprehension and Application on Its Basic Principles," Springer-Verlag, Inc., New York, NY, 2001.

McMurry, J.: "Fundamentals of Organic Chemistry," Thomson Learning Publications, Fresno, CA, 1997.

McMurry, J. and M. Castellion: "Fundamentals of Organic and Biological Chemistry, 2nd edition," Prentice-Hall, Inc., Upper Saddle River, NJ, 1998.

Morrison, R.T. and R.N. Boyd: "Organic Chemistry," 6th Edition, Prentice-Hall, Inc., Upper Saddle River, NJ, 1992.

Nomenclature of Organic Chemistry; Sections A, B, C, D, E, F and H combined; Pergamon Press, Oxford, 1979. (IUPAC nomenclature.)

Richey, H.G., Jr.: "Fundamentals of Organic Chemistry," Pentrice-Hall, Inc., Upper Saddle River, NJ, 1983.

Schmerling, L.: "Organic and Petroleum Chemistry for Nonchemists," Penn-Well Publishing, Tulsa, OK, 1981.

Smith, M.B. and J. March: "March's Advanced Organic Chemistry: Reactions, Mechanisms, and Structure," 5th Edition, John Wiley & Sons, Inc., New York, NY, 2001.

Wade, L.G. Jr.: "Organic Chemistry," 4th Edition, Prentice-Hall, Inc., Upper Saddle River, NJ, 1998.

WENDELL L. DILLING and MARCIA L. DILLING, The Dow Chemical Company, Midland, MI.

ORGANIC LIGHT-EMITTING DIODES (OLEDs).

The general phenomenon of *electroluminescence* (EL) encompasses many processes, including that of common inorganic light-emitting diodes (LED) devices, their organic counterparts, and even cathode ray tubes (CRTs), vacuum fluorescent displays (VFDs), and field emission displays (FEDs) (In the last three cases, the EL process is called *cathodoluminescence*). The term organic light-emitting diode (OLED) refers specifically to light-emitting diode devices (with p-n junctions) made from organic materials. See also **Inorganic Light-Emitting Diode Displays**; and **Cathode-Ray Tube**.

The term *organic electroluminescence* EL is often used incorrectly to refer to an OLED device; organic EL in the true sense of the term does not operate like a diode, because there is no p-n junction. (Instead, light is emitted from within an active layer, wherein the applied electric field has raised various transition metal dopants to excited energy levels, which emit light as they relax. For example the inorganic EL device uses manganese as a dopant in a zinc sulfide host, while organic EL devices have an organic base or even a transition metal-organic compound as the active layer.) The OLED devices currently being pursued, and which are the main subject here, are true diodes.

The ubiquitous device known as the LED (light-emitting diode), based on inorganic semiconducting materials, was introduced commercially in 1969. In LED devices, electrons and holes are electrically injected into a p-n junction device, where they recombine and emit light; this same general principle applies in an OLED device, except that organic materials are used in place of semiconductors. The emission of light from organic materials subjected to an electric field has been known since at least the early 1960s. OLED technology dates back to work on conducting polymers in England in the early 1970s and at the University of Pennsylvania in 1977 on "synthetic metals." Concerted efforts in industry did not occur until the end of the 1980s, fueled primarily by a convergence of positive developments. New materials, processes, and electronics have been developed; more researchers have entered the field; and new prototypes have shown promising performance. These factors clearly point to the likelihood of viable commercial OLED products in the near future. Such products have the potential to offer very attractive features: monolithic light emission (no backlight required) over a large area, a low voltage, and full color.

Unlike inorganic LED devices, OLED devices are created with transparent organic materials using large-area, thick-film deposition processes. There is no need to grow a crystal, saw it, polish it, or dope it. It is an inherently large-area process, not a chip-oriented process. And, unlike inorganic LED devices, which must be mounted as individual lamps to create a large-area, high information content display, OLED devices can be made on a single, solid substrate. Most OLED devices are fabricated on a transparent anode (indium tin oxide, or ITO) layer, which is itself deposited on a glass substrate. One type of device (the earliest) uses small molecules that are vacuum-deposited to form a hole transport layer, on top of which a layer of emissive material and metallic cathode are deposited. An alternative approach uses fluorescent polymers such as poly (*p*-phenylenevinylene) (PPV) and polyfluorene, prepared in solution and spin-casted.

Ink jet printing techniques have been used to pattern polymeric OLED devices.

Steady technical progress has been made in both the small-molecule and polymer approaches. A major advance in small-molecule OLED devices was announced in a 1985 U.S. patent from researchers at the Eastman Kodak Company[1]. Since that time, researchers at Kodak have amassed a very significant patent portfolio so that Kodak has become the dominant intellectual property holder in small-molecule OLED devices. Many companies have licensed these patents and have gone on to make further technical breakthroughs.

Equally dramatic progress in polymer-based OLED devices was announced by researchers at Cambridge University[2] in 1990; in 1993 they[3] patented this technology. Devices based on this technology consist of a series of polymer layers deposited by spin- or knife-coating. Ownership of this and subsequent patents was transferred to Cambridge Display Technology (CDT), which is now the major intellectual property player in polymer OLED devices.

In 1997 a 256 × 64 pixel, monochrome car radio display, using technology licensed from Kodak was made. In 1997 and 1998 Color passive matrix OLED devices in a 320 × 240 pixel format was demonstrated Monochrome, active matrix, 320 × 240 pixel prototypes were produced. In total, more than 80 companies are investigating OLED devices.

Laboratory devices based on small-molecule and polymer technologies have excellent technical performance today. Neither technology seems to have a major performance edge over the other. However, the early commercial products perform rather modestly compared to what has been accomplished in the laboratory, where, for example, small full-color televisions have been demonstrated. Many major companies have moved rapidly to start or expand research and development programs and prototype product development. This spreading interest in OLED technology stems from the perceived potential advantages that it offers:

Flexible substrates (displays shaped to product design; continuous coating); emissive devices (back lights and color filters not needed); ease of fabrication; design latitude; design simplicity; fast switching speed (video display capability); low weight; high brightness (useful in bright environments); wide color gamut; low voltage (compatible with TFT active matrix drivers); monolithic light emission with 140 to 160° viewing angle (vertical and horizontal); and rugged.

Although the results so far have been achieved quite rapidly, at least from the perspective of historical flat panel display development, there is still much work to be done. Lifetime, particularly packaging for long life, remains a key issue with these technologies. The techniques for making large, full-color video displays are just in the concept stage. Also, there are a number of business challenges relating to displacing the cathode ray tube (CRT) and liquid crystal display (LCD) with such a different technology as the OLED device. These areas of future concern are as follows: Operating lifetimes are too short; operation in extreme conditions has not been proved; improvements in color and video capability needed; people are unfamiliar with the technology; manufacturing infrastructure not yet established; and claimed that it will be inexpensive, but this is unproven.

Many researchers feel confident that when enough resources are available, these problems will be solved. Manufacturing any radically new type of display is sure to be a much longer process than originally planned, but it is rare that so many companies with similar technologies show positive preliminary results so quickly.

As noted previously, OLED devices emit light through the recombination of holes and electrons in a junction region. This physical phenomenon lends itself to a number of basic device designs, which are discussed in this section.

Monochromatic OLED Devices

The simplest organic LED device that one might construct is shown schematically in Figure 1. This structure contains one type of hole transport layer and one type of electron transport layer, so it is monochromatic. Three layers are shown as constituting the light-emitting heart of the device: hole and electron transport layers and a recombination zone, shown in black. All of these three elements need not be present. The recombination region need not be a separate material but could lie in either of the transport layers. Further, the mobility of one of the injected charges could be significantly higher than the mobility of the other. This would allow the use of a single charge transport layer, with the recombination occurring near one of the electrodes. Devices on the today make use of these possibilities.

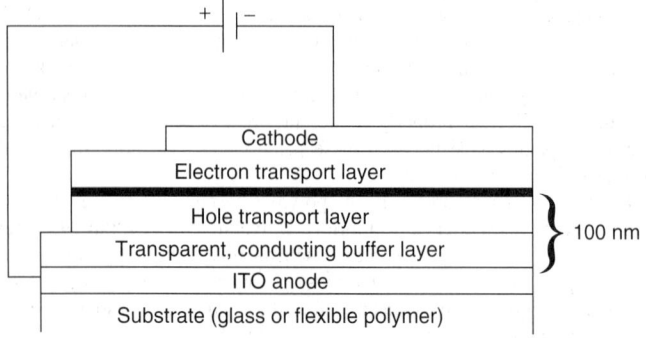

Fig. 1.   Monochromatic OLED Device Structure.

Although the specific materials and fabrication processes for this type of device will be discussed later, it is worth noting a few features at this time. The cathode is normally an opaque metal, so the light is emitted through the transparent indium tin oxide (ITO) anode and substrate. In Figure 1 the cathode and anode are drawn such that they both differ in size from the transport layers in order to illustrate one of the marked advantages of

OLED devices. Because the in-sheet electrical conductivity of the transport layers is very low (organic materials typically conduct far more poorly than inorganic semiconductors), light will be emitted only from the region under the cathode in Figure 1. A patterned cathode gives patterned light emission, with each pixel being electrically addressable independently. Further, the techniques for depositing the various layers in the device can yield uniform film properties over significant area. These two facts taken together mean that it is surprisingly easy to fabricate extended OLED devices, which can function as monochrome displays (with centimeter-scale diagonal dimensions in today's products, and tens of centimeters diagonal in the laboratory). Patterned cathodes can be deposited through a mask in a single step; one need not tile a large number of point LEDs together to obtain the display.

White OLED devices can be considered monochromatic because they emit "one" color of light—white (which is, of course, composed of all colors). They can be made in a variety of ways, of which the easiest is doping a hole or electron transporting polymer layer with red, green, and blue fluorescent dyes chosen in concentrations so that the emission is white. Very bright white OLED devices have been made in this fashion.

Full-color OLED devices, which can produce a broad gamut of colors by combining different proportions of red, green, and blue light, present much larger technical challenges. Thus, the first product marketed by Pioneer was a monochrome green display. Newer generations of Pioneer's products are displays with separate red, green, and blue areas. They are not full-color; they are tri-color. The issues can be appreciated by examining Figure 2, which is a schematic representation of a single full-color pixel. To span the entire color spectrum, it is necessary to use separate blue, green, and red subpixels, for which the voltage must be controlled individually. The materials in each subpixel differ from each other, at least in doping and probably in the small molecule or polymer itself. This means that the depositions of the three light-emitting layers cannot be done simultaneously; they must be done sequentially through masks. For large-area displays (such as computer monitors and televisions), manufacturing problems can arise with the registration of individual masks to high positional accuracy. Several companies have demonstrated full-color, 5-inch-diagonal passive matrix displays, and Sanyo has demonstrated a full-color active matrix display, but commercial products are not yet available.

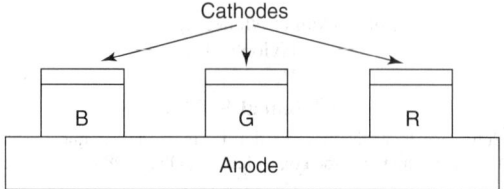

Fig. 2.   Full-Color OLED Pixel.

An approach proposed by Universal Display Corp., Princeton, New Jersey, (http://www.universaldisplay.com/) is to locate three primary color devices in a vertical stack in order to produce full color. The SOLED, or Stacked OLED, device could improve pixel density (resolution) when it is eventually implemented in a manufacturing line. The potential drawback is that light from the bottommost device must pass through two other devices on its way to the viewer's eye. An alternative structure used in some color TFEL systems is to pair up red and green on one level and to stack the blue on top in a two-layer device.

All full-color, high information content displays demonstrated so far have been made using small-molecule technology. Small molecules are deposited by vacuum vapor deposition, a process in which the use of very fine masks with very accurate and precise placement is a common manufacturing skill. Polymer OLED devices, however, are usually made by spin-coating the materials onto the substrate, a technique that is not compatible with masking. The differences between various display materials and device fabrication processes are discussed in the next two sections.

OLED devices are changing rapidly as development proceeds in more than 80 companies as well as many universities. However, it is worth tabulating the current specifications of OLED devices, both small-molecule and polymer based. These are presented in Tables 1 and 2.

TABLE 1. PERFORMANCE PARAMETERS FOR THREE SMALL-MOLECULE OLED DEVICE

	Blue	Green	Red
Operating voltage (V)	10	8	9
Luminous efficacy (lm/W)	0.56	3.9	1.3
Luminance (cd/m^2)	355	1,980	77

TABLE 2. PERFORMANCE PARAMETERS FOR THREE POLYMER OLED DEVICE

	Blue	Green	Red
Operating voltage (V)	4.13	4.07	3.5
Luminous efficacy (lm/W)	1.29	7.3	1.3
Luminance (cd/m^2)	340	1,900	300

One set of materials does not perform significantly better than the other. The major distinction is in operating voltage, where small-molecule voltages are about twice as great as polymer voltages. While smaller voltages are clearly desirable, it is possible that further research will diminish this gap.

The luminances in both tables are certainly adequate for OLED devices to be used as computer monitors or televisions. Much higher luminances have been observed in the laboratory, with a few reports of 100,000 cd/m^2, which is adequate for some outdoor lighting in daytime. However, lifetimes at such luminances are measured in seconds; such bright devices with long operating lifetimes are not yet on the development horizon.

The lifetime of an OLED device is an important concern, and state-of-the-art parameters are changing rapidly. At the luminances listed in Tables 1 and 2, It is common today for experienced laboratories to observe lifetimes in excess of 10,000 hours. These are generally acceptable lifetimes in a practical sense, but there are two further issues. First, in the particular case of monitors, even 10,000 hours may be too short; a computer monitor operating 10 hours per day would require only a bit over three years to accumulate 10,000 hours. Second, lifetime as defined in the literature is usually the time required for the device's luminance to fall to 50% of its initial value. For some applications this luminance falloff would be intolerable. The Pioneer display mentioned previously corrects for the temporal falloff in lifetime by correcting the drive current electronically.

Sato et al. have published a study of the degradation of small-molecule OLED devices. At the time of their study (1997), devices with initial luminances greater than about 300 cd/m^2 had lifetimes of less than 10,000 hours. For initial luminances of 500 cd/m^2, the lifetime was only about 4,000 hours, and the lifetime dropped to about 300 hours for an initial luminance of about 1,500 cd/m^2. A similar compilation of data obtained in 1999 would likely show improved lifetimes at all luminance levels.

The color performance of OLED devices matches well with the established NTSC and PAL standards for high definition television.

Organic Light-Emitting Molecules

Small Molecules. The selection of organic molecules suitable for OLED devices involves considering several variables. High electroluminescence efficiency is correlated with high fluorescence efficiency, although this is not a very restrictive requirement because many organic molecules have high fluorescence efficiency. More important is the fact that practical devices must operate at a low voltage (well below 10 volts). To obtain applied electric fields large enough for useful electroluminescence, the organic films must be very thin (a few tens of nanometers). They must also be very uniform in thickness and composition to assure uniform light intensity over the entire device. And finally, ease of manufacturing and cost considerations come into play.

Also, the emitted light must be bright. The brightness is determined by the rate of electron-hole recombination and is proportional to the current, after the threshold has been reached. This implies a high electrical conductivity for the transport layers so that, for a reasonable voltage of less than 10 V, emission is not limited by the number of charges available for recombination. For inorganic semiconducting materials, it is easy to find materials with useful electrical conductivities that also satisfy all other device variables. Organic materials are significantly less conductive, however, and present a serious challenge.

Effective conduction also relies on the electrodes' ability to efficiently inject electrons at one end and holes at the other end of the device. Thus, the anode work function should match the energy of the highest occupied molecular orbital (HOMO, roughly equivalent to the valence band) of the hole conduction layer. Similarly, the cathode work function should match the energy of the lowest unoccupied molecular orbital (LUMO, roughly equivalent to the conduction band) of the electron conduction layer.

Full-color light-emitting devices require the availability of blue, green, and red emitters, each of which satisfies all of the conditions discussed previously, and each of which can be deposited into the device without affecting any of the previously deposited materials. A fairly complete color palette can be constructed from small-molecule emitters.

Finally, a practical device must emit bright light for tens of thousands of hours of operation. Small-molecule OLED devices usually contain amorphous electron-transporting and hole-transporting layers, which must remain unchanged (that is, must not crystallize) through the normal exposure to high electric fields, high photon fluxes inside the device, and operating temperatures significantly higher than room temperature.

A large number of conjugated organic molecules that satisfy all of the criteria above well enough to qualify for commercial device applications have been found[4,5]. Figure 3 shows examples of the molecular structure of the electron- and hole-transporting layers for a prototype Kodak device. This device produces a fairly pure green light.

The performance of small-molecule OLED devices can be improved by doping the electron transport layer, the hole transport layer, or both with photoluminescent dyes. This has been shown to widen the color gamut, increase efficiencies, and reduce line widths of the emissive bands[5]. In the case of Alq (or one of its derivatives), successful dopants include perylene (blue), coumarin-6 (green), or DCM2 (red).

Although it is not strictly a p-n device, a new small-molecule technology from Opsys warrants mentioning here. Opsys has developed a class of materials called organolanthanide phosphors (OLPs) that rely on light emission from a lanthanide ion embedded in an organic shell (thus, they are organic EL devices, not OLED devices). The reason this development is significant is that it offers the potential for very stable devices through the separation of the light-emitting components from the more fragile organic components. Stability is a primary concern, as short lifetime continues to be a drawback of organic EL and OLED devices.

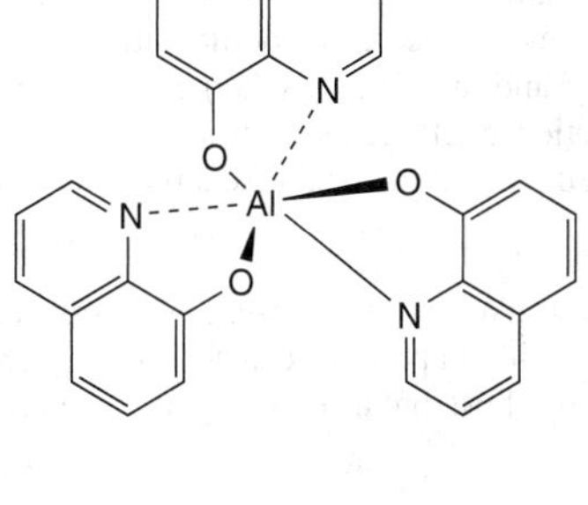

tris(8-hydroxyquinolinato)aluminum (Alq)
Electron transport material

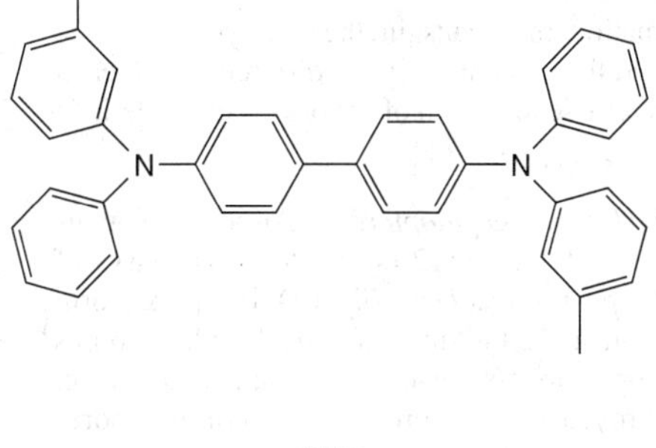

TPD
Hole transport material

Fig. 3.   Electron- and hole-transporting molecules.

Polymers. Organic polymers are composed of finite-length chains of atoms called *segments*, which are attached to a carbon-carbon backbone. The bulk polymer is made up of a three-dimensional, irregular array of polymer chains. As with the organic molecules described above, most commercial polymers are poor electrical conductors. However, those that do conduct have electrical conductivity far superior to that of small-molecule materials, rivaling that of semiconductors for current flow along the polymer backbone. Polymer materials are generally applied to the OLED device as monomers by a spin-on process. The monomers are then polymerized through light (usually ultraviolet), heat, or both. See also **Colloid System**.

The selection of a suitable polymer for OLED applications involves issues similar to those discussed for small molecules, but there are some differences. First, it is the polymer film, rather than the monomer molecules from which it is made, that must have the desired electroluminescence qualities. Second, unlike amorphous films of organic molecules, polymer films are unlikely to crystallize with exposure to electric fields, large photon fluxes, or high temperatures.

However, there are two common difficulties with polymer films that do not occur with small-molecule films. The first is exposure to temperatures higher than the glass transition temperature (Tg), the temperature at which the polymer chains have enough energy to make significant changes in their position or orientation. Upon cooling back below Tg, the polymer film will have a different microscopic structure, which will almost certainly degrade or destroy performance. Tg is typically $50-100\,°C(122-212\,°F)$, but has been pushed above $200\,°C(392\,°F)$ in some materials. Second, as polymer films age, they often lose free volume, which is called *shrinkage*. The speed and magnitude of this effect depend on the specific polymer and its state of crosslinking, but, overall, shrinkage causes polymer OLED devices to have a limited shelf life.

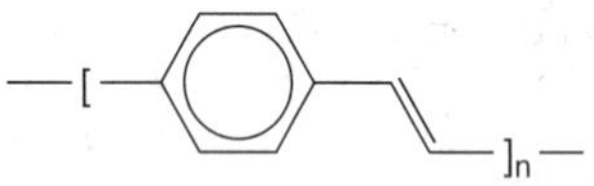

Fig. 4.   Structure of *p*-phenylenevinylene (PPV).

The structure of one commonly used polymer, *p*-phenylenevinylene (PPV), is shown in Figure 4. PPV or a comparable material will be used in the orange-emitting polymer devices being released by the PolyLED business unit of Philips and has also been employed successfully by CDT. See also family of articles catalogued under **Flat Panel Display Technology**, including **(Inorganic) Light-Emitting Diode Displays**; and **Semiconductor**.

For additional reading, refer to Flat Panel Display Technology entry.

Stanford Resources, Inc., San Jose, CA.

ORGANISMS (Dioecious). Organisms that produce only one type of reproductive gamete; they are usually classified as either male or female. Diecious organisms are the opposite of monecious organisms, which bear both male and female gametes. In botanical usage, however, the word diecious may be applied to trees that bear distinct male and female flowers or cones even though both may be on the same tree. Most higher animals, including all of the arthropods and chordates, are diecious, but most of the higher plants are monecious. Most of the flowers produced by the angiosperms bear both male and female organs.

ORGANOGENY. The formation of organs in the embryo. After the formation of the germ layers and their initial stages of differentiation, each layer or its subordinate parts gives rise to certain of the organs characteristic of the adult.

ORIENTAL FRUIT MOTH (*Insecta, Lepidoptera*). An inconspicuous gray moth with a wingspan of about $\frac{1}{2}$-inch (12 millimeters). The larva of this species, *Grapholitha* or *Laspeyresia molesta* (Busck), is a pink worm with a brown head and up to $\frac{1}{2}$-inch (12 millimeters) long. The larva bores into twigs and new shoots of apple, apricot, peach, pear, plum, and quince. The twigs and shoots are destroyed in short order. The larva also bores into the stem end of fruit and eats the pulp. This damage also increases the susceptibility of the tree to disease. The insect was probably imported from the Orient about 1915 on nursery stock. It is well established in the eastern United States and has been found on peach in some of the western states.

In some regions, early maturing varieties of peach and apricot may be planted, in which case the fruit often can be picked before the pest attacks the fruit. Damage can be prevented in this manner even in heavily infested areas. The soil around an infested tree should be cultivated to a depth of 1 to 4 inches (2.5 to 10 centimeters) about 1 to 3 weeks before blooming time. This cultivation will kill many of the overwintering larvae in the soil.

When the fruit moth becomes a serious pest, a dust impregnated with a light-grade mineral oil can be applied. An effective formulation is: 60% sulfur, 35% 300-mesh talc; and 5% light-grade mineral oil, by weight. The dust should be applied at 5-day intervals, starting about 20 days before the peaches are picked. The oil dusts act as irritants and not as poisons.

ORIFICE. An orifice is an opening having a closed perimeter through which a fluid may discharge. The orifice may be open to the atmosphere, which is the case of free discharge, or it may be partially or entirely submerged in the discharged fluid. The standard orifice is the sharp-edged orifice shown in Fig. 1, but other types, such as the well-rounded orifice, the partially rounded orifice, the Borda mouthpiece, and the short tube orifice, have their special uses. An orifice may be very small, as in the case of those used for leak ports or for calibration, or large, as illustrated by sluice gates in a dam. The head of water on the orifice is measured from the water level surface to the center line of the orifice. Should the head above the orifice be so small as to be less than approximately the vertical dimension of the orifice, the following remarks will not apply, as this would come under the special case of large orifices under low heads.

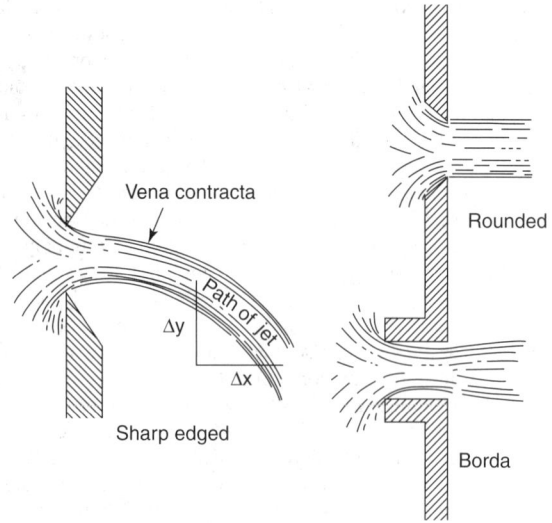

Fig. 1.   Orifices.

The streamlines in water approaching a sharp-edged orifice converge on the orifice from all directions, and so continue to converge for approximately one-half of the orifice diameter downstream. The jet contracts to a section somewhat smaller in diameter than the orifice, after which it increases in size. The contracted section is known as the *vena contracta*. The ratio of the cross section of the jet at the vena contracta to the area of the orifice is known as the contraction coefficient. Friction in the orifice slows the velocity to a somewhat lower value than the ideal free spouting velocity, which is $\sqrt{2gh}$. The ratio of actual to spouting velocity is the velocity coefficient. Since the discharge is the product of velocity and area, the discharge coefficient is the product of velocity and contraction coefficients. It has a numerical value of 0.61 for the average sharp-edged orifice. The discharge from an orifice of area *a* is

$$Q = .61a\sqrt{2gh}$$

The path taken by a jet discharging freely horizontally under head of *h* is parabolic in shape due to the pull of gravity acting on a particle having, originally, horizontal motion only. The equation

$$x^2 = 4C_v^2 hy$$

gives a curve of the center of the path of the jet, C_v is the velocity coefficient which averages 98% for sharp-edged orifices. Suppression of the contraction of a jet increases the discharge from an orifice. An orifice on the side wall of a tank near the bottom has a higher coefficient of contraction than one which is located farther away from the bottom. Similarly an orifice with the upstream edges rounded has a higher coefficient of contraction than one with sharp edges. The discharge may be as much as 30% greater for well-rounded orifices. Orifices which are submerged, orifices which are squared instead of circular, and orifices in which the water approaches with a high velocity, cannot be treated by the equation given above without corrections being made for these special conditions.

The foregoing discussion relates to the flow of water through an orifice. Orifices are much in use for measuring flows of vapors and gases. The method employed is to place an orifice of some type in the pipe or duct carrying the fluid. By means of a manometer, or pressure gauge, the upstream and downstream pressures are measured, and the discharge can be determined from those readings coupled with the known area of the orifice. The flow of a gas through an orifice depends on the area of the orifice, the upstream pressure, the temperature, and a factor which involves gas constants, such as the ratio of the specific heats at constant pressure and constant volume, and the ratio of the upstream and downstream pressures. The formula for the weight of gas flowing is $C_1 C_2 aP/\sqrt{T}$ (pounds per second). C_1 is the constant just mentioned, C_2 the velocity of approach correction, P the upstream pressure, a the area of the orifice, and T the absolute temperature. Many steam flow meters are based on the principle of the orifice. A sharp-edged, or thin-plate, orifice is clamped between the flanges at some joint of a flanged steam line. Pressure leads are taken from upstream and downstream sections to an instrument which is a pressure-measuring device, but which may be calibrated to read steam flow. See also **Flow Measurement (Liquids and Gases)**.

ORIGINS PROGRAM. Have you ever looked up at the night sky, marveling at the vastness of the Universe and your own connection to it?

It is hard to communicate the full sense of wonder that floods through us at such a moment, but we all understand. At least once, the dimly glittering night sky has stopped us in our tracks, bringing quiet contemplation of how the Universe came to be and what our relationship is to everything within it.

NASA's Origins Program seeks to answer two enduring human questions:

Where do we come from? To answer this, we need to understand the astronomical, physical, planetological, and biological processes necessary to generate and sustain life on earth. Knowing "where we come from" means understanding how the great chain of events unleashed after the Big Bang culminated in us and in everything we observe today. It is the story of our cosmic roots, told in terms of all that precedes us; the origin and development of galaxies, stars, planets, and the chemical conditions necessary to support life. See also **Cosmology**.

Are we alone? To answer this we need to understand the building blocks of life and the conditions necessary for life to arise. We need to search our solar neighborhood to see if such conditions exist elsewhere. We need to search for signatures of life. Knowing our uniqueness–"whether we're alone" in the cosmos–depends on our search for life-sustaining planets and on our understanding of its glorious diversity here on Earth. Only by seeing the innumerable possibilities on our home planet can we be sure that we'll recognize life if and when we find it somewhere else.

Ever since humans became capable of thought and reason, we have pondered these questions. Our ancestors, huddled around their ancient campfires, must have wondered about such mysteries. The questions are abstract, and profound in their implications; yet seem so natural that the youngest children gathered in modern classrooms ask them today.

Our generation is privileged to live in an era in which advances in science and technology allow us to investigate these intriguing questions. While the questions can be simply stated, the scientific and technical foundations needed to answer them are challenging.

Over the course of the next two decades, the Origins Program will develop the sophisticated telescopes and technologies that will bring us the information we seek. Although the questions are challenging, our generation is privileged to have the technological ability to reveal the possibilities for the first time. Just as the Greeks were known for democracy, the Egyptians for pyramids, and the Romans for roads, our civilization may well be remembered for discovering life beyond our own planet, forever changing our perception of the Universe and our place within it.

Science

Approximately 15 billion years ago in cosmic history, the first galaxies took shape from vast clouds of early chemical elements.

In the furnace of stars, life-sustaining chemicals such as carbon and oxygen came into being. Then, in awe-inspiring blasts from dying stars, life's chemicals blew out into space, only to condense anew into stars like our sun and planets like Earth.

Through the mixing of these vital chemicals and energy, the living Universe blossomed with the earliest self-replicating organisms and the profusion of life on our planet. Seeing similar chemical conditions wherever we look in the cosmos, the hope of finding life somewhere else rises inevitably within us.

To seek answers to the two defining questions of the Origins Program, scientists have outlined four goals that will speed us on our way to discovery:

Goal One. To understand how galaxies formed in the early universe. One of the biggest mysteries in science is how the Universe went from a uniform, relatively smooth structure to the clumpy, galaxy-strewn expanse we observe today.

Our telescopes are beginning to look farther and farther back in cosmic time, but we have almost no information on the span stretching from 100 million years to 1 billion years after the Big Bang. In many senses, we can think of this age as "the cosmic dark ages." For Origins, it is also one of the most interesting times of all, because that is when the first galaxies began to form from a vast sea of tiny particles. Consider it a time for the seeds of our cosmic roots.

Gravity's Role in Galaxy Formation

To understand how galaxies formed, the Origins Program will be taking a look at gravity in great detail. In the early Universe, uniformly distributed matter began to gather as gravity weakly acted upon it. Small variations, or ripples, began to appear as matter accumulated in different regions and began to grow. The result? The first galaxies were born.

By accurately mapping the amount, distribution, and chemical content of gaseous matter in the early Universe, Origins missions will tell us more about how this process took place. Detecting light from the Universe's very first generation of stars will also tell us when the first structures began to shine, generating the chemicals necessary for life.

How Galaxies Produce Chemicals for Stars, Planets, and Living Organisms

Life as we know it depends on the complex chemistry of organic matter, and yet the very early universe was made only of hydrogen, helium, and deuterium. None of the atoms necessary for life–carbon, for instance–had yet been formed.

One of the triumphs of 20th-century science is understanding that the elements needed for building planets and for supporting life are forged in the fiery furnace of stars. The chemical elements produced by these stars are gravitationally bound to the galaxies in which the stars "live."

By looking at the chemical composition of galaxies over cosmic time, we can see how their store of heavy elements grows throughout the ages. We want to understand how such chemical enrichment takes place and how new stars and planets can form the increasingly available materials. The answer will tell us if the development of a giant galaxy like our own Milky Way is essential to the eventual emergence of life. See also **Galaxy**.

Goal Two. To understand how stars and planetary systems form and evolve. Great clouds of gas and dust give birth to the stars, as neighboring atoms pull together in a gravitational bond. This bond grows stronger and stronger as more material packs in. See also **Star**.

So massive and dense does the central core become that nuclear fusion bursts on the scene, and a young star begins to shine. The energy is so intense that it blows most of the remaining gas away, leaving a disk of dusty material from which planets form. These "leftover" particles orbit the new star, knocking and gravitating together to form new worlds.

Our observations have shown that this birthing process is going on all the time. We're lucky, in fact, to have so many examples of star and planetary systems, all at various stages of formation. The Origins Program wants to study these transformations in much greater detail, hoping to answer perhaps the most intriguing question of all: whether our solar system (with its life-sustaining Earth) is a common outcome of planet formation or very rare.

Young Planetary Systems in Formation

Most stars have a twin nearby–or even multiple companion stars. As we know quite well, our own star (the sun) seems to be an exception. That's good news for us, of course, because our planet Earth does quite well with its stable, single sun for life-giving energy.

But why are we so different? The answer probably stems from initial conditions in the gas cloud from which our solar system emerged. That's why we want to study the life histories of young stars and planetary systems all the way back to their origins. Measuring the temperature, density, and chemical conditions of the planetary disks will help us understand how, and how often, solar systems like our own emerge.

Greater analysis of dust disks at later stages will also tell us more about the role of gravity in planet formation. As small dust particles coagulate into larger and larger grains to form planets, they increasingly draw in nearby materials as they orbit around their parent stars. This depletion of surrounding material produces detectable gaps in the disks, giving us a "high sign" that faraway planets are in formation.

Mature Planetary Systems

What really drives the Origins quest forward is our desire to know if, somewhere out there, another Earth exists in a mature planetary system. The best way to find out is to take a census of nearby stars and any planetary systems around them.

Origins will therefore survey about a thousand of the closest stars to us, as well as a significant sample of more distant ones. In our inventory of the galactic "neighborhood," we want to take a look at the orbital characteristics and physical properties of planets to find out what they're like. Mass, temperature, and atmospheric composition will be particularly important.

We will also try to study the parent stars in greater detail to find out how their characteristics might influence the kinds of planets that end up forming. We might even be able to say when the earliest planets began to take shape in our own home galaxy, the Milky Way.

By far the most fascinating search, however, will be our survey of planets that are about the same mass as Earth. Analyzing their characteristics will help us determine how frequently habitable worlds like our own are born. With that knowledge, we should begin to know whether life is exceedingly rare... or a cosmic imperative instead.

Goal Three. To determine whether habitable or life-bearing planets exist around nearby stars.

What does it mean to look for an Earth-like planet? The question is actually more complex than you might think. Our own home planet has been around for about 4.7 billion years, and it has not always been the world we know today. Think of the ice age... or the dinosaur age... or much further back—some 2 billion years ago—when living material on Earth had only just begun to pump oxygen into the atmosphere. While we would not have survived back then, Earth was abundant in life forms that could.

To find a habitable planet, then, we have to expand our mental horizons to encompass our planet's past as well as its future—and all of the possibilities in between. We have to think about what would happen if Earth were slightly different—larger or smaller, warmer or colder, with different gravitational or chemical conditions—and what effect that would have on the possibility for life.

Worlds Where Life Can Thrive

Based on our only example (Earth!), life seems to need a couple of key ingredients: liquid water, key chemicals such as carbon, and a source of energy for the complex chemical reactions that take place in all living organisms.

Of the 100 billion stars in our galaxy alone, there are certainly enough stable stars out there to provide energy for life's chemical transformations. We also find plenty of carbon and other necessary chemicals just about everywhere. Based on those two factors, the odds look good that life is possible somewhere else, but that still leaves the question of liquid water. One of the most important steps is to figure out what kinds of planets are likely to have water in a flowing, life-enabling form, and then to find out how commonly they form.

For a habitable planet, location is everything. You do not want to be too close to the sun, or water would boil away. Too far away, and water would freeze. A stable, more circular orbit is also important, because wild swings toward and away from the sun would not support a stable supply of liquid water over time. Those simple limits help us narrow down our search to the "Goldilock's Zone"—that is, a "geographical" band around stars where planets would neither be too hot nor too cold, but rather "just

right." The Origins census of nearby stars will look for planets in this comfort zone, giving us a sense of how many other life-supporting worlds might exist.

Identifying Life-bearing Worlds

No matter how different conditions on other world might be, we do know one thing: life and its environment are inextricably linked. Life changes a planet's condition as it takes in food and energy and releases waste products. A changing planetary environment, due to biological, geophysical, or climatic activity, in turn causes life to adapt, resulting in the rich diversity in plants and animals that we encounter all around the world.

Exploring this continuous cause-and-effect dance on Earth will help us comprehend the intertwining relationship between life and host planets everywhere. Nowhere is this relationship more apparent than in the observed characteristics of an atmosphere, especially combined with the planet's temperature and orbital location. Origins will particularly look for the presence of carbon dioxide, ozone, water, and chemical combinations that would not tend to occur in nature without biological activity.

Of course, it is not enough to look for chemicals that only indicate the presence of life as we find it throughout Earth's history. See also **Earth**. The Origins Program will therefore be developing a catalog of all the possible chemical signatures for habitable planets and for life on them. In this research, it will be very important to identify how atmospheric gases produced by geological activity differ from those produced by life. After all, we would not want to mistake a barren planet, bursting with gaseous chemicals, for one abundant in life. Reconstructing the environmental and biological history of Earth, while identifying the extreme environments in which life has flourished here, will give us a good sense of the widest environmental limits in which life is possible.

Goal Four. To understand how life forms and evolves.

Life on Earth covers the gamut, from the most primitive life forms to the most complex. With the help of microscopes, we have peered into everything from the world of single-celled organisms to the 100-trillion cells that make up the human body. No matter which living system we study, we find that each cell is largely made of proteins, and that each protein itself is a complex molecule made of millions and millions of atoms, arranged just so.

It turns out that six atoms are particularly important to life: hydrogen, carbon, nitrogen, oxygen, phosphorus, and sulfur. Ninety-eight percent of the material in all living organisms is made of these atoms. Among them, carbon is the most important of all, since it has a special chemical property that likes to bind with other atoms. Carbon is therefore the "glue" that holds life's large and complex molecules together. That is why we tend to talk about "carbon-based life"—it is the only life we know. See also **Carbon**; **Hydrogen (Fuel)**; **Nitrogen**; **Oxygen**; **Phosphorus**; and **Sulfur**.

While we can generally say that life is based on a few select atoms and the presence of moisture and energy, which hardly accounts for life in all its glory. How living systems first emerged from these basic conditions remains a fundamental mystery that the Origins Program will explore.

How Matter is Organized Into Living Systems

The first thing we want to understand is how organic (carbon-based) molecules formed on our planet. Some scientists propose that comets, meteorites, and cosmic dust may have brought them to Earth. Others propose that they formed in the early atmosphere, in hydrothermal vents in the ocean, or in geothermal environments on land. The geological record of Earth, as well as the composition of bodies in our solar system, may provide key insights into the source of organics.

The next critical step is trying to understand the all-important leap from basic organic materials to organized systems that could process energy and nutrients and transform them for life's biological processes. Origins laboratory research will study chemical reactions under conditions similar to Earth's early environment to understand how early biological structures may have emerged. Of particular interest is how the ancient counterparts of modern cells developed from simple structures, and came to process energy, metabolize, and transfer information to succeeding generations (genetics).

Of course, Earth's biological processes may not be the same as those on another world, as chemical conditions on other planets might have produced substantially different organisms. Having no other life-sustaining world for comparison, we do not know which of the properties we observe are necessary for life, and which are specific to life on our own planet. Therefore, the Origins Program will create laboratory models that exhibit "life-like" properties and use a variety of chemical "ingredients" for their make-up. In that way, we will be more confident about detecting life, however different, on another world.

Limits to Life in Different Planetary Environments

The abundance of life in the universe also depends on the ability of living systems to adapt to their environments. We know life on Earth thrives in a number of different places, and can find evidence of past life in fossil records when conditions on our planet were vastly different than today. By studying the Earth's history and micro-environments, we will better understand not only the conditions in which life can operate, but also the environments in which it can not.

In order to establish the limits for life, Origins research will look at self-replicating molecules and other systems, seeking clues for the way in which living systems adapt and pass on genetic information that enables future generations to survive. The one thing we know is that life is hardy, and finds a niche in wide range of environments, even those that seem impossibly extreme and hostile. Identifying the environmental limits to life will improve our ability to analyze the potential for life on other worlds.

Because the diversification and survival of early life on Earth depended on the ability of microbial communities to live in harmony with their environment, Origins researchers will also take a look at how these communities cooperate and compete to harvest energy and nutrients. These studies are important because microbial life accounts for most of the living material on Earth. It is likely that another life-sustaining world would be filled with these simple forms of life too. Understanding the biosignatures (the life signs, or markers) left by these microbial communities in planetary rocks and atmospheres will help us identify the chemical signature for life that we hope to observe on a distant world someday.

Astrobiology is the scientific discipline that will make all of these studies possible. See also **Astrobiology**.

Missions

For the first time in history, humanity is on the verge of having the technological capability to explore age-old questions about our cosmic origins and the possibility of life beyond Earth.

Even with the best talents and technologies available, our plans are ambitious and daring. To collect faint light from the first-ever galaxies and from Earth-sized planets around distant stars, we would essentially need telescopes the size of Texas. We have not achieved that capability here on Earth, let alone in space, and even if we could do it, the effort would triple the national debt.

That is why the Origins Program has embarked on a series of closely linked missions that build on prior achievements. As each Origins mission makes radical advances in technology, innovations will be fed forward, from one generation of missions to the next. By the end of the decade, we will have combined the very best imaging, formation flying, and other visionary technologies, giving us the power of enormous telescopes at a fraction of the cost.

Support for these missions is provided by the Interferometry Science Center (ISC), *http://isc.caltech.edu/* a science operations and analysis service sponsored by the Origins theme and operated by the California Institute of Technology. The ISC facilitates timely and successful execution of projects that use interferometry, a key technology in the Origins Program.

While the majority of Origins efforts are focused on developing space-based observatories above Earth's atmosphere, investigations on the ground pave the way for future achievements in orbit.

Space-based Observatories

The Origins missions form a family, in which each generation passes on a rich technological heritage to those that come after. Much like in human families, each mission has something unique to contribute, yet is closely tied to the others to form a supportive web. Origins currently has four chronological generations that move technology and knowledge forward:

The Precursor Missions

Hubble Space Telescope (HST). Our first-ever, long-term space observatory, HST reveals stunning views of the Universe with 10-times better resolution than any ground-based telescope.

Since 1990, NASA's Hubble Space Telescope (HST) has given us over 14,000 images of the Universe, allowing us to see stars exploding, galaxies colliding, planets forming, and other spectacular wonders that occur throughout our dynamic, evolving, ever mysterious Universe.

A cooperative program between NASA and the European Space Agency (ESA), the HST is the world's first long-lived, space-based observatory. See also **Hubble Space Telescope (HST)**.

Far Ultraviolet Spectroscopic Explorer (FUSE). FUSE looks at the Universe in a whole new light by studying objects in the ultraviolet portion of the spectrum, which is unobservable with other telescopes.

For hundreds of years, astronomers were only able to explain the Universe through the visible light that their eyes could see. NASA's Far Ultraviolet Spectroscopic Explorer (FUSE) gives astronomers a new tool for their exploration: a space telescope that studies the far ultraviolet light that is both invisible to us and is largely filtered out by Earth's atmosphere.

Far ultraviolet light in the Universe can help us understand more about conditions right after the Big Bang, the dispersion of chemical elements in galaxies, and the composition of interstellar gas clouds from which stars and planets form. A complement to other Origins missions.

FUSE's focus on the far ultraviolet permits astronomers to study the many important atoms, ions, and molecules that cannot be investigated otherwise.

FUSE was developed for NASA by the Johns Hopkins University, which has the primary responsibility for all aspects of the mission. Collaboration also comes from the Canadian and French space agencies, which share in observing time.

FUSE was launched on June 24, 1999, with a projected operational life of three years. For further information on the FUSE Mission see: *http://fuse.pha.jhu.edu/*

Space Infrared Telescope Facility (SIRTF). Able to see infrared (heat) radiation, SIRTF will peer through the veil of gas and dust that obscures most of the Universe from view.

Giant clouds of gas and dust block most of the Universe from view. NASA's Space Infrared Telescope Facility (SIRTF) will lift "the cosmic veil, " looking through these clouds to reveal stars forming in the heart of dusty galaxies, brown dwarfs, and even galaxies that existed near the beginning of time. It will also be able to characterize the disks of gas and dust around stars from which planets eventually form.

Because infrared radiation measures heat, astronomers have to cool the telescope to near absolute zero (-460 degrees Fahrenheit) so that the telescope can observe distant places in the Universe without interference from the heat of the near-Earth environment. With proper shielding from the sun, SIRTF will be launched into an Earth-trailing solar orbit that allows the telescope to cool rapidly. Rather than carrying large amounts of onboard cryogen (coolant), SIRTF also pioneered an innovative "warm launch" architecture that allows the telescope to cool in the frigid vacuum of space. This innovative design significantly reduced mass and launch costs.

SIRTF is the final element of NASA's Great Observatories Program, a series of four space-borne observatories designed to study the universe over many different wavelengths. The observatory is also a bridge to the Origins Program, providing information that will help us understand the formation and development of galaxies, stars, and planets. SIRTF is scheduled for launch July 2002, with an operational life of 2.5 years, minimum. See also **Chandra X-Ray Observatory**; and **Space Infrared Telescope Facility (SIRTF)**:

Stratospheric Observatory for Far Infrared Astronomy (SOFIA). The world's largest airborne telescope, SOFIA will make observations that are impossible for even the largest and highest of ground-based telescopes.

The Stratospheric Observatory for Infrared Astronomy (SOFIA) is a Boeing 747-SP aircraft that carries a 2.5 meters (8.2 feet) reflecting telescope. The largest airborne telescope in the world, SOFIA will make observations that are impossible for even the largest and highest of ground-based telescopes. Another advantage SOFIA offers is the opportunity for teachers and the media to experience science in action, on-board as guests.

One of SOFIA's primary goals will be to study the properties of the interstellar medium—clouds of gas and dust that lie between stars in a galaxy. These clouds are important, because new stars and planets will eventually form from them. SOFIA will measure the infrared (heat) emissions from the clouds, seeking to understand their chemical makeup. Particularly interesting is the presence and abundance of carbon, which cools the interstellar medium and alters its subsequent chemical evolution.

Carbon chemistry is important to study, because it is the basis of life as we know it. SOFIA will also study star and planet formation, along with other important Origins questions.

SOFIA is a joint project between NASA and the German Space Agency. The observatory is being developed and operated for NASA by a team of industry experts led by the Universities Space Research Association (USRA). SOFIA will be based at NASA's Ames Research Center at Moffett Federal Airfield near Mountain View, California.

SOFIA is scheduled for launch in 2004, with an operational life of 20 years or more. For further information on the SOFIA Mission see: *http://sofia.arc.nasa.gov/*

The First Generation Missions

Our seeking continues with missions that all serve as technological parents of second-generation Terrestrial Planet Finder.

StarLight. StarLight's technologies will enable future spacecraft to detect Earth-sized planets around other stars.

StarLight's two small telescopes can achieve the resolution of a telescope mirror 125 meters (137 yards) in diameter — wider than a football field is long!

StarLight's formation flying technologies will control the distance between the two spacecraft to within less than 1 centimeter (0.4 inch) and the angle (bearing) to within 3 arc minutes (873 μradians).

StarLight is scheduled for launch in 2005, with an operational life of 12 months or more. For further information on the StarLight Mission see: *http://starlight.jpl.nasa.gov/*

Space Interferometry Mission (SIM). With a pinpoint accuracy several hundred times greater than any previous mission, SIM will begin identifying stars that have planetary systems around them.

Out of all the stars in the night sky, we wonder which ones might have planets swirling around them. NASA's Space Interferometry Mission (SIM) will continue the search by identifying stars that "wobble", that is, stars that are pulled back and forth as orbiting planets move from one side of the star to the other. If SIM sees such a gravitational tug, it infers the presence of planets. While this method of finding planets is indirect, it paves the way for Terrestrial Planet Finder and other missions that will eventually image the distant worlds first discovered by SIM.

While ground-based observatories have identified large, gaseous planets around other stars, only SIM's incredible precision will allow us to begin detecting the extremely tiny star wobble caused by an Earth-sized planet. In its efforts to find Earthlike planets that lie closer to their parent stars, SIM will also pioneer a technique to block out the star's light so that the tiny, faint, orbiting planets can be seen. Its ability to measure the position of stars several hundred times more accurately than any previous program will also allow to astronomers to determine the size and age of the Universe.

In an Earth-trailing solar orbit, SIM will receive continuous solar illumination, avoiding the occultations that would occur in an Earth orbit. (While this orbit is similar to SIRTF's, SIM is an optical telescope and thus does not need to block out heat from the sun.)

SIM is scheduled for launch in 2009, with an operational life of five years or more. For further information on the SIM Mission see: *http://sim.jpl.nasa.gov/*

Next Generation Space Telescope (NGST). Nearly four times the size of Hubble's mirror yet ultra-lightweight, NGST will study the very first stars and galaxies to emerge in the Universe.

The Next Generation Telescope (NGST) will look back to an extremely important period in the early history of the Universe when the first stars and galaxies began to form. While we have a fairly good understanding of the Universe in other periods, we have no observations during this time when the Universe was between one million and a few billion years old. NGST's studies will help us understand the shape and chemical composition of the universe, the evolution of galaxies, and the nature of unseen "dark matter."

NGST will study infrared (heat) emissions from this early time, seeing objects 400 times fainter than those currently studied with large ground-based telescopes or the current generation of space-based infrared telescopes. At eight meters (26.2 feet), NGST's primary mirror is more than three-and-a-half times as large as Hubble's, giving it much more light gathering capability. Its tennis-court sized sunshade will help eliminate the heat from sun, which is necessary for reducing heat "pollution" from the surrounding environment.

NGST is managed for NASA by the Goddard Space Flight Center. Contributions come from a number of industry, academic, and government partners.

NGST is scheduled for launch in 2009 (approx.), with an operational life of 5 to 10 years or more. For further information on the NGST Mission see: *http://ngst.gsfc.nasa.gov/*

The Second Generation Mission

The culmination of a decade's work, this mission will combine preceding technologies to begin revealing whether life is a cosmic imperative.

Terrestrial Planet Finder (TPF). Flying four advanced telescopes in formation, TPF will give us the first "family portraits" of other planetary systems, and maybe even a picture of a planet where life might exist.

NASA's Terrestrial Planet Finder (TPF) will study all aspects of planets, from their formation to their final characteristics. In addition to measuring the size, temperature, and placement of Earth-sized and other planets, TPF will look for gases such as carbon dioxide, water vapor, ozone, and methane that would indicate that a far-away planet could, or even does, support life.

TPF will find the tiny, faint planets around distant stars by reducing the glare of their parent stars a hundred-thousand times, taking pictures of planetary systems as far away as 50 light years. With pictures a hundred times more detailed than those of the Hubble Space Telescope, TPF will also allow us to study the black hole at the center of the Milky Way and other exciting phenomena in the universe.

TPF is scheduled for launch in 2012 (approx.), with an operational life of six years or more. For further information on the TPF Mission see: *http://tpf.jpl.nasa.gov/*

The Third Generation Missions

For now, these missions remain just a vision because the required technology is not on the immediate horizon. Today's missions, however, put us on the path toward such monumental achievements.

Life Finder (LF). Once we identify any habitable planets, Life Finder would fly telescopes at even larger distances to detect chemicals that actually reveal biological activities, that is, the presence of life.

If earlier Origins missions lead us to discover another world with life-sustaining conditions, it does not necessarily mean that life has actually emerged there. Life Finder (LF) will seek to determine if a distant planet with the right living conditions actually has an abundance of living creatures!

Life Finder will be even more sensitive than Terrestrial Planet Finder, but the principle of characterizing a planet's conditions is the same. If a planet harbors life, biological activity on the planet will impact the atmosphere, just as it does on our own home planet. When we analyze the radiation coming from the planet, we can look for much finer dips in the energy. These dips indicate the presence of methane and other chemicals we do not expect to find in nature unless biological activity is pumping it into the atmosphere.

We will also have to keep in mind the history of Earth in relation to a new world. The simplest life forms existed on Earth well before an abundance of oxygen appeared in the atmosphere, which in turn allowed multicelled organisms to flourish. Origins astrobiology research will help us expand our knowledge of "life signs" that would appear at different stages in a planet's history, as well as signs that would appear given a planetary chemistry that is not exactly the same as our own. With these insights, we will give ourselves the best possible chance of recognizing life if and when we find it somewhere else.

Planet Imager (PI). If we found a planet with life, we would not leave it to our imagination! To create a picture would require a number of telescopes flying in formation to achieve the power of a telescope 360 kilometers (225 feet) wide.

Just imagine knowing that another Earthlike planet is out there... and that for the first time we know exactly where it is! That could never be the end of the story, because our thirst for knowing about that distant planet would be enormous. Would it have continents like ours? Oceans as large? Clouds? Or would it be foreign enough to rival the creations of our very best science-fiction writers?

This new earth, this Terra Nova, would be at the forefront of our imaginations. We'd never be content just knowing it exists, we would want to see it for ourselves! The Planet Imager (PI) mission will produce pictures of single planets at much higher resolution than any preceding mission. Instead of seeing a planet as a single dot, we would strive for a larger image, consisting of as many pixels as possible. The greater the number of pixels, however, the more complex the mission.

We do have an idea of what we would need to accomplish our goal: an array of interferometers that each carried NGST-sized telescopes (about eight meters (26.2 feet). They would have to fly in exquisitely precise formation... over distances of 6,000 kilometers (3730 miles) or more! Right now, we're not even close to having the technology to accomplish the task, but the generations of Origins missions leading up to Planet Imager will pave the way.

Ground-based Observatories

Reaching toward incredibly capable space observatories must begin by stretching here at home. The following ground-based efforts provide the technological bedrock for future Origins missions.

Two Micron All-Sky Survey (2MASS). The Two Micron All-Sky Survey (2MASS) used two 1.3 meter (51-inch) telescopes at Mount Hopkins, Arizona, and Cerro Tololo, Chile to conduct the most thorough census ever made of our Milky Way galaxy and the nearby universe. The work was completed in February 2001, but data processing will continue through 2002. Catalogues produced from 2MASS data contain more than 300 million stars and galaxies, including previously undetected star-forming regions, as well as galaxies behind the disk of the Milky Way. The 2MASS data will help scientists prepare for future infrared space missions, including the Space Infrared Telescope Facility (SIRTF).

For further information on the Two Micron All-Sky Survey (2MASS) project: See *http://www.ipac.caltech.edu/2mass/*

Keck Interferometer. By connecting the twin Keck telescopes and combining incoming light, Keck will function as a single, much larger and more powerful telescope.

To expand the capabilities of the world's largest telescopes even farther, the Origins Program is in the process of connecting the twin Keck telescopes. By combining the light paths from each twin, astronomers will gain the capability of a single telescope the size of the distance between them. That's about the equivalent of an 85-meter (279-foot) mirror; almost the size of a football field.

With the addition of four proposed 1.8-meter (5.9-foot) "outrigger" telescopes that are located nearby, Origins astronomers eventually hope to have the ability to simulate a telescope with mirrors anywhere between 25 and 140 meters (82 and 459 feet).

For further information on the Keck Interferometer Array project: See *http://huey.jpl.nasa.gov/keck/*

Palomar Testbed Interferometer (PTI). By combining light from telescopes at the Palomar Observatory, PTI takes the first crack at developing technologies that combine light coming in from distant objects in the cosmos.

Located near San Diego, the Palomar Testbed Interferometer (PTI) is developing some of the technologies needed for the Keck Interferometer and the Space Interferometry Mission (SIM). PTI uses multiple telescopes to measure interference fringes created when light gathered by the telescopes is combined and processed. This technique enables scientists to measure the positions and distances between stars with great accuracy.

PTI's dual-star tracking system, the first of its kind, will help provide measurements that detect a star's "wobble," allowing us to infer the presence of planets orbiting around the star. PTI has also been used to make scientific observations of giant and supergiant star, as well as binary stars.

For further information on the Palomar Testbed Interferometer (PTI) project: See *http://huey.jpl.nasa.gov/palomar/*

The Large Binocular Telescope Interferometer (LBTI). Two 8-meter (26.2-foot) class telescopes on Mount Graham, Arizona, will be linked to create an infrared interferometer capable of imaging distant galaxies and other faint objects over a wide field-of-view.

Because of its unique geometry and relatively direct optical path, the LBTI will offer science capabilities that are different from other interferometers. It will be cable of providing high-resolution images of many faint objects over a wide field-of-view, including galaxies in the Hubble Deep Field with 10 times the Hubble resolution.

Nulling techniques will enable the LBTI to study emissions from faint dust clouds around other stars. These dust clouds reflect light and give off heat, and so interfere with the search for planets. By helping the characterize these emissions, the LBTI will provide critically needed data

for the design of the Terrestrial Planet Finder, a later mission that will study planets orbiting nearby stars.

For further information on The Large Binocular Telescope Interferometer (LBTI) project: See *http://medusa.as.arizona.edu/lbtwww/lbt.html*

For more information on the Origins Program: See *http://origins.jpl.nasa.gov/index.html*

See also **Astrobiology**; and **Universe (The)**.

NASA/Jet Propulsion Laboratory, Pasadena, CA.

ORIOLE *(Aves, Passeriformes)*. Brightly colored birds of the family *Oriolidae*, found in all parts of the Old World. The North American birds to which this name is applied belong to the family *Icteridae* and are more closely related to the blackbirds than to the true orioles. The orchard oriole, *Icterus spurius*, and the Baltimore oriole, *I. galbula*, are the most widely known species of the latter group.

ORION (The hunter). This constellation is on the whole, the richest and most impressive of all of the constellations. The "belt and sword" of Orion are frequently referred to in both ancient and modern literature. Despite the wide area covered, the physical characteristics of many of the stars are so similar that there is considerable evidence in support of the theory that they have a common origin.

The star Betelgeuse must be considered an exception to this class, for it is quite different from the other bright members of the constellation. Its color is a distinct yellow-orange as contrasted with the blue-white tint of the typical "Orion star."

The middle "star" of the sword is not a star at all, but is a huge emission nebula. This is one of the very few nebulae that can be seen to any degree of satisfaction through a small instrument. Of course, up to a certain point, the larger the instrument, the finer the view.

Several of the bright stars in the constellation are multiple objects, and the possessor of a 4-inch telescope will find the stars of Orion worth more than a passing glance. (See map accompanying entry on **Constellations**.)

ORNITHOLOGY. The scientific study of birds. Oölogy is a special division dealing with the eggs of birds. See also **Birds**.

OROGENY. Literally, the process of formation of mountains. The term came into use in the middle of the nineteenth century, when the process was thought to include both the deformation of rocks within the mountains, and the creation of the mountainous topography. Only much later was it realized that the two processes were mostly not closely related, either in origin or time. Today, many geomorphologists and a few geologists use orogeny for the formation of mountainous topography; most geologists regard this process as postorogenic and epeirogenic. By present geological usage, orogeny is the process by which structures within mountain areas were formed, including thrusting, folding, and faulting in the outer and higher layers, and plastic folding metamorphism, and plutonism in the inner and deeper layers. This usage has practical advantages; only in the very youngest, late Cenozoic mountains is there any evident casual relation between rock structure and surface landscapes. Little such evidence is available for the early Cenozoic, still less for the Mesozoic and Paleozoic, and virtually none for the Precambrian—yet all the deformational structures are much alike, whatever their age, and are appropriately considered as products of orogeny. *Tectogenesis*, a synonym of orogenesis, in its present meaning, is the process of by which mountainous areas are formed, e.g., folding or thrusting, without implication of the formation of mountainous topography. *(American Geological Institute, by permission.)*

ORPIMENT. This mineral, like realgar, its frequent associate, is an arsenic sulfide. Orpiment, however, is the trisulfide corresponding to the formula As_2S_3. It is monoclinic, usually in foliated, granular, or powdery aggregates; hardness, 1.502; specific gravity, 3.49; with a resinous to somewhat pearly luster; color, various shades of lemon yellow; translucent to nearly opaque. Orpiment is found in association with realgar, although a somewhat rarer mineral. It is believed to be formed from the alteration of other arsenic-bearing minerals. It occurs in Czechoslovakia, Romania, Macedonia, Japan, and in the United States in Utah, Nevada, and Wyoming. The name orpiment is derived from a corruption of the Latin *auripigmentum*, meaning golden paint, because of its color as well as the belief that it contained gold.

ORTHOCLASE. See **Feldspar**.

ORTHOGONAL ANTENNAS. In radar, a pair of transmitting and receiving antennas, or a single transmitting-receiving antenna, designed for the detection of a difference in polarization between the transmitted energy and the energy returned from the target.

ORTHOGONAL FUNCTION. The word *orthogonal* means, in general, *perpendicular*, thus the three axes of a rectangular Cartesian coordinate system are orthogonal in pairs, hence mutually orthogonal. Conditions for orthogonality are readily expressible in vector notation, for if the scalar product of two vectors in two dimensions vanishes, $\mathbf{A} \cdot \mathbf{B} = 0$, the two vectors are perpendicular to each other or orthogonal. The concept is easily generalized to n-dimensional space by assuming two quantities with components $A_i, B_i (i = 1, 2, \ldots, n)$ and they are then orthogonal if

$$\sum_{i=0}^{n} A_i B_i = 0$$

If the vector space involved has an infinite number of dimensions and if the components A_i, B_i are continuously distributed so that the index i becomes a continuous variable, the two functions are orthogonal if

$$\int_a^b \mathbf{A}(x) \cdot \mathbf{B}(x) \, dx = 0$$

The limits of integration are needed to specify the range of x for which the functions are defined. They may be finite or infinite.

Arbitrary functions may be made orthogonal by the Schmidt process. Suppose such a set $f_1(x), f_2(x), \ldots$ is defined over the range $a \leq x \leq b$, so that

$$\int_a^b f_i(x) f_j(x) \, dx = 0; i \neq j$$

Then presumably

$$\int_a^b f_i^2(x) \, dx = c_i^2, \text{ a constant}$$

It is frequently convenient to redefine the functions so that they are *normalized*. Thus, if $c_i \phi_i(x) = f_i(x)$, then

$$\int_a^b \phi_i(x) \phi_j(x) \, dx = \delta_{ij}$$

and the functions $\phi_i(x)$ are said to form an *orthonormal* set. If x is complex, generalization of this procedure is possible.

A set of functions $f_i(x)$ is *complete* if an arbitrary function $F(x)$ satisfying the same boundary conditions as the functions of the set can be expanded as

$$F(x) \sum_{i=1}^{\infty} A_i f_i(x)$$

the A_i being constant coefficients. If the set is an orthonormal one, the expansion coefficients are readily found by integration

$$A_i = \int F(t) f_i(t) \, dt$$

with appropriate limits to the definite integral. Similar, but more complicated integrals result when the functions are not normalized. These procedures are generalizations of those used in a Fourier series expansion.

ORTHOGONAL GROUP. See **Lie Group**.

ORTHOGONAL POLYNOMIALS. If it is desired to fit a polynomial function

$$y = \alpha_o + \alpha_1 x + \alpha_2 x^2 \ldots$$

to observed data by the method of least squares, ordinary multiple regression techniques can be used, but it is often more convenient to use orthogonal polynomials, especially if the degree of the required polynomial is unknown. These polynomials are such that, if the polynomial is of degree r, then

$$P_0(x) = 1 \text{ and } \sum_i P_r(x_i) P_s(x_i) = 0$$

for all $r \neq s$. For equally spaced points, the values of the polynomials have been fairly extensively tabulated; rewriting the equation as $y = \beta_0 + \beta_1 P_1(x) + \beta_2 P_2(x) + \ldots$, the estimates b of the β's are given by $b_r = \sum y_i P_1(x_i) / \sum [P_r(x_i)]^2$, and the sum of the squares of the residuals by $\sum y^2 - \sum \{b_r \sum [P_r(x_i)]^2\}$. This technique has the advantage that the estimate of any one of the β's is not affected by the fitting of other terms; this does not apply to direct estimates of the α's.

ORTHOGONAL SQUARES. It may be possible to superimpose two $n \times n$ Latin squares in such a way that any letter of the first occurs just once with every letter of the second. The resulting array is called a Graeco-Latin square, and the two squares are said to be orthogonal. In certain cases further squares may be superimposed in an orthogonal manner, up to a maximum of $(n - 1)$ squares. The letters of these squares, together with the rows and columns, define $(n + 1)$ sets of $(n - 1)$ contrasts which together account for all the contrasts between the n^2 cells of the square. A complete set of $(n - 1)$ orthogonal squares has been shown to exist for $n = p^s$ where p is prime; no Graeco-Latin square exists for $n = 6$, but contrary to an old conjecture, a square exists for $n = 10$. See also **Latin Square**.

ORTHOGRAPHIC PROJECTION. A method of representing solid objects on a plane surface, using parallel rays or projectors, in contrast to a cone or rays as used in perspective. The rays are perpendicular to the plane of representation, in contrast to oblique representation or projection, in which the rays are parallel, but at an angle to the plane of representation. In auxiliary projection, the plane of representation is inclined with respect to the principal axis of the object, although the rays are perpendicular to the plane of representation, and usually perpendicular to an inclined face of the object.

Six principal orthographic views are possible, although usually only four: the top view or plan, the front view or front elevation, and the left-side and right-side views or side elevations are employed. The usual arrangement of views in American practice is shown in Fig. 1; in Europe, a view arrangement in which the top view is below the front view, and the so-called right-side view is to the left of the front or top view, referred to as first angle projection, is extensively used.

See also **Perspective**.

(a) Orthographic projection

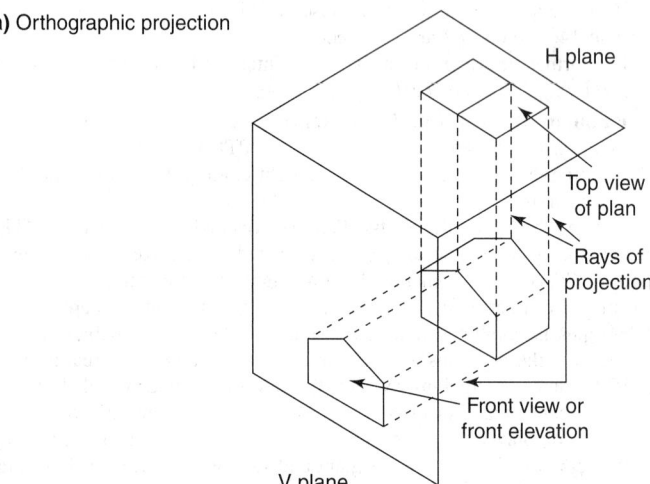

(b) Orthographic views

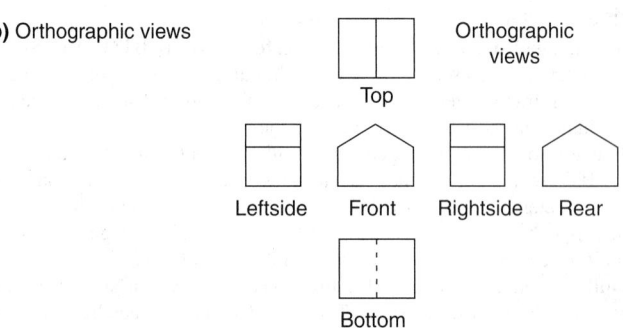

Fig. 1.   **(a)** Orthographic projection. **(b)** Orthographic views.

ORTHOKERATOLOGY.   Orthokeratology (ortho-K) is a nonsurgical system of treating myopia (nearsightedness) and astigmatism with a sequential series of specialized contact lenses that gradually flatten the cornea. The procedure uses rigid gas permeable contact lenses with very high oxygen permeability to gradually change the curvature of the front surface of the eye. These lenses are worn for a specific period of time then changed for new lenses with a slightly different shape. This repetitious method of fitting and changing the curves allows the anatomical shape of the cornea to slowly change, reducing the dependence on contact lenses or eyeglasses. Orthokeratology produces no permanent change in the structure of the cornea.

Although the procedure has been in existence for many years, recent improvement in lens technology and the availability of the corneal topographer, used for mapping the surface of the eye, have made the process more precise and dependable. The procedure works best for up to four diopters of myopia and two diopters of astigmatism. Refractive errors above this level may be reduced but not totally corrected.

To understand how orthokeratology works, it is first necessary to understand the visual function of the eye. See also **Visual Function (Eye)**.

If the cornea is too steep or if the eye is too long from front to back, light rays are focused in front of the retina, resulting in myopia. If the front of the cornea is unevenly curved, it causes the light rays to fall on different spots on the retina, resulting in astigmatism.

The objective of ortho-K is to correct nearsightedness and/or astigmatism by reshaping the cornea in order to correct for imperfections.

Ortho-K contact lenses, called corneal molds, are fitted in progressive stages to gradually reshape the cornea towards less curvature and a more spherical shape. Similar to a contact lens, the mold works as you wear it, whether awake or asleep. The program length usually varies between three and six months, depending on the degree of visual error and the rates of change. Over this period of time, you gradually decrease the time you wear the mold to establish the minimum wear time to maintain the corneal shape and good functional vision.

During the first phase of the program, lasting several days to several weeks, the visual changes occur quite rapidly, requiring frequent examinations and progressive lens changes. The second phase of the process, called the holding phase, lasts from three to six months and is designed to train the corneal tissue to retain its shape over a 24-hour period. In the final phase, lenses called retainer lenses are prescribed to stabilize the new corneal shape. The schedule for wear of these lenses is determined on an individual basis, but usually begins with full-time and then is reduced to short periods of the day or overnight. The cornea will return to its original shape if the retainer is not work on an ongoing basis.

Orthokeratology is a controversial procedure, according to some eye doctors. See also **Astigmatism**; **Cornea**; **Myopia (Nearsightedness)**; and **Retina**.

Vision Rx, Inc., Elmsford, NY.

ORTHOPTERA.   Insects of many forms with many common names, constituting one of the larger orders. The principal members of the group are the grasshoppers, crickets, mantises, stick insects, and cockroaches. Locusts are grasshoppers. Some authorities place these major forms in separate orders, but the general tendency is to group them together.

The order is characterized by biting mouthparts and gradual metamorphosis (Paurometabola). When wings are present there are two pairs, the front wings somewhat thickened as tegmina and the hind pair broad and membranous, folding beneath the tegmina when at rest.

These insects have a wide range of habits and occupy many terrestrial habitats. The locusts are important crop pests in some areas and cockroaches are common household pests.

The main families of Orthoptera include:

Blattidae	Roaches and cockroaches
Gryllidae	Crickets, tree crickets, mole crickets
Locustidae (formerly *Acrididae*)	Grasshoppers or locusts
Mantidae	Praying mantis
Phasmidae	Walking-sticks, walking-leaves
Tettigoniidae	Long-horned grasshopper, green meadow grasshoppers, katydids, cave and camel crickets

ORTHOSCOPIC SYSTEM.   An optical system corrected for both distortion and spherical aberration. Also called rectilinear system.

ORTHO-STATE.   1. In diatomic molecules, such as hydrogen molecules, the ortho-state exists when the spin vectors of the two atomic nuclei are in the same direction (i.e., parallel), whereas the para-state is the one in which the nuclei are spinning in opposite directions.

2. In helium the ortho-state is characterized by a particular mode of coupling of the electron spins. See also **Helium**.

ORTHOTOMIC SYSTEM.   An optical system that contains only rays which may be cut at right angles by a properly constructed surface.

ORYX.   See **Antelope**.

OSCILLATION.   An oscillation is one complete period of vibratory or periodic motion, for example, the whole succession of states that takes place before the motion begins to repeat itself. For example, one oscillation of a pendulum bob is a complete excursion from where it started back to its original position with the *same* velocity (magnitude and direction). The time of one oscillation is called one period and the number of oscillations per second (the reciprocal of the period) is called the frequency. The definition cannot be applied strictly to nonperiodic motions. In such motions the period is usually taken as the time between successive zeros of the displacement.

There are three types of oscillations: (1) Oscillations that continue in a circuit or system after the applied force has been removed, the frequency of the oscillations being determined by the parameters in the system or circuit, commonly referred to as shock-excited oscillations. (2) The oscillation of some physical quantity of a body or system when the externally applied forces consist either of those which do no work, or of those which are derivable from a potential that is invariant during the time under consideration, or both. (3) That type of oscillatory motion into which a suitable system not subject to external driving forces is capable of being excited by a displacement from an equilibrium position.

The frequency of oscillation is the number of complete oscillations of a given system per unit time, commonly symbolized by v or f. The frequency is the reciprocal of the period, the time for one complete oscillation. Sometimes the angular frequency, symbolized by ω, is used for greater convenience in manipulating trigonometric functions. The angular frequency has the unit radian per unit time and is equal to $2\times$ (frequency).

Modes of Oscillation. In the case of standing waves, the boundary conditions restrict the possible frequencies of oscillation to a discrete set of values. The whole set constitutes the modes of oscillation of the system. The oscillations of a string clamped at both ends, and the oscillations of sound waves in a closed or open pipe are examples of cases where boundary conditions impose modes.

A degenerate oscillating system is a vibrating system with several degrees of freedom in which the frequencies associated with two or more degrees of freedom may be equal in magnitude.

OSCILLATOR.   As applied to electronic equipment, a circuit that generates a periodic signal. The waveform may be sinusoidal or nonsinusoidal. As the term is most often used, however, it refers to a source signal which has a waveform closely approximating a sine wave. The oscillator is the heart of radio transmitters used for communication, navigation, radar, and similar functions requiring a source of radio-frequency energy, since it generates the high-frequency carrier signal essential for achieving the desired function. The early oscillators used in radio consisted of inductance-capacitance circuits to which a surge of electrical energy was applied by the breakdown of a spark gap or an arc. This energy then surged back and forth between the inductance and capacitance until it was all dissipated as radiation and as circuit losses. This type of oscillator produced damped oscillations and is no longer used. Modern oscillators use vacuum tubes or transistors in various circuit arrangements. Since both of these electronic elements provide amplification, they can be used as oscillators by feeding back some of the output energy to the input circuit so that the device effectively drives itself, i.e., furnishes its own input signal. The various oscillator circuits in common use effect this function by different means. Shown in the figures are various circuits employing transistors and vacuum tubes which find use as practical oscillators.

For extremely stable frequency characteristics, crystal-controlled oscillators, one form of which is shown in Fig. 1, are most often used. In these

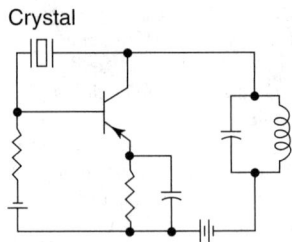

Fig. 1.   Crystal oscillator.

use is made of the frequency selective properties obtainable from a quartz crystal by means of the piezoelectric effect. An extremely sharp resonance characteristic is obtained thereby which permits feedback of signal from output to input inadequate to sustain oscillation only over a very narrow range of frequencies. This results is extremely good frequency stability; a variation of 1 part in 20,000,000 in representative of this type of oscillator.

Where continuously adjustable frequency output is needed some type of self-controlled oscillator is needed. The Hartley oscillator, shown in Fig. 2 is one of the simplest. The energy is fed back from output to input circuits through the inductive coupling of the two sections of the coil. The frequency is determined by the inductance and capacitance values in the tuned circuit.

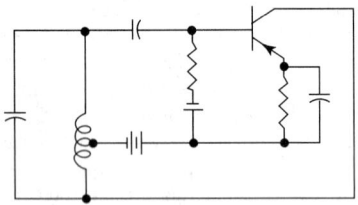

Fig. 2.   Hartley oscillator.

A very similar circuit is the Colpitts oscillator shown in Fig. 3. It differs from the Hartley circuit only in the manner in which energy is fed back from the output to the input, the coupling being accomplished by a capacitance voltage divider rather than a tapped coil.

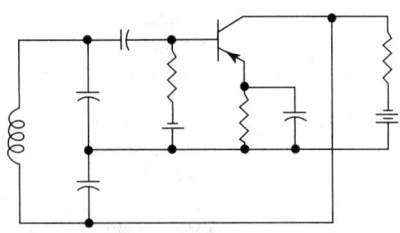

Fig. 3.   Colpitts oscillator.

A transistor blocking oscillator circuit is shown in Fig. 4.

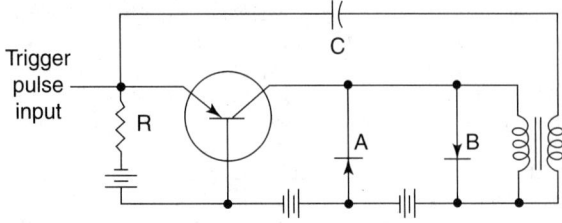

Fig. 4.   Transistor blocking oscillator circuit.

OSCILLOMETRY.   The measurement of capacitance produced by the dielectric constant and conductivity of a liquid sample in a cell. The cell is all glass, constructed as two concentric cylinders with bonded inner and

outer metal surfaces to form the plates of a capacitor, thus the sample contacts only the glass walls. See Fig. 1.

A frequency of 5 MHz is used for small capacitive reactances of the cell walls.

A dry cell (air filled) is connected to an oscillator circuit which is tuned to a resonant frequency of 5 MHz, by adding parallel capacitance inside the measuring instrument. The sample is then inserted into the cell with a resulting increase in the cell capacitance and calibrated parallel capacitance in the instrument removed until a return to the resonant condition is indicated. Thus the amount of capacitance removed is equal to the capacitance change of the cell.

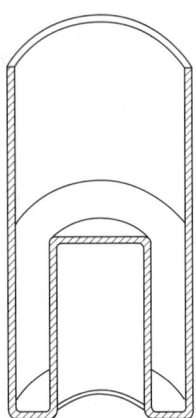

Fig. 1.   Sectional view of cell used in oscillometry.

Chemical titrations may be performed as point plotted capacitance change or by observing frequency change indicated on a meter or recording device. Dielectric constant measurements are determined by calibration using known standards or by mathematical equations derived using simulated cell constants. Solution conductivity effects a change in capacitance in the cell which cannot be distinguished from changes due to the dielectric constant. Generally conductivity measurements are made using known standards and the assumption that the dielectric constant is unchanging.

OSCILLOSCOPE.   An instrument primarily for making visible the instantaneous value of one or more rapidly varying electrical quantities as a function of time or of another electrical or mechanical quantity (IEEE). An *oscillograph* is an oscilloscope capable of recording as well as displaying the aforementioned measurements and is a form of $X - Y$ recorder. For decades the oscilloscope has been a workhorse in the development, testing, and troubleshooting of electrical and electronic equipment. A number of refinements (improving the performance, making the instrument more convenient to use, etc.) have been made, but the basic principles remain. History records that the first known use of a cathode-ray tube, the primary element of the oscilloscope, to display changing phenomena occurred as early as 1897, which was the same year that the tube was used to prove that cathode rays were electrified and would respond to a magnetic field.

An oscilloscope consists of several basic elements: (1) cathode-ray tube (CRT), (2) time-base generator, (3) vertical deflection channel, and (4) power supply. Oscilloscope construction over the years has taken full advantage of modern solid state, semiconductor electronics, as well as improvements in phosphors which are used in CRTs for other applications. Some CRT-based terminals and workstations are, in essence, highly sophisticated oscilloscopes.

See also **Spectrum Analysis**.

OSCULATING ORBIT.   The ellipse that a satellite would follow after a specific time t (the epoch of osculation) if all forces other than central inverse-square forces ceased to act from time t on. An osculating orbit is tangent to the real, perturbed, orbit and has the same velocity at the point of tangency.

OSCULATING PLANE.   The plane, if it exists, through a point P of a space curve which has contact of higher order with the curve than any other plane through the point; the limiting position of the plane through P, P', P'' as the points P', P'' approach P along the curve. See also **Curve (Space)**.

OSMIUM. Chemical element, symbol Os, at. no. 76, at. wt. 190.2, periodic table group 8, mp 3,015° to 3,075 °C, bp 4,927° to 5,127 °C, density 22.6g/cm³ (solid), 22.8g/cm³ (single crystal) (20 °C). Elemental osmium has a close-packed hexagonal crystal structure. Compact osmium is a bluish-white metal and is not attacked by acids. Discovered by Tennant in 1804. The seven stable isotopes of osmium are ^{184}Os,^{186}Os through ^{190}Os, and ^{192}Os. Electronic configuration $1s^2 2s^2 2p^6 3s^2 3p^6 3d^{10} 4s^2 4p^6 4d^{10} 4f^{14} 5s^2 5p^6 5d^6 6s^2$. Ionic radius Os$_4$+ 0.65 Å. Metallic radius 1.3377 Å. Oxidation potential Os + 4H$_2$O → OsO$_4$ + 8H$^+$ + 8e$^-$, −0.85 V.

Chemical Properties. Finely divided Os oxidizes in air, producing the poisonous and volatile tetroxide. The compact metal is not attacked by nonoxidizing acids. The finely divided metal dissolves in fuming HNO$_3$, aqua regia, and alkaline hypochlorite solutions. When fused with Na$_2$O$_2$ or KNO$_3$ and KOH, the metal is converted to the corresponding water-soluble osmate, K$_2$OsO$_4$. The brown or black insoluble osmium(IV) oxide, OsO$_2$, can be made by heating Os with a limited amount of O$_2$ or with osmium(VIII) oxide. This compound forms a brown to black-blue dihydrate that can be prepared by reducing a solution of the tetroxide or by hydrolyzing a solution of sodium hexachloroosmate, Na$_2$OsCl$_6$.

Osmium(VIII) oxide, the most important compound, is formed in one of the reactions unique to the platinum metals. Its ease of formation and volatility make it useful in a purification step for the refining or analysis of Os. The tetroxide is readily formed by heating the metal in air or distilling an osmium-containing solution from HNO$_3$. Although an aqueous solution of osmium(VIII) oxide is neutral to litmus, it is a weak acid with first dissociation constant of about 8×10^{-13}. Osmium(VIII) oxide is soluble in water, alcohol, and ether. The compound is widely used as a stain for tissues. When an alkaline solution of osmium(VIII) oxide is reduced with alcohol or KNO$_2$, an osmate(VI) is formed. Potassium osmate(VI) is formed by adding an excess of KOH to such a solution, resulting in the precipitation of violet crystals of K$_2$OsO$_4$ × 2H$_2$O. The osmate(VI) ion is probably better written as OsO$_2$(OH)$_2$.

Osmium(II) chloride can be prepared by heating osmium(III) chloride in vacuum at 500 °C. This dark-brown compound is insoluble in HCl or H$_2$SO$_4$. NHO$_3$ or aqua regia oxidizes it to the tetroxide. Osmium(III) chloride is best made by decomposing ammonium hexachloroosmate(IV), (NH$_4$)$_2$OsCl$_6$, in a Cl$_2$ stream at 350 °C. The brown hygroscopic powder sublimes above 350 °C, and at about 560 °C it disproportionates into the tetrachloride and dichloride. Osmium(IV) chloride is formed from the elements at 650–700 °C. The black compound slowly dissolves in water, eventually forming the dioxide. The free acid, H$_2$OsCl$_6$, is stable in solution and can be made by refluxing osmium(VIII) oxide with HCl and alcohol. The ammonium salt can be precipitated by adding NH$_4$Cl to such a solution. This salt is reduced to the metal when heated in H$_2$. The potassium salt is well known. Both are brownish-red solids yielding orange solutions in water.

Recent studies have established the reaction product of Os metal and F$_2$ at 300 °C to be the hexafluoride, OsF$_6$. This yellow volatile solid had previously been described as an octafluoride. Osmium(VI) fluoride melts at 33.4 °C and boils at 47.5 °C. OsF$_6$ can be reduced to a pentafluoride and a tetrafluoride. The pentafluoride is a blue-gray crystalline solid that melts at 70 °C to a green viscous liquid and boils at 226 °C. The tetrafluoride distills at about 290 °C. Potassium hexafluoroosmate(V), KOsF$_6$, can be made by reacting KBr, osmium(IV) bromide, and bromine trifluoride. The white powder dissolves in water to form a colorless solution that hydrolyzes to yield some osmium(VIII) oxide. On addition of 1 equiv of KOH to a fresh solution, an orange color develops, O$_2$ is evolved, and yellow crystals of potassium hexafluoroosmate(IV), K$_2$OsF$_6$, form.

Os forms many complexes with nitrite, oxalate, carbon monoxide, amines, and thio ureas. The latter are important analytically. Osmium forms the interesting aromatic "sandwich" compound, osmocene. A *metallocene* is described under **Ruthenium**. See also **Chemical Elements**; and **Platinum and Platinum Group**.

Proton nuclear magnetic resonance (NMR) is a widely used tool for researching biomolecules. Although much too detailed to report here, Zai-Wei Li and Henry Taube (Stanford University) reported in 1992 that they have had success in analyzing for certain molecules by using a dihydrogen osmium complex on a versatile 1H NMR recognition probe.

Osmium Isotopic Ratios in Paleogeology. As pointed out in 1983 by J.M. Luck and K.K. Turekian (Yale University), one of the most creative concepts regarding the cause of the many paleontologic extinctions at the Cretaceous-Tertiary boundary is the one put forward by Alvarez et al.in 1980 involving the impact of a large asteroid or comet with the earth at that geologic time period. The Alvarez hypothesis stemmed from the finding of an exceptional chemical signature (high iridium concentration) at the Cretaceous-Tertiary boundary at Gubbio, Italy, a signature that was later found in other marine as well as continental sections which bracket the Cretaceous-Tertiary boundary. Because of the radioactive decay of rhenium-187 (4.6×10^{10} years), the osmium-187/osmium-186 ratio changes in planetary systems as a function of time and the rhenium-187/osmium-186 ratio. For a value of the ^{187}Re/^{186}Os ratio of about 3.2, typical of meteorites and the earth's mantle, the present ^{187}Os/^{186}Os ratio is about one. The earth's continental crust has an estimated ^{187}Re/^{186}Os ratio of about 400. Thus for a mean age of the continent of 2×10^9 years, a present ^{187}Os/^{186}Os ratio of about 10 is expected. Marine manganese nodules show values (6 to 8.4), which are compatible with this expectation if an allowance for a 25% mantle osmium supply to the oceans is made. The Cretaceous-Tertiary boundary iridium-rich layer in the marine section at Stevns Klint, Denmark, yields a ^{187}Os/^{186}Os ratio of 1.65 and the one in a continental section in the Raton Basin, Colorado, is 1.29. The investigators conclude that the simplest explanation is that these represent osmium imprints of predominantly meteoritic origin.

As reported in 1992 by M.F. Horan, J.W. Morgan and J.N. Grossman (U.S. Geological Survey), and R.J. Walker (University of Maryland), "Rhenium and osmium concentrations and the osmium isotopic compositions of iron meteorites were determined by negative thermal ionization mass spectrometry. Data for the IIA iron meteorites define an isochron with an uncertainty of approximately ±31 million years for meteorites ~4500 million years old. Although an absolute rhenium-osmium closure age for this iron group cannot be as precisely constrained because of uncertainty in the decay constant of ^{187}Re, an age of 4460 million years ago is the minimum permitted by combined uncertainties. These age constraints imply that the parent body of the IIAB magmatic irons melted and subsequently cooled within 100 million years after the formation of the oldest portions of chondrites. Other iron meteorites plot above the IIA isochron, indicating that the planetary bodies represented by these iron groups may have cooled significantly later than the parent body of the IIA irons."

Additional Reading

Anderson, D.L.: "Composition of the Earth," *Science,* 367 (January 20, 1989).

Cherfas, J.: "Proton Microbeam Probes the Elements," *Science*, 1150 (September 28, 1990).

Considine, D.M. and G.D. Considine: "Van Nostrand Encyclopedia of Chemistry," 4th Edition, Van Nostrand Reinhold, New York, NY, 1984. (A classic reference.)

Davis, J.R.: "Metals Handbook," 2nd Edition, ASM International, Materials Park, OH, 1998.

Greenwood, N.N. and A. Earnshaw: "Chemistry of the Elements," 2nd Edition, Butterworth-Heinemann, Inc., Woburn, MA, 1997.

Horan, M.F., et al.: "Rhenium-Osmium Isotope Constraints on the Age of Iron Meteorites," *Science,* 1118 (February 28, 1992).

Krebs, R.E.: "The History and Use of Our Earth's Chemical Elements: A Reference Guide," Greenwood Publishing Group, Inc., Westport, CT, 1998.

Lewis, R.J., Sr.: "Hawley's Condensed Chemical Dictionary," 13th Edition, John Wiley & Sons, Inc., New York, NY, 1997.

Lide, D.R.: "CRC Handbook of Chemistry and Physics 2000-2001," 81st Edition, CRC Press, LLC., Boca Raton, FL, 2000.

Luck, J.M. and K.K. Turekian: "Osmium-187/Osmium-186 in Manganese Nodules and the Cretaceous-Tertiary Boundary," *Science,* 222, 613–615 (1983).

Sinfelt, J.H.: "Bimetallic Catalysts," *Sci. American*, 253(3), 90–98 (1985).

Staff: "ASM Handbook—Properties and Selection: Nonferrous Alloys and Pure Metals," ASM International, Materials Park, OH, 1990.

Zai-Wei, Li and H. Taube: "Use of a Dihydrogen Osmium Complex as a Versatile 1 H NMR Recognition Probe," *Science,* 210 (April 10, 1992).

OSMOTIC COEFFICIENT. A factor introduced into equations for nonideal solutions to correct for their departure from ideal behavior, as in the equation:

$$\mu = \mu_x^0 + gRT \ln x_1$$

in which μ is the chemical potential, μ_x^0 is a constant, representing a standard value of the chemical potential, R is the gas constant, T the absolute temperature, x_1 is the mole fraction of solvent, and g is the osmotic coefficient.

OSMOTIC PRESSURE.

Pressure that develops when a pure solvent is separated from a solution by a semipermeable membrane which allows only the solvent molecules to pass through it. The osmotic pressure of the solution is then the excess pressure which must be applied to the solution so as to prevent the passage into it of the solvent through the semipermeable membrane.

Because of the similarity in the relations for osmotic pressure in dilute solutions and the equation for an ideal gas, van't Hoff proposed his bombardment theory in which osmotic pressure is considered in terms of collisions of solute molecules on the semipermeable membrane. This theory has a number of objections and has now been discarded. Other theories have also been put forward involving solvent bombardment on the semipermeable membrane, and vapor pressure effects. For example, osmotic pressure has been considered as the negative pressure which must be applied to the solvent to reduce its vapor pressure to that of the solution. It is, however, more profitable to interpret osmotic pressures using thermodynamic relations, such as the entropy of dilution.

A number of methods have been developed for measurement of osmotic pressure.

In the Berkeley and Hartley method, a porous tube with a semipermeable membrane such as copper ferrocyanide deposited near the outer wall, and a capillary tube attached to one end contains the pure solvent. The solution surrounds the tube and is enclosed in a metal vessel to which a pressure may be applied which is just sufficient to prevent the flow of solvent into the solution. Berkeley and Hartley also developed a dynamic method for measuring osmotic pressure.

Simple osmometers have also been developed by Adair particularly for aqueous colloidal solutions. A thimble-type collodion membrane is attached to a capillary tube and contains the solution. When equilibrium is established the difference in level inside and outside the capillary is measured. Capillary corrections are made. For organic solvents a dynamic type osmometer may be used. A membrane of large surface area is clamped between two half cells and attached to each half cell is a fine capillary observation tube. With such an apparatus, equilibrium is rapidly established between solution and solvent contained in the half cells. The volume of the half cell may be small (about 20 cubic centimeters). The level of the solvent is usually arranged to be a little below the equilibrium position, and the height of the solvent in the capillary as a function of time is measured. This procedure is repeated with the level of the solvent just above the equilibrium position. A plot is then made of the half sum of these readings.

Since the osmotic pressure is related to the concentration of dissolved solute particles, it is related to the lowering of the freezing point and elevation of the boiling point.

The relation between osmotic pressure and lowering of the freezing point and the elevation of the boiling point may be expressed by the relation:

$$\Pi = \frac{LT}{\bar{v}\, T_0^2} \Delta T$$

where L is the molar heat of fusion or of vaporization of the solvent, T the temperature at which the osmotic pressure is measured, T_0 the freezing point or the boiling point of the solvent, $\bar{v}$, the partial molar volume of the solvent, and Π, the osmotic pressure.

Moreover, the relation to lowering of the vapor pressure is

$$\Pi = \frac{RT}{\bar{v}} \ln \frac{p_0}{p}$$

or

$$\Pi = -\frac{RT}{\bar{v}} \ln x_0 = -\frac{RT}{v} \ln(1 - x)$$

For dilute solutions,

$$\Pi = cRT$$

where Π is the osmotic pressure, $\bar{v}$ *the partial molar volume of the solvent in the solution*, p_0 the vapor pressure of pure solvent, p the partial vapor pressure of the solvent in equilibrium with the solution, x_0 the mole fraction of the solvent, x the mole fraction of the solute, c the concentration of the solution in moles per liter, R, the gas constant, and T, the absolute temperature.

OSMOTIC PRESSURE (Cell). See Cell (Biology).

OSTEOARTHRITIS.

This is the most common joint disease. As much as 80% of the population has radiographic evidence of osteoarthritis by the age of 65. Although only about 60% of patients with radiographically detectable osteoarthritis have symptoms, 15 to 30% of all visits to general practitioners may be attributed to difficulty with ambulation, largely due to osteoarthritis.

Osteoarthritis is a degenerative joint disease and, among people over age 50 (especially women), is the most common musculoskeletal problem. Under age 40, the disease usually is the result of an injury, from sports or occupational strain or of congenital abnormalities. Most often, the disease develops in the weight-bearing joints, such as the hips, knees, and spine and, to some extent, in joints that are used repetitively, such as the fingers.

Cartilage is a firm, rather rubbery material that covers the end of each bone. The normal purpose of cartilage is to act as a cushion and thus provide a smooth surface between the bones. In osteoarthritis, however, the cartilage breaks down, its surface becoming thin and uneven. Eventually, the ends of the bones may be left unprotected. Then the bones grind against each other, causing pain. With the development of pain, many patients tend to use affected joints less, and this leads to muscle weakness and stiffness of joints. A spur also may form in the surrounding bone tissue.

The three major contributing causes of osteoarthritis are injury, time, and weight. Overuse of a joint with time increases wear; obesity contributes to the stress placed on the joints and damages them prematurely. Overweight persons are more apt to develop osteoarthritis of the hips and knees earlier than their thinner neighbors.

The most common symptom is aching pain when the affected joint is in use. Except in advanced cases, this pain will subside when the affected joint is rested. Osteoarthritis does not always become worse if the aforementioned precautionary measures are taken. Although the pain may become rather constant, the common symptoms of *rheumatoid arthritis*, such as prolonged morning stiffness, swollen lymph nodes, fatigue, and fever usually are not present in osteoarthritis. Physical examination of a patient with osteoarthritis usually reveals joint tenderness and sometimes swelling. When moved, the joints sometimes cause *crepitus* (popping, clicking, or grating sounds). Enlargement of bone ends also may be present, such as the finger ends (Heberden's nodes) or the middle joints (Bouchard's nodes). A radiographic examination is required to yield complete information for the physician.

Degenerative joint disease of the spine can take several forms, but usually affects the vertebral bodies or disks, or both. These deformities are most severe when they occur in the middle to lower cervical spine and in the lower lumbar region. In contrast with ankylosing spondylitis, spinal fusion is uncommon in osteoarthritis.

Osteoarthritis and Exercise. Todd Schmidt (Hughston Sports Medicine Foundation, Columbus, Georgia), reports, "Forty years after they had begun training, one group of former champion runners had a lower incidence of osteoarthritis than a nonrunning group of the same age. Comparisons of the spines, hips, knees, and feet of a group of runners averaging 25 miles (40 km) per week with a group of nonrunners showed no difference in cartilage thickness, spur formation, grinding, and joint stability. Another study found that former college runners were no more likely than anyone else to have symptomatic osteoarthritis." The study dealt with normal uninjured joints. Joints that have meniscus and cartilage damage or ligament instability are clearly at an increased risk for osteoarthritis. Because joints are designed for motion and shock absorption, running on an appropriate surface appears to be physiologic and nontraumatic. Further research along these lines is continuing.

Heritable Factors. There is substantial qualitative evidence that osteoarthritis occurs with greater frequency in some families. Research along these lines has been underway for several years. No discovery of a breakthrough nature, however, has been made as of the early 1990s. For those readers of scholarly interest in this subject, reference to the Knowlton paper (reference listed) is suggested.

More is known about the rare Marfan syndrome, which is considered to be an inherited connective tissue disorder. There are some 40 to 60 cases per million population. The most prominent clinical manifestations occur in the skeletal, ocular, and cardiovascular systems. Kainulatinen (Kuopio University Central Hospital, Finland), with reference to a recent study, observes: "The chromosomal localization of the mutation in Marfan syndrome is a first step toward the isolation and characterization of the defective gene and serves as a diagnostic test in families in which cosegregation of these markers with the disease has been confirmed."

Additional Reading

Bradley, J.D., et al.: "Comparison of An Antiinflammatory Dose of Ibuprofen, An Analgesic Dose of Ibuprofen, and Acetaminophen in the Treatment of Patients with Osteoarthritis of the Knee," *New Eng. J. Med.*, **87** (July 11, 1991).

Hamanishi, C.: "Advances in Osteoarthritis," Springer-Verlag, Inc., New York, NY, 1999.

Hosie, G. and J. Dickson: "Managing Osteoarthritis in Primary Care," Blackwell Science, Inc., Malden, MA, 2000.

Kainulainen, K., et al.: "Location of Chromosome 15 of the Gene Defect Causing Marfan Syndrome," *New Eng. J. Med.*, 935 (October 4, 1990).

Knowlton, R.G., et al.: "Genetic Linkage of a Polymorphism in the Type II Procollagen Gene (COL2A1) to Primary Osteoarthritis Associated with Mild Chondrodysplasia," *New Eng. J. Med.*, 326 (February 22, 1990).

Koopman, W.J.: "Arthritis and Allied Conditions: A Textbook of Rheumatology," 14th Edition, Lippincott Williams & Wilkins, Philadelphia, PA, 2000.

Liang, M.H. and P. Fortin: "Management of Osteoarthritis of the Hip and Knee," *New Eng. J. Med.*, 125 (July 11, 1991).

Moskowitz, R.W., J.A. Buckwalter, D.S. Howell, et al.: "Osteoarthritis: Diagnosis and Medical/Surgical Management," W.B. Saunders Company, Philadelphia, PA, 2001.

Shipman, P., et al.: "The Human Skeleton," Harvard University Press, Cambridge, MA, 1985.

OSTEOLOGY.   A division of vertebrate anatomy which deals with the skeletal system. The study of bones. Comparative osteology is the study and comparison of the bones of different races and different species of organisms.

OSTEOMYELITIS.   See **Bone**.

OSTRICH *(Aves, Struthioniformes; Struthio camelus).*   The only living representative of the suborder *Struthiones*, family *Struthionidae*. Its relatives inhabited wide areas in Asia, Europe, and Africa since the Eocene about 55 million years ago. Eight extinct species all belonged to the same genus as the surviving species. See Fig. 1.

Fig. 1.   Ostriches.

Attaining 3 meters ($9\frac{1}{2}$ feet) in height and weighing more than 150 kilograms (330 pounds), the male ostrich is the largest living bird. The head and about $\frac{2}{3}$ of the neck are sparsely covered with short, hair-like, degenerated feathers, which make it appear naked. The skin is variably colored, depending on the subspecies. The legs are particularly strong and long. The foot has two toes: a large strongly clawed third toe and weaker, generally clawless fourth (outside) toe. The first and second toes are absent. The feathers have no secondary or aftershaft. There are 50 to 60 tail feathers. The wing has 16 primary, 4 alular, and 20–23 secondary feathers. The wing feathers and rectrices have changed to decorative plumes.

Ostriches live on the open savannah, in the dry South African bushland or on the wide sand plains of deserts which have hardly any plants, and also dense bush. See Fig. 2. This adaptable grazer is at home even in steep rocky mountain country. Depending upon habitat and seasons, they eat various grasses, bushes, and forage on trees. Plants that store water help them through dry seasons but cannot meet all water requirements for a longer period of time. Without open water, they eventually die of thirst. They supplement their plant diet as much as possible with animal food, such as invertebrates and small vertebrates which they chase by rather ungracefully zig-zagging after them. Long periods of abundant food influence the reproductive readiness of the ostrich. Even in the most unfavorable weather, during long droughts, or during occasional local rains, the ostrich is a very adaptable and opportunistic breeder which can rear a few young in any season.

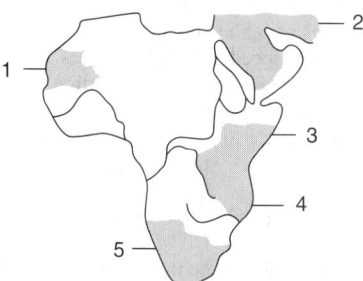

Fig. 2.   The ostrich (*Struthio camelus*) was distributed in these areas of Africa and Asia Minor until a few decades ago. Since then it has been exterminated in many areas. Subspecies: (1) North African ostrich, *Struthio camelus camelus*; (2) Arabian ostrich, *Struthio camelus syriacus*; exterminated; (3) Somali ostrich, *Struthio camelus molybdophanes*; (4) Maasai ostrich, *Struthio camelus massaicus*; (5) South African ostrich (*Struthio camelus australis*.)

The social life of ostriches is one of the most complex in the animal world. In rainless periods, when wandering, and in the common grazing grounds at watering places, they often form peaceful aggregations of up to 680 birds, but the individual flocks remain recognizable. Social contacts between birds of different groups are initiated by approaching the other in a submissive posture, with lowered head and tail down. Often a family of a herd adopts the chicks or young of another. Single males may join together and form "schools" of half-grown ostriches, with whom they wander about for days or weeks. For the communal sandbath, each flock seeks out a sandy depression.

Depending upon local circumstances and the composition of the population, ostriches live in monogamy or polygamy. Pair formation generally takes place within the large flocks. The "goading" of the old females and the animosities between the males often lead to displays and "dances" of the entire flock. The most common mating pattern is polygyny, in which a male generally lives with a "head female" and two "auxiliary" females. The head female tolerates the others and all lay their eggs into a common nest. In most cases, the experienced head female drives the auxiliary females away from the nest as soon s they have finished laying.

The male ostrich has far more to do during breeding than merely fertilizing the eggs. He is a genuine father. He scratches out the nest depression in a sandy spot, often in a dried-out river bed. Even before the eggs are going to be laid he guards the nest. Finally he sits down on it, the female lays her eggs before his breast, and he pushes them under his body with his neck and beak. The male ostrich incubates from the late afternoon until early morning; the female, therefore, does not have to sit for long since she only incubates in the hottest hours of the day.

When holding an ostrich egg, one may wonder how the chick could possibly get out of the shell without help. The shell is as thick as china. It weighs about $1\frac{1}{2}$ kilograms (3 pounds), about as much as 25–30 chicken eggs. Ostrich eggs are good to eat; they taste almost like chicken eggs. When refrigerated, an ostrich egg keeps up to a year and remains edible. It takes about 2 hours to hard-boil one.

Behavioral interactions between chicks and parents begin a few days before hatching, with pleasant-sounding contact calls from the chicks within the eggs. Actual hatching may take hours or even days. Soon after hatching the nestlings swallow the first small stones for grinding food, and already on the first day most chicks are ready to explore the vicinity of the nest. When in danger from a terrestrial enemy, such as a jackal, the adults try to decoy with conspicuous zigzag running, wing beating, and moot calls; at the appropriate time one parent calls the young, which meanwhile have cowered motionless, and leads them away.

One to three months after hatching, and sometimes much later in areas thinly populated by ostriches, the families rejoin the large flocks. It is believed that ostriches live thirty to seventy years.

Excellent eyesight and acute hearing are their most important senses, they often make the ostrich an unintentional but reliable sentinel for many African grazing animals, such as zebras, antelopes, and gazelles. They are capable of producing pleasant-sounding calls, moot and harsh guttural sounds, hissing and snorting, and a far reaching territory and courtship song. The volume and quality of these calls are not inferior to those of song birds, although the vocal organ, the syrinx, is very primitive. The calling of ostrich males during courtship resembles the distant roaring of lions.

Ostriches are excellent runners. The large bird, nearly 3 meters ($9\frac{1}{2}$ feet) high, easily takes steps of 3.5 meters ($11\frac{1}{2}$ feet) high when running. They are capable of attaining speeds of 50 kilometers per hour (31 miles per hour) for 15–30 minutes without obvious exertion or signs of exhaustion. Other wild animals can only run fairly short stretches at such a high speed. Ostriches are said to be able to reach a top speed of about 70 kilometers per hour ($43\frac{1}{2}$ miles per hour). See also **Ratites**.

OTITIS MEDIA.   See Hearing and the Ear.

OUTCROP. Everywhere beneath the soil at greater or lesser depths there exist the solid continuous masses of rock that make up the earth's crust. Wherever rock appears protruding through the soil cover it is spoken of as an outcrop.

OUTPUT UNIT. In computer terminology, a unit which delivers information from the computer to an external device or from internal storage to external storage.

OUZEL (or Ousel). *Aves, Passeriformes.* 1. The ring ouzel, *Turdus torquatus*, a mountain bird (Aves) of central and northern Europe. A thrush. 2. The blackbird of Europe and eastern Asia, also a thrush. 3. The water ouzel, or dipper. The name water ouzel applies to several species of the Northern Hemisphere which haunt rapid streams. Although not a swimming bird it wades into deep water, where it is said to make its way by using both feet and wings. The North American species, *Cinclus mexicanus*, is characteristically a western mountain bird. Its dull gray color is more than offset by its interesting habits and by a glorious song, heard all too rarely. See also **Thrush**.

OVAL OF CASSINI. A higher plane curve, also known as a cassinoid or cassinian, defined as the locus of a point which moves so that the product of its distance from two fixed foci ($\pm a$, 0) equals k^2, a constant. Its equation in Cartesian coordinates is $(x^2 + y^2 + a^2)^2 - 4a^2x^2 = k^2$. The curve cuts the X-axis at the two real points $\pm c$, where $k = c^2 - a^2$ and at two other points, real if $\sqrt{2}a > c$ but imaginary if $\sqrt{2}a < c$. Under the same conditions, it cuts the Y-axis in two real points or in four imaginary points. Thus, when $k < a$, there are two separated ovals each enclosing a focus and the curve is said to be *bipartite*. If $k > a$, the two ovals merge into one. When $k = a$, the curve is known as the lemniscate of Bernoulli.

Cassini ovals are useful in the study of optics. Another curve of similar shape is the Cartesian oval.

OVARIES.   See Gonads; Hormones; and Pituitary Gland.

OVERCAST.   See Sky Cover.

OVERFOLD. A term used by structural geologists to define an overturned anticline resulting in a recumbent fold often simulating a normal stratigraphic pile of strata for which it may be mistaken.

OVERLAP. 1. In some magnetic amplifier and rectifier circuits, the undesirable flow of current for a period after the supply voltage has gone to zero (and perhaps reversed) caused by energy stored in some circuit inductance.

2. In a magnetic amplifier, the simultaneous conduction of the half wave elements during certain intervals of supply voltage cycle.

3. In geology, the gradual burial of the land mass or mountain slopes from which the sediments were derived.

OVERRANGE. Of a system or element, any excess value of the input signal above its upper range-value or below its lower range-value. See also **Range and Span (Instrument)**.

OVERTHRUST. A low-angle fault of the thrust or compressional type frequently resulting in the translocation of a mass of rocks several miles. Because of the low angle of the thrust as well as the possible inversion of great piles of formations, this type of faulting or failure of the lithosphere has led, and still leads, to considerable technical difficulties in determining the stratigraphic history in regions where low angle faulting predominates. See Fig. 1.

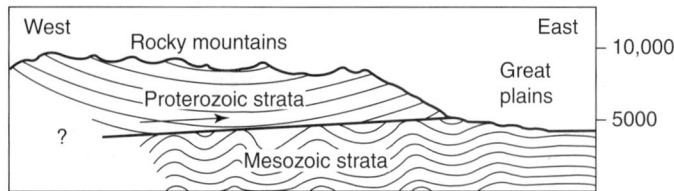

Fig. 1.   Diagrammatic structure section showing how the great body of Proterozoid strata has been thrust-faulted from the west for long distances over upon Late Mesozoic strata in Glacier National Park, Montana. The Mesozoic beds were folded by action of the overriding block. Length of section about 25 miles (40 kilometers). Vertical scale, much exaggerated.

OVERVOLTAGE. The excess of observed decomposition voltage of an aqueous electrolyte over the theoretical reversible decomposition voltage.

OVINAE.   See Goats and Sheep.

OVIPAROUS REPRODUCTION. The production of eggs that hatch after expulsion from the body. Frogs, birds, and many forms of reptiles and fish, as well as many insects and other invertebrate animals, have oviparous reproduction. See also **Fishes**.

OVIPOSITOR. An organ used by some of the arthropods to deposit their eggs. It consists of a maximum of three pairs of appendages formed to transmit the egg, to prepare a place for it, and to place it properly. In some of the insects the organ is used merely to attach the egg to some surface, but in many parasitic species (see also **Hymenoptera**) it is a piercing organ as well. It is used by the grasshoppers to force a burrow in the earth to receive the eggs and by cicadas to pierce the wood of twigs for a similar purpose. Both long-horned grasshoppers and sawflies cut the tissues of plants by means of the ovipositor. None of these examples is quite as remarkable as the ichneumon flies (parasitic *Hymenoptera*) which have a slender ovipositor several inches long, used to drill into the wood of tree trunks. These species are parasitic in the larval stage on the larvae of woodboring insects, hence the egg must be deposited in the burrow of the host.

The sting of wasps and bees is also an ovipositor, in this case highly modified and associated with poison glands. Some roach-like fish have an ovipositor as a tubular extension of the genial orifice in the breeding season for depositing eggs in the mantle cavity of the pond mussel.

OVOVIVIPAROUS REPRODUCTION. The condition where the young are born alive, but they hatch from eggs within the body of the mother just before being born. There is no placental connection between the young and the mother to furnish food before the birth. Sharks, rattlesnakes, certain snails, some minnows, and flesh flies are examples of the great variety of animals which may have this type of reproduction. It should not be confused with viviparous reproduction, where there is a placental connection between mother and embryo before birth, nor oviparous reproduction, where the eggs are laid before hatching. See also **Fishes**.

OWLS (*Aves, Strigiformes*). The *Strigiformes* comprise an order of birds with hooked beaks and talons of the birds of prey, but differing from the hawks and related forms in having the eyes directed forward and in their nocturnal habits. The common name for this order is owl. There are about 130 or more species. The head and neck of the owl is modified so that it can see nearly 270 degrees by turning its head. The ears and eyes are highly specialized. Although the eye cannot be turned in the socket, the owl can see small darting objects with very little light, substituting head movement for eye movement. It is reported that an owl can catch an insect with no more light than a candle would produce at 1,170 feet (351 meters). The eye is surrounded by a small disk of radiating feathers. The ear openings are unusually large and well adapted to detecting the slightest of sounds. The legs are strong and the claws are sharp. When landing from flight, the owl spreads its tail to serve as a brake.

Owls of various species utter different kinds of sounds, ranging from whistles, to hoots, and some sounds described as laughing. These birds nest in trees, on rocks, or on the ground, depending upon species. The egg is shaped like a sphere. The young are covered with white down. The adults have soft thick plumage that contributes to silencing their flight. A barn owl is shown in Fig. 1.

Fig. 1.   Barn owl. (*A.M. Winchester.*)

North American species range in size from the 6-inch (15 centimeters) Acadian owl to the 2-foot (0.6 meter) great horned owl (*Bubo virginanus*). There is a similar range of sizes in the other parts of the world — from the pigmy owls of the Old World to the great hawk-owl of Australia. The owl is represented on all continents.

Although owls are characteristically predacious, they do not disdain insects as food. The smaller species catch mice and other small animals; the larger owls prey upon animals as large as rabbits. The great horned owl will sometimes maraud a poultry yard. See also **Strigiformes**.

For several years the most famous of North American owls has been the spotted owl, whose habitat is found in the forests of the Northwest. This owl has been the subject of extensive environmental disputes.

OXALIC ACID. $H_2C_2O_4 \cdot 2H_2O$, formula weight 126.07, white solid, mp 101 °C (crystals), 180 °C (anhydrous), sublimes at 150 °C, soluble in H_2O, or alcohol, or ether.

Oxalic acid may be obtained (1) from some natural products, e.g., wood sorrel, and other members of the oxalis family, as potassium hydrogen oxalate, the bark of certain species of eucalyptus (sometimes containing 20% calcium oxalate in the cells and cell walls), (2) by reaction of acid with an oxalate, e.g., calcium oxalate plus H_2SO_4 as above. Sodium oxalate is made (1) by heating sodium formate at 300 °C in a vacuum, with the evolution of hydrogen gas, (2) by reaction of CO_2 plus metallic sodium at 360 °C, (3) by heating wood powder ("sawdust"), particularly of coniferous woods, with NaOH (addition of potassium hydroxide enables one to use a moderate temperature, i.e., 220 °C), (4) by oxidation of sucrose or starch with HNO_3. Oxalic acid is a dibasic acid, that is, two series of salts are known; a third series is also known, thus, sodium oxalate, $Na_2C_2O_4$, sodium binoxalate, $NaHC_2O_4$, sodium tetroxalate, $NaH_3(C_2O_4)_2$.

Oxalic acid is used (1) in the preparation of oxalates, e.g., titanium potassium oxalate, and esters, (2) in the purification of certain chemicals, e.g., glycerol, stearates, (3) in bleaching straw, (4) in ink and rust remover, (5) in the leather and textile industries, (6) in the manufacture of dyes, (7) in the preparation of glyoxalic and glycolic acids by regulated reduction.

A common test for oxalic acid is as follows: Heat solution with resorcinol in a test tube and on cooling add carefully a layer of H_2SO_4. A blue ring at the junction of the two layers indicates oxalic acid.

Representative esters of oxalic acid are: Methyl oxalic acid $COOH \cdot COOCH_3$, mp 37 °C, by 108 °C at 12 mm pressure; dimethyl oxalate $(COOCH_3)_2$, mp 54 °C, bp 163 °C; diethyl oxalate $(COOC_2H_5)_2$, mp −41 °C, bp 186 °C.

OX BOW LAKE.   The type of lake developed on the flood plains, especially in the delta regions, of large rivers. Ox bow lakes represent the detached or truncated former meanders of the main stream, which, from time to time, straighten their courses by forming cutoffs or shortcuts.

OXIDATION AND OXIDIZING AGENTS.   Many years ago, the term *oxidation* signified a reaction in which oxygen combines chemically with another substance. The term now has a much broader meaning and includes any reactions in which electrons are transferred. Oxidation and reduction always occur simultaneously (redox reactions), and the substance which gains electrons is termed the *oxidizing agent*. For example, cupric ion is the oxidizing agent in the reaction: Fe (metal) $+ Cu^{++} \rightarrow Fe^{++} + Cu$ (metal). Here, two electrons (negative charges) are transferred from the Fe atom to the Cu atom. Thus the Fe becomes positively charged (is oxidized) by loss of two electrons, while the Cu receives the two electrons and becomes neutral (is reduced). Electrons may also be displaced within the molecule without being completely transferred away from it. Such partial loss of electrons likewise constitutes oxidation in its broader sense and leads to the application of the term to a large number of processes which at first inspection might not be considered to be oxidations. Reactions of a hydrocarbon with a halogen, for example: $CH_4 + Cl_2 \rightarrow CH_3Cl + HCl$, involves partial oxidation of the methane. Also, when a halogen addition to a double bond is made, this is regarded as an oxidation.

Dehydrogenation is also a form of oxidation, when two hydrogen atoms, each having one electron, are removed from a hydrogen-containing organic compound by a catalytic reaction with air or oxygen, as in oxidation of alcohols to aldehydes. See also **Dehydrogenation**.

Oxidizing agents are widely used throughout the chemical and petro-chemical industries. It is also interesting to note that, while a primary thrust in food processing is to prevent oxidation (associated with rancidity and spoilage in foods), some powerful oxidizing agents are required by some processes to perform the function of bleaching. See also **Bleaching Agents**.

As one example, after a crude fat or oil is refined to remove its impurities, it must be further treated by bleaching to remove coloring materials that are typically present. Sulfuric and metaphosphoric acids and hydrogen peroxide have been used for this purpose. Calcium hypochlorite is used in bleaching sugar syrup prior to crystallization. Not all effective bleaching agents can be used because of their toxicity and inability to remove them completely. The aim is to remove all traces of the bleaching agent during subsequent processing so that they do not occur in the final product. Calcium peroxide, acetone peroxide, and benzoyl peroxides are other bleaching agents sometimes used in the food industry.

The use of oxidizers, such as potassium bromate, potassium iodate, and calcium peroxide as dough modifiers in the baking industry dates back many years. However, their mechanism has not been fully explained. Among authorities, there are at least two major viewpoints. It has been proposed that oxidizers inhibit proteolytic enzymes present in flour. It has also been proposed that the number of −S−S− bonds between protein chains is increased, forming a tenacious network of molecules. This action leads to a tougher, drier, and more extensible dough. The need for oxidizers is less with well aged flours and in connection with supplemented flours that have been effectively *brominated* at the mills. When used properly, oxidizers contribute to improve appearance, brighter crumb, and better texture of breads. Oxidizers do not appear to interfere with the generation of gas by yeast or leavening chemicals. However, oxidizers used to excess can destroy the desirable properties of the dough and end-products.

OXIDATION NUMBER. In its original and restrictive sense, the number of electrons which must be added to a cation to neutralize the charge. The concept has been extended to anions by assignments of negative oxidation numbers. Moreover, it has been further extended, first to all atoms or radicals joined by electrovalent bonds, and then to covalent compounds in which the shared electrons are distributed equally. For the broadest use of the concept, the expression "oxidation state" is often used.

OXIDATION POTENTIAL. The potential drop involved in the oxidation (i.e., ionization) of a neutral atom to a cation, of an anion to a neutral atom, or of an ion to a more highly charged state (e.g., ferrous to ferric).

OXO PROCESS. The general name for a process in which an unsaturated hydrocarbon is reacted with carbon monoxide and hydrogen to form oxygen-function compounds, such as aldehydes and alcohols. In a typical process for the production of oxo alcohols, the chargestock comprises an olefin stream, carbon monoxide, and hydrogen. In a first step, the olefin reacts with CO and H_2 in the presence of a catalyst (often cobalt) to produce an aldehyde which has one more carbon atom than the originating olefin: $R \cdot CH: CH_2 + CO + H_2 \rightarrow R \cdot CH_2 \cdot CH_2 \cdot CHO$. This step is exothermic and requires a cooling cycle. The raw aldehyde existing from the oxo reactor then is subjected to a higher temperature to convert the catalyst to a form for easy separation from the reaction products. The subsequent treatment also decomposes unwanted byproducts. The raw aldehyde then is hydrogenated in the presence of a catalyst (usually nickel) to form the desired alcohol: $R \cdot CH_2 \cdot CH_2 \cdot CHO + H_2 \rightarrow R \cdot CH_2 \cdot CH_2 \cdot CH_2OH$. The raw alcohol then is purified in a fractionating column. In addition to the purified alcohol, byproducts include a light hydrocarbon stream and a heavy oil. The hydrogenation step takes place at about $150\,°C$ under a pressure of about 100 atmospheres. The olefin conversion usually is about 95%.

Among important products manufactured in this manner are substituted propionaldehydes from corresponding substituted ethylenes, normal and iso-butyraldehyde from propylene, iso-octyl alcohol from heptene trimethylhexyl alcohol from di-isobutylene.

OXYGEN. Chemical element, symbol O, at. no. 8, at. wt. 15.9994, periodic table group 16, mp $-218.4\,°C$, bp $182.96\,°C$, critical temperature $118.8\,°C$, critical pressure 49.7 atmospheres, density 1.568 g/cm^3 (solid), 1.429 g/L $(0\,°C)$. Solid oxygen has a cubic crystal structure. Oxygen at standard conditions is a colorless, odorless, tasteless gas. Oxygen is slightly soluble in H_2O (4.89 parts oxygen in 100 parts H_2O at $0\,°C$), the solubility decreasing with increasing temperature (2.6 parts oxygen in 100 parts H_2O at $30\,°C$; 1.7 parts oxygen in 100 parts H_2O at $100\,°C$). Oxygen is slightly soluble in alcohol. Molten silver dissolves up to $10\times$ its volume of oxygen, but easily gives up the gas upon cooling. There are three stable isotopes, ^{16}O through ^{18}O. Three radioactive isotopes have been identified, ^{14}O, ^{15}O, and ^{19}O, with short half-lives measured in seconds and minutes. In terms of abundance in igneous rocks in the earth's crust, oxygen ranks first, with an average composition by weight of 46.6%. In terms of abundance in seawater, oxygen also ranks first, with an estimated 4 billion tons of oxygen per cubic mile of seawater. In terms of cosmic abundance, oxygen ranks eighth. For comparisons, assigning a value of 10,000 to silicon, the figure for oxygen is 220,000 and that for hydrogen, estimated the most abundant, is 3.5×10^8. Of dry air in the earth's atmosphere, 23.15% is oxygen by weight; 20.98% by volume. In the atmosphere, the oxygen is mixed with nitrogen, argon, the rare gases, CO_2, and H_2O vapor. Oxygen first was identified by Priestly in 1774 when he was experimenting with mercuric oxide. In the same year, Scheele also identified the element. Oxygen is required for burning and combustion, although the conditions of combustion vary widely. For example, phosphorus burns in air at the low temperature of $34\,°C$ when ignited. The temperature if ignition for ether in air is $340\,°C$, for ethyl alcohol in air, $560\,°C$, kerosene in air, about $300\,°C$, and hydrogen in air, about $600\,°C$. The oxidation process may occur with the rapidity and violence of an explosion, or may be as slow as the rusting of iron. Nearly all known species of living things require oxygen in some form, either free or chemically bound. First ionization potential 13.614 eV; second, 34.93 eV; third, 54.87 eV. Oxidation potentials $H_2O_2 \rightarrow O_2 + 2H^+ + 2e^-$, -0.68 V; $3H_2O \rightarrow \frac{1}{2}O_2 + 2H_3O^+(10^{-7}\,M) + 2e^-$, -0.815 V; $3H_2O \rightarrow \frac{1}{2}O_2 + 2H^+ + 2e^-$, 1.229 V; $4H_2O \rightarrow H_2O_2 + 2H_3O^+ + 2e^-$, -1.77 V; $3H_2O \rightarrow O(g) + 2H^+ + 2e^-$, -2.42 V; $HO_2^- + OH^- \rightarrow O_2 + 2H_2O + 2e^-$, 0.075 V; $4OH^- \rightarrow O_2 + 2H_2O + 4e^-$, -0.401 V; $3OH^- \rightarrow HO_2^- + H_2O + 2e^-$,

-0.87 V; $OH^- \rightarrow OH + e^-$, -1.4 V. Other physical properties of oxygen are given under **Chemical Elements**.

Allotropic Forms. The three known allotropic forms of oxygen are (1) the ordinary oxygen in the air, with two atoms per molecule, O_2, (2) ozone, O_3, with three atoms per molecule, and (3) the rare, very unstable, nonmagnetic, pale-blue O_4. The latter breaks down readily into two molecules of O_2.

When oxygen is subjected to the silent electric discharge, activated atomic oxygen is produced. Atomic oxygen displays an afterglow upon cessation of the current, and the oxygen is notably active with hydrogen bromide, forming bromine; with H_2S, forming sulfur, SO_2, sulfur trioxide, and H_2SO_4; with CS_2, forming carbon monoxide, CO_2, and SO_2, and, strangely, reduced molybdenum trioxide to a white oxide not reducible with hydrogen. The concentration of atomic oxygen obtainable by the silent electric discharge through oxygen is estimated at 20%.

The normal electron distribution of the electrons of the oxygen atom is $1s^2 2s^2 2p_x^2 2p_y^1 2p_z^1$, with 2 unpaired electrons in the $2p$ orbitals. The covalent or partly covalent compounds of oxygen would be expected to have $90°$ bonding angles. But in many cases they have values significantly greater (ca. $104°$ for R_2O and $105°$ for H_2O). This suggests the promotion of a $2s$ electron to a $2p$ orbital (i.e., $2p_y$ orbital), still leaving two unpaired electrons (a $2s$ and a $2p_z$ electron), and permitting partial sp^3 hybridization (which is incomplete because sufficient energy is not available) but producing bond angles between $90°$ and $109°$ for the sp^3 tetrahedral structure, with covalent-polar bonds.

The oxygen molecule is paramagnetic with a moment in accord with two unpaired electrons. In molecular orbital terms, the configuration is written

$$O_2[KK(z\sigma)^2(y\sigma^*)^2(x\sigma)^2(w\pi^*)^4(v\pi^*)^2]$$

in which KK designates the complete $1s$ shells of the two atoms, which are nonbonding, the term $(z\sigma)^2$ denotes the bonding effect of one pair of $2s$ electrons, one from each of the O atoms, $(y\sigma^*)^2$ denotes the antibonding effect of the second pair, the $(x\sigma)^2$ term represents the s-bond formed by one pair of p-electrons, $(w\pi^*)^4$ represents the 2 π-bonds formed by the other two pairs of p-electrons, while the $(v\pi^*)^2$ term denotes the last pair of p-electrons, which go into the next π subshell (two orbitals) with unpaired spins, and are antibonding.

Ozone, O_3, obtained by electrical discharge through oxygen or high-current electrolysis of sulfuric acid, is considered on the basis of electron diffraction studies to have an $O-O-O$ bond angle of $127 \pm 3°$ and $O-O$ bond length of 1.26 ± 0.02 Å. Its structure is considered to resonate among several forms, chiefly

Ozone is a blue gas, of characteristic odor, formed when ordinary oxygen is subjected to electrostatic discharge, density 1.5 times that of oxygen gas, mp $-251.4\,°C$, bp $-111.5\,°C$. It is explosive by percussion or under variations of pressure. Ozone reacts (1) with potassium iodide, to liberate iodine, (2) with colored organic materials, e.g., litmus, indigo, to destroy the color, (3) with mercury, to form a thin skin of mercurous oxide causing the mercury to cling to the containing vessel, (4) with silver film, to form silver peroxide, Ag_2O_2, black, produced most readily at about $250\,°C$, (5) with tetramethyldiaminodiphenylmethane $(CH_3)_2N \cdot C_6H_4 \cdot CH_2 \cdot C_6H_4 \cdot N(CH_3)_2$, in alcohol solution with a trace of acetic acid to form violet color (hydrogen peroxide, colorless; chlorine or bromine, blue; nitrogen tetroxide, yellow). In contrast to hydrogen peroxide, ozone does not react with dichromate, permanganate, or titanic salt solutions. Ozone reacts with olefin compounds to form ozonide addition compounds. Ozonides are readily split at the olefinin-ozone position upon warming alone, or upon warming their solutions in glacial acetic acid, with the formation of aldehyde and acid compounds which can be readily identified, thus serving to locate the olefin position in oleic acid, $C_{17}H_{33} \cdot COOH$, as midway in the chain $(CH_3(CH_2)_7CH:CH(CH_2)_7COOH$. Ozone is used (1) as a bleaching agent, e.g., for fatty oils, (2) as a disinfectant for air and H_2O, (3) as an oxidizing agent. See also **Aerosol**.

The protective effects of an ozone layer in the stratosphere of the earth have been known for many years. Ozone prohibits full penetration of ultraviolet radiation from the sun to the surface of the earth. Much research has been conducted and is still underway to determine the extent to which certain chemical pollutants may be destroying the ozone layer gradually and, among other factors, causing marked warming of the earth. This topic is discussed in considerable detail in the article on **Climate**.

Role of Oxygen in Water. The solvent properties of H_2O are due in great part to the dipole moment of its molecules (1.8 debye units) and its high dielectric constant (ca. 78). Its hydrogen atoms form hydrogen bonds with electronegative atoms such as fluorine, nitrogen, or oxygen. In fact, the H_2O molecules associate in H_2O by this mechanism. Also, the oxygen atoms of H_2O because of their residual negative charges are electrically attracted by cations, so that the H_2O molecules arrange themselves around cations, facilitating solution and ionization. In the same way, H_2O molecules surround anions by attraction of the positive ends of the dipoles. By these two processes, as well as the dissociation of water into oxonium and hydroxide ions, it forms hydrates with many compounds. Moreover, H_2O readily reacts with large numbers of compounds because of these properties. Thus the hydrolysis of covalent halides that have at least one lone pair of electrons is initiated by the donation of a proton by the H_2O, followed by splitting off of hydrogen chloride.

Oxides. Oxygen forms oxides with all the elements except some inert gases. Oxides are said to be normal when they contain no oxygen atoms that are bonded to each other, as in the peroxides. The normal oxides may be divided into three groups, basic, acidic, and neutral. The basic oxides, which react with or dissolve in H_2O to produce alkaline solutions, are formed by the alkali and alkaline earth elements (except beryllium) by the lighter Lanthanides and actinium, by silver(I), thallium(I) and lead(II). The oxides of the nonmetals and of the transition metals in their higher oxidation states are in general acidic. The oxides lying in the positions between the two groups exhibit both basic and acidic properties (amphiprotic or amphoteric) such as those of aluminum, tin(II) and iron(III), Al_2O_3, SnO, and Fe_2O_3.

The known facts about the structure of hydrogen peroxide, H_2O_2, are that the O—H distances are 0.97 Å, the O—O distances 1.47 Å, the HOO angles 94°, and the dihedral angle between the planes of the two O—H radicals 97°. The O—O bond is essentially a single one. In the liquid, H_2O_2 is somewhat more self-ionized than water. In water $pK_A = 11.75$, $pK_B = 17$. Its reactions may be oxidizing or reducing. Thus, it oxidizes Fe(II) to Fe(III), Ti(III) to Ti(IV) and SO_3^{2-} to SO_4^{2-}; but it reduces MnO_4^- (acid solution) to Mn^{2+}. Peroxides are known for the alkali and alkaline earth metals, as well as zinc, cadmium, mercury, thorium, uranium, plutonium, etc. However, not all compounds of formula MO_2 (where M is a metal atom) are peroxides; some are merely dioxides, as MnO_2, PbO_2, etc., others are superoxides, such as NaO_2, KO_2, RbO_2, CsO_2, CaO_4, SrO_4, and BaO_4. These last compounds contain the group O_2^-, as evident from their paramagnetism and crystal structure. Perhydroxyl, the free acid corresponding to the superoxides, is unstable ($H_2O_2 \rightarrow HO_2 + H^+ + e^-$, $E° = 1.5V$; $HO_2 \rightarrow O_2 + H^+ + e^-$, $E° = +0.13$ V). It is a moderately strong acid, $pK_A = 2.2$.

The peroxyacids containing —O—OH groups, are formed with all the transition elements in groups 4, 5, 6 of the periodic table, with main group elements 4 and 5 as well as elements of atomic numbers from boron to sulfur, inclusive. Representative peroxyacids are

peroxymonosulfuric acid, $H\!:\!\overset{\cdot\cdot}{\underset{\cdot\cdot}{O}}\!:\!\overset{:O:}{\underset{:O:}{S}}\!:\!\overset{\cdot\cdot}{\underset{\cdot\cdot}{O}}\!:\!\overset{\cdot\cdot}{\underset{\cdot\cdot}{O}}\!:\!H$ and peroxychromic acid,

$H\!:\!\overset{\cdot\cdot}{\underset{\cdot\cdot}{O}}\!:\!\overset{\cdot\cdot}{\underset{\cdot\cdot}{O}}\!:\!\overset{:O:}{\underset{:O:}{Cr}}\!:\!\overset{\cdot\cdot}{\underset{\cdot\cdot}{O}}\!:\!\overset{\cdot\cdot}{\underset{\cdot\cdot}{O}}\!:\!H$. The only peroxydiacids are formed by sulfur, phosphorus, carbon and boron, of which the most important is peroxy-

disulfuric acid $H\!:\!\overset{\cdot\cdot}{\underset{\cdot\cdot}{O}}\!:\!\overset{:O:}{\underset{:O:}{S}}\!:\!\overset{\cdot\cdot}{\underset{\cdot\cdot}{O}}\!:\!\overset{\cdot\cdot}{\underset{\cdot\cdot}{O}}\!:\!\overset{:O:}{\underset{:O:}{S}}\!:\!\overset{\cdot\cdot}{\underset{\cdot\cdot}{O}}\!:\!H$ although peroxy bridge compounds are also formed by certain transition element complexes, e.g., $[Co(NH_3)_5OOCo(NH_3)_5]^{4+}$ and $[Co(NH_3)_5OOCo(NH_3)_5]^{5+}$.

Industrial Oxygen. As with hydrogen, the electrolysis of water offers one approach to the production of pure oxygen. However, the economics are as unfavorable for oxygen production in this manner as for hydrogen. See also **Hydrogen**. For industrial oxygen production, air is the raw material. Using air, processes are of two major types: (1) liquid-oxygen processes wherein the oxygen is fractionally distilled from liquid air, and (2) gaseous-oxygen processes. See also **Cryogenics**.

Because of the relatively high energy costs of compressing and refrigerating involved in oxygen production, many processes have been developed and tested over the years, a high percentage of these later abandoned. An idea of the alternatives which face the process designer can be gathered from scanning the methods available specifically in the area of producing refrigeration for these processes: (1) Joule-Thomson effect only; (2) Joule-Thomson effect plus auxiliary refrigeration with an ordinary liquid-vapor cycle at moderate- or high-temperature levels, i.e., relative to liquid-air temperature; (3) Joule-Thomson effect plus approximately reversible expansion of the air or products in an expander; (4) refrigeration essentially due only to approximately reversible expansions of auxiliary fluid or fluids operating in liquid-vapor cycles, i.e., the cascade process; and (5) processes using an auxiliary nitrogen-liquefaction cycle.

Designers also face the choice of capacity of an oxygen plant. Costs per unit weight of oxygen made are lowered as the capacity of the plant goes up. For example, a plant with a capacity of 2000 tons (1800 metric tons) per day will produce oxygen at approximately 50% of the cost per unit weight as a plant with a 200-ton (180-metric ton) capacity per day.

The demand for industrial oxygen has created the need for several new plants during the past 20 years. Capacities for most recent plants range from about 1100 tons (990 metric tons) per day to 2500 tons (2250 metric tons) per day.

An oxygen pipeline system was established in western Europe in the late-1970s that is 592 miles (956 kilometers) long. The Eastern Network of this system serves 30 consumers in France, Luxembourg, and West Germany; the Northern Network serves some 40 additional users in France, Belgium, and the Netherlands.

Uses: In addition to the requirements by the chemical industry for oxygen as a reactant, either directly from the air or in purer, more concentrated form as from a separation plant, significant quantities of purified oxygen are used for welding and cutting metals. Oxygen of a purity of 99.5% is required for oxyacetylene and oxyhydrogen torches. When combined in proper proportions, acetylene and oxygen yield a flame with a temperature of about 3,480 °C. Oxyhydrogen flames are somewhat lower in temperature, but they are particularly useful for welding light-gage aluminum and magnesium allows and for underwater cutting. In welding applications, a reduction in purity of oxygen used from 99.5% to 99.0% will cut welding efficiency by over 10%. During the past several years, basic oxygen steelmaking has increased requirements for pure oxygen. In this process, nearly pure oxygen is introduced by means of a lance into molten iron and scrap. The oxygen combines with carbon and other unwanted elements and refines raw steel in much less time than the older open-hearth furnaces. The basic oxygen process exceeded the open-hearth process in terms of output to the United States in 1970 for the first time. On the total scale of consumption, relatively limited amounts of oxygen go into medical and life-support applications, as required for emergency situations in aircraft at high altitudes.

Role of Oxygen in Corrosion. Oxygen and oxidizing agents exert both a positive and negative influence on corrosion of metals. On the one hand, an oxidizing agent may form a protective oxide film on the surface of certain metals, aluminum being an excellent example, which essentially arrests corrosion by many external agents. On the other hand, the presence of oxidants may increase the rate of corrosion by supporting cathode reactions. As an example, Monel metal fully resists attack by oxygen-free 5% H_2SO_4 at room temperature. The corrosion rate rises, however, in almost direct proportion to oxygen content. A 20% oxygen content will cause a corrosion rate of about 150 mdd (milligrams of metal corroded per square decimeter per day). A concentration of 40% will increase the rate to about 250 mdd; a concentration of 80% to about 450 mdd. The oxygen need not be present in all of the acid contained in the metal vessel, but simply present in that concentration at the interface of metal, acid, and surrounding atmosphere. The effect of oxidizing salts on corrosion can be dramatic. Several factors, in addition to oxygen, affect corrosion, including the presence of other metals (electromotive-force displacements of one metal by another), temperature, acidity, and velocity. These factors are discussed further under **Corrosion**.

Oxygen Toxicity of Plants. As early as 1801, Huber and Senebier observed that grains develop more satisfactorily in an atmosphere containing a mixture of 3 parts nitrogen and 1 part oxygen than in an atmosphere containing 3 parts of oxygen and 1 part nitrogen. Considerably later, in 1878, Bert noted that the earlier observations also apply to the development of many plant species and are not peculiar to grains. Bert further suggested that excessive oxygen may slow down various reactions involving fermentation. It was not until much later, in the mid-1940s, that scientists (Dickens, Haugaard, and Stadie) further confirmed that enzymes are inactivated by oxygen excesses. They particularly stressed this fact in connection with enzymes that contain a sulfhydryl group in the active site. Machaelis (1946), Barron (1946), and Gilbert (1963) later pointed out that molecular oxygen alone acts in a rather sluggish manner in this regard and that, therefore, a special process or phenomenon must be involved. Molecular oxygen can be reduced only by accepting one electron at a time.

A number of scientists in the late 1960s through the mid-1970s pointed out that many sources in biological systems produce oxygen *free radicals*. For example, some oxidative enzymes which contain flavin as a prosthetic group proceed by a radical mechanism. When illuminated, chloroplasts produce superoxide ions and singlet oxygen. Because of its singlet configuration, the latter is not hindered in its interactions with biological materials. As pointed out by Griffiths and Hawkins, singlet oxygen can be formed from the ground state when energy, usually in the form of light, is supplied n the presence of a photosensitizer. The compounds that are photosensitized include many dyes and pigments, such as chlorophyll, flavins, and hematoporphyrins. The interaction between the sensitizer and oxygen results in the transfer of electrons, with the formation of superoxide ion. McCord and Fridovich (1969) discovered the enzyme superoxide dismutase. Their later findings show that aerobic organisms contain it, giving further credence to the proposal that all oxygen-metabolizing organisms from superoxide free radicals as a result of a univalent reduction of oxygen. As pointed out by Kon (1978), those free radicals that are toxic to the organism, by themselves or through interaction with other active forms of oxygen, are dismutated by the action of this enzyme.

There is a close relationship between oxygen toxicity and radiation on enzymes, DNA, and fats. Gerschman et al. (1954) showed that the same substances that afford protection against oxygen poisoning also increase resistance to radiation. Their results were further strengthened by experiments that demonstrated that additive nature of the two effects. Work on the effects that free radicals have on some of the polysaccharides used in food processing was commenced by Kon and Schwimmer and reported in 1977.

Environmental Aspects of Oxygen. Gaseous oxides, notably those of carbon, nitrogen, and sulfur which result from the combustion of fossil fuels and numerous industrial processes, comprise a large portion of the air pollution problem. These compounds are discussed under the specific elements and, in particular, are described under **Pollution (Air)**. In connection with the pollution of water in streams, lakes, ponds, rivers, etc., the content of dissolved oxygen in water is of prime concern. Dissolved oxygen must be available to support fish and other desirable living species in natural waters, and sufficient additional oxygen must be available in the water to effect biological degradation of both natural and manufactured materials which reach the water. The overuse of streams for disposal purposes in many instances has almost fully depleted the dissolved oxygen available for life support and hence has given rise to the term "dead" lakes or streams. Two terms are widely used: (1) BOD (biological oxygen demand) which is the requirement for dissolved oxygen in water to degrade or decompose organic matter within a measured time period at a given temperature, and (2) COD (chemical oxygen demand) which is the requirement for dissolved oxygen in water to combine with chemicals, essentially of an inorganic nature, which are introduced into a stream as the result of disposal operations. These aspects of oxygen are discussed under **Water Pollution**.

Earth's Oxygen Supply. The manner in which the earth's present oxygen system and reserves were formed has been the subject of much postulation for many years. Many of the details remain unclear and unconfirmed. In a theory proposed by Berkner-Marshall (1964, 1965), as the earth's atmosphere evolved, there was a slow buildup of the concentration of oxygen—proceeding from a trace to the present content

of 23.15% (weight). This theory also proposes that the oxygen content of the atmosphere fluctuated from time to time in a major and relatively rapid manner. There is speculation that these major alterations may have accounted for the extinctions of animal and life forms that took place at the ends of the Paleozoic and Mesozoic eras. For example, there was a great reduction in life in the latter part of the Permian period (Paleozoic era) when many kinds of strange reptiles and trilobites disappeared and seem to have left no descendents. Plant life declined greatly too during the late Paleozoic. From thousands of species in the Pennsylvanian period, there remained only a few hundred during the late Permian. Numerous explanations, particularly of a climatic nature, have been offered for these periods of reduction in life.

As pointed out by Van Valen (1971), photosynthesis does not produce a net change in oxidation. Except in bacterial photosynthesis, oxygen production is accompanied by a stoichiometrically equal quantity of reduced carbon. Thus, almost all of the oxygen is eventually used to oxidize reduced carbon. Predominantly, this oxidation occurs as the result of respiration in animals and plants. Further oxidation occurs as the result of forest fires. As observed by Borchert (1951), the only net gain in oxygen equals the amount of reduced carbon buried, as in the form of peat, black mud, and similar sediments. It has been estimated that most individual molecules of carbon remain reduced only for relatively short periods (months or years) because animals and plants have geologically very short lives. Plants respire and so oxidize some reduced carbon almost immediately. Other net sources of oxygen include nitrogen fixation and the photolysis of water in the upper atmosphere. Some investigators have considered these sources quantitatively unimportant, although Brinkmann (1969) suggests that this process would produce, over the earth's history (4.5×10^9 years), about seven times the present mass of oxygen in the atmosphere.

Numerous ways have been proposed to explain a net loss of molecular oxygen. Oxidation of volcanic gases, ferrous iron, sulfur, sulfide, and manganese, and the accretion of hydrogen from the solar wind are among these. Such processes are sometimes referred to as *oxygen sinks*. Estimates by Holland (1962) indicate that the net gain and net loss over geologic time are essentially in balance.

Van Valen has posed the question, "What can happen if photosynthesis is suddenly and drastically reduced?" Under such conditions, at a new steady state, production of oxygen and its consumption in the oxidation of carbon would be equal. But, before the new steady state occurs, would animals and decomposers use up much of the previously stored carbon in plants, thus creating a new loss of oxygen? Several investigators have observed that even if all the carbon in all organisms now alive were oxidized, this would decrease the atmospheric concentration of oxygen by less than 0.1% of its present value. And, further, still less than 1% of the present oxygen concentration would be used if all the reduced carbon available in soils and the like were reduced.

Much more detailed explanation of the stability of atmospheric oxygen is contained in the excellent review by Van Valen (1971).

As pointed out by Broecker (1970), the earth's oxygen supply is frequently included in lists of concerns over alterations in the environment, particularly as brought about by anthropogenic activities. Several investigators have made a number of observations which tend to invalidate any claims that oxygen is in danger of serious depletion. Broecker observes that each square meter of the earth's surface is covered by 60,000 moles of oxygen gas. Further, plants living in the ocean and on land produce about 8 moles of oxygen per square meter of surface each year. It is also observed that animals and bacteria destroy nearly all of the products of this photosynthetic activity—thus they use an amount of oxygen nearly equal to that generated by plants. Using the rate at which organic carbon enters the sediments of the ocean as a measure of the amount of photosynthetic product preserved each year, Broecker estimates this to be about 3×10^{-3} mole of carbon per square meter per year. This corresponds to approximately 1 part in 15 million of the oxygen present in the atmosphere. It is estimated, however, that this small amount of oxygen is probably being destroyed by a number of processes, including oxidation of reduced carbon, iron, and sulfur (weathering mechanisms). Broecker points out that the oxygen content of the atmosphere is thus well buffered, particularly in terms of relatively short time spans (100 to 1000 years).

Over a period of time, people have recovered about 10^{16} moles of fossil carbon and the fuels containing this carbon have been oxidized as sources of energy. Byproduct carbon dioxide from this combustion represent about 18% of the carbon dioxide content of the atmosphere. Two moles of atmospheric oxygen are used to liberate each mole of carbon dioxide from fossil fuel sources. Broecker points out that this process uses up only 7 out of every 10,000 available oxygen molecules. It is estimated that if these fuels are burned at an accelerating rate (5% per year), by the end of this century, only about 0.2% of available oxygen (20 molecules in every 10,000) will be used. It is estimated that if all known fossil fuels were ultimately burned, only 3% of available oxygen would be consumed. In terms of urban oxygen needs, particularly for automotive combustion needs, it is estimated that carbon monoxide levels in the atmosphere (in terms of physiological damage) would reach intolerable levels before the oxygen content of the atmosphere would have decreased by 2%.

The case of anthropogenic alterations of photosynthetic rates and its possible effects on oxygen supply has been covered previously by the observation that stoppage of all photosynthetic activity would require less than 1% of the present oxygen concentration.

Sverdrup et al. (1942) estimated that the oxygen content of deep sea water averages about 2.5 cubic centimeters at standard temperature and pressure per liter (0.1 mole per cubic meter). Thus, there are about 250 moles of oxygen gas in the deep sea for each square meter of earth surface. The oxygen content of the deep-sea waters is renewed about every 1000 years. The magnitude of this oxygen reservoir is tremendous. Broeker emphasizes this by observing that if the entire terrestrial photosynthetic product were dumped each year into the deep sea, the supply of deep-sea oxygen would last 50 years. But, if the waste products of 1 billion people were limited to 100 kilograms of dry organic waste per year, this would consume 0.01 mole of oxygen per square meter of earth surface and the deep-sea oxygen supply would last some 25,000 years.

In the summary of this report, Van Valen (1971) states, "There are three processes weakly concentration-dependent that keep changes in concentration of atmospheric pressure from being a random walk — inhibition of net photosynthesis by oxygen, the passage of hydrogen through the oxidizing part of the atmosphere before it escapes from the earth, and burial of reduced carbon in anaerobic water. A stronger regulator seems desirable but remains to be found. The cause of the initial rise in oxygen concentration presents a serious and unresolved quantitative problem."

And, in summary of his report, Broeker (1970) states, in part, "It can be stated with some confidence that the molecular oxygen supply in the atmosphere and in the broad expanse of open ocean are not threatened by human activities in the foreseeable future. Molecular oxygen is one resource that is virtually unlimited."

Additional Reading

Baukal, C.: "Oxygen Enhanced Combustion," CRC Press, LLC, Boca Raton, FL, 1998.

Berkner, L.V. and L.C. Marshall: in "The Origin and Evolution of Atmospheres and Oceans, "(P.J. Brancazio and A.G.W. Cameron, editors), pages 102–126, Wiley, New York, NY, 1964.

Berkner, L.V. and L.C. Marshall: *J. Atmos Sci.*, **22**, 225 (1965).

Borchert, H.: *Geochim. Cosmochim. Acta*, **2**, 62 (1951).

Brinkmann, R.T.: *J. Geophys. Res.*, **74m** 5355 (1969).

Brocker, W.S.: "Man's Oxygen Reserves," *Science*, **168**, 1537–1538 (1970).

Dickens, R.: "The Toxic Effects of Oxygen on Brain Metabolism and on Tissue Enzymes," *Biochem. J.*, **40**, 145, 170 (1946).

Gerschman, R., et al.: "Oxygen Poisoning and X-irradiation: A Mechanism in Common," *Science*, **119**, 623 (1954).

Gilbert, D.L.: "The Role of Pro-Oxidants and Anti-Oxidants in Oxygen Toxicity," *Radiation Res. Suppl.*, **3**, 44 (1963).

Griffiths, J. and C. Hawkins: "Mechanistic Aspects of the Photochemistry of Dyes and Their Immediates," *J. Soc. Dyers Colorists*, **89**, 173 (1973).

Holland, H.D.: in "Petrologic Studies: A Volume in Honor of A.F. Buddington, "(A.E.J. Engel, H.L. James, and B.F. Leonard, editors), pages 447–477, Geological Society of America, Washington, DC, 1962.

Kon, S. and S. Schwimmer: "Depolymerization of Polysaccharides by Active Oxygen Species Derived from Xanthine Oxidase Systems," *Food Biochem.*, **1**, 141 (1977).

Kon, S.: "Effects of Oxygen Fee Radicals on Plant Polysaccharides," *Food Technol.*, **32**, 5, 84–94 (1978).

Kruk, I.: "Environmental Toxicology and Chemistry of Oxygen Species: The Handbook Of Environmental Chemistry," Springer-Verlag, New York, Inc., New York, NY, 1997.

Lide, D.R.: "CRC Handbook of Chemistry and Physics 2000-2001," 81st Edition, CRC Press, LLC., Boca Raton, FL, 2000.

McCord, J.M. and I. Fridovich: "Superoxide Dismutase: An Enzymatic Function for Erythrocuprein," *J. Biol. Chem.*, **244**, 6046 (1969).

Michaelis, L.: "Fundamentals of Oxidation and Reduction, "in Currents in Biochemical Research (D.E. Green, editor), pages 207, Wiley, New York, NY, 1946.

Sawyer, D.: "Oxygen Chemistry," Oxford University Press, New York, NY, 1999.

Sundquist E.T. and W.S. Broecker, Eds.: "The Carbon Cycle and Atmospheric CO_2," American Geophysical Union, Washington, DC, 1985.

Sverdrup, H.U., Johnson, M.W., and R.H. Fleming: "The Oceans, Their Physics, Chemistry and General Biology," Prentice-Hall, Englewood Cliffs, New Jersey, 1942.

Van Valen, L.: "The History and Stability of Atmospheric Oxygen," *Science*, **171**, 439–443 (1971).

OXYGEN DEBT.   A term used to refer to the buildup of a need for oxygen through anaerobic respiration of muscle cells in a higher vertebrate animal during violent exercise. When the energy demands are too great to be satisfied by the aerobic respiration, the cells turn to anaerobic respiration. Lactic acid is an end product of such respiration; this acid tends to accumulate in the muscles and some of it diffuses out into the blood and accumulates in the liver. When the activity ceases, deep breathing continues and the extra oxygen is used to reconvert the lactic acid back to pyruvic acid and to carry the pyruvic acid on through the tricarbocyclic acid cycle. It may also be reconverted back to glucose and glycogen. We say that the muscles have built up an oxygen debt during the very active exercise and this is repaid in the continued deep breathing during rest following the exercise.

OXYGEN GROUP.   The elements of group 16 of the periodic classification sometimes are referred to as the Oxygen Group. In order of increasing atomic number, they are oxygen, sulfur, selenium, tellurium, and polonium. The elements of this group are characterized by the presence of six electrons in an outer shell. The similarities of chemical behavior among the elements of this group are less striking than hold for some of the other groups, e.g., the close parallels of the alkali metals or alkaline earths. With exception of oxygen, all elements of the group have a valence of 4+, in addition to other valences. All of the elements with the exception of polonium also have a valence of 2−. Unlike the alkali metals or alkaline earths, for example, the elements of the oxygen group are not so similar chemically that they comprise a separate group in classical qualitative chemical analysis separations. Tellurium and selenium do appear together among the rarer metals of the second group in terms of qualitative chemical analysis.

OXYTOCIN.   A polypeptide hormone which is secreted by the posterior lobe of the pituitary gland of mammals and other vertebrates. Oxytocin exerts a stimulating effect upon the muscles of the breast (milk-ejection) and those of the uterus of mammals. It is sometimes used medically to stimulate labor in cases of difficult childbirth and to time the onset of labor. See also **Central and Peripheral Nervous Systems.**

OZOCERITE.   Sometimes spelled ozokerite, this is a natural, brown to jet black mineral (paraffin) wax comprised mainly of hydrocarbons. The melting point is variable. The material is soluble in chloroform. When heated with sulfuric acid (20–30%) from 120–200 °C, ozocerite yields ceresine. Sometimes called earth wax, fossil wax, mineral wax, and native paraffin.

OZONOSPHERE.   The general stratum of the upper atmosphere in which there is an appreciable ozone concentration and in which ozone plays an important part in the radiation balance of the atmosphere. This region lies roughly between 10 and 50 kilometers (6.2 to 31 miles), with maximum ozone concentration at about 20 to 25 kilometers (12.4 to 15.5 miles). Also called *ozone layer*.

P

PACA. See **Rodentia**.

PACIFIC COD. See **Codfishes**.

PACIFIC EQUATORIAL WATER. An immense surface oceanic water mass, extending from Central America to the East Indies. It is broader on the east, extending from 20°N to 20°S latitude, but it narrows greatly on the western side of the Ocean. Its surface temperature range is 8–15°C (46.4 to 59°F), salinity range 34.6–35.15%, but both properties decrease with depth below the lower values.

PACIFIC HIGH. See **Atmosphere (Earth)**.

PACKING (Absorption Column). See **Absorption (Process)**.

PACOEMULSIFICATION. Phacoemulsification is a modified version of extracapsular cataract extraction (ECCE) and is the most common surgical procedure for removing cataracts. As in other forms of ECCE, phacoemulsification involves removing the eye's natural lens while leaving in place the back of the capsule, which holds the lens in place. The difference with phacoemulsification is that the cataract is broken into tiny pieces that are suctioned from the eye through a smaller incision than that required by other forms of cataract surgery. Healing and rehabilitation are faster with this procedure, and there is little, if any discomfort.

To understand how the phacoemulsification technique works, it is important to understand what a cataract is and how it interferes with vision. The eye works like a camera with two lenses. The first lens is the cornea, a clear membrane that covers the front of the eye. The second lens is the eye's natural crystalline lens, which is held in place by a capsule located behind the pupil. The cornea is responsible for about 70% of the eye's focusing power, while the natural lens fine-tunes the image.

When the natural lens becomes cloudy, usually because of the aging process, it keeps light rays from passing through or diffuses the light in such a way that vision becomes fuzzy or hazy. This cloudy lens is called a cataract. The object of cataract surgery is to remove this hazy lens and to replace it with a plastic prescription lens that is permanently implanted in the eye.

In phacoemulsification cataract surgery, the surgeon makes a very small incision, about 1/8th of an inch, in the white of the eye near the outer edge of the cornea. A small ultrasonic probe is inserted through this opening and, oscillating at 40,000 cycles per minute, is used to break up (emulsify) the cataract into tiny pieces. The emulsified material is simultaneously suctioned from the eye by the open tip of the same instrument. The hard central core of the cataract (the nucleus) is removed first, followed by extraction of the softer, peripheral cortical fibers that make up the remainder of the lens. The front (anterior) section of the lens capsule is removed along with the fragments of the natural lens. The back (posterior) portion of the capsule is left in place to hold and maintain the correct position for the implanted intraocular lenses.

After removal of the cataract, a prescription intraocular lens, or IOL, is permanently implanted in the lens capsule to replace the natural crystalline lens of the eye that was removed during the surgery. This lens is rolled inside a tiny hollow tube and inserted through the same incision that was used to remove the cataract. The folded lens is pushed out of the tube by a tiny plunger and, as it unfolds, is positioned by the surgeon in the center of the lens capsule. The new lens is held in place by microscopic, spring-like wires that are attached to the implant.

The tiny incision made during phacoemulsification surgery generally requires no stitches and heals itself in a few days. Antibiotic and steroid eye drops may be given to diminish inflammation, to prevent infection, and to keep the eye moistened for several days following surgery.

Phacoemulsification cataract surgery is one of the most effective surgical procedures performed in the United States today, and a large percentage of patients are very satisfied with the results. See also **Cataract**; and **Extracapsular Cataract Extraction (ECCE)**.

Vision Rx, Inc., Elmsford, NY.

PADDLEFISHES (*Osteichthyes*). Of the order *Chondrostei*, family *Polyodontidae*, the paddlefishes resemble to some extent some species of shark. Although classified as *Osteichthyes*, they are a strange combination of cartilage and bone. These fishes are named for the rather tremendous paddle that is attached to the nose, equal in length to about one-third the length of the body of the fish. The small barbels under the paddle are reminiscent of sturgeons.

Paddlefishes are known for very rapid growth and development. When fully grown, a paddlefish may measure up to 6 feet (1.8 meters) in length, with a paddle an additional 2 feet (0.6 meter) in length and about 4 inches (10 centimeters) in width. Biologists have not determined the use for the paddle because the fish obtains its food by swimming with its mouth wide open. The main diet is comprised of crustaceans and planktonic organisms.

Only two kinds of paddlefishes are known: (1) the *Psephurus gladius*, found in the Yangtse River valley; and (2) the *Polyodon spathula*, found in the Mississippi valley. It is considered to be an acceptable food fish. In the United States, the fish is sometimes called the spoon-bill, spoon-beaked sturgeon, and spoon-billed catfish, or simply spooner.

PAGET'S DISEASE. See **Bone**; and **Dermatitis and Dermatosis**.

PAINTS AND COATINGS. Traditionally, paints and other coatings have been considered in terms of protecting and decorating buildings, houses, furniture, automobiles, toys, boats, machines, and the like. These products represent a large industry estimated at about $15 billion per year worldwide. Nevertheless, these kinds of coatings account for only 10–15% of the total coatings industry.

One researcher[1] has classified coatings in two technically defined categories.

Type I—Products manufactured and sold with several coating layers. Examples are the aforementioned applications, but also include color photographic film, graphic arts films for printing, coated papers for printing, coated containers for food packaging, magnetic storage media for computers and audio/visual equipment, optical disks for digital data storage, adhesive tapes, wallpaper, and metallized films.

Type II—Coatings that become an integral part or a key intermediary for a device or piece of equipment. Sometimes, the term *core technology* is used for such applications. There is a wide variety of coating products in this classification. They would include photoresists for circuit board manufacture, thick- and thin-film coatings on integrated circuits, adhesives for bonding metals, phosphor coatings on electronic display screens, stain repellents on fibers, dyes on fibers, ceramic glazes, encapsulated time-release drugs, and thin-film photovoltaic cells.

The foregoing categories of coatings immediately dramatize the versatility and specialization of a huge industry that has developed essentially during the past three decades. With the continuing interest in materials composites, continued rapid growth is expected. Coating technology is

[1] E.D. Cohen, E.I. Du Pont De Nemours & Co., Parlin, New Jersey.

a complex undertaking fraught with numerous design and manufacturing problems. These are particularly difficult in scaling up laboratory procedures to high production rates on the factory floor. Consequently, a number of successful processes remain proprietary. The technology has demanded much of chemists and chemical and mechanical engineers, with particular emphasis on an understanding of fluid dynamics.

Although the two foregoing classifications may serve a very useful purpose at the scientific level, for the purposes of this article, the first part is devoted to traditional surface coating products that serve protective and decorative purposes. The second part, in less depth, addresses coatings that are used for special purposes as represented, (e.g., by uses in electronic devices).

PROTECTIVE AND DECORATIVE COATINGS[2]

Paint, coating, and finish are terms used to describe a wide variety of materials designed to adhere to a substrate and act as a thin, plastic-like layer. Paints are available for decorative, protective, and other purposes. They can be decorative by covering defects (being opaque), by changing color, or by providing a desired gloss or sheen. Protective uses include shielding metals from corrosion, protecting plastic from degradation caused by ultraviolet light, acting as a moisture barrier and providing mar and scratch resistance for wood or plastic surfaces. A paint can also be used for its special spectral properties (e.g., light absorbing for heating swimming pools; radar absorbing for military vehicles; etc.), or unusual physical properties (e.g., strippable coatings for the interior of paint spray booths, insulating coatings for electrical parts, etc.).

Paints are generally liquids before they are applied to a surface. When applied, they should completely replace the substrate/air interface with a substrate/paint interface (called wetting). The forces of attraction created by this wetting process are responsible for the paint's adhesion. The paint then dries and/or cures to form a hard film. Drying is the physical action of solvent leaving the film, while curing refers to a chemical reaction which connects polymer chains of the paint together.

Composition

Thousands of raw materials are used to manufacture coatings, but they can generally be classified into four categories: (1) binders, (2) pigments, (3) solvents, and (4) additives. Binders (or resins) are generally organic compounds, usually polymeric or oligomeric in nature, which provide a continuous matrix in the final film and have a major influence on the toughness, flexibility, gloss, chemical resistance and cure/dry properties of the coating.

Pigments. These are finely divided powders (particles between 0.1 and 50 micrometers in diameter) which are dispersed throughout the binder. In addition to reinforcing the final film, much as they do in composite plastics, they influence a coating's resistance to abrasion and corrosion, and they also are the major factor in the gloss, color, and opacity of a coating.

Solvents or Thinners. These substances have a major effect on a paint's application viscosity and also affect a coating's cure dry properties, and often its toxicity. Most solvents evaporate and leave the film during the drying process, although certain "reactive dilutents" are designed to react with the resin and become part of the binder system. Water is a popular solvent because of its low cost, low toxicity, and nonpolluting nature. Almost half of the coatings currently produced use water as a major solvent. Disadvantages of using water include the effect of humidity on the drying characteristics of the paint and the difficulty of making a water-resistant film from materials suspended in water. Major organic solvents include mineral spirits, ketones, acetates, alcohols, and xylene.

Additives. Among the more important classes of additives used in coatings are: (1) *surfactants*, which are used to suspend pigment and binder particles; (2) *thickeners* to obtain proper rheology (especially in latex paints); (3) *plasticizers*, which lower the glass transition temperature of the binder and increase the flexibility of the coating; (4) *antifoam agents* to prevent bubbles in aqueous paints; (5) *antiskin agents*, which prevent the formation of a dry layer on top of the paint while it is still in the can; (6) *preservatives*, such as biocides and mildewcides to protect the binder from microscopic organisms both before and after application;

(7) *ultraviolet light absorbers* to protect the binder and/or substrate from degradation due to sunlight; and (8) a variety of surface *conditioners* and *lubricants*, which help the film adhere to the substrate or protect the film by giving it a lubricated surface. Additives will often interact and coating formulators must be careful to watch for synergistic and antagonistic effects.

Ratio of Components. The ratio of the four aforementioned components (binders, pigments, solvents, and additives) greatly influences the properties of a coating. The volume fraction of the solid film (which is pigment) is referred to as the *pigment volume concentration*. The concentration where there is barely sufficient binder to fill in the voids between the pigment particles is referred to as the *critical pigment volume concentration* (or *CPVC*). When the composition of a coating is changed from below its CPVC to above it, the properties of the coating begin to dramatically degrade (except for opacity, which increases). The performance changes are primarily due to the pockets of air which form in the final film because of the shortage of binder. With the exception of flat architectural (house) paints, fillers, and certain primers, almost all coatings have a pigment volume concentration less than the CPVC. The CPVC is a function of a pigment combination's oil adsorption and particle size distribution.

The ratio of solvent to nonvolatiles is important since the volatile organic compound (VOC) composition of coatings is increasingly a target of government regulations. Usually the VOC of a coating is expressed in units of *mass per volume* and is calculated by multiplying the density of the coating by that fraction of the coating that is volatile. For coatings using water (or certain chloroalkanes) as a solvent, the effect of these "exempt" solvents is subtracted out before the calculation is made. Current VOC limits in the United States range from 250 grams per liter for some California architectural alkyds to 450 grams per liter for some furniture finishes.

Binder Classifications

Paints are often classified by the type of binder they include. The most common classifications (with percent of total coatings used in 1985) are: Latexes (31%); waterborne (10%); non-aqueous dispersions (2%); solventborne (55%); and one hundred percent solids coatings (2%). A small volume of paint is made with silane binders.

Latexes. These are dispersions of high-molecular-weight polymer particles in an aqueous medium. Since the polymer is in a suspended form, the viscosity of the mixture is almost exclusively a function of the viscosity of the continuous phase (i.e., water) and is not affected by the molecular weight of the polymer. This permits the use of higher-molecular-weight material than can be used with a solution-type approach. The film forms (i.e., the paint dries) when the water evaporates and the spherical latex particles are forced together, overcoming the steric and ionic forces which had been stabilizing them. Once the particles touch, the surface tension of the latex causes the individual particles to coalesce (i.e., the polymers in adjoining particles entangle), aided by the capillary forces created by the evaporating water. Latex particles are generally 0.1 to 0.5 micrometer in diameter, although particle sizes as much as an order of magnitude on either side of these values is used for specialized purposes.

The two most commonly used latex systems are acrylic systems (40% of usage), which perform very well, but are relatively expensive, and the vinyl-acrylic copolymer systems (57% of usage and growing), which do not perform as well, but are less expensive.

Latexes are usually considered separate from other waterborne binders because the method employed for synthesizing and suspending them is very different. For latexes, emulsion polymerization is used and no organic solvent is required to obtain or stabilize the emulsion. In contrast, most waterborne binders are synthesized in organic solvent solutions and then "let down" with water. Often, some quantity of organic solvent must remain or the resulting aqueous mixture will not be stable. Most waterborne resins need to be cured. Reactions commonly used include the oxidative polymerization of unsaturated aliphatic chains, the reaction with aminoplast resins, and the reaction of epoxies to form ether or ester bonds. Waterborne resin compositions include acrylics, polyesters (including alkyds), urethanes, phenolics, and epoxies, among others.

Non-Aqueous Dispersions (NADs). These are the solvent-borne analogues of latexes. They were used as automotive finishes in the 1970s, but their use is now declining. The use of NADs as an auxiliary binder, however, is increasing as it has been found that the addition of a small amount of NAD can improve a coating's drying characteristics and rheology.

[2] Some of this information was provided by The Sherwin-Williams Company, Cleveland, Ohio.

Solvent-Borne Coatings. These cover a wide variety of resins including alkyds. Alkyds are polyesters made from soya, linseed, or other oils and are a major factor in architectural, automotive, and industrial maintenance usages, among other applications. Acrylics are known for their exterior durability and are the major binder in automotive coatings. Epoxies are used mostly in automotive, industrial maintenance, metal container, and coil coatings. Polyurethanes are isocyanate-based binders and are used where excellent properties are required, such as in the magnetic media, magnet wire, industrial maintenance, and deck coatings fields. Polyesters, other than alkyds, are used in coil, metal furniture, metal container, appliance, and automotive coatings. Amino crosslinkers primarily are modified melamines and are important in metal container, automotive, coil, and metal furniture applications, among others.

In addition to drying, most solvent-borne coatings undergo some type of cure. Chemical reactions used for the cure of solvent-borne coatings include the oxidative crosslinking of unsaturated carbon bonds (alkyds and polyesters), the reaction of melamine derivatives to form ureas (polyesters, alkyds, acrylics, and amino resins), ether and ester formation by epoxies (epoxies, polyesters and acrylics), and urethan formation by isocyanates. Curing is especially important in low VOC coatings because the viscosity of a resin is a function of the molecular weight of the material. The demand for lower VOC (i.e., higher solids) coatings has resulted in a move toward lower-molecular weight and more reactive solvent-borne coatings. About one-quarter of the solvent-borne coatings currently used are considered "high solids" (i.e., they have a VOC of 350 grams/liter or less).

The thrust for higher solids also has been a factor in the growth of *100% solids* coating technologies. These technologies include polymerization initiated by ultraviolet radiation or an electron beam; the use of powdered coatings which coalesce when sufficient heat is applied; and vapor cure technology, where a reactive resin is crosslinked by exposure to a reactive vapor. While the capital investment required by these technologies is high, their use is expected to continue to grow because of their efficient use of material and the superior properties that can be obtained. Another means of using a *solventless system is hot melt coatings*. These are applied at high temperatures without solvent and dry by cooling them to room temperature.

Silanes. Silanes and silane derivatives dominate the small market for inorganic binders. These materials are used both in combination with organic binders and by themselves. As co-binders, they increase the chemical and moisture resistance of a film. When used by themselves, they form brittle, very chemical-resistant films. Silane coatings are usually more expensive than their organic counterparts and must be kept dry before application.

Pigment Classifications

Pigments used in the paint industry are commonly classified by function: (1) *Hiding* (or *prime*) pigments scatter light and are used to obtain opacity; (2) *extender* (or "inert") pigments are used to reinforce the binder, increase the pigment volume concentration, lower gloss, and lower the cost of a paint; (3) *colored* pigments which are used to tint a paint can be either inorganic or organic; (4) *metallic* pigments are used for corrosion prevention and appearance reasons; and (5) *protective* and other *functional pigments* can be used to add special features to a coating.

Hiding Pigment. The refractive index of hiding pigments must be sufficiently different from the refractive index of the binder (usually about 1.5) if light is to be effectively scattered. Two crystal structures of titanium dioxide (commonly referred to as titanium) are the most widely used hiding pigments in the paint industry with the rutile version being more popular than anatase because the refractive index of rutile (2.76) is higher than that of anatase (2.55). Rutile is also more thermally stable and photostable, although the chalking property of the anatase pigment has been used to advantage in making "self-cleaning" paints. Like most inorganic pigments, titanium dioxide is a naturally occurring compound that must be mined, crushed and processed before it is a suitable raw material for paint. The optimum diameter for light scattering is about 0.2 micrometer.

Other naturally occurring hiding pigments include zinc oxide (refractive index = 2.01) and zinc sulfide (refractive index = 2.37). At one time, lead carbonate (refractive index = 2.0) was a leading hiding pigment. In addition to hiding, zinc oxide is a fungistat (i.e., it inhibits the growth of fungi).

Pockets of air (refractive index = 1.0) encapsulated in plastic are also used as light-scattering pigments. The size of these synthetic pigments range from 0.6 to 20 micrometers, depending upon the number of 0.5-micrometer bubbles per particle.

Extender Pigment. The major classes of extender pigments are as follows:

Calcium carbonate (also called *whiting*). *Calcium carbonate* is inexpensive, has a low binder demand, and is not colored, but is acid sensitive.
Clay (or *kaolin*) covers a wide variety of materials which are inert and inexpensive, but it is more yellow than calcium carbonate.
Talcs are very inexpensive and easily suspended pigments.
Silicas (mostly silicon dioxides) are low cost, inert, and very hard.

Other extender pigments include barium sulfate, feldspar, diatomite, and mica.

Color Pigments. Chrome yellow is the leading inorganic color pigment. It is used primarily in traffic-marking paint for roads and highways. The yellow, red, and brown versions of iron oxide also are important and are used in a variety of industrial and architectural coatings.

There are hundreds of organic color pigments used in the paint industry, the vast majority of which are synthetic. Since many of these are vulnerable to ultraviolet light or chemical degradation, great care must be taken in choosing pigments that are suitable for a given paint and its intended usage.

The most commonly used black pigments are carbon blacks. In addition to being efficient light absorbers, some varieties of these small-particle-size materials impart electrical conductivity and thixotropy to paints. The leading inorganic black pigment is black iron oxide. This material is used primarily because it is easier to disperse than the carbon blacks. See also **Carbon Black**.

Metallic Pigments. The leading metallic pigment is zinc dust, which is used mostly in zinc-rich primers, where it acts as a *passivating agent*. Aluminum flake is used for the silvery metallic appearance that it imparts.

Anticorrosive Pigments. Several pigments are used primarily because of their anticorrosive properties. Chromates (zinc, strontium, and lead, if permitted) are the most effective anticorrosive pigments, but the chronic toxicity danger associated with them is a matter of serious concern. Barium metaborate, red lead, and borosilicates have been the principal nonchromate materials used. Other pigments used for special purposes include iron oxide for magnetic media and copper oxide in marine coatings to prevent barnacle and algae growth.

Tributyltin acetate $[(C_4H_9)_3SnOOCCH_3]$ gained widespread usage in marine paints. The loss of ship performance and efficiency resulting from the growth of barnacles, seaweeds, tubeworms, and other organisms on boat bottoms has been known since the time of the Phoenicians, when copper strips were fastened to hulls to prevent fouling. Various navies and ocean shippers worldwide have found tributyltin acetate (TBT) effective, particularly in tropical waters. Unfortunately, the ingredient in paint has been found to be toxic. One authority in ocean chemistry has observed that TBT is the most toxic compound man has introduced into the marine environment. In recent years, numerous studies and evaluations of TBT-based paints have been made. Some countries have regulations against its use. Details on the adverse growth of oysters, for example, have been reported. Some of these observations are covered in the Champ reference listed.

Coating Manufacturing Process

There are usually two steps in the paint-making process: (1) dispersing the pigment (called *grinding*) and (2) mixing in the raw materials not used in grinding (called *letting down*). Except for NADs, solvent-borne coatings are made by grinding the pigment into a binder/solvent solution. The polymer serves a dual purpose in the step: (1) It thickens the mixture, increasing the dispersing efficiency of energy put into the system, and (2) it adsorbs onto the surface of the pigment particles, stabilizing them in suspension. Once stabilized, the suspension is let down by stirring in the remaining raw materials. For most waterborne paints (including all latexes), the grinding step consists of dispersing the pigment in water. The binder is not included since it is not stable enough to withstand the grinding process. Instead, surfactants are used to help wet and stabilize the pigment. Paints prepared in this manner add the binder in the letdown.

Mills. Several types of machines (called *mills*) are used for grinding pigment. *Media mills* have a chamber where the pigment, binder and a solid media are all ground together. The grinding action of the media particles on one another provides the shear needed to breakdown the pigment agglomerates. Some media mills have chambers that hold an entire batch at one time, while others have smaller chambers through which a batch is passed in a continuous flow. Batch mills (those falling into the former category) include those using pebbles, ceramic beads, or steel shot as media. Continuous mills (where the batch flows through the chamber) usually use sand or ceramic beads as the media.

Roller mills have large, closely placed rollers capable of grinding very thick pigment suspensions. Adjoining rollers are turned in opposite directions and at high speeds. The point where adjoining roller surface separate is subjected to sufficient shear to pull the pigment agglomerate apart.

High-speed dispersers (HSDs) are a third general type of mill. HSDs use blades attached to rotating shafts to disperse the pigment, much as an egg beater is used to disperse flour. HSDs are currently the most widely used grinding equipment. HSDs or similar stirring devices are generally used for the letdown regardless of the type of mill used in the grinding step. See also **Ball, Pebble, and Rod Mills.**

Application Methods

Coatings can be applied in a variety of ways depending upon the nature of the substrate and the viscosity of the coating. Brush, roller and pressure pads are popular methods of application for architectural and industrial maintenance coatings. Advantages of these methods include lack of capital investment and the ability to apply coatings on site. Disadvantages include their labor intensity and their limitation to use *ambient cure* coatings. Air, and especially airless, spray equipment can apply coatings much faster than a brush or roller, but this method requires more equipment and is not as adaptable. Spraying is used for architectural, industrial maintenance, wood furniture, automotive refinish and other coatings.

Electrostatic Spraying. In this method, a paint with a negative charge is applied to a substrate with a positive charge. This is the method of choice for automotive, metal furniture, appliance, machinery, and metal container coatings. The equipment for electrostatic spray costs more than regular spray equipment, but the transfer efficiency can be much higher. A drawback is that electrostatic spraying can be used only if the object to be painted can hold an electrical charge and does not have deep crevices.

Electrodeposition. Another way of using electricity to paint is electrodeposition (ED). In this method, the object to be coated is dipped into a vat of charged aqueous coating. An opposite electrical charge is then applied to the object and the paint is attracted to the surface of the charged object. Having been painted, the object is removed from the vat, rinsed, and baked. Electrodeposition requires a very large capital investment, does not allow for color changes, works only for conductive (metal) objects, and is only suitable for coatings that are baked. Ambient cure paints do not have the long-term stability needed for ED. Nevertheless it is the greatly preferred application method for automotive primers and many other metal products because of its high transfer efficiency — desirable for both environmental and economic reasons.

Dip Tanks. This method, which does not use electricity, is especially suitable for small objects. The main disadvantage to dipping is the difficulty of getting an even coat without *drips*.

Roller Coaters and Sheet Coaters. These are often used in coil coating where very large, flat surface areas need to be painted quickly. In these methods, the objects to be painted are passed between a doctor blade and applicator rolls. These methods are suitable only for coatings that are to be baked.

Surface Preparation of the Substrate. This is extremely important for all methods of paint and coatings application. The failure of a paint system is often due *not* to the paint itself, but because of a failure in surface preparation. For example, an anticorrosive paint applied to a rusty surface will not be effective if the rust falls off taking the new paint with it. For wood and plastic surfaces, old paint or a weathered surface layer may have to be removed. For older metal objects, the removal of corrosion is often required. Sandblasting is one method to remove both the old paint and any corrosion. For new metal objects, a phosphate or chromate layer is often chemically bonded to the metal to provide a surface to which a coating can easily adhere.

Paints for Specific Functions

Paints are often separated into the types of jobs they perform. *Primers* are meant to be applied to bare substrates and then covered with a topcoat. As such, they must have good adhesion and good recoatability, but color and light stability are usually unimportant. When used over metal, primers are usually expected to provide corrosion protection. *Sealers* are similar to primers except they are used over porous substrates. Sealers eliminate the leaching of material from the substrate and also prevent paint components from migrating to the substrate. *Surfacers* are highly pigmented paints that are applied in a thick layer to mask surface irregularities and to allow good adhesion by a topcoat. *Fillers* are a type of surfacer that is used to fill holes. *Stains* are low-solids coatings applied to wood to accentuate its grain. Most interior stains require a topcoat.

Some finishes do not require a primer. Varnishes are clear, tough, and usually have a glossy finish. Exterior varnishes give wood some protection from sunlight. *Shellacs* are a type of varnish which offers the advantages of a quick dry and easy sanding, but which is sensitive to water and has a limited shelf life.

Topcoats are usually applied over an undercoat (i.e., primer, stain, etc.). A topcoat must protect its undercoat from environmental damage and should provide the appearance characteristics desired for the particular application. Topcoats are available in a variety of colors, textures, and glosses. One specialized topcoat is *lacquer*, which is a solution of resin in organic solvents. Lacquers dry, but do not cure. Because of their high VOCs, the use of lacquers is declining. *Enamel* is a term used to describe a glossy, opaque topcoat.

A unique two-coat, topcoat system is used in the automobile industry. A basecoat is applied to a primed surface to provide opacity, color, and a metallic appearance, while the clearcoat provides gloss and a mirrorlike finish (referred to as *distinctness of image*).

Various regulatory agencies in the United States and other countries have set limits on the use of low-volatile organic coatings. These are widely used in the chemical, petrochemical, and metallurgical industries to resist corrosion. Presently, high-temperature paints and coatings are largely silicone in nature. These are considered the best-performing coatings for smokestacks, boilers, mufflers, furnaces, incinerators, combustion chambers, and jet engines. High solids polyorganosiloxane polymers (silicone resins) are used to protect steel piping and other equipment where high temperatures accelerate deterioration of ferrous substrates. They also are used on storage tanks where appearance is an issue, as in the case of oil refinery tanks.

In addition to their high-temperature properties, high solids polyorganosiloxane polymers are resistant to numerous chemicals. For example, electric power generation frequently creates sulfuric acid as a byproduct. The process often involves extremely high temperatures as well. Silicone-based coatings currently appear to be the only materials capable of withstanding such conditions.

Special-purpose Coatings

These coatings include the use of materials (generic coatings) for purposes other than their contribution to protecting and decorating a structure or product. There is a multitude of such uses, and only a few can be described here because of space limitations.

Electronic Microstructure Fabrication. A microstructure may be defined as a pattern formed on or imbedded in the surface of some substrate material. Microstructure implies that the transverse dimensions of the patterns are in the microscopic and submicroscopic range, factors that determine the scale of electronic circuit integration. With the progress of electronic component miniaturization, such ranges have advanced from small-scale integration (SSI) to medium-scale integration (MSI) to large-scale integration (LSI) and to very-large-scale integration (VLSI). As of the early 1990s, such integration has progressed to millions and billions of transistors (e.g., on one integrated circuit). See also **Microelectronics**.

Patterns on microstructures may be formed in layers or insulators deposited on a surface or may consist of chemical or physical modification of shallow regions of the substrate. Traditionally, the most important use of microstructure fabrication has been the creation of large numbers of transistors, diodes, resistors, and capacitors fabricated with the interconnections that enable them to perform useful electronic functions on a single piece or "chip," usually silicon. Recent trends in miniaturization include two additional classifications (i.e., *application-specific* integrated circuits (ASICs) and *very-high-speed* integrated circuits (VHSICs).

Since the beginning of the concept of integrated circuits several years ago, the "yardstick" of dimension has been the micron (micrometer). A micrometer equals 1/1,000,000 of a meter, or 1/25,000 of an inch, or 1000 nanometers, or 10,000 angstroms. The wavelength of light, by comparison, extends from approximately 4000 to 7000 angstroms. Approximately 150 half-micron-wide lines would fit within the width of a human hair. The "resists" used in the fabrication of microminiaturized components fall into the "coatings" category in the parlance of modern coatings technology.

A *simplified* example will illustrate the process of microstructure fabrication. With reference to Fig. 1, an *n*-type region has been created by diffusion of a donor impurity into a surface of *p*-type silicon, forming a *p-n* junction diode. There is a metal contact to the *n*-region, and the contact line is insulated from the *p*-type surface by a layer of silicon dioxide. The diameter of the diode is on the order of 10 micrometers.

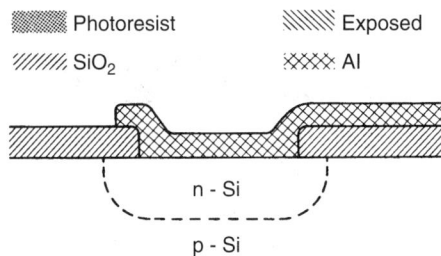

▨ Photoresist		▧ Exposed	
▨ SiO$_2$		✕✕✕ Al	

Fig. 1.   A diode fabricated on the surface of a wafer of silicon. An *n*-type region has been created by diffusing a donor impurity through an opening in a layer of SiO$_2$ on the silicon. Electrical contact is made to the *n* region by a deposited aluminum conductor. The SiO$_2$ insulates the silicon from the aluminum.

The fabrication begins with the application of a layer of photoresist to the oxidized surface of a silicon wafer. The photoresist is then exposed to light in the region where the diode is to be formed. Photoresist is a polymeric mixture that is deposited as a thin layer, perhaps 1 μm thick, upon an SiO$_2$ film on a silicon wafer. Irradiation with light in the near UV region of the spectrum modifies the chemical properties of the photoresist, and, in "positive" photoresist, makes it more soluble in certain developers. Thus, one step frequently employed in microstructure fabrication is the projection of the image of a mask onto the photoresist layer. It becomes possible to remove the exposed region of the photoresist by dissolving it with a suitable developer. The SiO$_2$ layer can then be removed from the areas that were exposed to light by hydrogen fluoride etches. The photoresist is resistant to HF etches and the SiO$_2$ in the unexposed areas is not affected by the etch. After etching, the remaining photoresist can be removed by a solvent, leaving a silicon substrate covered with SiO$_2$ only in the unexposed areas. The SiO$_2$ film acts as a barrier to the contact of impurities in a gaseous phase with the silicon. Thus, when the silicon wafer covered by the patterned SiO$_2$ film is exposed, to, for example, a gas containing phosphorus at high temperatures, the phosphorus, being very soluble in silicon, diffuses into the exposed areas rapidly. An idealized description of this sequence of process steps is shown in Fig. 2. The effect of this doping is very important in electronics, since phosphorus is a donor impurity and a region of n-type or electron conductivity is produced where it is present.

Many physical phenomena, however, obstruct the formation of the ideal structure depicted in Fig. 2. The technical literature is well supplied with papers devoted to each of the steps illustrated in Fig. 2. None is as straightforward as appears at first sight. It is instructive to discuss them further, since much of the essence of microstructure fabrication is revealed by examining them in detail.

Figure 2(**a**) suggests that the thickness of the photoresist and SiO$_2$ layers are independent of position. While the layer thickness is not an extremely critical process parameter, its control cannot be entirely neglected, as the time needed for the subsequent developing or etching steps depend on it. The wafers used in modern silicon technology have diameters of three or more inches, and maintaining uniformity of layers and process parameters across a wafer is not a trivial task. Also, very high standards of cleanliness must be maintained, as any particulate contamination will affect the resist adversely.

Figure 2(**b**) shows a well-defined boundary between the exposed and unexposed areas of the photoresist. In fact, the dimensions of the structures produced in modern microelectronics are comparable to the wavelength of the exposing light, so that diffraction prevents such sharp contrast from being achieved. Furthermore, high-resolution projection exposure schemes

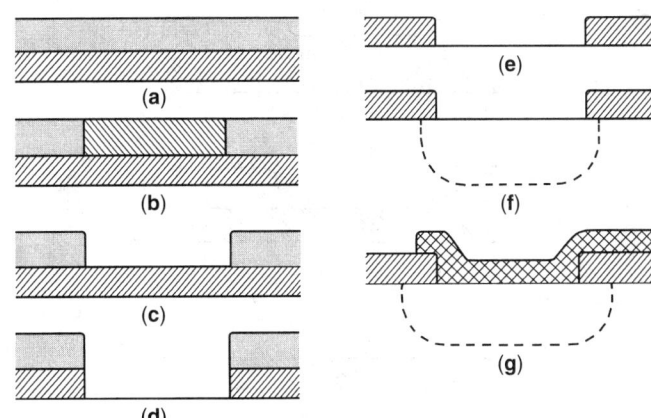

Fig. 2.   Process steps used to produce the structure shown in Fig. 1: (**a**) A film of SiO$_2$ has been formed by oxidizing the silicon and a layer of photoresist has been deposited on the SiO$_2$. (**b**) Shading shows a region of the photoresist that has been exposed to light and thereby made more soluble. (**c**) The exposed photoresist has been removed. (**d**) An etchant that reacts with the SiO$_2$, but not with the photoresist, has been removed. (**e**) Another solvent has been used to remove the unexposed photoresist. (**f**) Donor atoms have diffused into the silicon through the opening in the SiO$_2$ to produce an *n*-type region. (**g**) Additional masking steps, not shown, have permitted aluminum to be evaporated onto the diode in a pattern that forms a contact to the n region of the diode. (See Fig. 1 for legend.)

require that the light be monochromatic to avoid the problems of chromatic aberration in the lenses. The photoresist must be reasonably transparent to insure that its full thickness is exposed to the light. The silicon surface is, however, reflective, so that the interference of the incident and reflected light produces standing waves in the photoresist and nonuniform exposure of the photoresist in the vertical direction. Complicated effects of this kind are clearly important to microstructure fabrication. It must also be apparent that, as the amount of exposure received is a continuous function of position, the time of development required to remove a given region of photoresist will also be a continuous function, and that the profile of the developed photoresist will depend on the time of development. In particular, the size of the opening in the photoresist, to which Fig. 2 is oriented, will depend on the time.

Resists can also be exposed with focused electron beams, as used in electron microscopes, instead of light. The electrons, however, pass through the resist layer into the substrate, where they are scattered, and some eventually return from the substrate to the resist, exposing it at a distance from the intended opening. Great care is needed to allow for the backscattering phenomenon in calculating exposures for nearby openings. See **Electron Beam Lithography**.

Development of the photoresist proceeds somewhat as shown in Fig. 3, with simultaneous lateral and vertical removal of material. The tapered edge of the resist film may be a disadvantage, because the exact point at which the film is thick enough to protect the underlying SiO$_2$ layer during the succeeding etching step, and thus the size of the hole that will be produced in the SiO$_2$, is not clearly defined. Prolonging the development beyond the point shown in Fig. 3(**b**) allows continued lateral development and increase in the size of the opening. Also, however, achieving perfect adhesion of the photoresist film to the SiO$_2$ is difficult, and the developer may invade the interface between the two layers, producing the undesirable result shown in Fig. 3(**d**).

The developed photoresist, Fig. 2(**c**), is then used as a mask to etch the SiO$_2$ layer. Again, perfection is hard to achieve. Etching for too short a time will leave a certain amount of photoresist in the hole. Etching for too long a time can cause undercutting, as shown in Fig. 4(**c**). After removal of the photoresist, the wafer is exposed to a diffusant, affording additional opportunities for deviations from idealized behavior. Time and temperature of diffusion are important and can produce results resembling Figures 5(**a**) or 5(**b**). Diffusion can proceed rapidly along interfaces in certain cases, leading to junction profiles of the kind shown in Fig. 5(**c**). Preferential diffusion along crystalline defects can give rise to a profile resembling that shown in Fig. 5(**d**).

Next, a metal connection is to be made to the diffusion-doped region. Aluminum is frequently used for this purpose, as it has high electrical

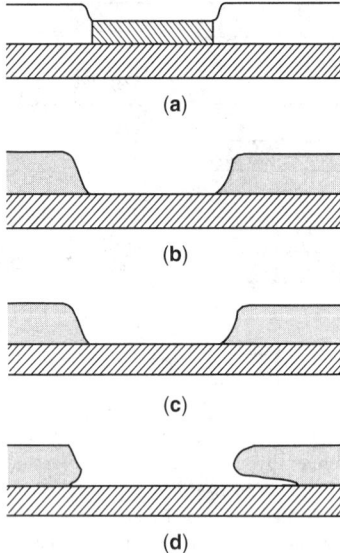

Fig. 3.   Development of the photoresist: (**a**) Although exposure increases the solubility of photoresist in the developer, there is only a finite ratio of dissolution of the exposed photoresist to that of the unexposed region. In addition, the exposure received is not a perfect step function at the boundary of the exposed areas. Thus, dissolution proceeds laterally as well as vertically. (**b**) The opening in the photoresist has penetrated to the surface of the SiO_2. (**c**) With continued development the opening continues to enlarge. (**d**) Poor adhesion of the photoresist to the SiO_2 has allowed the development to penetrate the interface. (See Fig. 1 for legend.)

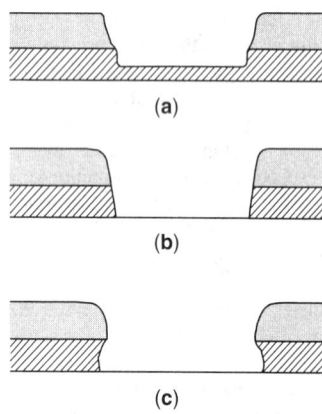

Fig. 4.   The opening in the photoresist is used to mask the etching of a hole in the SiO_2: (**a**) An early stage of development. (**b**) The opening in the SiO_2 has reached the silicon. (**c**) Prolonged etching can lead to removal of SiO_2 underneath the photoresist masks. (See Fig. 1 for legend.)

conductivity and does not enter the silicon and alter its properties. The aluminum is also evaporated through a mask that defines the shape and location of the conductor. Examples of the region of contact between the aluminum and the doped semiconductor are shown in Fig. 6. It is seen that the current will be forced to flow through a narrow constriction if the profiles are as shown in Fig. 6(**b**). The high current densities may cause electromigration of the aluminum atoms, leading to the open circuit shown in Fig. 6(**c**).

Also, silicon is somewhat soluble in aluminum. One can thus encounter the situation shown in Fig. 6(**d**), where enough silicon has been dissolved to allow the metal to completely penetrate the doped region, shorting the junction.

None of the problems illustrated in Figs. 2, 3, 4, 5, and 6 is insurmountable. A great many ingenious ways to avoid the difficulties described are known. Sometimes these are guided by physical or chemical knowledge; frequently they are empirical fixes.

The microstructure engineer must also be aware of constraints that have little to do with chemistry and materials science, but tend to be more closely related to mechanical technology. One of the most difficult of these is registration or alignment. A substrate is usually passed through several

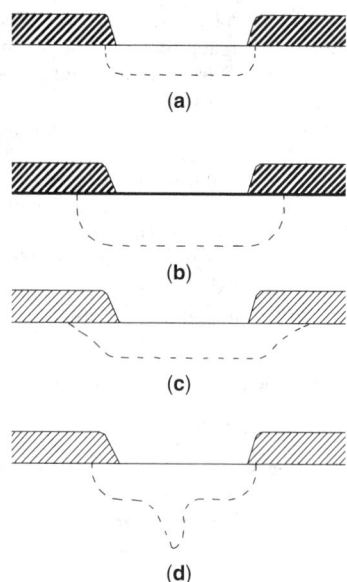

Fig. 5.   A donor impurity is diffused into the silicon from a gaseous phase. (**a**) A shallow n region has been created. (**b**) Continued diffusion, longer times, or higher temperatures increase the extent of the n region. (**c**) Surface diffusion has caused spreading of the n region along the SiO_2-silicon interface. (**d**) A crystal defect, such as a dislocation, has provided a path for anomalously high diffusion and led to penetration of the junction to unanticipated distance from the surface. (See Fig. 1 for legend.)

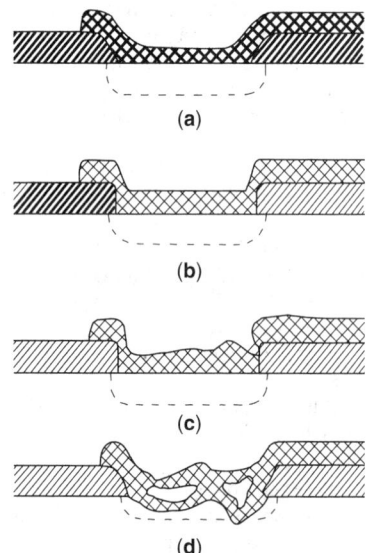

Fig. 6.   (**a**) Masking steps, not shown, have permitted the deposition of aluminum in selected areas to form a contact to the n region. (**b**) A sharp vertical profile in the SiO_2 opening may cause a reduction of the cross section of the aluminum conductor where it passes from the SiO_2 insulator to the silicon surface. (**c**) The high current density in the constriction shown in (**b**) has led to electromigration of aluminum atoms and opening of the conductor. (**d**) Solution of silicon in the aluminum has resulted in deformation of the metal-semiconductor interface and penetration of the aluminum through the n region. (See Fig. 1 for legend.)

process steps in order to form a desired structure. For example, fabricating a transistor may involve diffusing a base dopant through a window in SiO_2, subsequently diffusing an emitter dopant through a somewhat smaller opening in an SiO_2 layer, and finally, using masking layers to form contacts to these transistor elements. It is necessary to ensure that the emitter region is created in the correct position within the previously diffused base region. The mask that is used to define the emitter region lithographically must be precisely located with respect to the geometrical structure already established on the substrate by previous processing steps. This positioning is known as alignment. It may require that separate structures that can be easily located but have no electronic function be provided on the substrate.

Alignment all over a large substrate is made more difficult by dimensional changes that may take place during processing at high temperatures. Materials soften at high temperatures and can deform under the force of gravity or the stresses that accompany temperature gradients and contacts between different materials.

Further, economic factors also constrain the utilization of microstructure fabrication technology. These are the factors that control the cost of production, such as throughput, the rate at which substrates can be processed by the fabrication tools, capital investment required, and demands on operator time and skill. Electron beam exposure, for example, provides high resolution but uses expensive equipment that works slowly. Naturally, all of the elements of cost must be weighed against the value of the product produced.

Chemical Vapor Deposition. Deposition of tungsten, molybdenum, and their silicides by chemical vapor deposition (CVD) is of relatively recent interest in the microelectronics industry. These materials are useful for gates and interconnects in metal oxide semiconductors (MOS) devices. Aluminum, the widely used interconnect material, has a comparatively low melting point (600 °C) and a markedly different coefficient of thermal expansion (compared to silicon), so that over a period of years researchers have been seeking an alternative for aluminum.

In using CVD for microelectronics applications, the deposit thickness must be as uniform as possible. This can best be achieved by conducting the deposition in a surface-controlled, not a diffusion-controlled, regime. In this way, effects of reactor geometry on deposition rates are minimized and irregular-shaped substrates will tend to be uniformly coated.

An allied process is *metal-organic chemical vapor deposition* (MOCVD). In this process, two or more metal-organic chemicals (example: trimethylgallium) or one or more metal-organic sources and one or more hydride sources (example: arsine, AsH_3) are used to form the corresponding intermetallic crystalline solid solution. MOCVD materials technology is a vapor-phase growth process that is used to study the basic physics of novel materials and to grow complex semiconductor device structures, particularly for new optoelectronic and photonic systems. The process is reported in some detail by Dupuis.

Ion Implantation. In this process, alloying elements are introduced into a host material by accelerating the ions to a high energy level and allowing them to strike the surface of the host. The impinging atoms penetrate into the substrate material to a depth of one micrometer or less, depending upon atomic number and energy of the atom. Although it has a number of other applications metallurgically, the process has been of interest in the semiconductor industry primarily in connection with doping substrates with elements in Periodic Table Groups 13 and 15. Use of the process in fabricating a Schottky barrier gate used in a metal-semiconductor field-effect transistor (MESFET) is described in the article on **Semiconductors**. See also **Ion Implantation**.

Multidisciplinary Characteristics of Microstructure Fabrication. Upon observing the practice of microstructure fabrication, one cannot fail to notice a resemblance to certain aspects of modern metallurgy and chemical engineering. For example, the precipitates produced by metallurgical processing have a dimensional scale similar to that of electronic microstructures. Inhomogeneities on a scale of 0.01 ϕm to 10 ϕm control the desirable properties of a structure. This preoccupation with the properties of solids on a microscopic scale produces a common interest in techniques and in interactions with basic science. Thus, both microstructure fabrication and physical metallurgy: (a) rely on phenomena that take place in the solid state; (b) depend on analytical tools that are capable of chemical analyses with the highest possible spatial resolution; (c) involve the motion of atoms through solids, controlled by diffusion, solution, nucleation, and precipitation; (d) involve interface phenomena at the contact between different solids; and (e) are sensitive to crystal defects.

It must further be noted that both metallurgy and microstructure fabrication are practical disciplines, they are oriented toward the economic production of structures that have a useful role in commerce and industry. In this respect both are engineering rather than scientific disciplines. On the other hand, their deep probing of phenomena on an atomic scale and under unusual conditions produce new discoveries and lead to new concepts that enhance basic science.

This is not to say that microstructure fabrication is a branch of metallurgy. The detailed motivation of the two disciplines is rather different, metallurgy concentrating on the mechanical properties of solids, while microstructure fabrication controls the electronic properties of structures

made from magnetic and optical materials, semiconductors, metals, and insulators. The basic difference, of course, is that microstructure fabrication involves control of the fabrication process in detail at the dimensional level of the structure, while metallurgical processing exercises control at a much grosser level. The application of lithography, with the attendant use of exposure tools, clean rooms, resists, masks, and etchants is the province of microstructure fabrication. Crystalline defects are usually undesirable in microstructures; the metallurgist can frequently use them to advantage. Metallurgy also encompasses its extractive aspects. There is no doubt that microstructure fabrication is a distinct activity.

The technique of microstructure fabrication has grown up as an art in response to a continuous economic and functional motivation to push to the smaller and smaller. Chemistry, physics, and empiricism are combined into the creation of novel physical structures that have enormous economic impact, that, indeed, form the basis of whole new industries. Progress has been made by adaptation and invention to meet the needs of the moment. By and large, the art has grown rapidly, adapting and improving old methods to new situations and inventing as needed and possible. It is common experience that obtaining reproducible results requires very careful control of all aspects of the fabrication processes, such as temperatures, pressures and time. High standards of cleanliness and reagent purity must be enforced. Seemingly identical apparatuses and starting materials often yield different results. Recipes must be carefully followed, with little understanding of which aspects of a process are critical or what contaminants are important, and why.

The rapid development has outstripped basic understanding of the fundamental mechanisms underlying the techniques. It must be recognized, however, that the optimal exploitation of microstructural technology, the maximization of performance, yields, and utilization of available silicon area will depend on a detailed interpretation of each step in fabrication as a chemical or physical process. Furthermore, unanticipated phenomena are encountered and inject surprises into basic science. A new interdisciplinary field of applied science has emerged.

Photographic Color Film

One of the early examples of coating technology was for building up layers in color film. These films exemplify the use of multiple layers of coatings that are "built in" to the final product. In use, the camera lens also will be coated. See Fig. 7. See also **Photography and Imagery**.

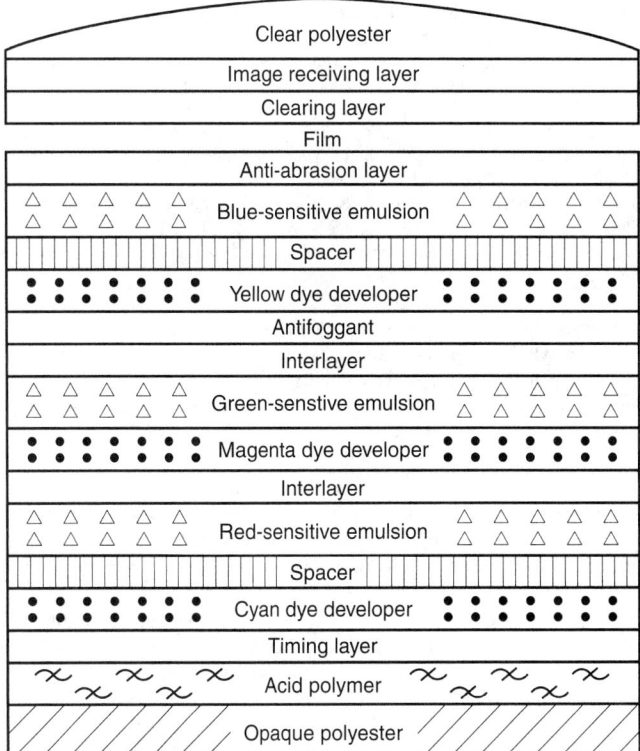

Fig. 7. Schematic sectional view of layers composing the "thickness" of a color photographic film. (*After Cohen.*)

Adhesives

Commonly, adhesives are applied by coating products and parts that must be held together. Some adhesives require application at the time of usage; others are pre-applied in semi-dry or dry form, but with a "sticky" surface. Adhesives range from natural products, such as glue to plastics and polymers. The epoxies are exemplary of modern adhesives. With the continuing replacement of many metal parts with plastics, adhesives are displacing welding and mechanical fasteners to a dramatic degree. In the final essence, most adhesives must form a binding coating on one form or other. See also **Adhesives**.

Additional Reading

Anderson, D.G.: "Coatings (Analysis of)," *Analytical Chemistry*, 87R (June 15, 1991).

Champ, N.A. and F.L. Lowenstein: "TBT: The Dilemma of High-Technology Antifouling Paints," *Oceanus*, 69 (Fall 1987).

Cohen, E.D.: "Coatings: Going Below the Surface," *Chem. Engineering Progress*, 19 (September 1990).

Cohen, E.D. and E.J. Lightfoot: "A Primer on Forming Coatings," *Chem. Engineering Progress*, 30 (September 1990).

Coyle, D.J., C.W. Macosko, and L.E. Scriven: "Fluid Dynamics of Reverse Roll Coating," *Amer. Inst. of Chem. Eng's J.*, 16 (February 1990).

Dupuis, R.D.: "Metalorganic Chemical Vapor Deposition of III—V Semiconductors," *Science*, **226**, 623–629 (1984).

Elliott, D.: "Microlithography: Process Technology for IC Fabrication," The McGraw-Hill Companies, Inc., New York, NY, 1986.

Fettis, G.: "Automotive Paints and Coatings," John Wiley & Sons, Inc., New York, NY, 1995.

Finzel, W.A.: "Use Low-VOC (Volatile Organic Compounds) Coatings," *Chem. Engineering Progress*, 50 (November 1991).

Freitag, W. and D. Stoye: "Paints, Coatings and Solvents," 2nd Edition, John Wiley & Sons, Inc., New York, NY, 1999.

Glass, J.E.: "Technology for Waterborne Coatings," Vol. 663, American Chemical Society, Washington, DC, 1997.

Herman, H.: "Plasma-Sprayed Coatings," *Sci. Amer.*, 112 (September 1988).

Lambourne, R.: "Paint and Surface Coatings: Theory and Practice," Prentice-Hall, Inc., Upper Saddle River, NJ, 1993.

Marrs, J.M.: "Ultraviolet Light for Photochemical Deposition," *Chem. Eng. Progress*, 31–34 (January 1986).

Satas, D. and A. Tracton: "Coatings Technology Handbook," 2nd Edition, Marcel Dekker, Inc., New York, NY, 2000.

Schweizer, P.M.: "Visualization of Coating Flows," *J. Fluid Mechanics*, 193, 285 (1988).

Scriven, L.E. and W.J. Suszynski: "Take a Closer Look at Coating Problems," *Chem. Engineering Progress*, 24 (September 1990).

Shumay, W.C., Jr.: "High-Performance Adhesives on the Line," *Adv. Material & Processes*, 82 (August 1987).

Staff: "Paints Related Coating and Aromatics," American Society for Testing & Materials, West Conshohocken, PA, 1998.

Sturge, J.M., V. Walworth, and A. Shepp: "Imaging Processes and Material: Neblette's," 8th Edition, John Wiley & Sons, Inc., New York, NY, 1997.

Tobiason, T.L.: "Choosing a Finish for Your Electronic Package," *Instruments & Control Systems*, 85 (May 1990).

Various: "6th International Coating Process Science and Technology Symposium Papers," *Amer. Inst. of Chem. Engineers*, New York, NY, 1992.

Weldon, D.G.: "The Failure Analysis of Paints and Coatings," John Wiley & Sons, Inc., New York, NY, 2001.

PAIR PRODUCTION.   The conversion of a photon into a negatron and a positron when the photon traverses a strong electric field, usually that surrounding a nucleus but occasionally that of an electron. The electric charge is conserved, since the negatron and positron have charges of equal magnitude and opposite sign.

PALEOBOTANY.   This is the study of ancient plants known today only through fossil remains. These are of two general types: prints and petrifactions. Much of our knowledge of fossil plants has been derived from the study of petrifactions in which the original structure of the plant has been replaced by mineral material.

Paleobotany yields much interesting information about the plants of prehistoric times, notably of the Carboniferous and Devonian periods. The best known fossil plants are those which grew during the Carboniferous period of the Paleozoic era. The many types of coal are goechemical compounds and mixtures of plants, including their spore cases, oxidized cellulose (mother of coal), lignite, and bacterial byproducts. The nature of the processes by which coal has been formed is such that coal is often not the best material to study to get a picture of the fossil plants

composing it. But found associated with the coal are mineral masses called coal balls. Within these, plant remains are often preserved with beautiful detail. Microscopic studies of peels and thin sections of coal balls have made it possible for the paleobotanist to reconstruct many of the Carboniferous plants.

PALEOCENE.   The earliest period in the Cenozoic Era of the geologic time-scale. It is also spoken of as basal Tertiary. The term was first proposed by Schimper in 1874. The greatest thickness of the formations of this system occur in Wyoming. The Paleocene began approximately 60 million years ago and lasted for about 10 million years. In the United States, the sediments are mainly of terrestrial origin, occurring as intermontane deposits of sands, gravels, and clays. The Paleocene is chiefly remarkable in that the dinosaurs have been replaced by the archaic or earliest types of mammals.

PALEONTOLOGY.   The study of fossils. A fossil is the evidence of the former existence of an organism, either animal or plant. Fossils may be classified according to their method of fossilization as follows: (1) Actual remains, sharks' teeth, ear bones of whales, chitin, etc. (2) Petrifications. Minute replacements in which the original organic matter has been completely or partially replaced by mineral matter. The principal replacing minerals are calcite, quartz, chert, and pyrite. (3) Molds and casts of interiors and exteriors. (4) Prints of leaves, jellyfish, etc., sometimes showing carbonized traces of organic matter, as in the case of some fossil plants. (5) Coprolites, fossil excrement. (6) Tracks, trails, and burrows.

A valuable reference is *"Principles of Paleontology"*, by D.M. Raup and S.M. Stanley, Freeman, San Francisco, 1979.

PALEOPHYTIC.   A paleobotanic division of geologic time. The term signifies the time period during which pteridophytes were abundant, occurring between the development of algae and the appearance of the first gymnosperms. See also **Paleobotany**.

PALEOZOIC.   The era of "ancient" life or the age of invertebrates. Subdivided from the base up into the following periods: Cambrian, Ozarkian, Canadian, Ordovician (lower Paleozoic), Silurian, Devonian (middle Paleozoic), Mississippian, Pennsylvanian, Permian (upper Paleozoic). The lower Paleozoic was characterized by: (a) the first known marine faunas; (b) dominance of trilobites; (c) rise of animals with hard shells (Cambrian); (d) rise of nautiloids, armored fishes, and corals; and (e) the first evidence of colonial life (Ordovician). The middle Paleozoic was characterized by: (a) the rise of the lung-fishes and scorpions (Silurian); (b) the first known land floras (Devonian), not very different from those of the Pennsylvanian; and (c) earliest evidence of a terrestrial vertebrate. The upper Paleozoic was characterized by: (a) ancient sharks and echinoderms (Mississippian); (b) primitive reptiles and insects (Pennsylvanian); (c) periodic glaciation and extinction of many Paleozoic groups during and after the Permian; and (d) rise of modern insects, land vertebrates, and ammonites (Permian). The Paleozoic Era began about 500 million years ago and lasted for approximately 300 million years.

PALLADIUM.   Chemical element symbol Pd, at. no. 46, at. wt. 106.4, periodic table group 10, mp 1554 °C, bp 2970 °C, density 12.02 g/cm^3 (solid), 12.25 g/cm^3 (single crystal) (20 °C). Elemental palladium has a face-centered cubic crystal structure. The six stable isotopes of palladium are ^{102}Pd, ^{104}Pd, through ^{106}Pd, ^{108}Pd, and ^{110}Pd. The seven unstable isotopes are ^{100}Pd, ^{101}Pd, ^{103}Pd, ^{107}Pd, ^{109}Pd, ^{111}Pd, and ^{112}Pd. In terms of earthly abundance, palladium is one of the scarce elements. Also, in terms of cosmic abundance, the investigation by Harold C. Urey (1952), using a figure of 10,000 for silicon, estimated the figure for palladium at 0.0091. No notable presence of palladium in seawater has been found. Discovered by Wollaston in 1803.

Electronic configuration $1s^2 2s^2 2p^6 3s^2 3p^6 3d^{10} 4s^2 4p^6 4d^{10}$. Ionic radius Pd^{2+} 0.50 Å (Wyckoff). Metallic radius 1.3755. First ionization potential 8.33 eV; second 19.8 eV. Oxidation potentials Pd → Pd^{2+} + 2e$^-$ 4 M HClO$_4$, −0.83 V, Pd + 2OH$^-$ → Pd(OH)$_2$ + 2e$^-$, −0.1 V. Further physical properties are given under **Platinum and Platinum Group**.

Pd has some similarities with both Ni and Ag and many with Pt. Pd dissolves more readily in acids than any other member of the platinum group of metals. In aqua regia, the metal dissolves quickly. Even the compact metal dissolves slowly in HCl. In finely divided form, it is quite soluble in all acids. When heated in air at red heat, the monoxide, PdO, is formed. Pd is similarly converted to the dihalides under the same conditions when it is exposed to F_2 or CL_2. The metal is not affected by H_2S.

The black compound, palladium(II) oxide is formed by fusing palladium(II) chloride with $NaNO_3$ to $600\,°C$ and then leaching out the salts with water. This strong oxidizing agent is easily reduced to the metal by H_2. The compound is insoluble in water and acids, including aqua regia. The hydroxide, $Pd(OH)_2$, is made by the hydrolysis of palladium(II) nitrate. The compound is soluble in acids, and water is evolved on heating, but even at $500–600\,°C$ some water still remains. At this temperature, the compound starts to lose O_2.

Palladium(III) oxide, P_2O_3, is made as a hydrate by careful oxidation of a solution of palladium(II) nitrate either by anodic oxidation or ozone treatment at $-8\,°C$. This unstable brown powder reverts to the monoxide in about 4 days. When heated, the compound loses water and may explode as it changes to the monoxide.

Palladium(II) chloride is formed by direct combination of the elements at $500\,°C$. It is the only stable chloride over $500–1500\,°C$. The red crystals are partly soluble in water and completely soluble in HCl. The fraction insoluble in water is probably a polymer. Palladium(II) chloride also is the product obtained by evaporation of a solution of Pd in HCl. Palladium(II) bromide can also be made from the elements.

When KI is added to a solution of palladium(II) chloride, an insoluble diiodide is precipitated. The dark red-black crystals are soluble in excess iodide with formation of the tetraiodide complex ion. Palladium(II) iodide evolves iodine at $100\,°C$, the decomposition to the elements being complete at $330–360\,°C$. The black compound, palladium(III) fluoride, is made by direct combination of the elements. On reduction, the brown difluoride is formed.

Divalent Pd forms many planar complexes with a coordination number of 4. The tetrachlorides are quite soluble. When a solution of palladium(II) chloride is oxidized with chlorite or chlorate ion, Pd(IV) is formed, which has a coordination number of 8. The addition of NH_4Cl to such a solution precipitates ammonium hexachloropalladate(IV) as a red compound. It is somewhat less stable than the platinum analog.

The soluble yellow-brown palladium(II) nitrate is formed by dissolving finely divided Pd in warm HNO_3 and then crystallizing the compound from this solution. The analogous sulfate is similarly formed from H_2SO_4. It crystallizes as a red-brown dihydrate. Both these compounds easily hydrolyze.

Palladium(II) sulfide is precipitated as a brown powder by adding H_2S to a solution of palladium(II) ion. When this sulfide is heated with sulfur at $400\,°C$, the insoluble disulfide is formed. The excess sulfur can be extracted with CS_2 to yield the gray-black crystalline palladium(IV) sulfide. This compound is not soluble in single acids but is soluble in aqua regia.

Some Pd complexes are important analytically or in the refining of Pd. The yellow dimethylglyoxime compound is quantitatively precipitated from a HCl solution of palladium(II) chloride by the addition of an alcoholic solution of dimethylglyoxime. Palladium(II) has a great affinity for nitrogen-containing ligands. The di- and tetramine find use in refining.

Palladium, as with other members of the platinum group, exhibits catalytic activity for various reactions. One of its best known uses is in conjunction with other platinum metals in the catalytic converters of present-day automobiles.

As reported by Chung-Chiun Liu et al. (*Science*, **207**, 188–189, 1980), a palladium-palladium oxide miniature pH electrode has been developed for pH measurement. The miniature wire-form electrode exhibits a super-Nernstian behavior and gives a mean pH response of 71.4 mV per [pH] (standard deviation, 5.3 mV). The electrode may find applications in biological, medical, and clinical studies.

See also **Chemical Elements**.

Additional Reading

Carter, G.F. and D.E. Paul: "Materials Science and Engineering," ASM International, Materials Park, OH, 1991.

Davis, J.R.: "Metals Handbook," 2nd Edition, ASM International, Materials Park, OH, 1998.

Greenwood, N.N. and A. Earnshaw: "Chemistry of the Elements," 2nd Edition, Butterworth-Heinemann, Woburn, MA, 1997.

Krebs, R.E.: "The History and Use of Our Earth's Chemical Elements: A Reference Guide," Greenwood Publishing Group, Inc., Westport, CT, 1998.

Lide, D.R.: "CRC Handbook of Chemistry and Physics 2000–2001," 81st Edition, CRC Press, Boca Raton, FL, 2000.

Parker, P.: "McGraw-Hill Encyclopedia of Chemistry," 2nd Edition, The McGraw-Hill Companies, Inc., New York, NY, 1993.

Sinfelt, J.H.: "Bimetallic Catalysts," *Sci. Amer.*, 90–98 (September 1985).

Staff: "ASM Handbook—Properties and Selection: Nonferrous Alloys and Pure Metals," ASM International, Materials Park, OH, 1990.

Stwertka, A. and E. Stwertka: "A Guide to the Elements," Oxford University Press, Inc., New York, NY, 1998.

LINTON LIBBY, Chief Chemist, Simmons Refining Company, Chicago, IL.

PALM OIL. See **Vegetable Oils (Edible)**.

PALM TREES (*Palmaceae*). This family of plants contains some 1,100 species, most of which are tropical. Many are of large size. A tall woody stem bearing at its top a crown of large compound leaves characterizes most of them, although some are short and bushy, while a few are vinelike. Only rarely does branching occur, although many do develop numerous basal offshoots. The leaves are either pinnately or palmately compound. The inflorescence is either a simple or a compound spike, surrounded by a spathe, which may become extremely large. The flowers are regular, with their parts in threes or multiples of three. Wind pollination generally occurs in palms. The fruit is a berry, a drupe, or a nut, which in many species has a fibrous mesocarp.

Many species of palms are planted extensively, because of their ornamental habit, in tropical and subtropical countries. Large numbers are also grown as greenhouse or conservatory plants; a few species are of great economic value.

Date Palm (Phoenix dactylifera). The mature trunk of this tree often attains a height of 75–100 feet (22.5–30 meters). From the base of this trunk, numerous adventitious roots extend into the soil to a depth of 20 feet (6 meters) or more. From the base of the stem, many basal offshoots develop. The pinnate leaves are from 10 to 20 feet (3 to 6 meters) long, with the individual pinnate sharp-pointed, linear, and from 1 to 3 feet (0.3 to 0.9 meter) long. These leaves remain attached for an indefinite time to the trunk, giving it a very ragged appearance. In cultivation, the old leaves are removed. The inflorescence is a large branched spike enclosed in a large tough spathe. See Fig. 1. In a mature tree, a single flower cluster may be composed of several thousands of flowers. The flowers are of two different sexes, borne on different plants, and are small, wax-white, and of firm texture. The pistillate (see also **Flower**) flowers have three carpels, each with a short curved stigma. The carpels are almost surrounded by the perianth, composed of three united sepals and three petals. The staminate flowers, bearing six stamens, are of larger size than the pistillate and more showy. The fruit is a drupe. See Fig. 2. After fertilization, only one of the three carpels develops, the other two being suppressed. At first the carpel is wax-white, but some time after fertilization, it becomes green

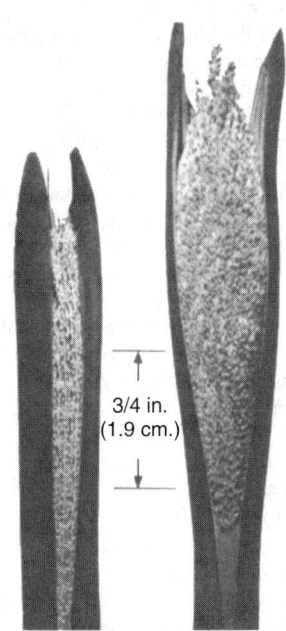

Fig. 1.   Flower clusters of the date palm as they appear when the spathe first opens. (Left) female; (right) male. (*USDA diagram.*)

Fig. 2.   A bunch of Barhee dates, showing wire spreader that serves to provide ventilation in the center. (*USDA photo.*)

and remains so during growth. As the drupe becomes mature, the color changes to yellow or red or a blending of these colors, according to the variety of date. When ripe, the color of the fruit varies from pale straw color through deep amber to a deep purple. The seed of the date contains an abundance of endosperm and a rather small embryo.

Many varieties of dates are grown in cultivation. Since cross-pollination is usually necessary to produce seeds, propagation by seeds cannot be used to increase the number of plants. For such plants would not remain true to the seed-bearing parent in type but would be affected by the pollen from the staminate plant; such crossing would cause the formation of new varieties with the possibility that less desirable forms might arise. Furthermore, about half the seedlings would be staminate, producing no fruit. To perpetuate desirable varieties, propagation is largely by means of the basal offshoots. These shoots are cut from the parent and planted, ensuring new plants identical with the parent.

Cultivation of the date has been carried on in Arabia and adjoining countries since prehistoric times. It is probable that the plant is native to this region, even though it is unknown in the wild state. Today the date palm is grown widely in regions having a sufficiently hot climate. The Mesopotamian valley is the chief producing region of the world. In the United States, the date can be grown successfully only in a very limited region in the southern parts of California, Nevada, and Arizona.

The principal product of the plant is the edible fruit. In the Orient, the fruit is often fermented, yielding a very potent alcoholic drink, and also vinegar. Various materials used in building houses, and in making baskets, ropes, and household articles are also obtained from the tree in regions where other sources of such materials may be lacking.

Another useful palm is the oil palm, *Elaeis guineensis*, a native plant occurring in large numbers in the forests of the west coast of Africa. The plant is also extensively cultivated elsewhere in Africa and in the East Indies, particularly in Sumatra. From the fruit of this palm, two valuable oils are obtained. The orange-yellow pericarp yields palm oil, much used in the making of soaps. The endosperm yields a white pleasantly flavored oil known as palm-kernel oil, also used in making soaps. In Africa, the fruits are eaten by the natives.

In India, *Phoenix sylvestris* is widely cultivated as a plant from which sugar is obtained. Many other palms are used as sources for sugar in regions where they grow.

Areca Palm (Areca catechu). This tree is native to Malaysia and is extensively cultivated in the southern Asian countries and in the East Indian islands. It is a slender unbranched tree that reaches a height of about 40 feet (12 meters) with a trunk diameter of from 3 to 4 inches (7.6 to 10 centimeters). It bears at its top a crown of 6 to 10 large leaves. The fruit is slightly more than an inch in diameter and has a fibrous rind

surrounding the very hard seed. This seed (the *betel nut*) is chewed by many people in the Asiatic countries. To prepare them for chewing, the fruits are gathered just before maturity, boiled, sliced, and dried. A piece of this dried fruit, together with a bit of lime, is wrapped in a betel leaf and the whole placed in the mouth. Chewing this preparation causes an abundant flow of saliva, which is colored brick red and stains the mouth and teeth. The betel leaves are obtained from an entirely different plant, the black pepper (*Piper betle*).

Coconut Palm (Cocos nucifera). To the natives of tropical islands, the coconut is of great value. It provides shelter, dishes, food, and drink. However, products of the tree are valuable throughout the world, well over 1 million tons of coconut oil being produced annually. The coconut tree is a striking plant having a columnar trunk some 60 to 80 feet (18 to 24 meters) in height, bearing at its top a crown of bright green pinnate leaves, each leaf ranging from 15 to 20 inches (38 to 50.8 centimeters) in length. The fruit is a large nut composed of three united carpels. The shell of the coconut has three distinct layers; the outer, or epicarp, is thin, smooth, and brown; the middle, or mesocarp, is thick and fibrous; it is the source of the fiber coir. Usually these two layers are removed before the coconuts are shipped, but not infrequently the entire fruit is exhibited as a curiosity by dealers in fruits. Within is the hard brown shell or endocarp. This contains the seed, which consists of an embryo, located under one of the germ pores at the end of the endocarp, and the endosperm, which is of two parts, one the familiar white meat of the coconut, the other the milk which partially fills the cavity of the seed. It is within the meat that the embryo is embedded. The thin brown skin immediately investing the meat and sticking to it when it is removed from the shell is the inner seed-coat. The hard shell of the coconut is much used as a dipper or as a vessel for storage of various substances.

The leaves of the tree yield a fibrous material used by the native islanders and are also used at times as a thatch for shelters. The trunk of the tree is sometimes used by cabinet makers for ornamental work.

Commercially, the most valuable product of the coconut is copra, the dried meat of the nut. This is obtained by splitting the nut and drying the meat, preferably by the sun. Artificial drying is done in regions of great humidity. From the meat is obtained coconut oil, which forms about 63% of the meat. The natives obtain this oil by various means. The hot sun of the tropics may cause it to dry out of the pounded meat. Crude presses are often used to squeeze the oil from the dried meats. Or again the crushed dried meats may be placed in hot water, the melted oil rising to the top and being skimmed off. Any one of these methods is sufficient to supply the moderate needs of the natives, but utterly inadequate to meet the requirements of the civilized nations.

For modern treatment, copra is shipped to large factories. Here it is cleaned and ground, then heated and pressed. The meat is then ground up once more, cooked in water, and pressed by powerful hydraulic presses. By this method, nearly all the oil contained in the copra is extracted.

The principal use of the oils is in the making of soap, especially in soaps that will produce a copious lather. Considerable quantities are also used in making shortenings. Some oil finds use in salad oils and in confectionery.

The copra cake remaining after the oil is expressed may be used as a stock feed.

Seychelles Palm. This tree is found on Seychelles Island in the Indian Ocean northeast of Madagascar. It is of interest because the tree does not bear fruit until it is about 100 years of age. The tree is straight with a bulbous trunk, marked by numerous holes. Roots penetrate into these holes, providing stabilization of the tree against gales. The leaves are large and fan-shaped, ranging from 10 to 12 feet (3 to 3.6 meters) in length. The fan portion covers an area of about 7 by 15 feet (2.1 to 4.5 meters). A single tree will have from 20 to 30 leaves. The leaves are so constructed that rain drains down to the base of the tree. Once the fruit commences, it takes from 10 to 12 months to ripen. The fruit contains the largest seed known in the plant world, the nuts weighing from 30 to 40 pounds (13.6 to 18.1 kilograms) at maturity. The nut is brown, smooth, with a thorny shell. The shell is easily torn away. The fruit is called *durian* and is egg-shaped with a very thick rind. The edible part is soft and creamy, rather custard-like in texture.

The tree was discovered on the aforementioned island in 1742. The nuts can be found on the market throughout Southeast Asia.

Raffia Palm (Raphia pedunculata). This African palm tree has leaves of great length, sometimes as much as 25 feet (7.5 meters). The leaf epidermis may be removed in strips. These strips, known commercially as raffia, are

TABLE 1. RECORD PALM TREES IN THE UNITED STATES[1]

Specimen	Circumference[2]		Height		Spread		Location
	Inches	Centimeters	Feet	Meters	Feet	Meters	
Buccaneer palm (*Pseudophoenix sargentii*)	26	66	25	7.6	8	2.4	Florida
Coconut palm (1979) (*Cocos nucifera*)	60	152	93	28.3	27	8.2	Hawaii
Royalpalm (1995) (*Roystonea elata*)	50	127	99	30.2	18	5.5	Florida
Florida Silverpalm (1979) (*Coccothrinax argentata*)	19	48	27	8.2	8	2.4	Florida
Florida Silverpalm (1994) (*Coccothrinax argentata*)	21	53	25	7.6	7	2.1	Florida
California washingtonia (fanpalm) (1991) (*Washingtonia filifera*)	120	305	83	25.3	21	6.4	California
California washingtonia (fanpalm) (1991) (*Washingtonia filifera*)	100	254	101	30.8	22	6.7	California
California washingtonia (fanpalm) (1997) (*Washingtonia filifera*)	141	358	66	20.1	18	5.5	California
PALMETTOS							
Cabbage palmetto (1994) (*Sabal palmetto*)	69	175	60	18.3	14	4.3	Florida
Mexican palmetto (1995) (*Sabal mexicana*)	61	155	50	15.2	15	4.6	Texas
Mexican palmetto (1995) (*Sabal mexicana*)	61	155	45	13.7	20	6.1	Texas
Saw palmetto (1994) (*Serenoa repens*)	22	56	20	6.1	13	4	Florida
Saw palmetto (1987) (*Serenoa repens*)	27	69	21	6.4	8	2.4	Florida
YUCCAS							
Beaked yucca (1994) (*Yucca rostrata*)	48	122	16	4.9	9	2.7	Texas
Carneros (Spanish-dagger) yucca (1977) (*Yucca carnerosana*)	51	130	25	7.6	10	3	Texas
Faxon yucca (1991) (*Yucca faxoniana*)	91	231	18	5.5	9	2.7	Texas
Joshua tree yucca (1999) (*Yucca brevifolia*)	155	394	46	14	38	11.6	California
Mojave yucca (1987) (*Yucca schidigera*)	66	168	24	7.3	7	2.1	California
Moundlilly yucca (L1998) (*Yucca gloriosa*)	106	269	33	10.1	31	9.4	California
Schott yucca (1997) (*Yucca schottii*)	43	109	15	4.6	12	3.7	Arizona
Sooptree yucca (1996) (*Yucca elata*)	62	157	30	9.1	11	3.4	Arizona
Torrey yucca (1987) (*Yucca torreyi*)	86	218	23	7	6	1.8	New Mexico
Trecul yucca (1991) (*Yucca treculeana*)	24	61	30	9.1	9	2.7	Texas

[1] From the "National Register of Big Trees," American Forests (by permission).
[2] At 4.5 feet (1.4 meters).

used in basketwork and for tying. The leaves curl so that the raffia must be obtained before they become dry. The tree is extensively found on Madagascar.

For record palm trees in the United States, see Table 1.

Development of Palm Leaves. In a most unusual and interesting article, D.R. Kaplan (*Sci. Amer.*, **249** (1), 98–107, July 1983) describes in detail well beyond the scope of this encyclopedia the development of palm leaves. The compound leaves of most plants usually arise either from differential growth or from selective cell death. Not so with palm leaves. Their development pathway combines both of these processes. Representative of the content of the Kaplan article is the following: "Palm-leaf blades typically show one or the other of two major configurations that reflect differences in the distribution of growth during the course of dissection.[1] One of the two is characterized as *pinnate* because of its feathery

appearance. The other is characterized as palmate because it is fanlike, rather like a hand with its fingers spread. Pinnate fronds tend to have short petioles; palmate fronds have long ones. What makes the developmental mode of palm leaves distinctive is that at first their blade surfaces are thrown into a series of pleats known as plications. A process of tissue separation along certain of these pleats then cleaves the pleated surface into a series of leaflets. It is easy to see, in looking at a fully developed leaf of the palmate kind, that the segmented nature of the leaf originates from the pleating of the leaf blade. Indeed, as with a Japanese fan, the leaf can be closed by pressing the pleats together. Since each leaflet is V-shaped in cross section, it appears to have been cut out of a pleated surface."

It has been observed that some molecular biologists tend to extrapolate their findings of developmental controls that operate in lower plants to the higher plants, an approach which may lead to oversimplification of the

[1] Leaves of flowering plants have a large variety of shapes and sizes. A common variant is the dissected or compound leaf. See also **Leaf**. The blades of these leaves are cut into segments or leafleats. Kaplan observes that, in terms of comparative development, dissected leaves are of particular interest because they

clearly illustrate how different path of development can lead to leaves that are closely similar in appearance. The giant fronds of palm trees are the largest and most complex of all dissected leaves.

growth or developmental processes. Studies of palm leaf development tend to indicate that many more and different molecular control systems may exist than previously assumed.

Yuccas. These plants (more like shrubs than trees) are abundant in desert areas, particularly in the southwestern United States and in Mexico, although they are found in other similar climatic regions of the world. Yucca leaves, all basal in most species, are long, thick, and shiny green with dagger-like points. Thus, the Carneros yucca is sometimes referred to as Spanish dagger. The flowering stem may be several feet tall and bears many pendant bell-shaped white flowers. A few species have been urbanized and frequently will be found in areas with a mild climate.

An unusual feature of the yuccas is the manner by which they are fertilized. The shape of the flowers makes pollination extremely difficult and consequently the common pollinating insects avoid them. Pollination is accomplished by a symbiotic process, first discovered in 1872 by a Missouri state entomologist, C.V. Riley. The yucca moth (or Navajo yucca borer) deposits an egg in the pistil and stuffs the hole in which the egg is laid with pollen, thus fertilizing an ovary.

A yucca of particular interest is commonly referred to as the Joshua tree, which enjoys a protected area ($\frac{1}{2}$ mil acres) in southern California at the junction of the Mojave and Colorado deserts. The Joshua Tree National Monument is located 140 miles (225 km) east of Los Angeles. The record specimen (see accompanying table) is located in the San Bernardino National Forest (California). Most experts consider calling this plant a tree is uncongenial scientifically—because it does not produce annual growth rings, the trunk is fibrous, the bark is soft and corklike, and its branching is erratic, among other factors. It has been reported that early pioneers likened the appearance of the Joshua "tree" to the biblical Joshua waving his arms to encourage the pioneers to push forward. Further lore of this plant is reported by S.L. Keith in an article, "A Tree Named Joshua." *American Forests,* 38–41, July 1982.

Additional Reading

Heinerman, J.: "Aloe Vera, Jojoba, and Yucca," Keats Publishing, Inc., Chicago, IL, 1990.
Henderson, A., G. Galeano, and R. Bernal: "Field Guide to the Palms of the Americas," Princeton University Press, Princeton, NJ, 1997.
Howard, F.W., D. Moore, R. Giblin-Davis, and R. Abad: "Insects on Palms," Oxford University Press, Inc., New York, NY, 1999.
Irish, M., et al.: "Agaves, Yuccas, and Related Plants: A Gardener's Guide," Timber Press, Portland, OR, 2000.
Jones, D.L.: "Palms throughout the World," Smithsonian Institution Press, Washington, DC, 1995.

Web Reference

www.usda.gov (Search: Palm Trees).

PALMITIC ACID. $CH_3(CH_2)_{14}COOH$, formula weight 256.42, white crystalline powder, mp 64 °C, bp 271.5 °C, sp gr 0.849. The acid is insoluble in H_2O, moderately soluble in alcohol, soluble in ether. About 60% of the content of palm oil is palmitic acid.

Palmitic acid is present as cetyl ester in spermaceti from which, by hydrolysis, the acid may be obtained; it is present in bee's wax as the melissic ester; and in most vegetable and animal oils and fats, in greater or lesser amounts, as glyceryl tripalmitate or as mixed esters, along with stearic and oleic acids. Palmitic acid is separated from stearic and oleic acids by fractional vacuum distillation and by fractional crystallization. With NaOH, palmitic acid forms sodium palmitate, a soap. Most soaps are mixtures of sodium stearate, palmitate, and oleate.

Representative esters of palmitic acid are: methyl palmitate, $C_{15}H_{31}COOCH_3$, mp 30 °C, bp 195 °C at 15 mm pressure; ethyl palmitate, $C_{15}H_{31}COOC_2H_5$, mp 24 °C, bp 185 °C at 10 mm pressure; cetyl palmitate, $C_{15}H_{31}COOC_{16}H_{33}$, mp 54 °C; glyceryltripalmitate (tripalmitin), $C_3H_5(COOC_{15}H_{31})_3$, mp 65 °C, bp 310 °C, approximately.

As the glyceryl ester, palmitic acid is one of the constituents of many vegetable and animal oils and fats.

Palmitic acid finds use in the production of cosmetics, food emulsifiers, pharmaceuticals, plastics, and soaps. One commercial formulation contains 95% palmitic acid, 4% stearic acid, and 1% myristic acid; another preparation contains 50% palmitic acid and 50% stearic acid.

See also **Vegetable Oils (Edible)**.

PALMITOLEIC ACID.   Also called *cis*-9-hexadecanoic acid, formula $CH_{33}(CH_2)_5CH:CH(CH_2)_7COOH$. This is an unsaturated fatty acid found in nearly every fat, especially in marine oils (15–20%). At room temperature, it is a colorless liquid. Insoluble in water; soluble in alcohol and ether; mp 1.0 °C; bp 140–141 °C (5 millimeters pressure). Insoluble in water; soluble in alcohol and ether. Combustible. Palmitoleic acid is used in organic synthesis, and as a standard in chromatographic analysis. See also **Vegetable Oils (Edible)**.

PALYNOLOGY.   The study of pollen of seed plants and spores of other embryophytic plants, whether living or fossil, including their dispersal and applications in stratigraphy and paleoecology.

PAMPAS DEER.   See **Deer**.

PANCREAS.   An elongated glandular organ located in the midportion of the abdominal cavity. The organ secretes juices that aid in digestion, and certain of its cells produce the endocrine hormone *insulin*, which aids in the regulation of blood sugar levels. The pancreas is a long, soft, yellowish-gray gland which lies transversely on the posterior abdominal wall, its right end enclosed by the curve of the duodenum and its left end touching the spleen. The gland lies, for the most part, behind the stomach. It is about 6 inches (15 centimeters) long and weighs about 3 ounces (85 grams). The gland secretes a clear, watery, alkaline fluid (the *pancreatic juice*), which passes to the duodenum through the pancreatic duct. Pancreatic juice is one of the chief chemical agents in digestion, for it contains enzymes that break down starch into sugar, fats into glycerine and fatty acids, and proteins into peptones and amino acids. Additionally, the pancreas also secretes directly into the bloodstream the antidiabetic hormone, *insulin*. See also **Diabetes Mellitus**. Patients who must have the pancreas removed are able to live only so long as they receive injections of insulin. It is the hormone function of the pancreas that makes it a part of the endocrine system. See also **Endocrine System**; and **Hormones**.

Pancreatic juice flows into the intestine when acid material comes into contact with the duodenal mucosa. A hormone, *secretin*, is liberated from the mucous coat of the duodenum when an acid substance is in contact with the coat. This hormone enters the bloodstream and is conveyed to the pancreas in a few seconds. There, it stimulates the glandular cells to produce pancreatic juice. There is also nervous control of pancreatic secretion, so that even the thought of food may stimulate its secretion.

Pancreatitis remains a disease of considerable mystery. In the United States, it has been reasonably well established that cholelithiasis (gallstones) and chronic alcoholism together account for about 80% of cases. Other diverse causes include hyperlipidemia, ductal obstruction, viral infection, some drugs, and impaired pancreatic perfusion. In an excellent article (reference listed), W.H. Steinberg, (George Washington University) reviews the numerous hypotheses that attempt to explain at least some of the probable causes of acute pancreatitis. Numerous cases of acute pancreatitis remain *unclassified*—that is, *idiopathic* (cause unknown).

The symptoms of acute pancreatitis are excruciating pain in the region of the organ and radiating toward the back. Unlike the undulating pain of gallbladder disease, the pain is essentially continuous and may persist for several hours or days, but because of its severity, the patient usually seeks medical attention promptly. When the pain reaches a peak plateau, it is usually accompanied by nausea and vomiting and, in about half of the cases, fever of 100–101 °F (37.8–38.3 °C) will be present even though infection may not be obvious. If the condition is allowed to persist without medical attention for several hours, shock and obtundation may occur in about half of the cases. Through differential diagnosis, the physician will check out the symptoms and laboratory findings to make certain that the symptoms are not caused by other conditions associated with abdominal pain, such as acute cholecystitis, perforated duodenal ulcer, myocardial infarction, and sometimes pneumonia, which can produce severe pain in the epigastrium as well as fever.

In view of the severity of pancreatitis, the mortality rate of an acute attack is comparatively low (about 5% of cases). However, where acute suppurative or hemorrhagic pancreatitis is present, the mortality ranges from 50 up to 90%. Where acute pancreatitis is related to alcoholic intake, the patient is prone to developing chronic disease. Statistics indicate that between 10 and 20% of chronic alcoholics will be found to have chronic pancreatitis when examined during surgery or autopsy. Chronic pancreatitis is also evident in patients, notably males approaching 40 years of age, who consume inordinate quantities of protein and fat.

Post-Cardiac Surgery. In 1990, more than one-half million cardiac surgical operations were performed in the United States. Pancreatitis is an infrequent but well-recognized complication of cardiac surgery, including cardiac transplantations and pediatric procedures. It has been reported that ischemia and drugs, such as phenylephrine, norpinephrine, narcotics, steroids, and calcium chloride contribute to "post-pump" pancreatitis. The cause of these complications is poorly understood. In 1991, C. Fernéndez-Del Castillo (Harvard Medical School and Massachusetts General Hospital) studied 300 consecutive patients undergoing cardiac surgery with cardiopulmonary bypass. The results of that study: "Evidence of pancreatic cellular injury was detected in 80 patients (27%), of whom 23 had associated abdominal signs or symptoms and 3 had severe pancreatitis." Conclusions: "Pancreatic cellular injury, as indicated by hyperamylasemia of pancreatic origin, is common after cardiac surgery. The administration of large doses of calcium chloride is an independent predictor of pancreatic cellular injury and may be a cause of it."

Cancer of the Pancreas. Middle-aged and elderly men are the most common sufferers from cancer of the pancreas. The disease is only one-third as frequent in women. When the tumor originates in the head of the gland and blocks the bile duct, a steadily increasing jaundice appears, and increasing itching of the skin develops. When the cancer begins in the body of the organ, pain is the most common symptom. The pain is usually constant and of a deep, penetrating character that radiates through to the back, between the shoulder blades. If the cancer occurs in the tail of the gland, there may be no symptoms until the growth has spread to the liver, with resulting loss of weight and general impairment of bodily function. Treatment is removal of the whole gland, a surgical procedure made possible because of means available for controlling shock and the availability of substances which can substitute for the secretions of the gland.

As observed by B.A. Chabner and M.A. Friedman (National Cancer Institute), "Islet-cell carcinoma of the pancreas is a rare, slow-growing neoplasm with a broad range of endocrinologic manifestations". C.G. Mortel (Mayo Clinic, Rochester, Minnesota) tested the efficacy of several drugs in 105 randomly assigned patients with advanced islet-cell carcinoma. Conclusions of the study: "The combination of streptozocin and doxorubicin is superior to the current standard regiment of streptozocin plus fluorouracil in the treatment of advanced islet-cell carcinoma."

Sometimes agglomerations of glandular tissue that simulate cancer occur in the pancreas. These are called islet cell tumors and may not be malignant. Because of the possibility of their becoming malignant, the tumors should be removed. Sometimes this mass of glandular tissue secretes so much insulin that weakness and faintness occur. The patient may even lapse into coma because of the extent to which the blood is depleted of its sugar. In recent years, cases have been reported in which islet cell tumors are associated with peptic ulcerations of the jejunum.

For the preoperative localization of endocrine tumors, such as insulinomas (99% of cases), endoscopic ultrasonography is reported as a highly sensitive and specific procedure to be used after clinical and laboratory diagnosis has been established.

Transplantation. H. Katz (University of Minnesota) and a group of researchers studied a small number of patients with insulin-dependent diabetes mellitus after they had a pancreas-kidney transplantation and determined that carbohydrate metabolism was similar to that in nondiabetic subjects who received the same immunosuppressive agents after kidney transplantation. Pancreas transplantation provides a means of normalizing plasma glucose concentrations, thus offering a way to delay or prevent the long-term complications of diabetes. Whether long-term systemic delivery of insulin has deleterious effects on carbohydrate metabolism in humans remains unknown.

K. Pyseoqaki (University of Minnesota) and a group of researchers reported in 1992 that the transplantation of pancreatic islets rather than the whole pancreas has been introduced as a treatment to diabetes mellitus. The group studied five patients ranging in age from 12 to 37 years. If successful, islet transplantation would have advantages over transplantation of the entire pancreas. However, there are only a few reports of transient successful transplantation of cadaveric islets in humans. The study conclusions: "Intrahepatic transplantation of as few as 265,000 islets can result in the release of insulin and glucagon at appropriate times and in prolonged periods of insulin independence."

Additional Reading

Becker, K.L., W. Hung, C.R. Kahn, et al.: "Principles and Practice of Endocrinology and Metabolism," 3rd Edition, Lippincott Williams & Wilkins, Philadelphia, PA, 2001.
Castillo, C. Fernández-Del, et al.: "Risk Factors for Pancreatic Cellular Injury After Cardiopulmonary Bypass," *N. Eng. J. Med.*, 382 (August 8, 1991).
Chabner, B.A. and M.A. Friedman: "Progress Against Rare and Not-so-Rare Cancers," *N. Eng. J. Med.*, 563 (February 20, 1992).
Cruickshank, A.H. and E.W. Benbow: "Pathology of the Pancreas," 2nd Edition, Springer-Verlag Inc., New York, NY, 1995.
Doppman, J.L.: "Pancreatic Endocrine Tumors–The Search Goes On," *N. Eng. J. Med.*, 1770 (June 23, 1992).
Howard, J.M., R. Prinz, Y. Idezuki, and I. Ihse: "Surgical Diseases of the Pancreas," Lippincott Williams & Wilkins, Philadelphia, PA, 1997.
Katz, H., et al.: "Effects of Pancreas Transplantation on Postprandial Glucose Metabolism," *N. Eng. J. Med.*, 1278 (October 31, 1991).
Laufer, H. and R.G.H. Downer: "Endocrinology of Selected Invertebrate Types," John Wiley & Sons, Inc., New York, NY, 1988.
Lee, S.P., J.F. Nicholls, and H.Z. Park: "Billiary Sludge As a Cause of Acute Pancreatitis," *N. Eng. J. Med.*, 589 (February 27, 1992).
Lloyd, R.V.: "Endocrine Pathology," Springer-Verlag, Inc., New York, NY, 1990.
Lott, J.A.: "Clinical Pathology of Pancreatic Disorders," Vol. 2, Humana Press, Totowa, NJ, 1997.
Moertel, C.B., et al.: "Streptozocin-Doxorubicin, Streptozocin-Fluorouracil, or Chlorozotocin in the Treatment of Advanced Islet-Cell Carcinoma," *N. Eng. J. Med.*, 510 (February 20, 1992).
Moore, W.T. and R.C. Eastman: "Diagnostic Endocrinology," 2nd Edition, Mosby-Year Book, Inc., St. Louis, MO, 1996.
Owen, D.A. and J.K. Kelly: "Pathology of the Gallbladder, Biliary Tract and Pancreas," W. B. Saunders Company, Philadelphia, PA, 2000.
Pyzdrowski, K.L., et al.: "Preserved Insulin Secretion and Insulin Independence in Recipients of Islet Autografts," *N. Eng. J. Med.*, 220 (July 23, 1992).
Reber, H.A.: "Acute Pancreatitis–Another Piece of the Puzzle," *N. Eng. J. Med.*, 423 (August 8, 1991).
Rösch, T., et al.: "Localization of Pancreatic Endocrine Tumors by Endoscopic Ultrasonography," *N. Eng. J. Med.*, 1721 (June 25, 1992).
Russell, R.C.G., D. Carr-Locke, M.G. Sarr, and H.G. Beger: "The Pancreas," Vol. 2, Blackwell Science Inc., Malden, MA, 1998.
Skerrett, P.J.: "New Transplant Method Evades Immune Attack," *Science*, 1248 (September 14, 1990).
Steinberg, W.M.: "Acute Pancreatitis — Never Leave a Stone Unturned," *N. Eng. J. Med.*, 635 (February 27, 1992).
Trey, C. and C.C. Compton: "Fever Three Weeks after an Operation for Pancreatic Cancer," *N. Eng. J. Med.*, 318 (February 1, 1990).
Waters, D.L., et al.: "Pancreatic Function in Infants Identified as Having Cystic Fibrosis in a Neonatal Screening Program," *N. Eng. J. Med.*, 303 (February 1, 1990).

PANDAS *(Mammalia, Carnivora, Ailuridae).* These animals are of the family *Ailuridae*; they were once classified together with the raccoons. The Lesser Panda (*Ailurus fulgens*) (also called Red Panda) reaches a length of 79 to 112 centimeters (31 to 44 inches), has a tail length of 28 to 48.5 centimeters (11 to 19 inches), and weighs from 3 to 4.5 kilograms (6.5 to 10 pounds). Two subspecies are distinguished. The head is unusually large, and the snout is short and pointed. The ears are medium-sized and are tapered. The eyes are small and brown, and the nose is black. There are 38 teeth, with three premolars in the upper jaw and four lower premolars. The trunk is supported by short, bear-like legs. The toes have short, partially retractile claws. The soles of the hands and feet are covered with woolly hair, and the lesser panda walks in a plantigrade manner. The fur is soft and long, and the underfur is very thick. The chin and inside of the ears are white. A rust-red stripe beneath each eye separates the white snout from the white cheeks. The upper side of the body is rust-brown and yellowish, while the forehead tends toward a rust-yellow hue. A white spot is located above each eye. The tail is long and bushy and is reddish-brown with distinctly separate light reddish rings. The underside, and the rear of the ears and limbs, are gleaming black. See Fig. 1.

The habitats of the lesser panda are mountain forests and bamboo thickets from 1,800 to 4,000 meters (5,906 to 13,124 feet) along the southeastern slopes of the Himalayas. Its distribution extends from Nepal across Bhutan, Sikkim, northern Assam, northern Burma, and as far east as the western Chinese provinces Yunnan and Szechuan. The western lesser panda (*Ailurus fulgens fulgens*) inhabits the western part of the distribution, while Styan's lesser panda (*Ailurus fulgens styani*) inhabits Szechuan, Yunnan, and northeastern Burma. Styan's lesser panda has a larger body, higher forehead region, and a more intense rust-yellow coloration than the western lesser panda.

Fig. 1.   Pandas.

Fig. 2.   Giant panda. (*A.M. Winchester.*)

Lesser pandas feed chiefly on bamboo shoots, juicy grasses, roots, berries, and fruits, occasionally taking young birds, eggs, small rodents, and insects. They are crepuscular and nocturnal animals, which may be found equally often in trees or on the ground. On the ground the lesser panda runs about at a gallop with its back arched and tail raised; it places its hands inward with each step. The hairy soles reduce the danger of slipping on wet, smooth branches in the foggy, high-altitude forests; they also reduce heat loss on snow-covered and ice-encrusted rocky mountain ground. Similar hair covering on the soles is found in the giant panda and the polar bear.

Although lesser pandas can leap up to 1.5 meters (5 feet), they avoid doing so if they can climb to wherever they are going. The sharp claws are put to good use when the panda is climbing about. During the day this very heat-sensitive species sleeps in shady branches or in tree hollows, lying coiled up on its side. The tail is used as a pad or to cover the face. The lesser panda often stretches out on a tree limb and sleeps there. It washes itself in a catlike manner; first it licks the soles of the hands for some period of time, and then it rubs the hands across the forehead and the ears. Food is skillfully seized with the hands and is eaten bit by bit, often with the panda sitting upright.

Lesser pandas are rarely seen in their native habitat. They are asocial and typically live alone, but sometimes in pairs. Calls to conspecifics include shrill cries, a whistle which is often uttered repeatedly, a peeping, and a birdlike chirping. If the lesser panda is endangered it withdraws into a rock crevice or a tree. When it is greatly excited, the anal glands secrete a penetrating, musky liquid; generally males use the anal glands to mark tree branches as they forage for food or to mark their territory on the ground.

In spring the female bears 1 to 4 young after a 130-day gestation period. The newborn young is about 20 centimeters (7.8 inches) long, 5 centimeters (2 inches) of which is the tail. The upper side is pale rust-red, with a gray-white head; the characteristic markings of the face are not yet visible, but the black spot on the rear side of the ears can already be clearly seen. The prominent black hair of the legs and belly of the adult exist as a dark gray color in the young panda, and the ring pattern on the tail cannot yet be seen. After 2 to 4 weeks the eyes open. After the young are six weeks old they begin to attain adult coloration.

The Giant Panda (*Ailuropoda melanoleuca*) is the closest related species to the lesser panda. It is the only member of its genus, *Ailuropoda*, and it has not changed its appearance since the Pleistocene. The overall length from the nose to the base of the tail reaches 150 centimeters (59 inches), the tail length is 16 centimeters (6.3 inches), and the weight reaches 125 kilograms (275 pounds). The head is short, pug-nosed, and looks quite powerful because of the greatly developed chewing musculature and the fur in front of the ears. The nose is naked; the ears are round and look as if they were simply placed artificially on top of the head or glued there. The rear of the ears is very steep. The fur is short and thick, and in the rear it is felt-like and gleaming. The white areas often have a yellowish tinge. The feet are oriented to the inside, and the front and rear feet have 5 toes each. Much of the soles of the feet is hair-covered; the giant panda walks in a semi-plantigrade manner. It has the most powerful, broad chewing teeth of any carnivore. See Fig. 2. Distribution is restricted to a very small part of the Hsifan mountain region in western Szechuan.

The panda is most prevalent in steep, rocky, moist mountains with a rather mild subtropical climate where there is an ample bamboo supply; such areas are very difficult for a man to traverse. It is generally found at altitudes from 1,500 to 4,000 meters (4,921 to 13,124 feet), where there is often a great deal of snow. The Chinese name for the panda is *beshiung-chin* (white bear). Its black/white coloration appears very striking, as does that of the zebra, but in both cases this color effectively breaks up the outline, as it were; the contrast makes it difficult to recognize the panda in the wild. Pandas generally sit on their rear legs in moderately high branches; the dark parts of their fur approximate the color of the tree limbs, while the white portions are against the sky. Thus the panda is quite effectively camouflaged.

The panda moves through its home range using the same paths and tunnels again and again, sometimes using people-made paths as well. These paths connect feeding sites with rock or tree dens and sleeping spots. The sleeping den is padded with twigs and leaves. Pandas move in a bear-like walk, or they trot and make short leaps. See Fig. 2. They ford streams and rivers. Food and other objects can be grabbed with the clenched hands. Karl Max Schneider reports that vision is the most highly developed sensory modality of the panda, but its hearing, smelling, and taste are also developed. Its dampened roar is rarely heard, only when attacking and during courtship.

Additional Reading

Angel, H.: "Pandas," Voyageur Press, Stillwater, MN, 1998.
Entwistle, A. and N. Dunstone: "Priorities for the Conservation of Mammalian Diversity: Has the Panda Had Its Day?" Cambridge University Press, New York, NY, 2000.
Lumpkin, S. and J. Seidensticker: "Smithsonian Book of Giant Pandas," Smithsonian Institution Press, Washington, DC, 2002.
Maple, T.L.: "Saving the Giant Panda," Longstreet Press, Inc., Atlanta, GA, 2000.
Schaller, G., et al.: "The Giant Pandas of Wolong," University of Chicago Press, Chicago, IL, 1985.

Web Reference

www.nationalgeographic.com (Search: Panda from National Geographic Society homepage).

PANDEMIC.   See **Contagion**.

PANTHER.   See **Cats**.

PANTOGRAPH.   A type of trolley frequently used on electric locomotives to connect with the overhead trolley wire. The pantograph is a parallel mechanism and is also used in parallel rulers and drafting machines. The principle is best explained by the simple example of the parallel ruler. In this instrument, the straight edges are always parallel, but the perpendicular distance between them may be varied up to the length of the connecting links. Two straight edges are joined with pin joints by two connecting links having points of connection so spaced that the rulers and the connecting links form a parallelogram.

PANTOTHENIC ACID.   The designation *vitamin B₃* for this essential substance is now used only infrequently, as are other, earlier terms, such as *chick antidermatitis factor, Bios IIa,* and *antigray-hair factor.* Pantothenic acid is a constituent of coenzyme A, which participates in numerous enzyme reactions. CoA was discovered as an essential cofactor for the acetylation of sulfanilamide in the liver and of choline in the brain. CoA is particularly important in the initial reaction of the TCA cycle (citric acid cycle) of carbohydrate metabolism and energy production. These factors are described in greater detail in the entry on **Coenzymes**. Pantothenic acid is unique among the vitamin group, in that it was one of the first to be isolated, using as a basis a microbiological assay method. Even more unique is the fact that its structure was largely determined, using a highly quantitative biological yeast test, long before it was isolated or obtained in concentrated form. R.J. Williams and coworkers described it as an acid with an ionization constant lower than that of an alpha-hydroxy acid, but about right for a hydroxy acid in which the hydroxyl group was farther removed from the carboxyl group.

In 1901, Wildiers described Bios, an essential for yeast growth. In 1933, Williams isolated crystalline Bios from yeast and named it pantothenic acid. In 1938, Williams isolated pantothenic acid from liver; and, in 1939, Jukes determined liver antidermatitis factor (chick) to be identical with yeast factor. Also, in 1939, Woolley et al. demonstrated beta-alanine as a vital part of pantothenic acid:

$$H-O-\overset{\overset{\displaystyle H}{|}}{\underset{\underset{\displaystyle H}{|}}{C}}-\overset{\overset{\displaystyle CH_3}{|}}{\underset{\underset{\displaystyle CH_3}{|}}{C}}-\overset{\overset{\displaystyle OH}{|}}{\underset{\underset{\displaystyle H}{|}}{C}}-\overset{\overset{\displaystyle O}{\|}}{C}-\overset{\displaystyle H}{\underset{\displaystyle \ }{N}}-\overset{\overset{\displaystyle H}{|}}{\underset{\underset{\displaystyle H}{|}}{C}}-\overset{\overset{\displaystyle H}{|}}{\underset{\underset{\displaystyle H}{|}}{C}}-COOH$$

(Pantoic acid)   (Beta-alanine)

d (+) Pantothenic acid ($C_6H_{17}O_5N$)

In 1940, Harris, Folkers, et al. reported structure determination and synthesis and crystallization of pantothenic acid. In 1950, Lipmann et al. discovered coenzyme A; and, in 1951, Lynen characterized the coenzyme A structure.

Pantothenic acid participates as part of coenzyme A in carbohydrate metabolism (2-carbon transfer-acetate, or pyruvate), lipid metabolism (biosynthesis and catabolism of fatty acids, sterols, +phospholipids), protein metabolism (acetylations of amines and amino acids), porphyrin metabolism, acetylcholine production, isoprene production.

Distribution and Sources. Particularly high in pantothenic acid content are yeasts, animal glands and organs. Fruits have a low content.

High pantothenic acid content (2.0–10.0 milligrams/100 grams): Beef (brain, heart, kidney, liver), chicken (liver), cod ovary, groundnut (peanut), herring, lamb (kidney), pea (dry), pork (kidney, liver), royal jelly, sheep (liver), wheat bran and germ, yeast.

Medium pantothenic acid content (0.5–2.0 milligrams/100 grams): Avocado, bean (lima), beef, broccoli, carrot, cauliflower, cheese, chicken, clam, kale, lamb, lentil (dry), mackerel, mushroom, oats, pea, pork (bacon, ham), rice, salmon, soybean, spinach, walnut, wheat.

Low pantothenic acid content (0.1–0.5 milligram/100 grams): Almond, apple, banana, bean (kidney), cabbage, grape, grapefruit, honey, lemon, lettuce, lobster, milk, molasses, onion, orange, oyster, peach, pear, pepper (white and sweet), pineapple, plum, potato, shrimp, tomato, turnip, veal, watercress.

Pantothenic acid is produced commercially by synthesis involving the condensation of *d*-pantolactone with salt of *β*-alanine. Some of the dietary supplement forms include calcium pantothenate, dexpanthenol, and panthenol.

Precursors in the biosynthesis of pantothenic acid include *α*-ketoisovaleric acid (pantoic acid), uracil (*β*-alanine), and aspartic acid. Intermediates in the synthesis include ketopantoic acid, pantoic acid, and *β*-alanine.

Some of the unusual features of pantothenic acid noted by investigators include: (1) it promotes amino acid uptake; (2) it is potentiated by zinc in preventing graying of hair in rats; (3) it promotes resistance to stress of cold immersion; (4) there is a deficiency of pantothenic acid in tumors; (5) it is required for chick hatchability; (6) it is useful in treating vertigo, postoperative shock, and poisoning with isoniazid and curare; (7) it is useful in accelerating wound healing; and (8) it is useful in treating Addison's disease, liver cirrhosis, and diabetes.

Additional Reading

Williams, R.J.: "Pantothenic Acid," in The Encyclopedia of Biochemistry: (R.J. Williams and E.M. Lansford, Jr., editors), Van Nostrand Reinhold, New York, NY, 1967.

PAPAYA TREE. Of the family *Caricaceae*, the papaya or papaw tree (*Carica papaya*) is a tropical American tree with a straight barely branching trunk from 6 to 25 feet (1.8 to 7.5 meters) tall. On the upper portion of the stem is borne a crown of large compound, long-petioled leaves. The pale yellow, fragrant flowers are of two kinds, pistillate and staminate, borne on different plants. Papayas are, therefore, dioecious plants. The smooth-skinned fruits vary considerably in shape and size, those of wild plants being not much larger than eggs, while cultivated fruits are much larger. The orange-colored flesh is sweet and juicy and surrounds a central cavity, which contains the numerous seeds. Usually the fruit is eaten raw but may be used in salads or cooked. See Fig. 1.

Fig. 1.   Close-up of cluster of papayas (*Carica papaya*). (*Papaya Administrative Committee, Honolulu, Hawaii.*)

From the fruit and sap of the plant is obtained an enzyme, papain, which is used as a digestive aid, its action being much like that of pepsin. The leaves, cooked with meat, are said to tenderize meat.

The plant is widely introduced in many tropical lands. It is extensively grown in the Hawaiian Islands and to some extent in Florida and California. Propagation is mainly by seed.

A record tree, determined by the American Forests, is located in Homestead, Florida with a circumference of 20 inches (50.8 centimeters), a height of 25 feet (7.62 meters), and a spread of 4 feet (1.22 meters).

The name papaw, sometimes given to this fruit, is also applied to a North American tree, *Asimina triloba*, with which it should not be confused.

PAPERMAKING AND FINISHING.   One of the most important factors in the progress of civilization has been paper, a thin flat tissue composed of closely matted fibers obtained almost entirely from plant sources. In modern life paper finds a variety of uses, for writing, for containers, wrappers, wall covering, and—perhaps most important—in all forms of printing: newspapers, magazines, books.

The art of making paper seems to have been discovered first by the Chinese, who were making paper as early as the beginning of the Christian era. From China the process was carried to Arabia and thence to Europe. Paper was not an important article at first and, since it is not a very durable substance under ordinary conditions, could not compete with parchment or vellum as a medium for the written word. In the fifteenth century writing became more general and the demand for a cheaper material increased. Paper became an important product. At this time paper was made largely from vegetable fibers reclaimed from cloth (especially linen), as had been done since the invention of paper in China. This paper was made entirely by hand, as is done even today in the manufacture of certain expensive types of paper. In making handmade paper, a pulp is formed by soaking the vegetable fibers in water in a vat. From this vat the pulp is dipped out in a mold, the bottom of which is a fine screen. By a deft motion of this mold, the soft pulp is spread over the screen in a thin layer of matted fibers. The water in the pulp drains off, leaving a rather firm mass, which is turned out on a piece of felt. More pieces of half-dried pulp spread on felt are added. The whole pile is then pressed to squeeze out more of the water, press the fibers closer together and form a firm sheet. These are then removed from between the felts, pressed again, and dried. During the final treatment surface sizing is added to render a surface more suitable to receive ink. Sheets of hand-made paper are naturally of limited size and expensive.

To meet the great demand for paper, machine methods were developed. This increased demand for paper also led to the utilization of material that could be obtained in quantities much greater than rags. Out of this developed the vast pulp industry, which today converts vegetable material, mostly soft woods such as spruce and fir, as well as poplar, into a white felt-like mass of fibrous substance, known as pulp. See also **Pulp (Wood) Production and Processing**.

Wet End. A modern paper machine begins with a flow spreader or distributor, conveying a dilute fiber suspension (0.1–1% fibers) to a headbox which delivers a jet of the suspension or slurry through a slice (sluice) across the full width of the machine, almost 400 inches (~10.2 meters) in some large machines. In the headbox, the fibers are dispersed, and the flow is rectified as well as possible so that the jet is delivered onto a moving endless fine-mesh wire screen with uniform composition, flow rate, and velocity. The pressure in the headbox and its slice opening are adjusted so that the jet velocity matches the speed of the wire screen, which may be up to 4000 feet (~1220 meters) per minute for newsprint. The proper stock flow per unit width corresponds to the desired *basis weight* of the paper. (Basis weight is weight per unit area and varies with grades and sizes of papers.)

The dispersion of fibers in the headbox is brought about by subjecting the slurry or suspension to shear stresses, usually with turbulence. Various designs have been developed to accomplish this.

As shown in Fig. 1, the most common type of paper machine is the Fourdrinier, in which the moving wire screen is in the form of an endless conveyor belt stretched between two large rolls. The roll situated under the headbox slice is called the *breast roll*. The roll located generally at the end of the straight wire run is the *couch roll*. Drainage of the slurry through the wire screen is induced by several types of driving forces. In the early, slow-speed machines, the principal force was gravity. Later, the hydrodynamic action of table rolls, which support the wire and rotate with it, began to play an important part in drainage as speed increased. More recently, foils came into use, i.e., rigid stationary, hydrodynamically shaped elements which support the wire and exert a pumping action through the wire screen. Other means are perforated or slotted boxes with vacuum over which the wire runs. When only water is drained, they are called *wet boxes*. When applied toward the dry end of the wire screen, they also draw air through the wet paper mat and are called *suction boxes*. Other equipment configurations have been developed to meet these objectives. On all modern Fourdriniers (Fig. 2), a forming board located close to the breast roll is used to scrape off the water, drained initially by gravity, from the bottom of the wire.

An important development in paper forming is the use of a top wire dewatering unit placed above the wire of the Fourdrinier. The top wire units use various dewatering elements such as rolls, foils, and vacuum boxes. The top wire units add drainage capacity to an existing Fourdrinier plus improved symmetry through the thickness of the sheet similar to the twin wire formers. An example of this type of former is shown in Fig. 3(**a**) (*Valmet, SymFormer MB*). See also Fig. 4.

A more recent development is the *roll and blade gap former* (*Valmet SpeedFormer*) shown in Fig. 3(**b**). In this type of machine, the fiber suspension is confined between two wire screens, and water is removed through both wires either simultaneously or alternately. This two-sided drainage leads to greater symmetry of distribution of fines and other nonfibrous particles through the thickness of the sheet. A significant feature of twin-wire forming is the elimination of the free surface of the fiber-water suspension while the sheet is being formed. This greatly reduces the larger-scale disturbances (waves, streaks, and jumps) which occur at higher speeds on Fourdrinier wires.

Not all the fiber and other solid materials are retained by the forming wire. For this reason and because so much water is used in the papermaking process, the *white water* removed in the sheet-forming process is recirculated in the overall system. A large part of it is added directly to the high-consistency stock and fed back to the headbox, while a small portion goes into a *save-all* device, which recovers much of the solids from the white water. These extracted fibers and other solids are returned and added to the suspension. The clarified water is used in showers for

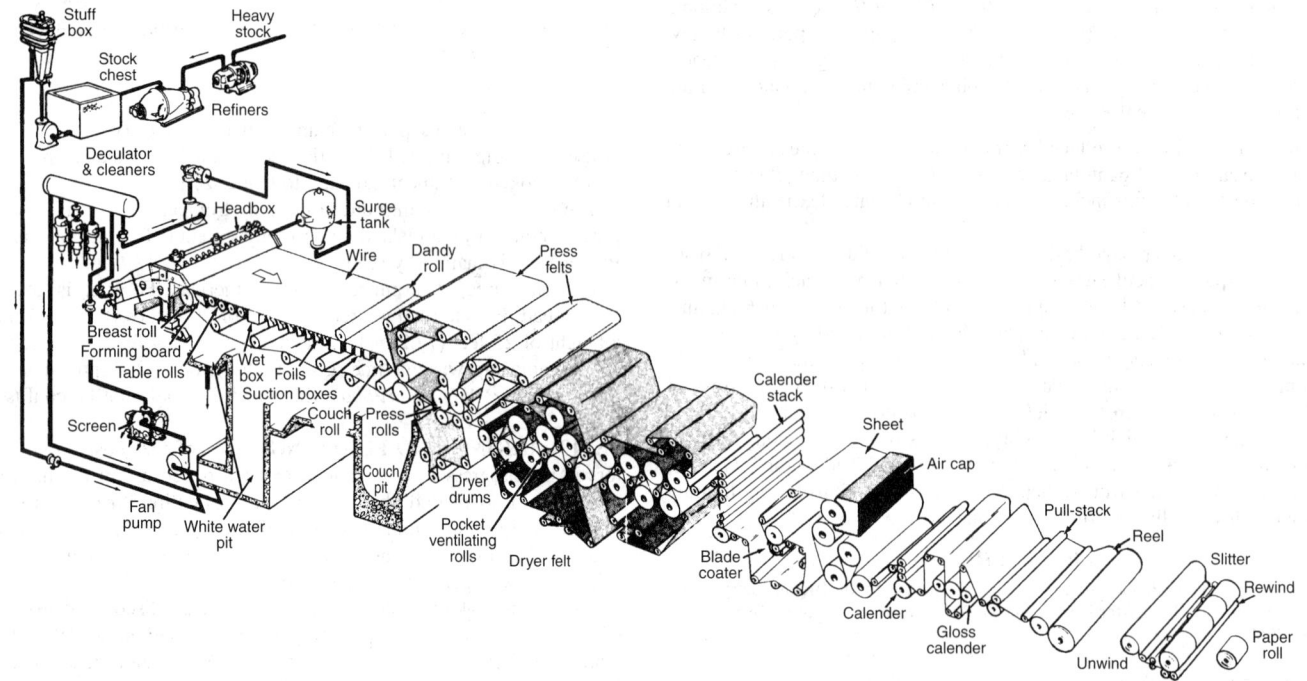

Fig. 1.   Fourdrinier machine for producing printing-grade paper. (*Beloit Corporation.*)

Fig. 2.   Fourdrinier machine located at Blandin Paper Company, Grand Rapids, Michigan. (*Beloit Corporation.*)

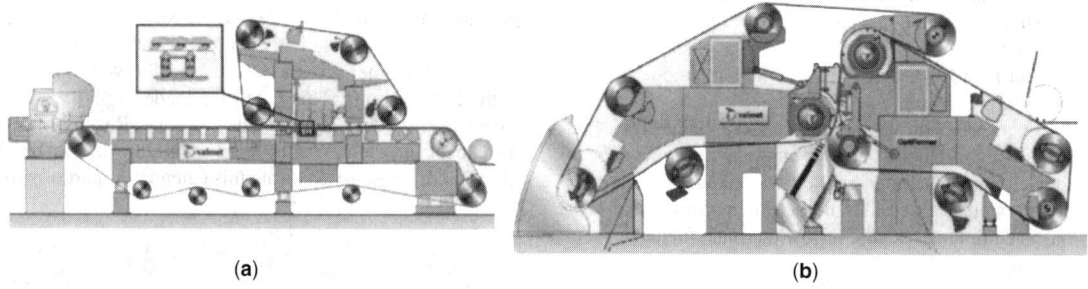

(a)   (b)

Fig. 3.   Paper formers: (**a**) top-wire former; (**b**) Roll and Blade gap former. (*Valmet Corporation.*)

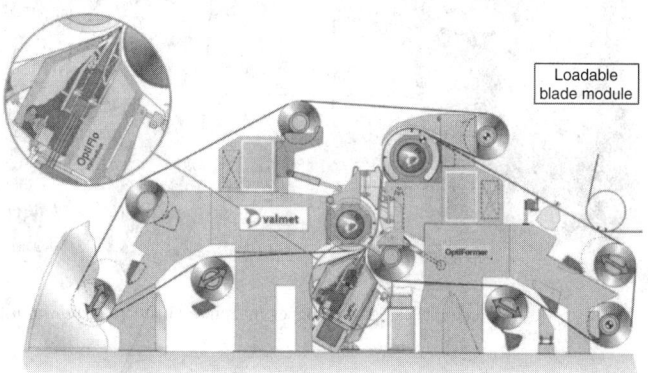

Loadable
blade module

Fig. 4.   Roll and blade gap. (*Valmet Corporation.*)

cleaning wires and felts and other purposes so that only a small amount of the reused water eventually is discharged.

Press Section. At the end of the forming system, the *paper web* is transferred from the wire to a *press felt*, a fine-textured, usually synthetic fabric. At this point, the web contains about 4 or 5 parts water to 1 part solids. The wet paper web and one or more press felts pass through two or more press-roll nips, where water is squeezed out. Pressing also compacts the paper mat. This increases the potential interfiber contact areas where bonds will be formed.

The early *plain press* used a pair of metal and rubber-covered solid rolls. The expressed water had to flow out of the nip in the upstream direction, parallel to the paper web, as in an old washing-machine wringer. Nip pressures were then limited by the damage to the wet web (crushing) caused by this lateral flow. Although the plain press was improved in many ways,

later development work led to the *fabric press* in which the felt contacting rolls are wrapped with a relatively coarse and incompressible mesh fabric. In another development (Beloit Ventanip press), the felt contacting rolls have narrow, closely spaced circumferential grooves. In both types, the lateral flow is virtually eliminated.

While the development of the modern presses has achieved high performance with simple constructions, the remaining problems of flow resistance and web rewetting leave room for improvement. It is generally recognized that mechanical removal of water is much less costly than drying. This has led to the development of presses with very long nips (*Beloit Extended Nip Press*) as shown in Fig. 5. These presses use a shoe typically of 10 in. (25 cm) width to replace one roll in the roll press configuration and use oil lubrication between the stationary shoe and a

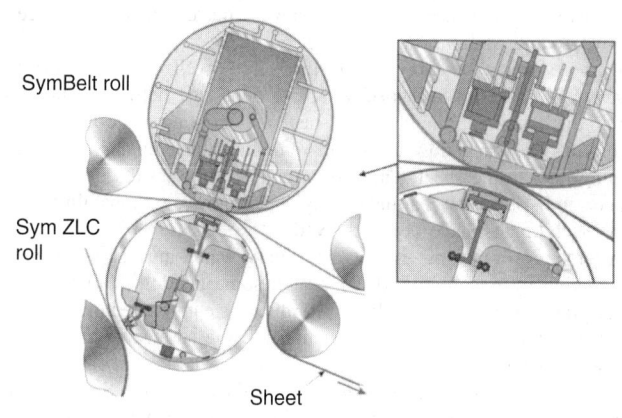

SymBelt roll

Sym ZLC
roll

Sheet

Fig. 5.   The basic belt nip press. (*Valmet Corporation.*)

moving impervious belt. An additional 20–40% of the remaining water in the sheet can be mechanically removed by these presses with their longer residence time under high pressure.

As the machine speed has continuously has increased the dynamic forces to the web have increased and it has forced to support the web with felts to avoid web breaks. Before 30's typically the web had open draw between press and wire section which was soon replaced with pick-up suction roll arrangement. In this arrangement web is picked up with a suction roll from wire and supported by press felt to the first nip. Most recent fast paper machines are using the same principle of suction roll and supporting fabric not only through press but also through dryer section.

Dryers. After water removal by pressing has been done to the extent practical with present technology, the paper web leaves the press section with 1–2 parts of water to 1 part fibers. Most of this remaining water, down to 5–10%, must be removed by evaporative drying. In the most common method, the paper web is passed over a series of staggered cast-iron drums internally heated by condensing steam at pressures ranging up to approximately 10.2 atmospheres. The paper web is held in contact with the rotating drums by means of dryer felts or fabrics under tension. The diameter of the dryer drums is typically 5–6 feet (1.5–1.8 meters). There may be as many as 100 of them in heavyweight paperboard machines. These dryer drums are shown in the panoramic view. See Fig. 1.

Ventilating devices that blow air of controlled temperature and humidity through the dryer felts into the spaces between adjacent dryers are used. Here the air is confined by the sheet and felt runs. These pocket ventilating systems, together with greater control of the flow patterns within the dryer hood (which usually encloses the entire dryer section) have led to significant improvements in cross-machine uniformity of paper drying. This results in paper and board of improved suitability for modern high-speed converting and printing operations.

Other types of dryers, including radiant heating, dielectric and microwave heating, and high-velocity, hot air impingement, have been developed. These devices are generally applied to drying coated paper where sheet contact to a solid surface may be detrimental during drying. Wider application has been limited because of low thermal efficiencies and high capital costs.

Size Press and Coaters. Many printing grades of paper and paperboard are coated with an aqueous suspension of pigments (such as clay) in adhesives (such as starch) to provide a smoother surface, control the penetration of inks, and improve the pick resistance, appearance, brightness, and opacity. These and other materials are also applied, such as *functional coatings*, to provide such features as water resistance, pressure sensitivity for carbonless copying, and a wide variety of other properties. The appropriate materials may be added to the papermaking furnish during some stage of stock preparation (called *internal sizing*). Application of sizing or coating to one or both surfaces of the formed and dried sheet, rather than as internal sizing, simplifies the sheet-forming process and provides better control of surface properties.

The principal methods of surface coating may be classified as roll, blade, and air-knife coating, according to the method used to apply and control the final coating-layer thickness and smoothness. One version of coater (*Beloit Billblade*) simultaneously coats and smooths both surfaces of the paper web by running it down through the nip between a blade and a roll while maintaining two puddles, one between the web and the roll and the other between the flexible blade and the web, thus eliminating the necessity for two coating stations.

A more recent development is the short dwell coater (*Beloit Short Dwell*). The coater consists of a captive pond just before the blade that limits the contact time between sheet and coating material as shown in Fig. 6. The back flow assists in removing the boundary layer of air coming in with the sheet. The shorter contact time results in less coating penetration. Superior coating quality and improved runnability, due to fewer web breaks, have been achieved.

After sizing or coating, the solvent, usually water, must be removed from the coating by evaporative drying. With some coating formulations and paper grades, drying can be done on ordinary steam-heated drums without damage to the coated surface, particularly if the surface of the first drum is smooth (sometimes chrome-plated). However, it is often desirable to do the initial drying with air impingement or radiant heating. Surface coating can be done on the machine as a step in the paper-machine operation, as shown in Fig. 1.

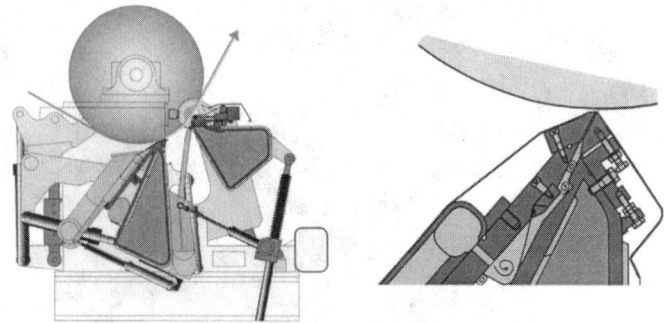

Fig. 6.   Jet coater head. (*Valmet Corporation.*)

Calenders and Winders. Nearly all paper grades are calendered after they have been dried to the desired final moisture content. Ordinary calendering involves passing the paper web through one or more nips between metal rolls with high linear pressures. The calendering process flattens out the paper structure by virtue of the high pressure and "irons" the sheet. Calendering causes bulk reduction, which often is not desired, and surface smoothing, which is desired. The results strongly depend upon moisture content, calender-roll temperature, roll pressure, and speed.

In *supercalendering*, an off-machine operation, the calender rolls consist of alternating chilled-steel and paper-filled rolls, i.e., paper disks clamped on a steel shaft. These roll fillers have to be replaced periodically. Very high pressures are used. The increased pressure and shear forces associated with deformation of the relatively soft paper roll and the very high roll pressures impart a smoother, glossier surface to the web than ordinary calendering with all-metal rolls. This type of calendering is frequently used on coated sheets to provide a glossy coated surface. Recently polymer covered rolls have been replacing paper-filled rolls and due to better durability have given an opportunity to locate this calender as part of papermachine. Fig. 7.

Fig. 7.   Multinip calender as part of paper machine. (*Valmet Corporation.*)

There are other process configurations for the various coating effects and specifications desired.

Other Types of Machines. Although the Fourdrinier machine is used for making almost all grades of paper and board, other designs are sometimes more advantageous. The *cylinder machine*, invented at about the same time as the Fourdrinier, consists of a rotating cylindrical mold covered with a wire screen and partially submerged in a vat. The stock flows into the vat, and a mat is formed on the cylinder under a hydraulic head difference between the stock level in the vat and the white-water level inside the cylinder. The wet mat is picked up by a felt running through the nip between a couch roll and the cylinder. The cylinder machine is used for making multiply board, employing several vats in series. Because of slow speed and other limitations, the cylinder machine is becoming obsolete. In recent years, several new types of machines have emerged.

The most recent multiply machines use headboxes with simultaneous delivery of two different stocks from the headbox slice (*Beloit Strataflo Headbox*), followed by top wire dewatering units (*Beloit Bel Bond*). A secondary headbox with additional top-wire dewatering follows the first unit on the forming wire. Other versions of multiply formers use mini-fourdiniers on top of the primary forming wire in various configurations.

ROBERT A. DAENE; (original preparer and formerly of the Beloit Corporation); revised and updated by Jipi Jaakkola, Paper Machine Product Manager, Valmet Corporation, Charlotte, NC.

PAPER WASP *(Insecta, Hymenoptera; Family Vespidae).* The hornets and yellow-jackets. Wasps that build their nests of coarse paper made by chewing fragments from weathered or partially decayed wood. Some build subterranean nests and others suspend the nest from the eaves of buildings or from the boughs of trees.

PARABOLA. A conic section obtained by a cutting plane parallel to an element of a right circular conical surface. It is the locus of a point that moves so that its distance from the *directrix* equals its distance from the focus; thus, the eccentricity is unity.

The standard equation in rectangular Cartesian coordinates is $y^2 = 2px$. The coordinates of its focus are $x = p/2$, $y = 0$ and its directrix is parallel to the Y-axis at $x = -p/2$. The straight line through the focus and perpendicular to the directrix is the *axis* of the parabola. The point where the parabola crosses the axis is the *vertex*. When the curve is placed in its standard position, the axis is the X-axis, and the vertex is the coordinate origin.

The *latus rectum* of the parabola has length $2p$. Its center is at infinity; hence, it is a noncentral conic.

The parabola has no asymptote. Its equation in polar coordinates is $r = 2a/(1 - \cos\theta) = \sec^2\theta/2$. If the tangents at the extremities of the latus rectum are taken as coordinate axes, the parabola is tangent to the new coordinate system, and its equation is $x^{1/2} + y^{1/2} = a^{1/2}$. The evolute of a parabola is a semicubical parabola.

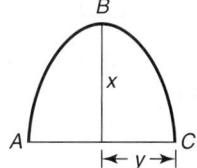

Fig. 1.   Points and dimensions of parabola.

With reference to Fig. 1,

the length of arc $ABC = \sqrt{4x^2 + y^2} + y^2 \dfrac{y^2}{2x} \ln \dfrac{2x + \sqrt{4x^2 + y^2}}{y}$;

and the area of section $ABC = \frac{4}{3}xy$.
See also **Conical Surface**; and **Conic Section**.

PARABOLA (Cubical). A higher plane curve represented by the equation $y = ax^3$. There is a point of inflection at the origin, and the X-axis is tangent to the curve at that point. See Fig. 1.
See also **Curve (Higher Plane)**.

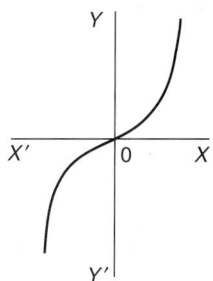

Fig. 1.   Cubical parabola.

PARABOLA (Semicubical). A higher plane curve represented by the equation $y^2 = ax^3$. It is also know as Neil's parabola. There is a cusp of the first kind at the origin, where the X-axis is a double tangent. It is the

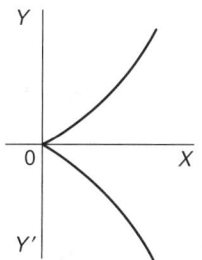

Fig. 1.   Semicubical parabola.

evolute of an ordinary parabola and a special case of a binomial curve, $y = C^{1-n}x^n$. See Fig. 1.
See also **Curve (Higher Plane)**.

PARABOLIC COORDINATE. In a limiting case of confocal paraboloidal coordinates, only two surfaces result. There are two families of parabolas with common foci at the origin of a rectangular coordinate system, one extending toward the positive and the other toward the negative Z-direction. Rotate the families about this axis to obtain paraboloids of revolution ($\xi, \eta = $ constant) and add planes from the Z-axis ($\phi = $ constant). The position of a point is then given by

$$x = \xi\eta \cos\phi$$
$$y = \xi\eta \sin\phi$$
$$z = \tfrac{1}{2}(\eta^2 - \xi^2)$$

See also **Coordinate System**.

PARABOLIC CYLINDRICAL COORDINATE. A curvilinear coordinate system similar to parabolic coordinates, with coordinate surfaces consisting of two families which are parabolic cylindrical ($\xi, \eta = $ constant) and a family of planes ($z = $ constant). A point in this system is given by

$$x = \xi\eta$$
$$y = \tfrac{1}{2}(\eta^2 - \xi^2)$$
$$z = z$$

See also **Coordinate System**.

PARABOLIC MIRROR. See **Spherical Aberration**.

PARABOLIC SPIRAL. A special kind of spiral, the equation of which in polar coordinates is $r^2 = a\theta$. It is also called Fermat's spiral. The windings of it get closer together as it approaches the pole. It has no asymptote. See Fig. 1.

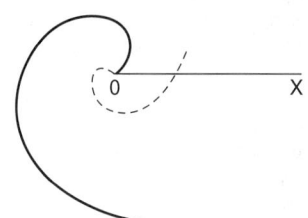

Fig. 1.   Parabolic spiral.

PARABOLOID. A noncentral quadric surface described by the equation

$$\frac{x^2}{a^2} \pm \frac{y^2}{b^2} = \frac{z}{c}$$

With the upper sign, the surface is an elliptic paraboloid. Sections parallel to the XZ- and YZ-planes are parabolas; sections parallel to the XY-plane at $z = k$ are ellipses if k/c is positive, but the locus is imaginary if k/c is negative. When $a = b$, the elliptic sections become circles, and the quadric is a surface of revolution which could be obtained by rotating a parabola about the Z-axis.

A hyperbolic paraboloid results when a minus sign is taken in the equation for the surface. Sections parallel to the XZ- and YZ-planes are parabolas; sections parallel to the XY-plane are hyperbolas. When placed in its standard position, as shown by the equation, the origin of the coordinate system is a saddle point for the hyperbolic paraboloid. No surface of revolution is obtained in this case. If one member of a confocal quadric system becomes a paraboloid, all three of the surfaces are paraboloids. The equation for this case may be taken as

$$\frac{x^2}{A-q} + \frac{y^2}{B-q} = \frac{2z}{C} - \frac{q}{C^2}$$

where $A > B$. When $q > A$ or $q < B$, the surfaces are elliptic paraboloids with vertex on the negative and positive Z-axis, respectively; when $A > q > B$, the surfaces are hyperbolic paraboloids.

See also **Quadric Surface**.

PARABOLOIDAL COORDINATE.   A curvilinear coordinate system of confocal paraboloids. If λ, μ, ν are the real roots of a cubic equation in a parameter describing such surfaces, they are elliptic paraboloids (λ, μ = const.); $-\infty < \lambda < b^2$; $a^2 < \nu < \infty$ and hyperbolic paraboloids ((μ = const.), $b^2 < \mu < a^2$ where $a > b > c$ are also constants).

As in the case of ellipsoidal coordinates, a convention is needed for the sign of the coordinates. They are related to rectangular coordinates by the equations

$$x^2 = \frac{(a^2 - \lambda)(a^2 - \mu)(a^2 - \nu)}{(b^2 - a^2)}$$

$$y^2 = \frac{(b^2 - \lambda)(b^2 - \mu)(b^2 - \nu)}{(a^2 - b^2)}$$

$$z = \frac{1}{2}(a^2 + b^2 - \lambda - \mu - \nu)$$

See also **Coordinate System**.

PARAFFINS.   See **Organic Chemistry**.

PARAFOVEAL VISION.   Vision in which the eye is so oriented toward the pertinent light source as to have the light fall upon some portion of the retina surrounding the fovea. Also called *scotopic vision*. See **Foveal Vision**.

The portion of the retina used in this type of vision contains receptors known as rods. Although these rods do not permit the sort of color-sensing vision possible with the cones in the central or foveal region of the retina, they have the useful property of responding to very low illuminance, particularly after dark adaptation is complete. Nighttime vision is performed primarily with the rods.

PARAKEETS.   See **Parrots and Cockatoos**.

PARALLAX (Astronomy).   Parallax effects are very important in astronomy. In transferring from one system of coordinates to another, parallax effects must always be applied when the location of the origin changes. For example, observations are taken from the surface of the earth, and geocentric parallax must be applied when transferring the observations to the center of the earth. Distances of many celestial objects are expressed in terms of parallactic angles. The angle subtended by an equatorial radius of the earth at the distance of the sun is known as solar parallax. The angle subtended by the radius of the earth at the distance of any other member of the solar system is known as the *horizontal parallax* of that object. In measuring the distances of the stars, the parallactic shift due to the revolution of the earth about the sun is employed; the angle subtended by one astronomical unit at the distance of the star is known as the stellar parallax of the object. As the characteristics of solar motion become more completely known, parallactic shifts, called secular parallax, due to this motion will undoubtedly be used for the determination of distances.

Stellar distances are given by

$$D = \frac{1}{\pi}$$

where π is the parallactic shift in seconds of arc and D is in parsecs.

See also **Dynamical Parallax**; **Secular Parallax**; **Solar Parallax**; **Spectroscopic Parallax**; and **Stellar Parallax**.

PARALLAX (Instrument).   As an observer moves about, the relative positions of distant objects seem to change. This apparent change, actually due to change of position of the observer, is technically known as parallactic shift. The amount of parallactic shift is inversely proportional to the distance of the object. In taking readings of instruments using scales and pointers, care must be taken to insure that the eye of the observer and the pointer are both in a line perpendicular to the plane of the scale. To facilitate this, some instruments have a mirror in the plane of the scale.

PARALLAX ERROR.   The error in measurement between two pairs of antenna caused by the fact that the center of the two baselines do not coincide. This error is a function of the distance of the target from the baseline, as well as its relative direction.

PARALLEL A/D CONVERTER.   This is the simplest design among the electronic analog-to-digital converters. A comparator is provided for each quantization level in an n-bit converter. See also **Quantization**. A digital representation of the input signal is obtained by appropriately decoding the output of the multiple comparators. Reference is made to the example and schematic diagram in Fig. 1. A 2-bit converter is shown. Comparators 1, 2, and 3 are biased with reference voltages of 0.75 V_r, 0.5 V_r, and 0.25 V_r. V_r is the full-scale reference voltage and is equal to the full-scale input range of the A/D converter. All the comparators that are biased at a level less than the level of the input signal provide a binary 1 output when the input signal is applied. In the example shown, an input signal of 0.6 V_r will result in a 1 output from comparators 2 and 3. The output of comparator 1 will be a 0. Through appropriate decoding, the binary digital representation of 1 0 is yielded. The binary state at the input and output of each logic block is included in the diagram.

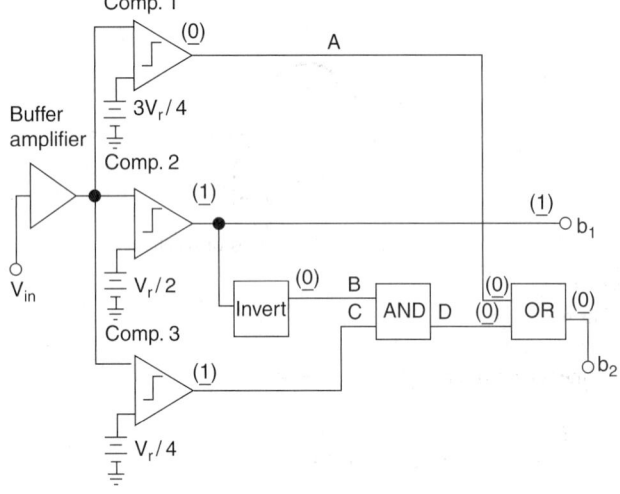

Input signal	A	B	C	D		b_1	b_2
$3V_r/4 < V_{in}$	1	0	1	0		1	1
$V_r/2 < V_{ir} < 3V_r/4$	0	0	1	0		1	0
$V_r/4 < V_{in} < V_r/2$	0	1	1	1		0	1
$V_i < V_r/4$	0	1	0	0		0	0

Fig. 1.   Parallel analog-to-digital converter.

Even though the example provides only 2-bit resolution, the addition of more comparators and decoding logic can increase the resolution. However, each additional bit of resolution requires a doubling of the necessary hardware. Consequently, the parallel A/D converter normally is used only for converters of fewer than 5 or 6 bits resolution. Not only must the hardware be increased, but increasing the resolution makes marked demands on the threshold stability of the comparators in the interest of avoiding output ambiguity and excessive errors.

Simplicity for low-resolution requirements and high speed are the principal advantages of the parallel A/D converter. The settling times of the comparators and logic delays are the main factors that determine the speed.

THOMAS J. HARRISON, IBM Corporation, Boca Raton, FL.

PARALLELEPIPED. A polyhedron with parallelograms as faces, thus, a six-sided prism with parallelograms as bases. The general case is an oblique parallelepiped where the lateral edges are not perpendicular to the bases. If the edges are perpendicular, it is called right; if the six faces are rectangles, rectangular; if the six faces are all squares, a cube. Sometimes the figure is also called a parallelepipedon and, incorrectly, parallelopiped or parallelopipedon.

The volume of a parallelepiped is conveniently written in vector analysis as the scalar triple product **[ABC]** where **A, B, C** are three vectors describing the edges of the parallelepiped. Its lateral area and volume can also be given by the same equations used for a prism.

PARALLELOGRAM. A quadrilateral with opposite sides parallel and equal in length. Its consecutive angles are complementary. If a, b are the lengths of its sides and C is the acute angle between them, the area of a parallelogram is $A = ab \sin C = bh$, where h is the distance between the bases, b. Its area is also given by $2A = d_1 d_2 \sin D$, where d_1, d_2 are the lengths of the two diagonals and D is the acute included angle. A diagonal divides the parallelogram into congruent triangles, the two diagonals bisect each other, and their point of intersection is the center of gravity of the figure.

If two vectors are drawn from the same origin and the same vectors are used to construct a parallelogram, the sum of the vectors equals the length of the diagonal drawn from the origin. This is the well-known parallelogram law of elementary physics, which describes the combination of forces, acceleration, velocity, etc.

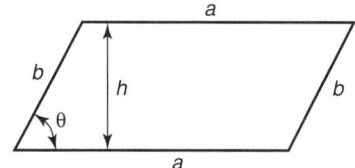

Fig. 1.   Principal parameters of parallelogram.

With reference to Fig. 1, area $A = ah = ab \sin \theta$, where a and b are the lengths of the sides of the parallelogram, h is the height, and θ is the angle between the sides. In a rhombus, which is an equilateral parallelogram, $A = 1ab$, where a and b are the lengths of the diagonals.

See also **Vector**.

PARALLEL SAILING. A term used to designate the problem of converting the distance traversed by a ship along a parallel of latitude into difference of longitude. This problem is solved under **Departure (Navigation)**.

PARALLEL-SERIAL A/D CONVERTER. This type of analog-to-digital converter is somewhat similar to the successive-approximation A/D converter because the conversion is accomplished in the form of a series of distinct steps. The speeds of parallel-serial A/D converters are not as high as those of parallel A/D converters, but practical attainment of resolution is considerably better, that is, 12- to 14-bit resolution at speeds over 100,000 samples/second.

A parallel-serial A/D converter is shown schematically in Fig. 1. Here the A/D converter is comprised of four 3-bit D/A converters (digital-to-analog) and four amplifiers. Each has a gain of 8 which corresponds to the 3-bit resolution of each section of the A/D converter, that is, $2^3 = 8$. A 3-bit conversion is accomplished by the seven comparators shown and in a fashion similar to that employed in parallel A/D converters. See also **Parallel A/D Converter**. The voltage that corresponds to the 3-bit conversion is subtracted from the input signal by digital-to-analog converter #1 (DAC 1). This difference signal then is amplified by a factor of 8. The next three bits are then determined in a similar manner in the next stage of the A/D converter. Inasmuch as three bits are determined simultaneously (nearly), five steps yield a 15-bit quantization of the input signal. To accomplish this in a successive-approximation A/D converter requires a 15-step procedure.

With reference to Fig. 1, assume an input voltage of 5.163086 V. Also assume a full-scale range of the A/D converter of 8.0 V. The operation

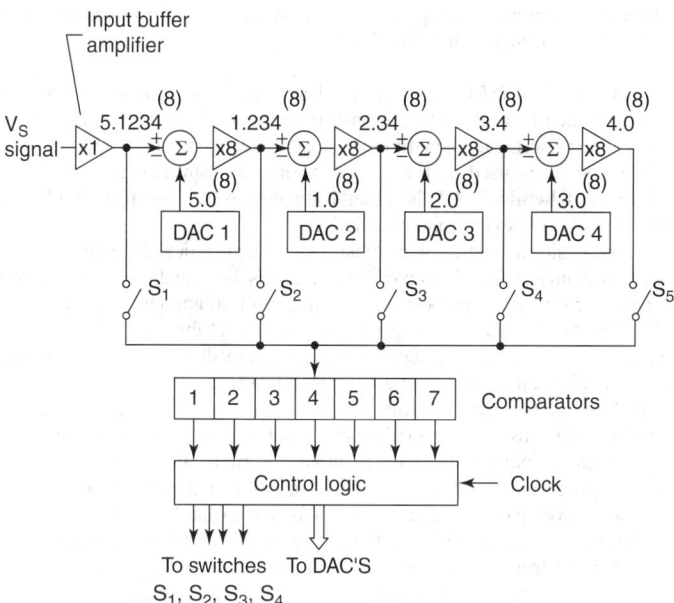

Fig. 1.   Parallel-serial analog-to-digital converter.

of this A/D converter configuration is best explained in terms of octal numbers. Thus, the input voltage is 5.1234^8. The superscript 8 shows that the number is expressed in octal code. The conversion process commences with the closing of switch S_1 and applying the input signal to the parallel comparators. The output of the comparators shows that the input is in excess of 5^8 but is less than 6^8. Hence, DAC 1 is set to 5^8 and a voltage of 0.1234^8 is applied to the first gain of 8 amplifier. By opening S_1 and closing S_2, the output of the amplifier, 1.234^8 V is applied to the parallel comparators. This parallel conversion yields the determination that the amplified difference voltage is in excess of 1^8 but less than 2^8 V. Hence, DAC 2 is set to 1.0^8 V to furnish a difference signal of 0.234^8 V. Other steps follow which involve the closing of switches S_3, S_4, and S_5, and these actions result in the octal result of 5.1234^8, or the 15-bit binary result 101001010011100.

In the device shown in Fig. 1, it is required that each stage of the converter settle completely before the subsequent 3-bit conversion is commenced. Further circuitry is added for error correction so that if a particular stage has not completely settled, with a resultant incorrect 3-bit conversion, the conversion in the following stage will provide a signal to correct the prior 3-bit representation.

Although the parallel-serial A/D converter has relatively high-speed conversion, the design requires a greater hardware complexity than that of a successive-approximation A/D converter with equal resolution. The parallel-serial A/D converter is not as fast as the parallel A/D converter, but it requires considerably less hardware for resolutions of more than 6 bits. Further, precise comparator adjustments, required in a high-resolution parallel A/D converter, are avoided with the parallel-serial A/D converter. See also **Analog-to-Digital Converter**.

THOMAS J. HARRISON, IBM Corporation, Boca Raton, FL.

PARALLEL SURFACES. The surface S' is said to be parallel to the surface S if the distance measured from S to S' along the normal to S at a point P is independent of the position of P on S. If P' is the point at which the normal to S at P intersects S', then the normals to S and S' at P and P', respectively, have the same directions. If S' is parallel to S, then S is parallel to S'.

PARALLEL TRANSMISSION. The system of information transmission in which the characters of a word are transmitted (usually simultaneously) over separate lines, as contrasted to serial transmission.

PARALYSIS. Loss of motor activity in some part of the body. Paralysis occurs when the nerve impulse has met with interference. This may occur in the brain centers, the spinal-cord centers or pathways, nerve trunks or individual nerves, or the nerve endings in the muscles. It may be temporary or permanent, depending on the cause. The cause may be due to a specific disease, to chemical or bacterial poisons, toxins, or injury to nervous tissue. See also **Hemiplegia**; and **Quadriplegia**.

PARAMAGNETISM. A physical characteristic of some matter, advantage of which is taken in certain instrumental systems and scientific apparatus. The paramagnetic qualities of oxygen, due to two unpaired electrons per molecule, is used as the basis for some oxygen analyzers. See also **Analysis (Chemical)**. Of the other common gases, only nitric oxide and nitrogen dioxide exhibit paramagnetism.

Three basic magnetic forces exist: (1) paramagnetic, (2) ferromagnetic, and (3) diamagnetic. Ferromagnetic materials are much more permeable than a vacuum and thus positive, aligning with an applied magnetic field. Diamagnetic materials are slightly less permeable, thus negative, aligning across the field. Para- and ferromagnetism diminish with temperature rise, whereas diamagnetism essentially is unaffected.

Precession of electron orbits in atoms and molecules induced by an applied field causes diamagnetism in all matter even when paramagnetism dominates. Unpaired electrons in atoms or molecules cause para- and ferromagnetism. Normally, electrons of opposite spin pair off, netting zero magnetism per pair. Unbalanced spin moment of unpaired electrons yield both para- and ferromagnetism. Paramagnetic matter has unpaired electrons in outer electron shells. Thermal agitation retards atomic or molecular alignment, and thus the net moment is weak. Ferromagnetic matter, such as iron, cobalt, and nickel, has unpaired electrons in the next-to-outer shell. Interatomic forces cause molecular alignment and permanent magnetism results. See also **Magnetism**.

PARAMETER. An arbitrary constant, as distinguished from a fixed or absolute constant. Any desired numerical value, subject in some cases to certain restrictions. Thus, in the set of equations $(x - a)^2 + (y - b)^2 = r^2$ for circles in the plane, there are three parameters, two describing the position of the center and one giving the radius, and for each choice of a, b, and r a corresponding circle is obtained. Similarly, the equations of a curve $x = \phi(t)$, $y = \psi(t)$ is said to be in parametric form and each choice for the parameter t produces a corresponding point on the curve.

Parameterization is the representation, in a dynamic model, of physical effects in terms of admittedly oversimplified *parameters*, rather than realistically requiring such effects to be consequences of the dynamics of the system.

PARAMETER (Statistics). An unknown constant appearing in the specification of a probability distribution, or in the specification of a model, e.g., the constants in a regression equation. The domain of permissible variation of the parameters defines the class of distribution or models of given form under consideration.

PARANA PINE. See **Araucarias**.

PARAQUE. See **Nightjars and Nighthawks**.

PARASITE. An organism living in close association with another organism, deriving its nourishment from the host, and harming the host organism in the process.

PARASITIC PLANTS. Normally, when one considers injuries to plants and trees by parasites, the causal factors that immediately come to mind are fungi, viruses, and certain kinds of insects and other animal pests. Some plants, known as *phanerogamic parasites*, also can be economically important in some regions on some valuable trees and crop plants. These are plants, such as the mistletoes, Spanish moss, and dodder, among others, which attach themselves to other plants and grow on them depriving the host plant of nutrients, light, and moisture—in essence inhibiting normal growth of the host plants through processes of deprivation and strangulation.

Parasitic plants that strangle the host include a number of climbing vines, of which bittersweet (*Celastrus scandens L.*) is an example. Parasitic plants which in essence suck or consume vital juices of a host plant include the mistletoes, whose sticky seed becomes attached to the bark of the host, germinating and thus utilizing the host as its growth medium and depriving the host of space. Most tree branches so "infected" ultimately expire. The dwarf mistletoes (*Arceuthobium*) operate in a similar fashion against the western yellow pine tree. An interesting description of the mistletoe is given by J.M. Haller (*American Forests*, 11–15, December 1978).

The damage wrought by the foregoing parasitic plants is generally confined to large trees used for shade or timber. Food crops are most seriously affected by the dodders (*Cuscuta* spp.), which can seriously injure such field crops as alfalfa, clover, lespedeza, onion, and sugar beet. The parasitic dodder vine does not simply entwine an affected host to deprive it of soil nutrition, available moisture, and light, but like the mistletoes, the dodder actually lives off the juices of the host plant. Common dodder (*C. gronovii Willd.*) and field dodder (*C. pentagona Engelm.*) will infest a variety of hosts. There are also specializing dodders, such as alfalfa, flax, and clover dodders. The most effective control is that of eliminating dodder seed from crop seed through use of special processing machinery.

PARASITISM. See **Symbiosis (Ecology)**.

PARA-STATE. 1. In diatomic molecules, such as hydrogen molecules, the para-state exists when the spin vectors of the two atomic nuclei are in opposite (i.e., antiparallel) directions, forming a singlet state, $S = 0$, whereas the ortho-state is the one in which the nuclei are spinning in the same direction.

2. In helium, the para-state is one group or system of terms in the spectrum of helium that is due to atoms in which the spin of the two electrons are opposing each other. Another group of spectral terms, the orthohelium terms, is given by those helium atoms whose two electrons have parallel spins. Because of the Pauli Exclusion Principle, a helium atom in its ground state must be in a para-state.

PARATHYROID GLANDS. Four glands with a combined mass scarcely greater than that of a large pea produce parathyroid hormone (PTH) which is one of three main substances required to control the amount of calcium in the extracellular fluid and in the bone marrow. The parathyroid glands are adherent to and sometimes embedded within the back part of the thyroid gland. All of these structures lie in the lower front portion of the neck.

The influence of the parathyroid glands on the regulation of calcium concentrations in the blood of mammals was first recognized by MacCullum and Voegtlin, who reported in 1909 that removal of the glands produced a marked drop in serum calcium levels—to the point of causing tetany, which is described later. That the glands also affect phosphate was demonstrated in 1911 by Greenwald, who produced a rapid fall in urinary phosphate by parathyroidectomizing dogs. In 1924 and 1925, Hanson and Collip independently prepared physiologically active extracts of beef parathyroid. Similarly prepared hormone substances are available today for treating acute hypoparathyroidism, but because patients soon become refractory to injections of the substance, long-continued use of the extract is no longer indicated. During the intervening 75 years, a great deal has been learned concerning calcium and phosphorus metabolism and of the role of the parathyroid glands. Current, quite successful therapies for parathyroid disorders were made possible only through an improved understanding of the endocrinological processes involved.

The second of the three main substances involved in calcium metabolism is *vitamin D*. This major component of the endocrine system regulates the metabolism of bone minerals. The synthesis of active vitamin D commences with ultraviolet radiation of 7-dehydrocholesterol in the skin, producing previtamin D_3. This is converted to vitamin D_3 (cholecalciferol). See also **Vitamin D**. The last of the three substances is *calcitonin*, a polypeptide which at one time was believed to be secreted by the parathyroid glands, but now known to be produced by the parafollicular cells of the thyroid gland.

As with the other endocrine glands (see also **Endocrine System**), disorders mainly develop as the result of over- or underproduction of hormones secreted.

Primary Hyperparathyroidism. In the great majority of patients, excessive PTH is produced as the result of a single adenoma (tumor) of the gland. Primary hyperparathyroidism also may originate from hyperplasia (abnormal increase in volume caused by growth of new cells), or carcinoma. Cancer of the parathyroid is a rare cause and only occurs in 3–5% of cases. Because primary hyperparathyroidism is increasingly diagnosed in an early stage, the more severe symptoms, such as renal stones, osteitis fibrosa cystica, anemia, constipation, and renal failure are not usually presented to the physician. Rather, the patient complains of generalized weakness, which may be accompanied by weight loss, pallor, and keratopathy (deterioration of the cornea). Definite diagnosis requires laboratory measurement of the serum calcium level. Significant bone changes may be revealed by radiologic examination. Diagnosis is extremely important because, if this is positive, correction of a serious disorder almost always involves surgery. Where there is only one adenoma, this will be removed and the remainder of the glands will be left intact. Some surgeons prefer that all four glands be exposed prior to making a decision. Preoperatively, the patient will be given large dosages of 1α-hydroxyvitamin D$_3$, during which time serum calcium levels are watched carefully. Postoperatively, the patient must be monitored carefully for any appearance of hypoparathyroid tetany. Usually with those patients who have a single adenoma, with a normal second gland, there will be a successful return to normal calcium levels.

In *ectopic hyperparathyroidism*, the symptoms may parallel those of primary hyperparathyroidism, but the source of PTH may arise from a malignancy. Some researchers have reported abnormal PTH in many cancer patients, regardless of the type of tumor or the presence or absence of metastases.

Hypercalcemia. The serum calcium level is regulated by gastrointestinal absorption of calcium, oseteoclastic resorption of mineral from bone stores (see also **Bone**), and tubular reabsorption of calcium from the glomerular filtrate (see also **Kidney and Urinary Tract**). The treatment of hypercalcemia is determined by the underlying cause. Drugs used include prednisone, mithramycin, and calcitonin.

Hypoparathyroidism. This condition may arise from a deficiency in the effective production of PTH by the parathyroid glands; or by a failure of tissues to react to it. Rarely, the condition may occur in newborn infants as a result of hemorrhage that partially destroys the parathyroid gland. It is most common in adults who have undergone surgical removal of the parathyroid tissues as an occasional consequence of removal of diseased thyroid tissue. The most common symptom is a tightening and a spasm of the muscles, most evident in the position assumed by the fingers and toes. An inability to straighten the fingers and toes is referred to as *carpopedal spasm*. There is usually considerable apprehension in the form of a sense of impending doom. There may be difficulty in inhalation, spasm of muscles in the larynx, vomiting, and abdominal pain. This is *tetany*. In persons who develop the disease more gradually, the symptoms are less severe. Treatment of the hypocalcemia resulting from hypoparathyroidism is by a combination of calcium and vitamin D therapy.

The reader who has a scholarly interest in more details concerning the biochemistry of parathyroid hormone and its possible relation to the hypercalcemia of cancer is referred to the references listed.

Additional Reading

Bilezikian, J.P.: "Parathyroid Hormone-Related Peptide in Sickness and in Health," *N. Eng. J. Med.*, 1151 (April 19, 1990).

Bilezikian, J.P., R. Marcus, and M. Levine: "The Parathyroids," 2nd Edition, Academic Press, Inc., San Diego, CA, 2001.

Eisenberg, B.: "Imaging of the Thyroid and Parathyroid Glands: A Practical Guide," Churchill Livingstone, Inc., Philadelphia, PA, 1997.

Epstein, K. et al.: "Transient Hypoparathyroidism During Acute Alcohol Intoxication," *N. Eng. J. Med.*, 721 (March 14, 1991).

Livolsi, V.A. and R.A. DeLellis: "Pathology of the Parathyroid and Thyroid Glands," Lippincott Williams & Wilkins, Philadelphia, PA, 1993.

Marcus, R., M.A. Levine, and J.P. Bilezikian: "The Parathyroids: Basic and Clinical Concepts," Lippincott-Raven, Philadelphia, PA, 1994.

Nussbaum, S.R., R.D. Gaz, and A. Arnold: "Hypercalcemia and Ectopic Secretion of Parathyroid Hormone By An Ovarian Carcinoma with Rearrangement of the Gene for Parathyroid Hormone," *N. Eng. J. Med.*, 1324 (November 8, 1990).

PARESTHESIA. Any abnormal sensation on the surface of the body. It may be described by the patient as burning, tickling, itching, pricking, etc. Paresthesias occur in nervous-system diseases involving the spinal cord, in pernicious anemia, and polyneuritis.

PARKINSON'S DISEASE. Parkinson's disease (PD) is a prototypal neurodegenerative disease—having a rich history of description, etiology, treatments, pathology, and neurochemistry. However, despite its classical clinical picture, its meaning and perception have varied for more than one hundred and eighty years.

Parkinson's disease, was recognized as a syndrome of tremor, stooped posture and shuffling gait by Dr. James Parkinson in 1817. The most prominent pathological characteristic, loss of neural pigments seen both grossly and by histopathology of the *substantia nigra* (SN) was noted in 1921 following viral (influenza) post-encephalitic Parkinsonism. Progress

in assessing degeneration in other pigmented brainstem nuclei, the identity of neurotransmitters, and the loss of dopamine (DA) and serotonin (5-HT) were recognized following the introduction of Rauwolfia alkaloids from India as antihypertensives in the 1950's and subsequent development of parkinsonism due to the neurochemical depletion as a side-effect of such treatment. Early surgical therapeutic attempts were nonspecific and included destruction of the spinal motor pathways, resulting in paralysis, for tremor reduction. Pallidotomy and thalamotomy (vide infra) were performed following the introduction of human stereotactic instruments and brain maps in the 1950's; these early semi-empiric procedures were effective in eliminating hyperkinetic symptoms such as tremor and rigidity, but were rapidly supplanted by the inadvertent introduction of dopa. Cotzias in 1967 reported three of sixteen patients who benefited from dopa supplementation. This however, was in an effort to reconstitute neuromelanins by its precursors of phenylalanine and then dopa, not in order to replace DA. This first trial resulted in the known side effects of dopa therapy such as induction of hallucinosis, choreo-athetotic "dyskinesias", and sleep disturbances and the effects of impure preparations such as fever, seizures and rashes in thirteen of the sixteen patients. Despite these distressing results, two patients showed dramatic improvement and interestingly had undergone pallidotomy by Cooper prior to the drug trial, presaging the synergism of pallidotomy and medical (dopaminergic, serotonergic and antiglutaminergic) combined treatments, (vide infra).

This new operative definition of Parkinson's disease based on response of early symptoms to dopaminergic medications did much to initiate circular reasoning regarding dopamine treatment and research between medical and scientific investigations of Parkinson's disease for the twenty years of 1967 to 1987.

Several factors were responsible for the resolution of the dopaminergic loss definition of Parkinson's disease during the late 1980's. Of primary importance was the failure of *L-dopa* replacement therapy and novel dopaminergics to affect the progression of the clinical or pathological disease. Fortuitously, a street-drug cogener of meperidine, MPTP was identified as an oxidative toxin, which resulted in rapid onset Parkinsonism in young addicts in San Francisco. The employment of this substance, which becomes an active toxin MPP+ following the action of monoamine oxidase B (found primarily in serotonergic and dopaminergic neurons) in primate models, lead to new understanding of brainstem glutaminergic influences of primary import in the akinetic symptoms of supervening levodopa failure in advanced Parkinson's Disease. Further interest was generated for toxic and oxidative (metabolic) stress etiologies and a broader focus on the underlying disease process. Lackluster mechanistic attempts to "replace dopamine" by structural surgical methods such as adrenal medullary chromaffin cell brain auto- grafts and fetal mesencephalic isolated dopamine neuron allografts (1984–1992) were quickly supplanted by renewed interest in conventional stereotactics directing lesions at targets identified by the new MPTP model such as the disinhibited and putatively hyperactive subthalamic nucleus, (STN). The STN's overamplified target, the posterior ventromedial portion of the globus pallidus interna (PVP/GPi) has since been the target of selective lesioning or disruptive high-frequency electrical stimulation using implanted electrodes (vide infra), with excellent therapeutic results.

Further characterization of metabolic deficits such as impaired activity of complex I—Coenzyme Q$_{10}$ NAD, cytochrome oxidase in mitochondria (with its maternal pattern of inheritance), the role of serotonergic losses in levodopa failure and in Parkinson's dementia, and clinical subtypes, were also undertaken. Current endeavors are aimed at enhancement of (dopaminergic) neuronal survival using trophic factors, e.g. glial-derived neurotropic factor GDNF, and genetic engineering for dopamine DA, serotonin 5HT and GDNF augmentation using viral vectors, transformed cell lines and neural stem cells.

Pathology

The pathological hallmarks of PD are depigmentation of the substantia nigra and other pigmented brainstem nuclei (e.g. locus coeruleus), progressive loss of dopaminergic neurons and subsequent loss of the other monoaminergic cells (i.e. serotonin and norepinephrine). The loss of dopaminergic neurons is variable and is not related to the severity of disease for the post-encephalitic form is associated with 95% DA loss or greater and is a mild clinical syndrome characterized by slow progression to akinesia; contrariwise classical idiopathic PD may demonstrate 50% to 90% DA cell loss with poor correlation to more progressive akinesia, although 80% DA cell loss is often cited as a threshold sufficient to induce the clinical presentation of PD. These DA-independent findings are exemplified by a recent animal model of low dose MPTP with the primate subjects showing up to 99% DA cell loss in the SN with 97% DA unavailability in the striatal targets, yet no signs or symptoms of parkinsonism. The explanation for this is centered on the noted overabundance of serotonin and it's compensatory role in this instance, possibly mimicking the post-encephalitic subtype of PD.

The pathognomic eosinophilic cytoplasmic inclusion called *Lewy body*, located primarily in pigmented neurons, is found in classical types of PD. This characteristic histopathological finding is absent in the post-encephalitic form and the MPTP model, and MPTP exposed human Parkinson patients, indicating the model as a milder or incomplete expression of idiopathic PD. In so-called *Lewy Body Dementia* (LBD) they are found in cortical and subcortical locations, their density possibly related to the aggressiveness or speed of disease progression as they must be considered transient. In Parkinson's Dementia Complex (PDD) it is the density of the Lewy bodies (LBD) found in the brain stem, which is the only known correlation to the severity of the dementia. In amyotrophic lateral sclerosis (ALS- also called motor neuron disease), if associated with PDD, Lewy bodies will be found in the cervical anterior horn cells. The distribution of Lewy bodies in all of these instances is highly consistent with the distribution of another marker, the neurofibrillary tangle (NFT). This product of cytoskeleton degeneration is common in neurodegenerative disease including post-encephalitic P.D., Alzheimer's dementia (ALZ), schizophrenia, and pugilism. Despite their apparent differing etiologies, the distribution of the NFTs which overlaps the LB's in PDD respects identical nuclei and cell fields consisting of phylogenetically old "reticular areas" and their projections, including hypothalamic and limbic areas. Moreover, the distribution of lesions in Wernicke's encephalopathy is an inclusive subset of this distribution.

Neurochemistry

The brainstem areas involved by LB's and NFT's, which share a common denominator in neurodegenerative diseases are marked by the tri-localization of melatonin (MLT), beta-endorphin (B-END), and glial derived neurotropic factor (GDNF) cell surface receptors. This commonality alludes to the protective role of intrinsic opioids (B-END), the importance of tryptophan-serotonin-melatonin metabolism (and its dual supportive physiological role with opioids), and an understandable dependence on trophic protective factors such as GDNF.

Monoamine loss in PD follows the pattern of early DA and 5-HT loss in most idiopathic cases. Prior to the introduction of L-dopa, cerebrospinal fluid (CSF) studies of patients with idiopathic disease showed an overall 60% to 70% DA loss and 40% to 55% reduction in serotonin. In the 15% of PD patients without tremor (Type B, vide infra) with predominant akinetic, gait and postural instability problems, the primary breakdown product of serotonin, 5-hydroxyindoleacetic acid (5-HIAA), is found in half the concentration as the classic tremor type idiopathic PD, Type A.

In Parkinson's disease progressive loss of 5-HIAA is correlated with advanced akinetic symptoms and DA medication failure. Of particular impact is the excitatory glutaminergic activity in advanced states which has an adverse effect on neuronal oxidative stress, possible white-matter degeneration and is ultimately responsible for the akinetic state (vide infra: Pathophysiology).

Relevant to the issue of neurochemistry in PD is a discussion of the biochemistry of metabolic oxidative stress in PD. As part of normal metabolism up to 3% of oxygen can be converted to reactive oxygen species (ROS) such as hydrogen peroxide or the hydroxyl radical. In neurodegenerative diseases, ROS are either raised beyond capacity of protective mechanisms such as superoxide dismutase (SOD), glutathione (GSH), or MLT or these latter systems may fail. In the former case, the first cytochrome oxidase reaction, complex I, is known to be poorly functioning in the mitochondrial respiratory chain of patients with idiopathic PD. The body's inability to mitigate ROS damage is implicated in cancer, aging and neurodegenerative disease.

Epidemiology

It is estimated that one million North Americans are affected by Parkinson's disease, with the incidence increasing with age and it is more common in males. However, with the demographics of aging, pollution, and stress, PD incidence as well as prevalence is felt to be increasing. One population study in Massachusetts (non-patients) found that greater than 15%

of those over 75 years old showed two or more symptoms of PD such as shuffling gait, poor facial expression, poverty of movement, stooped posture, etc., but not tremor. The racial distribution of PD is skewed being most common among fair Caucasians, less common in Asians and Latinos, and least common among Blacks, being rare in unmiscegenated Africans.

All links to central nervous system (CNS) damage including head trauma, high fever, encephalitis, nutritional or neurochemical depletion states (e.g. pellagra), toxic exposure (e.g. herbicides, insecticides), MPTP and other street drugs of abuse which damage monoaminergic systems (e.g. methamphetamine), and even emotional or 'silent trauma' (e.g. post traumatic stress disorder or prolonged untreated depression) have been linked as risk factors for PD. Clinically, such stressors as elective operative surgical interventions (e.g. coronary artery bypass grafts), motor vehicle accidents, death of a primary relative, and even career or financial tragedies are commonly noted as a date of onset of symptoms. Onset before age 45 is considered "young onset" or Narabayashi's "juvenile type" and has a stereo-typed presentation, course, and response to treatment, including induction of bilateral uncoordinated spontaneous movements referred to as dyskinesias.

Clinical Expression, Symptoms, Course and Prognosis

The premonitory and earliest symptoms and signs of PD include loss of sense of smell (anosmia), decreased eye blink rate and facial expression, monotone soft speech, stooped posture, a poverty of spontaneous movements, a feeling of chronic fatigue and then often a resting, "pill-rolling" tremor of the thumb and index finger. Sleep disturbance, a passive personality, and small handwriting (micrographia) with progressively decreasing amplitude of the letters and words in a sentence are characteristic. In the akinetic or Type B expression of the disease there is no tremor but antecedent stress (usually unrecognized) and depression are constant.

Most patients show asymmetric appearance and progression of appendicular hyperkinesias (i.e. tremor, cogwheel rigidity of the arms), but as the symptoms and disease progress, akinesia supervenes and the symptomatology becomes more axial evidenced by stooped posture, balance problems (postural instability), smaller, shuffling steps and such problems as being unable to roll over in bed. Many of the signs can be viewed as motor system loss of control over gravity which implicates the gamma motor system of importance in the pathophysiology of the disease. Undermedicated or under treated, the course of the disease can lead to death within 10 years, even in recently reported studies. In classic type, appropriately treated patients currently experience 10 to 15 years of quality functioning, whereas tremor dominant unilateral symptoms imply 20 or more years of functionality. In all cases, stereotactic surgical interventions may mitigate symptoms and disease for an additional 3 to 10 years. Patients without tremor at onset have a more aggressive progression of disabilities due to akinesia, with only 8 years of independent quality of life, on average. Estimates of from 20 to 80% are given for the occurrence of depression in PD, exceeding the later figure in advanced and akinetic patients. Depression is rare in highly asymmetrically affected patients such as tremor dominant type (representing 5% of cases) and in the young-onset type. Other non-motor signs and symptoms such as autonomic instability (e.g. episodes of profuse sweating, fluctuating blood pressure) also occur based upon poor hypothalamic modulation, possibly related to serotonin deficiency.

Dementia commonly supervenes in elderly, akinetic patients and is of the frontal type. Neurologic exam reveals frontal lobe release signs. Anosmia and a glabellar reflex are constant, and their absence challenges the diagnosis of PD. A snout reflex, palmomental reflex, and the inability to alternatively simultaneously open and close opposite hands confirm the clinical impression of frontal dysfunction. Neurocognitive testing (i.e. using the Wisconsin Card Sort or Stroop Test) will confirm the diagnoses as needed.

Frequently, the differential diagnoses of central hypothyroidism, vitamin B_{12} deficiency, and serotonin deficiency, will be noted as a triad, which when aggressively treated will result in dramatic improvements of these reversible forms of dementia often obfuscated by the constellation of co-morbid Parkinson's symptoms. Other differentials in this clinical scenario include so-called normal pressure hydrocephalus (NPH) with its triad of urinary incontinence, dementia, and gait problems, which can exist concurrently and confound the diagnosis whilst contributing to the parkinsonism. Parkinson's Plus Syndromes (PD Plus), which represent gross structural degenerations and atrophy of widespread CNS systems (beyond the limited pigmented, monoaminergic cell losses in PD) may appear to mimic PD in the first 2 to 5 years of recognition. However, the occurrence of intact sense of smell, poor response to pharmacologics, predominant axial symptoms, paucity of tremor, nuchal rigidity, or the inability to walk within less than 5 years of symptoms all weigh against PD and toward so called PD Plus, for which there is no satisfactory treatment at present.

Treatment

Treatment of PD includes pharmacologic replacement or substitution for losses of neurotransmitters (i.e. L-dopa or ergoloid-derived dopaminergic agonists for DA, selective serotonin re-uptake inhibitors (SSRIs) such as Prozac and B vitamin supplementation for 5-HT, and e.g. l-threo DOPS for norepinephrine) are employed. Other tactics are aimed at reducing glutaminergic influences using amantadine and/or magnesium. Alternative rational strategies focus on the use of antioxidants, such as lipoic acid, green tea, selenium, pycnogenol, and/or a vegetarian diet replete in these nutrients and minerals.

The clinical pharmacology of these treatments requires acumen and patient education. Many patients on L-dopa/carbidopa preparations may remain under-medicated as constipation or amino acid competition (especially dairy protein) block intestinal absorption and also active transport across the blood brain barrier. In decade-long treatment programs, DA therapy may be met with failure, partly due to the second hit of monoamine serotonergic loss or co-morbid conditions following a sequential pathophysiology, viz. serotonin loss reduces thyrotropin releasing hormone activity from the hypothalamus; resulting triiodothyronine deficiency blunts monoamine receptor sensitivity and also leads to gastric mucosal/intrinsic factor losses which result in vitamin B_{12} deficiency and increased oxidative stress and symptomatology.

Pathophysiology

Following losses of monoaminergic function in PD, the extrapyramidal motor system of the basal ganglia and brain stem become dysregulated.

Loss of DA inhibition to the putamen results in demonstrated neuronal hyperactivity of the globus pallidus (GPi) interna and further amplification from the disinhibited subthalamic nucleus. The inhibitory gabanergic pallidal outflow via the tegmental bundle of the ansa lenticularis restricts the activity of the midbrain locomotor center and the tegmental pedunculopontine nucleus (PPN). The resulting brainstem and reticulospinal inhibition results in axial akinetic symptoms, gait problems, postural instability, stooped posture, and impairment of anti-gravity activities. Projections from the GPi and brainstem to the thalamus and cortex are responsible for appendicular hyperkinesias such as tremor. See Fig. 1. Because of this mechanism of akinesia associated with an imbalanced neurotransmitter and hyperglutaminergic state, (glutamate being an neuronal excitatory transmitter and so-called "excitotoxin") then increased oxidative stress, the progression of the disease may follow the poorly treated or undermedicated clinical state.

Stereotactic Surgical Interventions

Beginning in 1948–55, Spiegel and Wycis, Narabayashi, Nashold, Cooper and others had begun to use precise instruments and brain maps allowing them free access with probes to therapeutically explore deep areas of the human brain. These techniques became widely used in the 1960's following Cooper's serendipitous discovery of the ventral lateral thalamus as an improved target for tremor. These techniques were all but forgotten following the widespread introduction of L-dopa in 1967–70. The advent of computerized CT and MR imaging along with advancing disease in a population of PD patients with so-called *levodopa failure syndrome* spurred new stereotactic therapeutic interventions including fetal brain allografts and adrenal autografts. During this era of increased interest in stereotactics and PD the utility of Leksell's posteroventral pallidotomy technique was rediscovered by Laitinen and Iacono. The ability of a conventional radiofrequency (heat) lesion within the pallidum to eliminate or improve both positive (e.g. tremor, dyskinesia) and negative (akinetic)

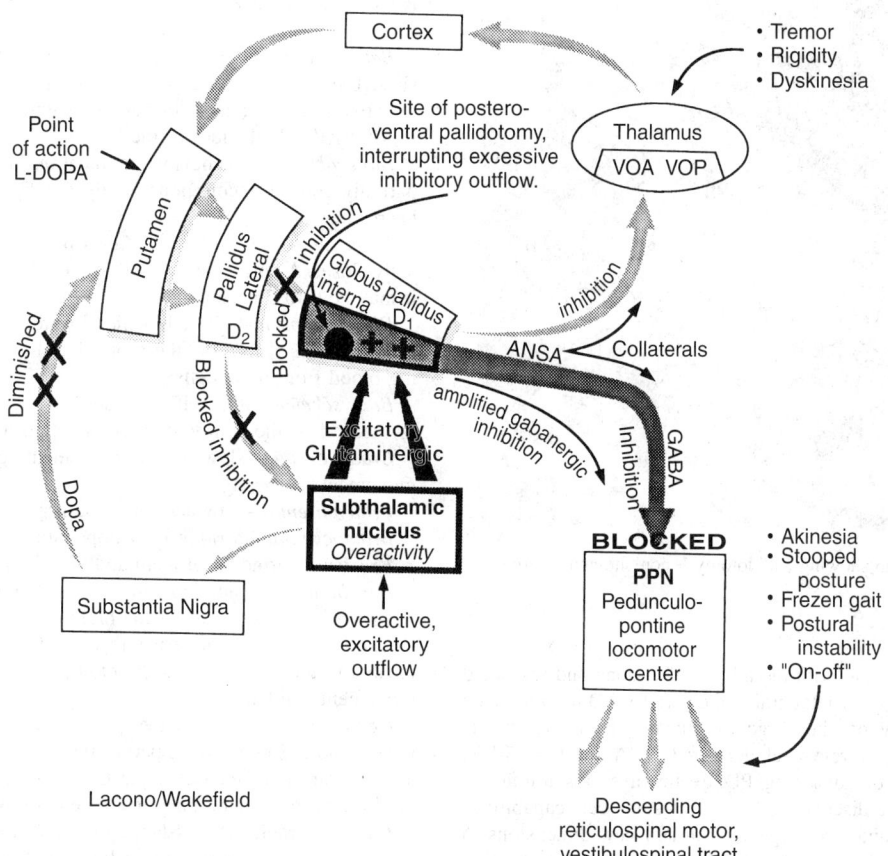

Fig. 1.   Proposed mechanism of the inhibition of the pedunculopontine locomotor nucleus (PPN). In this model, the inhibition is caused by the subthalamic amplification of the pallidal γ-aminobutyric-acid-inhibitory output and results in akinesia, gait freezing, and postural instability. In Type B patients, partial interruption of abnormal pallidal efferents produced by posteroventral pallidotomy (PVP) allows the reversal of brain stem inhibition, reversing akinesia. In Type A patients, the concomitant effects on the motor thalamus (by posteroventral pallidotomy via ansa lenticularis collaterals) accounts for the benefits of posteroventral pallidotomy in decreasing tremor, eliminating rigidity and dyskinesia, and reversing akinesia. Despite its appeal, the model in humans is limited. VOA, ventralis oralis anterior; VOP, ventralis oralis posterior.

symptoms was surprising and at first perplexing. Iacono (1994) proposed that partial interruption of abnormal pallidal efferents to the brainstem by posteroventral pallidotomy (PVP) allows reversal of brainstem inhibition of the PPN-locomotor center and reticulo-spinal projections reversing akinesia. See Figs. 1 and 2. This effect was found to be synergistic with medication therapeutic effects. Advances in technique have produced concomitant PVP interruption of subthalamic afferent amplification of the GPi and more direct measures to damage or suppress the subthalamic nucleus. Chronically implanted deep brain electrodes for high-frequency stimulation (DBS) of the refined thalamotomy target for tremor, of the pallidum, and STN for akinesia have been employed to create a reversible lesion effect. The PVP is successful in greater than 80% of operated patients and is able to reverse all symptoms of PD up to 90% of normal with 50% of patients benefiting for more than five years.

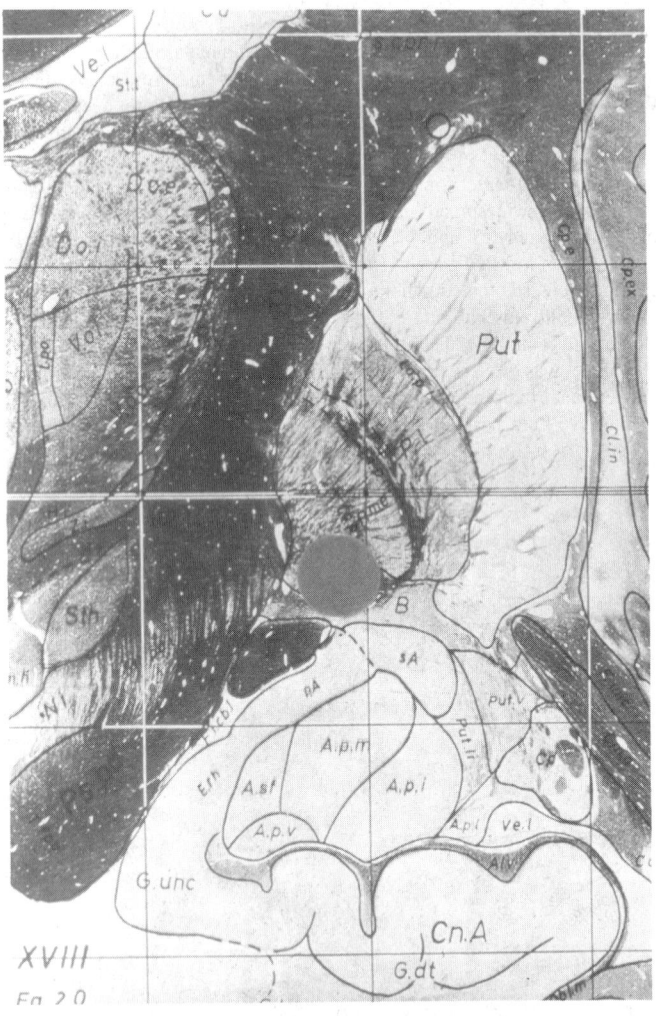

Fig. 2.   Coronal brain map showing pallidotomy lesion site and location.

Research and Cure

Despite hope of cure beginning with the advent of L-dopa and rekindled before the subsequent failures of fetal grafts, the cure for PD has remained evanescent. However, now on the verge of human genetic treatments utilizing transformed cells or retroviral delivered DNA for DA, 5-HT, or GDNF, the possibility of advancing PD treatment seems imminent. Moreover, the most recent discoveries of neural stem cell capabilities for neuronal repopulation and CNS regeneration add new dimensions to the treatment of PD. Optimism is generated by new research indicating that the adult human brain ordinarily repopulates up to 20% per year of neurons in the hippocampus (an area damaged in ALZ). Parenthetically, serotonin may represent the trophic stimulation for this stem cell derived regeneration. Stem cells in general have broad, flexible ontogenic potential, can migrate and adjust by differentiation to local demands. Utilizing, controlling and then understanding the enigma of embryonic development

and aging involving genetic expression is an endearing promise for the cure of neurodegenerative disease. See also **Blood-Brain Barrier**; and **Central and Peripheral Nervous Systems**.

An abridged glossary of terms used to describe Parkinson's disease would include:

Acetylcholine—a chemical which acts as a neurotransmitter. An imbalance between dopamine and acetylcholine results in some Parkinson's disease symptoms.

Action tremor—a tremor that increases when the hand is moving voluntarily.

Agonist—a drug which increases neurotransmitter activity by stimulating the dopamine receptors directly.

Akinesia—no movement.

Amantadine (Symmetrel)—an anti-Parkinson drug.

Anticholinergics—anti-Parkinson drugs that block the action of acetylcholine, thereby rebalancing it in relation to dopamine and reducing rigidity and tremor; e.g., Artane, Cogentin.

Antihistamines—drugs that are often used to relieve cold or allergy symptoms (i.e., Benadryl) but may also be effective in reducing tremor.

Ataxia—loss of balance.

Athetosis—slow, involuntary movements of the hands and feet.

Atrophy—wasting, shrinkage.

Autonomic nervous system—that part of the nervous system that is responsible for automatic functions, such as the heartbeat, digestion, salivation.

Axon—the long, hair like extension of a nerve cell that carries a message to the next nerve cell.

Basal ganglia—several large clusters of nerve cells deep in the brain below the cerebral hemispheres; crucial in coordinating motor commands. Include the striatum and the substantia nigra.

Benign Essential Tremor—A condition characterised by tremor of the hands, head, voice, and sometimes other parts of the body. Essential tremor often runs in families and is sometimes called *familial tremor*. It is sometimes mistaken for a symptom of Parkinson's. However, this is an action tremor and there is no rigidity or bradykinesia.

Beta-Blockers—Drugs which block the action of epinephrine at certain sites. Usually used to treat hypertension and heart disease, they may be effective in the treatment of benign essential tremor.

Bilateral—both sides of the body.

Biofeedback—a behavior modification in which patients are taught to partially control unconscious bodily functions, such as blood pressure or heart rate.

Blink rate—the number of times per minute that the eyelid automatically closes. A normal rate may be 10 to 30 per minute; for the parkinsonian it may be 0 to 5 per minute.

Blepharospasm—forced eyelid closure. See also **Blepharospasm**.

Blood-brain barrier—the protective membrane that separates circulating blood from brain cells.

Body scheme—the ability to identify body parts or to relate body parts to each other; the ability to sense one's position in space.

Bradykinesia—slowness of movement, gradual loss of spontaneous movement.

Bradyphrenia—slowness of thought processes.

Bromocriptine (Parlodel)—a dopamine agonist and anti Parkinson drug.

Bruxism—grinding of teeth and clenching of jaw muscles.

Buccinator—a muscle of the face and cheek.

Central nervous system—the brain and the spinal cord.

Cerebellum—a large structure consisting of two halves (hemispheres) located in the lower part of the brain; responsible for the coordination of movement and balance.

Cerebrum—consists of two parts (lobes), left and right, that form the largest and most developed part of the brain; initiation and coordination of all voluntary movement take place within the cerebrum. The basal ganglia are located immediately below the cerebrum.

Chorea—rapid, jerky, dance like movement of the body. May result from high doses of levodopa and/or long term levodopa therapy.

Cogwheel Rigidity—Stiffness in the muscles, with a jerky quality when arm and leg joints are repeatedly moved.

Constipation—Diminished ability of intestinal muscles to move feces (stool), often resulting in very hard stool. A common problem in Parkinson's.

Corpus striatum—a part of the brain that helps regulate motor activities.

Cortex — the outer layer of the cerebrum, densely packed with nerve cells.

Cryothalamotomy — a surgical procedure in which a supercooled probe is inserted into a part of the brain called the thalamus in order to stop tremors.

Dendrite — a threadlike extension from a nerve cell that serves as an antenna to receive messages from the axons of other nerve cells.

Delusions — a condition in which the patient has lost touch with reality and experiences hallucinations and misperceptions.

Dementia — loss of intellectual abilities.

Deprenyl (Eldepryl, Selegiline, Jumex) — anti-Parkinson drug. A drug that slows the breakdown of chemicals like dopamine by inhibiting the action of certain enzymes. It's increase effects of dopamine in the brain.

Dopa decarboxylase — an enzyme present in the body that converts levodopa to dopamine.

Dopa decarboxylase — inhibitors anti-Parkinson drugs that block the enzyme dopa decarboxylase.

Dopamine — a chemical substance, a neurotransmitter, found in the brain that transmits impulses from one nerve cell to another and regulates movement, balance, and walking. It is the substance that is lost in PD.

Dopamine Agonist — Drugs that mimic the effects of dopamine and stimulate the dopamine receptors

Dopaminergic — a chemical that works like, or has the same effect as, dopamine.

Drug holiday — a 3- to 14-day withdrawal of levodopa after long-term treatment when side effects of levodopa outweigh benefits; rarely done today because of the severe effects of drug withdrawal.

Drug Induced Parkinsonism — Parkinson's symptoms which have been caused by drugs used to treat other conditions, e.g., neuroleptic drugs, and reserpine, use to be used to treat hypertension.

Dyskinesia — an abnormal involuntary movement including athetosis and chorea. Can result from long-term use of high doses of levodopa.

Dysphagia — difficulty in swallowing. Dystonia a slow movement or extended spasm in a group of muscles.

Edema — tissue swelling due to excessive fluid.

Enzyme — a substance that speeds up a specific chemical reaction but that is not itself consumed in the reaction.

Euphoria — a feeling of well-being or elation; may be drug related.

Extensor (muscle) — any muscle that causes the straightening of a limb or other part.

Extrapyramidal system — the system of nerve cells, nerve tracts and pathways that connects the cerebral cortex, basal ganglia, thalamus, cerebellum, reticular formation, and spinal neurons; it is concerned with the regulation of reflex movements such as balance and walking. The extrapyramidal system is damaged in Parkinson's disease.

Festination — short, shuffling steps; involuntary speeding up of the gait. Quick forward steps. A symptom characterized by small, Flexor (muscle) any muscle that causes the bending of a limb or other body part.

Freezing — Temporary, involuntary inability to move.

Ganglion — a cluster of nerve cells.

Globus pallidus — The inner part of the lenticular nucleus. The lenticular nucleus and the caudate nucleus form the Striatum.

Gray matter — the darker-colored tissues of the central nervous system; in the brain, the gray matter includes the cerebral cortex, the thalamus, the basal ganglia, and the outer layers of the cerebellum.

Hormone — a substance secreted by a gland that is transported in the bloodstream to various organs in order to regulate or modify bodily functions.

Hypokinesia — Abnormally diminished motor activity.

Idiopathic — An adjective meaning "of unknown cause". The usual form of Parkinson's is idiopathic Parkinson's.

Incontinence — involuntary voiding of the bladder or bowel.

Intention Tremor — one occurring when the person's attempts voluntary movement.

Lenticular nucleus — This group of cells along with the caudate nucleus form the Striatum or Corpus Striatum.

Levodopa — the single most effective anti-Parkinson drug, which is changed into dopamine in the brain, usually combined with carbidopa (a dopa decarboxylase inhibitor) as Sinemet.

Levodopa-Induced Dyskinesias — A side effect of medication, which may occur with prolonged use. These abnormal, involuntary movements may be alleviated by reducing the amount of medication.

Lewy body — a pink-staining sphere, found in the bodies of dying cells, which is considered to be a marker for Parkinson's disease.

Livido Reticularis — A purplish or bluish mottling of the skin seen usually below the knee and sometimes on the forearm in persons under treatment with the drug amantadine (Symmetrel).

Micrographia — a change in handwriting with the script becoming smaller and more cramped.

Monoamine oxidase (MAO) — an enzyme that breaks down dopamine. There are two types of MAO "A" and "B." In Parkinson's disease, it is beneficial to block the activity of MAO B.

MPTP — a chemical produced during an attempt to make a synthetic narcotic. MPTP destroys the cells of the substantia nigra cells and produces a disease that mimics Parkinson's disease.

Myoclonus — jerking, involuntary movements of the arms and legs. May occur normally during sleep.

Neostriatum — Vital part of the brain comprised of two basal ganglia (caudate and putamen).

Neuroleptic Drugs — (Also called major tranquilizers) A class of drugs which act as dopamine antagonists (by blocking some dopamine receptors). They can aggravate symptoms of Parkinson's. This class includes Haloperidol (Haldol), and the phenothiazines, e.g., Compazine, Stelazine, Chlorpromazine, etc.

Neuron — a cell specialized to conduct and generate electrical impulses and to carry information from one part of the brain to another.

Neurotransmitters — chemical substances that carry impulses from one nerve cell to another; found in the space (synapse) that separates the transmitting neuron's terminal (axon) from the receiving neuron's terminal (dendrite).

Nigral — of or referring to the substantia nigra.

Nigrostriatal Degeneration — Degeneration of the nerve pathways from Substantia Nigra to the striatum. These pathways are normally rich in dopamine and are those affected in PD.

Norepinephrine — a neurotransmitter found mainly in areas of the brain that are involved in governing autonomic nervous system activity, especially blood pressure and heart rate.

On-off phenomena — abrupt changes in performance during the day caused by the taking effect or wearing off of anti-parkinson drugs. A change in the patient's condition, with sometimes rapid fluctuations between uncontrolled movements and normal movement, usually occurring after long-term use of levodopa and probably caused by changes in the ability to respond to this drug.

Orthostatic hypotension — a large decrease in blood pressure upon standing; may result in fainting. Orthostatic hypertension may occur spontaneously in PD or may be related to certain drugs.

Palilalia — A symptom of Parkinsonism, especially the postencephalitic form, in which a word or syllable is repeated and the flow of speech is interrupted.

Pallidotomy — a surgical procedure in which a part of the brain called the *globus pallidus* is lesioned in order to improve symptoms of tremor, rigidity, and bradykinesia.

Palsy — paralysis of a muscle or group of muscles.

Paraesthesia — Sensations, usually unpleasant, arising spontaneously in a limb or other part of the body, variously experienced as Òpins and needlesÓ or a feeling of warmth or coldness (thermal paresthesias).

Parkinson's Disease — That form of Parkinsonism originally described by James Parkinson' as a chronic, slowly progressive disease of the nervous system characterised clinically by the combination of tremor, rigidity, bradykinesia, and stooped posture, and pathologically by loss of the pigmented nerve cells of the Substantia Nigra in the brain.

Parkinson's Facies — A stolid masklike expression of the face, with infrequent blinking; it is characteristic of Parkinson's.

Parkinsonism — a term referring to a group of conditions that are characterized by four typical symptoms — tremor, rigidity, postural instability, and bradykinesia.

Pergolide (Permax) — an anti-Parkinson drug.

Peristalsis — wave like contractions that move food through the digestive tract.

Postural instability — impaired balance and coordination, often causing patients to lean forward or backward and to fall easily.

Postural Tremor — Tremor that increases when hands are stretched out in front.

Pyramidal pathway — a collection of nerve tracts that travel from the cerebral cortex through the pyramid of the medulla oblongata in the brainstem to the spinal cord. Within the pyramid of the medulla, fibers cross from one side of the brain to the opposite side of the spinal cord; the pyramidal pathway is intact in Parkinson's disease.

Range of motion — the extent that a joint will move from full extension to full flexion.

Resting tremor — a tremor of a limb that increases when the limb is at rest.

Retropulsion — the tendency to step backwards if bumped from the front or upon initiating walking, usually seen in patients who tend to lean backwards because of problems with balance.

Rigidity — increased resistance to the passive movement of a limb. A symptom of the disease in which muscles feel stiff and display resistance to movement even when another person tries to move the affected part of the body, such as an arm.

Sialorrhea — drooling.

Sinemet — Trade name for the antiparkinson drug that is a mixture of levodopa and carbidopa. This drug combination contains a ratio of levodopa 4 mg. or 10 mg. to carbidopa 1 mg. (Sinemet 100/25, Sinemet 250/25).

Sinemet CR — Controlled-release Sinemet. 200 mg. Levodopa with 50 mg. Carbidopa in a capsule contained in a matrix (outer layer) releasing the drug more slowly in the body. These capsules are not to be taken all at once, but rather in separate doses over the course of a day.

Spasm — a condition in which a muscle or group of muscles involuntarily contract.

Stereotactic Surgery — Surgical technique, that involves placing a small electrode in an area of the brain to destroy a tiny amount of brain tissue.

Striatonigral Degeneration — This is a degeneration of the nerve pathways travelling from the striatum to the Substantia Nigra. People with this degeneration also appear to have Parkinsonism. However, they respond differently to drug therapy than people with Parkinson's.

Striatum — part of the basal ganglia, it is a large cluster of nerve cells, consisting of the caudate nucleus and the putamen, that controls movement, balance, and walking; the neurons of the striatum require dopamine to function.

Substantia nigra — movement-control center in the brain where loss of dopamine-producing nerve cells triggers the symptoms of Parkinson's disease; substantia nigra means "black substance," so called because the cells in this area are dark. A small area of the brain containing a cluster of black-pigmented nerve cells that produce dopamine which is then transmitted to the striatum.

Sustention (postural) — tremor a tremor of a limb that increases when the limb is stretched.

Synapse — a tiny gap between the ends of nerve fibers across which nerve impulses pass from one neuron to another; at the synapse, an impulse causes the release of a neurotransmitter, which diffuses across the gap and triggers an electrical impulse in the next neuron.

Tardive Dyskinesia — This is a movement disorder associated with long-term use of neuroleptic drugs such as Chlorpromazine, Haloperidol, Loxapine, etc. Movements of a person with tardive dyskinesia are similar in appearance to those of a person with levodopa induced dyskinesias, but the causes of the two conditions are different.

Thalamotomy — Operation in which a small region of the thalamus is destroyed, achieved by stereotactic techniques. Tremor and rigidity in Parkinsonism and other conditions may be relieved by thalamotomy.

Thalamus — Anatomical term designating a mass of gray matter centrally placed deep in the brain near its base and serving as a major relay station for impulses travelling from the spinal cord and cerebellum to the cerebral cortex.

Toxin — A poisonous substance.

Tremor — Rhythmic shaking and involuntary movement of part(s) of the body as a result of sequential muscle contractions.

Tyrosine — the amino acid from which dopamine is made.

Unilateral — Occurring on one side of the body. Parkinson's symptoms usually begin unilaterally.

Vomiting Center — Term referring to an area of the brain where the nausea and vomiting reflex may be triggered by some medications.

"Wearing Off" Phenomenon — Waning of the effect of the last dose of levodopa, associated with abrupt reduction or loss of mobility.

White matter nerve tissue — that is paler in color than gray matter because it contains nerve fibers with large amounts of insulating material (myelin). The white matter does not contain nerve cells. In the brain, the white matter lies within the gray layer of the cerebral cortex. See **Gray matter**.

Glossary with permission from The Iacono Neuroscience Clinic: *http://Pallidotomy.com/index.html*

Additional Reading

Barbeau, A.: "The Pathogenesis of Parkinson's Disease: A New Hypothesis," *Can. Med. Ass. J.*, **87**, 802–807 (1962).

Cooper, I.S., G. Bravo: "Chemopallidectomy and Chemothalamectomy," Presentation at the meeting of the Harvey Cushing Society, Detroit, Michigan, (April 26, 1957).

Cotzias, G.C. et al.: "Modification of Parkinsonism — Chronic Treatment with L-dopa," *NEJM*, **280**, No 7, 337–345 (1969).

Coyle, J.T., P. Puttfarcken: "Oxidative Stress, Glutamate, and Neurodegenerative Disorders," *Science*, **262**, 689–695 (1993).

Fahn, S. et al.: "Monoamines in the Human Neostriatus: Topographic Distribution in Normals and in Parkinson's Disease and Their Role in Akinesia, Rigidity, Chorea, and Tremor," *J. Neurological Sciences*, **14**, 427–455 (1971).

Halliday, G.M. et al.: "Loss of Brainstem Serotonin and Substance P-containing Neurons in Parkinson's Disease," *Brain Research*, **510** (1990).

Horner, P.J., F.H. Gage: "Regenerating the Damaged Nervous System," *Nature*, **407**, 963–70 (2000).

Hornykiewicz, O.: "Dopamine in the Basal Ganglia: Its Role and Therapeutic Implications (Including the Clinical use of L-dopa)," *Br. Med. Bull.*, **29**, 172–178 (1973).

Iacono, R.P. and B.S.J. Nashold: "Stereotactic Neurosurgery, In Iacono Robert, P., M.D. Iacono, FACS, B.S.J. Nashold, ed. Textbook of Surgery: The Biological Basis of Modern Surgical Practice, W.B. Saunders, Philadelphia, PA, (1991).

Iacono, R.P. et al.: "Stereotactic Pallidotomy Results for Parkinson's Exceed Those of Fetal Graft," *The American Surgeon*, **60**, 776–782 (1994).

Iacono, R.P. et al.: "The Results, Indications, and Physiology of Posteroventral Pallidotomy for Patients with Parkinson's Disease," *Neurosurgery*, **36**, 1118–1127 (1995).

Iacono, R.P., et al.: "Chronic Anterior Pallidal Stimulation for Parkinson's Disease," *Acta Neuroch*, **137**, 106–112 (1995).

Iacono, R.P. et al.: "Concentrations of Indoleamine Metabolic Intermediates in the Ventricular Cerebrospinal Fluid of Advanced Parkinson's Patients with Severe Postural Instability and Gait Disorders," *J. Neural. Transm.*, **104**, 451–459 (1997).

Laitinen, L.V. et al.: "Leksell's Posteroventral Pallidotomy in the Treatment of Parkinson's Disease," *J. Neurosurg*, **76**, 53–61 (1992).

Lindner, M.D. et al.: "Implantation of Encapsulated Catecholamine and GDNF-producing Cells in Rats with Unilateral Dopamine Depletions and Parkinsonian Symptoms," *Exp. Neurol.*, **132**, 62–76 (1995).

Liu, H. etal: "A Comparative Study on Neurochemistry of Cerebrospinal Fluid in Advanced Parkinson's Disease", *Neurobiology of Disease*, **6**, 35–42 (1999).

Olanow, C.W.: "An Introduction to the Free Radical Hypothesis in Parkinson's Disease," *Ann. Neurol.*, **32**, S2–S9 (1992).

Parkinson, J.: "An Essay on the Shaking Palsy", Sherwood, Neely and Jones, London, (1817).

Spiegel, E.A. et al.: "Long Range Effects of Electropallidotomy in Extrapyramidal and Convulsive Disorders," *Neurology*, **8**, 738–743 (1958).

Svennilson, E. et al.: "Treatment of Parkinsonims by Stereotactic Thermolesions in Pallidal Region," *Psychiatr Neurol Scand*, **35**, 358–377 (1960).

Tasker, R.R.: Thalamotomy," *Neurosurg Clin. N. Am.*, **1**, 841–864 (1990).

Wichmann, T. et al.: "The Primate Subthalamic Nucleus. III. Changes in Motor Behavior and Neuronal Activity in the Internal Pallidum Induced by Subthalamic Inactivation in the MPTP Model of Parkinsonism," *J. Neurophysiol*, **72**, 521–530 (1994).

Web References

National Parkinson Foundation, Inc.: *http://www.parkinson.org/*
Parkinson Alliance.net: *http://www.parkinsonalliance.net/*
The American Parkinson's Disease Association: *http://www.apdaparkinson.com/*

R.P. Iacono, M.D., F.A.C.S.

PAROTITIS. Infection of the parotid glands. Parotitis may occur as a complication of any prolonged illness, especially after operations or illnesses in the aged and debilitated. Mumps is a form of parotitis resulting from infection with a virus that has a specific affinity for the parotid gland. See also **Mumps**.

PARROTFISHES *(Osteichthyes)*. Of the order *Percomorphi*, family *Scaridae*, the parrotfishes are brightly colored marine fishes with a prominent beak formed by the partial coalescence of the teeth — thus their name. They occur in tropical waters, chiefly about coral reefs. They are herbivorous and are known to erode tropical reefs. They usually remove a bit of coral along with vegetation. Because of their powerful teeth, they are able to grind bits of coral, and hence, this does not interfere with their digestive processes. In 1955, Dr. Howard Winn discovered that some of the parrotfishes construct an envelope made of secreted mucous for covering themselves at night. The exact mechanism required to form the envelope and the actual need for the envelope remain to be fully investigated. The size range of parrotfishes is great, varying from lengths of 18 inches (45 centimeters) and under, up to 6 feet (1.8 meters).

PARROTS AND COCKATOOS (Aves, Psittaciformes).

The parrots and related species, including cockatoos, macaws, and parakeets (paraquets) are all psittaciformes. These birds are characterized by a strong hooked beak, adapted for opening nuts and seeds. They have two toes directed forward and two toes directed backward.

The parrot is represented on all continents except Europe but is confined chiefly to tropical and subtropical areas. Parrots eat nuts and other fruits and seeds, although the kea parrot (*Nestor notabilis*) of New Zealand is reported to have acquired a taste for mutton. The brilliant colors of many parrots and their ready imitation of human speech have led to their being kept as cage birds for many centuries; hence, they are familiar well beyond their natural range.

There are many subsidiary forms of parrots. The nestor parrots of New Zealand and the neighboring islands are dark-colored birds, including the kea and the kaka. A curious member of the order is the owl-parrot or kakapo of New Zealand, which constitutes a distinct family. It is a flightless bird of owl-like appearance, barring its more brilliant colors, and is largely nocturnal in habits. The cockatiel is a small Australian parrot related to the cockatoos. The conures are small parrots of numerous species found from Mexico into South America. Their prevailing colors are green and yellow. The Carolina paraquet of North America is a member of this group. Broadtails are a group of parrots and parakeets found only in Australia, Norfolk Island, and Tasmania. The Budgerigar is an Australian parakeet, chiefly green with a blue tail and yellow face. In captivity, this species is also known as the Australian lovebird. Other names for the species include zebra, shell, and warbling grass-parakeet.

The cockatoo is a crested parrot whose beak is transversely ridged on the under surface of the hook. The tail is short and broad. Cockatoos occur in Australia and Oriental regions. The bill is sharp, short, and curved. The tongue is worm-like, tiny, and adapted for seizing and grasping various dietary items. Most cockatoos are white but also occur in various colors and with quite a range of sizes. They are frequently cherished by natives as pets. Numerous young birds are also taken for export as cage birds.

The cockatoo's nest is principally in holes of eucalyptus trees in the deep forest. The egg is white and is often deposited on bare wood chips or wood debris. The young are hatched naked and blind. They are fed by regurgitation of the parents for the first 3 months.

The black cockatoo (*Calyptorhynchus magnificus*) is the largest of the species and is a wanderer. With its strong bill, the bird cracks nuts that would require a hammer or vise for a human to break open. The bird's tongue assists much in extracting the kernel from the nut. The white cockatoo (*Cocatua galerita*) is a sulfur crested cockatoo of Australia. The white crested cockatoo (*C. alba*), found in the Molucca Islands, and the rose crested cockatoo (*C. moluccensis*), found on the island of Ceram, are representative of numerous varieties of cockatoos, embracing a wide range of colors and specific characteristics. See also **Psittaciformes**.

PARSEC.

The parsec is a unit of distance used for expressing distances between stars and other members of the sidereal universe. Technically, an object is at a distance of one parsec when it has a stellar parallax of $1''$ (one second of arc); or, in other words, one astronomical unit would subtend an angle of one second at the distance of one parsec. Thus, D(in parsecs) $= 1/\pi$.

$$1 \text{ parsec} = 3.26 \text{ light-years}$$

$$= 206,265 \text{ astronomical units}$$

$$= 1.924 \times 10^{13} \text{ miles}$$

$$= 3.084 \times 10^{13} \text{ kilometers}$$

$$= 3.084 \times 10^{18} \text{ centimeters}$$

Within recent years, in the discussion of distances between extragalactic objects, the parsec is not large enough to be convenient, and the terms kiloparsec (1,000 parsecs), and megaparsec (1,000,000 parsecs), are being used.

See also **Astronomical Unit**; and **Light-Year**.

PARSHALL FUME.

See **Flow Measurement (Liquids and Gases)**; and **Flume**.

PARSLEY.

See **Flavorings**.

PARTHENOGENESIS.

The development of eggs without fertilization. In some groups of animals, eggs normally develop in this way, either for a series of generations, interrupted occasionally by a normal fertilization, or as a special part of the reproductive process associated with the development of some young from fertilized eggs.

PARTIAL DIFFERENTIAL EQUATION.

Partial derivatives of the unknown function are involved. The general linear second-order partial differential equation in two variables is

$$A(x, y)\frac{\partial^2 f}{\partial x^2} + B(x, y)\frac{\partial^2 f}{\partial x \partial y} + C(x, y)\frac{\partial^2 f}{\partial y^2}$$

$$= D(x, y)\frac{\partial f}{\partial x} + E(x, y)\frac{\partial f}{\partial y} + G(x, y)f + H(x, y)$$

The two families of curves given by

$$A \, dy = [B \pm (B^2 - AC)]^{1/2}dx$$

having solutions λ, μ = constant, are the families of characteristic curves. When the differential equation is rewritten with λ, μ as independent variables, the equation is in normal form. The three cases arising are *elliptic, hyperbolic*, and *parabolic* partial differential equations. Another special case is Euler's equation, if A, B, C are constants and the right-hand side of the equation is zero. The general second-order equation can be obtained from it by a linear transformation of independent variable.

The boundary conditions appropriate to the various forms of the partial differential equations are different, but they are always needed to fix the functional form of the solution. Problems dealing with the physics of fields usually lead to partial differential equations. See also **Mathematical Physics (Equations of)**.

Some properties of the three special cases are: (1) *Elliptic*, where $A(x, y)C(x, y) > B^2(x, y)$ for all x, y. The characteristic curves become functions of the complex variable. Writing $\lambda = (u + iv)$, the other solution is $\mu = (u - iv)$, and the normal form of the equation is

$$\frac{\partial^2 \phi}{\partial u^2} + \frac{\partial^2 \phi}{\partial v^2} = P(u, v)\frac{\partial \phi}{\partial u} + Q(u, v)\frac{\partial \phi}{\partial v} + R(u, v)\phi$$

The specification of Dirichlet or Neumann conditions along a closed boundary assures a unique solution. The Laplace, Helmholtz, and Poisson equations are of this type.

(2) *Hyperbolic*, where $B^2(x, y) > A(x, y)$ for all x, y. The characteristic curves are all real, and the normal form is

$$\frac{\partial^2 \phi}{\partial \lambda \partial \mu} = P(\lambda, \mu)\frac{\partial \phi}{\partial \lambda} + Q(\lambda, \mu)\frac{\partial \phi}{\partial \mu} + R(\lambda, \mu)\phi$$

Unless the boundary coincides with a characteristic, unique solutions are obtained with specified boundary values and normal derivatives (Cauchy conditions). When the boundary is closed, the Cauchy conditions overdetermine the solution. The wave equation is an example of this type.

(3) *Parabolic*, where $B^2(x, y) = A(x, y)C(x, y)$ for all x, y. There is only one set of characteristic curves given by $Ady = Bdx$, having the solution λ = constant. The normal form is then

$$\frac{\partial^2 \phi}{\partial x^2} = P(x, \lambda)\frac{\partial \phi}{\partial \lambda} + Q(x, \lambda)\frac{\partial \phi}{\partial x} + R(x, \lambda)\phi$$

The *diffusion equation*, $\partial^2 \phi/\partial x^2 = a^2 \partial x/\partial t$, where t is the time, is an example of this type. Dirichlet conditions on a boundary open at the end toward increasing t result in a unique solution.

Methods of solving partial differential equations must be sought in texts on the subjects.

PARTIALLY BALANCED INCOMPLETE BLOCKS.

Partially balanced incomplete blocks form a very general class of experimental design in which not all treatments occur in every block. The designs are such that:

1. Each treatment is replicated r times.
2. Given any treatment, the remainders fall into groups of n_1, $n_2, \ldots, n_m$ such that every treatment of the ith group occurs λ_i times

in the same block as the given treatment, the n_i and λ_i being independent of the treatment at the start. Two treatments that occur in the same block λ_i times are called ith associates, and the groups of treatments are called associate classes.

3. Given two treatments which are ith associates, the number of treatments which are simultaneously jth associates of the first and kth associates of the second is denoted by p^i_{jk}. This number must be independent of the particular pair of ith associates at the start.

Under these conditions, estimates of the treatment differences and their standard errors can be obtained without difficulty.

PARTIAL PRESSURE. The pressure exerted by each component in a mixture of gases. In a mixture of perfect gases

$$p_i = \frac{n_i RT}{V}$$

The partial pressure of i is then the same as if component i occupies the same volume at the same temperature in the absence of the other gases. This is the Dalton law, which is treated more fully under that heading.

PARTICLE ACCELERATOR. See **Particles (Subatomic)**.

PARTICLES (Subatomic). For many years, the atom was traditionally described as having a central positively charged nucleus possessing considerable mass, but of minute dimension — this nucleus surrounded by a number of electrons in orbits at a relatively great distance from the nucleus. The number of electrons and their orbital arrangement determined the chemical properties of the atom, with the atoms of each chemical element possessing their own unique configuration. Recognition of the electron, the first elementary (presumably indivisible) particle, by J.J. Thomson and his associates in the 1890s ushered in an era of interest in *subatomic particles*. Ultimately, this led to the discipline of *high-energy physics*.

Organization of Matter — A Chronology[1]

A better understanding of the building blocks of nature has been a goal for many centuries, extending back to the period in Greek history of Anaxagoras of Ionia (500–428 B.C.), who held that "there was an infinite number of different kinds of elementary atoms, and that these, in themselves motionless and originally existing in a state of chaos, were put in motion by an eternal, immaterial, spiritual, elementary being, from which motion the world was produced." The concept of atoms appeared from time to time in medieval works, although the concepts expressed now seem vague. However, they seem to have been based on the idea that there could be a limit to the divisibility of matter and, consequently, the idea of a final indivisible particle out of which large pieces of matter could be built.

Atoms and Molecules. In the early 1800s, it became clear that chemical reactions could be most simply explained if each chemical element was thought of as composed of very small, identical entities characteristic of the chemical element. Thus there arose a rather well-defined idea of a chemical element composed of identical atoms, as distinguished from a compound composed of groups of different atoms combined into molecules. During the later part of the 1800s, the kinetic theory of gases made use of the idea of atoms and molecules in explaining the behavior of gases. During this period, few scientists still doubted the actual material existence and reality of atoms.

Electrons. It is perhaps rather curious that the idea of atoms became really well established only after it became clear that the atoms were not in any true sense indivisible, but that instead they probably had a complex structure that should be investigated. Since these investigations required equipment and methods that had been developed by physicists rather than chemists, the physicists took the lead and the work became known over a long period as *atomic physics*, or the physics of atomic structure. As mentioned previously, this era was inaugurated by Thomson, who first isolated and established the existence of electrons. He showed that electrons have only about 1/2000 the mass of the lightest known atom, hydrogen. He also showed that these particles, as indicated by their name, carry negative electrical charges. It was later shown by Millikan that all

[1] Some of the concepts mentioned in this brief historical review have long since been abandoned or altered.

the electronic charges are the same. Thus, the identification of electrons as small electrically charged pieces of matter, and as constituents of all matter, became firmly established.

Electrical Neutrality. Since it was clear that normal matter is electrically neutral, it had to be assumed that each atom contained a positive electrical charge, as well as negative electrons. J.J. Thomson developed the picture of a somewhat spherical, jelly-like mass of positive electricity in which electrons are located at various positions, bound by "quasi-elastic" forces.

A principal means of investigating the structure of atoms was the examination of light emitted by the material in the gaseous state. This light was found to consist of a number of discrete wavelengths, or colors. Each of these wavelengths was associated, in the early days of the last century, with a mode of vibration of the electrons in the positive jelly. In particular, Lorentz (University of Leiden) was able to show that such electrons, when placed in a magnetic field, would have their modes of vibration changed in a way that explained the findings of Zeeman, who had made early observations of the wavelengths of the light emitted by a radiating gas in a magnetic field.

During 1910–1911, Sir Ernest Rutherford suggested an experiment, carried out by Geiger and Marsden, in which alpha particles from a radioactive source were scattered from thin foils. The angles at which the alpha particles were scattered were found to be such as could best be described by the close approach of a heavy positively charged particle, the alpha particle, to another heavier and more highly positively charged particle, representing the scattering atom.

Nuclear Atom. From the results of the experiments, Rutherford concluded that the mass in the positive charge of an atom, instead of being distributed throughout the volume of a sphere of the order of 10^{-8} centimeter in radius, was concentrated in a very small volume of the order of 10^{-12} centimeter in radius. He thus developed the idea of a nuclear atom. The atom was pictured as a small solar system with the very heavy and highly charged nucleus occupying the position of the sun, and with electrons moving around it, as planets in their respective orbits.

Although this picture of nuclear atoms served to describe the alpha-particle scattering experiments, it still left many questions unsolved. One of these questions referred to the apparent stability of the atoms. An electron moving around the nucleus would tend to emit radiation, to lose its energy, and thereby to spiral into the nucleus. Why did it not do so? Why did the atoms all seem to be quite stable, and all to be of approximately the same size, even though some contain 90 or more electrons, while hydrogen contains only one?

Electron Motion Around the Nucleus. The first approach to a treatment of these problems was made by Niels Bohr in 1913 when he formulated and applied rules for "quantization" of electron motion around the nucleus. Bohr postulated states of motion of the electron, satisfying these quantum rules, as peculiarly stable. In fact, one of them would be really permanently stable and would represent the ground state of the atom. The others would be only approximately stable. Occasionally an atom would leave one such state for another and, in the process, would radiate light of a frequency proportional to the difference in energy between the two states. By this means, Bohr was able to account for the spectrum of atomic hydrogen in a spectacular way. Bohr's paper in 1913 may well be said to have set the course of atomic physics on its latest path.

Correlation with Chemical Properties. Out of the experimental work on the scattering of alpha particles and the theoretical work of Bohr, there grew a fairly definite picture of an atom that could be correlated with its chemical properties. The chemical properties were determined in the first place by the nuclear charge. The nucleus contained most of the atomic mass and carried an electric charge equal to an integral number of positive charges, each of the same magnitude as an electronic charge. This positive nucleus then accumulated around itself a number of electrons just sufficient to neutralize its positive charge and form a neutral atom.

Atomic Number. The number of positive charges or the number of negative electrons around the nucleus was designated as the atomic number of the atom. These showed a close parallelism with the arrangement of atoms in the periodic system. Through the formulation of a number of rules based upon Bohr's picture of quantized orbits, the periodic system of the elements could be understood. Hydrogen was given one electron, and helium two. The two electrons in helium constituted a "closed shell" which exhibited almost perfect spherical symmetry and chemical inactivity.

Thus, during the years after 1913, the feeling grew that the chemical properties of atoms could be pretty well understood. The idea that there were undiscovered elements, as indicated by gaps in the periodic system, was reinforced. These elements and more have since been discovered.

Quantum Mechanics. It was not until 1925 that Bohr's ideas were developed into a mathematical form complete enough and precise enough to permit their general application, under the name *quantum mechanics*. This development associated with the names of Dirac, Heisenberg, and Schrödinger, provided the basic laws which permit, in principle, the complete and quantitative description of an atom consisting of a heavy positively charged nucleus, and surrounded by enough electrons to make the whole system electrically neutral. See also **Quantum Mechanics**.

Electron Spin. One of the properties of electrons that became evident during the study of optical spectra of atoms was that of *electron spin*. The suggestion was made by Uhlenbeck and Goudsmit in 1925 that one of the features of such spectra could be understood if each electron had associated with it a quantity called spin, which is similar in many ways to angular momentum. Each electron also has a certain magnetic moment which affects the energy in the presence of a magnetic field.[2] This property also has been incorporated into the wave concepts of quantum mechanics.

[2] Even with acceptance of the spin concept, in terms of high-energy experiments, most scientists believed that spin effects would be observed only in low-energy atomic collisions. In the 1950s, C.L. Oxley (University of Rochester) noted large spin effects in high-energy collisions (several hundred million eV). Experimentation in this area, however, was rather limited until the late 1950s, when researchers (University of California Berkeley) proposed constructing polarized proton targets. In a technique involving low temperatures and strong magnetic fields, it was possible to cause electrons to spin in the same direction and, using another technique (microwave radiation) to cause neighboring protons also to spin in one direction. Interesting experiments followed at some laboratories, but by and large many high-energy physicists in the field considered spin as relatively unimportant and it would be even less important as collision energy levels were increased. This assumption has been disproved. Spin direction does seem to be important to collisions even at high energy levels.

In order to learn more about the role of spin in colliding protons, in 1973, using the Zero Gradient Synchrotron (ZGS) at the Argonne National Laboratory, Krisch and colleagues (University of Michigan) scattered beams of polarized protons from targets in which the protons were also polarized, i.e., the spinning was all in the same direction. During the series of experiments, it was found that when the beam and the target were polarized in the same direction, violent proton-proton collisions occurred with much greater frequency than where the beam and target were spinning in a like direction. Under the latter circumstance, it appeared that the particles would pass each other, but would not interact.

Some years later (in the late 1970s), this research group further investigated the spin-collision phenomenon of protons, but used a different accelerator, i.e., the Alternating Gradient Synchrotron (AGS) located at the Brookhaven National Laboratory. The latter apparatus made it possible to study the particles at much higher energy levels: (1) an energy level up to 18.5 GeV (beam and target both polarized; and (2) up to 28 GeV (with only the target polarized). Several unexpected findings were yielded by the experiments:

- Effects of spin appear to oscillate with an increase of collision energy of the protons.
- Spin directions of the particles continue to make a difference even at high energy levels. (Normally, one would reason that at high energy levels, the difference in spin directions and the effects of spinning would become smaller simply because the spin of the proton is believed to be constant and this would tend to be overwhelmed by the higher collision energy. This was not the case and provoked suspicion that much less is known about the proton than formerly believed.)

Out of clues gained from experiments with the AGS, one central question, as posed by researcher Krish in his 1987 paper (reference listed), is posed: What does the observed difference between scattering to the left and to the right mean? Krisch observed that perhaps (as some theorists suggest) both the violence and the energy (28 GeV) of the experiments were much too low for a fundamental theory, such as quantum chromodynamics (QCD), to apply. With higher energies, the scattering difference between left and right may soon be measured at the 70- to 800-GeV proton synchrotrons at Seupukhov, CERN, and Fermilab, and even higher levels of the proposed Superconducting Supercollider (SSC).

See also **Quantum Mechanics**; and **Quarks**.

Neutrons and Protons. By 1932, it had been established that atomic nuclei are made of comparatively small numbers of neutrons and protons. Even prior to the use of particle accelerators and the birth of high-energy physics, other experiments continued to "hint" at the need of additional subatomic particles to satisfy any theory that would unify scientists' understanding of the atom's infrastructure.

A quantum theory of nuclei was made possible by the discovery of the proton and the neutron. The nuclear interaction responsible for holding the nucleus together (against disruptive electrostatic repulsion of the protons) was found to be of an entirely new kind, much stronger than the electric interaction at short distances, but decreasing very much more rapidly with distance. The various complex nuclei differ in the number of protons and neutrons they contain.

By that time, the theory of the interactions between electrons and photons had developed to the point where the electrostatic repulsion or attraction between electrically charged particles could be understood in terms of the exchange of photons between them. In the lowest nontrivial approximation, it gave the Coulomb law for small velocities. The basic interaction was the emission and absorption of "virtual" photons by charged particles.

Pions and Other Particles. A similar mechanism could be invoked to explain the short-range nuclear interaction—i.e., it is due to the exchange of particles, which have nonzero masses which are a fraction of nuclear mass. These theoretical considerations predicted the existence of a set of three particles called pions, which were ultimately discovered.

Another kind of particle and another kind of interaction were discovered from a detailed study of beta radioactivity in which electrons with a continuous spectrum of energies are emitted by an unstable nucleus. The corresponding interactions could be viewed as being due to the virtual transmutation of a neutron into a proton, an electron, and a new neutral particle of vanishing mass called the neutrino. The theory provided such a successful systematization of beta decay rate data for several nuclei that the existence of the neutrino was well established more than 20 years before its experimental discovery. The beta decay interaction was very weak even compared to the electron-photon interaction.

Meanwhile, the electron was found to have a positively charged counterpart called the positron; the electron and positron could annihilate each other, with the emission of light quanta. The theory of the electron did in fact predict the existence of such a particle. It was later found that the existence of such "opposite" particles (antiparticles) was a much more general phenomenon than once surmised.

With intensification of particle physics research, many more particles were discovered and a classification of these particles into five families was proposed—the photon family, electron family, muon family, meson family, and baryon family. Most of these particles are unstable and decay within a time which is often very small by normal standards, but which is many orders of magnitude larger than the time required for any of these particles to traverse a typical nuclear dimension. There is a wide variety of reactions between them, but they could be understood in terms of three basic interactions—the *strong* (or nuclear), *electromagnetic*, and *weak* interactions.

The Nuclear Force. The nuclear forces and the interactions between pions and nucleons are strong; the electron-electron and electron-photon interactions are electromagnetic; the beta decay interactions are weak.

As mentioned previously, by 1932 it was known that nuclei are made of comparatively small numbers of neutrons and protons. A new force was discovered (in addition to the electromagnetic and gravitational forces) that held the positive protons and electrically uncharged neutrons together in the nucleus. This nuclear force was very strong, but of limited range. Its "quantum," the particle analogous to the photon in the electromagnetic field, was of nonzero rest mass. This particle, later called the π-meson or pion, was predicted by Yukawa in 1936 and discovered by Lattes, Occhialini, and Powell in 1947.[3] For a short time, it appeared

[3] Even though the meson was first predicted by the Japanese scientist Yukawa in 1935, the development of high-energy physics in Japan proceeded slowly from an experimental standpoint. It was not until 1975 that Japan established the National Laboratory for High Energy Physics (KEK). Rather than following traditional research approaches by way of building a proton synchrotron, for example, scientists at the University of Tokyo, in collaboration with physicists in the United States and Canada and later with CERN, decided to construct a meson facility with

that physicists had achieved a clear, simple, and correct theory of the fundamental constitution of matter. However, shortly thereafter, two new and unpredicted particles were reported. The first of these was another meson, somewhat like the pion but more massive. The second was a hyperon, i.e., a strongly interacting particle heavier than the neutron.

With the continuing discovery of more particles, investigators began to suspect that these particles were not in themselves fundamental or elementary, but that they had an internal structure. This paralleled the experiences of the 1800s when the large number of different types of atoms discovered suggested that atoms must have structure. Properties of particles also suggested an internal structure. For example, the neutron's total electric charge is indistinguishable from zero down to very fine limits, yet the neutron has a sizeable magnetic moment.

The discoverers of neptunium (1940), plutonium (1940), americium (1944), berkelium (1949), californium (1950), einsteinium (1952), fermium (1953), mendelevium (1955), and lawrencium (1961) gained much knowledge in the area of high-energy physics.

Research directed toward creating a nuclear bomb also contributed to an improved understanding of high-energy physics.

Exotic atomic nuclei may be described as structures that do not occur in nature, but are produced in collisions. These nuclei have abundances of neurons and protons that are quite different from the natural nuclei. In 1949, M.G. Mayer (Argonne National Laboratory) and J.H.D. Jensen (University of Heidelberg) introduced a spherical-shell model of the nucleus. The model, however, did not meet the requirements and restrains imposed by quantum mechanics and the Pauli exclusion principle. Hamilton (Vanderbilt University) and Maruhn (University of Frankfurt) reported on additional research of exotic atomic nuclei in a paper published in mid-1986 (see reference listed). In addition to the aforementioned spherical model, there are several other fundamental shapes, including other geometric shapes with three mutually perpendicular axes — prolate spheroid (football shape), oblate spheroid (discus shape), and triaxial nucleus (all axes unequal).

In 1964, M. Gell-Mann and G. Zweig (California Institute of Technology) independently pointed out that all the known hadrons (i.e., particles that interact via the strong nuclear force) could be constructed out of simple combinations of three particles (and their antiparticles). These hypothetical particles had to have slightly peculiar properties (the most peculiar being a fractional electric charge). Gell-Mann called these hypothetical particles *quarks* (referring to a sentence in James Joyce's work *Finnegan's Wake*, "Three quarks for Muster Mark"). The theory proposed postulated that three quarks bind together to form a baryon, while a quark and an antiquark bind together to form a meson. With supposition that the binding is such that the internal motion of the quarks is nonrelativistic (which requires the quarks be massive and sit in a broad potential well), then many quite detailed properties of the hadrons could be explained.

The purpose of the quark model was that of explaining the diversity of the hadrons; not to deal with the internal structure of any particle. But awareness of the model created a natural tendency among investigators to associate newly observed particles (among the poorly understood debris

a powerful superconducting muon channel. In his report on the evolution of meson science in Japan, Yamazaki (1986 reference cited) lists at least four advantages of opting for a meson facility: (1) It made possible the measurement of μ-e decay time spectra in a much wider time range (0 to 20 μsec) than previously possible without background, enabling muon-spin relaxation functions (mainly long-time behavior) to be determined precisely; (2) extreme external conditions pulsewise (pulsed RF, laser, high magnetic fields) could be applied; (3) because the time of muon arrival is uniquely defined, any time-dependent transient phenomena could be investigated; and (4) rare events could be selected from continuous backgrounds. Beyond the editorial scope here, Yamazaki describes in considerable detail the numerous accomplishments of KEK during the 1980s, including an improved understanding of nuclear structure from the viewpoint of quark structure. As mentioned, since nucleons and mesons are composed of quarks and since nucleons are densely packed in a nucleus, whether the nucleons in nuclei keep their free identities (mass, size, magnetic moment) is a rewarding problem to investigate. Recently, a new type of hypernuclear spectroscopy has emerged from KEK. Yamazaki defines meson science as an interdisciplinary since it uses "second generation" particles (muons and K mesons) for the creation and detection of exotic states in matter. To study this interesting frontier, scientists strongly sense the need for experimental facilities that will provide meson beams a hundred times as strong as those available today. Plans are underway along these lines. See also **Mesons**.

from particle experiments) with the hypothetical quarks. A number of properties of *partons* (a name given by Feynman, California Institute of Technology) were measured, including intrinsic spin angular momentum, and these were found to be consistent with the predictions of the quark model. Such observations, of course, added credence to the model.

In the 1960s, the quest for a grand unification theory — a theory that would explain all elementary particles and all forces acting between them — grew in intensity among most investigators who had the good fortune of discovering so many new particles, accompanied by the realization that the ultimate structure of matter was more complex than envisioned in the earlier years. The instrumental means for research (accelerators with higher and higher energies) were getting ahead of the theoretical aspects of the topic. Many particles resulting from collisions were found in the debris of experiments — their presence without plausible explanations. Many questions were posed — for instance, why four kinds of force, each with its own characteristic strength, the strengths differing by nearly 40 orders of magnitude (electromagnetism with its infinite range, the weak force for all practical purposes extending out only 10^{-15} centimeter)? For a while, prospects of a unified theory were dim, but a number of theories were proposed and given sufficient serious attention to warrant planning of experimental tests. As pointed out by Glashow (Nobel Prize, Physics, 1979, shared with Salam and Weinberg), in his Nobel Lecture, "In 1956, when I began doing theoretical physics, the study of elementary particles was like a patchwork quilt. Electrodynamics, weak interactions, and strong interactions were clearly separate disciplines, taught and separately studied. There was no coherent theory that described them all. Developments such as the observation of parity violation, the successes of quantum electrodynamics, the discovery of hadron resonances, and the appearance of strangeness were well-defined parts of the picture, but they could not be easily fitted together."

In the early years of investigation, the weak force and the electromagnetic force were regarded as indistinguishable. They were of the same strength and possessed the same infinite range, and they were transmitted by four bosons, all of which were massless. The forces manifested a symmetry, that is, they could be interchanged freely. It was believed that no matter which of these forces was applied, the net effect was the same. These views were later to be altered in the light of the process called *spontaneous symmetry breaking*.[4]

---

[4] Prior to 1956, it was believed that all reactions in nature obeyed the law of conservation of parity, so that there was no fundamental distinction between left and right in nature. However, Yang and Lee pointed out that in reactions involving the weak interaction between particles, parity was not conserved, and that experiments could be devised that would absolutely distinguish between right and left. This was the first example of a situation where a spatial symmetry was found to be broken by one of the fundamental interactions.

The principle of charge conjugation symmetry states that if each particle in a given system is replaced by its corresponding antiparticle, then it would not be possible to tell the difference. For example, if in a hydrogen atom the proton is replaced by an antiproton and the electron is replaced by a positron, then this antimatter atom will behave exactly like an ordinary atom — if observed by "persons also made of antimatter." In an antimatter universe, the laws of nature could not be distinguished from the laws of an ordinary matter universe.

However, it turns out that there are certain types of reactions where this rule does not hold, and these are just the types of reactions where conservation of parity breaks down. For example, consider a piece of radioactive material emitting electrons by beta decay. The radioactive nuclei are lined up in a magnetic field which is produced by electrons traveling clockwise in a coil of wire, as seen by an observer looking down on the coil. Because of the asymmetry of the radio-active nuclei, most of the emitted electrons travel in the downward direction. If the same experiment were done with similar nuclei composed of antiparticles and the magnetic field were produced by positron current rather than an electron current, then the emitted positrons would be found to travel in the upward, rather than in the downward, direction. Interchanging each particle with its antimatter particle has produced a change in the experiment.

However, the symmetry of the situation can be restored if we interchange the words "right" and "left" in the description of the experiment at the same time that we exchange each particle with its antiparticle. In the above experiment, this is equivalent to replacing the word "clockwise" with "counterclockwise." When this is done, the positrons are emitted in the downward direction, just as the electrons in the original experiment. The laws of nature are thus found to be invariant to the simultaneous application of charge conjugation and mirror inversion.

The first direct evidence that the proton has not only size, but structure was provided by an experiment at the Stanford Linear Accelerator Center (SLAC) in 1970. Previously it had been established that the proton is not a point-like particle, but has a finite size—a diameter of about 10^{-13} centimeter. Although it is only about 1/100,000 the size of an atom, it is still measurable. This is unlike certain other particles, notably the electron, for which no extension has been noted, so that the electron can be regarded as a mathematical point. In the experiment, electrons were raised to an energy of some 20 billion electron volts and struck protons and neutrons in the atoms of a stationary target. The angular distributions of the scattered electrons and of other particles created in the collisions were carefully monitored. Most of the electrons, as expected, passed through the target with little change in direction. An unexpected excess of widely scattered particles was produced, however—much greater than if the proton were diffuse and homogeneous. The excess of the widely scattered particles was attributed to a mass embedded within the proton, estimated at no more than $\frac{1}{50}$ the diameter of the proton. In later experiments, a target was illuminated by means of muons (like electrons but with a mass 200 times greater); and by a beam of neutrinos (which lack both mass and electric charge). The results of the original and later experiments were consistent and the deep scattering of particles was attributed to collisions between the incident leptons and some "hard" constituent of the proton.

Theories and postulations continue to be developed concerning the symmetry of nature. For example, in a 1986 paper, H.E. Haber (University of California, Santa Cruz) and G.L. Kane (University of Michigan) observe that *supersymmetry* could represent the next step in the quest for a few simple laws that explain the nature of matter. Physicists are seeking evidence to test the theory. As described, in supersymmetry, for every ordinary particle that exists there is a so-called "superpartner" having similar properties, with exception of the quantity referred to as *spin*, previously discussed here. In a 1985 paper, C. Quigg (Fermi National Accelerator Laboratory) further describes current theories and hypotheses pertaining to elementary particles and forces and observes that a coherent view of the fundamental constituents of matter and the forces governing them is beginning to emerge. A present goal of physicists is to merge disparate theories into a single comprehensive description of natural events. It is agreed among most high-energy specialists that to reach that goal, greater and greater energy must be brought to the experiments. The Superconducting Supercollider (SSC) would be one of these. The complex concept of CP invariance is described in a 1988 paper by R. Adair (Brookhaven National Laboratory). In this postulate, the claim is made that without CP invariance there would be no matter in the universe. Adair observes that if the approximate symmetry between matter and antimatter that has been observed were perfect, the universe would be elegantly simple but virtually empty of matter and of creatures made up of that matter who could contemplate that elegance. It is proposed that the existence of the universe as currently known comes from a flaw in a symmetry exhibited by a universal mirror (the CP mirror), i.e., a symmetry that requires that the

Time reversal invariance describes the fact that in reactions between elementary particles, it does not make any difference if the direction of the time coordinate is reversed. Since all reactions are invariant to simultaneous application of mirror inversion, charge conjugation, and time reversal, the combination of all three is called *CPT* symmetry and is considered to be a very fundamental symmetry of nature.

A relatively recent type of space-time symmetry has been introduced to explain the results of certain high-energy scattering experiments. This is *scale symmetry* and it pertains to the rescaling or "dilation" of the space-time coordinates of a system without changing the physics of the system. Other symmetries, such as chirality, are more of an abstract nature, but aid the theorist in an effort to bring order into the vast array of possible elementary particle reactions.

A feature of quantum field theory is that the quanta of the fields are initially massless. Spontaneous symmetry breaking offers a mechanism by which weak field quanta, for example, can acquire masses. Unification of weak and electromagnetic forces may be viewed thus in the following manner at short distances (high energies), the masses of the weak field quanta become unimportant and thus original symmetry is restored. Symmetry in this context refers to the properties of the equations of motion of particles in the field theories. Spontaneous symmetry breaking occurs when solutions of the equations do not display full symmetry. Some physicists have likened this to a ball moving on a roulette wheel, whose equations of motion are symmetrical about the axis of rotation even though it always stops in an asymmetric position.

outcomes of some events in nature should remain the same on changing matter to antimatter and viewing the result in a mirror.

Before discovery of the hadron particle (designated *psi* or *J*), and after much experimental and theoretical effort, physicists had about concluded that three massive, fractionally charged entities (quarks) were the primary building blocks of the universe. However, discovery of the psi particles in 1974 indicated a fourth quark was required. Previously, in the three-quark model, all mesons were made up of one quark and one antiquark; baryons, of three quarks; and all anti-baryons, of three antiquarks. Prior to 1974, all of the known hadrons could be accommodated within this basic scheme. Three of the possible meson combinations of quark-antiquark could have the same quantum numbers as the photon, and hence could be produced abundantly in e^+e^- annihilation. These three predicted states had all been found.

As pointed out by Richter (1977), the first proposal of a theory based on four quarks rather than three was published in 1964 by Amati and others. The motivation at that time was more esthetic than practical, and these models gradually expired for want of an experimental fact that called for more than a three-quark explanation. In 1970, Glashow explained in a paper that a fourth quark was required to explain the nonoccurrence of certain weak decays. The fourth or c quark was assumed to have a charge of $+\frac{2}{3}$, like the u quark, and also to carry +1 unit of a previously unknown quantum number, called *charm* by Glashow, which was conserved in both the strong and electromagnetic interactions, but not in the weak interactions. Discovery of the psi particles demonstrated a more compelling need for the fourth quark. Richter observes that the four-quark model of hadrons seemed to account, in at least a qualitative fashion, for all of the main experimental information that had been gathered about the psions, and by the early part of 1976, the consensus for charm had become quite strong.

In 1977, the *upsilon particle* was found as the result of energetic collisions between protons and copper nuclei. The upsilon particle has a mass three times greater than any other subatomic entity yet detected (early 1980s). Researchers on this experiment from Columbia University, the State University of New York (Stony Brook), and the Fermi National Accelerator Laboratory reported that, with a mass at its lower energy state equivalent to 9.0 GeV and masses in excited states equivalent to 10 and 10.4 GeV, the upsilon particle has been interpreted as consisting of a massive new quark (the fifth) bound to its antiquark. Confirming experiments were also conducted at the Deutsches Elektronen Synchrotron (DESY) located near Hamburg, Germany. The quantum attribute of the fifth quark was named "bottom." With a fifth quark reported, many physicists felt that finding a sixth quark ("top") was highly probable.

In 1979, the Nobel Prize (Physics) was awarded to Glashow and Weinberg (both of Harvard) and Salam, a Pakistani physicist, in recognition of the significance of the theory which unites the weak force with the electromagnetic force. But most scientists recognize this finding as only a milestone in a series of predictions that include the existence of new particles so massive that they cannot be expected to appear at the energies thus far available to physicists.

The chemists of the nineteenth century once thought that all material substances were comprised of only 36 elements.[5] Over the years, these expanded to over 100 elements. Fifty years ago, it was proclaimed that the elements were made of electrons, protons, and neutrons. Then, commencing in the 1940s, many other particles were found, as described here. Then for a while it seemed that elementary matter could be reduced to three particles—the quarks. But quarks have multiplied in number, with a sixth quark now seriously proposed. Will there be too many quarks? Perhaps hypothetical particles will be proposed from which the quarks are comprised. Possibly the ultimate answer will lie with the "mathematical groups that order the particles rather than in truly elementary objects."

In the late 1980s and thereafter, the *superstring* theory was drawing the attention of most theoretical physicists. Out of this theory may emerge the long sought concept that will account for all four fundamental forces. Quantum theories have been formulated for three of the four known forces of nature—strong, weak, and electromagnetic interactions. This comprises one of two basic foundations. The other, of course, is Einstein's general theory of relativity, which relates the force of gravity to the structure of space and time. A quantum theory of gravity is missing. Some scientists

[5] Line of thought suggested by L.M. Lederman (Columbia University) in a paper on "The Upsilon Particle," *Sci. Amer.*, **239**, **4**, 80 (1978).

currently feel that there is much potential for the string theory to bring this unification about. String theory was first proposed by Y. Nambu (University of Chicago) in 1970. Traditional models of elementary particles are based on quantum field theory, which involves dimensionless points and quantum numbers, but with no specified internal structure. In studying one of the alternatives to the foregoing, a dual resonance model, Nambu observed that it was equivalent mathematically to the interaction of bits of string. The dual resonance model had been constructed to show how one hadron should scatter off another. The theory envisions the strings about 10^{-35} meter long (10^{20} times smaller than the diameter of a proton). As observed by M. Green (Queen Mary College, University of London) in a 1986 paper (reference listed), string has extension; it can vibrate like a violin string. The harmonic, or normal, modes of vibrations are determined by the tension of the string. In quantum mechanics, waves and particles are dual aspects of the same phenomenon. Thus, each vibrational mode of a string corresponds to a particle. It is envisioned that two strings interact when they touch their tips together and are fused into one—or one string may split into two parts. Strings can absorb energy in a collision and may ripple and rotate (at the speed of light). The original string theory initially was limited to providing a satisfactory "explanation" for describing bosons (pi meson and rho meson), i.e., particles that have integral numbers of spin angular momentum units. Attempts to describe particles with half-integral spin, such as fermions (proton, neutron), failed. However, in 1976, a suggestion was made by Scherk, a French physicist, to the effect that a string model could represent a fermion, but with the proviso that a fermion was matched by a corresponding boson. Incidentally, this is the type of correspondence required by the principle of supersymmetry. One of several difficulties that lie ahead for the superstring theory is that of matching its mathematical concepts (26 dimensions—25 space and 1 time) with the real world of 4 dimensions. As one physicist has observed (E. Witten, Princeton University), it is a complete mystery what string theory is at a fundamental level.

State of the Science (Early 1990s)

In terms of comprehending the ultimate nature of matter, most scientists would agree that tremendous progress has been made over the past century and a half, but that a grand unifying theory of particles and their structures and interaction may continue to elude researchers for many years.

Number of Particle Families. How many families of matter may exist? Three, four, or more? An acceptable number among researchers today is three. Three family entities make up matter—the stars, the planets, molecules, and the atoms in the paper upon which this is printed. These fundamental particles are the "up" quark, the "down" quark, and the electron. Some other researchers are not quite so confident. One is reminded of the quotation from Jonathan Swift:

> So, naturalists, observe, a flea
>
> Hath smaller fleas that on him prey:
>
> And these have smaller still to bite'em
>
> And so proceed *ad infinitum*.

Within the limitations of contemporary knowledge, there are three families of fundamental particles, the approximate properties of which are given in Table 1. In 1991, G. Feldman (Harvard University) and J. Steinberger (CERN and 1988 Nobelist for discovery of the muon nutrino) observed, "Many questions remain unanswered. Why are there just three families of particles? What law determines the masses of their members, decreeing that they shall span 10 powers of 107? These problems lie at the center of particle physics today. They have been brought one step closer to solution by the numbering of the families of matter."

In their reference listed, Feldman and Steinberger describe how experiments at CERN and SLAC, using electron-positron collisions, showed that there are only three families of fundamental particles in the universe.

By contrast, as D. Cline (University of California) observed in 1988, "Several theorists think a new quark should exist in the vicinity of 246 GeV. One of the notable features of the standard model is its prediction that at high enough energies the various forces begin to unify. In particular, the electromagnetic force and the weak and strong nuclear forces should become a single 'grand unified' force. The forces should be unified at the incredible energy of 10^{15} GeV, considerably beyond what can ever be attained by an accelerator on earth. The extrapolation of measured values

TABLE 1. THREE FAMILIES OF FUNDAMENTAL PARTICLES

	Charge	Mass in Billions of Electron Volts (GeV)		
		Electron Family	Muon Family	Tau Family
Quarks	2/3	*Up* ~0.01 GeV	*Charm* ~1.5 GeV	*Top* Est. 89 GeV*
	−1/3	*Down* ~0.01 GeV	*Strange* ~0.15 GeV	*Bottom* ~5.5 GeV
Leptons	0	*Electron neutrino** <2 × 10^{-8} GeV	*Muon Neutrino** <2 × 10^{-4} GeV	*Bottom* <5.5 GeV
	−1	*Electron* 5.11 × 10^{-4} GeV	*Muon* 0.106 GeV	*Tau* 1.78 GeV

→ = Relative Increase in Mass →
(*Mass unknown)
Data source: G.J. Feldman and J. Steinberger (see reference listed).

of fundamental parameters from low energies to the grand-unified energy scale would require the existence of a new massive quark for consistency."

Nuclear Equation of State. In a late 1991 paper, H. Guthrod (Institute of Heavy-Ion Research, Darmstadt) and H. Stöcker (University of Frankfurt) are developing a nuclear equation of state in order to clarify the "new" states of matter and conditions that may occur inside a supernova and the organization of the universe. The authors observe, "Nuclear matter in its normal phase resembles a liquid. Increasing the temperature or density 'boils' nuclei into the hadron gas phase. Under extreme density but low temperature, nucleons could become 'frozen,' forming condensates, further heating or compression may produce the plasma phase, which would consist of free quarks and gluons. The gas and plasma phases may exist simultaneously over a wide region. Particles that have strange quarks, such as multistrange, metastable objects ('memos') and strangelets, may also form."

This approach parallels our consideration of the equation of state that applies to "ordinary" matter, such as gases, liquids, and solids that exist in macrostructured materials. See also **Equation of State**.

Particle Accelerators

Subatomic particles, such as electrons, positrons, and protons, can be accelerated to high velocities and energies, usually expressed in terms of center-of-mass energy, by machines that impart energy to the particles in small stages or nudges, ultimately achieving in this way very high-energy beams, measured in terms of billions and even trillions of electron volts. Thus, in terms of their scale, particles can be made to perform as powerful missiles for bombarding other particles in a target substance or for colliding with each other as they assume intersecting orbits. Because the particles are empowered with high energy, their smashing encounters are conducive to breaking the particles into their constituents. Instruments or machines used to arrange these particle encounters are known as *particle accelerators* and are very large, their dimensions frequently measured in terms of a few miles or kilometers. Thus, in a sense, it is ironic that the largest tools of science are required to seek knowledge concerning the smallest particles of matter that make up the universe and the earth. Theoretically, if at some future date sufficient energy can be imparted to two particles, their head-on collision will yield a complete fireball of disintegration, such that the absolute, indivisible particles of matter will be lain bare. A further division of matter then would be theoretically impossible. This is a goal of many particle physicists and perhaps achievable within a relatively few decades. Or, on the other hand, would then new theories be formulated to show that the true absolutes have indeed not yet been found?

Electromagnetic forces are used to accelerate particles, requiring that the particles must have an electric charge. Protons (+1) and electrons (−1) are commonly used as the media for particle-physics experiments, although not exclusively. The particles must be accelerated within a vacuum because otherwise they would collide with the molecules of air. The tube-shaped enclosures for the speeding particles are maintained under a vacuum of about 10^{-9} torr. The particles are set in motion by an electric field, which in the simplest configuration is an applied high voltage across a pair of electrodes. The positive electrode attracts electrons; the negative electrode attracts protons. It may be pointed out that an acceleration of this kind occurs in the ordinary television receiver. This basic principle, which applied to the early accelerators, remains the basis of the current operating and planned accelerators of the future. A simple electrode arrangement cannot sustain a potential of over a few million volts because of breakdown of insulation and arcing, and thus is confined to the simplest kind of accelerator where high energies are not required. In a practical sense, particles must be provided with energy in a large number of stages or "nudges." Thus, as in a *linear accelerator* (linac), the stages are strung out along a straight path, each stage requiring a radio-frequency oscillator which sets up an alternating electric field which is connected to each set of electrodes. Many radio-frequency cavities are formed along the line, and to provide the correct parity so that the particles will be accelerated rather than retarded the oscillators for the successive cavities must be synchronized. Effectively, an electromagnetic wave that travels continuously through the evacuated chamber is set up. It has been suggested by one physicist that the particle "rides the electrical wave as a surfer rides a water wave." In a less costly arrangement, known as the *synchrotron*, the particles are made to follow a circular or closed curve rather than a linear course. Groups of particles may circle a ring of this kind several million times while they are increasing their energy and velocity. Only one or a few radio-frequency cavities are required because energy is picked up each time the particle completes a revolution.

In addition to linear accelerators and synchrotrons, the accelerated particles of which ultimately are directed to strike a fixed target, there are *colliding-particle machines* in which particles are made to collide head-on. These machines are similar to synchrotrons except one bunch, or cluster of particles, travels in one direction while another bunch travels in the opposite direction. Where the colliding particles have the same rest mass and, after acceleration, have the same energy, the center-of-mass energy is the sum of the two beam energies. Thus, two beams with energies of 50 GeV (billion electron volts), for example, can provide the colliding force of one beam of 100 GeV against a fixed target. As will be shown later, head-on collision apparatus requires the use of storage rings, and systems are arranged in various configurations. Electron-positron storage rings are particularly efficient for creating new elementary particles from high-energy collisions.

The energy acquired by the particles in an accelerator is expressed in electron volts (eV), the amount of energy gained by any particle bearing a charge equal to that of an electron when it falls through a potential difference of 1 volt. Thus, 10^3 eV = 1 keV (kiloelectron volt); 10^6 eV = 1 MeV (megaelectron volt or one million eV): 10^9 eV = 1 GeV (gigaelectron volt or one billion eV); and 10^{12} eV = 1 TeV (teraelectron volt or one trillion eV). The large accelerators have been in the GeV range. Plans call for machines in the TeV range. Inasmuch as the mass and energy can be freely interconverted, the mass of a particle is usually expressed in terms of its energy equivalent in eV. The mass of the proton thus is 938 MeV.

High-energy machines require a supply not only of the type of particle desired, but also particles that have been preliminarily accelerated. Electrons are comparatively easy to generate as inputs to accelerators, as by the Cockcroft-Walton generator. Protons are obtained by ionizing hydrogen atoms.

As early as 1928, E.O. Lawrence constructed one of the first particle accelerators out of laboratory glassware. This was only a few inches in diameter. Principles remain essentially the same even today, but the size of accelerators has increased tremendously. They no longer are parts of laboratories, but rather the detection and laboratory aspects of modern accelerators are appended around the accelerator. Modern accelerators cost many millions of dollars and a major installation requires numerous scientists and scores of support people, aided by a number of digital computers. In the future, costs may become so high that several countries, working together, will have to support particle-physics research facilities, as already exemplified by the European Organization for Nuclear Research (abbreviation CERN for the former name, *Conseil Européen pour la Recherche Nucléaire*) with facilities located in Geneva, Switzerland, and adjacent land in France.

Particle Generators. Direct voltage for charging particles may be obtained in two basic ways: (1) by means of a cascade process, such as the cascade rectifiers or voltage-multiplying circuits; and (2) by charging up a terminal through actual transportation of the charge.

The Cockcroft-Walton generator is of the first type and consists of several stages of a voltage-doubling circuit together with an ion source and a suitably designed discharge tube. Although it is possible to use electrons, these accelerators are usually used as positive-ion sources and can provide

dc currents up to about 10 mA and energies up to about 1.5 MeV without special pressure tanks. With this type of accelerator, Cockcroft and Walton were able, in 1932, to induce the first nuclear reaction using artificially accelerated protons. The reaction was: $^1H + {}^7Li \rightarrow {}^8Be \rightarrow 2{}^4He$. See Fig. 1.

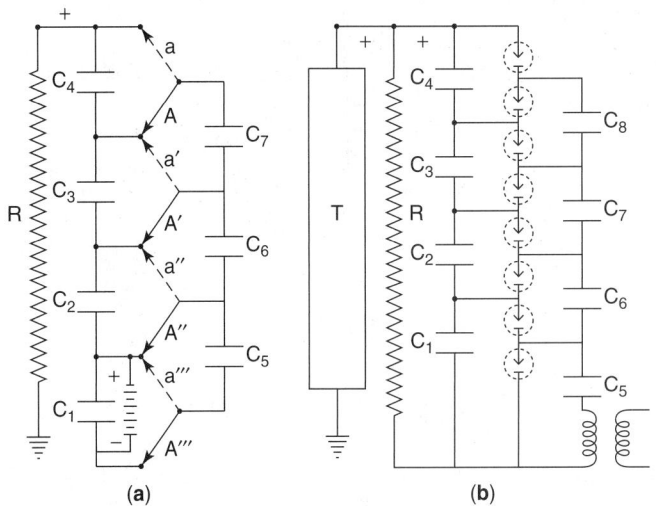

(a)   **(b)**

Fig. 1.   Cockcroft and Walton scheme in which rectified A.C. voltage is used to charge a series of capacitors to a high D.C. potential, which is then applied to the acceleration of charged nuclear particles. The principle of operation can best be described by the apparatus using mechanical switches shown in A. Condensers of equal capacity are represented by C_1, C_2, C_3, etc. If the switch blades are down, the battery B is connected across condensers C_1 and C_5. On moving the blades up (dashed lines), condenser C_5 transfers half its charge to C_2. When the blades are moved down again, C_2 transfers half its charge to C_6 and C_5 is recharged to full capacitance. A continuous up and down motion of the blades results in a transfer of charge up the condenser bank until every condenser is charged to the voltage of the battery. The total voltage applied to the discharge tube, symbolized here by the resistance R, is the sum of the voltage across the condensers C_1, C_2, C_3 and C_4 in series. In the actual apparatus (part B), an alternating voltage was applied through a transformer and the switching action was accomplished by the use of rectifier tubes.

The second method is employed in the *electrostatic*, or *Van de Graaff*, generator where a row of corona points sprays charge onto a moving belt that carries the charge to the field-free region inside a spherical metal terminal. Currents of about 1 mA for electrons, and up to about 500 μA for positive ions, are obtained in the range 1 to 5 MeV with a precision of about 0.1%. The whole apparatus is enclosed in a pressure tank and operated at about 10 atmospheres. In this form, the maximum energy is close to 10 MeV and is limited by breakdown between the terminal and its surroundings. However, double the energy can be attained by accelerating negative ions to the positively charged terminal; then, through electron-stripping, positive ions are created which can be accelerated again as they pass from the terminal to ground. Such tandem Van de Graaff generators, in two- and three-stage variations, can provide particles in the 10 to 30 MeV range, with high precision. See Fig. 2.

An electric field may also be produced by a time-varying magnetic field. The changing magnetic flux in the central core of a pulsed cylindrical electromagnet induces a transverse electric field that accelerates the particles. These travel in a doughnut-shaped vacuum chamber located between the poles of the magnet surrounding the core. The magnetic field between these poles keeps the particles traveling in a circle, but it must be carefully designed to keep the particle orbits within the vacuum chamber during each pulse. Although betatrons can accelerate positively charged particles, they have been used for electrons. The electrons can be extracted, but they usually bombard an inner target to produce beams of X-rays, which can be as intense as 1400 roentgens/minute at a distance of 1 meter from the target. Pulsing rates vary from 30 to 60 times per second.

Other types of accelerators use various forms of rf electric fields, at relatively low voltage, which are applied many times in a given direction to the particles and are prevented from influencing them when the rf field is reversed.

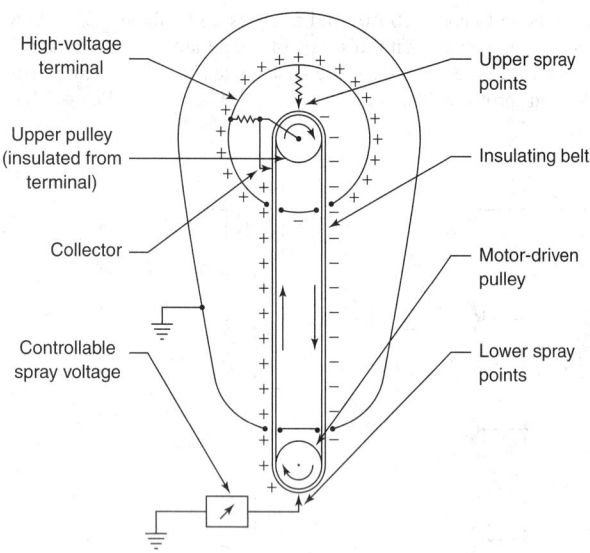

Fig. 2.   Diagram of Van de Graff electrostatic belt generator.

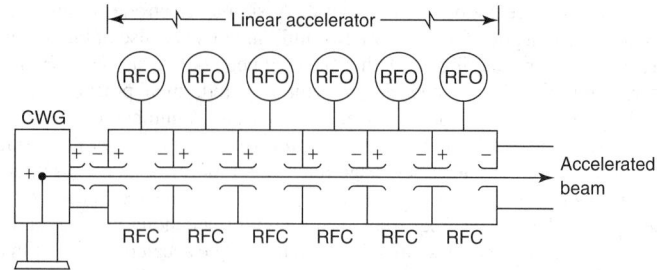

Fig. 3.   Principle of linear accelerator (linac). Partially accelerated electrons from a source, such as a Cockcroft-Walton generator, are further accelerated by stages as the electrons pass through radio-frequency cavities, powered by rf oscillators. Each particle receives a small "push" as it passes from one cavity to the next until the final desired accelerated beam is produced. The machine must be carefully synchronized. CSG = Cockcroft-Walton generator; RFO = radio-frequency oscillator; RFC = radio-frequency cavity.

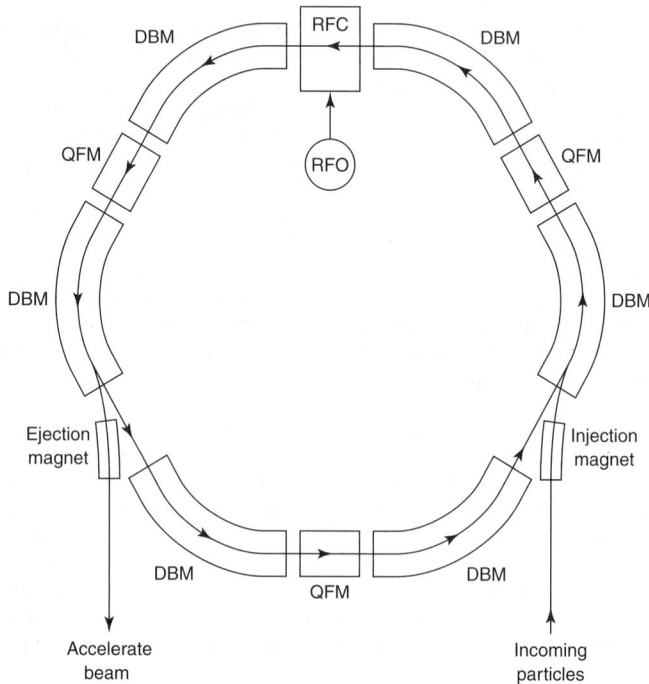

Fig. 4.   Principle of synchrotron. One radio-frequency cavity (there may be several) provides a small "push" each time a particle passes through it. Unlike the linear accelerator, the synchrotron requires only one or few rf cavities. Dipole bending magnets keep the particles on their proper course. Focusing magnets keep the particles in a narrow beam, thus preventing undesired scattering. Particles enter the machine through an injection magnet and leave through an ejection magnet. DBM = dipole bending magnet; QFM = quadrupole focusing magnet; RFC = radio-frequency cavity; RFO = quadrupole focusing magnet; RFC = radio-frequency cavity; RFO = radio-frequency oscillator.

Linear Accelerators. The linear accelerator (*linac*) has the advantage that the accelerated beam is easily extracted for experimental use. In principle, it is capable of producing well-focused beams of higher intensity than are available from circular machines of the synchrotron type. It does, however, require very high power levels at frequencies where conversion equipment is relatively expensive. For a given final energy, a linear accelerator will usually be materially more expensive than a synchrotron.

The rf fields used for acceleration are set up in a long cylindrical cavity whose axis is to be the axis of the accelerated beam. Hence, for acceleration the field pattern must have a major electric field component parallel to the axis. This requirement is satisfied by the TM$_{01}$ waveguide mode in which a paraxial electric field has its maximum strength at the axis and falls to zero at the cavity wall. Azimuthal magnetic fields lie in planes normal to the axis, have small values near the axis and increase to maximum values at the cavity walls. Usually the field pattern is maintained by coupling to these magnetic fields by loops or apertures excited by external power sources. Corresponding to the high rf magnetic field at the wall, paraxial currents flow in the walls and are responsible for a major function of the power loss in the system. When high electric fields are required on the axis to accelerate to high energy in reasonable distances, the wall currents are correspondingly high.

Both standing wave and traveling wave patterns can be used in linear accelerators. If traveling waves are used, the phase velocity of the waves must be made equal to the velocity of the particles accelerated; as the particle velocity increases, the phase velocity must also increase. But, phase velocities in simple waveguides always are greater than the velocity of light, and loading must be introduced to reduce the phase velocity to the desired value. This can be accomplished by the introduction at intervals of washer-shaped irises.

The operating principles of a linear accelerator are shown in Fig. 3.

Synchrotrons. A particle is made to follow a circular (or other closed curve) orbit by arranging a number of magnets in a ring. The principle is illustrated by Fig. 4. Two kinds of magnets are required. Dipole magnets (two poles) generate a uniform magnetic field. Spaced around the ring, these magnets bend the particle trajectory. To keep a concentrated beam of particles, quadrupole magnets (two north poles and two south poles), which have no effect on deflecting the particles, are used to focus them. Acting like lenses, these magnets form the particles into a narrow beam. Depending upon the size and general configuration of a synchrotron, radio-frequency cavities may be variously interspersed among the magnets where the actual acceleration occurs. Special magnets are used for injecting particles into the ring and for extracting the accelerated beam of particles.

The first substantial synchrotron was built as early as 1952. Known as the *Cosmotron*, it was installed at Brookhaven National Laboratory on Long Island, New York. The device achieved energies up to 3 GeV. Two years later, the *Bevatron*, with energies up to 6.2 GeV, was installed at the University of California at Berkeley. A shortcoming of these earlier designs was the magnet system, which provided inadequate focusing of

the beam. A system of strong focusing was introduced to later-generation synchrotrons. As pointed out by R.R. Wilson (1980). "The shape of the magnetic field can be described mathematically as being partly uniform (the dipole component) and partly a gradient in a direction transverse to the orbit of the beam (the quadrupole component). The quadrupole component was made stronger and was alternated in sign, so the oscillations of the particles around the desired orbit were more frequent, but of smaller amplitude. As a result of this alternating gradient, the aperture of the magnets and the bore of the vacuum chamber could be smaller. It is the invention of the synchrotron and of strong focusing that has made the very large accelerators of today economically feasible." Synchrotrons with this design are known as *alternating gradient synchrotrons* (AGS).

The operation of a synchrotron is cyclic, i.e., a bunch of particles will be injected, with the bending magnets precisely adjusted to cause the particles to closely follow the curvature of the evacuated chamber. But, as the particles increase in energy, the field strength in the bending magnets must also be increased. Upon achieving their desired or maximum possible energy level, the particles are extracted—possibly going directly to bombard a target, or to supply an even more powerful synchrotron for further acceleration. When experiments with the first group or bunch of particles have been completed, the magnetic field is reduced to its original level, after which the next bunch or group of particles is added. The term *synchrotron* stems from the fact that the particles automatically synchronize their motion with the rising magnetic field and the rising frequency of the accelerating voltage. It is interesting to note that some accelerators (linacs or synchrotrons), which in their day may have been regarded as most powerful, may later be used as preliminary particle accelerators, as feeders to larger machines of later designs. Thus, instead of dismantling the older machines, in some cases the cost of a more powerful machine can be reduced. An example is shown in Fig. 5. Another example at CERN in Geneva, Switzerland will be cited later. Once the particles in a synchrotron have reached full energy, they are nudged out of their orbit by a special ejection magnet.

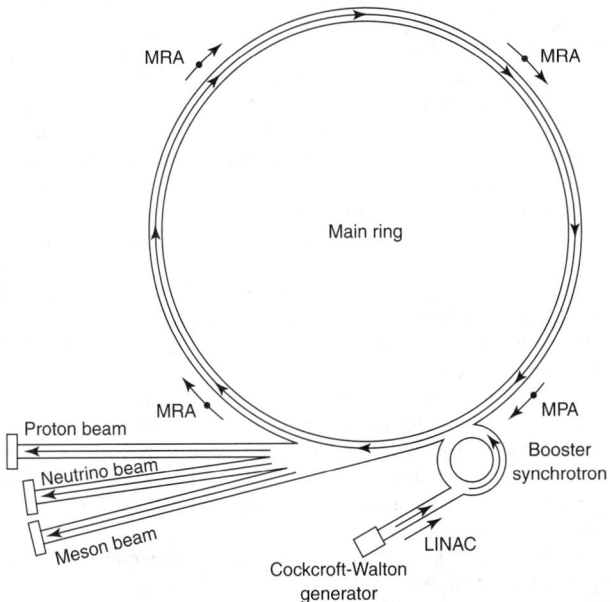

Fig. 5.   An early fixed-target accelerator comprised of a large main-ring synchrotron with four stages of acceleration (MRA); a booster synchrotron; a linear accelerator (linac), and a Cockcroft-Walton generator. Protrons are accelerated to 0.75 MeV in the Cockcroft-Walton generator; to 300 MeV in the linac; to 8 GeV in the booster synchrotron; and to 400–500 GeV in the main-ring synchrotron. Experiments are not limited to accelerated protons, but also can be conducted with beams of secondary particles (mesons and neutrinos) which are knocked out of the target by impacting protons.

The linacs and synchrotrons described thus far are for use with *fixed targets*. The advantages of colliding-particle machines will be described shortly. What, then, are the reasons for continuing to bombard fixed targets? The fixed-target configuration provides the creation of a variety of secondary beams (neutrinos, muons, pions and other mesons, antiprotons, and massive particles called hyperons). Fixed-target machines also are effective in furnishing particles to larger accelerators and storage rings. When particles are beamed at a fixed target, an interaction is assured. In contrast, since in a colliding-particle configuration the great majority of particles will simply pass each other without colliding, the number of events of interest may only occur at the rate of comparatively few per day.

Colliding-Particle Machines. As mentioned earlier, the effective beam energy essentially can be doubled when particles can be made to collide head-on with each other. This is shown dramatically by the graph of R.R. Wilson (1980) and given in Fig. 6. Wilson had made an effective comparison of the resolving power of an accelerator with that of a microscope. In a microscope, the ultimate limit to resolution is the wavelength of radiation that illuminates the specimen. Thus, a visual lightwave microscope is limited to distinguishing objects of 10^{-5} centimeter or larger. Since a particle can be described (quantum mechanics) as a wave, the wavelength is inversely proportional to the momentum of

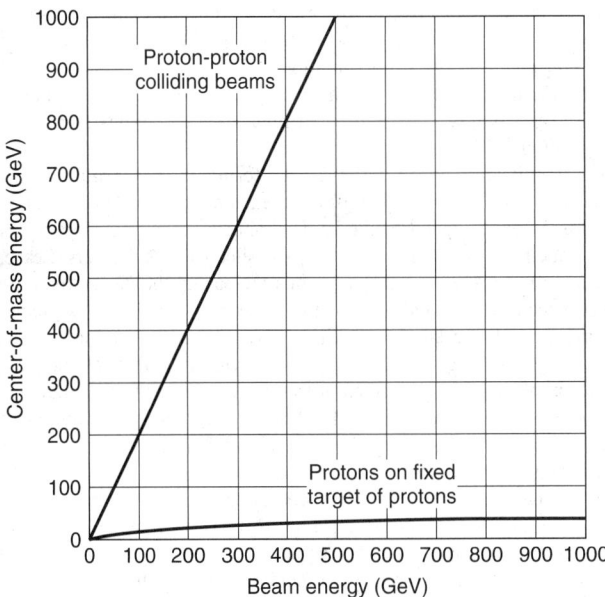

Fig. 6.   Comparison of effective collision energy. Colliding-beam devices double the beam energy effectiveness. In fixed-target accelerators, the center-of-mass energy is proportional to the square root of the beam energy (at low energy) and at higher-energy levels, this rises even more slowly because of relativistic effects. (*After R.R. Wilson.*)

the particle. Thus, one objective in accelerator design is improvement of resolution by reduction of particle wavelength. The early large accelerators had an effective resolution of about 10^{-16} centimeter (1/1000 the diameter of the proton).

When a particle and its antiparticle, such as an electron and a positron, or a proton and an antiproton, are used in head-on collision experiments, acceleration of the particles can be accomplished in one ring. This is because electrons and positrons, for example, behave in the same way in terms of their response to magnetic and electric fields. Thus, both particles can be injected into the same ring, one to follow an orbit in a clockwise direction; the other in a counterclockwise direction. Upon injection of a cluster of each type of particle, collisions occur at two points diametrically opposed. This arrangement provides maximum utilization of the equipment.

Other advantages of using a particle and its antiparticle, particularly the electron and the positron, is that when collisions occur, a less confusing splash of debris occurs.

When advantage of the single-ring, head-on collision approach cannot be taken, then *storage rings* are required. These rings are designed much like synchrotrons. Their primary purpose usually is not to accelerate the particles, but rather to maintain or "store" their energy while they continue to circulate in orbits in the ring. Just sufficient energy (usually furnished by a single radio-frequency cavity) is added to overcome losses, mainly due to synchrotron radiation. The storage rings are physically arranged tangentially to the main synchrotron ring so that precision transfer of particles from the storage rings to the main ring can be affected. Another concept was introduced in the early 1970s at the European CERN facility, where two interlaced rings that store counter-rotating proton beams are used. These rings cross over at eight points around their circumference. There are seven interaction zones that will accommodate detectors.

Scientists at the National Laboratories of the National Committee for Nuclear Energy (C.N.E.N.) at Frascati, Italy pioneered the first electron-positron ring in 1959. Each beam of the machine was 0.25 GeV and yielded a center-of-mass energy of 0.5 GeV. This ring was later moved to the Orsay laboratory outside Paris, France. The number of such rings grew quite rapidly.

Synchrotron Radiation

The electromagnetic radiation emitted as a result of continual acceleration toward the axis of rotation of charged particles moving in a magnetic field is generally known as synchrotron radiation because it was first observed during the operation of electron synchrotrons. The rate of emission of synchroton radiation varies with the fourth power of the particle energy, inversely with the radius of curvature, and inversely with the fourth power of the rest mass. Because of the rest mass relationship, no significant quantity of this radiation is formed from proton trajectories in magnetic fields, but an easily measurable loss of energy to synchrotron radiation occurs when electrons with high kinetic energies spiral through a magnetic field. Many microwave radiations observed in radio-astronomy measurements are believed to have been formed as synchrotron radiation.

At the beam energies that have been attained to date, synchrotron radiation is not a major factor in designing a proton accelerator, but it is the principal limitation on the energy of electron accelerators. Lessening the curvature of a synchrotron and thus increasing the circumference will reduce synchrotron radiation. For this reason, the radiation is practically negligible in a linear accelerator (linac). Synchrotron radiation energy loss in a storage ring must be made up by providing additional energy from one or more radio-frequency cavities. However, synchrotron radiation is not entirely wasted energy. The effect of the radiation is to damp out small excursions of the electrons away from their main trajectory, making a beam of electrons easier to control than a beam of protons.

Aside from its effects in high-energy experimentation, synchrotron radiation is of interest and value primarily as a source of tunable coherent x-rays. As summarized in a 1985 paper by Atwood, Halbach, and Kim (Lawrence Berkeley Laboratory), a modern 1- to 2-billion-eV synchrotron radiation facility (based on high-brightness electron beams and magnetic undulators) would generate coherent, laser-like, soft x-rays of wavelengths as short as 10 angstroms. The radiation would also be broadly tunable and subject to full polarization control. Radiation with these properties could be used for phase- and element-sensitive microprobing of biological assemblies and material interfaces as well as research on the production of electronic microstructures with features smaller than 1000 angstroms. These short wavelength capabilities, which extend to the K-absorption edges of carbon, nitrogen, and oxygen, are neither available nor projected for laboratory XUV lasers. Higher-energy storage rings (5 to 6 billion eV) would generate significantly less coherent radiation and would be further compromised by additional x-ray thermal loading of optical components.

Much interest and limited application of synchrotron radiation by medical professionals appeared in the late 1980s. Synchrotron radiation is intense and its brilliance extends continuously over a broad bandwidth from infrared to hard x-rays. For example at the Stanford Synchrotron Radiation Laboratory, the output in the x-ray region is more than 100,000 times that of the most powerful x-ray tube. The natural collimation of synchrotron radiation can be monochromatized by different crystals and by adjusting the angle at which the crystal intercepts the radiation, the energy of the monochromatic beam can be selected. The tunability and intensity of the monochromatic radiation are currently making transvenous angiographic applications possible. See also **X-Ray Scan and Other Medical Imagery**.

Third-Generation Synchrotrons. The first run of experiments at the new, $500-million European Synchrotron Radiation Laboratory (ESRF), located in the Dauphiné Alps, was scheduled to commence in the early 1990s. The facility will provide the brightest continuous x-rays in the world, but have been eclipsed by a new Japanese facility in the mid-1990s. The ESRF will generate x-rays two orders of magnitude more brilliant than any prior facility. Stronger x-ray sources are of an advantage to biologists, physicists, chemists, and other researchers because of their shortened exposure time and increased resolution. As pointed out by one American physicist who plans to use the facility, "We'll be able to get a picture in minutes, seconds, or fraction of a second and there will be much less specimen damage when the photons enter and leave a specimen in a millisecond."

A large variety of experiments was scheduled, representing numerous scientific disciplines. Another similar but slightly larger facility was scheduled for the Riso National Laboratory in Roskilde, Denmark, sometime in 1996.

The most powerful synchrotron now is at Harima Science Garden City, approximately 60 mi (96 km) west of Osaka, Japan. It will be known as the SPring-8 (Super Photon Ring-8 GeV). It is envisioned that it may be able to fulfill the x-ray crystallographer's dream—that is, x-ray holography. Hiromichi Kamitsubo, project head, states, "X-ray holography would enable the direct visualization, not just see complex x-ray diffraction patterns of the structure of materials—as if we were seeing them with a microscope." Three-dimensional holograms created in a photosensitive material require high-powered and coherent x-rays. The SPring-8 will have

an energy rating of 8 GeV, with a main ring circumference of 4708 ft (1435 m) and 80 beam lines. Estimated cost is $1 billion.

Superconducting Super Collider (SSC)

As mentioned earlier, the ultimate fate of the SSC will be determined by economic and political factors. Generally, the scientific community in the United States and worldwide, in fact, looks to the SSC for important information on the structure of matter. Specific goals include:

(1) *The origin of mass.* As explained by Quigg and Schwitters, the current model of the weak interactions, previously mentioned, suggests that a similar situation may be realized in the universe. The field involved is not a magnetic field, but rather what is called a Higgs field. A major objective of the SSC will be to clarify the exact nature of the Higgs field and its interactions with other matter. It is currently assumed that the Higgs field pervades the universe because the total density is minimized in its presence. To change the magnitude of the field significantly (or to manifest one of its particles), sufficient energy density must be supplied to overcome an assumed natural tendency of the field to return to its normal universal background value. The energy levels of the SSC is expected to be sufficient for this.

(2) *Families of particles.* Quigg and Schwitters observe that one theoretical approach to understanding particle families postulates new symmetries under which different families are interchanged. A facet of this concept involves combining family symmetry and gage symmetries of the strong, weak, and electromagnetic interactions into one all-encompassing symmetry, but hidden by the presence of suitable background Higgs fields. Implementing such symmetries could lead to a new class of "family interactions" mediated by heavy bosons that may be created at the levels made possible by the SSC.

(3) *Chirality* (a peculiar asymmetry in the weak interactions of the observed quarks and leptons) can be investigated.

(4) *Supersymmetry* will be investigated, and may lead to unified field theories that include gravity.

(5) *Compositeness* is a characteristic of some particles initially investigated at CERN. Better understanding of the cosmos through SSC experiments is also envisioned. The SSC will make it possible to simulate the conditions that prevailed about 10^{-15} second after the primordial Big Bang explosion when the temperature of the universe was estimated about 10^{17} K.

All of the foregoing factors are explained in considerable detail in the Quigg and Schwitters reference listed.

The new accelerator complex will be based on the accelerator principles and technology that were developed for construction of the Fermilab Tevatron, coupled with extensive experience with superconducting magnets gained over the past two decades. The energy of the SSC will be equal to 20 times that of the Tevatron collider, with a total collision energy of 40 TeV. Racetrack in shape with a circumference of 87 km (52 mi), the SSC will utilize 10,000 helium-cooled superconducting magnets. The SSC will utilize an injector system consisting of a linear accelerator followed by two circular accelerators. The diameter of the main ring used in the final acceleration phase will be about 30 km (18.7 mi), depending upon details yet to be developed pertaining to the magnets that will be used to guide the protons. The superconducting magnet system will require many hundreds of miles of cryogenic plumbing, including several hundred thousand vacuum joints to assure the establishment and maintenance of superconductivity conditions. The linear dimensions of the SSC will be roughly 15 times those of the Tevatron and about 4 times those of the electron-positron collider being constructed by CERN. As suggested by Quigg and Schwitters (Fermi National Accelerator Laboratory) in a 1986 paper, technological fallout from construction of the SSC would include: (1) large-scale industrialization of superconducting wire fabrication and cryogenic refrigerator manufacture, making such technologies available to future power distribution and transportation systems; (2) improved tunneling techniques for future application to public works and transportation projects; (3) large-volume storage of helium, a potentially critical and nonrenewable resource; and (4) computer control and mechanical alignment systems extending over very large areas.

The SSC may furnish answers to critical scientific questions.

Additional Reading

Adair, R.K.: "A Flaw in a Universal Mirror," *Sci. Amer.*, 50 (February 1988).
Alfassi, Z.B. and M. Peisach: "Elemental Analysis by Particle Accelerators," CRC Press, LLC., Boca Raton, FL, 1991.

Amato, I.: "New Superconductors: A Slow Dawn," *Science*, 306 (January 15, 1993).

Ando, M., C. Uyama, M. Ibaraki, and M. Osaka: "Medical Applications of Synchrotron Radiation," Springer-Verlag Inc., New York, NY, 1998.

Atutov, S.N.: "Trapped Particles and Fundamental Physics," Kluwer Academic Publishers, Norwell, MA, 2002.

Bertsch, G.F., L. Frankfurt, and M. Strikman: "Where Are the Nuclear Pions?" *Science*, 773 (February 5, 1993).

Bertschinger, E.: "Uniting Cosmology and Particle Physics," Freeman, Salt Lake City, UT, 1992.

Branco, G.C., Q. Shafi, and J.I. Silva-Marcos: "Recent Developments in Particle Physics and Cosmology," Kluwer Academic Publishers, Norwell, MA, 2001.

Breuker, H., et al.: "Tracking and Imaging Elementary Particles," *Sci. Amer.*, 58 (August 1991).

Brown, F.R. and N.H. Christ: "Parallel Supercomputers for Lattice Gauge Theory," *Science*, 1393 (March 18, 1988).

Brown, L.M., M. Dresden, and L. Hoddeson: "From Pions to Quarks: Particle Physics in the 1950s," Cambridge University Press, New York, NY, 1989.

Chanowitz, M.S.: "The Z Boson," *Science*, 36 (July 6, 1990).

Chupp, E.L.: "Transient Particle Acceleration Associated with Solar Flares," *Science*, 229 (October 12, 1990).

Cline, D.B.: "Beyond Truth and Beauty: A Fourth Family of Particles," *Sci. Amer.*, 50 (August 1988).

Cline, D.B., C. Rubbia, and S. van der Meer: "The Search for Intermediate Vector Bosons." (March 1982). A classic reference in The Laureates' Anthology, 133, Scientific American, Inc., New York, NY, 1990.

Conte, M. and W.M. MacKay: "An Introduction to the Physics of Particle Accelerators," World Scientific Publishing Company, Inc., River Edge, NJ, 1991.

Dawson, J.M.: "Plasma Particle Accelerators," *Sci. Amer.*, 54 (March 1989).

Dehmelt, H.: "Experiments on the Structure of an Individual Elementary Particle," *Science*, 539 (February 2, 1990).

Donoghue, J.F., E. Golowich, and B.R. Holstein: "Dynamics of the Standard Model," Cambridge University Press, New York, NY, 1994.

Dunning, F.B. and R.G. Hulet: "Atomic, Molecular, and Optical Physics: Charged Particles," Vol. 29, Academic Press, Inc., San Diego, CA, 1995.

Ericson, T. and W. Weise: "Pions and Nuclei," Oxford University Press, Inc., New York, NY, 1988.

Ezhela, V.V., B. Armstrong, and J.D. Jackson: "Particle Physics: One Hundred Years of Discoveries: An Annotated Chronological Bibliography ANNOTATED," Springer-Verlag Inc., New York, NY, 1996.

Feldman, G.J. and J. Steinberger: "The Number of Families of Matter," *Sci. Amer.*, 70 (February 1991).

Flam, F.: "CERN's New Detectors Take Shape," *Science*, 180 (April 10, 1992).

Flam, F.: "Neural Nets: A New Way to Catch Elusive Particles?" *Science*, 1282 (May 29, 1992).

Gottfried, K. and V.F. Weisskopf: "Concepts of Particle Physics," Oxford University Press, Inc., New York, NY, 1997.

Graham, D.: "Testing Physicists' GUTS (Grand Unified Theory)," *Technology Review (MIT)*, 10 (May–June 1988).

Gutbrod, H. and H. Stocker: "The Nuclear Equation of State," *Sci. Amer.*, 58 (November 1991).

Helliwell, J.R.: "Macromolecular Crystallography with Synchrotron Radiation," Cambridge University Press, New York, NY, 1992.

Hermann, A., et al.: "History of CERN," North-Holland AQ: 2 Different Publishers? Elsevier Science, New York, NY, 1990.

Hoddeson, L., M. Riordan, M. Dresden, and L.M. Brown: "Rise of the Standard Model: Particles Physics in the 1960s and 1970s," Cambridge University Press, New York, NY, 1997.

Lederman, L.M.: "Observations in Particle Physics from Two Neutrinos to the Standard Model," *Science*, 664 (May 12, 1980).

Lederman, L.M.: "The Tevatron," *Sci. Amer.*, 48 (March 1991).

Myers, S. and E. Picasso: "The LEP Collider," *Sci. Amer.*, 54 (July 1990).

Leader, E.: "Spin in Particle Physics," Cambridge University Press, New York, NY, 2001.

Martin, B.R. and G. Shaw: "Particle Physics," 2nd Edition, John Wiley & Sons, Inc., New York, NY, 1997.

Month, M. and M. Dienes: "The Physics of Particle Accelerators," Vol. 2, American Institute of Physics, College Park, MD, 1997.

Olive, K.A.: "The Quark-Hadron Transition in Cosmology and Astrophysics," *Science*, 1194 (March 8, 1991).

Peterson, I.: "Quantum Interference," *Science News*, 363 (December 2, 1989).

Peterson, I.: "Protons and Antiprotons Held in the Balance," *Science News*, 38 (July 21, 1990).

Peterson, R.J. and D.D. Strottman: "Pion-Nucleus Physics," American Institute of Physics, College Park, MD, 1997.

Peterson, I.: "Beyond the Z," *Science News*, 204 (September 29, 1990).

Polchinski, J.G.: "String Theory: Superstring Theory and Beyond," Vol. 2, Cambridge University Press, New York, NY, 1998.

Pool, R.: "The Hunting of the Quark — Computer Style," *Science*, 46 (April 3, 1992).

Quigg, C.: "Gauge Theories of the Strong, Weak and Electromagnetic Interactions," Perseus Publishing, Boulder, CO, 1997.

Rees, J.R.: "The Stanford Linear Collider," *Sci. Amer.*, 58 (October 1989).

Rice, T.M.: "Can Europe Keep up the Pace in Condensed Matter Physics?" *Science*, 482 (April 24, 1992).

Riordan, M.: "The Discovery of Quarks," *Science*, 1287 (May 29, 1992).

Rothman, T.: "Ambidextrous Universe: New Particles Blur Distinction Between Fermions and Bosons," *Sci. Amer.*, 26 (May 1989).

Rubbia, V.: "The European Strategy in Particle Physics," *Science*, 484 (April 24, 1992).

Ruthen, R.: "Quark Quest," *Sci. Amer.*, 32 (March 1993).

Ruthen, R.: "Attractive and Demure," *Sci. Amer.*, 30 (May 1993).

Sarkar, S.: "Big Bang Laboratory for Particle Physics," Cambridge University Press, New York, NY, 2002.

Schmidt, V.: "Electron Spectrometry of Atoms Using Synchrotron Radiation," Cambridge University Press, New York, NY, 1997.

Schramm, D.N. and G. Steigman: "Particle Accelerators Test Cosmological Theory," *Sci. Amer.*, 66 (June 1988).

Selvin, P.: "How Do Particles Put on Weight?" *Science*, 173 (January 8, 1993).

Shifman, M., M.A. Shifman, and B.L. Ioffe: "At the Frontier of Particle Physics: Handbook of QCD," World Scientific Publishing Company, Inc., River Edge, NJ, 2001.

Staff: "Particle Physics Phenomenology," World Scientific Publishing Company, Inc., River Edge, NJ, 1997.

Sundaresan, M.K.: "Handbook of Particle Physics," CRC Press, LLC., Boca Raton, FL, 2001.

Taubes, G.: "Are Neutrino Mass Hunters Pursuing a Chimera?" *Science*, 731 (May 8, 1992).

Waldrop, M.M.: "SLAC Feels the Thrill of the Chase," *Science*, 771 (May 10, 1989).

Wilcek, F.: "Anyons," *Sci. Amer.*, 58 (May 1991).

Willeke, K.: "Physics of Particle Accelerators: An Introduction," Oxford University Press, Inc., New York, NY, 2000.

Wilson, E.J.N.: "An Introduction to Particle Accelerators," Oxford University Press, Inc., New York, NY, 2001.

Yam, P.: "Spin Cycle: Rotating Nucleii Share A Few Moments of Inertia," *Sci. Amer.*, 26 (October 1991).

Yan, Y.T., J.P. Naples, and M.J. Syphers: "Accelerator Physics at the Superconducting Super Collider," Springer-Verlag Inc., New York, NY, 1995.

Zotter, B.W. and S. Kheifets: "Impedances and Wakes in High Energy Particle Accelerators," World Scientific Publishing Company, Inc., River Edge, NJ, 1998.

Pre–1988 References

Adair, R.: "The Great Design: Particles, Fields, and Creation," Oxford University Press, Inc., New York, NY, 1987.

Atwood, D., K. Halbach, and Kwange-Je Kim: "Tunable Coherent X-rays," *Science*, **228**, 1265–1272 (1985).

Barnett, R.M., H.E. Haber, and G.L. Kane: "Supersymmetry—Lost or Found?" *Nuclear Physics,*" **B267**(3, 4) 625–678 (April 21, 1986).

Bengtsson, T., et al.: "Nuclear Shapes and Shape Transitions," *Physica Scripta*, **29**(5) 402–430 (May 1984).

Black, J.K., et al.: "Measurement of the CP-Nonconservation Parameter e 1/e," *Physical Review Letters*, **54**(15) 1628–1630 (April 15, 1985).

Broglia, R.A., C.H. Casso: "Frontiers in Nuclear Dynamics," Plenum, New York, NY, 1985.

Court, G.R., et al.: "Energy Dependence of Spin Effects," *Physical Review Letters*, **57**(5), 507–510 (August 4, 1986).

Crosbie, E.A., et al.: "Energy Dependence of Spin-Spin Effects in p-p Elastic Scattering at 90°," *Physical Review*, **23**(3) 600–603 (February 1, 1981).

de Rujula, A.: "Superstrings and Supersymmetry," *Nature*, **320**(6064), 678 (April 24, 1986).

Eichten, E., et al.: "Supercollider Physics," *Reviews of Modern Physics*, **56**(4), 579–707 (October 1984).

Ellis, J.: "Hope Grows for Supersymmetry," *Nature*, **313**(6004), 626–627 (February 21, 1985).

Glashow, S.L.: "Toward a Unified Theory: Threads in a Tapestry," in "Nobel Lectures," Elsevier Science, Amsterdam and New York, NY, 1981.

Green, M.B.: "Unification of Forces and Particles in Superstring Theories," *Nature*, **314**(6010), 409–414 (April 4, 1985).

Green, M.B.: "Superstrings," *Sci. Amer.*, 48–60 (September 1986).

Haber, H.E. and G.L. Kane: "The Search for Supersymmetry: Probing Physics Beyond the Standard Model," *Physics Reports*, **117**(2, 3), 75–263 (January 1985).

Haber, H.E. and G.L. Kane: "Is Nature Supersymmetric?" *Sci. Amer.*, 52–60 (June 1986).

Hamilton, J.H., P.G. Hansen, and E.F. Zganjar, *Reports on Progress in Physics*, **48**(5) 631–708 (May 1985).

Hamilton, J.H.: "Magic Numbers, Reinforcing Shell Gaps and Competing Shapes in Nucleii," *Progress in Particle and Nuclear Physics*, **15**, 107–134 (1985).

Krisch, A.D.: "Collisions between Spinning Protons," *Sci. Amer.*, 42–50 (August 1987).

Lipkin, H.J.: "Colour Theory in a Spin," *Nature*, **324**(6092), 14–16 (November 6, 1986).

Martin, J.A., W. Greiner: "Potential Energy Surface Model of Collective States," in High-Angular Momentum Property of Nuclei (N.R. Johnson, Ed.) Harwood Academic Publishers, New York, NJ, 1983.

Mulvey, J.H.: "The Nature of Matter," Oxford University Press Inc., New York, NY, 1981.

Nadis, N.: "Anti-Proton Fishing," *Technology Review (MIT)*, 15 (July 1987).

News: "Antiprotons Captured at CERN," *Science*, **233**, 1383–1384 (1986).

News: "Bright Synchrotron Sources Evolve," *Science*, **235**, 841–842 (1987).

News: "CERN Panel Backs New Accelerator," *Science*, **235**, 1567 (1987).

News: "Soviets Plan Huge Linear Collider," *Science*, **238**, 16–17 (1987).

Quigg, C.: "Elementary Particles and Forces," *Sci. Amer.*, 84–95 (April 1985).

Quigg, C. and R.F. Schwitters: "Elementary Particle Physics and the Superconducting Super Collider," *Science*, **231**, 1522–1527 (1986).

Richter, B.: "From the Psi to Charm: The Experiments of 1975 and 1976," *Science*, **196**, 1286–1297 (1977).

Sachs, R.G.: "The Physics of Time Reversal," University of Chicago Press, Chicago, IL, 1987.

Scherk, J.: "An Introduction to the Theory of Dual Models and Strings," *Reviews of Modern Physics*, **47**(1), 123–164 (January 1975).

Schwartz, J.H., E. Witten: "Superstring Theory," Cambridge University Press, New York, NY, 1987.

Schwarzschild, B.M.: "Polarized Scattering Data Challenge Quantum Chromodynamics," *Physics Today*, **38**(8), 17–20 (August 1985).

Sutton, C.: "The Particle Connection," Simon and Schuster, New York, NY, 1984.

van der Meer, S.: "Stochastic Cooling and the Accumulation of Antiprotons," *Science*, **230**, 900–906 (1985).

Waldrop, M.M.: "String as a Theory of Everything," *Science*, **229**, 1251–1253 (1985).

Weinberg, S.: "The Discovery of Subatomic Particles," W.H. Freeman, New York, NY, 1983.

Wilson, R.R.: "The Next Generation of Particle Accelerators," *Sci. Amer.*, 42–57 (January 1980).

Yamazaki, T.: "Evolution of Meson Science in Japan," *Science*, **233**, 334–338 (1986).

Zweig, G.: "Quark Catalysis of Exothemal Nuclear Reactions," *Science*, **201**, 973–979 (1978).

PARTICULATES (Precipitation).   See **Electrostatic Precipitator**.

PARTITION.   A term in mathematics used in a number of ways: 1. A partition of a set S is its division into a number of subsets called *cells*, which must be *exhaustive* (that is, every element of S must belong to one of the subsets) and which must also be *disjoint* (no member of S can belong to more than one of the subsets). For some types of sets, the last requirement is better stated in the form that the intersection of any two sets is zero.

2. The partition of a positive integer is its expression as a sum of positive integers, the number of partitions of a given positive integer being the number of ways in which it can be so expressed. Special types of such partitions result when restrictions are imposed upon the process, such as limiting the number of integers in the sum or requiring that they be different.

3. The partition of a permutation may be described as follows. Consider a permutation

$$\begin{pmatrix} a_1, a_2, \ldots, a_n \\ b_1, b_2, \ldots, b_n \end{pmatrix}$$

of n objects, where the notation indicates that a_i is replaced by b_i. This permutation can also be expressed by cycles $(abc \ldots d)(ef) \ldots$, etc., indicating that a is to be replaced by b, b by $c, \ldots, d$ by a, e by f, f by e, and so forth. Suppose now that there are α cycles of degree (i.e., length) 1, β cycles of degree 2, etc. Then we can conveniently denote this property of the original permutation by the symbol $(1^\alpha 2^\beta 3^\gamma \ldots)$, which is called a partition of the permutation. This concept is useful in group theory, since every finite group can be represented as a group of mutations.

See also **Permutation**; and **Permutation Group**.

PARTITION FUNCTION.   An expression giving the distribution of molecules in different energy states in a system

$$Z = \sum q_r e^{-\varepsilon_r/kT}$$

where Z is the partition function, q_r the statistical weight of the rth state of energy ε_r, k is the Boltzmann constant, T, the absolute temperature, and the summation is taken over all the energy states of the system. The energy levels ε_r may be those attributed to rotation, translation, vibration, or electronic energies, etc.

PARTRIDGE (*Aves, Galliformes*).   Game birds of numerous species, related to the pheasants and turkeys. The francolins of Asia and Africa are included here. True partridges are similar to the quails. The latter, although more generally known in North America by the name quail, are members of the group. They are found over Europe, Asia, Africa, and North America. Aside from the common quail or bob-white of North America, the names quail and partridge are both applied to the several western species, and in the southern states even the bob-white becomes the partridge. To confuse the term still further the ruffed grouse is often called a partridge in the northern states.

The francolin is a bird of Africa and the Oriental region, related to the partridges. The spur-fowl is a long-tailed Indian and Ceylonese partridge, which resembles pheasants.

See also **Galliformes**; and **Quail**.

PARVOVIRAL ENTERITIS.   This is a serious disease of dogs, particularly of younger animals. Morbidity is high and mortality approaches 50%. At present the incubation period is unknown. Affected animals stop eating, may vomit, and become depressed and weak. The disease appears to be most common in kennels. Parvoviruses were first isolated from feces of asymptomatic dogs in 1970. The first report of parvoviruses being related to diarrhea in puppies was in 1977. Other species in which the virus has been associated with enteric disease include cats, rabbits, rodents, and calves. The disease features hemorrhagic enteritis that may involve most of the small intestine and, in some cases, the colon. Major microscopic changes in the intestinal tract have been confined primarily to the small intestine. Considerable research is underway to produce a more effective vaccine against this disease.

PASCAL, BLAISE (1623–1662).   Pascal was a French mathematician, physicist, and philosopher. He was home-schooled by his father who had unorthodox views on education and told him he was not going to study mathematics until he was fifteen. But Pascal was a prodigy and by the time he was twelve he had already worked on geometry by himself and discovered that the sum of the angles, of a triangle are two right angles. By the age of sixteen, he published a paper on conic sections, which was a groundbreaking theorem. At age nineteen, he invented the first digital calculator, which added and subtracted through use of a series of cogged wheels, in order to help his father with his work on collecting taxes.

Pascal is best known for Pascal's law. This principle states that fluid in vessels transmits pressure equally in all directions. Pascal also performed a series of experiments on atmospheric pressure and proved, air has weight and that air pressure can create a vacuum.

Pascal is also remembered for his theory of probability and Pascal's triangle can be used to calculate probabilities. Pascal is credited with the invention of both the syringe and the hydraulic press.

Pascal's scientific investigations led to valuable contributions for man but he is also remembered for his religious and philosophical writings. His Pensees contains "Pascal's wager" which claims belief in God is rational because, "If God does not exist, one will lose nothing by believing in him, while if he does exist, one will lose everything by not believing."

See also **Digital Computer Systems**; **Pascal Triangle**; and **Pressure**.

J. M. I.

PASCAL'S LAW.   See **Pressure**.

PASCAL TRIANGLE.   If the coefficients of $(x + y)^k$ in the binomial series are arranged as shown, successive coefficients can be obtained as a sum of two numbers in the preceding line. The second figure in each line is the value of k.

```
        1
      1   1
    1   2   1
   1   3   3   1
  1   4   6   4   1
 1   5  10  10   5   1
1   6  15  20  15   6   1
```

Other forms of the triangle are often shown, especially that in the shape of an isosceles triangle, with unity at the apex and unities along the sides, as shown below.

```
                              1           1
                         1         2    1
                    1         3    3         1
               1         4    6    4         1
          1         5    10   10   5         1
     1         6    15   20   15   6         1
1         7    21   35   35   21   7         1
1    8    28   56   70   56   28   8    1
1    9    36   84   126  126  84   36   9    1
1  10   45   120  210  252  210  120  45   10   1
```

The triangle can easily be extended by simple additions; hence the coefficient in a binomial expansion can be determined to any order with a minimum effort.

Pascal (1623–1662) also showed that the triangle could be used to find the number of combinations when selecting k objects from n objects, since this also equals the binomial coefficient,

$$\binom{n}{k}$$

See also **Binomial Series**; and **Probability**.

PASCHEN-BACK EFFECT. In a strong magnetic field, the anomalous Zeeman effect changes into a pattern similar to the normal effect, and this is known as the Paschen-Back effect. Each energy level with a given value of L, the electronic orbital angular momentum splits into $(2L + 1)$ components characterized by the magnetic quantum numbers $M_L = L, L - 1, \ldots, -L$, and each level with a given value of M_L splits into $(2S + 1)$ components with quantum numbers $M_s = S, S - 1, \ldots, -S$, where S is the resultant electron spin. The selection rules are $\Delta M_L = 0, \pm 1$, $\Delta M_s = 0$. Lines with $\Delta M_L = 0$ are plane polarized with electric vector parallel to the direction of the applied magnetic field; those with $\Delta M_L = \pm 1$ are plane polarized with components perpendicular to the field. See also **Atomic Spectra**; and **Hyperfine Structure**.

PASCHEN (Law of). The spark potential between electrodes in a gas depends on the length of the spark gap and the pressure of the gas in such a way that it is directly proportional to the mass of gas between the two electrodes, i.e., the sparking potential is a function of the pressure times the density of the gas. See also **Discharge (Gaseous)**.

PASCHEN SERIES. See **Energy Level**.

PASSERIFORMES *(Aves)*. An exceptionally large order of birds comprising more than a fifth of all living bird species. The length ranges from 7.5 to 110 centimeters (3 to 43 inches), and the weight is from 4.8 to 1350 grams (0.1 to 48 ounces). These birds have 4 toes (the babbling thrush is the only member of this order with a greatly regressed first toe). The toes all originate from the same level on the tarsus, and they are generally free to the base. There is always 1 toe (generally the largest) directed to the rear; this toe cannot be rotated forward. The claw of the rear toe is, with few exceptions, larger than that of the middle anterior toe. Passeriformes have a bony palate with a design only rarely seen in other orders. There is always a distinct sternal keel, but only traces of an appendix. The young hatch with their eyes closed; the inside of the mouth is brightly colored, often with dark spots inside the mouth as well as other juvenile developments which disappear later. The birds have a worldwide distribution with the exception of a very few remote oceanic islands and areas near the poles.

There are 4 suborders: 1. The Broadbills *(Desmodactylae)*, in which the flexor tendons of the third toe (the middle front toe) and those of the rear toe are joined together. The front toes are fused at their bases. 2. The Noisemakers *(Clamatores)*, in which the flexor tendons of the toes are separate; the lower syrinx has one or two tensor muscle pairs inserted on the half-rings of the trachea either in the middle, through the entirety, or only at its end. 3. The Lyre Birds *(Suboscines)* also have separate flexor muscles in the toes; the lower syrinx has two or three pairs of tensor muscles inserted at both ends of the tracheal half-rings. 4. The Songbirds *(Oscines)*, in which the flexor tendons of the toes are separate; the lower syrinx has four to nine pairs of tensor muscles inserted at both ends on the tracheal half-rings.

Passeriformes are land birds even though some of them may get food from water not far from shore. These birds evolved from ground-dwelling forms of tree-dwelling birds, as all have the typical perching foot. Their four toes are suitable for grasping branches, stalks, wires, etc. The grasp remains firm even when the bird is asleep, because the flexor tendon and its sheath rest inside one another, and each must be freed before the toes can extend and the bird can fly, hop off, or fall down. Only a few members of this order never fly at all.

Separate entries are included in this volume on the following passeriformes:

Bird of Paradise	**Finch**	**Raven**
Blackbird	**Gnatcatcher**	**Redstart**
Bluebird	**Grackle**	**Robin**
Bluethroat	**Jay**	**Shrike**
Bobolink	**Junco**	**Sparrow**
Bowerbird	**Kingbird**	**Starling**
Broadbills	**Lark**	**Swallow**
Bulbul	**Lyrebird**	**Tanager**
Bullfinch	**Magpie**	**Thrasher**
Bunting	**Manakin**	**Thrush**
Canary	**Martin**	**Tit**
Cardinal	**Meadowlark**	**Warbler**
Chatterer	**Myna**	**Waxwing**
Chickadee	**Nightingale**	**Weaverbird**
Cowbird	**Nuthatch**	**Wren**
Creeper	**Oriole**	
Crow	**Ouzel**	

Other species of interesting passeriformes are described in alphabetical order as follows:

Babbler — found in the Ethiopian and Indian regions, particularly those of the family *Crateropotidae*.

Beccafico — this is an Italian name translated "fig-eater" or "fig-pecker," said to apply to the European garden warbler, *Sylvia hortensis*. The English call this bird the "pretty chap." This small bird is a favorite among gourmets in Venice and elsewhere on the Tables of Italy, France, and Greece. The term Beccafico is also used for other warblers when used for food in these countries. An annual feast of the beccafico is called the Beccaficata.

Bee-Martin — the common kingbird of North America, *Tyrannus*. The name appears to be undeserved because the bird eats very few, if any, bees.

Bishop-Bird — any bird of several brightly colored species of African weaverbirds which make up the genus *Pyromeland*. The name also has been applied to some of the brightly colored birds of North America, especially by the early settlers of Louisiana.

Broadbill — an Oriental bird with a shallow, but very broad beak.

Brambling — a finch of the Old World, which nests in northern Europe and Asia. *Fringilla montefringilla*.

Calandra — a European lark noted for its song. The name is sometimes applied to other related species of the Old World.

Cassique — a South American bird related to Old World starlings. Several species.

Catbird — a common North American bird, *Damatella carolinensis*, related to the mockingbird and the thrashers. Although quietly colored in slate gray, it is a welcome resident because of its singing. The term catbird is also used to describe an Australian bowerbird.

Chaffinch — name for birds of several species found in Europe and western Asia.

Chat — name applied to several species of birds, usually designated by a compounded word, as the stone-chat and whinchat of Europe, the yellow-breasted chat of North America (*Icteria virens*), and several North African species. The European wheatear and hedge warbler are also called chats.

Chough — a Eurasian bird related to the crows and resembling them in form and color. A few other Asian birds are known as chough-thrushes.

Cock of the Rock — name for birds of several species found in tropical South America. The males are crested and brilliantly colored. See also **Chatterer**.

Cotinga — name for a group of Brazilian chatterers closely related to the bell birds.

Crossbill — a small seed-eating bird of the Northern Hemisphere whose mandibles cross at the tip when the mouth is closed. This adaptation enables the bird to open seeds and fruits, such as cones, very readily.

Dickcissel — small American bird, *Spiza americana*, related to the buntings. It is found in open country and is distinguished by its yellow breast and black throat patch.

Dipper — the water ouzel. See also **Ouzel (or Ousel)**.

Drongo — the king crow of southern Asia and Africa. Several species, mostly black, forming a family not closely related to the true crows.

Dunnock — the European hedge sparrow, *Prunella modularis*.

Fieldfare — a common thrush, *Turdus pilaris*, of northern Europe. See also **Thrush**.

Fire Eye — a common species of ant bird found in Brazil.

Flower Pecker — a brightly colored bird of the Oriental and Australian regions, related to the sun birds and having remarkable nests.

Forktail — a bird found in India and related to the European chats.

Grassquit — Jamaican name for small birds more commonly called buntings.

Hangnest — a group of birds whose nests are woven of vegetable fiber, grass, and hair and are suspended from small branches. The Baltimore oriole is a common North American species. Others occur from the southwestern states to Brazil.

Honey Creeper — species of small birds of tropical South America and the West Indies related to the warblers. These birds visit flowers in a fashion similar to hummingbirds but are incapable of hovering flight. One species is called the banana-quit.

Honey Eater — species of birds of the Australian region. They have long tongues with which they secure nectar from flowers. The group includes the parson bird, stitch bird, and several species called white eyes.

Honey Pecker — species of small brilliantly colored birds of the Oriental and Australian regions, related to the sun birds. One Australian species is called the diamond bird.

Huia — a New Zealand bird related to the starlings. The male has a short, straight beak, whereas that of the female is long and curved.

Manucode — a name applied to a few smaller birds of paradise, found in several islands of the Australian region. Derived from the generic name *Manucodiata,* which is a corruption of a Malay name.

Mavis — the European song thrush, *Turdus philomelus*, a bird that resembles the wood thrush of North America.

Munia — any bird of numerous species of weaver finches constituting the genus *Munia*. They are native to Africa and the Oriental region. The most common species is the rice bird, paddy bird, or Java sparrow, which is regarded as a pest in the rice fields. It has been valued as a cage bird in Europe.

Nutcracker — species of birds related to the crows and jays. The relatively few species are confined to the northern parts of the Northern Hemisphere. The Clarke nutcracker, *Nucifraga columbiana*, lives principally at higher altitudes in the mountains of western North America. The bird is associated with coniferous forests.

Ovenbird — the European willow wren, *Phylloscopus trochilus*, and other birds that build domed nests. Also, a North American warbler, *Seiurus aurocapillus*, sometimes called the golden-crowned thrush. Also, South American birds of several species, which build mud nests resembling old-fashioned ovens. Genus *Furnarius*.

Ox-Pecker — species of African birds of a group related to the starlings. The common name refers to their habit of climbing about the bodies of domestic cattle in search of ticks and other external parasites. They also visit wild animals for the same purpose.

Piping Crow — Australian birds of several species related to the crows and jays, but not true crows. They are black and white, whence comes their other name, Australian magpie. Unlike the true crows, these birds are quite musical and can be taught to whistle tunes and to speak. They are frequently maintained as cage birds.

Pipit — small quietly colored birds of numerous species related to the wagtails and warblers. They are widely distributed, but most species occur in the Old World. In North America, the common pipit, *Anthus spinoletta*, nests in the north and at high altitudes in the western mountains; the other known species, Sprague's pipit, *A. spraguei*, is a bird of the plains.

Pitta — species of small, brightly colored birds of the Old World, also called the ant thrushes, water thrushes, and ground thrushes. They are only superficially like the true thrushes.

Plant Cutter — several species found only in the temperate zones of South America. They have short, thick beaks with finely serrate edges. They are related to the chatterers.

Redpoll — a small bird related to the finches. It is named for its red crown. One species nests in the northern parts of the Northern Hemisphere and is known in Europe as the mealy redpoll. Europe has another species, the lesser redpoll, and Asia and North America have the related hoary redpoll, *Acanthis hornemanni*, which only occasionally enters the northern United States.

Rice Bird — a term used for the American bobolink. Also for the Java sparrow.

Rifle Bird — a bird of paradise found in Australia and New Guinea. The several species make up the genus *Ptilorhis*.

Rook — a European bird, *Corvus frugilegus*, related to the crows. Its black plumage is glossed with purple, and the face of the adult is usually naked and of a gray color.

Shama — a jungle bird of the Oriental region. The several species are found in the Malayan area, in the Philippines, in India, and on various Pacific islands. They are shy birds. The Indian species is maintained as a cage bird for its beautiful song.

Stonechat — a bird of central and northern Europe, *Saxicola torquata*. The male has a black head and back, a white collar, and reddish underparts. The female is of a brown coloration.

Towhee — a North American bird related to the finches and sparrows. The common eastern species, *Pipilo erythrophthalmus*, is a black and white bird with red-brown sides. It is seen chiefly on the ground and nests chiefly beneath tangled thickets. From its call, the bird is also known as the chewink. Four other species are found in the west, three congeneric with the eastern towhee, and a fourth, the green-tailed towhee. *Oberholseria chlorura*, is placed in a related genus.

Tree Creeper — a small dull-colored bird with a long curved beak. It seeks its prey, consisting of insects and small creatures, in the crevices of bark, moving about the trunks and branches of trees in almost any position. The group is represented in North America by the brown creeper, *Certhia familiaris*, and several subspecies.

Tree Pie — a bird of the Oriental region related to the magpies. The colors of these birds include shades of brown, black, and gray. The beaks are relatively short. In habits, the birds resemble the magpies.

Troupial — a name derived from the French and applied variously to members of the family *Icteridae*, including the orioles, blackbirds, and New World grackles. The name has been used by various writers for the grackles and other birds as mentioned. Also spelled troopial.

Wheatear — a bird, *Oenanthe oenanthe*, related to the thrushes and blue-birds. The bird nests in the northern part of Europe and in Alaska and is widely distributed in the Old World and occasionally in the United States during its southern migrations.

Whinchat — a small European bird related to the bluebirds and thrushes. The bird nests in the far north and winters in Africa.

Woodhewer — a small brown bird of a family found only from Mexico to southern South America. The family includes more than 200 species, mostly limited to the temperature parts of the continent. Among them are the ovenbird previously described.

Yellowbird — the American goldfinch, *Carduelis* (*Astragalinus, Spinus*) *tristis*, and the yellow warbler, *Dendroica aestiva*. Only the male of the former species is yellow, and it has the crown and wings black and the tail marked with black. The yellow warbler is more generally yellow in both sexes. The name yellowbird is not commonly used.

Zosterops—a small bird of the Old World tropics with white rings usually around the eyes. All of the numerous species are birds of small size and quiet colors.

See also **Birds**.

PASSIVE NETWORK (Electronic).   A grouping of resistors, capacitors, or resistors and capacitors required to accomplish the purpose of an electronic circuit. In many different circuit applications, prepackaged units replace clusters of comparably rated low-power resistors and capacitors. Applications include "pull ups" and "push down" transitions among logic circuits, line- and sense-amplifier terminations, decoupling, light-emitting display (LED) drives, ac coupling, supply filtering, line matching, and resistors for current limiting and pulse separating, among other uses. Such networks are available in a variety of packaging formats. See also **Capacitor (Electrical)**; and **Resistor (Circuit)**.

PASSIVITY.   When iron is immersed in concentrated nitric acid, there is no visible reaction (Keir, 1790), although dilute nitric acid results in a marked reaction with iron. Upon removal of the iron from the concentrated nitric acid and immersion in copper sulfate solution, the iron is not plated by copper, although this occurs with ordinary iron. Iron in such a condition is described as passive iron, and the phenomenon is known as passivity. See also **Iron Metals, Alloys, and Steels**.

PASTEUR, LOUIS (1822–1895).   Pasteur was a French chemist and microbiologist who made important contributions to biology, medicine, chemistry, and industry.

As a small child, Pasteur showed traits of becoming a scientist. He was fascinated by the local chemist, that made medicine for sick customers. He patiently and carefully observed the things the chemist did and then went home and made drawings of the herbs and roots the man used. Even before finishing high school, Pasteur's study of chemical crystals won attention of the scientific world. Most of what Pasteur is famous for is his work concerning the effects of microbes. He found that living organisms, microbes, cause fermentation. His discovery was important both for theoretical science and for industry. Pasteur's studies showed microbes could be killed by heat. His discovery made winemaking a more scientific process. Pasteur applied the same idea to milk. The process of keeping milk free from bacteria is named pasteurization after him.

During the 1800s the theory of spontaneous generation was raging in the scientific circles. Pasteur's work proved that food and other organic matter does not spontaneously generate microbes and settled the controversy.

Pasteur also discovered a vaccine to prevent rabies and another vaccine to prevent anthrax. Pasteur's greatest achievement was the founding of the science of microbiology.

See also **Fermentation**; **Grapes and Wines**; and **Rabies**.

J. M. I.

PASTEURIZER.   See **Heat Transfer**.

PATCH (Computer Program).   A section of coding inserted into a computer routine to correct an error or to alter the routine. Often, it is not inserted into the actual sequence of the routine being corrected, but rather it is placed elsewhere, with an exit to the patch and a return to the routine provided. Also, the act of altering a routine by using a patch. See also **Program (Computer)**.

PATENT LOG.   A term applied to any one of a large group of instruments for recording the speed of a ship through the water and, also, the distance run through the water in a given interval of time.

The screw of the ship itself is, in a sense, a patent log, for the speed of the ship through the water is proportional to the revolutions per minute of the screw, and the distance run is proportional to the total number of revolutions in a given time. However, the distance that the ship will move for a single revolution depends upon a number of variable factors, such as the trim of the vessel, the speed of the vessel, the state of the sea, etc.

The earliest, simplest, and, perhaps, the most reliable of the various types of patent logs is the so-called taffrail log. This instrument consists of a spinner, which is towed astern of the ship, well beyond the turbulence produced by the screw. The revolutions of the spinner are transmitted to a recording mechanism, which was originally at the stern, or taffrail, of the ship. In modern installations, the recording dials may be located on the bridge or wherever they will be of most use to the navigating staff. The dials show the speed of the ship at any instant, and also the distance run after the indicator was set to zero. The instrument must be continually watched to see that it is not fouled by seaweed or debris thrown overboard from the ship. Furthermore, it must be frequently checked by the log chip and line, or some other method, to be certain that the blades have not been bent by objects floating in the water.

The principle of the Pitot tube is used in another type of patent log. The tube itself is below the ship, at the turning center, and operates dials similar to those of the other instruments.

See also **Course**; and **Navigation**.

PATH.   An edge train in which each internal vertex is of degree two and each terminal vertex is of degree one. See also **Vertex**.

PATHFINDER MISSION TO MARS.   On July 4, 1997, Mars *Pathfinder* landed safely on the surface of Mars. Designed under the new "faster, cheaper, and better" *Discovery* program philosophy, the lander deployed and navigated a small rover named "*Sojourner* " onto the Ares Valles landing site and began collecting data from its onboard scientific instruments. Designed primarily as an entry, descent and landing demonstration, *Pathfinder* returned 2.3 billion bits of new data, including over 17,000 images, 16 chemical analyses of rocks and soil, and 8.5 million individual temperature, pressure and wind measurements. *Sojourner* traversed approximately 100 meters (330 feet) clockwise around the lander exploring about 200 square meters (2,153 square feet) of area. See Fig. 1. The mission captured the imagination of the public, garnered front page headlines during the first week of mission operations, and went on to became one of NASA's most popular missions. A total of about 56.6 million people visited the *Pathfinder* Web Pages during the first month of the mission, with 4.7 million people visiting the Web Pages on July 8, 1997 alone, making the *Pathfinder* landing by far the largest Internet event in history up to that time.

Mission Summary.   The Mars *Pathfinder* mission was the second mission launched under the National Aeronautics and Space Administration's (NASA) *Discovery* Program. The *Discovery* missions were developed for small planetary missions with a maximum three-year development cycle and a cost cap of $150 million (Fiscal Year 1992) for development that focused on engineering, science, and technology objectives. Originally conceived as an engineering demonstration of key technologies and concepts for use in future missions to Mars, the primary objective was to demonstrate a low-cost cruise, entry, descent, and landing system that could safely place a variety of science instruments on the surface of Mars. For *Pathfinder*, the cost of the mission was $171 million (Fiscal Year 1996), the *Delta II* launch vehicle was an additional $55 million, the development and operations of the rover cost an additional $25 million, and $14 million was allotted for operations.

Mission and Spacecraft Overview

The *Pathfinder* spacecraft or flight system consisted of three major components: the cruise stage, the entry decent subsystem, and the lander, which consisted of the science instruments and the rover.

Cruise Phase.   The cruise phase of the Mars *Pathfinder* mission began with the successful launch atop a *Delta II* rocket from the Kennedy Space Center in Florida on December 4, 1996. See Fig. 2. Once in earth orbit, the spacecraft was given a final boost with the help of a solid-fuel rocket motor called a *Payload Assist Module (PAM-D)*. This 'kick-stage' gave the spacecraft just the right amount of velocity increase it needed to escape Earth's gravity and enter its own orbit around the Sun. Once spent, the third stage was jettisoned.

At separation from the upper stage, the spacecraft was in Earth's shadow and spinning at 20 rpm. An onboard sequence of events was activated once the separation microswitch detected the separation. The Deep Space Network (DSN) initiated spacecraft acquisition and lockup activities using a 34-meter (112-foot) antenna located in the California desert. **See Antenna**. As soon as acquisition occurred, the engineering telemetry broadcast by the spacecraft was received on the ground at a rate of 40 bits per second (b/s). This telemetry consisted of a combination of real time engineering data and stored data from launch, separation, and Earth/Sun acquisition. See Fig. 3.

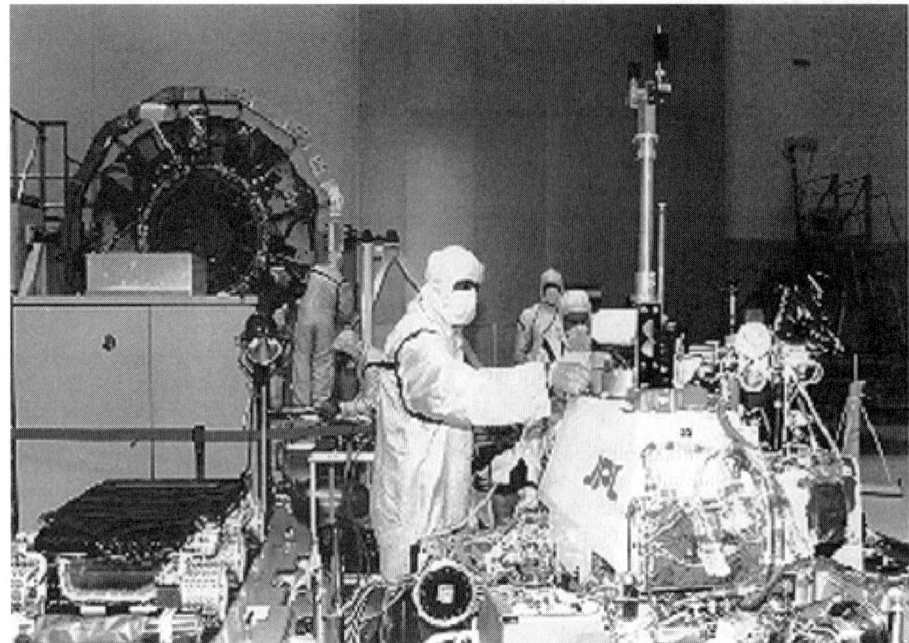

Fig. 1.   Mars *Pathfinder*, rover, and cruise stage being unpacked at the Kennedy Space Center. (*Image courtesy of KSC/NASA.*)

Fig. 2.   Mars *Pathfinder* launch onboard a Delta II on December 4, 1996. (*Image Courtesy of KSC/NASA.*)

The spacecraft automatically determines its orientation in space by first determining the location of the Sun with respect to the spin axis of the spacecraft using a Sun sensor located on the top of the cruise stage. This procedure, known as Sun acquisition, was supposed to provide the spacecraft with the information it needed to reduce the spin rate from 20 rpm to a nominal 2 rpm. But due to some difficulties during launch, it was soon discovered that two of the five sensors had been damaged with an unknown, foreign substance. A software patch was developed which corrected the problem and by using the data from the three working sensors, engineers were able to slow the spacecraft down. Once the spacecraft had cleared the moon's orbit and safely spun down to 2 rpm, the star scanner was activated. After star identification had been confirmed, the *Attitude and Information Management (AIM)* computer calculated the spacecraft's orientation and position, and started its seven-month trip to Mars.

Mars *Pathfinder* used an Earth-Mars transfer orbit. The total flight time from Earth to Mars took seven months. See Fig. 4 for a view of the interplanetary trajectory, as it would look from above the Sun. During the seven-month cruise to Mars, a number of activities were performed to maintain the health of the entry vehicle, lander and rover. Navigation was required to maintain the flight path, and the various spacecraft subsystems were monitored and adjusted as needed to keep them operating at peak efficiency.

Cruise activities began once the spacecraft was safely out of Earth orbit. After the attitude was established and the spacecraft was determined to be healthy, the flight team began a two-week initial characterization and calibration period. Systems included the solar array and battery, thermal control, attitude determination and control, and the communication subsystems. The primary spacecraft activities during the first month of cruise were to collect and downlink relevant engineering telemetry and tracking data, initial spacecraft health checks and calibrations, and attitude maneuvers to maintain the correct Earth/Sun geometry. One health check of the Rover and Science Instruments occurred on December 19, 1996.

Measurements of the spacecraft range to Earth and the rate of change of this distance were collected during every *DSN* station contact and sent to the navigation specialists of the flight team for analysis. They used this data to determine the true path the spacecraft was flying, and to determine corrective maneuvers needed to maintain the desired trajectory. The first of four *Trajectory Correction Maneuvers (TCMs)* was scheduled on January 4, 1997 to correct any errors collected from launch. The magnitude of this maneuver was less than 75 meters (246 feet) per second. Navigation was an ongoing activity that continued until the spacecraft entered the atmosphere of Mars on the 4th of July.

After TCM-1, the flight team transitioned from a "spacecraft checkout mode" to a more routine "spacecraft monitoring mode". DSN tracking coverage was reduced from three contacts a day to three per week to allow other spacecraft like Mars *Global Surveyor* and *Galileo* to use the DSN time. Spacecraft health and performance telemetry was downlinked at 40 b/s or greater during each tracking pass.

A key activity that took place during cruise was the designing and building of command sequences that dictated to the spacecraft how it was to perform each of the activities required. Each cruise command sequence was generated and tested, and then uplinked approximately once every four weeks during one of the regularly scheduled DSN passes. The uplink generation process required 14 days for planning, sequence generation, verification, and commanding.

Two more trajectory correction maneuvers were performed in early February and early May to further reduce any navigation guidance errors. TCM-2 required less than 10 meters (33 feet) per second, and TCM-3 was smaller still, less than 1 meter (3.3 feet) per second. These two maneuvers

Fig. 3. Mars *Pathfinder* in cruise configuration. The red panels are the solar cells that will supply power during the seven month cruise. (*Image courtesy of JPL/NASA.*)

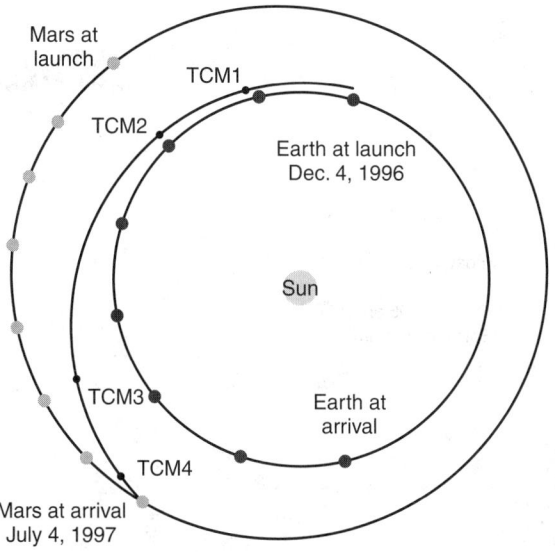

Fig. 4. Mars *Pathfinder* Cruise Trajectory.

further reduced any guidance error detected from navigation measurements during cruise.

Starting 45 days prior to entry, tracking was increased to three passes a day and the flight team stepped up its preparation for atmosphere entry and landing. A final health and status check of the instruments and rover was performed on June 4, 1997. A fourth and final trim maneuver was performed on June 24, requiring less than 0.50 meters (1.65 feet) per second due to the accuracy of the previous maneuvers. On June 30, the spacecraft performed a turn to the entry attitude, where it remained until atmosphere entry. The roll thrusters increased the spacecraft spin rate from 2 to 10 rpm for entry. At that time, the cruise phase ended and the flight team transitioned to the entry, landing, and surface operations phases. See Fig. 5.

During the final day of approach, the navigation team produced orbit solutions on a regular basis, and adjustments were made to the computer programs that determine when the parachute should be deployed. At 6 hours out, the final adjustments were made, and the flight team made final preparations for atmosphere entry.

Entry, Decent, and Landing Phase. The fast-paced approach of *Pathfinder* to Mars began with venting of the heat rejection system's

cooling fluid about 90 minutes prior to landing. See Fig. 6. This fluid is circulated around the cruise stage perimeter and into the lander to keep the lander and rover cool during the seven month cruise phase of the mission. Its mission fulfilled, the cruise stage was then jettisoned from the entry vehicle about one-half hour prior to landing at a distance of 8,500 kilometers (5,100 miles) from the surface of Mars. Several minutes before landing, the spacecraft began to enter the outer fringes of the atmosphere about 125 kilometers (75 miles) above the surface. Spin stabilized at 2 rpm, and traveling at 7.5 kilometers (4.5 miles) per second the vehicle entered the atmosphere at a shallow 14.8-degree angle. A shallower entry angle would result in the vehicle skipping off the atmosphere, while a steeper entry would not provide sufficient time to accomplish all of the entry, descent and landing tasks. A *Viking*-derived aeroshell (including the heatshield) protected the lander from the intense heat of entry. At the point of peak heating the heatshield absorbed more than 100 megawatts of thermal energy. The Martian atmosphere slowed the vehicle from 7.5 kilometers per second to only 400 meters per second (900 miles per hour).

Entry deceleration of up to 20 gees, detected by on-board accelerometers, set in motion a sequence of preprogrammed events that are completed in relatively quick succession. Deployment of the single, 8-meter (24-foot) diameter parachute occurred 2 minutes and 14 seconds after atmospheric entry at an altitude of 9.4 kilometers (6 miles) above the surface. The parachute was similar in design to those used for the *Viking* program but had a wider band around the perimeter, which helped to minimize swinging.

The heatshield was pyrotechnically separated from the lander 20 seconds later and dropped away. See Fig. 7. The lander soon begins to separate from the backshell and "rappels" down a metal tape on a centrifugal braking system built into one of the lander petals.

The slow descent down the metal tape places the lander into position at the end of a braided Kevlar tether, or bridle, without off-loading the parachute or placing excessive loads on the backshell. The 20-meter (66-foot) bridle provides space for airbag deployment, distance from the solid rocket motor exhaust stream and increased stability. Once the lander was lowered into position at the end of the bridle, the radar altimeter was activated and began a timing sequence for airbag inflation, backshell rocket firing and the cutting of the Kevlar bridle.

The lander's Honeywell radar altimeter acquired the surface about 28 seconds prior to landing at an altitude of about 1.6 kilometers (1 mile). The airbags were inflated 18 seconds later before landing at an altitude of 355 meters (less than 1/4 of a mile) above the surface. See Fig. 8. The airbags had two pyro firings, the first of which cut the tie cords and loosened the bags. The second firing, 0.25 seconds later, and 4 seconds before the rockets fired, ignited three gas generators that inflated the three

Fig. 5. Artist renditions of Mars *Pathfinder* as it enters the Martian atmosphere. (*Image courtesy of JPL/NASA.*)

Cruise stage separation
(8500 km, 6100 m/s)
landing - 34 min

Entry
(125 km, 7600 m/s)
landing - 4 min

Parachute deployment
(6-11 km, 360-450 m/s)
landing - 2 min

Heatshield separation
(5-9 km, 95-130 m/s)
landing - 100 s

Lander separation
bridle deployment
(3-7 km, 65-95 m/s)
landing - 80 s

Radar ground aquisition
(1-5 km, 60-75 m/s)
landing - 32 s

Airbag inflation
(300 m, 52-64 m/s)
landing - 8 s

Rocket ignition
(50-70 m, 52-64 m/s)
landing - 4 s

Bridle cut
(0-30 m, 0-25 m/s)
landing - 2 s

Mars *Pathfinder*

entry, descent and landing

Friday, July 4, 1997

landing at 10:05 am PDT
(Earth receive time)

Deflation /
petal latch firing
landing + 15 min

Airbag retraction /
lander righting
landing + 115 min

Final retraction
landing + 180 min

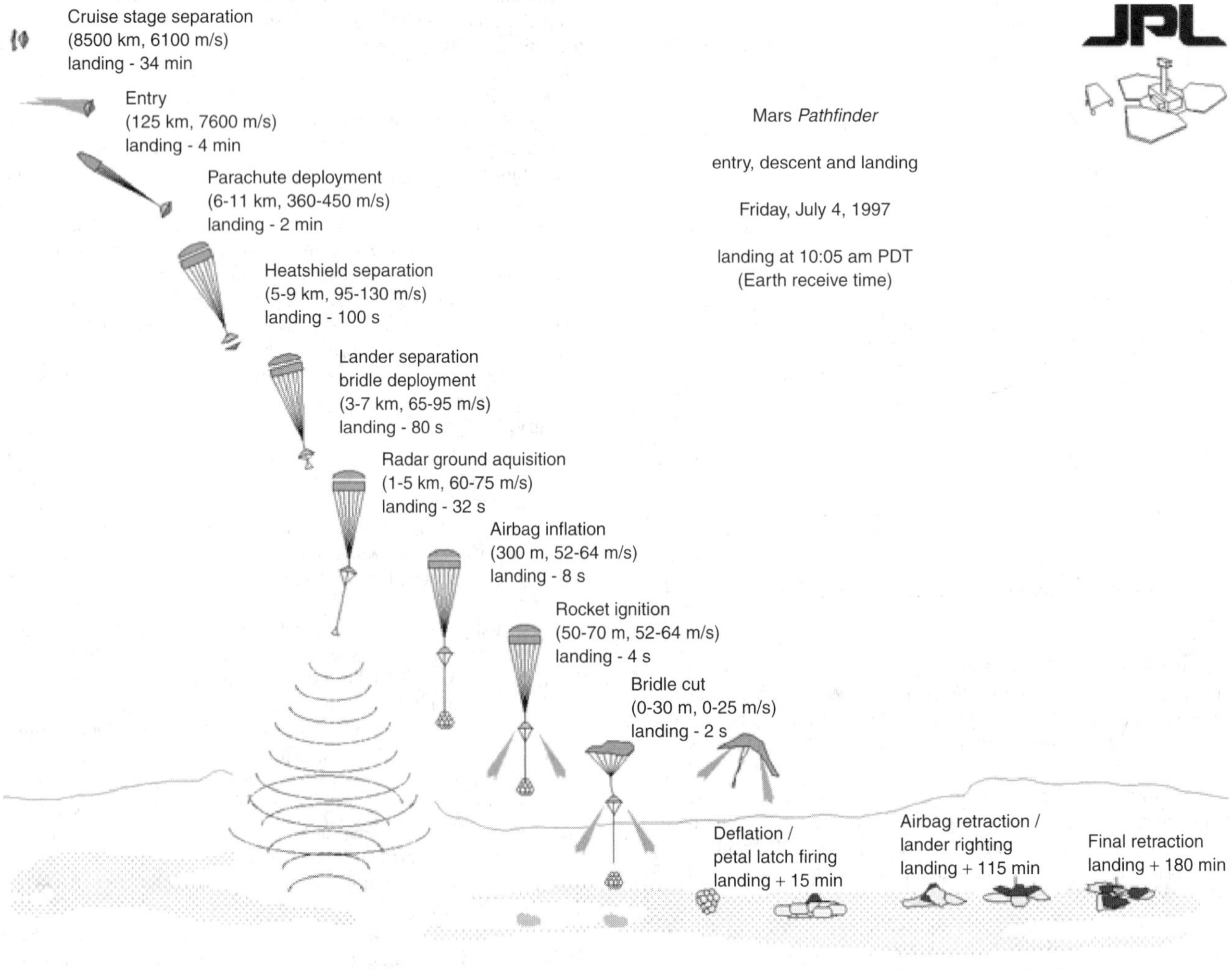

Fig. 6. Entry, Decent, and Landing schematic for July 4, 1997. (*Image courtesy of JPL/NASA.*)

Fig. 7.   Artist rendition of *Pathfinder* entering the Martian atmosphere. (*Image courtesy of JPL/NASA.*)

5.2-meter (17-foot) diameter bags to a little less than 1 psi. in less than 0.3 seconds.

The conical backshell above the lander contains three solid rocket motors each providing about a ton of force for over 2 seconds. The computer in the lander activates them. Electrical wires that run up the bridle close relays in the backshell which ignite the three rockets at the same instant.

The brief firing of the solid rocket motors at an altitude of 98 meters (323 feet) was intended to essentially bring the downward movement of the lander to a halt some 12 meters, +/−10 meters (40 feet, +/−30 feet) above the surface. In reality, the rockets fired approximately 21 meters (69 feet) above the surface. The bridle separating the lander and heatshield were then cut from the lander, resulting in the backshell driving up and into the parachute under the residual impulse of the rockets, while the lander, encased in airbags, fell to the surface. See Fig. 9.

Because it was possible that the backshell could be at a small angle at the moment that the rockets fire, the rocket impulse imparted a large lateral velocity to the lander/airbag combination. In fact the impact could have been as high as 25 meters per second (56 mph) at a 30-degree grazing angle with the terrain. It was expected that the lander could have bounced at least 12 meters (40 feet) above the ground and soared 100 to 200 meters (330 to 660 feet) between bounces. Tests of the airbag system verified that it was capable of much higher impacts and longer bounces. In reality, an onboard instrument calculated at least 15 bounces with the first bounce up to 12 meters (40 feet) high without any airbag rupture.

Once the lander had settled on the surface, pyrotechnic devices in the lander petal latches were blown to allow the petals to be opened. The latches locking the sturdy side petals in place were necessary because of the pulling forces exerted on the lander petals by the deployed airbag system. In parallel with the petal latch release, a retraction system began slowly dragging the airbags toward the lander, breaching vent ports on the side of each bag, in the process deflating the bags through a cloth filter. The airbags were drawn toward the petals by internal lines extending between attachments within the airbags and small winches on each of the lander sides. It took about 64 minutes to deflate and fully retract the bags. See Fig. 10.

There is one high-torque motor on each of the three petal hinges. If the lander had come to rest on its side, it would have to be righted to the base petal by opening a side petal with a motor drive to place the lander in an upright position. Once upright, the other two remaining petals would have been opened.

About three hours was allotted to retract the airbags and deploy the lander petals, but on Mars the whole operation only took 87 minutes because *Pathfinder* came to rest on the base petal. At this time, the lander's X-band radio transmitter was turned off for the first time since before it was launched on December 4, 1996. This saved battery power and allowed the transmitter electronics to cool down after being warmed up during entry without the cooling system. It also allowed time for the Earth to rise well above the local horizon so that it would be in a better position for communications with the lander's low-gain antenna.

Science Instruments and Objectives

The Mars *Pathfinder* project landed a single vehicle on the surface of Mars, which included a microrover, *(Sojourner),* and several science instruments. See Fig. 11. *Sojourner's* mobility provided the capability of "ground truthing" the local landing area and investigating the surface with three additional science instruments: A stereoscopic imager with spectral

Fig. 8.   One of the many airbag test performed prior to lift-off. Each lobe of the airbags consists of six spheres, with four lobes, one for each of the pedals. The airbags total 16 feet from the ground to the top. (*Image courtesy of JPL/NASA.*)

Fig. 9.   Artist renditions of the landing of Mars *Pathfinder* on the surface of Mars. (*Image courtesy of JPL/NASA.*)

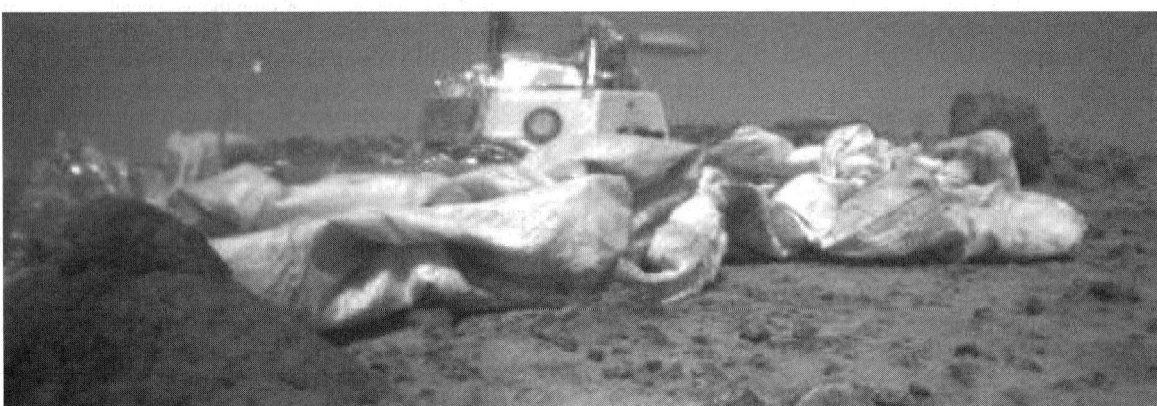

Fig. 10.   Rover view of the lander on the surface of Mars. Notice how far the airbags retracted. (*Image courtesy of JPL/NASA.*)

filters on an extendible mast (IMP), an Alpha Proton X-Ray Spectrometer (APXS), and an Atmospheric Structure Instrument/Meteorology package (ASI/MET). See Fig. 12. These instruments allowed for investigations of the geology and surface morphology at submeter to a hundred meters scale, the geochemistry and petrology of soils and rocks, the magnetic and mechanical properties of the soil as well as the magnetic properties of the dust, a variety of atmospheric investigations and rotational and orbital dynamics of Mars.

Landing downstream from the mouth of a giant catastrophic outflow channel (Ares Vales) offered the potential for identifying and analyzing a wide variety of materials in the crust, from the ancient heavily cratered terrain to intermediate-aged ridged plains to reworked channel deposits. Examination of the different surface materials allowed first-order scientific investigations of the early differentiation and evolution of the crust, the development of weathering products and the early environments and conditions that have existed on Mars.

Surface Morphology and Geology at Meter Scale. The Imager for Mars *Pathfinder* (IMP) examined Martian geologic processes and surface-atmosphere interactions similar to what was observed at the *Viking* landing sites. See Fig. 13. Observations of the general landscape, surface slopes and the distribution of rocks were obtained by panoramic stereo images at various times of the day. IMP was also designed to monitor any dust or sand deposition, erosion or other surface-atmosphere interactions. A basic

understanding of the surface and near-surface soil properties was obtained by the rover and lander imaging of rover wheel tracks, holes dug by rover wheels, and examining any surface disruptions caused by airbag bounces or retractions.

Petrology and Geochemistry of Surface Materials. The Alpha-Proton X-Ray Spectrometer (APXS) and the visible to near-infrared spectral filters on the IMP determined the dominant elements that made up the rocks and other surface materials of the landing site. A better understanding of these materials provided answers concerning the composition of the Martian crust, and secondary weathering products (such as different types of soils). These investigations provided a calibration point for orbital remote sensing observations such as Mars *Global Surveyor*. The IMP was able to obtain full multi-spectral panoramas of the surface and underlying materials exposed by the rover.

Magnetic Properties and Soil Mechanics of the Surface. Magnetic targets were distributed at various points around the spacecraft. Multi-spectral images of these targets identified the magnetic minerals that make up the airborne dust. Using the IMP images, it was possible to identify the mineral composition of the rocks. Detailed examination of the wheel-track images also gave a better understanding of the mechanics of the soil surrounding the landing site.

Atmospheric Structure as Well as Diurnal and Seasonal Meteorological Variations. The Atmospheric Structure Instrument/Meteorology

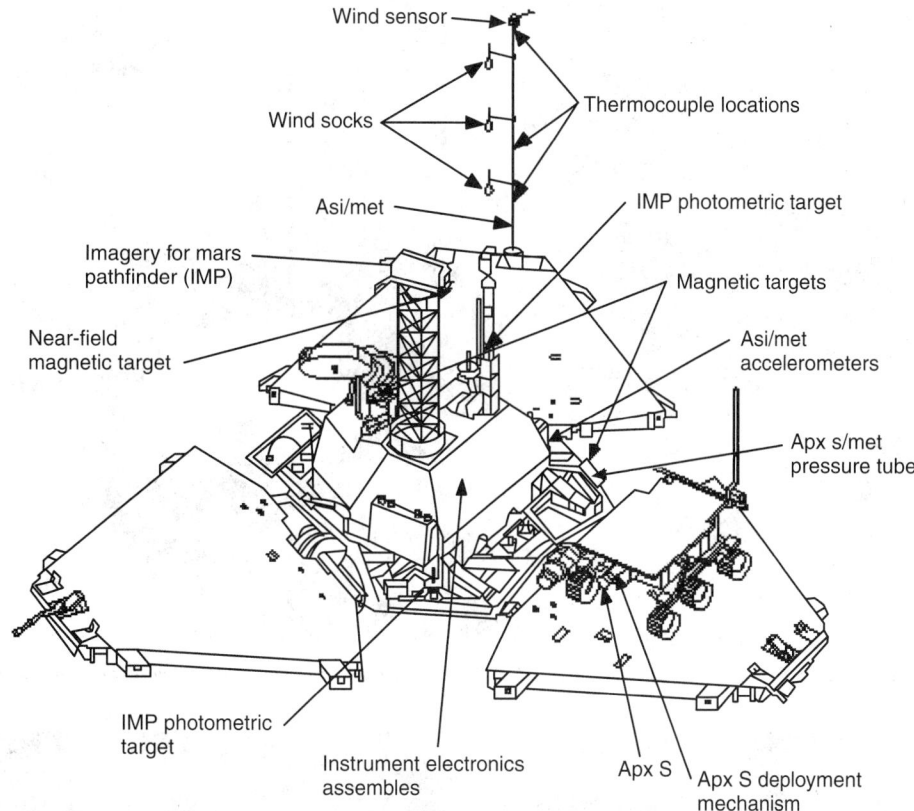

Fig. 11.   Computer drawing of the lander components. (*Image courtesy of JPL/NASA.*)

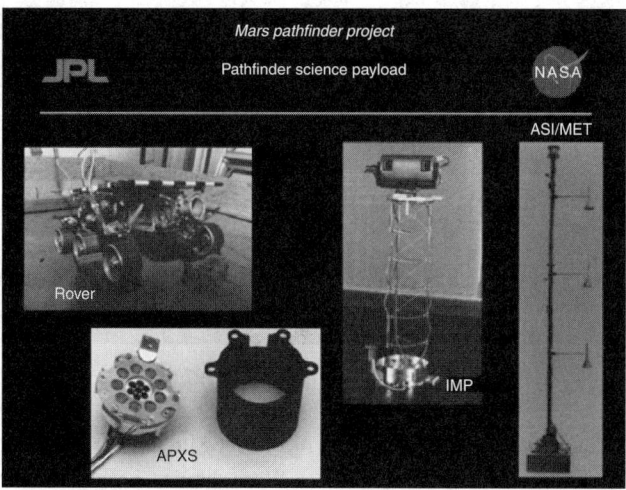

Fig. 12.   Mars *Pathfinder* instrument package. Imager for Mars *Pathfinder* (IMP), Alpha Proton X-Ray Spectrometer (APXS), an Atmospheric/Meteorology. (ASIMET.)

(ASI/MET) experiment was able to monitor the temperature and density of the atmosphere during Entry, Descent and Landing (EDL). In addition, three-axis accelerometers were used to measure atmospheric pressure during entry. Once on the surface, meteorological measurements such as pressure, temperature, wind speed and atmospheric opacity were obtained on a daily basis. Thermocouples mounted on a one meter (3.3 foot) high mast examined temperature profile with height. Wind direction and speed were measured by a wind sensor mounted at the top of the mast, as well as three windsocks interspersed at different heights on the mast. Understanding this data was important for identifying the forces that act on small particles carried by the wind. Regular sky and solar spectral observations using the IMP monitored windborne particle size, particle shape, distribution with altitude and the abundance of water vapor.

Rotational and Orbital Dynamics of Mars. The Deep Space Network (DSN), by using two-way X-Band and Doppler tracking of the Mars *Pathfinder* lander once it was on the surface, was able to address a variety of orbital and rotational dynamics questions. Spacecraft ranging involves

sending a code to the lander and measuring the time required for the lander to echo the code back to the Earth-based station. By dividing this time by the speed of light, results can be accurate within 1 to 5 meters (3 to 16 feet) of the distance from the station to the spacecraft. As the lander moves relative to the tracking station, the velocity between the spacecraft and Earth causes a Doppler shift in frequency. Measuring this frequency shift provided an accurate measurement of the distance from the station to the lander. After a few months of observing these features, the Mars *Pathfinder* lander location was determined within a few meters. Once the exact location of *Pathfinder* had been identified, the orientation and precession rate of the pole can be calculated and compared to measurements made with the *Viking* landers 20 years ago. Measurement of the precession rate allowed direct calculation for the moment of inertia. Measurements similar to these are used on earth to determine the makeup of the earth's interior.

Surface Science Phase. After receiving data indicating the health of the spacecraft and a successful landing, commands were sent to the spacecraft

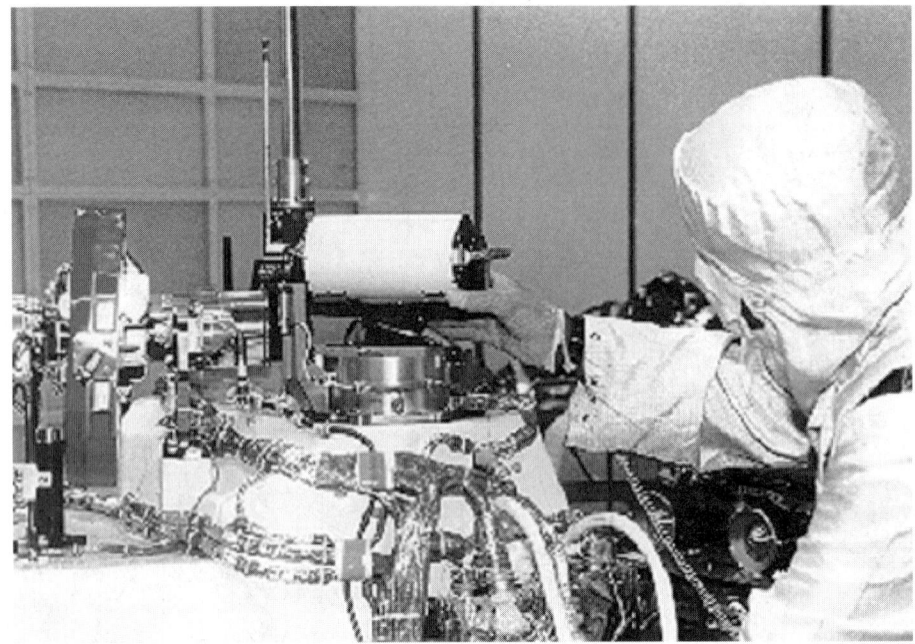

Fig. 13.   Figure 7 Imager for Mars *Pathfinder* (IMP) being tested before launch. (*Image courtesy of KSC/NASA.*)

Fig. 14.   Mission success panorama acquired on July 4, 1997. (*Image courtesy of JPL/NASA.*)

to unlatch the IMP camera and the high gain antenna. The first task of the lander was to determine the location of the Sun. To do this, the IMP scanned the horizon for the brightest spot on the horizon. Once the Sun's location was determined, the high gain antenna was directed towards Earth and the first images were received around 4:30 p.m. PDT. The first received data included the mission success panorama, stereo images of both rover ramps, and spacecraft engineering data which included the health status of the spacecraft and the status of the airbag retraction. See Fig. 14.

After examining the imagery, it was determined that the airbags had not fully retracted from the rover petal and that it would not be safe to deploy the rover petals. Commands were sent up to reclose the rover petal, retract the airbag further, and then redeploy the rover petal. After careful examination of a second set of images, the ramps were determined to be safe and the rover was commanded to stand up. A full panorama of the landing site was also returned on the first day of operation and the rover was driven down the rear ramp the following day (Sol 2). After it was determined to be safe to deploy the IMP camera, the camera mast was deployed to a height of 0.8 meters (2.6 feet) at the end of Sol 2. After some minor communication errors between the rover modem and the lander, the rover deployed the APXS at the surface for its first soil sample. See Fig. 15.

Lander Site Location

When the first images had arrived, five prominent horizon features and two small craters were identified in both lander horizon and *Viking Orbiter* images. This enabled the lander to be located within 100 meters (330 feet) of other surface features at 19.13 °N, 33.22 °W in the U.S. Geological Survey reference frame.

Characteristics of the landing site were determined to be consistent with its prelanding predication of a flat, level flood plain composed primarily of materials left behind by the Tiu and Ares catastrophic floods. The surface is composed of pebbles, cobbles and boulders that closely resemble depositional surfaces found from catastrophic floods on Earth. Two nearby peaks identified as "Twin Peaks," appear to be streamlined hills in IMP images; this is consistent with prelanding predictions of *Viking Orbiter* images of the region. Rocks identified in the Rock Garden are imbricated in the direction of the predicted flow; again agreeing with prelanding predictions Troughs are also visible throughout the scene and have been interpreted to be erosional features produced by the turbulent flood waters. Large boulders can be found perched on top of smaller rocks (i.e. Yogi), consistent with deposition by a flood. Except for later eolian activity, the site appears little altered since it formed up to a few billion years ago.

A variety of soil types have also been identified at the *Pathfinder* landing site. See Fig. 16. These soils appear to be composed of poorly crystalline

Fig. 15.   Sojourner on the surface of Mars. Rock to the left is Barnacle Bill. (*Image courtesy of JPL/NASA.*)

Fig. 17.   End of day IMP image of the rover in the Rock Garden. (*Image courtesy of JPL/NASA.*)

Fig. 16.   End of day image of the rover. (*Image courtesy of JPL/NASA.*)

characterized as a light yellowish brown, with clay-sized silicate particles and a small amount of a magnetic mineral (believed to be maghemite). The present interpretation for the maghemite formation is that iron was dissolved out of crystal materials by water and that the maghemite is a freeze-dried secondary precipitate.

Observations from wheel tracks and soil mechanics experiments illustrate that the subsurface consisted of a variety of different materials with different physical properties. See Fig. 18. Rover tracks observed in bright drift material preserved individual cleat marks indicating that they are compressible deposits of very fine-grained dust. Several cloddy deposits found at the landing site appear to be composed of poorly sorted dust, sand-sized particles, lumps of soil, and small rock granules and pebbles.

ferric-bearing materials. Elemental compositions of soil units measured by the APXS are similar in composition to those measured at both of the *Viking* landing sites. Due to the distance between the *Pathfinder* site and the two *Viking* landing sites, the similarities in soil compositions suggest that the compositions are influenced by globally distributed airborne dust.

Rocks that have been identified at the *Pathfinder* site are primarily dark gray and partially covered with coatings of bright dust and/or weathered surfaces. See Fig. 17. From the rock chemistry measured by the APXS they appear to be similar to basalt, basaltic andesites, and andesites found on Earth. Rover close-up and IMP images display rocks with a variety of different morphologies, textures and fabrics. Some of the rocks have been hypothesized to be conglomerates composed of rounded pebbles embedded in a finer matrix. Rocks such as these may be the source of numerous rounded pebbles and cobbles that were identified throughout the site. If these rocks are conglomerates, their formation suggests that running water was present to smooth and round the pebbles and cobbles over long periods of time. The rounded materials would then be deposited into a finer grained sand and clay matrix and lithified before being carried to the Ares site. This suggests a warmer and wetter past in which liquid water was stable on the surface.

The magnetic properties experiment identified the airborne magnetic dust that was deposited on most of the magnetic targets. The dust is

Fig. 18.   Rover on Merimaid Dune. (*Image courtesy of JPL/NASA.*)

The atmospheric opacity was determined to be 0.5 and changes slightly higher at night as well as early in the morning due to clouds. The sky is

Fig. 19.   Clouds observed at the Ares Valles landing site. (*Image courtesy of JPL/NASA.*)

a light yellowish brown color composed of micron-sized particles and water vapor. See Fig. 19. The upper atmosphere, above 60 kilometers (36 miles) altitude, was determined to be very cold and different from warmer measurements obtained by both *Viking* landers. The differences in the measurements may be attributed to seasonal variations at the time of landing. The multiple peaks in the landed pressure measurements and the entry and descent data are indicative of dust uniformly mixed in a warm lower atmosphere.

The meteorology measurements at the site identified diurnal and higher order temperature fluctuations. The barometric minimum was reached at the site on Sol 20 indicating the maximum extent of the winter south polar cap. Temperatures changed abruptly with time and height in the morning; these observations suggest that the warming of the cold morning air by the Sun created upward moving small eddies. Winds were fairly consistent, and dust devils were detected repeatedly throughout the mission.

Daily Doppler tracking and less frequent two-way ranging during communication sessions between the spacecraft and Deep Space Network antennas resulted in a solution for the location of the lander and the direction of the Mars rotation axis. Combined with earlier results from the *Viking* landers, the estimated precession constant constrains the core radius of Mars to be between 1,300 and 2,000 kilometers (780 to 1200 miles).

From all of the scientific results that have been completed so far, early Mars appears to have been very similar to an early Earth. Some of the materials that make up the crust may be similar to terrestrial continental crust materials. The rounded pebbles, cobbles suggest a possible conglomerate, which supports water rich early Mars. This would imply that the early environment of Mars was warmer and wetter than today and liquid water may have been in equilibrium. Further Mars missions may be able to answer these questions.

R.C. ANDERSON, JPL.

PATHOGENESIS.   The pathway followed in the development of a disorder or disease—from early clinical findings and onset through the many stages that may follow.

PATHOGENIC.   Capable of producing disease. The term most commonly is applied to bacteria, viruses, or fungi possessing this property.

PATHOLOGY.   The study of disease, particularly by laboratory methods, including the bacteriology, virology, parasitology, etc. of pathogenic organisms.

PATHOLOGY (Plants).   See **Botany**.

PATINA.   1. The geochemically altered surface of any discrete object such as a mineral, pebble, or rock. 2. A film, usually green, formed on copper and bronze after long atmospheric exposure. 3. A term used by archeologists to describe the altered surface of artifacts.

PATTERN RECOGNITION.   It is difficult to make a case for a general unified theory pertaining to pattern recognition. Rather, it may be described as a field of technical interests comprised of various concepts, tools, and techniques. The general objective of pattern recognition is that of classifying an unknown pattern—to place it into one of several classes of patterns. The need for automatic recognition of patterns rises constantly in terms of information processing systems involved in the scientific, medical, business, military, etc., areas. There are numerous instances, of course, where patterns are easily identified by human visual recognition. In such cases, pattern recognition techniques assist in improving the communication between humans and machines. In many other situations, however, patterns are very difficult to recognize quickly through human perception. Detailed examination of electrocardiograms would be an example, but by no means one of the more difficult areas. During the past decade, one of the most rapidly developing and exciting applications for pattern recognition has been in connection with machine vision in robotic systems. See also **Machine Vision (Recognition and Applications)**; and **Robots and Robotics**.

Pattern recognition technique may be divided into three parts: (1) first, *measurements* must be made of the object or event that is to be recognized; (2) measurements will contain relevant as well as much irrelevant data, and thus, pertinent features must be *extracted* that will better characterize the pattern classes; and (3) once features have been established, the input pattern is *classified* to one of a series of pattern categories.

Hundreds and even thousands of measurements may be required to convert a physical pattern to electric signals suitable for feature extraction. Elimination of noise is a particularly critical problem. In character recognition, locating a character on a line and isolating it from touching characters are not trivial problems. Even more difficult is the segmentation of speech patterns.

The objective of the extraction of features is to provide an intermediate process between measurement and classification that will make the design and implementation of the classifier feasible. The formulation of a set of adequate features is largely empirical and intuitive, based on knowledge of the particular pattern recognition problem and characteristics of the various techniques of classification. Features should emphasize differences between classes and deemphasize differences within classes. It is advantageous for features to be invariant for irrelevant variations in the patterns, which might be translation skew and sizes. Statistically independent features can simplify the design and implementation of the classifier. Generally, a feature cannot be evaluated with any certainty as an individual, but must be tested as a member of a set of features. For example, the two features shown in Fig. 1 taken together result in overlapping classes for the given samples. However, if either one of these features is considered by itself, the classes have significant overlap, as can be seen by projecting the samples on either feature axis. This illustrates the requirement of evaluating combinations of features, which aggravates the difficulty of feature selection. This also indicates the relevancy of context in the general sense to pattern recognition, whether it be the relationships among letters to form words or strokes to form letters.

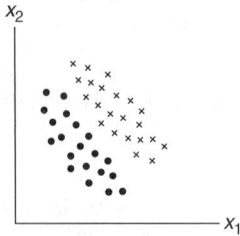

Fig. 1.   Statistically dependent features.

Applications.   Character recognition provides an automatic means of entering machine-printed and hand-printed alphanumeric data into systems. Machine-printed characters, including special fonts for character recognition, have been recognized without the use of features by template matching or correlation techniques operating on a binary array representation of the character. Features are also used in machine-print application and are generally required for the recognition of hand-printed characters. A few of the many features that have been investigated are strokes, line endings, line intersections, loops and bays with various sizes, orientations, and locations. See also **Magnetic Ink Character Recognition (MICR)**; and **Optical Character Recognition (OCR)**.

Fingerprint classification is normally concerned with identifying fingerprint samples as belonging to various basic types. Generally, some scheme of tracing the ridges in the fingerprints produces directed line segments, line endings, and line intersections as features. From these features, pattern types, such as arches, loops, and whorls, are detected.

The automatic interpretation of aerial photographs and other remote sensor images has application in navigation, earth resources studies, and military reconnaissance. Geometric features, such as dots, straight lines, parallel lines, and x-detectors are of interest in this application. Features providing texture information can be determined from the spatial frequency of the optical density. The analysis of particle tracks in bubble chambers is another form of automatic interpretation of photographs. Events are recognized with the aid of features describing the curvature and end-points of the tracks and intersections of tracks.

Electroencephalogram classification and speech recognition are examples of waveform patterns in which frequency spectrum analysis is important. The energy in selected frequency bands and certain aperiodic waveshapes are features that have been investigated for classifying electroencephalograms. In speech recognition, features such as format frequencies are derived from the energy-frequency-time spectrum called the speech spectrogram. See also **Voice and Sound Production**.

Particularly the approach to classification in many difficult cases requires rather sophisticated mathematical approaches beyond the scope of this article.

Additional Reading

Bezdek, J.C., J. Keller, R. Krisnapuram, and M.R. Pal: "Fuzzy Models and Algorithms for Pattern Recognition and Image Processing," Kluwer Academic Publishers, Norwell, MA, 1999.

Casacuberta, F. and A. Senfeliv: "Advances in Pattern Recognition and Applications," World Scientific Publishing Company, Inc., River Edge, NJ, 1994.

Chen, C.H., L.F. Pau, and P.S. Wang: "Handbook of Pattern Recognition and Computer Vision," 2nd Edition, World Scientific Publishing Company Inc., River Edge, NJ, 2000.

Pal, S.K. and S. Mitra: "Neuro-Fuzzy Pattern Recognition: Methods in Soft Computing," John Wiley & Sons, Inc., New York, NY, 1999.

Pal, N.R.: "Pattern Recognition in Soft Computing Paradigm," World Scientific Publishing Company Inc., River Edge, NJ, 2001.

Theodoridis, S. and K. Koutroumbas: "Pattern Recognition," Academic Press, Inc., San Diego, CA, 1998.

Yu, F.T.S. and S. Jutamulia: "Optical Pattern Recognition," Cambridge University Press, New York, NY, 1998.

PAULI EXCLUSION PRINCIPLE. The statement that any wave function involving several identical particles must be antisymmetric (must change sign) when the coordinates, including the spin coordinates, of any identical pair are interchanged. If the particles in a system can be considered as occupying definite quantum states, it follows from the principle that no more than one particle of a given kind can occupy a particular state; hence the name, exclusion principle. The principle applies to fermions, but not to bosons. Since electrons, protons, and neutrons are fermions, the Pauli exclusion principle must be used in the assignment of particles to quantum states in theories of atomic and nuclear structure.

PAULING, LINUS CARL (1901–1994). Dr. Linus Pauling was a famous American chemist. He was born in Portland, Oregon and his father was a pharmacist. When he was four, his family moved to Condon, Oregon. On the south edge of town, there was a creek and Pauling sent much time exploring the creek bed and collecting minerals. Eventually Pauling would establish structures for these minerals at the California Institute of Technology.

Pauling enrolled in Oregon Agricultural College (Oregon State University) and was majoring in chemical engineering, but when his mother became ill, he quit school and began working. Because he had show outstanding promise, he was offered a position to be an instructor of quantitative analysis at the college. He taught the class and graduated with a degree in chemical engineering. He attended Cal Tech graduate school.

His Ph.D. dissertation was on the crystal structure of different minerals. Pauling received a Guggenheim fellowship to study at the University of Munich where he began applied the concept of quantum mechanics to chemical bonding. Then he took an assistant professorship in chemistry at Cal. Tech and published his paper "The Nature of the Chemical Bond." In 1931, he received the Langmuir Prize of the American Chemical Society for his noteworthy work. In 1933, he was made a member of the National Academy of Sciences. He was 32 years old at that time and he was the youngest appointment ever made.

His serious interest molecular biology began about 1935. He was intrigued by the question of how protein molecules were constructed. As a professor at the California Institute of Technology, he was known for giving "baby toy lectures" because he made models of molecules out of string, rod- and-ball structures, and plastic bubbles in different colors, shapes, and sizes. One day, working with paper, he sketched atoms and chemical bonds and folded them in different ways and discovered the basic structure of the protein molecule.

Pauling, with the help of C.D. Coryell, analyzed the effects of the oxygenation of hemoglobin molecules by measuring their magnetic susceptibility. In 1936, along with A.E. Mirsky, Pauling developed a theory of native, denatured, and coagulated proteins. And in 1950, he and R.B. Corey described the structure of several molecules including muscles, fingernails, hair, and other tissues.

During World War II, Pauling chose not to work on the Manhattan Project for the development of the atomic bomb. He became worried about the radiation the atomic bomb produced and helped organize the Pasadena Federation of Atomic Scientists, a group of scientists working for safe control of nuclear power. He also joined the Emergency Committee of Atomic Scientists, known as the Einstein Committee since it was chaired by Albert Einstein. The aim of the committee was to educate people about atomic weapons. In 1947, he was awarded the presidential Medal of Merit by President Truman for his work on crystal structure, nature of chemical bond, and his efforts to bring about world peace.

In 1954, Pauling received the Nobel Prize in chemistry for his research into the nature of the chemical bond and its application to the elucidation of the structure of complex substances. On October 10, 1962 he was awarded the Nobel Peace Prize for his efforts towards nuclear test ban treaty.

Pauling, in recent years, researched the chemistry of the brain and mental illness, the cause of sickle-cell anemia, and the effects of large doses of Vitamin C on the common cold and on cancer. On August 19, 1994, Pauling, himself, died of cancer at the age of 93.

See also **Electronegativity**.

J. M. I.

PAULI, WOLFGANG ERNST (1900–1958). Pauli was an Austrian theoretical physicist. After WWII, he became an American citizen. When just 20 years of age he wrote "The Theory of Relativity." Later he wrote articles on "Quantum Theory" and "Principles of Wave Mechanics." He is most remembered for formulating the "Pauli exclusion principle". This principle says that two electrons in an atom can never exist in the same state. This is important concept for modern physics. Pauli was awarded the Nobel Prize in physics in 1945 for this discovery.

See also **Pauli Exclusion Principle**; and **Quantum Mechanics**.

J. M. I.

PAULOWNIA. See **Catalpa Trees**.

PAWPAW. See **Laurel Family**.

PEA. See **Bean**.

PEACH-TREE BORER (*Insecta, Lepidoptera; Synanthedon*). Any of three species of moths whose larvae burrow in the sapwood and inner bark of peach trees, sometimes killing younger trees. The moths belong to the family *Aegeriidae*, characterized by the long body and narrow wings, more or less free from scales. All three species attack other fruits as well as peaches.

The larvae of all species are destroyed by digging them out of their burrows and by burning badly infested boughs or entire trees.

PEACH TREES. See **Rose Family**.

PEACH-TWIG BORER *(Insecta, Lepidoptera).*   The larva of a small moth, *Anarsia lineatella*, which bores in the young twigs and fruit of peaches and other fruit trees. It is an important pest, chiefly in the western part of the United States, where various spraying programs have been found effective in controlling it.

PEA FAMILY.   See Leguminosae.

PEAFOWL *(Galliformes, Pavoninae).*   Related to the subfamily of turkeys *(Meleagridinae)*, peafowl are among the largest gallinaceous birds. They are very long-legged. The extraordinary long tail coverts of the male have strong shafts and magnificent colors and extend far beyond the tail, which has 20 feathers. They form a train with ocellate designs that is erected and spread like a fan to display during courtship. The fan is supported by hidden tail feathers that are much shorter. The feathers have no aftershaft. There are no feathers around the eyes, but rather a crest on the crown. The legs of the male have no spurs. The colors of the female are plain; she has no train and her tail has 18 feathers. There is one genus *(Pavo)* with two species: (1) the *Indian peafowl (Pavo cristatus)*, short-legged with a marked sex dimorphism, and (2) the green peafowl *(Pavo muticus)*, which has a longer and slimmer neck and longer legs, giving it a larger, more pompous appearance. Sex dimorphism is slight. Its crown carries a long bundle of erect, shiny green feathers, resembling a sheaf of ears and having narrow, spiky vanes. The plumage of both sexes is predominantly green. Like the female of the Indian peafowl, the green peafowl female has no train. See Fig. 1.

Fig. 1.   Male green peafowl. *(Pavo muticus.)*

In the wild, the peafowl prefer dense jungle on hilly territory near water. There they live polygamously in small family bands. In the early morning and evening, these large birds seek food in open fields. Wherever natives protect the peacock as the symbol of the god Krishna, it becomes very tame. Then it roams the fields even during the daytime, selecting high trees in the villages for sleeping quarters. The peacock enjoys a special reputation in India as an exterminator of cobras. In fact, it enjoys eating young cobras. As a result, this poisonous snake soon disappears from peacock-inhabited territories.

Peacock calls frequently warn game animals of the presence of tigers and leopards. This is understandable. The bird itself is one of the most common prey of the big cats. Its predilection for snakes and its warning calls have made it a highly valued and popular bird in its home country, probably long before it was introduced as an ornamental bird in parks and gardens of other parts of the world.

In India, the mating season of the peacock essentially coincides with the rainy seasons. The cock surrounds himself with a "harem" of two to five hens. The cock never courts the hen directly, but promptly turns his back on her as soon as she approaches. This, in turn, causes the hen to run around to face him. This curious behavior may be repeated several times. Finally, the hen lies down before the cock, whereupon he folds his fan and treads upon her, in the usual fowl-like manner. Ethologists have various explanations for the peacock's spectacular courtship behavior. One naturalist has observed that the peacock's magnificent fan is used to attract hens from long distances. In the opinion of another naturalist, the spreading fan indicates "feeding" to the hen, even though there is no food. Among many gallinaceous birds, the cock does indeed offer food to the hen and then mates with her.

Peacock chicks practice fan spreading frequently. Their small wings vibrate, and their tiny tail feathers are erected as if a fan were already spread above. Even hens, particularly young hens, sometimes assume this "pompous" attitude. It is mainly during the mating season that the peacock utters its loud, "meowing" call. Both sexes utter this call, but it is more frequently uttered by the males. The Indian natives interpret these calls as "minh-ao" which means "there will be rain." Peacocks do call more frequently before imminent thunderstorms.

The hen hides her nest in the underbrush, but occasionally she uses hollows between the main branches of strong trees, abandoned nests of birds of prey, or even buildings. She lays 3 to 5 creamy-white, thick-shelled eggs, which she incubates for 28 days. After birth, the peacock chicks like to be close under the mother's tail. They grow slowly, and small feather crowns appear after they are 1 month old. But not before they are 3 years old do the young cocks' trains reach their full size. The train can grow for up to 6 years and may reach a length of 160 centimeters (5.2 ft).

The peacock is probably the oldest known ornamental bird. More than 4000 years ago, it was introduced to the cultures of Mesopotamia via trade routes and from there to the Mediterranean nations. Its magnificent coloring, its adherence to its habitat, the ease with which it can be bred, and the way it gets along with other birds have made it the ideal ornamental bird, particularly in large parks. Besides, it destroys very little plant life and is insensitive to fluctuations in the climate.

Many experts consider the green peafowl to be the most beautiful and most impressive of all gallinaceous birds. Green peafowl do not "meow," but instead utter loud trumpet tones that sound like "hah-o-han." They are considered keen-eyed, watchful, and cautious jungle inhabitants. Rarely does an enemy that is sneaking up on them escape their attention. The green peafowl is not resistant to winter weather and requires frost-free shelter during the cold season. Moreover, the green peafowl have an extraordinarily lively, fierce, and courageous temper. The males do not get along well with each other either in "flying cages" or in parks. They become aggressive toward people. When crossbred with the Indian peafowl, a beautiful breed, the Spalding peacock, emerges. They can be raised from hybrids. More adapted to people, these birds are known to live up to 20–30 years.

The subfamily, *Congo peafowl*, is a link between the peafowl and the guinea fowl.

PEAKING CIRCUIT.   A circuit used to extend the frequency response of a video amplifier. In a *shunt peaking circuit* (Fig. 1), the impedance of the load circuit of an amplifier stage is raised when a small inductance, included in series with the load resistor, is caused to resonate with the circuit capacitance. In *series peaking*, a small inductance is included in series with the connection to the input of the next stage. At the higher frequencies, this inductance resonates with the input capacitance of the succeeding stage, thus improving the frequency response. Both series and shunt peaking may be included in the same amplifier stage.

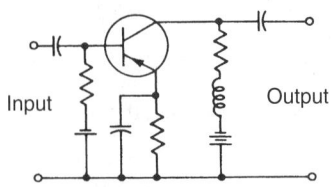

Fig. 1.   Shunt peaking circuit.

PEAKING TRANSFORMER.   A transformer which is supplied, through a large series impedance, with a voltage of many times the magnitude necessary to produce normal flux densities in the core. Twice each cycle, therefore, the flux travels rapidly from one saturation region to the other, resulting in a sharp pulse of induced voltage in a secondary winding. The secondary winding is sometimes made of high-permeability material with a smaller cross-sectional area, to increase the saturation effects resulting in a sharper pulse. Used to provide firing pulses for ignitrons and thyratrons, it is sometimes also called an impulse transformer. See also **Transformer**.

PEAK LOAD. See **Gas and Expansion Turbines**; and **Hydroelectric Power**.

PEANUT-GROUNDNUT *(Arachis hypogaea; Leguminosae)*. Generally known as the *groundnut* outside the United States, this plant is a native legume of South America that is now widely cultivated in warm climates throughout the world.

The plant is a bush annual with pinnately compound leaves and rather showy yellow flowers. After fertilization, the flower stalk elongates greatly and bends downward so that the ovary is pushed into the ground. There it develops into the familiar peanut or groundnut with its two or more seeds. See Fig. 1.

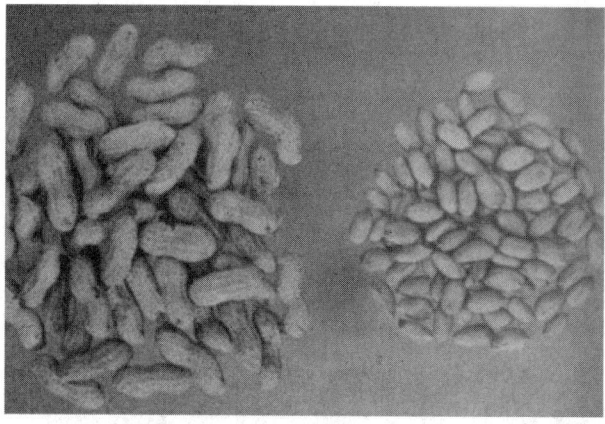

(a)

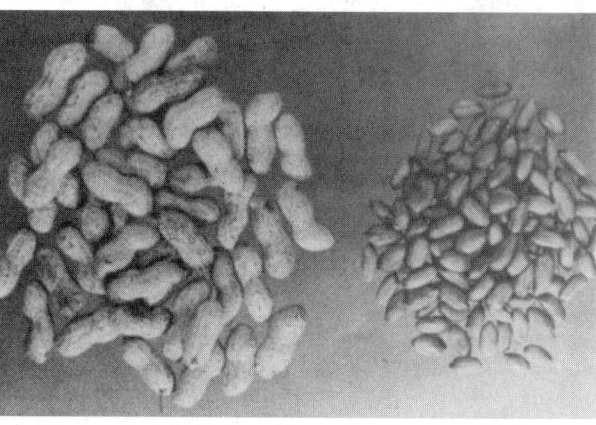

(b)

Fig. 1. Mature and immature peanut (groundnut) pods, showing fruiting habit of the plant. (*USDA photo.*)

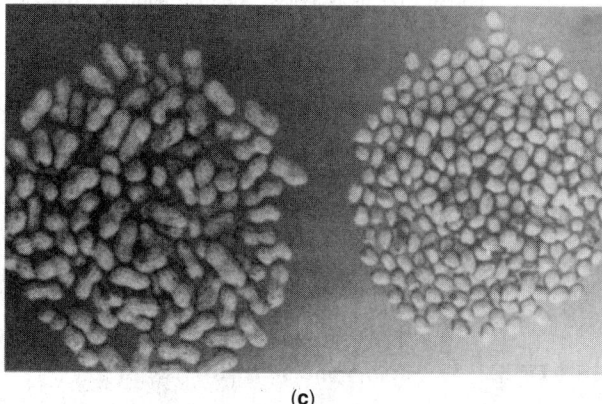

(c)

Fig. 2. Varieties of peanuts (groundnuts), showing pods at left and shelled nuts at right: (**a**) Virginia Bunch variety; (**b**) Holland Jumbo variety; (**c**) Dixie Spanish variety. (*USDA photos.*)

Large crops are raised annually in the United States. Continuing research on the peanut, because of its economic importance, is conducted at the Tifton, Georgia, experimental station of the U.S. Department of Agriculture.

Large crops are raised annually. Much of the crop (Fig. 2) is roasted in the shell or shelled and salted, and so marketed. Large quantities are ground for peanut butter or crushed for peanut oil. In some areas, boiled peanut are well regarded as snack food. See also **Leguminosae**.

Harvesting practices are among the most critical in peanut culture because of their far-reaching effect on yield, quality, and total value of the crop. In modern harvesting procedures, digging, shaking, and winnowing are accomplished in a single operation. Special machines permit peanuts to be combine harvested directly from the windrow. (A windrow is a row of any crop lined in a row prior to separation.) Whole plants interlaced as a ribbon are brought into the combine, where pods are removed, cleaned, stemmed, and conveyed into a bulk bin on the combine. See Figure 3.

PEANUT OIL. See **Vegetable Oils (Edible)**.

PEARL. A gem formed by bivalve mollusks, particularly by several marine species known as pearl oysters. Pearls are formed as a protection against the irritation caused by foreign objects, either parasites or bits of gravel, which lodge inside the shell. A fold of soft tissue envelops the foreign particle and deposits layer after layer of nacre on it, similar to the mother-of-pearl lining the shell.

PEARSON DISTRIBUTIONS. A family of distributions devised by Karl Pearson in 1895 and subsequently. If f is the frequency and x the variate, the family is defined by

$$\frac{1}{f}\frac{df}{dx} = \frac{a - x}{b_0 - b_1 x + b_2 x^2}$$

Pearson distinguished 12 types, according to the values of the constants. Type J is the form

$$f = \frac{x^{p-1}(1 - x)^{q-1}}{\beta(p, q)}$$

and is also known as the *Beta distribution*.

Fig. 3. Peanut (groundnut) combine with a capacity ranging from 0.5 to 2 acres (0.2 to 0.8 hectare) per hour. Note parking jack in foreground, which is designed to keep the combine at correct height for attachment to tractor. (*Lilliston Corporation.*)

Type III is of the form

$$f = \frac{x^{p-1}e^{-x}}{\Gamma(p)}$$

and is also known as the *Gamma distribution*.

The normal distribution is also one of the Pearson system. The other types are less important. Taken as a class, the distributions provide a very flexible set, to one of which many practically occurring distributions may be satisfactorily fitted.

PEAR TREES. See **Rose Family**.

PEAT. See **Coal**.

PEAT BOG. See **Bog**; and **Swamp**.

PEA WEEVIL (*Insecta, Coleoptera*). A beetle which attacks growing peas in the pod. The most effective control measures are to avoid planting infested seed or to fumigate it with carbon disulfide before planting.

PECAN TREE. See **Hickory and Wingnut Trees**.

PECCARY. See **Suines**.

PECLET NUMBER (symbol Np_e). A nondimensional number arising in problems of heat transfer in fluids. It is the ratio of heat advection to heat diffusion and may be written $Np_e = Ul/k$ where U is a characteristic velocity; l is a characteristic length; and k is the thermometric conductivity. Also, $Np_e = N_{Re}N_{Pr}$ where N_{Re} is the Reynolds number and N_{Pr} is the Prandtl number.

PECTINS. A *pectic substance* is a group designation for those complex carbohydrate derivatives that occur in or are prepared from plants and contain a large proportion of anhydrogalacturonic acid units, which are thought to exist in a chainlike combination. The carboxyl groups of polyglacturonic acids may be partially esterified by methyl groups and partly or completely neutralized by one or more bases. The general term *pectin* (or *pectins*) designates those water-soluble pectinic acids of varying methyl ester content and degree of neutralization which are capable of forming gels with sugar and acid under suitable conditions. The term *protopectin* is applied to the water-insoluble parent pectic substances which occur in plants and which upon restricted hydrolysis yield pectin or pectinic acids. *Pectic acids* is a term that is applied to pectic substances mostly composed of colloidal polygalacturonic acids and essentially free from methyl ester groups. The salts of pectic acids are either normal or acid *pectates*. The term *pectinic acids* is used for colloidal polygalacturonic acids containing more than a negligible proportion of methyl ester groups. Pectinic acids, under suitable conditions, are capable of forming gels with sugar and acid, or, if suitably low in methosyl content, with certain metallic ions. The salts of pectinic acids are either normal or acid *pectinates*.

Pectins occur commonly in plants, particularly in succulent tissues, and are characterized by the polygalacturonic acids that are fundamental to their structure. The pectins are important emulsifying, gelling, stabilizing, and thickening agents used in the preparation of numerous food products. About 75% of the pectins produced are used in making fruit jams, jellies, marmalades, and similar products. Additional uses include the preparation of mayonnaise, salad dressings, malted milk beverages, frozen dessert mixes, and frozen fruits and berries (to prevent leakage upon thawing), among others. The addition of a dilute pectin solution to milk coagulates the casein. In many food products, the use of pectins as stabilizers is preferred, since they blend better into the flavor complex than do many gums, starches, or a number of carbohydrate derivatives. Pectin jellies do not melt at temperatures below 49 °C, a distinct advantage over gelatin gels that require refrigeration. Pectins also have a number of nonfood uses, including pharmaceuticals and cosmetics.

The location of various pectin substances in plant tissues is well established. Pectins make up most of the middle lamella in unripe fruit and are to be found in the cell walls and in small proportion in all plant tissues. The genesis and fate of pectins in plant tissues have not been fully determined.

Citrus peel, apple pomace from juice manufacture, and beet pulp left over from the manufacture of sucrose are common commercial sources of pectins. After some preliminary purification of the raw material, the extraction is usually performed with hot dilute acid (pH = 1.0–3.5 in a temperature range of 70–90 °C). The pectin is then precipitated from the extract with ethanol or isopropanol, or with metal salts (copper or aluminum). The metal ions have to be subsequently removed by washing with water or acid ethanol. Specific formulas for denatured ethanol for use in pectin manufacture are used. The precipitates are purified, dried, and pulverized to form the yellowish-white powder of commerce.

Pectin substances in solution behave as typical colloids. See also **Colloid System**. Dry, purified pectins are light in color and soluble in hot water to the extent of 2–3%. The pH of pectin solutions is usually 2–3.5.

The proportion of sugar which pectin will form into a firm jelly determines the *jelly grade* of the pectin. In a jelly, jam, or marmalade, the proportions of total solids of sugars, the pH, and the proportion and nature of the pectin used will determine the extent of jellification obtained. The use of pectin in fruit jams and related products is approved by most food regulators because the addition is believed to compensate for an incidental natural deficiency.

In pectic acids, all carboxyl groups are free, or at least not present as the methyl ester. Under suitable conditions, pectins will form jellies with sugar and acid, whereas the low-ester pectins will form *gels* with traces of polyvalent ions. The general structure of pectin is:

Additional Reading

Fishman, M.L. and J.J. Jen: "Chemistry and Function of Pectins," American Chemical Society, Washington, DC, 1986.

Quilici-Timmecke, J.: "New Nutrients against Cancer: Modified Citrus Pectin, Soybeans, Lycopene and Other '90's Cancer Fighters," Keats Publishing, Inc. Chicago, IL, 1998.

Walter, R.H.: "The Chemistry and Technology of Pectin," Academic Press, Inc., San Diego, CA, 1997.

Wood, W.A. and S.T. Kellogg: "Biomass: Lignin, Pectin, and Chitin," Vol. 161, Academic Press, Inc., San Diego, CA, 1988.

PECTORAL FIN. See **Fishes**.

PEDIATRICS. That branch of medicine concerned with the prevention, diagnosis, and treatment of the diseases and disorders of children.

PEDICULOSIS. Infestation of the body with lice. Infestation of the scalp, most prevalent in women, is caused by *Pediculus humanus* var. *capitis* (head lice). The condition is accompanied by severe pruritus (itching) and sometimes is complicated by secondary infections, often arising from contamination of the skin as the result of aggressive scratching. Left unattended, the hair becomes matted with exudation of the lice. Treatment includes gamma benzene hexachloride shampoo or benzyl benzoate (20% solution). Occasionally, a second application will be required within one week. Physicians usually prescribe a systemic antibiotic to allay secondary infection if such is indicated. In body infestation, the cause is *Pediculus humanus* var. *corporis* (body lice) which tend to live within the crevices of clothing, but which attack the skin, forming macules and wheals. Pruritus can be severe. Scratching may initiate secondary infection. The best solution is destruction of infested clothing, but if garments must be retained, they should be boiled once or several times and ironed with a hot iron. Pediculosis pubis (sometimes called *crabs*) results from infestation of the pubic area with *Phthirus pubis*. Uncommonly, the condition also may be found on the eyelashes. Examination reveals lice attached to the skin and lice eggs attached to the hair shafts. Macules (bluish in color) are formed when the lice suck blood from the skin—in the pubic area and sometimes on the thighs. Gamma benzene hexachloride lotion or cream can be used for treatment. This agent should not be used on the face or eyelids. Physicians usually prescribe 0.5% physotigmine for local application in the case of eyelash infestation.

PEDIGREE METHOD. See **Plant Breeding**.

PEDOLOGY. The study of the origin and classification of soils. The investigation of the regolith as a fundamental natural resource or the basis of terrestrial plant life.

PEEWIT. See **Waders, Shorebirds, and Gulls**.

PEGASUS. Also referred to as the *Flying Horse*, this is one of the most ancient of the northern constellations. The constellation is situated in a southeasterly direction from Andromeda and is on the meridian at midnight in September. The three brightest stars in the grouping are α-, β-, and γ-Pegasi. Together with α-Andromedae, these stars form a figure known as the *square of Pegasus*. (See map accompanying entry of **Constellations**.)

PEGMATITE. The term *pegmatite*, derived from the Greek word meaning "joined together," was first applied by Haüy in 1822 to a peculiar interpenetrating growth of quartz and feldspar sometimes called graphic granite from its resemblance to written characters, particularly those of the Hebrew language. Pegmatite is also used to designate those coarse-grained dikes and sheets, chiefly of granite or syenite, that are apophyses of stocks or batholiths, or of the residual magma, during their congelation. The individual minerals may often reach great size. Granite pegmatites are chiefly composed of alkali feldspar and quartz with some muscovite or biotite but may carry such minerals as tourmaline, topaz, beryl, fluorite, apatite, garnet, lepidolite, etc. See also **Mineralogy**.

PELAGIC. See **Ocean Resources (Living)**.

PELECANIFORMES *(Aves)*. Among the many bird groups whose members live on or at the water, the Pelecaniformes are of especially striking appearance because of the peculiar structure of their feet. Their toes are joined by more or less well-developed webs, but the web in their case includes, in contrast to ducks and geese, the hind toe, which is directed forward and to the inside. This "paddle foot" is found in all members of this order.

Birds of this order are medium sized to very large, and feed exclusively on animal food. Most species obtain their food from the sea. There are a total of 6 well- differentiated families, with 7 genera.

The Tropic-Birds (*Phaethontidae*, genus *Phaethon*) with 3 species; the Pelicans (*Pelecanidae*, genus *Pelecanus*) with 7 species; the Cormorants (*Phalacrocoracidae*, genus *Phalacrocorax*) with 28 species; the Anhingas or Darters (*Anhingidae*, genus *Anhinga*) with 2 species; the Gannets (*Sulidae*, genera *Morus* and *Sula*) with 9 species between them; and the Frigate-Birds (*Fregatidae*, genus *Fregata*) with 5 species.

The Tropic-Birds (family *Phaethontidae*, with only 1 genus, *Phaethon*) are birds of the high seas with a pigeon-like flight. The length is 80–100 centimeters (31–39 inches), of which the body length is 30–45 centimeters (12–18 inches). The wing span measures 92–109 centimeters (36–43 inches) and the weight is 300–750 grams ($10\frac{1}{2}$–$26\frac{1}{2}$ ounces). They are predominantly white with a very long central tail feather. The short legs are far back on the body. Tropic-birds can hardly walk, but they can dig and scrape. There are 3 species: The Red-Billed Tropic-Bird (*Phaethon aethereus*), which reaches a length of 100 centimeters (39 inches); the White-Tailed Tropic-Bird (*Phaethon lepturus*), which reaches a length of 80 centimeters (31 inches); and the Red-Tailed Tropic-Bird (*Phaethon rubricauda*), which is the largest species, with a length of 100 centimeters (39 inches).

The largest birds of this order are generally known as the Pelicans (family *Pelecanidae*). The length is 170–180 centimeters (67–71 inches) and the wingspan is up to almost 300 centimeters (118 inches) in the Dalmatian pelican. The weight ranges from 7–14 kilograms (15–31 pounds). They appear clumsy, but because of the air in the bones and the skin, they are relatively light birds with large bodies, long broad wings, fairly long necks, and gigantic beaks. Between the branches of the lower mandible there is a distensible skin pouch; the upper mandible serves merely as a flat lid to cover it. The tongue is minute, the legs are short, the feet are large, and the 4 toes are connected by webs.

Pelicans float high on the water and they carry their wings slightly raised; since they have no wing pockets, the beak rests on the slightly curved neck. In flight the head is drawn back onto the shoulders. The flight is light and elegant; gliding often alternates with wing beats. The food consists exclusively of fish, which are scooped up in the bill; only the brown pelican is a plunge diver. They are sociable birds that fly in small groups or larger flocks, mostly in a diagonal line with respect to the direction which they are travelling. They search for food together, and often nest in very large colonies of up to several thousand birds. They breed on a base of reeds and branches, and on shore often on just a few feathers.

The Cormorant family (*Phalacrocoracidae*) has only 1 genus, *Phalacrocorax*. The length is 48–92 centimeters (19–36 inches), and the weight is 0.7–3.4 kilograms ($1\frac{1}{2}$–8 pounds). They are equally well adapted for flying and for swimming. Relatively clumsy on land, they are still not as helpless as some other members of the order. They lie rather deeply in the water when swimming, since the air spaces in their bones are very small. There are no external nares, and the edges of the beak are somewhat toothed. The head and neck have powerful muscles for closing the beak, which in part originate from special long, sesamoid bones behind the back of the head, and which are needed to maintain a grip on fish that have been caught. The plumage usually has crests in the breeding season. They are distributed worldwide, with 29 species, more species than in all the other Pelecaniformes together.

Two species of freshwater Pelecaniformes must be considered as a separate family because of a number of peculiarities of their body structure, although they are in many respects close to the cormorants. These are the Anhingas or Darters (family *Anhingidae*, genus *Anhinga*). The length is 90 centimeters (35 inches). The bill is straight and pointed, and is finely toothed on both cutting edges. The neck has 20 vertebrae and, when the bird is sitting or flying, it is bent into an S or even a G shape. The tail is long, stiff, and rounded at the end. They walk and swim underwater when diving. There are 2 species: The American Anhinga (*Anhinga anhinga*), with no subspecies; the Old World Anhinga (*Anhinga rufa*), which has 3 subspecies.

Anhingas live on freshwater fish and other aquatic animals. Like herons, they have dagger-like beaks and long G-shaped necks. But while herons wade in shallow water and stalk or wait for their prey above the water surface, anhingas hunt for their prey underwater. A special hinge and muscle arrangement on the eight and ninth cervical vertebrae enables them to thrust the head rapidly forward so that the prey is stabbed and stunned.

Like the plumage of the cormorants, the anhinga's is water permeable. This adaptation reduces buoyancy and enables anhingas to submerge silently without attracting the attention of prey animals or enemies. They often swim with only the head and the thin neck projecting above the surface, and when doing so they really resemble a swimming snake; hence they have another name, the snake-bird.

In the air they soar and glide like pelicans, skillfully using thermal air currents. Anhingas are very well adapted to both air and water; they are gliders as well as spear fishers.

The Gannets or Boobies are predominantly black and white seabirds (family *Sulidae*). The length is 70–100 centimeters ($27\frac{1}{2}$–39 inches), and the weight is 1.5–3.5 kilograms (3–8 pounds). The strong beak is pointed and conical, with fine serrations on the cutting edge near the tip. The bare parts of the face, throat, and feet are often colored. The strong feet have well-developed webs. Gannets breathe through a specially constructed palate. There are 2 genera with 9 species.

All gannets obtain their food by diving. Where they are numerous, this method of feeding makes a fascinating spectacle.

The Frigate-Birds (family *Fregatidae*) exceed the members of all the other families in aerial mastery. Almost half of their weight consists of the breast muscles and feathers; their load per unit of wing surface is extraordinarily small.

There is only 1 genus, *Fregata*, with a length of 75–112 centimeters ($29\frac{1}{2}$–44 inches), a wing span of 176–230 centimeters (69–$90\frac{1}{2}$ inches), and a weight of up to 1.5 kilograms (3 pounds). The beak is long and bent into a hook at the tip. The wings are narrow and the lower arm and hand bones are strongly elongated. The tail is deeply forked, often spread and then closed again in flight. The small feet are almost without webs. They are restricted to tropical and subtropical seas; they live mainly where flying fish are common, in water of at least 25 °C. (77 °F.).

There are 5 species. The Magnificent Frigate-Bird (*Fregata magnificens*) is the largest. Its length is 103–112 centimeters ($40\frac{1}{2}$–44 inches), the wing span is 230 centimeters ($90\frac{1}{2}$ inches), the beak is 12 centimeters ($4\frac{1}{2}$ inches), and the weight is 1.5 kilograms (3 pounds). The males have a white breast band and brownish upper. The Ascension Frigate-Bird (*Fregata aquila*) weighs 1.2 kilograms ($2\frac{1}{2}$ pounds). The female has a somewhat greenish gloss, while the males are brownish on the upper breast, nape, and wing band. In the Christmas Frigate-Bird (*Fregata andrewsi*), both the male and female have the same coloring. The Lesser Frigate-Bird (*Fregata ariel*); has a length of 82 centimeters (32 inches). The Great Frigate-Bird (*Fregata minor*); reaches a length of 95 centimeters (37 inches).

Although frigate-birds sometimes fly out over the sea, they tend to breed all year round and, as a rule, stay near their home islands. When a frigate-bird is seen at sea, land is generally not far away. Their ability to find their way is such that these birds are used to send messages in the South Sea islands, as are carrier pigeons elsewhere in the world.

Frigate-birds generally build their nests in low shrubs or trees, and only rarely on bare ground. The nesting colonies are generally close to those of other seabirds, particularly terns or gannets, which frigate-birds rob of their prey or young. See also **Pelicans and Cormorants**.

PELE'S HAIR.   A fibrous, basic, natural glass (tachyllite). The congealed liquid lava blown out of volcanoes. Type locality, the Hawaiian Islands.

PELICANS AND CORMORANTS (*Aves, Pelecaniformes*). These birds are widely distributed and of relatively few species. Millions of pelicans, cormorants, and boobies inhabit the islands off the coast of Peru, as evidenced by the views of Fig. 1. The birds almost completely cover the islands when they come in to roost in the evening. They deposit a tremendous amount of droppings (guano), which dries and hardens quickly in the prevailing dry climate. As many as 750 tons of guano per acre per year may be deposited. Food for this large number of birds is provided by abundant fish life in the surrounding cold waters.

The common pelican of North America (*Pelecanus erythrorhynchos*) is chiefly white with orange beak and feet during the breeding season. The bird is large with a long neck, short legs, and webbed toes. See Fig. 2. The beak is very long, and the lower mandible bears a flexible pouch, which can be distended to accommodate a large amount of food. In addition to fish, the pelican consumes other aquatic forms. The pelican lives in colonies. Some species build nests on the ground; others in low bushes or trees. The birds fly at high elevations and often soar in formation. The pelican has practically no voice. The pelican winters along the sea coast, and in summer, it may be found around inland lakes. The white pelican of Africa and Asia is much like the American pelican except the plumage has a delicate pink coloration. Possibly the most attractive pelican is the black

Fig. 1.   Guano birds. (*R.C. and G.E. Murphy.*)

Fig. 2.   American white pelican. Bill shows horny growth that appears in breeding season (*New York Zoological Society.*)

and white pelican of Australia (*P. conspicillatus*). Other species display brown and gray plumage. See also Fig. 3.

The cormorant is a large bird of slender build with a moderately long neck and slender beak, slightly hooked at the tip. The feet are webbed, and the birds are strong swimmers and divers. They live entirely on fish. The several species are widely distributed, the common cormorant (*Phalacrocorax*) occurring in eastern North America, Europe, Asia, and northern Africa. In Japan and China, cormorants are kept in captivity to be used for fishing. A ring or strap around the bird's neck keeps it from swallowing the fish that it catches, although some are said to be so well trained that they bring fish to their keepers without this check. The birds are eaten in some parts of the world.

The darter (*Anhinga*) is a long-necked diving bird with a long, sharp beak, found in all continents. Resembling the cormorants, the darter is also known as the snake bird or snake neck, and certain species are called wryneck and water turkey. The frigate bird (*Fregata*) is a marine bird related to the cormorant. It is of slender, long-winged build and powerful in flight. The bird lives chiefly on fish sometimes forcibly obtained from other birds. Also related to the cormorant is the gannet (*Sula*), a large fishing

Fig. 3. Formation of pelicans over coastal waters of northern Florida. (*G.D. Considine.*)

bird found on the seacoasts in the higher latitudes of both hemispheres. It breeds in dense colonies on steep rocky cliffs. See also **Pelecaniformes**.

PELLAGRA.   See **Niacin**.

PELL'S EQUATION.   See **Number Theory**.

PELORUS (or Dumb Compass).   An instrument once widely used on board ship for taking bearings of external objects. The pelorus consists fundamentally of a circular plate, heavily ballasted and mounted in gimbals. This plate has two pairs of indicators, one pair parallel to the keel of the ship and the other perpendicular to the keel. Concentric with the circular plate, and capable of being rotated independently of each other about a vertical axis, are a graduated dial plate and an alidade or arm for reading angles. The dial plate is graduated in a manner similar to the compass card and may be clamped to the circular plate in any desired position. The alidade carries sighting vanes, and the line through these vanes passes through the axis of rotation of the instrument, carrying indicators at each end, which may be read on the dial plate.

PELTIER EFFECT.   See **Thermocouple**.

PELTON WHEEL.   See **Hydroelectric Power**.

PELVIC FIN.   See **Fishes**.

PELVIS.   Literally, a basin. Commonly applied to the bony pelvis of man, which is a compact pelvic girdle comparable to that of vertebrates generally (skeletal system). The human pelvis is properly the basin-like abdominal cavity containing the viscera, supported by the bony pelvis, which is composed of the hip bones on either side, and in front, and the sacrum and coccyx. The pelvis rests upon the lower extremities and supports the spinal column. Within the pelvis are found the rectum, bladder, and generative organs. There are certain sexual differences in the pelvis. The female pelvis is lighter, more slender, with a cavity that is larger, less funnel-shaped, and shorter. When the female pelvis resembles the male type or is deformed by bony disease such as rickets, childbirth is interfered with and either made difficult or impossible by the vaginal route.

The pelvis of the kidney is the principal cavity, which receives the urine from all subordinate divisions and discharges it to the ureter.

PEMPHIGUS VULGARIS.   This is classified as a *vesiculobullous* (watery blisters and sacs or cysts containing fluid and affecting mucous membrane) disease. The disease is not commonly seen in patients before the fourth to sixth decade of life. Localized lesions in the mouth may be the only early symptoms. Untreated, the disease progressively involves the mucous membrane surfaces of the mouth, esophagus, vagina, and cervix. When the disease becomes extensive, large areas of the skin

may become denuded, leading to the loss of protein and fluid from those areas. Prior to the administration of glucocorticosteroids, serious denudation led to debilitation and even death. The physician will administer glucocorticosteroids with discretion because of the side effects associated with them. It is important that means be taken to prevent infection of denuded areas and thus systemic antibiotics are frequently administered. The etiology of the vesiculobullous diseases is poorly understood.

PENCIL BEAM.   Emission, from an antenna, having the form of a narrow conical beam.

PENCIL-BEAM ANTENNA.   A unidirectional antenna, so designed that cross section of the major lobe by planes perpendicular to the direction of maximum radiation are approximately circular, and having a very small angular cross section.

PENCIL FISHES.   See **Characids**.

PENDULUM.   In considering a gravity pendulum, let a rigid body of mass M swing on an axis located at distance r above its center of mass, and with respect to which axis the moment of inertia of the body is I. This motion is mathematically not simple, but if the amplitude of swing is small, certain terms in the differential equation of motion may be disregarded and the remaining equation readily solved. The solution gives as the period of the complete oscillation

$$T = 2\pi\sqrt{\frac{1}{Mgr}}$$

in which g is the acceleration of a freely falling body. Huygens found experimentally, what may be proved theoretically, that for any given period of oscillation greater than a certain minimum there are two different distances r for which I/r, and hence T, has the same value. If these two distances are laid off on opposite sides of the center of mass, the two points resulting are "conjugate points" (center of suspension and center of oscillation). Denoting the whole distance between these points by l, the period of swing when the pendulum is suspended at either of them is

$$T = 2\pi\sqrt{\frac{l}{g}}$$

This is the same as the period for an "ideal simple pendulum," i.e., a single particle of mass m suspended by a weightless thread of length l, for which $r = l$, and $I = ml^2$. Kater utilized this principle in his well-known reversible pendulum.

It was Huygens who first adapted the pendulum to regulate a mechanism for keeping time and thereby gave us the common clock.

PENDULUM CLOCK.   A clock that uses a pendulum as its frequency standard. Galileo discovered the principle of *isochronism* of the pendulum in 1582, observing experimentally that a pendulum's frequency, or period of swing, seemed independent of its arc. Not until 1641, however, did he begin to adapt the pendulum to clocks. In 1656, the Dutch scientist, Christian Huygens (who 20 years later developed the hairspring to regulate the oscillations of the spring-driven balance wheel) substituted a pendulum for the foliot bar in the then-existing weight-driven clocks. Theoretically, Huygens recognized that the period of the swing of the pendulum, unlike the oscillation rate of the foliot bar, would be practically independent of the clock's drive system. In Huygens' clock, built in 1657 by Salomon Coster, a pendulum was weight-driven through a verge and foliot escapement. However, the pendulum was required to maintain an arc of some 40 degrees.

Huygens recognized that a pendulum with such a large arc is not isochronous and determined that its bob should swing in a cycloidal, or more U-shape, curve. His English contemporary William Clement solved the problem by inventing the anchor escapement that permits the use of a small arc of 3 or 4 degrees. Only within such a small arc is a pendulum practically isochronous and, therefore, most accurate. The small arc also minimizes power requirements and made practical narrower clock cases.

A later development, the so-called deadbeat escapement invented in England by George Graham in 1715, perfected the anchor escapement and was employed in many observatory pendulum clocks for more than 200 years.

The first electrically driven pendulum clock was built in 1843 by Alexander Bain, a Scot. The bob swung between two permanent magnets. In 1873, George Airy, British Astronomer Royal, designed the first electric pendulum clock to compensate for variations in barometric pressure—which otherwise affect the period of swing. His clock became recognized as the standard time reference throughout the world until 1922, when the Shortt electric pendulum clock replaced it as the primary standard at the Royal Greenwich Observatory outside London.

In 1898, James Rudd, an Englishman, had built the first so-called free pendulum clock that used two pendulums: a master, or "free," pendulum that drove no mechanism; and a slave pendulum that drove an anchor escapement. The Shortt clock was an adaptation of this system. Its slave pendulum always ran slow as compared to the master, with a maximum difference of only $\frac{1}{240}$ of a second reported at the Royal Greenwich Observatory.

The U.S. National Bureau of Standards (NBS) also used the Shortt pendulum clock as its primary standard. Quartz crystal clocks, which in the 1930s and early 1940s could not maintain comparable accuracies for periods longer than three months, replaced the Shortt pendulum as the primary NBS standard during 1946–50. The accuracy of the Shortt clock ranged to 1 part in 4,000,000, equivalent to a variation of 0.020 second per day.

After the replacement of pendulum clocks as primary standards, a new Riefler electric pendulum eliminated the slave pendulum by substituting a diaphragm and pin-hole on the master pendulum. A pencil beam fixed on the diaphragm activates a photo cell that periodically penetrates the pin-hole as the pendulum swings, creating a precise frequency standard without burdening the pendulum with a mechanical escapement. Proposals for a similar system had been impractical previously because of the relatively short life of earlier vacuum tubes. However, electric free pendulum clocks, which must be set on ultrastable mountings and operate in a controlled-atmosphere environment, are too large and delicate for consumer use. See also **Clock**.

W.O. Buennett, J.J. Carpenter, F. Dostal, and E. Van Haaften, New York, NY.

PENDULUM (Foucault). See **Foucault Pendulum**.

PENEPLAIN. Meaning nearly a plain. A physiographic term implying a broad flat erosional surface which has been finally developed regardless of the structure, relative hardness, and solubility of the rocks of the region. In the case of widespread unconformities, the plain of erosion that truncates the subjacent deformed rocks and underlies the superjacent formations is an ancient peneplain. Local topographic features which rise above the peneplain are called monadnocks, after the type, Mt. Monadnock, in New England.

PENGUIN (*Aves, Sphenisciformes, Spheniscidae*). Flightless marine birds which inhabit mainly the coasts of the Antarctic, but they are also common in the southern temperate cool zone. Northward, one species is found as far as the Galápagos Islands, which lie on the Equator; they also occur on the subtropical coasts of South America, South Africa, and Australia.

The shape of their bodies makes penguins excellent swimmers. They can remain in the water for a long time without being harmed by the cold, yet they can move quickly and with agility on land. Their adaptation to life in the water is developed to a high degree, similar to that shown by the seals among mammals.

Penguins generally swim on or just below the surface of the sea; under water they reach a speed of 36 kilometers (22 miles) per hour. At higher speeds they dive alternately up and down like dolphins; this enables them to breathe at regular intervals without interrupting their forward movements, and to reduce friction by lubricating the plumage with air bubbles. The heat-conserving air-cushion under their plumage gives their bodies a buoyancy which is not altogether favorable for prolonged diving. Only rarely do they dive for longer than 2 to 3 minutes without surfacing.

On land or on ice, penguins walk upright, although in snow they often move by sliding forward on their bellies. The rock-penguins and their relatives hop in an upright posture with both feet simultaneously. At the edge of the ice, penguins often pick up speed under water so that they shoot out of the waves and land securely on the ice upright on both feet.

Warm-blooded animals that live in cool regions or in cold water must conserve their heat. Since penguins are relatively small, the ratio between body surface and body volume, even in a large penguin, is less advantageous than it is in any other warm-blooded vertebrate that spends much time in the water; it is therefore particularly important for penguins to counteract the heat loss at sea. This is done in part through elevation of the metabolic rate; penguins are much more lively in the water than on land and thus produce more metabolic heat, thereby causing the body temperature to remain constant. A further help is the 2–3 centimeter ($\frac{3}{4}$-1 inch) thick layer of subcutaneous fat, particularly in the polar penguins; further, their thick, waterproof plumage and the air trapped beneath it form a very effective protection.

Such heat insulation is very useful in a cold climate and is essential in enabling penguins to survive in and near the Antarctic. The heat insulation of the Antarctic adelie penguins is so effective that while they are incubating snow can accumulate on them without melting. In warmer areas, and sometimes even in the particularly cold ones, heat insulation becomes a problem; on land penguins are always in danger of overheating, especially when they fight, run, or are very active in some way. Of the two insulating layers the fat is the lesser problem, for it is penetrated by blood vessels. These can dilate so that more blood reaches the outermost skin layer, where it cools off. The protective plumage, however, allows less heat exchange. Penguins can erect the feather tips somewhat, but the downy under layer still holds back much of the heat. The plumage is very resistant to disturbances by wind.

Large penguins lay only one egg; the rock hoppers on Tristan da Cunha often lay three eggs, but most species have a clutch of two eggs and only occasionally one or three. The weight of the egg ranges from $1\frac{1}{2}$ to 4% of the body weight. The incubation period lasts from 33 to 62 days.

The young of the same clutch hatch at the same time or, at most within a day of each other; they are covered with a sparse downy coat and are carefully brooded until, at the age of 6 to 10 days, they begin to regulate their own body temperatures. They keep their thick, almost woolly, down plumage until they are almost fully-grown. They are fed by both parents, who regurgitate food for them; within a short time they almost reach the size of a fully grown penguin. The down is replaced by a new plumage and the young, without parental help or guidance, go into the water.

Most penguins are sociable. They breed in groups or in large noisy colonies; they take to the water in flocks and seek out the feeding grounds in large swarms. One of their largest and most densely populated habitats lies at the edge of Antarctica; on several Antarctic and sub-Antarctic islands, breeding colonies of hundreds of thousands, or even of millions, of penguins have been found. In lower latitudes their numbers are generally fewer.

Penguins feed on floating (planktonic) and swimming animals, particularly small fish, small floating crabs, and squids. Although a few species limit themselves to certain animals, as the black-footed penguin apparently does to fish or the gentoo penguin to small crabs and the largest penguins to fish and squids, nevertheless they usually take whatever is most abundant.

The largest living penguins are the Giant Penguins (genus *Aptenodytes*). There are two species, the Emperor Penguin (*Aptenodytes forsteri*; see Fig. 1), which reaches a length of 115 centimeters (45 inches) and weighs up to 30 kilograms (66 pounds), and the King Penguin (*Aptenodytes patagonica*), which reaches a length of 95 centimeters (37 inches) and weighs up to 15 kilograms (33 pounds).

Although the emperor penguin is only a little larger than the king penguin, it weighs almost twice as much. It is a bird of the high Antarctic and has long, dense plumage and very large fat stores. The king penguin, on the other hand, is a bird of the sub-Antarctic and temperate-cold zone. Its body is more slender and its plumage is less dense. During most of the year it has little subcutaneous fat. The vividly colored patches on the sides of the head of both species are shown off prominently during courtship; if they are colored dark experimentally, the bird fails to attract a partner and hence does not breed. Neither species builds a nest; instead they carry the egg on top of the feet and walk around with it.

The Adelie Penguins (genus *Pygoscelis*) reach a length of 72.5–75 centimeters (28.5–29.5 inches). There are three species: the Adelie Penguin (*Pygoscelis adeliae*), which reaches a length of 70 centimeters (27.5 inches) and a weight of 5 kilograms (11 pounds); the Chinstrap Penguin (*Pygoscelis antarctica*), which reaches a length of 68 centimeters (27 inches) and a weight of 4.5 kilograms (10 pounds); and the Gentoo Penguin (*Pygoscelis papus*), including the subspecies the Great

Fig. 1.   Adult Emperor penguin with young. (*New York Zoological Society.*)

Northern Gentoo Penguin (*Pygoscelis papua papua*), the Southern Gentoo Penguin (*Pygoscelis papua ellsworthii*), and the Macquarie Gentoo Penguin (*Pygoscelis papua taeniata*). All penguins of this genus have long, curved tail feathers, which sweep behind them like brooms as they walk. The adelie penguin lives farthest south on the coasts of the Antarctic continent and the barren islands which surround it. See Fig. 2.

Fig. 2.   Penguins from Adelie, Antarctica

The Crested Penguins (genus *Eudyptes*) reach a length of 70–72 centimeters (27.5–28 inches). There are four species. The Macaroni Penguin (*Eudyptes chrysolophus*) has two subspecies: *Eudyptes chrysolophus chrysolophus*, which reaches a length of 70 centimeters (27.5 inches) and a weight of 4.2 kilograms (9 pounds), and the Macquarie Island Macaroni Penguin or Royal Penguin, *Eudyptes chrysolophus schlegeli*, which reaches a length of 62 centimeters (24 inches). The Rock Hopper Penguin (*Eudyptes crestatus*) reaches a length of 55 centimeters (22 inches) and a weight of 2.5 kilograms (5.5 pounds). The Erect-Crested Penguin (*Eudyptes atratus*) reaches a length of 67 centimeters (26 inches) and a weight of 3.6 kilograms (8 pounds). The Fiordland Penguin (*Eudyptes pachyrhynchus*); which reaches a length of 55 centimeters (22 inches) and a weight of 3 kilograms (6.5 pounds).

The macaroni penguin is the most southern representative of this genus; the others live in warmer water. Using its sharp claws, the rock hopper can climb on rocks when it allows the waves to throw it on land. It lives on many islands with either temperate-warm or temperate-cold climates, and is also found in unexpectedly large numbers on Heard Island in the sub-Antarctic. Macaroni penguins which breed with adelie penguins on the edge of the Antarctic have feathers that are on longer than those of the erect-crested penguins which live on the much warmer islands around New Zealand or the rock hoppers of Tristan da Cunha and New Amsterdam.

The Black-Footed Penguins (genus *Spheniscus*) are small to medium-sized, reaching a length of 50–71 centimeters (20–28 inches). There are four species: the Black-Footed Penguin (*Spheniscus demersus*), which reaches a length of 70 centimeters (27.5 inches) and a weight of 2.9 kilograms (6 pounds); the Magellan Penguin (*Spheniscus magellanicus*), which reaches a length of 70 centimeters (27.5 inches) and a weight of 4.9 kilograms (11 pounds); the Peruvian or Humboldt Penguin (*Spheniscus humboldti*), which reaches a length of 65 centimeters (25.5 inches) and a weight of 4.2 kilograms (9 pounds); and the Galápagos Penguin (*Spheniscus mendiculus*), which reaches a length of 53 centimeters (21 inches) and a weight of 2.2 kilograms (5 pounds).

In this genus the beak is high and strong at the base with longitudinal grooves; it is used for digging. These penguins are short-tailed and have a smooth plumage. In front of the eyes and on the chin they have red to black bare spots. Their webs are often white spotted. They rub their beaks and necks on one another in a greeting ceremony unlike other penguins.

The Yellow-Eyed (or Yellow-Crowned) Penguin (*Megadyptes antipodes*) is about the size of the gentoo penguin. Its length is 75 centimeters (29.5 inches) and reaches a weight of 5.2 kilograms (11.5 pounds). It has a fairly long beak. There is a black and yellow crown patch, and behind it is a golden-yellow stripe across the upper ear coverts. The nape has slightly elongated feathers. The eyes are pale yellowishgreen. It resides on New Zealand and a few nearby islands.

The Dwarf Penguins (genus *Eudyptula*) are even smaller than the Galápagos penguins. They reach a length of 40–42 centimeters (16–16.5 inches). There are two species with several subspecies. The Little Penguin (*Eudyptula minor*) has three subspecies: a southern subspecies (*Eudyptula minor minor*), reaching a length of 40 centimeters (16 inches) and a weight of 2.5 kilograms (5.5 pounds); a northern subspecies (*Eudyptula minor novaehollandiae*), which reaches a length of 41 centimeters (16 inches), and a weight of 2.2 kilograms (5 pounds); and the Chatham Island Little Penguin (*Eudyptula minor iredalei*), which reaches a length of 39 centimeters (15 inches) and a weight of 2.1 kilograms (4.5 pounds). The White-Flippered Penguin (*Eudyptula albosignata*) reaches a length of 40 centimeters (16 inches), and a weight of 2.4 kilograms (5.2 pounds).

All dwarf penguins have very similar ways of life. Their nests are generally well hidden. They dig holes up to 2 meters ($6\frac{1}{2}$ feet) long or use holes dug out by shearwaters, as well as those found in natural rock or earth caves, under rocks or plants. They come ashore only after sunset. They spend the night on shore, usually in their breeding areas, where they live the whole year. At dawn they return to the sea. See also **Sphenisciformes**.

Web References

Adelie Penguins (Pygosceli adeliae): *http://www.siec.k12.in.us/~west/proj/penguins/ adelie.html*

African Penguins (*Spheniscus demersus*): *http://www.siec.k12.in.us/~west/proj/penguins/africanpen.html*

Chinstrap Penguins (*Pygosceli Antarctica*): *http://www.siec.k12.in.us/~west/proj/ penguins/chinstrap.html*

Emperor Penguins (*Aptenodytes forsteri*): *http://www.siec.k12.in.us/~west/proj/penguins/emperor.html*

Erect–Crested Penguins (*Eudyptes sclateri*): *http://www.siec.k12.in.us/~west/proj/ penguins/erect.html*

Fiordland Penguins (*Eudyptes pachyrhynchus*): *http://www.siec.k12.in.us/~west/ proj/penguins/fiord.html*

Galapagos Penguins (*Spheniscus mendiculus*): *http://www.siec.k12.in.us/~west/proj/ penguins/galap.html*

Gentoo Penguins (Pygosceli papua): *http://www.siec.k12.in.us/~west/proj/penguins/ gentoo2.html*

Humbolt Penguins (*Spheniscus humboldti*): *http://www.siec.k12.in.us/~west/proj/ penguins/humbolt.html*

King Penguins (*Aptenodytes patagonicus*): *http://www.siec.k12.in.us/~west/proj/ penguins/king.html*

Little Blue (Fairy) Penguins (*Eudyptula minor*): *http://www.siec.k12.in.us/~west/ proj/penguins/little.html*

Magellanic Penguins (*Spheniscus magellanicus*): *http://www.siec.k12.in.us/~west/ proj/penguins/magell.html*

Macaroni Penguins (*Eudyptes chrysolophus*): *http://www.siec.k12.in.us/~west/proj/ penguins/mac.html*

Rockhopper Penguins (*Eudyptes crestatus*): *http://www.siec.k12.in.us/~west/proj/ penguins/rock.html*

Royal Penguins (*Eudyptes schlrgeli*): *http://www.siec.k12.in.us/~west/proj/penguins/ royal.html*

Snares Island Erect-crested Penguins (Eudyptes robustus): *http://www.siec.k12.in.us/~west/proj/penguins/snare.html*

Yellow-Eyed Penguins (*Megadyptes antipodes*): *http://www.siec.k12.in.us/~west/proj/penguins/yellow.html*

PENICILLIN.   See **Antibiotic**.

PENIS.   See **Gonads**.

PENNING DISCHARGE.   A direct-current discharge where electrons are forced to oscillate between two opposed cathodes and are restrained from going to the surrounding anode by the presence of a magnetic field. It is sometimes referred to as a pig discharge since the device was originally used as an ionization gage (Penning ionization gage). It is used as a plasma-beam source by permitting the plasma to stream out along the magnetic field through a hole in one of the cathodes.

PENNING EFFECT.   An increase in the effective ionization rate of a gas due to the presence of a small number of foreign metastable atoms. For instance, a neon atom has a metastable level at 16.6 volts and if there are a few neon atoms in a gas of argon which has an ionization potential of 15.7 volts, a collision between the neon metastable atom with an argon atom may lead to ionization of the argon. Thus, the energy which is stored in the metastable atom can be used to increase the ionization rate. Other gases where this effect is used are helium, with a metastable level at 19.8 volts, and mercury, with an ionization level at 10.4 volts.

PENNSYLVANIAN PERIOD.   A period of the Paleozoic era. A systematic term first proposed by H.S. Williams in 1891. Type locality, Pennsylvania. The period began about 250,000,000 years ago. The term Pennsylvanian is roughly equivalent to the more general term Upper Carboniferous. In Britain this system is referred to as the Coal Measures. During this period there were many oscillations of sea level with relatively rapid alternations of marine and terrestrial sediments and freshwater swamp deposits in which were formed important coal beds. The principal occurrence of the formations of this system are in the Allegheny Plateau region of the eastern United States, westward to the Ohio River.

PENSTOCK.   Some considerable surface distance usually separates the intake works and turbines in a medium- or high-head hydroelectric project. Even where an open canal is employed to carry the water from the forebay of a diversion dam to intake works located near the plant, there is still a considerable span to be abridged by a closed water conduit of the pressure type. This water conduit is called the penstock and is always circular in form because that shape is best adapted to withstand internal pressure.

PENTADACTYL APPENDAGE.   The form of vertebrate appendage which is regarded as the fundamental terrestrial limb from which all of the specialized appendages of animals above the fishes have been derived.

The two pairs of vertebrate limbs, pectoral and pelvic, are similar in structure and both are attached to the girdles of corresponding name. Each girdle consists, in the primitive state, of three pairs of bones meeting at the articulation of the limbs. On each side of the body, one bone extends toward the back and two toward the middle of the body below. The skeletal structure of the limbs is made up of regions movable to one another and includes a single bone in the segment next to the body, followed by two bones. Then follows a group of small bones, and last five divergent series of moderately long bones, which extend into the digits. Of the more constant bones in the pectoral girdle of vertebrates with limbs, the dorsal bone is the scapula, and the two ventral bones are an anterior clavicle (of dermal origin) and a posterior coracoid. This girdle may contain a procoracoid between the last two bones. The bones of the pectoral appendage (forelimb) are, in the order described above, the humerus, the ulna and radius, the carpals, the five metacarpals of the hand, and the five series of phalanges in the digits. In the pelvic girdle of vertebrates with limbs, the dorsal bone is the ilium, and the two ventral bones are an anterior pubis and a posterior ischium. The bones of the pelvic appendage (hind limb) are the femur, the tibia and fibula, the tarsi, the metatarsals, and the phalanges.

Specialized appendages, such as the wings of birds, flippers of marine mammals, and the legs of the hoofed species, show either a simplification or a slight increase in complexity of the skeletal system and in some cases a consolidation by webbing of the digits or a still more compact fleshy

union between them. In all of these cases, however, the basic pentadactyl structure is still present.

PENTLANDITE.   The mineral sulfide of iron and nickel corresponding to the formula $(Fe, Ni)_9S_8$. It is isometric, appears in granular masses; hardness, 3.5–4; specific gravity, 5.0; color, bronze-yellow; opaque. Occurs with pyrrhotite, millerite, and nickeline. The best known deposit of pentlandite is at Sudbury, Ontario, Canada, where it is associated with a nickel-bearing pyrrhotite.

PENTOSES.   See **Carbohydrates**.

PENUMBRA.   See **Eclipse**.

PEPPER (*Piper nigrum; Piperaceae*).   The pepper plant, *Piper nigrum*, is a woody climbing shrub, which is indigenous in India. It is aided in climbing by the adventitious roots formed at the nodes. The ovate leaves are evergreen. The flowers are minute, without petals, and borne in slender spikes. The fruit is a bright red berry less than $\frac{1}{4}$ inch diameter. Each berry contains a single seed. On drying, the pericarp becomes black and wrinkled.

The berries are gathered before they are ripe and dried, usually by the sun. The dried berries are separated from the stem and ground, producing black pepper. If the pericarp is removed from the berry, leaving the seed and endocarp, the ground product is known as white pepper. The pericarp may be removed by using mature berries and soaking them to soften the pericarp. Or machines may rub off the dried pericarp, a method used in western countries. Pepper is one of the most extensively used of all spices. White pepper is less pungent than black, and hence not so conspicuous when used in cooking. Pepper is grown mostly in India and in the Malaysian regions. Propagation is usually by cuttings, which begin to fruit within four or five years, after which they bear continuously but somewhat irregularly for many years.

Related to *Piper nigrum* is *Piper Betle*, a perennial creeping vine native to Java, but widely grown in tropical Asia. From its leaves is prepared a chew with the nut of the Areca palm. *Piper Cubeba* is another species, also native of Java and the Molucca Islands, which yields a volatile oil, which is used medicinally.

PEPPERIDGE TREE.   See **Tupelo Trees**.

PEPTONE.   A secondary protein derivative that is water-soluble, not coagulated by heat, and not precipitated on saturation of its solutions with ammonium sulfate.

PERCHED.   1.   In hydrology, a term describing the ground water table when it is separated from an underlying layer of ground water by an unsaturated layer of rock.

2.   In glaciology, a term describing a glacial boulder, erratic, resting on a prominent topographic position.

3.   In a physiography, a term describing a stream whose bed is separated from the top of the ground water table by a dry, unsaturated zone.

PERCHES AND DARTERS (*Osteichthyes*).   Of the order *Percomorphi*, suborder *Percoidea*, family *Percidae*, there are several well known species, many of which are considered fine food fishes. The yellow perch (*Perca flavescens*) is a freshwater food fish that lives in lakes and streams in the United States from Iowa to South Carolina and northward into Canada. It comprises a significant part of Great Lakes fishery operations. Yellow perches are relatively small, a large fish weighing about a pound with a length of 14 to 16 inches (35 to 40 centimeters). In the mid-1800s, a yellow perch weighing $4\frac{1}{2}$ pounds (2 kilograms) was recorded. *Perca fluviatilis* (European perch) is quite similar to the American yellow perch, but a bit larger, sometimes weighing up to 6 pounds (2.7 kilograms). Occurring as far east as Siberia in European fresh waters, it is widely distributed but not found in southern Italy, northern Scandinavia, or Spain. Some perches also are found in the Baltic Sea.

Stizostedion vitreum (walleye) is a favorite sporting fish and one of the valuable food fishes in Lake Erie. Most walleyes are in the 1 to 4 pounds (0.45 to 1.8 kilograms) range, but one was recorded at 25 pounds (11.3 kilograms). *Stizostedion canadense* (sauger) appears much like the walleye but is somewhat smaller.

Darters appear to be confined to temperate North American waters east of the Rocky Mountains. They are very quick, small, and prefer a bottom habitat. The majority do not exceed 4 inches (10 centimeters) in length, rarely attaining a length of 8 to 9 inches (20 to 27 centimeters). Some are brilliantly colored.

The *Acerina cernua* is a small species of perch occurring in the slow gravelly streams of England. It is commonly known as the ruffe or pope.

PERCUTANEOUS. Penetrating through the skin as contrasted with entry through breaks and lesions of the skin. For example, some hepatitis viruses invade the body through the skin.

PERFECT FLUID. See **Fluid**.

PERFECT GAS. A perfect gas may be defined by the following two laws: The Joule law: the energy per mole, U, depends only on the temperature; the Boyle law: at constant temperature, the volume V occupied by a given number of moles of gas varies in inverse proportion to the pressure.

By combination of these two laws we obtain the equation of state for perfect gas:

$$pV = nRT \qquad (1)$$

where R is the gas constant, T, the absolute temperature. (It is also called the *perfect gas law*.)

The perfect gas is an abstraction to which any real gas approximates according to the nature of the gas and the conditions. For a given temperature and composition, the perfect gas condition is approached when the density tends to zero. From a molecular point of view, the perfect gas laws correspond to the behavior of a system of molecules whose interactions may be neglected in expressing the thermodynamic equilibrium properties. However, even at a low density, the transport properties depend essentially on the interactions.

The thermodynamic properties of a perfect gas are, of course, especially simple. For example, the difference between the molar heat capacities at constant pressure and constant volume is equal to the gas constant R,

$$C_p = C_v = R \qquad (2)$$

The value of R is 0.08205 liter-atm. degree^{-1} mole^{-1}, which in cgs units is equal to 8.314×10^7 g cm^2 sec^{-2} degree^{-1} mole^{-1}. This relationship, Formula (2), applies only approximately to real gases.

However, the way in which either C_p or C_v depends on the temperature can be calculated only from statistical mechanics.

PERIAPSIS. The orbital point nearest the center of attraction.

PERIASTRON. That point of the orbit of one member of a binary star system at which the stars are nearest to each other. That point at which they are farthest apart is called *apastron*.

PERICYNTHIAN. That point in the trajectory of a vehicle that is closest to the moon.

PERIDERM. A collective term for the tissues composing the outer bark of older stems of woody plants. When the stem increases in diameter as a result of the production of secondary vascular tissues by the cambium, the epidermis is no longer adequate as a protective layer. Cells of the cortex become meristematic and begin to divide, producing a new protective tissue. The dividing cells are called cork cambium or phellogen, while most of the cells produced become cork. These cork cells are small and rectangular; the protoplasm soon disappears leaving only the suberin-impregnated cell walls. Later, even the outer cells of the phloem become cork cambium, so that the bark of old trees contains only phloem and periderm.

Roots form a similar periderm. In this case, it is the pericycle cells that become cork cambium.

PERIDOTITE. The term peridotite is derived from peridor, the French word for olivine.

It is a coarse-grained igneous rock related to gabbro, which consists of olivine and proxene in varying proportions. Certain peridotites contain spinel, chromite, or mica as accessories.

Rocks consisting essentially of olivine alone are known as dunites, the name coming from the occurrence of this rock in the Dun mountains of New Zealand. In the United States, this mineral is found in North Carolina, South Carolina, and Georgia, where corundum is associated with the dunite in commercial quantities. The olivine of peridotites alters readily to the mineral serpentine, often to such an extent that the rock itself is called a serpentine. As mentioned above, the peridotites may contain chromite or other valuable minerals, often to such an extent that they may be commercially exploited, for nickel, platinum, and precious garnet.

Kimberlite from which diamonds are secured is commonly called a mica peridotite but is more closely related to the lamprophyres. See also **Kimberlite**.

PERIFOCUS. The point on an orbit nearest the dynamical center (focus). The pericenter is at one end of the major axis of the orbital ellipse.

PERIGEE. That orbital point nearest the Earth when the Earth is the center of attraction. That orbital point farthest from the Earth is called *apogee*. Perigee and apogee are used by some writers in referring to orbits of satellites, especially artificial satellites, around any planet or satellite, thus avoiding coinage of new terms for each planet and moon.

PERIHELION. The point in the orbit of any member of the solar system at which the object is closest to the sun. Since the orbits are all conic sections with the sun at one focus, perihelion must lie on the line of apsides of the conic. The point is diametrically opposite to aphelion.

In orbits of satellites and of other objects related to primaries other than the sun, terms similar to perihelion are used to indicate points in the orbit closest to the primary. For example, the point in the moon's orbit closest to the earth is known as perigee; the corresponding point in an orbit of a satellite of Jupiter is known as perijove; and a similar point in a double star orbit is known as periastron. See also **Orbit (Astronomy)**.

PERIMETRY. Perimetry is a visual field test of the eye that checks for problems of peripheral (side) vision. Peripheral vision is used primarily for detecting objects and for directing central vision so it is possible to see those objects in detail. A loss of peripheral vision results in a condition called tunnel vision and can lead to legal blindness. Severe glaucoma causes the loss of peripheral vision

The cells that make up the retina are responsible for the ability to see detail, brightness, and color. There are two types of photoreceptor cells in the cornea: rods and cones. The rods specialize in work at low light levels, and the cones provide sharp vision, color, and contrast discrimination. People with achromatopsia have defective cone cells and must rely on their rod photoreceptors for vision.

In normal eyes, there are about six million cone photoreceptors, located mainly in the macula at the center of the retina. These cells are primarily responsible for sharp, straight-ahead vision and also for the ability to distinguish colors. There are 100 million rod receptors, located mostly at the periphery of the retina. The rods are more sensitive to light than cones are, but rods are not able to differentiate among colors, nor can they perceive shades of gray, black, and white.

There are different variations in the severity of symptoms among individuals with achromatopsia. The rarest and most severe is called complete rod monochromatism, where there is a total lack of cone function. People with this disorder are extremely sensitive to light, even in normally lit rooms. They also have symptoms of poor visual acuity and nystagmus, which is involuntary movement of the eyes. Other less severe variations of the disorder are known as incomplete rod monochromatism and blue cone monochromatism. The type depends on which cones are affected.

A perimetry test is easy and comfortable. Sometimes a doctor administers the test, but usually a trained technician administers it. To take the test, you sit with your head in a chin rest at the edge of a large, bowl-shaped instrument with a fixed spot in the center, usually a yellow or green light.

There are two types of perimetry tests. In the Goldman kinetic perimeter test, the patient stare directly at the spot as the technician moves objects or lights of different size and brightness from the side. When the object or light is seen, the patient pushs a button.

The threshold static automated perimetry, which is the other type of perimeter test, uses stationary objects of light that blink on and off in various parts of the visual field.

With either method, each eye is tested independently. The maps of visual sensitivity, made by either of these methods, are very important in diagnosing diseases of the visual system.

The perimetry test usually takes no more than 30 minutes, and the results are available immediately. See also **Peripheral Vision**.

Vision Rx, Inc., Elmsford, NY.

PERIOD.   See **Geologic Time Scale**.

PERIODIC FUNCTION.   One for which a constant a exists so that $f(z) = f(z + na)$, where n is an integer. The function repeats itself periodically with the fundamental period a. Typical examples are $\sin z$, $\cos z$, with periods $2n$; e^z, with fundamental period, $2\pi i$. Functions may be doubly periodic such that the periods are of the form $(na + n'a')$, both n and n' integers but not both zero. See also **Elliptic Integral**.

PERIODIC TABLE OF THE ELEMENTS.   When the chemical elements are arranged in a matrix on the basis of increasing atomic numbers, a pattern of periodicity among the physical and chemical characteristics emerges. See Fig. 1. By no means is the resulting matrix perfect, but the resemblance of characteristics among groups of elements arranged in this manner is indeed both striking and illuminating. Attempts to classify the elements date back to the early work of de Chancourtois (1862) and Newlands (1863), but the discovery of the relationship between atomic-number groupings and characteristics was made by Dmitri Mendeleev in 1869. One year later, Lothar Meyer independently showed the periodicity of the elements in terms of atomic volumes. Meyer defined the latter characteristic as the atomic weight divided by the specific gravity of the element in the solid state.

Although there have been numerous refinements to Mendeleev's early tabulation, fortified by the discovery and isolation of several elements then unknown the fundamental principles of the matrix are the same. The conventional table is shown in the upper right in Fig. 2. The information also can be presented in polar fashion as shown in Fig. 2. It is interesting to note that as one proceeds clockwise around the circle the atomic numbers appear consecutively and that 18 sectors of the circle become the bases for families or groups of elements.

Notation for designating the grouping of the elements was changed and officially accepted in the mid-1980s. The new notations are used in Fig. 2. They are summarized in Table 1.

TABLE 1. REVISED CHEMICAL ELEMENT GROUP NOTATION VERSUS PRIOR SCHEMES

Revised Notation	Elements in Revised Grouping	Prior IUPAC Form	Prior CAS Version
1	H, Li, Na, K, Rb, Cs, Fr	IA	IA
2	Be, Mg, Ca, Sr, Ba, Ra	IIA	IIA
3	Sc, Y, La, Ac	IIIA	IIIB
4	Ti, Zr, Hf, 104	IVA	IVB
5	V, Nb, Ta, 105	VA	VB
6	Cr, Mo, W, 106	VIA	VIB
7	Mn, Tc, Re, 107	VIIA	VIIB
8	Fe, Ru, Os	VIIIA	VIII
9	Co, Rh, Ir	VIIIA	VIII
10	Ni, Pd, Pt	VIIIA	VIII
11	Cu, Ag, Au	IB	IB
12	Zn, Cd, Hg	IIB	IIB
13	B, Al, Ga, In, Tl	IIIB	IIIA
14	C, Si, Ge, Sn, Pb	IVB	IVA
15	N, P, As, Sb, Bi	VB	VA
16	O, S, Se, Te, Po	VIB	VIA
17	F, Cl, Br, I, At	VIIB	VIIA
18	He, Ne, Ar, Kr, Xe, Rn	VIIIA	VIIIA

Lanthanides and Actinides not included in group numbering.

Thus, the members of the alkali metals (Group 1), alkaline earths (Group 2), halogens (Group 17), and so on, all bear resemblance, one element to the other, within any given group. There are two significant breakpoints in any representation of periodicity, namely, commencing with atomic number 57 (lanthanum) and atomic number 89 (actinium). Attempts to place the elements which follow — in the one case, atomic numbers 59 through 71, and in the other case, atomic numbers 90 through 103, in the underlying geometric matrix (whether tabular or circular) do not succeed. These separate groups are known as the *Lanthanides* (rare earths) and the *Actinides*, respectively. Upon completion of the Lanthanide series (with lutetium, atomic number 71), the orderly geometry resumes with hafnium, atomic number 72 (Group 4) and continues through actinium, atomic number 89. The probable positions of elements 104 and 105 are indicated in Groups 4 and 5, respectively.

An amazing result of Mendeleev's pioneering classification was the prediction of elements yet to be discovered. Mendeleev found that he

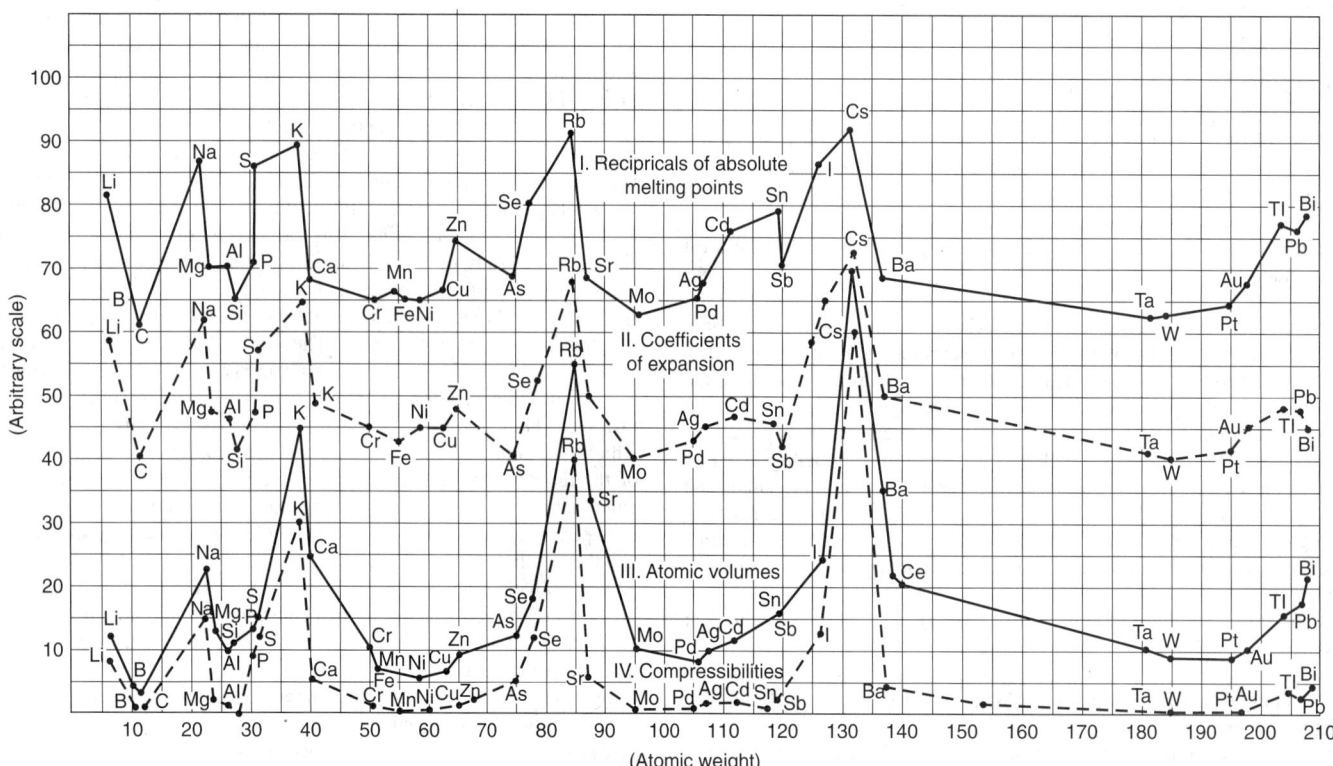

Fig. 1.   Pattern obtained when various parameters are plotted against increasing atomic weight of chemical elements.

Conventional representation of periodic table

1																	18
1 H	2											13	14	15	16	17	2 He
3 Li	4 Be											5 B	6 C	7 N	8 O	9 F	10 Ne
11 Na	12 Mg	3	4	5	6	7	8	9	10	11	12	13 Al	14 Si	15 P	16 S	17 Cl	18 Ar
19 K	20 Ca	21 Sc	22 Ti	23 V	24 Cr	25 Mn	26 Fe	27 Co	28 Ni	29 Cu	30 Zn	31 Ga	32 Ge	33 As	34 Se	35 Br	36 Kr
37 Rb	38 Sr	39 Y	40 Zr	41 Nb	42 Mo	43 Te	44 Ru	45 Rh	46 Pd	47 Ag	48 Cd	49 In	50 Sn	51 Sb	52 Te	53 I	54 Xe
55 Cs	56 Ba	57 La	72 Hf	73 Ta	74 W	75 Re	76 Os	77 Ir	78 Pt	79 Au	80 Hg	81 Ti	82 Pb	83 Po	84	85 At	86 Rn
87 Fr	88 Ra	89 Ac															

Lanthanides

58 Ce	59 Pr	60 Nd	61 Pm	62 Sm	63 Eu	64 Gd	65 Tb	66 Dy	67 Ho	68 Er	69 Tm	70 Yb	71 Lu

Actinides

90 Th	91 Pa	92 U	93 Np	94 Pu	95 Am	96 Cm	97 Bk	98 Cf	99 Es	100 Fm	101 My	102 No	103 Lw

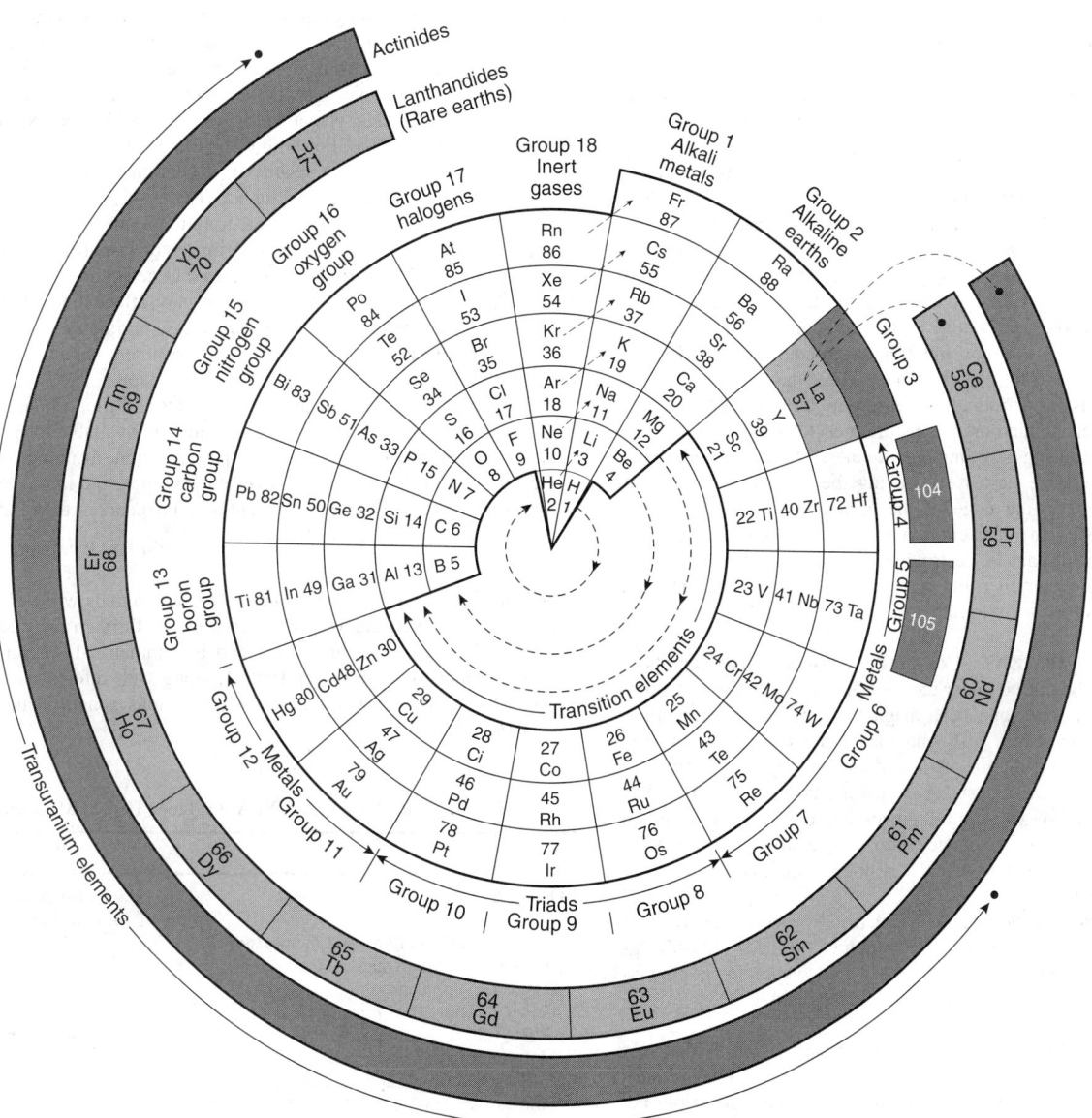

Fig. 2.   Polar representation of periodic relationships of the elements. (*Source: Omnibix, U.S.A.*) At upper right is shown the conventional representation.

could maintain geometric logic of his table only if he allowed for some blank spaces in his table. He further reasoned that elements later would be discovered that would occupy these vacant positions and, thus, Mendeleev predicted the existence of gallium, scandium, and germanium. In fact, Mendeleev gave a preliminary name to scandium, calling it eka-boron and predicted the probable properties of the element. The element later was isolated by Lars Fredrik Nilson in 1879. Mendeleev lived to see his prediction confirmed.

In retrospect, with a much fuller understanding of the underlying electronic and particle structure of the elements, most aspects of the periodicity of the elements come as no surprise, but the fact remains that Mendeleev, Meyer, and others made these striking observations without benefit of over 100 years of additional knowledge. The periodicity of the elements is demonstrated by the accompanying graph which plots atomic weights along the abscissa versus an arbitrary ordinate for various observed physical characteristics. See also **Chemical Elements**.

PERIOD LUMINOSITY LAW.   It has been observed that the cepheids of populations I and II form distinct sequences, where their absolute magnitudes are very closely related to their periods. Hence, upon observing the period and apparent magnitude (m) of a cepheid, one can read off the absolute magnitude (M) and then calculate its parallax (π) by means of

$$m - M = 5 \log D - 5$$

where D is in parsecs. The term $m - M$ is often referred to as the *distance modulus*. It is interesting to note that these pulsating stars also follow the relation

$$P\sqrt{\rho} = \text{constant}$$

where P is their period of pulsation, and hence allow us to calculate their mean density (ρ).

See also **Cepheids**.

PERIODONTITIS.   Sometimes called pyorrhea or gingivitis, this condition is caused by degenerative changes in the periodontium and is characterized by purulent discharge or inflammation. Periodontal disease usually begins at the edge of the gums, along sites that frequently are inadequately cleaned when brushing the teeth. The microorganisms that cause periodontitis generally differ from those that produce dental caries. See also **Caries, Cariology, and Dentistry**. Whereas caries results from microorganisms that are nurtured from the food supply of the patient, those organisms that cause periodontitis find their nutrients in the tissue. Thus, a person who consumes little carbohydrate may be free of cavities, but could have periodontal disease. This is demonstrated in some countries where there is little to eat and little dental care available. There may be few cavities among the population, but periodontal disease may be prevalent.

In periodontal disease, microorganisms produce toxic products, which irritate the tissues, causing swelling and redness. In time, underlying connective tissue fibers are destroyed, leaving the gum tissues weak. A space is formed between the edge of the gum and tooth (*periodontal pocket*) which provides an ideal environment for the continued growth of bacteria, the latter ultimately destroying soft tissue and bone. Serum and blood provide an even richer diet at the gingival margin. The process, without attention, continues until teeth are lost.

PERIPHERAL EQUIPMENT (Computer System).   The auxiliary machines which may be placed under the control of a computer. Examples include card readers, card punches, magnetic tape units, high-speed printers, scanners, diagnostics, test, and measuring devices. Peripheral equipment may be used on-line or off-line, depending upon the computer design, task requirements, and costs. A device is said to be on-line when it is under the control of the central processing unit, and information reflecting current activity is introduced into the data-processing system as it occurs. A device is off-line when it is not in direct communication with the central processing unit.

With the continuing developments in electronic microminiaturization and the wide acceptance of the personal computer (PC) for industrial as well as business applications, the trend has been toward integrating as many functions as possible into the computer unit per se.

PERIPHERAL VISION.   Peripheral vision, or side vision, is that part of vision that detects objects outside the direct line of vision. For instance, when you read a word on a page, you are using your central vision, but it is your side vision that tells you if the word is at the beginning or end of a sentence, or at the top or bottom of a page. Your peripheral vision also tells you where to look if someone enters the room or if a car is approaching from the side. Like most people, you are probably not aware of the limitations that would exist without peripheral vision, because you are constantly moving your eyes in order to focus with your central vision.

The difference between central and peripheral vision becomes apparent when you understand the visual function of the eye. The eye works like a camera with two lenses: the cornea at the front of the eye and the natural crystalline lens behind the pupil. The cornea is responsible for about 70 percent of the eye's focusing power, while the natural lens fine-tunes the image before it is focused on the retina at the back of the eye. The retina works like the film in a camera, receiving light rays and sending them through the optic nerve to the brain where they are converted into images.

There is a small area in the center of the retina called the macula, which is less than 1/4 inch in diameter, that is responsible for sharp, clear central vision and ability to perceive color. The densely packed photoreceptor (light-sensitive) cells in the macula control the eye's central vision and are responsible for the ability to read, drive a car, watch television, see faces, and distinguish detail. The rest of the retina handles peripheral vision that enables the eyes to see objects off to the side while looking forward.

There are two types of photoreceptor cells in the cornea; rods and cones. The rods specialize in work at low light levels, and the cones provide sharp vision and discrimination. The macula contains a high concentration of cones, which accounts for the sharper focus of straight-ahead vision, particularly in bright light. Most of the rods are located in the periphery of the retina. This is why you can often see faint objects more clearly if you do not look directly at them. A dim star, for instance, is best seen when your eyes are not aimed directly at it.

When you see something out of the corner of your eye, its image focuses on the periphery of your retina and you are unable to distinguish the color of what you are seeing. And, because there are fewer rods, the ability to resolve the shapes of objects at the periphery of vision is limited. Normal peripheral vision, called visual field, for one eye is approximately 150 degrees from side to side. For both eyes, it is approximately 180 degrees.

Loss of peripheral vision results in a condition called tunnel vision, which is like looking through a tunnel. Tunnel vision can be caused by several disorders including glaucoma, retinitis pigmentosa, and strokes.

Most eye examinations include a perimetry or visual field test to check peripheral vision. The perimetry test is used to detect and monitor damage from glaucoma and other conditions that may affect the visual pathway from the eye to the brain. In medical terms, perimetry is a systematic measurement of visual field function. An instrument called a perimeter is used to plot the central or peripheral field of vision.

Loss of peripheral vision can actually be more difficult than the loss of visual acuity. However, many low vision aids are available for people who have lost all or a portion of their peripheral vision. Training and optical devices, such as prisms, mirrors, reverse telescopes, and minus lenses, can improve awareness of the environment and independent travel ability.

Vision Rx, Inc.; Elmsford, NY.

PERISSODACTYLA (*Mammalia*).   Hoofed animals with an odd number of toes, in which the axis of the foot passes through the middle digit. This digit is larger than the others and is symmetrical. Organization of the *Perissodactyla* is shown in Table 1, along with references to specific entries in this volume which describe the various genera in the order of *Perissodactyla*.

TABLE 1. PERISSODACTYLA (Odd-toed Hoofed Mammals)

	In This Volume
EQUINES	See also **Horses, Asses, and Zebras**.
Horses (*Equus caballus* and *przewalskii*)	
Asses (*E. hemionus* and *asinus*)	
Zebras (*E. burchelli*. ...)	
Grevy's Zebra (*Dolichohippus*)	
TAPIRINES	See also **Tapir**.
Malayan Tapir (*Tapirus indicus*)	
Central American Tapir (*T. bairdii*)	
South American Tapirs (*T. terrestris*)	
Mountain Tapir (*T. roulini*)	
RHINOCEROTINES	See also **Rhinoceros**.
One-horned Rhinoceroses (*Rhinoceros*)	
Asiatic Two-horned Rhinoceroses (*Didimoceros*)	
African Black Rhinoceros (*Diceros*)	
Ceratotheres (*Ceratotherium*)	

PERISTALTIC MOVEMENT.   See **Digestive System (Human)**.

PERITONITIS. A widespread inflammation of the peritoneum, the membranous lining of the abdominal cavity. The surface area available to infection is large, thus usually producing profound symptoms. The large, moist, warm conditions prevailing in the peritoneum make an ideal environment for the rapid growth and development of bacteria. There are several avenues of access to the peritoneum by bacteria, including a perforation of the gastrointestinal tract, such as may occur with a gastric or duodenal ulcer that may break through, or the rupture of an inflamed and distended appendix. In these instances, the colon bacillus is frequently the causative agent. Other conditions favoring entrance of bacteria into the peritoneum include typhoid fever, obstructions of the bowel, serious dysentery, diverticulitis, and severe gallbladder conditions. Less frequently, bacteria may gain entry via the female genital tract. Pelvic peritonitis results from infection of the Fallopian tubes, most often with the gonococcus. Tuberculous peritonitis occurs in the course of tuberculous infection of the bowel or the genitourinary tract. Pneumococcus peritonitis is seen as a complication of nephrosis. A form of chemical peritonitis may occur if gastric juices, such as bile or pancreatic juices, spill over into the abdominal cavity. In relatively few instances, the peritoneum infection may take the form of local abscesses rather than be of a generalized nature.

Peritonitis may also result from penetrating trauma of the abdominal wall, such as caused by knife or gunshot wounds. Such externally inflicted injury carries with it not only hemorrhage and traumatic shock, but also the possibility of tetanic peritonitis and infection by anaerobic bacteria.

Prior to the availability of sulfonamide drugs and antibiotics, peritonitis was a most dreaded condition. The condition is most serious and requires immediate attention, but because of available treatment properly and timely administered, some of the former grave aspects of the condition have been diminished. The patient with peritonitis is acutely ill, with fever, rapid pulse and respirations, and a distended, tender, rigid abdomen. Treatment varies with cause and will almost always include chemotherapy with or without surgical intervention.

A. C. V.

PERIWINKLE *(Mollusca, Gasteropoda).* Marine snails with a thick conical spiral shell. They live in shallow water, in the tidal zone, and along the shore. Some of the many species are very widely distributed, and the genus *Littorina* is represented in all parts of the world. An edible species has been introduced into the United States and is now common on the Atlantic coast north of Delaware Bay.

PERLITE (or Pearlstone). An unusual form of siliceous lava composed of small spherules of about the size of bird shot or peas. It is grayish in color with a soft pearly luster. The spherules often show a concentric structure and are believed to be formed as a result of a peculiar spherical cracking developed while cooling. They may be confused with oölites, which are classified as concretions.

PERMANENT MEMORY. In computer terminology, storage of information that remains intact when the power is turned off. Also called *nonvolatile storage.*

PERMEABILITY. As applied in magnetism, *absolute permeability* is the ratio B/H, where B is magnetic flux induction and H is magnetizing force. *Relative or specific permeability* is $\mu_s = B/(\mu_0 H)$, where μ_s is specific permeability and μ_0 is the permeability of free space. See also **Magnetism**.

PERMEABILITY (Soil). See **Hydrology**.

PERMEAMETER. An electromagnet arranged for magnetizing a specimen, and for allowing measurement of the flux through the specimen and the magnetizing force at the surface. Various types of permeameter have been designed, aiming at ease of operation, accuracy of results, or use with high-coercivity materials.

PERMEATION. As applied to gas flow through solids, the passage of gas into, through, and out of a solid barrier having no holes large enough to permit more than a small fraction of the gas to pass through any one hole. The process always involves diffusion through the solid and may involve various surface phenomena, such as sorption, dissociation, migration, and desorption of the gas molecules.

PERMIAN PERIOD. The name of a geologic period. Type locality. Province of Perm, Russia. The formations of this system were first studied and described by R.I. Murchison, in 1841. The Permian period began about 230 million years ago and lasted for about 30 million years. Only the upper Permian appears to be represented in North America. Eastern North America was undergoing uplift and erosion during this period, while the seas invaded the West from the Arctic and Gulf. Increasing uplift of the continents and mountain building culminated in the Appalachian Revolution (U.S.) and the Hercynian Revolution (Europe). Increasing aridity and widespread equatorial continental glaciation are disclosed by tillites in Australia, Tasmania, New Zealand, India, South America, and Massachusetts.

PERMUTATION. Given n distinguishable objects or elements, each different arrangement of the elements is a permutation. The number of permutations is $n(n-1)(n-2)\cdots 3\cdot 2\cdot 1 = n!$ Several different symbols, such as $_nP_n$, $P_{n,n}$ or $P(n, n)$ are used to indicate this result. If the n things are taken r at a time $(r < n)$,

$$_nP_r = n(n-1)(n-2)\cdots(n-r+1) = \frac{n!}{(n-r)!}$$

When n_1 of the elements are all alike of the first kind, n_2 of the second kind, etc., so that $n_1 + n_2 + \cdots n_m = n$, the number of permutations is

$$\frac{n!}{n_1! n_2! \cdots n_m!}$$

This result also applies if the elements are separated into m parts with n_i elements in the i th part. If the number of elements in each of the m parts is not specified, but each part must contain at least one element, the number of permutations is

$$\frac{n!(n-1)!}{(n-m)!(m-1)!}$$

This number is increased to

$$\frac{(m+n-1)!}{(m-1)!}$$

if empty parts are permitted.

A *combination* is an arrangement of objects or elements, where the order of arrangement is not distinguished. Thus, given the three letters a, b, c, the possible permutations are (abc), (acb), etc., six in number, but there is only one combination. With symbols as before,

$$_nC_r = \frac{n(n-1)(n-2)\cdots(n-r+1)}{r!} = \frac{n!}{r!(n-r)!}$$

This number is identical with the $\binom{n}{r}$, the coefficient in the binomial series. Moreover, $_nC_r =_n P_{n-r}$; $_nP_r = r!_nC_r$.

The number of combinations of n different elements into m specified parts, with empty parts allowed, is m^n; of n identical elements into m different parts with empty parts is

$$\frac{(n+m-1)!}{n!(m-1)!}$$

but when at least one element is in each part, the number is

$$\frac{(n-1)!}{(m-1)!(n-m)!}$$

Finally, the total number of combinations of n things taken $1, 2, 3, \ldots, n$ at a time is $\Sigma_{i=1}^n \,_nC_i = 2^n - 1$.

PERMUTATION GROUP. Its elements, $n!$ in number, are the various permutations or rearrangements of a standard arrangement of n symbols of objects. A typical element is

$$S = \begin{pmatrix} s_1 & s_2 & \cdots & s_n \\ 1 & 2 & \cdots & n \end{pmatrix}$$

meaning that the operation S replaces 1 by s_1, 2 by s_2, etc. If another element is indicated by T, then ST, the rearrangement designated by T followed by S, the resulting permutation is also in the group.

A permutation sending s_1 into s_2, s_2 into s_3, etc., and finally s_n into s_1 is called a *cycle* on n letters. It is usually written as $(s_1, s_2, \ldots, s_n)$.

The degree of a cycle equals the number of symbols permuted. A cyclic permutation of degree two is a *transposition*. Any permutation may always be written as a product of transpositions, either even or odd in number. The permutation is then said to be even or odd.

The group of all permutations of *n* letters or objects, of order *n*!, is called the *symmetric group*. The even permutations of *n* objects form a *subgroup* of the symmetric group. Its order is *n*!/2 and it is called the *alternating group*.

Every group is *isomorphous* to some permutation group. It is easy to find a representation of a permutation group by using a permutation matrix. Each row and column of such a matrix has but one non-zero element and that is unity. The row and column thus designate the initial and final locations of the object permuted.

See also **Permutation**.

PEROVSKITE. The mineral perovskite is calcium titanate, essentially $CaTiO_3$, with rare earths, principally cerium, proxying for Ca, as do both ferrous iron and sodium, and with columbium substituting for titanium. It crystallizes in the orthorhombic system, but with pseudo-isometric character; fracture subconchoidal to uneven; brittle; hardness, 5.5; specific gravity, 4; luster, adamantine; color, various shades of yellow to reddish-brown or nearly black; transparent to opaque. It is found associated with chlorite or serpentine rocks occurring in the Urals, Baden, Switzerland, and Italy. It was named for Von Perovski.

PERSEIDS. The name given to the most reliable of all meteor showers. Although the Leonids provide some very brilliant displays about three times each century, their appearance in the intervening years is not at all striking. The Perseids make their appearance during August of each year. Because of the fact that the Perseid showers never are as striking as the Leonids, we should not expect to find them referred to as frequently in the ancient writings, however, and we find mention of the Perseids as far back as 830 A.D. The first determination of the radiant point in the constellation of Perseus was apparently made in 1834.

In appearance, the members of the Perseid shower are as striking as those from any other radiant point. Coming as they do during the month of August, when the nights are warm, they are seen by large numbers of people, who are always impressed by the relatively slow motion, distinctly reddish appearance, and trails, frequently of several seconds' duration, which characterize the members of this swarm. See also **Meteor Shower**.

PERSEUS. A rich and brilliant constellation of the northern sky. Because Perseus lies right in the Milky Way, it presents many beautiful fields for the opera glass or the small telescope, particularly the field of the bright star α Persei. The star Algol (β Persei) is the famous eclipsing variable star whose striking changes in light intensity caused the Arabs to name it the demon star.

In Perseus is to be found a famous double-star cluster, one of the finest objects in the entire sky for an observer with a small telescope. On a clear moonless night, using a relatively low power on the instrument, an observer will be well rewarded for his search for this wonderful object. See map accompanying entry on **Constellations**; and **Galaxy**.

PERSIMMON TREES. Of the family *Ebenaceae* (ebony family), there are several species of persimmon trees. The fruits of these trees also are called persimmons. Especially esteemed for their fruits are *Diospyros virginiana*, the American persimmon, and *D. kaki*, the Japanese persimmon. *Diospyros virginiana* is a large American tree, 60–100 feet (18 to 30 meters) high, with rather thick ovate-oblong leaves and pale yellow axillary flowers. The fruit is a large globular berry an inch or more in diameter, orange-yellow in color, and very astringent until fully ripe. See Fig. 1. The astringent quality is due to the presence of much soluble tannin, which is gradually formed into an insoluble compound as the fruit ripens, so that the mouth-puckering quality is nearly lost. Frost action has been considered by many to be the cause of the change in the fruit. The American persimmon is hardy as far north as Rhode Island. The Japanese persimmon is a smaller tree, seldom growing more than 40 feet (12 meters) tall, and is less hardy. Its fruits are larger than those of the American tree, and of reddish color. Both trees have a hard dark wood.

See Table 1.

Also of the ebony family is the ebony tree (*Diospyros ebenum*), a tree native of India and Ceylon. It is a large tree with entire leathery leaves

Fig. 1.   Persimmon fruit (Diospyros virginiana.) (*USDA photo.*)

TABLE 1. RECORD PERSIMMON TREES IN THE UNITED STATES[1]

Specimen	Circumference[2]		Height		Spread		Location
	Inches	Centimeters	Feet	Meters	Feet	Meters	
Common Persimmon (1999) (*Diospyros virginiana*)	88	224	132	40.2	30	9.1	Massachusetts
Common Persimmon (1987) (*Diospyros virginiana*)	136	345	66	20.1	85	25.9	Arkansas
Common Persimmon (1999) (*Diospyros virginiana*)	96	244	121	36.9	42	12.8	Georgia
Common Persimmon (1995) (*Diospyros virginiana*)	95	241	120	36.6	40	12.2	South Carolina
Common Persimmon (1995) (*Diospyros virginiana*)	85	216	132	40.2	37	11.3	South Carolina
Texas persimmon (1965) (*Diospyros texana*)	68	173	26	7.9	32	9.8	Texas

[1]From the "National Register of Big Trees," American Forests (by permission).
[2]At 4.5 feet (1.4 meters).

and axillary flowers. The wood of the tree is divided sharply into a soft white sapwood of little value and a hard very dark heartwood. The latter is much used for inlay work, for black piano keys, for musical instruments, and for handles of various instruments. Many other species of *Diospyros* have dark woods used as a substitute for true ebony. The wood of several other trees, especially that of the pear tree, are frequently stained to imitate ebony.

PERSONAL EQUATION.

In making measurements of any character, every observer, no matter how skilled he may be, is bound to make certain errors. These errors are of two kinds: accidental errors which will be small in the case of a good observer and which will be distributed in accordance with the laws of probability; and systematic errors or errors which are always in the same direction and of approximately the same magnitude. As an example of systematic errors we may cite the case of the observation of the transit of a star across the reticle of a meridian circle. In this case a good observer will always press the chronograph key either slightly too early or slightly too late, depending upon the observer.

The value of the systematic error is known as the personal equation of the observer. Personal equation must be determined empirically for each observer under a variety of different observing conditions. For a good observer, the personal equation remains remarkably constant over long periods of time and may be applied directly to any observation.

See also **Error**.

PERSPECTIVE.

A method of representing solid objects on a plane surface, as they would appear to an observer's eye when viewed from a given point. The projectors or visual rays converge from the object to the eye, forming a cone or pyramid of rays. The intersection of these rays with the picture plane results in a perspective drawing. Two types of perspective representation are in common use—parallel and angular. In the former, all horizontal lines parallel to the picture plane, and all vertical lines, appear horizontal and vertical; all other parallel lines will intersect, if extended, at one or more common points, called vanishing points. In angular perspective, vertical lines appear vertical; all other parallel lines will intersect, if extended, at one or more vanishing points. See Fig. 1.

See also **Orthographic Projection**.

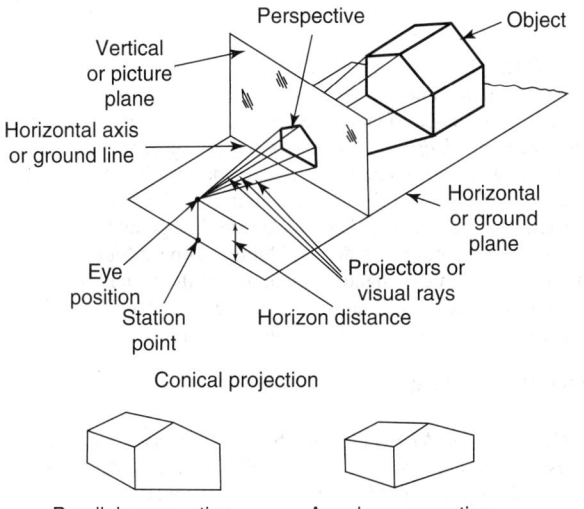

Fig. 1. Types of perspective.

PERSPIRATION.

A secretion formed by the sweat glands of the skin of some of the mammals. It consists chiefly of water with small quantities of other materials in solution. The dissolved substances include fatty acids, urea, sodium and potassium chloride, phosphates, lactic acid, and cholesterin. By evaporation at the surface of the skin, the sweat plays an important part in the regulation of body temperature when the surrounding air is too warm for adequate radiation. It is also important as an excretory medium.

PERTHITE.

An alkali feldspar comprising parallel or subparallel intergrowths. The potassium-rich phase, usually microcline, seems to be the host from which the sodium-rich phase, usually albite inclusions, exsolved. The exsolved areas typically form blebs, films, lamellae, small strings, or irregular veinlets and usually are visible to the naked eye.

PERTURBATION.

1. A small contribution to a physical quantity, such that the problem into which the quantity enters can be solved exactly or in a far simpler manner than otherwise if the perturbation is neglected. The form in which a perturbation is most frequently used in both classical and quantum mechanics is a small additional energy, called the *perturbation energy*.

2. Any departure introduced into an assumed steady state of a system. The magnitude of the departure is often assumed to be small so that product terms in the dependent variables may be neglected; the term *perturbation* is therefore sometimes used as synonymous with small perturbation. The perturbation may be concentrated at a point or in a finite volume of space; or it may be a wave (sine or cosine function); or, in the case of a rotating system, it may be symmetric about the axis of rotation.

3. In molecular spectra, perturbations cause the displacement of a band from its regular position in the band system (vibrational perturbation) or the displacement (and/or weakening) of corresponding lines in the different branches of a band (rotational perturbation). A perturbation observed in the spectrum is indicative of the presence of a perturbation (shift) of one of the energy levels involved due to interaction with another level of the same, or nearly the same, energy.

4. A much more frequent use is made of perturbation methods in quantum mechanics. The mathematical complexity of many quantum mechanical problems is such that one cannot hope to obtain exact solutions. However, good predictions can sometimes be obtained by means of perturbation theory, if one can assume that the actual system differs only slightly from a simpler system for which the problem can be solved, and the neglected difference can be dealt with as a perturbation of this simpler unperturbed system. The effect of a weak electromagnetic field on an atom, for instance, can be dealt with as a perturbation, and the transition probabilities between the energy states of the unperturbed atom can be calculated by means of perturbation theory. A weak interaction between two particles can be dealt with as a perturbation in the collision process of the two particles. The perturbation methods can be time-independent or time-dependent, according to whether the unperturbed states are described by time-independent wave functions or time-dependent ones. If the strength of a weak interaction between two systems is proportional to a constant parameter, the wave functions and energy values of the wave equation can be expanded in powers of this constant. The zero-order approximation is given by the unperturbed wave functions and energy values which are independent of this parameter. These determine the first-order approximations together with that part of the wave equation that is linear in the coupling parameter. By successive approximations one obtains expressions for second, third, and higher order perturbations in the wave function and the energy. If an unperturbed energy state is degenerate, that is, if two or more states have the same unperturbed energy, the effect of the perturbation has to be taken into account first between these degenerate states.

PERTURBATION (Astronomy).

Within the solar system, it fortunately happens that the attraction of one body is dominant, i.e., the attraction of the sun upon any planet is far greater than the attractions of all of the other planets combined. In the case of satellites, the attraction of the primary is preponderant. In such cases, a close approximation to the true orbit may be obtained by neglecting the attractions of other objects and obtaining a preliminary orbit by the methods of solution of the two-body problem.

Using the Keplerian ellipse thus obtained, it is possible to find at any instant, to a high degree of approximation, the distance of the object from other members of the solar system and hence the attractions of these other objects. The effects of these attractions in changing the motion of the object under consideration may be computed, a second approximation to the true position may be obtained, and this can be used to obtain more accurate values to the attracting forces. Usually, the second approximation is sufficiently accurate for all practical purposes.

The influences that the attractions of the other members of the solar system have on the motions of the object under consideration are known as perturbations. In the case of nearly circular orbits, as in the case of the motions of the planets about the sun, it is possible to obtain, in the form of infinite series, an analytic expression for the perturbations. Such a solution is known as general perturbations. If the orbit is highly eccentric, as in the case of many comet orbits, it is not possible to obtain any general analytic expression, and the perturbations are known as special perturbations.

Due to perturbations, the orbits of objects have slow, steady changes in the elements, known as secular perturbations, and also relatively short oscillations about an average value, which are known as periodic perturbations. In actual practice, it is customary to list both the secular and periodic perturbations in Tables from which the accurate position of the planet at any desired instant may be computed. In the case of the perturbations of the moon, one of the most complicated of all perturbation problems, E.W. Brown's Tables of the moon fill three quarto volumes totaling more than 360 pages.

Perturbations have played an important part in astronomical discovery. After the planet Uranus had been discovered and accurately observed, the preliminary orbit, plus perturbations from all known objects, did not accurately represent the observed positions. These deviations could only be explained on the basis of another planet outside the orbit of Uranus; accurate computations were made, and the planet Neptune was discovered. Deviations of observed positions of Neptune from those computed from the orbit stimulated the search that led to the discovery of the planet Pluto. In the case of the planet Mercury, a perturbation in the longitude of perihelion in the orbit could not be explained on the basis of gravitational theory. This led to a fruitless search for a planet between Mercury and the sun. The perturbation was later explained on the basis of the theory of relativity and was one of the early triumphs of this theory.

See also **Kepler's Laws of Planetary Motion**; and **Orbit (Astronomy)**.

PERTUSSIS (Whooping Cough). A disease primarily of young children, caused by a small aerobic, gram-negative coccobacillary *Bordetella pertussis*, which adheres to ciliated epithelial cells in the respiratory tract. This attachment is followed by ciliostasis and subsequent loss of the ciliated cells. Humans are the only known reservoir of the organism. Occasionally asymptomatic infections have been identified, but there is no evidence that these are important in the spread of the disease and there are no chronic carriers.

The disease, still under mandatory reporting to the Centers for Disease Control (Atlanta, Georgia) was prominent prior to 1950, during which year about 800 cases per 100,000 population were reported. With a few intermediate peaks, the incidence declined to a plateau in the early 1960s of less than 10 cases per 100,000. The incidence of the disease has lessened since then and, in 1984, less than one case per 100,000 population was reported—with the majority of reported cases involving children less than one year old. This decrease in incidence appears attributable to the widespread vaccination program introduced in the late 1940s.

The disease is spread by droplets from infected patients. The incubation period is from 12 to 20 days, when minor upper respiratory symptoms are found. These persist for about two weeks, during which period the disease is most contagious. Toward the end of this period, a dry hacking cough appears and becomes progressively worse. After one or two weeks, the cough becomes paroxysmal and prolonged coughing attacks are followed by the characteristic "whoop" which is produced by forced inspiration through a partially closed glottis. Frequently, efforts to expel thickened sputum will induce vomiting.

The appearance of fever suggests a secondary bacterial infection. Otitis media and pneumonia are the most common complications and most noninfectious complications are extremely rare.

Diagnosis is symptomatic and confirmed by isolation of the causative organism in culture.

Treatment of an active case of pertussis is generally supportive. Cough medicines are of no value, nor is passive immunization. The organism is sensitive to a variety of antibiotics, including erythromycin, tetracycline, chloramphenicol, and probably trimethoprin-sulfamethoxazole. Early treatment during the catarrhal stage will shorten the course of the clinical illness, but if withheld until the paroxysmal stage, antibiotics will have no effect. Even though there is no clinical benefit, however, patients in the coughing stage should receive antibiotics to render them noninfectious to others.

Prevention of *B. pertussis* infection relies upon use of vaccines. Because this is a killed whole bacterial preparation, containing both toxins and antigens, it has various side effects that have been persistently criticized. Attempts to produce a "purer" vaccine have not been wholly successful. An extract vaccine was associated with fewer side effects than the whole cell vaccine, but doubts have been expressed concerning its efficacy. The whole cell vaccine is commonly given together with tetanus and diphtheria toxoids as a series of injections at one or two month intervals, starting at 6 to 12 weeks of age.

Additional Reading

Birkebaek, N.H., M. Kristiansen, T. Seefeldt, et al.: "Bordatella Pertussis and Chronic Cough in Adults," *Clin. Infect. Dis.* **29**, 1239–1242 (1999).

Decker, M.D., K.M. Edwards, M.C. Steinhoff, et al.: "Comparison of 13 Acellular Pertussis Vaccines: Adverse Events," *Pediatrics* **96**(suppl), 557–566 (1995).

Evans, A.S. and P.S. Brachman: "Bacterial Infections of Humans: Epidemiology and Control," 3rd Edition, Plenum Medical Book Company, New York, NY, 1998.

Gangarosa, E.J., A.M. Galazka, L.M. Phillips, et al.: "Impact of Anti-vaccine Movements on Pertussis Control: The Untold Story," *Lancet* **351**, 356–361 (1998).

Guris, D., P.M. Strebel, B. Bardenheir, et al.: "The Changing Epidemiology of Pertussis in the United States: Increasing Reported Incidence in Adolescents and Adults, 1990–1996," *Clin. Infect. Dis.* **28**, 1230–1237 (1999).

Halperin, S.A., D. Schiefele, L. Barreto, et al.: "Comparison of a Fifth Dose of a Five-component Acellular or a Whole Cell Pertussis Vaccine in Children Four to Six Years of Age," *Pediatr. Infect. Disease J.* **18**, 772–779 (1999).

Orenstein, W.A., S. Hadler, and M. Wharton: "Trends in Vaccine-Preventable Diseases," *Semin. Pediatr. Infect. Dis.* **8**, 23–33 (1997).

Plotkin, S.A. and W.A. Orenstein: "Vaccines," 3rd Edition, W. B. Saunders Company, Philadelphia, PA, 1999.

Peter G.: "1997 Red Book: Report of the Committee on Infectious Diseases," 24th Edition, American Academy of Pediatrics, Elk Grove Village, IL, 1997.

Staff CDC: "Pertussis Vaccination: Use of Acellular Pertussis Vaccines Among Infants and Young Children: Recommendations of the Advisory Committee on Immunization Practices (ACIP)," *MMWR* **46**(RR-7), 1–25 (1997).

Staff CDC: "Pertussis—United States, January 1992–June 1995," *MMWR* **44**, 525–529 (1995).

Staff CDC: "Transmission of Pertussis From Adult to Infant—Michigan, 1993," *MMWR* **44**, 74–76 (1995).

Staff Institute of Medicine: Adverse Effects of Pertussis and Rubella Vaccines. National Academy Press, Washington DC, 1991.

Staff Institute of Medicine: "Adverse Events Associated with Childhood Vaccines: Evidence Bearing on Causality," National Academy Press, Washington, DC, 1994.

Web Reference

Centers for Disease Control and Prevention: *http://www.cdc.gov/health/diseases. htm#P* and *http://www.cdc.gov/nip/publications/pink/pert.pdf*

R. C. V.

PETALITE. The mineral petalite, lithium aluminum silicate ($LiAlSi_4O_{10}$), is monoclinic, although crystals are rare, this mineral usually occurring in cleavable, foliated masses, whence the name petalite, from the Greek meaning a *leaf*. Its hardness is 6–6.5; specific gravity 2.39–2.46. It is brittle with subconchoidal fracture; perfect basal cleavage; luster, vitreous, colorless to white or gray but may be greenish or reddish; transparent to translucent. Petalite occurs in granite pegmatites with sodium-rich feldspar, quartz, and lepidolite; has been found in Sweden; the former U.S.S.R.; on the Island of Elba; and in the United States at Bolton, Massachusetts, and Peru, Maine. It is interesting to note that lithium was first discovered in this mineral. See also **Lithium**.

PETIOLE. See **Leaf**.

PETRELS AND ALBATROSSES (*Aves, Procellariiformes*). These birds essentially comprise the procellariiformes, an order of marine birds known for their powerful flight. The feet are webbed, and the horny sheath of the beak is composed of several parts. Among the petrels are the fulmars and shearwaters, as well as several forms that bear the name petrel. There are many species, and many of them bear one or more names in sailors' vernacular. The Cape pigeon (*Daption capensis*) is a well-known petrel of the Southern Hemisphere, and the little stormy petrel of the North Atlantic (*Hydrobates pelagicus*), under the name Mother Carey's chicken, is probably the most familiar of all to ocean travelers. The name puffin is commonly applied to the Atlantic species of petrel; and the name shearwater to those of the Pacific. Together, they constitute the genus *Puffinus*.

The albatross is a large marine bird known from its habit of following ships for many hours without alighting. There are several species, belonging to *Diomedea* and allied genera. See Fig. 1. These birds are commonly called sea gulls although the term albatross is sometimes reserved for the species chiefly found in the South Seas. Of a large number

Fig. 1.   Royal albatross. (*Sketch by Glenn D. Considine.*)

of species, these birds display similar characteristics. One egg is hatched per time in a nest constructed of sticks, usually on the ground. They participate in spectacular courtship dancing, making loud shrieks during the performance.

The Laysan albatross is found in the environs of the mid-Pacific islands and summers in the Aleutians. Unlike most species, this albatross does not follow ships. The blacktailed albatross lives along the Pacific coasts, feeding on squid and fish. The feathers are dark, and the bird often rests on the surface of the water. The range is wide — from Japan and the Pacific islands and along the North American coast from Mexico to Alaska. See also **Procellariiformes**.

PETROCHEMICALS.   Chemicals derived from petroleum and, more specifically, substances or materials manufactured from a component of crude oil or natural gas. See Fig. 1 on p. 2686. In this sense, ammonia and synthetic rubber made from natural gas components are petrochemicals. Many of these chemicals are described in separate articles in this encyclopedia. Check alphabetical index.

Among the most important petrochemicals manufactured are:

Acetic acid	Ethylene dichloride	Phenol
Acetone	Ethylene glycol	Polyethylene
Acrylonitrile	Ethylene oxide	Polypropylene
Benzene	Formaldehyde	Polyvinyl chloride
Cumene	Isopropyl alcohol	Styrene
Cyclohexane	Maleic anhydride	Toluene
Ethylbenzene	Methanol	Vinyl chloride
Ethylene	Phthalic anhydride	Xylenes

The chemical unit operations, such as distillation, extraction, and various separation operations, and the chemical unit processes, such as alkylation, dehydrogenation, hydrogenation, and isomerization, are essentially identical to those operations used in the manufacture of chemicals from other sources.

In order to save materials transportation costs, a petrochemical plant frequently will be located adjacent to a petroleum refinery or gas processing plant. Short pipelines can be used in place of leasing long pipelines or having to depend upon tank car shipments by rail or truck. This also contributes to the overall safety of production. A representative petrochemical plant adjacent to a refinery is shown in Fig. 2 on p. 2687.

An excellent report of the petrochemical industry is prepared annually by the *Oil and Gas Journal* and *Hydrocarbon Processing* magazine.

PETROGENESIS.   That branch of petrology that deals with the origins of rocks. Practically a synonym for petrology unless, as is usually the practice, confined to the igneous (and possibly the metamorphic) rocks.

PETROLEUM.   A natural oil, ranging in color through black, brown, and green, to a light amber shade. It is often termed crude oil, and consists principally of hydrocarbons, that is, compounds of carbon and hydrogen, but varying amounts of oxygen-, nitrogen-, and sulfur-bearing compounds are almost invariably present. The term *mineral oil*, which was and is often used as a synonym for petroleum, is inadequate, for most geologists believe that it was derived from organic material resulting from reactions of organic materials such as plants and animals buried in sedimentary rocks.

The more important of these geologic formations in which petroleum is found are the Tertiary period of the Cenozoic era (50% of the world's oil production comes from these rocks, including regions in California and the Gulf Coast of the United States. Russia, Venezuela, Malaysia, Iran, and Iraq); the Cretaceous period of the Mesozoic era (including the East Texas, Kuwait, and Bahrein fields); the Jurassic period of the Mesozoic era (including the Arkansas and Rocky Mountain regions of the United States, and Saudi Arabia); and the Mississippian period of the Paleozoic era (including the West Texas, Pennsylvania, and Mid-Continent regions of the United States, and the Alberta, Canada, fields).

Petroleum oils vary considerably in composition, even when closely associated geographically. In some areas of the United States, for example, crude oils near the surface may have quite a different chemical composition from those found in deeper strata. Depth alone, however, does not correlate significantly with composition.

Analysis of typical crude oils found in representative areas of the United States are given in Table 1. It may be generalized that crudes found in the eastern and midwestern sections of the United States are predominantly sweet and paraffinic; those found along the Gulf Coast usually are naphthenic; those occurring in the inland southwest are sour and naphthenic; and those found along the west coast are asphaltic. The waxy, sweet paraffinic oils found in Pennsylvania first became prominent because of the high quality of lubricating oils and greases that could be made from them. The severe stresses imposed by the bearings and close-fitting reciprocating surfaces of machinery led to the development of refining processes and the discovery of additive materials whereby many other crude oils also can be transformed into excellent lubricants. Even Pennsylvania oils require special refining and additives to meet present quality specifications.

Analyses of some crude petroleums found outside the United States are given in Table 2, which illustrates the variety of crudes existent, but is not intended to give a full representation of worldwide petroleum source compositions.

API Gravity.   This parameter (API stands for American Petroleum Institute), expressed in "degrees," is mathematically related to specific gravity and can be determined with a hydrometer. The specific gravity of water (arbitrarily defined as unity) is 10.00 when expressed as degrees API. API gravity usually, although not infallibly, indicates the gasoline and kerosine contents of the crude. As an example, the Mississippi, Texas, New Mexico, and Louisiana crudes have API gravities between approximately 35 and 40; as do the Arabian, Iranian, and Colombian crudes. The gasoline content (that fraction boiling below about 400 °F (204 °C) of these crudes ranges from about 25% to over 35% by volume. The kerosene portions of such "light" crudes also are usually high. In contrast, Wyoming sour crude with an API gravity of 17.9 contains but 6% gasoline and about 40% asphalt. California crude has an even greater content of residuum and almost no gasoline.

Sulfur Content.   The amount of sulfur in crude is important in terms of handling the crude within the refinery and the undesirable effects of sulfur in finished products. High-sulfur crudes require special materials of construction for refinery equipment because of their corrosiveness. Certain refinery processes require desulfurization of sour charge stocks prior to use as a feedstock, not only because of their corrosiveness, but also because of the effect of sulfur-bearing compounds on expensive catalysts. From the standpoint of the consumer, sulfurous gasoline has an unforgettably offensive odor unless specially sweetened and it may corrode the fuel system and engine parts, as well as pollute the atmosphere after it has been burned.

Other factors indicated in the data of Tables 1 and 2 include: **Pour Point** — defined as the lowest temperature at which the material will pour and a function of the composition of the oil in terms of waxiness and bitumen content; **Salt Content** — which is not confined to sodium chloride, but usually is interpreted in terms of NaCl. Salt is undesirable because of the tendency to obstruct fluid flow, to accumulate as an undesirable constituent of residual oils and asphalts, and a tendency of certain salt compounds to decompose when heated, causing corrosion of refining equipment; **Metals Content** — heavy metals, such as vanadium, nickel, and iron, tend to accumulate in the heavier gas oil and residuum fractions where the metals may interfere with refining operations, particularly by poisoning catalysts. The heavy metals also contribute to the formation of deposits on heated surfaces in furnaces and boiler fireboxes, leading to

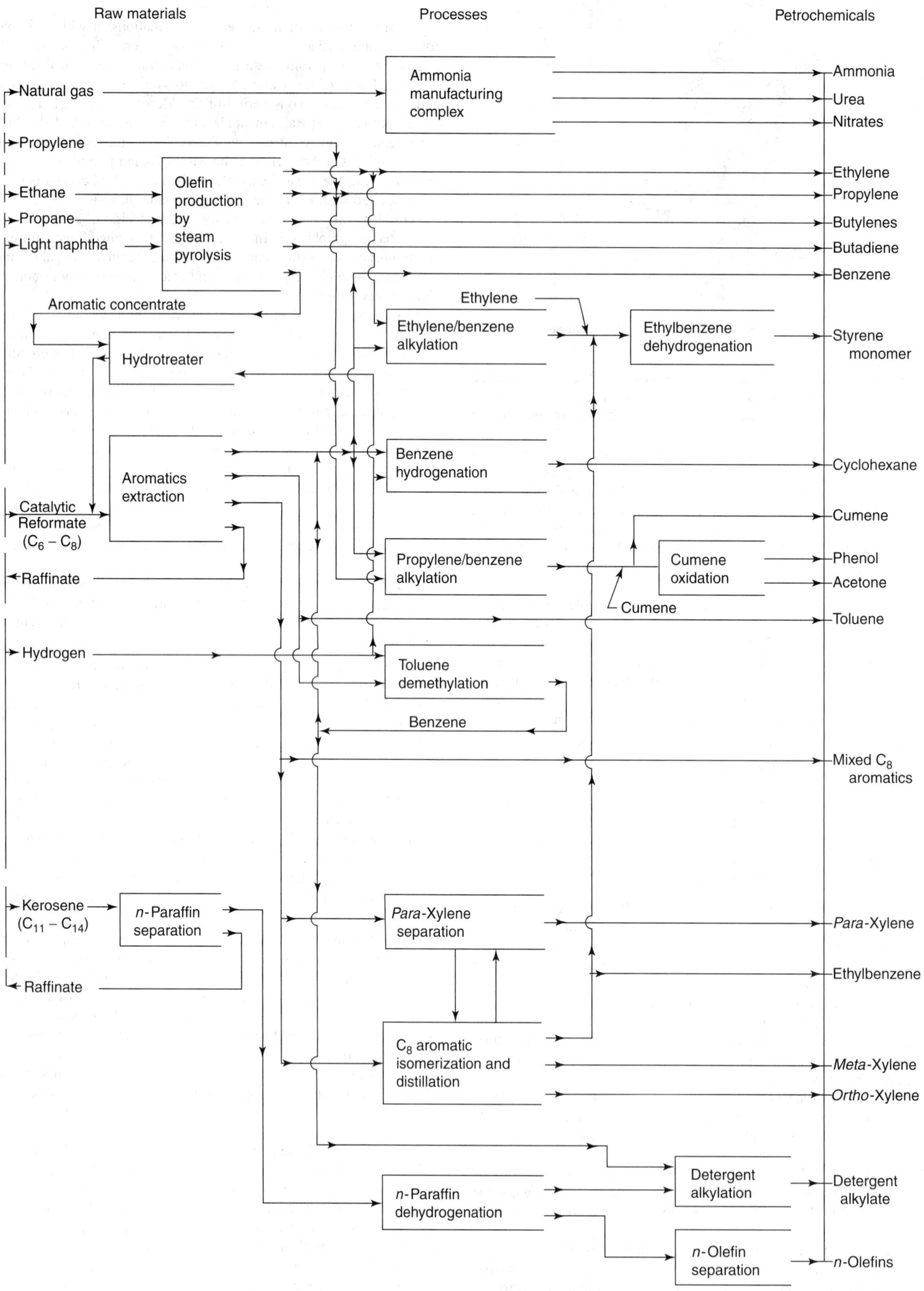

Fig. 1.   Interlocking processes and flow of materials in a representative petrochemical complex. (*UOP, Inc.*)

Fig. 2. Portions of a representative solvent-producing plant, using petrochemicals as raw materials.

permanent failure of equipment, interference with heat-transfer efficiency, and increased maintenance.

Natural Gas, Oil Shales, and Tar Sands. Natural gas is not formally defined as a component of crude petroleum, although natural gas commonly exists in the same geological formations, often directly in contact with crude petroleum. However, a large percentage of natural gas wells are not associated with producing oil wells. See also **Natural Gas**.

The oils derived from oil shales are not true petroleum, although they are petroleumlike products after being subjected to specialized chemical processing. Shales are sedimentary rocks that have a relatively high content of a bituminous substance called *kerogen* and 30–60% organic matter and fixed carbon. Kerogen, although not a definite chemical compound, yields an oily substance when heated (retorted) in the absence of air. Extraction of oil shale with ordinary solvents produces no oil, and their solubility in solvents is low. This evidence supports the conclusion that the "oil" is the result of a chemical change, i.e., the thermal cracking or fragmenting (pyrolysis) of the molecules that make up kerogen. See also **Oil Shale**.

Tar sands is an expression commonly used in the petroleum industry to describe sandstone reservoirs impregnated with a very heavy viscous crude oil which cannot be produced through a well by conventional production techniques. Two other terms, *bituminous sands* and *oil sands*, are gaining favor. The heavy viscous petroleum substances impregnating the "tar sands" are called asphaltic oils. See also **Tar Sands**.

Petroleum processing and petroleum end-products are described in the article on **Petroleum Refining** immediately following this article.

Origin and Geology of Petroleum

Among the general theories for explaining the origin of petroleum, the most widely accepted is the *organic theory*, which can be quickly summarized. Over millions of years, rivers flowed to the seas, carrying large volumes of mud and sand to be spread out by currents and tides over the sea bottoms near the gradually changing shorelines. New deposits were distributed, layer upon layer, over the floors of the seas. Because of the increasing weight of these accumulations, the sea floors slowly sank, building up a thick series of mud and sand layers. High pressure and chemical forces ultimately converted these layers into sedimentary rocks of the type that often contain petroleum — the sandstones, shales, limestones, and dolomites. The organic theory further stipulates most importantly that tiny marine organisms were buried with the silt. In an airless environment and under high pressures and elevated temperatures, these carbon- and hydrogen-containing minuscule life-forms were converted over

TABLE 1. ANALYSIS OF REPRESENTATIVE U.S. CRUDE OILS

Property	McComb, Mississippi	Southwest Texas	East Texas	Wyoming (Sour)	New Mexico	N. Kenia Peninsula, Alaska	San Ardo, Calif.	Ospelousas, Louisiana	Velma, Okla.
Total sulfur, wt%	0.07	0.45	0.2	3.33	1.0	1.04	1.93	0.08	1.13
Pour point, °C	15.6	−1.1	12.8	−20	−3.9			4.4	
°F	60	30	55	−5	25			40	< −30
Gasoline, vol%	35.5	32.0	29.0	6.3	37.8	14.4	1.9	26.1	22.3
Kerosene, vol%	18.1	12.1	10.1	9.1		18.0	16.1	18.9	17.3
Diesel fuel, vol%	14.6	38.0	13.8	14.0		18.4	10.6	22.9	8.5
Gas oil, vol%	28.1	12.6		30.7	41.2	22.3	23.3	27.9	31.9
Asphalt bottoms, vol%	3.7	5.3	47.1	39.9	20.8	26.9	48.1	4.2	20.0
Metals in gas oils, ppm									
Nickel	0.06						0.15		
Vanadium	0.08						<0.1		
Salt, lb/1000 bbl	4	<0.5	31	0.6	14	76		5	78

TABLE 2. ANALYSES OF REPRESENTATIVE WORLD CRUDE OILS

Property	Arabian	Minas, Central Sumatra Topped	Putomayo, Colombia	Gulf Nigeria	Zulia, Venezuela	Iran	Kuwait
Total sulfur, wt%	3.05	0.2	0.49	0.16	1.69	1.12	2.62
Pour point, °C	−36.1	−17.8	7.2	−6.7	< −15	15	< −15
°F	−33	0	45	20	<5	5	<5
Gasoline, vol%	29.1	11	34.1	24.9	18.9	32.2	25.5
Kerosene, vol%	16.0	16	9.3	26.5	14.1	18.3	13.7
Gas oils, vol%	12.5	14	40.7	19.3			
Residuum, vol%	42.4	59	15.9	29.3			
Metals in gas oils, ppm							
Vanadium	0		25	7			
Nickel	0		11	5			
Iron	3						
Salt, lb/1000 lb	12		trace	5			

an extremely long time span into hydrocarbons. This theory, of course, requires acceptance of the concept of drastically altered shorelines, because obviously oil deposits are found in many parts of the world long distances from the present coastlines.[1]

Geologists find it particularly difficult to trace the history of a given hydrocarbon deposit, because the oil and gas may have moved as the result of numerous seismic events, again occurring over a very long time span. A past requisite for commercially exploitable hydrocarbon deposits has been prior movement and concentration of large quantities of hydrocarbons in various forms of *traps*. In contrast with oil shales and tar sands, natural gas and petroleum flow relatively easily in permeable underground structures and, consequently, tend to concentrate, greatly assisting the economic exploitation of these materials.

The movement of petroleum from the place of its origin to the traps where accumulations are found is believed to have occurred in an upward direction. This movement took place as the result of the tendency for oil and gas to rise through the ancient seawater with which the pore spaces of the sedimentary formations were filled when originally laid down. An underground porous formation or series of rocks which occur in some shape favorable to the trapping of oil and gas must also be covered or adjoined by a layer or rock that provides a covering or seal for the trap. A seal of this type, frequently called a *cap rock*, stops further upward movement of petroleum through the pore spaces.

As oil and gas gathered in the upper part of a trap, because of differences in weight of gas, oil, and salt water, these fluids also separated vertically, much in the same manner as if these materials were all present in a bottle. Thus gas, if any is present, is found in the highest parts of the trap, followed by oil (and oil with gas) below the gas, and finally salt water below the oil. Experience has indicated that the salt water seldom was completely displaced by oil or gas from the pore spaces, even within the trap. Even in the midst of oil and gas accumulation, pore spaces within the trap may contain from 10 to 50% or more of salt water. It appears that the remaining water (termed *connate water*) fills the smaller pores and also exists as a coating or film, covering the rock surfaces of the larger pore spaces; thus oil and/or gas are apparently contained in water-jacketed pore spaces. The geological structures called traps are petroleum reservoirs, i.e., they are the oil and gas fields that are explored and produced. All oil fields contain some gas, but the quantity may range widely. See also *Natural Gas*.

Types of Oil Accumulations

A composite diagram showing different types of oil and gas accumulations is given in Fig. 1.

Structural Traps. The attitude of the rocks, whether they are folded, fractured, displaced, or otherwise disturbed, is called their geologic structure. Traps that are due to geologic structure are known as *structural traps*.

[1] In a more recent, alternate theory, Thomas Gold (Cornell University) suggests that, in contrast with the organic sediments theory, the prime source of natural gas is primordial, abiotic methane rising from deep within the earth's mantle. This is discussed in further detail in the article on **Natural Gas**.

A common structural trap, the *anticline*, is an upward bulge in the rock layers which forms an arch capable of holding oil under its apex. The buoyancy of oil and gas carries them upward through porous rock layers into the apex until they are trapped by an impermeable layer. Anticlinal type of folded structure is shown in Fig. 2. Reservoirs formed by folding of the rock layers or strata usually have the shapes shown in Fig. 2(a) and (b). These traps were filled by upward migration of oil and/or gas through the porous strata or beds to the location of the trap. Further movement was arrested by a combination of the forms of the structure and the seal or cap rock provided by the formation covering the structure.

Examples of domal structures are the Conroe Oil Field in Montgomery County, Texas and the Old Ocean Gas Field in Brazoria County, Texas. Another example of a reservoir formed by an anticlinal structure is the Ventura Oil Field in California.

Another type of structural trap is the *fault trap*. A fault is a fracture in the earth's crust along which movement has occurred such that a porous rock layer is offset by a nonporous layer. The oil moving along a porous stratum is dammed or blocked by an impermeable shale or limestone. See Fig. 3. Examples of fields of this type occur along the Mexia fault zone of East-Central Texas.

The *salt dome* is another interesting form of structural trap. This type of trap is found along the Gulf Coasts of Texas and Louisiana and in western Colorado and Utah. This type of structure resulted from the upward thrust of a great mass of salt far below the earth's surface. When a salt dome rose through a layer of oil-bearing sedimentary rock, oil may have been trapped in anticlines above the dome, or in structures similar to faults along its flanks. One famous example of a salt-dome reservoir is the Spindletop field, near Beaumont, Texas. It was "brought in" in 1901 by a 100,000-barrel-a-day gusher, giving birth to the modern petroleum industry. Another example of a salt-dome field is the Sugarland Oil Field in Fort Bend County, Texas. See Fig. 4.

Stratigraphic Traps. Petroleum geologists also seek another kind of trap, the *stratigraphic trap*, which results when a porous layer is "pinched" or phased out between two nonporous layers. Caught in an underground envelope of impermeable rock, the oil accumulates to form a reservoir. This type of trap may have formed from buried beaches or sandbars. The famous East Texas field is a "strat" trap. See Fig. 5. Because structural features such as anticlines and faults are often more obvious and easier for the geologist and geophysicist to detect, more fields thus far have been found in structural traps than in their stratigraphic counterparts. Stratigraphic traps are usually discovered only after exhaustive studies have been made of rock samples from outcrops and from core samples from wells drilled over large areas.

The serpentine plug, shown in Fig. 6, is an interesting type of trap, an example of which is the Hilbig Field in Bastrop County, Texas. As illustrated, a porous serpentine plug has formed a reservoir within itself by intruding into nonporous surrounding formations.

Lens-Type Traps. These form in limestone and sand. In this type of trap the reservoir is sealed in its upper regions by abrupt changes in the amount of connected pore space within a formation. A trap formed in sand is shown in Fig. 7(a). An example is the Burbank Field in Osage

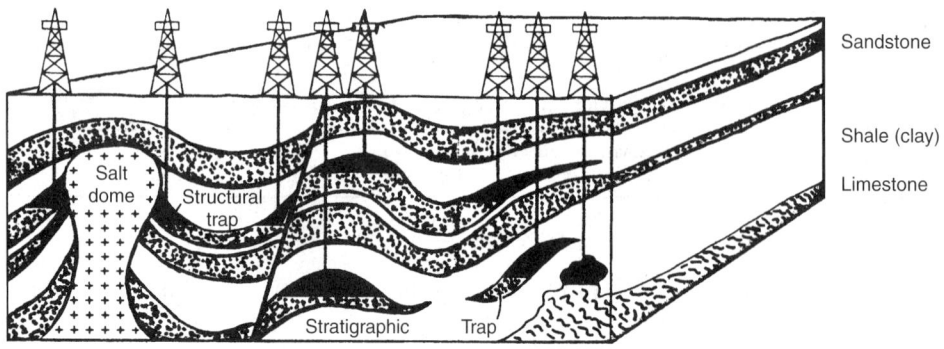

Fig. 1.   Composite diagram showing different types of oil and gas accumulations (shown in solid black). *Structural traps* where petroleum deposits may have accumulated are often found along the edges of salt domes or along fault lines. *Stratigraphic traps* may exist where reservoir rock is "pinched off" by denser strata. These accumulations are the "pools" or "reservoirs" which, singly or in groups, compose an oil or gas "field." The pores of the reservoir rock contain oil or gas or both, always accompanied by briny water. The fluids tend to be layered, with gas at the top of the trap, oil in the middle, and water underneath. Most of the petroleum which ever existed has been obliterated, either by attenuation in the earth's crust, or by exposure to heat and pressure high enough to break down its chemical bonds. The accumulations that do exist have endured against long odds. Additional detail on oil and gas reservoirs will be found in Figs. 2 through 7.

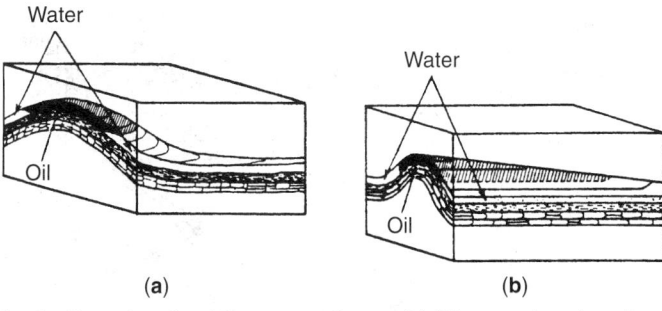

Fig. 2.   Examples of anticline structural traps: (**a**) Oil accumulates in a dome-shaped structure. The dome is circular in outline. (**b**) Anticlinal trap that is long and narrow, differing from the dome configuration.

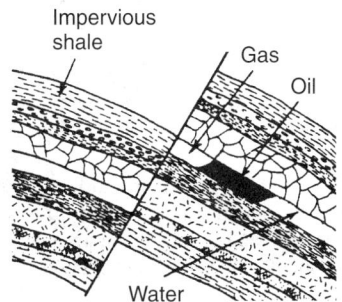

Fig. 3.   Example of a fault structural trap. The oil is confined in traps like this because of the tilt of the rock layers and faulting.

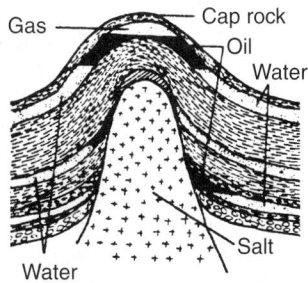

Fig. 4.   Example of a salt-dome structural trap. One of the earliest and greatest reservoirs of this type was Spindletop, brought into production in 1901 and located near Beaumont, Texas.

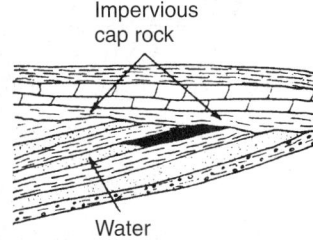

Fig. 5.   Example of a stratigraphic trap. This unconformity represents the condition where upward movement of oil has been halted by the impermeable cap rock laid down across the cutoff (possibly by water or wind erosion) surfaces of the lower beds. This type of reservoir is found in the great East Texas field.

County, Oklahoma. This type of trap may occur in sandstones where irregular deposition of sand and shale occurred at the time the formation was laid down. In these cases, oil is confined within the porous parts of the rock by the nonporous parts of rock surrounding it. A lens-type trap formed in limestone is shown in Fig. 7(b). In limestone formations there are frequent areas of high porosity with a tendency to form traps. Examples of limestone reservoirs of this type are found in the limestone fields of West Texas.

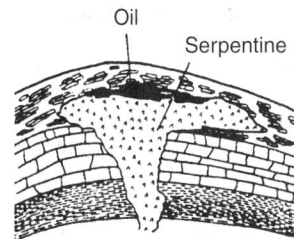

Fig. 6.   Example of a serpentine plug as found in the Hilbig Field in Bastrop County, Texas.

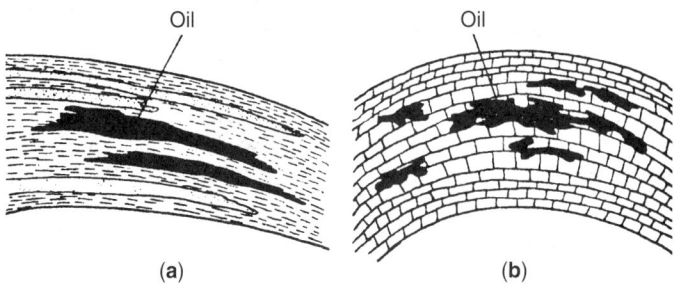

Fig. 7.   Lens-type traps: (**a**) sandstone; (**b**) limestone.

Reef-Type Traps. These have accounted for some of the most important production in recent years. The reef is generally considered a type of stratigraphic trap. The reef was formed under the right combination of conditions by the remains of millions of small underwater animals. Building their limestone residences on top of those built by their ancestors, the tiny creatures produced columns or mounds, often reaching several hundred feet high. If eventually surrounded by impermeable rock layers, the reef could become a trap for oil and gas.

Petroleum Production — Geophysics

In the petroleum industry, the word *production* is generally interpreted as the obtaining of crude oil, natural gas, and other related natural hydrocarbons, whether located in reservoirs under the land or under the oceans. Production generally does not relate to the manufacture of final end-products, such as naphtha, gasoline, diesel fuel, lubricating oils, etc. Products obtained by the processing of natural hydrocarbons are commonly referred to as *refined*. In a sense, then, petroleum production is analogous to other natural raw material mining, such as coal and metals mining.

The fundamental phases of petroleum production include: (1) the initial *exploration* required to find heretofore undiscovered oil and gas reservoirs; (2) *primary* and *secondary recovery* methods, which make use of both naturally occurring (or *primary*) reservoir energy and the application of *secondary* energy sources, such as the injection of gas or water; and (3) *enhanced oil recovery* used to increase ultimate oil production beyond that achievable with primary and secondary methods. Enhanced oil recovery (EOR) methods increase the proportion of the reservoir by improving the "sweep" efficiency, reducing the amount of residual oil in the swept zones (increasing the displacement efficiency), and reducing the viscosity of thick oils.

Petroleum Exploration

The exploration required to identify previously undiscovered oil and gas reservoirs is grossly affected by the economics and politics of the worldwide oil markets (supply and demand). The experience of one year cannot easily be extrapolated to that of subsequent years.

Although the petroleum geologist has for several years employed highly sophisticated instrumentation and computer technology, the fact remains that the geologist, in many respects, is still a *detective*, often commencing with the barest of clues and following the search hopefully to a successful conclusion. Decisions based on uncertain information are normal in petroleum exploration. How to capitalize on the creativity of the geological and geophysical scientists further complicates decision making.

Modern petroleum exploration geophysics is based on three important earth properties: (1) density, (2) magnetization; and (3) acoustic response.

Gravity Surveys. Gravitational pull is an effective way to measure the density of the strata lying far below the earth's surface. The *gravimeter* came into oil exploration use in about 1900. If there are dense rocks near the surface, a spring-loaded weight in the instrument will weigh more — obviously since denser rocks exert greater gravitational pull than those which are less dense. Usually, the denser rocks are older and their presence near or far from the surface indicates an underlying shallow or deep basement. Thus, a structural "high," such as an anticline, will often appear as a high reading on the gravimeter. Conversely, a structural "low," such as a syncline, will yield opposite effects. The low density of a salt dome will be reflected on the gravimeter as having a lesser gravity value. When using a gravity meter offshore, it is necessary to place the instrument so that it remains level, regardless of the movement of the vessel. The rise and fall of the ship and other effects of its travel must be measured by other equipment and taken into account when the gravity data are processed.

Magnetic Surveys. The natural magnetic properties of rocks can be useful in the early stages of petroleum exploration. More than three centuries ago, it was found that a freely suspended magnetic needle will move away from a level position in response to the presence of nearby iron. In 1879, this principle was applied in the first *magnetometer*, a simple device combining a magnetized needle with a compass. Both types and depths of rocks far underground affect how much and in which direction the needle will move.

Using a magnetometer, earth scientists can measure the depths at which basement rocks lie. These rocks are usually from the Precambrian period, dating back over 600 million years. They contain large amounts of magnetite, a naturally occurring iron oxide having magnetic properties. Where petroleum-bearing sedimentary rocks overlie these basement rock formations, the magnetometer can be used to indicate variations in the thickness of these sedimentary layers. This information can provide the geologist with a valuable clue as to the probability that those sediments, bent by the underlying basement rocks, may provide the right conditions for the accumulation of recoverable quantities of oil or natural gas.

It is not always necessary to place instruments on the ground to measure magnetic attraction of the subterranean rock layers. Magnetometers can also be used from aircraft flying high above the surface to determine quickly areas of general interest. Use of airborne magnetometers also eliminates the need for and the problems sometimes associated with personnel actually getting on the site.

Seismic Surveys. Gravity and magnetic measurements are considered preliminary. The principal detailing tool used is the seismograph.[2]
See also **Earth Tectonics and Earthquakes**.

The type of experiment used to make a seismic survey in petroleum exploration will be determined by the type of information required, the types of rock strata anticipated, environmental considerations, the type of terrain, and, of course, the economic costs involved.

In petroleum exploration, reflection seismic surveying is the method predominantly used. The seismograph records variations in the way rocks reflect sound waves sent downward from a surface source. The reflected sound waves vary with the type, depth, density, and dip of the rocks encountered. The returning sound waves made from a series of points along the survey path can then be displayed graphically to form a seismic record for interpretation by earth scientists. These principles are diagrammed and explained in Fig. 8.

Four sources of energy may be used to generate the sound waves: (1) controlled explosives; (2) vibrators; (3) weight dropping (where heavy weights are dropped to create the waves); and (4) compressed gases (producing bursts of energy using compressed air or propane). In exploring onshore sources, controlled explosives and vibrators are most often used. In offshore work, compressed air guns are most frequently used. These sources have been shown to cause virtually no damage to marine life.

When *controlled explosives* are used, the most common method is to place carefully measured charges in shallow "shot holes" a few inches

[2] The first "seismograph" is recorded to have been an inverted bronze urn with a pendulum inside and used by the Chinese in about 160 A.D. to announce earthquakes. Seismographs were "reinverted," so to speak, during World War I to locate heavy German artillery. In 1920, seismography for use in petroleum exploration was first demonstrated.

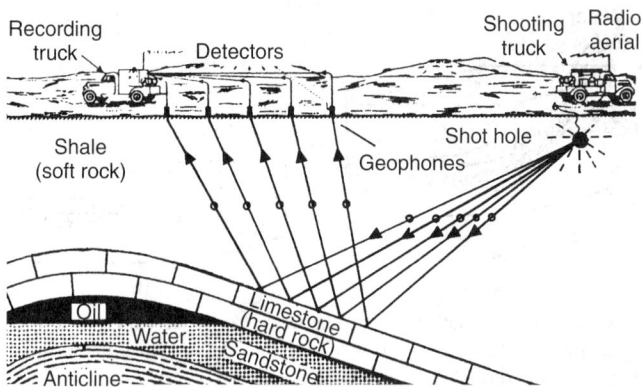

Fig. 8.   Onshore seismic field operation, using the shot hole energy method. Energy from a controlled explosion is directed toward underlying rock structures and is reflected to indicate shape of formation below — in this case an anticline. Geophones placed on the ground surface by the surveying crew measure and record the reflected acoustic energy. With the survey truck safely away from the shot hole, the explosives are detonated by an assigned radio frequency from the truck.

in diameter drilled from a truck-mounted drill. These charges are then detonated to produce the sound waves needed. In less accessible areas, a portable drill may be used and, in certain environmentally sensitive areas, the charges may be mounted on stakes above the ground to minimize plant disturbance.

Truck-mounted vibrators are frequently used. Metal plates, used singly or in groups, are pressed against the ground and vibrated briefly creating a frequency sweep similar to that used in radar, to send a brief acoustic signal into the ground. The vibrators cause little disturbance to the surrounding area and are often used in sensitive locations, in cities and along highways, where dynamite charges would be less acceptable. See Fig. 9.

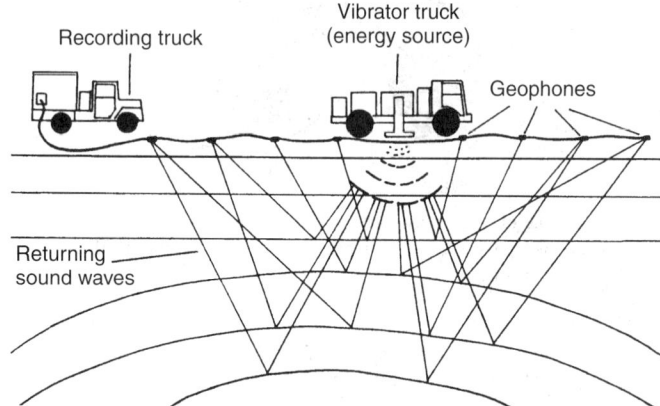

Fig. 9.   Seismic survey using a truck-mounted vibrator as the energy source for creating reflected acoustic sounds. This method is used in sensitive areas near cities and along highways because disturbance to the environment is minimal. (*American Petroleum Institute.*)

The principles of seismic surveying are the same regardless of the type of equipment used. As the waves strike the various strata, some are reflected back to the surface, where they are picked up by sensing devices called *geophones* (on land) and *hydrophones* (offshore). See Fig. 10. These sensors convert acoustic information into electrical signals, whence the data are recorded on magnetic tape for processing at a computer center.

Seismic Data Processing. Continuing advances in technology, such as fiber optics, have enabled the capabilities of seismic surveys to be markedly increased so that, where required, a 3-dimensional picture of the rocks in the subsurface can be obtained. Advanced processing has also enabled discrete anomalies in seismic data from individual rock horizons to be analyzed. Under certain conditions, the presence of hydrocarbons can be directly detected. This analysis, sometimes referred to as "bright spot" technology, has been responsible for numerous discoveries in the Gulf of Mexico in recent years.

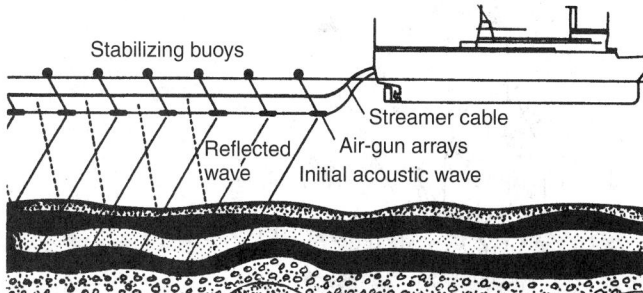

Fig. 10.   In offshore exploration, vessels are specifically designed for seismic surveying. Instead of using the same equipment that is used on-shore, such as "surface shaking" machines and geophones, offshore seismic crews use chambers containing compressed gases or fluids to generate the acoustic signals and hydrophones to pick up the returning sounds. Air-gun arrays are trailed in the water behind the vessel as it plies along a predetermined survey line. The crew activates the chambers at set intervals from onboard controls connected by cables to the air guns. Other cables contain arrays of hydrophones to detect sounds that echo off the underlying strata. The vessels are kept on course through the use of radar, loran, and satellite navigation equipment. (*After Exxon.*)

Raw data gathered from seismic surveys must be processed to compensate for and to remove a variety of distortions—unwanted noises created by weathered near-surface rocks, normal time delays, and echoing by rebounding acoustic waves—to provide the clearest possible image of the strata below. Computers can restore these distortions in a fraction of the time that was formerly required to adjust the data painstakingly by hand. Advanced techniques not only permit presentations in three dimensions, but also in color, and to create contour maps and models of subterranean features. However, even with the use of sophisticated tools, there remains a large measure of uncertainty. History has shown repeatedly that a prospective area rejected by one petroleum firm has been accepted by another and proved to be successful.

After thorough analysis of seismic and other exploration data, the next steps are management decision making and approval to proceed with exploratory drilling. The ultimate exploratory tool is the *drill bit*.

Exploratory Drilling

Onshore Drilling. When Col. Edwin Drake brought in the first commercial oil well in 1859, he struck oil at a depth of 59 feet, 8 inches (18.2 meters). See Fig. 11. Most early wells were less than 400 feet (122 meters) deep. Shallow oil and gas wells were fully exploited many years ago. Deep producing wells today often exceed depths of 25,000 feet (7,620 meters) and dry holes have been drilled to a depth in excess of 31,000 feet (9,449 meters). In an average year, wildcat wells reach a depth of about 6,000 feet (1,829 meters). Depth, however, is only one of the factors that makes the search for petroleum difficult in modern times. Increasingly, drilling must be done in remote places where it is costly to bring in materials and labor. For example, onshore locations, such as those found on the North Slope of Alaska, can result in drilling costs that are 10 times as high as they would be in the lower 48 states.

The Rotary Drill. This is the most commonly used method and consists of a rotary drill, a power source, a derrick and lifting and lowering devices, and a bit attached to a length ("string") of tubular high-tensile-strength steel. See Fig. 12. The drill string passes through a rotary table that turns it and thus provides the torque needed for the drilling operations. The weight applied to the formation is also a critical factor.

During the drilling operation, a special *drilling mud* (mixture of clay, water, and chemical additives) is pumped down through the hollow drill string and bit into the borehole. The fluid is forced up the borehole and through the area between the drill string and the casing (the "annulus") to the surface. There it is cleaned and recirculated into the well. The fluid helps to cool and lubricate the bit, control the pressures within the well, provide a protective and stabilizing coating to some permeable formations, and brings the rock cuttings up the borehole to the surface. The consistency of the fluid is carefully monitored and adjusted to compensate for pressure changes within the well, as the bit penetrates the various rock strata.

"Spudding" is the actual start of drilling a well and is akin to the first shovel of dirt at groundbreaking. A large bit, frequently from 18 to 38 inches (46–97 cm) in diameter, is used to drill a hole to a depth of from

Fig. 11.   Colonel Edwin Drake (right) and Peter Wilson, a druggist who endorsed a $500 bank loan for Drake, confer in front of the world's first commercial oil well near Titusville, Pennsylvania. Initially, Drake rigged a large wheel powered by steam to raise and lower a cable and iron bit. Later connected to a crude drill pipe and pump, this well produced about 35 barrels a day. (*ca. 1861.*)

10 to 100 feet (3–30 meters). The hole is then lined with a conductor pipe ("casing"). The space between the casing and the drilled hole (the "borehole") is filled with cement.

Drilling is then resumed using a smaller bit and, after the borehole reaches several hundred feet (meters), the bit and drill string are hoisted out of the well and another length of pipe ("surface casing") is lowered into the borehold and cemented in place. Besides preventing the generally unconsolidated surface formation from sloughing into the hole, the casing also protects the freshwater strata ("aquifers") from being contaminated by the drilling mud.

As the drilling proceeds, additional casings of concentrically smaller diameter are lowered into the well and sealed in place until the final depth ("target zone") is reached. During the drilling, the drill bit and string must be removed from the well whenever the bit becomes dull and requires changing or cores are taken from the well. The coring process involves a special cylindrical rock bit, generally with a diamond-encrusted face and a cylinder ("core barrel") into which the core passes and is retained for recovery at the surface. These cores are analyzed to determine the type of rocks penetrated and their porosity, permeability, chemical analysis and possible hydrocarbon content. See Fig. 13.

Drilling Geometry. Deviation surveying was introduced into oil-well drilling technology in 1929. Before that time, it was generally assumed that a hole properly started as a vertical hole would remain essentially vertical. In many instances, this was not a realistic assumption because many "vertical" holes were found to be quite crooked. Crooked holes not only caused operational problems, but also resulted in false indications of depth. Since the early 1930s, drilling contracts usually have specified a maximum deviation of 3 to 5 degrees. The problem of drilling a straight hole usually is simpler with uniform materials, such as limestone, and more difficult when laminar formations of sandstone and shale are encountered. Often of even greater concern than a crooked hole is an irregular, "jagged" hole that does not have a graceful bending contour in the vertical. The presence of abrupt changes in angle interferes with the casing program and ultimately with production. Although the mechanics involved in causing nonvertical drilling are not fully understood, much has been learned through experience and great improvements in drilling have been made. For one thing, deviation results from flexibility of the drill string (drill collars), the forces acting upon the string causing it to bend. A relatively

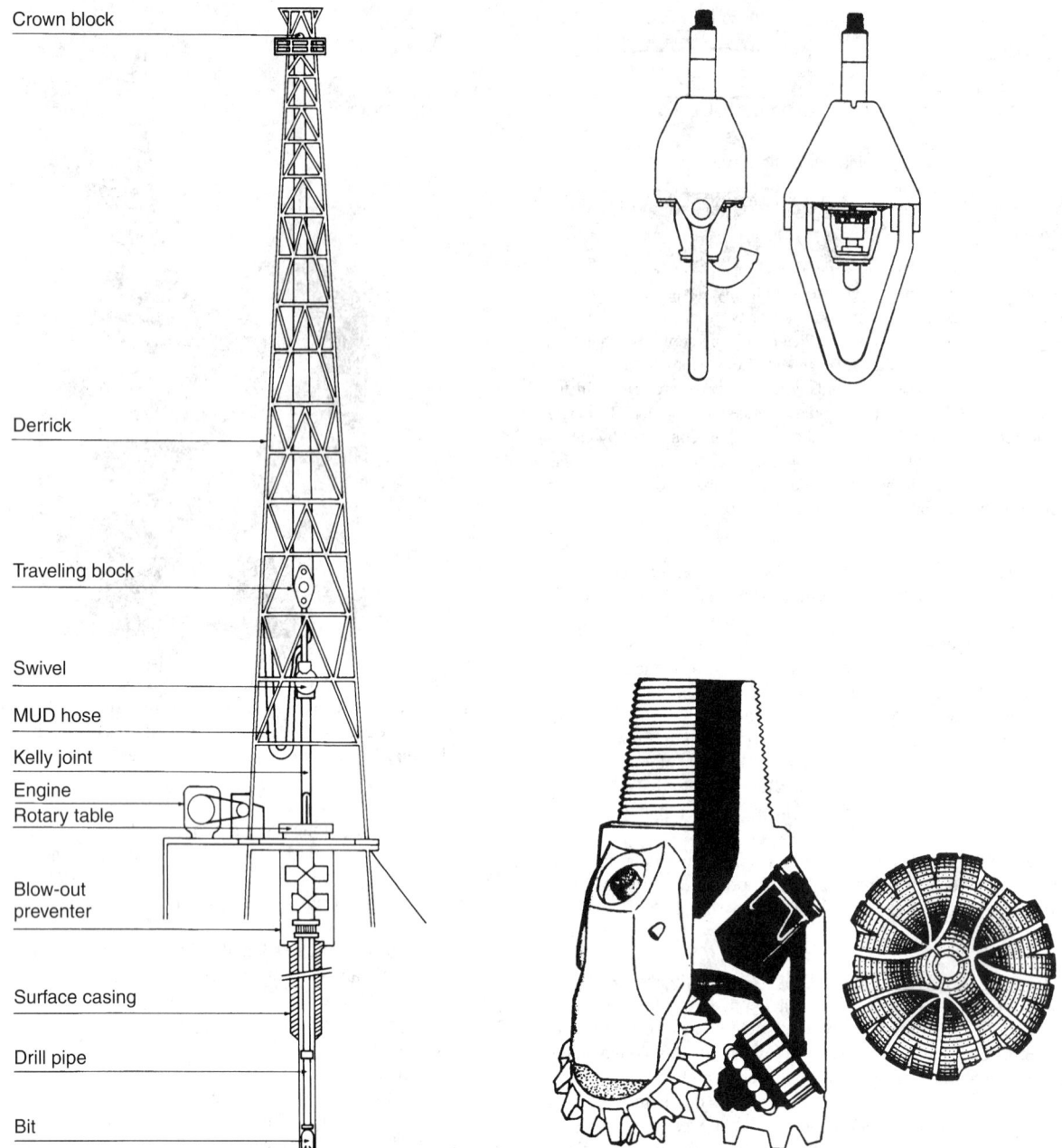

Fig. 12.   A rotary rig has four systems. The *rotary system* consists of a turntable, a swivel, a square or hexagonal pipe length called a "kelly," which transmits rotary motion from the turntable to the drill pipe, and the drill "string" itself. A *circulating system* of pumps, hoses and other apparatus keeps mud circulating through the well. The *hoisting system* includes the derrick, a drawworks, hoisting blocks, and other equipment needed to lift and lower heavy pipe joints and casing. The *power system* usually consists of diesel engines and generators, set apart from the rig, which provide power for the electric motors that drive the rotary, hoisting, and pumping equipment. The elevated floor allows installation of blowout preventer beneath the platform.

Shown in the upper right of this view is the *swivel* (front and side view) which permits the drill pipe to rotate while mud is pumped down to clean the hole. Shown in the lower right is a three-cone drill bit, with cutaway showing the bearings on which it rotates; and at the extreme lower right, the face of a diamond bit revealing openings through which fluids may pass. (*American Petroleum Institute; Exxon Corp.*)

simple change to square collars, as shown in Fig. 14, has brought about marked improvement.

Of course, it is frequently desirable to utilize a controlled directional drilling technique. There are several reasons, as indicated by Fig. 15. The three most commonly patterned directional holes are shown in Fig. 16. The planned course of direction depends upon several factors, including rig capacity, hole size, mud program, types of formation, and the casing program. Meticulous surveying is required to achieve the desired results. Several types of drilling tools may be required.

Offshore Drilling. When exploration moves offshore, standard drilling equipment obviously must be supplemented by some sort of structure that provides a stable platform for operations. The structure also must be movable, given the odds against a single wildcat finding commercial quantities of petroleum.

The first offshore exploration in the 1930s in the Gulf of Mexico was conducted from rigs on barges which could be towed to drilling sites and submerged to rest on the bottom during operations. These were forerunners of the twin-hulled submersible rig, which has an upper hull housing crew quarters and working spaces and a lower hull providing the buoyancy needed to move the unit. See Fig. 17(**a**) on p. 2695.

The jack-up or self-elevated rig, introduced in the 1950s, is a barge with movable legs, which can be lowered to the sea floor and the barge jacked into drilling position above the water. Jack-ups are used in water depths up to about 300 feet (91 meters). See Fig. 17(**b**) on p. 2695.

In deeper waters, exploration is conducted from floating rigs, including submersibles and drill ships. Drill ships with conventionally shaped hulls of seagoing vessels are not so stable as semisubmersibles (semis) in rough waters, but can be moved from location to location much faster.

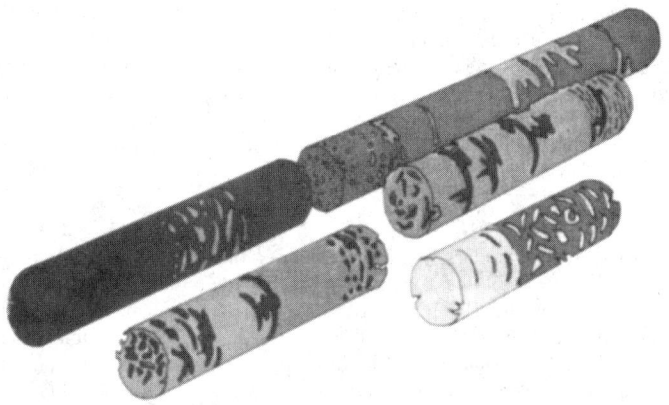

Fig. 13.   Rock core samples cut by a diamond-faced core bit reveal underlying structure and the possible presence of hydrocarbons. (*Exxon Corp.*)

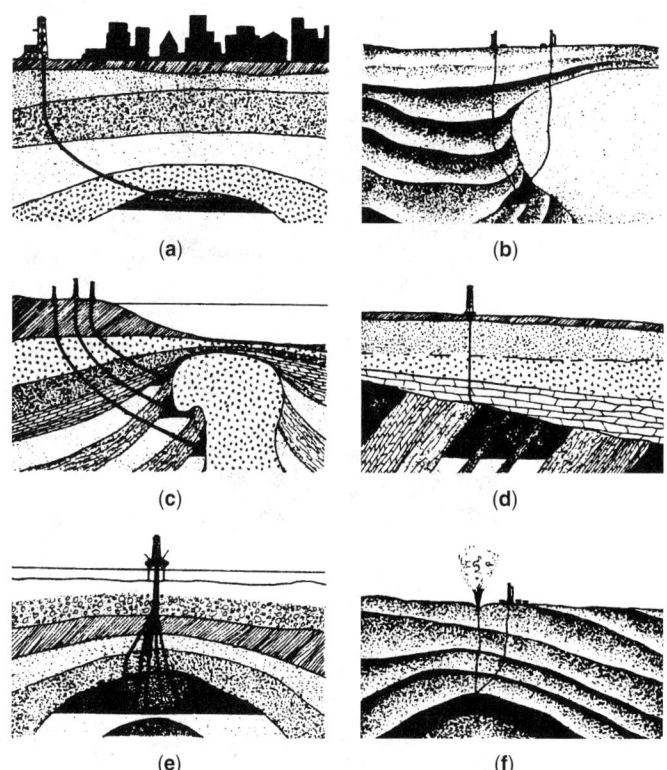

Fig. 15.   Some applications for controlled directional drilling: (**a**) Reaching formations which lie below inaccessible locations, such as towns, rivers, and lakes. (**b**) Formations sometimes are found below the overhanging cap of a salt dome. A well may be drilled around this cap, or through the salt and deflected into the productive formation. (**c**) Formations below harbors or the ocean floors sometimes can be reached from rigs located on the shore. (**d**) Directional drilling into the intersection of several oil sands from a single wellbore. Obviously, a straight hole would be less effective in this type of situation, (**e**) Offshore drilling is usually most economic when several directional wells can be drilled from a single platform. As many as 20 or more wells can be drilled from a small area. (**f**) Drilling of a relief well to intersect a wild, cratered well near the source of pressure. Mud and water can be pumped in to kill the blowout. This technique, first used in 1934, helped to establish the importance of directional drilling. (*Petroleum Extension Service, The University of Texas at Austin.*)

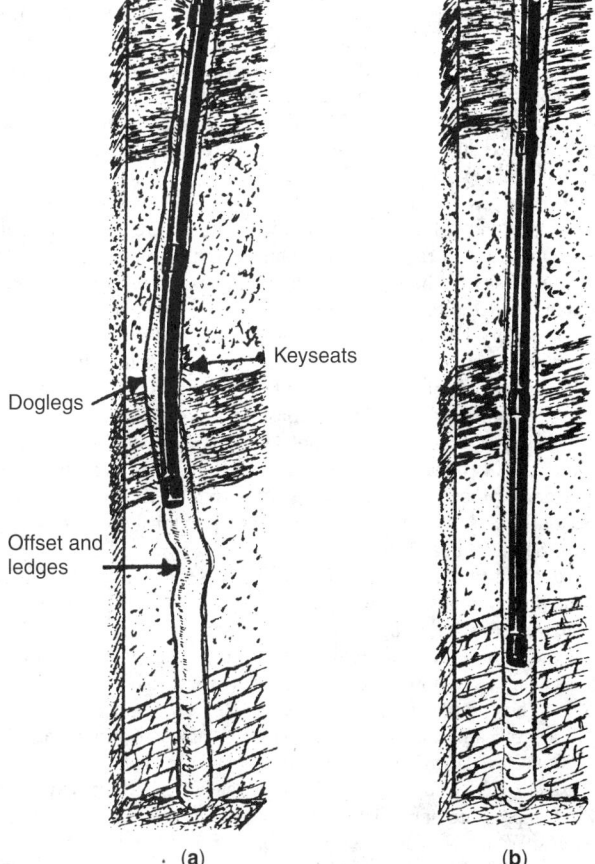

Fig. 14.   Example of a crooked hole (**a**) drilled without a square collar, and a relatively straight hole (**b**) drilled with a square collar. (*Drilco.*)

A critical requirement of all floating rigs is the ability to maintain position over the wellhead while drilling proceeds. Semis and drill ships use either multiple anchors or "dynamic positioning" systems to keep on station. A dynamic positioning system uses thruster engines which, responding to signals from acoustic beacons on the sea floor, automatically make the adjustments required to maintain the rig in position. Hydraulic devices keep a constant tension on the drill string to prevent the up-and-down motion of the sea from being transmitted to the drill bit.

Semis and drill ships find limited use in arctic waters where ice covers the sea most of the year.

In offshore operations, exploration wells are almost always plugged and abandoned even when they strike petroleum. Their sole function is to find oil or gas and to delineate the reservoir. The operator uses this information to pick a location for a permanent production platform from which development wells will be drilled to recover as much petroleum as economically possible. In onshore operations, however, successful exploration wells also become producers.

Measuring Well Characteristics

At selected intervals during the drilling, generally before the casing is run or when formations with hydrocarbon indications are encountered, measurements may be taken of the characteristics of the borehole and surrounding strata. Wire line logging tools are used.

In the early days of the industry, little was known about "downhole" geophysics, that is, the physical characteristics of the subsurface strata, how they might be measured, and what could be learned from such measurement. Since the Drake well, more than 3 million wells have been drilled in the United States. An estimated 27,000 fields have been found. With each new discovery, additional data become available and patterns begin to emerge. It was not until the late 1920s that technological changes were introduced that would have a lasting impact on "logging" (recording) the characteristics within the well during exploratory drilling. The first well-logging device (electrical resistivity log), invented by Conrad Schlumberger (France), was introduced into the United States. In 1934, a second development, the *spontaneous potential* (SP) *curve*, was introduced.

The *resistivity log* was lowered by cable into the borehole, with the drill bit and drill stem removed, thus enabling the recording of the electrical resistivity of the rock layers that the bit had penetrated. The record helped to identify the hydrocarbon content of the reservoir rock (since both oil and gas have different resistivities than does salt water). It was also used to correlate rock horizons between wells, which proved to be an invaluable tool in subsurface mapping.

The *SP Curve* recorded the differences in natural electrical potential between the fluids in the adjacent formations and those within the uncased borehole. This curve was soon accepted as an indicator of the porosity of

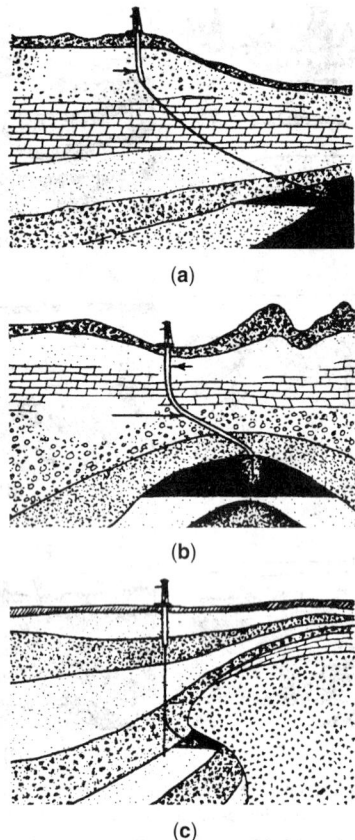

Fig. 16. Principal types of directional drilling patterns: (a) The most widely used directional drilling pattern is one in which the initial deflection is obtained at relatively shallow depth. Then, surface casing is set and cemented through the deviated section of hole. From that point, the angle is maintained as a straight line to the target zone. (b) This pattern is also initially deflected at shallow depth. Surface casing is set and cemented. Drilling continues on a straight line to a point where the hole is gradually returned to vertical. After intermediate casing is set, drilling is continued to final depth. This type hole is used when undesirable formations must be penetrated and isolated with an intermediate casing string. (c) In this pattern, deflection commences at a greater depth. Drift angle is maintained on a straight line to the target. This type hole may be used for exploratory drilling from a dry hole. Normally, the deflected part of this hole is not protected by casing during drilling operations. (*Petroleum Extension Service, The University of Texas at Austin.*)

the rock strata and as a means of locating the boundaries of rock beds. See Fig. 18.

Since then, a number of downhole measurements have been devised that use radioactive, acoustic, and electrical methods.

Presently, televiewers are used to look at rock features in boreholes and computers are programmed to compare, synthesize, and integrate the new range of measurements, thus providing more reliable information and definition of rock properties and formation fluids thousands of feet (meters) below the surface.

Until recently, these measurements could be taken only when the drill bit and string were removed from the borehole. However, in the mid-1980s, a new dimension was added, namely, *measurement while drilling* (MWD) instruments. These devices are mounted above the drill bit and around the drill string to provide a continuing source of data on downhole characteristics. This advancement reduces the drilling downtime previously required when measurements were taken.

Testing. Modern wire line logs will indicate with a good degree of accuracy the potential of a hydrocarbon zone. If the zone is sufficiently promising to warrant further study, a formation test will be undertaken.

Generally, a drill stem test is carried out—either in the open hole or after the hole has been cased. However, the case hole test is the most reliable.

Basically, the drill stem test involves attaching a tubing assembly to the end of the drill pipe, isolating the test zone with rubber packers, and

perforating the zone. The tool is then opened so that the fluids or gas in the formation can flow up the drill pipe for metering at the surface. During this process, extensive pressure measurements are taken, which can help to indicate the extent of the reservoir and the rate at which the hydrocarbons could be recovered. Prior to describing how a well is finally completed (if the hole is not dry!), it is in order to describe the forces utilized to transfer the oil from the reservoir to the surface.

Well Drive Systems

It is convenient to classify oil and gas reservoirs in terms of the type of natural energy and forces available to produce the oil and gas. At the time oil was forming and accumulating in reservoirs, pressure and energy in the gas and salt water associated with the oil were also being stored, which would later be available to assist in producing the oil and gas from the underground reservoir to the surface. Obviously, since oil cannot lift itself from reservoirs to the surface, it is largely the energy in the gas or the salt water (or both), occurring under high pressures with the oil, that furnishes the force to drive or displace the oil through and from the pores of the reservoir into the wells.

In nearly all cases, oil in an underground reservoir has dissolved in it varying quantities of gas that emerges and expands as the pressure in the reservoir is reduced. As the gas escapes from the oil and expands, it drives oil through the reservoir toward the wells and assists in lifting it to the surface. Reservoirs in which the oil is produced by dissolved gas escaping and expanding from within the oil are called *dissolved-gas-drive reservoirs*. See Fig. 19.

Often more gas exists with the oil in a reservoir than the oil can hold dissolved in it under the existing conditions of pressure and temperature in the reservoir. This extra gas, being lighter than the oil, occurs in the form of a cap of gas over the oil. This condition was previously illustrated by Figs. 3 and 4. See also Fig. 20. Such a gas cap is an important additional source of energy because, as production of oil and gas proceeds and as the reservoir pressure is lowered, the gas cap expands to help fill the pore spaces formerly occupied by the oil. Where conditions are favorable, some of the gas coming out of the oil is conserved by moving upward into the gas cap to further enlarge the gas cap. As compared with the dissolved-gas drive, the *gas-cap drive* is more effective, yielding greater recovery of oil. The gas-drive process is typically found with the discontinuous, limited, or essentially closed reservoirs of the types previously shown in Figs. 6 and 7(a). See also Fig. 20.

Where the formation containing an oil reservoir is quite uniformly porous and continuous over a large area, as compared with the size of the soil reservoir per se, very large quantities of salt water exist in surrounding parts of the same formation, often directly in contact with the oil and gas reservoir. This condition is demonstrated by previously shown Figs. 2, 3, 4, and 5. These large quantities of salt water occur under pressure and provide a large additional store of energy to assist in producing oil and gas. A situation like this is termed *water-drive reservoir* and is shown in Fig. 21. The energy supplied by the salt water comes from expansion of the water as pressure in the petroleum reservoir is reduced by production of oil and gas. Water will compress, or expand, to the extent of about one part in 2500 per 100 psi (6.8 atmospheres) change in pressure. Although this effect is slight with reference to small quantities, the phenomenon becomes of importance when changes in reservoir pressure affect large volumes of salt water that are often contained in the same porous formation adjoining or surrounding a petroleum reservoir.

The expanding water moves into the regions of lowered pressure in the oil- and gas-saturated portions of the reservoir caused by production of oil and gas, and retards the decline in pressure. In this way, the expansive energy in the oil and gas is conserved. As shown in Fig. 4, the expanding water also moves and displaces the oil and gas in an upward direction out of the lower parts of the reservoir. By this natural process, the pore spaces vacated by oil and gas produced are filled with water, and oil and gas are progressively moved toward the wells.

The water drive is generally the most efficient oil-production process. Oil fields in which water drive is effective are capable of yielding recoveries ranging up to 50% of the oil originally in place, if (1) the physical nature of the reservoir rock and of the oil are conducive to the process, (2) care is exercised in completing and producing the wells, and (3) the rate of withdrawal of products is optimal.

When pressures in an oil reservoir have fallen to the point where a well will not produce by natural energy, some method of artificial lift must be

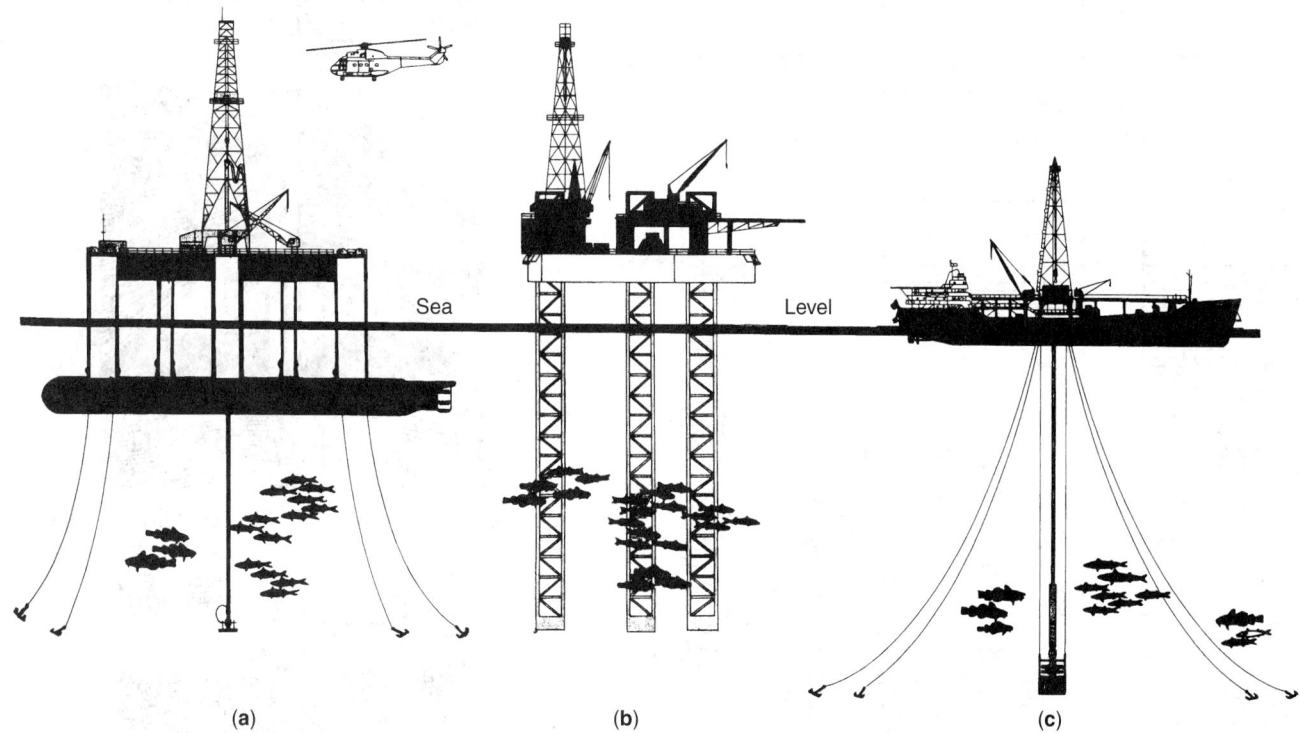

Fig. 17. Offshore drilling schemes: (**a**) Big, pontoon-mounted "semisubmersible" rig that is indispensable for exploratory drilling in rough water; (**b**) self-elevating or *jack-up* drilling rig widely used in water depth of less than 300 feet (91 meters); (**c**) turret mooring allows a drill ship to head into prevailing winds and currents while positioned over the well. Helicopters and boats are required to transport personnel to and from offshore well sites. (*Exxon Corp.*)

used. Oil-well pumps are of three general types: (1) pumps located at the bottom of the hole run by a string of rods, (2) pumps at the bottom of the hole run by high-pressure liquids, and (3) bottom-hole centrifugal pumps. Another method involves the use of high-pressure gas to lift the oil from the reservoir.

Well Completion

Production casing must be set through which the oil and/or gas can be brought safely to the surface. The "pay zone" (productive area) is then sealed off with cement. With the production casing in place, hollow charges are fired through it into the production formation and the drilling mud is gradually displaced, so that the hydrocarbons can flow into the well-bore and up to the surface. There, a "Christmas tree" (an assembly of valves and special connectors) is attached to the top of the production casing. This device controls the flow of oil or gas into the gathering pipelines. See Fig. 22.

Deepwater Production

Two basic types of platforms may be used in deepwater production—*fixed leg* and *compliant*. Each has its advantages and limitations. These facilities can provide all the functions required for drilling, completing, producing, and maintaining conventional wells or a combination of conventional and subsea wells.

Nearly all offshore fields have been developed with fixed-leg platforms. In 1947, for example, a fixed-leg platform weighing 1200 tons was in operation in 20 feet (6 meters) of water out of sight of land. Twenty years later, such platforms weighing in excess of 6500 tons were in use in 340 feet (104 meters) of water. See Fig. 23. To date (1986), the tallest fixed-leg platform, weighing 58,000 tons, is located in 1025 feet (312 meters) of water in the Gulf of Mexico. This platform, completed in 1979, has a total height from seabed to top of the derricks of 1265 feet (386 meters)—taller than the Empire State Building in New York City.

This type of design, known as the "steel jacket," accounts for most of the hundreds of platforms that dot the Gulf of Mexico. For larger oil fields, such as the North Sea, platforms must withstand severe environmental forces, handle large volumes of oil, gas, and water, support heavy equipment, and accommodate 200 to 300 production workers. Here, a favored type is the concrete "gravity platform," so called because its own immense weight pins it to the sea bottom and no piles are needed to secure it. Rigid platforms are impractical in waters much more than 1000 feet

(305 meters) deep. An alternative for deeper water is a "compliant" structure, such as a guyed tower, a slender steel tower held in place by a radial array of anchor cables. Heavy weights attached to the cables lie on the bottom some distance away; these keep the cables taut under normal sea conditions and lift gradually in storms, to allow the tower to tilt slightly to absorb wave forces.

When a platform is not a practical way to develop an offshore field, the operator may "complete" the well using a submerged production system, in which case the Christmas tree and other wellhead equipment are installed on the sea bottom and pipelines are connected to carry off the petroleum, either to shore, to a nearby platform, or to a vessel or storage buoy moored in the area. Divers can be used to make the necessary connections. For deepwater applications, the industry is continuing to develop remote, diverless techniques for installation and maintenance of these completions.

Enhanced Oil Recovery

Enhanced oil recovery (EOR) methods increase ultimate oil production beyond that achievable with primary and secondary methods. This is accomplished by increasing the proportion of the reservoir affected. EOR methods are of three broad groups: (1) thermal, (2) miscible, and (3) chemical.

Thermally Enhanced Recovery. Because oil becomes thinner and flows more easily when it is heated, considerable effort has been devoted to the development of techniques that introduce heat into a reservoir to improve recovery of the heavier, more viscous crude oils. Hot water flooding has been tried, but it is seldom used today because it contains too little heat energy and is very slow to warm the oil and rock surrounding an injection well. More heat is needed for efficiency.

Steam contains the extra heat energy that is required and it has been widely used by the petroleum industry since the mid-1960s to stimulate the production of thick oils. Two techniques, steam stimulation and steam flooding, are currently used.

Steam stimulation or steam soaking uses a well as both injector and producer. High-pressure steam is injected directly into the production zone for several days to weeks. After this period, the reservoir area around the well is allowed to soak in the new heat energy for an additional period. During this time, most of the steam condenses to hot water. After the soak period, the well is brought back into production to recover the heated (thinner) oil and hot water near the wellbore. Because natural driving forces are relied upon to move the oil through the reservoir during the production

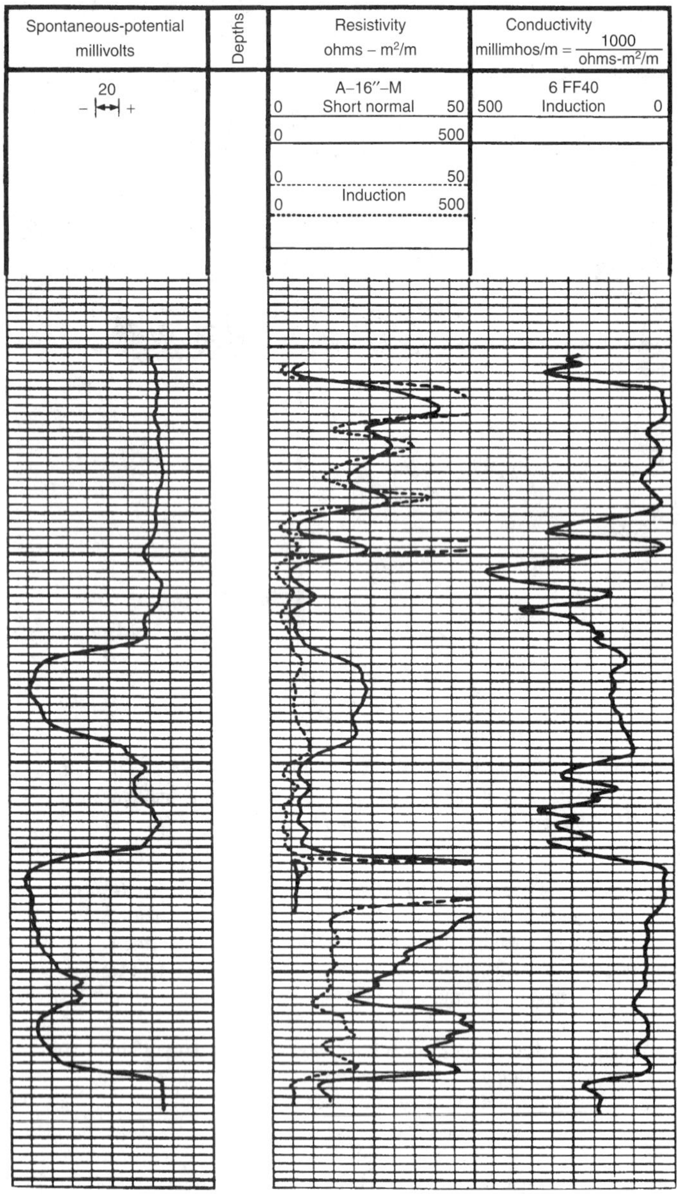

Spontaneous-potential millivolts	Depths	Resistivity ohms – m²/m		Conductivity millimhos/m = $\frac{1000}{\text{ohms-m}^2/\text{m}}$
20 − ⊢•→ +		A–16″–M Short normal 0 50		6 FF40 500 Induction 0
		0 500		
		0 50 Induction		
		0 500		

Fig. 18. Oil well logging provides valuable information on the "down-hole" characteristics of oil and gas wells. Spontaneous potential (SP), electrical resistivity/conductivity logs are frequently recorded simultaneously.

Fig. 19. Dissolved-gas drive. (*Texas Mid-Continent Oil and Gas Association.*)

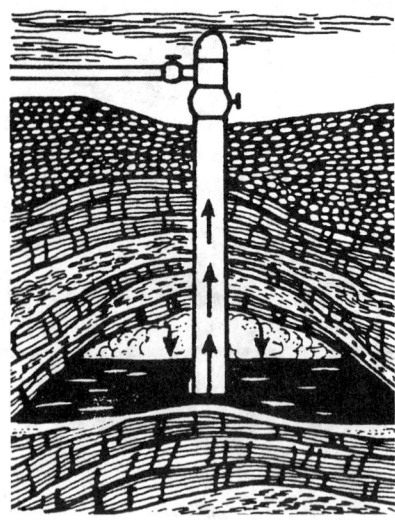

Fig. 20. Gas-cap drive. (*Texas Mid-Continent Oil and Gas Association.*)

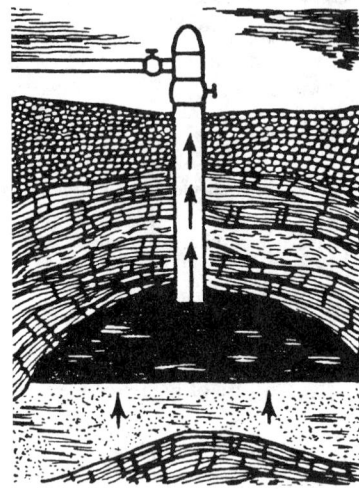

Fig. 21. Water drive. (*Texas Mid-Continent Oil and Gas Association.*)

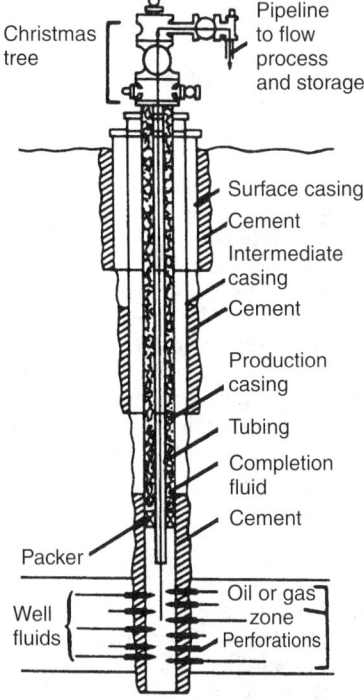

Fig. 22. Completed oil well showing the flow of oil into and up the well to the pipeline connection at the Christmas tree. (*American Petroleum Institute.*)

Fig. 23. Fixed-leg offshore drilling and production platform of the "steel jacket" variety. (*Exxon Corp.*)

phases, steam stimulation generally increases the *rate* of recovery rather than the *amount* of oil that ultimately may be recovered. This technique is particularly adapted to certain California fields containing heavy crude oils, as well as fields in the Orinoco oil belt of Venezuela and the Cold Lake area of Alberta, Canada.

Steam flooding is more sophisticated and difficult than steam stimulation. This technique uses separate injection and production wells to improve both *rate* and *amount* of production.

Miscible Recovery. Oil and water do not mix and they do not flow with equal facility through a porous rock. Over the years, many miscible flood processes have been tested, the most successful of which have been: (1) hydrocarbon miscible recovery; (2) carbon dioxide miscible flooding; and (3) chemically enhanced recovery.

Depending on the composition of the oil and the reservoir temperature and pressure, light hydrocarbons in liquid form, such as liquefied petroleum gas (LPG) and including propane, butane, and ethane, may be miscible with crude oil. Where conditions are right, natural gas can be used to drive a bank of injected light hydrocarbon liquids through the reservoir to form a miscible flood. The disadvantage of this method is that it involves the prolonged use of valuable hydrocarbon liquids, some of which may never be recovered. The use of natural gas alone has received consideration in special situations where high pressure, combined with low reservoir temperature, may result in miscibility.

Carbon dioxide miscible recovery is a preferred method. Carbon dioxide may not be initially miscible with crude oil, but when it is forced into an oil reservoir, some of the smaller, lighter hydrocarbon molecules in the contacted crude will vaporize and mix with the CO_2, forming a wall of enriched gas (CO_2 plus light hydrocarbons). If the temperature and pressure of the reservoir are suitable, this wall of enriched gas will mix with more of the crude, forming a "bank" of miscible solvents capable of efficiently displacing large volumes of crude oil. Carbon dioxide is found in underground deposits and can be produced through wells similar to gas wells. But its production and transportation to the oil reservoir can add

significantly to the cost. A project has been proposed that would involve the construction of a long CO_2 pipeline from a Colorado CO_2 well to a large oil field in Texas. The economic feasibility of the project remains to be proved.

The use of chemicals to coax more oil out of the ground has been investigated for many years. Chemically enhanced methods are of three major types: (1) polymer flooding; (2) surfactant flooding; and (3) alkaline flooding.

Experts estimate that prudent but aggressive application of enhanced recovery technology to known reservoirs in the United States could result in the production of 20 to 30 billion barrels of oil that might otherwise be lost forever. The amount of oil that might be recovered from known oil fields worldwide is estimated to be in the range of 100 to 200 billion barrels. See Fig. 24.

Petroleum Reserves

In 1988, the U.S. Geological Survey reduced its prior estimate of the oil and gas remaining to be discovered in the United States by 40%. As pointed out by R.A. Kerr (reference listed), "If the new estimates hold up, it would solidify a new realism in the agency's view of energy resources. In 1972, the agency claimed that there were 450 billion barrels of oil and 2100 trillion cubic feet of gas left to be found—Figures that the USGS itself soon characterized as four times too high." A revised estimate in 1981 showed 83 billion barrels and 594 trillion cubic feet left undiscovered. Even this estimate was challenged as too high by some industry experts.

The foregoing is exemplary of how assumptions of remaining oil and gas reserves are made and of the lack of confidence that *any* figures appeared to enjoy as of the 1990s. Generally, there is a consensus that, given current rates of consumption, oil reserves should be visualized as becoming exhausted in terms of decades rather than centuries!

Everyone recognizes that crude oil exists in finite amounts, but no one really knows how much of it remains in the earth. Since E.L. Drake drilled his first well in 1859, a vast industry has been established and oil became the predominant worldwide fuel. But petroleum geologists freely admit that definitive knowledge about the resource base that made this growth possible remains *elusive*. At any given time, the amount of oil available for consumption depends primarily on two factors: (1) the producibility of already discovered reserves; and (2) the production policies of governments in countries where those reserves exist. Over the long run, however, it is the amount of recoverable oil left in the ground, including those volumes yet undiscovered, that will be decisive. Estimates of the remaining petroleum resource base vary widely—in fact, too widely to record in this encyclopedia. Experts disagree about the size and producibility of individual reservoirs and about the total national and world reserves associated with already discovered fields. They are even further apart when assessing the world's undiscovered potential. This is not surprising when it is remembered that, in this current unscientific arena, the experts are making judgments—frequently educated guesses—about hydrocarbons contained in porous rocks many thousands of feet under the earth's surface. Often, they have little more to go on than the data from a few widely dispersed 8-inch-diameter holes in existing fields—and still less information in the case of fields expected to be discovered in the future. Consequently, in an attempt to be pseudoscientific, experts generally refer to three classes of reserves: (1) proved; (2) probable; and (3) potential. The chart shown in Fig. 25 represents a well-accepted approach to reserves analysis.

Petroleum Exploration and Production Progress

Some of the major milestones achieved in petroleum exploration and production technology are summarized in Table 3.

Additional Reading

Abelson, P.H.: "Hydrocarbon Energy Revisited," *Science*, 1433 (September 29, 1989).

Ahmed, T.H.: "Reservoir Engineering Handbook," 2nd Edition, Butterworth-Heinemann, Inc., Woburn, MA, 2001.

Arnold, K. and M. Stewart, Jr.: "Surface Production Operations: Design of Oil-Handling Systems and Facilities," Vol. 1, 2nd Edition, Butterworth-Heinemann, Inc., Woburn, MA, 1998.

Arnold, K. and M. Stewart, Jr.: "Surface Production Operations," Butterworth-Heinemann, Inc., Woburn, MA, 1986.

Bethke, C.M., et al.: "Supercomputer Analysis of Sedimentary Basins," *Science*, 261 (January 15, 1988).

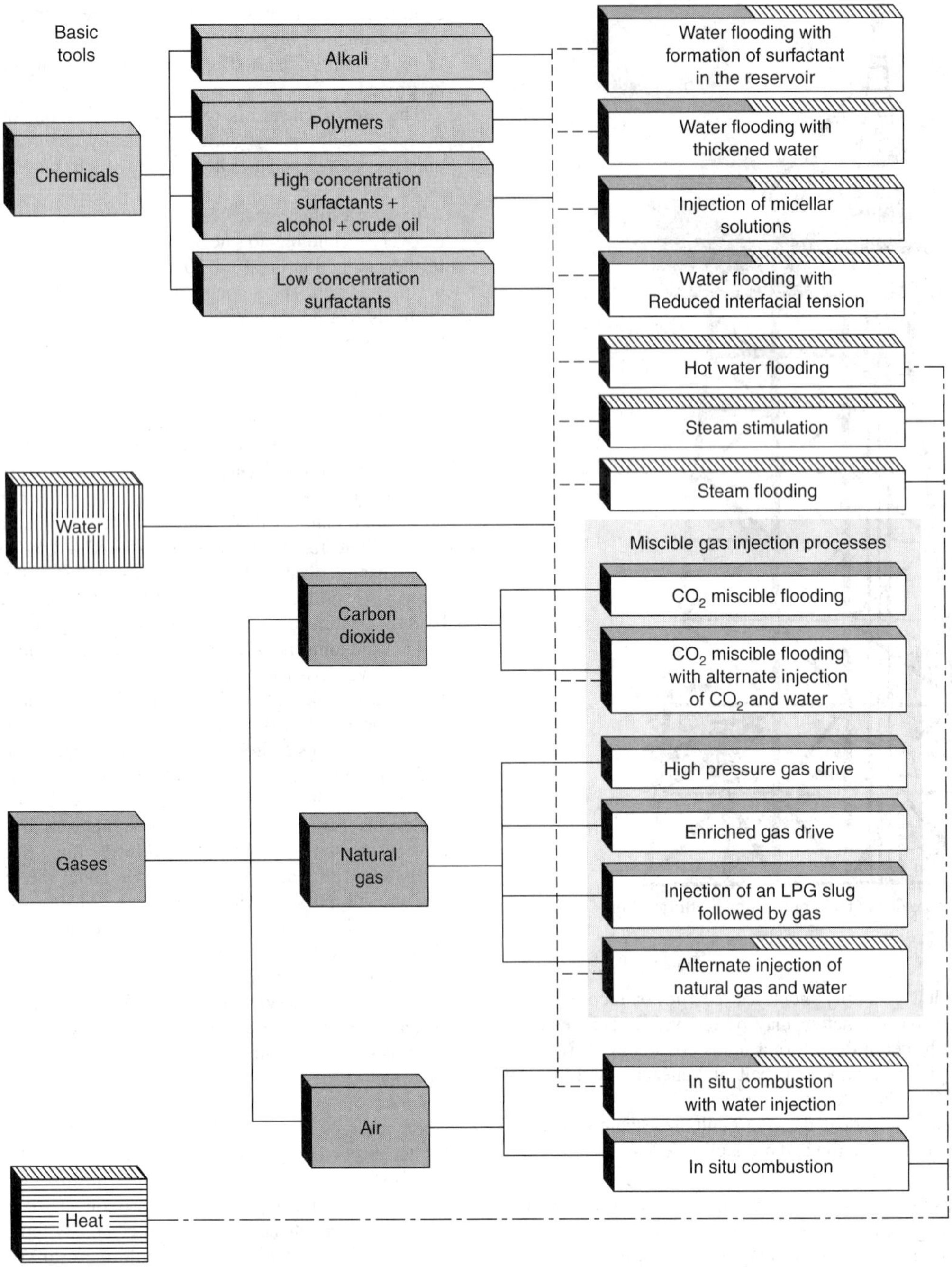

Basic
tools

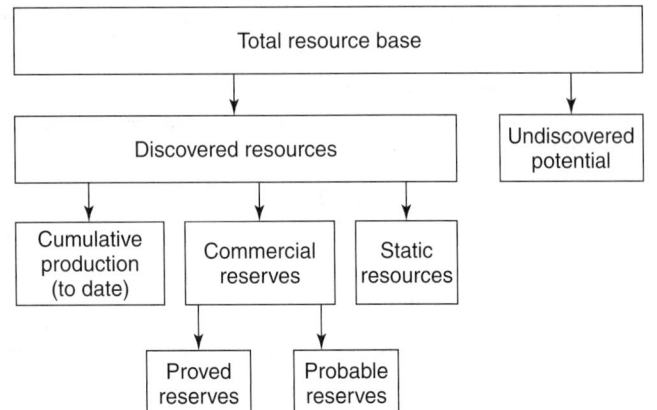

Fig. 24.   Enhanced oil recovery techniques employ heat, gases, chemicals, and water—singly or in combinations—to reduce the factors that inhibit oil recovery and to augment reservoir energy. (*Exxon Corp.*)

Fig. 25.   Commonly used approach for handling statistics in connection with estimating oil reserves.

Dalen, B.: "Computer Control Reduces Offshore Costs, Safety Risks," *Instrumentation Technology*, 22 (December 1989).

Dawe, R.A.: "Modern Petroleum Technology: 2 Volume Set," 6th Edition, John Wiley & Sons, Inc., New York, NY, 2000.

Devereaux, S.: "Practical Well Planning and Drilling Manual," PennWell Publishing Company, Tulsa, OK, 1998.

Economides, M.J., B.N. Murali, and L.T. Watters: "Petroleum Well Construction," John Wiley & Sons, Inc., New York, NY, 1998.

Elliott, D.H.: "Is There Any Oil and Natural Gas (Antarctica)?" *Oceanus*, 32 (Summer 1988).

Esmaeili, H.: "The Legal Regime of Offshore Oil Rigs in International Law," Ashgate Publishing Company, Brookfield, VT, 2001.

Gluyas, J. and R. Swarbrick: "Petroleum Geology," Blackwell Science, Inc., Malden, MA, 2001.

Hapgood, F.: "The Quest for Oil," *National Geographic*, 226 (August 1989).

Hyne, N.J.: "Nontechnical Guide to Petroleum Geology, Exploration, Drilling, and Production," PennWell Publishing Company, Tulsa, OK, 1995.

Kerr, R.A.: "Oil and Gas Estimates Plummet," *Science*, 1330 (September 22, 1989).

Lee, D.B.: "Oil in the Wilderness—An Arctic Dilemma," *National Geographic*, 858 (December 1988).

TABLE 3. MILESTONES IN PETROLEUM EXPLORATION AND PRODUCTION TECHNOLOGY

1853	Dr. Albert Gesner manufactures kerosene from coal.
1859	Edwin Drake completes the first successful well drilled in the search for oil at Titusville, Pennsylvania, striking oil at $69\frac{1}{2}$ feet. By the start of the 20th century, crude oil and/or natural gas were being produced in 20 states; in 1984, 33 of the 50 states had some oil or gas production.
1865	First oil pipeline, 2 inches in diameter and 32,000 feet long, laid at Oil Creek, Pennsylvania, to transport oil from the field to the Oil Creek Railroad.
1883	Dr. I.C. White proposes the theory that oil and gas deposits could be found in geological anticlines.
1896	First "offshore" wells drilled from piers extending into California waters.
1899	Threllfall and Pollock devise the first gravity meter.
1901	"Spindletop" oil field is discovered on a salt dome near Beaumont, Texas; proves the value of the rotary drilling rig and popularizes the use of drilling mud.
1914	Reginal Fessenden patents the reflections seismograph.
1924	Electric well logging first used in United States; refraction seismograph graph used to discover Orchard, Texas, salt dome; first geophysical discovery using magnetic torsion-balance.
1939	First airborne magnetometer developed.
1942	Fluid formation identified using electric logging.
1954	First oil and gas lease sale in federal offshore area held; through 1984, more than 100 such sales had been held, resulting in the leasing of some 38 million acres (4 percent of the federal offshore area); and federal revenues from that leasing had exceeded $77 billion.
1968	The Prudhoe Bay, Alaska, field is discovered some 250 miles above the Arctic Circle — the largest U.S. discovery ever made — containing some 10 billion barrels of oil and 26 trillion cubic feet of natural gas.
1972	First land remote sensing satellite (Landsat) launched; information from such satellites is playing an increasingly important role in identifying from space potential deposits of oil, natural gas and other minerals.
1974	Record-depth exploratory well — a natural gas well — drilled to 31,441 feet in Oklahoma.
1979	World's tallest fixed-leg platform — 1,265 feet tall and weighing 59,000 tons — installed in 1,025 feet of water in the Gulf of Mexico.
1984	Exploratory well drilled in world record water depth — 6,942 feet — off the coast of New England.

Source: American Petroleum Institute.

Longwen, W.: "China's Exploration and Development of Offshore Oil and Gas," *Oceanus*, 32 (Winter 1989/1990).

Lynch, M.C.: "Preparing for the Next Oil Crisis," *Chem. Eng. Progress*, 20 (March 1988).

Lyons, W.C.: "Standard Handbook of Petroleum and Natural Gas Engineering," Vol. 1, 6th Edition, Butterworth-Heinemann, Inc., Woburn, MA, 2001.

Lyons, W.C.: "Standard Handbook of Petroleum and Natural Gas Engineering," Vol. 2, 6th Edition, Butterworth-Heinemann, Inc., Woburn, MA, 1996.

Miall, A.D.: "Principles of Sedimentary Basin Analysis," 3rd Edition, Springer-Verlag, Inc., New York, NY, 1999.

McCain, W.D.: "The Properties of Petroleum Fluids," 2nd Edition, PennWell Publishing Company, Tulsa, OK, 1990.

Nelson, R.C.: "Chemically Enhanced Oil Recovery: The State of the Art," *Chem. Eng. Progress*, 50 (March 1989).

Pate-Cornell, M.E.: "Organizational Aspects of Engineering System Safety: The Case of Offshore Platforms," *Science*, 1210 (November 30, 1990).

Rutledge, G.: "Arctic Oil," *Chem. Eng. Progress*, 6 (October 1989).

Schmidt, R.L.: "Thermal Enhanced Oil Recovery: Current Status and Future Needs," *Chem. Eng. Progress*, 47 (January 1990).

Selly, R.C.: "Elements of Petroleum Geology," 2nd Edition, Morgan Kaufmann Publishers, Orlando, FL, 1997.

Short, J.A.: "Introduction to Directional and Horizontal Drilling," PennWell Publishing Company, Tulsa, OK, 1993.

Staff: "U.S. Oil and Gas Outlook Brightens," *Chem. Eng. Progress*, 6 (December 1989).

Staff: "International Petroleum Encyclopedia," PennWell Publishing Company, Tulsa, OK, 2000.

Staff: "Mobil's Arnold Stancell and the Pursuit of Oil," *Chem. Eng. Progress*, 70 (April 1990).

Tearpock, D.J. and R. Bischke: "Applied Subsurface Geological Mapping," 2nd Edition, Prentice Hall, Inc., Upper Saddle River, NJ, 2002.

Twiss, R.J. and E.M. Moores: "Structural Geology," W. H. Freeman Company, New York, NY, 1995.

Van Der Pluijm, B.A. and S. Marshak: "Earth Structure: An Introduction to Structural Geology and Tectonics," The McGraw-Hill Companies, Inc., New York, NY, 1997.

Web Reference

American Association of Petroleum Geologists: *http://www.aapg.org/*

PETROLEUM FLY. An unusual insect that lives in pools of crude petroleum in the California oil fields. The larvae will live in fresh water, brine, or crude oil. The insect is of the family *Ephydriaae*.

PETROLEUM REFINING. Because crude oils exhibit important differences from the standpoint of processing and final end-products from one geographic resource to the next (see Tables 1 and 2 of the preceding article on **Petroleum**), very few petroleum refineries operate on exactly the same basis. The principles of the operations performed are remarkably similar, but there are basic differences in the amount of throughputs for given products. The latter varies from one geographic area to the next and with the season of the year. Modern petroleum refineries are designed with considerable built-in flexibility, so that the production of petroleum products can be customized for specific market demands. For example, the production of greater amounts of heating fuels during the fall and winter season as contrasted with larger production of gasolines and diesel fuels during summer months, when the demand is greatest. In contrast with a few decades ago, modern refineries are integrated to produce numerous products and often will also serve nearby petrochemical plants where petroleum products become starting ingredients for a vast variety of chemicals used in many industries, such as plastics, solvents, polymers, fibers, among many others.

A representative integrated petroleum refinery is shown in Fig. 1.

The major processing units fundamental to the manufacture of fuel products from crude oil include: (1) crude distillation; (2) catalytic reforming; (3) catalytic cracking; (4) catalytic hydrocracking; (5) alkylation; (6) thermal cracking; (7) hydrotreating; and (8) gas concentration. Refineries also will use numerous auxiliary processes, such as treating units to purify both liquid and gas streams, waste-management and pollution-control systems, cooling-water systems, units to recover hydrogen sulfide (or elemental sulfur) from gas streams, desalters, electric-power stations, steam-producing facilities, and provisions for storage of crude oil and products.

Petroleum refining and petrochemical production is a 24-hour, 365-day operation with a very minimum of time planned for downtime. Unless one has visited a refinery firsthand, it is very difficult to comprehend the size and complexity of the equipment used. See Figures 2, 3, and 4.

Crude Distillation. To minimize corrosion of refining equipment, a crude-oil distillation unit generally is preceded by a *desalter*, which reduces the inorganic salt content of raw crudes. Salt concentrations vary widely (from nearly zero to several hundred pounds, expressed as NaCl per 1,000 barrels). The crude unit functions simply to separate the crude oil physically, by fractional distillation, into components of such boiling ranges that they can be processed by appropriately selected equipment in a long train of processing operations which follow. Although the boiling ranges of components (or fractions) vary between refineries, a typical crude distillation unit will resolve the crude into the following fractions:

By distillation at atmospheric pressure,

1. A light straight-run fraction, consisting primarily of C_5 and C_6 hydrocarbons. These also will contain any C_4 and lighter gaseous hydrocarbons that are dissolved in the crude.
2. A naphtha fraction having a nominal boiling range of $200°-400°F$ ($93°-204°C$).
3. A light distillate with boiling range of $400°-540°F$ ($204°-343°C$).

By vacuum flashing

1. Heavy gas oil, having a boiling range of $650°-1,050°F$ ($343°-566°C$).
2. A nondistillable residual pitch.

In the atmospheric-pressure distillation section of the unit, the crude oil is heated to a temperature at which it is partially vaporized and then introduced near, but at some distance above, the bottom of a distillation column. This cylindrical vessel is equipped with numerous trays through

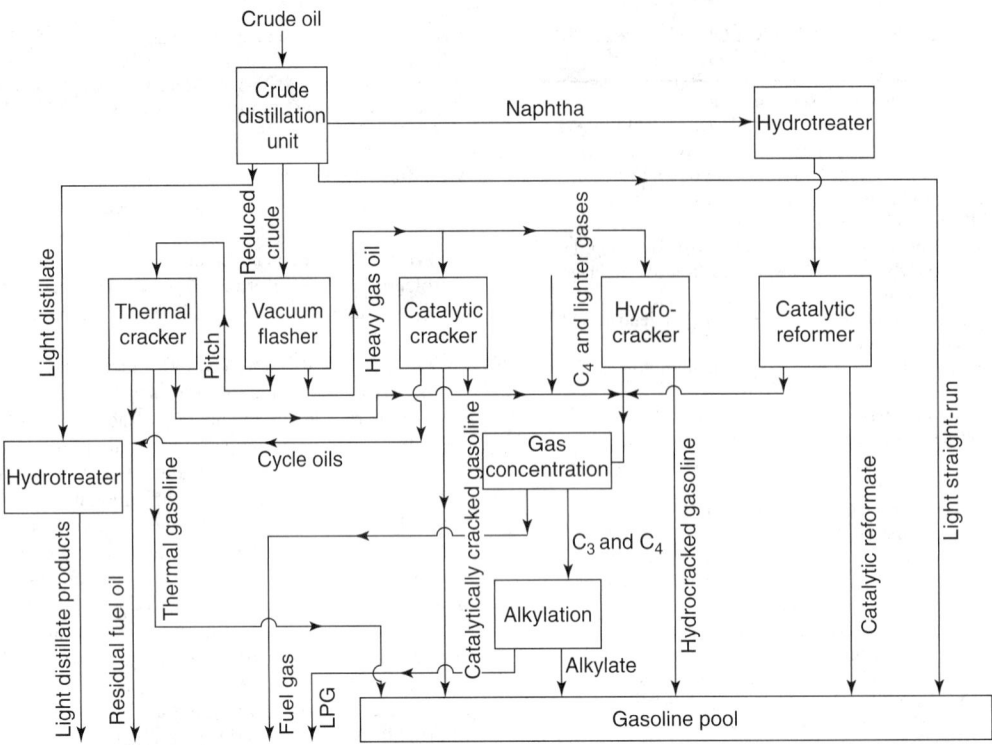

Fig. 1. Principal flow of materials in an integrated petroleum refinery for producing various fuels and raw materials for petrochemical plants.

Fig. 2. Aerial view of small portion of a Texas petroleum refinery, showing the hydrocracking unit just left of center of view.

containing the C_6 and lighter hydrocarbons are withdrawn from the top, while a liquid stream boiling at about 650 °F (343 °C) is taken from the bottom. The portion taken from the bottom is called *atmospheric residue*.

This residue is further heated and introduced into a vacuum column operated at an absolute pressure of about 50 millimeters of mercury, a vacuum maintained by the use of steam ejectors. A flash separation is made to produce heavy gas oil and nondistillable pitch.

The crude oil and atmospheric residue are heated in tubular heaters. Oil is pumped through the inside of the tubes contained in a refractory combustion chamber fired with oil or fuel gas in such manner that heat is transferred through the tube wall in part by convection from hot combustion gases and in part by radiation from the incandescent refractory surfaces.

The light straight-run gasoline contains all hydrocarbons lighter than C_7 in the crude and consists primarily of the native C_5 and C_6 families. After stabilization to remove the C_4 and lighter hydrocarbons (which are routed to a central gas-concentration unit), the stabilized C_5/C_6 blend is treated to remove odorous mercaptans and passed to the refinery gasoline pool for final product blending. The unleaded octane number (Research Method Number) is less than 70, and thus blending or further processing is required to improve its antiknock qualities. Isomerization can be used to improve octane rating, as well as the addition of lead alkyls.

Naphtha to become a suitable component for blending into finished gasoline pools must be further processed. The octane number will range from 40 to 50. Prior to introduction into a catalytic-reforming unit, most naphtha feedstocks are hydrotreated in the interest of prolonging the life of the reforming catalyst.

Gas oil separated from the crude by vacuum distillation, plus portions of light distillates, is the feedstock to catalytic cracking units. The main function of catalytic cracking is to convert into gasoline those fractions having boiling ranges higher than that of gasoline. Remaining uncracked distillates (cycle oils) are used as components for domestic heating fuels (generally after hydrotreating) and to blend with residual fractions to reduce their viscosity to make acceptable heavy fuel oils. In some refineries, cycle oils are hydrocracked to complete their conversion to gasoline.

The aforementioned processes are described in more detail in separate entries: **Alkylation; Cracking Process; Distillation**; and **Hydrotreating**.

Petroleum Terminology

An abridged glossary of terms used to describe petroleum products and processes would include:

Additives, Diesel Fuel — Chemicals for reducing smoke emissions and for cold weather conditioning.

Additive, Gasoline — In some instances, several functions can be combined in one chemical compound to provide "multifunctional" additives. Increasingly stringent control of automotive emissions in several countries has stimulated the development and use of "extended-range" detergents which are designed to promote peak engine performance by maintaining engine cleanliness. Some gasoline additives include:

Antiknock compounds to increase octane number.

Antioxidants to provide gasoline storage stability.

Antirust agents to prevent corrosion in gasoline-handling systems.

Detergents to control carburetor and induction system cleanliness.

Dyes to indicate kind of antiknock compounds used and to identify brand and grade of gasoline.

Viscosity index improves make lube oil more effective over a wide range of temperatures.

Alkylation — A refinery process for chemically combining isoparaffin with olefin hydrocarbons. The product, *alkylate*, has a high octane value and is blended with motor and aviation gasoline to improve the antiknock value of the fuel.

Base Oil — A refined or untreated oil used in combination with other oils and additives to produce lubricants.

Blending — The process of mixing two or more oils having different properties to obtain a final blend having the desired characteristics. This can be accomplished by off-line batch processes or by in-line operations as part of continuous-flow operations.

Bright Stock — High-viscosity, fully refined and dewaxed lubricating oils produced by the treatment of residual stocks and used to compound motor oils.

Catalytic Cracking — A refinery process that converts a high-boiling range fraction of petroleum (gas oil) to gasoline, olefin feed for alkylation, distillate, fuel oil, and fuel gas by use of a catalyst and heat.

Fig. 3.   Erection of 9.5-meter (31-foot) diameter, 263-ton vacuum tower at a petroleum refinery in Kuwait. (*The Fluor Corp.*)

Fig. 4.   Erection of a 675-ton hydrocracking reactor at Shauaiba, Kuwait, refinery, the world's first all-hydrogen refinery. (*The Fluor Corp.*)

which hydrocarbon vapors can pass in an upward direction. Each tray contains a layer of liquid through which the vapors can bubble, and the liquid can flow continuously by gravity in a downward direction from one tray to the next one below. As the vapors pass upward through the succession of trays, they become lighter (lower in molecular weight and more volatile with lower boiling temperature). The liquid flowing downward becomes progressively heavier (higher in molecular weight and less volatile with higher boiling temperature). The countercurrent action results in fractional distillation or separation of hydrocarbons based upon their boiling points. A liquid can be withdrawn from any preselected tray as a net product. Thus, the lighter liquids, such as naphtha, exit from trays near the top of the column, whereas heavier liquids, such as diesel oil, exit from trays near the bottom of the column. Thus, the boiling range of the net product liquid depends upon the tray from which it is taken. The vapors

Catalytic Reforming—A catalytic process to improve the antiknock quality of low-grade naphthas and virgin gasolines by the conversion of naphthenes (such as cyclohexane) and paraffins into higher-octane aromatics (such as benzene, toluene, and xylenes). There are approximately ten commercially licensed catalytic reforming processes, including fully regenerative and continuously regenerative designs.

Cetane Number—The cetane number (C.N.) of a fuel is the percentage by volume of normal cetane in a mixture of cetane and alpha-methylnaphthalene which matches the unknown fuel in ignition quality when compared with a standard diesel engine under specified conditions. The C.N. scale ranges from 0 to 100 C.N. for fuels equivalent in ignition quality to alpha-methylnaphthalene and cetane, respectively. For routine-testing, secondary reference fuels having cetane values of about 25 and 74 are blended in any desired proportion.

Clear Octane—The octane number of a gasoline before the addition of antiknock additives.

Cloud Point—The aniline cloud point is a measure of the paraffinicity of a fuel oil, a high value indicating a straight-run paraffinic oil and a low value indicating an aromatic, a naphthenic, or a highly cracked oil.

Coking—Distillation to dryness of a product containing complex hydrocarbons, which break down in structure during distillation, such as tar or crude petroleum. The residue is called coke.

Cracking—A process carried out in a refinery reactor in which the large molecules in the charge stock are broken up into smaller, lower-boiling, stable hydrocarbon molecules, which leave the vessel as overhead (unfinished cracked gasoline, kerosenes, and gas oils). At the same time, certain of the unstable or reactive molecules in the charge stock combine to form tar or coke bottoms. The cracking reaction may be carried out with heat and pressure (thermal cracking) or in the presence of a catalyst (catalytic cracking).

Cycle Stock—Unfinished product taken from a stage of a refinery process and recharged to the process at an earlier period in the operation.

Deasphalting—Process for removing asphalt from petroleum fractions, such as reduced crude. A common deasphalting process introduces liquid propane, in which the nonasphaltic compounds are soluble while the asphalt settles out.

Desulfurization—The removal of sulfur or sulfur-bearing compounds from a hydrocarbon by any one of a number of processes, such as hydrotreating.

Distillate—That portion of a liquid which is removed as a vapor and condensed during a distillation process. As fuel, distillates are generally within the 400° to 650°F (204° to 343°C) boiling range and include Nos. 1 and 2 fuel, diesel, and kerosene.

End Point—The temperature at which the last portion of oil has been vaporized in ASTM or Engler distillation. Also called the final boiling point.

Equilibrium Volatility of a Gasoline—The volatility of a gasoline is determined by the Reid vapor pressure and the ASTM distillation data. The Reid vapor pressure is the vapor pressure of a gasoline at 100°F (37.8°C) under specified conditions. The distillation curve of a fuel indicates the temperatures at which the various amounts of a given sample are distilled under specified test conditions. However, gasoline will completely evaporate in the presence of air at a temperature much lower than the end-point of the distillation curve. According to O.C. Bridgeman (U.S. National Bureau of Standards, Research Paper 694), the volatility of a gasoline is the temperature at which a given air-vapor mixture is formed under equilibrium conditions at a pressure of one atmosphere, when a given percentage is evaporated. According to this definition, one gasoline is more volatile than another for any given percentage evaporated if it forms the given air-vapor mixture at a lower temperature. Distillation temperature curves, for a given text sample, plot amount of sample distilled over (percentage of sample) at the time a given temperature has been reached.

Fire Point—The lowest temperature at which a fuel ignites and burns for at least 5 seconds under specified test conditions.

Flare—A device for disposing of gases by burning.

Flash Point—The lowest temperature at which a flash appears on the fuel surface when a test flame is applied under specified test conditions. This property is an approximate indication of the tendency of the fuel to vaporize.

Flue Gas Expander—A turbine used to recover energy where combustion gases are discharged under pressure to the atmosphere. The pressure reduction drives the impeller of the turbine.

Fractions—Refiner's term for the portions of oils containing a number of hydrocarbon compounds but within certain boiling ranges, separated from other portions in fractional distillation. They are distinguished from pure compounds which have specified boiling temperatures, not a range.

Fuel Oils—Any liquid or liquifiable petroleum product burned for the generation of heat in a furnace or firebox or for the generation of power in an engine. Typical fuels include clean distillate fuel for home heating and higher-viscosity residual fuels for industrial furnaces.

Gas Oil—A fraction derived in refining petroleum with a boiling range between kerosene and lubricating oil.

Heating Oils—A trade term for the group of distillate fuel oils used in heating homes and buildings as distinguished from residual fuel oils used in heating and power installations. Both are burned-fuel oils.

Heavy Ends—The highest-boiling portion of a gasoline or other petroleum oil.

Hydrocracking—The cracking of a distillate or gas oil in the presence of catalyst and hydrogen to form high-octane gasoline blending stock.

Hydrogenation—A refinery process in which hydrogen is added to the molecules of unsaturated (hydrogen-deficient) hydrocarbon fractions. It plays an important part in the manufacture of high-octane blending stocks for aviation gasoline and in the quality improvement of various petroleum products.

Hydrotreating—A treating process for the removal of sulfur and nitrogen from feedstocks by replacement with hydrogen.

Isomerization—A refining process which alters the fundamental arrangement of atoms in the molecule. Used to convert normal butane into isobutane, as alkylation process feedstock, and normal pentane and hexane into isopentane and isohexane, high-octane gasoline components.

Kinematic Viscosity—The absolute viscosity of a liquid (in centipoises) divided by its specific gravity at the temperature at which the viscosity is measured.

Knock—The sound or "ping" associated with the autoignition in the combustion chamber of an automobile engine of a portion of the fuel-air mixture ahead of the advancing flame front.

Lead Susceptibility—The increase in octane number of gasoline imparted by the addition of a specified amount of tetraethyl lead.

Low-Sulfur Crude Oil—Crude oil containing low concentrations of sulfur-bearing compounds. Crude is usually considered to be in the low-sulfur category if it contains less than 0.5% (weight) sulfur. Examples of low-sulfur crudes are offshore Louisiana, Libyan, and Nigerian crudes.

Lube Stock—Refinery term for fraction of crude petroleum suitable in terms of boiling range and viscosity to yield lubricating oils when further processed and treated.

Mercaptans—Compounds of sulfur having a strong, repulsive, garlic-like odor. A contaminant of "sour" crude oil and products.

Octane Number—The octane rating of a motor fuel is defined in terms of its knocking characteristics relative to those of blends of isooctane (2,3,4-trimethylpentane) and n-heptane, and a rating of 100 to isooctane. The octane number of an unknown fuel is numerically equal to the volume percent of isooctane in a blend with *n*-heptane which has the same knocking tendency as the unknown fuel when both the unknown and the reference blend are run in a standard single-cylinder engine operated at specified conditions. Motor Method octane numbers are measured at more severe engine conditions and are numerically lower than those determined by the milder Research Method. The difference between the two numbers is termed *sensitivity*.

Polymer—A product of polymerization of normally gaseous olefin hydrocarbons to form high-octane hydrocarbons in the gasoline boiling range. Polymerization is the process of combining two or more simple molecules of the same type, called monomers, to form a single molecule having the same elements in the same proportions as in the original molecule, but having different molecular weights. The combination of two or more dissimilar molecules is known as copolymerization—and the product is called a *copolymer*.

Pour Point—This property is defined as the lowest temperature at which the fuel will pour and is a function of the composition of the fuel.

Normally, the pour point of a fuel should be at least 10 to 15 degrees below the anticipated minimum use temperature.

Presulfide — A step in the catalyst regeneration procedure which treats the catalyst with a sulfur-bearing material such as hydrogen sulfide or carbon bisulfide to convert the metallic constituents of the catalyst to the sulfide form in order to enhance its catalytic activity and stability.

Process Unit — A separate facility within a refinery, consisting of many types of equipment, such as heaters, fractionating columns, heat exchangers, vessels, and pumps, designed to accomplish a particular function within the refinery complex. For example, the crude processing unit is designed to separate the crude into several fractions, while the catalytic reforming unit is designed to convert a specific crude fraction into a usable gasoline blending stock.

Raffinate — In solvent refining, that portion of the oil that remains undissolved and is not removed by the selective solvent.

Refinery Pool — An expression for the mixture obtained if all blending stocks for a given type of product were blended together in production ratio. Usually used in reference to motor gasoline octane rating.

Refluxing — In fractional distillation, the return of part of the condensed vapor to the fractionating column to assist in making a more complete separation of the desired fractions. The material returned is called *reflux*.

Residual Fuel Oils — Topped crude petroleum or viscous residuums obtained in refinery operations. Commercial grades of burner-fuel oils Nos. 5, and 6 are residual oils and include Bunker fuels.

Riser Cracking — Applied to fluid catalytic cracking units where the mixture of feed oil and hot catalyst is continuously fed into one end of a pipe (riser) and discharges at the other end where catalyst separation is accomplished after the discharge from the pipe. There is no dense phase bed through which the oil must pass because all the cracking occurs in the inlet pipe (riser).

Road Octane — A numerical value based upon the relative antiknock performance of an automobile with a test gasoline as compared with specified reference fuels. Road octanes are determined by operating a car over a stretch of level road or on a chassis dynamometer under conditions simulating those encountered on the highway.

SAE Numbers — A classification of motor, transmission, and differential lubricants to indicate viscosities, standardized by the Society of Automotive Engineers. They do not connote quality of the lubricant.

Smoke Point — The smoking tendency of a fuel is indicated by this value, which is the maximum height of a specified type of flame in a given wick lamp that results in no visible smoke.

Solvent Extraction — The process of mixing a petroleum stock with a selected solvent, which preferentially dissolves undesired constituents, separating the resulting two layers, and recovering the solvent from the raffinate (the purified fraction) and from the extract by distillation.

Sour Crude — Crude oil which (1) is corrosive when heated, (2) evolves significant amounts of hydrogen sulfide on distillation, or (3) produces light fractions which require sweetening. Sour crudes usually, but not necessarily, have high sulfur content. Examples are most West Texas and Middle East crudes.

Specific Gravity — The specific gravity of a petroleum fuel is the ratio of the weight of a given volume of the product at 60 °F to the weight of an equal volume of distilled water at the same temperature, both weights corrected for air buoyancy. The relation between API gravity scale and specific gravity is: °API = 141.5/(sp gr 60/60 °F) − 131.5.

Stability — In petroleum products, the resistance to chemical change. Gum stability in gasoline means resistance to gum formation while in storage. Oxidation stability in lubricating oils and other products means resistance to oxidation to form sludge or gum in use.

Sweet Crude — Crude oil that (1) is not corrosive when heated, (2) does not evolve significant amounts of hydrogen sulfide on distillation, and (3) produces light fractions which do not require sweetening. Examples are offshore Louisiana, Libyan, and Nigerian crudes.

Tricresyl Phosphate (TCP) — Colorless to yellow liquid used as a gasoline and lubricant additive and plasticizer. Formula, $PO(OC_6H_4Ch_3)_3$.

Tetraethyl Lead (TEL) — A volatile lead compound which is added to motor and aviation gasoline to increase the antiknock properties of the fuel. $Pb(C_2H_5)_4$. The use of this compound has diminished in recent years because of pollution regulations.

Thermal Cracking — A refining process which decomposes, rearranges, or combines hydrocarbon molecules by the application of heat without the aid of a catalyst.

Topped Crude — A residual product remaining after the removal, by distillation or other processing means, of an appreciable quantity of the more volatile components of crude petroleum.

Unsaturates — Hydrocarbon compounds of such molecular structure that they readily pick up additional hydrogen atoms. Olefins and diolefins, which occur in cracking, are of this type.

Vacuum Distillation — Distillation under reduced pressure, which reduces the boiling temperature of the material being distilled sufficiently to prevent decomposition or cracking.

Vapor Lock — The displacement of liquid fuel in the feed line and the interruption of normal motor operation, caused by vaporization of light ends in the gasoline. Vaporization occurs when the temperature at some point in the fuel system exceeds the boiling points of the volatile light ends.

Virgin Stock — Oil processed from crude oil which contains no cracked material. Also called straight-run stock.

Visbreaking — Lowering or breaking the viscosity of residuum by cracking at relatively low temperatures.

Viscosity — This is generally expressed in terms of the time required for a given quantity of fuel to flow through a capillary tube under specified conditions. Kinematic viscosity v is viscosity divided by mass density, or $v = \mu\rho$. The unit in cgs units is called the *stoke*; a customary unit is the *centistoke* ($\frac{1}{100}$ of a stoke). The value of the kinematic viscosity in (cm^2/seconds) can be obtained from the indications in seconds t of various viscometers by:

$$\text{Saybolt Universal, when } 32 < t < 100 = 0.00226t - 1.95/t$$

$$\text{Saybolt Universal, when } t > 100 = 0.00220t - 1.35/t$$

$$\text{Saybolt Furol, when } 25 < t < 40 = 0.0224t - 1.84/t$$

$$\text{Saybolt Furol, when } t > 40 = 0.0215t - 0.50/t$$

Yield — In petroleum refining, the percentage of product or intermediate fractions based on the amount charged to the processing operation.

Zeolitic Catalyst — Since the early 1960s, modern cracking catalysts contain a silica-alumina crystalline structured material called zeolite. This zeolite is commonly called a molecular sieve. The admixture of a molecular sieve in with the base clay matrix imparts desirable cracking selectivities.

Crude Oil Pipelines

Transportation of crude to petroleum refineries is essentially a part of the refining operation. Petroleum movement can be measured in barrels, tons moved, barrel-miles, ton-miles, etc. Because oil pipelines are the most economical means of moving large volumes of petroleum overland for long distances, ton-miles (movement of one ton over one mile) is the preferred unit. About 740 billion ton-miles per year of crude oil movement by all transportation modes is required to meet the needs of the United States. About half of this quantity is moved by pipeline; water carriers account for most of the remainder. Major interstate crude oil pipelines are shown in map (Fig. 5).

As domestic crude oil is produced, pipeline gathering systems collect and move it to central locations by means of low-pressure, small-diameter pipelines. Usually, these gathering lines feed into pipeline working tanks where the oil is held until it is ready for shipment by a crude oil *trunk line*. Gathering systems include pumping stations, meters, and samplers. About 90% of the gathering lines in the United States are 6 inches (diameter) or smaller. About 34% of all oil pipeline mileage in the United States consists of gathering lines. Crude oil trunk lines, larger in diameter, receive crude oil directly from gathering systems and also from barges and tankers. Crude oil trunk lines range from 8 to 56 inches in diameter and account for 30% of total oil pipeline mileage in the nation. These trunk lines generally originate in the major oil producing areas of the nation and terminate at the main refinery complexes. The largest of these refinery complexes is in the Gulf Coast area of Texas and Louisiana. Other major complexes are located in the St. Louis-Chicago area, northern Ohio, the East and West Coasts.

Petroleum Processing Progress

Some of the major milestones achieved in petroleum processing technology are summarized in Table 1.

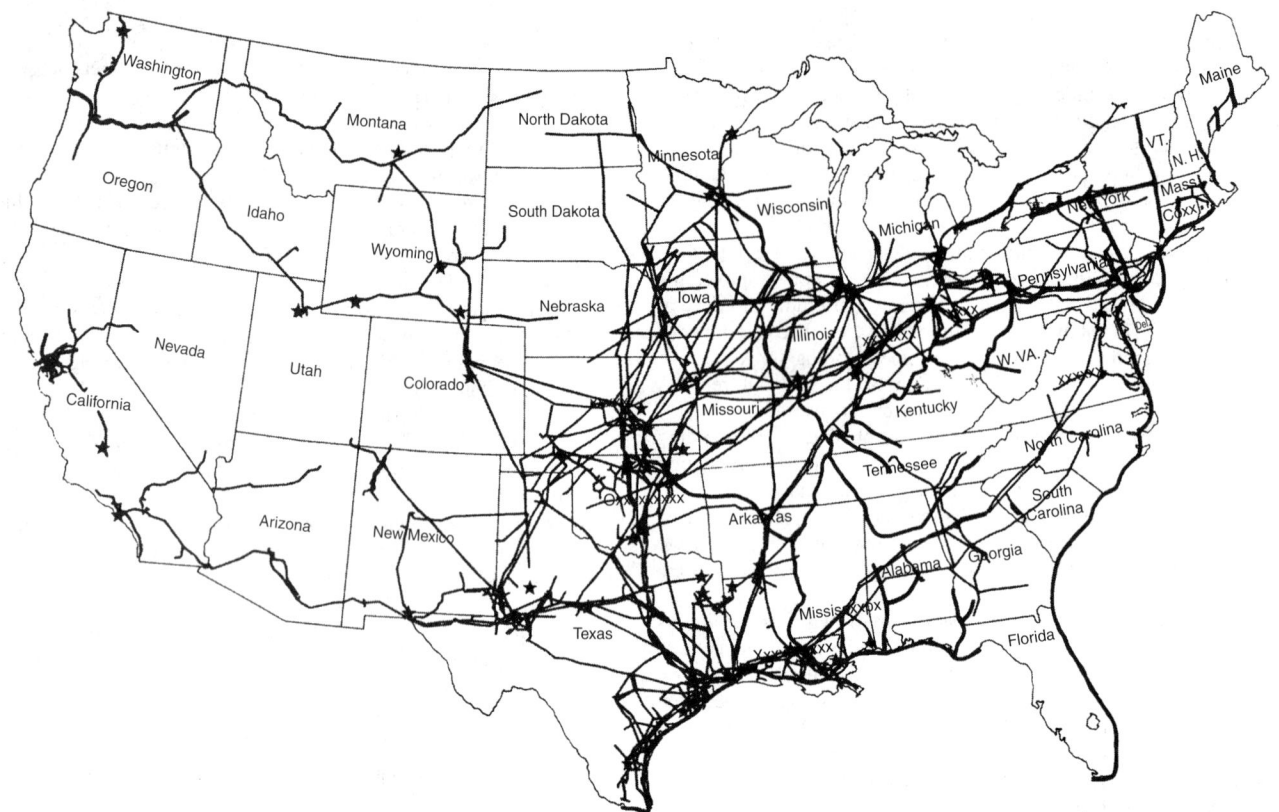

Fig. 5. Major interstate pipelines and waterways in the United States. Pipelines in Alaska not shown. Asterisks indicate principal refineries. (*American Petroleum Institute.*)

TABLE 1. CHRONOLOGY OF MAJOR PETROLEUM INDUSTRY ADVANCEMENTS

Challenge	Action
1. Better gasoline quality and greater quantity needed	1. Thermal cracking processes (about 1910)
2. Need to improve odor and stability of gasoline and kerosene	2. Refining with chemical solutions and synthesis and use of oxidation inhibitors; started in late 1920s
3. Better gasoline quality and greater quality needed	3a. Discovery of tetraethyl lead (1921)
	b. Polymerization of light olefins to make "poly" gasoline by catalysis (mid-1930s)
	c. Catalytic cracking invented and improved (late 1930s)
4. Combat grade aviation gasoline testing above 100 octane needed for World War II	4. Alkylation of light olefins with light isoparaffins by catalysis; discovered in 1932; commercialized in early 1940s
5. More aromatic hydrocarbons needed, especially toulene for TNT; benzene, toluene, and other high-octane aromatics needed for combat grade aviation gasoline and for chemical synthesis	5a. Catalytic reforming to make toluene from petroleum naphthas (early 1940s), using non-noble metal catalyst
	b. Extractive distillation of toluene from reformate with phenol and other materials (early 1940s)
	c. Extraction with SO_2, suggested in 1907 to purify kerosene, applied to secure aromatics from reformate (early 1940s)
	d. Alkylation of propylene with benzene using solid H_3PO_4 catalyst to make cumene (early to mid-1940s)
6. Butadiene needed for synthetic rubber in wartime	6. Thermal and catalytic processing applied to petroleum distillates, "quickly" butadiene program (early 1940s)
7. More isobutane for alkylation in wartime aviation-gasoline program	7. Isomerization of n-butane (early 1940s)
8. Improve quality of straight-run gasoline	8. Catalytic reforming using noble-metal catalyst (1949)
9. Remove catalyst poisons and sulfur compounds from gasoline and naphtha	9. Catalytic hydrotreating (early 1950s)
10. Increase supplies of pure aromatic hydrocarbons and aromatic concentrates	10. Liquid-liquid solvent extraction processes using aqueous glycols and improved contacting means (1952)
11. Purify kerosenes and light and heavy distillates	11. Modified catalytic hydrotreating (mid-1950s)
12. Improve quality of light hydrocarbons used in gasoline	12. New catalytic isomerization processes using noble-metal catalysts, converting C_4, C_5, and C_6 n-paraffins to isoparaffins (mid-1950s)
13. Increase production of light fuels and gasoline; reduce production of heavy fuels	13. Development of catalytic hydrocracking processes having great flexibility (1959-1960)
14. Ethylbenzene for styrene manufacture	14. Catalytic alkylation process developed, uniting benzene directly with dilute ethylene in refinery gases (1958)
15. Separation of normal paraffins from mixtures with isoparaffins	15. Molecular sieves used as solid adsorbents (1959, but not commercialized until late 1960s)
16. Increase benzene supply and decrease toluene	16. Hydrodealkylation of toluene; produce naphthalene from alkyl naphthalenes (early 1960s)

TABLE 1. (*Continued*)

Challenge	Action
17. Synthesize cyclohexane for nylon	17. Catalytic hydrogenation of benzene (early 1960s)
18. Improve quality of heavy fuel oils	18. Hydrodesulfurization of heavy fuels, also by hydrocracking (mid-1960s)
19. Biodegradable synthetic detergents	19. Development of processes of dehydrogenate *n*-paraffins to *n*-olefins and alkylate benzene with them (mid-1960s)
20. Increase production of *p*-xylene	20. Isomerization of C_8 aromatics to *p*-xylene (late 1950s)
21. Improve supplies of individual pure xylene isomers	21. Adsorptive separation of *p*-xylene in high yield and purity, making possible separation of other isomers by precise fractionation (early 1970s)
22. Utilization of metals-containing heavy petroleum fractions	22. Development of hydroprocessing techniques to effectively convert to synthetic crude oils
23. Increase supply of liquid fuels in the face of declining petroleum reserves	23. Development of processes for producing liquid and gaseous fuels from coal, shale oil, and tar sands

Source: Universal Oil Products Company.

Additional Reading

Abraham, O.C. and F.G. Prescott: "Make Isobutylene from Tertiary Butyl Alcohol," *Hydrocarbon Processing*, 51 (February 1992).

Ansari, R.M. and M.O. Tade: "Nonlinear Model-Based Process Control: Applications in Petroleum Refining," Springer-Verlag, Inc., New York, NY, 2000.

Chang, E.J. and S.M. Leiby: "Ethers Help Gasoline Quality," *Hydrocarbon Processing*, 41 (February 1992).

Chaput, G., et al.: "Pretreat Alkylation Feed," *Hydrocarbon Processing*, 51 (September 1992).

Dawe, R.A.: "Modern Petroleum Technology: 2 Volume Set," 6th Edition, John Wiley & Sons, Inc., New York, NY, 2000.

Desai, P.H., et al.: "Enhance Gasoline Yield and Quality," *Hydrocarbon Processing*, 51 (November 1992).

Devlin, J.F., L.A. Edwards, and J.E. Crosby: "Fluid Coker Benefits from Advanced Control," *Hydrocarbon Processing*, 55 (June 1992).

Elliott, J.D.: "Maximize Distillate Liquid Products," *Hydrocarbon Processing*, 75 (January 1992).

Gary, J.H. and G.E. Handwerk: "Petroleum Refining: Technology and Economics," 4th Edition, Marcel Dekker, Inc., New York, NY, 2001.

Johansen, T., K.S. Raghuraman, and L.A. Hackett: "Trends in Hydrogen Plant Management," *Hydrocarbon Processing*, 119 (August 1992).

Jones, D.S.: "Elements of Petroleum Processing," John Wiley & Sons, Inc., New York, NY, 1995.

Magee, J.S. and G.E. Dolbear: "Petroleum Catalysis in Nontechnical Language," PennWell Publishing Company, Tulsa, OK, 1998.

McKetta, J.J.: "Petroleum Processing Handbook," Marcel Dekker, Inc., New York, NY, 1992.

Meyers, R.A.: "Handbook of Petroleum Refining Processes," 2nd Edition, The McGraw-Hill Companies, Inc., New York, NY, 1996.

Monfils, J.L., et al.: "Upgrade Isobutane to Isobutylene," *Hydrocarbon Processing*, 47 (February 1992).

Nierlich, F.: "Oligomerize for Better Gasoline," *Hydrocarbon Processing*, 45 (February 1992).

Reynolds, B.E., E.C. Brown, and A. Silverman: "Clean Gasoline Via Vacuum Residuum Hydrotreating and Residium Fluid Catalytic Cracking," *Hydrocarbon Processing*, 43 (April 1992).

Speight, J.G.: "Petroleum Chemistry and Refining," Taylor & Francis, Inc., Philadelphia, PA, 1997.

Speight, J.G. and B. Ozum: "Petroleum Refining Processes," Vol. 85, Marcel Dekker, Inc., New York, NY, 2001.

Speight, J.G.: "The Chemistry and Technology of Petroleum," 3rd Edition, Marcel Dekker, Inc., New York, NY, 1999.

Staff: "Refining Handbook, '92," *Hydrocarbon Processing*, 133 (November 1992).

Staff: "What's Ahead in 1993?" *Hydrocarbon Processing*, 33 (December 1992).

Wagner, E.S. and G.F. Froment: "Steam Reforming Analyzed," *Hydrocarbon Processing*, 69 (July 1992).

Wauguier, J.P.: "Petroleum Refining: Crude Oil, Petroleum Products, Process Flowsheets," Gulf Publishing Company, Houston, TX, 2000.

Wheatcroft, G.: "Present and Future Refinery Information Systems," *Hydrocarbon Processing*, 101 (May 1992).

Web Reference

American Petroleum Institute: *http://api-ec.api.org/intro/index_noflash.htm*

PETROLOGY. That branch of geology that deals with the rocks forming the lithosphere or "crust" of the earth. The term is derived from the Greek, meaning rock and reason, hence, a more comprehensive term than petrography, which deals only with systematic descriptions of rocks, including their mineral composition, texture, structure, and occurrence. Petrology includes petrography and also methods of classification as founded on systematic description and genetic theories, both experimental and theoretical. Important branches of petrology are geochemistry and geophysics.

PETROLOGY (Structural). See **Structural Geology**.

PETZVAL CONDITION. To eliminate the aberration of curvature of field, at least two lenses must be used which are so related that they satisfy the Petzval condition; $f_1 n_1 + f_2 n_2 = 0$, where the subscripts refer to the two lenses, respectively, n is the refractive index and f is the focal length.

PETZVAL SURFACE. By changing the separation of a doublet lens, a condition free from astigmatism may be found, but the field will not then be flat. The single, paraboloidal surface over which point images of point objects are formed is called the Petzval surface. See also **Curvature of Field (Optics)**.

PEWTER. See **Antimony**.

PHAGE. See **Virus**.

PHAGOCYTE. A cell of the multicellular animal's body capable of ingesting foreign particles. In the sponges and coelenterates, and to a limited extent in other more complex animals, digestion is carried on wholly or in part by this process. In many animals bacteria and particles of dead tissues or cells are engulfed by such cells. Phagocytes are amoeboid in action and some of them move about in the tissue to which they belong by this means. Others are in fixed tissues, carrying on their amoeboid processes at the free end only. Phagocytes may be a part of the endodermal lining of the enteric cavity or wandering cells derived from endoderm, or they may be mesodermal. In the connective tissues and blood, mesodermal phagocytes occur, and in the lining of the circulatory system cells may act as phagocytes.

Large phagocytes are called macrophages, a term which includes wandering cells of the connective tissues. The opposite, microphage, is rarely met. In the blood of vertebrates all white cells (leukocytes) are phagocytic to some extent, the lymphocytes least of all.

PHALANX (plural Phalanges). The small bones of the toes and fingers of vertebrates.

PHALAROPE. See **Waders, Shorebirds, and Gulls**.

PHANTOM CIRCUIT. Two metallic communication circuits can be made to do the work of three by the addition of certain equipment. Since the third circuit has no wires definitely set aside to its use, it is called a phantom circuit.

PHARYNGITIS. A disease of some part of the throat and sometimes referred to simply as sore throat, tonsilitis, or strep throat, depending upon the specific site and severity of the condition. Pharyngitis is one of the most commonplace of upper respiratory tract infections, yet some cases can be complex and difficult to precisely diagnose immediately. It is estimated that from 70 to 80% of the cases arise from a variety of viruses; from 20 to 30% of cases from Group A streptococci, but other microorganisms can less frequently be involved. Simple acute pharyngitis is self-limiting and usually appears rather suddenly with a feeling of dryness and soreness in the throat. There may be a constant desire to clear the throat, pain on swallowing, headache, a dry, harsh cough, and an elevated temperature of 100–102 °F (37.8–39.9 °C). A generalized feeling of fatigue is usually present. Occasionally, there is pain in the ears, and if the infection spreads to the voice box, hoarseness will result. In children, the symptoms are usually more pronounced. The disease runs its course in a few days to a week. If the pharyngitis is thought to be caused by an organism sensitive to an antibiotic, such may be prescribed. Where the cause is definitely diagnosed as Group A streptococci, antibiotics are frequently administered as a preventive measure and to limit droplet spread of the organisms. In as much as rheumatic fever may follow an infection with Group A streptococci, early and effective treatment of respiratory tract infections, including pharyngitis, is a precaution against initial attacks of rheumatic fever. During untreated epidemics of streptococcal pharyngitis, surveys have shown that the incidence of rheumatic fever among those infected may reach 2%. Groups C and G streptococci also may cause pharyngitis, but such infections are not known to predispose rheumatic fever or other complications, such as glomerulonephritis.

Chronic pharyngitis is the result of repeated attacks of acute pharyngitis. Enlarged tonsils and adenoids may cause the condition, as well as constant breathing through the mouth. This form is frequently associated with chronic colds, sinusitis, and nasal infections.

In *septic sore throat* (strep or streptococcal throat), the symptoms are those of an acute pharyngitis, but more severe. A pseudomembrane often appears in the throat. The patient's temperature may rise as high as 105 °F (40.6 °C). The lymph nodes in the upper part of the neck enlarge and become sore. Antibiotics may be used.

Trench mouth (Vincent's angina) is characterized by a pseudomembrane, grayish or yellow-gray in appearance and often spreading from the gums. It is an acute inflammatory disease of the gums, which may be accompanied by pain, bleeding, and offensive breath, as well as fever.

Pharyngeal diphtheria is described in the entry on **Diphtheria**.

Peritonsillar abscess (quinsy sore throat) is caused by an abscess in the tissue surrounding the tonsil. The first symptoms are those of acute pharyngitis, but after a few days, the pain becomes localized in one side of the throat. Swallowing or expectorating becomes extremely painful. The abscess is a complication of strepotococcal tonsilitis and usually occurs in adolescents and young adults. Treatment of this condition consists of parenteral penicillin and surgery. A tonsilectomy is usually prescribed from 4 to 6 weeks after the initial phases of the condition, although earlier tonsilectomy has also been reported as providing excellent results.

Gonococci can produce pharyngitis in persons who practice orogenital sexual activity. Penicillin and tetracycline regimens generally are effective — spectinomycin is not. Although pneumococci and staphylococci commonly reside in the mouth and do cause diseases in other parts of the respiratory tract, they are seldom implicated in pharyngitis. In malnourished infants, a rare but serious invasive gangrene (*cancrum oris*) can occur. This condition is caused by a combination of bacteria and spirochetes. The spirochete *Treponema pallidum* may cause pharyngitis in primary or secondary syphilis.

PHARYNX. A muscular portion of the alimentary tract (digestive system) of invertebrates of various phyla, between the mouth and the esophagus. In some species, it acts as a suctorial organ. In vertebrates, the pharynx is a region of the endodermal part of the gut just behind its union with the ectoderm of the oral cavity. It is associated with the respiratory system. In humans, the pharynx is important as the source of ductless glands and is located under the soft palate and in back of the tongue. See also **Digestive System (Human)**.

PHASE DIAGRAM (Metallurgy). A graphical representation defining the phase fields of a multiphase system, such as an alloy, in a coordinate system using the temperature and the compositions of the phases as coordinates. A phase diagram may be an equilibrium diagram, but it may also sometimes show the boundaries of the phase field under nonequilibrium conditions corresponding to specific conditions of heating or cooling. See iron carbon equilibrium diagram in entry **Iron Metals, Alloys, and Steels**.

Additional Reading

Frick, J.P.: "Woldman's Engineering Alloys," 9th Edition, ASM International, Materials Park, OH, 2000.
Gupta, K.P.: "Phase Diagrams of Ternary Nickel Alloys: Part I and II," ASM International, Materials Park, OH, 1990.
Kassner, M.E. and D.E. Peterson: "Phase Diagrams of Binary Actinide Alloys," ASM International, Materials Park, OH, 1995.
Massalski, T.B.: "Binary Alloy Phase Diagrams," 2nd Edition, ASM International, Materials Park, OH, 1990.
Massalski, T.B.: "Binary Alloy Phase Diagrams Materials Network User," ASM International, Materials Park, OH, 1996.
Rogi, P. and J.C. Schuster: "Phase Diagrams of Ternary Boron Nitride and Silicon Nitride Systems," ASM International, Materials Park, OH, 1992.
Subramanian, P.R., D.J. Chakrabarti, and D.E. Laughlin: "Phase Diagrams of Binary Copper Alloys," ASM International, Materials Park, OH, 1994.
Staff: "ASM Handbook, Vol. 3, Alloy Phase Diagram," 10th Edition, ASM International, Materials Park, OH, 1992.
Staff: ASM International "Superalloys: A Technical Guide," 2nd Edition, ASM International, Materials Park, OH, 2002.
Villars, P., A. Prince, and H. Okamoto: "Handbook of Ternary Alloy Phase Diagrams," ASM International, Materials Park, OH, 1995.

PHASE DIAGRAM (Statistics). A diagram showing two time-series x_1 and x_2 plotted as ordinate and abscissa. If the fluctuations of these two variates keep in step, then the line joining the plotted points will trace a definite pattern, for example, somewhat like an ellipse for oscillatory series.

PHASE DIFFERENCE. With reference to industrial and scientific instruments, the Scientific Apparatus Makers Association defines phase difference as:

1. Between sinusoidal input and output of the same frequency, the phase angle of the output minus the phase angle of the input.
2. Of two periodic phenomena (e.g., in nonlinear systems), the difference between the phase angles of their two fundamental waveforms.

Phase difference is regarded as part of the transfer function which relates output to input at a specified frequency. Measurement of phase difference in the complex case is sometimes made in terms of the angular interval between respective crossings of a mean difference line, but values so measured will generally differ from those made in terms of the fundamental waveforms.

PHASE INVERTER. The push-pull amplifier has numerous advantages over single-ended operation but it does require special circuit provisions to accomplish the push-pull energization of the tubes or transistors. Since the usual amplifier has low-level stages operating single-ended, this requires some method of getting two signals equal in amplitude and 180° out of phase to drive the input of the push-pull stage. One method is to use a transformer to couple the output of the single-ended stage to the push-pull input by center-tapping the secondary. However, for resistance coupling, a phase inverter may be used. Various circuit arrangements may be used to accomplish this function; one typical circuit using transistors is shown in Fig. 1. With reference to the figure, part of the output signal of the upper transistor is coupled back into the input of the lower device. Since the output voltage of the amplifier stage is 180° out of phase with respect to the input, in the normal operating frequency range, the input to the second transistor is 180° from that of the first, hence their outputs are likewise displaced. If the tap on the output resistor of the first transistor is adjusted correctly, just enough voltage is applied to the input of the second to

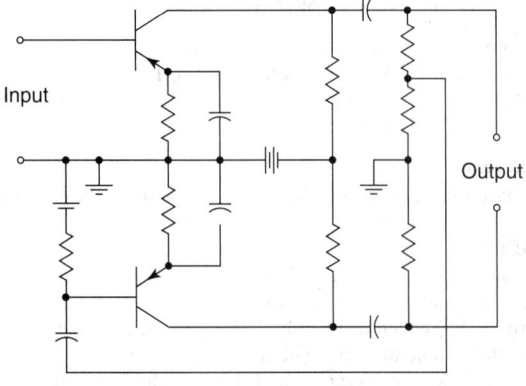

Fig. 1. Phase inverter.

cause its output to be equal in amplitude to that of the first. These equal and opposite voltages may then be applied to the inputs of the two devices in the succeeding push-pull amplifier stage.

PHASE RULE.   The phase rule, due to Gibbs, gives the number F of intensive variables which can be fixed arbitrary in a system in equilibrium. This number is also called the variance or the number of degrees of freedom of the system. It is given by

$$F = 2 + (C' - R) - P$$

where C' is the number of components, R, the number of independent chemical reactions, and P, the number of phases.

In terms of the number of independent components, C, Equation (1) may also be written

$$F = 2 + C - P$$

If $F = 0$ the system is invariant. We cannot fix either temperature or pressure arbitrarily. Equilibrium can only be established at isolated points. An example is the *triple point* at which pure solid, liquid and vapor are in equilibrium ($F = 2 + 1 - 3 = 0$).

If $F = 1$ the system is *monovariant*. We can, for example, fix the temperature, but the equilibrium pressure is then fixed. This is the situation for a system containing one component and two phases.

If $F = 2$ the system is *bivariant*. Within certain limits both pressure and temperature can be given arbitrarily. This is the situation for $C = 1$ and $P = 1$, or $C = 2$ and $P = 2$.

PHASE-SHIFTING CIRCUIT.   Any circuit containing resistive and reactive components used to produce a desired difference between the phase of the output voltage or current and the input voltage or current at some desired frequency.

PHASE-SHIFT OSCILLATOR.   An oscillator produced by connecting any network having a phase shift of an odd multiple of 180° (per stage) at the frequency of oscillation, between the output and the input of an amplifier. When the phase shift is obtained by resistance-capacitance elements, the circuit is an R-C phase-shift oscillator.

PHEASANT (*Aves, Galliformes*).   The pheasant is a game bird native to the Old World, especially to southeastern Asia and the high altitudes of China and Tibet. The males of many species are gorgeously colored and have much longer tails than the females. Pheasants are raised in large numbers for game both in Europe and in North America. In some parts of the United States, the ring-necked pheasant (*Phasianus colchicus*) became well established after a number of years of careful protection. See Fig. 1 The introduction of other species has met with more limited success.

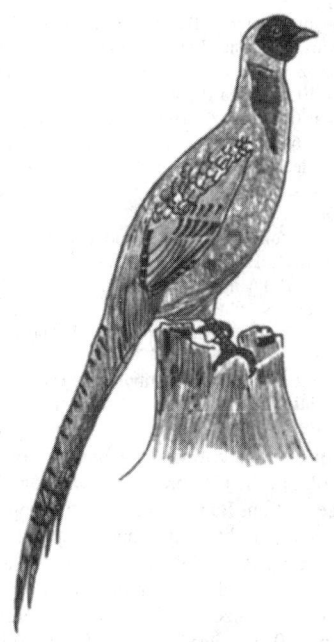

Fig. 1.   Ring-necked (Southern green) pheasant. (*Sketch by Glenn D. Considine.*)

Many species are called pheasants, in addition to the true pheasants of the genus *Phasianus*. Other forms are the tragopans or horned pheasants, the blood pheasants, the monals, the fire-backed pheasants, eared pheasants, golden pheasants, jungle fowls, and argus pheasants. Some of the Indian species bear the names kallege and pukra. The ruffed grouse of North America is sometimes referred to as a pheasant in the southern states, but this has no scientific foundation. The monal is brightly colored and is found in the higher forests of the mountains of Asia. The tragopan is a large game bird found in wooded country at high altitudes in China and northern India. Guinea fowl comprise several species of African birds related to the pheasants. All have some dark plumage with light spots, and brightly colored bare skin about the head and neck. The common Guinea fowl, *Numida meleagris*, is among the common domesticated species. See also **Galliformes**.

PHENACITE.   The mineral phenacite is a beryllium silicate corresponding to the formula Be_2SiO_4. It is hexagonal but the crystals are usually rhombohedral in habit. It has a conchoidal fracture; is brittle, hardness, 7.5–8; specific gravity, 3; luster, vitreous; colorless to yellowish or reddish, sometimes brown; transparent to translucent. Phenacite is found in pegmatites with topaz, quartz and microcline, and occurs also in emerald-bearing mica schists of the Ural Mountains. It is found also in France, Norway, Switzerland, Africa, Brazil, and Mexico; and in the United States, in Oxford County, Maine; Carroll County, New Hampshire; and in Chaffee and El Paso Counties in Colorado. It derives its name from the Greek meaning deceiver, as it resembles quartz and topaz with which it is associated. It is sometimes spelled phenakite. It has been used as a gem.

PHENOCLAST.   A textural term proposed by R.M. Field in 1916 for coarsely graded clastic sedimentary rocks in which the largest or "show" particles or fragments are referred to as phenoclasts, regardless of their shape or composition. The term implies that the larger constituents of the glomerate have been derived from prelithified rock. Rounded fragments are called pebbles or spheroclasts, which when lithified by means of matrix (sand and clay) and cement form a conglomerate. Angular fragments are called anguclasts (Field), which when lithified by means of matrix (sand and clay) and cement form a breccia.

PHENOCRYST.   A textural term proposed by Iddings in 1892 for macroscopic crystals which are relatively much larger than the crystalline matrix of the igneous rock in which they occur. Rocks which have phenocrysts are called porphyritic. The term phenocryst is derived from the Greek, meaning show, and crystal.

PHENOL.   1.   A class of aromatic organic compounds in which one or more hydroxy groups is attached directly to the benzene ring. Examples are phenol itself (benzophenol), the cresols, xylenols, resorcinol, naphthols. Although technically alcohols, their properties are quite distinctive.

2.   Phenol (carbolic acid; phenylic acid; benzophenol; hydroxybenzene), C_6H_5OH. Phenol is a white, crystalline substance that turns pink or red if not perfectly pure, or if under influence of light; absorbs water from the air and liquefies. It has a distinctive odor and a sharp burning taste. It is toxic by ingestion, inhalation, and skin absorption, and is a strong irritant to tissue. When in a very weak solution, phenol has a sweetish taste; specific gravity 1.07; mp 42.5–43 °C; bp 182 °C; flash point 77+ °C. Soluble in alcohol, water, ether, chloroform, fixed or volatile oils, and alkalies.

Most of the phenol used in the United States is made by the oxidation of cumene, yielding acetone as a byproduct. The first step in the reaction yields cumene hydroperoxide, which decomposes with dilute sulfuric acid to the primary products, plus acetophenone and phenyl dimethyl carbinol. Other processes include sulfonation, chlorination of benzene, and oxidation of benzene. The compound is purified by rectification.

Major uses of phenol include production of phenolic resins, epoxy resins, and 2,4-D (regulated in many countries); as a selective solvent for refining lubricating oils; in the manufacture of adipic acid, salicylic acid, phenophthalein, pentachlorophenol, acetophenetidine, picric acid germicidal paints, and pharmaceuticals; as well as use as a laboratory reagent. Special uses include dyes and indicators, and slimicides.

High-boiling phenols are mixtures containing predominantly meta substituted alkyl phenols. Their boiling point ranges from 238–288 °C they set to a glass below −30 °C. They are used in phenolic resins, as fuel-oil sludge inhibitors, as solvents and as rubber chemicals.

Phenol is regarded as a dangerous chemical. Refer to *Dangerous Properties of Industrial Materials*, 10th Edition, Sax and R.J. Lewis, Editors, Wiley, New York, 1999.

PHENOLICS.

These are products of the condensation reaction of phenol and formaldehyde.[1] Water is the byproduct of this reaction. Substituted phenols and higher aldehydes may be incorporated to achieve specific resin properties, e.g., flexibility, reactivity, or compatibility with elastomers and other polymers. A variety of phenolic resins can be produced by adjusting the formaldehyde:phenol molar ratio and the resinification temperature and catalyst. Single-stage (*resole*) resins are produced with an alkaline catalyst and a molar excess of formaldehyde. The reaction is carefully controlled to allow the production of low-molecular-weight, noncrosslinked resins. Single-stage resins complete the curing reaction in a heated mold with no additional catalyst. A three-dimensionally crosslinked, insoluble, and infusible polymer is formed.

Two-stage resins (*novolacs*) are produced by the acid-catalyzed reaction of phenol and a portion of the required formaldehyde. The resin product is brittle at room temperature. It can be melted, but it will not crosslink. Novolacs can only be cured by the addition of a hardener, almost always formaldehyde supplied as hexamethylene tetramine. Upon heating, the latter compound decomposes to yield ammonia and formaldehyde.

Phenolic resins are available in flake, powder, and liquid forms. A wide variety of industrial applications has been developed, including foundry molds and cores; plywood and particle board; brake and clutch linings; glass, cellulose, and foam insulation; grinding wheels and coated abrasives; adhesives and glues, rubber tackifiers; coatings; varnishes; and electrical and decorative laminates. Resole and novolac resins are combined with a variety of fillers, reinforcements, and additives to produce phenolic engineering plastics. See also **Paints and Coatings**.

Application development for phenolics has been spurred by weight and cost savings inherent in metal replacement and parts consolidation. Thermoplastics have been replaced by phenolics where creep resistance and thermal stability are required in downsized parts or applications in hostile environments.

PHENYLKETONURIA.

Also known as phenylpyruvic oligophrenia or PKU, this is an inherited disorder which involves about 1% of the mentally retarded population. Persons with the disease are born without the enzyme phenylalanine hydroxylase, which is necessary for the conversion of phenylalanine, an amino acid present in practically all protein foods, into tyrosine. This leads to greatly increased concentrations of phenylalanine in the blood, some of which is then converted to phenylpyruvic acid, phenyllactic acid, and *o*-hydroxyphenylacetic acid. All of these metabolites are excreted in the urine of phenylketonurics in higher than normal amounts. Varying concentrations of other unusual metabolites deriving from phenylalanine, tyrosine, and tryptophan are also present.

The mental defect in phenylketonuria is usually severe and is apparent by six months of age. Most patients are idiots, a few are imbeciles, and rare individuals have borderline intelligence. Seizures, eczema, and albinism may be present, and life expectancy is greatly decreased.

That phenylalanine accumulation in the tissues is primarily responsible for the biochemical abnormalities in phenylketonuria is established by demonstrations over the past 20–25 years that a low-phenylalanine diet may prevent or reverse these changes. More important, if such a diet is instituted in the first months of life and continued for 3 to 4 years, intellectual impairment may be prevented. Reversal of the mental defect in older patients is not certain.

Pathways of Research. In 1991, C.R. Scriver (McGill University, Montreal, Canada) reviewed the research on phenylketonuria, which had commenced as early as 1934. Scriver pointed out four phases of this research that took place over a half-century of study.

Phase I — In 1934, Følling identified a clinical entity which he called "imbecillatas phenylpyruvica." Over the two decades that followed, it became known that the mental retardation associated with the disease was related to persistent postnatal *hyperphenylalaninemia* (i.e., high phenylalanine levels). It also was recognized that the metabolic phenotype was the consequence of deficient activity of phenylalanine hydroxylase (an enzyme). Further, it was determined that the enzyme abnormality would be

[1] Data furnished by *Occidental Chemical Corp.*

found in a pair of recessive mutations inherited from healthy parents. From these findings, certain conclusions were drawn, to the effect that there was (a) an ultimate cause (mutation) and (b) a proximate cause (ingestion of the essential amino acid phenylalanine).

Phase II — L.S. Penrose (University College, London) renamed the disease (phenylketonuria) and emphasized (a) association between human genetic variation, (b) chemical imbalance, and (c) abnormal mental function. Penrose also speculated that the course of phenylketonuria could "be influenced by the deliberate alteration of body metabolism" — that is, by medical treatment. Penrose also observed the nonrandom distribution of the disease and noted that about 2% of Europeans carry the gene for the disease, wondering why such a gene was so prevalent.

Considerable research followed and, as reported by Scriver, the results included the following: (a) dietary treatment to control serum phenylalanine levels enjoyed modest success; (b) a set of enzymes (and genes) and a catalytic cofactor were identified; (c) a better understanding of why all untreated patients with phenylketonuria have hyperphenylalaninemia, but not all persons who have hyperphenylalaninemia necessarily have phenylketonuria; and (d) much additional knowledge was gained from screening tests. As observed by Scriver, "Therapy for the disorder became not only an epitome of the application of human biochemical genetics, but also a model for so-called genetic medicine and for public health. In addition, maternal hyperphenylalaninemia became a paradigm of metabolic teratogenesis, which if not dealt with successfully could nullify all the gains made to prevent mental retardation associated with phenylketonuria mutations."

Phase III — Commencing in the 1980s, S.L. Woo (Baylor College of Medicine), according to Scriver, "prepared a complementary DNC (cDNA) for the rat liver enzyme, then cloned a human cDNA, and eventually isolated and characterized the human phenylalanine hydroxylase gene. The gene resides on the long arm of chromosome 12 and has about 90,000 base pairs of DNA, has 13 exons, and is decorated with a suite of highly informative DNA markers." The genetic architecture is indeed complex.

Phase IV — In 1991, sponsored by the National Institute of Child Health and Human Development, the Howard Hughes Medical Institute, and the Deutsche Forschungsgemeinschaft, Yoshyuki Okano (Baylor College of Medicine) and colleagues conducted a study of 258 patients with phenylketonuria from Denmark and Germany for the presence of eight mutations previously found in patients from these countries. The conclusions of this research: "Our results strongly support the hypothesis that there is a molecular basis for phenotypic heterogeneity in phenylketonuria. The establishment of genotype will therefore aid in the prediction of biochemical and clinical phenotypes in patients with this disease."

Additional Reading

Addison, G.M., D.M. Isherwood, R.A. Harkness, and R.J. Pollitt: "Practical Developments in Inherited Metabolic Diseases," Kluwer Academic Publishers, Norwell, MA, 1986.

Kaurman, S.: "Tetrahydrobbioterin: Basic Biochemistry and Role in Human Disease," Johns Hopkins University Press, Baltimore, MD, 1997.

Koch, R., F. De la cruz, and L.D. Platt: "Genetic Disorders and Pregnancy Outcome," CRC Press, LLC., Boca Raton, FL, 1997.

Levy, H.L.: "Molecular Genetics of Phenylketonuria and Its Implications," *Amer. J. Human Genetics*, **45**, 667 (1989).

Lyonnet, S., et al.: "Molecular Genetics of Phenylketonuria in Mediterranean Countries: A Mutation Associated with Partial Phenylalanine Hydroxylase Deficiency," *Amer. J. Human Genetics*, **44**, 511 (1989).

Okano, Y., et al.: "Molecular Basis of Phenotypic Heterogeneity in Phenylketonuria," *N. Eng. J. Med.*, 1232 (May 2, 1991).

Romano, V.: "Advances in Phenylketonuria Research," S. Karger Publishers, Inc., Farmington, CT, 1993.

Scriver, C.R.: "Phenylketonuria — Genotypes and Phenotypes," *N. Eng. J. Med.*, 1280 (May 2, 1991).

Woo, S.L.: "Molecular Basis and Population Genetics of Phenylketonuria," *Biochemistry*, **28**, 1 (1989).

pH (Hydrogen Ion Concentration).

A measure of the effective acidity or alkalinity of a solution. It is expressed as the negative logarithm of the hydrogen-ion concentration. Pure water has a hydrogen ion concentration equal to 10^{-7} moles per liter at standard conditions. The negative logarithm of this quantity is 7. Thus, pure water has a pH value of 7. The pH scale usually is considered as extending from 0 to 14. When a strong acid fully dissociates (or ionizes) in water, a $1 N$ solution of this acid will have a pH value of 0.0. Conversely, a $1 N$ base fully ionized in water will have a pH value of 14. Both hydrochloric acid and sodium hydroxide come close to meeting these stipulations. Because of the logarithmic nature of the pH scale, there is a tenfold change in hydrogen- and hydroxyl-ion concentration per unit change of pH. Thus, a slightly acidic solution having a pH of 6 will contain ten times as many active hydrogen ions as a solution of pH 7. See also **pK**.

Effective acidity or alkalinity is stressed in pH measurement — not the total hydrogen present. Sulfuric acid and boric acid both contain significant amounts of hydrogen. Nearly all the hydrogen in sulfuric acid dissociates in the presence of sufficient water to become free hydrogen ions. On the other hand, when boric acid is added to water, it dissociates very little into free hydrogen ions. The pH of a 0.1 N sulfuric acid solution will be about 1.3 whereas for the same concentration, boric acid will have a pH of about 5.3. Thus, sulfuric acid is called a strong acid; boric acid a weak acid. In all materials, of course, dissociation increases with temperature, thus the same solution will have a somewhat different pH at a lower temperature than at a higher temperature. Pure water is neutral at a temperature of 25 °C, having a concentration of 1×10^{-7} hydrogen ions and 1×10^{-7} hydroxyl ions and, consequently, a pH of 7. Dissociation is less at 0 °C, at which temperature the hydrogen-ion concentration is 0.34×10^{-7}, or a pH of 7.47 (slightly basic rather than neutral). But, at a temperature of 100 °C, dissociation is greater. The hydrogen ion concentration is 8×10^{-7} and the pH is 6.10 (or slightly acid). The pH of various substances is given in Table 1.

TABLE 1. PH VALUES OF VARIOUS SUBSTANCES (AT 25 °C)

Material	pH
Seawater	7.75 to 8.25
Soils	3 to 10
Plant tissues and fluids	About 5.2
Animal tissues and fluids	About 7.0 to 7.5
Blood	7.35–7.5
Urine	5.0–7.0
Milk	6.5–7.0
Gastric juice	1.7
Pancreatic juice	7.8
Intestinal juice	7.7
Internal tissue fluids:	
Minimum, below which acidosis ensues	7.0
Maximum, above which tetany ensues	7.8
Hydrochloric acid (1N)	0.1
Hydrochloric acid (0.1N)	1.08
Hydrochloric acid (0.001N)	3.00
Sulfuric acid (1.0N)	0.32
Sulfuric acid (0.1N)	1.17
Acetic acid (1N)	2.37
Lemon juice	2.0–2.2
Acid fruits	3.0–4.5
Fruit jellies	3.0–3.5
Sodium hydroxide (1N)	13.73
Sodium hydroxide (0.1N)	12.84
Ammonia (10% NH_3)	11.8
Limewater, $Ca(OH)_2$ saturated	12.4
Trisodium phosphate, 2%	11.95

Buffer solutions can be added to resist changes in pH despite the addition of acid or base to the solution. This is explained under **Buffer (Chemical)**.

pH is measured in two basic ways: (1) colorimetrically, usually where high accuracy is not required and manual methods suffice; and (2) electrometrically. Color changes are based upon various organic dyes which alter their color within a relatively narrow range of pH values. Numerous dyes are required to cover the full pH range. Electrometric methods are used both in the laboratory and on-line for process control. They are continuous and easily adapted to automatic control systems. The possibility that a thin glass membrane of special composition could develop a potential in relation to hydrogen ion concentration was described as early as 1909 by the German chemist, Fritz Haber. Little progress was made until the middle-1920s. Glass electrodes are now the standard approach to electrometric pH measurement, after periods of trial with quinhydrone and antimony electrodes. The glass electrode responds in a predictable fashion throughout the 0 to 14 pH range, developing 59.2 millivolts per pH unit at 25 °C, values which are consistent with the classical Nernst equation. Contrary to earlier pH electrodes, the glass electrode is not influenced

by oxidants or reductants in solution. With suitable temperature compensation, pH measurements can be made up to 100 °C and higher. In pH measurement, a second or reference electrode is required to complete the circuit. After trials with numerous electrodes (the hydrogen electrode is the standard) for practical plant and laboratory applications, the mercury-mercurous chloride (calomel) electrode is widely used. There is also some use of the silver-silver chloride reference electrode.

pH control systems are widely used in waste control and neutralization systems, in pulp and paper manufacture, in food processing, and in the manufacture of numerous organic chemicals. pH measurement is very important in the medical field.

PHILLIPSITE. The mineral phillipsite is a zeolite, a hydrous silicate of potassium, calcium, and aluminum, corresponding to formula $(K,Na_2Ca)(Al_2Si_4)O_{12} \cdot 4{-}5H_2O$. It is monoclinic, forming penetration twins, and sometimes crosses resembling orthorhombic or tetragonal forms. It also may occur in radial groups. Phillipsite is a brittle mineral; hardness, 4–4.5; specific gravity, 2.2; luster, vitreous; color, white to light red; translucent to opaque. Like other zeolites, it is found in veins and cavities in basalts, and sometimes in more acidic rocks. It is believed to be a low-temperature mineral. Phillipsite is found in Italy, especially in the lavas of Vesuvius and Monte Somma, and in the basalts of Germany, Ireland, and Australia. It has been reported from Greenland. This mineral was named in honor of the British mineralogist William Phillips.

PHLEBOTOMUS FEVER. Also known as *sandfly fever* or *Papataci fever*, this is an acute, nonfatal virus disease transmitted by the sandfly (*Phlebotomus papatasii*). The disease is common in many of the tropical and subtropical areas, appearing mainly in the hot dry weather in the Mediterranean and Middle East, India, Asia, and parts of South America. It has been reported in communities living at altitudes up to nearly 5000 feet (1500 m) above sea level. In most areas where antimalarial residual insecticide spraying has been used, the vector has disappeared and, with it, the disease. Recent failure to maintain the antimalarial campaigns has followed in former endemic areas with return of the vector and the disease. The causal agent is a small, unclassified virus. The fly becomes infective within 6 to 8 days after ingesting human blood in which the virus is circulating. The fly transmits the disease by biting. No race is immune and both sexes are attacked at all ages.

An attack of phlebotomus fever closely resembles one of dengue, but without the rash, saddleback fever, or glandular involvement. The onset is sudden. There is a rise in temperature, which remains elevated for 1 to 3 days; it is remittent and ends by crisis accompanied by intense sweating. The conjunctiva are infected and photophobia and lachrymation are common. There is very severe headache and bone and joint pain can be very troublesome.

There is no laboratory method of making a certain diagnosis. The clinical picture should be recognized during an epidemic, but may be easily confused with atypical dengue and influenza. Treatment is entirely symptomatic.

R. C. V.

PHLEBOTOMY. Intentional removal of blood from the circulatory system. The therapy for hemochromatosis is an example. See also **Liver**.

PHLOEM. That part of a plant through which foods move from one part of the plant to another. In stems, the phloem forms a considerable portion of the bark, and is found outside the cambium.

PHLOGOPITE. The mineral phlogopite is a magnesium-bearing mica, with but little iron, corresponding essentially to the formula $K(Mg, Fe)_3(AlSi_3)O_{10}(F, OH)_2$. Fluorine is sometimes present. This mica is monoclinic like muscovite, biotite and lepidolite, forming prismatic crystals, occasionally very large, and occurring also in scales and plates. Its cleavage is basal and highly perfect with elastic laminae; hardness, 2–2.5; specific gravity, 2.76–2.90; luster, pearly to submetallic; color, yellowish-brown, green, white and colorless; transparent to translucent; may exhibit asterism, probably due to minute inclusions. Phlogopite is more nearly a characteristic of metamorphic than igneous rocks although occasionally occurring in the latter if they are rich in magnesia and with but little iron. Phlogopite is found especially in Rumania, Switzerland, Italy, Finland, Sweden and Madagascar where it occurs in the crystalline limestones in huge crystals. In the United States it occurs in New York State at Edwards, Hammond, DeKalb, Monroe and, in New Jersey, at Franklin. In Canada it is found at many places in Ontario and Quebec.

The name phlogopite comes from the Greek word meaning like fire, referring to the copper-like reflections often observed in the reddish-brown varieties.

Phlogopite is in demand commercially by the electrical industry for use as an insulator.

PHOENICOPTERI *(Aves)*.   The Flamingos (family *Phoenicopteridae*) are long-legged water birds highly adapted to taking small water animals. Because of their overly long legs, they were formerly grouped with the *Ciconiiformes*, but some zoologists considered them to be related to the *Anatidae*. It is, therefore, appropriate to set up this family as a separate order.

Their length reaches 80–130 centimeters (31–51 inches) from the tip of the beak to the tip of the tail, but it may reach 190 centimeters (75 inches) if measured to the tips of the toes of the extended legs; the weight is 2500–3500 grams ($5\frac{1}{2}$–$7\frac{1}{2}$ pounds). The long neck is curved, with 19 cervical vertebrae. The feathers have an aftershaft; there are 12 primaries, and 12 to 16 rectrices. The skeleton, muscles, and air sacs are formed as in storks. The voice is goose-like. There are 3 genera with 5 species, all rather similar to one another: the Greater Flamingo (*Phoenicopterus ruber*), with 2 subspecies, the American Flamingo (*Phoenicopterus ruber*), and the European Flamingo (*Phoenicopterus ruber roseus*); the Chilean Flamingo (*Phoenicopterus chilensis*); the Lesser Flamingo (*Phoeniconaias minor*); the Andean Flamingo (*Phoenicoparrus andinus*); and the James' Flamingo (*Phoenicoparrus jamesi*).

The food of all flamingos consists of small swimming crustacea, algae, and unicellular organisms that they sift out of the water with the beak, which has been transformed into a filtration apparatus. The lower mandible is large, and at its cutting edge, looks as if it is inflated; the upper mandible, on the other hand, is small and fits on the lower one like a lid.

The type of food and the manner of feeding of flamingos assume an abundance of prey of fairly uniform size, particularly since the birds live in large flocks of up to several 100,000 individuals. Such conditions are found in salt lakes and brackish coastal lagoons of warm areas. Their manner of feeding readily differentiates flamingos from the related orders *Ciconiiformes* and *Anseriformes*, to which they do, however, show numerous similarities. See also **Flamingo**.

PHOENIX.   A southern constellation located near Cetus, also sometimes spelled Phenix.

PHOLIDOTA.   A small order of mammals containing only the scaly anteaters or pangolins of the Malay archipelago, Africa, and southeastern Asia. They are peculiar animals which are covered with overlapping horny scales of large size and, in this respect, somewhat resemble the armadillo. Pangolins are slender animals with a long tail and short legs bearing powerful claws. Like other anteaters they have a sharp snout and long sticky tongue and live chiefly on termites. The most common species is *Manis pentadactyla*. For protection, these animals roll themselves into a ball with the outer protective scales discouraging predators.

PHON.   The unit of loudness level of sound, numerically equal to the sound pressure level in decibels, relative to 0.0002 mircobar, of a simple 1000 cycle per second tone judged by listeners to be equivalent in loudness.

PHONOMETER.   An instrument for measuring the intensity or frequency of sounds.

PHONONS.   Many of the thermal and vibrational properties of solids can be explained by considering the material to be a volume made up of a gas of particles called *phonons*. This particle description is a method of taking into account the actual motion of the atoms and molecules in the solid. Since each atom possesses energy due to its thermal environment, and since there are forces between the atoms that keep the solid together, each atom tends to oscillate about its equilibrium position. The formal mathematical development, obtained through solving the equations of motion of the array of individual atoms and molecules, indicates that the thermal energy of the solid is contained in certain combinations of particle vibrations which are equivalent to standing elastic waves in the sample and are called normal modes. Each normal mode contains a number of discreet quanta of energy $E = \hbar$ where ω is the frequency of the mode (or wave) and $\hbar$ is Planck's constant divided by 2π. Each of these quanta is called a phonon (in analogy with the light quanta or photon whose energy-frequency relationship is identical). Phonons are considered only as particles, each having an energy $E = \hbar$, a momentum q, and a velocity $v = \partial\omega/\partial q \sim \omega/q$. Analogous to the energy levels of electrons in a solid, phonons can have only certain allowed energies.

The phonon is of importance to many phenomena: electron mobility, optical absorption, electron spin resonance, electron tunneling, and superconductivity. The phonon spectrum represents a detailed picture of the forces that hold solids together. Thus, it is clear why the phonon has been and will continue to be of fundamental importance in solid-state physics.

See also **Acoustics**.

PHOSGENE.   See **Chlorinated Organics**.

PHOSGENITE.   This mineral is a chlorocarbonate of lead, $Pb_2(CO_3)$ Cl_3, crystallizing in the tetragonal system, associated with other lead minerals of secondary origin, e.g., cerussite and anglesite; hardness, 2–3; specific gravity 6.133; prismatic to tabular crystals, also massive and granular, adamantine luster; color, white, gray, brown, green or pink; transparent to translucent. Some specimens show yellowish fluorescence under ultraviolet light.

Found in the United States in California, Colorado, Arizona, and New Mexico. Magnificent crystals up to 5 inches (12.5 centimeters) in diameter have been found at Monte Poni, Sicily; as fine crystals in England at Derbyshire and Matlock; and in Poland, Russia, Tasmania, Australia, and Namibia.

PHOSPHATE ROCK.   See **Fertilizer**.

PHOSPHINE.   See **Phosphorus**.

PHOSPHOLIPIDS.   These compounds belong to a group of fatty acid compounds sometimes referred to as complex lipids. The simplest are esters of fatty acids with glycerol phosphate and are called *phosphatidic acids*. There are also phosphatidylcholines or lecithins, phosphatidyl-ethanolamines, phosphatidylserines, and phosphatidylinositols. The latter may have one or more additional phosphate groups attached to the inositol. A similar series also exists containing an aldehyde attached to the 1-position of the glycerol, in the form of an α, β-unsaturated ether. These are commonly referred to as *plasmalogens*.

The percentage of phospholipid content of tissues varies little under normal physiological conditions, thus giving rise to the term *element constant*, in contrast to the triglycerides, which have been called the *element variable*.

Phospholipids are considered to be involved in the transport of triglycerides through the liver, especially during mobilization from adipose tissue. Conditions which could be interpreted as interfering with phosphatidylcholine formation, such as deficiency of choline or its precursors, result in a pronounced increase in liver triglycerides.

Mitochondrial phospholipids play a role in electron transport and oxidative phosphorylation, two mechanisms by which the cell accomplishes the final oxidation of the metabolites to produce energy. Phospholipids also are linked in the transport of ions, especially sodium, across membranes.

In summary, phospholipids (phosphatides) comprise a group of lipid compounds that yield, upon hydrolysis, phosphoric acid, an alcohol, fatty acid, and a nitrogenous base. They are widely distributed throughout nature.

PHOSPHORIC ACID.   Generally the term *phosphoric acid* refers to orthophosphoric acid, H_3PO_4. Anhydrous orthophosphoric acid is a white, crystalline solid, which melts at 42.35 °C. It forms a hemihydrate, $2H_3PO_4 \cdot H_2O$, which melts at 29.32 °C. Although it is possible to produce almost any desired concentration, it is common practice to supply the material as a solution containing from 75% H_3PO_4 (melting point, 17.5 °C) to 85% H_3PO_4 (melting point, 21.1 °C). When phosphoric acid is heated to temperatures above about 200 °C, water of constitution is lost. A series of acids is formed by the dehydration, ranging from pyrophosphoric acid, $H_4P_2O_7$, to metaphosphoric acid, $(H_3PO_4)_n$. Salts of the dehydrated acids are used for the preparation of certain types of liquid fertilizers and have been used in some detergents. However, to counter the effects of "phosphate pollution," there has been a serious cutback in this latter use of the phosphates. See also **Fertilizer**.

One, two, or three of the hydrogens in phosphoric acid may be neutralized, leading to a series of products which range widely in their hydrogen ion concentration (pH): monosodium phosphate, NaH_2PO_4, with a pH of 4.0; disodium phosphate, Na_2HPO_4, with a pH of 8.3 (approximate); and trisodium phosphate, Na_3PO_4, with a pH of 12.0. Other phosphorous acids of little commercial importance are hypophosphorous acid, H_3PO_2; orthophosphorous acid, H_3PO_3; and pyrophosphorous acid, $H_4P_2O_5$.

Manufacture of Phosphoric Acid. The major sources of H_3PO_4 traditionally have been mineral deposits of phosphate rock. Mining operations are extensive in a number of locations, including the United States (Florida), the Mediterranean area, and Russia, among others. The major constituent of most phosphate rocks is fluorapatite, $3Ca_3(PO_4)_2 \cdot CaF_2$. The supply of high-grade phosphates, the raw material of choice for producing high-purity phosphoric acid by the wet process, is rapidly decreasing in some areas.

Two major methods are utilized for the production of phosphoric acid from phosphate rock. The *wet process* involves the reaction of phosphate rock with sulfuric acid to produce phosphoric acid and insoluble calcium sulfates. Many of the impurities present in the phosphate rock are also solubilized and retained in the acid so produced. While they are of no serious disadvantage when the acid is to be used for fertilizer manufacture, their presence makes the product unsuitable for the preparation of phosphatic chemicals.

In the other method, the *furnace process*, phosphate rock is combined with coke and silica and reduced at high temperature in an electric furnace, followed by condensation of elemental phosphorus. Phosphoric acid is produced by burning the elemental phosphorus with air and absorbing the P_2O_5 in water. The acid produced by this method is of high purity and suitable for nearly all uses with little or no further treatment.

Basic reactions of the wet process are

$$3\ Ca_3(PO_4)_2 \cdot CaF + 10\ H_2SO_4 + 20\ H_2O$$
$$\longrightarrow 10\ CaSO_4 \cdot 2\ H_2O + 6\ H_3PO_4 + 2\ HF$$

Numerous side reactions also occur. Phosphate rock and sulfuric acid, together with recycled weak liquors, are carefully metered to a large, stirred reactor, providing retention for 4–8 hours. Conditions in the reaction are carefully controlled to maintain preselected conditions. Temperatures (77–83 °C) are controlled by removing excess heat of reaction with a vacuum cooler, or by blowing air through the phosphoric acid slurry. The slurry contains precipitated gypsum and is sent to a filter. The gypsum is washed with water in several countercurrent steps, and weak liquor is returned to the reaction stage. For most uses, the acid requires further concentration, normally done in vacuum evaporators. Merchant-grade acid is generally concentrated to about 54% P_2O_5 (75% H_3PO_4). See Fig. 1.

Effluents and gypsum disposal pose problems. Fluorine is evolved at various steps in the process and scrubbers are required to reduce release to the atmosphere. Gypsum is frequently piled in diked areas or dumped into abandoned mines. Wastewater from these plants is heavily contaminated with fluorine, phosphates, sulfates, and other compounds. It is commonly impounded in large ponds, where a portion of the contaminants may precipitate or be lost by other processes. The cooled effluent from the ponds is recycled to the production unit. Any excess water must be treated with lime before it can be allowed to enter streams.

Developments of recent years include plants designed to precipitate the calcium sulfate in the form of the hemihydrate instead of gypsum. In special cases, hydrochloric acid is used instead of sulfuric acid for rock digestion, the phosphoric acid being recovered in quite pure form by solvent extraction. Solvent-extraction methods have also been developed for the purification of merchant-grade acid, which normally contains impurities amounting to 12–18% of the phosphoric acid content. Processes for recovering part of the fluorine in the phosphate rock are in commercial use.

Although more costly to operate, the electric-furnace process produces phosphoric acid of high purity. A mixture of coke, silica, and phosphate rock is formed into nodules by heating in a nodulizing kiln, and the resulting lump material is transferred to the electric furnace, where it is heated with an electric current introduced by means of graphite electrodes.

Fig. 1. View of three-stage evaporation process used for concentrating wet-process phosphoric acid. (*Swenson.*)

The entire charge is melted, and elemental phosphorus is volatilized. The slag is tapped off intermittently while the phosphorus vapor is condensed. The phosphorus is then burned in air and the P_2O_5 is absorbed in water. Reactions are

$$2\ Ca_3(PO_4)_2 + 6\ SiO_2 + 10\ C \longrightarrow P_4 + 10\ CO + 6\ CaSiO_3$$
$$P_4 + 5\ O_2 \longrightarrow 2\ P_2O_5$$
$$P_2O_5 + 3\ H_2PO \longrightarrow 2\ H_3PO_4$$

PHOSPHORS AND PHOSPHORESCENCE. A large variety of substances become luminescent when stimulated or excited by suitable radiation, or by emissions, such as cathode rays or beta-rays. This phenomenon is complex and exhibited in various aspects. In some cases, the light is emitted only so long as the exciting emission is maintained, in which case it is called *fluorescence*. See also **Illumination**. In other cases, the luminescence persists after the excitation is removed and it is then called *phosphorescence*. It has long been known, for example, that zinc sulfide, under certain conditions, glows brightly for a time after exposure to daylight or lamplight, but the luminosity decays rapidly and disappears, usually within a few minutes. The electroluminescent phosphor of zinc sulfide-zinc selenide-copper has the property that the wavelength of the emitted radiation increases with increasing selenium content. The white luminescence of some television tubes is obtained from a combination of cadmium-zinc sulfide phosphors, one that is blue-emitting and the other yellow-emitting. Also, in some color television tubes, the blue-emitting and green-emitting phosphors are of the sulfide type, but earlier use of sulfides for the red-emitting "dots" on the tube surface were replaced by rare-earth red-emitting phosphors. One composition used is prepared by combining about 4% europium oxide and 65% yttrium oxide, with various vanadium compounds and calcining the mixture. Rare-earths also have been used in producing phosphors for high-pressure mercury-arc lamps. These phosphors increase the proportion of red light emitted by reducing the green, blue, and ultraviolet portions.

Quantitatively, phosphorescence may be defined as luminescence that is delayed by more than 10^{-8} seconds after excitation. It may be associated with transitions from a higher excited state to a lower one, the energy going into a radiationless rearrangement of the system. If the lower state is metastable, its lifetime may be considerable before it finally decays by a highly forbidden radiative transition to the ground state. In the case of zinc sulfide, the process depends upon the ionization of activator atoms, the freed electrons being trapped and only released slowly for recombination.

See also **Luminescence**.

PHOSPHORUS. Chemical element, symbol P, at. no. 15, at. wt. 30.9738, periodic table group 15, mp 44.1°C (α-white), bp 280°C (α-white), sp gr 1.82 (white), 2.20 (red).

Four allotropes of phosphorus are known, the hexagonal β-white, stable only below $-77°C$, the cubic α-white (mp 44.1°C), the violet, and the black (which is thermodynamically the most stable). The α-white form is usually taken as the standard state. The violet is obtained by continued heating at 500°C of a solution of phosphorus in lead. When α-white phosphrus is heated to 250°C in the absence of air, a red variety (mp 590°C) is obtained which is believed to consist of a mixture of the α-white and violet allotropes, although the studies of the violet component in the mixture have shown that at least four polymorphic forms of red (violet) phosphorus exist.

White phosphorus is considered to be made up largely of P_4 molecules, as is the liquid and vapor up to 800°C, where dissociation becomes appreciable. The P_4 molecule is a tetrahedron, with single covalent bonds between the P atoms, and each having an unshared pair of electrons. White phosphorus is much more reactive than red or violet.

Black phosphorus has a graphite-like structure and has a similar electrical conductivity.

There is one stable nuclide, ^{31}P. Six radioactive isotopes have been identified, ^{28}P through ^{30}P and ^{32}P through ^{34}P, all with short half-lives, measured in terms of seconds, minutes, or days. See also **Radioactivity**. In terms of terrestrial abundance, phosphorus ranks 10th with an estimated average content of igneous rocks being 0.13% phosphorus. The element ranks 19th in abundance in seawater, there being an estimated 325 tons of phosphorus per cubic mile (70 metric tons per cubic kilometer) of seawater. In terms of cosmic abundance, phosphorus is ranked 15th among the elements. The element was first identified by Hennig Brandt in Germany in 1669 during an experiment in which he was distilling urine with sand and coal. White phosphorus is very toxic.

First ionization potential 11.0 eV; second, 19.81 eV; third, 30.04 eV; fourth, 51.1 eV; fifth, 64.698 eV. Oxidation potentials $H_3PO_2 + H_2O \rightarrow H_3PO_3 + 2H^+ + 2e^-$, 0.59 V; $P + 3H_2O \rightarrow H_3PO_3 + 3H^+ + 3e^-$, 0.49 V; $P + 2H_2O \rightarrow H_3PO_2 + H^+ + e^-$, 0.29 V; $H_3PO_3 + H_2O \rightarrow H_3PO_4 + 2H^+ + 2e^-$, 0.20 V; $PH_3(g) \rightarrow P + 3H^+ + 3e^-$, 0.04 V; $P + 2OH^- \rightarrow H_2PO_2^- + e^-$, 1.82 V; $P + 5OH^- \rightarrow HPO_3^{2-} + 2H_2O + 3e^-$, 1.71 V; $H_2PO_2^- + 3OH^- \rightarrow HPO_3^{2-} + 2H_2O + 2e^-$, 1.65 V; $HPO_3^{2-} + 3OH^- \rightarrow PO_4^{3-} + 2H_2O + 2e^-$, 1.05 V; $PH_3(g) + 3OH^- \rightarrow P + 3H_2O + 3e^-$, 0.87 V.

Other physical characteristics of phosphorus are given under **Chemical Elements**.

Because of its reactivity, phosphorus does not occur in nature in the elemental form. Phosphate rock is the principal source of phosphorus and phosphorus compounds. Very large deposits of phosphate rock occur and are worked in the Bone Valley area of Florida, as well as deposits in Tennessee, Idaho, and South Carolina. Large deposits are mined in Northern Africa (Morocco and Tunisia). Very significant reserves have been found on several of the Pacific Islands, the reserves on Christmas Island estimated at some 30 million tons (27 million metric tons) and those on Nauru Island in excess of 100 million tons (90 million metric tons). There also are large active mining operations in the Mediterranean area as well as in the former Soviet Union. Known reserves assure a supply for several centuries. The mineral apatite, $Ca_3(PO_4)_2 \cdot CaCl_2$ or $\cdot CaF_2$, found in Quebec, Virginia, Brazil, and the South Pacific also contains high percentages of phosphorus, up to 20% P_2O_5. The main constituents of most phosphate rocks is fluorapatite, $3Ca_3(PO_4)_2 \cdot CaF_2$. These rocks contain 30–37% P_2O_5.

Most phosphorus raw materials are converted into phosphorus and phosphorus compounds, such as phosphoric acid, on an extremely high-tonnage basis. Percentagewise, relatively little elemental phosphorus is produced for consumption as an end product. See also **Fertilizer**. Phosphorus is important both to plant and animal nutrition. Traditionally, phosphorus compounds have been key components of cleaning compounds and detergents although there have been trends to reduce or eliminate phosphates from high-consumption items. See also **Detergents**.

Production of Elemental Phosphorus. The tricalcium phosphate in phosphate rock, mixed with coke and silica, is thermally reduced to yield P_2 vapor. The phosphorus vapors condense to a liquid and the carbon monoxide produced is returned for burning in the furnace. The process requires much heat and, in addition to the heat provided by the combustion of the coke and the heating value of the recycled carbon monoxide, an electric arc also is used. The reaction takes place in very large furnaces at a temperature of 1,300–1,500°C and at atmospheric pressure. A 70-MW furnace will produce 44,000 short tons (39,600 metric tons) of P_4 per year, equivalent to 100,000 tons (90,000 metric tons) of P_2O_5 (if converted to acid). Although there are numerous intermediate and side reactions, the overall reaction is: $Ca_3(PO_4)_2 + 5C + 3SiO_2 \rightarrow P_2 + 5CO + 3Ca \cdot SiO_3$. Byproduct ferrophosphorus alloy and calcium silicate slag are tapped from the furnace periodically. Maximum furnace efficiency occurs when the SiO_2/CaO weight ratio is about 0.8. This ratio also assures a minimum melting-point eutectic for the melt and thus lengthens furnace life. This process was originally developed by Readman in England in 1888. The first 1,500-kW furnace in the United States was installed at Niagara Falls, N.Y. in 1896 because of the availability of low-cost energy. For many years the proximity of Tennessee brown stone (a phosphate rock) to the low-cost power of the Tennessee Valley Authority made a good economic combination. Worldwide production of phosphorus by this process is about $\frac{3}{4}$-billion short tons (0.675 billion metric tons) per year (installed capacity). In the United States, about 80% of the phosphorus produced is immediately converted to the oxide and thence to phosphoric acid. The remaining 20% has gone into alloys, organic intermediates for oil and fuel additives, pesticides, plasticizers, and pyrotechnics. In addition to use in detergents, cleaning compounds, and degreasing formulations, phosphoric acid has been consumed in the preparation of liquid fertilizers, water-treatment, pharmaceutical, and chemical products. Phosphorus-containing fertilizers, such as single superphosphate, wet-process orthophosphoric acid, triple superphosphate, ammonium phosphate, and nitrophosphates do not require elemental phosphorus (or the resulting pure P_2O_5) in their preparation, but are manufactured by directly reacting phosphate rock with requisite chemicals, such as H_2SO_4 or HNO_3. See also **Fertilizer**.

Chemistry and Compounds. Like carbon, phosphorus is covalently bound to its neighboring atoms in all of its compounds, except perhaps for some metallic phosphides. Indeed, the chemistry of carbon and that of phosphorus are somewhat similar as might be expected from the diagonal relationship of these elements in the periodic table.

Probably the major difference between carbon and phosphorus is that the former element is quite closely restricted to the use of s- and p-orbitals, because of the relatively high energy of d-orbitals in the case of second-period elements; whereas, phosphorus, being a third-period element, can use d-orbitals in bonding. For both carbon and phosphorus, the most common hybridization for σ-bonding is approximately the tetrahedral sp^3. However, in order to form π-bonds, carbon must go to lower hybrids: sp^2 and sp. Phosphorus, on the other hand, does not do this but can employ d-orbitals for π-bonding. This difference between carbon and phosphorus in sigma bond strength, and the ease with which phosphorus uses its d-orbitals for attachment of attacking nucleophilic groups, can be used to explain why catenation is common in carbon compounds, while at the same time phosphorus compounds containing long chains of connected phosphorus atoms have not yet been synthesized.

The known coordination numbers exhibited by phosphorus within the molecule-ions containing this element are 1, 3, 4, 5, and 6 which, to at least a first approximation, exhibit the symmetry of p, p^3, sp^3, sp^3d and sp^3d^2 hybridization, respectively. A very large number (several thousand in each case) of triply- and quadruply connected phosphorus compounds are known; but there are only a few compounds of higher coordination number in which d-orbitals are involved in the σ-bond base structure. These are the halogen compounds PF_5, PCl_5, PBr_5, PCl_2F_3, PBr_2F_3 and the pentaphenyl compound $(C_6H_5)_5P$, in which the phosphorus is quintuply connected to its neighboring atoms, and the PF_6^- and PCl_6^- anions, in which the phosphorus has a six-fold coordination. The singly connected phosphorus atoms appear only in compounds occurring at very high temperatures. Although singly connected phosphorus is not known under ordinary conditions, interpretation of diatomic spectra has given considerable information.

Several generalities can be stated concerning the phosphorus compounds that are stable under normal conditions:

1. In those compounds in which phosphorus shares electrons with three neighboring atoms, there are three σ-bonds, with little or no π-character, from the phosphorus.
2. In those compounds in which phosphorus shares electrons with four neighboring atoms there are four σ-bonds, with an average of about one π-bond per P atom.
3. When electrons are shared with five or six neighboring atoms, there is less than one full σ-bond for each connection between the phosphorus and a neighboring atom with apparently very little π-bonding.

These generalities are obviously dependent to a considerable extent upon the specific atoms connected to the phosphorus and, indeed, it is possible that the observed differences between the triply and quadruply connected phosphorus atoms may be attributed primarily to the individual ligands. Fluorine appears to contribute nearly as much shortening (assuming that the tabulated values for the fluorine bond length are correct) to the P—F connection in the triply connected as in the quadruply connected phosphorus compounds. On the other hand, chlorine shows essentially no shortening, whether attached to either triply or quadruply connected phosphorus.

Phosphine. PH_3, and its substitution products, have a pyramidal structure. The P—H bond length is 1.42 Å, and the H—P—H angle is 93°. Hypophosphites, containing the radical

are produced by alkaline hydrolysis of white phosphorus. The barium salt yields hypophosphorous acid, H_3PO_2, upon acidification with sulfuric acid. In this acid, only one H is capable of ionization, suggesting the experimentally confirmed formula $H_2P(O)OH$. Hypophosphorous acid and its salts are reducing agents, although their reaction rates are somewhat low, which is usually explained by an equilibrium between H_3PO_2 and its hydrate form, H_5PO_3.

Phosphorus Halides. PX_3, P_2X_4 and PX_5 are formed by direct reaction of the elements, though the pure substances require special methods. Mixed halides are also known. The trihalides are covalent pyramidal compounds, the X—P—X bond angles being generally between 98° and 104°. They all undergo hydrolysis, the rate being roughly inversely proportional to the sum of the atomic numbers of the halogen atoms.

The halogen derivatives of pentavalent phosphorus may be grouped on the basis of their structure into three classes, the pentahalides, oxyhalides and related compounds, and the fluorophosphoric acids. The pentahalides (except the pentaiodide, which is unknown) are produced by reaction of the elements, or, in the case of mixed halides, by reaction of a halogen and a phosphorus trihalide in correct proportions. Their structure in the vapor state has been determined to be a trigonal bipyramid and their bonding is covalent. In the solid, however, phosphorus pentachloride, PCl_5, is $[PCl_4^+][PCl_6^-]$ and phosphorus pentabromide, PBr_5, is $[PBr_4^+]Br^-$. Various mixed halides are known. The five halogen atoms are not in equivalent positions. One may be ionized, with the other four forming sp^3d orbitals or there may be a transition state, to explain the nonequivalence of the exchange between the three equatorial chlorine atoms and the two apical ones in PCl_5 in carbon tetrachloride solution. The pentahalides react with excess water to yield phosphoric acid and hydrohalic acids, but with less water to form phosphorus oxyhalides instead of phosphoric acid.

Phosphorus oxyhalides have the tetrahedral structure

$$\ddot{X}:\ddot{X}:\overset{..}{\underset{..}{P}}:\ddot{O}:$$

These compounds, particularly $POCl_3$ and $POFCl_2$, readily form complexes with metal halides. Closely analogous to the oxyhalides are the thiohalides of general formula PSX_3 and the phosphorus nitrilic halides, $(PNX_2)_n$, the chloride of the latter being obtained by partial ammonolysis of PCl_5, and existing as cyclic or polymeric structures of alternate nitrogen and phosphorus atoms.

Phosphorus Oxides and Oxyacids. The principal oxides are related to the acids which they yield when dissolved in H_2O in the following manner:

Trioxide, P_2O_3	Hypophosphorous acid, H_3PO_2
Tetroxide, P_2O_4	Phosphorous acid, H_3PO_3
Pentoxide, P_2O_5	Hypophosphoric acid, $H_4P_2O_6$
plus $3H_2O$	Orthophosphoric acid, $2H_3PO_4$
plus $2H_2O$	Pyrophosphoric acid, $H_4P_2O_7$
plus $1H_2O$	Metaphosphoric acid, $2HPO_3$

Normally when the term "phosphoric acid" is used, it is with reference to orthophosphoric acid, H_3PO_4. Anhydrous orthophosphoric acid is a white crystalline solid that melts at 42.35 °C. It forms a hemihydrate, $2H_3PO_4 \cdot H_2O$, which melts at 29.32 °C. Although practically any desired concentration can be produced, it is common to supply the material as a solution containing from 75% H_3PO_4 (mp −17.5 °C) to 85% H_3PO_4 (mp 21.1 °C). When phosphoric acid is heated to above 200 °C, the water of constitution is lost. Thus, a series of acids is formed by the dehydration, ranging from pyrophosphoric acid, $H_4P_2O_7$, to metaphosphoric acid, $(HPO_3)_n$. Salts of the dehydrated acids are used for the preparation of certain kinds of liquid fertilizers and are present in numerous cleaning compounds. The dehydrated acids can form water-soluble complexes with many metals, such as calcium. One, or two, or three of the hydrogens of phosphoric acid may be neutralized. When one hydrogen is replaced with sodium, for example, the product is slightly acidic; while replacement of all three hydrogens yields a highly alkaline product. The acidity of the solutions is: NaH_2PO_4, a pH of 4.0; Na_2HPO_4, a pH of about 8.3; Na_2PO_4, a pH of 12.0. Although of interest scientifically, the other acids of phosphorus, hypophosphorous acid, H_3PO_2, orthophosphorous acid, H_3PO_3, and pyrophosphorous acid, $H_4P_2O_5$, are not important commercially.

There are two main processes for the industrial production of phosphoric acid, H_3PO_4, from phosphate rock: (1) the *wet process* which involves the reaction of phosphate rock with H_2SO_4 to yield phosphoric acid and insoluble calcium sulfates. Several of the impurities present in the rock dissolve and remain with the product acid. These are not important when the acid is used for fertilizer manufacture. However, the impurities are deleterious to the manufacture of phosphorus chemicals. For a purer product, (2) the *furnace process* is used, wherein the phosphate rock is combined with coke and silica, producing elemental phosphorus as previously described. Oxidation of the phosphorus produces P_2O_5 which, when combined with H_2O, yields H_3PO_4.

Phosphorus sesquioxide, P_4O_6, produced by controlled oxidation of white phosphorus, is hydrolyzed in the cold to produce phosphorous acid, H_3PO_3, a colorless solid (mp 73 °C). Only two of its H atoms are capable of ionization, thus compounds such as M_3PO_3 do not exist. This fact leads to the formula $HP(O)(OH)_2$. Phosphorous acid is a somewhat stronger acid than phosphoric, and both it and the phosphite ion (HPO_3^{2-}) are strong reducing agents.

Hypophosphorous acid was discussed earlier in this entry.

Metaphosphorous acid, HPO_2, is produced by atmospheric combustion of PH_3, but in aqueous solution it hydrates to H_3PO_3.

Phosphorous acid is used in solution, and is usually a reducing agent. That is, in air it changes to phosphoric acid, with hot concentrated sulfuric acid it yields phosphoric acid plus SO_2, with copper sulfate it yields finely divided copper metal, with silver nitrate it yields finely divided silver metal, with permanganate after some time it yields manganous. Occasionally it is an oxidizing agent, e.g., with zinc plus dilute H_2SO_4 it yields phosphine.

Phosphorus tetroxide, P_2O_4, is obtained along with red phosphorus by heating P_4O_6 at 290 °C in a closed tube. It is believed to have the formula, P_8O_{16}, and to consist of trivalent and pentavalent phosphorus. Hypophosphoric acid, $H_4P_2O_6 \cdot 2H_2O$, cannot be produced directly from the tetroxide. The acid decomposes into phosphorous and phosphoric acids on heating, and must be prepared by indirect methods, such as treatment of white phosphorus with an HNO_3 solution of $Cu(NO_3)_2$.

Phosphorus(V) oxide is the chief product of atmospheric oxidation of phosphorus, whence it is obtained as the β-allotrope, of formula P_4O_{10}. Several other allotropes are obtained by various thermal treatments of the β-form, differing in structure and physical properties. The compound hydrates rapidly to form various phosphoric acids. With an excess of H_2O, orthophosphoric acid, $(HO)_3PO$, is formed, mp $42.3\,°C$. It is a triprotic acid, yielding $H_2PO_4^-$, HPO_4^{2-} and PO_4^{3-} ions. The other crystalline phosphoric acid, pyrophosphoric acid, is believed to have the formula $(HO)_2P(O)-O-P(O)-(OH)_2$. Its acid solution undergoes hydrolysis to the orthoacid. It is tetraprotic, yielding the ions $H_3P_2O_7^-$, $H_2P_2O_7^{2-}$, $HP_2O_7^{3-}$ and $P_2O_7^{4-}$.

Classification of Phosphates. Audrieth and Hill have proposed a classification on the basis of structure, that is, to divide them into glassy phosphates and crystalline phosphates, the latter class being subdivided into (1) linear phosphates and polyphosphates, and (2) cyclic phosphates. Three important members of class (1) are the orthophosphates, containing the PO_4^{3-} ion, the pyrophosphates, having the $P_2O_7^{4-}$ ion, and the triphosphates, having the $P_3O_{10}^{5-}$ ion. The general structural unit of the linear phosphates is the tetrahedron, containing a phosphorus atom surrounded by four oxygen atoms covalently linked to it. Such tetrahedra are linked through a common oxygen atom to form linear polyphosphates, the $P_2O_7^{4-}$ ion having two such tetrahedra, and the $P_3O_{10}^{5-}$ ion, three. Moreover, the tetrahedra may also be double linked through oxygen atoms to form cyclic structures, as in the trimetaphosphate ion, $P_3O_9^{3-}$, which has three such tetrahedra and the tetrametaphosphate ion, $P_4O_{12}^{4-}$, which has four tetrahedra. In general, heating acid salts of simpler phosphates produces polyphosphates (loss of H_2O), while alkaline hydrolysis reverses the process. The glassy phosphates, produced by fusion and rapid cooling of metaphosphates, appear to be true glasses, containing anions with molecular weights well into the thousands.

The widely diverse functionality of phosphates makes them of exceptional importance to technologists, particularly in the food processing field. Phosphoric acid finds many direct uses as an acidulant. It has three available hydrogens which can be replaced one by one with alkali metals, forming a series of *orthophosphate salts* with pH levels ranging from moderately acid (pH = 4) to strongly alkaline (pH = 12). This wide pH range makes phosphates very useful for adjusting the pH of food and chemical systems to almost any desired level. Heating orthophosphates converts them to condensed phosphates containing two, three, or more phosphorus atoms per molecule. The *condensed phosphates*, or *polyphosphates*, have many properties that the orthophosphates do not enjoy. They are polyelectrolytes, and have dispersing or emulsifying properties. They can sequester or chelate metals, such as calcium, magnesium, iron, and copper, rendering these metals nonreactive. This functionality is useful for controlling oxidative rancidity and color formation, as both are catalyzed by metal ions.

Condensed phosphates containing two atoms of phosphorus are *pyrophosphates*. Sodium acid pyrophosphate (SAPP) is used as a leavening acid in baking, and is particularly useful because of the way it can be modified to give different rates of reaction. Pyrophosphates are good sequestrants for iron and copper, which often catalyze oxidation in fruits and vegetables. Thus, the use of a pyrophosphate effectively prevents the discoloration of such foods during preparation and storage.

Condensed phosphates containing three atoms of phosphorus are *tripolyphosphates*, the most important of which is sodium tripolyphosphate (STPP). This compound reacts with the protein in meat, fish, and poultry to prevent denaturing or loss of fluids. This property is sometimes called "moisture binding." STPP also solubilizes protein, which aids in binding diced cured meat, fish, and poultry. It also emulsifies fat to prevent separation.

The chain length of phosphates can be increased further by melting and chilling to form a glass. Glassy sodium phosphates are generally called *sodium hexametaphosphates* (SHMP). SHMPs have excellent sequestering power toward calcium and magnesium. They are used in meat treatment as a partial replacement for STPP to improve solubility in strong pickling brine or to prevent hardness precipitation in very hard water. Considerably more information on the use of phosphates in food processing will be found in the *Foods and Food Production Encyclopedia*, D.M. Considine (editor), Van Nostrand Reinhold, New York, 1981. See also **Fertilizer.**

Peroxyphosphoric Acids. Two are known—peroxymonophosphoric acid, $HOOP(O)(OH)_2$, prepared by treatment of P_4O_{10} with hydrogen peroxide, and peroxydiphosphoric acid, $(HO)_2(O)POOP(O)(OH)_2$, prepared from metaphosphoric acid and peroxide or by electrolysis of an alkali hydrogen phosphate solution (cf. preparation of peroxydisulfuric acid). They and their salts are strong oxidants.

Fluorophosphoric Acids. H_2PO_3F, HPO_2F_2, $(POF_3$ is not a protic acid), "HPF_6" are obtained by replacing one or more hydroxyl groups with fluorine. They are strong fuming acids, like H_2SO_4 in most properties except its oxidizing power. Their salts are also known, extending as far as completely fluorinated MPF_6 where M is an alkali metal. The solubilities of the monofluorophosphates parallel those of the sulfates; while those of the di- and hexafluorophosphates parallel those of the perchlorates.

Upon acidification or neutralization of solutions containing phosphate anions with other anions, such as those of molybdenum and tungsten, complexes are formed which can readily be crystallized as salts, called phosphomolybdates, phosphotungstates, etc.

Oxygen-nitrogen Compounds of Phosphorus. These may be classified as aquo, aquo ammono, or ammono derivatives. The first group includes the various acids and oxides already discussed. The second group includes the amido phosphoric acids in which one or more of the hydroxy groups of the acid are substituted by amino groups. Thus there is amidophosphoric acid (*orthophosphoric acid* is understood when the substituted acid is not specified), amidopyrophosphoric acid, diamidophosphoric acid, and triamidophosphoric acid. The substitution of all —OH groups of phosphoric acid gives phosphoryl triamide, $OP(NH_2)_3$. Related compounds are phosphoryl amide imide, $OP(NH_2):NH$ and phosphoryl nitride $(OPN)_x$. The second group also includes imidodiphosphoric acid,

$$HN \begin{matrix} PO(OH)_2 \\ \\ PO(OH)_2 \end{matrix}$$

diamidotriphosphoric acid,

$$\begin{matrix} PO(OH)_2 \\ HN \\ PO(OH)_2 \\ HN \\ PO(OH)_2 \end{matrix}$$

and still longer chain acids. Finally, many other derivatives are possible because of the stability of the $N\equiv P$ arrangement. This gives rise to the phosphonitrilic acids, $[NP(OH)_2]_x$, as well as to many ammono derivatives containing only the two elements. The latter include phosphonitrilamide, $[NP(NH_2)]_x$, phospham $[NP=NH]_x$, and phosphoric nitride, $[P_3N_5]_x$. The phosphonitrilic chlorides, already discussed, are derivatives of phosphonitrilamide.

Organophosphorus Compounds. Most of the industrially important organic compounds of phosphorus commence with one of the basic inorganic phosphorus compounds, such as PCl_3, $POCl_3$, P_2S_5, and P_2O_5, reacted with an appropriate organic intermediate. Ester intermediates, such as alkyl phosphoryl chlorides, are made by the addition of primary alcohols to $POCl_3$. Triaryl phosphate plasticizers and gasoline additives, such as tricresyl phosphate (TCP), can be prepared from PCl_5 and an appropriate phenolic compound. Alkyl diaryl phosphates can be made from $POCl_3$ and corresponding phenols. A number of thiophosphate esters contain PS plus an ethyl or methyl group and a substituted aryl group and are based on $PSCl_3$ or P_2S_5. These compounds are finding use for pesticide control. Dialkyl dithiophosphates may be prepared from P_2S_5 and appropriate intermediates. They are finding application as flotation-agents, oil-additives, and insecticides. It is much more difficult to prepare organophosphorus compounds containing a C—P bond than it is to form the esters. Numerous organic phosphorus compounds are found in nearly all life processes and remain to be better understood before they can be synthesized. Classes of compounds of this type include the phosphoglycerides required for fermentation, the adenosine phosphates needed in photosynthesis and muscle activity, and the very complex phosphorus-containing groups identified in the nucleotides. Of structural interest are the catenation compounds, as illustrated below, which contain many cyclic phosphates and oxygen-linked chains. Compounds of this type include tetrachlorodiphosphine, Cl_2PPCl_2, tetraphenyldiphosphine $(C_6H_5)_2PP(C_6H_5)_2$, diphosphobenzene, $C_6H_5PPC_6H_5$, and tetramethyl hypophosphate, $(CH_3O)_2(O)PP(O)(OCH_3)_2$.

Inorganic Macromolecules. After the accelerated activity in the development of new polymers that took place during the past 30 or 40 years, some researchers observed some lessening of polymer research and polymer achievements during the 1970s. Not all authorities agreed, but most agreed that the time had arrived when polymer chemistry and applications deserved evaluation both in terms of the past and the future. Synthetic polymers generally have a number of relatively negative features — flammability (derived from their organic nature); a tendency to melt, oxidize, and char at high temperatures in regular atmospheric conditions (again, a result of their organic nature); and a tendency to become stiff and brittle at low temperatures. Many also have a tendency to soften, swell and dissolve in a number of common substances, such as gasoline, jet fuel, hot oil, and numerous other hydrocarbons. In terms of medical applications, most organic polymers tend to initiate a clotting reaction of the blood and many tend to cause toxic, irritant, and sometimes carcinogenic responses. Observers also noted that most polymer research in some way initiated with petrochemicals.

In the early 1970s, a number of investigators decided to shift emphasis and to look at a number of inorganic elements including silicon, phosphorus, sulfur, boron, and some metal atoms that might make up the backbone of a polymer. It was reasoned that the presence of some of these materials in the backbone might remedy some of the aforementioned shortcomings. It should be pointed out that as early as the 1940s, silicone, or poly(organosiloxane) polymers were developed and have proved highly satisfactory in many applications. Peters et al. (1976) reported on a new class of thermally stable polymers, which are based upon alternating siloxane and carborane units. In 1965, Allcock et al. (Pennsylvania State University) synthesized the first poly(organophosphazenes). Since that time, well over 60 new polymers have been made and presently constitute a substantial class of new elastomers. They appear to solve some of the biomedical problems previously mentioned. As pointed out by Allcock (1976), all linear, high polymeric polyphosphazenes have the general structure shown by (a), below. It is interesting to note that over a century ago researchers in Germany and Britain found that phosphorus pentachloride will react with ammonia or ammonium chloride to yield a volatile, white solid. It is now known that this product has the form of (b). Later experimentation showed that the compound would melt under strong heating to form a transparent rubbery material. Involving a ring-opening polymerization of the cyclic trimer, a poly(dichlorophosphazene), shown in (c), is formed. Mainly for the reason that the compound hydrolyzes slowly in the presence of atmospheric moisture to form a crusty mixture of ammonium phosphate and phosphoric acid, the substance was not given serious thought for many years.

Allcock and associates, during the 1960s, subjected the cyclic trimer to new procedures and, after considerable research, were successful in developing polymers that have molecular weights up to and sometimes exceeding 3 to 4 million. The investigators found that the introduction of different substituent groups had a marked effect on the properties of the polymers. See (d). The further detailed research is beyond the scope of this book, but interesting details can be found in the Allcock reference. In summarizing one of their reports, the researchers stated: "Polyphosphazenes are emerging as a new class of macromolecules that have an obvious future as technological elastomers, films, fibers, and textile treatment agents. However, they also possess almost unique attributes for use in biomedicine as reconstructive plastics or as drug-carrier molecules. Moreover, their possible value as 'pseudo-protein' model polymers is an exciting prospect."

Phosphorus Ylides. In 1979, G. Wittig received a share of the Nobel Prize for Chemistry in recognition of his development of the use of phosphorus-containing compounds as important reagents used in organic synthesis. According to the selection committee, Wittig's most important achievement was "the discovery of the rearrangement reaction that bears his name. In the Wittig reaction an organic phosphorus compound with a formal double bond between phosphorus and carbon is reacted with a carbonyl compound. The oxygen of the carbonyl compound is exchanged for carbon, the product being an olefin. This method of making olefins has opened up new possibilities, not the least of which is the synthesis of biologically active substances containing carbon-to-carbon double bonds. For example, vitamin A is synthesized industrially using the Wittig reaction." As early as 1919, some 30 years prior to Wittig's work, the first phosphorus ylide was described by Staudinger. This was diphenylmethylenetriphenylphosphorane, formed by pyrolysis of a phosphazine precursor. During that period, Staudinger conceived the possibility of olefin synthesis by condensation of this ylide with a carbonyl compound and visualized a four-membered phosphorus-oxygen heterocyclic (oxaphosphetane) as the intermediate. Staudinger's work was accomplished during a period when practical application of the concept was in doubt. At that time, the Lewis theory of electronic structure was new. The exact bonding of phosphonium salts was somewhat veiled and controversial. Attempts of an olefin synthesis were put aside. For more detail, see **Wittig Reaction**.

Additional Reading

Allcock, H.R.: "Polyphosphazenes: New Polymers with Inorganic Backbone Atoms," *Science*, **193**, 1214–1219 (1976).

Burges, R.J.: "Choose the Right Alloys for Fertilizer Acids," *Chem. Eng. Progress*, 82 (November 1992).

Considine, D.M. and G.D. Considine: "Foods and Food Production Encyclopedia," Van Nostrand Reinhold, New York, NY, 1982.

Corbridge, D.E.: "Phosphorus: An Outline of Its Chemistry, Biochemistry and Technology," 5th Edition, Elsevier Science, New York, NY, 1995.

Corbidge, D.E.: "Phosphorus 2000: Chemistry, Biochemistry and Technology," Elsevier Science, New York, NY, 2000.

Dillon, K.B., F. Mathey, and J.F. Nixon: "Phosphorus: The Carbon Copy," John Wiley & Sons, Inc., New York, NY, 1998.

Emsley, J.: "The 13th Element: The Sordid Tale of Murder, Fire and Phosphorus," John Wiley & Sons, Inc., New York, NY, 2000.

Greenwood, N.N. and A. Earnshaw: "Chemistry of the Elements," 2nd Edition, Butterworth-Heinemann, Inc., Woburn, MA, 1997.

Kent, J.A.: "Riegel's Handbook of Industrial Chemistry," 9th Edition, Chapman & Hall, New York, NY, 1992.

Krebs, R.E.: "The History and Use of Our Earth's Chemical Elements: A Reference Guide," Greenwood Publishing Group, Inc., Westport, CT, 1998.

Lewis, R.J. and N.I. Sax: "Sax's Dangerous Properties of Industrial Materials, 10th Edition, John Wiley & Sons, Inc., New York, NY, 1999.

Lide, D.R.: "CRC Handbook of Chemistry and Physics 2000-2001," 81st Edition, CRC Press, LLC., Boca Raton, FL, 2000.

Peters, E.N., et al.: *Rubber Chemistry*, 1976.

Somerville, R.L.: "Reduce Risks of Handling Liquefied Toxic Gas (Phosgene)," *Chem. Eng. Progress*, 64 (December 1990).

Stevenson, F.J. and M.A. Cole: "Cycles of Soils: Carbon, Nitrogen, Phosphorus, Sulfur, Micronutrients," 2nd Edition, John Wiley & Sons, Inc., New York, NY, 1999.

Vedejs, E.: "1979 Nobel Prize for Chemistry," *Science*, **207**, 42–44 (1980).

PHOSPHORUS (In Biological Systems).   Phosphorus is required by every living plant and animal cell. Deficiencies of available phosphorus in soils are a major cause of limited crop production. Phosphorus deficiency is probably the most critical mineral deficiency in grazing livestock. Phosphorus, as orthophosphate or as the phosphoric acid ester of organic compounds, has many functions in the animal body. As such, phosphorus is an essential dietary nutrient.

The biological roles of phosphorus include: (1) anabolic and catabolic reactions, as exemplified by its essentiality in high-energy bond formation, e.g., ATP (adenosine triphosphate), ADP (adenosine diphosphate), etc., and the formation of phosphorylated intermediates in carbohydrate metabolism; (2) the formation of other biologically significant compounds, such as the phospholipids, important in the synthesis of cell membranes; (3) the synthesis of genetically significant substances, such as DNA (deoxyribonucleic acid) and RNA (ribonucleic acid); (4) contributing to the buffering capacity of body fluids, cells, and urine; and (5) the formation of bones and teeth. Like calcium, the majority of the phosphorus in the vertebrate body is contained in the hard tissues; in the adult, approximately 80–86% of the total body phosphorus is contained in the bones and teeth, with the balance found in the soft tissues and body fluids.

In a very interesting dissertation by F.H. Westheimer (*Science*, **235**, 1173–1177, 1987), the role of phosphates in living substances is described. As pointed out by Westheimer, phosphate esters and anhydride dominate the living world but are seldom used as intermediates by organic chemists. Phosphoric acid is specially adapted for its role in nucleic acids because it can link two nucleotides and still ionize; the resulting negative charge serves both to stabilize the diesters against hydrolysis and to retain the molecules within a lipid membrane. A similar explanation for stability and retention also holds for phosphates that are intermediary metabolites and for phosphates that serve as energy sources. Phosphates with multiple negative charges can react by way of the monomeric metaphosphate ion PO_3^- as an intermediate. No other residue appears to fulfill the multiple roles of phosphate in biochemistry. Stable negatively charged phosphates react under catalysis by enzymes; organic chemists, who can only rarely use enzymatic catalysis for their reactions, because they need more highly reactive intermediates than phosphates.

Most of the coenzymes are esters of phosphoric or pyrophosphoric acid. The main reservoirs of biochemical energy, adenosine triphosphate (ATP), creatine phosphate, and phosphoenolpyruvate are phosphates. Many intermediary metabolites are phosphate esters, and phosphates or pyrophosphates are essential intermediates in biochemical syntheses and degradations. The genetic materials DNA and RNA are phosphodiesters.

Phosphorus deficiencies are not common in humans and most other species, but they have been observed in ruminants. Symptoms of the deficiency are loss of appetite and a depraved appetite (termed "pica") where the animal chews and consumes extraneous items, such as wood, clothing, bones, etc. Vitamin D deficiency may accentuate a marginal lack of phosphorus in the diet.

Cereals and meats are the major sources of phosphorus in human diets. Phosphorus deficiencies in most regions have not been a serious problem in human nutrition. Insofar as food is concerned, the primary value of phosphorus fertilizers is that they generally increase the total food production; not the content of phosphorus in the food per se.

Experimental phosphorus deficiency can be induced by feeding diets low in this element and by including excesses of calcium, strontium, barium, beryllium, and other cations that precipitate phosphates in the intestinal tract. In this situation, bone formation ceases, and the following histological bone changes have been noted in experimental animals: (1) a thickening of the epiphyseal plate and the formation of a typical rachitic metaphysis; (2) wide osteoid borders of trabecular bone and a considerable rarefaction of the shaft; and (3) irregular or complete cessation of calcification of the zone of provisional calcification of the cartilage matrix. Rickets can be produced in the laboratory by feeding a diet high in calcium and low in phosphorus, and containing little or no vitamin D.

The nutrient requirement for phosphorus depends upon the particular species and the physiological status of the animal. During growth, lactation, gestation, and egg laying, a higher phosphorus content of the diet is generally required in poultry than for the maintained adult. The availability

of phosphorus in the diet varies with its chemical form and the animal species in question. Diets high in foods of plant origin may contain a considerable portion of phosphorus in the form of phytic acid, which is the hexaphosphoric acid ester of inositol. When the acid occurs as salts of calcium, magnesium, sodium, etc., it is referred to as phytin. Phytate phosphorus usually is less available than inorganic phosphate to such species as rat, chicken, dog, pig, and human. However, a phytase has been shown to be present in the intestine and intestinal secretions of some animals, and the formation of this enzyme is dependent, in part, on the presence of vitamin D. Through the action of phytase, some of the phytate phosphorus would be made available for absorption.

Experimentation has indicated that, under normal dietary conditions and calcium intake, food phytate is of no nutritional concern in humans. The microbial population of the ruminant also elaborates a phytase enzyme that makes phytate phosphorus readily available in this class of animals. Phytates may be of nutritional consequence for another reason — dietary calcium can be bound in an unavailable, insoluble complex, thereby decreasing the absorption of this element.

Many studies have involved determination of the availability of phosphorus from other organic and inorganic sources. In chicks, orthophosphates, superphosphates, and phosphate rock products are good sources of phosphorus, whereas metaphosphate and pyrophosphate are relatively unavailable to the species. Most organic phosphorus sources, such as casein, pork liver, and egg phospholipid are found to be as available as inorganic phosphorus. Commonly used phosphorus supplements in human or animal nutrition or both are steamed bone meal, ground limestone, dicalcium phosphate, and defluorinated rock phosphates. Phosphorus dietary supplements include magnesium phosphate (dibasic and tribasic) manganese glycerophosphate and manganese hypophosphate, potassium glycerophosphate, sodium ferric pyrophosphate, sodium phosphate (mono-, di-, and tri-), and sodium pyrophosphate. Of course, phosphate compounds are not always added in the interest of augmenting phosphorus, but for the other elements which may be contained in the compound.

Absorption of Phosphate.   The phosphate ion readily passes across the gastrointestinal membrane. The rate of absorption of phosphate at various intestinal sites in rats has been observed to be most rapid in the duodenum, followed in decreasing order by the jejunum, ileum, colon, and stomach. When transit time is considered, most of the phosphorus is absorbed by the ileum.

The triangular relationship between calcium, phosphorus, and vitamin D is described briefly in entry on **Calcium (In Biological Systems).**

Plasma Phosphate.   Once absorbed, phosphorus enters the blood and the majority is present therein as orthophosphate ions. At an ionic strength of 0.165 and at 37 °C (98.6 °F), calculations show that the proportional concentration of the orthophosphate ions in plasma for $H_2PO_4^-$ is 18.6×10^{-30}; for HPO_4^{2-} is 81.4×10^{-30}; and for PO_4^{3-} is 8×10^{-30}. About 12% of the phosphorus present is bound to proteins. During egg laying in birds, the concentration of nonionized phosphorus compounds in plasma is greatly increased. The administration of diethylstilbestrol (regulated in some countries) to cockerels results in the formation of a plasma phosphoprotein, which forms relatively firm complexes with calcium. The function of the phosphoprotein appears to be one of phosphorus transport; in laying birds, the phosphoprotein is incorporated in egg yolk.

The approximate average plasma phosphorus levels for several species, in milligrams per 100 milliliters of plasma, are: pigs, 8.0; sheep, cattle, and goats, 6.0; horse, 2.3. Erythrocytes contain considerably more phosphorus than plasma, mostly in the form of organic esters. Some of the latter are acid soluble and hydrolyzable by intracellular enzymes.

Plasma phosphate appears to be homeostatically controlled. The primary organ concerned appears to be the kidney, although the skeleton also may play a role. Parathyroid hormone, by way of its direct action on the kidney and bone, is a significant hormonal factor.

Phosphate Excretion.   The excretion of body phosphorus occurs via the kidney and intestinal tract, the distribution between these pathways varying with species. For example, relatively small amounts of phosphorus are endogenously excreted into the feces of rat, pig, and human, but in the bovine, perhaps 50% or more of the fecal phosphorus may be from endogenous sources.

The amount of phosphorus excreted in the urine varies with the level of ingested phosphorus and factors influencing phosphorus availability and utilization. It has been shown that in the dog, when plasma phosphate is normal or low, over 99% of the filtered ion is reabsorbed, presumably in the upper part of the proximal tubule. Increased plasma concentrations of alanine, glycine, and glucose depress phosphate reabsorption.

Phosphate of Hard Tissues. Body phosphorus contained in the intracellular matrix of bone and teeth is of the general form of hydroxyapatite, $Ca_{10}(PO_4)_6(OH_2)$, this calcium phosphate salt providing the characteristic hardness of ossified tissue. Phosphate ions are also adsorbed onto the surface of bone crystals and exist in the hydration layers of the crystals. Early theories of calcification placed special emphasis on the role of alkaline phosphatase and organic esters of phosphoric acid. As part of the theory, it was stipulated that, with the hydrolysis of phosphate esters at the site of calcification, the K_{sp} for bone salt would be exceeded. Although phosphatase may have a function in bone formation, as in the synthesis of organic matrix, its role as earlier depicted has been revised. Later research emphasized the specific and characteristic properties of collagen and other substances, such as chondroitin sulfate. This is related to the local mechanism of calcification; the other component of calcification is the humoral mechanism whereby an adequate supply of calcium, phosphate, and other ions is made available to the calcifying site. A later theory proposed that either the functional groups on collagen are anionic, initially binding Ca^{2+}, or that the first reaction is with phosphate or phosphorylated intermediates. The first-held moiety of bone salt (Ca^{2+} or phosphate) subsequently attracts or binds the other component, providing the aggregation or "seed" for subsequent crystal growth. Since an ATPase-type enzyme has been demonstrated in cartilage, suggesting that ATP may be intimately involved in the calcification mechanism, another proposal is along the line that pyrophosphate is transferred from ATP to free amino groups of collagen, leading to nucleation and followed by combination with calcium and bone salt formation. Or, the ATP provides energy which increases the calcification mechanism.

Dietary inorganic phosphates have been shown to protect experimental animals against dental caries. Orthophosphates were effective cariostats, but $Na_4P_2O_7$ and $Na_5P_3O_{10}$ were not. Dicalcium phosphate, $CaHPO_4$, did not decrease dental caries unless a high level of NaCl was also included in the diet.

Toxicity. Although many phosphorus-containing compounds are vital to life processes, as previously described, there are also many phosphorus compounds that are quite toxic — elemental phosphorus, for example. While the elemental form is dangerous because of its low combustion temperature, its absorption also has an acute effect on the liver. The long and continued absorption of small amounts of phosphorus can result in necrosis of the mandible or jaw bone (sometimes called "phossyjaw"). Chronic phosphorus poisoning occurs particularly through the lungs and gastrointestinal tract. The most common symptom is the necrosis of the jaw already mentioned, but this is also usually accompanied by anemia, loss of appetite, gastrointestinal weakness, and pallor. Other bones and teeth may be adversely affected.

Phosphine is a very toxic gas. Inhalation of phosphine causes restlessness, followed by tremors, fatigue, slight drowsiness, nausea, vomiting, and, frequently severe gastric pain and diarrhea. Although most cases recover without after-effects, in some cases, coma or convulsions may precede death.

Phosphorus-halogen compounds are quite toxic.

Phosphorus in Soils. When phosphorus fertilizers are added to soils deficient in available forms of the element, increased crop and pasture yields ordinarily follow. Sometimes the phosphorus concentration in the crop is increased, and this increase may help to prevent phosphorus deficiency in the animals consuming the crop, but this is not always so. Some soils convert phosphorus added in fertilizers to forms that are not available to plants. On these soils, very heavy applications of phosphorus fertilizer may be required. Some plants always contain low concentrations of phosphorus even though phosphorus availability from the soil may be good. See also **Fertilizer**.

PHOSPHORYLATION (Oxidative). This is an enzymic process whereby energy released from oxidation-reduction reactions during the passage of electrons from substrate to oxygen over the electron transfer chain is conserved by the synthesis of adenosine triphosphate (ATP) from adenosine diphosphate (ADP) and inorganic orthophosphate. Since ATP is the major source of energy for biological work, and since most of the net gain of ATP in the animal cell derives from oxidative phosphorylation, research in the area has been very active.

Oxidative phosphorylation was discovered simultaneously and independently in 1939 by Kalckar (Denmark) and by Belitzer (the former U.S.S.R.). It was recognized by these workers that aerobic phosphorylation was different from and independent of phosphorylation supported by glycolysis. In addition, they found that the stoichiometry of phosphate esterification (ATP synthesis) and oxygen utilized was two or more, or that the reduction of one atom of oxygen to form water may be accompanied by the "activation" of two or more molecules of phosphorus (P_i), thus leading to an expression of the efficiency of the energy-conserving system. The efficiency expression is known as the P/O ratio, i.e., the ratio of molecules of P_i esterified per atom of oxygen utilized.

The quantitative importance of the ATP synthesized at the expense of energy liberated during electron transfer in the mitochrondrion is realized when one follows the conservation of energy during the metabolism of a molecule, such as glucose in the cell. The oxidation of one mole of glucose to carbon dioxide and water is accompanied by the release of 673,000 calories. In order to degrade the glucose molecule to a form that can be metabolized further by mitochondrial enzymes, the glycolytic enzymes consume two molecules at ATP and also synthesize two molecules of ATP in the presence of oxygen, a net energy conservation of zero. The mitochondrion may then degrade the pyruvate supplied by glycolysis to carbon dioxide and water, yielding a net total of 38 molecules of ATP, mostly at the level of the electron transfer process. Thirty-eight molecules of ATP per molecule of glucose results in between 260,000 and 380,000 calories conserved, between 39 and 56% of the total energy released in the complete oxidation of glucose, the remainder being released directly to heat. Inasmuch as the mitochondrion is approximately 50% effective in conserving energy from its major substrate, it is indeed an efficient machine. See also **Cell (Biology)**.

PHOSPHORYLATION (Photosynthetic). Photosynthetic conversion of light energy into the potential energy of chemical bonds involves an electron transport chain, and the phosphorylation of ADP (adenosine diphosphate)

$$ADP + P \xrightarrow[\text{Chlorophyll}]{\text{Light}} ATP$$

as intermediate stages. The process of phosphorylation defined by the foregoing equation was discovered simultaneously by Arnon and coworkers for green plant chloroplasts and by Frenkel, working with Geller and Lipmann, for bacteria in 1954. For both systems, the heart of the mechanism is the creation of a very oxidizing and a very reducing component, utilizing the energy of the photoexcited stage of one of the pigment (chlorophyll) molecules. This process will be designated a photoact. The redox components are both members of a photosynthetic electron transport chain, bound to the membranes of the chloroplasts (for green plants) or chromatophores (for bacteria). The photoact can be considered as electron transport against the thermochemical gradient, i.e., away from the member which is a better electron acceptor (high oxidation-reduction potential), through the excited chlorophyll, then to the member which is a better electron donor (low oxidation-reduction potential). Subsequent steps consist of ordinary, dark electron transport with the thermochemical gradient. The energy in at least one of these redox reactions is conserved as ATP by a phosphorylation reaction analogous to that found in oxidative phosphorylation by mitochondria. See also **Phosphorylation (Oxidative)**.

In bacteria, the photoact proper is accomplished by a special kind of bacteriochlorophyll, amounting to only 3% of the total present. It is unique in having a peak in absorption at 870–890 micrometers, or further into the infrared than the remaining 97% of the chlorophyll molecules. It is unique not by virtue of a difference in structure, but because of its environment — most probably in close association or complexing with cytochrome molecules. Since its absorption extends to longer wavelengths, it is an energy trap, and the function of the bulk of the bacterio-chlorophyll is that of capturing light and transmitting it to this active center.

Components of the electron transport chain in bacteria have been shown to include b- and c-type cytochromes, ubiquinone (fat-soluble substitute quinone, also found in mitochondria), ferredox (an enzyme containing nonheme iron, bound to sulfide, and having the lowest potential of any known electron-carrying enzyme) and one or more flavin enzymes. Of these a cytochrome (in some bacteria, with absorption maximum at 423.5 micrometers, probably c_2) has been shown to be closely associated with the initial photoact. Some investigators were able to demonstrate, in chromatium, the oxidation of the cytochrome at liquid nitrogen temperatures, due to illumination of the chlorophyll. At the very least this implies that the two are bound very closely and no collisions are needed for electron transfers to occur.

In both bacterial chromatophores and green plant chloroplasts, the existence of photo-induced high-energy intermediates or states leads to reversible confirmational changes in the structures of the membranes, and to gross swelling and shrinking. These are observed by changes in light scattering, viscosity, and sedimentation properties, and by electron microscope studies. The mechanisms may include ion transport, followed by water diffusion, internal pH changes leading to conformation changes of proteins, or possibly something resembling a contractile protein.

PHOT. A photometric unit of illuminance or illumination equal to 1 lumen per square centimeter.

PHOTOCHEMISTRY AND PHOTOLYSIS.

When certain substances are subjected to light, a chemical change results. Such reactions comprise *photochemistry*. The production of an image on a photographic plate is an example. Photosynthesis in the green leaf of a plant is another. Where the change involves chemical decomposition of the radiated material, the process is termed *photolysis*. As used in this context, the term light includes both visible light and ultraviolet radiation. One of the better known and most extensive examples of photolysis is the production of ozone, O_3, in the upper atmosphere, a reaction critical to life on earth because ozone acts as a filter of the middle- and far-ultraviolet radiations which destroy living organisms. The oxygen molecule, O_2, absorbs solar ultraviolet radiation with a wavelength of 190 nanometers, and the energy absorbed breaks the molecule to the atomic state. The released oxygen atoms may combine with oxygen molecules present to form ozone, or the freed oxygen atoms may recombine to form O_2. Thus, there is a continuing combination of processes in dynamic equilibrium, that is, the synthesis and the photolysis of ozone.

Similarly, oceanic nitrite (NO_2^-) photolysis by natural light produces detectable concentrations of nitric oxide (NO). This latter forms in the oceans during daylight and disappears rapidly at sunset when recombination occurs: $NO_2 \rightarrow NO + O \rightarrow NO_2$.

Isomerism can also be induced photochemically, although such processes are less well understood and probably require the presence of additional free radicals. The cytotoxic metabolite bilirubin can cause brain damage in infants with neonatal jaundice; this is prevented by exposing the child to intense blue light. The bilirubin is photochemically converted in the skin to metastable geometric isomers, which can be transported in the blood and excreted in bile.

Two major instances of photochemical reactions that have reached deeply into modern civilization are the photosensitive silver and uranium salts and dyes which are the basis of photography and the manufacture of Vitamin D by the ultraviolet irradiation of ergosterol.

Photochemical reactions are highly specific and their products are quite different from those of thermochemical reaction processes.

Sunlight in the near infrared, visible, and near ultraviolet regions possesses considerable energy; utilization of this through photochemical reactions could make a considerable contribution to energy resources. Since biosynthesis itself is relatively inefficient in conversion of solar energy, emphasis has been placed upon the fabrication of *artificial* photochemical systems. One of the more promising approaches has involved application of photoelectric chemical cells or catalysts of semiconductor materials.

The absorption of light by semiconductors creates electron-hole pairs ($e^- h^+$) which can be separated because their components diffuse in different directions. The energies of these moieties can be stored by several mechanisms or used in photocatalysis or photosynthesis for nitrogen fixation, formation of amino acids, methanol, etc. The efficiencies of such conversions depend almost entirely upon the semiconductor material, and as yet these efficiencies are too low for significant application. Currently the most promise is demonstrated by the use of titania on a platinum substrate or single crystals of strontium titanate. See also **Photoelectric Effect**.

Fundamental Considerations. In photochemical reactions, light supplies the energy necessary for the activation of the reacting molecules (Grotthus, 1818, and Draper, 1839). Sometimes the light waves absorbed by a body produce only an increase in temperature, sometimes fluorescence as in the cases of eosin and fluorescein, and sometimes chemical change. The reaction of hydrogen and chlorine in light was studied by Bunsen and

Roscoe (1862), and they discovered that the amount of chemical change is proportional to the intensity of the light and to the length of time of exposure to the light. The first law of photochemistry (Draper-Grotthus) states that light that is absorbed causes chemical change. The energy of light is measured in quanta, and according to the Stark-Einstein law,

$$E = Nhc/\lambda$$

where N is Avogadro's constant, h is Planck's constant, c is velocity of light, λ is wavelength of light; that is, each molecule that takes part in a chemical reaction induced by exposure to light absorbs one quantum of radiation causing the reaction. Photochemical processes are of two kinds: primary and secondary. The primary process in a photochemical reaction is limited by the Einstein law to the absorption of one quantum by a molecule or atom. A knowledge of the spectrum of the reactants is necessary to determine what happens in this process. The molecule may be disrupted into fragments or an electron may be excited from a lower orbit to a higher one. Which of these events takes place can often be determined by spectroscopic studies. The secondary process deals with the fate of the molecular fragments or of the excited molecules. The excited molecule may emit its extra energy as light, causing fluorescence; it may lose it by transferring it to other molecules as thermal energy; or it may cause a chemical reaction. On the other hand, the molecular fragments may either recombine to give the original reactant or cause further chemical reactions. The study of the quantum yield (which is the number of molecules reacting divided by the number of quanta absorbed), is used as a means of formulating the secondary processes. If the quantum yield is less than one, fluorescence, deactivation or recombination of fragments must take place. If the quantum yield is unity every photon absorbed decomposes one molecule. When the quantum yield is greater than unity (and in some reactions it may be as high as a million) chain reactions are involved. The classical example of such a reaction is the combination of hydrogen and chlorine. The primary reaction is Cl_2 and light $\rightarrow$ 2Cl. The chain propagation reactions are

$$Cl_2 + h\nu \longrightarrow 2Cl$$
$$Cl + H_2 \longrightarrow HCl + H$$
$$H + Cl_2 \longrightarrow HCl + Cl$$

creating a cycle which is only stopped by

$$Cl + Cl \longrightarrow Cl_2$$
$$H + H \longrightarrow H_2$$

Since the last two processes are slow compared to the two before them, one quantum of light can bring about a combination of a million molecules of hydrogen and chlorine.

See also **Photosynthesis**.

The existence of microbes capable of utilizing light energy to drive the synthesis of cellular components was firmly established by Winogradsky and Molisch by the late 19th century, but it was largely through the work of Van Neil in the 1930s on the physiology of the purple sulfur bacteria that a clearer picture of their photosynthetic processes started to emerge. The photochemical reaction center (RC) is a now well-defined physical entity which exists as an energy sink able to convert the energy of excitation into an electron transfer event. From the use of this RC, three broad patterns of microbiological photosynthesis are known. See Table 1.

The chlorophylls are the pigments responsible for initiating primary photochemistry in the cyanobacteria and higher plants and absorb light energy at 870 nm converting appropriate species to the single excited state and rapidly transferring an electron to a single molecule. Chlorophyll is remarkably similar to hemoglobin in that iron is replaced by magnesium as the chelated metal. Any other transition metals in that position would cause quenching of the initial photochemical reaction.

The carotenoids are largely responsible for the color of the green and purple bacteria and function in both light absorption and energy transfer processes as well as extending the usable light wavelengths and protecting against harmful photooxidation.

In green and purple photosynthetic bacteria, most of the bulk bacterial chlorophyll is inactive in photochemistry and does not undergo photooxidation when excited. Energy is transferred randomly between adjacent pigment molecules until trapped by the RC or lost as fluorescence or heat.

TABLE 1. BROAD PATTERNS OF MICROBIOLOGICAL PHOTOSYNTHESIS*

Bacterial Group	Type of Photosynthesis	Pigment in Primary Photoactivation	Electron Donors	Products	Carbon Sources
Eubacteria	Anoxygenation	Bacterial chlorophyll	H_2, H_2S, S, Organics	ATP + NO, D, P(H)	CO_2 + Organics
Eubacteria	Oxygenation	Chlorophylls	H_2O, H_2S	ATP + NO, D, P(H)	CO_2 + Organics
Archeobacteria	Halogenation	Bacterial rhodopsin	Do not participate	ATP	Organics

(*After Keely and Dow.)

But the photochemistry of bacterial activity must also be associated with photoeffects in the surrounding media, and in this context a rapid increase in the concentration of hydrogen peroxide has been observed when natural surface and ground waters are exposed to sunlight. The hydrogen peroxide is photochemically generated from organic constituents present in the water. Humic materials are believed to photochemically reduce oxygen to give the superoxide anion, which subsequently disproportionates to hydrogen peroxide. Since both hydrogen peroxide and peroxide radical are known to affect biological systems, they may be important factors in the photochemical reactions of photosensitive bacteria and other natural life forms.

Laser Chemistry

Lasers generate a high-intensity output of monochromatic photon energy, and studies of the photochemical reactions induced by this have created a virtual subdivision of photochemistry known as "laser chemistry." While the output of a laser can heat, anneal, burn, cut, or be used instrumentally as a spectral source, we are concerned here only with those chemical effects attributable to the photon output at wavelengths between near infrared and near ultraviolet, i.e., between about 12 and 0.2 microns.

When atoms or molecules are excited conventionally by elevated temperatures or pressure, they can follow several reaction paths yielding a variety of byproducts in addition to the desired substance. Since the basis of a chemical reaction is to weaken or break or make specific chemical bonds to yield the final product, energy ideally should be selectively introduced at the particular level necessary to accomplish this. The high energy and monochromaticity of laser output are ideal for imposition of the specific energy changes that induce or catalyze chemical changes.

The absorption of a quantum of energy by an atom or a molecule takes it from a low energy state to a higher one, and the jump will affect the different properties of the atom or molecule depending upon the amount of energy in the quantum. When absorbed, a quantum of visible or ultraviolet radiation raises an electron to a higher orbit; on the other hand, a quantum of infrared radiation will alter energy levels on an atomic basis.

A laser can supply a precise amount of energy to an atom or molecule, thus effecting a transition from one excited state to a higher excited one. Once known of the energy level displacement required to effect a chemical reaction is available, the laser can provide the specific energy for the specific excitation required. However, the energy input must be related to the total energy dissipation, for, if excess energy leads to ionization or dissociation, a continuum of allowed energy levels will be developed rather than the required discrete levels. If the energy is thus fragmented, the required reaction will proceed only weakly, if at all. Excess energy may be redistributed in two ways. It is either transferred from the excited vibrational state to one or more other vibrational states of the molecule, or it is transferred directly into rotational and translational states. The first mode of energy translation proceeds appreciably more rapidly than the second. Time is a further controlling factor in laser chemistry. The reaction must proceed in time either shorter than, or equal to, that required for transfer of vibrational energy from one state to another in the same molecule; molecule dissociation or atom ionization must take place before there is any depletion of energy by molecular or atomic collisions. Where a reaction proceeds within the lengthy period required for transfer of energy from the initial vibrational state to the much lower rotational and translational states, one cannot hope for laser action to effect a significant degree of reaction specificity, since the effect is basically a thermal one.

A goal of early laser-induced photochemistry was the initiation of specific chemical reactions, to fabricate specific chemicals, or to separate isotopes. Although the specificity of laser-induced photochemistry is important, equally significant is the ability of the laser to confine excited regions to microscopic areas. Thus, one of the better known capabilities of the laser beam is that even a low-powered laser can produce highly intense spots of light of submicrometer dimensions.

In many photochemical reactions, a specific excitation wavelength leads to a specific set of molecular fragments and ultimately products. This has enabled the dissociation of a large variety of molecular gases which can then deposit on, dope, or etch a semiconductor wafer. Further, laser excitation can photodeposit metals, insulators, and semiconductors by decomposing one or more photosensitive gases. Direct writing with lasers through such reactions as:

$$CF_3Br \xrightarrow{h\nu} CF_3 + Br; \text{ or } Al(CH_3)_3 \xrightarrow{h\nu} Al + 3\ CH_3$$

is just beginning to be applied to the fabrication of complete microelectronic devices and the modification of actual circuits.

Operation of visible-light and ultraviolet lasers costs more per photon produced than does infrared laser operation. Partly because of this, appreciable interest has centered over the past several years on unimolecular reactions driven by infrared lasers. But absorption of a single infrared photon will raise a molecule only one step in the energy ladder, and, to be dissociated, the molecule will require the absorption of many infrared photons in sequence. The carbon dioxide laser can supply this requirement cheaply and efficiently. A mole of photons (6.02×10^{23}) costs only a few cents in the infrared, but several dollars in the visible and near ultraviolet ranges. This has aided continued study of multiple-photon infrared laser excitation. Much study has gone into an exciting and fundamental reaction and its implication. When sulfur hexafluoride (SF_6) is irradiated by infrared laser light, it decomposes to the pentafluoride (SF_5) and fluorine (F). When the laser is tuned to the vibrational absorption of $^{32}SF_6$ in a mixture with $^{34}SF_6$, only the $^{32}SF_6$ decomposes, leaving the residual gas enriched some 3000-fold in $^{34}SF_6$. Changing the frequency of the irradiating light slightly from emission at 10.61 μ to emission at 10.82 μ selectively decomposes the $^{34}SF_6$ molecule. This method of isotope separation by lasers is being extensively studied for the separation of fissionable U^{235} from nonfissionable U^{238}.

Laser-induced processes are expected to increase in number and expand in application, but the principal obstacle to large-scale introduction of the laser into the chemical industry is an economic one. Laser photos are still much more expensive than those from thermal sources, and the initial application will undoubtedly be directed to those specialty chemicals and isotopes whose current cost far exceeds that of large volume chemicals.

A typical example of this is in the preparation of extraordinary divalent carbon intermediates (carbenes) by application of ultrafast laser techniques. Thus, photoexcitement of diphenyldiazomethane (DPDM) to an excited singlet state breaks the $C=N_2$ bond, releasing diphenylcarbene (DPC) in a singlet state and ultimately allows its stabilization in the triplet ground state.

The increasing number of known photochemical reactions is still very small in comparison with those in ground state chemistry, and our understanding of all the factors controlling photochemical reactions is quite primitive. In some cases it is the ease of conversion to the ground state that is significant, while in others it is the energy hypersurface surrounding the excited state that dictates the energy pathways through which the free electrons will move back toward the ground state, and hence the nature of the photochemical products.

Laser Femtochemistry

As explained by A.H. Zewail (California Institute of Technology), "Femtochemistry is concerned with the very act of the molecular motion that brings about chemistry, chemical bond breaking, or bond formation on the femtosecond (10^{-15} second) time scale. With lasers it is possible to record snapshots of chemical reactions with sub-angstrom resolution. This strobing of the transition-state region between reagents and products provides real time observations that are fundamental to understanding the dynamics of the chemical bond."

A longstanding problem in chemistry has been the development of a better understanding of the transition state between reagents and products. Several investigative approaches, including thermodynamics, kinetics, and synthesis, have been used to systematize masses of experimental data to ascertain the rates and mechanisms of reactions. Over the past few decades, advanced understanding of molecular reactions have stemmed from the use of molecular beams, chemiluminescence, and, more recently, laser techniques. In laser-molecular beam research, a laser is used (1) to excite one of the reagent molecules, thus influencing reaction probability, or (2) to initiate a unimolecular process simply by providing energy to a molecule. In what has been called a "half-collision" unimolecular process, the fragmentation of the excited molecule can be determined. As pointed out by Zewail, "During the last three decades many reactions have been studied and these methods, with the help of theory, have become the main source of information for deducing the nature of the potential surface of a reaction."

Unimolecular reactions, as typified by

$$ABC^* \longrightarrow [A \cdots BC]^{\ddagger *} \longrightarrow A + BC,$$

and bimolecular reactions, as shown by

$$A + B \longrightarrow [ABC]^{\ddagger} \longrightarrow AB + C,$$

have been studied in terms of their *time scale*.

In an excellent paper, commenting on the progress made from the "picosecond" era to the "femto-second" era of research, Zewail observes, "Prior to femtochemistry, molecular beams and picosecond lasers were combined, which led to studies of collision-free energy redistribution in molecules and state-to-state rates of reaction, but the time resolution was still not sufficient to directly view the process of bond breaking or bond formation. However, in femtochemistry, the 'shutter speed' has reached the 10^{-15} regime, so it is now possible to observe chemistry as it happens — the transition-state region between reagents and products. $\cdots$ The strobing of these ultrafast molecular reactions, stemming from the happy marriage between ultrafast lasers and chemistry, is what forms the central theme in real time femtochemistry."

In an extension of this technology, A.S. Moffat reported in 1992 two additional goals of development: (1) the use of lasers to *control* chemical reactions, and (2) development of "tunable" lasers that can vary the wavelength of the light they generate. Thus, the light source can be tailored exactly to the vibrational frequency of the bond targeted.

Time-Resolved Photoacoustic Calorimetry

As defined by K.S. Peters and G.J. Snyder (University of Colorado), time-resolved photoacoustic calorimetry is an experimental technique that measures the dynamics of enthalpy changes on the time scale of nanoseconds to microseconds for reactions initiated by absorption of light. As pointed out, "When the reaction is carried out in water, it is also possible to obtain the dynamics of the corresponding volume changes. This method has been applied to a variety of biochemical, organic, and organometallic reactions."

Although pulsed time-resolved photoacoustic calorimetry is in an early phase of development (1994), it is proving to be a powerful technique for understanding the dynamics of enthalpy and volume changes for ground- and excited-state species. Peters and Snyder, developers of the technique, observe, "For reactions in water, the problem is directly approached through temperature-dependent studies. For reactions in organic solvent, there must be further investigations into the magnitude of the effect. At some future date, time-resolved photoacoustic calorimetry will be extended onto the 1 ns time scale .... The technique should find wide ranging applications to problems in chemistry and biochemistry that include solid-state reactions and dynamics of proteins in membranes."

A schematic representation of the instrumental technique used is shown in Fig. 1.

Additional Reading

Andrews, D.L.: "Lasers in Chemistry," 3rd Edition, Springer-Verlag, Inc., New York, NY, 1997.
Dunning, T.H., Jr., et al.: "Theoretical Studies of the Energetics and Dynamics of Chemical Reactions," *Science*, **453** (April 22, 1988).
Eisenthal, K.B., et al.: "Divalent Carbon Intermediates: Laser Photolysis," *Science*, **225**, 1439–1445 (1984).
Gust., D. and T.A. Moore: "Mimicking Photosynthesis," *Science*, **35** (April 7, 1989).

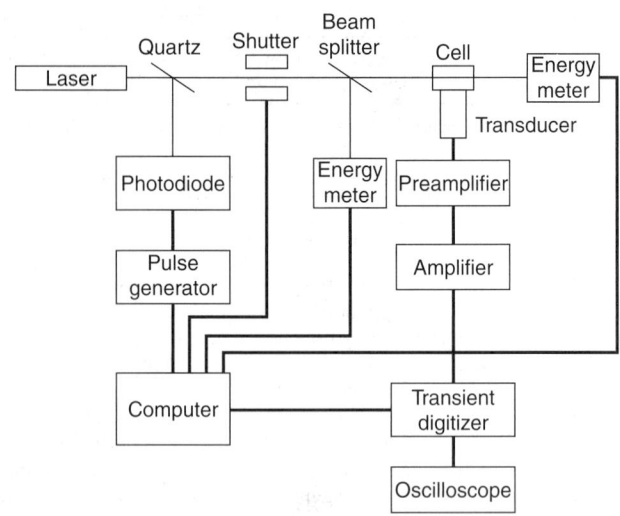

Fig. 1.   Schematic representation of the time-resolved photoacoustic calorimeter. Solid lines indicate light path; heavy solid lines represent signal paths. (*After Peters and Snyder.*)

Horspool, W.H. and Pill-Soon Song: "CRC Handbook of Organic Photochemistry and Photobiology," CRC Press, LLC., Boca Raton, FL, 1998.
Osgood, R.M. and T.F. Deutsch: "Laser Induced Chemistry for Microelectronics," *Science*, **227**, 709–714 (1985).
Murov, S.L., I. Carmichael, and G.L. Hug: "Handbook of Photochemistry," 2nd Edition, Marcel Dekker, Inc., New York, NY, 1993.
Neckers, D.C., D.H. Volman, and G. Von Bunau: "Advances in Photochemistry," Vol. 24, John Wiley and Sons, Inc., New York, NY, 1999.
Neckers, D.C., D.H. Volman, and G. Von Bunau: "Advances in Photochemistry," Vol. 25, John Wiley and Sons, Inc., New York, NY, 1999.
Neckers, D.C., D.H. Volman, and G. Von Bunau: "Advances in Photochemistry," Vol. 26, John Wiley & Sons, Inc., New York, NY, 2001.
Peters, K.S. and G.J. Snyder, "Time-Resolved Photoacoustic Calorimetry: Probing the Energetics and Dynamics of Fast Chemical and Biochemical Reactions," *Science*, 1053 (August 26, 1988).
Ramamurthy, V. and K.S. Schanze: "Organic, Physical and Materials Photochemistry," Marcel Dekker, Inc., New York, NY, 2000.
Staehelin, L.A. and P. Aentzer: "Photosynthetic Membranes and Light Harvesting Systems," in Encyclopedia of Plant Physiology, Vol. 19, Springer-Verlag, New York, NY, 1986.
Truhlar, D.G. and M.S. Gordon: "From Force Fields to Dynamics: Classical and Quantal Paths," *Science*, **491** (August 3, 1990).
Wayne, C.E. and R.P. Wayne: "Photochemistry," Oxford University Press, Inc., New York, NY, 1996.
Zewail, A.H.: "Laser Femtochemistry," *Science*, 1645 (December 23, 1988).

R.C. VICKERY, D.Sc., Blanton/Dade City, FL.

PHOTOELASTICITY.   A term that refers to certain changes in the optical properties of isotropic, transparent dielectrics when subjected to stresses. A block of glass, free of optical flaws, exhibits a "forced" double refraction when subjected to compression or tension parallel to one of its dimensions. If the block is placed between crossed Nicol prisms, the field remains dark so long as the glass is in its normal condition, but as stress is applied, colored fringes appear which are characteristic of the internal deformations of the glass.

PHOTOELECTRIC CONSTANT.   A quantity equal to h/e where h is the Planck constant, and e, the electronic charge, and which multiplied by the frequency of any radiation exciting photoemission gives the potential difference corresponding to the quantum energy absorbed by the escaping photoelectron.

$$h/e = 4.1349 \times 10^{-7} \text{ erg} \cdot \text{sec} \cdot \text{emu}^{-1}$$

$$= 1.3793 \times 10^{-17} \text{ erg} \cdot \text{sec} \cdot \text{esu}^{-1}$$

PHOTOELECTRIC EFFECT.   Changes in electrical characteristics of substances due to radiation, generally in the form of light. Radiation of sufficiently high frequency (short wavelength), impinging on certain substances, particularly, but not exclusively, metals, causes bound electrons to be given off with a maximum velocity proportional to the frequency of the radiation, i.e., to the entire energy of the photon. The Einstein photoelectric law, first verified by Millikan, states:

$$E_k = h\upsilon - \omega$$

where E_k is the maximum kinetic energy of an emitted electron, h is the Planck constant, υ is the frequency of the radiation (frequency associated with the absorbed photon), and ω is the energy necessary to remove the electron from the system, i.e., the photoelectric work function for the surface of the emitting substance. An inverse photoelectric effect results from the transfer of energy from electrons to radiation. For example, in an x-ray tube, there is observed the transfer of energy from electrons accelerated by the anode voltage to radiation emitted by the target. This radiation exhibits a continuous spectrum at lower voltages, upon which are superimposed, at higher voltages, intense lines characteristic of the anode material.

Two principal aspects of the photoelectric effect are described here: (1) Photoconductivity; and (2) photovoltage.

Photoconductivity is the phenomenon evidenced by the increase in electrical conductivity of a material by the absorption of light or other electromagnetic radiation. Although insulating or semiconducting materials exhibit this effect to some degree, there are relatively few materials that give sufficiently large changes of conductivity with illumination for application of the principle in useful devices. The principle can be explained briefly by using a cadmium sulfide photoconductor as an example. As in the case of luminescence, the band-type of energy level diagram is useful. See Fig. 1. Transition 1 represents absorption of a photon of energy at least equal to that of the band gap, giving rise to a free electron and a free hole. Transition 2 represents absorption at a local crystalline imperfection (defect or impurity), also producing a free electron, but with a hole trapped in the vicinity of the imperfection. While these carriers are "free" in the crystal, the conductivity can be greatly enhanced, so that the conductivity in the light can be a million times that in the dark. Recombination of the carriers may occur via transition (3), which is a "direct" electron-hole recombination across the band-gap, or via step 4, an electron recombining with a center containing a hole, so that they no longer contribute to the conductivity.

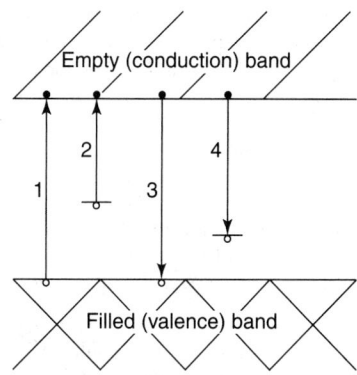

Fig. 1.   Simplified band model for photoconduction processes.

For a material in which one type of carrier predominates (i.e., electrons), the change in conductivity with illumination can be given as:

$$\Delta\sigma = \Delta n e \mu + e n \Delta \mu \qquad (1)$$

where $\Delta\sigma$ = conductivity change, Δn is the change in free carrier density, e is the electronic charge, μ is the carrier mobility, and $\Delta\mu$ is the change in carrier mobility. Usually, the first term in Equation (1) predominates.

Photoconductivity gain, G, may be defined as the number of interelectrode transits that can be made by an electron until the photogenerated hole is eliminated by recombination. For the case treated here, namely where one type of carrier predominates, the gain, G, can be stated as:

$$G = \tau\mu^{VL-2} \qquad (2)$$

where τ is the carrier lifetime, μ is the mobility, V is the applied voltage, and L is the spacing between electrodes. Since the specific sensitivity, S, varies as the product of carrier lifetime and mobility, $S\alpha\mu\tau$, we can state that

$$G\alpha(VL^{-2}S) \qquad (3)$$

Commercially, photoconductive devices are used as (1) Detectors of radiation; (2) switches which are sensitive to light and which can actuate relays; and (3) in combination with other photoelectronic materials, such as electroluminescent materials, as image intensifiers. Germanium and silicon devices of the *p-n* junction phototransistor type have long been used in computer detectors; lead sulfide has been used in photocells for infrared detection; cadmium sulfide or cadmium selenide have been used in photocells for detection of light in the visible range; zinc oxide and selenium devices have been used in photocopying machines; antimony sulfide has been used in television pickup tubes. There are numerous other applications.

Photovoltage or the *photovoltaic effect* may be defined as the conversion of light photons to electrical voltage by a material. Becquerel, in 1839, was the first to discover that a photovoltage was developed when light was shining on an electrode in an electrolyte solution. Nearly half a century elapsed before this effect was observed in a solid, namely, selenium. Again, many years passed before successful devices, such as the photoelectric exposure meter, were developed. Radiation is absorbed in the neighborhood of a potential barrier, usually a *p-n* junction, or a metal-semiconductor contact, giving rise to separated electron-hole pairs, which create a potential. An equivalent circuit for a photovoltaic cell is shown in Fig. 2, where R_{SH} and R_S are the internal shunting and series impedances; I_J is the junction current; R_L is the load resistance; I_S is a constant-current generator; and V_L and I_L are the voltage and current developed across the load. With R_L optimum, the maximum conversion efficiency, $\eta_{\max}$ can be given by:

$$\eta_{\max} = \frac{100V_{mp}I_{mp}}{P_{in}}$$

in which case, V_{mp} and I_{mp} are the voltage and current across R_L, and P_{in} is the radiant input power.

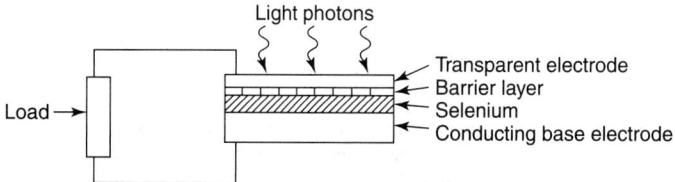

Fig. 2.   Diagram of photovoltaic cell.

Photovoltaic cells have found numerous applications in electronic and aerospace applications, notably in satellites (solar cells) for instrument power. See also **Solar Energy**. Materials used, in order of decreasing theoretical efficiency, include gallium arsenide (24%); indium phosphide (23%); cadmium telluride (21%); silicon (20%); gallium phosphide (17%); and cadmium sulfide (16%). See Fig. 3. In the past, disadvantages of photovoltaic cells have included: (1) high susceptibility to radiation damage; (2) high cost; and (3) requirement for auxiliary battery power when a source of radiation for the cells is not available. Intensive research is underway in the mid-1970s to improve the production of known materials, i.e., upping the practical efficiency to approach the theoretical efficiency — as well as decreasing production costs; and to find new materials and combinations. Such findings are extremely important to certain of the proposed approaches for utilizing solar energy on a large scale. Much of this research is currently of a proprietary nature. The scope of the problem can be visualized from published Figures giving ranges of 0.2 to 0.5 volts per cell with current output of a cell with an area of two

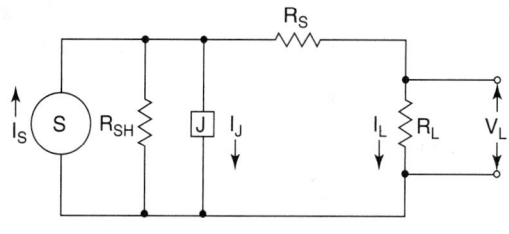

Fig. 3.   Photovoltaic cell equivalent circuit.

square centimeters amounting to only about 0.05 ampere. Consequently, without significant materials and efficiency improvements, extremely large numbers of cells are needed for any large-scale, practical solar energy application. Also, for maximum utility, a means must be found to orient the cells wherever possible so that the incident light will be perpendicular to the face of the cells.

See also **Photoemission and Photomultipliers**; **Photometers**; and **Photon**.

PHOTOEMISSION AND PHOTOMULTIPLIERS.

Photoemission is the ejection of electrons from a substance as a result of radiation falling on it. Photomultipliers make use of the phenomena of photoemission and secondary-electron emission in order to detect very low light levels. The electrons released from the photocathode by incident light are accelerated and focused onto a secondary-emission surface (called a dynode). Several electrons are emitted from the dynode for each incident primary electron. These secondary electrons are then directed onto a second dynode where more electrons are released. The whole process is repeated a number of times depending upon the number of dynodes used. In this manner, it is possible to amplify the initial photocurrent by a factor of 10^8 or more in practical photomultipliers. Thus, the photomultiplier is a very sensitive detector of light.

The major characteristics of the photomultiplier with which the user is generally most concerned include: (1) sensitivity, spectral response, and thermal emission of photocathodes; (2) amplification factor; and (3) noise characteristics and the signal-to-noise ratio.

Many different types of photocathodes are used in photomultipliers. With a selection of various cathodes, it is possible to cover the range of response from the soft x-ray region (approximately 5 to 500 Å) to the near infrared (approximately 12,000 Å). Materials and combinations used include cesium-oxygen-silver; cesium-antimony; cesium-antimony-bismuth; sodium-potassium antimony; sodium-potassium-cesium-antimony; copper-iodine; and cesium-iodine. The thermal emission at 25 °C of copper iodide and cesium iodide tends to run less than the other materials.

The amplification factor in a photomultiplier depends upon the secondary emission characteristics of the dynode and to some extent on the design of the multiplier structure. Important secondary-emission surfaces used in commercial photomultipliers are of two types: (1) alkali metal compounds, e.g., cesium-antimony; and (2) metal oxide layers, e.g., magnesium oxide on silver-magnesium alloy. The alkali metal compounds have higher gain at low primary electron energy (of the order of 75 volts). The metal oxide layers show less fatigue at high current density of emission (i.e., at several microamperes per square centimeter or higher).

The multiplier structures may be divided into two main types: (1) dynamic; and (2) static. The dynamic multiplier in its simplest form consists of two parallel dynode surfaces with an alternating electric field applied between them. Electrons leaving one surface at the proper phase of the applied field are accelerated to the other surface where they knock out secondary electrons. These electrons, in turn, are accelerated back to the first plate when the field reverses, creating still more secondary electrons. Eventually, the secondary electrons are collected by an anode placed in the tube; if they are not, a self-maintained discharge occurs. In practice, dynamic multipliers have been replaced by static ones mainly because the latter have better stability and are easier to operate.

Static multipliers may be either magnetically or electrostatically focused. In a magnetic type, primary electrons impinging on one side of a dynode cause the emission of secondary electrons from the opposite side. These electrons are then focused onto the next dynode by means of the axial magnetic field. The more common types of electrostatic multiplier structures use focusing from one stage to the next. Deposition of thin semiconductor secondary emission surfaces onto insulating substrates has been used in designing rugged miniature multiplier structures. In unfocused electrostatic structures, there is less sensitivity to stray electric and magnetic fields.

Dark noise in photomultipliers is caused by: (1) leakage current across insulating supports; (2) field emission from electrodes; (3) thermal emission from the photocathode and dynodes; (4) positive ion feedback to the photocathode; and (5) fluorescence from dynodes and insulator supports. Careful design can eliminate all but item (3). Associated with the photocurrent from the photocathode is shot noise. There is also shot noise from secondary emission in the multiplier structure.

A major use for photomultipliers has been in the scintillation counter where in combination with a fluorescent material, it is used to detect nuclear radiation. They also have been used in star and planet tracking for guidance systems as well as in star photometry and quantitative measurements of soft x-rays in outer space. Additional uses include facsimile transmission, spectral analysis, process control, and wherever extremely low-light levels must be detected. For applications in photometers, see also **Photometers**.

PHOTOGRAMMETRY.

The technology of obtaining reliable measurements by means of photographs as required in surveying and map-making. Frequently photographs are obtained from aircraft, but terrestrial-based cameras serve for some applications.

PHOTOGRAPHY AND IMAGERY.

Imagery is the representation (pictorial, graphical, etc.) of a subject by sensing quantitatively the patterns of electromagnetic radiation emitted by, reflected from, or transmitted through a subject of interest (object, body, scene, etc.). Imagery is not wavelength-limited, but is achievable (theoretically if not practically) with all bands of the electromagnetic spectrum—gamma rays, x-rays, ultraviolet radiation, visible light, infrared radiation, radar and radio waves.

Fundamentally, an excellent reflective surface, such as the unrippled surface of a clear, uncolored pool or pond of water, provided early people with images of themselves and their surroundings.[1] Under these near-perfect conditions, the observer had a "picture" of outstanding fidelity, including full *color* and *motion*, not to mention the accompanying environmental enhancements of other human senses. This *real* picture, however, lacked several attributes. Except in the observer's brain, the "picture" could not be *stored* and thus *retrieved*, nor could it be transported (transmitted), nor could it be reproduced in the form of complete and accurate copies so that others could respond to the "picture" as seen by the original observer. Further, the data content of the "picture" could not be magnified or in other ways processed to reveal information unobservable by the unassisted human eye.[2]

The *fidelity* of image recall by people is not precise and ranges with the clarity and intensity of the original perception of the image and with the passage of time. So-called "eyewitnesses" in court hearings, for example, recalling facial and dress characteristics of offenders, often are unreliable. The ability of humans to distinguish colors, for example is high, embracing a thousand or more shades, but the ability to remember a given shade is notably deficient. The chances of the average individual in a paint store selecting a matching color for the woodwork at home is next to impossible for most people. Lighting at home and at the paint store, rarely identical, also interferes with the color recollection process.

Verbalizing an image also is extremely difficult for the average individual. Some poets and writers are excellent in this regard, creating a reverence for many of their works.

The several main elements of imaging technology include:

1. *Capturing the image*—Cameras and sensors by any other name (radar, infrared (IR), ultraviolet (UV), radio (RF), sonar, x-ray, etc.) that are used to detect, sense, and otherwise make an image that becomes part of a database.
2. *Storing and retrieving the image*—In the form of prints on paper (albums, reports, printed matter, etc.) or on magnetic tape or disks and other electronic formats, including optical computers.
3. *Displaying the image*—Largely determined by its storage format. For many years, still or motion picture projectors were and still are used for displaying film formats. Image data in digital format commonly are displayed by some form of electronic screen, cathode-ray tubes, liquid crystal, and other flat-screen formats.
4. *Transmitting the image*—Using digital format over conventional short- or long-distance communication networks. Pictures, mainly for the news media, have been transmitted over telephone lines for

---

[1] Narcissus, in Greek mythology, was the son of the river god Cephissus. Narcissus, when first seeing his image reflected in a fountain, was overtaken with admiration for himself, ultimately precipitating his demise and transformation into the flower appropriately named narcissus.

[2] The manner by which the human brain processes, stores, and retrieves "picture" or "graphic" images is extremely complex and poorly understood. See also **Central and Peripheral Nervous Systems**.

several decades and are referred to as *telephotos*. The modern telefax (FAX) machine with an electrooptical scanner and an electrooptical printer is the widely accepted contemporary method. In the early 1990s, there is a strong trend toward the use of photonics (use of pulses of light over glass fibers).

5. *Designing the image* — A specialized application of imaging technology wherein a designer works up the image of a future product in a CAD/CAM (computer-aided drafting and manufacturing) system by rotating and otherwise maneuvering a created image at a computerized workstation.

6. *Analyzing and enhancing the image* — Image definition and clarity can be improved by assigning "shades of gray" classifications to black and white images as well as assigning color codes to differentiate kinds of objects (trees, water, soil, etc.) and, in some cases, to create actual colors of natural terrain. This is widely used in astronomy and aerial surveillance technology.

Photography

Traditional *photography*, as initially conceived and as commonly practiced, depends upon visible light and uses an optical light-gathering and focusing system (camera) and a light-sensitive medium (film emulsion) to record (store) the image — a *photo-image*. The subsequent availability of infrared, ultraviolet, and x-ray sensitive films extended the capabilities of traditional photography well beyond its dependence upon visible light. The word *photography* derives from the Greek roots *photos* (light) and *graphos* (to draw). Coining of the term is usually attributed to Herschel, although this has not been proved conclusively. Herschel did use the term in a memo dated January 17, 1839 and in a technical paper given on March 14, 1839.

Early History of Photography. The concept of a camera obscura (dark chamber) was first described by Giovanni Battista Porta in 1558 when he put a lens in a hole in the shutter of an otherwise fully darkened room. His objective was to drawn an image of the outside by means of tracing a pattern projected by the lens onto the screen rather than attempting to draw the scene simply by looking at it. Because there was a great desire to capture scenes and subjects on paper, canvas, and other media, but a scarcity of artistic and drawing abilities among many of the populace, the principle of the camera obscura persisted for some 250 years, with improvements and refinements of the optical system used.

During the sixteenth and seventeenth century, it became known to technically curious persons that a number of substances would change color when exposed to light. These observations provided the first hint that perhaps an image could be captured permanently and thus save a lot of time and labor and also provide a means of relatively quickly producing duplicates of any given subject. There were two main problems: (1) finding a suitable medium that would respond within a reasonably short time span to the projected image; and (2) finding some way to hold or fix an image without permitting the medium to follow a full course of development and thus completely obliterate the captured image. The properties of silver salts were discovered in 1725 by Johann Heinrich Schulze at the University of Altdorf. He found that chalk, when moistened with a solution of nitric acid and silver nitrate, became darker upon exposure to light. Early experiments involved contacting objects with the silver medium to produce silhouettes. The first image obtained through use of a camera-like technique was achieved in France by Joseph Nicéphore Niepce. His first success came in 1813, using paper which he had soaked in sodium chloride, followed by immersion in a silver nitrate solution, upon which silver chloride was precipitated throughout the paper. The crude image obtained was a negative and persisted for only a short period because he had not developed a required fixative. Niepce then turned his efforts to an asphalt process, known as *heliogravure*, which he used for copying prints. The process was quite insensitive, but an image (exposure of about 8 hours in direct sunlight) of some buildings is regarded as the *first photograph*. This old photograph is now part of the Gernsheim collection in the United States.

Daguerreotype. The first practical process of photography was invented by Louis J.M. Daguerre of Paris in 1837, although the details of the process were not published until 1839. The process was used chiefly for portraiture and became obsolete within a few years after the introduction of the wet collodion process in 1851. Although the daguerreotype process was the original, modern photography is based on the negative-positive methods introduced the same year by William H. Fox-Talbot of England. This was known as the *calotype* process. In the daguerreotype process

a light-sensitive layer of silver iodide is formed on a silver plate by contact with iodine. After exposure in the camera, a positive image is produced when the image is exposed to mercury and heated. The mercury, by attaching itself to the unexposed portions, forms a positive image. The silver iodide remaining was removed at first with a solution of sodium chloride (salt) which was soon replaced, however, with sodium thiosulfate (hypo), the properties of which had been discovered by Herschel in 1819. The daguerreotype image so produced is very weak. In 1840 Fizeau described a process of toning with gold which greatly increased the strength of the image and was generally adopted.

At first, from 5 to 10 minutes' exposure was required on open landscapes and street scenes. The invention of a fast, large-aperture portrait lens by Petzval in 1841 and the discovery by Goddard in London (1840) of the superior sensitivity of silver bromide reduced the time of exposure to a few seconds.

Problems were encountered in preparing positive prints from the calotype negatives because of reproduction of the grain of the paper that contained the negative. Attempts were made to wax or oil paper negatives, but these were essentially unsuccessful. De Saint-Victor attempted to coat plates with albumin (egg white). Upon hardening of the albumin, the plates were bathed in silver nitrate, causing precipitation of silver iodide within the film of albumin. This was not successful because the sensitivity of the plates was greatly lowered.

Early Emulsions. The use of collodion in photographic emulsions dates from 1851 when Frederick Scott Archer published details of his wet collodion process. Although this process is no longer in general use, it can be used in making the half-tone negatives required in photoengraving. In the collodion process, a clean glass plate is first coated with collodion containing potassium iodide and potassium bromide. It is next sensitized by immersion in a solution of silver nitrate. It is then placed in a plate holder — specially designed for the handling of the wet plate — and the exposure made. After exposure it is developed in a solution of ferrous sulfate and fixed in potassium cyanide, or in hypo, washed and dried. The wet collodion process, as it is used by the photoengraver, results in a negative of high density and extreme contrast, high resolution and with an extremely fine grain. These characteristics render wet collodion well adapted to the requirements of photoengraving. Much later, the wet collodion process was essentially replaced by the gelatino-bromide emulsions of similar characteristics.

Collodion printing-out paper was introduced by Obernetter of Munich in 1867 and was for many years the favorite printing process of the portrait and professional photographer. It was in general use until the early years of the present century when it was gradually replaced by developing-out paper.

Gelatin Emulsions. Not true emulsions, but suspensions of minute silver halide crystals dispersed in a protective colloid medium (gelatin), the suggestion of replacing collodion with gelatin was first made by R.L. Maddox in 1871. The first plates made by Maddox were not very sensitive, but their advantages far outweighed their defects, leading to further developments by Charles Bennett in England in the late 1870s, and the first mass production attempts by George Eastman in 1880. One of the several contributions of Eastman to photography was his early recognition of making and marketing gelatin dry plates on a large scale, eliminating the need for the photographer to prepare his own plates, as well as the need for developing and fixing the plates immediately thereafter. There soon followed the concept of strip film, making it unnecessary to change plates after each exposure. Eastman avoided the grain problem by using a coating that enabled the stripping of the thin layer of gelatin from the paper support. Later, in 1889, he replaced the paper support with a transparent plastic support (nitrocellulose), thus making it possible to produce prints without the need of stripping the gelatin layer from the support. Eastman's goals were to make it easy for the masses of people to take photographs in a simplified manner and, through mass production, market equipment at a price within grasp of the public.

Gelatin is a preferred photographic colloid because the sensitizing bodies in the gelatin make possible emulsions with great sensitivity and speed. Gelatin is an excellent emulsifying agent and is readily transformed, from gel to a liquid or the reverse, by changes in temperature. The latter property makes coating of supports and emulsion processing and working feasible. The strong protective action of gelatin lowers the rate of reduction of unexposed silver halide crystals in developers so that image formation is readily obtained.

Silver halides employed in emulsions are the chloride, the bromide and the iodide. Negative emulsions are composed of silver bromide with a small amount of silver iodide. Positive emulsions for films and paper contain silver chloride, or mixtures of silver chloride and silver bromide in varying amounts, according to the tone, speed, and contrast desired.

In photomicrographs of negative emulsions, the crystals of silver bromide appear as flat triangular or hexagonal plates with rounded corners. Some globular, needle-shaped or diamond shaped crystals may also be observed, as in Fig. 1. The thickness of the flat plates is approximately one-tenth of their diameter. The size of silver bromide crystals range from less than one to four micrometers. Crystals of silver bromide, as used in positive emulsions, are quite uniform and seldom exceed 0.5 micrometer in diameter. Multi-layered emulsions contain approximately 1 billion (10^9) crystals per square centimeter for low-speed emulsions to 1.0×10^{-8} centimeter for high-speed negative emulsions.

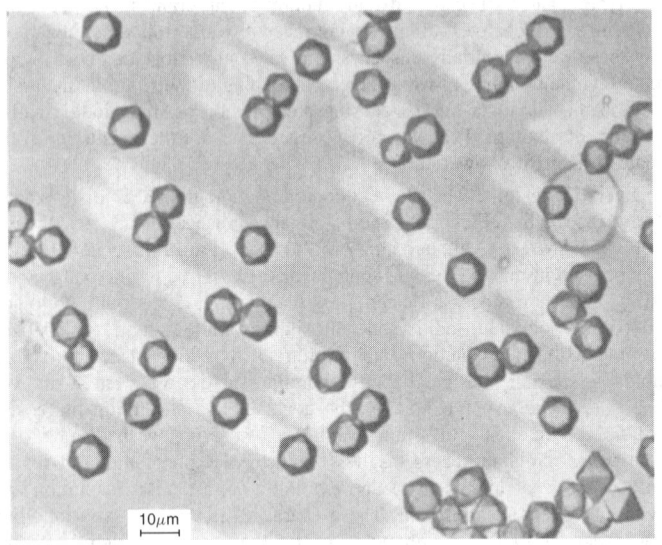

Fig. 1. Using the carbon replica technique, this is an electron micrograph of octahedral silver bromide grains. (*Photo by Dr. Donald L. Black, Eastman Kodak Company.*)

The characteristics of an individual emulsion are primarily dependent on two factors, the size-frequency distribution of the crystals and the composition of the silver halide crystals. For instance, Fig. 2 illustrates silver bromide grains with slightly rounded corners due to the presence of a silver complexing agent. The chief problems of the emulsion-maker are

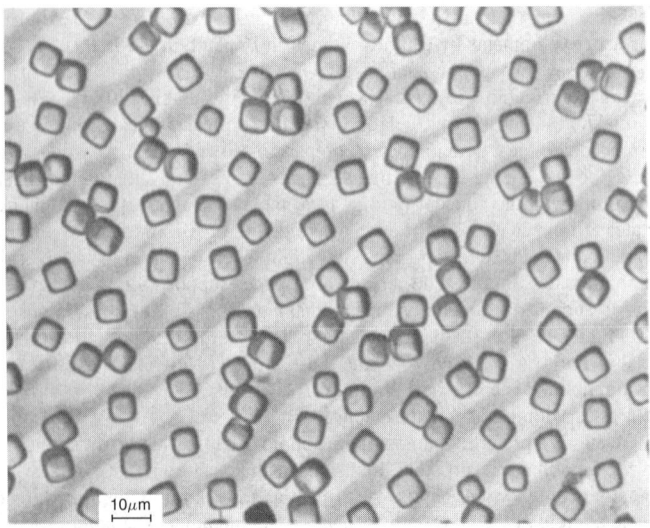

Fig. 2. Using the carbon replica technique, this is an electron micrograph of cubic silver bromide grains in which the corners have been slightly rounded due to the presence of a silver complexing agent. (*Photo by Dr. Donald L. Black, Eastman Kodak Company.*)

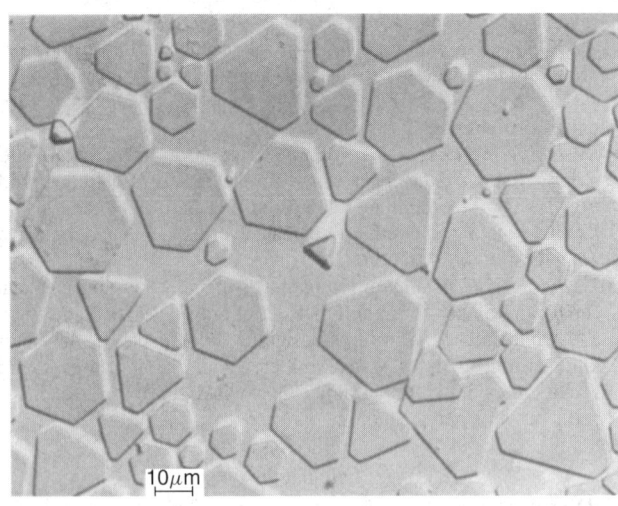

Fig. 3. Using the carbon replica technique, this is an electron micrograph of an Eastman Kodak T-Grain type emulsion of silver bromide that contains some iodide. (*Photo by Dr. Donald L. Black, Eastman Kodak Company.*)

the production of uniform suspensions of silver halide crystals with proper size-frequency distribution and the correct composition in gelatin, and the ability to reproduce results. In an attempt to meet these needs along with increased sharpness for high-speed negative color films, *Eastman Kodak* has developed T-Grain type emulsions of silver bromide which also contain some iodide. Illustrated in Fig. 3, these new emulsions exhibit flat grains with relatively sharper edges. As a result, when compared to traditional high-speed color negative films they are capable of producing more clearly defined images.

Classification of Emulsions

1. *Printing-out emulsions.* These emulsions produce images on exposure without development. They are used largely for making portrait proofs which are distinguished by their red or purplish color. Emulsions of this type differ from others in that they usually contain silver nitrate, some free silver, silver salt of an organic acid and a weak free acid. These are known as P.O.P. Proof Papers.

2. *Developing-out emulsion.* Emulsions for development have an excess of alkaline halides. By varying the composition of the silver halide and treatment, developing-out emulsions may be prepared which are suitable for either negative or positive purposes.

a. *Negative emulsions.* Negative emulsions are prepared by adding a small amount of a soluble iodide to the bromide used in making the silver halide. The mixed crystals of silver-bromiodide formed are more sensitive to light and produce emulsions with greater speed than silver bromide alone. Negative emulsions are referred to as neutral emulsions if precipitation of the silver halide is carried out in a gelatin solution with an excess of soluble bromide. They are referred to as ammonia emulsions if the precipitation takes place in a gelatin solution with an excess of soluble bromide in the presence of ammonia or ammoniacal silver. The latter method produces emulsions with coarser grains, which have the highest sensitivity.

b. *Positive emulsions.* Positive emulsions are prepared by precipitating silver halides containing chloride or mixtures of chloride and bromide in gelatin. The size of the crystals formed are smaller than those of negative emulsions and have a lower sensitivity. Positive emulsions are divided into four classes, according to the composition of the silver halides and their properties.

- *Chloride emulsions.* Because of their slow speed chloride emulsions are used largely for contact printing.
- *Bromide emulsions.* Bromide emulsions are very sensitive and fast. They are used for projection printing exclusively.
- *Chlor-bromide emulsions.* In chlor-bromide emulsions the amount of silver chloride is greater than that of silver bromide. These emulsions are somewhat faster than chloride emulsions and used for contact or slow projection printing. Chlor-bromide emulsions produce warm-toned silver images with a brown or brown-black color.

- *Brom-chloride emulsions.* Brom-chloride emulsions contain more silver bromide than silver chloride. They are faster than chlor-bromide emulsions and used for projection printing where black images and speed printing are desired. Image tones of brom-chloride emulsions are not as warm as chlor-bromide images nor as cold as bromide images.

Manufacture of Commercial Emulsions. Although the details are proprietary, the basic procedures of manufacturing commercial emulsions are known. A portion of the gelatin in the formula is swelled by soaking in water and later dissolved with heat. Mixtures of soluble bromides and iodides, or chlorides, are placed in water solution and added to the gelatin solution. Precipitation of silver halides is accomplished by slowly adding a solution of silver nitrate, while stirring, to the mixture. The relative concentration of the solutions, the rate of addition and temperature during mixing, are factors which control the formation, size and dispersion of the crystals in gelatin. The emulsion is then heated or "ripened" at $40-80\,°C$ to recrystallize the silver halides and readjust the size-frequency distribution. Following ripening, more gelatin is added and the emulsion is chilled so it will set quickly. The emulsion is then placed in a press and forced through a screen to break it into shreds or noodles, which are washed, in cold running water to remove the potassium nitrate formed, the excess soluble halides, and certain soluble byproducts of the reaction. Chloride emulsions are often prepared without washing or with only a limited washing. After washing, the emulsion is drained, remelted, and additional gelatin and certain agents, such as fog preventatives, are added. The emulsion is then heated, or "after-ripened," to form sensitizing nuclei on the silver halide crystals. This operation increases the sensitivity and contrast of the emulsion and is necessary for the preparation of high-speed negative emulsions. Certain preservatives, or stabilizers, are added so the emulsion can be stored in refrigerated rooms until needed. Before coating the emulsion is melted and sensitizing dyes, hardening agents, wetting agents, etc., are added. After thorough mixing, filtering and heating to coating temperature, it is placed in a coating machine. Supports, as film, paper, or glass, with substratum coatings are fed through machines at proper rates so they become coated with emulsions in uniform layers of desired thickness. The coated supports pass over chill boxes to set the emulsion and then through a series of drying compartments where the rate of drying is carefully controlled so as not to change the sensitivity on the surface. Following drying, the coatings are inspected under proper safelights and the film or paper is cut to desired size and packaged.

Numerous variations in the manufacturing process make possible a wide range of film characteristics, including film speed and spectral sensitivity. Film, unlike the human eye, can extend beyond the visible region of the spectrum. High-speed film can capture the details of a fast-moving object, seen only as a blur by the eye. By extended exposure, film can capture images entirely too faint to be seen by the eye. The three main types of film emulsions for black-and-white photography are: (1) *ordinary* (color-blind; sensitive to blue light only); (2) *orthochromatic* (sensitive to all but red light); and (3) *panchromatic* (sensitive to light of all colors). Ordinary and orthochromatic films generally offer greater contrast than most panchromatic emulsions. However, the response of panchromatic emulsions can be modified by use of color filters. Film is available in several sizes and formats. Obviously, a delineation of film specifications is beyond the scope of this encyclopedia. Some excellent references are listed at the end of this entry.

Color Films. The trichromatic theory of vision was first proposed by Thomas Young, a British physicist in 1801. He was the first to propose that the retina of the eye incorporates three different types of receptors, responding to blue, green, and red light, respectively. The theory was elaborated upon to the extent that color perception is based upon the stimulation of two receptors, with light stimulating both red and green receptors seen as yellow light; light equally stimulating all three types of receptors seen as white, etc. Young concluded that it should be feasible to match any color of the spectrum through the proper mixing of blue, green, and red light. Although not essentially interested in color photography, Maxwell effectively demonstrated the principle by way of specially-prepared lantern slides before the Royal Institution in London in 1861. Maxwell had demonstrated the *additive color principle* (mixing of blue, green, and red light).

Practical color photography on a massive amateur scale, of course, could not depend upon the preparation of three separate photographs and the use of three projectors, but rather dictated a process that would combine the three records on one plate. In 1907, the Lumière Autochrome plate was developed. This was comprised of a very coarse mosaic of potato starch grains, one third of which was dyed blue; another third, green, and the remaining third, red. An emulsion layer was exposed, with the light first passing through the mosaic. In the *Kodacolor* system of 1928, filters in the camera were used instead of color mosaics. A major problem of the mosaic and filter approaches was that of loss of light as it passed through one or the other media, greatly reducing sensitivity and loss of brightness of a projected image. See also **Additive Color Process**.

In the *subtractive color system*, the phenomenon of absorption is involved. A dye that will absorb red light will, in turn, reflect green and blue light, thus appears a greenish-blue (cyan); a dye that will absorb green light appears a bluish-red (magenta); and a dye that absorbs blue light appears yellow. Thus, cyan, magenta, and yellow are the three primary subtractive colors. A mixture of all three dyes in proper portion will absorb all primary light and thus appear black. Most processes of color photography make use of a subtractive synthesis to yield prints or transparencies. See also **Subtractive Color Process**.

Color-separation negatives are photographic negatives that record the relative intensities of the primary colors used in the analysis necessary to reproduce a subject by means of color photography. In three-color photography, for example, the separation negatives are records, in terms of silver densities, of the amounts of red, green and blue light received at the camera from the subject.

A set of color-separation negatives may be prepared by photographing the subject three times on separate color-sensitive emulsions so that each is a record of one of the primary colors. A panchromatic emulsion is generally employed with a set of tricolor filters, the colors of the primaries. It is only necessary, however, to obtain the color records on separate negatives so it is also possible to use for each record any combination of color filter and emulsion sensitivity that will record one of the primary colors. A set of color-separation negatives may be made by exposing (1) each one in turn in a camera, (2) by the use of a color camera that will expose them simultaneously, or (3) in a tripack.

It is common practice to balance a set of color-separation negatives, by altering the exposure and development times, so that a gray scale will be recorded equally on each negative. The particular densities desired are dependent on the method of color synthesis to be employed.

The majority of color is by use of integral tripacks. There are three layers of photographic emulsion in the tripack, one layer sensitive to red light, another layer to green light, and another to blue light. They are coated, one on top of the other. Since silver iodobromide emulsions usually selected for film emulsions are sensitive to blue light, sensitivity to the green and red light must be conferred by sensitizing dyes. Although this sensitivity can be obtained, the dyes do not negate the emulsion's natural sensitivity to blue light. Thus, those layers that are sensitive to green and red light must be protected from blue light. This is accomplished by inserting a yellow filter layer that will absorb the blue light. Chloride emulsions on the other hand are sensitive only to ultraviolet light. Whereas they do not require a yellow-filter layer, they have to incorporate a filter for exclusion of ultraviolet light. There are a number of dyes that may be used in dye-transfer systems, but for tripacks it is necessary to select only those dyes that will be formed during the development process. In 1912, the German scientist, Rudolf Fischer, discovered the role of couplers. In his early version of a color film, he placed three layers of emulsion one atop another as previously described and he also incorporated a coupler in each layer to cause the development of a particular color. Fischer's concept was brilliant, but the actual process failed because the couplers and sensitizers tended to wander from layer to layer.

In 1931, Leopold Godowsky, a violinist, and Leopold Mannes, a pianist, and both avid amateur photographers made crude experiments in a home laboratory on a type of color film that ultimately became *Kodachrome*, released by Eastman in 1935. In the *Kodachrome* process, the couplers are laced in the developers instead of in the emulsions. Phenols are usually the couplers that form cyan dyes; nitriles or pyrazolones form magenta dyes; and esters, ketones, or amides form yellow dyes. There are many hundreds of couplers and, consequently, there is continuing improvement in color film. Space here does not permit a detailed description of such important matters as the negative-positive system, reversal systems for transparencies, color corrections, etc., but these areas are well covered in some of the listed references.

Direct Positive Images. Even in the early days of black-and-white photography and the early work of Daguerre and Fox-Talbot, it was realized that there would be a great advantage gained from a system that would initially produce a positive rather than a negative image. As early as the late 1930s, Hippolyte Bayard and Robert Hunt proposed systems, but these did not produce satisfactory results. The *chemical transfer* process was developed by A. Rott in Belgium in 1939 and found application in the document-copying field. In 1947, E.H. Land demonstrated a camera which produced a finished black-and-white print without need for a negative and one that was available to the photographer within a very short period, approximately one minute. This was the first model of the *Polaroid* camera. In chemical transfer, a normal emulsion is used. Immediately after exposure and while within the camera, it is developed in a solution containing combined developer-fixer agents. The emulsion is in contact with a special positive white paper, not light sensitive, on which the finished image is printed. The developing reagent is of a jellylike consistency and in early models was contained in pouches or pods, one for each picture. The exposed grains develop in the normal fashion. The unexposed grains are dissolved by the fixing agent. Thus, in the unexposed areas, the dissolved halide is silver which forms on the nuclei in the receiving sheet. In connection with partially exposed areas, the developing grains and the receiving sheet nuclei compete for the silver. Thus, a negative image is formed on the original film or paper, whereas a positive image appears on the receiving sheet. Subsequent to the first Polaroid camera, models were developed to provide a permanent negative as well as print, with the processing time reduced to seconds.

To achieve an instant color film that could rival 35 mm color quality, current generation *Polaroid Spectra* system film utilizes two different color chemistries for greater control of the self-developing image formation process. Composed of 18 microscopically thin coated layers in a rectangular format for both horizontal and vertical composition, this film is able to produce photographs of improved color separation, saturation and brilliance. This is a result of combining images created in three chemical sandwiches in the film negative. Each sandwich is sensitive to red, green, or blue light and consists of a photosensitive emulsion and a related image dye.

For the *Spectra* film, the blue-light sensitive sandwich has been radically altered by the utilization of thiazolidine dye release. This required the creation of new molecules and a new dye-release mechanism involving only a minute quantity of silver. By utilizing this hybrid imaging system, chemical interaction and molecular cross talk between the red-, green-, and blue-sensitive sandwiches have been reduced, which results in greater color definition.

This new material also affords substantial improvement in yellow dye saturation and in recording pastels. By controlling chemical crosstalk between the red and green layers, *Spectra* film is further able to produce more brilliant greens, which is a difficult photographic accomplishment because of the low reflectivity of green in nature. This material also reproduces a broader range of hues and tints when compared with earlier-generation self-developing color films. The transparent support through which the image is viewed is thinner than previous *Polaroid* films and enhances image quality.

Infrared Films. The first photographs by infrared radiation appear to have been made about 1880 by Sir William Abney, using a specially prepared collodion emulsion. Abney is reported to have photographed a boiling tea-kettle, but efforts by others to repeat his work were not particularly successful. In 1903, the first real infrared sensitizer, Dicyanine, was discovered. While the sensitizing action extended to a wavelength of 960 nanometers in the infrared, the exposure was too long for general photography and Dicyanine-sensitized plates were used chiefly in infrared spectroscopy. The discovery of more efficient sensitizers in the early twenties, beginning with Kryptocyanine and neocyanine and followed by the penta- and tetra-carbocyanines, has made it possible to prepare films and plates whose sensitivity in the infrared is such that they can be used for general photography, including aerial and motion-picture photography.

Infrared-sensitive films and plates may be divided into two classes: (1) materials of relatively high speed to the extreme red and infrared, i.e., from approximately 700 to 900 nanometers (nm), and (2) materials sensitive to much longer wavelengths but of lower sensitiveness. The former are used for general photography, for aerial photography and cinematography; the latter for spectroscopy in the infrared and other scientific applications requiring sensitivity to wavelengths longer than about 900 nm. All infrared-sensitive materials are sensitive to violet and blue and to the extreme visible red, as well. Photographs made without a filter resemble those made on an ordinary blue-sensitive material. For most purposes it is sufficient to use an orange or light-red filter which will absorb blue and violet light. In this case, the picture is made partly by infrared and partly by the extreme red. The result, however, is generally only slightly different from that obtained with infrared radiation alone. For true infrared photographs, a visually opaque filter transmitting the infrared only must be used. No filter, however, is required when photographing hot bodies such as an electric flatiron, hot castings, or high-pressure boilers, provided that these show no visible glow.

All infrared-sensitive materials must be loaded and developed in total darkness, as safelight screens, even those for panchromatic films and plates, transmit the infrared freely.

Certain precautions are necessary when making pictures with infrared-sensitive materials. The bellows and shutter blades of some cameras, although perfectly safe for ordinary photographic films and plates, transmit the infrared and fog infrared-sensitive films. The slides of some film and plate holders transmit the infrared sufficiently to cause fog. Although some modern lenses are corrected for the infrared, with most, the focal distances for the visible and for the infrared are different. Usually, it is sufficient to extend the lens a distance equal to 2% of the focal length beyond the visual focus. Even this may often be ignored when using a lens of short focal length or a small diaphragm.

Infrared photographs of landscapes are quite different from those made in the usual way. Green foliage is reproduced light and blue sky and water almost black. The shadow portions of the subject are dark and without detail. The general effect is that of a photograph made by moonlight, particularly if the print is made rather dark. As a matter of fact, most night scenes in motion pictures are really infrared photographs.

Since infrared radiation is not scattered by atmospheric haze, as is light, distant objects are rendered sharper and more distinctly in infrared photographs. Objects invisible to the eye because of the intervening haze are often reproduced sharply in an infrared photograph. In fact, one of the most important applications of infrared photography is in photographing distant objects, whether from the ground or the air. Infrared photographs, however, cannot be made through dense fog.

The scientific applications of infrared photography are numerous and important.

Aerial Photography in the Infrared. Extensive use of color infrared film (CIR) has been made in the field of aerial photography for such applications as crop sensing and inventorying, flood assessments, etc. See also **Satellite (Scientific and Reconnaissance)**. Both normal color film and CIR consist of three separate layers of emulsion on a clear base material. It will be recalled that in normal color film one emulsion layer is sensitive to blue light, one to green light, and one to red light. The images recorded on the three emulsion layers of normal color film combine in the final image to form colors which closely match those of the original subject. CIR film, sometimes referred to as "false color film," also produces combinations of blue, green, and red in the final image; but the blue color results from exposure by green light; the green color from exposure by red light; and red color by exposure of the infrared sensitive layer by infrared energy. Therefore, the images are called false color images. See Fig. 4. Actually, all three layers of CIR film are also sensitive to blue light. For this reason, the film is always exposed through a minus-blue (yellow) filter which eliminates blue light before it reaches the film. The infrared energy needed to expose the infrared sensitive layer is reflected energy, not heat energy. Heat energy does not enter into the image forming process of CIR film.

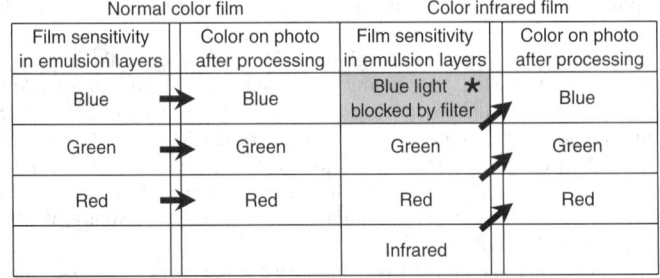

| Normal color film | | Color infrared film | |
Film sensitivity in emulsion layers	Color on photo after processing	Film sensitivity in emulsion layers	Color on photo after processing
Blue	Blue	Blue light ★ blocked by filter	Blue
Green	Green	Green	Green
Red	Red	Red	Red
		Infrared	

Fig. 4. Film sensitivities and final image color of normal color and color-infrared films. *Blue light absorbed by yellow filter. (*U.S. Geological Survey*.)

One of the most important features of CIR film is the manner in which vegetation is recorded. Healthy green plants appear in shades of red, because healthy plants reflect sunlight strongly in the photographic infrared region (therefore strongly exposing the infrared sensitive layer) while simultaneously reflecting relatively little energy in the visible region (therefore offering little exposure to the green and red sensitive layers). For all practical purposes, living healthy vegetation is the only natural source of high-infrared reflection coupled with low visible reflection. Because of the unique reflectance characteristic of healthy vegetation, the film was originally used by the military to differentiate between real vegetation and painted camouflage material.

Another unique and variable aspect of CIR photography results from the fact that plants do not reflect strongly in the photographic infrared when they are severely stressed or have died, and as a result, no longer appear red on the photographs, in contrast with normal, healthy vegetation. The reasons behind this phenomenon are complex, yet the ability to distinguish between healthy and stressed or dead vegetation by using CIR is very important for vegetation analysis. This characteristic is particularly useful in determining crop damage due to flooding.

Generalizations about the photographic appearance of other features commonly found on the agricultural landscape can also be made. Clear water usually appears very dark blue or black, but muddy or turbid water appears light to medium blue. This is useful for satellite tracking of pollution situations. Fresh grain stubble appears very light or almost white, whereas clean plowed fields of dark soil usually appear dark blue.

Other Applications of Infrared Film. Among several other scientific uses are:

1. *In medical photography.* For the study of the following: diseases and conditions affecting the venous pattern not revealed by light; the progress of healing beneath certain scabs; the eye, to determine atrophy; histological specimens, to reveal structures below the surface and invisible to the eye. Thermography has been used to detect tumors, the skin temperature often being as much as 1 to 2 °C higher than that of the surrounding skin.
2. *In industry.* For the study of irregularities in the dyeing and weaving of textile fibers, the interior of furnaces, the detection of carbon in lubricating oils, infrared spectroscopy of metals and alloys.
3. *In astronomy.* For detection of nebulae and stars otherwise invisible because of astronomical haze or because their radiation lies chiefly in the infrared; in infrared spectroscopy, for the determination of the composition, the temperature, and the movement of stars and nebulae. See also **Infrared Astronomy**.
4. *In criminology.* For deciphering writing or printing that has been crossed out with other inks to render it illegible; for obtaining copies of charred documents, detecting erasures, revealing finger prints, identifying blood and other stains, uncovering secret writings, etc.

Ultraviolet Films. In the near-ultraviolet region, photography is the same as with visible light. However, at shorter wavelengths, many materials are not transparent to ultraviolet radiation. For example, glass is not transparent at wavelengths shorter than about 3,000 Å. To produce a photograph at these shorter wavelengths, a quartz lens (transparent to about 1,800 Å) or a fluorspar lens (transparent to about 1,200 Å) is required. Inasmuch as gelatin also absorbs radiation of wavelength less than about 2,200 Å, photography in such regions requires specially-prepared plates where a minimal amount of gelatin is used. In some cases, the plates can be coated with a thin film of oil or other substance which fluoresces when exposed to ultraviolet. For particular work, the fact that air absorbs short wavelengths also must be considered and best results will be obtained in a vacuum. Even with these problems, however, spectra have been recorded down to about 50 Å.

Ultraviolet photography and spectroscopy have found particular usefulness in the study of combustion processes. Some of the very short-lived chemical species occurring during combustion can be observed in the near-ultraviolet region. Along with infrared, ultraviolet techniques also have been used for detecting the retouching of paintings. See also **Ultraviolet Astronomy**.

Cameras and Optical Systems

The term *camera obscura* was defined earlier in this entry. As portable cameras (rather than whole, darkened rooms) were developed as aids for sketching and through the use of such portable devices by Niepce. Fox-Talbot, and Daguerre, the word camera alone became a part of photographic terminology. Cameras essentially are comprised of the following principal components: (1) a light-tight enclosure for the sensitive material, (2) a lens, or other means of forming an image, (3) a holder for the sensitive material, and (4) a means of controlling the time during which light is permitted to reach the sensitive material (i.e., a shutter). A camera may have, in addition to these essentials: (1) a finder to show what will be included in the picture; (2) means for changing the sensitive material, i.e., roll holder, film or plate holder, film cartridge, or film pack; (3) means of focusing the image, e.g., a collapsible bellows, or a tube with a sliding or screw motion, combined with a focusing scale, or focusing screen; (4) a diaphragm to control the amount of light admitted by the lens; (5) a range finder to assist in focusing the image; (6) a depth-of-focus scale; (7) movements, or swings, of the lens or of the film holder or of both which enable the sensitive material to be placed at different angles to the optical axis of the film; and (8) an exposure meter reading the light reflected from the subject at a site close to the lens or, most recently, from behind the lens (see below).

Cameras, other than those designed for specialized fields of photography, such as photoengraving, photolithography, photomicrography, etc., may be divided into seven classes: (1) box, (2) folding roll film, (3) hand cameras with ground-glass focusing, (4) miniature cameras, (5) reflex cameras, (6) twin-lens cameras, (7) view cameras.

The obsolete box camera used roll film; had a single lens with a maximum opening of about f/14; a shutter with one "instantaneous" speed (usually from $\frac{1}{25}$ to $\frac{1}{40}$ second); two reflecting or one direct-vision finder; and, in some designs, an adjustable diaphragm. It was designed to meet the needs of those wishing a simple and inexpensive camera.

35-Millimeter Cameras. The 35 mm camera was invented in the early 1920s by Oskar Barnack, the head of the Design Development Department at the Ernst Leitz Optical Works in Wetzlar, Germany. He needed samples of single frames of exposed film for his motion picture camera experiments. The enlarged results from his motion picture films encouraged him to continue work on the development of a small format still camera. He doubled the normal 35 mm cine frame size to 24 × 36 mm and thus created the now standard miniature film format. The first production model was the Leica I[3] which was introduced at the Leipzig Fair in 1925.

Modern 35 mm cameras are available in many levels of sophistication, varying from complex models with built-in microprocessors to the very simple, what might be called "guess focus" variety. Nearly all 35 mm camera makers are now offering built-in exposure metering and it is this feature, along with focusing methods, automation and interchangeability of lenses, which broadly differentiate the types of cameras available.

Four basic methods of focus are in present use, the simplest being a focusing scale around the lens barrel, which is marked in feet or meters. This is a secondary system and is available no matter what the primary system of focus may be. Another method of focus is the *rangefinder*, which consists of a prism and a pellicle mirror yielding a "split image" across the base leg of the device. These are very accurate for wide-angle lenses and normal lenses, but suffer somewhat in accuracy with the use of lenses of telephoto design.

The most popular method of focus in modern 35 mm cameras is the *single lens reflex* (SLR), which consists of a movable mirror in the optical path of the lens. This mirror reflects the image upward at 90 degrees, where it appears on a viewing screen. Immediately above the screen is a pentaprism that reflects the image of the screen twice again, correcting it from left to right so that the image being viewed appears without inversion or distortion. This image is then magnified in the eyepiece, where accurate focus is accomplished. This is essentially "ground glass" focus and similar systems are also found in some roll film camera designs.

Automatic Focusing Methods. During the development of autofocus cameras, three principal approaches have been employed by camera manufacturers. The methods can be categorized as *active* or *passive*. An active system will allow autofocus regardless of the ambient light levels, while a passive system will function only when the subject is sufficiently illuminated. Even though present-generation cameras have become highly sophisticated in their approach to autofocus range and accuracy, the basic principles still apply.

In one system developed by Canon, an infrared-light-sensitive photocell "looks" at a concentrated portion of the subject (seen as a small oval

---

[3] The words *Leica* and *Leitz*, used in this article are registered trademarks of E. Leitz, Inc.

in the viewfinder) while a beam of infrared light scans the scene. When the photocell detects the beam, a computer microprocessor in the camera trigonometrically calculates the distance and, using an electric motor and gear trains, sets the focus of the lens. See Fig. 5.

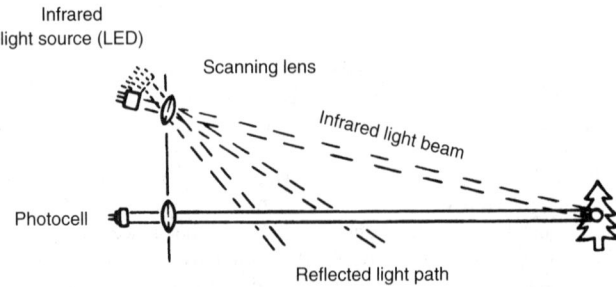

Fig. 5.   Active automatic focusing system employing infrared radiation.

Similar in principle to the function of a standard optical range-finder, triangulation or telemetering is intrinsically a more complex procedure. The triangulation technique depends heavily on precision optics, close-tolerance mechanical parts, and meticulous factory calibration. Triangulation measurements involve a variety of moving elements — optics, mirrors, photocells, cams, and linkages — all required in scanning the subject, making brightness comparisons from two segments of the triangulation geometry and translating the information to a lens-setting apparatus by means of additional mechanical devices.

In another active system (*Polaroid*), distance is computed by an electronic-based direct measurement system using ultrasonic waves. This sonar system is patterned after the center-weighted integrating photometer. The system is controlled by five key elements: (1) an electrostatic transducer which transmits and receives signals, (2) a crystal oscillator clock, (3) a return signal detector, (4) an accumulator, and (5) a focus motor. The transducer emits an inaudible sound (chirp) only about a millisecond long

and is divided into four ultrasonic frequencies: 60 kHz, 57 kHz, 53 kHz, and 50 kHz. As the sound signal is transmitted, the oscillator clock begins counting. The clock times the distance measurements of chirp as well as the necessary electrical timing functions. As the chirp leaves the transducer, the crystal oscillator clock sends regularly spaced timing signals to an accumulator until the echo returns to the transducer. The timing signals sequentially fill 128 empty positions in the accumulator, corresponding to the number of depth-of-field zones into which the focusing range has been divided. After the pulses have been sent, the transducer readies itself to become the "receiver" for the returning echo. Simultaneously, the detector waits to signal the counter to stop upon receipt of the first sound. Once received, the travel time of the signal is determined and data from the accumulator are directed to the lens rotation motor at one of 32 focus positions from 1 meter to infinity. A schematic diagram is given in Fig. 6. See also Fig. 7.

In another (inactive) system (Honeywell Visitronic), the measurement of available reflected light is used. This optical-based system (produced by Honeywell and Canon) is currently incorporated into a variety of rangefinder cameras, including models manufactured by Chinon, Fujica, Konica, Minolta, and Yashica. This autofocus system is similar to conventional rangefinding systems, but instead of visually aligning split or double images, a small portion of a scene is "viewed" on one set of photocells while a scanning mirror seeks an electronically measured brightness match for another set of photocells. When the images correlate, the lens is focused at the proper distance. See Fig. 8. In the triangulation system used, a pair of lenses is located on each side of the camera's optical viewfinder. The lenses produce mirror-to-mirror base lengths ranging from 39 to 60 millimeters, depending upon the camera system. When the autofocus action is initiated, one mirror scans the focus range, searching for an electronically measured brightness match. When correlation within certain preset limits occurs, the lens is then focused to the geometrically calculated distance setting. For most effective operation, a system of this type requires good lighting and contrast conditions. In an effort to overcome low-light limitations, one manufacturer (Fuji) has equipped its AF camera with a miniature spotlight that projects a narrow beam of white light as far as the maximum flash distance, approximately 4.5 meters.

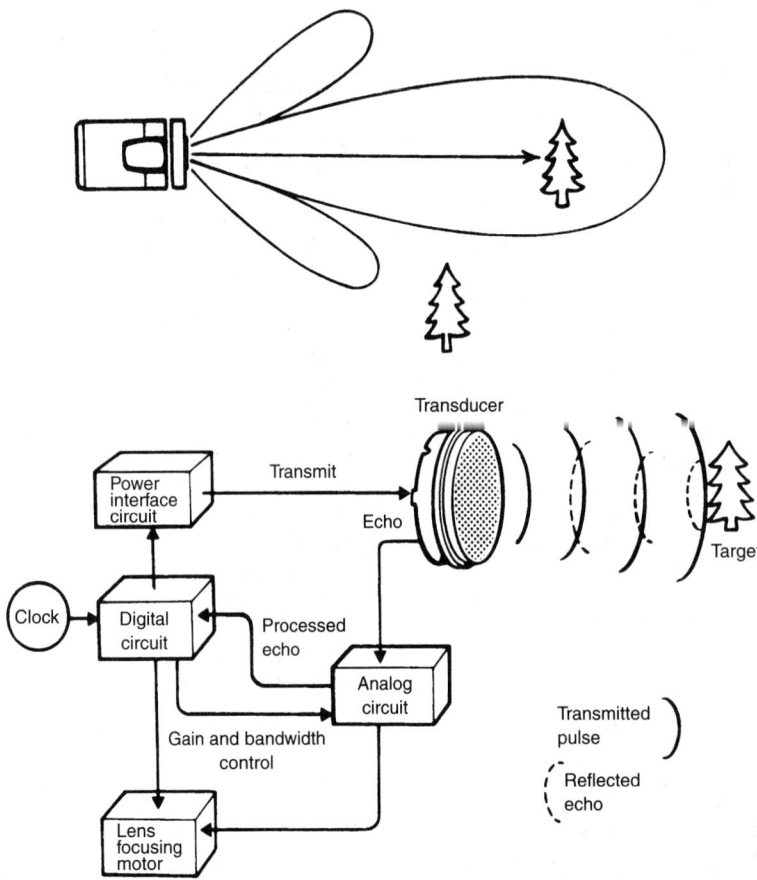

Fig. 6.   Schematic diagram of operating principle and flow of energy in the echo-ranging system. (*Polaroid.*)

While autofocus systems were originally developed to aid amateur photographers in achieving sharp pictures, new developments in these systems are now incorporated in sophisticated professional cameras. For instance, the *Nikon F4*, when used with its AF Nikkor interchangeable lenses, provides for high-speed autofocus even under changing conditions. This system is based on the Nikon AM200 focus sensing module, which consists of a one-piece optical block with 200 high-sensitivity charge-coupled device (CCD) sensors in diagonal arrays and software that guides the camera's internal computer. Refinements to this system include an automatic self-cleaning system that wipes the module's optical surface each time the camera is turned on. Another refinement is an infrared filter that automatically moves into place for available-light photography and moves out of the way for autofocus flash photography.

The *Nikon F4* user has a choice of single-servo or continuous-servo autofocus modes and automatically activated focus tracking. During single-servo autofocus, the shutter cannot be released until the subject is correctly focused, and, once in focus, it remains locked for as long as the shutter release button is pressed lightly. In continuous-servo autofocus, the camera continues focusing for as long as the shutter release button is pressed lightly and the reflex mirror is in the viewing position. This feature allows the photographer to track moving subjects and to release the shutter at any time, regardless of focus conditions. Focus tracking enhances the performance of the continuous-servo mode. When coupled with slow-speed operation of the camera's built-in motor drive and continuous-servo autofocus, the *Nikon F4's* computer system will monitor the rate of movement for moving subjects. If the subject is moving at a constant speed, the *Nikon F4* automatically switches to focus tracking mode. In this mode, it can predict the subject's position at the moment of exposure and rapidly focus the lens to the predetermined position. This function overcomes what is commonly referred to as shutter lag time, which is the time between the camera's mirror raising and the shutter actually being released. During this time, a moving subject can move out of its original position and be rendered unsharp on the film. If the subject's movement is not constant, the system automatically switches to standard continuous focus AF operation, and an LED viewfinder display alerts the photographer to the changed status.

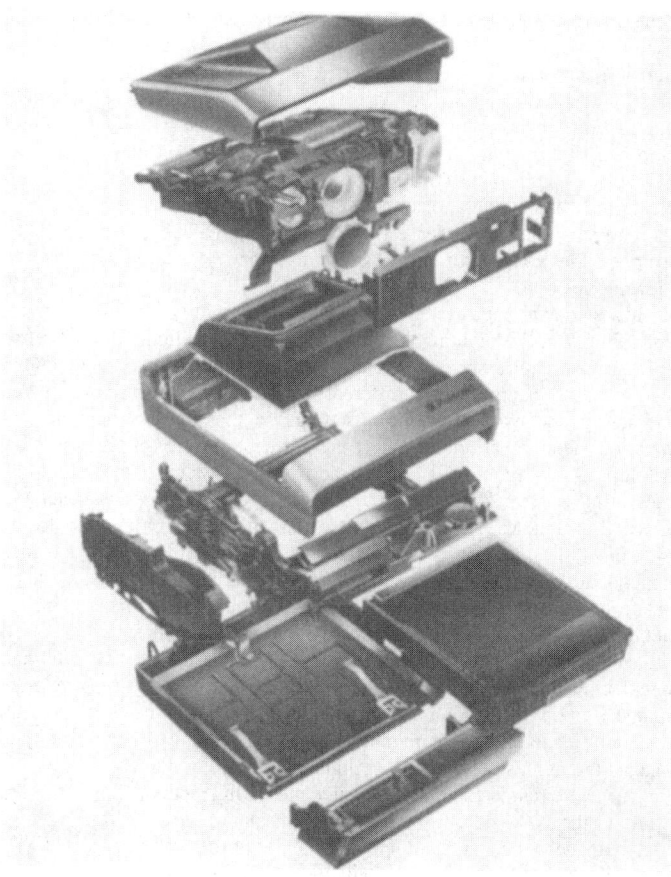

Fig. 7. *Polaroid's Spectra* system camera incorporates advanced electronic and optical controls for precise focus and exposure. This exploded view shows the modular construction common to state-of-the-art camera designs. (*Polaroid.*)

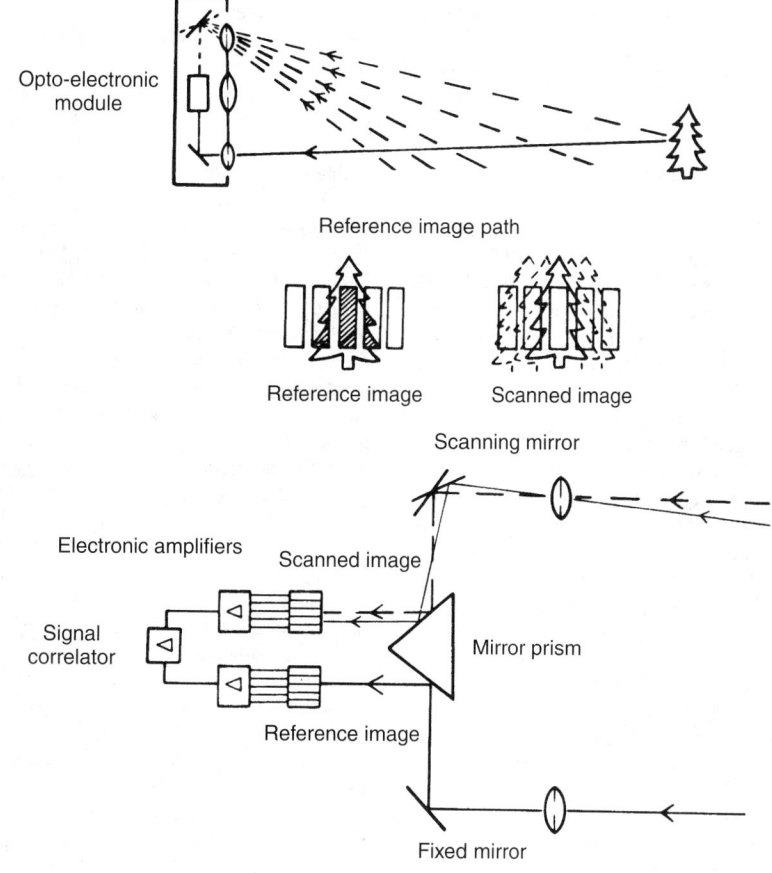

Fig. 8. Operating principle of the *Vistronic* system. (*Honeywell.*)

In situations where existing light, contrast, or subject detail is insufficient for autofocus operation, a Nikon autofocus electronic flash can be used. These units emit a patterned beam of infrared light that the camera can use for focus operation. The *Nikon F4* is shown in Fig. 9.

Fig. 9.   The *Nikon* F4 is an example of a microprocessor-controlled, autofocus, single-lens reflex 35-millimeter camera. (*Nikon Inc.*)

Of the automatic focusing systems just described, there are, of course, the inevitable relative advantages and limitations, a discussion of which is beyond the scope of this encyclopedia.

Exposure Metering Systems. These systems of the built-in type vary from the selenium cells, which require no battery power to the cadmium sulfide cells or silicon photodiodes. Selenium cells have poor sensitivity in low light levels and are seldom used in automated cameras. Cadmium sulfide cells suffer from "memory" and partial paralysis under extremely bright, high-light conditions. The silicon photodiode with transistor amplification offers the most superior performance. For both hand-held and built-in metering systems, light-emitting diodes (LEDs) have largely replaced the formerly used D'Arsonval-type meters. Usually the LED is used in a Wheatstone bridge circuit serving as a simple signal device to show when the bridge is balanced. More expensive types are offering full numeric readout of exposure values in F steps and shutter speeds. Performance is quite impressive with high sensitivity and accuracy. Such systems, of course, are totally dependent upon battery power.

The first of the automatic cameras used ingenious mechanical linkages which transformed exposure data from cadmium sulfide cells directly to in-finder meter readings while simultaneously setting lens and shutter readings automatically. The camera operator was required only to focus and press the shutter release after programing the film speed into the camera. Exposure calculations were essentially eliminated. Complexity and some lack of ruggedness developed a desire for automatic cameras that use electronic techniques and printed circuitry similar to that used in computers.

The use of microprocessors as integral parts of camera control is growing at an impressive rate.

Generally, the operating principles are about as follows. Actual exposure values are read by a single photodiode. Analog output of the "correct" exposure is translated into the Gray Code, which is stored in the central processing unit (CPU) portion of the chip. This is "counted down" at the instant of exposure. The CPU also processes this information from Gray Code to binary to numeric for the full informational readout made visible in the eyepiece of the camera before actual exposures are made. The pulse generator and multiplexer operating at approximately 30 Hz supply the oscillator frequency for the CPU, strobing for the LEDs, and timing for the mirror-aperture-shutter-aperture- mirror sequence. In order, these functions are: (1) determine correct exposure; (2) report exposure calculation in the camera's eyepiece via LEDs in numeric writeout, showing both F stop and shutter speed; (3) set these reported values on both shutter and diaphragm at the instant of exposure; (4) supply the time sequence for electromechanical operations: (a) raise the mirror; (b) close the aperture to the correct F stop; (c) release the shutter at the calculated sweep speed; (d) open the aperture to full F stop value; and (e) return the mirror to viewing position.

However, automatic systems cannot deal successfully with every lighting condition. Special lighting situations such as backlighting or contrast sidelighting require special measuring techniques. Various manufacturers have attacked this problem in different ways. Most of them provide manual measuring and exposure control settings so that the photographer can override the automation and select the proper exposure for the conditions at hand. An even more advanced and useful system is the multimode measuring system developed by E. Leitz for their present-generation Leica R5 cameras. This system offers largefield integrating or selective measuring. In many cases the integrating method is the correct and most dependable one if there are no extreme light and color contrasts, no heavy shadows, and when the bright and dark portions in the image plane are about even. Under these conditions the camera's exposure meter will register the result of the entire image area. Since usually the important detail is in the center of the frame, this measurement is center-weighted in the Leitz system. Unusual photographs, especially news photos, are generally made under uncommon and difficult lighting conditions such as against the light, side light, or spotlighted scenes. For making photographs under these circumstances, the selective measuring method allows a central circle in the viewfinder to be used to delineate a narrow measuring angle that may be aimed at the important image portion of the subject. The camera's central processing unit will consider in its exposure determination only that portion which lies within the circle, regardless of what occurs within the rest of the viewfinder area. The system allows for automatic or manual exposure so that the desired picture composition can be attained even if the area of dominant interest does not correspond to the center of the film frame.

Leitz is able to provide these two different exposure-measuring modes by an ingenious combination of traditional optics applications combined with contemporary fabrication techniques. The main reflex mirror in the *Leica R10* camera carries a 17-layer coating on its surface that allows 30 percent of the light falling on it to pass through to a secondary mirror which is hinged to the top back edge of the main one. The front surface of this molded plastic secondary mirror consists of 1345 concave reflectors in a spiral pattern radiating out from the bottom edge. Each of the pinhead-size surfaces acts as an individual lens and as a diffuser because of the surface texture of the material. Further, the entire array behaves like an off-axis Fresnel mirror since the optical axis of each of these concave reflectors is angled a given amount more toward the center with each successive ring of the spiral. As a result, the entire image area as seen in the camera's viewfinder is projected onto the silicon photodiode located on the bottom front edge of the mirror box. The optical properties of the Fresnel mirror cause emphasis to be placed in the central areas of the image. In order to obtain the selective readings described above, a small lens is shifted over the photodiode by a control ring mounted concentric with the shutter speed dial on the top plate of the camera. This lens and a masking device limit the light reaching the photodiode. This narrow angle is the equivalent of the image area covered by the central focusing aid on the viewscreen. Hence, the photographer can use this central circle to aim the camera for accurate spot readings of the most important part of the subject matter at hand. The *Leica R* is shown in Fig. 10.

As metering systems have advanced for single-lens reflex cameras, rangefinder cameras have kept pace. Many entry-level amateur 35 mm rangefinder cameras now exhibit computer controlled exposure systems, autofocus, bi-focal lens systems, and built-in exposure-controlled electronic flash. In addition, one top-of-the-line camera, the *Leica M6*, features interchangeable lenses and through-the-lens metering coupled to the large-base rangefinder system originally developed by Leitz in the early 1950s as an aid to exact focusing in low light or other difficult photographic situations. See Fig. 11. While the single-lens reflex camera has become the preferred tool of the professional photographer and of the advanced

Fig. 10.   *Leica R5* with optional Motor Winder R and Handgrip. (*E. Leitz, Inc.*)

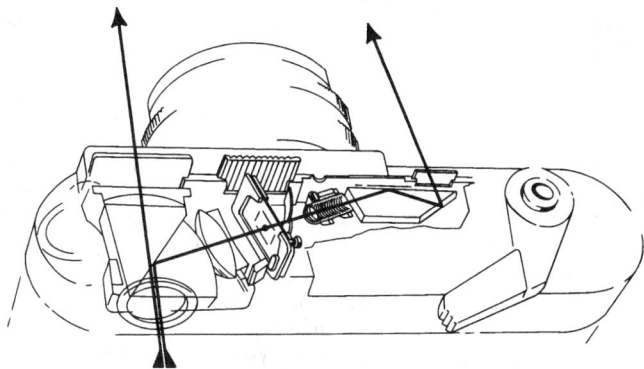

Fig. 11.   Schematic diagram of large-base (49.9-mm measuring base) viewfinder with split-image and double-image rangefinder, variable lens framing marks, and automatic parallax compensation. (*Schematic by Brenda Oras from E. Leitz, Inc. data.*)

amateur, many photojournalists still use the rangefinder camera, because of its pinpoint focus accuracy with wide-angle lenses, as well as its very quiet shutter mechanism. This allows the camera to be used in places and situations where the noisier reflex designs would be unacceptable.

In the *Leica M6*, illustrated in Fig. 12, light enters the lens and is reflected by the white area on the center of the shutter blind into the photocell located above and behind the lens flange inside the camera's dark chamber. Then either the shutter or aperture is adjusted depending on the scene, until both LEDs in the viewfinder light up to indicate correct exposure. Fine correction and intentional over- or underexposure in half or full stops are possible and are indicated in the viewfinder. As a result the photographer has complete manual control in unusual lighting conditions.

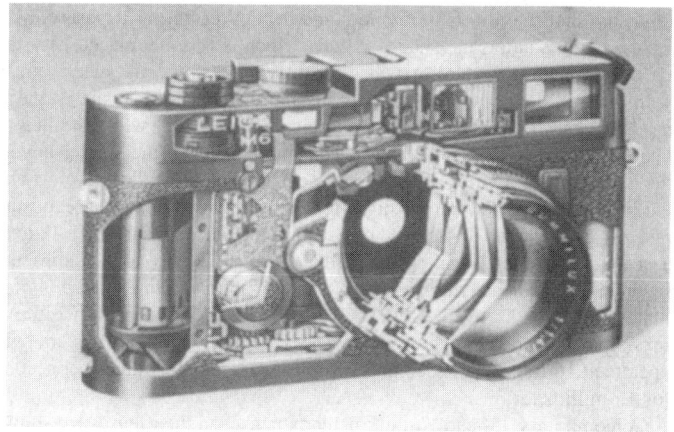

Fig. 12.   Cutaway view of *Leica M6* 35-mm rangefinder camera. (*E. Leitz, Inc.*)

Medium-Format System Camera. Although the foregoing descriptions are concerned with 35-mm cameras, similar advances are being made in medium-format camera systems as well. These systems, using 120 size roll film, particularly have benefited from advances in electronic exposure control. For example, the *Hasselblad 205TCC*, illustrated in Fig. 13, offers a wide range of metering options that give the photographer full control over the tone and contrast of the exposed picture. The camera incorporates four metering modes, including normal automatic, zone, differential, and meter-coupled manual. The exposure system is based around a built-in spotmeter. Also, the camera incorporates a feature that allows the photographer to program certain values different from the one's factory programmed into the camera's central processing unit.

Fig. 13.   View of left side of the *Hasselblad 205TCC* camera shows the controls for the camera's multimode exposure metering system, which allows the photographer to regulate the tone and contrast of the finished picture. (*Victor Hasselblad Inc.*)

In the automatic exposure mode, the *Hasselblad 205TCC* functions as an aperture priority camera, with the camera's electronics choosing the correct shutter speed after the photographer has selected a lens aperture. This mode continuously monitors light values.

In the differential mode, the camera stores the light value of an area of the subject as selected by the photographer. Then, by using the spotmetering capability of the camera, the photographer can measure light values from different parts of the subject and continuously compare them to the reference value first set. This method continuously shows the contrast difference between the initial and other areas of the subject.

The zone mode, which is based on the fact that different colors and shades of color reflect different amounts of light just as gray tones do, gives the photographer a simple way of controlling the image rendition on the film. In this mode, the light values are continuously metered, and provision exists for the locking and storing of a light value at a selected moment. These continuous zone indications are shown in the viewfinder display of the camera. This mode aids the photographer in analyzing a scene and determining what exposure will render all the subject's tones within the film's latitude.

In the manual mode, the *Hasselblad 205TCC* provides for continuous metering of the light value and completely manually controlled setting of the exposure. This mode also provides for continuous indication of the difference in light values between the preset exposure and the exposure calculated by the camera's central processing unit.

In addition to the above four modes, the camera provides an automatic flash function when used with a so-called dedicated electronic flash unit, such as the *Hasselblad Proflash*. When such a unit is connected to the dedicated flash socket and switched on, the *Hasselblad 205TCC* automatically shifts to its flash mode for exposure control based on timing the duration of the flash to properly expose the film.

Like their 35-mm counterparts, medium-format cameras are based on component parts systems that offer many options in terms of interchangeable lenses, viewfinders, film backs, close-up devices, and other assorted accessories. The *Hasselblad 205TCC* system, illustrated in Figs. 14 and 15, is an example of this building-block approach to professional camera systems.

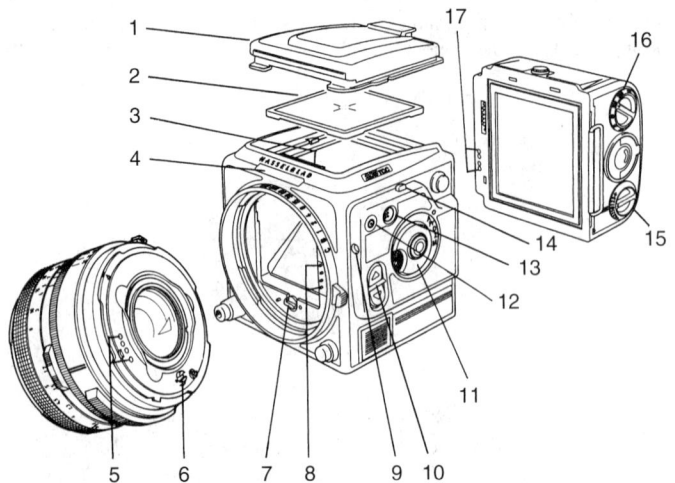

1 Focusing hood cover
2 Acute-matte* focusing screen
3 Liquid crystal display (LCD)
4 Display illumination window
5 TCC connectors
6 Lens drive shaft
7 Drive shaft
8 TCC-connection bracket
9 Selftimer indicator
10 Adjustment buttons
11 Mode selector dial
12 PC socket
13 Dedicated flash connector
14 Display illumination button
15 Film contrast dial
16 Film speed dial
17 System connectors

*Designed by Minolta

Fig. 14. The building-block system concept of the *Hasselblad 205TCC* camera. The type of modular construction is typical for professional medium-format camera systems. (*Victor Hasselblad Inc.*)

Twin-Lens Camera. This is a ground-glass focusing camera with two lenses of the same focal length mounted side by side, one of which forms an image on the ground-glass focusing screen while the other, mounted in a shutter, is used in making the picture. The twin-lens camera has many of the advantages of the reflex, but is simpler mechanically.

View Camera. This type of camera is used chiefly by professional and advanced amateur photographers. It is designed for use on a tripod and does not have either a view-finder or a focusing scale. Ordinarily it is made only in sizes 4×5 inches ($\sim 10 \times 12.5$ centimeters) or larger—5×7 inches ($\sim 12.5 \times 17.5$ centimeters) and 8×10 inches (20×25 centimeters) being two of the usual sizes—and has a long bellows extension with vertical and horizontal adjustments to the back, and in some cases to the front, for dealing with difficult subjects such as tall buildings, cramped interiors and industrial products requiring unusual perspective.

Cameras designed for professional portrait photography are similar to the view camera except that they are much larger and heavier. Ordinarily they are not made in sizes smaller than 8×10 inches and are fitted with a sliding carriage so that the loaded film holder is placed on one side and the slide withdrawn before the image is focused. When everything is ready, the shutter is closed and the carriage shifted to place the film in position for the exposure. This reduces the time elapsing between focusing and the exposure and assists greatly in making life-like portraits. The camera is mounted on a substantial stand, which enables it to be raised or lowered as required.

An entirely new control system for lens aperture and shutter speeds is now available in view cameras with formats up to 8×10 inches (20×25 centimeters). Such cameras feature internal metering with full automatic control settings, plus automatic aperture open and closing and shutter open-close as the film holder is inserted or withdrawn. In the darkroom there is increasing demand for computer technology, with digital clocks and timers and multifunctional timers being used to control enlargers and developing systems.

Digital technology also is changing the way still images are being created. Even though chemistry-based film technologies continue to offer the highest-quality images, digital imaging systems have improved to the point where photojournalists find them to be useful tools for producing news photographs because of their instant imaging capability. For example, Kodak markets a digital camera system that is based on a *Nikon F3* 35-mm camera body with a 1024-by-1280–pixel CCD installed at the film plane. This CCD is connected by cable to a battery-powered digital unit capable of storing a maximum of 156 digitized color images, using computer-based hard drive technology. Once stored, these images are easily accessed for viewing and can be transmitted over telephone lines or by other electronic means for use at other locations.

Motion Picture Cameras. The contemporary motion picture camera may be divided into two general mechanical categories: (1) those in which the film moves intermittently across the film plane, being at rest during the actual exposure (intermittent movement); and (2) those in which the film moves continuously during pull down and exposure (continuous movement).

In a motion picture camera with intermittent movement (Fig. 16), the film is drawn from the roll at the top and delivered to the take-up reel at the bottom by a sprocket and idler wheel system. This may be driven by a spring motor and escapement mechanism, or by an electric motor powered by rechargeable batteries or the A.C. power line. Many of the D.C. motors are quartz crystal oscillator-controlled for speed accuracy. The film passes around the drive sprocket, through the film gate, past the aperture, around the opposite side of the sprocket and finally is wound onto the take-up spool. The supply spool (unexposed film) and the take-up spool (exposed film) are driven in the directions shown by the same mechanism as the sprocket. However, they are geared and clutched in such a way as to allow them to rotate at different rates to compensate for the increase or decrease in circumference of the film loads wound on them. The spools turn slower as the amount of film wound on them increases; and faster when the film load decreases. Since the sprocket and spools operate continuously while the pull-down claw produces an intermittent movement of the film in the film gate between exposures, it is necessary to provide the camera with an eccentric mechanism to produce the reciprocating motion of the claw. It is also necessary to leave a loop on either side of the film gate to prevent damage to the perforations on the film. In some advanced camera designs, a registration pin is driven by the eccentric mechanism. This pin enters the film perforations during the time when the film is stopped in the film gate to ensure that the film is motionless and in register during exposure.

A rotating shutter, the shaded part of which is opaque, placed between the lens and the aperture, exposes the film. The shutter is synchronized with the pull-down mechanism, and the registration mechanism, if one is used, so that the film is stationary during exposure, i.e., while the lens is uncovered by the shutter. When the opaque portion of the shutter covers the lens, the perforations of the film are engaged by the pulldown claw and the film is transported into position for the next exposure. At an exposure rate of 24 pictures, or frames, per second, actual exposure is 1/48th of a second in cameras using 16 millimeter film. Twenty-four frames per second is the standard running rate for sound motion pictures in the United States and Canada; twenty-five frames per second is standard in Europe. In simple cameras for "home movies" without sound, the camera will operate at 16 frames per second for 8-millimeter and 18 frames per second for 16-millimeter.

Contemporary 35- and 16-millimeter cameras and their lenses are shown in Figures 17, 18, and 19.

In all cameras, the density of the exposure may be controlled by changing the size of the lens diaphragm opening. In addition, many professional motion picture cameras are provided with variable-angle shutters to regulate exposure.

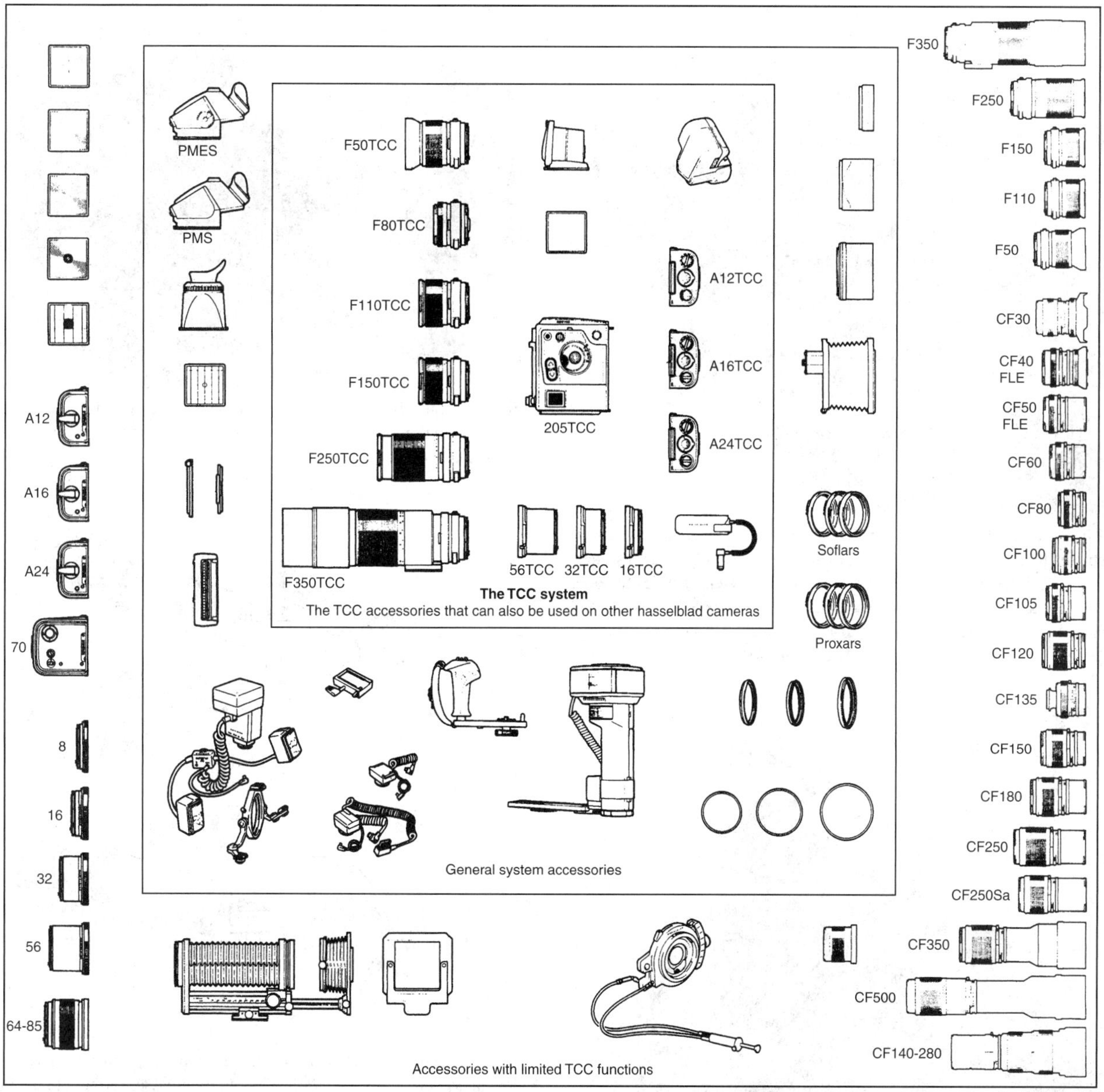

Fig. 15. Range of lenses, film magazines, viewfinders, focusing screens, electronic flashes, and other accessories of the total *Hasselblad 205TCC* camera system. (*Victor Hasselblad Inc.*)

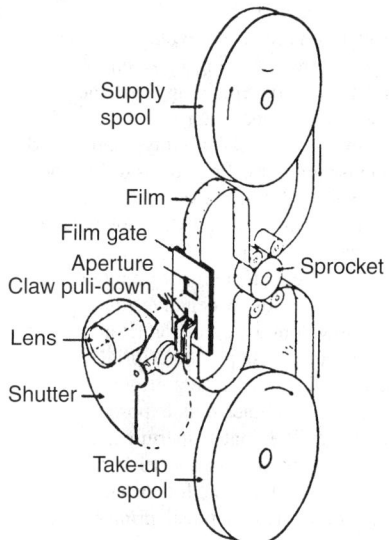

Fig. 16. Schematic diagram of motion picture camera with intermittent movement.

Some professional cameras and many amateur cameras use built-in exposure meters either to aid the operator in setting the correct exposure for the film being used in the camera or to set the exposure for that film automatically. These mechanisms are usually coupled to the lens diaphragm adjustment so that the lens is set to pass the correct amount of light to the film for proper exposure.

Simple cameras are fitted with wire frame or nonadjustable optical masks for viewfinders so that the camera may be lined up on the subject. Most professional cameras use beam splitters or mirror shutter designs so that the taking camera lens forms part of the viewing system, as shown in Fig. 20. This latter system eliminates the parallax problems of the separate viewfinder systems and allows for more accuracy in setting focus.

Continuous-movement cameras use the principle of optical compensation, which allows the film to move continuously past the exposing aperture at speeds up to 18,000 frames per second while optically arresting movement with a rotating glass block or prism so that steady and sharp images result. These cameras are generally only used for high-speed or special-effects cinematography.

The widest camera film is 65-millimeter. The size of the film projected from this format is 70-millimeter. The additional 5 millimeters are for the film sound track. The projected image is 65-millimeter. A 100-foot (30+ meters) roll of standard 35-millimeter film, as used in theaters, requires 67 seconds to run through the projector at sound speed (24 frames

Fig. 17.   The *Arriflex 535* 35-millimeter motion picture camera. The instrument has a viewfinder that pivots 270° and still provides a completely upright image. The camera incorporates integral timecode, a numbering system that ties together the film, sound, and video. The camera also offers programmable film speed and aperture controls that permit an operator to change the running speed without affecting depth of field or exposure. (*Arriflex Corporation.*)

Fig. 18.   *Carl Zeiss* fixed focal length and zoom, color-matched lenses for the *Arriflex* range of 35-millimeter motion picture cameras. (*Arriflex Corporation.*)

Fig. 19.   This 16-millimeter motion picture camera is used for sound and silent filming in the studio or on location. (*Arriflex 16SRII Camera, Arriflex Corporation.*)

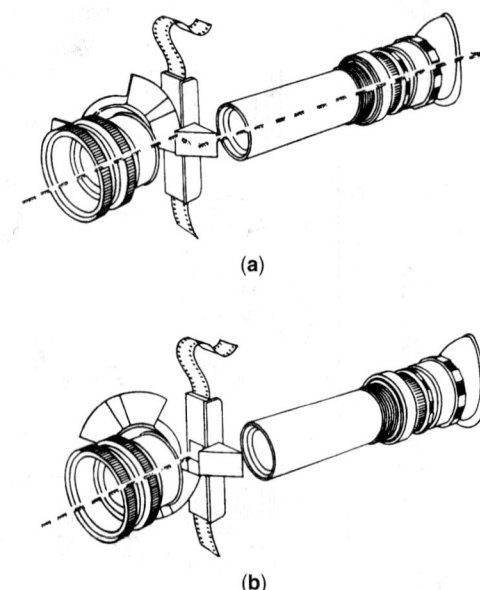

Fig. 20.   Shutter viewfinder: (**a**) shutter "closed" — all light passes through the viewing system; (**b**) shutter "open" — all light passes through to the film. (*Arriflex Corportion.*)

per second), while a roll of 16-millimeter film of the same length requires 2 minutes, 47 seconds at sound speed; and 8-millimeter requires 5 minutes, 33 seconds at sound speed.

Processing. Professional motion picture films are almost always processed by continuous machinery. Amateur films may be processed in this way, or short lengths may be processed by hand in special light-tight containers. In the former method, the strand of film is conducted at a pre-set steady speed through a number of developing tanks which are maintained at a constant temperature, a rinsing tank, one or more fixing tanks, and several washing tanks. Upon emerging from the washing tanks, the film is squeegeed to remove surplus water and then conducted through a series of rollers into a temperature-and-humidity-controlled drying cabinet. Both color and black-and-white films, regardless of whether they are positive or negative type, are developed in the same manner. Generally, the processing time for positive films is longer than for negative films. Color films take longer than black-and-white.

Positive prints for projection are made on motion picture positive release print film, which corresponds to negative film in size but is coated with a slow, fine-grain, low-contrast emulsion. Printing is usually done by the contact method on either a step-contact printer or a continuous contact printer. In contact printing, the camera original and the release print stock are brought together in contact and exposed by moving together at a uniform speed across an illuminated aperture in the printer. Optical printing is used for special effects and for reducing a large format to a smaller one (16-millimeter to 8-millimeter, for instance), or to enlarge a smaller format to a larger one. In the optical printer a lens system is used to focus the image from the camera original onto the print stock which is running through another part of the printer in synchronization with the original.

The exposure of the release print film in either type of printer is controlled by varying the voltage to an incandescent lamp, or by an adjustable diaphragm or variable back shutter, any of which methods will control the amount of light reaching the exposing aperture.

In addition to making corrections for exposure density variations, modern printers of contact or optical type can be programmed to correct for color variations of the original film as well. These color-correction controls may be used to compensate for color imperfections in the original, or they can be used to create special effects. Color analyzers are used to determine what corrections are needed for a given camera original. These analyzers produce a punched paper tape which is used to program the printer for the color corrections as the film is printed. The analyzer and printer are set up for additive color printing, which involves mixing of color illumination through various dichroic filters to produce a print with both color and density correction. The color-density correction is based on the primary and complementary colors of light. This system changes the color balance of a camera original when it is printed by adding or increasing the filtration of the complementary color. The printer works in the additive process by using a series of dichroic filters, which act like mirrors to separate a single light source into three color beams — red, green, and blue. It is this mirror system that filters and refines the color beams to eliminate the unusable wavelengths of color and to retain the pure colors for the printing process. Some modern analyzer/printer combinations use computer techniques instead of punched paper tapes for control of the printing operation. In either method, the control information is retained so that the results may be repeated for multiple prints from the original.

Projectors. Sound projectors are employed for viewing the completed release prints of a given motion picture. These projectors vary in size from large theater units, using carbon arcs for a light source, to small table models, which use incandescent lamps for illumination. The operating principles are the same regardless of the illumination source.

Just as in the camera, the mechanism must be arranged so that the film is moved intermittently past the film aperture. However, for proper sound reproduction, the film must be moved continuously past the sound head. The arrangement used in a modern 35-millimeter projector is shown in Fig. 21. Film is drawn from the supply spool by the top feed sprocket which rotates at a speed required to supply 24 frames of picture to the film gate per second. An intermittent sprocket, or in some designs, a pull-down claw, moves the film into the projector aperture and stops it so that it may be projected by the lens. Behind the film gate, a two-bladed rotating opaque shutter allows light to pass through the film frame in the projector aperture. This shutter has two openings arranged so that each frame is projected twice before the intermittent sprocket advances the film to the next frame. This serves to eliminate picture flicker on the screen and creates an illusion of continuous motion. The film is only moved in the gate when the shutter is closed to prevent blurred motion on the screen. The film is then advanced through the film gate by the bottom constant-speed sprocket. It then passes the sound head where the audio portion of the motion picture is picked up and fed to the audio reproduction circuits of the projector. Finally, the film is wound onto the take-up spool. As in the camera, loops are formed in the film to prevent damage to the film perforations. Generally, modern projectors are driven by ac electric motors and employ solid-state audio circuits.

The audio portion of a motion picture film is carried on the sound track, a strip of 0.1-inch (~2.5 millimeters) wide on a 35-millimeter film. In recording the sound track, two systems are generally used. In the *variable-area system*, the transmitted light amplitude is a function of the amount of unexposed area in the positive release print. In the *variable-density system*, the transmitted light amplitude is an inverse function of the amount of exposure in the positive release print. For reproduction of the sound track, light is passed through the film onto a photocell in the projector. The amount of light that impinges on the photocell is directly proportional to the unexposed portion of the sound track in the variable-area system, or to the inverse function of the density in the variable-density system. When the film is continuous motion, the light undulations that fall upon the photocell produce voltage variations in the projector's audio circuits that correspond to the original sound recording. See also **Television (TV)**.

This is an appropriate juncture to note that, while 16-millimeter is still the preferred format for professional documentary film production, and 35-, 65-, and 70-millimeter formats are likewise the mainstays of theatrical and even broadcast productions, videotape in its various sizes and recording formats is becoming ever more important as a distribution medium. For instance, many prime time television dramas, action shows and mini-series are originally produced on 35-millimeter motion picture film for reasons of picture and sound quality coupled with equipment portability on location. The finished print is then transferred to one of the one-inch or two-inch broadcast videotape formats for airing. In addition, theatrical films are transferred to one-half-inch formats for distribution to the home video recorder/player market. See also **Television (TV)**.

Electronic Imagery

Electronic Imagery, instead of using chemical means (emulsions), takes advantage of the sensitivity of various electronic detectors to different bands of the electromagnetic spectrum. The energy received is transduced by these sensors into an electronic or electrical effect (change of resistance, current, emf, the emission of electrons, etc.), from which effects an option of ways to process and display the information is available. The most common form of electronic imagery is found in television. Image orthicons, vidicons, and the more recent TV cameras using charge-coupled devices are described in entries on **Television**; and **Charge-Coupled Device**. See also **Cathode-Ray Tube**; and **Computer Graphics**.

Electronic imagery is particularly attractive for situations where image information must be transmitted over long distances where digitized signals offer greater accuracy and reliability — and where the incoming information is immediately compatible with digital data processing and computing equipment. Electronic imagery also has made certain imaging tasks possible, such as radar imaging, where traditional photographic means do not suffice.

While techniques are available in traditional photography to enhance raw information, these methods are largely of a qualitative, aesthetic nature rather than of a quantitative, scientific nature. In handling tiny pixels (one of the dots or resolution elements making up a digitized picture) of information, it becomes possible to computer program the processing of image information as it is received or after it is retrieved from tape or other electronic storage medium. The pixels can be measured one at a time at a rapid rate for brightness, sense (detection of obviously bad information), and other quantities, and over a wide scale of selections (for example, black = 0; medium gray = 32; white = 63) so that groupings of input information can be made to provide better contrasts (in pattern and

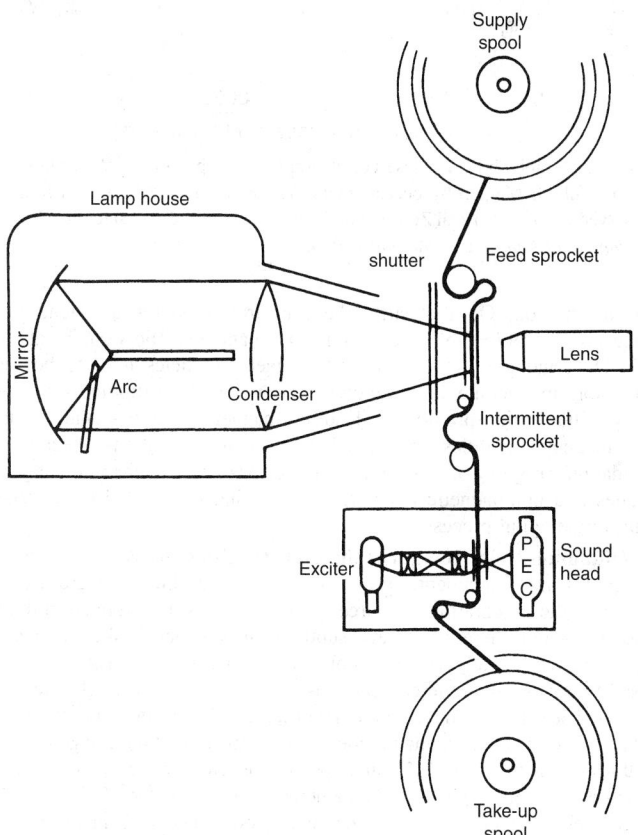

Fig. 21.   Principle of a 35-millimeter projector.

blackness or color) when the information is all regrouped and reassembled for display. Accomplishments of this nature are probably best exemplified by the image intensification and color enhancement of pictures returned to earth from various space explorations. This is discussed further in this entry, but also see entries on specific planets.

With the flexibility of modern data processing and computing equipment, a vast array of programming techniques can be applied to the handling of pixels similar to the handling of any other kind of information. Also, with electronic imagery, a full reconstitution of a scene need not always be a primary objective. For example, as color may be related to chemical composition (discussed later) an astrochemist may call for the proportion of certain "colored" pixels in an entire scene or part of a scene and thus make at least a preliminary judgment as to the composition of rocks or soil in a scene without having to see the entire scene reconstituted.

Whereas the final results of traditional photography are usually in the form of prints or transparencies (with attendant problems), electronic imagery can be projected at will on cathode ray tubes and, if desired, combined with computer graphics—or conventional photos can be made from digitized information.

Relating Color to Physical/Chemical Properties of Materials

Color of materials is of particular interest to geologists, mineralogists, oceanographers, astrophysicists, and astrochemists, who have found in recent years that color as recorded in images can lead to specific information pertaining to the chemical content and physical parameters of materials. Only a few specific examples can be given here.

Ocean Water. The possibility of using remote sensing of ocean water color to determine the general composition of the water has been recognized for many years. The physical processes of absorption and scattering relate the upwelling radiance just beneath the sea surface to the constituents of the water. Except for waters in close proximity to coastlines and the confluences of rivers and the sea, biological constituents play a dominant role in these processes. The most important constituent appears to be phytoplankton, microscopic plant organisms that photosynthesize and constitute the bottom link in the ocean food chain. These plankton contain chlorophyll (the dominant photosynthetic pigment), which absorbs strongly in the blue and red regions of the visible spectrum. Hence, increasing concentrations of phytoplankton (chlorophyll a) have the effect of changing the color of water to green hues from the deep blue of its pure state. By selecting a number of frequencies, the sensing of the chlorophyll content can be put on a quantitative basis. An instrument known as the Coastal Zone Color Scanner (CZCS) was installed in the Nimbus-7 satellite, launched in October 1978. There are six wavelength bands utilized: (1) 433–453 nm; (2) 510–520 nm; (3) 540–560 nm; (4) 660–680 nm; (5) 700–800 nm; and (6) 10,500–12,500 nm. When the data are processed, so-called false-color images are produced, each color representing one of the aforementioned frequency bands. See Fig. 22. Phytoplankton supports all higher life forms in the sea. Information like this may lead to improved methods for managing and exploiting fisheries. Massive concentrations of phytoplankton blooms (red tide) also may be detected well in advance by this technology. The excellent correlation between samples taken from the ocean research vessel (R.V. *Athena II*) and the CZCS instrument on the Nimbus satellite are shown in Fig. 23.

Color Values of Planetary Images. The *Viking Orbiter* and *Lander* cameras returned color information about many of the surface features of Mars, as well as the colors of the atmosphere and of a Martian sunset. The planet has a salmon-colored sky, surface materials that range from medium-warm umbers to brighter shades of orange and reddish yellows, and rocks that range from nearly black breccia-type blocks to the more characteristics reddish yellow rocks of recognizable volcanic origin. From orbit the colors are even more diverse and include color-shaded values of white from clouds, fogs, and surface ice, and many brightness variations of the reddish color that was noted of Mars some 3000 years ago. See also **Mars**.

Aside from their interest purely as the result of curiosity, detailed analysis and study of the colorations of the Martian features are of much assistance to scientists in extending the data returned by the chemical analytical equipment which was part of each of the two landers' sampling equipment. Because surface color is so consistent in many regions of the planet, the interest in color is to look for changes as a result of disturbances in reference to the passage of time. In the area of magnetic properties, spectral analysis and color produced a primary component

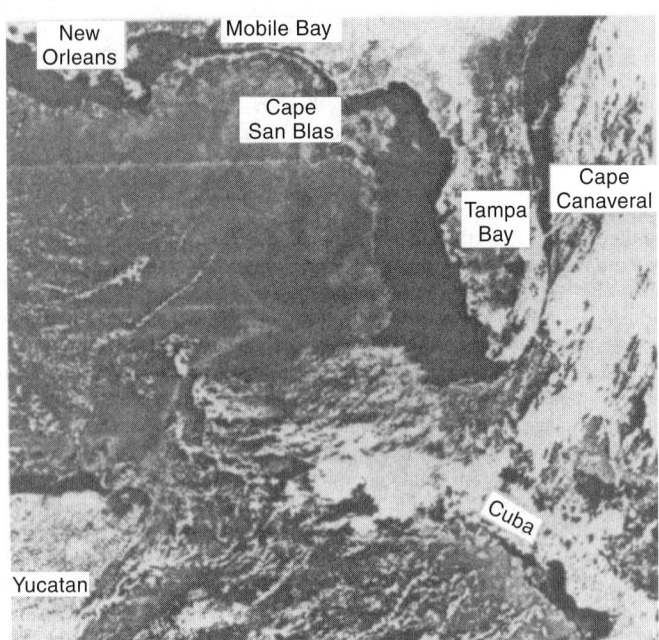

Fig. 22. Black-and-white reproduction of a false-color coded map of phytoplankton pigments in the Gulf of Mexico. Black area represents highest concentrations. (*National Oceanic and Atmospheric Administration.*)

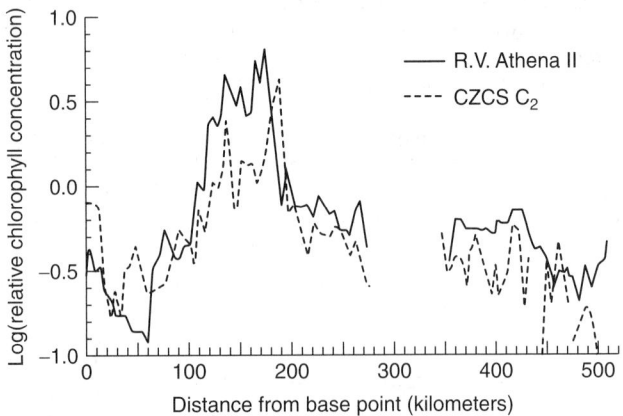

Fig. 23. Graph showing close correlation between phytoplankton counts taken in the Gulf of Mexico by oceanographic research vessel *R.V. Athena II* and as observed from the Coastal Zone Color Scanner in the *Nimbus 7* satellite. (*National Oceanic and Atmospheric Administration.*)

of information. On the earth, the four iron products that tend to be magnetically attractable are iron metal, magnetite (Fe_3O_4), maghemite (Fe_2O_3), and pyrrhotite. Iron and magnetic particles tend to be black in color; maghemite is a yellowish brown; and pyrrhotite is a brassy grey. Thus color provides a lead to magnetic qualities and, in turn, the magnetic qualities are related to planetary evolution—because the oxidation progress of iron, for example, passes through both magnetite stages and nonmagnetic hematite (or limonite, a hydrated oxide) stages during the aging process.

Obtaining Color Images of the Planets. Using the *Viking* missions as examples, the fundamental principles of photographic color are involved through the integration of three primary colors to produce full-color pictures—but there are many subtle factors which make the task of resolving completely accurate color very difficult. The *Viking* Orbiters used twin slow-scan vidicon cameras, but acquired only single frames by utilizing a shutter. The vidicons recorded the image on a photosensitive plate, which is then scanned a line at a time to convert the image to digital data for transmission (1056 lines per frame, each made up of 1182 tiny picture elements—pixels). The cameras' 475 millimeter telescopic lenses could resolve features no smaller than a football field from the minimal orbital altitude of 1512 kilometers (940 miles). The cameras alternated and each could recycle in 4.5 seconds. See Fig. 24. Each camera contained a

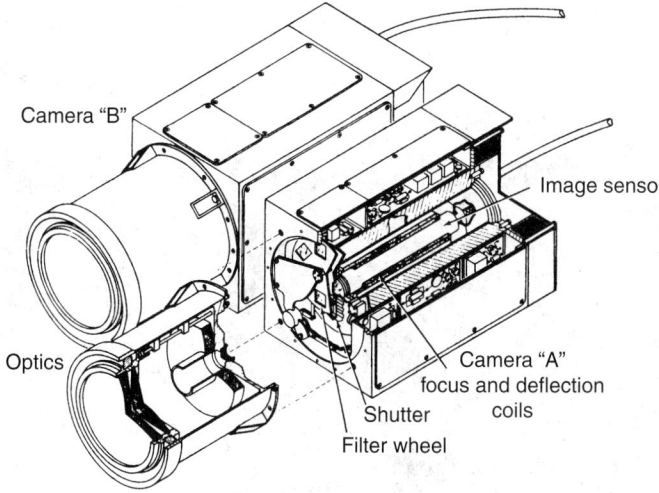

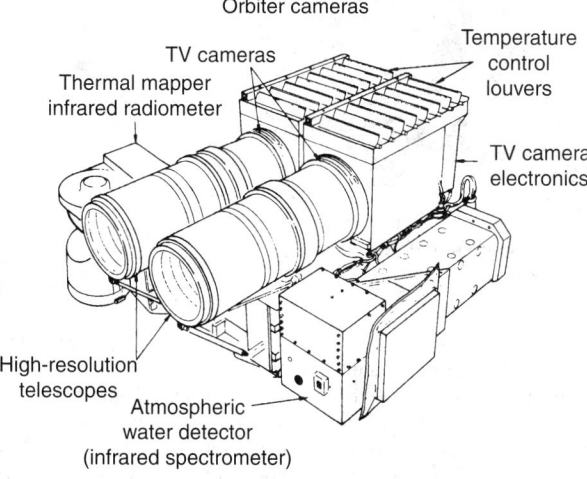

Orbiter science platform

Fig. 24. *Orbiter* camera and science platform used on *Viking* mission to Mars. (*NASA Jet Propulsion Laboratory, Pasadena, California.*)

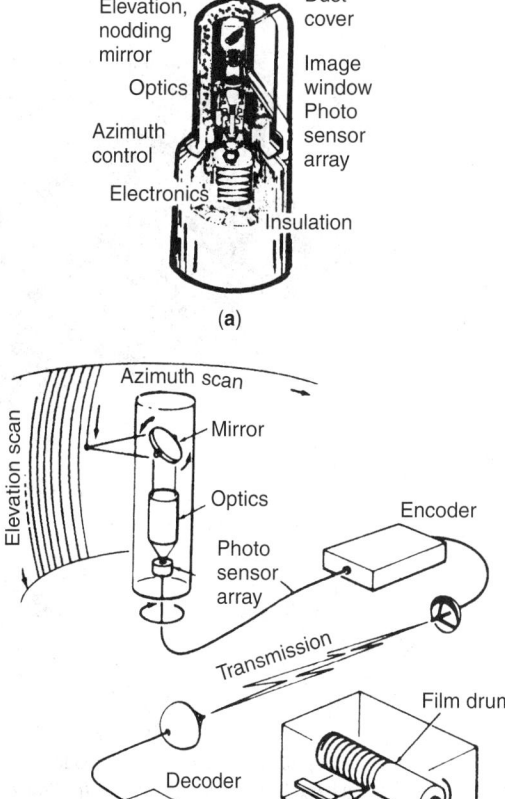

(a)

(b)

Fig. 25. (**a**) Schematic of Mars *Lander* photo sensors; (**b**) complete system, including reconstruction of information after transmission to Earth. (*NASA Jet Propulsion Laboratory, Pasadena, California.*)

filter wheel fitted with six filters; a clear filter to provide broad sensitivity across the near-ultraviolet and visible wavelengths; a violet filter sensitive only to the near-ultraviolet and violet (for cloud and ice enhancement); a minus-blue filter to yield a reverse effect to that of the filter; and three filters for color reconstruction (red, green, blue). The actual color pictures are constructed at the project office on earth—not on the spacecraft. The camera simply acquires three individual pictures in quick succession, each utilizing one of the color filters. The three pictures are combined at the project office as individual frames, enhanced as needed to improve contrast and color balance, and frequency mosaics are prepared.

The cameras on the *Viking* landers are technically classified as facsimile cameras. The principle of operation is similar to that of equipment used to transmit wirephotos by radio or telephone, and is quite a bit slower than vidicon photography. The operation is not unlike that of the Orbiter vidicons just described, in that a scanning technique is used. However, the scene is scanned directly rather than via a photosensitive plate before the lines and picture elements (pixels) are encoded as digital information for radio transmission.

The reconstruction equipment at the project office essentially reverses the camera process by converting the digital data back into an image on film with a unique artificial light beam produced by an argon/krypton laser. The fundamental principle involved in processing orbiter and lander data is similar, but it should be emphasized that the reconstruction process is also used for the preparation of orbiter pictures and for the visual presentation of data that are not imagery produced. Orbital thermal mapping and water-vapor mapping data can also be illustrated as color imagery. A schematic of the system is shown in Fig. 25. Some general concept of the quality of the reconstructed lander image is given in Fig. 26, which is reproduced in black and white from a color image.

The sequence of operations is as follows:
On the surface of Mars

1. Nodding mirror vertically scans Martian scene
2. Reflects scene through lens onto photo sensors
3. Sensors generate signal directly proportional to density of incident light
4. Signal output sampled at rate synchronized with mirror
5. Samples space vertically as picture elements (pixels). Each pixel—one resolution size on a side (square), 512 pixels per line
6. Mirror completes elevation scan. Returns to start position
7. Camera revolves one line in azimuth; starts next line (or rescans line two or more times for 3-color)
8. Digitized pixels relayed to Lander transmitter
9. Data transmitted to earth (direct or orbiter-relayed)

On Earth (project office)

1. Data are processed; recorded on magnetic tape
2. Data for complete scene played into reconstruction equipment via digital computer. Laser is initiated
3. Red, blue, green beams separated out of laser beam
4. Individual beams reflected into light modulators
5. Modulators are computer controlled to alter beam intensities according to digitized pixel intensity data
6. Three color beams recombined into laser beam and focused on rotating mirror
7. Duplicates action of camera mirror; puts image on conventional color film; unit moves one line width as each line is completed

Color Analysis of Lunar Images. Although it has been several years since the last images of the moon were returned to earth by the Apollo missions, detailed studies continue in some laboratories. Detailed chemical maps of the lunar surface have been constructed by scientists (Andre et al., 1977) who have applied a new weighted-filter imaging technique

Fig. 26.   The general character of images received from the *Viking Lander* on the Martian surface is apparent from this black-and-white reproduction of the reconstituted view. White structure in foreground is part of the *Viking Lander* vehicle. (*NASA Jet Propulsion Laboratory, Pasadena, California.*)

to Apollo 15 and Apollo 16 x-ray fluorescence data. The data quality improvement is amply demonstrated by (1) modes in the frequency distribution, representing highland and mare soil suites, which were not evident prior to data filtering, and (2) numerous examples of chemical variations which are correlated with small-scale (about 15-kilometer) lunar topographic features.

Radar Imagery

Although radar has been a useful tool in air traffic control, weather reporting, and military weaponry for several decades, the use of radar imagery for scientific research and geological survey and oceanographic investigations is somewhat more recent. One of the first scientific applications, of course, was in connection with astronomy. See also **Radio Astronomy**. The acoustic analog of radar, sonar, has been applied for many years and is finding increasing scientific applications. See also **Sonar**.

Geological Mapping of Land and Ocean Surfaces. When the Seasat satellite was put into orbit around the earth in June 1978, it was equipped with a payload of active microwave sensors consisting of an altimeter, a scatterometer, and an imaging synthetic aperture radar (SAR). The objective of the mission was a proof-of-concept demonstration of the capability to monitor the ocean surface and near-surface features, such as surface waves, internal waves, currents, eddies, surface wind, surface topography, and ice cover. The imaging radar, which was operated in the synthetic aperture mode, provided, for the first time, synoptic radar images of the earth's surface (both ocean and land areas) obtained from an orbiting platform. The resolution of these images is about 25 meters (82 feet). As pointed out by Elachi (1980), the success of this complex sensor was a major technological advance, and it opened up a new dimension in the capability to observe, monitor, and study the Earth's surface. The SAR imaging sensor is an active system, using its own energy to illuminate the surface and to generate an image from the backscatter echoes. Thus, it is not dependent upon illumination from the sun. The radar energy also penetrates cloud cover. Consequently, the system is not constrained by weather conditions. The illumination angle and direction can be controlled and selected, whereas in optical systems these parameters are constrained by the sun's location.

In the SAR approach, the Doppler information in the returned echo is used simultaneously with the time delay information to generate a high-resolution image of the surface being illuminated by the radar. The radar

usually "looks" to one side of the moving platform and perpendicular to its line of motion, thus eliminating right-left ambiguities. In Elachi's description, it is pointed out that points equidistant from the radar are located on successive concentric spheres. The intersection of these spheres with the surface gives a series of concentric circles centered at the nadir point. See Fig. 27. The backscatter echoes from objects along a certain circle will have a well-defined time delay.

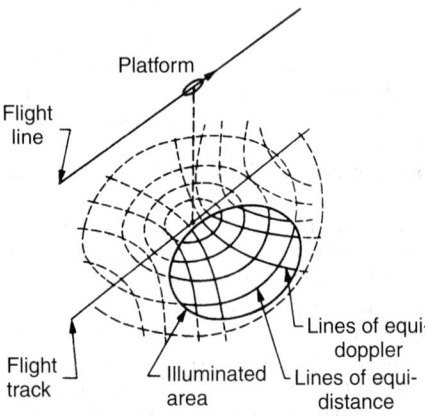

Fig. 27.   The constant time delay and Doppler contour lines for the radar imaging coordinate system. Each point on the surface can be uniquely identified if the energy in the appropriate time delay bin and Doppler shift bin is filtered out of the received echoes. (*Jet Propulsion Laboratory, Earth and Space Sciences Division, Pasadena, California.*)

The brightness in the radar image is a representation of the surface backscatter cross section, which is a function of the surface slope, surface roughness at the scale of the observing wavelength, and surface complex dielectric constant. Geologic interpretation of the radar image is based upon two general types of information: (1) geometric patterns and shapes, and (2) image tone and texture. The sensitivity of the amplitude of the radar echo to changes in the surface topography is very high in comparison to the optical and infrared albedo. A change in the surface slope of a few

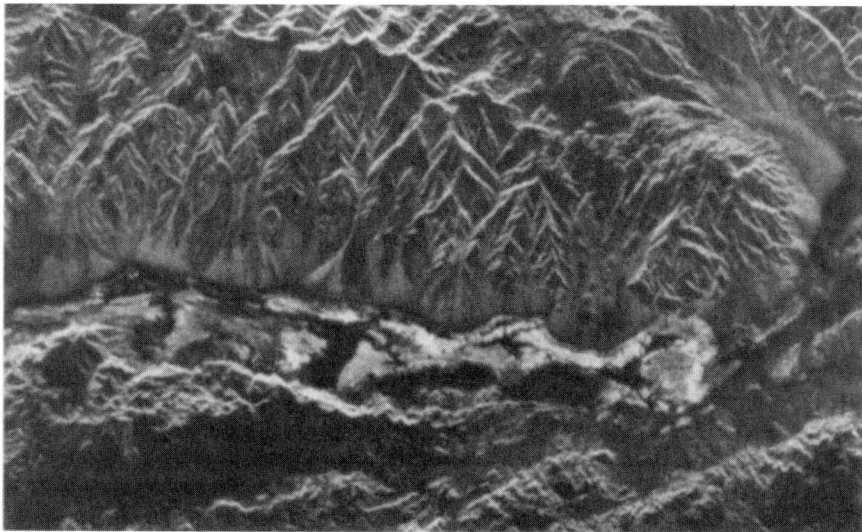

Fig. 28.   Radar image of Death Valley, California. Length of view represents a distance of about 90 kilometers (56 miles). (*Jet Propulsion Laboratory, Earth and Space Sciences Division, Pasadena, California.*)

degrees can easily change the amplitude of the radar echo by a factor of two or more. A radar image of Death Valley, California, is shown in Fig. 28.

Because the radar sensor basically provides a Doppler time-delay history of each point target, thousands of computational operations are required to generate a single image element. This processing requirement, combined with the desire to have large swath mapping with high resolution, requires extremely fast processing hardware which is just at the limit of current technology.

See also **Satellites (Scientific and Reconnaissance)**.

Color Image on Silicon Wafer. V.V. Doan and M.J. Sailor (University of California, San Diego), in mid-1962, reported that an electrochemical etch of silicon produces a microporous material that Photoluminescences in the visible region of the electromagnetic spectrum. The researchers attribute the action to quantum confinement effects that arise from isolated, nanometer-size silicon features that are produced during the etching. Images get generated on an n-Si wafer appeared colored under white light. Images appeared red-orange under UV radiation. The researchers also photoetched a diffraction grating into the substrate to demonstrate simultaneous encoding of gray-scale images into thin-film interference, luminescence, and diffraction phenomena. It is too early at this juncture to forecast possible practical uses for such devices.

Medical and Biochemical Imaging

Numerous mentions of imaging technology in the medical and biochemical research area were covered in this encyclopedia. Consult alphabetical index. As will be found from such references, outstanding progress has been made in both areas during the past three decades. However, at a meeting of the American Psychological Society (Chicago, Illinois) in 1993, several professionals observed that actually only a "good start" has been made, with reference to imaging techniques yet to be developed. The human heart and brain are two areas that will require extensive, innovative imaging approaches. Scientists are interested particularly in producing real-time imagery, and a beginning has been made pertaining to the use of magnetic resonance imaging (MRI) (e.g., to peer into the brain of a patient with obsessive-compulsive disorder). On the other hand, imaging the heart remains difficult because of its constant motion that tends to blur contemporary imaging devices. Some researchers have established the symbiotic effects of combining MRI with electroencephalography (EEG) techniques. One of the important questions posed at the meeting, "How does the brain coordinate activity to produce consciousness?" See also **Central and Peripheral Nervous Systems**. The long-neglected technique of *transillumination* has been reborn. History records the first use of this methodology by a British physician in the mid-1800s for exploring possible testicular tumors. With considerable refinement of the procedure and instrumentation used, a promising application may be in mammography.

R.A. Alfano (City University of New York) observed, "We can see inside tissue with enough clarity that we're starting to see 'fingerprint' patterns that distinguish between healthy and abnormal tissue."

Additional Reading

NOTE: Because of their large numbers, general books on photography are not included here. References that support some of the more specialized areas of photography and imagery, as reported in this entry, are included.

Alper, J.: "Echo-Plannar MRI: Learning to Read Minds," *Science*, 556 (July 30, 1993).

Andre, C.G., et al.: "Lunar Surface Chemistry: A New Imaging Technique," *Science*, **197**, 986–989 (1977).

Barger, M.S. and W.B. White: "The Daguerreotype," Smithsonian Institution Press, Washington, DC, 1991.

Barger, M.S. and W.B. White: "Daguerreotype: Nineteenth-Century Technology and Modern Science," Johns Hopkins University Press, Baltimore, MD, 2000.

Barinaga, M.: "Biology Goes to the Movies," *Science*, 1204 (November 30, 1990).

Bentley, J.: "Coloring the Invisible World," *Technology Review (MIT)*, 54 (July 1991).

Beardsley, T.: "Sharper Image: Picosecond Photography May Reveal Tumors," *Sci. Amer.*, 32 (October 1991).

Becher, P.: "Emulsions: Theory and Practice," 3rd Edition, Oxford University Press, Inc., New York, NY, 2001.

Benaron, D.A.: "Optical Time-of-Flight Absorbance Imaging of Biological Media," *Science*, 1463 (January 22, 1993).

Booth, S.A.: "Video To Go: Camcorders," *Popular Mechanics*, 38 (January 1991).

Cipra, B.A.: "Image Capture by Computer," *Science*, 1288 (March 10, 1989).

Corcoran, E.: "Not Just a Pretty Face: Compressing Pictures with Fractals," *Sci. Amer.*, 77 (March 1990).

Corcoran, E.: "Body Heat: Quantum-Well Infrared Photodetectors," *Sci. Amer.*, 123 (October 1991).

Cornwell, T.J.: "The Applications of Closure Phase to Astronomical Imaging," *Science*, 263 (July 21, 1989).

Crease, R.P.: "Biomedicine in the Age of Imaging," *Science*, 554 (July 30, 1993).

Doan, V.V. and M.J. Sailor: "Luminescent Color Image Generation on Porous Silicon," *Science*, 1791 (June 26, 1992).

Drury, S.A.: "Guide to Remote Sensing: Interpreting Images of the Earth," Oxford University Press, Inc., New York, NY, 1990.

Elachi, C.: "Spaceborne Imaging Radar: Geologic and Oceanographic Applications," *Science*, **209**, 1073–1082 (1980).

Elachi, C.: "Radar Images of the Earth from Space," *Sci. Amer.*, 54–61 (December 1982).

Elachi, C., J. Cimion, and M. Settle: "Overview of the Shuttle Imaging Radar-B: Preliminary Scientific Results," *Science*, **232**, 1511–1516 (1986).

Gibbons, A.: "Remote Sensing in War's Aftermath: Kuwait Oil Fires," *Science*, 920 (May 17, 1991).

Grimm, T. and M. Grimm: "The Basic Book of Photography," 4th Edition, Penguin USA, New York, NY, 1998.

Hedgecoe, J.: "The Photographer's Handbook," 3rd Edition, Alfred A Knopf, Inc., Westminster, MD, 1992.

Holloway, M.: "The Plankton Stalkers," *Sci. Amer.*, 20 (April 1992).

Huang, et al.: "Optical Coherence Tomography," *Science*, 1178 (November 22, 1991).

Izatt, J.A., et al.: "Ophthalmic Diagnostics Using Optical Coherence Tomography," *SPDIE Proceedings*, 1877 (1993).

Jenkins, F.A., Jr., K.P. Dial, and G.E. Goslow, Jr.: "A Cineradiographic Analysis of Bird Flight," *Science*, 1495 (September 16, 1988).

Kaufman, W.: "Radar Imaging: Forest X-Ray," *Amer. Forests*, 46 (September–October 1990).

Kerr, R.A.: "Do NASA Images Create Fantastic Voyages? Scientists Distort Images of Planetary Bodies by Accident or Design; Now They Are Being Scolded for Misleading the Public," *Science*, 1637 (March 27, 1992).

Lam, D., Man-Kit, and B.W. Rossiter: "Chromoskedasic Painting," *Sci. Amer.*, 80 (November 1991).

Lillesand, T.M.M. and R.W. Kiefer: "Remote Sensing and Image Interpretation," 4th Edition, John Wiley & Sons, Inc., New York, NY, 1999.

London, B. and J. Upton: "Photography," 6th Edition, Addison Wesley Longman, Inc., Redding, MA, 1997.

Mollet, H. and A. Grubenmann: "Formulation Technology: Emulsions, Suspensions, Solid Forms," John Wiley & Sons, Inc., New York, NY, 2001.

Newhall, B.: "Daguerreotype in America," 3rd Edition, Dover Publications, Inc., Mineola, NY, 1999.

Ourmazd, A., et al.: "Quantifying the Information Content of Lattice Images," *Science*, 1571 (December 22, 1989).

Pappas, D.L., et al.: "Atom Counting at Surfaces," *Science*, 64 (January 6, 1989).

Peterson, I.: "Needle Imaged in Animal-Tissue Haystack," *Science News*, 325 (May 25, 1991).

Pool, R.: "Molecular Photography with an X-ray Flash," *Science*, 295 (July 15, 1988).

Pool, R.: "Making 3-D Movies of the Heart," *Science*, 28 (January 4, 1991).

Richards J.A. and D.E. Ricken: "Remote Sensing Digital Image Analysis," 3rd Edition, Springer-Verlag, Inc., New York, NY, 1999.

Richelson, J.T.: "The Future of Space Reconnaissance," *Sci. Amer.*, 38 (January 1991).

Roberts, L.: "Mapping by Color and X-rays," *Science*, 425 (April 28, 1989).

Romer, G.B., J. Delamoir: "The First Color Photographs," *Sci. Amer.*, 88 (December 1989).

Silverman, J., J.M. Mooney, and F.D. Shepherd: "Infrared Video Cameras," *Sci. Amer.*, 78 (March 1992).

Staff: "Odyssey (Reviews of Photos in First 100 Years of National Geographic Magazine)," *Natl. Geographic*, 322 (September 1988).

Staff: "New Projectors," *Hughesnews*, 1, Culver City, California (February 21, 1992).

Staff: "Imaging Technologies, Inscribing Science," *Camera Obscura*, 28 (1992).

Vager, Z., R. Naaman, and E.P. Kanter: "Coulomb Explosion Imaging of Small Molecules," *Science*, 426 (April 28, 1989).

Stroebel, L.D., J. Compton, and I. Current: "Basic Photographic Materials and Processes," 2nd Edition, Butterworth-Heinemann, Inc., Woburn, MA, 2000.

Vander Voort, G.F.: "Metallography," *Advanced Materials & Processes*, 71 (January 1990).

Van Sant, T., et al.: "The Earth—From Space: A Satellite View of the World," Spaceshots, Inc., Manhattan Beach, CA, 1990.

Waters, A.J., M.J. Bader, J.R. Grant, G.S. Forbes, et al.: "Images in Weather Forecasting: A Practical Guide for Interpreting Satellite and Radar Imagery," Cambridge University Press, New York, NY, 1997.

Wilkie, D.S. and J.T. Finn: "Remote Sensing Imagery for Natural Resources Monitoring: A Guide for First-Time Users," Columbia University Press, New York, NY, 1996.

Zwingle, E., H.E. Edgerton, and B. Dale: "'Doc' Edgerton—The Man Who Made Time Stand Still," *Natl. Geographic*, 464 (October 1987).

WESLEY F. TREE, The College of Wooster, Wooster, OH.

PHOTOIONIZATION.   This process, which is also called the atomic *photoelectric effect*, is the ejection of a bound electron from an atom by an incident photon whose entire energy is absorbed by the ejected electron. This statement means that photoionization cannot occur unless the energy of the photon is at least equal at the ionization energy of the particular electron in the particular atom; any excess of energy in the photon above this value appears as kinetic energy of the ejected electron.

PHOTOLUMINESCENCE.   See **Luminescence**.

PHOTOMETERS.   Instruments for the measurement of luminous intensity, luminous flux density, and illumination. In usual terminology, only instruments that respond to the central portion of the electromagnetic spectrum, i.e., the ultraviolet, visible, and infrared regions, are called photometers. Essentially, a photometer is comprised of a transducer, which transforms electromagnetic waves (photons) into an electric current, and a current-measuring readout device. In the simplest form, the instrument

could be a voltage-generating photocell connected to a microammeter. Photographic exposure meters and light meters that measure ambient illumination are usually of this type. The latter are furnished with green filters, which correspond to the relative spectral sensitivity of the human eye. Photoresistors also are used for this purpose. Photoresistors require a voltage source (battery) in the circuit. The microammeter reads out the change in resistance caused by the illumination. These devices are more sensitive, but not of high precision because of fatigue effects of photoresistors. For precision work, photomultiplier tubes are usually the transducer selected. Some specific types of photometers and spectrophotometers include:

Atomic-Absorption Photometer.   This instrument operates on the very specific spectral absorption of an atomized sample rather than emission. The equipment is comprised of a stabilized hollow-cathode lamp (one for each element to be analyzed), a flame with sample nebulizer, a monochromator, and a photometer.

Brightness Meter.   A special type of reflection meter for evaluating the brightness of paper and similar products by measuring the diffuse reflectance in the blue range of the spectrum. Actually, these meters quantify the yellow characteristics of the paper.

Circular Dichrograph.   An instrument similar to a spectropolarimeter. Instead of a change in angle of optical rotation versus wavelength, the instrument records the difference in dichroic absorption versus wavelength.

Color-Difference Meter.   A specially designed reflection meter for assessing small color variations.

Colorimeter.   An instrument for routine chemical analysis. Compounds or ions which absorb light in the visible part of the spectrum (400 to 800 nanometers) or which are convertible by specific reagents to such compounds can be analyzed with a colorimeter. The instrument typically incorporates an incandescent light bulb as light source, filters to separate the spectral region, a cuvette to contain the sample solution, and a photometer. See also **Colorimetry**.

Densitometer.   An instrument used to measure the attenuation of a beam passing through, or reflected from the surface of solid samples.

Ellipsometer.   An instrument for determining the thickness of very thin films of monomolecular dimensions. Essentially, the instrument is a polarization interferometer that utilizes a photometer as a readout device.

Flame Photometer.   See *Atomic-Absorption Photometer* in this entry.

Fluorimeter (also Fluorometer or Fluorophotometer).   In this instrument, the sample is excited by a light beam of suitable short wavelength. The remitted fluorescent light is picked up by a photometer, usually placed 90 degrees from incidence. A filter or monochromator is provided which excludes the exciting waveband and transmits the fluorescent light. See also **Fluorometers**.

Footcandle Meter.   A color-corrected illumination meter calibrated in footcandles.

Glossmeter.   An instrument for measuring specularly reflected light from the surface of a flat sample. The angle of incidence and the angle of light pickup are identical and opposed from the normal to the surface. A typical glossmeter consists of a light source and simple optics to direct a defined beam onto the sample. In the opposite direction, there is a light detector connected to a readout meter.

Hemoglobinometer.   A specialized colorimeter for determining hemoglobin in blood.

Light-Scattering Photometer.   In one type, suspended particles are determined (counted). Another type is used to determine the molecular weight of macromolecules dispersed in solution. The former type operates on the basis of a nephelometer; the latter is of much higher precision and uses monochromatic light.

Lux Meter.   Essentially a footcandle meter calibrated in international lux units. (1 footcandle = 10.8 lux.)

Nephelometer.   An instrument for determining particle size or particle concentration by measuring the amount of light transmitted or scattered by the suspended particles. Quantitative determinations are made by comparing a given sample with a known standard. See also **Nephelometry**.

Opacimeter.   A reflection meter specifically designed to evaluate the opacity of thin sheets, such as paper, by measuring the diffuse reflectance over a white and a black surface in turn.

Optical-Emission Spectrometer.   Similar to flame photometer (atomic-absorption photometer) except that an electric spark rather than a flame is used to vaporize (atomize) unknown samples.

Polarimeter. An instrument for determining the concentration of optically active compounds in solution by determining the angle of rotation of plane-polarized light passing through the sample. See also **Polarimetry**.

Reflection Meter. A photometer arranged to pick up diffusely reflected light from the surface of a flat sample. The spectral evaluation of the reflected light permits quantitative color evaluation as seen by the human observer.

Refractometer. An instrument for determining the refractive index of solutes. Most of these instruments use photometric readout systems.

Saccharimeter. A polarimeter calibrated in "sugar degrees" for analyzing the concentration of sugar solutions.

Spectrofluorimeter. A fluorimeter with two separate monochromators. One serves to scan through the spectrum of the exciting light source; the other scans the emitted fluorescent light.

Spectrophotometer. An instrument comprising a light source, means of monochromatizing the light, a sample space, and a photometer. These instruments normally determine concentration of a solute by measurement of light attenuation, the logarithm of absorption being proportional to the concentration. If the instrument is designed to operate in the infrared region, it is known as an infrared spectrophotometer. If in the ultraviolet region, an ultraviolet spectrophotometer, etc.

Additional Reading

Decusatis, C.: "Handbook of Applied Photometry," Springer-Verlag, Inc., New York, NY, 1998.

Heranshaw, J.B.: "The Measurement of Starlight: Two Centuries of Astronomical Photometry," Cambridge University Press, New York, NY, 1996.

Swatland, H.J.: "Computer Operation for Microscope Photometry," CRC Press, LLC., Boca Raton, FL, 1997.

PHOTON AND PHOTONICS.

In common usage, a photon is a quantum of electromagnetic energy. The energy of a photon is hv, where h is the Planck constant, and v is the frequency associated with the photon. The term photon usually refers to a plane-wave quantum of electromagnetic energy, for which the momentum is hv/c, and the component of angular momentum in the direction of the momentum is $\pm\hbar$, where c is the velocity of light and $\hbar$ is $h/2\pi$.

The word *photonics* entered the scientific vocabulary in the mid-1980s to describe a communications transmission system that converted digital information into pulses of light that traveled over an optical fiber cable (fiber optics/light wave communication). The first crude cables were used in the mid-1970s. An exploratory system was established in a network between three buildings in downtown Chicago. Since then, the growth of optical fiber networks worldwide has been no less than dramatic. However, photonics also has a broader connotation and parallels in its hardware and systems aspects the well-established microwave technology, thus, a *photonics technology*. See also **Optical Fiber Systems**.

The existence of the photon was first suggested by Planck's famous research, about 1900, into the distribution in frequency of blackbody radiation. Planck arrived at agreement with the experimental distribution only by making the drastic (for that period in science) assumption that the radiation exists in discrete amounts with energy $E = hf$, where f is the frequency of the radiation and h is Planck's constant, $6.626 \, (10)^{-27}$ erg sec. Confirmation of the existence of these quanta of electromagnetic energy was provided by Einstein's interpretation of the photoelectric effect (1905). Einstein made it clear that electrons in a solid absorb light energy in the discrete amounts hf. The full realization that the photon is a particle with energy and momentum was provided by the Compton effect (1922), an aspect of the scattering of light by free electrons. Compton showed that features of the scattering are understood by balancing energy and momentum in the collision in the usual way, the light considered as a beam of photons each with energy hf and momentum hf/c.

The modern point of view is that, for every particle that exists, there is a corresponding field with wave properties. In the development of this viewpoint, the particle aspects of electrons and nuclei were evident at the beginning and the field or wave aspects were found later (this was the development of quantum mechanics). In contrast, the wave aspects of the photon were understood first (this was the classical electromagnetic theory of Maxwell) and its particle aspects only discovered later. From this modern viewpoint, the photon is the particle corresponding to the electromagnetic field. It is a particle with zero rest mass and spin one.

For a photon moving in a specific direction, the energy E and the momentum q of the particle are related to the frequency f and wavelength λ of the field by Planck's equation $E = hf$ and the de Broglie equation $q = h/\lambda$. As for all massless particles, the energy and momentum are related by $E = cq$ and the photon can only exist moving at light speed c. Another property of all massless particles is that, given the momentum, the particle can exist in just two states of spin orientation. The spin can be parallel or antiparallel to the momentum, but no other directions are possible. The photon state with the spin and momentum parallel (antiparallel) is said to be right- (left-) handed and is a right- (left-) hand circularly polarized wave. In analogy with the neutrino, one can say that the state has positive (negative) helicity and can call the right-handed particle the antiphoton, the left-handed particle the photon. There is an operation, CP conjunction, that converts a photon state into an antiphoton state and vice versa. It is possible to superpose photon and antiphoton states in such a way that the superposition is unchanged by CP conjugation and so gives a type of photon that is its own antiparticle. The photons produced by transitions between states of definite parities in atoms or nuclei are their own antiparticles in this sense. As for all particles with integer spin, the photon follows Bose-Einstein statistics. This means that a large number of photons may be accumulated into a single state. Macroscopically observable electromagnetic waves, such as those resonating in a microwave cavity, for example, are understood to be large numbers of photons all in the same state. The photon, among all particles, is unique in having its states be macroscopically observable in this way.

Additional Reading

Ackerman, E., et al.: "A 3 to 6 GHz Microwave/Photonic Transceiver for Phased-Array Interconnects," *Microwave J.*, 60 (April 1992).

Cusack, J.: "Photonics at Rome Laboratory," *Microwave J.*, 72 (February 1992).

Fujimoto, J.G. and M.S. Patterson: "Advances in Optical Imaging and Photon Migration," Optical Society of America, Washington, DC, 1998.

Howe, H.: "Let There Be Light," *Microwave J.*, 24 (January 1992).

Joannopoulos, J.D., R.D. Meade, and J.N. Winn: "Photonic Crystals," Princeton University Press, Princeton, NJ, 1995.

Kaminow, I.P. and T.L. Koch: "Optical Fiber Telecommunications IIIA," Morgan Kaufmann Publishers, Orlando, FL, 1997.

Polifko, D. and H. Ogawa: "The Merging of Photonic and Microwave Technologies," *Microwave J.*, 75 (March 1992).

Pradhan, T.: "The Photon," Nova Science Publishers, Inc., Huntington, NY, 2001.

Render, D.J.: "Photonics—Fast Track for Tomorrow's Communications," *AT&T Technology*, II, (1), 1987. *A classic reference.*

Sakoda K.: "Optical Properties of Photonic Crystals," Springer-Verlag, Inc., New York, NY, 2001.

Soukoulis, C.M.: "Photonic Band Gaps and Localization," Kluwer Academic Publishers, Norwell, MA, 1993.

Zmuda, H. and E.N. Toughlian: "Adaptive Microwave Signal Processing: A Photonic Solution," *Microwave J.*, 58 (February 1992).

Web Reference

Optical Society of America: *http://www.osa.org/*

PHOTON ENGINE.

A projected type of reaction engine in which thrust would be obtained from a stream of electromagnetic radiation. See **Ion Engine**. Although the thrust of this engine would be minute, it may be possible to apply it for extended periods of time. Theoretically, in space, where no resistance is offered by air particles, very high speeds may be built up.

PHOTONEUTRON.

A neutron emitted from a nucleus in a photonuclear reaction.

PHOTONUCLEAR REACTION.

A nuclear reaction induced by a photon. In some cases the reaction probably takes place via a compound nucleus formed by absorption of the photon followed by distribution of its energy among the nuclear constituents. One or more nuclear particles then "evaporate" from the nuclear surface, or occasionally the nucleus undergoes photofission. In other cases the photon apparently interacts directly with a single nucleon, which is ejected as a photoneutron or photoproton without appreciable excitation of the rest of the nucleus.

PHOTOPERIODISM.

This term is applied to the reaction of plants to the daily length of the period of illumination. It is one of the most noteworthy of the reactions of plants to an environmental factor. In most parts of the world, marked seasonal variations occur in the length of the daylight period. In temperate zones the length of the daylight period varies from about 8 or 10 hours at the winter solstice to about 14–16 hours at the summer solstice. At higher latitudes the annual variation in day length is greater; at lower latitudes less. In arctic and subarctic regions the length of day varies from 24 hours on the longest summer days to zero hours on the "shortest" days; in the equatorial zone day lengths approximate 12 hours the year round.

Although the length of the photoperiod (number of hours of illumination per day) also has effects upon the vegetative development of plants its most significant influences are upon flowering and other phases of the reproductive development of plants. Plants fall into three fairly well defined categories: (1) "long-day" species, which flower more or less readily in a range of photoperiods longer than a certain critical period, developing only vegetatively under shorter photoperiods; (2) "short-day" species, which flower more or less readily in photoperiods shorter than a certain critical period, developing only vegetatively at all longer photoperiods; and (3) "indeterminate" species, which exhibit no critical photoperiod, developing both vegetatively and reproductively over a wide range of photoperiods. The length of the critical photoperiod differs according to species, but for many plants of both the long-day and short-day types, lies in the range of 12 to 14 hours. Examples of short-day species are dahlias, chrysanthemums, asters, cocklebur, and salvia; of long-day species, radish, beets, dill, spinach, lettuce, and grains; of indeterminate species, tomato, cotton, sunflower and buckwheat. Both long-day and short-day varieties may exist even within the same species. Some varieties of soybeans, for example, are short-day plants, while others are long-day plants.

In temperate regions the season of blooming of a plant is largely determined by its photoperiodic reaction. In general, short-day plants bloom in the early spring or early fall; long-day plants in the late spring or summer. The geographical distribution of some kinds of plants is at least partly controlled by their photoperiodic reaction. A species cannot maintain itself in a climate in which it is impossible for the cycle of reproductive processes to be completed. Pronounced long-day species, for example, would not as a rule be found in tropical regions.

Practical applications of the principles of photoperiodism have been made in the growing of floricultural greenhouse crops. Short-day species such as chrysanthemums, can be brought into bloom earlier in the fall by decreasing the length of their daily exposure to light. Likewise, the time required for long-day floricultural species to attain the flowering stage during the winter months can be shortened by increasing the day length with artificial illumination.

See also **Photosynthesis**; **Plant Breeding**; and **Plant Growth Modification and Regulation**.

PHOTOPHOBIA.   See **Vision and the Eye**.

PHOTORECEPTOR.   A sensory organ that responds to the stimulus of light waves. Eyes are the most familiar organs of this kind but many one-celled animals are sensitive to light and some of the more complex forms have the surface of the body sensitive to it. True eyes serve for the formation of a visual image, whereas the more simple photoreceptors merely indicate the luminosity of the animal's surroundings. In some cases, adjustment to light is necessary and the simple light-sensitive organ is adequate for the initial step in the animal's orientation. Eyes serve the very different purpose of enabling the animal to perceive objects about it and are not primarily associated with its adjustment to light. Man's skin is sensitive to light in an entirely different way, which becomes evident only in the degree of pigmentation. See also **Vision and the Eye**.

PHOTOREFRACTIVE KERATECTOMY (PRK).   As with other refractive eye surgery procedures for correcting nearsightedness, the goal of the PRK procedure is to flatten the central zone, or visual axis, of the cornea so that light rays passing through the cornea and the inner lens will be focused properly on the retina.

This procedure came into use following the approval by the Food and Drug Administration (FDA) of the Summit Excimer laser system in October 1995, and of the VISX system for performing the PRK procedure in March 1996. The approval, with restrictions, was granted only after intensive FDA evaluation of the PRK procedure for a period of almost 10 years.

PRK Procedure

Step 1: Eye preparation. Before the procedure begins, a nurse or technician talks to the patient about any immediate health problems that

may affect readiness for the procedure. Antibiotic and anesthetic eye drops are then placed in the eye to numb it and prevent infection. The eye is swabbed with a sterile solution. The eyelid is then propped open with a lid retainer, and a paper or plastic "mask" is placed over the eye to keep eyelashes out of the way. The final step before the procedure begins is marking the cornea with a blue "dye ring, "which serves as a reference point for the surgeon throughout the procedure. Because the cornea is numb, most patients experience little if any discomfort during these pre-operative preparations.

Step 2: Creating the flap. The next step in the PRK procedure is to remove the epithelium, the ultra-thin, film-like protective outer covering of the cornea, from the central zone (visual axis). This can be accomplished with either the Excimer laser or a surgical instrument.

Step 3: The Excimer laser. The Excimer laser is then used to ablate, or vaporize, layers of cornea tissue in order to "sculpt" the cornea to the shape needed to achieve the degree of vision correction required. The sculptured area is less than the diameter of an average pencil eraser. Preoperative tests and the evaluation process determine the sculpture pattern. The data that was gathered is entered into a computer that is built into the Excimer laser system. The computer then calculates the sculpting pattern and directs the application of the laser under the guidance of the eye surgeon. The surgeon continually monitors the progress of the procedure through a microscope that is a part of the laser system.

The patient is asked to focus on a fuzzy red light inside the laser. As the doctor activates the laser, there is a "popping" or "tacking" sound. In addition, there is a slight odor similar to that of hair burning, but there is no discomfort for the patient. The number of laser pulsations will depend on the nature of the refractive vision problem that is being corrected. This phase of the procedure takes only a minute or so.

Step 4: Post-operative measures. When the procedure is complete additional antibiotic drops are placed in the eye, and it may be covered with a plastic shield. For a short while after the procedure has been completed, the eye is numb from the anesthetic drops. As the numbness wears off, the patient may experience some sensitivity to light and scratchy or dry sensation as though something is in the eye. This feeling usually goes away within a few hours. The patient must not drive home following the procedure.

The patient returns to the doctor's office the next day for a post-operative examination. The doctor checks the eye to see if the cornea is healing properly. Vision is checked and, for most patients, will range from 20/20 to 20/40 depending on the number of corrections received. For some patients, vision may continue to improve for several weeks before stabilizing.

The patient may experience some discomfort for the first 24 to 48 hours following surgery. Vision may be blurred for three to five days while the epithelium that was removed from the central corneal zone heals. The patient may also experience glare, halos, shadows, and some ghost images for a few weeks. These are normally transient phenomena and should lessen with time and then fade away. See also **Lasers (Eye Surgery)**; **Myopia (Nearsightedness)**; and **Refractive Eye Surgery**.

Vision Rx, Inc., Elmsford, NY.

PHOTOSPHERE.   The intensely bright portion of the sun visible to the unaided eye. The photosphere is that portion of the sun's atmosphere which emits the continuum radiation upon which the Fraunhofer lines are superimposed. In one sun model, the photosphere is thought to be below the reversing layer in which Fraunhofer absorption takes place. In another model, all strata are considered equally effective in producing continuous emissions and line absorption.

PHOTOSYNTHESIS.   This is the most important of all biological processes. With negligible exceptions the existence of the entire biological world hinges upon this process. From a few simple inorganic compounds and from the sugar made in photosynthesis are erected all of the complex kinds of molecules essential to the construction of the bodies of plants and animals or to maintenance of their existence. Some of these subsequent synthetic processes occur in the plant body, others in the bodies of animals after they have ingested plant materials as foods. Likewise, the energy used by plants and animals represents sunlight energy that has been entrapped in sugar molecules during photosynthesis. The entire organic world runs by the gradual expenditure of the energy capital accumulated in photosynthesis.

Under suitable conditions of temperature and water supply, the green parts of plants, when exposed to light, abstract and use carbon dioxide from the atmosphere and release oxygen to it. These gaseous exchanges are the opposite of those occurring in respiration and are the external manifestation of the process of photosynthesis by which carbohydrates are synthesized from carbon dioxide and water by the chloroplasts of the living plant cells in the presence of light. For each molecule of carbon dioxide used, one molecule of oxygen is released. A summary chemical equation for photosynthesis is:

$$6\ CO_2 + 6\ H_2O \xrightarrow{\text{light}} C_6H_{12}O_6 + 6\ O_2$$

In this process, the radiant energy of sunlight is stored as chemical energy in the molecules of carbohydrates and other compounds that are derived from them.

All photosynthetic organisms, except bacteria, use water as the electron or hydrogen donor to reduce various electron acceptors, and from the water they evolve molecular oxygen. Anaerobic bacteria cannot endure such oxygen, but derive their sustenance through slightly different photosynthetic routes:

$$2\ H_2S + CO_2 \xrightarrow{\text{light}} (CH_2O) + H_2O + 2S$$

or

$$2\ CH_3CHOHCH_3 + CO_2 \xrightarrow{\text{light}} (CH_2O) + CH_3COCH_3 + H_2O$$

Photosynthesis takes place in chlorophyll-containing cells only when carbon dioxide, water, and light are available, and when a suitable temperature prevails. Although carbon dioxide constitutes, on the average, only 0.03% of the atmosphere, land plants are entirely dependent upon this source for the carbon dioxide used in photosynthesis. It has been shown experimentally that an increase in the carbon dioxide concentration of the atmosphere results in an increased rate of photosynthesis. On the other hand, a deficiency of water results in a reduced rate of photosynthesis. In nature, sunlight is the source of radiant energy used in photosynthesis, although plants will also photosynthesize under artificial light sources of suitable quality and intensity.

The total radiant energy received at the earth's surface is 1–2 gram calories/square centimeter/minute, depending upon altitude, or approximately 1 hp/10–20 square feet. For crop plants in the field, a maximum of 2–3% of this energy remains stored in the plants at the end of the growing season. During that time about 20% more is actually used in photosynthesis and lost by respiration of the plant, the remainder of the energy being dissipated by re-radiation, transmission through the leaves, and evaporation of water from the plant.

The intensity, quality and daily duration of illumination all have influence on the amount of photosynthesis accomplished per day. Clearly, the longer the daily period of illumination, the more photosynthesis will be accomplished by a plant in the course of a day. The minimum light intensity at which a measurable rate of photosynthesis occurs varies according to species, but is seldom less than 1% of full midday summer sunlight. Under natural conditions, maximum rates of photosynthesis are attained in single leaves of many species at 25–35% of full sunlight intensity, and in some shade species at even lower intensities. For equal intensities, more photosynthesis appears to occur in the orange–short red and blue parts of the spectrum than in the green and yellow. This is because the chlorophyll pigments of the leaves absorb light energy at wavelengths of 6600 and 4250 micrometers. Radiation is most intense in the green and, if this radiation were absorbed, the plant could not utilize it and would overheat.

The range of temperatures most suitable for relatively rapid rates of photosynthesis is not the same for all kinds of plants. In general, it is higher in tropical than in temperate species, and higher in temperate species than in those of subarctic regions. Increase in temperature results in an increase in the rate of photosynthesis up to an optimum which varies with the variety of plant, but which, for most temperate zone species, lies within the range of 20–30 °C. With increase above the optimum, the rate of photosynthesis progressively decreases.

In the vascular plants, photosynthesis occurs chiefly in the leaves. Carbon dioxide diffuses into the intercellular spaces of the leaf from the atmosphere via the stomates, and then dissolves in the moist walls of the mesophyll cells. In solution, the carbon dioxide diffuses to the surface of the chloroplasts, which are the actual seat of the photosynthetic process. The first major step in photosynthesis is the absorption of radiant energy by the plant pigments in the chloroplasts, with the generation of electrons. The plant pigment consists of two closely similar pigments, chlorophyll and chlorophyll *b*, which are porphyrin-derived complexes of magnesium and which, upon excitation by radiant energy, become electron donors. See also **Chlorophylls**. The chloroplast is a complex, self-replicating organelle that possesses its own DNA and is able to synthesize at least a few of the proteins needed for its own functioning. It is filled with membranous thylakoid sacs which are specifically designed to harness the energy available in the excited electron and to carry out the light phase of photosynthesis. In this, the light energy captured is converted into the chemical energy of adenosine triphosphate (ATP) and nicotinamide-adenine dinucleotide phosphate (NADPH). See also **Adenosine Phosphates**; and **Coenzymes**. Hydrogen atoms are removed from water and used to reduce NADP, leaving behind molecular oxygen. Simultaneously, adenosine diphosphate (ADP) is phosphorylated to ATP:

In the second, or dark, reaction phase, NADPH and ATP provide the

$$Water + NADP^+ + PO_4 + ADP \xrightarrow{\text{light}} Oxygen + NADPH + H^+ + ADP$$

energy to reduce carbon dioxide to glucose and are themselves oxidized or decomposed:

$$CO_2 + NADPH + H^+ + ATP \longrightarrow Glucose + NADP^+ + ADP + PO_4$$

Peter Mitchell (Nobel Prize, 1978) of Great Britain was the first to realize, and to propose in his chemi-osmotic theory, that the energy required for the ADP-ATP reaction could be derived by an accretion of protons in the thylakoid sac to the point at which the electrochemical gradient across the membrane could effect the proton transport required as the driving force for this reaction. See also **Phosphorylation (Photosynthetic)**.

In most plants, the water used in photosynthesis is absorbed by the roots from the soil, whence it is translocated to the leaves. Except for a small portion used in respiration, the oxygen liberated in the process diffuses out of the leaf into the atmosphere, mostly through the stomates. See also **Ascent of Sap**; and **Stomate (or Stoma)**.

Carbohydrates other than hexoses are synthesized in the leaves, apparently as a result of secondary reactions following photosynthesis. Sucrose invariably accumulates in actively photosynthesizing leaf cells. This more complex sugar is built up from the molecules of the simpler hexoses. In most plants, insoluble starch also accumulates in leaf cells during photosynthesis. This carbohydrate is synthesized by the condensation of numerous glucose molecules. The sucrose and starch contents of leaves decrease at night as a result of the continued translocation from the leaves to other parts of the plant. The sucrose is probably translocated as such, but the starch must first be converted into simpler, soluble sugars before it can move out of the leaves. Synthesis of starch is not restricted to the green parts of plants; a familiar example of this is the accumulation of starch in potato tubers. Starch in the nongreen cells is made from glucose, which comes from the leaves or other photosynthetic organs. Starch occurs in cells in the form of small grains, the type of grain formed in each kind of plant being more or less characteristic of that species.

For many years, the nature and location of the complex of proteins (sometimes referred to as the "engine" of photosynthesis) were poorly understood. During the 1980s, much more was learned as the result of research carried out by Johann Deisenhofer (Howard Hughes Medical Institute), Robert Huber, and Harmut Michel (Max Planck Institute), and for this work the investigators were awarded the 1988 Nobel Prize for chemistry. The protein complex, called the membrane-bound proteins, are difficult to define structurally because they do not crystallize readily and thus could not be subjected to x-ray crystallography. However, over a period of three years, the researchers were able to create crystals and thus were able to determine precisely the position of some 10,000 atoms in the protein complex.

With this better understanding, researchers are able to depict how plants, algae, and rhodopseudomonads carry out synthesis. Also, it has been hypothesized that such membrane-bound proteins may have a functional role in some diseases, such as cancer and diabetes.

Finally, it must be realized that photosynthesis is not the sole prerogative of the higher plants. More than half the photosynthesis on the earth's surface is carried out in the oceans by phytoplankton.

Additional Reading

Amato, J.: "A Shady Strategy for Photosynthesis," *Science News*, 246 (October 20, 1990).

Anderson, B., J. Barber, and H. Salter: "Molecular Genetics of Photosynthesis," Oxford University Press, Inc., New York, NY, 1996.

Blankenship, R.E.: "Molecular Mechanisms of Photosynthesis," Blackwell Science, Inc., Malden, MA, 2001.

Blaxter, J.H.S. and A.J. Southward: "Advances in Marine Biology," Academic Press, Inc., San Diego, CA, 1993.

Bogorad, L. and I.K. Vasil: "The Photosynthetic Apparatus," Academic Press, Inc., San Diego, CA, 1991.

Charles-Edwards, D.A.: "Mathematics of Photosynthesis and Productivity," Academic Press, Inc., San Diego, CA, 1981.

Coleman, G. and W.J. Coleman: "How Plants Make Oxygen," *Sci. Amer.*, 50 (February 1990).

Darnell, J., H. Lodish, and D. Baltimore: "Molecular Cell Biology," 4th Edition, W. H. Freeman and Company, New York, NY, 1999.

Falkowski, P.G. and J.A. Raven: "Aquatic Photosynthesis," Blackwell Science, Inc., Malden, MA, 1996.

Hall, D.O. and K. Rao: "Photosynthesis," 6th Edition, Cambridge University Press, New York, NY, 1999.

Herring, P.J., et al.: "Light and Life in the Sea," Cambridge University Press, New York, NY, 1990.

Hogan, J.: "1988 Nobel Prize for Chemistry," *Sci. Amer.*, 33 (December 1988).

Holden, C.: "Picture-Perfect Plankton," *Science*, 681 (February 7, 1992).

Kirk, J.T.O.: "Light and Photosynthesis in Aquatic Ecosystems," 2nd Edition, Cambridge University Press, New York, NY, 1994.

Metzler, D.: "Biochemistry," 2nd Edition, Academic Press, Inc., San Diego, CA, 2002.

Miller, K.R.: "A Particle Spanning the Photosynthetic Membrane," *J. Ultrastruct., Res.*, **54**, 1, 159–167 (1976).

Miller, K.R.: "The Photosynthetic Membrane," *Sci. Amer.*, **241**, 4, 102–113 (1979).

Ort, D.R.: "Oxygenic Photosynthesis: The Light Reactions," Kluwer Academic Publishers, Norwell, MA, 1996.

Pessarakli, M.: "Handbook of Photosynthesis," Marcel Dekker, Inc., New York, NY, 1996.

Raghavendra, A.S.: "Photosynthesis," Cambridge University Press, New York, NY, 2000.

Sherman, K., L. Alexander, and B. Gold: "Large Marine Ecosystems: Patterns, Processes, and Yields," AAAS Books, Waldorf, MD, 1992.

Stoecker, D.K.: "Photosynthesis Found in Some Single-Cell Marine Animals," *Oceanus*, 49 (Fall 1987).

Stryer, L. and J.L. Tymoczko: "Biochemistry Extended, Chapters 1–34," 5th Edition, W. H. Freeman and Company, New York, NY, 2002.

Yunus, M. and Dr. U. Pathre: "Probing Photosynthesis: Mechanisms, Regulation, and Adaptation," Taylor & Francis, Inc., Philadelphia, PA, 2000.

Zilsnov, V.K.: "Living Marine Resources," *Oceanus*, 29 (Summer 1991).

See also references list at the end of entry on **Photochemistry and Photolysis**.

R.C. VICKERY, D.Sc., Blanton/Dade City, FL.

PHOTOTHEODOLITE. An instrument or device incorporating one or more cameras for taking and recording angular measurements. The phototheodolite, sometimes in conjunction with radar equipment, is used to track rockets and to measure and record attitude, altitude, azimuth and elevation angles, etc.

PHOTOVOLTAIC CELL. A transducer that converts electromagnetic radiation into electric current. The solar cells used on satellites and space probes are photovoltaic cells employing a semiconductor such as silicon, which releases electrons when bombarded by photons from solar radiation.

PHREATIC. The term proposed by Daubree in 1887 for the waters of the ground water reservoir, as distinct from the underground waters above the water table, called vadose.

PHRENIC NERVES. The nerves which control the movement of the diaphragm.

PHTHALIC ACID. $C_6H_4(COOH)_2$, formula weight 166.13, mp 208 °C (ortho), 330 °C (meta and iso), the ortho form sublimes and the meta and iso forms decompose with heat, sp gr 1.593 (ortho). Phthalic acid is very slightly soluble in H_2O, soluble in alcohol, and slightly soluble in ether. The solid form is colorless, crystalline. Because of their chemical reactivity and versatility, phthalic acid derivatives find wide use as starting and intermediate materials in important industrial organic syntheses. A common starting material is phthalic anhydride which is formed when phthalic acid loses water upon heating. See also **Phthalic Anhydride**; and **Terephthalic Acid**.

Orthophthalic acid is made by the oxidation of naphthalene (1) with H_2SO_4 fuming heated, in the presence of mercuric sulfate — SO_2 is also formed and recovered; (2) with air in the presence of vanadium pentoxide at 450 to 520 °C. Orthophthalic acid also is formed when benzene compounds containing carbon ortho-substituted groups are oxidized. Orthophthalic acid is used in the manufacture of indigo and other dyes.

PHTHALIC ANHYDRIDE. $C_6H_4(CO)_2O$, formula weight 148.11, mp 130.8 °C, bp 284.5 °C, sp gr 1.527. Phthalic anhydride is very slightly soluble in H_2O, soluble in alcohol, and slightly soluble in ether. The compound is a high-tonnage chemical and is widely used in a variety of industrial organic syntheses. Although phthalic anhydride may be derived directly from phthalic acid by heating and dehydration, it usually is prepared on a large scale by (1) oxidizing naphthalene, or (2) from the petroleum derivative, orthoxylene. Phthalic anhydride, in addition to its use as a raw and intermediate material for syntheses, finds wide application in the chlorinated form as a compounding ingredient for plastics. The chlorine content is approximately 50%. The compound provides increased stability and improved resistance of plastics to high temperatures.

Representative reactions of phthalic anhydride include: (1) phthalic anhydride reacts with phosphorus pentachloride to form phthalyl chloride which, upon rearrangement, can be transformed to unsymmetrical phthalyl chloride; (2) both forms of phthalyl chloride react with zinc plus acetic acid to form unsymmetrical phthalide, or with benzene plus aluminum chloride to form unsymmetrical-diphenylphthalide (phthalophenone); (3) phthalic anhydride reacts with NH_3 to form phthalimide $C_6H_4(CO)_2NH$; (4) phthalimide reacts with KOH in alcohol to form potassium phthalimide; (5) treatment of potassium phthalimide with an alkyl halide (e.g., ethyl chloride) forms an alkyl phthalimide (e.g., ethyl phthalimide); (6) ethyl phthalimide, when heated with fuming HCl, yields the primary amine $C_2H_5NH_2$ (ethyl amine) in a reaction used for the production of many primary amines and known as Gabriel's synthesis; (7) ethyl phthalimide, when treated with sodium hypochlorite, forms sodium anthranilate which upon treatment with an acid yields anthranilic acid; (8) phthalic anhydride reacts with phenol to form phthaleins, such as phenolphthalein, when in the presence of concentrated H_2SO_4; (9) phthalic anhydride reacts with resorcinol to form resorcinolphthalein (fluorescein); (10) fluorescein reacts with bromine to form tetrabromo-fluorescein, the potassium salt of which is eosin (a red dye for wool and silk); (11) phthalic anhydride reacts with N-diethyl-meta-aminophenol to form N-diethyl-meta-aminophenolphthalein (rhodamine) which is a red dye. See also **Phthalic Acid**; and **Terephthalic Acid**.

PHUGOID OSCILLATION. In a flightpath, a long period longitudinal oscillation consisting of shallow climbing and diving motions about a median flightpath and involving little or no change in angle of attack.

PHYLLOXERAN (Phylloxerid). *Insecta, Homoptera.* A sucking insect related to the plant lice and scale insects. The many species make up a subfamily, which, with the adelgids, constitutes the family *Phylloxeridae*. They differ from the aphids in that all females lay eggs and form the scales in their more complex structure, including the four wings of the winged stages.

The most important phylloxerid is a species that attacks grapevines, working on the leaves and roots. It once threatened to ruin the vineyards of France and has destroyed millions of acres of vines. The use of roots of certain American grapes which are not seriously harmed by the pest has greatly lessened the danger from its attack. Tender varieties are grafted onto the resistant roots.

PHYLUM. See **Taxonomy**.

PHYSICAL DOUBLE STAR. Two stars in nearly the same line of sight and at approximately the same distance from the observer, as distinguished from an optical double star (two stars in nearly the same line of sight but differing greatly in distance from the observer). If the stars revolve about their common center of mass, they are called a *binary star*. See also **Double Star**; and **Binary Stars**.

PHYSICAL METEOROLOGY. That branch of meteorology which deals with optical, electrical, acoustical, and thermodynamic phenomena of atmospheres, their chemical composition, the laws of radiation, and the explanation of clouds and precipitation. As generally accepted, it does not include mathematical theory of the motions of the atmosphere and the forces responsible therefore (which matters fall in the field of dynamic meteorology). Also called *atmospheric physics.* Subdivisions of physical meteorology include atmospheric electricity, cloud physics, precipitation physics, atmospheric acoustics, and atmospheric optics.

PHYSIOGRAPHY (or Geomorphology). The description and interpretation of the surface features or topographic pattern of the Earth. The scientific interpretation of scenery. The science of physiography is one of the major subdivisions of the earth sciences. The term is sometimes loosely used as synonymous with geography, hence the recent tendency to use geomorphology in its place. Since the scenery of any region is fundamentally the present stage of its geologic history, it naturally follows that a discussion of the origin of the topographic or scenic features must include not only an account of the processes of erosion and deposition which are now active, or have been active in the region, but also the manner in which the agents of erosion have been affected or controlled by the stratigraphy and structure.

PHYSIOLOGY. A division of biological science that deals with the normal functions of the living body. General physiology is a science that treats of the underlying physical and chemical foundations of vital processes. Physiology in the usual sense is concerned with the more evident vital processes themselves, analyzed to some extent in terms of physics and chemistry.

PHYTOPLANKTON. See **Ocean Resources (Living).**

PI. Consider the differential equation

$$\frac{d^2x}{dt^2} = -x,$$

with initial conditions

$$x(0) = 1, \frac{dx}{dt}\bigg|_0 = 0,$$

which describes a simple oscillating system in one of its extreme positions. The time taken for it to reach its opposite extreme position (that is, the smallest positive value of t for which $dx/dt = 0$) is denoted by the Greek letter π, called pi and approximately equal to 3.14159. It is easy to show that π is also equal to the ratio of the circumference of a circle to its diameter. See also **Circle (Geometry).**

Interesting background information on efforts made to calculate the value of pi out to hundreds of decimal places is given in *Science,* **193,** 836 (1976). As of 1987, pi had been computed to an unprecedented level of accuracy—more than 100 million decimal places. It is interesting to note that early in the present century, an Indian mathematical genius (Srinivasa Ramanujan) developed ways of calculating pi with extraordinary efficiency. Ramanujan's methods are now incorporated in computer algorithms, thus yielding pi in millions of digits. Detail can be found in article by J.M. and P.B. Borwein *Sci. Amer.,* 112–117 (February 1988).

PICARD METHOD OF SUCCESSIVE APPROXIMATIONS OR ITERATION. A numerical method for solution of a differential equation. If the given equation is $y' = f(x, y)$ subject to the condition that $y = y_0$ when $x = x_0$, the solution may be written in the form of an integral equation

$$y = y_0 + \int_{x0}^{x} f(x, y) \, dx$$

An approximate solution is

$$y_1 = y_0 + \int_{x0}^{x} f(x, y_0) \, dx$$

Iteration yields more exact solutions:

$$y_2 = y_0 + \int_{x0}^{x} f(x, y_1) \, dx; \ldots$$

$$y_n = y_0 + \int_{x0}^{x} f(x, y_{n-1}) \, dx$$

PICEA. See **Spruce Trees.**

PICIFORMES *(Aves).* This order of birds are more or less closely adapted to arboreal life (inhabiting or frequenting trees); however, of the 6 families included in the order, only the true woodpeckers are real tree climbers, and have stiff supporting tails. The length ranges from 8 to 60 centimeters (3 to $23\frac{1}{2}$ inches), and the weight is between 6 and 300 grams (0.2 and 10.5 ounces). The beak is strong; it is especially powerful and colorful in the toucans, while the true woodpeckers have a chisel-shaped beak. These birds have 2 toes directed toward the front and 2 (the first and fourth) directed toward the rear. The members of this order have various skeletal features in common. There are 14 cervical vertebrae. All of the thoracic vertebrae are unattached, and there are 5 complete ribs. There are also four notches on the rear edge of the sternum. These birds have other common features in their musculature, digestive system, and feather pattern. The woodpeckers' food consists of insects, fruits, seeds, and, for the honey guides, beeswax (the latter being a unique diet, at least among the birds). The white eggs are laid in holes. The young are blind when they hatch, and in most species they are naked as well. There are a total of 383 species, distributed over the whole world, with the exception of Madagascar, Australia, New Zealand, and the South Sea Islands.

Several structural characteristics allow us to separate the 6 families within this order into two suborders. (1) The Jacamars (*Galbuloidea*) have a syrinx that is expanded into a drum. The preen gland is bare, and the appendices are well developed. There are 2 carotid arteries. These birds breed in self-excavated ground holes in South and Central America. There are two families: The jacamars (*Galbulidae*), with 15 species (Fig. 1); and the Puffbirds (*Bucconidae*), with 31 species. (2) The Woodpeckers (*Picoidae*) do not have a drum-like syrinx; there are no appendices, and the preen gland is usually covered with feathers. Only the left carotid artery is present. The nestlings have ankle swellings. Almost all of the members of this suborder breed in tree holes, although a few families (honey guides) are brood parasites. These birds are distributed over all parts of the world, with the exception of Australia. There are four families: the barbets (*Capitonidae*), with 76 species; the honey guides (*Indicatoridae*), with 17 species; the toucans (*Ramphastidae*), with 40 species; and the woodpeckers (*Picidae*), with 209 species. See also **Woodpeckers and Toucans.**

Fig. 1. The jacamars (family *Galbulidae*) are slim, middle-sized tree birds. The beak is fine and curved slightly downward. The short feet have two climbing toes. The first segments of the second and third toe are fused, and these two toes are directed forward, while the first and fourth toes are directed toward the rear. The plumage generally is loose and has a metallic green iridescence that is reminiscent of the plumage of many hummingbirds. The preen gland is bare, and the tongue is long and thin. The contour feathers have a short secondary shaft. The short wings have ten primaries. There are 10 to 12 tail feathers. The great jacamar is shown here.

PICKEREL. See **Pike.**

PICKUP (Sensing). A device that converts a sound, scene, measurable quantity and other forms of intelligence into corresponding electric signals, as in a microphone, phonograph pickup, and television camera. A pickup is a transducer only when energy conversion is also involved, as in a microphone or phonograph pickup.

PICOLINES. See **Pyridine and Derivatives.**

PICTORIAL REPRESENTATION.   A method of representation based upon oblique representation, in which the effect of perspective representation is obtained. In the practical application of pictorial representation, three principal axes are selected, and the actual lengths of the edges of the object are laid off along these axes, resulting in a drawing that is not correct from the standpoint of either orthographic or perspective representation. Since the relative proportions of the object are retained, however, the differences do not detract from the value of the representation. In some forms of representation, however, one or more edges may be drawn to a reduced scale, termed scale reduction, to more closely simulate perspective representation.

See also **Orthographic Projection**; **Perspective**; and **Photography and Imagery**.

PIDDOCK (*Mollusca, Lamellibranchiata*).   A bivalve mollusk that bores in soft rock and floating wood. Especially a European species of the genus *Pholas*, commonly used as bait and in some localities regarded as a delicacy. The family to which these animals belong is near that containing the shipworm. Also spelled piddick.

PIERCE OSCILLATOR.   An oscillator that includes a piezoelectric crystal connected between the input and the output of a three-terminal amplifying element. Very similar to a Colpitts oscillator.

PIEZOELECTRIC EFFECT.   The interaction of mechanical and electrical stress-strain variables in a medium. Thus, compression of a crystal of quartz or Rochelle salt generates an electrostatic voltage across it, and conversely, application of an electric field may cause the crystal to expand or contract in certain directions. Piezoelectricity is only possible in crystal classes which do not possess a center of symmetry. Unlike electrostriction, the effect is linear in the field strength.

The directions in which tension or compression develop polarization parallel to the strain are called the piezoelectric axes of the crystal. Thus the axis of a hexagonal quartz crystal indicated by the arrows in Fig. 1 is known as an "X-axis," and a plate cut, as shown, with its faces perpendicular to this direction is an "X-cut"; while one cut with its faces parallel to the lateral faces of the crystal is a "Y-cut."

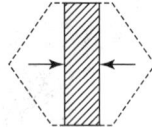

Fig. 1.   Hexagonal quartz crystal showing X-axis.

The magnitude of the piezoelectric polarization is proportional to the strain and to the corresponding stress, and its direction is reversed when the strain changes from compression to tension. The principal piezoelectric constants of a crystal are the polarizations per unit stress along the piezoelectric axes. While these constants are much greater for Rochelle salt than for quartz, the latter is better adapted to some purposes because of its greater mechanical strength. It is also stable at temperatures over $100\,^\circ$C.

If a quartz plate is subjected to a rapidly alternating electric field, the inverse piezoelectric property causes it to expand and contract alternately. As an elastic body, the plate has a certain natural frequency of expansion and contraction in the direction of the field, and if the field is made to alternate with the same frequency, the plate responds with a vigorous resonant vibration. This reacts, through the direct piezoelectric property, to augment the electric oscillations. A circuit arranged for this purpose, as in Fig. 2, is known as a piezoelectric or crystal oscillator, the crystal itself, P, being the piezoelectric resonator; T is the oscillation transformer, and C a variable condenser. This device has been much used as a frequency control in radio transmitters. Both X-cut and Y-cut quartz plates are subject to changes of frequency with temperature, due to change of elastic modulus; but certain planes in the crystal have been found, oblique to both X and Y,

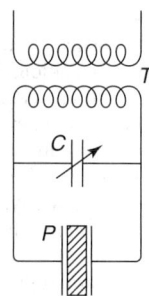

Fig. 2.   Piezoelectric oscillator circuit.

such that plates cut parallel to them are nearly free from the temperature effect.

See also **Accelerometer**; **Acoustics**; and **Microphone**.

In addition to natural quartz, Rochelle salts, and tourmaline, synthetic crystals, such as ethylenediamine tartrate (EDT), dipotassium tartrate (DKT), and ammonium dihydrogen phosphate (ADP) have varying suitability as piezoelectric elements. While Rochelle salt has a greater piezoelectric effect than any other crystal, it has the disadvantage of a greater sensitivity to temperature change than quartz. EDT has an advantage over quartz when used in frequency-modulated oscillators because of the wide gap between its resonant and antiresonant frequencies. See also **Quartz**.

PIEZOMETER RING.   A hollow ring surrounding a pipe to which it is connected by several symmetrically spaced small holes so that the pressure in the ring is the average of the various values obtained at the holes in the pipe. A piezometer, or other pressure-measuring device, is connected to the ring to measure this average pressure.

PIG.   See **Swine**.

PIGEONS AND DOVES (*Aves, Columbiformes*).   The birds of this order are almost exclusively characterized by the name pigeon or dove, although a few related species of the Old World are known as sand grouse. The extinct dodo also belonged here.

Pigeons and doves (Fig. 1) have the beak swollen at the tip and covered with soft skin at the base, about the nostrils. In North America the group is represented by the band-tailed, *Columba fasciata*, and red-billed pigeons, *Columba flavirostris*, of the western part of the continent, and by several species of doves of similar distribution. The turtle dove, *Streptopelia*, is the only widely distributed species, although the extinct passenger pigeon, *Eclopistes migratorius*, remains a memory of one of the most abundant and widely distributed birds. Numerous other species are found in all of the faunal regions of the world. They bear the names dove or pigeon, which have no exact scientific distinctness, with the exception of an Australian species called the wongawonga. The dodo of Mauritius and the solitaire of Rodriguez Island, both now extinct, were giant flightless pigeons. The last of these birds disappeared in the late seventeenth and in the eighteenth centuries, respectively. See Fig. 2.

Fig. 1.   Mourning dove. Soft olive-brown above, buff-gray below; white tips to outer tail feathers.

The sand grouse comprise a small group of birds related to the pigeons, but in some ways resembling game birds. They are found chiefly in Africa and Asia, but extend to Europe and Madagascar. As the name suggests, they frequent open ground. Their flight is powerful, hence some species migrate over considerable distances.

See also **Columbiformes**; and **Poultry**.

Fig. 2.   Model of a dodo. (*A.M. Winchester.*)

PIGMENTATION (Animals).   The accumulation of colored materials in living things, which is partly or wholly responsible for the characteristic coloration of different species. Pigments also serve special purposes in the body. In these functions the presence of color may be entirely incidental to the chemical composition of the material.

Pigments are important in visual organs. Here impervious black or dark brown deposits insulate the sensitive nerve endings against all light except that which is transmitted by the lens or cornea. In the eyes of some arthropods the pigment is redistributed to admit more light when the surrounding illumination is dim than when it is bright. A similar result is gained in the vertebrate eye by the muscular adjustment of the pigmented iris to change the size of the pupil through which light is admitted. Still another pigment, the visual purple (rhodopsin) is found in the rods of the retina of the vertebrate eye. It is bleached by light and resumes its color in darkness; it is associated with the sensitivity of the eye.

Pigments in the superficial layers of the body are also useful in some animals, independently of the relations discussed under coloration. A familiar example is the protective pigment deposited in the human skin as a protection against ultraviolet light. The deposition normally follows excessive exposure to sunlight or to other sources of ultraviolet, and the deposits are lessened when exposure is reduced. These deposits are in the form of granules of melanin in the cells of the innermost layer of the epidermis and in branching cells called melanoblasts in the underlying dermis. The pigment loses its granular form and becomes diffuse as the epidermal cells move toward the surface. Melanoblasts are possibly active in the formation of the pigment granules. The pigmentation of hair is not thoroughly understood but both granular and diffuse pigments have been reported.

A definite relation also exists between the normal illumination of the body and pigmentation in other animals, but the nature of the relation is not always known and a definite value to the animal need not exist. A familiar example is the dark upper surface and light lower surface of fishes, whether the upper surface is dorsal, as in most species, or lateral, as in the flatfishes. Lack of pigment in fishes of subterranean waters is closely associated.

Pigmentation of insects has been shown in several cases to respond to light. Lessened illumination may result in deeper colors, and some observers have secured the same result by moderate increase of light. Extreme changes, however, have resulted in decreased pigmentation in some experiments. Humidity and temperature also affect the depth of pigmentation in some insects, and may modify the pattern.

Incidental colors like the pink flush of human skin results from the presence in the body of the respiratory pigments, hemocyanin and hemoglobin, and waste products. Protein wastes deposited in the superficial tissues of insects are one source of color and the bile pigments of vertebrates, formed by the liver in the modification of hemoglobin, are a source of color in some organs.

The entire subject is closely associated with the chemistry of the living organism on the one hand and with coloration and mimicry on the other.

PIGMENTATION (Plants).   The distinctive green color of leaves and other plant organs results from the presence in such organs of two pigments called chlorophyll *a* and chlorophyll *b*. In the higher plants these pigments occur only in the chloroplasts. These pigments play so important a role in the fundamental process of plant life, photosynthesis, that their chemical reactions are discussed at length in that entry. The chlorophylls are not water-soluble but can be readily dissolved out of leaf tissues with alcohol, acetone, ether, or other organic solvents. The resulting solutions exhibit the phenomenon of fluorescence; they are deep green when held between an observer and the light, but deep red when viewed in reflected light. By suitable treatments it is possible to obtain pure crystals of chlorophyll from such solutions. Most leaves contain considerably more chlorophyll *a* than chlorophyll *b*, often two to three times as much. In the organs of the higher seed plants, with rare exceptions, chlorophyll is synthesized only upon exposure to light. Leaves of grass that develop under a board, for example, contain no chlorophyll. In the leaves of mosses, ferns, and gymnosperms, however, chlorophyll develops in the dark as well as in the light.

Invariably associated with the chlorophylls in the chloroplasts are the yellow pigments, the carotenes and the xanthophylls. These pigments are not, however, restricted in their occurrence to the chloroplasts, but may also be present in nongreen parts of the plant where they commonly occur in chromoplasts. Collectively, these pigments, together with certain others which are closely related chemically, are called the carotinoids. Carotene refers to a class of orange-yellow pigments. They are especially abundant in the roots of carrots. These compounds are of considerable importance because they are the precursors of vitamin A, one molecule of β-carotene being split into two molecules of vitamin A by a simple hydrolytic reaction.

Lycopene, a red pigment of this class, is responsible for the red color of the fruits of tomato, pepper, rose, and some other species. The commonest xanthophylls found in leaves are lutein and zeaxanthin, although others also occur. Another xanthophyll is fucoxanthin, which imparts to brown algae their distinctive color. None of the carotenoids is water soluble, but all of them can be extracted from plant tissues with suitable organic solvents.

Most of the red, blue, and purple pigments of plants belong to the group of anthocyanins. In general, the anthocyanins are red in an acid solution and change in color through purple to blue as the solution becomes more alkaline. Red pigmentation resulting from the presence of the anthocyanins is found in flowers, fruits, bud scales, young leaves and stems, and sometimes even mature leaves as in those of the red cabbage. Blue and purple pigmentation due to the presence of anthocyanins occur principally in flowers and fruits. The anthocyanins are diglucosides of the compounds pelargonidin, cyanidin, delphinidin and apigenidin. These compounds are closely similar in structure, all having the double ring benzopyrylium.

Another group of cell sap water-soluble pigments is the *anthoxanthins*. These pigments are also chemically related to the glucosides. Anthoxanthins often occur in the plant in a colorless form but under suitable conditions their typical yellow or orange color becomes apparent. Some yellow flowers, such as yellow snapdragons, owe their color to the presence of anthoxanthins, but the color of the majority of kinds of yellow or orange flowers is due to carotenoid pigments.

The autumnal coloration of leaves in temperate regions is one of the most spectacular accompaniments of the march of the seasons. Both carotenoid and anthocyanin pigments play an important role in autumnal leaf coloration which is not, contrary to popular opinion, a result of the action of frost. Brilliant development of the anthocyanin pigments in the fall is, however, favored by dry weather during which cool, but not frosty, nights alternate with clear days. During the late summer and early fall the chlorophyll in the leaves gradually decomposes. In many species this simply results in unmasking the yellow carotenoid pigments already present, accounting for the yellow autumnal pigmentation of such species as birch, sycamore, aspen, and tulip trees. In other species synthesis of anthocyanins occurs more or less concomitantly with the disintegration of the chlorophyll; this accounts for the reds or purplish reds characteristic in the autumnal coloration of such species as many oaks, maples, sumacs, and dogwood.

Except in flowers, white is an uncommon color in the externally visible parts of plants, and results from the complete absence of pigments. In some species white streaks or other markings are of common or regular occurrence in leaves, and in the leaves of some species such as roses completely white leaves or even entire branches bearing only white leaves sometimes occur. Such branches cannot be propagated because no photosynthesis can take place in the absence of chlorophyll. As long as such branches remain attached to a plant bearing green leaves they can obtain necessary food from the branches bearing normally pigmented leaves.

See also **Annatto Food Colors**; **Carotenoids**; **Chlorophylls**; **Colorants (Food)**; and **Photosynthesis**.

PIGMY ANTEATER.   See **Edentata**.

PIGMY ANTELOPES.   See **Antelope**.

PIGNOLIA NUT.   See **Conifers**.

PIKA.   See **Rabbits and Hares**.

PIKE (*Osteichthyes*).   Of the order *Haplomi*, family *Esocidae*, the pikes are freshwater food and game fishes. They are most abundant in the rivers and lakes of the northern United States and Canada. The northern pike (*Esox lucius*) attains a weight of about 40 pounds (18.1 kilograms) and a length up to 4 feet (1.2 meters) and is one of the principal game fishes of the north country. This species is sometimes called the northern pickerel, but it differs from the closely related pickerels in having the cheeks scaly, but the lower half of the opercula bare. The walleyed "pike" is of a different order (*Percomorphi*) and is not a pike per se. Similarly, the sandpike is related to the perches and is not a pike. See also **Perches and Darters (Osteichthyes)**.

The esocids have shovel-like bills something like a duck's and equipped with sharp teeth. Pikes are carnivorous with a diet mainly of smaller fish, but including small birds, mammals, and frogs. The northern pike is found throughout North America north of Ohio. However, it is not found in the far northwest. The esocids are quite sensitive to environmental changes. The muskellunge (*Esox masquinongy*) found in the Great Lakes and environs is larger than the northern pike, reaching a weight in excess of 70 pounds (31.8 kilograms). Chain pickerels frequent waters from Nova Scotia on the east as far south and west as Texas. This fish attains a length of about 24 inches (61 centimeters). All of the aforementioned fishes are considered good food fishes. See Fig. 1.

Fig. 1.   Pike.

PILCHARD.   See **Herring**.

PILEUS.   See **Clouds and Cloud Formation**.

PILLBOX ANTENNA.   A cylindrical parabolic reflector enclosed by two plates perpendicular to the cylinder, so spaced as to permit the propagation of only one mode in the desired direction of polarization.

PILL BUG (*Crustacea, Isopoda*).   A small oval terrestrial crustacean, commonly found in moist areas at the surface of the ground. They hide in crevices among rocks, under wood, and even invade basements. These forms are also commonly known as sow bugs and wood lice. The pill bugs are properly the species with the power of rolling up into a ball so nearly spherical that it will roll on a slight incline.

PILLOW LAVA.   Effusive volcanic rocks, generally of basic composition, which are characterized by pillow-like or bun-like structures formed during the concomitant movement and congelation of the lava. Most pillow lavas are basalts. Frequently the "pillows" have a skin of rock glass called tachylyte. The evidence of the rapidity of the chilling suggests that pillow lavas owe their peculiar structure to having flowed into a body of water or as having originated as aquatic lava flows.

PILOTAGE (or Piloting).   A term used to describe that type of navigation in which the positions and motions of a ship are determined by reference to fixed objects on the earth. The landmarks may be natural, such as hills, points of land, small islands, lakes, rivers, etc., or they may be artificial, such as lighthouses, light vessels, beacons, buoys, prominent buildings, water towers, railroads, highways, and power transmission lines. Two general types of pilotage are recognized by sea navigators; inshore or harbor piloting, and offshore or coast piloting. Before a ship proceeds up a channel, or into a bay or harbor, the pilot must have a clear mental picture of the locations of all available landmarks, range points, etc. A stranger to the region must obtain this image from a thorough study of charts, pilot directions, and similar publications for the region. With the mental picture thoroughly developed, the pilot then guides his ship in much the same manner as an individual finds his way about a city. Aviators use this type of pilotage when operating in the vicinity of a base with which they are thoroughly familiar, or when proceeding by contact flying. When a ship is off the coast, with but few recognizable landmarks available, or when a pilot is flying over unfamiliar terrain where prominent landmarks are few and far between, the pilot obtains lines of position from the available objects and then applies geometric constructions to obtain fixes.

The bearing of an object, taken with the pelorus of the compass, is the most frequently used line of position in pilotage. If two or more landmarks are available, and are spaced so that their bearings differ by more than 30°, the lines of position will intersect in a fix that is known as a cross-bearing position. A single object will provide only one bearing, but a fix can be obtained if the distance of the object can be measured. In this case, the two lines of position will be the bearing line and a circle centered on the object, the circle having a radius equal to the distance. Among the various methods for finding the distance, we have: the angular height of the object, measured with the sextant if the linear height is known; the use of stadia lines in binoculars; the difference in time of reception of audible and radio signals; and the time for the echo of a signal from the ship, such as a whistle blast, to return from the shore. None of these methods is as accurate as a bearing line, and they should not be used when other methods for obtaining a fix are available.

One object may be used for obtaining a running fix by taking two bearings of the object, at different times, provided that the distance and direction run between observations is accurately known. At 1015, the navigating officer of a ship, which is proceeding at 18 knots on heading 060°, sights a lighthouse about 12 miles away that bears 032° off the port bow. He immediately instructs the helmsman to be very careful with his steering, and proceeds to make a careful study of the current and tide tables of the region. From the predicted currents, he finds that the ship is making good a course of 058°, with ground speed 17.8 knots. He then starts a graphical construction (see Fig. 1) by drawing the 1015 line of position through the lighthouse. This line bears 028°, since $060° - 032° + 028°$. He then selects a point A on this line, usually the point closest to his dead-reckoning position, and draws the course line in direction 058°. At 1045, the lighthouse is 070° off the port bow, and a 1045 line of position is drawn through the lighthouse on bearing 350°. The 1015 line must now be advanced to 1045 by measuring off the distance made good in 30 minutes (8.9 miles) along the course line and drawing a line parallel to the 1015 line through the estimated position at 1045. The point of intersection of the 1045 line and the advanced 1015 line is the running fix at 1045. A line drawn through this fix in direction 058° will represent the probable course of the ship. This will intersect the 1015 line in the estimated position at that time. When the ship arrives at point C, the ship will have the lighthouse on the port beam. Measurements indicate that the ship, at 1045, bears 170° distant 7.2 miles from the light, and that the light will be on the port beam at 1054 and will be 6.7 miles distant at that time.

Proper selection of the second bearing will give a fix without plotting the lines. If, in the above case, the pilot had noted the time when the relative bearing of the light was 064° off the port bow (twice the first value), the triangle $A'FL$ would be isosceles, with the side $A'F$ equal to FL. The

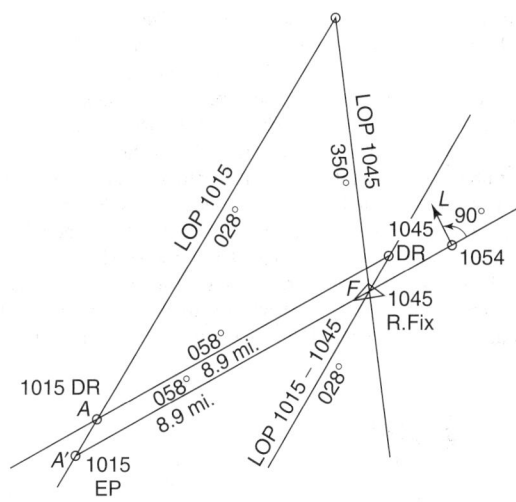

Fig. 1.   Pilotage. Scale diagram.

side $A'F$ is the distance run in the interval required to "double the angle on the bow," and the bearing and distance of the light at the time of the second observation is obtained without any plotting. An experienced pilot is familiar with a number of similar short-cuts for locating his positions by taking frequent bearings of available objects. Continual use of these, together with careful steering and a thorough knowledge of the speed of the ship and the set and drift of the current, will provide the pilot with a series of successive positions of his ship. In many cases, the scattering of the positions will indicate that the predicted currents are in error. By proper allowance for these abnormal conditions, the ship may be saved from disaster. Similar methods are available to the air navigator; they may be used to check wind directions and speeds, and may provide the pilot with accurate positions of his plane. Either visual or radio bearings may be used in flight and a series of running fixes obtained.

The contour of the sea bottom may be used for determination of position of a ship. If a ship is held on steady heading, and if frequent soundings are taken, the depths may be plotted on a sheet of tracing paper, using the distance scale of a chart. The position of the ship is found by placing this tracing over the chart, with the line of soundings parallel to the course line, and moving it about until the observed values correspond with the depths shown. At present, a set of accurate contour maps of the ocean floor is available and is used with a self-recording fathometer in the method of pilotage by depth. A similar type of air pilotage will be available when the absolute altimeter is completely developed and ready for general use.

See also **Course**; and **Navigation**.

PINEAL GLAND.   A small gland attached to the posterior part of the brain, behind and above the third ventricle. The gland is about the size of a small vitamin capsule and is located in the central part of the brain, as is the pituitary gland, but a little higher up. In humans, extensive calcification of the pineal body begins during the second decade of life and by the sixth decade over 70% of all pineals show X-ray evidence of calcification. However, this does not necessarily imply loss of functional activity inasmuch as recent evidence indicates that the human pineal may retain functional activity over the entire life span.

The product of the secretion of the pineal gland is a hormone known as *melatonin*. This substance causes marked skin blanching or lightening by its action on the pigment cells. This effect is the opposite of that produced by the pituitary melanocyte-stimulating hormone.

The functions of the pineal gland are still obscure—with some differences of opinion among endocrinologists concerning whether or not the gland is a part of the overall endocrine system. See also **Endocrine System**. Recent studies support the view that the pineal contains and probably secrets certain humoral substances and may influence the secretion of others. Other observations suggest that the pineal may serve as a *biological clock*, which helps to regulate certain endocrine rhythms.

Pineal tumors are among the rarest tumors. Processes destroying the pineal gland are characterized by precocious sexual development in a manner that is suggestive of an overactive pituitary gland. However, if the tumor is truly a pineal neoplasm, there may be a depression of gonadal function. The precocious sexual development that results from pineal destruction appears to be limited to boys about 2 to 3 years of age. The boys manifest definite signs of masculinity of an adult nature. The sex organs become adult in size and function, and public hair appears similar to that of a mature male. There has been no satisfactory explanation of why nearly all cases of sexual precocity associated with pineal tumors have occurred in boys.

The surgical removal of pineal tumors is not often advisable because of the high mortality rate. Radiation therapy may result in a temporary cure.

PINEAPPLE.   The fruit of a perennial herb of the genus (*Ananas*) of the *Bromeliaceae* family. Many species of *Ananas* are grown for their ornamental beauty. The main species of commercial value for its fruit is *A. comosus*. This plant is native to tropical America. The species is also found extensively in West Africa and botanists are not in full agreement as to whether the species is also native to that region or was introduced there from tropical America many decades ago. The pineapple was described in the diary of Columbus and his party in 1493, at which time plants cultivated by the native Indians on the Island of Guadeloupe in the Lesser Antilles were noted. Later, explorers with Columbus also noted the pineapple in an Indian settlement along the coast of Panama. Reports of pineapples growing in Brazil date back to 1519. Some authorities postulate that the fruit was carried by sailors from Brazil to Malaysia and China. They also carried pineapples back to Spain and England, where they were called the fruit pineapple because it looked like a pine cone. Historians report that transporting pineapples from tropical America to Europe and other regions was an arduous task. It is reported that out of one shipment of pineapples, only one fruit arrived at its destination unspoiled. Progeny of this specimen were prized by European gardeners, who constructed special greenhouses for their growth. There is controversy as regards how the pineapple reached the Hawaiian Islands. Some authorities observed that it was introduced there by a Spanish adventurer (not named) in the late 1700s. Other authorities observe that there is a form of pineapple on the islands of Hawaii that has been growing there for as long as the oldest inhabitants can recall, but its origin is unknown. There is no mention of the pineapple in the literature of the Egyptians, Arabs, Greeks, or Romans.

Cultivation of the pineapple is widespread. In descending order of production are China, the United States (notably Hawaii), Brazil, Thailand, Philippines, Mexico, Malaysia Peninsula, Ivory Coast, Ecuador, South Africa, Bangladesh, Australia, India, and Kenya.

The pineapple fruit generally weighs within the range of 2 to 3 pounds (0.9 to 1.4 kilograms), although some larger varieties may weigh up to 5 and 10 pounds (2.3 to 4.5 kilograms). The fruit is broadly conical in shape, with the larger fruitlets at the bottom, tapering to smaller fruitlets at the top. See Fig. 1. In all, the fruit consists of from 100 to 200 berry-like fruitlets. These are carried on the core or central axis of the fruit. The core, of course, is a continuation of the stem of the plant. Close observation of the outside of the fruit will show that the fruitlets follow a spiral-like pattern around the axis of the fruit. On most fruits, two series of spirals (clockwise and counterclockwise) can be noted. The slope of one spiral is somewhat less than that of the other. End-to-end length of the fruit (bottom to crown) ranges from 5 to 10 inches (12.5 to 25 centimeters). In addition to the particular variety or cultivar affecting dimensions and weight, these characteristics also are a function to some extent of the planting density, lower densities promoting growth of larger fruits. Characteristic of the mature fruit is the so-called blossom cup. This will be found to be an oval cavity located at the center of each fruitlet.

When growing in the field, the plant achieves a height of 2.5 to 3 feet (0.75 to 0.9 meter). The plant, as shown by Fig. 1, has rather rigid, sword-like leaves, which take the form of an elongated rosette. The fruit occurs in the center of the plant, out of the crown of which a smaller grouping of similarly stiff, but shorter sword-like leaves appear (appearing something like a junior version of the total plant).

The pineapple plant has the desirable feature of an ability to store water in its leaf axils, as well as in leaf tissue especially developed for this purpose. These features contribute to the drought-resistance of the plant.

Varieties.   Although there are hundreds of varieties of pineapple, only a comparative few are commercially important.

These include:

Abagaxi or *Abachi* — Conical, elongated fruit. Good taste. Grown in west Africa.

Fig. 1.   Close-up of mature pineapple (*Ananas Cosmosus*) growing in Hawaii. (*USDA Soil Conservation Service photo.*)

Baronne de Rothschild — Medium-size fruit seldom exceeding 3 pounds (1.4 kg). Good quality, but not exceptionally sweet. Grown in west Africa. Yellow-white flesh.

Carbazoni or *Smooth Cayenne* — Cylindrical shape. Weight ranges from 5 to 10 pounds (2.3 to 4.5 kg). Grown mainly in Puerto Rico for local consumption. Known mainly for its large size.

Cayenne or *Smooth Cayenne* — Cylindrical shape. Weight ranges 3 to 5.5 pounds (1.4 to 2.5 kg). Yellow flesh. High acid and sugar content. The most widely planted variety. Grown in Hawaii, west Africa, Australia, the Philippines, and South Africa. Used for fresh market and canning.

Pernambuco or *Pernambuca* — Cylindrical shape. Weight ranges from 3 to 4 pounds (1.4 to 1.8 kg). Yellow-white, tender flesh. Mild flavor. Grown mainly in Brazil for fresh market.

Queen — Weight ranges from 2 to 3 pounds (0.9 to 1.4 kg). Rich yellow, crisp flesh with mild flavor. Less acidic and less juicy. Grown in South Africa, Australia, and Malaysia for fresh market and canning. Keeps well when mature.

Red Spanish — Squarish shape. Weight ranges from 3 to 5 pounds (1.4 to 2.3 kg). Pale yellow, fibrous, aromatic, spicy, and acidic flavor. Tough shell is excellent for shipping. Grown mainly in the Caribbean and Florida. When fully ripe, the fruit is orange.

Sugarloaf — Globular shape. Yellow-white, sweet, and rich flesh. Grown in Mexico and Cuba, mainly for fresh market.

Propagation. Several different methods of propagation are possible. These include: (1) Crown suckers; (2) ratoon or side suckers; (3) basal slips or suckers; and (4) disk or stem cuttings.

Crown suckers are shoots that grow from the top of the fruit. They are taken off when fully developed and planted in a nursery bed. The lower leaves are removed before insertion. They normally produce fruit in 12 to 18 months from planting.

Ratoon or *side suckers* arise from buds situated low down on the main stem and should be detached at a point close to the parent stem. They should be rooted in a nursery bed before planting. Fruiting takes place in about 12 months from planting.

Basal slips or *suckers* are produced from immediately below the fruit in some varieties. These are rooted in a nursery bed. After planting, they will fruit in from 12 to 15 months.

Disk or *stem cuttings* are stems which have been stripped of their leaves and cut transversely into short sections from 2 to 3 inches (5 to 7.5 centimeters) long, or 9-inch (23-centimeter) lengths of stem can be sliced longitudinally into halves. These pieces are planted just below the soil level in a nursery bed. The short pieces are planted in an upright position, while the sliced halves are placed horizontally, with the rounded side facing upward. Dormant buds will eventually develop from the pieces of stem, arising from the axils of the removed leaves. See Fig. 2. Each bud will produce roots and may be detached from the stem when these are 0.5 to 1-inch (1.3 to 2.5-centimeters) long. These rooted shoots are established in nursery beds until they are large enough for planting in the field. Up to 9 or 10 buds can be obtained from a 9-inch (23-centimeter) stem, but plants produced by this method will not fruit in less than 15 to 18 months from planting.

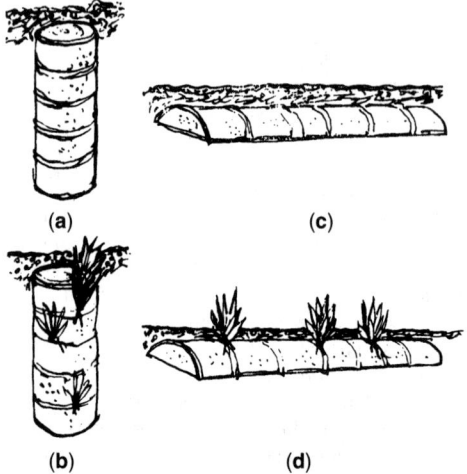

Fig. 2.   Propagation of pineapple by stem cuttings: (**a**) short length of stem is buried in soil of cutting bed; (**b**) buds arising from stem; (**c**) longer stem cut into halves and inserted in cutting bed in horizontal position; (**d**) shoots arising from dormant buds along the stem. (*After Tindall.*)

Most commercial planters in Hawaii use slips or shoots from a parent plant (removed from the plants and dried after harvest). These are inserted into the ground through mulch paper. Beds are usually about 3 feet (0.9 meter) apart, with two rows per bed about 2 feet (0.6 meter) apart. Range of numbers of plants per area are: 15,000 to 18,000 plants/acre; 37,050 to 44,460 plants/hectare. Handled in this matter, fruits will ripen within 18 months of planting. The type of propagation and planting technique used varies with the size of the operation and the philosophy of the grower.

Soils and Fertilization. Pineapples will grow on a wide range of soils and will tolerate dry conditions, although the largest fruits are obtained from plants grown in fertile soils, properly cultivated. Nitrogen deficiency is evidenced by yellowing of leaves, sometimes followed by a red coloration on young fruit.

In large operations, prior to planting (frequently in the fall, but can be in spring or early summer), the old pineapple plants are crushed and disked into the soil. Following further working of the earth to improve drainage and aeration, an asphalt-impregnated paper is placed along the rows. Marks are made to indicate plant locations. Growers who use this system claim that the paper provides a number of functions — retention of moisture, maintenance of higher soil temperature, and inhibition of weed growth. Frequently, in the last soil preparation operation, prior to laying down the paper, a soil fumigant will be used.

On large pineapple plantations, fields are laid out to allow the smooth passage of spraying and other production equipment. In typical Hawaiian operations, fields will range from 100 to 600 acres (40 to 240 hectares), with the fields bisected by roads that may be 130 feet (39 meters) apart, over which mechanized equipment with booms up to 65-feet (19.5 meters) in length are used for spraying and harvesting. Almost completely a hand operation prior to the mid-1950s, pineapple production has been mechanized in a number of aspects and research directed toward mechanizing the more difficult operations of selecting and picking has been under way for a number of years.

Harvesting. Pineapples can be picked in varying degrees of ripeness, depending upon the final use intended, that is, for processing or for the fresh market. The points on the pineapple where the flowers blossomed are termed "eyes." An ideal time to harvest the fruit is when the eyes in the lower or basal part of the fruit have turned yellow. Fruit at this stage of maturation can be maintained for up to 4 weeks if maintained under proper refrigeration. However, if the required cool temperatures cannot be provided, the plant is susceptible to a number of market diseases. To extend the life of fruit during the marketing cycle, growers will frequently harvest the pineapples before any of the eyes turn yellow. Such fruits are known as "green." Although these fruits will ripen and can be retained longer, they will not develop desired flavor to the full. For the fresh market, the crown leaves are always left on the fruit. They will not become any sweeter once picked.

Packing and Storing. Pineapples are packed by hand into shipping containers that contain from 8 to 15 fruits, according to size of fruit. Pineapples are subject to chilling injury and are not adapted to long storage. The usual storage life is from 2 to 4 weeks. Hawaiian pineapple harvested at the $\frac{1}{2}$-ripe stage can be held about 2 weeks at 45° to 55°F (7.2 to 12.8°C) and still have about 1 week of shelf-life remaining. Ripe fruit should be held at about 45°F (7.2°C). Mature-green fruit is particularly susceptible to chilling injury at temperatures below 50°F (10°C). Hawaiian pineapples are commercially treated with a fungicide. A temperature of about 48°F (8.9°C) is suggested for South African pineapples. For all pineapples, continuous maintenance of storage temperature is considered equal in importance to the specific temperature. Pineapples that are subjected to a temperature that is too low take on a dull hue, develop water soaking of the flesh, develop darkening of the core, and are particularly subject to decay when removed from storage. Black rot (*Penicillium fusarium*) is the most serious decay of pineapples during transit and marketing.

Processing. Of the pineapples produced in Hawaii, 92.4% of the fresh weight is processed and 7.6% goes to the fresh market. Of the processed pineapple, 59.3% is canned as fruit slices or chunks; 37.2% is canned juice; and 3.5% is made into frozen concentrate. Of the fresh market sales, 23.9% of the Hawaiian pineapple is consumed in Hawaii, 75.7% is shipped to mainland United States; and 0.4% is shipped to markets of other countries.

In the terminology of the pineapple processors, the wet suspended solid removed from the centrifuge in the process of juice manufacturing is called "centrifuged pineapple juice underflow." This is a viscous fluid of pH 3.5, with an appearance of apple sauce. Scientists at the University of Hawaii have been studying the possibilities of deriving useful byproducts from this underflow. Traditionally, some pineapple solid wastes have been used directly as cattle feed. Appreciable amounts of the underflow are produced daily. For example, from just one Hawaiian processing plant, during the processing season, this may amount to some 7200 gallons (273 hectoliters)/day.

PINE TREES. Members of the family *Pinaceae* (pine family), these trees are of several species. Pines are of the genus *Pinus*. It should be noted that other genera in the *Pinaceae* family include the silver firs, the Douglas fir, the spruces, hemlocks, true cedars, and larches. Also, it should be noted that the term pine is applied to a few trees not in the genus *Pinus* as, for example, the Black Pine (*Podocarpus*), the Brown Pine (*Podocarpus*), the Japanese Umbrella Pine (*Sciadopitys*), the King William Pine (*Athrotaxis selaginoides*), the Norfolk Island Pine (*Araucaria excelsa*), and the Tasmanian Huon Pine (*Dacrydium franklinii*).

Important species of pine trees (Pinus), not listed in Table 1, include:

Arolla pine	*Pinus cembra*
Austrian black pine	*P. nigra nigra*
Bhutan or Himalayan white pine	*P. wallichiana*
Bosnian pine	*P. leucodermis*
Calabrian-Corsican pine	*P. nigra maritima*
Crimean black pine	*P. nigra caramanica*
Japanese black pine	*P. thunbergii*
Japanese red pine	*P. densiflora*
Japanese white pine	*P. parviflora*
Jelicote or spreading-leaf pine	*P. patula*
Korean pine	*P. koraiensis*
Lacebark pine	*P. bungeana*

Loblolly pine	*P. taeda*
Macedonian pine	*P. peuce*
Mexican white pine	*P. ayachuite*
Montezuma pine	*P. montezumae*
Mountain pine	*P. mugo*
Pinaster pine	*P. pinaster*
Piñon pine	*P. cembroides, monophylla, etc.*
Scrub pine	See Mountain pine.
Shore pine	*P. contorta*
Siberian pine	*P. echinata*
Stone pine	*P. pinea*
Umbrella pine	See Stone pine.

All pine trees are conifers, evergreen, and prefer lots of sunlight. Pines are also extensively described in the entry on **Conifers**. See Table 1.

Highlights on the distribution of pine trees in North America include: the bigcone pine is found in the mountains of California; the bristlecone pine is found in the western part of the United States (Oregon, Colorado, Utah, Nevada, Arizona, the eastern part of California and the northern part of New Mexico); the lodgepole pine is found in the Pacific Northwest and in parts of the Rocky Mountains; the red pine is found east, north, and just west of the Great Lakes region to south of the Hudson Bay and in northeastern Canada to the Gulf of Saint Lawrence; the western white pine is found in the Pacific Northwest; the pitch pine ranges from New Brunswick in eastern Canada south and considerably inland along the east coast of the United States to Georgia and parts of Florida; the eastern white pine is found from Newfoundland and southward along the Atlantic coast to about New Jersey, inland and westward considerably west of the Great Lakes region and inland and south into northern Georgia. The loblolly pine ranges from southern New Jersey along the Atlantic coast to Florida and west to the Gulf states into Texas. It ranges inland from the coast for 250 to 350 miles (402 to 563 kilometers). The longleaf pine is found in the southern United States from Virginia to nearly the tip of Florida and westward into the Gulf states with patches of population in eastern Texas; the Monterey pine ranges from central California northward to the Monterey Peninsula. It is also found on some of the Channel Islands off the California coast and on Guadeloupe Island off Lower California; the knob-cone pine is found from southern Oregon to northern California and through the Coast and Cascade Ranges and the Sierra Nevada Mountains. It is a common tree along parts of the Sacramento River; the piñon pine is found in New Mexico and environs. A species update report on the longleaf pine by W. Voigt, Jr. is contained in *American Forests*, 43–45, May 1986.

Probably the most common pine in Europe is the Scots pine, which ranges from the Mediterranean northward to Siberia. Authorities believe that it was the only pine species to survive the Ice Ages. Pines found in the Alps include the Arolla pine of Switzerland and the Austrian black pine and further east, the Crimean black pine. The Corsican pine is found on Corsica as well as in southwestern Italy. The stone or umbrella pine is considered a classical landscape tree in the Mediterranean region. The pinaster pine prefers the seacoast and it is interesting to note that in the late 1700s, some 12,000 or more acres of sand dunes were planted with these trees on the Bay of Biscay and that the third and fourth generation of these trees are surviving well. The scrub pine found in the Alps is a favorite with some rock gardeners. The Bosnian pine is also a favorite of European gardeners.

Among the Asian pines, there are the Japanese pines—Japanese red pine, and Japanese black pine, now garden favorites in other parts of the world. The white pine, however, is not so hardy and is confined to arboretums. Other Asian pines include the Korean pine (there is an excellent specimen in the National Arboretum in Washington, DC), the lacebark pine of China, the much rarer Chinese white pine, and the umbrella pine of Japan.

There is insufficient space here to detail all species of pine trees. However, the bristlecone pine merits special attention because authorities now believe that at least one living specimen in the White Mountains of California is nearly 5,000 years old.

The range of the bristlecone pine was mentioned previously. The tree requires an altitude of from 7,000 to 12,000 feet (2100 to 3600 meters). About 100 years are required for the tree to reach maturity. The cone takes two years to mature. At Indio, California, stands "The Patriarch," considered the oldest living thing and a national monument. The American Forests estimated the tree to be about 4,600 years old. Many trees no taller

TABLE 1. RECORD PINE TREES IN THE UNITED STATES[1]

Specimen	Circumference[2]		Height		Spread		Location
	Inches	Centimeters	Feet	Meters	Feet	Meters	
Apache pine (1998) (*Pinus engelmannii*)	127	323	108	32.9	44	13.4	Arizona
Apache pine (1998) (*Pinus engelmannii*)	121	307	112	34.1	38	11.6	Arizona
Arizona pine (1998) (*Pinus ponderosa var. arizonica*)	153	389	127	38.7	57	17.4	Arizona
Austrian pine (1991) (*Pinus nigra*)	129	328	114	34.7	49	14.9	Washington
Bishop pine (1986) (*Pinus muricata*)	172	437	112	34.1	40	12.2	California
Bolander's pine (1983) (*Pinus contorta var. bolanderi*)	58	147	76	23.2	18	5.5	California
Border pinyon (1999) (*Pinus discolor*)	64	163	32	9.8	37	11.3	Arizona
Chihuahua pine (1997) (*Pinus leiophylla var. chihuahuana*)	121	307	87	26.5	34	10.4	Arizona
Colorado bristlecone pine (typ.) (1985) (*Pinus aristata var. aristata*)	132	335	76	23.2	39	11.9	New Mexico
Colorado bristlecone pine (typ.) (1986) (*Pinus aristata var. aristata*)	138	351	72	21.9	33	10.1	New Mexico
Coulter pine (1996) (*Pinus coulteri*)	209	531	80	24.4	78	23.8	California
Digger pine (1998) (*Pinus sabiniana*)	214	544	95	29	90	27.4	California
Eastern white pine (1999) (*Pinus strobus*)	200	508	150	45.7	53	16.2	Michigan
Foxtail pine (1982) (*Pinus balfouriana*)	316	803	76	23.2	34	10.4	California
Intermountain bristlecone pine (1978) (*Pinus aristata var. longaeva*)	473	1201	47	14.3	41	12.5	California
Jack pine (1995) (*Pinus banksiana*)	116	295	56	17.1	61	18.6	Minnesota
Jeffrey pine (1984) (*Pinus jeffreyi*)	307	780	197	60	90	27.4	California
Knobcone pine (1976) (*Pinus attenuata*)	135	343	117	35.7	66	20.1	California
Limber pine (1988) (*Pinus flexilis*)	275	699	58	17.7	46	14	Utah
Loblolly pine (1993) (*Pinus taeda*)	188	478	148	45.1	83	25.3	Arkansas
Lodgepole pine (1999) (*Pinus contorta var. latifolia*)	132	335	155	47.2	31	9.4	*Idaho*
Longleaf pine (1999) (*Pinus palustris*)	127	323	120	36.6	66	20.1	Georgia
Mexican pinyon pine (1982) (*Pinus cembroides*)	111	282	66	20.1	44	13.4	Texas
Monterey pine (1998) (*Pinus radiata*)	204	518	95	29	90	27.4	California
Parry pinyon pine (1976) (*Pinus quadrifolia*)	86	218	53	16.2	42	12.8	California
Pinyon (two-leaf) pine (1982) (*Pinus edulis*)	213	541	69	21	52	15.8	New Mexico
Pitch pine (1998) (*Pinus rigida*)	142	361	112	34.1	75	22.9	Georgia
Pitch pine (1999) (*Pinus rigida*)	169	429	99	30.2	40	12.2	New Hampshire
Pond pine (1998) (*Pinus serotina*)	112	284	119	36.3	60.5	18.4	Florida
Ponderosa pine (typ.) (1997) (*Pinus ponerosa var. ponderosa*)	293	744	227	69.2	68	20.7	California
Ponderosa pine (typ.) (1997) (*Pinus ponerosa var. ponderosa*)	294	747	223	68	59	18	California
Red pine (1993) (*Pinus resinosa*)	124	315	124	37.8	60	18.3	Michigan
Red pine (1998) (*Pinus resinosa*)	120	305	126	38.4	48	14.6	Minnesota
Rocky Mountain ponderosa pine (1982) (*Pinus ponderosa var. scopulorum*)	241	612	194	59.1	64	19.5	Montana
Sand pine (1997) (*Pinus clausa*)	97	246	91	27.7	42	12.8	Florida
Scotch pine (1983) (*Pinus sylvestris*)	186	472	64	19.5	76	23.2	Michigan

TABLE 1. (*Continued*)

Specimen	Circumference[2]		Height		Spread		Location
	Inches	Centimeters	Feet	Meters	Feet	Meters	
Shore pine (typ.) (1992) (*Pinus contorta var. contorta*)	138	351	101	30.8	37	11.3	Washington
Shortleaf pine (1999) (*Pinus echinata*)	139	353	88	26.8	68	20.7	Mississippi
Sierra lodgepole pine (1997) (*Pinus contorta var. murrayana*)	238	605	124	37.8	42	12.8	California
Singleleaf pinyon pine (1991) (*Pinus monophylla*)	164	417	45	13.7	40	12.2	California
Slash pine (typ.) (1992) (*Pinus elliottii var. Ielliottii*)	130	330	138	42.1	55	16.8	Florida
South Florida slash pine (1997) (*Pinus elliottii var. densa*)	138	351	68	20.7	64	19.5	Florida
Southwestern white pine (1974) (*Pinus strobiformis*)	185	470	111	33.8	62	18.9	New Mexico
Spruce pine (1998) (*Pinus glabra*)	125	318	149	45.4	53	16.2	Georgia
Spruce pine (1997) (*Pinus glabra*)	160	406	112	34.1	66	20.1	Louisiana
Sugar pine (1993) (*Pinus lambertiana*)	442	1123	232	70.7	29	8.8	California
Table mountain pine (1984) (*Pinus pungens*)	97	246	94	28.7	46	14	North Carolina
Torrey pine (1993) (*Pinus torreyana*)	245	622	126	38.4	130	39.6	California
Virginia pine (1998) (*Pinus virginiana*)	111	282	101	30.8	56	17.1	Kentucky
Washoe pine (1997) (*Pinus washoensis*)	243	617	145	44.2	64	19.5	California
Western white pine (1991) (*Pinus monticola*)	394	1001	151	46	52	15.8	California
Whitebark pine (1980) (*Pinus albicaulis*)	331	841	69	21	47	14.3	Idaho

[1]From the "National Register of Big Trees." American Forests (by permission).
[2]At 4.5 feet (1.4 meters).

than 10 to 20 feet (3 to 6 meters) may be from 500 to 900 years old. On the summits where these trees often grow, snow may remain for several summers, thus hindering growth, while winds blowing particles of sand from the desert below tear at the tree's bark and branches. In Colorado, some people refer to the wood of this tree as "wind timber." Studies indicate that some trees may only grow an inch during a century. The branches are usually split, twisted, and shattered. Growth of this nature is referred to as "crooked wood" by people in the Swiss Alps.

The bark of the bristlecone is shallow, furrowed with a red-brown color. The twig is light orange, becoming dark with foliage at the tip. The leaf or needle occurs in groups of five and are 1 to $1\frac{1}{2}$ inches (2.5 to 3.8 centimeters) long. It is dark green, curved, glossy, and seemingly brushed forward. It may be accompanied by an exudation of resin.

The male flower is orange red; the female, purple. The cone is about 3 to $3\frac{1}{2}$ inches (7.6 to 8.9 centimeters) long, prickly, bristle-like, with thick scales. The seed is about $\frac{1}{4}$ inch (0.6 centimeter) long, light brown in color. It is widely sown by the wind. The heartwood is light brown. The sapwood is pale, medium soft, brittle, and of fairly light weight (35 pounds/cubic foot; 561 kilograms per cubic meter).

At one time, the redwood tree was considered to be the oldest living thing. In the mid-1950s, a small group of bristlecone pines was discovered in the White Mountains. The redwoods add a new ring each year, usually of about the same size, because these trees live in areas where there is rainfall unfailingly each year. With the bristlecone, the rings are microscopically narrow because of the chronic draught conditions in the White Mountains. Scientists at the University of Arizona developed a procedure for matching samples of wood, both from living and dead trees and thus, with the aid of a computer, have been able to build up a continuous series of rings. There may be as many as 1,100 rings in the space of 5 inches (12.7 centimeters). These rings serve as a sensitive rain gage for the region and are of much interest to meteorologists and climatologists. These scientists expect to be able to prepare a weather history of the region for at least 10,000 years. Studies of the bristlecone pine rings also uncovered a surprise in connection with the use of carbon-14 dating techniques.

This method, based upon the amount of carbon in the atmosphere, has been subjected to revisions as the result of these studies. It was found that the atmospheric carbon level, assumed to be constant, actually holds true for only the last 3,400–3,600 years. The errors in dating structures, artifacts, etc., prior to about 1500 B.C. have had to be revised. For example, it is now estimated that Stonehenge is at least 1,000 years older than originally believed. This, in turn, has caused revisions in the thinking concerning the technology required at Stonehenge. An interesting report on the antiquity of the bristlecone pine is contained in *American Forests*, 26–42, September 1978. Author is R. Grant.

Longleaf and slash pine are the principal sources of turpentine. The sapwood contains about 2% oleoresin, heartwood from 7 to 10%, and stumpwood about 25%. The majority of oleoresin is obtained from the sapwood of living trees. However, this is not the sap of the tree. Oleoresin yields about 20% oil of turpentine and 80% rosin, the two commodities collectively known as naval stores.

Pines are widely used in the United States for making kraft paper, paperboard, and book paper. About half of this wood comes from the southern states. The fiber length of longleaf pine is about 3.5 millimeters ($\frac{1}{4}$ inch); and that of jack pine and lodgepole pine a little over 2 millimeters ($\frac{1}{8}$ inch). Along with Douglas fir and Sitka spruce, these pines make up the principal softwoods (as contrasted with the hardwoods, birch, cottonwood, and willow) for pulp and paper production. The term *jack pine* is sometimes rather loosely used. Usually, the term is synonymous with lodgepole pine, but the term is also applied to *P. banksiana* of central Canada (mainly used for telephone poles), and to black, prickly pine, and to certain species of spruce. Engineering constants of various commercial pine woods are given in Table 2.

Additional Reading

Ciesla, B.: "The Digger; California's Oddball Pine," *Amer. Forests*, (January–February 1987).

Dusek, K.H.: "Update on our Rarest Pine (Torrey)," *Amer. Forests*, 26 (November 1985).

Jone, S.: "White Pine Pest (Sawfly)," *Amer. Forests*, 22 (December 1981).

TABLE 2. MOISTURE AND WEIGHT OF VARIOUS PINE WOODS

Common Name of Species	Green Condition			Air-Dried to 12% Moisture	
	Moisture Content %	Weight/ Cubit Foot (pounds)	Weight/ Cubic Meter (kilograms)	Weight/ Cubic Foot (pounds)	Weight/ Cubic Meter (kilograms)
Eastern white pine	73	36	576	25	399
Loblolly pine	81	53	847	36	576
Lodgepole pine	65	39	625	29	466
Longleaf pine	63	55	879	41	657
Shortleaf pine	81	52	833	36	576
Sugar pine	137	52	833	25	399
Western white pine	54	35	561	27	431

Source: U.S. Forest Products Laboratory.

Kingsbury, L.: "New Beginning for the Western White Pine," *Amer. Forests*, 27 (December 1984).

Taylor, A.: "Mission in the Pines (Search for Blister-Rust Cankers)," *Amer. Forests*, (July 1985).

Additional interesting reading on pine trees includes: "White Pine Pest (Sawfly)," by S. Jones, *American forests*, 22–25 (December 1981); "Mission in the Pines (Search for blister-rust cankers)," by A. Taylor, *American Forests*, 27–29 (July 1985); "Update on Our Rarest Pine (Torrey)," by K.H. Dusek, *American Forests*, 26–29 (November 1985); and "New Beginning for the Western White Pine," by L. Kingsbury, *American Forests*, 30–33 (December 1984).

PINHOLE IMAGE. If a small opening is made in one side of a darkened room or box, an inverted picture of objects outside appears upon the wall opposite the opening. Such a picture differs from a true image in that it is not formed by light from a given point of the source diverging and being reconverged at the corresponding image-point, as by a lens, but is an effect of the rectilinear propagation of light. The only spot on the screen reached by light from a given point of the source is that in direct line with the opening. For this reason, pinhole images are of low intensity. On the other hand, they are free from the distortions to which lens images are subject, and with sufficient exposure, very good photographs can be made by means of them. The pinhole image also affords an excellent means of viewing eclipses of the sun.

PINK BOLLWORM (*Insecta, Lepidoptera*). A widely-distributed enemy of cotton, which probably originated in India or Africa. The adult is a small gray-brown moth, *Pectinophora* (platyedra) *gossypiella*, and the caterpillar is one-half inch long and is pinkish above. The larvae work in the flowers and bolls, causing imperfect development and destroying seeds and lint. It also attacks other plants, including the hollyhock and okra.

This species is found in the western part of the cotton-growing areas of the United States. Vigorous measures have been taken to eliminate it, for no adequate methods of control have been discovered.

PINK-EYE. See **Conjunctivitis**.

PINO. The column of smoke and ashes emitted by an explosive volcano, usually in the beginning or in the early stage of an eruption. The term is of Italian origin signifying the cauliflower-shape of the cloud as observed during the eruptions of Vesuvius.

PINWORM (*Nemathelminthes, Nematoda*). A small roundworm that lives in the alimentary tract of man, chiefly in the large intestine. The female is about $\frac{2}{5}$-inch (1 centimeter) long and the male somewhat smaller. Eggs are taken into the mouth in water or from the hands, or on raw vegetables. The entire life cycle takes place in the one host. Pinworms are usually not harmful but they may cause nervous symptoms.

PION. See **Muon**.

PIPEFISHES (*Osteichthyes*). Of the order *Solenichthys* (tube-mouthed fishes), family *Syngnathidae*, pipefishes have been aptly described as a "pipestem cleaner suddenly come to life." In this comparison, however, the cleaner would have to be equipped with a bony-plate armor. As with seahorses of the same family, pipestem fishes display an independent movement of their eyes. Each eye appears to operate independently.

Pipefishes range in length from about 1 inch to about 18 inches (2.5 to about 46 centimeters) maximum. There are some 150 species of pipefishes. Although essentially found in marine waters of both the Atlantic and the Pacific, some species can tolerate both salt and fresh waters. There are a few fresh water species. They prefer inshore, shallow water. Pipefishes are quite similar in numerous respects to their close relatives, the seahorses and it is believed that seahorses developed from a primitive pipefish. See also **Seahorses (Osteichthyes)**.

PIPE SNAKES. See **Snakes**.

PIRANHA. See **Characids**.

PIRANI GAGE. A thermal conductivity vacuum gage in which an increase of pressure from the zero point causes a decrease in the temperature of a heated filament of material having a large temperature coefficient of resistance, thus unbalancing a Wheatstone bridge circuit (or the circuit is adjusted to maintain the filament temperature constant).

PISCES (Constellation). Also referred to as the Fishes. (See map accompanying entry on **Constellations**.)

Pisces is a large constellation, which is of importance principally because it is the twelfth sign of the zodiac. There are relatively few interesting objects in the constellation although the brightest star (Alpha) is a close double, which may be resolved in instruments larger than a four-inch. In spite of the fact that Pisces is the twelfth sign of the zodiac, the vernal equinox is located in this constellation at present. This is because precession has caused the vernal equinox itself to move back an entire "sign" along the ecliptic since the time when the names were first assigned.

PISCIS AUSTRINUS. A southern constellation, somewhat resembling a fish, located between Aquarius and Grus.

PI SECTION. This is a type of network in which the elements are arranged in π shape, i.e., a shunt element across the circuit at each end of a series element.

PITCH. See **Coal Tar and Derivatives**.

PITCHING MOMENT. Rotation of an airplane about a lateral axis passing through the center of gravity is known as *pitch*. Nosing-up (positive) and diving (negative) motions are the result of moments acting around this axis. These moments are produced by thrust, wing lift and drag, parasitic drags, and tail surface forces. For airplane trim these moments must be in equilibrium, whereas for longitudinal stability an increase of angle of attack caused by an external gust or a momentary deflection of the elevator must be decreased by inherent diving moments of the airplane. Or, when a decrease in angle of attack is produced, it must be countered by a stalling moment, which will bring it back to its original angle of attack or trim. The relation between the moment and the moment coefficient is expressed in the following equation:

$$M = C_m q S c$$

Where q = dynamic pressure equivalent to the air speed = $pv^2/2$
   S = wing area
   C = wing chord
   C_m = pitching moment coefficient of the airplane

For airplane trim, the expression M equals zero, so that C_m must equal zero. At any other angle than for trim C_m is not zero, but increases either negatively (for diving motions) or positively (for stalling motions).

PITCH (Music). See **Musical Sound**.

PI THEOREM. A principal theorem in dimensional analysis that may be stated as follows: Suppose we have a dimensionally homogeneous relation B $(\alpha, \beta, \gamma, \ldots)$ in n dimensional variables, $\alpha, \beta, \gamma, \ldots$, valid for certain system of m fundamental units. The equation may then be put in the form $F(\pi_1, \pi_2, \ldots) = 0$, where the π's are the $n - m$ independent products of the variables $\alpha, \beta, \gamma, \ldots$, which are dimensionless in the fundamental units.

PITOT TUBE. A pitot-tube air-speed indicator consists of two elements: (1) A dynamic tube, which points upstream and determines the dynamic pressure; and (2) the static tube, which points normal to the air stream and determines the static pressure at the same point. The tubes are connected to the two sides of a manometer or inclined gage so as to obtain a reading of velocity pressure, which is the algebraic difference between the total pressure and the static pressure. See also **Manometer**; and **Manometer (Barometer)**. The relationship between air velocity and velocity pressure is:

$$v = \sqrt{2gH}$$

where v = velocity, feet per second; g = acceleration due to gravity; and H = velocity head or pressure, feet of air. The pressure differential created is quite small with relation to air velocity. At 100 feet (30 meters) per minute, the velocity pressure is only 0.0625 inches (1.6 millimeters) of water. Consequently, the instrument is not generally used for measuring velocities less than 1,000 feet (300 meters) per minute.

The principle of the pitot tube, in addition to aerospace applications, can be used as a liquid flow-measuring device, but because of its tendency to clog, cannot be used with liquids that have suspended solid matter. The device is useful for flow measurements in laboratory and research applications. See also **Airspeed Indicator**; and **Bernoulli Law**.

PITUITARY GLAND. A small organ of the endocrine system, the size of an average pea—with about the same weight. The gland is larger in women than in men, particularly in those women who have borne children. The gland is joined to the undersurface of the brain by a thin stalk and is protected by a bony structure that surrounds the gland. Because of its shape, the bony structure is called the "turkish saddle" (*sella turcica*).

Only within relatively recent years has the role of the pituitary become better understood. The pituitary gland is the most important organ in the regulation of growth, milk production, and in the control of several other endocrine glands. In turn, the pituitary is regulated to some extent by many of the other endocrine glands, as well as by the hypothalamus, which lies immediately above it. See also **Central and Peripheral Nervous Systems**; and **Endocrine System**.

It has long been known that severe pituitary disturbances, such as tumors, influence the function of other endocrine glands. The pituitary can upset the body's hormone balance so severely as to cause mental as well as physical illness. Tumors on the gland can usually be removed surgically, with good chance of relieving the emotional disturbances of which they are the indirect cause. Physical disorders caused by pituitary overactivity may be managed by surgical removal.

The pituitary is made up chiefly of two distinct parts called lobes—an *anterior lobe* and a *posterior lobe*. There is also a middle portion, the *pars intermedia*, that constitutes only a minor fraction of the entire gland. Under a microscope, this simple division of the pituitary appears far more complex. The supply of incoming nerve fibers is large; it has been estimated that approximately 50,000 nerve fibers enter into this organ, being confined almost exclusively to the posterior lobe. The blood supply, which is arranged in a circular pattern to avoid even the smallest temporary breakdown, is also extensive. It serves the gland by bringing food, gases, and hormones and by conveying the secretions of the pituitary to other parts of the body.

The pituitary produces a number of hormones, each endowed with the ability to produce some specific effect in one or more organs of the body, especially other endocrine organs. The hormones produced by the anterior lobe differ from those made by the pars intermedia. Most of the pituitary hormones are protein in nature.

The functions controlled by each part of the pituitary are entirely different. The largest number of hormones are produced by the anterior lobe; hence; it performs most of the functions of the entire gland. The posterior lobe does not in itself manufacture any hormones, although it does receive and store hormones made by the hypothalamus.

The pituitary has been called the "master gland" because it is believed to be the endocrinological center of the body. The anterior lobe of the pituitary regulates the growth and proper functioning of other endocrine organs by complex processes. For example, it produces a hormone, called the thyrotrophic hormone, which acts on the thyroid to stimulate its production of thyroid hormone. See also **Thyroid Gland**. In addition to the thyrotrophic hormone, the anterior pituitary is believed to secrete several other hormones. These include the adrenocorticotrophic hormone (ACTH), the follicle-stimulating and luteinizing components which make up the gonadotrophic hormone, the luteotrophic hormone, or lactogenic hormone, and the growth hormone.

Gonadotrophic hormone. The anterior lobe produces active principals that are effective stimulators of the *gonads*. The hormones that act on the sex organs are termed gonadotrophic hormones. The sexual organs in both male and female have a double function, reproduction and the production of sex hormones. The anterior lobe of the pituitary, by manufacturing and secreting the gonadotrophic hormones, controls the production of these hormones in the ovaries of the female and the testes of the male. In addition to these functions, the gonadotrophins, directly and indirectly, stimulate the development of the sex organs and the maintenance of their structure.

The testes, under the influence of the gonadotrophins, manufacture the male hormones. These, in turn, exert their action on the other parts of the body, chiefly the organs of reproduction. When the testicular tubules have developed under the influence of the male hormones, maturation of the spermatozoa also is stimulated by the gonadotrophin from the anterior lobe of the pituitary. Failure to produce male hormones results in immature appearance and lack of development of the accessory sex organs; in the previously normal adult, loss of the male hormones results in changes in appearance and degeneration of the accessory sex organs.

In women, the ovaries, under control of the hormones from the anterior lobe of the pituitary, produce the female hormones. The maturation of ova in the ovaries is stimulated by the gonadotrophins. The female hormones act on the reproductive organs and are responsible for the proper growth and function of the uterus, vagina, and other reproductive organs. It is not uncommon to observe disturbance in sexual characteristics of individuals with defective pituitary function.

Deficient pituitary activity is in many cases reflected in the lack of development of the sex organs, which may remain infantile. When accompanied by obesity, the condition is known as the *adiposogenital syndrome*. Other disturbances also can arise. The rate of production of gonadotrophins by the pituitary is influenced by the production of sex hormones. The effects are mutual and the two glands, the pituitary and the ovaries or testes maintain an exact balance in hormone production.

Adrenocorticotrophic hormone. A substance called adrenocorticotrophic hormone acts on the adrenal glands and is produced in the anterior lobe of the pituitary. Abbreviated, this substance is ACTH. This substance stimulates the production of most of the cortical hormones, but especially *hydrocortisone*. If the production of ACTH is below normal, the adrenal cortex diminishes in size and the production of most cortical hormones falls to low levels. Methods to measure ACTH directly and inadequate. ACTH has been isolated from pituitaries of cattle and pigs and is available in pure form. ACTH also is synthesized. The hormone has been found useful as therapy for a variety of disorders, as well as for the diagnosis of some conditions. Although the principal effect of ACTH is to stimulate the adrenal cortex to greater secretion, it also may perform some of the functions of the adrenal glands when the adrenals are absent.

ACTH is effective in the management of certain hematological diseases, as well as in conditions of stress. It can be used in the treatment for certain spasms involving the head, trunk, and arms of infants. It can also be used for treating young children subject to convulsions caused by diabetes. ACTH may be used in the treatment for severe allergic manifestations associated with dermatitis. Allergic reactions to the hormones, however, may occur.

The role of the pituitary gland in concert with the thyroid gland in the regulation of metabolism is described under **Thyroid Gland**.

Pituitary Hormones in Brain. During the last few decades, research into the hormone-generating facilities of the pituitary gland, particularly as these may involve the brain, has been accelerated. Through the use of radioimmunoassay, bioassay, and immunocytochemical techniques, peptides and protein hormones usually considered as being of pituitary origin have been detected within the central nervous system. Investigation continues into determining if these hormones are generated in the pituitary gland and then transported to the brain: or if they are generated elsewhere and mimic pituitary hormones. Some researchers have tentatively concluded that they are synthesized within the central nervous system (*neurosecretory cells*) and that their regulation may differ somewhat from that of their pituitary counterparts. But, other researchers suggest that pituitary hormones may be transported directly to the brain to modify brain function. A better understanding of this problem could lead to answers to a number of questions as regards such functions as memory, sleep, pain, orgasm, endocrine feedback loops, cerebral blood flow, cerebral vascular permeability, cerebrospinal fluid dynamics, epilepsy, headache, acupuncture, and mental illness.

Disorders of the Pituitary Gland

The pituitary gland is involved in a complex group of disorders that do not necessarily have a neatly classified group of causative factors.

Cushing's Disease and Cushing's Syndrome. Once considered rare, but now seen in some frequency, Cushing's disease is recognized by obesity of the abdomen, face, and buttocks, but not of the limbs. The skin about the face and hands is redder than normal. Hair grows profusely, and women may grow mustaches and beards. Bones become brittle and suffer a considerable loss of mineral components. Sexual functions may fall to a low level or become suppressed altogether.

When the manifestations of Cushing's disease occur in patients with excessive production of adrenal cortical hormones of the adrenal glands, the condition is called *Cushing's syndrome.* The adrenal hormone production is excessive also in Cushing's disease, as a result of overstimulation of the adrenals by pituitary hormone (ACTH) produced in excess by the tumor.

Some physicians in the early 1990s now stress that pituitary adenomas are not rare as once believed, but are relatively common, often being found at autopsy in a wide range (up to 25%) of persons who were not suspected as having the disease. Some 40% of tumors thus identified contain prolactin. Currently, surgery is the only effective treatment for gonadotroph-cell adenomas.

As reported by L. Daneshdoost (University of Pennsylvania School of Medicine) and co-researchers, "Adenomas that arise from the gonadotroph cells of the pituitary gland account for a substantial percentage of pituitary macroadenomas in men, but they are rarely recognized in women."

Endogenous (produced or synthesized within an organism) Cushing's syndrome remains difficult to diagnose as of the early 1990s. Specific tests have not proved reliable. As pointed out by D.N. Orth (Vanderbilt University Medical Center), "Cushing's syndrome is either adrenocorticotropin-dependent or independent. The dependent type is due to hypersecretion of adrenocorticotropin by a pituitary adenoma (Cushing's disease) or by a nonpituitary tumor (ectopic adrenocorticotropin syndrome)." These two causes are difficult to differentiate. Although too technical to describe here, a group of researchers representing several institutions reported in September 1991 about a new diagnostic test that may be effective in distinguishing Cushing's disease from ectopic adrenocorticotropin syndrome. This can prove to be of large significance because the therapy for each of the two syndromes differs considerably.

Atrophy of Anterior Lobe. Degeneration of the anterior lobe in adults results in a disease sometimes called *Simmonds'* or *Simmonds-Sheehan disease.* The disorder is characterized by extreme appearance of aging. Axillary and pubic hair are lost, there is a loss of teeth, and hair of the head becomes gray and sparse. The skin is wrinkled and the face has a wizened appearance. All of the metabolic functions of the body are affected, and eventually the mental functions decline. The condition occurs most often in women and nearly always arises after postpartum hemorrhage or shock and excessive loss of blood. The condition gradually deteriorates over a period of years. The pituitary atrophy is believed to be the result of lack of oxygen reaching the gland during the shortage of blood. The disease has been confused with *anorexia nervosa.* For the latter, hormonal treatment is secondary to psychiatric and dietary treatment. Amenorrhea (absence of menstruation) is a constant feature of Simmonds-Sheehan disease, but is not always present in anorexia nervosa.

Fröhlich's Syndrome. A lesion of the hypothalamus may affect the anterior lobe of the pituitary, resulting in Fröhlich's syndrome or *dystrophia adiposogenitalis.* The patient is excessively fat and the sexual organs are infantile. In early childhood, the disease causes dwarfism. The victim is mentally lazy and possesses a voracious appetite. When the disease develops in adulthood, male patients become effeminate, with soft skin and feminine distribution of fat in the breast region and thighs. In female patients, the obesity is extreme; it is not uncommon to see patients with this disorder weighing 300 pounds (136 kilograms).

The obesity is not a direct result of tumor in the pituitary, because the pituitary gland has no relationship to obesity, a fact often misunderstood. The obesity is a result of the same tumor's affecting the adjacent hypothalamus. Pituitary insufficiencies result in the immaturity of the sexual organs. Hypothalamic disease results in a disturbance of the appetite control center, with resulting obesity. The disease should not be confused with the typical obesity of childhood and adolescence. Fröhlich's syndrome is very rare and most obese children do not have this condition, nor any detectable glandular disturbance; rather, they are obese because of dietary habits. See also **Hormones.**

Dwarfism and Giantism. The pituitary gland is susceptible to the growth of tumors that may make the gland over- or underactive. Decreased function of the pituitary results in retarded growth. The growth hormone (*somatotrophic* hormone) exerts its major effect upon the size of the organs and the skeleton, which in cases of decreased pituitary function remains small. The condition that results is called *pituitary dwarfism,* or *infantilism.* Teeth grow slowly if insufficient growth hormone is produced, and the development of permanent teeth is considerably delayed. Untreated pituitary dwarfs do not grow over 3 to 4 feet (0.9 or 1.2 meter) in height and remain sexually immature. Specific therapy for such patients is the administration of pituitary growth hormone of primate or human origin. However, even with human growth hormone, refractory states may develop, resulting in poor growth. Laron dwarfism (growth hormone insensitivity)—that is, the lack of growth hormone receptors—may be alleviated by the administration of insulin-like growth factor 1 (IGF-1).

In late 1990, A.L. Rosenbloom (University of Florida) studied a small group (50 patients) with Laron dwarfism in southern Ecuador. This was a highly inbred group, and a genetic link was indicated. A marked predominance in females was indicated, but explained by the fact that there is early fetal death of most affected males.

Pituitary dwarfs do not achieve normal endocrine function. For dwarfed girls to develop breast tissue and to menstruate, they must be treated with estrogen. However, treatment with estrogen may stop growth of the bones before the patient has attained acceptable height. Therefore, it is wise to delay therapy with the female sex hormones as long as possible.

Although many forms of dwarfism are the result of pituitary or thyroid insufficiency, some are genetically determined. Since 1860, 49 cases of a dwarfism, known as the Ellis-van Creveld syndrome, have been verified in the Amish community of Pennsylvania—all dwarfs descended directly or indirectly from one ancestral couple. These dwarfs range in height from 40 to 60 inches (102 to 152 centimeters); they have six fingers on each hand, the extra finger on the outside beyond the little finger. Sometimes there is a sixth toe. Many of the infants have heart abnormalities and a weakness or deficiency of cartilage in the chest. One-fourth of dwarfed children with such defects dies within two weeks of birth; however, others achieve near-normal life spans. There is no mental retardation or loss of intelligence.

The most familiar form of dwarfism is *achondroplasia.* Persons in this category have large heads with saddle or scooped-out noses, short extremities, and sway backs. Advances have been made in the treatment of this condition with somatotrophin, but supplies of the substance are limited.

On occasion, the anterior lobe or the entire pituitary gland may be enlarged and the production of hormones may increase above normal range. This, in turn, causes excessive growth. If the condition develops while the bones are in the process of growing, the result is *giantism*; individuals with this condition may grow to over 8 feet (2.4 meters) in height. Prevention of giantism is relatively simple if diagnosis is made early. To close the epiphyses (open ends of the bones, which are still growing) of probable giants, estrogen is used in girls; and both estrogen and testosterone are used in boys. This treatment does not affect later gonadal function adversely. The epiphyses of these patients should be studied at 4-to-6-month intervals to determine whether growth is stopping and if therapy can be discontinued. X-ray examination of the hands and wrist provides good indication inasmuch as the epiphyses in the wrists are the last to close.

Acromegaly. In later life, after bones have ceased to grow, an overactive pituitary causes excessive stimulation of the growth centers which results in the disease known as *acromegaly.* This condition is characterized by an abnormal development of feet and hands. The jaw is prominent and large, as are the bones of the skull. The face may become angular and irregular, and the general appearance is that of a primitive man. The fully-developed disease is readily discerned by the layman; the early disease is difficult to detect. In this condition, the pituitary gland usually is enlarged by a tumor. Steroid therapy is given, depending in part on the extent of the condition and the level of circulating growth hormone in the blood. Visual acuity must be carefully monitored during this therapy. Acromegaly also occurs to a slight degree in some women during pregnancy, but regresses after delivery.

Additional Reading

Backer, K.L., J.P. Bilezikian, W. Hung, et al.: "Principles and Practice of Endocrinology and Metabolism," 3rd Edition, Lippincott Williams Wilkins, Philadelphia, PA, 2001.

Berkow, R. and M.H. Beers: "The Merck Manual," 17th Edition, Merck Company, Inc., Whitehouse Station, NJ, 1999.

Christy, N.P.: "Pituitary-Adrenal Function During Corticosteroid Therapy," *N. Eng. J. Med.*, 266 (January 23, 1992).

Daneshdoost, L., et al.: "Recognition of Gonadotroph Adenomas in Women," *N. Eng. J. Med.*, 589 (February 28, 1991).

deGroot, L.J. and J.L. Jameson: "Endocrinology: 3 Volumes," 4th Edition, W. B. Saunders Company, Philadelphia, PA, 2000.

Dowset, R.J. and B. Fowble: "Radiotherapy for Acromegaly," *N. Eng. J. Med.*, 612 (August 30, 1990).

Griffin, J.E. and S.R. Ojeda: "Textbook of Endocrine Physiology," 4th Edition, Oxford University Press, Inc., 2000.

Klabanski, A. and N.T. Zervas: "Diagnosis and Management of Hormone-Secreting Pituitary Adenomas," *N. Eng. J. Med.*, 822 (March 21, 1991).

Kostyo, J.L. and H.M. Goodman: "Handbook of Physiology: A Critical, Comprehensive Presentation of Physiological Knowledge and Concepts: Section 7: The Endocrine System: Hormonal Control of Growth," Vol. 5, Oxford University Press, Inc., New York, NY, 1999.

Melmed, S.: "Acromegaly," *N. Eng. J. Med.*, 966 (April 5, 1990).

Molitch, M.E.: "Gonadotroph-Cell Pituitary Adenomas," *N. Eng. J. Med.*, 626 (Ferbuary 28, 1991).

Monson, J.P.: "Challenges in Growth Hormone Therapy," Blackwell Science, Inc., Malden, MA, 1999.

Moran, A., et al.: "Gigantism Due to Pituitary Mammosomatotroph Hyperplasia," *N. Eng. J. Med.*, 322 (August 2, 1990).

Motta, M.: "Comprehensive Endocrinology," 2nd Edition, Raven Press, New York, NY, 1991.

Neal, J.M.: "How the Endocrine System Works," Blackwell Science, Inc., Malden, MA, 2001.

Oldfield, E.H., et al.: "Petrosal Sinus Sampling with and without Corticotropin-Releasing Hormone for the Differential Diagnosis of Cushing's Syndrome," *N. Eng. J. Med.*, 898 (September 26, 1991).

Orth, D.N.: "Differential Diagnosis of Cushing's Syndrome," *N. Eng. J. Med.*, 957 (September 26, 1991).

Pinchera, A., M. Serio, and X. Bertagna: "Endocrinology and Metabolism," The McGraw-Hill Companies, Inc., New York, NY, 2001.

Rosenbloom, A.L., et al.: "The Little Women of Loja—Growth Hormone-Receptor Deficiency in an Inbred Population of Southern Ecuador," *N. Eng. J. Med.*, 1367 (November 15, 1990).

Schlaghecke, R., et al.: "The Effect of Long-Term Glucocorticoid Therapy on Pituitary-Adrenal Responses to Exogenous Corticotropin-Releasing Hormone," *N. Eng. J. Med.*, 226 (January 23, 1992).

Takasu, N., et al.: "Exacerbation of Autoimmune Thyroid Dysfunction After Unilateral Adrenalectomy in Patients with Cushing's Syndrome Due to an Adrenocortical Adenoma," *N. Eng. J. Med.*, 1708 (June 14, 1990).

Walker, J.L., et al.: "Effects of the Infusion of Insulin-like Growth Factor in a Child with Growth Hormone Insensitivity Syndrome (Laron Dwarfism)," *N. Eng. J. Med.*, 1483 (May 23, 1991).

Williams, R.H., D.W. Foster, H.M. Kronenberg, and P.R. Larsen: "Williams Textbook of Endocrinology," 9th Edition, Harcourt Brace Company, San Diego, CA, 1999.

Web References

Cushing's Syndrome/Cushing's Disease and CRH: *http://neurosurgery.mgh.harvard.edu/e-f-942.htm*

Cushing's Syndrome: *http://www.ninds.nih.gov/healthandmedical/disorders/cushingsdoc.htm*

The Pituitary Gland: Location and Functions: *http://www.umm.edu/endocrin/pitgland.htm*

PIT VIPER. See **Snakes**.

PITYRIASIS ROSEA. See **Dermatitis and Dermatosis**.

PIXEL. An individual, identifiable element of a picture. For example, a large astronomical photographic plate may contain as many as 100,000 or more individual picture elements (pixels). In terms of a digitized picture, one of the dots or resolution elements making up the picture as a pixel.

pK. A measurement of the completeness of an incomplete chemical reaction. It is defined as the negative logarithm (to the base 10) of the equilibrium constant K for the reaction in question. The pK is most frequently used to express the extent of dissociation or the strength of weak acids, particularly fatty acids, amino acids, and also complex ions, or similar substances. The weaker an electrolyte, the larger its pK. Thus, at 25 °C for sulfuric acid (strong acid), pK is about −3.0; acetic acid (weak acid), pK = 4.76; boric acid (very weak acid), pK = 9.24. In a solution of a weak acid, if the concentration of undissociated acid is equal to the concentration of the anion of the acid, the pK will be equal to the pH.

PLACENTA. See **Embryo**.

PLAICE. See **Flatfishes**.

PLAIT POINT. The point at which two conjugate solutions of partially miscible liquids have the same composition, so that the two layers become one.

PLANCK LAW. The fundamental law of the quantum theory, expressing the essential concept that energy transfers associated with radiation such as light or x-rays are made up of definite quanta or increments of energy proportional to the frequency of the corresponding radiation. This proportionality is usually expressed by the quantum formula $E = hv$, in which E is the value of the quantum in units of energy and v is the frequency of the radiation.

The constant of proportionality, h, is known as the elementary quantum of action or, more commonly, as the Planck constant.

PLANCK, MAX (1858–1947). Planck was a German physicist who in 1900 proposed the quantum theory of electromagnetic radiation. The basic concept of the quantum theory is that radiant energy is a continuous stream of discrete packets of energy called *quantum*. A quantum is the smallest amount of energy possible. In 1918 he was awarded the Nobel Prize for his discovery of the quantum theory of energy. He also is remembered for providing the mathematical, Planck's constant.

Planck and Albert Einstein developed a close friendship. During World War II, however, Planck did not flee Germany. Planck served in many German scientific associations including the Prussian Academy of Science and the Kaiser Wilhelm Society of Berlin which, was later renamed the Max Planck Society.

Even with all of his fame and influence within Germany, Planck could not save his son's life when he was accused and executed for participation in the July 1944 plot to assassinate Hitler.

See also **Black Body**; **Energy**; **Fokker-Planck Equation**; **Planck Law**; **Planck Radiation Formula**; **Quantum**; and **Quantum Mechanics**.

J. M. I.

PLANCK RADIATION FORMULA. The relationship

$$E_\lambda \, d\lambda = \frac{hc^3}{\lambda^5} \frac{d\lambda}{e^{hc/k\lambda T} - 1}$$

where $E_\lambda \, d\lambda$ is the intensity of radiation in the wavelength band between λ and $\lambda + d\lambda$, h is the Planck constant, c is the velocity of light, k is the Boltzmann constant and T is the absolute temperature. This formula describes the spectral distribution of the radiation from a complete radiator or black body. $hc^3 = C_1$ is known as the First Radiation constant, with $ch/k = C_2$ as the Second Radiation constant. C_2 has the value 1.43879 centimeter-degree. This radiation formula can be written in other forms, such as in terms of wavenumber instead of wavelength. Also it may be written in terms of energy density instead of radiation intensity. The value of the First Radiation constant will depend on the particular form of the radiation formula used. See also **Black Body**; and **Wien Laws**.

PLANE (Geometry). A surface on which any two points may be connected by a straight line. One straight line does not determine a plane but a plane is determined by: a straight line and a point not in the line; three points not in a straight line; two intersecting lines; two parallel lines.

The general equation of a plane is $Ax + By + Cz + D = 0$, with A, B, C, not all zero. Thus the locus of every first-degree equation in x, y, z is a plane. Other forms of its equation are $x/a + y/b + z/c = 1$, where a, b, c are the x-, y-, z-intercepts; the normal form is, $\lambda x + \mu y + vz = p$, where λ, μ, v are direction consines of the normal from the origin to the plane and p is the length of this normal.

Figures on a plane surface are studied in both plane and analytic geometry. See, for example, **Conic Section; Curve; Polygon; Quadrilateral;** and **Triangle.**

PLANE SAILING. A term applied to the solution of various problems in the sailings, in which the earth is considered as a plane surface. The particular subject of plane sailing will be found discussed under the topics **Dead Reckoning.**

PLANE TABLE. A plane table is a surveying instrument used for locating and mapping topographical features. A drawing board, accurately made, and arranged so that it may be mounted on a tripod by an adjustable head which allows leveling of the board, is an essential feature of the plane table. Spirit levels are attached to the table in mutually perpendicular directions. The compass, the ruler, and a means for getting a line of sight, such as a telescope or open sights, complete the outfit. A ruler combined with a telescope or with slit sights is called an alidade. When the plane table is used for a survey, it is not necessary to take notes of angles or lengths of lines, since they are plotted, at the time of the survey, on the sheet of paper which covers the plane table. Obviously the plane table is not suitable for use in bad weather. When a survey is to be made with this instrument, the table is set up so that some convenient point on the paper is over a selected spot on the ground. The table is leveled and rotated horizontally until it is in azimuth. This is accomplished by means of the compass or by sighting back on a known point. It is then clamped in this position and the ruler is brought to the point selected on the paper and swung about it so that the line of sight that parallels the ruler bears on a distant point whose location is desired. A line is drawn in that direction, and after the distance to that point is measured, the length of line is plotted to some scale suitable to include the area being mapped on the surface of the plane table.

PLANETARIUM. A representation of the astronomical system. (1) A mechanized model reproducing the motion of the planets around the sun. (2) An optical instrument that projects images of the celestial bodies in their relative brightness and size just as they occur in nature on a hemispherical dome of a darkened auditorium. The planetarium is used mainly in education and for practical demonstration of the coordinated and relative positions and motions of the celestial objects, including artificial satellites, as well as for the training of astronauts in celestial navigation. The instrument makes it possible to go backward or forward in time to show the true panorama of the heavens as seen from any point on the earth; or from a space capsule; at any time in the past, present, or future.

The first projection-type planetarium was designed by Zeiss in 1923. Since that time a large number of projectors have been installed in science halls in major cities throughout the world. The projection instrument, as shown in Fig. 1, is about $16\frac{1}{2}$ feet high and comprises approximately 29,000 individual parts of some 2.00 types. About 150 projectors, which are mostly aspherical condensers and Tessar lenses or tele-objectives, are used. A special 1,000-watt incandescent lamp illuminates the 16 projectors in each of the two spheres. A diurnal event can be shown 120 to 480 times as fast as it would occur in nature. The movement of the celestial bodies over the period of an entire year, can be compressed into a time span ranging from several seconds to several minutes. The 25,800-year precessional revolution of the fixed-star system, which is caused by the slow gyroscopic movement of the earth, can be compressed into just 4 minutes. The variation of the sky during a trip around the earth from pole to pole only $6\frac{1}{2}$ minutes.

The various speeds for diurnal and annual movement are achieved by connecting the motors singly or together in the same or opposite directions. Diurnal, annual, and precessional movement are automatically coupled in that order. The diurnal movement is transmitted at correct time ratio by gears to the annual movement and from there to the precessional movement. Altogether there is a transmission ratio of about 1:156,000,000,000 between the rotation of the motor for the slowest diurnal movement and for the precessional rotation. The complex mechanism required for the movement of the planets in their elliptical orbits about the sun is controlled by various gear drives built into the planetarium projector.

The literature on planetariums is thin. Reference to "Geared to the Stars," by H.C. King and J.R. Millburn (University of Toronto Press, Toronto, Canada, 1978), is suggested.

Fig. 1.   Planetarium. (*Carl Zeiss.*)

W.E. DEGENHARD. Carl Zeiss, Inc., New York.

PLANETARY BOUNDARY LAYER. That layer of the atmosphere from a planet's surface to the geostrophic wind level including, therefore, the surface boundary layer and the Ekman layer. Above this layer lies the free atmosphere. Also called *friction layer*, or *atmospheric boundary layer*.

PLANETARY CIRCULATION. 1. The system of large-scale disturbances in a planet's troposphere when viewed on a hemispheric or worldwide scale.

2. The mean or time-averaged hemispheric circulation of a planetary atmosphere; also called *general circulation*.

PLANETS AND THE SOLAR SYSTEM. The word *planet*, which comes from a Greek root meaning "wanderer," was used prior to the fifteenth century to designate those celestial objects (other than meteors and comets) that were observed to be in motion relative to the stars. Before the fifteenth century, seven objects were listed as planets: Sun, Moon, Mercury, Venus, Mars, Jupiter, and Saturn. With the advent of the Copernican heliocentric hypothesis for the structure of the universe, the sun and moon were removed from the list and Earth added. Since the application of the telescope to astronomy, three major planets have been added (Uranus, Neptune, and Pluto) as well as over 1,000 small planets or asteroids. The term, as it is used at present, applies to an opaque object that shines by reflected sunlight and travels about the sun or a star in an orbit.

In spite of the fact that many of the planets are larger than Earth, their distance is so great that some of them appear to the naked eye as bright stars. The only certain method for distinguishing a planet from a star without the use of a telescope is to watch it carefully for a considerable period (frequently several days are required), and if the object is a true planet, it will move relative to the stars. For a quick method of identification, it may be said that usually a planet does not appear to twinkle as do the stars, but this rule is not infallible. With a telescope, a planet may be immediately distinguished from a star (with the exception of the planet Pluto or the asteroids) because of the fact that a planet will show an appreciable disk, whereas the stars appear as points of light.

The planets are classified in two general ways. Mercury and Venus are frequently referred to as the inferior planets, and the others are called the superior planets. Another system of classification considers Mercury, Venus, Earth and Mars as the minor or terrestrial planets, while Jupiter, Saturn, Uranus, Neptune, and Pluto are called the major (or gaseous, or Jovian) planets.

That group of objects, including the planets, asteroids, and comets, which is moving through space with the sun is known as the *solar system*. Each of the planets and its family of satellites is described in a separate entry in this encyclopedia. Also, there are separate entries on **Asteroid; Comet; Moon (Earth's); Planets (Motions); Sun (The); Voyager Missions to Jupiter and Saturn**; and **Pathfinder Mission to Mars**. The origin of the solar system is described from a theoretical standpoint in entry on **Cosmology.** The earth as a planet is described in the entry on **Earth.**

The orbital characteristics of the nine planets are given in accompanying Table 1. The physical characteristics of the planets are given in Table 2. The characteristics of the satellites of the planets (a total of over 30) are given in Table 3. More specific information on the satellites of the more recently explored planets is given in separate articles on these planets.

Although the Earth's moon is described in detail in the entry on **Moon (Earth's)**, additional convenient statistics for the moon are given here in Table 4.

Satellites serve a useful purpose for astronomers, since the mass of a planet can be determined accurately only if the planet has one or more satellites. By application of the rigorous expression for the harmonic Keplerian laws of planetary motion, the mass of any planet and satellite may be found in terms of the mass of Earth-Moon system after the distance of the planet from the satellite and its period of revolution are known. The problem of the determination of the masses of the satellites themselves is a more difficult problem. The mass of the moon can be determined in terms of the earth's mass by means of the so-called barycentrix parallax. Approximate values of the masses of the satellites of Jupiter can be obtained by the mutual perturbations they exert on each other. In the case of Saturn, the masses of the satellites may be approximately determined from their mutual perturbations and an approximate check is provided by the positions of the divisions in the ring. As the result of the *Voyager* findings, refinements in past statistical data are being formulated.

Some satellites revolve about their primaries in the retrograde sense, i.e., in the direction contrary to that in which all other planets and satellites are revolving and rotating. This retrograde motion can be fully explained on the basis of modern celestial mechanics.

The influences that satellites exert on their primaries are very slight. The tidal forces they exert have some slight effect upon the rotation periods of the primaries but such effects are so small as to be beyond observational measurement. The tidal effects the planets exert upon the satellites, on the other hand, are in many cases so large that the satellites rotate in approximately the same period as that in which they revolve.

Extrasolar Planets. In 1987, Canadian astronomers (Dominion Astrophysical Observatory, Victoria, B.C.) announced that their survey of 16 nearby solar-type stars had revealed clear indication of low-mass companions around two of the stars and possible evidence of low-mass companions around 5 others. As a scale of reference, "low mass" was considered one to be 10 times that of Jupiter (a large mass in comparison with the planets

TABLE 1. ORBITAL CHARACTERISTICS OF THE PLANETS

Characteristics	Mercury	Venus	Earth	Mars	Jupiter	Saturn	Uranus	Neptune	Pluto
Mean distance from sun:									
kilometers (millions)	57.91	108.21	149.60	227.94	778.3	1427	2869	4498	5900
miles (millions)	35.99	67.24	92.96	141.64	483.64	887	1783	2795	3666
astronomical units	0.387	0.723	1.0167[a]	1.524	5.203	9.539	19.182	30.057	39.440
Approximate distance from Earth:									
Maximum:									
kilometers (millions)	219	259		399	965	1654	3154	4682	7562
miles (millions)	136	161		248	600	1028	1960	2910	4700
astronomical units	1.46	1.73		2.67	6.45	11.06	21.08	31.3	50.5
Minimum:									
kilometers (millions)	80	40		56	591	1197	2584	4307	4296[b]
miles (millions)	50	25		35	367	744	1606	2677	2670
astronomical units	0.53	0.27		0.37	3.95	8.00	17.27	28.79	28.72
Orbital eccentricity[c]	0.2056	0.0068	0.0167	0.0934	0.0484	0.0543	0.0460	0.0082	0.2481
Angular momentum[d]	0.02	0.07	1.00	0.13	722	293	64	94	1.2
Inclination to ecliptic, degrees	7.003	3.4	0[e]	1.850	1.309	2.493	0.773	1.779	17.146
Period of rotation, sidereal days	58.82	224.59R	1.0[f]	1.03	0.41	0.43	0.45R	0.66	6.41
Approximate planetary day	8.4 weeks	32.1 weeks	23 hours, 56 min., 4.09 sec.	24 hours, 43 min.	9 hours, 50 min.	10 hours, 19 min.	10 hours, 48 min.	15 hours, 50 min.	6 days. 9 hours. 50 min.
Sidereal period, mean days	87.97	224.70	365.25636	686.98	4332.4	10,759.3	30,684.49	60,188.31	90,710.07
in terms of earth years	0.241	0.615		1.881	11.861	29.457	84.008	164.784	248.346

[a]The astronomical unit is defined as the distance equal to that of the geometrical mean distance of the earth from the sun. Refinements in measurements have altered the value slightly from unity.

[b]The minimum distance of Pluto, although the furthest planet from the sun, can be less than that of Neptune under certain circumstances because the orbits these two planets cross.

[c]Eccentricity is a number which defines the shape of an ellipse. It is the ratio of the distance from center to focus to the semimajor axis. The orbit of the earth is nearly circular.

[d]The angular momentum of a moving body, such as a planet revolving around the sun, is the production of the mass, the square of the distance from the center of motion, and the rate of angular motion.

[e]By definition, inclination to ecliptic is 0°. Inclination of equator to ecliptic is 23.45°.

[f]Actually slightly less than 24 hours, as indicated below.

R = retrograde motion. Motion in an orbit opposite the usual orbit direction of solar-system bodies, that is, motion from east to west around a center.

TABLE 2. PHYSICAL CHARACTERISTICS OF THE PLANETS

Characteristic	Mercury	Venus	Earth	Mars	Jupiter	Saturn	Uranus	Neptune	Pluto
Mean semidiameter:									
kilometers	2433	6051.4	6371	3380	69,758	58.219	23,470	22,716	1750^a
miles	1512	3760.4	3959	2100	43,348	36,177	14,584	14,116	1087
in terms of Earth = 1	0.382	0.950	1.0	0.531	10.949	9.138	3.684	3.566	0.275
Apparent, seconds of arc	5.45	30.50		8.94	3.43	9.76	1.80	1.06	0.11
Mass, kilograms	3.181×10^{23}	4.883×10^{24}	5.979×10^{24}	6.418×10^{23}	1.901×10^{27}	5.684×10^{26}	8.682×10^{25}	1.027×10^{26}	1.08×10^{24}
in terms of Earth = 1	0.053	0.817	1.0	0.107	317.946	95.066	14.521	17.176	0.181
Mean density, (grams/cubic centimeter	5.431	5.256	5.519	3.907	1.337	0.688	1.603	2.272	1.65
in terms of Earth = 1	0.98	0.95	1.0	0.71	0.24	0.125	0.29	0.41	0.30
Mean gravity,									
centimeters/sec^2	357.8	887.4	980.7	374.0	2601.0	1117.0	1049.0	1325.0	221.0
feet/sec^2	11.74	29.11	32.18	12.27	85.33	36.65	34.41	43.47	7.25
in terms of Earth = 1	0.36	0.90	1.0	0.38	2.65	1.14	1.07	1.35	0.23
Escape velocity,									
kilometers/second	4.173	10.365	11.179	5.028	60.238	36.056	22.194	24.536	5.023
miles/second	2.593	6.441	6.947	3.124	37.432	22.405	13.791	15.247	3.121
miles/hour	933.5	2318.8	2500.9	1124.6	13,475.5	8065.8	4964.8	5488.9	1123.6
Solar constant, calories/cm^2/min	12.8	3.7	1.920	0.83	0.071	0.021	0.005	0.002	0.001
Temperature (day),									
Kelvin	683^b	720°	287^b	$190{-}240^b$	$11,000^d$	223°	123°	123°	63°
approximate°C	410	447	14	−83 – −33	10.704	−50	−150	−150	−210
approximate°F	770	837	57	−117 – −27	19.300	−58	−238	−238	−346
Oblateness^e	0.029	0?	0.0034	0.005	0.066	0.103	0.07	0.08	0.156
Albedo^f	0.076	$0.59{-}0.76^g$	0.36	0.152	0.54	0.57	0.65	0.68	0.13

^{a}Recent investigations have indicated that Pluto is much smaller than previously estimated. New estimates are given here.
^{b}Surface temperature.
^{c}Upper atmosphere temperature.
^{d}Inner layer temperature.
^{e}The departure of a planet from spherical form because of centrifugal force of rotation. If equatorial diameter is a and polar diameter is b, oblateness $= (a - b)/a$.
^{f}A measure of the light-reflection power of a surface compared with an ideal white matte surface which absorbs no light.
^{g}Range of various estimates.

TABLE 3. SATELLITES OF THE PLANETS

Planet and Satellite	Mean Semidiameter Kilometers	Miles	In terms of Moon = 1	Rotation Period (days)	Mean Distance from Primary Body Kilometers (thousands)	Miles (thousands)	Orbit Inclination to Ecliptic of Planet (degrees)	Orbit Eccentricity	Apparent Stellar Magnitude	Mass (kilograms)	Date of Discovery	Equilibrium Temperature (°K)
EARTH												
Moon	1783.3	1108.1	1.0	27.3	383.403	238.857	18–29	0.055	−12.3	7.35×10^{22}	Antiquity	394
MARS												
Phobos	5.4–14.4	3.4–9.6	0:003	0.32	9	5.59	1.1	0.021	11.5	2.7×10^{16}	1877	319
Deimos	1.0–9.0	0.6–5.6	0.0005–0.005	1.26	23	14.29	1.6	0.003	13	1.8×10^{15}	1877	319
JUPITER												
Jo	1818	1130	1.02	1.77	422	262.2	0.0	0.0	5.5	7.9×10^{22}	1610	173
Europa	1533	953	0.86	3.55	671	417	0.5	0.0	5.7	4.8×10^{23}	1610	173
Ganymede	2608	1621	1.46	7.16	1070	665	0.2	0.001	5.1	1.54×10^{23}	1610	173
Callisto	2445	1519	1.37	16.69	1880	1168	0.2	0.01	6.3	7.35×10^{22}	1610	173
Amalthea	120	75	0.067	0.4	181	112.5	0.4	0.003	13	8.3×10^{18}	1892	173
Himalia	85	53	0.048	?	11.470	7127	27.6	0.158	13.7	—	1904	173
Elara	30	19	0.017	?	11,740	7295	24.8	0.207	16	—	1905	173
Pasiphae	13.5	8.4	0.008	?	23,300	14,479	145R	0.38	16	—	1908	173
Sinope	11.5	7.1	0.006	?	23,700	14,727	153R	0.28	18	—	1914	173
Lysithea	9.5	5.9	0.005	?	11,710	7277	29.0	0.13	18	—	1938	173
Carme	12	7.5	0.007	?	22,350	13,888	164R	0.21	18	—	1938	173
Ananke	8.5	5.3	0.005	?	20,700	12,863	147R	0.17	19	—	1951	173
1979-J1	15–20	9.3–12.4	0.01	?	—	—	—	—	—	—	1979	173
1979-J2	35–40	21.7–24.9	0.02	?	—	—	—	—	—	—	1979	173
SATURN^a												
Titan	2525	1600	1.4	15.95	1222	759	0.3	0.03	8.3	1.2×10^{23}	1655	128
Iapetus	900	559	0.5	79.33	3560	2212	14.7	0.03	11.0	2.3×10^{21}	1671	128

TABLE 3. (*Continued*)

Planet and Satellite	Mean Semidiameter			Rotation Period (days)	Mean Distance from Primary Body		Orbit Inclination to Ecliptic of Planet (degrees)	Orbit Eccentricity	Apparent Stellar Magnitude	Mass (kilograms)	Date of Discovery	Equilibrium Temperature (°K)
	Kilometers	Miles	In terms of Moon = 1		Kilometers	Miles (thousands)						
Rhea	765	475	0.45	4.4	527	327.5	0.4	0.0	10.0	1.8×10^{21}	1672	128
Tethys	525	326	0.32	?	295	183.3	1.1	0.0	10.5	4.9×10^{20}	1684	128
Dione	560	348	0.32	2.7	377	234.3	0.0	0.0	10.7	5.4×10^{20}	1684	128
Mimas	200	124	0.11	?	186	115.6	1.5	0.02	12.1	3.7×10^{19}	1789	128
Enceladus	250	155	0.14	1.37	238	147.9	0.0	0.0	11.6	7.4×10^{19}	1789	128
Hyperion	180	110	0.14	?	1481	920	0.4	0.1	13.0	6.8×10^{19}	1848	128
Phoebe	100	62	0.056	?	12,390	8035	150R	0.16	14.5	1.9×10^{19}	1898	128
Janus	185	115	0.1	?	160	99.4	0.0	0.0	—	1.2×10^{20}	1966	128
URANUS												
Titania	1200	746	0.67	?	438	272.2	0.0	0.0	14.0	2.1×10^{21}	1787	90
Oberon	1100	684	0.62	?	586	364.1	0.0	0.0	14.2	1.1×10^{21}	1787	90
Ariel	1000	621	0.56	?	192	119.3	0.0	0.0	15.2	5.0×10^{20}	1851	90
Umbriel	650	404	0.36	?	267	165.9	0.0	0.0	15.8	1.4×10^{20}	1851	90
Miranda	400	249	0.22	?	130	80.8	3.4	0.02	17.0	3.0×10^{19}	1948	90
NEPTUNE												
Triton	2500	1554	1.4	5.9	30.07AU	30.07AU	160R	0.0	13.6	1.46×10^{23}	1846	72
Nereid	350	217	0.2	?	354	220	27.5	0.76	19.0	5.0×10^{19}	1949	72
PLUTO												
Chiron[b]	1050–1500	650–930	~1.4	6.4	39.44AU	39.44AU	0.0	0.0	—	—	1978	63

[a]Saturn is known to have additional satellites. See article on **Saturn**.
[b]Very tentative. See article on **Pluto**.
AU = astronomical unit.
R = retrograde revolution.

TABLE 4. ADDITIONAL LUNAR STATISTICS

Distance of Moon from Earth	
greatest	406,697 km (252,710 miles)
least	356,700 km (221,643 miles)
mean	384,403 km (238,857 miles)
Equatorial horizontal parallax at mean distance:	57′03″
Apparent angular diameter:	
minimum	29′21″
maximum	33′30″
mean	31′05″
Eccentricity of orbit	$\frac{1}{18}$
Diameter of Moon	3,476 km (2,160 miles)
Volume	$\frac{1}{49}$ that of Earth
Mass	$\frac{1}{81}$ that of Earth
Mean density	$\frac{3}{5}$ that of Earth (3.34g/cm^3)
Surface gravity	$\frac{1}{6}$ that of Earth
Velocity of escape	2.4 km (1.5 miles) per second
Approximate temperature of soil	
at noon	100 °C 212 °F
at midnight	−150 °C −238 °F
Revolution and Rotation	
synodic month (from one new moon to the next)	29^{d}12^{b}44^{m}2.8
sidereal month (true period of revolution around Earth)	27^{d}7^{b}43^{m}11.5
period of axial rotation	27^{d}7^{b}43^{m}11.5
Period of revolution of nodes	18.6 years
Daily retardation in crossing meridian	
minimum	38 minutes
maximum	66 minutes
average	50$\frac{1}{2}$ minutes
Average velocity of Moon around Earth	
linear	3680 km (2287 miles) an hour
angular	13°.2 a day or 33′ an hour

(Moon moves in one hour a distance about equal to its own diameter)

of the solar system). Currently, astronomers are considering the evidence with caution. Numerous extrasolar planets have heretofore been claimed. As pointed out by Gatewood (Allegheny Observatory, Pittsburgh, Pennsylvania), a highly publicized case was made in 1984 with regard to the observation of a companion to the dim red star van Biesbroek 8, which to date has not been observed again. Nevertheless, many astronomers regard the 1987 Canadian observations as exciting. An important aspect of the Canadian observations is that they did *not* see brown dwarfs (star-like objects that just miss being massive enough to ignite by thermonuclear fusion). Current astrophysical theory indicates the threshold for thermonuclear burning is about 80 Jupiter masses. It is pointed out that anything approaching that mass would have stood out in the survey like a searchlight. Astronomers are quick to point out, however, that the fact that brown dwarfs are rare (or nonexistent) indicates that current knowledge of the star formation process is indeed wanting. As observed by Levy (University of Arizona Lunar and Planetary Laboratory), "What's amazing to me is that the lower limit for star formation seems to be so close to the lower limit for nuclear burning. It's not obvious why that should be. But then, in astrophysics there are a lot of coincidences that people still don't understand."

Some scientists as of 1994 suggested that an additional planet beyond Pluto in the solar system may exist, but the majority of researchers at least tentatively have concluded that the case for a tenth planet is a weak one. The infrared sky survey (IRAS) launched in 1983 has revealed nothing to support the existence of a tenth planet. The strongest argument for a tenth planet is the fact that the present model of the solar system appears to be somewhat flawed. This is evidenced by the fact that the positions of the outer planets cannot be predicted reliably beyond a span of about ten years. As one investigator observes, "If you sent a probe to Pluto at the moment, you'd miss it." But others observe that such estimations arise from feeding insufficiently precise information into the model of the system. Discovery of Pluto's moon, Charon, permitted a determination of the planet's mass and of its satellite. Robert Harrington (U.S. Naval Observatory) continues with research toward filling out the tenth-planet theory. After investigating the effects of numerous combinations of data, a best estimate of a tenth planet required to satisfy the model would be a body with a mass about 3.5 times that of Earth, following an oval orbit tilted at about 30° to the plane of the solar system and lying about three times as far from the sun as Neptune. Further, it is expected that the planet may be located in the constellation Centaurus (deep in the southern sky), thus explaining why most astronomers who observe the skies of the Northern Hemisphere may not have found it.

Additional Reading

Atreya, B.K., J.B. Pollack and M.S. Matthews: "Origin and Evolution of Planetary and Satellite Atmospheres," University of Arizona Press, Tucson, AZ, 1997.

Barnes-Svarney, P.: "A Growing Solar Family," *Technology Review (MIT)*, 21 (May/June 1991).

Bennett, J., M. Donahue, and N. Schneider: "The Solar System," 2nd Edition, Addison Wesley Longman, Inc., Redding, MA, 2001.

Black, D.C.: "Worlds Around Other Stars," *Sci. Amer.*, 76 (January 1991).

Eberhart, J.: "Straightening the Magnetic Tilts of Planets," *Science News*, 294 (May 12, 1990).

Eberhart, J.: "Panel Prods NASA to Seek Unknown Planets," *Science News*, 21 (January 12, 1991).

Eberhart, J.: "Why Three Planets Radio the Sun," *Science News*, 63 (January 26, 1991).

Freedman, R.A. and W.J. Kaufmann: "Universe: The Solar System," W.H. Freeman and Company, New York, NY, 2001.

Garlick, M.A.: "The Story of the Solar System," Cambridge University Press, New York, NY, 2002.

Gore, R.: "The Planets," *National Geographic*, 4 (January 1985).

Greeley, R. and R. Batson: "The NASA Atlas of the Solar System," Cambridge University Press, New York, NY, 1996.

Hanson, R.B.: "Planetary Fluids," *Science*, 281 (April 20, 1990).

Ingersoll, A.P.: "Atmospheric Dynamics of the Outer Planets," *Science*, 308 (April 20, 1990).

Jones, B.W.: "Discovering the Solar System," John Wiley & Sons, Inc., New York, NY, 1999.

Kasting, J.F., O.B. Toon, and J.B. Pollack: "How Climate Evolved on the Terrestrial Planets," *Sci. Amer.*, 90 (February 1988).

Kerr, R.A.: "Which Way is North? Ask Right-Handed Astronomers," *Science*, 999 (November 24, 1989).

Lewis, J.S.: "Physics and Chemistry of the Solar System," 2nd Edition, Morgan Kaufmann Publishers, Orlando, FL, 1997.

Lunine, J.I.: "Origin and Evolution of Outer Solar System Atmospheres," *Science*, 141 (July 14, 1989).

Matthews, R.: "Planet X: Going, Going...But Not Quite Gone," *Science*, 1454 (December 6, 1991).

Pasachoff, J.M. and D.H. Menzel: "Field Guide to the Stars and Planets," Vol. 15, 3rd Edition, Houghton Mifflin Company, New York, NY, 1992.

Powell, C.S.: "A Cosmic Unveiling: Newborn Stars Have Some Secrets About How Planets Form," *Sci. Amer.*, 26 (December 1989).

Price, F.W.: "Planet Observer's Handbook," 2nd Edition, Cambridge University Press, New York, NY, 2000.

Rubin, A.E.: "Disturbing the Solar System: Impacts, Close Encounters, and Coming Attractions," Princeton University Press, Princeton, NJ, 2002.

Stern, S.A. and J. Mitton: "Pluto and Charon: Ice Worlds on the Ragged Edge of the Solar System," John Wiley & Sons, Inc., New York, NY, 1998.

Taylor, S.R.: "Solar System Evolution: A New Perspective: An Inquiry into the Chemical Composition, Origin, and Evolution of the Solar System," 2nd Edition, Cambridge University Press, New York, NY, 2001.

Weissman, P.R., T. Johnson, and L.-A. McFadden: "Encyclopedia of the Solar System," Academic Press, Inc., San Diego, CA, 1998.

Web References

JPL Solar System Exploration Site: *http://sse.jpl.nasa.gov/index.html*
The Planetary Photojournal: *http://photojournal.jpl.nasa.gov/*
The Nine Planets: *http://seds.lpl.arizona.edu/billa/tnp/intro.html*
Views Of The Solar System: *http://www.hawastsoc.org/solar/eng/homepage.htm*

PLANETS (Motions). The apparent motions of the planets on the celestial sphere have been observed, recorded, and speculated about ever since mankind has existed on the earth. The motions as seen from the earth are complicated by the fact that the earth is moving about the sun in the same direction as the planets, but with a different rate.

Apparent Motions Relative to the Sun. In Fig. 1, we have S representing the sun, E representing the earth (assumed fixed for convenience), P'_1, P', P'_2, P'_3, P'_4 indicating various positions of a planet whose orbit lies between the earth and the sun (inferior planet), and P'_1, P, P_2, P_3, P_4, positions of a planet whose orbit is outside that of the earth (superior planet). The angle between the sun and the planet (e.g., SEP' or SEP) is defined as the elongation of the planet. For an inferior planet the elongation may have any value from 0 to SEP'_2 or SEP'_4, whereas for a superior planet the elongation varies either east or west from 0 to $180°$. With elongation 0 we have the planet in the aspect of conjunction.

Inferior planets: It will be noted that the elongation is 0 both at P'_1 and also at P'_3 and hence there are two conjunctions for an inferior planet.

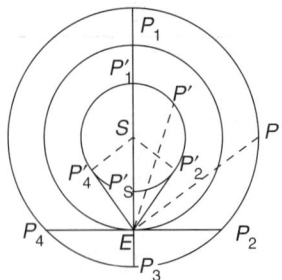

Fig. 1. Motions of planets as seen from Earth relative to the sun.

To distinguish between them P'_1 is known as superior conjunction and P'_3, inferior conjunction. An inferior planet moves more rapidly in its orbit than does the earth and accordingly from superior conjunction the planet moves out with increasing eastern elongation (evening object) to the point P'_2 (greatest eastern elongation). It then moves in with decreasing elongation, passes the sun and becomes a morning object at inferior conjunction (P'_3) and moves out with increasing western elongation to P'_4 (greatest western elongation) and thence back to superior conjunction again. Hence these planets apparently oscillate back and forth across the direction of the sun. They do not ordinarily pass either between the earth and the sun or directly behind the sun because of the fact that their orbits are not in the plane of the ecliptic.

Superior Planets: It must be remembered that these planets are moving more slowly in their orbits than is the earth. These planets apparently move from conjunction at P_1 slowly out to the west of the sun (morning objects) to P_2 where the western elongation is $90°$ and the aspect is western quadrature. From this point the increase in western elongation increases rapidly to $180°$ at P_3 (aspect opposition) from which point the elongation becomes east (evening object) and decreases rapidly to $90°$ eastern elongation at P_4 (eastern quadrature). The decrease in eastern elongation then slows down as the planet moves slowly back to conjunction again.

Apparent Motions of the Planets Relative to the Stars. In Fig. 2, we have S representing the position of the sun, E representing the (assumed circular) orbit of the earth, with successive positions marked, P the orbit of a planet with successive positions marked with numbers corresponding in date with the marked positions for the earth, and an outer circle representing directions on the celestial sphere, V being the direction of the vernal equinox, A the direction of the autumnal equinox and the direction of increasing longitude (or right ascension) indicated by arrows.

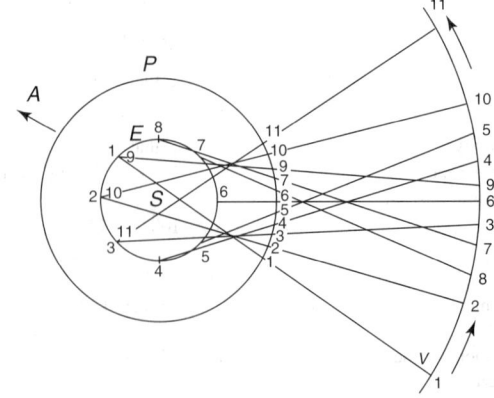

Fig. 2. Motion of planets as seen from Earth relative to the stars.

The successive positions of the planet as seen from the earth relative to the stars, are obtained by drawing lines, representing the lines of sight from successive positions of the earth through corresponding positions of the planet. By examining the successive directions, as indicated by numbers on the outer circle, it will be evident that the general trend of the planetary motion is in the direction of increasing right ascension. Such motion in increasing right ascension is known as direct motion. It will be noticed, however, that in the vicinity of opposition of the planet, the motion reverses for a period (numbers 6, 7, and 8) and the planet moves in the direction of decreasing right ascension. Such motion, in the direction of decreasing right ascension, is known as retrograde motion.

Since the diagram is plotted in the plane of the ecliptic with the planet also assumed in this plane, the reversal in direction appears only in longitude. However, there will also be changes in direction of motion both in celestial latitude and declination, giving the appearance on the celestial sphere of loops in the motion of the planet. These loops were observed by the early astronomers and it was to account for those on the assumption of the geoconcentric universe that the epicycles of the Ptolemaic system became necessary. For planetary motions referred to the sun alone, see **Orbit (Astronomy)**.

See also **Kepler's Laws of Planetary Motion**.

PLANKTON.   See **Ocean Resources (Living)**.

PLANTAIN (*Plantago* sp.; *Plantaginaceae*).   There are some 200 widely distributed species of plantains, many of which are ubiquitous weeds. Some species are stemless plants, the petioled leaves forming a rosette that covers a considerable area of ground, from which it excludes other desirable plants. These are the species that cause unsightly patches in lawns. These weeds thrive so well in cultivated areas that some of the American Indians call them "white-man's-footsteps."

Plantago major is one of these, with broad ovate leaves on long petioles. *Plantago lanceolata* has lanceolate erect leaves. The flowers are borne in elongated spikes, the pistil maturing before the stamens. Plantains are wind-pollinated plants, although insects occasionally visit them for their pollen. The fruit is a capsule, the upper half of which comes off when mature, through the development of a circumferential line of dehiscence. The seeds of plantains are often fed to caged birds. In some species the seeds imbibe water and become mucilaginous. One species, *Plantago psyllium*, native in southern Europe, is sometimes used in medicine under the name Psyllium seed.

Plantain also refers to a tropical fruit, *Musa paradisiaca*, a close relative of the banana. Plantains are always eaten cooked, or made into flour. They have furnished food for all tropical peoples for centuries.

PLANT BREEDING.   In general, there are three ways of effecting plant improvement: (1) introduction of varieties new to a given region or district; (2) plant breeding proper, that is, hybridization in all of its various phases; and (3) artificially introducing new genes or altering in some way the genetic structure of a given plant. Methodologies (1) and (2) are the predominant methods used, but method (3) is now in an advanced technological stage and expected to make serious inroads on the other methods during the next few years. See Chilton reference listed at end of article.

The term *variety*, equivalent to *cultivar*, is defined in the International Code of Nomenclature of Cultivated Plants as "an assemblage of cultivated plants which is clearly distinguished by any characters (morphological, physiological, cytological, chemical, and others) and which when reproduced (sexually or asexually) retains its distinguishing characters."

The agronomic value of a variety depends upon many characteristics, the most important of which are (1) high-yield ability, (2) high response to improved cultivation methods (e.g., fertilizers), (3) high quality of product, (4) resistance to diseases and pests, (5) resistance to adverse environmental factors (frost, drought, lodging, etc.), and (6) suitability for mechanized cultivation and harvesting methods.

In the assessment of new varieties, several steps may be distinguished: (1) The first is the evaluation carried out at the breeding station by the breeder. Normally, the breeder handles many experimental lines. By observation and by using various testing methods over several years, the breeder selects the most promising and develops them into new varieties. (2) The trials in the breeding station alone are not sufficient for objective determination of the agronomic value of new varieties. In a second step, the varieties are tested in several localities outside the vicinity of the breeding station. These trials are conducted either by the breeder or by a public or private agency. (3) Then, in a third step, the new varieties are tested for adaptability at a large number of locations with a wide range of soil and climatic conditions. In most countries, these trials are conducted by a neutral agency, to ensure objective comparison of the new varieties with commercial varieties already available. The object is to make certain that only those varieties having a higher agronomic value than the best existing varieties are released to producers for planting. (4) In a fourth and continuing step, the released varieties are bested periodically to confirm their performance against specifications and expectations. Variety testing procedures are described in detail in the Harrington reference listed at the end of this entry.

Background

There are few examples of commercially exploited plants today that are identical reproductions of the plants which appeared on the earth several thousand years ago, when and before animals and people first learned of their value as food sources. It is not uncommon for technical discussions of the nature of present-generation food plants to commence with a review of the presumed (possible, probable, etc.) nature of the precursors of present species and varieties. For example, some authorities believe that cereal seeds were first planted by people in the foothills of the Zagros Mountains in the Near East some 9000 years ago. In some instances, plant scientists are aided by the findings of archaeological digs and carbon dating techniques, making it possible to refine conjecture into refined scientific speculation. For example, some years ago, small-eared corn (maize) was found in the Bat Cave of New Mexico, dating back some 5600 years. By researching the records of early explorers and pre-Colombian Indian cultures of Mexico, Honduras, and other Central American countries, the record has been pieced together, hinting strongly that contemporary corn (maize) plants relate back to the cereal grass teosinte, which flourishes in these areas and which hybridizes with corn easily.

Up until about 150 years ago, most of the food plants used commercially were the result of natural developmental processes, those varieties most suited for a given region surviving, and unsuited types essentially becoming extinct. Authorities believe, however, that people intervened in the developmental improvement of plants a thousand or more years ago. Astute growers, strictly through the application of keen powers of observation, coupled with common sense, most likely continued to plant the seeds of varieties that were healthy and that yielded well and, by the reverse process, refrained from planting the seeds of unhealthy, poor-yielding, and otherwise unattractive plants. With the open-pollinated plants, it was only natural that from time to time new varieties would arise strictly by natural mechanisms and, again, astute growers would observe superior performance and gather seeds for planting. Such "unscientific" processes continue even in present times. An example is the finding, in 1956, of a few rice plants that remained standing after a typhoon had flattened an entire area. These plants were found by two peasants in an eastern Kwantung (China) rice field and became the parents of the well-known dwarf variety Ai Tze-Tzang which possesses excellent anti-lodging properties (does not bend over when being harvested, thus making cutting difficult).

A great boost, to both natural (or accidental) and purposeful intervention in the development of improved varieties, was the advent of globe-circling explorers and later intensive trade between all of the continents. This permitted the introduction of species and varieties from different parts of the world to specific agricultural locales. There were many failures, but also notable successes, of which a few examples are: (a) coffee from Ethiopia achieved its highest development in Central and South America; (b) Bahian cacao thrived in west Africa; (c) African grasses proved to become a main support for the Latin American cattle industry; (d) in Indonesia, African oil palm realized its most advanced development; and (e) the famed introduction of Amazonian rubber into Malaysia. Records indicate that a new group of rices, *Champa*, was introduced into Fukien (China) about 1000 A.D. from southeast Asia.

Probably the next most important step toward plant improvement commenced with the observations of Gregor Johann Mendel, the Austrian monk, made when he bred garden peas as a hobby during the late 1850s. Serious biologists took many years to appreciate the quality of Mendel's work, which was essentially unrecognized for nearly 35 years. See also **Heredity**. European biologists in the early 1900s found that they had independently duplicated what Mendel had found much earlier. Of course, an understanding of the genetic mechanism greatly accelerated interest in plant breeding and enabled breeders (as also with animal breeding for improvement) to conserve time by avoiding a number of mistakes and taking some shortcuts. Even guided by the principles of plant genetics, aided by computers and sophisticated data processing systems to keep massive genealogical records sorted out, plant breeding still retains the trappings of a numbers game in which chance, along with science and experiment, play a leading role in producing failures and successes. For example, in one recent year, a leading hybrid corn (maize) developer tested more than 4700 new inbred lines; planted more than 208,000 yield test plots; and made 1.4 million hand pollinations — in the interest of releasing fewer than 20 new commercial hybrids.

Hybridization

This technique is used to combine in a single variety the desirable characters of two or more varieties. The parental varieties may be closely related or quite unrelated, such as from different parts of the world, or from different species. Frequently, a cross gives rise to plants which are beyond the range of the parents for a given character. For example, they are earlier than the earlier maturing parent, or taller than the tallest parent. Such transgressive segregates may enable the breeder to attain objectives more readily or completely than might be anticipated. Sometimes a single cross is all that is required to attain the objective. Sometimes the first cross has to be followed with others. A group of two or three single crosses may have to be made, to be followed with others, particularly when dependence upon one cross is considered a high risk.

The crossing program cannot be planned beyond stating the objectives until a survey of potential parental material has been made. This requires collecting material and testing it thoroughly. Catalogs (like those of the Food and Agriculture Organization, United Nations) have been established so that plant breeders know where seeds for breeding uses can be obtained. The greatest amount of work done along these lines to date has related to the cereals, such as corn (maize), rice, and wheat.

In the case of newly introduced varieties, information supplied with the seed must be verified under the target conditions where the breeding is to be done. For example, a plant variety from Canada may be considered resistant to a given disease in Canada, but prove to be quite susceptible to races of that same disease in Argentina. Great care must be exercised in selecting parents that have the desired characters. Sometimes there are several varieties, say, A, B, and C, each possessing a certain character which the breeder wants in combination with the desired attributes of D, a standard variety. It may be that the use of only one of these three varieties is a more risky proposition than using all of them. In this situation, D is crossed with each of A, B, and C. Sufficient hybrid seeds are obtained to produce a total F_2 population[1] the same size as would be used if only one of the three crosses were made, that is, the group of three crosses have a single objective and the statistical estimate of the number of F_2 required is the same as for a single cross.[2]

The amount of F_2 seed needed in a given cross depends upon the genetic width of the cross, that is, the number of gene differences between the parents, the number of important characters concerned, the presence of linkage and its aid or hindrance, the importance of the breeding problem, and the assistance and facilities available. If a breeder is attempting to solve a very important problem, such as obtaining a combination of yield with resistance to a destructive disease, and if this objective involves a wide cross within a given species, but no linkage is known to be concerned, there should be an F_2 population of at least 10,000 to 20,000 plants. If no exacting commercial quality is required, the lower number is probably sufficient. If, however, the parents of the proposed cross are both satisfactory, except that one is disease-resistant, but not high in yield, and the other is susceptible to disease, but high in yield when disease is not present, fewer than 10,000 F_2 plants could suffice.

If the attainment of the desired combination of characteristics is blocked by high linkage (association) of the desired resistance with susceptibility to some other disease of lesser, but not negligible importance, it would be advisable to grow far more than 20,000 F_2 plants so as to be able to secure desirable crossover plants, that is, those with the unfavorable linkage broken and a favorable linkage obtained in its place, and which, at the same time, are desirable in the characters not affected by this linkage.

Where unfavorable linkage is so strong that an exceedingly large F_2 population is indicated to obtain crossovers which are otherwise desirable, a less costly method is to have just sufficient F_2 plants to ensure obtaining a few crossovers and then backcross these to the more desirable of the two parents. This procedure requires less work, but takes a year longer, unless more than one generation is produced in a given year.

---

[1] The terms F_1, F_2, F_3, etc., indicate the first, second, third, etc. generations of a cross.

[2] Much more detail is given in the Harrington reference listed.

Making the Crosses. The parental varieties, previously checked for uniformity, are each sown at several dates to secure flowering. For example, to cross variety P with variety Q, which flowers about 15 days earlier, a row of P is sown on each of date 0, 5, 10, and 15 and the one row of Q on each of dates 10, 15, and 20. Date 0 means the earliest date of sowing the breeding material of the crop concerned at a given breeding station. Thus Date 0, for example, may be November 1 for wheat and April 1 for rice in lower Egypt. With rice, the dates of flowering of the two varieties to be crossed may differ by as much as a month or two and the successive sowings of parental material are made accordingly. Where artificial manipulation of exposure to light is possible, reducing the hours of daylight for the late-maturing variety to induce it to flower earlier, or reducing the hours of light for the early variety to retard its flowering may be helpful. The variety used as a male parent should have at least one noticeable dominant character not possessed by the female parent, to aid in checking the F_1 plants for authenticity.

Sufficient heads or panicles are emasculated to furnish the required number of F_1 seeds. The removal of the anthers is usually done one or two days before the pollen is ripe when the anthers are still greenish and about half-grown. With rice, the emasculation may be done the prior evening if the temperature is not too high; otherwise it should be done shortly before crossing. Only 10 to 16 of the best developed florets on the central part of a wheat or barley head, for example, are emasculated, the other florets being clipped or pulled off. The emasculated head is enclosed in a plastic envelope for protection from outside pollen, and then tagged. The number of heads to emasculate in an ordinary cross or group of related crosses is about thirty if about 50% of seeds are expected to set from the crossing and an F_2 population of 20,000 plants is desired.

When the stigmas are receptive (generally a few days after emasculation), the pollinating is done and the plastic envelopes are returned to the heads. On each tag is recorded the name of the male parental variety, the date, and the pollinator's initials. The pollen is collected immediately before pollinating, usually in the morning. For a given cross, pollen is collected from at least five or six heads and, when possible, only the pollen from one plant used as a male is used on one plant used as a female.

In harvesting a cross, all the plastic enveloped heads with their attached labels are put in one or two large paper envelopes. A crossed head is discarded if the other heads on the same plant appear to be untypical of the variety used as the female. Two weeks or more after they are harvested, the crosses are threshed. The success of each cross is recorded as to the number of seeds obtained from each head, the number of florets pollinated, the identity of the operators, etc. This information is invaluable for reference when making other crosses. These procedures emphasize the delicacy and patience of the work and the great amounts of statistics that must be maintained.

Growing the F_1 Seeds. These seeds are sown in a good location as free as possible from insects and diseases. Parental variety material is grown close by under similar spacing for checking the hybridity of the F_1 plants. A knowledge of the dominance or partial dominance of obvious plant characters is desirable.

Each F_1 plant is harvested separately, keeping any of doubtful parentage apart. The F_1 plants are threshed individually and the seed examined on a laboratory table. Any that are doubtful in the field and are doubtful on the table are discarded. All others, including reciprocals, are bulked for each cross unless there is some particular reason for keeping their identities separate. This bulked seed is the F_2 seed of the cross.

Further Generations. Essentially, the procedures used to produce the F_2 generation are used for the F_3 and F_4 generations, except that fewer plant selections are needed for F_4. Some lines will appear to be fairly uniform, and if these appear otherwise desirable they are promoted to preliminary yield tests in F_5. Many valuable varieties have started as bulked F_4 progeny of an individual F_3 plant. However, sufficient uniformity usually is lacking in F_4 progenies and most breeders wait until F_5 or F_6 to consider a line as reasonably uniform and *pure breeding*.

Selection from segregating progenies is only continued to F_6 in wide crosses, such as those between the different 28 chromosome species of *Triticum*.

The foregoing procedures generally constitute what is known as the *pedigree method*, the quickest and also the most expensive of the hybridizing methods. With this method, a new variety can be selected and well tested by the time it has reached its ninth generation after the cross.

Other procedures include the *mass method*, which makes use of the fact that the change from homogeneous heterozygosity of F_1 (where each plant resembles every other one, but all are high heterozygous) to the heterogeneous homozygosity of F_{10} (where, if no natural crossing or mutation occurs, the plants are a heterogeneous population showing all combinations of parental characters, and each plant is reasonably pure breeding) occurs naturally during a period of ten years in a normally, self-pollinated crop. Usually it is considered that after ten generations of self-fertilization nearly all the individual plants are homozygous, although they differ from each other in many ways. Selecting is done in F_{10} or F_{11}. See also **Heterozygous**.

During the years from F_9 to F_{10}, while heterozygosity is being replaced by homozygosity, the phenomenon of natural selection operates continuously to increase the proportions of the adapted plants and reduce the numbers of those which competed at a disadvantage. However, experiments have shown that the plants that survive best in a mixture are not necessarily the best for future crop use. It has been shown that many of the plants most useful for crops are crowded out to virtual extinction when in a varietal mixture. This is a significant disadvantage of the mass method. The ten years and more of time required is also an extreme disadvantage.

There are also modified mass methods and fanned mass methods, but the details of these procedures are too detailed for the scope of this book. These methods are described in some of the references listed.

Crossing Methods. The *backcross method* is used chiefly for adding one or two simply inherited attributes to an otherwise satisfactory variety. Suppose that L is such a variety and K is a variety that is undesirable except that it possesses the character or characters lacking in L. A cross is made between K and L and the F_1 of this cross is crossed with L. Variety L is called the *recurrent parent* and K is known as the *nonrecurrent parent*. The size of the population required in the first segregating generation is much smaller than for an ordinary cross. In ordinary crosses, the number of important genes in which the parents differ may be ten or more. In a backcross program, the number of genes desired from the nonrecurrent parent may be only one or two, less often several. The method is most useful and most easily used when only one character dependent upon a single gene is to be transferred.

Multiple-cross methods involve mixing the germ plasm of a number of varieties in a simple or complex manner for the purpose of adding together genes not available in any two varieties. The *triple-cross* is frequently used. It consists of crossing a variety P with the F_1 or a later generation segregate of a cross, $K \times L$. It is done to add to $K \times L$ some character or characters desired from P. It achieves the most effective mixing of genes when the further cross is made on the F_1 of the first cross. The prominent rust-resistant wheat variety Apex was produced in this manner.

The *double-cross* is also frequently used. It consists in crossing two simple crosses, such as $K \times L$ and $M \times N$, when they are both F_1, or as later-generation segregates. The cross of the two F_1s is simplest and mixes the genes of the two varieties quite effectively. However, the renowned rust-resistant wheat variety Thatcher was produced from crossing desirable later-generation segregates of two single crosses.

Complex or *successive crossing* is practiced by many breeders. This consists in crossing single, triple, or double crosses among themselves, or with additional parental varieties at any stage from F_1 to F_5, or later. Such a procedure requires careful record keeping and much concentration on the part of the breeder for knowing the material sufficiently well to practice intelligent selection of parents for the successive crosses.

Multiple or *composite crosses* involve elaborate pyramiding of the germ plasm of many varieties into one cross. This has been practiced to a limited extent. The procedure is to select varieties from each of plasm of which one or more characters are desired — then to proceed to make single crosses, then double crosses, then quadruple crosses, etc. A disadvantage of this method is the part that natural selection plays. Plants best able to compete in a mixture may not be the best in pure stands. In some instances, the best competitors are the poorest yielders under field conditions.

It is sometimes necessary to resort to crosses of species of different chromosome numbers in order to attain a desired combination of characters. For example, in the breeding of stem rust-resistant bread wheat, genes for a superior resistance were obtained from *durum* and *dicoccum* varieties and incorporated by hybridization into *vulgare* wheat. These are very wide crosses and require large populations for the experimental programs. Usually a desirable commercial variety cannot be expected as a direct result of a wide interspecific cross. It is generally necessary to make a further

cross of a narrow type, or to backcross to the standard variety parent, in order to obtain a variety that satisfies both the producer and the market.

Crosses involving *artificially changed* germ plasm are sometimes made. Hybridization is used to follow up the creation of chromosome doubling, new chromosome combinations, new gene arrangements, and gene changes, which have been induced by treatment with various chemicals such as colchicine and by irradiation of various kinds.

Plant Introduction

Plant introduction can be defined as the orderly transfer of a cultivated species or variety to a *new place*, following the usual procedures of quarantine, evaluation, multiplication, and distribution. When well organized, plant introduction is a powerful tool for agriculture food production, particularly in the developing countries. Biologically, plant introduction involves the adaptation of germ plasm to a new environment. As adaptation is the key to failure or success, in a program of plant introduction the essential factor is the introduction of as many variants of a species as possible to fit into the conditions of the new habitat.

To introduce only one genotype of a crop, as is sometimes done, is to offer a very narrow margin of adaptability. Most failures in plant introductions are due to insufficient variability in the introduced materials. This is becoming more important as genetic materials are introduced for specific purposes, such as disease resistance, which makes it imperative to test the widest possible range of variability in a crop. The interaction of genetic characteristics and environment as expressed in the physiology of the plant (growth, flowering, fruit setting, etc.) can be foreseen only in wider terms. The effect of temperature and length of day could be assumed rather simply, provided that general guidelines are prepared for the plant introduction program. Another consideration is selecting as many sites as possible for a new introduction. The interaction of the plant and its environment has to be thoroughly tested under different conditions, especially of climate and soil.

The main constraints on free exchange of germ plasm are quarantine regulations. These are established by countries to protect their crops or animals and usually are in a continuous state of change and expansion. To conserve time, all persons engaged in such transfers must observe all governmental rulings from the start. Otherwise, inevitable and costly delays result.

Genetic Resources

Experienced scientists are required to continuously search for previously unfound genetic materials and, through the supervision of the Food and Agriculture Organization (United Nations), a score or more such persons operate in the field throughout the world. Once found, such materials, sometimes of a very early origin, must be preserved not only for experiments scheduled for fairly early use, but also for the long term. Comparatively few scientists until the last couple of decades stressed the great need of finding and preserving germ plasm directly descended from the ancient precursors of modern plant species, to be used for future breeding experimentation. Such materials (plants) still can be found growing in the wild state in out-of-the-way, sparsely inhabited places on earth. But with population pressures coupled with general expansionistic movements of people, such areas are rapidly diminishing and thus genetic resources that could prove invaluable in plant breeding work and genetic research at some future date are rapidly vanishing.

The Russian agronomist and geneticist N.I. Vavilov was one of the first scientists to recognize this problem and to take concrete action toward establishing a national collection of genetic resources in Leningrad. Vavilov directed the All-Union Institute of Plant Industry during the period 1920–1940. In 1969, the institute was renamed to N.I. Vavilov All-Union Institute of Plant Industry.

In the United States, as early as 1819, there was an awareness of the importance of collecting plants materials and storing them, long before the appearance of any genetic concepts. At that time, the Secretary of the Treasury requested consuls representing the United States in other nations to send useful plant materials back to the United States. A section of Seed and Plant Introduction was established by the U.S. Department of Agriculture in 1898. A National Seed Storage Laboratory was constructed in Fort Collins, Colorado in 1958. The principal charge of the laboratory is the long-term storage of seed. Research on the physiology of germination, dormancy, and longevity of seeds is also conducted at this facility.

As early as 1936, H.V. Harland and M.L. Martini of the USDA included the following observation in the *Agriculture Yearbook*: "In the great laboratory of Asia, Europe, and Africa, unguided barley breeding has been going on for thousands of years. Types without number have arisen over an enormous area. The better ones have survived. Many of the surviving types are old. Spikes from Egyptian ruins can often be matched with ones still growing in the basins along the Nile. The Egypt of the Pyramids, however, is probably recent in the history of barley. In the hinterlands of Asia, there were probably barley fields at an earlier time. The progenies of these fields, with all of their surviving variations, constitute the world's priceless reservoir of germ plasm. It has waited through long centuries. Unfortunately, from the breeder's standpoint, it is now being imperiled. When new barleys replace those grown by the farmers of Ethiopia or Tibet, the world will have lost something irreplaceable."

By the end of World War II, what might be called *genetic erosion* had advanced in much of Europe, the United States, Canada, Japan, Australia, and New Zealand. In the early 1960s, more of an organized warning from knowledgeable scientists was heard and, in 1961, the Food and Agriculture Organization (United Nations) convened a technical meeting on plant exploration and introduction. Since that time, a series of actions has aided greatly in protecting what remains of genetic resources. In 1973, the International Board for Plant Genetic Resources was established by the United Nations. See also **Green Revolution**.

NOTE: See also references listed at ends of articles on **Cell (Biology)**; **Genetics and Gene Science**; **Industrial Biotechnology**; **Molecular Biology**; and **Protein**.

Additional Reading

Abelson, P.H.: "Plant Gene Expression Center," *Science*, 1465 (September 27, 1991).
Allard, R.W.: "Principles of Plant Breeding," 2nd Edition, John Wiley & Sons, Inc., New York, NY, 1999.
Beardsley, T.M.: "Doing Agricultural Genetics in the Marketplace," *Sci. Amer.*, 24 (April 1990).
Benfey, P.N. and N.H. Chua: "Regulated Genes in Transgenic Plants," *Science*, 174 (April 14, 1989).
Bennetzen, J., W.F. Blevins, and A.H. Ellingboe: "Cell-Autonomous Recognition of the Rust Pathogen Determines Rpl-Specified Resistance in Maize," *Science*, 208 (July 8, 1988).
Borojevic, S.: "Principles and Methods of Plant Breeding," Elsevier Science, New York, NY, 1990.
Chahal, G.S. and S.S. Gosal: "Principles and Procedures of Plant Breeding: Biotechnological and Conventional Approaches," CRC Press, LLC., Boca Raton, FL, 2002.
Chilton, Mary-Dell: "A Vector for Introducing New Genes into Plants," *Sci. Amer.*, 50–59 (June 1983).
Crawford, M.: "Agricultural Research Service (U.S.)," *Science*, 719 (February 12, 1988).
Crawford, M.: "Plan to Map Crop Genes," *Science*, 1137 (March 3, 1989).
Crawford, R.: "Gene Mapping Japan's Number One Crop," *Science*, 1611 (June 21, 1991).
Dziezak, J.D.: "Biotechnology Enzyme Firm Embraces Innovation," *Food Technology*, 117 (September 1990).
Erickson, D.: "Genetically Engineered Plants Head for the Harvest," *Sci. Amer.*, 81 (May 1990).
Gasser, C.S. and R. T. Fraley: "Genetically Engineering Plants for Crop Improvement," *Science*, 1293 (June 16, 1989).
Gibbons, A.: "Biotechnology Takes Root in the Third World," *Science*, 962 (May 25, 1990).
Hamer, J.E.: "Molecular Probes for Rice Blast Disease," *Science*, 632 (May 3, 1991).
Haring, V., et al.: "Self-Incompatibility: A Self-Recognition System in Plants," *Science*, 937 (November 16, 1990).
Harrington, J.B.: "Cereal Breeding Procedures," FAO Development Paper 28, Food and Agriculture Organization (United Nations), Rome, 1970.
Janick, J.: "Plant Breeding Reviews," Vol. 21, John Wiley & Sons, Inc., New York, NY, 2001.
Jensen, N.F.: "Plant Breeding Methodology," John Wiley & Sons, Inc., New York, NY, 1998.
Kiernan, V.: "appropriate Biotech," *Technology Review (MIT)*, 11 (August/September 1989).
Lea, P.J. and R.C. Leegood: "Plant Biochemistry and Molecular Biology," John Wiley & Sons, Inc., New York, NY, 1994.
Levings, C.S., III: "The Texas Cyctoplasm of Maize: Cytoplasmic Male Sterility and Disease Susceptibility," *Science*, 942 (November 16, 1990).
Lindow, S.W., N.J. Panopoulos, and B.L. McFarland: "Genetic Engineering of Bacteria from Managed and Natural Habitats," *Science*, 1300 (June 16, 1989).
Manuel, J.: "North Carolina Regulates Biotech," *Technology Review* (MIT), 20 (July 1990).
Moffat, A.S.: "Bumpter Transgenic Plant Crop," *Science*, 33 (July 5, 1991).
Moses, P.B. and N.-H. Chua: "Light Switches for Plant Genes," *Sci. Amer.*, 88 (April 1988).
Rhodes, C.A., et al.: "Genetically Transformed Maize Plants from Photoplasts," *Science*, 204 (April 8, 1988).
Richards, A.J.: "Plant Breeding Systems," 2nd Edition, Chapman & Hall, New York, NY, 1997.
van Harten, A.M.: "Mutation Breeding: Theory and Practical Applications," Cambridge University Press, New York, NY, 1998.
Wallace, D.H. and W. Yan: "Plant Breeding and Whole-System Crop Physiology: Improving Adaptation, Maturity and Yield," CAB International, New York, NY, 1998.

PLANT GROWTH MODIFICATION AND REGULATION. In the entry on **Cell (Biology)**, the highly specialized nature of various proteins, enzymes, coenzymes, etc., to effect changes during the life cycle of an animal or plant are pointed out. Those chemical substances having the most to do with plant growth and form are given the general term *plant hormones*. A plant hormone, or *phytohormone*, may be defined as an organic compound produced naturally in plants, which controls growth or other functions at a site remote from its place of production, and which is very active in minute amounts. Three chemically quite different types of compounds apparently act as plant hormones: the *auxins*, the *gibberellins*, and the *kinetins*. In addition, the growth of roots is dependent upon vitamins of the B group which are synthesized in leaves and transported thence to the roots, thus qualifying as hormones.

Auxins. The best-studied hormones are those belonging to the class of auxins. These are defined as organic substances which promote growth along the longitudinal axis, when applied in low concentrations to shoots of plants freed as far as practical from their own inherent growth-promoting substances. Auxins generally have additional properties, but this one is critical.

Natural auxins have been identified in a number of instances. Indole-3-acetaldehyde occurs in a number of etiolated seedlings and in pineapple leaves; indole-3-acetonitrile has been isolated from cabbage and its presence indicated in a number of plants. One of the most widely occurring auxins is indole-3-acetic acid, which has been isolated in pure form from fungi and from corn (maize) grains. Its presence has been conclusively demonstrated by biochemical and chromatographic tests in a wide variety of flowering plants, including mono- and dicotyledons.

Many synthetic auxins have been produced, including 2,4-dichlorophenoxyacetic acid or 2,4-D; naphthalene-1-acetic acid; and 2,3,6-trichlorobenzoic acid, among others. Used as a herbicide, 2,4-D is described in **Herbicide**.

By definition, these synthetic compounds are not hormones, although they are sometimes loosely referred to as hormone-type compounds.

An auxin is formed in fruits, seeds, pollen, root tips, coleoptile tips, young leaves, and especially in developing buds. The auxin travels away from the site of production in shoots by a special transporting system, depending on oxygen, which moves it in a predominantly polar direction from apex toward base. Movement in the opposite direction, i.e., from base toward apex, takes place to a variable extent depending upon the tissue and the plant. In the course of the polar transport, a large part of the auxin becomes bound and is no longer transportable. The transport is rather specifically inhibited by related compounds, particularly 2,3,5-triiodobenzoic acid, 2,4-D, and other synthetic auxins, which are transported either more slowly or to a much lesser extent in the polar system. Auxin applied artificially to intact plants can travel rapidly upward by penetrating into the conducting tissues of the wood, where it is carried upward in the transpiration stream.

In its normal polar, downward movement, the auxin stimulates the cells below the tip to elongate and sometimes to divide. Specific tissues, notably the cambium, are caused to divide laterally by auxin coming from the developing buds, which accounts for the wave of cell division occurring in tree trunks in the spring. Stimulation of other stem cells to divide leads to the production of root initials, which grow out as lateral roots. Cells of the young ovary are commonly caused to multiply and enlarge so that an apparently normal fruit is produced without requiring pollination (*parthenocarpic fruit*). This latter phenomenon indicates that the growing seeds normally secrete an auxin to which enlargement of the fruit is due, a conclusion which has been directly confirmed by bioassay in several fruit types.

Gibberellins can also cause enlargement of fruit. On reaching the lateral buds, however, auxin inhibits their elongation into shoots, and this accounts for *apical dominance*, i.e., suppression of the growth of lateral buds by the terminal bud of a shoot. Auxin also inhibits the falling off of leaves or fruits, which normally occurs when they are mature or aged by the formation of an *abscission layer* of special cells whose walls come apart. That the leaves or fruits do not absciss earlier is due to their steady production of auxin, which prevents formation of these cells. In the root, auxin inhibits elongation except in very low concentrations, but its level therein is usually low. Auxin can be transported for a short distance from the root apex toward the base, but the transport is not fully polar and in the more basal parts of the root the transport is slight.

When the shoot is placed horizontal, auxin is transported toward the lower side, causing accelerated growth there and hence upward curvature (*geotropism*); in the root, this causes decreased growth on the lower side and hence downward curvature. However, in the downward geotropic curvature of roots, other phenomena appear to enter in, and the complexities are not yet fully resolved. When shoots are illuminated from one side, auxins accumulate on the shaded side and, therefore, the plant curves toward the light (*phototropism*). Both geotropic and phototropic auxin movements have been confirmed with carboxyl-^{14}C-labeled compounds. The first observed effect when auxin is applied is the acceleration of the streaming of cytoplasm, but acceleration of growth begins in 7 to 14 minutes at about 23 °C.

In plants that flower on short days, auxin may inhibit flowering; in plants that flower on long days, however, if close to the transition from the vegetative to the flowering state, auxin may promote flowering. In hemp and some of the squashes, auxin modifies the sexuality of the flowers toward femaleness. In the special case of pineapple, auxin directly causes flowering in an unusually clear-cut and quantitative response.

The principal uses of synthetic auxins are to promote the formation of roots on stem cuttings, to prevent abscission, especially of apples and pears, to induce flowering in pineapples, and occasionally to produce seedless fruits. The largest use, however, is that of weed killing. This action depends upon the fact that, at concentrations from 100 to 1000 times those concentrations occurring naturally, the auxins are highly toxic. Monocotyledonous plants, however, are usually resistant. In years past, 2,4-D has been favored in North and South America, whereas 2-methyl-4-chlorophenoxyacetic acid (*methoxone*) has been popular in Europe. However, as of the early 1980s, the regulatory status of these compounds in various countries is under study and may be subject to change.

Some chemically related compounds antagonize the action of auxins, for example, relieving the inhibition of root growth caused by 2,4-D. In contrast, 2,3,5-triiodbenzoic acid synergizes the action.

Gibberellins. These compounds were originally isolated from a parasitic fungus that causes excessive leaf elongation in rice plants. The mechanisms and applications of this group of compounds are described in **Gibberellic Acid** and **Gibberellin Plant Growth Hormones**.

Kinetins. Considerably less is known about this class of compounds. The first one to be discovered, produced by autoclaving yeast nucleic acid, was 6-furfurylaminopurine. Somewhat later, zeatin was isolated from immature corn (maize) kernels. The kinetins promote cytokinesis and protein synthesis, thus causing amino acids to accumulate where kinetins are synthesized (or externally applied) and maintaining the chlorophyll content of yellowing leaves. The kinetins antagonize auxin in apical dominance, releasing lateral buds from inhibition by a terminal bud or by applied auxin. It is believed that through the same mechanism, the kinetins promote the development of buds and leaves on tissue cultures. Their action is primarily local, and if there is transport in vivo, it probably occurs mainly in the transpiration stream (where amino acids are also often found).

Ethylene. The production of ethylene in fruit tissue and in small amounts in leaves may justify its consideration as a hormone, functioning in the gaseous state. Cherimoyas and some varieties of pear produce 1000 times the effective physiological concentration. Ethylene formation is closely linked to oxidation and may be centered in the mitochondria. Its effects are to promote cell-wall softening, starch hydrolysis, and organic acid disappearance in fruits — the syndrome known as *ripening*. Ethylene also decreases the geotropic responses of stems and petioles.

Daminozide Growth Modifier. Daminozide, the chemical name of which is 2,2-dimethylhydrazide, was developed in the early 1960s as a modifying or regulating agent for the growth process of several food plants. The action varies with each plant. For example, on apple, the compound accelerates the start of flower budding, restricts nonproductive vegetative growth, and

assist in fruit drop control. It is also claimed that the compound accelerates fruit coloring and helps to retain the firmness of the fruit. For some of these and other similar reasons, the compound has been used effectively for certain varieties of grape (particularly Concord), for peanuts (groundnuts), for tomatoes, nectarines (except Cherokee), and peaches. Other commercial designations are Alar, B-Nine, Kylar, and Sadh.

Ethephon Growth Modifier. This compound, (2-chloroethyl)phosphonic acid was developed in the United States in the mid-1960s and is used effectively on a number of fruit and vegetable crops for controlling a variety of factors. These include loosening fruit and causing earlier ripening of the fruit (apple, blackberry, blueberry, cherry, cranberry, filbert, tangerine, and walnuts); for encouraging uniform ripening and increasing yield (pepper and tomato); to improve color as well as accelerate maturity (cranberry); and to decrease time required for degreening in citrus fruits, particularly lemon. Other commercial designations for this compound include Cepha, Ethrel, and Florel.

Maleic Hydrazide. This compound, 1,2-dihydro-3,6-pyridazinedione, is also used as a growth regulator, herbicide, and plant modifier. It is used in the treatment of tobacco plants; as a post-harvest sprouting inhibitor; and as a sugar content stabilizer in sugar beets.

PLAQUE. See **Caries, Cariology, and Dentistry**.

PLASMA (Blood). See **Blood**.

PLASMA DISPLAY PANELS. Mechanisms of Operation. The plasma display panel (PDP) can be thought of as a descendant of the neon lamp, which was invented in 1915 by Georges Claude in France. The term *plasma* refers to a gas that consists of electrons, positively charged particles known as anions, and neutral particles. The PDP has sometimes been referred to as a gas-discharge display because it operates by passing electricity through neon gas, causing it to become "charged" temporarily; light is produced when the gas spontaneously discharges. The displays operate at high voltages, low currents, and low temperatures, resulting in long operating lifetimes.

Plasma display panels use glow discharge reactions. Four important reactions can occur in a plasma discharge: ionization, excitation, metastable generation, and Penning ionization. Ionization of the Ne atom generates electrons and neon anions (Ne$^+$) that can cause the generation of an avalanche in the gas. The avalanche begins near the cathode and grows toward the anode as it generates a very large number of electron–neon-ion pairs. The ionization of the gas results in a visible glow. Ions travel back to the cathode, ejecting secondary electrons, which start new avalanches. A feedback loop is set up, and depending on the applied voltage, the current will be maintained at a point either above or below the threshold.

The ions, electrons, and neutral Ne atoms are in constant motion in the plasma, creating a number of chemical reactions that occur in this gaseous phase. One of these reactions, called excitation, causes some of the Ne atoms to transform into an unstable condition known as the excited state, or Ne*, where it remains for only about one billionth of a second before returning to its original form (Ne), the ground state. As it returns to its ground state, the energy that had been absorbed during the excitation process is released in the form of light energy. This light may be visible (orange) or invisible (ultraviolet light) to the naked eye. The ultraviolet light emission can be used in conjunction with color phosphors to produce full-color PDPs.

The process of excitation and light emission takes place so quickly that the eye perceives the plasma as producing orange light continuously, as long as the plasma is activated by the electric field. Figure 1 depicts this operation.

The process of light emission for a monochrome (orange) PDP occurs in the following way Neon gas in a sealed glass envelope receives a high-voltage electrical charge. The gas becomes a mixture of electrons, anions, and neutral atoms; this is the plasma. Some neon atoms are raised to a high-energy, unstable excited state (Ne*). After a nanosecond, Ne* converts to ground-state Ne, resulting in the production of orange light.

The main advantage of the PDP over nearly all other display devices is that it can be made into a display panel, with diagonal sizes of 20 to 60 inches (51 to 152 centimeters) currently in the production, or advanced prototype stages that are not thicker than 4 inches (10 centimeters), including drive electronics. PDPs larger than 50 inches (127 centimeters) are in the production or advanced prototype stages. Moreover, these large

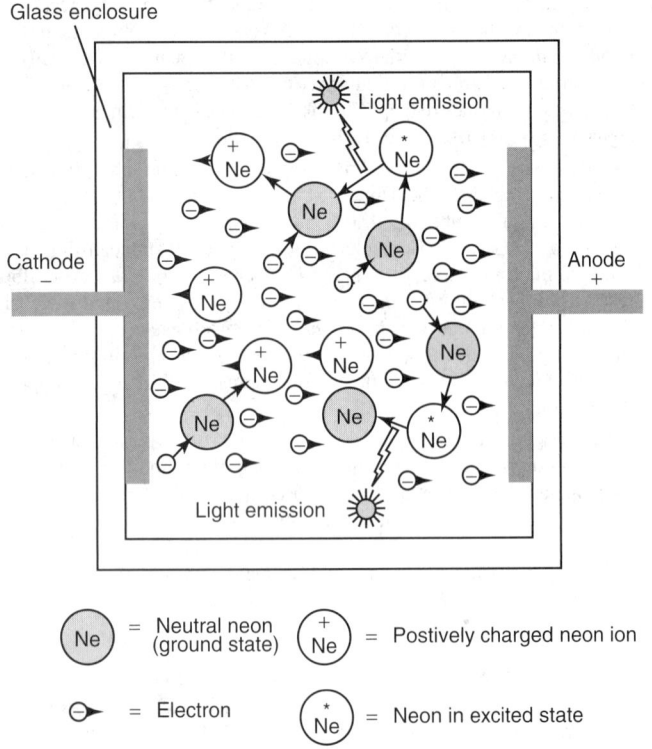

Fig. 1. Light emission processes in plasma display panels.

panels can provide high information content and full-color images. This makes the PDP the ideal technology for the fabled wall-mounted television, which can be used in the home as well as in numerous commercial and industrial environments as a replacement for large, bulky CRT monitors and rear projectors.

PDP Characteristics

The central characteristic of a PDP is the highly nonlinear electro-optic response curve. When the addressing voltage is driven above a certain "threshold" level, the ionization increases dramatically. As in all electro-optic response curves, the threshold allows multiplex addressing; the steepness of the plasma curve allows a large number of lines to be addressed without active matrix elements. Another characteristic of PDPs is that the discharge in a PDP can be switched *on* and *off* in a few microseconds, allowing for a very fast display. This property is proving to be useful in the developing video applications for large PDPs.

Another characteristic of some types of PDPs is inherent memory. The memory is located in the plasma panel itself, rather than in an external device, such as random-access memory. Inherent memory is a curious characteristic. Although at first glance memory might appear to be an advantage in every type of display, it is only helpful when paired with other features, such as easy erasing and low power consumption. It appears to be particularly useful in light-emitting displays. In the case of PDPs, it is very helpful, because it allows a display to be very bright even when the display contains a large number of lines. The duty cycle limitations imposed on multiplexed liquid crystal displays are not seen in PDPs with memory. Essentially, the memory effect means that the display operates with a duty cycle of 1.

There are two main types of PDPs, A.C. and D.C., differentiated by their driving technique and the required structural differences in the panel. Subcategories and hybrid types have evolved, so that now there are numerous variations on PDP technology, including many approaches to providing a full-color display.

Monochrome PDPs are characteristically very long lived. Some estimates put the properly designed A.C.-PDP lifetime at 350,000 hours. Some monochrome PDPs have actually been operating continuously for 20 years. The first color PDPs had to be replaced because they were not as long lived, but progress has been rapid in the past five years, and the lifetime of color PDPs is no longer a major issue.

Because chemically stable rare gases are used exclusively in the panels, PDPs are affected by temperature only to the degree at which driving

and addressing circuits are affected. In the case of D.C.-PDPs with added mercury, shortened lifetimes can be observed if the display is operated at a low temperature for a long period of time.

D.C. Operation

The first type of D.C.-PDP was the famous NIXIE tube. From 1950 to 1965, this display, manufactured by Burroughs, was the major type of electronic digital display used in measuring instruments and other control-oriented applications. The NIXIE tube had a common anode and ten cathodes, each shaped in the form of a digit and contained in a tube that resembled the vacuum tube used in radios and early televisions. The tube was filled with neon, and the selected digit was displayed by applying 100 volts between the appropriate cathode and the common anode.

Although they were crude and unattractive, the NIXIE tubes ushered in the era of the digital display as a replacement for needle point gauges. After the NIXIE tube, segmented and character-type D.C.-PDPs were developed. These displays have been used in millions of pieces of equipment, such as cash registers and ticket machines. The main drawback of these displays was their limitation to a single orange color.

The basic structure of a D.C.-driven PDP cell is shown in Figure 2. The cell is made with inexpensive soda lime glass plates, the same type of glass used in windows, except that it is much flatter and free of surface defects. The back, or rear, plate has a thin, coating of a metal or a transparent conductive coating such as tin oxide or indium–tin oxide (ITO). This coating acts as the cathode or negative electrode. The top, or front, plate also has a conductive coating, but this material must be transparent so that the light emission can be seen. This coating, which is typically tin oxide or ITO, will act as the anode or positive electrode. The two plates are sealed together at 450°C using a glass solder material known as frit to seal the plates, but a small opening (sometimes a tube is used) is left in the seal.

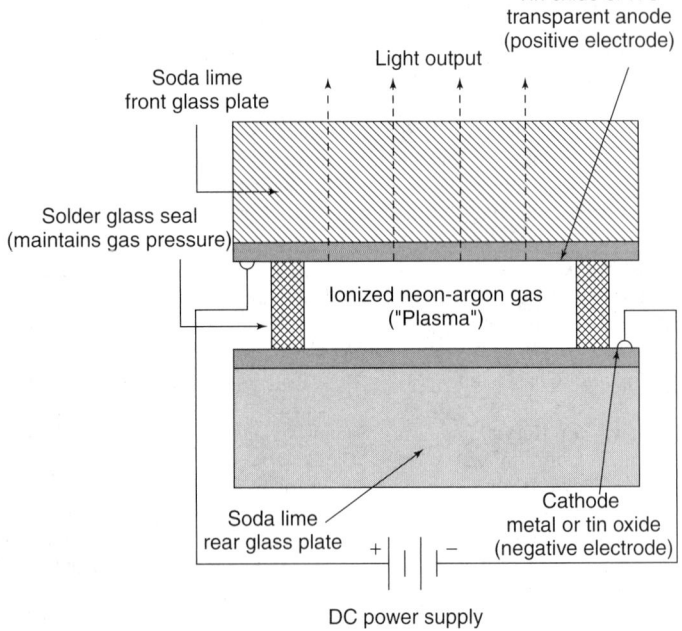

Fig. 2. Basic structure of DC plasma display panel.

The tube is used to remove air from the space between the plates and to refill (or backfill) the space with the neon gas. After filling with the gas, the opening or tube is sealed off. The process of simultaneously sealing and filling with gas is very similar to the process used to manufacture cathode ray tubes (CRTs). The positive terminal of a D.C. power supply is connected to the anode, and the negative terminal to the cathode. When a voltage of 100 to 120 volts is applied from the power supply, light emission occurs. In practice, the construction of a dot matrix or graphic D.C.-PDP is somewhat more complicated. The electrodes are patterned to form a row (horizontal) and column (vertical) arrangement. See Figure 3. The structure includes barrier ribs, which maintain cell separation and isolate the glow to one particular dot. In addition, the electrodes are not coated with an insulator, as is required in A.C.-PDPs, and each column electrode

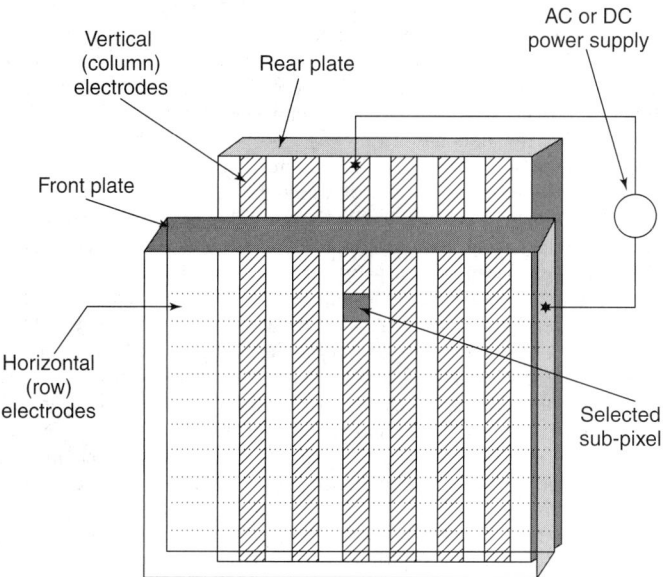

Fig. 3.   Diagram of electrode matrix panel in an AC or DC plasma display panel.

- Low contrast ratio
- Low luminance (lower than A.C.-PDP)
- Complex structure, because ribs adopt cell configurations
- Short lifetime (shorter than A.C.) due to exposed electrodes
- Background glow in some designs

A.C. Operation

A simple alternating current (A.C.) PDP has a construction that is similar to a D.C.-PDP as shown in Figure 4. The cell is also made with inexpensive soda lime glass plates. The back, or rear, plate has a thin coating of a metal or a transparent conductive coating such as tin oxide or ITO, while the top or front plate has a transparent conductive coating of tin oxide or ITO. The important difference from a D.C.-PDP is that the conductive coatings on both plates are covered by an insulating, dielectric film, followed by a coating of magnesium oxide, which acts as a cathode.

typically has a ballast resistor attached to it. The voltage drop across the resistor allows only one discharge to be started along that column at any one time. Mercury is added to a D.C.-PDP to protect the electrodes from sputtering damage. At low temperatures, the mercury vapor condenses to a liquid, leaving the electrodes unprotected.

In operation, data pulses (electrical signals) are supplied to the vertical, or column, electrodes according to the image stored in memory. The scan pulses are supplied to the horizontal, or row, electrodes by scanning the rows sequentially, one at a time. Once ionization occurs, the brightness of the PDP is directly proportional to the current passing through the plasma. The current continues even when the voltage is decreased below the threshold voltage (V_{th}). This is known as the holding, or sustaining, voltage. If the voltage is further decreased to a level known as the extinguishing voltage (V_{ex}), light emission will cease.

There is a memory effect in a D.C.-PDP, realized by the fact that there are two stable states on the current–voltage curve. The threshold voltage (V_{th}) is sharply defined in that no ionization and, hence, no light emission will occur until V_{th} is reached. The extinguishing voltage (V_{ex}) is the level at which the emission stops. Depending on the gas used, its pressure, and the geometry of the cell, V_{th} typically ranges from 65 to 120 volts, and V_{ex} typically ranges from 70% to 90% of V_{th} for D.C.-coupled operation.

To reduce the manufacturing cost and to extend the life of D.C.-PDPs, other complex driving schemes are employed. One commonly used technique, known as priming, creates a faint background glow that essentially reduces the contrast ratio of a D.C.-PDP from 15:1 to about 6:1. Without priming particles, the initiation of discharge can take about 100 microseconds, and some of the isolated pixels tend to flicker off. Higher driving voltages could reduce this time, but then more expensive, higher voltage drivers would be required. Priming can reduce the discharge time by a factor of 10.

Monochrome D.C.-PDPs have evolved as reliable, rugged, and fairly low-cost devices for applications suitable for displays with limited grayscale. The display markets have evolved away from monochrome, however, forcing nearly all manufacturers to drop their monochrome PDPs and concentrate on developing color PDPs. Some developers of color PDPs have made color D.C.-PDP prototypes, but so far only A.C. color PDPs have been developed for mass markets. With the shift away from D.C. plasma technology at Matsushita, there are essentially no developers left working on color D.C.-PDPs.

Advantages of D.C.-PDPs:

- Simplified driving circuitry
- Long lifetime for monochrome displays
- Rise time is relatively shorter in D.C.-PDPs than in A.C.-PDPs

Disadvantages of D.C.-PDPs

- High voltage drivers needed

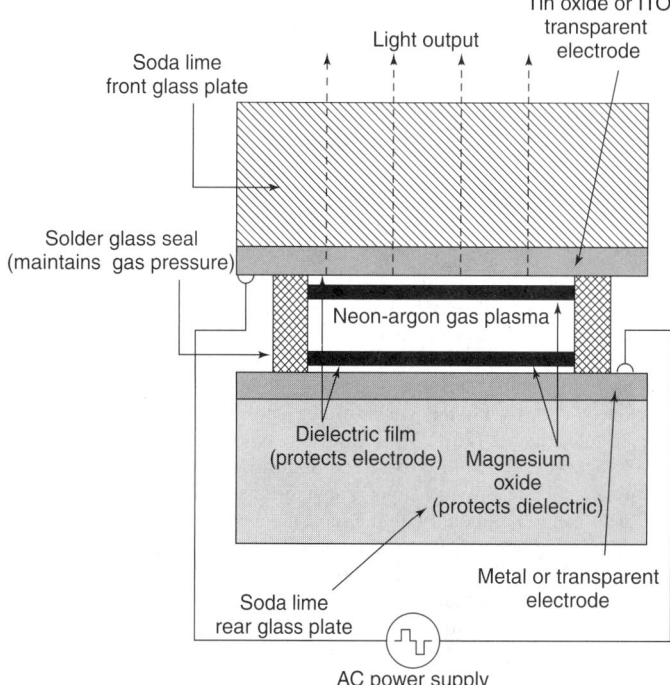

Fig. 4.   Basic structure of an AC plasma display panel.

The purpose of the dielectric film, which is typically a lead oxide glass, is to create a capacitor in series with the cell. The magnesium oxide film protects the dielectric film from damage caused by a bombardment of charged particles in the plasma and reduces the operating voltage of the panel. It has a low (95 V) normal cathode fall, is chemically stable, is transparent to visible light, and has a low sputtering rate. The two plates are sealed together at 450 °C (842 °F) using a frit, leaving a small opening or tube in the seal. The tube is then used to remove air from the space between the plates and to backfill that space with the Ne gas. After filling with the gas, the opening or tube is sealed off. Again, the process of simultaneously sealing and filling with gas is very similar to the process used to manufacture CRTs. One terminal of an A.C. power supply is connected to the electrode on the front plate, and the other terminal is connected to the electrode on the rear plate. When a voltage of about 200 volts is applied from the power supply, light emission occurs.

In practice, the construction of a dot matrix or graphic A.C.-PDP has electrodes that are patterned to form the same row (horizontal) and column (vertical) arrangement. The structure also includes barrier ribs, which maintain cell separation and isolate the glow to one particular dot or pixel. The A.C.-PDPs achieve memory operation by using current limiting capacitors instead of resistors. The thin dielectric (lead oxide glass) coating forms the capacitor over the row and column electrode array; this is protected by a thin film of magnesium oxide. The two thin-film layers and the gas cell are equivalent to three capacitors in series. If a high current source were available, the cell would conduct very high currents. However, the current is only allowed to flow until the capacitive charge is dissipated. The thin-film dielectric coating also protects the electrodes from contamination and erosion from the charged particles in the plasma.

An A.C. sustain voltage, which is a 20-microsecond pulse, is applied to the cell. Since the sustain voltage level is small, no discharge is initiated by the application of the sustain voltage alone. For cells already in the *on* state, the sustain pulse refreshes the wall charge and maintains the status of the cell. When the wall voltage changes, a pulse of light is given off by the gas discharge. For a cell in the *off* state, the sustain voltage is not great enough to initiate a discharge, so it is maintained in an *off* state. Writing and erasing an A.C.-PDP require a complex set of operations. Writing is facilitated when another pulse with a voltage greater than the sustain voltage is applied to the cell. While the write pulse is delivered to the cell after the sustain pulse, an erase pulse must be delivered before the sustain pulse in order to turn a pixel off by bringing the wall voltage to 0 volts.

The long history of experience and the rugged nature of A.C.-PDPs have made them the most popular for fully militarized flat panel display systems. The panels are used in a wide range of systems, from compact battlefield computers used for fire control to 1.5-meter (diagonal) displays used in war rooms. This experience has given the A.C.-PDP a solid reputation as a long-lived, highly reliable display system that can be used in many commercial applications.

Perhaps the most attractive aspect of PDP technology is that the structure of the display lends itself to the use of macroscopic manufacturing techniques, such as screen-printing. Screen-printing is a very old process that is used routinely to print on T-shirts and other materials. It is possible to screen-print substrates up to several feet across, and this is done for relatively coarse designs, such as large circuit boards. Equipment for screen-printing is relatively inexpensive, and the throughput can be quite high. The characteristics of screen-printing are well matched to the manufacturing process of large-screen, color PDPs.

The minimum feature size needed to build a high-resolution display obviously varies with the viewing area and the format of the display. For 10 to 14 inches (25 to 36 centimeters) color displays with 640×480 pixels or more, the ability to screen-print all the necessary structures is marginal at best. But for larger viewing areas, 19 to 42 inches (48 to 107 centimeters) or more (with the same number of pixels), the match with screen-printing becomes more acceptable. Screen-printing areas larger than 42 inches (107 centimeters) becomes more difficult in terms of maintaining uniformity. Displays with a resolution of 66 lines per inch are readily made with thick-film processing, although thick-film processing has a practical limit of about 0.005-inch (0.127 mm) spaces.

The new emphasis on large-area, color PDPs for television has pushed the screen-printing technique to a level such that it is now possible to make displays larger than 40 inches (102 centimeters)(diagonal). The technique uses a screen-printing process to apply thick-film pastes, which act as conductors, resistors, and dielectrics. When greater precision is required, PDPs use a more complex, thin-film deposition technique.

The A.C.-PDPs require a driving voltage of about 200 volts, leading to special problems for the integrated circuit drivers. However, on the positive side, the high voltage drivers are not required to deliver high currents. Due to the construction of the panel, the capacitance of an A.C.-PDP is about 1,000 times lower than it is for a comparable electroluminescent display, which operates at a similar voltage level. Nevertheless, driver circuits can add several hundreds of dollars to the manufacturing cost of an A.C.-PDP.

To produce grayscale in an A.C.-PDP, the approach is to modulate the frame time. The entire display is addressed multiple times in a frame. A given pixel will be "hit" for a certain number of the subframes, depending on the intensity level it requires. For example, in a 16-level grayscale scheme, a full *on* pixel would be addressed in each of the 16 subframes making up a single frame. It is possible to produce 256 gray levels in an A.C.-PDP.

Color A.C. Plasma Display Panels

There are several approaches to achieving color operation in a PDP, but all of them use the ultraviolet light generated by the plasma discharge, rather than the orange glow of the plasma directly. A fluorescent material, such as zinc sulfide or zinc oxide, placed in the vicinity of the discharge converts the ultraviolet light into visible light. This is the same principle employed in the ordinary fluorescent tube lamp used in offices around the world. If the fluorescent material, called a phosphor (it does not contain phosphorous, but the base is composed of zinc oxides or sulfides), is doped with a small

amount of a rare earth or other compound, it can emit light of various colors, depending on the specific compound selected. By using red, green, and blue phosphors, multicolor and even full color (16.777 million colors) can be achieved by forming arrays of these phosphors on the inner surface of the panel. This is similar to the way in which color CRTs for televisions and computer monitors are made.

Controlling the intensities of the red, green, or blue phosphor deposited on the wall of each discharge cell allows full color representation. Rare earth materials (from the group IIIb, Lanthanide Series of elements in the periodic table) are used as activators by most of the high-performance, ultraviolet-light-sensitive phosphor powders. A depletion of activators in the ground state causes the phosphor output to saturate with respect to the ultraviolet light intensity. A short decay time constant is desirable in the phosphor to avoid tailings from moving images. Increasing activator density or reducing the adsorption coefficient can reduce saturation. The National Television Standards Committee (NTSC) has determined three primary colors. It is important that these coincide with products manufactured for television use. It is also important that the intensity balance and the intensity persistence balance coincide with the NTSC standards.

Phosphor deposition can be done several ways. One method uses thick-film printing with careful adjustments of binder contents as well as printing conditions. A sandblasting technique and a photosensitive phosphor paste are other possible phosphor deposition processes. Alternately, a phosphor is deposited on top of a layer; when this is irradiated with UV light, it becomes tacky. The tackiness allows the phosphor powder to adhere to the material. Capsulated color filters have been installed on the entire surface of the PDP to improve color reproducibility.

Fujitsu was one of the pioneers in the development of color PDPs. This company developed and produces the surface-discharge color A.C.-PDP panel with a common electrode structure; it employs a pair of parallel transparent (tin oxide) display electrodes on the front (viewing side) glass plate. These electrodes are coated with a dielectric film of lead oxide glass, followed by a thin layer of magnesium oxide. Meanwhile, the back glass plate has metallic silver address electrodes coated with the red, green, and blue phosphors, each separated by a barrier rib made of lead oxide glass. The barrier ribs keep the gas discharge confined to each red, green, and blue cell so that no crosstalk occurs. After the two glass plates are sealed together, the cells are filled with a neon/xenon gas mixture. The structure is shown in Figure 5.

The surface-discharge color A.C.-PDP works in the following way. When a voltage is applied between the electrodes, a surface-discharge is generated on the magnesium oxide layer, which results in the generation of ultraviolet light from the plasma. When this light strikes the phosphor layer in the cell enclosed by the barrier ribs, the atoms in the phosphor are transformed into an unstable condition known as the excited state. They remain in this state for only about one billionth of a second before returning to their original form known as the ground state. As they return to the ground state, the energy that had been absorbed during the excitation process is released in the form of light energy. This light, which is visible to the naked eye, passes through the front plate to the viewer. By controlling the intensity of the light emitted from the three primary color cells, all the colors of the spectrum can, in principle, be produced.

These color A.C.-PDPs have a reduced load to capacitance between adjacent cells so the pixel pitch can be shortened. Optical crosstalk is avoided by using a dielectric barrier rib of the appropriate height. The common electrode configuration provides a simpler row and column electrode arrangement for using the thick-film-through-hole method of printing. Thus, this panel is relatively easy to make in large sizes because of its simple structure, which is entirely fabricated by screen-printing technology. The key steps in the process are as follows:

(1) Pair of electrodes on front plate are coated with dielectric lead oxide glass and magnesium oxide.
(2) Silver electrode on back plate is coated with color phosphor.
(3) Barrier ribs keep discharge confined to each primary color.
(4) UV light is generated from surface discharge.
(5) UV light strikes color phosphor to produce colored light.

The use of three primary color phosphors (red, green, and blue) gives all colors of the spectrum, with appropriate scanning of the address and surface-discharge electrodes.

The most recent development in PDP design is Fujitsu's alternate lighting of surfaces, or ALiS, which uses an interlace approach to driving

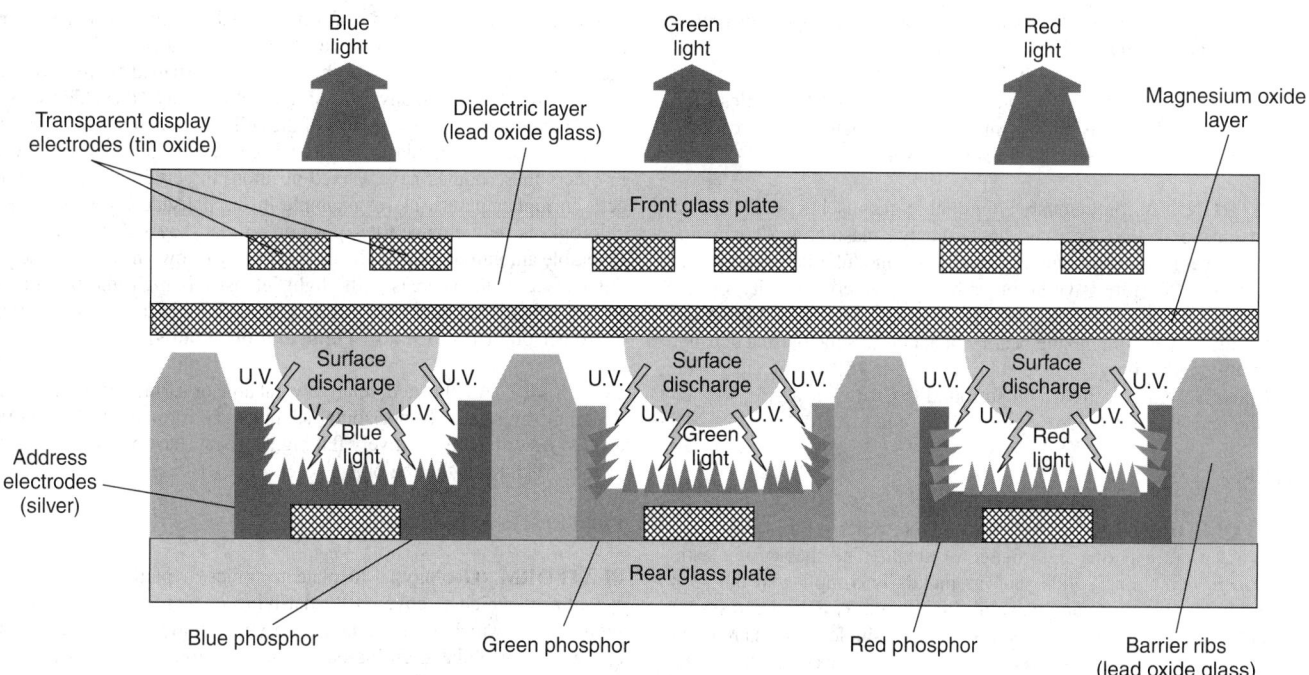

Fig. 5.   Structure of Fujitsu's color PDP using surface discharge (*Courtesy of Fujitsu Microelectronics.*)

the PDP pixels, effectively doubling the number of scan lines for the same number of electrodes. Rather than using dedicated pairs of electrodes to address a scan line, ALiS uses each adjacent pair of electrodes, scanned in interlace format.

For video and noncomputer applications, pixel size increases for a PDP to dimensions compatible with screen-printing, especially as the screen measurements increase above 36 inches. The pixel size for a 36-inch-wide picture with 640 horizontal pixels is about 1 millimeter. The key is that the screen is not meant to be viewed from a distance of 2 feet (0.61 meter), but rather from 8 to 12 feet (2.44 to 3.66 meters).

Photometric issues are quite detailed. Brightness is determined by the luminance of the display, which is the measure of luminous intensity. A typical desktop CRT monitor operates at a level of about 150 nits (1 nit $= 1$ cd/m^2). This is bright enough to provide a pleasant-looking display in a typical office environment. It would not be bright enough for viewing in bright sunlight, nor would a consumer be happy with this luminance level on a home television set, which typically has a luminance of at least three times this value.

The contrast ratio is a comparison of the luminance of a pixel in the on state to one in the *off* state; the higher the contrast ratio, the better. The current 42-inch (107 centimeter) PDP specifications refer to contrast ratio as 400:1 peak under dark room conditions. A 50-inch (127 centimeter) PDP with a contrast of 300:1 for dark areas and 120:1 for the same display in a bright area has been announced. Generally, references to contrast by specifications must be interpreted as referring to contrast in dark settings unless otherwise specified.

The current contrast is adequate for use in the home television market, but contrast improvement needs to be made for use in high ambient light conditions. Contrast could be improved first by increasing emission intensity from the display discharges and then by minimizing background light by either reducing emissions from control discharges or by hiding auxiliary discharge emissions. The driving method has been improved the frequency of the pilot discharge and background emissions has been cut, and the black display luminance has been lowered. Another way to improve contrast would be to use a neutral density filter or a circular polarizing filter that would reduce ambient light. These filters also reduce output luminance, requiring a higher phosphor output. However, by selecting a red-transmitting filter in front of a red cell and a green-transmitting filter in front of a green cell, this effect is reduced by a factor of three.

The saturation of light output and reduction of efficiency as the discharge current is increased are significant issues with PDPs. These occur because neutral Xe atoms easily absorb photons radiated from excited Xe atoms. An increase in the number of photons and the frequency of the sustain pulse reduces efficiency.

Increasing resolution in a display of the same size means reducing the size of each discharge cell. This in turn increases diffusion losses of charged particles to the cell walls. To compensate for this loss, increased ionization is needed. In the normal glow discharge mode, only one in forty electrons is effective for ionization. Because there are many unnecessary excitation levels, the excitation of atoms to a desired level is already a loss-provoking process. Adjustments to electron temperatures in plasmas by a mode different than the glow discharge method allow more efficient ionization and excitation of gases. If size can be increased, these issues will be resolved. See also **Television (TV)**, **Computer Graphics**, and the family of articles catalogued under **Flat Panel Display Technology**.

For additional reading, refer to Flat Panel Display Technology entry.

Stanford Resources, Inc., San Jose, CA.

PLASMA FREQUENCY.   The oscillation frequency of plasma electrons about an equilibrium charge distribution is called the plasma or Langmuir frequency and is

$$\omega_p = \sqrt{\frac{4\pi n_e q^2}{m}} = 5.7 \times 10^5 n_e^{1/2} \text{(radians/sec)}$$

where n_e is the electron density, m is the electron mass, and q is the electronic charge in esu. Its value is

$$(5.7 \times 10^{-5} \text{ radian cm}^{-3/2} \text{ sec}^{-1}) n_e^{1/2}$$

PLASMA OSCILLATIONS.   A plasma is capable of supporting various modes of vibration and can propagate several types of waves. These can be classified in terms of their oscillation frequencies. The study of plasma oscillations is the general branch in plasma physics that includes the classification of the various modes of vibration that a plasma is capable of supporting and of the several types of waves that can be propagated in a plasma. Usually these collective effects are classified by their oscillation frequency.

PLASMA (Particle).   1. An assembly of ions, electrons, neutral atoms and molecules in which the motion of the particles is dominated by electromagnetic interactions. This condition occurs when the macroscopic electrostatic shielding distance (Debye length) is small compared to the dimensions of the plasma. Because of the large electrostatic potentials which would result from an inhomogeneous distribution of unlike charges, a plasma is effectively neutral. Thus there are equal numbers of positive and negative charges in every macroscopic volume of a plasma. Also, because a plasma is a conductor, it interacts with electromagnetic fields. The study of these interactions is called *hydromagnetics* or *magnetohydrodynamics*. See also **Magnetohydrodynamic Generator**.

2. A collection of electrons and ions, usually at a high enough temperature so that the ionization level is about 5% and at densities such that the Debye shielding distance is much smaller than the macroscopic dimensions of the system. See also **Fusion Power**.

PLASMATRON. A continuously controllable gas-discharge tube which utilizes an independently generated gas-discharge plasma as a conductor between a hot cathode and an anode. Continuous modulation of the anode current can be effected by variation either of the conductivity or the effective cross-section of the plasma. The first method is based upon the modulation of the electronionizing beam which controls the plasma density, and hence its conductivity. The second method makes use of the gating action of positive-ion sheaths which surround the wires of a grid, located between the anode and cathode.

PLASTER. See **Gypsum**.

PLASTIC DEFORMATION. When a metal or other solid is plastically deformed it suffers a permanent change of shape. The theory of plastic deformation in crystalline solids such as metals is complicated but well advanced. Metals are unique among solids in their ability to undergo severe plastic deformation. The observed yield stresses of single crystals are often 10^{-4} times smaller than the theoretical strengths of perfect crystals. The fact that actual metal crystals are so easily deformed has been attributed to the presence of lattice defects inside the crystals. The most important type of defect is the dislocation. See also **Creep (Metals)**; **Crystal**; **Hot Working**; and **Impact**.

PLASTICS. These are materials formed from resins through the application of heat, pressure, or both. Most starting materials prior to the final fabrication of plastic products exhibit more or less plasticity — hence the term plastic. However, the great majority of plastic end-products are quite *nonplastic*, i.e., they are nonflowing, relatively stable dimensionally, and are hard. There are scores of different kinds of plastics. They fall into two broad categories: (1) *thermoplastic resins*, which can be heated and softened innumerable times without suffering any basic alteration in characteristics; and (2) *thermosetting resins*, which once set at a temperature critical to a given material cannot be resoftened and reworked. Since most plastic fabrication methods, such as casting, molding, or extruding, involve heat, the thermosetting materials must be properly and accurately formed during any thermal cycling that exceeds the critical temperature.

The principal kinds of thermoplastic resins include: (1) acrylonitrile-butadiene-styrene (ABS) resins; (2) acetals; (3) acrylics; (4) cellulosics; (5) chlorinated polyethers; (6) fluorocarbons, such as polytetra-fluorethylene (TFE), polychlorotrifluoroethylene (CTFE), and fluorinated ethylene propylene (FEP); (7) nylons (polyamides); (8) polycarbonates; (9) polyethylenes (including copolymers); (10) polypropylenes (including copolymers); (11) polystyrenes; and (12) vinyls (polyvinyl chloride). The principal kinds of thermosetting resins include: (1) alkyds; (2) allylics; (3) the aminos (melamine and urea); (4) epoxies; (5) phenolics; (6) polyesters; (7) silicones; and (8) urethanes.

The relatively recent development of electrically conductive plastics is reviewed by R.B. Kaner and A.G. MacDiarmid (*Sci. Amer.*, 106–111, February 1988).

Numerous plastics are described in terms of manufacture and chemical and physical properties throughout this volume. Consult the alphabetical index.

PLASTIDS. Pigments in plants are often located in special bodies called plastids. There are many kinds of plastids: leucoplasts, those which contain no pigment and which are therefore colorless; chloroplasts, those which contain chlorophyll (by far the commonest kind); and chromoplasts, colored plastids that do not contain chlorophyll.

Leucoplasts occur in parts of stems and roots where light fails to penetrate. They absorb glucose and change it to starch.

Chloroplasts occur in cells exposed to light. They are indispensable to photosynthesis. In the algae the shapes of these bodies are many; in a large number of cases the plastid is a thick cup-shaped body occupying the greater part of the volume of the cell; in other algae the plastids have a central mass from which radiating plates of arms extend outward to the cell wall; spiral, net-shaped and ring-shaped plastids are not uncommon in this group of plants. In some algae and in nearly all higher plants, the chloroplasts are small subspherical or lens-shaped bodies, varying in number from one to many in a single cell. Always the chloroplasts are found embedded in the cytoplasm of the cell. In many plants the continuous movement of the cytoplasm in the cell carries the plastids along with it; in others these bodies have a fixed position. In certain algae and in many cells in higher plants, as for example in the palisade layer of leaves, the chloroplasts may change their position so that they will receive the most favorable amount of light. If the light intensity is low, they will present their flat surface to it; whereas if the light intensity is high, the plastid rotates so that it is placed edgewise to the light. Chloroplasts contain chlorophyll and other pigments. See also **Pigmentation (Plants)**.

PLATEAU. An elevated, relatively flat area or surface of wide extent and underlain by relatively horizontal sedimentary formations. Less correctly used to describe any relatively flat high-level surface of erosion regardless of the structure of the region.

PLATELETS. See **Blood**.

PLATFORM (Geology). In plate tectonics, a platform is that part of a continent which is covered by flat-lying or gently tilted strata, mainly sedimentary, which are underlain at varying depth by a basement of rocks that were consolidated during earlier deformations. Platforms are parts of the cratons. See also **Craton**. In geomorphology, a platform is any level or nearly level surface, e.g., a terrace or bench, a ledge or small space on a cliff face, or a flat and elevated piece of ground, such as a tableland or plateau, a peneplain, or any beveled surface. The term also sometimes indicates a small plateau. In discussions of a coastline, a platform is a flat or gently sloping underwater erosional surface extending seaward or lakeward from the shore. Specific types include a *wave-cut platform* or an *abrasion platform*.

A *platform beach* is a looped bar or dike of sand and gravel formed on a wave-cut platform. These are found, for example, on Madeline Island along the Wisconsin shore of Lake Superior. A *platform reef* is an organic reef, generally small but more extensive than a patch reef, with a flat upper surface. Platform reefs are common off the coast of Australia; they are also called *table reef*.

See also **Ocean**.

PLATINUM AND PLATINUM GROUP. Chemical element, symbol Pt, at. no. 78, at. wt. 195.09, periodic table group 10 (formerly 8 — transition metals), mp 1,772°C, bp 3,727 to 3,927°C, density 21.37 g/cm^3 (solid), 21.5 g/cm^3 (single crystal) (20°C). Elemental platinum has a face-centered cubic crystal structure. The five stable isotopes of platinum are ^{192}Pt, ^{194}Pt through ^{196}Pt, and ^{198}Pt. The seven unstable isotopes are ^{188}Pt through ^{191}Pt, ^{193}Pt, ^{197}Pt, and ^{199}Pt. In terms of earthly abundance, platinum is one of the scarce elements. Also, in terms of cosmic abundance, the investigations by Harold C. Urey (1952), using a figure of 10,000 for silicon, estimated the figure for platinum at 0.016. No notable presence of platinum in seawater has been found.

Electronic configuration

$$1s^2 2s^2 2p^6 3s^2 3p^6 3d^{10} 4s^2 4p^6 4d^{10} 4f^{14} 5s^2 5p^6 5d^9 6s^1$$

Ionic radius Pt^{2+} 0.52 Å. Metallic radius 1.3873 Å. First ionization potential 8.96 eV. Oxidation potentials Pt → Pt^{2+} + 2e$^-$, ca. −1.2 V; Pt + 2OH$^-$ → Pt(OH)$_2$ + 2e$^-$, −0.16 V.

Platinum is one member of a family of six elements, called the *platinum metals*, which almost always occur together. Before the discovery of the sister elements, the term *platinum* was applied to an alloy with Pt as the dominant metal, a practice that persists to some degree even today. The major properties of the platinum metals are given in Table 1. See also **Iridium**; **Osmium**; **Palladium**; **Rhodium**; and **Ruthenium**.

Occurrence. These metals occur in both primary and secondary deposits. The primary deposits are generally associated with Ni-Cu sulfide ores. The Sudbury ores of Canada and the deposits of the Bushveld complex of South Africa are of this type. Native platinum occurs as a primary deposit in the Ural Mountains of the former U.S.S.R. and also in the Choco district of Colombia. Weathering and erosion of these deposits have resulted in the formation of secondary, or placer, deposits of native Pt in riverbeds and streams. One nugget of Pt found in the Urals weighed over

TABLE 1. REPRESENTATIVE PROPERTIES OF PLATINUM GROUP METALS

Property	Iridium	Osmium	Palladium	Platinum	Rhodium	Ruthenium
Atomic volume, cm^3/g-atom	8.54	8.43	8.88	9.09	8.27	8.29
Atomic radius	1.355	1.350	1.373	1.335	1.342	1.336
Crystalline form	fcc	hcp	fcc	fcc	fcc	hcp
Lattice parameters, Å, a	3.8389	2.7341	3.8902	3.9310	3.804	2.7041
b	—	4.3197	—	—	—	4.2814
Thermal conductivity at 20°C, (cal)(cm)/(s)(cm^3)(°C)	0.14	—	0.168	0.166	0.21	—
Electrical resistivity at 0°C, micro-ohm-cm	5.3	9.5	10.8	10.6	4.5	7.2
Thermal expansivity, °C × 10^6 at 20°C	6.6	6.6	12.4	9.0	8.3	9.6
Hardness, Mohs scale	6.5	7.0	4.8	4.3	—	6.5
Specific heat, cal/g-atom at 20°C	0.031	0.031	0.0584	0.031	0.059	0.057
Heat of fusion, kcal/mole	6.3	7.0	4.0	4.7	5.2	6.1
Heat of vaporization, kcal/mole	134.7	150	90	122	118.4	135.7

fcc = face-centered cubic
hcp = hexagonal close-packed

25 pounds (11.3 kilograms). Most of the world's platinum comes from Canada, the former U.S.S.R., and South Africa. Minor amounts have been found in Alaska, Colombia, Ethiopia, Japan, Australia, and Sierra Leone.

Because of their unique properties and in spite of their high initial cost, the platinum metals find many applications in industry. Since used platinum metals retain a large portion of their initial value, many scrap materials are a major source of recoverable platinum metals. Practically every application of platinum generates scrap in some form, which is eventually returned to the platinum refiner for recycling. Although there are ample mine reserves, they soon would be depleted without constant scrap recycling.

Refining Processes. The refining procedures are a good introduction to the complex chemistry of the platinum metals. Some of these methods are still the best analytical techniques available for the separation of the metals. South African ore is smelted to form a Cu-Ni matte containing small amounts of the platinum metals (0.18%). The matte is melted, cast into anodes, and electrolytically dissolved. The contained Cu is deposited at the cathode, the Ni remains in the H_2SO_4 electrolyte, and the Pt metals are contained in the anode slimes. The resulting Cu is refined and the $NiSO_4$ solution purified and crystallized. The anode slimes are treated by roasting to remove sulfur and leached with dilute H_2SO_4 and air to remove Cu and Ni. The leached slimes are treated with aqua regia. The aqua regia solution is evaporated to concentrate the solutions and expel the excess HNO_3. The residue from this treatment contains Rh, Ir, Ru, Os, and Ag. The solution contains Pt, Pd, and Au.

Platinum is first removed by precipitating as ammonium hexachloroplatinate(IV) [$(NH_4)_2PtCl_6$] by the addition of a saturated solution of NH_4Cl. The precipitate is washed, dried, and calcined to form platinum sponge about 98% pure. The sponge is purified by redissolving in aqua regia and evaporating the solution to dryness with NaCl. The resulting sodium hexachloroplatinate is dissolved in H_2O and boiled with $NaBrO_3$ to convert impurities, such as Ir, Rh, Pd, and base metals, to valence states which produce readily filterable hydroxides. The Pt left in solution is free of impurities. It is then treated with NH_4Cl, and the pure ammonium hexachloroplatinate precipitate is calcined at 1000°C to pure Pt sponge.

The first aqua regia solution is treated with $FeSO_4$ to precipitate the gold. Pd is precipitated by oxidizing the solution with HNO_3 and adding NH_4Cl. Ammonium hexachloropalladate(IV) is formed (analogous to the Pt compound). This salt is purified by dissolving in NH_4OH, filtering off the impurities, and reprecipitating the Pd by the addition of HCl. The insoluble complex $Pd(NH_3)_2Cl_2$ is formed, which when calcined and reduced in H_2 yields pure Pd sponge.

The insolubles from the first aqua regia treatment are fused with a flux of litharge, soda ash, borax, and carbon in a gas-fired furnace at 1000°C for 1 hour. This procedure converts silica, alumina, and some base metals to slag. The precious metals are retained in the lead phase. The lead portion is heated with HNO_3, which dissolves the Pb and Ag. The Pb is precipitated as a sulfate and then the Ag as a chloride. The residue is treated with concentrated H_2SO_4 at 300°C. Rh will dissolve, leaving Ir, Ru, and Os as insolubles. The Rh solution is treated with Zn powder, precipitating

an impure Rh. The impure Rh is heated in an atmosphere of Cl_2. Many impurities form volatile chlorides at this temperature and are expelled. Rh forms a polymeric trichloride, which is insoluble in aqua regia. The rhodium trichloride is digested in aqua regia for several hours, then filtered, dried, and calcined, yielding a commercial grade of Rh sponge.

The residue insoluble in H_2SO_4 is fused with Na_2O_2, poured into thin slabs and cooled. Ir is oxidized in the fusion of IrO_2, which is insoluble in H_2O. Ru and Os form soluble sodium salts and are separated from the Ir by filtration. The insoluble IrO_2 is dissolved in aqua regia, and ammonium hexachloroiridate(IV) is precipitated by the addition of NH_4Cl. Calcining yields pure Ir sponge.

The filtrate from the dissolution of the Na_2O_2 fusion contains $NaRuO_4$ and $NaOsO_4$. Ethyl alcohol is added to the solution, causing the precipitation of RuO_2, which is separated by filtration.

The Ru is purified by distilling with Cl_2. Volatile ruthenium tetroxide is collected. A saturated solution of NH_4Cl is added, causing the precipitation of ammonium hexachlororuthenate(III). The precipitated salt is calcined in H_2, yielding commercial Ru sponge.

The filtrate from the alcohol precipitation of Ru contains the Os. The solution is neutralized with HCl and is treated with powdered Zn, reducing the Os to the metallic state. Osmium tetroxide is formed by roasting the impure Zn in a current of O_2. The volatile OsO_4 is trapped in an aqueous solution of KOH. Ethanol is added to the solution, precipitating potassium osmate(VI), which is mixed with an excess of NH_4Cl and calcined in an atmosphere of H_2. The resulting Os sponge is leached to remove KCl, leaving a commercial-grade Os sponge.

The refining of secondary scrap follows much the same procedures with minor variations. For example, solid metallic Pt and especially the Rh and Ir alloys of Pt are very difficult to dissolve in aqua regia. Therefore the scrap generally is alloyed with Cu, Ni, Pb, or Zn before dissolution with acids.

Uses. These metals, in various forms, currently are used as catalysts for a wide variety of reactions. Products include high-octane gasoline, HNO_3, H_2SO_4, HCN, vitamins, antibiotics, H_2O_2, cortisone, alkaloids, and fuel-cell chemicals. These catalysts also are used to remove trace impurities, e.g., acetylene in ethylene or O_2 in H_2, or noxious constituents of partial combustion, e.g., automobile exhausts. Although substitutes are being sought, Pt is by far the best catalyst for pollution control of auto exhausts. In the future, the catalytic converters currently installed in automobiles will become a significant source of platinum metals. See also **Catalysis**.

The corrosion resistance of the Pt metals has made the Pt crucible and the Pt electrodes commonplace laboratory tools. The glass industry makes use of large amounts of Pt and its alloys for manufacturing very pure glass. Synthetic fibers often are extruded through spinnerettes made of Pt alloys. The large use of Pt metals in dental and medical devices, in jewelry, and for decorative purposes is based on the corrosion resistance and general appearance of these metals.

Because of their high melting points and stability, Pt alloys have found applications in thermocouples, resistance thermometers, potentiometer windings, electrodes, insoluble anodes, high-temperature furnace winding, crucibles that can withstand corrosive materials at high temperature, and generally as materials of construction that will not contaminate products at very high temperatures. Often Pt and Pd are alloyed with Rh, Ir, Ru, or Os to increase their strength, hardness, and corrosion resistance.

Platinum metals, in particular Pd, find extensive use in the electrical industry. Most of these metals are used as contacts, particularly in telephone relays, where their resistance to oxidation and sulfidization results in circuits of reliability and stability. Alloys of Pt find use as grids for electronic tubes, in electrodes for aircraft spark plugs, for contact metal in printed and solid-state circuits, and in pressure-rupture disks.

In the medical field, *cis*-dichlorodiammineplatinum(II) has been available for cancer therapy (*Science*, **192**, 774–775, 1976).

Platinum Compounds. Platinum forms many di- and tetravalent compounds. The latter valence is more common and more stable. Pt in compact form is inert to all mineral acids except aqua regia. Under oxidizing conditions, fused alkalies will attack Pt to some extent. Molten halides, carbonates, and sulfates have little effect on the metal. Concentrated boiling H_2SO_4, fused cyanides, and fused alkaline sulfides will attack the finely divided metal. Pt is vigorously attacked by Cl_2 at elevated temperatures. In hot aqua regia or HCl containing chlorate ion or H_2O_2, the metal slowly dissolves, yielding a solution of hexachloroplatinic acid, H_2PtCl_6.

Platinum(II) hydroxide is made by adding KOH to a solution of platinum(II) chloride. The unstable black powder is easily oxidized by air and must therefore be handled in an inert atmosphere. In hot alkali or HCl, it disproportionates into the platinum(IV) compound and the metal. Very careful dehydration results in the formation of a gray powder that approaches the composition of platinum(II) oxide. Platinum(II) oxide can also be made by combining the elements at 420–440 °C at an O_2 pressure of 8 atm.

When a solution of hexachloroplatinic(IV) acid is boiled for some time with NaOH, all the chloride ions are replaced by hydroxide ions. The resulting sodium hexahydroxoplatinate(IV), $Na_2Pt(OH)_6$, is soluble in the basic solution, but it can be precipitated as hexahydroxoplatinic(IV) acid, $H_2Pt(OH)_6$, by the addition of acetic acid. The hydroxide ions of the salt are replaced by the corresponding ions of mineral acids when the compound is dissolved in acid. Hexahydroxoplatinic acid can be dehydrated to yield compounds corresponding to the tri-, di-, and monohydrate of platinum(IV) oxide. The last water molecule cannot be removed without some destruction of the dioxide.

Brown-black, insoluble, anhydrous, platinum(IV) oxide is made by fusing hexachloroplatinic(IV) acid with $NaNO_3$ at about 500 °C. The alkali salts are washed out with H_2O to free the fine insoluble residue of platinum(IV) oxide. This compound is known as *Adam's catalyst*.

When Pt is heated to 500 °C in the presence of Cl_2, yellow-green, insoluble platinum(II) chloride is formed. At a pressure of 1 atm of Cl_2, the compound is stable from 435 to 581 °C. It can also be made by heating hexachloroplatinic(IV) acid in Cl_2 at about 500 °C. Platinum(II) chloride is soluble in HCl as tetrachloroplatinic(II) acid. It forms many salts that are water-soluble. These salts can be made by reducing a hot solution of the corresponding hexachloroplatinate(IV) with oxalic acid or SO_2. Platinum(III) chloride has a narrow range of stability. It can be made by contacting Pt or a platinum chloride with 1 atm of Cl_2 at 364–374 °C. This dark-green to black compound is practically insoluble in cold concentrated HCl but does dissolve on warming, forming a mixture of tetrachloroplatinic(II) and hexachloroplatinic(IV) acids. Anhydrous platinum(IV) chloride is very difficult to prepare. This brown soluble solid can be made by heating hexachloroplatinic(IV) acid in Cl_2 at 360 °C. The most common Pt compound, hexachloroplatinic(IV) acid, is readily made by dissolving Pt in aqua regia, followed by several evaporations with additional HCl to destroy nitrosyl compounds. The acid crystallizes as a hexahydrate. It is difficult to stop the evaporation at just this point, and slight local overheating causes excess loss of water. The sodium salt is quite soluble, and the compound is resistant to hydrolysis in basic solution, allowing the bromate hydrolysis to precipitate base metals and other Pt metals as their hydroxides. The Pt remains in solution. The insolubility of ammonium hexachloroplatinate(IV) often is used in refining Pt. Its slight solubility can be overcome sufficiently by mass action to allow its use as a gravimetric procedure for the determination of Pt. This yellow compound decomposes at red heat, yielding pure Pt sponge. The insolubility of the potassium salt is used for the gravimetric determination of potassium.

A series of di-, tri-, and tetrabromides is well known. Platinum(II) iodide is precipitated as a black insoluble compound by the addition of 2 equiv of iodide to a hot solution of platinum(II) chloride. The black, insoluble, graphitelike substance, platinum(III) iodide, is made by combining the elements in a sealed tube at 350 °C.

In contrast with Pd, Pt does form a Pt(IV) iodide. When a concentrated solution of hexachloroplatinic(IV) acid is treated with a hot solution of KI, this brown-black substance is precipitated. The compound is somewhat unstable and light-sensitive. It dissolves in excess KI to form the complex salt, also rather unstable.

Pt forms a nonvolatile tetrafluoride, a pentafluoride, and a volatile hexafluoride. The dark-red PtF_6 melts at 56.7 °C and is very reactive. It even reacts with O_2 at 21 °C to form dioxygenylhexafluoroplatinate(V), O_2PtF_6.

When sulfur and Pt sponge are ignited, some platinum(II) sulfide is formed. The naturally occurring mineral is called cooperite. When heated in air or H_2, the products are metallic Pt and S. Platinum(IV) sulfide can be made by heating ammonium hexachloroplatinate(IV) or Pt and S at 650 °C. When precipitated by H_2S from chloroplatinic acid, the compound may exist as $PtS_2 \cdot H_2S$.

Divalent and tetravalent Pt probably form as many complexes as any other metal. The platinum(II) complexes are numerous with N_2, S, halogens, and C. The tetranitritoplatinum complexes are soluble in basic solution. Tetranitritoplatinum(II) ion is formed when a solution of platinum(II) chloride is boiled, at about neutral pH, with an excess of $NaNO_3$. The ammonium salt may explode when heated. Generally, platinum-metal nitrites should be destroyed in solution. They never should be heated in the dry form. Platinum(II) complexes most often have a coordination number of 4. Many compounds have been prepared with olefins, cyanides, nitriles, halides, isonitriles, amines, phosphines, arsines, and nitro compounds.

Platinum(IV) has a coordination number of 6. It forms complexes with halides, nitrogen and sulfur compounds, and other donors but to a lesser extent than platinum(II).

Additional Reading

Carter, G.F. and D.E. Paul: "Materials Science and Engineering," ASM International, Materials Park, OH, 1991.

Greenwood, N.N. and A. Earnshaw: "Chemistry of the Elements," 2nd Edition, Butterworth-Heinemann, Inc., Woburn, MA, 1997.

Krebs, R.E.: "The History and Use of Our Earth's Chemical Elements: A Reference Guide," Greenwood Publishing Group, Inc., Westport, CT, 1998.

Lide, D.R.: "CRC Handbook of Chemistry and Physics 2000–2001," 81st Edition, CRC Press, LLC, Boca Raton, FL, 2000.

Meyers, R.A.: "Handbook of Chemicals Production," The McGraw-Hill Companies, Inc., New York, NY, 1986.

Schweitzer, P.A.: "Corrosion Resistance Tables," 3rd Edition, ASM International, Materials Park, OH, 1991.

Parker, P.: "McGraw-Hill Encyclopedia of Chemistry," 2nd Edition, The McGraw-Hill Companies, Inc., New York, NY, 1993.

Staff: "ASM Handbook—Properties and Selection: Nonferrous Alloys and Pure Metals," ASM International, Materials Park, OH, 1990.

Stwertka, A. and E. Stwertka: "A Guide to the Elements," Oxford University Press, Inc., New York, NY, 1998.

LINTON LIBBY, Chief Chemist, Simmons Refining Company, Chicago, IL.

PLATYHELMINTHES. The flatworms, a major division of the animal kingdom containing the most primitive of the triploblastic *Metazoa*. The phylum includes both free-living and parasitic species. Among the latter are the flukes and tapeworms, some of which are serious parasites of humans.

The phylum is characterized by the following details of structure: (1) The body is bilaterally symmetrical and flattened. (2) The ectoderm is ciliated in free-living forms but forms a cutical in the parasitic species. (3) The mesoderm forms a compact tissue between the various organs, known as a parenchyma. (4) The alimentary tract, when present, has a single opening. (5) The nervous system is a network in which a brain and longitudinal nerve cords are developed. (6) The excretory system consists of large hollow cells with a group of cilia extending into the cavity, known as flame cells, connected with tubes.

PLAYA. The flat interior part of an undrained basin, on which accumulates fine clastic sediments and chemical precipitates. Playas are formed within desert basins due to intermittent interior drainage, which, during cloudburst, forms intermittent lakes in which the sediments are deposited. A region remarkable for playas is the Great Basin of the western United States, which covers all of Nevada and Utah. Commercially important mineral salts derived from playa deposits are: gypsum, sodium carbonate, the soluble chlorides, and borates.

PLECOPTERA. The stone flies. An order of insects with aquatic early stages. The mouth is formed for biting but is usually poorly developed in the adult. They have four wings, which fold flat over the back when at rest. Sometimes abundant in the vicinity of water.

PLECOSTOMUS *(Osteichthyes).* Of the suborder *Siluroidea*, family *Loricariidae*, the plecostomus is a heavy-bodied loricariid (catfish) and is found present in many of the tanks maintained by tropical-fish fanciers. The average length is from 4 to 5 inches (10 to 12.5 centimeters), although the plecostomus can attain a length of about 20 inches (51 centimeters). The fish functions to clean small organisms from the sides of fish tanks and blends well in most tropical fish communities. The fish is considered an excellent food, when baked, by various Indian tribes in South America.

The loricariids occupy various habitats in South and Central America. Some are found in waters comparable to the European trout zones, something that would not be expected of these plump fishes. Others live in slowly flowing rivers. Interestingly, the gill openings of the loricariids are always on the lower side of the body, where they would always come in contact with mud or gravel. This is nonetheless advantageous for these fishes: they can maintain their suction (grip) on stones even in small waterfalls. Loracariids generally eat plants and refuse, although they also take small bottom-dwelling organisms. Because of their special gill apparatus, they can breathe and feed simultaneously. They have rows of horny teeth on large cupped lips, forming a good rasping surface. The catfishes scrape algae and other growth from stones and wood lying in the water. The teeth are also used to chew dead fishes or other animal cadavers in the water. In aquariums, it has repeatedly been observed that large loricariids will attack some living fishes, adhering to them and chewing their skin. Whether this results from a lack of appropriate food or is natural behavior is not fully understood. Like most catfishes, loricariids are also nocturnal or dusk-active creatures. They can decrease the amount of light entering their eyes with a distensible, whitish, drop-shaped structure on the upper part of the eye. The brighter the incoming light, the smaller the pupil, and the more this structure can distend over the eye.

Most plecostomids inhabit fast-flowing water, and they migrate into the clear Andes streams at an altitude of about 13,000 feet (3960 meters). They use the sucking mouth for propulsion while migrating, gaining a sucking hold in very rapidly flowing spots, and pushing forward bit by bit with jerky movements. By this means, they can get through places where the water is roaring past with an extremely powerful current. They prefer concealed sites when resting near the shore or behind stones. They also adhere to tree trunks that have fallen into the water. Such a loricariid will hold on to the stone even when lifted out of the water.

Almost all loricariids have a brown basic coloration with some gray or reddish colors. The body has irregular black or dark brown spots arranged in different ways, resulting in rather beautiful patterns. The dorsal fin is higher than the body itself and forms an impressive sail when erect.

Other loricariids include the antenna armored catfishes (genus *Ancistrus*); the loricaria (genus *Loricaria*); and the very small *Otocinclus* loricariids, which range from 1.5 to 3 inches (3.5 to 8 centimeters) in length.

PLEIADES. A very famous group of bright stars in the constellation of Taurus. Probably no one group of stars in the entire sky has received so much notice in classical literature and mythology as has the Pleiades. The Great Pyramid, which was undoubtedly designed for astronomical purposes, is so oriented that in 2170 B.C., when the Pleiades were on the meridian at midnight on the first day of spring, they could be seen through the south passageway. References to this group of stars are to be found in both the Old and New Testaments. One of the seven stars is, at present, distinctly fainter than the other six, and there are many myths regarding this so-called "lost pleiad." In fact, the myths occur in so many different ancient literatures that there is a well-established theory that at one time all seven of the stars were of approximately the same brightness.

The Pleiades is an open star cluster, with the various members all moving through space together. Long exposure photographs of this group indicate that the space surrounding the stars is filled with a luminous nebulosity. See Fig. 1.

Fig. 1.   The Pleiades. *(Lick Observatory.)*

PLENUM. Pressures slightly above atmospheric are known as plenums. Such pressures usually occur in air or gas systems as the result of the action of fans or blowers. The plenum is measured in small units of pressure, such as ounces per square inch, or in inches head of a liquid on a differential manometer.

PLEURISY. Inflammation or infection of the pleura, often producing a characteristic sharp, piercing sensation, which appears during the inspiratory phase of breathing as the inflamed pleural surfaces rub together. The friction produced by this approximation of the parietal and visceral pleura can be distinguished by the stethoscope as a sound simulating the creaking of leather and is termed a friction rub. Pleurisy is often accompanied by effusion, the outpouring of a thin fluid, which distends the pleural space. The appearance of fluid causes the disappearance of both the rub and the pain. Where expansion of the lung is sufficiently cut down by a massive accumulation fluid, dyspnoea will develop. Although once established, pleurisy may run a course independent of the primary course, pleurisy generally is the secondary manifestation of pneumonia, lung tumor, tuberculosis, abscess, rib fracture, or chest wounds. So-called primary pleurisy in which the disease commences in the pleura proper is rarely seen. Treatment is directed to the primary cause and frequently antibiotics are used. The latter also act to alleviate the pleurisy per se. In mild cases, the fluid formed will be reabsorbed, but if the fluid becomes infected, an empyema will make spontaneous reabsorption difficult and at best slow. This sometimes will cause the formation of fibrous adhesions, which may not permit the lung to re-expand upon removal of the fluid. The latter situation usually requires decortication procedure, that is, the lung is freed from the pleural peel.

PLEXUS. A network. 1. In many animals the processes of nerve cells join to form a plexus or nerve net. This is the characteristic form of nervous system in the coelenterates and persists with modifications in the flatworms. The nerves of the radially symmetrical echinoderms also take on this form. A plexus underlies the ectoderm of these animals and deeper in the body other nerve fibers form plexuses of limited extent. In the vertebrates nerves branch and rejoin in some parts of the body. The brachial plexus, made up of the spinal nerves which enter the arm, and the solar plexus above the stomach are examples. Almost a hundred such plexuses have been named in the human body.

2. A network of blood vessels. The choroid plexuses of the brain are the most commonly mentioned examples of this group. They are the very thin and highly vascular roof plates of the most anterior and the most posterior cavities of the brain, which expand into the interior of the cavities. Other vascular plexuses are found elsewhere in the body.

PLIOCENE. The last major subdivision of the Tertiary in the geologic time-scale. Term proposed by Charles Lyell in 1832 after the type locality in the Paris Basin. The Pliocene Period began approximately 8,000,000 years ago and lasted for about 6,000,000 years. In the United States the principal marine deposits occur on the Pacific Coast, but the outline of the continental margins was approximately what it is at the present day. There was continued mountain-building during the period, and the interior continental terrestrial deposits were relatively thin and unimportant. There was considerable volcanic activity in the Rocky Mountain region, with great extrusions of rhyolitic lavas in the Yellowstone Park.

PLOTTER (Curve).   A digital computer output device, which draws curves of one or more variables as a function of one or more variables. The function is essentially that of an x-y recorder. Data from the digital computer are translated into plotter-actuating signals and converted to incremental plotter movements to produce a drawing, chart, or other graphic portrayal. Two configurations are common. In a flat-bed plotter, the paper is fixed to a flat surface and one or more pens are mounted on a carriage capable of moving in two dimensions, normally designated x and y. In a drum plotter, the paper is affixed to a cylindrical drum whose rotation provides one dimension (usually) of motion. A pen mounted above the drum surface and capable of moving axially with respect to the drum provides the x-dimension. Control is also provided to raise or lower the pen from or to the paper surface. The information from the computer is decoded into fixed incremental movements of the drum and or pen carriage. Each plotter command transmits information which is decoded into the number of pen-carriage increments to be moved (y-axis) and the number of drum increments to be moved (x-axis). Specified bit combinations also control the raising or lowering of the pen to the paper.

PLOTTING SHEET (Navigation).   A small-area plotting sheet is an approximation to a mercator plotting sheet, and may be used for solving problems in dead reckoning and for plotting lines of position.

In constructing the graticule for the sheet, the fundamental assumption is made that, within the limit of errors inherent in the problems to be solved, the distance between successive parallels of latitude, separated by 1°, is constant all over the sheet. This is equivalent to assuming that the same scale of distance may be used all over the sheet. The ratio between the linear distance, X, between successive meridians and the linear distance, Y, between successive parallels, is that of middle-latitude sailing: $X = Y \cos L_m$, in which L_m is the average latitude of the region covered by the sheet.

Two types of small-area plotting sheet are in common use: (1) the fixed meridian type, and (2) the fixed parallel type. The method for completing the graticules and plotting a point is shown in Figs. 1 and 2, respectively. In the Figures the heavy lines are those printed on the published forms, and the light lines are those drawn to complete the graticule. Dotted lines in the figure are construction lines which need not appear on the completed sheet. The sheets in the figure are completed for $L_m = 48°$ and the point, P, is in latitude 48°17'N and longitude 50°23'W.

In the older or fixed meridian type, Fig. 1, a diagonal line is drawn making an angle L_m with the base line. The distance, X, is that between the fixed meridians, and the distance, Y, is the length of the diagonal, measured between the meridians. Elementary trigonometry shows that $X = Y \cos L_m$. This distance, Y, is then transferred to one of the fixed meridians and becomes 1° of latitude for the small-area plotting sheet. This length, Y, is also equivalent to 60 nautical miles and is used as a scale of distance all over the sheet. The central parallel on the sheet is that of L_m. The central meridian is that of the middle of the region covered by the particular problem to be solved. With this type of sheet the longitude scale is constant, no matter for what region the sheet is drawn. The latitude spacing, and hence the scale of distance, is different for different sheets.

In the more modern, and far more frequently used, fixed parallel type, Fig. 2, the diagonal line making an angle of L_m with the base is drawn. Distances equal to that between the fixed parallels are laid off along this diagonal and lines are drawn through these points parallel to the latitude scale of the sheet. These become the successive meridians. It will be seen that, as above, $X = Y \cos L_m$. However, on this style of sheet the distance Y is fixed, no matter for what region the sheet is prepared, and a scale for distance ruled on celluloid, or other permanent material, may be used on all sheets. Furthermore, this same scale may be used for measurement of longitude by placing it along the diagonal and then projecting the desired value, parallel to the latitude scale, to the proper latitude.

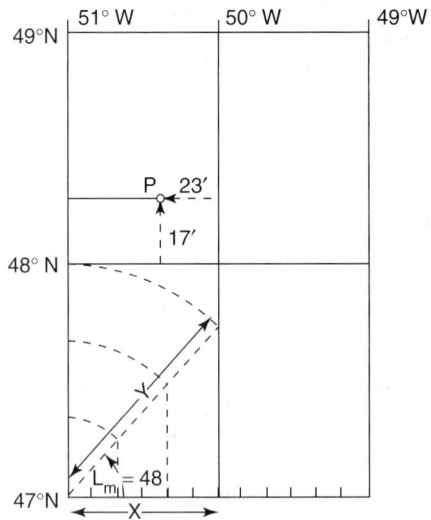

Fig. 1.   Fixed meridian type.

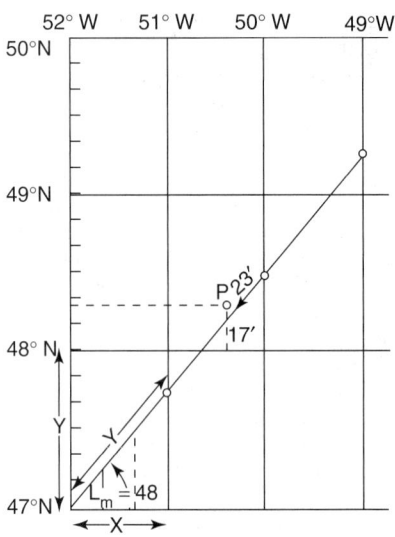

Fig. 2.   Fixed parallel type.

Plotting sheets of both types and for various scales are published by the U.S. Government, and by several publishing firms. Most of the forms have a compass rose at the center of the sheet so that parallel rulers and dividers may be used, instead of protractor and scale, if the user prefers. However, printed forms are by no means necessary, for the graticule can quickly be drawn on blank paper with the aid of a protractor and scale. Using a blank sheet requires about 1 minute to obtain the completed graticule, whereas with the printed forms this time is cut in half.

Either type of sheet can be used, without introducing errors greater than those inherent in the problems for which they are designed, for any mid-latitude between the equator and 60° and for areas of about 300 miles square. For low latitudes the area can be correspondingly increased.

See also **Course**; and **Navigation**.

PLOVER.   See **Waders, Shorebirds, and Gulls**.

PLUM CURCULIO (*Insecta, Coleoptera*).   A weevil, *Conotrachelus nenuphar*, which damages plums and other stone fruits, apples, pears, and quinces in the eastern half of the United States. The insects hibernate in fencerows and rubbish in orchards and pupate in the ground; hence, the destruction of their hiding places and thorough cultivation of the soil in late July and early August are useful measures of control. Spraying just after the petals drop and again after 10 days is effective. For control of the pest on peaches, special methods are necessary which differ in various peach-growing regions.

PLUME MOTH (Insecta, Lepidoptera).

A small moth whose wings are deeply split to form two to six slender fringed lobes. In the family *Pterophoridae* most species have the front wings split for about $\frac{1}{3}$ of their length to form two short lobes and the hind wings deeply divided into three lobes. One genus has the wings entire. Members of the *Orneodidae* have six slender plumes to each wing.

PLUMERIA TREE.

Of the family *Gentianaceae* (gentian family), there are some seven species of plumerias found from the West Indies southward to northern South America. The tree was named for Charles Plumier, a French botanist who was traveling in America in the late 1600s. It is found today in Hawaii and Mexico as well as the Caribbean region. It is highly regarded for its blossoms, which are used for a variety of decorative items, including leis. The flowers are fragrant and rugged, withstanding long periods after cutting. The tree blooms continually from spring until fall. The flowers are frequently white with yellow centers, but some have a reddish or golden yellow coloring, depending upon species. The blooms have a funnel shape.

Plumerius rubra may attain a height of 15 feet (4.5 meters) and bears pink and red flowers. The leaf is about 18 inches (46 centimeters) in length and 6 inches (15 centimeters) in width, with narrow pointed ends and conspicuous veining. The plant may be potted from cuttings and makes a very desirable plant for gardens and patios. The shrub is found mainly in Mexico and south through Venezuela. There are some 18 species of *Amsonia walt*, principally found in North America and Japan. The leaf is alternate, its flower is blue and white, the corolla being cylindrical in shape with 5 erect lobes. The shrub was named for Dr. Amson of Virginia. Leis are made from the flowers.

PLUM TREES.   See Rose Family.

PLUTONIUM.

Actinide radioactive metal. Atomic number 94. Symbol Pu. This element does not occur in nature except in minute quantities as a result of the thermal neutron capture and subsequent beta decay of ^{238}U; all isotopes are radioactive; atomic weight tables list the atomic weight as [242]; the mass number of the second-most-stable isotope ($t_{1/2} = 3.8 \times 10^5$ years). The most stable isotope is ^{244}Pu ($t_{1/2} = 7.6 \times 10^7$ years). Electronic configuration $1s^2 2s^2 2p^6 3s^2 3p^6 3d^{10} 4s^2 4p^6 4d^{10} 4f^{14} 5s^2 5p^6 5d^{10} 5f^6 6s^2 6p^6 7s^2$. Ionic radii Pu^{4+} 0.86 Å; Pu^{3+} 1.01 Å (Zachariasen). Oxidation potentials in acid solution $Pu \rightarrow Pu^{3+} + 3e^-$, 2.03 V; $Pu^{3+} \rightarrow Pu^{4+} + e^-$, -0.982 V; $Pu^{4+} + O_2 + 3e^- \rightarrow PuO_2^+$, $-1.17V$; $PuO_2^+ \rightarrow PuO_2^{2+} + e^-$, -0.91 V. Oxidation potential in alkaline solution $Pu^{3+} + 4H_2O \rightarrow Pu(OH)_4 + 4H^+ + e^-$, 0.4 V; $Pu(OH)_4 \rightarrow PuO_2^+ + 2H_2O + e^-$, -1.0 V; $PuO_2^+ + 2OH^- \rightarrow PuO_2(OH)_2 + e^-$, -0.8 V.

The isotope of major importance is ^{239}Pu ($t_{1/2} = 2.44 \times 10^4$ years). The importance of this isotope stems from its property of being fissionable with slow neutrons, together with the fact that the problem of its mass production has been solved. The first isotope to be produced was ^{238}Pu ($t_{1/2} = 86.4$ years).

Processes for the isolation and purification of plutonium, including the enrichment of spent nuclear reactor fuels, are described in the entry on **Nuclear Power**. These processes take advantage of Pu's several oxidation states, each of which has different chemical properties. The processes may involve carrier precipitation, solvent extraction, and ion exchange.

Plutonium is of major importance because of its successful use as an explosive ingredient in nuclear weapons and the role it plays in the industrial applications of nuclear power. Exemplary of the energy available from plutonium: (1) One pound (0.45 kilogram) $\cong$ 10 million kilowatts; (2) one kilogram (2.2 pounds) = 22 million kilowatts; (3) one kilogram (2.2 pounds) = 20,000 tons of chemical explosive. Plutonium has the important nuclear property of being readily fissionable with neutrons. ^{238}Pu was used in the Apollo lunar missions to power seismic and other experimental instruments placed on the lunar surface. Because comparatively large quantities of plutonium are produced in reactors, the amount available for various applications has increased considerably during recent years. It is estimated that as of the late 1980s, nuclear reactors throughout the world are producing in excess of 20,000 kilograms of Pu per year. Within a few years, there will be an accumulation of some

300,000 kilograms of Pu or more. The element is available for purchase by qualified potential users.

In a typical fast breeder nuclear reactor, most of the fuel is ^{238}U (90 to 93%). The remainder of the fuel is in the form of fissile isotopes, which sustain the fission process. The majority of these fissile isotopes are in the form of ^{239}Pu and ^{241}Pu, although a small portion of ^{235}U can also be present. Because the fast breeder converts the fertile isotope ^{238}U into the fissile isotope ^{239}Pu, no enrichment plant is necessary. The fast breeder serves as its own enrichment plant. The need for electricity for supplemental uses in the fuel cycle process is thus reduced. Several of the early liquid-metal-cooled fast reactors used plutonium fuels. The reactor "Clementine," first operated in the United States in 1949, utilized plutonium metal, as did the BR-1 and BR-2 reactors in the former Soviet Union in 1955 and 1956, respectively. The BR-5 in the former Soviet Union, put into operation in 1959, utilized plutonium oxide and carbide. The reactor "Rapsodie" first operated in France in 1967 utilized uranium and plutonium oxides.

Plutonium was the second transuranium element to be discovered. The isotope ^{238}Pu was produced in 1940 by Seaborg, McMillan, Kennedy, and Wahl at Berkeley, California by deuteron bombardment of uranium in a 150-cm cyclotron. Plutonium exists in trace quantities in naturally occurring uranium ores. The metal is silvery in appearance, but tarnishes to a yellow color when only slightly oxidized. A relatively large piece will give off sensible heat as the result of alpha decay. Large pieces are capable of boiling water.

Chemical Properties. Plutonium has the oxidation states (III), (IV), (V), and (VI), and a complex chemistry in aqueous solutions, as can be judged from such a multiplicity of states. A large number of solid compounds corresponding to these states have been made, and they are in general similar in formulas and properties to the corresponding compounds of uranium and neptunium. An important difference, especially as regards ranges of stability of these compounds, arises as a result of the much greater stability of the (III) and (IV) states of plutonium. This also leads to differences in the aqueous solution chemistry of plutonium as compared to uranium and neptunium. The pentavalent state, like that of uranium, but unlike that of neptunium, is unstable in aqueous solution with respect to disproportionation.

The ionic species corresponding to the four oxidation states of plutonium vary with the acidity of the solution. In moderately strong (one-molar) acid the species are Pu^{3+}, Pu^{4+}, PuO_2^+, and PuO_2^{2+}. The ions are hydrated but it is not possible at present to assign a definite hydration to each ion. The potential scheme of these ions in one-molar perchloric acid is the following:

$$Pu \xrightarrow{+2.03\text{ V}} Pu^{3+} \xrightarrow{-0.982\text{ V}} Pu^{4+} \xrightarrow{-1.17\text{ V}} PuO_2^+ \xrightarrow{-0.91\text{ V}} PuO_2^{2+}$$

with -1.043 V from Pu^{4+} to PuO_2^{2+} and -1.023 V from Pu^{3+} to PuO_2^+.

The potentials are in volts relative to the hydrogen–hydrogen-ion couple as zero.

The values given for the potential scheme in one-molar acid may be altered extensively by a change in hydrogen ion concentration (pH) or as a result of the addition of substances capable of forming complex ions with the plutonium species. Among such substances are sulfate, phosphate, fluoride, and oxalate ions, and various organic compounds, especially those known as chelating agents. The tetrapositive and hexapositive ions are complexed appreciably even by nitrate and chloride ions. The stability of the complex formed with a specified anion increases in the order: PuO_2^+, Pu^{3+}, PuO_2^{2+}, Pu^{4+}.

The hydrolysis of the ions follows a similar order; Pu^{4+} begins to hydrolyze even in tenth-molar acid and in hundredth-molar acid forms partly the hydroxide, $Pu(OH)_4$, and partly a colloidal polymer of variable but approximate composition $Pu(OH)_{3.85}X_{0.15}$, where X is an anion present in the solution. Further reduction of the acidity results in the hydrolysis of PuO_2^{2+} near pH 5, of Pu^{3+} at about pH 7, and of PuO_2^+ at about pH 9.

The plutonium ions in aqueous solution possess characteristic colors: blue-lavender for Pu^{3+}, yellow-brown to green for Pu^{4+}, and pink-orange for PuO_2^{2+}.

Plutonium monoxide occasionally appears on the surface of metal exposed to atmospheric oxidation, but is prepared more conveniently by treating the oxychloride with barium vapor at about 1,250 °C. The oxide is classified with the interstitial compounds rather than with the typical metal oxides.

The so-called sesquioxide ($PuO_{1.5-1.75}$) is a typical mixed oxidation state oxide, similar to those formed by uranium, praseodymium, terbium, titanium, and many other metals. Its composition shows continuous variation with changes in temperature and pressure of oxygen above the oxide.

Plutonium dioxide (yellow-green to brown, cubic) is the most important oxide of the element. Almost all compounds of plutonium are converted to the dioxide upon ignition in air at about 1,000 °C.

The important halides and oxyhalides of plutonium are PuF_3 (purple, hexagonal), PuF_4 (brown, monoclinic), PuF_6 (red-brown, orthorhombic), $PuCl_3$ (green, hexagonal), $PuCl_4$ (green-yellow, tetragonal), $PuBr_3$ (green, orthorhombic), PuI_3, $PuOF$, $PuOCl$, $PuOBr$, $PuOI$.

All of the halides except the hexafluoride and the triiodide may be prepared by the hydrohalogenation of the dioxide or of the oxalate of plutonium(III) at a temperature of about 700 °C. With hydrogen fluoride the reaction product is PuF_4, unless hydrogen is added to the gas stream, in which case the trifluoride is produced. With hydrogen iodide the reaction product is $PuOI$, and the other oxyhalides may be formed by the addition of appropriate quantities of water vapor to the hydrogen halide gas. Plutonium triiodide is produced by the reaction of the metal with hydrogen iodide at about 400 °C. The hexafluoride is produced by direct combination of the elements or by the reaction $2PuF_4 + O_2 \rightarrow PuF_6 + PuO_2F_2$ at high temperature. The hydrides of plutonium include PuH_2 (black, cubic) and PuH_3 (black, hexagonal).

Plutonium forms several binary compounds that are of interest because of their refractory character and stability at high temperatures. These include the carbide, nitride, silicide, and sulfide of the element.

The monocarbide is formed by reacting the dioxide in intimate mixture with carbon at about 1,600 °C. The mononitride may be obtained by heating the trichloride in a stream of anhydrous ammonia at 900 °C; it is prepared more easily, however, by reacting finely divided metal with ammonia at 650 °C. Although the lower temperatures are favorable to the production of higher nitrides, none are obtained, in contrast to the uranium-nitrogen system in which compositions up to $UN_{1.75}$ are easily realized.

The disilicide is formed when a slight stoichiometric excess of calcium disilicide is heated with plutonium dioxide in vacuum at about 1,550 °C. The disilicide is only moderately stable in air and burns slowly to the dioxide when heated to about 700 °C.

Plutonium "sesquisulfide" may be prepared by prolonged treatment of the dioxide in a graphite crucible with anhydrous hydrogen sulfide at 1,340°–1,400 °C, or by the reaction of the trichloride with hydrogen sulfide at 900 °C.

Handling Precautions. Care must be taken in the handling of plutonium to avoid unintentional formation of a critical mass. Plutonium in liquid solutions is more apt to become critical than solid plutonium. The shape of the mass also determines criticality. Plutonium's chemical properties also increase handling difficulty. Metallic plutonium is pyrophoric, particularly in finely divided form. Because of the high rate of emission of alpha particles, and the physiological fact that the element is specifically absorbed by bone marrow, plutonium, like all of the transuranium elements, is a radiological poison and must be handled with special equipment and precautions. To assure the safety of personnel, plutonium operations are normally handled in an essentially closed system, such as a *glovebox*. In addition, shielding is required when certain isotopes, including [240]Pu and [241]Pu, are present in appreciable quantity. Because research continues on the hazards and toxicity of plutonium, specific toxicity data should be sought from current authoritative literature, including government (U.S., UK, France, etc.) publications. As of the early 1980s, permissible body burden was established at 0.6 microgram; lung burden at 0.25 microgram. Chemical toxicity is trivial compared with radiation effects. The permissible levels for plutonium are the lowest for any of the radioactive elements.

The behavior of actinides in natural waters has great relevancy to the safe long-term storage of radioactive wastes. The enhanced solubility of plutonium and other actinides in the water of Mono Lake, California was studied by a group of scientists with the U.S. Geological Survey (Denver, Colorado). J.M. Cleveland and associates found that the solubility of plutonium in Mono Lake water is enhanced by the presence of large concentrations of indigenous carbonate ions and moderate concentrations of fluoride ions. In spite of the complex chemical composition of this water, only a few ions govern the behavior of plutonium, as demonstrated by the fact that it was possible to duplicate plutonium speciation in a synthetic water containing only the principal components of Mono Lake water. See reference listed.

Practical Utilization. Since the potential reserves of [235]U are limited, some point will be reached where this power source no longer will be competitive with fossil fuels, synthetic fuels, solar power plants, etc.—unless the development of means for the practical utilization of plutonium can be achieved. An important element of nuclear fuel cost is the credit received from the sale or future utilization of plutonium after its recovery from spent fuel. The plutonium credit is realistic only if the plutonium is used for power production, since, at present, there are few commercial uses envisioned where it would yield a similar economic return.

See also **Chemical Elements**.

NOTE: References pertaining to plutonium in nuclear reactors and nuclear wastes are listed at end of entry on **Nuclear Reactor**.

Additional Reading
Albright, D. and F. Berkhout: "Plutonium and Highly Enriched Uranium, 1996: World Inventories, Capabilities, and Policies," Oxford University Press, Inc., New York, NY, 1996.
Cleveland, J.M., T.F. Rees, and K.L. Nash: "Plutonium Speciation in Water from Mono Lake, California," *Science*, **222**, 1323–1325 (1983).
Lewis, R.J. and N.I. Sax: "Sax's Dangerous Properties of Industrial Materials," 10th Edition, John Wiley & Sons, Inc., New York, NY, 1999.
Lide, D.R.: "CRC Handbook of Chemistry and Physics 2000–2001," 81st Edition, CRC Press, LLC, Boca Raton, FL, 2000.
Kent, J.A.: "Reigel's Handbook of Industrial Chemistry," 9th Edition, Chapman Hall, New York, NY, 1992.
Krebs, R.E.: "The History and Use of Our Earth's Chemical Elements: A Reference Guide," Greenwood Publishing Group, Inc., Westport, CT, 1998.
Parker, P.: "McGraw-Hill Encyclopedia of Chemistry," 2nd Edition, The McGraw-Hill Companies, Inc., New York, NY, 1993.

Classical References
Kennedy, J.W., Seaborg, G.T., E. Segrè, and A.C. Wahl: "Properties of 94(239)," *Phys. Rev.*, **70**, 7/8, 555–556 (1946).
Seaborg, G.T.: "The Chemical and Radioactive Properties of the Heavy Elements," *Chem. Eng. News*, **23**, 2190–2193 (1945).
Seaborg, G.T., E.M. McMillan, J.W. Kennedy, and A.C. Wahl: "Radioactive Element 94 from Deuterons on Uranium," *Phys. Rev.*, **69** (7/8), 366–367 (1946).
Seaborg, G.T. and A.C. Wahl: "The Chemical Properties of Elements 94 and 93," *J. Amer. Chem. Soc.*, **70**, 1128–1134 (1948).
Seaborg, G.T. (editor): "Transuranium Elements," Dowden, Hutchinson & Ross, Stroudsburg, Pennsylvania, 1978.

PLUTO (Planet). Ninth and outermost known and confirmed planet from the sun, Pluto is estimated to have a diameter considerably less than half that of earth. The mass is estimated at about 0.2 that of earth. Discovery of the planet on March 13, 1930 by the Lowell Observatory occurred on the anniversary both of Percival Lowell's birth and the discovery of Uranus. See Fig. 1. The discovery marked the culmination of a search for a planet outside the orbit of Neptune that had been carried on

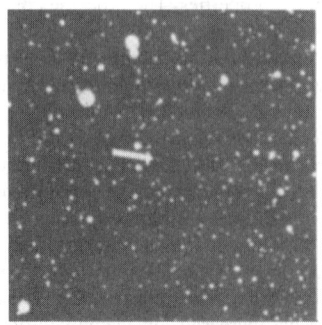
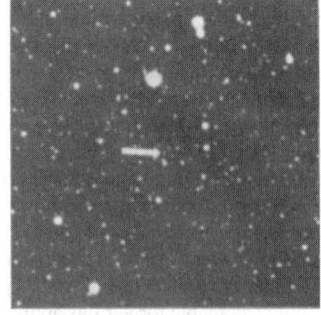

January 29, 1930   January 23, 1930

Fig. 1.   Pluto shows just to the right of the arrow. Small sections of the discovery plates. (*Lowell Observatory.*)

for many years at the observatory, at Flagstaff. Arizona. The circumstances that led to the belief that such a planet existed are similar to those that led to the discovery of Neptune. After the orbit of Neptune was computed and the motion carried back through the years, it was found that the planet had been observed several times previous to its announcement as a planet, the early observers having recorded it as a star. These early observations were of great value in making an accurate determination of the orbit, and when all perturbations due to known objects had been computed and applied, certain unexplainable differences between observed and computed positions appeared. On the basis of these perturbations, Lowell made the necessary laborious computations to determine the positions of a possible planet that might be causing the attractions, and he predicted Pluto. There is considerable doubt in the minds of many astronomers as to whether it is Pluto that is actually producing the perturbations in the orbit of Neptune or whether the perturbations may not be due to accidental errors in the observations of Neptune itself. Whatever the case, it is certain that the computations of Dr. Lowell stimulated the search, and that the planet was found as a result in the approximate position predicted.

The name Pluto was selected for the new planet and the first two letters of the name, combined in monogram form **P** are used as its symbol. These *two* letters are particularly fortunate in being both the first letters in the name Pluto and the initials of Percival Lowell.

The orbit of Pluto is the most eccentric of all the orbits of the major planets, and the inclination to the plane of the ecliptic is also the largest. The mean distance of the planet from the sun is slightly less than 40 astronomical units. Due to the large value of the eccentricity (0.25), the planet is more than 50 astronomical units from the sun at aphelion and within 30 at perihelion. The latter figure is less than the distance of Neptune from the sun, and so, at times, the planets Pluto and Neptune pass each other. However, the large inclination of the orbit of Pluto makes a collision virtually impossible, the closest approach of the two planets being about 38.4×10^7 kilometers.

Pluto appears as a very faint star of about the fifteenth magnitude, with a yellowish color, in contrast to the greenish appearance of its nearer neighbors in the solar system. It is only within the last few years that spectroscopic data have revealed a few specifics concerning the nature of Pluto's surface. At one time it was believed that Pluto was similar in size and mass to Mars and Earth, but these estimates have been revised downward. In observing the infrared reflection of Pluto through two very narrow band filters, astronomers from the University of Hawaii working at Kitt Peak National Laboratory in 1976 found the response was exactly as expected for methane ice. With a surface of frozen methane, Pluto must be colder than 50 K (methane condenses at this temperature under low pressure). Methane gas previously had been found in the atmospheres of Jupiter, Saturn, Uranus, and Neptune, but this was the first finding of solid methane on a planet. Some scientists believe that Pluto may be the only planet that closely exhibits the pristine state presumed to have existed some 4.6 billion years ago when the solar system was formed. Now many astronomers believe that Pluto is comparatively small, icy, with low density–more like the satellites of the outer planets than like the planets themselves. Revised estimate place the diameter of Pluto at about 3500 kilometers (2175 miles), making it the smallest planet. For some years, it had been suspected by some astronomers that possibly Pluto is not a true planet, but rather at escaped satellite of Neptune.

A satellite (Charon) of Pluto was first reported in 1978 and it was noted that a series of mutual eclipses might be observable beginning in 1979. An improved orbit determination revised that estimate to the early 1980s. As reported by Binzel (University of Texas) and colleagues at the University of Hawaii and California Institute of Technology, these eclipse events are observable from Earth for only a short period every 124 years (when Pluto's heliocentric motion causes the plane of the satellite's orbit to sweep across Earth's orbit). These conditions, of course, offer a rare means to gain information of this distant planet-satellite system. The investigators use the term "eclipse" broadly to refer both (1) the satellite passing behind the planet, and (2) the satellite passing in front of the planet.

The first eclipses were detected in January and February 1985, further confirming the existence of the satellite.

Shortly after midnight on June 9, 1988, two astronometers (Massachusetts Institute of Technology) in NASA's high-flying observatory 3,500 miles south of Hawaii watched as Pluto eclipsed a small star for 80 seconds. It gave the scientists an opportunity to see what kind of a shadow Pluto casts. Because the shadow was fuzzy, this indicated that the

planet has an atmosphere. The researchers estimated Pluto's temperature at $-415\,°F$ ($-248\,°C$), suggesting a methane (or possibly, argon, carbon monoxide, neon, nitrogen, or oxygen atmosphere).

Although still speculative, some tentative consensus on the nature of Pluto is commencing to develop. Based largely upon studies of the planet's flickering light as the result of Charon's fortuitous eclipses of the planet in recent times, from data that may prove to be more valuable than that which may be obtained with the Hubble Space Telescope. See also **Hubble Space Telescope (HST)**. Pluto is considered a small body of rock and ice only about two-thirds the size of Earth's moon. According to data extrapolation, Pluto's south polar region is extremely bright and presumed to be coming from frozen methane and other chemical ices. The north polar cap, by contrast, is much smaller and less brilliant. This is puzzling because Pluto has a 248-year orbit around the sun. During recent years, the planet's south pole is approaching the end of a century-long summer. During this period, the north pole has been in shadow. Further, Pluto recently passed its closest approach to the sun. Thus, why is there so much ice on the southern cap and less on the northern cap?

Currently, scientists are modeling the planet's atmosphere. Included are astronomers R. Binzel and E. Young (Massachusetts Institute of Technology). These researchers observe that, just as the planet starts to move away from the sun and its atmosphere starts to condense, darkness sets in at the south pole. Thus, the south pole is the most likely place for the deposition of ice when the atmosphere condenses. Further, although the south pole has experienced full sunshine for a century, it nevertheless bears a residual ice cap from its decades in frigid conditions. Perhaps its high reflectivity interferes with sunlight absorption and thus slows the evaporation of the frost and the consequent building of layer upon layer of frost rather than promoting the slow evaporation of the frost.

One astronomer, when commenting on the unusual characteristics of the Plutonian system—the spin of the planet and a satellite almost as large as the planet itself—observed that perhaps these characteristics are not so unusual after all. As the result of computer simulations, the scientist speculates that the early solar system may have included numerous Pluto-like icy bodies, and that the current system may have by some freak occurrence escaped becoming part of the Oort cloud of "icy snowballs," of the kind that form the nucleii of comets. The Plutonium system may have been influenced by gravitational resonance with Neptune. Perhaps the system is one of a kind? Perhaps not!

Additional Reading

Binzel, R.P., et al.: "The Detection of Eclipses in the Pluto-Charon System," *Science*, **228**, 1193–1194 (1985).

Binzel, R.: "Hemispherical Color Differences on Pluto and Charon," *Science*, 1070 (August 26, 1988).

Binzel, R.P.: "Pluto," *Sci. Amer.*, 50 (June 1990).

Kerr, R.A.: "Pluto's Orbital Motion Looks Chaotic," *Science*, 986 (May 20, 1988).

Kerr, R.A.: "Geophysicists Take a Tour Around the Solar System: The Tiniest Planet Shines in the Best Portrait Ever," *Science*, 1635 (June **19**, 1992).

Lunine, J.I.: "Origin and Evolution of Outer Solar System Atmospheres," *Science*, 141 (July **14**, 1989).

Pasachoff, J.M. and D.H. Menzel: "Field Guide to the Stars and Planets," Vol. 16, 3rd Edition, Houghton Mifflin Company, New York, NY, 1992.

Powell, C.S.: "A Rare Glimpse of a Dim World," *Sci. Amer.*, 24 (August 1992).

Rothman, T.: "A Computer Finds that Pluto's Orbit is Chaotic," *Sci. Amer.* 20 (October 1988).

Shulman, S.: "The Seasons of Pluto," *Technology Review (MIT)*, 9 (July 1989).

Stern, S.A., L. Cesana, and D.J. Tholen: "Pluto and Charon," University of Arizona Press, Tucson, AZ, 1997.

Stern, S.A. and J. Mitton: "Pluto and Charon: Ice Worlds on the Ragged Edge of the Solar System," John Wiley & Sons, Inc., New York, NY, 1999.

Sussman, G.J. and J. Wisdom: "Numerical Evidence that the Motion of Pluto is Chaotic," *Science*, 432 (July 22, 1988).

Web References

JPL Solar System Exploration Site: *http://sse.jpl.nasa.gov/index.html*

Pluto-Kuiper Express...to explore Pluto/Charon and the fringes of our Solar System: *http://pluto.jhuapl.edu/*

The Planetary Photojournal: *http://photojournal.jpl.nasa.gov/*

The Nine Planets: *http://seds.lpl.arizona.edu/billa/tnp/intro.html*

Views Of The Solar System: *http://www.hawastsoc.org/solar/eng/homepage.htm*

PNEUMATIC. Related to, or pertaining to, air or derivatives to other gases.

PNEUMATIC CONTROLLER. Although electronic and digital control systems have made serious inroads in the field of process and manufacturing control since the early 1970s, a high percentage of installed instrumentation in industry is pneumatic. Pneumatic controllers still retain certain inherent advantages over other kinds of control systems and, therefore, pneumatic systems are expected to continue an important role over many years in the future.

The function of a pneumatic controller is basically the same as that of an electric or hydraulic controller, the primary difference being that compressed air is used as the controlling medium instead of electricity or hydraulic pressure. Because both pneumatic and hydraulic controllers utilize fluids within mechanical-type systems, they have many design similarities. In hydraulic systems, a jet-pipe is one of the fundamental detectors; in pneumatic systems, the baffle-nozzle is the principal detector. The baffle-nozzle is also commonly called the flapper-nozzle or orifice-nozzle system. A device of this type is shown in Fig. 1. Input motion from some measured variable, such as the movement of a bourdon tube in a pressure- or temperature-measurement system, is applied to a simple pivoted baffle to change the clearance X between a flat surface on the baffle and nozzle. The nozzle normally exhausts directly to atmosphere. Under fixed conditions, air flows through the orifice and out of the nozzle through clearance X. As X is increased from zero, the nozzle back pressure P_n decreases. The upper limit of the value for nozzle pressure is determined by the air-supply pressure when X equals zero. The lower limit is determined by the ratio of resistance to airflow, which is established between the fixed orifice and the nozzle when X is very large. Between these two extremes, various nozzle pressures are established. This relationship is shown in Fig. 2. The slope of the curve is called "nozzle sensitivity," or gain. Detectors of this type exhibit high gains. For a commonly used 0.010-inch diameter orifice and a 0.025-inch (~0.6 millimeter) diameter nozzle combination, a motion of the baffle of 0.001 inch (~0.03 millimeter)

creates a change in nozzle pressure in excess of 8 pounds/square inch (0.54 atmosphere) in the central portion of the curve.

In some designs, the flat baffle of Fig. 1 is replaced by a ball as shown in Fig. 3. An advantage of this arrangement is elimination of the need to align the baffle surface parallel with the lip of the nozzle. However, slight friction may result if the ball rubs against its guide. Another configuration is the "free vane" shown in Fig. 4. Two nozzles share a common restriction and have a common centerline. The baffle is moved parallel with the nozzle lips. The sensitivity of this type of detector is somewhat less—a baffle motion of 0.005 inch (~0.13 millimeter) producing a change in nozzle back pressure of about 1 pound/square inch. The direct-operated pneumatic ball pilot, shown in Fig. 5, is still another design configuration for a pneumatic detector.

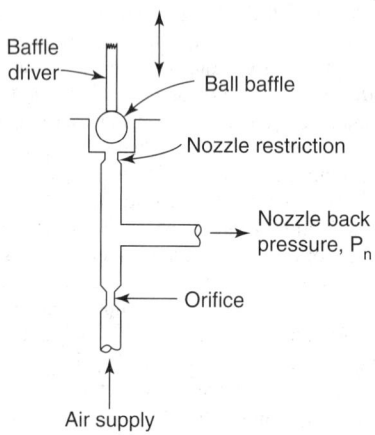

Fig. 3.   Ball-nozzle used in pneumatic controller.

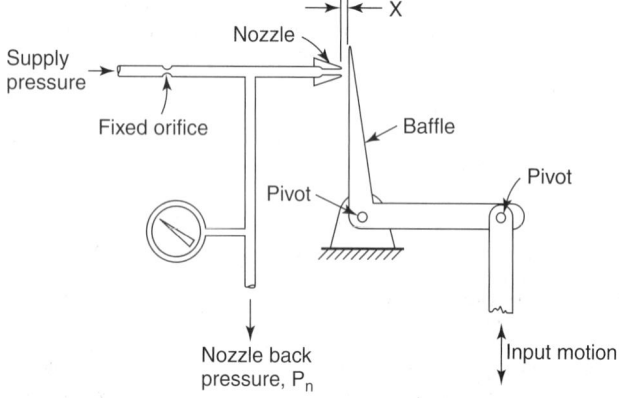

Fig. 1.   Baffle-nozzle detector used in pneumatic controller.

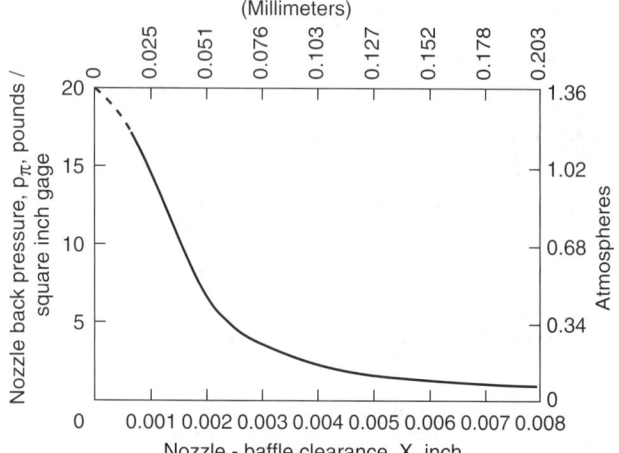

Fig. 2.   Baffle-nozzle characteristic relation between X and P_n for steady state. Conditions are based upon an orifice diameter of 0.01 inch (0.254 mm) and a nozzle diameter of 0.025 inch (0.635 mm).

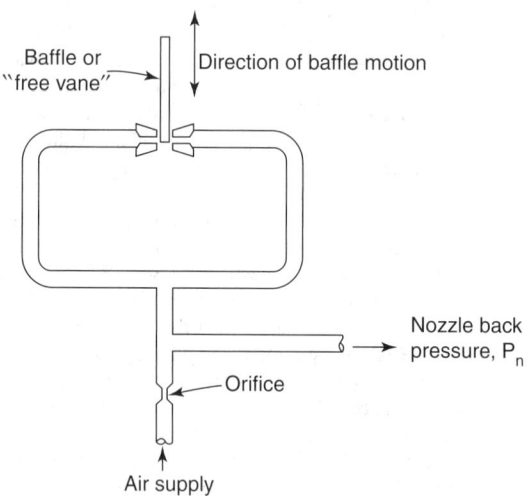

Fig. 4.   Free-vane type detector used in pneumatic controller.

In many pneumatic controllers, a pneumatic pilot relay or pilot valve is used to increase gain. A relay also increases airflow capacity when a change of output pressure is needed. Thus, the dynamic response is improved when the pneumatic instrument is connected to a large volume or to a long transmission line. Gains usually range between 3 and 10. A relay of this type is shown in Fig. 6. Reverse-acting relays also are used to cause a decrease in pneumatic signal pressure with an increase in value of the process variable being measured and controlled. Pneumatic controllers are available with practically all of the modes of control—two-position, proportional, proportional-plus-integral, proportional-plus-derivative, and proportional-plus-integral-plus-derivative actions—that are available in electric controllers. Because of the wide acceptance and preference for

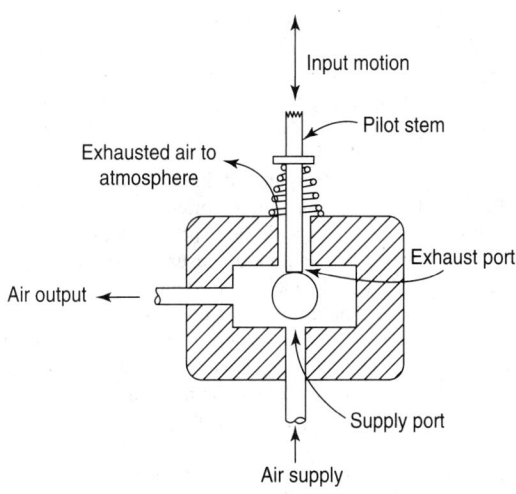

Fig. 5.   Direct-operated pneumatic ball pilot used in pneumatic controller.

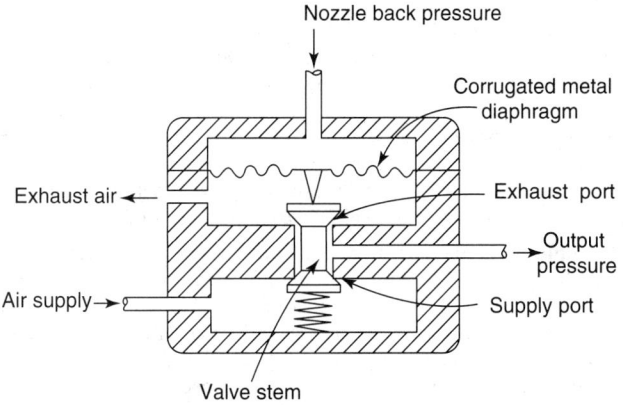

Fig. 6.   Direct-acting pneumatic relay (bleed type) used in pneumatic controller.

pneumatically-operated final controlling elements (valves, dampers, etc.), a pneumatic controller makes an all-pneumatic system possible.

The systems previously described sense a change in position. Modern pneumatic controllers also incorporate force-balance systems. The measured variable, when changing, creates differences in force (pressure) rather than actual small movements of an element in space (as in a position-balance system). Pneumatic controllers are frequently designed to be housed within a recorder or an indicator case.

PNEUMATIC-PROBE PYROMETER.   A thermometer for high-temperature gases, in which the gas is sucked through a nozzle and then cooled. Reliance is place principally on knowledge of the law of gas expansion through the nozzle and on measurement of pressure and mass flow rate of the gas.

PNEUMATOLYSIS.   Pneumatolysis is the process of alteration of existing rocks and mineral deposits, or the formation of new ones, by means of gases or vapors emanating from the magma.

PNEUMOCYSTOSIS.   Although of uncertain affinities, *Pneumocystis carini* is usually considered a protozoan belonging to the class *Sporozoa*. The "cysts" have been found in a number of common animals, and human isolates have been reported from all over the world, but the epidemiology is not fully understood and the natural reservoir is unknown. The mode of transmission is believed to be inhalation and it is suspected that healthy people may carry these organisms as saprophytes in their lungs for some time and develop *Pneumocystis* pneumonia during the course of some other illness.

Pneumocystis infection assumes two patterns: An infantile endemic was seen among severely malnourished children in war-stricken Europe and later in Viet Nam. This was characterized by a chronic, insidious respiratory illness. The second form is an acute, or subacute pneumonitis in patients whose cell-mediated immunity is compromised by a malignancy or therapy involved in organ transplantation. *Pneumocystic pneumonia* is also the predominant opportunistic infection in persons suffering from AIDS.

The pathologic foci of the disease are the pulmonary interstitium, with thickening of the alveolar septa and the alveoli.

Pneumocystis infection may be rather abrupt with rapid development of fever, hacking and non-productive cough, tachypnea, and progressive dyspnea. Pleural cavitation and lymphedema is unusual. Diagnosis entails demonstration of the organism—a 4–6 micrometer cyst with 6 to 8 merozoites. Examination of sputum has, however, only a 5% chance of success and transtracheal aspirates demonstrate the organisms in only 15% of cases. Open lung biopsy is the ultimate approach and has a 95% or better success rate in providing material containing the organisms.

Treatment can be with pentamidine isethionate, but this has major disadvantages in its side effects of hypotension, tachycardia, nausea, vomiting, etc. The major therapeutic regime is in the fixed drug combination of trimethoprim-sulfamethoxazole.

R.C. VICKERY, M.D., D.Sc., Ph.D., Blanton/Dade City, FL.

PNEUMOKONIOSES.   Diseases of the lungs produced by inhalation of dusts, particularly those containing silica, asbestos and other inorganic material, or certain vegetable substances, notably sugar cane waste and raw cotton dust (brown lung).

Silicosis occurs in industries in which the air is polluted by silica dust, e.g., pottery, metal grinding, sandblasting and mining in rock. The inhaled silica gives rise to the production of diffuse fibrosis in the lungs; moreover it facilitates the growth of the tubercle bacillus so that tuberculosis is a possible complication. A special form of silicosis, called anthracosis (black lung), occurs in coal miners who are exposed to a mixed dust, mainly of coal, with a small proportion of silica.

Asbestosis is much less common but more serious than silicosis, since once contracted it is more rapidly fatal, and is associated with a liability to lung cancer.

Another form of pneumokoniosis is an acute and often fatal form which results from inhalation of beryllium, much used in the manufacture of fluorescent lamps.

Corticosteroid therapy has had encouraging results but the primary consideration in these diseases is the environmental protection of exposed workers.

PNEUMONIA.   An acute inflammation of the lungs (parenchyma—alveolar spaces and/or interstitial tissue). Involvement of an entire lobe is *lobar pneumonia*; of parts of the lobe, *segmental pneumonia*; of the bronchi, bronchopneumonia. Pneumonia is identified in accordance with cause—*bacterial* or *nonbacterial* (fungal, protozoan, mycoplasmal, viral, etc.)

Bacteria causing pneumonia include *Streptococcus pneumoniae* (pneumococci); *Staphylococcus aureus*; Group A hemolytic streptococci (*Klebsiella pneumoniae*—Friedländer's bacillus); *Hemophilus influenzae*; and *Francisella tularensis*. Other bacterial pathogens, such as tubercle bacillus, as well as viruses, rickettsias, and fungi may cause pneumonia. When pneumonia is due to certain gram-negative bacilli, such as *Escherichia coli*, treatment may be complicated when using antimicrobial drugs for inhibiting gram-positive bacteria. Treatment with immunosuppressive agents also may cause complications. In some populations, viral pneumonia may be more common than bacterial pneumonias.

Certain factors will predispose pneumonia; these include the common cold, other acute viral respiratory infections, acute and chronic alcoholism, malnutrition, debility, exposure to bad weather and environmental conditions, coma, bronchial tumor, foreign matter in the respiratory tract (such as aspiration of vomitus), immunosuppressive agents, and hypostasis. *Pneumocystis carinii* pneumonia results from AIDS.

Pneumococcal Pneumonia.   The pneumococcus accounts for about 60% of all pneumonias in adults. Typical symptoms, as described by Sir William Osler in 1892, remain lucid and accurate today: Abruptly, or preceded by a day or two of indisposition, the patient has a severe chill, lasting from 10 to 30 minutes. In no acute disease is an initial chill so constant or so severe. The fever rises quickly. There is pain in the side, often of an agonizing character. A short, dry painful cough soon develops and the respirations are increased in frequency. When seen on the second or third day the patient presents an appearance which may be quite pathognomonic (specifically characteristic). The patient lies flat in bed, often on the affected side; the face is flushed, particularly the cheeks; the breathing is hurried; the alae nasi (cartilaginous flap on the outer side of either nostril) dilate with every inspiration; the eyes are bright, the expression is anxious, and there is a frequent short cough which makes the patient wince and hold his/her side.

The expectoration is blood-tinged and extremely tenacious. The temperature rises rapidly to 104° or 105 °F (40–40.6 °C). The pulse is full and bounding and the pulse-respiration ratio much disturbed. Examination of the lung shows the physical signs of consolidation—blowing breathing and fine rales. After persisting for 7 to 10 days, the crisis occurs, and with a fall in the temperature the patient passes from a condition of extreme distress and anxiety to one of comparative comfort.

Not all patients present a full menu of symptoms as just described. Prior or current use of antipyretics may result in only a modest fever; the initial rigor may be absent, but recurrent chills are common, and, in the elderly, the additional symptoms of confusion or stupor may be present. The description by Osler portrays the inflammatory response of the alveoli of the lungs. See also **Respiratory System**. There is exudation of fluid (edema, congestion). The pneumococci survive the edema and, in fact, the condition is favorable to the proliferation of the microorganisms. Within several hours, polymorphonuclear leukocytes arrive at the alveoli, but apparently phagocytize (envelope and destroy) only a comparatively small number of the pneumococci. It is believed that phagocytosis at this stage depends upon the capsular polysaccharide (serotype) of the organism. It is at the time of crisis, as previously mentioned, that type-specific anticapsular antibodies arrive and destroy the pneumococci en masse. This is the general course of untreated pneumococcal pneumonia, a rather risky period for some patients. With the availability of antibiotics, prognosis has been markedly enhanced. Penicillin is the usual antibiotic of choice, but erythromycin, chloramphenicol, and others are used. Some strains of pneumococci have developed resistance to penicillin at usual dosage levels. Thus far it has been clinically safe to raise the dosage levels where required to destroy the pathogens. However, in a case in South Africa, reported in 1977, a strain of pneumococci was encountered that was fully resistant to penicillin. In such instances, which to date are rare, vancomycin, rifampin, or bacitracin can be used. See also **Antibiotic**.

In 1978, a new pneumococcal vaccine was introduced. This is effective against 14 serotypes of pneumococci. Although the vaccine is not for the general population, it is indicated for persons considered at high risk.

The natural residence of the pneumococcus is in the nasopharynx (cavity behind the nasal cavities). The bacteria, although always present in most humans, do not usually cause illness in this location because of the very effective defense mechanisms present. An infection of the upper respiratory tract, however, may prepare a pathway for the pneumococci, mainly because such infections increase the volume and lower the viscosity of secretions, making it easier for the pneumococci to spread.

Of treated patients (age 2 to 50), 90% to 95% survive. Patients who are treated within the first 5 days of pneumococcal pneumonia usually respond favorably. Response to antibiotic therapy is often prompt, but fever may persist for a few days in about half of the patients. Antibiotics should be administered for a period at least 48 hours after the disappearance of fever.

Other Pneumonias

Staphylococcal Pneumonia. Caused by *Staphylococcus aureus*, this pneumonia is frequently a complication of influenza, but can be primary. It is not uncommon among hospitalized patients as a superinfection accompanying debility, surgery, tracheostomy, coma, or immunosuppressive therapy. Staphylococcal pneumonia is a life-threatening disease requiring prolonged, high-dose parenteral antibiotic therapy. Oxacillin or nafcillin may be administered, with cephalothin an excellent alternative. Erythromycin or clindamycin may be useful in patients allergic to penicillins and cephalosporins, but if such patients have advanced disease or bacteremia, vancomycin is preferred because it is bactericidal.

Streptococcal Pneumonia. Infrequently seen as a complication of measles and influenza, streptococcal pneumonia is caused by hemolytic streptococci of Lancefield's Group A. Before the advent of modern chemotherapy, this type of pneumonia sometimes followed so-called "strep throat" or scarlet fever. The disease was common among the military in both World Wars. Penicillin G is the antibiotic of choice; cephalosporins for the penicillin-allergic patient.

Klebsiella **Pneumonia.** Gram-negative bacillary pneumonias mainly occur in hospital, although *Klebsiella pneumoniae* will be seen in nonhospitalized persons as well. These bacilli reach and infect lower respiratory areas in three ways—direct expansion of pharyngeal flora, usually by aspiration; contaminated fluid droplets from respirators and ventilatory equipment can introduce the infection; bacteremic spread to the lung, as may be caused by the presence of organisms in drugs injected intravenously. Predisposition for this type of pneumonia often is a severe underlying disease, such as chronic pulmonary disease, heart disease, and alcoholism. *Pseudomonas* lung infections, in contrast, usually are associated with systemic immunosuppression and leukopenia. Symptoms include fever, chills, and malaise, with cough, sputum production, dyspnea, and pleuritic chest pain. In immunosuppressed patients, these symptoms may not be present or they may be overlooked. X-rays cannot establish an etiologic diagnosis. Treatment consists of antimicrobial chemotherapy along with drainage of sequestered fluid in cavitary lesions or of pleural fluid. The prognosis for gram-negative bacterial pneumonia depends largely on the presence of an underlying disease. In immunocompromised hosts, mortality rates may range up to 80%. *Klebsiella* pneumonias have a lower mortality rate, ranging up to 50% for severe cases. These infections often cause severe lung tissue damage. *Pseudomonas* pneumonia produces multiple and widespread abscesses.

Pneumonia Resulting from *Hemophilus influenzae*. In connection with the viral influenza pandemics of 1889 and 1918, cases were complicated by pneumonia caused by *H. influenzae*. This bacillus is now an infrequent cause of pneumonia, occurring mainly in patients with chronic bronchitis and bronchiectasis. Most patients recover unless treatment is delayed to the point where bronchiolitis and pneumonia have advanced to produce severe cyanosis and anoxemia. In applying antibiotic therapy, the physician must be aware of penicillinase-producing strains.

Mycoplasmal Pneumonia. At one time, this pneumonia was designated as "primary atypical pneumonia" until it was found that the disease is caused by the pleuro-pneumonia-like microorganism, *Mycoplasma pneumoniae*. These are among the smallest known free-living organisms. Unlike bacteria, mycoplasmas lack a rigid cell wall. Eight species have been found in humans, but only three have been implicated in human disease. It is estimated that from 4% to 10% of cases admitted to hospital for pneumonia suffer with the *M. pneumoniae* infection. The peak seasonal incidence is fall and early winter. The disease is spread by infected respiratory secretions. Epidemics are rare except in military populations. The infection characteristically spreads slowly throughout a family or in other situations where people live in close contact. The incubation period is 8 to 10 days. The illness usually commences with sore throat, followed by cough, headache, malaise, chills, and fever. Cough is frequently nonproductive. Fine or medium rales are noted. Symptoms usually disappear within 1 to 3 weeks after onset, even without treatment. Ear involvement sometimes occurs (10% to 20% of patients). Nonlobar pneumonia with a subacute onset and nonpurulent sputum usually suggests the diagnosis of this type of pneumonia. The antibiotics of choice are erythromycin or a tetracycline. Chemotherapy does not prevent shedding, which may persist from 3 to 8 weeks.

Viral Pneumonia. This illness usually results from exposure of nonimmune individuals to infected persons shedding virus. It is estimated that the disease may account for about 75% of all acute pulmonary infections in some populations (schools, offices, military establishments, etc. where people associate closely). A number of agents—influenza and parainfluenza viruses; adenoviruses; respiratory syncytial virus; rhinoviruses; coxsackie-, echo-, and reovirus; cytomegalovirus; herpes simplex—may cause the infection. Symptoms usually are mild; pulmonary involvement is not always detected. However, severe and sometimes fatal cases may result, particularly with influenza A virus. Prognosis varies widely with the nature of the causative virus, the patient's age, and presence or absence of underlying diseases. Physicians generally recommend prophylactic vaccination with influenza A and B preparations, particularly for patients over 50 years of age. Amantadine is sometimes administered to moderate the course of the disease.

Associated Pneumonias. As in the case of influenza just described, pneumonia can be associated with a number of other infections, including plague, tularemia, Q fever (rickettsial pneumonia), psittacosis, and legionellosis. These illnesses are described in separate articles in this encyclopedia.

Pneumocystis carinii **Pneumonia.** Identified as a protozoan, *Pneumocystis carinii* has been known for many years to cause pneumonia in infants with *immune deficiencies* and in children (1 to 4 years old) who have acute lymphatic leukemia. Within the last several years, the disease has been recognized with increasing frequency among patients who are undergoing immunosuppressive treatment. The disease is suspected in any patient receiving immunosuppressive therapy or having an immunologic deficiency who develops progressive pulmonary infiltration and respiratory insufficiency. Although the parasite may occur in the body essentially unnoticed, when untreated an active infection will progress to death.

Only since the discovery and spread of AIDS has the disease become prominent. *Pneumocystis carinii* pneumonia, along with Kaposi's sarcoma, are the principal clinical manifestations of the acquired immunodeficiency syndrome. In the case of the AIDS patient, *P. carinii* is considered an opportunistic infection along with several other such infections (mucosal candidiasis, progressively ulcerating perianal herpes simplex infection, and disseminated cytomegalovirus infection).

Additional Reading

Berkow, R. and M.H. Beers: "The Merck Manual," 17th Edition, Merck & Company, Inc., Whitehouse Station, NJ, 1999.

Cimolai, N.: "Serodiagnosis of the Infectious Diseases: Mycoplasma Pneumoniae," Kluwer Academic Publishers, Norwell, MA, 1999.

Godfrey, S.: "Pneumonia," Blackwell Science, Inc., Malden, MA, 1996.

Jarvis, W.R.: "Nosocomial Pneumonia," Marcel Dekker, Inc., New York, NY, 2000.

Marrie, T.J.: "Community-Acquired Pneumonia," Kluwer Academic Publishers, Norwell, MA, 2001.

Rello, J. and K.V. Leeper: "Severe Community Acquired Pneumonia," Kluwer Academic Publishers, Norwell, MA, 2001.

Stevens, D.L. and E.L. Kaplan: "Streptococcal Infections: Clinical Aspects, Microbiology, and Molecular Pathogenesis," Oxford University Press, Inc., New York, NY, 1999.

Web Reference

Centers for Disease Control and Prevention: *http://www.cdc.gov/health/diseases.htm#S*

PNEUMONIA (Legionellosis). See **Legionellosis**.

PODICIPEDIFORMES. In this order of birds (*Aves*) are found the grebe and dabchick. The grebe is a swimming bird with lobed toes, short legs, short neck, and a sharp beak, which in some species is quite long. The grebe is found in temperate regions of both hemispheres and members of the same species may have a very wide range. The little grebe (*Podiceps ruficollis ruficollis*) is an aquatic bird of the Old World. The pied-billed grebe (*Podilymbus podiceps*) is of the New World. Little grebes are also called dabchicks. See also **Grebe**.

PODOCARPS. Medium-to-large evergreen shrubs and trees that are the main conifers of the Southern Hemisphere. They are members of the family *Podocarpaceae* and many belong to the genus *Podocarpus*. The main source of lumber and possibly the only native conifer of South Africa is the yellow-wood tree (*Podocarpus latifolius*). Most podocarps bear a nutlike fruit, which is embedded in a colored berry and, technically, these are considered the "cones." The leaves are narrow, ranging from 2 to 5 inches (5 to 12.5 centimeters) in length. The leaves have a rather luxuriant yellow-green color and glossy surface. The brown "pine" and black "pine" of Australia are podocarps, as is the white "pine" (*P. dacrydioides*) of New Zealand. This tree, which attains a height of 200 feet (60 meters) or more, is the largest of the podocarps.

Podocarps exhibit considerable variation, in that, in addition to the aforementioned species, there are some podocarps that resemble the yews; and others that resemble the cypresses. A subfamily of *Podocarpaceae*, known as the *Dacrydium* branch, includes trees with scale-like leaves reminiscent of some cypresses or junipers. In this family branch are the New Zealand rimu (*D. cupressinum*), a small, weeping-type tree, and the Tasmanian Huon "Pine" (*D. franklinii*), which has been described as a cypress-like weeping willow. In Chile, a yew-like podocarp, Prince Albert's yew (*Saxegothaea conspicua*) flourishes. Another yew-like podocarp is *P. andinus* of Chile. There is one podocarp native to Japan, the only known native podocarp of the Northern Hemisphere.

POINCIANA. Sometimes referred to as the *flame tree* or *flamboyant tree*, the poinciana (*Delonix regia*) is regarded by some authorities and beholders as the world's most beautiful tree. The tree is native to Madagascar, but is found in ornamental plantings in several tropical and semitropical regions, such as the Caribbean islands, Central and South America, including Mexico, Hawaii, and southern Florida and the Florida keys. When in Cuba at the turn of the century, Theodore Roosevelt observed, "The tropical forest was very beautiful, and it was a delight to see the strange trees, the splendid royal palm and a tree which looked like a flat-topped acacia, and which was covered with a mass of brilliant scarlet flowers." However, aside from its beauty, the poinciana is deficient in most other respects. The tree produces no fruit and the wood is considered inconsequential and inferior, both for working and as a fuel.

Poincianas grow at a very fast rate, estimated at about twice the growth rate of a Chinese elm. The tree can reach a height of about 20 feet (6.1 meters) in five years. In urban plantings, where the climate is suitable, the poinciana frequently is used, often as a replacement for other trees that die from disease or are accidentally destroyed. The average poinciana achieves a height of from 25 to 30 feet (7.6–9.1 meters), with a trunk diameter ranging between 1.5 and 2 feet (0.5–0.6 meter). The shape of the tree is reminiscent of the African mimosa, with the crown spread out to form a flat, slightly rounded head.

The flowers have brilliantly red petals and appear at the extreme tips of leaf-bearing twigs. When fully open, the flowers measure 4 to 5 in. (10–13 cm) across. As observed by J.M. Haller (*American Forests*, 49–58, June 1982), "Gorgeous when examined singly, doubly so when seen against the feathery green of the leaves, the flowers must be reckoned among the most beautiful of all arboreal flowers. The seed pods are as picturesque as the flowers and, like them, help establish the tree as a member of the great legume family. Warped like the pods of the honey locust, they are two to three times longer than those; it is always surprising to find so small a tree producing such enormous seed containers." The average pod measures $20 \times 1.5 \times \frac{3}{8}$ in. ($51 \times 4 \times 1$ cm). One of the five flower petals is flecked with red and yellow, rather than being solid red as are the others. This variegated petal always has an upward orientation and is on top, thus guiding insects directly into the nectar cup at the pistil's base.

The record tree is located in Florida with a circumference of 102 inches (259 centimeters), a height of 61 feet (16.59 meters). and a spread of 57 feet (17.37 meters).

POINSETTIA. See **Euphorbiaceae**.

POINT. 1. An element of geometry that has position but no extension. 2. An element of geometry defined by its coordinates, such as the point (1,3). 3. An element that satisfies the postulates of a certain space. 4. In positional notation, the character, or the location of an implied symbol, which separates the integral part of a numerical expression from its fractional part. For example, it is called the *binary point* in binary notation and the *decimal point* in decimal notation. If the location of the point is assumed to remain fixed with respect to one end of the numerical expressions, a fixed-point system is being used. If the location of the point does not remain fixed with respect to one end of the numerical, but is regularly recalculated, then a *floating-point system* is being used. A *fixed-point system* usually locates the point by some convention, while a floating-point system usually locates the point by expressing a power of the base.

POINT SOURCE. No finite source of radiation is a true point, but any source viewed from a distance sufficiently great compared to the linear size of the source may be considered as a point source. Point source is a term also used in describing the origin of air or water pollutants — as contrasted with an area or regional source.

POISSON DISTRIBUTION. The Poisson distribution is a discrete distribution with one parameter, whose probability function is given by $P(r) = e^{-\mu}\mu^r/r! \, r = 0, 1, 2, \ldots$. The mean and variance are both equal to μ, and are best estimated from the sample mean. For large μ, the distribution approaches normality. The binomial distribution $(q + p)^n$ approaches the Poisson distribution as a limiting form when $n \rightarrow \infty$ and $p \rightarrow 0$ in such a way that $np = \mu$ remains constant. If events occur in such a way that the probability of an occurrence in a small interval of space or time dt is $\lambda \, dt + O(dt^2)$, independently of other intervals, the numbers of events in equal finite intervals follow the Poisson distribution. For this reason such a series of events is known as a Poisson process.

POISSON EQUATION. An inhomogeneous analogue of the Laplace equation, a partial differential equation of the form

$$\nabla^2 \phi = f(x, y, z)$$

It occurs in (1) electrostatics, where ϕ is potential due to a charge distribution of volume density ρ and $\phi = -4\pi\rho$; (2) thermal conductivity, where ϕ is the temperature in a homogeneous medium of thermal conductivity k and in which $A(x, y, z)$ calories of heat are generated per unit of volume and time, so that $f(x, y, z) = -A/k$. When no heat is generated in the medium, $A = 0$, or when there is no charge, $\rho = 0$; hence, Laplace's equation results in each special case.

See also **Electromagnetic Phenomena**; and **Laplace Equation**.

POISSON, SIMEON DENIS (1781–1840). Poisson was a French Mathematician. His important works include a series on definite integrals and his advances in the Fourier series. He published almost 400 mathematical works including applications to electricity, magnetism, and astronomy. The Poisson distribution established a law governing the distribution of rare and randomly occurring events.

Poisson taught at Ecole Polytechnique from 1802 until 1808. Then he became an astronomer at Bureau des Longitudes. In 1809 he was appointed the chair of pure mathematics at the Faculte des Sciences.

See also **Electromagnetic Phenomena**; **Gravitation**; **Poisson Distribution**; **Poisson Equation**; and **Poisson's Ratio**.

J. M. I.

POISSON'S RATIO. If a rod of elastic material is stretched with sufficient force it can be elongated. The unit elongation (elongation per unit of length) is the strain, and may be denoted by s. At the same time the lateral dimensions will contract, the unit lateral contraction being c. The ratio c/s, which is constant for a given material within the elastic limit, is known as Poisson's ratio. For materials in which there is no directionality to elasticity, the value of Poisson's ratio was demonstrated by that celebrated mathematician to be 0.25. The value of 0.30 is generally used for steels, although recent careful determinations indicate 0.28 is a better average value. For aluminum alloys 0.33 is generally used. For values of Poisson's ratio up to 0.50, stretching results in a net increase in volume. At 0.50 the volume remains constant, as in the case of plastic deformation of metals. See also **Elasticity**.

POLAR BEAR. See **Bears**.

POLAR COORDINATES. If r is the distance from the origin of a rectangular Cartesian coordinate system to a point (x, y, z) and if the direction angles of a line drawn from the origin to the point are α, β, γ, then the polar coordinates of the point are given by

$$x = r\cos\alpha; \quad y = r\cos\beta; \quad z = r\cos\gamma; \quad r^2 = x^2 + y^2 + z^2$$

This system is generally called spherical polar coordinates.

If the point lies in a plane determined by a pair of the coordinates, the XY-plane, for instance, then $z = 0$, and with the usual symbols, $\alpha = \beta = (\pi - \theta)$

$$x = r\cos\theta; \quad y = r\sin\theta; \quad \theta = \tan^{-1} y/x$$

The coordinate origin is called the *pole*; the X-axis is the *polar axis*; the angle θ is the *polar* or *vectorial angle* (sometimes the *azimuth* of the point); r is the *radius vector*. Complex numbers are often plotted in this way, the vectorial angle then being called the *amplitude, argument,* or *phase*, and the radius vector is the *modulus*. See also **Argand Diagram**.

POLAR DISTANCE. Angular distance from a celestial pole; the arc of an hour circle between a celestial pole, usually the elevated pole, and a point on the celestial sphere, measured from the celestial pole through 180 degrees. If the declination, d, and the celestial pole are of the same name, the polar distance is 90 degrees $-d$, but if of contrary name, it is 90 degrees $+d$.

POLAR FRONT THEORY. Originated by the Scandinavian school of meteorologist, a theory whereby a polar front, separating air masses of polar and tropical origin, gives rise to cyclonic disturbances, which

intensify and travel along the front, passing through various phases of a characteristic life history. See also **Fronts and Storms**.

POLARIMETER. An instrument for determining the degree of polarization of electromagnetic radiation, specifically the polarization of light.

POLARIMETRY. The basic principles of polarimetry as a method of quantitative chemical analysis were established over 150 years ago. The method is simple and nondestructive. A polarimeter measures the angle of rotation of linearly polarized light upon passage of the light from the unknown sample. Saccharimetry represents the polarimetric analysis of sugar and is a specialized area with its own form of instrumentation and well-established procedures of international acceptance. Polarimetry in other fields is less standardized, but is extensively used for the qualitative determination of numerous alkaloids, steroids, pharmaceutical, and organic chemical products. See also **Saccharimeter**.

Polarimetric instruments are operative with asymmetric molecules in the direct measurement of circular dichrosm (i.e., the difference of absorption of the left and right circularly polarized light as it passes through the sample). The technique is analogous to absorptiometry.

When plane polarized light passes through an anisotropic medium, the refractive indices of the two beams which emerge, which are right-hand and left-hand polarized, respectively, are not the same. This causes a phase difference between the two component beams and the resultant beam is rotated in its plane of polarization as it emerges from the medium.

Molecules of inherent structural asymmetry are anisotropic; they are *optically active* and exhibit *optical rotation* in solution. The typical optically active center is a carbon atom with four different substituents. In addition, any structural dissymmetry that results in a spatial left- and right-handedness will cause optical activity. Compounds of these types of come in a right-hand (R) and left-hand (L) form. When equal amounts of these two forms are mixed (racemic mixtures) there is no optical rotation because the activity of the two forms exactly cancel. Internal compensation of optically active centers in complex molecules is also found. Left- and right-handed optical isomers were first studied by Pasteur well over 100 years ago, and extensive surveys are found in most organic chemical texts.

Visual Polarimeters. A typical visual polarimeter is shown in Fig. 1. Light source may be a sodium or a mercury arc (less usual is the cadmium arc for the 509-millimicron and 644-millimicron lines). A filter isolates the emission line for monochromatic illumination. (While an instrument does not produce absolutely monochromatic light, the term is used by spectroscopists to describe light within a very narrow wavelength range, such as 0.2 millimicron.) The light then passes a polarizer prism system. This is usually a Nicol prism (a prism made of calcite that is cut and recemented in such a way that the incident light is split into a linear polarized beam which is transmitted, while the second beam is reflected and absorbed). The polarized beam is then passed through the analyzer, which is essentially identical with the polarizer. One of these two elements (usually the analyzer) can be rotated, and it is provided with a graduated circle for the precise read-out in angular degrees. By using a large circular scale and a vernier, a precision of $0.002°$ can be obtained in research-type polarimeters.

The principle of measurement is straightforward. If the two "Nicols" are oriented identically with respect to their optic axes, maximum light is passed. When they are crossed (90°), the intensity is at minimum (following a $\sin^2$ law). A refinement in all commercial visual polarimeters is that the observation of the crossed analyzer position is made easier by a half-shade field. See Fig. 2. Because the human eye is a comparative, rather than absolute, light-measuring device, very much better precision can be obtained by comparing two adjacent fields, rather than attempting to evaluate the brightness of a single field. The half-shade fields are created by an auxiliary prism, and the details of the optical arrangement can be found in the literature. Here we are only concerned with the operational features. The zero position of the instrument is that angle at which the two (or three) segments of the observed field are *equally* dim. Between the polarizer and the analyzer, a space is provided to accept the sample. The sample is placed in a tube that has precisely ground ends corresponding to the light path. End windows are held to the tube by gasketed fittings.

Routine polarimetric determinations are simple enough. First the polarimeter is balanced to zero degrees with the solvent. Then the solution is placed into the instrument, the instrument is rebalanced, and the angle α read off the scale. Nevertheless, when many measurements are taken, this

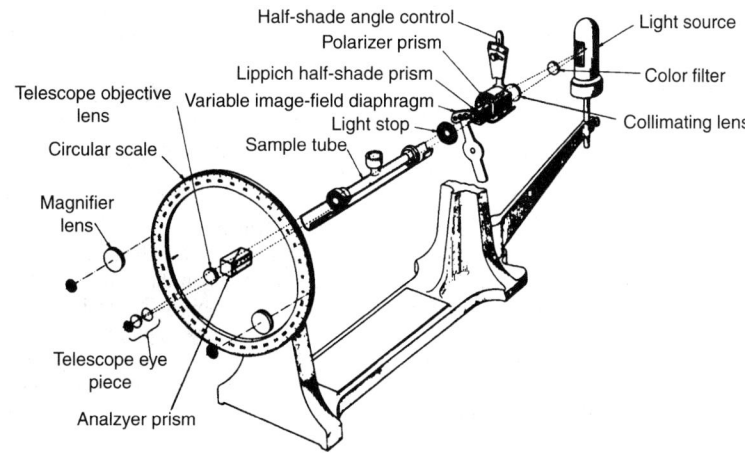

Fig. 1.   Schematic diagram of a visual polarimeter

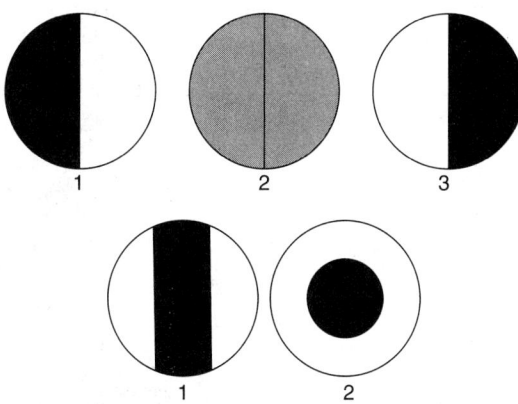

Fig. 2.   Aspect of half-shade fields. Top diagram shows conventional split fields with (2) at the balance point. Bottom diagrams show special field configurations when at the off-balance point.

becomes somewhat tedious. For the assessment of the half-shade field, the operator's eyes must be dark-adapted. Extended work in a darkened room peering through the eyepiece at an almost black field is tiring. The precision of visual polarimetric measurements will tend to increase rapidly at first, as the observer's eyes become adapted, but then it will decrease gradually because of fatigue. These facts justified the introduction of photoelectric polarimeters.

Photoelectric Polarimeters. In one design, there is no split half-shade field, but rather the analyzer is mechanically flip-flopped over an adjustable angle. At balance (and only at balance), the two extreme positions of the analyzer yield equal (low) intensity. A sensitive photomultiplier-photometer serves as a null indicator. The analyzer prism is manually rotated until a minimum deflection results on the large photometer scale and then the angle of rotation is read visually. In another design, Faraday cells are used. In the Faraday effect, a magnetic field induces optical rotatory power in liquids and solids by its influence on the atomic electron configuration. By using an electromagnet surrounding a glass rod, or a suitable crystal or solution, an alternating optical rotation can be introduced. This is analogous to the flip-flop just described. In the second design, a rotating polarizer is used, followed by the sample cell and the Faraday modulator. The emerging light passes through the analyzer and a photomultiplier pickup. The latter actuates a servosystem which drives the polarizer to balance. When the polarizer is exactly crossed with the analyzer, then, and only then will the alternating polarization introduced by the Faraday modulator have equal magnitude. This is the null point for the servosystem which thus establishes the balance automatically. The operator places the sample into the instrument (after it has been set to zero with the solvent) and then reads off the value of rotation from a magnified scale which allows estimation to 0.0025°. Several other designs have been developed for various measurements in optical rotatory dispersion. A number of spectropolarimeters, both automatic and recording, are available.

Kinetic Polarimetry. Polarimetry is particularly well suited for kinetic studies. The reason lies in the cyclic nature of the phenomenon, which allows the measurement of small changes in the angle of rotation with equal precision in the presence and absence of large background values. Moreover, subtle changes in structure, which are common in enzyme reactions, are often strongly reflected in rotatory power.

Spectropolarimetry. The determination of a rotatory dispersion curve is a prerequisite to the establishment of a sensitive polarimetric technique. This is analogous to colorimetry, where a complete spectrophotometric curve is required to establish the absorption peak best suited for routine work at a fixed wavelength. Similarly, in polarimetry, the gain in sensitivity may be enormous when working at an extremum (peak or trough). Thus, spectropolarimetry plays a role even though the analytical technique is simply polarimetry at a fixed wavelength. It is important to make a judicious choice in wavelength in an industrial assay.

POLARIS (α *Ursae Minoris*). The principal star of the constellation Ursa Minor. The fact that Polaris is located in the "Little Dipper" is about the only claim to recognition that this constellation has. Polaris has been the closest star to the pole of rotation of the celestial sphere for the past three millenniums and has been used by navigators throughout written history. The antiquity of the use of this star is attested to by the fact that it is found represented on the earliest known Assyrian tablets. At present, Polaris is slightly over 1° from the pole of rotation and hence revolves about the pole in a small circle about 2° in diameter. Twice, and only twice, during every 24 hours Polaris accurately defines the true north azimuth.

At present, the star is universally used by navigators and surveyors for the purpose of determining true azimuth and also astronomic latitude. Tables have been computed and are to be found in the Ephemerides of the various governments for the purpose of reducing the actual position of the star at any particular instant to the actual position of the pole of rotation.

POLARISCOPE. An instrument for detecting polarized radiation and investigating its properties.

POLARIZATION.   1.   The process of bringing about a partial separation of electrical charges of opposite sign in a body by the superposition of an external field.

2.   A vector quantity representing the dipole moment per unit volume of a dielectric medium. In rationalized units, the electric induction in a dielectric is given by $\mathbf{D} = \varepsilon\mathbf{E}$, which can be written

$$\mathbf{D} = \varepsilon_r\varepsilon_0\ \mathbf{E} = \varepsilon_0 E + (\varepsilon_r - 1)\varepsilon_0\mathbf{E}$$

where ε_r is the relative permittivity or dielectric constant (κ) of the medium. The term

$$(\varepsilon_r - 1)\varepsilon_0\mathbf{E}$$

is the additional induction attributable to the matter of the dielectric, and is called the polarization of the dielectric. The coefficient $(\varepsilon_r - 1)$ is the "electric susceptibility" of the dielectric, and is often written as χ_e. In unrationalized systems,

$$\chi_e = \frac{\varepsilon_r - 1}{4\pi}$$

3. The process of confining the vibrations of the magnetic (or electric) field vector of light or other radiation to one plane.

4. The formation of localized regions near the electrodes of an electric cell during electrolysis, of products which modify (usually adversely) the further flow of current through the cell.

POLARIZED LIGHT. Whenever ordinary light is reflected from a glass plate, a varnished table-top, or other polished dielectric surface, we find upon suitable examination that a much larger part of the reflected beam is vibrating at right angles to the plane of incidence than in that plane; whereas in the incident beam there was no evidence of any preferential direction of vibration. A little experimenting shows that at a certain angle of incidence (the "polarizing angle," different for different dielectrics), the component vibrating in the plane of reflection is practically extinguished, all vibration being confined to the plane at right angles to this. The light is then said to be plane-polarized. The effect is more conveniently produced by a Nicol prism or by one of the polarizing films that polarize by transmission with less loss of light.

When the light passes through two such polarizers in succession, as in a polariscope, the fraction of it finally emerging depends upon the angle between the transmission planes of the polarizers, and varies all the way from nearly 100% to zero (see also **Malus Cosine-Squared Law**). The same effect may be produced by two reflections at glass plates turned to reflect in different planes but at the same (polarizing) angle. It seems probable that when the polarizing films above mentioned have been further perfected and cheapened, this intensity-reducing effect will be turned to account in reducing automobile headlight glare.

Metallic reflectors do not produce plane-polarization, but when plane-polarized light falls on a polished metal, its vibration is in general changed from a rectilinear to an elliptic one, and the light is said to be elliptically polarized.

When plane-polarized light traverses a crystal exhibiting double refraction, such as calcite, at right angles to its axis, it is transformed into elliptically polarized, or even circularly polarized, light.

If plane-polarized light is passed through quartz along its axis, or in any direction through one of the many optically active liquids such as turpentine or sugar solution, it undergoes optical rotation. That is, its vibration plane is twisted around through an angle that steadily increases with the distance traversed in the substance. Different substances have very different rotatory power, and some rotate one way and some the other. The sacharimeter is especially adapted to the study of this effect in liquids, especially sugar solutions.

POLAROGRAPHIC ANALYZERS. Polarography is an electrometric method of chemical analysis that is based on the current-voltage relationship at a special type of electrode. In most electrometric processes, it is desirable to use electrodes of relatively large surface area and, in some cases, to stir the solution, and thus avoid an effect termed "concentration polarization." The cause of this effect is the formation of a region in the solution that differs in composition from the main body of the solution. There is a depletion layer in cases where the ions of the solution are discharging on the electrode, and an excess layer in cases where the electrode is ionizing. Under such conditions, the value of the current at the electrode will not be determined solely by the usual factors: impressed electromotive force, electrode potential, and concentration of the ions in question in the body of the solution. It will also be affected by the rate of diffusion of these ions, from the body of the solution into the depletion layer. Since this rate of diffusion depends in turn upon the concentration of the ions in question in the body of the solution, it can be used as a measure of that concentration by constructing a cell in such a way as to maximize the effect of "concentration polarization." This is done by using a *microelectrode* (an electrode having a very small area of contact with the solution) as the electrode at which the ionic reaction to be measured is to occur, and a large or normal size electrode as the other one in the cell. As a further step, there is added to the solution an excess of an electrolyte that is inert to the electrochemical reaction, so that the diffusion effect on the ion under analysis will not be masked by its migration effect, that is, by its role in carrying current through the cell.

The conditions described for the polarized electrode are met by the dropping mercury electrode, shown in Fig. 1. The size of the polarized electrode, which is merely a forming drop of mercury, is certainly small. It has the further advantage that it is constantly being replaced so that no solid deposit can form on its surface and so contribute an unwanted polarization effect to the one being measured. (Since the polarographic process involves the reduction of the ions, there is a deposition of metal on the mercury drop in all cases in which they are reduced to the metal, the effect of which is overcome by the constant replacement of the drop with fresh mercury.) This advantage is not processed by the other electrode, the pool of mercury in the bottom of the cell. For that reason, among others, the mercury pool electrode is not used as the other electrode in many instruments, being replaced by a standard calomel electrode connected electrically to the solution by a salt bridge, as described earlier in this entry. Moreover, the simple galvanometer and slidewire arrangement for measuring current and potential difference, respectively, is replaced and supplemented by more sensitive and easier-operated measuring and control devices, often automatic in operation and provided with recorders for producing the graphs of the current-potential difference directly. In fact, this instrumentation is so sensitive that it records the small fluctuations in the graph that occur during the formation of the drop of mercury and the beginning of the next one, the characteristic polarographic waves shown in Fig. 2. As can be seen from this figure, the polarographic method makes it possible to measure more than one concentration at a time, as represented by the two waves shown.

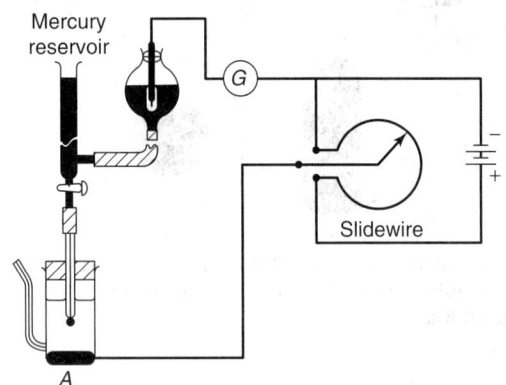

Fig. 1.   Dropping mercury cathode assembly.

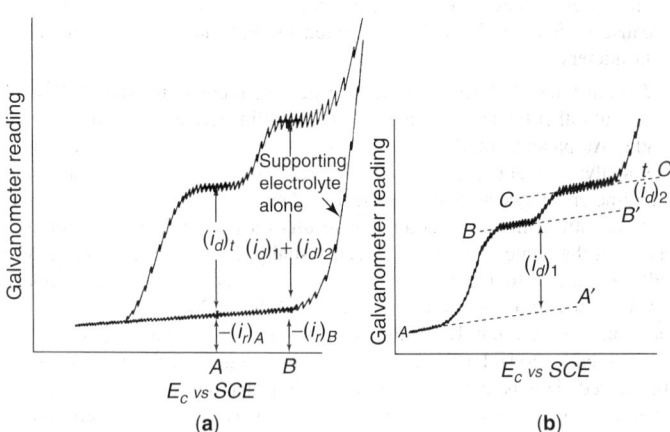

Fig. 2.   Measuring a diffusion current: (**a**) exact method; (**b**) extrapolation method.

In addition to its advantage of permitting the determination of more than one ionic concentration in the same run, polarography also is most useful in determining very small ionic concentrations, of the order of millimoles per liter or lower.

POLAR RESEARCH. The term *polar* is applied to the regions of both the North and South Poles of the Earth. These regions lie within the Arctic and Antarctic Circles, respectively. In many ways, the two polar regions are similar—low mean annual temperatures; oceans with the presence of sea-ice cover; the presence of ice sheets, glaciers, and ice shelves; alternating 6-month periods of continuous daylight and darkness at the poles, as well as auroral phenomena and geomagnetic disturbances. On the other hand, the two polar regions are also quite different. With exception of the Greenland Ice Sheet (1.8×10^6 square kilometers; 0.7 million square miles), the central Arctic region has few large land masses and comprises the Arctic Ocean (14×10^6 square kilometers; 5.4 million square miles). In the Arctic region, there are significant land-dwelling forms of fauna and flora and indigenous human populations, as well as newcomers to the region from the south. In contrast, the Antarctic region is a continent that is surrounded by the Southern Ocean (36×10^6 square kilometers; 13.9 million square miles). On this oceanic area (about 10% of the world's oceans) are found the *Antarctic Ice Sheet*, which in reality represents two ice sheets that have been butted together—the large *East Antarctica Ice Sheet* and the smaller *West Antarctica Ice Sheet*. The larger ice sheet (13.5×10^6 square kilometers; 5.2 million square miles) is known to be at least 4,000 meters (13.124 feet) thick in some locations. It contains about 85% of the world's ice. Living forms, by comparison with the Arctic, are sparse.

The Polar Areas in Perspective

Exploration of the Earth's polar regions has fascinated many people over the years, ranging from those persons who simply were curious and who sought high adventure, to opportunists, to scientists and historians. In retrospect, polar explorations represented tremendous challenges to people and equipment. Expeditions to the Arctic date back to the 1600s; to the Antarctic, the late 1700s. An abbreviated chronology of principal polar exploration events is given in the accompanying *editorial inset*. During the last few decades, the sheer adventurism of the polar regions has waned because of great improvements in transportation and communication, the establishment of bases, improved housing, clothing etc. even though the regions embrace some of the most rugged environments encountered on Earth. The curiosity of the regions from a geographic standpoint has abated because of the availability of good maps and geodetic information. The regions are now fascinating and intriguing from a scientific standpoint because, as briefly described in this article, much information about the Earth and even the solar system can be gleaned from polar research.

Mainly catalyzed by the International Geophysical Year (IGY), which extended over 1957–1958, scientific interest in the polar regions has increased over the last 40 years. This research is essentially targeted toward five topics—the atmosphere, the hydrosphere, the cryosphere, the lithosphere, and the biosphere. Over 60 important research stations, operated by 16 different countries, are located in the polar regions. These stations are supported by dozens of additional minor research stations. See Table 1.

TABLE 1. POLAR RESEARCH STATIONS[a]

Country	Number of Stations	
	Arctic	Antarctic
Russia[b]	6	7
Argentina		10
United States	3	5
Canada	5	
United Kingdom		5
Australia		3
Norway	1	2
New Zealand		2
Denmark	2	
Finland	2	
Japan		2
Chile		2
South Africa		1
France		1
Poland		1
Sweden	1	
Total	20	41

[a]Does not include numerous minor research stations.
[b]Not officially published as of 1994.

Arctic Research and Policy Act of 1984. This U.S. Congressional Act was designed to advance Arctic research in the national interest. Some of the research fields that require attention are weather and climate; national defense; renewable and nonrenewable resources; transportation; communication and space disturbance effects; environmental protection; health, culture, and socioeconomics; and international cooperation. As reported by the Polar Research Board of the National Research Council, a research framework recommended by the U.S. Arctic Research Commission includes, in order of priority, integrated investigations to understand: (1) the Arctic Ocean, including the marginal seas, sea ice, and seabed, and how the ocean and atmosphere operate as coupled components of the arctic system; (2) the coupled atmosphere and land components and how their interaction governs the terrestrial environment; and (3) the high-latitude upper atmosphere and its extension into the magnetosphere with emphasis on predicting and mitigating effects on communications and defense systems.

Targets for research in the Antarctica, as fostered by the nations that participated in the IGY, follow along somewhat similar lines.

The Antarctic Treaty. The text of this treaty was drawn up in 1959 by the governments of Argentina, Australia, Belgium, Chile, the French Republic, Japan, New Zealand, Norway, the Union of South Africa, Russia, the United Kingdom, and the United States.

The Preamble of the Treaty reads:

> Recognizing that it is in the interest of all mankind that Antarctica shall continue forever to be used exclusively for peaceful purposes and shall not become the scene or object of international discord;

Acknowledging the substantial contributions to scientific knowledge resulting from international cooperation in scientific investigation in Antarctica;

Convinced that the establishment of a firm foundation for the continuation and development of such cooperation on the basis of freedom of scientific investigation in Antarctica as applied during the international Geophysical Year accords with the interests of science and the progress of all mankind;

Convinced also that a treaty ensuring the use of Antarctica for peaceful purposes only and the continuance of international harmony in Antarctica will further the purposes and principles embodied in the Charter of the United Nations;

The Treaty was ratified by all participating nations by 1961.

Polar Climate

It has been established for many years that the polar regions serve as large heat sinks and in so doing provide an unusual degree of stability to the global climate. Even after several decades of research, the polar atmosphere-ice-ocean system is poorly understood. Weather forecasting for the polar region per se, for example, is difficult, and it is now recognized that the system is much more sensitive and reactive than previously thought. There is an important feedback system made up of several meteorological elements, including the nature and extent of snow cover, which in turn affects albedo (solar energy reflectivity from polar surfaces) and the temperature. Particular interest in recent years has been directed to the possible modulating effects on global climate by the Antarctic, notably over much shorter time spans (10 to 100 years), as contrasted with the recognized longer-term (1,000–100,000 years) effects. For example, large-scale ice sheet surges could increase the albedo of the Southern Ocean to the point where global temperatures would be reduced, triggering glaciation in the Northern Hemisphere. However, not all scientists agree with this hypothesis, noting that evidence to date indicates that the Antarctic ice sheet has been quite stable for over 50,000 years.

Considerable research is going forth to show a correlation with past polar climates and global climates as a whole. Studies involving sedimentology, paleontology, palynology (study of pollen and spores, living or fossil) paleosols (a buried soil horizon of the geologic past), glacial geology, periglacial (frost action) features, and ice cores are being studied by a number of polar investigators.

As described in the entry on **Climate**, there has been much concern in recent years over the increasing content of carbon dioxide in the atmosphere (mainly from combustion of fossil fuels), which theoretically should cause a warming of global climate. As pointed out by Washburn (1980), the polar regions clearly constitute an excellent focus for monitoring changes in atmospheric carbon dioxide and climate. It has been predicted that breaking up of blocking ice shelves, as a result of Antarctic

THE POLAR REGIONS IN PERSPECTIVE

THE ARCTIC

Arctic exploration dates back to 1587 when John Davis (England) surveyed the Davis Strait to Sanderson's Hope (72°12′N). Other famous names involved in Arctic exploration during the 1600s and 1700s included Barents and van Heemskerck (both of Holland) who discovered Bear Island and touched the tip of Spitsbergen (79°49′N). Henry Hudson (England) reached the north of Spitsbergen (80°23′N) in 1607. In 1616, William Baffin and Robert Bylot (England) explored Baffin Bay to Smith Sound. Vitus Bering (Russia) proved that Asia and America were separate continents in 1728. In 1771, Samuel Hearne (Hudson's Bay Co.) traveled overland from Churchill to the mouth of Coppermine River. James Cook (Britain), in 1778, traveled through the Bering Strait to Icy Cape, Alaska and North Cape, Siberia. In 1789, Alexander Mackenzie (Britain) traveled from Montreal to the mouth of the Mackenzie River.

Considerable exploration of the Arctic commenced in the 1800s. William Scoresby (Britain) reached a location north of Spitsbergen (81°30′N). In the period 1820–1823, Ferdinand von Wrangel (Russia), joining the ventures of James Cook, confirmed the separation of Asia and North America. In 1845, in search for the Northwest Passage, Sir John Franklin (Britain) encountered disaster at Lancaster Sound. In 1888, Nansen (Norway) crossed Greenland's ice cap, and, in a futile attempt to reach the North Pole, reached only as far as Franz Josef Land. Salomon A. Andree (Sweden) in 1896, made an unsuccessful attempt to reach the pole by balloon. (The frozen bodies of his team were not discovered until 1930 on White Island, 82°57′N.) Very serious efforts to reach the North Pole commenced in the 1900s. After a number of unsuccessful attempts, Robert E. Peary reached the pole (90°N) on April 6,1909. In 1925, Roald Amundsen (Norway) and Lincoln Ellsworth (U.S.) reached 87°44′N in an attempt to fly to the pole from Spitsbergen. Adm. Richard E. Byrd and Floyd Bennett (Both of the U.S.) flew over the pole for the first time on May 9, 1926. In that same year, Amundsen, Ellsworth, and Umberto Nobile (Italy) flew from Spitsbergen over the pole to Teller, Alaska. They traveled in the dirigible *Norge*. In 1928, Nobile again crossed the pole on May 24, but the airship crashed on the following day. Amundsen later lost his life in a plane crash in an attempt to rescue Nobile.

Comdr. William R. Anderson, on August 3, 1958, in the nuclear-powered submarine *U.S. Nautilus*, led the first team to cross the North Pole beneath the Arctic ice. The vessel sailed from Portsmouth, New Hampshire, headed between Greenland and Labrador through Baffin Bay, then west through Lancaster Sound and McClure Strait to the Beaufort Sea. Traveling submerged for much of the voyage, the submarine made 850 miles (1360 km) from Baffin Bay to the Beaufort Sea in 6 days.

On August 16, 1977, the Soviet nuclear-powered icebreaker *Arktika* reached the pole and became the first surface ship to break the Arctic ice pack to gain the pole. On April 30, 1978, Naomi Uemura (Japan) was the first person to reach the pole alone by dog sled. In April 1982, Sir Ralph Flennes and Charles Burton (Britain) were the first persons to circle the Earth from pole to pole. The trek required about 3 years, at a cost of about $18 million On May 2, 1986, American and Canadian explorers reached the North Pole, assisted only by dogs. Thus, they became the first party to reach the pole without some type of mechanical support since the 1909 venture of Peary. The 500-mile (800 km) trip was completed in 56 days.

THE ANTARCTIC

Exploration of the Antarctic region did not commence until the late 1700s. Captain James Cook (Britain) reached 71°10′S during the period 1773–1775. In 1819–1821, Bellingshausen (Russia) discovered Palmer Peninsula, but did not know then that he was on a continent. In 1823, James Weddell (Britain) discovered the Weddell Sea. Charles Wilkes (U.S.) was the first person to posit the existence of the continent of Antarctica (1840). The Ross Ice Shelf was discovered by James Clark Ross (Britain) in 1841–1842. In 1895, a party led by Leonard Kristensen (Norway) landed on the coast of Victoria Island and were the first persons to explore the main continental mass. A British expedition was the first to spend an entire winter in Antarctica (1899).

Robert F. Scott (Britain) discovered Edward VII Peninsula in 1902–1904, reaching a point (82°17′S) east of McMurdo Sound. In 1908–1909, Ernest Shackleton (Britain) used Manchurian ponies in Antarctic sledging. He reached 88°23′S and discovered a route onto the plateau by way of the Beardmore Glacier, thus pioneering a land path to the South Pole. In 1911, Roald Amundsen with a team of 4 men and dogs reached the South Pole on December 14. Shortly thereafter, Capt. Scott reached the pole from Ross Island (January 18, 1912). There were 5 men on Scott's team, all of whom perished. However, they previously had found Amundsen's tent and had reached the pole. Hubert Wilkins (Britain) was the first person to fly over the pole (1928). In 1929, Richard Byrd established Little America on the Bay of Whales. On an airplane flight (November 29, 1929), Byrd and his party crossed the pole by air, having covered a distance of 1600 miles (2560 km). A second expedition to Little America was led by Byrd in 1934–1935, during which an area of some 450,000 square miles (1.17 mil square km) was explored. The party wintered at a weather station (80°08′S).

In 1935, Lincoln Ellsworth flew south along the coast of the Palmer Peninsula and then crossed the continent to Little America, making four landings on unprepared terrain in very bad weather. In 1939–1941, the U.S. Navy launched Operation "Deep Freeze," led by Adm. Byrd. This project established five coastal stations fronting the Indian, Pacific, and Atlantic oceans as well as three interior stations. During this project, more than 1 million square miles (about 2.5 million square km) were explored and mapped. During the International Geophysical Year (1957–1958), scientists from twelve countries participated in Antarctic research. A network of some sixty stations on the continent and sub-Antarctic islands was used to study the oceanography, glaciology, meteorology, seismology, geomagnetism, ionosphere, cosmic rays, aurora, and airglow of the Antarctic region. During that period, V.E. Fuch led a 12-man expedition on the first crossing of Antarctica by land, a distance of 2158 miles (3472 km) in 98 days.

In 1958, several U.S. scientists traveled by tractor from Ellsworth Station on the Weddell Sea to discover a huge mountain range (5000 feet; 1524 meters high) above the ice sheet and some 9000 feet (2743 meters) above sea level. Named the Dufek Massif, this range had been identified earlier by the U.S. Navy, but not fully confirmed. In 1959, a treaty was signed by twelve nations, suspending territorial claims in the Antarctic for a period of 30 years and reserving the continent for research. In 1961–1962, scientists discovered a trough (Bentley Trench) that runs from the Ross Ice Shelf (Pacific side) into Marie Byrd Land and around the end of the Ellsworth Mountains toward the Weddell Sea. In 1962, the first nuclear power plant commenced operation at McMurdo Sound. The longest nonstop flight made in the area of the South Pole was accomplished on February 22, 1963 by a crew (U.S.) working out of the McMurdo Station. In 1964, a British survey team revisited Cook Island by helicopter, this being the first visit there since its discovery in 1775. In 1964, a New Zealand team completed one of the last and most important surveys of the continent, mapping the mountain area from Cape Adare and westward some 400 miles (640 km) to Pennell Glacier. See Fig. 1 for flags of countries carrying on active research.

---

warming, could lead to ice sheet advances possibly sufficient to elevate the sea level by as much as 6 meters (19.6 feet) over a 200-year time span. However, even though carbon dioxide concentrations have been increasing, recent trends in polar regions have been cooling instead of warming.

The polar regions are advantageously located for a number of investigations, such as studying the effects of the upper atmosphere on the lower atmosphere. The Antarctic continent is well situated because of its near-spherical ice surface offset from the geographical axis (SCAR, 1979). Also studied are energy transfer processes in the high-latitude magnetosphere-ionosphere system and the electrical coupling between the upper and lower atmosphere. Studies of radio communication and solar-terrestrial physics also take advantage of the polar location.

Pollution of Polar Atmospheres

In connection with research that has been going forward since 1981, investigators have noted unexplained high-frequency radar echoes from the polar mesosphere at an altitude of approximately 90 km (56 mi). Also, they have observed noctilucent clouds (highest clouds in Earth's atmosphere) in increasing numbers. Experiments at Poker Flat, Alaska, indicated receipt of 50-MHz echoes from the very top of the mesosphere (mesopause). During more recent years, echoes of frequencies as high as 900 MHz have been observed at other radar station locations. J. Ulwick (University of Utah — Stewart Radiance Laboratory) noted that the echoes are coming from electrons in the mesosphere, and their high frequency suggests that the electrons are interacting with dense layers of particles. "The electrons are sticking to something up there like paint," observed

Fig. 1.   The South Pole showing flags of all Antarctic Treaty countries carrying out active scientific research on the continent. (*Source: British Antarctic Survey.*)

R. Goldberg (Goddard Space Flight Center). Atmospheric methane in polar regions has increased about two-fold since 1900, and noctilucent clouds have become nearly ten times brighter. Inasmuch as these clouds (droplets of ice crystals) must be seeded to form, some scientists propose that proton hydrates are responsible. Other researchers suggest charged aerosols as the medium. Cloud sampling has been carried out through the use of over 30 rockets, and much information remains to be processed. Should methane be detected, G. Thomas (University of Colorado at Boulder) notes that pollution even at these very high altitudes may indicate anthropogenic origins.

Antarctic Ozone "Hole." Dramatic losses in the ozone layer over Antarctica, displaying definitive variations, were first noted in 1977 by members of the British Antarctic Survey team. The dips in ozone levels were so large that several research groups, including the British team, as well as scientists from the United States and other countries with a direct interest in Antarctica, were somewhat reluctant to report them immediately, believing that there might have been something seriously wrong with the data. Not until May 1985 did the British group publicly report that a massive decrease in ozone concentrations over Antarctica had been occurring during the Antarctic springtime (September and November) over an 8-year period. Routinely, over the period 1957 to 1973, past records showed that October ozone concentrations in the affected area averaged about 320 parts per billion (ppb). A routine measurement in 1984 showed a concentration of less than 200 ppb, indicating a depletion loss of about 40%. Interim October measurements have confirmed this trend. These ozone losses have been confirmed by numerous measurement methodologies, including balloon-borne, high-altitude aircraft, and rocket sampling.

The possible very adverse effects of a continuously diminishing ozone layer have thrust the study of the ozone "hole" into the forefront of scores of scientific investigations. The ozone layer dampens the passage of ultraviolet radiation from the sun through Earth's atmosphere. Excessive UV radiation can injure organisms of nearly all kinds, promoting skin cancers in humans and retarding the growth of phytoplankton, which is a major key in the food chain of the Antarctic.

In the early phases of this scientific investigation, numerous hypotheses were developed to explain the causes and effects of the depletion of the ozone layer. For example, was this phenomenon related to the sun? Or had the missing ozone been transported elsewhere by some poorly understood meteorological process? These and other postulations did not withstand the rigors of careful examination. Although some form of anthropogenic pollution was suspected, specific connections had to await the development of an understanding of the complex chemistry that occurs. Progressively, a step-by-step series of chemical reactions involving the halogens and notably chlorine appeared to be the essentially reactive and destructive ingredients, and this led to the indictment of widely used, vaporous

chemicals that easily enter Earth's atmosphere. Although other compounds ultimately may be also play a role, the scientific consensus today targets the family of chlorofluorocarbons (CFC 11 and CFC 12), used mainly in refrigerating and air-conditioning systems. Accepted as common knowledge today, this is indicative of the outstanding scientific progress made within a time span of less than two decades. Consequently, regulatory actions and targets have been established by numerous governments throughout the world (e.g., the Montreal Protocol), and it is also of interest to note that refrigerant chemical manufacturers have been searching for CFC substitutes.

Chemistry of Ozone Layer and "Hole." It is very difficult to duplicate in the laboratory the physical and chemical conditions that prevail in the clouds over Antarctica and to show how these conditions differ between the winter and summer seasons. The presence or absence of stratospheric clouds is an important determinant of what chemistry takes place, particularly when pollutants of any nature, natural or anthropogenic, enter into the atmosphere. Explanations to date largely represent theories arising out of "blackboard" chemistry.

One may ask, "How does ozone get into the stratosphere in the first place? Under what may be considered "normal" conditions, the oxygen molecule (O_2) is quite stable. However, when this molecule is subjected to ultraviolet (UV) radiation of just the right frequency (energetic solar radiation), the O_2 molecule can be broken to create two free atoms of oxygen (O). Each of these atoms, in turn, can combine with an oxygen molecule to form ozone (O_3), which is comprised of three oxygen atoms.

It is interesting to note that, although ozone accounts for less than 1 part per million (ppm) of the gases in Earth's atmosphere, it absorbs most of the UV radiation from the sun.

Destruction of ozone occurs mainly in polar stratospheric clouds (PSCs), which contain water ice crystals, sulfuric acid particles, and nitric acid trihydrate particles, among others. The latter, in essence, may be considered *pollutants*, although all are not of anthropogenic origin. Various natural causes contribute, such as the ejecta from volcanic eruptions. But, without regard to source, tiny particles form nuclei for condensation reactions, including those that determine the fate of ozone. Thus, any pollution can indirectly contribute to the sum total of ozone depletion, although certain materials, such as those containing chlorine and bromine, greatly accentuate the process.

Why does the ozone depletion maxima occur in October? Theories to explain this involve numerous factors, such as atmospheric circulation and, more directly, the type of cloud and the manner in which it is formed. PSCs may result from either slow or rapid cooling, factors that reflect the meteorological cycles of the Antarctic.

How is ozone destroyed? A chlorine molecule, for example, can combine with a molecule of ozone to yield a molecule of chlorine monoxide and a molecule of oxygen. Once formed, two chlorine monoxide molecules may form a *dimer*, which is a compound comprised of two chlorine monoxide molecules. Near-UV radiation can catalyze the dimer to form two chlorine atoms and one oxygen molecule. Thus, ozone disappears.

Nearly all of the reactions proposed are reversible, depending upon alterations in the fundamental conditions. For example, if shielded from UV radiation, ozone can be regenerated.

As described by Molina, "Chlorofluorocarbons (CFCs) are unaffected by rain and by the chemical reactions that cleanse most other gases in the troposphere. The CFCs slowly rise into the upper stratosphere, above the ozone layer, where UV is strong enough to break the molecules apart, releasing chlorine atoms that react very rapidly with ozone. Occasionally, these chlorine atoms combine with other chemicals to form relatively stable 'chlorine reservoirs,' which in turn decompose, periodically returning the free chlorine atom to the stratosphere. Each chlorine atom released by the decomposition of a CFC molecule is capable of destroying tens of thousands of ozone molecules before it returns to the Earth's surface."

While all of the physicochemical details of ozone destruction have not been worked out, the proposition that CFCs are a primary cause of ozone depletion now has few doubters. Earlier contradictory theories essentially have been ruled out as of 1993.

The principal driving force for more research pertaining to the ozone "hole" has come from the damaging effects of ozone depletion worldwide. But what are the effects specifically on Antarctica itself?

Considerable research in this regard has been going on at the Palmer Station, located at the tip of the Antarctic Peninsula, about 965 km (600 mi) south of Cape Horn. Observations have tended to indicate that the "hole" tends to run in 2-year cycles. In one year, for example, ultraviolet radiation may be double or one-half that of a prior year. D. Karentz (University of California at San Francisco) notes that the increase in radiation is sudden and occurs at a time when organisms are emerging from the dark winter period and thus have not had time to adapt to the sun being up. Another scientist has proposed the analogy—"like a Norwegian going to the Mediterranean over Christmas vacation."

Thus, the effect on Antarctic organisms not only is one of UV dosage, but one of timing as well. Not all organisms adapt as well and as readily as people.

A major concern is the effect of UV on phytoplankton synthesis. Researchers at Palmer Station have found that this process can decline by as much as 15–20% in the uppermost meter of water, diminishing exponentially with depth, with little if any decline at depths of 10 to 15 meters. Because of UV damage to their DNA, the microorganisms are unable to divide. But researchers also have found that phytoplankton, like some other organisms, introduce a mechanism for repairing DNA damage. The tolerance to UV ranges widely from one species of phytoplankton to the next.

Krill, a mainstay of the Antarctic ecosystem, feed on phytoplankton. Just what effect, if any, genetic changes in the phytoplankton may have on the reduction of krill remains unknown. One scientist has observed, "I am positive there will not be a collapse of the southern ecosystem." Also, it is most likely that larger marine animals, such as the penguin, will not be affected directly by the effects of the ozone "hole."

Ozone Depletion in the Arctic Region. The Airborne Arctic Stratosphere Expedition, conducted by the U.S. National Aeronautics and Space Administration and the National Oceanic and Atmosphere Administration, assisted by researchers from Norway, Great Britain, Germany, and Russia, during January and February 1989, did not reveal an ozone hole comparable to that over Antarctica. However, one researcher observed what he termed "highly perturbed" chlorine chemistry, a probable precursor of an ultimate ozone hole. The scientists were surprised to find chlorine monoxide and chlorine dioxide in abundance more than 50 times higher than normal over a wide range of altitudes. Nitrogen compounds that inhibit ozone destruction were absent. R.T. Watson (NASA) noted, "Ozone loss would be expected if the mass of cold air over the Arctic (known as the Arctic polar vortex)[1] persisted into spring, when the sun rises after the polar winter." The data strongly backed up the concerns that resulted in the *Montreal Protocol* that targets cutting emissions of chlorofluorocarbons in half by the year 1998.

In a subsequent mid-1991 paper, W.H. Brune (Pennsylvania State University) and colleagues from other institutions in the United States

and United Kingdom, summarized the potential for ozone depletion in the Arctic polar stratosphere in the following terms: "The nature of the Arctic polar stratosphere is observed to be that of the Antarctic polar stratosphere.... Most of the available chlorine (HCl and ClONO₂) was converted by reactions on polar stratospheric clouds to reactive ClO and Cl₂O₂ throughout the Arctic polar vortex before midwinter. Reactive nitrogen was converted to HNO₃ and some, with spatial inhomogeneity, fell out of the stratosphere. These chemical changes ensured characteristic ozone losses of 10–15% at altitudes inside the polar vortex where polar stratospheric clouds had occurred. These local losses can translate into 5–8% losses in the vertical column abundance of ozone. As the amount of stratospheric chlorine inevitably increases by 50% over the next two decades (by 2011), ozone losses recognizable as an ozone 'hole' may well appear." See also **Climate**.

Ironical Sources of Pollution. Not all scientific research activity in the polar regions is directed toward determining the anthropogenic sources of pollution throughout the world that affect the well-being of the Arctic and Antarctic. But, when pollution stems from research activities within this region, it is particularly disturbing. There are several research stations in these regions, including the presence of well over a thousand staff and support personnel in one Antarctic station. An Environmental Defense Fund report shows that at least some research stations in the past have polluted the pristine environment with polychlorinated biphenyls (PCBs), used fuel, and emissions from burning waste. A somewhat-questioned report indicated that PCB concentrations in McMurdo Bay ranged from 18 to 340 ppb, as compared with little presence of PCBs in Galveston Bay and 70 ppb in the Oakland Bay.

Polar Atmospheres as Telltales. In addition to ozone depletion, the polar atmosphere may serve in assessing and forecasting other air contaminants on Earth. For example, Mayewski (University of New Hampshire) and colleagues reported in 1986 that an ice core in south Greenland (covering the period 1869–1984) was analyzed for oxygen isotopes and chloride, nitrate, and sulfate concentrations. It revealed that an "excess" (non-sea salt) sulfate concentration had tripled since approximately the 1900–1910 period; nitrate concentration had doubled since about 1955. The investigators suggest that the increases may be attributable to the deposition of these chemical species from air masses carrying North American and Eurasian anthropogenic emissions. This ice core record was derived from a site that is devoid of any locale-specific contamination. It is the longest and most detailed record of anthropogenically introduced sulfate and nitrate available to date. The increase in "excess" sulfate in the record (by a factor of 3) is close to the estimated increase (by a factor of 2.5) previously determined from a calculation of the export of sulfate from North America eastward. This observation also ties in with observations of Arctic and Antarctic haze.

In commenting specifically on the Arctic atmosphere and climate. Washburn (University of Washington) and Weller (University of Alaska) observed that at high latitudes, the upper atmosphere (with magnetosphere and ionosphere) serves as a window to space, in that particles from the sun are focused primarily on polar regions. There, they give rise to magnetospheric and ionospheric phenomena that have a direct impact, for example, on important space processes. These phenomena can strongly impact communications and defense capabilities—charged particles precipitating into the ionosphere, causing auroras, can interrupt rf communications during magnetic storms. Thus, communications with satellite and radar systems become vulnerable. High background noise induced by the aurorta can adversely affect optical and infrared sensors used by surveillance spacecraft; electric currents induced in long conductors, such as telephone cables, power lines, and pipelines, can produce deleterious effects. Thus the need for a better understanding of the polar atmospheres. Because of the prolonged polar nights, the polar regions become excellent laboratories for studying the aurora and for probing the long tail of the magnetosphere. See also **Climate**.

Washburn and Weller also report that any temperature increase due to the greenhouse effect (CO₂ buildup) is intensified in the Arctic because of the melting of snow and ice and the accompanying reduced *albedo* and changes in energy balance. Some recent numerical models have indicated temperature increases that would affect the distribution of sea ice, navigation in the Arctic Ocean, the northward extent of agriculture and the tree line, and, as land-based glaciers melt, perhaps some global rise of sea level, important to low-lying coasts. Although still considered speculative, some investigators believe that a temperature rise of 6 °C

---

[1] Polar vortex is defined in article on **Atmosphere (Earth)**.

in winter and 1 °C in summer could occur in Arctic Alaska. A change of this magnitude is regarded with mixed reactions. The temperature increases most likely would benefit oil exploration and agriculture with but slight effect on the timber industry, but adversely affect a number of other industries because of thawing of permafrost. Despite large-scale climatic influences and the significance of Arctic weather for regional forecasting and for understanding global weather, important knowledge remains lacking because Arctic observing stations are widely scattered and air-sea-ice interactions are physically complex.

Polar Hydrosphere

The polar oceans influence about three-quarters of the Earth's oceans. Research is being directed toward a better understanding of the bottom sediments of the Arctic Ocean and the Southern Ocean, and on improved data pertaining to ice and seawater exchanges between the Arctic and Atlantic Oceans, which occur between Greenland and Svalbard. Although it is the world's largest ocean current, the Antarctic Circumpolar Current (or West Wind Drift) is still poorly understood. See also **Antarctic Convergence**. Even though the bottom water originating in the Antarctic constitutes over half of the bottom water of the global oceans, its exact mode and place of origin are not fully understood.

Considering the prevailing cold temperatures of Antarctica, it is interesting to mention briefly Lake Vanada in the Wright Dry Valley, where water temperatures near the bottom of the lake run as high as +25 °C (77 °F). The mean annul air temperature in this region is −20 °C (−4 °F). Once believed to be due to volcanic heat, researchers have since concluded that the unexpected high water temperatures derive from a snow-free ice cover that transmits and traps solar radiation, which becomes available to heat the dense saline water that remains at the bottom.

The growth, drift, and decay of sea ice are closely related to the circulation of polar oceans. As reported by Hibler (U.S. Army Cold Regions Laboratory) and Bryan (Geophysical Fluid Dynamics Laboratory, Princeton), this is especially true in the Greenland and Norwegian seas in winter, where warm currents flowing northward encounter rapidly cooling atmospheric conditions together with sea ice advancing southward. These investigators describe a diagnostic ice-ocean model of the Arctic, Greenland, and Norwegian seas that has proved useful in examining the role of ocean circulation in seasonal sea-ice simulations. The model includes lateral ice motion and three-dimensional ocean circulation. (In past studies, the ocean has been approximated by a motionless mixed layer of fixed depth.) The ocean portion of the model is weakly forced by observed temperature and salinity data. Simulation results show that including modeled ocean circulation in seasonal sea-ice simulations substantially improves the predicted ice drift and ice margin locations. Simulations that do not include lateral ocean movement predict a much less realistic ice edge. Additional work of this type is underway in the Marginal Ice Zone Experiment (MIZEX) by CRREL (U.S. Army Cold Regions Research and Engineering Laboratory, Hanover, New Hampshire).

As early as 1978, the NORSEX Group (Norwegian Remote Sensing Experiment) conducted three field investigations in Norwegian waters. This group, consisting of scientists from Canada, Norway, the United States, and Switzerland conducted three important experiments: (1) investigation of sea surface temperature and wind in the Norwegian and Barents seas, as derived from the scanning multichannel microwave radiometer (SMMR) on the *Nimbus 7* satellite; (2) similar measurements as gained from improved remote sensing methods and concentrating on the Norwegian Coastal Current; and (3) specific investigation in the marginal ice zone north of Svalbard. Details of the latter experiment are outlined by the NORSEX group (1983). In their conclusions, the group observed that microwave signatures from measurements at 5 to 100 GHz are excellent for remote sensing of sea ice. The large contrast in emissivity between water and all ice types at low frequencies facilitates the retrieval of data on total ice concentration. At high frequencies, the contrast between first and multiyear ice can be used to separate types of ice. The *Nimbus 7* SMMR was found to locate the ice edges accurately to within 10 kilometers.

Sea level appears to have risen 10 to 15 centimeters during the last century. Part of this rise may be due to thermal expansion of the oceans. See also **Ocean**. The remainder generally has been attributed to the melting of polar ice. However, as reported by Meier (U.S. Geological Survey), studies of the current mass balance of the Antarctic ice sheet, which makes up about 85% of the total glacier ice area on Earth, suggest that a negative mass balance is not likely and that this ice sheet may be subtracting water

from the world's oceans. Observed long-term changes in glacier volume and hydrometeorological mass balance models have yielded important data on the transfer of water from glaciers, excluding those in Greenland and Antarctica, to the oceans. The average observed volume change for the period, 1900–1961, has been scaled to a global average by using seasonal amplitude of the mass balance. These data have been used to calibrate other models to estimate the changing contribution of glaciers to sea level for the period, 1884–1975. Although the error band is large, the Greenland and Antarctica glaciers appear to account for a third to half of observed rise in sea level, approximately that fraction not explained by thermal expansion of the ocean.

In an interesting observation, Washburn and Weller stress that the acoustical characteristics of polar seas, as affected by differing water masses and the background of noise of ice movement, internal waves, and other influences, require further study. Acoustic characteristics of these waters are of key interest in the detection of submarines, tanker movements, oil exploration activities, and even on studies of whales.

Isotopic Composition Studies. During the 1980s and 1990s, much attention was directed toward the use of short-lived radionuclides, such as ^{3}H, ^{14}C, ^{210}Pb, ^{226}Ra, and ^{230}Th, in studies of ocean circulation, notably in the polar regions. As described by Piepgras and Wasserburg (California Institute of Technology), these nuclides have half-lives which are short compared with the time scales of the processes studied. They have been used in conjunction with numerous hydrographic measurements, including tracer studies of oceanic circulation paths, mixing rates, and the chemical behavior and distribution of associated stable elements in seawater.

A problem of interest to oceanographers is that of determining mixing rates in and between the oceans. Studies with ^{14}C have indicated that at least 1500 years are required for the exchange of deep water with the mixed layer. A longer time may be required for the exchange of deep waters between ocean basins. A minimum time of about 150 years to mix the world oceans is obtained by assuming that the Pacific Ocean is emptied by the flow through the Drake Passage and mixed with the Atlantic. The Antarctic Circumpolar Current, which controls interocean mixing, flows through this passage. The isotopic composition of neodymium has been determined in seawaters from the Drake Passage. By using a box model to describe the exchange of water between the Southern Ocean and the ocean basins to the north, together with the isotopic results, an upper limit of approximately 33 million cubic meters per second is calculated for the rate of exchange between the Pacific and the Southern Ocean.

The Polar Cryosphere

This is that part of the Earth's surface that is permanently frozen; the zone of the Earth where ice and frozen ground are formed. Studies of the immense ice sheets in the Antarctic and Greenland can provide a better understanding of the ice sheets that once covered large parts of North America, Europe, and Eurasia at various times during the Pleistocene (10^4 to 1.8×10^6 years ago). Much research in recent years has concentrated on development of a better understanding of the complex Earth-sun relationships that influence climatic change, notably in the polar regions. The nature of the climatic change responsible for the growth and decay of the Pleistocene glaciers are still problematical. The moisture sources and mechanisms permitting the growth of the Northern Hemisphere ice sheets also remain to be established.

The Antarctic Ice Sheet. When Bellingshausen (Russia) confirmed that Antarctica was an ice-covered continent (and not a frozen ocean) in 1820, interest in the Antarctic ice sheet commenced. Numerous expeditions managed ultimately to establish the lateral extent of the ice sheet, but the first estimate of its depth was not made until 1911, when Meinardus (Germany), using clever calculations based upon air moving to and fro over the edge of the ice sheet, claimed the thickness to be about 2100 meters. This early estimate was remarkably close to the depth of 2200 meters estimated by a team during the IGY program. IGY teams also refined the general shape of the ice sheet, the main features of which are the large East Antarctic ice mass and the much smaller West Antarctic lobe (nearly as large as Greenland). The center of the ice sheet does not correspond with the South Pole, but is located at 83 °S, 53 °E and because of its location is sometimes referred to as the *Pole of Relative Inaccessibility*. See Fig. 2. Progress in describing the ice sheet was made during the 1980s and continues. This largely embraces technological advances in satellite observations, remote-sensing devices, and computer simulations.

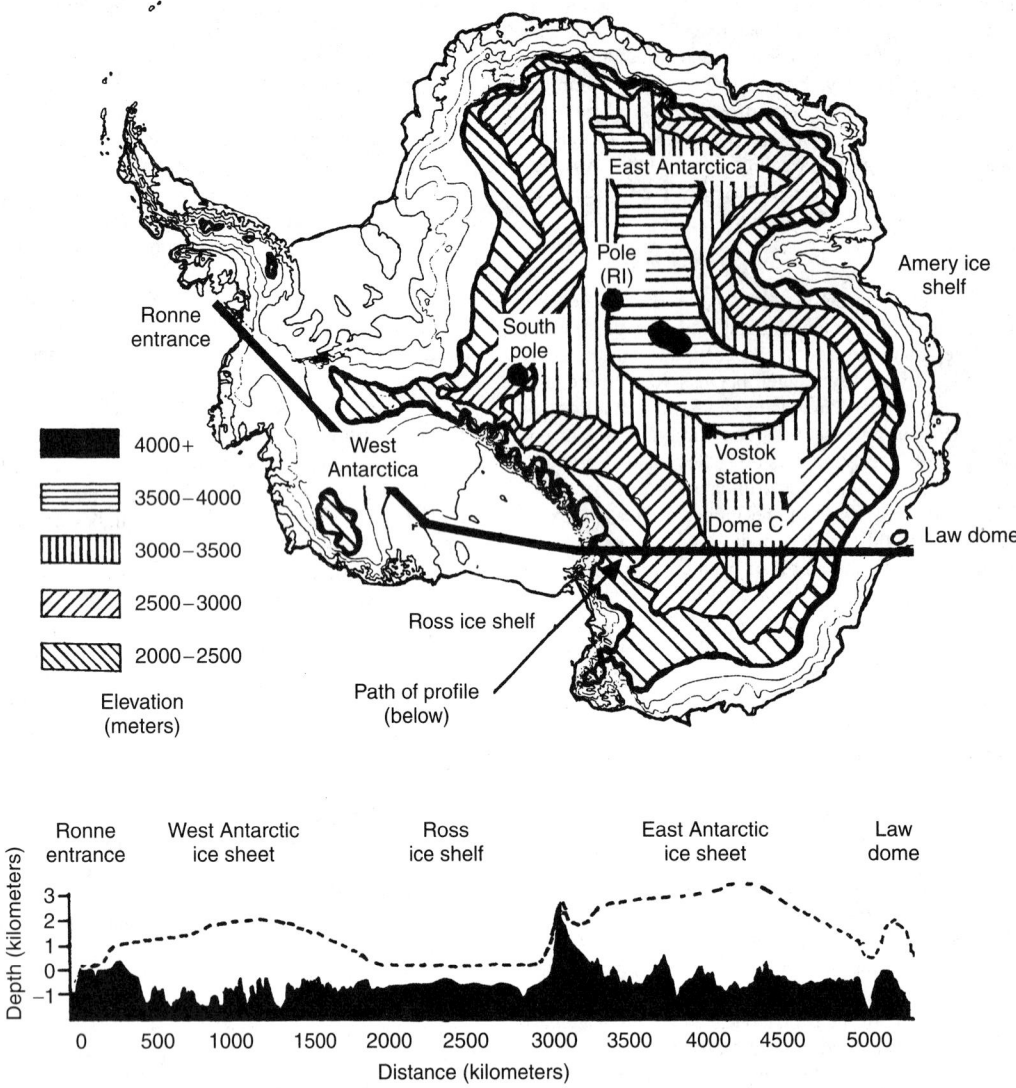

Fig. 2.   Surface of the ice on the Antarctic continent. Pole (RI) indicates the Pole of Relative Inaccessibility (83 °S, 53 °E), located essentially at the center of the ice sheet. The pole (RI) is equally remote from all surrounding coasts. The bedrock (shown in black) and the ice level (dashed line) is shown by profile below map. The line (shown in solid black) is path along which the profile is constructed. Map is based upon data available from the International Geophysical Year (IGY) gathered during the period, 1956–1957 and by the International Antarctic Glaciological Project (IAGP), which has been in continuous operation since 1969. This latter project is comprised of teams from Australia. France, Japan, the United States, and the former U.S.S.R., with a group from the United Kingdom. Data on the bedrock were derived from radar measurements taken through the ice, both from surface ships and airplanes. The technique used is sometimes referred to as *radioglaciology*. (*After Radok.*)

Radok (Cooperative Institute for Research in Environmental Sciences) indicates that the ice sheet poses three broad research tasks: (1) topography (surface and bottom)—where more details are needed pertaining to the surface elevation and thickness of the ice; (2) clarification of the ice regime—temperature at various depths and balance between what is gained as snow and what is lost as icebergs and meltwater; and (3) a better understanding of the structural and chemical properties of the ice. With better information along these lines, models can be constructed which describe the history of the ice sheet and predict its future behavior.

Radioglaciology has proved the most effective means for mapping surface topography, a technique first pioneered by the U.S. Army Signal Corps in 1957. Data are collected from aircraft, satellites, surface ships, and balloons. Internal echoes received from a variety of internal layers indicated that the ice sheet is not uniform. Echoes may be caused by changes in density and crystal structure (possibly from melting or deformation), as well as concentrations of impurities and layers of increased acidity, which could result from very old volcanic activity. Doppler radar has also been used in more recent years. Measurements of this effect on signals from satellites passing over a point on the ice sheet give the position of that point with a precision of a few

meters. A computer-generated view of the Antarctic ice sheet, made from measurements taken during the IGY and IAGP programs, is shown in Fig. 3.

Details on the structural and chemical properties of the ice are obtained principally from drilling cores of ice, usually to a depth of a few hundred meters. A French team has drilled a 900-meter hole on Dome C and the Russians are now drilling a hole of a few thousand meters in depth at the Vostok Station. See also **Earthquakes**; **Seismology**; and **Plate Tectonics**. A representative ice core sample is shown in Fig. 4.

Many clues to the past are found in ice cores drilled from the Antarctic sheet. Among such findings are evidence of global pollution—radioactive fallout from hydrogen bomb tests in the atmosphere (mid-1950s to early 1960s); volcanic eruptions, such as Krakatau (1883) and Guning Agung (1963), among others. Trapped air in the cores represents a measure of the atmospheric composition many years ago. Fischer and Oeschger (University of Bern) report that air entrapped in bubbles of cold ice has essentially the same composition as that of the atmosphere at the time of bubble formation. Measurements of the methane concentration in air extracted by two different methods from ice samples from Siple Station in West Antarctica have allowed the reconstruction of the history of the

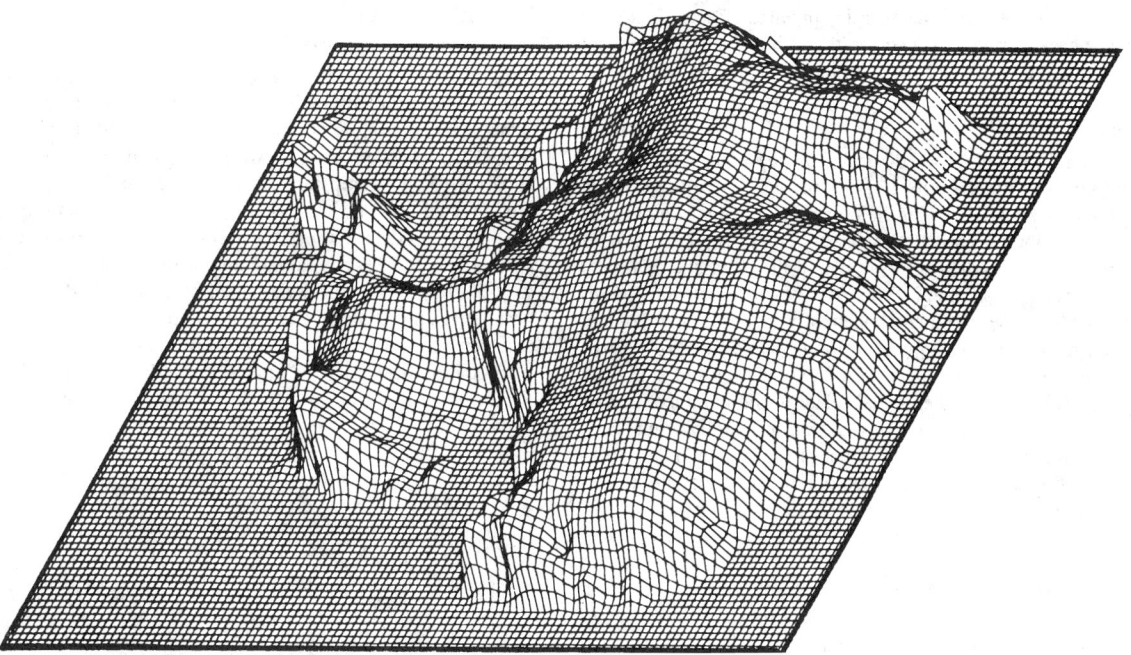

Fig. 3. Computer-generated view of topography of Antarctica ice sheet prepared from data collected during the International Geophysical Year and the International Antarctic Glaciological Project. Average thickness of the ice is estimated at 2,200 meters (7,220 feet.)

Fig. 4. Black-and-white facsimile of a section of an ice core from the East Antarctic sheet. The individual crystal separations indicated result from the use of polarized light. For examination, core sections usually are thin (1 mm). (*After Radok.*)

increase of atmospheric methane for the past 200 years. Many experts feel that it is a bit premature to extrapolate ice core findings to pollution of the atmosphere as derived from anthropogenic causes. In connection with CO_2, some modeling studies have suggested that a CO_2-induced climatic change would include a polar warming several times the global average warming (variously estimated to be from 1 to 4 K for a doubling in CO_2). This would predict a marked change in polar atmospheric temperature and the warming should be observed in the polar regions prior to the tropics. A significant decrease in overall ice extent during the mid-1970s, previously suggested to reflect warming induced by carbon dioxide, has not been maintained. The extent of ice in the Weddell Sea region rebounded after a large decrease concurrent with a major oceanographic anomaly (the Weddell polynya).[2] The decrease of the mid-1970s was preceded by an increase in ice extent from 1966 to 1972, further indicating the presence of cyclical components of variation. As pointed out by Zwally (Goddard Laboratory for Atmospheric Sciences), these cyclical components of variation may obscure any long-term trends that might be caused by a CO_2-induced warming.

As described in article on **Meteoroids and Meteorites**, a number of meteorites have been found on the Antarctic ice sheet, precipitating a controversy as to their source (lunar or martian). Further, scientists are now analyzing Antarctic cover for the presence of dust from meteoroids. Most of these bodies burn up in the Earth's atmosphere, but may leave dust in the atmosphere that ultimately settles down and is preserved in the Antarctic's emerging science storehouse.

Arctic Cryosphere. As in Antarctica, ice cores are revealing fascinating scientific information. Sponsored by the U.S. Arctic Research Commission, the Arctic cryosphere is subject to intensive investigation. The glacial history of large parts of the Arctic remains controversial. The fact that former widespread shelf ice in the Canadian Arctic Archipelago was only recently identified, (1984) illustrates how little is known about this history. The cryosphere is an important resource. In most of the Arctic, water stored in snow and glacial ice; then it is released as runoff to rivers and lakes during seasonal thawing is much more significant than rain as a source of fresh water. It has been suggested that shelf ice can be a transportation resource in some places inasmuch as sections may break off from their parent glaciers and drift around the Arctic Ocean as ice islands (several kilometers in diameter). They may last for years and form platforms for research programs.

The extent of Pleistocene ice sheets in the polar regions is at present unknown. Hughes (1977) and other investigators have suggested that possibly an ice sheet covered a large part of the Arctic Basin, representing greater glaciation than previously believed for that region.

Researchers are studying the sea ice that covers most of the Arctic Ocean all year. The power required to move pack ice is tremendous. The mechanics of floating ice are poorly understood and are being researched. Pack ice also is a problem to the exploitation of petroleum and mineral resources in polar regions.

Permafrost (defined in the entry on **Biome**) is considered part of the cryosphere. In polar regions, permafrost extends to depths exceeding 500 meters (1641 feet) and presents difficult engineering problems, as in

---

[2] A polynya is a region in a frozen sea that stays anomalously free of ice.

the case of construction of the Trans-Alaska pipeline. Recently, it has been found that subsea permafrost also exists offshore and will present difficulties for recovering hydrocarbons from the polar region coastlines.

Polar Lithosphere

Investigations of the lithosphere are described in the entries on **Earthquakes, Seismology, and Plate Tectonics; Ocean**; and **Volcano**. Since the lithosphere obviously also exists under the polar regions, investigations in those regions can contribute to a better understanding of the Earth's interior. Bushnell and Craddock point out that 98% of the Antarctic continent is ice-covered, and thus its geology is poorly known. It is quite possible that the sediments of the continental shelf of Antarctica may contain large hydrocarbon resources. Inadequate exploration of date hardly justifies speculation. The problems of this type of exploration and study are well summarized: "Scientific problems that relate to the continental margins bordering the Arctic Ocean must be assessed differently from those of other continental margins. So little is known of the area, and the working conditions are difficult. We know at least one order of magnitude less about Arctic marine geology and geophysics than about any other areas. Not even the broad tectonic framework of the area has been adequately described.

After passage of the U.S. Arctic Research and Policy Act of 1984, renewed evaluation of our knowledge of the Arctic lithosphere and recommendations for research were made. Summarizing from the Washburn and Weller paper (1986): It is essential to acquire data from deep cores from the Arctic ocean bottom and paleomagnetic data are essential to determine its tectonic and sedimentary history. The sedimentary history may also provide clues to changes in climate and sea-ice cover. The interpretation of even the uppermost sedimentary layers of the lithosphere is still open to debate. The continental shelves are highly significant sources of oil and gas. Offshore exploration for these resources presents many geologic and engineering problems. The stratigraphy and structure of some Arctic lands are very complex, and in Alaska this includes juxtaposition by extensive faulting and lateral displacements of terrains of quite different origin. Oil, gas, coal, and mineral resources of the Arctic are large, especially the energy resources of Alaska. In 1983, the Prudhoe Bay field had a proven crude oil reserve about a quarter that of the entire United States. Coal resources of the Alaska North Slope range from an identified 150 billion to a hypothetical 4 trillion short tons, reflecting limited surface exposures and scattered drilling. Alaska's coal resources as a whole have been estimated as making up 50% of the total in the United States and 15% of world resources.

Other mineral resources of the Arctic region include barite, beryllium, cobalt, copper, fluorite, lead, silver, tin, tungsten, and zinc. The overall potential is unknown. Only 4.5% of the bedrock geology of Alaska has been mapped on a scale of 1 mile (1.6 km). Thus, the potential of the arctic region not only is of interest to science, but to economists and resource planners as well.

Polar Biosphere

There are relatively few studies of life in the polar regions that could be considered in an advanced state. One of these pertains to penguins, which spend most of their life at sea, but nest on land. See also **Penguin**. Quite a lot of information is also available on seals. See also **Sea Lions and Seals**. Less information is known of lichens, mosses, and aquatic algae, and insects, which are sparse, particularly in Antarctica. One wingless insect (*Belgica Antarctica*) is known. This is a midge confined to the Antarctic Peninsula. Various species of *Collembola* and mites are also known. In recent years, there has been considerable research directed toward Antarctic krill (*Euphausia superba*). Krill is a critical link in the Antarctic food chain for fishes, whales, seals, penguins, and some other birds. See also **Crustaceans (Edible)**. The Arctic accounts for about 10% of the world's fish catch—the cod family, capelin, and herring among the most important. The Bering Sea is the world's most productive for the Pacific (walleye) pollock. The Kotzbue Sound salmon fishery is particularly important for Alaska. More interest needs to be shown concerning endangered species, such as the bowhead whale and the Arctic cisco (fish). The bowhead whale is an example of a conservation problem for which critical data are lacking. Data are also needed for the shelf ecosystem of the Bering Sea and for most other Arctic shelf ecosystems on which renewable marine resources depend.

Polar Engineering

Whether for the exploration of minerals, scientific expeditions, or planning military operations in the polar regions, numerous and very practical engineering problems arise. A number of years ago, the U.S. Army established the Cold Regions Research and Engineering Laboratory (CRREL), operated by the Army Corps of Engineers, headquartered at Hanover, New Hampshire. Principal research and development efforts are directed to: (1) studies of snow and ice (both freshwater ice and sea ice); (2) determining the properties of frozen ground, including permafrost; and (3) developing and testing equipment and structures, such as vehicles, roads, airfields, pipelines, etc. CRREL is also concerned with environmental protective measures for the Arctic. For examples, researchers are developing a pioneering study of the climate and biology of northern Alaska. Typical projects handled by CRREL are illustrated in Figs. 5 through 7.

Fig. 5.   Special drilling and coring equipment is required to collect samples of sea ice. Cores produced by this large-diameter auger are used to determine mechanical properties of ice from pressure ridges. (*U.S. Army Corps of Engineers.*)

Fig. 6.   Field data on stream flow under ice covers is required to validate models developed to predict winter river hydraulics. These models are used to predict the effect of an ice cover on erosion, sediment transport, and ice jam formation. (*U.S. Army Corps of Engineers.*)

Fig. 7.   Specially equipped ice breakers are used in polar research and incorporate instrumentation and cold research laboratories to study the geophysical properties of sea ice. The boom extending from the ship supports a radar for measuring ice thickness. (*U.S. Army Corps of Engineers.*)

Additional Reading

Abelson, P.H.: "The Arctic: The Key to World Climate," *Science*, 873 (February 17, 1989).

Alley, R.B. and I.M. Whillans: "Changes in the Antarctic Ice Sheet," *Science*, 959 (November 15, 1991).

Anderson, J.B.: "Antarctic Marine Geology," Cambridge University Press, New York, NY, 1999.

Anderson, J.G., D.W. Tooley, and W.H. Brune: "Free Radicals within the Antarctic Vortex: The Role of CFCs in Antarctic Ozone Loss," *Science*, 39 (January 4, 1991).

Beardsley, T.M.: "Arctic Angst: No Arctic 'Ozone Hole,' But Conditions Could Lead to One," *Sci. Amer.*, 26 (April 1989).

Beardsley, T.M.: "Low Zone: The Infamous Hole has Influence Beyond Antarctica," *Sci. Amer.*, 26 (October 1989).

Bindschadler, R.A. and T.A. Scambos: "Satellite-Image Derived Velocity Field of an Antarctic Ice Stream," *Science*, 242 (April 12, 1991).

Brigham, L.W.: "The Soviet (Russian) Antarctic Program," *Oceanus*, 87 (Summer 1988).

Brune, W.H., et al.: "In Situ Northern Mid-Latitude Observations of ClO_3, O_3, and BrO in the Wintertime Lower Stratosphere," *Science*, 558 (October 28, 1988).

Brune, W.H., et al.: "The Potential for Ozone Depletion in the Arctic Polar Stratosphere," *Science*, 1260 (May 31, 1991).

Bushnell, V.C. and C. Craddock: "Antarctica Map Folio Series," *Folio 12, American Geographical Society*, New York, NY, 1970.

Cooper, A.K., P.F. Barker, and G. Brancolini: "Geology and Seismic Stratigraphy of the Antarctic Margin," American Geophysical Union, Washington, DC, 1995.

Craig, H., et al.: "The Isotopic Composition of Methane in Polar Ice Cores," *Science*, 1535 (1988).

Curtin, T.B., N.M. Untersteiner, and T. Callaham: "Arctic Oceanography," *Oceanus*, 58 (Winter 1990–1991).

Davies, T.D.: "New Evidence Places Peary at the Pole," *National Geographic*, 44 (January 1990).

Drewry, D.J.: "The Challenge of Antarctic Science," *Oceanus*, 5 (Summer 1988).

El-Sayed, S.Z.: "The BIOMASS Program," *Oceanus*, 75 (Summer 1988).

Elzinga, A.: "Changing Trends in Antarctic Research," Kluwer Academic Publishers, Norwell, MA, 1993.

Feeney, R.E.: "Food Technology and Polar Exploration," *Food Techy.*, 70 (May 1989).

Fogg, G.E.: "A History of Antarctic Science," Cambridge University Press, New York, NY, 1993.

Fowler, A.N.: "Antarctic Logistics," *Oceanus*, 80 (Summer 1988).

Frederick, J.E. and H.E. Snell: "Ultraviolet Radiation Levels During Antarctic Spring," *Science*, 438 (July 22, 1988).

Friedmann, E.I. and A.B. Thistle: "Antarctic Microbiology," John Wiley & Sons, Inc., New York, NY, 1994.

Galimberti, D.: "Antarctica: An Introductory Guide," Zagier and Urruty Publications, Miami Beach, FL, 1991.

Gordon, A.L.: "The Southern Ocean and Global Climate," *Oceanus*, 39 (Summer 1988).

Hempel, G.: "Antarctic Science: Global Concerns," Springer-Verlag, Inc., New York, NY, 1994.

Herbert, W.: "Commander Robert E. Peary—Did He Reach the Pole," *National Geographic*, 386 (September 1988).

Hibler, W.D. and K. Bryan: "Ocean Circulation: Its Effects on Seasonal Sea Ice Simulations," *Science*, **224**, 489–491 (1984).

Hodgson, B.: "Land of Isolation No More—Antarctica," *National Geographic*, 2 (April 1990).

Horgan, J.: "Antarctic Meltdown," *Sci. Amer.*, 19 (March 1993).

Jacobs, S.S. and R.F. Weiss: "Ocean, Ice, and Atmosphere: Interactions at the Antarctic Continental Margin," American Geophysical Union, Washington, DC, 1998.

Joyner, C.C.: "The Antarctic Legal Regime and the Law of the Sea," *Oceanus*, 22 (Summer 1987).

King, J.C. and J. Turner: "Antarctic Meteorology and Climatology," Cambridge University Press, New York, NY, 1997.

Lee, D.B.: "Oil in the Wildnerness—An Arctic Dilemma," *National Geographic*, 858 (December 1988).

LeMasurier, W.E. and J.W. Thomson: "Volcanoes of the Antarctic Plate and Southern Oceans," American Geophysical Union, Washington, DC, 1990.

Mayewski, P.A., et al.: "Sulfate and Nitrate Concentrations from a South Greenland Ice Core," *Science*, **232**, 975–977 (1986).

Mech, L.D.: "Life in the High Arctic," *National Geographic*, 750 (June 1988).

Meier, M.F.: "Contribution of Small Glaciers to Global Sea Level," *Science*, **226**, 1418–1421 (1984).

Meriwether, J.W., Jr.: "Atmospheric Sciences in Antarctica," American Geophysical Union, Washington, DC, 1990.

Mitchell, B.: "Undermining Antarctica," *Technology Review (MIT)*, 49 (February 1988).

Molina, M.J.: "The Antarctic Ozone Hole," *Oceanus*, 47 (Summer 1988).

Mount, G.H., et al.: "Observations of Stratospheric NO_2 and O_3 at Thule, Greenland," *Science*, 555 (October 28, 1988).

Ousland, B.: "The Hard Way to the North Pole," *National Geographic*, 124 (March 1991).

Peltier, W.R.: "Global Sea Level and Earth Rotation," *Science*, **240**, 895 (1988).

Piepgras, D.J. and G.J. Wasserburg: "Isotopic Composition of Neodymium in Waters from the Drake Passage," *Science*, **217**, 207–214 (1982).

Radok, U.: "The Antarctic Ice," *Sci. Amer.*, 98 (August 1985).

Roberts, L.: "Does the Ozone Hole Threaten Antarctic Life?" *Science*, 288 (April 21, 1989).

Schoeberl, M.R. and D.L. Hartmann: "The Dynamics of the Stratospheric Polar Vortex and Its Relation to Springtime Ozone Depletions," *Science*, 46 (January 4, 1990).

Scott, Sir Peter: "The Antarctic Challenge," *National Geographic*, 538 (April 1987).

Sherman, K. and A.F. Ryan: "Antarctic Marine Living Resources," *Oceanus*, 59 (Summer 1988).

Smith, W.O., Jr.: "Polar Oceanography: Physical Science," Academic Press, Inc., San Diego, CA, 1990.

Solomon, S., et al.: "Observations of the Nighttime Abundance of Chlorine Dioxide in the Winter Stratosphere Above Thule, Greenland," *Science*, 10 (October 28, 1988).

Stauffer, B.: "The Greenland Ice Core Project," *Science*, 1766 (June 18, 1993).

Steger, W.: "North to the Pole," *National Geographic*, 288 (September 1986).

Stolarski, R.S.: "The Antarctic Ozone Hole," *Sci. Amer.*, 30 (January 1988).

Stone, R.: "Signs of Wet Weather in the Polar Mesosphere?" *Science*, 1488 (September 27, 1991).

Sun, M.: "NSF and Antarctic Wastes," *Science*, 897 (August 19, 1988).

Tolbert, M.A., M.J. Rossi, and D.M. Golden: "Antarctic Ozone Depletion Chemistry: Reactions of N_2O_5 with H_2O and HCl on Ice Surfaces," *Science*, 1018 (May 20, 1988).

Toon, O.B. and R.P. Turco: "Polar Stratospheric Clouds and Ozone Depletion," *Sci. Amer.*, 68 (June 1991).

Vesilind, P.J.: "Antarctica," *National Geographic*, 556 (April 1987).

Walton, D.W.H., Editor: "Antarctic Science," Cambridge University Press, New York, NY, 1987.

Washburn, A.L. and G. Weller: "Arctic Research in the National Interest," *Science*, **233**, 633–639 (1986).

Whitworth, T., III: "The Antarctic Circumpolar Current," *Oceanus*, 53 (Summer 1988).

Young, O.R.: "Global Commons—The Arctic in World Affairs," *Technology Rev. (MIT)* 52 (February/March 1990).

Zwally, H.J., C.L. Parkinson, and J.C. Comisco: "Variability of Antarctic Sea Ice and Changes in Carbon Dioxide," *Science*, **220**, 1005–1012 (1983).

Web References

American Geophysical Union (AGU): *http://www.agu.org/*
Byrd Polar Research Center: *http://www-bprc.mps.ohio-state.edu/polarpointers/PolarPointers.html*
National Institute of Polar Research: *http://www.nipr.ac.jp/welcome.html*
National Science Foundation: Polar Research: *http://www.nsf.gov/home/polar/start.htm*
Scott Polar Research Institute: *http://www.spri.cam.ac.uk/*
The International Commission on Polar Meteorology: *http://www.nerc-bas.ac.uk/public/icd/icpm/*

POLECAT. See **Mustelines**.

POLE FIGURE. A diagram used in metallurgy to show the preferred orientation of crystals in a metal. A pole figure is prepared by plotting, on a statistical basis, the positions in space of the poles of a specific crystallographic plane using a stereographic projection as the basis of the representation. The data for a pole figure is normally obtained using x-ray diffraction techniques.

POLE (Mathematics). 1. The origin of a polar coordinate system.

2. A nonessential singularity of an analytic function. Let $w = u + iv$ be a single-valued function of the complex variable $z = x + iy$, and u, v be real single-valued functions of x and y. Then $z = z_0$ is a pole of order k, provided that $(z - z_0)^k w(z)$ is analytic and not zero at $z = z_0$. The number k is an integer, greater than unity, and is the order of the pole. Singular points of this kind are called *nonessential* because they may be effectively removed if $w(z)$ is multiplied by $(z - z_0)^k$. They are called poles because a three-dimensional plot of w, x, y shows that w becomes infinite at the singular point and thus looks like a pole of infinite length erected on the plane of $z = x + iy$.

3. The intersection of an axis of rotation or of symmetry with a surface, often spherical.

POLIOMYELITIS. More than one generation of people has reached adulthood with little awareness of this devastating virus infectious disease. It was essentially brought under control in many of the industrialized countries of the world during the mid-1950s and early 1960s with the introduction of the Salk vaccine (1953) and the later introduction of the Sabin vaccine. Because there has been laxity on the part of some parents to have their offspring vaccinated against polio, it is appropriate here to review briefly the cruel nature of the disease, both as a reminder of its potency, and as an outstanding example of achievement by the science of medicine. To ensure the use of vaccine, many regions and communities in advanced countries throughout the world are making vaccination of youngsters mandatory.

In the Salk vaccine, viruses that have been killed are present. The Sabin vaccine is a live virus, but so weakened that it will not produce the disease, but will permit development of protective antibodies.

Nature of Poliomyelitis. At one time called *infantile paralysis*, poliomyelitis is a disease involving inflammation of the gray matter in the spinal cord. Transport of the virus causative agent is via the nerves, either within the axon or by way of perineural cells. The disease is particularly feared, because it may result in paralysis of any part of the body, leaving the victim crippled for life, although this occurs in a minority of cases. There are four principal types of poliomyelitis, only one of which is the paralytic type. The abortive, nonparalytic, and encephalitic types account for over half of the cases diagnosed during epidemic periods. It is believed that mild cases go unrecognized as poliomyelitis due to their resemblance to colds or intestinal disorders. Even mild experiences with the disease appear to offer future immunity to the virus.

Paralytic poliomyelitis may occur in several forms, depending upon the extent of viral involvement and the area of the central nervous system which is affected. *Spinal poliomyelitis* involves the muscles of the extremities or trunk, and is the form most frequently encountered. *Bulbar poliomyelitis* involves the cranial nerves arising from the brain stem, and the vital centers of circulation or respiration are affected. *Bulbo-spinal poliomyelitis* is usually severe and is associated with respiratory impairment and with paralysis involving both the spinal cord and brain stem. From 10 to 25% of paralytic cases seen during an epidemic are of the bulbar or bulbo-spinal type.

The disease dates back to antiquity. Egyptian skeletons (circa 3700 B.C.) show evidence of bone malformations indicative of involvement with polio during childhood. The disease was not confirmed as an infectious disease until 1907 when this fact was demonstrated by O.I. Wickman, a German physician. Although there were mild epidemics during colonial times, the disease was virtually unknown in the United States until 1916, when an epidemic occurred in New York City. At that time, about 9,000 children contracted the disease, with about 2,000 fatalities. Although infantile paralysis is rarely seen in infants under one year of age, and is again rare although more severe among persons over 40, there is no safe age and all persons should be immunized.

Other than rest and comfort, there appears to be no means of altering the course of poliomyelitis if the disease has developed. Treatment measures are directed toward preventing extension of the disease to other neuromuscular units of the body. Application of moist, hot packs over the affected arms, legs and back is the best method for the relief of pain and prevention of muscle spasms, which, if not prevented, may cause deformity of the limbs. Rehabilitation of poliomyelitis patients combines use of physical medicine with psychological and vocational adjustment in an effort to achieve maximal function and prepare the patient physically, mentally, socially, and vocationally for fullest possible life compatible with abilities. A great variety of equipment and techniques has been devised for all stages of muscle weakness.

Eradication Program

In May 1988, the WHO (World Health Organization) made a commitment to eradicate poliomyelitis worldwide by the year 2000. The general strategy followed that which was used earlier and concluded successfully in 1977 with the eradication of smallpox. Poliomyelitis, however, is inherently more difficult to eradicate than smallpox. Among the epidemiological characteristics in which the two diseases differ are the asymptomatic illness that is characteristic of most poliovirus infections and the ability of the poliovirus to spread by enteric transmission, both of which make the identification and containment of cases more difficult. In contrast, smallpox was clinically obvious and eradication quite easy to confirm.

Differences between the vaccines also are important. Smallpox vaccine is heat stable, one dose is required for protection lasting several years, and vaccination leaves a readily visible scar. In contrast, trivalent oral poliovirus vaccine (TOPV) loses substantial potency after 1 day at $37\,^\circ C$, and multiple doses are required for full protection. Another difference is that properly administered smallpox vaccine has been a highly effective immunogen, whereas seroconversion rates after one to four doses of TOPV have been suboptimal in developing countries. Confirming that poliovirus has stopped being transmitted requires far more sophisticated tests and facilities. Even in the face of these difficulties, the most promising evidence that poliomyelitis can be eradicated has come from the Americas.

As of 1994, all of North America, southern Central America (not including Mexico), Greenland, Iceland, Norway, Sweden, Finland, the United Kingdom, Venezuela, Chile, Argentina, Japan, and Australia were classified in Stage A of the program. That is, no indigenous cases of poliomyelitis has been reported for at least the prior 3 years. All countries had immunization coverage of at least 80 percent, with a full course of vaccine among children reaching their first birthday.

Stage B countries or areas have immunization coverage exceeding 50 percent and report fewer than ten cases of the disease per year. These areas include France, Germany, Italy, parts of the Balkans, extreme western Russia, Saudi Arabia, and the Republic of South Africa.

Stage C countries or areas have immunization coverage exceeding 50 percent and report ten or more cases of poliomyelitis per year. These areas include large portions of the world (Mexico, most of northern South America, Spain, northern Africa, and most of Russia, China, and Malaysia).

Stage D countries or areas have immunization coverage of 50 percent or less (or an unknown coverage) or report ten or more cases per year (or have an unknown number of cases). These areas include most of central Africa, Madagascar, Pakistan, Thailand, Burma, and the Malaysian Peninsula. It is highly likely that the "last" case of poliomyelitis will occur in one of the latter areas.

The WHO program is indeed a highly ambitious undertaking, but past results with smallpox eradication is driving the expectations for success. Cost of the program is in the several millions of dollars. Private assistance of many millions of dollars has come from such organizations as Rotary International.

Further Research Needs. As reported by P.F. Wright (Vanderbilt University) and colleagues at other institutions, the attenuated poliovirus strains derived by Sabin are present in TOPV used throughout the world. Three limitations of the Sabin strains are identifiable:

1. Thermolability, probably the most amenable to immediate improvement.
2. Rare vaccine-associated cases of poliomyelitis have been reported. Such cases lower confidence and increase liability risks. Genetic instability of the attenuating mutations is readily demonstrable in vaccine recovered from children, particularly mutation in the type 3 strain.
3. Although the immunogenicity of the vaccine in many countries, including the United States is effective, this is not always the case in developing countries. Even four doses of vaccine [as recommended by the Expanded Programme on Immunization [EPI]] may not be sufficient to achieve seroconversion rates that will block the spread of the virus.

Major progress has been made in recent years toward understanding the poliovirus at the molecular level, with complete sequence analysis of the three serotypes, the creation of full-length copies of the infectious DNA, the recognition of key attenuating mutations, and the identification of the crystallographic structure of the virus.

Additional Reading

Prevots, D.R., R.W. Sutter, P.M. Strebel, et al.: "Completeness of Reporting for Paralytic Poliomyelitis, United States, 1980–1991: Implications for Estimating the Risk of Vaccine-associated Disease," *Arch. Pediatr. Adolesc. Med.* **148**, 478–485 (1994).

Staff: American Academy of Pediatrics. "Poliomyelitis Prevention: Revised Recommendations for Use of Inactivated and Live Oral Poliovirus Vaccines," *Pediatrics* **103**, 171–172 (1999).

Staff: CDC. "Notice to Readers: Recommendations of the Advisory Committee on Immunization Practices: Revised Recommendations for Routine Poliomyelitis Vaccination," *MMWR* **48**, 590 (1999).

Staff: CDC. "Progress Toward Global Poliomyelitis Eradication, 1997–1998," *MMWR* **48**, 416–421 (1999).

Staff: CDC. "Paralytic Poliomyelitis—United States, 1980–1994," *MMWR* **46**, 79–83 (1997).

Strebel, P.M., R.W. Sutter, S.L. Cochi, et al.: "Epidemiology of Poliomyelitis in the United States: One Decade After the Last Reported Case of Indigenous Wild Virus-associated Disease," *Clin. Infect. Dis.* **14**, 568–579 (1992).

Sutter, R.W., E.W. Brink, S.L. Cochi, et al.: "A New Epidemiologic and Laboratory Classification System for Paralytic Poliomyelitis Cases," *Am J Public Health* **79**, 495–498 (1989).

Travis, J.: "Good News, Bad News for Polio," *Science*, 1467 (September 11, 1992).

Wright, P.F., et al.: "Strategies for the Global Eradication of Poliomyelitis By the Year 2000," *N. End. J. Med.*, 1774 (December 19, 1991).

Web References

Centers for Disease Control and Prevention: *http://www.cdc.gov/nip/publications/pink/polio.pdf*

The Global Polio Eradication Initiative: *http://www.polioeradication.org/*

The Polio Information Center Online (PICO): *http://cumicro2.cpmc.columbia.edu/PICO/PICO.html*

The Story of Polio: *http://www.pbs.org/storyofpolio/*

World Health Organization: *http://www.who.int/home-page/*

POLLACK. See **Codfishes**.

POLLINATION. The act of transference of pollen grains to the stigmatic surface of a flower, where the pollen grain will germinate, forming a slender pollen tube, the development of which leads to the process of fertilization.

Self-pollination must be a fact in certain flowers of the type known as cleistogamous—flowers which never open to allow pollen to be shed into the air or transferred by any means from flower to flower. In such flowers, which are found in many plants, as in several species of violet, the pollen grains may germinate while still in the anther, the pollen tubes growing out to the stigma, and then on to bring about fertilization. In other cleistogamous flowers the pollen is shed from the anthers and falls directly onto the stigma, and there develops. Self-pollination also occurs, though not of necessity, in many perfect flowers. In many cases the stigma is directly beneath the ripened anthers, so that pollen shaken from the anther is likely to fall onto the stigma. In some plants the stigma is always beneath the mature anther, while in other plants movement occurs so that as the part grows older the stigma bends over to a position beneath the anther. In such a case, if cross-pollination has not already occurred, self-pollination

may be effected. In other flowers it is the stamen that exhibits movements, curving or bending in such manner that pollination shall be accomplished. In still other plants the filaments gradually elongate so that the anthers are carried upward to the stigma.

Cross-pollination is insured by several means. In many plants, the stamens mature and shed their pollen long before the stigmas are receptive, while in other plants, the stigmas are mature before the pollen of the same flower has ripened. Either case necessarily insures cross-pollination. Equally certain is the occurrence of cross-pollination in those plants which bear unisexual flowers on different plants, or if on the same plant male flowers mature some time before the female.

Many plants, notably those in the Carrot and Composite families, often bring about cross-pollination in another way. The flowers grow in compact groups, either umbels or heads. As the floral organs, stamens and pistils, mature, they grow out in a way that brings about contact between nearby flowers, the stigmas of one flower touching the anthers of another. That this does often occur is seen in the Composite Family, where in many instances the styles curl back at maturity so that the stigmatic surface is brought directly against the pollen masses. Many plants in this family are self-sterile.

Another method that tends to effect cross-pollination is the occurrence of flowers of two, or three, or even four different kinds. The purple swamp loosestrife, *Lythrum salicaria*, shows one such case. Some of the flowers of its spike have anthers borne on long filaments, and they contain pollen grains of relatively large size. Other flowers have stamens with short filaments and contain small pollen grains; while a third flower type is intermediate in habit, the filaments being between the others in length and the pollen grains of intermediate size. Such plants are known as *heterostylous plants*.

The agents which are instrumental in carrying pollen from one flower to another are principally air-currents and insects. Plants pollinated by air-currents, or wind, are said to be *anemophilous*; those by insects are *entomophilous*.

As a rule wind-pollinated flowers are inconspicuous, and of small size; color, odor, and nectar, all associated with insect-pollinated flowers, are largely lacking. Pollen is produced in immense quantities, since much will be lost. The pollen grains are dry and dust-like, and so float buoyantly in the air. In many plants of this group the stigmas are much-branched, feathery objects, offering a considerable area of sticky surface to catch any pollen that falls on it. To further the ready discharge of pollen, which ordinarily occurs only when the atmosphere is dry, the flowers are often borne in pendulous catkins, or on slender flexible pedicels. Or they have stamens, the anthers of which are versatile, that is, attached at the middle and easily moved by any slight disturbance. Examples of wind-pollinated plants are found in the gymnosperms, all of which are so pollinated, in most grasses, in many hardwood trees, such as birches, alders, oaks, and beeches, and in the common cat-tail of the marshes.

Water-pollinated plants are not numerous. That a plant grows in water is not an indication that its flowers shall be pollinated by water. Indeed, only a very few water-plants are so pollinated.

Plants with flowers pollinated by animals are numerous, and with many variations seemingly calculated to assure cross-pollination. A few tropical plants are said to be pollinated by bats. The flowers have very fleshy petals, open during the evening hours, and are apparently sought by bats, which eat the petals, or perhaps seek any insects which may occur in the flowers. Quite possibly, in going from flower to flower, bats do bring about a transfer of pollen. Pollination by such animals is undoubtedly restricted to a very small number of plants.

Similarly birds, especially humming-birds and honey-suckers, are held to be the agents pollinating several tropical plants. In these the question arises as to whether the birds are seeking the numerous insects which are to be found in the flowers or after the nectar which is in the flowers. The flowers visited by birds are rather large, brilliantly colored, often scarlet. Unquestionably humming-birds do seek nectar in flowers, both in the tropics and in temperate regions. Pollination may well result from their visits.

A few flowers are said to be pollinated by snails or slugs. Plants so pollinated have dense masses of flowers borne on a fleshy stock. This mass of sterile tissue attracts these animals, which crawl over it in the search for food, and so are said to carry pollen from flower to flower. Many observers doubt that such animals ever bring about pollination.

All, however, recognize that insects are very important in the pollination of many flowers. In many cases the flowers show remarkable adaptations fitting them to be pollinated by certain insects. The insects that effect pollination are for the most part bees, flies, beetles, and moths and butterflies.

As certain features characterize wind-pollinated flowers, so also insect-pollinated flowers exhibit several common features. In them the pollen grains are usually somewhat adhesive, often sticking together into considerable masses. The surface of each grain is variously sculptured, with knobs, spines, or other protuberances definitely increasing the ability of the grain to stick to the insect body. As in wind-pollinated flowers, so here the pollen would be seriously impaired by water, from which it must be protected during periods when pollination would not occur. Many are the ways in which protection is obtained. In some plants the entire flower bends down at night, while in many others the petals close together over the stamens; often a passing cloud is sufficient stimulus to cause closing, which takes place with surprising speed. In many plants the flowers are located beneath the leaves, as in the common Jewel-weed or Touch-me-not, *Impatiens biflora*; in the common Iris, each stamen is located beneath the broad-petaloid stigma. Often the anther itself is so constructed as to afford considerable protection, the pollen frequently being shed through narrow slits which close tightly during periods of excessive moisture; or small pores may allow the pollen to escape when advantageous but protect it from water otherwise.

A most obvious characteristic of insect-pollinated flowers is color, which may be found in a single flower with large conspicuous petals, or may result from the massing together of many small flowers, as in the composite family. While it is generally assumed that bright color is an aid to pollination because it attracts insects, it should be recognized that insect vision is not necessarily like that of human beings. Often the color of a flower changes with age, many becoming gradually deeper toned, as in the common Lady's Slipper, while others as gradually fade out. Flowers opening at night are almost all white or very light-colored.

The odors of flowers are also assumed to attract insects. Every fragrant flower has an odor that is quite distinctive. The odors of flowers are of many types, from the foul rankness of the Skunk Cabbage and Carrion Flower to the delightful perfume of Verbena, Gardenia, the Roses, and many Lilies. In many cases the odor of the flower is very delicate, being scarcely detectable to many people; in others it is of such penetrating strength as to become objectionable. Often the odor is evident only during certain periods, some flowers being scentless by day but fragrant during the night, while others emit their odors only in broad daylight. All these differences in odors seem designed to attract special insects, which will accomplish pollination. The foul odor of carrion calls carrion-flies, and the sweet fragrance of night-blooming flowers attracts moths. After pollination the odor of the flower generally ceases, attraction of insects being no longer of any value.

Many insects undoubtedly visit flowers for the purpose of obtaining nectar, a sweet watery secretion formed in special glands called nectaries. These nectaries are variously located in the flower, usually deep down at the base of the corolla, so that any insect obtaining nectar must either have a mouth part of sufficient length to reach the nectar, or be strong enough to push its way into the flower, passing any obstructions that may serve to protect the nectar from less fortunate insects. Obstructions are found of several sorts. A very common means of excluding such crawling insects as ants, which in all probability would not be efficient pollinators, is by the presence of a barricade of hairs, especially in the throats of flowers having a tubular corolla. Long-tongued bees are able to push through these hairs enough to reach the nectar, while at the same time they are thoroughly powdered with pollen, which may be removed later in another flower. The existence of a sticky secretion over the outside of the flower or on the stem of the plant is an effective barrier to many crawling insects, as is also a waxy coating. Especially noteworthy are those flowers in which the nectar is located at the base of a long narrow tube or spur; often the nectar is present in quantities large enough to form a considerable volume. In such cases there is usually an insect with mouth parts just long enough to reach through the tube into the nectar supply, which is sucked up greedily. This becomes all the more remarkable when one considers that, in some flowers, the nectar is at the bottom of a tube which may be 3 or 4 inches (7.5 or 10 centimeters) long. In the case of one tropical Orchid a nectar-secreting sac over a foot long exists; in such cases insects, usually moths or butterflies, with correspondingly long sucking tubes are found. Often short-tongued insects succeed in obtaining the nectar illegitimately by biting a hole in the wall of the nectar-containing part of the flower, and obtaining

the nectar thereby. It is interesting to note, in connection with this problem of nectar-secreting flowers and insects, that the introduction of clover into Australia was not a success until honey-bees were also introduced. The native Australian bees were too short-tongued to reach the nectar and so pollination was not effected. As a consequence the clover crop soon died out, no seed being formed to perpetuate it. With the introduction of suitable bees the plant seeded abundantly.

It is interesting to note that in addition to the existence of colors, odors, and nectar, the structure of the flower and its position on the plant seem designed to facilitate the work of the insect in transferring the pollen. Many flowers are broad and flat, affording convenient support to the insect as it crawls about over the flower. Others, especially those visited by long-tongued moths, are borne in such a position as seems most suited to permit the insect to insert its tongue and obtain the nectar. The pollen is collected in quantities by many species of bees, who use it as a food for the developing young.

Several plants have developed a most striking relationship with insects. While the majority of flowers are not visited indiscriminately by all insects but only by certain species or genera, in these special cases the restriction is extreme, both the flower and the insect seeming to be modified especially to serve one another. One such case is seen in the edible fig, pollinated only by a small insect, *Blastophaga grossorum*, which lays its eggs in the ovaries of certain flowers of the fig. Another example of this insect-flower association occurs in a species of Yucca. The creamy-white flowers of this plant are borne in large panicles. They open during the evening and are visited by a small moth, *Pronuba fuccasella*, which seeks the abundant pollen of the flower. Of this pollen the moth makes a tiny ball which it carries away to another Yucca flower. In the ovary of this flower the female moth lays her eggs, while on the stigmatic surface it deposits its ball of pollen. The developing ovules serve as food for the growing larvae of the insect. However, many ovules grow to maturity to form viable seeds, which perpetuate the plant and so continue the food supply of the moth, which seems to be the sole agent capable of transferring the sticky pollen of the Yucca from plant to plant. Many other cases are known where pollination of the flower depends on the visit of certain insects, which do not, however, depend on the flower for existence. On the other hand, many elaborate devices, such as the keel mechanism in papilionaceous flowers, which ought to assure insect pollination, fail to do so. Garden peas are normally self-pollinated.

See also **Composite Family (Compositae)**; **Gymnosperms**; **Honey-bees**; and **Plant Breeding**.

Additional Reading

Dafni, A., M. Hesse, and E. Pacini: "Pollen and Pollination," Springer-Verlag, Inc., New York, NY, 2000.

Dafni, A.: "Pollution Ecology: A Practical Approach," Oxford University Press, Inc., New York, NY, 1993.

Delaplane, K.S. and D.F. Mayer: "Crop Pollination by Bees," CAB International North America, New York, NY, 2000.

Free, J.B.: "Insect Pollination of Crops," 2nd Edition, Academic Press, Inc., San Diego, CA, 1993.

Kearns, C.A. and D.W. Inouye: "Techniques for Pollination Biologists," University Press of Colorado, Boulder, CO, 1999.

Proctor, M., P. Yeo, and A. Lack: "The Natural History of Pollination," Timber Press, Inc., Portland, OR, 1996.

Thomson, J.D.: "Cognitive Ecology of Pollination: Animal Behavior and Floral Evolution," Cambridge University Press, New York, NY, 2001.

POLLUCITE. The mineral pollucite is rather rare. It contains cesium, aluminum, silicon, and oxygen, its chemical composition being approximately $(Cs, Na)_2(Al_2Si_4)O_{12} \cdot H_2O$. It is isometric, usually in cubic crystals or crystalline masses; conchoidal fracture; brittle; hardness, 6.5–7; specific gravity, 2.9; luster, vitreous on fresh surfaces; colorless and transparent. Found on the Island of Elba and in the pegmatites of Maine, and as masses 3–4 feet (0.9–1.2 meters) thick in South Dakota; at Varutrask, Sweden; in Italy; and in Kazakhastan, Russian. Pollucite and petalite were found in the granites of Elba and at first named pollux and castorite for the two famous brothers of Roman mythology, Castor and Pollux. Pollucite is derived from the Latin genitive *Pollucis*.

POLLUTION (Air). Prior to the Industrial Revolution (circa 1840s), the composition of "pure" air making up Earth's surrounding atmospheres essentially remained constant for several thousand years. True, certain natural phenomena, such as volcanic eruptions, may have altered the atmospheric composition over relatively short time spans. See also **Atmosphere (Earth)**.

By many orders of magnitude, the greatest alteration of the atmosphere occurred during the middle Precambrian period, between 2.9 and 1.8 billion years ago. It is generally accepted that prior to that time, the terrestrial atmosphere was chemically of a reducing nature—as contrasted with an oxidative environment of the present general composition required to support humans and other mammals and life forms which abound on the earth today. See also **Air**. Brought about by greatly accelerated plant growth, that earlier natural change represented the most dramatic pollution effect ever suffered by the earth's environment. During that period, the oxygen liberated by plant activity proved to be a very toxic substance for anaerobic life forms and eradicated most of the biotic community existing at that time. New types of life had to develop which were capable of survival in an oxidative environment. Geochemical processes took on new characteristics, based upon the slow oxidative degradation of both organic and inorganic materials.

Alteration of the earth's atmosphere as the result of human (anthropogenic) activities is extremely recent on the life scale of the earth. This altering process was essentially commenced when humans first discovered and started to use fire as a means of heating, cooking, etc. It is the *combustion* of organic fuels today that is the principal contribution to anthropogenic air pollution. For centuries the pollutants added to the atmosphere by humans were essentially insignificant in terms of the mass and the dynamics of the earth's atmosphere. Except on a local and sometimes regional basis, air pollution was no problem prior to the invention of the steam engine and, of course, the later invention of the internal combustion engine. Traditionally, air pollution has a direct relationship with increasing population and the growing sophistication of the population, which demands ever increasing quantities of energy and the manufacture of goods by processes which yield byproducts that require removal to some kind of sink, the earth's atmosphere being one of these sinks.

With few exceptions, air pollutants ultimately fall by gravity to the surface of the earth. On land, pollution of the soil and freshwater lakes and rivers and ultimately the groundwater occurs. Fallout on the seas and oceans also occurs, but unless radioactive, the effects are less easy to discern except on the long term. It is indeed difficult to separate air and water pollution. The relationship is explored in the article on **Water Pollution**. The winds contribute both to the spread and, in some instances, to the contribution of air pollutants. Frequently, as in the case of "acid rain," the precipitation of water (an excellent solvent) in the form of rain, snow, sleet, ice pellets, etc. causes entrainment of pollutants (gases, mists, particles, etc.). Thus the soils, rocks, lakes, and rivers are subject to the corrosive and biodestructive processes brought about by the presence of alien substances. Acid rain is described later in this article.

During the last few decades and including the early 1990s, the total amount of pollutants in the atmosphere has increased exponentially. Currently, many hundreds of different pollutants, largely from anthropogenic sources, rise into the atmosphere. Even though many of these substances are measured on a scale of parts per million (ppm) or even parts per billion (ppb), a majority of these substances cause ill effects on the health and well-being of air-breathing creatures as well as corroding and eroding structures.

The following *incomplete* list is given here to dramatize the complexity of the current air pollution problem, to demonstrate the probable impracticality of finding a few simple solutions, and to illuminate why air pollution measurement and control presently requires the skills of many hundreds of scientists and engineers. To accomplish these tasks, many excellent instrumental techniques are required, as exemplified by chromatography, laser radar, various forms of spectrometers, and particle analyzers, among other advanced analytical tools. The use of radioactive isotopes has been effective in many instances of pollution source tracing.

Partial List of Air Pollutants

Common gases—carbon monoxide (CO), carbon dioxide (CO_2), nitrogen oxides (NO_x), ammonia (NH_3), hydrogen sulfide H_2S), chlorine (Cl_2).

Volatile inorganics—sulfuric acid (H_2SO_4), hydrochloric acid (HCl), nitric acid (HNO_3), hydrogen peroxide (H_2O_2).

Volatile organics—hydrocarbons*, fluoroalkenes, alcohols, polychlorinated biphenyls (PCBs), ketones, aldehydes.

Free radicals—hydroxyl, sulfate.

Solid particles—carbonaceous and metal particles (lead, zinc, manganese, cadmium, chromium).

Formulated commercial products and byproducts—insecticides, pesticides, fungicides, herbicides, solvents, coatings, chlorinated fluorocarbons, petroleum and petrochemical products, plastics, fibers.

Radioactive substances—from weapon testing, radon in some enclosed spaces.

Pollution also affects the manufacture of certain materials and products. This is evidenced by the need for "clean rooms" in metrology standards laboratories and in the production of certain electronics materials and of component assembly operations. In addition to elaborate filtering systems, such rooms are held at a slightly positive pressure (above outside atmospheric pressure) to prevent the entry of raw air from the outside.

Air Pollution Settings. In developing preventive and remedial technologies for air pollution abatement, it is helpful to consider the fundamental settings in which pollution occurs.

(1) *Workplace pollution* usually represents the highest concentration and length of exposure to specific pollutants. It usually is the most obvious mode of air pollution and, consequently, the easiest to correct. In this setting, workers breathe specific pollutants on a day-to-day basis. Well-known and publicized examples would include: miners; chemical plant workers; farmers; textile workers; metal production workers; transportation-related personnel (who are exposed to high concentrations of carbon monoxide and other gaseous products of fossil fuel combustion as well as coal dust and other carbonaceous particles and volatile hydrocarbons); and insulation installers (who are exposed to airborne tiny particles and strands of glass, plastics, and natural minerals, such as asbestos and mica). In the industrial and manufacturing complex, which has continued to expand markedly during the last several decades, hundreds of specific examples of exposure to dangerous air pollutants could be recited.

(2) *Point-source pollution* extends beyond a specific entity within a plant, such as a particular machine or process, and comprises a source of pollution that may emanate from only one or a few particular facilities within a small area. Point-source pollutants become mixed in the atmosphere with pollutants from other sources. Hence, beyond the immediate vicinity of a given facility, sources are difficult to identify. Considerable success has been achieved by isotopic matching to specific coals and other fuels that may be burned at a given facility.

(3) *Confined-area pollution*, where there may be no natural circulation of air and no effective air purification effluent system. This situation is found in factories or mines and inside improperly ventilated garages, service stations, and vehicular tunnels.

(4) *Limited-area pollution* occurs on a small geographic scale, such as strips of land adjacent to major highways or close to a pollution point source.

(5) *Regional pollution* is pollution that occurs in the greater part of a city, valley, or basin and frequently is publicized in connection with cities like Los Angeles, London, and, in more recent years, other major cities of the world, including New York. Regional pollution is particularly affected by weather conditions, such as, for example, an inversion layer hanging over a natural basin. When pollution occurs on this scale, the beginnings of ecological damage (to trees, plants, natural life, etc.) are seen.

(6) *International pollution* or wide-area pollution is the occurrence of massive air pollution, usually extending over many years, to the extent that the atmosphere becomes severely overloaded with pollutants—that is, the atmosphere's ability to contain pollutants is exceeded. After holding pollutants for long periods, during which time prevailing winds transport the pollutants over long distances (from one country to the next, for example), the point is reached where pollutants "drop out" and contaminate the topography below. This is the type of pollution that has been the subject of debate, for example, between Canada and the United States—where pollutants from coal-burning power plants in the Midwestern states of the United States are transported to northeastern Canadian provinces (and northeastern U.S. states as well). Such pollution damages forests and lakes. See also "Acid Rain" later in this entry.

Other examples of international pollution occur in other parts of the world, but are less well understood at this time.

---

* Motor vehicle pollution is addressed in article on **Automotive Engineering**.

(7) *Worldwide pollution* is simply an extension of wide-area pollution and encompasses emissions that essentially become mixed with the entire atmosphere of the Earth. Even though the mixing time may be quite long, it is reasonable to assume that, over time, the ultimate effects of almost *any* pollution ultimately will affect the atmosphere on a worldwide basis. For example, as the result of nuclear events, particularly those that occurred prior to the ban on nuclear weapons testing, radioactive particles could be discerned over extensive regions of Earth. Depletion of Earth's ozone layer also is exemplary of how destruction of ozone can be caused by chlorine molecules, essentially without regard to where the pollutants are released into the atmosphere. The so-called ozone "hole" currently over Antarctica and a possible similar "hole" over the Arctic (now threatened to occur by about the year 2000) is discussed in the article on **Polar Research**.

The possible effects of increased carbon dioxide content of the atmosphere are discussed in the article on **Global Change**.

The Energy vs. Environment Conflict

Just as it is difficult to separate the topics of air and water pollution, so is it hard to separate the problems of pollution from energy generation and consumption. With exception of some of the nontraditional sources of energy, such as nuclear energy and the more direct utilization of solar energy (as contrasted with combustion), the needs of the earth's population for energy tend to follow a collision course with concerns over the environment. For example, until the nontraditional energy sources can be reduced to practical usage (of which economics is an important, if not scientific factor in the equation), coal, wood, biomass, and other organic fuels when burned are air polluters unless very costly measures are taken to treat the effluents. Even when the numerous chemical and electroprecipitation measures, among others, are taken, there remains the problem of increasing the carbon dioxide content of the atmosphere.

Energy has an impact on the environment by tending to worsen the environment. The environment has an impact on energy by requiring considerable energy to alleviate the degradation of environmental quality. Environmental concerns also tend to limit the energy options available. It is generally agreed that the standard of living of the technologically sophisticated and developed nations is at least partially the result of inexpensive energy. Generally, the societies in these countries have not readied themselves to very high energy costs, there being a realization that these increased costs will, by and large, come out of so-called discretionary income and, consequently, impact the standard of living in a negative way. The significance of the incremental addition of energy required to effect environmental protection is shown in Table 1.

The topic of the conflict between energy, environment, and economics is discussed in greater detail in the article on **Electric Power Production and Distribution**.

Principal Air Pollutants

The major air pollutants as identified by a number of countries in recent years in connection with pollutant regulatory programs are:

(1) Particulate matter; (2) nitrogen oxides (NO_x); (3) sulfur oxides (SO_x); (4) hydrocarbons; and (5) carbon monoxide (CO).

Particulates and Aerosols. These may be comprised of numerous mineral and organic materials and frequently result from such operations as milling, crushing, screening, grinding, and demolition operations—as well as quarries and cement plants. Soot and fly ash as well as heavy carbonaceous smoke, arising from fuel-burning operations and smudge pots, also may fall into this category of pollutants. Aerosols generally are considered to be very tiny spherical droplets of a liquid that may be as small as 0.01 micrometer in diameter. These small liquid particles and the larger liquid particles, including mists and sprays, along with dusts, permit numerous physical separating and isolating means that do not apply to gases and vapors. Recent investigations of particle size distribution of atmospheric aerosols have revealed a multimodal character, usually with a bimodal mass, volume, or surface area distribution and frequently trimodal surface area distribution near sources of fresh combustion aerosols. These modes are attributed to the following factors: (a) the course mode (2 micrometers and greater) is formed by relatively large particles generated mechanically or by evaporation of liquid from droplets containing dissolved substances; (b) the nuclei mode (0.03 micrometer and smaller) is formed by condensation of vapors from high-temperature processes or by gaseous reaction products; and (c) the intermediate or accumulation mode (0.1 to 1.0 micrometer) is formed by coagulation of nuclei. This evidence indicates that atmospheric particles tend to form a stable aerosol having a size distribution ranging from 0.1 to 1.0 micrometer in general. However, larger and smaller particles occur. The larger particles (greater than 1.0 micrometer) settle out, and the very fine particles (smaller than 0.1 micrometer) tend to agglomerate to form larger particles which remain suspended. The nuclei mode tends to be highly transient and is concentration-limited by coagulation with both other nuclei and also particles in the accumulation mode. Therefore, the particulate content of a source emission and the ambient air can be viewed as composed of two portions, i.e., the settleable and the suspended.

Control of emissions in both size ranges is required because both settleable and suspended atmospheric particulates have deleterious effects upon the environment. Significantly, it is the suspended particles from an upper level of about 2 to 5 micrometers and smaller that health experts consider most harmful to humans because particles of this size have been found to penetrate the body's natural defense mechanisms and reach most deeply into the lungs. Efforts to control particulate emissions to the atmosphere have historically been geared to maximizing the efficiency of control (by weight) of the overall particulate loading emanating from the generating process. This work has led to the empirical understanding that present systems can perform with high control efficiencies down to a particle size of about 2–3 micrometers. But, below this size, the control efficiency appears to decrease with decreasing particle size to a minimum between 1.0 and 0.1 micrometer; and then increases again. This relationship of control efficiency and particle size is highly significant to any strategy for

TABLE 1. AIR POLLUTION COSTS FOR REPRESENTATIVE TECHNOLOGIES (10^6 BTU/TON OF PRODUCT)*

Process Option	Primary Energy Source	Process Energy (10^6 Btu)	Air Pollution Control Energy (10^6 Btu)	Percent of Total for Air Pollution Control
Glassmaking				
side port regenerative furnace	natural gas	7.0	0.57	7.5
side port regenerative furnace with preheat of charge	natural gas	5.7	0.37	6.0
electric furnace	electric power	8.2	0.03	0.3
coal gasification	coal	8.6	0.9	9.5
direct coal firing	coal	7.0	0.65	8.5
Cement				
long kiln (conventional)	oil	5.6	0.07	1.2
suspension preheater with long kiln	oil	4.2	0.05	1.2
fluid bed	oil	5.0	0.1	2.0
Copper Production				
roast-reverb smelting (conventional)	gas, oil, or coal	22.0	5.3	19.4
flash smelting (90–95% sulfur recovery)	oil	10.0	7.8	43.8

*10^6 Btu = 252×10^3 Calories

controlling particulate air pollution, and serves to underscore the need to adequately measure and evaluate both ambient particulate air pollution and source emissions.

The particle diameters of some substances commonly found in the atmosphere or important in various manufacturing operations are given in Fig. 1. Various particle measurement techniques versus particle diameter are given in Fig. 2. The most suitable ranges of particle size versus types of gas cleaning equipment are given in Fig. 3. Modern instrumental techniques are shown in Fig. 4.

Source Identification of Airborne Particles. In recent years, ingenious methods of identifying the point source of airborne particles have been developed. If not specific point sources of pollution, rather small regional areas can often be identified. Source of particles can be important for numerous reasons, including the enforcement of regulation and also in sorting out, for example, the various distant sources that contribute to acid rain pollution.

As reported by Olmez and Gordon (University of Maryland), the concentration pattern of rare earth elements on fine airborne particles (less than 2.5 micrometers in diameter) is distorted from the crustal abundance pattern in areas influenced by emissions from oil-fired plants and refineries. The ratio of lanthanum (La) to samarium (Sm) is often greater than 20 (crustal ratio is less than 6). The unusual pattern apparently results from the distribution of rare earths in zeolite catalysts used in refining oil. Oil industry emissions have been found to perturb the rare earth pattern even in very remote locations, such as the Mauna Loa Observatory in Hawaii. Rare earth ratios are probably better for long-range tracing of oil emissions than vanadium (V) and nickel (Ni) concentrations because the ratios of rare earths on fine particles are probably not influenced by deposition and other fractionating processes. Emissions from oil-fired plants can be differentiated from those of refineries on an urban scale by the much smaller amounts of V in the latter. Pb in urban areas originates mainly from combustion of leaded gasoline. Arsenic (As), selenium (Se), and other chalcophile elements are usually associated with coal-fired plants or sulfide ore smelters (or both). In addition to the use of receptor models on an urban scale, they also have been used on a global scale to identify sources of Arctic haze, based upon manganese (Mn) and V ratios.

R.W. Shaw (U.S. Army Research Office, Research Triangle Park, North Carolina) commented that in analyzing particle samples, one must always consider that a particle sample could be produced by various sources. Chemical-element balance copes with this complication by positing (on the basis of the known ratios) several different elemental concentrations that could be generated by the suspected sources. It then finds the "mix" that best fits the actual concentrations on the collected samples. Such manipulations might reveal, for example, that 80% of the fine fraction of a sample is a byproduct of coal combustion and 15% comes from motor vehicles. Or it might show that 94% of the lead in a specimen is from motor vehicles, 4% from the burning of refuse, and 1% from the burning of coal.

Smoky Mountain Haze. A rather thorough study of what contributes to the increasing haze that exists in the Great Smoky Mountains National Park in Tennessee was reported by Shaw in 1987. The overall purpose of the study was to determine the contributions of natural emissions, as from vegetation, and motor vehicle and industrial emissions. The researchers also conducted surveys in the Georgia region (former Soviet block), the latter highly regarded for clear air and massive areas of evergreen forests. The Smoky Mountains haze is ascribed to sulfate particles in the air. It was tentatively concluded that the sulfate is a byproduct of coal burned at distant power plants. Surprisingly, the concentration of coal-derived particles in the Smoky Mountains was not much lower than concentrations found at several sites in an industrialized city, St. Louis, Missouri. Sulfate particles also were found in Georgia, but considered to be much older and the particles contained thirty times more sulfur than in sulfur dioxide gas. These studies, however, did not make a direct link with remote coal-burning plants. This was accomplished later in a 16-month study in the Ohio River valley, with sampling stations located in rural Kentucky, Indiana, and Ohio, far away from cities, roads, and power plant smoke plumes. About 50% of the fine-particle mass was found to be sulfate. The total concentration nearly equaled that found in the industrialized cities of the regions. The link was confirmed by a consistent association between sulfate concentration and the trace element selenium. No other probable sources of selenium exist in or near the Ohio River valley. The investigators (R.W. Shaw, et al.) concluded in their report that there is little doubt that fuel combustion is the main source of acid deposition and particulate fallout.

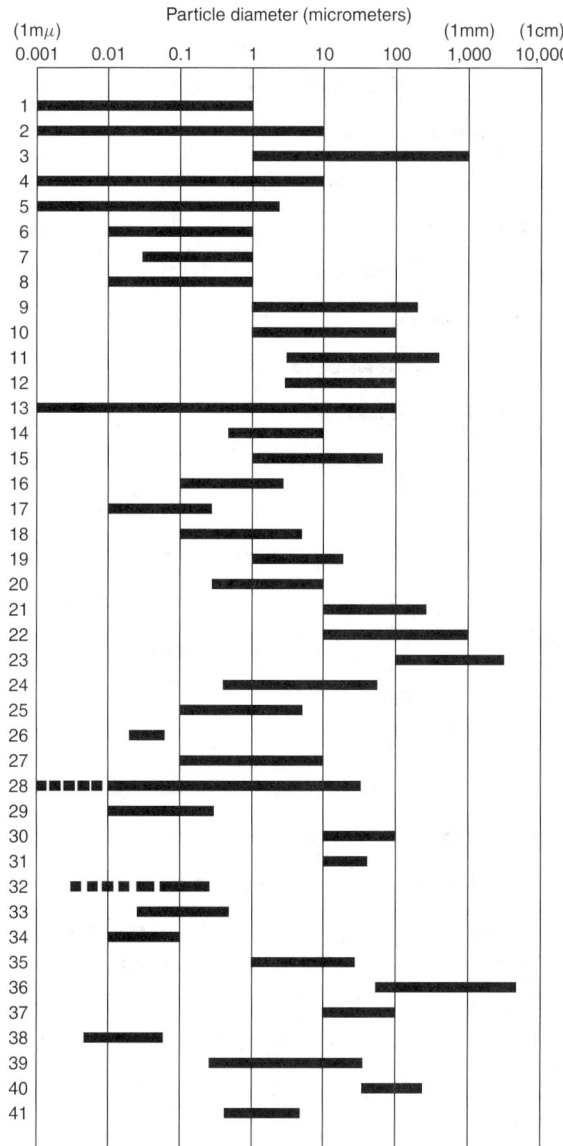

Fig. 1.   Range of diameters of particles and particle dispersoids of substances commonly encountered in the atmosphere or associated with various manufacturing operations. (*Stanford Research Institute.*)

Legend:

1. Gas dispersoids — solid fumes
2. Gas dispersoids — liquid mists
3. Gas dispersoids — solid dusts
4. Gas dispersoids — liquid sprays
5. Common atmospheric dispersoids
6. Smoke — rosin
7. Smoke — oil
8. Smoke — tobacco
9. Fly ash
10. Coal dust
11. Pulverized coal
12. Cement dust
13. Metallurgical dusts and fumes
14. Insecticide dusts
15. Milled flour
16. Fumes — ammonium chloride
17. Fumes — zinc oxide
18. Fumes — alkali
19. Sulfuric acid concentrator mist
20. Contact sulfuric acid mist
21. Ore flotation mist

22. Ground limestone fertilizer
23. Beach sand
24. Ground talc
25. Paint pigments
26. Colloidal silica
27. Spray dried milk
28. General atmospheric dust
29. Carbon black
30. Plant pollen
31. Plant spores
32. Nuclei (Aitken)
33. Nuclei (Sea salt)
34. Nuclei (Combustion)
35. Nebulizer drops
36. Hydraulic nozzle drops
37. Pneumatic nozzle drops
38. Viruses
39. Bacteria
40. Human hair (diameter)
41. Most severely lung-damaging dust

Shaw also describes two techniques that now make it possible to analyze particles without removing them from collection filters. To determine the

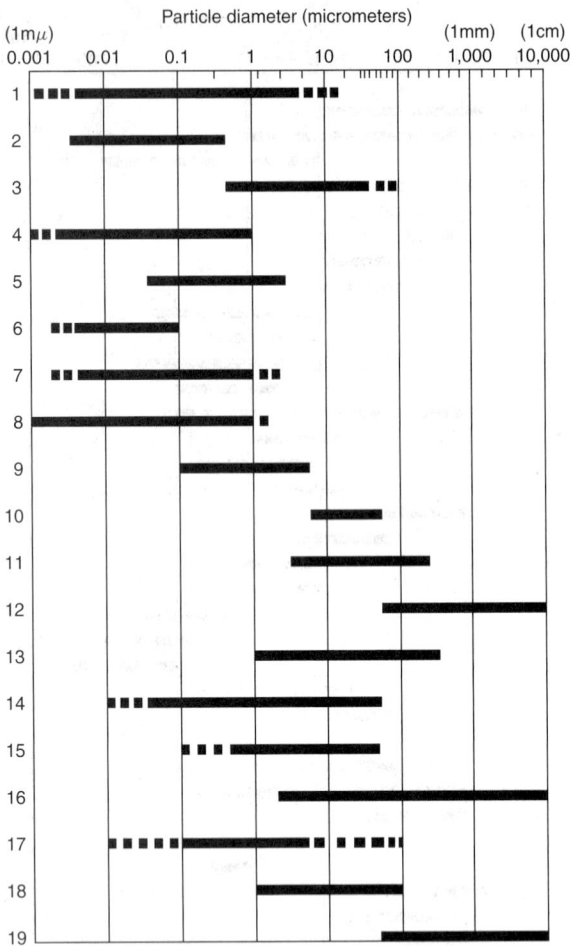

Fig. 2.   Various particle measurement techniques versus particle diameter. (*Stanford Research Institute.*)
Legend:

1. Electron microscope	11. Elutriation
2. Ultramicroscope	12. Sieving
3. Microscope	13. Sedimentation
4. Ultracentrifuge	14. Turbidimetry
5. Centrifuge	15. Permeability
6. X-ray diffraction	16. Scanners
7. Adsorption	17. Light scattering
8. Nuclei counter	18. Electrical conductivity
9. Impingers	19. Visible to eye
10. Electroformed sieves	

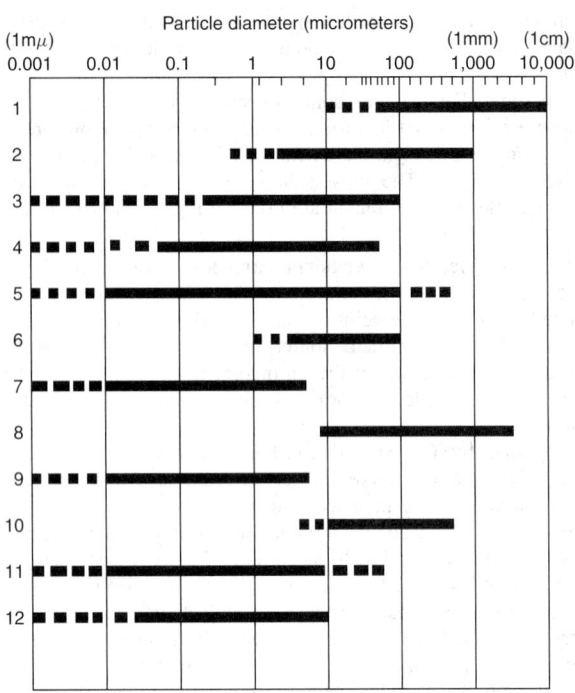

Fig. 3.   Types of gas cleaning equipment versus particle size diameters. (*Stanford Research Institute.*)
Legend:

1. Settling chambers	7. High-efficiency air filters
2. Centrifugal separators	8. Impingement separators
3. Liquid scrubbers	9. Thermal precipitators
4. Cloth collectors	10. Mechanical separators
5. Packed beds	11. Electrical precipitators
6. Plain air filters	12. Ultrasonic methods

mass of a sample, technicians insert the particle-laden filter between a source that emits beta particles and a detector that counts them. As the mass increases, the number of particles that can penetrate the sample decreases. To determine the atomic elements in a specimen, laboratory workers may also separately carry out x-ray fluorescence spectroscopy. X-rays passed through the sample cause each element to emit characteristic x-rays. The energy levels of the rays reveal the identity of the elements: the intensity of the x-rays (number emitted) reflects the concentrations.

Sulfur Oxides (Sulfur dioxide, SO_2, and sulfur trioxide, SO_3). The primary sources of these oxides SO_x are sulfur-bearing fuels — as used for heat and power, both industrially and residentially. Chemical and metallurgical plants of various kinds also emit SO_x as the result of processing activities, such as the manufacture of sulfuric acid, the roasting of ores, etc. In order of decreasing pollution, the fossil fuels contributing to SO_x pollution are: (a) Untreated coal; (b) untreated petroleum fuels, particularly those originating from so-called sour crude oils; and (c) natural gas. Thus, the preference for natural gas by many large fuel users, such as power plants. With only small variations in the cost of raw fossil fuels, there was an advantage in burning a naturally low-sulfur fuel as contrasted with installing elaborate SO_x removal or reduction systems. But, with a rapidly lessening natural gas supply and accompanying higher costs, it has become economically attractive to pay more for desulfurized coal

and petroleum fuels, as well as to install SO_x abatement equipment. The allowable sulfur content of oil and coal fuels varies from one community to the next, ranging from 0.50% by weight or less, up to 4% and slightly higher. Such regulations usually take into consideration new versus old fuel-burning equipment, the incidence of serious pollution in a given area, as well as economic impact and practicability. Logically, for some years to come, such regulations must represent a compromise of social, economic, and technological factors. See also **Electric Power Production and Distribution**.

The chemical nature of the oxides of sulfur is given in the entry on **Sulfur**. Treatment of SO_x effluents is described later in this article.

Nitrogen Oxides. These compounds result from all fossil-fuel combustion processes where air is used as the oxidant. Oxygen from the air and nitrogen combine at combustion flame temperatures to form nitric oxide, NO, according to $N_2 + O_2 \leftrightarrow 2NO$. The rate at which NO is formed and decomposed depends largely upon temperature. For the majority of stationary combustion processes, there is too short a residence time for the full oxidation of NO to NO_2, an estimated average of only 5 to 10% of this reaction occurring. Thus, it is important to observe that although NO_x emissions generally are given as "equivalent NO_2," the predominant NO_x in combustion gases is NO. Several factors affect the generation of NO_x pollutants. Factors which tend to decrease NO_x emissions are: (a) decrease in excess air for combustion; (b) decrease in preheat temperature; (c) decrease in the heat-release rate; (d) increase in the heat-removal rate; (e) increase in back-mixing; and (f) decrease in fuel nitrogen content. With exception of very large installations, coal appears to generate more NO_x than oil; and oil generates more NO_x than natural gas. Thus, as with SO_x, natural gas is the preferred fuel when properly burned to minimize NO_x.

The major sources of NO_x are the large fuel-burning operations as previously mentioned, automotive vehicles, and certain chemical plants, notably nitric acid manufacturing facilities. Research to date indicates that effective steps toward reducing the overall emission of NO_x can be effected from stationary combustion sources by: (a) using low excess air firing; (b) providing for two-stage combustion; (c) utilizing flue-gas recirculation; and (d) using water injection. These objectives, when reduced to terms

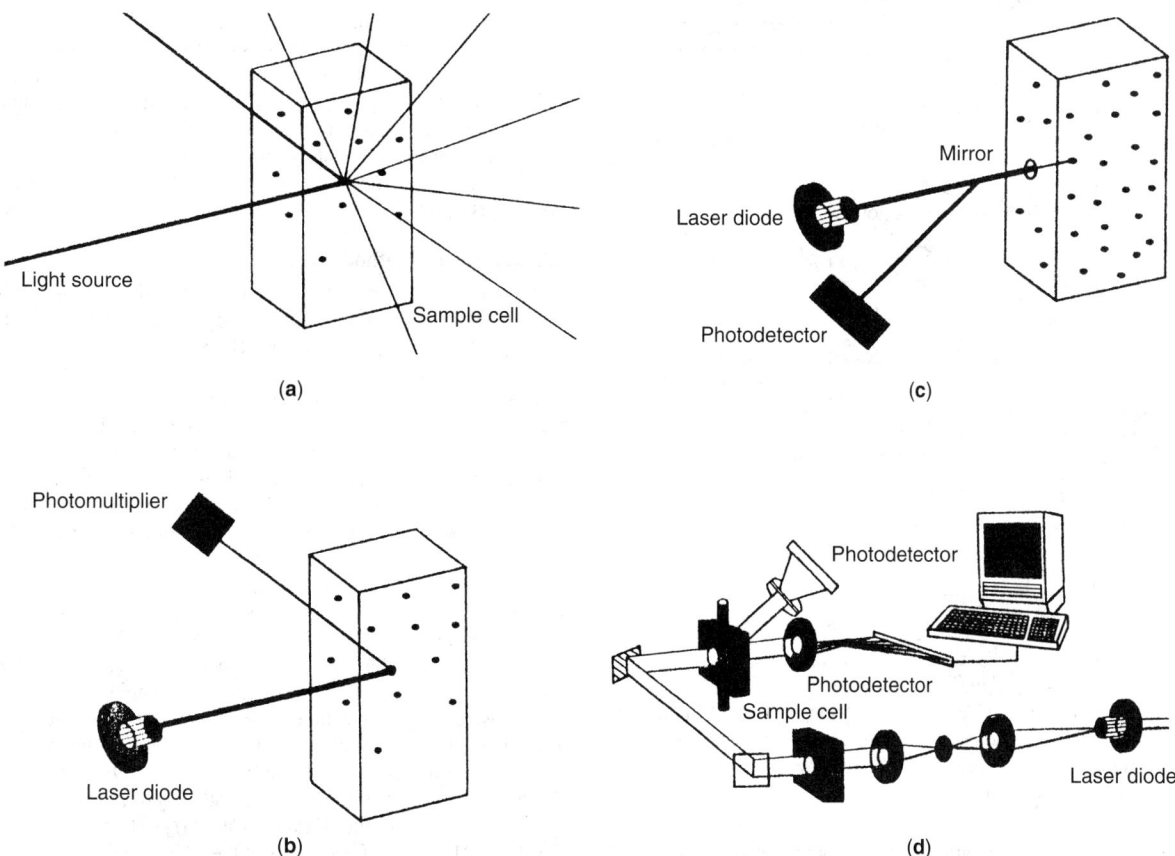

Fig. 4. Instrumentation for measuring air pollution particle dimensions. (**a**) Some fine-particle-size analyzers determine particle size by measuring the Doppler shift of light as it is scattered by moving particles. Smaller particles move faster, causing a greater Doppler shift in the light that they scatter. (**b**) In some conventional analyzers, light passes completely through an extremely dilute suspension and scatters in all directions. The detector measuring the Doppler shifts of the scattered light has no high-level signal to reference against the resulting low-level signal. The resulting low-level signals require amplification from photomultipliers, which can introduce noise errors. (**c**) In instrument shown (*Microtrac — Leeds & Northrup*), light travels to the sample via an optical wave guide. A mirror reflects some of the light, creating a high-level reference signal. Moving particles backscatter the light penetrating the mirror. The instrument combines the reflected and backscattered light to create a high-level signal sufficiently strong to be fed directly to a solid-state photodetector with a requirement for amplification. (**d**) Infrared light, emitted from a long-lived solid-state laser diode, scatters when it hits particles in the sample cell. Multi-train optics direct all scattered light onto solid-state photodetectors, which measure scattered-light angles and send signals to a computer control module. (*Leeds & Northrup.*)

of hardware, mean changes in the configuration, location, and spacing of burners, and the kinds of firing and combustion techniques used. Two-stage combustion is defined as firing all fuel below stoichiometric amounts of primary air in a first stage of combustion, followed by injecting air in a second stage, whereupon burnout of the fuel is completed. There is removal of heat between the two stages. The formation of NO in the first stage is limited because the available oxygen for combustion with nitrogen is limited. The removal of heat between stages kinetically limits the formation of NO when excess air is added to the second stage. Experience shows that a 90% reduction in NO_x emission can be achieved in this manner. By recirculating flue gas, both the peak flame temperature and oxygen content are lowered. Injecting low-temperature steam or water also provides a diluting effect. Although probably of limited value for electric utility boilers (because thermal efficiency is lowered), the water-injection technique may be one of the better ways to reduce NO_x emissions in connection with internal-combustion engines of the stationary type. The situation in the case of internal combustion engines for automotive vehicles is considerably more complicated — there is a wide range of loads on such engines, high performance is required at all loading conditions, and the combustion process from fuel to exhaust must be kept simple and hence low-cost. As of the early 1980s, there remained some differences in opinion as regards the use of catalytic converters to remove NO_x from automobile exhausts. See also **Catalytic Converter (Internal Combustion Engine); Combustion**; and **Petroleum**. The chemical characteristics of NO_x are described under **Nitrogen**.

Hydrocarbons. Extensive pollution of air occurs from the introduction of hydrocarbons either from (a) the incomplete combustion of hydrocarbon fuels in both stationary and vehicular engines; or (b) from paint spraying, solvent cleaning, printing, chemical and metallurgical, and other plants that use various fluids that have a high hydrocarbon content. Engine design and tuning are major factors in abating exhaust hydrocarbons. The intent is to fully burn the hydrocarbon content of the fuel. In a major city, industrial and commercial sources of organic solvent fumes (principally hydrocarbons) may average from 300 to 600 tons/day. For years, without legal restrictions, some operators found it more economical to permit vapor-laden air to escape to the atmosphere rather than to invest in solvent recovery equipment. Regulations coupled with higher costs of solvents have gone a long way toward eliminating this source of industrial pollution. Also, the chemical industry has successfully developed newer solvents, which are less volatile, and easier to handle and recover.

Carbon Monoxide. This pollutant is also associated with combustion operations, again being a product of incomplete combustion. Over the years, there has been a much greater awareness of carbon monoxide as a pollutant than the aforementioned gases because of its potent toxicity, dramatized by numerous deaths in earlier years as the result of keeping an automobile engine running in an enclosed space. Faulty residential heaters continue to take their toll of life and in recent years an important killer is the outdoor grill or hibachi with glowing coals taken into a camper or cabin as a means to temper the evening chill. Vehicular tunnel and large parking garage designers, of course, have practiced careful control over carbon monoxide concentrations for many years. See also **Carbon Monoxide**. The effects of carbon monoxide on human beings are shown in Fig. 5.

Other Pollutants. Some of these are gases; others fall into the particulate category. Of considerable importance are beryllium dust — very

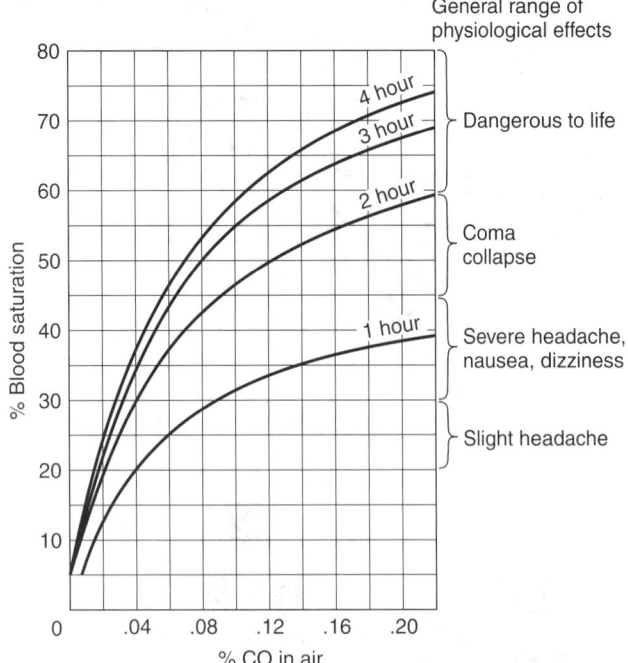

Fig. 5. Effects of carbon monoxide on human beings. This chart can be considered only as a general guide because the percent of CO blood saturation will vary with exertion, excitement, fear, depth of respiration, anemia, and general physical condition of the individual.

toxic and arising from ore preparation and metalworking operations, but of relatively limited extent because this metal is not a common structural material. Other contaminants include fluorides; metal fumes, such as arsenic, lead, and zinc; organic phosphates, notably from crop dusting and spraying; numerous kinds of organic vapors, including chlorinated hydrocarbons and hydrofluorocarbons (used in aerosol containers but suspect in connection with altering the ozone content of the upper atmosphere—see also **Aerosol**); radioactive fallout, such as ^{14}C, ^{137}Ce, and ^{90}Sr, arising from nuclear-device testing–no longer the major concern it was prior to the nuclear test ban accepted by most nations); and uranium dust. See also **Asbestos**; **Beryllium**; and **Pneumokonioses**.

Acid Rain

This term is used almost exclusively in connection with the effect of airborne pollutants on the natural health of forests and lakes. The term was coined by Angus Smith several years ago, when he referred to the effect of industrial emissions on precipitation over the British midlands. Unfortunately, the term does *not* fully or accurately describe what may be occurring in certain forests and lakes.

The topic of acid rain during the 1980s and early 1990s was one of controversy and of incomplete answers in terms of official policy and science—after an expenditure of many millions of dollars. In 1980, the National Acid Precipitation Assessment Program (NAPAP) was established and subsequently consumed thousands of scientific research hours and costly field investigations, including the use of numerous helicopter expeditions to northwestern mountain and lake areas of the United States and Canada. Thousands of hours of computer power were consumed.

Segments of the jigsaw puzzle were worked out in excruciating detail, but only a few of the larger pieces of the puzzle were put together. Bottom-line conclusions and recommendations needed for governmental and private forestry operators were not forthcoming. A 6,000-page report was generated by NAPAP. But, in terms of a summary, the findings can be condensed to the few following points:

1. Acid rain has adversely affected aquatic life in about 10 percent of eastern lakes and streams.
2. Acid rain has contributed to the decline of red spruce at high elevations by reducing that species' cold balance.
3. Acid rain has contributed to erosion and corrosion of buildings and materials.
4. Acid rain and related pollutants, especially fine sulfate particles, have reduced visibility throughout the Northeast and in parts of the West.

Ellis Cowling (Ecologist, North Carolina State University) observed in early 1991, "You can say no symptoms were found, and we looked hard"—but that is different from saying no problem exists. A. Johnson (University of Pennsylvania) has observed, "It is the marginal ecosystems, the tenuous ones on the edge, that are affected by acid rain at current levels. I think control measures are justified. They will probably go a small way toward restoring some of the ecosystems that have been altered. And they will largely prevent more harm from occurring." See also **Acid Rain**.

Air Treatment Methodologies

Numerous techniques have been used to reduce, if not fully eliminate, various forms of air pollution over the years. For example, in connection with particulates, see entry on **Electrostatic Precipitator**. Most methods, however, fall into what might be termed the category of wet-scrubbing processes. As of the early 1980s, there were nearly 100 such processes on the market. Progress in this field has been steady, but not characterized by major breakthroughs. The concentration of sulfur dioxide in stack gases emitted by steam generation plants is usually in the range of only 400 to 2,000 parts per million (ppm). However, the volume of gases produced by the utility industry results, for example, in the liberation of large tonnages of sulfur dioxide into the atmosphere.

Chemical scrubbing systems for SO_2 absorption fall into two broad categories: (a) Disposable systems; and (b) regenerative systems. Typical of systems in use for a number of years are those that use an aqueous slurry of an insoluble calcium compound, which can be discarded after use. Disposable SO_2-removal systems use aqueous slurries of finely ground materials, such as lime, limestone or dolomite, to produce a mixture of insoluble sulfites and sulfates. On passing through the scrubber, SO_2 from the waste gas dissolves to form sulfurous acid: $SO_2 + H_2O \rightarrow H_2SO_3$. The dissolved SO_2 reacts with the lime, $Ca(OH)_2$ or limestone, $CaCO_3$, to form insoluble calcium sulfite, $CaSO_3$: $Ca(OH)_2 + H_2SO_3 \rightarrow CaSO_3 + 2H_2O$; $CaCO_3 + H_2SO_3 \rightarrow CaSO_3 + H_2O + CO_2$. Unfortunately, SO_2 is less soluble (and hence less easily removed by scrubbing) in slightly acid solutions, so that it is extremely difficult in practice to operate a calcium-based system in such a manner that SO_2 removal is maximized while the quantities of calcium chemicals are minimized in order to approach stoichiometric conditions. As calcium-based slurry systems are usually operated at pH 6–10, disposal of the very large masses of used slurry presents a major problem. A typical power station using a calcium-based SO_2-removal slurry system will produce several hundred tons of spent slurry per day. A further disadvantage of lime or limestone systems is their marked tendency to precipitate insoluble calcium salts inside the scrubber. Unless the scale is removed, the scrubber shortly becomes inoperable.

Although chemically analogous to calcium-based systems, magnesium-based scrubbing systems possess several advantages. A slurry of finely divided magnesium hydroxide, $Mg(OH)_2$, is pumped through the scrubber to remove SO_2 from stack gases. Insoluble magnesium sulfite, $MgSO_3$, is formed: $Mg(OH)_2 + SO_2 \rightarrow MgSO_3 + H_2O$. Hydrated magnesium sulfite, $MgSO_3 \cdot 6H_2O$, can be disposed of as such, although it is usually heated to produce a rich stream of SO_2 and regenerate MgO. The SO_2 is compressed, liquefied, and stored in tanks for market; or catalytically oxidized to sulfur trioxide, SO_3, and treated with water to produce sulfuric acid, H_2SO_4. Alternatively, the SO_2 is mixed with hydrogen sulfide, H_2S, to produce elemental sulfur by the Claus process: $SO_2 + 2H_2S \rightarrow 3S + 2H_2O$. Absorption efficiency of SO_2 attainable in a magnesium system is good, and removal efficiencies from 90 to 95% have been claimed without difficulty at reasonable liquor recirculation and MgO feed rates. As with calcium systems, serious scaling occurs due to build-up of insoluble $MgSO_3$.

Scrubbing solutions containing sodium (or other alkali metals) compounds have been extensively studied for removal of SO_2. Justification for the use of sodium compounds includes: (a) complete solubility in water with no formation of scale; and (b) simple reactions with SO_2 : $Na_2CO_3 + SO_2 \rightarrow Na_2SO_3 + CO_2$; $2NaHCO_3 + SO_2 \rightarrow Na_2SO_3 + 2CO_2 + H_2O$; $2NaOH + SO_2 \rightarrow Na_2SO_3 + H_2O$. In one commercial process, a scrubbing solution of sodium sulfite is used, which readily absorbs SO_2 to form the bisulfite: $Na_2SO_3 + H_2O + SO_2 \rightarrow 2NaHSO_3$. In practice, only a portion of the Na_2SO_3 is converted to $NaHSO_3$ because the SO_2 absorption efficiency diminishes as the bisulfite concentration increases. The resulting solution is heated to decompose the bisulfite and thermally regenerate the sulfite. The gaseous SO_2 is compressed, liquefied and handled as previously mentioned under the magnesium system.

Ammonia-based chemicals appear to have some advantages over sodium systems. They are less costly, and regeneration by conventional means is possible, with the byproduct, ammonium sulfate, a marketable commodity for fertilizer.

Solutions containing ammonium sulfate, with or without the addition of ammonium hydroxide, have been widely used. The ammonium system can operate effectively only within a pH range of 4.0 to 7.0. As the pH value increases above 7.0, progressively more gaseous ammonia is liberated and this reacts in the gaseous phase with water vapor and SO_2 to produce a dense aerosol (white plume) which is difficult for scrubbers to remove. In an ammonia system, in order to regenerate the scrubbing solution, the ammonium bisulfite and sulfite mixture is heated to drive off gaseous SO_2: $2NH_4HSO_3 \rightarrow (NH_4)_2SO_3 + H_2O + SO_2$. Alternatively, the ammonium bisulfite/sulfite mixture can be treated with calcium hydroxide. Gaseous ammonia is evolved and trapped in water, which is then recirculated to the scrubber.

Sodium citrate also is used in an SO_2 removal system. The solution is buffered at a pH 3.0–3.7 by the citrate ion, sulfur dioxide is absorbed, and an equilibrium mixture of sodium bisulfite and citric acid is produced.

$$\begin{array}{c}CH_2COONa\\|\\HO-C-COONa\\|\\CH_2COONa\end{array} + 3\,SO_2 + 3\,H_2O \longrightarrow 3\,NaHSO_3 \;+\; \begin{array}{c}CH_2COOH\\|\\HO-C-COOH\\|\\CH_2COOH\end{array}$$

The bisulfite leaving the scrubber is then reduced with gaseous hydrogen sulfide, which precipitates elemental sulfur by a modified Claus reaction:

$$3\,NaHSO_3 + \begin{array}{c}CH_2COOH\\|\\HO-C-COOH\\|\\CH_2COOH\end{array} + 6\,H_2S \longrightarrow \begin{array}{c}CH_2COONa\\|\\HO-C-COONa\\|\\CH_2COONa\end{array} + 9\,H_2O + 9\,S$$

A formate system uses two reactions involving potassium formate, HCOOK, which is regenerated after recovery of elemental sulfur. This method has the advantage over other wet scrubbing methods in that no precipitation of insoluble intermediates occurs at any stage of the process. Disadvantages include the need to heat K_2CO_3 solution, at high temperature and pressures, with carbon monoxide to regenerate the potassium formate. The energy requirements thus are high.

While it has been demonstrated that solutions of NaOH, $NaHCO_3$, and Na_2CO_3 are effective for SO_2 removal, these solutions are not effective for removal of mixtures of NO and NO_2, particularly when the gas stream velocities are reasonably high. Under conditions where from 95 to 99% SO_2 may be removed, the solutions may only be effective in removing from 5 to 15% NO_x. The fundamental difference between SO_2 and NO_x removal is that NO_x gases (mixtures of NO, N_2O_3, NO_2) are approximately 1,000 to 2,000 times less soluble in water than SO_2 at any given temperature. It has been found that conventionally designed wet scrubbers often do not provide sufficient liquid-to-gas contact surface areas or residence times to permit the NO_x to dissolve in the scrubbing solution. Consequently several stages may be required. Concentrations of NO_x in the range of 20,000 to 40,000 ppm require from 6 to 12 stages.

If no SO_2 is present, a sodium-based process may be used to remove NO_x efficiently. In the Neville-Krebs process (patent applied for), removal efficiencies of 60 to 90% have been achieved from gas streams containing up to 1,500 to 2,000 ppm of NO_x passing through a 3-stage scrubber (1 cubic foot (0.028 cubic meter per stage) at 150 to 500 cubic feet (4.2 to 14.2 cubic meters) per minute.

The urea system is another relatively new system for removing NO_x from low-volume, slow-flowing waste gas streams. The system uses a slightly acid solution of urea, $CO(NH_2)_2$. Unfortunately, the cost of urea is quite high, particularly for a large installation.

Other systems using electron-donor compounds have been tried or are in development. Such compounds include tri-*n*-butyl phosphate, dimethyl-formamide, triethyleneglycol dimethyl ether, dimethylsulfoxide, hexam-ethylphosphoramide, diethyleneglycol dimethyl ether, tricresyl phosphate, and dioxane. Most of these compounds are expensive compared with inorganic compounds used in other scrubbing systems.

Thus, it is evident that as of the early 1980s, the panacea for stack gas treatment was yet to be realized. Difficulties in treating stack gases have provided incentives at the other end of the combustion cycle, namely, in the treatment of the fuels prior to combustion. Various means for treating coal are described in detail in the entry on **Coal**; and desulfurization of petroleum fuels is described under **Petroleum**.

Numerous other entries in this volume take the energy/pollution interface into consideration. These include: **Air Mass; Atmosphere (Earth); Catalytic Converter (Internal Combustion Engine); Climate; Combustion; Diesel Engine; Electric Power Production and Distribution; Energy; Fuel; Gas and Expansion Turbines; Geothermal Energy; Hydrogen (Fuel); Internal Combustion Engine; Natural Gas; Nuclear Power; Oil Shale; Tar Sands; Visibility**; and **Weather Technology**.

Additional Reading

Abelson, P.H.: "Asbestos Fiasco," *Science*, 1017 (March 2, 1990).

Abelson, P.H.: "New Technology for Cleaner Air," *Science*, 793 (May 18, 1990).

Abelson, P.H.: "Excessive Fear of PCBs," *Science*, 361 (July 26, 1991).

Alley, E.R., L.B. Stevens, and W.L. Cleland: "Air Quality Control Handbook," The McGraw-Hill Companies, Inc., New York, NY, 1998.

Arya, S.P.: "Air Pollution Meteorology and Dispersion," Oxford University Press, Inc., New York, NY, 1998.

Ashmore, M., L. Emberson, and F. Murray: "Air Pollution Impacts on Crops and Forests: A Global Assessment," World Scientific Publishing Company, Inc., River Edge, NJ, 2001.

Ayres, J., R. Richards, and R. Maynard: "Air Pollution and Health," Vol. 3, World Scientific Publishing Company, Inc., River Edge, NJ, 2002.

Barner, R.A. and A.C. Lasaga: "Modeling the Geochemical Carbon Cycle," *Sci. Amer.*, 74 (March 1989).

Baron, P.A. and K. Willeke: "Aerosol Measurement: Principles, Techniques, and Applications," 2nd Edition, John Wiley & Sons, Inc., New York, NY, 2001.

Barth, H.G.: "Particle Size Analysis," *Analytical Chemistry*, (June 15, 1991).

Bell, N. and M. Treshow: "Air Pollution and Plant Life," 2nd Edition, John Wiley & Sons, Inc., New York, NY, 2002.

Benedick, R.E.: "Ozone Diplomacy: New Directions in Safeguarding the Planet," Harvard University Press, Cambridge, MA, 1997.

Bohn, H.: "Consider Biofiltration for Decontaminating Gases," *Chem. Eng. Progress*, 34 (April 1992).

Boss, M.J. and D.W. Day: "Air Sampling and Industrial Hygiene Engineering," Lewis Publishers, Boca Raton, FL, 2000.

Boubel, R.W., D.L. Fox, and D.B. Turner: "Fundamentals of Air Pollution," 3rd Edition, Academic Press, Inc., San Diego, CA, 1994.

Brimblecombe, P.: "Air Composition and Chemistry," 2nd Edition, Cambridge University Press, New York, NY, 1995.

Clement, R.E.: "Environmental Analysis," *Analytical Chemistry*, 270T (June 15, 1991).

Colls, J.: "Air Pollution: An Introduction," Chapman Hall, New York, NY, 1996.

Conrad, J.: "An Acid-Rain Trilogy," *American Forests*, 21 (November–December 1987).

Cordasco, E.M., C. Zenz, and S.L. Demeter: "Environmental Respiratory Diseases," John Wiley & Sons, Inc., New York, NY, 1997.

Crawford, M.: "Scientists Battle Over Grand Canyon Pollution," *Science*, 911 (February 23, 1990).

Davenport, G.B.: "Understand the Air-Pollution Laws that Affect Chemical Process Industries Plants," *Chem. Eng. Progress*, 40 (April 1992).

de Nevers, N.: "Air Pollution Control Engineering," 2nd Edition, The McGraw-Hill Companies, Inc., New York, NY, 1999.

Downing, T.M.: "Preparing for New Smokestack Monitoring Regulations," *Instruments and Control Systems*, 47 (February 1992).

Ebert, L.B.: "Is Soot Composed Predominantly of Carbon Clusters?" *Science*, 1469 (March 23, 1990).

Fulkerson, W., R.R. Judkins, and J.K. Sanghvi: "Energy from Fossil Fuels," *Sci. Amer.*, 128 (September 1990).

Graedel, T.E. and P.J. Crutzon: "The Changing Atmosphere," *Sci. Amer.*, 58 (September 1989).

Hall, J.V., et al.: "Valuing the Health Benefits of Clean Air," *Science*, 812 (February 14, 1992).

Hesketh, H.E.: "Air Pollution Control: Traditional and Hazardous Pollutants," CRC Press, LLC., Boca Raton, FL, 1996.

Hobbs, P.V. and L.F. Radke: "Airborne Studies of the Smoke from the Kuwait Oil Fires," *Science*, 987 (May 15, 1992).

Hoffman, D.J.: "Increase in the Stratospheric Background Sulfuric Acid Aerosol Mass in the Past 10 Years," *Science*, 996 (May 25, 1990).

Holdren, J.P.: "Energy in Transition," *Sci. Amer.*, 156–163 (September 1990).

Holgate, S.T., J.M. Samet, R.L. Maynard, and H.S. Koren: "Air Pollution and Health," Academic Press, Inc., San Diego, CA, 1999.

Hutterman, A. and D. Godbold: "Effects of Acid Rain on Forest Processes," John Wiley & Sons, Inc., New York, NY, 1994.

Kennedy, I.R.: "Acid Soil and Acid Rain: Research Studies in Botany and Relate Applied Fields," 2nd Edition, John Wiley & Sons, Inc., New York, NY, 1992.

Koenig, J.Q.: "Health Effects of Ambient Air Pollution: How Safe Is the Air We Breathe?" Kluwer Academic Publishers, Norwell, MA, 2000.

Little, C.E.: "The California X-Disease," *Amer. Forests*, 32 (July 8, 1992).

Liu, D.H. and B.G. Liptbak: "Air Pollution," Lewis Publishers, Boca Raton, FL, 1999.

Lyons, C.E.: "Environmental Problem Solving: The 1987–88 Denver Brown Cloud Study," *Chem. Eng. Progress*, 6171 (May 1990).

Majewski, M.S., P.D. Capel, and R.J. Gilliom: "Pesticides in the Atmosphere: Distribution, Trends, and Governing Factors," CRC Press, LLC., Boca Raton, FL, 1999.

Matthews, S.W. and J.A. Sugar: "Is Our World Warming? Under the Sun," *National Geographic*, 66 (October 1990).

Mohnen, V.A.: "The Challenge of Acid Rain," *Sci. Amer.*, 30 (August 1988).

Nazaroff, W.W. and L. Alvarez-Cohen: "Environmental Engineering Science," John Wiley & Sons, Inc., New York, NY, 2000.

Nierenberg, W.A.: "Atmospheric Carbon Dioxide: Causes, Effects, and Options," *Chem. Eng. Progress*, 27 (August 1989).

Ondov, J.M. and W.R. Kelly: "Tracing Aerosol Pollutants with Rare Earth Isotopes," *Analytical Chemistry*, 691A (July 1, 1991).

Patrick, D.R.: "Toxic Air Pollution Handbook," Van Nostrand Reinhold Company, Inc., New York, NY, 1997.

Regens, J.L. and R.W. Rycroft: "The Acid Rain Controversy," University of Pittsburgh Press, Pittsburgh, PA, 1988.

Roberts, L.: "Learning from the Acid Rain Program," *Science*, 1302 (March 15, 1991).

Schifftner, K.C.: "Air Pollution Control Equipment Selection Guide," CRC Press, LLC., Boca Raton, FL, 2002.

Schneider, T.: "Air Pollution in the 21st Century: Priority Issues and Policy," Elsevier Science, New York, NY, 1998.

Schnelle, K.B., C.A. Brown, C. Carelli, and F. Kreith: "Air Pollution Control Technology Handbook: A Handbook Series for Mechanical Engineering," CRC Press, LLC., Boca Raton, FL, 2001.

Sher, E.: "Handbook of Air Pollution from Internal Combustion Engines: Pollutant Formation and Control," Academic Press, Inc., San Diego, CA, 1998.

Snow, R.H. and T. Allen: "Effectively Measure Particle-Size Classifier Performance," *Chem. Eng. Progress*, 29 (January 1993).

Staff: "ICI Plans U.S. Plant for CFC Substitute," *Chem. Eng. Progress*, 11 (February 1990).

Stradling, D.: "Smokestacks and Progressives: Environmentalists, Engineers, and Air Quality in America 1881–1951," Johns Hopkins University Press, Baltimore, MD, 1999.

Turco, R.P.: "Earth under Siege: From Air Pollution to Global Change," 2nd Edition, Oxford University Press, Inc., New York, NY, 2001.

Van Wormer, M.B.: "Use Air Quality Auditing as an Environmental Management Tool," *Chem. Eng. Progress*, 62 (November 1991).

Wallich, P.: "Dark Days: Eastern Europe Brings to Mind the West's Polluted Past," *Sci. Amer.*, 16 (August 1990).

Wark, K., C.F. Warner, and W.T. Davis: "Air Pollution: Its Origin and Control," 3rd Edition, Addison Wesley Longman, Inc., Redding, MA, 1997.

Wettestad, J.: "Clearing the Air: European Advances in Tackling Acid Rain and Atmospheric Pollution," Ashgate Publishing Company, Brookfield, VT, 2002.

POLLUX (*β Geminorum*). The brighter star of the "heavenly twins." These two stars are always considered together in ancient writings and in astrology, and as a matter of fact, are not mentioned individually in literature, but always as the constellation Gemini. The constellation is of very ancient lineage and is referred to throughout all classical literature. The two stars were always considered of good omen by all peoples, and were always referred to as twins. Ranking seventeenth in apparent brightness among the stars, Pollux has a true brightness value of 45 as compared with unity for the sun. Pollux is an orange, spectral type *K* star. Estimated distances from the earth is 40 light years. See also **Constellations**.

POLONIUM. Chemical element, symbol Po, at. no. 84, at. wt. 210 (mass number of the most stable isotope), mp $252\,^\circ$C, bp $960\,^\circ$C, sp gr 9.4. The element was first identified as an ingredient of pitchblende by Marie Curie in 1898. The element occurs in nature only as a decay product of thorium and uranium. Because of limited availability and high cost, relatively few practical uses for the element have been found. Meteorological instruments for measuring the electrical potential of air have used small quantities of the metal. It is interesting to note that when Mme. Curie first identified polonium, she found that an electroscope was a far better instrument for detecting the metal than spectroscopic means. Polonium-plated metal rods and strips have been used as static dissipators in textile coating equipment and in various electrical equipment. The alpha rays from the polonium ionize the air, causing it to conduct and draw off accumulations of static electrical charges. Polonium is a member of periodic group 16 (formerly 6a).

Three isotopes of polonium occur in the uranium ($4n + 2$) radioactive series: ^{218}Po (radium A), $t_{1/2}$ 3.05 min; ^{214}Po (radium C'), $t_{1/2}$ 1.6×10^{-4} s; and ^{210}Po (radium F), $t_{1/2}$ 138.4 days, and the most stable isotope of polonium. It is used as a source of a-radiation. The thorium ($4n$) series has two isotopes, ^{216}Po (thorium A), $t_{1/2}$ 0.16 s, and ^{212}Po (thorium C'), $t_{1/2}$ 3×10^{-7} s. The actinium ($4n + 3$) series also has two isotopes, ^{215}Po (actinium A), $t_{1/2}$ 1.83×10^{-3} s, and ^{211}Po (actinium C'), $t_{1/2}$ 0.52 s, which occurs in a 0.3% branched chain disintegration of ^{211}Bi (actinium C). Several other isotopes of polonium have been prepared, one of which occurs in the neptunium ($4n + 1$) series as ^{213}Po, $t_{1/2}$ 4.2×10^{-6} s.

Polonium exhibits the allotropy of the lower members of the chalcogen group, having a low-temperature, cubic form, α-polonium, and a high-temperature, rhombohedral form, β-polonium.

The tendency of the chalcogens to show increasing metallic character as one moves down the periodic table is quite marked for polonium; in fact, it resembles lead more than it does tellurium. Its compounds have a more ionic character in its lower oxidation states than do the tellurium compounds. The stability of the 6+ state is low, the existence of polonate(VI) ion being doubtful. The common oxidation states of the element are 2+ and 4+.

The halides, consisting of both dihalides and tetrahalides, are covalent and volatile, and they are not well characterized. The fluorides have not been established. The complex $PoCl_6{}^{2-}$ is known. Polonium compounds are usually colored, a fact that is useful in following their reactions. Thus, polonium(II) chloride, $PoCl_2$, formed by dissolving polonium(IV) oxide, PoO_2, in HCl is pink, and an oxidation by heating or treatment with chlorine yields yellow $PoCl_4$. Polonium(IV) bromide, $PoBr_4$, dark red, gives purple polonium(II) bromide, $PoBr_2$, on heating. $PoBr_4$ also gives ammonium polonium bromide, $(NH_4)_2[PoBr_6]$ with ammonia. Complex iodides $M_2[PoI_6]$ have been prepared.

Metallic polonium reacts with air readily on heating, to form PoO_2, which exists in a yellow face-centered form having fluorite structure at low temperatures, and a red tetragonal one on heating. Polonium(IV) hydroxide, $Po(OH)_4$, precipitated from polonium(IV) solutions by ammonia, exhibits only slight activity, and is thus not amphiprotic. On reaction of polonium with HNO_3, $Po(NO_3)_4$ is formed, and on treatment of polonium(IV) chloride, $PoCl_4$, with H_2SO_4, polonium(IV) sulfate, $Po(SO_4)_2$, is formed, both being ionic-type salts, as indeed are other oxyacid compounds. The sulfate, however, is quite reactive, being hydrated in solution, dehydrated on removal from solution, and forming a basic compound, $2PoO_2 \cdot SO_3$, on heating. H_2S precipitates black polonium(II) sulfide, PoS.

Additional Reading

Greenwood, N.N. and A. Earnshaw: "Chemistry of the Elements," 2nd Edition, Butterworth-Heinemann, Inc., Woburn, MA, 1997.

Krebs R.E.: "The History and Use of Our Earth's Chemical Elements: A Reference Guide," Greenwood Publishing Group, Inc., Westport, CT, 1998.

Lide, D.R.: "CRC Handbook of Chemistry and Physics 2000–2001," 81st Edition, CRC Press, LLC., Boca Raton, FL, 2000.

Parker, P.: "McGraw-Hill Encyclopedia of Chemistry," 2nd Edition, The McGraw-Hill Companies, Inc., New York, NY, 1993.

Stwertka, A. and E. Stwertka: "A Guide to the Elements," Oxford University Press, Inc., New York, NY, 1998.

POLYAMIDE-IMIDE RESINS. An injection-moldable, high-performance engineering thermoplastic, polyamide-imide is the condensation polymer of trimellitic anhydride and various aromatic diamines with the general structure:

Polyamide-imides are available[1] in unfilled; 30% glass fiber-filled; 30% graphite fiber-filled; and modified 40% glass-filled grades. The unfilled grade has the highest impact resistance, while the graphite fiber-filled grade has the highest modulus or stiffness. The modified version offers the lowest cost while still maintaining an impressive slate of properties. The resins can be molded into complex precision parts and also can be extruded and machined to close tolerances. The resins are used extensively in the aerospace industry, offering significant weight reduction by replacing metal parts. Aircraft usage includes jet engine components, compressor and generator parts, and electronic/electrical devices. Polyamide-imide resins are also used in the hydraulic/pneumatic equipment industry for wear surfaces, bushings, seals, vanes, and flow control devices. The automative and heavy equipment industries use this material as parts in transmissions, universal joints, and power-assisted devices. Internal combustion engines use many polyamide-imide structural-mechanical parts, such as valve train components.

The material is opaque and characterized by good dimensional stability, creep resistance, impact resistance, and superior mechanical properties that persist up to temperatures of about 500 °F (260 °C). Unfilled polyamide-imide has a tensile strength of about 27,000 psi (186 mPa); flexural strength of about 35,000 psi (241 mPa); compressive strength of about 32,000 psi (221 mPa); and an elastic modulus of about 750,000 psi (5172 mPa). Mechanical properties at 450 °F (232 °C) exceed those of many polymers at room temperature. At cryogenic temperature, the unfilled polymer has a tensile strength of about 31,500 psi (241 mPa) with 6% elongation at −196 °C (liquid nitrogen). Heat deflection is high (525 °F; 273 °C), while the coefficient of linear thermal expansion is low. When burned, polyamide-imide produces a char rather than drip and produces very little smoke. Electrical properties are attractive. Radiation resistance is good. The resin is virtually unaffected by aliphatic and aromatic hydrocarbons, halogenated solvents, and most acid and base solutions. It is attacked by high-temperature caustic, steam, and some acids. At 50% relative humidity, the material (70 °F; 21 °C) will absorb about 1% moisture in 1000 hours.

POLYARYLATES. These are clear, amorphous thermoplastics that combine clarity, high heat deflection temperatures, high impact strength, good surface hardness, and good electrical properties with inherent ultraviolet stability and flame retardance. No additives or stabilizers are required to provide these properties. Polyarylates are aromatic polyesters that are manufactured from various ratios of iso- and terephthalic acids with bisphenol A.[1] The resultant products are free-flowing pellets which can be processed by a variety of thermoplastic techniques in transparent and opaque colors. Because polyarylate's weatherability is obtained through polymer chemistry rather than additives (as with most UV-resistant polymers), the properties of polyarylate do not deteriorate significantly with time. (Over 5000 hours of accelerated weathering and actual Florida and Arizona aging resulted in virtually no change in performance with respect to luminous light transmittance, haze, gloss, yellowness, and impact.) The flammability characteristics are inherent. Properties include a high oxygen index, low smoke density, low flame spread, and low toxic gas formation. Because the flammability properties are achieved without additives, the resultant products of combustion are essentially limited to Commercially available as CO_2, CO, and water.

Polyarylates are offered in several glass-reinforced versions with loading available up to 40%. The glass fibers provide higher stiffness, improved tensile strength, and higher heat deflection temperatures. Polyarylates may also be mineral-filled and reinforced with other fibers, such as carbon. Alloys/blends with other polymers are also available.

These materials are useful in outdoor applications, such as high-intensity discharge lighting (traffic signals), automobile halogen headlamp lenses and bodies, and rear-end elevated automobile stop lights. High-temperature lighting and microwave cookware are suitable applications. Electronic/electrical connectors and housings are also important applications for the polymer.

Polyarylates have good optical properties. Luminous light transmission can range from 84% to 88% with only 1% to 2% haze. Refractive index is 1.61. An important feature of the polyarylate family is high heat resistance demonstrated by a 340 °F (171 °C) heat deflection temperature at 264 psi (1.8 mPa). The material exhibits good retention of properties at high temperature exposures: 270,000 psi (1380 mPa) at 300 °F (149 °C); and over 200,000 psi at 350 °F (177 °C).

Polyarylates are injection molded, using standard screw machines, as well as extruded, foam molded, and blow molded. Melt temperatures range from 600 to 680 °F (316 to 360 °C). Mold temperatures should be maintained between 200 and 300 °F (93 and 149 °C).

POLYBASITE. A mineral antimony sulfide of silver $(Ag,Cu)_{16}Sb_2S_{11}$, in which copper substitutes for silver to approximately 30 atomic percent. It crystallizes in the monoclinic system; hardness, 2–3; specific gravity, 6.3; color, black, dark ruby red in thin splinters with metallic luster; nearly opaque. From the Greek, meaning *many*, suggesting the many-metal basis.

Occurs in low-temperature silver deposits commonly associated with silver and lead minerals. Found in various Western States in the United States, and as superb crystals at Arizpe and Las Chiapas, Mexico; in Chile, Peru, Sardinia, Germany, and Australia.

POLYBENZIMIDAZOLES. These are heterocyclic polymers that have outstanding high thermal characteristics, the highest obtainable in commercial polymers. These materials also have superior ablative and hydrolytic stability as well as high compressive and dimensional stability. Polybenzimidazoles essentially are unaffected by solvents, acids, and bases. They are marketed in stock shapes and as finished parts. The materials are not available in resin form. Hoechst Celanese markets the products under the tradename *Celazole*.®

Parts are produced by a high-pressure sintering process wherein the melt polycondensation resin is densified and then coalesced. Metallurgical pressures at temperatures exceeding over 400 °C (750 °F) are required. Polybenzimidazoles have repeating benzimidazole groups in the polymer backbone. These materials were synthesized first in 1961. Currently, the products result from a melt polycondensation reaction of aromatic, bis-ortho-diamines (diphenylisophthalate). By way of compounding, fabrication and end-use performance characteristics can be customized to specific needs. Often the materials are preferred for tribological applications inasmuch as they have the desirable characteristics of low coefficient of friction, low abrasion, and good high-temperature dimensional stability.

Polybenzimidazoles have been used for seals, mechanical components, electrical connectors, valve seats, and as components of materials-handling equipment in the petrochemical, geothermal, chemical process, aerospace, defense, automotive, and electrical products industries. These materials frequently are procured as replacement parts for other materials in an effort to improve equipment performance.

POLYBUTYLENE RESINS. These materials (PBs) are semicrystalline polyolefin thermoplastics based on poly(1-butene) and include homopolymers and a series of poly(1-butene-ethylene) copolymers. The resins available commercially[1] are manufactured via stereospecific Ziegler-Natta polymerization of 1-butene monomer. The commercial products are based on isotactic (98% to 99.5%), high-molecular-weight (230,000 to 750,000) polymer. Five crystalline modifications of poly(1-butene) are known. Of these, the glass transition temperature ranges from about −4 °F (−20 °C) for the homopolymers to about −30 °F (−34:18C) for the high-ethylene copolymers.

PB resins generally are resistant to acids, bases, solvents, paraffinic and naphthenic oils, detergents, and various chemicals. Resistance decreases, however, at elevated temperatures. They have good moisture barrier and electrical insulation properties. They exhibit a broad range of flexibility: tensile moduli vary from 41,500 psi (286 mPa) for homopolymers to 7500 psi (52 mPa) for the high-ethylene polymers. The resins are

---

[1] Commercially available as *Torlon*™. (*Amoco Chemicals Co.*)

[1] Data furnished by *Celanese Engineering Resins*.

[1] Commercially available as *Duraflex*™. (*Shell Chemical Co.*)

particularly resistant to creep, environmental stress cracking, chemicals, and abrasion. PB resins are offered in a special pipe grade, film grades, and five general-purpose grades.

PB pipe can be fabricated by conventional single-screw extrusion technology using vacuum or pressure sizing for dimensional control. The pipes can be joined by thermal fusion or mechanical fittings. Applications include cold and hot water plumbing. Other uses include well, heat pump, and fire-sprinkler piping as well as specialty hoses. Large-diameter PB pipe finds uses in the transport of abrasive or corrosive materials at high temperature, as found in the mining, chemical, and power generation industries.

PB film is usually made by the blown film process, but also can be cast on chill rolls. Film applications include food and meat packaging, compression wraps, and hot-fill containers. The material can be formulated to provide a wide range of seal strengths for peelable or easy-opening packaging.

The ability of PB to accept high filler loading (up to 80%) has resulted in its use as a color, mineral filler, and flame-retardant concentrate carrier. Polybutylene is compatible with polypropylene in all proportions and as a modifier it provides enhancement of processibility, impact, and weld line strength in injection molding and extrusion. It also provides improved impact strength and heat stability in films and improved hand and bondability in fibers of polypropylene. A comparatively recent use of PBs is for hot-melt adhesives and sealants where high strength, high-shear adhesion failure temperature, and a long open time are needed. They are particularly suited for use with aliphatic tackifying resins.

POLYBUTYLENE TEREPHTHALATE POLYESTERS.   A semicrystalline thermoplastic polyester. Because of its rapid crystallization, injection molding is the preferred method of processing. The material has been used for many years in the connector industry because of its good chemical resistance, high-temperature capabilities, good electrical properties, and long flow lengths in thin sections. This material (PBT) typically is formed in a transesterification reaction between 1,4-butanediol and dimethylterephthalate. Unmodified PBT is translucent in thin sections and opaque white in thicker sections.

The glass transition temperature of PPBT is about 52°C (125°F). Melting point is about 230°C (440°F). Unreinforced PBT is obtainable in several molecular weights. Compounded resins are available with numerous types and levels of fillers and reinforcements. Glass fiber reinforcement has a wide spectrum of physical properties. These materials can be made flame-retardant through the use of additives.

Exceptional electrical properties and temperature resistance qualify the material for numerous electrical parts—connectors, coil bobbins, light sockets, terminal blocks, fuse holders, and motor parts. PBT provides weight reduction of final parts. PBT has replaced a number of thermoset materials and is particularly popular for appliance housings and fibers for paint brushes.

POLYCARBONATE.   This material is classified as an engineering thermoplastic, mainly because of its toughness.[1] Exceptional clarity and high heat-deflection temperatures are other outstanding properties of polycarbonate. The most successful commercial polycarbonates are based on bisphenol A. Polycarbonate is made in an interfacial process by the reaction of bisphenol A with carbonyl chloride. Molecular weight is controlled by the use of a phenolic chain stopper. The carbonate moiety in the molecular structure provides high toughness and ductility; the bisphenol A moiety contributes to the polymer's high heat-deflection temperature. Molecular weights for typical commercial polycarbonates vary from 22,000 to 35,000.

Polycarbonate is available in general-purpose and food grades. Glass-filled products are available when less mold shrinkage and higher modulus are required. Branched polycarbonate is available for applications requiring high melt strength. Special grades include elastomer- and polyolefin-modified polycarbonates for applications where better low-temperature toughness is required. Polycarbonate blends with polyesters and with ABS (Acrylonitrile-Butadiene-Styrene) resins for certain special needs. Aromatic ester carbonates have been developed for applications where exceptional heat-deflection properties are specified. Polycarbonate

is available in a variety of colors, ranging from clear tints, custom tints, and custom opaques to black.

Polycarbonate can be melt-processed using all of the traditional methods for thermoplastics. Coextrusion with other engineering resins is possible in making multilayer sheet, profiles, and extrusion-blow molded containers. Sheet stock, both mono- and multilayer, can be thermoformed using vacuum, fluid pressure, or matched dies. Sheet and profile stock also can be cold formed by punching, rolling, or machining. Very-high-molecular-weight polycarbonate has been solvent-cast into thin, tough films.

Polycarbonate finds application in a broad range of industries, including electrical/electronics, business machines, glazing, transportation, lighting, appliances, food service, medical, and recreational equipment. One of the largest uses is in the electronics and business machine field, including small connectors and gears to large covers for computers. A new growth area is for laser-read compact recording disks for superior high-density recording quality. A special high-purity polycarbonate has been developed for this application. Transportation applications include tail and sidemarker lights, head lamps, and supports for instrument panels. Polycarbonate alloys are used in the construction of automobile bumpers. Other uses include traffic light housings and signal lamps.

Polycarbonate is particularly useful for glazing. Outdoor lighting applications can be made essentially vandalproof. Ophthalmic lenses and safety glasses are also important uses. Hard coatings are available to increase durability.

Food contact service includes large water bottles and carboys, baby formula bottles, microwave ovenware, beverage mugs and pitchers, tableware, and food-storage containers. Medical applications are also important, where sterilization of containers by steam and gamma radiation is often a requirement. Polycarbonate performs well in this area. Recently introduced formulations can be sterilized by gamma radiation without objectionable color change.

High-impact properties and colorability of polycarbonate provide wide design choices for vacuum sweepers and other household appliances.

POLYCYCLO-HEXYLENE-DIMETHYLENE TEREPHTHALATE.   PCT is 1,4-cyclohexylene-dimethylene terephthalate and is a high-temperature, semicrystalline thermoplastic polyester. PCT possesses excellent thermal properties. Injection molding is the predominant method of processing glass fiber-reinforced grades of PCT. It is widely used for products where excellent thermal (heat-resistant) properties are needed, as exemplified by surface-mountable electronic components, automotive parts, and dual-ovenable cookware. The material also is used for flexible electronic circuitry. PCT-based polycarbonate is a polymeter that provides melt blends that exhibit excellent clarity, toughness, chemical resistance, flow, and gloss.

PCT is differentiated from other thermoplastic polyesters by its higher heat deflection temperature. Continuous use temperatures of up to 150°C (300°F) are possible.

Principal uses for this material are found in the electrical/electronics industries; automotive parts, such as alternator armatures and pressure sensors; optical uses, such as safety goggles; and garden vehicles, such as mower decks and shrouds, tractor hoods, grills, and fenders.

POLYEMBRYONY.   The development of more than one individual from a single fertilized egg cell. In this process the egg breaks up during its early development into several to many component parts, each of which becomes a complete animal. It takes place in the phylum *Bryozoa* in connection with the formation of colonies and has been reported in some of the parasitic insects (*Hymenoptera*). Since the host animal defends itself against the efforts of the female parasite to deposit her eggs in its body, the development of many young from each egg successfully placed in an obvious advantage. The process is akin to an asexual reproduction.

POLYESTER FIBERS.   The principal characteristics of these fibers are described in the entry on **Fibers**. Polyester fibers are defined as synthetic fibers containing at least 80% of a long-chain polymer compound of an ester of a dihydric alcohol and terephthalic acid. The first polyester fiber to be commercialized was prepared from the ester in which the dihydric alcohol was ethylene glycol; this fiber is the material used in the largest quantity by the textile industry. For some other commercial uses, the ester 1,4-dimethyldicyclohexyl terephthalate is also used.

The original process, still in use for making the polymer, employs dimethyl terephthalate (DMT) and ethylene glycol as raw materials. A

---

[1] Data furnished by *Dow Chemical Co.*

later process, using direct esterification of terephthalic acid (TPA) with ethylene glycol, also gained acceptance after the increased availability of highly purified TPA. With either process, the first step is the preparation of the intermediate diester, *bis*-hydroxyethyl terephthalate (bisHET), which then is further condensed to the polymer.

The basic process for making polyester fibers from the polymer is called *melt spinning*, i.e., heating the polymer above its melting point, forcing it through small holes in a metal plate, and then quenching the molten stream as it issues from the holes by means of a current of cool air. The spun yarn is weak and highly extensible because the polymer molecules are randomly oriented. To impart strength and dimensional stability the yarns must be drawn at temperatures above the glass-transition temperature of the material by pulling the yarn between two *godet wheels*, the second of which is rotating at a speed three to six times as fast as the first. The higher the draw ratio, i.e., the ratio of the two speeds, the more oriented the molecules become and the stronger the yarn.

The two main classes of polyester fibers are continuous-filament yarns and short-cut fibers, called staple. A wide range of deniers is available in continuous-filament yarns, varying from very fine deniers of about 20 up to 2000 for heavy industrial yarns. (Denier is the weight in grams of 9000 meters of yarn). The number of filaments in these yarns ranges from about 7 for the 20-denier yarns up to 384 for the heavy material. Staple fiber is produced in sizes ranging from 0.5 to 1.5 denier per filament. The finer deniers are used in making blends with cotton and rayon for apparel, while the coarser-denier yarns generally are used for carpets. Staple lengths vary from $1\frac{1}{4}$ to 6 inches (3.1 to 15.2 centimeters).

Fiber with no added delustrant is designated as *clear*. *Bright* fiber has about 0.1% titanium dioxide (TiO_2); *semidull* fiber has about 0.25% TiO_2; and *dull* fiber has up to 2% TiO_2. Other variations in physical properties and dyeing characteristics include optically brightened, high-modulus, high-shrink, high-tenacity, low-pilling, deep-dyeable, and cationic-dyeable fibers.

POLYETHER-ETHERKETONE.   Abbreviated PEEK, polyether-ether-ketone is a high-temperature resistant thermoplastic suitable for wire coating, injection molding, film, and advanced composite fabrication. The wholly aromatic structure of PEEK contributes to its high-temperature performance. Its crystalline character gives it important advantages, including resistance to organic solvents and dynamic fatigue, and retention of ductility on short-term heat aging. The material is available as dry, free-flowing granules and exhibits very low water absorption. Continuous service at temperatures up to 470 °F (243 °C) and intermittent use up to 600 °F (316 °C) are possible. PEEK has good resistance to aqueous reagents, with long-term performance in super-heated water at 500 °F (260 °C). It resists attacks over a wide pH range, from 60% sulfuric acid to 40% sodium hydroxide at elevated temperatures. Attack can occur with some concentrated acids. No organic solvent attack has been observed on molded parts, although a limited range of solvents will stress-craze highly stressed PEEK-coated wire. Radiation resistance is excellent. Typical applications include wire and cable, automotive engine parts, aerospace components, valve plates, valve linings, oil well data logging tools, bearings, woven monofilament, and film.

POLYETHERIMIDE.   This is an amorphous, high-performance thermoplastic. The material is characterized by high strength, rigidity, heat resistance, dimensional stability, and electrical properties, combined with broad chemical resistance and processibility. Unmodified polyetherimide is amber-transparent in color and exhibits inherent flame resistance and low smoke evolution without the use of additives. The material is commercially available in several grades — unreinforced and in 10, 20, 30, and 40% glass fiber reinforced formulations for general-purpose molding and extrusion. Also available are easy-flow and release grades, wear-resistant grades, carbon fiber-reinforced grades for high-strength and static dissipation, along with a family of high-heat grades. A relatively new family of polyetherimide blends is available for use in vapor-phase soldering environments and for high-impact applications.

Polyetherimide has a chemical structure based on repeating aromatic imide and ether units. High performance strength characteristics at high temperatures are provided by rigid imide units, while the ether linkages confer the chain flexibility required for good melt processibility and flow. Polyetherimide is resistant to a wide range of chemicals, including most hydrocarbons, alcohols, and fully halogenated solvents. It is resistant to

mineral acids and tolerates short-term exposure to mild bases. These resins are rated for 170–180 °C (338–356 °F) in continuous-use applications. Intermittent use at 200 °C (392 °F) is possible.

Resins are used in electronic/electrical applications (connectors, circuit boards that are vapor and wave solderable, microwave transparent radomes, integrated circuit chip carriers, miniature switches, explosion proof enclosures, lamp reflectors, and high-precision fiber optic components). Polyetherimide is used for medical components that require all forms of sterilization. Other uses are found in the transportation field, dual-ovenable cookware, as well as bearings, fasteners, and advanced composites.

POLYETHYLENE.   A thermoplastic molding and extrusion material available in a wide range of flow rates (commonly referred to as melt index) and densities. Polyethylene offers useful properties, such as toughness at temperatures ranging from −76 to +93 °C, stiffness, ranging from flexible to rigid, and excellent chemical resistance. The plastic can be fabricated by all thermoplastic processes.

Polyethylenes are classified primarily on the basis of two characteristics, namely, density and melt index. The former is the criterion used to distinguished the type; and the latter for the designation as to category (ASTM-D-1248). ASTM type I polyethylene (sp. gr. 0.910–0.925) is commonly referred to as low-density, conventional, or high-pressure polyethylene. ASTM type II polyethylene (sp. gr. 0.926–0.940) is commonly referred to as medium-density or intermediate-density polyethylene. ASTM type III polyethylene (sp. gr. 0.941–0.965) is commonly called high-density, linear, or low-pressure polyethylene. High-density type III polyethylene has been divided into two ranges of density: 0.941–0.959 (considered type III); and 0.960 and higher, commonly considered type IV. Within each density classification, products with different melt indexes are categorized numerically as follows: category 1 has a melt index (MI) greater than 25; category 2 has an MI greater than 10 to 25; category 3, MI > 1.0 to 10; category 4, MI > 0.4 to 1.0; category 5 has a 0.4 maximum.

Chemical Composition.   Polyethylene is formed from the polymerization of ethylene under specific conditions of temperature and pressure and in the presence of a catalyst, according to:

The reaction is exothermic and may form polymer from a molecular weight of 1000 to well over 1 million. The high-pressure process, which normally produces types I and II, uses oxygen, peroxide, or other strong oxidizers as catalyst. Pressure of reaction ranges from 15,000 to 50,000 psi (∼1,020–3,400 atmospheres). The polymer formed in this process is highly branched, with side branches occurring every 15–40 carbon atoms on the chain backbone. Crystallinity of this polyethylene is approximately 40–60%. Amorphous content of the polymer increases as the density is reduced.

The low-pressure processes, such as slurry, solution, or gas phase, can produce types I, II, III, and IV polyethylenes. Catalysts used in these process vary widely, but the most frequently used are metal alkyls in combination with metal halides or activated metal oxides. Reaction pressures normally fall within 50 to 500 psi (∼3.4–34 atmospheres). Polymer produced by this process is more linear in nature, with branching occurring about every 1000 carbon atoms. Linear polyethylene of types I and II is approximately 50% crystalline and types III and IV are as high as 85% crystalline.

Ethylene has been polymerized with other monomers, e.g., propylene, butene-1, hexene, ethyl acrylate, vinyl acetate, and acrylic acid, to develop such specific properties as environmental stress crack resistance, low-temperature toughness, and improved flexibility and toughness. High-molecular-weight (HDPE) and chlorinated polyethylenes have been developed to extend the property range of polyethylenes from extremely rigid to elastomeric.

Applications.   Polyethylene products include extruded films for food packaging (baked goods, frozen foods, produce); nonfood packaging (heavy-duty sacks, industrial liners, shrink and stretch pallet wrap); non-packaging (agricultural, diaper liners, industrial sheeting, trash bags);

extrusion coating of films, foils, paper, and paperboard; blow molding of bottles, drums, tanks, toys, and pails; injection molding of industrial containers, closures, housewares, toys; extrusion of electrical cable jacketing, pipe, sheet, and tubing; and rotational molding of tanks, drums, toys, and sporting goods.

Properties. Tensile strength, hardness, chemical resistance, surface appearance, and flexural modulus increase with an increase in density (from type I through type IV).

Polyethylene is translucent to opaque white in thick sections, opacity increasing with density. Relatively clear film can be extruded from polyethylene, especially if it is quenched rapidly. The plastic accepts pigmentation readily. Most coloring is performed using dry-blend techniques. Color dispersion devices are required to ensure thorough mixing of resin and pigment.

Mechanical properties of polyethylenes vary with density and melt index. Low-density polyethylenes are flexible and tough; high-density products are quite rigid and have creep resistance under load. Toughness is the primary mechanical property affected by melt index, with lower-melt-index polyethylenes having greater toughness. Under loads, polyethylene is subject to creep, stress relaxation, or a combination of both.

Excellent dielectric characteristics at all frequencies and high electrical resistivity have made polyethylene one of the most important insulating materials for wire and cable.

At no-load conditions, polyethylene has good heat resistance. However, small loads can cause distortion at relatively low temperatures. Dimensional stability of polyethylene is fair to good. Dimensional changes caused by crystallization during cooling usually occur in a non-uniform pattern, resulting in warpage. Narrower molecular weight distribution resins within given families result in less warpage. Types I and II polyethylenes produced by the low-pressure process offer significant improvement in heat distortion temperatures. This property is directly related to melting point and is much higher for low-pressure, low-density resins than for conventional LDPE resins. This allows molded parts to be exposed to significantly higher service temperatures, e.g., dishwasher parts, without undergoing distortion or warpage. Most shrinkage occurs within 48 hours after fabrication and for type I and type II materials is 0.01–0.03 inch/inch (centimeter/centimeter).

Rupture of molecular bonds by external and internal stress in the presence of certain compounds is referred to as environmental stress cracking. Small molecular fractures in the amorphous regions propagate until visible cracks appear. In time, the part may fail. Chemical agents which accelerate stress cracking in polyethylene include detergents; aliphatic and aromatic hydrocarbons; soaps; animal, vegetable, and mineral oils; ester-type plasticizers; organic acids; and aldehydes, ketones, and alcohols. There is no adequate test for stress cracking.

Deterioration occurs in uncolored polyethylene exposed to weather. Ultraviolet light causes photoactivated oxidation. Satisfactory weathering formulations contain 2–2.5% well-dispersed carbon black and stabilizers. The carbon black prevents ultraviolet light penetration.

Unmodified polyethylenes are flammable and are classified in the slow-burning category by the National Board of Fire Underwriters. Burning rate is approximately 1–1.5 inches (2.5–3.8 centimeters) per minute. The flammability of polyethylene may be retarded significantly by the addition of flame retardant compounds, such as antimony trioxide along with halogenated compounds.

At room temperature, polyethylene is insoluble in practically all organic solvents, although softening, swelling, and environmental stress cracking can occur. At high temperatures, some concentrated acids and oxidizing agents chemically attack polyethylene. Above 60 °C, the material becomes increasingly soluble in aliphatic and chlorinated hydrocarbons. Chemical resistance increases slightly as density is increased.

Polyethylene is water-resistant and is a good water vapor barrier. Less than 0.1% water is absorbed in a 2-inch (5-centimeter), 1/8-inch (3-millimeter) thick disk of polyethylene in 24 hours. Transmission of other gases is high when compared with that of most other plastics. Polyethylene is not satisfactory for retention of vacuum.

Fabrication. Polyethylene is readily fabricated by all methods of thermoplastic processing. The principal methods used are film and sheet extrusion, extrusion coating, injection molding, blow molding, pipe extrusion, wire and cable extrusion coating, rotomolding, and hot melt and powder coatings.

Decorating. Polyethylene parts are decorated by silk screening, hot stamping, or dry offset printing. For satisfactory printing, the surface must be oxidized by hot air, flame, chlorination, sulfuric acid-dichromate solution, or electronic bombardment. Hot air or flame methods are used with molded parts; flame or electronic methods with films. Inks specially made for polyethylene give best results. Roll-leaf hot stamping does not require pretreatment of the surface.

Design. Because of high mold shrinkage, parts must be carefully designed to minimize warpage. Wall cross-sectional thicknesses should be uniform throughout the part. Large flat areas should be avoided. Corners should be curved rather than square. Stiffening ribs should be less than 80% of the thickness of the wall to which they are attached. Thermoformed parts require liberal radii and draft angles. Slight undercuts can be incorporated when a female mold is used. Dimensional variations in a part made of polyethylene are difficult to predict. In general, greater tolerances should be allowed than with more rigid plastics.

Additional Reading

Gsell, R.A., H.L. Stein, and J.J. Ploskonka: "Characterization and Properties of Ultra-High Molecular Weight Polyethylene," *American Society for Testing & Materials*, West Conshohocken, PA, 1998.

Harris, J.M.: "Poly(Ethylene Glycol): Chemistry and Biological Applications," American Chemical Society, Washington, DC, 1997.

Peacock, A.J.: "Handbook of Polyethylene: Structures, Properties, and Applications," Marcel Dekker, Inc., New York, NY, 2000.

B.W. HEINEMEYER, The Dow Chemical Company, Freeport, TX.

POLYGON. A plane figure with n vertices and n sides, also called an n-gon. Depending on the value of n, the following names are used: 3, triangle; 4, quadrilateral; 5, pentagon; 6, hexagon; 7, heptagon; 8, octagon; 9, nonagon; 10, decagon, etc. If all sides are equal, the polygon is equilateral; if all angles equal, equiangular; if all sides and angles are equal, the polygon is regular. Let A be the angle between two sides, B the angle between lines connecting the center of an inscribed or circumscribed circle to two vertices, a the length of a side, R the radius of a circumscribed circle, r the radius of an inscribed circle, and S the area of a regular polygon with n sides, then

$$A = (n-2)\pi/n \text{ radians} \qquad B = 2\pi/n \text{ radians}$$
$$a = 2R\sin B/2 \qquad r = a/2 \cot B/2$$
$$R = a/2 \csc B/2 \qquad S = na^2/4 \cot B/2$$

A convex polygon has no side which produced will enter the polygon. Its angles are called salient and each of them is less than 180°. The term *polygon* usually means a convex polygon. A concave polygon has two or more sides which, when produced, enter the polygon. One or more of its angles are reentrant and greater than 180°. The diagonal of a polygon is a line joining two nonadjoining vertices. A spherical polygon is a part of a sphere bounded by arcs of a great circle.

The sides of a polygon are usually considered to be straight lines but the sides could be curved and then one should call the resulting figure a curvilinear polygon.

See also **Sphere**; and **Triangle**.

POLYHALITE. Polyhalite, K_2Ca, $Mg(SO_4)_4 \cdot 2H_2O$ is a late evaporate mineral associated with halite, sylvite and carnallite from the famous oceanic salt deposits at Stassfurt, Germany, and near Carlsbad, New Mexico. It is of triclinic crystallization, with color grading from gray to brick-red; hardness, 3–3.5; specific gravity 2.78; translucent with vitreous luster; very bitter taste. It is a source of potassium.

POLYHEDRON. A solid with faces formed from plane polygons. The intersections of faces are edges and the points where three or more edges meet are vertices. If the faces are congruent regular polygons and the polyhedral angles are congruent, the polyhedron is regular. There are only five regular polyhedra, which are called the Platonic solids. Their names and the nature of their faces are: tetrahedron, 4 equilateral triangles; hexahedron or cube, 6 squares; octahedron, 8 equilateral triangles; dodecahedron, 12 pentagons; icosahedron, 20 equilateral triangles. See Fig. 1. See also **Parallelepiped**; and **Prism (Mathematics)**.

An important equation for polyhedra, discovered by Descartes and Euler, is $V - E + F = 2$, where V is the number of vertices, E the number of edges, and F the number of faces. With this equation it is easy to see that there are only five regular polyhedra. The equation is more general, however, for it holds for any polyhedron with curved faces and edges,

Fig. 1. A polyhedron with 240 sides. A polyhedron of this type may be referred to as a solid starred small rhombicosidodecahedron and always has 122 vertexes and 360 edges. This is in accordance with Euler's formula. $E + 2 = V + S$, where E = edges, V = vertexes, and S = sides.

TABLE 1. AREA, VOLUME, AND EDGE RELATIONSHIPS OF POLYHEDRONS

Polyhedron	A/a^2	V/a^3
Tetrahedron	1.7321	0.1179
Cube	6.000	1.0000
Octahedron	3.4641	0.4714
Dodecahedron	20.6457	7.6631
Icosahedron	8.6603	2.1817

A = area of surface; V = volume; a = edge

provided it can be deformed continuously into the surface of a sphere. Such considerations are studied in topology. See Table 1.

See also **Topology**.

POLYMER. See **Colloid System**.

POLYMORPHISM. 1. A phenomenon in which a substance exhibits different forms. Dimorphic substances appear in two solid forms, whereas trimorphic exist in three, as sulfur, carbon, tin, silver iodide, and calcium carbonate. Polymorphism is usually restricted to the solid state. Polymorphs yield identical solutions and vapors (if vaporizable). The relation between them has been termed "physical isomerism." See "Allotropes" under **Chemical Elements**. See also **Mineralogy**.

2. The occurrence of individuals of distinctly different structure or appearance within a species. In many cases two such forms occur and the species is said to be dimorphic rather than polymorphic.

Polymorphism depends upon many different conditions in various groups of animals. The various forms may be adapted for different places in a life cycle, for special parts in a colonial or social organization, or for special stages in a metamorphosis. They may also result from the incidence of different environmental conditions due to seasons or to unusual climatic conditions.

POLYNOMIAL. A rational integral function, sometimes also called a multinomial, in n variables of the form

$$c_1 x_1^{a_1} \cdots x_2^{b_1} + c_2 x_1^{a_2} x_2^{b_2} \cdots x_n^{r_2} + \cdots + c_k x_1^{a_k} x_2^{b_k} \cdots x_n^{r_k}$$

For any term in this expression c_i is the coefficient, a_i is the degree with respect to x_1, b_i with respect to x_2, etc., and the total degree of that term is $a_i + b_i + \cdots + r_i$. The highest degree in x_i of any term whose coefficient is not zero is the degree of the polynomial in x_i, and the highest total degree of any term with nonvanishing coefficient is the degree of the polynomial. The coefficients, c_i, which are constants, may be real or complex.

If the terms of a polynomial all have the same degree, the polynomial is homogeneous. The expressions *quantic* and *form* are also used. If they have two, three, four, etc., variables they are binary, ternary, quaternary, n-ary forms or a binomial, trinomial, etc. If their degree is 1, 2, 3, etc., they are linear, quadratic, cubic, etc., forms or equations.

The commonest case is the nth-degree polynomial in one variable which may be written as

$$a_0 x^n + a_1 x^{n-1} + \cdots + a_{n-1} x + a_n$$

For such a polynomial, the fundamental theorem of algebra (see also **Factor**), states that there exists one and only one set of constants, $x_1, x_2, \ldots, x_n$, such that

$$(x - x_1)(x - x_2)(x - x_3) \cdots (x - x_n) = 0$$

The constants are the roots of the polynomial. They may be real or complex, but they need not all be different. If $k \leq n$ roots are equal to each other, the root is said to be k-fold or k-tuple. The following relations hold for the roots:

$$x_1 + x_2 + \cdots + x_n = -a_1/a_0;$$

$$x_1 x_2 + x_1 x_3 + \cdots + x_{n-1} x_n = a_2/a_0;$$

$$x_1 x_2 x_3 + \cdots + x_{n-2} x_{n-1} x_n = -a_3/a_0; \cdots$$

$$x_1 x_2 x_3 \cdots x_n = (-1)^n a_n/a_0.$$

The roots of linear, quadratic, cubic, and biquadratic or quartic equations may be obtained in terms of algebraic expressions containing the coefficients but polynomials of degree higher than four cannot be solved in this way. Approximate values of the roots of equations of any degree may be determined by graphical or numerical methods (see also **Approximate Calculation**). Properties of the roots can be found by Descartes' rule and by the theorems of Budan and Sturm.

Given a polynomial in one variable of order n, $f(x) = 0$, it is often useful to introduce a new variable which will change the roots of the original equation in a predetermined way. This process is called transformation of variable. Some possibilities are as follows, where the original variable, x, is replaced by the new variable indicated in order to produce the corresponding change in the roots: (*a*) $-x$, the signs of all roots are changed; (*b*) x/h, the roots are all multiplied by the constant h; (*c*) $1/x$ and multiply the equation by x^n, the new roots are reciprocals of the original roots; (*d*) $x \pm h$, the roots will be decreased by the constant amount h if the plus sign is used or increased by h with the minus sign.

Polynomials frequently occur as the solution of a differential equation. For examples, see also **Generating Function**.

POLYP. A smooth-coated tumor projecting from a mucous surface, named from the Greek "polypos" (octopus). Polyps range in size from almost invisible bumps (102 millimeters diameter) to mushroom-shaped bodies (1–3 cm diameter), attached to the surface by a vascular penduncle. Commonly found in the nose, bladder and gastrointestinal tract, polyps may occur almost anywhere in the body. Most polyps are nonmalignant and are surgically removed only when they interfere with a normal body function. The majority of colorectal polyps prove to be adenomas with malignant potential, however, and thus their detection and removal are important.

POLYPEPTIDE. A compound composed of two or more amino acids, similar in many properties to the natural peptones. The amino acids are joined by peptide groups

$$-NH-C{\overset{\textstyle O}{\diagdown}}$$

formed by the reaction between an $-NH_2$ group and a

$$-C{\overset{\textstyle O}{\diagdown}}_{OH}$$

group, whereby there is elimination of a molecule of water, and formation of a valence bond. They may be termed di-, tri-, tetra-, etc., peptide according to the number of amino acids present in the molecule.

The sequence of amino acids in the chain of a protein is of critical importance in the biological functioning of the protein, and its determination is very difficult. The chains may be relatively straight, or they may be coiled or helical. In the case of certain types of polypeptides, such as the keratins, they are crosslinked by the disulfide bonds of cystine. Linear polypeptides can be regarded as proteins. See also **Amino Acids**; and **Protein**.

R. C. V.

POLYPROPYLENE. A synthetic crystalline thermoplastic polymer, $(C_3H_5)_n$, with molecular weight of 40,000 or more. Low-molecular-weight polymers are also known which are amorphous in structure and used as gasoline additives, detergent intermediates, greases, sealant, and lubricating oil additives. They are also available as high-melting-point waxes.

Polypropylenes are derived by the polymerization of propylene with stereospecific catalyst, such as aluminum alkyl. These polymers are translucent, white solids, insoluble in cold organic solvents, softened by hot solvents. They maintain their strength after repeated flexing. They are degraded by heat and light unless protected by antioxidants. Polypropylenes are readily colored, exhibit good electrical resistance, low water absorption and moisture permeability. They have rather poor impact strength below $15\,°F$ ($-9.5\,°C$). They are not attacked by fungi or bacteria, resist strong acids and alkalies up to about $140\,°F$ ($60\,°C$), but are attacked by chlorine, fuming nitric acid, and other strong oxidizing agents. They are combustible, but slow burning.

Polypropylenes are available as molding powder, extruded sheet, cast film, textile staple, and continuous-filament yarn. They find use in packaging film; molded parts for automobiles, appliances, and housewares; wire and cable coating; food container closures; bottles; printing plates; carpet and upholstery fibers; storage battery cases; crates for soft-drink bottles; laboratory ware; trays; fish nets; surgical casts; and a variety of other applications.

See also **Fibers**.

POLYP (Zoology). One of the two types of individuals found in many species of coelenterates. The two are the polyp or hydroid and the medusa. Polyps are approximately cylindrical, elongated on the axis of the body. One end is usually attached and the other bears the mouth, surrounded by a circlet of tentacles. The wall is relatively thin, due to the thinness of the mesogloea. In the class *Hydrozoa*, polyps are often very simple, like the common little freshwater species of the genus *Hydra*. Actinozoan polyps, including the corals and sea anemones, are much more complex, due to the development of a tubular stomodaeum leading inward from the mouth and a series of radial partitions called mesenteries. Many of the mesenteries project into the enteric cavity but some extend from the body wall to the central stomodaeum.

POLYSTYRENES. General purpose (or crystal) polystyrene is a clear, water-white, glassy polymer commonly derived from coal tar and petroleum gas. Physical properties of this material can be altered by addition of modifying agents, such as rubber (for increased toughness), methyl or α-methyl styrene (for heat resistance, methyl methacrylate (for improved light stability), and acrylonitrile (for chemical resistance). In general, varying the level of modifying agent (e.g., comonomer) will alter the level of desired property improvement.

Special grades of polystyrene include impact polystyrene modified with ignition-resistant chemical additives. These were developed because of increased emphasis on product safety and used in many electrical and electronic appliances. The addition of flame-retardant chemicals does not make the polymer noncombustible, but increases its resistance to ignition and decreases the rate of burning when exposed to a minor fire source.

Chemistry. The polymerization of styrene is an exothermic chain reaction which proceeds by all known polymerization techniques. This reaction can be shown schematically as:

The exact nature of the beginning and end of such a polymer chain is not certain. In general, the polymer can be characterized by its average degree of polymerization, i.e., the value of n, or more precisely by the distribution of n values. The heat of polymerization is 17.4 ± 0.2 kcal/mole at $26.9\,°C$. The reaction may be initiated by heat or by means of catalysts. Organic peroxides are typical initiators. Styrene also will polymerize in the presence of various inert materials, such as solvents, fillers, dyes, pigments, plasticizers, rubbers, and resins. Moreover, it forms a variety of copolymers with other mono- and polyvinyl monomers.

It is a matter of general observation that with styrene, the polymerization-rate curves will exhibit three distinct phases, the nature of which can be determined by the polymerization conditions and the purity of the monomer: (1) an initial slow period at the beginning of the reaction, known as the *induction period*, which appears to be associated with the presence of an inhibitor or other impurity in the monomer; (2) a period of relatively rapid polymerization, which persists almost to the end of the reaction, and for which the rate is exponentially dependent upon temperature; and (3) a final slowing down in rate as the reaction approaches completion and the monomer becomes exhausted. This effect is particularly apparent at low temperatures with relatively impure monomers.

General Properties. The *specific gravity* of general purpose and impact polystyrene is 1.05. It can vary for copolymers. It is higher for some specialty grades. Density varies slightly with pressure, but for practical purposes, the polymer is noncompressible.

In terms of *heat-resistance*, deflection temperatures range from about 66 to $99\,°C$ (170 to $215\,°F$), depending upon the formulation. Continuous resistance to heat for polystyrene is usually 60 to $80\,°C$ (140 to $175\,°F$). Time and load have a significant influence on the useful service temperature of a part.

Polystyrene is nontoxic when free from additives and residuals. It has no nutritive value and does not support fungus or bacterial growth.

Dimensional stability of polystyrene resins is excellent. Mold shrinkage is small. The low moisture absorption (about 0.02%) allows fabricated parts to maintain dimensions and strength in humid environments.

General-purpose polystyrene is water white, and transmission of visible light is about 90%. Modifiers reduce this property, and translucence results. The refractive index is about 1.59; critical angle about 39. Polystyrene molecules do not have the same optical properties in all directions. When molecules become oriented in a given direction during fabrication, a double refraction occurs and a birefringence effect can be observed if the part is examined through a polarized lens under a polarized light source. Injection moldings often exhibit birefringence in a random pattern. This can be beneficial if the birefringence is in the direction of load.

In terms of *weatherability*, polystyrene does not exhibit ultraviolet stability and is not considered weather-resistant as a clear material. Continuous, long-term exposure results in discoloration and reduction of strength. Improvement in weatherability can be obtained by the addition of ultraviolet absorbers, or by incorporating pigments. The best pigmenting results are obtained with finely dispersed carbon black.

In terms of *chemical resistance*, polystyrene has a high resistance to water, acids, bases, alcohols, and detergents. Chlorinated solvents will mar the surface and, in the presence of an external load or high internal stresses, will cause failure. Aliphatic and aromatic hydrocarbons, in general, will dissolve polystyrene. Such foodstuffs as butter and coconut oil should be avoided. The chemical resistance depends upon chemical concentration, time, and stress.

Typical mechanical properties of polystyrene are given in the accompanying table. The long-term load-bearing strength of most polystyrene materials is about one-third of the typical tensile strength given in Table 1.

Uses. Packaging applications are the most extensive. Meat, poultry, and egg containers are thermoformed from extruded foamed polystyrene sheet. The fast-food market also accounts for a substantial amount of polystyrene for takeout containers where the insulation value of a foamed container is an advantage. Containers, tubs, and trays formed from extruded impact polystyrene sheets are used for packaging a large variety of food. Biaxially oriented polystyrene film is thermoformed into blister packs, meat trays, container lids, and cookie, candy, pastry, and other food packages where clarity is required.

Housewares is another large segment of the use of polystyrenes. Refrigerator door liners and furniture panels are typical thermoformed impact polystyrene applications. Extruded profiles of solid or foamed

TABLE 1. COMPRESSION MOLDED PROPERTIES OF POLYSTYRENE

Property	General Purpose psi (MPa)	Impact psi (MPa)
Tensile strength	5500–8000 (38–55)	2500–5000 (17–35)
Compressive strength	21000–16000 (145–110)	4500–9000 (31–62)
Flexural strength	9000–15000 (62–104)	5000–10000 (35–69)
Tensile (Young's) modulus	400000–500000 (2760–3450)	200000–400000 (1880–2760)
Impact strength, Izod, foot-pounds/inch	0.3–0.5	1–4
Hardness, Rockwell M	65–80	60
Elongation, %	0.8–2.0	5–50

impact polystyrene are used for mirror or picture frames, and moldings for construction applications.

General-purpose polystyrene is extruded either clear or embossed for room dividers, shower doors, glazings, and lighting applications. Injection molding of impact polystyrene is used for household items, such as flower pots, personal care products, and toys. General-purpose polystyrene is used for cutlery, bottles, combs, disposable tumblers, dishes, and trays.

Injection blow molding can be used to convert polystyrene into bottles, jars, and other types of open containers.

Impact polystyrene with ignition-resistant additives is used for appliance housings, such as those for television and small appliances. Structural foam impact polystyrene modified with flame-retardant additives is used for business machine housings and in furniture because of its decorability and ease of processing. Consumer electronics, such as cassettes, reels, and housings, is a fast growing area for use of polystyrenes. Medical applications include sample collectors, petri dishes, and test tubes.

In an effort to make homes and other buildings more energy efficient, the use of polystyrenes in extruded foam board with flame-retardant additives for walls and under slabs has experienced exceptional growth in recent years. Used as a sheeting material, extruded foam board complies with the requirements of the major building codes as well as federal and military specifications.

In general, polystyrene is used in applications where ease of fabrication and decorability are required. Polystyrene has excellent electrical properties, good thermal and dimensional stability, resistance to staining, and low cost. General purpose polystyrene is preferred where clarity is also of prime concern. Impact polystyrene is preferred where toughness is needed.

POLYSULFONE. This is a transparent, heat-resistant, ultrastable high-performance engineering thermoplastic.[1] It is amorphous in nature and has low flammability and smoke emission. It possesses good electrical properties that remain relatively unchanged up to temperatures near its glass transition temperature of 374 °F (190 °C). The molecular structure of polysulfone features the diaryl sulfone group. This group tends to attract electrons from the phenyl rings. Oxygen atoms para to the sulfone group enhance resonance and produce oxidation resistance. High resonance also strengthens the bonds spatially, fixing the grouping into a planar configuration. The polymer consequently has good thermal stability and rigidity at high temperatures. Ether linkages provide chain flexibility, thereby imparting good impact strength. The polymer resists hydrolysis and aqueous acid and alkaline environments, because the linkages connecting the benzene rings are hydrolytically stable.

Polysulfone is available in transparent and opaque colors in both molding and extrusion grades (unfilled). A special medical grade meets U.S.P. criteria. Two mineral-filled grades are available; one is designed specifically for plating using conventional techniques; the other is a combination of polysulfone compounds with glass fiber or beads as well as other fillers, such as Teflon.

Polysulfone is widely used in medical instrumentation and trays to hold instruments during sterilization. It is also used in food processing equipment, including piping, scraper blades, milking machines, steam Tables, microwave oven cookware, coffee makers, coffee decanters, and beverage dispensing tanks. Electrical/electronic uses include connectors, fuse and switch housings, coil bobbins and cores, TV components, capacitor film, and structural circuit boards. In chemical processing equipment, uses are found in corrosion-resistant piping, both transparent and glass fiber-bonded, tower packing, pumps, filter modules, and membranes.

Polysulfone has high resistance to acids, alkalies, and salt solutions, and good resistance to detergents, oils, and alcohols, even at elevated temperatures under moderate stress. It is attacked by polar organic solvents, such as ketones, chlorinated hydrocarbons, and aromatic hydrocarbons. Polysulfone can be used continuously in steam at temperatures up to 300 °F (149 °C). Maximum stress in water at 180 °F (82 °C) is about 2000 psi (14 mPa) for steady loads and up to 2500 psi (17 mPa) for intermittent loads. Polysulfone offers a good combination of electrical properties—dielectric strength and volume resistivity are high, while dielectric constant and dissipation factor are low. The latter two properties (which determine lossiness) remain relatively constant over a wide range of temperatures and frequencies (including microwave). Polysulfone can be plated by an electroless nickel or copper process.

POLYTROPIC PROCESSES. The expansion or compression of a constant weight of gas may assume a variety of forms, depending on the extent to which heat is added to or rejected from the gas during the process, and also on the work done. There are, theoretically, an infinite number of ways possible in which a gas may expand from an initial pressure p_1, and volume v_1 to a final volume v_2. All these expansions may be grouped generically as polytropic expansions, and all could be represented graphically on the PV plane by the family of curves $pv^n = C$. They are all, in theory, perfectly reversible. n may have any positive value, 0 to ∞, and having been selected numerically it defines the type of expansion. From the infinite number of possible polytropic expansions, it is worthwhile to isolate four that deserve special attention. When one of the four physical characteristics, to wit, pressure, temperature, entropy, or volume, remains constant, expansions of more than ordinary interest are denoted, since they are frequently employed in a practical way, in situations which can be subjected to thermodynamic analysis. The value of the exponent n of the polytropic family for each of these is:

isobaric	$n = 0$
isothermal	$n = 1$
isentropic	$n = \gamma$ (γ = ratio of specific heat at constant pressure to that at constant volume)
isometric	$n = \infty$

Note, however, that the first and last are limiting cases, since in the first the pressure remains constant, and in the fourth it approaches zero. Note also that the second applies strictly only to ideal gases.

These thermodynamic processes, as they occur in useful machines, are not often of the exact polytropic form desired. For example, an isentropic process, which is exemplified, at least theoretically, by expansion of the burned gases after the explosive combustion in the gasoline engine, is modified slightly by the interchanging of heat between gases and cylinder wall, whereas a true isentropic has no heat either added or rejected in this way. The particular polytropic curve that would suit these conditions of expansion would depart somewhat from the adiabatic form.

During a polytropic process conditions of the working medium are constantly varying, and analysis may be aimed at determining one of the following: the work done, the heat added, the variation of temperature, and the change of entropy. Some information may be obtained merely by comparing the value of the exponent n with certain other data. For example, if n lies between 0 and 1, the temperature rises during an expansion and falls during a compression; when n is greater than 1, the temperature falls during expansion and rises during compression. Also, when n is less than γ, heat must be added to obtain an expansion, whereas when it is greater than γ, heat must be expelled. From the above it will be noted that there is a certain range of polytropic expansion in which, although heat is added, the temperature falls. This may seem to some to be paradoxical, but it is readily explained. During these expansions work is being done by the gas at a rate greater than that at which heat is being added, with the result that the deficiency must be made up from within the gas. The only way that this may be accomplished is for the gas to cool and give up some of its internal energy.

The equations for work done and for heat added in the case of the general polytropic expansions are:

$$W = \frac{p_1 v_1 - p_2 v_2}{n - 1}$$

$$Q = (p_1 v_1 - p_2 v_2)\left(\frac{1}{n-1} - \frac{1}{\gamma - 1}\right)$$

Both of these are expressed in foot-pounds. Sometimes a substitution of a definite value of n in one or the other of these equations leads to an indeterminate; for example, with the isothermal,

$$W = \frac{p_1 v_1 - p_2 v_2}{1 - 1}$$

But since the equation of the isothermal for an ideal gas is

$$pv = C$$

$$p_1 v_1 = p_2 v_2$$

and the work equation becomes indeterminate,

$$W = \frac{0}{0}$$

By approaching the isothermal from a different angle, however, the equation

$$W = pv \, \log_e \frac{v_2}{v_1}$$

may be deduced for work done.

POLYURETHANES. These materials comprise a conglomerate family of polymers in which formation of the urethane group

$$
\begin{array}{cc}
\text{H} & \text{O} \\
| & \| \\
\text{N} & -\text{C} - \text{O}
\end{array}
$$

is an important step in polymerization. Because the urethane linkage usually is formed by reaction of hydroxyl and isocyanate groups, urethane chemistry is the chemistry of isocyanates. The high reactivity of isocyanates and knowledge of the catalysis of isocyanate reactions have made possible the simple production of diverse polymers from low- to moderate-molecular-weight liquid starting materials. Several isocyanates (tolylene diisocyanate, hexamethylene diisocyanate, dicyclohexylmethane diisocyanate, etc.) are used in preparing polyurethanes. All are low-viscosity liquids at room temperature with the exception of 4, 4'-diphenylmethane diisocyanate (MDI), which is a crystalline solid. The aromatic isocyanates are more reactive than the aliphatic isocyanates and are widely used in urethane foams, coatings, and elastomers. The cyclic structure of aromatic and alicyclic isocyanates contributes to molecular stiffness in polyurethanes.

Flexible and rigid urethane foams, probably the most familiar of the polyurethanes, are produced in very large quantities. Foam formulations contain isocyanates and polyols with suitable catalysts, surfactants for stabilization of foam structure, and blowing agents, which produce gas for expansion. The largest volume of flexible urethane foam is used as a cushioning material. Expanding uses for flexible foam include carpet underlays and bedding. Weight reduction programs in the transportation field also take advantage of polyurethane forams for seating and trim. Rigid foams find application in insulation for appliances. Thermoplastic urethane elastomers form a widely used family of engineering materials, which appear to combine the best properties of elastomers and thermoplastics. They are tough, have high load-bearing capacity, low-temperature flexibility, and resistance to oils, fuels, oxygen, ozone, abrasion, and mechanical abuse. Possible carcinogenic properties are being studied. See also **Elastomers**.

POMEGRANATE TREE. Of the family *Myrtaceae* (myrtle family), the pomegranate tree (often a shrub), *Punicum granatum*, is native to southeastern Europe and southwestern Asia. The tree has been cultivated since early times and is now grown extensively in tropical and subtropical regions in both hemispheres. The flowers are borne either singly or in small clusters in the axils of the leaves. They are perfect and have a bright red corolla of five to eight petals, and many stamens. The fruit is a many-seeded berry. The outer coat of each seed is the edible portion. It is soft and fleshy and red in color.

See Fig. 1.

Fig. 1.   Fruits of the pomegranate tree. (*USDA photo.*)

POPLAR TREES. Members of the family *Salicaceae* (willow family), some species of poplars are also known as cottonwoods and aspens. Poplars sometimes are classified into four groups as indicated in the following listing:

1. White Poplar (*Populus Alba*)
    Hybrid with Japanese balsam ('*Pyramidalis*' *Richardii*)
2. Balsam Poplars
    Black Cottonwood (*P. trichocarpa*)
        Hybrid with Chinese black poplar (× *generosa*)
    Balsam (*P. balsamifera*)
        Hybrid with Eastern cottonwood (*Candicans* or balm of Gilead)
    Simon Balsam (*P. simonii*)
    Japanese Balsam (*P. maximowiczii*)
3. Black Poplars
    Cottonwood (*P. deltoides*)
    "Black Poplar" (*P. nigra*)
    Carolina Black Poplar (*P. angulata*)
    Chinese Black Poplar (*P. lasiocarpa*)
4. Trembling Poplars
    American Aspen (*P. tremuloides*)
    European Aspen (*P. tremula*)
    Bigtooth Aspen (*P. grandidentata*)

Much interbreeding of the four groups has resulted in numerous improved and fast-growing species. Notably, interbreeding of the black poplars has produced a number of excellent hybrid black poplars, including: *P. canadensis*; *P. serotina*; *P. marilandica*; *P. regenerata*; *P. 'Italica'* (the Lombardy poplar) and several others.

The balsam poplar (*P. balsamifera*) is predominantly a Canadian tree, ranging from Labrador and Nova Scotia westward to Alaska, and northward to the tree limit. A few species are found beyond the Arctic Circle. In some locales, the trees are planted close together to form shelter belts. The tree is large, with the height ranging from 80 to 90 feet (24 to 27 meters) in adult trees. The bark is coarse with deep furrows. The leaf is deciduous and from 5 to 6 inches (12.7 to 15 centimeters) in length, egg-shaped, pale green, lighter underneath. The amber-colored twig is covered with blisters of resin which is quite fragrant. The seed is minute, hairy, and tufted. When released, it appears much like falling snow. The bud is fragrant. There are about 12 species, some being natural hybrids.

The black cottonwood (*P. trichocarpa*) is found mainly in the western part of North America, extending from California northward to British Columbia and Alaska. The tree reaches eastward from the West Coast into the Sierra Nevada mountains and is found at altitudes up to 10,000 feet (3,050 meters), but rarely below 3,000 feet (915 meters). The tree also occurs in the San Diego mountains. A healthy tree will range from 80 to 100 feet (24 to 30 meters) or more in height. The branches are slender and wide-spreading. The leaf is 5 to 6 inches (12.7 to 15 centimeters) long, 2 to 4 inches (5 to 10 centimeters) across, and darkly veined. The top of the leaf is a glistening olive-green, the underside is silver white. The bud is resinous, ovate, and fragrant, having a balsam odor. In older trees, the bark is dark brown and furrowed. Young trees have a smooth gray bark. Autumn coloration is a deep yellow.

The quaking aspen of America (*P. tremuloides*) extends over an extremely wide range—from Labrador to Mexico. Some trees are found as far north as the Bering Strait. Although vigorous, the life of the tree is relatively short, requiring much light. It is highly regarded as a shade tree. A related species is the bigtooth aspen (*P. grandidentata*) named for its large, toothed leaves. The European aspen (*P. tremula*) ranges north and eastward from western Europe into Russia.

The botanical explanation for the mysterious and intriguing quaking motion of the *quaking aspen* is that the aspen leafstalk is longer than the leaf itself and flattened opposite the plane of the leaf, thus becoming a sensitive pivot upon which the thin, papery leaves flutter almost magically, even in imperceptible breezes (Voynick, 1984). Prior to the present century, aspen thrived in a natural ecosystem in which wildfire regularly devastated mature stands of both aspen and conifers. With the shading canopy reduced or eliminated, sunlight stimulated prolific formation of root buds. Some

of the present largest and most beautiful aspen groves started on fire-blackened slopes before the turn of the century. Forest management in recent years has altered the aspen's natural forest ecosystem. Prevention and containment of wildfire has permitted both conifers and aspen to grow older and larger and thus shade more of the forest floor. In the long run, however, this is a condition that favors the reproduction of conifers to the detriment of the aspen. Some foresters now regard this condition as a serious long-term threat to the aspen. Two interesting articles on aspens will be found in *American Forests* magazine: Ciesla, B., "A Tree (Aspen) for All Seasons," October 1982; and Voynick, S.M., "Trouble in the Quakies," May 1984. The yellow poplar is described in "Rediscovering the Yellow Poplar" by D.A. Boerner-Ein in the July 1991 issue.

The yellow poplar is actually not a poplar, but a tulip tree and member of *Magnoliaceae* (magnolia family).

The wood from various species of poplar trees is valuable commercially. Lumber from the cottonwoods (*P. monilfera* and *P. deltoides*) is soft and of a yellowish-white color, possessing a fine, open grain. It is sometimes referred to as Carolina poplar or whitewood. The weight is about 30 pounds per cubic foot (481 kilograms per cubic meter). Easy to work, but not strong and prone to warping, the wood is used mainly for making paneling, packing boxes, and some general carpentry. Wood from the balsam poplar (*P. balsamifera*) is a weak and soft wood and used mainly for making containers and excelsior. It also makes an excellent pulpwood for paper production. Wood from the aspen (*P. tremula*) is widely used for making match sticks and excelsior, with limited use for inside construction. Easily bleached, the wood also is used for paper pulp.

See Table 1.

POPPY *(Papaver somniferum; Papaveraceae).* The poppy from which opium is obtained is an annual herb having a smooth branching stem 2–3 feet (0.6–0.9 meter) tall, large, dull, green, smooth leaves and solitary single flowers, varying from white to purple in color and rather showy. The flower consists of two sepals, which soon fall off when the flower opens, four petals, many stamens and a single pistil with a one-celled ovary. The fruit is a capsule, 1–2 inches (2.5–5 centimeters) in diameter,

TABLE 1. RECORD POPLAR TREES IN THE UNITED STATES[1]

Specimen	Circumference[2]		Height		Spread		Location
	Inches	Centimeters	Feet	Meters	Feet	Meters	
ASPENS							
Bigtooth aspen (1980) (*Populus grandidentata*)	140	356	102	31.1	64	19.5	Kentucky
Bigtooth aspen (1984) (*Populus grandidentata*)	105	267	132	40.2	67	20.4	Michigan
Quaking aspen (1991) (*Populus tremulides*)	122	310	109	33.2	59	18	Michigan
Quaking aspen (1998) (*Populus tremulides*)	127	323	114	34.7	32	9.8	Arizona
COTTONWOODS							
Black cottonwood (1995) (*Populus trichocarpa*)	320	813	158	48.2	110	33.5	Oregon
Eastern cottonwood (typ.) (1991) (*Populus deltoider var. deltoides*)	433	1100	85	25.9	121	36.9	Idaho
Fremont cottonwood (typ.) (1996) (*Populus fremontii var. fremontii*)	504	1280	92	28	108	32.9	Arizona
Meseta cottonwoodk (1986) (*Populus fremontii var. mesetae*)	190	483	60	18.3	60	18.3	Texas
Narrowleaf cottonwood (1973) (*Populus angustifolia*)	314	798	79	24.1	80	24.4	Oregon
Plains cottonwood (1967) (*Populus deltoides var. occidentalis*)	432	1097	105	32	93	28.3	Colorado
Rio Grande cottonwood (1997) (*Populus fremontii var. wislizeni*)	366	930	123	37.5	104	31.7	Texas
Swamp cottonwood (1990) (*Populus heterophylla*)	42	107	55	16.8	23	7	Mississippi
POPLARS							
Balsam poplar (1991) (*Populus balsamifera*)	165	419	128	39	57	17.4	Michigan
White popular (1992) (*Populus alba*)	263	668	93	28.3	86	26.2	Illinois

[1]From the "National Register of Big Trees," American Forests (by permission).
[2]At 4.5 feet (1.4 meters).

containing many small seeds, which escape through a ring of pores which form around the top of the capsule, beneath the persistent stigma.

To obtain opium, the unripe capsules are incised with a knife. From these cuts the milky juice oozes and dries to form a plastic gummy substance, which is scraped off and molded into a ball. This crude opium contains fragments of the plant tissues and considerable dirt. About 10% of the opium is the alkaloid morphine. When first prepared opium is brownish and easily molded. It gradually dries to a hard brittle substance, easily ground to a powder. Besides morphine it contains many other alkaloids.

Poppy seeds, which contain no harmful substances, are frequently used in bread and cakes. From them is expressed an expensive oil used in cooking and in making artist's paints.

POPULATION (Statistics). A set of observations is commonly regarded as a sample from a larger set, called the population. The population may be actually existing and of finite size, as in practical sampling inquiries; or it may consist of the infinite number of observations that might hypothetically be obtained under the same condition as the actual sample.

PORCUPINE. See **Rodentia.**

PORCUPINE FISHES (*Osteichthyes*). Of the order *Plectognathi*, family *Diodontidae*, porcupine fishes are puffers with spines. While swimming, the fish holds the spines pressed close to the body. The spines appear when the fish swallows air or water. There are about 15 types of porcupine and burrfishes known. The burrfishes (genus *Chilomycterus*) have shorter spines that normally are extended. Also sometimes called the spiny boxfish. By most people, the porcupine fish is usually seen as a decorative lamp made from a dried skin of the fish. These fishes also are called balloon fishes. A porcupine fish is shown in Fig. 1. See also **Puffers (Osteichthyes).**

Fig. 1. Porcupine fish.

PORGIES (*Osteichthyes*). Of the family *Sparidae*, the porgy is a valuable food fish found along the Atlantic coast from Cape Cod to South Carolina. This species is also called the scup or scuppaug. Some 14 species of porgies occur in the American Atlantic, including the *Stenotomus chrysops* (northern porgy or scup); the *Archosargus rhomboidalis* (the sheepshead); and the *Lagodon rhomboides* (the pinfish). The sheepshead is also a valuable sporting and food fish. Porgies are not found on the eastern side of the Atlantic. The *Calamus brachysomus* occurs in California waters, and the *Monotaxis grandoculis* is found in Hawaiian waters. The *Chrysophrys guttulatus* (bump-headed porgy) is an important fish in Australian waters. The Australians refer to the fish as a snapper. Possibly the largest of the sparids is *Cymatoceps nasutus*, a musselcracker that can weigh up to 100 pounds (45 kilograms) and is a favorite sporting fish among South Africans. See also **Fishes.**

PORIFERA. The sponges. A phylum of animals of low organization, related to some of the one-celled protozoans and much more primitive than any other multicellular group. Because of their loosely integrated structure the sponges are regarded as one of three major types of animal

organization, designated by the term *Parazoa*. This group lies between the *Protozoa*, also a single phylum, and the *Metazoa*, containing all of the other multicellular phyla. See Fig. 1 on p. 2817.

Sponges develop only two germ layers, the ectoderm and endoderm, but many different cells lie in the mesogloea between the two. The body wall is perforated by many canals leading to a central cavity, the paragaster. Some part of these passages is lined with collared flagellate cells, which produce currents of water flowing inward through the pores and out of a larger opening of the paragaster called the osculum. The body wall contains several kinds of specialized cells. The scleroblasts form hard supporting structures of various forms and materials, called spicules. Phagocytes ingest, digest, and transport food. Porocytes become perforated to form canals. The outer surface is covered with flattened cells called pinacocytes.

Three kinds of sponges are recognized, according to the plan of the canal system: (1) Ascon sponges have canals leading entirely through the body wall and collared cells (choanocytes) in the lining of the paragaster. (2) Sycon sponges have radial canals lined with choanocytes and opening into the paragaster. Between them inhalant canals lead inward from the outside but do not reach the paragaster. The two types of canals are connected by minute pores, called prosopyles, through which water must pass to reach the interior. (3) In the rhagon or leucon sponges the canal systems are more intricately branches and the choanocytes are located in small chambers.

Commercial sponges are the skeletal remains of species whose bodies are supported by fibers of a peculiar material, spongin. The organic matter is removed by maceration and washing.

The phylum is divided into three classes:

Class *Calcarea*. Sponges whose bodies contain calcareous spicules only. The choanocytes are large and all three types of canals systems are represented.

Class *Hexactinellida*. Spicules six-rayed and siliceous. Choanocytes small. Canal system of a simple rhagon type. The glass sponges.

Class *Demospongiae*. Spicules siliceous but not six-rayed or an association of siliceous (silicon) material and spongin. Rhagon type of canal system. The sponges of commerce are included here. The subfamily *Spongillinae* includes the only species of sponges found in fresh water. See also **Invertebrate Paleontology.**

POROSITY. Two common uses of this term are: (1) the property of containing pores, which are minute channels or open spaces in a solid; (2) the proportion of the total volume occupied by such pores.

PORPHYRIN. Any of several physiologically active nitrogenous compounds occurring widely in nature. The parent structure is comprised of four pyrrole rings, shown in I, II, III, and IV in Fig. 1 on p. 2817, together with four nitrogen atoms and two replaceable hydrogens, for which various metal atoms can be readily substituted. A metal-free porphyrin molecule has the structure shown in the diagram. Porphyrins of this type have been made synthetically by passing an electric current through a mixture of ammonia, methane, and water vapor. Some biochemists suggest that this phenomenon may account for the early formation of chlorophyll and other porphyrins which have been essential factors in the development of life.

The most important porphyrin derivatives are characterized by a central metal atom; hemin is the iron-containing porphyrin essential to mammalian blood, and chlorophyll is the magnesium-containing porphyrin that catalyzes photosynthesis. Other derivatives include the cytochromes, which function in cellular metabolism, and the phthalocyanine group of dyes. Porphyrins are described in considerable detail in a 7-volume set of books, *The Porphyrins*, Academic, New York, 1978.

PORPHYRY. Porphyry is a textural term applied to igneous rocks in which one or more of the mineral constituents present exists as well crystallized individuals in a ground mass that is relatively of much finer grain. The derivation of the word presents an interesting study. The gasteropods of the genus *Murex* were much used for obtaining a purple dye; the Greek name for both the animal and the dye is the same. A certain Egyptian rock which was once much used for building and ornamental purposes displays very prominent crystals in a purplish ground-mass and so the same Greek word was applied to it, then later came to mean all rocks of this general appearance. Modern use now restricts the term porphyry to the description of texture alone as in the case of the Egyptian rock.

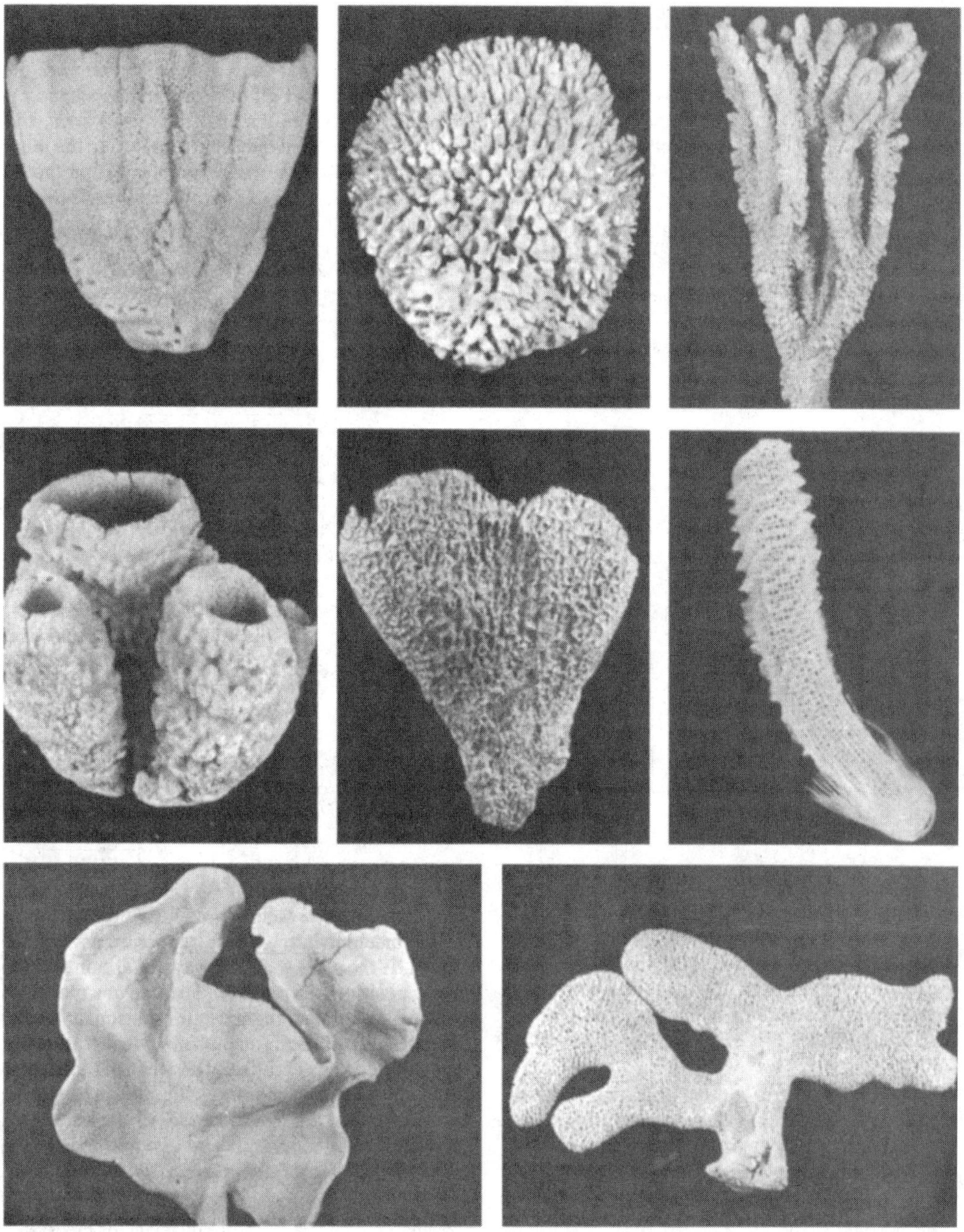

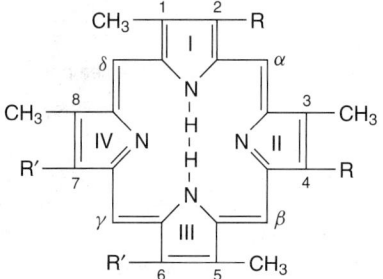

Fig. 1.   Varieties of sponges. (*A.M. Winchester.*)

Fig. 1.   Suggested structure of a metal-free porphyrin molecule.

PORPOISE.   See **Whales, Dolphins, and Porpoises**.

PORTLAND CEMENT.   See **Cement**; and **Gypsum**.

POSITION AND DISPLACEMENT MEASUREMENT.   The measurement of position is usually expressed in terms of a dimension, a grid, or a vector. Position is a very important quantity in terms of radar, navigational, astronomical, and surveying systems. Industrially, position measurement is important to the control of machine tools and many other production machines. See also **Numerical Control**. Descriptions here are confined to the latter applications.

To position a machine member with acceptable accuracy, it is necessary to establish the extent of the backlash or dead-band region for the positioning mechanism used. The measuring transducer and its attendant dead-band characteristics, when used with the machine member, will determine the amount of dead band or backlash to be included in the total control system loop.

The controlled member may have undesired movement with respect to the machine base — in a direction transverse to the controlled axis of travel. If the slide and tableways wear nonuniformly, variation in transverse position of a point on the table of a machine may cause a variation in the air gap of a magnetic slot transducer system. This same problem will result in misalignment of optical transducer systems if the table motion becomes crab-like after wear of the slides has progressed. Nondata components of both cyclic and random nature may be superimposed on the true data because of machine-induced vibrations on the transducer. Attention must be given to reduce these effects and to take the residue effects fully into account when designing a total positioning system.

Resolvers. The resolver, among the several shaft-type mechanical input transducers available, is perhaps the most versatile because of the many possibilities for its electric input. The electrical diagram of Fig. 1 discloses the data input or stator winding in the manner of a two-phase motor stator. The electrical characteristics of this device are such that the output voltage of the rotor winding varies in a sinusoidal manner with respect to the data readout position. Zero voltage will occur at the readout position that corresponds to chosen data. But zero voltage also may occur 180 shaft degrees from this position. Thus, there is the possibility of a false-data region. To eliminate this possibility, the designer chooses a gear ratio for use between the machine member and its transducer that will permit only 180 mechanical degrees of transducer shaft rotation during full-member travel.

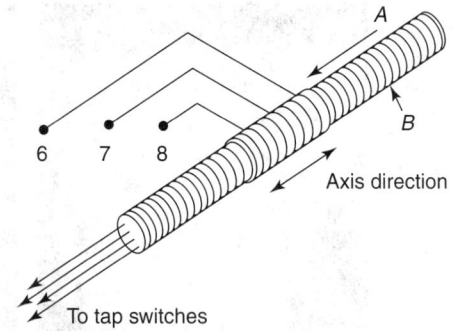

Fig. 3.   Inductive-bridge transducer for position measurement.

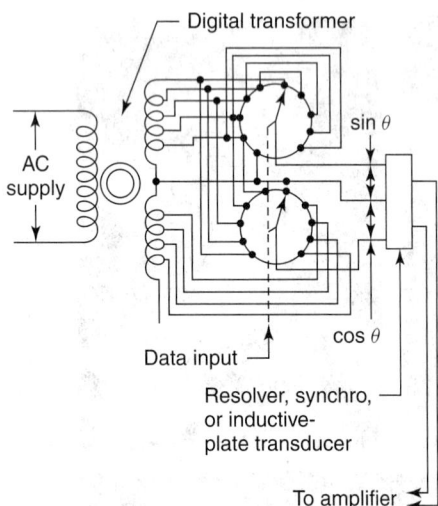

Fig. 1.   Digital transformer circuit for resolvers, synchros, and inductive plates.

Synchros. The term *synchro* is applied to a class of variable-coupling transformers in which the variable coupling is obtained by changing the orientation of the primary to the secondary through rotation of the movable element. Specifically, the synchro transduces a rotor-position angle into a voltage or a set of voltages unique to that angle. Unlike the resolver, a synchro stator has a three-phase winding configuration, while the rotor may use a single- or a three-phase winding.

Inductive Plates. As shown in Fig. 2, an inductive plate includes an etched stator winding that has been projected upon a dimensionally stable nonconducting surface by a photographic process. The rotor associated with this transducer is constructed in a like manner. Variations in inductor displacement are averaged over a large number of inductors by summing the voltages from a like number of coils located on the rotor plate. Thus, the reproduced rotor and stator inductors need not be printed with a positional accuracy equivalent to that of the final transducer. Essentially, this transducer is a 2-phase synchro or resolver whose windings have been projected onto a linear medium.

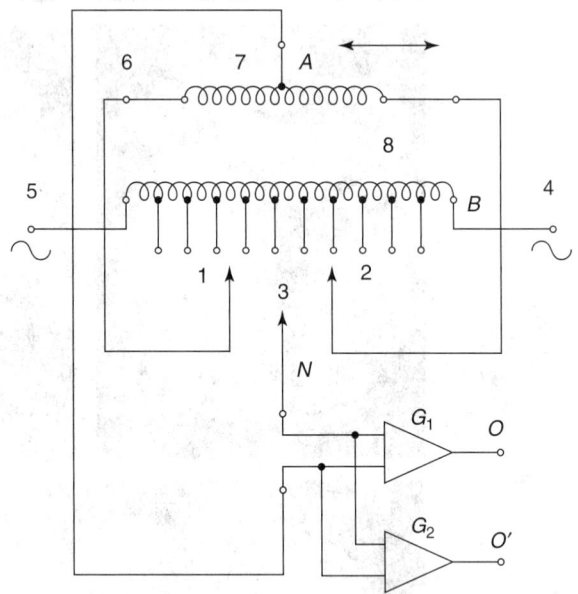

Fig. 4.   Schematic circuit of inductive-bridge transducer.

axis to be measured, and a movable member *A* that is approximately half the length of *B*. Selectable taps are placed on *B* in a successive decade with externally located inductors to provide a bridge configuration that may be externally unbalanced by placing *A* (coil) and *N* (point) across a pair of tap points—then moving the coil until equal voltage prevails between the two ends of the coil, as evidenced by the existence of a small voltage at *O* and *O'*. See Fig. 4. A disadvantage of this system is the relatively large number of wires that must be taken from the device through the machine to the control system. An advantage is the high output voltage per unit of displacement.

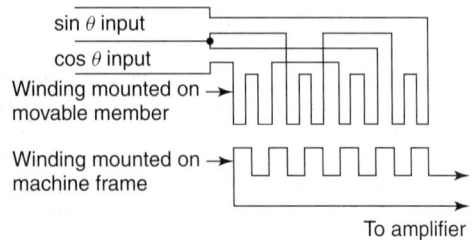

Fig. 2.   Schematic circuit for a position transducer utilizing inductive plates.

Inductive-Bridge Transducers. As shown by Fig. 3, operation is based upon the use of a fixed inductive member *B* that is slightly longer than the

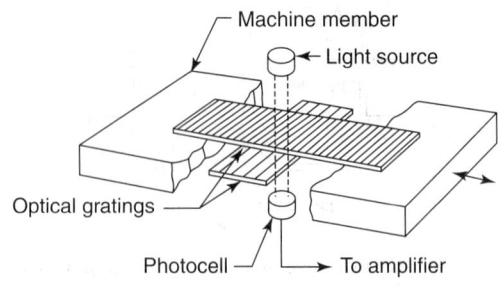

Fig. 5.   Optical grating-type position transducer.

Optical Gratings. A transducer of this type is shown in Fig. 5. The device requires an amplifier to raise the power level of the pulses from the associated photocell so that position readout displays or relays used in conjunction with other relay logic elements can be operated.

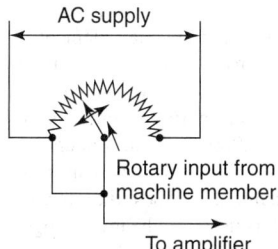

Fig. 6. Basic circuit of a multiturn potentiometer as used in a position transducer.

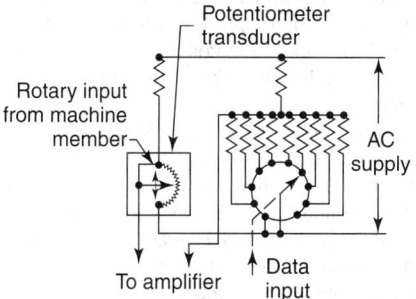

Fig. 7. Decade switch circuit for multiturn potentiometer.

Potentiometers. Multiple-turn potentiometers are used in machine control systems that do not require very high accuracy. Circuitry for such a transducer system is shown in Figs. 6 and 7.

Other Transducers. Their descriptions are beyond the scope of this volume; other position and displacement sensors include digital contactors, optical binary scales, magnetic scales, linear potentiometers, linear variable reluctance transducers, linear transformers, and linear variable differential transformers (LVDT).

Further details on this complex topic can be found in the *Process/Industrial Instruments and Controls Handbook*, 4th Edition (D.M. Considine, Editor), McGraw-Hill, New York, 1993.

POSITION ANGLE (Stellar). A term used in astronomy to denote the angle between the great circle joining any two celestial objects, and the hour circle through one of the objects. In measuring double stars, the position angle is the angle between the great circle joining the two stars and the hour circle through the brighter of the pair, the angle being measured from the north to the east through 360°.

POSITIONAL NOTATION. One of the schemes for representing real numbers, characterized by the arrangement in sequence of digits (symbols for integers) with the understanding that the successive digits are to be interpreted as the coefficients of successive integral powers of a number called the radix or base of the notation. The representation of a real number by the notation

$$A_n A_{n-1} \ldots A_2 A_1 A_0 \cdot A_{-1} A_{-2} \ldots A_{-m}$$

which is an abbreviation for the sum

$$\sum_{i=-m}^{\infty} A_i r^i$$

where the $\cdot$ is called the radix point, the A_i are integers ($0 \leq |A_i| \leq r$) called digits, and r is an integer greater than one called the radix (or base). The signs of all of the A_i are the same as the sign of the number represented. In the decimal number system, the radix is ten and the radix point is called the decimal point. In the binary number system, the radix is two and the radix point is called the binary point. For some purposes the system of notation has been broadened to include the case in which the radix assumes more than one value in a single number system. In this case the notation

$$A_n A_{n-1} \ldots A_2 A_1 A_0 \cdot A_1 A_{-2} \ldots A_{-m}$$

in an abbreviation for the sum

$$\left(\sum_{i=1}^{n} A_i \prod_{j=1}^{i} r_j \right) + A^0 + \left(\sum_{i=-m}^{-1} A_i \prod_{j=1}^{-1} \frac{1}{r_j} \right)$$

Several such systems have been used. The biquinary system uses a radix which is alternately two and five for successive values of j. The quinary vicenary system uses a radix which is alternately five and twenty for successive values of j.

POSITRON. The positron is one of many fundamental bits of matter. Its rest mass (9.109×10^{-31} kilogram) is the same as the mass of the electron, and its charge ($+1.602 \times 10^{-19}$ coulomb) is the same magnitude, but opposite in sign to that of the electron. The positron and electron are antiparticles for each other. The positron has spin $\frac{1}{2}$ and is described by Fermi-Dirac statistics, as is the electron.

The positron was discovered in 1932 by C.D. Anderson at the California Institute of Technology while doing cloud chamber experiments on cosmic rays. The cloud chamber tracks of some particles were observed to curve in such a direction in a magnetic field that the charge had to be positive. In all other respects, the tracks resembled those of high-energy electrons. The discovery of the positron was in accord with the theoretical work of Dirac on the negative energy of electrons. These negative energy states were interpreted as predicting the existence of a positively charged particle.

Positrons can be produced by either nuclear decay or the transformation of the energy of a gamma ray into an electron-positron pair. In nuclei that are proton-rich, a mode of decay that permits a reduction in the number of protons with a small expenditure of energy is positron emission. The reaction taking place during decay is

$$p^+ \longrightarrow n^0 + e^+ + v$$

where p^+ represents the proton, n^0 the neutron, e^+ the positron, and v a massless, chargeless entity called a neutrino. See also **Neutrino.** The positron and neutrino are emitted from the nucleus while the neutron remains bound within the nucleus. Although none of the naturally occurring radioactive nuclides are positron emitters, many artificial radioisotopes that decay by positron emission have been produced. The first observed case of positron decay of nuclei was also the first observed case of artificial radioactivity. An example of such a nuclear decay is

$$_{11}Na^{22} \longrightarrow \, _{10}Ne^{22} + e^+ + v \quad \text{(half-life} \sim 2.6 \text{ years)}$$

This decay provides a practical, usable source of positrons for experimental purposes.

The process of pair production occurs when a high-energy gamma ray interacts in the electromagnetic field of a nucleus to create a pair of particles—a positron and an electron. Pair production is an excellent example of the fact that the rest mass of a particle represents a fixed amount of energy. Since the rest energy ($E_{\text{rest}} = m_{\text{rest}} - c^2$) of the positron plus electron is 1.022 MeV, this energy is the gamma energy threshold and no pair production can take place for lower-energy gammas. In general, the cross section for pair production increases with increasing gamma energy and also with increasing Z number of the nucleus in whose electromagnetic field the interaction takes place.

The positron is a stable particle (i.e., it does not decay itself), but when it is combined with its antiparticle, the electron, the two annihilate each other and the total energy of the particles appears in the form of gamma rays. Before annihilation with an electron, most positrons come to thermal equilibrium with their surroundings. In the process of losing energy and becoming thermalized, a high-energy positron interacts with its surroundings in almost the same way as does the electron. Thus, for positrons, curves of distance traversed in a medium as a function of initial particle energy are almost identical with those of electrons.

It is energetically possible for a positron and an electron to form a bound system similar to the hydrogen atom, with the positron taking the place of the proton. This bound system has been called "positronium" and the chemical symbol Ps has been assigned. Although the possibility of positronium formation was predicted as early as 1934, the first experimental demonstration of its existence came in 1951 during an investigation of positron annihilation rates in gases as a function of pressure. The energy levels of positronium are about one-half those of the hydrogen atom, since the reduced mass of positronium is about one-half that of the hydrogen atom. This also causes the radius of the positronium system to be about twice that of the hydrogen atom.

In principle, positronium can be observed through the emission of its characteristic spectral lines, which should be similar to hydrogen's except that the wavelengths of all corresponding lines are doubled. Positronium is also the ideal system in which the calculations of quantum electrodynamics can be compared with experimental results. Measurement of the fine-structure splitting of the positronium ground state has served as an important confirmation of the theory of quantum electrodynamics.

It is possible for a positron–electron system to annihilate with the emission of one, two, three, or more gamma rays. However, not all processes are equally probable.

See also **Particles (Subatomic)**.

POSITRONIUM.   A quasi-stable system consisting of a positron and a negatron bound together. Its set of energy levels is similar to that of the hydrogen atom (electron and proton). However, because of the different reduced mass, the frequencies associated with the spectral lines are less than half of those of the corresponding hydrogen lines. The mean life of positronium is at most about 10^{-7} seconds, its existence being terminated by negatron-positron annihilation. See also **Positron**.

POTAMOGALE.   See Moles and Shrews.

POTASSIUM.   Chemical element, symbol K, at. no. 19, at. wt. 39.098, periodic table group 1 (alkali metals), mp 63.3 °C, bp 760 °C, density 0.86 g/cm^3 (20 °C). Elemental potassium has a body-centered cubic crystal structure. Potassium is a silver-white metal, can be readily molded, and cut by a knife, oxidizes instantly on exposure to air, and reacts violently with H_2O, yielding potassium hydroxide and hydrogen gas, which burns spontaneously in air with a violet flame due to volatilized potassium element, is preserved under kerosene, burns in air at a red heat with a violet flame. Discovered by Davy in 1807.

There are three naturally occurring isotopes, ^{39}K through ^{41}K, of which ^{40}K is radioactive with a half-life of 1.3×10^9 years. In ordinary potassium, this isotope represents only 0.0119% of the content. There are four other known isotopes, all radioactive, ^{38}K and ^{42}K through ^{44}K, all with relatively short half-lives measured in minutes and hours. In terms of abundance, potassium ranks seventh among the elements occurring in the earth's crust. In terms of content in seawater, the element ranks eighth, with an estimated 1,800,000 tons of potassium per cubic mile (388,000 metric tons per cubic kilometer) of seawater. First ionization potential 4.339 eV; second, 31.66 eV. Oxidation potential K → K$^+$ + e$^-$, 2.924 V. Other important physical properties of potassium are given under **Chemical Elements**.

Potassium does not occur in nature in the free state because of its great chemical reactivity. The major basic potash chemical used as a source of potassium is potassium chloride, KCl. The potassium content of all potash sources generally is given in terms of the oxide K_2O. The majority of potash produced comes from mineral deposits that were formed by the evaporation of prehistoric lakes and seas which had become enriched in potassium salts leached from the soil. In addition to natural deposits of potassium salts, large concentrations of potassium also are found in some bodies of water, including the Great Salt Lake and the Salduro Marsh in Utah, the Dead Sea between Israel and Jordan, and Searles Lake in California. All of these brines are used for the commercial production of potash.

The main potassium minerals are sylvite, KCl, sylvinite, KCl/NaCl, carnallite, KCl · MgCl$_2$ · 6H$_2$O, kainite, MgSO$_4$ · KCl · 3H$_2$O, polyhalite, K$_2$SO$_4$ · MgSO$_4$ · 2CaSO$_4$ · 2H$_2$O, langbeinite, K$_2$SO$_4$ · 2MgSO$_4$, jarosite, K$_2$Fe$_6$(OH)$_{12}$(SO$_4$)$_4$, leucite, K$_2$O · Al$_2$O$_3$ · 4SiO$_2$, alunite, K$_2$Al$_6$(OH)$_{12}$(SO$_4$)$_4$, microcline, K$_2$O · Al$_2$O$_3$ · 6SiO$_2$, muscovite, K$_2$O · 3Al$_2$O$_3$ · 6SiO$_2$ · 2H$_2$O, biotite, H$_2$K(Mg, Fe)$_3$(Al, Fe)(SiO$_4$)$_3$, and orthoclase, K$_2$O · Al$_2$O$_3$ · 6SiO$_2$. See also **Alunite; Biotite; Carnallite; Jarosite; Leucite; Muscovite;** and **Polyhalite**. The principal workable mineral deposits are in Stassfurt, Germany, Alsace, New Mexico, Saskatchewan, the former Soviet Union, Spain, Poland, Italy, the Atlantic Seaboard of the United States, and Utah. There are significant potassium reserves in many other parts of the world, notably in Canada and the former Soviet Union. World consumption of potash is about 18 million tons annually. Potassium metal is obtained by electrolysis of fused potassium hydroxide or chloride fluoride mixture in a specially designed cell.

Uses. Like so many of the chemical elements, the compounds of potassium are far more important than elemental potassium—by several orders of magnitude. The uses for metallic potassium are extremely limited, mainly because metallic sodium serves about the same needs and is much less costly. Sodium production, for example, exceeds potassium production by a factor of at least 1,000. A large amount of elemental potassium is used to produce the superoxide, KO$_2$ which finds application in gas-mask canisters. The compound also goes into the production of a sodium-potassium alloy, which is used as a heat-exchange medium. This alloy also has been used in magnetohydrodynamic power generation and as a catalyst for the removal of CO$_2$, H$_2$O, and oxygen from inert-gas systems. The handling precautions for potassium metal are similar to those for sodium metal. See also **Sodium**.

Chemistry and Compounds. Potassium is more electropositive than sodium in many of its reactions, as is consistent with its position in group 1. Its reaction with H$_2$O is more vigorous and it reacts violently with liquid bromine, and readily on heating with solid iodine.

Because of the ease of removal of its single 4s electron (4.339 eV) and the difficulty of removing a second electron (31.66 eV) potassium is exclusively monovalent in its compounds, which are electrovalent. (Some experimental work indicates that the potassium alkyls may be covalent, but even they form conducting solutions in other metal alkyls.)

Potassium solutions in liquid NH$_3$ react readily with the elements on the further right side of the periodic table to produce normal and poly compounds such as potassium sulfide, K$_2$S, and tetrapotassium plumbide, K$_4$Pb, in the first instance and K$_2$S$_6$ and K$_4$Pb$_9$ in the second. Ammoniates are not formed by potassium as readily as by sodium or lithium and solubility of salts exhibits a minimum at the cation: anion radius ratio of 0.75 (potassium fluoride, KF, 16 moles per kilogram, potassium chloride, KCl, 0.0177 moles per kilogram, potassium bromide, KBr, 2.26 moles per kilogram, potassium iodide, KI, 11.09 moles per kilogram). Potassium nitrate reacts in liquid ammonia with potassium amide, KNH$_2$ to form the azide, KN$_3$.

Like the other alkali metals, potassium forms compounds with virtually all the anions, organic as well as inorganic. Like sodium bicarbonate, the reactivity of potassium bicarbonate with many metallic oxides permits of the preparation of many compounds (such as the meta- and pyroarsenates) which are unstable in aqueous solution. For a general discussion of these reactions, and for a general picture of the inorganic salts of potassium, see the discussion of the compounds of sodium, which differ principally in their greater degree of hydration and greater number of hydrates. However, potassium, rubidium, and cesium coordinate with large organic molecules even though they do not with water. Potassium, like the others, coordinates with salicylaldehyde. It is believed to have two coordination numbers, 4 and 6. The tetracoordinate compounds of potassium (and sodium) are the most stable. The following reasons are given: (1) Increasing atomic number carries with it increasing electropositiveness and ease of ionization, which diminishes the tendency to coordinate. (2) The increasing distance of the nucleus from the coordinating electrons with increasing atomic volume makes it less likely that additional electrons will be held with ease. (3) On the other hand, there is an increase in the maximum coordination number with the elements of higher atomic number. These factors are in keeping with a maximum stability for the tetracoordinate compounds occurring with potassium.

Charge Density Waves in Potassium. Frequently, because of their relatively simple electronic structure, the alkali metals are selected as a basis for the study of the behavior of electrons in solids. As early as 1964, Overhauser (Purdue University) predicted the existence of "charge density waves," a phrase coined by Overhauser, in the potassium atom. This conclusion was the result of calculations made by Overhauser to the effect that K, in its lowest energy or ground state, does not exhibit a uniform distribution of its free electrons (which cause K to behave as a metal), but rather the electron density varies sinusoidally with a characteristic wavelength—and that this usually is not an integral multiple of the crystal lattice constant. This concept, of course, was not in agreement with the traditional conclusion that free electrons are uniformly distributed. The reasoning—the sinusoidal clumping lowers the electron energy, which in turn causes the lattice to distort and, as explained by Robinson (1986), this distortion is an attempt to reduce the huge electric fields generated by the separation between the positive charge of the K ions and the negative charge of the electrons.

At the time of Overhauser's work in 1964, experimental examples were not available and the concept was generally considered academic. Several years later, however, investigators working with layered materials (electrons essentially move in only two directions) and with linear conductors (motion is essentially in one direction) attributed a charge density wave phenomenon to what has been termed the Pierls instability. The latter effect, which involves lowering electron energy and lattice distortion, currently is not believed to apply to the simple, three-dimensional metals (K etc.). In summary, the Pierls instability and the Overhauser charge density waves concept appear to be similar, but different.

In 1985, Giebultowicz (National Bureau of Standards), Overhauser, and Werner (University of Missouri) conducted a neutron diffraction study and tentatively proved the Overhauser concept. Some solid-state physicists are seeking further evidence. If the concept is fully confirmed, some modifications in the thinking of how electrons behave in solids may be required.

Salt-Forming Properties. One major difference between potassium and sodium in their salt-forming properties is the much greater ability of potassium to form alums, although potassium does not form quite as many types of these compounds as do the higher alkali metals, or ammonium or monovalent thallium.

Potassium also differs from sodium, and especially from lithium, in the greater stability of its salts of polarizable polyatomic anions, such as peroxide, superoxide, azide, polysulfide, polyhalides, etc. The rubidium and cesium salts, on the other hand, are even more stable.

Among the other inorganic compounds of potassium are the following:

Bromate. Potassium bromate, $KBrO_3$, white solid, soluble, mp $434\,°C$, upon heating oxygen is evolved and the residue is potassium bromide; formed by electrolysis of potassium bromide solution under proper conditions. Used as a source of bromate and bromic acid.

Carbonate. Potassium carbonate, potash, pearl ash, K_2CO_3, white solid, soluble, formed (1) in the ash when plant materials are burned, (2) by reaction of potassium hydroxide solution and the requisite amount of CO_2. Used (1) in making special glasses, (2) in the making of soft soap, (3) in the preparation of other potassium salts (a) in solution, (b) upon fusion; potassium hydrogen carbonate, potassium bicarbonate, potassium acid carbonate, $KHCO_3$, white solid, soluble, (4) in vat dyeing and textile printing, (5) in titanium enamels, (6) in boiler water treating compounds, (7) in photographic chemical formulations, (8) in electroplating baths, and (9) as an important absorbent for CO_2 in the process industries.

Chlorate. Potassium chlorate, chlorate of potash, $KClO_3$, white solid, soluble, mp about $350\,°C$, powerful oxidizing agent, and consequently a fire hazard with dry organic materials, such as clothes, and with sulfur; upon heating oxygen is liberated and the residue is potassium chloride; formed by electrolysis of potassium chloride solution under proper conditions. Used (1) in matches, (2) in pyrotechnics, (3) as disinfectant, (4) as a source of oxygen upon heating. (Hazardous! Use of potassium perchlorate is recommended instead.)

Chloride. Potassium chloride, KCl, colorless or white crystals; strong saline taste. Occurs naturally as sylvite. Soluble in water; slightly soluble in alcohol. Sp. gr. 1.987; mp $772\,°C$; sublimes at $1500\,°C$; noncombustible; low toxicity. Used in fertilizers, as a source of potassium salts; pharmaceutical preparations; photography; spectroscopy; plant nutrient; salt substitute; laboratory reagent. See also **Fertilizer.**

Chloroplatinate. Potassium chloroplatinate, K_2PtCl_6, yellow solid, insoluble, formed by reaction of soluble potassium salt solution and chloroplatinic acid. Used in the quantitative determination of potassium.

Chromate. Potassium chromate, K_2CrO_4, yellow solid, soluble, formed by reaction of potassium carbonate and chromite at a high temperature in a current of air, and then extracting with water and evaporating the solution. Used (1) as a source of chromate, (2) in leather tanning, (3) in textile dyeing, (4) in inks.

Cobaltinitrite. Dipotassium sodium cobaltinitrite, $K_2NaCo(NO_2)_6 \cdot H_2O$, golden yellow precipitate, formed by reaction of sodium cobaltinitrite solution in acetic acid with soluble potassium salt solution. Used in the detection of potassium.

Cyanate. Potassium cyanate, KCNO, white solid, soluble, formed along with lead metal by reaction of potassium cyanide and lead monoxide solids upon heating. Source of cyanate.

Cyanide. Potassium cyanide, cyanide of potash, KCN, white solid, soluble, very poisonous, formed by reaction of calcium cyanamide and potassium chloride at high temperature. Used as a source of cyanide and for hydrocyanic acid, but usually replaced by the cheaper sodium cyanide. Also used in metallurgy, electroplating, extraction of gold from ores, as a pesticide and fumigant, in photography and analytical chemistry. Upon acidification, produces dangerous HCN gas.

Dichromate. Potassium dichromate, chromate of potash, $K_2Cr_2O_7$, red solid, soluble, powerful oxidizing agent, formed by acidifying potassium chromate solution and then evaporating. Used (1) in matches, (2) in leather tanning and in the textile industry, (3) as a source of chromate, (4) in pyrotechnics, (5) in colored glass, (6) as an important laboratory reagent, (7) in blueprint developing, and (8) in wood preservation formulations.

Hydroxide. Potassium hydroxide, caustic potash, potassium hydrate, KOH, white solid, soluble, mp $380\,°C$, formed (1) by reaction of potassium carbonate and calcium hydroxide in H_2O, and then separation of the solution and evaporation, (2) by electrolysis of potassium chloride under the proper conditions, and evaporation. Used in the preparation of potassium salts (1) in solution, and (2) upon fusion. Also used in the manufacture of (3) soaps, (4) drugs, (5) dyes, (6) alkaline batteries, (7) adhesives, (8) fertilizers, (9) alkylates, (10) for purifying industrial gases, (11) for scrubbing out traces of hydrofluoric acid in processing equipment, (12) as a drain-pipe cleaner, and (13) in asphalt emulsions.

Hypophosphite. Potassium hypophosphite, KH_2PO_2, white solid, soluble, formed (1) by reaction of hypophosphorous acid and potassium carbonate solution, and then evaporating, (2) by reaction of potassium hydroxide solution and phosphorus on heating (poisonous phosphine gas evolved).

Iodate. Potassium iodate, KIO_3, white solid, soluble, melting point $560\,°C$, formed (1) by electrolysis of potassium iodide under proper conditions, (2) by reaction of iodine and potassium hydroxide solution, and the fractional crystallization of iodate from iodide. Used as a source of iodate and iodic acid.

Manganate. Potassium manganate, K_2MnO_4, green solid, soluble, permanent in alkali, formed by heating to high temperature manganese dioxide and potassium carbonate, and then extracting with water, and evaporating the solution. The first step in the preparation of potassium manganate and permanganate from pyrolusite.

Nitrate. Potassium nitrate, saltpeter, niter, KNO_3, white solid, soluble, mp $333\,°C$, formed by fractional crystallization of sodium nitrate and potassium chloride solutions. Used (1) in matches, explosives, pyrotechnics, (2) in the pickling of meat, (3) in glass, (4) in medicines, (5) as a rocket-fuel oxidizer, and (6) in the heat treatment of steel. See also **Fertilizer.**

Nitrite. Potassium nitrite, KNO_2, yellowish-white solid, soluble, formed (1) by reaction of nitric oxide plus nitrogen tetroxide and potassium carbonate or hydroxide, and then evaporating, (2) by heating potassium nitrate and lead to a high temperature and then extracting the soluble portion (lead monoxide insoluble) with H_2O, and evaporating. Used as a reagent (diazotizing) in organic chemistry.

Oxides. See discussion later in entry.

Perchlorate. Potassium perchlorate, $KClO_4$, white solid, very slightly soluble, mp $610\,°C$, but above $400\,°C$ decomposes with evolution of oxygen gas and formation of potassium chloride residue; formed (1) by electrolysis of potassium chlorate under proper conditions, (2) by heating potassium chlorate at $480\,°C$ and then fractional crystallization. Used (1) as a convenient and safe (preferred to use of potassium chlorate) method of preparing oxygen by heating, (2) in the determination of potassium in soluble salt solution.

Periodate. Potassium periodate, KIO_4, white solid, very slightly soluble, mp $582\,°C$, formed by electrolysis of potassium iodate under proper conditions.

Permanganate. Potassium permanganate, permanganate of potash $KMnO_4$, purple solid, soluble, formed by oxidation of acidified potassium manganate solution with chlorine, and then evaporating. Used (1) as disinfectant and bactericide, (2) in medicine, (3) as an important oxidizing agent in many chemical reactions.

Persulfate. Potassium persulfate, $K_2S_2O_8$, white solid, slightly soluble, formed by electrolysis of potassium sulfate under proper conditions. Used (1) as a bleaching and oxidizing agent, (2) as an antiseptic.

Silicate. Potassium silicate, K_2SiO_3, colorless (when pure) glass, soluble, mp 976 °C, formed by reaction of silicon oxide and potassium carbonate at high temperature, similar in properties and uses to the more common sodium silicate.

Sulfates. Potassium sulfate, sulfate of potash, K_2SO_4, white solid, soluble. Common constituent of potassium salt minerals. Used (1) as an important potassium fertilizer, (2) in the preparation of potassium or potash alums; potassium hydrogen sulfate, $KHSO_4$, white solid, soluble; potassium pyrosulfate, $K_2S_2O_7$, white solid, soluble, formed by heating potassium hydrogen sulfate to complete loss of H_2O. See also **Fertilizer**.

Sulfides. Potassium sulfide, K_2S, yellowish to reddish solid, soluble, formed by heating potassium sulfate and carbon to a high temperature; potassium hydrogen sulfide, potassium bisulfide, potassium acid sulfide KHS, formed in solution by reaction of potassium hydroxide or carbonate solution and excess H_2S.

Sulfite. Potassium sulfite, $K_2SO_3 \cdot 2H_2O$; potassium hydrogen sulfite, $KHSO_3$; white solids, similar in properties and formation to the corresponding sodium sulfites.

Thiocarbonate. Potassium thiocarbonate, K_2CS_3, yellow solid, soluble, formed by reaction of potassium sulfide and CS_2.

Thiocyanate. Potassium thiocyanate, potassium sulfocyanide, potassium rhodanate, KCNS, white solid, soluble, mp about 170 °C, formed by fusing potassium cyanide and sulfur, and then crystallizing. Used as a source of thiocyanate.

In addition to the inorganic salts, potassium forms such binary compounds as a phosphide, K_3P, by direct union with phosphorus, a boride, KB_6, by electrolysis of fused fluorides and borates in the presence of a metal boride, a nitride, and the oxides. Of the latter, direction reaction of potassium and oxygen yields the superoxide, KO_2, a paramagnetic, orange-colored substance. The likelihood of KO_2 having a monomeric structure is supported by these properties, since the O_2^- ion would have an odd electron, which would confer paramagnetism and color upon the compound. The lower oxides of potassium, K_2O and K_2O_2, which are less stable in air than the superoxide, have been prepared, as have their hydrates. K_2O unites explosively with the oxygen of the air. One other oxide, K_2O_3, has been reported, but this appears to be a double salt of KO_2 and K_2O_2. The properties of potassium hydroxide are in keeping with its position in Group 1; thus its heat of solution is somewhat lower than that of rubidium hydroxide, RbOH, or cesium hydroxide, CsOH, and much higher than that of lithium hydroxide, LiOH, and NaOH.

The organic compounds of potassium include many oxycompounds, such as salts of organic acids, alcohols and phenols (alkoxides, phenoxides, etc.). A few potassium-carbon linked compounds have been reported, such as a phenylisopropyl potassium, $C_6H_5C_3H_7K$, and a carbonyl compound of unknown composition, $K_x(CO)_x$. The adduct of ethyl potassium and diethyl-zinc is a true salt, $K_2[Zn(C_2H_5)_4]$, potassium tetraethylzincate.

See also **Potassium and Sodium (In Biological Systems)**.

Additional Reading

Giebultowicz, T.M., A.S. Overhauser, and S.A. Werner: *Phys. Rev. Lett.*, **56**, 1485 (1986).
Greenwood, N.N. and A. Earnshaw: "Chemistry of the Elements," 2nd Edition, Butterworth-Heinemann, Inc., Woburn, MA, 1997.
Krebs, R.E.: "The History and Use of Our Earth's Chemical Elements: A Reference Guide," Greenwood Publishing Group, Inc., Westport, CT, 1998.
Lide, D.R.: "CRC Handbook of Chemistry and Physics 2000–2001," 81st Edition, CRC Press, LLC., Boca Raton, FL, 2000.
Parker, P.: "McGraw-Hill Encyclopedia of Chemistry," 2nd Edition, The McGraw-Hill Companies, Inc., New York, NY, 1993.
Robinson, A.L.: "Charge Density Waves Seen in Posassium," *Science*, **232**, 713 (1986).
Stwertka, A. and E. Stwertka: "A Guide to the Elements," Oxford University Press, Inc., New York, NY, 1998.

POTASSIUM AND SODIUM (In Biological Systems). Potassium and sodium play major roles in biological processes. Because of the numerous parallels between these two elements in metabolism, they are treated in a single entry, with appropriate distinctions made.

Potassium is required by both plants and animals. Although the total amount of potassium in most soils is usually rather high, the level of available or soluble forms of the element is frequently too low to meet the needs of growing plants. Deficiencies of plant-available potassium are more frequent in the soils of the eastern rather than of the western United

States. See also **Soil**. Potassium in the form of soluble potassium salts is a very common constituent of fertilizers. See also **Fertilizer**.

Many plants will not grow at normal rates unless the plant tissues, especially the leaves, contain as much as 1 or 2% potassium and, for some plants, even higher concentrations are required. Therefore, if a plant grows at all, it will nearly always contain sufficient potassium to meet the requirements of the people or animals that consume the plant. Potassium deficiencies do occur in humans and animals, but these are largely due to metabolic upsets and illnesses that interfere with the utilization of potassium in the body, or via excessive losses of potassium from the body, rather than due to inadequate levels of dietary potassium.

The general role of potassium fertilizers in improving human and animal nutrition is to help increase food and feed supplies rather than to improve the nutritional quality of the crops produced. Excessive use of potassium fertilizers may decrease the concentration of magnesium in crops. Sodium is essential to higher animals that regulate the composition of their body fluids and to some marine organisms, but it is dispensable for many bacteria and most plants except for the blue-green algae. Potassium, on the other hand, is essential for all, or nearly all forms of life. The importance of these cations for all forms of life has been related to the predominance of sodium and potassium in the ocean where primitive forms of life are thought to have originated and developed. During most of the period of evolvement of living organisms, there has been little change in the sodium and potassium content of seawater, either as to proportion or total amount. The body fluids of sea animals are, in most instances, similar to seawater in sodium and potassium level and ratio. In freshwater and terrestrial animals, the sodium and potassium level of body fluids is usually somewhat lower, and the ratio is likely to vary from the 40:1 ratio of seawater. Most fresh waters contain small and variable amounts of sodium and potassium, usually in a ratio of from 1:1 to 4:1.

Despite the higher level of sodium in natural water, potassium is universally the characteristic cation found within both plant and animal cells. Although sodium is not an absolute requirement for most plants and bacteria, it is found in these organisms and is essential to higher animals where it is the principal cation of the extracellular fluids. Sodium and potassium are important constituents of both intra- and extracellular fluids. Generally, the best external and internal medium for function of cells not adjusted to low salt levels is a medium involving a balance of sodium and potassium.

Beyond the osmotic effects depending on the sum of the concentration of the ions in the solution, Ringer found in 1882 that to maintain the contractility of an isolated frog heart, it was necessary to perfuse it with a medium containing sodium, potassium, and calcium ions in the proportion of seawater. It has since been recognized that the normal life activities of tissues and cells may depend on a proper balance among the inorganic cations to which they are exposed. Sodium is required for the sustained contractility of mammalian muscle, while potassium has a paralyzing effect. Thus, a balance is necessary for normal function. Other investigators have found that the antagonism among univalent and divalent cations observed by Ringer is demonstrable with various simpler or more complicated organisms or biological systems.

Excessive salt in soil, such as soils recently soaked with seawater, is toxic to most plants, although there are many plants, e.g., those of the salt marshes and the sea, which are adapted to a high salt concentration. Ingestion of seawater by man as the only source of water is eventually fatal because of the inability of the body to eliminate salt at a concentration comparable to that of seawater. This results in accumulation of salt, with severe toxic effects and eventually fatal results.

It is probable that potassium is absorbed by the plant roots from the soil by an active transport mechanism which carries it through the cell wall structure. Similarly, potassium and sodium if required, are accumulated by animals also by active transport. The actual cellular content of potassium and sodium is likewise controlled by transport mechanisms that specifically move potassium in and sodium out of the cell against the concentration gradient. The energy for this is derived from the metabolic processes of the cell. The nature of these transport mechanisms has not been fully determined.

Ions and Transport Mechanisms. Potassium differs from most other essential constituents of plant and animal cells in that it is not built into the cell as a part of an organic compound, but is rather an ion from a soluble inorganic or organic salt. Potassium ions may chelate with cellular constituents, such as polyphosphates. The ion is of the correct size to fit into the water lattice adsorbed by the protein in the cell. In general, the potassium and sodium ions are attracted to protein or other colloidal or structural units having a negative charge. Mucopoly-saccharides within the cell, on the cell surfaces and of the intercellular structures, are of particular importance in holding cations, such as potassium and sodium. Active centers of other configurational features of the proteins in the cell may be affected or altered by the potassium held by electrostatic or covalent binding. There are several enzyme systems activated by potassium.

In general, most of the sodium and potassium in the animal is in a dynamic state, being exchanged between different parts of the cell, between the cell and the extracellular fluid, and intermixing with ingested sodium and potassium in body fluids.

Most cellular constituents do not selectively bind potassium in preference to sodium. Myosin of muscle fibers, for example, will bind either. But, in contrast, the mitochondria and ribosomes are organized cellular organelles able to selectively take up or extrude potassium. This accounts for only a part of the potassium held in the cell.

In blue-green algae and some yeasts, sodium may in part replace cellular potassium. While potassium is usually the principal cation concerned with the maintenance of the osmotic pressure within the cell, sodium contributes appreciably to the total, and amino acids and other organic compounds may help make up any deficit, particularly in marine invertebrates.

The sodium content of the body extracellular fluids of marine invertebrates from the coelenterate through the arthropod phyla is approximately that of seawater. In freshwater and terrestrial invertebrates, the sodium of body fluids varies over a wide range and there is considerable variation among vertebrates. There are both fish and crustaceans so highly adaptable that they are able to live in either fresh or salt water.

Osmotic Pressure Regulation. The regulation of osmotic pressure within the cell and the control of the passage of water into or out of the cell is dependent to a considerable extent on the control of the potassium and sodium in the cell by the transport systems of the cell wall. The cell wall itself is of protein-lipid composition and is in general impermeable to the passage of water and inorganic salts. Recent studies of the cell walls with electron microscopes and with the use of other investigative techniques indicate that the cell wall contains pores connecting the cell contents with the extracellular fluid, or in some plants, with other cells. In cells having an endoplasmic reticulum, the intracellular vacuolar system may have openings through the cell wall communicating with the extracellular fluid. The ease with which water passes in or out of the cell in response to changes in external or internal osmotic pressure varies over an extreme range, from easy passage to rigid control, depending on the cell and its functions.

Phagocytosis and pinocytosis may bring salts and water, as well as other substances, into the cell.

In some unicellular organisms, osmotic equilibrium may be maintained by a contractile vacuole, which collects water; in other organisms, water may be excreted through the cell wall. The kidney and sweat glands of higher animals, gills of fish and salt glands of birds serve to excrete salt. Most animals, through control of sodium and potassium excretion and loss, are able to adapt to a wide range of intake.

The importance of sodium chloride in nutrition has been recognized from the beginning of history. Agricultural populations that lived on cereal grains, nuts, berries, and other vegetable foods poor in sodium, experienced a hunger for salt which led them to go to great lengths to obtain the mineral. This was particularly true if they lived in a hot climate with the attendant increased loss of salt in perspiration. Similarly, herbivorous animals will travel long distances to supply their need for additional salt. In contrast, peoples or animals subsisting on meat, milk and other foods receive quite appreciable amounts of sodium salts in the diet, and experience no special desire or hunger for salt. See also **Sodium Chloride**.

In plants, the meristematic tissues in general are particularly rich in potassium, as are other metabolically active regions, such as buds, young leaves, and root tips. Potassium deficiency may produce both gross and microscopic changes in the structure of plants. Effects of deficiency reported include leaf damage, high or low water content of leaves, decreased photosynthesis, disturbed carbohydrate metabolism, low protein content and other abnormalities.

Since potassium is found abundantly in most natural foods consumed by animals, deficiency is ordinarily no problem. With prolonged maintenance through parenteral (intravenous) feeding when normal oral feeding is not possible, potassium must be supplied.

Role of Kidney. Experimental potassium deficiency in rats results in stunted growth, loss of chloride with hypochloremic acidosis, loss of potassium and increase of sodium in muscle. In man, disease of the gastrointestinal tract, involving loss of secretions through vomiting or diarrhea, may result in serious loss of both sodium and potassium. Trauma, surgery, anoxia, ischemia, shock and any damage to or wasting away of tissues may result in loss of cellular potassium to the extracellular fluid and plasma, and the loss from the body through kidney excretion. Recovery with rapid uptake or potassium by the tissues may result in low plasma levels. Low extracellular potassium concentration may cause muscular weakness, changes in cardiac and kidney function, lethargy, and even coma in severe cases. There are no reserve stores of either sodium or potassium in the animal body, so any loss beyond the amount of intake comes from the functional supply of cells and tissues. See also **Kidney and Urinary Tract**.

The kidney is the key regulator of the sodium and potassium content of higher animals and makes possible adaptation to wide variations of intake. In the glomerulus of the kidney nephron (or individual unit), an ultrafiltrate containing the smaller molecules of plasma is normally produced. As this ultrafiltrate passes down the kidney tubule, 97.5% or more of the sodium is actively resorbed, along with nearly all of the potassium. The remaining 2.5% of the sodium is sufficient to account for even the maximum sodium excretion. Potassium is added to the filtrate in the distal tubule through exchange for sodium. Control of this exchange appears to be the principal mode of action of aldosterone, which thus exerts a final control over sodium excretion. Aldosterone is a steroid hormone from the adrenal cortex, secretion of which seems to result from lowering of the Na/K ratio in the blood. Water is passively resorbed with the electrolytes along the length of the tubule.

Water excretion is further controlled by the antidiuretic hormone from the posterior pituitary gland which acts to increase water resorption in the kidney through making the collecting tubule permeable to water for additional resorption beyond what took place in the tubule. The posterior pituitary gland secretes the hormone as a rapid and sensitive response to a rise in the osmotic pressure of the extracellular fluid. The osmotic pressure of the extracellular fluid is, of course, principally due to its sodium chloride content.

With low intake of sodium, excretion is reduced to a very low level to conserve the supply in the body. Potassium is not so efficiently conserved.

The kidney regulates the acid-base balance of the body by control over resorption of sodium ions, which may exchange for hydrogen ions in the kidney tubule. Since most dietaries are of acid-ash, the urine is usually more acid than the original plasma filtrate and much of the phosphate excreted is thus changed to the acid monosodium salt. Within the range of normal variability, with an alkaline ash diet, the urine may become alkaline, and in extreme instances, some sodium bicarbonate may be excreted.

The salts of the buffer pairs responsible for control of the pH of plasma and extracellular fluid involve sodium as the principal cation, while the cellular buffers involve potassium salts. See also **Acid-Base Regulation (Blood)**; and **Diuretics**.

Additional Reading

Benos, D.J. and D.M. Fambrough: "Amiloride-Sensitive Sodium Channels: Physiology and Functional Diversity," Academic Press, Inc., San Diego, CA, 1999.

Evans, J.M., T.C. Hamilton, S.D. Longman, and G. Stemp: "Potassium Channels and Their Modulators: From Synthesis to Clinical Experien," Taylor & Francis, Inc., Philadelphia, PA, 1997.

Young, D.B.: "Role of Potassium in Preventive Cardiovascular Medicine," Kluwer Academic Publishers, Norwell, MA, 2001.

POTATO FAMILY.   See **Solanaceae (Potato Family)**.

POTENTIAL ENERGY.   The negative of the work done by the forces of a conservative system when the particles of the system move from one configuration to another is the potential energy of the second configuration relative to the first configuration. This quantity is independent of the path followed by the particles in changing their configuration and is a function of the initial and final positions only.

An equivalent definition states that the potential energy is that particular function of the coordinates $V(x, y, z)$ whose negative gradient exists and is equal to the force, i.e.,

$$\mathbf{F} = -\Delta V$$

The existence of V implies a conservative force field.

If the force between two particles separated by a distance r is given by $\mathbf{F} = K/r^2$ the mutual potential energy of the particles when separated by a distance R is

$$-\int_{r0}^{r} \frac{K^2}{r} \, dr$$

where r_0 is the distance of separation in the initial or standard configuration. For convenience, r_0 is often taken as infinity, in which case the potential energy at infinity is considered to be zero and the potential energy of the final configuration is then K/r. Actually, the numerical value of the potential energy is arbitrary because the initial configuration can be chosen arbitrarily. Any constant can be added to the potential energy function and the condition $\mathbf{F} = -\Delta V$ will still be satisfied.

A particle on the surface of the earth is acted upon by a force mg, where m is the mass and g is the acceleration of gravity at the point. If the particle is raised a height h centimeters above the surface of the earth, where h is small in comparison with the radius of the earth, the potential energy of the particle with respect to the earth's surface becomes mgh. For example, a 10 gram mass at a distance of 100 cm above the ground has a potential energy of about $(10 \text{ gm}) \times (100 \text{ cm}) (980 \text{ cm/sec}^2) = 980,000$ ergs. If the mass were allowed to fall to the ground in a vacuum, the potential energy would be converted completely into 980,000 ergs of kinetic energy, thus exemplifying the conservation of energy.

POTENTIOMETER.   1.   An instrument used for the measurement or comparison of small potential differences or electromotive forces, based upon the "law of potential drop" (see also **Electric Circuits**). One of the simplest potentiometer circuits is shown in Fig. 1.

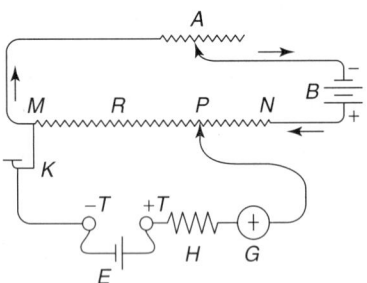

Fig. 1.   Potentiometer circuit.

Current from a battery B is sent through a resistance MN and is adjustable by means of a rheostat A. From one extremity M of this resistance is taken off a branch circuit containing the potentiometer terminals $+T$, $-T$, between which E, one of the electromotive forces to be compared, is connected. This circuit rejoins the main circuit at a point P which is adjustable so that the partial resistance MP or R can be varied, while MN as a whole remains constant. The $+$ and $-$ leads from E must be connected as shown, and the electromotive force of B must exceed E. The position of P is now adjusted until the galvanometer G shows no current, indicating that the potential drop from P to M just balances the electromotive force E. If two different electromotive forces E_1, E_2, are thus connected and balanced in succession, and if the corresponding values of the resistance MP are R_1, R_2, then since the current through MN is unaltered, the law of potential drop gives

$$\frac{E_1}{E_2} = \frac{R_1}{R_2}$$

In particular, one of the electromotive forces may be a standard cell of accurately known voltage; the other is thereby determined. In such case the standard cell should be safeguarded by a high resistance H, which is gradually reduced as the zero-current adjustment is approached; and the key K should be closed only for an instant. In some potentiometers the whole equipment, including galvanometer and standard cell, is contained in one compact case.

2.   The term potentiometer is also used to denote a three-terminal voltage-dividing network such as is used in volume controls for radio receivers, etc. Confusion may be avoided if the term voltage-divider is used for this purpose.

POT HOLE.   Under favorable conditions, where streams flow over the bedrock, swirling eddies will wash sand, gravel or pebbles around and around in the same place with the result that cylindrical holes called pot holes are worn, often to a considerable depth. These pot holes may be from a few inches to several feet in diameter and rarely as much as 40–50 feet (12–15 meters) deep. Similar features found on the seashore, the result of wave action, are called sea-mills. Pot holes also have been formed by water from crevasses and ice cliffs and glaciers.

POTOMETER.   A device for measuring transpiration, consisting of a small vessel containing water, and sealed so that the only escape of moisture is by transpiration from a leaf, twig, or small plant with its cut end inserted in the water. A similar device, the *phytometer*, consists of a vessel containing soil in which one or more plants are rooted and sealed.

POTTER WASP (*Insecta, Hymenoptera*).   A small wasp whose nest is built of mud in the form of a globular pot with a narrow neck. The several species belong to the genus *Eumenes*. All are solitary.

POULTRY.   Any of a variety of domesticated birds bred and raised for their meat and eggs. Of all meat (poultry and red meat), poultry accounts for about 20% of worldwide meat production. It is estimated that the production of poultry meat exceeds 21 million metric tons. Production of eggs worldwide exceeds 23 million metric tons annually.

A Chinese book, dated about 1400 B.C., refers to fowls as "creatures of the West." This statement is consistent with the views of early authorities, who believed that the domestication of various wild birds for use as food (flesh and eggs) commenced in Asia, notably in the western part of the continent in the Indian region. Mentions of fowls were made much later in the writings of Aristophanes and others as early as 500 B.C. They were mentioned frequently in the literature of Biblical and Roman times. But, unfortunately, little information found in the early literature serves as a key to the origin of the common domesticated birds as we know them today. Some early authorities traced back to the wild jungle fowl (*Gallus bankiva*), which still can be found wild in some regions of India. This correlation resulted from comparing the fundamental characteristics of modern breeds of chickens with the characteristics of the wild jungle fowl. More recently, however, some experts have pointed out resemblances between present fowls with a number of other forms of wild game from various regions of the Orient, such as *Gallus sonneratii* (Afghanistan), *G. fercatus* (Indonesia), *G. stanleyi* (Sri Lanka), and *G. giganteus* (Malaysian Peninsula), among others.

The genealogy of the domesticated birds that are known today as *poultry* is complicated by many factors. Birds are relatively easy to domesticate, and they interbreed and combine characteristics quite readily, thus accounting for the large numbers of varieties developed in nature. The varieties are mainly the result of human intervention, a process that has gone on for many hundreds of years, though admittedly quite unscientifically until the last century or so. Further, these birds are relatively small and are quite easy to transport, even during the earliest times when transport was crude. Thus, the blending of bird varieties from many regions was inevitable. These and other factors tended to discourage the development of breeds along simplistic, classical lines that are easy to trace.

The term *poultry*, once confined to birds easily raised by farmers, peasants, and natives in relatively small numbers, has been expanded to include any species of bird that is subject to commercialized domestication. Turkeys, for example, at one time were strictly considered wild game. Today, wild turkeys are difficult to find, but many millions produced each year are a major product of the poultry industry. The same situation would apply to swans if it were found commercially advantageous to industrialize their production and market them in huge quantities. Growing interest in the 1980s, for example, is being shown in commercializing production of quail as a common item for the marketplace.

Breeds of Poultry

Although tracing the ancestry of modern poultry is difficult, authorities recognize well over a dozen basic *classes* of chickens (or fowls as preferred by some authorities).[1] Each class has a number of breeds and varieties. A

[1] Some authorities prefer the designation fowl rather than chicken, because the latter word is sometimes used by people to designate a young bird as contrasted with an older bird.

breed of bird possesses certain fundamental and consistent qualities that reliably appear generation after generation and the results of which can be forecasted when there is mating between breeds. The similarities between the breeds of any given class are usually more obvious than any relationship between classes. Further, varieties within a given breed tend to be even more closely and obviously related. The *variety* is the first subdivision of a breed. When certain intrinsic qualities appear within a variety and then consistently reappear, it may be desirable from a classification standpoint to establish a *subvariety*. Usually a subvariety is spoken of as a *strain*. The well recognized classes of chickens or fowls are delineated in Table 1. Of the breeds and varieties listed, the American, Asiatic, English, and Mediterranean breeds are of principal importance in North America. The principal characteristics of these breeds are summarized in Table 2. Some concept of the range of popularity among the various breeds can be gleaned from Table 3.

Of the 200 or more officially recognized varieties of chickens, the majority are not of commercial importance. Of the breeds listed in Table 1, only a relatively few breeds are raised in significant numbers in North America; these include the White Leghorn, the White Plymouth Rock, the Rhode Island Red, the Barred Plymouth Rock, the New Hampshire, and the Cornish. There has been a strong movement for a number of years on the part of American poultry producers to pay progressively less attention to the traditional purebreds. The latter are mainly of interest in the development of new foundation stocks. The poultry industry differs from the livestock field in that less attention is paid to breed registration. The closest approach to this is The American Poultry Association (Crete, Nebraska), which publishes the "Standard of Perfection" that contains listings and specifications of a large number of varieties of poultry.

Breeding techniques in the poultry field parallel those used in the cattle, swine, and sheep industries, and also in the crop plant field, because the parameters of what can be accomplished are established by the fundamental laws of genetics. More experimentation is possible in the poultry field, as compared with large animals, for a number of reasons. The biological cycle is much shorter and the results of breeding efforts can be determined during a shorter time span. The investment in individual birds is much less than in other forms of livestock. The compatibility among breeds of birds is good. In the improvement cycle, foundation breeders furnish eggs to hatcheries. In turn, the hatchery sells chicks to the producer, who, in turn, raises the commercial egg layers, broilers, and market turkeys. New varieties are also developed by publicly funded agricultural experiment stations and universities, where proprietary interests are not a problem.

Breeding objectives include fertility, viability (general health), and feed conversion. There are also specific objectives as regards egg-type and meat-type birds. The heritability of characteristics ranges from a low of 8–10% (laying intensity for example) to as high as 50% (egg size) and higher. The interrelationship and interdependence of heritable qualities and bird management practices are always present and not always easy to sort out.

Body weight, which has a heritability factor of about 60%, varies widely among the basic breeds (contrast between the Brahmas and the bantams). This is an important factor because there is a strong correlation between body weight and size and rate of growth and feed utilization efficiency. Large body size is of particular importance to broiler and turkey breeders. Closely related to body weight is excessive abdominal fat in broilers. Traditionally, broiler breeders have selected for large body weight. Because of high heritabilities for abdominal fat, breeders probably have inadvertently selected for increased percentage of fat. It is estimated that at least 1% of the weight of broilers purchased by the processing plant is gizzard fat and sections of the leaf fat torn from the bird during the processing. This tissue goes to the offal plants and is processed as a byproduct and at a loss to the processor.

Egg production increases have been less amenable to genetic manipulation than increases in meat production. Two factors are of major consideration: (1) length of the laying period prior to molting. Early sexual maturity of a pullet is obviously desirable. Age at maturity is a heritable factor. Sexual maturity in normal Leghorns occurs at about 24 to 26 weeks (170 to 185 days) of age. Dual-purpose breeds (White Wyandotte) require 1 to 3 weeks longer. There are considerable variations among the various strains available. Environmental effects (lighting, feed management, diseases, etc.) also influence the age of maturity. To achieve an objective of 20 dozen or more eggs in the first pullet year, it is evident that laying should commence when the bird is about 5 months old. In judging a given flock of birds, it is usually assumed that the first 70–80% of the birds that commence laying will be the best layers in the long term.

TABLE 1. PRINCIPAL CLASSES OF CHICKENS[1]

AMERICAN CLASS

Buckeye, Dominique, Holland, Java, **Jersey White Giant, New Hampshire, Plymouth Rock, Rhode Island Red**, Rhode Island White, and **Wyandotte**. Composite breeds — originated in North and Central America. Medium-to-large size, with moderate, but colorful and widely varied plumage. These birds grow and develop at a moderate rate. The birds are slow, but not sluggish in their movements. Generally, they are a hardy stock with well-protected bodies. Although somewhat combative, the birds do not have a vicious personality. They are excellent winter layers and in general prolificacy are better than the Asiatics and somewhat inferior to the Mediterraneans. Reliable in terms of breeding efficiency, but not as prepotent as the Asiatics and Mediterraneans. Maternal qualities are excellent.

ASIATIC CLASS

Brahma, Cochin, Langshan

Presumed to be of Chinese origin — one of the oldest of the classes. They are raised for meat and winter egg production. Some of the birds, especially the Brahmas, are excellent for heavy roasters and capons. Probably better suited to small-scale than large-scale commercial operations. Plumage is heavy. Wide variety among breeds in terms of posture and stature. Skin is yellow and coarse. Rate of development is slow by comparison with American class breeds. These birds have an easy-going temperament and are not quickly frightened. They are known as good, but clumsy mothers. Generous plumage enables coverage of many eggs and chicks. By nature, they are not excellent egg producers, but eggs are medium to large size and of uniform characteristics.

MEDITERRANEAN CLASS

Ancona, Blue Andalusian. **Leghorn, Minorca**, Spanish

Originated in Italy, Spain, and other parts of the Mediterranean region. With exception of Minorca, the Mediterraneans are light to medium size and weight. The Minorca is heavy. Birds have a light, closely feathered plumage with a wide variation of color in different breeds. Development is rapid and exceeds the American and Asiatic classes. The birds are alert, nervous, and quite easily frightened. Although best suited to temperate climates, they can be adapted with relative ease to different environments. However, changes in environment affect productivity. Not rated too high in terms of mothering ability or interest. The birds are courageous, but not quarrelsome. The Mediterraneans are rated among the highest in terms of prolificacy. First eggs are laid at 6 months of age or less. Possess prepotent breeding qualities.

ENGLISH CLASS

Australorp, Cornish, Duckwing, Orpington, Redcap, Sussex

Originally developed mainly as meat producers, but with objective of combining meat- and egg-producing qualities. The birds are large and white, with a white skin, excepting the Cornish, which has a yellow skin.

POLISH CLASS

This is one breed with eight varieties. Essentially ornamental rather than food birds. However, they are good layers of white eggs. Generally regarded as nonsitters.

HAMBURG CLASS

A Dutch class with one breed and six varieties. Considered unsatisfactory for meat production. Small size of eggs limits market appeal.

FRENCH CLASS

Four breeds and five varieties. The birds are of the meat-type, although can be raised for reasonably satisfactory egg production.

CONTINENTAL CLASS

Of Belgian origin, the birds are quite small, quite prolific, and very hardy.

GAME AND GAME BANTAM CLASS

Sixteen varieties in this class are mainly raised for exhibition and fighting (in some countries).

ORIENTAL CLASS

Mainly birds for exhibiting.

ORNAMENTAL BANTAM CLASS

A few breeds and several varieties of birds, generally considered too small to be of interest as food-producing birds.

[1] Breeds listed in boldface type are the most important breeds in North America. There is also a Miscellaneous Class (mainly ornamental birds).

Rate of production, sometimes called intensity, has a heritability factor of only about 10%. Emphasis on bird management is more important than inheritance in this case and management of environment and feed is paramount. For example, laying production goes up when eggs are removed

TABLE 2. CHARACTERISTICS OF MAJOR BREEDS OF CHICKENS

Breed	Weight of Cock Pounds	Weight of Cock Kilograms	Weight of Hen Pounds	Weight of Hen Kilograms	Coloration Skin	Coloration Shank	Coloration Ear Lobe	Coloration Plumage	Coloration Egg	Type of Comb	Illustration
AMERICAN CLASS											
Jersey White Giant	13	5.9	10	4.5	yellow	yellow	red	white	brown	S	—
New Hampshire	8.5	3.9	6.5	2.9	yellow	yellow	red	red	brown	S	Fig. 1
Plymouth Rock	9.5	4.3	7.5	3.4	yellow	yellow	red	white	brown	S	Fig. 2
Rhode Island Red	8.5	3.9	6.5	2.9	yellow	yellow	red	red	brown	S&R	Fig. 3
Wyandotte	8.5	3.9	6.5	2.9	yellow	yellow	red	white	brown	R	—
ASIATIC CLASS											
Brahma (light)	11	4.9	9	4.1	yellow	yellow	red	*	brown	P	Fig. 4
Cochin	11	4.9	8.5	3.9	yellow	yellow	red	buff	brown	S	—
Langshan (black)	10	4.5	7	3.2	white	blue-black	red	dark	brown	S	—
MEDITERANNEAN CLASS											
Ancona	6	2.7	4.5	2	yellow	yellow	white	varies	white	S&R	—
Leghorn	6	2.7	4.5	2	yellow	yellow	white	white	white	S&R	Fig. 5
Minorca (white)	8	3.6	6.5	2.9	white	white	white	white	white	S	—
ENGLISH CLASS											
Australorp	8.5	3.9	6.5	2.9	white	dark	red	black	brown	S	Fig. 6
Cornish (white)	10	4.5	8	3.6	yellow	yellow	red	white	brown	pea	—
Orpington	10	4.5	8	3.6	white	white	red	white and buff	brown	S	—

* = Columbian pattern; S = single; R = rose. All breeds given in table, with exception of Asiatic class, do *not* have feathered shanks. Weights shown are for mature birds.

TABLE 3. RELATIVE POPULARITY OF BREEDS OF CHICKENS (In North America)

Breed or Type	Percent of Total	Trend Since Early 1960s
Crossmated	82.8	sharply up
White Leghorn	8.9	moderately down
Incrossmated	4.8	sharply down
White Plymouth Rock	0.6	sharply down
Rhode Island Red	0.4	sharply down
Barred Plymouth Rock	0.2	steady
New Hampshire	0.1	sharply down
Others	2.2	

Fig. 2.   White Plymouth Rock male chicken. (*USDA.*)

Fig. 1.   New Hampshire female chicken. (*USDA.*)

from the nest daily. Egg size has a heritability factor of about 50%. Egg size also relates to body size, age of pullets, and environment. Egg color is also a heritable factor. In some markets, white eggs are preferred; in others brown eggs are preferred. However, the astute marketer of eggs will not mix slightly tinted eggs with white eggs.

Meat-quality indicators include plumage color, skin and shank color, and rate of feather development. Feathers that are white or light in coloration are much sought in breeding broilers. Dark feathers are much more difficult to pluck clean when the broiler is dressed. It has been found that when an early-feathering male is mated with a late-feathering female, the male progeny will be slow-feathering and the pullets will be early-feathering. This helps the producer in selecting pullets at hatching time.

Production Technology. The production of poultry meat and eggs is of the most common food-producing operations found throughout the world. Poultry plays an important role in the nutritional requirements, from the most underdeveloped of countries and regions to the most advanced of nations. There is an exceptionally wide spectrum of production intensity, ranging from small flocks numbering well under 100 birds to highly

Fig. 3. Rhode Island Red female chicken. (*Texas Agricultural Extension Service.*)

Fig. 5. White Leghorn female chicken. (*Texas Agricultural Extension Service.*)

(a)

(b)

Fig. 4. Brahma chickens: (**a**) Dark variety female; (**b**) light variety female. (*USDA.*)

Fig. 6. Australorp male chicken. (*USDA.*)

efficient and specialized commercial operations where tens of thousands of birds may be involved. While the fundamentals of managing a flock in order to obtain the best possible production remain the same, the technology (housing, equipment, etc.) varies widely. In general, one may observe that the small poultry operation represents an interface between birds and people, whereas in the large operations, the interface involves the interaction of birds and machines. In any situation, it is extremely important to select good starting stock, matching the birds with the environmental needs. For a number of years, little attention was paid by most producers to the matter of spacing required by birds. Reduction in the production of poultry meat and eggs, as a result of crowding, is well documented. The mortality of poultry is increased by crowding due to cannibalism and the spread of virulent diseases. The type of housing, whether littered floor, slats, wire floors, or cages (of various configurations) influences the optimum space allotment. Temperature and humidity control are equally important. Although further research is required, experience over the years has shown that the optimum temperature for layers lies between 12.8 and 21.1 °C. Experience also has shown that an optimum humidity for the laying house lies between 50 and 75% relative humidity.

A number of semiautomated and automated bird-feeding systems have been developed during the past few decades. Two major design criteria are reduction of labor and reducing the wastage of feed. Such systems also take into consideration the bird's environment, so that there will be a smooth (without stress) interface between bird and machine. See Fig. 7.

Other Species

Turkeys. Native to the Western Hemisphere, the original habitat of the wild turkey ranged from Quebec and Ontario southward to Florida and west to Mexico and the Rocky Mountains. As vast areas of open hardwood forests were cleared in North America, the prime turkey habitat was destroyed. Fortunately, the species retreated to seldom-visited pockets of wilderness in Pennsylvania, West Virginia, South Carolina, Florida, and a few other states. The "American Standard of Perfection" recognizes only one breed of turkeys. The recognized varieties include the Broad-Breasted Bronze, the Broad-Breasted White, the White Holland, the Beltsville Small White, the Bourbon Red, the Narragansett, the Black Slate, and the Jersey Buff. Of these only three varieties are of major commercial importance in North America. See Table 4. Unlike chickens, most turkeys are bred as standardbreds, rather than hybridized. Improvements are gained through individual selections and mass-matings. The popular fast-growing,

broad-breasted birds are clumsy in their mating and this has led to increasing use of artificial insemination in breeding the birds.

In the early 1950s, the Broad-breasted Bronze variety was most popular, enjoying over 75% of total production, followed by the Small White (16%), and the Large White (nearly 5%). By the mid-1960s, the emphasis shifted to the Large White (48%), followed by the Broad-Breasted Bronze (41%), and the Small White (8.4%). Many producers object to the Small White because of the smaller quantity of meat produced with about the same investment in labor and housing as required by larger birds.

Geese. Generally, geese are very hardy and not susceptible to many of the common poultry diseases. They are excellent foragers, although selective, and can be put on good succulent pasture or lawn clippings as early as their first week of life. In North America, the Toulouse, Emden, and African geese are the most popular breeds raised for meat production. Other common breeds found are Chinese, Canada, Buff, Pilgrim, Sebastopol, and Egyptian. The weights of various breeds of geese are summarized in Table 5.

The *Toulouse goose* derives its name from the city of Toulouse in southern France. Because of its late maturity, it is sometimes called the Christmas goose. This breed has a broad, deep body and is loose-feathered, a characteristic which gives it a massive appearance. The plumage is dark gray on the back, gradually shading to light gray edges, with white on the breast, and to white on the abdomen. The eyes are dark brown or hazel, the bill pale orange, and the shanks and toes are a deep reddish orange.

The *Emden goose* was one of the first breeds of geese to be imported into the United States. The breed was first identified as the Bremen, named for the city in Germany from which early importations were made. The Emden is a pure white, springly goose. It is much more tight-feathered than the Toulouse and therefore appears more erect. The Emden is a fairly good layer, but production depends upon the breeding and selection of the flock. Egg production averages from 35 to 40 eggs per year per mature breeding goose. The Emden is usually a better sitter than most Toulouse

Fig. 7.   Artist's representation of a modern cage-layer system. The components are currently available: (A) Bulk feed bin; (B) straight-line auger feed delivery; (C) watering system; (D) floor stands; (E) egg deescalators; (F) cross egg conveyor which moves eggs to central location; (G) air inlet system; (H) cage-delivery system; (I) dropping board scrapers; (J) cages; (K) feed intake cups; (L) thermostatically-controlled fans; and (M) egg collectors. (*Chore-Time Equipment, Inc., Milford, Indiana.*)

TABLE 4. CHARACTERISTICS OF TURKEY VARIETIES

BRONZE VARIETY (See Fig. 8)

The subvariety, Broad-Breasted Bronze, is the most predominantly produced and is characterized by a uniform, amply-fleshed carcass.

Normal Weights

Adult male	36 pounds (16.3 kilograms)
Adult female	20 pounds (9.1 kilograms)

Coloration

Plumage	black with an iridescent sheen ranging from red to green to bronze
Throat wattle	red, but sometimes can appear bluish-white
Beard	black
Shanks and toes	in young birds, a dull black; in mature birds, a tarnished pink
Beak	light at tip and dark at base

WHITE HOLLAND VARIETY

Although somewhat more fertile than the Bronze variety, it is quite similar in many characteristics.

Normal Weights

Adult male	33 pounds (15 kilograms)
Adult female	18 pounds (8.2 kilograms)

Coloration

Plumage	pure white
Throat wattle	red, but sometimes pale pink
Beard	rather intense black
Shanks and toes	pale pink
Beak	light-pink horn

BELTSVILLE SMALL WHITE VARIETY (See Fig. 9)

Developed by the U.S. Department of Agriculture and named after the research center located in Beltsville, Maryland. The objective of development was to obtain a well-fleshed, smaller, light-weight bird. The variety is known for exceptional hatchability and egg production.

Normal Weights

Adult male	23 pounds (10.4 kilograms)
Adult female	13 pounds (5.9 kilograms)

Coloration

Plumage	pure white
Throat wattle	red, but sometimes pinkish-white
Beard	black
Shanks and toes	pale pink
Beak	light to medium pink

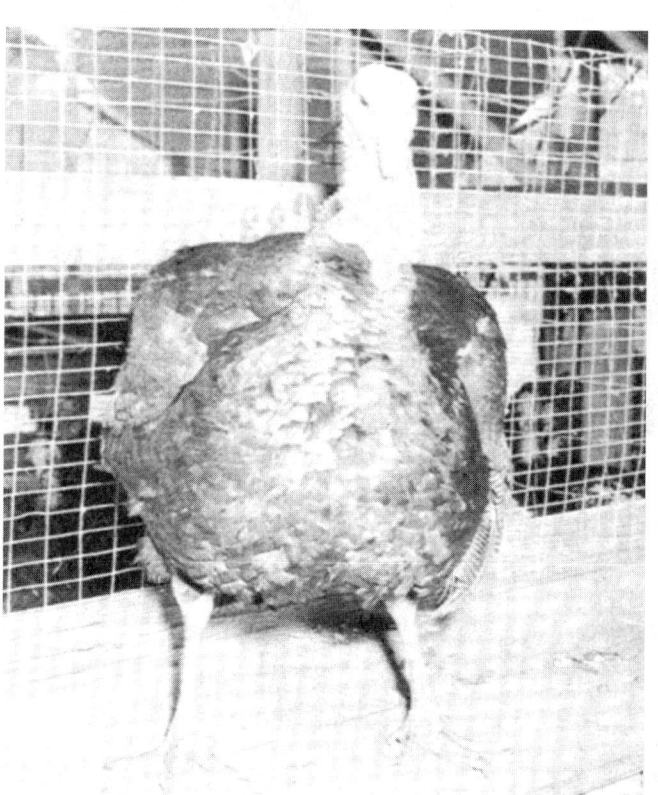

Fig. 8. Broad-breasted female turkey. (*Texas Agricultural Extension Service.*)

Fig. 9. Beltsville Small White turkey developed by the U.S. Department of Agriculture

and is one of the most popular breeds for marketing. The breed grows rapidly and matures early.

The *African goose*, shown in Fig. 10, is a handsome breed with a distinctive knob or protuberance on its head. Its carriage is more erect than that of the Toulouse, and its body more nearly oblong and higher from the ground. The head is light brown, the knob and bill are black, and the eyes are dark brown. The plumage is ash brown on the wings and back and is light ash brown on the neck, breast, and underside. The African goose is a good layer, grows rapidly, and matures early. However, it has not gained in market popularity as have the Emden or Toulouse.

The *Chinese goose*, of which there are two standard varieties, originated in China and probably came from the wild Chinese goose. See Fig. 11. It is smaller than the other standard breeds and more swanlike in appearance. Both the brown and white varieties mature early and are better layers than other breeds, usually averaging from 40 to 65 eggs per bird annually. The Chinese goose grows rapidly, is very attractive, and makes a desirable medium-size market bird.

The *Pilgrim goose*, shown in Fig. 12, is a medium-size goose that is well suited for marketing. A unique feature of this breed is that the males and females may be distinguished by color. In day-old goslings, the male is creamy-white and the female is gray. The adult male remains white and has blue eyes. The adult female is gray and white with dark hazel eyes.

The *Buff goose* is of fair economic value as a market goose and thus only limited numbers are raised. Other less-commercial geese are the *Canada*, the common wild goose of North America. The breed is difficult to keep in confinement. The wild gander is sometimes used to cross with domestic breeds, resulting in the so-called mongrel goose (a hybrid), which usually is sterile but has fine-quality flesh. The *Sebastopol* is a white ornamental goose, very attractive because of its soft, plume-like feathering. The *Egyptian* is a long-legged, but very small goose, kept primarily for ornamental purposes.

Breeding geese prefer to be outdoors. Except in extremely cold weather or in storms, mature geese seldom seek shelter. In northern climates, colony poultry houses, open sheds, or barns are provided for shelter. Geese make nests on the floor of the house or in coops, boxes, or barrels provided. Straw or grass hay is used for outside nests as well as for nests on the floor of a house. The producer must provide one nest for every three females and permit the geese to select their own nests. Geese generally start laying in February or March in the Northern Hemisphere and often lay until early summer. The incubation period for eggs is about 30 days. Geese require ample quantities of palatable drinking water. Geese are very selective and tend to pick out the most palatable of forages. They will reject alfalfa and narrow-leaf tough grasses and select the more succulent clovers and grasses. Geese cannot be raised satisfactorily on dried-out, mature pasture. Because geese will eat weeds without harming certain cultivated plants, they are frequently used as weeders.

Ducks. Breeds of ducks suitable for producing mainly table meat are the Alesburys and Pekins. Others, which are layers, such as Magpies, Buff Orpingtons, Blue and Black Orpingtons, Stanbridge Whites, and large utility Fawn and White Indian Runners are also useful for table purposes. The Campbell White and Khaki all give drakes and ducks which are

TABLE 5. WEIGHTS OF DIFFERENT BREEDS OF GEESE

Breed	Young Male		Adult Male		Young Female		Adult Female	
	Pounds	Kilograms	Pounds	Kilograms	Pounds	Kilograms	Pounds	Kilograms
Toulouse	20	9.1	26	11.8	16	7.3	20	9.1
Emden	20	9.1	26	11.8	16	7.3	20	9.1
African	16	7.3	20	9.1	14	6.4	18	8.2
Chinese	10	4.5	12	5.4	8	3.6	10	4.5
Egyptian	5	2.3	5.5	2.5	4	1.8	4.5	2
Canada	10	4.5	12	5.4	8	3.6	10	4.5
Sebastopol	12	5.4	14	6.4	10	4.5	12	5.4
Pilgrim	12	5.4	14	6.4	10	4.5	13	5.9
Buff	16	7.3	18	8.2	14	6.4	16	7.3

Fig. 10.   African geese. (*USDA.*)

quite useful eating—good in flavor and with good deep breast meat. See Figures 13, 14, 15.

Quails. Usually considered mainly for the sportsperson, in most regions of the world, serious interest is shown in the commercial domestication of these birds and they are widely available in the markets of major cities and towns. Sportspersons in North America recognize seven kinds of quails. Ornithologists recognize five genera, seven species, and nearly twenty races. In the eastern and southern parts of North America is the Bobwhite. Along the Mexican border are the Scaled or Blue Quail, the Massena Quail (also called Fool Quail), and the Masked Bobwhite. In California, there are the Desert Quail (sometimes called Gambel's Quail), the Valley Quail, and the Mountain Quail. The birds in these groups differ sufficiently in plumage and habitat and calls from those in other groups to be recognized at a glance by the experienced hunter. All are ground birds. They are small, averaging 11 inches (28 centimeters) in length. The overall shape is nearly round. Tails are short and strong; bills are heavy and hard. They feed on grasses and clovers and seeds of different kinds.

In Corio (State of Victoria, Australia), an experiment was commenced in 1977 toward intense breeding of *Coturnix* or European quail, which are also bred in large numbers in Italy, France, and the United States. Apart from the fairly rank-tasting native variety, Australians had not had frequent opportunity for tasting quail, which has been a delicacy in Europe for hundreds of years.

Other Species of Poultry. Relatively minor species of poultry include guinea fowl and pigeons (squabs). Native to Africa, *guinea fowl* were introduced into Europe and the British Isles during the Middle Ages. The three major varieties of domesticated guineas are the Lavender, the Pearl, and the White. The Pearl variety is the one most esteemed by guinea fowl fanciers.

Pigeons are found in numerous regions of the world and squab has been a gourmet item for hundreds of years. A favorable quality for producers of pigeons is their exceedingly fast rate of growth, reaching normal adult

Fig. 11.   White Chinese geese. (*USDA.*)

Fig. 12.   Pilgrim geese. (*USDA.*)

Fig. 13.   White Pekin drake. (*USDA.*)

Fig. 14.   Khaki Campbell drake. (*USDA.*)

weight within just a little over a month. Among the most popular varieties of pigeons (for eating) are the Homer, the Swiss Mondaines, and the White King.

Pigeons and doves, particularly homing pigeons, are described in greater detail in the entry on **Columbiformes**.

Hundreds of varieties of essentially ornamental and exhibit or show birds, closely related to poultry, but seldom eaten, include peafowls and swans. See also **Anseriformes** (swans); and **Galliformes** (peafowls).

Metabolic System of Fowls

The digestive tract of the fowl differs in several respects from that of the mammals. The fowl does not chew in the conventional sense. It has no teeth. Feed is picked up by the beak and forced into the *gullet* (esophagus) by a specialized tongue, the rear part of which incorporates a fork-like configuration. The tongue also assists the bird in taking water. The esophagus, capable of extensive expansion, provides for continuous flow of feed and water to the stomach, but is divided into an upper and lower chamber, with an intermediate chamber known as the *crop*. The crop, coupled with the expandable characteristics of the esophagus, enables the bird to consume feed at an accelerated rate, and provides a place for initial

softening and processing of the feed, largely as the result of an admixture of moisture with enzymes present in the feed. However, very little saliva as such is secreted.

From the lower gullet, feed passes to the glandular stomach, also known as the *proventriculus*, where gland-secreted gastric juices containing hydrochloric acid and pepsin break down proteins to peptones and other simpler compounds. The partially-processed feed then proceeds to the *gizzard*, also known as the *ventriculus*. This is a very muscular organ, equipped with a horny lining, that by way of repeated contractions and expansions, masticates or massages the feed contents. This action accomplishes the type of disintegration of the feed into small particles that is accomplished by the teeth in mammals. Sometimes, the horny lining of the gizzard erodes or sloughs off, leaving ulcerations in the lining. Usually, a well-balanced diet will prevent the condition from occurring. The gizzard also acts as an efficient mixer of gastric juices and pulverized feed.

Fig. 15.   White Runner drake, a variety of the Indian Runner. (*USDA*.)

The well-processed feed substances then enter the small intestine where further enzymes are secreted to finish the breakdown of proteins and to split sugars. Absorption of the nutrients is also performed, enriching the bloodstream with energy-containing and body-building materials. A peristaltic action also occurs within the small intestine to ensure proper flow of the residue to the *ceca* (large intestine) and rectum. The ceca serves as an intermediate storage or buffer chamber. Bile generated in the liver is discharged into the gizzard and the duodenal loop, which connects the gizzard with the small intestine. Amylase, trypsin, and lipase are generated by the pancreas and secreted into the duodenal loop to aid digestion of carbohydrates, proteins, and fats. Insulin, also generated by the pancreas, regulates sugar metabolism.

A laying hen will require about 2.5 hours to digest feed, whereas a nonlaying hen will require up to 8 or even 12 hours.

Physiology of the Hen. Unlike most animals, which have two functional ovaries (right and left), the hen has only one functional *ovary* (left). This is located in the body cavity near the backbone. At the time of hatching, the female chick has close to 4000 very small *ova* contained in the left ovary. These ova have the capability of developing into full-sized yolks once a pullet attains full sexual maturity. Approximately once every 24 hours, in a normal, healthy hen, one of these ova will be released to the *oviduct*, which is a long (20 to 30 inches; 51 to 76 centimeters), folded tube contained on the left side of the abdominal cavity. The tube is complex and is comprised of five regions, all of which participate in the completion of a whole egg. For example, the white part of the egg is formed in an albumen-secreting region of the oviduct.

Within a short period (about 30 minutes) after laying a completed egg, another ovum is released from the ovary to the oviduct for another cycle of egg production. The service of a cock is not required to commence the egglaying process of the pullet; nor to restart the process after the first molting period. Infertile eggs are preferred over fertilized eggs in the marketplace. Mating is required, of course, to produce fertile eggs for hatching.

A number of important physiological changes occur in the laying hen. During the course of a laying period, a fowl will consume surplus fat from the body, particularly from the skin. Thus certain parts of the bird's body become whiter as the fat disappears. Other important physiological changes include some alteration in blood flow and in characteristics of the vent and pelvic arches and feathers as well. These changes occur in phase and progress slowly throughout the laying period.

A laying hen has a large, moist vent. The vent is dilated and loose and unlike the hard, puckered vent of a nonlaying hen. The pelvic arches and keel configure so as to increase the abdominal capacity of the bird. As fat is removed from the skin, the skin becomes velvety and the abdomen is soft and pliable. In nonlayers or poor producers, there will be an underlying layer of hard fat.

At the end of the laying period (200 or more eggs), the bird goes into the molting phase, during which time feather replacement and other physiological changes occur. Poultry producers who retain pullet flocks for a second year of production often use molt as a sign for selecting the good layers. The bird sheds its feathers in stages—first from the head, secondly from the neck, thirdly from the breast, fourthly from the body, and last from the wings and tail. On the average, from 3 to 4 months are required to molt. The producer can ascertain the length of time a given bird has been out of egg production by counting the number of new primary feathers. There are ten of these feathers in each wing. About six weeks are required to produce the first new feather and two weeks for each additional new feather.

Much more detail on poultry and the production of poultry meat and eggs can be found in the "*Foods and Food Production Encyclopedia*," (D.M. Considine, editor), Van Nostrand Reinhold, New York (1981). For further background on various birds, see entries listing under **Birds**.

Additional Reading

Barbut, S.: "Poultry Products Processing: An Industry Guide," CRC Press, LLC., Boca Raton, FL, 2001.
Calnek, B.W.: "Diseases of Poultry," 10th Edition, Iowa State Press, Ames, IA, 1997.
Etches, R.J.: "Reproduction in Poultry," CAB International, New York, NY, 1996.
Gillespie, J.R.: "Modern Livestock and Poultry Production," 6th Edition, Delmar Thomson Learning, Albany, NY, 2000.
Hunton, P.: "Poultry Production," Elsevier Science, New York, NY, 1995.
Jordan, F.T.: "Poultry Diseases," 5th Edition, Harcourt Health Sciences, San Diego, CA, 2001.
Pattison, M.: "The Health of Poultry," 4th Edition, Addison-Wesley Longman, Inc., Redding, MA, 1994.
Sainsbury, D.: "Poultry Health and Management: Chickens, Ducks, Turkeys, Geese, Quail," 4th Edition, Blackwell Science, Inc., Malden, MA, 2000.

Web References

Breeds of Livestock, Chickens, Ducks, Geese & Turkeys: *http://www.ansi.okstate.edu/breeds/other/*
Poultry Science Virtual Library: *http://posc.tamu.edu/library/dother.html*
U.S. Poultry & Egg Association: *http://www.poultryegg.org/*

POUND, ROBERT (1919). Pound is a Canadian-born American physicist who pioneered many fruitful ideas and is especially remembered for co-discovering, with Purcell, nuclear magnetic resonance (NMR) and establishing it as one of physics' most valuable analytical techniques. NMR is used as an analytical technique in chemical research, medical diagnosis, and a number of other fields. For this he received and shared the Nobel Prize in Physics in 1952.

Pound worked with his associate, Glen A. Rebka, Jr., carrying out an experiment using the Mossbauer effect to measure the gravitational effects of electromagnetic radiation and to test the predictions of Einstein's theory of general relativity. Pound's experiments continued and results predicted the Red Shift discovery.

During WW II, Pound worked at the Submarine Signal Company and then at MIT's radiation laboratory helping to develop radar and microwave technology. After the war, he became a professor at Harvard in 1948 and stayed until his retirement in 1989. Among his many awards have been the Thompson Memorial Award of the Institute of Radio Engineers in 1948, The Eddington Medal of the Royal Astronomical Society in 1965, and the National Medal of Science in 1990.

See also **Gravitation**; and **Nuclear Magnetic Resonance (NMR) and Magnetic Resonance Imaging (MRI)**.

J. M. I.

POUR POINT.   See Petroleum.

POUT.   See Codfishes.

POWDER METALLURGY. Powder metallurgy (PM) embraces the production of finely divided metal powders and their union through the use of pressure and heat into useful articles. The temperatures required are below the fusion point of the principal constituent, and bonding depends on interdiffusion of the metal particles in the solid state. It is necessary to provide intimate contact between particles, hence reducing atmospheres are provided in the sintering process to prevent formation of oxide films. Readily oxidized powders such as aluminum require special technique.

Probably the most important applications of powder metallurgy are those in which a product is made which cannot be duplicated by other methods. There are many examples of this kind. The melting point of tungsten, 6,100 °F (3.371 °C), is much too high for ordinary melting and casting methods and the only way in which filaments for electric lights can be made is to draw them from rods of compacted and sintered tungsten powder. The cemented carbide cutting tools are another important product of refractory nature readily made by powder metallurgy.

Self-lubricating bronze bearings having controlled porosity are products that can be made only by powder metallurgy. The pores are impregnated with oil, and flow to the bearing surface is maintained by capillary action. Graphite is incorporated with the metal powder in one type of oil-less bearing. A material made from powdered copper and graphite is used for electric-current collector brushes, and tungsten-copper or tungsten-silver combinations are used for electric contact points. In contrast to these high-conductivity materials, a high-resistance element is produced from a mixture of copper and porcelain powders, combining a metal with a nonmetallic substance.

Advances in PM Technology

Particularly during the past decade, remarkable progress was made in PM technology. Major trends in the early 1990s included: (1) rapid solidification processing (RSP), (2) liquid-dynamic compaction (LDC); (3) self-propagating high-temperature synthesis (SHS); (4) greater use of intermetallics and additives in PM products; (5) advancements in PM injection molding; and (6) improvements in heat treating PM parts — not to mention the appearance of PM in products and structures traditionally made by other metallurgical processes, such as seamless tubing.

Rapid Solidification Processing. RSP holds high promise for producing engineering alloys with refined microstructures, improved chemical homogeneity, extended solute solubility, and possible retention of metastable phases. RSP usually involves cooling rates greater than 100 °C/second (212 °F/s). For high cooling rates, RSP products must have a large surface-to-volume ratio, and thus are commonly in the form of powder, flakes, or ribbon. To be commercially acceptable, such rapidly solidified particulates must be consolidated into fully dense, metallurgically bonded forms suitable for engineering applications. RSP properties are quite sensitive to heat treatment and the desired properties can easily be lost without careful control over the consolidation process. Among the consolidation methods currently in commercial or near-commercial use include hot extrusion, hot isostatic pressing, vacuum or inert-atmosphere pressing or sintering, and powder forging. Unfortunately, these processes require elevated temperatures for relatively long times, which may destroy the benefits achieved by RSP. A major problem involves the tenacious oxide that forms on the surface of many RSP materials, particularly aluminum, nickel, and stainless steels.

A shock wave moving through the medium at velocities in excess of that of sound appears to be one solution to this problem. The shock wave can greatly exceed the yield stress. Passage of the shock wave causes plastic flow, interparticle melting and bonding, and can produce a fully dense, metallurgically bonded product. Three methodologies have evolved for introducing a shock wave: (1) use of a gas gun incorporating propellants or compressed gas; (2) direct application of explosives; and (3) impact of a projectile accelerated by explosives.

Guns are available of several designs. In one configuration, a high-pressure burst of gas launches a projectile down an evacuated tube where the projectile imparts a shock wave by driving a punch into the powder bed. As pointed out by Wright, the gun may be in the form of a high-impact press in which a reusable piston is accelerated in an evacuated chamber by introducing a rapid burst of gas into the breach. The impact of the ram produces a pressure pulse.

Hitchcox (1986) describes a process being developed at the Massachusetts Institute of Technology, which uses high-velocity pulses of an inert gas to atomize a stream of molten metal. Semisolid droplets of the metal are collected as rapidly solidified "splats" on a chilled metallic substrate. (This liquid dynamic compaction (LDC) process is attractive from a cost standpoint.) Substrates can be flat surfaces, molds, or shaped containers. The splats build up rapidly, forming high-density bodies suitable for further processing. Because the splats are thin, they cool at relatively high rates (1000 °C/second; 1800 °F/s). It is claimed that the LDC process improves ductility and fracture toughness because oxides and powder particle boundaries are minimized. Although in an early stage of development, materials such as high-strength aluminum and superalloys and (FeCo)-Nd-B have been produced with the process. Grant (MIT) reports that rapidly solidified material may exhibit grain sizes as fine as 0.2 micrometer (8 microinches) after crystallization of glasses. The fine grain size allows superplastic forming of aluminum alloys, stainless steels, and other materials.

Self-Propagating High-Temperature Synthesis. SHS usually involves an exothermic reaction producing temperatures in excess of 2500 °C (4532 °F). In essence, a mixture of compressed powders is ignited with a heat source in air or an inert atmosphere and in an instant, a refractory compound or multicomponent material results. SHS eliminates the need for high-temperature furnaces as required by conventional processes. Processing time is shortened to seconds or minutes versus hours and days as required with normal sintering. The products are usually of a higher purity, some having less than 0.2% (wt) of unreacted elements. This is the result of vaporizing volatile contaminants during the "explosion." SHS has been used to produce borides, carbides, and other difficult materials and is considered to have much potential for making ceramic matrix composites with unique microstructures.

In SHS, there are fundamentally two types of reactions: (1) *thermite*, where oxidation-reduction produces multiphase products, such as cermets; and (2) *compound formation*, as resulting from the starting elements, such as $Ti + 2B = TiB_2$. A combination of the two types of reaction also can be used. SHS requires a strong exothermic reaction where the heat of reaction is at least 40 kcal/mole (168,000 Joules/mole). The adiabatic temperature must be greater than the melting point of the product in order to produce a liquid phase for enhancing diffusion. Sheppard also breaks the reactions into (1) propagating, and (2) bulk. *Propagating reactions* are initiated locally, so that a synthesis wave of reactants, or, conversely, chemical activators can be added to accelerate the reaction. Also, if a higher reaction temperature required, preheating of the reactants is practiced.

Examples of products made by the SHS process include borides, carbides, chalcogenides, hydrides, intermetallic compounds, nitrides, silicides, carbonitrides, sulfides, cemented carbides (cermets), and various heterogeneous mixtures (microcomposites).

PM Intermetallics and Additives. An example of improved materials for which PM technology may solve past metallurgical processing problems is found in turbine parts, where high-temperature performance and oxidation resistance is mandatory. Aluminides of iron, nickel, and titanium have received consideration for a number of years, not only because they appear to meet the two foregoing criteria, but also because of their relatively low density, high strength, and corrosion resistance. Conventional casting of these materials results in unacceptable inhomogeneities. This has led to the evaluation of several PM methodologies, including hot isostatic pressing (HIP), vacuum hot pressing (VHP), injection molding, transient liquid-phase sintering, reactive sintering, and hot extrusion. Of considerable promise, reflecting research at Rensselaer Polytechnic Institute, is *reactive sintering*. This process involves a transient liquid phase. The reaction takes place above the lowest eutectic temperature in the system, but still at a temperature at which the compound remains in the solid phase. Research has shown that a transient liquid forms at the lowest eutectic temperature and spreads through the compact during heating. Actually, the reaction is approximately spontaneous because heat is liberated due to the thermodynamic stability of the compound's high melting temperature. In terms of the reaction of nickel and aluminum powders, a temperature over 550 °C (1020 °F) is the optimum. The time required for processing is relatively short (about one-half hour). Densities over 97% (of theoretical) are obtained. Even with the presence of some residual porosity, the ductility and strength of the product are good, which properties are retained after subsequent high-temperature exposure.

Researchers at Case Western Reserve University and the NASA Lewis Research Center, both located in Cleveland, Ohio, have evaluated *hot extrusion* as a candidate process. In essence, the process consists of canning the powder (prealloyed aluminide powders [FeAl, NiAl, and Ni_3Al]) and then extruding the material at a temperature and area-reduction ratio sufficiently high to produce satisfactory material flow and efficient filling of interparticle spaces, the latter for eliminating porosity and to encourage grains to recrystallize dynamically.

A basic advantage of PM technology has been that of minimizing or eliminating machining in making a final part. Nevertheless, some machining operations may be required. Traditionally, the machinability of sintered PM steels, for example, is poor, mainly due to porosity, hardness,

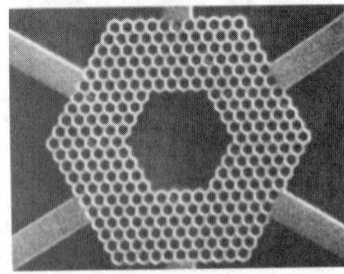
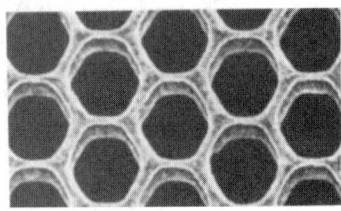

Fig. 1.   Representative cross sections of tiny (submicrometer) mechanical parts that can be produced by nanofabrication technology. (*Cornell University.*)

and low thermal conductivity, Porosity causes an interrupted cut and causes tool wear—with the possible results of both higher tool costs and poorer surface finish. In recent years, PM techniques have been improved by the incorporation of additive, notably manganese sulfide (MnS), to enhance machinability.

Powder metallurgy also is playing a major role in the pioneering but rapidly developing technology of *nanofabrication*. Melding the technologies of PM and electronic components manufacture are reducing operational minute machine parts to submicron levels. See Fig. 1.

Additional Reading

Alman, D.E. and J. Newkirk: "Powder Metallurgy Alloys and Particulate Materials for Industrial Application," The Minerals, Metals & Materials Society, Warrendale, PA, 2000.

Anderson, I.E.: "Boost in Atomizer Pressure Shaves Powder-Particle Sizes," *Advanced Materials and Processes*, 30 (July 1991).

Craighead, H.G.: "The National Nanofabrication Facility at Cornell University," Cornell University, Ithaca, New York, NY, October 1990.

Froes, F.H.: "Powder Metallurgy," *Advanced Materials and Processes*, 55 (January 1990).

German, R.M.: "Powder Metallurgy of Iron and Steel," John Wiley & Sons, Inc., New York, NY, 1998.

Keishi Gotoh, K. and H. Masuda: "Powder Technology Handbook," 2nd Edition, Marcel Dekker, Inc., New York, NY, 1997.

Hitchcox, A.L.: "Advances in Powder Metallurgy Cover Many Fields," *Advanced Materials and Processes*, 63–65 (December 1986).

Jenkins, I. and J.V. Wood: "Powder Metallurgy: An Overview," Ashgate Publishing Company, Brookfield, VT, 1991.

Kloecker, C.J.: "Hammers Take on Presses for Forging PM Steel," *Advanced Materials and Processes*, 37 (July 1991).

Marquis F.D.S.: "Powder Materials: Current Research and Industrial Practices," The Minerals, Metals & Materials Society, Warrendale, PA, 1999.

Scott, W.W., Jr.: "Engineering the Part," *Advanced Materials and Processes*, 4 (July 1991).

Staff: "Properties and Selection: Nonferrous Alloys and Special-Purpose Materials," ASM International, Materials Park, OH, 1991.

Staff: "ASM Handbook: Powder Metal Technologies and Applications," Vol. 7, ASM International, Materials Park, OH, 1998.

Staff: "Top Powder Metallurgy Parts Honored," *Advanced Materials and Processes*, 8 (August 1991).

Staff: "Forecast for Metals," *Advanced Materials and Processes*, 17 (January 1991); 17 (January 1992); 18 (January 1993).

Staff: Metallic and Inorganic Coatings, Metal Powders, and Sintered P/M Structural Parts," American Society for Testing & Materials, West Conshohocken, PA, 2001.

Suslick, K.S.: "Ultrasound 'Makes a Hit' with Metal Powder," *Advanced Materials and Processes*, 10 (September 1990).

Thummler, F. and R. Oberacker: "An Introduction To Powder Metallurgy," Ashgate Publishing Company, Brookfield, VT, 1994.

Web References

Institute of Materials Processing (IMP): *http://www.imp.mtu.edu/*
The Minerals, Metals, Materials Society: *http://members.tms.org/Staff.asp*

POWER.   In general, power is the time rate of doing work, as defined by the equation

$$P = \frac{dW}{dt}$$

where P is power, W is the work done, t is the time and therefore dW/dt is the time rate of doing work. Power may be expressed in units of work per unit time (e.g., foot-pounds per minute or ergs per second) or more arbitrarily, as in horsepower or watts. One horsepower is 33,000 foot-pounds per minute. One watt is 10^7 ergs (or 1 joule) per second.

POWER (Mean).   The instantaneous power in a device, or branch of a network is EI. The mean power is

$$\frac{1}{T} \int_0^T EI\,dt$$

where T is an integral number of periods, or T approaches infinity for nonperiodic currents. The mean power is also given by

$$P = \overline{I^2}R$$

where $\overline{I^2}$ is the mean square current, and R the resistance.

POWER FUNCTION.   An algebraic function, $y = ax^n$. It is often convenient to plot this function on logarithmic paper, where both abscissa and ordinate are graduated in divisions proportional to the logarithm of the number plotted. The result is thus $\log y = n \log x + \log a$ and the curve is a straight line of slope n and intercept $\log a$. The number n is called the exponent. A series of the form $a_0 + a_1x + a_2x^2 + \cdots$, either finite or infinite, is called a power series. See also **Curve Fitting**; and **Neyman-Pearson Theory**.

POWER GAIN.   1. The ratio of the power that a transducer delivers to a specified load, under specified operating conditions, to the power absorbed by its input circuit. If the input and/or output power consist of more than one component, such as multifrequency signal or noise, then the particular components used and their weighting must be specified. This gain is usually expressed in decibels.

2. Of an antenna, in a given direction, 4pi times the ratio of the radiation intensity in that direction to the total power delivered to the antenna.

POWER LOADING.   The ratio of the gross weight of a propeller-driven aircraft to its power, usually expressed as the gross weight of the aircraft divided by the rated horsepower of the power plant corrected for air of standard density. With turboprop engines, the equivalent shaft horsepower is used.

POWER SERIES.   An infinite series of the form

$$\sum_{n=0}^{\infty} a_n x^n = a_0 + a_1x + a_2x^2 + \cdots + a_nx^n + \cdots$$

where $a_0, a_1, \ldots, a_n, \ldots$ are constants and x is a variable, is called a power series.

For every power series $\Sigma a_n x^n$ there exists a constant l such that the series is absolutely convergent for all values of x such that $|x| < l$, and is divergent for all values of x such that $|x| > l$. This number l is called the limit of convergence of the power series, and the interval $(-l, l)$ is called the interval of convergence. The series may converge or diverge at the ends of this interval, for $x = l$ and $x = -l$.

If $\lim_{n \to \infty} |a_n/a_{n+1}|$ exists, it is the limit of convergence.

If $\lim_{n \to \infty} \sqrt[n]{|a_n|}$ exists, it is the reciprocal of the limit of convergence.

A power series is uniformly convergent within the interval $(-l', l)$, where $l' < l$ and l is the limit of convergence of the given power series.

The function defined as the sum of a power series $\Sigma a_n x^n$ is a continuous function of x at all points within the interval of convergence.

For the operations with power series, we have the following theorems:

Two power series may be added together for all values of x for which both series are convergent, i.e., for the smaller of the two intervals of convergence.

Two power series may be multiplied together for all values of x for which both series are absolutely convergent, i.e., for the smaller of the two intervals of convergence.

A power series may be differentiated term by term for all values of x within its interval of convergence.

A power series may be integrated term by term between any limits lying within the interval of convergence.

Power series are very useful in the analytical representation of functions by the expansion of functions in series, and for calculation by series.

See also **Series**.

POWER SOURCES AND SUPPLIES.

Because of the variety of ways in which instruments, computers, communications equipment, and control systems are used—from undersea to outerspace applications and from laboratory and factory installations to air dropping of equipment packages over remote terrain—practically every form of available energy has been considered to provide electric power for such equipment.

Batteries. Within their ampere-hour rating, batteries provide low ripple and a measure of voltage regulation. Most batteries can be recharged when necessary and are relatively compact and inexpensive, but have severe limitations on the amount of energy stored. The three types most commonly used as power sources are (1) lead acid, (2) nickel cadmium, and (3) mercury cells which obtain their power by electrochemical (ionic) exchange between elemental anodes, cathodes, and electrolytes. See also **Battery**.

Rotating Generators. Small rotating generators powered by such varied prime movers as hot gas turbines (propane or hydrocarbon fuels) or cold gas turbines (CO_2 or compressed air tanks) generally are quite expensive and are limited to military or remote mapping or monitoring applications.

Solar Cells. Silicon photovoltaic cells offer an inexpensive solution to low power battery charging on intermittent monitoring applications in remote, unattended locations. The number of manufacturers is limited. Approximately 0.2 V/cell is available, and current output is proportional to the area of each cell. The operation of the cells results from a conversion of the electromagnetic energy of the incident visible radiation into carrier (electron-hole) pair generation across a *p-n* junction or within the silicon crystal which produces an emf that varies somewhat with the wavelength of the incident radiation (light). See also **Solar Energy**.

Thermoelectric Cells. Liquified petroleum gas—LPG (propane or butane) power sources utilizing lead telluride cells as thermoelectric generators giving nearly 1 V/cell offer a compact though expensive remote power source. Lead telluride converts thermal power into electric power mainly owing to the setting up of a thermal gradient across a *p-n* junction formed by doping—resulting in an emf and current flow in the circuit.

Fuel Cells. These cells are described in detail in the entry on **Fuel Cells**. A power supply, generically speaking, is a device that converts electric power from one power source type to another power source type more suitable for use in the instrument or control system. The power supply contrasts with a power source that may be a battery, fuel cell, thermoelectric or thermionic cell, or even a prime source, such as a motor-generator set.

Power supplies are of two general types, dependent upon the degree of precision required by the load. On the one hand, when source and load precision or regulation are almost the same, all that is needed is a device to convert the form of electric energy—an example of which is a transformer-rectifier which converts A.C. energy to D.C. energy. Filtering can be added if required. On the other hand, most process and instrumentation control power requirements are incompatible with the variations inherent in the average industrial or utility power source. Consequently, this power must be controlled or regulated before use. Refer to Table 1 for a listing of power supply types available, the power conversion required, a typical application, and cost per watt. The following brief descriptions of these power supply types give their function and operation.

A.C. Voltage Regulator. This is a device that compensates for A.C. voltage fluctuations in the utility power source and provides essentially constant voltage within a specified regulation band of the preset output voltage. Major types include:

1. *Ferroresonant*: either flux coupled or non-flux coupled.
2. *Motor-driven Autotransformer* (also induction regulator).
3. *Electronic*:
    a. Shunt regulated buck boost autotransformer filtered:
        (1) Magamp type (tube, SCR, or transistor controlled).
        (2) Solid state (SCR) type.
    b. Add-subtract dissipation types:
        (1) Tube type (Class *A* or *AB*).
        (2) Solid state type (usually push-pull).

The *ferroresonant* types depend upon the fact that at line frequency the magnetizing inductance of one winding of the transformer forms a series resonant circuit with an external capacitor. The resulting, higher-than-normal current acts in such a way as to saturate a portion of the magnetic circuit, thereby giving the winding that surrounds this magnetic circuit better line voltage regulation than the input source. By properly choosing the circuit parameters, load voltage regulation can be achieved. These regulators can give 1% line and 1% load regulation with up to

TABLE 1. POWER SUPPLY CHARACTERISTICS

Type of Power Supply	Prime Source	Load Source	Conversion Required	Typical Applications
Autotransformer	A.C.	A.C.	voltage change	raise or lower line voltage
Isolation transformer	A.C.	A.C.	voltage and isolation	isolate load from line
A.C. voltage regulator	A.C.	A.C.	voltage regulation	close process control medical applications
Frequency changer	A.C.	A.C.	change frequency	motor drive; testing
Frequency regulator	A.C.	A.C.	regulate frequency	timing; gyro drive
Frequency converter	A.C.	A.C.	change and regulate frequency	timing; testing
Rectifier	A.C.	D.C.	change A.C. to D.C.	plating; battery charge
Transformer rectifier	A.C.	D.C.	change voltage, rectify, and isolate	plating; battery charge
D.C. power supply	A.C.	D.C.	change voltage, rectify, isolate, filter	amplifiers; radios
Regulated D.C. power supply	A.C.	D.C.	rectify, change, adjust, and regulate voltage	strain gage; computers
High-voltage D.C. power supply	A.C.	D.C.	rectify, change, and filter	paint spray; ionization
Regulated high-voltage power supply	A.C.	D.C.	rectify, change, filter, and regulate	photomultipliers; accelerators
D.C. current regulator	A.C.	D.C.	rectify voltage; regulate current	magnets; plating; forming
Inverter (rotary or vibrator)	D.C.	A.C.	convert D.C. to A.C.	motor and transducer drives
Static inverter	D.C.	A.C.	convert D.C. to A.C., solid state	motor and transducer drives
Sine wave inverter	D.C.	A.C.	D.C. to A.C., low distortion	emergency or remote power source
Regulated inverter	D.C.	A.C.	D.C. to regulated A.C.	emergency or remote power source
D.C.–D.C. converter	D.C.	D.C.	change voltage	battery-powered equipment power supply
Regulated D.C.–D.C. converter	D.C.	D.C.	change and regulate voltage	precision power source for amplifiers
High-voltage D.C.–D.C. converter	D.C.	D.C.	raise voltage to high voltage	photomultipliers; accelerators
D.C. regulator	D.C.	D.C.	lower voltage and regulate	amplifiers; computers

25% distortion in the less expensive versions. Cancellation of most of the unwanted harmonics, by adding windings and introducing various flux paths, can be achieved with a maximum distortion of 3% without degenerating regulation.

The *electronic shunt* regulated buck boost types consist of: (1) an autotransformer; (2) a low-dissipation regulated device, such as a saturating reactor or SCR pair across the input line with suitable harmonic reducing filters across the output; and (3) an electronic reference, sensing, and regulation circuit which can be either of the thermionic (tube) or solid state type. In either case, the regulation is held within approximately 1.0% rms and the distortion within 3% greater than the input at the line frequency. Transient response to line or load changes (the time required to recover to within a given regulation bandwidth) generally is less than 10 cycles of line frequency—as also is the case with ferroresonant types. Some adjustment of output voltage generally is provided.

Push-pull electronic amplifiers buck out the objectionable harmonics and poor input voltage regulation by means of an add-subtract transformer in series with the load. Whether tubes or transistors are used, considerable heat is generated within the supplies. Regulation is better than 0.1%, distortion is 1%, and transient response is in the microsecond region.

Frequency Changers. These devices convert power of one frequency into power of another frequency. In addition, voltage conversion and regulation are possible. Uses include (1) driving induction or synchronous motors at other than the available input frequency; and (2) testing control systems that utilize uncommon frequencies, such as aircraft, from commercial power frequency sources.

D.C. Rectifier. This is the most commonly used power supply. It uses an arrangement of electronic switches or rectifiers, which allow current flow in only one direction. The silicon rectifier—small, efficient, and inexpensive—is the most common type of rectifier in use today. When voltage transformation is required, a transformer is employed. The transformer increases the size, weight, and cost of a power supply, but also allows electrostatic isolation of the rectified D.C voltage from the input A.C. When filtering of the normal ripple from the normal half or full wave rectified output voltage is required, a D.C. filter is added, which generally consists of a large electrolytic capacitor. For very large currents, an inductance or choke usually is added unless closed-loop regulation circuitry is added.

Regulated D.C. Power Supply. See Table 2. In general, a regulated D.C. power supply consists of a transformer, rectifier, filter, and regulator. The first items have been discussed, but the method of regulation varies widely. For instrument and control systems, certain types are more useful than others. In general, low voltage (up to 150 V) systems use transistors as the main voltage or current regulating element with zener diode references and transistor error amplifiers to drive the main control transistors. When the transistors are D.C. controlled, the power supply is the so-called "linear" type, and the main regulating transistor (or transistors when higher currents are required) is referred to as a "pass" transistor. Regulation is the static D.C. error due to either line or load changes about a nominal or set output

condition. Since D.C. gain is high, very good D.C. regulation results. Since the regulating element is in series with the load, high-speed recovery from line or load transients results—generally on the order of tens of microseconds. Also, due to the high A.C. gain of the amplifier, excellent ripple reduction is achieved—generally of the same order (in percent) as the static dc regulation or better.

Current limiting is almost universally available on this type of supply. It is accomplished by sensing the output current, comparing it with a reference and, in effect, shutting off the amplifier and main regulating transistors, causing the output voltage to decrease at the load to limit the output current to the preset value. Some supplies actually reduce the load current under this condition. This is referred to as "fold back" current limiting. Actual current regulation can be accomplished and is described later.

In general, transistor regulated D.C. power supplies are available in several mechanical configurations. The most popular from the standpoint of volume is the "system" type which is usually contained in a metal box and has limited output voltage and current limit adjustments and no switches or meters. This type can be mounted near the load. The more expansive "laboratory" type has wider output voltage and current regulation adjustments and generally has better performance specifications.

Although the linear transistor regulated D.C. power supplies are the most popular, other types are available for many reasons.

First, for higher output power requirements nonlinear or switching-type D.C. power supplies are available. SCRs (silicon controlled rectifiers) are generally used because of their high current and high voltage capabilities both as primary regulating elements and as preregulators for a final transistor passing stage. When SCRs serve as the only regulating element, they generally switch on and off at the input line frequency rate, and regulation is achieved by varying the portion of the input sine wave that the SCR conducts (so-called phase control). Since the forward drop of the SCRs and/or possibly rectifiers is low in relation to the output voltage (except for 3- and 5-V supplies), the efficiency of this type of supply is somewhat higher than that of a linear regulator. D.C. regulation is somewhat worse as is transient response since it takes about 5 cycles of input line frequency for the regulator to recover to the regulation band. Ripple is related to the output filter size and is also higher than in the transistor regulators, although active filters can improve the ripple considerably at very little increase in dissipation.

Some of the switching-type power supplies utilize transistors as ON-OFF switches. These usually operate at ultrasonic frequencies. This results in a great reduction in overall size and weight as compared with other methods of regulation.

Where constant nonvarying loads must be protected against line variations the simplest and most reliable regulator consists of a ferroresonant ac transformer and a passive rectifier and filter. Line regulation of 1% is available, but load regulation is 3 to 10% for this type of supply which although inexpensive is rather large and heavy because of the ferroresonant transformer.

TABLE 2. REGULATED DC POWER SUPPLIES

Type of Power Supply	Input	Output Adjustment, %	Maximum Available Regulation, B2			Maximum Available Ripple, %	Regulation Means	Efficiency Average, %
			Line	Load	Current			
Line regulated	ferroresonant transformer	none	1.0	3–10	NA	1–3	ferroresonant transformer	75
Narrow range (slot)	transformer	±5	0.03	0.03	NA	0.03	transistor (usually Si)	40
Wide range	transformer	100	0.01	0.01	0.1	0.01	transistor (usually Si)	30
High wattage, low ripple	transformer	some-limited	0.1	0.1	0.1	0.03	SCR and transistor filter	70
High wattage, good regulation	transformer	100	0.01	0.01	0.1	0.01	SCR or transistor switch and transistor pass	50
High efficiency, small size	rectifier	±5	0.1	0.1	NA	0.2	switching transistor (high voltage)	65
High voltage (100–300 V), narrow range	transformer	±5	0.03	0.03	NA	0.03	high-voltage transistor	40
High voltage, wide range	transformer	100	0.01	0.01	0.1	0.01	high-voltage transistor, two-stage	30
High voltage (300–3000 V)	transformer	100	0.05	0.05	0.1	0.01	vacuum tube and SC control	25
High voltage, high current	transformer	100	0.1	0.1	0.1	0.3	SCR (usually primary)	60
High voltage (5000–50,000 V)	transformer	50	0.05	0.05	Opt.	0.05	tube	30

Current Regulators. The last type of D.C. regulator that can be used in control applications is the current regulator. Practically all standard D.C. voltage regulators can be reconnected as precision current regulators by utilizing a wire-wound power resistor in series with the load and controlling the voltage across this resistor, thus controlling the load current. However, this method uses a high proportion of the available power. The laboratory types have built-in current regulators that offer current regulation almost as good as the voltage regulation offered, although current ripple is somewhat higher than voltage ripple. For either configuration, precision current regulators are available "off the shelf."

Inverters. These are very useful devices for converting direct to alternating current. Control applications include audio frequency oscillators for linear variable differential transformers (LVDTs), servo amplifiers and control systems, and remote battery operated instrumentation systems requiring A.C. power. Older, cheaper types included the rotary type and a static type, which had a movable armature vibrator. Newer static types are generally transistor driven and, as with everything electronic every specification adds to the cost, so that frequency regulation, voltage regulation and output distortion control each contributes to the overall cost of the unit. As indicated in Table 1, most inverters are custom designed but a few are available off the shelf. The larger volt-ampere ratings include SCRs that also require auxiliary commutation circuitry, adding complexity and lessening reliability. Because all static inverters begin with square waves, the generation of and suppression of radio frequency interference are constant problems. In addition, the generation of low-distortion sine waves requires filtering except in the rotary types.

D.C.-D.C. Converters. D.C.-D.C. converters are useful for converting power at one D.C. voltage level to power at another D.C. voltage level. The usual technique is to "chop" the input D.C. at an ultrasonic frequency and transform or filter the resulting A.C. to the required level. For stepped-up voltage, a push-pull transistor stage is generally used. For stepped-down voltage a single transistor "chopper" suffices. Because of the high switching frequency, the size and weight of the magnetic components is minimal. Regulation, when required, is accomplished by modifying the chopped waveform (say by pulse width modulation) or by including a passing transistor in the lowest current path. For remote power supply applications requiring battery or solar cells as a power source, the D.C.-D.C. converter offers a very small and efficient method of producing precisely controlled voltage for amplifiers or transducers.

For some applications requiring "at the load" precision regulation with a wide variation of input D.C. voltages, e.g., from a car battery, a D.C. regulator consisting only of a series pass transistor and a reference amplifier is available, sometimes in a single small package that can be mounted on the amplifier or transducer chassis. These devices are obtainable in a variety of package sizes with power-handling capabilities ranging from milliwatts to 100s of watts. The lower-power devices often are monolithic integrated circuits. The high-power devices often are hybrid designs.

Thermionic Regulation and Switching Devices. Most of the power supply descriptions have included solid state devices as rectifiers and regulators. However, for many applications particularly high voltage (>250 V) regulated supplies, vacuum tubes are used as both passing regulators and amplifiers. Furthermore, in the medium to high kilovolt region, vacuum tube rectifiers are often used, although selenium controlled rectifiers offer some advantages over tubes. But selenium controlled rectifiers have a "wear out" mode due to pinhole melting of cadmium selenide, which causes high resistivity areas on the cells, increasing the forward voltage drop for a given current density.

Grid-controlled mercury vapor rectifiers (thyratrons) are still used in power supplies but have very little advantage over silicon controlled rectifiers except in the >1000-V rating, and of course they have a higher forward voltage drop and lower overall efficiency. See also **Zener Diode**.

POWER TRANSMISSION RATIO (Acoustic).
The ratio of the average acoustic energy transmitted normally through a surface to the average acoustic energy incident normally upon that surface.

POYNTING-ROBERTSON EFFECT.
The effect upon the motion of a micrometeoroid or other very small particle due to the absorption and emission of radiation. The particle absorbs radiation from the sun only in one direction, but re-radiates energy in all directions. This effect produces a drag upon the particle that is directly tangential to the orbit of the particle about the sun, and thus decreases the orbital angular momentum. Since this angular momentum varies as the square root of the orbital radius, decreases in angular momentum are accompanied by a decrease in orbital radius, so that the particle follows a spiral path directing steadily closer to the sun. Although the solar radiation pressure upon the particle opposes this effect, it is not sufficient to offset it completely.

PPB.   Parts per billion. One part per billion is a frequently used dimension for expressing the composition and analysis of substances—as found in air, water, food substances, etc. Instrument developments and other assay techniques perfected during the past decade or so have made the determination of such minute quantities a practical possibility for many materials. One part per billion is approximately equivalent to 1 drop in a 10,000-gallon (37,850-liter) tank.

PPM.   Parts per million. One part per million is a common dimension for expressing the composition and analysis of substances—as found in air, water, raw materials, food substances, etc. One part per million is approximately equivalent to about 1/32 ounce (1 gram) in 1 ton of substance. One gram is exactly one-millionth of a metric ton.

PRAIRIE DOG.   See **Squirrels and Other Sciuromorphs**.

PRAIRIE HARE.   See **Rabbits and Hares**.

PRAIRIE WOLF.   See **Canines**.

PRANDTL NUMBER.   A dimensionless number equal to the ratio of the kinematic viscosity to the thermometric conductivity (or thermal diffusivity). For gases, it is rather under one and is nearly independent of pressure and temperature, but for liquids the variation is rapid. Its significance is as a measure of the relative rates of diffusion of momentum and heat in a flow and it is important in the study of compressible flow and heat convection. See also **Heat Transfer**.

PRASEODYMIUM.   Chemical element symbol Pr, at. no. 59, at. wt. 140.91, second in the Lanthanide Series in the periodic table, mp 934°C, bp 3,512°C, density 6.769 g/cm^3 (20°C). Elemental praseodymium has a close-packed hexagonal crystal structure at 25°C. The pure metallic praseodymium is silver-gray in color, the luster dulling rapidly upon exposure to air and forming a nonadherent oxide which hastens the process of oxidation. When pure, the metal is soft and workable with ordinary tools. Processing and handling require storage under a nonreactive liquid or inert atmosphere or vacuum. Finely-divided praseodymium is pyrophoric, burning at a red heat. There is only one isotope of the element in nature ^{141}Pr. It is not radioactive and has a low acute-toxicity rating. Fourteen artificial isotopes have been produced. Of the light (or cerium-group) rare-earth metals, praseodymium is the fourth most plentiful and ranks 59th in abundance of the elements in the earth's crust, exceeding tantalum, mercury, bismuth, and the precious metals, excepting silver. The element was first identified by C.A. von Welsbach in 1885. Electronic configuration

$$1s^2 2s^2 2p^6 3s^2 3p^6 3d^{10} 4s^2 4p^6 4d^{10} 4f^2 5s^2 5p^6 5d^1 6s^2$$

Ionic radius, Pr^{3+} 1.01 Å, Pr^{4+} 0.90 Å. Metallic radius, 1.828 Å. First ionization potential, 5.42 eV; second, 10.55 eV. Other important physical properties of praseodymium are given under **Rare-Earth Elements and Metals**.

Primary sources of the element are bastnasite and monazite, which contain from 4 to 8% praseodymium. Plant capacity involving liquid-liquid or solid-liquid organic ion-exchange processes for recovering the element is in excess of 100,000 pounds Pr$_6$O$_{11}$ annually. Metallic praseodymium is obtained by electrolysis of Pr$_6$O$_{11}$ in a molten fluoride electrolyte, or by a calcium reduction of PrF$_3$ or PrCl$_3$ in a sealed-bomb reaction.

For many years, praseodymium has been a component of light rare-earth mixtures used in mischmetal, a pyrophoric alloy used in cigarette-lighter "flints." Mixtures of cerium, lanthanum, neodymium, and praseodymium, as oxides and fluorides, are used in the cores of arc carbons for the production of light of greater intensity. Similar mixtures of rare-earth oxides, including praseodymium, are used in optical glass polishing formulations. Mixtures of the lanthanide compounds, including about 5% praseodymium, find application as catalysts in petroleum cracking processes. A mixture containing 10% Pr, 30% Nd, and 60% La is used for cracking crude oil and comprises the largest single use of the element as well as of all other Lanthanide elements. Use of elemental praseodymium as a colorant for glass was one of the early applications. The color ranges from clear yellow to green and finds use in sunglasses, protective glasses for industry, art objects of glass, tableware, and optical filters. In the manufacture of ceramic tile, a praseodymia-zirconia yellow stain is used. Metallurgically, the most important intermetallic compound is $PrCo_5$, which has unsurpassed permanent magnetic properties. The compound has a very high resistance to demagnetization and has a high magnetic saturation value. PrNi5 has been used for adiabatic magnetization cooling of samples down to the milli-Kelvin range for low-temperature research. Investigations continue into further electronic and optical uses of the element and its compounds.

Additional Reading

Greenwood, N.N. and A. Earnshaw: "Chemistry of the Elements," 2nd Edition, Butterworth-Heinemann, Inc., Woburn, MA, 1997.
Krebs, R.E.: "The History and Use of Our Earth's Chemical Elements: A Reference Guide," Greenwood Publishing Group, Inc., Westport, CT, 1998.
Lide, D.R.: "CRC Handbook of Chemistry and Physics 2000-2001," 81st Edition, CRC Press, LLC., Boca Raton, FL, 2000.
Parker, P.: "McGraw-Hill Encyclopedia of Chemistry," 2nd Edition, The McGraw-Hill Companies, Inc., New York, NY, 1993.
Stwertka, A. and E. Stwertka: "A Guide to the Elements," Oxford University Press, Inc., New York, NY, 1998.

PRAWN (*Crustacea, Decapoda*).   Small marine crustaceans closely related to the shrimps. They differ from the lobsters and crabs in the compressed body and in the use of the abdominal appendages for swimming. See also **Aquaculture**; and **Crustaceans (Edible)**.

PRAYING MANTIS.   See **Mantis**.

PREAMPLIFIER.   This is a class-A voltage amplifier, which receives the signal from a microphone, pick-up, television camera tube, or other device supplying a low signal level and amplifies it so it can supply the input for additional amplifier circuits. Thus a preamplifier is commonly used in a radio or television studio to amplify the audio or video signal before feeding it into a mixer, line to the transmitter, or other amplifying equipment at the studio. See also **Amplifier**.

PREANTENNA.   A sensory organ of the pair borne by the first segment of the body in the *Onychophora*. Most arthropods have the first body segment developed in the embryo but lacking in the adult, hence their antennae are appendages of a more caudal segment. The similar organs of *Onychophora* are named preantennae to distinguish them from the true antennae of the other classes.

PRECESSION.   An effect manifested by a rotating body when a torque is applied to it in such a way as to tend to change the direction of its axis of rotation. If the speed of rotation and the magnitude of the applied torque are constant, the axis, in general, slowly describes a cone, its motion at any instant being at right angles to the direction of the torque.

A familiar example of precession is an ordinary top. If the axis of spin is not exactly vertical, the force of gravity exerts a torque tending to overturn the top; but instead of tipping over, it "wobbles" with a precessional motion about the vertical through the pivot-point. The gyroscope exhibits similar behavior. A hoop or a coin can roll on edge across the floor because, whenever it tends to tip either way, precession swerves its plane and changes its path, so that it automatically steers itself as a bicycle is steered by the rider.

Precession is due to the fact that the resultant of the angular velocity of rotation and the angular velocity produced by the torque is an angular velocity about a line that makes an angle with the permanent rotation axis. This angle lies in a plane at right angles to the plane of the couple producing the torque. The permanent axis must turn toward this line, since the body cannot continue to rotate about any line that is not a principal axis of maximum moment of inertia. That is, the permanent axis turns in a direction at right angles to that in which the torque might be expected to turn it. If the rotating body is symmetrical and its motion unconstrained, and if the torque on the spin axis is at right angles to that axis, the axis of precession will be perpendicular to both spin axis and torque axis. Under these circumstances the period of precession is given by

$$T_p = \frac{4\pi^2 I_s}{QT_s}$$

in which I_s is the moment of inertia and T_s the period of spin about the spin axis, and Q is the torque. In general, the problem is more complicated.

PRECESSION (Astronomy).   The slow, rotary motion of the axis of a rotating body in space. The term is, perhaps, most frequently used in connection with the earth, but all rotating bodies may exhibit the effect. The earth is an oblate spheroid with the minor axis as the axis of rotation. Hence, if we subtract from the earth a sphere with radius equal to the minor axis, we shall have left a shell of continually increasing thickness as we pass from the pole to the equator. Such a rotating shell, with the greatest amount of mass in the plane perpendicular to the axis of rotation, is a characteristic gyroscope.

The axis of rotation of the earth is inclined at an angle of 66.5° to the plane of the ecliptic. The forces of gravitational attraction of the sun and the moon tend to pull the equatorial shell into the plane of the ecliptic and hence a torque is applied tending to change the direction of the axis of rotation. This causes the axis to describe a cone in space (i.e., produces precession).

As the axis of rotation describes a cone, the poles of rotation describe circles in space. The radii of these circles, expressed in angular measure on the celestial sphere, are 23.5°. The effect of this motion of the poles is to cause the vernal equinox (one point of intersection of the ecliptic with the equator) to move along the ecliptic. The motion is slow, about 26,000 years being required for the vernal equinox to make a complete circuit of the ecliptic. The motion is not perfectly regular because of the fact that both the sun and the moon are in different planes and are moving relative to each other, causing a variation in the torque applied to the earth. The variations in the torque produce a slight irregularity in the motion of the poles known as nutation.

The motion of the equator among the stars causes slow, and slightly irregular, changes in the equatorial coordinates of the stars. Since celestial longitude is measured from the vernal equinox, the motion of this point produces a change in this coordinate of the stars. The motion of the vernal equinox also changes its location among the constellations along the zodiac, the "sign of Aries" (i.e., the vernal equinox) now being located in the constellation of Pisces instead of in Aries. At present the north pole of rotation of the celestial sphere is close to the star Polaris (α Ursae Minoris); but 12,000 years hence, the star Vega (α Lyrae) will be close to the pole of rotation and will be known as the pole star.

PRECIPITATION AND HYDROMETEORS.   Hydrometeors are any products of condensation or sublimation of atmospheric water vapor, whether formed in the free atmosphere or at the earth's surface; also, any water particles blown by the wind from the earth's surface. Hydrometeors are classified in a number of different ways, among which are the following. (1) Liquid or solid water particles formed, and remaining suspended, in the air. These include damp haze, cloud, fog, ice fog, and mist. (2) Liquid precipitation, including drizzle and rain. (3) Freezing precipitation, including freezing drizzle and freezing rain. (4) Solid (frozen) precipitation, including ice crystals, ice pellets, hail, snow, and the variations thereof. (5) Falling particles that evaporate before reaching the ground; namely, virga. (6) Liquid or solid water particles lifted by the wind from the earth's surface, including drifting and blowing snow, and blowing spray (i.e., spray lifted from the sea surface by the wind and blown about in such quantities that the horizontal visibility is restricted). (7) Liquid or solid water deposits on exposed objects, including dew, hoarfrost, glaze, and rime.

Precipitation, in meteorology, is any or all of the forms of water particles, whether liquid or solid, that fall from the atmosphere and reach the ground. Thus a major class of hydrometeor, as distinguished from clouds, fog, dew, and frost, is that precipitation must "fall" in reaching the ground. *Precipitable water vapor* is the total atmospheric water vapor contained in a vertical column of unit cross-sectional area extending between any two specified levels. It is commonly expressed in terms of the height to which that water substance would stand if completely condensed and collected in a vessel of the same unit cross section. In actual rainstorms, particularly thunderstorms, amounts of rain very often exceed the total precipitable water vapor of the overlying atmosphere. This results from the action of convergence, which brings into the rainstorm the water vapor from a surrounding area that often is quite large. Nevertheless, there is a general correlation between precipitation amounts in given storms and the precipitable water vapor of the air masses involved in those storms.

Condensation in the Atmosphere

Clouds and precipitation are visible evidence that water vapor in the atmosphere condenses into liquid and solid water. Moisture in cloud form may be re-evaporated into the air, but rain, snow, and allied forms of precipitation actively lessen the total water content. This moisture loss by precipitation, considering the world at large, is replaced by equivalent evaporation of moisture into the atmosphere.

Nuclei upon which condensation can take place are absolutely necessary for cloud formation and precipitation. Non-ionized and pollution-free air will become up to 400% supersaturated before condensation occurs. It also is known that clouds sometimes form before air becomes 100% saturated.

Lowering of the temperature of air below its dew point, or adding water vapor beyond the holding capacity of the air, is a second requirement for the formation of clouds. At high temperatures, air is able to hold up to 3% water vapor before becoming saturated; but at low temperatures, this amount may be as low as .01%. When air is cooled, therefore, it slowly loses its capacity to hold water vapor, and that excess vapor condenses into cloud droplets (or ice crystals). Cooling is achieved by one of four methods:

1. By contact with surfaces cooler than the air. This method of cooling is effective in forming fogs, dew, or hoarfrost; but it forms no precipitation except drizzle. If clouds or precipitation are to result, it is necessary that cooling proceed beyond the dew point of the air in question.

2. By lateral mixing of one air parcel with another at a lower temperature. This method is virtually non-effective in forming any extensive clouds, or fogs, and it does not cause precipitation.

3. By vertical mixing of a layer of air whose lapse rate is less than its adiabatic rate. In this method of cooling, the top layer of air is cooled and the base is warmed. The transfer of heat is due to the establishment of an adiabatic lapse rate in the thoroughly mixed layer in contrast to its previously nonadiabatic rate. Cooling at the top of the layer, if there is sufficient water vapor present, will produce a cloud layer (and considerable cloudiness of weather), but very little precipitation.

4. By lifting of whole layers of air or parcels of air, a process that occurs on windward sides of sloping terrain, along frontal surfaces, and in convection currents. Nearly all precipitation is caused by this method of cooling. Lifted parcels or layers of air cool dry-adiabatically while unsaturated, until the temperature and dew point coincide at a level known as the lifting condensation level. This level, at which condensation into cloud droplets or ice crystals occurs, depends upon the temperature and dew point of the lifted air; it is high for dry air and low for nearly saturated air.

Sublimation in the Atmosphere

When saturation is reached in the atmosphere at temperatures less than 0 °C and an additional cooling occurs, however small, and ice particle nuclei are present, water vapor collects directly on the ice nuclei to form an ice crystal cloud. Continued growth causes the ice cloud particles to grow and become precipitable snowflakes. This process is sublimation. Particles upon which ice crystals may grow by this process are known as *sublimation nuclei*.

The *Bergeron-Findeisen theory* offers an explanation of the process by which precipitation particles may form within a mixed cloud (composed of both ice crystals and liquid water drops).

The basis of this theory is the fact that the equilibrium vapor pressure of water vapor with respect to ice is less than that with respect to liquid water at the same sub-freezing temperature. Thus, within an admixture of these particles, and provided that the total water content were sufficiently high, the ice crystals would gain mass by sublimation at the expense of the liquid drops, which would lose mass by evaporation. Upon attaining sufficient weight, the ice crystals would fall as snow and very likely become further modified by accretion, melting, and/or evaporation before reaching the ground.

The most important cloud ingredient specified by this hypothesis is an aggregation of cloud drops that have undergone supercooling. This is a common feature, mainly of cumuliform clouds in temperate latitudes. The current intensive research regarding the introduction of the necessary ice crystals into the cloud has served to increase interest and activity in cloud-seeding methods.

Saturation, Supersaturation, and Supercooling

Saturation. If a free water surface is introduced into a box from which all gases have been removed, molecules of water will emerge from the liquid surface until the number of molecules escaping from the liquid equals the number returning to the liquid. When a balance is maintained between those evaporating or escaping from the liquid water and those condensing or impinging on and remaining in the liquid, the space within the box is said to be saturated with water vapor. It can hold no more water molecules and, if more are added, these will condense into liquid. If air is admitted into the box, there will be no change whatsoever in the rate at which water molecules leave and impinge on the liquid. Component parts of the atmosphere, therefore, have no bearing on the number of molecules of water vapor present. Saturation of space above any liquid water surface depends solely on the number of molecules of water in the air as a vapor when a balance is achieved between the vapor molecules condensing on the liquid and the liquid molecules evaporating into the space above. This molecular balance depends almost entirely on the temperature of the liquid and the water-vapor molecules. The higher the temperature, the more molecules of water vapor can escape from the water surface before they begin to condense and return to the surface.

In meteorological practice, because air is the medium in which weather phenomena take place, it is useful to relate saturation of water vapor to the air carrying it. Saturated air is used, therefore, as a term, even though the presence of the air does not directly affect the number of molecules of water vapor at saturation. It is customary to speak of saturation specific humidity with particular reference to the weight of air carrying the water-vapor molecules. These saturation specific humidities vary from a few tenths of a gram per kilogram of cold air to 30–40 grams per kilogram of very warm air.

Dew point is the temperature to which a given parcel of air must be cooled at constant pressure and constant water vapor content in order for saturation to occur. When this temperature is below 0 °C, it is sometimes called the *frost point* (i.e., the highest temperature at which atmospheric moisture will sublimate in the form of hoarfrost on a cooled polished surface). The dew point may alternatively be defined as the temperature at which the saturation vapor pressure of the parcel is equal to the actual vapor pressure of the contained water vapor. For example, let the temperature of the air be 20 °C and the relative humidity 60%. Then, since the maximum vapor pressure of water at 20 °C is 17.4 millimeters, the actual water vapor pressure is 0.6 of this, or 10.4 millimeters. The temperature at which 10.4 millimeters is the maximum vapor pressure of water is 12 °C. Hence, if the air is cooled to 12 °C, it will reach saturation, and under suitable conditions, dew will form; 12 °C is the dew point. Likewise, if the dew point is known to be 12 °C when the air is 20 °C, it follows that the relative humidity is 60%.

Isobaric heating or cooling of an air parcel does not alter the value of that parcel's dew point, so long as no vapor is added or removed. The dew point of the atmosphere can be determined directly by any of several types of dew point (or frost-point) hygrometers, or by the dew cell, but it is more commonly determined with the aid of the psychrometric calculator or Tables after the direct reading of a psychrometer. See Fig. 1.

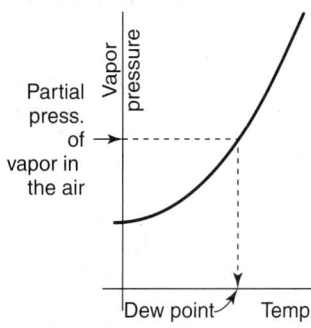

Fig. 1. Graphical location of dew point from vapor pressure curve. Project the known vapor pressure in the air over the curve and down to the temperature axis.

Supersaturation. This exists in a given portion of the atmosphere when the relative humidity is greater than 100%, i.e., when it contains more water vapor than is needed to produce saturation with respect to a plane surface of pure water or pure ice. Such supersaturation develops because, frequently, there is no "plane surface of pure water (or ice)" available. In the absence of water surfaces and in the absence of condensation nucleii or any wettable surfaces, phase change from vapor to liquid cannot occur due to the free energy barrier imposed by the surface free energy of the embryonic droplets, which would then have to form by spontaneous nucleation. Humid air, purified of all foreign nuclei, can be expanded in experimental cloud chambers to relative humidities of the order of 400% without any condensation taking place. Cloud condensation occurs in the atmosphere at relative humidities near 100% only because there is an abundance of condensation nuclei.

Supercooled Liquid in the Atmosphere. Drops of rain and cloud droplets are often cooled to temperatures well below 0 °C and remain in liquid form. They are then said to be supercooled. If such drops and droplets are disturbed, they freeze almost instantly, partially or entirely, depending upon their temperature and size.

The supercoolability of a substance whose solid form is crystalline (as opposed to amorphous matter) stems from the unique energy transformations necessary for the formation of the first crystal nucleus; whereafter, all adjacent liquid immediately becomes solid unless or until the latent heat released elevates the system's temperature sufficiently to arrest the process. It should be noted that the reverse process is not possible; a crystalline solid cannot be "superheated;" therefore, a substance's melting point is very conservative, only slightly dependent upon pressure.

Supercooled clouds are quite common. In extreme cases, they have been observed at temperatures as low as −40 °C. The smaller and more pure the water droplets, the more likely is supercooling. Passage of aircraft through supercooled drops and droplets in the atmosphere results in rapid and severe icing.

Cloud Process

In addition to the processes described here, see also **Clouds and Cloud Formation**.

Accretion. In terms of cloud physics, accretion is the growth of a precipitation particle by the collision of a frozen particle (ice crystal or snowflake) with a supercooled liquid droplet that freezes upon contact.

Agglomeration. The process in which precipitation particles grow by collision with, and assimilation of, cloud particles or other precipitation particles. Regardless of air motions within the cloud, collisions always result from different rates of particle motion due to size differences. The assimilation may result from coalescence when both particles are water droplets or by accretion if one particle is an ice crystal and the other supercooled liquid water. See also **Agglomeration**.

Coagulation. In cloud physics, coagulation is generally used synonymously with accretion. Less frequently, it refers to any process by which a cloud's numerous small cloud drops are converted into a smaller number of precipitation particles. When so used, the term is employed in analogy to the coagulation of any colloidal state.

Coalescence. In cloud physics, coalescence means the merging of two water drops into a single larger drop. This process is believed to be of importance in the production of rain in "warm clouds," i.e., clouds that are warmer than 0 °C throughout, and hence are wholly incapable of supporting an ice-crystal precipitation. Physical details of the coalescence process are still obscure, but the relative velocity of impact, the relative and absolute sizes of colliding drops, electric charge of the drops, and external electric fields must enter the picture. The fraction of all collisions between water drops of a specified size that results in actual merging of the two drops into a single larger drop is called *coalescence efficiency*. It is clear that not every collision of two drops results in coalescence; that under certain conditions, the coalescence efficiency is less than unity.

Orography. Generally, this term means any weather phenomena caused by the flow of air over prominent features of the terrain. *Orographic lifting* is the lifting of an air current caused by its passage up and over mountains, whereby it undergoes certain changes: (1) So long as the air is unsaturated, it cools at the dry-adiabatic rate of approximately 5.5 °F (3 °C) per 1,000 ft. (304.8 meters). (2) As soon as the air is saturated, it cools at the pseudo-adiabatic rate, which depends on the amount of water being condensed by cooling, but is, in any event, less than the dry-adiabatic rate. (3) Clouds, and rain or snow (orographic precipitation) are often formed in this orographically lifted air. Thunderstorms occur if the air is unstable. The Indian monsoon rains and the rains of the United States and European north-west coasts are large-scale orographic phenomena.

Specific Types of Hydrometeors and Precipitation

Cloudburst. In popular terminology, any sudden and heavy fall of rain. An unofficial criterion sometimes used specifies a rate of fall equal to or greater than 100 millimeters (3.94 inches) of rain per hour.

Condensation Trail. These trails are cloudlike streamers called contrails that are observed to form behind an aircraft, particularly jet-engined, flying at high altitudes in very cold air. There are two types of contrails: (1) Exhaust trails are formed when the air in the wake of the aircraft is mixed with and saturated by the water vapor that is a combustion product of the aircraft's engines. Exhaust trails may linger for many miles to the rear of an aircraft, painting its track in the sky. (2) Aerodynamic trails are formed in saturated or nearly saturated air that is cooled in passing over the surfaces of an aircraft and cooled by adiabatic expansion. Aerodynamic trails are more rare than exhaust trails and are of short duration. They are observed most frequently at low altitude levels.

Dew. Classified as a hydrometeor, dew is water condensed onto grass and other objects near the ground, the temperatures of which have fallen below the dew point of the surface air due to radiational cooling during a prior time (usually night), but are still above freezing. Hoarfrost forms if the dew point is below freezing. If the temperature falls below freezing after dew has formed, the frozen dew is known as "white dew."

The conditions favorable to dew formation are: (1) a radiating surface, well-insulated from the heat supply of the soil, on which vapor may condense; and (2) a clear, still atmosphere with low specific humidity in all but the surface layers, to permit sufficient effective terrestrial radiation to cool the surface, and with high relative humidity in the surface air layers, or an adjacent source of moisture, such as a lake.

Dew plays an important role in the propagation of certain plant pathogens, such as late potato blight, which require dew-covered leaves for certain stages of sporulation. Also reliant upon dew is the optical effect known as *heiligenschein*, a diffuse white ring surrounding the shadow cast by the observer's head upon a dew-covered lawn when the solar elevation is low.

Drizzle. This is a small falling droplet of water whose dimensions across are usually less than 0.5 millimeter. Drizzle droplets are buffeted by gusts of wind whereas rain tends to travel in a more or less straight line, although the path may be slanted. Drizzle falls predominantly from low stratus clouds and is a phenomenon associated with vertical atmospheric stability in the low levels.

Fog. See also **Fog and Fog Clearing**.

Fog Tracks. Linear regions of condensation, produced in air or other gases that are supersaturated with water vapor, by the passage of electrified particles. Fog tracks are useful in following the courses and collisions of such particles, as in an experimental cloud chamber.

Frost. This is the condition that exists when the temperature of the earth's surface and earthbound objects falls below freezing (0 °C or 32 °F). Depending upon the actual values of ambient-air temperature, dew point, and the temperature attained by surface objects, frost may occur in a variety of forms, including a freeze and hoarfrost. If a frost period is sufficiently severe to end the growing season or delay its beginning, it is commonly referred to as a "killing frost."

Frost, like snow, is the result of the sublimation of water vapor in saturated air. If there is excessive radiation from solid objects, as on a clear night in late fall, the air coming in contact with these objects may be chilled below the sublimation point, and spicules of ice may grow out from the cold surfaces. The process is entirely similar to the artificial formation of metallic crystals, from vaporized metals, on the interior of a glass tube communicating with the vaporizing furnace. In either case, the size of the crystals is a matter of time and the supply of saturated vapor. Frost is often observed around cracks in a wooden sidewalk, because of the damp air escaping from the ground below. The objects upon which frost forms most readily are those of low specific heat and high thermal emissivity, such as blackened metals; hence, the marked accumulation of frost on the heads of rusty nails. The apparently erratic occurrence of frost in adjacent localities is due partly to differences of level, the lower areas becoming colder; but also largely to differences in absorptivity and specific heat of the ground, which, in the absence of wind, greatly influences the temperature attained by the superincumbent air. It should be understood that vegetation is not damaged by frost itself, but by cold air; the appearance of frost merely indicates that the temperature has dropped below the freezing point. The formation of white frost on the indoor surface of window panes indicates low relative humidity of the indoor air; otherwise water would first condense in small drops and then freeze into clear ice.

Hail. Classified as a hydrometeor, hail is precipitation in the form of balls or irregular lumps of ice, always produced by convective clouds, nearly always cumulonimbus. Single units of hail, called hailstones, range in size from that of a pea to that of a grapefruit (i.e., from less than $\frac{1}{4}$ inch (6 millimeters) to more than 5 inches (13 centimeters) in diameter). The largest hailstone observed in the United States is believed to be one that fell at Potter, Nebraska, on July 6, 1928; it measured 17 inches (43.2 centimeters) in circumference and weighed about 1.5 pounds (0.7 kilograms).

Hailstones may be spheroidal, conical, or generally irregular in shape, the spheroidal form being the most common. When broken, the stones reveal a structure of concentric alternate layers of clear and opaque white ice. This characteristic layered inner structure has been explained in terms of the "multiple incursion theory," which pictures cyclical ascent and descent of the hailstone into alternate above- and below-freezing regions of the cloud. Thus, thunderstorms, which are characterized by strong updrafts, large liquid water content, large cloud-drop sizes, and great vertical height, are favorable to hail formations. It has been suggested that descent alone from great heights through an updraft exhibiting water-content stratification is probably sufficient to yield the foliated structure. In either case, the hailstones grow, basically, by accretion of supercooled water drops upon the growing ice particles; and the nature of the ice depends upon such things as the rate of accretion, drop size, and temperature.

The destructive effects of hail storms upon plant and animal life, buildings and property, and aircraft in flight render them a prime object of weather modification studies.

Haze. Classified as a hydrometeor, haze is a fine dust or salt particles dispersed through a portion of the atmosphere. The particles are so small that they cannot be felt or individually seen with the naked eye, but they diminish horizontal visibility and give the atmosphere a characteristic opalescent appearance that subdues all colors.

Many haze formations are caused by the presence of an abundance of condensation nuclei, which may grow in size and become mist, fog, or cloud. Distinction is sometimes drawn between "dry haze" and "damp haze," largely on the basis of differences in optical effects produced by the smaller particles (dry haze) and the larger particles (damp haze). Dry haze particles, with diameters of the order of 0.1 micrometer, are small enough to scatter shorter wavelengths of light preferentially, though not according to the inverse fourth-power law of Rayleigh scattering. Such haze particles produce a bluish color when the haze is viewed against a dark background, for dispersion allows only the slightly bluish scattered light to reach the eye. The same type of haze, when viewed against a light background, appears as a yellowish veil, for here the principal effect is the removal of the bluer components from the light originating in the distant light-colored background. Haze may be distinguished by this same effect from mist, which yields only a gray obscuration, since, in the mist, the particle sizes are too large to yield appreciable differential scattering of various wavelengths.

Any small liquid droplet contributing to an atmospheric haze condition is known as a "haze droplet." In certain industrial areas, such droplets may be entirely non-aqueous (largely hydrocarbons), but there is little doubt that most haze droplets are water solutions of some type. Near seacoasts, droplets of sea-salt solutions are responsible for haze; and even far inland, some haze conditions have been shown to be due to salt-solution droplets probably produced by sea-salts. A combination of smoke and haze, or a very light smoke condition resembling haze is known as "smaze."

Arctic haze is a condition of reduced horizontal and slant visibility (but unimpeded vertical visibility) encountered by aircraft in flight (to above 30,000 feet; 9144 meters) over arctic regions. When viewed away from the sun, it appears grayish-blue; into the sun, it appears reddish-brown. It has no distinct upper and lower boundaries, and produces none of the optical phenomena that would be expected if it were composed of ice crystals. Color effects suggest particle sizes of two microns or less. See also **Polar Research**.

Ice-crystal haze is a type of very light ice fog, usually associated with the precipitation of ice crystals, and observable at times to altitudes as great as 20,000 feet. Observed from the ground, it may be dense enough to hinder observation of celestial bodies, sometimes even the sun. Looking down from the air, however, the ground is usually visible and the horizon only blurred.

Hoarfrost. Classified as a hydrometeor, hoarfrost is a deposit of interlocking ice crystals (hoar crystals) formed by direct sublimation on objects, usually those of small diameter freely exposed to the air, such as tree branches, plant stems and leaf edges, wires, poles, etc. Also, frost may form on the skin of an aircraft when a cold aircraft flies into air that is warm and moist, or when it passes through air that is supersaturated with water vapor. The deposition of hoarfrost (also known as "white frost") is similar to the process by which dew is formed, except that the temperature of the befrosted object must be below freezing. It forms when air with a dew point below freezing is brought to saturation by cooling. In addition to its formation on freely exposed objects (air hoar), hoarfrost also forms inside unheated buildings and vehicles, in caves, in crevasses (crevasse hoar), on snow surfaces (surface hoar) and in air spaces within snow, especially below a snow crust (depth hoar).

Hoarfrost may be distinguished from two other ice deposits, rime and glaze, by the following characteristics: *Rime* is a white or milky and opaque granular deposit of ice, which is denser and harder than hoarfrost. *Glaze*, is a generally clear and smooth coating of ice, denser, harder, and more transparent than either rime or hoarfrost. Both rime and glaze are formed when supercooled water drops strike an object at a temperature below freezing. Such formation on terrestrial objects is known as an "ice storm," on aircraft, as "aircraft icing." See also **Aircraft Icing**. When rime forms on ice particles in the atmosphere, snow pellets result; when glaze forms on these ice particles, the result is hail.

Ice Crystal. A type of precipitation composed of slowly falling, very small, unbranched crystals of ice, which often seem to float in the air; it is popularly referred to by such names as "frost snow," "ice needles," or "diamond dust." It may fall from a cloud or from a cloudless sky; it is visible only in direct sunlight, or in an artificial light beam, and does not appreciably reduce visibility. Ice crystal precipitation often produces sun pillar and other halo phenomena.

Ice Pellets. A type of precipitation consisting of transparent or translucent pellets of ice, 5 millimeters or less in diameter. They may be spherical, irregular, or (rarely) conical in shape. Ice pellets usually bounce when hitting hard ground, and make a sound upon impact. They include two basically different types of precipitation.

1. Transparent, globular, solid grains of ice, which have formed from the freezing of raindrops or the refreezing of largely melted snowflakes when falling through a below-freezing layer of air near the earth's surface. When raindrops enter a layer of intensely cold air, they become supercooled, i.e., cooled below the freezing point, but without freezing. In this state, they are highly unstable, and upon coming in contact with any object, even with a speck of dust, they suddenly freeze into pellets of sleet. Larger objects, such as twigs or telephone wires, when touched by these supercooled drops, receive a coating of ice, which may result in a landscape of marvelous beauty when the sun appears, but which often proves very destructive because of the heavy weight of the ice. Sleet is quite commonly mixed with snow or with rain, but it is always a cold-weather product, and is not as sometimes believed, a small form of hail.

2. Translucent particles, consisting of snow pellets encased in a thin layer of ice. The ice layer may form either by the accretion of droplets upon the snow pellet, or by the melting and refreezing of the surface of the snow pellet.

Mist. According to international definition, mist is a hydrometeor consisting of an aggregate of microscopic and more-or-less hygroscopic water droplets suspended in the atmosphere. It produces, generally, a thin grayish veil over the landscape, but reduces visibility to a lesser extent than does fog. The humidity with mist is often less than 95%.

Rain. The most common type of atmospheric precipitation, in the form of liquid water drops, ranging in diameter from 0.5 millimeter to approximately 5.0 millimeters (or larger), and usually falling with a velocity ranging from 5 to 8 meters per second. A typical rain drop might have a diameter of 1–2 millimeters, while the largest drops (observed in heavy thunderstorms, and flattened on their undersides as a result of aerodynamic effects), may have equivalent spherical diameters of 5–8 millimeters. Drops larger than this become quite unstable and break up as a result of microturbulence in the air through which they fall.

Rain is to be distinguished from the only other form of liquid precipitation, *drizzle*, in that drizzle drops are generally less than 0.5 millimeter in diameter, are very much more numerous, and reduce visibility much more than does light rain. Also, unlike rain, drizzle drops may appear to float while following air currents.

Shower. Precipitation characterized by the suddenness with which it starts and stops, by rapid changes of intensity, and, usually, by rapid changes in the appearance of the sky. In weather observing practice, showers are always reported in terms of the basic type of precipitation that is falling, i.e., rain showers, snow showers, sleet showers, and so on.

Air-mass showers are produced by local convection within an unstable air mass. They are most frequently within a moist air mass that is sufficiently unstable so that daytime heating at the surface can produce well-developed cumulus clouds. They are not associated with a front or instability line.

Silver Thaw. After a period of cold weather and below-freezing temperatures, a mass of warm air passing over the region will cause frost or glaze to form on objects that are still at a low temperature. This condition is known as a silver thaw. A silver thaw usually lasts only a few hours, because the warm air soon warms all exposed objects above the freezing point.

Sleet. See "Ice Pellets" previously described in this entry.

Snow. Classified as a hydrometeor, snow is precipitation composed of white or translucent ice crystals, chiefly in complex branched hexagonal form and often agglomerated into snowflakes. Snow appears white only because of the multitude of reflecting surfaces. The individual crystals are of transparent ice.

Single snow crystals, when freshly formed, are often almost perfect and exhibit an endless variety of detail. They are commonly flat, six-sided polygons, stars, or spangles, often of very complicated and beautiful design, but always with the 60° and 120° angles characteristic of the hexagonal system. Sometimes they are needle-like with a hexagonal head, like a pin. The finer spicules are sometimes suspended high in the atmosphere, and are the cause of halos. Snow differs from frost chiefly in being formed in the air instead of upon solid objects near the ground, the crystallization nuclei being particles of dust.

Partly melted crystals often cling together to form snowflakes of varying size, and may melt into raindrops before reaching the ground. Snowflakes made up of clusters of crystals or crystal fragments may grow as large as three to four inches in diameter, frequently building themselves into hollow cones falling point downward. In extremely still air, flakes with diameters as large as ten inches have been reported.

Snow may also fall in the form of snow grains and snow pellets. *Snow grains*, as a form of precipitation, is the solid equivalent of drizzle, being very small (less than 1 millimeter) opaque particles of ice. Unlike snow pellets, they neither shatter nor bounce when they hit a hard surface, and usually fall in very small quantities. *Snow pellets* (sometimes called "soft hail" or "tapioca snow") is a form of precipitation consisting of white, opaque, approximately round ice particles, having a snow-like structure, and measuring about 2 to 5 millimeters in diameter. Snow pellets are crisp and easily crushed, rebounding when they fall on a hard surface, and often breaking up. They most often fall with the suddenness and changes of intensity characteristic of showers, and often together with or before snow.

In the United States, a heavy snowfall is commonly followed by intense cold, partly because of the low absorptivity of snow for solar radiation, and partly because of the cyclonic character of the snowstorm, which brings a change of wind to the north on the westward or following side of the storm.

Virga. A hydrometeor in the form of wisps or streaks of water or ice particles falling out of a cloud, but which evaporates before reaching the earth's surface as precipitation. Virga is frequently seen trailing from altocumulus and altostratus clouds, but also is discernible below the bases of high-level cumuliform clouds from which precipitation is falling into a dry subcloud layer. Virga typically exhibits a hooked form in which the streaks descend nearly vertically just under the precipitation source, but appear to be almost horizontal at their lower extremities.

Measurement of Hydrometeors and Precipitation Phenomena

In weather observation activities and meteorological forecasting, several variables concerning hydrometeors, precipitation, and associated cloud formation and other processes are very important. Several of these instruments are described as follows:

Atmometer. This is the general name for an instrument that measures the evaporation of water into the atmosphere. Several types may be mentioned: (1) The *evaporation pan* is a cylindrical container made of galvanized iron or monel metal, 10 inches deep and 48 inches in diameter. It is filled with water to a depth of 8 inches, and periodic measurements are made of the changes of the water level. (2) The *clay atmometer* is a porous porcelain container connected to a calibrated reservoir filled with distilled water. Evaporation is determined by the depletion of water in the reservoir. (3) The *Piché evaporimeter* consists of a graduated tube, closed at one end, which is filled with distilled water and then covered with a larger circular piece of filter paper held in place by a disc-and-collar arrangement. In operation, the instrument is inverted so that the distilled water is in contact with the filter paper. Evaporation is determined by noting the change in level of the meniscus of the water. (4) The *radio atmometer* is designed to measure the effect of sunlight upon evaporation from plant foliage. It consists of a porous clay atmometer whose surface has been blackened so that it absorbs radiant energy.

Cloud-Detecting Radar. A type of weather radar designed specifically for the detection of clouds rather than precipitation. It is capable of detecting clouds in multiple layers up to heights of about 50,000 feet above the radar.

Cloud-Height Indicator. A general term for instruments that measure the height of cloud bases (i.e., the lowest levels of the atmosphere at which the water or ice particles comprising a cloud appear). Cloud-height indicators may be classified according to their principle of operation:

1. Height determination by the principle of triangulation is represented by two instruments: (a) The *ceilometer* is an automatic, recording, cloud-height indicator. The photoelectric cell pickup detector of one type, the fixed-beam ceilometer, scans continuously to detect in a cloud an illuminated spot directed by a stationary projector. In the other type, the rotating-beam ceilometer, the projector rotates rapidly through 360° while the detector is fixed vertically. (b) The *ceiling light* projects a narrow beam of light onto a cloud base, the height of which is then determined by means of a clinometer.
2. Height determination by the principles of pulse techniques is represented by the *pulsed-light cloud-height indicator* and the vertically-directed *cloud-detection radar*. With these instruments, the time required for a pulse of energy to travel from a radiator located on the ground to the cloud base and back to the ground again is measured electrically. The height of the cloud is computed from this transit time and a knowledge of the propagation velocity of the pulse.

In addition to these instruments, a small balloon known as a ceiling balloon is used to determine the height of a cloud base. The height can be computed from the ascent velocity of the balloon and the time required for its disappearance into the cloud.

Dew Cell. An instrument used to determine the dew point. It consists of a pair of spaced bare electrical wires wound spirally around an insulator and covered with a wicking wetted with a water solution containing an excess of lithium chloride. An electric potential applied to the wires causes a flow of current through the lithium chloride solution, which raises the temperature of the solution until its vapor pressure is in equilibrium with that of the ambient air. A modification of the dew cell, the hygristor, is used for upper-air measurements.

Drosometer. An instrument for measuring the amount of dew formed on a given surface. One type consists of a hemispherical glass vacuum cup exposed to the atmosphere. Dew forming on the glass surface automatically collects in the bottom of the cup, which is weighed at the end of the exposed period. Another type consists of a block of wood whose surface has been treated so that dew forms on it in characteristic patterns. Photographs are supplied with each instrument to enable the observer to match the dew formation with a set of standards corresponding to a dew "fall" of from 0.01 to 0.45 millimeter.

Evapotranspirometer. An instrument for measuring the rate of evapotranspiration (i.e., the combined processes of evaporation of liquid or solid water plus transpiration from plants). It consists of a vegetation soil tank so designed that all water added to the tank and all water left after evapotranspiration can be measured.

Hygrothermograph. A recording instrument combining, on one record, the variation of atmospheric temperature and humidity content as a function of time. See Fig. 2. The most common hygrothermograph is a hair hygrograph (a recording hair hygrometer) combined with a thermograph.

Nephometer. A general term for instruments designed to measure the amount of cloudiness. An early type consisted of a convex hemispherical

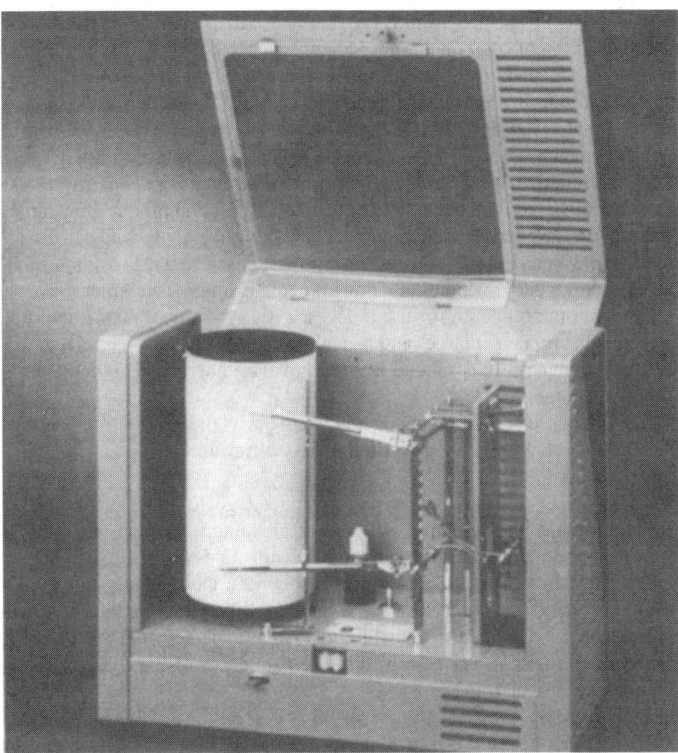

Fig. 2. Hygrothermography for measuring both temperature and humidity. A time record is linked on the slowly rotating drum. Digital instrumentation also is available.

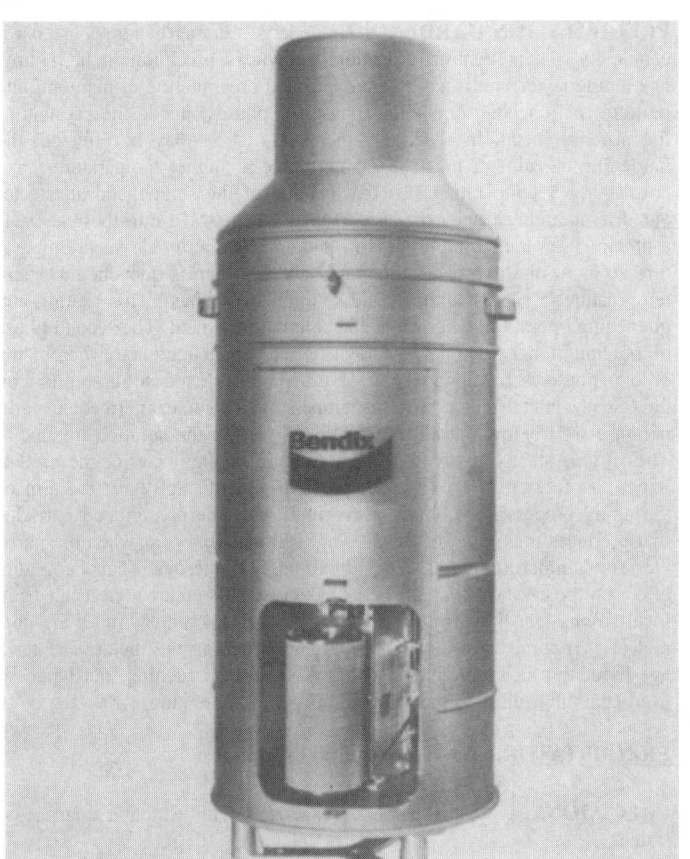

Fig. 3. Universal recording precipitation gage.

mirror mapped into six parts. The amount of cloud coverage on the mirror was manually noted by the observer.

Nephoscope. An instrument for determining the direction of cloud motion. With the direct-vision nephoscope, the observer notes the motion of the cloud by looking directly at it through the instrument, and aligning a grid so that the cloud appears to move parallel to its major axis. With the mirror nephoscope, the cloud motion is indicated by the azimuth at which the cloud image, seen in a mirror, leaves the mirror.

Rain Gage. These gages may be classified according to their principle of operation: (a) A recording rain gage automatically records the amount of precipitation collected, as a function of time. One of these recording instruments, frequently used at climatological stations, consists of a receiver in the shape of a funnel, which empties into a bucket mounted on a weighing mechanism. See Fig. 3. The weight of the catch is recorded on a clock-driven chart as inches of precipitation. (b) A nonrecording rain gage indicates, but does not record, the amount of precipitation captured. The nonrecording rain gage of the U.S. Weather Bureau consists of a receiver in the form of a funnel (8 inches; 20.3 centimeters in diameter), an overflow can upon which the receiver fits, a measuring tube into which the small end of the funnel fits, and a measuring stick graduated in proportion to the difference in area between the receiving piece and the measuring tube. The ratio of the area of the receiver to that of the measuring tube is such that tenfold magnification is obtained for ease in determining small amounts. (c) The *rain-intensity gage* measures the instantaneous rate at which rain is falling on a given surface. In the Hudson-Jardi design, water from the rain collector enters a chamber containing a float and outlet valve, and shaped in such a fashion that the height of the float is directly proportional to the rate of rainfall. The motion of the float is recorded either electrically or mechanically.

The term *ombrometer* refers in general to any rain gage, but specifically to a rain gage capable of measuring very small amounts of precipitation.

Snow Gage. These instruments are designed to measure the vertical depth of snow. (a) The *snow sampler* is a hollow metal tube in which snow is collected, melted, and weighed. (b) The *snow stake* is a wood scale, calibrated in inches and used in regions of deep snow to measure its depth. It is bolted to a wood post or angle iron set in the ground. (c) The *radioactive snow-gage* automatically and continuously records the water equivalent of snow on a given surface, as a function of time. A small sample of a radioactive salt is placed in the ground in a lead-shielded collimator that directs a beam of radioactive particles vertically upwards. A Geiger-Müller counting system (located above the snow level) measures the amount of depletion of radiation caused by the presence of the snow. (d) The *snow bin* is simply a box in which snowfall is collected and measured. (e) The *snow board* is a sheet of thin white board, about 16 inches (40.6 centimeters) square, with a layer of cotton flannel tacked to its upper surface. This surface retains falling snow better than metal surfaces for snowfall measurement purposes. (f) The *snow mat* is a special device used to mark the surface between old and new snow. It consists of a piece of white duck 28 inches (71.1 centimeters) square, having in each corner triangular pockets in which are inserted slats placed diagonally to keep the mat taut and flat.

See the section on "Acid Rain" in the entry on **Pollution (Air)**. See also **Atmosphere (Earth)**; **Clouds and Cloud Formation**; **Fog and Fog Clearing**; **Fronts and Storms**; and other entries listed under **Meteorology**.

Additional Reading

Cotton, W.R. and R.A. Anthes: "Storm and Cloud Dynamics," Academic Press, Inc., San Diego, CA, 1997.

Collier, C.G.: "Applications of Weather Radar Systems: A Guide to Uses of Radar Data in Meteorology and Hydrology," 2nd Edition, John Wiley & Sons, Inc., New York, NY, 1996.

Cotton, W.R. and R.A. Anthes: "Storm and Cloud Dynamics," Academic Press, Inc., San Diego, CA, 1992.

Garfield, J. and O. Sohnel: "Precipitation: Basic Principles and Industrial Applications," Butterworth-Heinemann, Inc., Woburn, MA, 1993.

Jaenicke, R.: "Dynamics and Chemistry of Hydrometeors: Final Report of the Collaborative Research Centre 233 "Dynamik Und Chemie Der Meteore". Collaborative Research Centres," John Wiley & Sons, Inc., New York, NY, 2001.

Sumner, G.N.: "Precipitation Process and Analysis," John Wiley & Sons, Inc., New York, NY, 1988.

Upgren, A. and J. Stock: "Weather: How It Works and why It Matters," Perseus Publishing, Boulder, CO, 2000.

Williams, J.: "The Weather Book," 2nd Edition, Random House, Inc., New York, NY, 1997.

PETER E. KRAGHT, Certified Consulting Meteorologist, Mabank, TX.

PRECIPITATION HARDENING.   A large number of alloys are hardenable by a heat treating procedure known as precipitation hardening. Hardening is accomplished by the controlled precipitation of many minute particles of a second crystalline phase (or phases) inside the crystals of the primary metal. In order that the precipitation may be effected, the hardening constituent must be more soluble at higher temperatures than it is at lower temperatures, so that heating of the solid metal at an elevated temperature causes the second phase to dissolve into the matrix. If a precipitation hardening alloy is heated and held at an elevated temperature so as to dissolve the hardening phase and then is quenched to room temperature, a supersaturated solid solution is obtained. This heating and quenching operation is known as the solution treatment. The second phase of precipitation hardening is known as the aging treatment wherein the second phase is precipitated out of the supersaturated solid solution by holding the metal either at room temperature or some intermediate temperature well below the temperature employed in the solution treatment. The various stages involved in the formation of the nuclei of the precipitation particles may be very complex. In general, however, the aim of the aging process is to obtain a distribution of the precipitated particles that produces maximum hardness. This will usually occur when the particles are submicroscopic in size and extremely numerous. Their hardening effect on the crystal lattice of the matrix crystals is believed to result from local strains that they produce in the matrix. These latter hinder the normal easy motion of dislocations, thereby hardening the metal. The term age hardening is synonymous with precipitation hardening, but when so used generally refers to metals aged at room temperature.

PRECIPITATOR.   See **Electrostatic Precipitator**.

PRECISION.   1.   The degree of exactness with which a quantity is stated.

2.   The degree of discrimination or amount of detail; e.g., a 3-decimal digit quantity discriminates among 1,000 possible quantities. A result may have more precision than it has accuracy; e.g., the true value of pi to 6 significant digits is 3.14159; the value 3.14162 is precise to 6 Figures, given to 6 Figures, but is accurate only to about 5.

Double precision. The retention of twice as many digits of a quantity as a given computer normally handles; e.g., if a computer, whose basic word consists of 10 decimal digits is called upon to handle 20 decimal digit quantities, then double precision arithmetic must be resorted to.

Triple precision. The retention of three times as many digits of a quantity as the computer normally handles; e.g., a computer whose basic word consists of 10 decimal digits is called upon to handle 30 decimal digit quantities.

PRECURSOR.   In biological systems, an intermediate compound or molecular complex present in a living organism which, when activated physiochemically, is converted to a specific functional substance. Sometimes the prefix "pro" is used to indicate that a compound in question plays the role of a precursor. Examples from the history of vitamin and other essential chemical developments include: ergosterol (pro-vitamin D2), which is activated by ultraviolet radiation to form vitamin D; carotene (pro-vitamin A) is a precursor of vitamin A; prothrombin forms thrombin upon activation in the blood-clotting mechanism.

PREDICTION (Statistics).   In general, prediction is the process of forecasting the magnitude of statistical variates at some future point of time. In statistical contexts the word may also occur in slightly different meanings; e.g., in a regression equation expressing a dependent variate y in terms of dependents x's, the value given for y by specified values of x's is called the "predicted" value even when no temporal element is involved.

PREGNANCY (or Cyesis; Gestation).   The condition of being with child. The duration of pregnancy in humans is usually about 280 days, 9 calendar or 10 lunar months, dating from the time of the last menstrual period. Presumptive signs of pregnancy are absence of the menstrual periods, nausea or vomiting in the morning (morning sickness), enlargement of the breasts with pigmentation of the nipples, and enlargement of the abdomen during the last half of pregnancy. Absolute signs of pregnancy are palpation of the fetal body, movement of the child, and sound of the fetal heart.

Diagnosis of pregnancy in early stages usually depends upon detecting the presence of certain hormones in the urine which will produce characteristic changes in animals into which it is injected.

An extrauterine or ectopic pregnancy is one in which the fertilized ovum lodges and develops outside the uterine cavity. This usually takes place somewhere along the fallopian tubes and, more rarely, free in the abdominal cavity. Such ectopic gestations almost always terminate spontaneously early in the course of the pregnancy, and produce the clinical picture of ruptured ectopic pregnancy. Abdominal pain and shock due to massive hemorrhage into the abdominal cavity make this condition a serious emergency, which demands immediate surgical treatment.

A phantom pregnancy, or pseudocyesis, is an hysterical manifestation in which the abdomen enlarges and resembles the enlargement associated with a true pregnancy. It is treated by treating the underlying psychoneurosis with psychotherapy.

The physiology of pregnancy is discussed in the entries on **Embryo**; and **Embryology**. See also **Artificial Insemination**.

PREHENSION.   The flexion of an appendage to grasp an object by folding around it. There are two types of grasping organs among animals: forcipate and prehensile. The human hand illustrates both. Objects may be taken between the thumb and fingers as between the jaws of a forceps, or they may be grasped between the fingers and the palm by the prehensile folding of the digits. Less versatile organs are capable of one or the other type of action, as in the case of the forcipate chela of the lobster and the prehensile tails of some monkeys.

PREHNITE.   Prehnite is a hydrous silicate of calcium and aluminum, $Ca_2Al_2Si_3O_{10}(OH)_2$, crystallizing in the orthorhombic system. Usual occurrence as intergrown crystals of reniform, stalactitic character, and as rounded groups of such crystals; hardness, 6–6.5; specific gravity 2.90–2.95; luster, vitreous to pearly; color, various shades of light green to gray or white; translucent. Though not a zeolite it is found associated with them and with datolite and calcite, in veins and cavities of basic rocks, sometimes in granites, syenites, or gneisses. It is found in Austria, Italy, the Harz Mountains, France, Scotland, and the Republic of South Africa, where it was originally discovered. Magnificent crystal casts after an unknown mineral have been found in a single large cavity in the basaltic rocks near Bombay, India. In the United States well-known localities are Somerville, Massachusetts; Farmington, Connecticut; Paterson, New Jersey; and Keweenas County, Michigan. Named for Colonel Prehn, its discoverer, who was an early Dutch Governor of the Cape of Good Hope colony.

PRESBYOPIA.   Presbyopia is the condition that exists when the natural crystalline lens of the eye loses some of its ability to change shape in order to focus on near objects. Sometime after the age of 40, most people eventually develop presbyopia, usually signaled by a need for reading glasses or bifocal lenses.

The eye functions much like a camera with two lenses. The first lens is the cornea, a clear membrane that covers the front of the eye. The second lens is the eye's natural crystalline lens, which is located behind the pupil. The cornea is responsible for about 70% of the eye's focusing power, while the natural lens "fine-tunes" the image before it is focused on the retina at the back of the eye. The natural lens accomplishes this fine-tuning function by changing shape to accommodate both near objects and those that are further away. Muscles called ciliary muscles are attached to the lens and are responsible for its ability to change shape.

People who are nearsighted may not need corrective lenses for reading as soon as other people do, because their eyes naturally focus more easily on objects that are close. Even when the crystalline lens in a nearsighted eye loses some of its flexibility, the flatter cornea compensates and may continue to offer sharp close-up vision.

Because presbyopia affects near vision, the usual correction is with reading glasses or with bifocals for those who also require distance vision correction. Several innovations in bifocal eyeglasses and contact lenses have occurred in recent years that allow presbyopes (those people with presbyopia) to see better, look better, and be more comfortable. Progressive or "no-line" multifocal eyeglasses graduate from distance to reading power without the noticeable lines that exist in standard bifocals or trifocals. Although progressive lenses may require a greater period of adjustment, these are the most versatile of all multifocal designs because of the continuous range of focus Several versions of multifocal contact lenses are also now available that provide near, intermediate, and distance vision in one contact lens. See also **Vision and the Eye**.

Vision Rx, Inc., Elmsford, NY.

PRESSURE. If a body of fluid is at rest, the forces are in equilibrium or the fluid is in static equilibrium. The types of force that may act on a body are shear or tangential force, tensile force, and compressive force. Fluids move continuously under the action of shear or tangential forces. Thus, a fluid at rest is free in each part from shear forces; one fluid layer does not slide relative to an adjacent layer. Fluids can be subjected to a compressive stress, which is commonly called *pressure*. The term may be defined as force per unit area. The pressure units may be dynes per square centimeter, pounds per square foot, torr, mega-Pascals, etc. Atmospheric pressure is the force acting upon a unit area due to the weight of the atmosphere. Gage pressure is the difference between the pressure of the fluid measured (at some point) and atmospheric pressure. Absolute pressure, which can be measured by a mercury barometer, is the sum of gage pressure plus atmospheric pressure.

Pascal's law states that the pressure in a static fluid is the same in all directions. This condition is different from that for a stressed solid in static equilibrium. In such a solid, the stress on a plane depends upon the orientation of that plane. A liquid in contact with the atmosphere is sometimes called a free surface. A static liquid has a horizontal free surface if gravity is the only type of force acting.

Imagine a body of static fluid in a gravitational field. The mass of the fluid is m (in grams) and the weight of the fluid is mg (as dynes) where g is the local gravitational acceleration. Figure 1 shows a large region of any static fluid with a very small or infinitesimal element. Figure 2 indicates the element in detail. The vertical distance z is measured positively in the direction of decreasing pressure (up); dA is an infinitesimal area; p is the pressure acting on the top surface; and $(p + dp)$ is the pressure acting on the bottom surface. The pressure difference is due only to the weight of the fluid element. Let r represent density, which is mass per unit volume (as grams per cubic centimeter). Thus the weight of the element is $\rho g dz dA$.

Considering the element as a free body, an accounting of forces in the vertical direction gives:

$$dp \, dA = -\rho g \, dz \, dA; \quad dp = -\rho g \, dz \tag{1}$$

As z is measured positively upward, the minus sign indicates that the pressure increases with an increase in height. This fundamental equation of fluid statics can be applied to all fluids. In integral form, Equation (1) becomes:

$$\int_1^2 \frac{dp}{g} = \int_1^2 dz = -(z_2 - z_1) \tag{2}$$

where 1 refers to one level and 2 refers to another level. The functional relation between pressure p and the combination ρg must be established before Equation (2) can be integrated. There are two major cases: (a) incompressible fluids, in which the density ρ is a constant; and (b) compressible fluids, in which the density ρ varies.

Liquids can be considered as incompressible in many cases. For small differences in height, a gas might be regarded as incompressible. For an incompressible fluid, with constant g, Equation (2) becomes:

$$p_2 - p_1 = -\rho g \, (z_2 - z_1) \tag{3}$$

The term $(z_2 - z_1)$ may be called a static "pressure head," and it can be expressed in feet or inches of water, or some height of any liquid. For example, barometric pressure can be expressed in inches of mercury.

A manometer is a device that measures a static pressure by balancing the pressure with a column of liquid in static equilibrium. Many types of manometers are used. See also **Manometer**. The common mercury barometer is essentially a manometer for measuring atmospheric pressure; a mercury column in a glass tube balances the weight of the air above the mercury. Figure 3 illustrates a manometer in which the left leg is open to the atmosphere; the liquid has a specific weight (weight per unit volume) $\rho_2 g$. In the other leg is a liquid of specific weight $\rho_1 g$. Starting with the left leg, the gage pressure p_A is:

$$p_A = h_2 \rho_2 g$$

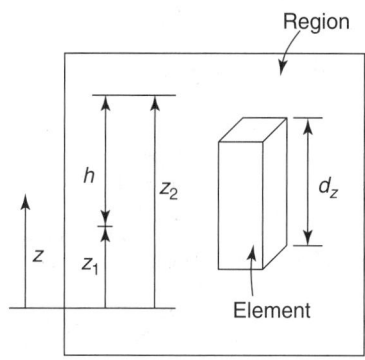

Fig. 1.   Large region of any static fluid.

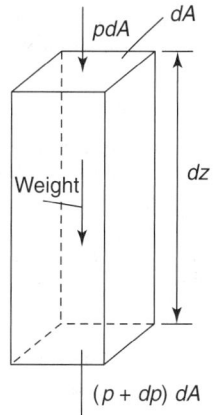

Fig. 2.   Vertical forces on infinitesimal element.

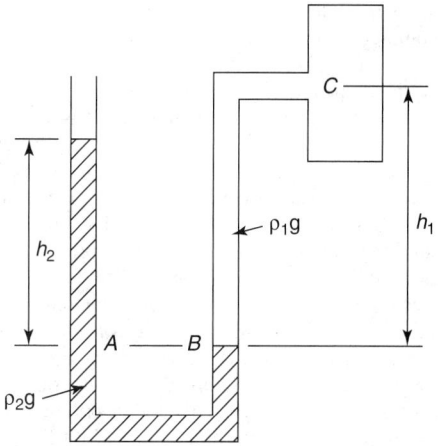

Fig. 3.   Manometer.

Since the fluid is in static equilibrium, the pressure p_B at point B equals the pressure at point A. Thus:

$$p_A = p_B = h_2 \rho_2 g$$

The pressure p_C at point C is less than that at B. Thus:

$$p_B - p_C = h_1 \rho_1 g$$

Then the gage pressure at point C is:

$$p_C = g(h_2 \rho_2 - h_1 \rho_1)$$

When a body of any kind is partly or fully immersed in a static fluid, every part of the body surface in contact with the fluid is pressed on by the fluid. The pressure is greater on the areas more deeply immersed. The resultant of all these fluid pressure forces is an upward or buoyant force. The pressure on each part of the body is independent of the body material.

Archimedes' principle states that the buoyant force equals the weight of the displaced fluid.

Equation (3) is for the special case of an incompressible fluid. As an example of a compressible fluid, consider an isothermal or constant-temperature layer of gas. The equation of state for such a gas can be written:

$$p = \rho R T_1 \tag{4}$$

where T_1 is the given absolute temperature and R is a gas constant or gas factor depending upon the gas. Assuming a constant g, Equation (2) gives:

$$\frac{RT_1}{g} \int_1^2 \frac{dp}{p} = -(z_2 - z_1)$$

$$z_2 - z_1 = \frac{RT_1}{g} \log_e \frac{p_1}{p_2} \tag{5}$$

Equation (5) is sometimes called a "barometric height" relation. For an isothermal atmosphere, a measurement of the temperature T_1 and the static pressure (as with a barometer) at two different levels will provide data for the calculation of the height difference.

Other pressure designations include:

Vacuum. A gage pressure below atmospheric.

Hydrostatic Pressure. The pressure at a point below a liquid surface due to the height of fluid above it.

Tons-on-Ram. The force that acts over a given area as in various types of hydraulic machinery.

Partial Pressure. The pressure exerted by one component in a system, usually one gas or vapor in a mixture.

Internal Pressure. The effect of the attractive forces of the molecules of a substance, which is called pressure because its result is the same as that of an added external pressure. In liquids, its effect appears as the ability of liquids to stand substantial negative pressures without rupture.

Cohesion Pressure. A term in Van der Waal's equation introduced to take care of the effect of molecular attraction. It is usually expressed as a/V^2, where a is a constant and V is the volume of the gas.

Pressure Measurement. Liquid-column elements, such as the manometer, are commonly used for pressure measurement. See also **Manometer**. A variety of diaphragm and other elastic elements is used to measure pressure. A metallic diaphragm element is primarily a device for measuring relatively low pressures. It consists of a single diaphragm or of one or more capsules connected together, so that upon pressure application, each capsule deflects. The total deflection is the sum of the deflections of all capsules. A variety of bellows elements is similarly used in pressure gages. One of the most common forms of pressure gage makes use of a bourdon-spring element. See also **Bourdon Tube**. Gages for medium-to-high vacuums usually incorporate an electronic type transducer. See also **Vacuum Gages**. Electrical transducers, such as strain gages, moving-contact resistance elements, inductance, reluctance, capacitative, and piezoelectric devices also are used in pressure detection systems. Pressure not only is important as a key variable for direct measurement, but differential pressures are commonly measured in connection with various flowmeters that use a differential-producing element, such as an orifice plate, to measure flow. Manometers and other pressure sensors are also used in liquid-level measuring devices.

High-Pressure Technology. Until the mid-1970s, the limit to most high-pressure experimentation was confined to about 300 kilobars.

As of 1988, the maximum pressure created in the laboratory by the diamond anvil pressure cell approximates 5 million atmospheres. Theoretical estimates, however, forecast that diamond is stable up to 23 million atmospheres with respect to any phase transition. Although plastic deformation would limit its capability, predictions for the diamond anvil cell are for pressures somewhere between 5 and 23 million atmospheres. See also **Diamond Anvil High Pressure Cell**.

PRESSURE SUIT.   A garment designed to provide pressure upon the body so that respiratory and circulatory functions may continue normally, or nearly so, under low-pressure conditions, such as occur at high altitudes or in space without benefit of a pressurized cabin. A pressure suit is distinguished from a pressurized suit, which inflates, although it may be fitted with inflating parts that tighten the garment as ambient pressure decreases.

PRESSURE WAVE.   1. In meteorology, a short period oscillation of pressure such as that associated with the propagation of sound through the atmosphere; a type of longitudinal wave. These waves are usually recorded on sensitive microbarographs capable of measuring pressure changes of amounts down to 10E-4 millibar. Typical values for the period and wavelength of pressure waves are 1/2 to 5 seconds and 100 to 1500 meters, respectively. Pressure waves produced by explosions in the upper atmosphere are of value in determining the high-altitude temperatures and winds.

2. A wave or periodicity which exists in the variation of atmospheric pressure on any scale, usually excluding normal diurnal and seasonal trends. Such waves can persist for an indefinite length of time only if they coincide approximately with the free oscillations of the atmosphere. Waves of a period longer than that associated with the passage of large-scale weather disturbances are difficult to isolate, since they usually have such a small amplitude that they can be extracted from the data and only by means of precise statistical methods.

PRESTRESSED CONCRETE.   Prestressed concrete resulted from the desire to overcome the disadvantage of the low tensile strength of concrete. See also **Concrete**. By means of high strength steel wires, cables, or rods—particularly the first two—a concrete member is pre-compressed. Then, when the structure receives its load, the compression is relieved on that portion which would normally be in tension. Thus, a beam is prestressed so that under load the concrete on the side normally in tension has no tensile forces acting on it.

There are two general methods of prestressing, namely pretensioning and post-tensioning. Pretensioning consists of pouring concrete around wires kept under tension until the concrete has gained sufficient strength. The wires are then cut, and compressive forces are thereby imparted to the concrete through bond between the steel and concrete. Post-tensioning consists of jacking bond-free cables against the ends of an already hardened concrete section, and then anchoring the ends of the cables.

Prestressed concrete structures require less concrete and steel, but the steel used is a more costly high-strength wire. Prestressed concrete is especially well adapted to combination with precast concrete.

PREVOST LAW OF EXCHANGES.   In an evacuated enclosure, with walls maintained at constant temperature, objects within will reach a condition of thermal equilibrium at which they will attain, and remain at, the temperature of the walls. Each body is constantly exchanging heat energy with its surroundings, the net result of which exchange tends to equalize the temperature of the body and its surroundings. Cold bodies radiate less heat than they receive from warmer surroundings and thus rise in temperature to the equilibrium value.

PRIESTLEY, JOSEPH (1733–1804).   Priestley was an English chemist who researched relationships among plants, air, and animals. After meeting Benjamin Franklin he became interested in science and the two men became lifelong friends. Priestley started doing chemical experiments as a hobby, but it soon became a passion. He had little scientific education but his observations were very keen.

Priestley lived near a brewery and his curiosity about how it operated and about the gases involved lead him to discover a gas (carbon dioxide) was heavier than air. He found water and this heavy "air" made a great drink and in 1773 he was awarded a medal by the Royal Society for his invention of soda water. In 1774, he announced the results of his experiment, which described the unusual properties of a new "air", this was in fact, the discovery of oxygen. His experiments with "air" and gases were important for leading to the first ballooning flights.

Priestley also researched relationships among plants, air, and animals. He observed the respiration of plants, by which they take in carbon dioxide and produce oxygen. His observation helped others understand the process. He observed "green matter", which now we know as photosynthesis.

He was a strong religious and political leader and was persecuted for his support of the American Revolution. He came to America in 1794 and spent his last years experimenting in his laboratory.

See also **Balloon**; and **Oxygen**.

J. M. I.

PRIMARY AMEBIC MENINGOENCEPHALITIS.   Free-living amebae (*Naegleria* and *Acanthameba* sp.), ubiquitous in warm fresh water ponds and lakes, are the etiologic agents of this rare, rather recently identified disease. The infection is acquired by swimming in (and probably inhaling) infested waters. The amebae enter through the nasal mucosa, penetrating to the brain via the olfactory nerve sheaths, producing an acute purulent meningitis, and then spread rapidly beyond the leptomeninges to brain tissue, where they produce an acute hemorrhagic, necrotizing encephalitis.

The incubation period is approximately five days, when nuchal rigidity, headache, pyrexia, and nausea occur and progress to bizarre behavior, coma, and death within 4 to 7 days. Amebae with large nuclei are found in the brain tissue at autopsy.

Infections have been seen in South Australia, Eastern Europe, and the United States (Florida and Virginia). Patients are typically young children and low-age adults. Treatment is generally unsuccessful, but one patient in the United States and one in Australia survived with heroic treatment involving intrathecal amphotericidin.

R. C. V.

PRIMARY BODY.   The celestial body or central force field about which a satellite or other body orbits, or from which it is escaping, or towards which it is falling. The primary body of the moon is the earth; the primary body of the Earth is the sun.

PRIMARY CIRCULATION.   In meteorology, the prevailing fundamental atmospheric circulation on a planetary scale which must exist in response to radiation differences with latitude, to the rotation of the planet, and to the particular distribution of land and oceans; and which is required from the viewpoint of conservation of energy.

PRIMARY COLORS.   Three colors which, when suitably mixed, will produce all the other colors, as well as white and black. The colors generally used are an orange-red, green and blue-violet. These colors are sometimes called the additive primaries to distinguish them from the three subtractive or minus-colors, cyan, magenta, and yellow.

PRIMARY ELEMENT.   In an instrumentation or automatic control system, the primary element is that system element that quantitatively converts measured variable energy into a form suitable for measurement. For transmitters not used with external primary elements, the sensing portion is the primary element. Other terms used for the same function include sensor and detector. See also **Sensor (Measurement)**.

PRIMARY RADAR.   Radar using reflection only, in contrast with secondary radar which uses automatic retransmission on the same or a different radio frequency.

PRIMATES.   Most primates are arboreal animals and all species either grasp by opposing the thumb to the fingers or show similarity to grasping appendages of this type in the anatomy of the hand. With exception of the marmosets, all primates have nails on at least part of the digits. The lemurs retain a claw on the second toe and the marmosets have a nail only on the great toe. The brain is more highly developed in the primates than in any other animals. The major categories of mammals are described in this volume. See Table 1 for general organization of the primates and for reference to other entries in this encyclopedia.

PRIME MERIDIAN.   1.   The meridian of longitude 0 degrees, used as the origin for measurement of longitude. The meridian of Greenwich, England, is almost universally used for this purpose.

2.   Any meridian in any coordinate system used as an origin for measurement of longitude.

PRIME NUMBER.   See **Number Theory**.

PRINCIPAL COMPONENTS (Statistics).   Given a multivariable complex with variables $x_1, x_2, \ldots, x_n$ it is possible to transform to n new variables which are (1) linear functions of the x's and (2) uncorrelated among themselves. These variables are called principal components. Except in degenerate cases they are unique apart from sign. The principal components have optimal properties in the sense that one will have the largest

TABLE 1. PRIMATES (Top Mammals)

	In This Encyclopedia
TUPAIOIDS	**See Moles and Shrews.**
Tree-Shrews (*Tupaia*, . . .)	
Pen-tailed Tree-Shrews (*Ptilocercus*)	
LORISOIDS	**See Lorisoids.**
Lorises (*Lorisidae*)	
The Slow Loris (*Nycticebus*)	
The Slender Loris (*Loris*)	
Pottos (*Periodicticus*)	
The Angwantibo (*Arctocebus*)	
Bush-Babies (*Galagidae*)	
Common Bush-Babies (*Galago*)	
Pigmy Bush-Babies (*Galagoides*)	
Needle-clawed Bush-Babies (*Euoticus*)	
LEMUROIDS	**See Lemur.**
The Aye-Aye (*Daubentoniidae*)	
Small Woolly Lemurs (*Cheirogalaginae*)	
Mouse-Lemurs (*Microcebus*)	
Dwarf Lemurs (*Cheirogaleus* and *Phaner*)	
Large Woolly Lemurs (*Lemurinae*)	
Weasel-Lemurs (*Lepilemur*)	
Gentle Lemurs (*Hapalemur*)	
Common Lemurs (*Lemur*)	
—Ruffled Lemur	
—Black Lemur	
—Brown Lemur	
—Mongoose-Lemurs	
—Red-bellied Lemur	
Silky Lemurs (*Indriidae*)	
Sifakas (*Propithecus*)	
The Avahi (*Lichanotus*)	
The Indri (*Indri*)	
TARSIOIDS	**See Tarsioids.**
HAPALOIDS	**See Marmoset.**
Marmosets (*Callithricidae*)	
Pigmy Marmosets (*Cebuella*)	
Maned Marmosets (*Leuntocebus*)	
Plumed Marmosets (*Callithrix*)	
Ruffed Marmosets (*Hapale*)	
Bald Marmosets (*Marikina*)	
White Tamarins (*Mico*)	
Black Tamarins (*Tamarin*)	
Moustached Tamarins (*Tamarinus*)	
Pinches (*Oedipomidas*)	
Goeldi's Marmoset (*Callimiconinae*)	
Titis (*Callicebinae*)	
CEBOIDS	**See Monkeys and Baboons.**
Half-Monkeys (*Pithecinae*)	
Douroucoulis (*Aotes*)	
Sakiwinkis (*Pithecia*)	
Bearded Sakis (*Chiropotes*)	
Uacaris (*Cacajao*)	
Hand-Tailed Monkeys (*Cebinae*)	
Squirrel-Monkeys (*Saimiri*)	
Capuchin Monkeys (*Cebus*)	
Woolly Monkeys (*Lagothrix*)	
Woolly Spider Monkeys (*Brachyteles*)	
Spider Monkeys (*Ateles*)	
Howler Monkeys (*Alouatta*)	
SIMIOIDS	**See Monkeys and Baboons.**
Colobine Monkeys (*Colobinae*)	
Guerezas (*Colobus*)	
Languars (*Presbytis*,. . .)	
Snub-nosed Monkeys (*Rhinopithecus*)	
Proboscis Monkey (*Nasalis*)	
Long-tailed Monkeys (*Cercopithecinae*)	
Guenons (*Cercopithecus*)	
Allen's Swamp Monkey (*Allenopithecus*)	
Military Monkeys (*Erythrocebus*)	
Mangabeys (*Cercocebus*)	
Dog-faced Monkeys (*Cynopithecinae*)	
—Barbary Ape	
—Rhesus	
—Bearlike Monkey	
The Black Ape (*Cynopithecus*)	
Baboons (*Papio*)	

(continued)

TABLE 1. (*Continued*)

	In This Encyclopedia
The Gelada (*Theropithecus*)	
Drills (*Mandrillus*)	
ANTHROPOIDS	**See Anthropoids**.
Lesser Apes (*Hylobatidae*)	
The Siamang (*Symphalangus*)	
Gibbons (*Hylobates*)	
Greater Apes (*Pongidae*)	
Gorillas (*Gorilla*)	
The Chimpanzees (*Pan*)	
Orangutans (*Pongo*)	
Humans (*Hominidae*)	**See Mammals**.

possible variance among linear functions of the original x's, a second will have the largest variance among linear functions which are uncorrelated with the first, and so on.

The determination of the principal components depends on the calculations of the eigenvalues of the variance-covariance matrix of the x's.

Principal component analysis is to be sharply distinguished from factor analysis, to which it bears some formal resemblance. Principal components are simple variate-transformations; factor analysis imposes a model on the data.

PRINCIPAL PLANES (Lens).   Two planes so located in a thick lens or lens system such that if object distances are measured from the first principal plane and image distances are measured from the second principal plane, the thin-lens formula will hold. The intersections of the principal planes with the optical axis are the principal points of the lens or system of lenses. See Fig. 1.

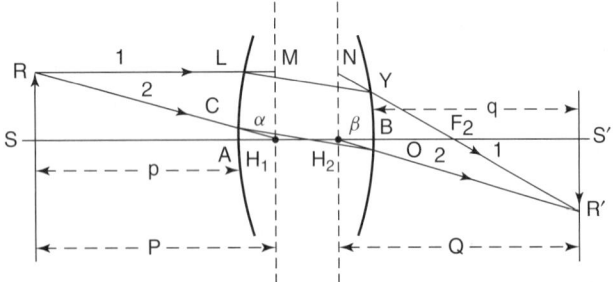

Fig. 1. H_1 and H_2 are the principal points; MH_1 and NH_2 the principal planes of a thick lens. In locating images by a graphical method, principal rays may be used as for thin lenses if the incident light falls on an imaginary thin lens coincident with the first principal plane, the lens being shifted to coincide with the second principal plane for emergent light.

PRINTED CIRCUITS AND CIRCUIT BOARDS.   A printed circuit is any circuit formed by depositing a conducting material on the surface of an insulating sheet. This may be achieved by the use of electrically conducting ink, electroplating, or other methods.

Circuit boards represent a mid-phase in the progress of electronic (and some electric) circuitry — ranging from the "spaghetti" wiring that was characterized by individually soldered or wire-wrapped connections and stuffed into the bottom of a chassis (as commonly encountered in radio receivers, control instruments, and numerous other equipment of just a few decades ago), to the highly sophisticated integrated circuits of the present era. See also **Integrated Circuit (IC)**. In the pre-circuit board days, components were mounted and interwired individually. The appearance and ease of trouble-shooting ranged from chaotic and difficult to rather neat and easy, depending upon care exerted by the designers and assemblers. Because assembly was essentially manual, sufficient working space for manipulating components and connections had to be made. Thus, equipment, on present standards, was excessively heavy and bulky. Earlier wiring was simplified to quite a degree by means of wire harnesses.

As components became smaller with the entry of solid-state devices and as economic pressures toward automating assembly became greater, the concept of circuit boards and printed circuitry received a rather rapid

response. During the interim, circuit boards (sometimes called circuit modules, cards, etc.) have greatly improved. Multilayer circuitry helped to increase the component density of boards. Both the materials and technology for making circuit boards have improved at a steady rate.

Coombs listed a number of advantages of printed circuits over former conventional loose wiring methods, including: (1) weight reduction by as much as 10 to 1; (2) better control over volume required for wiring; (3) cost savings as the result of standardization and automation in production; (4) increased reliability mainly as the result of reducing human errors; (5) easier inspection and troubleshooting; and (6) easier part identification. Disadvantages pointed out include: (1) difficulty in repairing; (2) heat dissipation problems; (3) design regimentation and restriction, making it difficult to make improvements in circuitry design, among others. The design and production of circuit boards is volume sensitive in terms of production costs and time.

Single-board computers appeared during the mid-1970s and created a large expansion in the use of circuit boards. Among boards required were analog input and output boards, core memory boards, digital input/output boards, floppy disk controller boards, hard disk interface boards, keyboard controller boards, math boards, optical isolator boards, PROM boards, RAM boards, synchro-to-digital boards, and so on. One of the major problems that arose, as the result of so many suppliers, was the lack of a standard data bus, a problem that is now approaching resolution.

There is a considerable blending of circuit board and integrated circuit production technology, and much of the latter technology has been transferred from earlier experience in producing simpler circuit boards.

PRISM (Mathematics).   A polyhedron, with two bases which are equal polygons in parallel planes and additional faces, called lateral, which are parallelograms. The intersections of the lateral faces are the lateral edges of the prism. Special cases are: right, lateral edges perpendicular to the bases; regular, a right prism with regular polygons for bases; triangular, quadrangular, etc., if the bases are triangles, quadrilaterals, etc. A prism is truncated if it is cut by a plane oblique to its base. See also **Parallelepiped**.

The lateral area of a prism, $A = ep$, where e is the length of a lateral edge and p is the perimeter of a section made by a plane perpendicular to the lateral edges. Its volume, $V = bh$, is the product of its base by its altitude.

PRISM (Optics).   A transparent solid, cut at precise angles for various optical purposes. Incident light may pass directly through a prism, or may emerge after one or more internal reflections; in some cases the light is polarized.

The common triangular prism, familiar in older forms of spectroscope, receives light upon one face and passes it through another after two refractions, resulting in a total deviation Δ dependent upon the angle of the prism and its refractive index for the light used. See Fig. 1. If the light is incident at angle i on the first prism face, and if the prism angle is α and the refractive index is n, the total deviation after passage through the prism in a plane at right angles to the prism edge is given by

$$\Delta = i - \alpha + \arcsin\left[n \sin\left(\alpha - \arcsin\frac{\sin i}{n}\right)\right]$$

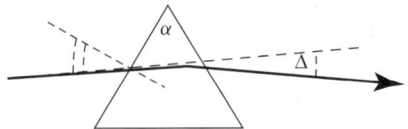

Fig. 1.   Ray of light passing through triangular prism.

A little experimenting shows that this deviation has a minimum value when the light traverses the prism symmetrically, entering and emerging at the same angle with the corresponding faces. This angle of minimum deviation is easily shown to be

$$\Delta_{\min} = 2 \arcsin\left(n \sin \frac{1}{2}\alpha\right) - \alpha$$

from which the refractive index may be obtained, by experiment, as

$$n = \frac{\sin \frac{1}{2}(\Delta_{\min} + \alpha)}{\sin \alpha/2}$$

The effect of a prism on heterogeneous light may be deduced from the formulas for refractive dispersion. Prisms are used in binoculars, in monochromatic illuminators, and in many other optical instruments.

The *Amici* prism is a *direct-vision* prism, that is, a prism combination by which a beam of light is dispersed into a spectrum without mean deviation. Such prisms are sometimes used in direct-vision spectroscopes.

The principle will be clear from the following example. Assume an inverted prism of crown glass with an angle of 40°, used with an erect prism of flint glass. See Fig. 2. Yellow sodium light (5,893 Å) is deviated by the crown-glass prism through +22°32′ (upward). For the flint-glass prism to produce an equal negative (downward) deviation it must have an angle of 33°40′. (Each prism is supposedly set for minimum deviation.) Together they produce no deviation for this wavelength. But if light of 7,682 Å (red) is used, the deviation of the crown glass is +22°16′, while that of the flint is −22°10′, giving a net deviation of +6′. And for the wavelength 4,047 Å (violet), the deviations are, respectively, +23°12′ and −23°44′, giving −32′. There is thus, between the ends of the visible spectrum, a separation of 38′. Additional pairs of prisms may be used to increase the dispersion.

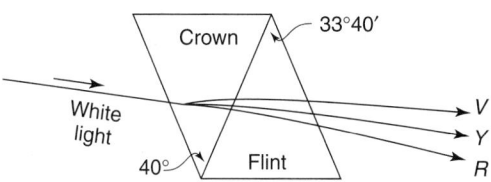

Fig. 2.   The deviation due to the crown-glass prism is neutralized by that of the flint-glass prism, for the middle of the yellow spectrum only.

A *Littrow* prism is a 30–60–90-degree prism silvered on the side opposite the 60° angle. A single lens can then be used as both collimator and telescope.

A *Dove* prism has the property of inverting a beam of light. Three faces are polished, and the size and index of refraction must be properly correlated if the beam is not to be displaced laterally. Rotation of the prism about the axis of the beam rotates the beam at twice the rate of rotation of the prism. See Fig. 3.

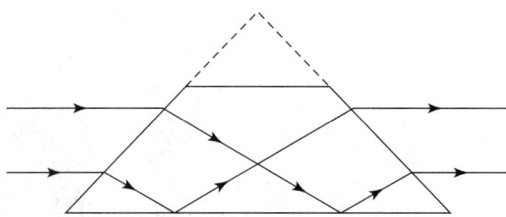

Fig. 3.   Dove prism.

A *constant-deviation* prism refracts any required wavelength of light in a specified direction to the incident beam. An example is the *Pellin-Broca* prism consisting of two 30° prisms, connected by a 45° total-reflecting prism. See Fig. 4.

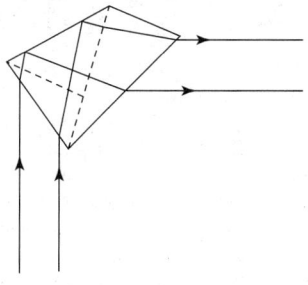

Fig. 4.   Pellin-Broca prism.

A total-reflecting prism is arranged to provide total reflection of light incident upon it in a specified direction. Most such prisms are 45–45–90 prisms in which the light enters normally to a face opposite to a 45° angle, is totally reflected by the hypotenuse face, and leaves by the third face, thus having been totally reflected through 90°. The beam of light must be sufficiently parallel so that all of the light strikes the reflecting face outside of the critical angle, or the reflection will not be total. The *Porro* prism is of this type, and two of them are used in each telescope of prism binoculars. The *roof*-prism is a total-reflection prism in which the surface opposite the right angle has been replaced by two surfaces at right angles (roof), with their common element parallel to the hypotenuse of the triangle. The roof prism turns the beam through 90° like a total-reflection prism, and also inverts the beam like a Dove prism.

The *Nicol* prism is one of the best-known devices for producing plane-polarized light. It consists of two pieces of Iceland spar (pure calcium carbonate) cut as shown in Fig. 5. The optic axis of each is approximately indicated by the double arrow, and they are cemented together with colorless Canada balsam along the plane *MN*. If the incident beam *IP* is unpolarized, it suffers double refraction at *P*, dividing into an ordinary component *PO′* and an extraordinary component *PE′*. The refractive index of Iceland spar for the ordinary ray (sodium light) is 1.658 and for the extraordinary it is 1.486, while that of Canada balsam for both is 1.53. The ordinary ray therefore encounters at *O′* a less refractive medium, and, the incidence being at an angle larger than the critical angle, it is totally reflected (*O′O*); while the extraordinary ray incident at *E′* encounters a more refractive medium, therefore cannot suffer total reflection and most of it passes on along *E′Q*, emerging along *QE* completely plane-polarized with its vibration plane in the plane of the paper. Modifications of this prism, having different shapes and using other cements, have been designed for special purposes. However, Nicol prisms cannot be used in ultraviolet light as the Canada balsam is not transparent to these shorter wavelengths. The *Rochon* and *Wollaston* prisms are made of quartz or calcite and "cemented" with glycerine or castor oil. See Fig. 6.

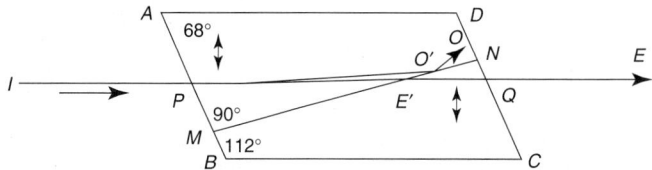

Fig. 5.   Nicol prism. Double-headed arrows indicate direction of optic axis.

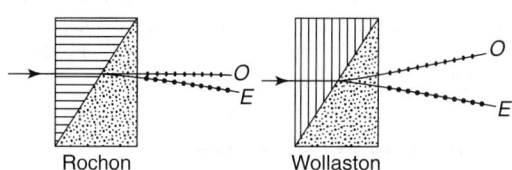

Fig. 6.   Rochon and Wollaston prisms.

The direction of the optic axis of each crystal is indicated by the shading. The Rochon prism transmits the ordinary ray without deviation, the ray being achromatic. The light should travel through the prisms in the directions indicated.

A *Cornu-Jellet* prism is made by splitting a Nicol prism in a plane parallel to the direction of vibration of the transmitted light and removing a wedge-shaped section. When the two pieces are joined together again, the planes of vibration of the light transmitted by the two halves make a small angle with each other.

A *Cornu-double* prism is designed to utilize the ultraviolet-transmitting properties of quartz without introducing its double refraction. This is accomplished by cementing together two 30° prisms, one of right-handed and the other of left-handed quartz.

A *corner cube* prism (trihedral), shown in Fig. 7, is used to reflect a beam back on itself very precisely in two dimensions by means of two total internal reflections. Typical uses are the coincident alignment

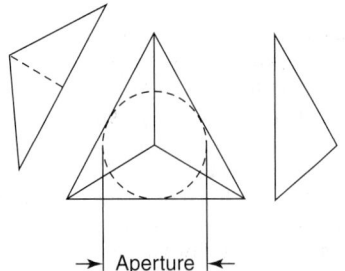

Fig. 7.   Corner cube prism (tribedral).

of laser systems and reticles; for determination of the infinity focus for autocollimators or other telescopic equipment; and as a reference for indirect sighting and triangulation systems.

A *polarizing* prism, shown in Fig. 8, utilizes uniaxial birefringent crystals of quartz or calcite, which are double refracting, so that two separate beams are created (ordinary and extraordinary). Each beam is totally polarized and orthogonal with respect to the other. In most designs, one beam exits from the prism in a direction parallel to the light beam that enters, while the other beam emerges obliquely or impinges on a light-absorbing blackened face.

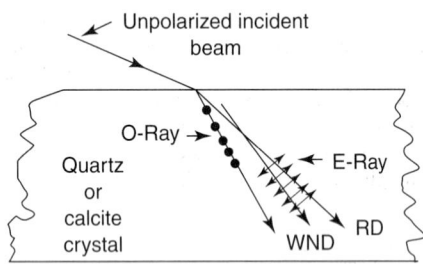

Fig. 8.   Polarizing prism.

A *Glan air-spaced prism* (Glan-Taylor), shown in Fig. 9, is designed for use where maximized transmission is desired throughout the spectral range of the calcite being used. Increased transmission occurs because of the special angle of the optic axis and the resulting incidence of the transmitted beam at about Brewster's angle at the air-spaced interface. The prism has exceptional performance when extinction is required. Two crossed Glan air-spaced prisms will transmit less than 10^{-5} of the incident beam. For high-power laser application, one of the beams is allowed to exit from the top face of the prism (double-beam type) rather than being absorbed (single-beam type).

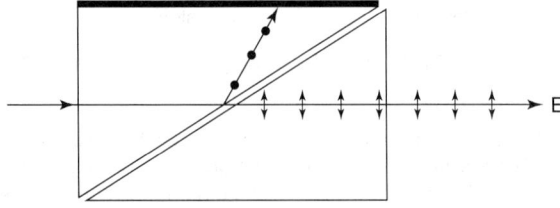

Fig. 9.   Glan air-spaced prism. (*Glan-Taylor.*)

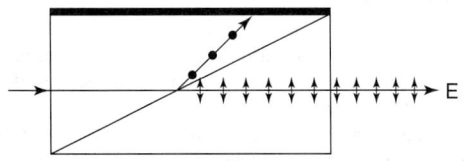

Fig. 10.   Glan Thompson prism.

A *Glan Thompson prism*, shown in Fig. 10, is a polarizing prism that finds use in high-precision instruments. Since the elements of the prism are cemented together, the prism should not be used in the path of high-power beams. The prism has a wide field of view as compared with the Glan air-spaced prism.

PROBABILITY.   Suppose that a trial may have one of two mutually exclusive outcomes — say "success" and "failure" — and that we associate with it a number p, $0 \le p \le 1$, called the probability of a success. The calculus of probability is concerned with the manipulation of such numbers, and is largely based on the following results:

1. If two trials with probabilities p_1, p_2 are independent, in that the outcome of either has no effect on that of the other, the probability that both will result in successes is $p_1 p_2$.
2. If the probabilities of a set of mutually exclusive events are $p_1, p_2, p_3, \ldots$, the probability that one will occur is $p_1 + p_2 + p_3 + \cdots$.

The application of statistical techniques usually results in a statement couched in terms of probabilities, but the application of probability theory to the real world encounters formidable logical difficulties. Two main schools of thought may be distinguished.

Fig. 1.   Demonstration of probability. A Galton Board, named after Sir Francis Galton who constructed the first device of this kind, can be used to illustrate Gaussian (normal or bell-shaped) distribution. Tiny spheres that roll easily are placed in an upper hopper. They fall by gravity to the receiving slots below, but in falling must pass an array of hexagonally shaped obstacles. Theoretically, at each obstacle, probability indicates that one-half of the spheres will be diverted to the right; the other half to the left. Thus, the distribution follows the format of Pascal's triangle. The distribution at the bottom of the diagram will fully fill out the distribution curve, once the hopper is completely emptied. See also **Pascal Triangle**.

1. In the frequency theory, the trial is supposed capable of indefinite repetition, either real or hypothetical, and the probability is identified with the limiting frequency of successes in this infinite sequence. However, the existence of a limiting value is difficult to establish theoretically, and impossible to establish empirically because of the infinite length of the sequence.

2. The other school identifies probability with the strength of belief that a success will occur. The main difficulty here is the subjective nature of the definition.

For demonstration of probability, see Fig. 1.

PROBABILITY DISTRIBUTION. If a quantity may take any one of a set of values with different probabilities, it is said to possess a probability distribution. Formally, this is the same as the frequency function, where the relative frequencies are interpreted as probabilities.

PROBABLE ERROR. In the theory of observational error, the so-called probable error of an observation was defined as $\pm 0.6745\sigma$ where σ is the standard deviation of the distribution from which the observation was drawn. The coefficient derives from the fact that if the population is normal (Gaussian) a deviation of $\pm 0.6745\sigma$ from the mean covers half the distribution. Likewise, in a sample of n observations the probable error of the mean was $\pm 0.6745\sigma/\sqrt{n}$.

In statistics the expression is obsolete and has been replaced by the standard error, which is simply the standard deviation of the sampling distribution of the statistic concerned. Thus the standard error of the mean is $\sigma/\sqrt{n}$. Significance is customarily attributed to the difference between an observed value of a statistic and the parameter which it estimates if that difference exceeds twice (or more stringently, three times) the standard error.

SIR MAURICE KENDALL, International Statistical Institute, London.

PROBOSCIS. A protruding or protrusible organ associated with the mouth and therefore at the front of the head in most animals. It is used in feeding and in some cases for other purposes.

PROCELLARIIFORMES (Aves). These narrow-winged flying acrobats of the high seas belong to quite a different order of birds than the gulls and terns: they are the petrels, albatrosses, and their relatives, a group which ornithologists call Tube-Noses because of the peculiar shape of their nostrils.

The nostrils are horny tubes generally found on the culmen and less often on the sides of the beak, which is straight with a hooked tip. The beak-sheath is composed of several separate horny plates. There are large nasal glands for salt excretion. The long gullet and proventriculus secrete an oil. There are 15 cervical vertebrae, and the furcula is movable and located on the keel of the sternum. The pelvis is penguin-like, and is fused with the synsacrum. The knee joint has a projection of the tibial crest. There are 3 forward toes that are connected by webs, and the hind toe is degenerated. These birds have an outstanding flight capacity and great endurance.

They are inhabitants of the high seas, particularly of the Southern Hemisphere, and they only go on land to breed. They lay 1 egg, and have a long incubation period. The young grow very slowly. This order includes both the largest and the smallest sea birds; the length is 14–135 centimeters (5.5–53 inches), and the weight is 20–8000 grams ($\frac{1}{2}$–$17\frac{1}{2}$ pounds). Today there are 4 families: the Albatrosses (*Diomedeidae*); the Petrels and Shearwaters (*Procellariidae*); the Storm-Petrels (*Hydrobatidae*); and the Diving-Petrels (*Pelecanoididae*). There are 22 genera and 92 species.

The paired nose tubes may vary in form and length; they join round or oval openings with large nasal or olfactory cavities. The significance of the tubes is not known; many suggestions have been made regarding their purpose, but none are altogether convincing. Evidently the olfactory sense, otherwise of low efficiency in birds, is well developed in the tube-nosed swimmers, but this does not explain the peculiar nostrils. All members of this order fly low over the sea, so possibly the nose tubes keep spray out of the inner nasal cavities. All tube-noses also have large nasal glands which secrete a saturated salt solution, and so the tubes could possibly serve to keep this solution away from the eyes and the skin of the gape.

Another peculiarity of these birds is the flesh colored, oily liquid that most species secrete from special cells of the proventriculus. When in danger, breeding birds and the young as well can regurgitate this stomach-oil and spray it at an aggressor for a distance of several meters (feet). When this oil cools it solidifies to a wax-like consistency; in this form it is often found around the nests in cold regions. The birds can use this oil when preening their plumage and possibly they apply it directly from the nose tubes. It is also possible that the stomach-oil accounts for the strong musky smell, peculiar to all tube-noses, which can even persist on skins which have been in museums for over 100 years.

While the tubular nostrils and the stomach-oil occur only in tube-noses, their particularly slow reproduction and the long growth period of their young are not so unique in the bird world. All *Procellariiformes* lay only 1 egg. As a rule, it is relatively large, weighing 6–10% of the mother's body weight in large species and 10–25% or more in the smaller ones. Incubation and the growth periods last longer in tube-noses than in all other birds of similar size.

All *Procellariiformes* are birds of the high seas, adapted in various ways for feeding on or just beneath the surface of the water. They can spend days, weeks, and even months away from land. Two-thirds of the living species live in the Southern Hemisphere, which must be regarded as the main area of development of this order. From the temperate-cool zone of westerly winds, they have spread south as far as the coasts of the Antarctic and north over the equator as far as Arctic latitudes. See also **Petrels and Albatrosses**.

PROCESS CONTROL. The use of instruments and control devices and systems to measure and manipulate one or many variables to assure the safe and efficient operation of machines and equipment required in manufacturing. Processes may be categorized by the kinds of materials which they handle: (1) *fluids* (gases and liquids), (2) *bulk solids*, (3) *sheeted and webbed materials*, and (4) *discrete pieces*. The chemical and so-called process industries (petroleum refining, food processing, petrochemical manufacture, some of paper and textile manufacturing, and some of the metallurgical industries) essentially are concerned with fluids and bulk solids. These are materials that flow in pipelines or are handled by various forms of bulk conveyors. Principal variables with which these industries are concerned include temperature, pressure, flow, liquid level, solid level (as in bins, silos, and hoppers), specific gravity; density, viscosity, consistency, and chemical composition. Industries concerned with sheeted and webbed materials include the paper, textile, printing, plastics, and parts of the metals industries. Here, the major variables of concern include temperature, thickness, dimensional widths and lengths, linear speed, and weight/area (such as basis weight). A discrete-piece handling industry is typified by most of the automotive and aircraft manufacturing facilities that involve machine tools, transfer machines, assembly operations, and parts inspection and testing. Major variables include dimension (linear and angular), position, hardness, vibration, electrical properties, thermal properties, and structural soundness (such as by x-ray inspection). It is difficult to find one industry that is purely one type. It is obvious, however, that the form of instrumentation and control is affected importantly by the particular variables that must be measured and controlled.

PROCESS (Control System). The term *process* when used as a part of control-system terminology may be defined as the collective functions performed in and by the equipment in which the variable(s) is (are) to be controlled. Equipment as embodied in this definition should be understood not to include any automatic control equipment. The process may also be referred to as the *controlled system*.

PROCTITIS. Inflammation and infection of the rectum due to a variety of causes. Symptoms may include discharge, pain on defecation, bleeding, and rectal fullness. A major cause of proctitis is inflammatory bowel disease. See also **Colitis and Other Inflammatory Bowel Diseases**. Amebiasis may be a cause. See also **Amebiasis**. Other causative factors include the virus *Lymphogranuloma inguinale*, syphilis, chancroid, and malignancy. Gonococcal proctitis is increasingly seen in homosexual men, in which cases the gonococcal infection not only involves the urethra, but the anal canal as well.

PROCYON (α Canis Minoris). A bright, nearby star, which receives its name from the fact that in its nightly journey across the sky, it is close to the star Sirius. These two "dog stars" are referred to in the most ancient literatures, and were objects of veneration and worship both by the Babylonians and the Egyptians.

Procyon, like the other "dog star" Sirius, has as a faint companion a white dwarf, a type of star whose common proper motion is 1.25″/yr.

Ranking eighth in apparent brightness among the stars, Procyon has a true brightness of 7.3 as compared with unity for the sun. Procyon is a yellow-white, spectral type F star. Estimated distance from the earth is 11.3 light years. See also **Constellations**; and **Star**.

PRODUCT.   The result obtained when two or more quantities, such as numbers, functions, or equations are combined by multiplication. In vector and tensor analysis, the concept of a product must be generalized for there are several kinds of products and multiplication does not always obey the commutative law.

In matrix algebra, if A is a rectangular matrix of order $(m \times h)$ and B of order $(h \times n)$, the product $C = AB$ is of order $(m \times n)$ and its elements are $C_{ij} = \rho A_{ik} B_{kj}$. It is not necessary that the commutative law be obeyed. Another matrix combination is the direct product. If A and B are square, of order m and n, respectively, the direct product is of order $(m \times n)$ and defined by the relation

$$A \times B = [A_{ij}B_{rs}]$$

The index pairs (i, r) and (j, s) refer to row and column, respectively. They are customarily arranged in dictionary order so that (j, s) precedes (j', s') if $j < j'$, $s < s'$ or if $j = j'$, $s < s'$, etc.

In group theory, *multiplication* means any defined combination law and the result is a product. For example *multiplication* might be defined as addition and then the product is that result commonly known in algebra as a sum.

The product of two infinite series is called a Cauchy product. An expansion of the form

$$u_1 u_2 u_3 \cdots u_n \cdots = \Pi_{k=1}^{\infty} u_k$$

is an infinite product. The partial products form a sequence $\{p_n\}$ such that $p_1 = u_1$; $p_2 = u_1 u_2$; $p_3 = u_1 u_2 u_3$; $\ldots$, $p_n = \Pi_{k=1}^{n} u_k$. The convergence of infinite products is studied by methods similar to those used for infinite series. An infinite product of the form $(1 + u_1) \times (1 + u_2) \cdots$ may be converted into an infinite series $1 + x_1 + x_2 + x_3 + \cdots$, where $u_n = x_n/(1 + x_1 + x_2 + \cdots + x_{n-1})$. The product converges only if the positive infinite series, $u_1 + u_{2x} + u_3 + \cdots$ converges.

PRODUCT MODULATOR.   A modulator whose output is proportional to the product of the carrier and the modulating signal. The desired result can be achieved by sampling the modulating wave briefly at regular intervals at the carrier rate, and applying the ensemble of samples to the input of a band-pass filter having a center frequency coincident with the carrier frequency. One fundamental property of a product modulator is that the carrier is normally suppressed.

PRODUCT-MOMENT.   If the distribution function of n variates $x_1, x_2, \ldots, x_n$ is given by $F(x_1, \ldots, x_n)$ the product-moment, joint- or multivariate-moment of order $r, s, \ldots, u$ is the mean value of $x_1^r x_2^s \ldots x_n^u$, namely:

$$\int_{-\infty}^{\infty} \int_{-\infty}^{\infty} \ldots \int_{-\infty}^{\infty} x_1^r x_2^s \ldots x_n^u dF(x_1, \ldots, x_n)$$

PROFILE.   1.   Of a variable, a curve representing corresponding values of two or more variables which may occur. A profile accounts for the correlation from point to point on the curve and has some possibility, not necessarily specified, of actual occurrence.

2.   The contour or form of a body, especially in a cross section; specifically, an airfoil profile.

3.   See also **Differential (Mathematics)**.

PROGRAM (Computer).   1. The complete plan for the computer solution of a problem, more specifically the complete sequence of instructions and routines necessary to solve a problem. 2. To plan the procedures for solving a problem. This may involve among other things the analysis of the problem, preparation of a flow diagram, preparing details, texting, and developing subroutines, allocation of storage locations, specification of input and output formats, and the incorporation of a computer run into a complete data processing system.

Internally stored program. A sequence of instructions, stored inside the computer in the same storage facilities as the computer data, as opposed to external storage on punched paper tape and pinboards.

Object program. The program which is the output of an automatic coding system, such as an assembler or compiler. Often the object program is a machine language program ready for execution, but it may well be in an intermediate language.

Source program. A computer program written in a language designed for ease of expression of a class of problems or procedures, by humans; e.g., symbolic or algebraic. A generator, assembler translator or compiler routine is used to perform the mechanics of translating the source program into an object program in machine language.

PROGRAM GENERATOR (Computer System).   A program that permits a computer to write other programs automatically. Generators are of two types: (a) the *character-controlled generator*, which operates like a compiler in that it takes entries from a library of functions, but unlike a simple compiler in that it examines control characters associated with each entry, and alters instructions found in the library according to the directions contained in the control characters; and (b) the *pure generator*, which is a program that writes another program. When associated with an assembler, a pure generator is usually a section of program which is called into storage by the assembler from a library and which then writes one or more entries in another program. Most assemblers are also compilers and generators. In this case, the entire system is usually referred to as an *assembly system*.

See also **Assembler (Computer System)**.

PROGRAMMABLE CONTROLLER.   NEMA (National Electrical Manufacturers Association) defines a programmable controller as: *a digital electronic device that uses a programmable memory to store instructions and to implement specific functions, such as logic, sequence, timing, counting, and arithmetic operations to control machines and processes.* The programmable controller as of the late 1980s was a mainstay of industrial automation.

Perspective.   As recently as the early 1960s, industrial control systems had been constructed from traditional electromechanical devices, such as relays, drum switches, and paper tape readers. This was particularly true of the discrete-piece manufacturing industries, such as the machinery, parts, automotive, aircraft, and electronics industries, as contrasted with the fluid processing industries. See also **Automation**. Although many of the earlier devices still are used today and many of the problems associated with using them have been eliminated due to technological advances in their design, such approaches continue to suffer from some inherent problems. Relays are susceptible to mechanical failure, they require large amounts of energy to operate, and they generate large amounts of electrical noise. Extreme care had to be taken in the design of relay-based control systems because it was not uncommon for the outputs to "chatter," i.e., to turn on and off rapidly when they changed states. The logic of the circuit was dictated by the hard wiring of contacts and coils. In order to make changes, as was required when production patterns changed, more time was required to rewire the logic than was needed originally.

In the late 1960s, the need to design more reliable and more flexible control systems became apparent. For example, the automotive industry was spending millions of dollars for rewiring control panels in order to make relatively minor changes to the control systems at the time of the annual model changeovers. In 1968, a team of automotive engineers wrote a specification for what they called a "programmable logic controller." What they specified was a solid-state replacement for the relay logic. The machine would use solid-state outputs and inputs, instead of control relays, to control the motors starters and sense push buttons and limit switches. The first commercially successful programmable controller was introduced in 1969. By present standards, the instrument was a massive machine containing thousands of electronic parts. It should be stressed that the first machine was designed long before microprocessors became available. The early programmable logic controller (PLC) used a magnetic core memory to store a program that was written in a graphical language (relay logic), a scheme long established in connection with conventional relay systems. See Figs. 1 and 2.

In the late 1970s, the microprocessor became a reality and greatly enhanced the role of the PLC, permitting it to evolve from simply replacing relays to the sophisticated control system it has become today.

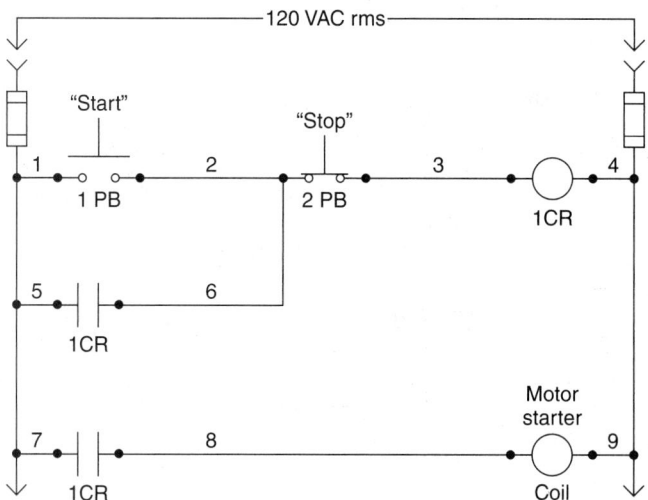

Fig. 1. Typical motor control circuit. When the pushbutton labeled START (1PB) is pressed, the control relay (1CR) is energized. A contact from 1CR is then closed and is used to "seal" 1CR "on" after 1PB is released. Another contact from 1CR is used to energize the motor starter coil, turning the motor on. When the STOP pushbutton (2PB) is pressed, it deenergizes 1CR, which "unseals" 1CR, and deenergizes the motor starter coil which stops the motor. Implementing this motor control circuit requires nine wires, not counting the power supply. The equivalent PRC Ladder Diagram Program is shown in Fig. 2.

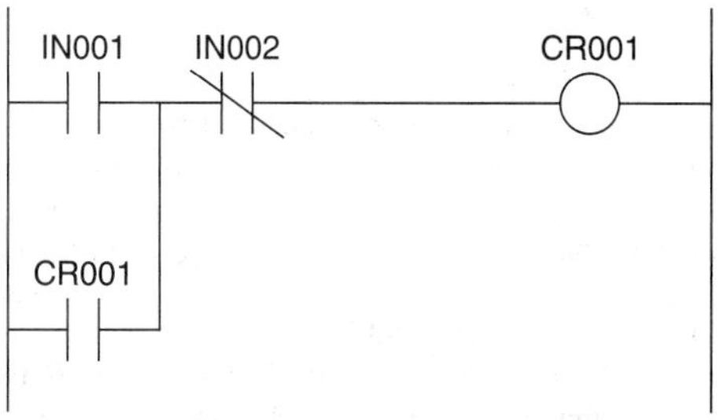

Fig. 2. Ladder Diagram Program used to control the motor circuit shown in Fig. 1. In this case, all the inputs and outputs are assigned variable names, such as IN001 for the START input and CR001 in place of the control relay 1CR. This diagram is then drawn on a program loader and entered in the PLC's user memory. The PLC's processor then solves the logic that is stored in memory. Only six wires are needed between the PLC's output and the motor starter coil and between the PLC's inputs and the pushbuttons (not counting power supply wiring).

Programmable controllers now have the ability to manipulate large amounts of data, perform mathematical calculations, and communicate with other intelligent devices, such as robots and computers. Concurrent with the increased capability and flexibility of the PLC was the expansion into many other industrial applications, including the control of machine tools, material handling systems, food-processing operations, and use in the continuous process control field.

Characteristic Functions of a Programmable Controller[1]

Seven of the most important characteristics of a programmable controller include:

1. *Field-programmable by the user.* This characteristic allows the user to write and change programs in the field without rewiring or sending the unit back to the manufacturer for this purpose.
2. *Contains preprogrammed functions.* PLCs, when procured, are already programmed with at least logic, timing, counting, and memory functions that the user can access through some type of control-oriented language.

---

[1] The abbreviation for programmable controller, PLC or PC, is optional. The editors here have selected PLC to avoid confusion that frequently arises from using PC, which is also the common abbreviation for personal computer.

3. *Scans memory and input/output (I/O) in a deterministic manner.* This feature allows the control engineer to precisely determine how the machine or process will respond to the program.
4. *Provides error checking and diagnostics.* A PLC will periodically run internal tests on its memory, processor, and I/O systems to ensure that what it is doing to the machine or process is what it was programmed to do.
5. *Can be monitored.* A PLC will provide some form of monitoring capability, either through indicating lights that show the status of inputs and outputs, or by an external device that can display program execution status.
6. *Packaged appropriately.* Modern PLCs are designed to withstand the temperature, vibration, and noise found in most factory environments.
7. *General-purpose suitability.* Generally, a PLC is not designed for a specific application, but it can handle a wide variety of control tasks effectively.

A simplified model of a PLC is shown in Fig. 3. The input converters convert the high-level signals that come from the field devices to *logic-level signals* that the PLC can read directly. The logic solver reads these inputs and decides what the outputs should be, based on the user's program logic. The output converters take the logic-level signals output from the logic solver and convert them into the high-level signals that are needed by the various field devices. The program loader is used to enter and/or change the user's program into the memory and to monitor the execution of the program.

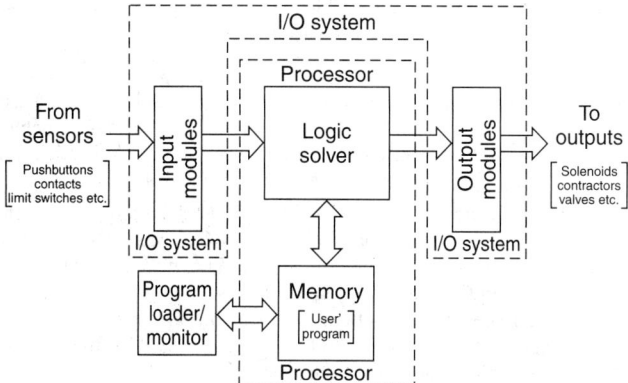

Fig. 3. Simplified block diagram of a programmable controller illustrates the basic functionality of the PLC. The control engineer (user) enters the control program on the program loader. The latter writes the program into the memory of the processor. The logic solver reads the states of the sensors through the input modules, then uses this information to solve the logic stored in the user memory (program) and also writes the resulting output states to the output devices through the output modules.

Memory. A PLC's memory can be of two different types—volatile or nonvolatile. Volatile memory loses its contents when power is removed, whereas a nonvolatile memory does not. PLCs will use nonvolatile memory for a majority of the user's memory because the program must be retained during a power-down cycle, meaning that the user will not have to reload the program every time power is lost. It is important that all nonvolatile memory in a PLC use some form of error checking in order to assure that the memory has indeed not changed. Types of memory currently used include (1) battery-backed up CMOS RAM, (2) EPROM, and (3) EEPROM. See also **Memory (Electronic)**.

Central Processing Unit (CPU). How the CPU is constructed will determine the flexibility of the PLC (whether or not the PLC can be expanded and modified for future enhancement) as well as the overall speed of the PLC. The speed is expressed in terms of how fast the PLC will scan a given amount of memory. The measure, called the scan rate, is typically expressed in milliseconds per thousand words of memory. Faster PLCs will typically cost more than the slower models. Thus, it is important to choose a PLC with a scan time appropriate to present and planned use.

It is important to note that many of the commercially available PLCs specify their scan time using contacts and coils only. A real program that uses other functions, such as timers, counters, and mathematical functions,

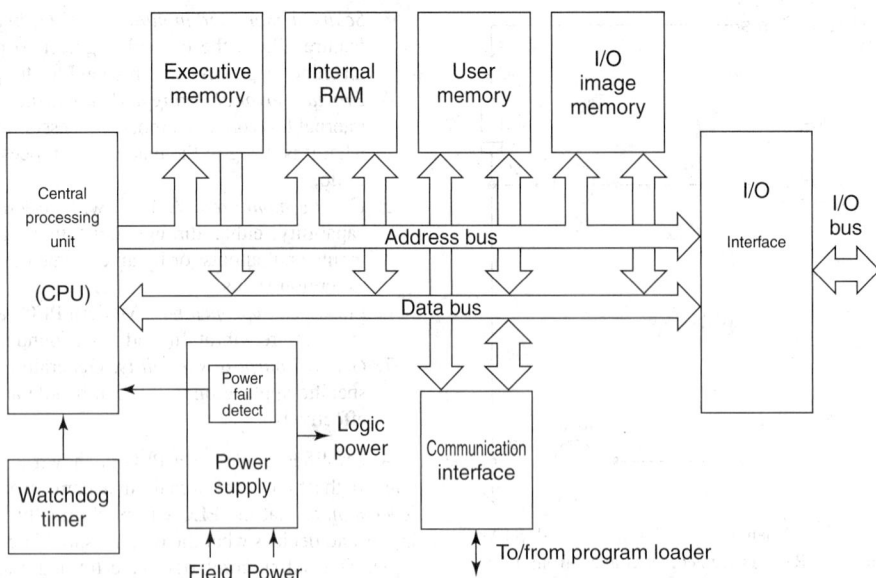

Fig. 4.   Detailed block diagram of the processor section of a programmable controller. The central processing unit (CPU), typically a microprocessor, executes a program written by the manufacturer of the PLC that is stored in the Executive Memory. This executive program that the CPU executes gives the CPU the ability to interpret the user's program. The CPU does not operate on the I/O directly. Rather, it works with an image of the I/O that is stored in the I/O image memory. The I/O interface is responsible for transferring the image outputs to the I/O system and reading the inputs from the I/O system and writing them into the image memory. A "watchdog" timer is provided to time how long it takes the CPU to execute the user's program. If this time exceeds a predetermined value, the watchdog timer causes the processor to fault. If the CPU fails and does not execute the user's program, the watchdog timer will ensure that at least a fault will be indicated and that the processor will shut down in a safe manner.

may take considerably longer to execute. Also in considering a PLC, one should include the scan time of the I/O, the scan time of the memory, and any additional time overheads the processor requires. A detailed block diagram of the processor section of a PLC is given in Fig. 4.

Processor Software. The hardware of the PLC is not too different from that of a lot of computers. What makes the PLC special is the software. The executive software is the program that the PLC manufacturer provides internal to the PLC that executes the user's program. The executive software determines what functions are available to the user's program, how the program is solved, how the I/O is served, and what the PLC does during power up/down and fault conditions.

Executive Software. A simplified model of what the executive software does is shown in Fig. 5. Specific PLC designs perform the basic functions shown somewhat differently. This can make a large difference in program execution. For example, some PLCs may perform diagnostics only at a single point in the executive program, while others may perform diagnostics "on-line," i.e., while the user's program is being solved.

Close attention must be given to how the PLC runs diagnostic tests and what it does during failures. Ignoring this aspect of the PLC can result in an unsafe system.

Multi-Tasking. In a later development, PLCs capable of executing multiple tasks with a single processor appeared. Multi-tasking takes several forms, of which two are: (1) time-driven, and (2) event-driven. In a time-driven system, the user writes programs and assigns I/O for each task. The user will then configure the processor to run each task on periodic time intervals. This type of system is shown graphically in Fig. 6. This feature allows the *time-critical* portion of the control system, such as the portion that controls high-speed motions or machine fault detection, to run many times per second, while allowing the *non-critical* portions, such as servicing indicator lights, to run much slower. Because only the time-critical logic and I/O need quick solutions, versus the entire user's program, faster throughput can be achieved.

Event-driven multi-tasking (also called interrupt-driven) is similar. In this case, the user defines a particular event, such as an input changing state or an output turning off, that causes each task to be run.

In either the time-driven or event-driven case, it is important to recognize the priority of tasks. Some multi-tasking systems allow any task to access any variable, such as an I/O point. Thus, caution must be used when programming multiple tasks that access the same variables. It may be difficult to determine which task is writing which variable while trying to debug a program.

User Software. This is the software that the control engineer writes and stores in user memory in order to perform the required control over the machine or process. User software can contain both configuration data and language programs. The configuration data contain information that tells the processor what its environment is and how it should execute the language problem. The configuration process typically consists of assigning I/O points to particular I/O racks, telling the processor how much memory and I/O it has, assigning specific memory for tasks, determining fatal versus nonfatal faults, and many other items interactively on a program loader.

In as much as the modern PLC is required to do more and more in terms of operator interfacing, communications, data acquisition, and supervisory control, more is required of the language that implements these functions. Therefore, it is crucial that the various aspects of the language be considered.

Information/Output Systems

Direct I/O, as the name implies, is the brute force way of getting I/O to and from the PLC's processor. There is one input signal and one output signal corresponding to the number of inputs and outputs the processor supports. This approach is typically used in the very small PLCs that have all the I/O circuits in the same package as the processor (sometimes called internal I/O). Cost is the principal advantage of internal direct I/O. Some flexibility, however, is lost because the processor must be changed in order to change the I/O.

Parallel I/O Systems. In a parallel system, a parallel I/O bus emanates from the processor's I/O interface and individual I/O modules are plugged into this bus. The I/O module contains the necessary circuitry to decode the bus signals and convert these signals into voltage levels that can drive the necessary loads. I/O modules will typically drive multiple loads. This multiplicity of I/O points is called the *modularity* of the I/O system. Most commercially available I/O systems have modularities of 2, 4, 8, 16, or 32 I/O points per module. Adding more I/O points on a module will commonly reduce the cost per I/O point and reduce the amount of space required to install a given number of I/O points. See Fig. 7.

The failure of one I/O module with many points of I/O on it can be disastrous if it controls many critical devices, such as those causing motions or controlling emergency stops. Thus, it is good practice to split up the critical I/O between the high-density modules.

Serial I/O Systems. Parallel systems are limited in the distance over which one can extend the I/O bus, typically less than 50 feet (15 meters). If the machine should be 100 feet (30 meters) long, one would have to use

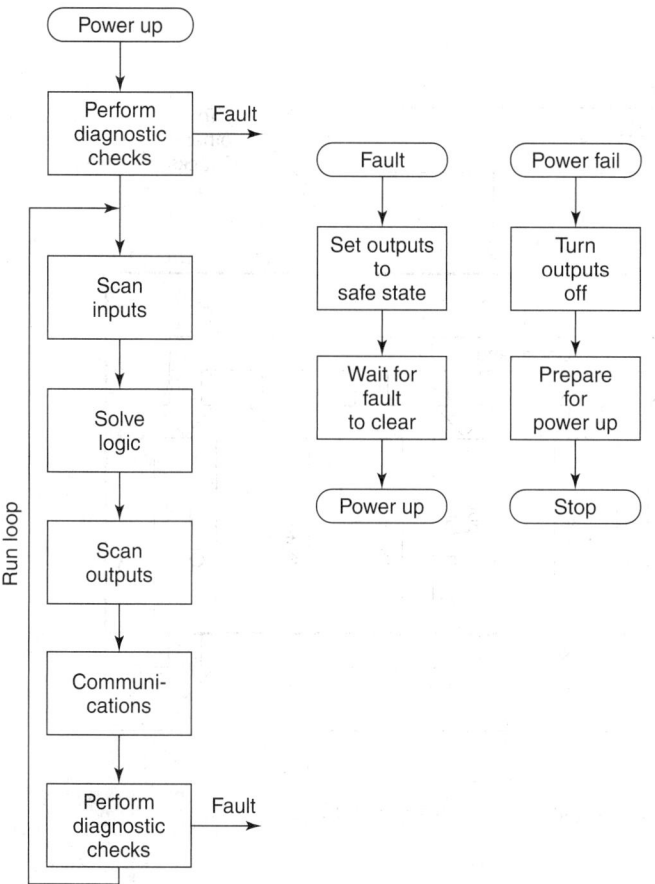

Fig. 5.   The Executive Program shown here controls the functionality of the programmable controller. It controls the actions of the CPU to perform the indicated actions. Diagnostic checks must be run at power-up as well as during the run loop, which is executed while the PC is controlling the process. When faults are detected, the outputs must be set to a predetermined "safe" state the user usually has the choice of turning all of the outputs off or leaving them in their last state. When advance warning of power failure is given by the power supply, the executive program shuts down the CPU in a controlled manner. During the run loop, inputs are only scanned once. This allows the entire user program to operate from a consistent set of inputs because they are only determined prior to executing the user's program and do not change state in the middle of the run loop.

two PLCs. Serial I/O systems solve this problem by transmitting the I/O information over a serial data link capable of being extended over longer distances (1000–10,000 feet; 300–3050 meters). A serial bus emanates from the processor and is connected to a parallel bus through a serial-to-parallel converter. Since a single serial bus contains fewer wires than does the wiring to the loads, large wiring cost savings can be realized by using serial systems. See Fig. 8.

Care must be taken when using serial I/O systems in time-critical applications because two I/O buses have to be scanned—both the serial and parallel bus—instead of one, thus making them slower than straight parallel systems. Some, but not all, serial systems may "desynchronize" themselves from the logic scanning, thus making it more difficult to predict I/O responses to fast-changing signals.

I/O Circuits. An I/O module performs signal conversion and isolation between the internal logic-level signals inside the PLC and the field's high-level signals. There are several different types of I/O circuits available that are capable of driving almost any conceivable load and sensing the status of a wide variety of sensors. Most of these I/O circuits fall into one of five categories, as shown in Fig. 9.

Programmable Controller Communications

The communications aspects of a PLC can severely limit or greatly enhance the applicability of these devices.

Point-to-Point Communications. Most PLCs have at least one communication port built in, i.e., the program loader interface. However, only a few manufacturers release the information needed in order to communicate over this interface. Even so, these ports typically use unusual protocols that can require considerable effort to implement. Some manufacturers of peripheral equipment, such as color graphics displays, have converted their equipment to talk to some PLCs directly, thereby saving the expense of writing specific communications software.

Most PLCs also provide some form of ASCII communications. Some PLCs have separate I/O modules for this purpose, while some others allow the user to reconfigure the program loader port for this purpose. With ASCII communications, it is possible to talk to a wide variety of devices, such as color graphics terminals, intelligent push-button stations, bar code readers, servomotor controllers, etc. Usually such communications can be made over telephone links.

Network Communications. Most PLC manufacturers provide some type of network allowing for communication between their own PLCs. With these networks, it is possible to distribute PLCs physically, but yet have them work in unison by using the network's communication functions. Most of these networks provide three basic functions: (1) reading variables, (2) writing variables, and (3) program upload and download. However, because these networks are designed to provide communication

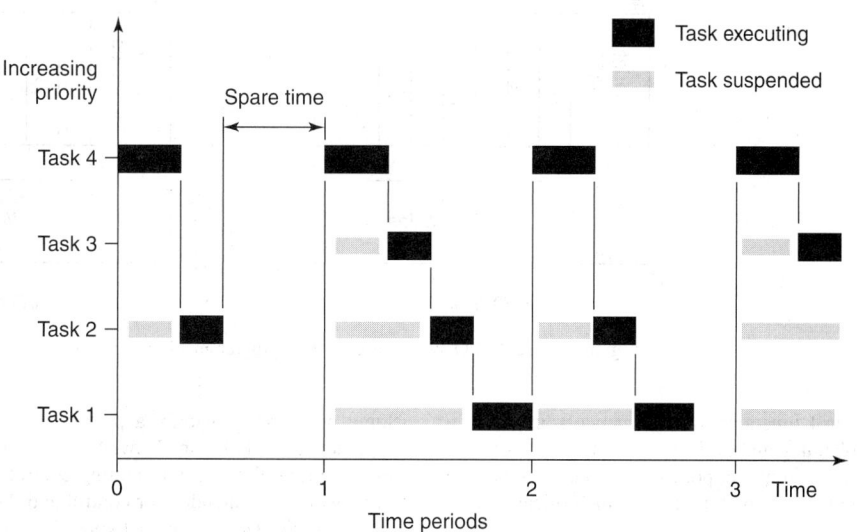

Fig. 6.   In a time-driven multi-tasking system, tasks are scheduled to run on predetermined time intervals. In the example shown here, tasks 4 and 2 are scheduled to run every period while tasks 3 and 1 are scheduled to run every other time period. The higher-priority tasks always execute before the low-priority tasks. During period #1, all four tasks are scheduled to run. However, there is not enough time for task 4 to finish executing. Its execution is suspended until the spare time in period #2 is available. Care must be taken to ensure that there is enough spare time for all the tasks to execute. Some multi-tasking systems will provide an indication that not enough time exists to execute all the programmed tasks, thus making it easier to debug the programs.

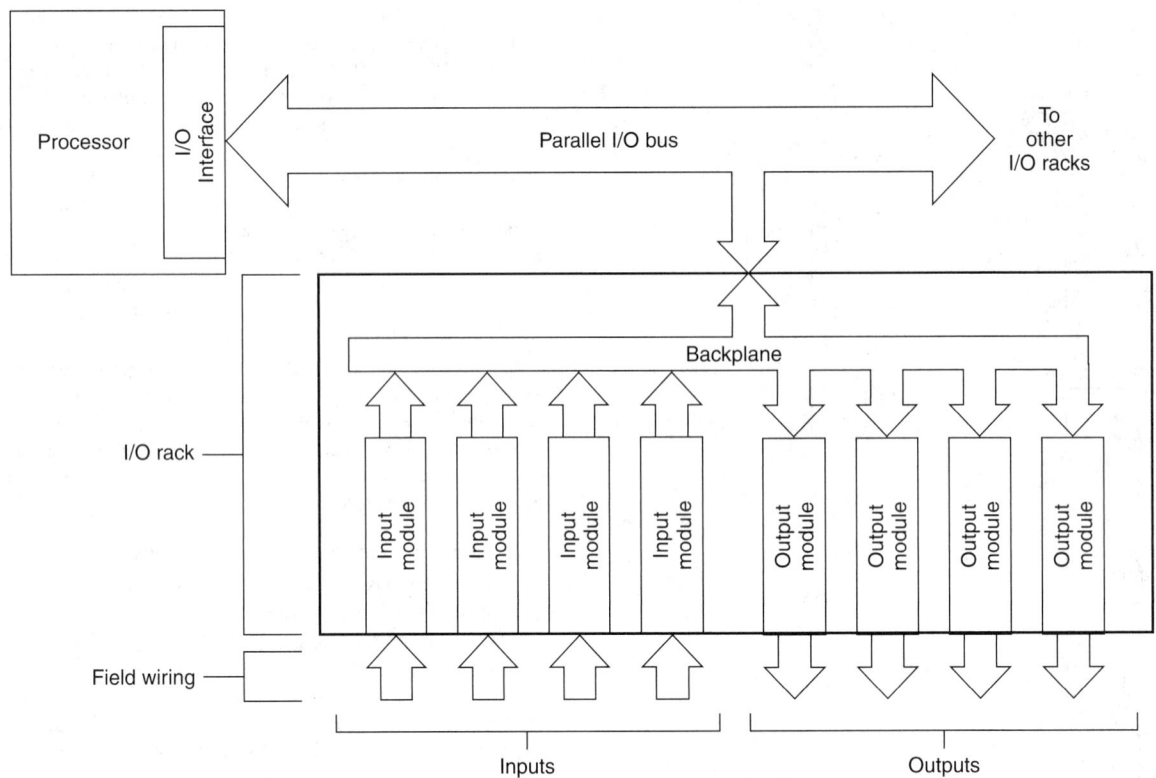

Fig. 7.   Block diagram of a parallel I/O system used with a programmable controller.

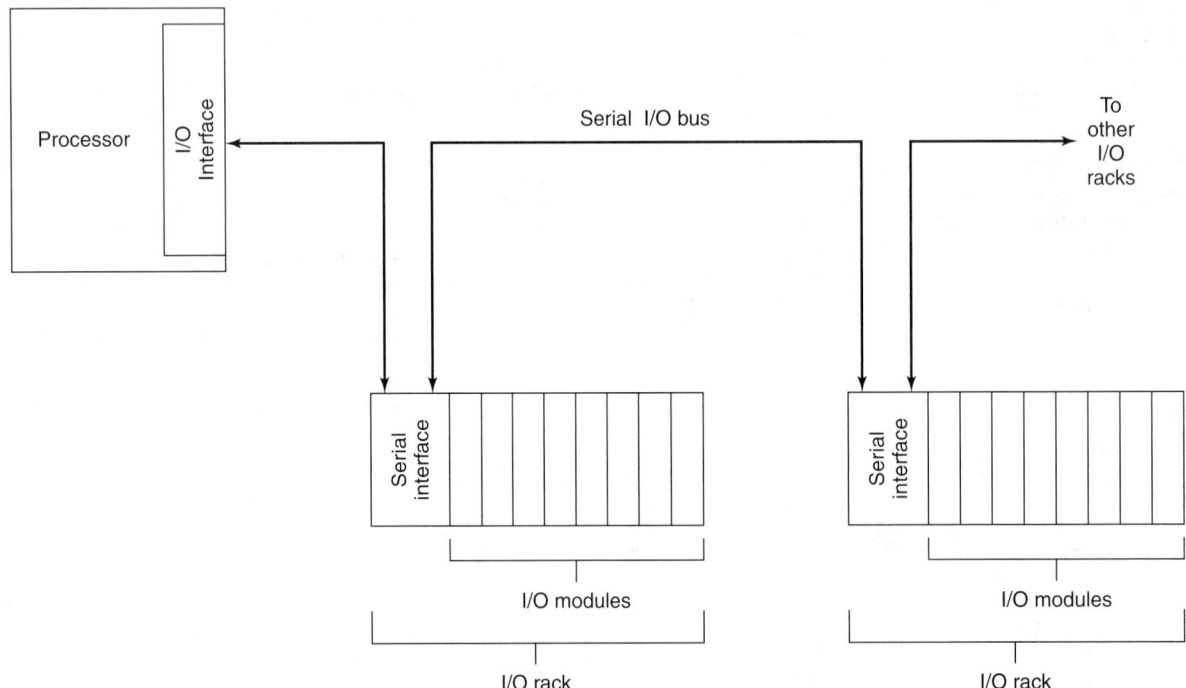

Fig. 8.   Block diagram of a serial I/O system used with a programmable controller.

functions, not necessarily control functions, using a network inside of a control loop requires careful planning and evaluation. Some networks have difficulty transporting information from two points on the network in such a manner as to allow the use of this information in a time-critical control application.

Certain factors must be considered when evaluating a network for a control application, including: (1) *Response time* — the length of time required from an input changing state on one node of a network to a second node receiving notice that the input has changed. This is a critical parameter when trying to implement control of a process over a network.

Some networks give only a probabilistic response time based on some hypothetical installation. However, one should always know the precise response time limits before putting control information on a network. Not all networks are intended for control and thus must be evaluated carefully. (2) *Error checking* — any network that is used for transferring control information should utilize extensive error checking on the information sent over it. Both ends of the network, the sender and the receiver, should be capable of detecting errors. Both ends should also perform specific and known error recovery mechanisms, such as retransmission or, at a minimum, be able to notify both the sender and receiver that there was

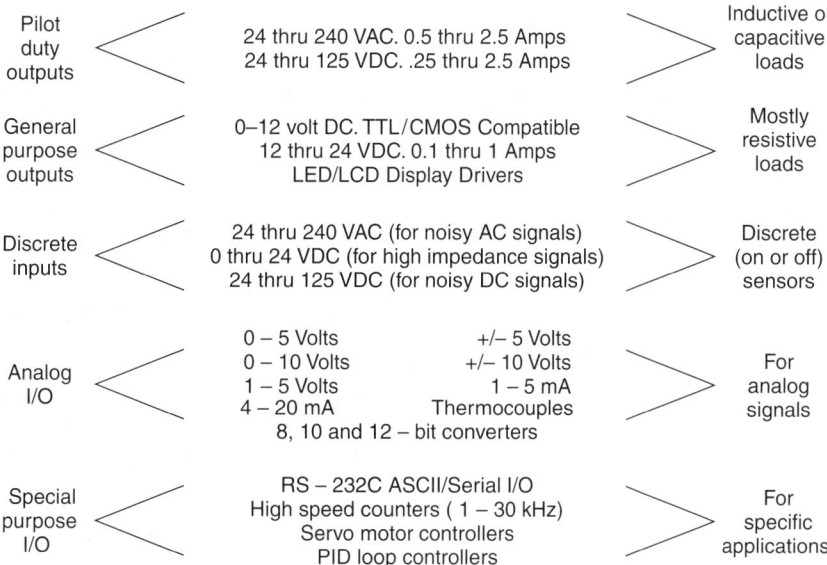

Fig. 9.   Examples of commonly available I/O circuits used with programmable controllers.

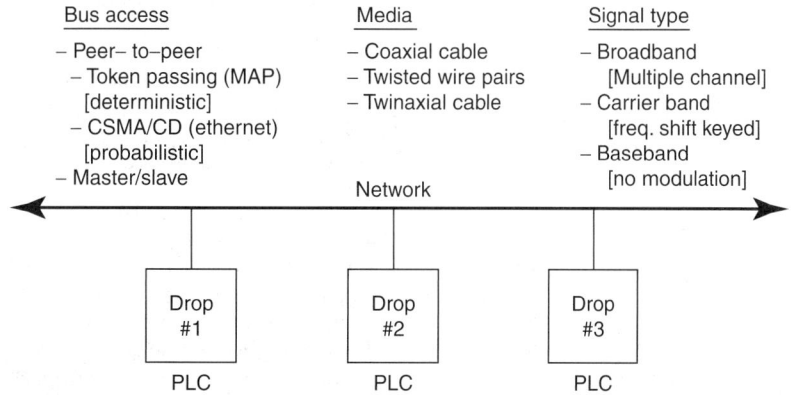

Fig. 10.   Simplified network block diagram with typical features. In master/slave networks, only the master can initiate communications. In peer-to-peer networks, any drop on the network can initiate communications. Although peer-to-peer networks typically offer faster communications, they are sometimes difficult to use for control because they require that a large number of variables be sent between a large number of drops. This causes the number of communication paths to increase exponentially. A network that is used for control should have guaranteed response times and known error recovery methods. A master/slave system can be easier to maintain if variables in the drops change—because only the master and the drop in which the variable changes need be updated. Some networks alleviate this problem by allowing variables to be accessed by names instead of addresses. The software tools that are provided to communicate with can be the most important factor to consider. The media and signal type affect the noise immunity of the network. Coax and twin axial based networks offer good noise immunity, but may be more expensive than some twisted wire networks. Broadband allows for multiple communication channels on the same network (much like a TV has multiple channels), but is very expensive. Baseband is the lowest cost, but does not offer the noise immunity that modulated signals provide.

an error so the control engineers can program their own recovery scheme. (3) *Access mechanisms*—because a network usually contains only one channel over which all PLCs must talk, some method for determining who has access to the network at any given time must be used. Two of the more popular access mechanisms are (a) master/slave, and (b) peer-to-peer. See Fig. 10.

On a *master/slave system*, there is only one master PLC. The master sends commands out to the other slave PLCs, and they respond appropriately. The slaves on the network never initiate their own commands—they always respond to what the master commands them to do.

The *peer-to-peer* mechanism allows any PLC on the network to initiate messages. However, as in the case of humans talking, if everybody talks at once, nothing intelligible can be heard. Peer-to-peer networks need some mechanism for determining access between all the PLCs—not just between the master and the slave. Various mechanisms for determining access have been implemented, such as token-passing and carrier-sense-multiple-access-with-collision detection (CSMA/CD). More details regarding these systems are given elsewhere in this encyclopedia. Check alphabetical index.

RALPH E. MACKIEWICZ, SISCO, Inc., Warren, MI.

PROGRAMMING FLOWCHART (Computer).   Refers to a graphical representation for the definition of a program, in which symbols are used to represent data, flow, operations, equipment, etc. A digital computer program may be charted for two primary reasons: (1) ease of initial program design, and (2) program documentation. By coding from a flowchart, instead of coding without any preliminary design, the programmer usually conserves time and effort in developing the program. In addition, the flowchart is an effective means for transmitting an understanding of the program to someone else.

A programming flowchart is comprised of function blocks with connectors between these blocks. A specific function box may represent an input/output operation, a numerical computation, or a logic decision. The program chart shown in Fig. 1 is of a program that reads values from the process, converts those values to engineering units, limit-checks the converted values and, if there is a violation, prints an alarm message on the process operator's typewriter.

Various levels of detail are presented in programming flowcharts. A functional block in a low-level flowchart may represent only a few computer instructions, whereas a functional block in a high-level flowchart may represent many computer instructions. The high-level flowchart is used mainly for initial program design and as a way of informing a

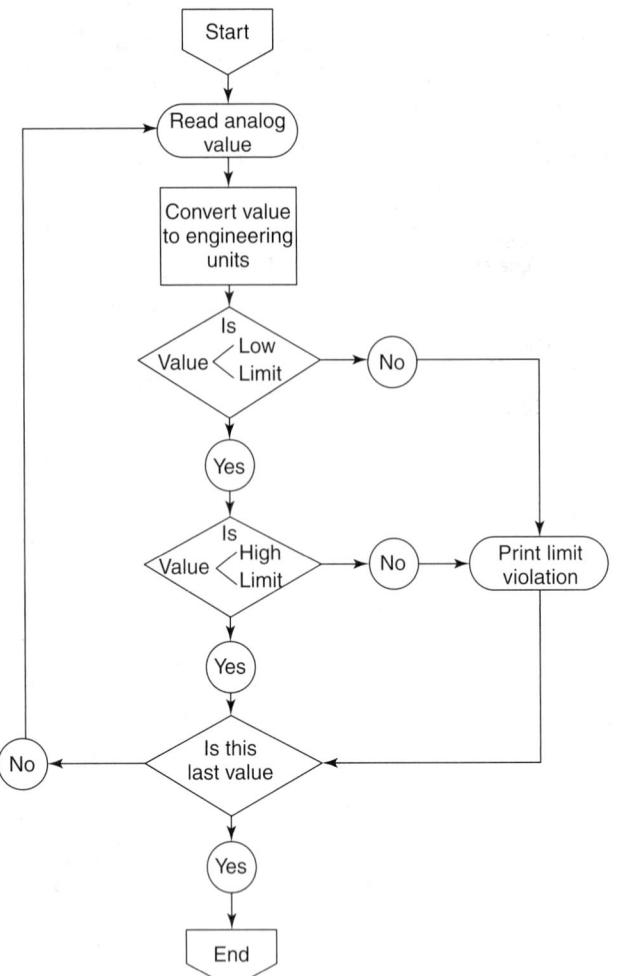

Fig. 1.   Representative computer programming flowchart.

nonprogrammer of what the program does. The low-level chart usually appears in the last stage of flowcharting before a program is actually coded. It is used for documentation that may be needed for later modifications and corrections.

THOMAS J. HARRISON, International Business Machines Corporation, Boca Raton, FL.

PROGRAMMING LANGUAGE.   A language that computer programmers use in writing instructions for a computer to execute (i.e., carry out). Also known as a computer language. There are many kinds of programming languages. Depending on whether the instructions in a programming language closely parallel the primitive instructions that are directly built-in into a computer, one may distinguish between two general classes of programming languages: low-level languages and high-level languages.

Low- level languages. Instructions in a low-level language specify primitive operations that a physical computer is designed to perform directly. Machine languages and assembly languages are low-level languages. A machine language uses the binary coding internal to a physical computer. Each instruction is a sequence of 0s and 1s that denotes a primitive operation of a physical computer. For example, an instruction to add the value at some memory address (say, 12) to a register named R3 (a high-speed storage device within a computer's central processing unit) might be written in a hypothetical machine language as

$$01101011000000001010$$

where 01101 denotes addition, 011 denotes the register R3, and 000000001010 denotes the memory address 12.

Most instructions in an assembly language also denote primitive operations of a computer, but are written using symbols more meaningful to human programmers. For example, the above machine language instruction

might be written in a hypothetical assembly language as

$$\text{ADD R3 X}$$

where ADD denotes addition, R3 denotes the register, and X denotes a memory address. An assembly language program must first be translated into a machine language in order for a computer to execute it. The translator itself is a computer program called an assembler. Clearly, both machine languages and assembly languages are machine-dependent: computers of distinct internal designs (different architectures) usually support different set of instructions for their machine and assembly languages. To a certain extent, a programmer needs to know the internal structure of a physical computer in order to gain proficiency in its assembly language or machine language. Low-level languages are generally considered difficult to use for large programming projects. Their instructions are primitive, and even a program of modest size and complexity may require a fairly large number of instructions and significant programming effort.

High-level Languages. The vocabulary of a high-level language generally includes common mathematical symbols, such as $+$ and $-$, and a small number of key words, mostly from the English language. A computer programmer uses such a vocabulary and follows rigid syntactical rules to write instructions in a high-level language. High-level languages are machine-independent. The mathematical symbols and key words are not tied to the primitive operations or the internal structure of a specific physical computer. They denote arithmetical and logical operations and programming concepts that are commonly understood without reference to a specific computer. High-level languages are easier to use for large programming projects than low-level languages. A statement in a high-level language generally denotes an abstract operation equivalent to several primitive operations (low-level instructions) that are built into a computer. The use of commonly understood abstract operations and programming concepts in a high-level language relieves a programmer from concerns with the low-level details within a computer and enables a programmer to focus on figuring out a logically correct solution to a given problem and expressing the solution in a high-level language.

Clearly, a computer program written in a high-level language cannot be directly recognized and executed by any physical computer, but can be translated into a machine language and then executed on a computer. The translation can be totally automated: the translator itself is a computer program, called a compiler. Alternatively, a program written in a high-level language can be directly executed by another computer program, called an interpreter. An interpreter is essentially a computer built in software, i.e., a computer program designed to recognize and execute the instructions of other computer programs. Compilers and interpreters are referred to as language processors.

There are an enormous number of different high-level languages and their dialects. Based on the computing models adopted by these high-level languages, one may distinguish between three broad families of languages: imperative, functional and logic-based. The languages in the same family are based on a similar computing model, and share similar features that distinguish them from those in other families. Mastery of one representative language in a family makes it easy to learn other languages in the same family.

Imperative Languages. Fundamental to understanding of an imperative language is the concept of a variable, which is a named container of a value. Program statements (instructions in an imperative language) can retrieve and use the value of a variable. Program statements can also change the value of a variable by placing a new value into the variable, overwriting the old value. When a program in an imperative language is executed, the program statements are executed successively, causing the values in variables to be successively used and changed, with the goal of achieving a desired program result. Most high-level languages in use today belong to the imperative family, including the general-purpose languages C, C++, Java, Pascal and Basic, the language COBOL (Common Business Oriented Language) for business data processing and the language FOTRAN (Formula Translation) for use in science and engineering. As an examples, the following statements in the C++ language obtains two integers from the user (one who executes a program to solve a problem) and shows to the user the GCD (greatest common divisor) of the two integers if the two integers are positive. To compute the GCD, the statements use the fact that, given two positive integers, if we repeatedly replace the larger number by the result of subtracting the smaller from the larger, eventually the two will have the same value, which is the GCD.

```
1.    int x, y;
2.    cin >> x >> y;
3.    while (x! = y)
4.        if (x > y)     x = x − y;
5.        else           y = y − x;
6.    cout >> x;
```

Line 1 declares that there are two integer variables named x and y. Line 2 obtains two integers from the user and places them into x and y. Line 3 dictates that, as long as x and y are not equal, lines 4 and 5 are to be repeatedly executed. In lines 4 and 5, if x is greater than y, the value of x is replaced by that of $x-y$; otherwise the value of y is replaced by that of $y-x$. In short, lines 4 and 5 subtract the smaller from the larger of the two values stored in x and y. Such an operation is to be performed repeatedly until x and y have the same value (as specified by line 3). As an example, if the user gives the values 21 and 35 to the program when line 2 is executed, the following table illustrates how the values of x and y are changed by line 4 and 5 as they are repeatedly executed.

x	21	21	7	7
y	35	14	14	7

Line 6 shows the value of x to the user. Since line 6 is executed after lines 4 and 5 have been repeatedly executed so that x and y have the same value, which is the GCD, the value shown to the user by line 6 is the desired result: the GCD of the two input integers.

Although a typical imperative language defines many versatile operations that a programmer can use directly, often a programmer needs an operation that is not built into the language. Modern imperative languages provide a means for a programmer to define a new operation in terms of the existing operations available to the programmer. Such a new operation is variously called a subprogram, a procedure or a function. Defining such a new operation is called procedural abstraction. Besides, most modern languages include features to support a programming concept called data abstraction, which allows grouping of related data and operations as integral parts of an entity called object. Data abstraction is the basis of object-oriented programming (OOP), which is a way to structure a computer program. OOP has been widely accepted in recent years and is supported by such widely used modern languages as C++ and Java. In object-oriented programming, a major task in program design is to identify the objects in the application domain, the relationship between the objects (e.g., an objects may be a specialization of another object), and the interactions among the objects.

Functional Languages. Functional languages are based on the theory of functions in mathematics. A function takes a number of given values as parameters and yields a single value as the result. A program is defined as a function, which may call other functions to compute the values for parameters. Execution of a program is application of a function on given parameter values. As an example, we may define a function to find the GCD (result) of two positive integers x and y (parameters) as follows. The definition uses the fact that if two positive integers are unequal then their GCD is equal to the GCD of the smaller of the two and the result of subtracting the smaller from the larger.

$$gcd(x, y) = \begin{cases} x & \text{if } x = y \\ gcd(x - y, y) & \text{if } x > y \\ gcd(x, y - x) & \text{if } x < y \end{cases}$$

Using 21 and 35 as parameters, the function gcd is evaluated as follows.

$$gcd(21, 35) = gcd(21, 14) = gcd(7, 14) = gcd(7, 7) = 7$$

It is worth noting that recursion (defining a function using the function itself) is used to accomplish the task of repeatedly subtracting the smaller from the larger of the parameters. LISP (LISt Processing) and Scheme, a dialect of LISP, are the most well known languages that adopt functional programming features. As an example, the above function gcd is defined below in the language Scheme

```
(define (gcdxy)
    (cond      ((= xy)x)
        ((> xy)      (gcd(−xy)y))
        ((< xy)      (gcdx(−yx)))
```

After the function has been defined, the function call $(gcd\ 21\ 35)$ in Scheme will yield the answer 7.

Logic-based languages. Logic-based languages are based on predicate logic. A program in a logic language is a set of facts and rules of inference, by which new facts can be deduced from existing facts. Logic-based languages are considered declarative, as opposed to procedural, because a program specifies what a problem is instead of providing a procedure for solving a problem. As an example, one may define the predicate that the GCD of two given positive integers is an integer X as follows.

1. For all positive integers X, the GCD of X and X is X
2. For all positive integers X, Y and Z, if $X > Y$ and Z is the GCD of X-Y and Y, then Z is the GCD of X and Y
3. For all positive integers X, Y and Z, if $X < Y$ and Z is the GCD of X and Y-X, then Z is the GCD of X and Y

A query is a theorem to be proved from known facts and rules of inference. By using the above rules (2 and 3 repeatedly), the query that the GCD of 21 and 35 is 7 can be proved to be true. It can also be deduced that in order for the query that the GCD of 21 and 35 is A (some unknown number) to be true, A must be 7.

The most well known logic-based programming language is Prolog. As an example, the three rules specified above are defined in Prolog below.

$gcd(X, X, X)$
$gcd(X, Y, Z):\text{-}\ X > Y, X1\ is\ X - Y, gcd(X1, Y, Z)$
$gcd(X, Y, Z):\text{-}\ X < Y, Y1\ is\ Y - X, gcd(X, Y1, Z)$

Using these rules, a Prolog interpreter will be able to deduce that the query $gcd(21, 35, 7)$ (i.e., the GCD of 21 and 35 is 7) is true. It can also deduce that in order for the query $gcd(21, 35, A)$ (i.e., the GCD of 21 and 35 is some unknown number A) to be true, A must be 7.

See also **Computer Operating System**.

SAMUEL C. HSIEH, Ball State University, Muncie, IN.

PROGRAM RELOCATION (Computer System). A process of modifying the address in a unit of code, i.e., a program, subroutine, or procedure, to operate from a different location in storage than it was originally prepared for. Program relocation is an integral part of nearly all programming systems. The techniques allow a library of subroutines to be maintained in object form and made a part of a program by relocation and appropriate linking. When a program is prepared for execution by relocating the main routine and any included routines to occupy a certain part of the storage on a one-time basis, the process usually is termed *static relocation*. The resulting relocated program may be resident in a library and be loaded into the same location in the core each time it is executed. Dynamic storage-allocation schemes also may be set up so that each time a program is loaded, it is relocated to an available space in the core. This process is termed *dynamic relocation*. Time-sharing systems may temporarily stop execution of programs and store them on auxiliary storage and later reload them into a different location for continued execution. This process also would be termed dynamic relocation. See also **Time Sharing**.

Software relocation commonly refers to a method whereby a program loader processes all the code of a program as it is loaded and modifies any required portions. Auxiliary information is carried in the code to indicate which parts must be altered. Inasmuch as all code must be examined, this method can be time consuming. In *hardware relocation*, special machine components are used, as a base or relocation register, to alter addresses automatically at execution time to achieve the desired results. In dynamic relocation situations, this is a fast method. Coding techniques also are used for relocation. The resulting code is caused to be self-relocating. The code executes in any core location into which it is loaded. Index register may be used to make all references to storage. Double indexing is required to do both the relocation and the normal indexing operations. Or, the code may actually modify all addresses as it executes to provide the correct reference. This method is slow compared with other methods and requires additional storage.

THOMAS J. HARRISON, International Business Machines Corporation, Boca Raton, FL.

PROGRESSION. A rather simple type of sequence or series, the most common ones being known as arithmetic, geometric, and harmonic.

1. *Arithmetic progression.* The form is a, $a + d$, $a + 2d$, $a + 3d, \ldots$ where a is the first term and d is the constant difference between

two successive terms. If there are n terms in the series, its sum $S_n = n(a + l)/2 = n[2a + (n - 1)d]/2$, where l is the last term. If any three of the five quantities a, d, n, l, S_n are known the other two may usually be found from the simultaneous equations $l = a + (n - 1)d$ and the sum, S_n. If three numbers are members of an arithmetic progression, the middle one is the arithmetic mean. A plot of the magnitude of the terms in an arithmetic progression as ordinate against the number of terms n as abscissa will be a straight line of slope d and intercept a on the Y-axis.

2. *Geometric progression.* Its form is a, ar, ar^2, ar^3, ... and its last term is $l = ar^{n-1}$. If written as $a(1 + r + r^2 + r^3 + \cdots)$ its sum to n terms is $a(1 - r^n)/(1 - r)$. When $r < 1$, the infinite series converges and its sum is $a(1 - r)$. The series diverges for other values of r. The n-th root of the product of n positive quantities x_i is the geometric mean $G = (x_1 x_2 x_3 \cdots x_n)^{1/n}$.

3. *Harmonic progression.* The sequence $a, b, c, \ldots$ is a harmonic progression if the reciprocals $1/a, 1/b, 1/c, \ldots$ form an arithmetic progression. The harmonic mean between two numbers is the middle term of a harmonic progression whose first and last terms are the given numbers. The harmonic mean between a and b is given by $H = 2ab/(a + b)$. If A, G, and H are respectively the arithmetic, geometric, and harmonic mean of two numbers then $G^2 = AH$. The sum of its terms is a harmonic series. There is no general method of finding the sum. It is usually done by taking the reciprocal series and solving the resulting arithmetic series. If the series is

$$\sum_{n=1}^{\infty} 1/n^r, \text{ with } r \text{ real}$$

it converges for $r > 1$, but diverges for $r \leq 1$. This is often called the hyperharmonic series.

Series of these three types are convenient as comparison series.

PROGRESSIVE MULTIFOCAL LEUKOENCEPHALOPATHY.

A disease thought to be caused by an opportunistic virus in patients with impaired immune responses. Papoviruses have been implicated as the most common causal agents. The infection may occur in patients suffering from lymphoproliferative disorders and in those therapeutically immunosuppressed. Multifocal demyelinating lesions, often asymmetrical, result in progressive alteration of personality and intellect, sensory deficit, cortical blindness, and impairment of consciousness. The disease begins insidiously and usually terminates in death within six months.

No treatment has been found effective.

R. C. V.

PROLAPSE.

The falling or protrusion of an organ or structure, due to lack of support, usually secondary to weakness of its retaining ligaments or surrounding muscles. Prolapse of the rectum is a protrusion of a part of the rectal wall externally. Prolapse of the uterus is a protrusion of the uterus and other pelvic organs through the muscular floor of the pelvis. See also **Gonads**.

PROMETHIUM.

Chemical element symbol Pm, at. no. 61, at. wt. 145 (mass number of the most stable isotope), fourth in the Lanthanide Series in the periodic table, mp $1042\,°C$, bp $3000\,°C$ (estimated), density 7.26 g/cm^3 ($20\,°C$). Elemental promethium has a double hexagonal closepacked crystal structure at $25\,°C$. The pure metallic promethium is silverwhite in color, is soft, and can be cast or machined. The naturally occurring isotope [147]Pm is radioactive with a half-life of 2.52 years. Consequently, the element must be handled within a shielded area. Eighteen artificially produced isotopes, ranging from [140]Pm to [146]Pm and from [148]Pm to [158]Pm have been identified, all with very short half-lives. Many of the properties of promethium remain classified by the United States Atomic Energy Commission, or are known by other proprietary sources. Although first identified as an element by J.A. Marinsky, L.E. Glendenin, and C.D. Coryell in 1947, the element was not available on more than a gram-scale for several years. Electronic configuration

$$1s^2 2s^2 2p^6 3s^2 3p^6 3d^{10} 4s^2 4p^6 4d^{10} 4f^4 5s^2 5p^6 5d^1 6s^2$$

Ionic radius 0.98 Å. Other important physical properties of promethium are given under **Rare-Earth Elements and Metals**.

[147]Pm is extracted from the wastes of uranium or plutonium reactors, the most important source of the element. [146]Pm and [148]Pm also are derived from reactor wastes. In 1970, [147]Pm became available in kilogram quantities. [147]Pm has been under intensive study as a heat and power source; however, before it can be used for this, [146]Pm and [148]Pm, which produce penetrating gamma radiation, must be eliminated. The desirable property of [147]Pm is that it decays by beta emission only, at a low energy level compared with most fission products, and thus requires only light to moderate shielding. [147]Pm has been used to activate luminescent phosphors. Beads (Microspheres®, 3 M Company) containing [147]Pm mixed with a phosphor provide a long-lived, reliable green light and were used by astronauts to assist in docking and other maneuvers in outer space. Commercial applications of [147]Pm as a power source include beta-voltaic cells for surgical implant with heart pumps and pacemakers.

Additional Reading

Greenwood, N.N. and A. Earnshaw: "Chemistry of the Elements," 2nd Edition, Butterworth-Heinemann, Inc., Woburn, MA, 1997.

Krebs, R.E.: "The History and Use of Our Earth's Chemical Elements: A Reference Guide," Greenwood Publishing Group, Inc., Westport, CT, 1998.

Lide, D.R.: "CRC Handbook of Chemistry and Physics 2000–2001," 81st Edition, CRC Press, LLC., Boca Raton, FL, 2000.

Parker, P.: "McGraw-Hill Encyclopedia of Chemistry," 2nd Edition, The McGraw-Hill Companies, Inc., New York, NY, 1993.

Stwertka, A. and E. Stwertka: "A Guide to the Elements," Oxford University Press, Inc., New York, NY, 1998.

PRONGHORN ANTELOPE *(Mammalia, Artiodactyla)*.

Also sometimes referred to as the American Antelope, actually this animal is not classified as an antelope *(Antelopine)*, but rather it falls into a special, small class of the *Artiodactyla* (even-toed hoofed animals), known as the *Antilocaprines*. The *Antilocapra americana* is a plains animal of the western half of North America. It is distinguished by the erect horns, hooked at the tip, and bearing a short branch in front. The pronghorn differs from true antelopes in the branching of the horn sheaths and in the periodical shedding and renewal of these sheaths. The horns are constructed something like those of a giraffe. However, it is much like an antelope in general appearance. The pronghorn is approximately 3 feet (0.9 meter) high, 5 feet (1.5 meters) long, with large eyes, short tail, erect ears, and slender legs. The coloration is a mottled brown, with a chestnut mane. Its recognition mark is a brilliant white rump patch. The pronghorn is known for its fast speed and running endurance. The animal, although timid, is described as curious. The diet is almost entirely grass. It is hunted for sport and food. The flesh is dry, but considered good. The animals faced extinction in the early 1900s, but as the result of government protection, have staged a strong comeback on reservations.

The pronghorn is the only animal in the *Antilocaprines* family. Nothing like the pronghorn is known elsewhere and it is assumed that the species is of North American origin. The animal is considered by some authorities as a leftover from pre-glacial times, when there were probably several more species. The animal also has been described as a possible experiment by nature in the development of an antelope, which was given up without carrying the process to completion.

PROPAGATION CONSTANT.

This is a characteristic of a transmission line and indicates the effect of the line on the wave being transmitted along the line. It is a complex quantity having a real term, the *attenuation constant*, and an imaginary term, the *wavelength constant* or *phase constant*. See also **Line (Mathematics)**. The attenuation constant is a measure of the reduction in amplitude as the wave travels along a matched line while the wavelength constant is a measure of the phase shift which it undergoes. These relations may be expressed by the following equations:

$$\gamma = \alpha + j\beta$$
$$I = I_s e^{-\gamma x} = I_s e^{-\alpha x} e^{-j\beta x}$$
$$E = E_s e^{-\gamma x} = E_s e^{-\alpha x} e^{-j\beta x}$$

where γ is the propagation constant, α, the attenuation constant, β, the wavelength constant, x the distance from the input of the line, I_s and E_s the current and voltage at the input of the line, and I and E the corresponding quantities at a distance x from the input, and e is the number $2.718\ldots$, the base of natural logarithms.

Because of the attenuation on lines used for communication purposes it is necessary to insert amplifiers or repeaters at intervals to build the signal back to suitable levels. For sound transmission work the phase shift is usually not important, but for television and picture transmission it is extremely important and necessitates correcting circuits.

PROPAGATION (Direction of).
At any point in a homogeneous, isotropic medium, the direction of time-average energy-flow. In a uniform waveguide, the direction of propagation is often taken along the axis. In the case of a uniform lossless waveguide, the direction of propagation at every point is parallel to the axis, and in the direction of time-average energy-flow.

PROPANE.
$CH_3 \cdot CH_2 \cdot CH_3$, formula weight 44.09, colorless gas, mp $-187.1°C$, by $-42.2°C$, sp gr 0.585 (at $-45°C$). The gas is slightly soluble in H_2O, moderately soluble in alcohol, and very soluble in ether. Although a number of organic compounds which are important industrially may be considered to be derivatives of propane, it is not a common starting ingredient. The content of propane in natural gas varies with the source of the natural gas, but on the average is about 6%. Propane also is obtainable from petroleum sources.

Liquefied propane is marketed as a fuel for outlying areas where other fuels may not be readily available and for portable cook stoves. In this form, the propane may be marketed as LPG (liquefied petroleum gas) or mixed with butane and pentane, the latter also constituents of natural gas (1.7% and 0.6%, respectively). LPG also is transported via pipelines in certain areas. The heating value of pure propane is 2,520 Btu/ft^3 (283 Calories/m^3); butane 3,260 Btu/ft^3 (366 Calories/m^3); and pentane 4,025 Btu/ft^3 (452 Calories/m^3). Propane and the other liquefied gases are clean and appropriate for most heating purposes, making them very attractive where they are competitively priced.

PROPER MOTION (Star).
The individual motion of a star relative to the other stars. Up to the early part of the eighteenth century, the belief was current that the stars were all fixed on a sphere commonly known as the celestial sphere. Since this sphere was apparently rotating about the earth, and possessed other motions such as precession and nutation, all of the stars had certain motions in common. In 1718, Edmund Halley, while reducing his observations of the positions of the stars, noted that the positions he obtained for Sirius and Arcturus differed in position relative to the other stars from the positions given by Ptolemy. Since the differences were greater than could be ascribed to errors in observation, Halley concluded that these two stars were actually not fixed on the celestial sphere but were in motion.

Since Halley's time, many stars have been observed to have proper motion, and many long programs are under way to study these motions. The standard method of procedure is to compare positions of the stars at two epochs as widely separated as possible. Visual observations made with extreme precision with a meridian circle may be used for this purpose, but the photographic methods are far more fruitful since a large number of star positions may be obtained on a single plate. The longer the interval of time between the observations, the more accurate is the determination of the proper motion, and also the smaller the proper motion which can be detected.

Proper motion can be determined only in angular units, and the results are usually expressed in terms of seconds of arc per year. If the parallax of the star is known, the velocity in linear units may be computed in accordance with methods discussed under space velocity of a star. The largest known proper motion, 10.25″ per year, was found by Barnard for a tenth magnitude star. Such a star would require about 200 years to change its position by an amount equal to the apparent diameter of the moon. There are only about 50 stars known to have proper motions greater than 2″ per year, and not more than 1,000 with values greater than $\frac{1}{2}″$ per year. Hence, we should not expect the constellation Figures to have altered appreciably in the 5,000 years since they were first described.

PROPOLIS.
A material gathered by honey-bees and used for closing crevices in the hive and for filling in sharp angles and attaching loose parts. It is also applied as a varnish to the combs and to the smooth surfaces in the hive. The substance is composed chiefly of resins gathered from plants and has an aromatic fragrance much like that of the leaf buds which furnish some of these resins.

PROPORTIONAL COUNTER.
A detector of ionizing radiations that operates in a voltage region intermediate between an ionization chamber and a Geiger counter. For this counter, the size of the output pulse is proportional to the number of ions formed in the initial ionizing event. Because its operating voltage is lower than for a Geiger counter, avalanche ionization is limited to that portion of the counter in the immediate vicinity of the primary ionization and does not spread along the entire central wire electrode. As a result, its gas amplification is constant for all pulses at any one voltage. The gas amplification is the number of additional ions produced by each electron produced in the initial ionizing event as it travels to the central wire. See also **Geiger Counter**.

PROPORTIONAL LIMIT.
The maximum unit stress that can be obtained in a structural material without causing a change in the ratio of the unit stress to the unit deformation is called the proportional limit.

PROPRIOCEPTIVE STIMULATION.
Stimulation originating within the deeper structures of the body (muscles, tendons, joints, etc.) for sense of body position and movement and by which muscular movements can be adjusted with a great degree of accuracy and equilibrium can be maintained.

PROSTAGLANDINS.
A group of physiologically active compounds (PGs) derived from fatty acids with 20 carbon atoms (approximate formula, $C_{20}H_{36}O_5$). The compounds originally were isolated as lipid-soluble extracts from sheep and human prostates. Later studies have shown that prostaglandins are found in most mammalian tissues. There are numerous prostaglandins, individually named by the substituents present on the cyclopentane ring that is part of the parent molecule, prostanoic acid. Thus, they are identified as PGA$_1$, PGE$_1$, PGI$_2$ (prostacyclin), etc. The chemical structure and metabolic functions of the prostaglandins have been established, in most cases, with considerable accuracy. Some have been synthesized. Each prostaglandin has specific effects. The compounds participate in pulmonary circulation and hypertension, with varying vasodilator and vasoconstrictor effects. Prostaglandins of the E series have been implicated as a cause of hypercalcemia—they resorb fetal bone in vitro, urinary prostaglandin metabolites are elevated in certain hypercalcemic patients with malignancy, and clinically very important, in certain cancer patients with hypercalcemia. Chemical improvement has been seen after treatment with indomethacin, which inhibits prostaglandin synthesis. Prostaglandins also are implicated in systemic mastocytosis, due partly to marked overproduction of prostaglandin D$_2$. Prostacyclin plays an important role in platelet function, acting as an effective antiaggregating agent. The prostaglandins are involved in the biochemical pathways that participate in bronchial asthma. PGs are synthesized ubiquitously in the body from unsaturated fatty acid precursors with high rates of production by the seminal vesicles and renal medulla. The metabolism of prostaglandins occurs mainly in the lungs, renal cortex, and liver, with the metabolites excreted in the urine. The most prolific source of natural prostaglandins is a marine organism (gorgonian sea whip) found in great numbers in coral reefs, notably in the Caribbean area. Intermediates and chemical analogs derived from this organism are sometimes referred to as *syntons*.

PROSTATE GLAND.
See **Gonads**.

PROTACTINIUM.
Chemical element, symbol Pa, at. no. 91, at. wt. 231.036, radioactive metal of the Actinide Series, mp is estimated at less than 1600°C. All isotopes are radioactive. The most stable isotope is ^{231}Pa with a half-life of 3.43×10^4 years. The latter is a second-generation daughter of ^{235}U and a member of the actinium ($2n + 3$) decay series. See also **Radioactivity**. Electronic configuration

$$1s^2 2s^2 2p^6 3s^2 3p^6 3d^{10} 4s^2 4p^6 4d^{10} 4f^{14} 5s^2 5p^6 5d^{10} 5f^2 6s^2 6p^6 6d^1 7s^2$$

Ionic radii Pa^{4+} 0.91 Å; Pa^{3+} 1.06 Å. See also **Chemical Elements**.

The probable existence of protactinium was predicted as early as 1871 by Mendeleev to fill up the space on his periodic table between thorium (at. no. 90) and uranium (at. no. 92). He termed the unconfirmed element *ekatantalum*. In 1926, O. Hahn predicted the properties of the element in considerable detail, including descriptions of its compounds. In 1930, Aristid v. Grosse isolated 2 milligrams of what then was termed ekatantalum pentoxide and showed that element 91 differed in all reactions with comparable amounts of tantalum compounds with exception of precipitation by NH_3. However, credit for the discovery of protactinium generally is attributed to Lise Meitner and Otto Hahn in 1917.

Protactinium-231 yields actinium-227 by α-particle emission and has a half-life of 3.43×10^4 years. Its other isotopes include two isomers of mass number 234: uranium X_2 with a half-life of 1.17 minutes, and uranium Z with a half-life of 6.7 hours, the former being an excited state which undergoes de-excitation to give the latter. Other nuclear species have mass numbers 225–230, 232, 233, 235 and 237.

Protactinium (of mass number 231) is found in nature in all uranium ores, since it is a long-lived member of the uranium series. It occurs in such ores to the extent of about $\frac{1}{4}$ part per million parts of uranium. An efficient method for the separation of protactinium is by a carrier technique using zirconium phosphate which, when precipitated from strongly acid solutions, coprecipitates protactinium nearly quantitatively. Then the protactinium is separated from the carrier by fractional crystallization of zirconium oxychloride.

Isotopes of protactinium can also be produced artificially, i.e., by the nuclear reactions of other elements with such particles as deuterons, neutrons, and alpha-particles. Thus, when thorium is bombarded with deuterons of various high energies, five of the reactions are: $^{232}Th(d,4n)^{230}Pa$, $^{232}Th(d,6n)^{228}Pa$, $^{232}Th(d,7n)^{227}Pa$, $^{232}Th(d,8n)^{226}Pa$, and $^{230}Th(d,3n)^{229}Pa$.

Quantitative methods of obtaining protactinium start from the carbonate precipitate from the treatment of the acid extract of certain uranium ores. After this carbonate precipitate is dissolved, the protactinium remains in the silica gel residue, from the solution of which it is obtained on a manganese dioxide carrier. An alternate method effects final separation of the protactinium by formation of a complex compound, protactinium-cupferron, and its extraction with amyl acetate.

The methods of purification include the use of ion exchange resins, the precipitation of protactinium peroxide and the extraction of aqueous solutions of protactinium salts by various organic solvents.

Protactinium metal is prepared: (1) by reducing the tetrafluoride with metallic barium at about $1,500\,°C$; (2) by heating the halide, usually the iodide, under a high vacuum; and (3) by bombardment of the oxide under high vacuum with 35-keV electrons for hours at a current strength of 0.005–0.010 Amperes.

As early as 1965, investigators at Los Alamos (Fowler et al., 1965) reported that protactinium metal is superconductive below 1.4 K. In 1972, researchers at Harwell (Mortimer, 1972) reported no superconductivity of the metal down to approximately 0.9 K. An exchange of information to resolve the differences in data was conducted over the next few years (Fowler, 1974; Hall et al., 1977). Smith, Spirlet, and Mueller (1979) reported that differences in experimental research were due to problems with the crystal structure of the metal and sample purity that arise when dealing with radioactive material. These investigators observed very-high-purity protactinium, produced by the Van Arkel procedure, and observed an extremely steep superconductivity transition at 0.42 K in protactinium in the presence of rather high self-heating. The superconducting transition temperature and upper critical magnetic field of protactinium were measured by alternating-current susceptibility techniques. Inasmuch as the superconducting behavior of protactinium is affected by its $5f$ electron character, it has been further confirmed that protactinium is a true actinide element.

The predominant oxidation state of the element is (V). There is some evidence that the (IV) state is obtained under certain reduction conditions. When the pentapositive form is not in the form of a complex ion it may exist in solution as PaO_2^+. The compounds are very readily hydrolyzed in aqueous solution yielding aggregates of colloidal dimensions, thus showing marked similarity to niobium and tantalum in this respect. These properties play a dominant role in the chemical properties of aqueous solution, because the element is so easily removed from solution by hydrolysis and adsorption. Protactinium coprecipitates with a wide variety of substances, and it seems likely that the explanation for this lies in the hydrolytic and adsorptive behavior.

The element is difficult to maintain in aqueous solution in the form of simple salts. Solubility data seem to indicate that such amounts as can be dissolved probably do so entirely by formation of complex ions. Fluoride ion strongly complexes protactinium, and it is due to this that protactinium compounds are in general soluble in hydrofluoric acid.

Protactinium oxide may be prepared from the hydrated oxide or the oxalate by ignition. The product is a dense white powder with a very high melting point; the ignited material is not hygroscopic and maintains a constant weight upon exposure to the air. The formula Pa_2O_5 has been determined indirectly, and there is evidence for the existence of $PaO_{2.25}$ (air oxidation) and PaO_2 (reduction of P_2O_5 by H_2).

Volatile protactinium pentachloride has been prepared in a vacuum by reaction of the oxide with phosgene at $550\,°C$ or with carbon tetrachloride at $200\,°C$. Reduction of this at $600\,°C$ with hydrogen leads to protactinium(IV) tetrachloride, $PaCl_4$, which is isostructural with uranium(IV) tetrachloride, UCl_4. The pentachloride can be converted into the bromide or iodide by heating with the corresponding hydrogen halide or alkali halide.

The volatile fluoride protactinium(V) fluoride, PaF_5, or possibly protactinium(V) oxyfluoride, $PaOF_3$, is formed at relatively low temperatures such as $200\,°C$ from the action of agents such as bromine tri- or pentafluoride, BrF_3 or BrF_5, on one of the protactinium oxides. At higher temperatures, treatment of Pa_2O_5 with hydrofluoric acid and hydrogen yields PaF_4.

The reduction of protactinium to the (IV) state in aqueous solution can be accomplished by reducing agents, such as zinc amalgam, and polarographically.

Additional Reading

Fowler, R.D., et al.: *Phys. Rev. Lett.*, **15**, 860 (1965).

Fowler, R.D., et al.: "Proceedings of the 13th International Conference on Low Temperature Physics" (K.D. Timmerhaus, et al.), Plenum, New York, NY, 1974.

Greenwood, N.N. and A. Earnshaw: "Chemistry of the Elements," 2nd Edition, Butterworth-Heinemann, Inc., Woburn, MA, 1997.

Hall, R.O.A., J.A. Lee, and M.J. Mortimer: *J. Low Temp. Phys.*, **27**, 305 (1977).

Krebs R.E.: "The History and Use of Our Earth's Chemical Elements: A Reference Guide," Greenwood Publishing Group, Inc., Westport, CT, 1998.

Lide, D.R.: "CRC Handbook of Chemistry and Physics 2000–2001," 81st Edition, CRC Press, LLC., Boca Raton, FL, 2000.

Mortimer, J.J.: Harwell Report AERE-R 7030 (1972).

Parker, P.: "McGraw-Hill Encyclopedia of Chemistry," 2nd Edition, The McGraw-Hill Companies, Inc., New York, NY, 1993.

Smith, J.L., J.C. Spirlet, and W.C. Miller: "Superconducting Properties of Protactinium," *Science*, **205**, 188–190 (1979).

Stwertka, A. and E. Stwertka: "A Guide to the Elements," Oxford University Press, Inc., New York, NY, 1998.

PROTEIN. Along with the carbohydrate and lipid[1] components of the animal diet, protein substances are a major source of nutrition and energy for the living system. Because of his high regard for the proteins, but well before they were really understood, the Dutch chemist Gerardus Mulder (1802–1880) pioneered the use of the term *protein*, derived from the Greek word meaning "to come first." Although proteins furnish energy to the body and thus can be considered body fuels, as are the carbohydrates and fats, the major nutritional roles of the proteins reside in other functions, usually of a highly specific nature. Thus, there are structural, contractile, process-activating, and transport proteins, among others, which essentially are responsible for the chemical workability of the animal system.

Considering the research tools available, the amount of qualitative and quantitative information pertaining to proteins collected over several decades of effort has been tremendous. The data amassed have been highly beneficial to the medical and health sciences, notably in terms of dietary requirements and protein deficiency diseases, to biologists, and, of course, to organic chemists. Past protein research has led to the development of many useful protein substances for industry and commerce. Scientists stretched the limitations of their available instrumental techniques (crystallography, electron microscopy, chromatography, electrophoresis) in their efforts to better understand protein structure and protein function. With the advent of molecular biology (studying proteins at the molecular level), the potential for learning more pertaining to structure and of what proteins do and how they behave in living organisms increased, conservatively speaking, by an order of magnitude or more. As of the later 1980s, protein science has progressed just a little beyond the initial efforts to reduce protein studies to the molecular level. Highlights are summarized in the latter portion of this article. There are several keys to expanding protein knowledge, two of the most important of which are continued mapping of organism genomes, notably mapping the human genome; and the continuing development of improved instrumental and procedural techniques. In

---

[1] Fats, oils, fatty acids, phospholipids, and sterols.

using this newly acquired knowledge to manipulate protein structures, the term "protein engineering" is sometimes used. Protein engineering largely lies in the future.

Protein Requirements

In the growing animal body, a significant portion of proteins consumed is required for the creation of new tissue. This results in an increasing requirement for proteins in the diet of humans, for example, up to about the age of 20 years, at which time the protein requirement tends to level off to a fairly stable figure. After body maturity, the portion of proteins needed for tissue maintenance is greater than the need for new tissue building. It must be emphasized, however, that immediately at the commencement of life both new tissue building and tissue maintenance take place and even as the body grows older, the two needs continue—only the proportions between the two roles change.

Proteins, on a weight basis, are second only to water in their presence in the human body. If the factor of water is discounted, then about 50% of the body's dry weight is made up of numerous protein substances, distributed about as follows: 33% in muscles; 20% in bones and cartilage; 10% in skin; the remaining 37% in numerous other body tissues. With exception of the urine and bile in the normal healthy individual, all other body fluids contain from small to relatively large portions of protein substances.

Chemically, proteins are distinguished from other body substances in that all proteins contain nitrogen. Some contain sulfur, phosphorus, iron, iodine, cobalt, and other elements, some of which are generally not thought of as components of the life process, but which nevertheless do play extremely important roles (e.g., as catalysts), even if present only in very minute quantities.

In considering the importance of proteins to building and maintaining body functions, it must be emphasized that proteins consumed essentially are raw materials that contain the building blocks for the creation of different proteins. These building blocks are the amino acids of which the protein molecules consumed are constructed and of which the proteins restructured in the body (after consuming or metabolizing the raw materials) are also constructed. Thus, the desirability of proteins for the diet is based upon the best combination of amino acids present. Therefore, some foods are desirable from a protein nutrition standpoint not only because, with relation to their carbohydrate and fat content, they contain a high percentage of protein, but also because they contain most or all of the amino acids needed to form new proteins within the body. See also **Amino Acids**.

Examples of this situation (desirable versus less desirable proteins) popularly cited are the soybean proteins and the grain proteins. With exception of the sulfur-bearing amino acids, notably methionine, the amino acid balance of soybean proteins is reasonably good. With exception of the amino acid lysine, the amino acid balance of grain proteins is reasonably good. By mixing protein substances from these two sources, an excellent source of protein for the human diet is obtained, this explaining growing trends toward fortification of wheat and other cereal flours with soy flour. There are scores of examples of this type which are representative of the trend toward so-called *fabricated foods*.

From years of experience in studying the dietary needs of humans, nutritionists and biologists established the hen egg as having the most perfect balance of amino acids in a natural protein substance. Against this standard, other foods can be rated in their performance. In naming the following food substances in order of their diminishing chemical score, it should be stressed that these foods are arranged only in terms of this one nutritional criterion: fish (70), beef (69), cow's milk, whole (60), brown rice (57), polished white rice (56), soybeans (47), green leaves (45), brewer's yeast (44), groundnuts (peanuts) (43), whole grain maize (corn) (41), cassava (manioc) (41), common dry beans (34), white potato (34), white wheat flour (32). The foregoing food items were selected *randomly* to provide a sense of the spectrum of foods from this one particular standpoint. The Figures represent only the chemical balance of amino acids present and not the total amount of protein available as a weight percentage of food intake, or from the standpoint of protein utilization, once ingested.

In looking at a number of food substances, again a random selection, from the standpoint of total protein (with no regard to quality) in an average serving, the following amounts of protein (grams) are present: fried chicken breasts (27.8); canned tunafish (24), cooked round roast of beef (24), roasted leg of lamb (22), oven-cooked pork loin (21), dry cooked soybeans (13), whole milk (1 cup) (9), canned red beans (7.5), cheddar cheese (1 ounce = 28 grams) (7), fresh cooked lima beans (6.5), egg (medium size) (6), vanilla ice cream (6), fried crisp bacon (5), baked potato (3), cooked broccoli (2.5), cooked oatmeal (2.5), enriched white bread (1 slice) (2), cooked green snap beans (1), lettuce ($\frac{1}{4}$ head) (1), and reconstituted frozen orange juice (1).

Consequences of Protein Deficiency. Because proteins are so important to numerous and very complex bodily functions, years of research have just commenced to provide some understanding of most of the mechanisms involved. As would be expected, recognition of the extreme manifestations of protein deficiencies has taken place, at least to the extent of providing new guidelines for assisting millions of inadequately fed people in several regions of the world. As further experience is gained in researching the gross problem, the important subtleties of protein performance within the body will become more apparent.

Exemplary of a better understanding and appreciation of protein nutrition is a comparison of the 1945 report of the Food and Agriculture Organization (United Nations) with more recent findings, recommendations, and nomenclature used. In the first *World Food Survey*, the terms *undernourishment* and *malnourishment* were used throughout the report. The general interpretation of undernourishment was taken to mean an inadequate caloric intake, i.e., insufficient energy input to support normal body functions and activities, with body weight loss the inevitable result. Similarly, *malnourishment* was taken to mean a deficiency of one or all of the protective nutrients, such as proteins, vitamins, and minerals. During the last few years, inasmuch as these two problems are so interrelated, the term *protein-calorie malnutrition* (PCM) has come into wide use. PCM of early childhood, particularly in regions that are a part of some of the less developed countries, is quite widespread. PCM apparently is manifested in minor ways at first, but when prolonged very severe syndromes become evident. These include the conditions known as *kwashiorkor* and *marasmus.*

Kwashiorkor usually occurs in the second or third year in the life of a child. Edema is the principal symptom. The condition arises from a combination of circumstances, but the primary cause appears to be a weaning diet that is both inadequate and indigestible and, notably, is lacking of protein. The principal calories are supplied by carbohydrate. The condition is accelerated by repeated infections of a bacterial, parasitic, or vital nature. Without treatment, the disease is fatal in most cases.

Nutritional marasmus is a severe manifestation of PCM and is a condition that usually occurs during the first year of life. Again, it arises from a combination of conditions, frequently widespread in many regions, of feeding an overly diluted formula of cow's milk, thus reducing the protein input well below minimum needs. The condition is accelerated by filthy surroundings and contaminated bottles. Characteristic of the syndrome are a wasting of muscle and subcutaneous fat, a body weight that may be only 60% of standard, and diarrhea. Children who have access to human milk usually are protected against marasmus and diarrheal disease.

A more recent finding and term now used for a protein deficiency syndrome is *PCM-plus*, or *infantile obesity*. This is a condition that occurs among the more affluent populations where an infant is bottlefed, where hygiene is adequate, and where funds are adequate. Overfeeding of an improperly balanced formula can cause the condition. The condition does not occur with breast feeding because the volume of intake is regulated by the infant's appetite and thirst.

Sources of Proteins

The two basic categories of protein sources for the animal diet are other *animals* (living or dead) and *plants*. Thus, in the animal category as a source of human and pet protein foods, there are what might be called terminal sources or nonreplenishing sources, in which the living animal is killed and disassembled into its protein-containing parts. The most common examples including the meaty flesh and organs of beef cattle, pigs, sheep, and horses and goats, as well as the more occasional sources of meat, such as deer, elephant, hippopotamus, etc., depending upon availability and regional eating preferences. To these sources are added the flesh and organs of birds (chickens, ducks, turkeys, pheasants, etc.) and of fish caught in saline and fresh waters. In the overall animal protein category, one also would include those less conventional and essentially unexplored categories, such as earthworms and single-cell proteins (produced by microorganisms) and algae. Renewable or repeating protein sources from living animals, of course, include the milk from dairy cows and buffaloes and the eggs from hens, from which hundreds of high-protein foods (cheese, for example) are prepared. And, to this category, must be added the excellent source of protein provided by human milk to the nursing infant.

Plants, of course, also require protein to build and maintain their life processes and, consequently, are protein sources for the animal diet. In the case of herbivores, plants are essentially the exclusive source of proteins, energy, and all other dietary elements.

In terms of percentage of protein content of basic sources, the animal sources far excel the plant sources. For example, the protein content of some typical unfortified foods is as follows: 20–30% for cooked poultry and meats; 19–30% for cooked or canned fish; 25% for cheese; 13–17% 17% for cottage cheese; 16% for nuts; 13% for whole eggs; 7–14% for dry cereals; 8.5–9% for white bread; 7–8% for cooked legumes; and about 2% for cooked cereals.

Of course, in achieving the higher protein contents of meat from poultry and cattle, a rather costly two-step production process is involved, wherein the animal first converts plant proteins (as from grasses) into animal protein. In a sense, the animal both converts and concentrates the protein source for humans. Several economic factors enter into the picture — the utilization of land, the costs of labor, the additional costs of feed materials, and the costs related to a greater time span of production, among others. As a case in point, an animal must be fed between 3 and 10 pounds (1.4 and 4.5 kilograms) of grain to produce 1 pound (0.45 kilogram) of meat. All of these factors in recent years, particularly in consideration of protein shortages in many regions of the world, have given rise to conflicting opinions pertaining to the ever-increasing production and consumption of meat, not only in several of the western nations of the world, but in the developed nations of the Orient as well. A few authorities have suggested that the western countries should cut back on meat production, thus making more land, skills, etc. available to increasing vegetable protein production to the level where a generous excess supply would be available to underdeveloped countries as well as amply supplying the protein needs of the developed countries. Quickly, these arguments penetrate not only into technological and economic factors, but psychological considerations as well — because any moves of this type necessarily require drastic changes in eating habits, and to bring them about successfully would require much more governmental regulation and policing than any system of private enterprise is likely to tolerate. Further, attitudes tend to swing rather widely from times of grain surplus to times of grain shortage.

Fortunately, as of the early 1980s, it appeared that protein-processing techniques were providing a very satisfactory compromise, even though the industry is just getting underway toward a large-scale operation. Protein meat extenders, for example, wherein meat and vegetable protein are blended to produce an edible product that retains much of what is desired of meats, including their good protein content, are finding acceptance. The wide acceptance of vegetable protein in analogue meat products has many hurdles to overcome, but it appears that a solid start has been made. The hurdles not only include acceptability in the marketplace, but also some justifiable resistance on the part of cattle and poultry producers. For many reasons, the transition, if it ultimately takes place, will occur over quite a long period of time. Because of continuing economic inflation, the earlier cost advantages that tended to favor blends of meat and vegetable proteins have become less significant.

An early impetus to soy protein foods was given when the United States introduced soy protein products into its overseas donation program in 1966 as a component of foods formulated to meet special needs of certain population groups. Chief among these were children in developing nations, especially the weanling infant and preschool child whose requirements for growth put special demands on diet composition. Pregnant and lactating mothers also had dietary needs frequently not met in countries where food supplies were marginal. Beyond these needs, there were nutrient deficiencies in large population groups, which could be best overcome by enrichment or fortification of commonly eaten foods.

Shortages in the domestic supply of nonfat dry milk, which developed in 1965, stimulated the development of high-protein formulated foods which would serve as supplements in the diets of the children or in the emergency feeding of adults. These formulations had to pass rigid specifications, one of the principal criteria being the recommended daily dietary allowance for protein, vitamins, and minerals. The U.S. Department of Agriculture and the U.S. Agency for International Development developed the guidelines and designed various formulated foods. Among these formulations were Corn-Soy Milk (CSM), Corn-Soy Blend (CSB), and Wheat-Soy Blend (WSB).

Further impetus was given to protein blends in foods when such products were introduced into the domestic food assistance program in the United States. Soy protein foods were introduced into school lunch and breakfast programs for which federal assistance has been given in the form of a subsidy administered by the federal government. Soy-fortified foods also were distributed to needy families through a family food distribution program.

Textured soy protein products in their use as meat alternatives have become increasingly popular in school lunch programs since their introduction in 1971. A soy-modified macaroni was introduced into the family food assistance program a number of years ago.

Less Conventional Sources of Protein. In addition to the traditional animal sources of protein already described and the very large amounts of vegetable protein derived from the soybean, other sources of protein on a large scale for the future are under intense study. Among these are (1) oil-seed crops, such as rapeseed and cottonseed; (2) leaf proteins; (3) algae; and (4) single-cell protein.

Rapeseed, one of the five most widely produced oilseeds, is cultivated mainly in India, Canada, Pakistan, France, Poland, Sweden, and Germany. Past objections to using rapeseed as a source of edible protein has been its content of deleterious glucosinolates. Considerable research has been conducted in Sweden to develop a rapeseed protein concentrate. The first full-scale production plant using a new process was installed in Alberta, Canada. The plant, with a capacity of 5000 tons/year produces a material containing 65% protein. Rapeseed is rich in essential amino acids, with exception of methionine, which soybeans also lack.

Cottonseed offers an attractive source of protein provided that certain objectionable ingredients can be removed. One of these is gossypol, a substance in cottonseed gland that is harmful to humans. A process developed by the U.S. Department of Agriculture has been designed to turn out a satisfactory edible cottonseed protein product. Employing solvent-extraction techniques, the first plant was built in Texas. Cottonseed flour extrudes easily and can be water-extracted to produce a nearly 100% protein isolate. The product has been used as a bland extender and fortifier for processed meats, baked goods, candies, and cereals. Research of a different approach has been used in Central America. In this approach, iron compounds are used to tie up the gossypol in nontoxic form without having to remove it.

Leaf Protein Concentrates. Laboratories in Hungary, Japan, the United Kingdom, and the United States, among other countries, have been engaged in perfection of a leaf protein concentrate process, with emphasis upon increasing yields and palatability and reducing flavor problems and cost. To date, alfalfa appears to be most attractive as a source of leaf protein. Alfalfa will produce more protein per unit of land than most other crops — up to 2800–4000 pounds/acre (3136–4480 kilograms/hectare). It has been estimated that the raw material costs for edible protein from alfalfa would be about 50% that for soybean meal. Several processes have been worked out, ranging from a green curd containing 52% protein to a white powder containing about 90% protein.

Single-Cell Protein. The advantages of single-cell protein (SCP) made from growing microorganisms are several: (1) SCP is independent of agricultural or climatic conditions; (2) SCP doubles in mass rapidly for high production rates and fast genetic experimentation; (3) the crop is free of surface-area limitations, and (4) the protein in microbial cells is generally of a high nutritional quality. Many of the processes proposed and tested, some with limited operating experience, commence with hydrocarbon feedstocks — gas oil and normal paraffin substrates. Two objections have been raised. The first is the possibility that carcinogenic polyaromatic materials present in gas oil may be passed along to the final protein product. The second is an adverse public reaction. A more recent, third objection is the proposition that perhaps technology should be concentrating on manufacturing fuels from farm products rather than food from petroleum products.

Some of the more recent SCP process concepts start with other materials, such as ethanol, acetic acid, starches, sugars, and cellulosic products that may be more available and particularly so in the protein-needy developing countries.

Algae have the highest intrinsic rates of photosynthesis and growth found among green plants. Human food and animal feed are being produced from algae. In Japan, a full plant-scale production harvests algae from open ponds to yield green powder extract that can be used for animal or human consumption. The genus *Chlorella* has perhaps received the most research to date.

Conservation Sources of Protein. Tightening pollution restrictions have forced cheese makers in many regions to end a long-time practice of dumping whey (with its high biological oxygen demand) as a liquid waste. Although many of these manufacturers are now evaporating or spray-drying whey to produce a whole-solids product, several fractionation techniques have been devised to separate a concentrated protein. In the United States, whey as a byproduct of cheese making totals well over 30 billion pounds (13.6 billion kilograms) per year. From 6.5 to 7% of the whey is solids, of which 0.9% is protein. Some authorities believe that whey and other milk-based protein ingredients offer a high growth potential among all of the non-soybean sources.

Fish protein concentrate is regarded by some authorities as having a high long-term potential. A major restraint is competition for the whole fish. As fish food sources become increasingly competitive, fishes currently considered "trash" fishes from a fresh marketing viewpoint may ultimately become more desirable for table use. Animal-feed fish meal also will be a strong contender for available fish. In terms of processes required for preparing fish-protein concentrate, extraction processes using single or mixed solvents of isopropanol, ethylene dichloride, ethanol, and hexane already have been developed. Experiments with enzymatic processing also are underway.

Chemical Nature of Proteins

In defining a protein structurally, it is first necessary to define a peptide. Peptides are compounds made up of two or more amino acids covalently bound in an amide linkage. The characteristic amide linkage, in which the carboxyl group of one amino acid joins with the amino group of the next amino acid, is called a peptide bond. A peptide is a chain of amino acid residues. Provided that the chain is not circular or blocked at either of the ends, the peptide has an N-terminal amino acid, bearing a free amino group, and a C-terminal amino acid, bearing a free carboxyl group. This is illustrated as follows:

$$H \text{---} NHCHRCO \text{---} OH + H \text{---} NHCHR''CO \text{---} OH$$
$$\downarrow -2H_2O$$
$$H \text{---} NHCHRCO \text{---} NHCHR'CO \text{---} NHCHR''CO \text{---} OH$$

N-terminal   Nonterminal   C-terminal

Usually a form of shorthand is used to represent the structure of a peptide. For example, H-Val-Gly-Ala-OH, represents a peptide where abbreviation for each amino acid is given in terms of three letters each (Val = valine; Gly = glycine; Ala = alanine). Abbreviations for other amino acids are given in entry on **Amino Acids.** The H denotes the amino terminal (N-terminal) and the suffix OH denotes the carboxyl terminal (C-terminal). Peptides may consist of from two to eight amino acid residues and thus are known as dipeptides, tripeptides, or oligopeptides (eight), depending upon the number of residues contained. A peptide consisting of ten or more amino acid residues and with a molecular weight in the range of $1-5 \times 10^3$ is called a polypeptide. Emil Fischer, father of protein chemistry, proposed early in the twentieth century that proteins are peptide in nature. Actually, no sharp demarcation exists between large polypeptides and small proteins. Examples of small proteins include insulin (hormone protein), protamine, and some components of histone (basic proteins of chromosomes).

Almost all proteins are comprised of amino acid residues, more than 100 in number, and their molecular weight may range from 10^4 to 10^7. A few examples include: Insulin (6×10^3); ribonuclease (13×10^3); lysozyme (eggwhite) (15×10^3); chymotrypsinogen (21×10^3); ovalbumin (43×10^3); serum albumin (66×10^3)—all of the foregoing being single peptide chains. Multiple chains include: Hemoglobin (68×10^3); gamma globulin (IgG) (160×10^3); fibrinogen (340×10^3); urease (460×10^3); thyroglobulin (640×10^3); myosin (850×10^3); hemocyanin (octopus)

($2,800 \times 10^3$); hemocyanin (snail) ($8,900 \times 10^3$); and tobacco mosaic virus ($40,000 \times 10^3$). Proteins of huge molecular weight (millions) are enormous aggregates of protein subunits, each of which may be so large (molecular weight = $1.5-10 \times 10^4$) in most instances. The independent peptide chains that constitute a protein molecule are often held by the disulfide bridges of cystine residues. From the diagram below, it will be seen that in a single chain the bridges may hold together two quite distant points in terms of the linear amino acid sequence, forming a large loop structure:

$$
\begin{array}{l}
H \text{------} NHCHCO \text{------} \\
\qquad\qquad | \\
\qquad\qquad CH_2 \\
\qquad S \\
\qquad | \\
\qquad S \\
\qquad\qquad CH_2 \\
\qquad\qquad | \\
HO \text{------} COCHNH \text{------}
\end{array}
$$

Although more than 200 amino acids have been found in living organisms, only 20 alpha-amino acids of the L configuration have been found serving as the building units for proteins and related peptides. These 20 amino acids occur in varying proportions in different proteins. Some proteins are fully lacking in one or more of them. Some amino acids occur only in some of the proteins. For example, hydroxyproline has been found only in collagen and elastin (proteins of animal connective tissue) and in gelatin derived from collagen.

Numerous classifications of proteins have been proposed over the years. In terms of function, there are:

a. *Structural proteins.* Proteins that support the skeletal structures, maintain the form and position of organs, impart the structural rigidity to walls of containers for biological fluids, and often form part of the external tissues. In keeping with their functions, they are insoluble in many liquids, especially body fluids, and are otherwise relatively resistant to biochemical reactions. The proteins of nails, horn, hoofs, and hair are familiar examples.

b. *Contractile proteins.* Those substances that have the property of undergoing a change in configuration, which results in a change in length or shape. Thus they give the organism the power to move itself, its parts, or other objects. The proteins of muscles are prominent examples.

c. *Process-activating proteins.* As used here, the term *process* includes the biochemical reactions, which are catalyzed by enzymes, and in some of which the cytochromes play an intermediate role; it also includes the endocrine reactions activated by the hormones, some of which are proteins.

d. *Transport proteins.* Proteins which transport an essential substance or factor, from that part of the organism where it becomes available from a source external to the organism to the point where it is used. Examples are many of the chromoproteins, such as hemoglobin, or the blue hemocyanins (from mollusks) which contain copper instead of iron as does hemoglobin, or the chlorophyll-protein complexes of plants.

Another basis of classification is that of solubility, which has been applied to proteins from all sources, plant and animal. (a) Thus the albumins were soluble in water and coagulable by heat. They included serum albumin, egg albumin, lactalbumin (from milk), leucosin (from wheat), and legumelin (from legumes, chiefly peas). (b) The globulins are soluble in neutral salt solutions and in strong acids and alkalies. They include blood globulin (which has been separated by electrophoresis into alpha, beta, and gamma fractions, and is further discussed later in this entry), ovoglobulin (from egg yolk), edestin (from hempseed), phaseolin (from beans), arachin (from peanuts), and amandin (from almonds). (c) The glutelins, such as glutenin from wheat, are soluble in dilute acids and alkalies, and insoluble in neutral salt solutions. (d) The scleroproteins are quite insoluble, and the structural proteins (group I mentioned above) belong to this group. All these groups, and several others not included here, are simple proteins, i.e., they consist only of polypeptide chains of amino acids. The many conjugated proteins must then be classified upon the basis of their nonprotein portions: glycoproteins which contain carbohydrate groups, lipoproteins which contain lipid groups, chromoproteins which contain metal-containing complexes that are usually colored, as hemoglobin contains heme.

Still another classification places proteins into three major categories: (a) Simple proteins; (b) conjugated proteins; and (c) derived proteins. The last classification embraces all denatured proteins and hydrolytic products of protein breakdown and no longer is considered a general class.

A relatively simplistic concept of a protein structure is indicated in Fig. 1. The molecular weight for the hemoglobins is on the order of 68,000. They are conjugated proteins and consist of four heme groups and the globin portion. The heme group is a porphyrin in which the metal ion coordinated is iron, which may be Fe^{3+} or Fe^{2+}, but only in the latter case (ferrohemoglobin) can the molecule bond molecular oxygen and be effective in respiration, i.e., by forming oxyhemoglobin. The globin portion of the molecule consists of four polypeptide chains. These chains are designated as alpha, beta, gamma, etc. according to their amino acid composition. Normal adult hemoglobin consists of two alpha chains and two beta chains. The composition and conformation of the beta chain are shown in the diagram, together with the point of attachment of the heme groups: Note that they are attached to histidine groups. It has been learned that the central iron atom in heme, which is chelated to the porphyrin ring by four bonds, is attached to the polypeptide chain in adult human hemoglobin by three imidazole ligands of the globin chain, which belong to the histidines at positions 58,87, and 89 of the alpha chain. See also **Hemoglobin**.

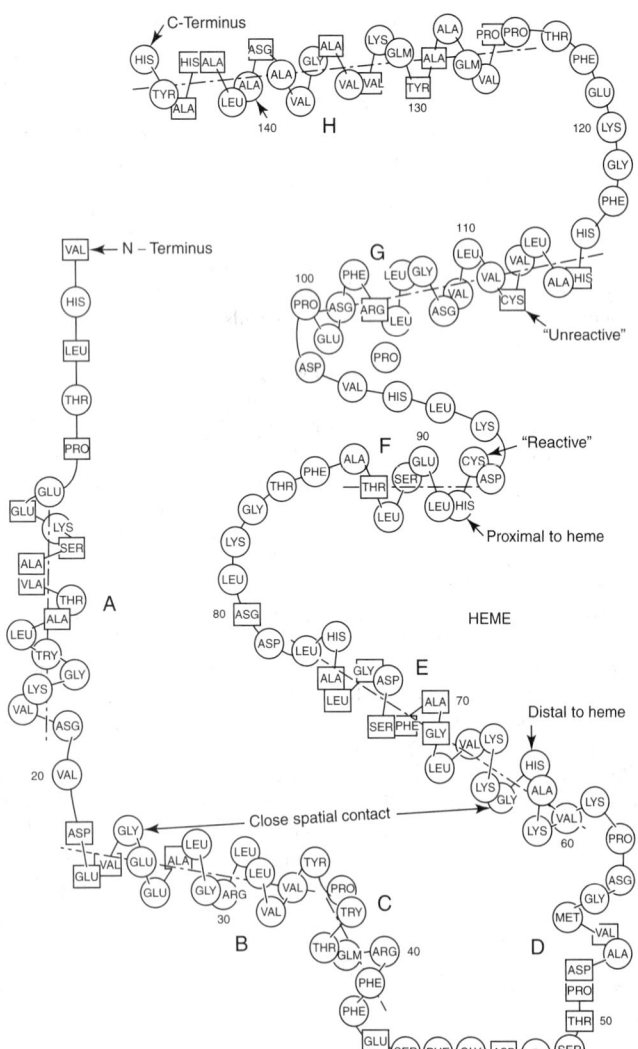

Fig. 1. Simplified representation of the beta chain of human hemoglobin A.

Besides hemoglobin, other proteins of blood are of considerable importance. They are the plasma proteins, serum albumin and fibrinogen, and the globulins. Serum albumin is responsible for the major part of the osmotic pressure of human plasma. Its molecular weight is on the order of 68,000. It is a typical globular protein, having nearly one-half

helical character. Although not as nearly symmetrical as hemoglobin or myoglobin, it has a symmetry indicated by its molecular dimensions of 150 Å long and 38 Å wide. It is the smallest, and most abundant, of the plasma proteins; for this reason, and also because of its relatively low isoelectric point, it undergoes migration rapidly in an electric field. See also **Electrophoresis**. By this method it may be separated into two types of molecules, similar in composition except for the presence of a single cysteine residue in one and not the other. However, it contains cystine residues, which form seventeen disulfide bridges cross-linking the polypeptide chain, i.e., the molecule of serum albumin consists of a single polypeptide chain.

Another plasma protein to be discussed here is fibrinogen, which is the chief substance involved in the process of blood clotting. Its molecular weight is on the order of 330,000. It contains all twenty of the amino acids described in that entry as the most general in proteins, although it is relatively low in cysteine, and highest in the acidic amino acids (aspartic and glutamic acids). The process of clotting occurs in three major steps. In the first the substance prothrombin, a blood glycoprotein containing about 5% carbohydrate as glucosamine and a hexose sugar, is converted to the clotting enzyme thrombin. (The latter is unstable, and hence must be formed when needed.) The conversion process is catalyzed by the calcium ion and a group of substances known as thromboplastins. In the second step the enzyme thrombin catalyzes the transformation of fibrinogen to an activated form, called profibrin, with an altered pattern of electric charge. This change is considered to be due to the liberation of two short-chain polypeptides (one bearing 18 amino acid residues and the other 20), and a corresponding change in the character of the remainder of the fibrinogen molecule, which collectively constitute the substance profibrin. In the third step, this mixture of substances undergoes spontaneous polymerization to form the substance fibrin, which has been shown in electron microscope photographs to consist of a network of striated fibers. This polymerization occurs in stages, and in some views of the process they are divided into two steps, polymerization and clotting, the former being regarded as the formation of linear polymers, and the latter as their cross-linking by an enzymatic reaction whereby disulfide bonds are formed.

Animal organisms generally require effective assistance of intestinal flora, as in ruminants to assimilate inorganic nitrogen into a very wide variety of foreign substances, called antigens.

The life span of individual proteins in living organisms is relatively short — about 4 months for hemoglobin and but a week or two for serum albumin. The aged proteins are digested by proteolytic enzymes of tissues, such as cathepsin. A significant portion of the recovered amino acids may be available for the biosynthesis of new proteins, but another part is catabolized and the nitrogen is excreted as urea in mammals, uric acid in birds, reptiles, and insects, or ammonia in organisms of lower classes. For the maintenance of nitrogen balance and for growth, the human organism requires a daily intake of from 70 to 80 grams of proteins. The present world population thus requires 3×10^7 tons of animal proteins and 10×10^7 tons of plant proteins annually.

Living organisms can synthesize their own proteins from amino acids. In terms of the ability to carry out the de novo synthesis of amino acids, however, there are wide variations among different organisms. Plants, for example, can synthesize amino acids from nitrogen in the form of ammonium salts or nitrate and other simple compounds. The annual production of cereal and vegetable proteins so assimilated from inorganic nitrogen over the world is estimated to be about 10×10^7 tons. Of this, 4×10^7 tons are provided by wheat; 2×10^7 tons by rice; and 1×10^7 tons by corn and other sources. Lactic acid and some other microorganisms require preformed amino acids for growth, lacking some ability to synthesize. However, some microorganisms perform well with ammonium sulfate and carbohydrate as the sole sources of nitrogen, sulfur, and carbon. In some cases, they accumulate particular amino acids in a process referred to as amino acid fermentation.

Animal organisms generally require effective assistance of intestinal flora, as in ruminants, to assimilate inorganic nitrogen into body protein. This accounts for the human needs of a daily requirement of 70–80 grams of protein. However, over half of the protein-constituent amino acids can be derived from other amino acids by their own enzymic reactions. Thus, amino acids are classified as essential or nonessential. Amino acid requirements vary with the physiological state of the animal, age, and possibly with the nature of the intestinal flora.

The Food and Agricultural Organization (FAO) established the following essential amino acids in the ratios indicated:

	Percent of Crude Protein
Isoleucine	4.2
Leucine	6.2
Lysine	4.2
Methionine	2.2
Phenylalanine	2.8
Threonine	2.8
Tryptophan	1.6
Valine	5.0

A distribution of amino acids in dietary proteins can be obtained accordingly by taking both animal and plant proteins at a ratio of 1:3–4. Although plant proteins are lower cost, they are markedly deficient in some essential amino acids. Their protein efficiency is low without addition of deficient amino acids. Enrichment of human and animal diets with free amino acids, such as lysine, methionine, threonine, and tryptophan, as a substitute for animal proteins, has proved successful.

Excesses of amino acids are not harmful — with few exceptions. An imbalance of amino acids can result in a few instances. For example, a rat that feeds on eggwhite proteins with threonine or isoleucine added in high concentrations can experience an undesirable imbalance.

Several industries are based upon proteins as exemplified by the keratins of wool, feather, or horn; the fibroin of silk; the collagenous tissues as leather; proteins in milk, wheat, soybean, egg, and numerous other natural substances. Making cheese from milk casein and flavor seasonings from plant or fish proteins are old processes. Gelatin derived from collagen has been used widely in processed foods and as an adhesive material and in photography. Gluten in cereal is a protein. Major proteins in meat are *myosin*; in egg, *ovalbumin*; in rice, *oryzenin*; in soybean, *glycinin*; and in corn, *zein*.

Protein Quality and Evaluation. Protein quality relates to the efficiency with which various food proteins are used for synthesis and maintenance of tissue protein. Food industry evaluators of protein nutritional quality must operate on several levels of awareness. In particular, manufacturers of processed foods must measure the biological value of the protein content of a variety of processed foods for several reasons: (1) to comply with various governmental regulations; (2) to satisfy nutrition labeling regulations; and (3) an accurate knowledge of protein effectiveness is required in developing new food products and in controlling sources of protein ingredients. Protein quality is also very important in the formulation of animal feedstuffs.

In a number of countries, including the United States, the stipulated measurement of protein quality is the so-called protein efficiency ratio (PER), which may be defined as the *gain in weight* divided by the *weight of protein consumed* by experimental laboratory animals. As of the early 1980s, the AOAC (Association of Official Analytical Chemists) method, defined in 1975, is only one of the codifications of PER work since the concept was first proposed in 1919. More specifically, the PER is the ratio of the weight gained by a group of ten weanling rats fed a diet containing about 10% protein, to the weight of protein consumed over a 28-day period. No sample to be studied should contain less than 1.8% nitrogen according to the AOAC method, and the diet should supply 1.6% nitrogen. Since the samples are not analyzed for protein, but rather for nitrogen, and since protein efficiency ratio rather than nitrogen efficiency ratio is reported, it is important to be clear about whether one of the specific nitrogen factors or the conventional 6.25 figure is used to calculate the protein in the final diet.

Advances in Protein Chemistry

For many years, research was directed to a better understanding of the structure of proteins, notably based upon X-ray crystallography. Remarkable structural details were evidenced. But this avenue of research tended to regard proteins as being static in nature, whereas more recent findings show that indeed proteins are dynamic and that, if they were rigid, they simply could not function. The internal motions that underlie their workings are best explored in computer simulations. As pointed out by Karplus and McCammon, it is now recognized that the atoms in a protein molecule are in a state of constant motion. Thus, what the crystallographer finds is at best a representation of a protein's average structure. The chemical bonds between the atoms along the polypeptide chain in a protein

act much like springs. There are also weaker forces between unbonded atoms, including forces that prevent more than one atom from occupying the same point in space at any given time. Thus, in a protein consisting of many atoms, the total force acting on any one atom at any given time depends upon the positions of all the others. Not surprising, the solution of Newton's equations of motion for determining the positions and velocities of all the atoms in a protein requires a high-speed computer. Such calculations constitute what is called a *molecular-dynamics simulation*.

In summarizing their recent research, Karplus (Harvard University) and McCammon (University of Houston) observe that, from future research, much will be learned regarding how to calculate the rates of enzymatic reactions, and the binding of small molecules to large ones, as well as the role of flexibility and fluctuations in the function of macromolecules. For example, it should become possible to determine how particular solvent conditions and amino acid sequences produce certain patterns of protein fluctuations. Such information will become useful in applying new genetic technologies in a practical way.

Protein research of this kind is important because all enzymes are proteins. They catalyze the speed of essential reactions in living systems, including the synthesis of proteins themselves. Knowledge of the dynamics of proteins will assist in better understanding those proteins that transport small molecules, electrons and energy to specific parts of an organism where they are needed. Those proteins of a structural nature, which make up fibrous tissue and muscle, also will be better understood.

The molecular biology of proteins has progressed markedly since the development of detailed knowledge of DNA and RNA, as discussed in the entry on **Genetics and Gene Science**.

As pointed out by Phillips, the level of understanding of enzyme (protein) action has been achieved for many enzymes through the use of chemical, crystallographic, and spectroscopic methods. Gene science, however, has enormously advanced protein studies. By using cellular machinery for protein synthesis, proteins can be manufactured with any primary structure and then introducing whatever changes seem useful in the chemical constitution of naturally occurring proteins. With further knowledge, at some future date it most likely will be possible to design and manufacture fully novel proteins with new and useful properties.

Currently, the most useful advances are being made by the detailed modification of existing protein structures. A very small change, often involving only a single base, is made of the DNA coding for the protein. This is followed by use of natural cellular machinery (frequently bacteria) to synthesize the modified protein. The method is known as *site-directed mutagenesis*, which was first used in 1982. Classical chemical modification of protein structure still is used, but the site-directed mutagenesis approach is usually more straightforward and reliable.

Phillips has projected an imaginary oligopeptide with side chains grouped in accordance with their properties to illustrate intricacies of structure and regions of specializing functions. See Fig. 2.

Much progress is being made in connection with *fibronectins*, those adhesive proteins that act as biological organizers by holding cells in position and guiding their migration. Studies are now revealing the molecular bases for the functions of fibronectins. As observed by Hynes, within the complex architecture of a multicellular organism most normal cells remain reasonably stationary. They are anchored to basement membranes and connective tissue, which is made up mainly of a fibrous mesh of proteins and other substances. In the adults of most species, only a few cell types will routinely move through this extracellular matrix. It is known that during embryonic development and wound healing, some cells migrate extensively and usually unerringly. The question is asked — how can the organization of these cells be both fixed and dynamic? Glycoproteins (those with attached sugars) may be part of the answer. Of these glycoproteins, the fibronectins are currently the best understood. These molecules have several functions — they can assemble into fibrils, bind to cells, and link cells to other kinds of fibrils in the extracellular matrix. Fibronectin, of course, is a critical component of the blood clotting function. Several lines of research are now being followed in fibronectin studies. These are well described and illustrated in the Hynes reference. It has been suggested that, inasmuch as cancer most frequently involves metastasis (migration of tumor cells to unrelated tissues elsewhere in the body), there may be some connection with fibronectins, because their currently best understood role is that of keeping cells in place and when they move they control their migration.

W.R. Schaffer (University of California, Berkeley) and colleagues have been investigating what are known as *isoprenoids*. These compounds

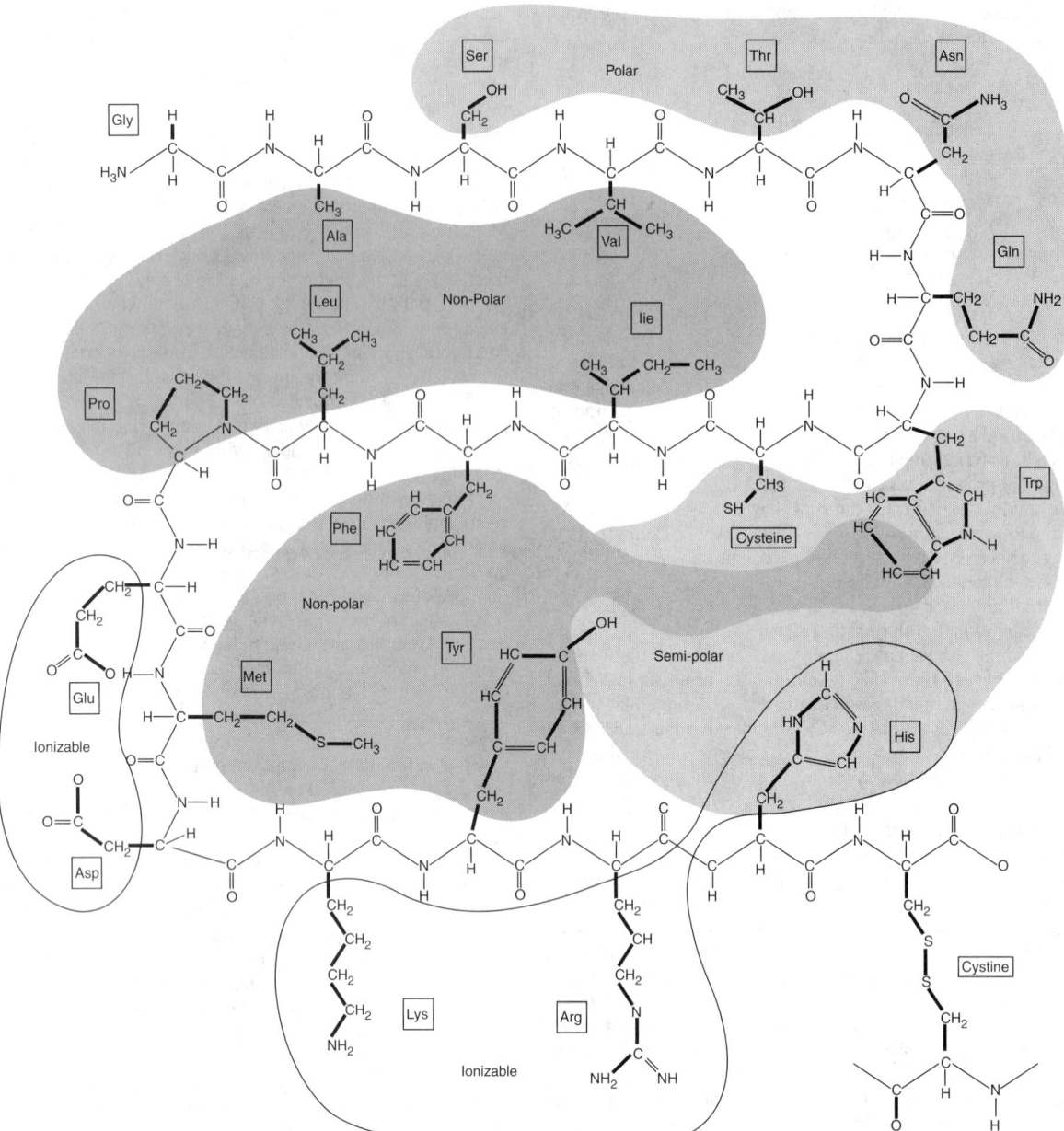

Fig. 2.   Facsimile depiction of an imaginary oligopeptide with side chains grouped in accordance with their properties as proposed by Phillips (1987) in an excellent summary of "Protein Engineering" in the new and exceptional publication, *Scientific & Technology Review (The University of Wales)*. All twenty amino acids are represented as shown by three-letter abbreviations in the boxes on the diagram. Polar, semipolar, nonpolar, and ionizable portions of the hypothetical oligopeptide are indicated by shaded and dotted areas. Also, note disulfide bridge shown. (*After Phillips.*)

are structurally related lipophilic molecules that perform a wide variety of essential cellular functions. These lipids include such functionally diverse molecules as cholesterol, ubiquinone, dolichols, and chlorophyll, yet isoprenoids are derived from a common precursor, mevalonic acid. These studies may lead to a better understanding of the Ras oncogenic proteins.

In recent years, molecular biologists have found that proteins, in their various roles (binding of receptors, assembling into cellular structures, catalyzing metabolic reactions, etal) depend largely on their three-dimensional structure. For quite some time, biologists have been successful with their techniques for sequencing the amino acids that make up a given protein. In contrast, progress was slow toward determining how the protein chain of components folds into a three-dimensional structure.

It was not until quite recently that neural computers have been used to solve the protein-folding problem. Some proteins have been described in the past as being contorted into "tangled" structures, sometimes likened to a twisting telephone cord. Attempts to predict a folding pattern would require the computation of each part of a chain and its effect on adjacent parts of the chain, a computation process of great magnitude. Further, protein

crystals are difficult to develop, thus eliminating or at least reducing the effectiveness of x-ray crystallography. Nuclear magnetic resonance (NMR) also has been used, but unfortunately tends to be limited to the smaller proteins and requires much computer time. Although several thousand proteins have been amino acid sequenced, only a few hundred structures have been determined.

In 1988, researchers T.J. Sejnowski and N. Olan (Johns Hopkins University) reasoned that a computer (NETtalk), that had been designed to pronounce written English words might be applied to the protein structure problem—because NETtalk also depended upon deciphering that occurs at the junctions of numerous separations in a word (as it may appear hyphenated) and that this analysis may be similar to the occurrence of a multi-hyphenated structure exhibited by proteins. The researchers explain that a learning rule modifies the network so that eventually the network will produce the correct phoneme a large percentage of the time. Further work along these lines resulted in a network that could correctly predict over 64% of a test sequence.

S. Brunak and R.M.J. Cotterill (Technical University of Denmark) pursued the approach further, based upon data inputs from NMR and x-ray diffraction. The neural network approach remains very active so that encoding of the intricacy of interconnections may be achievable.

In addition to studying the structure (folding) of proteins for fundamental knowledge, the study of enterotoxins has the additional incentive where life-threatening diseases are concerned, particularly toward the development of improved vaccines.

In research conducted at the University of Groningen (Netherlands) over a 14-year period, scientists succeeded in developing a pure crystal of the cholera toxin. Over 25,000 diffraction measurements of pure crystals, it became possible to generate a computer image of the cholera toxin. It has been observed that all bacterial toxins act in the same manner — one component is an enzyme that performs the invasive function and another component performs destruction once it enters the cell.

Research also has indicated that *E. coli* and diphtheria toxin perform in a similar manner. Active research programs currently are being conducted at Harvard University and the University of California, Los Angeles.

Similar structural determination studies are going forward to determine enzyme structures. In 1991, S. Taylor, D. Knighton, J. Sowadski, and colleagues (University of California, San Diego) announced their development of the three-dimensional structure of a protein kinase.

As aptly put by Doolittle (1985) — "If DNA is the blueprint of life, then proteins are the bricks and mortar."

Additional Reading

Abbott, N.L. and T.A. Hatton: "Liquid-Liquid Extraction for Protein Separations," *Chem. Eng. Progress*, 31 (August 1988).

Angeletti, R.H.: "Proteins: Analysis and Design," Academic Press, Inc., San Diego, CA, 1998.

Barton, G.J.: "Protein Structure and Prediction," Blackwell Science, Inc., Malden, MA, 2002.

Bollag, D.M., S.J. Edelstein, and M.D. Rozycki: "Protein Methods," 2nd Edition, John Wiley & Sons, Inc., New York, NY, 1996.

Bohr, H.G.: "Neural Network Prediction of Protein Structures," Springer-Verlag, Inc., New York, NY, 2001.

Bowie, J.U., et al.: "Deciphering the Message in Protein Sequences: Tolerance to Amino Acid Substitutions," *Science*, 1306 (1990).

Branden, C. and J. Tooze: "Introduction to Protein Structure," 2nd Edition, Garland Publishing, Inc., New York, NY, 1998.

Brown, W.E. and G.C. Howard: "Modern Protein Chemistry: Practical Aspects," CRC Press, LLC., Boca Raton, FL, 2001.

Builder, S.E. and W.S. Hancock: "Analytical and Process Chromatography in Pharmaceutical Protein Production," *Chem. Eng. Progress*, 42 (August 1988).

Clore, G.M. and A.M. Gronenborn: "Structures of Larger Proteins in Solution: Three- and Four-Dimensional Heteronuclear NMR Spectroscopy," *Science*, 1390 (June 7, 1991).

Considine, D.M. and G.D. Considine: "Foods and Food Production Encyclopedia," Van Nostrand Reinhold Company, Inc., New York, NY, 1982.

Copeland, R.A.: "Proteins: Masterpieces of Polymer Chemistry," *Today's Chemist*, 53 (June 1992).

Creighton, T.E.: "Protein Function: A Practical Approach," 2nd Edition, Oxford University Press, Inc., New York, NY, 1997.

DeGrado, W.F., Z.R. Wasserman, and J.D. Lear: "Protein Design, a Minimalist Approach," *Science*, 622 (1989).

Deutscher, M.P. and J.N. Abelson: "Guide to Protein Purification," Vol. 182, Academic Press, Inc., San Diego, CA, 1990.

Fersht, A.: "Structure and Mechanism in Protein Science: A Guide to Enzyme Catalysis and Protein Folding," W. H. Freeman Company, New York, NY, 1999.

Gennadios, A. and C.L. Weller: "Edible Films and Coatings from Wheat and Corn Proteins," *Food Techy.*, 63 (October 1990).

Gierasch, L.M. and J. King: "Protein Folding," *Amer. Assn. for the Adv. of Science*, Waldorf, MD, 1990.

Hall, A.: "The Cellular Function of Small GTP-Binding Proteins," *Science*, 635 (August 10, 1990).

Hoffman, M.: "New 3-D Protein Structures Revealed," *Science*, 382 (July 26, 1991).

Hoffman, M.: "New Role Found for a Common Protein 'Motif'," *Science*, 742 (August 16, 1991).

Hoffman, M.: "Playing Tag with Membrane Proteins," *Science*, 650 (November 1, 1992).

Hynes, R.O. and K.M. Yamada: "Fibronectins: Multifunctional Modular Glycoproteins," *J. of Cell Biology*, **95**(2), Part I, 369–377 (November 1982).

Hynes, R.O.: "Molecular Biology of Fibronectin," *Ann. Rev. of Cell Biology*, **1**, 67–90 (1985).

Hynes, R.O.: "Fibronectins," *Sci. Amer.*, 42–51 (June 1986).

Karplus, M. and J.A. McCammon: "Dynamics of Proteins: Elements and Function," *Ann. Rev. of Biochemistry*, **52**, 263–300 (1983).

Karplus, M. and J.A. McCammon: "The Dynamics of Proteins," *Sci. Amer.*, 42–51 (April 1986).

Kinoshita, J.: "Net Result: Folded Protein," *Sci. Amer.*, 24 (April 1990).

Knighton, D.R., et al.: "Crystal Structure of the Catalytic Subunit of Cyclic Adenosine Monophosphate-Dependent Protein Kinase," *Science*, 407 (July 26, 1991).

Lesk, A.M.: "Introduction to Protein Architecture: The Structural Biology of Proteins," Oxford University Press, Inc., New York, NY, 2000.

Linder, M.E. and A.G. Filman: "G Proteins," *Sci. Amer.*, 56 (July 1992).

Marx, J.L.: "New Family of Adhesion Proteins Discovered," *Science*, 1144 (March 3, 1989).

Nakai, S. and H.W. Modler: "Food Proteins: Processing Applications," Vol. 2, John Wiley & Sons, Inc., New York, NY, 1999.

Neurath, H.: "Protein Science," Cambridge University Press, New York, NY, 1991.

Otting, G., E. Liepinsh, and K. Wuthrich: "Protein Hydration in Aqueous Solution," *Science*, 974 (November 15, 1991).

Patthy, L.: "Protein Evolution," Blackwell Science, Inc., Malden, MA, 1999.

Phillips, D.C.: "Protein Engineering," *Review (Univ. of Wales)*, 46 (March 1987).

Richards, F.M.: "The Protein Folding Problem," *Sci. Amer.*, 54 (January 1991).

Radousky, H.B., G. Hammond, Z. Xu, et al.: "Gene Families: Studies of DNA, RNA, Enzymes and Proteins," World Scientific Publishing Company, Inc., River Edge, NJ, 2001.

Schaffer, W.R., et al.: "Enzymatic Coupling of Cholesterol Intermediates to a Mating Pheromone Precursor and to the Ras Protein," *Science*, 1133 (September 7, 1990).

Sikorski, Z.E.: "Chemical and Functional Properties of Food Proteins," CRC Press, LLC., Boca Raton, FL, 2001.

Skolnick, J. and A. Kolinski: "Simulations of the Folding of a Globular Protein," *Science*, 1121 (November 23, 1990).

Smith, D.M.: "Meat Proteins," *Food Techy.*, 116 (March 1988).

Utermann, G.: "The Mysteries of Lipoprotein (a)," *Science*, 904 (1989).

Villafranca, J.J.: "Current Research in Protein Chemistry: Techniques, Structure, and Function," Academic Press, Inc., San Diego, CA, 1990.

Walker, J.M.: "Protein Protocols Handbook," 2nd Edition, Humana Press, Totowa, NJ, 2002.

Walsh, G.: "Proteins: Biochemistry and Biotechnology," 2nd Edition, John Wiley & Sons, Inc., New York, NY, 2002.

Whiting, R.C.: "Ingredients and Processing Factors that Control Muscle Protein Functionality," *Food Techy.*, 104 (April 1988).

Wuthrich, K.: "Protein Structure Determination in Solution by Nuclear Magnetic Resonance Spectroscopy," *Science*, 45 (1989).

PROTEIN HYDROLYSATE.   Solutions of protein hydrolyzed into its constituent amino acids.

PROTEROZOIC (Algonkian).   The next to the oldest of the five eras of the Earth's history. Separated from the Archeozoic (the oldest Era) by a profound unconformity. The formations of the Proterozoic contain a preponderance of red sandstones and shales suggesting increasing aridity. Tillites also prove that continental glaciers existed in Eastern Canada, Australia, Tasmania, Norway, South Africa and India. The only undoubted forms of life appear to have been low forms of marine plants called calcareous algae. The sedimentary formations of the Proterozoic, especially in North America, contain important ores of iron and copper. Length of time since the beginning of the Proterozoic, 1,000 million years.

PROTIUM.   The lighter isotope of hydrogen, with a single proton and electron, and constituting 98.51% of ordinary hydrogen is termed protium.

PROTON.   The proton is the atomic nucleus of the element hydrogen, the second most abundant element on earth. Positively charged hydrogen atoms or "protons" were identified by J.J. Thomson in a series of experiments initiated in 1906. Although the structure of the hydrogen atom was not correctly understood at that time, several properties of the proton were determined. The electric charge on the proton was found to be equal but opposite in sign to that of an electron. The traditionally accepted proton mass is 1836 times the electron rest mass, or 1.672×10^{-24} grams.[1]

An estimate of the size of the proton and an understanding of the structure of the hydrogen atom resulted from two major developments in atomic physics: the Rutherford scattering experiment (1911) and the Bohr model of the atom (1913). Rutherford showed that the nucleus is vanishingly small compared to the size of an atom. The radius of a proton

---

[1] Particularly since the early 1970s, physicists have been seeking a grand unification theory to explain all the elementary particles of matter and all the forces acting between them. Although this goal continues to be elusive, work toward that end is producing many new findings and revised concepts. In the main part of this entry, the traditional viewpoints on the proton are described. Some of the more recent postulations are given toward the end of the entry.

is on the order of 10^{-13} centimeter as compared with atomic radii of 10^{-8} centimeter. Thus, the size of a hydrogen atom is determined by the radius of the electron orbits, but the mass is essentially that of the proton.

In the Bohr model of the hydrogen atom, the proton is a massive positive point charge about which the electron moves. By placing quantum mechanical conditions upon an otherwise classical planetary motion of the electron, Bohr explained the lines observed in optical spectra as transitions between discrete quantum mechanical energy states. Except for hyperfine splitting, which is a minute decomposition of spectrum lines into a group of closely spaced lines, the proton plays a passive role in the mechanics of the hydrogen atom. It simply provides the attractive central force field for the electron.

The proton is the lightest nucleus, with atomic number one. Other singly charged nuclei are the deuteron and the triton, which are nearly two and three times as heavy as the proton, respectively, and are the nuclei of the hydrogen isotopes deuterium (stable) and tritium (radioactive). The difference in the nuclear masses of the isotopes accounts for a part of the hyperfine structure called the isotope shift.

In 1924, difficulties in explaining certain hyperfine structures prompted Pauli to suggest that a nucleus possesses an intrinsic angular momentum or "spin" and an associated magnetic moment. The proton spin quantum number I is $\frac{1}{2}$, and the angular momentum is given by $[I(I + 1)h^2/(2\pi)^2]^{1/2}$, where h is Planck's constant. The intrinsic magnetic moment is 2.793 in units of nuclear magnetons (0.50504×10^{-23} erg/gauss), which is about a factor of 660 less than the magnetic moment of the electron.

Two types of hydrogen molecule result from the two possible couplings of the proton spins. At room temperature, hydrogen gas is made up of 75% orthohydrogen (proton spins parallel) and 25% parahydrogen (proton spins antiparallel). Several gross properties, such as specific heat, strongly depend upon the ortho or para character of the gas.

See also **Particles (Subatomic)**.

PROTON–PROTON REACTION.   A thermonuclear reaction in which two protons collide at very high velocities and combine to form a deuteron. The resultant deuteron may capture another proton to form tritium and the latter may undergo proton capture to form helium. See also **Carbon Cycle (Nuclear)**. The proton–proton reaction is now believed to be the principal source of energy within the sun and other stars of its class. A temperature of the order of five million degrees Kelvin and high hydrogen (proton) concentrations are required for this reaction to proceed at rates compatible with energy emission by such stars.

PROTOZOA.   The one-celled animals, constituting a major division of the animal kingdom. They occur in soil or water and many species live as parasites or symbionts in the bodies of other animals. Some protozoans are colonial and in some colonies a division of labor occurs, accompanied by structural specialization of the individuals. Some species are widely distributed. See Table 1.

Since the body of a protozoan is a single cell it has the subordinate structures of the cell in addition to other specialized parts. It contains one or more nuclei surrounded by cytoplasm. The body may be naked and without permanent form or held in a definite shape by a delicate surface membrane called a pellicle. Some species secrete a shell or test and some form internal hard parts. See also **Ocean Resources (Living)**.

The structures that perform special functions are called organelles, since they resemble the multicellular organs of other animals in function but are simpler than the cell itself in structure. Among the more evident and important are the external organelles for locomotion and for securing food. In the various forms of protozoans these are pseudopodia, cilia, flagella, cirri, membranelles, undulating membranes, or tentacles. Some are associated with a depression in the surface of the body called the cytosome through which food is ingested. Within the body the cytoplasm is differentiated into a clear layer of ectoplasm at the surface and an inner granular endoplasm containing the nucleus. Here also masses of food with a little water form the food vacuoles in which digestion takes place. One or more pulsating or contractile vacuoles are interpreted as excretory organelles or osmoregulators. They fill periodically with clear liquid and then discharge. Slender rod-like defensive structures, the trichocysts, lie in the ectoplasm of some species and are discharged when the animal is irritated. See Fig. 1.

In the bodies of the species which carry on a type of nutrition like that of green plants colored bodies (chloroplasts) containing chlorophyll are

TABLE 1. CLASSIFICATION OF THE PROTOZOA

Subkingdom Unicellulates (Protozoa)

Class Flagellata
   ORDER Chrysomonadina
      Families: *Chromulinidae; Rhizochrysidae; Ochromonadidae; Coccolithophoridae*
   ORDER Cryptomonadina
   ORDER Phytomonadina
      Families: *Chalmydomonadidae; Volvocidae*
   ORDER Euglenoidina
      Families: *Englenidae; Peranemidae*
   ORDER Dinoflagellata
      Families: *Ceratidae; Noctilucidae; Gymnodinidae*
   ORDER Protomonadina
      Families: *Eumonadidae; Craspedomonadidae* (choanoflagellates); *Trypanosomatidae* (trypanosomes)
   ORDER Diplomonadina
   ORDER Polymastigina
      Families: *Trichomonadidae; Calonymphidae; Prysonymphidae; Hypermastigidae*
   ORDER Opalinia

Class Rhizopoda
   ORDER Amoebina
   ORDER Testacea
   ORDER Foraminifera
   ORDER Heliozoa
      SUBORDER Centrohelidia
   ORDER Radiolaria
      SUBORDER Acantharia
      SUBORDER Spumellaria
      SUBORDER Nassellaria
      SUBORDER Phaeodaria

Class Sporoza
   ORDER Gregarinida
      SUBORDER Schizogregarinida
      SUBORDER Gregarines (Eugregarinida)
   ORDER Coccidia
      SUBORDER Schizococcidia
         Families: *Eimeridae; Haemosporidae*

Class Ciliata
   ORDER Holotricha
      SUBORDER Gymnostomata
      SUBORDER Trichostomata
      SUBORDER Hymenostomata
      SUBORDER Astomata
   ORDER Peritricha
      SUBORDER Sessila
      SUBORDER Mobilia
   ORDER Spirotricha
      SUBORDER Heterotricha
      SUBORDER Hypotricha
      SUBORDER Oligotricha
         Family: *Tintinnidae*
      SUBORDER Entodiniomorpha
   ORDER Chonotricha
   ORDER Suctoria

Class Cnidosporida
   ORDER Myxosporidia
   ORDER Actinomyxidia
   ORDER Microsporidia

Class Halosporida

Class Sarcosporidia

Class Piroplasmida
      Families: *Theilerida; Babesidae*

found in the cytoplasm. These organisms have been interpreted both as fission although sexual reproduction also takes place at intervals through the process of conjugation.

Protozoans are economically important chiefly as the causes of several serious diseases.

Additional Reading

Amos, W.B. and J.G. Duckett: "Prokaryotic and Euraryotic Flagella," Cambridge University Press, New York, NY, 1982.

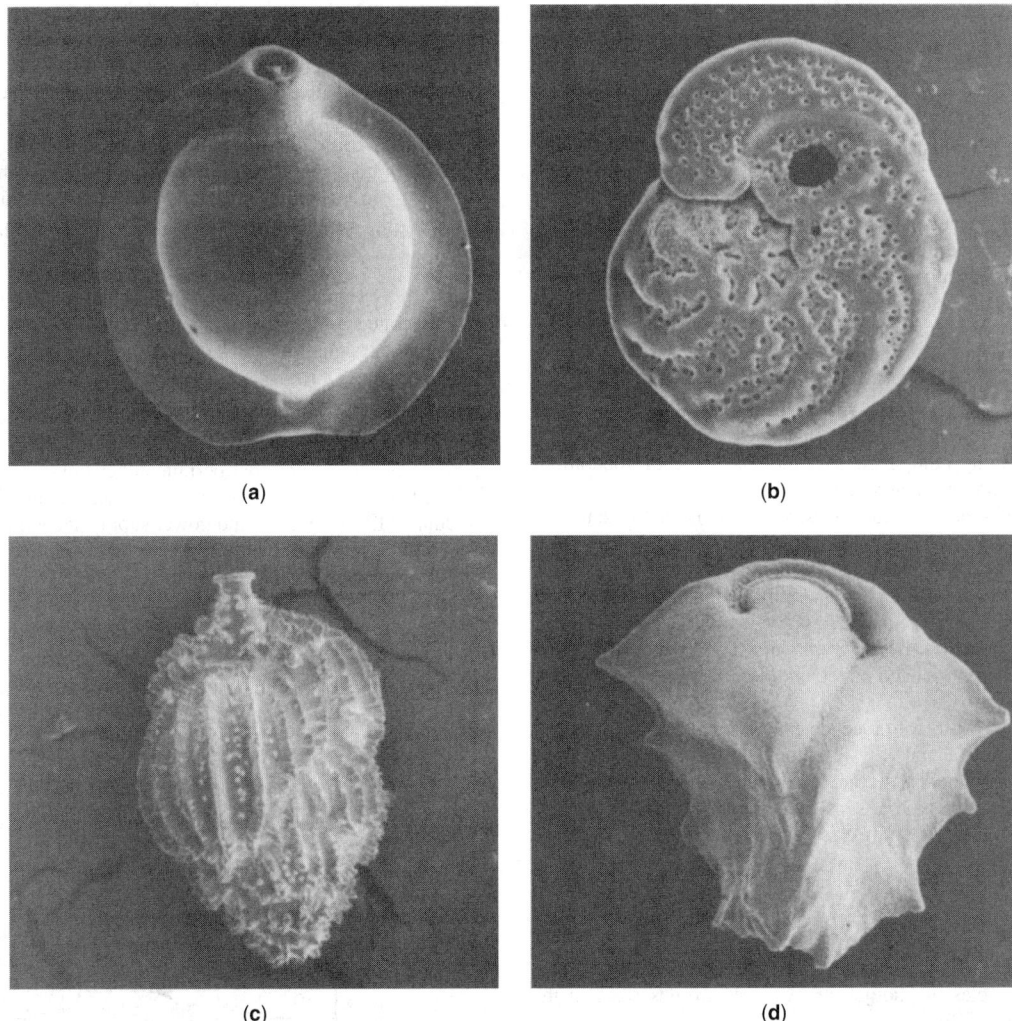

(a)

(b)

(c)

(d)

Fig. 1. Oceanologists and climatologists have found that the abundance of benthic foraminifera may be a key to present and past deep ocean water flows, such as the Atlantic Deep Water and the Antarctic Bottom Water and, in turn, indicators of past ice ages. Benthic foraminifera form calcitic shells that are preserved in deep-sea sediments. The above are scanning electron micrographs of several species of benthic foraminifera that presently live on the sea floor around the Rio Grande Rise. (a) *Pyrgo* and (b) *Planulina wuellerstorfi* are associated with the North Atlantic Deep Water. (c) *Uvigerina peregrina* is widespread in the deep Pacific Ocean, and occurs on the Rio Grande Rise beneath the Circumpolar Deep Water, a water mass that originates in the Pacific, then flows into the western South Atlantic Ocean. (d) *Ehrenbergina*, which is found in the South Atlantic Ocean. (*Woods Hole Oceanographic Institution.*)

Bell, G.: "Sex and Death in Protozoa: History of an Obsession," Cambridge University Press, New York, NY, 1989.

Bloodgood, R.A.: "Ciliary and Flagellar Membranes," Kluwer Academic Publishers, Norwell, MA, 1990.

Capriulo, G.M.: "Ecology of Marine Protozoa," Oxford University Press, Inc., New York, NY, 1990.

Curds, C.R.: "Protozoa and the Water Industry," Cambridge University Press, New York, NY, 1992.

Dentler, W., P.T. Matsudaira, and G. Witman: "Cilia and Flagella," Harcourt Brace & Company, San Diego, CA, 1995.

Gray, P.: "The Encyclopedia of the Biological Sciences," Krieger Publishing Company, Melbourne, FL, 1981.

Jahn, T.L. and E.C. Bovee: "How to Know the Protozoa," 2nd Edition, The McGraw-Hill Companies, Inc., New York, NY, 1978.

Laybourn-Parry, J.: "A Functional Biology of Free-Living Protozoa," University of California Press, Berkeley, CA, 1984.

Patterson, D.J.: "Free-Living Freshwater Protozoa: A Colour Guide," John Wiley & Sons, Inc., New York, NY, 1996.

Pennak, R.: "Fresh-Water Invertebrates of the United States: Protozoa to Mollusca," 3rd Edition, John Wiley & Sons, Inc., New York, NY, 1990.

Rietschel, P. and K. Rohde: "The Unicellular Animals," in "Grzimek's Animal Encyclopedia," Vol. 1, Van Nostrand Reinhold, New York, NY, 1974.

PROTRACTOR. 1. A muscle that draws some part of the body forward or draws out or extends a part.

2. A mathematical instrument consisting of a graduated arc for measuring or plotting angles.

PROUSTITE. This ruby-silver mineral crystallizes in the hexagonal system; its name is a product of its scarlet-to-vermilion color when first mined. It is a silver arsenic sulfide, Ag_3AsS, of adamantine luster. Hardness of 2–2.5; specific gravity of 5.55–5.64. Usual crystal habit is prismatic to rhombohedral; more commonly occurs massive. Conchoidal to uneven fracture; transparent to translucent; color, scarlet to vermilion red. Light sensitive; must be kept in dark environment to maintain its primary character. A product of low-temperature formation in most silver deposits. Notable world occurrences include the Czech Republic and Slovakia, Saxony, Chile and Mexico. Found in minor quantities in the United States; the most exceptional occurrence at the Poorman Mine, Silver City District, Idaho where a crystalline mass of some 500 pounds (227 kilograms) was recovered in 1865. It was named for the famous French chemist, Louis Joseph Proust.

PROXIMITY AND OBJECT DETECTORS. Generally, in position and motion control commonly encountered in automation systems, control action is taken to establish a position or series of positions that lie along a trajectory or path. In these cases, the exact coordinates of position are usually paramount. By contrast, *object detection* does not always require the accurate measurement of an object's position, but rather the prime purpose is one of detecting the *presence* or *nonpresence* (absence) of an object.

The needs for object detectors are several: (1) conveyor-associated applications, such as jam detection or protection, empty line detection, automatic routing; (2) safety and accident avoidance—to detect human hands and fingers where machines are manually loaded or unloaded and during maintenance procedures; (3) inspection of products—containers filled to the proper levels? labels properly placed or missing? incorrect closures? open flap detection and folding and wrapping imperfections, detection of web breaks, and plating and coating imperfections; (4) counting—to detect missing parts and to measure throughput; (5) sorting—by size, color, and other parameters; (6) as hopper level detectors and as feed cutoff controllers; and (7) a host of miscellaneous applications, such as edge guidance, remote door openers, and overhanging roof detectors. In industry, as well as in the lay world, these same kinds of devices play a major role as security monitors, warning of the unauthorized presence of persons and actions in designated secure areas.

It is interesting to note that object detection technology, although continuously refined from an engineering standpoint, dates back in principle for several decades and was one of the forerunners of modern industrial automation. Whereas today, robots seem to be in the forefront in terms of public recognition, it was not too many years ago that automated conveyor lines, packaging equipment, and the like were the operations proudly shown to visitors by plant managers.

Very few of the reasonably viable physical phenomena for detecting objects have been overlooked—a wide potpourri of detection methods is commercially available. A survey (1985) indicated that inductive and photoelectric detectors predominate, but other physical realms include electrical, electromagnetic, electromechanical, optoelectronic, radiant energy, air flow, and sonic approaches. Also very popular among industrial users are capacitive, Hall effect, Wiegand effect, and, to a lesser extent, magnetostrictive approaches.

A rather clean categorization of object detectors is: (1) contacting types, where the sensor makes actual physical contact with the object; and (2) noncontacting types, where the object only need be in the vicinity of the sensor. Depending on the type of sensing technology used, the vicinity can range from a millimeter or so up to several hundred feet (meters), although these are extreme cases.

Electromechanical (limit) contact switches are described in the article on **Limit Switch**.

Photoelectric Detectors

The fundamental principles of photoelectric devices are described in the article on **Photoelectric Effect**.

The principal aspects of the photoelectric effect of interest industrially are: (1) *photoconductivity*, evidenced by the increase in electrical conductivity of a material upon the absorption of light (or other electromagnetic radiation); and (2) the *photovoltage* or *photovoltaic effect*, wherein the energy of photons is converted to electrical voltage by the substance receiving the radiation. Photoconductivity is the basis of the operation of photocells and phototransistors utilized in industrial photoelectric switches. Principal applications for photovoltaic cells have been in aerospace applications, satellites for power, and the solar energy field.

Scope of Usage. Photoelectric controls respond to the *presence* or *absence* of either opaque or translucent materials at distances from a fraction of an inch (a few mm) up to 100 or even 700 feet (30 to 210 m). Photoelectric controls need no physical contact with the object to be triggered—important in some cases, such as those involving delicate objects and freshly painted surfaces. Some of the more common applications include thread break detection, edge guidance, web break detection, registration control, parts ejection monitoring, batch counting, sequential counting, security surveillance, elevator and conveyor control, bin level control, feed and/or fill control, mail and package handling, and labeling, among many others.

Photoelectric Control System Configuration. A self-contained control includes a light source, a photoreceiver, and the control base function, which amplifies and imposes logic on the signal to transform it into a usable electrical output. A *modular control* uses a light source-photoreceiver combination or reflective scanner separate from the control base. Self-contained retroreflective controls require less wiring and are less susceptible to alignment problems, while modular controls are more flexible in permitting remote positioning of the control base from the input components and hence are more easily customized.

Photoelectric controls are further classified as nonmodulated or modulated. *Nonmodulated* devices respond to the intensity of visible light. Thus, for reliability, such devices should not be used where the photosensor is subject to bright ambient light, such as sunlight. *Modulating*

controls employing light-emitting diodes (LEDs), respond only to a narrow frequency band in the infrared. Consequently, they do not recognize bright, visible ambient light.

Controls typically respond to a change in light intensity above or below a certain value of threshold response. However, certain plug-in amplifier-logic circuits cause controls to respond to the rate of light change (transition response) rather than to the intensity. Thus, the control responds only if the change in intensity or brightness occurs very quickly (not gradually).

Operating Mode. Both modulated and nonmodulated controls energize an output in response to:

1. A light signal at the photosensor when the beam is not blocked (light-operated, LO).
2. A dark signal at the photosensor when the beam is blocked (dark-operated, DO).

Although some controls have built-in circuitry that determines a fixed operating mode, most controls accept a plug-in logic card or module with a mode selector switch that permits either light or dark operation.

In addition to a light source, light sensor, amplifier (in the case of modulated LED devices), and power supply, a complete system includes an electrical output device (in direct interface with logic level circuitry—the output transistor of a DC-powered modulated LED device or of an amplifier-logic card).

Scanning Techniques. There are several ways to set up the light source and photoreceiver to detect objects. The best technique is that one which yields the highest signal ratio for the particular object to be detected, subject to scanning distance and mounting restrictions. Scanning techniques fall into two broad categories: (1) thru (through) scan, and (2) reflective scan.

In *thru (direct) scanning*, the light source and photoreceiver are positioned opposite each other, so light shines directly at the sensor. The object to be detected passes between the two. If the object is opaque, direct scanning will usually yield the highest signal ratio and should be the first choice. See Fig. 1.

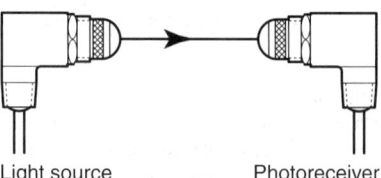

Light source   Photoreceiver

Fig. 1. In direct (or *thru*) scan configuration, the light source is aimed directly at the photoreceiver. Sometimes the configuration is referred to as the *transmitted beam system*. (Micro Switch.)

In *reflective scanning*, the light source and photoreceiver are placed on the same side of the object to be detected. Limited space or mounting restrictions may prevent aiming the light source directly at the photoreceiver, so the light beam is reflected either from a permanent reflective target or surface, or from the object to be detected, back to the photoreceiver. There are three types of reflective scanning: (1) *retroflective scanning*, (2) *specular scanning*, and (3) *diffuse scanning*.

Retroflective Scanning. With retroflective scanning, the light source and photosensor occupy a common housing. The light beam is directed at a retroreflective target (acrylic disk, tape, or chalk)—one that returns the light along the same path over which it was sent. See Fig. 2. Perhaps the most commonly used retrotarget is the familiar bicycle-type reflector. A large reflector returns more light to the photosensor and thus allows scanning at a further distance. With retrotargets, alignment is not critical. The light source-photosensor can be as much as 15° to either side of the perpendicular to the target. Also, inasmuch as alignment need not be exact, retroreflective scanning is well suited to situations where vibration would otherwise be a problem.

Specular Scanning. The specular scan technique uses a very shiny surface, such as rolled or polished metal, shiny plastic, or a mirror to reflect light to the photosensor. See Fig. 3. With a shiny surface, the angle at which light strikes the reflecting surface equals the angle at which it is reflected from the surface. Positioning of the light source and photoreceiver must be precise. Mounting brackets, which firmly fix the

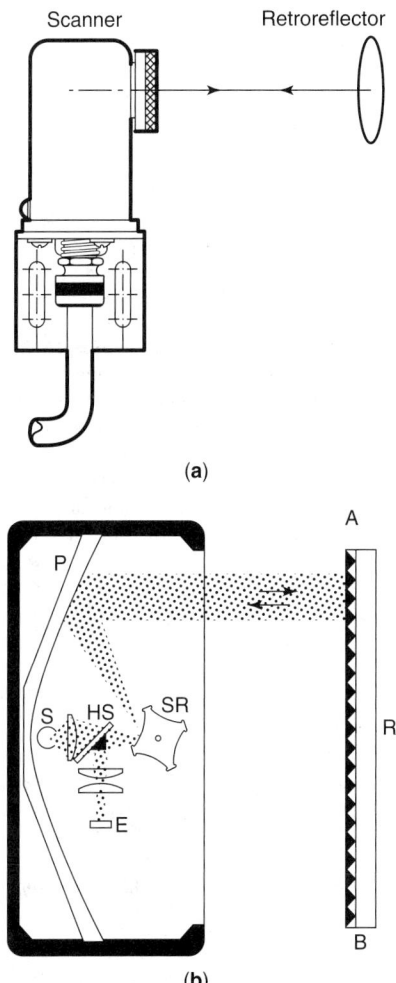

(a)

(b)

Fig. 2.   (**a**) Reflected beam (retroreflective scan) system in which the light source and photoreceiver are contained in a single enclosure. This simplifies wiring and avoids critical alignment of the source and sensor. (*Micro Switch.*) (**b**) By adding a rotating-mirror wheel (SR), a parabolic reflector (P), and a semitransparent mirror (HS), a parallel-scanning beam can be obtained. This beam moves at high speed from A to B, thus forming a "light curtain," any interruption of which is detected and signaled by a relay, S. E is photoreceiver. (*Sick Optik Elektronik.*)

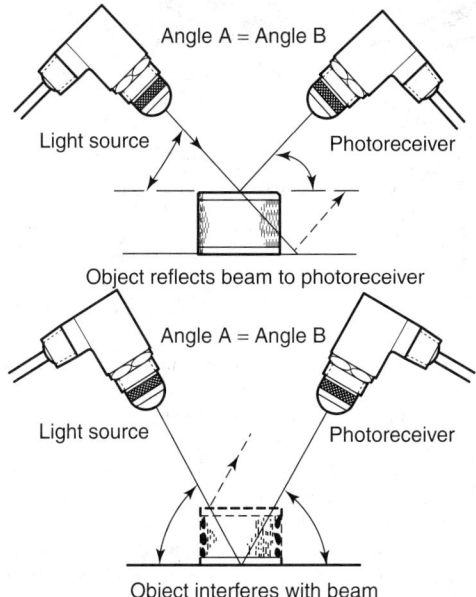

Fig. 3.   Specular scan technique uses a very shiny surface, such as rolled or polished metal, shiny plastic, or a mirror to reflect light to the photosensor. (*Micro Switch.*)

light source-photoreceiver relationship, must be used. Also, the distance of the reflecting surface from the light source and photoreceiver must be consistently controlled. The size of the angle between the light source and photoreceiver determines the depth of the scanning field. With a narrower angle, there is more depth of field. With a wider angle, there is less depth of field. For a fill-level detection application, for example, this means that a wider angle between the light source and photoreceiver allows detection of the fill level more precisely.

Diffuse Scanning. Nonshiny (matte) surfaces, such as kraft paper, rubber, and cork, absorb most of the incident light and reflect only a small amount. Light is reflected or scattered nearly equally in all directions. In diffuse scanning, the light source is positioned perpendicularly to a dull surface. Emitted light is reflected back from the target to operate the photoreceiver. See Fig. 4. Because the light is scattered, only a small percentage returns. Therefore, the scanning distance is limited (except with some high-intensity modulated LED controls), even with very bright light sources. It is often difficult to obtain a sufficient signal ratio with diffuse scanning when the surface to be detected is almost the same distance from the sensor as another surface (for instance, a nearly flat or low-profile cork liner moving along a conveyor belt). Contrasting colors can help in such situations.

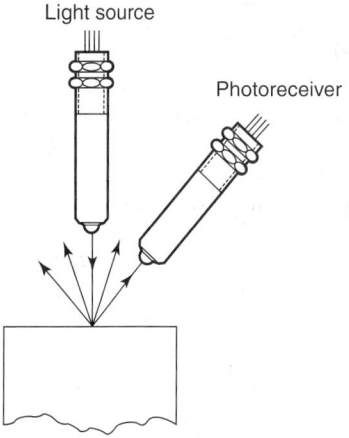

Fig. 4.   Diffuse scan is used in registration control and to detect material (corrugated metal, for example) with a slight vertical flutter — which may prevent a consistent signal with specular scan. Alignment is not critical in picking up diffuse reflection. (*Micro Switch.*)

Diffuse scanning is used in registration control and to detect material (corrugated metal, for example) with a slight vertical flutter — which might prevent a consistent signal with specular scanning. Alignment is not critical in picking up diffuse reflection.

Light Sources and Sensors. Early photoelectric control systems used incandescent light sources and traditional photocells — a combination still used. A photocell changes its electrical resistance with the amount of light that falls on it. A number of photocells have been used over the years for different applications (photoelectric controls, copying machines, TV pickup tubes, etc.). Widely used for photoelectric controls are cadmium sulfide and cadmium selenide cells. During recent years, phototransistors and photodiodes have become available as sensors — and LEDs have been used as light (infrared) sources. There are several advantages in using the more recent hardware.

Photocells. There are at least four parameters that are important in the operation of a photocell: sensitivity, speed of response, light history effect, and effect of temperature.

Phototransistors. A phototransistor produces a collector current that is a function of both base current and light. Since the base lead of a phototransistor is usually left unconnected, only variations in light intensity produce variations in current output. There are several differences between the phototransistor and the photocell. (1) Current output of a phototransistor is largely independent of the voltage across it, whereas that of a photocell is not. As a result, controls designed to work with photocells will not necessarily work well with phototransistors, and vice versa. (2) The response of phototransistors is affected by changes in temperature, but in a way opposite to that of photocells; the higher the temperature, the higher the current output. (3) Phototransistors have a polarity, which must be observed; photocells do not. (4) Phototransistors respond to light much faster than photocells, but typically have a lower sensitivity.

Photodiode response is narrower than that of the phototransistor, making the diode more effective in blocking stray light from incandescent, sun, or other sources.

LED Sources. The useful life of an LED is estimated at 100,000 hours, which is at least ten times that of an incandescent lamp. However, incandescent lamps are still frequently used because they have a spectrum from the ultraviolet to the visible to the infrared, allowing a wide range of colored targets to be detected. LEDs have the advantage that they can be modulated directly, whereas incandescent lamps require a mechanical chopper. Silicon phototransistors and photodiodes are excellent matches for infrared LEDs because their greatest sensitivity peaks almost match precisely at the transmitter's (LED) wavelength.

Use of Fiber Optics with Photocells. Fiber-optic bundles can be added to existing photoelectric switches to provide object sensors, and these can be combined to implement logic functions. Such systems are useful for applications that require several sensing inputs and one or more outputs to interface with microcomputers. Program selection permits use of the LO or DO mode and allows operation of any channel for a predetermined time, thus avoiding sequential channel operation in fixed time frames. These systems frequently find application where a programmable controller is not warranted because of cost or complexity. Input can be from a relay or switch contacts, transducers, memory devices, CMOS, or TTL. The output section provides a channel signature for each emitter and detector pair, resulting in the capacity of actuating one or more output devices.

Applications of Photoelectric Controls. As illustrated by Fig. 5, the applications for photoelectric controls in automated systems seem to be limited only by the ingenuity of the control and system engineer. Most of these applications can be served by other types of proximity sensors, but there are exceptions.

Magnetic Proximity Switches

There are four principal types of magnetic proximity switches: (1) variable-reluctance-type sensors, the operation of which depends on the interruption of a fixed magnetic field (circuit) by a ferrous actuator; (2) magnetically actuated dry reed or mercury switches; (3) Hall-effect sensors; and (4) Wiegand-effect sensors.

Variable-Reluctance Sensors. The principle of operation of variable-reluctance position (presence) sensors is shown schematically in Fig. 6. These transducers convert motion (rotating, sliding, oscillating) into electrical control signals. As shown in Fig. 6(a), with no actuating object in the vicinity of the sensor (pole piece plus coil plus magnet), the path of magnetic flux is undisturbed. As an object approaches and passes near the pole piece, the flux path is distorted. This system is often used in connection with rotating equipment for speed measurement (tachometry) where discontinuities, such as gear teeth, shaft keyways, drilled holes in steel plates, etc., alter the magnetic flux in proportion to rpm. Sensors are available in active and passive forms. Passive sensors require no external electric power. The output signal is an alternating current, the waveform of which is a function of the actuator, usually sinusoidal. The amplitude and frequency of the output signal are both proportional to the surface speed of the actuator as it passes the sensor's pole piece. The active configuration requires a DC power supply. The output signal is a pulse train whose amplitude is constant over the operating range for a fixed supply voltage level. Active magnetic sensors provide usable output signals at very low actuator speeds and at relatively large air gaps between the sensor pole piece and the actuator. They produce a logic-level output signal directly compatible with digital instrumentation.

Magnetically Actuated Dry Reed Switches. Generally consisting of a thin reed (wire) contained in a hermetically sealed container (encapsulated), this type of switch is both inexpensive and rugged. Whenever an activating magnet approaches the critical range of the switch, a contact closure is made. Life expectancy usually is in excess of 20 million operations at contact ratings of about 15 VA. These switches generally can operate loads

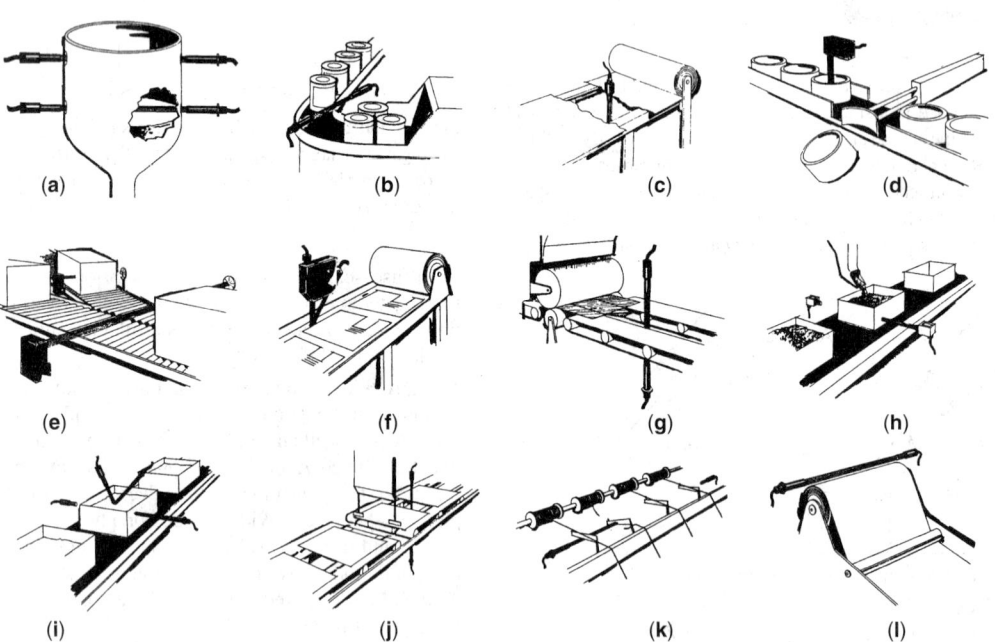

Fig. 5.   Representative applications of photocell detectors in automated systems: (**a**) Two light source-photoreceiver pairs are used to keep hopper fill level between high and low limits. (**b**) Counting products is a common application of photoelectric controls. Counting batches or groups of cans or other items prior to packaging or group processing is also common. (**c**) A photoelectric control operating on reflected light is a simple way to detect a web break. An alternative is to put a light source above the web, and a photoreceiver below. (**d**) Dark caps are checked for white liners by a photoelectric scanner. The scanner activates a mechanism that rejects caps without the liners. (**e**) To prevent collisions where two conveyors merge, each conveyor is monitored by a control that powers the other conveyor when its own conveyor is cleared. (**f**) A tubular light source and photoreceiver in a specially designed bracket detect registration marks to initiate any related operation, such as printing, cutoff, or folding. (**g**) Gluing, buffing, or flattening can be done efficiently by controlling the pressure rollers or buffer with a photoelectric light source and photoreceiver that detect the product to be processed. (**h**) Using logic for one-shot pulse output, a photoelectric control slows a conveyor and fills the carton that has interrupted the light beam. (**i**) Two light source-photoreceiver pairs work together to check fill level. The box-detecting pair turns on, or enables, the fill inspection pair—thereby preventing the inspection pair from mistaking the space between boxes as an "improper fill." (**j**) Light source and photoreceiver placed near a guillotine are used to detect products and operate the blade for cutting the link between products. (**k**) Thread break detection made possible when the photoelectric beam is interrupted by a lightweight flag riding on the taut thread. (**l**) The size of a paper or fabric roll can be controlled by positioning a light source and a photoreceiver so the roll diameter blocks the beam. (*Micro Switch.*)

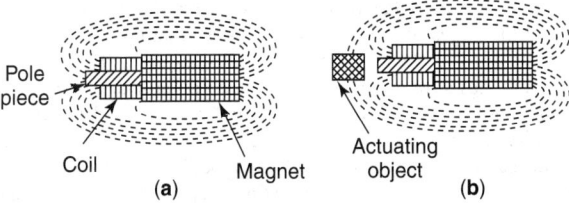

Fig. 6. Schematic representation of the action of a variable-reluctance object sensor: (**a**) A sensor with no actuating object in the magnetic field, (**b**) actuating object in field alters the voltage generated at coil terminals. The voltage is proportional to the rate of change of magnetic flux.

directly. Since the actuating magnet (powerful alloy magnets now usually used) can be installed on a rotating or reciprocating object, the switch can be used in a wide variety of applications for counting, positioning, and synchronizing. Contact closure speeds can be up to 100 per second. At one time, mercury switches with flexible electrodes that can be attracted by the proximity of a magnet were more widely used than at present.

Inductive Proximity Sensors

In an inductive proximity sensor, an electromagnetic field (radio frequency, rf) is generated by an oscillator circuit. When a metal object enters the effective field generated by the sensor, a countercurrent (eddy current) is set up in the metal object. This causes a voltage drop in the oscillator. This drop is sensed by the detector, which triggers the output. The output can be used for many industrial control purposes. See schematic diagram of a typical sensor given in Fig. 7. The sensors are available in three basic configurations: (a) cylindrical—a two-piece mounting clamp with socket head screws allows the installed sensor to be moved to the desired position; (b) threaded—the installed sensor can be rotated to the desired position and held in place with two flat nuts; and (c) rectangular—the sensor has slots in mounting base, which allow adjustment to desired position after installation. Nominal detection ranges from about 2 mm to about 20 mm, depending on specific design. Some units are adjustable from 10 to 50 mm.

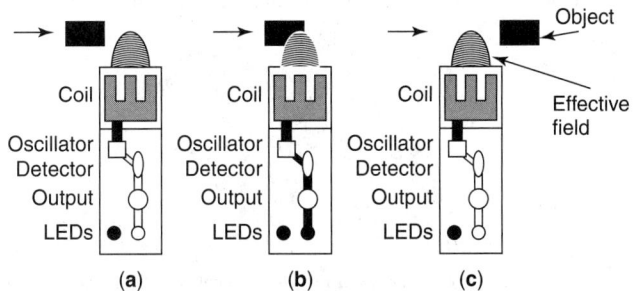

Fig. 7. Inductive proximity sensor: (**a**) Object approaching sensor, (**b**) object in field of sensor, (**c**) object beyond sensor. (*Cutler-Hammer.*)

Wiegand-Effect Switches

A Wiegand wire is a small-diameter wire that has been selectively work-hardened so that the surface and the core of the wire differ in magnetic permeability. When subjected to a magnetic field, the wire emits a well-defined pulse that requires little signal conditioning. This pulse induces a voltage in the surrounding sensing coil. The wire is insensitive to polarity and emits a pulse whether the magnetic field is flowing from north to south, or vice versa. A Weigand proximity senses the presence or absence of ferromagnetic material.

Capacitive Proximity Sensors

The basic element of a capacitive proximity sensor is a high-frequency oscillator containing a capacitor, one of the plates of which is built into the end of the sensor. When oscillating, a field is created around this free capacitor plate. When an object is placed in the field, the amplitude of the oscillator output changes. These oscillations are rectified and smoothed by an integrating coupling stage. The resulting DC signal is fed to a trigger circuit, which switches an output transistor. The sensing zone or envelope of the sensor is influenced by the physical properties of the object being sensed in the following ways: (1) Nonconducting materials, such as glass

and plastics, are sensed by a change in dielectric characteristics. Since this change is small, the sensing ranges are necessarily limited. (2) Conducting materials are sensed by a change in dielectric characteristics as well as by an additional disturbance of the noise field caused by terminal conductivity. (3) Materials containing both conducting and nonconducting properties, especially if grounded, are sensed by a combination of the foregoing characteristics, as well as by absorption. These conditions produce the greatest switching distance for a given sensor. Because capacitive sensors are so markedly affected by material characteristics, an estimate of the switching range requires knowledge of the medium to be detected.

Magnetostrictive Limit Switch

Of limited application to date, the principle of magnetostriction was discovered in 1858 and represents a change in dimension of a material as the magnitude or direction of magnetization in a crystal is changed. The principle was introduced in the limit switch field in the mid-1970s. The active element is a helical spring that is subjected to axial extension or compression caused by mechanical displacement. The magnetostrictive transducer is considered to have good potential where high-resolution information is required over a substantial range.

Ultrasonic Switches and Other Novel Approaches

The presence of objects can be detected by return echoes from reflecting materials in a form of sonar. Similarly, rf reflection (radar) can be applied. A number of the large, well-established suppliers serving the proximity switch field have been joined by scores of newer, smaller firms that are busily engaged as of the late 1980s in applying "smart" electronics to new and old concepts.

One firm has developed a lacquer detection system (inside coating of cans). The system is based on the principle that ultraviolet light is reflected by uncoated metal surfaces, such as aluminum or steel, but is absorbed by lacquer. The system measures the amount of light reflected from a can and, through a series of statistical routines performed by a microprocessor, the system is able to distinguish areas of metal exposure as small as 0.5 square inch (3.2 square centimeters), depending on location. The detector-ejector is built as a unit and located above the conveyor systems. Speeds up to 1500 cans/minute are achieved. Also performed by the microprocessor are: (1) electronic compensation for UV light source aging, (2) eject control to adjust ejector timing to the actual conveyor speed, (3) counting of tested, ejected, and knocked-over cans, and (4) printer control and data supply to a host computer.

Impact of Machine Vision Systems. Normally, one equates machine vision systems with considerable complexity and relatively high cost. But, there are opportunities for simpler, lower-cost vision systems that accomplish some of the tasks traditionally assigned to photoelectric and proximity switch systems. The cost differential, however, still remains quite large. See also **Machine Vision (Recognition and Applications)**.

PRURITUS. Itching of the skin, usually due to irritation of peripheral nerves.

PSEUDOCODE. An arbitrary code not directly understandable by a computer. Also called *interpreter code*.

PSEUDOMORPH. In mineralogy and geology, a mineral, having the crystal form of one species and the chemical composition of another. Typical pseudomorphs are malachite in the form of cuprite, barite in the form of quartz, limonite in the form of pyrite. In such cases of pseudomorphism the evidence seems to be that there has been a complete chemical and molecular change but without any change of the original outward form. See also **Mineralogy**.

PSEUDOMORPHISM. See **Mineralogy**.

PSEUDOSCALAR. A scalar quantity that changes its sign when the coordinate system, to which the quantity is referred, is changed from a right-handed to a left-handed one, or vice versa. An example is the scalar product of a polar vector and a pseudovector. See also **Vector Multiplication**.

PSEUDOSCORPION. A small animal, resembling the scorpions slightly in appearance. The abdomen is rounded and has no slender posterior part and no sting, but the pair of chelate pinchers at the anterior end of the body resembles those of the scorpions. Pseudoscorpions are commonly found in leaves at the surface of the ground and under the bark of decaying logs. They are only a few millimeters long. The relatively few species make up the order *Chelonethida* of the class *Arachnida*.

PSEUDOVECTOR.

The vector product of two vectors does not completely satisfy the formal requirements of a vector, for it changes sign if the coordinate system is changed from a right-handed to a left-handed one, or vice versa. It is a typical example of a pseudovector (also called an axial vector). Thus the vector for an element of area, represented by the vector product $d\mathbf{S} = d\mathbf{x} \times d\mathbf{y}$ is not determined with respect to direction unless an arbitrary convention is established for the positive side of the surface element. The three components of a pseudovector are actually the components of a three-dimensional antisymmetric tensor of second rank. Physical quantities which are pseudovectors include angular momentum (vector product of momentum and radius vector); moment of a force (vector product of force and distance); linear velocity (vector product of angular velocity and radius vector).

In the following relations, symbols ($\pm$) refer to true and pseudoquantities, S for scalar and $\mathbf{V}$ for vector; the primed and double primed vectors are so designated merely to indicate that they are not identical with the unprimed vectors: $S_+\mathbf{V}_\pm = \mathbf{V}'_\pm$; $S_-\mathbf{V}_\pm = \mathbf{V}'_\mp$; $\mathbf{V}_\pm \cdot \mathbf{V}'_\pm = S_+$; $\mathbf{V}_+ \cdot \mathbf{V}_- = S_-$; $\mathbf{V}_\pm \times \mathbf{V}'_\pm = \mathbf{V}''_-$; $\mathbf{V}_+ \times \mathbf{V}'_- = \mathbf{V}''_+$; $\nabla S_\pm = \mathbf{V}_\pm$; $\nabla \cdot \mathbf{V}_\pm = S_\pm$; $\nabla \times \mathbf{V}_\pm = \mathbf{V}'_\mp$.

PSILOMELANE.

Psilomelane is a massive black mineral, essentially a basic oxide of barium with divalent and quadrivalent manganese, corresponding to the formula $BaMn^{2+}Mn^{4+}O_{16}(OH)_4$. It crystallizes in the monoclinic system, but is found only in massive, botryoidal or reniform to earthy habits; hardness, 5–6, less in earthy varieties; specific gravity, 6.45; color, black to gray; opaque; submetallic to dull luster. It is a product of secondary weathering of manganese carbonates and silicates. Of widespread occurrence, usually associated with pyrolusite. Major world occurrences include Michigan in the United States, Scotland, Sweden, France, Germany, and India. It is a major source of manganese. The word psilomelane is derived from the Greek words meaning smooth and black, in reference to the smooth black surfaces so often exhibited.

PSI PARTICLE.

Discovery of this subatomic particle in 1974 was announced independently by Ting (Brookhaven National Laboratory) who named it the *J particle* and by B.D. Richter (Stanford) who named it the *psi particle*. The discovery of this particle resolved a number of important problems in particle physics. Intensive research on the psi particle was carried out by Richter and the Stanford group during 1975 and 1976 and is reported firsthand by Richter (*Science*, **196**, 1286–1297, 1977). As pointed out by Richter, the four-quark theoretical model became much more compelling with the discovery of the psi particles. The long life of the psi is explained by the fact that the decay of the psi into ordinary hadrons requires the conversion of both c and $\bar{c}$ into other quarks and antiquarks. See also **Particles (Subatomic)**.

PSITTACIFORMES *(Aves)*.

The parrots and related species, including cockatoos, macaws, and parakeets (paraquets) are all members of this order. These birds are characterized by the crooked bill, adapted for opening nuts and seeds. It is popularly believed that all parrots are extremely colorful and inhabit the tropical jungles. Neither belief holds good always. While it is true that the more colorful forms are indeed found in tropical regions and in South America, and most particularly in territories of New Guinea and northern Australia, which are thought to be the original homeland of the parrot stock, it is also true that some species live above the tree line in snow and frost. Moreover, not all parrots are colorful; there are many forms with green plumage as a means of camouflage and many species are dark; some are even black. Today Europe is the only continent where there are no parrots.

The length of this order ranges from 10 centimeters (4 inches) in the pygmy parrot to 100 centimeters (39 inches) in the blue macaw. The fourth (lateral) toe is reversed, like the first, so that the two function opposed to the second and third. Thus a pair of pincers is formed which serves in climbing and grasping objects. The legs are short; the upper mandible is articulate and can be raised; the lower can slide (a useful property in shelling nuts and comminuting seeds and fruits). The underside of the upper mandible usually has hard filing ridges across its width which serve to sharpen the edges of the lower mandible so that the bird can get better grasp of its food and can grate hard shells more easily. The tongue is usually thick, has strong muscles, and many touch and taste papillae. The lories (subfamily *Trichoglossinae*) are brush tongued and thus thoroughly adapted to visiting flowers on whose pollen and nectar they feed. Almost all species can use their beaks as a "third foot" when climbing.

The sexes, as a rule, are easily distinguishable. With the exception of the keas, parrots are usually monogamous. Parrots usually breed in caves; their clutch numbers from 1 to 10 white eggs. As a rule brooding, which takes from 18 to 30 days, is done exclusively by the female. The young are crop-fed, usually by both parents, the food thus being already softened and supplied with vitamins from the glandular stomach. There are specific breeding seasons and sometimes there are 2 broodings.

Parrots comprise only 1 family (*Psittacidae*), which is subdivided into 7 subfamilies. 1. Keas (*Nestorinae*); 2. Vulturine Parrots (*Psittrichasinae*); 3. Cockatoos (*Kakatoeinae*); 4. Pygmy Parrots (*Micropsittinae*); 5. Lories (*Trichoglossinae*); 6. Owl parrots (*Strigopinae*); and 7. True Parrots (*Psittacinae*). There are 79 genera, 326 species, and 816 subspecies.

The most primitive of the still extant parrot forms is the Kea (subfamily *Nestorinae*). Seeing these birds at a distance for the first time, one would think that they are crows, judging from their motions. The habitat of the keas is also somewhat unusual for parrots; they breed in the New Zealand Alps above the tree line.

They are as large as crows: the length is 50 centimeters (20 inches) and the bill is long and narrow, the tooth-like serrations are undeveloped, and it has no filing ridges. They are birds of the dawn and dusk. Their food consists of vegetable and animal matter. The clutch numbers 2–4 eggs, the incubation period is 29 days.

There are 3 species, one of which is already extinct: (1) the Kaka (*Nestor meridionalis*); (2) the Kea (*Nestor notabilis*); (3) the Slender-Billed Kea (*Nestor productus*) has been extinct since the middle of the nineteenth century.

The Vulturine Parrot (*Psittrichas fulgidus*) is the only representative of the subfamily *Psittrichasinae*. See Fig. 1. It has bristle-like feathers on the neck. Its size is that of a crow; the length is 50 centimeters (20 inches). Its habitat is mountainous woodlands at altitudes of from 800 to 2,000 meters (2,645–6,562 feet); it feeds on sprouts, buds, berries, and fruits.

Fig. 1. Vulturine parrot: Kea (*Nestor notabilis.*) (Sketch by Glenn D. Considine).

The Cockatoos (subfamily *Kakatoeinae*) are the largest parrots of the Indo-Australian region and at least some species are known to every bird lover. Cockatoos are as large as pigeons or ravens; the length is 32–80 centimeters (13–31 inches). Their feather crest is diagnostic, serving as a signaling device, and it sometimes differs in color from the rest of the plumage. Their food consists mostly of plant matter. Their clutch numbers from 2 to 4 eggs and the incubation period is 20–30 days, during which male and female relieve each other; the young are fledged at the age of 60–70 days. There are 5 genera with 17 species and 48 subspecies.

The Palm Cockatoo (*Probosciger aterrimus*) is the largest cockatoo, with a length reaching 80 centimeters (31 inches).

The Black Cockatoos (*Calyptorhynchus*) have a length of 50–65 centimeters ($19\frac{1}{2}$–$25\frac{1}{2}$ inches). Among the 4 species are: (1) the Yellow-Tailed Cockatoo (*Calyptorhynchus funereus*), which has a length of 60 centimeters ($23\frac{1}{2}$ inches); usually its clutch consists of 2 eggs; (2) The White-Tailed Cockatoo (*Calyptorhynchus baudinii*).

The Gang-Gang-Cockatoo (*Callocephalon fimbriatum*) has a length of 35 centimeters ($13\frac{1}{2}$ inches). The sexes differ markedly: the male has a red head and crest, while the female is blackish-gray. The clutch numbers 2 eggs.

The White-Billed and Black-Billed Cockatoos (*Kakatoe*) have a length of 32–50 centimeters ($12\frac{1}{2}$–$19\frac{1}{2}$ inches). They generally have light, usually white plumage. The division of this group into 2 genera (*Plyctolophus* and *Kakatoe*) is generally accepted. *Plyctolophus* comprises the species with black bills, while the *Kakatoe* describes those having light-colored bills. Among them are the species that bird fanciers like to keep: (1) The Yellow-Crested Cockatoo (*Kakatoe galerita*) has a length of 50 centimeters ($19\frac{1}{2}$ inches). (2) The White Cockatoo (*Kakatoe sulphurea*) has a length of 50 centimeters ($19\frac{1}{2}$ inches). (3) The Rose Cockatoo (*Kakatoe moluccensis*) has a length of 35 centimeters ($13\frac{1}{2}$ inches). (4) The Pink Cockatoo (*Kakatoe leadbeateri*) has a length of 38 centimeters (15 inches). (5) The Galah (*Kakatoe roseicapilla*) has a length of 37 centimeters ($14\frac{1}{2}$ inches). (6) The Bare-Eyed Cockatoo (*Kakatoe sanguinea*) has a length of 40 centimeters ($15\frac{1}{2}$ inches). (7) The Slender-Billed Cockatoo (*Kakatoe tenuirostris*) has a length of 40 centimeters ($15\frac{1}{2}$ inches).

The Cockatiel (*Nymphicus hollandicus*) reaches a length of 30 centimeters (12 inches), and the clutch consists of 4–7 eggs. See Fig. 2.

Fig. 2. Cockatoo: Cockatiel (*Numphicus hollandicus.*) (Sketch by Glenn D. Considine).

The Pygmy Parrots (subfamily *Micropsittinae*) are the smallest species among parrots. Their length is 10 centimeters (4 inches) and their weight is 13 grams (0.5 ounce). Little is known so far about the life pattern of these dwarfs among parrots. The toes are very long and thin; the bird has short, strong, stiff tail feathers that serve as a prop against the bark (as in the woodpecker) when it climbs around the tree trunk. Pygmy parrots feed on fruits, tree termites, and the sap of trees such as the *Albizzia procera* (related to the true acacia); the food is heavily mixed with saliva, which is secreted from the large salivary glands. They breed in nests of tree-inhabiting termites. There are 6 species and 23 subspecies; among them the most common are *Micropsitta pusio* and the Yellow-Capped Pygmy Parrot (*Micropsitta keiensis*).

The Lories (subfamily *Trichoglossinae*) are the most colorful of all parrots. Red is predominant among their conspicuously bright, poster-like colors. Only a few species are less colorful or even black. They range in size from that of a sparrow to that of a pigeon; the length is 12–35 centimeters ($4\frac{1}{2}$–$13\frac{1}{2}$ inches). Since they feed on blossoms, berries, and soft fruits, the filing ridges on the underside of the upper jaw are degenerate. The tip of the tongue is shaped like a paintbrush. Sex differences are slight. They have a clutch of 2–4 eggs, the incubation period lasting 21–26 days. The young stay in the nest for about two months. They occur in New Guinea and neighboring islands, and occasionally in Australia. There are 14 genera with a total of 61 species and about 150 subspecies.

The subfamily Owl Parrots (*Strigopinae*) has only 1 species, the Owl Parrot (*Strigops habroptilus*). It is about the size of a crow, the length reaching 60 centimeters ($23\frac{1}{2}$ inches). Its plumage is soft and the feathers around the beak are bristle-like, as they are in owls and goatsuckers. Its wings are short and rounded and it is virtually flightless. It feeds on plants and its beak is thick and without serrations. The owl parrot is nocturnal. Its clutch numbers two.

Since the owl parrot is virtually flightless it must climb, but it is able to come down in a slanted gliding flight of about 100 meters (328 feet). Its habitat is mountainous forests on the southern island of New Zealand. Owl parrots stamp down genuine paths in their habitats. They feed on mosses, leaves, sprouts, berries, and, if possible, fungi; they will also dig up roots and rhizomes of ferns. Indigestible cellulose fibers are formed into balls

and regurgitated. The bird usually does not tear out grass blades and the berry-like *Carmichaelia* branches, but crushes them on sight.

The seventh and last subfamily, the True Parrots (*Psittacinae*), has the greatest number of species. There are 5 tribes: the Rosellas (*Platycercini*); *Loriini*; *Loriculini*; *Psittacini*; and *Araini*.

The best known and most frequently kept parrot, the budgerigar, belongs to the rosellas (tribe *Platycercini*). Its larger and usually very colorful relatives are among the most popular aviary-kept birds. They range abundantly from sparrow to magpie in size. The length is 18–38 centimeters (7–15 inches). The beak often has serrations, and is short and thick; the tail is staggered and long. The bird has a well-developed uropygium (the fleshy and bony prominence at the posterior extremity of a bird's body, which supports the tail feathers). Both sexes have the same or very similar colors. They usually inhabit the steppes and are poor climbers; their food consists of grass seeds. They breed once or twice a year, mostly in hollows, and have a clutch size of 4–6 eggs. The incubation period is from 18 to 21 days. There are eleven genera in this tribe with 29 species and 67 subspecies: (1) the Night Parrots (*Geopsittacus*); (2) the Ground Parrots (*Pezoporus*); (3) the Red-Fronted New Zealand Parakeets (*Cyanoramphus*); (4) the Horn Parakeets (*Eunymphicus*); (5) the Budgerigars (*Melopsittacus*); (6) *Neophema*; (7) the Red-Backed Parrots (*Psephotus*); (8) *Northiella*; (9) the Red-Capped Parrots (*Purpureicephalus*); (10) the *Rosellas*, in a narrower sense (*Platycercus*); and (11) the Swift Parrots (*Lathamus*). Two of the genera occur in New Zealand and on neighboring islands, and all others in Australia and Tasmania.

The Wax-Billed Parrots (*Loriini*) do not form as homogeneous a group as the rosellas. There are large and small species, and short and long-tailed ones, and even the shapes of their beaks differ considerably. Some of the forms are not even known by name to parrot lovers.

The wax-billed parrots range from the size of a sparrow to that of a magpie; the length is 13–54 centimeters (5–21 inches). See Fig. 3. The beak in most of the species has a wax-like sheen and smoothness; red beaks occur frequently, at least at the upper mandibles of the males (with the exception of the rosy-faced lovebird, the black-collared lovebird, and the gray-headed lovebird); while the staggered tails are longer than the wings, the wedge-shaped, rounded, or clipped ones are shorter. They occur in Australia, Asia, and Africa; there are 10 genera, 50 species, and 168 subspecies: (1) King Parrots (*Alisterus*); (2) the Red-Winged Parrot (*Aprosmictus*); (3) *Polytelis*; (4) *Lorius*; (5) *Psittacula*; (6) the Lovebirds (*Agapornis*); (7) *Mascarinus*; (8) *Tanygnathus*; (9) Racket-tailed Parrots (*Prioniturus*); and (10) Masked Parakeets (*Prosopeia*).

Fig. 3. Lovebird: Rosy-faced lovebird (*Agapornis roseicollis*) (Sketch by Glenn D. Considine).

The tribe *Loriculini* inhabits Southeast Asia and the adjoining island groups that stretch toward Australia. The *Loriculini* reach a length of 10–16 centimeters (4–6 inches). The beak is slender and longer than it is high. As with the wax-billed parrots, the beak has a smooth, shiny surface and is red or black. The tail is as long as the wings or shorter, and the upper tail coverts reach as far as the tip of the tail. They have a clutch of 2–3 eggs. There is 1 genus with 10 species and 31 subspecies; among them is the Malay Lorikeet (*Loriculus galgulus*).

All *Loriculini* usually scurry gracefully on the ground, and can climb deftly among branches and boughs. When sleeping, and occasionally when looking for food, they suspend themselves head down from branches rather like bats. In the wild their food consists predominantly of fruits, nectar from blossoms, and pollen. Like most species of lovebirds, the *Loriculine* also carry nesting materials under their feathers.

The tribe *Psittacini* includes the gray parrots and blunt-tailed parrots, known as good imitators. See Fig. 4. They range from the size

Fig. 4.   Gray parrot (*Psittacus erithacus.*) (Sketch by Glenn D. Considine).

of a chaffinch to that of a crow; the length is 16–50 centimeters (6–19 1/2 inches). Their tails are blunt or somewhat rounded. The beak is not smooth and wax-like. It may be black, gray, brownish, or yellowish-white, but rarely red. They occur in South and Central America, Africa, and Madagascar. There are 13 genera with 67 species and 143 subspecies: (1) the Vasas (*Coracopsis*); (2) *Psittacus* (Gray Parrot, *Psittacus erithacus*); (3) *Poicephalus* (Yellow-billed Senegal Parrot, *Poicephalus senegalus*); (4) *Pionites*; (5) *Pionus* (White-crowned Parrot, *Pionus senilis*); (6) *Deroptyus* (Red-Fan Parrot, *Deroptyus accipitrinus*); (7) The Blunt-Tailed Parrots (*Amazona*); (8) *Graydidasculus* (Short-tailed Parrot, *Graydidasculus brachyurus*); (9) *Pionopsitta*; (10) *Hapalopsittaca*; (11) *Gypopsitta* (Vulturine Parrot, *Gypopsitta vulturina*); (12) *Touit*; (13) *Triclaria* (Blue-Bellied Parrot, *Triclaria malachitacea*).

The tribe *Araini* inhabits North and South America. Besides small forms, we also find among them the giants of the parrots, the macaws. Their sizes range from that of a titmouse to that of a pheasant; the length is 12–98 centimeters (5–38 1/2 inches). The tail is almost always staggered, and the tail feathers are narrowed or the tips are pointed; the eye rings are often bare. The bill is short, thick, never red. The most conspicuous characteristic of most *Araini* is their loud, penetrating voice. Since the parrots of this group can be quite destructive to their enclosures, they are not very popular with bird lovers.

There are 19 genera (one of them extinct) with 84 species and more than 180 subspecies: (1) the Parrotlets (*Forpus*); (2) *Nannopsittaca*; (3) *Brotogeris* (Orange-flanked Parakeet, *Brotogeris phyrrhopterus*); (4) The *Bolborhynchos* (Rufous-Fronted Parakeet, *Bolborhynchos ferrugineifrons*); (5) *Psilopsiagon*; (6) *Amoropsitta* (Sierra Parakeet, *Amoropsitta aymara*); (7) *Myiopsitta* (Green Parakeet, *Myiopsitta monachus*); (8) *Microsittace* (Chilean Parakeet, *Microsittace ferruginea*); (9) *Enicognathus* (Slender-billed Parakeet, *Enicognathus leptorhynchus*); (10) *Pyrrhura* (White-Eared Parakeet, *Pyrrhura leucotis*); (11) *Ognorhynchus* (Yellow-Eared Parakeet, *Ognorhynchus icterotis*); (12) *Leptosittaca* (Golden-Plumed Parakeet, *Leptosittaca branickii*); (13) *Nandayus* (Black-headed Parrot, *Nandayus nenday*); (14) *Aratinga* (Sun Parakeet, *Aratinga solstitialis*); (15) *Conuropsis* (Carolina Parakeet, *Conuropsis carolinensis*); (16) *Cyanoliseus* (Burrowing Parrot, *Cyanoliseus patagonus*); (17) The *Rhynchopsitta* (Thick-billed Parrot, *Rhynchopsitta pachyrhyncha*); (18) the Macaws (*Ara*); (19) the Blue Macaws (*Anodorhynchus*).

Almost all macaws are inhabitants of the forests. With their large beaks they are able to crack open very hard nutshells, after first filing down the thickness of the shell at one place. See Fig. 5. The beak is also an

adaptive feature of their locomotor pattern, which is chiefly climbing. As in almost all parrots it is used as a "third foot." See also **Parrots and Cockatoos.**

PSITTACOSIS (Parrot fever).   A disease caused by *Chlamydia psittaci*, an obligate intracellular parasite similar to a virus, but classified as a bacterium because it possesses both RNA and DNA, a discrete wall membrane similar to that of Gram-negative bacteria, and a primitive enzyme system. The life cycle of the organism is complex and not yet fully understood. The infectious elementary body is DNA-rich and relatively stable. It initiates the infection upon entering a susceptible cell by phagocytosis and some hours later reorganizes to form the noninfective large, labile, metabolically active reticulate body. This has an increased RNA content and divides several times by binary fission, ultimately producing new elementary bodies. The whole cycle is completed in 36 to 48 hours.

Psittacosis is a serious hazard to people who keep pet birds or who work in the poultry industry. The infecting agent is harbored by avians and, although an infected bird does not usually act as sick, it is lethargic, has ruffled feathers, and usually diarrhea. *C. psittaci* is present in the bird's nasal secretions, excreta, and feathers and is transmitted to humans by inhalation of such dried sources. Transmission between humans is extremely rare.

Birds admitted to a number of countries are first quarantined for a period. In the United States, this quarantine period is thirty days, during which time the birds receive chlortetracycline in their feed. Occurrence of the disease in humans in the United States ranges between about 40 and 100 cases per year, although in 1974 and again in 1984, the numbers of cases reported to the Centers for Disease Control reached 175, with most of the cases coming from turkey processing operations.

C. psittaci enters the body through the respiratory tract and is rapidly disseminated through the blood stream to the reticuloendothelial system. The principal lesion is then found in the lung, but others occur in the liver and spleen. The pulmonary lesion is principally an interstitial pneumonitis accompanied by airspace involvement.

In humans, the incubation period is generally one to two weeks. Symptoms include high fever, headache, myalgia, chills, and coughing. Clinically, psittacosis presents in two major forms. The first is pneumonitis with extensive focal or lobar pneumonia. The second has features suggestive of a toxic or septic condition. The severity of clinical signs may diminish in the second week of illness, but the course of psittacosis is often prolonged and relapses are not uncommon.

Laboratory diagnosis is usually made by detecting a rise in complement-fixing titer to a group antigen. In specialized facilities, a definitive diagnosis can also be made from sputum, by the isolation of *C. psittaci* in tissue culture.

Tetracycline is treatment of choice and may be administered orally or intravenously. Chloramphenicol is an alternative agent. Before the advent of chemotherapy, the case fatality rate was about 20%; now it is less than 5%.

R. C. V.

PSORIASIS.   See **Dermatitis and Dermatosis**.

PSYCHROMETRIC CHART.   The semiempirical relation giving the vapor pressure in terms of the barometer and psychrometer (hygrometer) readings is:

$$e = e' - \left[3.67 \times 10^{-4} \left(1 + \frac{t' - 32}{1571} \right) \right] p(t - t')$$

where t is the dry-bulb temperature, t' the wet-bulb temperature, e' the saturation vapor pressure at t', p the barometric pressure, and e the vapor pressure; all pressure units being the same. The temperatures are in °F.

A psychrometric chart is a nomogram constructed to provide convenient determination of the properties of air-water vapor mixtures, such as humidity, dew point, and water-vapor pressure from temperatures obtained with a psychrometer. A form of psychrometric chart plotted for °F is shown in Fig. 1. Dry-bulb temperature is represented by the vertical lines; wet-bulb temperature by the diagonal lines; dew-point temperature by the horizontal lines; and relative humidity by the curved lines.

To find the relative humidity value, determine the wet-and-dry-bulb temperatures and follow the lines to their junction on the chart. For

Fig. 5.   Blue-and-yellow macaw (*Ara ararauna.*) (Sketch by Glenn D. Considine).

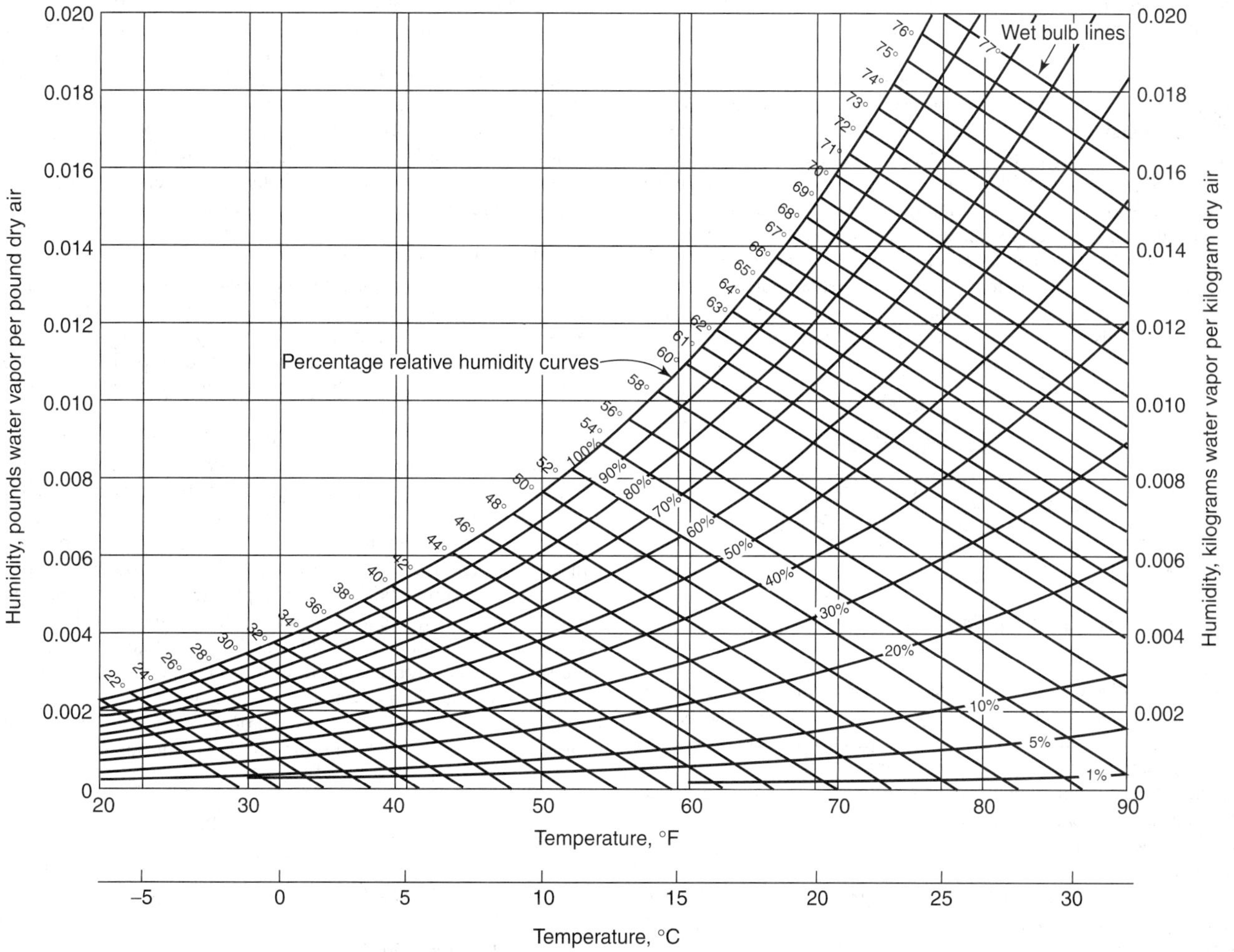

Fig. 1. Psychrometric chart. Wet-bulb temperature lines and percentage relative humidity curves. Temperature range 20 to 90 °F (−6.7 to 32.2 °C). Pressure reference = 29.92 inches of mercury (Sea Level); 1013.2 millibars; 101.325 kiloPascals.

example, assume that the wet-bulb temperature is 50 °F and the dry-bulb temperature is 55 °F. Find the wet-bulb temperature at the left termination of the diagonal lines. Follow the diagonal line toward the lower right to its junction with the dry-bulb temperature vertical line. The curved line at this junction indicates the relative humidity. In this example, the curved line indicates 70% relative humidity. Values falling between lines are found by interpolation. Lines for temperature in degrees Celsius are included on chart.

To determine the dew-point temperature of the water-vapor mixture, follow the horizontal line from the junction of the wet-and-dry-bulb temperature lines to its junction with the 100% humidity curve. The dew point is read from the temperature °F scale and in this example is 45.2 °F. See also **Humidity**.

Psychrometric charts are useful for most observations. The charts are referenced to a specific pressure.

PTARMIGAN *(Aves, Galliformes; Lagopus).* Birds *(Aves)* related to the true grouse but with the feet and legs fully clothed with feathers. They are found in the far northern parts of Europe, Asia and North America and at high altitudes, above timber line, as far south as Colorado. They are largely mottled gray and brown in the summer but assume white plumage in winter. See Fig. 1. The change is not, however, always complete but is somewhat conditioned by the climate. The red grouse and the willow grouse or ripa of northern Europe are closely related to the ptarmigans.

The hoatzin is a peculiar South American bird of possible, but doubtful relationship to the ptarmagin. The bird lives along streams of the Amazon valley and eats fruit and other vegetation. The young have a clawed digit on the margin of the wings that they use to grasp boughs in climbing. See also **Galliformes**; and **Grouse**.

Fig. 1. Ptarmigan, showing seasonal change of plumage. (*A.M. Winchester.*)

PTEROPSIDA. Members of the *Pteropsida* have large leaves, and definite gaps in the vascular cylinder, where a vascular strand, or leaf trace, passes from the stele to the leaf. It is also characteristic of the *Pteropsida* that the sporangia are located on the lower surface of the leaf. Most of the vascular plants of today, including all ferns, gymnosperms and angiosperms, belong to this group.

PTERYGOTA. One of two subclasses into which the class *Insecta* is divided. The members of this subclass are either winged or closely related to winged forms. Some show evidence of derivation from groups that are typically winged. The great majority of insect orders belong here, only three falling into the suborder *Apterygota*.

PTOMAINE. A class of amines formed by the action of bacteria on proteins or by the metabolism of amino acids, which are broken down into toxic products.

PUFFERS *(Osteichthyes).* Of the order *Pectognathi*, family *Tetraodontidae*, a puffer, if pulled out of water, instantly reacts by swallowing air and inflates itself much as a balloon. If returned to the water, a once-expanded puffer may require five minutes or so to expel the air and return to normal. The puffer also can swallow water in the same fashion. The prime difference between a puffer and a porcupine fish is the presence of a solid beak and fused teeth in the latter. There are about 90 species of carnivorous puffers that are widely distributed throughout temperate and tropical waters. Although they can attain a size of 36 inches (0.9 meter) the usual fish does not exceed 18 inches (46 centimeters). These fishes are consumed as a food item (fugu) in Japan. However, because puffers contain a dangerous poison (tetrodotoxin), a Japanese cook requires a special license, assuring proper training in the preparation of fugu. Mortality from food poisoning of this source can run as high as 60%. See Fig. 1.

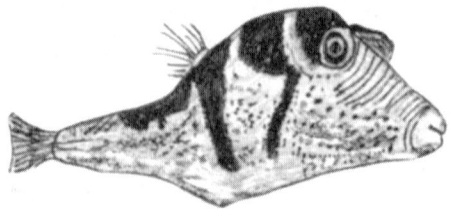

Fig. 1.   Puffer, *Canthigaster valentini.*

Puffers are highly territorial, and need recesses into which they can retreat. This behavior is not immediately evident in species that engage in a great deal of swimming. If two puffers meet, they usually become aggressive. If larger fishes appear, or if several males court a single female, the puffers pump themselves full of water in order to appear larger. This behavior is particularly impressive in the puffer *Carinotetradon somphongsi*; if two of these puffers meet, a thin comb rises in the middle of their backs, and a ridge also rises on the belly. The fishes' coloration and markings almost entirely disappear. The opponents usually stand upside down and point their belly sides at each other, in order to appear as impressive as possible.

PULLEY.   See **Machine (Simple)**.

PULP (WOOD) PRODUCTION AND PROCESSING. Pulps can be defined as fibrous products derived from cellulosic fiber-containing materials and used in the production of hardboard, fiberboard, paperboard, paper, and molded-pulp products. With suitable chemical modification, pulps can be used in the manufacture of rayon, cellulose acetate, and other familiar products. Pulps can be produced from any material containing cellulosic fiber; but in North America and several other regions of the world, wood is the predominant source of pulp. This description is confined to the production and processing of wood pulp.

Wood is a cellular substance chemically composed of roughly 70% holo-cellulose, 25% lignin, and 5% water and ethyl alcohol-benzene soluble extractives. These percentages are based on oven-dry wood.

The chemical composition and physical character of wood vary from species to species, within species grown in different geographical locations, and within a given tree, depending upon the location of the fiber cell in the tree. Both lignin (noncarbohydrate) and holocellulose (carbohydrate) are polymeric substances. Holocellulose is composed of approximately 70% alpha cellulose and 30% hemicellulose, the long-chained alpha cellulose being characterized by nonsolubility in alkali; whereas the shorter-chained hemicellulose is alkali-soluble, the degree depending upon the alkali concentration. Lignin concentration in wood substance is greatest in the middle lamella (the zone around each individual fiber cell), decreasing in concentration through the cross section of the fiber, and reaching a concentration of about 12% at the inner layer of the fiber adjacent to the fiber cavity, or lumen. It is the middle-lamella material (lignin and hemicellulose) that cements the fiber cells together, thus giving rigidity to the fibrous wood structure.

The objective of wood pulping is to separate the cellulose fibers one from another in a manner that preserves the inherent fiber strength while removing as much of the lignin, extractives, the hemicellulose materials as required by pulp end-use considerations. Wood pulp to be used for the manufacture of hardboard, for example, requires only the removal of

water-soluble wood sugars and sufficient fiberization, i.e., separation of fibers, to permit effective felting of the fibers in a sheet-forming operation. In a subsequent operation in which the felted fiber sheet is subjected to high pressure and heat, the lignin in the fiber mass softens and flows, ultimately acting as a bonding agent cementing the fibers together into a coherent hardboard. At the other extreme, wood pulp to be used for rayon manufacture must be of a high alpha-cellulose content (~88–93%), have extremely low amounts of noncarbohydrate material, and be well fiberized to permit uniform reactions during chemical processing.

Pulping Processes

Wood is converted to pulp by mechanical and chemical actions, which constitute the pulping process. Their selection depends upon the type of wood supply available and the pulp qualities desired. Pulps can be characterized on the basis of the unbleached pulp yields achieved by the pulping process used, i.e., the yield of oven-dry (OD) pulp obtained from oven-dry debarked wood.

Five major types, or classes, of pulps, related to pulp yield ranges normally considered to define each class of pulp, are shown in Fig. 1. Pulp yield is a direct indication of degree of chemical action (delignification and chemical attack on carbohydrate and other nonligneous material). Also shown in this figure are the degrees of defibration effected by chemical and mechanical action utilized to produce the pulp, although this representation is not strictly correct. For example, in producing a full chemical pulp, wood chips are subjected to chemical action (digestion or cooking) in a pressure vessel. When digestion is completed, the cooked and softened chips retain the same physical form as the raw chips originally charged to the digester. But they separate into essentially discrete fibers as a result of mechanical action occurring upon sudden release of the chips from the pressure vessel into a receiving tank, which ordinarily is at atmospheric pressure.

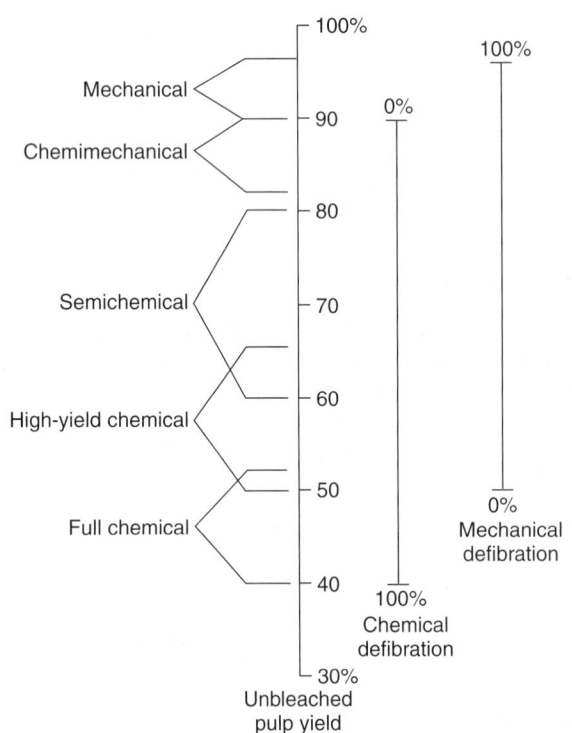

Fig. 1.   Wood pulp characterized on basis of yield.

At the other extreme, no chemicals are used in the production of mechanical pulp, and defibration is effected by subjecting wood to a mechanical grinding or attrition action. In this instance, the defibration is aided by some small degree of chemical change and solubilization of wood substance occasioned by heat generated by the grinding operation.

The pulps listed in Fig. 1 are characterized on an unbleached basis as produced by processes conventionally called *pulping processes*. In many instances, these pulps must be further treated chemically to remove residual lignin, hemicellulose, and color bodies before they can be considered suitable for use in specific applications. This further treatment is called *bleaching*, and the bleaching operation is actually an extension of the pulping process.

Customarily, pulping processes and bleaching processes are considered separately, although the choice of bleaching process is highly dependent upon the pulping process used. With this distinction between pulping and bleaching in mind, it will be understood that the pulping processes that are briefly described here pertain only to the production of *unbleached* pulps.

The soda, kraft, and sulfite pulping processes are used to prepare full chemical pulps. The soda process, which uses sodium hydroxide as the cooking chemical for delignification purposes, has largely been superseded by the kraft process, which is characterized by its use of sodium hydroxide and sodium sulfide as active delignification agents in the chip-cooking phase of the process.

Chip-digestion parameters are digester pressure and temperature, digestion time to and at maximum temperature, amount of active alkali used per unit weight of OD wood (percent active alkali), percentage ratio of sulfide to active alkali (percent sulfidity), and weight ratio of cooking liquor (including chip moisture) to OD wood weight. No two kraft pulp mills use the same set of parameter values. Such values must be frequently adjusted, even within a given mill, because of variations in incoming wood and pulp-quality requirements.

Kraft processes are applicable to nearly all species of wood, and effective means of recovering spent cooking chemicals for recycle in the process have been developed. Some sodium and sulfur losses do occur and are replenished in the cooking-liquor system by adding sodium sulfate at the recovery boiler, where it is converted to sodium carbonate and sulfide. In order to maintain a proper sulfur-to-sodium ratio in the recovered chemicals, other chemicals, such as sodium carbonate, sodium sulfite, and sulfur, are sometimes used for chemical makeup.

In contrast to the highly alkaline (pH 11–13) kraft processes, sulfite pulping processes are acidic in nature and are of two general types: (1) the *acid sulfite processes* utilize calcium, sodium, magnesium, or ammonium bisulfite in combination with free or excess sulfur dioxide as cooking chemicals (pH 1.7–2.3). (2) The *bisulfite processes* use sodium, magnesium, or ammonium bisulfite (pH 3.5–5.5) for chip digestion.

Several sulfite processes are multistage and use various combinations of acid sulfite and bisulfite cooking stages and can even use the alkaline kraft cook as one of the multistages. Although spent calcium acid sulfite cooking liquor can be incinerated, there is no recovery of calcium or sulfur. The sodium and magnesium bases can be recovered with or without sulfur recovery, and spent ammonium base liquor can be burned with recovery of sulfur as an option.

High-yield chemical pulps can be produced by the soda, kraft, or sulfite processes, in which chemical use and digestion time and/or temperature are suitably reduced to effect a milder cook than used for full chemical pulps. Mechanical defibrators are used to complete the separation of wood fibers not accomplished by the chemical action.

Semichemical pulps are usually prepared by the neutral sulfite semi-chemical (NSSC) process, although modifications of the full chemical processes can be used. Active pulping chemicals are (in the sodium-base NSSC process) sodium sulfite buffered with sodium bicarbonate (pH 7.0–9.0) and (in the ammonium-base NSSC process) ammonium sulfite with ammonium hydroxide used as a buffer. Defiberization is usually accomplished by attrition mills of the disk type.

Mechanical pulps are produced by two basic processes: (1) *stone groundwood pulp* (SGW) is produced by the defibration action of natural or artificial grindstones rotated at moderate speeds (200–300 rpm) against bark-free bolts of roundwood axially aligned across the peripheral face of the stone in the presence of water. By air-pressurization of the grinder, a *pressurized groundwood pulp* (PGW) of improved quality can be produced. (2) *Refiner mechanical pulp* (RMP) is produced by the attrition action upon raw wood chips of an open (atmospheric) discharge disk refiner. By preheating the chips in a pressurized vessel via direct steaming at temperatures of 120 °C or higher and fiberizing the heated chips in either a pressurized or atmospheric disk refiner, *thermomechanical pulp* (TMP) is produced.

Chemimechanical pulps (CMP) are produced by processes in which roundwood or chips are treated with weak solutions of pulping chemicals, such as sulfur dioxide, sodium sulfite, sodium bisulfite or sodium hydrosulfite, followed by mechanical defibration. By presteaming chemically treated chips before attrition, *chemithermo-mechanical pulps* (CTMP) are produced. The mild chemical action, augmented by heat, softens wood lignin and promotes easier defibering with less fiber damage than achieved by the purely mechanical processes.

Wood Pulping Operations. The preceding description of pulps and pulping processes were given as a background to the following descriptions of the various operations involved in the preparation of wood pulp. The pulping system of a typical kraft linerboard mill, as indicated in the simplified flow diagram in Fig. 2, is illustrative of that required for the preparation of both full and high-yield chemical pulps. Linerboard normally is two-layered. The base, or primary sheet, is formed from a high-yield chemical pulp (50–54% yield) and the top, or secondary sheet, is formed from a full chemical pulp, either unbleached (48–50% yield) or bleached (46–48% unbleached yield), laid upon the wet primary sheet on the sheet-forming wire.

Pulp, paper, and paperboard mills are characterized by high capital investment costs and use of high tonnage and rugged but precisely engineered machinery capable of continuous operation with minimum maintenance. A modern kraft linerboard mill with a capacity of 1000 short tons per day (900 metric tons) will have an installed cost, excluding woodlands, of from $275,000 to $325,000 (1986 dollars) per daily ton of board produced. Indication of machinery sizes will be given in the following paragraphs.

Wood-Chip Preparation. As indicated in Fig. 2, pulping operations begin with receipt of wood at the mill site. Pulpwood is supplied in log form (roundwood) or chips in accordance with specifications set by the pulp mill. Roundwood is usually received with bark on and in lengths and diameters suitable for proper handling in the wood-preparation equipment at the mill. It has been customary for mills to specify multiple lengths of pulpwood, i.e., 4 feet (1.2 meters) and 8 feet (2.4 meters) as standard receipts, but there is a trend to the procurement of tree-length logs, up to 70 feet (21 meters) in length, either exclusively or in combination with short logs. Another trend has been to the use of chips already prepared, except perhaps for final screening, by independent suppliers or by satellite wood yards operated by the pulp mill itself.

Although linerboard mills formerly used only softwoods (coniferous) for pulping, continued improvements in pulping and board-making technology have permitted the inclusion of up to 20% or more of hardwoods

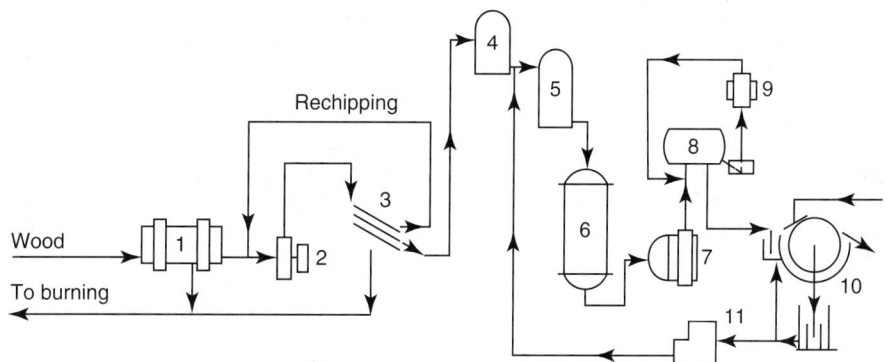

Fig. 2. Flow diagram of kraft pulp mill: (1) debarking, (2) chipping, (3) screening, (4) steaming, (5) impregnating, (6) digesting, (7) fibrilizing, (8) screening, (9) fiberizing, (10) washing, (11) chemical recovery.

(deciduous) in the wood furnished to the mill, with improved utilization of woodlands as a beneficial result. Softwood and hardwood species are processed through the chipping operation and stored separately; they are either blended into the digester and cooked together, or they are processed separately and the respective pulps blended just ahead of the linerboard machine.

Former practice was to store pulpwood receipts in either a debarked or unbarked condition in stacks or random piles in the wood yard and to reclaim the yard wood for processing into chips just a few hours in advance of chip needs at the digester. A common practice today is to convert the wood into chips immediately after pulpwood receipt and to place the chips, usually by belt or air conveyance, in chip piles built up on concrete or asphalt pads. Separate piles are provided for softwood and hardwood chips, and storage capacities of 40,000 cords or greater can be maintained.

Pulp logs are conveyed to the debarking area, where they are cut to proper length, if necessary, and sorted. Accepted logs are mechanically fed into one end of a large horizontal, cylindrical drum, usually consisting of one or more sections constructed of spaced steel plates, channels, or bars mounted in carrying rings and supported on trunnions and driven by ring gears or suspended from an overhead structure by heavy chains, one or more of which are motor driven. This *barking drum rotates* at a speed of 5–8 rpm, and as the logs tumble about in passing from the intake to the discharge end of the drum, bark removal is effected by the logs rubbing against each other or against the bars or plates constituting the drum shell. Provision can be made for introduction of steam into the feed-end for log de-icing when needed.

Bark removal also can be accomplished by use of a *ring barker* or *hydraulic barkers* which employ high-pressure water jets for bark stripping. Bark removed is collected, shredded in a hog or hammer mill, and used as fuel in steam boilers, where it contributes about 3863 kJ/kg (9000 Btu/pound) of dry solids.

Debarked wood is conveyed to a chipper for conversion into chips of proper length for chemical treatment in a subsequent cooking operation. Chip length of 0.5–1.0 inch (12.7 to 25.4 millimeters) is conventional.

Chip Digestion. This cooking operation is accomplished in either a batch or continuous digester. A chip digester is essentially a large pressure vessel provided with suitable raw-chip and cooking liquor feed ports and a cooked-pulp discharge port. It is equipped with means for heating and maintaining its contents to and at a specified temperature for the required periods of time.

Batch digesters are vertical, stationary, cylindrical pressure vessels into which chips and cooking liquor are charged under atmospheric conditions. Heating of the digester, after sealing of the feed ports, is effected by direct steam addition or by continual withdrawal of liquor through screened ports and reintroduction of the liquor, after passage through external heat exchangers, onto the top (and sometimes into the bottom) of the chip mass within the vessel. Often, a combination of the direct and indirect heating methods is used. Modern batch digesters are typically 4000–6000 cubic feet (113–170 cubic meters) in volume, with height-to-diameter ratios of 3.5–5.5, and pre-cook pulp capacities of 10–12 tons (9–18 metric tons).

Continuous digesters have been developed as part of the highly successful effort to convert pulp and papermaking from a series of strictly batch operations into an integrated series of continuous operations. A number of successful types of continuous digesters range from horizontal and inclined tube (single or multiple) designs, in which the chip charge is moved through the digester by mechanical screw or bucket conveyors, to vertical digesters, in which chip movement is effected by gravity. See Fig. 3.

Screened chips are conveyed from storage to a chip supply bin in the digester house. The chip bin is designed so that low pressure steam recovered from the hot, spent cooking liquor can contact the chips, preheat them and expel most of the air from the chip interior. If hardwood and softwood chips are to be cooked together, they are blended by weight proportion during the transfer to the chip bin. The chips drop by gravity from the bin to a chip meter, either a twin-screw or a multi-pocket rotary feeder, the speed of which determines chip and cooking liquor flow rate to the digester and pulp discharge rate.

Metered chips drop into a low-pressure rotary feeder valve, through which they are introduced to a steaming vessel maintained at a pressure of about 15–55 psi gage (1 to 3.5 atmospheres). There the chips are further preheated, the remaining air expelled from the chip interior, and chip

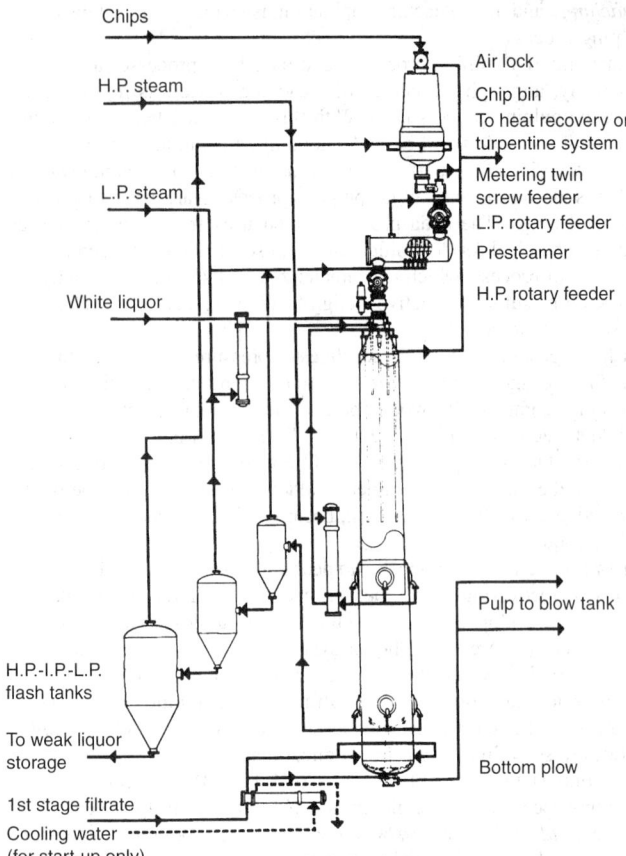

Fig. 3.   Continuous digester system (*Ingersoll Rand Co.*)

moisture is leveled in preparation for impregnation of cooking liquor. Since cooked chips are continuously removed from the bottom of the digester, chips pass downward in the digester, replacing those discharged. The time of passage through the cooking zone is normally 90–120 minutes. As cooked chips reach the bottom zone of the digester, the hot, spent cooking liquor is displaced with cooler filtrate from the pulp washers, and removed via extraction strainers to a heat recovery system in which steam is generated to be used to precondition the chips being fed to the digester. As the partially cooled chips move further down the digester, they are plowed to a central well in the bottom of the digester, and are mixed with more filtrate from the pulp washers for dilution and final cooling. Mechanical forces exerted in the transfer of chips from the digester to the blow tank effect fiberization of the chips, the degree of which depends upon cooking conditions. The fibrous material in the blow tank is called pulp, and separate blow tanks are normally used to collect the several types of pulp produced alternately in the digester.

Pulp Screening and Washing. Pulp (brown stock) discharged to the blow tank is in admixture with black liquor, a water solution of spent and residual cooking chemicals and dissolved wood substance, and is at a consistency of from 10 to 18%. The term *consistency* has a meaning peculiar to the pulp and paper industry and refers to the percentage ratio of washed, dry (either oven- or air-dried) fiber to total fiber slurry weight. The fiber bundles left in the pulp after blowing must be fiberized, i.e., separated into discrete fibers, and the black liquor removed in order for the pulp to be refined (a conditioning of individual fibers) and formed into a fiber sheet on the linerboard machines.

Pulp is diluted with filtrate from the pulp washer to a consistency of about 4.5% in the lower portion of the blow tank and fed to fibrilizers, which serve the purposes of metal trapping, fiber-bundle breaking, rough screening, and pumping.

Removal of the black liquor from screened brown stock is usually accomplished on rotary-drum vacuum filters, arranged for multistage countercurrent washing, as shown in Fig. 4.

Refining is accomplished by disk mills, equipped with different plate designs or patterns than those used for defibration. During the refining operation, cellulose fibrils, which wind spirally around the fiber at various

Fig. 4. Line of three brown stock washers, 9.5 feet (3 meters) in diameter and 16 feet (3 meters × 5 meters) long, equipped with multiport circumferential valve. First stage washer is shown in foreground. (*Ingersoll-Rand Co.*)

positions in its cell wall, are loosened, the cell wall swells due to water absorption, and the fiber is conditioned for sheet formation and inter-fiber binding in the paper- or board-making operation.

Chemical Recovery. Economic and environmental control factors dictate that chemical and heat values of black liquor solids be carefully conserved, and the recovery system of the modern kraft pulp mill has developed into a highly sophisticated system with still more improvement in efficiency continually being sought. See also **Papermaking and Finishing**.

HENRY F. SZEPAN (retired) and DUNBAR G. TERRY (retired),
Ingersoll-Rand Co., Impco Division, Nashua, NH.

PULSAR. A neutron star with a strong magnetic field, first detected in 1967 as sources of intense nonthermal pulsed radio emission (hence sometimes called *radio pulsars*).

At the end of the life of a massive ($>8M_\odot$) star, following the production of iron in the core by nucleosynthetic processes, the collapse of the core produces a supernova ejection of the envelope and leaves a very compact, degenerate object as a remnant. This remnant, having a radius of about 10 kilometers and a mass of order $1M_\odot$, has a central density about equal to an atomic nucleus ($\approx 10^{15}-10^{16}$ g cm^{-3}) and consists of nucleons and more exotic states of matter. This is a neutron star. Also present may be a strong magnetic field, either as a remnant from the precollapse core or generated immediately following the supernova, which can reach strengths of order 10^{13} gauss. In such a strong field, electrons can radiate via synchrotron processes at radio, optical, and even x-ray wavelengths. The emission is confined to the magnetic polar cap, which if it is inclined to the rotational axis will produce a pulse of emission as the region sweeps through the line of sight to a distant observer. This is the mechanism presumed responsible for the radio pulsars. The mechanism for accelerating the electrons to the high energies observed in these regions is not completely understood at present. It appears to be at least in part due to strong electric fields generated by the rotation of the magnetized star ripping electrons from the surface and accelerating them to high energies nearby. The emission of radiation and the torquing of the star by the magnetic field produce a slowing down of the rotation rate of the star, the rate of which can be linked to the magnetic dipole moment of the neutron star.

When such stars occur in binary systems in which matter is accreting onto the neutron star, the magnetic field acts to funnel the gas toward the polar region. The matter emits x-rays as it falls onto the surface of the accreting star, which will be pulsed due to the rotation of the neutron star. Many such systems are known, the best studied being Hercules X-1. The interaction between the magnetosphere of the accreting pulsar and the

accretion disk, which forms in the plane of the binary system, can also cause *quasi-periodic oscillations* (QPO's), due to instabilities generated by the shearing between the trapped and circulating gas. This shows up as noisy pulsing on timescales of milliseconds in the intensity of x-ray binaries at high energy.

A new class of objects, the millisecond pulsars, contains the most rapidly rotating objects known in the cosmos. These rotate about 10 to 20 times faster than the shortest period radio pulsar, the Crab pulsar (period of about 33 msec) and appear to have unusually weak magnetic fields (less than 1010 gauss). Their rapid rotation has been explained as resulting from spin-up of a weakly magnetized neutron star in a loose binary system, which may have since disrupted. Their weak magnetic fields are inferred from their slow rate of spin-down.

Some pulsars are members of binary systems in which the separation of the stars is great enough that no mass transfer is occurring between the members. The prototype of this class is PSR 1913 + 16. The changes in the orbital parameters of these systems, because of their masses and exceptionally well-determined orbital parameters, allow for tests of general relativity, since the periods and eccentricities change from the emission of gravitational radiation. They also show that the formation of compact objects through supernovae does not necessarily disrupt the binary.

There are now about 400 known pulsars, found largely through radio surveys at Aricebo, Parkes in Australia, and the National Radio Astronomy Observatory at Green Bank. Of these, fewer than a half-dozen are associated with known supernova remnants, a problem that remains a major question in the origin and development of such systems. The two best studied radio pulsars, the Crab and Vela pulsars, are imbedded in radio remnants of known age, but the vast majority are isolated systems. See also **Cosmology**.

S. N. S.

PULSE. 1. Commonly, a pulse is a variation of a quantity whose value is normally constant; this variation is characterized by a rise and decay, and has a finite duration.

2. In physics and related sciences, a pulse is a waveform whose duration is short compared to the time scale of interest, and whose initial and final values are the same. The word "pulse" normally refers to a variation in time; when the variation is in some other dimension, it should be so specified, such as "space pulse." This definition is broad so that it covers almost any transient phenomenon. The only features common to all pulses are rise, finite duration, and decay. It is necessary that the rise, duration, and decay be of a quantity that is constant (not necessarily zero) for some time before the pulse and has the same constant value for some time afterwards. The quantity has a normally constant value and is perturbed during the pulse. No relative time scale can be assigned.

3. In animal physiology, the pulse is the expansion and elongation of the arterial walls, produced passively by changes in intra-arterial pressure during contraction (systole) and relaxation (diastole) of the heart. The pulse is usually felt in the radial artery at the wrist, but it may be felt in any artery lying near the surface. The heart, with systole, forces blood into the arterial circulation. This blood is accommodated partly by moving the entire arterial column on at greater velocity, and partly by distending the arterial walls. The increase in pressure and distention of the vessel walls is transmitted from one segment of the artery to the next as the pulse wave. The ability of the vessel wall to distend is dependent upon its elasticity. In old age, when the vessel becomes sclerotic and inelastic, it offers increased resistance and this results in elevation of the blood pressure.

The examination of the pulse in disease is one of the oldest customs in medicine. Variations in the rate, rhythm and force are significant, particularly in heart disease. The normal rate is 60–80 in adults, and 80–140 in children. With fever the rate increases as a general rule. A disproportionately slow pulse with a high fever is of diagnostic significance in certain infections, notably typhoid, typhus and so-called virus pneumonia. An increase in the pulse rate is a normal reaction to emotional stimuli, being dependent in this case upon a release of epinephrine (adrenalin) by the adrenal glands.

PULSE (Botany). A general name for leguminous plants (peas, beans, etc.) or their seeds.

PULSE GENERATOR.   A large portion of electronic engineering effort is devoted to the development of circuits and systems that operate in the time domain. Switching, pulse, and digital circuits predominate computing, navigation, and data communication systems. In the laboratory, the oscilloscope as the detector and the pulse generator as the signal source largely have replaced the voltmeter and sinusoidal oscillator. Many electronic devices or circuits utilize sharp pulses of current or voltage as a basic part of their operation. Such pulses must frequently be of very short time duration (microseconds or nanoseconds) and accurately spaced in time. Others need not be repeated at regular intervals, but are initiated by some signal and are single narrow pulses occurring each time a signal is received.

In essence, a pulse generator is a highly versatile and controllable switch. Two parameters of interest which must be controlled are: (1) the pulse repetition frequency or switching rate, and (2) the pulse duration, or length of time the switch is closed (or open). The rise time, or speed of switching, also is an important parameter. In addition to the characteristics of the switch, specific applications require particular characteristics for the energy source switched. The output impedance, open-circuit voltage, and available current all must be known and specified to fit a given pulse generator to a specific requirement.

In electronic equipment, pulse repetition rates range from nearly dc to over 100 MHz—with durations to less than 1 nanosecond. For computation and data transmission systems, most applications can be served by relatively low power outputs. Radar and certain magnetic data storage systems may require pulses of very high energy. Because of these wide variations, a universal pulse generator is not practical.

For purposes of illustration, a few specific instruments are described briefly. A low cost unit for the student laboratory is shown in Fig. 1. This instrument produces pulses ranging from 0.1 microseconds to 1 second over a repetition rate ranging from dc to 2 MHz. The device contains an internal prf (pulse repetition frequency) oscillator with continuous range from 3 kHz to 1.2 MHz. Both pulse polarities are available simultaneously and the output circuits are dc coupled, the latter a necessity when pulses of long duration must be produced. A block diagram of the instrument is shown in Fig. 2. An oscillogram of a 1 microsecond pulse into 50 ohms with delayed sync pulse is shown in Fig. 3. This device may be used with a pulse amplifier to translate performance to 1-ampere output levels.

A modular pulse generator is shown in Fig. 4. This instrument is a hybrid system that offers much flexibility. The duration ranges, rise times, and prf ranges are similar to those of the instrument previously described. An input circuit module, one of several in one package, either serves as a prf oscillator or processes an external driving signal from dc to 2 MHz to produce a standardized system-synchronizing pulse. A second module produces pulses or delayed 0.1 microsecond synchronizing pulses in the range from 0.1 to 1 second with rise times of 15 nanoseconds.

A single input module and three pulse/delay modules form a double-pulse generator, while five pulse/delay modules provide a triple pulse. A third module, timed from a pulse/delay module, produces pulses with linear and independently variable rise and fall times over a range from 0.1 microsecond to 1 second. A fourth module produces up to 16-bit pulse words. A power amplifier with up to 0.4 ampere output of a limited duty ratio is the fifth module. Several pulse configurations of which this device is capable of generating are shown in Fig. 5.

Fig. 5(a). Waveform that appears at the "adder No. 1" terminal with one prf unit driving two pulse/delay units at 10 kHz. Amplitudes and durations of positive and negative pulses can be adjusted independently.

Fig. 5(b). One prf unit, one pulse/delay unit, and one pulse shaper are used to form this train of triangular waveforms. The prf unit is set for 5 kHz. The positive-going ramp rises linearly to 20 volts in 50 microseconds, while the negative-going ramp falls to zero in 10 microseconds. Rise and fall times are independently variable.

Fig. 5(c). Pulse train produced by one prf unit and three pulse/delay units operating at 100 kHz. The amplitude and duration of the positive pulse are controlled by one pulse/delay unit whose delayed output triggers a second pulse/delay unit. This second unit provides the delay between the positive and negative pulses, and its delayed output triggers the third unit to produce the negative pulse.

Fig. 5(d). This train of ramp pulses is produced with one prf unit, one pulse/delay unit, and one pulse shaper. Prf is 100 and the zero-volt level is adjusted by the main chassis "pulse dc component" control.

Fig. 5(e). Pattern produced when a word generator is connected between prf unit and the first pulse/delay unit in the situation shown in Fig. 5(a), with switches set as shown.

In another unit, the tone-burst generator, the instrument operates as a coherent gate for an externally introduced signal. This instrument is used for testing and calibrating sonar transducers and amplifiers, the measurement of room acoustics, and in automatic gain-control circuits. The device also is useful in the synthesis of time ticks on standard-time radio transmissions and in psychoacoustic instrumentation. A binary scaler is used to establish both the number of cycles in a burst and the time duration between bursts.

The device incorporates (1) a switch that holds the gate open for preliminary alignment of external equipment; (2) trigger controls, which allow control of the relative phase of the gate and input signal; (3) the ability to use separate input signals for the gate timing and gated signals; and (4) a timed mode, for very long periods between bursts. A typical waveform is shown in Fig. 6. The tone-burst generator also is useful with pulse and aperiodic signals. If pulses are applied to its input, the device can perform as a word generator, or a frequency divider.

PULSEJET ENGINE.   A type of compressorless jet engine in which combustion takes place intermittently, producing thrust by a series of explosions, commonly occurring at the approximate resonance frequency of the engine. Often called a *pulsejet*.

PULSE RADAR.   A type of radar, designed to facilitate range measurement, in which the transmitted energy is emitted in periodic short pulses. Also called *pulsed radar*. The distance to any target a detectable echo can be determined by measuring one-half the time interval between transmitted pulse and received echo and multiplying this number by the speed of light. This is by far the most common type of radar.

PUMA.   See **Cats.**

Fig. 1.   Unit pulse generator with power supply.

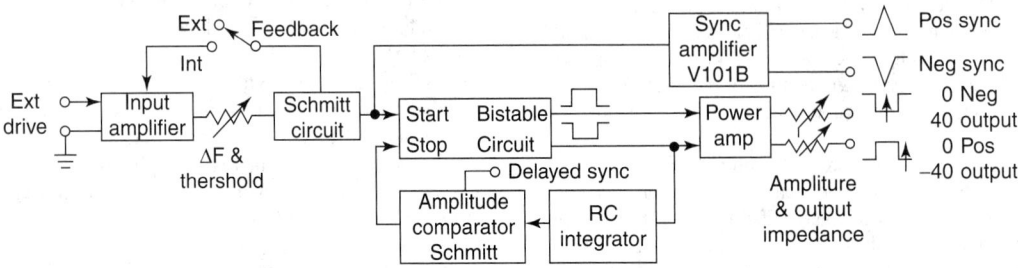

Fig. 2.   Block diagram of circuit used in unit pulse generator.

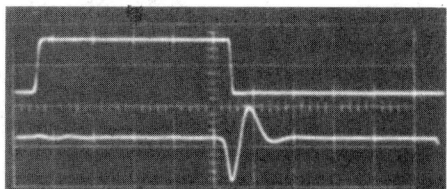

Fig. 3. Oscillogram made from unit pulse generator.

Fig. 4. Modular pulse generator.

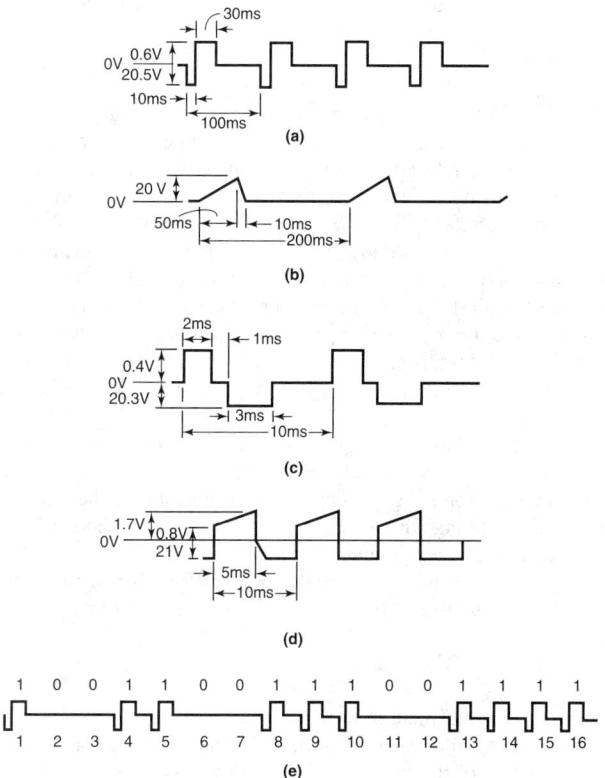

Fig. 5. Pulse configurations available with modular pulse generator.

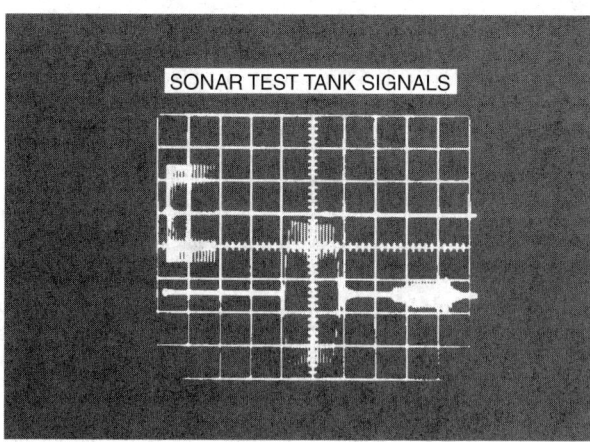

Fig. 6. Typical waveform produced by tone-burst generator with a 15 kHz signal turned on for 15 cycles and off for one-half second. Upper trace shows input to sonar projector; lower trace shows output from projector and subsequent echo return from wall of test tank.

PUMICE. Rhyolitic lavas with a high gas content, when suddenly discharged by volcanic action, congeal in the form of a highly vesicular natural glass called pumice. When ground, mixed with an appropriate binder and pressed into cakes it is the "pumice stone" of commerce which is used as a light abrasive.

PUMP (Liquid). The function of a pump is to add to the pressure existing in a liquid an increment sufficient for the required service. This service may be the production of a velocity, or the overcoming of friction or external pressure. Principal types of pumps include: (1) reciprocating pumps–(a) direct-acting steam (simplex and duplex), (b) power-driven, single-acting (simplex and triplex); (2) centrifugal pumps–(a) single and multistage, (b) volute and turbine types; (3) rotary pumps–(a) gear and screw pumps, (b) propeller pumps, (c) lobe pumps; and (4) jet pumps–(a) steam jet injectors and ejectors, (b) water jet ejectors.

Very widely used is the centrifugal pump, which is a velocity machine, i.e., its pumping action requires first, the production of a liquid velocity; second, the conversion of velocity head to pressure head. The velocity is given by the rotating impeller, the conversion accomplished by diffusing guide vanes in the turbine type, and in a volute casing surrounding the impeller in the volute type. With few exceptions, all single-stage pumps are of the volute type.

The specific speed of a centrifugal pump is $NQ^{1/2}/H^{3/4}$. Ordinarily N is expressed in rotations per minute, Q is gallons per minute, and head H in feet. The specific speed of an impeller is an index to its type. Impellers for high heads usually have low specific speeds, while those for low heads have high specific speeds. The specific speed is a valuable index in determining the maximum suction head that may be employed without danger of cavitation or vibration, both of which adversely affect capacity and efficiency.

Allied to the centrifugal pump in several ways are the axial-flow pumps. An axial-flow pump, sometimes called a propeller pump, develops most of its head by the propelling action of the vanes in the liquid. It has a single-inlet impeller with the flow entering axially into a guide case. Where the head is developed partly by centrifugal action and partly by vane propulsion, the pump is called a mixed-flow pump.

PUMPKIN. See **Cucurbitaceae**.

PUNKIE *(Insecta, Diptera)*. Small biting midges, also called sandflies. They are found in abundance at certain times along streams in the eastern mountains of the United States and at some parts of the seashore.

PUPA. The third stage of insects that undergo complete metamorphosis. The pupa is a more or less inert stage but in some insects it retains the power of locomotion to a high degree. The pupae of mosquitoes are an example; they swim as freely as the larvae when disturbed. In contrast, the pupae of butterflies and moths can merely move the abdominal segments and those of many flies are quite rigid. Among these inactive pupae some, like those of the beetles and wasps, have the legs and wings free and are called exarate. Others have the appendages closely attached to the body and are said to be objected. Those of *Diptera* in some cases are enclosed in a hardened larval skin, the puparium, and are called coarctate pupae. The pupae of butterflies, often brightly colored and strangely shaped, are called chrysalids (singular, chrysalis or chrysalid).

PUPPIS. A constellation located near Canis Major and once, with Carina and Vela, was part of a superconstellation known as Argo Navis.

PURINES. Derivatives of the dicyclodiureide of malonic and oxalic acids. The dicyclodiureide is uric acid and the parent compound is purine: so that uric acid is 2,6,8-trioxypurine or the keto form of

2,6,8-trihydroxypurine. Caffeine, theobromine, and theophylline are other important purine compounds.

Uric acid ($C_5H_4O_3N_4$) is a white solid, insoluble in cold water, alcohol or ether, sparingly soluble in hot water. Uric acid is a weak dibasic acid thus forming two series of salts, most of which are very slightly soluble in water (lithium urate soluble).

Uric acid is found in the urine, blood, and muscle juices of carnivorous animals (herbivorous animals secrete hippuric acid), in the excrement of birds, serpents and insects, and is an oxidation product of the complex nitrogenous compounds of the animal organism.

Purine Metabolism. Purines are major building blocks for the nucleic acids, DNA and RNA. Adenine, also a purine, plays several important roles—as a cofactor component in energy metabolisms and in enzymatic reactions in which the coenzymes NAD^+ and $NADP^+$ are involved. The end product of purine metabolism is uric acid. It has been well established for many years that biochemical shortcomings in purine metabolism are the principal cause of gout. An average adult male will excrete between 200 and 600 milligrams of uric acid in the urine per day, representing about two-thirds of the total uric acid production in the body. Less than 10–20% of uric acid can be accounted for directly as dietary intake. When insufficient uric acid is excreted, a condition known as *hyperuricemia* will result. When the concentration of uric acid nears the saturation threshold, precipitation in tissues commences. Increased amounts of uric acid may be produced as the result of faulty enzyme activity or other abnormal factors that may occur in the purine metabolism system. Hyperuricemia may be evidenced by the development of an acute, extremely painful, swollen, inflamed joint, frequently at the base of the great toe (podagra). This condition is most commonly encountered in obese, overindulgent people. Usually this condition persists for several days to several weeks without treatment. The condition may recur periodically. The condition responds well to the administration of colchicine. See also **Alkaloids**. Treatment also includes removal of carbohydrates from the diet for a few days, as well as deprivation of alcohol and certain medications, such as thiazide diuretics. See also **Gout**.

Abnormalities in purine metabolism also may create a purine nucleoside phosphorylase (PNP) deficiency, which ultimately may surface as *hypoplastic anemia.*

Purine, uric acid, and other associated compounds play a role in organic synthesis of industrial products.

PURKINJE EFFECT.

With a good level of illumination, the spectral sensitivity of the normal eye is greatest in the yellow-green region. As the illumination is reduced, the maximum sensitivity shifts toward the blue. This shift is called the Purkinje effect.

PURPLE LIGHT.

The faint purple glow observed on clear days over a large region of the western sky after sunset, and over the eastern sky before sunrise. The purple light first appears, in the sunset case, for example, at a solar depression of 2°; at that time, it extends from about 35° to about 50° elevation above the solar point, and has an azimuthal extent of between 40° and 80°. Maximum intensity of the glow typically occurs at the time the sun is about 4° below the horizon. Increasing depression of the sun causes the top of the purple light to descend steadily toward the western horizon. The effect disappears at solar depression angles near 7°, being replaced in the western sky by the bright segment. See also **Atmospheric Optical Phenomena**; and **Twilight**.

PURPURA.

A heterogeneous group of disorders characterized by the formation of purple patches on the skin and the mucous membranes. In *vascular purpura*, there usually is cutaneous hemorrhage, sometimes associated with mucosal bleeding. Clinical tests of platelet number and function, as well as tests of procoagulant function, are normal.

In *hereditary hemorrhagic tolangiectasia*, small red focal lesions, caused by dilation of capillaries, arterioles, or venules, appear on the finger pads, buccal (mouth) mucosa, the tongue, and lip borders. Arteriovenous shunts (anomalous passages) may appear in the liver and lungs. Platelets may be abnormal, but coagulation tests are usually normal. There may be gastrointestinal bleeding for which iron administration may be indicated.

Although rare today, skin, gingival, and mucosal bleeding can occur in scurvy. See also **Ascorbic Acid (Vitamin C)**.

Cutaneous hemorrhages also may be precipitated by corticosteroids. Amyloidosis (intercellular deposit of amyloid in tissue) may present purpura lesions, particularly of the neck and upper chest regions.

In *Schönlein-Henoch purpura*, a vasculitis caused by allergy, there are raised purpura lesions that itch, with subcutaneous edema. There may be gastrointestinal bleeding and acute glomerulonephritis. See also **Kidney and Urinary Tract**. This disorder is most common in children. Drug reactions may produce the vasculitic form of purpura in adults. Notably, aspirin and phenacetin should be avoided. Other symptoms may include fever and arthritis. This disorder mimics the rash found in meningococcemia.

In *senile purpura*, there are cutaneous hemorrhages on the back, hands, wrists, and upper arms—less frequently, the calves. There is no serious bleeding and no treatment is required.

In *autoerythrocyte purpura*, which is rare, there are very large, painful subcutaneous hematomas (massive clots of extravasated blood). Mainly this occurs in middle-aged women and can be debilitating. The disease is considered to have psychiatric vectors. Drugs and low-allergin diets generally have been ineffective in treatment of the disorder.

PUSTULE.

A sore (abscess, bubo, pimple, etc.) with which is associated the formation and/or discharge of pus. See also **Suppuration**.

PYCNOGONIDA.

The sea spiders, a small number of species constituting a class of *Arthropoda*. All are marine, crawling about on plants and sessile animals. The body is very small and the legs very long; hence the animals seem like clusters of legs attached to each other. They have a sucking mouth.

PYCNOMETER.

A device for measuring densities of liquids. It is a container, usually in the form of a bottle or a pipette-like tube, the capacity of which is accurately known and which may be completely filled with the liquid. The difference in weight when filled and when empty, together with the known volume of the liquid, gives the density. The pipette form has a mark to show how far to fill it, and is bent into a V-shape to facilitate immersion in a temperature bath. A familiar design is the "specific gravity bottle," a small flask with a ground and perforated stopper, and sometimes provided with a thermometer. In one of the most precise forms the stopper has a conical top with the capillary leading to the apex, and both neck and stopper are covered by a tight-fitting ground-glass cap to prevent evaporation.

A preliminary step necessary to precise work with the pycnometer is the determination of its two volume constants; that is, the constants of the linear equation expressing the capacity as a function of the temperature. This is done by filling with distilled water and weighing accurately several times at each of two temperatures near the ends of the range for which the pycnometer is to be used. The bottle form is also adapted to the precise measurement of densities of solids. See also **Specific Gravity**.

PYELITIS.

See **Kidney and Urinary Tract**.

PYRAMID.

A polyhedron with one base, which is any polygon and additional faces, called lateral, which are triangles with a common vertex. The intersection of the lateral faces are the lateral edges of the pyramid. Special cases are: regular, the base is a regular polygon with center determined by its axis, the perpendicular from vertex to base; triangular, quadrangular, etc., if the base is a triangle, quadrilateral, etc. A tetrahedron is a triangular pyramid with four triangular faces and, because of its symmetry, any one of them may be taken as the base. The slant height of a regular pyramid is the altitude of any of its lateral faces. A pyramid is truncated if it is cut by a plane oblique to its base, provided all the lateral edges are cut. If the is plane is parallel to the base, the result is the frustum of a pyramid.

The volume of a regular pyramid $V = \frac{1}{3}$ (area of base × height). The lateral area of a regular pyramid $= \frac{1}{2}$ (perimeter of base × slant height). The volume of any pyramid $V = \frac{1}{3}$ (area of base × distance from vertex to plane of base). The volume of a frustum of any pyramid $V = h/3(A_1 + A_2 + \sqrt{A_1 + A_2})$, where A_1 and A_2 are the areas of bases made by the parallel planes.

See also **Polygon**; **Polyhedron**; and **Triangle**.

PYRARGYRITE. An antimony-bearing silver mineral corresponding to the formula Ag_3SbS_3. It crystallizes in the hexagonal system, commonly in rhombic prismatic forms. It displays a rhombohedral cleavage; fracture, conchoidal to uneven; brittle; hardness, 2.5; specific gravity, 5.24; luster, adamantine to submetallic; color, deep red, but being light sensitive alters readily to black. In thin fragments deep red by transmitted light, otherwise practically opaque; streak, purplish red. Pyrargyite occurs with proustite, other silver minerals, and galena, and sphalerite. It is found in the Harz Mountains, in the Czech Republic and Slovakia, Bolivia, Chile, Mexico, and in the United States in Colorado, Idaho, and Nevada. In Canada it is found in the Cobalt region of the Province of Ontario. It derives its name from the Greek words meaning fire and silver.

PYRHELIOMETER. An actinometer which measures the intensity of direct solar radiation, consisting of a radiation sensing element enclosed in a casing which is closed except for a small aperture, through which the direct solar rays enter, and a recorder unit. See also **Actinometer**.

PYRIDINE AND DERIVATIVES. Pyridine is a slightly yellow or colorless liquid; hygroscopic; bp, 115.5 °C; fp, −41.7 °C; unpleasant odor; burning taste; slightly alkaline in reaction; soluble in water, alcohol, ether, benzene, and fatty oils; specific gravity, 0.978; flash point (closed cup), 20 °C; autoignition temperature, 482 °C. Pyridine, a tertiary amine, is a somewhat stronger base than aniline and readily forms quaternary ammonium salts.

Pyridine and derivatives of pyridine occur widely in nature as components of alkaloids, vitamins, and coenzymes. These compounds are of continuing interest to theoretical physical, organic, and biochemistry and to industrial chemistry. Pyridine and derivatives have many uses, e.g., herbicides and pesticides, pharmaceuticals, feed supplements, solvents and reagents, and chemicals for the polymer and textile industries.

Structure and Nomenclature. The pyridine group consists of a six-membered, heterocyclic, aromatic compound with one nitrogen atom in the ring. The parent compound of this group is pyridine I with ring positions numbered as shown. Alternative denotations of the 1, 2, and 4 positions in the ring are alpha, beta, and gamma, respectively.

The behavior of pyridine in substitution reactions can be understood on the basis of its resonance structures (Ia–d) and on the basis of the electron-density distribution at the various ring positions as derived from molecular-orbital-theoretical calculations. An example of the published pi-electron density distribution is shown in II. The resonance energy of pyridine is 35 kcal/mole (versus 39 kcal/mole for benzene).

Electrophilic substitution occurs at the 3 and 5 positions, but usually requires drastic conditions because the species actually being attacked is a pyridinium ion. For example, nitration of pyridine with KNO_3 and concentrated H_2SO_4 at 300 °C gives a 15% yield of 3-nitropyridine. Electrophilic substitution in the pyridine ring is facilitated by the presence of electron-donating substituents.

Nucleophilic substitution occurs in the 2, 4, and 6 positions of pyridine under relatively mild conditions. As an example, amination of pyridine with sodium amide in N,N-dimethylaniline at 180 °C gives 2-aminopyridine in good yield.

Homolytic (free-radical) substitution may occur in any of the 2 to 6 positions of pyridine. Thus, the reaction of pyridine with benzene-diazonium salts gives a mixture of 2-, 3-, and 4-phenylpyridine.

Many pyridine derivatives difficult to make directly from pyridine are readily accessible starting from pyridine N-oxide, made by oxidation of pyridine with hydrogen peroxide in acetic acid. As but one example, the nitration of pyridine N-oxide gives 4-nitropyridine N-oxide in high yield. Reduction of the D-oxide to the parent pyridine nucleus is readily effected by hydrogenation or reagents, such as PCl_3 or triphenyl phosphine.

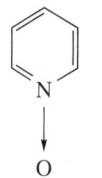

Pyridine N-oxide

Trivial names for the methylpyridines are the *picolines*; the dimethylpyridines are the *lutidines*; and the trimethylpyridines (and in older literature the ethyldimethylpyridines) are the *collidines*. The refractive indices for these alkyl pyridines and for pyridine itself fall in the range: $n_D^{20} \sim 1.50$–1.51.

Production of Pyridine and Homologues

Coke Manufacture By-products. In United States practice, coking of coal is done almost exclusively by the high-temperature (900–1200 °C) process. For many years, the major source of the pyridines was the chemical-recovery coke oven. The volatiles produced in the coke oven are only partially condensed. The noncondensed gases are passed through a scrubber (the ammonia saturator) containing sulfuric acid. After removal of crystals (ammonium sulfate), a solution of ammonium sulfate and pyridinium sulfates is obtained and treated with ammonia to liberate and contained pyridine bases (~70% is pyridine itself). See also **Coal Tar and Derivatives**. The balance of the pyridine bases is extracted from the crude coal tar, i.e., the condensed, main portion of the volatilization products from coking. The crude tar contains approximately 0.1–0.2% pyridine bases. Further separation of the pyridines involves a rather complex series of extractions, distillations, and crystallizations.

Synthetic Methods of Manufacture. Due to rising demand, production of the pyridine bases by large-scale synthesis passed the volume of tar bases extracted from coal tar in the 1960s. By the early 1970s, capacity in the United States for the synthetic manufacture of pyridine, the picolines, and 2-methyl-5-ethylpyridine (MEP) was in the tens of millions of pounds. All of these products can be made by condensation reactions of aldehydes and ammonia. MEP is no longer made in the United States.

When acetaldehyde and ammonia in a 3:1 mole ratio are fed over dehydration-dehydrogenation catalysts, such as PbO or CuO on alumina, ThO_2, or ZnO or CdO on silica-alumina, or CdF_2 on silicamagnesia at 400–500 °C and atmospheric pressure, an equimolar mixture of 2- and 4-picolines can be obtained in 40–60% yields. When a mixture of acetaldehyde, formaldehyde, and ammonia in about 2:1:1 mole ratio is passed over such catalysts, pyridine and 3-picoline are produced; their ratios are usually 1:0.8, but the amounts of pyridine can be increased by changes in the feed.

The lowest-cost synthetic pyridine base, 2-methyl-5-ethylpyridine, is made in a liquid-phase process from paraldehyde (derived from acetaldehyde) and aqueous ammonia in the presence of ammonium acetate at approximately 102–190 atmospheres and 220–280 °C in 70–80% yield. Minor byproducts include 2- and 4-picoline.

A new synthetic method for preparing 2-methylpyridine has been commercialized. This process involves the acid/base-catalyzed condensation of acetone with acrylonitrile to make 5-oxo-hexanenitrile. Then the nitrile is converted to a 2-methylpyridine by catalytic cyclization/dehydrogenation:

$$CH_3COCH_3 \ + \ CH_2{:}CHCN \longrightarrow$$
acetone   acrylonitrile

$$CH_3COCH_2CH_2CH_2CN \longrightarrow$$
5-oxo-hexanonitrile   (2-methylpyridine)

Much recent work has been done on the synthesis of pyridines from alkynes and nitriles over cobalt catalysts. For example, 2-vinylpyridine has been obtained in good yield from acetylene and acrylonitrile using a cyclopentadienyl-cobalt catalyst. Pyridine has also been obtained from cyclopentadiene and ammonia over a silica/alumina catalyst.

In the synthetic processes, mixtures of products are often obtained. Variation in the supply/demand balance of the alkyl pyridine isomers has led to much research on processes which may alleviate such imbalances, including development of the catalytic hydrodealkylation of alkyl pyridines to pyridine as well as the alkylation of pyridine.

Major Uses of Pyridine Derivatives

The applications of these compounds are wide-ranging and new uses are proliferating. The following examples are a selection of important commercial products, but hardly a complete listing.

Herbicides. A major outlet for pyridine (20–30 million lb/yr worldwide) is in the manufacture of the desiccant herbicides and aquatic weed killers, such as 1,1'-ethylene-2,2'-dipyridilium dibromide, known as Diquat;® and 1,1'-dimethyl-4,4'-dipyridilium dichloride (or dibromide or dimethylsulfate), known as Paraquat.®

1,1'-ethylene-2,2'-dipyridilium
dibromide

1,1'-dimethyl-4,4'-dipyridilium
dichloride

4-Amino-3,5-dichloro-6-fluoro-2-pyridyloxyacetic acid, tradenamed Starane,® is a herbicide used to control broadleafed weeds and brush species, and certain deep-rooted perennial weeds. 3,5,6-Trichloro-2-pyridyloxyacetic acid, tradenamed Garlon,® is a herbicide used for vegetation management, such as in rights-of-way.

4-amino-3,5-dichloro-6-fluoro-
2-pyridyloxyacetic acid

3,5,6-trichloro-2-pyridyloxyacetic acid

A new class of herbicides, the pyridylosy-phenoxyalkanoic acids, is typified by n-butyl 2-[4-(5-trifluoromethyl-2-pyridyloxy)phenoxy] propionate, tradenamed Fusillade,® active against annual and perennial grasses:

n-butyl-2-[4-(5-trifluoromethyl-2-pyridyloxy)phenoxy]propionate

The newest class of pyridine herbicides with pre- and post-emergent activity, the pyridinesulfoneamides, is typified by N-(2-chloro-3-pyridinesulfonyl)-N'-[2-(4-chloro-5,6-dimethylpyrimidyl] urea.

N-(2-chloro-3-pyridinesulfonyl)-N'-{2-4-chloro-
5,6-dimethylpyrimidyl]urea

2-picoline (2-methyl pyridine) is the source of 2-chloro-6-trichloromethylpyridine, known as N-Serve,® which is useful as a fertilizer additive for reduction of nitrogen losses in the soil due to bacterial

oxidation. 2-Picoline also is the starting material for the production of 4-amino-3,5,6-trichloropicolinic acid, a powerful broad spectrum herbicide for broad-leaved plants, known as Tordon.® 3,6-Dichloropicolinic acid, tradenamed Lontrel® and Format® in different formulations, is used for the postemergence control of broadleafed weeds.

2-chloro-6-
trichloromethylpyridine

4-amino-3,5,6-
trichloropicolinic acid

3,6-Dichloropico-
linic acid

2,3-Lutidine (2,3-dimethylpyridine) is the starting material for the herbicide 2-[4,5-dihydro-4-methyl-4-(1-methylethyl)-5-oxo-1H-imidazol-2-yl]-3-pyridinecarboxylate (compound with 2-propanamine), tradenamed Arsenal.®

2-[4,5-dihydro-4-methyl-4-(1-methylethyl)-5-oxo-1H-imidazol-2-y1]3-
pyridinecarboxylate (compound with 2-propanamine)

Pesticides. The compound 2-picoline is a component of 1-[(4'-amino-2'-n-propyl-5'-pyrimidinyl)methyl]-2-picolinium chloride hydrochloride, known as Amprolium,® a broad-spectrum coccidiostat. A newer coccidiostat is 3,5-dichloro-4-hydroxy-2,6-lutidine and known as Clopidol.®

1-[(4'-amino-2'-n-propyl-5'-pyrimidinyl)methyl]-
2-picolinium chloride hydrochloride

3,5-dichloro-4-
hydroxy-2,6-lutidine

The acaricide O,O-diethyl-O-(3,5,6-trichloro-2-pyridyl) thiophosphate, known as Dursban,® is used to control ectoparasites. The similar, O,O-dimethyl-O-3,5,6-trichloro-2-pyridyl) thiophosphate, tradenamed Reldan® and Tumar,® is a nonsystemic insecticide/acaricide. Di(n-propyl)isocinchomerate, known as MGK Repellent 326® is used in fly repellents and is made by oxidation of 2-methyl-5-ethylpyridine and esterification of the isocinchomeronic acid obtained. Nicotine (sulfate) (Black Leaf 40®) is used as an agricultural insecticide, as an external parasiticide, and as an anthelminthic, and is obtained by extraction of tobacco wastes (not by synthesis).

O,O-diethyl-O-(3,5,6-trichloro-2-pyridyl) thiophosphate

Di (n-propyl) isocinchomerate

Nicotine

3-(2-Methylpiperidino)propyl-3,4-dichlorobenzoate, tradenamed Pipron,® is a foliar fungicide for the control of powdery mildew:

3-(2-methylpiperidino)propyl   3,4-dichlorobenzoate

Dimethyl 3,5,6-trichloro-2-pyridyl phosphate, known as Fospirate® or Dowco® 217, is an insecticide useful in antiflea collars for dogs and cats. The compound 4-aminopyridine, known as Avitrol® 100, and 4-nitropyridine-N-oxide, known as Avitrol® 200, are useful as bird repellents. See also **Pesticide**.

Dimethyl-3,5,6-trichloro-2-pyridyl phosphate

4-aminopyridine

4-nitropyridine-N-oxide

Pharmaceuticals. A wide variety of pyridine compounds, with varying, and often multiple, drug action are used commercially. A few examples are given. A number of **antihistamines** contain the pyridine moiety in their structure, as exemplified by chlorpheniramine maleate (2-[p-chloro-α-(2-dimethylaminoethyl)benzyl] pyridine acid maleate); doxylamine succinate (2-[α-(2-dimethylamino)ethoxy-α-methylbenzyl]-pyridine acid succinate); and pyrilamine maleate (2-(2-dimethyl-aminoethyl-2-p-methoxybenzyl) aminopyridine acid maleate). These products are synthesized, e.g., from the appropriate benzylpyridines or aminopyridines.

Chlorpheniramine maleate

Doxylamine succinate

Pyrilamine maleate

Cetylpyridinium chloride is used as a germicide and antiseptic, e.g., in mouthwashes; it is made by quaternization of pyridine with cetyl chloride.

Cetylpyridinium chloride

Isonicotinehydrazide, also known as isoniazid, is an important antitubercular drug made by oxidation of 4-alkylpyridine (or 2,4-lutidine) or by hydrolysis of 4-cyanopyridine to isonicotinic acid (pyridine 4-carboxylic acid) and reaction of an ester or the acid chloride of the latter with hydrazine.

Isonicotinehydrazide

Meperidine hydrochloride (1-methyl-4-carbethoxy-4-phenylpiperidine), also known as Demerol,® is an important narcotic and analgesic. It is not made from piperidine, but rather by ring-closure reactions of appropriate precursors.

Meperidine hydrochloride

Cephapirin sodium, tradenamed Bristocef,® Cefadyl,® Today® (and others) is a cephalosporin C antibiotic:

Cephapirin sodium

Nalidixic acid (1-ethyl-7-methyl-1.8-naphthridine-4-one-3-carboxylic acid), many tradenames (e.g., Nalidicron®), is an antibacterial. *Bisacodyl* [4,4'-(2-pyridylmethylene)diphenol diacetate], tradename Dulcolax,® is a laxative.

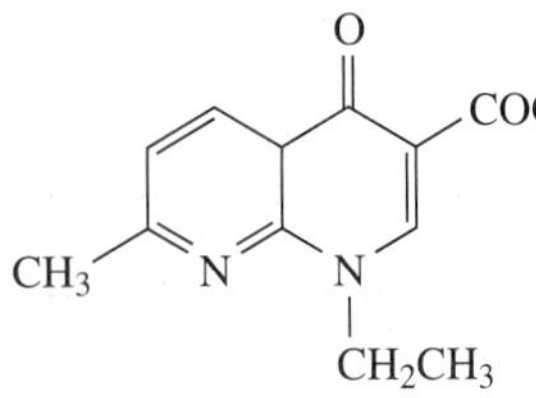

Nalidixic

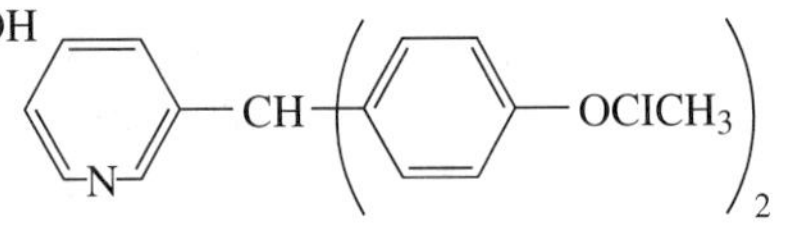

Bisacodyl

Nifedipine   [1,4-dihydro-2,5-dimethyl-3,5-dicarbmethoxy-4-(2-nitrophenyl)pyridine], tradename Procardia,® is used in the treatment of angina.

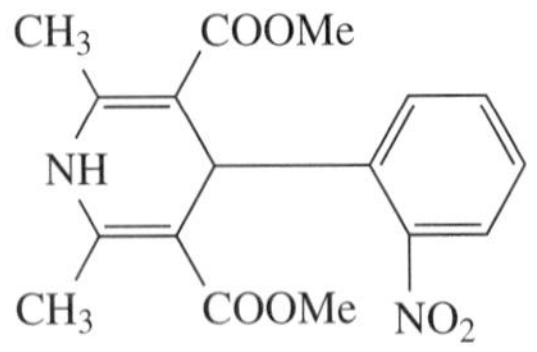

Nifedipine

Nicotinic acid and nicotinamide, members of the vitamin B group and used as additives for flour and bread enrichment, and as animal feed additive among other applications, are made to the extent of 24 million pounds (nearly 11 million kilograms) per year throughout the world. Nicotinic acid (pyridine-3-carboxylic acid), also called *niacin*, has many uses. See also **Niacin**. Nicotinic acid is made by the oxidation of 3-picoline or 2-methyl-5-ethylpyridine (the isocinchomeric acid produced is partially decarboxylated). Alternatively, quinoline (the intermediate quinolinic acid) is partially decarboxylated with sulfuric acid in the presence of selenium dioxide at about 300 °C, or with nitric acid, or by electrochemical oxidation. Nicotinic acid also can be made from 3-picoline by catalytic ammoxidation to 3-cyanopyridine, followed by hydrolysis.

Nicotinamide is prepared by partial hydrolysis of the nitrile, or by amination of nicotinic acid chloride or its esters. Some of the compounds mentioned in the foregoing are shown below.

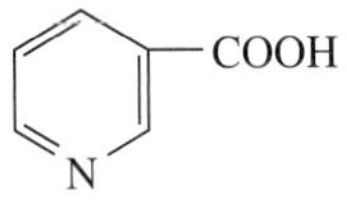

Nicotinic acid

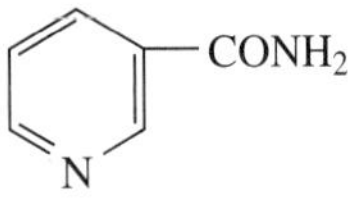

Niacinamide

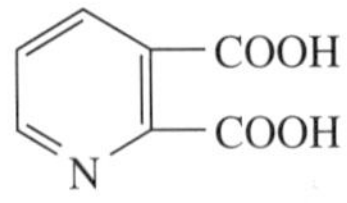

Quinolinic acid

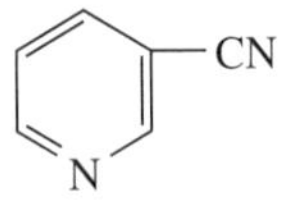

3-cyanopyridine

Several esters of nicotinic acid are used as vasodilators.

Nikethamide is a respiratory and heart stimulant, used beneficially against overdoses of barbiturates and morphine. Also known as Coramine,® this compound (*N,N*-diethylnicotinamide) is made by reaction of nicotinic acid esters or the acid chloride with diethylamine. Its formula is shown below.

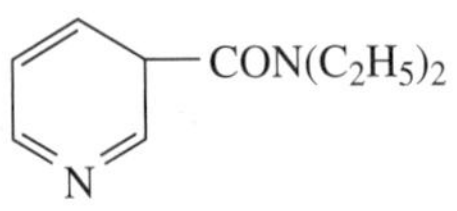

Nikethamide

Pipadrol is a central nervous system stimulant. This compound, α,α-diphenyl-2-piperidinemethanol, is made by condensation of 2-pyridyl-magnesium chloride with benzophenone and catalytic hydrogenation of the pyridine ring of the resultant carbinol. Its formula is shown below.

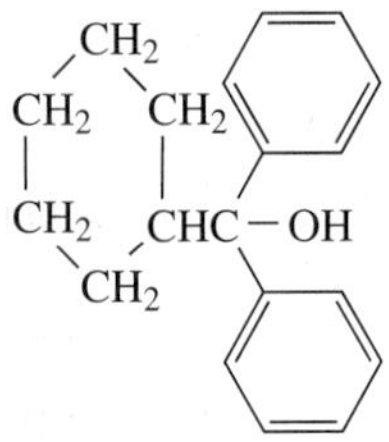

Pipadrol

Piperocaine hydrochloride is used as a local anesthetic. This compound (*d, l*-(2-methylpiperidino)propyl benzoate hydrochloride) is made by reaction of 2-methylpiperidine with 3-chloropropyl benzoate. Its formula is shown below.

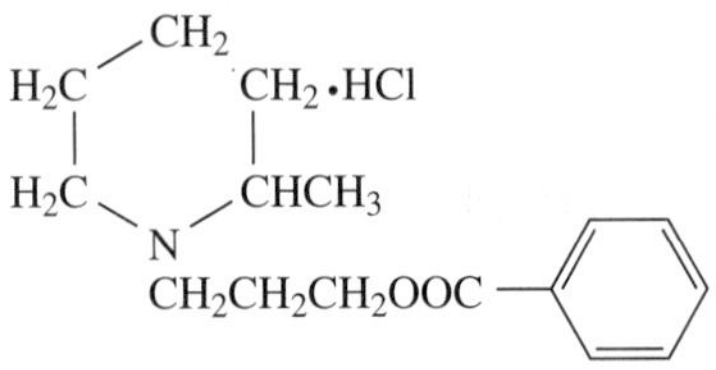

Piperocaine hydrochloride

Pyrithione (zinc salt of) is used as a component of antidandruff shampoos and as a bactericide in soap and detergent formulations. This compound (2-mercaptopyridine *N*-oxide) exists in equilibrium with N-hydroxy-2-pyridinethione and is a fungicide and bactericide, prepared by reaction of 2-chloropyridine *N*-oxide with sodium hydrosulfide and sodium sulfide. This compound is also known as Omadine.® Its formula is shown below.

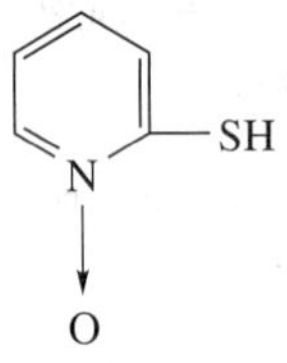

Pyrithione

Sulfapyridine is used to treat dermatitis herpetiformis and also has been used by veterinarians against pneumonia, shipping fever, and foot rot of cattle. This compound (2-sulfanylamidopyridine) is made by condensation of 2-aminopyridine with the appropriate sulfonyl chloride. Its formula is shown below.

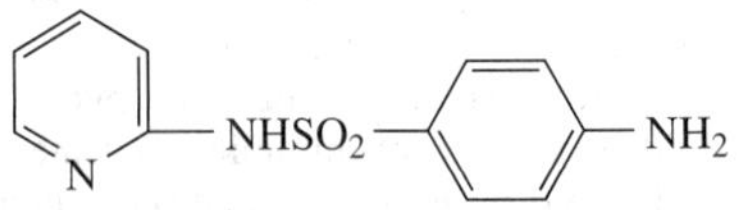

Sulfapyridine

Vitamin B6 is described in detail under **Vitamin B$_6$**. This is 2-methyl-3-hydroxy-4,5-di(hydroxymethyl)pyridine or pyridoxol. World demand of this compound is estimated at about 5 million pounds (about 2.3 million kilograms) per year. Commercial production is by synthesis, starting, for example, with the base-catalyzed condensation of cyanoacetamide and ethoxyacetylacetone. The formula for pyridoxol is shown below.

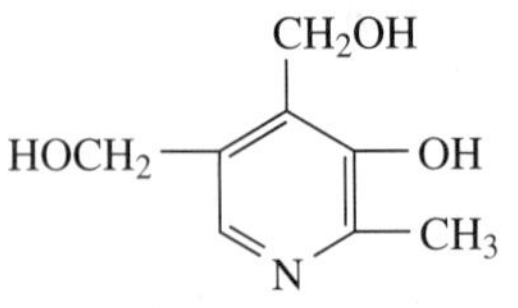

Pyridoxol

Methyridine or 2-(2-methoxyethyl)pyridine, also called Mintic,® is used as an anthelmintic. *Piroxicam*, also known as Feldene,® is a relatively new anti-inflammatory for the treatment and relief of arthritis. See formulas below.

Methyridine   Piroxicam

Pyridinol carbamate has been used as an anti-inflammatory/anti-arteriosclerotic. This compound 2,6-pyridinedimethanol-bis-(N-methyl-carbamate) is also known as Anginin.® See the formula below.

Pyridinol carbamate

Pyrithioxin is a neurotropic agent that reduces the permeability of the blood-brain barrier to phosphate. This compound, 3,3′-dithio-dimethylene-bis-(5-hydroxy-6-methyl-4-pyridinemethanol), is also known as Life® and Bonifen.® Its formula is shown below.

Pyrithioxin

Textile Chemicals. Pyridine derivatives find a number of quite different applications in the textile and related fields.

Stearamidomethylpyridinium chloride is used in waterproofing textiles. It is made by reacting pyridine hydrochloride with stearamide and formaldehyde. *Vinylpyridines* are used as components of acrylonitrile copolymers to improve the dyeability of polyacrylonitrile fibers. The commercially important products are 2-vinylpyridine; 4-vinylpyridine; and 2-methyl-5-vinylpyridine. Formulas are shown below.

Stearamidomethylpyridinium   2-vinylpyridine
chloride

4-vinylpyridine   2-methyl-5-vinylpyridine

2-Vinylpyridine is used in the terpolymer latex component of tire cord dips to improve the bonding of textile to rubber. Rubber tires built with steel cord, however, do not require vinylpyridine latex-based adhesives for the steel belt. Therefore, the consumption of vinylpyridines may be affected in the future.

Other. The pyridines and methylpyridines and their mixtures are used as chemical processing aids (e.g., acid acceptors, solvents) and as industrial corrosion inhibitors.

Piperidine, the hydrogenation product of pyridine, is used as an intermediate for drugs and for making rubber-vulcanization accelerators, e.g., piperidinium pentamethylenedithiocarbamate (also known as Accelerator 552®). On a commercial scale, piperidine (hexahydropyridine) is prepared by the catalytic hydrogenation of pyridine, e.g., with nickel catalysts at

from 68 to 136 atmospheres pressure and at 150–200 °C, or under milder conditions with noble-metal catalysts. Pyridine derivatives can be similarly reduced to substitute piperidines. See formulas below.

Piperidine   Piperidinium pentamethylendithiocarbamate

4-N,N-Dialkylaminopyridines have found use as catalysts for acylation reactions.

There are developing applications for linear and crosslinked poly vinylpyridines in photovoltaic cells and batteries, electron beam resists, as catalysts and reagents (e.g., in pollution control).

The hindered-amine light stabilizers for polymers are piperidine derivatives. An example of these products is bis-(2,2,6,6-tetramethyl-4-piperidinyl)sebacate, tradenamed Tinuvin 770,® useful as a light stabilizer for polyolefins and styrenics.

Bis-(2,2,6,6-tetramethyl-4-piperidinyl)sebacate

There is growing evidence of developing high-technology uses of pyridines, particularly as quaternary salts, as components of electrolytic capacitors, photoconductors, rechargeable batteries, complex-coated electrodes for photosensors, electrochromic display elements and photoresist matrix resins.

HANS DRESSLER, Koppers Company, Inc., Monroeville, PA.

PYRITE. The mineral pyrite or iron pyrites is iron disulfide, FeS_2, its isometric crystals usually appearing as cubes or pyritohedrons. It has a slightly conchoidal to uneven fracture; brittle; hardness, 6–6.5; specific gravity, 5; metallic luster; color, pale to normal brass-yellow; streak, greenish-black; opaque. Arsenic, nickel, cobalt, copper, and gold may be found in small quantities in pyrite, auriferous pyrite being sometimes a very valuable ore. Pyrite is the commonest of the sulfide minerals, and is of worldwide occurrence. It is found associated with other sulfides, or with oxides, in quartz veins, in sedimentary and metamorphic rocks, in coal beds, and as the replacement material in fossils. There are many well-known pyrite localities, among which are the Rio Tinto mines in Spain, where copper-bearing pyrite is obtained from huge deposits. Magnificent crystals and crystal groups occur at Ambasaguas (Logrono) in Spain; Quirivulca, Peru; and from the Island of Elba. In the United Stated pyrite is found in California, New York, and Virginia in workable deposits. The name pyrite is derived from the Greek word meaning fire, because of the sparks that result when pyrite is struck with steel.

PYROCLASTIC. Pertaining to clastic rock material formed by volcanic explosion or aerial expulsion from a volcanic vent. The word also pertains to rock texture of explosive origin. However, it is not synonymous with "volcanic."

PYROGENETIC MINERALS. A term for the primary magmatic minerals of igneous rocks as distinguished from those minerals which are the result of special and later processes such as come under the head of pneumatolytic, hydrothermal, etc.

PYROLUSITE. The mineral pyrolusite, manganese dioxide (MnO_2), crystallizes in the tetragonal system, but may be only pseudomorphous after manganite. It is found massive or in indistinct crystalline aggregates, often acicular, and as dendritic growths on fractured rock surfaces and as inclusions within moss agates and other chalcedony varieties of quartz. Hardness, 6–6.5 (crystals), 2–6 (massive); specific gravity, 5.06; luster, metallic; color, steel gray to black; streak black; opaque. Pyrolusite is found as replacement deposits and as residual and sedimentary masses. Psilomelane is its usual associate. European localities for pyrolusite are in Bohemia, Saxony, the Harz Mountains, England, and elsewhere. Other deposits occur in India and Brazil. In the United States it is found in Arkansas and Michigan. It is an ore of manganese. It is from this latter use that it derives the name pyrolusite, from the Greek words meaning *fire* and *to wash*.

PYROMETER. An instrument for the measurement of temperatures; generally applied to instruments measuring temperatures above 600 °C.

PYROMETRIC CONES. Small cones that differ in the temperatures at which they soften on heating. They are made of clay and other ceramic materials and are used in the ceramic industries to show furnace temperatures within ranges. In practice, three or four of the cones which have softening points at consecutive temperature ranges are used, and the increase in kiln temperature is judged from the progressive deformation of the cones.

PYROMETRY. High-temperature thermometry, the technique of measurement of temperatures, generally above 600 °C, at a distance.

PYROMORPHITE. The mineral pyromorphite is lead chlorophosphate with a formula corresponding to $Pb_5(PO_4)_3Cl$. The phosphorus is sometimes replaced by arsenic and the lead by calcium. It occurs in prismatic, sometimes hollow, hexagonal crystals or may appear in massive forms. It is brittle; hardness, 3.5–4; specific gravity, 7.04; luster, resinous; color, green, yellow-green, yellow, brown, and less often gray or white; translucent to opaque.

Pyromorphite is a secondary mineral associated with other lead minerals, but is seldom found in large quantities. It has probably resulted from the action of waters bearing phosphoric acid upon the preexisting lead minerals. Localities for pyromorphite are in the Ural Mountains, Saxony, France, Spain, Cornwall and Cumberland, England; in Scotland, Zaire, and Australia. In the United States pyromorphite has been found in Chester and Montgomery Counties, Pennsylvania; in Davidson County, North Carolina, and in the Coeur d'Alene mining district of Idaho. The name is derived from the Greek words meaning fire and form.

PYRON. A unit of radiant intensity of electromagnetic radiation equal to 1 calorie per square centimeter per minute.

PYROPHANITE. Manganese analogue of ilmenite.

PYROPHYLLITE. The mineral pyrophyllite is a hydrous silicate of aluminium corresponding to the formula $Al_2Si_4O_{10}(OH)_2$. Monoclinic with a basal cleavage, it is usually, however, in foliated, radiated lamellar, or fibrous masses, sometimes compact. It is a soft mineral with a greasy feel; hardness, 1–2; specific gravity, 2.65–2.9; luster, pearly to dull; color, white, greenish, grayish, yellowish, and brownish; translucent to opaque. It is found making up schists or in foliated masses in the Ural Mountains, in Switzerland, Sweden, Brazil, and in the United States in Pennsylvania, North Carolina, Georgia, and California. It is used to some extent for the same purpose as is the mineral talc, and also for making slate pencils, hence the name pencil stone sometimes applied to pyrophyllite.

PYROXENE. This is the name given to a closely related group of minerals, all of which show a distinct cleavage angle of 87° or 93° parallel to the fundamental prism. Chemically the pyroxenes are metasilicates corresponding to the formula $RSiO_3$, where R may be calcium, magnesium, iron, or less commonly manganese, zinc, sodium, or potassium. Rarely titanium, zirconium, or fluorine may be present. A general formula is $ABSi_2O_6$, where A is Ca, Na, Mg, or Fe^{2+}, and B is Mg, Fe^{3+}, or Al. Sometimes the Si is replaced by Al. The pyroxenes crystallize in the orthorhombic, and monoclinic systems, like the amphiboles, the chief difference between the two groups being the cleavage angles, which for amphibole are 56° and 124°. Pyroxene crystals tend to be short, stout, complex prisms as opposed to the long, slender, and simpler amphiboles.

The pyroxenes are common in the more basic igneous rocks, both intrusive and extrusive, and may be developed by the metamorphic processes in gneisses, schists, and marbles.

For descriptions of members of the pyroxene group, see also **Acmite-Aegerine**; **Augite**; **Diallage**; **Diopside**; **Enstatite**; **Hypersthene**; **Jadeite**; and **Spodumene**.

PYRRHOTITE. The mineral pyrrhotite, sometimes called magnetic pyrites, is a sulfide of iron with varying amounts of sulfur. Analyses indicate formulae $Fe_{1-x}S$. Pyrrhotite exists in two modifications: it is monoclinic below, and hexagonal above 138 °C (280 °F). It is a brittle mineral; hardness, 3.5–4.5; specific gravity, 4.53–4.97; luster, metallic; color, reddish bronze-yellow when fresh, otherwise tarnished; streak, grayish-black; magnetic. It may carry nickel, generally as pentlandite, when it becomes a valuable nickel ore as at Sudbury, Ontario. Pyrrhotite is commonly associated with the basic igneous rocks like gabbro, and norite, and occurs with chalcopyrite, magnetite, and pyrite. Besides being apparently of magmatic origin, it has been found as contact metamorphic and as vein deposits. Austria, Italy, Saxony, Bavaria, Switzerland, Norway, Sweden, and Brazil have deposits of more or less importance, and in the United States it has been found associated with andalusite crystals at Standish, Maine; also at Brewster, New York; Lancaster County, Pennsylvania, and elsewhere. At Ducktown, Tennessee, it is found together with copper and zinc minerals. It is mined for its nickel content, in the form of admixed pentlandite, in Sudbury, Ontario.

Pyrrhotite derives its name from the Greek word *pyrrhos*, meaning reddish, in reference to the color of the fresh ore.

PYRROLE AND RELATED COMPOUNDS. Pyrrole (monoazole, C_4H_5N or C_4H_4NH), contains a ring of 1 nitrogen and 4 carbons, with 1 hydrogen attached to nitrogen and to each carbon:

Pyrrole is a colorless liquid, boiling point 131 °C, insoluble in water, soluble in alcohol or ether. Pyrrole dissolves slowly in dilute acids, being itself a very weak base; resinification takes place readily, especially with more concentrated solutions of acids; and on warming with acid a red precipitate is formed. Pyrrole vapor produces a pale red coloration on pine wood moistened with hydrochloric acid, which color rapidly changes to intense carmine red. Pyrrole may be made (1) by reaction of succinimide

with zinc and acetic acid, or with hydrogen in the presence of finely divided platinum heated, (2) by reaction of ammonium saccharate or mucate $COONH_4 \cdot (CHOH)_4 \cdot COONH_4$ with glycerol at 200 °C by loss of carbon dioxide, ammonia, and water.

When pyrrole is treated with potassium (but not with sodium) or boiled with solid potassium hydroxide, potassium pyrrole C_4H_4NK is formed, which is the starting point for *N*-derivatives of pyrrole, since reaction of the potassium with halogen of organic compound and with carbon dioxide, readily occurs. When pyrrole is treated with magnesium metal and ethyl bromide in ether, pyrrole magnesium bromide plus ethane is formed, which may be used as the starting point for C-derivatives of pyrrole, since reaction with sodium alcoholates readily occurs (with separation of magnesium oxybromide).

The pyrrole nucleus has been shown to be present in the complex substances chlorophyll (the green coloring matter of plants), hematin (the red coloring matter of blood), and in the coloring matter of bile.

PYTHAGOREAN SCALE. A musical scale such that the frequency intervals are represented by the ratios of integral powers of the numbers 2 and 3.

PYTHAGOREAN THEOREM. Also known as the hypotenuse theorem, the square of the hypotenuse of a right-angle triangle is equal to the sum of the squares of the other two sides. This is illustrated in Fig. 1. Numbers so related (for example, 3, 4, 5) are referred to as Pythagorean numbers. A generalization of the Pythagorean theorem, so as to include oblique triangles, is the Law of Cosines, "The square of any side of a triangle is equal to the sum of the squares of the other two sides minus twice the product of these two sides times the cosine of their included angle." See also **Direction Cosine**.

PYTHON. See **Snakes**.

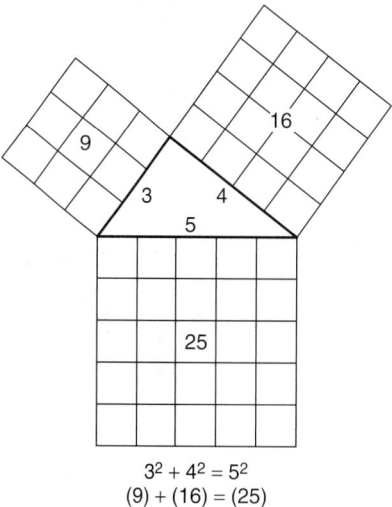

$$3^2 + 4^2 = 5^2$$
$$(9) + (16) = (25)$$

Fig. 1. Demonstration of Pythagorean theorem.

Q

Q. A figure of merit of an energy-storing system equal to

$$2\pi\left(\frac{\text{average energy stored}}{\text{energy dissipated per half cycle}}\right)$$

which is equal to $\omega L/R$ for an inductor, where R is the equivalent series resistance. For a capacitor, Q is $1/\omega CR$, again the ratio of reactance to effective resistance. For a medium, Q is the ratio of displacement current density to conduction current density. The basic equation may also be expanded to include series and parallel resonant circuits, for which cases appropriate approximate equations may also be developed.

The term Q (or more commonly Q-value) is also used as a synonym for nuclear disintegration energy.

QUAD. One quadrillion Btus (British thermal units). This amount of energy may be envisioned by considering it the equivalent of 8 billion gallons (U.S.) of gasoline; or about one year's supply for 10 million automobiles.

QUADRANT. An instrument formerly used in surveying and astronomy for measuring angles and altitudes, consisting of a graduated arc of 90 degrees (180 degrees in range) equipped with a sighting device and a movable index or vernier and usually a plumb line or spirit level for fixing the vertical or horizontal direction. The instrument has been superseded by the sextant.

QUADRATE BONE. A small bone of roughly quadrangular shape in the vertebrate skull. In the reptiles and lower forms it is included in the articulation of the lower jaw with the skull, and a similar condition persists in the birds. In the mammal the incus or anvil, one of the three small bones of the middle ear, is regarded as the homologue of this bone.

QUADRATIC EQUATION. An algebraic equation of the second degree in one or more variables. If there are two variables, the resulting curves are conic sections; if three variables, quadric surfaces are obtained. The roots of the quadratic equation in one variable, $ax^2 + bx + c = 0$, are given by the quadratic formula, which was apparently known to the Greek mathematicians, for in geometric terms it is found in Euclid's "Elements" (ca. 300 B.C.). This equation, also called the Hindu method for quadratics, is

$$2ax = -b \pm \sqrt{b^2 - 4ac}$$

Its discriminant, $D = b^2 - 4ac$. If $D > 0$, the roots of the equation are real and unequal; if $D = 0$, they are real and equal; if $D < 0$, both roots are imaginary.

TABLE 1. POSSIBLE CASES

I	Δ	Nature of Curve
0	$A\Delta < 0$	ellipse; circle if $B = 0$, $A = C$
	$A\Delta > 0$	imaginary locus
	0	a point
>0	$\neq 0$	hyperbola
	0	two intersecting straight lines
<0	$\neq 0$	parabola
	0	$A \neq 0$; $H = D^2 - 4AF = 0$, one line;
		$H > 0$, two parallel lines;
		$H < 0$, imaginary locus
		$A = 0$; $J = E^2 - 4CF = 0$, one line;
		$J > 0$, two parallel lines;
		$J < 0$, imaginary locus

The general form of the quadratic equation in two variables is $Ax^2 + Bxy + Cy^2 + Dx + Ey + F = 0$ and its curve is a conic section. The nature of the curve is determined by the invariant of the equation, **I**, and the discriminant, Δ.

$$\mathrm{I} = (B^2 - 4AC)$$

$$\Delta = \frac{1}{2}\begin{vmatrix} 2A & B & D \\ B & 2C & E \\ D & E & 2F \end{vmatrix}$$

The possible cases are shown in Table 1.

See also **Algebraic Equations**.

QUADRATRIX OF DINOSTRATUS. A higher plane curve, the equation of which in polar coordinates is $r = a\phi \csc \phi$. The curve has an infinite number of asymptotes. It was studied by Dinostratus, a Greek mathematician who lived about 350 B.C. and who showed that it could be used to divide an angle into an arbitrary number of equal parts and thus applied to the quadrature of a circle.

Another quadratix, known by the name of *Tschirnhausen*, has the equation $x = b\cos ny/2b$.

See also **Curve (Higher Plane)**.

QUADRATURE. Given a differential equation of the form $F(x, y, y' = dy/dx) = 0$, it is reduced to quadrature if it may be written

$$y = \int f(x)\, dx$$

When the integral cannot be evaluated in terms of known functions, numerical, graphical, or mechanical methods may be used. These are known as approximate or mechanical quadratures.

QUADRATURE (Astronomy). An arrangement of the earth, sun, and another planet or the moon, in which the angle subtended at the earth between the sun and the third body, in the plane of the ecliptic, is 90°. The first and third quarters of the moon are positions of quadrature.

QUADRIC SURFACE. Any of the surfaces represented in Cartesian coordinates by an algebraic equation of the second degree in three variables

$$Ax^2 + By^2 + Cz^2 + 2Fyz + 2Gxz + 2Hxy + 2Pr$$
$$+ 2Qy + 2Rz + D = 0$$

Every plane section of a quadric surface is a conic section.

There are sixteen special cases, which in their standard forms can all be written as $f(x, y, z) = k$, as shown in Table 1. The conical and cylindrical surfaces, as well as the planes, are called degenerate quadrics. They are often called *conicoids*.

The 16 cases can also be classified by means of certain invariants of the general equation. Consider its discriminating cubic,

$$U(u) = u^3 - a_1u^2 + a_2u - a_3 = 0$$

where $a_1 = (A + B + C)$; $a_2 = (BC + CA + AB - F^2 - G^2 - H^2)$;

$$a_3 = \begin{vmatrix} A & H & G \\ H & B & F \\ G & F & C \end{vmatrix}$$

the discriminant of the second degree terms; Δ, the discriminant of the equation, a determinant of the fourth order obtained from a_3 by adding one more row and column composed from the elements P, Q, R, D. If the roots of the cubic are u_1, u_2, u_3 the cases are given in Table 2, where the

TABLE 1. CHARACTERISTICS OF QUADRIC SURFACES

Designation	$f(x, y, z)$	k	Surface
1	$x^2/a^2 + y^2/b^2 + z^2/c^2$	1	ellipsoid
2	Same as 1.	−1	imaginary ellipsoid
3	Same as 1.	0	imaginary conical surface
4	$x^2/a^2 + y^2/b^2 - z^2/c^2$	1	hyperboloid of one sheet
5	Same as 4.	−1	hyperboloid of two sheets
6	Same as 4.	0	conical surface
7	$x^2/a^2 + y^2/b^2$	1	elliptic cylindrical surface
8	Same as 7.	−1	imaginary cylindrical surface
9	Same as 7.	0	pair of imaginary planes
10	Same as 7.	−2z	elliptic paraboloid
11	$x^2/a^2 - y^2/b^2$	1	hyperbolic cylindrical surface
12	Same as 11.	0	pair of intersecting planes
13	Same as 11.	2z	hyperbolic paraboloid
14	y^2	4ax	parabolic cylindrical surface
15	Same as 14.	a	two parallel lines, real or imaginary
16	Same as 14.	0	two coincident planes

TABLE 2. CHARACTERISTICS FOR u_1, u_2, u_3

Δ	a_3	u_1	u_2	u_3	Surface
≠0	≠0	≠0	≠0	≠0	1, 2, 4, 5
≠0	0	≠0	≠0	0	10, 13
0	≠0	≠0	≠0	≠0	3, 6
0	0	≠0	≠0	0	7, 8, 9, 11, 12
0	0	≠0	0	0	14, 15, 16

numbers in the last column are the same as those used to designate the surfaces in Table 1.

When squared terms only appear in the equation, so that its form is $Ax^2 + By^2 + Cz^2 = 1$, one speaks of a central quadric. Thus the surface is symmetric about the coordinate origin. It could be an ellipsoid or a hyperboloid. The other quadric surfaces: a paraboloid and the degenerate quadrics are non-central quadrics. A system of three quadric surfaces having the same foci is a confocal quadric. Sections of these surfaces through their axes are confocal conics. The quadrics composing the system are represented by the equation

$$x^2/(A - q) + y^2/(B - q) + z^2/(C - q) = 1$$

where $A > B > C$ and q is a variable parameter. When q is negative or less than C, the surface is an ellipsoid; if $B > q > C$, a hyperboloid of one sheet; if $A > q > B$, a hyperboloid of two sheets; if $q > A$, the quadric is imaginary. The three surfaces are mutually orthogonal and are used as a curvilinear coordinate system called ellipsoidal coordinates. If two of the constants A, B, C become equal, quadrics of revolution result; if all three are equal, the quadric is a sphere.

See also **Algebraic Equations**; **Conical Surface**; **Conic Section**; **Ellipsoid**; **Hyperboloid**; and **Paraboloid**.

QUADRILATERAL. A polygon with four vertices and four sides. Special cases of it are named as follows: trapezium, no two sides parallel; trapezoid, two sides only are parallel; parallelogram, opposite sides are parallel; rectangle, all of its angles equal 90°; square, a rectangle with equal sides; rhomboid, all of its angles are oblique; rhombus, a rhomboid with equal sides.

See also **Polygon**.

QUADRIPLEGIA. Paralysis of both arms and both legs. When there is severe injury to the cord high in the neck, the patient usually dies quickly. If the patient survives, both arms and legs are usually paralyzed. Lower in the neck, a few muscles of the upper arm may be spared while the forearm and hand may be paralyzed as well as the legs. If damage is below this level, the legs only are paralyzed (*paraplegia*).

QUADRUPLE POINT. The temperature at which four phases are in equilibrium.

QUADRUPOLE. Let a collection of electric or magnetic charges be distributed around a point; for example, the center of mass of a system of atoms, molecules, or nuclei. The potential at a distance r from this point may be represented by an infinite series of terms in inverse powers of r. The term in the inverse first power is the Coulomb potential, the inverse second power term is the dipole potential, the inverse third power term is the quadrupole potential, etc. A typical example is an array of four charges of equal magnitude so spaced that they coincide with the vertices of a parallelogram. Charges located on opposite vertices are of the same sign; the distance of separation between charges is taken to be of the order of molecular or infinitesimal dimensions.

QUADRUPOLE MOMENT. When the radiation field due to a set of moving electric or magnetic charges is expanded in a series of powers of the product of the charges times space coordinates, the sum of the quadratic terms is the quadrupole moment.

QUADRUPOLE RADIATION. Radiation emitted by a quadrupole. Since the selection rules were deduced by analogy between the behavior of a classical electric dipole and the emission of radiation by quantum transitions, then there are arrangements of two dipoles, e.g., a linear arrangement in which the positive charges coincide and are located at the center of two negative charges separated by a distance $2x(x = $ distance from center), so $(\Sigma e_i x_i = 0)$ when the quadrupole moment $(\Sigma e_i \cdot x_i 2)$ is not zero but varies as $1/r^3$, and therefore acts as a source of radiation.

QUAGGA. See **Horses, Asses, and Zebras**.

QUAHOG. See **Mollusks**.

QUAIL *(Aves, Galliformes)*. Small compactly built game birds related to the partridges. See Fig. 1. Every continent has species of this name, but those of North and South America and those of the Old World belong to different divisions of the family. The widely distributed bob-white, *Colinus virgianus*, is the best known North American species. Some of the western species are known both as partridges and as quails, notably the valley quail of California, *Lophortyx californica*, and in the south the bob-white also receives the name partridge. Quails are swift-footed but their wings are short and they fly only short distances. Although their colors are mostly very quiet, they are beautiful birds. See also **Galliformes**; and **Poultry**.

Fig. 1. Quail (partridge, bob-white) *Colinus virginianus*. Above, mottled, reddish-brown and gray; under parts, white barred with black. White spots on head in male; yellow in female.

QUALITY CONTROL (Statistical). Maintaining product quality in accordance with acceptable standards has been a major role for industrial instrumentation since its inception decades ago. With the ever-growing interest in speeding up production, one becomes increasingly aware of the fact that rejects as well as acceptable products can be produced at very high rates. What constitutes product quality? Apparently, in the long run, it is that degree of excellence that the ultimate consumer demands at an affordable price. Obviously, manufacturers must compare cost versus quality acceptance in the marketplace. Based upon years of experience, the manufacturer learns what affordable quality really means in terms of market acceptance. Astute competitors also learn these basic facts.

With the foregoing knowledge, the manufacturer established quality standards, which then are reduced to engineering specifications for each product. But knowing in advance that there will be variations from exact specifications, acceptable variations must be established. These variations are then reduced to plus or minus tolerances.

The tolerable variations and the intolerable variations must be determined and, in an ideal situation, immediately fed back to production machines or processes. Adjustments made would again, ideally, affect every part or unit of substance being produced. But, because the target of acceptance is bracketed (±), a strictly go/no-go type of sorting does not suffice. Thus, at least as early as a century ago, the concept of statistical quality control (SQC) was introduced. The general intent of SQC is that of sampling units and parts being produced and essentially determining trends in deviation from production as continuously (affordable and achievable) as possible. Since the early part of this century, the literature on SQC theory has continued to grow, and, because a full understanding of the concept involves rather intricate mathematics, most of the principles have been applied. But since the 1960s and continuing, the emphasis on SQC has shifted from essentially the exertion of manual controls and interpretation to the present semiautomation of data collection and interpretation. The mathematics have improved as the result of computer technology, including the development of algorithms and other shortcuts, and the interface between SQC and management has been streamlined by modern display technology, notably computer graphics, as well as by imbedded computer calculations.

Basic Assumptions of SQC

The basic assumptions of SQC are as follows:

1. Variations are inherent in any process. Inherent variations are not the primary target of SQC. If there are *only* inherent variations, the process is said to be in statistical control. It is assumed that inherent variations affect all measurements and that, over a period of time, these variations will stabilize.
2. Other variations are designated as correctable — that is, they are in the realm of statistical control.
3. A key element in the SQC concept is the familiar normal distribution (bell curve), shown and explained later.
4. In some interpretations of SQC, a further objective is considered. The acceptable tolerance limits (+ and −) are not the target to seek; those data within the + and − range should be further correlated so that the true "aim" of +0 is achieved as closely and as often as possible. For example, a manufacturer may specify that all products must fall within a given ± tolerance, but that the majority should be closer to the midrange of the tolerance — that is, as "near perfect" as possible. This differs from simply tightening the ± tolerance because it still allows less perfect units to escape full rejection.

SQC Glossary of Terms

An abridged glossary of terms used in SQC can be helpful. The following symbols are used in the next several paragraphs

N = number of data points
R = range
R = average range
s = standard deviation
x_i = value of specific data point
USL = upper specification limit
LSL = lower specification limit

Capability and Control (Concept of). A process is in control if the only sources of variation are common causes. The mean spread of such a process will appear stable and predictable over time. See Fig. 1.

The fact that a process is in control does not imply that it will yield only good parts — that is, parts within specification limits. Control only denotes a stable process. The size variation due to the common causes may be so large that some parts are outside the specification limits. Under these conditions, the process is said to be *not* capable. Capability is the ability of the process to produce parts that conform with engineering specifications.

Cp (Inherent Capability of Process). Cp is the ratio of the tolerance to 6σ. The formula is

$$C_p = \frac{\text{USL} - \text{LSL}}{6\sigma}$$

The Cp ratio is used to indicate whether a process is capable.

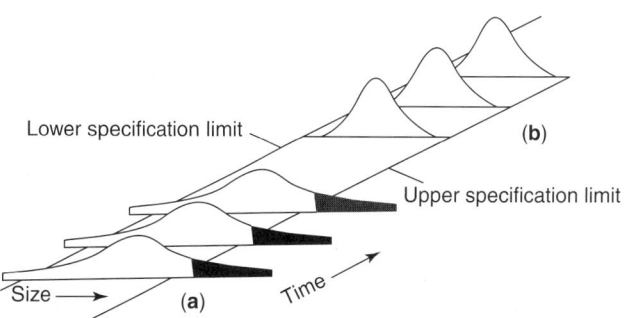

Fig. 1. Demonstration of process variations. (**a**) In control, but not capable. (**b**) Variations from common causes are excessive. (*Moore Products.*)

1.33 or greater	Process is capable.
1.0 to 1.3	Process is marginally capable; should be monitored.
1.0 or less	Process is *not* capable.

It should be noted that Cp does not relate the mean to the midpoint of the tolerances. If the mean is not at the midpoint, out-of-tolerance parts may still be probable, even if the process is capable. See Fig. 2.

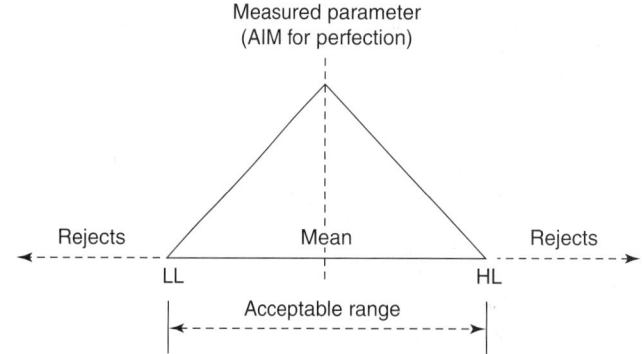

Fig. 2. Parts or units are acceptable if they barely achieve either of the specified limits (Low, LL; High, HL). In the quality control program, the manufacturer can gear the operation such that a majority of units will lie in the middle of the acceptable range. This is the basis for the AIM mode of operation.

Cpk (Capability in Relation to Specification Limits). Cpk relates the capability of a process to the specification limits — that is, Cpk equals the lesser of

$$\frac{\text{USL} - \text{mean}}{3\sigma} \quad \text{or} \quad \frac{\text{mean} - \text{LSL}}{3\sigma}$$

The Cpk value is useful in determining whether a process is capable and is producing good parts based on the specification limits. Values for the Cpk index have the following meaning:

Greater than 1.0	Both of the 6σ limits fall within the specification limits. The process is both capable and producing good parts (99.73 percent or greater).
1.0	At least one of the 6σ limits falls directly on the specification limits.
Less than 1.0	The mean is outside of the specification limits.

Capability Ratio. This ratio is the inverse of Cp — that is,

$$\frac{6\sigma}{\text{USL} - \text{LSL}}$$

The value of this ratio can be thought of as the portion of the part tolerance consumed by 6σ. A common interpretation for the various values of the capability ratio is as follows:

50 or less	Desirable
51 to 70	Acceptable
71 to 90	Marginal
91 and greater	Unacceptable

Common Cause. A source of random variation that affects all of the individual measurements in a process. The distribution is stable and predictable. (See also *Special Cause* in this glossary.)

```
      TEST CHART                 OVER SCALE        0
      PROC. "Q"    94.495%       +.019000          0
      SKEWNESS      +.00         +.017000          0
      KURTOSIS     +3.19         +.015000          0
                                 +.013000          1
                                 +.011000         51   X
      HI LIMIT     +.010000      +.009000        129   XX
      ----------------           +.007000        300   XXXXX
      ----------------           +.005000        960   XXXXXXXXX
      SAMPLE "N"    25442        +.003000       1590   XXXXXXXXXXXXXXX
            MEAN   -.003000      +.001000       2840   XXXXXXXXXXXXXXXXXXXXXXXXXX
      STD DEV      +.004346      -.001000       4350   XXXXXXXXXXXXXXXXXXXXXXXXXXXXXX
      +3 SIGMA     +.010040      -.003000       5000   XXXXXXXXXXXXXXXXXXXXXXXXXXXXXXXXX
      -3 SIGMA     -.016041      -.005000       4350   XXXXXXXXXXXXXXXXXXXXXXXXXXXXXX
            RANGE  +.032000      -.007000       2840   XXXXXXXXXXXXXXXXXXXXXXXXXX
                                 -.009000       1590   XXXXXXXXXXXXXXX
      LO LIMIT     -.010000      -.011000        960   XXXXXXXXX
      ----------------           -.013000        300   XXXXX
      ----------------           -.015000        129   XX
       +999999          12       -.017000         51   X
       -999999          13       -.019000          1
       -??????          22       UNDER SCALE       0
      TOTAL PARTS   25489
      TOTAL ON      H                          FXT    STATS    PRINT    CAL
                                  Date/Time    #1      ON       OFF     MODE
```

Fig. 3.　Facsimile of computer-generated histogram.

Control Limits (LCL and UCL). The upper and lower control limits are values used to determine whether or not a process is in statistical control. The values are inherent to the process and should not be confused with specification limits.

Histogram. A chart that plots individual values versus the frequency of occurrence and is used for statistical data analysis. Note that earlier, these diagrams were created manually. The diagrams now can be displayed on a CRT (computer graphics) and, of course, stored in memory. See Fig. 3.

Individual. A single measurement of a particular process characteristic.

Kurtosis. An indication of whether the data in a histogram have a normal distribution. Specifically, it is a measure of the "flatness" or "peakness" of a curve. The formula for kurtosis is

$$\sum_{i=1}^{\eta} \frac{(X_i - \overline{X})^4}{4s^4}$$

Values for kurtosis have the following meaning:

3　　　　　　　Normal distribution.
Less than 3　　Leptokurtic curve — that is, the curve has high peak (data are concentrated close to the mean).
Greater than 3　Platykurtic curve — that is, the curve has low peak (data are dispersed from the mean).

Mean X. The value of the middle individual when the data are arranged in order from lowest to highest.

Mean X. The arithmetic average value of the data. The process mean is the average of all the process data. The subgroup mean averages just those values in the subgroup.

Median $\widetilde{X}$. The value of the middle individual when the data are arranged in order from lowest to highest. If the data have an even number of individuals, the median is the average of the two middle values.

Mode. The most frequently occurring value — that is, the highest point on a histogram.

Normal Distribution. Data often are summarized graphically as a means of better understanding and analyzing the variation. A plot of the frequency of occurrence of a particular variable is one of the common tools used for analysis. If a definite pattern emerges from the data, this plot is referred to as a distribution.

Many distributions have been identified and named since their pattern of variation is repeatable and certain mathematical characteristics can be defined for each distribution. One of the most commonly occurring patterns of distribution is the normal distribution. It describes many natural and other phenomena encountered in the field of statistics.

The mean and the standard deviation define a specific normal distribution. Knowing the values of the normal distribution, the total spread of expected outcomes can be predicted. It is important to note that 99.73 percent of the population lies between −3 standard deviations (often called −3σ) from the mean and +3 standard deviations (+3σ) from the mean. This fact is the basis for many statistically calculated indications of the status of a process, including capability and control limits.

It is equally important to note that if the distribution of a process is not normal, a number other than 99.73 percent of produced parts or units will fall within 6 standard deviations. Because of this, many of the statistical control indicators calculated from a non-normal distribution will not describe that distribution accurately. Often a distribution is sufficiently close to being normal, however — that is, the errors are insignificant. See Fig. 4.

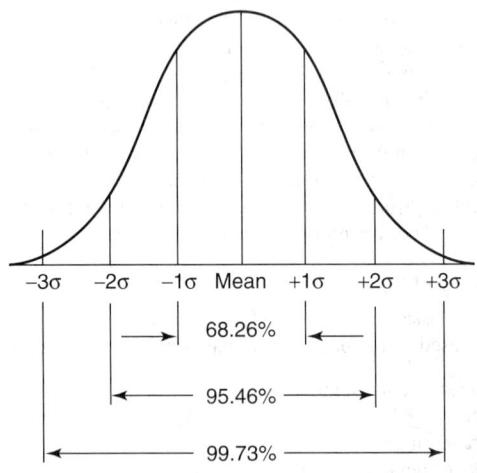

Fig. 4.　Normal distribution chart. This is basic to the formal statistical quality control (SQC) concept. The curve is useful for many other statistical purposes in other fields, such as variations in a population of people (clothing sizes, for example). The concept dates back to Laplace and Gauss and is sometimes referred to as gaussian distribution.

Pareto Chart. This type of chart ranks events according to a specified parameter (Fig. 5), such as rejected individuals by frequency of occurrence.

Process Quality. The process quality or process capability coefficient denotes the area under the normal curve that falls within the specification limits. It is expressed as a percentage. The calculation is a multistep procedure.

1. Calculate the distance of the mean from the specification limits in terms of standard deviations:

$$L_U = \frac{\text{USL} - \overline{X}}{s} \qquad L_L = \frac{\overline{X} - \text{LSL}}{s}$$

```
            GAGED    8550          x x x x x x x x x x x x x x x x x x x x x x x x x x x x
         ACCEPTED    7835   91.6   x x x x x x x x x x x x x x x x x x x x x x x x x
         REJECTED     715    8.4   x x x
```

```
        LABELS     OVER          UNDER         SCALE AS % OF REJECTED
        N . N    400   55.9    315   44.1   x x x x x x x x x x x x x x x x x x x x x x x x x x
        P   P    915   44.1    315   44.1   x x x x x x x x x x x x x x x x x x x x x x x x x
        3   0    500   69.9                 x x x x x x x x x x x x x x x x x x x x
        A   A    100   14.0    390   54.5   x x x x x x x x x x x x x x x x x x
        G   G    333   46.8    111   15.5   x x x x x x x x x x x x x x x x x
        M   M    201   28.1    203   28.4   x x x x x x x x x x x x x x x x
        O   O    190   25.2    180   25.2   x x x x x x x x x x x x x x
        E  DIA E  312   43.6                x x x x x x x x x x x x
        O   O     68    9.5    199   27.8   x x x x x x x x x x x
        L   L    180   25.2     41    5.7   x x x x x x x x x
        I   I                  170   24.9   x x x x x x x
        H   H    100   14.0     34    4.8   x x x x x
        F   F     43    6.0     47    6.6   x x x x
        B   B     20    2.9     24    3.4   x x
        J   J                    1     .1
        K   K      1     .1
```

```
        TEST ON      REJECT      XR      STATS     PRINT     TEST
        Date/Time                 0      ON        OFF       MODE
```

Fig. 5.   Facsimile of computer-generated pareto diagram. Listings include total parts gaged, parts accepted, and parts rejected, with the parameters for which parts are rejected in descending order of frequency. This type of diagram stresses the relative importance of corrections to be made.

(For this equation to be valid, the values for L_U and L_L must be greater than or equal to 0. If either is negative, the corresponding area is calculated using $A = 1 - A'$, where A' is obtained using $L' = -L$.)

2. Determine the area outside of the specification limits, expressed as a fraction of the normal curve:

$$A_{\text{upper}} = \frac{1}{\sqrt{2}} \cdot e^{-(L_U^2/2)} \cdot [B_1 Y_U + B_2 Y_U^2 + B_3 Y_U^3 + B_4 Y_U^4 + B_5 Y_U^5]$$

$$A_{\text{lower}} = \frac{1}{\sqrt{2}} \cdot e^{-(L_U^2/2)} \cdot [B_1 Y_L + B_2 Y_L^2 + B_3 Y_L^3 + B_4 Y_L^4 + B_5 Y_L^5]$$

where
$$Y_U = \frac{1}{1 + R \cdot L_U}$$
$$Y_L = \frac{1}{1 + R \cdot L_L}$$
$$R = 0.2316419$$
$$B_1 = 0.31938153$$
$$B_2 = -0.356563782$$
$$B_3 = 1.781477937$$
$$B_4 = -1.821255978$$
$$B_5 = 1.330274429$$

(When a tolerance limit is unused, that is, has a U prefix, the area under the normal curve that is outside that limit does not enter into the process quality calculation.)

3. Then the process quality is found from the equation:

$$\text{Process quality} = [1 - (A_{\text{upper}} + A_{\text{lower}})]\,100\%$$

Range R. The difference between the highest and lowest values in the group. The average of subgroup ranges is denoted by $\overline{R}$.

Run Chart (xR Chart). A chart that displays the most recent individual measurement x and the absolute difference from the previous measure R on a consecutive basis.

Sample. A synonym of *subgroup* in process control applications. However, sometimes "sample" is used to denote an individual reading within a subgroup. Because this can be confusing, "subgroup" is the preferred term.

3 Sigma. $\pm 3\sigma$ are the two specific points on a normal distribution centered about the mean. 99.73 percent of the population will fall between these values. Since this is essentially the entire population, $+3\sigma$ and -3σ represent the probable range of variation:

$$+3\sigma = X + 3s$$

$$-3\sigma = X - 3s$$

Skewness. An indication of whether the data in a histogram have a normal distribution. Specifically, skewness is a measure of symmetry. Values for skewness have the following meaning:

0   Symmetrical distribution.

Greater than 0   Positive skewness—that is, the distribution has a "longer tail" to the positive side. The median is greater than the mode.

Less than 0   Negative skewness—that is, the distribution has a "longer tail" to the negative side. The median is less than the mode.

The formula for skewness is

$$\sum_{i=1}^{\eta} \frac{(X_i - \overline{X})^3}{3s^3}$$

Another version of this formula is

$$\frac{\eta^2 \sum_{i=1}^{\eta} X_i^3 - 3\eta \sum_{i=1}^{\eta} X_i \sum_{i=1}^{\eta} X_i^2 + 2\left(\sum_{i=1}^{\eta} X_i\right)^3}{\sqrt{\eta \sum_{i=1}^{\eta} X_i^2 - \left(\sum_{i=1}^{\eta} X_i\right)^2}}$$

Special Cause. A source of nonrandom or intermittent variation in a process. The distribution is unstable and unpredictable. It is also referred to as an assignable cause. (See also **Common Cause** in this glossary.)

Specification Limits (LSL and USL). The upper and lower specification limits (engineering-blueprint tolerances) are values that determine whether or not an individual measurement is acceptable. They are engineering tolerances established external from the process and should not be confused with control limits.

Standard Deviation. A measure of the variation among the elements in a group. The formula for the standard deviation is

$$s = \sqrt{\sum_{i=1}^{\eta} \frac{(X_i - \overline{X})}{\eta - 1}}$$

Another version of this formula is

$$\sqrt{\frac{\sum_{i=1}^{\eta} X_i^2 - \left[\left(\sum_{i=1}^{\eta} X_i\right)^2 \Big/ \eta\right]}{\eta - 1}}$$

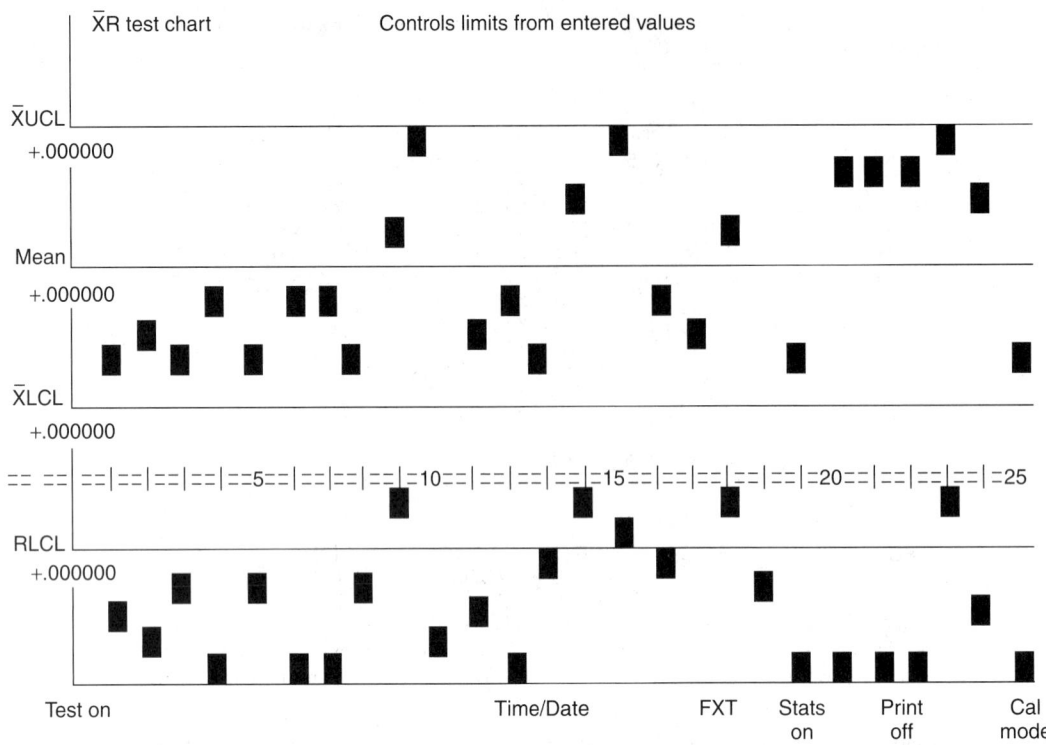

Fig. 6.   Facsimile of computer-generated $\widetilde{X}R$ chart.

Subgroup. A group of individual measurements, typically 2 to 10, used to analyze the performance of a process.

Variation Concept. Variation occurs in manufacturing because the process conditions are never exactly identical for any two parts or units. If variation did not exist, quality would not be a problem. Every piece or unit would be identical—so in the case of parts and pieces, all components would assemble and test correctly. Inasmuch as differences do exist, the goal of SQC is to reduce the variation as much as possible, thereby improving product consistency and quality.

$\overline{X}R$ Chart. A chart that displays the subgroup mean and range on a consecutive basis. Upper and Lower control limits are plotted to help analyze the process. See Fig. 6.

Note that SQC data are plotted in various columnar forms.

System Approach to SQC

Modern instrumentation and computer technology have eliminated or can eliminate manually entered and manual charting by way of tremendously speeding up SQC data acquisition and the numerous computations required. SQC can be made more encompassing and penetrating. Histograms in electronic storage, and thus easily called up on graphic displays, essentially eliminate paperwork. The remaining areas—namely, the interpretation and corrective action aspects—can be further speeded up through the application of further automation technology. Much of the hardware is available to create SQC systems of varying extent and complexity. In applying advanced instrument and computer technology to SQC systems, there will be a tendency, as occurred in the past in attacking a new area for instrumentation, to overdo in some instances. The established guidelines still apply—that is, do *not* collect more data than are required. A flood of essentially useless information taxes the hardware, the software, and the personnel who are confronted with it.

SQC is very easy to identify with controlling parts. It is somewhat more difficult to implement as the parts become subassemblies and the latter, in turn, become finished products ready for shipment. It is in the areas of subassembly and final assembly testing where much remains to be learned and achieved.

SQC in the Process Industries

The fundamentals of SQC also are applied in the continuous-process industries, but, because of the marked differences in the characteristics of the end products and of the variables that are measured and controlled, there are some differences in the application of SQC. This has led to the general acceptance of the term *statistical process control* (SPC) in these industries.

Examples of Statistical Quality Control

For several years, *quality control* has generally connoted statistical quality control. It is a tool of industrial management comparable with production control and cost control. When effectively applied, quality control commences with raw materials, as received or processed into parts and components. A target of most manufacturers is that of inspecting parts and subassemblies at several points during manufacture in order to make necessary corrections in the early stages of production rather than in the later, more costly phases.

All items or products fall into three classifications. There are (1) those items which have measurable characteristics or variables, (2) those items which are acceptable or rejectable because of either good or bad attributes, and (3) those which *can* have an infinite number of defects, but which *usually* have only a few observable ones.

As an example of the inspection for variables, consider an inspector who is weighing boxes of powder. The specification states that the net weight must be 340 grams (about 12 ounces) plus 10 grams minus nothing. Every half-hour, the inspector takes a sample of five boxes from the packaging machine, weighs the contents, and records the values. Then he computes the average of the five and notes the range of the sample (the difference between the lightest and the heaviest contents). Either from previous data or from six such samples, he computes the average range and multiplies by a constant to find the control limits within which the average weight of each sample must lie. He multiplies the average range by another constant and learns what extreme individual weights might be expected if he were to weigh several hundred boxes. If these lightest and heaviest anticipated weights lie within the specified Figures, the machine is *capable* of doing its work and, if the average of each sample lies within the control limits, the machine is *actually* doing its work. In this example the grand average of all samples should remain very close to 345 grams.

When a process is not capable of working within the tolerances, either the process or the tolerances must be changed. If a machine or process does not work within its control limits, there is some assignable cause of variation, which must be found and eliminated.

As an example of the inspection for attributes, consider an inspector who is checking small electrical devices for "grounds" or electrical paths between the windings and the metal frame. This defect could result from either faulty material or careless workmanship in assembly. During each

2-hour period, the inspector takes 100 units from the assembly line and makes a quick check on each to determine the presence or absence of a ground fault. He records the total number of items tested and the number of defectives found. This enables him to compute the average percent defective for the process over a period when the rate of production is steady. By the application of statistics, he then computes control limits within which the percent defective for each group of 100 items should lie. Any group showing more or less defective items must be investigated because it indicates an assignable cause of deviation. This cause should be discovered whether it makes the process better or worse.

As an example of the indeterminate sample size where the possible number of defects is large, consider the "squawk sheet" for a large airplane. The final inspectors of the finished plane fill out this sheet which lists all the faults that are observed. The statistical department analyzes these sheets to determine whether the over-all assembly process is getting better or worse. It also notes the frequency with which various faults occur. The total number of faults must follow a random fluctuation within certain limits, which are computed by statistical methods. Lack of this random fluctuation may reveal improper performance of the inspectors. These limits may shift slowly with time if the quality of the assembly work changes.

A few of the special terms used in statistical quality control are the following: *Average Outgoing Quality Level* (AOQL) is the proportion of acceptable items (or sometimes the complementary proportion of defectives) which emerges after inspection and any necessary rectification. *Acceptable Quality Level* is the proportion of effective units in a batch which is regarded as desirable by the consumer of the batch; the complement of the proportion of defectives which he is willing to tolerate. *Consumer's Risk* is the risk a consumer takes that a lot of a certain quality q will be accepted by a sampling plan. It is usually expressed as a probability of acceptance and depends, of course, on q as well as the sampling plan itself. *Producer's Risk* is the risk a producer takes that a lot will be rejected by a sampling plan even though it conforms to requirements. *Acceptance Number*, in sequential sampling, is the number of defective items (dependent on the sample number) which, if attained, requires the acceptance of the batch under examination. It is usually accompanied by a *rejection number*, which is the number of defectives requiring rejection of the batch. If the number of defectives at any stage is above the acceptance and below the rejection number sampling is continued. The *number of allowable defects* in a sample is the critical number (designated by the particular scheme of sampling inspection) such that, if this number is exceeded, the whole of the remainder of the inspection lot must be examined or the lot rejected out of hand. The term *batch variation* applies to the variations in a product that is made or examined in batches, and distinct from one which is produced or examined continuously. The batch variation may be made up of variation within each batch, due to the ordinary process of manufacture, and variation between batches, which may also be due to the quality of raw materials used.

QUANTIC. A homogeneous algebraic function of two or more variables, in general containing only positive integral powers of the variables, and usually a polynomial in several variables. Quantics are classified into quadratic, cubic, quartic, quintic, etc., according to degree, and into binary, ternary, quaternary, etc., according to the number of variables involved.

QUANTILE. If a set of n observations forming a frequency distribution are arranged in order of magnitude and every (n/p)th observation is marked off, these observations are called quantiles. Special cases are (1) $p = 2$, the central observation, called the median; (2) $p = 4, 5, 6, 10$ giving the quartiles, quintiles, sextiles, deciles; (3) $p = 100$, giving the centiles or percentiles.

QUANTIZATION. When the range of a variable is divided into a finite number of distinct, nonoverlapping subranges, each of which is assigned a value within the subrange, the variable is said to be quantized. The process of quantizing is known as *quantization*. The distinct subranges sometimes are called *quanta*. The magnitude of the subranges need not necessarily be equal. Where the widths are nominally equal, they may be termed *quantization units*. Where they are not equal by intent, the term *quantization steps* is used. The degree used in temperature measurement is a quantization unit. A slide rule, which performs quantization in terms of a nonlinear device, where the subranges (considering the range of numbers 1 and 10) are unequal, involves quantization steps.

It should be stressed that quantize is not synonymous with digitize or digitalize. As related to digital data acquisition and instrumentation systems, electronic devices that perform quantization are analog-to-digital converters (encoders). Interpreted broadly, quantization also would apply to the assignment of discrete values to a quantity obtained by reading the position of a pointer on a calibrated scale. Most quantizers of analog-to-digital converters are linear. Thus, the discrete subranges are properly called quantization units. Inasmuch as most of this equipment is based upon binary representations, the common term for the quantization unit is the bit. In a 10-bit converter, the least significant bit is the quantization unit, which is equal to $V_{fs}/2^{10}$, where V_{fs} is the full-scale input to the A/D converter.

QUANTIZATION OF SIGNALS. A process in which the range of values of a wave is divided into a finite number of smaller subranges, each of which is represented by an assigned or "quantized" value within the sub-range. "Quantized" may be used as an adjective modifying various forms of modulation, for example, quantized pulse-amplitude modulation.

QUANTIZED PULSE MODULATION. Pulse modulation, which involves quantization of signals. This is a generic term, including pulse-numbers modulation and pulse-code modulation as specific cases.

QUANTUM. As stated in the entry on **Quantization**, an observable quantity is said to be quantized when its magnitude is, in some or all of its range, restricted to a discrete set of values. If the magnitude of the quantity is always a multiple of a definite unit, then that unit is called the quantum (of the quantity). For example, the quantum or unit of orbital angular momentum is $h(= h/2\pi)$, and the quantum of energy of electromagnetic radiation of frequency, v is hv, where h is Planck's constant. In field theories, a field (or the field equations) is quantized by application of a proper quantum-mechanical procedure and this results in the existence of a fundamental field particle, which may be called the field quantum. Thus the photon is a quantum of the electromagnetic field and in nuclear field theories, the meson is considered to be the quantum of the nuclear field.

QUANTUM CHEMISTRY. The use of the principles of quantum mechanics for the resolution of problems in chemistry, notably in connection with the electronic structure of molecules. Some authorities attribute the beginnings of this field to James and Coolidge who, as early as 1933, theorized on the molecular structure of hydrogen and, in these efforts, demonstrated that the Schrödinger equation (primarily proposed for atoms) could be applied to molecules. Over a period of years, studies by other researchers followed. In 1968, for example, Kolos and Wolniewicz carefully investigated the dissociation energy of the hydrogen molecule. But, as early as 1960, some investigators in quantum chemistry turned their attention to methylene (CH_2) as the molecular target of choice. The predictive powers of computational quantum chemistry have since been demonstrated in connection with other molecules. Schaefer (1986) suggests, however, that methylene is the paradigm for computational quantum chemistry. In his paper, three important roles for quantitative theory are outlined: (1) theory precedes experiment, (2) theory overturns experiment, as resolved by later experiments, and (3) theory and experiment work together to gain insight that is afforded independently to neither.

Additional Reading

Goddard, W.A.: "Theoretical Chemistry Comes Alive: Full Partner with Experiment," *Science*, **227**, 917–923 (1985).

Levine, I.N.: "Quantum Chemistry," Prentice-Hall, Inc., Upper Saddle River, NJ, 1999.

Lowdin, Per-Olov, E. Brandas, J. Sabin, and Mike Zerner: "Advances in Quantum Chemistry," Vol. 39, Academic Press, Inc., San Diego, CA, 2001.

Roberts, M.W.: "Chemistry in Two Dimensions," *Review (University of Wales)*, 58 (Autumn 1987).

Schaefer, H.F.: "Methylene: A Paradigm for Computational Quantum Chemistry," *Science*, **231**, 1100–1107 (1986).

Stucky, G.D. and J.E. MacDougall: "Quantum Confinement and Host/Guest Chemistry: Probing a New Dimension," *Science*, 669 (February 9, 1990).

Veszpremi, T. and M. Feher: "Quantum Chemistry: Fundamentals to Applications," Kluwer Academic Publishers, Norwell, MA, 1999.

Warren, W.S., Rabiz, H., and M. Dahleh: "Coherent Control of Quantum Dynamics: The Dream is Alive," *Science*, 1581 (March 12, 1993).

Wasserman, E. and Schaefer H.F. : "Letters — Methylene Geometry," *Science*, **233**, 829–830 (1986).

QUANTUM EFFICIENCY.   A measure of the efficiency of conversion or utilization of light or other energy, being in general the ratio of the number of distinct events produced in a radiation sensitized process to the number of quanta absorbed (the intensity-distribution of the radiation in frequency or wavelength should be specified). In the photoelectric and photoconductive effects, the quantum efficiency is the number of electronic charges released for each photon absorbed. For a phototube, the quantum efficiency is defined as the average number of electrons photometrically emitted from the photocathode per incident photon of a given wavelength. In photochemistry, the quantum efficiency or yield is the ratio of the number of molecules transformed to the number of quanta of radiation absorbed.

QUANTUM ELECTRODYNAMICS.   A quantized field theory of the interaction between electrons, positrons and radiation based on the quantized form of the Maxwell equations and the Dirac electron theory. The theory is characterized by its remarkably accurate predictions (see also **Positronium**) and its meaningless results. The latter arise from divergent integrals that appear in the development of the theory by perturbation techniques based on expansion in powers of the fine structure constant. These divergences may be pictured in terms of the model of a vacuum as consisting of an infinite sea of negative energy electrons, since the introduction of a charge into this distribution causes infinite currents to be induced.

In 1948, techniques introduced by Schwinger and Feynman enabled these difficulties to be avoided, without being removed. Their relativistically covariant development of the theory allowed such infinite terms to be treated unambiguously, and in particular terms which are to be understood as electrodynamic contributions to the charge and mass of a particle were put in a form which is invariant under Lorentz transformations. The program of charge renormalization and renormalization of mass then enabled such terms to be related to the experimentally observed charge and mass of the particle. See also **Quantum Mechanics**.

QUANTUM MECHANICS.   The wave theory of light as originally developed by Maxwell in the 1860s became well established, but it did not accommodate certain phenomena. For example, experiments on thermal radiation uncovered gross disagreement or contradiction with classical theories. The equilibrium distribution of electromagnetic radiation (i.e., emission and absorption of radiation at constant temperature) in a hollow cavity could not be explained on the basis of classical electrodynamics (Maxwell's equations plus the laws of motion of particles). Thermal radiation is a certain function of the temperature (T) of the emitting body. When dispersed by a prism, thermal radiation forms a continuous spectrum. It was found that the energy distribution of the radiation had a regular dependence on its wavelength. Furthermore, the energy E_v as a function of the temperature of the material did not depend upon the structure of the cavity or its shape. On these bases, it was shown that the energy E_v should have a functional dependence upon frequency v, at temperature T, in the form:

$$E_v = v^3 F\left(\frac{cT}{v}\right)$$

All attempts to find the correct form of the function F on the basis of classical theory failed. The classical theory led to the now well-known "ultraviolet catastrophe," since the contribution of high frequencies caused the energy to assume an infinite value. The difficulty was removed by a hypothesis of Planck, according to which the energy of a monochromatic wave with frequency n can only assume those values which are integral multiples of energy hv, i.e., $E_n = nhv$, where n is an integer referring to the number of "photons." Thus the energy of a single photon of frequency v is:

$$E = hv \qquad (1)$$

The finiteness of Planck's constant h and its resulting implications laid the foundations of quantum theory. Quantum theory, like the special theory of relativity, was discovered through experiments on electromagnetic phenomena and their theoretical interpretations.

The fundamental equation of quantum mechanics, Eq. (1), implies, on the one hand, that energy of radiation stays concentrated in limited regions of space in amounts of hv and, therefore, behaves like the energy of particles. On the other hand, it establishes a definite relationship between the frequency v and the energy E of an electromagnetic wave. This dual behavior of light corresponds, in one way, to experimental situations of the interference properties of radiation, for the description of which one uses the wave theory of light. In another way, it corresponds to the properties of exchange of energy and momentum between radiation and matter, which require for their explanation the particle picture of light. Thus, the dual behavior of light has necessitated the quantum description (quantization) of the electromagnetic field. A unified point of view was formulated quantitatively by de Broglie, according to which all forms of energy and momentum related to matter will manifest a dual behavior of belonging to a wave or particle description of the physical system, depending upon the type of experiment performed.

The most interesting example of a quantum mechanical object is the photon itself. By using the relativistic and quantum mechanical definition of the photon energy, we can obtain a quantitative formulation of the concepts just described. The relativistic form of the total energy of a particle with rest mass m and momentum ρ is:

$$E = c(\rho^2 + m^2 c^2)^{1/2} \qquad (2)$$

We set $m = 0$ and obtain the relativistic definition of the energy of a photon:

$$E = c\rho \qquad (3)$$

Hence the first unification of relativity and quantum theory originated from the combination of Eqs. (1) and (3) in the form

$$c\rho = hv \qquad (4)$$

By using $v\lambda = c$ for the plane electromagnetic wave, we obtain the fundamental statement of quantum mechanics:

$$\lambda \rho = h \qquad (5)$$

valid for all particles with or without mass, where

$$\lambda = \frac{\lambda}{2\pi}; \quad = \frac{h}{2\pi} \qquad (6)$$

These assumptions of quantum theory have laid the foundations of new physical and philosophical concepts for the process of measurement in physics and the definition of physical reality.

It is necessary to develop a dynamic theory to describe the wave character of material particles. In the case of particles with mass, one has the possibility of comparing their kinetic energies with their rest masses. If the kinetic energy is small compared to rest energy, then we can formulate a nonrelativistic theory. However, with the photon there exists no possibility for the formulation of a nonrelativistic theory. There are important advantages in entering quantum mechanics via the photon:

1. The energy of a photon is a quantum mechanical quantity, $E = hv$.
2. It has provided a natural basis to postulate the wave-particle relation, $\lambda = h\rho$.
3. The wave aspects of the photon are completely described by charge-free Maxwell equations. Therefore, it is natural to try to reconcile Planck's hypothesis with the wave theory of light.

During 1979, scores of distinguished scientists reviewed the accomplishments of Albert Einstein who was born a century earlier (March 14, 1879). Many papers describing and reviewing the works of Einstein were prepared in honor of the centenary of Einstein's birth. See also **Gravitation**; and **Relativity and Relativity Theory**. Among Einstein's accomplishments, three were cited by Viktor Weisskopf, one of the speakers at the Pontifical Academy of Science (Vatican City) on November 10, 1979 at a special session devoted to Einstein. The relation between waves and particles was given as one of these, the other two, topically related, were special and general relativity. In commenting on Einstein's interest in wave-particle duality, Weisskopf made the following observations.

... Einstein's idea started a truly revolutionary development in physics: quantum mechanics. It opened up wide new horizons and clarified many outstanding problems in our view of the structure of matter. Quantum mechanics is based on the idea of wave-particle duality. Einstein first applied this idea to the nature of light, but it was soon applied to the nature of elementary entities such as electrons and other constituents of matter.

The idea was that all these entities exhibit both wave and particle properties. This double nature taxes our imagination: few things differ as much as a beam of particles and a running wave. In a beam of particles, matter is concentrated in small units, whereas a wave spreads continuously over space. Still, wave and particle properties are observed for electrons and other fundamental entities.

The wave nature of electrons explains so many previously unexplained facts for the following reason. If waves are confined to a finite region of space, they form characteristic shapes and patterns that are specific to the nature of the confinement. [Figure 8 in the entry on **Chemical Elements** shows waves in space confined to the neighborhood of a central point.] Only those and no other patterns can develop in this sort of confinement. But this is just the confinement that electrons suffer when they are confined around the atomic nucleus by electric attraction. The electron waves in atoms must assume some of these patterns. The simple patterns are "lower" than the more complex ones; they are lower in energy. Indeed, the electrons in an atom assume the lowest possible patterns.

This is the explanation of the stability of atoms—it takes energy to change to the next higher pattern. For example, the energy of molecular collisions in air is not sufficient to change the electron patterns in oxygen. Thus oxygen survives unchanged the many millions of collisions in air.

The typical shapes of the electron patterns determine the specific properties of atoms. For example, in the oxygen atom the electrons fill the lowest patterns up to the fourth one. The resulting pattern combination is characteristic for oxygen and is responsible for its properties; it determines how oxygen combines with other atoms (forming water with hydrogen, for example) and how the atoms fall into a symmetrical crystalline order when they form solids, such as ice crystals.

The electron patterns are the primal shapes of nature. Fundamentally, all of nature's shapes can be traced to such patterns. Even the properties of living substances are based on them—in particular, the properties of the molecules that carry the hereditary code. In the final scientific analysis, the stability of electron wave patterns causes the same flowers to bloom every spring and makes children similar to their parents.

Einstein started this great development as early as 1905 by an almost unimaginable act of vision, when he concluded that the concept of such an electromagnetic wave does not suffice to explain important properties of light. He drew the revolutionary conclusion that there must exist light-particles, the photons. The particle-wave duality was born. Einstein recognized the fertility of his idea, but he was never completely satisfied with the conceptual basis of quantum mechanics. The lack of complete causality and the frequent use of probability instead of certainty were always a matter of deep concern for him.

The next great development in physics was again an outgrowth of Einstein's ideas. Dirac was not satisfied with the fact that early quantum mechanics did not fit into the framework of relativity theory. The velocities of electrons in ordinary atoms are so small compared to the speed of light that the neglect of relativity theory did not matter much. But what about wave mechanics of particles that move much faster? Dirac was able in 1927 to unite relativity with quantum mechanics.

In so doing, Dirac discovered a new symmetry in nature, the matter-antimatter symmetry. He discovered it, not by experimenting, but solely by putting together the two great ideas of Einstein: the space-time unity of relativity and the wave-particle duality of quantum mechanics. Dirac saw that for every particle there must be an antiparticle with opposite charge. Although in our own environment we find only negatively charged electrons and protons with positive charge, which is ordinary matter, Dirac concluded that nature must also admit the opposite side. Such anti-matter, he predicted, would not be stable in the presence of ordinary matter; it would annihilate when in touch with it; in a sort of explosion where the masses would be transformed into energy—a direct manifestation of Einstein's equivalence of mass and energy.

A few years later the antielectron was found, and almost 30 years later, the antiproton. Antimatter indeed exists in nature, as Dirac predicted from Einstein's work. This theoretical prediction was one of the greatest intellectual achievements of science. Today, beams of antimatter are produced in many laboratories; they run in carefully evacuated tubes in order not to hit any ordinary matter until they reach their target, where they annihilate with the target substance.

Also, at the aforementioned Pontifical Academy Session on Einstein, P.A.M. Dirac observed:

> By 1905, the wave theory of light based on Maxwell's equations was well established, but certain phenomena would not fit in. It seemed that emission and absorption of light occur discontinuously. This

led Einstein to the view that the energy is concentrated in discrete particles. It was a revolutionary idea, very hard to understand, as the successes of the wave theory were undeniable. It seemed that light had to be understood sometimes as waves, sometimes as particles, and physicists had to get used to it. The idea was incorporated into Bohr's theory of the hydrogen atom and forms an essential part of it.

The statistics of an assembly of light particles was studied by Bose, who found that ordinary statistics was not applicable. The laws for the new statistics were formulated jointly by Bose and Einstein. By studying an atom in statistical equilibrium, Einstein saw the necessity for the phenomenon of stimulated emission of radiation. This effect is, in the first place, extremely small, but it can be very much enhanced with a suitable apparatus, because of the new statistics. This led to the laser, a useful tool in present-day technology, which we owe to Einstein.

The appearance of waves connected with particles was shown by de Broglie to be applicable to all particles, not just those having the velocity of light. De Broglie worked out the mathematical relations between waves and particles, using only the requirements of special relativity. He found that the waves move faster than light. However, they cannot be used to transmit signals faster than light, which is an important feature of special relativity.

De Broglie's theory was extended by Schrödinger and led to wave mechanics, which is fundamental for modern atomic theory. Here again, we have a long line of development of physics, originated by Einstein.

In 1926, the Schrödinger equation described the motion of the de Broglie phase waves under the influence of an externally applied potential, and the physical significance of the phase wave ψ was recognized particularly by Born by identifying $\psi^*(q_k)\psi(q_k)d\tau$ with the probability of finding the system in the element of configuration space $d\tau$ between q_1 and $q_1 + dq_1$, etc. Independently in 1925 Heisenberg developed a calculus of observable quantities, representing dynamical variables such as momentum, position, etc., by means of matrices, the time rate of change of a variable X being given by $i\,X = XH - HX$ where H is the Hamiltonian of the system. This formulation (matrix mechanics) of quantum theory is equivalent to the Schrödinger formulation (wave mechanics). However, it emphasizes the role played by the observer in the measurement of a physical quantity, and the fact that natural limits imposed on measurements he makes must be incorporated into a theory which purports to describe such measurements. Thus in particular to specify the momentum p and corresponding position x of a particle is strictly speaking not legitimate since the very measurement of the one will lead to an unpredictability of the other given by the Heisenberg indeterminacy relation $\Delta x \Delta p <$. Dynamical variables which cannot be measured simultaneously with arbitrary precision are thus represented by matrices, or, more generally, operators, which may not commute, while a system in the state ψ has a definite value for the dynamical variable A if ψ is an eigenfunction of the operator A, i.e., $A\psi = a\psi$ (a = number). Thus if A and B do not commute ψ cannot be at once an eigenfunction of both A and B. A system in an eigenstate of energy is thus described by the equation $H\psi = E\psi$ where H is the Hamiltonian of the system. In the Schrödinger representation (wave mechanics) ψ is regarded as a function of position and time and the momentum p appearing in the Hamiltonian is represented by the operator—i grad, which automatically yields the commutation relation $p_i x_j - x_j p_i = -i\delta_{ij}$. In the Heisenberg representation (matrix mechanics) the position and momentum are represented by matrices which satisfy this commutation relation, and ψ by a constant vector in Hilbert space, the eigenvalues E being the same in two cases.

The Hamiltonian of a particular system is formally identical with that of the classical theory, the simplest, for one particle of mass m moving in a potential V, being

$$H = \frac{p^2}{2m} + V = E$$

which in the Schrödinger representation gives the Schrödinger equation

$$\left(-\frac{\hbar^2}{2m}\nabla^2 + V\right)\psi = E\psi$$

In the presence of a magnetic field B derived from the vector potential A it is necessary to replace

$$\mathbf{p} = -i\hbar\nabla$$

by

$$\mathbf{p} - \frac{e}{c}\mathbf{A} = -i\hbar\left[\nabla - \frac{ie}{hc}\mathbf{A}\right]$$

and in addition to note the contribution $-\mu \cdot B$ to the energy arising from the magnetic moment μ of the particle. The vector μ is itself an operator, being, for an electron, $(e/mc)\mathbf{S}$ where $\mathbf{S}$ is the electron spin, $\mathbf{S} = (h/2)\sigma$ the components of σ being the Pauli spin operators.

The value e/mc for the electron gyromagnetic ratio was first postulated by Uhlenbeck and Goudsmit and later shown to be a consequence of the Dirac electron theory.

Nonrelativistic quantum mechanics, extended by the theory of electron spin and by the Pauli exclusion principle, provides a reliable theory for the computation of atomic spectral frequencies and intensities, of cross sections for scattering or capture of electrons by atomic systems, of chemical bonds and many properties of solids, including magnetic properties, although with much more complicated systems it has not always proved possible to develop with adequate accuracy the consequences of the theory. Quantum mechanics has also had a limited success in nuclear theory although in this field it is possible that a more fundamental system of mechanics is required.

Relativistic quantum mechanics is a generalization of nonrelativistic quantum mechanics in which the quantum equations of motion satisfy the principle of relativity. In its simplest form the equation of motion of a particle is the Klein-Gordon equation, but since this neglects the spin properties of the particle, the best verified form of relativistic quantum mechanics is provided by the Dirac electron theory. In general the relativistic quantum theory of a system may always be derived, in accordance with present knowledge, from a Lagrangian, the relativistically covariant properties of the resulting equations of motion being automatically assured if the Lagrangian is invariant under Lorentz transformations.

See also **Field Theory**; **Gravitation**; and **Relativity and Relativity Theory**.

Additional Reading

Adar, R.K.: "A Flaw in a Universal Mirror," *Sci. Amer.*, 50 (February 1988).

Ahimony, A.: "The Reality of the Quantum World," *Sci. Amer.*, 46 (January 1988).

Batalin, I.A., Isham, C.J., and G.A. Vileovisky, Editors: "Quantum Field Theory and Quantum Statistics," Hilger, Bristol, U.K., 1987.

Canright, G.S. and S.M. Girvin: "Fractional Statistics: Quantum Possibilities in Two Dimensions," *Science*, 1197 (March 9, 1990).

Chakaraborty, T. and P. Pietilainen: "The Fractional Quantum Hall Effect," Springer-Verlag, New York, NY, 1988.

Clarke, J. et al.: "Quantum Mechanics of a Macroscopic Variable: The Phase Difference of a Josephson Junction," *Science*, 992 (February 26, 1988).

Cohen-Tannoudii, C., Dupont-Roc, J., and G. Grynberg: "Photons and Atoms: Quantum Electrodynamics," Wiley, New York, NY, 1989.

Davies, P., Editor: "The New Physics," Cambridge University Press, New York, NY, 1989.

Dirac, P.A.M.: "Einstein," Einstein Session of the Pontifical Academy, Vatican City (November 10, 1979). Reprinted in *Science*, **207**, 1161–1162 (1980).

Eisenstein, J.F. and H.L. Stormer: "The Fractional Quantum Hall Effect," *Science*, 1510 (June 22, 1990).

Ellis, P.M. and Y.C. Tang, Editors: "Trends in Theoretical Physics," Addison—Wesley, Redwood City, CA, 1990.

Fayer, M.D.: "Elements of Quantum Mechanics," Oxford University Press, Inc., New York, NY, 2000.

Freedman, D.H.: "A Chaotic Cat Takes a Swipe at Quantum Mechanics," *Science*, 626 (August 9, 1991).

Greiner, W.: "Quantum Mechanics: An Introduction," 2nd Edition, Springer-Verlag, Inc., New York, NY, 2001.

Imry, Y. and R.A. Webb: "Quantum Interference and the Aharonov-Bohm Effect," *Sci. Amer.*, 56 (April 1989).

Mehra, J. and H. Rechenberg: "The Fundamental Equations of Quantum Mechanics 1925–1926: The Reception of the New Quantum Mechanics 1925–1926," Springer-Verlag, Inc., New York, NY, 2000.

Mehra, J. and H. Rechenberg: "The Historical Development of Quantum Theory: The Completion of Extensions of Quantum Mechanics—1926–1941," Springer-Verlag, Inc., New York, NY, 2001.

Penrose, R. and C.J. Isham, Editors: "Quantum Concepts in Space and Time," Oxford University Press, New York, NY, 1986.

Pool, R.: "Quantum Chaos: Enigma Wrapped in a Mystery," *Science*, 803 (February 17, 1989).

Pope John Paul II: "Einstein," Einstein Session of the Pontifical Academy, Vatican City (November 10, 1979). Reprinted in *Science*, **207**, 1165–1167 (1980).

Powell, C.S.: "Can't Get There from Here: Quantum Physics Puts a New Twist on Zeno's Paradox," *Sci. Amer.*, 24 (May 1990).

Ruthen, R.: "Quantum Pinball: A Quantum System Can Be Observed Without an Observer," *Sci. Amer.*, 36 (November 1991).

Schwinger, J.S. and C. Clarice Schwinger: "Quantum Mechanics: Symbolism of Atomic Measurements," Springer-Verlag, Inc., New York, NY, 2001.

Styer, D.F.: "The Strange World of Quantum Mechanics," Cambridge University Press, New York, NY, 2000.

Trefil, J.: "Quantum Physics' World," *Smithsonian*, 66 (August 1987).

Waldrop, M.M.: "Viewing the Universe as a Coat of Chain Mail: New Calculations Have Pointed the Way to Quantum Gravity and Suggested a Novel Structure for the Sub-sub-Microscopic World," *Science*, 1510 (December 14, 1990).

Weisskipf, V.F.: "Einstein," Einstein Session of the Pontifical Academy, Vatican City (November 10, 1979). Reprinted in *Science*, **207**, 1163–1164 (1980).

Ziock, K. and D. Pocanic: "Introduction to Quantum Mechanics," Cambridge University Press, New York, NY, 2002.

QUANTUM NUMBER. A number assigned to one of the various quantities that describe a particle or state. Many different characteristics of atomic and nuclear systems, as well as of those entities that are introduced as a part of particle physics, are described by means of quantum numbers. The quantum numbers arise from the mathematics of the eigenvalue problem and may be related to the number of nodes in the eigenfunction. Any state may be described by giving a sufficient set of compatible quantum numbers. In the customary formulations, each quantum number is either an integer (which may be positive, negative, or zero) or an odd half-integer.

Quantum Number (Magnetic). A quantum number that describes the component of the angular momentum vector of an atomic electron or group of electrons in the direction of an externally applied magnetic field. The values of these components are restricted, i.e., quantized. The symbol for the magnetic quantum number is m.

Quantum Number (Orbital). A quantum number characterizing the orbital angular momentum of an electron in an atom or of a nucleon in the shell-model description of the atomic nucleus. The symbol for the orbital quantum number is l.

Quantum Number (Principal). A quantum number that, in the old Bohr model of the atom, determined the energy of an electron in one of the allowed orbits around the nucleus. In the theory of quantum mechanics, the principal quantum number is used most commonly to describe the atomic shell in which the electrons are located. In a somewhat general way, it is related to the energy of the electronic states of an atom. The symbol for the principal quantum number is n. In x-ray spectral terminology, a K-shell is identical to an $n = 1$ shell, and an L-shell to an $n = 2$ shell, etc.

Quantum Number (Spin). A number that describes that part of the total angular momentum of the electron that is due to the rotation of the electron on its own axis. This contribution is quantized, having only a single value $\frac{1}{2}$ in terms of $h/2\pi$ units of angular momentum (h = Planck's constant). The magnetic quantum number associated with the spinning electron can have two values, either $+\frac{1}{2}$ or $-\frac{1}{2}$. The spin angular momentum of the electron can then couple in more than one way with its orbital angular momentum to provide a basis for the occurrence of many multiplet lines in atomic spectra. The symbol for the spin quantum number is s.

QUANTUM NUMBER (Isospin). A nuclear quantum number based upon the concept that the proton and the neutron are different states of the same elementary particle, the nucleon. The nucleon is assigned an isospin quantum number of $\frac{1}{2}$, and its two possible orientations, $\tau_z = +\frac{1}{2}$ and $-\frac{1}{2}$, and assigned to the proton and neutron, respectively. The isospin vectors of all the nucleons in a given nucleus combine, as do angular momentum vectors, to give a total isospin vector T, which is equal to $\Sigma\tau$ and has an orientation T_z, which is equal to $\Sigma\tau_z$. The symbolic space in which these orientations occur is called isospin space or isospace; its Z-axis corresponds to the direction of observable charge. Nuclei with a common value of T_z have the same proton excess, equal to $2T_z$, by the relationship

$$T_z Z = Z(+\tfrac{1}{2}) + N(-\tfrac{1}{2}) = \tfrac{1}{2}(Z - N)$$

where T_z is as defined above, N is the number of neutrons in the nucleus, and Z is its number of protons. Both T and T_z are integral for even values of A, and half-integral for odd values. Isospin is equally often called isobaric spin, and sometimes isotopic spin. No universal agreement appears to have been reached as yet as to which term to use.

QUANTUM STATISTICS. The classical Maxwell-Boltzmann statistics treated statistical mechanics by assuming that distinguishable particles are distributed among n cells ($r < n$) in such a way that each of the n^r arrangements is equally probably. However, if quantization of the energy states is considered, then the principle of the indistinguishability of the particles must be considered. If the particles are indistinguishable there are $\binom{n + r - 1}{r}$ distinguishable arrangements and if these are taken as equally probable there result Bose-Einstein statistics. As a particular case, if not more than one particle may appear in any cell there are $\binom{n}{r}$ equally probable arrangements; and these form the basis of Fermi-Dirac statistics. Particles described by these statistics are sometimes called fermions and bosons, respectively. No particle has been found to be neither a fermion nor a boson. All known fermions have total angular moments $(n + \frac{1}{2})h$, where n is zero or an integer, and all known bosons have angular momenta nh. At sufficiently high temperatures, where a large number of energy levels are excited, both quantum statistics reduce to the classical Maxwell-Boltzmann statistics. The basis of the two quantum statistics is the observation that any wave function that involves identical fermions is always antisymmetric with respect to interchange of the coordinates, including spin, or any two of the fermions, whereas for identical bosons, the wave function is always symmetric.

QUANTUM THEORY (Gravitation).   See **Gravitation**.

QUANTUM THEORY OF HEAT CAPACITY.   A theory that explains on the basis of energy quantization the decrease of specific heats at low temperatures to values below their classical values.

QUANTUM THEORY OF RADIATION.   The energy of radiation emitted or absorbed is concentrated in quanta or photons each with an energy $E = hv$ in ergs, where h is Planck's constant and v is the frequency of the radiation in cycles per second. See also **Planck Law**.

QUANTUM THEORY OF SPECTRA.   The present theory of spectra, which is based on an idea that there exist in each atom or molecule certain permitted energy levels. An atom or molecule absorbs or radiates energy as it moves from one energy level to another. The frequency (v) of the radiation associated with such change of energy level is given by

$$E_1 - E_2 = hv$$

E_1, E_2 are the energy levels and h is the Planck constant.

QUAQUAVERSAL.   A term used by structural geologists to describe a dome in which the formations dip outward in all directions.

QUARANTINE.   The temporary limitation of freedom of exposed or affected persons who may be in a position to transmit certain communicable diseases to others. The term was also used in connection with various space programs in which returning astronauts were confined to special quarters for a limited period out of fear that harmful organisms may have been carried back to earth from lunar missions.

In the agricultural and food fields, quarantines are established to regulate the transportation of cattle and other livestock to and from various locations in an effort to prevent the spread of various cattle diseases. Similarly, regulations are imposed on the transport of seeds and living and cut plants from one location to the next in the interest of halting the migration of insects, viruses, parasites, and other plant pathogens from invading unaffected regions. See also **Plant Breeding**.

QUARTZ.   The mineral quartz, oxide of the nonmetallic element silicon, is the commonest of minerals, and appears in a greater number of forms than any other. Its formula is SiO_2. Quartz commonly occurs in prismatic hexagonal crystals terminated by a pyramid. This pyramid is due to the equal development of two rhombohedrons, and may be observed in cases where one rhombohedron predominates. Cleavage is not observed; the fracture is typically conchoidal; hardness is 7; specific gravity, 2.65; luster, vitreous to greasy or dull; colorless to white, pink, purple, yellow, blue, green, smoky brown to nearly black; transparent to opaque.

There are two distinct modifications of quartz, depending upon the temperature at which they were formed. The low-temperature variety is formed below 573 °C and is the more common sort, being found in veins, geodes, etc. It is called low-quartz. The high-temperature modification

is formed between 573 °C and 870 °C, and is found chiefly in granites and granite or rhyolite porphyries. This is called high-quartz. Above 870 °C tridymite is the stable form of SiO_2. The differences between high- and low-quartz are entirely crystallographic, low-quartz having a vertical axis of threefold symmetry and three horizontal axes of twofold symmetry, while high-quartz has a vertical axis of six-fold symmetry and six horizontal axes of twofold symmetry. It is usual to separate the many kinds of quartz into (1) crystalline or vitreous varieties, actual crystals or vitreous crystalline masses, and (2) cryptocrystalline varieties, mostly compact nonvitreous sorts, but which may show a crystalline structure under the microscope.

1. *Crystalline or Vitreous*: Rock crystal, colorless crystals or masses. Amethyst, clear violet or purple, either crystals or masses. Rose quartz, usually massive but rarely in crystals, delicate shades of pink or rose, sometimes red. Citrine or yellow quartz, sometimes called false or Spanish topaz, light to deep yellow. Smoky quartz, smoky brown to almost black, often called cairngorm stone from Cairngorm, Scotland. Milky quartz, often showing delicate opalescence, transparent to nearly opaque, often with a greasy luster. Aventurine quartz incloses glistening scales of mica or hematite. Rutilated quartz incloses needle-like prisms of rutile called "fleches d'amour." Other acicular minerals such as actinolite, tourmaline, and epidote, may also be thus inclosed; Cat's Eye shows a peculiar opalescence, probably due to inclosed masses of some fibrous mineral. Tiger's Eye is a siliceous pseudomorph after crocidolite of a golden yellow brown color.

2. *Cryptocrystalline*: The following cryptocrystalline varieties of quartz are treated under their own headings: agate, basanite, bloodstone, carnelian, chalcedony, chert, chrysoprase, flint, heliotrope, jasper, moss agate, onyx, plasma, prase, sard, and sardonyx. Quartz readily forms pseudomorphs after various minerals or structures. Silicified wood is a quartz pseudomorph after the organic material of which it originally consisted. Quartz is often pseudomorphic after calcite, barite, and fluorite. Quartz is an essential constituent of many igneous rocks, for example, granites, granite porphyries, and felsites, as well as quartz diorites and their surface equivalents, the dacites. In the metamorphic rocks quartz Figures very largely in the gneisses and schists, and, of course, in quartzite. In the sedimentary rocks most sandstones are composed chiefly of grains of quartz, and quartz forms veins and nodules in limestones.

Of the many places that have yielded fine specimens of quartz, a few include: the Swiss Alps, the Piedmont of Italy, the Island of Elba, Dauphiné in France, Cumberland in England, Banffshire in Scotland, and Madagascar, Uruguay, Mexico and Brazil. Magnificent rose quartz crystals occur at the Arassuahy-Jequitinhonha District, Minas Gerais, Brazil. In the United States the following localities are well known: Paris, Maine, especially for rose quartz; Herkimer County, New York, for small but very brilliant crystals found in the Cambrian dolomites or in the soil. Amethyst County, Virginia, furnishes amethysts, as do Lincoln and Alexander Counties, North Carolina. Other localities for amethyst and smoky quartz are South Dakota, in the Black Hills; the Pikes Peak district, Colorado; Yellowstone National Park, Wyoming; Jefferson County, Montana; and in Canada in the Province of Ontario in the Thunder Bay region. The word *quartz* is believed to have been originally of German origin. Besides the use of the different varieties of quartz for jewelry and other ornamental purposes, this mineral has extensive industrial uses in the ceramic arts, optical and other sorts of scientific instruments, abrasive, scouring, polishing materials, and for refractories.

Certain mineral classes of low symmetry possess no center of symmetry, and their axes, known as polar axes, have different properties at their terminal ends. Quartz belongs to one of those classes. When quartz is exposed to an exerted compressive or mechanical stress along one of these polar axes, electrical charges are developed on that axis; a negative charge is produced at one end, a positive charge at the opposing end. Conversely, when quartz crystals are subjected to an applied electric field along a polar axis, mechanical strains will be developed in those crystals. This phenomenon is known as piezoelectricity. Plates or disks cut perpendicular to such polar axes and properly oriented with established specifications are subject to mechanical vibrations (oscillations) at predetermined frequencies under an applied electric field. Those frequencies are designed to coincide with and stabilize the circuit frequency of radio transmitters and receivers. This property is utilized extensively in the control of frequency oscillations in the field of radio telemetry. Between January 1942 and V-J Day over 70 million such units were manufactured for the United States armed forces, and consumed over 4 million pounds (1.8 million kilograms) of radio grade quartz. The excessive demand for natural quartz of required quality to produce those wafers resulted in the development of a new industry, synthesizing quartz to meet the demand.

See also **Piezoelectric Effect**.

ELMER B. ROWLEY, F.M.S.A., formerly Mineral Curator,
Department of Civil Engineering, Union College,
Schenectady, NY.

QUARTZ CRYSTAL (Piezoelectric). See **Piezoelectric Effect**.

QUARTZITE. A hard, tough, and compact metamorphic rock composed almost wholly of quartz sand grains that have been recrystallized to form a particularly massive siliceous rock. The term is also used for non-metamorphosed quartzose sandstones and grits whose clastic grains have been firmly cemented by silica that has grown in optical continuity around each grain.

QUARTZ PORPHYRY. One of the hypabyssal or effusive rocks chemically related to the granite or alkali family but rich in silica, which occurs as quartz phenocrysts in a crypto- or microcrystalline ground mass.

QUASARS. Quasars, a contraction of "quasi-stellar radio sources," are a class of objects defined by having spectra of extremely high redshift and optical appearance that is almost stellar. Present theory indicates that the quasars are the farthest objects visible in the universe. They must be extremely luminous to appear as bright as they do on earth in the optical, radio, and x-ray parts of the spectrum given the extreme distances that are assigned to them on the basis of Hubble's law. See also **Cosmology**.

Quasars were found in the course of an investigation of "radio stars." See also **Radio Stars**. The first such object was found in 1960 as a result of an unusually accurate radio position deduced at Caltech's Owens Valley Radio Observatory with the interferometer there and its identification with an optical object observed with the 5-m Hale Telescope of the Palomar Observatory. The work was carried out by Allan R. Sandage and Thomas Matthews. The object was 3C 48, the 48th object in the 3rd catalogue of radio sources compiled by Cambridge University. In the next couple of years, two other similar objects were detected.

Another unusual object, 3C 273, was identified with an optical object as a result of its occultation by the moon three times within a few months' period in 1962, a series observed by Cyril Hazard of the Australian National Radio Observatory at Parkes, N.S.W., Australia. The 13th magnitude object they identified appeared to optical telescopes as a bluish star with a faint jet emanating from it. The radio image had two peaks, one corresponding with the star-like object and the other with the jet.

The spectra of these objects showed emission lines similar to spectra of a gas at medium temperature, but the wavelengths of the lines did not correspond to those of known elements. In 1963, Maarten Schmidt of Caltech and the Mt. Wilson and Palomar Observatories realized that the spectrum of 3C 273 corresponded to that of hydrogen displaced to the red by 16%. His colleague Jesse Greenstein immediately realized that the spectrum of 3C 48 corresponded to that of hydrogen displaced to the red by 37%.

Since Hubble's law implies that high redshifts correspond to large distances, these redshifts imply that 3C 273, 3C 48, and their fellow quasi-stellar radio sources are among the farthest objects known. To supply the amount of energy that reaches the earth from them, they must be extremely copious emitters, among the most intense sources in the universe.

Sandage soon found a class of radio-quiet quasi-stellar objects that also corresponded to bluish objects. Both radio-emitting and radio-quiet quasi-stellar sources are included in the class we now call "quasars," a name resisted for a time by the Mt. Wilson and Palomar astronomers. Quasar spectra have since been studied by many observatories.

The discovery that quasars vary in overall intensity in a period of hours, days, or weeks indicates that the region of strongest emission must be smaller than light-hours, light-days, or light-weeks, respectively. The "energy problem" asks how the quasars can be such copious emitters in such a small volume. The problem has not been completely solved, though most astronomers currently believe that there is probably a giant black hole present in the midst of the quasar. The energy, in this model, as matter swirls around the giant black hole is released before the matter enters the black hole. Black holes are a way of providing a lot of energy in an extremely small space.

Some scientists had thought that the energy problem was so severe that they questioned whether the quasars were really at the distances assigned from the Doppler shift. This would be a serious challenge to a basic tenet of extragalactic astronomy, for if the Doppler shift does not give the true distance in this case, then all applications of Hubble's law would be in doubt. At the time, black holes were not in vogue as explanations for a variety of energetic objects, and no obvious way of producing all the energy was available.

One of the methods investigated to provide a giant redshift for quasars that are relatively close by was to assume that there is a gravitational redshift caused by massive objects in the gas that emits the observed quasar spectra. But theoreticians could never make a satisfactory model for such objects. And recent work shows that the redshift in the "fuzz" that apparently surrounds some quasars does not vary with distance from the quasar. This is in contradiction with this gravitational-redshift model and so rules it out.

Almost all astronomers now accept that the quasars are at cosmological distances, that is, the distances assigned by application of Hubble's law to their Doppler shifts. The leading dissenter is Halton Arp of the Mt. Wilson and Las Campanas Observatories, who has collected a variety of photographs of quasars that appear to be linked with galaxies because of apparent bridges, by proximity, or because of symmetry considerations. Statistical arguments as to the probability of chance alignments then come into play, and no resolution to the controversy currently exists.

The work of Alan Stockton of the Institute for Astronomy of the University of Hawaii has come down strongly on the side of quasars being at their cosmological distances. Stockton considered a complete sample of fields selected in advance around 27 nearby quasars, removing one of the major statistical objections to Arp's work. He used the 2.2-m telescope at the University of Hawaii's outstanding Mauna Kea site to search for galaxies in these fields, and found 29 galaxies associated with 17 of the fields. These galaxies for the most part have redshifts identical to those of the quasars, proving to a high statistical accuracy that the quasars are also at those distances.

Other evidence indicating that quasars are indeed at cosmological distances concerns the discovery of an apparent sequence of active nuclei of galaxies, including Seyfert and N galaxies. Also, the object BL Lacertae, long known as a variable star, turned out to have a redshift of about 7%, making it one of the nearest quasars. Other "Lacertids," similar objects, have since also proved to have redshifts of the same order. Still nearer quasars, with redshifts down to about 4%, have also since been found. If these lower-redshift objects had been discovered before the quasars of higher redshifts, astronomers would not have thought it so strange that quasars could be so far away or so powerful, and the energy problem might not have been bothersome.

Astronomers are now in agreement that quasars are probably events of some sort in the midst of galaxies, with giant black holes as the probable central object. There is probably a continuum of active galactic nuclei, with intensities of different degrees. Quasars are probably merely the most extreme members of the class.

Several quasars are now thought to be rejuvenated, gaining a new supply of fuel from interaction with a galaxy. The remnants of the galaxy can sometimes be barely resolved from the quasar image. These quasars are usually relatively near to us; the more distant quasars are seen in an earlier stage in the universe when they had their original supply of fuel.

In 1979, Dennis Walsh of the University of Manchester, Robert F. Carswell of Cambridge University, and Ray J. Weymann, then of the University of Arizona, realized a pair of quasars separated by only a very small distance in the sky were probably a pair of images of the same quasar formed by a "gravitational lens." Such a gravitational lens would be the consequence of Einstein's general theory of relativity, as light and radio waves from a distant quasar passed nearby a massive intervening object. Spectra taken with the Multiple Mirror Telescope in Arizona showed that the spectra of the pair of objects are all but identical. Observations made by Stockton at Hawaii and Jerome Kristian and colleagues at Palomar have detected the intervening galaxy and its surrounding cluster of galaxies, apparently confirming the gravitational-lens theory. High-resolution radio maps made with the VLA (Very Large Array of radio telescopes of the National Radio Astronomy Observatory) give images that agree with this interpretation. Several other multiple quasars have since been found, giving additional credence to the gravitational-lens model.

The International Ultraviolet Explorer spacecraft has taken quasar spectra in the ultraviolet region of the spectrum. The Einstein Observatory, the x-ray telescopes aboard NASA's High-Energy Astronomy Observatory-2, also observed many quasars, which turn out to be powerful x-ray emitters. Even long exposures of otherwise blank fields turned out to have faint quasars in them. The redshifts of the quasars could then be measured by optical means. The Einstein Observatory results indicated that most or all of the X-ray background is supplied by these faint quasars.

Additional Reading

Appenzeller, T.: "Latest Quarry in the Quasar Quest," *Science*, 1094 (September 6, 1991).

Arp, H.C.: "Quasars, Redshifts and Controversies," Cambridge University Press, New York, NY, 1999.

Courvoisier, T. and E. Robson: "The Quasar 3C 273," *Sci. Amer.*, 50 (June 1991).

Cowen, R.: "Enigmas of the Sky: Partners or Strangers?" *Science News*, 181 (March 24, 1990).

Cowen, R.: "Quasar Erupts with Relativistic Flair," *Science News*, 39 (January 19, 1991).

Crawford, M.H.: "The Abnormally Normal Quasar," *Science*, 1116 (December 1, 1989).

Horgan, J.: "Points of View: Quasars and Radio Galaxies May Be Two of a Kind," *Sci. Amer.*, (April 20, 1989).

Kembhavi, A.K. and J.V. Narlika: "Quasars and Active Galactic Nuclei: An Introduction," Cambridge University Press, New York, NY, 1998.

Peterson, I.: "Quasar Light Points to Younger Galaxies," *Science News*, 389 (June 23, 1990).

Rees, M.J.: "Dead Quasars in Nearby Galaxies?" *Science*, 817 (February 16, 1990).

Sanders, D.B., Scoville, N.Z., and B.T. Solfer: "The Birth of a Quasar?" *Science*, 625 (January 29, 1988).

Williams, S.: "Twin Quasars Found," *Science*, 1179 (March 9, 1990).

QUASSIA. A family of trees (*Simaroubaceae*) closely related to the mahogany family. Several Quassias are found in the West Indies, which is also the preferred region of the closely related Tree of Heaven. Among the quassias is the Spanish cedar (*Cedrela odorata*), poorly named because its only resemblance to a cedar is the scent of its wood, popularly used over the years in making cigar boxes. Another is the Chinese counterpart (*Cedrela sinensis*). The Chinese cedar is also known as the *toon*. The tree is becoming more popular as a tree for shade and street planting, with beautiful yellow autumnal coloring. The flowers are unscented; the leaves are onion-flavored. A rare genus is the *Ehretia*. The species *E. dicksonii*, however, is growing into favor for planting. It is a short broad tree. It flowers in June and the leaves are large, some 10 inches (25 centimeters) in length.

QUATERNARY. The second period of the Cenozoic era, following the Tertiary, considered to cover the last 2 or 3 million years. The period consists of two epochs, the Pleistocene and the Holocene. The name was originally assigned as an era rather than as a period designation, with the epochs considered to be periods; and it is still sometimes so used in geologic literature.

QUATERNARY AMMONIUM COMPOUNDS. See **Amines**.

QUATERNION. A generalization of the complex number is called a hypercomplex number. It has the form $Q_0 + Q_i e_i$, where Q_0 and Q_i are real and e_i is related to $\sqrt{-1}$. If $i = 1, 2, 3$ so that the hypercomplex number has four components, it is a special kind of four-dimensional dyadic and called a quaternion. In that case,

$$e_i^2 = 1; \quad e_i e_j = \varepsilon_{ijk} e_k, \quad i \neq j$$

where ε_{ijk} is a set of 27 quantities, which can take only the values O, ± 1. A given ε_{ijk} vanishes if two of its indices are identical; it equals $+1$, if the subscripts are obtained by an even permutation of (1, 2, 3) and -1, if by an odd permutation.

Quaternions add like ordinary complex numbers but their multiplication is not, in general, commutative. Thus,

$$PQ = (P_0 + P_i e_i)(Q_0 + Q_j e_i)$$

$$= (P_0 Q_0 - P_i Q_i + P_0 Q_i e_i + Q_0 P_i e_i + \varepsilon_{ijk} P_i Q_j e_k)$$

which does not equal QP unless $P_i = kQ_i$, where k is a real number.

It is often useful to write a quaternion in the form $Q = Q_0 + \mathbf{Q}$, where Q_0 is a scalar and $\mathbf{Q}$ is a vector with components Q_1, Q_2, Q_3. More generally let $\mathbf{u}$ be a unit vector, then $Q = Q_0 + q\mathbf{u}$, showing the analogy between a quaternion and a complex number, for Q_0 is the real part of Q and q the imaginary part.

Consider in classical mechanics a rigid body rotated through an angle θ about a line through the origin of a rectangular coordinate system with direction angles a, b, c. The coordinates of a point before and after the rotation are conveniently given in terms of the Euler-Rodrigues parameters. They are $Q_0 = \cos\theta/2$, $Q_1 = \sin\theta/2 \cos a$, $Q_2 = \sin\theta/2 \cos b$, $Q_3 = \sin\theta/2 \cos c$. They form the components of a quaternion and $Q_0^2 + Q_1^2 + Q_2^2 + Q_3^2 = 1$. These parameters are also related to the Euler angles, the Pauli spin matrices of quantum mechanics, and the rotation group in group theory.

QUEEN SNAKE. See **Snakes**.

QUENCHING. Immersion of hot metals in liquid baths in order to effect rapid cooling. In steel heat-treating practice quenching oils give slower and brine solutions give faster cooling rates than water. Dilute caustic solutions are sometimes used for rapid cooling rates comparable to brine. These baths are maintained at or near room temperature. For special quenching procedures requiring baths held at moderately elevated temperatures, molten metals, such as lead, and fused salts may be used.

Quenching of ordinary steel is for the purpose of hardening it. Certain non-hardenable steels, for example austenitic stainless steel, and many non-ferrous metals may be cooled rapidly from elevated temperatures for other reasons.

QUERCUS. See **Oak Trees**.

QUETZAL (*Pharomachrus mocino*). The best known of all trogons, the quetzal was much revered by the ancient civilizations of the Mayas and the Aztecs and still is the hearaldic emblem of Guatemala, whose monetary unit is named for it. Red and green are beautifully contrasted in the plumage of the bird. Added to this are the extraordinarily elongated upper tail coverts of the male, forming a magnificent train that can measure up to 3 feet (approximately 1 meter) in length. See Fig. 1 on p. 2908.

According to Indian tradition, the quetzal played a part in the struggle against the Spanish conqueror Pedro de Alvarado. This legendary bird, whose very existence occasionally has been doubted, inhabits mountainous, wooded areas from southern Mexico to Panama. There are also related forms in South America. The bird is rare in Guatemala today. Not only ruthless persecution, but also deforestation of the jungle have dramatically reduced its area of habitation. It has found provisional refuge in barely accessible mountain forests, especially in Honduras and Costa Rica. There and also in the state of Chiapas in southern Mexico, the quetzal is found at very high altitudes.

After two weeks, the nestlings are covered profusely with feathers except for the head. They are fed insects almost exclusively for the first few days after hatching; later on, they are given various fruits, small vertebrates, and even snails. Grown-up quetzals, however, feed mainly on fruits. After leaving their nests, the young remain with their parents for a considerable time. It also takes several months for their colors to develop. The full length of the upper tail coverts of the male is not reached until the third year.

Quetzal feathers had a special significance even in pre-Columbian times. Among the Mayas and the Aztecs, no ordinary person was allowed to possess them. Only the highest dignitaries were permitted to wear them as plumes on top of their feather wreaths.

QUEUE (Computer System). When events occur at a faster rate than they can be handled in a computer system, a waiting line or queue must be formed. The elements of a *queue* typically are pointers (addresses) which refer to the items waiting for service. The items may be tasks to be executed or messages to be sent over communication facilities, for example. The term is also used as a verb, meaning to place an item in a queue. Several methods of organizing queues are used. The sequential queue is common. As new elements arrive, they are placed at the end; as elements are processed, they are taken from the front. This is the first-in—first-out (FIFO) organization. In the push-down queue, the last one in is the first one out (LIFO). The multipriority queue processes from the front, but in terms of priority of the elements waiting. Essentially, this is a modified sequential queue containing subsequences and is sometimes referred to as priority in, first out (PIFO).

T. J. H.

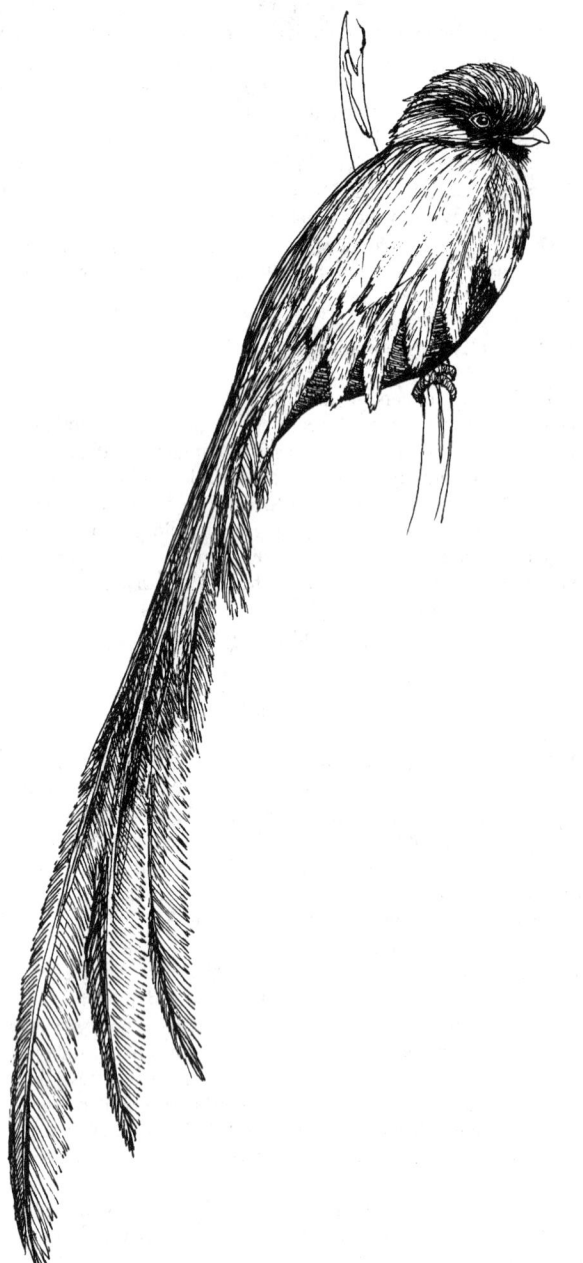

Fig. 1.   **Quetzal** (*Pharomachrus mocino*), a Central American bird of beautiful green colorations. Underside is bright red.

QUEUEING PROBLEM.   The problem of queues, or congestion, arises in a variety of fields where there is a service to be offered and accepted. In general the problem is concerned with the state of a system, e.g., the length of the queue (or queues) at a given time, the average waiting time, queue discipline and the mechanism for offering and taking the particular service. The analysis of queueing problems makes extensive use of the theory of stochastic processes.

Queueing problems arise in many fields of technology. They were first formulated from the traffic congestion in long distance telephoning. More recently they have become critically important in deciding upon the number of duplicate facilities required in automatic telephone exchanges. They also occurred in inventory control, in plant designing where machines occurred in multiple, in traffic work such as landing fields for aircraft, and toll stations for automobiles. Some solutions can be formulated mathematically. More complicated situations have to be solved by simulation on a computer.

QUEVENNE SCALE.   See **Specific Gravity**.

QUICK-FREEZING.   See **Freeze-preserving**.

QUICKSAND.   A thick mass or bed of fine sand, as at the mouth of a river or along a seacoast, that consists of smooth rounded grains with little tendency to mutual adherence. It is usually thoroughly saturated with water flowing upward through the voids, forming a soft, shifting, semi-liquid, highly mobile mass that yields easily to pressure and tends to suck down and readily swallow heavy objects resting on or touching its surface.

QUICKSILVER.   See **Mercury**.

QUILL.   The portion of a feather that bears none of the slender lateral branches. The hollow shaft attached to the skin of the bird. Also, the thickened and barbed spines of the porcupines, which are modified hairs.

QUILLWORTS.   These curious plants, species of the genus Isoetes, grow, as a rule, under water. They have short thick corm-like stems and slender dichotomously branching roots. The leaves are slender and somewhat grass-like and are crowded on the upper surface of the short stem. Within the basal portion of the leaf are borne sporangia of two kinds. One, containing large spores or megaspores, is called a megasporangium. The other is a microsporangium and contains numerous very small spores. The megaspores give rise to small multicellular gametophytes on which the archegonia are formed. The microspores become minute multicellular bodies in which are formed the small sperms. A sperm swims to the egg and unites with it to form a zygote. From this a new quillwort is formed. The quillworts are living relics of a once important group. They are of no commercial importance.

QUINCE TREES.   See **Rose Family**.

QUININE.   See **Alkaloids**; and **Malaria**.

QUINTUPLE POINT.   The temperature at which five phases are in equilibrium.

QUOTA SAMPLE.   A form of sample in which its constitution is predetermined according to assigned characteristics. For example, a sample of 100 human beings might be required to have 50 males and 50 females, 20 children and 80 adults, 30 graduates and 70 nongraduates, and so on. The object of quota sampling is to attain an accurate representation of the constitution of the parent population. Complicated forms are difficult to attain and in any case, as contrasted with random sampling, it is also difficult to assign probabilistic limits to the errors involved. See also **Sampling (Statistics)**.

QUOTIENT.   The result obtained when division is applied to numbers or functions. Thus a quotient equals a numerator divided by a denominator. The indicated quotient of two numbers is frequently expressed as a fraction and written as $a : b$, a/b, or $a \div b$ and is also called a ratio.

The quotient group has a special meaning in group theory.

See also **Group**.

R

RAASE. See **Viverrines**.

RABBIT FEVER. See **Tularemia**.

RABBITFISH (*Chimaera monstrosa*). One of seventeen species of Chimaeridae, the rabbitfish is a bottom dweller of the northeastern Atlantic, from Iceland and Norway to northern Africa, as well as of the western and central Mediterranean. The habitat chiefly is the continental shelf at depths of about 200 meters (650 feet), although it is occasionally found at considerably greater depths. The coloration is silver-gray, with a violet sheen on the upper side and a dark marble pattern underneath. The unpaired fins have black seams. The rabbitfish achieves a length of about 1.5 meters (5 feet). The fish is so named because of its slight resemblance to a rabbit in the eye, mouth, and throat area. See Fig. 1.

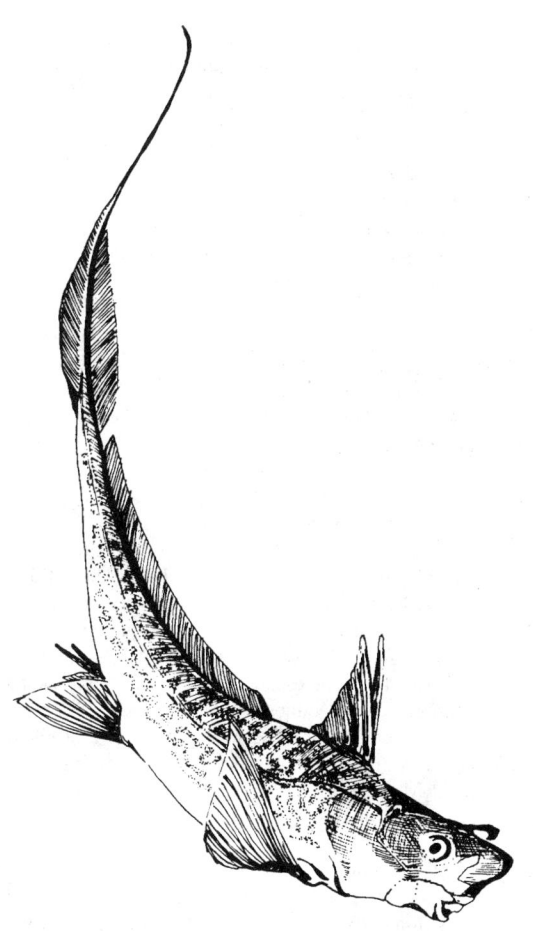

Fig. 1. Artist's sketch of the rabbitfish (*Chimaera monstrosa*).

The fish feeds on various crustaceans, mollusks, echinoderms and small floor-dwelling fishes. Although the fish is not suitable as a source of food for people, the liver, which amounts to about one-third of the total body weight, is a valuable source of oil. Fishermen dealing with rabbitfishes must proceed cautiously because the stinging spine of the dorsal spine is poisonous and lethal injuries have been reported from accidents by contacting them.

RABBITS AND HARES. At one time, these leaping animals were classified with other gnawing animals in the order of *Rodentia*. They are now placed in their own order of mammals, i.e., *Lagomorpha* (leaping mammals). Nevertheless, in a general sense, rabbits are rodents with long ears, large hind legs, and small front legs. Some species burrow and others occupy similar retreats which they do not make for themselves. Many members of this group are called hares. There is no sharp distinction between the terms except in their established application to certain species. Differences in the young sometimes are used to distinguish between the rabbits and the hares. Rabbits are born naked, blind, quite helpless and in a fur-lined nest. The young are smaller than the baby hares. The young hares are born in small, open dens on top of the ground, are fully haired and with eyes open. They hop about almost immediately after birth.

Rabbits are found on all continents, although they were introduced into the Australian region. In New South Wales the introduced stock threatened to crowd out even the settlers by its destruction of vegetation. Millions of the animals have been killed per year and the exportation of their hides has somewhat offset their destructiveness. Sudden reduction in numbers is caused by epidemics. Myxomatosis, a virus disease, is quickly fatal to rabbits and the artificial creation of this disease in Australia has reduced the enormous numbers of European rabbits, *Oryctolagus* (*Lepus*), which existed in the absence of natural enemies.

In North America the common or cottontail rabbit, *Sylvilagus floridanus*, and a few closely related species are widely distributed. One of these species is the brush rabbit, *S. bachmani*, of the Pacific northwest, and two others are the southern marsh rabbit or pontoon, *S. palustris*, and swamp rabbit or cane-cutter, *S. aquaticus*. The large western species are sometimes called hares but more commonly rabbits. The snowshoe rabbit, *Lepus americanus*, also known as the white rabbit or varying hare, lives in the north and in the mountains as far south as Virginia and Colorado. The white-tailed jack rabbit or prairie hare, *L. townsendi*, ranges from the Mississippi River to eastern California. This species becomes white in winter in the northern part of its range. Other species of jack rabbit are found farther west. See Fig. 1.

The flesh of rabbits is excellent, and in the more heavily settled parts of the country they are an important game animal. The fur is thick and soft but the hides are weak, hence they are used chiefly for linings, for cheaper fur garments, and for making felt. After shearing and dyeing rabbit fur reaches the market as northern seal.

Rabbits are bred extensively in captivity as pets, as laboratory animals for use in medicine and bacteriology, to some extent for food, and for the study of heredity. Since they are very prolific they have been among the most useful mammals to the geneticist.

Rabbits and hares have a varied diet, but prefer tender shoots, vegetables, buds, and small leaves. Sometimes they chew their food with a characteristic lateral motion of the jaw. Some of the larger animals can jump from 10 to 12 feet (3.0 to 3.7 meters) in one hop, although their normal gait is a short series of hops. The animals are active in late evening and often during the night.

The pika is a small rodent related to the rabbits. The pika lives chiefly at high altitudes, ranging from 11,000 to 19,000 feet and is found only in the Northern hemisphere. Two species of the genus *Ochotona* occur in the mountains of western North America and about two dozen in the Old World. All are compactly built, with small ears and a rudimentary tail. In the Old World, they are also called tailless hares or mousehares.

More details on rabbits and hares can be found in the *Food and Food Production Encyclopedia* (D.M. Considine, editor), Van Nostrand Reinhold, New York, 1982.

Fig. 1.   Jack rabbit. (*A.M. Winchester.*)

RABIES.   Viruses that are not usually found in humans, but that exist in some domestic and wild animals, are known as *viral zoonoses*. The effects of these viruses on humans may differ markedly from their effects in the reservoir animal. The *rhabdoviruses*, the causative agents of rabies, are among the viral zoonoses. The rabies virus is a bullet-shaped membrane envelope covering a coiled nucleocapsid structure containing single-stranded RNA. The virus measures about 80×180 nanometers (0.80×0.18 micrometer). Rabies is found throughout the world, wherever there are domestic and wild animals. The animals infected and incidence of the disease vary considerably from one region to another.

Rabies is still a major problem in developing countries. Although fewer than a thousand fatal cases are reported to the World Health Organization per year, some 15,000 deaths from the disease are said to occur in India, while in Central America 250 cases are reported per year. Many island communities are free of rabies. These include Great Britain, Australia, New Zealand, Hawaii, Taiwan, and Japan.

With a widely varying rate of occurrence worldwide, medical reporting and therapeutic measures also differ. Although once considered to be a reasonably well understood infection, studies are showing that much remains to be learned, particularly about the phenomenon of latent rabies.

Progression of the Rabies Infection.   In the general course of the disease, the rabies virus replicates in muscle cells near the site of the bite. There may be an incubation period of 20 to 60 days (even as long as 14 months) before symptoms of serious infection are manifested. In other cases, the disease may develop rapidly and result in death in a period as short as 3 to 4 days, particularly if left unattended without application of supportive measures. The virus spreads by way of nerves to the central nervous system. There is further replication of the virus in the brain in most cases before the virus spreads to other body tissues. Salivary glands are a common target, meaning that the patient can shed the virus in saliva and thus be infective to others.

Normally, after expiration of the incubation period, the patient will go through a prodomal phase of one or two days, during which time there will be fever and pain in the vicinity of the bite. There may be other, less specific symptoms, including irritability, nervousness, and sometimes a sensation of impending death. There follows an excitation stage, characterized by hyperventilation, hyperactivity, disorientation, and sometimes seizures. In furious rabies, most patients develop hydrophobia—a combination of inspirational muscle spasm, with or without painful laryngopharyngeal spasm, associated with terror. Initially provoked by attempts to drink water, this reflex can be excited by a variety of stimuli, including a draft of air, water splashed on the skin, or ultimately the sight, sound, or even mention of water. The spasm is violent and jerky, the neck and back are extended, the arms thrown up and the episode may end in generalized convulsions with cardiac or respiratory arrest. This is followed by a few

days of lethargy and varying degrees of paralysis, mainly in areas of the body that are innervated by the cranial nerves. The somatic muscles, bladder, and bowels may be affected. As the infection proceeds to the heart and respiratory muscles, the condition of the patient deteriorates rather rapidly and without cardiopulmonary support, death may shortly occur. The treatment of rabies is essentially supportive. Intensive cardiopulmonary support assists in prolonging the patient's life. Because of the high mortality, rabies is a disease for which exhaustive preventive measures are indicated.

Fortunately, there are by far many more instances of suspicion of rabies infection than actual cases. When a person knows or suspects exposure to rabies virus, a number of actions should be taken rapidly but carefully. Thorough washing of a wound with soap and water immediately after a bite or wound is mandatory. Next, the species of animal, knowledge of whether or not the animal has been vaccinated against rabies, the type and location of the bite or scratch, and the immediate history of rabies in the given geographical area must be considered. Where there is reasonable suspicion of a possible infection, rabies immune globulin for post-exposure prophylaxis may be ordered. Authorities do not all agree that treatment should be delayed pending proof of an infectious bite because of the costs or discomfort of treatment. Most physicians prefer the immune globulin to equine antirabies serum because the latter may cause serum sickness in many individuals. In most countries, the very old Pasteur treatment is no longer used. Vaccine cultured in duck embryos has essentially replaced the Pasteur vaccine (made from extract of brain of virus-injected rabbits) since the early 1960s. The duck embryo sometimes causes local reactions, but is much less painful than the Pasteur treatment. A daily dose of the vaccine for 22 days is required. Pre-exposure prophylaxis (for veterinarians and animal handlers), involves a shorter course of duck embryo vaccine administration, usually without administration of immune globulin. In the United States, about 20,000 persons are vaccinated against rabies each year.

Within the last few years, a human diploid inactivated rabies vaccine has been developed. This requires as few as 5 injections and, to date, there is little evidence of side-effects.

Sources of Rabies Virus.   Rabies virus is almost always transferred to humans by way of the animal's saliva, predominantly as the result of a bite, but some persons may contact the saliva as the result of the animal licking them.

It is sometimes difficult to assess potential exposure to wild animals. Many people will allow a wild animal to crawl on them, kiss them, or will feed them with a medicine dropper or baby bottle. When questioned after the fact, very few people can remember whether they had direct contact with the animal's saliva. Thus health officials in such cases must assume that exposure occurred and recommend a complete series of rabies injections.

Enzootic rabies (from wildlife) is of growing relative importance as the result of effective control measures taken by most cities and counties as regards immunization of domestic pets. From the Midwest to the far western United States, skunks are major reservoirs of rabies virus. Particularly in Florida and Georgia, raccoons are the major source. In the Appalachians, foxes are the principal source. In Alaska, in addition to the arctic fox, dogs, coyotes, wolves, and other carnivores that feed on fox carcasses are carriers of the rabies virus. In Europe, foxes are a major carrier of rabies virus among wildlife. Mongooses are a problem in Puerto Rico and Trinidad. Vampire bats, particularly as they infect cattle, are a major problem in Mexico. In Egypt, rabies in dogs, cats, and jackals is endemic. In India, jackals, in Iran, wolves, and in southeastern Asia and northern Africa, wild dogs are major rabies reservoirs.

Although dogs account for 90% of rabies cases, wild animals still account for thousands of cases worldwide each year. In the United States, skunks, raccoons, gray foxes, and Arctic and red foxes are the principal sources of rabies bites. For example, an epidemic of raccoon rabies has been progressing steadily up the East Coast from Florida since 1950 and by 1993 had worked its way north to eastern Pennsylvania. Skunks are found throughout the mid-continent, from Texas to the Dakotas and Montana. The skunk population of California is high, mainly along the coast and inland from Los Angeles north to the Oregon border. The gray fox population is more limited, affecting southeastern Arizona and central Texas. The population of the Arctic fox is found in the western half of Alaska and the extreme north of Maine.

During a period of over 30 years, scientists at the Centers of Disease Control in Atlanta have been researching self-vaccination methods for these wild animals. In 1970, Swiss researchers successfully placed baits in chicken heads. By 1985, machine-made baits of various composition found wide use in Europe and Canada and may be adopted on a wide scale in the United States. Attempts to launch widespread appeals to the Mexican population to immunize their pets have found some success.

Latent Rabies. In 1991, J.S. Smith (U.S. Centers for Disease Control, Atlanta, Georgia) and colleagues reported on studies of rabies viruses isolated from three individuals who had died from the disease and prior to their death had not revealed possible sources of their infections. All three patients had immigrated to the United States from Laos, the Philippines, and Mexico. The viral isolated in each case matched the antigenic or genetic characteristics of a rabies variant found in specimens from rabid animals obtained from or near the country in which the patient lived before immigrating to the United States. None of these variants were found among the isolates collected from rabid animals in the United States. Inasmuch as these patients had lived in the United States for nearly 5 years before onset of clinical manifestations of rabies, the study panel assumed that rabies occurred after long incubation periods. Thus the findings emphasize the need for careful questioning of patients and their family members who have lived in areas outside the United States in which rabies is endemic. Evidently, rabies virus can persist for long periods without producing clinical signs.

In some countries where rabies is not endemic, it is common to postpone post-exposure rabies treatment if the offending animal appears healthy at the time of the attack and can be observed and remains well for a period of 10 days. This may be a relatively safe procedure in regions where rabies is not endemic, but it is hazardous in a country in which canine rabies is hyperendemic.

As pointed out by T. Hemachudha (Queen Saovabha Memorial Institute, Bangkok, Thailand) and colleagues, "We have cared for one patient with rabies in whom treatment was delayed for five days while the animal responsible for the bite was being observed. The patient had the first signs of rabies two weeks after the start of treatment. Seven other patients with rabies received no treatment or only partial treatment after exposure, since the animals that bit them remained healthy for more than two weeks. Four dogs and one cat outlived the patients they bit. All the victims and their family members denied any other possible exposures to rabies. Unfortunately, none of these animals were available for study.

"Most animals that bite humans in Thailand are semidependent and semirestricted. Thus, observation of such animals is unlikely to be successful, and any attempt at observation only delays treatment. Furthermore, human rabies in Thailand is known for its short incubation periods, with the first symptoms developing in 71 percent of cases within one month. We therefore do not observe animals without first starting treatment of the patient with tissue-culture vaccine and equine or human rabies immune globulin where indicated. The vaccinations are discontinued if the animal remains well after two weeks. By the end of the second week, the level of neutralizing antibody in the patient has reached an arbitrary protective level, providing protection against exposure in the future. We suggest that observing a dog or a cat that has bitten a person in a country such as Thailand may be considered a form of 'Siamese roulette.'"

Reasons for Delaying Rabies Therapy. With numerous unknown factors pertaining to human rabies, notably its incubation period, some of the reasons given for a "waiting period" prior to commencing treatment include:

1. Exposure is seemingly insignificant, such as superficial bites by bats or other small mammals,
2. exposure to airborne particles (aerosols),
3. ignorance or fear of rabies treatment,
4. patient is too ill to be interviewed, thus ruling out important details until it is too late,
5. in the case of latent rabies, the initial cause may have occurred months or years before. Only recently has it been possible to make rapid clinical tests for the disease.

Additional Reading

Baer, G.M.: "The Natural History of Rabies," 2nd Edition, CRC Press, LLC., Boca Raton, FL, 2000.
Childs, J.E., L. Colby, J.W Krebs, et al.: "Surveillance and Spatiotemporal Associations of Rabies in Rodents and Lagomorphs in the United States, 1985–1994," *Journal of Wildlife Diseases*, **33**(1), 20–27, 1997.
Conzelmann, K.K.: "Non-segmented Negative-strand RNA viruses: Genetics and Manipulation of Viral Genomes (Review)," *Annual Review of Genetics* **32**, 123–162, 1998.
Dean, D.J., M.K. Abelseth, and P. Atanasiu: "The Fluorescent Antibody Test. In Meslin, F.X., M.M. Kaplan, and H. Koprowski (Eds.), Laboratory Techniques in Rabies (pp. 88–95). World Health Organization, Geneva, Switzerland, 1996.
Kuwert, C., et al.: "Rabies in the Tropics," Springer-Verlag, New York, NY, 1985.
Metze, K. and W. Feiden: "Rabies Virus Ribonucleoprotein in the Heart," *N. Eng. J. Med.*, 1814 (June 20, 1991).
Smith, J.S., et al.: "Unexplained Rabies in Three Immigrants in the United States," *N. Eng. J. Med.*, 205 (January 24, 1991).
Staff: "Morbidit and Mortality Weekly Report," issued weekly by the Massachusetts Medical Society, Waltham, Massachusetts.
Winkler, W.G. and K. Bögel: "Control of Rabies in Wildlife," *Sci. Amer.*, **86** (June 1992).

Web Reference

Center for Disease Control and Prevention: *http://www.cdc.gov/health/diseases.htm*

R.C. VICKERY, M.D., D.Sc., Ph.D., Blanton/Dade City, FL.

RACCOONS *(Mammalia, Carnivora)*. These animals are of the family *Procyonids*, the organization of which is shown in Table 1. The most primitive procyonid species are the Ring-Tailed Cats or Cacomistles (*Bassariscus*). Their total length ranges from 61.5 to 100 centimeters (24 to 39 inches), tail length from 31 to 53 centimeters (12 to 21 inches), and weight from 870 to 1300 grams (1.9 to 2.9 pounds). The head is flattened and has a long, tapered snout. The ears are large, oval, and thin, with well-developed pouches. The slender, graceful body is borne by short legs. The ring-tailed cat walks in a semi-plantigrade manner; the heels of the feet have dense fur. The prominent head markings include the whitish lips, cheeks, and eyebrow regions, a small, dark spot on each side of the snout and in front of the base of the ears, the black nose, and the large eyes with their dark borders. The soft, rather long fur is tan on the upper side of the body, with brownish or brown hues as well, and in some spots even black. The underfur is lead-colored and the belly is whitish to whitish-brown. The black-and-white-banded tail is particularly striking; it is quite bushy. Including the black tip of the tail, there are seven to nine stripes in the tail; they are not closed on the underside of the tail. Modern species are barely distinguishable from the fossil forms found in the Upper Tertiary, and for this reason the ring-tailed cat could be called a "living fossil."

TABLE 1. GENERAL ORGANIZATION OF THE RACCOONS (PROCYONIDS)

Ring-tailed cats or cacomistles (*Bassariscus*)
   North American ring-tailed cat (*Bassariscus astutus*)
   Central American ring-tailed cat (*Bassariscus sumichrasti*)
Olingos (*Bassaricyon*)
Raccoon (*Procyon*)
   North American raccoon (*Procyon lotor*)
   Guadalupe Islands raccoon (*Procyon minor*)
   Cozumel Island raccoon (*Procyon pygamaeus*)
   Crab-eating raccoon (*Procyon cancrivorus*)
Coatimundis (*Nasua*)
   White-nosed coati (*Nasua narica*)
   Nelson's coatimundi (*Nasua nelsoni*)
   Ring-tailed or red coati (*Nasua nasua*)
   Mountain coati (*Nasua olivacea*)
Kinkajou (*Potos*)

Two species are distinguished: (1) North American Ring-Tailed Cat (*Bassariscus astutus*), with 14 subspecies; and (2) Central American Ring-Tailed Cat (*Bassariscus sumichrasti*), with 5 subspecies. The latter differs from the North American species in having dark brownish fur, a blackish snout and feet, slightly tapered ears, and rings which are not as distinct toward the tip of the tail. The claws are also different, being partially retractile in the North American species and completely nonretractile in the Central American ring-tailed cat.

The North American ring-tailed cat is distributed in southern Oregon, the southwestern U.S.A., and into Mexico (including Baja California) as far south as Veracruz and Oaxaca. The Central American species is found from southern Mexico southward to western Panama.

Both species live on small mammals, birds and their eggs, arthropods, and reptiles. Domestic poultry that roost in trees are sometimes taken as well. The plant material taken varies with season and location; the most common plant items include juniper berries, persimmons, wild plums, the fruits of the opuntia and saguaro cactus, wild figs, legumes, and fresh corn.

The Olingos are often confused with the kinkajous. Their total length is from 75 to 95 centimeters (29.5 to 37 inches), tail length from 40 to 48 centimeters (16 to 19 inches), and weigh from 970 to 1500 (2 to 3.3 pounds). The head is rounded but flattened on top, with small, rounded ears and a tapered snout. The eyes are fairly big; they protrude and have a cinnamon-colored iris and a narrow, vertical pupil. The body is bony, with a very slender trunk and rather long limbs. The manner of walking is semi-plantigrade. The fingers and toes have sharp, greatly bent claws. Coloration of the long, loose hair is gray-brown with blackish hues on the upper side, and yellowish on the underside and insides of the limbs. A yellowish band extends across the neck to the base of the ears. The tail is very long, with fairly long hair, and 11 to 13 dark rings, often indistinct, which are not closed on the underside of the tail. Unlike the kinkajou's, their tail is not prehensile.

The olingo is distributed from northern Nicaragua across northwestern South America to Peru and northern Bolivia. The name "olingo" is a Panamanian word; in some regions it is known as cautaquil or cusacusa. It inhabits the tropical rainforest at an altitude of about 1800 meters (5905 feet). The species is not nearly as prevalent as the kinkajou, which inhabits the same ecological niche.

The olingo feeds chiefly on fruits but occasionally hunts insects and warm-blooded animals. Olingos are primarily nocturnal, living alone or in pairs, usually in treetops, and only rarely coming to the floor.

The procyonid family received its name from its most famous member, the Raccoon (*Procyon*). Its total length ranges from 60.3 to 105 centimeters (24 to 41 inches), tail length from 19.2 to 40.5 centimeters (7.5 to 16 inches). Weight varies with species from 1.5 to 22 kilograms (3.3 to 48.5 pounds). The head is broad, with a tapered snout and upright, rounded ears of medium size. The body looks plump with its long, thick fur. The legs are relatively long and have short hair and toes which can be spread greatly. The claws are long and sharp. The head has short fur, and a white-edged black "mask" extends from the cheeks across the eyes and snout, becoming somewhat lighter across the nose and running across the forehead in a thin band. The forehead and the sides of the snout and chin are white; the nose is black. The chief color is iron-gray, with yellow-brown and rust hues mixed into the region around the nape of the neck. The underside has short fur with white instead of black tips, and is not as thick as the fur on the upper side. Five to seven dark, narrow rings alternate with broader gray to light brown rings on the tail; the tip of the tail is always dark. See Fig. 1.

Fig. 1.   Racoon. (*A.M. Winchester.*)

There are 7 species with 32 subspecies of raccoons, of which the Raccoon or North American Raccoon (*Procyon lotor*), with 25 subspecies,

is the most familiar. There are 5 small species (*Procyon insularis*, *Procyon maynardi*, *Procyon gloveralleni*, *Procyon minor* and *Procyon pygmaeus*) which are found only on islands off Florida and Mexico. The South American Crab-Eating Raccoon (*Procyon cancrivorus*) has 5 subspecies distributed from Costa Rica and Panama across most of South America to northern Argentina. It differs from its northern relative in its coarser, thinner fur; the hair in the nape of the neck is directed forward. Since the crab-eating raccoon lacks underfur, it looks more slender and long-legged than the North American species. The claws are straighter, broader, and blunter; the lips, chin, and throat are gray-white, while the back is ash-gray to ochre or reddish with black tips on the fur.

Raccoons prefer forested terrain and stay in the vicinity of ponds, lakes, streams, and swamps; they also occur in mangrove forests along subtropical and tropical coastal plains, and on the edges of savannas and semi-arid regions as long as they have an ample water supply. Raccoons are not found at altitudes above 2500 meters (8202 feet), in pure evergreen forests, or in arid regions lacking water.

The diet of the raccoon changes with the season, and raccoons make very good use of what is available at a particular time. Animal prey includes insects, young small mammals, earthworms, crustaceans, snails, mussels, reptiles (especially their eggs), amphibians, and fishes. Birds are less often taken. In swampy regions and long fresh-water lagoons, raccoons take young muskrats from their nest, and in some areas this has almost led to the disappearance of these rodents. Plant materials comprise over half the annual diet component in the North American raccoon; it feeds on wild fruit, berries, grasses, leaves, rinds, beechnuts, and similar foods. This raccoon also feeds in stands of young corn, melons, sweet potatoes, young sugar cane, and fruit. In Canada and the northern U.S.A., where the raccoon endures varying periods of cold weather in a semihibernation state, it builds up a fat store from the great supply of acorns. During the winter the raccoon loses up to 50% of its fall weight. The crab-eating raccoon, as its common name indicates, has a much more specialized diet. Its molar teeth are broader and have prominent ridges better suited for masticating tough material.

Raccoons move across land at a slow amble, with the head lowered, back arched, and tail dangling downward. They can run at a gallop for short time, and reach a maximum speed of 24 kilometers per hour (15 miles per hour). Trees are climbed at the normal pace or even at this gallop. Raccoons are solitary. If two feeding competitors meet, they threaten each other by growling and lowering the head, baring their teeth, and laying their ears back. The fur of the nape of the neck and the shoulders becomes erect. This bluff usually has the desired effect of frightening both of them away from each other, and generally no fight ensues. Individual territories overlap considerably and they are defended by their owners (and thus in the truest sense should not strictly be considered as territories). The raccoon population density in a particular area is dependent upon the food supply, the number of trees suitable for nests, and predation pressure.

Raccoons are chiefly nocturnal animals. They climb readily and usually nest in hollow trees. In some sections of the United States, coon hunting at night with specially bred and trained dogs is regarded as a sport. Raccoon fur is among the better grades, although not the finest or most beautiful. It is used in making coats.

The Coatimundis (*Nasua*) have two very striking characteristics: (1) a moveable, trunklike snout, which protrudes beyond the lower jaw, and (2) a long, vertically carried tail. Their length is from 74 to 134 centimeters (29 to 53 inches), including a tail length from 36 to 68 centimeters (14 to 27 inches); they weigh from 3 to 6 kilograms (6.5 to 13 pounds). The head is long. The ears are rounded, short, and hidden within the fur. The fur on the head and legs is short; elsewhere on the body it is thick and coarse. It is quite stiff but has underfur, which is woolly and curly. Coloration varies considerably from cinnamon-brown or reddish-brown to brown gray. In one animal it may even change with successive molts! A group of coatimundis may contain both dark and light individuals; coloration differences are also independent of sex and age. The coloration on the under side is yellowish to dark brown. There is a face mask. The forehead and top of the head are yellowish-gray; the lips in North and Central American specimens are white. There is a whitish spot above and behind the eye and at the base of the long whiskers. A stripe extends along both sides of the nose from the eyes to the tip of the nose. The snout, chin and throat are whitish-yellow, and the insides of the ears are pale yellow. The paws are dark brown to blackish. The claws on the hand are long, powerful, blunt, and slightly curved, while those on the feet are short, greatly arched, and very sharp. The tail generally has an indistinct ringed marking.

Four species with seventeen subspecies are distinguished: (1) The White-Nosed Coati (*Nasua narica*), which generally has a light color, comprises three subspecies distributed from the southwestern U.S.A. to Panama. (2) Nelson's Coatimundi (*Nasua nelsoni*), a much smaller species with shorter, softer, silkier fur and smaller teeth, is found on the island of Cozumel off the Mexican province of Quintana Roo. (3) The Ring-Tailed or Red Coati (*Nasua nasua*), with eleven subspecies (which are difficult to distinguish), is found as far south as Argentina and along coastal plains up to altitudes of 3000 meters (9843 feet). (4) The Mountain Coati (*Nasua olivacea*) is another small species, with a slender head, long snout, and small, short-crowned, sharp-ridged teeth. Its coloration is olive-brown to rust-brown, with blackish underfur in specimens from Colombia and Venezuela but whitish underfur in specimens from Ecuador. Its tail is yellowish-gray and has black rings. Distribution is in the mountain forests and clearings in western Venezuela and the Colombian-Ecuadorian Andes from 2,700 to 3,100 meters (8858 to 10,171 feet).

Within this great range, the coatimundis have a highly adaptable nature, living in tropical lowlands, dry, high-altitude forests, in oak forests, mesquite grassland, and even on the edges of forests. During the last few decades, the white-nosed coati has penetrated further into the U.S.A. and has become a stable part of the animal life in southern Arizona, southwestern New Mexico, and southwestern Texas.

Coatimundis feed chiefly on invertebrates but also prey on lizards and small rodents; birds are caught infrequently. Like the raccoon, the coatimundi rolls its prey under the thick soles of its forefeet. This quickly kills prey which bite and sting, and the rolling process also removes harmful spines and other chitinous parts. Vertebrates are pressed to the ground with the paws and are killed by a bite to the head. The coatimundi eats large fruits, scraping the meat of the fruit out with the claws.

The Kinkajou (*Potos flavus*) is the only member of its genus. It differs from all the other raccoonlike species by its prehensile tail. Length is from 81 to 113 centimeters (32 to 44 inches), of which tail length is from 39.5 to 55.5 centimeters (15.5 to 22 inches). Weight ranges from 1.8 to 4.6 kilograms (4 to 10 pounds). The head is round, the ears short and round. The snout is blunt. The protruding eyes have chestnut-brown irises and round pupils. The trunk is long and the limbs are short. The fingers and toes are covered with membrane for one-third of their length. They have curved, sharp claws. The soles, like those of the olingo, are short and thickly covered with hair. The tail is about the same length as the body length; it is round in cross section and uniformly covered with short hair, and tapers toward the tip. The fur is very thick, soft, short, and gleaming. The upper side of the body is olive-brown, yellowish-brown, or reddish-brown to sandy, often covered with a bronze sheen. The middle of the back is darker, and the underside is yellow-brown, light tea-colored, or even golden-yellow. There are fourteen subspecies of the kinkajous; they differ in skull and tooth characteristics, coloration, and body size.

The kinkajou feeds chiefly on plant materials, primarily fruits such as wild figs, zapote, guava, avocado, and mango. It also takes soft-shelled nuts and legumes; insects are eaten less often. The narrow, greatly extensible tongue is used to pull out the soft fruit meat and to lick nectar, insects, and the honey of wild bees. The kinkajou often eats bird eggs and sometimes eats young birds as well. During the day the kinkajou sleeps in a coiled position on its side; the front feet cover the eyes. It sleeps in a tree hollow or in a thick, cool network of leaves and vines.

RACE RUNNER (*Reptilia, Sauria*). Slender lizards, reaching a length of about 10 inches, including the long tapering tail. They are found throughout the United States with the exception of the most northern part. One species is known as the swift, *Cnemidophorus sexlineatus*.

RACON. A transponder for interrogation by a primary radar.

RADAR. The use of electromagnetic energy for the detection and location of reflecting objects. Radar operates by transmitting an electromagnetic signal and comparing the echo reflected from the target with the transmitted signal. The first demonstration of basic radar effects was by Hertz in the late 1880s, when he verified Maxwell's electromagnetic theory. Hertz showed that shortwave radiation could be reflected from metallic and dielectric bodies. Although the basic principle of radar was embodied in Hertz's experiments, the practical development of radar did not arrive for another 50 years. Practical models of radar appeared in the late 1930s. The rapid advance in radar technology during World War II was aided by the many significant contributions of physicists and other scientists pressed into the practical pursuit of a new technology important to the military. In addition to its military application, radar now finds extensive use in air and ship navigation, air traffic control, rainfall observation, tornado detection, hurricane tracking, surveying, radar astronomy, and highway patrol activities. See also **Radar Astronomy**; "Radiosonde" in the entry on **Wind and Air Velocity Measurements**; and "Weather Radar" in the entry on **Weather Technology**. The contributions of radar to other nondefense uses are described later.

The measurement of distance, or range, is probably the most distinctive feature of radar. Range is determined from the time taken by the transmitted signal to travel out to the target and back. The distances involved may be as short as a few feet, or as long as interplanetary distances. If the target is in motion relative to the radar, the echo signal will be shifted in frequency by the doppler effect and may be used as a direct measurement of the relative target velocity. A more important application of the doppler shift is to separate moving targets from stationary targets (clutter) by means of frequency filtering. This is the basis of MTI (moving target indication) radar.

Radar antennas are large compared to the wavelength so as to produce narrow, directive beams. The direction of the target may be inferred from the angle of arrival of the echo. Radar antenna technology has profited from the theory and practice of optics. Both the lens and the parabolic mirror have their counterparts in radar, and the analysis of antenna radiation patterns follows from diffraction theory developed for optics. The greater versatility of materials in the radar frequency region, however, offers more flexibility in implementing many of the principles of optics not practical in the visual portion of the spectrum.

In defense system radar of the early 1980s, systems were being developed which were so quiet that anti-radiation missiles were less likely to home in on the radar's beam. These new radars have two antennas, one to transmit low-energy beams continuously; the other to listen for returns. Conventional radars differ by transmitting high-energy pulses so that one antenna can alternately transmit and receive. A new antenna technique reduces the radar's side lobes — the secondary patterns of energy that enemy missiles can home on.

The external appearance of a radar is dominated by the antenna. Most radars use some form of parabolic reflector. The radar antenna can also be a fixed array of many small radiating elements (perhaps several thousand) operating in unison to produce the desired radiation characteristics. Array antennas have the advantage of greater flexibility and more rapid beam steering than mechanically steered reflector antennas because the beam movement can be accomplished by electrically changing the relative phase at each element of the antenna. High power can be radiated since a separate transmitter can be applied at each element. The flexibility and speed of an array antenna make it necessary in some instances to control its functions and analyze its output by automatic data processing equipment rather than more simple formats involving display tubes.

Two obstacles to the advent of phased array radars — high cost and weight — are being overcome with innovations in technology and manufacturing. Tiny diode phase shifters now operate on the same power as their bulkier ferrite counterparts. The many wires that required individual connections and testing are giving way to thick-film fabrication, in which circuits are silkscreened onto aluminum wafers. It is now possible to place radiators, phase shifters, and power dividers onto single substrates — building blocks that can be assembled into larger sections before undergoing initial tests.

Radars are generally found within the microwave portion of the electromagnetic spectrum, typically from about 200 MHz (1.5 meters wavelength) to about 35,000 MHz (8.5 millimeters wavelength). These are not firm bounds. Some radars operate outside these limits. The well-known British CH radar system of World War II, which provided warning of air attack, operated in the high-frequency region in the vicinity of 25 MHz. Experimental radars have been demonstrated in the millimeter wavelength region, where small physical apertures are capable of narrow beam widths and good angular resolution. Radar principles, of course, have been applied at optical frequencies with lasers for the measurement of range and detection of small motions, using the doppler effect.

The detection performance of a radar system is specified by the radar equation, which states:

$$P_{rec} = \frac{P_t G}{4\pi R^2} \times \sigma \times \frac{1}{4\pi R^2} \times A$$

$$\begin{pmatrix} \text{Received} \\ \text{Power} \end{pmatrix} = \begin{pmatrix} \text{Power Density} \\ \text{at a Distance} \\ R \end{pmatrix} \times \begin{pmatrix} \text{Target} \\ \text{Backscatter} \\ \text{Cross Section} \end{pmatrix}$$

$$\times \begin{pmatrix} \text{Space} \\ \text{Attenuation on} \\ \text{Return Path} \end{pmatrix} \times \begin{pmatrix} \text{Antenna} \\ \text{Collecting} \\ \text{Area} \end{pmatrix}$$

where P_t is the transmitted power; G is the transmitting antenna gain; R is the range; σ is the backscatter cross section; and A is the effective receiving aperture of the antenna. The wavelength λ of the radar signal does not appear explicitly in this expression, but it can be introduced by the relationship between the gain and effective receiving area of an antenna, which states:

$$G = \frac{4\pi A}{\lambda^2}$$

The detection capability and the measurement accuracy of a radar are ultimately limited by noise. The noise may be generated within the radar receiver itself, or it may be external and enter the receiver via the antenna, along with the desired signal. External noise is generally small at microwave frequencies, but it can be a significant part of the overall noise if low-noise receiving devices, such as the maser and the parametric amplifier are used.

Since the effects of noise must be considered in statistical terms, the analysis and understanding of the basic properties of radar have benefited from the application of the mathematical theory of statistics. The statistical theory of hypothesis testing has been applied to the radar detection problem where it is necessary to determine which of two hypotheses is correct: The output of a radar receiver is due to (1) noise alone, or (2) signal plus noise. One of the results is the quantitative specification of the signal-to-noise ratio required at the receiver for reliable detection. Also derived from hypothesis testing based on the likelihood ratio or a *posteriori* probability are concepts for ideal detection methods with which to compare the performance of practical receivers. The statistical theory of parameter estimation has also been applied with success to analyze the accuracy and theoretical limits of radar measurements.

Reliable detection of targets requires signal-to-noise power ratios of the order of 10 to 100 at the receiver, depending upon the degree of error that can be tolerated in making the decision as to the presence or absence of a target. Even larger values are generally needed for the accurate measurement of target parameters. Although these values may seem high, for comparison, the minimum signal-to-noise ratio of quality television signals is usually of the order of 10,000.

The rms error δT in measuring the time delay to the target and back (range measurement) can be expressed as:

$$\delta T = \frac{1}{\beta \sqrt{2E/N_0}}$$

where β is defined as the effective signal bandwidth; E is the total energy of the received signal; and N_0 is the noise power per unit cycle of bandwidth assuming the noise has a uniform spectrum over the bandwidth of the receiver. The square of β is equal to $(2\pi)^2$ times the second central moment of the power spectrum normalized with respect to the signal energy. For a simple rectangular pulse, E/N_0 is approximately equal to the signal-to-noise (power) ratio. To obtain an accurate range measurement, E/N_0 and the signal bandwidth must be large. A similar expression applies to the accuracy of the measurement of doppler frequency if the rms time delay error is replaced by the rms frequency error and the effective bandwidth is replaced by the effective time duration of the signal. Thus, the longer the signal duration and the greater the ratio E/N_0, the more accurate is the doppler frequency measurement. Likewise, the angular measurement accuracy also depends on the ratio E/N_0 and the effective aperture size.

In addition to noise, radar can be limited by the presence of unwanted interfering echoes from large nearby objects, such as the surface of the ground, trees, vegetation, sea waves, and weather. Although these "clutter" echoes may be troublesome in some applications, they are sometimes echoes of interest, as, for example, in ground mapping and meteorological applications.

Radar as Catalyst of Technological Progress. In 1940, the Radiation Laboratory (Massachusetts Institute of Technology) was established to investigate microwave frequencies for radar. Scientists were agreeably surprised to find that the usable frequency spectrum could be extended by some three orders of magnitude. Power sources were developed that were capable of delivering several megawatts of power at 3,000 megahertz and kilowatts up to 24,000 megahertz. It was found that good radar resolution required equipment with bandwidths of several megahertz as contrasted with the few kilohertz required for voice radio circuits. A whole technology was required to build servomechanisms for driving highly precise and large engineering structures. Also required were improved pulse techniques and cathode ray tubes and delay line storage devices for operation with the early electronic computers. New concepts in circuitry and components were required, several of which made television practical just a few years later. In the late 1940s, the western world was threatened by a new kind of attack from the Soviet Block, which exploded its first atomic device in 1948. This caused a vitally renewed interest in radar and associated technology, which had cooled a bit after the close of World War II. Particularly targeted among the new interests were effective means for coupling radar with the rapidly developing digital computer technology of that period. Radar developments, in turn, accelerated computer technology in a sort of technological symbiosis.

Great interest was shown in developing radars that could detect low-flying aircraft. A solution proposed at that time involved the use of large numbers of radars operating in concert and yielding both high-and low-altitude surveillance information. So much data from such systems required data analyzers in the form of computer systems. In the late 1940s, the Massachusetts Institute of Technology built the MIT Whirlwind, the first reliable and fast computer designed for real-time usage. Air defense equipment at that time had not been transistorized, but it turned out that the application required the speed and reliability only obtainable with transistors. These, however, did not become available until the early 1950s. Later developments include a core memory (1955). The British were also having similar problems in updating their radar nets. Missile guidance systems placed a new load on radar technology, as did later needs for satellite-tracking radars. The needs of radar also catalyzed the development of modern signal-processing techniques, including pulse compression and matched filtering, as well as the development of acoustic wave technology and charge-coupled devices. Radar developments led to several nondefense uses, such as radio and radar astronomy, microwave spectroscopy, and the instrument technology required for earth resource meteorological, and navigational satellites. See also **Satellites (Scientific and Reconnaissance)**.

Radar in the 1990s. Throughout the "Cold War" period extending over several decades, most advancements in radar technology stemmed from weapons development. A technology referred to as gallium arsenide radio-frequency (RF) wafer-scale integration was developed shortly before the dissolution of the former Soviet block, but is now an important advancement for all radar uses. This technology has resulted in the reduction of radar size, weight, and cost. The original incentive was for its adaptation to stealth aircraft. See Fig. 1.

Although most stealth technology has been applied to aircraft, some efforts also have been directed to seacraft, as exemplified by Sweden's *Smyge*, a small attack seacraft that was launched in 1990. Antennas and mast are hull-integrated. When not in use, missiles and guns retract into a "stealth cupola." The hull is angular, the bridge is low-profile, and the hull is smooth and constructed of fiberglass-reinforced plastic.

By contrast, other developments have been underway to develop an effective anti-stealth radar that will, to an effective extent, negate the "hiding" characteristics of stealthcraft. One of these techniques is over-the-horizon backscatter radar (OTH-B). This radar uses the principle first developed by Marconi, namely using ionospheric reflections to cover some 4.8 mil square nautical miles over a distance of 1800 nautical miles. The system operates at frequencies from 5 to 28 MHz ($\lambda = 60$ to 11 m). The system uses a long antenna: 3630 ft (1115 m), employing steel beams and cables that range in height from 35 to 135 ft (10.6 to 35 m). Such a transmitter may be powered by twelve 10 kW tubes per sector, emitting 360 kW of radiated RF power. The receiving antenna may be an array some 4980 ft (1518 m) long and 64 ft (20 m) high. An experimental installation was located in Maine in 1990 and covers ranges from 500 to 1800 nautical miles (925 to 3330 km), thus reaching the coast of Cuba, as well as Haiti, the Dominican Republic, and Puerto Rico. Plans are underway to

Radar imaging from spacecraft has been used extensively in recent year to explore the planets and has been particularly effective where the planet's atmosphere interferes with optical observations. See also **Venus**. Radar imaging of Earth is playing a major role toward understanding the pceams and terraom in a number of global-change research programs. See also **Global Change**.

Additional Reading

Bierman, H.: "Microwave and mm-Wave Technology," *Microwave J.*, **44** (June 1991).

Curlander, J. and R. McDonough: "Synthetic Aperture Radar," John Wiley & Sons, Inc., New York, NY, 1992.

Goldman, S.J.: "Phase Noise Analysis in Radar Systems," John Wiley & Sons, Inc., New York, NY, 1990.

Harper, J.D. and J.W. Downs: "A New Resin System for Radomes," *Microwave J.*, **94** (November 1992).

Kaufman, W.: "Radar Imaging: Forest X-Ray," *Amer. Forests*, **46** (September–October 1990).

Mott, H.: "Polarization in Antennas and Radar," John Wiley & Sons, Inc., New York, NY, 1999.

Oliver, C.J., and S. Quegan: "Understanding Synthetic Aperture Radar Images," Artech House, Inc., Norwood, MA, 1998.

Stiglitz, M.R., and C. Blanchard: "Over-the-Horizon Backscatter Radar," *Microwave J.*, **32** (May 1990).

Thomas, L.: "Radar Investigations of the Middle Atmosphere," *Review (University of Wales)*, **47** (Spring 1989).

Wehner, D.R.: "High-Resolution Radar," Artech House, Inc., Norwood, MA, 1994.

RADAR ASTRONOMY.   See **Radio Astronomy**.

RADAR BEACON.   A beacon transmitting a characteristic signal on radar frequency, permitting a craft to determine the bearing and sometimes the range of the beacon. A racon returns a coded signal when triggered by the proper type of radar pulse; a ramark continuously transmits a signal that appears as a radial line on the plan position indicator.

RADAR MILE.   A time unit of 10.75 microseconds duration; the time it takes for the signal emitted by a radar to travel from the radar to a target one mile distant and return to the radar.

RADIAL DISTRIBUTION FUNCTION.   The radial distribution function for a liquid is defined as the function $\rho(r)$ where $4\pi r^2 \rho(r)\,dr$ is the average number of molecules with centers at distances between r and $r + dr$ from some selected molecule. If the liquid is isotropic it is the average number density at distance r from the selected molecule. The radial distribution function may be computed from measurements of x-ray diffraction patterns and it is of central importance in the kinetic theory of liquids.

RADIAL VELOCITY (Star).   That component of the space motion of a star that is directed toward the sun is known as the radial velocity of the star, or the velocity of the star in the line of sight. It is measured by spectroscopic methods, employing the Doppler-Fizeau principle, and is determined directly in linear units (i.e., kilometers per second, or miles per second).

Since a comparison spectrum must be available for measurement of the Doppler displacement of the stellar spectral lines, a slit spectrograph must be used for an accurate determination of radial velocity. This instrument is wasteful of light, and only one star can be observed at a time. For these reasons, the number of stars for which accurate radial velocities are known is small relative to the total number of stars. The objective prism may be used to determine approximate radial velocities for a large number of stars. In one of the applications of this instrument, a comparison spectrum is obtained by interposing a neodymium screen between the prism and the photographic plate. This produces a few absorption lines on the stellar spectrum relative to which the stellar lines themselves may be measured. Another application of the objective prism to this problem utilizes the fact that the Doppler displacement for a line in the red is greater than that for a line in the violet. Hence, the length of the spectrum between these extremes will be changed by an amount proportional to the radial velocity. Although the results obtained by the use of the objective prism are only approximate, they may be used for statistical study of stellar motions.

From a study of the variations in radial velocity of certain stars, known as spectroscopic binaries, the relative orbits of these objects may be

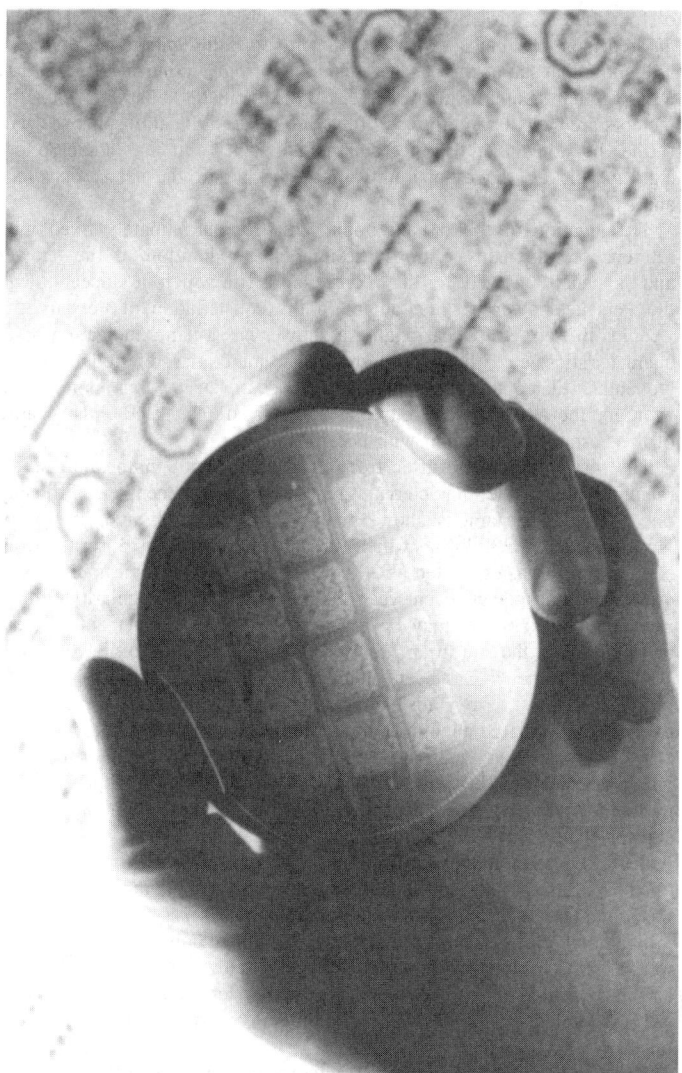

Fig. 1.   World's largest gallium arsenide microwave circuit. Units like this are well suited for radar because of their inherent speed, allowing use of higher frequencies and smaller antennas. (*Westinghouse Electric Corporation.*)

construct similar systems, one in Minnesota, one on the U.S. West Coast, and one in Alaska. The system is planned, not only for defense, but also for detecting small aircraft engaged in drug smuggling. The United Kingdom also is planning to make a similar installation at St. David's airfield in Pembrokeshire.

Radomes are primarily used to protect antennas and electronic systems from weather. A major requirement of a radome is radar transparency (ability to minimize attenuation of the radar signal). Research is continuing to develop the ideal radome material. The radome designer has a number of resin systems from which to choose. These include polyesters, vinyl esters, epoxies, polyimides, polybutadienes, phenolics, cyanate esters, and silicones.

Microwave and mm-wave technology played a prominent role during the Desert Storm campaign. Although most of this technology remains classified, an excellent summary is given in the Bierman reference listed.

Radar systems used by highway patrol officers have been threatened by the more recently developed laser speed gun, but laser gun jammers are also being developed. Ironically, it has been proposed that jammers would permit drivers to dial in whatever speed they want the speed gun to register.

Automobile designers currently are developing an all-weather radar that will warn drivers of obstacles obscured by rain or fog. A saucer-sized antenna, operating at high-resolution millimeter wavelengths, would fit behind the grille. Processors would convert signals to a head-up display on the windshield. A release date by the year 2000 has been announced for this feature.

determined. Since the first order relation

$$\frac{\Delta\lambda}{\lambda} = \frac{v}{c}$$

(where λ is the local rest wavelength, $\Delta\lambda$ is the observed shift, v is the frequency and c is the velocity of light) does not differentiate between motion of the source and the observer, one must correct for the revolution and rotation of the earth and, in many cases, for the space motion of the sun.

See also **Spectroscopic Binaries**.

RADIANCE.   In radiometry, a measure of the intrinsic radiant intensity emitted by a radiator in a given direction. It is the irradiance (radiant flux density) produced by radiation from the source upon a unit surface area oriented normal to the line between source and receiver, divided by the solid angle subtended by the source at the receiving surface. It is assumed that the medium between the radiator and receiver is perfectly transparent; therefore, radiance is independent of attenuation between source and receiver.

If the radiant source is a perfectly diffuse radiator (that is, emits exactly according to Lambert law), then its radiance is equal to its emittance per unit solid angle. The radiance of a light source is termed *luminance* (formerly, brightness).

RADIANT.   1. Pertaining to the emission or the measurement of electromagnetic radiation.

2. In astronomy, the apparent location on the celestial sphere of the origin of the luminous trajectories of meteors seen during a meteor shower. For convenience, the common meteor showers are named for the constellations of stars in which their radiants appear.

3. In describing auroras, a projected point of intersection of lines drawn coincident with auroral streamers; that is, the point from which the aurora seems to originate.

RADIANT ENERGY THERMOMETER.   An instrument that determines the black body temperature of a substance by measuring its thermal radiation. The substance need not be thermally black over the whole spectrum, since it is possible to limit the measurement to those frequencies where it is black.

RADIANT HEATING.   See Infrared Radiation.

RADIANT POINT (Meteor).   If the paths of all the meteors observed from a single station on a given night are plotted on a chart of the sky, it will usually be found that a number of them seem to be coming from a certain particular point in the sky. Such a point is known as a meteor radiant point, and the group of meteors associated with the radiant point is known as a meteor shower. It will further be noticed that, among the meteors belonging to the shower, those at the greater distance from the radiant point will have the longer trails.

This observed effect is merely due to the perspective view of a number of meteors actually entering the atmosphere of the earth in parallel paths. Fig. 1 represents the cause of the radiant point. The circular segment AA represents the surface of the earth with the observer at O; CC represents the upper part of the atmosphere of the earth where the meteors first become visible, and BB the lower atmosphere where the meteors burn out and disappear; ab, cd, ef, and gh represent the actual parallel paths

of four meteors through this layer of atmosphere, and ab', cd', ef', and gh' represent the paths as observed from O. Examination of the figure will show that the apparent paths all radiate from a point in the direction R, the radiant point, which is a direction parallel to that in which the meteors are actually entering and traveling through the atmosphere. It will further be noted that the meteors more distant from the radiant point, e.g., ab' and gh', have apparently longer trails than the nearer ones, cd' and ef'.

The location of the radiant point remains approximately fixed with reference to the constellations throughout the duration of the shower and is usually named for the constellation in which it appears; e.g., the Perseid shower has its radiant in the constellation of Perseus, the Leonids in Leo, etc. Occasionally, a shower has a name indicating other characteristics; e.g., the Leonid shower is sometimes referred to as the November meteors because the shower occurs during that month each year, and the Andromedes are frequently referred to as the Bielids because of their established relation with Biela's Comet.

The various showers differ from each other, both in the number of members and in the characteristics of the individual members. Probably one of the most famous showers on record is the Leonid shower of November 12, 1833, during which the number of meteors observed from some stations was estimated as 200,000 per hour for several hours.

Many of the showers occur year after year with define regularity of date. Such periodic showers may be explained by huge numbers of meteors traveling about the sun in an orbit which intersects the orbit of the earth. Such a phenomenon has been referred to as a "flying gravel bank," but such a descriptive term is misleading because of the fact that few, if any, of the meteors are large enough to be considered as gravel pebbles. In some cases, the meteors are distributed with fair uniformity all along the orbit, in which case the showers will recur on successive years with approximately the same frequency and appearance. Such is the case with the Perseid shower, which may be observed during the latter part of July and the early part of August each year. In other cases, the meteors are concentrated in one or more large swarms with a few scattered members in between along the orbit. This is the case with the Leonid shower.

In a number of cases, the orbits of meteor radiant points have been found to agree with orbits of comets. In some cases, the comets are still observed as comets, and in other cases, the comet itself no longer appears. At present, work is being done in applying new and more powerful techniques, notably radar.

See also **Bielids**; **Leonids**; and **Meteoroids and Meteorites**.

RADIANT POWER.   The intensity of a beam of radiation. It is proportional to the number of photons passing through a plane of unit area perpendicular to the beam and in unit time.

RADIATION.   1. The emission and propagation of energy through space or through a material medium in the form of waves; for instance, the emission and propagation of electromagnetic waves, or of sound and elastic waves.

2. The energy propagated through space or through a material medium as waves; for example, energy in the form of electromagnetic waves or of elastic waves. The term radiation, or radiant energy, when unqualified, usually refers to electromagnetic radiation; such radiation commonly is classified, according to frequency, as radio-frequency, microwave, infrared, visible (light), ultraviolet, x-rays, and γ-rays. Radiation may also be designated as monochromatic, when it has, ideally, one wavelength or actually a narrow band of wavelengths; or as heterogeneous, when it has two or more narrow bands of wavelengths (or particles of two or more narrow energy ranges); or as homogeneous, when it has only one narrow band of wavelengths (or consists of essentially monoenergetic particles).

3. Corpuscular emissions, such as alpha- and beta-radiation or rays of mixed or unknown type.

See also **Electromagnetic Phenomena**.

RADIATIONAL COOLING.   In meteorology, the cooling of the earth's surface and adjacent air, accomplished (mainly at night) whenever the earth's surface suffers a net loss of heat due to terrestrial radiation. See also **Atmosphere (Earth)**.

RADIATION BELT.   An envelope of charged particles trapped in the magnetic field of a spatial body. See **Van Allen Radiation Belts**.

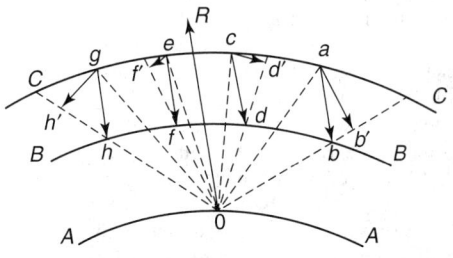

Fig. 1.   Explanation of radiant point.

RADIATION FOG.   See **Fog and Fog Clearing**.

RADIATION FORMULA (Planck).   See **Planck Radiation Formula**.

RADIATION HARDENING (Electronics).   Many electronic components are sensitive to the effects of radiation (charged particles, such as electrons, protons, ions, neutrons, photons, etc.). In the majority of applications, low levels of radiation are not a problem. In military and aerospace applications, however, equipment must be designed to withstand heavy radiation that may be encountered, for example, by a satellite when passing through the Van Allen belts surrounding the Earth and by other electromagnetic energy and atomic particles that appear to abound in the universe (from multitudes of galactic sources, for example). Military electronics also must be built to withstand the radiation effects of a nuclear event. In the United States, the Sandia National Laboratories (Albuquerque, New Mexico) has devoted a number of years of study to radiation hardening. As shown in Fig. 1 when a single, heavy ion shoots through a semiconductor, it may cause a memory bit to change state.

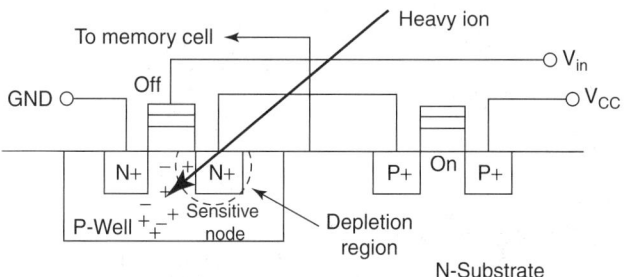

Fig. 1.   Shown in cross section of an integrated circuit, ionized path that results from heavy-ion penetration. In a cross-coupled CMOS memory cell, charges that move along this path cause the inversion of a stored bit.

A radiation hardened device should function normally when exposed to several types of ionizing radiation: (1) *photons*—x-rays, gamma rays; (2) *charged particles*—electrons, protons, alpha particles, beta particles, ions; and (3) *neutrons*.

The energy absorbed by a semiconductor (or other material) from radiation is measured in units of *rads*. Since the amount of energy absorbed by each material differs for a given exposure (measured in *roentgens*, R), the material must be specified. Thus, in the case of silicon (Si) substrates: 1 rad (Si) = 100 ergs/g (Si).

Radiation exposure may occur over a long period of time. *Total Dose* is the term used in connection with electronic circuits that may be used in outer space over long periods; or in a terrestrial situation (equipment near a nuclear reactor). This total dose, for example, may be 100 krad over a 20 year period. *Dose Rate* is the term used for situations where a device may be subjected to pulses of ionizing photon radiation over an extremely short interval (a few hundreds of nanoseconds), but with an amplitude on the order of 10^8 to 10^9 rad (Si). This is called *transient radiation*, denoted by the derivative (rate of change) of gamma with respect to time in seconds, or gamma dot ($\dot{\gamma}$). Dose rate = rad (Si)/s.

Particle levels are measured by concentration and as a time integral of concentration: Thus,

$$\text{Flux} = \frac{\text{Particles}}{\text{cm}^2 \text{ s}}; \quad \text{Fluence} = \frac{\text{Particles}}{\text{cm}^2}$$

Numerous methods have been used to radiation-harden silicon polar devices, including junction isolation, oxide isolation, and dielectric isolation. These methods are designed to eliminate so-called *soft errors* (sporadic, unexpected losses of data) as well as the worst case involving the inversion of a stored bit.

An important advantage of gallium arsenide (GaAs) over silicon is its much greater ability to withstand the effects of radiation. GaAs devices can tolerate a total dose of 10^8 rads, or 10^8 rads/s transient-dose. These Figures better those of silicon by three to four orders of magnitude.

Shielding, of course, can be used to protect against radiation, but this approach requires additional weight and space.

RADIATION (Infrared).   See **Infrared Radiation**.

RADIATION LAWS.   1.   The four physical laws which, together, fundamentally describe the behavior of black-body radiation: (a) the Kirchhoff law is essentially a thermodynamic relationship between emission and absorption of any given wavelength at a given temperature; (b) the Planck law describes the variation of intensity of black-body radiation at a given temperature, as a function of wavelength; (c) the Stefan-Boltzmann law relates the time rate of radiant energy emission from a black body to its absolute temperature; (d) the Wien law relates the wavelength of maximum intensity emitted by a black body to its absolute temperature.

2.   All the more inclusive assemblage of empirical and theoretical laws describing all manifestations of radiative phenomena; e.g., Bouguer law and Lambert law.

See also **Black Body**; **Bouguer and Lambert Law**; **Planck Law**; and **Wien Laws**.

RADIATION MEDICINE.   That branch of medicine dealing with the effect of radiation, specifically high-energy radiation such as X-rays, gamma rays, and energetic particles on the body and with the prevention or cure of physiological injuries resulting from such radiation.

RADIATION PATTERN.   A graphical representation of the radiation of an antenna as a function of direction. Cross sections in which radiation patterns are frequently given are vertical planes and the horizontal plane, or the principal electric and magnetic polarization planes. Also called *antenna pattern*, *lobe pattern*, or *coverage diagram*.

Two types of radiation patterns should be distinguished. They are: (a) the free-space radiation pattern which is the complete lobe pattern of the antenna and is a function of the wavelength, feed system, and reflector characteristics, and (b) the field radiation pattern which differs primarily from the free-space pattern by the formation of interference lobes whenever direct and reflected wave trains interfere with each other as is found in most surface-based radars. The envelope of these interference lobes has the same shape, but, for a perfectly reflecting surface, it has up to twice the amplitude of the free-space radiation pattern.

RADIATION PRESSURE.   That electromagnetic radiation exerts a pressure upon any surface exposed to it was deduced theoretically by Maxwell in 1871, and proved experimentally by Lebedew in 1900 and by Nichols and Hull in 1901. The pressure is very feeble, but can be detected by allowing the radiation to fall upon a delicately poised vane of polished metal.

It may be shown by the electromagnetic theory, by the quantum theory, or by thermodynamic reasoning, making no assumption as to the nature of radiation, that the pressure against a surface exposed in a space traversed by radiation uniformly in all directions is equal to $\frac{1}{3}$ the total radiant energy per unit volume within that space. For black-body radiation, in equilibrium with the exposed surface, the energy density is, in accordance with the Stefan-Boltzmann law, equal to $(4\sigma/c)T^4$; in which σ is the Stefan-Boltzmann constant, c is the velocity of light, and T is the absolute temperature of the space. One-third of this energy density is equal to $2.523 \times 10^{-15}T^4$ (ergs/cm.^{3}), which is therefore the pressure in bars. For example, at the boiling point of water ($T = 373.2°$), the pressure amounts to only 0.00005 dyne/cm.^{2} or about 3 pounds per square mile. Such feeble pressures are, nevertheless, able to produce marked effects upon minute particles like gas ions and electrons, and are of importance in the theory of electron emission from the sun, of cometary matter, etc.

In acoustics, radiation pressure is the unidirectional pressure force exerted at an interface between two media due to the passage of a sound wave.

RADIATION (Quantum Theory).   See **Quantum Theory of Radiation**.

RADIATION RESISTANCE.   1.   The quotient of the power radiated by an antenna to the square of the effective antenna current referred to a specified point.

2.   The acoustic impedance of a plane wave in a given medium, equal to the product of the density of the medium and the velocity of the wave divided by the area of the wave front.

RADIATION SICKNESS.   A syndrome following intense acute exposure to ionizing radiation. It is characterized by nausea and vomiting a few hours after exposure. Further symptoms include bloody diarrhea, hemorrhage under the skin (and internally), epilation (hair falling), and a decrease in blood-cell level.

RADIATION THERAPY.   See **Cancer and Oncology**.

RADIATION (Thermal).   See **Thermal Radiation**.

RADIATION THERMOMETRY.   Radiation thermometry is a means for measuring the temperature of an object without making physical contact with it. It is a practical application of the Planck law and Planck radiation formula. Planck's thermal radiation law predicts very accurately the radiant power emitted per unit area per unit wavelength and per unit solid angle by a blackbody, or complete thermal radiator. The radiant power in this format is called the radiance, $L(\lambda, T)$, and its relation to the other variables involved can be written:

$$L(\lambda, T) = c_1/\{\lambda^5\{\exp[c_2/\lambda T] - 1\} \quad \text{Watts/m}^2 - \mu m - sr$$

Where λ is the wavelength, T is the Absolute temperature, c_1 and c_2 are the first and second Planck radiation constants, sr is the solid angle and exp [ ] is the exponential function.

Advantages of the Method.   In radiation thermometry, the sensor does not have to be in thermal equilibrium with the object. Thus very high temperatures can be measured and measurements can be made very rapidly, limited only by the speed of response of the detector and its electronics. Planck's radiation law is the basis for the International Temperature Scale of 1990 (ITS-90) at temperatures above the freezing point of silver (961.78 °C-the silver point). The realization of the temperature scale above the silver point with carefully designed metrological instruments, while virtually without upper limit, is possible to within a precision of $\pm 0.1\,^\circ$C or better.

However, radiation thermometry is not limited to only high temperatures. Modern instruments are commercially available that can measure well below $-18\,^\circ$C (0 °F). Typical industrial precision lies in the range of ± 0.5 to 1% of absolute temperature.

Advantages and limitations of radiation thermometry are summarized in Table 1. Because of the wide selection of instruments offered, determining the best-suited instrument for a given application can be difficult. Two of the major criteria are (1) wavelength passband response, and (2) target size. Speed of response also may be a primary factor in selection. Factors that also contribute to difficulty of selection include lack of standards and precise terminology. Critical parameters of wavelength passband width, target size or field of view and calibration uncertainty are not always stated explicitly in the commercial literature.

TABLE 1. RELATIVE ADVANTAGES AND LIMITATIONS OF RADIATION THERMOMETERS

Advantages
Can measure:
very high temperatures
moving objects
large areas
inside vacuum or pressure vessels
inside semi-transparent objects
Does not contact (hence mar) object of measurement
Instrument not physically exposed to temperature it measures (as are devices which require physical contact)
Rapid response
High differential sensitivity

Limitations and Disadvantages
Relatively high cost:
initial
installation
requires maintenance
Application engineering required to solve some problems
No uniform calibration Tables

Perspective.   Early radiation thermometers, called radiation and optical pyrometers, were radically different from one another both in design and

use. The simple radiation pyrometer was compact, designed for fixed installation, and served as a transducer only. In contrast, the portable optical pyrometer was a complete measuring system and a much more sophisticated instrument. Developments in integrated circuits, transducer or detector devices, and optical technology have had a profound impact on both fixed and portable instrument design.

Types of Radiation Thermometers

A convenient classification of commercial radiation thermometers is:

1. Wideband instruments
2. Narrowband instruments
3. Ratio (two-color) thermometers
4. Optical pyrometers
5. Fiber optic instruments

Both portable and fixed-installation instruments are available in each class. See Table 2.

TABLE 2. PRINCIPAL TYPES OF COMMERCIAL RADIATION THERMOMETERS

Type	Temperature Range, °C
Wideband	
fixed	1–4000
portable	0–2000
Narrowband	
fixed	−50–2500
portable	0–2500
Ratio (two color)	
fixed	1000–2500 and 300–1200
portable	1000–2500
Optical	
portable	800–2500
Fiber optic	
fixed	100–2500
portable	250–800

Wideband Instruments.   These are the simplest and least expensive of the radiation thermometers. They are available for responding to radiation with wavelengths from 0.3 μm to between 2.5 and 20 μm, depending upon lens or window material used. These instruments also are called *broadband* or *total radiation pyrometers* because of their relatively wide wavelength response and the fact that they measure a significant fraction of the total radiation emitted by the object of measurement. Historically, these devices were the earliest fixed or automatic units. They still find wide application. The characteristics of four specific commercial instruments are given in Table 3.

TABLE 3. CHARACTERISTICS OF FOUR GENERAL-PURPOSE WIDEBAND RADIATION THERMOMETERS

Temperature Range Limits, °C	Waveband Limits μM
500–1800	0.4–2.6
600–1900	0.4–2.6
0–1000	7–20
825–1800	0.4–2.6

Narrowband Instruments.   These instruments usually have a carefully selected, relatively narrow wavelength response, often selected to meet the requirements of a very specific application. The detector, lens, window, and filter(s) are selected to provide the particular wavelength response desired. Optical pyrometers can be considered a subset of this class. See Table 4. See also Fig. 1.

Ratio Thermometers.   Ratio or so-called two-color radiation thermometers measure radiation in two different wavebands and "compute" temperature from the ratio of the two measurements. Changes in the sight path, which affect the signals in both wavebands equally, or variations in the apparent target size do not affect the temperature reading. However, contrary to popular conception, ratio thermometers can be sensitive

TABLE 4. CHARACTERISTICS OF SOME GENERAL-PURPOSE NARROWBAND RADIATION THERMOMETERS

Temperature Range Limits, °C	Mean Effective Wavelength, μM
600–3000	0.9
300–1000	1.6
100–1500	2.3
–40–300	11.0
500–3000	0.9
80–1500	2.2
0–500	11.0
1100–1700	0.6
600–2500	0.9
–50–600	11.0
800–1700	0.9
0–1000	11.0
800–1700	0.9
250–1500	2.2
0–1000	11.0
250–1000	1.9
500–2000	1.0
600–3000	0.8–1.1

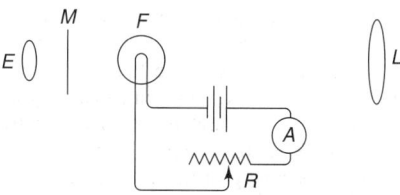

Fig. 1. Diagrammatic arrangement of the parts of an optical pyrometer. The hot body is viewed through a telescope, whose objective L produces at F a real image of the glowing surface. At point F is placed a lamp filament, which is thus viewed through the eyepiece E against the hot surface as a background. A monochromatic filter M is interposed before both, so that their brightness is compared in one spectral region only. The current in the filament is so adjusted by means of rheostat R that the filament becomes invisible against the bright background. The ammeter, A, then gives the current, from which the temperature may be deduced; or the ammeter scale may be graduated to read temperatures directly. In another type, the balance is secured by keeping the current constant and introducing an absorbing wedge between the filament and the objective, as in a wedge photometer. In still others, the temperature is determined, not by the total brightness, but by the relative brightness at two selected wavelengths.

to emissivity changes of the object. Many common industrial materials, notably metals subject to emissivity changes during processing, change differently at different wavelengths. Thus, the ratio of emissivities changes. Ratio thermometers are sensitive to the *ratio* of emissivities in the two wavebands and very sensitive to any changes or errors in setting the correct ratio.

A ratio thermometer is essentially two radiation thermometers contained within a single housing. Several internal components, such as the lens and detector, may be shared. The unique characteristic of the ratio thermometer is that the output from the two thermometers, each having a separate wavelength response, is ratioed. See Table 5.

Emissivity is a source of reduced radiation intensity reaching the thermometer (compared to the calibration on a blackbody source at the

TABLE 5. CHARACTERISTICS OF SOME RATIO (TWO-COLOR) RADIATION THERMOMETERS

Temperature Range Limits, °C	Wavebands Centers, μM	Equivalent Wavelength, μM
175–1250	1.65 and 2.2	6.6
750–1750	0.81 and 0.45	5.5
1000–2200	0.55 and 0.70	2.6
700–3500	0.95 and 1.05	10.0
800–1900	0.75 and 0.88	5.1
1200–2500	0.64 and 0.88	2.3
800–2200	0.71 and 0.81	5.8

same temperature) and it is argued that if the spectral emissivity in one waveband changes the same amount as in the second band, the ratio thermometer will be unaffected. This is a reasonable argument for some materials, but not oxidizing metals, since emissivity, as a rule, is not a strong function of an object's temperature, but is mostly affected by the material's composition, phase, and surface roughness. Oxidizing metals, however, change emissivity rapidly and quite differently at various temperatures. Pure dielectrics such as silica dioxide behave in a similar manner.

Optical Pyrometers. These instruments utilize a unique method of measurement, i.e., a photometric match is made between the brightness of the object and an internal lamp as the basis of measurement. Optical pyrometers are sensitive only in a very narrow wavelength range. The most popular instruments are manually operated, i.e., the operator performs the photometric match visually.

The manual optical pyrometer or visual optical pyrometer is the earliest and most respected portable radiation thermometer system available. It enjoys a reputation, unique among radiation thermometers, for outstanding accuracy. The instrument occupies a special historical place in radiation thermometry because it was the first instrument widely accepted in both research and manufacturing and it demonstrated that excellent temperature measuring performance is possible with properly designed and maintained radiation thermometers.

Optical pyrometers differ from other radiation thermometers in both the type of reference source used and the method of achieving the brightness match between the object and reference. Figure 2 shows the arrangement of parts of an optical pyrometer. The combination of filter characteristics and the response of the average human eye produce a net instrument wavelength response band that is very narrow and centered near 0.65 μm. By adjusting the current through the lamp or varying the intensity of the object radiation, the operator can produce a brightness match over at least a portion of the lamp filament, according to relative target size. Under matched conditions, the two "scenes" merge into one another, or the filament apparently vanishes.

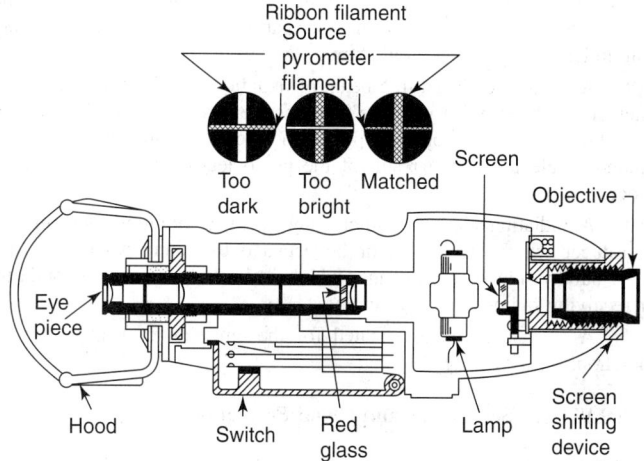

Fig. 2. Sectional view of optical pyrometer telescope.

The temperature range of optical pyrometers is limited at the lower end by the need for an incandescent image of the filament to about 800 °C. The upper temperatures are limited only by applicational needs. Temperature indications on manual units are analog scales or meters.

Fiber Optic Thermometers. This technology enables near-infrared and visible radiation to be transmitted around corners and away from hot, hazardous environments to locations more suitable for the electronics associated in modern thermometers. Fiber optics also makes possible measurements in regions where access is restricted to larger instruments and where large electric or radio frequency fields would seriously affect an ordinary sensor. Fiber optic transmission devices have helped to solve a number of applications, such as hot turbine blade temperature measurement in gas turbine engines and the temperature of hot metal inside induction heating coils or inside vacuum vessels. Conceptually, a fiber optic system differs from an ordinary thermometer system by the addition of a fiber optic light guide, with or without a lens. The optics of the light guide define the field of view of the instrument, while the optical transmission properties of the fiber optic elements form an integral part of the thermometer spectral response. Present fiber optics transmit well to wavelengths as long as about 2.0 μm and thus the thermometers are limited to measuring temperatures upward from about 200 °F (93 °C). Fiber optics must be maintained in a clean condition just as an ordinary thermometer lens.

Additional Reading

Carlson, D.R.: "Temperature Measurement in Process Control," *Instrument Technology*, 26 (October 1990).

Cleaveland, P.: "Temperature Monitoring and Control," *Instruments and Control Systems*, 31 (June 1987).

Considine, D.M., Editor: "Process/Industrial Instruments and Controls," 4th Edition, McGraw-Hill, New York, NY, 1993.

DeWitt, D.P. and G.D. Nutter Eds.," Theory and Practice of Radiation Thermometry", Wiley Interscience (John Wiley & Sons, Inc.), New York, NY (1988).

Richmond, J.C. and D.P. Dewitt, Eds., "Applications of Radiation Thermometry", ASTM Special Technical Publication 895, American Society for Testing and Materials, Philadelphia, PA (1985).

Sakuama, F. and S. Hattori: "Establishing a Practical Temperature Standard by Using Silicon Narrow-Band Radiation Thermometer," 6th Intl. Temp. Symposium, Washington, DC. (March 1982).

Schoenstein, P.G.: "Infrared Sensor Tracks Temperature," *Instruments and Control Systems*, 38 (April 1991).

Siskovic, C.: "New Developments Expand Use of Fiber Optic IR Thermometers," *Instruments and Control Systems*, 37 (June 1989).

Warren, C.: "Spectral Response and IR Temperature Measurement," *Sensors*, 16 (January 1992).

R. PEACOCK, LTV Steel Company, Inc. Independence, OH.

RADIATIVE CORRECTION.   Difference between the theoretical values of some property of a dynamical system as computed from the quantized field theory of the system and from the corresponding unquantized field theory. Applied particularly to the theory of electrons, positrons and the electromagnetic field.

RADIATOR.   Four uses of this term are as follows. 1.   A body that emits energy quanta or certain material particles; more commonly a body that emits electromagnetic radiation.

2.   A substance placed in a beam of radiation, which as a result of the interaction of the beam with the substance, emits radiation of a different type. For example, a metal foil placed in a beam of γ-radiation will emit secondary electrons as a result of the photoelectric and pair production processes.

3.   A radiating element, which may be (a) a vibrating element in a transducer which can cause, or be actuated by sound waves, or (b) a basic subdivision of an antenna, which in itself is capable of radiating or receiving radio-frequency energy.

4.   A surface especially heated for the emission of heat energy by radiation.

RADIATUS.   See Clouds and Cloud Formation.

RADICAL (Mathematics).   An indicated root of a number, usually a principal root; thus, the radical symbol $\sqrt[n]{a}$ means the principal n-th root of a. Operations with radicals are expressed by the formulas:

$$\sqrt[n]{ab} = \sqrt[n]{a}\,\sqrt[n]{b}$$

$$\sqrt[n]{\frac{a}{b}} = \frac{\sqrt[n]{a}}{\sqrt[n]{b}}$$

if a and b are positive. Such functions or equations containing them are also called irrational.

RADIOACTIVE GAS.   1.   In atmospheric electricity, any one of the three radioactive inert gases, radon, thoron, and actinon, which contribute to atmospheric ionization by virtue of the ionizing effect of the alpha particles which each emits on disintegration. These three gases are isotopic to each other, all having atomic number 86.

2.   Any gaseous material containing radioactive atoms.

RADIOACTIVE WASTES.   See Nuclear Power Technology.

RADIOACTIVITY.   The spontaneous disintegration of the nucleus of an atom with the emission of radiation. This phenomenon was discovered by Becquerel in 1896 by the exposure-producing effect on a photographic plate by pitchblende (uranium-containing mineral) while wrapped in black paper in the dark. Soon after this, it was found that uranium minerals and uranium chemicals showed more radioactivity than could be accounted for by the uranium content. About the same time, radioactivity of thorium minerals and thorium chemicals was also discovered.

The excess radioactivity of mineral over chemical uranium led Pierre and Marie Curie to experiment with the mineral. To detect the presence of radioactivity the discharge of a charged gold-leaf electroscope was used. A quantitative estimation of the amount of radioactivity was made by observing the rate of drop of the gold leaf. By chemically separating the uranium mineral into fractions and examining each fraction by the electroscope, they found in the bismuth element fraction the first new radioactive element to be discovered. It was named polonium (1898). They found that polonium disappeared rapidly, half of its radioactivity vanishing in about six months. The fraction containing barium element was also found by them to be radioactive. Repeated fractional crystallizations of the chloride and bromide solutions made possible the recovery by them of practically pure salt of the second new radioactive element. It was named radium (1898).

Radium is chemically similar to barium; it displays a characteristic optical spectrum; its salts exhibit phosphorescence in the dark, a continual evolution of heat taking place sufficient in amount to raise the temperature of 100 times its own weight of water 1 °C every hour; and many remarkable physical and physiological changes have been produced. Radium shows radioactivity a million times greater than an equal weight of uranium and, unlike polonium, suffers no measurable loss of radioactivity over a short period of time (its half life is 1620 years). From solutions of radium salts, there is separable a radioactive gas; radium emanation, radon, which is a chemically inert gas similar to xenon and disintegrates with a half life of 3.82 days, with the simultaneous formation of another radioactive element, Radium A (polonium-218).

Definitions[1]

Decay. The diminution of a radioactive substance due to nuclear emission of alpha or beta particles, gamma rays or positrons.

Decay Constant. A constant λ that relates the instant rate of radioactive decay of a radioactive species to the number of atoms N present at a *given time t*:

$$-(\partial N / \partial) = \lambda N$$

Where N is the number of atoms present at time *zero*:

$$N = N_0 e^{-\lambda t}$$

Decay Product. A *nuclide* that results from the radioactive disintegration of a radionuclide that is formed either directly or as the result of progressive transformations in a radioactive series. The nuclide thus produced is sometimes called the daughter or *daughter* element.

Half-Life (Radioactive Element). The average time required for one-half of the atoms in a sample of radioactive element to decay. The half-life $t_{\frac{1}{2}}$ is given by

$$t_{\frac{1}{2}} = (\ln 2)/\lambda$$

Types of Radioactivity

Beginning in 1899 and continuing through the next two decades, E. Rutherford and his associates conducted a rather thorough study of the radiations emitted by radioactive substances. During this study the radiations were found to be of three types, called alpha, beta, and gamma radiations. In kind, they resemble anode rays, cathode rays, and x-rays, respectively. In this behavior toward electrical and magnetic fields, the resemblance is qualitatively complete: (1) Alpha rays are positively charged particles of mass number 4 and slightly deflected by electrical and magnetic fields. (2) Beta rays are negatively charged electrons, and strongly deflected by electrical and magnetic fields. (3) Gamma rays are

[1] Data pertaining to radioisotopes is provided in quantitative detail in the "Handbook of Chemistry and Physics," 80th Edition, CRC Press, LLC., Boca Raton, FL. 2000.

undeflected by electrical and magnetic fields, and of wavelength of the order of 10^{-8} to 10^{-9} centimeters.

Alpha Ray Emission. Alpha rays have a definite velocity and a definite range for each radioactive nuclide. The velocity is from 5–7% that of light. *Range* is defined as the distance traversed in a homogeneous medium before absorption. The penetrating power of alpha rays is the smallest of the three kinds of rays, the beta rays being of the order of 100 times, and the gamma rays 10,000 more penetrating. The alpha rays are twice-ionized nuclei of helium (He^{2+}). Ramsay and Royds (1909) experimentally demonstrated that accumulated alpha particles, quite independently of the matter from which they have been expelled, consist of helium. They sealed radon in a glass tube with a wall so thin that the alpha particles passed through the wall into a surrounding vessel and after six days the optical spectrum of helium was observed. Helium itself does not diffuse through such a wall. Therefore, alpha particles on losing their positive charge become ordinary helium. This is the first instance of the production of a known element during radioactive transformation. The loss of a single alpha particle by an atom leaves the residual atom four units less in mass number, and two units less in atomic number. The shooting of alpha particles was visibly registered by Crooke's spinthariscope in which the tip of a wire, coated by a tiny amount of radium salts, was placed near a screen coated with zinc blende. Viewed in the dark with a magnifying eyepiece, each alpha particle striking the zinc blende target was observed to produce a visible scintillation. The detection and counting of single alpha particles was accomplished by Rutherford and Geiger (1908), by the deflection of an electrometer needle upon the arrival of each alpha particle in a gas at low pressure in an electric field somewhat below the sparking point.

Beta Radiation. Beta rays are electrons. They have varying velocities almost up to that of light. The loss of a single negatron by an atom leaves the residual atomic nucleus the same in mass number and one unit greater in atomic number, while the loss of a positron or an orbital electron capture leaves the residual atomic nucleus the same in mass number, and one unit less in atomic number.

Double-Beta Decay. In a scholarly paper, M. Moe (University of California, Irvine) and S. Rosen (Los Alamos National Laboratory) describe what is considered to be a very rare radioactive event, namely that of double-beta decay. Searching since 1971, Moe and colleagues observed a double-beta event directly in 1987. In an event of this type, using a selenium atom, which contains 48 neutrons and 34 protons, as an example, two of the neutrons decay simultaneously into two protons. During the process, two beta rays and two antineutrinos are generated. Application of an external magnetic field will cause the paths of the ejected electrons to spiral, but in different directions. A double spiral of this type, when observed, is indicative of a double-beta event. The atom remaining has gained two additional protons and lost two neutrons, as compared with its original state (i.e., prior to the double-beta decay event). The selenium has been transformed to krypton.

The neutrino problem is described in the article on **Particles (Sub-atomic)**. The double-beta decay event may contribute to the solution of that problem. In their introductory to the aforementioned article, the authors observe, "The future of fundamental theories that account for everything from the building blocks of the atom to the architecture of the cosmos hinges on studies of this rarest of all observed radioactive events."

Gamma Radiation. Gamma rays are photons of electromagnetic radiation. This radiation is much more penetrating than alpha or beta particles. The presence of gamma rays from 30 milligrams of radium can be observed in an electroscope after passing through 30 centimeters of iron (Rutherford). For the protection of the operator, radium is kept in lead outer containers or screened by lead sheets.

The naturally occurring radioactive elements at the upper end of the periodic table of elements form a number of series, the elements of each series existing in radioactive equilibrium, unless individual elements are separated chemically away from the series. These series include the Uranium Series, the Thorium Series, and the Actinium Series. See Tables 1, 2, and 3. These arrangements are useful in showing the decay chain (i.e., the parent-daughter) relationships of radioactive elements, including such concepts as radioactive equilibrium. Other naturally occurring radioactive elements also exist, including, for example, ^{40}K, ^{87}Rb, and ^{148}Sm.

Ionizing Radiation. This type of radiation is of major importance because it represents a biological and environmental hazard. Radioactive

isotopes contribute to this potential danger. The extent of damage varies immensely with the dose (exposure over a long or short period of time) and with the source material. The principal ionizing radiations are summarized in Table 4.

Artificially Induced Radioactivity

In addition to the radionuclides already discussed, there are also the great numbers of artificially produced radioactive elements. They are represented in the Neptunium Series and in various collateral series, because, in addition to the three main natural, and the one artificial, disintegration series of radioelements, each has been found to have at least one parallel or collateral series. The main series and the collateral series have different parents, but they become identical when, in the course of disintegration, they have a member in common. Collateral with the natural uranium series is an artificial series discovered in the United States by M.H. Studier and E.K. Hyde. Its parent is ^{230}Pa formed by the bombardment of thorium with alpha particles or deuterons of high energy. The decay scheme of the series has been found to be

$$^{230}Pa \xrightarrow{\beta-} ^{230}U \xrightarrow{\alpha} ^{226}Th \xrightarrow{\alpha} ^{222}Ra \xrightarrow{\alpha} ^{218}Rn \xrightarrow{\alpha} ^{214}Po \longrightarrow$$
Uranium Series

The loss of the alpha particle by the emanation, ^{218}Rn, leads to the formation of ^{214}Po, which is identical with radium C′ of the Uranium Series; the subsequent decay of the collateral series thus becomes identical with that of the main Uranium Series at this point.

Another collateral Uranium Series has for its progenitor ^{226}Pa which is found among the products of bombardment of thorium with 150-MeV deuterons. The decay scheme is represented by:

$$^{226}Pa \xrightarrow{\alpha} ^{222}U \xrightarrow{\alpha} ^{218}Th \xrightarrow{\alpha} ^{214}Ra \xrightarrow{\alpha} ^{210}Bi \longrightarrow$$
Uranium Series

Still other collateral series are the following:

$$^{228}Pa \xrightarrow{\alpha} ^{224}Ac \xrightarrow{\alpha} ^{220}Fr \xrightarrow{\alpha} ^{216}At \xrightarrow{\alpha} ^{212}Bi \longrightarrow$$
Thorium Series

$$^{232}Pa \xrightarrow{\alpha} ^{228}U \xrightarrow{\alpha} ^{224}Th \xrightarrow{\alpha} ^{220}Ra \xrightarrow{\alpha} ^{216}Rn \xrightarrow{\alpha} ^{212}Po \longrightarrow$$
Thorium Series

$$^{227}Pa \xrightarrow{\alpha} ^{223}Ac \xrightarrow{\alpha} ^{219}Fr \xrightarrow{\alpha} ^{215}At \xrightarrow{\alpha} ^{211}Bi \longrightarrow$$
Actinium Series

$$^{239}U \xrightarrow{\beta} ^{239}Np \xrightarrow{\beta} ^{239}Pu \xrightarrow{\alpha} ^{235}U \longrightarrow$$
Actinium Series

$$^{239}U \xrightarrow{\alpha} ^{225}Th \xrightarrow{\alpha} ^{221}Ra \xrightarrow{\alpha} ^{217}Rn \xrightarrow{\alpha} ^{213}Po \longrightarrow$$
Neptunium Series

See Table 5 on p. 2925.

Frederic and Irene Joliot-Curie found in 1933 that boron, magnesium, or aluminum, when bombarded with α-particles from polonium, emit neutrons, proton, and positrons, and that when the source of bombarding particles was removed, the emission of protons and neutrons ceased, but that of positrons continued. The targets remained radioactive, and the emission of radiation fell off exponentially just as it would for a naturally occurring radioelement. The results of this work may be stated in two equations as follows:

$$^4He + ^{27}Al \longrightarrow ^{30}P + n$$
$$^{30}P \xrightarrow{\beta+} ^{30}Si$$

The first of these equations shows that the result of the nuclear reaction in which aluminum is bombarded with α-particles is the emission of a neutron and the production of a radioactive isotope of phosphorus. The second equation shows the radioactive disintegrations of the latter to yield a stable silicon atom and a positron. Continuation of this line of investigation by several research groups confirmed that radioactive nuclides are formed in many nuclear reactions.

Generally, if any two isobars differ in charge by $\pm e$, one had a higher ground-state energy than the other and is beta radioactive. Any nuclide that

TABLE 1. THE URANIUM SERIES

Radioelement	Corresponding Element (2)	Symbol	Radiation	Half-life
Uranium I	Uranium (92)	^{238}U	α	4.51×10^9 yr
Uranium X_1	Thorium (90)	^{234}Th	β	24.1 days
Uranium X_2^*	Protactinium (91)	^{234}Pa	β and I.T.	1.17 min
99.87% | 0.13%				
Uranium II	Uranium (92)	^{234}U	α	2.48×10^5 yr
Uranium Z	Protactinium (91)	^{234}Pa	β	6.66 hr
Ionium	Thorium (90)	^{230}Th	α	7.5×10^4 yr
Radium	Radium (88)	^{226}Ra	α	$1.62^3 10^3$ yr
Ra Emanation	Radon (86)	^{222}Rn	α	3.82 days
Radium A	Polonium (84)	^{218}Po	α and β	3.05 min
99.96% | 0.04%				
Radium B	Lead (82)	^{214}Pb	β	26.8 min
Astatine-218	Astatine (85)	^{218}At	α	2 sec
Radium C	Bismuth (83)	^{214}Bi	β and α	19.7 min
99.96% | 0.04%				
Radium C$'$	Polonium (84)	^{214}Po	α	1.5×10^{-4} sec
Radium C$''$	Thallium (81)	^{210}Tl	β	1.32 min
Radium D	Lead (82)	^{210}Pd	β	19.4 yr
Radium E	Bismuth (83)	^{210}Bi	β and α	2.6×10^6 yr
~100% | ~10^{-5}%				
Radium F	Polonium (84)	^{210}Po	α	138.4 days
Thallium-206	Thallium (81)	^{206}Tl	β	4.23 min
Radium G (end product)	lead (82)	^{206}Pb	None	Stable

*Undergoes ismeric transition (I.T.) to form uranium Z(^{234}Pa); the latter has a half life of 6.66 hr, emitting β radiation and forming Uranium II ^{234}U.

can be formed from a nuclear reaction and is not one of the known stable nuclides is radioactive. Nuclides having higher atomic number (Z) than the nearest stable isobar decay to it through positron emission (β^+ decay) or orbital electron capture. Nuclides with lower Z than the nearest stable isobar decay to it through negatron emission (β^- decay). Occasionally, as for ^{64}Cu, a radioactive nuclide is located between two stable isobars and can decay to either of them, in this case either ^{64}Ni or ^{64}Zn. The simplest radioactive nuclide is the neutron, which has a half life of 12 minutes, and decays into a proton, a negatron, and a neutrino.

Energy-level diagrams for nuclear transformations are usually drawn to show the relative energies of levels of an entire neutrally charged atomic system. Since the nuclear charge increases in magnitude by e if a beta radioactive nucleus emits a negatron, one additional external electron must be added to maintain a neutral atom. On the other hand, the nuclear charge changes by $-e$ during positron emission; therefore, in order to maintain a neutral atom, an electron must also be lost by one of the atomic shells. Thus, for negatron decay the total energy difference between initial and final energy states is only the sum of the negatron kinetic energy and the neutrino energy. For positron emission, however, the atom loses a minimum energy equal to twice the rest energy of an electron, $2m_0c^2$. These energy relationships are shown schematically in Fig. 1. During orbital electron capture the nucleus loses a single positive charge merely by taking an electron from one of its own atomic shells. The only energy loss is that energy emitted as X radiation during rearrangement of the atomic shells following electron capture and the energy carried away by the neutrino. See Fig. 1 on p. 2925.

A table of nuclides showing mass number and isotopic abundance is given in the entry on **Chemical Elements**.

TABLE 2. THE THORIUM SERIES

Radioelement	Corresponding Element	Symbol	Radiation	Half-life
Thorium	Thorium	^{232}Th	α	1.39×10^{10} yr
Mesothorium I	Radium	^{228}Ra	β	6.7 yr
Mesothorium II	Actinium	^{228}Ac	β	6.13 hr
Radiothorium	Thorium	^{228}Th	α	1.90 yr
Thorium X	Radium	^{224}Ra	α	3.64 days
Th Emanation	Radon	^{220}Rn	α	54.5 sec
Thorium A	Polonium	^{216}Po	α	0.16 sec
~100% | 0.014%				
Thorium B	Lead	^{212}Pb	β and α	10.6 hr
Astatine-216	Astatine	^{216}At	α	3×10^{-4} sec
Thorium C	Bismuth	^{212}Bi	β and α	60.5 min
66.3% | 33.7%				
Thorium C	Polonium	^{212}Po	α	3×10^{-7} sec
Thorium C''	Thallium	^{208}Tl	β	3.1 min
Thorium D (end product)	Lead	^{208}Pb	None	Stable

Exotic Nuclei and Their Decay. As reported by J.C. Hardy (Chalk River Nuclear Laboratories, Atomic Energy of Canada, Ltd.), recent advances in nuclear accelerators and experimental techniques have led to an increasing ability to synthesize new isotopes. As isotopes are produced with more and more extreme combinations of neutrons and protons in their nuclei, new phenomena are observed, and the versatility of the nucleus is increased as a laboratory for studying fundamental forces. Hardy reports that, among the newly discovered decay modes are: (1) proton radioactivity, (2) triton, two-proton, two-neutron, and three-neutron decays that are beta-delayed, and (3) ^{14}C emission in radioactive decay. Precise tests of the properties of the weak force have also been achieved.

The fundamental usefulness of exotic nuclei and their decay assures a continuing interest in the field. New heavy-ion accelerators will ensure that this interest is matched by an ever-increasing capability to synthesize new isotopes and provide the nuclear laboratory with renewed flexibility. Hardy further emphasizes that applications uniquely suited to the decay modes of exotic nuclei, are starting to appear and indeed are sophisticated. Many of the new forms provide greater detail than can be obtained with stable nuclei. As an example, beta-delayed proton decay has been used to time the life of excited nuclear states. The technique is sensitive to lifetimes in the range of 10^{-16} second, a span in which it has few competitors, and has been applied to a number of different nuclei.

Useful Applications for Isotopes

Although care must always be exercised to avoid undue exposure to various radioisotopes, these materials have found wide acceptance in analytical chemistry, medicine, radiocarbon and other radioelement dating (geology, archaeology, etc.), and other special situations—for example, as a fuel source in spacecraft.

The radionuclides commercially available and most commonly used for a number of the foregoing applications include: antimony-125; barium-133, 207; bismuth-207; bromine-82; cadmium-109, 115m; calcium-45; carbon-14; cerium-141; cesium-134, 137; chlorine-36; chromium-51; cobalt-57, 58, 60; copper-64; gadolinium-153; germanium-68; gold-195, 198; hydrogen-3 (tritium); indium-111, 114m; iodine-125, 129, 131; iron-55, 59; krypton-85; manganese-54; mercury-203; molybdenum-99; nickel-63; phosphorus-32, 33; potassium-42; promethium-147; rubidium-86; ruthenium-103; samarium-151; scandium-46; selenium-75; silver-110m; sodium-22, strontium-85; sulfur-35; technetium-99; thallium-204; thulium-171; tin-113, 119m, 121m; titanium-44; ytterbium-169; and zinc-65.

Radioisotopes in Chemical Analysis

There is a wide range of applications for methods of analysis that are based upon the energies and intensities of the radiations emitted by radioactive nuclides. These techniques sometimes are termed *radiometric methods of analysis*. The methods are not restricted to the determination of substances initially radioactive, since there is wide use of methods involving the irradiation of stable nuclides to produce radioactive ones, followed by measurement of their radiations, from which the composition of the original stable substance can be inferred. This method is *radioactivation analysis*. Another method for the use of measurements of radioactivity in the analysis of stable substances is that of *tracer techniques*, that is, by the addition to them of radioactive nuclides, which can then be used to follow the course of various reactions or processes. There are various ways of introducing the radioactive nuclides, which are discussed later in this entry.

All methods of radiometric analysis involve, of course, the use of various radiation detection devices. The devices available for measuring radioactivity will vary with the types of radiations emitted by the radioisotope and the kinds of radioactive material. Ionization chambers are used for gases; Geiger-Müller and proportional counters for solids; liquid scintillation counters for liquids and solutions; and solid crystal or semi-conductor detector scintillation counters for liquids and solids emitting high-energy radiations. Each device can be adopted to detect and measure

TABLE 3. THE ACTINIUM SERIES

Radioelement	Corresponding Element	Symbol	Radiation	Half-life
Actinouranium	Uranium	^{235}U	α	7.07×10^8 yr
Uranium Y	Thorium	^{231}Th	β	25.6 hr
Protactinium	Protactinium	^{231}Pa	α	3.25×10^4 yr
Actinium	Actinium	^{227}Ac	β and α	21.7 yr

98.8% | 1.2%

Radioactinium	Thorium	^{227}Th	α	18.2 days
Actinium K	Francium	^{223}Fr	β	21 min
Actinium X	Radium	^{223}Ra	α	11.7 days
Ac Emanation	Radon	^{219}Rn	α	3.92 sec
Actinium A	Polonium	^{215}Po	α and β	1.83×10^{-3} sec

~100% | ~5×10⁻⁴%

Actinium B	Lead	^{211}Pb	β	36.1 min
Astatine-215	Astatine	^{215}At	α	~10^{-4} sec
Actinium C	Bismuth	^{211}Bi	β and α	2.16 min

99.96% | 0.32%

Actinium C	Polonium	^{211}Po	α	0.52 sec
Actinium C''	Thallium	^{207}Tl	β	4.76 min
Actinium D (end product)	Lead	^{207}Pb	None	Stable

TABLE 4. PRINCIPAL TYPES OF IONIZING RADIATION

Name	Symbol	Location in Aton	Relative Rest Mass	Charge
Proton (H^1)$^+$	p	Nucleus	1	+1
Neutron	n	Nucleus	1	0
Electron	e	Outer shells	0.00055	−1
Beta⁻ (electron)	β	Emitted during decay processes	0.00055	−1
Beta⁺ (positron)	β^+		0.00055	+1
Alpha (He^4)$^{++}$	α		4	+2
Gamma[a] (photon)	γ	Emitted during decay processes	0.0	0

[a]X-rays of equal energy are identical, but of extranuclear origin. Although only the gamma or x-rays are electromagnetic in character and thus "radiations" in the classical sense, the distinction between radiations and ionizing (particles) is often not made. X-rays are distinguished from gamma rays only with respect to their origins. Gamma rays result from nuclear interactions or decays. X-rays result from transitions of atomic or free electrons, produced artificially by bombarding metallic targets with energetic electrons. As pointed out by Mel and Todd (see references), it is sometimes difficult to make a clear distinction between ionizing and nonionizing electromagnetic radiations, particularly in condensed phases. The ionization potential of gaseous elements, that is, the energy required for removal of the first, electron, ranges from 3.9 eV (cesium) to 24.6 eV (helium). Although ultraviolet and even visible light in special cases can ionization, the general assumption is that the more energetic x- or gamma radiation is needed to insure ionization. Neutrons also lead to ionization, but for other reasons. Ionization, of course, is not the only interaction of high-energy particles and radiations with matter. The excitation of atomic electrons into higher-energy states always accompanies ionization as well.

radioactive material in another state, e.g., solids can be assayed in an ionization chamber. The radiations interact with the detector to produce a signal.

Since many radionuclides decay with gamma rays, many measurements are being made by gamma-ray scintillation spectrometry. Usually, a crystal detector, such as a sodium iodide crystal, is connected to a spectrometer. As described above, the gamma-rays interact with the crystal to produce light pulses which are converted to electrical pulses by a multiplier phototube. The pulse height analyzer of the spectrometer sorts out the gamma-rays of various energies. From this operation, a spectrum of the radionuclide's gamma-rays can be obtained to the *photopeaks* of full-energy pulses and the continuum of lower-energy pulses associated with the decay of the radionuclide. The photopeak, or photopeaks, in the gamma-ray spectrum can be used to identify and quantitatively measure the radionuclide.

Radioactivation Analysis. The principle of this technique is that a stable isotope when irradiated by neutrons, by charged particles such as protons or deuterons or by gamma rays, can undergo a nuclear reaction to produce a radioactive nuclide. After the radionuclide is formed, and its radiations have been characterized by radiation detection devices, calculations can be made of the elements contained in the sample before irradiation.

An important reaction used quite widely for this purpose is irradiation by neutrons and measurement of the energies of radiations emitted. The source of the neutrons may be a nuclear reactor, a particle accelerator, or an isotopic source, that is, a sealed container in which neutrons are produced by alpha rays emitted by a source such as radium, sodium-24(^{24}Na), yttrium-88(^{88}Y), etc., and arranged so that the alpha rays react-with a substance such as beryllium which in turn emits neutrons. The neutrons react with stable nuclides in the sample to produce radioactive ones. Thus

TABLE 5. THE NEPTUNIUM SERIES

Element (2)	Symbol	Radiation	Half-life
Curium (96) ↓	^{245}Cm	α	9300 yr
Plutonium (94) ↓	^{241}Pu	β	13.2 yr
Americium (95) ↓	^{241}Am	α	458 yr
Neptunium (93) ↓	^{237}Np	α	2.20×10^6 yr
Protactinium (91) ↓	^{233}Pa	β	27.4 days
Uranium (92) ↓	^{233}U	α	1.62×10^5 yr
Thorium (90) ↓	^{239}Th	α	7340 yr
Radium (88) ↓	^{225}Ra	β	14.8 days
Actinium (89) ↓	^{225}Ac	α	10.0 days
Francium (87) ↓	^{221}Fr	α	4.8 min
Astatine (85) ↓	^{217}At	α	1.8×10^{-2} sec
Bismuth (83) 96% | 4%	^{213}Bi	β and α	47 min
Polonium (84)	^{213}Po	α	4.2×10^{-6} sec
Thallium (81)	^{209}Tl	β	2.2 min
Lead (82) ↓	^{209}Pb	β	3.3 hr
Bismuth (83) (end product)	^{209}Bi	None	Stable

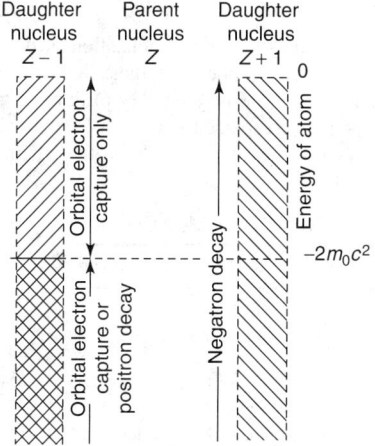

Fig. 1.   The energy regions for which negatron emission, positron emission, and orbital electron capture are energetically possible.

ordinary sodium undergoes a nuclear reaction with neutrons as follows:

$$\text{Na}^{23} + n \longrightarrow \text{Na}^{24} + \gamma$$

The ^{24}Na decays with a half-life of 15 hours to yield gamma rays and β-particles

$$\text{Na}^{24} \longrightarrow \text{Mg}^{24} + e^- + \gamma$$

Moreover the energies of these β-particles (electrons) are known to be 1.39 MeV and that of the gamma-rays 1.38 MeV so that the measured values of these magnitudes are characteristic of substances containing sodium. (Measurement of the γ-radiation is the usual procedure.) At least 70 of the elements can be activated in this way, by the capture of thermal neutrons, i.e., by *neutron activation analysis*. An activation analysis follows a procedure similar to that shown in Fig. 2. In almost all analyses, the sample materials are not treated before the bombardment, but are placed directly into the bombardment capsule or container. The length of the bombardment interval is usually determined by the half-life of the radionuclide used for the element of interest and the flux of nuclear particles.

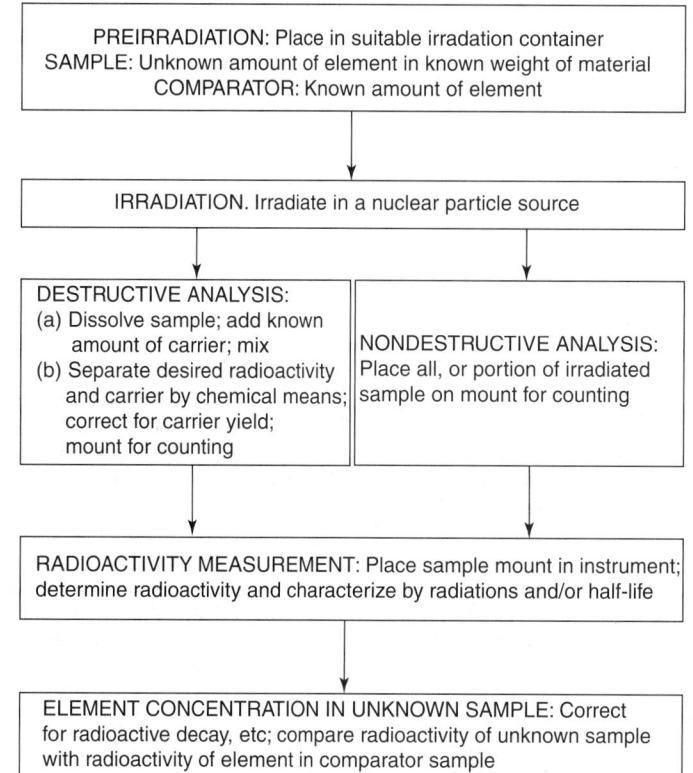

Fig. 2.   Representative program followed in activation analysis.

The post-bombardment processing of the activated sample may follow either a nondestructive assay of the radioactivity in the sample (gamma-ray scintillation spectrometry is used most often for this) or a chemical processing of the sample prior to the radioactivity assay. Techniques involving either precipitation, electrodeposition, solvent extraction, and ion exchange or some combination of these form the basis of the radio-chemical separation techniques used in activation analysis.

Neutron activation has been successfully applied to a great variety of determinations of small concentrations of elements present in alloys, for example, vanadium and manganese in iron. Other metallurgical applications include the determination of some 70 elements: including the metals, aluminum, antimony, arsenic, barium, bismuth, cadmium, calcium, cerium, cesium, and so on alphabetically down the list. Minerals and soils have also been extensively analyzed by the method. However, it is not restricted to trace quantities or to inorganic substances, one interesting application being the determination of phosphorus, oxygen and nitrogen in organic phosphorus compounds. Sodium has been determined in blood plasma, and numerous other biochemical determinations have been made accurately. See also **Neutron**.

Tracer Analysis. This method is readily performed with radioactive isotopes because their ease of detection by measurement of their radioactivity makes them effective means of "tagging" their stable isotope counterparts (e.g., ^{24}Na(sodium-24) to tag ordinary sodium). Since compounds are usually involved, rather than elements, one merely synthesizes enough of the radioactive compound to tag the compound under analysis. The tagged compound may be followed through any analytical scheme, industrial system, or biological process. It is essential that a compound be tagged with an atom, however, which is not readily exchangeable with similar atoms in other compounds under normal conditions. For example, tritium could not be used to trace an acid if it were inserted on the carboxyl group where it is readily exchanged by ionization with the solvent.

Radiometric methods employing reagent solutions or solids tagged with a radionuclide have been used to determine the solubility of numerous organic and inorganic precipitates, or as a radioreagent for titrations involving the formation of a precipitate. In this type of application it is necessary to establish the ratio between radioactivity and weight of radionuclide plus carrier present. This may be established by evaporating an aliquot to dryness, weighing the residue, and measuring the radioactivity.

Closely related to tracer analysis is the method of *isotopic dilution analysis*. Here, instead of checking the effectiveness of a method from known amounts of an element in the sample, and of its radioactive isotope, one knows only the amount of radioactive isotope added, and by precipitating or otherwise separating the total amount of that element present, and then measuring its radioactivity, one determines its amount, and hence the amount present in the original sample.

Radioactive tracer methods lend themselves well to research applications in studying entire processes in science and industry, and in the biological as well as the physical sciences.

Industrial Applications for Radioisotopes

Radioisotopes are widely used in the measurement of process variables, including the level of liquids and solids in tanks, silos, and other vessels, the density and specific gravity of fluids and solids, the thickness of sheets and coatings, the moisture content of soils and other solids, the mass flow of materials in pipelines or on belts, and the determination of chemical composition of raw materials, in-process materials, and end-products. Representative examples of these applications are given in Table 6.

Density, level, and thickness measurements all depend upon the determination of the number of radiations per unit time penetrating the sample and producing a measurable signal in the radiation detector. When the amount of matter between the source and the detector increases, there usually is a decrease in the signal. The following relationship demonstrates the exponential nature of the attenuation of beta or gamma radiation:

$$\frac{I}{I_0} = Be^{-\mu\rho t}$$

where I_0 = initial radiation intensity

I = radiation intensity through absorbing material

B = a buildup factor dependent on the energy and collimation of the source, and on ρ and t. B accounts for radiation which has been "scattered" or changed in direction by interactions that do not stop the radiation.

μ = absorption coefficient, dependent on composition of absorbing material, energy of radiation, and source detector geometry

ρ = specific gravity of absorbing material

t = thickness of absorbing material

Beta radiation has a finite range, whereas in theory x- and gamma rays are exponentially attenuated. It should be noted that x-radiation of each energy exhibits sharp changes in absorption coefficients for certain absorbers.

Where density and specific gravity measurements are made of liquids and slurries, major signal generation changes can result from temperature changes. Thus, compensation for temperature changes should be built into the instrumentation. Major errors also may result from the presence of air bubbles, grease deposits, and pipewall corrosion.

In a typical level-measuring gage, a gamma-emitting radioactive source is mounted in a shielded holder on one side of the vessel and a suitable detector is mounted on the other side. Several radiation receivers, mounted at vertical increments, are required to provide a range of levels. Greater accuracy and flexibility can be obtained with a motor-driven level gage. Here the radiation source and detector cell move up and down in unison as controlled by a motor-driven drum. Potentiometers provide output voltages; one indicates full range in feet; a second potentiometers indicates inches within range. A digital readout also may be provided.

Moisture measurements can be made by using a source of fast neutrons. If such a source (Sb-Be, Pu-Be, ^{252}Cf) is located on or in a medium containing hydrogen, some of the neutrons will be slowed (moderated or thermalized) by collisions with the hydrogen atoms. The number of slow neutrons per unit area per unit time can be determined with a detector selectively sensitive only to slow neutrons. A representative detector of this type is a ^{3}He or BF$_3$-filled proportional or Geiger-Mueller detector. The relative response of the detector will provide a measure or the hydrogen content of the medium surrounding or adjacent to the source. Thus, if water is the only or principal hydrogen-containing variable constituent of the medium, the technique can be used to measure moisture content.

Pressure measurements involve the interaction of alpha radiations with a gas, which results in the formation of positive and negative ions. The latter can be collected and measured as electric current. The number of ions produced in a gas by alpha particles depends upon the density and composition of the gas. Where either of these factors is known the other can be inferred from these measurements. Several vacuum gages employ this principle.

For the measurement of thickness and coverage (coatings) of very thin materials, the absorption of alpha radiation may be used as a measure of weight/unit area. For moderately thick materials, the beta radiation emitted by ^{90}Sr (strontium-90), ^{85}Kr (krypton-85), or ^{14}C (carbon-14) is used. For greater thicknesses, bremsstrahlung or gamma sources are used.

The uniformity of mixing may be determined by mixing radioactive ^{24}Na in sodium chloride as the tagged compound and, after mixing, determining the uniformity of the presence of the isotope in the samplings. The mixing patterns in fluid catalytic cracking units have been ascertained by tagging catalysts with ^{51}Cr (chromium-51), ^{46}Sc (scandium-46), and ^{144}Ce (cerium-144).

Radioisotope techniques also find application in determining diffusion rates in the study of metals, porous bodies, liquids, and gases. Tritiated water is used in studies of the permeability of thin, flexible plastic sheets. The thin sample sheets are placed over a dish containing tritiated water (^{3}H, heavy hydrogen in the water) solidified by gelatin. Then methane is

TABLE 6. REPRESENTATIVE RADIOISOTOPE PROCESS INSTRUMENTATION APPLICATIONS

Variable Measured	Representative Applications	Radioisotope Frequently Used
Mass flow in pipes	Slurries and viscous or corrosive materials	^{60}Co, ^{124}Sb
Mass flow on belts	Conveyance of solids, such as coal, woodchips, gravel, etc.	varies with material
Level	Slurries, viscous materials, solids, crushed rock, aircraft and rocket fuels, etc.	^{137}Cs, ^{60}Co, ^{226}Ra and daughter
Density — fluids	Coal slurries, sewage sludges, granular materials, black and green liquor (paper pulp processing), chemicals	^{137}Cs, ^{60}Co, ^{226}Ra and daughter
Density — solids	Compacted soil density for roads, footings, dams, asphaltic concrete	^{137}Cs, ^{60}Co, ^{226}Ra and daughter
Thickness	Thin plastic sheets, films	^{14}C
	Light paper and plastics	^{85}Kr
	Heavy paper, thin metal, rubber	^{90}Sr-^{90}Y
Moisture	Soils and solids	Neutron sources: ^{226}Ra-Be, ^{210}Po-Be, ^{124}Sb-Be, ^{239}Pu-Be, ^{252}Cf, ^{241}Am-Be

passed over the upper surface of the septum and into a counter. Pulses of the counter resulting from the tritium diffused in the methane are measured. Similar techniques have been used for studying flow patterns of underground water supplies of for tracing cross flow between oil wells. In one application, water tagged with ^{131}I (iodine-131) is used in the study of the subsurface flow of water in secondary oil recovery—to determine the path, velocity, and carrying strata of the water.

Three main techniques are used for applying radioisotopes to the measurement of flow rate: (1) peak timing, (2) dilution, and (3) total count. In peak timing, a gamma emitter (^{60}Co or ^{124}Sb) is injected quickly at a point close to the section of the pipe in which the velocity is to be determined. The time of passage of the peak of the tracer wave is determined by two detectors located at a known distance apart and external to the pipe. The dilution technique is based on the fact that the concentration of mixed tracer in a line resulting from the continuous bleeding of a tracer at a known rate into the line will be inversely proportional to the relative flow rates of the line and the bleeder, assuming that the tracer will uniformly mix with the flowing liquid. In the total-count method, flow rate is based upon measurement of the total counts from a radioactive tracer that has been added to the flowing stream. The total count bears a simple inverse relation to the flow rate.

Radioisotopes also are used for detection of interfaces in pipelines. Since some pipelines handle many different stocks, ranging from crude oil to finished petroleum products to chemicals, etc., effective control requires a knowledge of the precise instant when the interface between two materials passes the control point. This is obtained by adding to the interface a radioactive nuclide, which emits a strong beam of radiation, such as ^{124}Sb. The half-life of the radioisotope used must be quite short to insure that the activity will have decayed to a safe level before it reaches the ultimate user of the transported material.

Metallurgical Applications. Some of the outstanding industrial applications of radioactivity have been in metallurgy. The wear of piston rings in internal combustion engines is important in the selection of alloys for this service. Since the rate of wear is relatively slow, a long period of test would be required to give effects measurable by ordinary methods. However, if the piston ring is made radioactive by irradiation, and the lubricating oil is then tested for the radioactivity of particles of metal abraded from the ring, the more sensitive radiation counters will show measurable and comparable results after relatively short test periods.

Another metallurgical application of radioactivity is in determining the distribution of metals in alloys. A case in point is the molybdenum alloys, where the distribution of the molybdenum between the various solid phases is important to the properties of the alloy. Such studies can be made by polishing a sample of the alloy, etching its surface with properly chosen reagents, and examining its grain structure under the microscope. This process can often be shortened by enriching the molybdenum with one of its radioactive isotopes before adding this metal to the melting furnace. Alloys containing such radioactive elements can be photographed directly through the microscope, whereupon the radioactive areas of the metal show dark on the film.

A related metallurgical application is the use of radioactive nuclides, which emit high energy gamma rays and have a reasonably long half-life, such as ^{60}Co, 113-day ^{182}Ta or 100-day ^{88}Y, to take radiographs of metallic articles. This replaces the x-ray technique for the detection of holes, inclusions and other internal defects in castings and other parts. A similar technique is employed in the beta-ray gauges used in checking the uniformity of the thickness of sheets of material and various articles, such as paper, plastics, steel, textiles, rubber, glass, roofing, flooring, and cigarettes. Here the highly penetrating gamma rays would not be appreciably absorbed by the thin, nonmetallic material; therefore, radioactive nuclides emitting beta rays are used instead. This type of application constitutes an extensive use of radioactive radiations for testing purposes.

Radioisotopes in Medicine

Radiopharmaceuticals are almost ideal diagnostic tools because radioisotope tracers do not alter body physiology, and they permit external monitoring with minimal instrumentation. Presently, there are three major areas of nuclear medicine: (1) physiological function studies, (2) radionuclide imaging procedures, and (3) therapeutic techniques.

Physiologic Function Testing. An example of this application is the assay of thyroid hormone levels in the blood which, in turn, can aid in the assessment of thyroid function. The radioactive iodine uptake test, which involves the administration of a dose of ^{131}I (iodine-131) to the patient, is also a valuable procedure in assessing thyroid function. At present, the technique is best reserved for problem cases rather than used as a primary screening test. The main disadvantage of this test is the effect of the dietary intake of iodine, which reacts in various ways in different individuals.

The gamma camera, with computer-assisted data analysis, is used together with ^{131}I-hippuran to measure renal function. The renogram is of most clinical value in the assessment of ureter impairment in pre-and postoperative patients with carcinoma of the cervix and other pelvic and gynecological tumors.

Radionuclides also are useful in assessment of hematological status to detect anemia and iron deficiency, and in studying radioactivity in feces in order to detect significant blood loss through the gastrointestinal tract. Although considerable development remains, radioisotopes show promise of facilitating differentiation between well-vascularized and ischemic tumors and organs.

Radionuclide Imaging Procedures. Brain tumors can be detected by external counting of radionuclides, a procedure introduced into general clinical use in the late 1950s. A significant advance in brain tumor imaging was the introduction of the gamma camera, which permitted more rapid studies with multiple views, as well as dynamic cerebral blood flow assessment.

The ^{85}Sr (strontium-85) scanning of metastatic bone disease is another important tumor localization technique. Since metastatic lesions of bone are frequently associated with new bone formation, there is usually a significant localization of several radioisotopes in the general vicinity of the metastasis. Early metastatic lesions often go undetected on roentgenographic examination because a 30–50% change in bone density is required to produce visible changes on X-ray examination. Bone scans, however, are generally positive quite early in the development of metastasis. Patients with prostatic carcinoma and carcinoma of the breast are most often candidates for study with this technique.

The liver scan, using radioactive colloids, utilizes a slightly different approach to tumor detection. In this scan, the radioisotope concentrates in the normal tissue, and the tumor appears as a nonradioactive, or "cold" area. This procedure is often an indicated procedure in the cancer patient because of the frequency of liver metastases, and because the liver is not easily visualized using routine radiographic techniques. There are limitations to the approach. Lesions that are smaller than two centimeters in diameter generally go undetected because of limitations of resolution of scanning devices. A number of other disease conditions also may interfere with localization of radiocolloids, producing defects on the liver scan that are indistinguishable from neoplastic disease.

Lung scans also are useful in checking for changes before and after radiation treatment of carcinoma of the lung. Although not widely used at present, a technique for detecting bronchial obstruction has been developed using inhalation of radioactive aerosols. A liver-pancreas scan also can be performed, although interpretation of pancreatic scans is often difficult because of normal variation in size and shape and in trace concentration. When the scan appears to be within normal limit range, however, the presence of disease is unlikely.

Thyroid scans with ^{131}I are useful in determining the activity of thyroid nodules in the intact thyroid gland. A nonradioactive, "cold" nodule indicates a higher risk of thyroid carcinoma, but the scan alone is not recommended as a technique of selecting patients for surgery. After removal of a thyroid carcinoma, a scan of the neck may demonstrate areas of increased activity in the cervical lymph nodes and other organs, indicating metastatic disease.

Scintigram techniques of the kidney can be helpful in distinguishing between cysts and neoplasms, and salivary gland scanning can be useful in confirming abnormality in the salivary gland where tumor is suspected. Lymph node scans with the radiocolloid injected subcutaneously on the dorsum of the feet can be used as screening procedures for lymph flow.

The search continues for a general tumor-scanning agent. Although several radionuclides have been found to localize tumors of widely different types and regions of the body, current interest is in the use of ^{67}Ga-citrate, which is undergoing a wide clinical trial and may prove to be useful in the localization of lymphomas as well as some adenocarcinomas. Medical imaging techniques are discussed in several articles of this encyclopedia. Consult alphabetical index.

Therapeutic Techniques. Probably the most prominent therapeutic use of radiopharmaceuticals is radioactive iodine in the treatment of metastatic thyroid cancer. [131]I has a half-life of about 8 days and emits gamma and beta rays. When iodine salts are taken into the body, most of the dose is concentrated in the thyroid gland. A dose of radioactive iodine salt similarly concentrates in the thyroid gland. When there is a cancer in the thyroid gland, or the gland is overactive (hyperthyroidism), the excessive tissue may be destroyed by the radiation from the radioactive iodine that has been administered. Although removal of metastatic thyroid cancer is not always achieved with ^{131}I therapy, significant palliation can occur. In some instances of lung metastasis and lymph node metastasis in the neck, patients may show no evidence of recurrence, even many years after treatment.

Radiophosphorus is used in the treatment of patients with a number of diseases. This element has a half-life of about 14 days and emits beta rays. It is taken up by the body in the greatest quantity by those tissues which manufacture blood cells. In polycythemia vera, a condition in which too many red blood cells are formed, the radiation from this isotope often brings about a sufficient suppression of the blood cell-making tissues to alleviate some of the symptoms of the disease. Leukemia patients, in whom there is an excessive production of white cells, are offered added comfort and, in some instances, prolongation of life by the use of radiophosphorus. This element also may be used in treatment of metastatic cancer to the bone and, although the treatment is not used in an attempt to eradicate cancer, it can result in significant palliation of pain in some patients.

[198]Au (gold-198), in the form of a suspension in water, has found increasing use in the treatment of certain types of cancer. Isotopes of gallium, sodium, arsenic, and other elements have been tested for possible uses in medicine and show some promise. Other methods for using nonsealed sources include arterial therapy of liver cancer, endolymphatic therapy of lymph node cancer, and intracavitary therapy of pleural and peritoneal cancer. The basic principle back of the internal use of all radioactive isotopes depends upon the concentration of the isotope in some particular tissue. The search for elements that are concentrated in each of the organs by the selective abilities of the tissues, or elements that concentrate in tumor tissue as contrasted with normal tissues, is the key to all techniques in radiodiagnosis.

[60]Co (cobalt-60), a gamma and beta ray emitter with a half-life of about 5.3 years, is used in cancer therapy. Small pieces of radioactive metallic cobalt, made radioactive in a nuclear reactor, are placed into a proper shielding device and the radiation from them used in place of a high-powered x-ray machine or radium implantations in treating patients with localized cancer. Thousands of ^{60}Co irradiators are in use. Among the most significant advances in radiotherapy since 1925 has been the development of supervoltage equipment and the ^{60}Co megavoltage units. The latter are the most suitable, since they are compact, have high activity source in a small volume, are flexible and adapt to many geometric patterns for therapeutic use, and are easy and economical to maintain. With the ^{60}Co isotope, supervoltage is now made available throughout the world, and at a very low cost when compared with the cost of radium per se, or of x-ray generators. [137]Cs (cesium-137) units also are in use, but do not have the same therapeutic usefulness as the ^{60}Co, and the activity cannot be concentrated as easily as with cobalt.

See also **X-Ray Scan and Other Medical Imagery**.

Relative Biological Effectiveness (RBE). The roentgen is a measure of the intensity of ionizing radiation in air. One roentgen corresponds to the creation of 1 esu of charge in 1 ml of standard air. The roentgen can be considered as a measure of energy dissipation in air, and its definition has been extended to cover ionization or energy dissipation in other media. One roentgen corresponds to a radiation field dissipating 83.8 ergs/g of air, which dissipates approximately 93.8 ergs/g of body tissue. Ionization in tissue is a measure of physical damage. Thus, allowable radiation exposures for human tissue are expressed in roentgens. Since the same number of roentgens from various types of radiation produce different amounts of body damage, a term called roentgen equivalent man (rem) is used in stating allowable radiation exposure values. The rem $= R \times$ rbe, where rbe, known as *relative biological effectiveness*, has the values given in Table 7.

Radioactive Dating Techniques

Age determinations using radioactive nuclides may be looked upon as processes that are the inverse of half-life measurements. If a radionuclide

TABLE 7. VALUES OF RELATIVE BIOLOGICAL EFFECTIVENESS

Type of Radiation	rbe
X- and gamma radiation	1
Beta rays	1
Alpha rays	20
Fast neutrons[a]	10
Thermal or slow neutrons	5

[a]Having energies in the range 0.1 to 10 MeV. Above 10 MeV, the rbe increases rapidly.

of known half life exists within an object, the age of that object can be determined either by measuring the number of radionuclides that remain or the number of product nuclides of the radioactive decay. In these determinations it is assumed that, if we know the half life of the radionuclide, an elapsed time t, or age, for the object can be found by using the formula $t = (\ln N_1/N_2)/\lambda$, where λ is the decay constant of the radionuclide and N_1 and N_2 are the amounts of the radionuclide present at the beginning and the end of the interval spanning the time t.

In any use of radioactive dating or age determining processes, a basic assumption is, in general, that the concentration of the radioactive element is changed during the life of the sample only by its natural decay process, and that the accuracy of the determination depends primarily, therefore, upon the accuracy with which the half-life of that radionuclide is known.

Ages of specimens may sometimes be determined by other methods that the measurement of radioactivity, as by combination of radioactive measurements with mass spectroscopic determinations.

D.Q. Bowen (The University College of Wales, Aberystwyth) stresses that, in studies of paleoclimatology (global warming periods, etc.), "The last 130,000 years is especially important because it includes geological analogues which may be useful for predicting future changes in climate, and against which the predicted trace-gas induced global warming may be evaluated."[2]

Preliminary climate modeling suggests that natural trends will eventually overcome the predicted global warming, but better dating of past changes is required to refine such models. Two such dating methods are: (1) *thermoluminescence dating* of sediments, and (2) *amino acid geochronology* of fossil mollusks. All physical and biological sciences involved in research into the Quaternary Period[3] of the last 2.4 million years were revolutionized by the reinterpretation of Emoiliani's (1955) classic work on oxygen isotope variability in marine microfossils. For the reader with a scholarly interest in this topic, the Bowen reference listed contains a depth of understanding of radioactive decay dating in this area.

Age of Rocks. In the table of nuclides given under **Chemical Elements**, there are listed a number of naturally occurring radionuclides with long half-lives. From these known half-lives, the geological age of a rock may be calculated. One method of making this estimate is based upon the amount of radionuclide and its daughter nuclide contained in the rock. This method is based upon various assumptions, which may be stated as follows:

1. Since the rock was formed, the parent nuclidic content of the mineral has been changed only by radioactive decay.
2. All the decay products produced by the parent nuclide have been retained since the mineral was formed.
3. The geological separation of the parent and daughter elements at the time of formation of the mineral was sufficient to make the determination of the decay products unambiguous. For example, if a uranium mineral does not exclude all lead at the time it is formed, the isotopic abundance of the lead at the time of formation cannot be calculated with certainty.
4. The radioactive decay scheme of the parent nuclide is well known.

---

[2] The most appropriate past analogue for predicated global warming is from the mid-12. Global temperatures at that time simulate the predicted warming at high latitudes, whereas analogues based on the warmest interglaciations of the past 2 to 4 million years only give appropriate warming in middle latitudes.

[3] The Quaternary Period is subdivided into the Pleistocene Epoch and, commencing at 10,000 years ago, the Holocene Epoch. The Holocene is synonymous with the "Post-glacial or present interglaciation."

Another method uses the decay of cosmogenic isotopes that are produced in the atmosphere and then incorporated into terrestrial reservoirs. Examples of this approach include standard ^{14}C and ^{10}Be dating.

The contributions of modern chemistry, including the availability of separated isotopes, the extension of the range of mass spectrometers, and the developments of new chemical methods, which make possible the determination of microgram quantities, have extended the range of application of radioactive age measurements. This extension has been either to minerals that contain relatively little of the parent element, but maintain a good separation of the parent and daughter elements when they are formed; or to minerals containing radioactive elements that have a very low natural abundance, such as ^{40}K, or a very long half-life, such as ^{87}Rb. Although these extensions have in turn introduced certain new problems and forced some compromises, they have made possible certain conclusions about geological questions and have opened new avenues for research.

A number of possible radioactive dating methods exist, but each method is practical, of course, only if the appropriate radionuclide exists in the mineral. One series of possible dating methods is based on the decay of natural uranium and natural thorium. If the rock has retained the helium produced by the decay of ^{238}U, for example, 8 helium atoms should exist for each nuclide of ^{238}U that has decayed through its complete chain to ^{206}Pb, since 8 alpha particles result from this chain. From a measure of the ratio of the amount of helium to the amount of ^{238}U in the rock, a calculation may then be possible of the age of the rock. In this method, corrections must be made for the decay of ^{235}U and of ^{232}Th, both of which are the initiating nuclides for a natural chain of radioactive nuclides. Because the half lives of ^{232}Th and ^{238}U are different, another method for determining the age of a rock containing both these nuclides is the measurement of the ratio of the amount of ^{206}Pb to the amount of ^{208}Pb, which are the ultimate decay products of the ^{238}U and ^{232}Th chains, provided neither of these isotopes of lead existed in appreciable quantity prior to formation of the rock. A related measurement is the ratio of radiogenic lead (either ^{206}Pb or ^{208}Pb) to nonradiogenic lead (^{204}Pb), which can be assumed to have been of primordial origin. Another correction that may be necessary, especially if the rock comes from a high altitude, is a determination of the amount of helium that has been produced as a result of spallation reactions caused by very high-energy cosmic radiation. Other radioactive age-dating systems are those of potassium-argon (which consists of the decay of ^{40}K to ^{40}Ar, by electron capture, a process with a half-life of 1.27×10^{10} years) and rubidium-strontium which consists of the decay of ^{87}Rb to ^{87}Sr, by electron emission, a process having a half-life of 4.7×10^{10} years.

One conclusion drawn from radioactive measurements is that the pre-Cambrian history of the earth's crust extends beyond 2,700 million years. The pegmatites that have been found to be this old are located in North America and Australia, and they probably exist on all the continents. The oldest rocks in the United States that have been measured are on the south rim of the Bridger Mountains near the Wind River Canyon in Wyoming. These ancient pegmatites intrude geologic formations of sedimentary and volcanic rocks that, themselves, are the result of even more ancient processes than those in which they were formed. Thus, a period of the order of 3,000 million years or more is available for geologic processes that have formed the crust seen today.

Next, the facility to measure the absolute age of micas in igneous intrusives of pre-Cambrian sediments provides a method of correlating these sediments wherever they occur in much the same fashion that fossil correlation of more recent sedimentary formations is possible. A method that is independent of the lithologic characteristics and the general structure of the sediments will provide a crucial test of the validity of these criteria, which have been all that was available to the geologist. Further, any attempts to look for more subtle evidence of such things as changes in the composition of the atmosphere or origins of life itself must be fitted into a time scale of the pre-Cambrian.

Radiocarbon Dating. This is a method of estimating the age of carbon-containing materials by measuring the radioactivity of the carbon in them. The validity of this method rests upon certain observations and assumptions, of which the following statement is a brief summary. The cosmic rays entering the atmosphere undergo various transformations, one of which results in the formation of neutrons, which in turn, induce nuclear reactions in the nuclei of individual atoms of the atmosphere. The dominant reaction is

$$n + {}^{14}\mathrm{N} \longrightarrow {}^{14}\mathrm{C} + p$$

in which the neutrons react with the nuclei of nitrogen atoms of mass number 14 (which make up the nitrogen molecules that constitute nearly $\frac{4}{5}$ of the atmosphere) to form carbon atoms of mass number 14 and protons (p). The ^{14}C atoms are radioactive, having a half life of about 5730 years. The largest rate of formation of ^{14}C atoms from cosmic rays is at 30,000–50,000 feet (9,144–15,240 meters) above sea level and at higher geomagnetic latitudes, although formation occurs at varying rates throughout the entire atmosphere. The ^{14}C atoms react with oxygen in the atmosphere to form carbon dioxide, which is mixed with the nonradioactive carbon dioxide in the atmosphere, and with it gains worldwide distribution by various processes. The radioactive $^{14}\mathrm{CO_2}$ enters the carbon cycle in which plants take up carbon dioxide from the atmosphere to form carbohydrates, which enter through plant foods into the composition of animals. In another world-wide process, also of exchange nature, carbon dioxide is dissolved in seawater and then, under changing conditions of acidity and temperature, is partially evolved from the seawater again. As a result of these and other processes, the ^{14}C formed in the atmosphere by cosmic rays tends to become distributed throughout all the nonradioactive carbon, not only in the atmosphere, but in the biosphere, the hydrosphere, and even the upper levels of the lithosphere (there are many carbonate-containing minerals).

Obviously this wide distribution of the ^{14}C formed in the atmosphere takes time; it is believed to require a period of 500–1000 years. This time is not, however, a deterrent to radiocarbon dating because of two factors; the long half-life of ^{14}C and the relatively constant rate of cosmic-ray formation of ^{14}C in the earth's atmosphere over the most recent several thousands of years. These considerations lead to the conclusion that the proportion of ^{14}C in the carbon reservoir of the earth is constant, and that the addition by cosmic ray production is in balance with the loss by radioactive decay. If this conclusion is warranted, then the carbon dioxide on earth many centuries ago had the same content of radioactive carbon as the carbon dioxide on earth today. Thus, radioactive carbon in the wood of a tree growing centuries ago had the same content as that in carbon on earth today. Therefore, if we wish to determine how long ago a tree was cut down to build an ancient fire, all we need to do is to determine the relative ^{14}C content of the carbon in the charcoal remaining, using the value we have determined for the half life of ^{14}C. If the carbon from the charcoal in an ancient cave has only $\frac{1}{2}$ as much ^{14}C radioactivity as does carbon on earth today, then we can conclude that the tree which furnished the firewood grew 5730 ± 30 years ago. See also discussion on the age determination of the bristlecone pines in the White Mountains of California in the entry on **Pine Trees**.

As pointed out by Muller (1978), there are well-documented differences between the ages of materials determined by dating with radioisotopes and the ages determined by other means, such as tree-ring counting. In addition to systematic effects, there are statistical errors due to the limited number of atoms observed. Both types of errors can be considered to be fluctuations in n, the number of atoms observed. A relationship can be derived between the magnitude of these fluctuations and the resulting error in the estimation of age of the sample:

$$n = ke^{-t/\tau} \text{ or } t = \tau \ln(k/n)$$

where τ is the mean life of the isotope, t is the age of the sample, and k is the initial number of radioactive atoms in the sample multiplied by the efficiency for detecting them. If n has errors associated with it of $+\delta n_1$ and $-\delta n_2$, then the corresponding values of t will be:

$$t = \tau \ln \frac{k}{n^{\delta n_1}_{\delta n_2}} = \tau \ln(k/n) {}^{+}_{-} \left| \frac{\ln(1 - \delta n_2/n)}{\ln(1 + \delta n_1/n)} \right|$$

Muller has shown that for $n = 1$, inverse Poisson statistics gives $n_1 = 1.36$ and $n_2 = 0.62$ and thus the foregoing equation becomes:

$$t = \tau \ln (k)^{+0.96\tau}_{-0.86\tau}$$

Further details of this method can be found in aforementioned reference.

Determination of the ratio of two oxygen isotopes has been effective in fixing the age of fossil sediments and can provide information about ice formation and, possibly, water temperatures. The lighter isotope ^{16}O evaporates preferentially and thus precipitation and hence ice in glaciers and polar caps should be enriched with ^{16}O relative to seawater. Thus, fluctuations in the amount of water locked up as ice can be determined from variations in the oxygen isotope ratio of fossils locked up in deep-sea sediments. And, because this ratio also varies with water temperature, thermal information also can be gleaned. Kennett (University of Rhode Island) has employed this technique in determining when significant amounts of ice first formed at the poles. This research has indicated that the Antarctic ice cap formed only about 16 million years ago, after Australia had split off and moved away from Antarctica, leaving the latter continent isolated at the pole and surrounded by the fast-moving circumpolar current.

In 1986, J.I. Hedges and colleagues (see reference) reported on how dissolved and particulate organic material transported by rivers can provide a continuous record of physical and biological processes at work within the drainage basin. Rivers also contribute a potentially important quantity of organic matter to the ocean, where the dissolved component may exhibit an appreciable residence time. Although the magnitude of the global river contribution is known, the dynamics of organic materials within terrestrial ecosystems and their effects on the composition of the corresponding marine reservoirs are poorly understood. This is particularly true for rivers draining topical rain forests, which account for about 40% of the total riverine organic carbon discharged into the ocean. Hedges and a team studied the Amazon River System using organic carbon-14 dating methods. They found that coarse and fine suspended particulate organic materials and dissolved humic and fulvic acids transported by the Amazon River all contain bomb-produced carbon-14, indicating relatively rapid turnover of the parent carbon pools. However, the carbon-14 contents of these coexisting carbon forms are measurably different and may reflect varying degrees of retention of soils in the drainage basin.

Dating by Accelerator Mass Spectrometry. The cyclotron is mainly used as a source of energetic particles. The cyclotron also can be used as a very sensitive mass spectrometer. Alvarez and Cornog (1939) were the first researchers to use a cyclotron in this manner. This was in connection with their discovery of the true nuclear properties of ^{3}He and tritium. Within the last few years, Muller and associates at the Lawrence Berkeley Laboratory have used this method in a search for integrally charged quarks in terrestrial material. For radioisotope dating, the cyclotron is tuned to accelerate the isotope of interest and the sample is introduced into the ion source, preferably as a gas. For radioisotope dating, the greatest gains over radioactive counting techniques apply to the longer-lived species, which have lower decay rates. It has been estimated that the cyclotron can be used to detect atoms or simple molecules that are present at the 10^{-16} level or greater. For ^{14}C dating, the Berkeley investigators indicate that one should be able to go back 40,000 to 100,000 years with 1-to-100 microgram carbon samples; for ^{10}Be dating, 10–30 million years with from 1 cubic millimeter to 10 cubic centimeter rock samples; and for tritium dating, 160 years with a 1-liter water sample. Over 50 cyclotrons are in operation today that could perform radioisotope dating and, although the instruments are costly, the cost for a dating determination experiment may not be much higher than for decay dating technology.

Other isotopes with which an accelerator mass spectrometer may be effective include ^{26}Al, ^{36}Cl, ^{53}Mn, ^{81}K, and ^{129}I. Chlorine-36 has a half-life of 300,000 years and may be used for dating water in underground reservoirs. ^{10}Be is produced in the atmosphere at the rate of 1.5×10^{-2} atom cm^{-2}sec^{-1} by cosmic rays that break up oxygen and nitrogen nuclei. ^{10}Be has been used in studies of both seafloor spreading and manganese nodule formation. Although tritium (^{3}H) has a short mean life of 17.8 years, tritium dating has been important in cosmic-ray physics, hydrology, meteorology, and oceanography. For example, if one desires to know how long an underground water reservoir may require for refilling, the age of the water can be determined by tritium dating methods.

In 1986, researchers at the Research Laboratory for Archaeology and the History of Art, University of Oxford, reported on how the radioactive carbon-14 isotope can be separated from other atoms in a sample by use of accelerator mass spectrometry, thus making it possible to derive more accurate chronologies from much smaller archaeological or anthropological specimens. For details, consult Hedges/Gowlett reference listed.

In 1992, K.R. Ludwig and colleagues of the U.S. Geological Survey reported on their use of mass-spectrometric ^{230}Th-^{234}U-^{238}U dating of the Devil's Hole calcite vein (Derbyshire, England), which contains a long-term climatic record, but requires accurate chronological control for its interpretation. Mass-spectrometric U-series ages for samples from core DH=11 yielded ^{230}Th ages, with precisions ranging from less than

1000 years to less than 50,000 years for the oldest samples. The ^{234}U/^{238}U ages could be determined to a precision of approximately 20,000 years for all ages. Tentatively, the researchers have concluded, "Overall, the U-series ages form a remarkably self-consistent suite of age determinations. Because this consistency is both internal (from replicate samples) and external (from the stability of the overall age-distance trend), it seems highly unlikely that the dates have been significantly corrupted by open-system processes, such as uranium gain or loss or alpha-recoil phenomena. The apparent ideality of the U-Th system in the vein material is probably the result of continuous submergence in water that showed limited secular variations of its physical and chemical properties."

Non-Radioactivity Dating Techniques

Several methods in addition to those involving radioactivity have been used to estimate the ages of various materials and objects.

Obsidian Hydration Rate. Obsidian (rhyolitic volcanic glass) can be used as a key to age determinations for both archeological and geological purposes. As pointed out by Friedman and Long (1976), the method depends upon the fact that obsidian absorbs water from the atmosphere to form a hydrated layer, which thickens with time as the water slowly diffuses into the glass. The hydrated layer can be observed and measured under a microscope on thin sections cut normal to the surface. To convert the measured hydration thickness to an age, the equation relating thickness to time must be known. This requires not only the form of the equation (functional dependence), but also the constants in it. Prior to the early 1960s, age could be related to hydration thickness only if combined with known history of a region or through the use of carbon-14 techniques. In the mid-1960s, Friedman and associates conducted actual experimental hydration experiments on obsidian, exposing the materials (taken from the Valles Mountains in New Mexico) to a temperature of 100 °C and steam at a pressure of 1 atmosphere over a 4-year period. An equation of the form, $T = kt^{1/2}$ was developed, where T = thickness of hydration layer, t = time, and k is a constant. Investigators have developed a procedure for calculating hydration rate of a sample from its silica content, refractive index, or chemical index and a knowledge of the effective temperature at which the hydration occurred. The effective hydration temperature (EHT) can either be measured or approximated from weather records. The investigators concluded that if the EHT can be determined and measured for the hydration of a particular obsidian, it should be possible to carry out absolute dating to ±10% of the true age over periods as short as several years and as long as millions of years.

Manufactured Glass Objects. Other investigators (Lanford, 1977) have extended the principles applying to obsidian to manufactured glass, which extends back for thousands of years and thus can be useful to archeologists. However, as observed by Lanford, one cannot use the same optical method for measuring the thickness of hydration layers as used with obsidian. The hydration of the two materials differs. Also, glass that is less than a few hundred years old would generally have hydration layers thinner than the wavelength of visible light. The optical method is destructive in that it requires removal of a slice of glass from an object, something much discouraged by art historians and dealers. The Lanford method involves a resonant nuclear reaction between ^{15}N and ^{1}H for measuring the distribution of hydrogen in solids. With this technique, complete depth profiles of the surface hydration layer can be obtained in a fully nondestructive manner. Lanford summarizes by observing that this method of hydration dating need not be limited to glass. Since most silicates are unstable against slow reactions with atmospheric water, many may develop surface hydration layers suitable for dating and authenticating. The glazes on pottery are chemically similar to glass, and it may be possible that a dating method for glazed pottery based upon these procedures can be developed.

Amino Acid Racemization. This dating method is based upon the incorporation of L-amino acids, exclusively, into proteins by living organisms. As pointed out by the researchers Masters and Zimmerman (1978), given sufficient periods of time over which proteins are preserved after synthesis, a number of spontaneous chemical reactions take place. Among these is racemization, which converts L-amino acids into their enantiomers, the D-amino acids. The different amino acids racemize at various rates, and these rates (as with all chemical reactions) are proportional to temperature. One of the fastest racemization rates known is that of aspartic acid, with a half-life of 15,000 years at 20 °C. It follows, then, that the older a fossilized material may be, the higher will be its

D-aspartic acid content or D/L Asp ratio. Once the k_{asp} is known for a given fossil locality, the age of a specimen can be calculated from the D/L ratio.

This method was used in the examination of an Eskimo who died 1600 years ago. The body was discovered in a frozen state on St. Lawrence Island, Alaska in 1972 and remained frozen until it was brought to Fairbanks in 1973. Examination of the female individual revealed that she had a skull fracture, probably resulting from instant burial caused by a landslide. Aspartic acid racemization analysis of a tooth from the mummy yielded an age at death of 53 ± 5 years, which correlated well with earlier estimates based upon morphological features. This method is an example of the need to preserve mummies (Alaskan, Egyptian, and Peruvian, among others) for application of new dating techniques as they develop.

The racemization dating technique fell out of favor with many paleoanthropologists in the 1980s, but staged a comeback of acceptance in the early 1990s. This resurgence was the result of analyzing samples of African ostrich egg fragments collected from the Border Cave located on the east coast of South Africa. These or close-by sites also contain reputedly human bones. If the age of these bones is determined to be in the range of 100,000 years, this would provide an additional bit of independent evidence for the theory that modern humans came from Africa. In the Border Cave determination, researchers found that amino acid racemization (AAR) time dating compared relatively closely with electron-spin resonance dating carried out at Cambridge University. AAR techniques revealed an age of 80,000 to 100,000 years for the egg-shells, whereas electron-spin resonance dating of human bones shows 60,000 years.

Geochemical Methods. These methods usually involve a combination of chemical analysis of materials coupled with the curiosity of a detective. This is not a singular methodology, but incorporates numerous disciplines and a lot of past experience with the materials in question. The authentication of ancient marble sculptures prior to their procurement by the J. Paul Getty Museum is an example. In this case, an expert geochemist with past experience with dolomite marble studied a phenomenon known as *dedolomitization*, wherein after many centuries the exposed surface of the marble changes into calcite, or calcium carbonate. Thus, this provides a key to age as well as to source of the original material. See Margolis reference listed.

After using numerous dating techniques over a span of several years, by the late 1980s, the much publicized "Shroud of Turin" was finally explained to the satisfaction of most investigators. The answer was put forth by W.C. McCrone, a microscopist who specializes in authenticating art objects. McCrone examined fibers and other materials lifted from the surface of the cloth with adhesive tape. He determined that light-colored portions of the figure were comprised of a gelatin-based medium speckled with particles of red ocher and that fibers from the dark areas (representing blood) contained stains not of blood, but rather were made up of particles of vermilion. The vermilion was found to be of the type developed for artists in the Middle Ages. Also, the practice of painting linen with gelatin-based temperas was common during the late 13th and the 14th century. Final conclusion: The "shroud" had been forged by some unknown 14th century artist.

The use of thermoluminescence (i.e., heating small particles to a high temperature and then analyzing the emanations spectroscopically) can be a key to an object's age. In a paper by P. Lang presented at the 1988 Pittsburgh Conference and Exposition on Analytical Chemistry and Applied Spectroscopy, modern methods involving Fourier transform spectrometry or Raman spectrometry are described. As early as 1818, Sir Humphrey Davey presented a paper along these lines to the Royal Society on the colors used by the ancients in painting.

Tree-ring counting as a measure of age is described under Pine Trees. However, it is of note to mention an August 1990 report on how tree rings were used to reveal the age of the "oldest road," which was found in a peat bog in 1970. The road is located in southwestern England. It dates back 4000 years.

See also articles on **Mass Extinctions**; and **Meteorides and Meteorites**.

Additional Reading

Abelson, P.H.: "Isotopes in Earth Science," *Science*, 1357 (December 9, 1988).

Alpen, E.L.: "Radiation Biophysics," 2nd Edition, Morgan Kaufmann Publishers, Orlando, FL, 1997.

Appenzeller, T.: "Roving Stones: A Landmass Was Wandering Over Three Billion Years Ago," *Sci. Amer.*, 19 (February 1990).

Avignone, F.T., III and R.I. Brodzinski: "A Review of Recent Developments in Double-Beta Decay," 21 (A. Faessler, Editor) Pergamon Press, 1988.

Badash, L.: "The Age-of-the-Earth Debate," *Sci. Amer.*, 90 (August 1989).

Barnes, D.M.: "Probing the Authenticity of Antiquities with High-Tech Attacks on a Microscale," *Science*, 1374 (March **18**, 1988).

Beardsley, T.: "Fallout: New Radiation Risk Estimates Prompt Calls for Tighter Controls," *Sci. Amer.*, 35 (March 1990).

Bowen, D.Q.: "The Last 130,000 Years," *Review (University of Wales)*, 39 (Spring 1989).

Bower, B.: "Eggshells Help Date Ancient Human Sites," *Science News*, 215 (April 7, 1990).

Brooks, A.S. et al.: "Dating Pleistocene Archeological Sites by Protein Diagenesis in Ostrich Eggshell," *Science*, 60 (April 6, 1990).

Chesley, J.T., A.N. Halliday, and R.C. Scrivener: "Samarium-Neodymium Direct Dating of Fluorite Mineralization," *Science*, 949 (May 17, 1991).

Cobb, C.E., Jr.: "Living with Radiation," *National Geographic*, 403 (April 1989).

Eisenbud, M. and T.F. Gesell: "Environmental Radioactivity from Natural, Industrial, and Military Sources," 4th Edition, Morgan Kaufmann Publishers, Orlando, FL, 1997.

Elliott, S.R., A.A. Hahn, and K.M. Moe: "Direct Evidence for Two-Neutrino Double-Beta-Decay in $_{82}$Se," *Physical Review Letters*, **59**, 18, pp. 2020–2023 (November 2, 1987).

Friedman, I. and W. Long: "Hydration Rate of Obsidian," *Science*, **191**, 347–352 (1976).

Gaisser, T.K.: "Gamma Rays and Neutrinos as Clues to the Origin of High Energy Cosmic Rays," *Science*, 1049 (March 2, 1990).

Greiner, W. and A. Sandulescu: "New Radioactivities," *Sci. Amer.*, 58 (March 1990).

Hamilton, D.P.: "U.S. Faces Uncertain Medical Isotope Supply," *Science*, 603 (July 31, 1992).

Hardy, J.C.: "Exotic Nuclei and Their Decay," *Science*, **227**, 993–999 (1985).

Hedges, J.I., et al.: "Organic Carbon-14 in the Amazon River Systems," *Science*, **231**, 1129–1131 (1986).

Hedges, R.E.M. and J.A.J. Gowlett: "Radiocarbon Dating by Accelerator Mass Spectrometry," *Sci. Amer.*, 100–107 (January 1986).

Horgan, J.: "The Shroud of Turin," *Sci. Amer.*, 18 (November 1988).

Huntley, B. and I.C. Prentice: "July Temperatures in Europe from Pollen Data 6000 Years Before Present," *Science*, **687** (August 5, 1988).

Knoll, G.F.: "Radiation Detection and Measurement," 3rd Edition, John Wiley & Sons, Inc., New York, NY, 1999.

Lanford, W.A.: "Glass Hydration: A Method of Dating Glass Objects," *Science*, **196**, 975–976 (1977).

L'Annunziata, M.F.: "Handbook of Radioactivity Analysis," Harcourt Brace & Company, San Diego, CA, 1998.

Lowenthal, G. and P. Alrey: "Practical Applications of Radioactivity and Nuclear Radiations," Cambridge University Press, New York, NY, 2001.

Ludwig, K.R., et al.: "Mass-Spectrometric ^{230}Th-^{234}U-^{238}U Dating of the Devil's Hole Calcite Vein," *Science*, **284** (October 9, 1992).

Margolis, S.V.: "Authenticating Ancient Marble Sculpture," *Sci. Amer.*, **104** (June 1989).

Marshall, E.: "Racemization Dating: Great Expectations," *Science*, **790** (February 16, 1990).

Masters, P.M. and M.R. Zimmerman: "Age Determination of an Alaskan Mummy: Morphological and Biochemical Correlation," *Science*, 201, 811–812 (1978).

Moe, M.K. and S.P. Rosen: "Double-Beta Decay," *Sci. Amer.*, 48 (November 1989).

Monastersky, R.: "Coral Corrects Carbon Dating Problems," *Science News*, **356** (June 9, 1990).

Muller, R.A., Stephenson, E.J., and T.S. Mast: "Radioisotope Dating with an Accelerator: A Blind Measurement," *Science*, 201, 347–348 (1978).

Pasachoff, N.: "Marie Curie and the Science of Radioactivity," Oxford University Press, Inc., 1997.

Pszczola, D.W.: "Food Irradiation," *Food Technology*, **92** (June 1990).

Rusting, R.: "A Clock in the Trees: Tree Rings Reveal the Precise Age of the Oldest Road," *Sci. Amer.*, 30 (April 1990).

Smith, F.A.A.: "A Primer in Applied Radiation Physics," World Scientific Publishing Company, Inc., River Edge, NJ, 1999.

Staff: "Rocks of Ages," *Technology Review (MIT)*, 80 (April 1990).

Strauss, S.: "Archeology's Dating Game," *Technology Review (MIT)*, 8 (October 1987).

Swisher, C.C., III and D.R. Prothero: "Single-Crystal ^{40}Ar/^{39}Ar Dating of the Eocene-Oligocene Transition in North America," *Science*, **760** (August 17, 1990).

Theodorsson, P.: "Measurement of Weak Radioactivity," World Scientific Publishing Company, Inc., River Edge, NJ, 1996.

RADIO ALTIMETER.   See **Altimetry**.

RADIO ASTRONOMY.   Observations of astronomical objects made with wavelengths longer than about one centimeter (3.3 GHz, rf frequencies) are generally classified as radio observations. The term also extends into the millimeter wavelengths, although different detectors are required for this wavelength range. The field began in 1932, with the accidental discovery of radio emission from the galactic plane by Karl Jansky, working at Bell Laboratories. Further observations were made following the invention of the first parabolic radio telescope by Reber in 1940. Following World War II, groups at Cambridge and Manchester in the UK, CSIRO in Australia, and in the Netherlands, began the development of a variety of methods for observing the universe at radio wavelengths, often using war surplus equipment. Discrete radio sources, the Crab Nebula, a supernova remnant, and the radio galaxies Centaurus A and M 87 in Virgo, were identified by the end of the decade. Solar radio emission was discovered during the war and imaging work began shortly thereafter, and long-wavelength radio emission from the Jovian magnetosphere was discovered in the 1950s.

Types of Radio Telescopes

Single-Dish Measurements. The resolution of a telescope depends on the wavelength of the incoming light, λ and the aperture, D, by $\theta \approx \lambda/D$. Thus, to achieve reasonable accuracy in location of a source, it is necessary for a radio telescope to be significantly larger than an optical instrument. Specifically, the aperture must be tens to hundreds of meters in order to achieve resolutions even remotely comparable with optical telescopes (for which the wavelengths are more than 100 times smaller). The largest single disk is the Arecibo telescope, which is a fixed reflector constructed in a natural crater in the mountains of Puerto Rico (with an aperture of some 300 meters). See Fig. 1. In order to scan the sky, the feed, mounted at the focus and suspended above the aperture, is steered over the surface of the reflector. Only a small swatch of sky, bounded by declination (celestial latitude), can be seen with this instrument as it transits across the beam. The telescope has also been used for radar experiments on solar system objects (planets, comets, and asteroids), and conducts surveys of pulsars and mapping of neutral hydrogen in our own and other galaxies.

A large, steerable 100-meter telescope is located near Bonn, Germany. The 300-foot ($\sim$91-meter), partially steerable (north-south radio) telescope installed in 1962 at Green Bank, West Virginia, collapsed of metal fatigue on November 15, 1989, after having served for just over a quarter century. In late 1990, the National Science Foundation announced a contract for rebuilding the facility, which was expected to be operable by the mid-1990s. The new telescope would be fully steerable. The major discovery made by the former instrument at Green Bank was that of locating a pulsar in the Crab nebula.

Limitations of the types of telescopes just described center on the typically arcmin sizes of the beams at centimeter wavelengths. Especially along the galactic plane, there are often enough sources located within the beam that the observation is "confusion limited" — that is, it proves impossible to separate out individual sources in order to determine specific fluxes and sizes.

Interferometers. Light signals contain two essential pieces of physical information — amplitude and phase. At radio wavelengths, it is especially easy, because of the long integration times allowed, to recover both with an array of telescopes. While each of the instruments individually has a large beam, the combined angular resolution of the array can be made as small as the physical spacing of the elements of the array will allow. The idea derives from coherence of the incoming wavefront. If a signal is received by one element of the array at a time t, another element some distance away will receive the signal either earlier or later, depending upon the direction of the source. Accurate timing of the arrival of the signals then permits the combination of the two or more elements, which will produce fringes whose spacing depends upon the size of the source and of the array. In the earliest versions of interferometers, the antennas were in fixed configurations on the earth, and the source location was determined by the transit of the celestial coordinates over the central beam of the telescope, such as in the Mills Cross. With the advent of steerable large antennas, it is now possible to track sources across the sky, increasing the time on source and therefore the signal to noise ratio (SNR).

Signals from the individual antennas of the array are either combined directly in a central correlator, which compares the phases of the signals

Fig. 1.   Aerial view of radio telescope at Arecibo, Puerto Rico. Control room complex is in lower center; factory and service building is in lower left; helicopter landing pad is in far left center. All facilities are clustered around the 100-foot (305-meter) diameter reflector bowl. The central platform is suspended from three concrete towers. (*The National Astronomy and Ionosphere Center, Cornell University. Photo by Russell C. Hamilton.*)

and produces a flux for each baseline in the array, or by imposing timing signals on the data for later digital analysis. Beam characteristics and the sensitivities of each antenna in the array are calibrated by frequent observation of standard sources of known intensity and spectrum.

The technique for producing images of sources by creating (using individual phased telescopes) kilometer-size effective areas is called *aperture synthesis*. In this method, the sky moves over the telescope, whose individual antennas produce tracks on the celestial sphere that sample many parts of the sky with different time delays between the telescopes. This is because, to a celestial source, the telescopes of an array represent a flat surface rotating on the earth at some projection angle (depending on latitude of the telescope) to the sky. Thus, by tracking the source for a time, it is possible to obtain information with many different relative angles to the different parts of the source, and to produce, as seen by the celestial sphere, a nearly filled aperture with a resolution equal to that of a single disk the size of the array. Among the first such instruments were the 3-km array at Cambridge, constructed by Ryle and collaborators following the introduction of aperture synthesis methods in the 1950s; the Westerbork Radio Synthesis Telescope in the Netherlands; and the Green Bank interferometer at NRAO (National Radio Astronomy Observatory).

A new era in radio astronomy was opened in late 1970s with the opening of the Very Large Array (VLA) operated by NRAO. This was the first movable, steerable, fully two-dimensional radio synthesis telescope to operate on a regular basis. It consists of 27 identical movable 25 meter radio telescopes, mounted in a Y-configuration, located on the Plains of San Agustine, near Socorro, New Mexico. The array can be changed from about 3 kilometers to over 30 kilometers by physically transporting the telescopes to different stations along the three arms (24 stations per arm), and can reach a resolution of about 0.1 arcsec, better than most optical telescopes. Because of the ability to vary the ground based spacing of the antennas in the array, the VLA is capable of producing radio images which have nearly circular beams at essentially all positions on the sky accessible from central New Mexico.

The highest angular resolution achieved to date uses *Very Long Baseline Interferometry* (VLBI). In this technique, radio observers on different continents simultaneously observe a single portion of sky. The signals are later digitally combined. Resolutions of milliarcsec have been obtained, permitting imaging of radio emitting regions on physical size scales as small as light days. A continental size version of the VLA, the *Very Long Baseline Array*, is being constructed by NRAO using about a dozen identical antennas spread out among sites from Hawaii to Puerto Rico. A project under discussion for the next decade is a space-based orbiting array of satellite radio telescopes.

Arrays of millimeter telescopes are also possible. Several are currently operating, at Hat Creek in California, Nobayama in Japan, and the IRAM interferometer, currently (1987) under construction in the French Alps. A design study is underway by NRAO for a VLA-like millimeter array, and submillimeter arrays are also being planned. The sizes of these short-wavelength arrays are more severely limited than at longer wavelengths by the turbulence and transparency of the earth's atmosphere and the accuracy required for the telescope surfaces and detectors. They require high, dry, radio-quiet sites.

Types of Radio Emission

The standard unit of radio source strength, the monochromatic flux density, is the Jansky; $1 \text{ Jy} = 10^{-26} \text{ W m}^{-2} \text{ Hz}^{-1}$. The strongest radio sources are about 1 Jy, and the weakest sources thus far detected are at about the microjansky level.

Continuum Emission: Thermal. When protons and electrons undergo near misses in a plasma, photons are emitted as a consequence of the electromagnetic interaction between the charges. The rate of emission increases with increasing temperature and increasing electron density. This radiation is called *free-free emission* or *bremsstrahlung*, indicating that it is due to unbound electrons and protons. The spectrum has a characteristic form. The flux density for the optically thin frequencies is almost independent of frequency, but drops steeply at high frequencies (near kT/h where k and h are the Boltzman and Planck constants, respectively, and T is the source temperature). Thermal spectra turn over at the low frequencies due to optical depth effects internal to the emitting plasma, *self-absorption*, and have a characteristic low frequency spectrum of $F_v \sim v^2$, where F_v is the flux density. The source brightness temperature, the temperature a blackbody emitter would need at the same

wavelength to achieve the same surface brightness, is essentially the same as the plasma kinetic temperature.

Continuum Emission; Synchrotron. The most important emission mechanism in radio astrophysics is *nonthermal* continuum radiation from relativistic electrons spiraling in cosmic magnetic fields. This emission produces a characteristically *power law shape*, typically $F_v \sim v^{-1}$, and very high brightness temperatures. The low-frequency spectral cutoff is due to self-absorption, but with a slope different than for thermal radiation, $F_v \sim v^{2.5}$. The high-end cutoff is due to the aging of the highest energy electrons, the lifetime of which depends on the magnetic field strength B and the electron energy E, both of which determine the characteristic frequency of the radiated photons. Synchrotron radiation, because it is not thermal in origin, gives information about the most energetic acceleration processes accessible in astrophysical objects. It is both linearly and circularly polarized, the linear polarization providing information about the direction of the magnetic field. Active sources of such emission are solar flares, mass outflows from very high luminosity hot stars, nova and supernova remnants, and especially active galactic nuclei, radio galaxies, and quasars. A low-energy form of this emission is detected from planetary magnetospheres.

21-cm Line of Neutral Hydrogen. The ground state of HI is split due to hyperfine interactions between the nucleus (proton) and electron, with collisions inducing upward transitions that decay on very long time scales. The prediction that this line would be observable from the diffuse gas of the interstellar medium was published in 1945 by Van de Hulst, who calculated its wavelength as 21 cm (1420 MHz). Ewen and Purcell detected the line in emission in 1951, with the first absorption measurement being by Hagen and McClain in 1954. The importance of this radiation is that, because it is discrete, it provides information about both the abundance and velocity of the emitting gas. Using high frequency resolution, of order 1 to 5 KHz, velocity resolutions of a few kilometers per second can be achieved. As a result, the emission from the diffuse interstellar medium, from the most abundant element in the universe, can be mapped at radio wavelengths at which the galaxy is optically thin. The entire galaxy can be seen in this line, including the regions of the galactic center which are totally obscured at optical and ultraviolet wavelengths.

Recombination Lines. When an electron recombines with an ion, it cascades through the atomic energy levels toward the ground state, emitting photons with each transition. The radiation from the highest levels, near the continuum (the Rydberg states) can be studied at radio wavelengths and provide information about many species heavier than hydrogen, like carbon and helium. In addition, the observation of radio recombination lines of hydrogen gives information about the ionized gas in the same way as the 21-cm line yields a picture of the neutral component. This can be compared directly with continuum radiation (free-free emission) from the same gas and help determine the ionization fraction and dynamics of the ionized medium.

Molecular Lines. Emission from hyperfine structure lines in molecular species, which occur in the millimeter and short centimeter wavelength portion of the spectrum, permit determination of abundances and dynamics for the coldest component of the interstellar medium. Dark clouds are especially productive sources of molecular line emission, which is found in association with regions of star formation. Molecules as complex as $HC_{11}N$ have been detected at millimeter frequencies, although the most important probes of the dense cores of the interstellar clouds in which the emission typically arises are molecules like CO, CS, NH_3, H_2CO, and CN.

Masers. These are a special class of molecular transition, in which the populations of excited levels have been enhanced through collisions and absorption of higher energy photos. The lines produced are both narrow and intense, often having brightness temperatures of millions of degrees. Masers have been detected both from stars and from interstellar clouds. The most important transitions are SiO, H_2O, OH, and NH_3.

See also **Telescope (Astronomical-Optical)**.

Additional Reading

Burke, B.F. and F. Graham-Smith: "An Introduction to Radio Astronomy," Cambridge University Press, New York, NY, 1996.

Emerson, D.T.: "Multi-Feed Systems for Radio Telescopes," The Astronomical Society of the Pacific, San Francisco, CA, 1995.

Jackson, N., and R.J. Davis: "High-Sensitivity Radio Astronomy," Cambridge University Press, New York, NY, 1997.

Kraus, J.D.: "Radio Astronomy," 2nd Edition, Cygnus-Quasar Books, Powell, OH, 1986.

Malphrus, B.K.: "The History of Radio Astronomy and the National Radio Astronomy Observatory: Evolution toward Big Science," Krieger Publishing Company, Melbourne, FL, 1996.

Pacholczyk, A.: "Radio Astrophysics," W.H. Freeman, New York, NY, 1970.

Raimond, E. and R.E. Genee: "Westerbork Observatory, Continuing Adventure in Radio Astronomy," Kluwer Academic Publishers, Norwell, MA, 1996.

Robertson, P.: "Beyond Southern Skies: Radio Astronomy and the Parkes Telescope," Cambridge University Press, 1992.

Sullivan, W.T. III: "The Early Years of Radio Astronomy," Cambridge University Press, New York, NY, 1984.

Thompson, A.R., J.M. Moran, and G.W. Swenson, Jr.: "Interferometry and Synthesis in Radio Astronomy," 2nd Edition, John Wiley & Sons, Inc., New York, NY, 2001.

Wilson, T.L. and S. Huttemeister: "Tools of Radio Astronomy: Problems and Solutions," Springer-Verlag, Inc., New York, NY, 2000.

Web References

Arecibo Observatory: *http://www.naic.edu/*
Basics of Radio Astronomy: *http://www.jpl.nasa.gov/radioastronomy/*
National Radio Astronomy Observatory (NARO): *http://www.nrao.edu/*
The Astronomical Society of the Pacific: *http://www.aspsky.org/*Radio Astronomy and SETI — Big Ear Radio Observatory: *http://www.bigear.org/*

S. N. S.

RADIOBIOLOGY.   The study of the effects produced on living organisms by radiation.

RADIO COMMUNICATION.   Transmitted radio waves at all frequencies may travel in either of two general directions. One wave closely follows the surface of the earth, whereas the other travels upward at an angle which is dependent upon the position of the transmitting antenna. The former is known as the *ground wave*; the latter as the *sky wave*. At the low frequencies (up to approximately 1,500 kHz), the ground-wave attenuation is low, and signals travel for long distances before they disappear. Above the broadcast band, ground-wave attenuation increases rapidly. Long-distance communication is carried on mostly by means of the sky wave.

The sky wave leaves the earth at an angle that may have any value from 3 to 90 degrees, and travels in almost a straight line until the ionosphere is reached. This region, which begins about 70 miles (~113 kilometers) above the earth's surface, contains large concentrations of charged gaseous ions, free electrons, and uncharged, or neutral, molecules. The ions and free electrons act on all passing electromagnetic waves and tend to bend these waves back to earth. Whether the bending is complete (and the wave does return to the earth) or only partial, depends upon several factors: (1) the frequency of the radio wave; (2) the angle at which the wave enters the ionosphere; (3) the density of the charged particles (ions and electrons) in the ionosphere at that particular moment; and (4) the thickness of the ionosphere at the time.

Extensive experiments indicate that, as the frequency of a wave increases, a smaller entering angle is necessary in order for complete bending to occur. Consider waves A and B in Fig. 1. Wave A enters the ionosphere at a small angle (ϕ) and, hence, little bending is required to return it to earth. Wave B, subject to the same amount of bending, does not return to earth because its initial entering angle (θ) was too great.

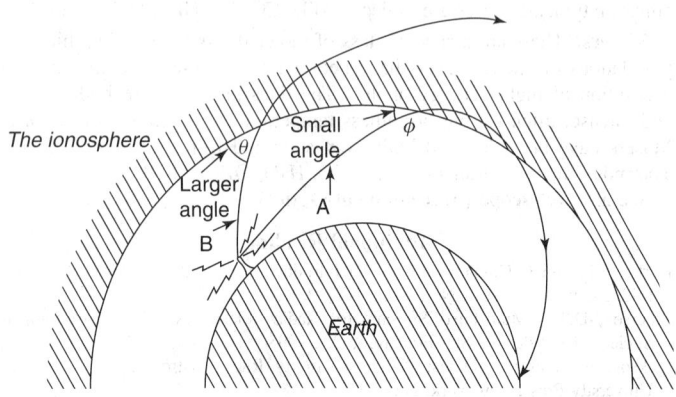

Fig. 1.   At the higher frequencies, a radio wave must enter the ionosphere at small angles if it is to be returned to earth.

Obviously, wave B would not be useful for communicating between points on earth.

By raising the frequency still higher, the maximum allowable incident angle at the ionosphere becomes smaller, until finally a frequency is reached where it becomes impossible to bend the wave back to earth, no matter what angle is used. For ordinary ionospheric conditions, this frequency occurs at about 35 to 40 MHz. Above these frequencies, the sky wave cannot be used for radio communication between distant points on earth. Only the direct ray is of any use. Television bands starting above 40 MHz fall into this category. By direct ray (or rays), it is meant that the radio waves travel in a straight line from transmitter to receiver. At television frequencies, the ionosphere is no longer useful, so the former sky waves must be concentrated into a path leading directly to the receiver. It is this restriction on the use of the direct ray that limits the distance in which high-frequency communication can take place.

There are present, at times, unusual conditions that cause the concentrations of charged particles in the ionosphere to increase sharply. At these times, it is possible to bend radio waves of frequencies up to 60 MHz. The exact time and place of these phenomena cannot be predicted and hence they are of little value for commercial operation. They do explain, however, the distant reception of high-frequency signals that may occur.

At the frequencies employed for television, reception is possible only when the receiver antenna directly intercepts the signals as they travel away from the transmitter. These electromagnetic waves travel in essentially straight lines, and the problem is resolved by finding the maximum distance from the transmitter where the receiver can be placed and still have its antenna intercept the rays. The distance, called the *line-of-sight distance*, may be computed quite simply. In Fig. 2, the height of the transmitting antenna is h_t; the radius of the earth is R; and the distance from the top of the transmitting antenna to the horizon is d. These distances form a right triangle. By applying the Pythagorean Theorem, using a value of 4,000 miles for the radius of the earth, the formula is:

$$d = 1.23\sqrt{h_t}$$

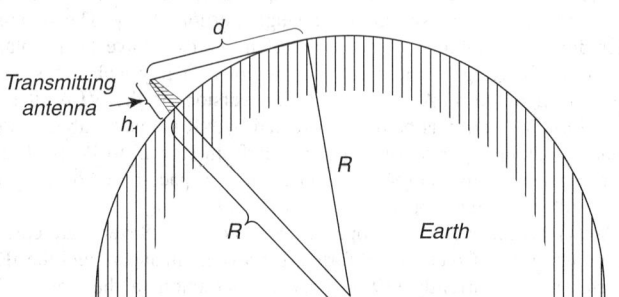

Fig. 2.   Computation of the line-of-sight distance for high-frequency radio waves.

where d is the line-of-sight distance from the top of the transmitting antenna in miles, and h_t is the height of the transmitting antenna in feet. The relationship between d and h_t for various values of h_t is shown in Fig. 3.

The ground coverage for any transmitting antenna will increase with height. Likewise, the number of receivers capable of receiving the signals obviously will increase. The foregoing equation can be used for computing the distance from the top of the transmitting antenna to the horizon. By placing the receiving antenna some distance in the air, it would be possible to cover a greater distance before the curvature of the earth interferes with the direct ray. This condition is shown in Fig. 4. By geometrical reasoning, the maximum line-of sight distance between the two antennas can be arrived at from the distances shown in Fig. 4.

$$d_1 = 1.23\sqrt{h_1} = \text{maximum distance from the transmitting antenna to the horizon}$$

$$d_2 = 1.23\sqrt{h_r} = \text{maximum distance from the receiving antenna to the horizon}$$

$$d = d_1 + d_2 = \text{maximum distance from the transmitting antenna to the receiving antenna}$$

$$= 1.23\sqrt{h_t} + 1.23\sqrt{h_r} = 1.23(\sqrt{h_t} + \sqrt{h_r})$$

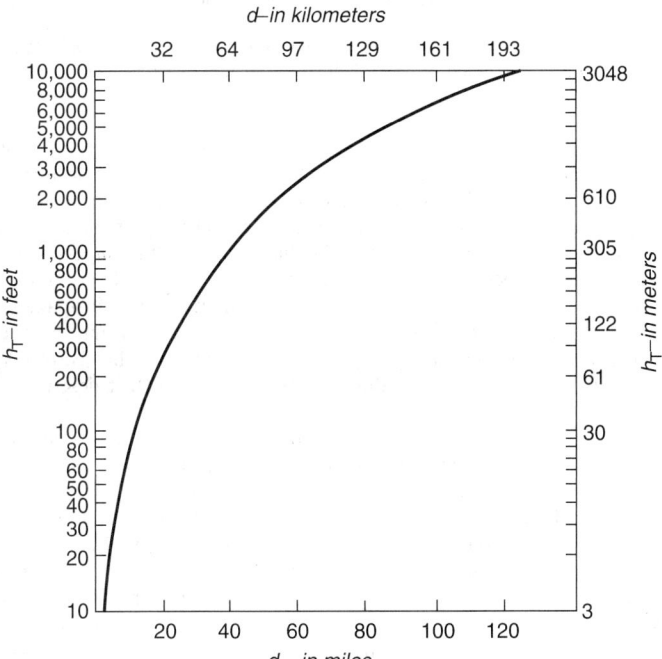

Fig. 3. Relationship between height of transmitting antenna and distance from receiving antenna.

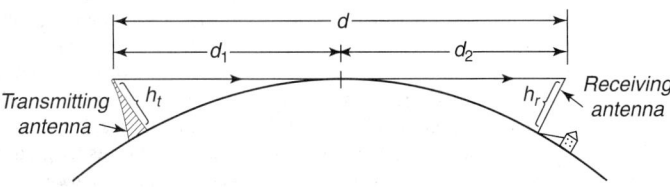

Fig. 4. Increase in line-of-sight distance from receiving antenna to transmitter can be achieved by raising both structures as high as possible.

where h_t is the height of the transmitting antenna, feet; h_r is the height of the receiving antenna, feet; and d is the maximum line-of-sight distance between the two antennas, miles.

The foregoing equations are for the geometrical line of sight. However, electromagnetic waves are bent slightly as they move across the contact point at the horizon, and this increases the television line-of-sight distance by about 15% over the geometric line of sight. Thus, for a geometric line of sight of 30 miles, the television line-of-sight distance is about 34.5 miles.

While the foregoing computed distances apply to the direct ray, there are other paths that waves may follow from the transmitting to the receiving antennas. These waves are undesirable, because they tend to distort and interfere with the direct-ray image on the screen. One type of indirect wave is produced by reflection from surrounding objects. Another may arrive at the receiver by reflection from the surface of the earth. This path is shown in Fig. 5. At the point where the reflected ray strikes the earth, phase reversals up to 180 degrees may occur. This phase shift places a wave at the receiving antenna, which generally acts against the direct ray. The overall effect is a general lowering of the resultant-signal level and the appearance of annoying ghost images.

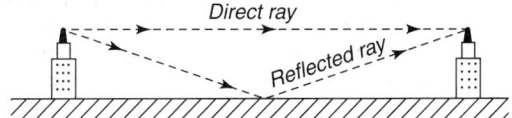

Fig. 5. Reflected radio wave, arriving at the receiving antenna after reflection from the earth, may lower the strength of the direct ray.

There are two compensating conditions that reduce the problem of ground reflection. One is the weakening of the wave strength by absorption at the point where it grazes the earth, and the other results from the

added phase change caused by the fact that the length of the path of the reflected ray is longer than that of the direct ray. Thus, there are two phase shifts affecting the reflected signal: (1) One at the point of reflection from the earth; and (2) one that is the result of the longer signal path. These two phase shifts are additive, so that the total phase shift is nearer to 360 degrees. The worst possible phase shift would be 180 degrees, because this would mean that the direct and the reflected waves are canceling. Since 360 degrees is equivalent to no phase shift, the fact that the individual phase shifts are additive reduces the problem of signal cancellation considerably.

The height of the antenna is one important factor that determines the quality of the reproduced picture in television. Another is the manner in which the antenna is held, that is, either vertically or horizontally. The position of the antenna is determined by the nature of the electromagnetic wave itself.

All electromagnetic waves have their energy divided between an electric field and a magnetic field. In free space, these fields are at right angles to each other. Thus, if one visualizes these fields and represents them by their lines of force, the wave front would appear as shown in Fig. 6. The squares represent the wave front, and the arrows represent the direction in which the forces are acting. The direction of travel of these waves in free space is always at right angles to both fields. As shown in the figure, if the lines of the electric field are vertically directed upward and those of the magnetic field are horizontally directed to the right, then the wave travel is forward.

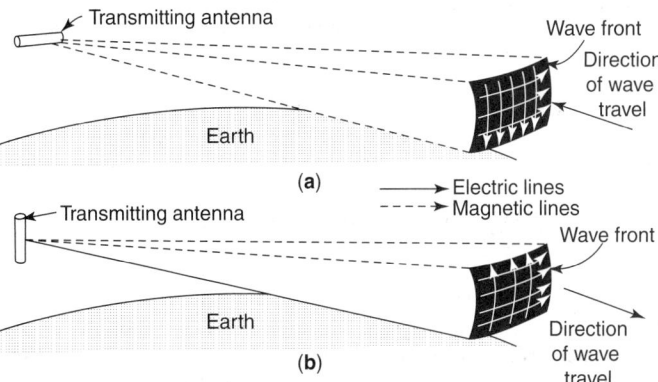

Fig. 6. Components of an electromagnetic wave, showing their relationship with the direction of travel of the wave form.

In radio, the polarity of a radio wave has been taken to be the same as the direction of the electric lines of force. Hence, a vertical antenna radiates a vertical electric field (the lines of force are perpendicular to the ground), and the wave is said to be vertically polarized. A horizontal antenna radiates a horizontally-polarized wave. In most cases, the signal that is induced in the receiving antenna is greatest if this antenna has the same polarization as the transmitting antenna.

There are different characteristics for horizontally- and vertically-polarized waves. For antennas located close to the earth, vertically-polarized rays yield a better signal. When the receiving antenna is raised about one wavelength above ground, this difference generally disappears and either vertical or horizontal antennas may be employed. When the antenna is at least several wavelengths above ground, the horizontally-polarized waves give a more favorable signal-to-noise ratio. In television, the wavelengths are short, and the antennas are placed several wavelengths in the air. Horizontally-polarized waves have been accepted as standard for the television industry. All television receiving antennas are mounted in the horizontal position.

Television receivers are described under **Television**. A radio receiver consists of an antenna suitably arranged to receive the impressed voltages of the electromagnetic waves that reach it from the transmitting antenna. Since the receiver must be selective, so that it will reproduce the output of only one of several possible transmitters, means must be provided to tune it to the frequency of the selected transmitter. The received signal is very weak and it must be strengthened by being passed through one or more stages of radio and audio-frequency amplification. Between the radio and audio amplification occurs the action of detection, or demodulation, separating the carried wave from the electrical wave, which mirrors the original voice or sound wave. Finally, the amplified voice current is fed into a speaker that re-creates the original sound.

The precursor of current AM radio receivers was the superheterodyne, developed by Armstrong in the early 1920s. This basic circuit converts all incoming radio-frequency signals to a common carrier frequency. This is accomplished in the first detector (mixer or converter). The signal from the antenna is fed by a tunnel coupled circuit to the mixer stage (in more elaborate designs, a tuned radio-frequency stage may be inserted between the antenna and the mixer). In the mixer stage, the incoming signal is heterodyned with a locally-generated signal so a beat frequency signal (b.f. or intermediate frequency) is produced. This new frequency signal is radio-frequency, ranging from around 450 kHz to several MHz, depending upon the purpose for which the receiver is designed. The intermediate frequency has exactly the same modulation as the original signal. In some broadcast receivers, the mixer combines the functions of mixer and oscillator.

The development of television led to extensions of the beat frequency principles involved in the radio receiver, but did not require radical redesigns. The main difference between the sound receiver and the picture receiver in a television set is in the width of the bands which must be handled, television requiring a band several megahertz wide while sound requires only a few kilohertz. This dictated that radio frequency channels had to be capable of selecting between stations, yet also pass very wide sidebands. The greater design change in radio receivers was initiated with the introduction of (FM) frequency modulation. In such sets, the incoming frequency modulated signal is amplified, converted to the intermediate frequency and this further amplified as in the AM set. However, since the modulation is present as a frequency variation and the loudspeaker responds to an amplitude variation, it is necessary to change from one to the other upon detection. This is accomplished in a discriminator (a frequency-sensitive detector) and then the resulting audio signals, which are amplitude variations, are amplified in a standard audio amplifier. The radio and intermediate frequency channels of the FM set must be fairly wide band (about 200 kHz) and the audio frequency response b.f. must be good to about 15 kHz to realize the full benefits of this type of modulation.

Radio Frequency Allocation

That part of the electromagnetic spectrum that is used for all forms of communication is commonly referred to as the *radio spectrum*. This span ranges from (a) waves of very low frequency, less than 10 kilohertz (kHz), and length of several kilometers to (b) waves of very high frequency, up to 300 gigahertz (gHz), and length of about one millimeter. Considering the radio spectrum as a *natural resource* for use by various countries and peoples of the earth as a whole, control and cooperation in allocating the spectrum into numerous bands for specific use are mandatory.

In addition to allocating portions of the spectrum for specific purposes, any given segment of the spectrum can be used in various geographical areas simultaneously through regulation of amplification and distance between transmitters. Thus, if locations are sufficiently far distant and power is limited, a given frequency can be assigned to several locations. Noise of various kinds is also a part of the radio spectrum. Poorly controlled communications systems are a source of noise when transmissions vary from their assigned frequency; all manner of machine-generated radiation, as from power tools, sewing machines, etc., when not properly shielded; and natural causes, notably lightning, are among noise sources.

In the case of a radio station transmitting 50 kilowatts of power, the density of energy at a distance of some 19 miles (30 kilometers) will be about 0.000004 watt per square meter. But inasmuch as the power is not uniformly radiated from the transmitter, but is directed toward the horizon, the effective signal is more likely to be about a millionth of a watt.

Several other entries in this encyclopedia pertain to radio communications. See also **Amplifier**; **Antenna (Communications)**; **Channel Frequency**; **Detection (Radio)**; **Fading (Communications)**; **Loudspeaker**; **Modulation**; **Navigation**; **Satellites (Communication and Navigation)**; and **Telephony (Telecommunications)**.

Additional Reading

Gibilisco, S.: "Handbook of Radio and Wireless Technology," McGraw-Hill Professional Book Group," New York, NY, 1998.
Gosling, W.: "Radio Spectrum Conservation: Radio Engineering Fundamentals," Butterworth-Heinemann, Inc., Woburn, MA, 2000.
Mazda, F.F.: "Principles of Radio Communication," Butterworth-Heinemann, Inc., Woburn, MA, 1996.
Rohde, U.L. and J.C. Whitaker: "Communications Receivers: DSP, Software Radios, and Design," 3rd Edition, McGraw-Hill Professional Book Group, New York, NY, 2000.

RADIO DUCT.   A rather shallow, almost horizontal layer in the atmosphere through which vertical temperature and moisture gradients are such as to produce an index of refraction lapse rate of greater than -48 N-units per 1000 feet. Strong temperature, or moisture inversions, or both are necessary for the formation of radio ducts. The resulting *superstandard propagation* is such as to cause the curvature of rays traveling through it to be greater than that of the Earth. See also **Superstandard Propagation** described in this entry. Radio energy, which originates within the duct, and leaves the antenna at angles near the horizontal may thus be *trapped* within the layer. See also **Radio Energy**; and **Skip Effect** described in this entry.

The effect is similar to that of a mirage (it is sometimes called radio mirage), and radar targets may be detected at phenomenally long ranges if both target and radar are in the duct. The greater the elevation angle between radar and target, the less the possibility of serious distortion due to transmission through ducts. Ducts may be surface based or elevated, with thickness ranging from a few tens of feet up to a maximum of 1000 feet. Elevated ducts are generally associated with subsidence or frontal inversions and are seldom found above 15,000 to 20,000 feet.

Radio Energy. Electromagnetic radiation of greater wavelength (lower frequency) than infrared radiation, that is, of wavelength greater than about 1000 microns (0.01 centimeter). The high-frequency end of the radio energy spectrum is known as microwave radiation.

Skip Effect. A phenomenon in which sound or radio energy may be detected only at various distance intervals from the energy source as the result of the presence of an energy reflecting or refracting layer in the atmosphere. For long radio waves, the ionosphere acts as the reflecting layer. For shorter wavelengths, the effect may be produced by strong superstandard propagation in elevated layers of the troposphere. Skip effects make it possible on occasion to detect targets at distances far greater than the normal radio horizon, while closer targets remain undetected.

Standard Propagation. The propagation of radio energy over a smooth spherical earth of uniform dielectric constant and conductivity under conditions of standard refraction in the atmosphere, i.e., an atmosphere in which the index of refraction decreases uniformly with height at a rate of 12 N-units per 1000 feet. Standard propagation results in a ray curvature due to refection which has a value approximately one-fourth that of the Earth's curvature, giving a radio horizon which is about 15% greater than the distance to the geometrical horizon. This is equivalent to straight-line propagation over a fictitious Earth whose radius is four-thirds the radius of the actual Earth.

Substandard Propagation. The propagation of radio energy under conditions of substandard refraction in the atmosphere, that is, refraction by an atmosphere or section of the atmosphere in which the index of refraction decreases with height at a rate of less than 12 N-units per 1000 feet. Substandard propagation produces a less-than-normal downward bending or even an upward bending of radio waves as they travel through the atmosphere, giving closer radio horizons and decreased radar and radio coverage. It results primarily when propagation takes place through a layer in which moisture remains constant or increases with height.

Superstandard Propagation. The propagation of radio waves under conditions of superstandard refraction (superrefraction) in the atmosphere, that is, refraction by an atmosphere or section of the atmosphere in which the index of refraction decreases with height at a rate of greater than 12 N-units per 1000 feet. See **Standard Propagation**, and **Substandard Propagation**. Superstandard propagation produces a greater-than-normal downward bending of radio waves as they travel through the atmosphere, giving extended radio horizons and increased radar coverage. It results primarily from propagation through layers near the earth's surface in which the moisture lapse rate is greater than normal, or the temperature lapse rate less than normal, or both. A condition in which warn dry air moves out over a cool water surface is an example of superrefraction. A layer in which the downward bending is greater than the curvature of the earth is called a radio duct. Frequently, the general term, anomalous propagation, is used for superstandard propagation.

See also **Anomalous Propagation**; and **Atmosphere (Earth)**.

RADIO GALAXY.   See **Galaxy**; and **Radio Astronomy**.

RADIOMETER.   An instrument for detecting and, usually, measuring radiant energy.

RADIOMETRY.   The science of measurement of radiant energy. In practice, there is no clear distinction between radiometry and photometry, although photometry usually refers to measurement in the visible and near-visible range.

RADIONUCLIDE.   See **Radioactivity**.

RADIOSONDE.   A balloon-borne instrument for the simultaneous measurement and transmission of meteorological data, consisting of: transducers for the measurement of pressure, temperature, and humidity; a modulator for the conversion of the output of the transducers to a quantity that controls a property of the radio frequency signal; a selector switch, which determines the sequence in which the parameters are to be transmitted; and a transmitter, which generates the radio frequency carrier. A pilot balloon carries the instrument aloft, and a small parachute lowers it to earth again when the balloon bursts in the upper atmosphere. By means of a small actuating device and a very light-weight radio-transmitting set, signals are automatically transmitted at regular intervals during the flight to a special recording receiver on the ground. These signals are then translated into readings of pressure, temperature, and humidity at the various altitudes.

See also **Weather Technology**.

RADIO STARS.   The name used in the early 1960s for the objects that are now called quasars. The first of these was announced at the December 1960 meeting of the American Astronomical Society, as a result of a combination of work in identifying the position of the 48th radio source in the 3rd Cambridge catalogue of radio sources, 3C 48, to a sufficiently high accuracy to enable it to be found optically. The work was carried out by Allan Sandage, of the Mt. Wilson and Palomar Observatories, with the 5-m Hale telescope, and Thomas Matthews, then of Caltech's Owens Valley Radio Observatory, with the interferometer there.

By 1963, three such "radio stars" were known to have optical objects coinciding with small-angular-size radio sources. An additional source, 3C 273, was identified with an optical object as a result of successive lunar occultations. One of the pair of radio sources in 3C 273 coincided with an apparently faint star and the other with a faint jet emanating from it.

The spectra of these radio stars included spectral lines, but the lines were in emission, contrary to normal stellar spectra, and did not appear to coincide with spectra of any known element. In 1963, Maarten Schmidt of the Mt. Wilson and Palomar Observatories realized that the lines in 3C 273 coincided with the spectrum of hydrogen redshifted by 15.8%. Jesse Greenstein, also of Caltech and the Mt. Wilson and Palomar Observatories, then realized that the spectrum of 3C 48 coincided with the hydrogen spectrum redshifted by 37%. The objects were then called "quasi-stellar radio sources," which was eventually contracted to "quasars." Allan Sandage soon found a class of radio-quiet quasi-stellar sources, which are considered to be included in the quasar class.

The extremely sensitive radio telescopes and arrays now in use have recently discovered radio emission from a variety of normal stars of various spectral types. The name "radio stars" is reserved for abnormally intense emitters, though, just as the name "radio galaxies" refers only to galaxies emitting extraordinary amounts in this part of the spectrum.

See also **Quasars**.

JAY M. PASACHOFF, Williams College, Williamstown, MA.

RADIO TELESCOPE.   A device for receiving, amplifying, and measuring the intensity of radio waves originating outside the Earth's atmosphere or reflected from a body outside the atmosphere. A radio telescope usually includes a source of radiation of known power for calibration of the received signal. The term radio telescope is not restricted to devices incorporating a paraboloidal dish antenna. A radio telescope can use any antenna or combination of antennas which will accept the radiation being studied. See also **Radio Astronomy**.

RADIUM.   Chemical element symbol Ra, at. no. 88, at. wt. 226.025, periodic table group 2 (alkaline earths), mp 700 °C, bp 1,140 °C,

density 5 g/cm^3 (20 °C). Radium metal is white, rapidly oxidized in air, decomposes H_2O, and evolves heat continuously at the rate of approximately 0.132 calorie per hour per mg when the decomposition products are retained, and the temperature of radium salts remains about 1.5 °C above the surrounding environment. Radium is formed by radioactive transformation of uranium, about 3 million parts of uranium being accompanied in nature by 1 part radium. Radium spontaneously generates radon gas at approximately the rate of 100 mm^3 per day per gram of radium, at standard conditions. Radium usually is handled as the chloride or bromide, either as solid or in solution. The radioactivity of the material decreases at a rate of about 1% each 25 years. All isotopes of radium are radioactive. See also **Radioactivity**. The first ionization potential of radium is 5.227 eV; second, 10.099 eV. Other important physical properties of radium are given under **Chemical Elements**.

One year after the discovery of X-rays by Röntgen (1895), Henri Becquerel investigated the relationship between the phosphorescence of various salts after their exposure to sunlight and the fluorescence in an operating x-ray tube. One of the salts under investigation was potassium-uranium sulfate. After exposure to sunlight, Becquerel noted that the salt not only emitted visible light, but also rays similar to x-rays that were able to penetrate the heavy black paper and thin metal foils within which his photographic plates were wrapped. During a period of cloudy weather, Becquerel stored the salt and photographic plates in a closet, awaiting further sunny days. Later, when he inspected the package, he noted that a very intense image had been developed on the photographic plate even though it had not received much prior exposure to sunlight. By further experiments, Becquerel confirmed that the intense image was derived directly from the presence of the salt, regardless of any exposure to sunlight. This constituted the first demonstration of radioactivity. Through further investigations, Ernest Rutherford demonstrated that both alpha and beta radiations were emitted by the salt. Rutherford learned that the alpha rays were easily absorbed by thin sheets of paper, whereas the beta rays acted in the same manner as observed by Becquerel. Later Mme. Marie Curie found that thorium produced about the same intensity of radioactivity as uranium. Further tests disclosed that the uranium ore with which she was working (pitchblende) exhibited more radioactivity than could be accounted for by its uranium content alone. Subsequently, Mme. Curie and her husband, Pierre Curie, successfully separated two previously unknown elements, radium and polonium. Thus, both radium and polonium were identified as chemical elements in 1898. It was found that each of these elements was over a million times more radioactive than uranium.

Radium gained prominence not only from its scientific interest, heralding a whole new area of physics and chemistry, but from its wide use in therapeutic medicine, as an ingredient (very dangerous) of luminous paints, and in various instruments for inspecting structures, such as metal castings. Commercially, radium generally is marketed as the bromide or sulfate and is extremely radioactive in these forms. Use of radium in medical technology has largely been replaced by other sources of radioactivity.

Radium occurs in pitchblende, and in carnotite along with uranium. Radium was first obtained from the uranium residues of pitchblende of Joachimsthal, the Czech Republic and Slovakia, later from carnotite of southwestern Colorado and eastern Utah. Richer ores have been found in Republic of Congo and in the Great Bear region of northwestern Canada.

In an interesting narrative, Landa (1979) describes and depicts what the author calls the "first nuclear industry," one that flourished during the first third of the 20th century. At that time, the prized element was *not* uranium, but rather it was radium. Radium initiated the concept of radiotherapy for the treatment of cancerous tumors. Even at the peak of radium production no more than a few hundred grams per year were purified. At one time the price approached $180,000 per gram (in early 1900s dollars). In one large extraction plant located in Denver, Colorado (National Radium Institute), carnotite ore was processed by a direct-dissolution method. During the three years (1914–1916) this plant was in operation, only 8.5 grams of Ra had been purified at an average cost of about $38,000 per gram. Pitchblende deposits from the Haut Katanga district of the Belgian Congo were discovered, but because of World War I, were not commercially exploited until 1921. Whereas it previously had required between 300 and 400 tons of a typical American carnotite ore to produce one gram of Ra, less than ten tons of the Kantanga ore were required. In 1931, an extraction plant at Port Hope, Ontario, designed to process uranium found in outcrops along the shores of Great Bear Lake in northwestern Canada, was commissioned. The hazards of radium production and handling were slow to surface and were not seriously recognized until the 1920s after a report had been prepared which implicated the ingestion of Ra in jaw necrosis in radium dial painters and lung cancer in uranium miners. Several workers, possibly including Marie Curie, died of conditions that were probably the result of long-term radiation exposure. By the 1930s, numerous precautions were initiated, including the design of tunnels at the Great Bear Lake mines to minimize radon hazards to miners.

The radium isotope of mass number 226 occurs in the uranium $(2n + 2)$ alpha-decay series. Its half-life is 1,620 years, and it yields radon-222 by α-disintegration. Other naturally occurring isotopes of radium are ^{228}Ra in the thorium series, half-life 6.7 years, producing actinium-228 by β-decay, which yields by β-decay thorium-228, which in turn yields ^{224}Ra, half-life 3.64 days, giving radon-220 by α-decay. Another naturally, occurring isotope of radium is found in the actinium series; it is ^{223}Ra, half-life 11.7 days, giving radon-219 by α-decay. In the neptunium series there is ^{225}Ra, half-life 14.8 days, undergoing β-decay to actinium-225. Other isotopes of radium include those of mass numbers 219, 221, 225, 227, 229, and 230.

Chemically related to barium, radium is recovered from its ores by addition of barium salt, followed by treatment as for recovery of barium, usually as the sulfate. The sulfates of barium and of radium are insoluble in most chemicals, so they are transformed into carbonate or sulfide, both of which are readily soluble in HCl. Separation from barium is accomplished by fractional crystallization of the chlorides (or bromides, or hydroxides). Dry, concentrated radium salts are preserved in sealed glass tubes, which are periodically opened by experienced workers to relieve the pressure. The glass tubes are kept in lead shields.

In many of its chemical properties, radium is like the elements magnesium, calcium, strontium and barium, and it is placed in group 2, as is consistent with its $6s^2 6p^6 7s^2$ electron configuration. Its sulfate ($K_{sp} = 4.2 \times 10^{-15}$) is even more insoluble in water than barium sulfate, with which it is conveniently coprecipitated. Like barium and other alkaline earth metals, it forms a soluble chloride ($K_{sp} = 0.4$) and bromide, which can also be obtained as dihydrates. Radium also resembles the other group 2 elements in forming an insoluble carbonate and a very slightly soluble iodate ($K_{sp} = 8.8 \times 10^{-10}$).

Additional Reading

Greenwood, N.N. and A. Earnshaw: "Chemistry of the Elements," 2nd Edition, Butterworth-Heinemann, Inc., Woburn, MA, 1997.

Hawley, G.G. and R.J. Lewis: "Hawley's Condensed Chemical Dictionary," 13th Edition, John Wiley & Sons, Inc., New York, NY, 1999.

Krebs, R.E.: "The History and Use of Our Earth's Chemical Elements: A Reference Guide," Greenwood Publishing Group, Inc., Westport, CT, 1998.

Landa, E.R.: "The First Nuclear Industry," *Sci. Amer.*, **180** (November 1982).

Lewis, R.J. and N.I. Sax: "Sax's Dangerous Properties of Industrial Materials," 10th Edition, John Wiley & Sons, Inc., New York, NY, 1999.

Lide, D.R.: "CRC Handbook of Chemistry and Physics 2000–2001," 81st Edition, CRC Press, LLC., Boca Raton, FL, 2000.

Williams, P.L. and J.L. Burson: "Industrial Toxicology: Safety and Health Applications in the Work Place," John Wiley & Sons, Inc., New York, NY, 1997.

RADIX POINT. The index which separates the digits associated with negative powers from those associated with the zero and positive powers of the base of the number system in which a quantity is represented; i.e., binary point, decimal point.

RADOME. A dome used to cover the antenna assembly of a radar installation, to protect it from wind and weather. The term may refer to either a surface or airborne installation. Radomes must be made of a material that is transparent to radio energy.

RADON. Chemical element symbol Rn, at. no. 86, at. wt. 222 (mass number of the most stable isotope), periodic table group 18 (inert gases), mp $-71\,°C$, bp $-61.8\,°C$. First ionization potential, 10.745 eV. Density 9.72 g/1($0\,°C$, 760 torr), 7.5 $\times$ more dense than air. The gas has been liquefied at $-65\,°C$ and solidified at $-110\,°C$. Radon was first isolated by Ramsay and Gray in 1908. Prior to acceptance of the present designation, radon was called niton or radium emanation. See also **Radioactivity**.

^{222}Rn is formed by the alpha disintegration of ^{226}Ra. Actinon, its isotope of mass number 219, is produced by alpha disintegration of ^{223}Ra (AcX) and is a member of the Actinium Series. Similarly, thoron, its isotope

of mass number 220, is a member of the thorium series. Since the name "radon" may be considered to be specific for the isotope of mass number 222 (from the radium series), the term "emanation" is sometimes used for element number 86 in general. Other isotopes of radon include those of mass numbers 209–218 and 221.

A fluorine compound of radon has been formed by reaction of the elements under higher temperature and pressure, similar to the conditions for forming xenon fluorides. Radon forms a hydrate at atmospheric pressure at $0\,°C$. It forms a compound with phenol, $Rn \cdot 2C_6H_5OH$ that is stable enough to give a sharply defined melting point at $50\,°C$. At low temperatures and pressures, HCl, hydrobromic acid, H_2S, SO_2, and CO_2 all add considerable percentages of radon; the HCl product, although possibly not a compound in the classical sense, being stable enough for its use as a method for separating radon from other gases.

Ionizing Radiation from a Natural Source. Since the mid-1980s, there has been a growing concern among some pollution experts with the topic of *indoor air pollution*, and as a major part of this problem, the presence in homes and other structures of ionizing radiation in the form of radon gas. It is beginning to appear that radon's earlier designation as radium emanation was not inappropriate. Most health hazards faced by the citizens of industrialized nations are anthropogenic. Not so in the case of radon. Numerous experts have concluded that the largest dose of ionizing radiation an average person may receive during a lifetime is the radioactive radon gas that emanates from rock formations as the uranium they contain decays. Some epidemiologists and environmental scientists believe that the gas can cause lung cancer in people who have done nothing more hazardous than to live in houses built over such rock formations. Others do not agree and caution has been exercised to avoid the tremendous costs that could be involved in taking remedial actions.

Radon issues continuously from the ground since uranium is present in virtually all rocks and soils. Radon dissipates quickly in open air, but when trapped inside a building, the gas can accumulate in concentrations tens, hundreds, and even thousands of times higher than out of doors.

It is important to note that radon per se is not the direct radiation hazard, but rather it is certain daughters (radioactive-decay products of the radon—mainly isotopes of polonium) that contribute the major radiation dose to lung tissue. These isotopes are chemically reactive. They can stick, either in elemental form or adsorbed onto minute airborne particles, to the lining of the bronchial passageways, whence they eradicate the surrounding tissue.

Dangerous radon concentrations were first noted in Sweden. Radiation measurements were made in well-insulated houses and immediately from that experience it was postulated that energy-efficient houses, designed to minimize the ventilation rate and thus conserve heating or cooling losses, were in effect "radon traps." Later investigations have discounted

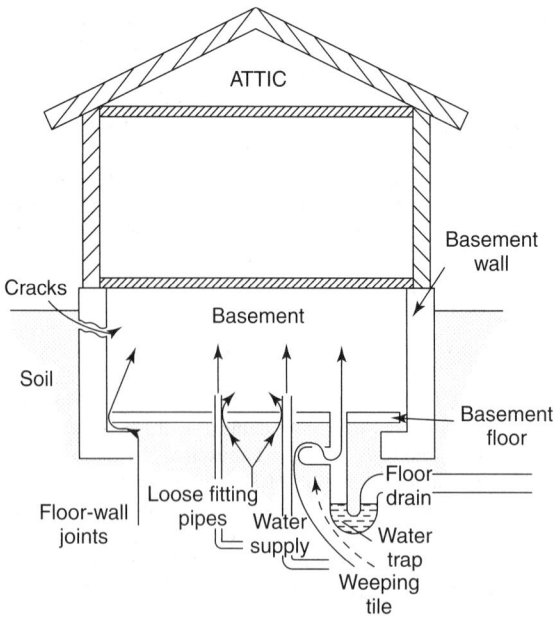

Fig. 1. Ports of entry for radon gas into average residence. (*National Indoor Environmental Institute, Plymouth Meeting, Pennsylvania.*)

the importance of ventilation. Attention was shifted from "tight" houses to ways in which radon may enter a building — in an effort to explain the wide disparity of data collected from thousands of structures. In these studies, building materials per se were rarely implicated as major sources of radon. Pathways for radon to enter buildings from underneath are illustrated in Fig. 1.

Because there are so many unanswered questions, most experts are attempting to quell any alarm complex that might arise among the public and to continue to collect data before drawing tentative conclusions. The U.S. Environmental Protection Agency is conducting a comprehensive survey of domestic radon levels. Radon concentrations are being measured in a random nationwide sampling of residential buildings, during which survey consistent instrumentation techniques will be used.

Additional Reading

Abelson, P.H.: "Uncertainties About Health Effects of Radon," *Science,* **353** (October 19, 1990).
Abelson, P.H.: "Mineral Dusts and Radon in Uranium Mines," *Science,* 777 (November 8, 1991).

Web References

National Safety Council Radon Information Page: *http://www.nsc.org/ehc/radon.htm*
NEHA — National Radon Proficiency Program: *http://www.radongas.org/*
Radon information from the U.S. EPA: *http://www.epa.gov/iaq/radon/*
The High Radon Project-Lawrence Berkeley National Laboratory: *http://eande.lbl.gov/IEP/high-radon/hr.html*
USGS Radon Information: *http://sedwww.cr.usgs.gov:8080/radon/radonhome.html*

RAFFIA. See **Palm Trees**.

RAFFINATE. See **Extraction (Liquid-Liquid)**.

RAFFINOSE. See **Sweeteners**.

RAILS, COOTS, AND CRANES (*Aves, Gruiformes*). The *Gruiformes* is an order of wading and swimming birds of varied form, with lobed toes or with neither lobes nor webs, but feet never fully webbed. The rail is a long-legged marsh bird of wide distribution. It has a moderate to long and slender beak, rather small wings, and short tail. North America has several species, including the clapper, *Rallus longirostris*; the Virginia, *R. limicola*; and the sora rail, *Porzana carolina*. The corncrake, *Crex crex*, is a Eurasian species which reaches North America occasionally. It is called the land rail and is related to the Carolina rail of North America. It has a meadow or marsh habitat, and a rasping call. In New Zealand, rails are represented by the large weka rails, which do not fly although they have wings.

The courlan is a large Brazilian bird that resembles the rail in appearance and habits. It is also called the limpkin. There are two species. One *Aramus pictus*, ranges from Florida through the Antilles and Central America; the *A. Scolopaceus* lives in tropical South America. The bird measures some 26 inches (66 centimeters) in length and the bill is twice as long as the head. Dietary favorites include snails, frogs, and small reptiles. There are fairly large numbers of limpkins in the Florida Everglades and around Lake Okeechobee. The limpkin usually nests near the water in bushes or on a platform of vegetation. The cry is piercing and mournful.

The bustard-quail, *Trunix*, is a small bird related to the pigeons and rails, as well as to the gallinaceous birds. Sometimes called hemipodes, bustard-quails are widely distributed in the Old World. The hemipode is unusual in that the females are larger and of brighter coloration than the males. Also, the males incubate the eggs and care for the young.

The bustard is a large bird and is of numerous species found chiefly in Africa, although some occur in Europe and Asia. The bustard is chiefly terrestrial in habits, but is a powerful flier. One of the African species is called the hubara and those of India are known as floricans. The bustards are related to the rails and cranes.

Coots are of several species occurring in Europe, Asia, North America, and Africa. They are waders and swimmers, with lobed toes. The plumage is dull in the adult, in contrast with the conspicuous white of the beak and part of the head. In the American species, *Fulica americana*, the beak is ivory white. Although sometimes eaten, the coots do not rank with the ducks as food and game birds.

Seriemas are peculiar South American birds of several species. They resemble the secretary bird, although in anatomical characteristics they are like the cranes. They have long legs and moderately long necks, with a broad beak, slightly hooked. These birds live in open country and eat small animals and insects.

Cranes are large birds with long legs and neck. They are superficially like the larger herons and the name crane is sometimes inaccurately applied to the latter birds, especially to the great blue heron. Cranes are found in Europe, Asia, North America, and Africa. The sun bittern is a South American bird, *Europyga helias*, of moderate size and related to the cranes. See Fig. 1.

Fig. 1. Black-necked crane (*Grus nigricollis*). This bird breeds in the most remote steppes of central and eastern Tibet, from Ladakh to as far as Kuku-Nor. This relatively restricted breeding territory is consistent with an equally small wintering ground in southeast China and North Vietnam. At one time, thousands of black-necked cranes could be counted at their wintering habitat, but their numbers in recent years have dwindled.

The trumpeter is a long-legged and long-necked bird of South America. The few species are characteristically terrestrial in habits, living in the forests and flying poorly. They live in flocks and are said to be tamed in Brazil for the protection of domestic fowls, with which they live contentedly. The word also appears in the names of the trumpeter-hornbills of Africa and the trumpeter swan of North America, both species of other orders.

The kagu, *Rhinochetus jubatus*, is about the size of a domestic fowl with longer legs and beak and a long drooping crest. It is found on the island of New Caledonia and is related to the sun bittern. This nocturnal bird is nearly extinct. It is a flightless ground bird that sleeps under roots of trees. The bird is dark gray, pale gray underneath. The legs are orange in color and very strong. The voice of the kagu is loud and calling is in early morning and at dusk. A London zookeeper once likened the disposition of the kagu to a playful puppy, taking its tail in its mouth and running around in circles and tossing leaves in the air and running to catch them. See also **Gruiformes**.

RAILWAYS (High-Speed). Essentially during the last two decades, several countries have developed high-speed railway travel to a degree far beyond that achieved in North America and, in fact, well in excess of what most advanced engineering planners would have envisioned a half century ago. European countries, notably France, Germany, Sweden, and Japan have been quite successful with their high-speed programs and are aggressively continuing to improve them. The practical, sociological needs for high-speed rail travel are well documented in the literature and need not be repeated here. Suffice it to say that, with continuing population pressures, the future heavy reliance on motor vehicle highways for moving people and goods appears untenable for any progressive nation. Failure to create an alternate pathway could lead to the construction of 12 to 15 traffic lanes in each direction, with accompanying huge investments in land acquisition and road construction costs, not to mention the costs to be borne by the environment. Most experts believe that some form of electronically guided highway for non-polluting vehicles may be in the technological picture, but ground movement of masses of people and of goods will require some form of transport that is much more akin to the traditional railway rather than the highway.

Advanced Approaches in Railway System Design

Over the past quarter century, the engineering approaches taken to overcome the limitations of the traditional installed railway system that stemmed from the fundamental developments of the 1850–1950 time span fall into three general classes:

1. *Technologically Minor Improvements to Existing Systems* — In the short run, this approach may be viewed, although questionably, as a minimal-cost program. The U.S. Northeast Corridor rail system is typical of the "fix-up" type program.

    Actions taken include: (a) smoothing out some of the curves; (b) restoring the roadbed and track and their maintenance to the high standards that prevailed in the early times when rail traffic was heavy; (c) simplifying track patterns to eliminate dangers of inadvertent switching; (d) improving rolling stock design, but with relative little attention given to reducing aerodynamic effects; (e) making marginal improvements in contemporary motive power; and (f) improving control systems and communications, largely in the interest of safety assurance at higher speeds. Most experts agree that fixing up an old system in the absence of major technological change simply will not suffice for future travel requirements between major cosmopolitan areas.

2. *Technologically Major Improvements to Existing Systems* — These systems do not depend wholly upon modernizing old concepts, but incorporate major technological changes in various aspects of existing systems. Sweden's *ABB Fast-Train* is an example of this kind of change. Cars (wagons) still are pushed (pulled) by locomotives along standard rails, but they feature mechanical innovations in rolling stock, such as car body tilting and soft suspension bogies, that contribute to increased speed while enhancing passenger comfort and safety at high speeds.

3. *Radically Innovative Systems* — Although the term railway persists for convenience, these comparatively new systems do not employ steel tracks or wheels in the conventional sense. Rather, the train is magnetically levitated and the motive power is furnished by a linear motor that responds to coils that are strung along the complete length of the train's pathway. Thus, the train no longer depends upon friction between locomotive wheels and track to accomplish motion or between car wheels and accompanying mechanical brakes to stop motion. Although there are other designs, bogies equipped with rubber-tired wheels can be used to position the train within its guidepath. The guidepath appears more like a channel or trough than a track-and-tie roadbed. These systems for obvious reasons represent the largest initial investment in the quest for high-speed land transportation, but may not be practical or even necessary for all future travel needs, except for links between major cities. Operating and maintenance costs for these radically different systems remain to be worked out.

Swedish X2000 High-Speed Train

Prior to the introduction of the X2000 train in the late 1980s, the run between Stockholm and Gothenberg was 4 hours. The X2000 has reduced that time to 2 hours, 55 minutes.

Stockholm–Gothenburg (457 km [284 mi])

	Conventional train	X2000 train
Top speed	160 km (99.4 mi)/h	200 km (124.3 mi)/h
Average speed	115 km (71.5 mi)/h	155 km (96.3 mi)/h

The X2000 runs on traditional tracks. Electric locomotives are used, one at each of the train to eliminate turn-around time at terminals. Three-phase thyristor-controlled engines produced 4400 horsepower and feature modern diagnostic electronics. The train has an aerodynamic fiberglass nose and is shaped to withstand collision with an elk, a common problem on Sweden's railroads. Body-tilting technology is used to permit banking up to $6\frac{1}{2}°$ in turns. Rubber components in each bogie permit axles to follow curves more freely. High speeds through curves place tough demands on bogie design. The X2000 bogies feature wheel axles that move individually in curves. Track forces are reduced, allowing a substantial increase in operating speeds on straight track as well as through curves. For improved passenger comfort, the passenger cars are designed with active tilt technology, which reduces the effects of centrifugal forces felt while passing through curves.

Car body tilting has been conceived and developed to increase passenger comfort.

The X2000 can be adapted to a variety of capacity requirements. One or two power units can be combined with up to ten intermediate coaches, providing a capacity of up to 600 passengers.

Asea Brown Boveri (ABB), the builder of the X2000, observe that current systems that operate trains in the 120–200 km (75–125 mi) per hour speed range can reach speeds of 160–250 km (99–155 mi) per hour. See Figs. 1, 2, and 3.

Fig. 1.   The Swedish X2000 high-speed train reaches a top speed of 200 km (124.3 mi) per hour on the run from Stockholm to Gothenburg. (*Swedish State Railways.*)

It has been estimated that the X2000 train uses one-ninth the energy of a passenger aircraft. During the Stockholm–Gothenburg run, the X2000 uses 10,000 kWh. With a maximum of 240 passengers on board, that means 40 kWh per passenger. A plane uses 60,000 kWh, which means about 375 kWh per passenger on board a 160-seat aircraft. While the X2000 emits no harmful pollutants, an aircraft on the same run will emit 50 kg of carbon monoxide and 12.6 tons of carbon dioxide.

The French TGV Atlantique Train. This train is now in its second generation. The train commenced runs between Paris and Lyon over a decade ago. The newer train, generally comparable with the first-generation design, runs between Paris and Le Mans. It incorporates a decade of experience and has a cruising speed of nearly 200 km (186 mi) per hour, which makes the more recent TGV approximately 30 km (18 mi) per hour faster than the earlier version. Exquisitely maintained and improved (but conventional) track and rail beds are used. Locomotives are electric. The axle load has been reduced to 17 tons, as compared with 11 tons for conventional American and German trains. Key to the improved TGV is a brushless synchronous motor that generates twice the horsepower, yet weighs 10% less than the earlier train. The new *Atlantique* uses eight rather than twelve motors, but can haul ten trailer cars, as contrasted with eight cars on the earlier version. This reduces power costs per passenger by more than 15%. At 12,000 horsepower, the more recent version can handle a 5% gradient without reducing speed. In the new TGV, pneumatic shock absorbers replace springs that suspended the body over the bogies. Car-to-car dampers are used to lessen vibration. The braking system is microprocessor controlled. High-voltage power electronic equipment is housed in cooled canisters. See Fig. 4. The locomotive cab features an extensive data-processing network.

Other high-speed rail trains in Europe and Japan could be described. These would include the Japanese *Shinkansen* "bullet train," which has

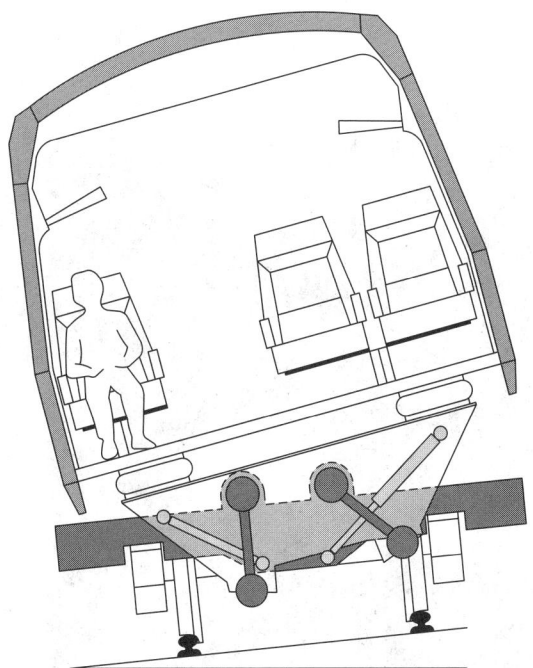

Fig. 2. Carbody tilt technology has been introduced for increased passenger comfort in curves. It has been found that the optimal compensation for lateral acceleration is 80%, which is achieved at tilting angles up to a maximum of 8.0°. For greater accuracy and faster reaction times, active tilt technology is utilized. An accelerometer placed in the front bogie transmits information to hydraulic tilting cylinders on each of the passenger coach bogies. All tilt equipment is fitted under the passenger coach floor in order not to intrude on passenger space.

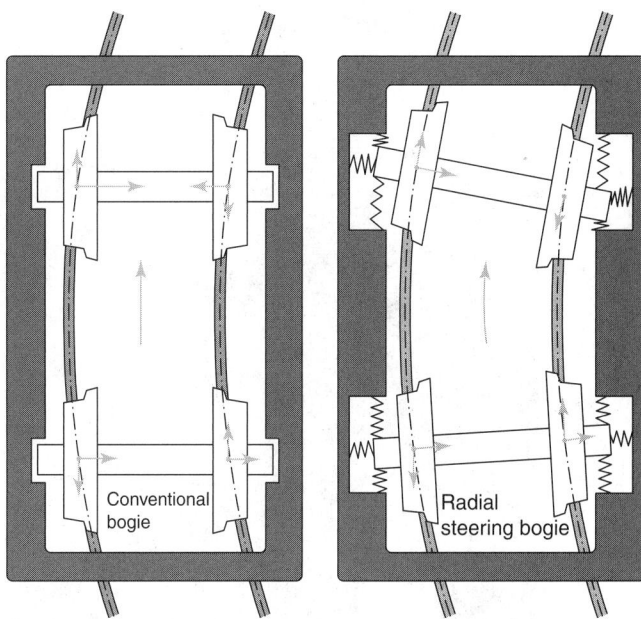

Fig. 3. The key to increased speeds lies in ABB's bogie design, featuring reduced dynamic forces. As opposed to traditional bogies, the ABB fast-train bogie concept allows wheel axles to respond to curves. The large creep-forces that arise when traveling through curves automatically steer the individual wheel axles, which are suspended in rubber elements. This technology allows for up to 40% higher speeds through curves while maintaining safety.

been operational since 1964. The train has a top speed of 201 km (126 mi) per hour.

Japan and Germany appear to be taking the lead in making the next revolutionary step in mass land transportation, notably magnetically levitated trains. There is considerable scientific and political unrest in the United States regarding the comparatively little commercial and governmental support that the United States is devoting to the maglev concept.

Other rail concepts not previously mentioned include single-rail overhead systems, which have been in limited operation for many decades in what are called *monorail* systems. The engineering principles parallel those found in vertical overhead conveyor systems widely used in parts and automotive assembly factories. Also, they are a common feature of some amusement parks and for highly localized conveyance of people and packages, as found in a few airline terminals. In the past, these systems have been mentioned frequently as convenient systems to parallel major highways, particularly in congested urban areas.

ACELA Express High-Speed Train

When conceived in 1995, the project was first known as the American Flyer, but Acela–derived from the words "excellence" and "acceleration"–took over as the project took further shape, as a new brand to reflect the main benefits which Amtrak aims to offer.

America's North-East Corridor, which links Boston, New York and Washington, is one of the most densely populated and fastest-growing areas of the country. The corridor stretches from Washington, DC at the southern end, through Baltimore, Wilmington, Philadelphia, New York City, and the Connecticut coast to Boston. With the complete rebranding of its train services in the North-East Corridor, Amtrak has introduced the Acela Express, a new fleet of 20 high-speed trains.

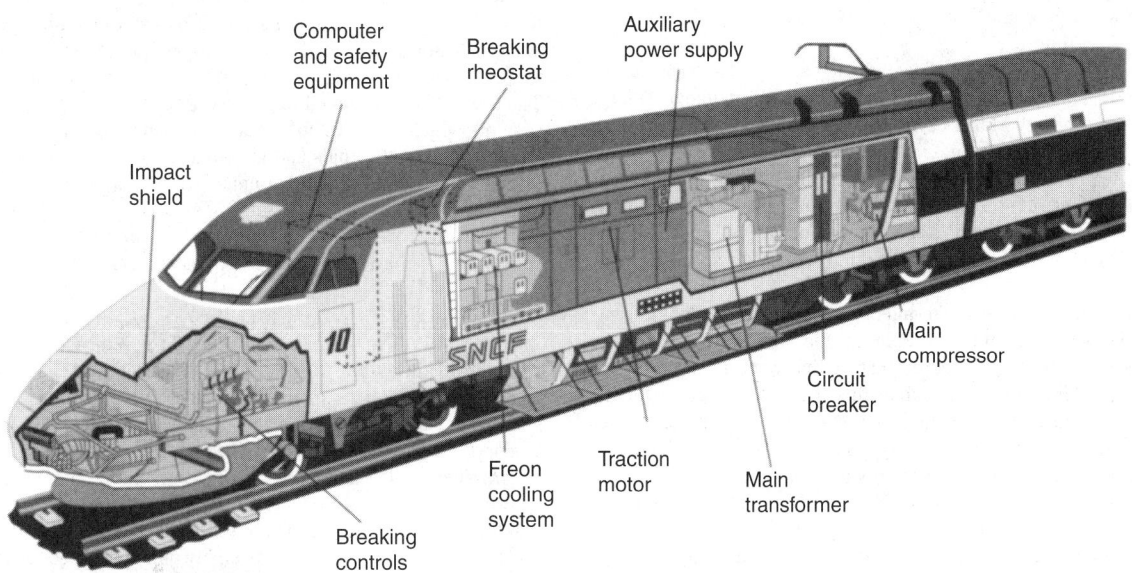

Fig. 4. Schematic diagram of forward locomotive used on the French TGV *Atlantique*, which runs between Paris and Le Mans.

Fig. 5.   Amtrak ACELA Express (*National Railroad Passenger Corporation*).

On November 16, 2000 Amtrak made history with the launch of Acela Express, the first high-speed rail service in North America, zipping across the Northeast at a top speed of 150 mph. Setting a North American speed record for scheduled passenger rail service, Acela Express made a non-stop run between Washington and New York in just 2 hours and 28 minutes. Previous rail service between these points averaged three hours. On the second leg of the trip, from New York to Boston, the travel time was just 3 hours and 15 minutes compared to previous times of four hours and 30 minutes. Acela Express is the fastest train in Amtrak's 29-year history. See Fig. 5.

Infrastructure

The Acela project also involves a series of improvements to existing stations, including major railheads in New York, Wilmington and Baltimore. A new station at Route 128 south of Boston was also built. New stations are planned at Metropark and Trenton in New Jersey as well as at BWI Airport south of Baltimore.

Installation of the 25 kV overhead catenary to carry the electrical supply for the new trains had to be installed. Engineers faced a number of large physical barriers in their work. In Connecticut, wires had to be laid across two lifting and three swing bridges, each of which carry the line across busy estuaries, which see regular ship and yacht traffic.

Balfour Beatty Construction and its joint venture partner, Mass Electric Construction Company, adopted a novel method to help the system overcome the problems inherent in energizing the moving sections of the bridges.

Fixed contact wires made of silver and copper are held in a longitudinal aluminium tube, which is then secured to the bridge structure. The transition between normal and fixed catenary at each end of the bridge is carried in a portal frame. Undersea cables link these structures. Vertically hinged contact wires have been installed on the line's swing bridges, which are designed to rotate clear when the bridge is opened. The wires have also been given extra protection against severe winter weather as a result of a pre-service testing programme undertaken in the National Laboratory for Climate Control in Switzerland.

The consortium of Alstom and Bombardier were contracted to design and manufacture the trainsets, each of which will be of a configuration familiar in high-speed railways around the world. Each trainset having a

power car an each end and a series of intermediate trailers. In this case four business class, one first class and a Café Car.

Alstom provided the trains' electrical equipment and electromechanical parts, including the bogies, while Bombardier manufactured the body shells and assembled the trains at its plants at La Pocatiere (Quebes), Plattsburgh (New York), and Barre (Vermont).

Each of the new trains carries a total of 304 passengers and the first to be fully compliant with the Americans with Disabilities Act. They incorporate larger windows than existing rolling stock, larger enclosed storage compartments for luggage, and the toilet facilities have been designed specially for ease of use by wheelchair users. There are 32-conference tables spread throughout the train and six public phones. Each seat is equipped with an electrical outlet to plug in laptop computers and two channels for audio entertainment. The Café car offers upscale menu selections.

As the trains are designed to have a top speed of 150 mph, a greatly upgraded signalling system is needed to ensure maximum levels of safety. Alstom, working alongside its recently acquired partner, North American systems specialist GRS, provided the control systems for the Acela project. The trainsets are equipped with a two-frequency, nine-aspect cab signalling system, which receives information transmitted through the rails in the form of electrical signals and is displayed to the traincrew in the cab. The driver is supervised by an Automatic Train Control (ATC) system, and the trains are also governed by Amtrak's own ACSES (Amtrak Civil Speed Enforcement System) equipment, which automatically adjusts a train's speed within lineside speed restrictions.

Ride Quality

Acela Express is more than high-speed; it was designed to be the quietest, smoothest riding train ever built for American passenger rail service.

Acela Express passenger cars feature tilt technology to allow higher speeds and a smoother ride through the many curves on the Northeast Corridor. Sensors on the lead locomotive detect upcoming curves and progressively tilt each car as the train enters the curve. The system determines when and how far to tilt each coach and constantly monitors conditions to provide the most comfortable ride possible.

Acela Express provides an extremely quiet passenger environment that is in marked contrast to the noise and vibration of passenger compartments of airplanes. Tests show that the typical noise level in the seating area registers at 63–65 dB. Interior treatments and car insulation materials were selected for their sound absorption qualities. Motion sensor-operated glass doors at the ends of the passenger cars open and close quietly and shield undesirable train noise from entering the seating area.

The Acela Express trucks (wheel assemblies) are taken from the French TGV and are designed exclusively for high-speed operation.

Seats for the *Acela Express* have been specially designed to achieve a high level of comfort for all customers, regardless of their size or shape. The seating configurations are two-by-two in Business class and two-by-one in First Class; there are no middle seats. The seats incorporate the best features identified through extensive consumer testing of items such as seat cushion hardness, adjustable backrests, tray table configuration and foot rest operation, as well as seat covering material and color. The seating areas for disabled customers provide a place for wheel chairs and a facing seat for a companion.

Windows in the Acela Express passenger cars are double the size of those found in Amtrak's current fleet, providing an open feel and allowing plenty of natural light to enter the car interior.

Magnetically Levitated and/or Magnetically Propelled Vehicles

Although, as of the early 1990s, the scientific principles were well understood, much scientific and engineering progress remains to perfect high-speed magnetically levitated and magnetically propelled vehicles, including trains. Notably, these include further developments of magnetic materials and the perfection of system geometry. Pertaining to the latter, it has been observed that the ultimately successful vehicle will combine some of the design principles of the railroad, the automobile, and the airplane.

A vehicle may utilize electromagnetic forces in one of two ways (i.e., for *levitation* and for *propulsion*). Most current design consideration is given to combining the two principles in the same vehicle. Usually, when the term *maglev* (magnetic levitation) is used, it refers to using both principles in combination by a single vehicle.

1. *Levitation* is an age-old term meaning "a rising or lifting of a person or thing by means held to be supernatural."

   A vehicle may use either of two principles to keep the vehicle completely free from touching what normally would be considered a basis for support (i.e., counteracting gravitational force).

   a. *Electrodynamic Suspension (EDS)* — The repulsive force of two opposing magnets, one located at the bottom of the guidepath, which normally would be considered "ground level," and one located on the bottom of the vehicle. Such a system is inherently stable and, as pointed out by researcher R.D. Thornton (see reference listed), "The current induced in the guideway will increase as the gap shrinks, thereby increasing the repulsive force and providing steady suspension. Since the vehicle's magnetic field can be constant, it can be supplied by superconducting magnets, allowing an airgap of 13.1 to 15.2 cm (2 to 6 in)." Consequently, available, low-temperature superconductors are adequate for EDS systems. Greater efficiency can be achieved by replacing the continuous sheets in the guideway with improved coil designs that use less current. Further research is required to shield people from the magnetic fields, both inside and outside the vehicle. See Fig. 6(a).

   b. *Electromagnetic Suspension (EMS)* — The attractive force of two opposing magnets — one located on the "ceiling" of the guidepath channel and the other along the top of the vehicle. The magnetic attraction must overcome gravitational forces. Again, as pointed out by Thornton, "Attractive systems are unstable unless the current in the magnets can be varied widely and rapidly, as is possible with normal magnets. Without a way of controlling the current, the attractive force increases as the gap decreases, further narrowing the gap until it finally closes. No maglev system that requires magnets with normal conductors can operate with an air gap greater than about 9.5 mm ($\frac{3}{8}$ in) without unacceptable power consumption, vehicle weight, and guideway cost. If superconductors still to be developed were available, the system could be feasible. All known superconductors must be operated with essentially constant current and thus cannot be controlled in a way that is necessary for maintaining a stable gap." See Fig. 6(b).

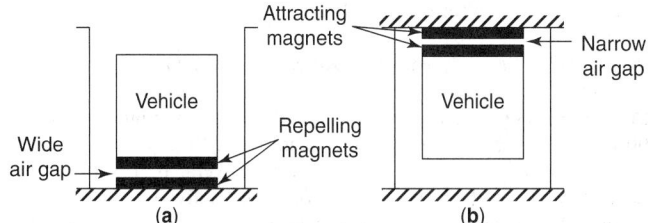

Fig. 6. (**a**) Magnetic repulsion, as used in electrodynamic suspension (EDS); (**b**) magnetic attraction, as used in electromagnetic suspension (EMS).

2. *Propulsion*, considered for use in the next generation, operates on the principle of the linear synchronous motor (LSM). According to Thornton, "A magnetic field travels along the guidepath, acting on superconducting magnets attached to the vehicle. By keeping the vehicle motion synchronous with the traveling field, the propulsive force can be forward, backward, or even straight up or down. In a typical design, the same superconducting coils that create the lift also create the reaction field for the LSM." A feedback control system changes the polarity of the field, so that a guideway section (marked "S") in Fig. 7 becomes "N" and vice versa.

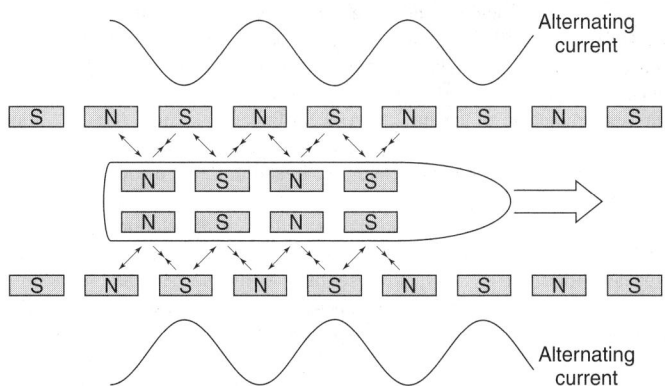

Fig. 7. A linear synchronous motor uses alternating current to generate a magnetic wave that travels along with the vehicle. The fields, which vary with time, interact with magnets on the vehicle to push and pull the vehicle. Optional arrangements also may be used. (*Japanese MLU002.*)

In another version, referred to as a *magneplane*, the guidepath takes the form of a trough (something like a bobsled run). Magnetic levitation of the vehicle would lift a 50-ton vehicle to a maximum altitude of about 15 cm (6 in) above the shallow bowl-like track. Propulsion could be by magnetic means, as previously described, or by propellers or jets.

Additional Reading

Kerson, R.: "Magnetic Trains," *Technology Review (MIT)*, **13** (April 1989).
Lynch, T.: "High Speed Rail in the U.S.: Super Trains for the Millennium," Gordon and Breach Science Publishers, Newark, NJ, 1998.
Thornton, R.D.: "Why the U.S. Needs a Maglev System," *Technology Review (MIT)*, **31** (April 1991).
Stix, G.: "Riding on Air," *Sci. Amer.*, **104** (February 1992).
Stix, G.: "Air Trains," *Sci. Amer.*, 102 (August 1992).
Strohl, M.P.: "Europe's High Speed Trains: A Study in Geo-Economics," Greenwood Publishing Group, Inc., Westport, CT, 1993.
Wachs, M.: "U.S. Transit Subsidy Policy: In Need of Reform," *Science*, 1545 (June 10, 1989).

Web References

Amtrak *Acela Express* : *http://www.acela.com/*
High Speed Rail (HSR): *http://www.o-keating.com/hsr/index.htm*
High Speed Train Links Page: *http://www.trainweb.org/railwaytechnical/hst-01.html*
High Speed Trains for Canada: *http://www.fraserinstitute.ca/publications/books/essays/chapter3.html*
Railway-Technology.com: *http://www.railway-technology.com/*
Railway Technology—Finland Tilting Trains (HSR): *http://www.railway-technology.com/projects/finland/*
Railway Technology — Swiss Tilting Trains (HSR): *http://www.railway-technology.com/projects/sbb/*

RAIN. See **Precipitation and Hydrometeors**.

RAINBOW. See **Atmospheric Optical Phenomena**.

RAIN FOREST. A tropical forest, where annual rainfall is at least 100 inches (254 cm). The region is characterized by tall, lush evergreen trees and by a vast variety of life forms. Several of the world's rain forests have been damaged by anthropogenic activities, and others are severely threatened. Many rain forests are situated in underdeveloped nations that are short of commerce, causing some governments to exploit the timber and other assets of the forests as a major means of bettering their economic position. This is still another instance of the triangular conflict between energy needs, environmental protection, and economics. See also **Biome**.

RAIN (Hydrology). See **Hydrology**.

RAIN (Runoff). See **Drainage Systems**.

RAISIN. See **Grapes and Wines**.

RAMAN SPECTROMETRY. This form of spectrometry is based upon the Raman effect which may be described as the scattering of light from a gas, liquid, or solid with a shift in wavelength from that of the usually monochromatic incident radiation. Discovered by the Indian physicist, C.V. Raman in 1928, it has also been called the Smekal-Raman effect, the former investigator having made some earlier theoretical predictions about it. If the polarizability of a molecule changes as it rotates or vibrates, incident radiation of frequency v, according to classical theory, should produce scattered radiation, the most intense part of which has unchanged frequency. This is Rayleigh scattering. In addition, there should be Stokes and anti-Stokes lines of much lesser intensity and of frequencies $v \pm v_k$, respectively, where v_k is a molecular frequency of rotation or vibration. The anti-Stokes line is always many times less intense than the Stokes line and this fact is satisfactorily explained by the quantum mechanical theory of the effect. The vibrational Raman effect is especially useful in studying the structure of the polyatomic molecule. If such a molecule contains N atoms it can be shown that there will be $3N - 6$ fundamental vibrational modes of motion only ($3N - 5$ if the molecule is a linear one). Those accompanied by a change in electric moment can be observed experimentally in the infrared. The remaining ones, if occurring with a change in polarizability, will be observable in the Raman effect. Thus both kinds of spectroscopic measurements are usually required in a complete study of a given molecule.

Like infrared spectrometry, Raman spectrometry is a method of determining modes of molecular motion, especially the vibrations, and their use in analysis is based on the specificity of these vibrations. The methods are predominantly applicable to the qualitative and quantitative analysis of covalently bonded molecules rather than to ionic structures. Nevertheless, they can give information about the lattice structure of ionic molecules in the crystalline state and about the internal covalent structure of complex ions and the ligand structure of coordination compounds both in the solid state and in solution.

Both the Raman and the infrared spectrum yield a partial description of the internal vibrational motion of the molecule in terms of the normal vibrations of the constituent atoms. Neither type of spectrum alone gives a complete description of the pattern of molecular vibration, and, by analysis of the difference between the Raman and the infrared spectrum, additional information about the molecular structure can sometimes be inferred. Physical chemists have made extremely effective use of such comparisons in the elucidation of the finer structural details of small symmetrical molecules, such as methane and benzene. But the mathematical techniques of vibrational analysis are not yet sufficiently developed to permit the extension of these differential studies to the Raman and infrared spectra of the more complex molecules that constitute the main body of both organic and inorganic chemistry.

The analytical chemist can use Raman and infrared spectra in two ways. At the purely empirical level, they provide "fingerprints" of the molecular structure and, as such, permit the qualitative analysis of individual compounds, either by direct comparison of the spectra of the known and unknown materials run consecutively, or by comparison of the spectrum of the unknown compound with catalogs of reference spectra.

By comparisons among the spectra of large numbers of compounds of known structure, it has been possible to recognize, at specific positions in the spectrum, bands which can be identified as "characteristic group frequencies" associated with the presence of localized units of molecular structure in the molecule, such as methyl, carbonyl, or hydroxyl groups. Many of these group frequencies differ in the Raman and infrared spectra.

When a transparent medium was irradiated with an intense source of monochromatic light, and the scattered radiation was examined spectroscopically, not only is light of the exciting frequency, v, observed (Rayleigh scattering), but also some weaker bands of shifted frequency are detected. Moreover, while most of the shifted bands are of lower frequency, $v - \Delta v_1$, there are some at higher frequency, $v + \Delta v_1$. By analogy to fluorescence spectrometry (see below), the former are called *Stokes bands* and the latter *anti-Stokes bands*. The Stokes and anti-Stokes bands are equally displaced about the Rayleigh band; however, the intensity of the anti-Stokes bands is much weaker than the Stokes bands and they are seldom observed. This article deals only with the more intense Stokes bands. The geometric arrangement for observing the Raman effect is shown diagrammatically in Fig. 1. See also **Infrared Radiation**.

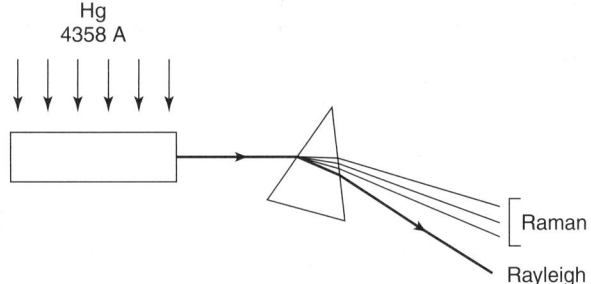

Fig. 1. Optical system to observe a Raman spectrum. The Rayleigh scattering is at the wavelength of the exciting line.

Additional Reading

Long, Fayer, M.D.: "Ultrafast Infrared and Raman Spectroscopy," Marcel Dekker, Inc., New York, NY, 2001.

Laserna, J.J.: "Modern Techniques in Raman Spectroscopy," John Wiley & Sons, Inc., New York, NY, 1996.

Lindon, J.C., G.E. Tranter, J. John, and L. Holmes: "Encyclopedia of Spectroscopy and Spectrometry," Academic Press, Incorporated, San Diego, CA, 2000.

McCreery, R.L.: "Raman Spectroscopy for Chemical Analysis," John Wiley & Sons, Inc., New York, NY, 2000.

Nakamoto, K.: "Infrared and Raman Spectra of Inorganic and Coordination Compounds," Vol. 2, 5th Edition, John Wiley & Sons, Inc., New York, NY, 1997.

Nakamoto, K. and J.R. Ferraro: "Introductory Raman Spectroscopy," Academic Press, Inc., 1994.

Ogilvie, J.F.: "The Vibrational and Rotational Spectrometry of Diatomic Molecules," Academic Press, Inc., San Diego, CA, 1998.

Pelletier, M.J.: "Analytical Applications of Raman Spectroscopy," Blackwell Science, Inc., Malden, MA, 1999.

Schrader, B.: "Infrared and Raman Spectroscopy: Methods and Applications," John Wiley & Sons, Inc., New York, NY, 1995.

Socrates, G.: "Infrared and Raman Characteristic Group Frequencies: Tables and Charts." 3rd Edition, John Wiley & Sons, Inc., New York, NY, 2001.

RAMARK. A fixed radar frequency facility that continuously emits a signal so that a bearing indication appears on a radar display. See also **Radar Beacon**.

RAMIE FIBERS. Pericyclic fibers, which are both long (up to nearly 8 inches (20 centimeters) in length) and of unusual strength, obtained from a perennial asiatic plant, *Boehmeria nivea*, of the family *Urticeae* (nettle family). The plant is a perennial that grows to a height of about 7 feet (2.1 meters) and produces several crops of canes annually. In the Orient, these fibers are removed from the plant, almost always manually, and the cortex is scraped off by drawing the stems over a coarse knife against which they are pressed. The fibers are dried and marketed in this condition as "China grass." Later, repeated washing and drying remove the gummy substances that hold the fibers together. Ramie fibers are considerably coarser than those of flax, and have great tensile strength, but are not extensively used because of their inability to withstand twisting for the formation of ropes.

RAMJET ENGINE. A type of jet engine with no mechanical compressor consisting of a specially shaped tube or duct open at both ends, the air necessary for combustion being shoved into the duct and compressed by the forward motion of the engine, where the air passes through a diffuser and is mixed with fuel and burned, the exhaust gases issuing in a jet from the rear opening. The ramjet engine cannot operate under static conditions. Often called a *ramjet*. Also called *Lorin tube*.

RAMP A/D CONVERTER. This type of analog-to-digital converter quantizes an analog-input signal through conversion of the signal into a time-duration pulse. The latter is measured by a counter and a constant-frequency pulse generator. A ramp A/D converter is shown schematically in Fig. 1. Considering an n-bit converter, a "start convert" signal will reset the n-bit counter and initiate the operation of a function generator. The latter produces a linear ramp output signal $V_R = kt$, where k is a constant and t is time. A comparator then will continuously compare the reference ramp signal with the value of the unknown input signal V_S. The resulting output of the comparator represents a binary 0 where V_R is greater than V_S and binary 1 where V_R is less than V_S. While the comparator output is 1, pulses from a clock pulse generator will be counted by the n-bit counter. At such time the input signal V_S and the ramp signal V_R are equal, the comparator output changes to a binary 0. Also, by action of the AND gate, clock pulses are inhibited from entering the counter. The time during which the comparator output remains in the 1 state is proportional to the magnitude of the input signal. Also, the count in the counter, at the instant the comparator changes state, will be proportional to the time interval that the comparator output is 1. Thus, the count in the counter will be a digital representation of the input signal.

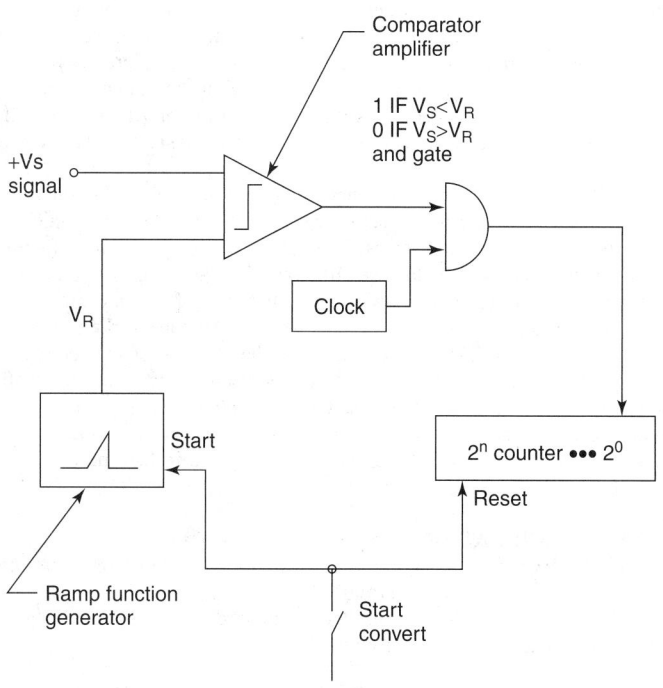

Fig. 1.   Ramp-type analog-to-digital converter.

This type of A/D converter is one of the simplest forms of electronic A/D converters and may be used for conversions of less than 10 to 12 bits at speeds that do not exceed several thousand conversions per second. Higher speeds can be obtained at lower resolution. Logic speed is the main factor in determining the speed of the device. It will be noted that the speed of the converter equals the clock frequency divided by the number of quantizing intervals. For example, 4,096 pulses must be counted for a full-scale input signal in a 12-bit converter. At a clock frequency of 10 MHz, this necessitates 1,024/10 MHz, or approximately 0.4 ms, consequently providing a converter speed of 2,500 samples per second. Added to simplicity is the advantage of excellent differential linearity, accounting for the very wide use of this type A/D converter, particularly in such applications as the generation of histograms as encountered in the field of nuclear experimentation.

See also **Analog-to-Digital Converter**.

THOMAS J. HARRISON, International Business Machines Corporation, Boca Raton, Florida.

RAMSDEN CIRCLE. If a telescope is focused for infinity, and pointed toward a bright sky, while a sheet of white paper is held near the eyepiece, a sharp, bright circle of light (the exit pupil), called the Ramsden circle, can be found. The diameter of this circle divided into the diameter of the objective lens gives the magnification of the telescope.

RANDOM ERROR. Errors that are not systematic, are not erratic, and are not mistakes. Such random errors are caused by disturbed elements in the measuring instrument and usually are of an approximately normal or Gaussian distribution. Such random errors are sometimes called *short-period errors*.

RANDOMIZATION. A set of objects is randomized when arranged in a random order. By slight extension, a set of treatments applied to a set of units is said to be randomized when the treatment applied to any given unit is chosen at random from those available and not already allocated.

RANDOM NUMBER. An expression formed by a set of digits selected from a sequence of digits in which each successive digit is equally likely to be any of the digits.

RANDOM SELECTION. A method of selecting sample units such that each possible sample has a fixed and determinate probability of selection. Ordinary haphazard or seemingly purposeless choice is generally insufficient to guarantee randomness when carried out by human beings. Therefore, devices, such as tables of random sampling numbers, are used to remove subjective biases inherent in personal choice.

RANDOM VARIABLE. A variable which can take any one of a given set of values with assigned probability. In statistics, a particular value of a random value is often referred to as a *variate-value*, and sometimes "variate" is used as synonymous with "random variable."

RANDOM VIBRATION. An oscillation whose instantaneous magnitude is not specified for any given instant of time. The instantaneous magnitudes of a random oscillation are specified only by probability distribution functions giving the fraction of the total time that the magnitude, or some sequence of magnitudes, lies within a specified range. A random vibration whose instantaneous magnitudes occur according to the Gaussian distribution is called Gaussian random vibration. Wide-band vibration amplitude is usually expressed as root-mean-square acceleration in gravitational units of acceleration g. The parameter used to specify the frequency distribution of a random vibration is power spectral density (g^2 per cycle per second), sometimes called *acceleration density* or *acceleration spectral density*.

RANDOM WALK. The path traversed by a particle or other entity, which moves in steps, each step being determined by chance either in regard to direction or in regard to magnitude or both. Cases most frequently considered are those in which the particle moves on a lattice or points in one or more dimensions, and at each step is equally likely to move to any of the nearest neighboring points. The theory of random walks has many applications, e.g., to the migration of insects, to sequential sampling, and in the limit, to diffusion processes. An additive random walk process is a stochastic process with independent increments, that is to say, a process $\{(x_t)\}$ is additive if, for $t_1 < t_2 > \cdots < t_n$, the differences, $x_{t2} - x_{t1}, x_{t3} - x_{t2}$, etc., are independent. The expressions *differential process* and *process with independent increments* are equivalent, but are usually confined to the case when the parameter process may also be said to be additive; a synonym in this case is *random walk process*.

RANGE AND SPAN (Instrument). With reference to industrial and scientific instruments, the Scientific Apparatus Makers Association defines *range* as the region between the limits within which a quantity is measured, received, or transmitted, and expressed by stating the lower and upper range-values. Examples would include:

(a) 0 to 150 °C   (b) −20 to +200 °F   (c) 20 to 150 °F

Unless otherwise modified, input range is implied.

TABLE 1. USE OF RANGE AND SPAN TERMINOLOGY

Typical Ranges	Name	Range	Lower Range-Value	Upper Range-Value	Span	Supplementary Data
0 +100	—	0 to 100	0	+100	100	—
20 +100	Suppressed zero range	20 to 100	20	+100	80	Suppression ratio = .25
−25 0 +100	Elevated zero range	−25 to +100	−25	+100	125	—
−100 0	Elevated zero range	−100 to 0	−100	0	100	—
−100 −20	Elevated zero range	−100 to −20	−100	−20	80	—

The following compound terms are used with suitable modification in the units: Measured variable range; measured signal range; indicating-scale range; chart-scale range; and so on. For multirange devices, this definition applies to the particular range that the device is set to measure.

Range-Limit (Lower). The lowest quantity that a device can be adjusted to measure.

Range-Limit (Upper). The highest quantity that a device can be adjusted to measure.

Range-Value (Lower). The lowest quantity that a device is adjusted to measure.

Range-Value (Upper). The highest quantity that a device is adjusted to measure.

Range (Elevated-Zero). A range where the zero value of the measured variable, measured signal, etc., is greater than the lower range-value. See Table 1.

The zero may be between the lower and upper range-values, at the upper range-value, or above the upper range-value. The terms *suppression, suppressed range, or suppressed span* are frequently used to express the condition in which the zero of the measured variable is greater than the lower range-value. The term *range, elevated-zero* is preferred.

Range (Suppressed-Zero). A range where the zero value of the measured variable is less than the lower range-value. Zero does not appear on the scale. An example is a range from 20 to 100. The terms *elevation, elevated range,* or *elevated span* are frequently used to express the condition in which the zero of the measured variable is less than the lower range-value. The term *range, suppressed-zero* is preferred.

Suppression Ratio. Of a suppressed-zero range, the ratio of the lower range-value to the span. Example:

Range: 20 to 100
Suppression Ratio: 20/80 = 0.2

Instrument Span

Span is defined as the algebraic difference between the upper and lower range-values. Examples:

(a) Range: 0 to 150 °C   (span is 150 °C)
(b) Range: −20 to +200 °F   (span is 220 °F)
(c) Range: 20 to 150 °F   (span is 130 °F)

The following compound terms are used with suitable modifications in the units: Measured variable span; measured signal span; and so on. For multi-range devices, this definition applies to the particular range that the device is set to measure. See accompanying table.

Span Error. The difference between the actual span and the ideal span, usually expressed as a percent of ideal span.

RANGEFINDER (Camera). See **Photography and Imagery.**

RANGE MARKER. The index marks displayed on radar indicators to establish the scale or facilitate determination of the distance of a target from the radar. On the plan position indicator scope, for example, range markers take the form of concentric circles with the position of the radar at the center. Also called *distance marker.* See also **Azimuth Marker.**

RANGE MARKS. Two prominent objects, either natural or artificial, which are located along a line that has some particular value for navigators, are known as range marks. Range marks are used for so many different purposes in navigation that it would be futile to attempt to list them all. However, one example may be of interest. A channel is entered on track 200°, followed for 1,100 yards (1,006 meters), then the channel turns and track must be altered to 296°. The turning point is marked by a black and white striped buoy. A lighthouse on shore is so placed that its bearing from the buoy is 200°. Accordingly, when a ship has the lighthouse and the buoy in line, the bearing of the buoy from the ship is 200°, and to follow the first leg of the channel the ship simply keeps the lighthouse "ranging" on the buoy. As the ship approaches the buoy the pilot watches the shore and when a red and white striped target ranges on a white church spire, he alters heading and holds the target on the spire to follow the 296° leg of the channel. Currents may force the pilot to head quite differently from the directions of the channel as given on his chart, but the range marks give a line of position and, so long as the ship is on the proper line, the pilot knows he is proceeding in safe water. See also **Course**; and **Navigation**.

RANGE (Probability). The range of a probability distribution (or of the associated variate) is the name given to the limits between which the probability takes nonzero values. The range of a sample is the difference between the highest and lowest observations. The range is of itself an elementary measure of dispersion and, in terms of the mean range in repeated sampling, it may afford a reasonable estimate of the population standard deviation. The *effective range* is the range after the removal of a limited number of outlying observations at either or both ends of the original range. The removal may have to be a matter of subjective judgment and inferences based on effective range are of somewhat doubtful value, and it yields at best a rough measure of dispersion; in fact the term itself is not a good one.

RANK CORRELATION. Suppose we have a sample of n pairs of ranked observations, (x_i, y_i), $i = 1$ to n. Two measures of correlation between the samples are in current use.

1. Spearman's ρ is the ordinary correlation between the ranks regarded as variate values. It may readily be calculated as

$$1 - \frac{6 \Sigma (x_i - y_i)^2}{n(n^2 - 1)}$$

2. Kendall's τ may be defined as $1 - 4s/n(n-1)$, where s is the smallest number of interchanges of neighboring members needed to transform one ranking into the other. Both ρ and τ take values in the range −1 to +1.

ρ is somewhat easier to calculate than τ, but τ has some practical and many theoretical advantages. The (ordinary) correlation between the two in a population in which all rankings are equally possible is high, being greater than 0.98 for $n \geq 5$ — the value of τ being approximately $\frac{2}{3}$ that of ρ except in extreme cases.

Sir Maurice Kendall, International Statistical Institute, London.

RANK (Mathematics). The rank of a matrix is the order of the non-zero determinant of greatest order that can be selected from the matrix by taking out rows and columns. The concept rank facilitates, for instance, the statement of the condition for consistency of simultaneous linear equations:

Fig. 1.   Dry lake mineral bed near Mountain Pass, California contains over one million pounds of neodymium and nearly one-half million pounds of praseodymium, both elements once regarded as "rare earths" and of limited scientific curiosity. During recent years, the rare earths have become significant materials in the electronic, chemical, metallurgical, glass, cryogenic, nuclear, and ceramic refractory industries. Lanthanum, another rare-earth element, is more abundant than lead.

m linear equations in n unknowns are consistent when, and only when, the rank of the matrix of the coefficients is equal to the rank of the augmented matrix. In the system of linear equations,

$$x + y + z + 3 = 0$$
$$2x + y + z + 4 = 0$$

the matrix of the coefficients is

$$\left\| \begin{matrix} 1 & 1 & 1 \\ 2 & 1 & 1 \end{matrix} \right\|$$

and the augmented matrix is

$$\left\| \begin{matrix} 1 & 1 & 1 & 3 \\ 2 & 1 & 1 & 4 \end{matrix} \right\|$$

The rank of both is two, because the determinant

$$\left| \begin{matrix} 1 & 1 \\ 2 & 1 \end{matrix} \right|$$

is not zero. Hence these equations are satisfied by some set of values of x and y and z.

RAOULT'S LAW.   The vapor pressure of a substance in solution is proportional to its mole fraction. See also **Vapor Pressure.**

RAPE PLANT.   See **Brassica.**

RAPESEED OIL.   See **Vegetable Oils (Edible).**

RARE-EARTH ELEMENTS AND METALS.   Sometimes referred to as the "fraternal fifteen," because of similarities in physical and chemical properties, the rare-earth elements actually are not so rare. This is attested by Fig. 1, which shows a dry lake bed in California that alone contains well in excess of one million pounds of two of the elements, neodymium and praseodymium. The world's largest rare earth body and mine near Baotou, Inner Mongolia, China is shown in Fig. 2. It contains 25 million tons of rare earth oxides (about one quarter of the world's human reserves). The term *rare* arises from the fact that these elements were discovered in scarce materials. The term *earth* stems from the fact that the elements were first isolated from their ores in the chemical form of oxides and that the old chemical terminology for oxide is earth. The rare-earth elements, also termed Lanthanides, are similar in that they share a valence of 3 and are treated as a separate side branch of the periodic table, much like

Fig. 2.   Open-pit operation at Baiyunebo mine. (*The Chinese Society of Rare Earth.*)

the Actinides. See also **Actinide Series**; **Chemical Elements**; **Lanthanide Series**; and **Periodic Table of the Elements**.

The properties of the Rare-earth elements are given in Tables 1 and 2. Pronunciation of the elements is as follows: Cerium (*sear' ium*), dysprosium (*dis pröz' ium*), erbium (*ur' bium*), europium (*yoo rö pium*), gadolinium (*gado lin' ium*), holmium (*hol' mium*), lanthanum (*Ian' tha num*), lutetium (*loo tee' shium*), neodymium (*neo dim' ium*), praseodymium (*pra zee o dim' ium*), promethium (*pro mee' thium*), samarium (*sa mar' ium*), scandium (*scan de'ium*) terbium (*tur' bium*), thulium (*thoo' lium*), ytterbium (*i tur' bium*), and yttrium (*it' rium*). The lanthanides are further described by individual alphabetical entries for each element.

C.A. Arrhenius, in 1787, noted an unusual black mineral in a quarry near Ytterby, Sweden. This was identified later as containing yttrium and rare-earth oxides. With the exception of promethium, all members of the Lanthanide Series had been discovered by 1907, when lutetium was isolated. In 1947, scientists at the Atomic Energy Commission at Oak Ridge National Laboratory (Tennessee) produced atomic number 61 from uranium fission products and named it promethium. No stable isotopes of promethium have been found in the earth's crust.

TABLE 1. ATOMIC AND THERMAL PROPERTIES OF RARE-EARTH ELEMENTS

Atomic Number	21	39	57	58	59	60	61	62	63	64	65	66	67	68	69	70	71
Symbol	Sc	Y	La	Ce	Pr	Nd	Pm	Sm	Eu	Gd	Tb	Dy	Ho	Er	Tm	Yb	Lu
Element	Scandium	Yttrium	Lanthanum	Cerium	Praseodymium	Neodymium	Promethium	Samarium	Europium	Gadolinium	Terbium	Dysprosium	Holmium	Erbium	Thulium	Ytterbium	Lutetium
Estimated abundance:																	
ppm	6	33	30	60	8.2	28	0	6.0	1.2	5.4	0.9	3.0	1.2	2.8	0.5	3.0	0.5
g/ton	—	28–70	5–18	20–46	3.5–5.5	12–24	0	4.5–7	0.14–1.1	4.5–6.4	0.7–1	4.5–7.5	0.7–1.2	2.5–6.5	0.2–1	2.7–8	0.8–1.7
Atomic constants:																	
Atomic weight,	44.96	88.91	138.91	140.12	140.91	144.24	(145)	150.36	151.96	157.25	158.93	162.50	164.93	167.26	168.93	173.04	174.97
Metallic radius, Å, (CN=12)	1.641	1.801	1.879	(+3) 1.846 (+4) 1.672	1.828	1.821	1.811	1.804	(+2) 2.042 (+3) 1.798	1.801	1.783	1.774	1.766	1.757	1.746	(+2) 1.939 (+3) 1.741	1.735
Volume, cm³/g atom	15.04	19.89	22.60	(+3) 21.43 (+4) 15.92	20.80	20.58	(20.24)	20.00	(+2) 28.98	19.90	19.31	19.00	18.75	18.45	18.12	(+2) 24.84 (+3) 17.98	17.79
Density, g/cm³	2.989	4.469	6.146	6.770	6.773	7.008	7.264	7.520	5.244	7.901	8.230	8.551	8.795	9.066	9.321	6.966	9.841
lb/in.³	0.108	0.161	0.222	0.244	0.244	0.253	0.262	0.271	0.189	0.285	0.297	0.308	0.317	0.327	0.336	0.251	0.355
Crystal structure at 25°C	hcp	hcp	dhcp	fcc	dhcp	dhcp	dhcp	rhom	bcc	hcp	hcp	hcp	hcp	hcp	hcp	fcc	hcp
Unpaired 4f electrons	0	0	0	1	2	3	4	5	6	7	6	5	4	3	2	1	0
Number of isotopes:																	
Natural	1	1	2	4	1	7	0	7	2	7	1	7	1	6	1	7	2
Artificial	11	14	19	15	14	7	15–18	11	16	11	17	12	18	12	17	10	14
Lattice constants, Å:																	
a	3.309	3.648	3.774	5.161	3.672	3.658	3.65	3.629	4.583	3.634	3.605	3.592	3.578	3.559	3.538	5.485	3.505
c	5.268	5.732	12.171		11.833	11.797	11.656	26.207		5.781	5.697	5.650	5.618	5.585	5.554		5.549
Ionic radius, Å:																	
+2								1.19	1.17							1.00	
+3	0.745	0.900	1.045	1.010	0.997	0.983	0.97	0.958	0.947	0.938	0.923	0.912	0.901	0.890	0.880	0.868	0.861
+4				0.80	0.78						0.76						
Color of 3⁺ ion (in solution)	Colorless	Colorless	Colorless	Colorless	Green	Reddish violet	Pink	Yellow	Pale pink	Colorless	Almost colorless	Yellow	Pink	Reddish violet	Green	Colorless	Colorless
Electronegativity	1.28	1.177	1.117	(+3) 1.123 (+4) 1.43	1.130	1.134	1.139	1.145	(+2) 0.98 (+3) 1.152	1.160	1.168	1.176	1.184	1.192	1.200	(+2) 1.02 (+3) 1.208	1.216

Absorption bands, 3+ ion, Å	None	9750	3600 6825 7800	3642 3792 4870 5228 6525	2870 3611 4508 5370 6404	3504 3650 9100	3694 3780 4875	2729 2733 2754 2756	3625 3745 3755 3941 4020	5485 5680 7025 7355		3540 5218 5745 7395 7420 7975 8030 8680	4445 4690 4822 5885	2105 2220 2380 2520	None	None	None
Thermal properties:																	
Melting point: °C	1663	819	1545	1529	1474	1412	1365	1313	822	1074	1042	1021	931	798	918	1522	1541
°F	3025	1506	2813	2784	2685	2574	2489	2395	1512	1965	1908	1868	1708	1468	1684	2772	2806
Boiling point at 1 atm.: °C	3402	1196	1950	2868	2700	2567	3230	3273	1529	1794	3000	3074	3520	3443	3464	3345	2836
°F	6156	2185	3542	5194	4892	4653	5846	5923	2784	3261	5432	5565	6368	6229	6267	6053	5137
Heat of fusion ΔH_f kcal/g atom	5.26	1.831	4.015	4.756	4.063	2.643	2.579	2.390	2.201	2.060	1.84	1.706	1.646	1.305	1.482	2.724	3.370
Heat of sublimation ΔH_s at 25 °C, kcal/g atom	102.20	36.35	55.50	75.79	71.89	69.41	92.90	95.00	41.90	49.40	83	78.30	84.99	101.0	103.0	101.5	90.30
Heat capacity ΔC_p at 25 °C, cal/(g atom) (°C)	6.41	6.38	6.45	6.72	6.50	6.62	6.91	8.87	6.62	7.05	6.52	6.55	6.55	6.43	6.48	6.34	6.09
Coefficient of expansion, per °C × 10^{-6}	9.9	26.3	13.3	12.2	11.2	9.9	10.3	9.4	35.0	12.7	11	9.6	6.7	6.3	12.1	10.6	10.2
Nuclear properties:																	
Thermal neutron capture, barns/atom	108	37	125	170	64	1100	46	40,000	4,300	5,600		50	11.6	0.73	8.9	1.31	17

*Table compiled by Molybdenum Corporation of America, White Plains, N.Y. (Joseph G. Cannon); edited by Rare-Earth Information Center, Energy and Mineral Resources Research Institute, Iowa State University, Ames, Iowa (Karl A. Gschneidner, Jr. and N. Kippenhan). Data from S.R. Taylor, Abundance of Chemical Elements in the Continental Crust: A New Table, *Geochim. Cosmochim. Acta*, vol. 28, pp. 1273–1285, 1964; E.T. Teatum, et al., Compilation of Calculated Data Useful in Predicting Metallurgical Behavior of Elements in Binary Alloy Systems, *Univ. Calif., Los Alamos Sci. Lab. Rep.* LA-4003, pp. 11–12. Dec. 24, 1968; Clifford A. Hampel, "Rare Metals Handbook," 2d ed., chaps. 1 and 35, Van Nostrand Reinhold Company, New York, 1961; O.A. Songina, "Rare Metals: Scandium, Yttrium, Lanthanide and Actinides," chap. 6, trans. from Russian (1970), 3d ed (1964), U.S. Dept. of Interior and The National Science Foundation, Washington, DC.; Karl A. Gschneidner, Jr., "Solid State Physics," vol. 16, "Physical Properties and Interrelationships of Metallic and Semimetallic Elements," pp. 275–426, Academic, New York, 1964; Clifford A. Hampel, "The Encyclopedia of the Chemical Elements," Van Nostrand Reinhold Company, New York, 1968; R. Hultgren, R.L. Orr, and K.K. Kelley, supplement to "Selected Values of Thermodynamic Properties of Metals and Alloys," Wiley, New York, 1963; Data from Department of Mineral Technology and Lawrence Radiation Laboratory, The University of California, Berkeley, Calif. (data and revision published periodically). Data from Karl A. Gschneidner, Jr. and Leroy Eyring, eds., "Handbook on the Physics and Chemistry of the Rare Earths, Vol. 1," North-Holland, Amsterdam, (1979).

TABLE 2. MECHANICAL, ELECTRICAL, AND OXIDE PROPERTIES OF RARE-EARTH ELEMENTS

Atomic Number	21	39	57	58	59	60	61	62	63	64	65	66	67	68	69	70	71
Symbol	Sc	Y	La	Ce	Pr	Nd	Pm	Sm	Eu	Gd	Tb	Dy	Ho	Er	Tm	Yb	Lu
Element	Scandium	Yttrium	Lanthanum	Cerium	Praseodymium	Neodymium	Promethium	Samarium	Europium	Gadolinium	Terbium	Dysprosium	Holmium	Erbium	Thulium	Ytterbium	Lutetium
Mechanical properties†																	
Yield strength:																	
$\quad$ kg/mm^2	17.6	4.3	12.8	2.9	7.4	7.2	N.A.	6.9	N.A.	1.5	N.A.	4.4	22.6	6.1	N.A.	0.7	N.A.
$\quad$ 1,000 psi	28.0	6.1	18.2	4.1	10.5	10.2	N.A.	9.8	N.A.	2.1	N.A.	6.3	32.1	8.7	N.A.	1.0	N.A.
Elongation, %	5.0	34	7.9	22	15.4	25	N.A.	17	N.A.	37	N.A.	30	5	11.5	N.A.	43	N.A.
Tensile strength:																	
$\quad$ kg/mm^2	26.0	13.2	13.3	11.9	15.0	16.7	N.A.	15.9	N.A.	12	N.A.	14.2	26.4	13.9	N.A.	5.9	N.A.
$\quad$ 1,000 psi	37.0	18.8	18.9	16.9	21.3	23.8		22.6		17.1		20.2	37.5	19.8		8.4	
Vickers hardness, 10-kg load, kg/mm^2	—	41	38	29	37	35	63	40	17	42	38	44	46	42	48	17	44
Elastic properties (values in parentheses estimated):																	
Compressibility, cm^2/kg $\times 10^{-6}$	1.73	3.98	3.23	4.96	3.39	3.09	(2.96)	2.60	11.76	2.59	2.52	2.44	2.37	2.23	2.21	7.26	2.06
Shear modulus, kg/cm^2 $\times 10^6$	0.297	0.260	0.152	0.122	0.150	0.169	(0.183)	0.199	(0.079)	0.226	0.232	0.259	0.269	0.289	(0.310)	0.101	0.276
Young's modulus, kg/cm^2 $\times 10^6$	0.759	0.648	0.392	0.306	0.387	0.431	(0.471)	0.510	0.186	0.569	0.582	0.643	0.665	0.672	0.754	0.314	0.697
Poisson's ratio	0.279	0.246	0.288	0.248	0.289	0.279	(0.278)	0.282	0.167	0.254	0.255	0.238	0.237	0.250	0.217	0.207	0.261
Electrical properties at 25°C:																	
Resistivity, $\mu\Omega$-cm	56.2	59.6	61.5	74.4	70.0	64.3	75	94.0	90.0	131	115	92.6	81.4	86	67.6	25	58.2
Hall coefficient, V-cm/(A)(Oe) $\times 10^{12}$	-0.13	-0.77	-0.35	$+1.81$	$+0.71$	$+0.97$	N.A.	-0.2	$+24.4$	-4.48	-4.3	-2.7	-2.3	-0.34	-1.8	$+3.77$	-0.54
Work function, eV	3.5	3.1	3.5	2.9	2.7	3.2	3.1	2.7	2.5	3.1	(3.1)	(3.1)	(3.1)	(3.1)	(3.1)	(2.6)	(3.1)

Magnetic properties:

Element oxide	Moment, theoretical for 3+ ion, Bohr magnetons	Susceptibility, emu/g atom × 10^6	Curie temperature, °C	Neel temperature, °C	Formula	Color	Molecular weight	Melting point, °C	Melting point, °F	Density g/cm³
	0	295	None	None	Sc2O3	White	137.92	2403	4357	3.88
	0	191	None	None	Y2O3	White	225.81	2410	4370	5.03
	0	101	None	None	La2O3	White	325.82	2300	4172	6.58
	2.5	2430	None	−260.6	CeO2	Buff	172.12	2210	4010	7.22
	3.6	5320	None	None	Pr6O11	Black	1021.79	2183	3961	6.83
	3.6	5650	None	−253	Nd2O3	Light blue	336.48	2233	4051	7.31
	N.A.	N.A.	N.A.	N.A.	Pm2O3	White	342	2320	4208	7.60
	1.6	1275	None	−258	Sm2O3	Cream	348.70	2269	4116	7.11
	3.5	33,100	None	−184	Eu2O3	Pale pink	351.92	2291	4156	7.29
	7.95	356,000	+20	None	Gd2O3	White	362.50	2339	4242	7.61
	9.7	193,000	−53	−43	Tb4O7	Dark brown	747.69	2303	4117	7.87 (Tb2O3)
	10.6	99,800	−185	−97	Dy2O3	Cream	373.00	2228	4042	8.16
	10.6	70,200	−254	−143	Ho2O3	Cream	377.86	2330	4226	8.41
	9.6	44,100	−253	−188	Er2O3	Rose	382.52	2344	4251	8.65
	7.6	26,100	−248	−215	Tm2O3	Light green	385.87	2341	4246	8.90
	4.5	71	None	None	Yb2O3	White	394.08	2355	4271	9.21
	0	17.9	None	None	Lu2O3	White	397.94	2427	4401	9.41

*Table compiled by Molybdenum Corporation of America, White Plains, N.Y. (Joseph G. Cannon); edited by Rare-Earth Information Center, Energy and Mineral Resources Research Institute, Iowa State University, Ames, Iowa (Karl A. Gschneidner, Jr. and N. Kippenhan). Data from S.R. Taylor, Abundance of Chemical Elements in the Continental Crust: A New Table, *Geochim. Cosmochim. Acta*, vol. 28, pp. 1273–1285, 1964; E.T. Teatum, et al., Compilation of Calculated Data Useful in Predicting Metallurgical Behavior of Elements in Binary Alloy Systems, *Univ. Calif., Los Alamos Sci. Lab. Rep.* LA-4003, pp. 11–12, Dec. 24, 1968; Clifford A. Hampel, "Rare Metals Handbook," 2d ed., chaps. 1 and 35, Van Nostrand Reinhold Company, New York, 1961; O.A. Songina, "Rare Metals: Scandium, Yttrium, Lanthanide and Actinides," chap. 6, trans. from Russian (1970), 3d ed. (1964), U.S. Dept. of Interior and The National Science Foundation, Washington, DC.; Karl A. Gschneidner, Jr., "Solid State Physics," vol. 16, "Physical Properties and Interrelationships of Metallic and Semimetallic Elements," pp. 275–426, Academic, New York, 1964; Clifford A. Hampel, "The Encyclopedia of the Chemical Elements," Van Nostrand Reinhold Company, New York, 1968; R. Hultgren, R.L. Orr, and K.K. Kelley, supplement to "Selected Values of Thermodynamic Properties of Metals and Alloys," Wiley, New York, 1963; Data from Department of Mineral Technology and Lawrence Radiation Laboratory, The University of California, Berkeley, Calif. (data and revisions published periodically). Data from Karl A. Gschneidner, Jr. and Leroy Eyring, eds., in "Handbook on the Physics and Chemistry of the Rare Earths, Vol. 1 (metals) & 3 (oxides)," North-Holland, Amsterdam, 1978, 1979.

†Highest reported value for metal at room temperature after 10–50% reduction in area or annealed or as-cast; purity unknown.

N.A.—not available.

Natural mixtures of these elements have been used commercially since the early 1900s. Mischmetal is the source of the hot spark in cigarette lighter flints. The mixed rare-earth fluorides are burned in the cores of carbon electrodes to create the intense sunlike illumination required by motion-picture projectors and searchlights. The mixed rare-earth oxides, which contain primarily cerium dioxide, are used to grind and polish almost all optical lenses and television faceplates. In the late-1940s, it was discovered that the rare-earth metals effectively control the shape of carbon in normally brittle cast iron, resulting in ductile or nodular iron. During the 1950s, interest in several of the pure elements (europium, gadolinium, dysprosium, samarium, and erbium) was stimulated because these elements have the highest thermal-neutron-absorption properties among the elements. These elements have found application in control rods and as burnable poisons. Yttrium metal was fabricated into tubing and mill products because it is almost transparent to thermal neutrons and has a unique stability at high temperature in contact with liquid uranium, potassium, and sodium. Nuclear aircraft and submarine propulsion programs were the main impetus for these efforts. Radioactive promethium has been used as a power source for pacemakers.

Early in the 1960s, mixtures of the rare-earth elements were incorporated with synthetic molecular-sieve catalysts, resulting in increased petroleum refining efficiency. Various rare-earth compounds have been found to act as catalysts in several chemical processes, such as hydrogenation. Rare-earth mixed oxides and especially CeO_2 are being used in auto exhaust catalysts. In 1964, a new red phosphor for color television was discovered. Relatively large quantities of highly purified europium and yttrium oxides were needed as commercial color television production started. Rare-earth phosphors are also being used in color monitors for computers, X-ray screens, fluorescent lamps, UV-conversion phosphors, dental and surgical lasers, electro- and thermoluminescent devices, and fiber optics.

Permanent magnets having properties several times superior to any other known materials were developed in 1967. Praseodymium, yttrium, samarium, lanthanum, and cerium are alloyed with cobalt in the range RCo_5 to R_2Co_{17}, where $R = a$ rare-earth element. The new family of permanent-magnet materials is bringing about improvements in power generation and electronic communications. Conventional applications now include watches, electric motors, computer printers, automotive devices, frictionless bearings, and loudspeakers. Novel applications for the powerful magnets include magnetic earrings and use in medical treatments.

About 15 years later a new family of iron-neodymium-based permanent magnets were discovered. In 1983 several research groups in the United States and Japan announced the discovery of a new compound with the probable composition, $R_2Fe_{14}B$ ($R = a$ light rare earth lanthanide, predominantly neodymium). These materials exhibit extremely powerful magnetic qualities as compared with traditional magnet materials and about 10–20% stronger than the Sm-Co magnets. One shortcoming, is the loss of desirable magnetic qualities at elevated temperatures above about 200°F. Applications for the new magnetic materials span a wide range from the very sophisticated applications, such as found in nuclear magnetic resonance imaging systems, down to the inexpensive magnets used in toys and around the home. The $Nd_2Fe_{14}B$ materials are made either by rapid solidification or by powder metallurgy techniques. Some investigators have found that the addition of 6% cobalt increases the Curie temperature 100 K while others found that small additions of Dy (less than a 10% substitution of the Nd) are added to raise the Curie temperature. The crystal structure of $Nd_2Fe_{14}B_1$ is tetragonal, an anisotropic structure that contributes to the high coercivity. The relatively low concentrations of light lanthanides and boron, and the fact that the magnetic moments in iron and the lanthanides align parallel (ferromagnetically) to each other, allow the magnetization to remain high. See also **Magnetism**.

Metallurgical Uses. During the 1960s, the rare-earth metals were established as reactive and refining metals in the iron and steel industry. As alloying elements, lanthanum and yttrium improve the high-temperature oxidation and corrosion properties of superalloys. Other metallurgical applications of the rareearths include welding solders, brazing alloys, nonferrous alloys (such as magnesium and aluminum), and dispersion hardening of complex alloys (eg. Y_2O_3 dispersed in nickel based alloys). In the 1990s complex $LaNi_5$-based alloys were developed for use in rechargeable metal hydride batteries, which are slowly replacing NiCd batteries in many applications because of their superior performance and because the $LaNi_5$-based materials are more environmentally friendly. Another fairly recent development is the use of mixed rare earths as fertilizers in agriculture. The Chinese have been quite successful in improving various crop yields, but this application has seen only limited use in other countries.

Miscellaneous applications of rare-earth compounds, complexes and alloys include: MRI (magnetic resonance imaging) contrasting agents in the medical field; high temperature oxide superconductors; magnetic recording alloys and magnetic bubble devices; electronic components, capacitors and semiconductors; optical glasses and fiber optics; magnetic cooling and refrigeration; corrosion inhibitors; dying and printing textiles; cosmetics; oxidizer in self-cleaning ovens; fuel cell cathodes; and Y_2O_3 (and other R_2O_3)-stabilized ZrO_2 as oxygen sensors, electrolytes, structural ceramics, and synthetic jewelry.

The Institute for Physical Research and Technology sponsors a Rare-Earth Information Center at Iowa State University. Ames, Iowa, which provides a comprehensive service to science and industry by cataloging the vast amount of technical information generated about these elements each year.

Occurrence. Rare-earth minerals exist in many parts of the world; the overall potential supply is essentially unlimited. As a group, these elements rank fifteenth in abundance, somewhat more plentiful than zinc. Rare-earth minerals generally are classified as sources for *light* (La through Gd) or *heavy* (Y plus Tb through Lu). Typical mineral distributions are given in Table 3.

Until 1964, monazite, a thorium-rare-earth phosphate, $REPO_4Th_3$ $(PO_4)_4$, was the main source for the rare-earth elements. Australia, India, Brazil, Malaysia, and the United States are active sources. India and Brazil supply a mixed rare-earth chloride compound after thorium is removed chemically from monazite. Bastnasite, a rare-earth fluocarbonate mineral; $REFCO_3$, is a primary source for light rare earths. From 1965 to about 1985, an open-pit resource at Mountain Pass, California, has furnished about two-thirds of world requirements for rare-earth oxides. In the early 1980s the Chinese started to produce rare earths from their Baiyunebo mine, which contains both bastnasite and monazite, located about 100 miles from Baotou, Inner Mongolia. In the early 1990s the rare earth production from this mine exceeded that mined in Mountain Pass. In 1998 the Mountain Pass operation was temporarily closed because of a blocked wastewater pipe. Once the governmental delays due to environmental issues are approved, production will resume. The main source for yttrium and heavy rare-earths is a by-product of uranium mining in the Elliott Lake Region, Ontario. Some xenotime, found in Malaysia, is processed in Japan and Europe.

A highly generalized flowsheet of the production of some of the rare-earth oxides is shown in Fig. 3. Crushed and finely ground bastnasite contain about 70% rare-earth oxides is roasted under oxidizing conditions to convert soluble trivalent cerium compounds to insoluble tetravalent CeO_2. The roasted product is leached with HCl, which dissolves the remaining rare earths (La, Pr, Nd, Sm, Eu, Gd), leaving behind a concentrated cerium product. The solution is passed through liquid-liquid organic solvent extraction (SX) cells, resulting in a primary separation of La-Nd-Pr from Sm-Eu-Gd. Further SX separates a pure lanthanum solution and a concentrated Nd-Pr solution, which another SX circuit separates. Europium is reduced to a divalent state in solution and precipitated. A final SX system separates and purifies gadolinium and samarium. Pure elements are usually precipitated as oxalates and calcined to oxides.

In connection with production of the heavy rare earths, monazite, containing about 55% rare-earth oxides and 5% thorium, is treated in one of two ways: (1) finely ground particles are leached with hot H_2SO_4, which dissolves thorium and the rare earths, leaving an insoluble residue; or (2) finely ground particles are reacted with hot caustic (NaOH), which dissolves the phosphate, creating a solution of trisodium phosphate, which may be recovered as a by-product. The thorium and rare-earth hydrate cake is then dissolved in H_2SO_4. Thorium sulfate is selectively precipitated by pH adjustment. Separation of the other rare earths in solution is usually completed by selective absorption on ion-exchange resins and elution from ion-exchange columns. After thorium is removed from the H_2SO_4 solution, the rare earths remaining are precipitated, using NaOH, forming a double salt, $NaRESO_4 \cdot xH_2O$, known as pink salt. This salt is dissolved in HCl, treated to remove impurities, and evaporated until the hydrated $RECl_3 \cdot 6H_2O$ can be cast.

TABLE 3. RARE EARTH CONTENT OF SEVERAL PRIMARY SOURCE MINERALS

Rare earth element	Bastnasite Calif. (%)	Bastnasite China[a] (%)	Monazite China[b] (%)	Monazite Australia[c] (%)	Monazite Brazil and India (%)	Xenotime Malaysia (%)	Uranium residues Ontario, Canada (%)	Ion-adsorption Clays (China) Longman (%)	Ion-adsorption Clays (China) Xunwu (%)	Loparite, Russia (%)
La	32.0	22.8	23.4	23.9	22.8	0.5	0.8	2.2	29.8	25.0
Ce	49.0	49.8	45.7	46.0	45.7	5.0	3.7	1.1	7.2	50.5
Pr	4.4	6.2	4.2	5.0	5.0	0.7	1.0	1.1	7.1	5.0
Nd	13.5	18.5	15.7	17.4	18.9	2.2	4.1	3.5	30.2	15.0
Sm	0.5	1.0	3.0	2.5	3.0	1.9	4.5	2.3	6.3	0.7
Eu	0.1	0.2	0.1	0.05	0.1	0.2	0.1	0.1	0.5	0.1
Gd	0.3	0.7	2.0	1.5	1.7	4.0	8.5	5.7	4.2	0.6
Tb		0.1	0.1	0.04	0.2	1.0	1.2	1.1	0.5	—
Dy		0.1	1.0	1.2	0.5	8.7	11.2	7.5	1.8	0.6
Ho		0.1	0.05	0.1	2.1	2.1	2.6	1.6	0.3	0.7
Er	0.1		0.5	0.2	0.1	5.4	5.5	4.3	0.8	0.8
Tm		0.1	0.5	0.01	—	0.9	0.9	0.6	0.1	0.1
Yb			0.5	0.1	0.1	6.2	4.0	3.3	0.6	0.2
Lu			0.1	0.04	—	0.4	0.4	0.5	0.1	0.2
Y	0.1	0.5	3.0	2.4	2.0	60.8	51.4	64.1	10.1	1.3

[a]Baiyunebo iron ore mine, Inner Mongolia
[b]Guangdong/Guangxi Provinces
[c]Western Australia

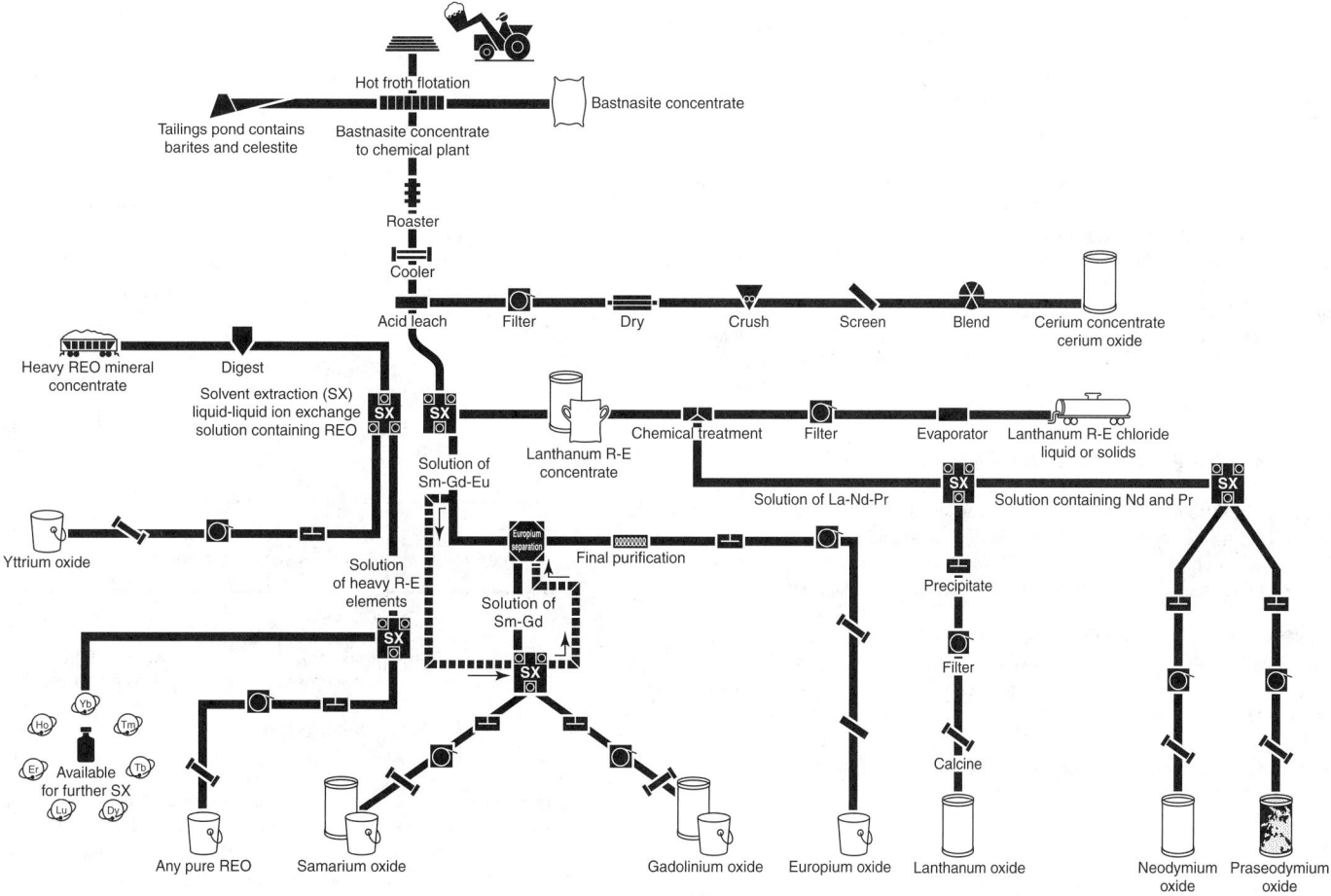

Fig. 3.   Separation and purification of rare-earth elements. REO = rare-earth oxides; SX = solvent extraction.

In the case of the Canadian yttrium-heavy rare earth concentrate, this is leached with HNO_3, causing all rare earths to go into solution. Solvent extraction separates yttrium from the other heavy rare earths, each of which can eventually be separated by further solvent extraction. In the case of xenotime, this is leached with hot H_2SO_4 and separation of yttrium and the heavy rare earths completed in ion-exchange columns. The liquid-liquid organic solvent extraction cycle is complete within 5–10 days and is a continuous process. The resin ion-exchange cycle requires 60–90 days

and is a batch process. Both processes result in pure rare earth oxides and chemicals.

Mischmetal is produced commercially by electrolysis. The usual starting ingredient is the dehydrated rare earth chloride produced from monazite or bastnasite. The mixed rare earth chloride is fused in an iron, graphite, or ceramic crucible with the aid of electrolyte mixtures made up of potassium, barium, sodium, or calcium chlorides. Carbon anodes are immersed in the molten salt. As direct current flows through the cell, molten mischmetal builds up in the bottom of the crucible. This method is also used to prepare lanthanum and cerium metals.

Additional Reading

Anderson, D.L.: "Composition of the Earth," *Science*, **367** (January 20, 1989).

Carter, G.F. and D.E. Paul: "Materials Science and Engineering," ASM International, Materials Park, Ohio, 1991.

Gschneidner, K.A., Jr. and L. Eyring: "Handbook on the Physics and Chemistry of Rare Earths," Vol. **1-28**, Elsevier Science B.V., Amsterdam, 1979–2000.

Gschneidner, K.A., Jr., B.J. Beaudry, and J. Capellen: "Rare Earth Metals," pp. 720–732 in "Metals Handbook, Properties and Selection: Nonferrous Alloys and Special Purpose Materials," 10th Edition, Vol. **2**, ASM International, Materials Park, OH, 1990.

Gschneidner, K.A., Jr.: "Physical Properties of the Rare Earth Earth Elements," pp. 4–112 to 4–121 in "CRC Handbook of Chemistry and Physics 1996–1997," 77th Edition, CRC Press, LLC., Boca Raton, FL, 1997.

Hammond, C.R.: "The Elements," pp 4–1 to 4–34 in "CRC Handbook of Chemistry and Physics, 1996–1997," 77th Edition, CRC Press, LLC., Boca Raton, FL, 1997.

Lewis, R.J. and G.G. Hawley: "Hawley's Condensed Chemical Dictionary," 13th Edition, John Wiley & Sons, Inc., New York, NY, 1999.

Lewis, R.J. and N.I. Sax: "Sax's Dangerous Properties of Industrial Materials," 10th Edition, John Wiley & Sons, Inc., New York, NY, 1999.

Lide, D.R.: "CRC Handbook of Chemistry and Physics 2000–2001," 81st Edition, CRC Press, LLC., Boca Raton, FL, 2000.

Mayers, R.A.: "Handbook of Chemicals Production Processes," The McGraw-Hill Companies, Inc., New York, NY, 1986.

White, R.M.: "Opportunities in Magnetic Materials," *Science*, 229, 11 (1985).

K.A. GSCHNEIDNER, JR., and B. EVANS
Rare-Earth Information Center, Institute for Physical Research and
Technology, Iowa State University, Ames, IA.

RARE-EARTH MAGNETS. See **Magnetism**.

RARE GASES. See **Inert Gases (The)**.

RASH. A skin eruption, the lesions of which may vary in size, location, and color. A rash may result from sensitivity to various drugs (penicillin rash) or various foods (tomatoes, strawberries, etc.), and may disappear upon withdrawal of the irritating substance. Rashes of various kinds also occur with certain diseases, such as measles, chickenpox, scarlet fever, rubella. With recover from the primary disease, the rash will disappear. What may be called a rash in some instances will occur in connection with a number of skin diseases. See also **Dermatitis and Dermatosis**.

RASPBERRY. See **Rose Family**.

RASPBERRY-CANE BORER (*Insecta, Diptera; Pegomya rubivora*). The larva bores in the canes of raspberry plants, ultimately killing them. The adults girdle the tender growth and cause it to wilt. If wilted canes are removed to a few inches below the girdling the development of the larvae is prevented. The red-necked Cane Borer *Agribus ruficollis* (Coleoptera) causes enlargements on swellings of the raspberry or blackberry canes.

RASPBERRY FRUIT-WORM (*Insecta, Coleoptera*). The larva of a small beetle, *Byturus unicolor* in America, and *B. tomentosus* in Europe, which lives on the inside of the fruit on raspberries. It eats the receptacle but is often found in the fruit itself.

RASTER. In television, a predetermined pattern of scanning lines which provides substantially uniform coverage of an area.

RAT. See **Rodentia**.

RAT-BITE FEVER. Two almost identical diseases are known as rat-bite fever. One is caused by a spirochete, *Spirillum minus*; the other is caused by a pleomorphic Gram-negative bacillus, *Streptobacillus moniliformis*. Both may be transmitted by the bite of infected rats, which are the apparent natural reservoir. Other rodents may also be reservoirs. There is severe inflammation around the bite, accompanied by headache, chills and arthralgia, followed by a relapsing type of fever lasting for several weeks when untreated. For treatment, penicillin or tetracycline is effective. The disease is commonest in Japan and India, where it is usually due to the spirillum, as well as in the United States, where, as the so-called Haverhill fever, it may occur rarely in epidemic form, and is due to the streptobacilli.

R. C. V.

RATE-OF-CLIMB INDICATOR. An instrument for installation in aircraft. One type of indicator comprises an enclosed volume of air connected to atmospheric (static) pressure through a constriction. As the altitude changes, the enclosed pressure lags that outside, and the pressure difference is measured in terms of rate of change of altitude (or rate of climb). Because any change of temperature causes a proportional change in pressure of an enclosed volume of air, such as used in the indicator, the container must be a good thermal insulator so as to prevent all but very gradual temperature changes; or proper correction for this must be made. Other corrections must be introduced for both static pressure and temperature where precise rate-of-climb information is required.

RATIO CONTROLLER. A controller or control system that maintains a predetermined ratio between two or more variables. In a ratio control system, two or more controllers are used, each with its own measured variable and output primary signal that is modified by individual ratio settings. Typical of industrial needs for ratio control are: cement kiln speed versus slurry flow control; propane gas versus airflow mixing controls; natural gas flow versus bottled gas flow mixing controls; steam flow versus airflow in boiler control; liquid blending process (very common in the chemical and petrochemical industries); and, of major importance, fuel flow-air flow ratio in a combustion control system.

A fuel-air ratio control system is shown in Fig. 1. The fuel-flow transmitter sets the setpoint of the air-flow controller through the ratio station. As the fuel flow changes, the setpoint of the air controller is changed automatically to a new value so that an exact ratio is maintained between air flow and fuel flow. Thus, no matter how the fuel flow changes, the correct amount of air for optimum combustion conditions is assured. Basically, the ratio station is simply a multiplier with the multiplying factor selected in accordance with the needs of the process. For flexibility, the multiplying factor can be changed manually, even remotely from a control console.

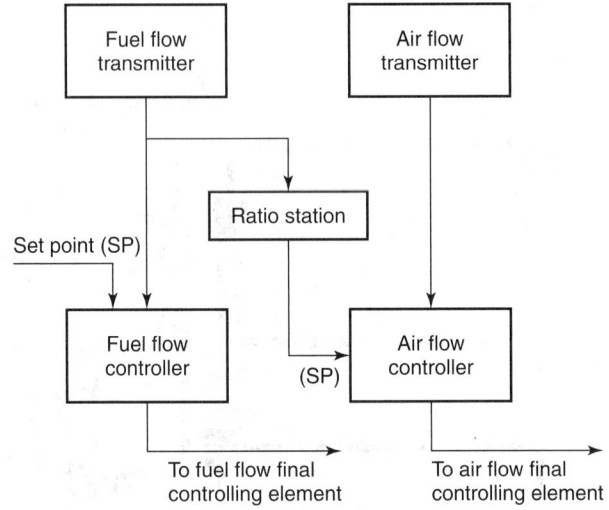

Fig. 1.   Ratio control system for fuel-air ratio.

More complex ratio control systems may include five or ten separate flows all to be proportioned in accordance with a final formula specification. In addition to mixing two or more fluids, ratio control systems find wide application in connection with the proportioning of bulk solids, such as the ingredients of feedstock and cereal; or in ore beneficiation, the blending of raw ores of various mineral contents best suited for subsequent processing.

RATIO DETECTOR. A frequency-modulation discriminator that utilizes the ratio of two intermediate-frequency voltages whose relative magnitudes are a function of frequency, rather than the difference of those voltages as in the case of the Armstrong discriminator circuit.

RATITES. A group of flightless birds (*Aves*) among which the order *Struthioniformes* (the ostrich) is the best known and most important. Other orders include *Rheiformes* (*Rheidae*; Rheas); *Casuariiformes*

(*Casuariidae*; Cassowaries); *Casuariiformes* (*Dromaiidae*; Emus); and *Apterygiformes* (*Apterygidae*; Kiwis). In many of their characteristics, they are more primitive than most other living birds. Therefore, it was once thought that ratites had split off from all other birds at a time when birds had not yet "invented" flight. However, if this were so, many of the structural features of ratites would not be understandable. For example, all of them have a wing skeleton not fundamentally different from that of flying birds. Their wings also still bear flight feathers and coverts, hence are degenerated wings rather than degenerated forelegs, as was the case in the bipedal walking dinosaurs. Therefore the ratites undoubtedly stem from flying ancestors and have evidently lost their ability to fly as their body size increased. This has led to considerable changes in the bones, muscles, and plumage.

Their characteristics include: degenerated breast muscles; a retrogressed keel of the sternum; an almost absent wishbone (furcula); a simplified wing skeleton and musculature; flight and tail feathers which have retrogressed or have been converted to decorative plumes; strong legs; leg bones without air chambers except in the femur; no separation of pterylae; apteria, a loss of feather vanes which means that oiling the plumage is not necessary; and no preen gland. See also **Cassowaries**; **Emu**; **Kiwi**; **Ostrich**; and **Rhea**.

RATO (From rocket-assisted take-off). 1. A take-off in which a rocket or rockets, commonly of the solid-fuel type, are used to provide additional thrust. Hence, RATO bottle, Rato bottle, rato unit, etc., a rocket so used.

2. A RATO bottle or unit; the complete apparatus on an aircraft, comprising rockets, ignition system, etc., for assisted take-off. See also **JATO**.

RATTLESNAKES. See **Snakes**.

RAVEN *(Aves, Passeriformes).* Large black birds with more or less iridescent luster. They are closely related to the common crow. The common raven, *Corvus corax*, is found in Europe, Asia, and North America. One variety, the American raven, extends from Canada to Guatemala and from the Rockies to the Pacific. Another, the northern raven, is found from Alaska to Greenland, southward into the northern tier of states, and in the mountains to Carolina. Two species of white-necked ravens are known, one in the southwestern deserts of the United States and the other in Africa. All of these birds eat carrion, eggs, insects, small animals, and to a limited extent vegetable matter.

The ravens are among the largest of the passeriformes. The common raven is about 25 inches (64 centimeters) in length. General characteristics of ravens include: blue-black glossy coloration; bill is large and strong; tail is wedged-shaped; nests are preferably on cliffs; they have a soaring type of flight something like the hawk and are capable of interesting acrobatics during flight; the eggs are green and are usually five to seven in number. See also **Blackbird**; and **Crow**.

RAWIN SYSTEM. See **Wind and Air Velocity Measurements**.

RAYLEIGH LAW. For small magnetization, the induction may be approximated by

$$B = \mu_0 H + \nu H^2 + \cdots$$

yielding

$$\mu = \mu_0 + \nu H$$

where μ is the normal permeability and μ_0 the initial permeability.

RAYLEIGH NUMBER. The quantity R defined for the fluid-filled space between two parallel horizontal planes as

$$R = \frac{\alpha(\theta_1 - \theta_2)g d^3}{\nu k}$$

where α is the coefficient of thermal expansion of the fluid, $\theta_1 - \theta_2$ is the difference of temperature between the bottom plane and the top plane, g is the acceleration due to gravity, d is the separation of the planes, ν is the kinematic viscosity, k is the thermal conductivity. Convection currents appear only when the Rayleigh number exceeds a critical value. For rigid planes, the critical Rayleigh number is of order 1,700.

A Rayleigh number may be defined for any system of natural (or free) convection and, with the Prandtl number, sets the condition for dynamical

similarity of geometrically similar flows. The Grashof number is also used for the same purpose. See also **Grashof Number**.

RAY TRACING (Optical). It is not practicable to set up completely accurate equations to describe an image in terms of the object and optical surface. However, it is possible to trace a ray from a point on an object through an optical system which has only spherical (or plane) interfaces, with complete accuracy. See Fig. 1.

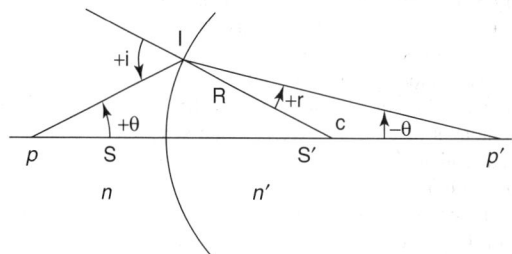

Fig. 1.   Demonstration of optical ray tracing.

For rays from a point on the optical axis, the four conventional equations are

$$\sin i = \frac{R + S}{R} \sin r$$

$$\sin r = \frac{n}{n'} \sin i$$

$$\theta' = r + \theta - i$$

$$S' = R - R \frac{\sin r}{\sin \theta'}$$

More complicated methods are needed for skew rays.

RAZORBILL. See **Waders, Shorebirds, and Gulls**.

RAZOR FISH. See **Wrasses**.

REACTION CURVE. The time response of a component or system is defined as the reaction curve in process control terminology.

REACTION ENGINE. An engine that develops thrust by its reaction to a substance ejected from it; specifically, such an engine that ejects a jet or stream of gases created by the burning of fuel with the engine. Also called *reaction motor*.

A reaction engine operates in accordance with Newton's third law of motion, i.e., to every action (force) there is an equal and opposite reaction. Both rocket engines and jet engines are reaction engines.

REACTION RATE (Chemical). See **Chemical Reaction Rate**.

READOUT. As a verb, to output information from a computer, generally to a display device. As a noun, readout refers to any of the several forms of display that may be used—moving indicators, dials, lights, printed or punched tape or cards, cathode-ray tube displays, etc. Although the term may be used in connection with information that automatically goes into some form of storage without a display interface, readout generally is considered output information that is visible or "read out."

REALGAR. The mineral realgar is a monosulfide of arsenic corresponding to the formula AsS. It is monoclinic, showing short prismatic crystals, or may be in granular or compact masses. It is a soft sectile mineral; hardness, 1.5–2; specific gravity, 3.5; luster, resinous; color, red to orange-yellow; transparent to translucent. Realgar occurs associated with other arsenic minerals and with gold, silver, and lead ores, although not in great quantities. It has been found as a hot-spring deposit and in volcanic sublimations. Realgar has been found in Macedonia, Japan, Switzerland; and in the United States, in Yellowstone National Park, as a hot-spring deposit, and in Utah and Nevada. The name realgar is derived from the Arabic words *rahj al ghar*, which means the powder of the mine.

REAL-TIME COMPUTING. The computer was used first mainly to solve scientific problems and automate record keeping. In these applications, the problem description and solution format are presented to the computer in the form of a program, after which data are typically furnished on which the program operates. The results of program runs are reports distributed for use by people. Other than the time required for execution, there are no real-time constraints on tasks of this nature.

However, some applications require the computer to respond to external events and to perform computations and control functions within specified, often very brief, time limits. Systems of this type are referred to as response- or real-time-oriented systems. The time intervals involved may range from several seconds to several microseconds.

Airline-reservation systems exemplify systems that must respond within a few seconds with certain information. The problem in this case is one of large amounts of data files (reservation schedules), constant changing of the data, and their use and modification by many sources (agents at various locations).

In process control applications, the primary task of the system may be to control information output as a result of sensor information read in. Petroleum refinery and chemical plant control, engine and transmission production, and performance testing of products, gas-distribution control—all typify uses for real-time computing. The tasks range from simple control algorithm calculations to complex process optimization and resource scheduling. See also **Process Control**.

REAMER TEMPERATURE SCALE. A temperature scale in which, under a pressure of 1 atmosphere, the ice point is 0 degrees and the boiling point of water is 80 degrees.

REBOILER. See **Heat Transfer**.

RECALESCENCE. A phenomenon exhibited by iron and some other ferromagnetic metals. If iron is heated white hot and allowed to cool, it will, at a certain temperature, suddenly evolve enough heat to halt the cooling and even produce a momentary heating. This is easily exhibited by stretching an iron wire against the tension of a spring and arranging a lever index to show slight changes in length. The wire is first heated by an electric current. As it cools and contracts, the index will at a certain point give a perceptible jerk, and then resume its steady motion of contraction. The effect is due to an exothermic change in the crystalline structure. The reverse phenomenon, exhibited on heating, is called "decalescence." For cast iron, the recalescence point is a little below 700 °C. Pure iron has two such points, at 780 °C and 880 °C.

A somewhat analogous effect is exhibited by some amorphous solids upon devitrification, which takes place when the temperature becomes high enough for the substance to crystallize. Noncrystalline sodium silicate, for example, has such a transition point near 500 °C, where it suddenly begins to glow.

RECIPROCAL. Given a number or fraction a, its reciprocal is $1/a$. It is a special case of division, where the numerator is unity. Reciprocals are particularly important in vector and matrix algebra, for division is only defined there as multiplication by a reciprocal.

RECIPROCAL VECTOR SYSTEM. From the properties of the quadruple product of vectors (see also **Vector Multiplication**), the following relation is found to hold for any four vectors $\mathbf{r}$, $\mathbf{a}$, $\mathbf{b}$, $\mathbf{c}$:

$$\mathbf{r[abc]} = \mathbf{[rbc]a} + \mathbf{[rca]b} + \mathbf{[rab]c}$$

Which may also be written in the equivalent form

$$\mathbf{r} = \mathbf{r} \cdot \mathbf{a}'\mathbf{a} + \mathbf{r} \cdot \mathbf{b}'\mathbf{b} + \mathbf{r} \cdot \mathbf{c}'\mathbf{c}$$

The system of three vectors

$$\mathbf{a}' = \frac{\mathbf{b} \times \mathbf{c}}{\mathbf{[abc]}}; \quad \mathbf{b}' = \frac{\mathbf{c} \times \mathbf{a}}{\mathbf{[abc]}}; \quad \mathbf{c}' = \frac{\mathbf{a} \times \mathbf{b}}{\mathbf{[abc]}}$$

is reciprocal to the three non-coplanar vectors $\mathbf{a}$, $\mathbf{b}$, $\mathbf{c}$. The unit vectors $\mathbf{i}$, $\mathbf{j}$, $\mathbf{k}$ form a system which is its own reciprocal. Conversely, a system which is its own reciprocal is a set of mutually perpendicular unit vectors, forming either a right-handed or left-handed Cartesian coordinate system.

RECIPROCITY THEOREM (Acoustical). In an acoustic system comprising a fluid medium having bounding surfaces $S_1, S_2, S_3, \ldots$, and subject to no impressed body forces if two distributions of normal velocities v_n' and v_n'' of the bounding surfaces produce pressure fields p' and p'', respectively, throughout the region, then the surface integral of $(p''v_n' - p'v_n'')$ over all the bounding surfaces $S_1, S_2, S_3, \ldots$, vanishes. If the region contains only one simple source, the theorem reduces to the form ascribed to Helmholtz; viz., in a region as described, a simple source at A produces the same sound pressure at another point B as would have been produced at A had the source been located at B.

RECIPROCITY THEOREM (Electric-Network). In an electric network composed of passive bilateral linear impedances, the ratio of an electromotive force introduced in any branch to the current measured in any other branch, called the transfer impedance, is equal in magnitude and phase to the ratio that would be observed if the positions of the electromotive force and the current were interchanged. When altering the location of an electromotive force in a network, the branch into which the electromotive force is to be introduced must be opened, while the branch from which it has been removed must be closed.

RECIPROCITY THEOREM (Electroacoustical). For an electroacoustic transducer satisfying the reciprocity principle, the quotient of the magnitude of the ratio of the open-circuit voltage at the output terminals (or the short-circuit output current) of the transducer, when used as a sound receiver, to the free-field sound pressure referred to an arbitrarily selected reference point on or near the transducer, divided by the magnitude of the ratio of the sound pressure apparent at a distance, d, from the reference point to the current flowing at the transducer input terminals (or the voltage applied at the input terminals), when used as a sound emitter, is a constant called the "reciprocity constant" independent of the type of constructional details of the transducer. The reciprocity constant is given by

$$\left| \frac{M_0}{s_0} \right| = \left| \frac{M_s}{s_s} \right| = \frac{2d}{\rho f} \cdot 10^{-7}$$

where M_0 is the free-field voltage response as a sound receiver, in open-circuit volts per microbar, referred to the arbitrary reference point on or near the transducer; M_s is the free-field current response in short-circuit amperes per microbar, referred to the arbitrary reference point on or near the transducer; s_0 is the sound pressure produced at a distance d centimeters from the arbitrary reference point in microbars per ampere of input current; s_s is the sound pressure produced at a distance d centimeters from the arbitrary reference point in microbars per volt applied at the input terminals; f is the frequency in cycles per second; ρ is the density of the medium in grams per centimeter^3; d is the distance in centimeters from the arbitrary reference point on or near the transducer to the point at which the sound pressure established by the transducer when emitting is evaluated.

RECOGNITION (Pattern). See **Pattern Recognition**.

RECOIL PARTICLE. A particle that has been set into motion by a collision or by a process involving the ejection of another particle. The direction and magnitude of the recoil are determined by the conservation of momentum. Examples are Compton recoil electrons, recoil nuclei in alpha decay, and fission fragments.

RECOMBINATION. The process by which a positive and a negative ion join to form a neutral molecule or other neutral particle. In the literature of atmospheric electricity, this term is applied both to the simple case of the capture of free electrons by positive atomic or molecular ions, and also to the more complex case of the neutralization of a positive small ion by a negative small ion, or a similar (but much more rare) neutralization of large ions.

Recombination is, in general, a process accompanied by emission of radiation. The light emitted from the channel of a lightning stroke is recombination radiation. The much less concentrated recombinations steadily occurring in all parts of the atmosphere where ions are forming and disappearing does not yield observable radiation. The intermediate ions are forming, and disappearing does not yield observable radiation.

RECOMBINATION ENERGY. The energy released as heat or light when two oppositely charged ions join to form a neutral atom or molecule, or two dissociated atoms combine to form a stable molecule.

RECRUDESCENT.   A medical term indicating the return of clinical features of a prior disease. For example, Brill-Zinsser disease is recrudescent epidemic typhus. Recrudescence implies a longer period between incidences (sometimes years) as contrasted with relapse experienced prior to the full recovery from a specific episode.

RECTIFICATION.   See **Distillation**.

RECTIFIERS.   Alternating current is ideal for generating and transporting energy alone because no net electronic or electrolytic charge or material transport is required. In contrast, whenever electrical energy is stored in batteries, energizes vacuum tube amplifiers, or is used for electrochemical separation or particle acceleration, the permanent and irreversible transport of charges is mandatory — hence a direct current is required. Rectifiers provide the physical means that achieve electric rectification, comprising all the elements which connect a complete alternating current circuit to a complete direct current circuit, without being part of either circuit. See Fig. 1. One rectifier may consist of a plurality of rectifier diodes, their mode of interconnection referred to as a rectifier circuit.

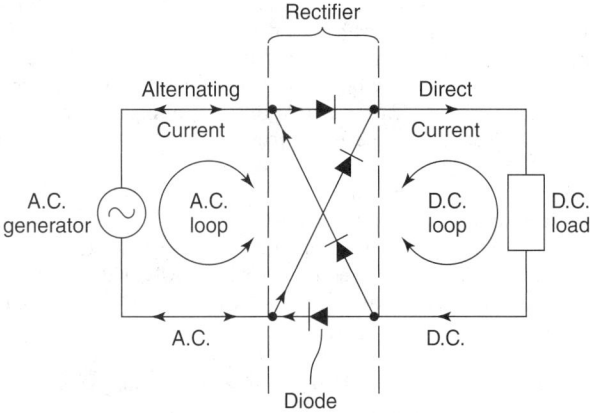

Fig. 1.   Rectifier circuit: single-phase bridge. Four rectifier diodes are connected together to achieve full rectification of both (negative and positive) waves of a single-phase alternating current. The alternating current flowing in the closed loop (A.C. circuit) on the left-hand side is converted into a direct current flowing in the closed loop (D.C. circuit) on the right-hand side.

Rectifier Diodes.   These are unilaterally, conducting component devices with two terminals, similar to resistors because they are passive (not generating electric energy) and nonreactive (not able to store energy), but differing from resistors in that they are nonlinear. Their differential resistance varies over a wide range, depending only upon the direction and magnitude of the current through the device, i.e., they are not time-dependent.

Rectifier Switching.   Rectification implies the concept of switching i.e., introducing a circuit element which has a resistance, which varies instantaneously over such a wide range that it may (mathematically) be considered as a discontinuity.

Controlled Rectifiers.   These are similar to rectifier diodes, except that they have two states of forward conductivity: (1) forward blocking the same resistance characteristics as in the reverse direction, which is the same as in a diode; and (2) forward conducting (same as in a diode). The control element (gate, grid) allows switching from the forward blocking to the forward conducting state. This control is very rapid and requires very little energy. Hence, the rectifier output can be controlled with a high amplification and with high speed. Gas tube controlled rectifiers are thyratrons. Semiconductor controlled rectifiers (SCR) are also known as thyristors. SCR more popularly refers to silicon controlled rectifier.

Electronic Rectifiers.   These devices are based on the Edison effect. Thermally emitted electrons from a hot metal oxide cathode are propelled across a short gap in a high vacuum to a cold metallic anode. The high velocity of thermal electrons and the absence of a gas, generating positive ions, assure ideal conditions for electric rectification, i.e., a flow of pure electrons. Because of their high speed, the high-frequency response is very good. The potential field of the driving voltage accelerates the

electrons — hence the rectifier is essentially a linear resistor in the forward direction and a good insulator in the reverse direction.

1. *Thyratron.* In this rectifier tube containing a low-pressure gas instead of a vacuum, the resistance in the current-carrying stage is very low, but not linear, as in the vacuum tube. This makes the device useless for variable resistance control (amplification), but much more efficient for rectification. A major resistance in the current flow direction is eliminated. This reduction of resistance is caused by impact ionization of the gas — hence the thyratron responds less rapidly than the vacuum tube. The ionized gas is usually a pure element (e.g., mercury, hydrogen, krypton) to avoid chemical reactions between ions and electrodes. Current is carried both as positive ions and negative electrons. Thus, reversing the voltage on the gap results in a definite period of current reversal (ion sweep-out). This severely limits the useful operating frequency. A control grid allows one to select the time at which the thyratron becomes conductive ("firing"). Once conduction is initiated, the grid field is neutralized by the ionized gas and hence the grid control has no effect.

2. *Mercury-Arc Rectifier (Excitron).* This device is similar to the thyratron except the cathode is a pool of liquid mercury. An arc is sustained between the cathode and an exciting anode. Conduction to the main anode is controlled both by anode voltage and a control grid.

3. *Ignitron.* Replacing the heated solid cathode of a diode by a pool of liquid mercury results in an ignitron. To initiate conduction, a localized spot on the liquid surface (cathode spot) is forcibly over-heated (ignited). Selecting the firing time (controlled rectifier) is achieved by energizing an ignitor which creates a cathode spot at a definite time. Ignitors are silicon carbide rods dipping into the mercury. They are subjected to a brief current pulse whereupon a surface arc occurs at the mercury-silicon carbide interface. The liquid is locally driven to a very high temperature by forced electron emission and ion impact, resulting in violent evaporation and ionization of the mercury. Once conducting the ignitron has the properties of a gas-filled tube (thyratron). It also requires the same deionization time and ion sweep-out current. Excess mercury vapor precipitates on the cold walls of the tube, flowing back into the liquid cathode. The ignitron is particularly applicable to very high power. It has the advantages of high efficiency and reliability and small size.

Semiconductor Rectifiers.   Crystallized semimetals, such as selenium, copper oxide, germanium or silicon and some organic compositions can be used to make devices which rectify electric currents. Semiconductors carry current by excess electrons or electron vacancies (carriers) moving in the solid crystal lattice. The transfer of charges is very rapid and driven by very small potential differences. The polarity of a semiconductor is not determined by the material itself, but by relatively few impurity nuclei ("doping") substituted in the crystal. Impurities (compared to the base material) have either an excess (n-type, negative, conducting by "electrons") or a deficiency (p-type, positive, conducting by "holes") of nuclear charges and hence electron shells. In the rigid lattice, the nuclei and normal electron shells are immobile. Excess or defect carriers are freely mobile. The density of these majority carriers is determined by the relative content of impurities in the crystal. The background of the lattice with its mutually neutralizing nuclei and electron shells does not contribute to the conduction or to the distribution of potentials.

Semiconductors conduct both by majority carriers (e.g., holes in p-type) and minority carriers (e.g., electrons in p-type) if such are injected by a junction with the opposite polarity material, e.g., majority carrier electrons coming from n-type material injected into p-type material. Injected minority carriers are ultimately trapped and recombined with majority carriers. However, they transfer a major quantity of charge from one zone to another, depending upon the lifetime of these minority carriers.

Semiconductor rectifier diodes contain one thin, flat wafer consisting of a single crystal, e.g., silicon. The wafer is brazed to metallic electrodes (anode and cathode) on its two opposing flat sides. The rim of the wafer is insulated (by oxidation, insulating resin or fused glass). Within the wafer, a junction is established by heavy doping with p-type impurities on its anode face and n-type impurities on the cathode face. The junction consists of an intermediate, thin, flat zone in which the density of both p-type and n-type dopants is very low.

Applying a forward bias (p-positive, n-negative) to the device injects majority carriers from both zones through the junction into the opposite zone. Attracted by the opposite potential, they effect a total transfer of available charges; i.e., a current flows with only a low driving potential difference. An unlimited number of carriers can flow across the small potential (energy level) barrier between the zones. Carriers are replenished by metal-to-semiconductor brazed joints on both faces of the wafer.

Reversing the bias at the junction (*n*-positive, *p*-negative) reverses the flow of carriers. Majority carriers from both zones are displaced away from the junction, a potential wall is created by depleting the crystal of its mobile carriers. The immovable charges of the lattice bound nuclei which are now uncompensated by the displaced mobile charges, create a high potential wall. A very small reverse current flows, sustained only by the thermally generated minority carriers of both zones which are swept across the junction. When applying a very high potential (e.g., 1,000 volts), these highly accelerated carriers generate more carriers by avalanche multiplication due to impact with the lattice. Above a certain voltage, so many carriers are generated that the reverse characteristic remains at a constant voltage at any current level. Semiconductor diodes have a very low forward voltage drop (e.g., 1 volt) and the forward resistance is not constant, but decreases with increasing current. The reverse current is negligibly small, except at very high voltage where the reverse resistance is negligible.

Varying the semiconductor material (mainly germanium and silicon), the impurity content, the distribution of impurities across the junction, the area of the junction, and its peripheral configuration (planar, mesa, cut wafer) allows for a multitude of possible designs, each preferred for certain applications. Silicon wafers with diffused impurities (e.g., boron and phosphorus) are used for high-power, low-frequency rectifier diodes. Planar junctions on the surface of a solid wafer, in which the impurities are diffused under a layer of protective oxide, yield diodes with good high-frequency and low-noise response. Junctions are also made by recrystallizing silicon from an alloy melt (alloyed junctions) or by epitaxially depositing pure silicon, with measured impurities, from the vapor phase upon a solid wafer.

Semiconductor-controlled rectifiers are similar to transistors (rather than diodes). They consist of four layers of semiconductor material forming three closely adjacent junctions (against two in a transistor). Only three layers are connected to outside electrodes. Reverse and forward blocking characteristics are similar to diode reverse characteristics. Forward conducting characteristics are similar to diode forward characteristics.

Rectifier Circuits. There are numerous arrangements for the circuitry of rectifiers. A half-wave rectifier utilizes only half of the input alternating waveform. A full-wave rectifier utilizes the full waveform. Within these general classes, rectifier circuits are named according to their types of connections. Half-wave types include the star, half-wave wye, zigzag, and fork rectifiers; while full-wave types include the full-wave bridge, full-wave delta, and full-wave wye rectifiers.

RECTILINEAR CHART.   A method of representing change by means of a curve drawn on chart paper having rectilinear, or right-angled uniformly spaced coordinates, known as Cartesian coordinates. In general, the independent variable is plotted along a horizontal, and the dependent variable along a vertical axis.

See also **Coordinate System**.

RECURRENCE TIME.   In stochastic processes, the time elapsing between the point when a system is in a certain state and the first subsequent time when it again attains that state.

REDBIRD.   See Cardinal.

REDBUD.   The eastern redbud (*Cercis canadensis*), a small to medium-size tree sometimes achieving a height of 50 to 60 feet (15–18 meters), occurs in the eastern United States and is particularly prevalent in parts of Appalachia, including eastern Kentucky. The redbud flowers early in the spring; the blossoms are of a rose color. Seed pods which appear later in summer definitely identify the tree with *Leguminosae* (pea family). The western redbud (*C. occidentalis*) occurs on the west coast, notably in parts of California. This is a somewhat smaller tree than its eastern counterpart. The height generally averages about 30 feet (9 meters) or less. The flowers are of a similar rose coloration, with the peak in blooming occurring during late March and early April. The flowers appear considerably before development of the leaves. The western redbud most commonly occurs along the eastern slopes of the north Coast Range; also at lower elevations along the western slopes of the Sierra Nevada range. The tree is often associated with the digger pine (*Pinus sabiniana*) and blue oak (*Quercus douglasii*). Traditionally, the wood of the redbud has been used by Indian tribes in basket making. Incidentally, the derivation of *Cercis* is from the Greek *kerkis* (a weaver's implement). There are at least seven species of *Cercis* known throughout the world, including the Middle East, Asia, and southern Europe (see Table 1 for record redbud trees).

As described by B. Ciesla (*American Forests*, 22–27, April 1981), "One of the species of *Cercis* occurs in the Holy Land. According to legend, this species once had a white blossom. After Judas betrayed Jesus, he hung himself from a redbud. The blossoms blushed pink from shame and have remained so ever since. This ancient legend is the source of another of the redbud's common names, the Judas tree."

RED CLAY.   The most common of deep-sea sediments. A ferruginous clay formed from the alteration products of volcanic ash and other aeolian sediments, including meteoric material. Manganese concretions develop in these muds which are deposited exceedingly slowly and contain little or no organic or calcareous matter, due to the solvent power of the sea water under great pressure.

REDFISH (*Osteichthyes*).   Also known as the *ocean perch*, because it looks much like a perch, the redfish (*Sebastes marinus*) is an important edible fish. The fish attains a length up to about 19.5 inches (50 centimeters). See Fig. 1. Its distribution extends from the White Sea and Greenland to Scotland and western Norway, and in the western Atlantic Ocean, the species is distributed along the cold Labrador Current to the latitude of New York. Redfish are caught in very large quantities and usually are sold in the fillet form. The female produces young which are about 5 millimeters ($\frac{1}{4}$ inch) in length, and which initially live in the open water, feeding on plankton. Once the young are about 6 millimeters long, they begin a bottom-dwelling life, feeding on crustaceans and fishes. Ocean perch or redfish are found mainly on rocky ground at depths of from 300 to 2000 feet (90 to 600 meters). There is also a deepsea species, *Sebastes marinus mantellus*, which inhabits deep channels and grooves. The meat of the deepsea species is not firm and cannot be kept as long as that of the principal species.

Not all marine biologists are in agreement with the present classification of the *Sebastes* group. Because of its wide range, there appear to be some racial variations. Color differences in fish taken from various grounds have been observed and loosely associated with depth. When exploitation of redfish extended into the Gulf of Saint Lawrence, the most striking variation was in eye diameter.

TABLE 1. RECORD REDBUD TREES IN THE UNITED STATES[1]

Specimen	Circumference[2]		Height		Spread		Location
	Inches	Centimeters	Feet	Meters	Feet	Meters	
California redbud (1997) (*Cercis occidentalis*)	74	188	29	8.8	35	10.7	California
Eastern redbud (typ.) (1997) (*Cercis canadensis var. canadensis*)	78	198	42	12.8	40	12.2	Texas
Texas redbud (1992) (*Cercis canadensis var. texensis*)	72	183	30	9.1	33	10.1	Texas

[1] From the "National Register of Big Trees," American Forests (by permission).
[2] At 4.5 feet (1.4 meters).

Fig. 1.   Redfish.

Spawning time for redfish is during the spring months and into mid-summer, varying with region. During the period of spawning, the decks of vessels bringing in heavy catches will be covered with immature fry or larvae, pressed from the parents' bodies. Many of these will live for hours when placed in a bucket of sea water, but survival is unsuccessful for these premature young and tiny fish. The female redfish produces up to 135,000 young each year. The species is ovoviviparous, that is, the offspring are extruded alive and, although they possess some motility, they are at the whim of wind and wave for some time.

Since the mid-1950s, breaded and precooked fish sticks made from rectangular strips of redfish meat have increased in popularity.

See also **Fishes**.

RED HARDNESS.   Most metals, including tool steels, lose much of their hardness at red heat—about 1,000 °F (538 °C) and above. Tool materials such as high-speed tool steel, cemented carbides, and diamond, which retain a considerable part of their hardness at these temperatures are said to have high red hardness or hot hardness.

RED MARROW.   See **Bone**.

RED MUD.   A reddish brown deep-sea mud composed of aeolian terrigenous dust or loss deposited off the seaward end of a large delta or off desert coast lines.

RED MULLET.   See **Mullets**.

REDSHANK.   See **Waders, Shorebirds, and Gulls**.

RED SHIFT.   The displacement of spectral lines toward the red, of particular interest in the spectra of galaxies and quasi-stellar sources. It was first observed (for galaxies) by V.M. Slipher, and was interpreted as a Doppler effect, i.e., as being due to the velocity of recession (in the line of sight) of the galaxies. For relatively low velocities, the ratio of the change in wavelength to the normal wavelength would be $\Delta\lambda/\lambda = v/c$ where v is the velocity of the object and c is the velocity of light. For higher velocities, at which the effects of relativity become important, this expression becomes

$$\frac{\Delta\lambda}{\lambda} = \left(\frac{1 + v/c}{1 - v/c}\right)$$

Except for some nearby galaxies, all galaxies exhibit red shifts, and the velocities calculated by the above relations are proportional to their distances from us, so that Hubble proposed the law

$$d = v_r H$$

where d is the distance of the galaxy, v_r, its velocity of recession, and H is Hubble's constant. This relation, of course, is in line with the original interpretation of the red shifts as due entirely to velocities, and leads directly to the conclusion that the universe is expanding.

Red shifts have been observed for many galaxies since Slipher's early work. Red shift is described further in the entries on **Cosmology** and other astronomical subjects.

RED SNAPPER.   See **Snappers**.

RED SPIDER.   See **Mite**.

REDSTART *(Aves, Passeriformes)*.   1. Birds of Europe, northern Africa, and palaearctic Asia. The common redstart, *Phoenicurus phoenicurus*, also called the firetail, has the tail and rump chestnut above. This and the allied species are related to the thrushes.

2.   The American, *Setophagea ruticilla*, and painted, *S. picta*, redstarts of North America are warblers. Both are variable in color according to age and sex, adult males showing black and white with red or orange markings. The painted redstart extends into Mexico. See also **Warbler**.

REDTOP.   See **Grasses**.

REDUCED MASS.   In treating any two-body problem, the most satisfactory coordinate frame in which the laws of motion may be applied is an inertial system, i.e., a system that is not accelerated with respect to the fixed stars. The center of mass system of two bodies, having masses M and m and acted on only by mutual forces, is such an inertial system. When the equations of motion are transformed to center of mass coordinates, it is found that they are identical with equations in a system having its origin fixed at M if the mass m is replaced by the reduced mass $\mu = Mm/(M + m)$. If $M \gg m$, the reduced mass is closely approximated by m. See also **Center of Mass**.

REDUCING MOTION.   A motion in which a given displacement of rectilinear, rotary, or curvilinear character is converted by the apparatus to a similar motion in which the displacements are at all times proportional to the original motion, but on a smaller scale. The multiplying lever, the large and small pulley, and the inclined plane are a few examples of the many common elements of mechanisms which may be adapted to reducing motions. These are not necessarily exact reducing motions, for frequently very close approximations serve just as well as exactly similar motion. The pantograph is typical of another group of reducing motions, in which the apparatus is of a more specialized character, and not merely some adaptation of a general element of mechanism. See also **Pantograph**.

REDUCTION POTENTIAL.   The potential drop involved in the reduction of a cation to a neutral form, as in the electrolytic deposition of metals; or the reduction to a less highly-charged ion, as in the reduction of ferric to ferrous ions.

REDUNDANCY (Structural).   In a structure of the type of a truss, redundancy refers to the condition in which there are more members than would be needed to produce stability if the joints are considered to be pin-connected. Redundant members may be used for the purpose of producing a more rigid structure than that which would be obtained with just enough members to satisfy the conditions for static equilibrium. A flag pole held in a vertical position by 4 guy wires which are spaced equidistant around the pole is not a redundant system, since the wires are incapable of carrying compression and only 2 wires act at one time. If the guy wires were replaced by stiff members capable of carrying either tension or compression the system would be redundant, as but two are necessary. A truly redundant structure is incapable of analysis by statics alone because the distribution of load between the redundant members and the other members depends upon the elastic properties of the members. The distribution of stress between the stiff members in the above example would depend upon their size and elastic properties. The analysis of such structures can be made by the methods for the solution of indeterminate structures.

The degree of redundancy or indeterminancy (see also **Indeterminate Structure**) of a structure is a number which represents the difference between the number of unknown conditions which must be satisfied and the number of equations of static equilibrium which are applicable.

REDWING.   See **Blackbird**.

REDWOOD (Coast).   Of the family *Taxodiaceae* (swamp cypress family), genus *Sequoia*, the coast redwood (*Sequoia sempervirens*) is not to be confused with the Giant Sequoia (or "Big Tree" or Sierra redwood), which is also of the family *Taxodiaceae*, but of a separate and exclusive genus, *Sequoiadendron*, the full name being *S. giganteum*. See also **Giant Sequoia**.

The coast redwood grows along a 500-mile (805 kilometers) strip of mountainous coast of northern California and southwestern Oregon. See

Fig. 1. Narrow strip of foggy coast line, ranging from mid-California to southwestern Oregon, where coastal redwoods are found. The southernmost trees are found west of the Santa Lucia Mountains a few miles south of Monterey (near Big Sur). The trees follow the Pacific coastline northward to some miles beyond the Oregon border.

Fig. 1. Although it has been transplanted successfully to five continents, the redwood of California grows inland "as far as the fog flows," according to an old saying, or about 30 miles (48 kilometers). The *Sequoia sempervirens* is the tallest tree species. The tallest known living specimen, as of the mid-1970s, was estimated to be 367.8 feet (112.1 meters) high and is located in Redwood National Park, north of Eureka, California. Although not the tallest, an excellent specimen (known as the "Founders' Tree") located in the Founders' Grove, Humboldt Redwoods State Park, California, is shown in Fig. 2.

Of the coast redwoods and the Sierra redwoods only the coastal variety is commercially valuable. Further, all of the Sierra redwoods are located in state and federal reserves. About 1,400,000 acres (566,566 hectares) of coast redwood forest land is privately owned, much of it by the major lumber companies who must practice sound forest management programs to insure a lasting supply. The commercial forest is primarily in the steeply rugged upland areas which produce trees smaller than those in the park-like groves. Groves of the superlative trees, occurring in the rich, alluvial flatlands, have always comprised a relatively small percentage of the total coastal redwood forest. It is estimated that about one-third of the original number of the truly magnificent old trees, existing when logging first began, are preserved in present parks and reserves. The some fifty redwood parks total about 175,000 acres (70,821 hectares), of which nearly half is dense, old-growth redwood. The remainder is in young growth, mixed stands of redwood and fir, or open areas.

During the summer months along the coastal strip, dense fog from the Pacific Ocean keeps the trees dripping with moisture. Precipitation runs as high as 100 inches (2.5 meters) during spring and winter months. However, redwoods are not found without summer fog as well. It is estimated that it requires about 300 years for the coastal redwood to mature. The redwood forests are estimated to have formed several thousand years ago. The "General Sherman" tree is estimated at about 3,500 years of age.

The coastal redwoods grow from seeds or sprouts. Some of the young trees (seedlings) spring up from the tiny seeds that drop from the small redwood cones. Or, unlike most conifers, the redwood has the ability to grow new trees from stump sprouts. When a coastal redwood is harvested for lumber, as many as a dozen sprouts are likely to spring from the parent stump. Roots also may sprout.

Fig. 2. The Founders' Tree, located in the Founders' Grove, Humboldt Redwoods State Park, California. (*Save-the-Redwoods League.*)

Ecology of the Redwood. The redwood forest is a unique example of the plant community. There is no other forest like it anywhere else in the world. The old growth forest with its tall, spire-like and stately redwoods has been termed an "unforgettable sight" by numerous tree authorities. The lush green carpet of ferns, trillium, and other small plants of the undergrowth contribute to the air of fantasy of the redwood forest. Redwoods have a number of interesting species of trees growing with them. Some of the major trees are Douglas fir (*Pseudotsuga menziessi*), Grand fir (*Abies grandis*), western hemlock (*Tsuga heterophylla*), and Sitka spruce (*Picea sitchensis*). These trees, along with the western cedar (*Thuja plicata*), and the Port Orford cedar (*Chamaecyparis lawsoniana*) comprise the softwood, or coniferous associates of redwoods throughout most of the range.

There are a number of hardwood or broadleaf trees associated with the redwoods. The greatest in number are Tanoak (*Lithocarpus densiflora*) and Madrone (*Arbutus menziessi*). Both grow quickly and move in where other trees are cut down. Both sprout from the stump when cut. Also found along the stream beds and in few numbers are red alder (*Alnus oregona*) and big-leaf maple (*Acer macrophyllum*). The leaves of the maple turn brilliant yellow in the fall. Another important broadleaf tree growing with the redwoods is the California laurel (*Umbellularia californica*), also known as Oregon myrtle, pepperwood, and bay tree. Other trees are the western dogwood (*Cornus nuttallii*), Oregon oak (*Quercus garryana*), and black oak (*Quercus kellogii*).

California huckleberry (*Vaccinium ovatum*) and salal (*Gaultheria shallon*) are the two most common shrub associates of redwood. Blue Blossom (*Ceanothus thyrsiflorus*) is also often very common, as well as the beautiful flowering rhododendron (*Rhododendron macrophyllum*), and the western azalea (*Rhododendron occidentale*). There are many flowering herbs and smaller plants, including redwood sorrel, trillium, deer-foot, mountain iris, alum root, and wild ginger. Horsetails also grow here. There are several ferns, the most common being the sword fern. Other ferns include maidenhair, chain fern, bracken, and gold fern.

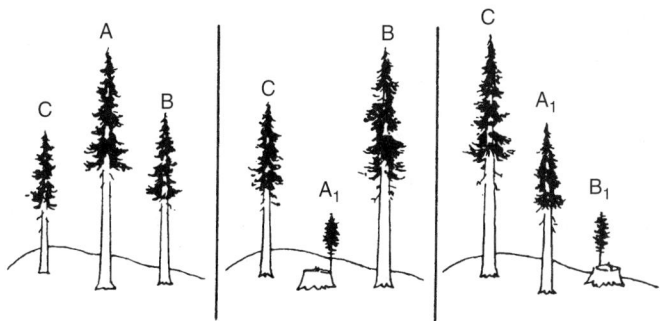

Fig. 3.   Steps in selective cutting.

Selective Cutting. A redwood forest has trees of all ages and sizes. Figure 3 illustrates how selective cutting works in such a stand. The full-grown redwood (A) is flanked by two younger trees (B) and (C), as shown in the diagram at the left. The mature tree is chosen by a forester for cutting. This releases (B) and (C) for faster growth, as indicated in the middle diagram. When (B) reaches its full growth it is harvested, as shown by the diagram at the right, allowing (C) to attain full growth faster and allowing (A₁) to develop into a healthy, full-size tree. The growth cycle is commenced again by the young sprout (B₁) which finds plenty of space in which to grow. New seedlings also spring up in the open spaces.

Timbering Operations. Before felling a redwood, the woods crew first determines the direction of fall to make sure the falling log will not damage trees left standing. The next step is bulldozing a bed of earth to cushion the shock of the falling tree and prevent damage to the trunk. To fell a tree, the logger uses a motor-driven chainsaw to make an undercut on the side of the trunk facing the prepared bed. A second, shallower cut, called the back cut, is made on the opposite side of the trunk above the undercut. Wedges are driven into this notch to topple the tree. The logging crew saws the log into shorter, more easily handled lengths (an operation called bucking); removes branches and peels loose bark. In some instances, mechanical or hydraulic barkers are used at the mill.

In the early days, teams of oxen once dragged redwood logs on skids from the woods to the sawmill. Small, greased logs, laid side by side through the woods, formed the skid road over which the logs were hauled. Tractors are now used. These are equipped with logging arches and skid the logs to the truck landing, being careful not to damage standing trees and young seedlings. Examples of loading devices are (1) The split yoke with cable suspended from two spar trees; and (2) the mobile boom loader. Logs are usually sorted for size and lumber quality at the truck landing before they are hauled to the sawmill.

The logging trucks which have replaced most of the old logging railroads to haul logs from woods to the mill may carry as much as 120,000 pounds (54,432 kilograms) in one load. The trip to the mill may begin on logging roads bulldozed through the forest, and end on a forest highway built and maintained by the lumber company. Length of the haul may be a few miles, or nearly 100 miles (161 kilometers). Most logs will be taken to a sawmill for manufacture into redwood lumber; some may be routed to plywood mills.

Many logs hauled from the forest to the mill are dumped in the millpond for storage. The millpond, however, is gradually being replaced by the cold deck, where the logs are stored in huge stacks. Logs taken from the millpond or cold deck are carried by the log chain to the headrig, the first step in the production of lumber. The headrig bandsaw cuts the logs into large planks, which are called *cants*. These, in turn, are passed through edging and trimming machines and cut to various widths and lengths. Waste material is taken to other plants to be made into lumber by-products, such as fence pickets and fiberboard. See Fig. 4.

From the trimmer, each piece of lumber is conveyed along the green chain where it is graded according to certain wood characteristics. Lumber free of knots and other defects is made into quality siding for home construction. Other grades are sorted out for fencing, decking, and numerous other end-uses. Fork lifts carry the rough lumber from the green chain to an outdoor storage yard, where it is stacked on stickers. These are cross-sticks, which allow air to move freely through the drying stacks. The *air seasoning* process may take 6 to 12 months or more because of the large amounts of moisture in green redwood. A piece of redwood one inch (2.5 cm) thick, one foot wide and 20 feet (6 meters) long may contain 70 pounds (32 kilograms) of moisture.

After air drying, top grades of lumber are ready for *kiln drying*. Carefully controlled temperature, humidity, and air circulation in the kilns permit the removal of additional moisture from the wood. Since all lumber shrinks when dried, kiln drying of quality grades is necessary before redwood can be remanufactured into finished lumber products. Dried lumber is taken to the planing mill where the edges and ends are trimmed and the face planed smooth. It is here that many patterns of redwood siding and paneling are milled. After milling, the lumber is stored until it can be shipped. The wood's natural beauty and durability, as well as its easy workability, make it in strong demand for siding and paneling, general garden uses, farm structures, and numerous commercial and industrial applications. Redwood is shipped worldwide.

See also **Conifers.** Diagrams and information regarding timbering operations are from "The Story of the Redwood Lumber Industry," prepared by California Redwood Association, San Francisco, California. Technical characteristics are based upon data of U.S. Forest Products Laboratory.

Plantings of Redwoods in Other Areas. As of the late 1980s, the coast redwood has been successfully planted outside its California-Oregon native range in at least five other states — Georgia, Hawaii, South Carolina, Virginia, and Washington — as well as in a score or more other countries. As pointed out by J. Kuser (*American Forests*, 30–31, December 1981), "One tree, planted in 1867 at Homestead, South Canterbury (New Zealand) was 115 feet (35 meters) high and 103 in. (262 cm) in diameter by 1967 and contained 1387 cubic feet (39.3 cu meters) of wood ... In England, a grove of redwoods planted in 1858 at Leighton, Montgomeryshire, contained 19,455 cubic feet (551 cu meters) of wood/acre by 1958, making it the heaviest stand of timber in Europe. Several redwoods grow in the New Forest near Lyndhurst tower surrounded by European beeches. One young redwood there reached a height of 80 feet (24 meters) in just 21 years. A stand of redwood at Dartington, Devon, grew to 70 feet (21.3 meters) in 20 years." In the former U.S.S.R., redwood has grown well along the Black Sea coast of the Caucasus Mountains. Selected coast redwoods growing outside the U.S. west coast include:

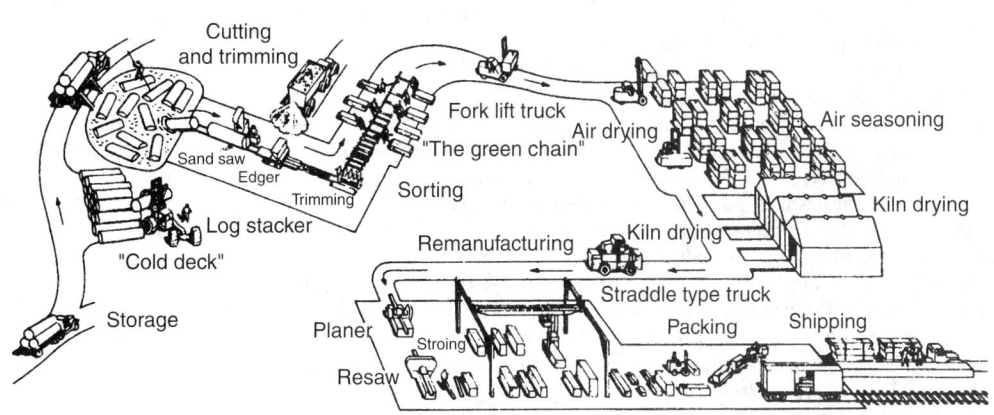

Fig. 4.   Principal operations in readying redwood for market.

TABLE 1. RECORD REDWOOD TREES IN THE UNITED STATES[1]

Specimen	Circumference[2]		Height		Spread		Location
	Inches	Centimeters	Feet	Meters	Feet	Meters	
Coast redwood (1993) (*Sequoia sempervirens*)	867	2202	313	95.4	101	30.8	California
Coast redwood (1998) (*Sequoia sempervirens*)	893	2268	307	93.6	97	29.6	California
Coast redwood (1998) (*Sequoia sempervirens*)	950	2413	321	97.8	80	24.4	California

[1]From the "National Register of Big Trees," American Forests (by permission).
[2]At 4.5 feet (1.4 meters).

Abbeville, South Carolina	~96 feet (29.3 meters) high; ~141 in. (358 cm) in circumference. Planted in 1849.
Williamsburg, Virginia	~81 feet (24.7 meters) high; ~97 in. (246 cm) in circumference. Planted in 1939.
Mexico City (Chapultepec Park)	~100 feet (30.5 meters) high. Planted in 1958.
Santiago, Chile	~60 feet (18.3 meters) high; ~75 in. (191 cm) in circumference. Planted in 1944.
Powerscourt, Wicklow, Ireland	~130 feet (39.6 meters) high; ~167 in. (424 cm) in circumference. Planted in 1866.
Vernets les Bains, France (Pyrenees)	~165 feet (50.3 meters) high. Planted in 1965.
Black Sea Coast, former USSR	~125 feet (38.1 meters) high; ~128 in. (325 cm) in circumference. Planted in 1966.
Rotorua, New Zealand	~180 feet (54.9 meters) high; ~226 in. (574 cm) in circumference. Planted in 1969.
Sourflats, Goudeveld, S. Africa	~124 feet (37.8 meters) high. Planted in 1974.

Genetic Research. Concerted efforts to genetically improve the coast redwood commenced in the early 1960s at the University of California, Berkeley, assisted by a grant from a leading redwood timber producing firm in Eureka, California. Redwood has 66 large chromosomes; all other known wild conifers have between 20 and 24. In further contrast with most conifers, some redwood trees have from one to six accessory chromosomes. To investigate this wealth of redwood chromosomes, early research began with redwood cell and tissue cultures. Searches were made of the redwood forests, with the objective of selecting a single outstanding tree from each of 200 separate areas. A goal was to avoid inbreeding and another to find many superior trees for future breeding. Characteristics such as growth rate, branch size, and stem form are strongly influenced by various components of the environment, as well as by the genetic constitution of the trees. After many years of research, the first selected pedigreed redwood was produced in 1977. This early phase of redwood genetic research is well detailed by W.J. Libby and B.G. McCutchan (*American Forests*, 38–39, August 1978).

Authorities generally agree that redwood is a prime candidate for domestication, since its growth rate is among the highest known (250 to 400 cubic feet/acre; 17.5–28 cubic meters/hectare) per year—this is over five times the national average for fully stocked stands. The process of genetic improvement can continue. Controlled crosses can be made at 30–40 year intervals, and another search begun for offspring for third-generation clones. Genetic theory indicates that this process can continue for at least ten generations. Although clones with unusual properties will be occasionally discovered and released to nurseries for urban and decorative plantings, the main effort concentrates on the production of clones for use as a renewable timber source. Other areas of the world with similar climate may also plant selected redwood clones on a large scale. Work can also proceed on selecting redwood for adaptation to different environmental conditions. As observed by Libby and McCutchan, "If domestication of redwood proceeds as we think it will, most of the original native redwoods and their sprouts will be crowded out and replaced by interplanted selected

trees. Although many fine old-growth groves are now preserved in parks and reserves, they do not constitute an adequate sample of the variability of the native species. Gene conservation is thus our final consideration. It should be possible to preserve for all time an appropriate sample of the predomesticated redwood species, by rooting cuttings from existing trees sampled from redwood's entire native range, and repropagating them at intervals as necessary. Thus, this technique of rooting cuttings, which will hasten the change of redwood from a wild to a domestic species, also can and should be used to preserve the variability redwood once had in the wild."

Indicative of a genetic deficiency or mutation of some kind, albino redwoods are uncommon, but not rare. Numerous sites have been mapped in the coastal mountains between San Francisco and big Sur where albinos have been found. Most albino redwoods, as reported by D.F. Davis (*American Forests*, 40–42, August 1982), are small sprout groups or modest shrublike growths under 5 feet (1.5 meter) tall found near the base of a parent tree or stump. A few individual albinos ranging from 20 to 30 feet (6–9 meters) in height have been spotted. The tallest albino ever recorded stands 80 feet (24 meters) high. Many oldtimers of the redwood region have never seen an albino. Rangers, natural scientists, and foresters who know of one or more albino-redwood sites are often reluctant to mention them to nonprofessional for fear of vandalism. (see Table 1 for record redwood trees)

REDWOOD (Dawn). Of the family *Taxodiaceae* (swamp cypress family), the dawn redwood (*Metasequoia glyptostroboides*) has been known only since 1941. Prior to that time, it was not believed that any taxodiums lived in Asia, but there had been unconfirmed reports of a form of swamp cypress in southern China, near Canton. The fossil remains of a *Metasequoia glyptostroboides* were found in Tokyo in 1941 and, at that time, the conclusion was drawn that the fossil specimen represented an extinct form for the region. However, also in 1941, three additional trees of this genus were located in eastern Szechwan near Chungking in the People's Republic of China. After considerable searching, many such trees were located. The first specimens were collected in 1944 and a thorough search completed in 1946. Seeds were successfully germinated in Britain and the United States and plantings were commenced worldwide in 1948. The tree grows fast, is pointed, and has what are described as rather twiggy branches. Perhaps removed to new habitats, the tree will broaden as do the swamp cypresses. Older trees in their natural habitat have very deeply marked and fluted trunks of a rich-red color.

REENTRANT (Computer System). Usually used with subroutines or procedures, reentrant is a method of program coding which permits the code to be used concurrently by different calling programs. One copy of the code is resident in storage and can be used by several programs simultaneously. This conserves the storage that would be required for multiple copies of the code. The technique is commonly used in multiprogramming situations.

REENTRY VEHICLE. Any payload-carrying vehicle designed to leave the sensible atmosphere and then return through it to Earth. This term applies both to return vehicles from orbital or space payloads and to boostglide vehicles.

REFERENCE ELLIPSOID. An ellipsoid of revolution used as a datum for geodetic measurements.

REFINING (Petroleum).   See **Petroleum**.

REFLECTANCE.   Reflectance is a term used singly and in combination to denote various quantitative expressions for the reflection properties of surfaces for electromagnetic radiation, usually light. When used without qualification, the term reflectance usually refers to radiant reflectance, which is the ratio of the reflected radiant flux to the incident radiant flux.

Luminous reflectance is the ratio of luminous emittance to the illuminance of a reflecting surface.

Spectral reflectance is the radiant reflectance for a specified wavelength of the incident radiation flux.

Specular reflectance is the ratio of the radiance measured by reflection to that measured directly.

REFLECTION.   When an emission, such as radiation or sound, traveling in one medium encounters a different medium, part of it in general passes on and undergoes refraction, while part is reflected. Even water waves exhibit reflection upon meeting an obstacle, and some of the characteristics of the process are conveniently observed by watching surface ripples. In all cases of "regular" reflection, in which the direction of propagation is sharply defined after reflection, the change takes place in accordance with a very simple law; that is, the reflected and incident wave trains travel in directions making equal angles with the normal to the reflecting surface and lie in the same plane with it. These angles are called, respectively, the angle of reflection and the angle of incidence. For normal incidence, both of these angles are zero. Rough surfaces reflect in a multitude of directions, and such reflection is said to be "diffuse." Only part of the emission or of the energy associated with it is reflected; the ratio of that part to the whole incident emission is called the "reflectivity" of the surface.

Various phenomena may accompany reflection under appropriate circumstances. Sometimes there is a change or even a reversal of phase (see also **Vibration**); the reflected wave train may be polarized (see also **Polarized Light**); or the incident and reflected waves may, through their interference, produce stationary waves. If the incident waves are of complex character, the reflection may be selective, due to the difference in reflectivity for the different components.

REFLECTION GRATING.   A diffraction grating ruled on a reflecting surface such as speculum metal or a glass-chromium-aluminum surface.

Rowland was the first to rule them on concave metal surfaces thus eliminating the need of lenses since the surface behaves like a concave mirror. As a typical example, the radius of curvature of the ruled surface might be 21 feet (6.4 meters). If the slit and grating are then placed on this Rowland circle, the spectrum is recorded on photographic plates suitably placed around the circle. This is called the *Paschen-Runge* mounting. Its advantages include the possibility of using more than one slit and source simultaneously, the photography of several orders at the same time, and the lack of moving parts. Its chief disadvantage is the large area needed. In the *Rowland* mount, grating and camera are attached to the opposite ends of a rigid bar of length equal to the radius of curvature of the grating. The grating and camera then slide in tracks perpendicular to each other, with the slit above the junction of the two tracks. The *Abney* mount is similar but grating and camera are fixed while the slit moves. The *Eagle* mount is compact, requiring less space than those previously described. Here the grating is moved toward or away from the camera and, at the same time, is rotated so that it, as well as the photographic plate and the virtual position of the source, lie on the Rowland circle. Another advantage in this case is the decrease in astigmatism, which produces increased intensity at the photographic plate. A completely stigmatic image is produced by the *Wadsworth* mount. Slit and camera are fixed, the grating is attached to a moveable arm of length equal to half the radius of the Rowland circle but light from the slit must be collimated by a spherical or paraboloidal mirror.

REFLECTIVITY.   The fraction of the incident radiant energy reflected by a surface that is exposed to uniform radiation from a source that fills its field of view.

REFLECTOMETER.   Two instruments commonly designated by this term are: 1. In optics, an instrument for measuring the reflectance of reflecting surfaces. 2. In electronics, a directional coupler containing matched calibrated detectors in both arms of the auxiliary line, or a pair of single-detector couplers oriented so as to measure the power flowing in both directions in the main line.

REFLECTOR.   In general, any substance, surface or device which exhibits reflection of radiation or sound. Two specific types of reflectors are: 1. A scattering substance surrounding the core of a nuclear reactor, used for the purpose of reducing the loss of neutrons due to leakage, and therefore, making the dimensions of the reactor smaller. Common reflectors are water, graphite and beryllium. 2. In antenna terminology, a parasitic element located in a direction other than the general direction of the major lobe of radiation. An example is an antenna wire placed behind a dipole to improve its directional characteristics and gain.

REFLUX.   See **Distillation**.

REFRACTION.   The term refraction properly applies to the change of direction which radiation, especially light, or sound experiences on passing obliquely from one medium to another in which its velocity of propagation is different. The physical nature of the effect can be visualized by considering a regiment marching in columns of platoons across a boundary between smooth turf and freshly plowed ground. If the line of march is perpendicular to the boundary, the platoons are simply slowed up and thus crowded more closely together; but if it is oblique, one end of each platoon is retarded sooner than the other, and the file swings around to a direction nearer the normal. See Fig. 1. A train of waves is similarly affected as it passes into a new medium with change of velocity.

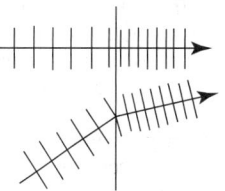

Fig. 1.   Change of wavelength and (in general) of direction of wave train upon entering new medium.

It is easy to show that if i is the angle of incidence and r the angle of refraction at such a boundary (Fig. 2), the refraction is governed by a simple relation known as *Snell's law*:

$$\sin i = n \sin r$$

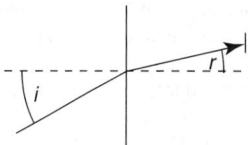

Fig. 2.   Angles of incidence (i) and refraction (r).

in which n has the same value for various angles of incidence and refraction. This constant n, known as the refractive index, depends upon the character of the wave train and of the two media. Physically it represents the ratio of the velocity of the disturbance in the first medium to that in the second. For light passing from one medium to another in which its velocity is greater, so that $n < 1$, we may, for a sufficiently large angle of incidence, encounter the curious phenomenon known as total reflection.

Refraction often occurs in a single medium due to variations in its properties resulting from changes in conditions through the portion of the medium traversed by the radiation or sound. The twinkling of stars is caused by differences in refractive index in the atmosphere resulting from differences in temperature. Sound also exhibits this temperature refraction.

Specific refraction is a relationship between the refractive index of a medium at any definite wavelength and its density, of the form

$$r = \left(\frac{n^2 - 1}{n^2 + 2}\right)\left(\frac{1}{\rho}\right)$$

in which r is the specific refraction of the medium, n is its index of refraction at any definite wavelength, and ρ is its density. The relation does not always give a constant value of r as the density is varied, and hence must be considered as an approximation.

Molar refraction is the product of the specific refraction by the molecular weight. The form of this relationship is

$$R = Mr$$

in which R is the molar refraction, M is the molecular weight, and r is the specific refraction. The direct form of this relationship is

$$R = \left(\frac{n^2 - 1}{n^2 + 2}\right)\left(\frac{M}{\rho}\right)$$

in which R is the molar refraction, n is the index of refraction for any chosen wavelength, M is the molecular weight, and ρ is the density.

An empirical relationship between molar refraction, density, and molar volume, that applies to many liquids over a considerable range of temperatures is that of Eykiman:

$$R = \frac{M(n^2 - 1)}{\rho(n + 0.4)} = \frac{V(n^2 - 1)}{(n + 0.4)}$$

in which R is the molar refraction at a given wavelength, n is the index of refraction at that wavelength, M is the molecular weight, ρ is the density, and V is the molar volume.

Atomic refraction is the product of the specific refraction of an element by its atomic weight.

Standard refraction is the refraction that would occur in an idealized atmosphere in which the refractive index decreases uniformly with height at the rate of 39×10^{-6} per kilometer. Standard refraction may be included in ground wave calculations by use of an effective earth radius of 8.5×10^6 meters, or $\frac{4}{3}$ the geometrical radius of the earth.

REFRACTION (Astronomical). In any type of astronomical observation, the light from the distant object must pass through the atmosphere of the earth and suffer a change of direction known as refraction. The amount of change of direction depends upon two fundamental factors: the relative refractive index of the atmosphere and the angle that the ray from the distant object makes with the normal to the surface of the atmosphere. Since the normal to the atmosphere is the direction of the astronomical zenith, the amount of refraction will depend upon the altitude of the object, being greatest when the altitude is least, or when the object is on the horizon. The effect of refraction is to make the altitude of an object appear greater than it would be if no atmosphere were present.

To calculate the amount of astronomical refraction, the index of refraction of the atmosphere is needed, and, unfortunately, this quantity varies with meteorological conditions. Various theoretical methods for computing the amount of astronomical refraction have been proposed, but none of them are very satisfactory for altitudes of less than 20°. A fair approximation to the true value may be obtained from the expression

$$R = \frac{983B}{460 + T} \cotan h$$

in which B is the reading of the barometer in inches, T is the temperature of the air in degrees Fahrenheit, h is the apparent altitude of the object, and R is the amount of refraction in seconds of arc. More accurate values may be obtained by using refraction Tables such as those published in Bowditch American Practical Navigator. These Tables give the amount of refraction in terms of observed altitude, and various meteorological conditions such as temperature and barometric pressure. This refraction must be subtracted from any observed altitude. In case changes due to refraction in other spherical coordinates than altitude are desired, the astronomical triangle must be solved.

Sudden and irregular changes in astronomical refraction are produced by varying meteorological conditions, and cause effects of twinkling in the stars.

REFRACTIVE EYE SURGERY. Refractive surgery is a general term that refers to any surgery that changes the shape of the cornea or the way the eye focuses light internally. The goal of refractive surgery is to reduce or eliminate nearsightedness (myopia), astigmatism, or farsightedness (hyperopia) and lessen a person's dependence on eyeglasses and contact lenses. There are several refractive procedures being performed today.

For a person to see clearly, light rays must be focused by the cornea and lens to fall precisely on the retina. The retina begins the process of using the optic nerve to convert those light rays into images the brain understands, similar to how a camera takes a picture. The cornea and lens in the eye serve as the camera lens. The retina is like the film. If the image is not in focus, the end result is a blurred image. When this happens, eye care professionals term it a refractive error.

In general, candidates for laser surgery are more than 18 years of age, have not had a significant increase in their prescription in the past 12 months, have a healthy cornea, and have a diagnosed refractive error. People with certain medical conditions such as eye disease or pregnancy may not prove to be good candidates.

It is important to note, however, that despite the successful results in a majority of laser surgery patients, it does not restore "perfect" vision in every case. It can not correct the condition called *presbyopia*; a part of the normal aging process of the eyes that require the use of reading glasses.

To make an informed decision on what type of refractive surgery is best, patients are advised to contact their eye care professional for a thorough evaluation and details on each option.

Types of Refractive Surgery. Popular today and receiving a lot of publicity is laser eye surgery. The Excimer laser, a highly precise type of laser system, has been adapted specifically for surgical use on the human eye. It is now widely used in optical practices throughout the world to correct refractive errors. Today, after more advances and improvements, the laser offers precision with better control and safety.

The numerous types of refractive surgery are described here, in alphabetical order.

Clear Lens Extraction (CLE). The CLE procedure is virtually the same as for the removal of cataracts. It involves removing the eye's natural crystalline lens and replacing it with a plastic prescription lens implant. The only difference is that the natural lens being replaced by the CLE procedure is clear; a cataract lens is cloudy.

Implantable Contact Lens (ICL). With the ICL procedure, a prescription lens resembling a contact lens is surgically inserted between the iris and the natural lens of the eye. This procedure can be used to correct virtually the same refractive vision problems that external contact lenses are used for, but results to date indicate that ICL works best for high degrees of farsightedness and nearsightedness.

Intracorneal Lens Implants. A tiny plastic lens, similar to a contact lens, is implanted inside the cornea within the corneal tissue to correct problems of nearsightedness. This procedure is in the investigation stage, as is a similar procedure for the treatment of presbyopia, the vision problem that results from the loss of flexibility of the natural lens because of the normal aging process.

Intrastromal Corneal Ring Segment (ICRS). The ICRS procedure involves the surgical implantation of two tiny plastic arcs (ring segments) in the peripheral area of the cornea. This serves to flatten the cornea to the degree required to correct a myopic (nearsighted) condition. The thickness of the ring segments determines the degree of flatness. If necessary, the segments can be removed and replaced with larger or smaller implants or removed permanently.

Laser In-Situ Keratomileusis (LASIK). LASIK offers a number of advantages over the other forms of refractive surgery techniques because it is performed using corneal tissue to protect the laser-flattened part of the eye. To undergo LASIK surgery, the patient's eye is numbed with eye drops, and an eyelid holder is placed between eyelids to prevent blinking. The laser is computer-controlled and programmed by an eye care professional to customize the surgery to a patient's own unique corneal shape and refractive error. Next, the surgeon uses a specialized instrument called a *microkeratome* to make a protective flap in the cornea which is folded out of the way to expose the underlying corneal tissue. Patients sometimes feel pressure at this point. The laser then reshapes the cornea, and the protective flap is folded back in place where it bonds and seals itself naturally without the need for stitches.

As a result, there is less surface to heal, less risk of postsurgery discomfort, and less need for postoperative medications. Vision returns faster, usually within the same day. LASIK surgery is highly effective. Most LASIK patients are able to pass a driver's license test without corrective lenses.

Laser Thermal Keratoplasty (LTK). The LTK procedure uses a Holmium laser to shrink the peripheral area of the cornea in order to make the shape of the cornea steeper and correct mild to moderate cases of farsightedness. It is also being used to a limited degree to treat some cases of astigmatism.

Photorefractive Keratectomy (PRK). PRK is the first laser corneal refractive surgery. An outpatient corneal surgery, PRK can reduce or correct mild to moderate myopia. A laser is used to precisely reshape the cornea with intense, focused light. No incisions are made in laser surgery. The PRK laser is computer-controlled and programmed by an eye care professional to customize the surgery to a patient's own unique corneal shape and refractive error. The laser light flattens the front surface of the cornea by removing microthin layers of tissue. Removing the layers, however, disrupts the surface and consequently, patients must normally wear a protective contact lens for a few days following surgery to help heal the surface. PRK is known to produce excellent results. The speed of recovery is a little slower than the recovery period following LASIK procedure.

Most of the people who have had PRK report they no longer need to wear glasses or contacts. A great portion of patients who had PRK can see 20/20 or better without corrective lenses. Ninety-five percent can see at least 20/40, which is the measurement required for a driver's license.

Radial Keratotomy (RK). Radial Keratotomy consists of a surgeon making small incisions called *keratotomies*, in the cornea to change its shape. RK surgery is an outpatient procedure, which can usually be completed in approximately 30 minutes.

Patients receive medication to help them relax and are given eye drops to numb the eye. First, the surgeon marks the part of the cornea a person actually sees through, to avoid making incisions in this area. Next, the surgeon measures the thickness of the cornea to determine how deep to make the incisions. Last, the surgeon makes the incisions to help flatten the cornea and reduce or eliminate the refractive error.

Patients may feel some pain following RK surgery, but are given eye drops and medications to help. For some days, patients' eyes may feel gritty, sensitive to light, and appear red. Antibiotic eye drops help guard against infection. In comparison to other refractive surgery being done today, the cornea after RK surgery heals more slowly because of the small, but sometimes deep, incisions made.

Though rare, RK surgery may cause cataract development, pain, infection, rupture of the incisions from eye injury or trauma, and vision loss.

Vision Rx, Inc., Elmsford, NY.

REFRACTIVE INDEX. The phase velocity of radiation in free space divided by the phase velocity of the same radiation in a specified medium. Because of the Snell law (see also **Refraction**) the refractive index may also be defined as the ratio of the sine of the angle of incidence to the sine of the angle of refraction.

The absolute index for all ordinary transparent substances is greater than 1 (see Table 1); but there are some special cases (X-rays and light in metal films, which are discussed below) for which the index of refraction is less than unity. Since the absolute index for air exceeds unity by less than 0.0003, the relative indices for solids and liquids in air are very nearly equal to their absolute indices. It should be noted that since the refractive index varies with the wavelength, any exact statement of its value must specify the wavelength to which it refers; in Tables it is usually given for sodium light of frequency 5,893A. See also **Dispersion (Radiation)**.

TABLE 1. ABSOLUTE INDEX

Substance	Absolute Index
Air	1.0002926
Bromine	1.661
Carbon dioxide	1.00045
Diamond	2.419
Glass	1.5 to 1.9
Glycerine	1.4729
Helium	1.000036
Ice	1.31
Rock salt	1.516
Water (20 °C)	1.333

Various relationships have been used to express the refractive index. Thus there are several semi-empirical relationships expressing the refractive index of a medium as a function of wavelength:

$$n^2 = 1 + \frac{A_1 \lambda^2}{\lambda^2 - \lambda_1^2}$$

where A_1 is a constant characteristic of the material and λ_1 is an idealized absorption wavelength of the medium. When $\lambda_1 \ll \lambda$, this reduces to

$$n^2 = 1 + A_1 + \frac{A_1 \lambda_1^2}{\lambda^2}$$

or

$$n = A + \frac{B}{\lambda^2} \quad \text{(Cauchy Formula)}$$

A better approximation is

$$n = A + \frac{B}{\lambda^2} + \frac{C}{\lambda^4} + \cdots$$

For equations for specific and molar refraction, see also **Refraction**.

As a consequence of his electromagnetic theory of light, Clerk Maxwell obtained the relation between the dielectric constant of a medium and its refractive index n:

$$\varepsilon = n^2$$

This relation holds only under rather restrictive conditions, such as measurement with light of long wavelength, absence of permanent dipoles in the substance, etc.

For strongly absorbing media, such as metals, the customary refractive index must be replaced by the complex refraction index $n(1 - i\kappa)$ where κ is called the absorption index. Then the reflectivity for normal incidence is

$$R = \frac{(n-1)^2 + n^2 \kappa^2}{(n+1)^2 - n^2 \kappa^2}$$

The refractive index for metals varies over a much larger range than for conventional dielectrics, e.g., sodium at $\lambda = 0.546$ micrometer, $n = 0.052$, while for silicon at $\lambda = 0.589$ micrometer, $n = 4.24$. The refractive index is sometimes defined as the relative dielectric constant of a medium, $\sqrt{\varepsilon/\varepsilon_0}$. This expression is invariably a function of the wavelength of the radiation.

REFRACTIVITY. Two uses of this term are: (1) In general, the property of refraction, or a quantitative relationship by which it is expressed, which is commonly some function of the index of refraction. (2) The quantity $(n - 1)$, which enters many optical formulas, is sometimes called "refractivity." Here, n is the refractive index.

REFRACTOMETERS. Several types of instruments, called refractometers, have been devised for measuring the refractive index of any substance. See also **Photometers**. Special forms are used for solids, for liquids, and for gases. Solid and liquid refractometers usually depend upon the principle of total reflection and the fact that the sine of the critical angle is equal to the refractive index for light passing from the more to the less refractive medium. The critical angle is what is measured, or deduced from other measured angles.

Suppose that a specimen of the solid or liquid to be tested is brought into optical contact with one face of a glass prism (or "block") of known, higher refractive index and known angle, and that a slightly convergent pencil of light, entering the test substance, is directed at grazing incidence upon the interface between it and the prism. Those rays incident at less than 90° to the normal of the interface enter the prism; the others do not, and the boundary between is sharply defined. The resulting half-pencil traverses the prism and emerges from the other face where the direction of its cutoff edge can be observed (Fig. 1). The angle between the cutoff boundary of the pencil and the first prism face, inside the prism, is the critical angle, and can be easily calculated from the observations and the known data. This is the principle of the Pulfrich refractometer.

Another form, due to Abbe, is used for liquids. A film of the liquid is enclosed between two similar glass prisms (Fig. 2), and the total reflection at the interface observed. Any spectrometer, with a pair of good prisms (preferably right-angled) mounted on the prism table, can be used in this way.

Rayleigh utilized an interference method for measuring the indices of gases. Using a collimator to render the rays parallel (Fig. 3), the

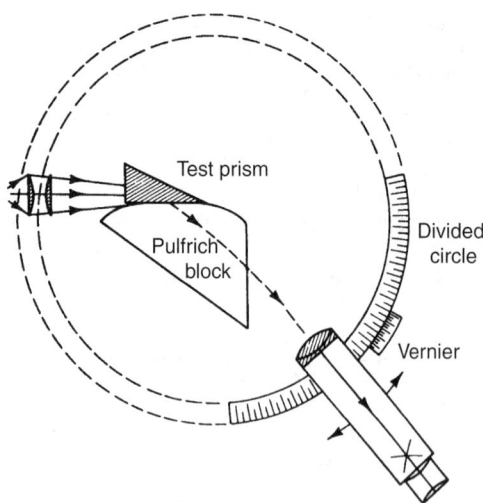

Fig. 1.   Pulfrich refractometer.

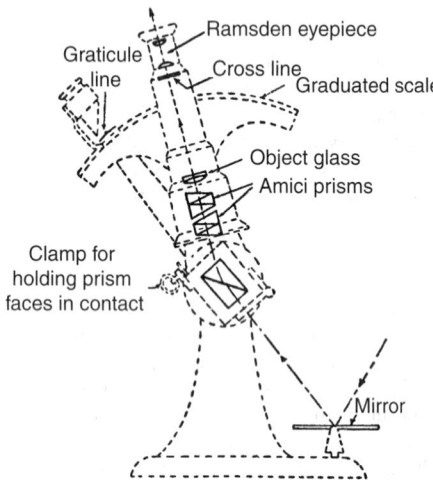

Fig. 2.   Optical system of Abbe refractometer.

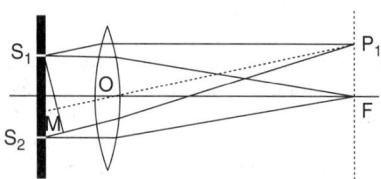

Fig. 3.   When a double slit S_1S_2, illuminated by light from a distant narrow source, is placed in front of the objective of a telescope, interference fringes are observed in the focal plane P_1F.

stream of light entering one of the slits of the apparatus for Young's experiment is passed through a tube of the gas to be tested. The resulting retardation in phase causes a shift of the interference fringes, the amount of which gives the retardation and hence the refractive index of the gas relative to the air outside the tube. See also **Refraction**; and **Refractive Index**.

A chemical analytical technique is based upon measurement of refractive index. The velocity of light in a material, and therefore the refractive index, depends upon several physical properties of the sample. Theoretical studies have indicated that the refractive index is related to the number, charge, and mass of vibrating particles in the material through which the light is passing. Further, it has been possible to relate refractive index to density and molecular weight for classes of compounds that have a relatively constant number of vibrating particles per unit weight. The number of vibrating particles in a compound is determined by the atoms in the structure and by the type of electronic bonding. Correlations of this sort have been particularly successful for the analysis of hydrocarbon mixtures.

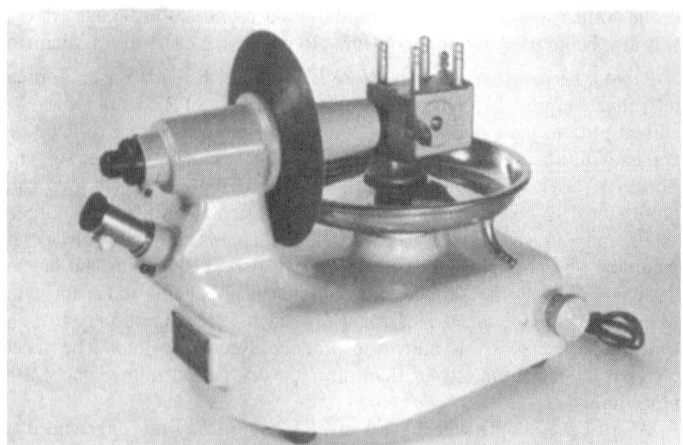

Fig. 4.   Abbe refractometer equipped with sodium are for studies of turbid liquids. (*Gaertner Scientific Corp.*)

Several techniques have been developed and applied in the petroleum industry. See Fig. 4. See also **Analysis (Chemical)**.

REFRIGERATION.   A process of cooling or freezing a substance to a temperature lower than that of its surroundings and maintaining that substance in a cold state. Refrigeration can be accomplished by arranging heat transfer from a warm body to a colder body through processes such as convection or thermal conduction. Other, more erotic methods include the exploitation of thermoelectric properties of semiconductors, the magnetothermoelectric effects in semimetals, or the diffusion of ^{3}He atoms across the interface between distinct phases of liquid helium having high and low concentrations of ^{3}He in ^{4}He, among other methods. See also **Thermoelectric Cooling**.

Mainly because of two factors, (1) energy conservation and cost and (2) the desirability of phasing out the use of chlorofluorocarbons (CFCs) in an attempt to slow the deterioration of Earth's ozone layer, refrigeration technology is under intense scrutiny as of the early 1990s.

In an effort to accelerate refrigeration technology, a group of U.S. electric utilities established in 1992 the "Super Efficient Refrigerator Program" (SERP), which will award approximately $30 million to that U.S. manufacturer who first succeeds in producing the most efficient home refrigerator. Targets that must be met include:

1. Consumes as little as 400 kWh per year, compared with 1993 federal efficiency standards of 704 kWh for comparable refrigerators.
2. The refrigerator will be moderately priced.
3. The system will require no chlorofluorocarbons.

In this article, after a review of the common contemporary means of refrigeration, some new or revised innovative methods for refrigeration and cooling systems will be described. The chemistry of CFCs is described in article on **Polar Research**.

Contemporary Refrigeration Systems

Most commercial refrigeration systems operate on a cyclic basis. A refrigerator operating in this manner may be considered a heat pump, for it continuously extracts heat from a low-temperature region and delivers it to a high-temperature region. It is rated by its *coefficient of performance*, which may be defined as the ratio of the heat removed from the cold region per unit of time to the net input power for operating the device, in symbols $K = Q_t/P$. Vapor-absorption and thermoelectric refrigeration systems have lower coefficients of performance than vapor-compression refrigerators, but they have other characteristics that are superior, such as quietness of operation and compactness. See also **Heat Pump**.

A *vapor-compression refrigerator* consists of a compressor, a condenser, a storage tank, a throttling valve, and an evaporator connected by suitable conduits with intake and outlet valves. See Fig. 1. The refrigerant is a liquid that partly vaporizes and cools as it passes through the throttling valve. Among the common refrigerants are ammonia, sulfur dioxide, and various halides of methane and ethane.

CFCs have been used widely in all kinds of cooling systems, including automobile air-conditioning systems. But for their atmospheric pollution problems, they are excellent refrigerants.

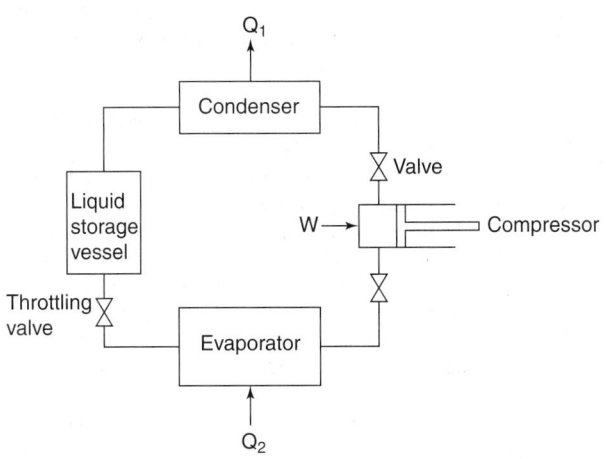

Fig. 1.   Vapor-compression refrigeration system.

Nearly constant pressures are maintained on either side of the throttling valve by means of the compressor. The mixed liquid and vapor entering the evaporator is colder than the near-surround; it absorbs heat from the interior of the refrigerator box or cold room and completely vaporizes. The vapor is then forced into the compressor, where its temperature and pressure increase as the result of compression. The compressed vapor then pours into the condenser, where it cools down and liquifies as the heat is transferred to cold air, water, or other fluid medium in the cooling coils. Comparative tests have shown that the coefficient of performance of vapor-compression refrigerators depends very little on the nature of the refrigerant. Because of mechanical inefficiencies, the actual value may be well below an ideal value, which ordinarily lies between 2 and 3.

In a *vapor-absorption refrigeration system*, there are no moving parts. The added energy comes from a gas or liquid fuel burner or from an electrical heater, as *heat*, rather than from a compressor, as *work*. See Fig. 2. The refrigerant used in this example is ammonia gas, which is liberated from a water solution and transported from one region to another by the aid of hydrogen. The total pressure throughout the system is constant and therefore no valves are needed.

Heat from the external source is supplied to the generator, where a mixture of ammonia and water vapor with drops of ammoniated water is raised to the separator in the same manner as water is raised to the coffee in a percolator. Ammonia vapor escapes from the liquid in the separator and rises to the condenser, where it cools and liquefies. Before the liquefied ammonia enters the evaporator, hydrogen, rising from the absorber, mixes with it and aids in the evaporation process. Finally, the mixture of hydrogen and ammonia vapor enters the absorber, where water from the separator dissolves the ammonia. The ammonia water returns to the generator to complete the cycle. In this cycle, heat enters the system not only at the generator, but also at the evaporator, and heat leaves the system at both the condenser and the absorber to enter the atmosphere by means of radiating fins.

No external work is done, and the change in internal energy of the refrigerant during a complete cycle is zero. The total heat $Q_a + Q_c$ released to the atmosphere per unit of time by the absorber and the condenser equals the total heat $Q_g + Q_c$ absorbed per unit of time from the heater at the generator and from the cold box at the evaporator. Thus $Q_e = Q_a + Q_c - Q_g$, and therefore, the coefficient of performance is $K = Q_e/Q_g = [(Q_a + Q_c)/Q_g] - 1$.

The vapor-absorption refrigerator is free from intermittent noises, but it requires a continuous supply of heat. Once very popular for households, refrigeration systems of this type are now most frequently found in camping facilities and some rural areas where commercial electric power may not be easily available.

In a *dilution refrigeration system*, the properties of helium are used advantageously for attaining very low temperatures. Below a temperature of 0.87 K, liquid mixtures of ^{3}He and ^{4}He at certain concentrations separate into two distinct phases. One is a concentrated (^{3}He-rich) phase floating on the other, denser (^{4}He-rich) phase with a visible interface between them. The concentrations of ^{3}He in the two phases are functions of temperature, approaching 100% in the concentrated phase and about 6% in the dilute phase at 0 K. The transfer of ^{3}He atoms from the concentrated to the dilute phase, like an evaporation process, entails a latent heat, an increase in entropy, and a lowering of temperature. The main features of a recirculating dilution refrigeration system are shown in Fig. 3. The pump forces helium vapor (primarily ^{3}He) from the still into the condenser, where it is liquefied at a temperature near 1 K in a bath of rapidly evaporating ^{4}He, through a flow controller that consists of a narrow tube of suitable diameter to obtain an optimum rate of flow, and then through the still

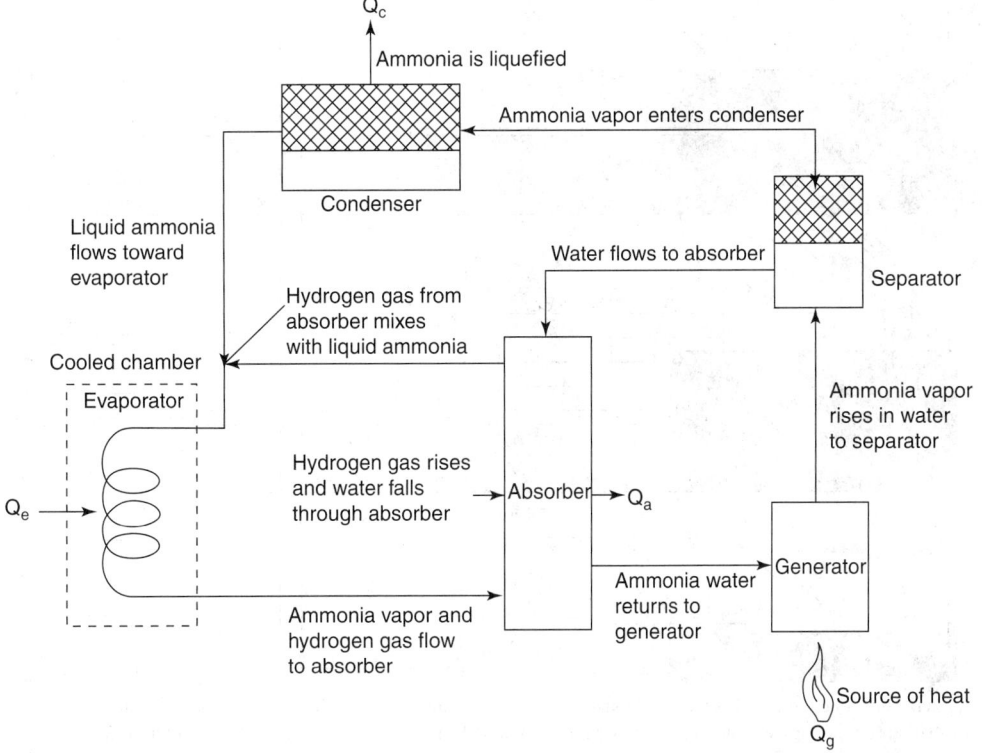

Fig. 2.   Vapor-absorption refrigeration system.

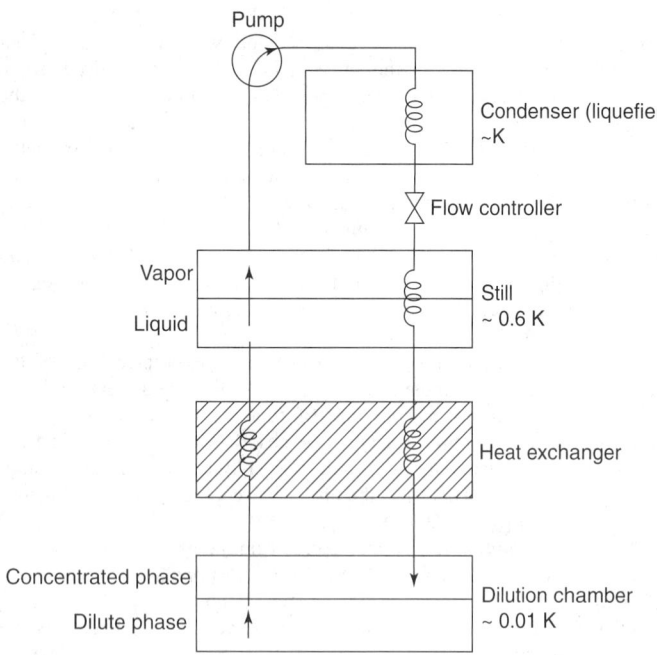

Fig. 3.   Dilution refrigeration system.

where its temperature is further reduced to about 0.6 K. The liquefied ^{3}He next passes through a heat exchanger so as to reduce its temperature to nearly that of the dilution chamber, by giving up thermal energy to the counter-flowing dilute phase, before entering the concentrated phase therein.

The diffusion of ^{3}He atoms from the concentrated into the dilute phase within this chamber can produce steady temperatures of very low values (0.01 K or less). Liquid ^{3}He from the dilute phase then passes through the heat exchanger to the still, where it is warmed to transform the liquid to the vapor phase that goes to the pump, thus completing the cycle. Modified versions of this system have been constructed, sometimes with an added

single-cycle process for temporarily producing temperatures lower than previously mentioned. The low-temperature limit in any system of this type is governed largely by two important sources of inefficiency that cannot be completely eliminated — heat leakage, especially severe because of the extreme range of temperatures, and recirculation of some ^{4}He with ^{3}He.

See also **Helium**; and **Thermodynamics**.

Cool-Storage Concept. In a majority of buildings and nearly all residences, automobiles, and other forms of transportation, comfort cooling systems operate on command — that is, the air conditioner is "on" or "off" in obedience to the thermostat. Thus, the air conditioner starts and stops a number of times during a 24-hour period. Studies have shown that the performance of cooling equipment operates at highest efficiency when used continuously. It has been estimated by authorities that about 25% of the total power consumption of an air-conditioning unit can be attributed to discontinuous operation.

Such inefficient operation essentially can be alleviated by converting to a "cool-storage" system. In this concept, a large pool of cooling medium is created during off-peak hours. The medium is stored in fluid or semi-fluid (slushy) form in a holding tank, which becomes the source of very cold water that can be circulated through a structure. There are definite parallels with circulating hot water or steam heat.

The principle of cool storage was used many years ago when motion picture theaters introduced air-conditioning to the general public. Precooling of the theater prior to opening and during off-peak load hours saved on electricity costs and also cut down on the investment in air-conditioning equipment.

In establishing a cool storage system, there is a choice of cooling media that may be used: (1) cold water just above the freezing point; (2) ice (commonly used with blowers in theaters, shops, and railway cars in the earlier days of air conditioning); (3) so-called "slippery ice," which is comparable to the mixtures containing calcium magnesium acetate used for deicing aircraft; and (4) eutectic salt mixtures. Slippery ice has the desirable property of a flowing slush that does not cling to metal parts.

Many years ago, groups of owners of large adjacent buildings would buy their steam from a district steam plant for heating purposes. This same concept also can be applied to a district cooling system. See Figs. 4 and 5.

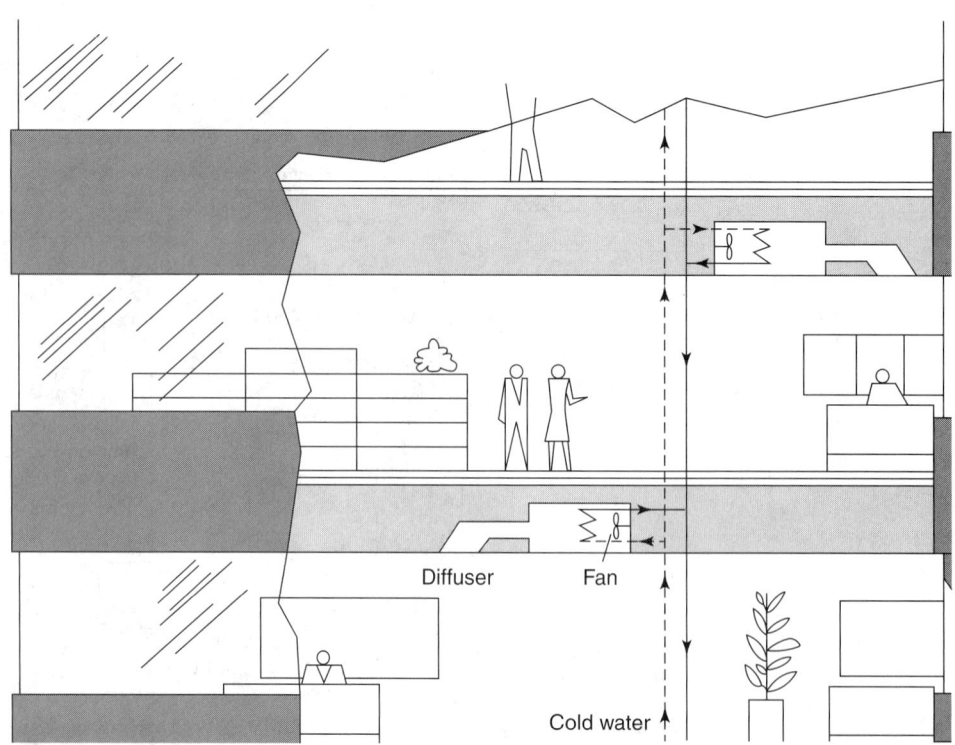

Fig. 4.   The cool storage system shown here differs from conventional air conditioners mainly by the inclusion of a storage tank. The tank contains a thermal medium (water, ice, or eutectic salt) that stores cooling generated by the refrigeration unit. When cooling is required, a water solution from the storage tank is circulated in a pipe system that runs throughout the building. The storage capacity effectively decouples the refrigeration process from the building load, allowing building operators to generate cooling during "off-peak" hours of the local electric utility at a time when rates are lower. (*Source: Electric Power Research Institute.*)

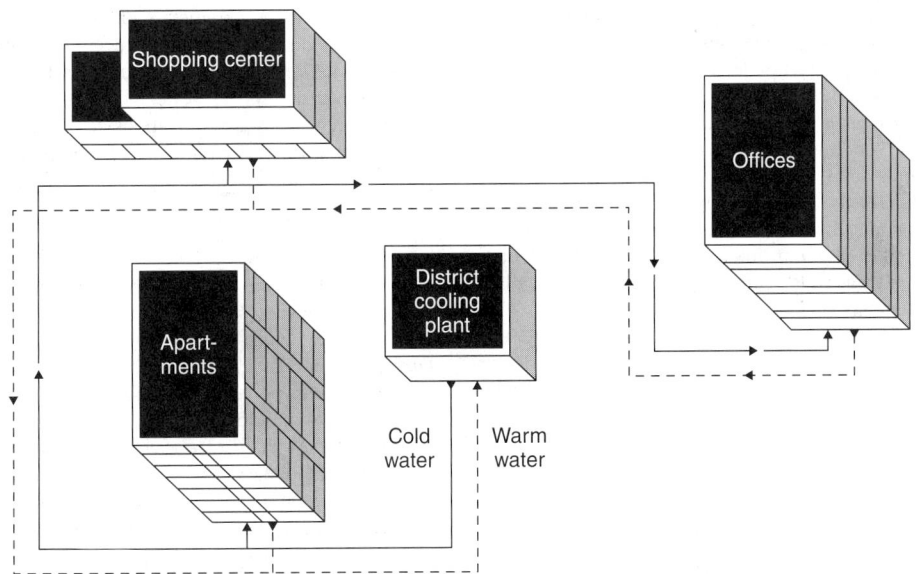

Fig. 5.   District cooling. The basic components of cool storage systems also exist in district cooling systems, which use a central plant to cool nearby buildings. Although some systems of this kind are already in use in the United States, they are much more common in Europe. Some authorities believe that district cooling would be a profitable business for many utilities because they could sell cooling while tapping generation capacity that typically goes unused in the off-peak hours. (*Electric Power Research Institute.*)

Magnetic Refrigeration

The basic physical principle behind magnetic refrigeration (or cooling) is the ability of a ferromagnetic material near its magnetic ordering temperature (called the Curie temperature) to heat-up when magnetized (i.e. placed in a magnetic field), and to cool-off when demagnetized (i.e. the magnetic field is removed). This phenomenon is known as the magnetocaloric effect, which was discovered in 1881. It was first utilized in the mid-1930's to reach temperature below 1 K by cooling a paramagnetic salt [$Gd_2(SO_4)_3 \cdot 7H_2O$] in a large magnetic field to as low a temperature as possible by pumping on liquid He (about 1.5 K), and when salt was thermally isolated and the magnetic field removed the sample cooled to well below 1 K due to the magnetocaloric effect. This one step process is known as adiabatic demagnetization, and it is still used today to reach temperatures in the microkelvin range (the record stands at 0.000025 K).

In order to make magnetic refrigeration practical one must use a continuous cyclic process in which heat must be rejected when the ferromagnet experiences a magnetic field increase (just as the heat is rejected during the compression of a gas in a conventional gas cycle refrigerator). The cooling is achieved by removing the magnetic field experienced by the magnetic refrigerant (this is equivalent to the gas expansion step in the gas cycle refrigerator). One of the first continuously operating magnetic refrigerators, which was used to cool from 1 to 0.2 K, was built and tested in the mid-1950's. Over the next 40 years progress was slow, but in 1997 Zimm and co-authors announced the successful testing of a proof-of-principle continuous near room temperature magnetic refrigerator, which showed for the first time that magnetic refrigeration is a viable technology and competitive with vapor compression cooling. They used gadolinium metal as the magnetic refrigerant and water as the heat transfer medium and were able to cool down to almost 0 °C (273 K), while the heat was rejected at a temperature of 38 °C from the hot heat exchanger. They achieved a maximum coefficient of performance, COP, (the cooling power divided by the power required to operate the apparatus) of 16 using a maximum field change of 0 to 5 T at a frequency of 0.17 Hz. For smaller field changes the COP was less, about 4 for a 0 to 1.5 T magnetic field change. Gas compression refrigeration has COPs ranging from 2 to 4.

Since magnetic refrigeration uses a non-toxic solid refrigerant, it is an environmentally friendly technology because it replaces the commonly used gaseous refrigerants such as ozone-depleting chemicals (chloro-fluorocarbons), hazardous chemicals (ammonia), and greenhouse gases (hydro-chloro-fluorocarbons and hydro-fluorocarbons). Furthermore, since magnetic refrigeration is expected to be 20 to 30% more efficient than conventional cooling systems this will reduce the amount of energy consumed and in turn reduce the amount of CO_2 (also a greenhouse gas) released into the atmosphere.

Magnetic refrigeration and cooing is expected to be used in household air conditioning and refrigerators/freezers, automotive and aircraft climate control systems, large scale building air conditioning, supermarket chillers, frozen food processing plants, liquefaction of natural gas (methane) and hydrogen gas.

Additional Reading

Althouse, A.D., C.H. Turnquist, and A.F. Bracciano: "Modern Refrigeration and Air Conditioning," Goodheart-Willcox Publisher, Tinley Park, IL, 2000.

Bounds, T.W. Ellis, and B.T. Kilbourn, eds, The Minerals, Metals & Materials Society, Warrendale, PA, 1997.

Gschneidner, K.A., Jr. and V.K. Pecharsky, "*Magnetic Refrigeration*," pp. 209–221 in "Rare Earths: Science, Technology and Applications III," R.G. Bautista, CO.

Kreith, F., Z. Lavan, and P. Norton: "Air Conditioning and Refrigeration Engineering," CRC Press LLC., Boca Raton, FL, 1999.

Pecharsky, V.K. and K.A. Gschneidner, Jr. "Magnetocaloric Effect and Magnetic Refrigeration," *J. Magn. Magn. Mater.* **200**, 44–56 (1999).

Stoecker, W.F.: "Industrial Refrigeration Handbook," McGraw-Hill Professional Book Group, New York, NY, 1998.

Tishin, A.M. "Magnetocalorid Effect in the Vicinity of Phase Transitions," pp. 395–524 in "Handbook of Magnetic Materials," vol. 12, Elsevier Science B.V., Amsterdam, 1999.

Trott, A.R. and T. Welch: "Refrigeration and Air-conditioning," Butterworth-Heinemann, Inc., Woburn, MA, 1999.

Wang, S.K.: "Handbook of Air Conditioning and Refrigeration," 2nd Edition, McGraw-Hill Professional Book Group, New York, NY, 2000.

Whitman, B.C., B. Johnson, and J. Tomczyk: "Refrigeration and Air Conditioning Technology," 4th Edition, Delmar Publishers, Albany, NY, 1999.

Zimm, C., A. Jastrab, A. Stenberg, V. Pecharsky, K. Gschneidner, Jr., M. Osborne, and I. Anderson, "Description and Performance of a Near-room Temperature Magnetic Refrigerator," *Adv. Cryo. Engin.* **43** 1759–1766 (1998).

Web Reference

Air-Conditioning and Refrigeration Institute: *http://www.ari.org/*

REGELATION.   The phenomenon that occurs when two pieces of ice are rubbed together, the pressure causing the ice to melt at the surfaces of contact while the temperature drops, and, on relieving the pressure, the two surfaces freeze together, producing one mass of ice. This phenomenon is due to reduction of the freezing point of water (melting point of ice) under increased pressure. At very high pressures, the relationship changes, and the melting point of ice increases steadily with increasing pressure.

REGENERATION (Zoology).   The development of a tissue or part of the living body to replace a similar structure that has been damaged or destroyed.

A conspicuous degree of regeneration is possible in some of the simpler animals, including sponges, coelenterates, and worms. When cut into pieces, the fragments undergo a reorganization of their materials to form complete individuals of smaller sizes. The process is not unlimited, however, for abnormalities of regeneration take place in some groups when the mutilation is of a certain type. In experiments with flatworms (see also **Platyhelminthes**), for example, C.M. Child has found that halves of worms or a segment from the middle of the body develop into complete animals, but a head produces only another head and so perishes. T.H. Morgan, in experiments with a species of earthworm, found that the amputation of a limited part of the anterior end was followed by complete regeneration but that the removal of more segments resulted in the formation of a minimum number like the original extreme anterior end. Starfishes undergo the regeneration of amputated arms very readily and mollusks and arthropods are capable of some restoration of lost parts. Insects and crustaceans develop new appendages if the loss occurs before the completion of their growth.

Among the most remarkable cases of regeneration are those of the bryozoans and sea cucumbers. The animal (polypide) breaks down within the body wall (zooecium) in the former group to become a disorganized mass called the brown body. From the zooecium a new animal is formed. Sea cucumbers, under extremely irritating stimuli, sometimes discharge the entire intestine, along with the defensive Cuvierian organs. The tract is later replaced.

In complex animals, including man, regeneration is limited to the replacement of parts subject to wear and easy loss, such as hair and nails, and to the renewal of damaged tissues, such as skin. Even the renewal of tissues is limited, some kinds undergoing normal and complete regeneration while others are repaired by the formation of scar issue of different origin but cannot be replaced.

REGISTER (Computer). A hardware device used to store a certain amount of bits or characters and usually provided for a particular purpose. A register usually is constructed of semiconductor devices, such as transistors, and usually contains approximately one word or one byte of information.

Address Register. A register for the temporary storage of an address in a computer.

General-purpose Register. A register that may be utilized for several purposes, such as accumulation, address indexing, shifting, etc.

Index Register. A register that contains a quantity that may be used to modify addresses. See also **Index Register**.

Instruction Register. A register that holds the identification of the instruction word to be executed next in time sequence following the current operation. The register is often a counter that is incremented to the address of the next sequential storage location, unless a transfer or other special instruction is specified by the program.

Program Register. A register in which the current instruction of the program is stored. Contrasted with control register.

Shift Register. A register in which the characters may be shifted one or more positions to the right or left. In a right shift, the rightmost character(s) are lost. In a left shift, the leftmost character(s) are lost.

Storage Register. A register in the storage of the computer, in contrast with a register in one of the other units of the computer.

REGRESSION. In statistics, this term has two somewhat different meanings, although the analysis in both cases is identical.

1. In a bivariate distribution, say of x, y, there will be a relation between the values of x and the mean of the values of y for given x. This is the regression of y on x. For bivariate normal variation the regression is linear. Likewise, there will be a regression of x on y, which in general is different from the regression of y on x. From this viewpoint the regression relationships can be considered as generalizations to stochastic situations of the functional relations of mathematics: the regression of y on x shows the dependence of the mean of a distribution of y-values for assigned values of x.

2. From a more general viewpoint x need not be a random variable and the stochastic variation lies solely in y, so that the relationship is of the type $y = f(x) + \varepsilon$ where ε (and therefore y) is a random variable.

In both cases the parameters of the relationships are usually estimated by the method of least squares, i.e. by minimizing the sum of squares of the residuals ε. This is optimal if the ε have a normal distribution with constant variance for all values of x.

There are various generalizations. In a multivariate complex one variable y may be regressed on a number of others, for example in the linear form

$$y = \beta c + \sum_{i=1}^{p} \beta_i x_i + \varepsilon$$

which expresses the way in which the mean of y varies according to assigned values of x. Again the x's need not themselves be random variables but could, for example, be predetermined in an experiment. Further generalizations include the case where other functions of the x's appear, e.g. powers; or where the ε's are not independent from one observation to another.

The goodness of fit of a regression equation is judged by the variance of the random element ε as a proportion of the variance of y, small values meaning a good fit. Alternatively, use is made of the complementary quantity $R^2 = 1 - \text{var}\varepsilon/\text{var } y$, known as the square of the multiple correlation coefficient.

As in the case of partial correlation (see **Correlation**) attempts are sometimes made to compute partial regressions, the object of which is to measure the dependence of the y-variable on certain x's when the effect of other x's on y has been removed. The terminology has been confused by the fact that "partial regression" is sometimes applied to the individual coefficients in a complete regression on all variables, as compared with "total regression" of y on any particular x.

The so-called "diagonal regression" is not a regression at all. It is a function connecting the two variables when they are both subject to errors of observation and is properly to be considered as part of the analysis of functional relationship.

Sir Maurice Kendall, International Statistical Institute, London.

REGULUS (α Leonis). Ranking twenty-first in apparent brightness among the stars, Regulus has a true brightness value of 120 as compared with unity for the sun. Regulus is a blue-white, spectral type B star and is located in the constellation Leo, a zodiacal constellation. Estimated distance from Earth is 75 light years. According to best authority, the present name of the star was given by Copernicus.

REHYDRATION.   See **Dehydration (Chemical)**.

REINDEER.   See **Deer**.

REINDEER MOSS.   See **Lichen**.

RELAPSING FEVER. A group of acute infectious diseases caused by spirochetes of the genus *Borrelia*, which are transmitted to humans by several species of ticks and lice. Relapsing fever has been known all over the world, but the chief centers of spread are Russia, Poland, and the Balkan states. An African form is also known and in the western United States, the tick-borne form is also seen in the mountains during summer. Epidemics of relapsing fever and typhus are often associated, and occur in periods of depression following war when overcrowding, famine, and poor hygienic conditions are prevalent.

The incubation period is usually 7–10 days. The disease is characterized by paroxysms of acute fever lasting several days with intervals between the attacks when the temperature is normal and the patient is apparently well. The diagnosis is easily established by finding the organism in specimens of the patient's blood during a paroxysm of fever.

The treatment of choice is tetracycline, administered orally several times per day for about 10 days. This eliminates the organisms from the bloodstream and prevents relapses. In some cases, the physician may prefer chloramphenicol. The mortality rate of relapsing fever, if untreated and if occurring during an epidemic, may be as high as 40%. Appropriate treatment reduces this statistic to 2–5%.

R. C. V.

RELATIVE COORDINATES. Any coordinate system which is moving with respect to an inertial coordinate system. In practice, atmospheric motion is always referred to a relative system fixed to the surface of the Earth.

Referred to a relative system, various apparent forces arise in Newton's laws owing to motion of the system, such as centrifugal force and coriolis force.

RELATIVE HUMIDITY.   See **Humidity**.

RELATIVITY AND RELATIVITY THEORY.   Relativity is a principle that postulates the interdependence of matter, space, and time in the universe, for various frames of reference. A theory that utilizes such a principle is called a relativity theory. The basic concepts of modern relativity theory are largely ascribed to the work of Albert Einstein. Both main branches of pre-Einstein physics had relied on an absolute space. To Newton, this served as the agent responsible for a particle's resistance to acceleration; to Maxwell — in the guise of an "aether" — it was the carrier of electromagnetic stresses and waves. Relativity may be defined briefly as the abolition of absolute space. Special relativity (Einstein, 1905) abolished it in its Maxwellian sense. General relativity (Einstein, 1915) abolished it in its Newtonian sense as well.

During 1979, scores of distinguished scientists reviewed the accomplishments of Albert Einstein, who was born a century earlier (March 14, 1879). Many papers describing and reviewing the works of Einstein were prepared in honor of the centenary of Einstein's birth.

P.A.M. Dirac, who addressed the Einstein session of the Pontifical Academy of Science (Vatican City) on November 10, 1979, pointed out three main innovations introduced to scientific thought by Einstein: (1) special relativity; (2) the relation between waves and particles; and (3) general relativity.

Prior to Einstein's profound observations and analyses, scientists relied on absolute space. Einstein built upon and revised the earlier insights that were expounded by his predecessors. Prior to Einstein, two principal guidelines were regarded as true: (1) that electricity, magnetism, and light are the same — that light is a wave of electric and magnetic forces; and (2) that there exist atoms and molecules made up of electrically charged particles as constituents. These guidelines presented certain inconsistencies. Einstein recognized these contradictions and sought a resolution of them. As it turned out, Einstein did not propose minor modifications of the prior thinking, but rather he introduced whole new avenues of thought, which since their presentation have guided scientists in their basic approaches to the physics of the universe.

Viktor Weisskopf, in his address to the Pontifical Academy, clearly summarized the contradictions of pre-Einstein physics: (1) "According to the laws of electromagnetism an electrically charged particle cannot move faster than light because it would produce infinitely strong electric forces. But matter is made of charged particles." (2) "The second contradiction is quite different. It concerns the surprising stability of atoms and molecules and their characteristic features. An oxygen molecule in air suffers a million times a million collisions every second, but remains unchanged in all its specific properties as an oxygen molecule. Ordinary mechanics is totally inadequate to explain this stability and specificity of systems made of charged particles, such as atoms that consist of electrons moving around atomic nuclei like planets around the sun."

Einstein's revisionary concepts of space, time, and energy have served as a resolution to the first contradiction; Einstein, by introducing the wave-particle duality to physics, gave decisive impetus to the solution of the second contradiction.

Special Relativity.   This theory was developed by Einstein on the hypothesis that the velocity of light is the same as measured by any one of a set of observers moving with constant relative velocity. According to Newtonian theory and the Galilean transformation, the mechanical motion of an object with respect to an inertial system could be predicted from a knowledge of the forces acting on it and the initial conditions, independently of any knowledge of the motion of the inertial system itself. Einstein extended this to optical phenomena, postulating that these also could be described without knowledge of the velocity of the laboratory with respect to the universe.

Before looking at the theoretical background of relativity, it is in order to mention some of its more striking practical implications. According to special relativity, for example, a rod moving longitudinally at speed v through an inertial frame is shortened, relative to that frame, by a factor $\gamma = (1 - v^2/c^2)^{-1/2}$ where c is the speed of light. This factor increases with v. When v is as large as $\frac{1}{7}c$, γ is only 1.01, but at higher speeds it grows rapidly and becomes infinite when $v = c$. The rate of a clock moving at speed v is decreased by the same factor γ; this is one aspect of Einstein's revolutionary prediction that time is not absolute and that, for example, after journeying at high speed through space, one could, upon return, find the world aged very much more than oneself. In fact, time and space become merged in a four-dimensional continuum in which neither possesses more absoluteness than, e.g., the x-separation between points in a Cartesian plane, which depends on the choice of axes. According to special relativity, time- and space-separations between events similarly depend on the choice of motion of the observer. The mass of a body moving at speed v is also increased by the factor y and thus becomes infinite at the speed of light. But the single most important result of special relativity, in Einstein's opinion, was the equivalence of mass m and energy E according to the formula $E = mc^2$. Although the original impact of special relativity was mainly theoretical and philosophical, technology since 1905 has made such vast studies (nuclear power, particle accelerators, Mössbauer effect, among others) that today special relativity is one of the most practical and, at the same time, best verified branches of physics.

Continuing with Weisskopf's observations, with reference to the two aforementioned contradictions:

> The solution of the first contradiction is embodied in so-called special relativity theory. In it, electromagnetism and mechanics are unified in one great conceptual system. To achieve this, our ordinary concepts of space and time had to be thoroughly changed. The simultaneity of events at different places has become a relative relation depending on the state of motion. Two events that appear to happen at the same time for one observer who does not move appear to happen at different times for a moving observer. The course of time also depends on the state of motion. Incredible as it may seem at first, this fact has been shown clearly in experiments with some fast-moving entities; their course in time was shown to be retarded compared to the course in time of the same entities remaining at rest. In a famous experiment with decaying particles, the fast-moving ones lived much longer than the same sort of particles at rest. Finally, any form of energy acquires a mass, and every mass is a form of energy. A moving body appears heavier than a body at rest because its energy of motion adds to its mass. In some modern particle accelerators electrons acquire masses more than 20,000 times their original mass when they are accelerated, an effect that is clearly observed when they collide with an obstacle. All these properties affect the motion of fast electrons in electric and magnetic fields. Indeed, many practical applications of electronics are based on these properties.

In commenting on special relativity, Dirac commented at the Pontifical Academy:

> With special relativity, Einstein showed that such commonplace ideas as space and time need to be modified. The traditional views do not provide an adequate basis for an accurate description of physical processes. They have to be replaced by a picture in which space and time are intimately related and are united in a four-dimensional continuum. Elementary notions of kinematics and of dynamics are altered.
>
> People sometimes say that special relativity was discovered by Lorentz or Poincaré and refer to work that was published by these authors before Einstein published his famous paper on relativity in 1905. But these statements give only part of the truth and not the main part. Lorentz and Poincaré believed in the ether. They obtained some of the relativity equations working within the framework of the ether,[1] which was always at the back of their minds. Einstein destroyed the ether, and so the framework on which the others built was gone. He introduced a new symmetry principle between space and time. For Einstein the symmetry principle was all-important. This was his great achievement and here he stands alone. Symmetry principles are

---

[1] EDITOR'S NOTE: Prior to acceptance of the relativity theory, the either was a postulated material substance which was assumed to fill all space, and to penetrate freely among the ultimate particles of which all matter is constituted. Existence of ether could not proved by the famous Michelson-Morley experiment in Cleveland, Ohio (Case Institute of Technology) in 1887 and doubtless was a building block toward Einstein's theory of 1905 and plays a dominant role over others.

now very important in a large part of physics. Many of the symmetry principles in use nowadays are only approximate, and they are broken. The symmetry principle introduced by Einstein connecting space and time is an exact principle in physics and plays a dominant role over others.

Two types of argument can be made in support of Einstein's principle. The first is experimental; all experiments devised to discover the frame of Maxwell's "aether," such as the Michelson-Morley experiment, failed to give positive results, although such results would have been well within range of observability. The second argument is theoretical, and rests on the unity of physics. For example, mechanics involves matter, which is electromagnetically constituted; electromagnetic apparatus involves mechanical parts. If, then, physics cannot be separated into strictly exclusive branches, it would seem unlikely that the laws of different branches should have different transformation properties.

Consider two observers O and O' and a light signal emitted at their coincidence. If each observer remains at the origin of a Cartesian reference system and sets his clock to read zero when the signal is emitted, the events on the light front must satisfy both the following equations:

$$x^2 + y^2 + z^2 - c^2t^2 = 0$$
$$x'^2 + y'^2 + z'^2 - c^2t'^2 = 0 \tag{1}$$

where primes distinguish the space and time coordinates used by observer O' from those used by observer O. Suppose the two observers arrange their corresponding y- and z-axes to be parallel, and their x-axes to coincide. In classical mechanics, with this configuration of reference systems, the so-called Galilean transformation equations:

$$x' = x - vt; \quad y' = y; \quad z' = z; \quad t' = t \tag{2}$$

relate the corresponding coordinates of any event. But under this transformation, the two equations (1) are not equivalent. Einstein showed that for these equations to be equivalent, the transformation equations must necessarily be

$$x' = \gamma(x - vt); \quad y' = y; \quad z' = z; \quad t' = \gamma(t - vx/c^2) \tag{3}$$

where $\gamma((1 - v^2)/c^2)^{-1/2}$ These are the well known *Lorentz equations*, which constitute the mathematical core of the special theory of relativity. They replace equations (2), to which they nevertheless approximate when v is small. The most striking of equations (3) is the last. It implies that events with the same value of t do not necessarily correspond to events with the same value of t', which means that *simultaneity is relative*. Setting $x = 0$ in that equation also shows that the clock at the origin of O goes slow by a factor γ in the frame of O'. But, setting $x = vt$, we see that the clock at the origin of O' similarly goes slow in the frame of O. Setting $t = 0$ in the first of equations (3) we see that a rod, fixed in the frame of O' along the x'-axis, appears shortened by a factor γ in the frame of O; this phenomenon too can be shown to be symmetric between the frames.

Another important property of equations (3) is that they leave invariant the differential quadratic:

$$ds^2 = dx^2 + dy^2 + dz^2 - c^2 dt^2 \tag{4}$$

which leads to the possibility of mapping events in a four-dimensional pseudo-Euclidean *space-time* in which an *absolute interval ds* exists, and in which the language and results of four-dimensional geometry can thus be applied. For example, a uniformly moving particle is described simply by a straight line in this space-time.

It was the first task of special relativity to review the existing laws of physics and to subject them to the test of the relativity principle by seeing whether they were invariant under Lorentz transformations. Any law found lacking had to be modified accordingly.

Since Newton's laws of mechanics are invariant under the transformation (2) and *not* (3), it was necessary to amend these laws so as to make them "Lorentz invariant." It was found possible to do this by retaining the classical laws of conservation of mass and momentum, but postulating that the mass of moving bodies increases by the factor γ, a fact amply borne out by modern particle accelerators. This led to the theoretical discovery of the equivalence of mass and energy—most spectacularly exemplified by atomic fission.

In contrast to Newton's theory, Maxwell's vacuum electrodynamics was already compatible with Einstein's theory. In other words, Maxwell's equations already were "Lorentz invariant" and needed no modification. Nevertheless, relativity has considerably deepened our understanding of Maxwell's theory. Other branches like kinematics, optics, hydrodynamics, thermodynamics, nonvacuum electrodynamics, among others, all underwent slight modifications to make them Lorentz invariant. Only Newton's inverse square gravitational theory proved refractory; several Lorentz-invariant modifications of it were proposed but none were entirely acceptable.

General Relativity. In speaking to the Pontifical Academy on November 10, 1979, Dirac made the following observations:

> Einstein provided a geometric picture of gravitation and thereby started an entirely new direction for physics. Previously, there were two pictures of physical forces in general use: action at a distance and action through a field. With Newtonian gravitation both pictures are possible. With electric and magnetic forces the action at a distance concept is useful, but action through a field provides a more complete picture since it allows electromagnetic waves. With Einstein, gravitation is interpreted in terms of the curvature of space, and only the field picture is possible.

> There were some small differences between the predictions of Einstein's and Newton's theories, which provided several lines of work for astronomers and physicists and enabled them to make checks. First, there was the motion of the planet Mercury, which was anomalous according to Newton but was brilliantly explained by Einstein. Then there was the bending of light passing close by the sun, which can be observed during a total eclipse of the sun. Observations were made in 1919 that confirmed Einstein's theory of general relativity (1915). These observations have been repeated many times since and his theory is always confirmed.

> The theory of general relativity also predicts effects concerning the shifting of spectral lines of light emitted in a gravitational field. The opportunities here are usually not very good for accurately testing the theory, but the results have supported the theory as well as could be expected.

> Besides all these astronomical and physical developments following from general relativity, there has been an extensive stimulus to mathematical work. The simple kind of curve space that Einstein used, Riemannian space, which can be embedded in a flat space with a higher number of dimensions, proved so successful with gravitation that people have wondered whether more elaborate kinds of curved space might not similarly account for the other fields of physics, in particular the electromagnetic field. Einstein himself worked on this problem for many years.

> But these efforts have not had any definite success. Whereas Einstein's original curved space was brilliantly successful, the more complicated spaces, on which an extensive amount of work has been done, have not led to results of physical importance so far.

> There is also the problem of cosmology, the understanding of the universe as a whole. This is necessary for getting the boundary conditions at great distances in applications of Einstein's field equations. A cosmological model was first proposed by Einstein, but it was not satisfactory. Then a model was proposed by de Sitter, also not satisfactory. Then many other models were worked out, by Friedmann, Lemaitre, and others, based on Einstein's equations. This is a large subject that was initiated by Einstein's general relativity. The simplest acceptable model is one that was proposed jointly by Einstein and de Sitter, and it may very well be the one that is used in the future.

See also **Cosmology**.

At the Pontifical Academy on November 10, 1979, Weisskopf commented on general relativity:

> It [the general theory of relativity] was a new way to understand gravity, as a warping of space and time. The consequences of Einstein's third contribution [general relativity] are staggering. Many of its predicted consequences have been observed—for example, the bending of light beams by the sun. One of the most interesting consequences is what happens if a large star collapses after having used up all its internal energy sources. Then the space around it collapses too, and something is formed that astronomers call a black

hole, an entity that engulfs everything in its neighborhood and does not allow even light to escape. Objects that may indeed be such black holes have been observed....

Einstein's theory of gravity as a deformation of the space-time structure had an enormous influence on our ideas of the structure of the universe, its beginning, its evolution, and its extension. The modern view that the universe originated from an infinitely compressed hot assembly of primal matter in the big bang and the subsequent expansion of the universe are ideas that were spawned by Einstein's conception of space and time. This view of the origin of the world was recently supported by the observation of an unmistakable faint optical echo of that grand explosion, an echo that still fills the universe with infrared radiation.

Einstein eventually solved the gravitational problem in an unexpected way. He rejected Newton's absolute space as the cause of inertia on the ground that "It is contrary to the spirit of science to conceive of a thing which acts but cannot be acted upon." Einstein's general theory of relativity ascribes to the space-time continuum discovered by special relativity the role of an inertial guiding field (free particles and light follow geodesics) but allows this field to be affected (curved) by the matter in it.

This extension was made possible by the so-called *principle of equivalence*. To Newton, an inertial frame was, primarily, the frame of "absolute space" in which the stars were assumed to be fixed, and secondarily, any frame moving uniformly relative to absolute space. Thus an inertial frame exhibited its defining property, viz., that in it free particles move uniformly and rectilinearly (Newton's first law), only in the regions far from attracting masses. In 1907, Einstein changed this global definition to a local one; a local inertial frame is a freely falling, nonrotating reference system. (The meaning of "local" here is determined by the extent to which the nonuniformity of the gravitational field is negligible.) Within the limits of each such frame, Newton's laws of mechanics would be valid according to the classical theory; in particular, Newton's first law would be strictly satisfied. Einstein also made the generalization from mechanics to the whole of physics. His principle of equivalence asserted that all the laws of physics are the same in each local inertial frame. It is these frames, therefore, which are the proper province of the special principle of relativity. Special relativity thus becomes a local theory. In recompense, we need no longer go to the tenuous interstellar regions for its strict validity.

An elementary consequence of the principle of equivalence is the bending of light in a gravitational field. For, if light travels rectilinearly in the local inertial frame, and that accelerates freely in the gravitational field, the light path is evidently curved in the field. No property of light other than its uniform motion in an inertial frame has been used in this argument. This, in turn, suggests that one might ascribe the bending of light to an inherent space curvature, rather than to the nature of light. In much the same way, the characteristics motion of free particles in a gravitational field suggests that they follow "natural" paths (geodesics) in a curved space. Their motion is independent of everything except their initial position and velocity, owing to the equality of "gravitational mass" (the analog of electric charge) and "inertial mass" (the measure of a particle's resistance to acceleration). It is this that makes the principle of equivalence possible. It should be noted, however, that for the geodesic law to be possible, space and time must be welded into four-dimensional space-time. Free motions could not be represented by geodesics in a *three*-dimensional curved space. For a geodesic is uniquely determined by a point on it and a direction at that point. But an initial point and an initial direction do not uniquely determine a free path in a gravitational field. *That* depends also on the initial speed. In space-time, on the other hand, a direction is equivalent to a (vector-) velocity. And it *is* the case in Newton's theory that an initial point and velocity uniquely determine a free path in a gravitational field. Note that all this depends upon the equivalence of inertial and gravitational means. This equivalence is an unexplained and inessential coincidence in Newton's theory; it is the *sine qua non* of Einstein's.

Special relativity forces a four-dimensional metric surface (Eq. 4) on the events within an inertial frame. By patching together the structures of all the local inertial frames, we obtain the structure of the world of general relativity. Locally, it can be regarded as flat. But it is evident that, if the very pleasing geodesic law of motion is to hold, the presence of matter must impress a curvature on the space-time. For example, the planets move in patently curved paths around the sun (in four dimensions these are helicoidal rather than elliptical); for these paths to be geodesic, the space-time around the sun must be curved. Just how matter curves the

surrounding space-time is expressed by Einstein's field equations:

$$G_{ij} = -\frac{8\pi G}{c^4} T_{ij} \tag{5}$$

which look deceptively simple. Technically, they represent ten second-order partial differential equations for the metric of space-time. This metric enters the 16 components of the "Einstein tensor" G_{ij}, of which only 10 are independent, for $G_{ij} = G_{ji}$. G is the constant of gravitation; T_{ij} is the so-called energy tensor of the matter, and its components represent a generalization of the classical concept of density.

The exact solution of Eq. (5) has been possible only in a limited number of physical situations. For example, in 1916 Schwarzschild gave the exact solution for the space-time around a spherical mass m (e.g., the sun):

$$ds^2 = (1 - a/r)^{-1} dr^2 + r^2(d\theta^2 + \sin^2\theta \, d\phi^2)$$
$$\times -(1 - a/r)c^2 \, dt^2 \tag{6}$$

where $a = 2Gm/c^2$, r is a measure of distance from the central mass, t is a measure of time, and θ and ϕ are the usual angular coordinates. Note that when $m = 0$, Eq. (6) reduces simply to the flat space-time of Eq. (4), written in polar coordinates, and its geodesics would be straight lines (in space and time). But for Eq. (6), the geodesics in the plane $\theta = \pi/2$ are found to satisfy the equation:

$$\frac{d^2\mu}{d\phi^2} + \mu = \frac{Gm}{h^2} + 3\frac{Gm\mu^2}{c^2} \tag{7}$$

where $\mu = 1/r$ and h is a constant. This differs formally from the classical orbit equation only by the presence of the last term, which is very small. But as a consequence of that term, the solution of Eq. (7) is:

$$\mu = Gmh^{-2}(1 - e\cos\rho\phi),$$
$$\rho = 1 + 3G^2m^2h^{-2}c^{-2} \tag{8}$$

instead of the classical solution, which has $\rho = 1$. Now r is a function in 1 of period $2\pi/\rho$ instead of 2π, and therefore the orbital ellipse precesses. For the planet Mercury, for example, the secular precession predicted as 42″ (seconds of arc), and this agrees well with observation. In the space-time defined by Eq. (6) one also finds that light-signals which pass close to the central mass are bent by an angle twice as big as that predicted on a simple Newtonian corpuscular theory of light; and again observations bear out the relativistic prediction. The third "crucial" prediction, which has also been verified observationally, is the reddening of the light received from the surface of very dense stars.

Tests have involved the timing of radar signals past the limb of the sun. According to general relativity, these should be slowed by the field of the sun.

Worldwide Relativistic Sagnac Experiment. Hafele and Keating, in 1971, first observed that a portable clock transported slowly eastward once around the Earth's equator will *lag* a master clock *at rest* on the Earth's surface by approximately 207.4 nanoseconds, but that when transported *westward* will *lead* a master clock at rest by 207.4 nanoseconds. In their experiment, Hafele and Keating used commercial jet aircraft to transport an ensemble of cesium clocks eastward and then westward around the globe.

In 1984, Allan and Weiss (U.S. National Bureau of Standards) and Ashby (University of Colorado) took a different approach. They made observations of the effect by using electromagnetic signals instead of portable clocks. Global Positioning System (GPS) satellites having accurate atomic clocks aboard transmit electromagnetic signals that can be viewed simultaneously from remote stations on Earth and, thus, an around-the-world Sagnac experiment can be performed on this basis.

In their 1984 paper, these investigators observe that the Sagnac effect has the same form and magnitude whether slowly moving portable clocks or electromagnetic signals are used to complete the circuit. For slowly moving portable clocks, the effect can be viewed from a local nonrotating geocentric reference frame as being due to a difference between the second-order Doppler shift (time dilation) of the portable clock and that of the master clock whose motion is due to the Earth's rotation. For electromagnetic signals, the effect arises from a well-known consequence of the special theory of relativity — the *relativity of simultaneity* — which follows from the principle of the constancy of the speed of light.

The investigators point out that if one imagines two clocks fixed a small east-west distance (x) apart on the Earth, then viewed from the nonrotating frame they will move with approximately equal speeds; $v = \omega r$, where ω is the angular rotation rate of the Earth and r is the distance of the clocks from the rotation axis. If a clock synchronization process involving electromagnetic signals were carried out by Earth-fixed observers who ignored the Earth's rotation, then the two clocks would *not* be synchronous when viewed from the nonrotating frame. The magnitude of the discrepancy is approximately: $vx/c^2 = (2\omega/c^2)(rx/2)$. In general, such discrepancies depend on the path along which the light signals travel relative to the rotation axis. An acceptable way to avoid this problem is to synchronize the clocks in the nonrotating frame.

Thus, in synchronizing clocks on the surface of the rotating Earth by means of electromagnetic signals or portable clocks, it is necessary to apply a correction that arises from the Sagnac effect in order to avoid problems of nontransitivity of the synchronization process. The Consultative Committee for the Definition of the Second and the International Radio Consultative Committee have agreed that, in order to obtain consistently synchronized clocks on the Earth's surface at the subnano-second level, the correction term to be applied is:

$$\Delta t = 2\omega/c^2 = A_E = 1.6227 \times 10^{-21} \text{ sec/m}^2 \, A_E$$

where A_E is the projected area on the Earth's equatorial plane swept out by the vector whose tail is at the center of the Earth and whose head is at the position of the portable clock or the electromagnetic signal pulse. The A_E is taken as positive if the head of the vector moves in the eastward direction. If two clocks located on the Earth's surface are compared by using portable clocks or electromagnetic signals in the rotating frame of the Earth, then Δt must be subtracted from the measured time difference (east clock minus west clock) in order to synchronize the clocks so they will measure coordinate time on the Earth.

In the Allan-Weiss-Ashby experiment, signals from GPS satellite vehicles, 3, 4, 6, and 8 were used in simultaneous common view between three pairs of Earth timing centers to accomplish the circumnavigation. The centers were the NBS (Boulder, Colorado), the Physikalisch-Technische Bundesanstalt (PTB) in Braunschweig (W. Germany), and the Tokyo Astronomical Observatory (TAO). A typical geometrical configuration of ground stations and satellites, with the corresponding projected area, is given in Fig. 1. The size of the Sagnac effect calculated from the expression given earlier varies from about 240 to 350 nanoseconds, depending upon the location of the satellites used in a circumnavigation carried out at a particular moment. Sufficient data were collected to perform ninety independent circumnavigations. The actual mean value of the Sagnac residual over the 90-day period was 5 nsec, which is less than 2% of the magnitude of the calculated total Sagnac effect. Even though the atomic clocks used in the experiment were among the best in the world, they are perturbed by natural random processes. The net time dispersion for the experiment attributable to these perturbations on the three clocks is about

2.5 nsec. The remainder nonnull result is explained by uncertainties in the propagation delays and in the ephemerides of the GPS satellites.

The researchers conclude that the theoretical predictions have been well verified by observation. Although the Sagnac effect had been observed frequently in laboratory settings, the experiment described here is probably the first such experiment to be conducted on a large scale. More detail can be found in the D.W. Allan et al. paper presented at the Netherlands conference in August 1984 (reference listed). A more concise version can be found in *Science*, 69–70 (April 5, 1985).

Additional Reading

Adar, R.K.: "A Flaw in a Universal Mirror," *Sci. Amer.*, **50** (February 1988).

Allan, D.W. and M.A. Weiss: Paper presented at 34th Annual Frequency Control Symposium, Fort Monmouth, New Jersey, 1980.

Allan, D.W., et al.: Paper presented at Conf. on Precision Electromagnetic Measurements, Delft, Netherlands, August **18**, 1984.

Allan, D.W., M.A. Weiss, and N. Ashby: "Around-the-World Relativistic Sagnac Experiment," *Science,* **228**, 69–70 (1985).

Bertotti, B., F. de Felice, and A. Pascolini: "General Relativity and Gravitation," Kluwer Academic Publishers, Norwell, MA, 1984.

Brush, S.G.: "Prediction and Theory Evaluation: The Case of Light Bending," *Science*, **1124** (December 1, 1989).

Carmeli, M.: "Group Theory and General Relativity," Imperial College Press, London, UK, 2000.

Chagas, C.: "Einstein," Einstein Session of the Pontifical Academy, Vatican City (November 10, 1979). Reprinted in *Science*, **207**, 1159–1161 (1980).

Cline, D.B. and F.E. Mills: "Unification of Elementary Forces and Gauge Theories," Gordon & Breach Publishing Group, Newark, NJ, 1980.

Cohen, I.B.: "Newton's Discovery of Gravity," *Sci. Amer.*, 166–179 (March 1981).

Davies, P.: "The New Physics," Cambridge University Press, New York, NY, 1989.

Dirac, P.A.M.: "General Theory of Relativity," Princeton University Press, Princeton, NJ, 1996.

Dirac, P.A.M.: "Einstein," Einstein Session of the Pontifical Academy, Vatican City (November 10, 1979). Reprinted in *Science*, **207**, 1161–1162 (1980).

Einstein, A.: "Relativity: The Special and the General Theory," Taylor & Francis, Inc., Philadelphia, PA, 2001.

Ellis, P.J. and Y.C. Tang: "Trends in Theoretical Physics," Addison-Wesley, Longman, Inc., Reading, MA, 1990.

Falomir, H., R.E. Gomboa and F.A. Schaposnki: "Trends in Theoretical Physics," American Institute of Physics, College Park, MD, 1998.

Freedman, D.Z. and P. van Nieuwenhuizen: "The Hidden Dimensions of Space-time," *Sci. Amer.*, **252**(3), 74–81 (1985).

French, A.P.: "Einstein: A Centenary Volume," Harvard University Press, Cambridge, MA, 1979.

Geroch, R.: "Mathematical Physics," University of Chicago Press, Chicago, IL, 1985.

Gibbons, A.: "Putting Einstein to the Test—In Space," *Science*, **939** (November 15, 1991).

Glick, T.F.: "The Comparative Reception of Relativity," Kluwer Academic Publishers, Norwell, MA, 1987.

Goldberg, S.: "Understanding Relativity," Birkhäuser, Boston, Massachusetts, 1984.

Hafele, J.C. and R.E. Keating: *Science*, **177**, 168 (1972).

Hamilton, D.P.L.: "Laser Interferometry Gravitational Observatory," *Science*, **635** (May 3, 1991).

Hawking, S.W. and W. Israel:"General Relativity," Cambridge University Press, New York, NY, 1979.

Hawking, S.W. and W. Israel: "Three Hundred Years of Gravitation," Cambridge University Press, New York, NY, 1990.

Hegstrom, R.A. and D.K. Kondepudi: "The Handedness of the Universe," *Sci. Amer.*, **108** (January 1990).

Held, A.: "General Relativity and Gravitation," Plenum, New York, NY, 1980.

Hey, T. and P. Walters: "Einstein's Mirror," Cambridge University Press, New York, NY, 1997.

Miller, A.I.: "Albert Einstein's Special Theory of Relativity: Emergence (1905) and Early Interpretation (1905–1911)," Springer-Verlag, Inc., New York, NY, 1997.

Pais, A.: "Subtle is the Lord: The Science and the Life of Albert Einstein," Oxford University Press, Inc., New York, NY, 1983.

Pool, R.: "Closing in On Einstein's Special Relativity Theory," *Science*, 1207 (November 30, 1990).

Pope John Paul II: "Einstein," Einstein Session of the Pontifical Academy, Vatican City (November 10, 1979). Reprinted in *Science*, **207**, 1165–1167 (1980).

Prugovecki, E.: "Principles of Quantum General Relativity," World Scientific Publishing Company, Inc., River Edge, NJ, 1995.

Pyenson, L.: "The Young Einstein: The Advent of Relativity," Hilger, Bristol, U.K., 1985.

Resnick, R.E. and D. Halliday: "Basic Concepts in Relativity and Early Quantum Theory," John Wiley & Sons, Inc., New York, NY, 1990.

Staff: "General Relativity and Gravitation: Proceeding of the 14th International Conference," World Scientific Publishing Company, Inc., River Edge, NJ, 1997.

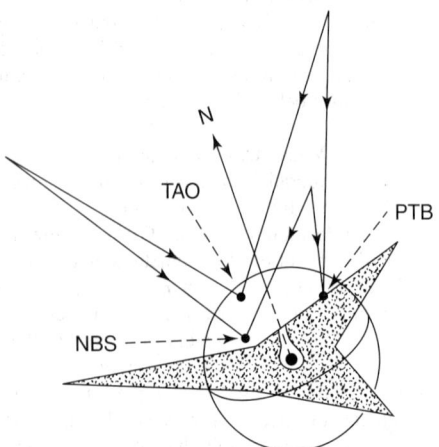

Fig. 1.   Diagram of Sagnac experiment performed by Allan, Weiss, Ashby, and colleagues in 1984. The paths of electromagnetic signals from three satellites are observed in common view to the three Earth-timing centers: NBS (National Bureau of Standards, Boulder, Colorado); PTB (Physikalisch-Technische Bundesanstalt, Braunschweig, W. Germany); and TAO (Tokyo Astronomical Observatory). (*After Allan, et al.*) NBS is now the National Institute of Standards and Technology.

Tipler, F.J.: "Essays in General Relativity," Academic Press, Inc., San Diego, CA, 1980.

Weisskopf, V.F.: "Einstein," Einstein Session of the Pontifical Academy, Vatican City (November 10, 1979). Reprinted in *Science*, **207**, 1163–1164 (1980).

Will, C.M.: "Theory and Experiment in Gravitational Physics," Cambridge University Press, New York, NY, 1993.

Will, C.M.: "General Relativity at 75: How Right Was Einstein?" *Science*, **770** (November 9, 1990).

RELAXATION BEHAVIOR.   All phenomena where consideration of equilibrium conditions alone would give an incomplete picture and where the study of the time-dependence of the approach to equilibrium is essential for an adequate comprehension of the effect.

RELAXATION FREQUENCY.   In general terms, the inverse of the relaxation time. A system is usually incapable of reacting to any periodic stimulus whose frequency is very much higher than the relaxation frequency for the effect concerned.

RELAXATION METHOD.   Originally used to calculate displacements in a structure subjected to known loads in engineering problems, it may be used to obtain eigenvalues and eigenvectors (see also **Eigenfunction**) of algebraic or differential equations. Initial values of the eigenvalues and eigenvectors are put into the simultaneous equations of the problem. If they were the correct ones, each equation of the system would vanish; otherwise, there would be a set of residuals. The initial solutions are then varied (relaxed) until the residuals are minimized.

RELAXATION PHENOMENON.   Any phenomenon in which a system requires an observable length of time in order to respond to sudden changes in conditions, forces, or effects which are applied to the system.

RELAXATION TIME.   In general, the time interval required for a system exposed to some discontinuous change of environment to undergo fraction $(1 - e^{-1})$, or about 63% of the total change of state which it would exhibit after an indefinitely long time.

Examples of processes exhibiting such exponential relaxation times are: the change in magnetic induction resulting from a change in magnetizing force; the change in magnetic moment on removal of a crystal from a magnetic field by which its electronic spins have been aligned; and the decay of stress in a Maxwellian fluid.

RELAY.   The electrical relay is a device that utilizes the variation of current in an electric circuit as a controlling factor in another. For example, a certain change of current in one circuit may cause current to begin to flow in another, by the operation of a relay connected between them. There are numerous types of electrical relays, as they have been widely used in industry, particularly in apparatus of an automatic or semi-automatic nature, or for the protection of electric power equipment, or for communication systems. Protective relays are highly specialized and developed to where they will detect any electrical abnormality, and open the circuit containing that abnormality in any required time interval. Suitable relays will detect overcurrent, undercurrent, overvoltage, undervoltage, overload, reverse current, reverse power, abnormal frequency, high temperature, grounds, and phase unbalance.

Usually the relay involves two circuits, the energizing circuit and the relay circuit (the latter variously called the trip circuit, the sounder circuit, etc.). Protective relays may close the trip circuit immediately, or after a definite time interval, or after an inverse time interval. If the trip circuit contacts are normally open, the relay is called circuit closing; if they are normally closed, the relay is called circuit opening.

The automatic protection of electric power circuits is necessary for safety and economy. Fuses and automatic circuit breakers are the device most used for opening the circuit. A relay must be used to operate the tripping circuit of the circuit breaker.

See also **Circuit Breaker**; and **Fuse (Electric)**.

Over a number of years, relays were extensively used in macrologic systems (relay or ladder logic) for the sequential programming of partially or extensively automated manufacturing systems. Relay logic, although still used, was essentially displaced by the programmable logic controller (usually referred to simply as programmable controller) for a number of reasons as explained in the article on **Programmable Controller**.

RELIABILITY.   This term is used in two different contexts. In connection with biological assay, Finney defined the reliability of an assay as the reciprocal of a function of the confidence interval (see also **Confidence Interval**) of the estimate of potency of the stimulus.

The term is also used in factor analysis, especially in connection with the statistical analysis of psychological and educational tests. The "reliability" of a result is conceived of as that part which is due to permanent systematic effects, and therefore persists from sample to sample, as distinct from error-effects which vary from one sample to another. The term has not spread to other sciences. In a slightly more specialized sense the noun "reliability" sometimes means a reliability coefficient.

RELIABILITY COEFFICIENT.   If a measured quantity y can be written in the form

$$y = a + b + c$$

where a is constant, b varies from one individual to another with variance σ_b^2, and c (representing errors of measurement, etc.) varies from one measurement to another on the same individual with variance σ_c^2, the coefficient of reliability is defined by

$$r^2 = 1 - \frac{\sigma_c^2}{\sigma_b^2 + \sigma_c^2}$$

r is in fact the intraclass correlation between measurements on the same individual.

REMANENCE.   The residual induction B_r when the magnetizing field is reduced to zero from a value sufficient to saturate the material.

REMORAS (*Osteichthyes*).   Of the order *Discocephali*, family *Encheneisdae*, remora are marine fishes whose anterior dorsal fin is modified to form an oval sucker on the top of the head. This sucker is used to attach the creature to boats, turtles, or other large objects by which the fish may be carried about without effort. From their frequent attachment to sharks, they are also called shark suckers or simply suckerfishes. The *Remilegia australis* is a whalesucker. At one time, it was felt that the remoras used their hosts essentially for transportation, but it is believed that they feed on parasitic crustaceans, which also become attached to the host. Thus, the function of the remoras may be similar to that of cleaner fishes. There are about eight species; the *Echeneis naucrates* (sharksucker) is striped and may attain a length up to 36 inches (91 centimeters); the *Remoropsis pallidus*, which favors swordfish or tuna as the host, is the smallest species, attaining a length of about 7 inches (18 centimeters). See photograph in entry on **Sharks**.

REMOTE CONTROL.   The ability to operate equipment, apparatus, and processes from a distance. A high percentage of industrial control systems fall into this category, where the indicating, recording, and controlling instruments will be located up to several hundred feet from the equipment that is being controlled. Commonly, the instrumentation will be placed on a control panel or console where the operator can take readings on all points of a large installation quickly and efficiently. Otherwise, several operators would be required to physically move about the plant. Without remote control, the coordination of complex processes would be chaotic. Frequently, instruments and controllers will be contained within a separate control house or control room, a facility often air conditioned for protection of the equipment and to provide comfort for the operator. In installations of this type, the signals of information flowing from a control house to the process and back are either electric or pneumatic. Without the use of boosters, however, pneumatic transmission is distance-limited.

There are varying degrees of *remoteness*. Where telecommunications are used, the ability to control remotely, as in the case of space vehicles and probes, is essentially limitless. Solid wire and microwave communications commonly are used to carry control signals and information over long distances as encountered, for example, in pipeline management. Telephone lines are commonly used for such industrial communications purposes. Some interesting arrangements of remote control have been developed in connection with the manipulation of robots for undersea and radioactive applications.

REPEATABILITY.   With reference to industrial and scientific instruments, the Instrument Society of America defines *repeatability* as the

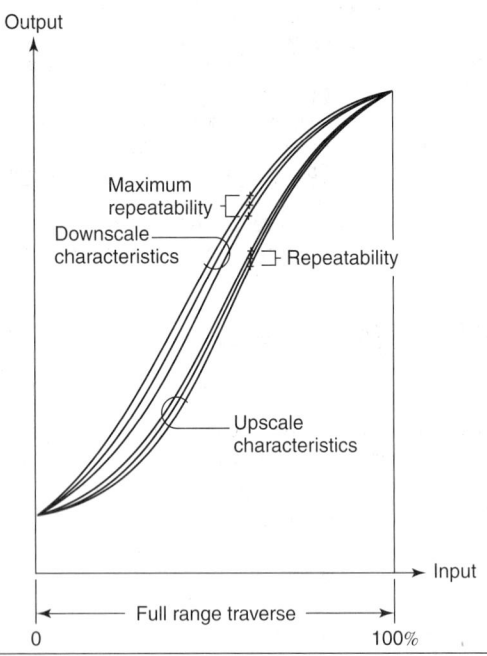

Fig. 1.   Graphic description of repeatability.

closeness of agreement among a number of consecutive measurements of the output for the same value of the input under the same operating conditions, approaching from the same direction, for full range traverses. See Fig. 1. Repeatability is usually measured as a *nonrepeatability* and expressed as repeatability in percent of span. Repeatability does *not* include hysteresis. See also **Reproducibility**.

REPOSE (Angle of).   The maximum angle with the horizontal at which an object on an inclined plane will retain its position without tending to slide. The tangent of the angle of repose equals the coefficient of static friction.

The term angle of repose is also used with the closely related meaning of the maximum angle with the horizontal at which loose material such as grain, sand, coal, or stone will retain its position without tending to slide. The moisture content and the distribution of the fine and coarse particles have a marked effect on the value of this angle. The angle of repose is an important factor in the design of retaining walls, earth dams, and embankments and is particularly valuable in the design of storage bins and bunkers since the allowable surcharge as well as the active horizontal pressure depends upon its value.

Angles of repose for some materials include: ashes (wet), 38–45°; bentonite (pulverized), 45°; cement (portland), 40°; clay (loose dry lump), 25–45°; coal (bituminous, slack, dry), 37°; coke (sized), 30°; earth (common loam, dry), 30–40°; feldspar (crushed), 45°; Fuller's earth, 23°; gravel (dry), 30–40°; malt, 33°; salt (cake), 36°; sand (dry, loose), 25–35°; sand (wet), 30–45°; slag (granulated), 45°; soda ash, 37°; and stone (crushed), 30–40°.

REPRODUCIBILITY.   With reference to industrial and scientific instruments, the Instrument Society of America defines *reproducibility* as the closeness of agreement among repeated measurements of the output for the same value of input made under the same operating conditions over a period of time, approaching from both directions. Reproducibility is usually measured as a *nonreproducibility* and expressed as reproducibility in percent of span for a specified period of time, but under certain conditions the period may be a short time during which drift may not be included. Reproducibility, in contrast to repeatability, does include hysteresis, drift, and also repeatability. See also **Repeatability**.

REPTILIA.   The reptiles, including the lizards and snakes, crocodiles, alligators and related forms, and turtles and tortoises. A class of the phylum *Chordata*.

The class is characterized as follows: (1) The skull articulates with the spinal column by a single process. (2) The mandibles are made up of several bones, joined with the skull by the quadrate bone. (3) The skin is covered with scales. (4) The heart is 4-chambered but the separation of the ventricles is incomplete. (5) The members are poikilothermal. (6) The members have extraembryonic membranes during development, a fundamental requirement of terrestrial life in the vertebrates; hence, the reptiles are grouped with the birds and mammals as *Amniota*.

Although the group is the lowest class of vertebrates to attain the capacity for entirely terrestrial life, many reptiles are now amphibious or aquatic. Even the aquatic species, however, come to the land to deposit their eggs.

Reptiles are economically important to a rather limited extent. The flesh of some turtles, lizards, and even snakes is used as food, and the skin of the alligator makes an excellent, if conspicuous, leather. In some regions, crocodiles have been known to kill human beings. Usually, it seems that the danger from them is very limited. Poisonous snakes also may destroy human life, but here again the danger is usually encountered only under special conditions and snake bites may be regarded as accidental.

There are a number of extinct groups, but the classification of the living groups is briefly as follows:

> Order *Prosauria*. A single living species, the tuatara of New Zealand, makes up this order. It is a lizardlike animal with a few primitive structural characteristics, including a well-developed pineal eye. The order also bears the name *Rhynchocephalia*.
> Order *Chelonia (Testudinata)*. The turtles, tortoises, and related species. Characterized by the shell, consisting of an upper carapace and a lower plastron, formed of bony plates with a horny sheath.
> Order *Crocodilia*. Large reptiles resembling lizards in form. The jaws are elongated. The thick skin is provided with bony plates. The crocodiles, alligators, garial, and related species.
> Order *Sauria*. The lizards and snakes. In some classifications these forms are included in separate orders, and in some the lizards constitute the suborder *Sauria* and the snakes the suborder *Serpentes* of the order *Squamata*. The order including the lizards is also named *Lacertilia*. All of these animals have elongate bodies. The skin bears horny scales and in some cases bony plates.

Numerous species of reptilia are described throughout this encyclopedia.

REPULSIVE FORCES.   Forces between bodies which tend to move them apart. The existence of such forces between molecules is shown by their collision diameters and similar properties; while in the case of crystals these forces, in equilibrium with forces of attraction, result in the formation of stable ionic systems.

REQUIEM SHARKS.   See Sharks.

RESIDUE.   If $f(z)$, a function of the complex variable, has a pole at $z = z_0$ so that it may be expanded in a Laurent series $\Sigma a_n (z - z_0)n$, then the coefficient a_{-1} is the *residue* of the function at the point z_0. Determination of the residues of a function makes it possible to evaluate an integral in the complex plane by the Cauchy theorem (see **Calculus**). However, expansion in a Laurent series is not always easy so the following procedures are often used.

1. Let the function be given in the form $w = f(z)/g(z)$ with a simple pole at $z = z_0$. If the denominator can be factored so that $g(z) = (z - z_0)F(z)$, its residue $R = f(z_0)/F(z_0)$.

2. If the denominator is not easily factored and there are n simple poles at $z_1, z_2, z_3, \ldots$ the sum of the residues is

$$S = \sum_{i=1}^{n} f(z_i)/g'(z_i)$$

where the prime means differentiation.

3. There are simple poles at $z = 0$, $z_1, z_2, \ldots$ and the function has the form $w = f(z)/zg(z)$. The sum of the residues is

$$S = f(0)/g(0) + \sum_{i=1}^{n} f(z_i)/z_i g'(z_i)$$

4. The function w has a pole of order k at $z = z_0$, then its residue is

$$R = \frac{1}{(k-1)!} \left[\frac{(z - z_0)^k f(z)}{g(z)} \right]_{z=z_0}^{(k-1)}$$

the $(k - 1)$th derivative to be evaluated at $z = z_0$.

Modifications of these methods can be made to care for other cases.

RESILIENCE. The resilience of a body measures the extent to which energy may be stored in it by elastic deformation. The implication of the word "stored" in the above definition is that this energy may be released in the form of mechanical work when the force causing the elastic deformation is removed, and that resilience is a property of a material within its elastic limit. The "modulus of resilience" is the maximum energy storage in a unit volume of the material. In practical units, it is the inch-pounds of energy stored in a cubic inch of the material stressed to the elastic limit. The modulus of resilience is directly proportional to the square of the stress, and inversely proportional to the modulus of elasticity. It is equal to the area under the stress-strain diagram up to the elastic limit, or

$$\frac{1}{2}\frac{\sigma^2}{E}$$

in which E is the modulus of elasticity and σ is the elastic limit.

RESINS (Natural). Complex compounds composed of carbon, hydrogen, and relatively small amounts of oxygen, which are secreted in various tissues of many plants. In the pine family, where resins are very common, they are secreted as oleoresins in resin canal cells, which break down finally, producing resin canals. These canals appear as longitudinal ducts in the sapwood and inner bark, connected laterally by resin canals in the compound wood rays, thus forming an extensive network. A common name given to the oleoresin in this group is pitch, the sticky juice that exudes from the plant wherever it is wounded. On exposure to the air the volatile oil in this pitch (oil of turpentine) gradually evaporates, leaving a clear hard glassy substance, the resin, which forms a protective coating over the wound.

Most resins have the same physical properties, being clear, translucent, and of a yellow or brownish color. Amber, a fossil resin, is a more or less familiar example. Resins are insoluble in water, but soluble in common organic solvents such as ether and alcohol. All resins burn with a sooty flame. Resins seem to be mainly of value to the plan in that they form protective coverings against the entrance of disease-producing organisms and also prevent excessive loss of water from the thin-walled tissues exposed in the wound.

Resins are separated into several classes. Many of the resins contain almost no volatile oil and are hard, without taste or odor. These are the varnish or hard resins. Other resins, when removed from the plant in which they are formed, and dissolved in volatile oils, form a thick semi-solid mass: these are the oleoresins. In still other cases the resin occurs in combination with a gum, forming a gum resin.

Hard Resins. Several of the hard resins, used mainly for making varnishes, are called copals. Most of them come from Africa and are either found in fossil form or obtained from living plants. Other copals come from Australia, New Zealand and East Indian Islands. The plants that form them are members of the legume and pine families. The African copals are products of several species of *Trachylobium*, fairly large trees growing in east Africa and Madagascar. The best resin from these trees occurs in a fossil form, often deeply buried in the ground—sometimes in regions where the trees no longer grow. These resins dissolve slowly and are used in making varnishes, which are very durable. A South American tree of large size, *Hymenaea courbaril*, also of the Leguminoseae, yields a very similar resin, which is also found in lumps in the ground around the trees, and used in varnishes.

Another copal is obtained from *Agathis australis*, a very large coniferous tree native in Australia and New Zealand, where it is known as the Kauri pine. Like the other copals, that from the Kauri pine is found in lumps buried in the ground. Most of these lumps are 1 or 2 inches in diameter, but some are much larger, weighing up to 100 pounds. Nearly all of this resin comes from the northern part of North Island of New Zealand. It is frequently called Kauri gum, though it is not a gum, but a true copal resin. Another group of hard resins, known as dammar resins, is obtained from many different trees growing in southern Asia and the East Indian Islands. These resins dissolve readily in alcohol, forming spirit-varnishes.

One of the commonest and most important of the hard resins is rosin, obtained by distilling the pitch, or turpentine, which is a product of several of the native pines of the southeastern United States. This rosin, also known as colophony, is a very important product of that region. Originally the turpentine was obtained by chopping a deep hollow in the base of the trunk of the tree and allowing it to fill up with the turpentine, which was then scooped out. This method was very destructive and wasteful, since much

of the oleoresin, turpentine, was lost during the process. The weakened trees were easily blown down.

Now turpentine is obtained by cutting V-shaped gouges in the bark and inserting metal gutters beneath the gouges. These gutters carry the turpentine to a cup placed underneath. As soon as the cut is made, turpentine begins to flow and continues to do so for two or three days, gradually slowing as the drying turpentine allows resin to accumulate and plug the wounds. A new flow is obtained by cutting off a narrow strip of bark from the upper edge of the cut. The process is continued as long as the pitch will flow, which is usually all summer and well along into the late fall. Each tree may be turpentined for 6 to 7 successive years or even longer before it ceases to be profitable.

The crude turpentine collected in the cups and the product that has dried on the wound of the tree are removed and carried to the still. Here the turpentine, to which a little water is added, is carefully heated to drive off the oil of turpentine present, together with the water added. The distillate is condensed by passing it through a coil around which cold water is flowing. It is collected in a barrel or any suitable container. The two substances, water and oil of turpentine, which make up the distillate, are immiscible and soon separate, the lighter oil of turpentine rising to the top and floating on the water, which is drawn off from the bottom. Oil of turpentine is often called spirits of turpentine, or, in the paint trade, turpentine. In medicine the word turpentine is reserved for the oleoresin which upon distillation yields oil of turpentine and rosin.

The residue remaining in the tank at the end of the distillation is skimmed to remove any impurities such as twigs, bits of bark and dirt, and run into vats to cool. Then it is put into barrels and allowed to harden, forming rosin.

Oil of turpentine is used principally as a solvent for paints and varnishes, because it mixes readily with the various substances used and also because it evaporates quickly, causing the paint or varnish to dry. It is also used in making such things as sealing wax and shoe polish. Very pure grades of turpentine (the oleoresin) are used medicinally.

Large quantities of rosin are used in sizing paper, which makes it take ink without spreading or blotting, gives it a smoother surface and makes it heavier. Rosin is also used in cheaper varnishes, in paints, and in soap making. It is furthermore used as an adulterant of the more expensive resins. Linoleum manufacturers use large amounts of rosin.

In early times, large quantities of crude turpentine were used to waterproof the rigging of the sailing vessels and to calk the seams of the hull.

Mastic is a hard resin exuding from the branches of one of the Pistachio trees, *Pistacia lentiscus*, native of Mediterranean Europe and southern Asia. Formerly, it was extensively used medicinally, for stomach troubles and dysentery, as well as other ailments. Now it is used in making varnishes and in lithographic work. Natives of the region in which it is found chew mastic, which has a pleasant taste.

Since turpentine is a mixture of a volatile substance, spirits of turpentine and a hard resin, it is one of the oleoresins.

Oleoresins. Canada balsam is one of the oleoresins. It is obtained from the bark of *Abies balsamea*, the common balsam fir of northern North America. Canada balsam, because its refractive index is so near that of glass, is much used in optical work and in preparing materials for examination with a microscope.

Little used today is Dragon's blood, an oleoresin obtained from the fruits of *Daemonorops draco*, a native palm of southeastern Asia and the Molucca Islands. The resins exudes from the surface of the ripening fruits. It is removed from them by boiling in water. The resin is then moulded into balls or long sticks. It is sometimes used in making varnishes and lacquers.

True lacquer, obtained from the juice of *Rhus verniciflua*, a sumac tree of southeastern Asia, is another oleoresin. To obtain the juice lateral cuts are made in the bark. The exuded sap is collected not only from these cuts but from small branches which are cut off and soaked in water. The juice is cleaned of any foreign substances by straining it through hemp cloth. By slow heating, either artificial or by the sun, the juice is evaporated and stored until used. Lacquer is a poisonous substance, causing intense irritation of the skin in many people. Others seem to be immune. Lacquer is usually applied over some soft wood, commonly soft pine, the pores of which have first been filled by rubbing in a paste of rice and resin, followed by a paste of soft clay and resin. The surface is then covered with cloth and layer after layer of lacquer put over that. Each layer is allowed to dry and rubbed down very smooth before the next layer is added. Any color to be added is mixed with the lacquer, with each colored layer covered by a clear layer before another is put on. The final product is a thick covering composed of many thin layers of lacquers. If this is carved the edges of the carving, on careful examination, will show the fine lines separating the different layers. Lacquering is a very old industry, having been carried on in China since the sixth century.

Certain resins occur in combination with fragrant volatile oils. One of these is benzoin, obtained from *Styrax benzoin* by cutting notches in the bark and allowing the resin to collect in them. It is used in making perfumes, in incense, and as a source of benzoic acid, used medicinally.

Another fragrant oleoresin is storax, obtained from *Liquidambar orientalis*, a medium-sized tree growing in southwestern Asia. The resin is obtained by boiling the bark and wood of young branches. It is used medicinally and also in incense.

Gum Resins. Gum resins include myrrh, which exudes from the trunk and branches of *Commiphora myrrha*, a tree growing in the region around the Red Sea. The lumps of resin are used medicinally, and also in making incense. Another gum resin is frankincense, obtained by cutting notches in the stem of *Boswellia carterii*, which grows in northeastern Africa and in Arabia. This resin is used in incense. *Asafoetida* is also a gum resin. See also **Gums and Mucilages.**

RESISTANCE.   The uses of this term in physics are in accordance with its general meaning of "that which tends to oppose motion."

1. Mechanical resistance is the opposition offered by a material body to forces that tend to produce motion. This mechanical resistance may arise from friction, from stresses set up in rigid anchors, or from inertia. Whenever the power dissipated in friction is proportional to the square of the velocity, mechanical resistance may be defined as the real part of mechanical impedance, the unit of which is the mechanical ohm.

2. Acoustic resistance is defined as the real component of acoustic impedance, the commonly used unit being the acoustic ohm. Acoustic flow resistance (D.C. acoustic resistance) is defined as the quotient of the pressure difference between the two surfaces of a sound-absorbing material by the volume current through the material.

3. Fluid resistance is the opposition offered by gases or liquids to the passage of bodies through them.

4. Electric conductors are believed to contain free electrons, the movement of which through the substance constitutes electric conduction. In this migration the moving particles evidently meet with some restraint, since heat is generated. Electrical resistance is the factor by which the square of the instantaneous conduction current must be multiplied to give the power lost by dissipation as heat or other permanent radiation of energy away from the electric circuit. (Consider the "radiation resistance" of an antenna.) The unit of electrical resistance is the ohm. The measure of the resistance of a given conductor is the electromotive force required per unit current, and is usually expressed in ohms. The resistance of a wire or other linear conductor of uniform cross section is proportional to the length l and inversely proportional to the cross section a: $R = \mathbf{r}/a$. The constant $\mathbf{r}$ is the "resistivity" of the substance, usually expressed in ohm-centimeters; and its reciprocal is the "electric conductivity." The dependence of resistivity of metals upon temperature is one of the major problems of electron physics.

Pieces of wire may be cut off at such lengths as to have definite resistances, and mounted with convenient connections to form a "resistance box," used in many electrical measurements. A "rheostat" is usually a rugged conductor, often with adjustable resistance, used to introduce a resistance load into a circuit. Resistances are commonly measured by means of some form of bridge, of which the Wheatstone bridge is most familiar.

Some typical resistivities are given in Table 1:

TABLE 1. RESISTIVITIES OF SOME COMMON MATERIALS
(In ohm-centimeters at 20 °C)

Aluminum	2.83×10^{-6}	Mercury	95.78×10^{-6}
Brass	7.0	Nickel	7.8
Copper	1.72	Platinum	10.0
German silver	33.0	Silver	1.63
Iron (pure)	10.0	Tin	11.5
Lead	22.0	Tungsten	5.51

For some purposes, it is convenient to express the resistivity as the resistance of one foot of wire of the given metal having a cross-section of one circular mil [a circular mil is the area of a circle 0.001 foot (0.0003 meter) in diameter]. This value may be obtained by multiplying the resistivity in ohm-centimeters by the factors 6.015×10^6.

The dependence of resistance of many metallic conductors upon temperature is expressed with fair approximation by the linear equation

$$R = R_0(1 + At)$$

in which R_0 is the resistance at 0 °C and t is the centigrade temperature. The temperature coefficient A is a constant characteristic of the metal.

5. There remains yet another type of resistance for consideration — heat resistance. This might be said to be the property of offering opposition to the flow of heat. This property is desirable in a heat insulator such as pipe covering, but undesirable in heat transfer equipment. The term resistivity is rarely used in connection with heat flow, as its reciprocal, thermal conductivity, is quite satisfactory, and enjoys the advantage of common usage.

RESISTANCE THERMOMETER.   The fact that the electrical resistance of a metal wire increases with rising temperature is the basis of a very useful class of thermometers. One has only to calibrate a given length of wire, as to its resistance in relation to its temperature, enclose it in a suitable protecting tube, and keep it connected with the resistance-measuring bridge, to have a resistance thermometer adapted to a variety of uses over a very wide temperature range. The metal nearly always employed is platinum. The variation of resistivity with temperature of platinum is very nearly linear, being closely approximated by the formula $\mathbf{r} = 0.000000037t + 0.000011$, in ohm-centimeters and centigrade degrees. Callendar found that for any given platinum resistance thermometer there is a slight systematic departure from this formula, characteristic of the particular sample of wire. It is best, therefore, to calibrate each instrument throughout the range for which it is intended.

Care must be taken, in mounting the platinum wire, that it does not come in contact with materials that will contaminate it at high temperatures. Compensation is also necessary for the change of resistance in the wires leading to the platinum spiral. This is commonly effected by balancing against these wires a pair of "dummy" wires similar to and laid alongside them. The instrument is usually provided with a suitably designed resistance bridge, such as the Callendar and Griffiths or the Mueller bridge; which for practical purposes should be portable and self-contained, with battery, galvanometer, balancing rheostat, etc., all in one case, and with a cable leading to the thermometer proper. The resistance thermometer is also called a resistance pyrometer.

Only a few of the pure metals have a characteristic relationship suitable for the fabrication of sensing elements used in resistance thermometers. The metal must have an extremely stable resistance-temperature relationship so that neither the absolute value of the resistance R_0 nor the coefficients a and b drift with repeated heating and cooling within the thermometer's specified temperature range of operation. The relationship may be expressed by:

$$R_t = R_0(1 + at + bt^2 + ct^3 + \cdots)$$

where R_0 = resistance at reference temperature (usually at ice pint, 0 °C), ohms

R_t = resistance at temperature t, ohms

a = temperature coefficient of resistance, ohm/ohm-degree C

b and c = coefficients calculated on the basis of two or more known resistance-temperature (calibration) points.

The material's specific resistance in ohms per cubic centimeter must be within limits that will permit fabrication of practical-size elements. The material must exhibit relatively small resistance changes for nontemperature effects, such as strain and possible contamination which may not be totally eliminated from a controlled manufacturing environment. The material's change in resistance with temperature must be relatively large in order to produce a resultant thermometer with inherent sensitivity. The metal must not undergo any change of phase or state within a reasonable temperature range. Finally, the metal must be commercially available with essentially a consistent resistance-temperature relationship to provide reliable uniformity.

Industrial resistance thermometers, often referred to as resistance temperature detectors (RTD), usually are made with elements of platinum, nickel, or copper. The entire resistance thermometer is an assembly of parts which include the sensing element, internal leadwires, internal supporting and insulating materials, and protection tube or case. A platinum industrial resistance thermometer assembly is shown in Fig. 1.

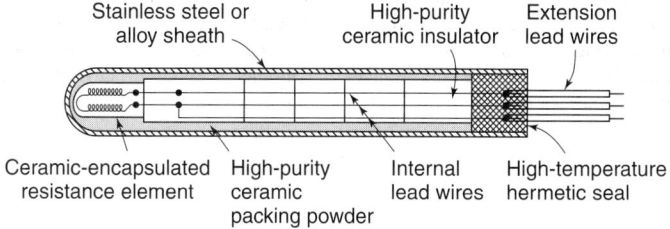

Fig. 1.   Platinum industrial resistance thermometer assembly.

RESISTANCE TRANSDUCER.   A large number of transducers for the measurement of several variables (temperature, pressure, flow, and so on) are designed around some aspect of Ohm's law, wherein the electrical resistance of a material or device can be caused to change when the material is subjected to some condition or situation that is the object of measurement.

As in the case of a potentiometer, the resistance of the device can be altered in accordance with the input from some external force to be measured by moving a contact and thus change the resistance value between two points. Inasmuch as resistance changes with the cross section of a conductor, that cross section can be altered as in the case of a strain gage. Resistance also is a function of temperature. Hence, resistance transducers can be used as temperature sensors. Similarly, resistance often is affected by humidity changes and by chemical composition changes.

Numerous transducers of the resistance type are described throughout this volume.

RESISTIVE-WALL AMPLIFIER.   An electron-beam amplifier in which the beam flows near a resistive wall. Gain is obtained through interaction between the stream-charge and the wall-charge, which is induced by the stream. The wall-charges act on the stream so as to cause larger and larger bunches to be formed, which result in exponential growth of the original signal with distance. Gain and gain-bandwidth are comparable to other forms of traveling wave tubes, and the stability is inherently greater.

RESISTIVITY (Specific Resistance).   A proportionality factor characteristic of different substances equal to the resistance that a centimeter cube of the substance offers to the passage of electricity, the current being perpendicular to two parallel faces. It is defined by the expression:

$$R = \rho \frac{l}{A}$$

where R is the resistance of a uniform conductor, l is its length, A is its cross-sectional area, and ρ is its resistivity. Resistivity is usually expressed in ohm-centimeters.

RESISTOR TRANSISTOR LOGIC.   A form of logic circuit using transistors and resistors. The abbreviation RTL is sometimes used to refer to this class of circuits. A circuit for performing the logical operation NOT

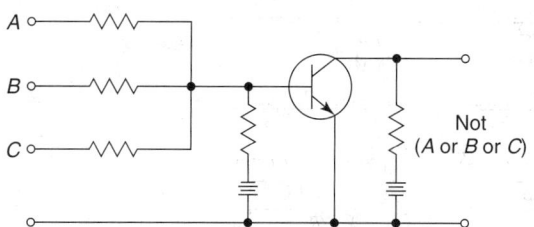

Fig. 1.   Resistor transistor logic (RTL) circuit effecting NOT-OR (NOR) function.

OR (NOR) is shown in Fig. 1. Although the circuit values are chosen to bias the transistor beyond cutoff when A, B, and C are all at ground potential (0 logic state), change to the positive voltage associated with 1 operation of either of the three inputs will result in the transistor being brought into full conduction. Its collector will then be close to ground potential corresponding to a 0 (NOT 1) output. Thus, the circuit effects the operation NOT (A OR B OR C).

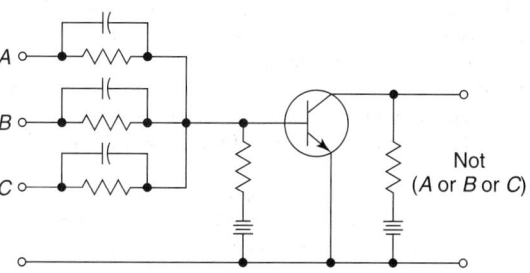

Fig. 2.   Resistor capacitor transistor logic (RCTL) circuit effecting NOT-OR (NOR) function.

A variation of this arrangement, known as *Resistor Capacitor Transistor Logic* (RCTL), is shown in Fig. 2. Operation is the same as RTL except that higher speeds of operation are possible because the switching time of the transistor from one state to the other is reduced markedly by the use of the capacitors.

THOMAS J. HARRISON, International Business Machines Corporation, Boca Raton, FL.

RESOLUTION.   A term used in a number of specific cases in science to denote the process of separating closely related forms or entities or the degree to which they can be discriminated. The term is most frequently used in optics to denote the smallest extension which a magnifying instrument is able to separate or the smallest change in wavelength which a spectrometer can differentiate. In this last sense, it is defined as the ratio of the average wavelength (wave number or frequency) of two spectral lines, which can just be detected as a doublet, to the difference in their wavelengths (or wave numbers or frequencies). The term resolution is also applied to such varied processes as the separation of a racemic mixture into its optically active components or as the breaking up of a vectorial quantity into components.

RESOLUTION (Computer System).   In systems where either the input or output of the subsystem is expressed in digital form, the resolution is determined by the number of digits used to express the numerical value. In a digital-to-analog converter, the output analog signal takes on a finite number of discrete values, which correspond to the discrete numerical input. The output of an analog-to-digital converter is discrete although the analog input signal is continuous.

In digital equipment, resolution is typically expressed in terms of number of digits in the input or output digital representation. In the binary system, a typical specification is that "resolution is x bits." As an example, if V_{fs} is the fullscale input- or output-voltage range, this specification states that the resolution is $V_{fs}/2^x$. If $x = 10$ and $V_{fs} = 5V$, the resolution is $5/2^{10}$, or $0.00488V$. It is also common to express resolution in terms of "parts." A 4-digit decimal converter may be said to have a resolution of "1 part in 10,000" and a 10-bit binary converter may be said to have a resolution of "1 part in 1,024." The term *least significant bit* (LSB) also is used. It may be stated, for example that the binary resolution is "$\pm\frac{1}{2}$ LSB." Also used is the term *least significant digit* (LSD). This term is used with relation to decimal or other nonbinary digital equipment.

RESOLVING POWER (Microscope).   This is given by the relation $d = 1.22\lambda/2$ N.A., where d is the linear separation of two points, λ is the wavelength used, and N.A. is the numerical aperture of the object lens. Most telescopes have large objective lenses in order to have large light-gathering power, and to have high resolution. This high resolution may produce resolved images too close together to be resolved by the human eye. Hence an eye-lens or ocular is included in the system for the purpose of magnifying the initial image so that the eye can see it as resolved. Note that no amount of magnification of the initial image can increase the resolving power of the telescope over the resolving power of the objective lens.

RESOLVING POWER (Telescope). The ability of a telescope to separate the images of the two stars of a double star, for example. Most studies of resolving power are based on the Rayleigh criterion of resolving power. Most telescopes have large objective lenses in order to have large light-gathering power, and to have high resolution. This high resolution may produce resolved images too close together to be resolved by the human eye. Hence an eye-lens or ocular is included in the system for the purpose of magnifying the initial image so that the eye can see it as resolved. Note that no amount of magnification of the initial image can increase the resolving power of the telescope over the resolving power of the objective lens. See also **Telescope (Astronomical-Optical)**.

RESONANCE. 1. Every physical system, in general, has one or more natural vibration frequencies characteristic of the system itself and determined by constants pertaining to the system. Thus a flexible string of length l and mass δ per unit length, and subjected to a tension f, will, if struck or plucked and left to itself, vibrate with frequencies equal to

$$\frac{1}{2l}\sqrt{\frac{f}{\delta}}$$

and to various integral multiples thereof (overtones). If such a system is given impulses with some arbitrary frequency, it will necessarily vibrate with that frequency even though it is not one of those natural to it. These "forced vibrations" may be very feeble; but if the impressed frequency is varied, the response becomes rapidly more vigorous whenever any one of the natural frequencies is approached, its amplitude often increasing many fold as exact synchronism is reached. This effect is known as resonance. The many uses of this conception in present-day physics stem from this initial use in mechanics or acoustics to denote a prolongation or reinforcement of sound by induced vibration. Such acoustical (and mechanical) resonance can often be represented by a differential equation of the form

$$M\frac{d^2x}{dt^2} + R\frac{dx}{dt} + Sx = A\cos\omega t$$

which permits a mathematical statement of velocity resonance and displacement resonance as given in the Table 1.

2. Electrical resonance is a condition that tends to produce relatively great currents in reactive circuits. There are two types, series resonance and parallel resonance, as explained in the following discussion. In an alternating-current circuit containing a coil and a capacitor in series, the impedance is given by

$$Z = R + j\left[\omega L \frac{1}{\omega C}\right]$$

where R is the resistance of the coil, ω is 2π times the frequency, L is the inductance of the coil and C is the capacitance. It can readily be seen that at some frequency the terms in the brackets will cancel each other, and the

impedance will equal the resistance alone. This condition, which gives a minimum impedance (and thus a maximum current for a fixed impressed voltage) and unity power factor, is known as series resonance. Where the resistance is small the current may become quite large. As the voltage drop across the capacitor or coil is the product of the current and the impedance of that particular unit, it may also become very large. The condition of resonance may even give rise to a voltage across one of these units which is many times the voltage across the whole circuit, being, in fact, **Q** times the applied voltage for the capacitor and nearly the same for the coil. This is possible since the drops across the coil and condenser are nearly 180° out of phase and thus almost cancel one another, leaving a relatively small voltage across the circuit as shown in Fig. 1a–b (circuit at left).

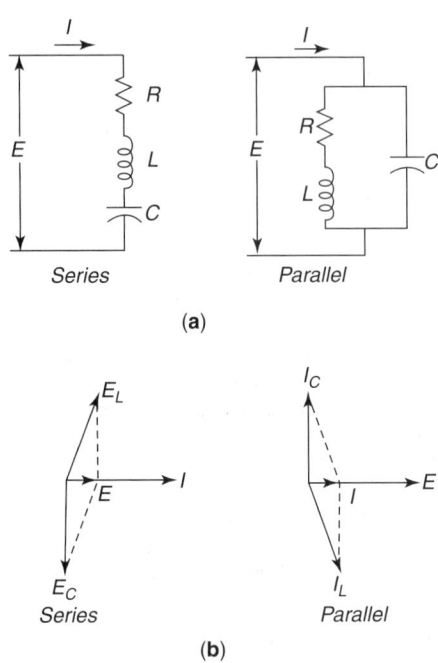

Fig. 1.   Typical (**a**) resonant circuits and (**b**) vector relations.

For a circuit composed of a coil in parallel with a capacitor the opposite effects of these two types of reactance will counteract one another at some frequency and produce unity power factor for the circuit. This is parallel resonance or anti-resonance (Fig. 1a–b, circuit at right). In such a circuit, the currents in the individual branches may be many times that in the line, since they are out of phase and combine vectorially to give the line current. The impedance of a parallel resonant circuit is very high, its behavior being almost identical with that of the current in a series circuit if the Q of the

TABLE 1. PROPERTIES OF RESONANT SYSTEMS

$$M\frac{d^2x}{dt^2} + R\frac{dx}{dt} + Sx = A\cos\omega t$$

	At Velocity Resonance	At Displacement Resonance	At the Natural Frequency
Frequency	$\dfrac{1}{2\pi}\sqrt{\dfrac{S}{M}}$	$\dfrac{1}{2\pi}\sqrt{\dfrac{S}{M} - \dfrac{R^2}{2M^2}}$	$\dfrac{1}{2\pi}\sqrt{\dfrac{S}{M} - \dfrac{R^2}{4M^2}}$
Amplitude of displacement	$\dfrac{A}{R\sqrt{\dfrac{S}{M}}}$	$\dfrac{A}{R\sqrt{\dfrac{S}{M} - \dfrac{R^2}{4M^2}}}$	$\dfrac{A}{R\sqrt{\dfrac{S}{M} - \dfrac{3R^2}{16M^2}}}$
Amplitude of velocity	$\dfrac{A}{R}$	$\dfrac{A}{R\sqrt{\dfrac{S}{M} - \dfrac{R^2}{4MS - 2R^2}}}$	$\dfrac{A}{R\sqrt{\dfrac{S}{M} - \dfrac{R^2}{16MS - 4R^2}}}$
Phase of displacement with reference to applied force	$\dfrac{\pi}{2}$	$\tan^{-1}\sqrt{\dfrac{4MS}{R^2} - 2}$	$\tan^{-1}\sqrt{\dfrac{16MS}{R^2} - 4}$

For values of R small compared to $\sqrt{SM}$ there is little difference between the three cases shown.

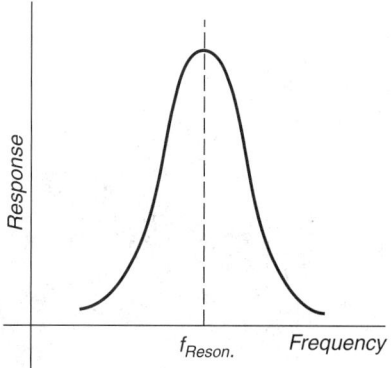

Fig. 2.   Typical frequency-response curve for resonant circuit.

parallel circuit is above 10. Figure 2 shows a typical frequency-response curve for resonant circuits, the ordinates being current for a series and impedance for a parallel resonant circuit. In series resonance, the condition for resonance is given exactly by the following expression, which is also approximately true for parallel resonance if the Q is high:

$$\omega L = 1/\omega C$$

From this equation the resonant frequency is found to be

$$f = \frac{1}{2\pi\sqrt{LC}}$$

Both types of resonance are widely used in communication circuits to select certain frequencies in preference to others. An example is the tuning circuit of the radio receiver.

3. Resonance phenomena are exhibited by all systems in motion, including molecular, atomic and electronic systems. The approach of quantum mechanics clarifies the behavior of such systems.

4. Another resonance phenomenon involving electrons or atomic nuclei is magnetic resonance.

Nonlinear Resonance. Provided the dissipation of the system is low, the curve of the response (e.g., rms value) to a driving force, of a double energy-storage, passive system containing at least one nonlinear energy storage element may be double-valued over a certain range, when plotted against the independent variable of driving force, or frequency, or one of the system parameters. Such a system is said to be in nonlinear resonance, if the operating point is on the upper leg of this double-valued response curve, and if the lowest-frequency component of the response is of the same frequency as the fundamental frequency of the driving force. If the circuit parameters, or the excitation, or both, preclude the existence of a double-valued response, the system is said to be in nonlinear resonance, if the operating point is in that range of the curve of response (e.g., root-mean-square value) versus driving force, in which the slope of the curve is greater than the slope in the vicinity of the origin.

RESONANCE RADIATION.   A process of reemission of radiation by gases and vapors. The process involves excitation, by incident photons, of atoms to higher energy levels, from which they may return to the ground state, or to other states. Therefore the radiation emitted, while characteristic of the particular atom, is not necessarily of the same frequency as that absorbed. Resonance radiation appears to be a species of fluorescence, except that it may take place with no change in frequency. For example, Wood found that sodium vapor, upon absorption of D-light (16,973 cm^{-1}), re-emitted the same frequency and therefore the name resonance radiation was adopted. But when Strutt excited sodium vapor by light of wave number 30,273 cm^{-1} (second line of the principal sodium series), the emitted radiation was D-light (16,973 cm^{-1}).

RESONATOR.   1. A device used to utilize or exhibit the effects of resonance. 2. A group of electrons which absorbs electromagnetic radiation of certain frequencies.

RESORPTION.   The absorption, or less commonly, the adsorption by a body or system of material previously released from absorption or adsorption by that same body or system.

RESPIRATION (Plants).   Plant cells, like all living cells, require energy to maintain the organization of materials and activities we associate with life. In plant cells, as in animal cells, this energy is provided by the oxidation of foods, commonly carbohydrates, in the process known as respiration. This oxidation is a sequence of enzyme-controlled reactions, in which the energy of the food is released a little at a time. The net result of the breakdown of a carbohydrate, commonly glucose, is the production of carbon dioxide and water, and the utilization of oxygen. The following equation expresses the overall result:

$$C_6H_{12}O_6 + 6\,O_2 \longrightarrow 6\,CO_2 + 6\,H_2O + \text{useful energy}$$

Thus oxygen is absorbed from the atmosphere and carbon dioxide is released.

The net results of this process are the opposite of those in photosynthesis, in which carbon dioxide is used and oxygen is produced. When green parts of plants are exposed to light, ordinarily photosynthesis proceeds much faster than respiration, so that respiration cannot be detected by the gaseous exchanges. In the dark, however, the green tissues as well as those lacking chlorophyll demonstrate the gas exchange characteristic of respiration.

Sometimes other foods, such as proteins, fats, or organic acids, may be oxidized in plant cells. In such cases the respiratory quotient (R.Q.) or respiratory ratio (ratio of amount of CO_2 released to amount of O_2 used) is not 1, as it is when carbohydrates are oxidized. Proteins and fats contain relatively less oxygen than do the carbohydrates, so more oxygen is required to bring about their complete oxidation, and the R.Q. is less than 1. When certain of the organic acids are respired, the R.Q. is greater than 1.

The sequence of chemical reactions in respiration results in a stepwise breakdown of the carbohydrate molecule. Phosphate groups are incorporated as a part of the sugar molecule, and are exceedingly important in the energy relationships of the cell. In certain types of phosphate-containing organic compounds, the phosphate is attached through "energy-rich" bonds. When these bonds are broken, the energy thus released may drive a chemical reaction or may perform useful work within the cell. Several of the steps in the respiratory process produce these energy-rich compounds. This represents a means of transferring energy within the cell, from respiration to other chemical processes or to any site where energy is required.

The separate reactions in the sequence of respiration are controlled by enzymes. A large number of enzymes are required for the whole process. A number of these depend for their activity upon coenzymes that are vitamins of the B complex, or derivatives of these compounds. For example, thiamine (vitamin B_1) serves as a coenzyme in the release of carbon dioxide.

The energy released in respiration may be used in a variety of ways. Protoplasm must be made continually, especially where there is rapid growth. The manufacture of protoplasm requires energy. The synthesis of many other compounds within the cell also requires energy. The accumulation of mineral ions by plant cells, the "pumping" of water by root pressure, and the maintenance of the semipermeability of the plasma membrane all require some of the respiratory energy. Unfortunately for the plant, the energy transfers are not completely efficient, and some of the energy is lost as heat. The release of heat is frequently demonstrated by placing germinating seeds in a thermos bottle and noting the increase in temperature of the seeds with a thermometer.

The foregoing discussion refers primarily to aerobic respiration, or respiration at the expense of atmospheric oxygen. *Anaerobic respiration* also may occur in plant cells. A familiar example is the fermentation which occurs when yeast cells act upon a sugar solution:

$$C_6H_{12}O_6 \longrightarrow 2\,CO_2 + 2C_2H_5OH$$

The sugar is partially oxidized, releasing carbon dioxide and alcohol and liberating some of the energy stored in the sugar. The process can occur in the absence of oxygen. Sometimes other products, such as acetic acid or lactic acid, result from similar processes. Some of the initial enzymes and chemical reactions in these processes are the same as those in aerobic respiration. In most plant cells aerobic respiration occurs in preference to anaerobic respiration if oxygen is available. This is advantageous to the plant because the aerobic reaction releases considerably more of the available energy of the food being oxidized.

RESPIRATORY SYSTEM.   The assemblage of organs by which air or water is brought into contact with tissues which can absorb part of its contents of oxygen.

Many animals absorb oxygen through the surface of the body. This is particularly true of small and simple forms, but the skin of the earthworms and that of some amphibians absorb all of the oxygen required by the animals, and any moist skin may absorb small quantities. The simplest modification to be introduced as a respiratory system is some extension of the surface to provide for the needs of a more bulky body. Tufted or thin plate-like structures called gills project into the water from the surface of many aquatic animals. Such structures are not adapted for air breathing because their epithelium must be moist for the ready passage of oxygen and their extensive surface favors drying when exposed to air.

In aquatic insects gills contain gas-filled tubes (*tracheae*) and are known as tracheal gills, but in most animals the blood or body fluids circulate through them. Gills of this kind are found in many annelid worms and in the crustaceans. In the latter group they are sometimes protected by a fold of the body wall. This fold encloses them in a chamber through which water is propelled by special appendages.

In many terrestrial arthropods, the respiratory system consists of air tubes or tracheae, metamerically arranged. In the primitive state each segment contains a pair of tracheae opening to the surface of the body separately through small pores called spiracles or stigmata. The openings are usually guarded by some closing device or by a grating formed from the cuticula. The tracheae have coiled chitinous filaments (*taenidia*) in their walls which keep them distended, and at their inner ends they communicate with finer tracheoles which lack these filaments. These fine tubules lead to the various tissues of the body, although oxygen probably passes from them into the body fluids, rather than directly to the cells. The gas-filled tubes form a closed system in aquatic insects. In these forms the oxygen content is renewed by diffusion from the surrounding water in tracheal gills, as mentioned above.

Spiders have a pair of lung books formed of many thin leaves in depressions in the abdomen. Blood circulates in these leaves and air between them.

The respiratory system of vertebrates is associated with the pharynx. In primitive chordates, cyclostomes, fishes, and larval amphibians the gill slits persist along this passage, so that water taken into the mouth may be expelled from the pharynx without being swallowed. Finely divided blood vessels in the walls of the pharynx or in special outgrowths known as gills along the walls of the pharyngeal clefts receive oxygen from the water as it flows over the surfaces. The gill surface is elaborately folded into numerous gill lamellae.

In some fishes and in terrestrial vertebrates generally, a saccular outgrowth of the ventral wall of the pharynx forms lungs for the reception of air. In the simplest forms the outgrowth branches to form two saclike lungs. In more complex lungs the surface is increased by ridges projecting into the cavities from their walls, and in the most highly developed organs of this type there are many minute chambers (alveoli) in a spongy mass of tissue containing muscle and elastic fibers. The original connection with the pharynx persists, leading into a single tube, the trachea, supported by cartilage rings. The principal branches of the trachea are the bronchi or bronchial tubes. They lead into finer bronchioles whose branches communicate with the alveoli.

Lungs in Humans. The lungs are the most important organs of respiration. They are paired structures, containing thousands of small sacs, the *alveoli*. See Fig. 1. The lungs are conical in shape. The right lung is composed of three lobes; the left has two. Each lobe is subdivided, in turn, into two or more bronchopulmonary segments, each segment representing the lung tissue supplied by one of the main branches of the lobar bronchi. In diseases, such as pneumococcal pneumonia, atelectasis, and lung abscess, the lesions are typically confined to a single lobe or segment.

The lungs are soft and spongy in texture; in the adult, they are gray, mottled with black, or even totally black in color, but in the infant they are pink. The dark color of the lungs in adults living in cities is the result of carbon deposits produced from the atmosphere.

Sounds produced by the lungs during respiration may be heard by the physician through a stethoscope. The value of listening to sounds in the chest area was known to Hippocrates in the fifth century B.C. The conventional type of stethoscope was invented by Dr. George Cammann of New York in the mid-1800s. Each lung is covered with a membrane called the *pleura*. This membrane extends over the inner chest wall and

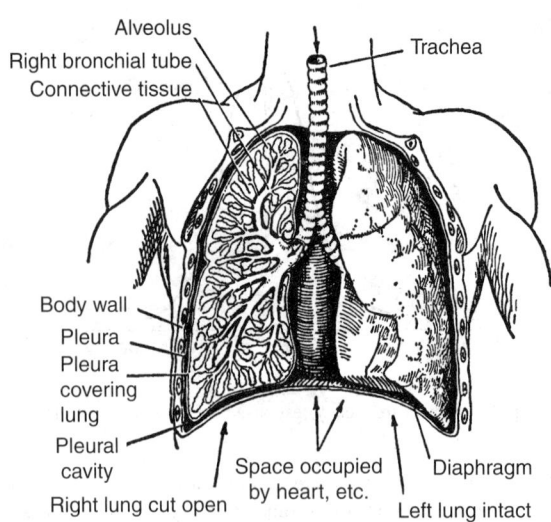

Fig. 1.   Human respiratory system.

down to the upper surface of the diaphragm. The two layers, therefore, are a closed sac, one on either side of the chest. The space enclosed by the pleural membrane is known as the *pleural cavity*. When a person is free from disease, no real space is present. Instead, these two coverings for the lungs lie next to each other, separated only by a thin coating of fluid, which permits the surfaces to slide easily over one another during the breathing processes. The lungs are housed in the thoracic cavity which is composed of twelve vertebrae, twelve pairs of ribs, the breastbone, muscles, and fibrous tissue. It is the expansion of this cavity, triggered by nerve impulses from the brain, that causes the lungs to expand and air to be drawn into the lungs. Collapse of the cavity causes expiration.

The pleural cavity is walled off from the abdominal cavity by a sheet of muscle called the *diaphragm*. When a person inhales, the diaphragm moves downward toward the abdominal cavity. Conversely, it moves upward when the person exhales. The up-and-down piston-like movements of this organ account for 60% of the air breathed. The diaphragm is attached in the rear to the spine, at the sides to the lower six ribs, and in front to the breastbone (*sternum*).

During strenuous exercise, the amount of oxygen used by the body increases to 10 or more times that normally used. An increase in carbon dioxide also occurs in the blood. This results in a gasping for breath, a feeling of fatigue, aching muscles, and other signs of shortage of air. These symptoms may appear to be caused by an inadequacy of the respiratory system, but actually they result from failure of the body to increase the amount of blood pumped into the heart. When the body adjusts to this condition, and panting and other symptoms cease, a so-called "second wind" occurs.

The thought that a person completely fills the lungs with fresh air when inhaling and completely empties them of used air when exhaling is erroneous. Only part of the air inhaled, which is about a pint at one time, enters the lungs. The remainder stays in the passageways that lead to the lungs. This is known as "dead air" inasmuch as it does not enter into an exchange of gases in the tissues of the body. When respiratory failure occurs, as in electrocution or drowning, the mechanical act of breathing often can be simulated artificially.

There are several diseases which involve the respiratory system. Some of these are described in separate entries in this encyclopedia. See also **Atelectasis**; **Bronchial Asthma**; **Bronchiectasis**; **Bronchitis**; **Common Cold**; **Cystic Fibrosis**; **Dyspnea**; **Emphysema**; **Influenza**; **Legionellosis**; **Pleurisy**; **Pneumokonioses**; and **Tuberculosis**.

Additional Reading

Bastian, G.F.: "The Respiratory System," 'Addison Wesley Longman, Inc., Reading, MA, 1997.

Beachey, W.: "Respiratory Care Anatomy and Physiology: Foundations for Clinical Practice," Mosby-Year Book, Inc., St. Louis, MO, 1997.

Lee, J.: "The Respiratory System," The Rosen Publishing Group, Inc., New York, NY, 2000.

Taylor, A.E.: "Clinical Respiratory Physiology," W.B. Saunders Company, Philadelphia, PA, 1996.

RESPONSE (Instrument). With reference to industrial and scientific instruments and control systems, the Instrument Society of America includes the following definitions:

Dynamic Response. The behavior of the output of a device as a function of the input, both with respect to time.

Ramp Response. The total (transient plus steady-state) time response resulting from a sudden increase in the rate of change in the input from zero to some finite value.

Step Response. The time response of an instrument when subjected to an instantaneous change in input from one steady-state value to another.

Time Response. An output, expressed as a function of time, resulting from the application of a specified input under specified operating conditions.

RESTITUTION COEFFICIENT (or Collision Coefficient). In a two-body collision involving particles 1 and 2, moving in the same straight line, the coefficient of restitution is defined by

$$e = \frac{v_2 - v_1}{u_1 - u_2}$$

where $u_1 > u_2$ are the velocities with respect to a primary inertial system before collision and $v_2 > v_1$ are the corresponding velocities after collision. For a completely elastic collision $e = 1$. For an inelastic collision $e < 1$.

REST MASS. The mass m_0 of an entity in the system in which it appears to be at rest. It is the mass in the classical, or Newtonian sense; that is, it does not include the additional mass which, according to the relativity theory is acquired by a particle or body when set in motion.

RETAINING WALL. A retaining wall is a structure for supporting loose material at an angle greater than its natural angle of repose. Usually, a retaining wall is of concrete construction, and used to retain a bank of earth. Four types of retaining walls are illustrated in Fig. 1. They are the simple trapezoidal section with vertical loaded face, the section with earth pressure aiding masonry weight, and the cantilever retaining wall.

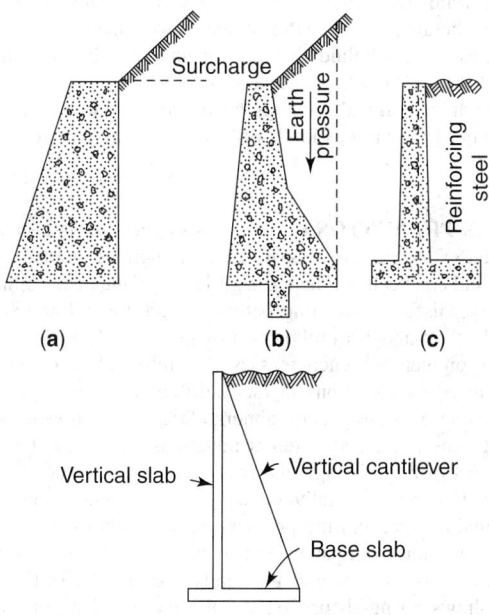

Fig. 1.   Types of retaining walls.

The successful retaining wall must be built so that it will neither slide on its base under horizontal pressure, nor tip over. The pressure of the earth is, of course, variable, depending on its composition and moisture content. If the earth back of the wall is well drained, so that its moisture content is a minimum, it will possess more cohesion, and exert less overturning pressure against the wall. The retaining wall may, in some cases, have to be designed to withstand pressures of a surcharged embankment having a slope of at least the natural angle of repose. Quite often the retaining wall must be designed to withstand pressure due to a surcharge caused by highway or railway traffic. The stability of a solid masonry section, such as that in (a) may be determined in much the same way as that for gravity masonry dams, except that earth pressure is not so directly and definitely computable as water pressure. The section shown in (b) will need less masonry because its shape causes it to make use of some of the vertical weight of earth as a stabilizing earth pressure. Also, it is keyed into the ground so that the resistance to sliding is greatly increased. A cantilever retaining wall, shown in (c), is the one with a minimum of masonry, but, because its resistance to overturning is obtained by bending action, internal moments are developed which require that this type of wall be reinforced with steel. A counterfort retaining wall (d) is made up of a continuous vertical slab which is connected to a continuous horizontal base slab or footing and supported at intervals by reinforced buttresses, called counterforts, which rest on the base slab. The earth pressure is carried by the vertical slab, which transfers this load by beam action to the counterforts. Since concrete is weak in tension, the sections of this type of wall must be properly reinforced with steel rods.

RETENTIVITY. That property (B_{rs}) of a magnetic material which is measured by the residual induction when a saturating magnetizing force is removed. See also **Hysteresis**.

RETICLE. A reticle is a set of two or more fine wires placed at the principal focus of a telescope lens. The reticle wires are usually sections of spider web or some equally fine fiber. In accordance with the fundamental principle of telescope construction this reticle must be in the principal focus of the eyepiece, and when the telescope is directed on an object both the image and the reticle will be in clear view in the eyepiece. Also called *reticule*.

RETICULUM (Constellation). A southern constellation located near Hydra.

RETINA. The retina is the light-sensitive membrane that covers the inside of the back of the eye. It receives the images that come through the cornea and lens. The retina contains light-sensitive nerve cells, the rods and cones, which send nerve impulses along the optic nerve to the brain by way of a network of connecting and integrating cells, some of which have very long fibers.

The rods and cones are specialized for different purposes. The rods perceive light, while the cone cells perceive both light and color. The retina has 20 times more rod cells than cone cells. Because there are fewer cone cells and because they need more light to function, it is difficult to discern colors in dim light.

At the center of the retina is an area called the *fovea*, which has no blood vessels. Light-sensitive cells, primarily cones, are packed tightly in the fovea, so that it has the highest visual resolution.

When conducting an eye examination, the examiner uses the ophthalmoscope, a hand-held device with a bright light, to examine the retina. The eye care professional looks for such abnormal conditions as damage to the tiny blood vessels that can be caused by high blood pressure or diabetes, detachment of the retina from the back of the eyeball, or blank spots on surface of the eyeball.

Heavy drinking and cigarette smoking, along with poor nutrition, can also lead to retinal damage, as can vitamin deficiency and lead poisoning.

Congenital Disorders of the Retina

Color vision deficiency. Some disorders that affect the retina are inherited. Perhaps the best-known disorder is color vision deficiency. True color blindness, the inability to detect any color so that the world is seen in black and white, is rare.

Color vision deficiency, is caused by defect in or a reduced number of cone cells. Most cases are inherited, but some result from injury or certain diseases of the retina or optic nerve. The two common types of hereditary deficiency are a reduced ability to discriminate light wavelengths in the middle (green) and long (red) segments of the light spectrum. A blue deficiency is much less common.

The deficiency is gender-based, so that it is eight times more common in men of white European origin than in women. About 8% of men and 1% of women of that ethnic background have either green or red deficiency.

The incidence is lower in Asians and Native Americans and lowest of all in the black population.

Color vision is tested with color plates. Most persons with color vision deficiency can function well in everyday life, but the condition can be dangerous for airline pilots or electricians, who work with color-coded wires.

Retinitis pigmentosa. This is a gradual degeneration of the rods and cones that can begin as early as adolescence, but most commonly occurs in middle age.

Tay-Sachs disease. Tay-Sachs affects not only the eye but also the brain, causing early death.

Other Retinal Diseases and Disorders

Age-related macular degeneration. An age-related retinal disorder is macular degeneration, gradual deterioration of the innermost part of the retina, the macula. There are two kinds of macular generation, wet and dry. In both kinds, there is a gradual breakdown of cells of the macula and the layer of cells below it, the retinal pigment epithelium.

The "dry" form is most common, accounting for about 90% of cases. Progressive breakdown of the cells causes a blind spot in the center of the eye. The "wet" form is caused by an overgrowth of blood cells in the area of the macula. Wet macular degeneration can be treated to some extent, but the dry form is currently untreatable.

Retinopathy. Retinopathy is a term used to describe a variety of disorders of the retina. One form is diabetic retinopathy, damage to the retina and the blood vessels that serve it when diabetes is not well controlled. The blood vessels can leak or burst, new vessels can grow on the surface of the retina, and there can be abnormal growth of fibrous tissue. Diabetic retinopathy is a major cause of loss of vision.

Hypertensive retinopathy results from high blood pressure, which causes the retinal blood vessels to become abnormally narrow.

Atherosclerosis of the retinal artery, the same kind of blood vessel blockage that leads to heart attack and stroke, can cause severe damage to the retina.

Retinal vein occlusion is a common cause of blindness. It happens when the central vein or artery of the retina is blocked.

Retinal tear and detachment. Retinal tear is a split in the retina that usually is caused by gradual degeneration. It is common in persons with severe nearsightedness, who have thinner-than-normal retinas. It can also be caused by a severe eye injury. A retinal tear often is followed by retinal detachment, in which the vitreous fluid of the eye collects between the delicate nerve membrane and the underlying layer of pigment.

Retinoblastoma. Retinoblastoma is a cancer that occurs early in childhood and has a genetic basis. It occurs in about one of every 20,000 births. It can be treated by radiation but most often requires surgical removal of the affected eye. If both eyes are affected, one may be removed, and the other may receive radiation therapy.

Retinal infections. The retina is also subject to infection.

One common infection is toxoplasmosis, caused by a parasite that is found in raw beef and cat feces. The infection often occurs before birth, and it causes progressive retinal damage over the years.

Other infections include toxocara canis, in which worm larvae lodge in the retina, causing severe damage, and onchocerciasis, another worm infestation.

Bacterial or viral infections elsewhere in the body may also be carried to the retina. They pose a special danger for persons with impaired immune systems. See also **Color Blindness**; **Macular**; **Ocular Hypertension**; **Retinal Detachment**; **Retinitis Pigmentosa**; and **Vision and the Eye**.

Vision Rx, Inc., Elmsford, NY.

RETINAL DETACHMENT.   Retinal detachment occurs when the light-sensitive inner surface of the back of the eye becomes separated from the outer layers. Quick treatment is needed when this happens to prevent loss of vision in the affected eye. The risk of retinal detachment is higher for persons who are highly myopic (nearsighted), who have thin retinas, and those who have had cataract surgery.

There are three forms of retinal detachment. Most often, retinal detachment occurs when a tear or hole develops in the retina. Fluid from the vitreous gel seeps into the opening, causing the inner surface of the retina to become detached. This kind of retinal detachment, called *rhegmatogenous*

or *mechanical detachment*, can occur in persons who have suffered injuries to the eye.

A less common form is *traditional retinal detachment*, in which there is fibrous tissue on the inner surface of the retina that contracts to pull the layers of the retina apart. Patients with diabetic retinopathy who have experienced a vitreous hemorrhage that produces scar tissue are vulnerable to this form of detachment. This can also occur after an eye injury or after surgery to repair the more common kind of detachment.

A third form is *exudative retinal detachment*, in which fluid oozes from the choroid, the layer of blood vessels between the retina and the sclera. This fluid then accumulates under the retina. An inflammation of the eye, uveitis or age-related macular degeneration can trigger this seepage.

The warning symptoms of retinal detachment include the occurrence, gradual or sudden, of flashes in one eye, usually at the edge of the field of vision and especially noticeable in the dark, lasting only a few seconds. Another symptom that may accompany these lightning-like flashes is a gradual or sudden appearance of floaters in one eye, which are spots or specks that can resemble insects, rings, or even hearts. The flashes occur when the light-sensitive cells of the retina are stimulated as the detachment begins, while the floaters result from the release of blood or pigment into the vitreous gel, the fluid that fills the eye. But sometimes the first symptom is the sudden occurrence of a dark drape, or curtain, that obscures part of the field of vision.

Quick action is needed to prevent detachment of the macula, the center of the retina, which causes loss of central vision. When the macula becomes detached, restoration of central vision may be impossible. If a drape or curtain has appeared, the position of the patient during treatment is important. If there is a descending drape, or lower detachment, the patient can remain upright, but if there is an upper detachment, the patient will be told to lie down to help prevent further detachment.

Retinal detachment is treated by specialized surgery. For mechanical detachment, the retinal surgeon will compress the sclera, which is the white of the eye, and use either a buckle or very small sponges to push the layers of the retina together again. Sometimes gas bubbles may be injected into the vitreous cavity to increase the push. Diathermy, which is heat treatment, may help reattach the retina. The patient may be told to stay in one position for several hours to facilitate the reattachment.

The tear or hole causing the detachment can be treated by bursts of laser light or cryotherapy, a freezing technique, to weld the retina together again. In some cases, retinal fluid that has accumulated beneath the detached segment of the retina will be drained surgically.

Complex microsurgical methods have been developed in recent years for treatment of the various kinds of retinal detachment. See also **Retina**.

Vision Rx, Inc., Elmsford, NY.

RETINITIS PIGMENTOSA.   Retinitis pigmentosa is a genetic condition that causes progressive loss of vision through the gradual deterioration of the rod and cone cells of the retina. It is one of the most common human inherited eye diseases, affecting between 50,000 and 100,000 persons in the United States and 1.5 million worldwide.

Retinitis pigmentosa encompasses a number of different conditions that are caused by mutations in many different diseases. As recently as the1990s, none of these genetic abnormalities were known. Now at least 6 different genes, and 100 different mutations, have been found to cause various forms of retinitis pigmentosa.

The first symptoms generally do not appear until adolescence, sometimes not until middle age, and the progression to blindness can take decades, with great variation from patient to patient. The first symptom usually is a loss of night vision, affecting visual ability in dim light. Testing of the field of vision shows a ring-shaped region of blindness that gradually extends over the visual field. The rods, which are found more in the periphery of the retina, are affected more than the cones. The cones are concentrated in the macula, the centermost part of the retina, so central vision usually is retained longer. An eye examination shows masses of branching black pigment in the areas of vision loss. That pigment gives the disease its name.

Until recently, there was no treatment for the condition. In 1993, however, a large-scale study sponsored by the National Eye Institute showed that daily doses of 15,000 units of vitamin A could slow the progression of the disease. An average person in the study who started taking the 15,000-unit dose daily retained some useful vision. The study also found that high doses of vitamin E had a negative effect. The disease progressed faster for patients taking 400 units of vitamin E each day. Normal dietary amounts of vitamin E had no effect.

The National Eye Institute recommends that persons with retinitis pigmentosa consult their eye professional about vitamin A supplementation, have blood levels of the vitamin measured before starting supplementation, and eat a balanced diet. In addition, retinitis pigmentosa patients should take vitamin A palmitate, the form used in the study. Vitamin A palmitate is better than beta-carotene, a natural precursor of vitamin A, because each person breaks down beta-carotene differently.

Identification of some of the genetic mutations has led to experiments on gene therapy, in which a virus is used to introduce a normal version of the affected gene. These experiments are in their early stage and are limited to work with laboratory animals. Some limited success has been reported, but human gene therapy for retinitis pigmentosa is at best a long-term prospect.

Vision Rx, Inc., Elmsford, NY.

RETINOBLASTOMA.   Retinoblastoma is the most common eye cancer of children, occurring in one of every 20,000 births. Some 300 cases are diagnosed every year in the United States, usually at about 18 months of age.

Retinoblastoma is often an inherited condition, although it can occur without a family history. It was the first cancer to be directly associated with a genetic abnormality, deletions, or mutations of one specific region of chromosome 14.

If genetic testing finds such a chromosomal abnormality in a family, there is a 45–50% chance that the parents will have another child with retinoblastoma. If there is no family history and no mutation is found, the risk is only 2–5%.

If retinoblastoma occurs in one child, young brothers and sisters should be observed for signs of the cancer. New retinoblastomas are rarely seen after the age of 7.

In about three-quarters of those cases, the first sign of retinoblastoma is an unusual whiteness of the pupil of the affected eye, which may lack vision. It can be accompanied by strabismus, an abnormal deviation of the line of sight of the eye. In some cases, both eyes are affected. Untreated, the cancer can spread from the eye to the orbit, the socket containing the eye, and then along the optic nerve to the brain.

Because other conditions such as congenital cataract and *Toxocara caris* can cause symptoms resembling retinoblastoma, the diagnosis usually is made by specialized blood tests, CAT scans, ultrasound evaluations, and an examination during which the young patient is anesthetized. In about one of every 20 cases, a biopsy may be needed for a definitive diagnosis.

Until recently, the standard treatment for retinoblastoma was enucleation, which is removal of the affected eye, or radiation therapy. If both eyes are affected, the one with the largest tumor can be removed and the other eye can receive radiation therapy. Enucleation successfully eliminates the cancer in 90% of the cases.

In recent years, the team of specialists treating retinoblastoma, an ophthalmic oncologist, pediatric oncologist, and radiation therapist, has favored radiation treatment, so the child can keep the eye. But evidence suggesting that radiation may increase the risk of other cancers has led to a number of studies focusing on a combination of chemotherapy and such modalities as laser therapy and freezing. The effectiveness of this treatment combination is being assessed.

Vision Rx, Inc., Elmsford, NY.

RETRACTOR.   A muscle which pulls an eversible or extensible part back to its normal resting position in the body. In the starfishes (see also **Asteroidea**) a pair of muscles in each ray, attached to the pouches of the stomach, serve as retractors for that organ, and in the mussels retractors draw the foot back into the mantle cavity when the animal closes its shell.

RETROGRADE MOTION.   1. Motion in an orbit opposite to the usual orbital direction of celestial bodies within a given system. Specifically, of a satellite, motion in a direction opposite to the direction of rotation of the primary. 2. The apparent motion of a planet westward among the stars. Also called *retrogression*. See also **Planet (Motions)**.

RETROREFLECTOR.   Any instrument used to cause reflected rays to return along paths parallel to those of their corresponding incident rays.

Also called *retroflector*. A type used for light is the retrodirective mirror. Another type of retroreflector, the corner reflector, is an efficient radar target.

RETROROCKET.   A rocket fitted on or in a spacecraft, satellite, or the like to produce thrust opposed to forward motion. See also **Space Vehicle Guidance and Control**.

REVEGETATION.   Generally refers to the purposeful seeding and planting of an area that once was covered with grass, trees, shrubs, forbs, etc., but which was denuded of vegetation because of a natural disturbance, such as a lightning-caused fire, or because of an artificial disturbance, such as mining and construction projects. A great deal of attention has been given to seeding and planting programs in connection with large areas that have become barren as the result of unplanned strip mining of the coal fields, notably in Appalachia.

Extraction of coal by surface mining has disturbed about two million acres of land in 26 states of the United States, approximately 50% of this disturbance taking place between 1965 and 1975. Ninety percent is on land owned by mining companies, farmers, and other private interests. Even though there has been extensive damage, most of the mining was done under then existing state laws and regulations. Regulations are becoming more rigid. Land disturbed by surface mining is generally an undeveloped land resource. Evidence indicates, however, that all but a small percentage of such land is capable of producing some tangible or intangible societal benefit. Mined sites, appropriately reshaped, can be developed for many uses. In the Midwestern and Appalachian coal fields, they have been developed, in relatively few instances to date, into successful and profitable production of agricultural and horticultural crops. Conversion of such areas into recreational and wildlife developments is a popular target. Industrial and residential sites have been established on areas disturbed by surface mining. Sites planted to trees yield various forest products. As of the early 1980s, some progressive mining companies have identified opportunities and have developed areas that provide social and economic benefits. Such reclamation and revegetation programs, coupled with superior surface mining methods (See also **Coal**), can yield vitally needed coal without extensive spoiling of the land.

Spoil Evaluation. In evaluating old sites that were mined by conventional methods, prior to the use of such techniques as haulback methods, (valley fill or head-of-the-hollow method, mountaintop leveling, etc.), it is necessary to determine both chemical and physical characteristics of the spoil. Chemical factors include toxic ion concentrations and availability of nutrients. The important physical features include particle size, texture, and color of the spoil. Spoil evaluations are often complicated by changes that occur in weathering. Each rock stratum has distinct chemical and physical characteristics. Many of these properties change when fragments of rock are exposed to moisture, light, air, and temperature variations. Release of iron, sulfur, manganese, and aluminum compounds may result in off-site pollution and failure of vegetation. The release of essential nutrients, such as phosphorus, potassium, calcium, and magnesium may, on the other hand, improve the growth of vegetation, reduce acidity problems, and increase the possibilities of reclamation.

Physically, the spoil is a heterogeneous mass of rock fragments derived from the rock strata above the coal. The size of rock fragments after mining and reshaping depends on the size of the basic particles and the strength of the cementing materials, which naturally hold small mineral particles together. Fine-grained shales strongly bonded together may break into hard, platy fragments that resist weathering. Coarse-grained sandstones weakly cemented together may disintegrate rapidly into sand. Spoil may be a dynamic material, more reactive than natural soils that have been exposed to the process of soil genesis for centuries. The chemical and physical properties of a spoil may change rapidly for several years. The rate of change will gradually lessen as characteristics of natural soils develop. Needless to say, however, time required for natural reclamation processes is exceedingly long and provides no relief to the immediate problem.

Evaluations of the chemical properties of a soil could be perplexing if an attempt were made to include all components that affect plant growth. Methods in use today rely on indicators for judging the interaction of many factors at one time. Soil reaction or hydrogen ion concentration (expressed as pH) is the most widely used indicator. Acidity alone may or may not affect the establishment of plants. Changes in acidity determine the concentration of toxic ions and nutrients in the soil solution. At pH levels below 4.0, two chemical changes may occur: Toxic ions, such as manganese, aluminum, and iron, become more available to plants and some essential nutrients become less so. When the pH of a spoil is between 5.5 and 7.0, the concentration of toxic ions in the soil solution decreases and more essential nutrients become available. A soil classification system for acidic spoils is given in Table 1. No system has been proposed for the alkaline or sodic spoils in the western U.S. coal fields. The same principles can be applied, but the toxic ions or cations may differ.

TABLE 1. SYSTEM FOR CLASSIFYING SPOILS (ACIDIC)

		Acidity	
Class Number	Description	pH Value	Extent on Area Sampled
1	Toxic	Less than 4.0	More than 75%
2	Marginal	Less than 4.0	50 to 75%
3	Acid[a]	4.0 to 6.9	50 to 75%
4	Calcareous	7.0 or more	More than 50%
5	Mixed	(Too varied to be classified as any above)	—

	Texture
Group	Description of Texture
A	Chiefly sand, sandstones, or sandy shales
B	Chiefly loamy materials and silty shales
C	Chiefly clay and clay shales

Combine acidity and textural classes to describe spoil type.

[a]Acid spoils may be subdivided into two classes: pH 4.1 to 5.4; and pH 5.5 to 6.9.
Source: Northeastern Forest Experiment Station, Forest Service, U.S. Department of Agriculture.

Other chemical analyses have been considered for identifying characteristics that result in specific nutrient deficiencies or toxicities. Phosphorus is often deficient (0 to 7 parts per million) in spoil material. Several laboratory analyses provide estimates of plant-available phosphorus. Total soluble salts in the soil solution may be important for the very strongly acid or alkaline spoils. This does not necessarily mean high concentrations of sodium salts, but salts of all anions and cations in the soil solution. Some consideration has been given to laboratory analyses that give concentrations of exchangeable aluminum, manganese, and hydrogen. These analyses can identify specific toxicities and thus permit the selection of tolerant plant materials. Research has been initiated to determine whether the heavy metals, such as copper, zinc, nickel, mercury, and cobalt, occur in concentrations toxic to plants.

Texture and clay mineralogy influence the degree of compaction that results from heavy equipment passing over the surface during mining and reshaping. There is evidence that compaction may limit plant growth by reducing water infiltration and nutrient release. Texture may also determine the rate of release of toxic ions or nutrients from soil particles. Smaller particles expose a larger surface area per unit volume to the forces of weathering than coarse fragments. This results in a more rapid release of chemicals. Surface color is important because it influences heat exchange at the spoil surface. Dark materials absorb solar energy so the spoil may attain temperatures lethal to plants. The degree of risk depends on the season of the year, the slope of the exposed surface, and the exposure of the site. The highest risk may occur during summer months on steep slopes of black or dark gray material facing south or west.

Preparation of the Site. Treatments to prepare the site for seeding or planting are determined after evaluation of site data and establishment of the land-management options and objectives. Grass and legumes often require more surface preparation than trees and shrubs. Uncompacted fill slopes and freshly reshaped surfaces may make acceptable seedbeds. Hard crusts form on many spoils; fine clay-size particles are consolidated by the drying action of wind and sun. This crust may be broken by rainfall, frost, or mechanical scarification. In many cases, ground cover will be denser if it is seeded into fresh spoil or where spoil surfaces have been broken by natural or mechanical scarification.

Treatment to create special surface configurations may be required on toxic spoils, on steep slopes, or in geographical locations where precipitation is low or unfavorably distributed throughout the year. Rows or depressions, made by machinery, trap precipitation, moderate wind velocities, reduce evapotranspiration rates, and moderate spoil temperature extremes. The orientation of the furrows with respect to direction of slope, exposure to sun, or prevailing winds may determine the effectiveness of the treatment. Amendments to modify acidity should be applied weeks or months before seeding. Scarification will mix the ameliorating material into the top 6 to 8 inches of spoil and create a neutralized layer for root development. On extremely rocky spoils where scarification is not practical, frost action and infiltrating water may carry the neutralizing materials below the spoil surface. The rate of application of neutralizing material depends on the type of vegetation to be seeded, the present and predicted acidity of the surface spoil, and the neutralizing capacity of the material used. Agricultural limestone is preferred for most treatments, but it can be used only in areas accessible to application equipment. No practical system has been developed for applying large amounts of agricultural limestone to steeply sloping land. Small quantities of lime can be applied with a hydroseeder to steep slopes. Finely ground limestone or hydrated lime may be mixed with water to form a slurry. Repeated applications may be necessary to achieve high rates per acre. The cost of repeated treatments could make this procedure impractical. Alkaline power plant fly ash and bottom ash are neutralizing materials, but they are usually not as effective as limestone. Fly ash contains various quantities of essential plant nutrients, such as phosphorus, zinc, molybdenum, and boron. However symptoms of boron toxicity have been observed on plants growing on sites treated with large amounts of fly ash. Rock phosphate provides a slowly available source of phosphorus and helps to neutralize acidity. Spoil acidity may react with rock phosphate to slowly release plant-available phosphorus.

Fertilization is generally recommended for grass and legume crops. Nitrogen and phosphorus have been used to accelerate the growth of trees on spoils. Tests show that most spoils are deficient in nitrogen. Phosphorus occurs in various concentrations, but deficiencies often limit plant growth. Potassium is usually adequate. There is little information about the concentrations of other nutrients. For most land-management objectives, the formulation of the fertilizer makes little difference when equivalent rates are applied. High-analysis fertilizers, ammonium nitrate, triple superphosphate, and ammonium phosphate are often preferred. Using low-analysis fertilizers on sites that require high rates of application may increase the total soluble-salt concentration to levels toxic to some plants.

Rates of fertilization vary with the seeded crop, the inherent fertility of the spoil, and the land-management objective. Nitrogen and phosphorus at 50 and 22 pounds per acre (56 and 246 kilograms per hectare) respectively, are sufficient to establish grass on many spoils. Higher rates of phosphate are important for legumes. More consideration is being given to retreatment at regular intervals. Multiple treatments are attractive because they reduce the chance of unacceptable cover and increase the probability of achieving land use objectives within the shortest time.

Much research on fertilizer application remains to be done. Research results thus far show that placing selected fertilizers in or near the planting hole increases the growth of black locust. This leguminous species responded to additions of nitrogen and phosphorus placed near seedlings on extremely acid spoil. In other trials, direct-seeded black locust made more rapid growth after surface applications of nitrogen and phosphorus. Five tons of lime per acre per foot of soil increased the growth of loblolly, shortleaf, Virginia, and a hybrid (pitch/loblolly) pine planted in extremely acid spoil. Pitch pine did not respond to the liming treatments. Ten tons of lime per acre-foot reduced the growth of the pines.

Municipal waste products have been considered for surface-mine reclamation. Waste applications could improve soil texture, add essential plant nutrients, and provide mulch for seed and seedlings. Shredded or composted waste could be applied to the surface, and mechanical methods could be used to incorporate it into the spoil. Sewage sludge can be applied in water slurry or in dried form. Mixtures of shredded or composted waste and sewage sludge offer another possibility. Thus far, use of these materials has been restricted to relatively small demonstration areas for public health reasons.

Seeding and Planting. For each of the major coal-producing regions in the United States, there is a group of preferred plant species. These may be grouped under four major categories: (1) grasses; (2) forbs; (3) trees; and (4) shrubs. The grasses and forbs may be further classified as (a) temporary; (b) semipermanent; and (c) permanent, depending on life expectancy of the plant. Temporary species give prompt and effective site protection by reason of quick growth, fibrous roots, and ability to endure unfavorable site conditions. Semipermanent species are perennials that will be ultimately replaced by permanent vegetation. Under favorable conditions, permanent species will persist for many years.

The grasses are a varied group of plant species well suited to surface-mine restoration. Experience has shown that many grasses are adapted to wide ranges of climate, spoil texture, nutrient regimes, and toxic ion concentrations. Germination is usually rapid, and growth often produces a crop, or at least site protection, during the first growing season. Annual grasses often grown as agricultural crops may be used to provide quick site protection. These temporary quick-cover crops also may serve as nurse crops for slower-developing perennials. Some species are better adapted to summer seeding; others should be seeded in the fall.

Leguminous forbs are considered by many to be essential components of ground-cover mixtures. The fixation of atmospheric nitrogen by the legumes benefits associated plants. This assists in maintaining a vigorous ground cover and may reduce the need for retreatment. The leguminous forbs generally are less tolerant of toxic ion concentrations and require more phosphorus than the grasses.

Forbs not classified as legumes are being evaluated in the western United States. The emphasis is on species that are components of natural vegetative cover and are common invaders of disturbed areas. Initial evaluations indicate that some species may be useful for surface-mine revegetation.

Inoculation with specific strains of rhizobium bacteria stimulates nodulation on leguminous forbs. Commercial inoculants are available for the important legume species. Rhizobium bacteria may not survive or produce effective nodules in acidic spoils with pH below 5.0.

Grass and legume ground covers reduce tree growth, but this is often unavoidable because some state laws and regulations require a herbaceous cover. Manipulation of the species composition and reduction of ground-cover density may minimize the adverse effects.

Trees are often planted by hand, using one of several planting tools. On selected spoils, machine planting is possible. Most planting stock is small, 1 to 2 years old, and bare-rooted. Spacing between trees varies by species, site characteristics, and end-objectives. Plantings may be mixtures of several species, or pure plantings of one species. The arrangement may be random or designed to protect seedlings from environmental extremes. There is increasing interest in the establishment of trees by direct seeding. Success has been achieved with several pine species in Alabama. Black locust is used in West Virginia and Kentucky, The planting of shrub species has not been emphasized in surface-mine revegetation. Shrubs have little tangible value; benefits accrue from site protection, wildlife food and cover, and aesthetics.

Costs of Refuse Disposal and Land Reclamation. A predominant advantage of so-called strip mining over underground mining over the years and particularly with reference to smaller, previously unestablished mining operations, has been one of economics. Obviously, as costs of reclamation of stripped areas and of improved surface mining techniques are added to strip mining costs, the differential becomes much less. Past despoilage of the land thus has resulted from comparative economics. While some progressive operators have reduced their profits in order to lessen the damage to the land, such voluntary action is simply more than can be expected on a massive scale. Concerns with land spoilage prevention and reclamation of spoiled lands, unfortunately, are coincident timewise with a growing demand for coal. This, in the early 1980s, hardly felt the impact of the numerous coal conversion programs most of which were still under test and development for the gasification and liquefaction of coal in terms of the new coal technology for supplanting waning supplies of natural gas and petroleum energy sources. The practical, interim solution to the energy/environment problem must be one of compromise.

Additional Reading

Holzworth, L.K. and R.W. Brown: "Revegetation with Native Species: Proceedings, 1997 Society for Ecological Restoration Annual Meeting," DIANE Publishing Company, Collingdale, PA, 2000.

Munshower, F.F.F.: "Practical Handbook of Disturbed Land Revegetation," Lewis Publishers, Boca Raton, FL, 1994.

Vogel, W.G.: "Manual for Training Reclamation Inspectors in the Fundamentals of Soils and Revegetation," DIANE Publishing Company, Collingdale, PA, 1998.

Note: See references listed at end of article on **Coal**.

The cooperation of W.T. Plass of the U.S. Dept. of Agriculture Forest Service, Northeastern Forest Experimentation, Princeton, WV in making information available for this summary is gratefully acknowledged.

REVELLE, ROGER R. (1909–1991). Once described by the New York Times as "one of the world's most articulate spokesmen for science" and "an early predictor of global warming," Roger Revelle was a giant in American science who accomplished enough during his eighty-two years to distinguish several lifetimes.

Revelle first made his mark in oceanography as a scientist, explorer, and administrator, and went on to become a senior spokesman for science, giving counsel in areas ranging from the environment and education to agriculture and world population. He was one of the first scientists to recognize the effects of rising levels of atmospheric carbon dioxide on the Earth's surface temperature.

Born in Seattle, Washington, Revelle was raised in Pasadena, California, and was identified as a gifted student early in his academic career. In 1925, Revelle entered Pomona College with an interest in journalism, but later turned to geology as his major field of study.

After receiving his bachelor's degree from Pomona in 1929, Revelle entered the University of California-Berkley to continue his studies in geology. In 1931, at the recommendation of his major professor, George Davis Louderback, he received a research assistantship in oceanography at the Scripps Institute of Oceanography in La Jolla, CA. While at Scripps, Revelle was awarded his Ph.D. and immediately appointed oceanography instructor at Scripps.

Revelle served in the U.S. Navy during World War II as the commander of the oceanographic section of the Bureau of Ships and became head of their geophysics branch in 1946. He returned to Scripps in 1948 and served as its director from 1951 to 1964.

In 1957, Revelle and Hans Suess, one of the founders of radiocarbon dating, demonstrated that carbon dioxide had increased in the air as a result of the use of fossil fuels in a famous article published in Tellus, a European meteorology and oceanography journal. Revelle's interest in atmospheric carbon dioxide engaged his attention for the remainder of his life. He brought the subject to the attention of the public as a member of the President's Science Advisory Committee Panel on Environmental Pollution in 1965. Under Revelle's leadership, the committee published the first authoritative U.S. government report in which carbon dioxide from fossil fuels was officially recognized as a potential global problem.

Revelle chaired the National Academy of Sciences Energy and Climate Panel in 1977, which found that about forty percent of the anthropogenic carbon dioxide has remained in the atmosphere, two-thirds of that from fossil fuel, and one-third from the clearing of forests.

Revelle influenced public opinion on the carbon dioxide issue through a widely read article published in *Scientific American* in August 1982. His research addressed issues such as the rise in global sea level and the relative role played by the melting of glaciers and ice sheets versus the thermal expansion of the warming surface waters. Revelle's international scientific contacts did a great deal to disseminate research findings and to foster discussion about the data, environmental and social effects of increased atmospheric carbon dioxide, and governmental policy and action. Revelle considered this work very important, once estimating that he spent twenty percent of his time keeping current with the issue.

In 1963, Revelle took a leave of absence from Scripps and formally switched fields from oceanography to public policy. He founded the Center for Population Studies at Harvard University, and spent more than a decade as director. His primary interests were applications of science and technology to world hunger. In 1976, Revelle returned to the University of California-San Diego where he received the title of Professor of Science and Public Policy and joined the Department of Political Science.

Revelle served on scores of academic, scientific, and government committees advising on a wide spectrum of topics. He was science adviser to the Secretary of the Interior, president of the American Association for the Advancement of Science, and a member of the NASA Advisory Council. In November 1990, Roger Revelle received the National Medal of Science from President George Bush. He remarked to a reporter: "I got it for being the grandfather of the greenhouse effect."

See also **Climate**; **Global Temperature**; **Global Warming**; and **Pollution (Air)**.

Web Reference

Scripps Institute of Oceanography Archives: *http://scilib.ucsd.edu/sio/archives /siohstry/revelle-biog.html*

REVERBERATION.   1. The persistence of sound in an enclosed space, as a result of multiple reflections after the sound source has stopped.

2.   The sound that persists in an enclosed space, as a result of repeated reflection or scattering after the source of the sound has stopped.

REVERBERATORY FURNACE.   A metallurgical furnace in which the charge, lying in a shallow hearth, is heated by flame passing over its surface and by radiation from a low roof.

REVERSE ACTING CONTROLLER.   A controller in which the absolute value of the output signal decreases as the absolute value of the input (measured variable) increases.

REVERSIBILITY (Principle of).   If all parts of a beam of light are reflected directly back on themselves, no matter how many reflections or refractions it has undergone, the light will travel back over the identical path (or paths) it followed before reversal.

REVERSIBLE AND IRREVERSIBLE PROCESSES.   Consider a system which undergoes the transformation *ABC*. See Fig. 1. The change is said to be *reversible* if there exists a change *CBA* such that:

1. The variables characterizing the state of the system return through the same values, but in the inverse order;
2. Exchanges of heat, matter and work with the surroundings are of the reverse sign and take place in the reverse order. Thus, for example, if in the trajectory *ABC* the system receives a quantity of heat Q, it must give up the same quantity in the inverse trajectory *CBA*.

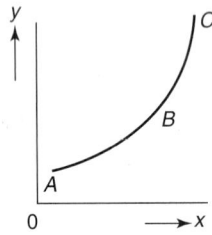

Fig. 1.   Depiction of reversible and irreversible processes.

All changes that do not satisfy those two conditions are termed *irreversible*. No changes that take place in nature are reversible. All are irreversible. But in many cases real processes can be derived approaching as nearly as we wish to reversible processes.

Thus let us consider a wide U-tube containing water in which initially the water level is higher in one arm than in the other. The level in this arm falls, passes through the equilibrium position, and rises again. If we could reduce friction and viscosity to zero, we should obtain on the limit a reversible change.

A reversible change may always be considered as a succession of equilibrium states. However, there are processes in nature that cannot be considered as reversible without suppressing them completely. For example, the equalization of temperatures, chemical reactions, and diffusion are processes of this kind.

REVERSION OF SERIES.   A given power series

$$y = a_0 + a_1 x + a_2 x^2 + \cdots$$

may be reverted to give an explicit representation of x as a function of y. The result is

$$x = z + c_1 z^2 + c_2 z^3 + \cdots$$

$$z = (y - a_0)/a_1; \quad c_1 = -a_2/a_1; \quad c_2 = -a_3/a_1 + 2(a_2/a_1)^2$$

$$c_3 = a_4/a_1 + 5 a_2 a_3/a_1^2 - 5(a_2/a_1)^3; \ldots \ldots$$

The method of undetermined coefficients can be used to obtain further coefficients in the series but the labor becomes great. A series can be reverted more elegantly by means of the theory of residues. Inversion is also used for this procedure, instead of reversion.

REVOLUTION.   1.   Motion of a celestial body in its orbit; circular motion about an axis usually external to the body. In some contexts, the terms revolution and rotation are used interchangeably but, with reference to the motions of a celestial body, revolution refers to motion in an orbit or about an axis external to the body, whereas rotation refers to motion about an axis within the body. Thus, the Earth revolves about the Sun annually and rotates about its axis daily.

2.   One complete cycle of the movement of a celestial body in its orbit, or of a body about an external axis, as a revolution of the Earth about the Sun.

REYE'S SYNDROME.   Sometimes fatal, Reye's syndrome is an encephalopathy usually accompanied by fatty infiltration of the liver. Statistics indicate that the disease is frequently found in persons under the age of 16 years who have had prior infections with influenza or other viruses. Less frequently, the disease is noted among a small percentage of middle-age adults.

The onset is sudden, with intractable vomiting a few days after viral illness. Sensorial impairment appears soon afterward and may progress to coma. Seizures may occur. The liver is usually enlarged. Reye's syndrome is sometimes a complication of chickenpox (varicella), depending upon the patient's immunologic competency.

During the 1978–1979 period, peak reporting of cases of Reye's syndrome occurred between December and March. Using data collected through nearly sixty World Health Organization laboratories as a measure of influenza activity, a temporal association was observed between the occurrence of Reye's syndrome and the reporting of isolates of the H1N1 influenza virus. Concurrent widespread influenza activity was reported in all regions where the incidence of Reye's syndrome was high. In the United States, Reye's syndrome outbreaks followed the occurrence of outbreaks of influenza A in each region, with the first cases of Reye's syndrome occurring in the western United States, followed by cases in the Southeast and Midwest. While influenza B has been epidemiologically associated with outbreaks of Reye's syndrome, statistics showed for the first time that influenza A outbreaks have been associated with temporal and geographic clustering of cases of Reye's syndrome in the United States. See also **Influenza**; and **Liver**.

Specific therapy is not available. Supportive measures include lactulose to control hyperammonemia, fresh frozen plasma to replenish clotting factors, mannitol or dexamethasone to lower increased intercranial pressure, and mechanical ventilation. Fatality rates as high as 23% of cases have been reported in some epidemics.

Although convincing causative evidence has not been uncovered, physicians strongly suggest that salicylates (such as aspirin) may be implicated and thus avoidance of these medicants in patients, especially children, with influenza or chicken pox, is indicated as a precautionary measure.

Web Reference

Reye's Syndrome: *http://my.webmd.com/*

REYNOLDS NUMBER.   A dimensionless number that establishes the proportionality between fluid inertia and the sheer stress due to viscosity. The work of Osborne Reynolds has shown that the flow profile of fluid in a closed conduit depends upon the conduit diameter, the density and viscosity of the flowing fluid, and the flow velocity.

Pipe Reynolds number, R_D, is the dimensionless ratio, $VD\gamma/g\mu_e$, where

V = Velocity in any units consistent with the rest of the equation
D = Inside diameter of the conduit (pipe)
γ = Specific weight in any units consistent with the rest of the equation
g = Acceleration of gravity, feet/second^2
μ_e = Absolute viscosity, pound-second/feet^2

The foregoing equation is inconvenient for commercial use because commonly used units of measurement are rarely consistent. Two alternative equation forms are:

$$(1) \quad R_D = \frac{6.32 \times \text{Rate of Flow (pounds/hour)}}{\text{Pipe Diameter (inches)} \times \text{Absolute Viscosity (centipoise)}}$$

$$(2) \quad R_D = \frac{52.77 \times \text{Rate of Flow (gallons/hour at flowing temperature)}}{\text{Pipe Diameter (inches)} \times \text{Kinetic Viscosity (centistokes)}}$$

Flow generally is considered to be fully laminar at Reynolds numbers below 1500; in transition between 1500 and 4000; and turbulent above 4000. Thus, the Reynolds number is a most useful tool in constructing piping system designs and in sizing flowmeters. Orifice or throat Reynolds number, $R_d = R_D/d/D$, where d = diameter of orifice bore (inches); D = inside diameter of conduit or pipe (inches).

Another general way of expressing Reynolds number is by the ratio

$$\frac{\text{Scale Velocity} \times \text{Scale length}}{\text{Kinematic Viscosity}}$$

This also is referred to as the Reynolds criterion. Reynolds numbers are only comparable when they refer to geometrically similar flows; and, provided that all the boundary conditions can be described by the scale velocity and scale length, flows of the same Reynolds number are dynamically similar. For an airfoil, it is

$$\frac{\text{Air Velocity} \times \text{Chord of the Airfoil}}{\text{Kinematic Viscosity of the Air}}$$

Reynolds numbers are of value in the various fields because tests of models are directly comparable to full-scale results of geometrically similar shapes if the Reynolds ratio for the model equals that of the actual or full-scale project. This has its practical applications in the field of hydrodynamics, in the study of water resistance of hulls or floats, and in the study of water velocities, levee problems, etc., of large rivers. It is used also to establish the best proportions of hydraulic turbines through the use of models. Much of the science of aeronautics rests upon experimental data obtained in wind tunnels. Dangerous inaccuracies might exist in drawing conclusions for actual construction from model tests, unless either the model were tested at a Reynolds number equal to that of the completed project, or due corrections and allowances were made for the Reynolds number. See also **Aerodynamics and Aerostatics**; and **Heat Transfer**.

RHEA *(Aves, Rheiformes)*. Large flightless birds of South America, which weigh considerably less and are much smaller than ostriches. Standing upright they reach 1.70 meters ($5\frac{1}{2}$ feet) and may weigh up to 25 kilograms (55 pounds). The head, neck, rump and thighs are feathered, and their plumage is soft and loose. There are 3 front toes, and the hind toe is absent. The tarsus has horizontal plates in front. The gut and particularly the caeca are very long. They are grass and leaf eaters.

There are 2 genera, each with 1 species and 7 subspecies. (a) The Common Rhea (*Rhea americana*) and (b) Darwin's Rhea (*Pterocnemia pennata*).

The Common Rhea reaches heights up to 170 centimeters (70 inches); the height of the back is 100 centimeters (39 inches), the wingspan reaches 250 centimeters (98 inches), the tarsal length is 30–37 centimeters (12–15 inches), and the beak length is 9–12 centimeters ($3\frac{1}{2}$–5 inches). The tarsus has about 22 horizontal plates in front. Albinos occur in this species. The eggs measure 135×95 millimeters (5.3×3.7 inches) and weigh 530–680 grams (19–24 ounces). They are elliptical with a shiny surface and are ivory or golden-yellow with black stripe-shaped pores; the color pales with time. The young at first are yellow with black longitudinal stripes on the back; after 2 years they resemble the parents.

Darwin's Rhea is smaller with a height at the back of 90 centimeters (35 inches). The tarsus is 28–30 centimeters (11–12 inches) and has about 18 horizontal plates. The eggs are 125×85 millimeters (4.9×3.3 inches) and weigh 500–550 grams (18–19 ounces). They are yellowish-green when laid and later become a pale yellow.

The home of the rheas is the grass steppes of South America. See Fig. 1. They avoid forests and mountains. Generally they live in groups of one cock and several hens within a distinct territory. After the breeding season, loose flocks of 50 or more may form. Good eyes and acute hearing allow them to detect enemies from far away; their fast legs, which can take strides of up to 1.5 meters (5 feet), carry them quickly out of danger. In

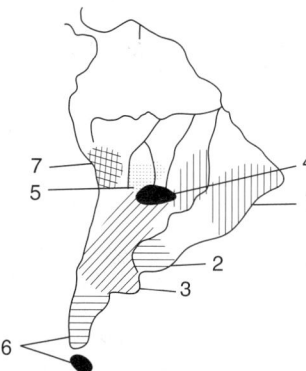

Fig. 1. Areas inhabited by the Rheas, including subspecies. Rhea (*Rhea americana*); (1) *Rhea americana americana*; (2) *Rhea americana intermedia*; (3) *Rhea americana albescens*; (4, 5) *Rhea americana araneiceps*; Darwin's Rhea (*Pterocnemia pennata*); (6) *Pterocnemia pennata pennata*; (7) *Pterocnemia pennata garleppi*.

an emergency they can escape by suddenly dodging aside. This maneuver is facilitated by the use of the wings, which, for a flightless bird, are remarkably long. While running, they raise one wing and lower the other; this has a steering effect, like the rudders of an airplane, and makes such sudden changes of direction possible.

The rheas' food consists of grasses and herbs (alfalfa, clover, serradella) and insects and other small animals. Because they prefer the same food plants as sheep, they are food competitors. However, their consumption of the burrlike seeds that tangle the wool of sheep also makes them useful. If they find enough juicy plants, they require very little water.

In the breeding season between September and December, earlier in the north than in the south, the cock expels all rivals from his territory. He only tolerates hens. The courtship display consists of his running around with his plumage erected, dodging sideways and swinging the neck from side to side. At this time he gives his deep call, which can be heard far away, "nan-du," which is responsible for one of its names, "nandu." The cock builds the nest, a simple depression lined with a few pieces of plants; only the cock incubates. Each hen lays her eggs outside the nest at two day intervals. Usually there are 15 to 20 eggs in a clutch; since each hen lays 10 to 15 eggs and several hens lay for their cock, it hardly matters if many eggs are lost. Nests containing up to 80 eggs have been found. But the cock, during incubation, cannot cover so many and hence he cannot hatch them all. The young hatch after about 40 days of incubation. After half a year, they reach adult size, and at 2 to 3 years they are capable of reproduction. See also **Ratites**.

RHEBOK. See **Antelope**.

RHEIFORMES. See **Rhea**.

RHENIUM. Chemical element, symbol Re, at. no. 75, at. wt. 186.2, periodic table group 7, mp 3180°C, bp 5627°C, density 21.04 g/cm³ (20°C). Elemental rhenium has a close-packed hexagonal crystal structure.

Rhenium is a platinum-white, very hard metal; stable in air below 600°C (at this temperature, the metal begins to generate a white, nonpoisonous, vaporous oxide, Re_2O_7); practically insoluble in HCl or hydrofluoric acid, but soluble in HNO_3 with the formation of perrhenic acid; forms sodium rhenate when fused with NaOH and nitrate. Discovered by Noddack and Tacke in 1925 in tantalite, wolframite, and columbite by the Moseley X-ray spectrographic method of analysis, and later found present in molybdenite, from which rhenium is obtained. Predicted by Mendeleev, in 1871, as an element to be discovered with properties resembling manganese, and named by him dvi-manganese.

There are two natural isotopes, ^{185}Re and ^{187}Re, of which the latter is radioactive with respect to beta decay, having a half-life of 5×10^{10} years reverting to ^{187}Os. Other radioactive isotopes include ^{177}Re, ^{178}Re, ^{180}Re, ^{183}Re, ^{184}Re, ^{186}Re, ^{188}Re, and ^{189}Re. The latter isotope has a long half-life, something less than 10^3 years; the half-lives of the remaining isotopes are comparatively short, measured in minutes, hours, or days.

Because Re and Os are highly refractory and siderophilic elements, the Re-Os isotope system is important in studies concerned with metal phases and high-temperature inclusions of meteorites. R.J. Walker (Carnegie Institute of Washington) and J.W. Morgan (U.S. Geological Survey) observe, "Potential applications of the system include dating meteorites, especially with respect to the chronology of the assembly and subsequent metamorphism of genetically disparate components, and providing estimates of the initial Os isotopic composition and Re/Os ratio of the early earth. This ratio is an important chemical tracer for understanding the formation of the earth's core and the chemical evolution of the mantle and crust."

In terms of abundance, rhenium ranks 75th among the elements, based upon estimated contents of the universe. Rhenium is not plentiful in the earth's crust, being essentially confined to association with the mineral molybdenite. First ionization potential, 7.87 eV. Oxidation potentials $Re + 4H_2O \rightarrow ReO_4^- + 8H^+ + 7e^-$, -0.15 V; $Re + 8OH^- \rightarrow ReO_4^- + 4H_2O + 7e^-$, 0.81 V. Other important physical properties of rhenium are given under **Chemical Elements**.

Rhenium is a minor constituent (100 ppb) of molybdenite-bearing porphyry copper ores, of which there is extensive mining in the United States and South America. Commercial rhenium is recovered from by-product molybdenum. As the result of roasting MoS_2 and MoO_3, the rhenium is concentrated to levels of 300–1,000 ppm. At the high process temperatures, the rhenium is oxidized and volatilized as rhenium heptoxide, Re_2O_7. This compound is recovered from flue gases by way of wet scrubbing and chemical separation techniques as the relatively crude ammonium perrhenate. The latter compound is reduced with hydrogen to produce rhenium metal.

Uses: Rhenium finds rather wide use as a catalyst in selective hydrogenation and other chemical reactions. It sometimes is used in conjunction with platinum in reforming operations. Additional processes for which rhenium has been tested and used as a catalyst include alkylation, de-alkylation, dehydrochlorination, dehydrogenation, dehydroisomerization, enrichment of water, hydrocracking, and oxidation. The outstanding feature of rhenium catalysts is their high selectivity, very important in hydrogenation reactions. Rhenium catalysts also resist such catalyst poisons as nitrogen, sulfur, and phosphorus. In terms of activity, rhenium commonly surpasses cobalt, molybdenum, and tungsten type catalysts and approximates palladium, nickel, and platinum catalysts.

Because of the very heavy ionic weight (250) of the perrhenate ion, it is one of the heaviest simple anions obtainable in readily soluble salts. It has found use as a precipitant for potassium and some other heavy univalent ions; also as a precipitant for such complex ions as $Co(NH_3)_6^{2+}$, and for the separation of alkaloids and organic bases. Perrhenate also is used in the fractional crystallization of the rare-earth elements.

When rhenium is added to other refractory metals, such as molybdenum and tungsten, ductility and tensile strength are improved. These improvements persist even after heating above the recrystallization temperature. An excellent example is the complete ductility shown by a molybdenum-rhenium fusion weld. Rhenium and rhenium alloys have gained some acceptance in semiconductor, thermocouple, and nuclear reactor applications. The alloys also are used in gyroscopes, miniature rockets, electrical contacts, electronic-tube components, and thermionic converters.

Because rhenium is very difficult to machine with carbide tools and other conventional methods, electrical-discharge machining (EDM), electrochemical machining (ECM), abrasive cutting, or grinding is recommended.

Rhenium can be consolidated by powder metallurgy techniques, inert-atmosphere arc melting, and thermal decomposition of volatile halides. In the powder metallurgy process, bars are pressed at 200 MPa, and this is followed by vacuum sintering at $1200\,°C$ and hydrogen sintering at $2700\,°C$. Rhenium is usually fabricated from sintered bar by cold working, following by annealing. Reductions of 10–20% can be taken with intermediate anneals for one to two hours at $1700\,°C$. Primary working is by rolling, swaging, or forging. Wire drawing is possible down to 2 mils for strip and wire. Because of its excellent ductility at room temperature, rhenium is suitable for forming of complex shapes (Knipple, 1979).

The cost and scarcity of pure rhenium preclude its extensive use for large structural components. Typical applications include wear-resistant electrical contacts and mass spectrometer cathodes. Rhenium also is being studied for use in the reactor core of space nuclear power systems as, for example, fuel-pin liners that prevent interaction with the fuel and also provide high neutron capture under some operating conditions. Rhenium coatings are used to enhance the heat resistance of carbon and graphite parts

in low-oxygen environments. Tungsten and molybdenum alloys containing rhenium have been used for heating elements, compact electromagnetic coils, high-temperature thermocouples, anti-friction and wear parts, and high-temperature elastic elements. Rhenium and rhenium-containing alloys are principally produced by powder metallurgy. The use of vacuum or inert-gas atmosphere melting or chemical vapor deposition processes tend to be cost prohibitive.

As observed by B.D. Bryskin (Sandvik Rhenium Alloys Inc., Elyria, Ohio), "Rhenium has poor oxidation resistance, but is chemically inert in most oxygen-free atmospheres. When it oxidizes in air, Re_2O_7 is emitted as a white smoke. When hot worked in air, rhenium is embrittled by grain boundary penetration of liquid-phase oxide. Consequently, hot deformation of Re is practical only in a nonoxidizing protective atmosphere (hydrogen or vacuum), or if the workpiece is encapsulated by an oxidation-resistant material. Rhenium is unique among refractory metals in that it does not form a stable carbide. The solubility for carbon is rather high, resulting in a eutectic melting point at about 2773 K ($2500\,°C$; $4530\,°F$) and 0.085% (wt) carbon." Further excellent application information is given in the Bryskin reference listed.

Chemistry and Compounds. Rhenium has a $5d^5 6s^2$ electron configuration and all oxidation states from 0 to 7+ are known, although the heptavalent is the most stable state.

The supposed compounds of Re($-I$), formulated as $M^I Re(H_2O)_4$, have been shown actually to be tetrahydrorhenates(III), e.g., potassium tetrahydrorhenate, $KReH_4$. These compounds are obtained by reducing potassium perrhenate, in a solution of water and ethanolamine, with the alkali metal. In addition to these compounds, however, rhenium differs from its congener manganese in that it forms more compounds that are in the higher valence group. Instead of the larger number of divalent compounds formed by manganese, the stable compounds of rhenium begin with the trivalent ones. The oxides include Re_2O_3, ReO_2, ReO_3, and Re_2O_7. The heptoxide differs from the corresponding manganese compound in its stability, as does the acid obtained by its reaction with water, perrhenic acid, $HReO_4$. Perrhenic acid is a strong acid ($pK_A = 1.25$), but not as strong an oxidizing agent as the permanganic acid. Rhenium dioxide is obtained by reduction of perrhenic acid or rhenium heptoxide, and thermal decomposition of the dioxan addition product of the latter yields rhenium trioxide, ReO_3.

Rhenium forms a number of halides and oxyhalides. Direct reaction by heating with chlorine yields $ReCl_5$, but heating that compound in a nitrogen atmosphere yields $ReCl_3$. $ReCl_4$ and $ReCl_6$ are also known. Direct reaction by heating with fluorine carries the halogenation further, to ReF_6 and ReF_7, which may be reduced to ReF_4. Direct reaction by heating with bromine yields $ReBr_3$. Reaction of the halides with oxygen, or the oxides or oxyacid salts with halogen-containing substances yields a considerable number of oxyhalogen compounds, e.g., $ReOF_4$, ReO_2F_2, ReO_2Cl_2, ReO_2Br_2, ReO_3F and ReO_3Cl. ReO_3F may also be prepared by the action of liquid HF on $KReO_4$ and ReO_3Cl may be prepared by reaction of ReO_3 and Cl_2.

The complexes of rhenium have not been studied as extensively as those of manganese. Among the known complex ions are $Re(NH_3)_6^{3+}$, $Re(CN)_6^{3-}$, $Re(CN)_8^{4-}$, $Re(CN)_8^{3-}$, $ReO_2(CN)_4^{2-}$, $ReCl_4^-$, ReF_6^{2-}, $ReCl_6^{2-}$, $ReBr_6^{2-}$ and ReI_6^{2-}.

The sulfides include ReS_2 and Re_2S_7.

Unlike manganese, rhenium is reported to form an alkyl compound, trimethyl rhenium, $Re(CH_3)_3$. Like manganese, it forms a dirhenium compound with carbon monoxide, $(CO)_5ReRe(CO)_5$, as well as hydrogen-halogen and alkyl-carbonyl compounds. It also forms a dicyclopentadienyl compound, $(C_5H_5)ReH$.

Additional Reading

Bryskin, B.D.: "Rhenium and Its Alloys," *Advanced Materials and Processes*, **22** (September 1992).

Greenwood, N.N. and A. Earnshaw: "Chemistry of the Elements," 2nd Edition, Butterworth-Heinemann, Inc., Woburn, MA, 1997.

Horan, M.F. et al.: "Rhenium-Osmium Isotope Constraints on the Age of Iron Meteorites," *Science*, 1118 (February 28, 1992).

Knipple, W.R.: "Rhenium," in "Metals Handbook," 9th Edition, Vol. 2, American Society for Metals, Metals Park, OH, 1979.

Lewis, R.J. and J.I. Sax: "Sax's Dangerous Properties of Industrial Materials," 10th Edition, John Wiley & Sons, Inc., New York, NY, 1999.

Lide, D.R.: "CRC Handbook of Chemistry and Physics 2000–2001," 81st Edition, CRC Press, LLC., Boca Raton, FL, 2000.

Meyer, C.: "Ore Metals Through Geologic History," *Science*, **227**, 1421–1428 (1985).

Savittskii, E.M. and M.A. Tylkina (editors): "The Study and Use of Rhenium Alloys" (translated from the Russian), Amerind Publishing Co., Pvt. Ltd. Available through U.S. Bureau of the Interior, or National Science Foundation, Washington, DC. (1978).

Sinfelt, J.H.: "Bimetallic Catalysts," *Sci. Amer.*, 90–98 (September 1985).

Staff: "RE Rhenium: Organorhenium Compounds: Binuclear Compounds," Springer-Verlag, Inc., New York, NY, 1996.

Staff: "ASM Handbook—Properties and Selection: Nonferrous Alloys and Pure Metals," ASM International, Materials Park, OH, 1990.

Staff: "Forecast for Metals," Advanced Materials and Processes, (Published annually in January issue.)

Walker, R.J. and J.W. Morgan: "Rhenium-Osmium Isotope Systematics of Carbonaceous Chondrites," *Science*, **519** (January 27, 1989).

RHEOLOGY.   The study of the response of materials to an applied force. Rheology deals with the deformation and flow of matter.

Heraclitus, a pre-Socratic metaphysician, recognized in the fifth century B.C. that $\pi\acute{\alpha}\nu\tau\alpha\rho\varepsilon\acute{\iota}$, or "everything flows." Long before Heraclitus, the prophetess Deborah, fourth judge of the Israelites, had sung that "the mountains flowed before the Lord" in celebrating the victory of Barak over the Canaanites (*Judges* 5:5). Reiner (1964) proposed the dimensionless quantity D (for *Deborah*), where:

$$D = \frac{\text{time of relaxation}}{\text{time of observation}} = \frac{\tau}{t} \tag{1}$$

The difference between solids and liquids is found in the magnitude of D. Liquids, which relax in small fractions of a second, have small D. Solids have a large D. A sufficient time span can reduce the Deborah number of a solid to unity, and impact loading can increase D of a liquid. Viscoelastic materials are best characterized under conditions in which D lies within a few decades of unity.

Force Balance Equation. When a force f is applied to a body, four things may happen. The body may be accelerated, strained, made to flow, or slide along another body. If these four responses are added, one can write an expression for motion in one direction:

$$f = m\ddot{x} + r\dot{x} + sx - f_0 \tag{2}$$

where m is the mass, r is a damping parameter related to viscosity, s to elasticity, and f_0 to the yield value. Evaluation of the coefficients m, r, and s involves the measurement of displacements x and their time derivatives in a manner which links these kinematic variables via an equation of state, such as given in Eq. (2), to stress σ (force per unit area) and its time derivatives.

Scope of Rheology. In contrast to the discipline of mechanics, wherein the responses of bodies to unbalanced forces are of concern, rheology concerns balanced forces which do not change the center of gravity of the body. Since rheology involves deformation and flow, it is concerned primarily with the evaluation of the coefficients r and s of Eq. (2). The coefficients account for most of the energy dissipated and stored, respectively, during the process of distorting a body. Most rheological systems lie between the two extremes of ideality—the Hookean solid and the Newtonian liquid.

Measurements of Viscosity and Elasticity in Shear (Simple Shear). Shear viscosity η and shear elasticity G are determined by evaluating the coefficients of the variables $\dot{x}$ and x, respectively, which result when the geometry of the system has been taken into account. The resulting equation of state balances stress against shear rate $\dot{\gamma}$ (reciprocal seconds) and shear γ (dimensionless) as the kinematic variables. For a purely elastic, or Hookean, response:

$$\sigma = G\dot{\gamma} \tag{3}$$

and for a purely viscous, or Newtonian response:

$$\sigma = \eta\dot{\gamma} \tag{4}$$

As a consequence, G can be measured from stress-strain measurements, and η from stress-shear rate measurements.

Elasticoviscous behavior is described in terms of the additivity of shear rates:

$$\dot{\gamma} = \frac{\sigma}{\eta} + \frac{\dot{\sigma}}{G} \tag{5}$$

whereas viscoelastic behavior is characterized by the additivity of stress, according to Eq. (2);

$$\sigma = G\gamma + \eta\dot{\gamma} \tag{6}$$

Relaxation. Numerous attempts have been made to fit simplified mechanical models to the two behavior patterns described by Eqs and (6). One can picture the elastic element as a spring-arrayed network parallel with the viscous element to give essentially a (Kelvin) solid with retarded elastic behavior, wherein:

$$\frac{\eta_k}{G_k} = \text{retardation time (sec)} \tag{7}$$

or as a (Maxwellian) series network which flows when stressed or relaxes, under constant strain:

$$\frac{\eta_m}{G_m} = \text{relaxation time (sec)} \tag{8}$$

and transient experiments may be designed to measure these parameters singly. In real systems, a single relaxation (or retardation) time fails to account for experimental results. A distribution of relaxation time exists.

Dynamic Studies. When Eq. (2) is written in the form:

$$\ddot{x} + 2k\dot{x} + \omega_1^2 x = 0 \tag{2b}$$

the equation suggests that the variation in stress should be cyclic. Rheometers are designed so that the system may oscillate in free vibration of natural resonant frequency ω_1, or else so that a cyclic shearing stress of the form $f_0 \cos \omega t$ is impressed on the sample over a frequency range which spans ω_1. In neither case is the material strained beyond its range of linearity. Equation (2b) represents a damped harmonic oscillator, providing that the coefficients are constant (i.e., providing that they do not depend on the strain magnitude). Not all systems meet this requirement in the strict sense, with the result that one of the first checks the experimenter makes is for linearity. Doubling the amplitude of oscillation should double the stress and should not change the phase relationships between the cyclic stress and the deformation.

Time-Temperature Equivalence (Steady-State Phenomena). The creep of a viscoelastic body or the stress relaxation of an elasticoviscous one is employed in the evaluation of η and G. In such studies, the long-time behavior of a material at low temperatures resembles the short-time response at high temperatures. A means of superimposing data over a wide range of temperatures has resulted which permits the mechanical behavior of viscoelastic materials to be expressed as a master curve over a reduced time scale covering as much as twenty decades (powers of ten).

Polymeric materials generally display large G values (10^{10} dynes/cm^2 or greater) at low temperatures or at short times of measurement. As either of these variables is increased, the modulus drops slowly at first, then attains a steady rate of roughly one decade drop per decade increase in time. If the material possesses a yield value, this steady drop is arrested at a level of G which ranges from 10^7 downward.

Dynamic Behavior. The application of sinusoidal stress to a body leads inevitably to the complex modulus G^*, where

$$G^* = G' + iG'' = G' + i\omega\eta' \tag{9}$$

where G' is the in-phase modulus ($\sigma/\dot{\gamma}$) which represents the stored energy, and G'' is the out-of-phase modulus ($\sigma/\dot{\gamma}$) representing dissipated energy (as its relation to η suggests); the variable against which G' and η' are determined is the circular frequency ω. Superposition of variable temperature data or variable frequency data provides a master curve of the type previously described for steady-state parameters.

Problems in Three Dimensions (State of Stress). The forces and stresses applied to a body may be resolved in three vectors, one normal to an arbitrarily selected element of area and two tangential. For the yz plane, the stress vectors are σ_{xx} and σ_{xy}, σ_{xz}, respectively. Six analogous stresses exist for the other orthogonal orientations, giving a total of nine quantities, of which three exist as commutative pairs ($\sigma_{rs} = \sigma_{sr}$). The state of stress, therefore, is defined by three tensile or normal components ($\sigma_{xx}, \sigma_{yy}, \sigma_{zz}$) and three shear or tangential components ($\sigma_{xy}, \sigma_{xz}, \sigma_{yz}$). The shear components are most readily applicable to the determination of η and G.

Strain Components. For each stress component σ there exists a corresponding strain component γ. Even for an ideally elastic body, however, a pure tension does not produce a pure γ_{xx} strain; γ components exist which constrict the body in the y and z directions.

The complete stress-strain relation requires the six σs to be written in terms of the six γ components. The result is a 6×6 matrix with 36 coefficients k_{rs} in place of the single constant. Twenty-one of these

TABLE 1. EXAMPLES OF NON-NEWTONIAN FLUIDS EXHIBITING DIVERSE RHEOLOGIC PROPERTIES

Pseudoplastic	Plastic	Trixotropic*	Rheopectic	Dilatant*
Catsup	Chewing gum	Silica gel	Bentonite sols	Quicksand
Printers ink	Tar	Most paints	Gypsum in water	Peanut butter
Paper pulp	Various slurries	Glue		Many candy compounds
		Molasses		
		Lard		
		Fruit juice concentrates		
		Asphalts		

*Some liquids may change from thixotropic to dilatannt or vice versa as the temperature or concentration changes.

coefficients (the diagonal elements and half of the cross elements) are needed to express the deformation of a completely anisotropic material. Only three are necessary for a cubic crystal, and two for an amorphous isotropic body. Similar considerations prevail for viscous flow, in which the kinematic variable is $\dot{\gamma}$.

Applications. The study of rheology is important to many sciences and technologies. Numerous instruments have been constructed to make manual, semiautomatic, and automatic measurements (and control) of such variables as viscosity, and consistency (flowability). These variables are particularly important in the food processing, chemical, and plastics industries, where many of the products and intermediates lie between Hookean solids and Newtonian liquids. Consistency measurement is of large importance in the paper manufacturing industry. See also **Papermaking and Finishing**; and **Pulp (Wood) Production and Processing**.

Some terms commonly used in industry include:

Newtonian Substance — Fundamentally, liquids or suspensions in liquids when subjected to a shear stress behave in two ways: (1) A Newtonian substance undergoes deformation, the ratio of shear rate (flow) to shear stress (force) is constant. See Fig. 1.

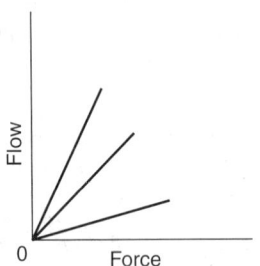

Fig. 1.   Behavior of Newtonian substances.

Non-Newtonian Substance — (2) In a non-Newtonian substance, the ratio of shear rate to flow is not constant. See Fig. 2.

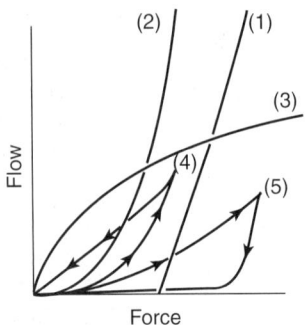

Fig. 2.   Behavior of non-New-tonian substances: (1) true plastic (sometimes called a Bingham body); (2) pseudoplastic; (3) dilatant; (4) thixotropic; and (5) rheopectic.

Dilatant — The initial flow of a dilatant substance under a low shear stress is at a high rate; further increases in shear stress result in a lower flow rate. A dilatant substance is sometimes called an inverted plastic or inverted pseudoplastic.

Pseudoplastic — A material of this type appears to have a yield stress beyond which flow commences and increases sharply with increase in stress. In practice, such substances are found to exhibit flow at all shear stresses, although the ratio of flow to force increases negligibly until the force exceeds the apparent yield stress.

Thixotropic — The flow rate of a thixotropic substance increases with increasing duration of agitation, as well as with increased shear stress. When agitation is stopped, internal shear stress exhibits hysteresis. Upon reagitation, generally less force is required to create a given flow than is required for first agitation.

Rheopectic — If certain thixotropic suspensions are rhythmically shaken or tapped, they will "set" or build up very rapidly, a phenomenon termed rheopexy. Apparent viscosity of a rheopectic substance increases with time (duration of agitation) at any constant shear rate.

See Table 1. See also **Colloid System**; **Fluid**; **Stokes Flow**; **Stokes Law for Viscosity**; **Viscoelasticity**; and **Viscosity**.

Additional Reading

Acierno, D. and A.A. Collyer: "Rheology and Processing of Liquid Crystal Polymers," Chapman & Hall, New York, NY, 1996.

Cheremisinoff, N.P.: "Introduction to Polymer Rheology and Processing," CRC Press, LLC., Boca Raton, FL, 1999.

Gupta, R.K.K.: "Polymer and Composite Rheology," Marcel Dekker, Inc., New York, NY.

Morrison, F.A.: "Understanding Rheology," Oxford University Press, Inc., New York, NY, 2000.

Tanner, R.I.: "Engineering Rheology," 2nd Edition, Oxford University Press, Inc., New York, NY, 2000.

Tanner, R.I. and K. Walters: "Rheology: An Historical Perspective," Elsevier Science, New York, NY, 1998.

RHEOPECTIC SUBSTANCE.   See Colloid System.

RHEOSTAT. A variable resistance used for operation or control of electrical equipment. Rheostats might be classified as metallic, carbon, and electrolytic types. The most common form is the metallic type, in which the resistance is in the form of a metal wire or ribbon, or cast grid, these being made of a metal having poor conductivity, and little deterioration from heating. The variable resistance of metallic rheostats is obtained by bringing out taps from different points of the resistance wire to the points of a multi-pointed switch which can be used to short-circuit different sections of the resistance. Laboratory rheostats are frequently coils of resistance wire wound closely on an insulating cylinder and provided with a sliding contact finger which will bear on the wires themselves, and which can be employed to short-circuit any desired number of turns of the resistance wire.

RHESUS MONKEY.   See Monkeys and Baboons.

RHEUMATIC DISEASES. Characteristically, and essentially by definition, the rheumatic diseases involve the joints. The functional characteristics of the various connective tissues that make up a joint are manifestations of the chemical structure of these elements. In a healthy, well-functioning (*diathroidal*) joint, there are essentially perfect biochemical processes going on to maintain ideal functionality. Connective tissue consists of large and complex molecules, such as collagen. See also **Bone**. In processes as complex and intricate as these, there are many potential opportunities for "natural" errors. Some genetic disorders of collagen have been identified. These reflect errors in the biosynthesis processes. In the makeup of a joint, there are other equally complex substances. The other major component of connective tissue matrix is the ground substance, composed of complex proteoglycans. Synovial membrane and synovial fluid are also complex biochemically. These varied substances, each characterized by its own biochemistry, explain in a very general way why there are so many kinds and degrees of rheumatic diseases.

The etiology of only a few rheumatic diseases is reasonably well understood. Much progress has been made in this area in recent years, with some fundamental findings dating back to the 1930s. Thus, over the years, the science of the rheumatic diseases has been revised considerably. In 1964, the American Rheumatism Association proposed a reclassification of the rheumatic diseases and has since been revising the classification as new information warrants. This classification is given in Table 1.

TABLE 1. CLASSIFICATION OF RHEUMATIC DISEASES[1]

1. *Polyarthritis of unknown etiology*, as exemplified by rheumatoid arthritis, ankylosing spondylitis, and Reiter's syndrome.
2. *Acquired connective tissue disorders*, as exemplified by systemic lupus erythematosus, scleroderma, polymyositis, and dermatomyositis.
3. *Rheumatic fever*.
4. *Degenerative joint disease*, where the primary form is of unknown cause, and where the secondary form may be associated with mechanical or metabolic disorders. Exemplified by osteoarthritis and osteoarthrosis.
5. *Nonarticular rheumatism*, exemplified by fibrositis, intervertebral disk and low back syndrome, myositis and myalgia, tendinitis and bursitis.
6. *Diseases frequently associated with arthritis*, such as sarcoidosis, relapsing polychondritis, ulcerative colitis, regional enteritis, Sjögren's syndrome, and familial Mediterranean fever.
7. *Infectious arthritis*, such as bacterial arthritis involving *Gonococcus, Meringococcus, Pneumococcus, Streptococcus, Staphylococcus, Salmonella, Brucella, Streptobacilus moniliformis*; or involving mycobacteria, such as *M. tuberculosis*; or involving fungi, parasites, and viruses.
8. *Traumatic and neurogenic disorders*, as exemplified by traumatic arthritis resulting from direct trauma, neuropathic arthropathy, and diabetic neuropathy, among others.
9. *Biochemical or endocrine disorders*, as exemplified by gout, Gaucher's disease, and Fabry's disease, among others.
10. *Neoplasms*, as found in synovioma, primary juxta-articular bone tumors, multiple myeloma, and others.
11. *Allergic and reactive disorders*, such as serum sickness and drug reactions.
12. *Inherited and congenital disorders*.
13. *Miscellaneous disorders*.

[1]Essentially as proposed by American Rheumatism Association (in abridged format).

The concept that immunologic alterations may play a role in rheumatoid arthritis was first suggested a number of years ago. In the 1930s, the abnormal agglutinating properties of the serum of several rheumatoid arthritis patients led to the discovery of the *rheumatoid factor*. See also **Immune System and Immunology**. Later, the abnormality of the humoral immune mechanisms was indicated by noting increased circulatory levels of the immunoglobulins, IgM, IgA, and IgG, in a high percentage of affected individuals.

Several articles in this encyclopedia relate to rheumatic diseases. Check alphabetical index for such key words as arthritis, bone, bursa and bursitis, gout, osteoarthritis, rheumatic fever, rheumatoid arthritis, scleroderma, spondylarthropathies, and systemic lupus erythematosus.

Additional Reading

Fassbender, H.G.: "Pathology and Pathobiology of Rheumatic Diseases," Springer-Verlag, Inc., New York, NY, 1998.
Hill, J.S., S.H. Gaston, and J.S. Gaston: "Rheumatic Diseases: Immunological Mechanisms and Prospects for New," Cambridge University Press, New York, NY, 1999.
Rosenbaum, R.R., S.M. Campbell, and J.T. Rosenbaum: "Clinical Neurology in Rheumatic Diseases," Butterworth-Heinemann, Inc., Wodurn, MA, 1996.
Silman, A.J. and M.C. Hochberg: "Epidemiology of the Rheumatic Diseases," Oxford University Press, Inc., New York, NY, 1999.

Web Reference

The National Arthritis Foundation: *http://www.arthritis.org.sg/New/arthritis.html*

RHEUMATIC FEVER. An inflammatory disease precipitated by a Group A streptococcal infection. The preceding infection, which may occur almost anywhere in the body, is key to diagnosis. It is important to affirm the infection because the disease can be mediated by early treatment of the infection. Rheumatic fever is predominantly found in persons between the ages of 5 and 15 years. Much remains to be learned concerning the manner by which Group A streptococci damage various tissues in the body, but it has been reasonably well established that antibodies directed against streptococcal antigens play an important role. It is also believed that a genetically determined aberration in the immune response is of major importance.

Principal manifestations of rheumatic fever are: inflammation of the heart (carditis), which occurs in 50–60% of cases; inflammation of joints (polyarthritis), which occurs in 70% of cases, subcutaneous nodules, which occur in most severe cases; erythema marginatum (characteristic skin rash), which occurs in about 5% of cases; and chorea, in about 5% of cases.

The lesions of rheumatic fever tend to favor connective tissue. Such lesions (Aschoff nodules) may be seen throughout the heart in some cases and can cause severe permanent damage. Ultimately the lesions regress and cause an area of fibrosis. Valvular lesions tend to undergo fibrous thickening as they heal, sometimes resulting in permanent valvular dysfunction.

The course of the disease ranges from a few weeks to three months (in 90% of cases) to six months (less than 5% of cases) or longer. Rheumatic fever is prone to recur, but such recurrences have been greatly reduced through antibiotic therapy. In moderate cases, heart valvular lesions will heal, but in some cases this condition may worsen. Recurrence retards the healing process. Patients who escape carditis in the initial attack usually also escape this involvement during recurrent attacks.

Four stages are recognized in the progressive development of rheumatic fever: (1) acute streptococcus infection; (2) a period of apparent recovery lasting from a few days to about 3 weeks; (3) appearance of the major symptoms of the disease as previously described; and (4) chronic manifestations of the disease in the form of damage, such as occurs in valvular stenosis. The disease usually runs a milder course in adults than in children. See also **Bacterial Diseases**.

Additional Reading

English, P.C.: "Rheumatic Fever in America and Britain: A Biological, Epidemiological, and Medical History," Rutgers University Press, Piscataway, NJ, 1999.
Massell, B.F.: "Rheumatic Fever and Streptococcal Infection: Unraveling the Mysteries of a Dread Disease," Harvard University Press, Cambridge, MA, 1997.
Taranta, A.V. and M. Markowitz: "Rheumatic Fever," 2nd Edition, Kluwer Academic Publishers, Norwell, MA, 1989.

RHEUMATOID ARTHRITIS. A chronic, inflammatory, systemic disorder that principally affects the joints and tendon sheaths. The disease is of varied characteristics, ranging from one of relatively minor aches and pains of one or more joints to a severely debilitating condition involving mutilative structural changes. Onset of the disease usually occurs during the third or fourth decade of life, but can appear at any age. About 1% of the population is affected in some way by rheumatoid arthritis. The disease appears in women at three times the rate of men. The etiology of the disease is difficult to determine—either generally for rheumatoid arthritis or for specific cases. There are no apparent biological markers for the disease and thus the usual routine laboratory tests are of little, if any value, to the physician in making a definite diagnosis of rheumatoid arthritis. The joints of the extremities are particularly prone to involvement and symmetry is often present. The joints most commonly affected, in general order of rate of occurrence, are the hands and wrists, elbows, shoulders, hips, knees, ankles and feet, and spine (cervical spine disease). Systemic involvement of the disease appears in a number of organs of the body, with each case having its own particular profile. Other systemic manifestations may include *Sjögren's syndrome* (eye involvement). This condition involves a sensation of grittiness of the eye, an accumulation of dried mucoid material, and frequently a lessening of eye fluid (tear) production. Other systemic manifestations include *lung involvement* (pleurisy with effusion); *heart* (pericardial effusion); *blood* (mild anemia); *neuromuscular* (muscle weakness); and *vasculature* (vasculitis).

Although some linkages of a genetic nature have been demonstrated in rheumatoid arthritis, there is a lack of convincing evidence particularly as regards the adult form of the disease. It also has long been suspected that bacteria and/or viruses may be causative agents, but research to date has not revealed well-documented evidence of such connections. The absence of a strong biological marker for the disease makes such connections very difficult to prove. Although *rheumatoid factors* are found in 80% of cases of adult rheumatoid arthritis, these factors are not specific to the disease. See entry on **Rheumatic Diseases**. Thus diagnosis of the disease in its early phases is essentially a matter of the physician's experience with numbers of cases of the disease and relating early physical evidence to the patient's history and description of complaints. X-rays taken in the early stages are not always helpful because, during the early period, the disease is essentially confined to the synovium and alterations that do not show up on X-ray films. Some months after the onset of the disease, x-rays may then indicate some loss of bone mass (*osteoporosis*), particularly of finger joints. X-ray examinations generally are most helpful in connection with the larger joints, such as the knee and hip, after the disease has progressed for a number of months or longer.

Because of the various profiles of the disease, generalization is difficult. Keeping this in mind, the most common form of rheumatoid arthritis is that which is only marginally noticeable during the early phases and that slowly progresses to more serious phases over a period of many months, even years. One qualitative confirmation of the disease is the nature of joint stiffness. For example, in most older people, there will be a certain amount of stiffness of joints experienced upon rising in the morning. This may persist for several minutes to close to an hour before the person feels "limbered up" for the day. No swelling follows such stiffness. If rheumatoid arthritis is present, certain joints will swell after brief periods of exercise.

Rheumatoid arthritis is frequently of a rather cyclic nature, i.e., alternating periods of remission and exacerbation—where the symptoms may essentially disappear or decrease to minimal proportions, but with periods of increasingly severe symptoms occurring between remissions. This is highly variable with patients and difficult, if not impossible, to forecast in advance. Usually patients who experience a relatively favorable course of the disease, that is, with long remissions, are usually less than 40 years of age and with involvement of just a few of the larger joints. The less favorable course of the disease is most frequently experienced by persons over 40 years of age and where the disease has commenced slowly, but where, after the initial phase of the disease, rheumatoid nodules have appeared at a rather rapid rate. Such patients also will usually be found to have high titers of rheumatoid factor and to have experienced weight loss, these factors indicating more serious systemic involvement.

After rheumatoid arthritis has progressed for a period, increased numbers of lining cells (most often found in recesses of the joint) will be found. It is believed that these cells may he due to proliferation of the cells per se or represent the presence of cells that have migrated to the joint from the blood. The surface of the cartilage will be covered with fibrin (an insoluble protein), and there may be edema of the sublining cells. The diseased synovial fluid contains various forms of debris, with consequent alteration of the fluid volume (increase) and of various physical characteristics (decrease of viscosity; increase in protein content, notably of large protein molecules). There is alteration of hyaluronic acid brought about by reactions between hyaluronic acid and proteins. There is demineralization of bone collagen, which may result from local decreases in pH and possibly chelation by organic acids. Other changes have been confirmed or postulated, the end result of which processes are bone loss or degeneration. Nodules (from a few millimeters to two centimeters in diameter and often occurring in clusters) will appear under the skin on the forearm, the tibia, the Achilles tendon, and other sites, varying with the particular course of the disease in a given patient. It is interesting to note that these nodules may be noted elsewhere (not confined to joints) in the body—the lungs, intestinal tract, heart, and dura, among others. Similar nodules are also seen in systemic lupus erythematosus and rheumatic fever and thus are not specific to rheumatoid arthritis.

In the treatment of rheumatoid arthritis, the regimen includes the administration of analgesics (for pain), anti-inflammatory agents, considerable dependence on physical therapy, sometimes in a hospital environment, and, less frequently, surgery.

Aspirin has been a mainstay for arthritis therapy for a number of years. Aspirin serves the dual function as an analgesic and anti-inflammatory agent. The drug, properly administered, assists in reducing swelling, relieving stiffness, and generally makes most patients, particularly in the initial phase of the disease, feel much better. Levels just below those that cause *tinnitus* (ringing in ears) and loss of hearing may be prescribed by some physicians. In some patients, gastrointestinal side effects usually can be alleviated by taking aspirin with milk or by using specially buffered or enteric-coated preparations. Some complications, such as interference with platelet formation, may rule out aspirin in some patients.

Nonsteroidal anti-inflammatory drugs, such as indomethacin and phenylbutazone, are prescribed in some cases. Although not fully free of gastric side effects in some patients, ibuprofen (Motrin®), fenoprofen, naproxen, tolmetin, and sulindac, among others, are sometimes used as substitutes for aspirin.

Although sometimes used, antimalarial drugs, such as chloroquine and hydroxychloroquine, are regarded by some authorities as of limited usefulness in rheumatoid arthritis and may cause retinopathy and irreversible decrease in vision. They are usually only used in patients who do not respond well to any of the drugs previously mentioned.

In highly persistent rheumatoid arthritis, soluble gold salts may be indicated. These compounds have been demonstrated in controlled tests to be effective in decreasing inflammation and contributing to improved frequency of remissions. However, tests also show that not all patients have a positive response to these compounds. The salts, such as gold sodium thiomalate or gold thioglucose, are usually administered intramuscularly. Several months may be required to determine the nature of response. Some patients develop a skin outbreak or stomatitis (inflammation of mouth) of a relatively minor and transient nature. Where such symptoms are serious and persist, another gold salt or a lower dosage may be attempted after a short period, or gold salt therapy may be ruled out altogether. This therapy requires the careful monitoring of a physician because, although they are uncommon, serious side effects may occur in some patients.

Released relatively recently for use in the United States and some other countries, D-penicillamine is now used for some severe cases of rheumatoid arthritis. The drug produces toxic reactions in some patients not unlike the reactions from gold salts as previously mentioned. The physician will usually start the patient on a low dosage, gradually increasing the dosage until the patient develops a tolerance for the drug.

Glucocorticoids, such as prednisone, are powerful anti-inflammatory agents. In some patients, suppression of the major symptoms of rheumatoid arthritis is quite marked. However, when prescribed in dosages sufficient to achieve such dramatic effects, there are long-term side effects of a serious nature, including increased susceptibility to infection, osteoporosis, osteonecrosis, cataracts, and gastrointestinal bleeding. Convincing evidence has not been forthcoming to the effect that the glucocorticoids actually alter the course of the disease, which leads to joint and other structural damage. Thus, these drugs are used with discretion, often confining their use to local injections, which usually lower the risk of side effects considerably.

In extreme cases of rheumatoid arthritis, where there are life-threatening implications and very serious crippling, cytotoxic agents, such as cyclophosphamide and azathioprine, may be used. These drugs are used infrequently because of serious side effects, including the induction of neoplasias and chromosomal abnormalities.

Physical therapy is an important adjunct in the overall management of prolonged cases of rheumatoid arthritis. Surgery to repair or replace affected joints and structures is used where there is a serious degree of crippling or debilitation. Complete joint replacement, including total hip replacement, is sometimes effected. In some patients, replacement of the hip joint with an alloy steel articulating ball mounted on a stem within a high-density polyethylene socket has proved quite successful, providing a joint that is stable, that allows satisfactory motion, and that is free from pain.

Additional Reading

Bird, H.A. and M.L. Snaith: "Challenges in Rheumatoid Arthritis," Blackwell Science, Inc., Malden, MA, 1999.

Baumgartner, H.: "Rheumatoid Arthritis," Thieme Medical Publishers, Inc., New York, NY, 1996.

Firestein, G.S., G.S. Panayi, and F.A. Wollheim: "Rheumatoid Arthritis: Frontiers in Pathogenesis and Treatment," Oxford University Press, Inc., New York, NY, 2000.

Goronzy, J.J. and C.M. Weyand: "Rheumatoid Arthritis,"S. Karger Publishers, Inc., Farmington, CT, 2001.

Harris, E.D. and R. Zorab: "Rheumatoid Arthritis," W.B. Saunders Company, Philadelphia, PA, 1997.

Lahita, R.G.: "Rheumatoid Arthritis: Everything You Need to Know," The Putnam Publishing Group, New York, NY, 2001.

Web References

Arthritis Resource Center: *http://www.healingwell.com/arthritis/*

Mayo Clinic: Rheumatoid Arthritis: *http://www.mayoclinic.com/home?id=5.1.1.21*

RHINITIS. Inflammation or infection of the mucous membrane of the nose. The common head cold is primarily a rhinitis. Chronic rhinitis, also termed dry rhinitis, usually is associated with some form of chronic debilitating disease, notably persistent kidney problems and diabetes. The principal complaint is dryness of the nose and accompanying encrustation. There is no atrophication of nasal bone and lining in the case of *rhinitis sicca*, but in *atrophic rhinitis* such alterations occur. The primary treatment of any rhinitis is directed to the general health and other causative disease factors, coupled with local medication for the stimulation of nasal mucosa to alleviate dryness. Rhinoviruses are described under **Virus**. See also **Common Cold**.

Increased secretion by nasal and bronchial mucous glands, in some cases, is believed to result from some of the mediators of immediate hypersensitivity as part of a complex of reactions occurring in the immune system. See also **Immune System and Immunology**.

RHINOCEROS *(Mammalia, Perissodactyla).* A large animal, *Rhinoceros*, of the Oriental region and Africa, with rather short legs and a long muzzle bearing one or two conical horns behind the nostrils. There

Fig. 1. (Top) Former and present distribution of the black rhinocerous (*Diceros bicornis*). This is the only species of rhinoceros that still occurs with some, but diminishing frequency in areas indicated by black triangles. (Bottom) The fights of the rhinoceros, after study, have been shown to be fair duels, which are performed according to specific rules. Serious injuries rarely occur, and often these fights are mere play.

are four species in the Oriental region and two in Africa. The common African species stands over 6 feet (1.8 meters) high at the shoulders and has an anterior horn over 3 feet (0.9 meter) long. Because of its thick skin, it is difficult to kill. These animals are described as generally dull and timid, but when brought to bay they can be very aggressive. Because of their great size, they can be dangerous when provoked. See Figs. 1 and 2.

The true rhinoceros of the Rhinocerotidae family during the Tertiary period was a widely distributed group of many species. During the Eocene, a hornless, small form with slender feet, probably not too different from the other odd-toed ungulates of the era, first appeared. The skull was low and flat, without any indication of horns. The molar teeth consisted of premolars and molars with low crowns and ridges across and on the sides. This basic structure, in spite of some variations, is the same as found in the later rhinoceros. Fossils of members of the subfamily Caenopodinae (*Eotrigonias, Caenopus*, etc.), which belong to the most primitive, oldest rhinoceros, and fossils of several such forms have been found in the Early Tertiary stratum of North America and Europe. These slender-footed, hornless, primitive rhinoceros still had a complete set of front teeth and molars.

Of the contemporary rhinoceros, the *Asiatic two-horned rhinoceros (Dicerorhiniae)* may be traced back approximately forty million years into the Oligocene. At first they occurred as small animals (*Dicerorhinus tagius*), which were less than the size of a tapir and soon split up into different lines. One line led to the well known, early glacial *Wooly Rhinoceros (Coelodonta antiquitatis)*. This was a cold-resistant species with a long-haired, thick coat. Knowledge of this is gained not only from bones, but also from complete bodies with skin and fur discovered in the Siberian permafrost soil. In addition, the people of the Early Stone Age have portrayed the animal on their cave drawings. The species was extinct by the end of the glacial period.

The contemporary *square-lipped rhinoceros* is a grass-eating animal of the steppe, which traces back to the earlier and middle glacial period of Europe. The larger *Merck rhinoceros (Dicerorhinus kirchbergensis)* from the same glacial period was a forest type. The only contemporary species of this group, the *Sumatran rhinoceros (Dicerohinus sumatrensis)* is much closer to the phylogenetically older forms than its glacial relatives, a fact which is frequently found in the inhabitants of the tropical prime forests. However, since the species still has front teeth and molars with low crowns, which are not suitable for crushing hard steppe grass, authorities consider the species as a slightly modified survivor from the Tertiary Period. The *great Indian rhinoceros*, which lives in South Asia today, can be traced back to the Tertiary (Miocene, approximately 25 to 10 million years ago). The *African rhinoceros* forms a separate branch which includes the *black rhinoceros*, an animal that originally fed on foliage. Another

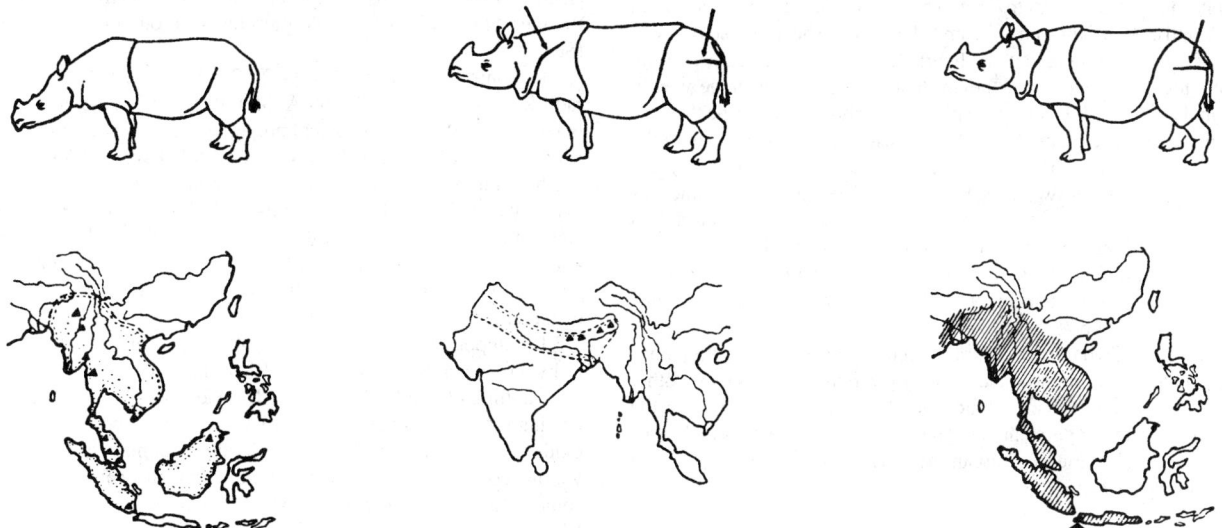

Fig. 2. Asiatic species of rhinoceros: (Left) Former and present distribution of the Sumatran rhino. This species now exists only in those few places marked by triangles on map. (Middle) Former and present distribution of the great Indian rhino. Presently, it is found only in a few protected areas, marked by triangles. (Right) Former and present distribution of the Javan rhino. Only a few animals exist today in the Udjong-Kulon Preserve in Java, noted by arrow. Skin folds at shoulder and base of tail are indicated by arrows in cases of the great Indian and Javan rhinos.

African species, the *square-lipped rhinoceros*, is a more highly evolved grass-eater.

A History of Extinction

Compared with the multitude of species that existed in earlier times, the surviving genera (great Indian, Javan, Sumatran, square-lipped, and black) appear rather stunted despite their size. They all live in remote habitats, seemingly because they have not been able to compete any longer with the other ungulates, especially the ruminants. Above all, however, human influence has basically changed wide areas of Africa and Asia, thus making them uninhabitable for the rhinoceros. Since humans first pursued animals, the rhinoceros has been hunted. The pictures found in the Early Stone Age caves of Pech-Merle, Rouffignac, Colombiàre, and Les Trois Freres tell an obvious story. They show that these animals have had mystical significance since early times.

The Rhino Horn — Fascination and Superstition. It is commonly held that rhinoceros horns consist of matted hair, a view that is not quite correct. The horns consist throughout of *keratin*, and they do not have a bony pith like the horns of cattle. Under a microscope, however, one can see that the individual rods are not coated with an individual protective layer as is real hair. They adhere densely together in layers and thus resemble neither the hair nor the horn of a ruminant, but rather the material of the hoof. This construction gives the nose horn a stiffness and quality similar to a ruminant's horn with a pith. The nose horn sits on a bony dome formed by the nasal bone; it may unravel in places, causing it to look like a growth of hair. If it is torn by accident, only a lightly bleeding area remains on the nose. Soon a new horn begins to grow. In young animals, a horn may be replaced completely.

Except for the elephants, we find the largest terrestrial mammals among the rhinoceros. However, these handsome mammals provide a classical example of the extent to which humans are responsible for the decrease and extinction of large mammals. *Superstition* played a dominant and especially destructive part in the disappearance of many rhinoceros species. The Chinese, as well as other Asiatic people, have believed that powdered rhinoceros horns make an aphrodisiac. Many centuries ago, the powder made from these horns was sold in East Asiatic pharmacies at a high price. Since the rhinoceros is easy to kill, it has been poached ever since. Poaching has essentially made Asiatic species extinct, and African rhinos are rapidly facing extinction. A number of scientists have carefully tested the potions made from rhino horns and no evidence whatever has been produced to back up the superstitious claims.

Protection of the rhinoceros by raising them in captivity has had some success. The first rhinoceros ever to be born in captivity was a Sumatran rhino, born in 1889 at the Calcutta, India zoo. At that time the Sumatran rhino was not as rare as it is now. Success in raising rhinos in a zoo atmosphere is difficult to achieve for many reasons. Although the females reach sexual maturity at three years, the bulls do not become sexually mature until they are seven to nine years old. The average gestation period in captivity is between 462 and 489 days. The birth starts with episodes of labor lasting about an hour; the actual birth requires only 15–30 minutes. A neonate great Indian rhinoceros has an average weight of 143 pounds (65 kg). It has folded skin like an adult, with all the "rivets" and protuberances. The plum-shaped head is especially conspicuous in the newborn. There is a flat, smooth, oval plate where the nose horn will grow later. The young rhino grows at a fairly rapid rate, gaining as much as $4\frac{1}{2}$ to $6\frac{1}{2}$ pounds (2–3 kg) per day. Thus, the weight at birth is multiplied tenfold within one year. Shortly after birth, the shoulder height is about 25 inches (63 cm) and, in the second year, this approximately doubles. At the age of $3\frac{1}{2}$ to 4 years, the female is fully grown, while the bull may continue growing for up to five years.

Unlike roaming elephants, rhinos rarely return to areas where they once were exterminated. To reintroduce them, they must be caught in other places, transported, and set free. Since we obtain our information on the rhino no longer from big game hunters, but rather from scientists and game wardens, much has been learned about their life. Studies on their behavior really began only in 1960.

Additional Reading

Cunningham, C. and J. Berger: "Horn of Darkness: Rhinos on the Edge," Oxford University Press, Inc., New York, NY, 1997.

Gould, E. and G. McKay: "Encyclopedia of Mammals," 2nd Edition, Academic Press, Inc., San Diego, CA, 1998.

Grzimek, B.: "Grzimek's Encyclopedia of Mammals," The McGraw-Hill Companies, Inc., New York, NY, 1990.

MacDonald, D.: "Encyclopedia of Mammals," Barnes & Noble Books," New York, NY, 1999.

RHINOSPORIDIOSIS. A disease due to a yeastlike organism (*Rhinosporidium seeberi*), which infects the mucous membrane of the nose producing nasal polyps and, in other locations, tumors on the cheek, conjunctivae, lacrima, uvula, penis, and skin. It is found most often in India and Sri Lanka, but has also been reported from Indonesia, Malaysia, the Philippines, Iran, South Africa, Italy, the United Kingdom, the southern United States, and South and Central America. The disease is most often seen in children and young adults, in men more than women, but it can occur at any age and there is no racial differentiation. Infection is most often seen in laborers who are frequently exposed to water in streams or pools and cases have occurred in men who dive to recover sand. This suggests that *R. seeberi* has a natural habitat in the water, growing as a parasite of either fish or water insects. The most striking feature in the stroma of the polyps is the appearance of sharply defined globular cysts, varying in size from 10 to 200 micrometers (diameter). There is a chronic inflammatory reaction and occasionally micro-abscesses occur.

Friable, highly vascular, sessile and pedunculated polyps may be disseminated hematogeneously to the urine, palate, lungs, liver, spleen and other organs.

Treatment is essentially, surgical.

A. C. V.

RHINOVIRUSES. See **Virus**.

RHIZOIDS. Filamentous outgrowths from the surface, or from epidermal cells, formed of one or many cells, which serve to hold the plants of mosses and hepatics or the prothallia of ferns to the substratum. Similar structures occur in the thallophytes.

RHIZOME (or Rootstock). A horizontal stem growing beneath the surface of the ground or at times, at the surface. Rhizome has all the characteristics of a stem, such as nodes, and internodes, leaves and branches. Often it is very much enlarged and contains much reserve food material. See also **Stem (Plant)**; and **Asexual Reproduction**.

RHODIUM. Chemical element, symbol Rh, at. no. 45, at. wt. 102.906, periodic table group 9, mp. 1,963 to 1,969 °C, bp. 3,627 to 3,827 °C, density 12.44 g/cm^3 for solid (20 °C). Elemental rhodium has a face-centered cubic crystal structure. The one stable isotope is ^{103}Rh. The seven unstable isotopes are ^{99}Rh through ^{101}Rh and ^{104}Rh through ^{107}Rh. In terms of earthly abundance, rhodium is one of the scarce elements. Also, in terms of cosmic abundance, the investigation by Harold C. Urey (1952), using a figure of 10,000 for silicon, estimated the figure for rhodium at 0.0067. No notable presence of rhodium in seawater has been found.

Electronic configuration is $1s^2 2s^2 2p^6 3s^2 3p^6 3d^{10} 4s^2 4p^6 4d^8 5s^1$. Ionic radii Rh^{3+} 0.75 Å, Rh^{4+} 0.65 Å. Metallic radius 1.345 Å. First ionization potential 7.7 eV. Other physical properties of rhodium will be found under **Platinum and Platinum Group**. See also **Chemical Elements**.

Rhodium was discovered by Wollaston (England) in 1803. Compact Rh is almost insoluble in all acids at 100 °C, including aqua regia. Hot concentrated H_2SO_4 will slowly dissolve the finely divided metal. When alloyed with 90% or more of Pt, it is soluble in aqua regia. The metal is attacked by fused bisulfates. Rh is soluble in molten Pb. This is the basis of the classic separation of Rh and Ir.

Rh compounds exhibit valences of 2, 3, 4, and 6. The trivalent form is by far the most stable. When Rh is heated in air, it becomes coated with a film of oxide. Rhodium(III) oxide, Rh_2O_3, can be prepared by heating the finely divided metal or its nitrate in air or O_2. The rhodium(IV) oxide is also known. Rhodium trihydroxide may be precipitated as a yellow compound by adding the stoichiometric amount of KOH to a solution of $RhCl_3$. The hydroxide is soluble in acids and excess base. When the freshly precipitated Rh(OH)$_3$ is dissolved in HCl at a controlled pH, a yellow solution is first obtained in which the aquochloro complex of Rh behaves as a cation. The hexachlororhodate(III) anion is formed when the solution is boiled for 1 hour with excess HCl. The solution chemistry of $RhCl_3$ is often very complex. Two trichlorides of Rh are known. The trichloride formed by high-temperature combination of the elements is a red, crystalline, nonvolatile compound, insoluble in all acids. When Rh is heated in molten NaCl and treated with Cl_2, Na_3RhCl_6 is formed, a soluble salt that forms a hydrate in solution. Rhodium(III) iodide is formed by the addition of KI to a hot solution of trivalent Rh.

Rhodium(III) sulfate exists in yellow and red forms. If $Rh(OH)_3$ is dissolved in cold H_2SO_4, the product is the yellow form, in which the sulfate is ionic. If this solution is evaporated in hot H_2SO_4, the product is a red, nonionic sulfate. When Rh is treated with F_2 at 500–600 °C, RhF_3 is slowly formed. This compound is practically insoluble in water, concentrated HCl, HNO_3, H_2SO_4, HF, or NaOH.

If a solution of $RhCl_3$ is treated with $NaNO_2$, the very soluble sodium hexanitritorhodate(III), $Na_3Rh(NO_2)_6$, is formed. The solubility of this compound in alkaline solution makes it useful for refining, as many base metals are precipitated as their hydroxides under these conditions. The analogous ammonium and potassium salts are relatively insoluble.

When H_2S is passed into a solution of a trivalent Rh salt at 100 °C, the hydrosulfide, $Rh(SH)_3$, is formed. This black precipitate is insoluble in $(NH_4)_2S$. Rh forms many complexes with NH_3, amines, cyanide, chloride, bromide, and numerous polynitrogen and polyoxygen chelating agents.

EDITOR'S NOTE: In 1982, J. Halpern (University of Chicago) reported that rhodium complexes containing chiral phosphine ligands catalyze the hydrogenation of olefinic substrates such as alpha-aminoacrylic acid derivatives, producing chiral products with very high optical yields. Elucidation of the mechanisms of such reactions leads to the conclusion that the stereoselection is dictated not by the preferred initial binding of the substrate to the chiral catalyst, but rather by the much higher reactivity of the minor diastereomer of the catalyst-substrate adduct corresponding to the less favored binding mode. In the Halpern 1982 reference listed, a relatively restricted class of asymmetric catalytic reactions, namely, the hydrogenation of alpha-acylaminoacrylic acid derivatives and related substrates, catalyzed by rhodium complexes containing chiral phosphine ligands, is discussed.

Additional Reading

Bauccio, M.L.: "ASM Metals Reference Book," 3rd Edition, ASM International, Materials Park, OH, 1993.

Carter, G.F. and D.E. Paul: "Materials Science and Engineering," ASM International, Materials Park, OH, 1991.

Considine, D.M. and G.D. Considine: "Encyclopedia of Chemistry," 4th Edition, Van Nostrand Reinhold Company, New York, NY, 1984. (A classic reference).

Davis, J.R.: "Metals Handbook," 2nd Edition, ASM International, Materials Park, OH, 1998.

Greenwood, N.N. and A. Earnshaw: "Chemistry of the Elements," 2nd Edition, Butterworth-Heinemann, Inc., Woburn, MA, 1997.

Halpern, J.: "Mechanism and Stereoselectivity of Asymmetric Hydrogenation," *Science*, **217**, 401–407 (1982).

Krebs, R.E.: "The History and Use of Our Earth's Chemical Elements: A Reference Guide," Greenwood Publishing Group, Inc., Westport, CT, 1998.

Lewis, R.J. and N.I. Sax: "Sax's Dangerous Properties of Industrial Materials," 10th Edition, John Wiley & Sons, Inc., New York, NY, 1999.

Lide, D.R.: "CRC Handbook of Chemistry and Physics 2000–2001," 81st Edition, CRC Press, LLC., Boca Raton, FL, 2000.

Meyers, R.A.: "Handbook of Chemicals Production Processes," The McGraw-Hill Companies, Inc., New York, NY, 1986.

Staff: "ASM Handbook — Properties and Selection: Nonferrous Alloys and Pure Metals," ASM International, Materials Park, OH, 1990.

LINTON LIBBY, Chief Chemist, Simmons Refining Company, Chicago, IL.

RHODOCHROSITE. Rhodochrosite, manganese carbonate, $MnCO_3$, is a rose-pink to red hexagonal mineral, occurring as small crystals, in cleavable masses, granular or compact. It is a brittle mineral; hardness, 3.5–4; specific gravity, 3.7 (pure $MnCO_3$, usually 3.4–3.6); luster, vitreous to pearly; color, various shades of pink, red and reddish-brown; transparent to opaque; streak, white. Rhodochrosite has a perfect rhombohedral cleavage. Rhodochrosite is a product of high-temperature metamorphic deposits; as a gangue mineral in ore veins of hydrothermal origin, and as secondary residual deposits from bodies of manganese or iron oxides, and

in sedimentary deposits precipitated like siderite by organic matter acting, in the absence of oxygen, upon bicarbonates.

Spectacular, gemmy red rhombohedrons, up to 3 inches (7.5 centimeters) on an edge occur at the Sweet Home Mine, Alma, Colorado, and the John Reed Mine, Alicante, Colorado. Exceptionally fine stalactitic formations are found in Catamarca Province, Argentina. Recently the most beautiful, large transparent gem red scalenohedron crystals ever found have come from Hotazel and the Kalahari manganese field in northern Cape Province, South Africa. This region encompasses one of the largest and richest known manganese deposits in the world. Other localities for this mineral are in Rumania, Saxony, Westphalia, and Cornwall, England. In the United States rhodochrosite is found at Franklin, New Jersey; Butte Montana; and in various localities in Colorado and Nevada. The name rhodochrosite is derived from the Greek meaning rose, and color.

RHODODENDRONS.   See **Heather Shrubs and Trees**.

RHODONITE.   The mineral rhodonite, manganese metasilicate, $(Mn,Fe,Mg)SiO_3$, crystallizes in the triclinic system forming large, irregular tabular crystals but usually occurring massive. Prismatic and basal cleavages excellent; fracture conchoidal to uneven; hardness, 5.5–5.6; specific gravity, 3.57–3.76; luster, vitreous to pearly on cleavage faces; color, red to brownish-red; rarely yellow to gray; streak, white; transparent to translucent. A variety containing much calcium is called bustamite. Zinc may replace the manganese in rhodonite; it is then known as fowlerite. Rhodonite is found in the Harz Mountains, Germany; in the Urals of the former U.S.S.R.; in Hungary, Italy, and Sweden. Bustamite from Mexico, Franklin and Sterling Hill, New Jersey, occurs with fowlerite.

Rhodonite has been occasionally used for an ornamental stone. Its name is derived from the Greek meaning a rose, because of the color.

RHOMBIC ANTENNA.   See **Antenna**.

RHO-THETA SYSTEM.   1. Any electronic navigation system in which position is defined in terms of distance, or radius ρ and bearing θ with respect to a transmitting station. Also called an *R-theta system*. 2. Specifically, a polar-coordinate navigation system providing data with sufficient accuracy to permit the use of a computer which will provide arbitrary course lines anywhere within the coverage area of the system. See also **Navigation**.

RHUBARB *(Rheum rhaponticum; Polygonaceae).*   The rhubarb plant is perennial from thick, short rhizomes. The large, somewhat triangular leaf blades are elevated on long fleshy petioles. The flowers are small, greenish-white and borne in large compound leafy inflorescences. The plant is principally grown for its fleshy petioles. These are stewed to yield a tart sauce used as filling for pies and tarts. The plant is indigenous to Asia.

RHUMB LINE.   Unless a ship is tacking or executing some other maneuver its course is generally constant for several hours at least. In such a case the ship is said to be following a rhumb line. The rhumb line, or loxodromic curve, may be defined as any curve on the surface of the earth such that the tangent of the curve at any point cuts the meridian through that point at a constant angle. In case this angle has any value other than 0° or 90°; it may be proved that the rhumb line is a spiral approaching one of the poles of the earth as a limit.

Obviously, the rhumb line course between any two points is the simplest course to follow, for once having set the course it will not have to be changed until the destination is reached. However, except in the particular cases where the two points are either on the same meridian or are both on the equator, the rhumb line will not be the shortest distance between the two points. The mercator chart was designed for the purpose of facilitating the laying down of rhumb-line courses. On a mercator chart, and only on this chart, the rhumb line appears as a straight line.

See also **Course**; and **Navigation**.

RHYOLITE.   The general term for a group of acidic igneous rocks, the effusive equivalent of the granites. It occurs as lava flows, breccias, and in volcanic necks and dikes. In the porphyritic varieties the phenocrysts are frequently quartz or othoclase feldspar imbedded in a highly felsitic or glassy ground mass. Rhyolites, including obsidian, frequently show flow, spherulitic, nodular, and lithophysal structures.

RIBBON PARACHUTE.   A type of parachute having a canopy consisting of an arrangement of closely spaced tapes. This parachute has high porosity with attendant stability and slight opening shock.

RIBOFLAVIN (Vitamin B₂).   Some earlier designations for this substance included vitamin G, lactoflavin, hepatoflavin, ovoflavin, verdoflavin. The chemical name is 6,7-dimethyl-9-d-l'ribityl isolloxazine. Riboflavin is a complex pigment with a green fluorescence. Riboflavin deficiency frequently accompanies pellagra and the typical lesions of both nicotinic acid and riboflavin deficiency are found in that disease. See also **Niacin.** Riboflavin, like nicotinic acid, forms an oxidation enzyme and, as such, acts as an oxygen carrier to the cell. The structure of riboflavin is:

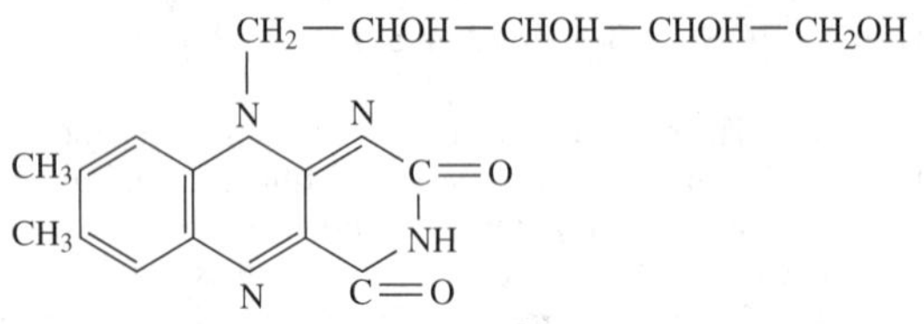

$$CH_2 - CHOH - CHOH - CHOH - CH_2OH$$

Disorders caused by a deficiency of riboflavin include anemia, cheilosis (a lip disorder); corneal vascularization, seborrheic dermatitis, and glossitis. Research leading to the current knowledge of riboflavin essentially commenced in 1917 when Emmet and McKim showed dietary growth factor for rats in rice polishings. In 1920, Emmet suggested the presence of several dietary growth factors in yeast concentrate, including the heat-stable component and B₁. The British Medical Research Council, in 1927, proposed that the designation B₂ be given to the heat-stable component. Warburg and Christian, in 1932, isolated yellow enzyme (containing riboflavin, FMN) from bottom yeast. In 1933, Kuhn isolated pure B₂ (riboflavin) from milk and recognized its growth-promoting activity. Several researchers (Kuhn et al.; Karrer et al.), in 1935, worked out the structure and synthesis of vitamin B₂, during which period it was named *riboflavin*. By 1954, Christie et al. had determined the structure and synthesized riboflavin dinucleotide (FAD).

Riboflavin has been shown to be a constituent of 2 coenzymes: (1) Flavin mononucleotide (FMN); and (2) flavin adenine dinucleotide (FAD). The structures are:

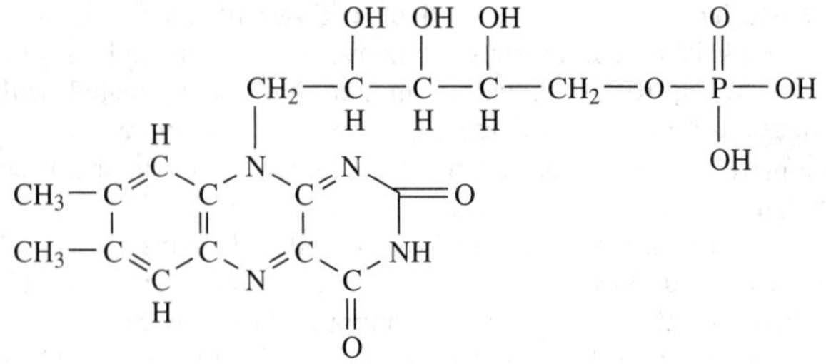

Flavin mononucleotide (FMN)

FMN was first identified as the coenzyme of an enzyme system that catalyzes the oxidation of the reduced nicotinamide coenzyme, NADPH (reduced NADP), to NADP (nicotinamide adenine dinucleotide phosphate). NADP is an essential coenzyme for glucose-6-phosphate dehydrogenase which catalyzes the oxidation of glucose-6-phosphate to 6-phosphogluconic acid. This reaction initiates the metabolism of glucose by a pathway other than the TCA cycle (citric acid cycle). The alternative route is known as the phosphogluconate oxidative pathway, or the hexose monophosphate shunt. The first step is:

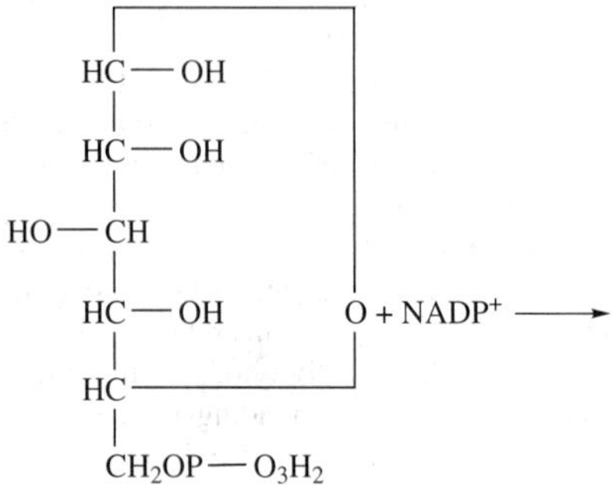

Glucose-6-phosphate

Most of the numerous other riboflavin-containing enzymes contain FAD. FAD is an integral part of the biological oxidation-reduction system where it mediates the transfer of hydrogen ions from NADH to the oxidized cytochrome system. FAD can also accept hydrogen ions directly from a metabolite and transfer them to either NAD, a metal ion, a heme

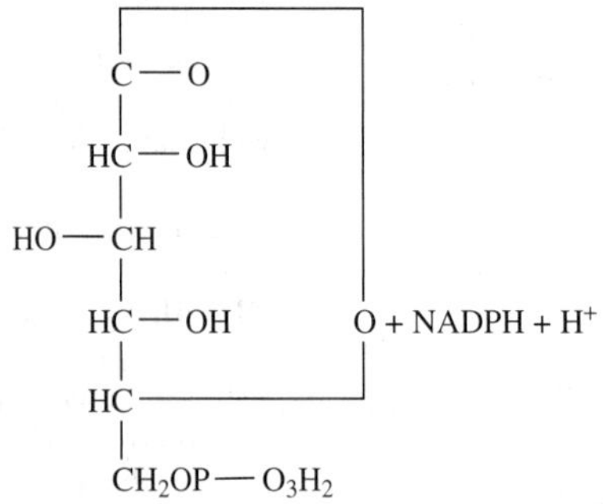

6-Phosphogluconolactone

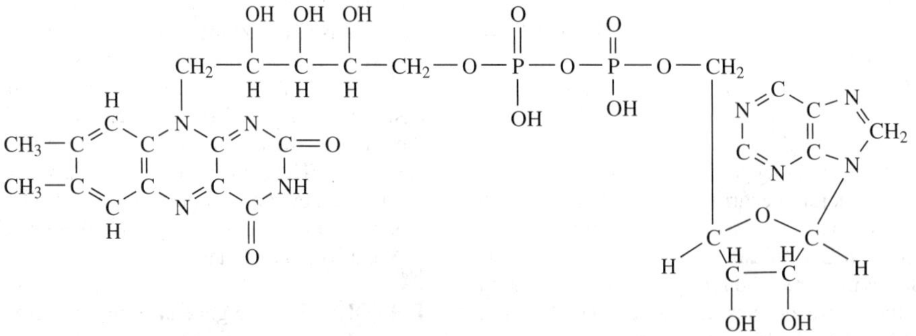

Flavin adenine dinucleotide (FAD)

derivative, or molecular oxygen. The various mechanisms of action of FAD are probably due to differences in protein apoenzymes to which it is bound. The oxidized and reduced states of the flavin portion of FAD are:

FAD
(oxidized)

FADH$_2$
(reduced)

In the biological oxidation-reduction system, reduced NAD (i.e., ADH) is reoxidized to NAD by the riboflavin-containing coenzyme FAD as shown by:

See also **Coenzymes**.

Distribution and Sources. Research indicates that all organisms require riboflavin. Endogenous sources exist in high plants, algae, some bacteria, and some fungi. All animals, some fungi and bacteria receive at least a partial supply of riboflavin from generation by intestinal bacteria. In the case of humans, there is a large dependence upon exogenous sources.

High riboflavin content (1000–10,000 micrograms/100 grams)
Beef (kidneys, liver), calf (kidneys, liver), chicken (liver), pork (heart, kidneys, liver), sheep (kidneys, liver), yeast (killed)
Medium riboflavin content (100–1000 micrograms/100 grams)
Almond (dry), asparagus, avocado, bacon, bean (kidney, lima, snap, wax), beef, beet greens, broccoli, Brussels sprouts, cashew, cauliflower, cheeses, chicken, chicory, corn (maize), cream, dandelion greens, eggs, endive, fish, goose, groundnut (peanut), kale, kohl-rabi, lamb, lentil (dry), milk, oats, parsley, parsnip, pea, pecan, pork, rice bran, soybean (dry), spinach, turkey, turnip greens, veal, walnut, wheat germ
Low riboflavin content (10–100 micrograms/100 grams)
Apple, apricot, artichoke, banana, barley, beet, berry (black-, blue-, cran-, rasp-, straw-), cabbage, carrot, celery, cherry, coconut, cucumber, date (dry), eggplant, fig, grape, grapefruit, lettuce, melons, onion, orange, peach, pear, pepper (sweet), pineapple, plum, potato, radish, raisin (dry), rice, sweet potato, tangerine, tomato, turnip

Commercial riboflavin dietary supplements are prepared (1) by the fermentation process (bacteria or yeast); and (2) by chemical synthesis from alloxan, ribose, and *o*-xylene.

Precursors in the biosynthesis of riboflavin include purines, pyrimidines, and ribose. Intermediate in the synthesis is 6,7-dimethyl-8-ribityllumazine. In plants, riboflavin production sites are found in leaves, germinating seeds, and root nodules. Storage sites in animals are heart and liver, with small amounts in the kidneys. Riboflavin in overdose is essentially nontoxic to humans.

Bioavailability of Riboflavin. Factors which tend to decrease the availability of riboflavin include: (1) cooking, inasmuch as riboflavin is slightly soluble in water; (2) in some plant foods, availability is lower than might be expected because of bound forms; (3) decreased phosphorylation in intestines prevents absorption; (4) exposure of foods to sunlight; (5) enzymes required for breakdown are not present; (6) presence of gastrointestinal disease; and (7) diuresis. Riboflavin availability is increased by storage in heart, liver, and kidneys and by the presence of very actively producing intestinal bacteria.

Antagonists of riboflavin include isoriboflavin, lumiflavin, araboflavin, hydroxyethyl analogue, formyl methyl analogue, galactoflavin, and flavin-monosulfate. Synergists include vitamins A, B$_1$, B$_6$, and B$_{12}$, niacin, pantothenic acid, folic acid, biotin, tetraiodothyronine (thyroxine), insulin, and somatotrophin (growth hormone).

Determination of Riboflavin. Bioassay includes observance of the growth rate of rats; microbiological — *L. caseli*, and *L. mesenteroides*. Physicochemical methods include fluorimetry, paper electrophoresis, and polarography.

Unusual features of riboflavin as recorded by some researchers include: (1) High levels in liver inhibit tumor formation by azo compounds in animals; (2) free radicals are formed by light or dehydrogenation: flavine $\rightleftharpoons$ semiquinone $\rightleftharpoons$ dihydroflavin+
; (3) free vitamin is found only in retina, urine, milk, and semen; (4) substitution of adenine by other purines and pyrimidines destroys activity of flavin adenine dinucleotide (FAD); (5) phosphorylation of vitamin in intestines allows absorption as flavin mononucleotide (FMN); (6) blood levels decrease during life in humans: (7) brain content remains constant; (8) available in plants as FMN and FAD; (8) very concentrated in bull semen.

RIBOSE. See **Nucleic Acids and Nucleoproteins**.

RIBS. In humans, the ribs number twenty-four, twelve on either side. They are attached to the vertebral column behind, and the first seven pairs are connected with the sternum in front and are called true ribs. The remaining five are called false ribs. The eighth, ninth and tenth are attached in front to the cartilaginous portion of the next rib above. The lower two, that is the eleventh and twelfth, are not attached in front at all and are called floating ribs. The spaces between the ribs are called intercostal spaces; they contain the intercostal muscles, nerves and arteries. The ribs form the greater part of the bony cage of the thorax; they preserve its outline and allow for easy motion in breathing, due to their elasticity.

RICE *(Oryza sativa; Gramineae)*. An annual grass, which grows wild in tropical Asia and Africa. Cultivated rice is probably derived from an Asiatic species, and is especially adapted to grow in swampy or very wet lowlands. It is a shallow rooted plant, the stems of which tiller abundantly and grow from 2–6 feet (0.6–1.8 meters) or more in height. The leaves are long and smooth. The inflorescence is a panicle the branches of which may occur singly or in pairs. See Fig. 1. The laterally compressed spikelets are one-flowered, and have a pair of small bristlelike glumes; the lemma is tough, parchment-like and sometimes awned; the palea resembles the lemma, but is somewhat smaller. A distinctive character of the flower is the presence of six functional stamens. Commonly, rice is self-pollinated. The grain or karyopsis is enwrapped in the palea, and frequently also in the lemma. In this condition, rice grain is known as paddy, or rough rice. The grain itself is smooth and shining, has a pair of longitudinal grooves on its surface, and a glassy endosperm. In structure it is very similar to wheat grain.

The milling of rice involves several processes. First, the outer coverings are removed by revolving stones and fans. After this, the outer seed-coats and embryo are largely removed by rubbing; the remaining grain is scoured and polished by rubbing on leather surfaces. Finally, the polished grain is given a coat of glucose and talc, and is ready for market.

Varieties. There are two major eco-geographic races of *O. sativa*: (1) indica and (2) japonica. The latter sometimes also is known as *sinica* or *keng*.

Indica is the major group grown throughout southeastern Asia and in most areas of the People's Republic of China. The majority of indica varieties raised in the monsoon tropics have evolved from combined natural and human selection processes. They are well adapted to conditions of low-soil fertility, uncertain weather, and poor water control. Most indicas have

Fig. 1. Rice panicle in bloom showing the anthers that have just dehisced and shed pollen on the stigma. (*USDA photo.*)

resistance to endemic diseases and insects, and they also compete well with weeds. They also have the dry cooking characteristics preferred by consumers in tropical and subtropical areas. But the features that enable the tropical types of indica to survive (i.e., tall and high tillering plants, late maturity, long and drooping leaves) also provide the basis for their weakness under modern agricultural practices. Improved fertilization, for example, will lead mainly to vegetative growth and lodging (tendency of a crop to bend over and avoid being cut by a machine) rather than significantly increased yield.

Japonica varieties are distributed widely in several areas of the Temperate Zone, such as the Yangtze valley of China, Korea, Japan, Europe, part of Australia, and the United States. The japonica varieties evolved in China more recently than the indicas and are the result of an intensive human selection process. In comparison with the indicas, the japonicas have darker and more upright leaves, a shorter and stiffer stalk, earlier maturity, and more thrifty vegetative growth. Japonicas respond well to improved cultural practices, especially the application of fertilizer, and are more resistant to lodging. See Fig. 2.

Fig. 2. Rice field in the United States showing a large, specially designed harvesting machine. (Allis-Chalmers.)

Japonicas, however, are not well adapted for the traditional cultural practices in tropical Asia because:

1. They require precise amounts of water.
2. They need weed and insect control.
3. Most are susceptible to virus diseases of the tropics.
4. Some react to high temperatures during early growth stages by flowering too early.

5. They lack the grain dormancy needed in the monsoon season.
6. The grains have a sticky cooking quality not desirable in the Orient.

Wild rice is found and harvested in some regions of the world. The variety *Zizania acquatica* grows in lakes and small streams in the upper midwestern United States and southern Canada. Wild rice is an annual grass that also can be grown in flooded soils similar to white rice. The grain of wild rice was a staple food for the American Indians in such regions and now is considered a favorite of the gourmet chef. Much of the wild rice grown in the United States comes from the Minnesota and Wisconsin lakes.

High-yielding types of rice tend to be raised under irrigated conditions. Both the quality of irrigation systems and the need for irrigation vary widely in the developing nations. Irrigation systems range from virtually complete, year-round supplies to occasional supplementations of rainfall. Most commonly, the systems supplement rainfall during the wet season and service only a limited area during the dry season.

Extensive research into the genetics of rice and toward breeding improved varieties has occurred, mainly since the early 1960s. The activities of the International Rice Research Institute (IRRI, Los Banos, Philippines) and the Indian Council of Agricultural Research are particularly well known. The use of high-yielding varieties of rice (along with wheat) formed the basis for what became known as the "green revolution." See also **Green Revolution**.

Culture. As previously mentioned, most rice is grown in tropical and subtropical regions of the world (between 30° North and 30° South latitude). Notable exceptions are Italy, Japan, Korea, Spain, and the United States. The growing season for rice ranges from 4 to 6 months, during which time the mean temperature should be no lower than 70 °F (21.1 °C). Without irrigation, rainfall in excess of 40 inches (102 cm) per year is required. But the crop does very well in irrigated and flooded areas where the weather may be hot and dry. Situations like this prevail in the Po valley of Italy, the Nile valley of Egypt, and the Sacramento valley of California, where rice is grown on flooded land. See Fig. 3. In many areas, such as California, the soil is submerged with 4 to 8 inches (10–20 cm) of water from seeding or shortly thereafter until a short time before the grain is mature. Water may be drained off once or twice for a few days for applications of fertilizer and chemicals to aid the control of algae, rice

(a)

(b)

Fig. 3. Rice fields in the United States. (**a**) Submerged field of young rice; (**b**) 6 to 8 weeks after submerging the land. (*USDA photo.*)

water weevils, and aquatic weeds. The chemistry of flooded soils differs markedly from dry soils:

1. Decrease in exchange of air (gases) between soil and the atmosphere, leading to low oxygen levels and high levels of hydrogen, methane, and various oxides of nitrogen in the soil;
2. An increase in soil pH from 0.7 to 1.5 units; and
3. An increased salt content of the soil.

Soil microorganisms under flooded conditions tend to be active forms of anaerobes. These forms require less energy, and thus soil organic matter decomposition takes place at a slower rate. A significant factor is that flooding makes the use of nitrate forms of nitrogen less efficient because of potential denitrification and loss of nitrogen. Nitrogen deficiency as well as those of phosphorus and potassium can be applied preplant, at planting, or within 2 to 3 weeks after planting.

In summary, of the cereal crops, rice ranks second in world tonnage and sixth in tonnage of cereal crops produced in the United States.

In terms of major rice-producing countries, China accounts for about 30%, followed by India (about 20%) and the southeastern Asian countries (about 50%). The United States accounts for less than 2% of world production.

Additional Reading

Crawford, R.: "Gene Mapping Japan's Number One Crop (Rice)," *Science*, 1611 (June 21, 1991).
Dover, M.J. and L.M. Talbot: "Feeding the Earth — An Agroecological Solution," *Technology Rev. (MIT)*, 26 (February 1988).
Dziezak, J.D.: "Romancing the Kernel: A Salute to Rice Varieties," *Food Technology*, 74 (June 1991).
Gasser, C.S. and R.T. Fraley: "Genetically Engineering Plants for Crop Improvement," *Science*, 1293 (June 16, 1989).
Hamer, J.E.: "Molecular Probes for Rice Blast Disease," *Science*, 632 (May 3, 1991).
Kulp, K. and J.G. Ponterotto: "Handbook of Cereal Science and Technology," 2nd Edition, Marcel Dekker, New York, NY, 2000.
Leath, M., Livezey, J., and K. McManus: "Government Programs for Rice: What They Mean to Producers, Processors, and Consumers," *Nat'l. Food Review*, 5 (July–September 1988).
Murray, T.D., D.W. Parry, N.D. Cattlin, and G.R. Dixon: "A Diseases of Small Grain Cereal Crops," Iowa State University Press, Ames, IA, 1998.
Pszczpola, D.E.: "U.S. Researcher Wins 1988 General Foods World Food Prize (Rice)," *Food Technology*, 50 (August 1988).
Staff: "Bumper Transgenic Plant Crop," *Science*, 33 (July 5, 1991).
Walsh, J.: "Second Chance for Rice Research Center (West Africa Rice Development Association)," *Science*, 969 (February 26, 1988).

Web References

International Rice Research Institute: *http://www.cgiar.org/irri/*
Research Resources: *http://www.carleton.ca/~bgordon/Rice/research_resources.htm*
Rice, Food Resource: *http://www.orst.edu/food-resource/g/rice.html*

RICHARDSON NUMBER. This number is a ratio used in meteorological evaluation of the degree to which atmospheric flow (winds) will be turbulent or nonturbulent. The ratio is formed by dividing the square of the shear of the horizontal wind in the vertical direction into the thermal stability of the atmosphere. The general form of the number is

$$R_i = \frac{\dfrac{g}{\theta}\dfrac{d\theta}{dh}}{\left[\dfrac{dV}{dh}\right]^2}$$

where g = gravitational acceleration
i = the meteorological potential temperature
V = the horizontal wind vector
h = height

Liberally interpreted, the denominator represents the turbulent generating forces and the numerator the turbulent damping forces. The expectancy would be that the borderline value would be unity. However, in actual cases, the breakover from laminal to onset of turbulent flow is in the general region of $R_i = 0.25$. See also **Atmospheric Turbulence**.

RICHTER, CHARLES FRANCIS (1900–1985). Richter was an American scientist who was a pioneer in seismology. He was born in Ohio, but moved to California when he was nine years old. He loved the outdoors and hiked and backpacked throughout California. He earned an A.B. degree in physics from Stanford University. Although he started graduate work at Caltech, he changed directions and took a job at the Carnegie Institution of Washington's seismological laboratory. Here, working with Beno Gutenberg, Richter in 1935, devised a scale to measure the magnitude of earthquakes. Richter's scale measured a quake's magnitude at its epicenter.

Richter and Gutenberg went to Caltech in 1937 and continued to work on earthquake research documenting that several hundred thousand earthquakes occur each year around the world. They mapped earthquake-prone areas. Richter is credited with convincing Los Angeles City to update its building codes and do retro-fit work to lessen damage from earthquakes.

See also **Earth Tectonics and Earthquakes**.

J. M. I.

RICHTER SCALE (Earthquake). See **Earth Tectonics and Earthquakes**.

RICKETS. See **Vitamin D**.

RICKETTSIAL DISEASES. This group of diseases is caused by very small ($1–2 \times 0.3$ micrometer) Gram-negative bacilli which are distinguished from other bacteria because of their obligate intracellular parasitism. The mode of entry into, and growth in, eukaryotic cells is variable and in this the symptoms manifested by their presence are influenced. As a whole, the group of diseases has been divided into three groups: (1) the *typhus group*, including endemic (murine) typhus, epidemic (louse-borne) typhus, and Brill-Zinsser disease; (2) the *spotted-fever group*, including Rocky Mountain spotted fever and rickettsial pox; and (3) a miscellaneous group, of which the main disease is Q fever. With the exception of Q fever, rickettsia enter the body by way of the bite of an insect, such as tick and louse. In Q fever, the microorganisms are inhaled as components of infected dust.

Typhus (Murine Endemic). In the United States, the disease is endemic in the southeastern and Gulf states. The reservoir of *Rickettsia typhi*, the causative agent, is the rat, in which the disease is a natural infection. Murine typhus occurs worldwide. The disease is transmitted from rat to human by a rat flea. When infected, the rat flea excretes rickettsias in its feces. In the absence of their normal host, the rat flea will attack humans and while feeding will drop rickettsia-containing feces. If the latter are rubbed into the bite site or other breaks in the skin, human infection commences. The incidence of murine typhus in the United States dropped to a low point of about 40 cases in 1969, rose to nearly 70 cases in 1977 and, in recent years, has cycled between 50 and 60 cases per year. As in past years, most cases are reported from Texas, although Maryland, Florida, Tennessee, Louisiana, California, and Hawaii also have reported occurrences. Statistics show that the highest incidence occurs during the summer and fall, with a majority of cases found in persons who work in various food depositories, which are, of course, attractive to rats. Effective rodent and flea control can reduce the incidence of the disease.

After an incubation period of 6–14 days, clinical features include headache, malaise, backache, and chills — to be followed later by shaking chills, fever, severe headache, vomiting, and nausea. The fever ranges between 102–103 °F (38.9–39.4 °C). In the untreated individual, there is a pattern of remittent fever, sometimes accompanied by tachycardia (fast heartbeat). On about the fifth day of fever, a rash consisting of irregular, discrete, pink macules occurs in the axilla and inner surface of the upper arms. Later, the lesions will appear on the trunk, thighs, and lower arms. Rarely are they found on the face, hands, or feet. The rash usually disappears in about one week. During the second week of infection. a dry cough may develop. In elderly people and persons of marginal health, more severe symptoms may be manifested. These include a greater severity of headache, stiff neck, aggravating backache, and, sometimes, mental confusion. These symptoms may be confused with those of meningitis.

Murine typhus is sometimes difficult to differentiate from Rocky Mountain spotted fever, particularly since the two diseases are frequently present in the same geographical region. Compared with other rickettsial diseases, murine typhus is relatively mild. The mortality rate of murine typhus is low, even in untreated cases. Therapy usually consists of oral doses of tetracycline.

Typhus (Louse-Borne Epidemic). Unlike murine typhus, this form of typhus, of which the causative agent is *R. prowazekii*, is very serious and has caused millions of deaths, particularly in eastern Europe and the Balkan countries after World Wars I and II. The body louse is the vector of transmission. The incubation period approximates one week. Clinical features include an abrupt commencement of headache, chills, prostration, and high fever, which may rise to $103-104\,°F$ ($39.4-40.0\,°C$). These symptoms, including fever, may persist for several days. As with murine typhus, a pink rash occurs on about the fifth day. Conjunctivitis frequently occurs. There may be a dry cough. In severe infections, renal failure and mental confusion may be present. The fever lowers in about two weeks in untreated patients.

As with murine typhus, louse-borne typhus is sometimes difficult to differentiate from Rocky Mountain spotted fever. In making the diagnosis, the physician will carefully note the differences in the evolution and character of the rash, as well as considering the epidemiologic setting.

Drugs of choice for the treatment of epidemic typhus include chloramphenicol and tetracycline. Doxycycline is also sometimes used. If treatment is commenced within a day or two of onset of symptoms, fever is usually reduced within 48 hours. The severity of epidemic typhus is age-related. The disease is mild and with few fatalities in children. The mortality in untreated adults ranges between 10 and 50%, even higher among adults over 60 years of age. With the availability of antibiotic therapy, the mortality rate has been reduced to a few percent of cases.

Prior vaccination against typhus reduces the severity and duration of the disease. This is a killed rickettsial vaccine derived from infected yolk sac tissue. In the United States, the rate of typhoid fever occurrence has remained relatively steady for the past decade. Underreporting does not appear to be a major problem affecting the statistics. In recent years, more than half of the cases have been acquired during travel to other countries. During World War II years, there were just over 4 cases per 100,000 population reported. Current incidence is about 0.2 case per 100,000 population.

Even at some risk, an effective way to prevent epidemic typhus is the mass delousing of the population with an insecticide, such as DDT or lindane powder, which have been banned by many countries except in emergency situations.

Brill-Zinsser Disease (Recrudescent Epidemic Typhus). This disease occurs after a prior infection with louse-born typhus and results from reactivation and multiplication of *R. prowazekii*, which have been dormant for a long period. The delayed mechanism for this activation is poorly understood. The disease is usually a mild form of the prior infection and is to be suspected in patients who present with fever and headache and who, upon interviewing, had epidemic typhus, particularly during the epidemic of World War II. The frequency of Brill-Zinsser disease is greatest in Eastern Europe, Poland, and the former U.S.S.R.

Rocky Mountain Spotted Fever. Caused by *R. rickettsii*, this disease is transmitted by tick vectors. In the eastern United States, the dog tick is usually responsible; the wood tick in the western states; and the Lone Star tick in the southwestern states. Ticks are infected by feeding on infected rodents and other small wild animals. The infection is non-lethal to the tick and is carried throughout the life of the tick. Although the disease was first identified in the Rocky Mountain area, this region now accounts for less than 5% of cases. The majority of cases in the United States are now found in the southeastern states, notably in the Piedmont region. About 90% of patients have onset of illness from early April to early September. A large number of cases occur in children. Mortality rates have fallen, but still remain above the overall mortality rate of 3.2%. In 1979, the incidence was 0.5 case per 100,000 population. Although not fully understood, it is now known that pet dogs can be accidental hosts to the rickettsiae.

The incubation period of the disease is approximately 2 to 7 days. Onset is usually abrupt and the clinical features are severe headache, nausea, rigor, and high fever. Severe abdominal pain may be present. Within one or two days, the fever rises to $103-105\,°F$ ($39.4-40.6\,°C$). A rash appears, usually in the regions of the wrists and ankles, but gradually spreads to include the trunk and face. Unlike the rickettsial diseases previously described, in Rocky Mountain spotted fever, lesions do appear on palms and soles, these serving as useful diagnostic indicators. A tenderness of the muscles is noted upon compression of limbs. In some cases, a dry cough may be present. Heartbeat increases with fever. Prior to appearance of the rash, the symptoms tend to parallel those of several acute bacterial or viral infections. The physician, during this period, will question the patient as regards possible recent exposure to areas where ticks may be found. Antimicrobial therapy should be commenced at the time the rash appears. Drugs of choice are chloramphenicol or tetracycline. Sulfonamides have not proved effective in the treatment of this disease. Commencement

of therapy prior to the sixth day of illness usually results in excellent prognosis.

Q Fever. This is an influenzalike disease first reported from Queensland (Australia) in 1935 and since then from Montana and other parts of the United States as well as North Africa, Switzerland, and Great Britain. Q fever is caused by the rickettsia *Coxiella burnetii*, which has its major effect on the lungs. The infection is nearly always airborne. The disease is particularly found in regions where cattle, sheep, and goats are produced. Livestock usually are infected by breathing dust containing the microorganism, but may be infected by ticks. Notably prone to infection are workers in dairies, abattoirs, and laboratory workers engaged in research on the rickettsiae. The incubation period ranges from 18 to 20 days. Clinical features include headache, chills, fever, myalgias, anorexia, and malaise. There is no rash. Fever may reach $104\,°F$ ($40\,°C$). Chest x-rays almost always indicate focal areas of pneumonitis. Complications include granulomatous hepatitis, jaundice, and abnormal liver function. In several instances, infective endocarditis has been reported. Early diagnosis is often difficult because the symptoms resemble those of numerous other acute febrile illnesses, such as salmonellosis, infectious hepatitis, brucellosis, leptospirosis, and infectious mononucleosis. In the early stages, indication by the patient of recent contact with livestock is an excellent clue.

The drugs of choice for treatment include tetracycline or chloramphenicol. Recovery with treatment is normally rapid. Some protection to persons of high risk to infection can be provided by killed vaccines made from *C. burnetti* grown in chick embryo culture. In rare instances, Q fever can be transmitted by infected milk.

Other Rickettsial Diseases. *Boutonneuse fever* is found along the coasts of the Mediterranean and Black Seas, as well as in the interior of Africa, where it is called *South African tick bite fever*. *Siberian tick typhus* is found in Mongolia and Siberia. *Queensland tick typhus* occurs in Australia. Caused by *R. tsutsugamushi* with a mite as vector, scrub typhus occurs in southeastern Asia, India, northern Australia, and the western Pacific Islands. *Trench fever* or *Volhynia fever* is transmitted by a louse and occurs in Mexico, northern Africa, Poland, and the former U.S.S.R. This disease is caused by *Rochalimaea quintana*. These rickettsial diseases essentially have symptoms that parallel those of the diseases previously described in this entry. Chloramphenicol or tetracycline are usually the drugs of choice in their treatment.

Additional Reading

Anderson, B., H. Friedman, and M. Bendinelli: "Rickettsial Infection and Immunity," Kluwer Academic Publishers, Norwell, MA, 1997.
CDC: "Morbidity and Mortality Report," Center for Disease Control, Atlanta, Georgia (Issued weekly). *http://www.cdc.gov/mmwr/*
Gear, J.H.S.: "Handbook of Viral and Rickettsial Hemorrhagic Fever," CRC Press, LLC., Boca Raton, FL, 1988.
Hattwick, M.A.W., R.J. O'Brien, and B.F. Hanson: "Rocky Mountain Spotted Fever: Epidemiology of an Increasing Problem," *Ann. Intern. Med.*, **84**, 732 (1976).
Maugh, T.H., II: "Rickettsiae: A New Vaccine for Rocky Mountain Spotted Fever," *Science*, **201**, 604 (1978).
Oster, C.N., et al.: "Laboratory-acquired Rocky Mountain Spotted Fever: "The Hazard of Aerosol Transmission," *N. Engl. J. Med.*, **297**, 859 (1977).
Turck, W.P.G., et al.: "Chronic Q Fever," *Q. J. Med.*, **45**, 193 (1976).
Woodward, T.E., et al.: "Prompt Confirmation of Rocky Mountain Spotted Fever," *J. Infect. Dis.*, **134**, 297 (1976).

Web References

National Center for Infectious Diseases: *http://www.cdc.gov/ncidod/diseases/index.htm*
National Foundation for Infectious Diseases: *http://www.nfid.org/factsheets/*

ANN C. DEBALDO, Ph.D., University of South Florida, Tampa, FL.

RIDGE (Meteorology). See **Atmosphere (Earth)**.

RIDGE (Ocean). See **Earth Tectonics and Earthquakes**; **Ocean**; and **Volcano**.

RIEBECKITE. The mineral riebeckite, essentially sodium iron silicate, $Na_2(Fe_3^{2+}, Fe_2^{3+})_5Si_8O_{22}(OH)_2$, is a monoclinic member of the amphibole group, usually in prismatic crystals. It has a prismatic cleavage; hardness, 5; specific gravity, $3.32-3.382$; vitreous luster; color dark bluish to black. It occurs in granites and syenites chiefly. It is found in Greenland, Portugal, Madagascar, and South Africa (crocidolite), and in the United

States at Quincy, Massachusetts; near Pikes Peak, Colorado, and the San Francisco Mountains, Arizona.

RIEKE DIAGRAM. A polar-coordinate load diagram for microwave oscillators, particularly klystrons and magnetrons. Constant-power and constant-frequency contours are plotted against the polar plot of load admittance, commonly called the Smith chart.

RIEMANN, (GEORG FRIEDRICH) BERNHARD (1826–1866). Riemann was a German mathematician whose early work was on the theory of functions of a complex variable and on the potential theory. He was a brilliant thinker and is remembered for a new non-Euclidean system of geometry that is important in modern physics. Riemann is credited with introducing the idea of finite but unbounded space.

See also **Geometry**; **Riemann-Papperitz Equation**; **Riemann Surface**; **Riemann Zeta Function**; and **Number Theory**.

J. M. I.

RIEMANN HYPOTHESIS. See **Number Theory**.

RIEMANNIAN GEOMETRY. See **Geometry**.

RIEMANN-PAPPERITZ EQUATION. A second-order linear differential equation, as studied in the Fuchs theorem, with three regular singular points in the finite plane at $x = a, b, c$. It may be symbolized by the Riemann P-function

$$y = P \left\{ \begin{array}{ccc} a & b & c \\ a' & b' & c' \; ; & x \\ a'' & b'' & c'' \end{array} \right\}$$

where a', a'', b', b'', c', c'' are the exponents at the singularities, y is the dependent variable and x, the independent variable.

The Riemann problem in the theory of linear differential equations consisted of the search for a function to satisfy a given P-function. It is shown that the transformation of variable

$$u = \left(\frac{x-a}{x-b} \right)^k \left(\frac{x-c}{x-b} \right)^l y$$

will shift the exponents of the differential equation to $a' + k$, $a'' + k$, $b' - k - l$, $b'' - k - l$, $c' + l$, $c'' + l$, without affecting the singular points. On the other hand, the general linear transformation

$$x = (Az + B)/(Cz + D); \quad (AD - BC) \neq 0$$

will shift the singular points to three new positions, say a_1, b_1, c_1, as determined by the transformation but the exponents will not be altered. Combination of these two transformations will thus result in a standard form of the linear differential equation of Fuchsian type. The standard form is usually taken as the Gauss hypergeometric equation. The solution of any given linear differential equation of second order with three or fewer singular points will therefore be a special case of solutions to the Gauss equation, for the given differential equation can always be converted into the latter by suitable transformation of variables as described.

See also **Gauss Hypergeometric Equation**.

RIEMANN SURFACE. A surface used in representing multivalued functions of the complex variable. One sheet is assigned to each branch of the function, each sheet is cut at the branch line, and all are joined together so that a closed contour may be traced by passing continuously along the sheets of the surface.

See also **Geometry**.

RIEMANN ZETA FUNCTION. An infinite series in the complex variable $z = x + iy$, with n an integer:

The function is analytic with a simple pole at $z = 1$. It is a higher transcendental

$$\zeta(z) = \sum_{n=1}^{\infty} n^{-z}$$

function, that is, one not defined as the solution of a differential equation.

RIESZ, FRIGYES (1880–1956). Riesz was born in Hungary. He studied at Budapest and did his doctoral dissertation on geometry. He is best remembered for being the founder of functional analysis. He provided a link between Lebesgue's work on real functions and the area of integral equations developed by Hilbert. The Riesz-Fischer theorem is fundamental in the Fourier analysis of Hilbert space. His findings in functional analysis had fundamental importance in early quantum theory.

Riesz was chair of mathematics in the University of Budapest, he was elected to the Hungarian Academy of Science and was awarded the 1949 Kossuth Prize.

See also **Functional Analysis**.

J. M. I.

RIFT VALLEY FEVER. Rift Valley fever (RVF) is an acute, fever-causing viral disease that affects domestic animals (such as cattle, buffalo, sheep, goats, and camels) and humans. The RVF is most commonly associated with mosquito-borne epidemics during years of heavy rainfall.

The disease is caused by the RVF virus, a member of the genus *Phlebovirus* in the family Bunyaviridae. This disease was first reported among livestock by veterinary officers in Kenya in the early 1900s. RVF is generally found in regions of eastern and southern Africa where sheep and cattle are raised. However, RVF virus also exists in most countries of sub-Saharan Africa and Madagascar.

The RVF virus primarily affects livestock and can cause disease in a large number of domestic animals (this situation is referred to as an "epizootic"). The presence of an RVF epizootic can lead to an epidemic among humans who are exposed to diseased animals. The most notable epizootic of RVF, which occurred in Kenya in 1950–1951, resulted in the death of an estimated 100,000 sheep. In 1977, the virus was detected in Egypt (probably exported there in infected domestic animals from Sudan) and caused a large outbreak of RVF among animals and humans. The first epidemic of RVF in West Africa was reported in 1987 and was linked to construction of the Senegal River Project. The project caused flooding in the lower Senegal River area and altered interactions between animals and humans resulting in transmission of the RVF virus to humans.

An epizootic of RVF is generally observed during years in which heavy rainfall and localized flooding occur. The excessive rainfall allows mosquito eggs, usually of the genus *Aedes*, to hatch. The mosquito eggs are naturally infected with the RVF virus, and the resulting mosquitoes transfer the virus to the livestock on which they feed. Once the livestock is infected, other species of mosquitoes can become infected from the animals and can spread the disease. In addition, it is possible that the virus can be transmitted by other biting insects.

Humans can get RVF as a result of bites from mosquitoes and possibly other blood-sucking insects that serve as vectors. Humans can also get the disease if they are exposed to either the blood or other body fluids of infected animals. This exposure can result from the slaughtering or handling of infected animals or by touching contaminated meat during the preparation of food. Infection through aerosol transmission of RVF virus has resulted from contact with laboratory specimens containing the virus.

The RVF virus can cause several different disease syndromes. People with RVF typically have either no symptoms or a mild illness associated with fever and liver abnormalities. However, in some patients the illness can progress to hemorrhagic fever (which can lead to shock or hemorrhage), encephalitis (inflammation of the brain, which can lead to headaches, coma, or seizures), or ocular disease (diseases affecting the eye). Patients who become ill usually experience fever, generalized weakness, back pain, dizziness, and extreme weight loss at the onset of the illness. Typically, patients recover within two days to one week after onset of illness.

The most common complication associated with RVF is inflammation of the retina (a structure connecting the nerves of the eye to the brain). As a result, approximately 1–10% of affected patients may have some permanent vision loss.

Approximately 1% of humans that become infected with RVF die of the disease. Case fatality proportions are significantly higher for infected animals. The most severe impact is observed in pregnant livestock infected with RVF, which results in abortion of virtually 100% of fetuses.

There is no established course of treatment for patients infected with RVF virus. However, studies in monkeys and other animals have shown promise for ribavirin, an antiviral drug, for use in humans. Additional studies suggest that interferon, immune modulators, and convalescent-phase plasma may also help in the treatment of patients with RVF.

Studies have shown that sleeping outdoors at night, in geographical regions where outbreaks occur, could be a risk factor for exposure to mosquito and other insect vectors. Animal herdsmen, abattoir workers, and other individuals who work with animals in RVF-endemic areas (areas where the virus is present) have an increased risk for infection. Persons in high-risk professions, such as veterinarians and slaughterhouse workers, have an increased chance of contracting the virus from an infected animal. International travelers increase their chances of getting the disease when they visit RVF-endemic locations during periods when sporadic cases or epidemics are occurring.

A person's chances of becoming infected can be reduced by taking measures to decrease contact with mosquitoes and other blood-sucking insects through the use of mosquito repellents and bed nets. Avoiding exposure to blood or tissues of animals that may potentially be infected is an important protective measure for persons working with animals in RVF-endemic areas.

A number of challenges remain for the control and prevention of RVF. Knowledge regarding how the virus is transmitted among mosquitoes and the role of vertebrates in propagating the virus must be answered to predict and control future outbreaks of RVF. Vaccines for veterinary use are available, but they can cause birth defects and abortions in sheep and induce only low-level protection in cattle. The human live attenuated vaccine, MP-12, has demonstrated promising results in laboratory trials in domestic animals, but more research will be needed before the vaccine can be used in the field. In addition, surveillance (close monitoring for RVF infection in animal and human populations) is essential to learning more about how RVF virus infection is transmitted and to formulate effective measures for reducing the number of infections. See also **Viral Hemorrhagic Fevers**.

Additional Reading

Fields, M.B.N., D.M. Knipe, P.M. Howley, and R.M. Chanock: "Fields Virology," 3rd Ed., Lippincott Williams & Wilkins, Philadelphia, PA, 1996.

Galasso, G.J., T.C. Merigan, and R.J. Whitley: "Antiviral Agents and Human Viral Diseases," 4th Ed., Lippincott Williams & Wilkins, Philadelphia, PA, 1997.

Gear, J.H.S.: "Handbook of Viral and Rickettsial Hemorrhagic Fever," CRC Press, LLC., Boca Raton, FL, 1988.

Love, C.B. and P.B. Jahrling: "Viral Hemorrhagic Fever," DIANE Publishing Company, Collingdale, PA, 1996.

Pattison, J.R., J.E. Banatvala, and A.J. Zuckerman: "Principles and Practice of Clinical Virology," 4th Ed., John Wiley & Sons, Inc., New York, NY, 2000.

Richman, D.D., R.J. Whitley, and F.G. Hayden: "Clinical Virology," Harcourt Brace & Company, San Diego, CA, 1998.

Voyles, B.A.: "The Biology of Viruses," Mosby-Year Book, Inc., St Louis, MO, 1993.

Centers for Disease Control and Prevention (CDC), Atlanta, GA.

RIFT VOLCANO.   See Volcano.

RIGEL (β Orionis).   Ranking seventh in apparent brightness among the stars, Rigel has a true brightness value of 40,000 as compared with unity for the sun. Rigel is a blue-white, spectral type B star and is located in the constellation Orion. Rigel is known as a blue supergiant star, with a total mass in the range of 36 times that of the sun. Although the estimated distance of this star from the earth is 800 light years, its brightness signifies its enormity. See also **Constellations**; and **Star**.

RIGHI-LEDUC EFFECT.   If heat is flowing through a strip of metal and the strip is placed in a magnetic field perpendicular to its plane, a temperature difference develops across the strip. This effect, discovered in 1887 independently by Righi and by Leduc, bears the same relation to the Nernst effect that the Ettingshausen effect bears to the Hall effect. It may indeed be regarded as analogous to the Hall effect, but with a longitudinal flow of heat replacing the electric current and a transverse temperature difference replacing the potential difference. If, to one looking along the strip in the direction of the heat flow, and with the magnetic field downward, the decrease of temperature is toward the right, the effect is said to be positive. It is positive in iron and negative in bismuth.

RIGHT ASCENSION.   Angular distance east of the vernal equinox; the arc of the celestial equator, or the angle at the celestial pole, between the hour circle of a point on the celestial sphere, measured eastward from the hour circle of the vernal equinox through 24 hours. Angular distance west of the vernal equinox, through 360 degrees, is sidereal hour angle.

RIGID BODY.   An aggregate of material particles in which the interaction forces of the particles are such that the distance between any two particles remains constant with time.

RIGID FRAME (or Continuous Frame).   An indeterminate structure in which continuity of action between the intersecting or adjacent members is obtained by means of moment-resisting joints (joints capable of resisting bending moment).

Rigid frames are usually constructed of structural steel or reinforced concrete, although some frames are built of wood and of prestressed concrete. In the steel structures, the joints are either riveted or welded whereas in the concrete structures continuity is obtained by running the main reinforcing rods through the joint.

Rigid frames are used as bridges, and as bents in mill and multiple-story buildings. The Vierendeel girder bridge is a rigid frame which is similar in outline to the usual bridge truss. However, the diagonals are omitted since the chords are designed for flexure. This truss is very useful for special cases of building framing. Rigid frame action is also utilized in the design of reinforced concrete culverts and sewers. See Fig. 1.

See also **Indeterminate Structure**.

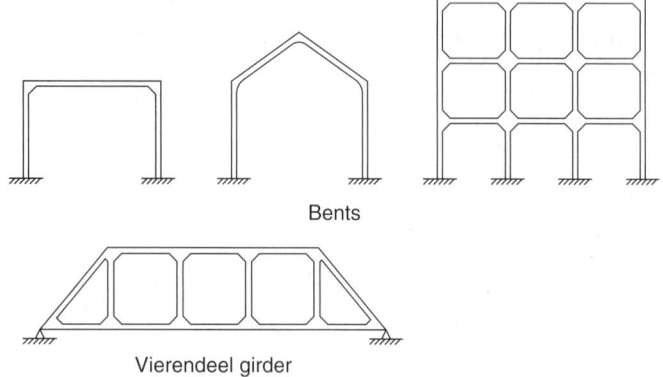

Bents

Vierendeel girder

Fig. 1.   Types of rigid frames.

RILL.   A deep, narrow depression on the lunar surface that cuts across all other types of lunar topographic features. (from German *rille*, meaning groove).

RIME ICING.   Rime aircraft icing is opaque, brittle, and granular. It is formed by the rapid freezing of small supercooled water droplets, allowing air to be trapped in. It is generally less hazardous than glaze icing, because it usually forms more slowly and is more conformal to the existing aerodynamic surface. It is the most frequent type, composing about 75% of icing reports.

RING AROUND.   Self-interrogation of a beacon due to insufficient isolation between receiver and transmitter, i.e., the beacon transmitter pulse passes through the receiver and retriggers the transmitter.

RINGED SNAKES.   See **Snakes**.

RING GALAXY.   See **Galaxy**.

RING (Mathematics).   A mathematical system for which two binary operations are defined, call them addition and multiplication, such that both operations are commutative and associative (these conditions are sometimes relaxed for multiplication) and multiplication is distributive over addition; also subtraction is always possible (cf. **Field (Mathematics)**). For example, the even integers ... − 4, −2, 0, 2, 4, ... form a ring.

RING OF FIRE.   See **Earth Tectonics and Earthquakes**; and **Volcano**.

RINGWORM.   See **Dermatitis and Dermatosis**.

RIOMETER.   An acronym for a relative ionospheric (sound) opacity meter, an instrument designed to determine the degree of absorption of high-frequency radio waves during the period of ionospheric storms. At times of sudden noise absorption in the ionosphere, this automatically operating device registers a drop in received noise power, which may be used to determine certain properties of the ionosphere.

RIPPLE.   Ripples are surface waves on a liquid whose wavelength is so short that the motion is effectively controlled by surface tension forces. This requires that the wavelength should be less than

$$\lambda_c = 2\pi\sqrt{\frac{\gamma}{\rho g}}$$

where γ is surface tension, and ρ, liquid density. For water, $\lambda_c = 1.7$ centimeters.

In electricity, ripple is the alternating-current component from a direct-current power supply, arising from sources within the power supply. Unless otherwise specified, percent ripple is the ratio of the root-mean-square value of the ripple voltage to the absolute value of the total voltage, expressed in percent.

RIPPLE MARK.   Corrugations developed in sands and muds by currents in the water which covers them. Ripple marks may be classified as due either to oscillation or translation. The former are the result of oscillation currents set up in the water by the wind; the result of ordinary water waves. The latter are the result of progressive, directional water currents, and the resulting ripple mark is essentially a subaqueous dune. Ripple mark is helpful in determining the conditions under which aqueous, clastic sediments are deposited. Ripple mark has also been used to help determine the depth of water in which the rippled sediments have been deposited. Ripple mark is also helpful in determining the original position of formations that have been subsequently deformed or overturned.

RIPPLE VOLTAGE.   When an ac is rectified, the resultant current or voltage consists of pulsating dc. In the case of simple half-wave rectifiers, this output is a series of half-sine waves spaced by equal intervals of no output while for full wave, single-phase rectifiers it consists of half-sine waves with no appreciable space between them. Poly-phase rectifiers give outputs which, while they do not vary as markedly as this, consist of a series of adjacent portions of sine waves. In every case the output may be considered as made up of a smooth dc component and an ac or ripple component. The first is the value read by a dc instrument in the circuit and the latter is the component which will produce objectionable hum in communication equipment supplied by the voltage. To reduce this ripple component various filter systems are used, the amount of filtering depending upon the equipment being supplied by the rectifier-filter combination. In many industrial applications the output of the rectifier alone is satisfactory, in communications the output must contain an extremely small amount of ac. This is measured in terms of percent ripple, which is the A.C. component of voltage divided by the D.C. component of voltage and multiplied by 100.

RISE TIME.   The time required for the output of a system (other than the first order) to make the change from a small specified percentage (often 5 or 10) of the steady-state increment to a large specified percentage (often 90 to 95), either before or in the absence of overshoot. See figure that accompanies entry on **Response (Instrument)**. If the term is unqualified, response to a unit step stimulus is understood; otherwise the pattern and magnitude of the stimulus should be specified.

RISK/BENEFIT ANALYSIS.   A nonstandard, rather poorly defined procedure for weighing the pros and cons of materials, machines, and devices in terms of their impact upon society. The concept can be applied to a wide range of consumer products, ranging from food additives to drugs, cosmetics, transportation safety devices, fireproof materials, insulation, aerosol packages — items of commerce that in some way have been identified with a threat to human life, health, or more generally, to the environment and ecology of various regions. Risk/benefit analysis usually is not brought into play until some chemical or physical quality of a product is *found to be* or is *believed to be* a serious threat. Frequently both sides of an advocacy contest will apply risk/benefit analysis, quite understandably sometimes with built-in biases in one direction or the other. Although scientific findings are often introduced as parts of such analyses, the procedure is essentially a judgmental art and not a science.

RITCHEY-CHRETIEN TELESCOPE.   A two-mirror telescope combining an oblate spheroidal primary and an ellipsoidal secondary, and resulting in a large field free of coma. See also **Telescope (Astronomical-Optical)**.

ROADRUNNER.   Turacos and Cuckoos.

ROBBER FLY (*Insecta, Diptera*).   A predacious fly of the family *Asilidae*. Many of these flies are large and all capture living insects as prey, including bees of all kinds. Most of the included species have smooth slender bodies but some are quite hairy and one group is characterized by stout form and dense vestiture. These last mimic bumblebees closely and furnish an apparent case of aggressive mimicry, since the bumblebees are said to be among their victims.

On balance, the robber fly is classified as an economically beneficial insect for food producers.

ROBERVAL PRINCIPLE.   A number of low-capacity scales incorporate a principle discovered by Gilles Personne de Roberval, a French mathematician, in 1670. A good scale must weigh as accurately when the load is placed on any of the four corners as it does when the load is placed directly over the center of the platform or platter. A scale that does not perform in this manner is either out of adjustment or is of inherently poor design. Any error of this type is known as shift error. The Roberval principle is one means for avoiding shift error.

The principle is applied in the counter scale shown in Fig. 1. Parallelogram *ABCD* always maintains the weight and load platters in a horizontal plane. Stems *AD* and *BC* are integral parts of the parallelogram. These stems always remain vertical regardless of load position. Links *CE* and *ED* prevent tipping of the platter and aptly are termed check links. Although the parallelogram may assume different relative positions and dimensions in scales of various designs, the check links are always present, since the Roberval principle applies only to a completed parallelogram.

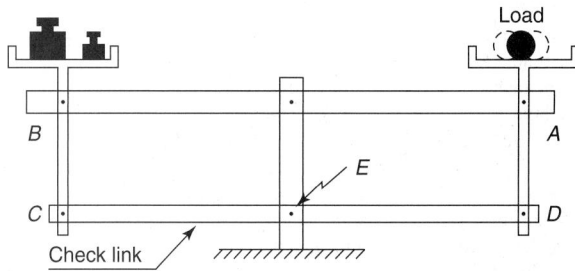

Fig. 1.   Counter scale that utilizes the Roberval principle to avoid shift error.

The principle can be proved by kinetics or by statics. Proof by statics is based upon the fact that the rigid arms extending from the connecting stems may be considered to be cantilever beams. Because the extending arms are rigidly fixed to the connecting stems, the force system is identical to that of the cantilever beam shown in Fig. 2. The system resolves itself into two couples, one comprised of W_1 and W_2 and the other by R_1 and R_2. The lateral position of W_1 along the extending arm does affect the values of R_1 and R_2, but since R_1 always equals R_2, it follows that W_2 always equals W_1 regardless of position. Since it is the value of W_2 that tends to disturb the equilibrium of the balance, the value and not the point of application of the force along the extending arm determines the equilibrium

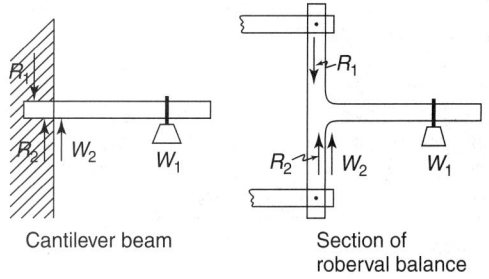

Fig. 2.   Static proof of Roberval principle.

of the balance. Thus, so long as the weight of the mass suspended from the left arm equals the weight of the mass suspended from the right arm, the relative lateral positions of these masses have no significance insofar as the point of balance is concerned.

ROBIN *(Aves, Passeriformes).* 1. The European redbreast, *Erithacus rubecula,* a warbler, and a related species of the Canary Islands. 2. The American robin, *Turdus migratorius,* a thrush. 3. In Australia, a species related to the wheatear; in New Zealand, other birds of the same group.

See also **Passeriformes**; and **Thrush**.

ROBITZSCH ACTINOGRAPH. A pyranometer developed by M. Robitzsch. Its design utilizes three bimetallic strips, which are exposed horizontally at the center of a hemispherical glass bowl. The outer strips are white reflectors, and the center strip is a blackened absorber. The bimetals are joined in such a manner that the pen of the instrument deflects in proportion to the difference in temperature between the black and white strips.

ROBOTS AND ROBOTICS. A robot may be defined as a "reprogrammable, multifunctional manipulator to move materials, parts, tools, or specialized devices, such as welders and paint sprayers, through variable programmed motions to accomplish a variety of tasks."[1] It follows that *robotics* is the technology of designing, applying, and maintaining robots.

Generally, a robot is a specifically designed piece of production hardware that is used predominantly to execute what may be termed "handling" operations — that is, such tasks as "picking up," "setting down," "putting into place," "locating and positioning," and "transferring" objects from one location to another. For these duties, the province of the robot is geometric (usually three-dimensional) in nature and the robot is programmed in the geometric terms of positions (coordinates) and motion (pathway). When robots are used in this manner, they must be capable of the same order of dimensional and positional precision that also applies to the machines that they serve.

Although each robot may operate under its own specific preprogrammed control schedule, it is common in large manufacturing facilities to lump groups of machines and the robots that serve them into *cells,* which adds another layer of computer control.

It is interesting to note that, after the late 1970s, when robots were introduced with great flair by the nontechnical news media, intense interest in robots by managers of discrete-piece manufacturing was created in their quest to cut production costs. Thus, during the 1980s, many hundreds of costly robots were procured and installed in large manufacturing facilities, as exemplified by the automotive, aircraft, and other parts manufacturing and assembly industries. Unfortunately, in numerous cases, the results did not meet the expectations. Two principal factors contributed to the disappointments and to the long periods of "debugging." One factor simply was undue haste to use a new technology that would offset international competition, and this led to the second factor — namely, basing this transfer of a new and highly specialized technology on shallow studies and failing to sufficiently develop detail and test plans for integrating robotics into existing facilities. These difficulties were present not only in retrofit situations, but also in connection with entirely new production facilities. It was found that robots cannot be treated as simple "add-on" hardware.

As of the present time (1993), robotics is better understood and now is viewed with an improved perspective, as the result of over a decade of experience. The unrealistic "rush" to robotize and computerize in discrete-piece manufacturing generally has yielded to more meticulous economic and practical scrutiny by users and potential users. Today, robotics is on a much firmer foundation and obviously will continue to progress.

Chronology of Robotics

The image of robotics over past centuries was burdened by an aura of mystique and romanticism that eventually led to an "overworked" comparison of robots with people — an image that still persists in the general press, but to a much lesser degree. During the late 1700s, for example, there existed an exceptional fascination with androids, which essentially were mechanical "people," so to speak. In 1774, Jaquetdroz

[1] Adapted from the *Robot Institute of America.*

exhibited three "mechanical marvels" — a musician, a writer, and a draftsman. These and other charmingly attired mechanical people were presented at court like visiting dignitaries. During that period, Diderot in his *Encyclopedie* observed that in the construction of machines, engineers should look to monsters for inspiration, but that instead the eighteenth century engineers looked to man and built beautiful automatons.

As recently as 1923, this general approach was carried on by Karel Čapek a Czech playwright who used the Czech word *robot* (for worker) to describe humanoid creations in his play, "Rossum's Universal Robots." In the 1940s, the science-fiction writer Isaac Asimov coined the word *robotics.*

Early industrial use of robots, which resembled machines more than people, dates back some 20 to 30 years. Although these robots performed well for certain applications (mainly for handling large and heavy loads under adverse conditions), their use did not commence to bloom until the 1970s.

In the discrete-piece manufacturing industries (metalworking and automotive, for example), many of the operations now performed or assisted by robots were formerly accomplished by what is referred to as *fixed* or *hard* automation. Such systems, still widely employed, use limit switches, relays, photoelectric sensors, and other electromechanical and magnetic devices for controlling the motion- and position-related geometric variables. The principal constraint of fixed or hard automation is a lack of *flexibility.* Transfer lines, for example, commonly used in the automotive and machinery industries and which were tailored to earlier continuous production and assembly line concepts, were representative of fixed automation. In connection with model changes (usually on an annual basis) and with manufacturers who produced only limited runs of many different products, the lack of flexibility was a major problem. Fixed systems required costly retooling when going from one product to the next.

This shortcoming was considerably relieved by the introduction of programmable relay logic systems (adjustable plugboard memories) and, in the late 1960s, by the entry of the *programmable controller.* These tools brought a degree of flexibility and universality to pre-robot industrial automation systems. See also **Programmable Controller**.

Classes of Robots

Because there are numerous types of robots, they are difficult to place in neat groups. Some useful classifications are by their: (1) axes of motion; (2) control system; (3) programming (and teaching) system; (4) load capacity and power requirements; (5) dynamic properties, including stability, resolution, repeatability, and compliance; (6) end-effectors (grippers) used; (7) workplace configuration; and (8) appropriate applications for a given robot design.

Axes of Motion. A robot is movable from one factory location to another (like a machine) as may be dictated by factory layout changes or major alterations in job assignment. However, for any given task that will be repeated again and again over a long period (usually months or more), a robot will be *firmly fastened* to the operating floor (in some cases, to the ceiling). This establishes a firm *geometric location of reference,* a very important, unchangeable position that will *geometrically relate* precisely with associated fixed machinery. In the case of a work cell, each of several robots will precisely relate geometrically with each and every machine that the robot serves. For moderate changes in use, the average robot will incorporate within its design sufficient operating flexibility and adjustments obviating the need to alter the installed robot's *fundamental reference location.* In the case of a so-called "smart robot," final very small changes in the positioning of the arm will be made by inputs from a machine vision or tactile system. Intentionally designed, movable robots are special cases (described later).

Once installed, a robot's ability to move parts and materials will be established by the built-in axes of motion, sometimes called *degrees of freedom.* The axis of motion refers to the separate motion a robot has in its manipulator, wrist, and base. The robot designer will usually select one of four different systems of geometric coordinates for any given need. These coordinate systems are: (1) revolute (jointed arm); (2) Cartesian (X, Y, Z); (3) cylindrical (rectilinear); and (4) spherical (polar).

Where revolute (jointed-arm) coordinates are used, the robot arm is constructed of several rigid members connected by *rotary joints.* As illustrated by Fig. 1, three independent motions are permitted. In some robots, these members are analogous to the human upper arm, forearm, and

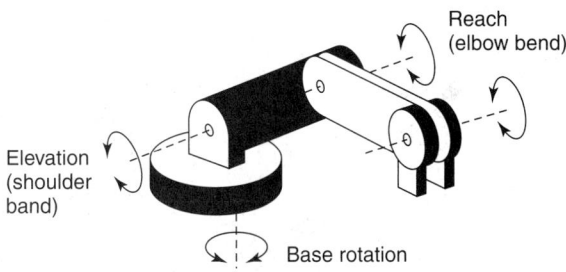

Fig. 1. Jointed-arm manipulator incorporating revolute coordinates.

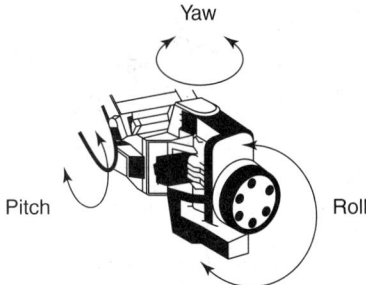

Fig. 2. Wrist assembly on robot arm for orienting the end-effector in accordance with requirements of workpiece.

hand, while the joints are, respectively, equivalent to the human shoulder, elbow, and wrist. The arm incorporates a wrist assembly for orienting the end-effector (gripper) in accordance with the workpiece. See Fig. 2. These three articulations are pitch (bend), yaw (swing), and roll (swivel). In some applications, fewer than six articulations may suffice, depending upon the orientation of the workpiece and the machine(s) which the robot is serving.

Where Cartesian coordinates are used, all robot motions travel in right-angle lines to each other. There are no radial motions. Consequently, the profile of a Cartesian robot's *work envelope* is a rectangular shape. See Fig. 3. Some systems utilize rotary actuators to control end-effector orientation. Robots of this type generally are limited to special applications. A robot also can incorporate rectilinear-Cartesian coordinates. In one example, a continuous-path extended-reach robot offers the versatility of multiple robots through the use of a bridge and trolley construction that enables it to have a large rectangular work envelope. When ceiling-mounted, a device of this type may service many stations with several functions, thus leaving the floor clear. The X and Y motions are performed by the bridge and trolley; the vertical motions are performed by telescoping tubes.

In a Cartesian coordinate system, the location of the center for the coordinate system is the center of the junction of the first two joints. Except for literally moving the robot to another factory location, this center *does*

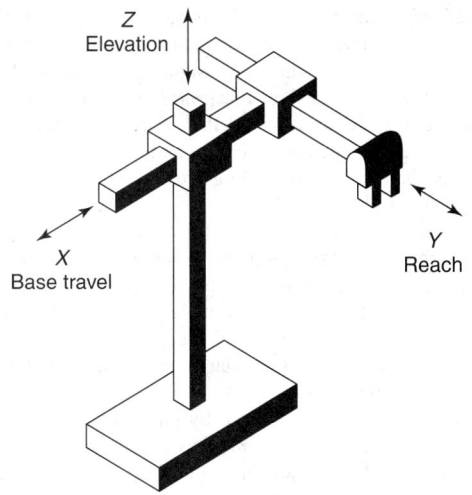

Fig. 3. A manipulator incorporating Cartesian coordinates.

not move. In effect, it is tied to the "world" as if anchored in concrete. If the X measurement line points toward a column in the area where the robot is placed, the X line will *always* point toward that same column no matter what way the robot turns while performing its programs. These are known as the *world coordinates* for a given robot installation. See Fig. 4.

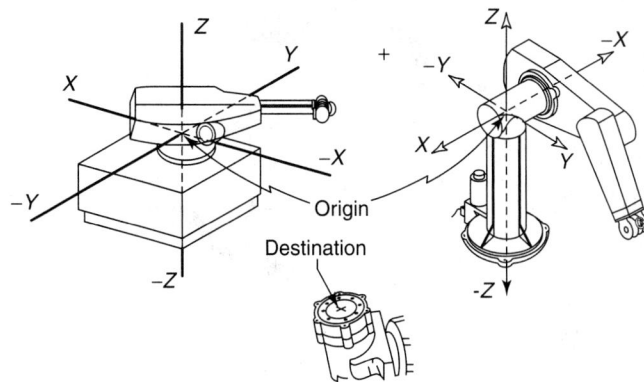

Fig. 4. World coordinate system of a robot using Cartesian coordinates. (*Westinghouse.*)

However, in the operation of a robot, having an origin for a measurement reference is not sufficient. We also need a reference to the point where we are measuring. This measurement is made from the origin of the coordinate system to a point that is exactly in the center of the circle on which the tool (end-effector) is to be mounted. This system moves with the tool and is aptly called the tool coordinate system. See Fig. 5.

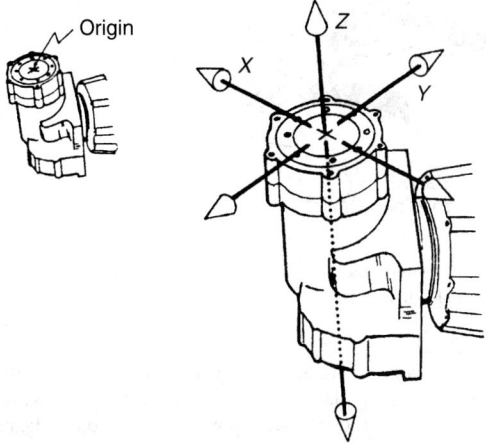

Fig. 5. Tool coordinate system of robot using Cartesian coordinates. In this system, the X and Y lines lie at right angles flat on the tool mounting surface. The Z line is the same as the axis of rotation for the joint, i.e., it points directly through the tool in one direction and through the wrist in the other direction. This system is *not* tied to the "world." While the origin of this system is thus allowed to move around, the destination (where it measures to) is left to the discretion of the user. Sometimes the tool coordinate system is actually used to measure where the tip of the tool lies relative to where it is mounted. Sometimes it is used to measure where one position in space lies relative to some other point in space. (*Westinghouse.*)

Robots incorporating cylindrical coordinates have a horizontal shaft that goes in and out, and rides up and down, on a vertical shaft that rotates about the base. See Fig. 6. Additional rotary axes sometimes are used to allow for end-effector orientation. Cylindrical-coordinate robots are often well suited where tasks to be performed or machines to be serviced are located radially from the robot and where no obstructions are present. A robot incorporating cylindrical coordinates has a working area or envelope that is a portion of a cylinder.

Robots using spherical (polar) coordinates may be likened to a tank turret, i.e., they are comprised of a rotary base, an elevation point, and

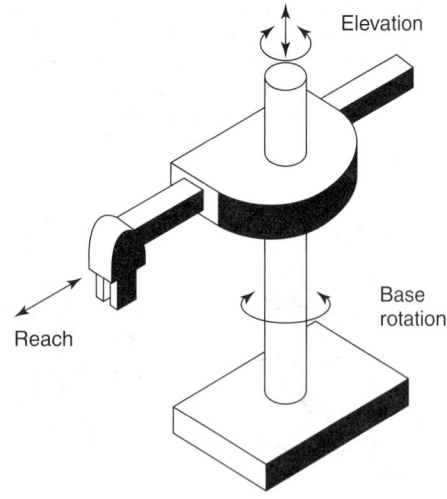

Fig. 6.   A manipulator incorporating cylindrical coordinates.

a telescoping extend-and-reach boom axis. See Fig. 7. Up to three rotary wrist axes (pitch, yaw, and roll) may be used to control the orientation of the end-effector. The arm moves in and out and is raised and lowered through an arc while rotating about the base. The end-effector moves in a volume of space that is a portion of a sphere.

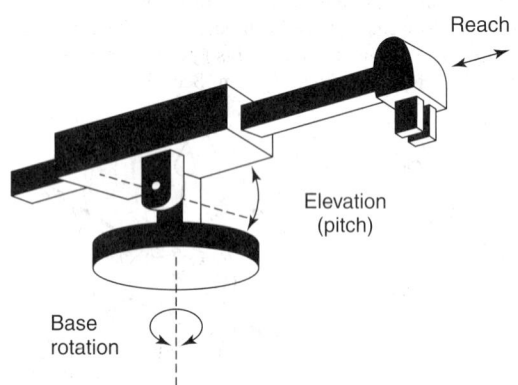

Fig. 7.   A manipulator incorporating spherical polar coordinates. Operation is similar to that of a tank turret.

The *work envelope* of a robot is that area in *space* that the robot can touch with the mounting plate on the end of its arm. Actually, the envelope will be somewhat larger, depending upon the dimension of the end-effector that is fastened to the tool mounting plate. See Fig. 8.

Robot Control Systems. Robots may be classified as (1) non-servo controlled; or (2) servo controlled.

In a non-servo controlled robot, the directional controls are fully off or fully on, causing essentially constant speed movement along an axis in one direction, or in the reverse direction. Some form of limit switch is used to stop the movement at the desired point. Non-servo controlled robots are the least complex of all robots. They move in an open-loop fashion between two exact end points on each axis, or along predetermined paths in accordance with fixed sequences. Such robots can operate over an infinite number of points enclosed within their operational envelope. Non-servo controlled robots are given start and end points on each axis which must be passed. There is little or no control of end-effectors between these points. Technically, controlled trajectory is possible on a non-servo controlled robot only if the unit is given the coordinates of all points lying between the start and end parameters. This specific type of programming will allow the system to perform motions, such as straight lines and circles.

Non-servo controlled robots are sometimes called *limited-sequence* robots. The number of limb articulations is usually few. Because of their control characteristics, the non-servo controlled robot is also sometimes called a "bang-bang" device or a *pick-and-place unit* (a common designation for robot used for picking up an object, transporting it to a

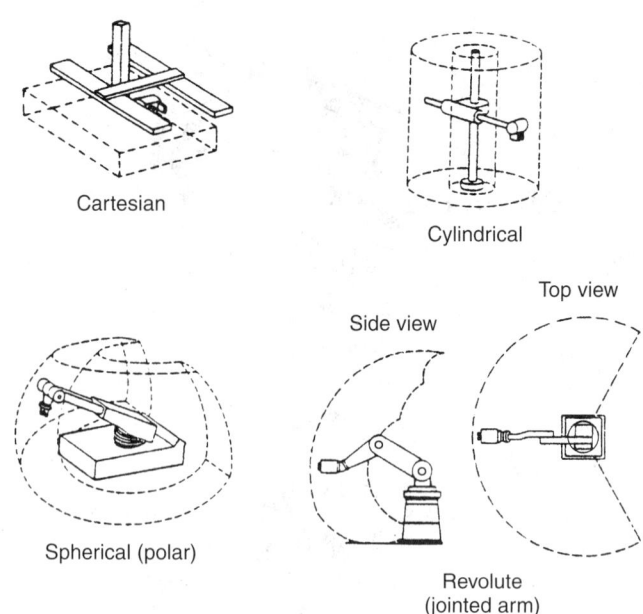

Fig. 8.   Work envelope of a robot is that volume in space that the robot can touch with the tool mounting plate on the end of its arm.

predetermined location, and placing the object at that location). Robots of this type are capable of high speed and usually have good repeatability.

The servo controlled robot incorporates one or more servomechanisms that enable the arm and gripper to alter direction in mid-air, without having to trip a mechanical switch. Servo controlled robots generally have larger program and memory capacity than their non-servo counterparts.

In a system of this type, upon start of program execution, the controller addresses the memory location of the first command position and also reads the actual position of the various axes as measured by a position feedback system. These two sets of data are compared and their differences (error signals) are amplified and transmitted as command signals to servo valves for the actuator of each axis. Thus, a servo controlled robot is a *closed-loop* system.

This closed loop refers to the robot system per se and does not embrace the total machine system that is served by the robot. The latter would require feedback of measurements made on the product of the handling operation and adjustments of the total system (machine plus robot) required to assure desired control of the total system. This requires additional inspecting, gaging, and other hardware and, in more sophisticated systems, may require *machine vision*.

Servo valves, operating at constant pressure, control flow to the manipulator's actuators. As the actuators move the manipulator's axes, feedback devices, such as encoders, potentiometers, resolvers, and tachometers (see also **Position and Displacement Measurement**), send position (and in some cases, velocity) data back to the controller. These feedback signals are compared with the desired position data and new error signals are generated, amplified, and sent as command signals to the servo valves. This process continues until the error signals are effectively reduced to zero, whereupon the servo valves reach *null*, and flow to the actuators is blocked and the axes come to rest at the desired position. The controller then addresses the next memory location and responds to the data stored there. This may be another positioning sequence for the manipulator or a signal to an external device.

Generally, the memory capacity of a servo controlled robot will be sufficient to store up to 4000 and more points in space. Specific program select and sequence activity points for a given operating scheme. Programs can be varied to maintain the scheme while changing the activity points. Both *continuous-path* and *point-to-point* capabilities are possible. In this regard, robot system control is similar to numerical control of machine tools. (See also **Numerical Control**.)

Accuracy can be varied, if desired, by changing the magnitude of the error signal, which is considered *zero*. This can be useful in "rounding the corners" of high-speed continuous motions. Programming is accomplished by manually initiating signals to the servo valves to move the various axes into a desired position and then recording the output of the feedback devices into the memory of the controller. This process is repeated for the entire sequence of desired positions in space.

A servo controlled robot can be one of two types: (1) point-to-point; and (2) continuous path.

With a *point-to-point robot*, there are two main commands: (1) attitude of all limbs at the start of the move; and (2) the new attitude of those limbs when a particular move has been completed. While making the move as fast as possible and while moving all limbs simultaneously to fulfill a given command, there is no precise definition of the paths which the robot limbs will traverse. Thus, the term, point-to-point. In programming a robot of this type, the system designer must consider all possible *intervening points* between start and destination. For example, the robot may have to clear an object that may fall in its "direct line" path, or it may be desired for the robot arm to approach its destination at the best angle (for instance, in picking an object up from a pallet). Point-to-point robots can do any job performed by a limited-sequence robot previously described and, with sufficient memory capacity, these robots can handle jobs, such as palletizing, stacking, and spot welding, among others.

Continuous-path robots are required for applications where it is required to control, not only the start and finish points of each robotized steps, but also to control the path traversed by the robot hand as it travels between these two extremes. Seam welding is an example of where a robot wields a welding gun and moves it along some complex contour at the correct speed to produce a strong and neat weld. Theoretically, the continuous-path robot is an extension of the point-to-point concept because the curved path is made up of numerous straight-line segments. This requirement, of course, calls for a very substantial memory.

Programming Robots. With the availability of sophisticated electronic hardware and system software during the past decade, programming represents one of the most advanced segments of robot technology. Early in the development of automated systems, at a time when robots and other automation techniques were largely associated with *replicating the skills of human operators*, the guidance of robots was essentially a "copy cat" technique. The "playback" concept was used exclusively. With certain robots, this technique is still used.

In the playback method, the robot is programmed through a procedure known as "teaching." When one considers the numerous variables and complexities that can be encountered in applying robots, the need for a short cut to programming becomes evident. For example, in any reasonable volume of factory space, there are literally many thousands (depending upon resolution needed) of points that may become part of a robot program, particularly in the case of continuous-path systems. If Cartesian (X, Y, Z) coordinates are used, the storing of three coordinate values for each point of travel rapidly adds up to a lot of data to be stored. Further, in planning a robotic system, the designer who does not have a short cut method must visualize just how the system will operate in three-dimensional space and express design objectives in terms of *very long* lists of coordinate positions.

In the early days of robotics, of course, the designer did not have computer graphics and computer systems with large memories that approached even in a small way the abilities of present minicomputers and microprocessors. Thus, early designers developed a clever, innovative "teach-and-playback" methodology for robot programming. In the *teach mode*, the robot is directed through various movements in sequence, this accomplished by actually manually guiding the robot through its complete act so to speak. During the period, the sequence of movements is recorded in memory. In the *playback mode*, the robot simply repeats, as desired, the sequence of movements, as taught, from the memory system.

Levels of Programming for Robots. Ranging from simple to complex, there are three levels of robot programming:

Level 1 Programming — Manual, lead-by-hand teaching and front-panel programming. Applications include some spot welding and pick-and-place tasks.

Level 2 Programming — Programs are written in simple robot programming languages, i.e., techniques that assist programmers in entering motion, branching, coordinate-transformation, and signal instructions. Such systems may also provide a number of the lead-by-hand and control-panel operations that are characteristic of Level 1 programming. These programming capabilities permit running of *user programs* that are more complex than those at Level 1. Representative applications are found in some palletizing and arc welding tasks.

Level 3 Programming — These are the most modern and expanding methods for robot programming. The programming languages incorporate extended capabilities, including structured constructs, full arithmetic functions, external robot-path modifications, and supervisory computer-communications support. Since these systems support the functions and features found on the two lower programming levels, Level-3 systems can handle Level-2 applications as well as modifying the robot arm's path, based upon data transmitted from external sensing devices, including machine vision in the more complex systems.

Shown in Fig. 9 is a microprocessor-based, operator-friendly programming system. In this system, a lightweight hand-held teach pendant is used to lead the robot through its required moves. A controlled-path motion feature of the controller automatically coordinates all six axes to move the robot from one point to the next in world coordinates at the programmed velocity. An average of 3000 points can be programmed and stored within the control memory.

Fig. 9.   Microprocessor-based, operator-friendly programming system. Menu-driven keyword approach to programming provides a simple interface with the robot. Keyboard commands permit creation and editing of application data. While teaching, the status of the robot's program and operating statistics, input/output or available memory may be examined. The communications interface provides a bridge to an FMS (flexible manufacturing system). The interface allows communication with host computers as well as intelligent sensors. (*Cincinnati Milacron*™.)

Large manufacturers and users of robots are aware of the importance of a day of lost production. Thus, manufacturers are concerned with robot downtime attributed to the normally slow process of entering program logic statements for machine, peripheral and sensor interfaces, conditional and unconditional program branching, looping, register manipulation, digital input/output, indirect addressing, timers, macro instructions and other commands. This input is essentially the same for all robots performing a similar task and can represent up to 80% of total programming requirements. This problem essentially can be solved, by installing a system that permits the logic portion of a program to be generated *off line* before robots are installed.

One approach is shown in Fig. 10. The personal computer workstation shown consists of four modules, which are supplied in either of two basic combinations to fit a user's needs: (1) a turnkey system, including a 32-bit CPU (central processing unit), CRT, keyboard, dual disk drives and printer, plus an Edit/Store software package and an interface unit; (2) an interface unit and a Store software package for use with certain personal computers. The latter option provides upload/download, storing, copying and printing capabilities as well as the ability to view programs on a remote CRT screen and change program names. At the personal computer workstation, the operator can create and modify programs off line, even while the robot is working in the plant (or before installation). The user can edit stored programs and enter, store and display programming or editing comments, notes, and reminders.

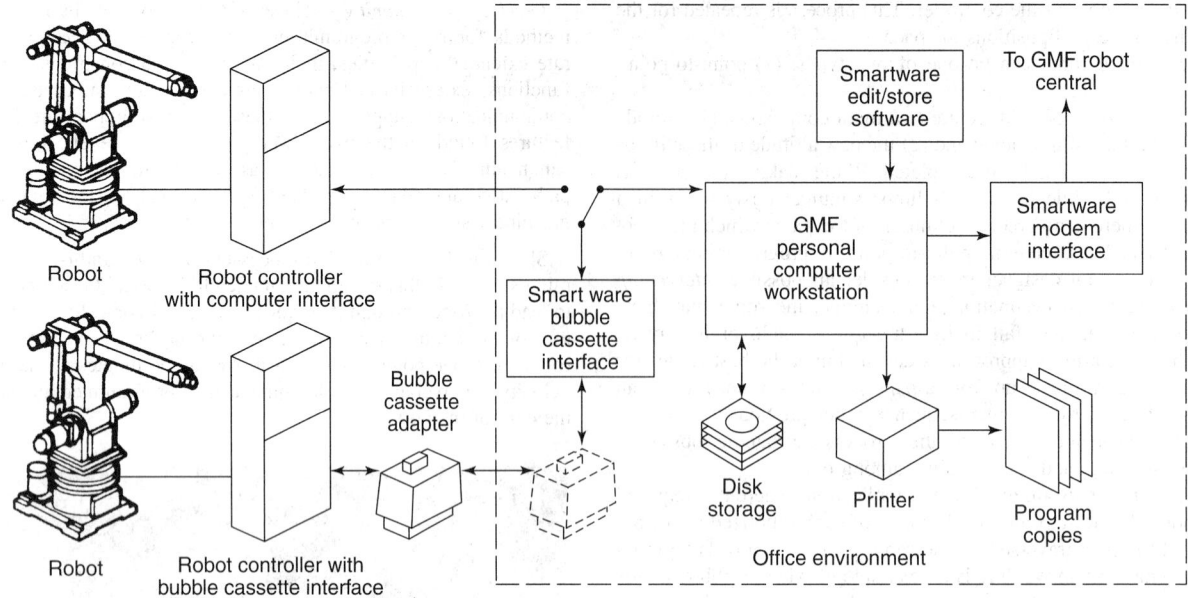

Fig. 10. Off-line personal computer (PC) workstation robot programming system. (*SmartWare*™, GMF Robotics Corp.)

A six-axis robot, the motion control of which uses 15 microprocessors, 6 optical encoders, and 6 stepper motors is shown in Fig. 11. Six degrees of freedom are required to minimize the demands on the construction of the robot workstation. In the interest of cost and speed, a distributed processing system is used. Stepper motors are used because of their relatively low cost and ease of control. Problems of accuracy and motor control are solved by software.

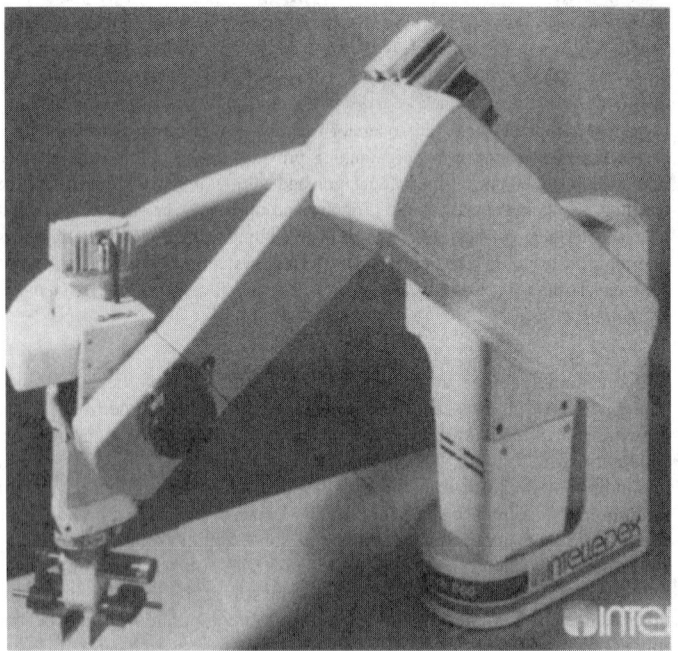

Fig. 11. A six-axis robot with a humanlike (anthropogenic) joint configuration, giving it exceptional dexterity and allowing it to approach any point in the work envelope from any angle, thus providing easy adaptability to existing workstations. (*Intellidex*™.)

Some of the tasks assigned to software include calibrating the arm, control of end-effectors (grippers) and safety devices, and image processing analysis by an integrated machine vision system. (See also **Machine Vision (Recognition and Applications)**.) The software is also responsible for motor control functions and software control of motion. Although the robot is built with as much mechanical accuracy as possible, considering cost restraints, the arm-parameter calibration, for example, can make up for

certain manufacturing tolerances. Nearly 20 sources of mechanical error have been identified. These errors, however, are constant for a given robot; they may vary from robot to robot. Once these constants are determined for a specific robot, they can be used by the software to perform calculations that accurately position the robot.

The robot shown in Fig. 12. is an electric-drive, overhead gantry configuration that incorporates tactile sensor technology. As designed, the machine is a fully self-contained workstation that allows either stand-alone or integrated assembly line adaptation. It has a high level of dexterity of the kind required for the assembly of disk drives, printed circuit boards (PCBs), wire harnesses, and telecommunication and electromechanical devices. The system has the potential of automatic laser inspection. The intelligent gripper incorporates both optical sensing and a strain gage that allows force-sensing thresholds to be programmed in three orthogonal axes. Parts that are out of tolerance or that are not properly oriented are detected quickly, enabling the robot to respond as required. Intelligent action based on the specific "force signatures" programmed into the gripper recognize and accommodate specific assembly tasks.

Load Capacity and Power Requirements of Robots. A recent survey of manufacturers of robots in the United States, Europe, and Japan indicated that models available (not total units installed) were designed to handle loads ranging from about one pound (0.5 kg) upwards to about 2300 pounds (1043 kg). These Figures, of course, do not bracket all robots ever made, either in terms of very heavy or very light loads. Applications included in the survey were found in die casting, forging, plastic molding machine tools, investment castings, spray painting, welding, and machining, among many others. It is well established that the use of robots in light manufacturing and inspection operations is expanding, particularly in the electronics manufacturing industry.

Many electric robots utilize direct current stepping motors. (See also *Servomotors.*) Hydraulically powered robots usually employ hydraulic servo valves and analog resolvers for control and feedback. Digital encoders and well-designed feedback control systems can provide hydraulically actuated robots with an accuracy and repeatability generally associated with electrically driven robots. Pneumatically driven robots normally are found in light service, limited-sequence, and pick-and-place applications.

Electric Drives for Robots. Electric motors provide the greatest variety of choices for powering manipulators in the low- and moderate-load range, and for low-speed, high-load operations. They are relatively easy to control and a number of control techniques can be used. Electric motors are not as responsive as hydraulic systems and are considerably stiffer than pneumatic systems, unless the latter are operated under very high pressure.

Motors generally operate at speeds that far exceed those desirable for manipulator joints. Thus, speed reducers are required. Although speed reducers have the advantage of amplifying available torque and in

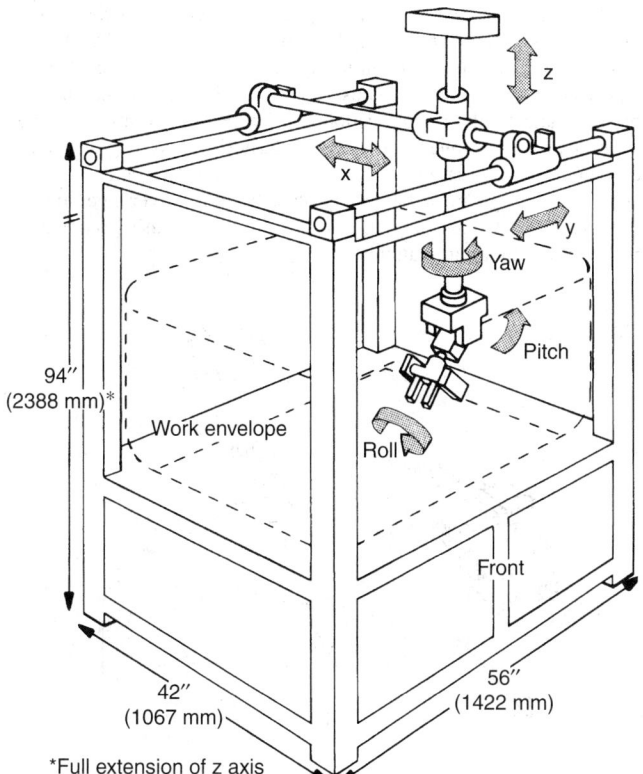

94″ (2388 mm)*

Work envelope

Yaw

Pitch

Roll

Front

42″ (1067 mm)

56″ (1422 mm)

*Full extension of z axis

Fig. 12. Intelligent, high-precision parts-assembly robot that incorporates both optical (vision) and force (tactile) sensors for feedback. Modular work-station concept of overhead gantry, as shown at right, facilitates integration into assembly flow. The six-axis configuration consists of three linear axes (X, Y, Z) and three rotary axes (yaw, pitch, and roll). All are driven by direct current servomotors. Accuracy and repeatability are ±0.002 and ±0.001 inch, (0.05 and 0.03 mm) respectively. Resolution is specified as 0.0005 inch (0.013 mm).

A teach pendant is provided for on-line programming and reprogramming to accommodate production line changes. A hand-held tool allows production personnel to control all robot movements, time delays, high-level commands for self-calibration, and to call up complex subroutines. Supporting the teach pendant, including on-line and off-line programming through an IBM PC compatible computer, is a special programming language developed by the manufacturer. Programs can be written and tested interactively on line or written off line and downloaded from disk for implementation. (*Adaptive Intelligence Corp.*)

preventing or inhibiting back driving, they are usually a major source of inefficiency and error. In demanding applications, the most important factors in selecting a motor include the power-to-weight ratio and torque-speed characteristics. The ability to accelerate and decelerate the working load quickly is a very desirable attribute. Also required is the ability to operate at variable speeds.

The development of microstepping techniques, pioneered by the office automation and instrument industries, contributed to the knowledge of an effective use and control of electrical stepping motors. For example, microstepping had led to the use of stepping motors in applications requiring incremental rotary motion of only a fraction of the particular motor's primary step angle. Hybrid step motors combine the fast response of variable reluctance motors with the detent torque of permanent-magnet step motors. Step motors are easily used in closed-loop servo systems and also may be operated in a synchronous mode at their slow speed.

Because electric robots do not require a hydraulic power supply, they conserve floor space and decrease factory noise. In an electric manipulator, the motors generally connect to the joints through a mechanical coupling, such as a leadscrew, pulley block, spur gears, or harmonic drive. This is because electric motors generally produce much less force or torque than a hydraulic actuator of the same size and thus require a mechanical impedance matcher between them and the joint if they are to overcome the loads that are encountered in a typical manipulator. A hydraulic actuator usually can drive a joint directly.

Permanent-magnet direct current motors have proved a good choice for medium- and small-size manipulators. They are generally more efficient, less costly, lighter, and smaller than wound-field motors. These servoed motors have been and continue to be a good choice for many applications. The brushless, electronically commutated versions have long lives. Printed circuit motors have high torque relative to their rotor inertia and thus have fast response time. These motors are capable of driving at low speeds without the need for speed reducers.

Hydraulic Actuators for Robots. Hydraulic actuators are either (1) linear piston actuators, or (2) a rotary vane configuration. Hydraulic systems are relatively easy to control because the low compressibility of hydraulic fluids results in the systems being very stiff. The high power-to-weight ratio makes the hydraulic actuator an attractive choice for moving moderate-to-high loads at reasonable speeds. Hydraulic systems are characterized by fast response time, high natural frequencies, and low signal noise levels. A major disadvantage of hydraulic systems is their need for an energy storage system, including pumps and accumulators. Several years ago, for machine tools, a switch was made from hydraulic to electric drives mainly because of better reliability and leakage problems associated with hydraulics. In some applications, such as paint spraying, hydraulics do not present an explosion hazard.

Dynamic Properties of Robots. Included among these properties are: (1) stability; (2) resolution; (3) repeatability; and (4) compliance. Considering these factors, the design of a robot is innately complex because of the manner in which these properties interrelate. Stability is associated with the oscillations in the motion of the tool. The fewer the oscillations, the more stable the operation of a robot. Lack of stability increases wear and sometimes inconsistencies in performance. Resolution is function of the design of the robot control system and specifies the smallest increment of motion by which the system can divide the working space. Repeatability is the ability of the robot to reposition itself to a position to which it was previously commanded or trained. The compliance of a manipulator is indicated by its displacement relative to a fixed frame in response to a force (torque) exerted on it. High compliance means the tool moves a lot in response to a small force (the manipulator is then said to be spongy or springy). If it moves very little, the compliance is low (the manipulator is said to be stiff).

These matters are dealt with in considerable detail in the Considine reference listed.

End-Effectors (Hands or Grippers). The end-effector is the device fastened to the free end of a manipulator. It provides the means for the manipulator to interact with its surroundings. The function of the end-effector is first to grasp an object or a tool, then hold it while the manipulator moves—thereby also moving the object—and finally to release the object.

Four main methods are used for holding parts or tools: (1) mechanical clamping; (2) vacuum suction; (3) magnetic attraction; and (4) plug-in or detent fittings. Scoops, ladles, and sticky fingers using adhesives are among numerous specialized types of end-effectors. A montage of representative grippers is given in Fig 13.

In the robot shown in Fig. 14, the end-effector, instead of taking the form of a gripper for grasping pieces, is an inspection sensor used to test tubes in steam generators.

Workplace Configuration and Environmental Factors

Robots must cope with the same environmental hazards as other industrial instruments and controls on the factory floor. Among the most important of possible adverse conditions are: (1) high temperature; (2) dusty, dirty, corrosive, and sometimes potentially explosive and fire prone atmospheres; (3) shock and vibration; and (4) electromagnetic interference, among others. In turn, robots contribute their share to the environmental problems, not the least of which is noisy operation in numerous instances. Further, safe shutdown of robots (fail-safe protection) is required because a meandering, heavily loaded robot can constitute a dangerous threat to both personnel and equipment. All workers who are in the vicinity of a robot must know and respect the full robot's working envelope which in most cases is impractical to screen. As the use of "stock" robots becomes more popular, increasing care and attention to their safe operation must be given.

Packaging. In many applications, it is acceptable to package the robot as a self-contained entity, but there are advantages in mounting the electronics separately. In extreme shock conditions, it may be desirable to mount the control console on a shock-absorbing pad in a remote location as protection against hostile atmospheres. If the power supply of a robot can be separated from the robot's arm, then it is possible to introduce

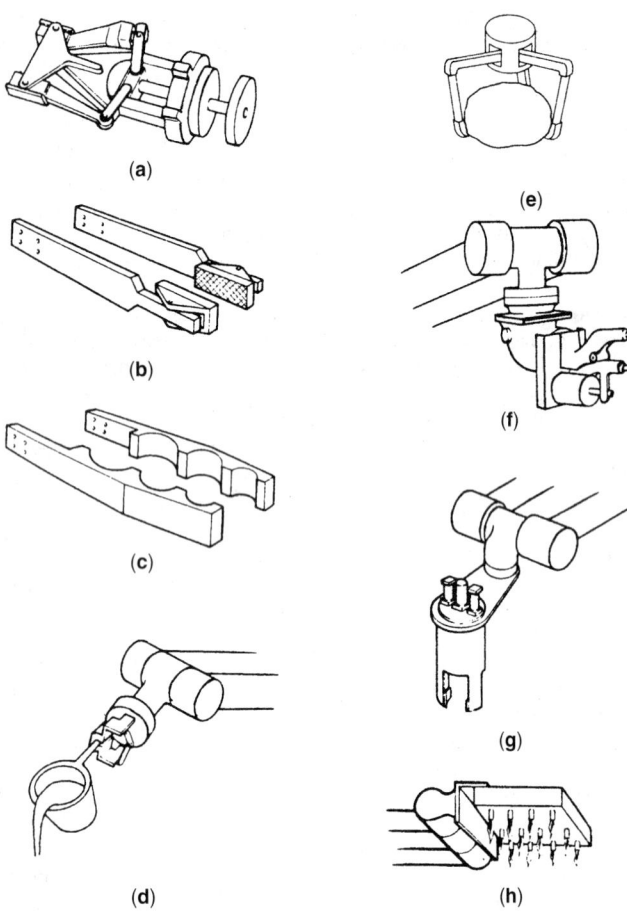

only the arm into an explosive atmosphere, such as a paint room, or some potentially dangerous area as may be found in a nuclear facility. Following the design practices for inherent safety is always worthwhile. Particularly for protection against abrasive dust, the joints of the robot may be booted. Nonflammable fluids for lubrication and hydraulics represent good robot design. Where atmospheric air is particularly dirty, cooling air should be well filtered and enter enclosures to provide positive internal pressure. Robot logic design must be well protected from power line spikes and noise pickup that may enter through any of the robot's communication links with surrounding equipment.

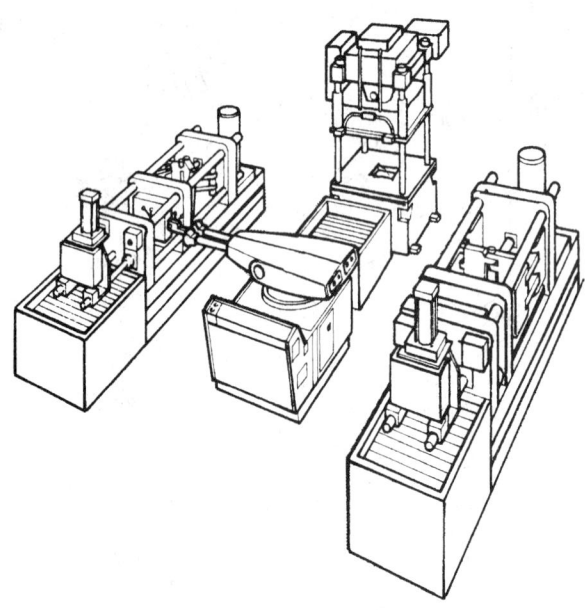

Fig. 15.   Die casting installation to unload, quench, and dispose of parts. In this installation, quite exemplary of earlier robotics, the work is arranged around the robot. (*Westinghouse.*)

Fig. 13.   Representative end-effectors (hands or grippers) used by robots for various applications: (**a**) *Standard hand* — inexpensive, all-purpose design. Will accept a wide variety of custom fingers, tailored to the parts to be manipulated or moved. Parts should be of moderate weight. Simple linkages provide both finger action and force multiplication needed to grip object with just the right force. (**b**) *Self-aligning pads* for fingers are valuable for assuring a secure grip on a flat-sided part. "Cocking" of the part is unlikely when pads like these are used. (**c**) *Multiple fingers* for grasping different-size parts. Thus, a particular finger design need not be restricted to parts within a given size range. (**d**) *Ladle* for moving hot fluid materials, such as molten metal. (**e**) *Three-finger gripper.* (**f**) *Sportwelding gun.* (**g**) *Pneumatic nut-runner.* Similar configurations can be used for holding drills and impact wrenches. (**h**) *Heating torch* as may be used to bake out foundry molds by playing the torch over the surface. Scores of standard and special robot end-effectors are used.

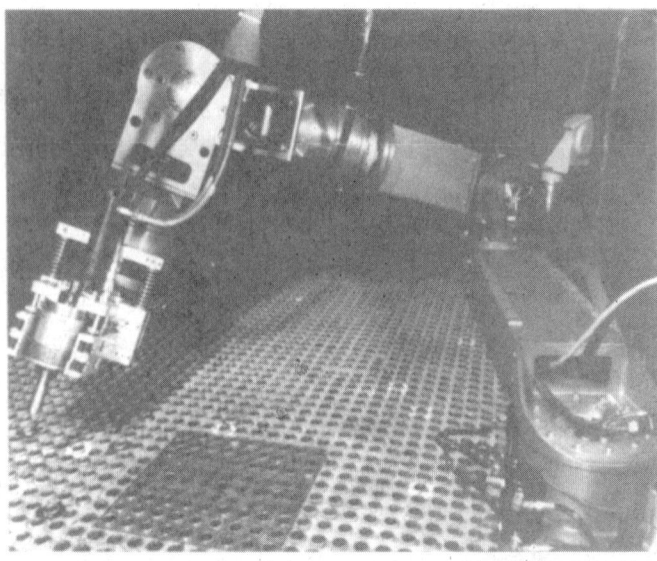

Fig. 14.   Robot used for inspecting tubes in a steam generator. (*Westinghouse.*)

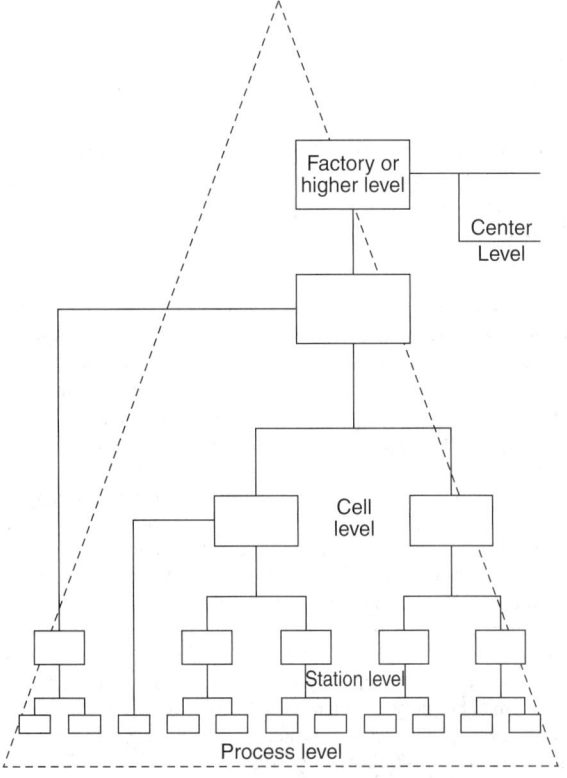

Fig. 16.   Pseudopyrimidal hierarchy where communications are predominantly vertical rather than horizontal.

The Robot Setting

There are four basic situations pertaining to the flow of work and the location of the robot: (1) work may be *arranged around* the robot, (2) work may be *brought* to the robot, (3) work may *travel past* the robot, and (4) the *robot may travel* to the work. Arrangement (3), of course, is a variant of (2).

Work Is Around the Robot. In the early installations of robots, the work was usually arranged around the robot. This arrangement causes the least commitment of space and the least disruption of plant procedures. The system is still used and frequently found in such operations as forging and trimming, press-to-press transfer, plastic molding, and investment casting. See Fig. 15.

Work Is Brought to the Robot. Flexibility has been enhanced through the use of computer control. For example, the robot can be made to track a workpiece that is being carried by a conveyor and thus the robot performing its task as the work passes by. The versatility of such a system can be extended to accommodate variations in conveyor speed. Automatic robot welding systems are of this variety. This is illustrated in the article on **Automation**. The configuration is sometimes called "stationary-base line tracking."

In a contrasting situation, the robot(s) may be mounted on some form of transport system, such as a rail and carriage, which moves parallel to the line at line speed. This type of system requires a powerful drive system so that the robot can be returned to its starting point from the other end

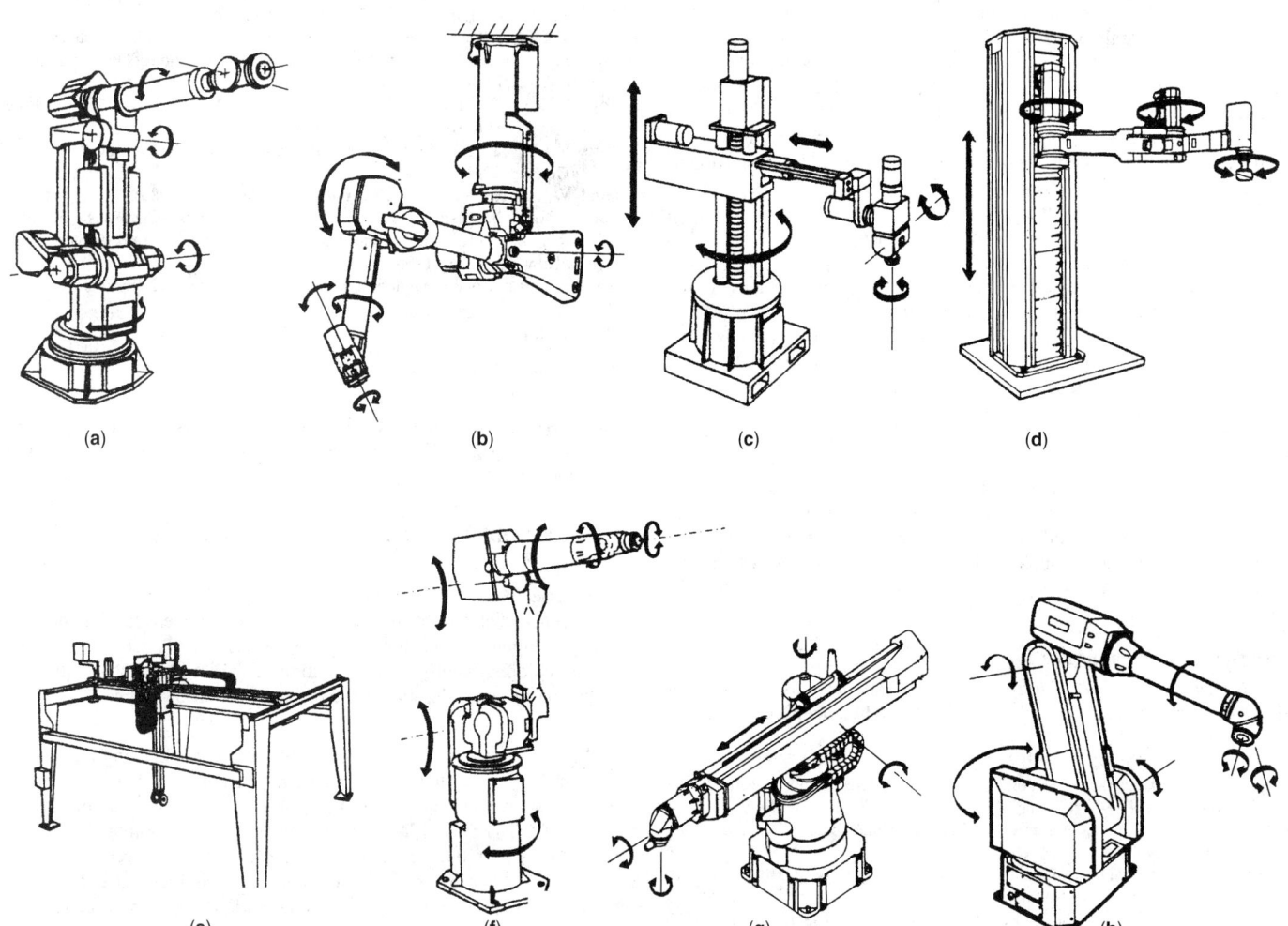

(a)   (b)   (c)   (d)

(e)   (f)   (g)   (h)

Fig. 17.   Representative industrial robots. (*Courtesy of GM Fanuc Robotics Corporation.*)

(**a**) Spot welding, heavy part or tool handling, parts transfer, palletizing, material removal. Payload 120 kg (264 lb). Six axes of motion; floor-or wall-mounted; repeatability ±0.5 mm (0.02 inch); base rotation 300°; vertical travel 2731 mm (107.5 inches); reach 2413 mm (95 inches). (*S-420*)

(**b**) Arc welding of large parts on conveyors and fixtures. Payload 5 kg (11 lb). Six axes of motion; overhead-mounted; repeatability ±0.1 mm (0.004 inch); base rotation 300°; reach 1309 mm (51.5 inch). (*ArcMate OH*)

(**c**) Material handling, machine loading, palletizing, mechanical assembly in severe environments. Payload 50 kg (110 lb). Three to five axes of motion; floor-mounted; repeatability ±0.5 mm (0.02 inch); base rotation 300°; vertical travel 550 or 1300 mm (21.6 or 51.2 inches); horizontal travel 500 to 1100 mm (19.7 to 43.3 inches). (*M-100*)

(**d**) Palletizing and machine loading. Payload 50 kg (110 lb). Four to five axes of motion; floor-mounted; repeatability ±0.5 mm (0.02 inch); vertical travel 1850 mm (72.8 inches); radius reach 1930 mm (76 inches); access to two or more conveyors. (*M-400*)

(**e**) Gantry robot for medium- to heavy-payload machine load and unload uses. Also palletizing, mechanical assembly, parts transfer. Cartesian coordinates. Area (shown) or linear configurations. Payload 50 kg (110 lb). Repeatability ±0.5 mm (0.02 inch); very large work envelope; two to four axes of motion for linear design; three to five for area design. (*G-500*)

(**f**) Multipurpose material handling; light-payload applications. Payload 10 kg (22 lb). Six axes of motion; floor-, ceiling-, or wall-mounted; repeatability 0.2 mm (0.008 inch); base rotation 300°; front reach 1529 mm (60.2 inches). (*S-10*)

(**g**) Laser robot for integration with a laser generator. For precision-path laser processing—welding, cutting, heat treating, and cladding. Payload 5 kg (11 lb). Five axes of motion; floor-mounted; antiback-lash drive; repeatability ±0.05 mm (0.002 inch); base rotation 200°; vertical travel 1968.5 mm (77.4 inches); horizontal travel 3964 mm (156 inches). Complete robotic laser cells available. (*L-100*)

(**h**) Industrial and automotive paint finishing of stationary or moving parts. Payload 7 kg (15.5 lb). Six or seven axes of motion; floor-or rail-mounted; repeatability ±0.5 mm (0.02 inch); maximum reach 2613 mm (103 inches), large work envelope. Robot also used for dispensing and applying antichip sealers and underbody deadeners. (P-155)

of its tracking range in the shortest possible interval. The designer also must guard against interference problems that may arise from adjacent stations.

Robot Travels to the Work. When machining cycles are quite long, a robot can be mounted on a track to enable it to travel among more machines than can be conveniently grouped around a stationary robot.

Work Cells. A robotic work cell may be defined as a cluster of two or more robots and several machine tools or transfer lines which are interconnected in such a way that they work together in unison. All of the necessary accessory equipment is embraced within the work cell. Criteria for justifying a robotic work cell include: (1) the system must be capable of performing required machining functions on a limited number of parts of sizes within predetermined limits; (2) one worker must be capable of operating the work cell with minimal skill needed; (3) selection of operator personnel from the shop floor must be possible; (4) manual parts fabrication in the work cell must be feasible in cases of system failure; (5) compliant end-effectors must be used because of inherent inaccuracies of available commercial robots; (6) part programming must be done on line; (7) safety sensors must be installed for protection of personnel, equipment, and work in process; (8) consistent quality must be maintained on parts; (9) productivity must be increased to make the system economical; and (10) early implementation must be made for quickest payoff.

As reported in a survey of productivity in the United States, it was noted that (1) of the time consumed in manufacturing a part, only 5% is actually used by a given machine tool; and (2) of that time, only 1.5% is used in making chips. Thus, 95% of the time is used for handling, record-keeping, and similar nonproductive activities. These observations have led to three criteria for assuring profitability from work cells: (1) work cells should not be isolated from the remainder of the plant; (2) the product should be made with automation in mind (from the start); and (3) the work cell should be designed for maximum product-making flexibility. Further studies have shown that maximum profitability is obtained when work cells can communicate with the remainder of the plant. A pyramidal hierarchy, where communications are exclusively vertical rather than horizontal, is the most desirable arrangement. See Fig. 16.

Overview of Contemporary Robots

Robots used for a number of different industrial applications are illustrated and described in Fig. 17. See also **Artificial Intelligence: Robotics**.

Additional Reading

Angeles, J.: "Fundamentals of Robotic Mechanical Systems: Theory, Methods, and Algorithms," Springer-Verlag, Inc., New York, NY, 1995.

Argote, L. and D. Epple: "Learning Curves in Manufacturing," *Science*, 920 (February 23, 1990).

Arimoto, S.: "Control Theory of Nonlinear Mechanical Systems," Oxford University Press, Inc., New York, NY, 1996.

Brooks, R.A.: "New Approaches to Robotics," *Science*, 1227 (September 13, 1991).

Cohen, S.S. and J. Zysman: "Manufacturing Innovation and American Industrial Competitiveness," *Science*, 1110 (March 4, 1988).

Considine, D.M., Editor: "Robots" in Process/Industrial Instruments & Controls Handbook, 4th Edition, The McGraw-Hill Companies, Inc., New York, NY, 1993.

Demiris, J. and A. Birk: "Robot Learning: An Interdisciplinary Approach," World Scientific Publishing Company, Inc., River Edge, NJ, 2000.

Dewdney, A.K.: "Insectoids Invade a Field of Robots," *Sci. Amer.*, 118 (July 1991).

Frosh, R.A. and N.E. Gallopoulos: "Strategies for Manufacturing," *Sci. Amer.*, 144 (September 1989).

Ghosh, B.K., T.J. Tarn, and N. Xi: "Control in Robotics and Automation: Sensor-Based Integration," Academic Press, Inc., San Diego, CA, 1999.

Gorinevsky, D.M., A. Formalsky, and A. Schneider: "Force Control of Robotic Systems," CRC Press, LLC., Boca Raton, FL, 1997.

Gupta, K.C.: "Mechanics and Control of Robots," Springer-Verlag, Inc., New York, NY, 1997.

Hodges, B.: "Industrial Robotics," 2nd Edition, Butterworth-Heinemann, Inc., Woburn, MA, 1992.

Holden, C.: "Juggling Robot," *Science*, 742 (February 15, 1991).

Jacak, W.: "Intelligent Robotic Systems: Design, Planning, and Control," Perseus Publishing, Boulder, CO, 2000.

Mansfield, E.: "Industrial Innovation in Japan and the United States," *Science*, 1769 (September 30, 1988).

Mason, M.T.: "Mechanics of Robotic Manipulation," MIT Press, Cambridge, MA, 2001.

Michie, D.: "Application of Machine Learning to Recognition and Control," *University of Wales Review*, 23 (Spring 1989).

Nof, S.Y.: "Handbook of Industrial Robotics," 2nd Edition, John Wiley & Sons, Inc., New York, NY, 1999.

Readman, M.C.C.: "Flexible Joint Robots," CRC Press, LLC., Boca Raton, FL, 1994.

Ryan, D.L.: "Robotic Simulation," CRC Press, LLC., Boca Raton, FL, 1994.

Sandler, B.Z.: "Robotics: Designing the Mechanisms for Automated Machinery," 2nd Edition, Academic Press, Inc., San Diego, CA, 1999.

Sheridan, T.B.: "Merging Mind and Machine," *Technology Review (MIT)*, 32 (October 1989).

Shirai, Y. and S. Hirose: "Robotics Research: The Eighth International Symposium," International Foundation of Robotics Research, Springer-Verlag, Inc., New York, NY, 1998.

Slutski, L.I.: "Remote Manipulation Systems: Quality Evaluation and Improvement," Kluwer Academic Publishers, Norwell, MA, 1997.

Smit, M.C. and M.W. Tilden: "Beam Robotics," *Algorithm*, 15 (March 1991).

Thro, E.: "Robotics: The Marriage of Computers and Machines," Facts on File, Inc., New York, NY, 1993.

Towhill, D.R.: "The Dynamic Analysis Approach to Manufacturing System Design," *Univ. of Wales Review*, 3 (Spring 1988).

Tyre, M.J.: "Managing Innovation on the Shop Floor," *Technology Review (MIT)*, 58 (October 1991).

Uttal, W.R.: "Teleoperators," *Sci. Amer.*, 124 (December 1989).

Van De Velde, W.: "Toward Learning Robots," MIT Press, Cambridge, MA, 1996.

Wolovich, W.A.: "Robotics: Basic Analysis and Design," Oxford University Press, Inc., New York, NY, 1995.

Yoerger, D.R.: "Robotic Undersea Technology," *Oceanus*, 32 (Spring 1991).

Web Reference

The Robotics Institute: *http://www.ri.cmu.edu/*

ROBUSTNESS.   In statistical tests, robustness is the property of being insensitive to small departures from the assumptions from which the test was derived.

ROCHE LOBE.   See **Binary Stars**; and **Neutron Stars**.

ROCHES MOUTONNÉES.   The term given by the Saussure, in 1796, to glaciated rock hills resembling the wigs which were fashionable during the late eighteenth century. Typical roches moutonnées have rounded surfaces sloping gently in the direction of the ice movement with steeper slopes on the lee side, due to the plucking action of the ice.

ROCK.   In geologic usage, rock is the material composing the outer part or "crust" of the earth. It is a popular concept that the term rock is restricted to hard or firm substances. To a geologist, however, a body of clay or of volcanic ash is as truly rock as is a mass of hard granite. In general, a rock consists of one or more definite minerals. However, natural glass (obsidian), which has no definite composition and structure, makes large rock masses, as in Yellowstone National Park. Rocks composed of definite minerals may be either simple or compound; for example, the purest marble is formed entirely of the one mineral calcite, whereas granite consists of feldspar and quartz mixed in varying proportions, usually with minor quantities of mica and other accessory minerals.

All rocks are divided into three major classes, according to mode of origin. Igneous rocks are those that have been formed by cooling and solidification of molten masses derived from within the earth. Because of their origin they are sometimes called the primary rock. In part they are lavas and other products of volcanoes; other masses solidify slowly below the earth's surface to form granite and other crystalline types of igneous rock whose mineral content depends chiefly on the chemical composition of the parent molten mass. Sedimentary rocks consist of material that formed a part of pre-existent rocks, and that was moved from its former position, deposited by the action of water, the atmosphere, or glacier ice, and subsequently converted into rock. Examples are conglomerate, sandstone, and limestone. Metamorphic rocks are those that were originally igneous or sedimentary, but have been changed either in mineral composition or in texture, or in both, so that their original characters have been radically altered. Slate, formed by alteration of shale, and marble, derived from pre-existent limestone, are familiar examples.

See also **Igneous Rock**; and **Metamorphic Rocks**.

ROCKET ENGINE. A reaction engine that contains within itself, or carries along with itself, all the substances necessary for its operation or for the consumption or combustion of its fuel, not requiring intake of any outside substance and hence capable of operation in outer space. Also called *rocket motor*. Chemical rocket engines contain or carry along their own fuel and oxidizer, usually in either liquid or solid form, and range from simple motors consisting only of a combustion chamber and exhaust nozzle to engines of some complexity incorporating, in addition, fuel and oxygen lines, pumps, cooling systems, etc., and sometimes having two or more combustion chambers. Experimental rocket motors have used neutral gas, ionized gas, and plasmas as propellants.

ROCKET PROPELLANTS. Chemical rocket propellant can be classified in several ways, including liquid or solid, monopropellant, bipropellant or tripropellant, cryogenic or storable, hypergolic or nonhypergolic, double-base, composite or composite double-base. Propellant classification frequently influences the classification of rocket engines—for example, monopropellant rocket, liquid rocket, solid rocket, and hybrid rocket (usually using a liquid oxidizer and a solid fuel). Propellants for chemical rockets serve two primary functions as contrasted to one function for nuclear, solar, electrical or laser heated rockets. In chemical propellant rockets, the propellant is both the energy source and the ejected mass or "working fluid."

Compared with the almost limitless number of chemical compounds that exist or can be formed, the number of chemical propellants in common use is relatively few. This situation arises from criteria including costs, source availability, toxicity, resistance to shock, and other requirements imposed by the vehicle application and the propulsion system design. Another practical reason is that extensive overlap of physical, chemical, and economic properties are displayed by many of the theoretically possible propellants. During the 1960s, the universities, industry, and government in the United States pursued extensive research programs for synthesizing new chemical compounds viewed as candidate propellants, which would increase the performance capability of chemical rockets. Although dozens of compounds were synthesized, few results reached the production line. This does not mean, however, that a scientific breakthrough in increasing the molecular energy of a propellant may not ultimately be obtainable.

The characteristics desired of a rocket propellant are several in number and can be divided into economic, safety, materials compatibility, engine-cycle needs and vehicle requirements.

In general terms, the engine-cycle needs ideally are: (1) a propellant or propellant combination that has a high heat of reaction per unit weight (also called heat of combustion). Most vehicles add a requirement for high heat of reaction per unit volume of propellant to minimize the vehicle size. (2) reaction products that are all gaseous, that have a very low molecular weight and that have a very high temperature of dissociation.

In addition to specific impulse, the vehicle requirements usually influence propellant selection in terms of storability, density, toxicity, and other hazards, and other application-sensitive factors, including exhaust plume properties and radar cross section and radiation emissions. Other factors being essentially equal, the higher the heat of reaction of a propellant (or combination), the more attractive the propellant. Sharp exceptions to this rule occur in some missiles because of volume limitations, the need for smokeless exhaust or similar restraints.

The heat released by a propellant is the difference in heat between the constituents and the end-products of combustion.

$$\Delta H_r^\circ = \left[\sum_{k,\ products} \eta_k (\Delta H_f^\circ)_k - \sum_{j,\ reactants} \eta_j (\Delta H_f^\circ)_j \right]$$

where ΔH_r° is the heat generated; ΔH_f° is the standard heat of formation of the constituent at reference temperature (298K); and η is the number of moles of each j reactant or k product. Large heat release is afforded by reaction products having large negative values, while the reactants should have positive, or at least small negative values, if possible. The heat of reaction is often noted in energy/weight units, such as kilocalories/gram.

Specific impulse, I_{sp}, the universally accepted measure of rocket engine performance, can also be used to indicate the performance of propellants. The most commonly stated expansion ration is $1,000 \rightarrow 14.7$ giving "sea-level specific impulse at 1,000 psi chamber pressure." Sometimes the expansion ratio is $1,000 \rightarrow 0.2$ to indicate specific impulse for high-altitude or space flight.

By definition, specific impulse I_{sp} is:

$$I_{sp} = \frac{F}{\dot{W}}$$

with the I_{sp} units being seconds; the short designation for units of thrust (force) per units of propellant mass flow per second.

For an ideal rocket with the nozzle exhaust pressure being the ambient pressure, the thrust, recognizing Newton's second law of motion, is:

$$F = \frac{\dot{W} c_e}{g}$$

where $\dot{W}$ is propellant flow rate in pounds per second; c_e is exhaust velocity in feet per second; and g is the gravitational constant in feet/second/second.

In practice, only about 10% of the elements on the periodic chart are adaptable to chemical rocket propellants. Propellants have made little use of elements other than hydrogen, carbon, nitrogen, oxygen, chlorine, fluorine, aluminum, boron, and beryllium.

Liquid propellants fall into two broad classes: (1) earth storable (monopropellants and bipropellants), and (2) cryogenic, depending upon whether they can be kept in the vehicle tankage for months and years, or must be used in a few hours or days. The theoretical performances of storable and cryogenic bipropellants combusting ideally at 1,000 psia chamber pressure and expanding to sea-level pressure without loss, assuming shifting chemical equilibrium of the combustion products during expansion in the engine exhaust nozzle, are listed in Tables 1 and 2. For comparison purposes, a few properties of the more common monopropellants are listed in Table 3. Water (not listed) as a source of hydrogen and oxygen via electrolysis has merit as a propellant in long-life satellites (5 years plus), equipped with solar electric cells.

Solid propellants fall into three general types: (1) double-base, (2) composite; and (3) composite double-base. Double-base propellants form a homogeneous cured propellant, usually a nitrocellulose-type of gun-powder dissolved in nitroglycerin plus minor percentages of additives. Both the major ingredients are explosives and both contribute to the functions of fuel, oxidizer, and binder. Composite propellants form a heterogeneous propellant grain with the oxidizer crystals and a powdered fuel (usually aluminum) held together in a matrix of synthetic rubber (or plastic) binder such as polybutadiene. Normally, composite propellants are less hazardous to manufacture and handle than double-base propellants. Composite double-base propellants are a combination of the two aforementioned types—usually a crystalline oxidizer (ammonium perchlorate) and powdered aluminum fuel held together in a matrix of nitrocellulose-nitroglycerin. The hazards of processing and handling this type of propellant are similar to those experienced with the double-base propellants. The characteristics of several common solid propellants are given in Table 4.

Ingredients are generally classified according to their function, e.g., fuel, oxidizer, binder, curing agent, burn-rate catalyst, etc. Ingredients used in small amounts are called additives and usually have functions other than the fuel, oxidizer, or binder. For example, an additive can reduce the viscosity of the propellant during mixing and casting (pouring) of the propellant, increase the burning rate of the propellant, or improve the storage stability. Often an ingredient serves or affects more than one function, the most diffused situation relating to composite double-base ingredients where the binder is a nitrocellulose-nitroglycerine complex with each of these two ingredients having its own fuel and oxidizer chemical elements. The binder contributes also as a fuel and in some propellant formulations, such as asphalt-base nonmetallized propellants, the binder is the fuel.

Ammonium perchlorate, NH_4ClO_4, is the most widely used crystalline oxidizer in solid propellants. Because of its characteristics, including compatibility with other propellant materials, specific impulse performance, quality uniformity and availability, it dominates the solid oxidizer field. Both ammonium and potassium perchlorate are only slightly soluble in water, a favorable trait for propellant use. Nitronium perchlorate is objectionably hygroscopic, is relatively incompatible with available binders, and detonates easily. All of the perchlorate oxidizers produce hydrogen chloride in their reaction with fuels. Their exhaust bases are toxic and corrosive to the extent that care is required in firing rockets, particularly the very large rockets, to safeguard operating personnel and communities in the path of exhaust clouds. Ammonium perchlorate is available in the form of small white crystals and close control of the size range and percentage of several

TABLE 1. STORABLE LIQUID BIPROPELLANT COMBINATIONS

Oxidizer	Fuel	Oxygen-Fuel Ratio by Weight for Maximum I_{sp}	Bulk Specific Gravity	Theoretical $I_{sp}{}^a$
Nitrogen tetroxide	50/50 Hydrazine/ Unsymmetrical dimethylhydrazine	2.0	1.19	289
Nitrogen tetroxide	Unsymmetrical dimethylhydrazine	2.6	1.17	286
Nitrogen tetroxide	Hydrazine	1.3	1.21	292
Red fuming nitric acid (15% NO_2)	Unsymmetrical dimethylhydrazine	3.4	1.28	266
Red fuming nitric acid (15% NO_2)	Kerosene-type fuel	5.6	1.37	257
Maximum density red fuming nitric acid (mixture of red fuming nitric acid and N_2O_4)	Unsymmetrical dimethyl-hydrazine	2.9	1.29	278
N_2O_4 with 15% NO	Unsymmetrical dimethylhydrazine	2.6	1.15	288
Hydrogen peroxide	Hydrazine	2.0	1.24	287
Hydrogen peroxide	Unsymmetrical dimethylhydrazine	4.2	1.22	284
Chlorine trifluoride	Hydrazine	2.8	1.64	295
Chlorine trifluoride	Unsymmetrical dimethyl-hydrazine	3.0	1.39	280
Chlorine pentafluoride	Hydrazine	2.7	1.47	313
Chlorine pentafluoride	Unsymmetrical dimethyl-hydrazine	2.9	1.34	297
Hydrazine	Pentaborane	1.3	0.80	328

^{a}1,000 → 14.7 psia, shifting equilibrium (chemical composition of exhaust gases changes during nozzle flow).

TABLE 2. CRYOGENIC LIQUID BIPROPELLANT COMBINATIONS
(At least one propellant is cryogenic)

Oxidizer	Fuel	Pounds Oxidizer/ Pound Fuel Stotchio Metric	Maximum I_{sp}	Bulk Specific Gravity Maximum I_{sp}	Theoretical $I_{sp}{}^a$
Liquid oxygen	Kerosene-type fuel	3.41	2.6	1.02	300
Liquid oxygen	Hydrazine	3.0	0.9	1.07	313
Liquid oxygen	Unsymmetrical dimethylhydrazine	—	1.7	0.98	310
Liquid oxygen	Ammonia	2.37	1.4	0.89	294
Liquid oxygen	Ethyl alcohol	2.09	1.8	0.99	290
Liquid oxygen	Methane	—	3.3	0.82	311
Liquid oxygen	Liquid hydrogen	7.95	4.2	0.29	290
Liquid fluorine	Liquid hydrogen	19.0	8.0	0.46	412
Liquid fluorine	Hydrazine	2.71	2.3	1.31	365
Liquid fluorine	Kerosene-type fuel	4.07	2.6	1.21	322

^{a}1,000 → 14.7 psia, shifting equilibrium (chemical composition of exhaust changes during nozzle flow).

TABLE 3. LIQUID MONOPROPELLANTS

Propellant	Specific Gravity at 68 °F	Theoretical $I_{sp}{}^a$	Exhaust (Average Molecular Weight)
Hydrogen peroxide	1.39	165	22.68
Hydrazine	1.01	199	12.77
Nitromethane	1.12	245	20.34
Ethylene oxide	0.89	199	20.50

^{a}1,000 → 14.7 psia, shifting equilibrium (chemical composition of exhaust gases changes during nozzle flow).

sizes present in a given quantity or batch is required, since particle size influences propellant processing and the physical and ballistic properties of the finished propellant.

Inorganic nitrates are relatively low-performance oxidizers as compared with the perchlorates. However, ammonium nitrate is used in some applications for economy and because of its smokeless and relatively nontoxic exhaust. Its main use is in low-burning rate, low-performance applications, such as gas generators for turbine pumps.

One or two crystalline high explosives, such as HMX (cyclotetramethylene tetranitramine) and RDX (cyclotrimethylene trinitramine), are sometimes included in a propellant formulation to achieve a specific performance characteristic. Depending upon the objectives, the percent can range from 5 to 50%.

The one prominent solid fuel is powdered aluminum, and it is used in a wide variety of composite and composite double-base propellant formulations, usually being between 14 and 22% of the propellant by weight.

Boron, even though it appears as one of the high-energy fuels and is lighter than aluminum, has not proven to be a practical fuel because it is so difficult to burn with high efficiency in combustion chambers of reasonable length. Beryllium burns much more easily than boron and improves the specific impulse of a solid propellant motor, usually by about 15 seconds, but as a powder or dust it is highly toxic to animals and humans. The technology with composite

TABLE 4. CHARACTERISTICS REPRESENTATIVE SOLID PROPELLANTS

Propellant Type	I_{sp} (Seconds)	Flame Temperature (°F)	Flame Temperature (°C)	Density Pounds/Cubic Inch	Density Grams/Cubic Centimeter	Metal Content (Weight %)	Burning Rate Inches/Second	Burning Rate Centimeters/Second	Hazards Class (Military)	Stress Strain (psi) −60°F (−51°C)	Stress Strain (%) 150°F (66°F)	Processing Method
DB	255	5,340	2449	0.057	1.58	0	0.45	1.1	7	4,600/2	490/60	Extruded
DB/AP/Al[a]	258	6,990	3866	0.069	1.91	25	0.78	2.0	2	2,750/5	120/50	Extruded
DB/AP-HMX/Al[a]	272	6,630	3666	0.067	1.85	20	0.55	1.4	7	2,375/3	50/33	Solvent Cast
PVC/AP[b]	239	4,810	2654	0.065	1.80	0	0.45	1.1	2	369/150	38/220	Solvent Cast
PVC/AP/Al[c]	253	6,120	3382	0.069	1.91	20	0.45	1.1	2	259/150	38/220	Solvent Cast
PBAN/AP/Al[c]	265	5,600	3093	0.063	1.74	19	0.55	1.4	2	520/16 (at −10°F)	71/28	Cast
PU/AP/Al[c]	263	6,000	3316	0.065	1.80	23	0.27	0.7	2	1,170/6	75/33	Cast
CTPB/AP/Al[c]	265	5,540	3060	0.063	1.74	19	0.45	1.1	2	325/26	88/75	Cast
HTPB/AP/AL[c]	264	5,540	3060	0.063	1.74	19	0.40	1.0	2	910/50	90/33	Cast
PBAA/AP/AL[c]	265	5,660	3127	0.063	1.74	20	0.32	0.8	2	500/13	41/31	Cast

Al	Aluminum	HTPB	Hydroxy-terminated polybutadiene
AP	Ammonium perchlorate	PBAA	Polybutadiene-acrylic acid polymer
CTPB	Carboxy-terminated polybutadiene	PBAN	Polybutadiene-acrylic acid-acrylonitrile terpolymer
DB	Double base	PU	Polyurethane
HMX	Cyclotetramethylene tetranitramine	PVC	Polyvinyl chloride

Notes: [a] AP/Al optimized with 40% DB as binder.
[b] AP/Al optimized with 20% binder.
[c] AP/Al optimized with 15% binder.

propellants using powdered beryllium fuel is sufficiently advanced for vehicle application, with space travel being the most likely application.

Theoretically, both aluminum hydride, AlH_3, and beryllium hydride, BeH_2, are attractive fuels because of their high heat release and gas volume contribution. Both are difficult to manufacture and both deteriorate chemically during storage due to loss of hydrogen. Because of these difficulties, coupled with relatively modest I_{sp} gains, these compounds remain experimental.

Hybrid rocket propellants are various combinations of solid and liquid propellants, usually a solid fuel and a liquid oxidizer. Sometimes, a third propellant, liquid hydrogen, is added, not for energy release, but as a low-molecular-weight working fluid. The main advantages of a hybrid rocket are: (1) use of liquid and solid propellant combinations offering the highest performance attainable with chemical rockets; (2) simplicity of a solid grain (usually fuel); (3) a liquid for nozzle cooling and thrust modulation (compared with a solid rocket); (4) restart capabilities; and (5) good storability and safe storage characteristics.

The chemical bond energy present in propellant molecules is the energy source used by chemical rocket engines to date. This source affords energy densities of approximately 3 kilocalories/gram in the liquid hydrogen/liquid oxygen combination, and up to about 5.7 kilocalories/gram with the lithium/fluorine combination. Theoretically, supplemental energy can be added to molecules or molecular fragments that, upon recombination or relaxation to their normal energy state, release significant amounts of energy. For example, 52 kilocalories/gram is theoretically released when two hydrogen atoms (free radicals) recombine to form hydrogen. Even higher energy densities, as much as 100 kilocalories/gram, are theoretically available from lightweight molecules, such as helium, that are in an excited state.

Metastable, in the sense of propellant ingredients, means that the "energized" molecule, atom, or molecular fragment, tends to promptly return to its normal state. Some molecular species distinctly assume a metastable state upon excitation with the lifetime at room temperature being 10^{-3} to 10^{-2} second as compared with less than 10^{-6} for nonmetastable excited species. Atoms subjected to excitation move into a more energetic state of translational motion of vibration or into a high-energy electron orbital state; diatomic molecules do likewise. Molecules containing more than two atoms can experience higher translational rotational motion, as well as higher electron orbital state.

Most of the research to date on metastable propellants has been with gaseous atoms and molecules. Obviously, energized ingredients in a condensed phase, solid or liquid, would be needed for most rockets. The primary objectives to be reached, if metastable ingredients are to

benefit rocket propulsion, are: (1) an efficient process for energizing the ingredients; and (2) a means of storing the ingredients for days at a time without appreciable energy loss. Actual use of metastable ingredients in a rocket is envisioned in the company of liquid hydrogen, or other low-molecular-weight working fluid.

Limited research has been conducted on two approaches to generating and storing (stabilizing) metastable propellant ingredients: (1) free radicals, specifically, atomic hydrogen; and (2) helides, which are excited states of helium. In the late-1950s, the U.S. National Bureau of Standards produced low concentrations of free radicals and stored them in inert matrices at very low temperatures. More recently, an approach has been taken to generate hydrogen atoms, immediately condensed at liquid-helium temperature, in the presence of a high density (70 to 100 kilogauss) magnetic field for the purpose of stabilizing the hydrogen atoms. Theoretically, the high-strength magnetic field is capable of aligning the spin of the electron of the hydrogen atom so as to prevent recombination into the hydrogen molecule.

Triplet helium has a theoretical energy level of 114 kilocalories/gram above the ground state. Assuming release of this energy and subsequent expansion through a rocket nozzle gives a specific impulse of 2,800 seconds. Techniques for generating activated helium and other noble gases are well known, but concentrating and storing these metastable species is quite another matter inasmuch as they revert to their ground state by collision processes. Experimental approaches to activating helium and trapping the helium molecules in a hydrocarbon wax have been reported.

The creation and use of metallic hydrogen (hydrogen derived from normal hydrogen subjected to about 2 megabars pressure) should release about 52 kilocalories/gram upon transitioning from the metallic to the normal solid form. The concept dates back to 1935, but interest has been renewed because some scientists believe that metallic hydrogen exists in some large planets.

Antimatter Rockets. Sufficient atomic particle research has been accomplished to warrant discussion of possible methods of applying energy available from particle mass annihilation to rocket propulsion. Complete conversion of matter to energy would allow exhaust velocities near that of light to be obtained from a propulsion device. Antimatter, by definition is matter made up of antiparticles, such as antineutrons, negatrons (antiprotons), and positrons (antielectrons). An annihilation property is known to exist between particles with one particle termed the antiparticle of the other.

Rocket design concepts envisioned for utilizing the reaction between atomic particles and antiparticles (matter and antimatter) are based upon the following postulations: (1) annihilation products can be accelerated using electrical and magnetic forces (consider the annihilation reaction of a neutrino with an antineutrino, yielding a proton and an electron); (2) annihilation products can be used indirectly to heat a working fluid for thermal expansion through a nozzle (consider the annihilation reaction of hydrogen and antihydrogen, leaving high-energy gamma rays); (3) antimatter possesses negative gravitational mass although its inertial mass may be positive. This could give rise to antigravity propulsion; and (4) annihilation products of ordinary quanta give rise to the possibility of a photon-expelling beam for the direct generation of thrust.

Before any form of antimatter rocket can exist, a lightweight method must be developed for producing antiparticles at a flow rate of grams/second in contrast with the few dozen of antiparticles produced in research laboratory generators. Also, a practical storage or containment method must arise inasmuch as antiparticles explode violently upon contact with normal matter. Reference 5 gives a performance estimate of an I_p of 3.06×10^7 seconds for a rocket propelled vehicle with a thrust/weight ratio of 10^{-7}.

Multiple Uses of Propellants. Propellants, both solid and liquid, are used in many secondary propulsion applications, including crew capsule ejection, attitude control and station-keeping of satellites, braking of reentry vehicles, extravehicular space operations—as well as being essential in rocket engine igniters, signal and illumination flares, and fuel-cell type electric generators. New developments, such as the high-powered gas dynamic laser[6] continue to broaden the field of applications.

Acknowledgment. Information for this entry as furnished by Mr. Donald M. Ross, Consulting Engineer, Lancaster, California and for confirmation of tabular performance data by Mr. Curtis C. Selph, Propellant Research Engineer, U.S. Air Force Rocket Propulsion Laboratory, Edwards, California, are gratefully acknowledged.

Additional Reading

Baryakhtar, V.H. and T. Rosendorfer: "Demilitarisation of Munitions: Reuse and Recycling Concepts for Conventional Munitions and Rocket Propellants," Kluwer Academic Publishers, Norwell, MA, 1997.

Cohen, W.: "New Horizons in Chemical Propulsion," *Astronautics and Aeronautics,* 2, **12**, 46–51 (1973).

Davenas, A.: "Solid Rocket Propulsion Technology," Elsevier Science, New York, NY, 1992.

Kent, J.A.: "Riegel's Handbook of Industrial Chemistry," 9th Edition, Chapman & Hall, New York, NY, 1992.

Quinn, L.P., et al.: "High Energy Storage Investigations," Rept. AFRPL TR-71-36, U.S. Air Force Rocket Propulsion Laboratory, Edwards, California, 1971.

Ross, D.M.: "Propellants," in "Energy Technology Handbook," (D.M. Considine, editor), McGraw-Hill, New York, NY, 1977.

ROCKETRY. Rocketry, based on mathematics, physics, and formulae, is an exacting science. One might trace humanity's interest in rocketry back to the fourth century when Aulus Gellius described a mechanical pigeon made to move by jets of steam escaping from its tail. Others might say the Chinese deserve credit for discovering rocketry. As early as 1232, the Chinese used rockets in their war against the Mongols who were besieging the city of Kai-fung-fu. Called *arrows of fire*, these primitive rockets were propelled by *black powder.*

The Tartars, who invaded Europe in the 13th century, also knew how to build primitive rockets. Moreover, they knew how to frighten their enemies. In 1241, the Tartars defeated a group of Polish knights by carrying a dragon head that spewed smoke and fire. More concerned with practical improvements than with psychological advantage, the English monk, Roger Bacon, developed an improved form of gunpowder during the first half of the 13th century. This improvement enabled rockets to be transformed into incendiary projectiles with a relatively long range.

Far beyond the borders of England, the Mongols used their "fire arrows" while capturing the city of Baghdad in 1258. More than 20 years later, the Arab, Hassan er-Rammah, completed a manuscript which included directions for making a torpedo that could be used against a land target. The manuscript resulted in nothing tangible. By the Seventh Crusade, however, the Arabs were ready to use rockets against their French invaders.

Throughout the 14th and 15th centuries, medieval rocket pioneers focused on using rockets as weapons. In 1420, Joanes de Fontana authored a sketch book which contained suggestions for making military rockets. Jean Froissart (1338–1410) proposed the use of tube-fired rockets.

Manuscripts ranged from those proposing rockets with crude parachutes to rockets that would dive and explode underwater. In 1577, the German armorer, Leonhart Fronsperger, wrote a book on firearms that historians believe resulted in the modern word "rocket." Fronsperger described a device called a "roget," made from powder with a base of saltpeter, sulfur and charcoal tightly wrapped in paper. Less than 100 years later, Colonel Friedrich von Geissler, a Polish field artillery commander, fired a 120-pound rocket near Berlin, using a mixture similar to that described by Fronsperger.

In 1687, Sir Isaac Newton published his *Principia Mathematica.* Newton's Third Law of Motion stated that for every action there is an equal and opposite reaction. Therefore, the rocket is a reaction device. Minute particles are thrown out through a nozzle as a propellant is burned in a suitable chamber. The reaction to the discharge makes the rocket fly in the opposite direction. Newton's discoveries would provide the classical physical basis for future generations of rocket theorists. Kazimierz Siemienowicz, a Polish general, also contributed to 17th century intellectual achievements in rocketry. In 1650, he published the theory of multistage rocketry in a book called *Artis Magnae Artileria.* The Age of Reason also provided a glimpse of the day when man might use rockets for civil purposes. In 1784 a Prussian master weaver, Ehrgott Friedrich Schaefer, drew up plans for a lifesaving rocket that sailors might use to throw a small anchor across the surf. About 20 years later, Claude Ruggieri, an Italian with an apparent interest in rocketry and fireworks as public entertainment, rocketed small animals into space. His payloads were recovered by parachute. Years later, Ruggieri's rockets were able to lift a full-grown ram.

Most major developments in rocketry in the 18th and 19th centuries, however, focused on military rather than civil applications One of the first major engagements with rockets involving Europeans took place in India in the year 1792 when the British cavalry was repulsed with 6.5-pound rockets mounted on bamboo sticks. The lessons on rocketry in India were not lost on the British. In 1807, Sir William Congreve (1172–1828), a British artillerist, developed the 3.5-inch diameter incendiary rocket, using black

Fig. 1.   The V-2 became one of the best known of all early missiles. The 46-foot rocket utilized alcohol and liquid oxygen as fuel and could carry a 1,650 pound warhead 225 miles. (*MSFC.*)

powder, an iron case, and a 16-foot guide stick. In 1805 Congreve had been allowed to bombard Boulogne with rockets fired from boats. The fires, which started in the town, burned a day and a night. In 1807, Congreve directed a rocket attack against Copenhagen in which approximately 25,000 rockets were fired, leaving much of the city burned. In 1813 at the Battle of Leipzig, the English bombarded a village held by five battalions of French infantry.

Czar Alexander I of Russia was among those who appreciated the role that Congreve's rockets had played at Leipzig. In 1815 the Russian, Alexander Zasyadko based his blueprint for a fighting rocket on a British model. Following Zasyadko's suggestion, the Russians ordered the formation of the nation's first rocket company of soldiers in 1826. The unit was attached to the St. Petersburg Rocket Works. Two years later, under Zasyadko's guidance, the Russians employed warhead rockets with occasional success against the Turks.

In 1814, the United States militia opposed the advance of British regulars on Washington. At the Battle of Bladensburg, less than 10 miles northeast of Washington, the British put their rockets into action. A special rocket squad, partially concealed in underbrush, fired the projectiles causing the Americans to retreat. Discourse and experimentation with rocketry continued in Britain. The British had depended on rockets in battle. However, they were not always ready to accept the possibility that they might be used for other purposes. Critics caricatured a rocket conceived by the Englishman Charles Golighty in 1841. A cartoon described his rocket as a "steam-horse on which one may ride from Paris to St. Petersburg."

William Hale's invention of the stickless rocket in 1846 was taken more seriously. Congreve's rocket had carried a 16-foot stick. Although the stick helped guide the rocket and kept it from moving erratically, it also added extra weight. Hale's rockets eliminated the stick. Instead, he inserted three metal vanes in the exhaust nozzle, slightly inclined so the exhaust itself caused the rotation.

Rockets were used to a limited extent in the American Civil War. On July 3, 1862, the Confederates, under Jeb Stuart, fired rockets at McClellan's troops. The Confederates also placed rocket batteries in service in Texas in 1863-64. Union troops formed the New York Rocket Battalion, and in 1864 Union rockets were fired in South Carolina at picket boats and other vessels in the harbor.

It is not possible to say exactly when man began to seriously consider the possibility that rockets might be used for space travel as well as for military purposes. Advances in technology were needed before rockets could use liquid propellants instead of solid propellants. The liquid propellants would increase man's capability to guide and control the rockets. One of the foremost rocket pioneers was Konstantin Tsiolkovskii, a Russian school teacher with a hearing affliction. Although he never built a rocket, his grasp of rocketry and space fundamentals has been described as extraordinary. In 1903, he published "Exploration of Cosmic Space by Rocket Devices," providing data on liquid propellants for rocket propulsion. The work, which was written in 1898, also considered the possibility of space travel.

By the time of Tsiolkovskii's work, other writers and intellectuals had begun to imagine the possibility of rockets that might allow men and women to leave the confines of Earth. In 1897, the German Kurd Lasswitz focused his imagination on the planet Mars. His writings were based on the assumption that if a life-form of higher intelligence existed on Mars, then the first space trip would be from Mars to the Earth. More well known than Lasswitz, the French author Jules Verne anticipated flights into outer space as well as a host of other 20th century scientific achievements. Verne's imagination would inspire an entire generation of rocket scientists including Dr. Wernher von Braun.

In 1906, Alfred Maul, a German engineer, employed rockets to take a camera to high altitudes. His 1912-model rocket carried an 8 inch by

Fig. 2.   The Redstone was a high-accuracy, liquid-propelled, surface-to-surface missile. The Von Braunteam developed and launched the first Redstone missile in August 1953. (*MSFC.*)

Fig. 3.   Redstone, known as Jupiter-C. The family of launch vehicles, developed by Von Braun and his team, also came to include the Juno II used to launch the Pioneer IV satellite on March 3, 1959. Pioneer IV passed within 37,000 miles of the moon before going into solar orbit. (*MSFC.*)

10 inch photographic plate stabilized by a gyroscope. During World War I, however, rocketry was again employed for military purposes. Rocket flares were used to illuminate trenches at night, and in 1916 the French fired rockets against German zeppelins, destroying at least one. A year later, the French fired incendiary rockets against hydrogen observation balloons.

As America entered the 20th century, the United States manifested an interest in designing aircraft rather than rockets. The Advisory Committee for Aeronautics, later named the National Advisory Committee for Aeronautics (NACA), was founded in 1915 to study the problems of flight and recommend practical solutions to basic aircraft design and construction.

While NACA engineers pursued their work in aeronautics in the early part of the century, other Americans, like Robert H. Goddard, dedicated themselves to new achievements in the area of rocketry. On March 16, 1926, Goddard launched a liquid-fueled rocket that traveled 184 feet in 2.5 seconds. See also **Goddard, Robert H**.

Interest in rocketry was also growing outside the United States. In 1928, Max Valier, a spaceflight popularizer, developed a rocket-propelled car in collaboration with Fritz von Opel, the automobile tycoon. The two developed the world's first rocket-powered car, which reached speeds of 180 miles per hour in Germany. Other novel applications of rocketry were also tried. In 1931, Friedrich Schmiedl used rockets to operate a postal service in Austria. Schmiedl used a 6-foot powder rocket that he manufactured. The rocket was designed to eject the mailbag by parachute shortly before crashing.

The foremost authority on rocketry, outside the United States, was Dr. Hermann Oberth, a Hungarian-born (Transylvanian) German. Near the end of 1923, he published his famous work, *The Rocket Into Interplanetary Space*, the genesis for considerable discussion regarding rocket propulsion. From 1926–30, Oberth experimented with liquid rocket propellants. He also worked as an advisor to the company making a film called, "Girl in the Moon," which increased interest in the potential of rocketry in Germany.

In the Spring of 1930, a young Wernher von Braun assisted Oberth in his early experiments in testing a liquid-fueled rocket stage with about 15 pounds of thrust. In September 1930, Oberth returned to a teaching post in Romania while Von Braun continued experiments under the sponsorship of the German Society for Space Travel. During July and August 1932, the society impressed officers of the German Army Ordnance by successfully firing a rocket to a height of 200 feet. As a result, the German Army formalized the rocket development program. In 1937 Von Braun was named technical director of the Peenemuende Rocket Center where he and his growing team of specialists worked on the development of the V-2 rocket, the first of which was launched in 1942. See Fig. 1. Some historians would

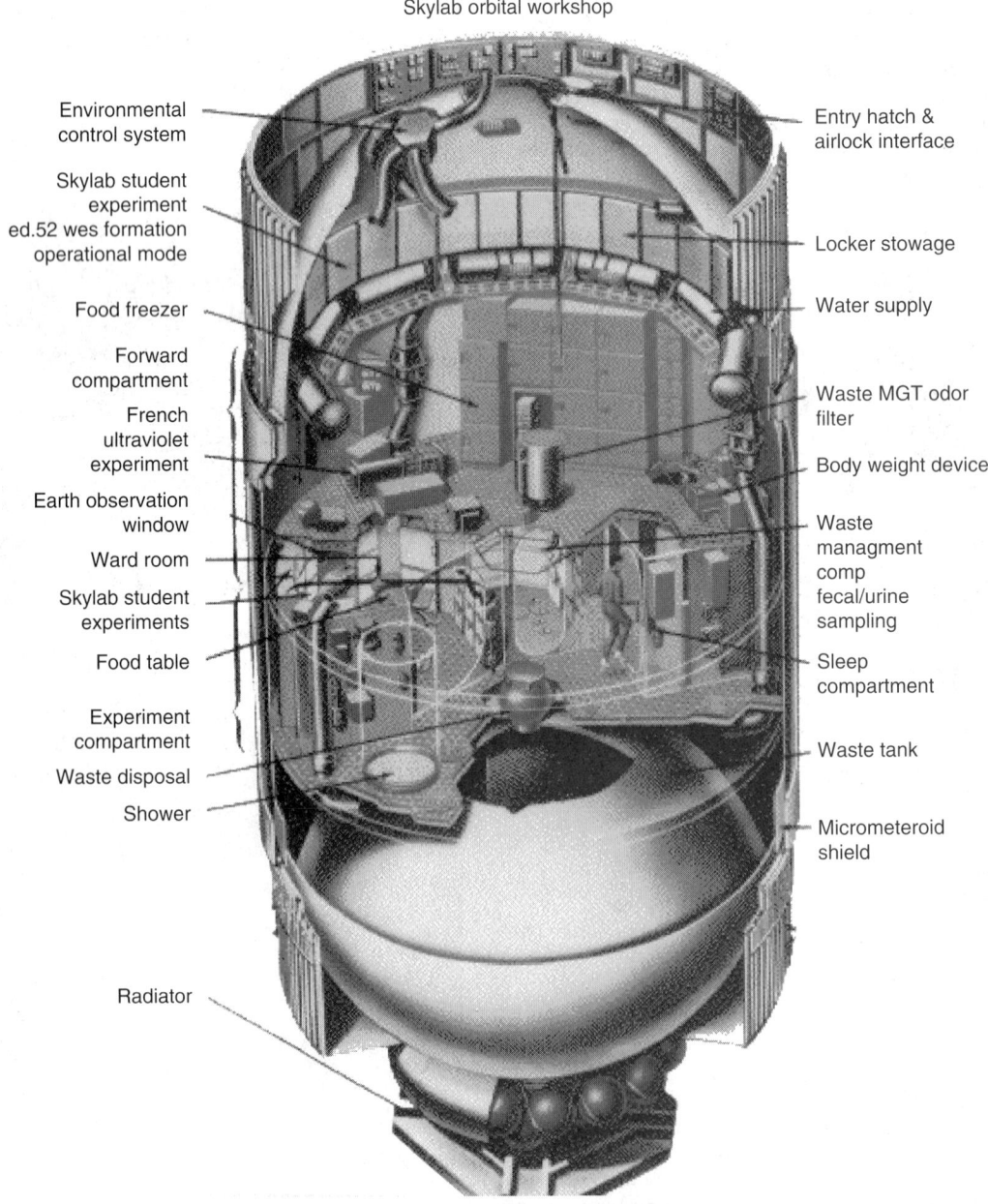

Skylab orbital workshop

Environmental control system

Skylab student experiment ed.52 wes formation operational mode

Food freezer

Forward compartment

French ultraviolet experiment

Earth observation window

Ward room

Skylab student experiments

Food table

Experiment compartment

Waste disposal

Shower

Radiator

Entry hatch & airlock interface

Locker stowage

Water supply

Waste MGT odor filter

Body weight device

Waste managment comp fecal/urine sampling

Sleep compartment

Waste tank

Micrometeroid shield

Fig. 4. *Skylab*, America's first Space Station. (*MSFC.*)

later estimate that by the end of World War II, the Germans had fired nearly 3,000 V-2 weapons against England and other targets.

As the war ended, the United States was interested in the technical capability of the Germans, and a team of American scientists was dispatched to Europe to collect information and equipment related to German rocket progress. Arrangements made under "Project Paperclip" enabled the German rocket specialists to come to the United States to initiate advances in American rocketry. Initially assigned to Fort Bliss, Texas, and later to Redstone Arsenal in Huntsville, AL, the German and American members of the team developed the Redstone Rocket, also known as "Old Reliable" because of its many diverse missions. See Fig. 2.

On October 4, 1957, the Russians launched a 23-inch-diameter ball designated *Sputnik*, the world's first artificial satellite. The news of the Soviet success created shock waves across the United States. The Soviet Union had inaugurated the Space Age. It had also presented American planners with the painful realization that there was no launch vehicle in the U.S. stable capable of orbiting anything approaching Sputnik's weight. Before the end of the year, the Soviets had also successfully launched *Sputnik 2* carrying the space dog, "Laika."

Only a few months after *Sputnik*, the Von Braun team used a modified Redstone, known as Jupiter-C, to launch the United States first artificial Earth satellite, *Explorer I* on January 31, 1958. Von Braun and his team were responsible for the Jupiter-C hardware. See Fig. 3. The satellite's principal scientific achievement was a major one—the discovery of the Van Allen Radiation Belts surrounding the Earth. The satellite dispatched information from space to Earth stations until May 23, 1958, when its

batteries were exhausted. The vehicle, however, continued to orbit for several years, reentering the Earth's atmosphere on March 31, 1970.

Soon after the success of *Explorer 1*, their research would begin focusing on the development of new launch vehicles, the Saturn family, to carry heavier payloads into space. But throughout 1958, the Nation's leadership in Washington had to face a series of decisions on how to manage the U.S. space program.

As a result of the debate in Washington, the National Aeronautics and Space Administration (NASA) came into being on October 1, 1958, with a broad charter for civilian aeronautical and space research. The new agency absorbed the existing National Advisory Committee for Aeronautics and made broad transfers from other government programs including the Von Braun team in Huntsville.

Launch vehicle development occupied much attention in NASA's early years leading to the Scout, Centaur, and Saturn launch vehicles.

On April 12, 1961, the Soviet, Yuri Gagarin became the first man in space.

Less than a month later, a Mercury-Redstone launched Alan B. Shepard, the first American astronaut, into space. Astronaut Gus Grissom followed him on an identical mission shortly thereafter. In 1962, a Mercury-Atlas launched Friendship 7, carrying John Glenn, who became the first American to orbit the Earth.

The Gemini Program, America's next major activity in space, used a modified Titan intercontinental ballistic missile. Its two-stage design permitted two men to ride into space together so they could conduct extra-vehicular activities (EVA). Astronaut Ed White made the first United States EVA in 1965, the same year that the Soviet cosmonaut, Aleksey Leonov, made the first Soviet space walk. On May 25, 1961, President Kennedy had announced a national goal of "landing a man on the Moon and returning him safely to Earth" within the decade. The Apollo program vastly expanded man's knowledge about the Moon and the Earth. Six Apollo expeditions explored the Moon, the last in December 1972. The lunar landings required an entirely new launch vehicle, the Saturn V. This new vehicle, developed by the Von Braun team at NASA's Marshall Space Flight Center, evolved in three phases. The Saturn I provided NASA with significant new payload lifting capabilities. The Saturn IB had even

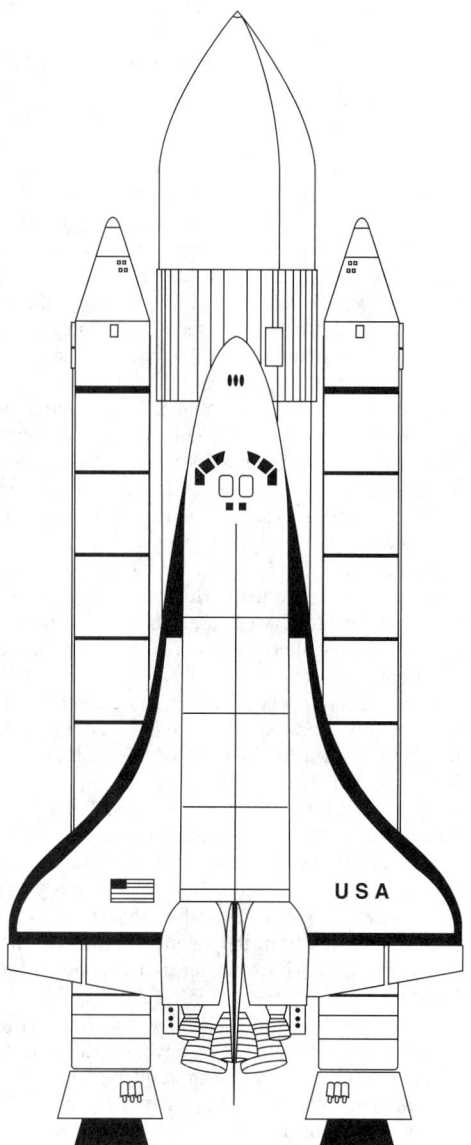

Fig. 5.   Illustration, showing the Space Shuttle. (*MSFC*.)

Fig. 6.   Launch of Space Shuttle Columbia. (*MSFC*).

more power, enough for orbital missions with the Apollo spacecraft. The Saturn V was clustered with five F-1 rocket engines burning liquid oxygen and kerosene and generating 1.5 million pounds thrust per engine.

In 1973, a two-stage Saturn V rocket also launched Skylab, America's first space station, designed to investigate long-duration spaceflight. See Fig. 4. Skylab's three different three-man crews, also launched into space by Saturn IB rockets, spent up to 84 days in Earth orbit and performed a variety of more than 100 experiments.

The final launch of a Saturn V rocket came on July 15, 1975, as part of the Apollo-Soyuz Test Project. Earlier that day, a Russian Soyuz spacecraft lifted off its launchpad at a Soviet launch site carrying three cosmonauts. Seven and one-half hours later, the U.S. Apollo spacecraft was launched with its crew of American astronauts. Rendezvous and docking of the two ships were accomplished on July 17. The two ships remained docked for two days, conducting joint.

The Apollo program demonstrated that men could travel into space, perform useful tasks there, and return safely to Earth. But space had to be more accessible. This lead to the development of the Space Shuttle. See Fig. 5.

With the launch of the Space Shuttle Columbia on April 12, 1981, the United States entered a new era in transportation for living and working in space. See Fig. 6. Described by NASA as the first "true aerospace vehicle," the space shuttle takes off like a rocket. The winged orbiter then maneuvers around the Earth like a spaceship. But unlike earlier manned spacecraft, which were good for only one flight, the shuttle orbiter and solid rocket boosters can be used again and again. Only the external tank is expendable on each mission. See also **Rocket Engine**; **Rocket Propellants**; **Satellites (Scientific and Reconnaissance)**; **Space Shuttle**; **Space Station**; and **Von Braun, Wernher**.

Web References

A Timeline of Rocket History: *http://history.msfc.nasa.gov/rocketry/index.html*
Biographical Sketch of Dr. Wernher Von Braun: *http://history.msfc.nasa.gov/vonbraun/index.html*
Milestones in Space Exploration: *http://history.msfc.nasa.gov/milestones/index.html*
NASA Histories on Line: *http://history.msfc.nasa.gov/otherlinks.html*
Saturn V History: *http://history.msfc.nasa.gov/saturnV/index.html*

MIKE WRIGHT, Marshall Space Flight Center, Huntsville, AL.

ROCK FLOUR.   A peculiar and distinctive white mud, the product of the grinding action of glaciers. When deposited in lakes, rock flour forms an important constituent of varves, or the type of annual cyclic stratification used by glaciologists in determining glacial time. When desiccated and transported by wind this material may form extensive deposits of loess.

RODENTIA (*Mammalia*).   Gnawing mammals characterized particularly by the two chisel-like incisor teeth in each jaw. These teeth oppose each other and are worn down by use to maintain a keen enamel edge on the front, while the softer dentine slopes away inward. Most species are small, but the beavers and capybara are moderately large animals. The rats and mice are the most familiar species, with squirrels, chipmunks, and woodchucks scarcely less known. Because of their vegetarian habits and their destructive gnawing, many rodents are serious crop pests, and as household nuisances and disease carriers they are all too common. The rabbits and their allies at one time were included under *Rodentia*, but now are placed in their own, separate order, *Lagomorpha*. See also **Rabbits and Hares**.

The general organization of *Rodentia* is given in Table 1. The common nomenclature applied to *Rodentia* is confusing, with such terms as rat, squirrel, gopher, etc. applying to numerous species. The principal species among the Sciuromorphs are described in this volume under **Beaver**; or **Squirrels and Other Sciuromorphs**. In the latter entry, there are descriptions of the chipmunk, gopher, marmot, prairie dog, spermophile, squirrel, and woodchuck. Mouse is a term used to describe pocket mice species as found under the Sciuromorphs and so-called ancient and modern mice under the Myomorphs. The term is loosely applied to numerous species of small burrowing and gnawing rodents with slender bodies, long tails, and either the front legs or both pairs short. The sewellel is a Sciuromorph although commonly called the mountain beaver or boomer. This is a broad, stout animal with short legs, found in mountain forests in the western part of the United States. Of the genus *Apolodontidae*, it is a burrowing species that feeds on bark, twigs, and leaves. The animal is more closely related to the squirrels than to the beavers.

Only a representative number of Myomorphs can be described here. The hamster is a burrowing animal found in Europe and Asia and known for its complex underground dwellings. The common hamster, *Cricetus*, attains a length of about 1 foot (0.3 meter). When numerous in an area, the animal can be a serious pest to farmers, damaging crops of almost all kinds. The fur is used and the flesh can be eaten. The animal is light brown above and black underneath. Hamsters are also used widely in medical laboratories. These are usually *Mesocricetus auratus*, of Syrian origin. Hamsters are extremely prolific, a female producing several litters per year. The gestation period is a brief 2 to 3 weeks. The young are born blind and naked, but they mature rapidly and are fully on their own in about 3 weeks. As a diet, the hamster prefers fruits, vegetables, and grain and occasionally small rodents or birds.

The dormouse is a small arboreal rodent of the Palaearctic and Ethiopian regions. The animal has a squirrel-like appearance, with long hairy, bushy tail. The common dormouse is *Muscardinus avellanarius* and is sometimes called a sleeper. Unlike the squirrel, the dormouse does not have cheek pouches. The animal is not much larger than an ordinary house mouse. The eyes are large and the head disproportionately large for the body size. Color is rusty-red above and white underneath. The animal hibernates during the colder season, during which time the body temperature may drop close to freezing temperatures. The heart beats but a few times per minute, the blood pressure is extremely low, and breathing is difficult to detect.

The muskrat is a moderately large amphibious rodent of North America. It is stoutly built, short-legged with partially webbed hind toes, and a compressed tail. The muskrat inhabits swamps and streams, making houses of sticks, while also burrowing in the banks. Two species are recognized, one a dark brown animal, *Ondatra rivalicia*, of the coastal part of Louisiana; the other *O. ziebethica*, found over most of the continent. Muskrat fur consists of a fine woolly undercoat and long glossy hairs. It is probably the best of the inexpensive furs and because of the ability of the animal to withstand extensive trapping is an important commercial fur. Several million pelts are taken annually in North America, many from fur farms. The fur is marketed in its natural colors, as well as in the form of Hudson seal after being plucked, clipped, and dyed black.

The lemming is a small animal of northern latitudes. The European lemmings resemble the woodchuck in form, but are much smaller. The American species, *Synaptomys*, looks like a short-tailed mouse. The common lemming, *Lemmus lemmus*, of northern Europe is noted for its occasional migrations. Many thousands of the animals take part in such migrations, crossing mountains, fording streams, and always pushing straight on until they enter the sea and are drowned.

The jerboa is a small jumping desert animal of Asia and northern Africa. It resembles a mouse, with long tufted tail and very long hind legs. The small forelegs are not used in locomotion. The Asiatic jerboas have five toes on the hind feet; the African, three. The ears vary considerably in shape. The fur is long, soft, and silky. The most common Asiatic species, also called the alagdaga, is *Alactaga indica*.

The vole is a meadow or field mouse, constituting the genus *Microtus*. The genus is limited to the Northern Hemisphere and in Asia does not extend south of the Himalayas. Voles are characterized by rootless molar teeth, formed of two rows of alternating triangular prisms. About 20 species occur in North America.

The wood rat is a moderately large rodent, more common in the western states, but represented by a few species in other sections. The wood rat is well and unfavorably known for its habit of invading houses and camps and carrying away anything edible. The wood rat differs from true rats in the shorter, furry tail and the larger eyes and ears. Ten species have been described, all in the genus *Netoma*. These animals also are called pack rats and trade rats, the latter from their habit of replacing what they take with some other object.

The jumping mouse is a small mouse-like rodent with very large hind legs and long tail. It differs from the kangaroo mouse in the absence of cheek pouches. Several genera have been recorded, all related to the jerboas of the Old World.

The kangaroo rat is a burrowing rodent of the western United States, with long tufted tail and long hind legs. Several genera of these animals are related to the pocket mice and kangaroo mice.

The gerbil is a small burrowing animal of Asia and Africa. Gerbils resemble rats, but have long hind legs and large eyes and move about by jumping. In these points, they resemble the jerboas, but they are less extreme.

TABLE 1. RODENTIA
(Gnawing Mammals)

In this Encyclopedia		In this Encyclopedia	
SCIUROMORPHS	See **Rodentia**.	Muskrats (*Ondatara rivalicia, ziebethica,* ...)	
Swellels (*Aplodontidae*)		Lemmings (*Lemmus lemmus,* ...)	
Squirrels (*Sciuridae*)	See **Squirrels and**	Selevinids (*Seleviniidae*)	
Typical Tree-Squirrel (*Sciurus,* ...)	**Other Scuivomorphs.**	Jumping-Mice (*Zapodidae*)	
The Chickaree (*Tamiasciurus*)		Striped Mice (*Sicista*)	
Palm-Squirrels (*Funambulus,* ...)		Jumping-Mice (*Zapus and Neozapus*)	
Oriental Tree-Squirrels (*Callosciurus,* ...)		Jerboas (*Dipodidae*)	
African Ground-Squirrels (*Xerus,* ...)		Black Rats (*Rattus rattus*)	
Northern Ground-Squirrels (*Marmota,* ...)		Roof Rats (*R. alexandrinus*)	
Flying-Squirrels (*Petaurista,* ...)			
Beavers (*Castoridae*)	See **Beaver**.	**HYSTRICOMORPHYS**	See **Rodentia**.
Pocket-Gophers (*Geomyidae*)		Old World Porcupines (*Hystricidae*)	
Pocket-Mice (*Heteromyidae*)		Crested Porcupines (*Hystrix*)	
Pocket-Mice (*Perognathus,* ...)		Noncrested Porcupines (*Acanthion*)	
Kangaroo-Rats (*Dipodomys*)		Sumatran Porcupine (*Thecurus*)	
Spiny-Rats (*Heteromys,* ...)		Brush-tailed Porcupines (*Atherura*)	
Scale-Tails (*Anomaluridae*)		The Rat-Porcupine (*Trichys*)	
Gliding Scale-Tails (*Anomalurus*)		New World Porcupines (*Erethizontidae*)	
Nongliding Scale-Tail (*Zenkerella*)		North American Porcupines (*Erethizon*)	
Gliding Mice (*Idiurus*)		The Bristly Porcupine (*Chaetomys*)	
Spring-Haas (*Pedetidae*)		Coendous (*Coendou*)	
		Mountain Porcupine (*Echinoprocta*)	
		Cavies (*Caviidae*)	
MYOMORPHS	See **Rodentia**.	Guinea-pigs (*Cavia,* ...)	
Ancient Mice (*Cricetidae*)		The Mara (*Dolichotis*)	
New World Mice (*Peromyscus,* ...)		Capybaras (*Hydrochoeridae*)	
Hamsters (*Cricetus,* ...)		Pacuranas (*Dinomyidae*)	
Sokhors (*Myospalax*)		Pacagoutis (*Dasyproctidae*)	
Malagasy Voles (*Nesomys,* ...)		The Paca (*Cuniculus*)	
The Crested Hamster (*Lophiomys*)		The Mountain Paca (*Stictomys*)	
Voles (*Microtus,* ...)		Agoutis (*Dasyprocta*)	
Sand-Rats (*Gerbillus,* ...)		The Acouchi (*Myoprocta*)	
Modern Mice (*Muridae*)		Hutias (*Capromyidae*)	
Old World Mice (*Mus,* ...)		Long-tailed Hutias (*Capromys*)	
Wading Rats (*Deomys,* ...)		Short-tailed Hutias (*Geocapromys*)	
The Shrew-Rat (*Rhynchomys*)		Zagoutis (*Plagiodontia*)	
Cloud-Rats (*Phloeomys*)		The Venezuelan Hutia (*Procapromys*)	
Australian Water-Rats (*Hydromys,* ...)		The Coypu (*Myopotamus*)	
Mole-Rats (*Spalacidae*)		Tucotucos (*Ctenomyidae*)	
Root-Rats (*Rhizomyidae,* ...)		Octodonts (*Octodontidae*)	
Blesmols (*Bathyergidae*)		The Rat-Chinchilla (*Abrocomidae*)	
Stand-Rats (*Bathyergus,* ...)		Chinchillas (*Chinchillidae*)	
Sand-Puppies (*Heterocephalus*)		Mountain-Chinchillas (*Lagidium*)	
Dormice (*Gliridae*)		Viscachas (*Lagostomus*)	
The Hazelmouse (*Muscardinus*)		Porcupine-Rats (*Echimyidae*)	
Squirrel-tailed Dormice (*Glis,* ...)		African Rock-Rats (*Petromyidae*)	
African Dormice (*Graphiurus*)		Cutting-Grass (*Thyronomyidae*)	
Spiny Dormice (*Platacanthomyidae*)		Gundis (*Ctenodactylidae*)	

A rat, in the strict sense, is an animal of the genus *Rattus*, a term that is widely applied to many small gnawing animals. The true rats are dull-colored animals with long scaly tails, short legs, small ears, and a pointed muzzle. The common brown rat, *R. norvegicus*, is a typical example. This species was introduced from Europe into the United States during Colonial times and has been a troublesome and often dangerous pest ever since. It is much more aggressive than the related black rat, *R. rattus*, and has almost crowded the latter out. The roof rat, *R. alexandrinus*, is a third species found in the southern United States. Still other species of the genus are found on other continents. Some of the closely related genera include the bandicoot rats, bush rats, bamboo rats, kangaroo rats, and cane rats. The name, American pouched rat, is sometimes applied to a group containing the pocket gopher.

The Hystricomorphs tend to number larger animals than found in the two foregoing families of *Rodentia*.

The porcupine is an animal with many modified hairs resembling quills in form. These hairs are sharp-tipped, finely barbed, rigid spines which penetrate flesh very readily and serve as an almost impregnable defense. North America has two species of porcupines, one ranging from the eastern half of the continent as far south as Virginia and the other, *Erethizon epixanthus*, in the far west, from Alaska to Mexico. The common eastern species, *E. dorsatum*, is also called the hedgehog. Both species are partial to the leaves, twigs, and bark of evergreen trees as food. The tree porcupines differ in having long prehensile tails. They are found in Mexico and South America. See Fig. 1. Still other species live in Eurasia, Africa, and the Oriental Region.

Guinea pigs are widely kept as pets and are useful to medical science as laboratory animals. Because of their high rate of reproduction, they have also been bred for studies in heredity. The guinea pig is one species of cavy. Among other species are the capybara, largest of the rodents and aquatic in habits, although most of the species are small and terrestrial. All cavies are native to South America. The domesticated guinea pig achieves a length of about 10 inches (25 centimeters) and a weight of about 2 pounds (0.9 kilograms). The animal has no tail. There are 4 toes on the forefeet; 5 toes on the hindfeet. The size and color vary considerably. In nature, the animal feeds on grass and vegetables. As a pet, the animal will consume some kinds of dog or cat food, but requires a lot of water. They can be bred 2 to 3 times per year. There are from 2 to 8 young per litter. Gestation period is from 63 to 75 days. The life span is from 6 to 7 years.

The capybara is a large South American species, *Hydrochoerus capybara*, of the cava family. These animals attain a length of 4 feet (1.2 meters) and a weight of almost 100 pounds (45 kilograms). They are semiaquatic in habits and while their food usually consists of water plants

Fig. 1.   Brazilian tree porcupine with prehensile tail. (*New York Zoological Society.*)

and other vegetation, they sometimes make inroads on cultivated crops. Locally, they are called capivaras and carpinchos.

The coypu is a large aquatic rodent of South America whose habits are like those of the muskrat. The fur is of commercial value. The animal is now protected by law because it was nearing extinction. The coypu achieves a length of $2\frac{1}{2}$ to 3 feet (0.8 to 0.9 meter). There are nine young in a litter. The hutia is a large arboreal rodent of the West Indies. It is related to the coypu, but resembles a rat, with exception that the muzzle is blunt and the tail is moderately long.

The agouti is a large rodent of the genera *Dasyprocta* and *Myoprocta* found in Central and South America and the West Indies, where they live chiefly in the forests. The natives hunt the animal for its flesh. There are about 15 species, all are brown with yellow coloration. They are about the size of an average American rabbit. The best known species is *Azaras* "acuchi" of Guinea. The animal feeds on nuts, fruits, and sugar cane. Nocturnal in habits, the animal can be quite destructive to banana plantations and sugar fields. The paca is a stoutly built rodent, *Agouti paca*, about 2 feet (0.6 meter) long, marked with rows of light spots on a fawn-to-blackish ground color. The animal is found through most of South America east of the Andes. It is closely related to the agoutis.

The tucotuco is a ratlike burrowing animal of South America. It has gray fur and red incisor teeth. The tail is only moderately long and is clothed with short fur. A closely related form with vestigial ears is known in Chile as the cururo.

The chinchilla is a small squirrel-like rodent related to the porcupines. The animal lives in communal burrows at high altitudes in the mountains of South America. The fur of the common chinchilla has commercial value. There are two genera, *Chinchilla* and *Lagidium*. The chinchilla is about 8 to 10 inches (20 to 25 centimeters) long with a tail of about 5 inches (13 centimeters) in length and with a black streak its full length. The fur is about 1 inch (2.5 centimeters) long, soft, pale gray, and dusky-looking. The diet consists of fruit, grain, moss, and herbs. There are two litters each year, usually with two young per litter. The gestation period is about 110 days. The Chilean chinchilla can be raised in captivity, but is very sensitive to excessive humidity. The animal was first introduced into the United States in about 1923 for breeding. The rare Peruvian royal chinchilla is practically extinct. This genus dates back to the age of the Incas, who prized the animal for food.

The viscacha is a large, stout burrowing animal, *Lagostomus trichodactylus*, of the South American pampas and related to the chinchillas. The contrast between these animals has been likened to that between the squirrels and woodchucks, the one gracefully built and the other a clumsy burrowing form.

ROE DEER.   See Deer.

ROENTGEN.   A unit of radiation, that quantity of X-rays or gamma rays which will produce, as a consequence of ionization, 1 electrostatic unit of electricity in 1 cubic centimeter of dry air measured at 0 degrees C and standard atmospheric pressure. See also **Radioactivity.**

ROENTGEN-EQUIVALENT–MAN (abbr rem).   A unit of radiation which when absorbed by a human being, produces the same effect as the absorption of 1 roentgen of high-voltage X-rays.

ROENTGEN-EQUIVALENT-PHYSICAL (abbr rep).   A unit measuring a purely physical effect of radiation by the number of ion pairs produced per unit volume of target material per time unit. One rep is equivalent to the absorption of 93 erfs per gram of tissue.

ROLLE THEOREM.   Let $f(x)$ be a function which vanishes at $x = a$ and at $x = b$, and which has a finite derivative $f'(x)$ at all points in the interval (a, b). Then $f'(x)$ vanishes at some point x_0 between a and b. This theorem is used in calculus to prove mean value theorems.

ROLLING FRICTION.   In the rolling of a wheel on a plane surface there is some distortion of the two surfaces in contact due to the normal force between the surfaces. Such distortion smears out the ideal line contact and effectively introduces a force with a component in opposition to the motion. This component of force is called the rolling resistance or friction F_r and is proportional to the normal force N. A coefficient of rolling resistance or friction μ_r can be defined by $\mu_r = F_r/N$, where F_r can be determined experimentally by observing the deceleration on a horizontal surface. See also **Friction (Mechanical).**

ROLL OUT.   In computer terminology, to read out of a storage device by simultaneously increasing by one the value of the digit in each column and repeating this r times (where r is the radix) and, at the instant the representation changes from $(r - 1)$ to zero: generating a particular signal, or terminating a sequence of signals, or originating a sequence of signals.

RONTGEN, WILHELM CONRAD (1845–1923).   Rontgen became one of the outstanding experimental physicists of his time. In 1869, he received his doctorate in mechanical engineering from the University of Zurich. He held various positions at higher institutions throughout Germany until, in 1888, he became the chair for the physics department at the University of Wurzburg.

In 1895, Rontgen was the first scientist to observe and record X-rays. While experimenting with a cathode ray discharge tube that was completely enclosed in cardboard so no light could escape, he observed that a sheet of barium platino-cyanide coated paper lying several feet away from the discharge tube was glowing in the dark. Rontgen realized the importance of this accidental discovery and he devoted six straight weeks of working in his laboratory investigating the X-ray. While investigating, he actually saw the image of his bones within his fingers and deduced that his X-rays must be a very short wavelength form of electromagnetic radiation. He published his work in a series of classic papers and the phenomenon of his X-rays had immediate use in hospitals in the diagnosis and treatment of fractures. His discovery could have been patented, but Rontgen gave it to humanity. Later, when he received a cash prize for receiving the first Nobel Prize (1901) in Physics, Rontgen gave the money to the University of Wurzburg.

Rontgen was truly a great scientist and great humanitarian. His discovery of the X-ray changed the world and today it has universal usage.

See also **X-Ray.**

J. M. I.

ROOT LOCUS.   The root locus method is a graphical technique applied in control system and feedback studies in an effort to determine the transient response of the system. The objectives of the method are similar to those of the frequency response techniques. In this method, the locus of roots of the system characteristic equation is plotted. In obtaining the locus of roots, the variable parameter is normally system gain. It is, however, possible to vary other system parameters to see their effect on system performance.

A typical root locus diagram is shown in Fig. 1, for a second-order or quadratic equation of the form $s^2 + 2\zeta\omega_n s + \omega_n^2$, where the roots are s_1, $s_2 = -\zeta_n \pm \omega_n \sqrt{\zeta^2 - 1}$. The roots which are in the form of $s = -\sigma \pm j\omega$ may be real or complex, depending on system parameters. When the damping ratio ζ is > 1, the roots are real and unequal and the system is overdamped. When $\zeta = 1$, the roots are real and equal and the system has critical damping. For $\zeta < 1$, the roots are imaginary and the system is underdamped, giving an oscillatory response. The limits for this ideal

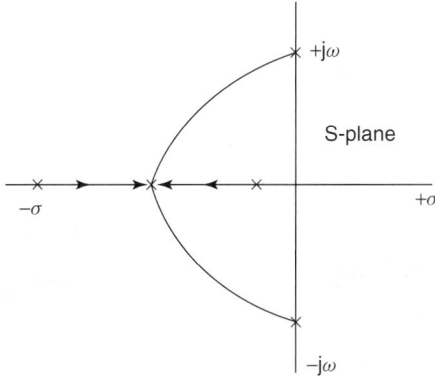

Fig. 1.   Root locus plot.

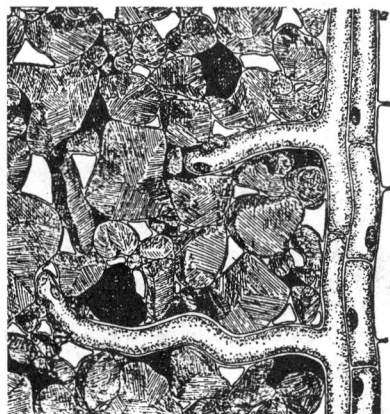

Fig. 1.   Root hairs penetrating the soil. Note the tiny rock particles, the bits of humus, the films of water, and the air spaces.

system are of course when $\zeta = 0$. In this case, the roots are imaginary, and the system will have a continuous oscillation. Thus, if the roots of the equation are known, it is possible to predict the transient response. In an actual system, the damping ratio ζ and resonant frequency ω_n are functions of system gain and other system parameters.

See also **Characteristic Equation**.

ROOT (Mathematics).   If a is a real number, n an integer, then x is the nth root if the product of x taken n times equals a. The number of n th roots is n but not all of them need be real. One of them is chosen as the principal nth root in accordance with the following rules: (a) if a is positive and n is even or odd, the one positive root; (b) a negative, n odd, the one negative root; (c) a negative, n even, any one of the complex roots. All of the roots, real or complex, may be found by the De Moivre theorem.

Tables of the positive real roots of numbers are usually given in mathematical handbooks. They may be calculated to any number of significant Figures with logarithms or desk calculating machines; approximately, by the binomial series; numerically, by several methods, usually that of Horner. For properties of the roots of an equation, see **Polynomial**.

See also **Cholesky Method of Solving Equations**; **Radical (Mathematics)**; and **Square and Square Root**.

ROOT-MEAN-SQUARE.   The root-mean-square current or voltage is the effective value of the quantity in an alternating-current circuit. The defining equation for current is:

$$I = \sqrt{\frac{1}{T} \int_0^T i^2 \, dt}$$

where I is the root-mean-square current, T is the interval of integration and i is the current equation as a function of time t. A similar equation defines the voltage.

In statistics, the root-mean-square of the variates $x_1, x_2, \ldots, x_n$ is the average defined as

$$\text{R.M.S.} = \sqrt{\frac{x_1^2 + x_2^2 + \cdots + x_N^2}{N}}$$

and in a frequency distribution, by $\sqrt{\Sigma x^2 f_x / \Sigma f_x}$.

ROOT (Plant).   In seed plants, the root is generally the first part of the plant to emerge from the germinating seed. The functions of the root are primarily anchorage of the plant, absorption of water and mineral salts in solution, and conduction of these to the stem, and also storage of food. Anchorage is obtained by the much-branched, far-reaching root system, which penetrates deeply into the ground and resists such forces as wind acting on the top of the plant. Commonly, one thinks of the root as the part of a plant found in the ground. While this is true in the majority of cases, in some plants the roots are found in the air.

Upon breaking through the seed coats of the germinating seed, the young roots turn downward and soon put out an abundance of minute hairs. These serve the twofold purpose of attaching the root firmly in the soil and of absorbing moisture and nutrients from the soil. The root continues to grow into the soil, elongating and branching. See Fig. 1.

Externally there are certain characteristics that distinguish a root from a stem, even when the latter grows underground. The root bears no leaves on its surface and is not separable into nodes and internodes, as is the case in the stem. The apex of the root is covered by a protective structure called the root-cap, a distinctive feature never found in stems. Just back of the tip, the surface of the root is usually provided with root hairs. Branching in roots is quite distinct from that in stems, the branch-roots, appearing at irregular intervals; frequently these branchroots are borne in longitudinal rows, a fact that is correlated with their origin. Branch-roots develop from the pericycle, a tissue deep in the root, and not from the sub-epidermal tissue as do stem branches.

Several types of root systems are recognized. See Fig. 2. In many plants, the first formed, or primary, root continues to grow downward to form a long root, which penetrates deep into the ground. Branchroots arising from this are commonly much shorter, and of smaller diameter. Such a root is called a tap root, and the entire root system in such cases is known as a tap-root system. Familiar examples are found in the dandelion, and in oak trees. Monocotyledonous plants rarely show this form of branching. In them and in many other plants, the primary root soon loses its individuality, the secondary roots becoming larger, and forming an extensive root system, in which no single root is distinguished from the others by its larger size and more obvious downward growth. Such a system is called a fibrous root system, and the individual roots are known as fibrous roots. Wheat has such a root system. In many plants, the roots become important places of storage of foods, the roots often being conspicuously enlarged because of this storage. Such roots are called fleshy roots. Their existence is of great value to man, since many of them become important crop plants, as, for example, carrots, parsnips, turnips and beets. In many plants, several roots are swollen with food so that a group or fascicle of storage roots is formed, as in the *Dahlia*.

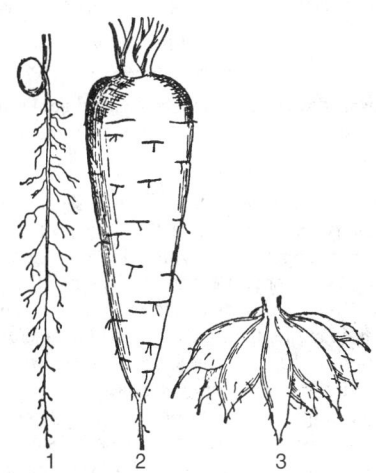

Fig. 2.   Different kinds of roots: (1) fibrous tap root of pea; (2) fleshy tap root of carrot; (3) fleshy foscicled roots of dahlia.

In many plants, especially in the tropics, the roots are formed in the air. Such aerial roots are often quite different from ordinary roots, being much coarser, less branched and lacking root hairs. In aerial roots the outer portion is frequently modified to a spongy tissue capable of absorbing moisture rapidly and retaining it tenaciously. See Fig. 3. This tissue, the velamen, is particularly well- developed in epiphytic orchids. Other plants have aerial roots which extend either downward from the branches, or from the lower part of the stem. Such roots are called, respectively, prop roots or brace roots. The terms are often used interchangeably. These roots soon penetrate the surface of the ground and become like ordinary roots. The Banyan tree shows the downward-growing prop roots particularly well, while in the Screw-Pine and in Corn plants, brace roots are well-developed. See Fig. 4. In many tropical plants, especially in those of large trees, the roots radiate out over the surface of the ground. Often these roots are conspicuously developed vertically, forming thin plates of considerable depth, but only a few inches thick; they are called buttress roots.

Fig. 3. A type of modified roots–aerial roots of poison ivy (*Rhus toxicodendron.*)

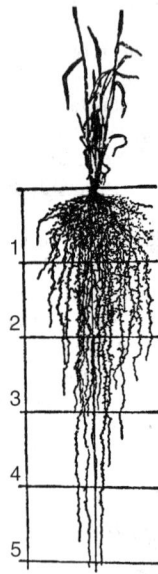

Fig. 4. Root system of wheat plant at blossoming time. Figures at left indicate depth in feet. (*Weaver.*)

In parasitic plants the roots may be curiously modified or replaced by absorptive organs which penetrate the tissues of the host plant until they reach the conducting system, from which they obtain their nutrient supply. In some plants, roots are entirely lacking, their functions being taken over entirely by other parts of the plant. In the saprophytic Orchid, *Corallorrhiza*, for example, the much-branched coarse underground stem, or rhizome, functions as a root; in the Bladderworts, *Utricularia*, floating in water, roots are entirely unnecessary and nonexistent. In the Spanish Moss, *Tillandsia*, growing epiphytically in tropical and subtropical America, roots are entirely lacking, absorption occurring over the surface of the plant.

Commonly, the extent of a root system is very much underestimated. See Fig. 5. Only when the entire plant is removed from the soil by careful and extensive digging is the great spread of the entire root system seen.

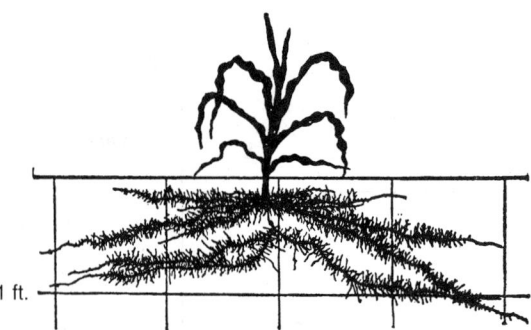

Fig. 5. Root system of Iowa Silver Mine corn 36 days old. (*Weaver.*)

Then it is found that the roots may penetrate the soil to a depth much greater than was supposed. In common Red Clover, for example, the roots may go downward to a depth of 8 or 9 feet, while in alfalfa depths of 15–20 feet or more are found. The lateral spread of roots is also often very great. In the common Squash the lateral roots may extend outward 10–15 feet, while other roots of the same plane penetrate the soil to depths of 4–6 feet, forming a very extensive system. The nature of the soil, and especially the amount of oxygen present in it, are important factors in determining the amount of branching and the extent of the root system. In general, porous well-aerated soil favors extensive branching.

Roots do not seek water, as popularly supposed from the fact that they seem to grow toward water. This tendency is explained by the fact that roots respond very definitely to certain external factors, such as gravity. Most roots react positively to gravity, that is, are positively geotropic, growing directly downward into the soil. If a young plant is placed so that its root is horizontal, in a short time it will be found that this root has changed its direction of growth and is growing downward again. Many roots are also affected by light, from which they turn away, being negatively phototropic. Temperature also affects roots, so that a root encountering a region having a temperature more favorable for its growth will increase more rapidly than other roots. Entirely like this is the response of roots to favorable moisture conditions, which lead to greater growth of the roots extending in that direction. It is such phenomena that lead to the statement that roots seek water. Often they do seem to seek water, especially when they penetrate joints in sewer pipes, sometimes at distances of 30–60 feet or more from the stem of the plant, and form great masses of much branched roots, which may completely clog the sewer pipe, and become a source of great inconvenience and expense. Other factors such as the nature of the soil and its contained minerals, also influence the growth of roots.

The internal structure of the root is quite constant and distinct from that of the stem. In the tip of the root an apical meristem, or zone of dividing cells is found. The cells of this region are more or less cubical in shape, thin-walled, and contain a dense protoplasm and a relatively very large nucleus.

Over the apical end of the root, and covering the actively dividing cells is the root cap. This is a mass of loosely aggregated cells arising in various ways. In some plants the cells of the root cap are formed from the meristem cells in general. In dicotyledonous plants there is frequently a special layer of cells, called the calyptrogen, covering the apical surface of the root. These cells divide and form cells that become the root cap. However formed, the root cap is continuously renewed throughout the life of the root, forming a protective cover over the tip. The outer and consequently older cells of this cap are loosely arranged. Their walls become considerably modified to form a mucilaginous mass, which greatly reduces the friction of the elongating root against the soil particles. This mucilaginous mass is often very conspicuous in the brace roots of corn.

Just back from the apical meristem of the root is the zone of elongation. In this region the cells show very characteristic changes. The most obvious of these is the increase in length that occurs in most of the cells. At the same time there appears in each cell a conspicuous central vacuole which enlarges greatly, the cytoplasm being pushed out to a thin peripheral layer in which is found the nucleus. In roots this elongation region is much larger than is the corresponding region in the stem, a fact presumably correlated with the denser medium through which it grows. Usually it is less than a centimeter long.

As the root continues elongating the cells gradually change to form a third zone not sharply set off from the second. In this, the zone of maturation, the cells gradually assume their final form. Externally the most conspicuous feature of this zone is the presence of the root hairs. A root hair is a slender outgrowth of a cell of the epidermis, or outermost single layer of cells. In size they vary from 0.1 to 10 millimeters long, with a diameter averaging about 0.01 millimeter; the number of them formed on the root surface varies from 200 to 400 or more, so that they cause a tremendous increase in the surface of the root. The life of a root hair is not long, being commonly only a few days, after which it disappears. Usually the wall of a root hair is very thin and modified externally to a pectic substance, which sticks closely to the soil particles and absorbs water readily therefrom. Within the wall is a thin peripheral layer of actively streaming cytoplasm. The central part of the hair is occupied by an evident vacuole, which is continuous with that in the basal portion of the cell from which the hair protrudes.

As the root grows older and the cells composing it mature, a very definite structural pattern appears. See Fig. 6. The outer portion, or cortex, is composed almost entirely of parenchyma cells, which are of irregular shape, loosely aggregated, and serve mainly for storage of materials.

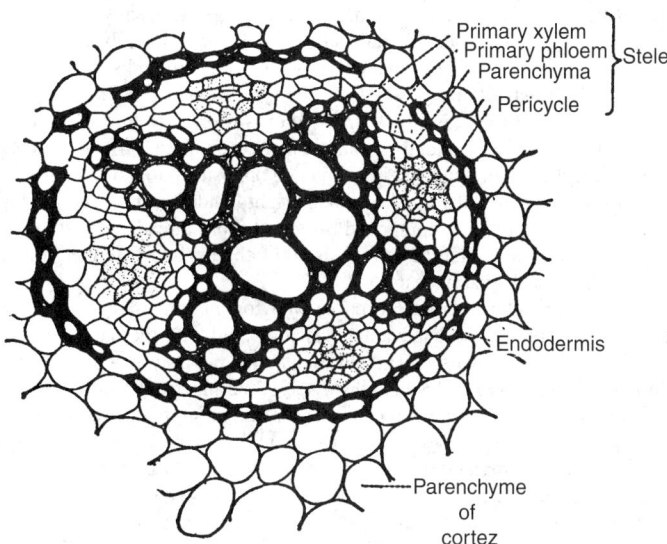

Fig. 7. A cross section of the stele of a young root of the buttercup (*Ranunculus acris.*)

meristem. (1) A central cylinder of alternate masses of xylem and phloem cells, surrounded by a sheath of slightly modified cells called the pericycle, these composing the central cylinder. (2) Outside this is the endodermis, with its cells showing characteristic thickenings of the walls. (3) Around the endodermis is the cortex, surrounded by (4) the outermost layer, the epidermis. All these tissues are collectively known as primary tissues.

With further growth of the root, secondary tissues appear. These are formed from a special group of cells known as cambium cells. Cambium cells first appear in the region inside the phloem patches, appearing in cross-sections of the root as crescent-like patches which gradually extend radially until they unite to form a continuous sheath around the xylem. By their divisions new cells are formed, both inside and outside the cambium band. Those inside gradually develop into secondary xylem cells, while those outside become secondary phloem cells. Continued development of the cambium band causes the primarily phloem cells and all other cells formed externally to be pushed outward and gradually crushed. Since growth often occurs periodically in the root, especially in regions having alternations of favorable and unfavorable seasons, annual rings are found in roots as in stems. As root enlargement occurs, the pericycle cells become meristematic and by their divisions form a mass of periderm cells around the central cylinder.

The functions of the root are primarily anchorage of the plant, absorption of water and mineral salts in solution, and conduction of these to the stem, and also storage of food. Anchorage is obtained by the much-branched far-reaching root system, which penetrates deeply into the ground and resists such forces as wind acting on the top of the plant.

The water of the soil, together with substances in solution, is taken into the plant largely through the portion of the root covered by root hairs. The outer wall of the latter soaks up water readily. The wall of the cell is permeable, permitting water to pass through easily. The outer surface of the cytoplasm of the cell is also a membrane, which is semipermeable, that is, readily permits certain substances such as mineral salts in solution to pass through, but does not allow organic substances to pass. A similar condition exists in the membrane of the cytoplasm that separates it from the cell sap in the vacuole. This cell sap has a high concentration of organic solutes dissolved in it, so that it has an osmotic pressure of 4 to 10 atmospheres, which is much greater than that of the soil solution. Since these two solutions are separated by a semipermeable membrane, it is but natural that water molecules should pass more freely into the cell than out, causing water to move from the soil into the plant. Once in the root hair, the water increases the turgor of the latter. This water then passes into the cortical cells and through them to the xylem. There it enters the vessels and moves up through the root into the stem. See also **Ascent of Sap**.

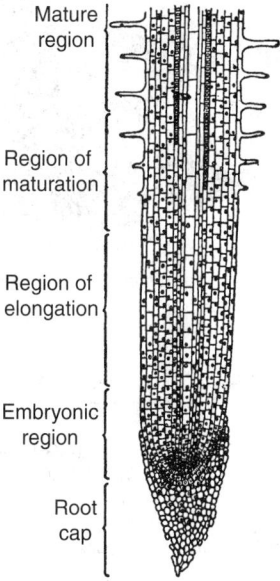

Fig. 6. Longitudinal section of onion root showing the regions of development.

Mature region

Region of maturation

Region of elongation

Embryonic region

Root cap

The innermost layer of cortical cells is very distinct, forming the endodermis. In this layer, the walls of certain cells become thickened and lignified; other cells, called passage cells, retain their thin walls and permit free flow of water to the xylem. In Monocots the walls are all thickened.

Within the endodermis is the stele (see Fig. 7), or central cylinder. The first cells to differentiate in this region are the water-conducting xylem elements, which arise in discrete groups. The number of these groups is usually constant for any given species of plant, cross-sections through the root at this stage showing a number of isolated groups of cells surrounded by unmodified cells. Between these groups of cells and somewhat farther from the center of the root other groups of cells become evident as the root grows older. These are the phloem cells. While the walls of xylem cells soon become thickened by the formation of ligno-cellulose against the primary wall, those of the phloem cells remain constantly thin. With continued growth of the root, the cells of the central portion gradually become changed into additional xylem cells, so that finally, in most roots, no unmodified cells remain in the center; that is, characteristically the root does not contain pith cells. Remaining outside the phloem and xylem cells are many only slightly changed cells, which form the pericycle. This is the region from which branch roots originate, and also the region which by the division of its cells forms the cork cambium which produces the cork in the outer bark of tree roots after the first few months of growth. The root therefore contains the following tissues, derived entirely from the modification of the cells originally resulting from the divisions of the apical

ROPE. A structure comprised usually of from three to six strands of fibrous material, twisted in such fashion that the twist of the rope offsets the twist in the strands, thus making it possible for the strands to hold together. Until about 1850, nearly all fiber rope was hemp (*Cannabis*

sativa), a soft fiber cultivated since ancient times. Currently, well over 90% of the cordage used is made up of hard fibers, notably abaca and agave. Some of the physical characteristics of both hard and soft fibers are given under **Bast Fibers**.

Twine is small rope, usually composed of two or more threads. Yarn is also termed *thread*. A strand is defined as two or more yarns or threads twisted together, with the twist in a direction opposite to that of the threads. In commercial twine, the twist is usually right-handed. Cord is defined as small commercial twine, comprised of two or more threads. The standard packaging of rope is 1,200 feet (equals 200 fathoms) and is termed a coil. Rope is also packaged in half-coils.

Several synthetic fibers are used as well for the manufacture of rope, but usually have a cost disadvantage as compared with the natural fibers. Among the commonly used synthetics are nylon and Dacron, Glass, Saran, and polyethylene synthetics also have been used in rope manufacture.

Natural fibers, such as manila and sisal, will lose strength when exposed to dry air over long periods, but this strength can be regained through reconditioning, provided the exposure temperature was not excessive. Exposure to air in excess of 250 °F (121 °C) can cause a strength reduction of from 10 to 20%. Above 300 °F (149 °C), manila and sisal fibers begin to char. Nylon can withstand higher temperatures, but in addition to cost, has the disadvantageous characteristic of stretching appreciably when loaded. Glass fibers are incombustible, withstanding high temperatures, but do lose strength beyond a temperature of 400 °F (204 °C). They are considered serviceable, however, up to about 1,000 °F (538 °C). Objections are poor flexing and abrasion properties, particularly when wet. Saran fibers are serviceable up to about 170 °F (77 °C), rapidly losing strength at higher temperatures. The melting point is 340 °F (171 °C). High-tenacity polyethylene is essentially unaffected by temperature changes below 220 °F (104 °C) and the monofilaments, if containing no plasticizer, will remain flexible down to about −100 °F (−73.3 °C). A disadvantage of polyethylene is its slippery characteristic, making it difficult to use with capstans or winches.

ROSACEAE.   See Rose Family.

ROSAT (Roentgen Satellite).   The X-ray observatory ROSAT was operated in a near-earth orbit from 1 June 1990 through February 1999. ROSAT stands for Roentgen satellite, in honor of Wilhelm Conrad Röntgen, who discovered the X-rays in Würzburg in 1895 and won the first Nobel Prize in Physics in 1901. The general aim of the project was to explore the whole X-ray sky with unprecedented sensitivity, angular resolution and coverage. See Fig. 1.

The ROSAT project began in 1975 with a successful proposal of the Max Planck Institute for Extraterrestrial Physics (MPE) in response of an announcement of opportunity for large projects of the German Ministry for Research and Technology. The basic technology, high resolution Wolter optics and imaging proportional counter (PSPC) was developed at MPE in the 1970s and tested by means of high altitude rocket flights. In the early 1980s the project was internationalized: NASA agreed to provide the launch with the Space Shuttle and a focal plane instrument. The UK contributed a small EUV telescope and financial support. After the Challenger catastrophe in 1986, NASA offered a launch with a Delta rocket, which took place on 1 June 1990. Table 1 summarizes the division of tasks between Germany and its partners.

The X-ray telescope on ROSAT had three image detectors, which were sitting on a carousel. Two of them were position-sensitive proportional counters (PSPC), providing colored (energy resolved) images with 20 arc sec resolution over a 2 degrees field of view. The third detector was a multichannel plate detector (high resolution imager, HRI), delivering black and white images of high angular resolution (5 arc sec) over a field of view of about 0.5⁰. See Fig. 2.

ROSAT saw first light two weeks after launch on 17 June 1990. After six weeks of calibrations and instrument verifications, ROSAT performed half-a-year all-sky survey, the first ever done with an imaging X-ray telescope, leading to the discovery of 80,000 X-ray sources. Thereafter, the satellite was used in a pointed mode for more than seven years. This program was completely open for guest observers. In total, 9,000 fields in the sky have been studied, with an average observation time of three hours. The longest pointing, the ROSAT Deep Survey, lasted two weeks. In total, a few thousand scientists from 26 countries have used ROSAT data, and more than 4,500 scientific publications have been written by them until

Fig. 1.   The ROSAT Satellite. (*Max Planck Institute, MPE.*) *http://heasarc.gsfc. nasa.gov/docs/rosat/gallery/rosat_sat.html.*

TABLE 1. INTERNATIONAL COOPERATION ON ROSAT

GERMANY	
Satellite	Dornier/MBB, now DASA
Ground station	DLR/GSOC
X-ray mirrors	Carl Zeiss
Focal plane assembly	MPE
Two position sensitive proportional counters	MPE
German ROSAT Data Centre	MPE
German ROSAT EUV Data Centre	Astron. Institut, University of Tübingen
Overall Project Management	DLR/DARA
UNITED STATES	
High resolution imager	Smithsonian Astrophysical Obs./GSFC
Launch (Delta 2)	NASA
Ground station back-up	NASA
US ROSAT Data Center	GSFC
US project management	NASA/GSFC
UNITED KINGDOM	
EUV Wide Field Camera	University of Leicester (UL) and the WFC Consortium
UK ROSAT Data Center	Rutherford Appleton Laboratory (RAL)
UK Guest Observer Centre & Survey Centre	UL
UK project management	RAL/UL

September 2000. The ROSAT archives will be a precious astrophysical database for many years to come.

The first X-ray picture of the **moon** ever taken was obtained by ROSAT in 1990. It shows the sunlit side glowing in X-rays from the solar corona that are reflected at the lunar surface. The lunar disk casts a shadow on the "X-ray sky background." ROSAT discovered **comets** as a new class of X-ray sources. Comets are cold objects ("dirty snowballs"), and

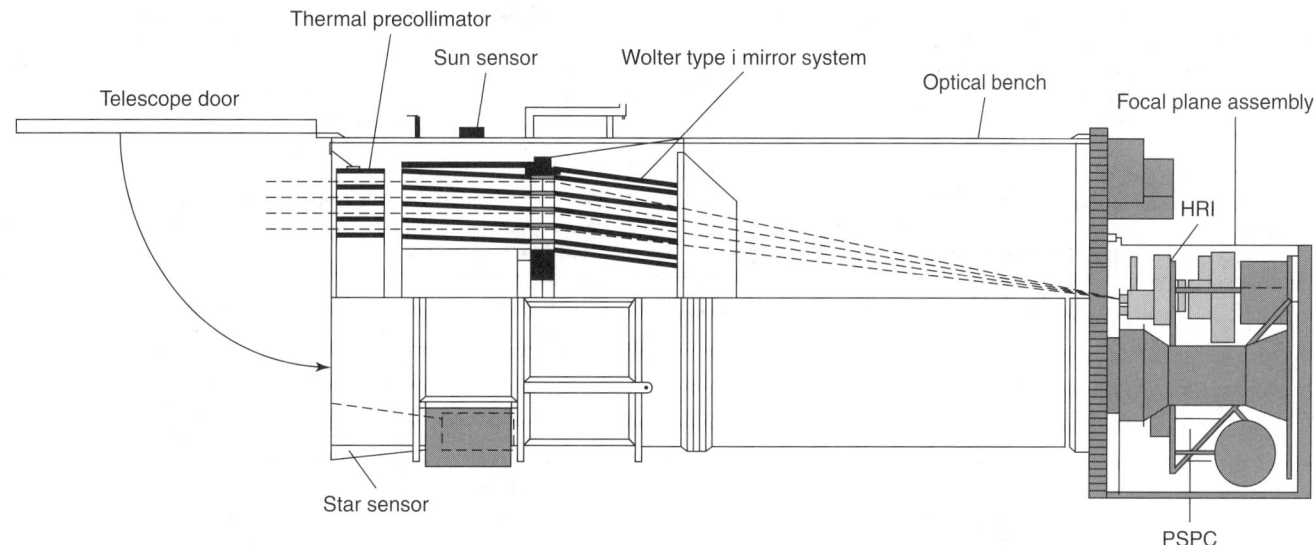

Fig. 2. Diagram of the ROSAT X-ray Telescope (the Wide Field Camera is not shown). X-rays would enter from the left and were focussed by the Wolter Type 1 mirror. The ROSAT mirror consisted of four nested parabola/hyperbola pairs. The double reflection is required for focusing in grazing-incidence optics of this type. The incident X-rays were focussed on the instruments carried by the focal plane assembly. (*Max Planck Institute, MPE.*) *http://heasarc.gsfc.nasa.gov/docs/rosat/gallery/rosat_diagram.html.*

therefore, it was a great surprise when Comet Hyakutake turned out to be a bright X-ray source. The total number of comets detected with ROSAT is a dozen. The data suggest that the X-rays are due to the transitions of excited carbon, nitrogen and oxygen atoms which are produced by charge exchange reactions between solar wind ions (e.g., C^{5+}, C^{6+}, O^{6+}, O^{7+}) and water molecules in the cometary coma.

ROSAT has given many new results in stellar astronomy, both on the coronal emission of cool **stars** and on the emission from shock heated gas in the strong stellar winds of hot and massive stars. A surprising results of the ROSAT All Sky Survey was the detection of many young stars (T Tauri Stars) far outside the known star-forming regions. They either have been ejected from their birthplace with high velocities or have been born in cloudlets which are not visible any more. ROSAT also discovered X-ray emission from **Brown Dwarves**. The ROSAT survey revealed many fewer **White Dwarves** than predicted. The reason is a pollution of the White Dwarf atmospheres by heavy elements that absorb soft X-rays very effectively.

In the beginning of the ROSAT mission a number of objects were discovered emitting extremely soft X-ray rays. They are very bright and show temperatures of a few hundred thousand Kelvin. It turned out that these sources are species that had been predicted to exist, but which had not been found before ROSAT. They are White Dwarves in binary systems, accreting matter from their companion at a rate just sufficient to sustain steady nuclear burning on the surface of the White Dwarf. This provides the unique opportunity to directly observe a thermonuclear reactor, which is usually deeply buried in the interior of a star.

The famous **Supernova 1987A** was the ROSAT First Light Target on June 16, 1990, but it was too faint at that time. It was first discovered in soft X-rays with ROSAT in 1992 and is steadily becoming brighter since*. In total, some 200 **supernova remnants** (SNR) have been found with ROSAT. In the Vela, SNR protrusions were found at the periphery of the shell type source; they are thought to be produced by fragments of the exploding star. ASCA observations revealed that they have different chemical compositions. A new SNR was found in the Vela complex which must be very young ($\sim$680 years), since it shows gamma ray emission from radioactive ^{44}Titanium (lifetime 90 years) detected subsequently by the Compton Gamma ray observatory. See also **Compton Gamma-Ray Observatory (CGRO).**

With ROSAT 34 **radio pulsars** were detected in X-rays showing an X-ray luminosity proportional to the rotational energy loss, suggesting that their X-ray emission is produced by high energy electrons accelerated in the pulsar magnetosphere. Four of these radio pulsars, including the Vela pulsar, exhibit an additional thermal spectrum corresponding to a temperature of $\sim$1 MioK, which is interpreted as the thermal radiation from the photosphere of the **neutron star**. A few point sources discovered near the centers of young Supernova remnants also show very soft

X-ray emissions, which must be attributed to photospheric emissions. The importance of these observations lies in the fact that the surfaces of these tiny stars (radius 10 km) have now become visible! Future X-ray spectroscopy should allow to measure their enormous gravity as well as their radii which contains important information on the physical properties of matter at supranuclear densities.

The ROSAT survey of the **Andromeda Nebula** yielded 500 X-ray sources. As in the Milky Way, the brightest of them are young supernova remnants or binary systems with neutron stars or black holes accreting matter from their companion. These bright source populations have been studied with ROSAT in many **galaxies**. In addition, the hot interstellar medium, which is heated by supernova explosions, has been investigated.

A small fraction of all galaxies have an **active galactic nucleus** (AGN) emitting huge amounts of energy in all spectral bands. Quasars represent the most powerful AGN. The irradiation is variable, indicating that it is emitted from a small region, generally not more than a few light weeks or months across. AGN often show jets, originating in their core. It is generally believed that the central engine is a supermassive black hole, swallowing matter at a high rate. The famous quasar 3 C 273* contains a black hole of $\sim$10^9 M$_{\odot}$ accreting one earth mass per second.

Because of their large luminosity AGN can be detected at large distances or redshifts. In the ROSAT deep survey about 1000 (!) sources per square degree in the sky were detected, mostly quasars and other classes of AGN. These observations showed that the X-ray sky background is at least to 80% produced by these faint sources, thus solving one of the oldest puzzles in X-ray astronomy.

ROSAT observations have shown that the hot plasma ($\sim$10^8 K), a typical **cluster of galaxies**, has a mass which is a few times larger than the total mass of all visible galaxies. The X-ray data also allow to determine the total gravitational mass of the cluster which turns out to be a few times larger than visible in X-rays and in optical light. The physical nature of the missing "dark matter" is still unknown. Many clusters show double structures, indicating the merging of two clusters or complicated inner structures. This must be due to earlier merging or interactive processes. Thus, one "sees" how these large structures evolve in the universe. However, the evolution of the cluster population with redshift is smaller than expected in a high density universe. Actually, the matter density inferred from cluster evolution is only about 30% of the "critical density" which is necessary to close the universe.

See also **X-ray Astronomy**.

Web References

X-Ray Astronomy ROSAT MPE: *http://wave.xray.mpe.mpg.de/rosat*
A large number of ROSAT images of the X-ray sky and about the ROSAT instruments can be found in ftp://ftp.xray.mpe.mpg.de/rosat/images/rosat/

JOACHIM TRUEMPER, Ph.D., Max Planck Institute.

ROSE CHAFER *(Insecta, Coleptera).*   Of the family *Scarabacidae,* this insect is particularly damaging to grape. The gray or fawn-colored, slender beetles, with long legs, consume both leaves and blossoms of grape. They are quite damaging to newly set grapes. The beetles are found in largest numbers during the first few weeks after bloom. The insect is also damaging to cherry (it eats the ripe fruit) and rose (it eats the leaves). Poultry also can be poisoned by eating rose chafers. Other plants attacked include apple, cabbage, clover, corn, garden beans, and peppers, as well as bush berries, such as blackberry, raspberry, and strawberry. Cultivation is the best protection. This should be practiced in May and June at the time when the insect is in the pupal stage. Relatively few rose chafers are found in areas where crops that require cultivation, such as corn and potato, are raised. See Fig. 1.

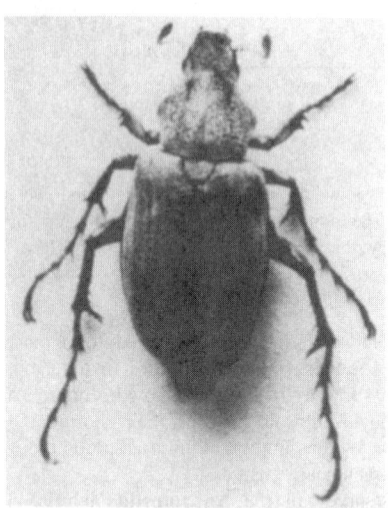

Fig. 1.   Rose chafer. *(USDA.)*

ROSE CURVE.   A higher plane curve of trochoidal type (see **Cyclic Curve**). Its equation in polar coordinates can be found from the general equation of the cyclic curve if one sets $a = (R - r)$, $n\theta = R\phi/2r$, and $k = 2a$. The result is $r = k \sin n\theta$. If n is an odd integer, there are n leaves, and if n is even, $2n$ leaves. Substitution of $\cos n\theta$ for $\sin n\theta$ rotates the curves by $45°$, so that one leaf lies on the positive OX-axis. The origin in each case is a multiple point, since nodes occur and there are n tangents. These curves are also sometimes called *trifolium, quadrifolium,* etc. See Fig. 1. (See also **Folium of Descartes**; and **Lemniscate of Bernoulli,** which are similar in appearance to the rose curves.)

See **Curve (Higher Plane).**

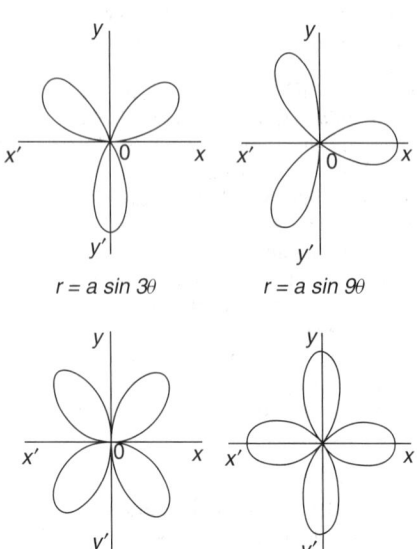

$r = a \sin 3\theta$ $r = a \sin 9\theta$

Fig. 1.   Rose curves.

ROSE FAMILY.   Included in this family (*Rosaceae*) are some 2,000 species of plants, a few of which are of tremendous value to man. Most of them are perennial plants, either trees or shrubs or herbs. Nearly all have alternate leaves, either simple or compound, with stipules present. The flowers are of many forms, and grow in racemes or cymes. Frequently the receptacle is more or less hollowed and often forms a part of the mature fruit. The flowers are commonly perfect and have their parts in multiples of five. The fruit may be either dry or fleshy, and in many species is an aggregate of many individual fruits. The flowers are usually conspicuous and insect pollinated. The most important economic members are found in three of the five or more suborders. One of these, the *Pomoideae,* has the carpels often united together and fused with the inner wall of the receptacle, which becomes fleshy; such a fruit is called a pome. In this subfamily are found the apple, pear, and quince.

Rosoideae, a second subfamily of *Rosaceae,* is large, containing many genera and species. Here the fruit is one-seeded and indehiscent, with usually many carpels borne on an enlarged central stalk or carpophore. In some genera, the floral axis encloses the carpels in the mature fruit. Shrubs or herbs with simple or compound leaves and various inflorescences comprise the group. Several are important plants extensively cultivated.

Drupoideae, the third major subfamily of *Rosaceae,* contains genera of considerable commercial importance. Members of this subfamily are trees or shrubs with simple leaves, and the fruit, a drupe, is usually one-seeded, the ovary wall or pericarp being differentiated into a fleshy outer portion, the exocarp and mesocarp, and an inner endocarp which is very hard or stony. In these plants, the bark contains a gum that frequently exudes in masses. The leaves, bark, and seeds are bitter due to the presence of a glucoside amygdalin, which, in the presence of an enzyme emulsin, produces prussic acid. The important economic members are the peach, the apricot, and related species, the plums and prunes, and the cherries. The almond is also a member of this subfamily.

Apple Tree (*Pyrus Malus*)

A medium-size tree, usually less than 40 feet (12 meters) in height. The wood of the tree is red, hard, and dense and sometimes is used for making tool handles. The flowers are borne on short lateral branches known as spurs, and are pink and pleasantly fragrant. They are self-sterile and insect-pollinated, particularly by the honeybee. The fruit is a pome, the fleshy part of which is regarded as modified stem tissue, the receptacle.

When unripe, the fruit contains much malic acid and starch. On ripening, the amount of malic acid decreases, while much of the starch changes to sugar. At the same time, characteristic aromas and flavors develop. When ripe a change occurs in the middle lamella of the cell wall, the substance dissolving away in part so that the cells become more or less separate. On further ripening, the flesh in certain varieties becomes mealy and loses most of its taste. Each of the five carpels that form the parchmentlike core of the apple contains two seeds.

The taxonomy of the commercial apple is obscured by centuries of selection and breeding experimentation and thus it is difficult to assign the numerous varieties to certain species. Just one example of the variation among the numerous types is shape, ranging from oblate to spherical and from conical to ovoid. See Fig. 1.

Champion specimens of apple trees in the United States are described, along with other trees of the rose family, in Table 1.

Propagation of the apple tree may be accomplished by seed planting, but with doubtful success since the new plant will usually bear small undesirable fruits. To perpetuate desirable strains, therefore, grafting or budding is usually practiced. Many people top-graft old apple trees; frequently a single tree may have a dozen or more varieties grafted thereon.

Temperature is a major factor limiting areas of apple production. The trees require a sufficient period of cold to induce dormancy. Without the cold period apples do not flower and fruit properly. Thus, although the trees will grow in semitropical-to-tropical climates, they will not set fruit satisfactorily and generally will not be vigorous. Conditions damaging to apples include long, hot summers and spring frost injury. The trees do not tolerate high humidity and excessive moisture very well. A temperature of less than $45°F$ ($7.2°C$) for at least 900 to 1000 hours (39 to 42 days) is needed to break dormancy. When the trees are not exposed to sufficient cold for the proper length of time, the leaf buds will not open in the spring. The flower buds require less cold than the leaf buds. When leaves do not appear shortly after blossoming, the fruit set will be small or none at all. Temperature requirements vary considerably with the particular variety of fruit.

Fig. 1.   The apple is a fruit of many variations. There are over 20 leading varieties (see Table 2), with a range of colors and shapes. One botanist estimates that over a thousand varieties of apple have been known, the majority of which have not been candidates for commercialization. (*USDA.*)

Although apples can be produced in many temperate regions, there are usually some shortcomings in any region. On the average, Washington State has a near-ideal climate for many apple varieties, but periodically the crop will be severely damaged by spring freeze conditions. In Michigan and New York, problems include cold injury in winter and spring, plus high humidity and moisture during summer.

The ranking of leading varieties of apples produced in the United States is shown in Table 2. Apples are usually grouped into summer and late-fall varieties.

Apples are used in several ways. Many are eaten raw; others used as sauce or made into pies and tarts. The high pectin content causes them to be much used in the making of jelly. Large quantities of apples, particularly cull fruit, are ground up and the juice pressed out to produce cider. Fermentation causes cider to acquire a considerable percentage of alcohol, it then being known as hard cider. By the action of acetic acid bacteria, cider may be changed to vinegar. From the crushed pulp, called pomace, remaining after the juice is expressed, commercial pectin is obtained.

Controlled-Atmosphere Storage. The use of controlled-atmosphere storage (CA) for apples has been increasing steadily since the early 1970s, when less than one-third of fresh-stored apples were stored by this method, compared with well over half of the fresh-stored apples as of the early 1980s. Apples in CA storage ripen, respire, and soften much more slowly than those in regular cold storage. Experience has shown that the storage life of the McIntosh variety can be doubled by CA. CA storage, a process first developed in England (Kidd and West) and further engineered at Cornell University (Smock), dates back to the early 1940s. Storage costs of the CA method are about twice those of traditional cold storage, a factor that has slowed its use to some extent, but increasing use reflects the outstanding advantages gained from it. Basically, in CA, the oxygen and carbon dioxide content of the atmosphere are adjusted in accordance with the requirements of the fruit. Oxygen level is usually reduced from the 20.9% content of oxygen in normal air to 2–3%, whereas the normal 0.03% carbon dioxide of air is increased in the process to a range of 1–8%, the exact values of oxygen and carbon dioxide depending upon variety of fruit stored and the storage temperature used. Generally, the temperature is held above 29 °F (−1.7 °C) and below 32 °F (0 °C). The relative humidity is usually maintained somewhat in excess of 90%. Packing of the fruit in closed spaces is done carefully to ensure good air circulation throughout the fruit. Initially, the fruit is quickly chilled. Varieties that respond well to CA include Delicious, McIntosh, Rome Beauty, Jonathan, Stayman, and Newton Pippin, with somewhat poorer results with Golden Delicious.

Productivity. Apple production has undergone a number of changes during the present century and particularly since World War II. The number of orchards has been markedly reduced, the remaining orchards and new orchards have been much larger, and, during the first half of the century, yields of apples per unit area of ground required have increased

three- to four-fold. Over the last several years, the trend has been to high-density plantings. Trees are much smaller than the apple tree that for many years was considered standard. The trend has been to semidwarf and dwarf trees. The smaller trees have a number of advantages, including their better utilization of sunshine and consequent production of higher-quality fruit. Smaller trees usually produce earlier than the standard trees. Even though the dwarf tree produces less fruit per tree, the number of trees per area can be substantially increased, so that yields per acre (hectare) are increased. The smaller trees, however, require greater care than the standard sizes if maximum production is to be achieved. A dwarf tree is expected to develop not taller than about 6 to 8 feet (1.8 to 2.4 meters); a semidwarf tree will be from 10 to 12 feet (3 to 3.6 meters) tall; a standard tree will average some 20 feet (6 meters) in height. The ultimate size of the tree is determined by the rootstock and by the soil richness, as well as the cultural practices (pruning, etc.) used by the orchardist.

Cultural Practices. Seeds of the tree rarely improve upon the parent and thus seedlings are chiefly used to produce stocks for grafting or budding. Standard trees are produced by planting seeds, by growing the seedling for 2 years, and by grafting the desired variety on them. Dwarfing rootstocks must be propagated vegetatively by rooting the shoots of specific rootstocks in stoolbeds. Dwarfing rootstocks have been studied extensively in Europe since about 1912, such studies commencing considerably earlier than in the United States. Quite early in this century, the East Malling Research station in Kent, England cataloged a series of rootstocks, ranging from very dwarf to near standard-size trees. Most of the common dwarfing rootstocks so developed carry the designation M (standing for Malling) and a number.

The rootstock with the greatest dwarfing effect in common use is the M9. Trees on this rootstock rarely grow larger than 8 feet (2.4 meters) in height. The M9, however, has a poor root system that requires good soil and frequent irrigation, and the trees on this rootstock must be staked or grown beside a trellis. Free-standing trees grown on M9 rootstock without support are frequently toppled over by heavy weight of fruit and by high winds. The next most dwarfing rootstock is M26. Trees on this rootstock are somewhat larger and the root system is stronger, but these trees also require staking. The M7 and M106 rootstocks produce semi-dwarf trees. These trees generally do not require staking, but the M7 rootstock still may be pushed over by heavy winds. M2 and M111 rootstocks produce more vigorous semi-dwarf trees and their root systems are less sensitive to diseases. Trees on M106, M2, and M111 rootstocks are relatively large and require a more laborious training method during early years.

Trees may be dwarfed by the *interstem method.* They usually require grafting twice. First, a piece of dwarfing rootstock is grafted onto a large root system of a semi-dwarf rootstock. The variety is then grafted on the top of the stem-piece, which becomes a so-called interstem. The length of the interstem determines the degree of dwarfing. Such trees are produced to

TABLE 1. RECORD TREES OF THE ROSE FAMILY IN THE UNITED STATES[1]

Specimen	Circumference[2]		Height		Spread		Location
	Inches	Centimeters	Feet	Meters	Feet	Meters	
APPLE							
Common apple (1993) *(Malus sylvestris)*	183	465	44	13.4	49	14.9	New Hampshire
Oregon crab apple (1989) *(Malus fusca)*	66	168	79	24.1	47	14.3	Washington
Pond apple (1989) *Annona glabra)*	125	318	44	13.4	47	14.3	Florida
Prairie crab apple (1994) *(Malus ioensis)*	38	97	46	14	68	20.7	Michigan
Southern crab apple (1996) *(Malus angestifolia)*	96	244	47	14.3	60	18.3	Maryland
Sweet crab apple (1976) *(Malus coronaria)*	70	178	37	11.3	35	10.7	Virginia
CHERRY							
Alabama black cherry (1995) *(Prunus serotina* var. *alabamensis)*	40	102	35	10.7	27	8.2	Florida
Bitter cherry (1997) *(Prunus emarginata)*	58	147	100	30.5	27	8.2	Washington
Bitter cherry (1999) *(Prunus emarginata)*	65	165	86	26.2	40	12.2	Washington
Black cherry (typ.) (1997) *(Prunus serotina* var. *serotina)*	210	533	134	40.8	70	21.3	Tennessee
Catalina cherry (1992) *(Prunus lyanii)*	120	305	41	12.5	55	16.8	California
Chokecherry, common (typ.) (1999) *(Prunus virginiana* var. *virginiana)*	168	427	74	22.6	88	26.8	Maryland
Chokecherry, western (1991) *(Prunus virginiana* var. *melanocarpa)*	54	137	73	22.3	18	5.5	Idaho
Escarpment cherry (1998) *(Prunus serotina* var. *eximia*	102	259	51	15.5	53	16.2	Texas
Hollyleaf cherry (1999) *(Prunus ilicifolia)*	54	137	50	15.2	56	17.1	California
Laurel cherry, carolina (1987) *(Prunus caroliniana)*	127	323	47	14.3	55	16.8	Florida
Laurel cherry, carolina (1996) *(Prunus caroliniana)*	128	325	47	14.3	49	14.9	Texas
Laurel cherry, english (1987) *(Prunus laurocerasus)*	96	244	32	9.8	52	15.8	Washington
Mahaleb cherry (1993) *(Prunus mahaleb)*	78	198	36	11	47	14.3	Washington
Mazzard cherry (1993) *(Prunus avium)*	263	668	80	24.4	80	24.4	Pennsylvania
Pin cherry (1999) *(Prunus pensylvanica)*	71	180	85	25.9	28	8.5	Tennessee
Sour cherry (1972) *(Prunus cerasus)*	119	302	68	20.7	75	22.9	Michigan
Southweatern black cherry (1996) *(Prunus serotina* var. *rufula)*	102	259	45	13.7	26	7.9	Arizona
Southweatern black cherry (1999) *(Prunus serotina* var. *rufula)*	100	254	39	11.9	42	12.8	Arizona
West Indies cherry (1989) *(Prunus myrtifolia)*	65	165	53	16.2	50	15.2	Florida
HAWTHORN							
Beautiful hawthorn (1993) *(Crataegus pulcherrima)*	25	64	46	14	31	9.4	Florida
Biltmore hawthorn (1982) *(Crataegus intricata)*	90	229	23	7	42	12.8	Virginia
Black hawthorn (1993) *(Crataegus douglasii)*	111	282	41	12.5	57	17.4	Washington
Blueberry hawthorn (1993) *(Crataegus brachyacantha)*	98	249	36	11	46	14	Texas
Cerra hawthorn (1997) *(Crataegus erythropoda)*	24	61	16	4.9	21	6.4	California
Cockspur hawthorn (1987) *(Crataegus crus-galli)*	60	152	40	12.2	48	14.6	Virginia

TABLE 1. (*Continued*)

Specimen	Circumference[2]		Height		Spread		Location
	Inches	Centimeters	Feet	Meters	Feet	Meters	
Cockspur hawthorn (1994) (*Crataegus crus-galli*)	71	180	33	10.1	30	9.1	Kentucky
Columbia hawthorn (1992) (*Crataegus columbiana*)	38	97	24	7.3	23	7	Idaho
Dotted hawthorn (1979) (*Crataegus punctata*)	97	246	38	11.6	38	11.6	West Virginia
Downy hawthorn (1972) (*Crataegus mollis*)	105	267	53	16.2	62	18.9	Michigan
Fanleaf hawthorn (1985) (*Crataegus flabellata*)	24	61	30	9.1	18	5.5	Virginia
Fanleaf hawthorn (1988) (*Crataegus flabellata*)	26	66	30	9.1	28	8.5	Virginia
Fleshy hawthorn (1991) (*Crataegus succulenta*)	51	130	21	6.4	30	9.1	West Virginia
Frosted hawthorn (1991) (*Crataegus pruinosa*)	64	163	30	9.1	36	11	Virginia
Frosted hawthorn (1991) (*Crataegus pruinosa*)	63	160	32	9.8	32	9.8	Virginia
Green hawthorn (1981) (*Crataegus viridis*)	61	155	40	12.2	45	13.7	West Virginia
Kansas hawthorn (1995) (*Crataegus coccinioides*)	33	84	30	9.1	36	11	New York
Littlehip hawthorn (1999) (*Crataegus spathulata*)	42	107	35	10.7	32	9.8	Georgia
May hawthorn (1993) (*Crataegus aestivalis*)	31	79	43	13.1	19	5.8	Texas
Oneflower hawthorn (1991) (*Crataegus uniflora*)	14	36	18	5.5	16	4.9	Florida
Oneseed hawthorn (1992) (*Crataegus monogyhna*)	111	282	37	11.3	58	17.7	Washington
Parsley howthorn (1996) (*Crataegus marshallii*)	46	117	27	8.2	33	10.1	Mississippi
Pear hawthorn (1991) (*Crataegus calpodendron*)	23	58	20	6.1	25	7.6	Illinois
Riverflat hawthorn (1989) (*Crataegus opaca*)	45	114	29	8.8	36	11	Mississippi
Scarlet hawthorn (1983) (*Crataegus coccinea*)	54	137	37	11.3	29	8.8	New York
Washington hawthorn (1987) (*Crataegus phaenopyrum*)	54	137	33	10.1	39	11.9	Tennessee
Washington hawthorn (1988) (*Crataegus phaenopyrum*)	54	137	36	11	38	11.6	Virginia
Washington hawthorn (1988) (*Crataegus phaenopyrum*)	62	157	30	9.1	32	9.8	Virginia
Yellow hawthorn (1999) (*Crataegus flava*)	36	91	15	4.6	19	5.8	Virginia
PEACH							
Peach (1986) (*Prunus persica*)	72	183	18	5.5	32	9.8	Virginia
Peach (1994) (*Prunus persica*)	50	127	37	11.3	25	7.6	Maryland
PEAR							
Common pear (1991) (*Pyrus communis*)	174	442	59	18	56	17.1	Washington
PLUM							
Allegheny plum (1991) (*Prunus alleghaniensis*)	37	94	37	11.3	24	7.3	Virginia
Allegheny plum (1991) (*Prunus alleghaniensis*)	52	132	24	7.3	32	9.8	Virginia
American plum (1993) (*Prunus americana*)	39	99	48	14.6	36	11	Florida

(*continued*)

TABLE 1. (*Continued*)

Specimen	Circumference[2]		Height		Spread		Location
	Inches	Centimeters	Feet	Meters	Feet	Meters	
Canada plum (1972) (*Prunus nigra*)	50	127	51	15.5	48	14.6	Michigan
Chickasaw plum (1987) (*Prunus angustifolia*)	51	130	32	9.8	32	9.8	North Carolina
Darling plum (1992) (*Reynosia septentrionalis*)	21	53	28	8.5	13	4	Florida
Flatwoods plum (1991) (*Prunus umbellata*)	18	46	34	10.4	24	7.3	Florida
Garden plum (1993) (*Prunus domestica*)	130	330	47	14.3	48	14.6	Oregon
Guiana plum (1998) (*Drypetes lateriflora*)	21	53	23	7	8	2.4	Florida
Hortulan plum (1999) (*Prunus hortulana*)	34	86	27	8.2	30	9.1	Missouri
Klamath plum (1972) (*Prunus subcordata*)	42	107	28	8.5	19	5.8	Oregon
Mexican plum (1999) (*Prunus mexicana*)	28	71	39	11.9	36	11	Mississippi
Pigeon plum (1994) (*Coccoloba diversifolia*)	84	213	49	14.9	22	6.7	Florida
Wildgoose plum (1991) (*Prunus munsoniana*)	56	142	20	6.1	20	6.1	Illinois
SERVICEBERRY							
Allegheny serviceberry (1997) (*Amelanchier laevis*)	78	198	101	30.8	36	11	Tennessee
Downy serviceberry (1986) (*Amelanchier arborea*)	108	274	60	18.3	53	16.2	Virginia
Roundleaf serviceberry (1989) (*Amelanchier sanguinea*)	54	137	34	10.4	42	12.8	Vermont
Utah serviceberry (1996) (*Amelanchier utahensis*)	268	681	38	11.6	46	14	California
Western serviceberry (1993) (*Amelanchier alnifolia*)	39	99	42	12.8	43	13.1	Washington

[1] From the "National Register of Big Trees," American Forests (by permission).
[2] At 4.5 feet (1.4 meters).

TABLE 2. LEADING VARIETIES OF APPLE PRODUCED IN THE UNITED STATES

Variety	Percent of Total
Delicious	37.9
Golden Delicious	16.4
Rome Beauty	8.1
McIntosh	7.8
Jonathan	5.1
Stayman	2.8
York Imperial	2.8
Newtown Pippin	2.6
Winesap	2.4
Cortland	1.8
Gravenstein	1.5
Northern Spy	1.3
Rhode Island Greening	1.2
Other varieties	8.3

Source: International Apple Institute.
Note: Other varieties include Baldwin, Ben Davis, Empire, Grimes, Idared, Lodi, Macoun, Milton, and Wealthy.

take advantage of the larger root and the dwarfing effect of the interstem. It requires 1 year longer to produce double-grafted trees. See Fig. 2.

Ultra-dwarf trees, sometimes referred to as meadow orchards, are being pioneered in England and Israel.

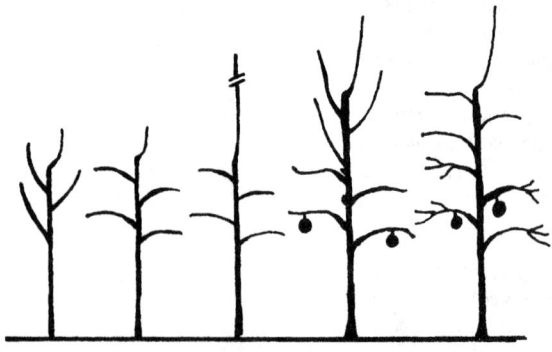

Fig. 2.   Development of a dwarf apple tree. (*After Faust.*)

Almond Tree (*Prunus amygdalus*)

The almond is related to peach and stone fruits and is a medium-size tree with pale pink or white flowers, probably native to western Asia and northern Africa. The fruit, a drupe, has the seed or kernel enclosed within a reticulated endocarp. There are two kinds of almond, bitter and sweet. Bitter almonds, used for flavoring, contain some hydrocyanic acid (HCN). The sweet almond is used as a dessert and for confections and yields almond oil.

There are numerous references to the almond in the Old Testament, suggesting that the crop has long been familiar in the countries of the Near East. Almost invariably, these references concern the esteem with which almonds were held, and their value as gifts. Gradually, plantings of the crop moved westward to the Mediterranean region and, for many years, Italy and Spain, were the two major and rival producers of almonds. There are reports of almonds being brought from Europe to the United States as early as 1840 for planting along the eastern coast. Climate and soil conditions were not favorable to the almond and, consequently, production in the United States had to await experimental plantings in California. In the 1870s, research and cross-breeding developed basic stock that resulted in the Nonpareil, NePlus, and other varieties which currently are popular in the West Coast almond industry.

Nutritional Aspects. The dry almond has a water content of 4.7% and a relatively high food energy content of 598 calories per 100 grams. Protein content is 18.6%; fat, 54.2%; and carbohydrate, 19.5%. Like most nuts, the almond is high in glutamic acid.

Almond flowers are self-incompatible and thus two or more compatible varieties must be selected for a planting to ensure pollination. In California, the Nonpareil variety, a consistent and heavy producer, accounts for about 55% of all plantings. Frequently, NePlus variety will be planted in rows with Nonpareil, for a total of 50 to 90 trees per acre (124 to 222 trees per hectare). The NePlus plantings will average about 7.5% of the total. To ensure a good crop, the grower usually will have honeybees brought in by a commercial beekeeper during the bloom period. Under favorable conditions, a well-managed orchard will yield approximately 1 ton (2000 pounds) of nut meats per acre (2240 kilograms per hectare). In recent years, a number of varieties have been introduced. Davey, Karpareil, Merced, and Thompson, for example, are also good pollinators for Nonpareil, and some authorities consider that these also produce a superior kernel.

Almonds are the first crop to bloom in spring. In California, this occurs in late February. From that point, the orchards must be virtually frost-free, rains should be at a minimum, and the days sufficiently warm (55–60 °F; 12.8–15.6 °C) to ensure full pollination by the bees. By mid-March, the trees are leafed out and small, fuzzy, gray-green nuts will be observed. As the weather warms, the crop matures rapidly. In early July, the hulls begin to split open, slightly exposing the shell within. The kernels then start to dry, the split widens, and the nuts are soon ready for harvest.

Almond Flavorings. Some food processors find it more convenient to use almond flavoring rather than almonds per se in various products. Bitter almond oil is a principal source of almond flavorings.

Bitter almond oil is obtained by cold expressing the fixed oils from previously ground bitter almond (*Prunus amygdalus*), apricot (*P. armeniaca*), and peach (*P. persica*). The kernels of these fruits contain the glucoside amygdalin, which when hydrolyzed by enzymes, will yield benzaldehyde and hydrogen cyanide. Bitter almond oil for use as a flavoring must be distilled to render it fully free of hydrogen cyanide (HCN; prussic acid). After pressing, the ground kernels are mixed with water to complete the aforementioned enzymatic hydrolysis. The mixture is then steam distilled to yield from 0.5% to 0.7% essential oil. Oil intended for food use is further treated to remove HCN by precipitation as an insoluble calcium ferrocyanide. The oil is used in both nonalcoholic and alcoholic beverages, in ice creams, candies, baked goods, gelatins and puddings, chewing gums, and maraschino cherries. Levels of usage in these products range from about 30 parts per million up to about 340 parts per million. Most countries generally regard the oil safe (GRAS) if it is free from prussic acid.

Apricot Tree (*Prunus armeniaca*)

The apricot is a stone fruit (drupe) and ranks fifth in terms of total worldwide deciduous fruit production. Of the genus *Prunus*, the apricot is *Prunus armeniaca*. This genus also includes almond, cherry, nectarine, peach, and plum. A drupe is a fleshy, one-seeded fruit that does not split open of itself, and with the seed enclosed in a stony endocarp, which is called a *pit*. As with other members of *Prunus*, the apricot tree has simple, alternate, serrate leaves. In the case of the apricot, they are ovate to round-ovate, abruptly short-pointed, closely serrated (toothed), thin, bright green, smooth above and below (usually), with stout stalks. See Fig. 3. Flowers are white, solitary, with no or very short stalks (peduncles) appearing from lateral buds of last year's growth or sometimes on short year-old spurs before the leaves appear. The fruit is short-stalked, somewhat flattened, nearly smooth when ripe mostly yellow, and overlaid more or less with red. The stone is flat and smooth, ridged on one edge. The tree is small, round-topped, with reddish bark. The eating quality of tree-ripened fruit rates high, but unlike most other stone fruits, is not widely available in the fresh market.

Fig. 3.   Apricot (*Prunus armeniaca*). (*USDA.*)

Background. Although the name of the species would indicate that its origin may have been Armenia, there is much more evidence that the apricot is native to China, as is the case of the peach and nectarine. A very early Chinese record (the Shan-hai-king) makes reference to a *sing* (believed to be apricot) in China as early as 2205 B.C. Field plant researchers also have found apricots growing wild in some of the mountain ranges of China. Pling recorded that the apricot reached Italy by about 100 B.C. No records have been found, however, that associate knowledge of the apricot with the period of the Greeks. There is not full agreement regarding the importation of apricot into England. Some authorities place the date at 1524; others say that the fruit was brought to England about 1620 by an armed privateer. The earliest mention of the fruit in North America is 1720, records indicating that the fruits were being grown in Virginia at that time. By 1835, 17 varieties of apricot were listed in British horticultural catalogs. By 1866, in North America, 26 varieties were listed by Downing. In 1879, the American Pomological Society recorded only 11 varieties. As with so many other fruits (and vegetables), the Spanish missionaries were responsible for introducing the apricot into southern California. A fine orchard of apricot trees was reported in Santa Clara, California as early as 1792.

Apricot trees are adapted to a variety of soils and climatic conditions, but the growing site must be relatively frost-free for the trees to produce fruit. The most frost-free sites are near large bodies of water, on the tops or sides of hills, or near the base of high hills or mountains. In the valleys, there is little air movement and the coldest air settles in the lowest places where damaging temperatures occur more frequently and for longer periods than on sites with good air drainage.

Apricot trees grow best in deep, fertile, well-drained soil, but they grow well in light, sandy soil when adequately fertilized and watered. Growers must avoid poorly drained soil, as well as sites where tomato, cotton, or brambles have grown because these crops harbor the *Verticillium* wilt fungus that causes "black heart" of apricot.

Apricot trees are vigorous and, unlike peach trees, are long-lived and hardy. They may grow to very large size, but it is desirable not to allow the trees to become larger than economical harvesting permits. Generally, apricot trees require less space than peaches and most plums. Planting distances range from 22 feet (6.6 meters) or somewhat less to 24 feet (7.2 meters and somewhat over). Planting density of all deciduous fruit trees has been the object of discussion of growers for many years.

Although a number of new varieties and cultivars of apricot have been introduced in recent years, the varietal situation applying to apricots is considerably simpler than is the case of peaches, for which thousands of varieties have been listed and hundreds of varieties have appeared during the last several decades.

Apricots are not commonly stored in commercial quantities, although they keep well for 1, 2, and 3 weeks within a temperature range of 31–32 °F (−0.6–0 °C). Fruit picked when firm enough to ship or store has about the same maturity as that commonly used for canning and lacks the character and full flavor of tree-ripened fruit. Apricots stored at 40–45 °F (4.4–7.2 °C) have less flavor and a more mealy texture after ripening than fruit stored at temperatures within the first-mentioned range. Relative humidity required is about 90%. After storage apricots will ripen at temperatures between 65 and 75 °F (18.3 and 23.9 °C).

Blackberry and Dewberry (*Rubus*)

These berries, as well as raspberry, are of the genus *Rubus* in the larger family *Rosaceae*. In these species, the fruit is an aggregate of many drupelets, which cling closely to the axis in the blackberries and dewberries, but slip free therefrom in the raspberries. An erect habit distinguishes blackberries from the prostate dewberries. All have a perennial root system and annual or biennial stems. Propagation is mainly by suckers and root cuttings. The fruits frequently are consumed fresh near their place of origin, but they are also cultivated on a commercial scale for shipment to distant markets and for processing.

The blackberry in one or another of its many varieties is native in most of the temperate parts of the northern hemisphere — in fact, brambleberries, of which blackberries are a part, inhabit almost the whole globe with exceptions of dry desert regions. The blackberries of North America fall into 5 major groups: (1) *Erect or nearly-erect bush*, such as the Early Harvest and Eldorado, which are found along the Atlantic Coast from Florida to Canada and westward to the prairie states; (2) *eastern trailing blackberries*, which inhabit about the same range; (3) the *southeastern trailing blackberries*, such as the Manatee and Advance, which range along the Atlantic and Gulf coasts from Delaware to Texas; (4) the *trailing blackberries*, from which the loganberry is derived, which occur on the Pacific Coast from southern California northward to Canada; and (5) the *semi-trailing evergreen* Black Diamond and Himalaya varieties found in Oregon, Washington and California, which were naturalized from European importations (Darrow, 1937). Possibly of more convenience, is the simpler classification in which erect blackberries (See Fig. 4) are referred to as blackberries and the trailing blackberries are called dewberries (Hedrick, 1948).

A variety or cultivar also may be described as being thorny or thornless.

Propagation. Erect blackberry plants can be propagated from root suckers or root cuttings. The latter method yields the greatest number of new plants. Trailing blackberries and some semitrailing varieties are propagated by burying the tips of the canes. These take root and form new plants. Thornless Evergreen and Thornless Logan varieties must be propagated by tipping, as the other methods give rise to thorny plants.

Pollination. Some self-unfruitful varieties of blackberries require cross-pollination. Other varieties, even though self-fruitful, may benefit from the pollen-distributing visits of insects. The flowers of blackberries are very attractive to honeybees, the primary pollinators, and small plantings are usually pollinated adequately by them. Control chemicals should be scheduled to avoid injury to pollinating insects.

Cherry Tree (*Prunus cerasus*)

Cherry trees are relatively small-fruited trees of medium size, indigenous to Europe. The principal cherries of commerce are (1) *Prunus avium* L. the sweet cherry; (2) *P. cervasus* L., the sour or tart cherry. Occasionally, the sour cherry hybridizes with unreduced pollen of sweet cherry to produce *Duke cherries*[1] (*P. gondouinii* Poit & Turp.). Other essentially nonfood species of cherry include *P. mahaleb*, of European origin and used as propagating stock; *P. mazzard*, also used as rootstock; *P. pennsylvanica*, the pin, wild red, or bird cherry; *P. bessey* and *P. pumilla*, both classified as sand or dwarf cherries; and *P. serotina*, the wild black cherry, highly valued for its wood in quality cabinet and furniture making. Some of the wild cherries can be used for making jellies, jams, wines, and cordials.

Botanically, cherries are more closely related to plums than to peaches or apricots of the *Prunus* genus. Some borderline *Prunus* species are difficult to classify definitely into one or the other group. Cherries generally differ from plums in that leaves emerging from buds are folded lengthwise, in contrast to being rolled in plums. Also, in cherries, the stone is more globular, the fruit and stone are smaller, and the flowers occur in corymbose (short, broad, more-or-less flat-topped inflorecence) rather than the umbelliferous (having a common point of attachment) clusters.

Cherry trees can become large trees, but in orchard practice relatively low-growing types are used and trees are trained low and wide by pruning. An example of a record sour cherry tree (not orchard-trained) is that selected by the "American Forests" and located in Calhoun County, Michigan. The tree is 9.9 feet (3 meters) in circumference at a height of 4.5 feet (1.4 meters); 68 feet (20.4 meters) high; with a spread of 75 feet (22.5 meters).

When fully hardened, sweet cherry trees are hardier than peaches, but more tender than most apple varieties. Cherry bark is somewhat like that of birch, but it is predominantly red or black. The leaves are oblong-ovate to obovate, large, and soft, with double teeth. Flowers are white, about 1 inch (2.5 centimeters) across, and come out with the leaves; they are in clusters. The cherry fruit is globular or oblong, long-stemmed, and white or deep red to black when ripe. See Fig. 5. These characteristics vary considerably with variety and type. Sweet cherry trees require cross-pollination from other varieties to set fruit.

Fig. 4.   Close-up of fruit of trailing variety of blackberry. (*USDA*.)

Fig. 5.   Sweet cherry (*Prunus avium*) ripening in a Michigan orchard. (*USDA*.)

---

[1] Duke cherries are tetraploid and intermediate between the parent species. Some of these are very similar to sour cherries in fruit characteristics, but have an upright growth habit.

Cherry trees are in the group of deciduous tree fruits that require a winter dormant period for proper development and fruit production. Thus, they are limited to temperate regions having sufficient winter cold to break the natural rest period. In their distribution northward in the Northern Hemisphere, they are limited by the duration and intensity of winter cold. If the trees are not exposed to sufficient cold, the buds do not open properly in the spring. For best growth, cherry trees require ample moisture in the soil of their root zone throughout the growing season. Since the trees develop large leaf areas, the total water requirement is relatively high. A minimum of 30 inches (75 centimeters) of precipitation or a combination of precipitation and irrigation must be made available. Several champion cherry trees in the United States are listed in Table 1.

Background. The first recorded mention of cherry is contained in Greek literature. Theophrastus in his "History of Plants" described the cherry as early as 300 B.C., and indicated that by then the fruit had been cultivated for a number of centuries. Early Chinese and Egyptian literature makes no mention of the fruit. Pliny suggested that Lucullus introduced the cherry to Italy in 65 B.C. Historical research in recent years, however, indicates that the cherry had been known in Italy at a much earlier date. Early herbalists and historians paid little attention to describing the different varieties of cherry or in giving them names. Gerarde, in 1596, wrote: "The Ancient herbalists have set down 4 kinds of cherry trees: (1) the first is great and wild; (2) the second is tame or of the garden; (3) the third hath sour fruit; and (4) the fourth is that which is called in Latin *Chamaecerasus,* or the dwarfe cherry tree. The later writers have found divers sorts more, some bringing forth great fruit, others lesser; some with white fruit, some with blacke, others of the colour of black blud, varying infinitely according to the climate and country where they are grown."

The cherry was one of the first fruits planted in the fields cleared and enriched by the American pioneers. From Canada to Florida, the colonists, although having different skills, were forced to turn to cultivation of the soil. Possibly, the French settlers were the first to plant cherries in Nova Scotia, Cape Breton, Prince Edward Island and in the early settlements on the Saint Lawrence River. Settlers in New England brought trees from England. A memorandum of the Massachusetts Company of March 16, 1629 stated: "Stones of all sorts of fruits, as peaches, plums, filberts, cherries, pear, apple, quince, kernells were to be sent to New England." For a long time after introduction in New York, the cherry in common with other fruits was grown as a species. Varieties and budded or grafted trees were probably unknown. The first nursery for cherry and several other fruit trees was established at Flushing, Long Island, New York in 1730.

The cherry quickly spread in Colonial days to the South and to the Franciscan missions in California. However, modern fruit growing on the Pacific Coast began in Oregon. Until 1847, the few cultivated fruits to be found in Oregon were seedlings, mostly grown by people associated with the Hudson Bay Fur Company. In 1847, Henderson Lewelling crossed the plains from Henry County, Iowa with a choice selection of grafted fruit trees. These he transported in boxes of soil, which he hauled in a wagon drawn by oxen. In this traveling nursery, Lewelling brought to Oregon cherries of the Bigarreau, English Morello, and probably several other types. The label of one of the cherries was lost and it was renamed the Royal Ann. Actually it was the Napoleon, which had been cultivated for more than 3 centuries. Later the work of a relative, Seth Lewelling, stimulated others to breed cherries, and their work included the well-known Lambert variety.

Varieties. *Sweet Cherries.* In most cherry-producing regions, sweet-cherry varieties are not as dependable as sour ones. They are more subject to difficulty in establishing the trees, and are subject to frost damage, cracking of fruit, brown rot, and loss of fruit from birds.

Sour or Tart Cherries. These varieties are also sometimes called *pie cherries or red cherries.* Sour cherries are separated into two distinct groups based on the color of the fruit juice: (1) the Amarelle or Kentish group, with colorless juice; and (2) the Morello or Griotte group, with reddish juice and usually dark-red fruit. In addition, a subspecies classification is sometimes given to the Marasca group, which is used for distilled or brined products. Further classification into Montmorency, Morello, Brusseler, and Vladimir groups has been suggested.

About 270 named cultivars (varieties) of sour cherries were described as early as 1914 by Hedrick. By 1950, only 12 more varieties had been added and, by 1970 only 5 additional varieties. Ten of the 17 varieties added since 1914 are mutations of Montmorency.

Numerous strains or mutations of Montmorency have been selected for productiveness, season of fruit maturity, or disease resistance. The fruits of these are similar to the Montmorency. Only three cultivars are of economic importance: (1) Montmorency; (2) English Morello; and (3) Early Richmond.

Duke Varieties of Cherries. Unless there is a known demand, these cherries are seldom grown on a large scale. The Duke varieties are neither sweet nor sour, but a blend of both. Most people find them too sour for eating fresh, but many prefer them for canning, freezing, and pie making.

Peach Tree *(Prunus persica)*

Native to China, peach trees are small and usually not very long-lived. They can be considered semi-hardy and, due to their habit of producing flowers very early in the growing season, before leaves appear, are frequently severely damaged by late heavy frosts. The calyx tube of the flower surrounds the pistil, but is not adnate thereto. The ovary is one-celled, but frequently contains two ovules when young, only one of them developing. Peach fruits are of two types. In one, the fleshy mesocarp slips readily from the stony endocarp. These are termed *freestone peaches.* In the other type, the two layers are closely adherent, giving *clingstone fruits.* Peach fruits are quite perishable when mature.

Clingstone and Freestone Types. Generally, the freestone-type peaches are sold on the fresh market, while clingstone peaches are usually processed. There are also some varieties that qualify as *semi-freestone* peaches. With all types there are varieties that have white flesh and others with yellow flesh. Generally, in North America, peaches with yellow flesh are in greatest demand.

Exemplary of the innovative thinking that goes into fruit breeding research is a study being made of freestone varieties at the Russell Research Center (Athens, Georgia). One scientist has observed that recognition of the relative levels of enzymes in freestone peaches compared with those of clingstone peaches may be helpful in improving the freestone peach. The freestone peach, as viewed by consumers is somewhat "ragged" when it comes out of the can, softening more than the clingstone peach. This softening problem is related to the "melting" characteristics of peach flesh. Freestones are characterized by a melting-type of flesh, while clingstones are frequently nonmelting. This characteristic tends to make the freestone quite difficult to process. Firming up the freestone would be very helpful to processors.

Scientists have found that the difference in texture may be due to more degradation of pectin in freestone than in clingstone fruit. Polygalacturonase is the enzyme involved in pectin degradation. In examining six varieties each of freestone and clingstone peaches, it was found that, at the unripe stage, all varieties of both types had low levels of water-soluble pectin and virtually no polygalacturonase activity. But in ripe freestone peaches there were high levels of soluble pectin and both endo- and exo-polygalacturonase. In the clingstone peach, only exo-polygalacturonase is present. The scientists found that endo-polygalacturonase is much more effective in degrading pectin than the exo-polygalacturonase. Thus, the conclusion—if pectin is the critical component in peach softening, the difference in enzyme composition accounts for the difference in softening characteristics. Further research, based upon this fundamental observation, may lead to later practical assistance to the processor.

Varieties. The selection of the variety of peach to plant is probably the most important decision a grower must make in developing a new orchard, or in replanting an old orchard. The selection process is not simplified by the large number of new varieties introduced each year. These emanate from a number of research teams that constantly are seeking improvements in the fruit for better quality, better resistance to diseases, insect, and environmental factors, and better shipping qualities, as well as making varieties available for different harvesting dates, among numerous other objectives. In a 50-year period of research, records indicate that some 700 new varieties of peaches have been introduced. Most of these varieties already have become obsolete.

The number of varieties introduced may seem large, but many are special-purpose varieties for adaptation by only one or a few of the several peach-growing regions. For example, the chilling requirement of a variety to break the rest period of its buds narrowly limits the area to which the variety may be adapted, particularly as regards the southern states. Florida requires very low-chilling varieties. In the North, winter hardiness of fruit buds is a limiting factor. Fruits of some varieties tend to develop enlarged sutures or elongated tips when grown in a warm climate. In the North where it is adapted, the same variety may produce perfectly round fruit. Maximum red color of fruit is developed when the temperatures just preceding maturity are cool. Thus, many varieties are not successfully grown in warm climates, such as California, because they lack sufficient exterior red color. Other varieties having adequate color for warm climates develop too much red color and are too dark in a cool climate, such as Michigan. Fruit of some varieties develops too much astringency when grown in the cool climates of western Oregon or Washington, where varieties with low astringency are needed. Resistance to bacterial leaf spot is important in varieties grown in Texas and the Southeast. It is less important elsewhere, such as California. Peach brown rot and leaf curl are diseases which also limit distribution (Weinberger).

Because of this constantly changing situation, specific varieties are not described here. The *Elberta*, although still a significant variety, has been taking on the role of a classic, with new varieties frequently compared against it.

Pear Tree *(Pyrus communis)*

Of Eurasian origin, the pear tree differs in several ways from the apple. See Fig. 6. Usually it is a taller tree, with a tendency to more upright growth. In contrast to the rough leaves of the apple, those of the pear are smooth and glossy. The wood is quite dense and sometimes used in the same manner as apple wood. The flowers are white. The fruit is more juicy and sweeter than is the apple, and the flesh especially when green contains an abundance of stone cells. Pears are used almost entirely as edible fruit, eaten either fresh or canned. The alligator pear (avocado) is not related to the true pear. Champion specimens of pear trees in the United States are described in Table 1.

Fig. 6.   Pear *(Pyrus communis)*. *(USDA.)*

Early cultivation of the pear probably predates recorded history by several centuries. Charred remains of the fruit have been found in the prehistoric lake dwellings of Switzerland. It is recorded that pears were cultivated in Greece at the time of Homer (850 B.C.), who referred to pears being grown in the garden of Alcinous and called them a "gift of the gods." Evidence shows that the Roman conquerors carried pears with them to the temperate parts of the Old World. Pliny, in his *Natural History*, names 41 varieties of pear. Thousands of varieties of pear are known. There were over 700 varieties in the Horticultural Society's garden in England as early as 1842. In 1866, T.W. Field of England cataloged 850 varieties of which he indicated 683 were of European origin. Pear seeds were mentioned in a communication between the Massachusetts Company and American colonists as early as 1629. It is believed that the first pear

tree in the United States was planted near Salem, Massachusetts in about 1630. This became known as the Endicott pear tree. Pear trees did very well in the eastern Colonial settlements. As with so many other fruits and vegetables, the pear was originally introduced into southern California by the padres for planting in their mission gardens. The first evidence of commercial pear sales in the West was during the period of the Gold Rush. However, commercial pear production in California did not commence until about 1914.

Varieties. More than 3000 varieties of pear are known in North America, but fewer than a dozen are of commercial importance.

Plum Tree *(Prunus avium)*

Of several different species, these small to medium trees have been cultivated since before the Christian era. The flesh of the fruit surrounds a hard pit in which there is a seed. There are more than 2000 varieties of plum, of which relatively few are of commercial importance. See Fig. 7. Champion specimens in the United States are listed in Table 1.

Fig. 7.   Plum *(Prunus domestica)*. *(USDA.)*

In a classic book, "The Plums of New York," Hedrick (1919) observed that, of all the stone fruits, plums furnish the greatest diversity of kinds. The trees are diverse in structure, some of the plums being shrublike plants with slender branches, others true trees with stout trunks and sturdy branches; some species having thin, delicate leaves and others coarse, heavy foliage.

In generalizing on the important plum trees of commerce, the trees are deciduous with simple and alternate oval leaves, veined, dull green to pale-green, serrated at the edges. The tree may be upright, spreading, drooping, or round-topped, depending largely upon climate, soil, and cultural handling. The bark of some varieties is birch-smooth, while that of other varieties is locust-rough. Flowers are white and generally appear with the leaves. The fruit is longer than broad, somewhat compressed, but may be smooth or roughened, with a groove or suture on the ventral side.

Nomenclature. At one time, the words "plum" and "prune" were used as synonyms for the fruits of hundreds of varieties comprising some 15 different species. A distinction in meaning evolved gradually. Botanically, all prunes are plums. In current usage, particularly in North America, *prune* signifies a variety that can be and is normally dried *without the removal of the pit*. The word refers to both the fruit in its fresh state and to the dried product. *Plum* designates a variety grown primarily for uses other than drying—mainly for fresh consumption, but also for canning, freezing, crushing, and jam- and jelly-making. Most plum varieties will ferment when dried with the pit. If they are dried after removal of the pit, the product is called 'dried plum' and not 'prune.' The *fresh prune*, which is grown extensively in the Pacific Northwest of the United States, does not fit into either category. Varieties of this prune are equally well suited and have been utilized in substantial volume for fresh use, canning, and drying. A more recent trend is to call these plums by the name "purple plum."

Species and Varieties. The two principal species of commercial plums are (1) *Prunus domestica*, also known as the European plum, varieties of which tend to be purple-to-black; and (2) *P. salicina*, also known as the Japanese plum, the varieties of which tend to be yellow-to-crimson.

Cultural Factors. Plums are propagated by budding on seedlings in the nursery, in a method similar to peach propagation. Peach rootstock does best in light, sandy, well-drained soils. It is sensitive to wet soil. About 50% or more of Italian prune trees in some regions are on peach rootstocks (in Idaho and foothill areas of California). Myrobalan rootstock is adaptable to a wide variety of soils and degrees of moisture, and its seedlings are compatible with a wide variety of Japanese—and European-type plums. Myrobalan is hardy, with deep roots, but is not particularly vigorous. Plums are sometimes budded on apricot where nematodes are a problem, or where there is soil of high alkalinity. Almond as a rootstock is best suited to warm, dry soils. Most Japanese plum varieties can be topworked on European varieties, producing durable trees.

Fresh Plums. Most of the fresh market plums grown in California are of the Japanese type. Tree growth habits vary considerably in character among the Japanese varieties, ranging from low and spreading to upright. Leaves are of medium size, pointed, and free from hair on the lower surface, a characteristic that distinguishes them from the European type. European plum trees are characteristically large, vigorous growers with large, thick leaves that are dark green above and pale green beneath. Fruit of the Japanese varieties is large and usually heart-shaped. It often has a pronounced apex, and is bright-red or yellow in skin color. The flesh is juicy and firm, either red or yellow. Most European plums marketed in fresh form belong to the Imperatice group, which are blue in color. The fruit is medium sized with firm flesh and thick skin.

Plums can be grown in most of the fruit production areas of California. Adequate growth can be attained on a fairly wide range of soil types. However, ample, good-quality irrigation water is required to produce commercially acceptable crops. For proper leaf and bud development, Japanese varieties require 700 to 1000 hours of winter chilling (temperatures below 45°F; 7.2°C). European varieties require 800 to 1000 hours of winter chilling. Japanese plums bloom in late February and March and are more susceptible to frost injury than European plums, which bloom in March and April.

Commercial plum production is reported in 30 California counties, extending from Butte in the north to Riverside in the south. The major production areas are concentrated in a few counties—Tulare, Fresno, Kern, and Placer. There has been a major shift in the geographical location of the California plum industry during the past 30 years. Plantings have declined substantially in the Sierra Nevada foothills region, but this trend has been offset by plantings in the lower San Joaquin Valley. Many varieties of plums have been developed in California during the over 100 years since introduction. New varieties have resulted from systematic breeding to develop superior hybrids, chance seedlings, and bud mutants. Characteristics, such as time of bloom and maturity, fruit size, color, eating quality, and intercompatibility with pollenizer trees determine which varieties are selected for commercial production.

Quince Tree (Cydonia oblonga)

A small, rather shrubby tree, the quince is much less important commercially than the pear or apple. The fruit is large, more or less hairy during growth, and hard and yellow at maturity. Each carpel contains several seeds invested with a mucilaginous pulp. It is not edible in the fresh state and thus does not qualify as a dessert fruit, although some varieties grown in the Near East are reputed to be delicious when consumed raw. The quince, like the persimmon and pomegranate, has an illustrious historical past and was greatly esteemed by ancient cultures, but it has fallen out of favor in the modern marketplace. For many years, the quince has been regarded in North America as a secondary fruit and used mainly as a raw material for jellies and preserves—and in recent years it has become increasingly difficult to find these products in the marketplace. There are only a few of what might be called fruit orchards in production in the United States today (in New York State). The literature, both from an agro-economics and scientific research standpoint, on the quince is extremely limited and generally of old vintage.

Believed to have originated in central and eastern Asia, the quince has been cultivated for over 2000 years. The quince is a small tree, usually not over 10 to 15 feet (3 to 4.5 meters) in height. The trunk is erect; the branches are crooked. Pink or white solitary flowers appear in spring. The tree has ovate leaves, tapering at one end and fruits resembling a large, yellow apple. In making flavoring substances, both the fruits and seeds can be used. Among flavorings made are a decoction (5%), an infusion (15% from seeds), and a fluid extract. For marmalades, jellies and preserves, the hard, acid flesh of unripe fruits, harvested in October, is preferred. A quince-seed mucilage is yielded by the seeds of ripe fruits and this can be used as a thickening agent in lieu of tragacanth gum. Quince flavoring is used on occasion in non-alcoholic beverages, ice creams, ices, and baked goods.

Raspberry (Rubus)

The raspberry is closely associated with the blackberry and dewberry. In these species, the fruit is an aggregate of many drupelets, which slip freely from the axis in the raspberry, but which cling close in the blackberries and dewberries. From a botanical standpoint, this quality of the raspberry separating from its receptacle is the primary difference between these species of berries.

Generally, the raspberries are perennial plants with either erect, procumbent or trailing stems; mostly armed with fine, slender, more-or-less stiff bristles and variously-shaped prickles; with stems or canes of most shrubby species of the temperate region biennial and reaching their full size the first year. In the second year, short lateral branches appear which bear flowers and fruits. After flowering and fruiting, the 2-year-old canes die and are replaced. The canes are either circular in cross section, or angled, or angled and furrowed. In some species, the tips of the canes bend over, touch the soil, strike root, and thus give rise to new plants.

The leaves are alternate, either simple, lobed or pinnately or palmately or pedately (with laterial division cleft) compound, mostly deciduous but in some species wintergreen or evergreen. Stalks of the leaflets resemble the canes. Stipules are always present at the base of the leaf stems. Flowers are always stalked and borne either solitary or in racemes or panicles. They are mostly bisexual, with both stamens and pistils. Flowers are white, rose, or pink. In the fruit, the pistils are transformed into small, more-or-less juicy and coherent drupelets. The ripe fruits are usually red, yellow or black, rarely green.

Raspberries grow best in cool climates. In the United States, they are not well adapted south of Virginia, Tennessee, or Missouri. Nor are they well adapted to areas in the Plains States or Mountain States where summers are hot and dry and winters are severe.

There is little evidence of formal cultivation of the raspberry in ancient times. The red raspberry did not draw the attention of Europeans until the 16th century. However, Hedrick (1925) observes that no doubt the raspberry crept into fields and was more or less cultivated from the very beginnings of agriculture in regions where it grows wild. Greek and Roman agricultural writers who lived prior to the Christian era do not mention the raspberry, although they have much to say about the tree fruits and grapes. Pliny, at the beginning of the Christian era, mentioned wild raspberries as coming from Mount Ida. The first well-documented reference to cultivated raspberries dates back to 1548 when Turner, the English herbalist, observed that they "growe in certayne gardines in Englande." Lawson in 1618 mentioned raspberries and currants growing along the border of gardens. In 1629, Parkinson discussed the "rapis berry" as useful for sickness, and mentioned several ways to use the berries and leaves for alleviating a number of human ills. The Horticultural Society of London listed 23 "sorts" of raspberries in the Catalogue published in 1826.

Hedrick, in referring to the culture of the red raspberry in North America observed that not until agriculture was well advanced with little land in waste could there have been a need for cultivated raspberries. Wherever its culture could have succeeded the native plant runs riot in waste places. It is one of the first plants to follow forest fires, to creep into newly cleared lands and become a weed in fence corners and neglected fields. Hedrick also mentioned that early explorers and settlers on the Atlantic seaboard often mentioned black raspberry as one of the delectable wild fruits of the country. It was found from New England to the Carolinas in the borders of woods, as a fringe about fields, around the stumps that dotted clearings, and came uninvited into the gardens.

It appears that the serious start toward domestication of the black raspberry started in 1850 by finding better methods of propagation. The red raspberry is propagated by suckers. The black raspberry throws no suckers and first was propagated laboriously by division. Bending the canes to the soil and covering them so that the tips will take root was later adopted from nature as the best means of manual culture.

Varieties and Culture. There are three main types of raspberries — red, black, and purple. They differ in several ways in addition to the color of the fruit. *Red raspberries* have erect canes. They are grown most extensively in the western states. *Black raspberries (blackcaps)* have arched canes that root at the tips. They are grown mostly in the eastern half of the United States and in Oregon. See Fig. 8. *Purple raspberries* are hybrids of red raspberries and blackcaps. They have the same growth characteristics as blackcaps and are propagated in the same way. They are grown extensively only in western New York State, although they are adapted to about the same regions as that for blackcaps. Some raspberries have yellow fruit; these *yellow raspberries* are variations of red raspberries and, except for fruit color, have all the characteristics of red raspberries. They are usually limited to home gardens.

Fig. 8.   Black raspberry (*genus Rubus*). (*USDA.*)

Strawberry (*Fragaria*)

The strawberry is a perennial herb of several species. The fruit is an enlarged, fleshy receptacle with numerous seeds embedded at the surface. What appear to be seeds to the casual observer are, in actuality, achenes. The seed is contained within a thin, dry ovary wall. See Fig. 9. The strawberry plant grows from a central stem called a crown whose terminal is a growing point. From this growing point, leaves, flowerbuds, and

Fig. 9.   A ripening cluster of strawberries. (*USDA.*)

runners develop. Runners are branches from the main stem. Branch crowns may develop following rapid runner development. Buds in the axils of the leaves produce flower clusters when temperatures are cool and days are relatively short. Different varieties produce clusters with many flowers, while others produce clusters with few flowers. Some varieties produce clusters that branch close to the crown, while others branch far out on the stem. Clusters with many flowers may produce a large number of berries, but the berries may be small. Natural propagation of the plant is by runners which form mainly after the blooming season. Shape is variable. See Fig. 10.

The strawberry is among the most widely adapted of fruit crops. Varieties have been selected that can be grown in at least the higher elevations in the tropical regions, and others are grown in northern latitudes where very severe winter conditions prevail. Although the strawberry can be grown as far north as most fruits, it is not truly hardy in the sense that the plant parts can withstand very low temperatures. As grown in cold climates, the vital plant parts are at or below ground level during the winter. Thus, they are protected by snow or other cover. Without such protection, the plants are very susceptible to winter killing. In commercial production, the practice of heavy mulching with straw or similar materials is followed in cold regions to ensure protection if there is little or no snow cover.

In the United States, varieties for the most southern latitudes differ in their growth response from those adapted to severe winters. The principal fruiting in the most southerly regions occurs during the winter and early spring months. Varieties adapted there must grow, flower, and fruit during the relatively short, cool days of winter.

Background. In the days of the Romans and Greeks, all plants of the genus *Arbutus* were included along with strawberry in one classification. The strawberry was first distinguished by Pliny, who called it *fragum*, a tribute to the sweet taste of the fruit. The present genus name *Fragaria* is Latin, although some botanists believe the name may have stemmed from *fragrans* (fragrant). The French *fresas* (*frayses* and modern *fraises*) and the Spanish *fresa* apparently came from the same source and the names definitely refer to the sweet odor of the fruit.

The unimproved or wild strawberry is native to many regions of the world. There are numerous varieties. As early as the 14th Century, it was reported that at least 1200 plants of small-fruited, indigenous European species were cultivated in the Royal Gardens at Louvre, France. At this time, the strawberry was commonly a part of French home gardens, a practice that commenced in England in the 15th century. The early American colonists were delighted to find *Fragaria virginiana* growing along the Atlantic coast and north to New England and west to the Rocky Mountains. American Indians in New England called this berry the "wuttahinmeash" and mixed it with meal in bread making. Roger

Oblate   Globose   Globose conic   Conic

Long conic   Necked   Long wedge   Short wedge

Fig. 10.   Range of shapes of strawberries. (*USDA.*)

Williams (1643) referred to the strawberry as the wonder of all fruits growing naturally in these parts. From observations of the colonists, it appeared that the Indians probably had cultivated the strawberry for several centuries prior to the colonial period.

A species, *F. chiloensis*, native to the south coastal region of Chile and the Cordillera of the Andes of South America, as well as the beaches and coastal mountains of western North America and the mountains of Hawaii, was taken to France from Chile by Captain Frezier in 1712. Records show that the species reached England by 1727. The original Chilean plants apparently were large-fruited selections that had been cultivated by the aborigines since before the arrival of the Spaniards. Some of these berries are still cultivated. Because only female plants were carried to Europe by Frezier, there was little effect of their introduction for about a century. Natural hybridization with *F. virginiana*, already in European gardens, occurred. As pointed out by Hedrick (1948), the rapid improvement that occurred can be regarded as one of the most remarkable phenomena recorded in pomology.

Varieties. The cultivated strawberry in America is a fruit that originated by hybridization from the wild species of eastern North America and the wild species of South America. The berries have a unique, tangy taste. They are highly valued as dessert fruit and are rich in vitamin C. Among the most important varieties, based upon total tonnage of fruit marketed, are the Tioga, Northwest, Shasta, Midway, Surecrop, Fresno, Blakemore, Florida Ninety, Pocahontas, Tennessee Beauty, Catskill, Albritton, Dabreak, Sparkle, Headliner, Raritan, Siletz, Hood, and Rodinson. Several other varieties are grown successfully. New varieties of strawberries appear from time to time.

The two general classes of strawberries are: (1) everbearers; and (2) one-crop varieties, sometimes called "June bearers." Everbearers produce fruit during spring, summer, and fall. The once-crop varieties produce fruit only in late spring and early summer.

Minor Fruits of the Rose Family

Medlar Tree (Mespilus germanica). A small tree or large shrub, the medlar is found in southern Europe and is native to that region. However, it is found in the United States, particularly in parts of New York State. In the wild state, the tree has thorns which cultivated trees do not bear. The fruit is tart and is used in making preserves. Flowers are large, solitary, and white. Frost assists in the ripening of the fruit and improves the flavor.

Hawthorn Tree (Crataegus). Of the same family (*Malaceae*), the hawthorn bears a fruit somewhat resembling a miniature apple. Over a thousand species of hawthorn have been described. The hawthorn is the hedgerow tree in northern Europe. The name means hedgethorn. The common species is *C. monogyna.* The common hedgerow hawthorn has attained a height of almost 50 feet (15 meters) and, in some cases, a girth of nearly 10 feet (3 meters). The trees are admired for the beauty of their foliage and can assume large proportions when permitted to develop as single trees. Champion specimens in the United States are listed in Table 1.

Crab Apple Tree (Pyrus angustifolia, etc.). There are approximately 15 species of these relatively small trees, found in the Temperate Zone of the Northern Hemisphere. Crab apples were valued in the Middle Ages for their juice, serving much as vinegar does in modern cooking. The trees are valued today for their flowers and foliage in gardens. Because they are considerably hardier and adaptable to poor soil, they can make fitting substitutes for cherry trees. Champion crab apple trees in the United States, as selected by American Forests are described in Table 1.

Shadbush or Serviceberry (Amelanchier). Usually small shrubs of value for coloring and foliage in gardens, but as indicated by the accompanying table, can become large trees. In England, the plant is known as the juneberry or snowy mespilus. Flowers are white and star-shaped. Usually in June, bunches of berries somewhat like black currants appear. Autumnal colors are soft red, orange, and brown. Champion serviceberry trees in the United States, as selected by The American Forests, are described in Table 1.

ROSIN.   See **Resins (Natural)**.

ROSSI X-RAY TIMING EXPLORER (RXTE). The Rossi X-ray Timing Explorer is a large X-ray space telescope capable of making observations over a large range of the X-ray spectral region (2–200 kiloelectron Volts (keV)) with unprecedented high time resolution (as little as one microsecond between readings). Its All Sky Monitor (ASM) collects a continuous daily record of hundreds of X-ray sources in the sky, searching for transients and recording slow variations in X-ray brightness. Its more sensitive Proportional Counting Array (PCA) and High Energy X-Ray Timing Explorer (HEXTE) instruments make in-depth observations of the X-ray variability of sources in our galaxy and of distant galaxies. Scientists compete for time on these instruments to schedule studies of well-known objects or to quickly observe transients uncovered by the ASM.

Compared to the night sky that is visible to the naked eye, the X-ray sky is incredibly dynamic: X-ray sources wink into being, reaching surprising brightness before gradually fading over weeks or months. Bright, persistent, but highly variable X-ray sources shine where there is only a faint optical counterpart. Even distant giant galaxies have tremendous X-ray output that can vary on short time scales (days) to the longest time scales yet observed. Scientists now understand that most cosmic X-rays are produced near the surfaces of white dwarfs and neutron stars and near the event horizons of black holes. White dwarfs, neutron stars and black holes are often collectively referred to as *compact objects*. Each is an end point of the stellar life cycle, during which a normal star like the Sun eventually sheds its outer, lighter layers, leaving behind a dense core that has collapsed due to its own gravity. These exotic objects are described by very extreme physics: matter compressed to unimaginable densities where even space and time behave very differently than expected. It is impossible to reach these conditions in any laboratory on Earth; the Rossi Explorer was designed to peer carefully at these objects to test the physics of extremes.

Compact objects often reside in a binary star system, in which X-ray source and a normal hydrogen-burning star like our Sun orbit each other. Gas is stripped off the normal star and usually forms a disk of material around the neutron star before eventually crashing onto its surface. This disk is referred to as an accretion disk. The Rossi Explorer has made great strides in the understanding of neutron stars in binary star systems. During X-ray outbursts of several such systems, Rossi has detected very rapid time and spectral variability. While the signal detected is not strictly periodic there is a strong preferred frequency in the X-ray output, often referred to as *quasiperiodic*. The rapid spectral changes and these high frequency quasi-periodic oscillations (QPOs) are actually probes of the inner accretion disk: this is the only window into the behavior of space and time near compact objects. Often, two distinct high frequency QPOs are observed in the same source. While both slowly adjust their frequency as the outburst progresses, the difference in their frequencies typically remains constant. The frequency difference between these twin peaks is believed to reveal the otherwise masked neutron star spin period. See also **Binary Stars**.

As the number of neutron stars with known spin periods increases, scientists can place new constraints on the mass, radius, and equation of state of neutron stars. Already, some theoretical equations of state for matter at neutron star densities are all but ruled out by RXTE observations. Watching the frequencies of QPOs change with time has produced the first evidence of some predictions of Einstein's theory of General Relativity in the strong field regime. The fact that there appears to be a "speed limit" for QPOs is taken as evidence for the innermost stable orbit around neutron stars. Other scientists have found evidence of Lense-Thirring precession (or frame dragging) around neutron stars and black holes in the behavior of kHz QPOs. See also **Neutron Stars**.

The Rossi Explorer has discovered dozens of new transients, peered intensely at numerous outbursts in over 30 systems, and detected many orbital and long term periods in systems where the period was not previously known. Also in the Rossi collection is the youngest known pulsar (about 700 years old) and the closest known microquasar, a stellar-sized black hole emitting jets of high-speed particles from its poles about 40,000 light years from Earth. Intensive study of one microquasar has uncovered an intricate interplay between the accretion disk of matter swirling into the event horizon of the black hole and the intense jets sometimes seen from these sources. When an instability disrupts the disk, a portion of the material is accelerated and ejected as a relativistically expanding cloud, causing an outburst that is seen in the radio and infrared regions of the spectrum. See also **Pulsar**.

Active Galactic Nuclei (AGN) are the extremely energetic central engines of distant galaxies. Intensive monitoring of AGNs over a wide range of time scales has revealed a surprising similarity between the characteristics of rapid brightness flickering in AGN and the much smaller X-ray binaries found in Earth's galaxy. Some scientists interpret the result as evidence that material falling into a black hole follows a simple scaling law: the flickering properties are proportional to the absolute brightness (and therefore the mass) of the black hole. If that is true, then the RXTE observations mean this rule applies over a very wide range of masses (from single black holes in our galaxy with masses a few times the mass of the Sun up to galactic cores, estimated to contain a few million times the mass of the Sun). The masses of black holes in X-ray binaries can

be accurately determined from the orbital motion of the two stars. Using the black hole flickering scaling rule, scientists can now estimate the total mass of the supermassive black holes in the cores of active galaxies from a measurement of their flickering properties. See also **Galaxy**.

The ASM has amassed a collection of long-term light curves that give scientists important insight on the dynamics of these objects at the longest time scales. Many X-ray binaries show a periodicity much longer than the orbital period of the two stars. This superorbital period is thought to be the precession of the disk of material around the compact object. Much like a spinning toy top wobbles as it moves across a table, the saucer-shaped accretion disk may slowly change its plane of orientation with time, effectively casting a shadow and blocking some of the X-rays that are headed toward Earth. Other systems show random long term variations thought to be caused by an instability in the accretion disk which can disrupt the flow of matter onto the compact object. The ASM is finding that many X-ray binary systems have long term light curves that can not be classified as simply periodic or random. Perhaps both disk instabilities and precession are at work in such systems.

The Rossi Explorer was launched by NASA aboard a Delta II rocket from Cape Canaveral, Florida, on December 30, 1995. The PCA was built by NASA Goddard Space Flight Center; the HEXTE was built by the University of California, San Diego; and the ASM was built by the Massachusetts Institute of Technology. See Fig. 1. The Rossi Explorer maintains a low-earth circular orbit at an altitude of 580 kilometers, corresponding to an orbital period of about 90 minutes, with an inclination of 23 degrees. NASA Goddard operates the satellite and manages the Rossi Explorer data archive. This X-ray mission is named in honor of Professor Bruno Rossi (1905–1993), an authority of cosmic rays whose experimental techniques at the Los Alamos Laboratory and at MIT gave birth to the field of X-ray astronomy. See also **X-Ray Astronomy**.

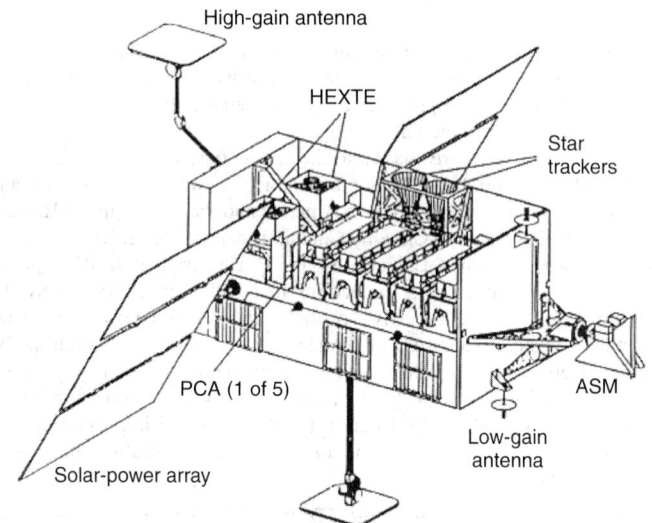

Fig. 1.   Rossi X-ray Timing Explorer (GSFC/NASA). Diagram location: *http://rxte.gsfc.nasa.gov/Images/xte/xte_spacecraft.gif.*

Web References

RXTE Homepage: *http://rxte.gsfc.nasa.gov/docs/xte/xte_1st.html*
RXTE Image Gallery: http://rxte.gsfc.nasa.gov/docs/xte/xhp_image.html

PATRICIA T. BOYD, Ph.D, University of Maryland, Baltimore and NASA Goddard Space Center.

ROTAMETER.   See **Flow Measurement**.

ROTATION AXIS.   A symmetry element possessed by certain crystals, whereby the crystal can be brought into a physically equivalent position by rotation about an axis which can be onefold, twofold, threefold, fourfold, or sixfold, according to whether the crystal can be brought into self-coincidence by the operations of rotation through 360, 180, 120, 90, or 60 degrees about the rotation axis. See also **Mineralogy**.

ROTATION (Dynamics).   A body is said to rotate when all of its particles move in circles about a common axis with a common angular velocity. This motion may be either free or constrained, as illustrated, respectively, by the earth turning on its axis, and by a flywheel or a pendulum.

If one twirls an umbrella about its handle, it tends to open. This is because the centrifugal forces exert torques tending to throw the stays outward on their pivots. Through any point of a rigid body there are at least three lines, mutually perpendicular, about which the body would rotate without any such centrifugal torque. It may be shown that the moment of inertia of the body with respect to any one of these lines is either a maximum or a minimum as regards all lines through the given point. They are called principal axes. In general there is only one line about which a free body will rotate permanently; it is the principal axis of greatest moment of inertia through the center of mass. A body constrained to rotate about an arbitrary axis will, when released, tend to change its motion so as to rotate about this permanent axis, but the adjustment is complicated by precession, so that the body may "wobble" like a badly thrown discus.

If a free body, at rest, is given a sudden push along some line not through the center of mass, it begins to rotate about some other line beyond the center of mass and perpendicular to the applied force. This line is the axis of instantaneous rotation. It is only a temporary axis, the rotation being at once transferred to an axis through the center of mass. The line mutually perpendicular to the instantaneous axis and to the line of the force passes through the center of mass, and its intersections with the other two lines are conjugate points, having the same relation as the center of oscillation and the center of suspension of a rigid pendulum. If the push is given in line with the center of mass, the axis of instantaneous rotation is at infinity, and the motion is then one of pure translation.

A torque applied so as to tend to change the axis about which a body is rotating results in the peculiar behavior known as precession. The angular momentum of a rotating body is the product of its angular velocity by its moment of inertia about the axis of rotation. The kinetic energy associated with rotational motion is equal, in absolute units, to $\frac{1}{2}$ the product of the moment of inertia by the square of the angular velocity — a formula analogous to that for kinetic energy of linear motion.

ROTATION (Earth).   See **Earth**.

ROTATION-REFLECTION AXIS.   A symmetry element possessed by certain crystals, whereby the crystal is brought into self-coincidence by combined rotation and reflection in a plane perpendicular to the axis of rotation. Rotation-reflection axes may be onefold, twofold, threefold, fourfold, or sixfold, according to whether the rotation which, with the reflection, brings the crystal into self-coincidence is through an angle of 360, 180, 120, 90, or 60 degrees.

ROTATORIA (*Rotifera*).   The wheel animalcules, minute animals with a circlet of cilia at one end of the body, whose movements in some species give the appearance of rotation to the entire disk. They live in water, even in the small quantities found in matted vegetation and temporary pools, and are adapted to withstand long dry periods in such situations. The group is a phylum of minor importance. See Fig. 1.

Although rotifers are minute their bodies are complex in structure. The body wall consists of an external cuticle and a syncytial ectodermal layer which bounds the internal cavity. There are no muscle layers but bands of muscle are present. The alimentary tract (digestive system) is tubular and includes a pharynx with an elaborate grinding apparatus, known as the mastax, an expanded stomach, with a ciliated (cilia) lining, and a short intestine. Near the anus is an expanded cloaca, which receives the ducts of the excretory and reproductive systems. The essential unit of the excretory system is the flame cell, like that of flatworms. The pair of ducts bearing these cells empty into a contractile vesicle or bladder. Reproduction in this phylum is complex, owing to the frequent occurrence of parthenogenesis and to the adjustment of the life cycle to fluctuating environmental conditions.

Some authorities regard the rotifers as a phylum and others associate with them two other forms of animals, making each of the three groups a class in the phylum *Rotifera*, also named *Trochelminthes*.

ROTIFERS.   See **Rotatoria**.

ROUNDING.   The process of approximating to a number by omitting certain of the end digits, replacing by zeros if necessary, and adjusting the

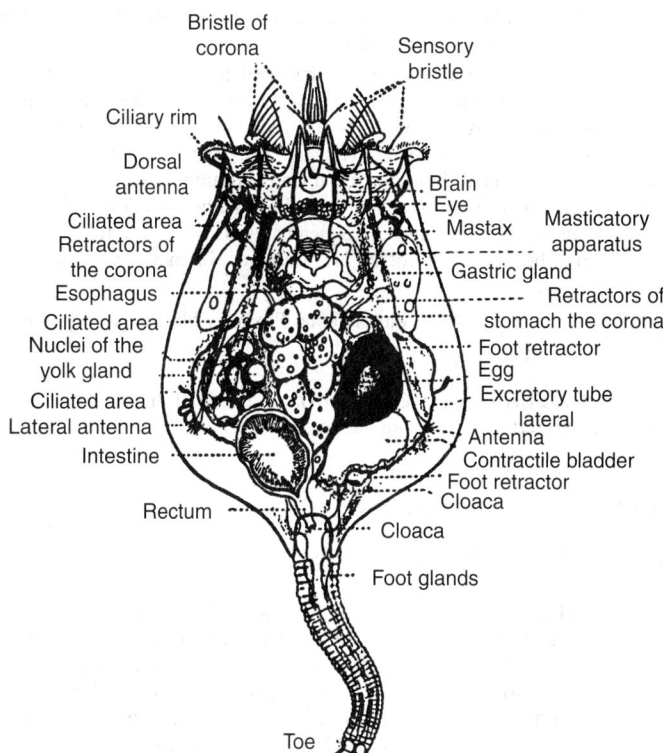

Fig. 1.   Sectional diagram of a rotifer. (*Wesenburg-Lund.*)

last digit retained so that the resulting approximation is as near as possible to the original number. If the last digit is increased by unity the number is said to be rounded up; if decreased by unity it is rounded down. Thus, if the number 0.04645 were to be rounded to three significant Figures, it would be rounded up to 0.0465, since the digit dropped is 5 (or greater).

ROUNDWORMS.   See **Nematodes**.

ROUTINE (Computer System).   A set of coded instructions arranged in proper sequence to direct a computer to perform a desired operation or sequence of operations. A subdivision of a program consisting of two or more instructions that are functionally related; therefore, a program.

Diagnostic Routine.   A routine used to locate a malfunction in a computer, or to aid in locating mistakes in a computer program. Thus, in general, any routine specifically designed to aid in debugging or trouble-shooting.

Executive Routine.   A routine that controls loading and relocation of routines and in some cases makes use of instructions not available to the general programmer. Effectively, an executive routine is part of the machine itself. Synonymous with monitor routine; supervisory routine and supervisory program.

Heuristic Routine.   A routine by which the computer attacks a problem not by a direct algorithmic procedure, but by a trial and error approach frequently associated with the act of learning.

Interpretive Routine.   A routine that decodes and immediately executes instructions written as pseudocodes. This is contrasted with a compiler that decodes the pseudocodes into a machine language routine to be executed at a later time. The essential characteristic of an interpretive routine is that a particular pseudocode operation must be decoded each time it is executed. Synonymous with interpretive code.

Service Routine.   A broad class of routines which are provided at a particular installation for the purpose of assisting in maintenance and operation of the computer as well as the preparation of programs as opposed to routines for the actual solution of production problems. This class includes monitoring or supervisory routines, assemblers, compilers, diagnostics for computer malfunctions, simulation of peripheral equipment, general diagnostics and input data. The distinguishing quality of service routines is that they are generally tailored so as to meet the servicing needs at a particular installation, independent of any specific production type routine requiring such services.

Tracing Routine. A diagnostic routine used to provide a time history of one or more machine registers and controls during the execution of the object routine. A complete tracing routine would reveal the status of all registers and locations affected by each instruction, each time the instruction is executed. Since such a trace is prohibitive in machine time, traces that provide information only following the execution of certain types of instruction, are more frequently used. Furthermore, a tracing routine may be under control of the processor, or may be called in by means of a trapping feature.

THOMAS J. HARRISON, International Business Machines Corporation, Boca Raton, FL.

ROVE BEETLE *(Insecta, Coleoptera).* A beetle of the family *Staphylinidae*, characterized by the long flexible abdomen which is exposed behind the short wing covers (elytra).

ROWAN TREE. See **Ash Trees**.

ROYAL ANTELOPE. See **Antelope**.

ROYAL JELLY. The food given by worker honeybees to the young larvae during the first 3 days of their existence and to the larvae of queens until they are fully developed. It is a thick white liquid formed in the stomach of the worker by partial digestion of honey and pollen, and is apparently a highly concentrated food. Queen cells are supplied with the material in excess of the needs of the larvae. If conditions within the colony deprive queen larvae of this abundance they fail to become as large, and in some cases they may even develop as intermediate forms between queen and worker. Such individuals may, however, have the instincts of queens and so may mate and lay fertile eggs. The change from royal jelly to a less concentrated food in the case of worker larvae apparently is responsible for the development of worker bees, since both queens and workers may develop from identical eggs. See also **Honey**.

RUBBER (Natural). *Natural rubber* is the name applied to the polymer *cis*-polyisoprene obtained chiefly from the *Hevea brasiliensis* tree.[1] Originally, the tree grew wild in the Amazon valley, but during the last part of the 19th century it was planted in well-organized plantations in tropical lands of the Far East and later in Africa. See Table 1. The average rubber trees stand about 40–50 feet (12–15 meters) high. For optimum growth, a tropical climate having 80 inches (203 centimeters) or more of annual rainfall is required.

TABLE 1. NATURAL RUBBER EXPORTED

Producing Area	Quantity Produced (1000 Long Tons)				
	1940	1960	1970	1975	1979
Malaysia	547	775	1304	1424	1595
Indonesia	543	587	755	788	866
Other Asian countries, including Oceania	283	400	484	525	704
Africa	16	149	207	197	141
Tropical America	26	15	10	12	19

Note: 1 long ton is slightly more than 1 metric ton (1016 kilograms).

Rubber comes from the tree as a milky white fluid, which is a colloidal suspension of rubber in a liquid consisting mostly of water. The tree is tapped by well-trained workers, who use a sharp-edged tool, and the cutting action goes at an angle of 30° from top left to bottom right. It is important that the rubber latex-bearing cells be cut, but that the blade not wound the inner cambium layer, as this would harm the tree. A cup is hung below the cut to collect the white, milklike latex, which contains about 35% rubber, the remainder being water, protein, resins, organic materials, and other plant substances.

[1] It is alleged that English scientist Joseph Priestly observed that the material could be used for rubbing out lead pencil marks and thus gave the material the name rubber. ("Introduction to the Theory of Perspective," Joseph Priestly, 1835.)

The yield of the *Hevea* tree can be increased by applying chemicals to the bark. These include 2-chloroethylphosphonic acid, which supplies small quantities of ethylene gas. This type of chemical is applied in a high-viscosity liquid form, usually mixed with palm oil or other diluent. The function of it is to stabilize the latex so that it continues to flow for a longer time and thus increases rubber yield. Because of the higher rubber yield per tapping operation, the cost of tapping labor is reduced. When stimulants are used, the tree is given a longer rest period between tappings to avoid diseases that would eventually kill it. See also Plant **Growth Modification and Regulation**.

During the 1980s, the annual yield for Malaysian plantations was about 1010 pounds/acre (1133 kilograms/hectare), which includes high-yielding trees which produce 1500 pounds/acre (1680 kilograms/hectare), as well as older, lower-yielding clones.

Properties of Natural Rubber. Chemically, natural rubber or *cis*-polyisoprene, has a broad molecular-weight distribution, ranging from several million to about one hundred thousand.

$$-H_2C \diagdown \underset{C=C}{\overset{H_3C \qquad H}{}} \diagdown H_2C - H_2C \diagdown \underset{C=C}{\overset{H_3C \qquad H}{}} \diagdown H_2C - H_2C \diagdown \underset{C=C}{\overset{H_3C \qquad H}{}} \diagdown H_2C -$$

Natural rubber is soluble in practically all aromatic and aliphatic hydrocarbons and particularly in halogenated hydrocarbons. When cements and solvent adhesives are made using natural rubber, methylethylketone (MEK) frequently is used to reduce viscosity. Although MEK is not a solvent, it tends to disperse large molecular particles, resulting in lower-viscosity dilution. Crude rubber is decomposed by heat and can be cyclized at 250 °C. It can easily be hydrogenated and reacts readily with halogens. The stress-strain properties of natural rubber are the best of all the elastomeric polymers. In vulcanized films made by the latex process, the tensile strength may exceed 6000 pounds per square inch (41 mPa), and ultimate elongation is as high as 700% or more.

Natural rubber is readily attacked by oxygen. Copper and manganese, if present in amounts greater than the specified 0.001%, greatly accelerate oxidation. There are, however, naturally occurring antioxidants in natural rubber that help preserve it until vulcanization. All vulcanized natural-rubber products contain added antioxidants to ensure satisfactory life.

Rubber burns quite readily and generates more than 10,000 cal/g. The specific gravity of rubber is 0.934, a property utilized in concentrating natural-rubber latex by the centrifuge process. The serum, which is mostly water and has a specific gravity of about 1.0, tends to separate readily from the rubber. The liquid concentrated latex is used in making foam rubber, dipped goods, adhesives, and carpet backing for nonwoven carpets. An industry has developed around this application, which involves spreading foamed latex onto the underside of carpeting, making an integral carpet-foam system.

Compounding and Vulcanization. Crude rubber in the raw state has few applications with the exception of crepe soles for shoes. To make commercial rubber products, the material must be mixed with a variety of chemicals and vulcanized into desirable end shapes. Charles Goodyear discovered in 1839 that adding sulfur to rubber and heating the mixture greatly enhances the physical properties of rubber. The material no longer becomes tacky in warm weather and in cold weather it does not become brittle. The material is much tougher, and the quality of products made this way results in service for a much longer period of time. In addition to sulfur, which crosslinks the large rubber molecules and makes it a giant organic molecule, zinc oxide, organic accelerators, antioxidants, reinforcing pigments, and other processing aids are used in compounding rubber for useful vulcanized products.

The function of the antioxidant is to improve service life of the product against such well-known degrading agents as oxygen, light, and nitroso compounds. One theory is that the antioxidant selectively reacts with the degrader, slowing down its reaction with the rubber molecule, which would result in scission and eventually poorer physical properties. During recent years, considerably aggravated by air pollution, degradation of vulcanized rubber by small quantities of ozone (a few parts per million) in the air has become a serious problem. Ozone has little noticeable effect on unstretched rubber, but even under slight stretch it causes cracks in the surface, which grow perpendicularly to the direction of extension. Hundreds of different antioxidants and antiozonants are employed, amine and phenol complexes being the basis of most. See also **Antioxidant**.

Accelerators act as catalysts of vulcanization, but, unlike most catalysts, they undergo chemical change during the reaction. Benzothiazyl disulfide is one of the oldest types, dating back to 1925, but it still accounts for the greatest use in the industry today. Besides the thiazole types, other popular accelerators are sulfenamides, aryl guanidines, dithiocarbames (extremely fast accelerators used mostly in latex compounds), and thiurams (also very fast and often used as a secondary accelerator to hasten the vulcanization rate). Accelerators also contribute to improved ageing properties of the end product.

Stearic acid is an activator of vulcanization, as is zinc oxide, both reacting to form zinc stearate, which enhances the activity of the organic accelerators. Zinc stearate is impractical to add directly to the rubber because its slippery, lubricating nature makes it difficult to mix in the batch.

With an accelerated system, a simple network structure with dialkenyl mono- and disulfide crosslinks and conjugated triene units as main-chain modifications is obtained:

With an unaccelerated sulfur-natural-rubber system, the poor crosslinking efficiency results in sulfur being incorporated into the rubber network as long polysulfide crosslinks, cyclic monosulfides, and vicinal crosslinks, which are very close together and act physically as a single cross-link:

It is theorized that between the complex network structure of the unaccelerated system and the simpler network structure of the accelerated system, structures made up of the two models represent natural-rubber vulcanizates made at various times and temperatures of cures, with different reactant concentrations, and showing the effects of other variants.

At any given degree of crosslinking, the tensile strength is highest with polysulfide bonds. High elongation at break is obtained by slightly decreasing the crosslinking action. If lower elongation is required, slightly excessive crosslinking is used, usually accompanied by higher tensile strength. Vulcanization of rubber decreases its solubility in solvents, and this property frequently is used as a qualitative measure of cure.

Vulcanization by sulfur accounts for practically all the commercial products. However, peroxide types of curing systems may be used, especially for some of the synthetic rubbers.

Ultrahigh-frequency (UHF) energy may be used for preheating and precuring rubber compounds for continuous vulcanization (CV) of rubber, containing carbon black, for such applications as weather stripping, tubing, hose, and, in some instances, tire tread compounds.

Carbon black is the major reinforcing pigment used, not only for natural rubber, but for practically all the synthetic rubbers. As much as 40–50 parts by weight, based upon 100 parts of rubber, is used in all tire-tread compounds. Carbon black greatly increases tensile strength at low elongations (modulus) and results in longer-wearing tires. Colloidal silica contributes some reinforcing properties to rubber, but not to the same degree as carbon black. See also Carbon Black.

Uses of Natural Rubber

Thousands of flexible products requiring top performance characteristics are made of natural rubber, e.g., huge earthmover tires, truck tires, tires for large aircraft, bridge supports, and surgeons' gloves. The treads of most passenger car tires in the United States consist mainly of styrenebutadiene synthetic rubber because of lower cost and lower temperature buildup during use.

The use of natural rubber in passenger car tires has increased in recent years due to the industry going from bias to radial types which, in North America, now account for 75% of the total. Higher degree of tack or cohesive bonding during the building of the radial tire, as compared with that of styrene-butadiene rubber, is largely responsible for this.

Because of its excellent high- and low-temperature properties, many products used in the arctic and tropical areas of the world are made from natural rubber. However, it is not suitable for applications where there is contact with naphtha, e.g., gasoline hoses, because the solvent swells the material. Almost all elastic bands are made from natural rubber. Because of its excellent tack properties, the material is used in solvent and latex form as the base for adhesives.

With the dependence of synthetic rubber on petroleum, natural rubber, which is produced by solar energy, may look increasingly attractive over the years ahead.

Processing Raw Materials

Field latex is bulked in large tanks at a factory adjacent to the rubber estate. If a high-solids latex is desired, the field latex is strained, stabilized with ammonia or other chemicals, such as soap and bactericide, and either centrifuged or creamed to 62–68% total solids.

Smoked Sheet. For making ribbed smoked sheet, the field latex is immediately mixed with dilute formic acid in long horizontal tanks. Because fresh latex is somewhat protected by a protein surface layer, it does not coagulate or gel immediately on addition of the acid. Within a few hours, however, the rubber particles in the latex gel form a spongy mass which is then run through a series of smooth metal rolls with clearance decreased from one set to the next, an arrangement that squeezes out the serum and densifies the wet rubber. Water is run over the wet coagulum to wash out non-rubber materials and dirt. The last unit consists of ribbed rolls, which imprint ribbed markings on the sheet. After drying in air for a few hours, the sheets are hung in a drying shed at 40–50°C until dry. Modern installations use efficient drying tunnels. Sheets are inspected by holding them over a strong light to determine clarity, color, presence of dirt and other factors. The rubber is classified by various grades. Sheets then are piled up and squeezed in a baling machine to form 250-pound (~113-kilogram) bales that measure 19 × 19 × 24 inches (~48 × 48 × 61 centimeters).

Crepe. Another popular type of commercial rubber, known as crepe, consists of two major classes—*pale crepe* and *thick blanket crepe*. Pale crepe is made by adding sodium hydrogen sulfite, $NaHSO_3$, to field latex to inhibit discoloration and softening during processing. Formic acid is used as the coagulant. The wet coagulum is passed through rolls with longitudinal grooves, which give the rubber a crepe-like appearance. Water running over the surface cleans out dirt and other nonrubber ingredients. Sheets are hung up to dry in circulating warm air. The quality of pale crepe is assessed on its whiteness and how good the finished rubber appears.

Blanket crepes are of lower quality and are made from wet slabs obtained usually from small landholders. These are creped, dried, and baled. Other types of crepe are made from coagulum left in collection cups and from dried skin remaining from the tapping incision. In addition to collecting latex, a tapper collects all dried and coagulated rubber that remains from the previous round, usually as skin in the cup or on the tapping panel.

Grading of Rubber. Commercial grades of natural rubber are classified into two main groups: (1) "Green Book International Grades," and (2) "Technically Specified Forms." The former depends on a visual grading system, the source of the rubber and the method of preparation. This system, dating back many years and kept current by the International Rubber Quality and Packing Conference Committee, consists of 35 grades under 8 major types, such as Ribbed Smoked Sheets, White and Pale Crepe, Estate Brown Crepes, Compo Crepes, Thin Brown Crepes (Remills), Thick Blanket Crepes, and Pure Smoked Blanket Crepe. Publisher of the "Green Book" is the Rubber Manufacturers Association, Inc., New York.

Technically Specified Rubbers (TSR), originated by the Malaysian Rubber Producers Association, classifies rubber not only on the basis of the source of rubber, but on its physical properties, such as dirt content, ash, and nitrogen content, volatile matter, plasticity, and, with the higher-quality grades, cure rate and color are standardized. This type of rubber is packaged in 75-pound (34-kilogram) bales, wrapped in transparent plastic, and the color of the printing on the bale identification strips indicate whether the source is latex grade, sheet material grade, blended grades, or field grades. An additional convenience of this rubber is that the bales can be charged into the mixing machine (Banbury) without removing the wrapper. This is in contrast with the Green Book grades, which are bonded together by a press, with the outside layer treated with soapstone to keep the bales from sticking together. These larger bales require cutting before they can be charged into the mixer.

Guayule

During the past few years, natural rubber from the desert shrub *Parthenium argentatum* has been under intensive study by scientists in the United States and Mexico as a possible domestic source of natural rubber. This plant grows wild in the arid areas of Mexico and the United States. In 1910, guayule produced 10% of the world's rubber, but lower-cost Hevea rubber from the Far East displaced it from the market. Rubber in the guayule plant is present in the roots and branches of the shrub and must be separated and purified by a flotation and solvent system. The purified product is equivalent in chemical properties to the Hevea rubber. An advantage of guayule is that it can be grown on semiarid land that is not suitable for other crops. Presently, agricultural experimentation on increasing rubber yield of the plant is underway. The U.S. government has passed legislation providing funds to help in developing an American-based guayule industry. Several large rubber product manufacturers have experimental plots planted with the shrub. The National Research Council, Washington, DC., published a report, "Guayule: An Alternative Source of Natural Rubber," in 1977.

Additional Reading

Bhowmick, A.K., M.M. Hall, and H.A. Benarey: "Rubber Products Manufacturing Technology," Marcel Dekker, Inc., New York, NY, 1994.
Hepburn, C.: "Rubber Technology," 3rd Edition, Butterworth-Heinemann, Inc., Woburn, MA, 1998.
Loadman, M.J.: "Analysis of Rubber and Rubber-like Polymers," Kluwer Academic Publishers, Norwell, MA, 1998.
Mark, J.E., F.R. Eirich, and B. Erman: "Science and Technology of Rubber," 2nd Edition, Academic Press, Inc., San Diego, CA, 1994.
Sethuraj, M.R. and N.M. Mathew: "Natural Rubber: Biology, Cultivation, and Technology," Elsevier Science, New York, NY, 1992.
Tinker, A.J. and K.P. Jones: "Blends of Natural Rubber: Novel Techniques for Blending with Specialty Polymers," Chapman & Hall, New York, NY, 1998.

THOMAS H. ROGERS, Consultant (Rubber and Plastics Industries), formerly Research Manager, Goodyear Tire and Rubber Company, Akron, OH.

RUBBER PLANT. See **Euphorbiaceae**; and **Latex**.

RUBBER (Synthetic). See **Elastomers**.

RUBELLA (German measles). This disease is caused by the rubella virus, a member of the togavirus family. Symptoms of the disease are mild — fever, a characteristic rash, viremia, and subsequent involvement of lymph nodes in the region of the neck. Incubation period is 18 days. Relatively uncommon complications include encephalitis, arthritis, and rarely, thrombocytopenia. Even with vaccination, reinfection may occur in 4% of the population. In the absence of a specific therapy for the disease, treatment is directed toward alleviating symptoms.

A principal concern with this disease is the major fetal damage that may occur when mothers in the first trimester of pregnancy are infected with rubella. Statistics show that from 20 to 35% of children born under such conditions will have one or more congenital defects, which usually involve the heart, eyes, ears, brain, and bones. Some of these abnormalities are not fully manifested until the child is several years old. The risk remains but is lower when the infection occurs during the second trimester.

Since 1969, when the first vaccine was licensed, vaccination programs in the United States and a number of other countries have greatly lowered the incidence of rubella. Nevertheless, nearly 12,000 cases were reported to the Centers for Disease Control (Atlanta, Georgia) in 1979. This number had dropped to 4,000 in 1980 and by 1984 to 752 cases. Although adolescents and young adults continue to have the highest age-specific incidences of rubella, reductions in incidence of more than 35% in 15-, 19-, and 20–24-year-olds have been showing up in the statistics.

Health authorities recommend routine immunization of all children between the ages of 15 months and 12 years of age. Frequently, combined measles, mumps, and rubella immunizations are given. With women of childbearing age whose serum test shows a lacking of hemagglutination inhibition antibody, vaccination at least 2 months prior to becoming pregnant is a seriously recommended precaution. Rubella vaccinations have been given in the United States since 1969 to well over 70 million people.

Additional Reading

CDC. Measles, mumps, and rubella — vaccine use and strategies for elimination of measles, rubella, and congenital rubella syndrome and control of mumps. Recommendations of the Advisory Committee on Immunization Practices (ACIP). *MMWR* 1998; 47(RR-8): 1–57.
CDC. Immunization of health-care workers. Recommendations of the Advisory Committee on Immunization Practices (ACIP) and the Hospital Infection Control Advisory Committee (HICPAC). *MMWR* 1997; 46(RR-18): 1–42.
CDC. Rubella and congenital rubella syndrome — United States, 1994–1997. *MMWR* 1997; **46**: 350–4.

Web Reference

Centers for Disease Control and Prevention: *http://www.cdc.gov/health/diseases.htm*

R. C. V.

RUBIACEAE. A family comprising some 4,500 species, particularly abundant in tropical regions. Some species are found in temperature regions, and a few in Arctic climates. The family includes trees, shrubs, and herbs having opposite entire or sometimes toothed leaves with stipules, the latter often large and conspicuous. The flowers are perfect, regular and epignous, and four- or five-parted. Few members of the family are important. Species of *Gardenia*, natives of tropical Old World regions, are frequently grown for their fragrant showy flowers. *Rubia tinctorium*, the madder plant, was formerly a very important source of the dye madder, also called alizarin. Now, however, the dye is prepared synthetically. Gambier, *Uncaria gambii*, a climbing plant native in tropical Asia and the Oceanic Islands, yields quantities of pyrogallol tannin, extracted with boiling water from the leaves and young shoots. This is used in tanning leathers, often mixed with other tannins. It is also used as an astringent in medicines.

Ipecac, *Cephaelis Ipecacuanha*, a native of South American tropics, is a shrubby plant, the roots of which are 6 millimeters thick. From the dried roots and the lower part of the stem the drug ipecac is obtained. Used in small doses, ipecac is a stimulant; in large doses it is an eliminant, causing vomiting, sweating, and elimination through the kidneys and bowels. It is a very efficient means for clearing an overloaded stomach. Quinine and coffee are two other important products from members of this family.

RUBIDIUM. Chemical element symbol Rb, at. no. 37, at. wt. 85.468, periodic table group 1, mp 38.9 °C, bp 686 °C, density 1.53 g/cm^3 (20 °C). Elemental rubidium has a body-centered cubic crystal structure.

Rubidium is a silver-white, very soft metal; tarnishes instantly on exposure to air, soon ignites spontaneously with flame to form oxide; best preserved in an atmosphere of hydrogen rather than in naphtha; reacts vigorously with H_2O forming rubidium hydroxide solution and hydrogen gas. Discovered by Bunsen and Kirchhoff in 1860 by means of the spectroscope.

There are two naturally occurring isotopes ^{85}Rb and ^{87}Rb, of which the latter is unstable with respect to beta decay ($t_{12} = 5 \times 10^{10}$ years) into ^{87}Sr. There are eight other known radioactive isotopes ^{81}Rb through ^{84}Rb, ^{86}Rb, and ^{88}Rb through ^{90}Rb, all with comparatively short half-lives, measured in terms of minutes, hours, or days. In terms of abundance, rubidium ranks 34th among the elements in the earth's crust. In terms of content in seawater, the element ranks higher (18th) with an estimated 570 tons of rubidium per cubic mile of seawater. First ionization potential 4.176 eV; second, 27.36 eV. Oxidation potential Rb → Rh$^+$ + e$^-$, 2.99 V. Other important physical properties of rubidium are given under **Chemical Elements**.

Rubidium occurs in lepidolite (lithium aluminosilicate, in amount up to 1% Rb), in certain mineral waters and rare minerals. Rubidium salts may be recovered from the mother liquor upon crystallization of (1) lithium salts, (2) potassium salts. Rubidium metal is obtained by electrolysis of the fused chloride out of contact with air.

Uses. The main uses of rubidium are in photocathodes and photoelectric cells. However, rubidium cells are inferior to cesium cells in their sensitivity and range. Although very small quantities are involved, rubidium gas cells now perform as secondary time standards, on the order of quartz crystal oscillators, inasmuch as they must be referenced to more accurate systems. The rubidium systems have a characteristic resonance at 6,835 MHz and, unlike other atomic frequency standards, require little power and are relatively compact. Portable rubidium atomic clocks were introduced by the U.S. Army in 1963. They weight as little as 44 pounds (20 kilograms) and occupy a volume of only about 1 cubic foot (0.028 cubic meter). The units operate on 110-V current, on the 24-V output of military vehicles, or both. Clocks of this type are used to synchronize radar nets, to assist in the accurate tracking of missiles and satellites, and to set precise radio broadcasting frequencies. Rubidium-vapor instruments also were developed as absolute-type magnetometers and introduced in 1958 by U.S. government scientists. The rubidium-vapor magnetometer uses a rubidium lamp, mounted in the tank coil of a radio-frequency oscillator. After collimating and filtering, the rubidium light is circularly polarized and then passed through a rubidium-vapor cell, after which it is focused on a sensitive photocell. Numerous combinations of amplifier parameters and various rubidium isotopes permit considerable range in the measurement of ambient magnetic fields. Inasmuch as the total world range is from 15,000 to 80,000 gammas, a system capable of this span finds use anywhere in the world.

Potential uses of rubidium include use as a fuel for ion-propulsion engines and as a heat-transfer medium.

Rubidium alloys easily with potassium, sodium, silver, and gold, and forms amalgams with mercury. Rubidium and potassium are completely miscible in the solid state. Cesium and rubidium form an uninterrupted series of solid solutions. These alloys, in various combinations, are used mainly as getters for removing the last traces of air in high-vacuum devices and systems.

Small quantities of rubidium are found in certain foods, including coffee, tea, tobacco, and several other plants. There is evidence indicating that trace quantities of the element are required by living organisms.

Chemistry and Compounds: Rubidium is more electropositive than potassium (or the lower alkali metals) as is consistent with its position in main group 1. It reacts more vigorously with H_2O, and ignites on exposure to oxygen.

Because of the ease of removal of its single $5s$ electron (4.159 eV) and the difficulty (27.36 eV) of removing a second electron, rubidium is exclusively monovalent in its compounds, which are electrovalent.

In its solutions in liquid NH_3, rubidium is, like the other alkali metals, a powerful reducing agent, so that in such solutions titrations of rubidium polysulfide with rubidium are made by electrometric methods. The solubility of rubidium salts in liquid NH_3 increases markedly with the radius of the anion (rubidium chloride, RbCl, 0.024 moles per kilogram, rubidium bromide, RbBr, 1.35 moles per kilogram, and rubidium iodide, RbI, 10.08 moles per kilogram). However, in water they exhibit minimum solubility at cation: anion radius ratio of 0.75

(rubidium fluoride, RbF, 12.5 moles/kilogram, RbCl 6.8 moles/kilogram, RbBr, 6.6 moles/kilogram, RbI 7.2 moles/kilogram).

As in the case of the other alkali metals, rubidium forms compounds generally with the inorganic and organic anions; for a general discussion of these compounds, see the entry on **Sodium**, because the sodium compounds differ principally in their greater extent of hydration and greater number of hydrates. However, rubidium coordinates with large organic molecules, such as salicylaldehyde, even though it does not with H_2O.

One respect in which rubidium and cesium are outstanding among the alkali metals is the readiness with which they form alums. Rubidium alums are known for all of the trivalent cations that form alums, Al^{3+}, Cr^{3+}, Fe^{3+}, Mn^{3+}, V^{3+}, Ti^{3+}, Co^{3+}, Ga^{3+}, Rh^{3+}, Ir^{3+}, and In^{3+}.

As in the case of potassium and cesium, rubidium forms a superoxide on reaction of the metal with oxygen. The compound is dark brown in color and paramagnetic, and hence believed to contain the O_2^- ion with an odd electron, and to have the formula RbO_2. On heating, it loses oxygen to form Rb_2O_3. Rubidium also forms a peroxide Rb_2O_2, and a normal oxide, Rb_2O, which is prepared by heating rubidium nitrite with metallic rubidium.

Rubidium hydroxide, RbOH, is the strongest, except for cesium hydroxide, CsOH (and francium hydroxide, FrOH), of the alkali hydroxides, as would be expected from its position in the periodic table. For the same reason, it has the next smallest lattice energy (146.6 kilocalories per mole).

The most numerous organic compounds of rubidium are those of oxy compounds, such as the salts of organic acids, the alcohols and phenols (alkoxides, phenoxides, etc.). An ethyl rubidium-zinc diethyl adduct has been reported, $RbZn(C_2H_5)_3$, which is certainly the true salt, rubidium triethylzincate, $Rb[Zn(C_2H_5)_3]$.

Additional Reading

Christensen, J.N., J.L. Rosenfeld, and D.J. DePaolo: "Rates of Tectonometamorphic Processes from Rubidium and Strontium Isotopes in Garnet," *Science*, 1465 (June 23, 1989).
Considine, D.M. and G.D. Considine: "Van Nostrand Reinhold Encyclopedia of Chemistry," 4th Edition, Van Nostrand Reinhold, New York, NY, 1984. (A classic reference.)
Greenwood, N.N. and A. Earnshaw: "Chemistry of the Elements," 2nd Edition, Butterworth-Heinemann, Inc., Woburn, MA, 1997.
Krebs, R.E.: "The History and Use of Our Earth's Chemical Elements: A Reference Guide," Greenwood Publishing Group, Inc., Westport, CT, 1998.
Lewis, R.J. and N.R. Sax: "Sax's Dangerous Properties of Industrial Materials," 10th Edition, John Wiley & Sons, Inc., New York, NY, 1999.
Lide, D.R.: "CRC Handbook of Chemistry and Physics 2000–2001," 81st Edition, CRC Press, LLC., Boca Raton, FL, 2000.
Staff: "ASM Handbook — Properties and Selection: Nonferrous Alloys and Special-Purpose Materials," American International, Materials, Park, OH, 1990.
Zhu, O., et al.: "X-ray Diffraction Evidence for Nonstoichometric Rubidium-C60 Intercalation Compounds," *Science*, 545 (October 25, 1991).

RUBY. See **Corundum**.

RUFF. See **Waders, Shorebirds, and Gulls**.

RUFFE. See **Perches and Darters**.

RULED SURFACE. A surface that can be generated by the motion of a straight line. The straight lines lying in the surface are the *generators of the surface*. The point into which the common perpendicular to two neighboring generators degenerates as these are brought into coincidence is the *central point* of the generator. The locus of the central points of all the generators of the surface is the *line of striction* of the surface. It is, of course, perpendicular to every generator of the surface.

See also **Coordinate System**; and **Surface**.

RUM. A spirit distilled directly from sugarcane products and usually produced in sugar-growing countries. The Chinese are known to have produced a spirit from sugarcane many centuries ago, and sugarcane was cultivated in Spain and on the Mediterranean islands as well as Madeira as early as the 3rd century A.D. The cultivation of sugarcane in the West Indies was not reported until the 15th century. Nevertheless, the origin of rum as it is known today is generally attributed to the West Indies and particularly with reference to the era of the pirates. Like gin, rum over the years has been called by a number of uncomplimentary names — Rumbullion, Rumbustion, and "kill-devil," among others. One early dictionary defined rum as "a great tumult or a strong liquor."

Although rum can be produced directly from sugarcane juice, it is traditionally and principally made from molasses (blackstrap), a by-product of the cane sugar industry. The molasses is mixed with water, yeast is added, and the mixture allowed to ferment in large tanks. Some of the variations in rums are due to the strain of yeast that is used, but also importantly by the distillation techniques employed. Depending upon local methodology, the fermentation process will span from a minimum of 2 days to nearly 2 weeks. In recent years, the Jamaican distillers have depended upon natural or "wild" yeast, with inoculation occurring directly from the vats or air. Generally, the procedure for making rum follows that for making whiskeys and gin.

Both the pot still and continuous stills are used to separate alcohol from the *wash*, a term equivalent to mash in the manufacture of grain spirits. See accompanying illustration. The distillate in either case is colorless. Caramel (burnt sugar) is the principal coloring agent used, although dyes may be used in the very dark rums. Continuous distillation produces a much more neutral, light rum with much less character than is obtainable with the pot still, where the separation of flavor- and aroma-containing components picked up from the molasses is less sharp. Consequently, ageing or maturing is required of the heavier, pot distilled rums.

Rums can be classified in a number of ways, possibly the most meaningful being the *light rum* and *full-bodied* rum categories.

Light rums, such as those produced in Puerto Rico and Cuba, are distilled to a proof range of 160° to 180° proof. Continuous stills are used. See Fig. 1. Ageing is not required. The lighter rums are preferred in the West Indies and Latin countries. They are also widely used in the United States for preparing cocktails. Bacardi rum, once exclusively Cuban, is now produced elsewhere as well.

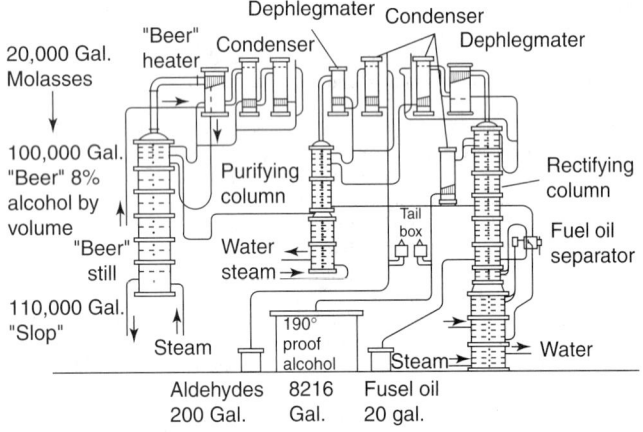

Fig. 1.   Continuous still of the type used in the production of light rums.

Heavy, or full-bodied, rums, such as those produced in Jamaica, are usually preferred in Western Europe and in the United Kingdom, particularly in the Midlands and northern England. These rums are of good quality and pungent. They are distilled to a proof range of 140° to 160° proof in pot stills. So-called "Continental Flavor" rum is particularly favored in Germany. It is highly flavored and aromatic. Commonly, the Germans mix this rum with neutral spirits to form Rum Verschnitt.

Rum for the United Kingdom is shipped from the West Indies in casks, ranging in size from 40 and 56 to 110 gallons, the latter containers called *puncheons*. The rum usually ranges between 130° and 145° proof, as received at the London docks. Law requires that it be matured for 3 years in wood before bottling, but some rums require a minimum of 6 years to reach acceptable good quality. Shipment of the rum in wood reduces the time required for further maturing in England. However, the casks, barrels, and puncheons are difficult to handle and the shipping method is under revision. Rum is shipped to France in metal containers and upon arrival is transferred to large wooden vats (small wood ageing is not specified by law), where it is matured for a minimum of 2 years. A few countries, such as Eire and New Zealand, require ageing in small wood for at least 5 years prior to bottling.

RUNOFF.   See Drainage Systems.

RUPTURE DISK. An intentionally designed weak spot within a pressurized system. The rupture disk is expected to fail before other more valuable equipment is damaged or destroyed, or before an explosive force can be created that will endanger human life. Pressure relief devices of an emergency nature may be required because of the generation of abnormal pressures which may result from: (1) faulty manual operation or automatic control of equipment; (2) presence of accidental fires or other unexpected sources of heat near the equipment; (3) sudden expansion or contraction of a liquid in a closed system; and (4) flow stoppage that may cause sudden pressure buildup and clogging of conventional pressure relief devices. See Fig. 1.

Fig. 1.   Solid-metal rupture disk.

Rupture disks are relatively simple in concept and virtually have no moving parts (except at time of rupture). The disks are designed to provide instant relief at a predetermined pressure and temperature rather than a gradual bleeding off of the excess pressure. The disks provide positive failure. Little can be done to alter the disks after installation in a pressure system to change their rupturing pressure. In selecting rupture disks, the following points are important: (1) type and thickness of metal; (2) mechanical methods of construction; (3) operating margin; (4) temperature extremes during operation and (5) types of loads the pressure system will impose on the disk during operation.

RUSAS DEER.   See Deer.

RUST FUNGI.   See Fungus.

RUTABAGA.   See Brassica.

RUTHENIUM. Chemical element, symbol Ru, at. no. 44, at. wt. 101.07, periodic table group 8 (platinum metals), mp 2,310°C, bp 3,900°C, specific gravity 12.41 (20°C). Elemental ruthenium has a close-packed hexagonal crystal structure. The seven stable isotopes are 96Ru, 98Ru through 102Ru, and 104Ru. The five unstable isotopes are 95Ru, 97Ru, 103Ru, 105Ru, and 106Ru. In terms of earthly abundance, ruthenium is one of the scarce elements. Also, in terms of cosmic abundance, the investigation by Harold C. Urey (1952), using a figure of 10,000 for silicon, estimated the figure for ruthenium at 0.019. No notable presence of ruthenium in seawater has been found. Ruthenium was discovered by Claus (Germany) in 1844.

Electronic configuration $1s^2 2s^2 2p^6 3s^2 3p^6 3d^{10} 4s^2 4p^6 4d^7 5s^1$. Ionic radius Ru^{4+} 0.60 Å. Metallic radius 1.3251 Å. First ionization potential 7.5 eV. Other physical properties of ruthenium will be found under **Platinum and Platinum Group**. See also **Chemical Elements**.

The chemistry of Ru is still poorly understood. The existence of at least eight valence states, coupled with the tendency to complex with many ions, often results in the presence of several different complexes in a given solution.

Ru metal is quite refractory. It is not significantly soluble in any single acid; even aqua regia has little effect. At room temperature, the metal does not react with O_2, but, when heated in air, a film of the dioxide appears. The metal is insoluble in fused sulfates. Molten alkali slowly dissolves the metal. The rate of attack is rapid under oxidizing conditions, and a molten mixture of NaOH and Na_2O_2 will readily dissolve the metal.

The finely divided metal is soluble in hypohalites if an excess of alkali is present. At red heat, the metal combines with Cl_2 to form the dichloride. Ruthenium(VIII) oxide is formed when an alkaline ruthenium solution is treated with a strong oxidant, such as chlorine, or bromate ion when the Ru is in acid solution.

Ruthenium(III) hydroxide is formed by the action of alkali on a solution of ruthenium(III) chloride. It is easily oxidized by air to the tetravalent state. The dioxide, RuO_2, forms when the metal is heated in air. Hydrous ruthenium(IV) oxide can be precipitated by adding alcohol to a less than 3-M NaOH solution of ruthenium(VIII) oxide, followed by boiling. Above 3-M NaOH, complete reduction is not obtained. The hydrous oxide that is soluble in concentrated HCl tends to occlude impurities.

The only known octavalent Ru compound is the tetroxide, RuO_4, which exists in a yellow and a brown form. The volatile and poisonous tetroxide melts at about $25\,^{\circ}C$ and sublimes readily. It may explode in contact with oxidizable substances or when heated above $100\,^{\circ}C$. It is formed by distillation from either an alkaline or acid solution under strongly oxidizing conditions. The tetroxide is moderately water-soluble. When dissolved in alkali, it initially forms a green solution of heptavalent perruthenate of the form $MRuO_4$, which further reduces to the orange ruthenate M_2RuO_4. The reduction to the hexavalent state is quicker in strong alkali. The ruthenates also are made by fusing finely divided metal with a mixture of alkali hydroxide and nitrate or peroxide.

Anhydrous ruthenium(III) chloride, $RuCl_3$, is made by direct chlorination of the metal at $700\,^{\circ}C$. Two allotropic forms result. The trihydrate is made by evaporating an HCl solution of ruthenium(III) hydroxide to dryness or reducing ruthenium(VIII) oxide in a HCl solution. The trihydrate, $RuCl_3 \cdot 3H_2O$, is the usual commercial form. Aqueous solutions of the trihydrate are a straw color in dilute solution and red-brown in concentrated solution. Ruthenium(III) chloride in solution apparently forms a variety of aquo- and hydroxy complexes. The analogous bromide, $RuBr_3$, is made by the same solution techniques as the chloride, using HBr instead of HCl.

Ruthenium(III) iodide, RuI_3, is a black, insoluble compound precipitated by the addition of iodide ion to a solution of $RuCl_3$.

Tetravalent ruthenium chloride, $RuCl_4$, and the hydroxychloride, $Ru(OH)Cl_3$, are intermediate products when $RuCl_3$ is prepared by evaporating the tetroxide in HCl. When the hydroxychloride in hot HCl is treated with Cl_2, it is converted to the tetrachloride. The anhydrous tetrachloride also is known. The tetrabromide and tetraiodide have not been isolated; attempts to prepare these compounds result in the formation of the respective trihalides.

The only pentavalent Ru compounds known are the fluorides; RuF_5 is made by combining the elements. The compound melts at $107\,^{\circ}C$ and boils at $313\,^{\circ}C$. The salt $NaRuF_6$ was recently made by mixing $RuCl_3$, with NaCl and treating the mixture with BrF_3.

Ru forms many complex ions. The nitrosyl compounds are frequently encountered by accident due to the great affinity of Ru for the nitrosyl group. Ruthenium(III) nitrosylchloride, $Ru(NO)\,Cl_3 \cdot 4H_2O$, is a by-product of most solutions of $RuCl_3$ in aqua regia or solutions containing HNO_3. It also is present in HCl solutions resulting from a KOH and nitrate fusion of the metal. The chloride and bromide are respectively raspberry and violet in solution. Alkaline chlorides form complex salts of the type $M_2Ru(NO)Cl_5$, which can be crystallized from solution. A black gelatinous precipitate of the nitrosylhydroxide, $RuNO(OH)_3$, is slowly formed when a solution of the nitrosylchloride is heated with a strong base. A series of nitrato- and nitro- derivatives of nitrosylruthenium also have been described and separated.

It is generally accepted that the disulfide is the only certain sulfide of Ru. It is formed by the action of H_2S on a solution of Ru or from the elements at about $1000\,^{\circ}C$. When ruthenium(IV) sulfide is treated with HNO_3, the sulfate is formed.

Dichlorodicarbonylruthenium(II), $Ru(CO)_2Cl_2$, is formed when $RuCl_3$ is heated above $210\,^{\circ}C$ in the presence of CO. It is a yellow, insoluble, volatile compound. The bromine and iodine analogs are similarly formed.

When finely divided Ru metal is heated at $180\,^{\circ}C$ under 200 atm of CO, pentacarbonylruthenium(0), $Ru(CO)_5$, is formed.

Ruthenium forms a large number of complex ions with amines.

Recently, a new group of organometallic sandwich compounds, called *metallocenes*, has been discovered. Ruthenocene is made in about 50% yield by reacting RuCl3 with cyclopentadienylsodium in tetrahydrofuran.

After refluxing and distilling the solvent, the light-yellow crystals of ruthenocene are sublimed. The compound, $Ru(C_5H_5)_2$, undergoes a large number of substitution reactions typical of aromatic systems.

Ruthenium is commonly used with other platinum metals as a catalyst for oxidations, hydrogenations, isomerizations, and reforming reactions. The synergetic effect of mixing ruthenium with catalysts of platinum, palladium, and rhodium has been found for the hydrogenations of aromatic and aliphatic nitro compounds, ketones, pyridine, and nitriles.

Additional Reading

Coles, D.G. and L.D. Ramspott: "Migration of Ruthenium-106 in a Nevada Test Site Aquifer," *Science*, **215**, 1235–1237 (1982).

Considine, D.M. and G.D. Considine: "Van Nostrand Reinhold Encyclopedia of Chemistry," 4th Edition, Van Nostrand Reinhold, New York, NY, 1984.

Davis, J.R.: "Metals Handbook," 2nd Edition, ASM International, Materials Park, OH, 1998.

Greenwood, N.N. and A. Earnshaw: "Chemistry of the Elements," 2nd Edition, Butterworth-Heinemann, Inc., Woburn, MA, 1997.

Krebs, R.E.: "The History and Use of Our Earth's Chemical Elements: A Reference Guide," Greenwood Publishing Group, Inc., Westport, CT, 1998.

Lewis, R.J. and N.I. Sax: "Sax's Dangerous Properties of Industrial Materials," 10th Edition, John Wiley & Sons, Inc., New York, NY, 1999.

Lide, D.R.: "CRC Handbook of Chemistry and Physics 2000–2001," 81st Edition, CRC Press, LLC., Boca Raton, FL, 2000.

Seddon, E.A. and K.R. Seddon: "The Chemistry of Ruthenium," Elsevier Science, New York, NY, 1984.

Sinfelt, J.H.: "Bimetallic Catalysts," *Sci. Amer.*, 90–98 (September 1985).

Staff: "ASM Handbook — Properties and Selection: Nonferrous Alloys and Special-Purpose Materials," American International, Materials Park, OH, 1990.

LINTON LIBBY, Chief Chemist, Simmons Refining Company, Chicago, IL.

RUTHERFORD, ERNEST (1871–1937). Rutherford was a British physicist who was born in the South Island of New Zealand and is famous for his pioneering work in nuclear physics and for his theory of the structure of the atom.

Rutherford was awarded a scholarship to be a research student at the University of Cambridge and began research under J.J. Thomson. He soon abandoned research on his radio wave detector to work on the power of X-rays to confer electric charge on gases but soon turned to researching the problem of the rays emitted by thorium. Rutherford found three kinds of radiation, which he named alpha, beta, and gamma. In collaboration with Frederick Soddy, he was able to isolate a substance, thorium X, and identify the phenomenon of radioactive half-life and formulated an explanation of radioactivity. Rutherford was awarded the 1908 Nobel Prize for chemistry for his work in radioactivity.

In 1907, Rutherford moved to the University of Manchester and in 1909 he discovered the atomic nucleus. In 1911 he announced his revolutionary idea on the nature of the atom and he developed a model of the atom showing it similar to the solar system. He proposed the idea that almost all the mass and all the positive electricity in an atom was densely concentrated in a tiny nucleus and the electrons circled around it like planets around the sun.

While at Manchester, Rutherford produced the first human "nuclear reaction" with the disintegration of a non-radioactive atom, dislodging a single particle. He became famous as the man who "split the atom." In 1919, Rutherford succeeded J.J. Thomson as Cavendish Professor of Physics at Cambridge. He was a leader of a research team encouraging others in the investigation of the nucleus.

See also **Radioactivity**; and **Proton**.

J. M. I.

RUTILE. A mineral, composed of titanium dioxide, which occurs in three distinct forms: as rutile, a tetragonal mineral usually of prismatic habit, often twinned; as octahedrite (anatase), a tetragonal mineral of pseudo-octahedral habit; and as brookite, an orthorhombic mineral. Both octahedrite (anatase) and brookite are relatively rare minerals.

Rutile has a sub-conchoidal fracture; is brittle; luster, metallic-adamantine; color, commonly reddish-brown but sometimes yellowish, bluish or violet; streak, brown; transparent to opaque. Rutile may contain up to 10% of iron.

Experiments in the artificial preparation of titanium dioxide appear to show that rutile is the most stable form and produced at the highest temperature, brookite at a lower temperature, and octahedrite (anatase) at a still lower temperature.

Rutile is found as an accessory mineral in many kinds of igneous rocks, and to some extent in gneisses and schists. In groups of acicular crystals it is frequently seen penetrating quartz as the "flèches d'amour" from Grisons, Switzerland, and Brazil. Rutile is found also in Austria, Italy, Norway, South Australia, and Brazil. In the United States it occurs in Vermont, Massachusetts, Connecticut, New York, Pennsylvania, Virginia, Georgia, North Carolina, and Arkansas.

Rutile derives its name from the Latin *rutilus*, red, in reference to the deep red color observed in some specimens when viewed by transmitted light.

RYDBERG CONSTANT.   A quantity that enters into the frequency or wave number formula for all atomic spectra. Bohr showed that, in terms of known constants, the Rydberg constant is given by

$$R = \frac{2\pi^2 m e^4}{ch^3(1 + m/M)}$$

where e is the charge on the electron, c is the velocity of light, h is Planck's constant, m is the mass of the electron, and M is the mass of the atomic nucleus. Since m/M is very small, R can vary only slightly for different elements. In the limit, as m/M approaches zero, a recommended numerical value of the constant is

$$R_\infty = 109{,}737.31 \text{ cm}^{-1}$$

with an estimated error limit of 3 based on 3 standard deviations in the last digit given. If R is multiplied by c, the dimension of the constant is frequency and so used in a formula for series in line spectra gives the frequency of the calculated lines rather than the wave number. See also **Atomic Spectra.**

RYE *(Secale cereale; Gramineae).*   Rye is an annual plant, which has a tendency to become perennial. It is a sturdy cereal grass having a much-branched root system which penetrates 4–6 feet (1.2–1.8 meters)

into the ground and tough slender stems which may grow as tall as 6 feet (1.8 meters). The leaves are like those of other cereal grasses, but have a definite bluish color, as does the stem. The inflorescence is a spike, with the individual spikelet three-flowered and occurring singly at each of the 20 or more joints of the rachis. See Fig. 1. Of the three

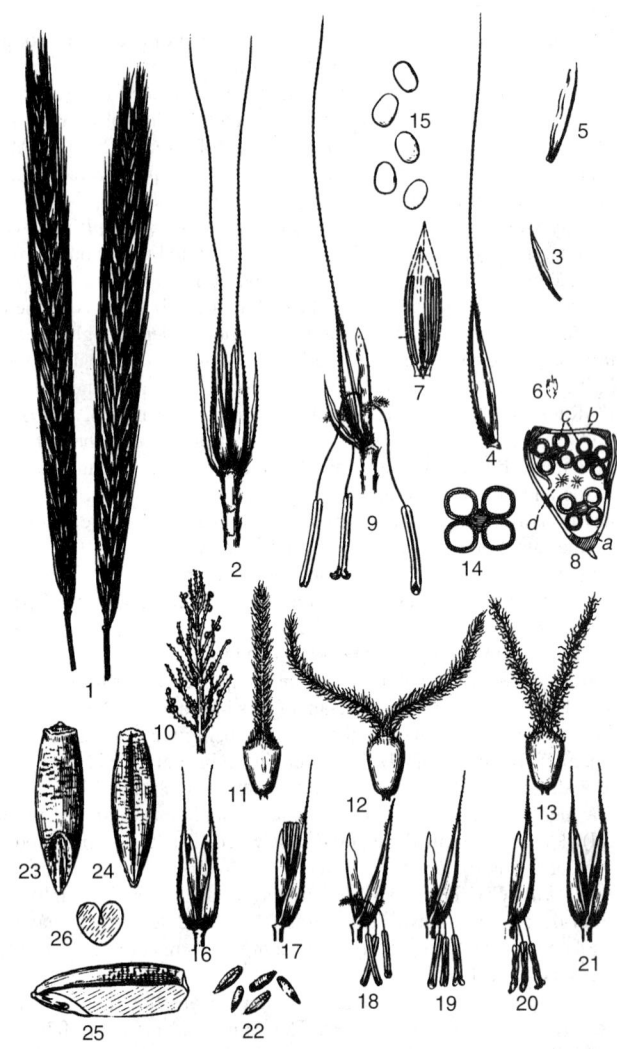

Fig. 2.   Inflorescence of rye:

(1) Two spikes, lateral view.
(2) Spikelets, dorsal view, attached to node of rachis.
(3) Glume, lateral view.
(4) Lemma, lateral view.
(5) Palea, lateral view.
(6) Lodicule.
(7) Diagrammatic lateral longitudinal view of floret at anthesis, showing position of gynoecium and androecium.
(8) Diagrammatic cross section of spikelet.
(9) Lateral view of floret at anthesis, with one subtending glume.
(10) Portion of stigma with adhering pollen grains.
(11) Through (13) Gynoecium, before, during, and after anthesis.
(14) Diagrammatic cross section of anther.
(15) Pollen grains.
(16) Two florets, lateral view, at beginning of anthesis.
(17) Through (20) Floret, lateral view, showing successive stages in anthesis.
(21) Two florets after anthesis.
(22) Caryopses (seeds).
(23) Caryopsis, dorsal view, showing embryo.
(24) Caryopsis, ventral section, showing endosperm and embryo.
(25) Caryopsis, lateral longitudinal (sagittal) section, showing endosperm and embryo.
(26) Caryopsis cross section.

(USDA diagram.)

Fig. 1.   Spikes of winter rye growing in Maine. (*USDA photo.*)

flowers in a spikelet, only the two lower ones mature, the third aborting. The two glumes are narrow, the lemma is broad, distinctly keeled, and has a long stiff terminal awn, while the palea is thin and blunt. Unlike most of the cereal grasses, rye must be cross-pollinated in order to set fruit abundantly. The fruit, a grain, is very similar to that of wheat in structure, and readily separated from the lemma and palea when mature. The grain is long and slender and of much darker color than wheat grains.

The principal parts of the rye plant are shown in Fig. 2.

Authorities estimate that rye has been cultivated for about 2000 years, although historical records on this plant are considerably less detailed than for other major cereal grasses. There is no evidence of rye in early records of the Greeks or of Swiss lake-dwellers. It is known that rye was grown and consumed extensively during the medieval period in Europe. During that time, rye became a major bread grain. During the past century, the popularity of rye bread has decreased in favor of wheat breads, but it does remain very popular in much of Europe and western Asia. In the United States, only about one-quarter of rye production is used in bread and related products. Approximately one-quarter of rye production is used in the manufacture of alcohol and distilled spirits, notably for rye whiskey. The majority of rye grain is used for livestock feed. Another major use is that of a green manure for crops and as pasture. When rye concentrate is used for feed, it usually is diluted with other cereal grains. Rye straw finds a number of uses, including packing and paper manufacture.

Varieties. There are winter and spring varieties of rye, but the largest portion of world production is from the winter varieties. The latter that do well in northern latitudes are not suited to conditions of the lower latitudes. For the southern United States, where considerable amounts of rye are produced, particularly during recent years, strains have been developed from the varieties that are grown in southern Europe and the Mediterranean region.

Culture. Rye should not be planted on the same land more often than once every three years if maximum yields and disease resistance are to be achieved. Although rye will germinate at a temperature as low as 33 °F (18.3 °C), the optimal planting temperature is 55°–65 °F (12.8°–18.3 °C).

Rye is harvested much as other small grains. A predominant amount of rye grown in the United States is harvested with a combine.

Production. Of the cereal crops, rye ranks eighth in world tonnage and seventh in tonnage of cereal crops produced in the United States. Continental Europe leads by far with over 90% of world production. North and Central America account for nearly 4%, Asia, 3.5%, and South America, 1.3%. Production of rye in Africa is negligible on a world scale.

RYEGRASS.   See **Grasses.**

S

SABIN VACCINE. See **Poliomyelitis**; and **Virus**.

SABLE. See **Mustelines**.

SACCHARIDE. See **Carbohydrates**.

SACCHARIMETER. An instrument for the measurement of sucrose solutions. A saccharimeter differs from a polarimeter in that the saccharimeter uses white light, whereas a polarimeter is operated with sharply monochromatic light. Consequently, a saccharimeter can only be used with sugar solutions (in which case the quartz compensates for the rotatory dispersion of sucrose). Conversely, a polarimeter is suitable for the measurement of optical rotation of any solution, including sugar. However, saccharimetric sugar determinations are the basis of internationally accepted *sugar degrees* (°S), and saccharimeters are appropriately calibrated. When a polarimeter is employed that utilizes monochromatic light rather than a quartz wedge, deviations from sugar degrees will be found in some solutions, and they may not be inconsequential. The International Sugar Scale assigns 100 °S to a pure sucrose solution of normal weight (26 grams in 100 milliliters of pure water) at 20 °C and a 200-millimeter light path, measured in a saccharimeter with white light and a dichromate filter. Of course, an exact numerical conversion from sugar degrees to angular degrees is possible with pure sucrose (100 °S corresponds to $\alpha = 34.6°$), but in practical, more or less impure solutions, the relationship is not exactly predictable, and the sugar scale is conventionally and legally binding. It is not surprising, then, that virtually all sugar laboratories use visual saccharimeters (also called polariscopes). They differ in vintage, the half-shade presentation, and in construction features, but they all use the quartz-wedge compensating principle, and they read out in sugar degrees. A typical instrument will have a split half-shade field or a triple field for observation. The quartz wedge is equipped with a fine scale that is read off a second observation tube. It is graduated in °S (e.g., −30 to +110 °S) and comes with a vernier, readable to 1/10 °S.

See also **Photometers**; **Polarimetry**; and **Polarized Light**.

SACCHARIN. See **Sweeteners**.

SACCHAROMYCES. See **Yeasts and Molds**.

SACCHAROSE. See **Carbohydrates**.

SACRUM. The portion of the spinal column of vertebrates, usually formed of several fused vertebrae, with which the pelvis is articulated.

SADDLE POINT. A point (x_0, y_0) on a surface $f(x, y)$ where $f(x, y_0)$ is a maximum at $x = x_0$ and $f(x_0, y)$ at the same time is a minimum at $y = y_0$. A familiar example is the hyperbolic paraboloid, which has a saddle point at the origin if its standard equation is taken as $x^2/a^2 - y^2/b^2 = 2cz$. A person walking toward the origin in the *XZ*-plane would be ascending a mountain peak while he would be descending into a valley if he walked in the *YZ*-plane. It is also called a *minimax* or a *col*.

See also **Game Theory**; **Paraboloid**; and **Surface**.

SAFETY (Intrinsic). See **Intrinsic Safety**.

SAFETY SWITCH. See **Limit Switch**.

SAFETY VALVE. The common form of the safety valve is the pop valve. It is held against its seat by a heavy spring and having a "huddling chamber" to make it open quickly and remain open until a predetermined

pressure drop (2–4% of the working steam pressure) has occurred. The A.S.M.E. Boiler Construction Code requires boilers having more than 500 square feet (~46 square meters) of heating surface, or those generating better than 2000 pounds (907 kilograms) of steam per hour, to have two or more safety valves. The safety valves should have sufficient relieving capacity to prevent more than 6% pressure rise at maximum rate or combustion. Required discharge capacity of a safety valve may be based either on the heat units in the fuel consumed or on the amount of steam generated.

In case more than one safety valve is used, the smaller one can be set to pop at the desired maximum pressure and the larger at 2 or 3 pounds per square inch (0.1 or 0.2 atmospheres) higher.

The relief valve is a form of safety valve, but usually intended for less severe service and of less importance from the safety viewpoint. Relief valves are applied to air, to water, and to steam lines, and also to tanks, heaters, and so on. Among them could be mentioned the back pressure valves and atmospheric relief valves.

SAFFLOWER *(Carthamus tinctorius; Compositae)*. This plant, a native of the East Indies, is now widely cultivated in tropical Asia and Egypt and to a lesser extent in southern Europe and elsewhere. It is a low annual plant with yellowish-red flowers that have tubular corollas.

SAFFLOWER SEED OIL. See **Vegetable Oils (Edible)**.

SAFFRON *(Crocus sativus; Iridaceae)*. *Crocus sativus* is a perennial herb, the native home of which is the eastern Mediterranean region. The stem is an underground flattened corm, the surface of which is covered by a few scaly leaves. At the top of the corm is a terminal bud, which develops into linear leaves 5–9 inches (12.5–22.5 centimeters) long and flowers. The flowers are white or lilac-tinted, with the perianth six-parted and with a very long tube, so that the ovary remains below the surface of the ground. The three stigmas are bright red. These, when dried, are known as saffron, an orange-yellow dye with a considerable percentage of volatile oil present. See also **Flavorings**.

SAGE. See **Antioxidant**; and **Flavorings**.

SAGEBRUSH *(Artemisia tridentata; Compositae)*. A number of other species of this genus are also called sagebrush, but *A. tridentata* is more prominent and of wider distribution than the others. This shrub has an extensive root system and may attain heights up to 7 feet (2.1 meters). The gray-green foliage and aromatic odor of this shrub are distinctive. The flowers occur in inconspicuous heads. This plant is distributed from the Black Hills to southern British Columbia to southeastern California to northern Arizona. It reaches its best development, however, in the Great Basin region, where it may occur over large areas in nearly pure stands. See also **Artemisia**. See Table 1 for record trees.

SAGITTARIUS (the archer). This large constellation is the ninth sign of the zodiac. Lying as it does in a particularly rich portion of the Milky Way, it contains a large number of star clusters and gaseous nebulae of great beauty visible in a moderate-sized telescope. From the large number of faint stars, cepheid variables, and globular clusters that seem to congregate in this region, we can deduce that the stellar galactic system has its greatest extension and hence its center in this direction. Long-exposure photographs indicate that large numbers of dark or obscuring nebulae lie in this portion of the Milky Way. The only way to penetrate these clouds is by means of radio astronomy. (See map accompanying entry on **Constellations**).

TABLE 1. RECORD SAGEBRUSH TREES IN THE UNITED STATES[1]

Sagebrush trees	Circumference[2]		Height		Spread		Location
	Inches	Centimeters	Feet	Meters	Feet	Meters	
Big sagebrush (1991) (Artemisia tridentata)	17	43	17	5.2	16	4.9	Washington
Big sagebrush (1995) (Artemisia tridentata)	20	51	13	4	17	5.2	Oregon

[1]From the "National Register of Big Trees," American Forests (by permission).
[2]At 4.5 feet (1.4 meters).

SAGITTA (the arrow). A northern constellation located next to Aquila.

SAIGA. See **Goats and Sheep**.

SAILFISH. See **Billfishes**.

SAILINGS (The). The position of a vessel at sea or in the air is defined by the latitude and longitude. The position at any particular instant is connected with any other position, either the one just left or the one toward which the vessel is proceeding, by means of the true course and distance.

Any given course and distance may be resolved into two components at right angles to each other: the northing or southing and the easting or westing, each expressed in nautical miles. The northing or southing may be immediately converted into difference of latitude, expressed in angular units, for the nautical mile is, by definition, approximately equal to a minute of arc along a great circle. However, the conversion from easting or westing, commonly known as departure, into difference of longitude, can be accomplished only after taking into account the shape of the earth and the approximate latitude of the ship.

The navigator is continually faced by one of two problems. (1) Given the difference of latitude and longitude between two points on the surface of the earth, to find the course and distance between them. (2) Given the course and distance followed by a ship, to find the difference of latitude and longitude between the point of starting and the destination. The different methods of solving these problems are known as the sailings and include plane, parallel, middle latitude, mercator, great-circle, and composite sailings.

See also **Course**; and **Navigation**.

SAILPLANE. A sailplane is a highly efficient glider. Being designed for the use of expert glider pilots, it is unsuited for primary training. It is characterized by very low sinking speed, nearly flat glide and perfection of construction. It is capable of rising flight on weak thermal air currents and is the type of aircraft employed for cross-country motorless flights of a sporting nature. The sailplane has high aspect ratio (about 20), careful streamlining, clean and smooth external surfaces, and minimum weight consistent with structural safety. The gliding angle in still air is often as flat as 22:1 and the sinking speed as small as 2 feet (0.6 meter) per second.

SAINT ELMO'S FIRE (or Corona Discharge). A brush-like, luminous, and often audible discharge from charged objects in the atmosphere. It occurs on ship masts, on aircraft propellers, wings, and other projecting parts, and on objects projecting from high terrain when the atmosphere is charged and a sufficiently strong electrical potential is created between the object and the surrounding air. Aircraft most frequently experience St. Elmo's fire when flying in or near cumulonimbus clouds or thunderstorms, in snow showers, and in dust storms.

SAITHE. See **Codfishes**.

SAKI. See **Monkeys and Baboons**.

SALAMANDER (Amphibia, Urodela). A vertebrate with a slender body, short legs, and a long tail. The moist skin of the amphibians limits them to protected habitats, either near water or under some protection on moist ground, usually in the woods. Some species are aquatic throughout life, some take to the water intermittently, and some are entirely terrestrial as adults. The salamanders resemble the lizards superficially, but they are easily distinguished from lizards by their moist skin, which is without scales. See also **Hellbender**.

SALICACEAE. See **Willow Trees**.

SALICYLIC ACID. Salicylic acid or $C_6H_4(OH)(COOH)$ is a white solid, melting points 159 °C, sublimes at 76 °C, insoluble in cold water, soluble in hot water, alcohol, or ether. With ferric chloride solution, salicylic acid solutions are colored violet (distinction from benzoic acid).

Salicylic acid may be obtained (1) from oil of wintergreen, which contains methyl salicylate, or (2) by heating dry sodium phenate C_6H_5ONa plus carbon dioxide under pressure at 130 °C and recovering from the resulting sodium salicylate by adding dilute sulfuric acid. Salicylic acid is a mild disinfectant and antiseptic and has been used as a food preservative. Salicylic acid and certain salicylates are used in medicine as antirheumatics.

SALINOMETER. An instrument for determining salt concentration (salinity), particularly one based upon electric-conductivity measurements. See also **Electrical Conductivity**.

SALIVA. See **Caries, Cariology, and Dentistry**.

SALK, JONAS EDWARD (1914–1995). Salk, an American physician, is famous for making a killed-virus vaccine that was effective in preventing poliomyelitis. His vaccine eliminated polio as a public health threat. For his vaccine he was awarded the Lasker Award of the American Public Health Association in 1956 and in 1958 he received the Bruce memorial Award of the American College of Physicians.

Salk began his work in virology and immunology in 1938 at New York University. Over the years, he worked with Thomas Francis, Jr. who was working on ways to kill influenza virus without destroying the virus's ability to stimulate production of antibodies. In 1947, Salk went to the University of Pittsburgh School of Medicine where he became the head of the Virus Research Center. Here he confirmed there were only three types of polio virus and then proceeded in developing his polio vaccine. In 1963, Salk became the director of the Salk Institute for Biological Studies at San Diego, California. He retired in 1975.

See also **Poliomyelitis**.

J. M. I.

SALK VACCINE. See **Poliomyelitis**; and **Virus**.

SALMON (Osteichthyes). The order Salmoniformes consists of eight suborders, with some quite unlike others. Most important commercially is the suborder Salmonoidei. These fishes are primarily migrating species and associated with the freshwater of the Northern Hemisphere. The three subfamilies of Salmonoidei include the ayus and smelts. Many of the salmon species are good eating not only because of their flavorful, fatty meat, but also because they lack those bones that in most fishes are embedded in the cartilaginous walls between the muscular segments. Salmonidae contains such familiar fishes as salmon, trout, and chars.

Atlantic Salmon. The species Salmon solar, prior to extensive pollution of certain waters, was one of the most prevalent fishes in the Atlantic drainage areas. Its distribution extended from Kara in northeastern Russia, along the coast of Europe to Douro in the northwestern part of the Iberian peninsula, and on to Iceland, the southern tip of Greenland, and across Newfoundland to Cape Cod in the northeastern United States. Salmon migrate extensively. The early part of the salmon's life is spent in the upper courses of large rivers. Then they migrate into the sea, where they grow relatively quickly, and then return to swim up rivers for spawning. During their stay in the oceans, salmon traverse great distances. Thus, salmon marked off the European coast have been recovered in waters of western Greenland. Generally, however, the salmon stays near the shore. Feeding grounds are primarily in the southern Baltic Sea and off northwestern Norway. When preying upon other fishes, they are found in the upper water levels to a depth of about 32 feet (10 meters), but may penetrate deeper.

Some studies indicate that their distribution at various depths depends upon daily and seasonal changes. During their period in the sea, salmon grow at a remarkable rate, often exceeding 2.2 pounds (1 kilogram) per month. They spend 1 to 3 years in the ocean before returning to the rivers to spawn. During this period, they have stored great quantities of fat, so much that their skin is orange-red. They leave freshwater when they are from 4 to about 8 inches (10 to 20 centimeters) long. After a year in the ocean, they measure nearly 20 to 26 inches (51 to 66 centimeters) in length and weigh from 3.3 to 7.7 pounds (1.5 to 3.5 kilograms). After 2 years, their length is from about 28 to 36 inches (71 to 91 centimeters), with a weight of from 9 to 17.5 pounds (4 to 8 kilograms). After 3 years, salmon are from 36 to 41 inches (91 to 104 centimeters) long and weigh from 17 to nearly 28 pounds (8 to 13 kilograms). Salmon probably reach a maximum age of 10 years. Occasionally, old males up to 41 inches (104 centimeters) in length and weighing as much as 80 pounds (36 kilograms) are caught. Females are generally smaller and rarely exceed a length of 39 inches (100 centimeters). Their greatest weight is 44 pounds (20 kilograms). See Fig. 1.

Fig. 1.   Atlantic salmon.

Salmon from various rivers meet at the feeding grounds. When the spawning season comes, they separate once again, and each salmon seeks out the river in which it was born. The exact manner in which salmon find their way back to their home river is not understood. It is only known for certain that their olfactory sense plays a crucial role in the second phase of their ascent up the river. This ascent takes place throughout the year in some rivers, while in others it is found only at certain seasons. Often, larger salmon are seen ascending a river at one period, while smaller ones are found at some other time. Four major types are distinguished—large and small summer salmon and large and small fall or winter salmon. When they meet at the river mouths, the salmon can be distinguished by the development of their gonads. The individual groups then seek out their various spawning sites. If these sites are far from the mouths of rivers, it is generally the large fall salmon that ascend the river. Their germ cells are still immature when they begin their ascent. The ascent is interrupted by the onset of frost; at this time, the salmon winter somewhere upriver and will reach their spawning sites in the following fall.

While the energetic large salmon are capable of swimming great distances upstream, the smaller species generally find spawning sites near the river mouths. Each day these salmon cover greater distances. In small Scottish rivers, daily distances of up to 34 miles (58 kilometers) have been recorded. It must also be recalled that the salmon only migrate for 5 to 6 hours per day, during which time they swim with great strength and endurance. Their highest speed has been estimated at 10 miles (13 kilometers) per hour. They can also swim through rapids. Against a current of 19.6 feet (6 meters) per second, a salmon can push ahead at a rate of over 3.3 feet (1 meter) per second. Small waterfalls are passed by jumping over them, and leaps of up to 10.8 feet (3 meters) high and 16.4 (5 meters) long have been observed. To jump out of the water, the salmon swim up through the water surface on a slant, and with a particularly

strong beat of the tail they gain additional acceleration at the surface. While jumping, salmon typically show a distinct lateral arching of the body. If the first attempt fails, the leaps are repeated constantly. This often results in skin injuries in stony waters, which sometimes become infected and cause death prior to reaching the spawning site.

During the entire migration, the salmon virtually cease feeding from the time they enter the river. During the journey, their fat reserves are converted into energy, and the orange-red hue of the skin disappears. As the germ cells mature, the salmon also alter their appearance. In the ocean, salmon have rather plain coloration, with a gray-green back, silvery sides, and white belly. X-shaped spots are above the lateral line and round black spots mark the head. However, when migrating upstream, a brilliant coloration develops. The back becomes considerably darker, while the sides take on a bluish shimmer and the stomach becomes reddish. Purple-red spots appear beside the black ones, and even the lower sides of the pectoral, anal, and caudal fins take on reddish hues. The color change is accompanied by a major anatomical modification of the lower jaw. The jaw points upward and develops cartilaginous growths from which a hook-shaped appearance develops.

The salmon-spawning period in central European waters occurs generally from mid-November to mid-December. In the north, the onset of spawning is in mid-September and in some groups may last until February. Spawning sites are located in regions with clear, cold, oxygen-rich, fast-flowing water and a clean gravel bottom. Salmon typically seek gravel banks in the upper water levels at a depth of about 1.5 feet (0.5 meter). Once the female arrives at the spawning site, she prepares the nest. The female digs up the floor with powerful rump and tail motions and constructs a depression some 4 to 8 inches (10 to 20 centimeters) deep and often well over 3 feet (1 meter) long. Bohemian fishermen have said that a horse could fit in one of these nests! During the entire period, there are usually several males in the vicinity, but they do not assist with construction of the pit. All salmon species carry out their spawning habits in a similar manner. After maneuvers of rivalry among males, the male and female salmon will swim side-by-side, pressing close together with mouths open just above the spawning pit, and the eggs and sperm are released. Eggs are released several times, interrupted by additional rivalry fights and more courtship display behavior. A single female lays a total of from 10,000 to 30,000 eggs. It lays about 500 to 2000 eggs per kilogram (2.2 pounds) of body weight. This is actually a low number when compared to some fishes, such as the carp. The yolk-rich, sticky eggs are from 5 to 7 millimeters large and, according to water temperature, will lie between the stones in the spawn pit for some 70 to 200 days. The young hatch in April or May. As long as the larvae still have yolk upon which to feed, they remain hidden in the pit. When that supply is exhausted they move into the water and initially feed on small crustaceans and insect larvae. As they get older, the young salmon increasingly feed on fishes and, at the end of their juvenile period, their diet is exclusively small fishes. The young stay in fresh water for 1 to 2 years (in the north, up to 5 years), after which they enter the oceans and take on their typical oceanic coloration.

The *effects of pollution* on the Atlantic salmon population have been marked and have been going on for decades. These fishes have almost completely disappeared from the Rhine, Weser, and Elbe Rivers in Germany. A few isolated salmon still appear in the lower Rhine, but they are inedible because their fat is contaminated with phenol and other wastes in the river. In Europe, the decline of the salmon began in the mid-1890s as the rivers became polluted. In England, industrialization began earlier, and thus the Thames, once a salmon river, became contaminated still earlier. The last salmon caught in the Thames was recorded in 1833. Until the early part of the twentieth century, the East Prussian coast was the most productive one for German commercial fishing. Large catches no longer are made. Currently, most European salmon are caught in Norway and Denmark. Canada has the greatest catch in the world. Salmon are generally smoked.

Inevitable dying after the first spawning is a rare phenomenon among vertebrates. Death of the adult salmon occurs in nearly all instances within 1 to 2 weeks after spawning. The immediate cause of the post-spawning death is not well understood. Physical exhaustion from the long migrations does not appear to be the principal cause of death, since salmon running up short streams may reach the spawning site in very good condition, but they still undergo degeneration that rapidly leads to death after shedding their sex products.

Pacific Salmon. There are several species: *chinook* (*Oncorhynchus tshawytscha*); the *coho* (*O. kisutch*); the *sockeye* (*O. nerka*); the *chum* (*O. keta*); and the *pink* (*O. gorbuscha*). These are generally known as the North American species. There is an additional Asian species (*O. masou*), the common name of which is *masu* salmon.

The various species of salmon vary greatly in size reached at maturity (Harry 1969). The chinook is the largest of the Pacific salmon, with a record weight of 126 pounds (57.2 kilograms), but with an average weight of about 20 pounds (9.1 kilograms). Coho salmon range in size up to 30 pounds (13.6 kilograms), with most of the fish being in the 8-to-12-pound (3.6-to-5.4-kilogram) category. Chum salmon weigh as much as 33 pounds (15 kilograms) and the average weight is about 8 pounds (3.6 kilograms). Sockeye salmon are usually between 5 and 7 pounds (2.3 and 3.2 kilograms), but weights up to 15 pounds (6.8 kilograms) have been recorded. Pink salmon usually weigh between 3 and 5 pounds (1.4 and 2.3 kilograms) and specimens up to 12 pounds (5.4 kilograms) have been taken. The masu salmon has an average weight of about 10 pounds (4.5 kilograms) and a maximum weight of 20 pounds (9.1 kilograms).

Chinook Salmon. This fish can be distinguished from the other species by the heavy black spotting on the back, the dorsal fins, and both lobes of the caudal fin, as well as by the black pigmented skin along the base of the teeth. Young chinook salmon can be recognized in freshwater by strongly developed parr marks (marks characteristic of young fishes). The chinook salmon enter the Sacramento-San Joaquin system of California to spawn, but are found only rarely in streams south of San Francisco Bay. Spawning occurs in rivers north to the Bering Sea and on the Asiatic side down to the Amur River, although rarely that far south, and in Japan in the rivers' of northern Hokkaido. In the North Pacific Ocean, chinook salmon are found generally to the north of 46° north latitude in the eastern half of the ocean, but west of the 180th meridian they occur about as far south as 42° north latitude. Chinook salmon generally ascend the larger streams to spawn, and they are abundant in such rivers as the Sacramento, Columbia, Fraser, and Yukon. The center of abundance of this species is the Columbia River, in which adults return to spawn during every month of the year. Each female chinook salmon carries from 2,000 to 13,000 eggs. After the eggs are deposited, they are protected by a cover of gravel. Hatching may take as long as 4 months, depending upon water temperature. From a commercial tonnage standpoint, the chinook salmon represents only about 4% of the total salmon catch of the major fishing countries (Canada, the United States, Japan, and the U.S.S.R.).

Coho Salmon. This fish is found in North American rivers and streams from Monterey Beach, California, in the south, to the Chukchi Sea the north. In Asia, they are rarely found in the Anadyr River, but occur in large numbers in Kamchatka and south almost to the Amur River. The species is also present on Sakhalin Island and in Hokkaido. During their fast-growing period in the ocean, coho salmon are distributed across the northern Pacific Ocean and in the Bering Sea. They are found as far south as California waters in the eastern half of the northern Pacific and in the western North Pacific, they are found as far south as 42° north latitude.

Adult coho salmon begin the freshwater migration between September and December, often coincident with a freshet (stream of freshwater flowing into the sea). This species enters the larger rivers, but also is common in the very small coastal streams throughout its range. Spawning takes place often in tiny tributaries only 3 to 4 feet (1 meter plus) wide. The young emerge from the gravel in the early spring. Coho salmon usually remain in fresh water for about 1 year after fry emergence and then begin their downstream journey to the sea. In Alaskan streams, coho commonly remain in fresh water for 2 years. Larger coho salmon feed principally on squid, small fish, and euphausiids. Adult coho salmon return to spawn late in the year following that in which they entered the ocean. In their last summer of ocean life, coho salmon grow very rapidly and commonly double their weight during this period. Of the commercial tonnage taken by the four major salmon fishing countries previously mentioned, the coho catch represents about 7% of the total. See also **Aquaculture**.

Chum Salmon. This fish is distributed along the North American coast from the Klamath River in California, north to the Arctic coast of Alaska and even in the Mackenzie River. In Asia, chum salmon are abundant in the Amur River and are found on the island of Sakhalin and in Hokkaido streams and south almost to Tokyo. Chum salmon are in the Pacific Ocean north of 45° north latitude in the eastern part of the ocean, and south almost to 36° north latitude in the western Pacific, as well as in the Bering Sea.

Chum salmon enter freshwater to spawn from July to January. To the north, in Alaska and northern British Columbia, the runs are primarily between July and early September, while south of Vancouver Island the runs are from October through January. Chum salmon usually spawn in smaller coastal streams or in the lower portions of larger rivers. However, they ascend several hundred miles (kilometers) up the Yukon to spawn and also spawn in the headwaters of some Asian rivers. Chum salmon frequently dig their nests in the same area as used by pink salmon. The young emerge from the gravel in March through May. Newly emerged chum salmon are a little over 1 inch (2.5 centimeters) long. In most North American streams, young chum salmon almost immediately move downstream to salt water, but in some Asian streams and the Yukon River, it may be several weeks before the young reach the ocean. Chum salmon usually mature at 4 years, but 3- and 5-year-old fish are common. Of the commercial tonnage taken by the four major salmon fishing countries, the chum salmon catch represents about 24% of the total.

Sockeye Salmon. This fish occurs rarely in North American coastal streams south of the Columbia River, but it can be found north to the Yukon River of Alaska, along the Asian coast from the Anadyr River, in the rivers of Kamchatka, where it is the principal species in the Kamchatka River, and south to northern Hokkaido, where it is very rare. Sockeye salmon are distributed in the ocean throughout the northern North Pacific and the Bering Sea.

The mature sockeye salmon male in freshwater becomes a brilliant red and the female is a dark red. This species is distinguished from others by having 28 to 40 long, slender, closely set gill rakers on the first gill arch. There are no black spots. Young sockeye salmon have oval parr marks, which extend only slightly below the lateral line. Sockeye salmon are typically lake-dwelling during the brief freshwater part of their life. After migrating from the lake they usually spend 2 to 3 years at sea. Adult sockeye may begin the upstream migration as early as May and in some areas the migration may extend into October. Sockeye salmon spawn in outlet and inlet streams of lakes and also along some lake shores. After leaving the protection of the gravel, the young move into lakes, where they remain for 1 to 3 years before migrating to the ocean at a length of from 3 to 6 inches (7.5 to 15 centimeters). The downstream migration takes place from April to June, usually under the protection of darkness. In the North Pacific, sockeye salmon feed heavily on amphipods, copepods, euphausiids, pterepods, fish, squid, and similar animals. In the ocean, growth is rapid and those fish that return to freshwater after spending 2 years in the ocean are about 21 inches (53 centimeters) long.

In the ocean, there is a considerable overlap of sockeye salmon from Kamchatka and North America. Maturing sockeye salmon in the high seas begin moving shoreward in May and June. The very important Bristol Bay (Alaska) runs travel an average of about 24 miles (39 kilometers) per day when heading toward their home stream.

Of the commercial tonnage taken by the four major salmon fishing countries, the sockeye salmon catch represents about 24% of the total.

Pink Salmon. This fish occurs occasionally in streams south of Puget Sound and from there northward into Arctic Ocean streams of Alaska. Pink salmon are also found in the Mackenzie River. On the Asiatic side of the Pacific, pink salmon have been reported from the Lena River and south into the streams of the Sea of Okhotsk, the Kurile Islands, Sakhalin, Hokkaido, and on the northeastern coast of Hondo. Pink salmon are found usually north of the 40th parallel across the entire north Pacific Ocean.

Pink salmon move from the ocean into the streams where they spawn from July to November. Adults usually migrate only a short distance from salt water and sometimes spawn in streams in areas that are affected by the tide. In larger rivers, pink salmon may move considerable distances upstream. The young migrate to salt water as soon as the yolk sac is absorbed, in March through early June, but principally in April. In many of the smaller coastal streams of Alaska, the young pink salmon emerge from the gravel and migrate to the sea in one night. Downstream movement begins as darkness approaches and ceases before morning. In the estuaries, schools of young fish migrate along the shore near the surface where the currents gradually carry them toward the ocean. Here their food consists of euphausiids, amphipods, pterepods, small fish, crustaceans, larval squid, and copepods. Pink salmon migrate from the stream to the estuaries at about 1.5 inches (3.8 centimeters) in length and, when they return to spawn after 15 to 17 months of life in the ocean, their average length is about 20 inches (51 centimeters). See Fig. 2.

Fig. 2. Male pink or humpback salmon.

Of the commercial tonnage taken by the four major salmon fishing countries, the pink salmon catch represents about 41% of the total.

Masu Salmon. Masu salmon are found only in streams on the Asiatic side of the Pacific Ocean, from the Amur to the Pusan River, in Sakhalin streams, and on the island of Hokkaido and Hondo. They are also occasionally taken on the western coast of Kamchatka.

Freshwater Salmon. An important sport fishery for salmon has been developed in the Great Lakes.

Salmon Products. Salmon are sold as fresh fish and are also canned, smoked, and fresh-frozen. Salmon eggs are used as bait for sport fishermen or are eaten as red caviar or salted as food. Salmon for the fresh market are usually troll caught in the ocean within a few miles of the port of landing. Coho salmon is the principal species used in the fresh market, but fresh chinook and pink salmon are also marketed. By far the greatest percentage of sockeye, pink, and chum salmon are processed in cans weighing 1 pound (0.45 kilogram) or less.

Commercial salmon fishing, as an industry, commenced in North America in the early 1860s, with the first commercial pack produced at Sacramento, California. The industry soon moved north where canneries on the Fraser River in British Columbia, on Puget Sound, and on Prince of Wales Island in Alaska were established by the late 1870s. The salmon catch in North American waters reached its first peak in the 1930s, after which production has been cyclic. Japanese salmon fisheries commenced in the early 1950s and developed rapidly. Somewhat later, the U.S.S.R. entered into serious commercial salmon fishing operations and is now a major factor.

Salmon Technology. Much research has been devoted to protecting and expanding the traditional salmon fisheries as well as developing sport fisheries for salmon in freshwater areas, such as the Great Lakes. Experimental stockings have been carried out with coho and chinook salmon in these lakes, notably Lake Michigan and Lake Superior.

In the area from British Columbia to California, and to some extent in Alaska, an increasing percentage of adult salmon found passage hindered or blocked by power, irrigation, or flood control dams. The Columbia River and its wild tributary, the Snake River, have increasingly become polluted by wastes as well as thermally. Fishways have been constructed over the low dams of the main Columbia and Snake Rivers, and these have been generally successful in passing adult salmon. Downstream migrating salmon pass through the turbines of the low dams or over the spillways with usually only a small percentage of loss.

In recent years, fish passage facilities have been provided at most dams, but at some high dams, fishway construction has not proved feasible and other methods have been provided for maintaining the runs. Chief among these has been the production from salmon and steelhead hatcheries. In recent years, research efforts have resulted in a great improvement in the success of rearing salmon in hatcheries. Many hatchery diseases can now be controlled, and nutritious diets have been developed for young salmon. Scientists have also given attention to improving survival of salmon eggs and fry by constructing spawning channels. An artificial spawning channel generally consists of a dam at the head to control the flow of water and an artificial channel of appropriate width and slope with gravel of optimum size for best egg and fry survival. Many of the conditions detrimental to the survival of salmon eggs and fry in nature can be controlled in such artificial spawning areas. The optimum number of adults can be allowed to spawn. Flood waters, which wash eggs and fry from the gravel, can be prevented. Predators can be controlled. Gravel can be selected to allow optimum circulation of water with its life-giving oxygen, and the amount of water needed to ensure best production of fry can be maintained. At Jones Creek, a tributary of the Fraser River in British Columbia, one of the first artificial spawning channels for salmon was constructed as early as 1954, where it was found that the survival of pink salmon eggs and fry

was from 4 to 6 times greater than that of the natural environments See also **Aquaculture**.

The smelt, closely related to the salmon, is described in the entry on **Smelt**.

Additional Reading

Eschmeyer, W.N., C.J. Ferraris, et al.: "Catalog of Fishes," *California Academy of Sciences*, San Francisco, CA, 1998.
Evans, D.H.: "The Physiology of Fishes," 2nd Edition, CRC Press, LLC., Boca Raton, FL, 1998.
MacDonald, D. and E. Knudson: "Sustainable Fisheries Management Pacific Salmon," Lewis Publishers, Boca Raton, FL, 1999.
Paxton, J.R., W.N. Eschmeyer, et al.: "Encyclopedia of Fishes," 2nd Edition, Academic Press, San Diego, CA, 1998.
Royston, A.: "Salmon," Heinemann Library, Woburn, MA, 2000.
Stead, S.M., L.M. Laird: "Handbook of Salmon Farming," Springer Publishing Company, Inc., New York, NY, 2001.
Stickney, R.R.: "Encyclopedia of Aquaculture," John Wiley & Sons, Inc., New York, NY, 2000.

SALMONELLA. See **Foodborne Diseases**.

SALMONELLOSIS. See **Foodborne Diseases**.

SALPINGITIS. Infection of the fallopian tubes. This common disease is most frequently due to a gonorrheal infection. More rarely, it may be due to tuberculosis, *streptococcus*, or *pneumococcus* infection. Gonorrheal salpingitis is not only the most prevalent but is the most disabling in its aftereffects. Salpingitis may not occur for months or even years after the original gonorrheal infection.

SALSIFY. See **Composite Family**.

SALT. A compound formed by replacement of part or all of the hydrogen of an acid by one (or more) element(s) or radical(s) that are essentially inorganic. Alkaloids, amines, pyridines, and other basic organic substances may be regarded as substituted ammonias in this connection. The characteristic properties of salts are the ionic lattice in the solid state and the ability to dissociate completely in solution. The halogen derivatives of hydrocarbon radicals and esters are not regarded as salts in the strict definition of the term.

In the classical concept of the process of neutralization, whereby an acid and a base in solution react to form a salt, the proton of the acid and hydroxyl ion of the base react to form water, leaving the cation of the base and the anion of the salt by recombination.

Upon evaporation of the solvent, the salt is obtained as such, frequently as crystals, sometimes with and sometimes without water of crystallization. A salt, when dissolved in an ionizing solvent, or fused (e.g., sodium chloride in water), is a good conductor of electricity and when in the solid state forms a crystal lattice (e.g., sodium chloride crystals possess a definite lattice structure for both sodium cations (Na^+) and chloride anions (Cl^-), determinable by examination with x-rays).

A broader definition than that confined to solutions is demanded in some fields of chemistry (e.g., in high temperature reactions of acids, bases, and salts). In the formation of metallurgical slags, at furnace temperatures, calcium oxide is used as base and silicon oxide and aluminum oxide as acids; calcium aluminosilicate is produced as a fused salt. Sodium carbonate and silicon oxide when fused react to form the salt sodium silicate with the evolution of carbon dioxide. In this sense:

$$\begin{bmatrix} \text{Oxide of any} \\ \text{element functioning} \\ \text{as a metal — that is,} \\ \text{as a base} \end{bmatrix} \text{plus} \begin{bmatrix} \text{Oxide of any} \\ \text{element functioning} \\ \text{as a nonmetal — that} \\ \text{is, as an acid} \end{bmatrix} \text{yields [Salt]}$$

Iron and sulfur when heated react to form the salt ferrous sulfide. In this sense:

$$\text{metal plus nonmetal yields salt}$$

Salts therefore, are prepared (1) from solutions of acids and bases by neutralization and separation by evaporation and crystallization; (2) from solutions of two salts by precipitation where the solubility of the salt formed is slight (e.g., silver nitrate solution plus sodium chloride solution yields silver chloride precipitate [almost all as solid], and sodium nitrate present in solution as sodium cations and nitrate anions [recoverable as sodium nitrate, solid by separation of silver chloride and subsequent evaporation of the solution]); (3) from fusion of a basic oxide (or its suitable compound—sodium carbonate above) and an acidic oxide (or its suitable compound—ammonium phosphate), since ammonium and hydroxyl are volatilized as ammonia and water. Thus, sodium ammonium hydrogen phosphate

$$\begin{array}{c} NH_4 \\ | \\ Na - PO_4 \\ | \\ H \end{array}$$

yields sodium metaphosphate, $NaPO_3$, upon heating. (4) Salts also are prepared from reaction of a metal and a nonmetal.

Reactions of salts as such in solution, without decomposition of cation or anion, are dependent upon the presence of the cation and the anion of salt.

An *acid salt* is a salt in which all of the replaceable hydrogen of the acid has not been substituted by a radical or element. These salts, in ionizing, yield hydrogen ions and react like the acids (e.g., $NaHSO_4$, $KHCO_3$, Na_2HPO_4).

An *amphiprotic* (also called *amphoteric*) *salt* is a salt that may ionize in solution either as an acid or a base, and react either with bases or acids, according to the conditions.

A *basic salt* is a salt contains combined base as $Pb(OH)_2 \, Pb(C_2H_3O2)_2$, a basic acetate of lead. These salts may be regarded as formed from the basic hydroxides by partial replacement of hydroxyl (e.g., $HO-Zn-Cl$). They react like bases and, when soluble, ionize to yield hydroxyl ions.

A *complex salt* is a saline compound having the structure of a combination of two or more salts and that is regarded as the normal salt of a complex acid. Complex salts do not split into a mixture of the constituent salts in solution, but furnish a complex ion that contains one of the bases (e.g., potassium molybdophosphate and potassium platinochloride).

A *double salt* is a substance consisting of two simple salts that crystallize together in definite proportions and exist independently in solution (distinction from complex salts). The alums are representative double salts.

An *inner salt* is a member of a special class of internal salts in which an acid group and a neutral group coordinate with metals to form a cyclic complex. These salts occur widely in analytical chemistry, (where they are formed between metallic ions and organic reagents) in dyestuffs, in life processes (chlorophyll and hematin belong to this class of compounds), and in many other fields.

An *internal salt* is a compound in which the acidic or basic groups that react to produce the salt linkage (which may or may not entail the formation of water) are in the same molecule. This particular salt linkage may consist of a polar or a nonpolar bond.

A *mixed salt* is a salt of a polybasic acid in which the hydrogen atoms are replaced by different metallic atoms or positive radicals.

A *pseudo salt* is a compound that has some of the normal characteristics of a salt, but lacks certain others, notably the ionic lattice in the solid state and the property of ionizing completely in solution. The absence of these properties is due to the fact that the bonds between the metallic and nonmetallic radicals are covalent or semicovalent instead of polar. Because these salts do not ionize completely, they are also called *weak salts*.

SALTATION. This term, as proposed by McGee, in 1908, is used by geologists to designate the particular mode of the stream transportation of clastic sediments by intermittent leaps or bounds. Probably an important factor in the ultimate transportation of the coarser fragments by streams and rivers.

SALT BRIDGE. A type of liquid junction used to connect electrically two electrolytic solutions. It consists commonly of a U-tube filled with a strong solution, and provided with porous plugs. It is used for such purposes as to connect electrolytic half cells in making measurements of electrode potential.

SALT (NaCl).   See Sodium Chloride.

SALTPETER.   Potassium nitrate. Sodium nitrate is often called Chile saltpeter, and calcium nitrate is sometimes called Norway saltpeter.

SALT RIVER.   Although many rivers of the world contain relatively high contents of salt, a salt river is usually identified as a river that penetrates into a huge deposit of rock salt and forms glistening white cliffs of salt in its course. One of these formations occurs in a broad bend of the Huallaga River in eastern Peru. The salt cliff extends some 8 kilometers in length and is about 100 meters in height. The Huallaga is a major tributary of the Amazon River. Several exposed salt deposits of this kind are found in the region of the Amazon headwaters. It is reasoned that during the formation of the Andes, compression forces caused rock salt to flow in a way somewhat analogous to the manner of flow of glacial ice. Salt deposits in the Andes region greatly affect the chemical composition of the Amazon tributaries and it is estimated that dissolved salt in these tributaries accounts for about half of the sodium and chlorine content in the Amazon River.

SALVE BUG (*Crustacea, Isopoda*).   A marine crustacean, *Aega psora*, parasitic on various fishes. It is elongate oval in form and is a little more than one-half inch long. Found on both sides of the Atlantic.

SAMARIUM.   Chemical element symbol Sm, at. no. 62, at. wt. 150.35, fifth in the Lanthanide Series in the periodic table, mp 1,073 °C, bp 1,791 °C, density 7.520 g/cm^3 (20 °C). Elemental samarium has a rhombohedral crystal structure at 25 °C. The pure metallic samarium is silver-gray in color, retaining a luster in dry air, but only moderately stable in moist air, with formation of an adherent oxide. When pure, the metal is soft and malleable, but must be worked and fabricated under an inert gas atmosphere. Finely divided samarium as well as chips from working are pyrophoric and ignite spontaneously in air, burning at 150–180 °C. There are seven natural isotopes of samarium ^{144}Sm, ^{147}Sm through ^{150}Sm, ^{152}Sm, and ^{154}Sm. Eleven artificial isotopes have been identified. The natural ^{147}Sm isotope is weakly radioactive with a half-life of 2.5×10^{11} years. The samarium isotope mixture is the second highest (after gadolinium) of all elements in terms of its thermal-neutron-absorption cross-section (5,800 barns at 0.025 eV). The cross-section of ^{149}Sm is about 40,000 barns. Samarium ranks 62nd in abundance of the elements in the earth's crust, exceeding tantalum, mercury, bismuth, and the precious metals, excepting silver. The element was first identified by Lecoq de Boisbaudran in 1879. Electronic configuration

$$1s^2 2s^2 2p^6 3s^2 3p^6 3d^{10} 4s^2 4p^6 4d^{10} 4f^5 5s^2 5p^6 5d^1 6s^2$$

Ionic radius Sm^{2+} 1.11 Å, Sm^{3+} 0.964 Å. First ionization potential 5.6 eV; second 11.1 eV. Other important physical properties of samarium are given under **Rare-Earth Elements and Metals**.

The principal sources of samarium are monazite (4.5% Sm_2O_3) and bastnasite (0.5% Sm_2O_3). Current demands for the element are met by the coproduction with europium and gadolinium from these minerals. The residues of uranium mining (Canada) also contain about 4.5% Sm_2O_3. Unlike the other light rare-earth metals, the salts and oxide of samarium do not reduce to metal using barium, calcium, or lithium, nor can electrolytic processes be used. The most effective reducing agent is lanthanum, which is mixed with Sm_2O_3 and heated under vacuum in a tantalum crucible. The samarium metal volatilizes and is condensed as powder or sponge on coiled tantalum or copper condenser plates. Subsequently, the samarium must be remelted under an argon or inert atmosphere before it is cast into graphite or tantalum molds.

Samarium has been alloyed with gadolinium and aluminum to produce nuclear reactor hardware that will absorb neutrons for short periods. The use of samarium in intermetallics, cermets, and other chemical forms for use in nuclear applications holds promise. Small quantities of Sm_2O_3 are used in optical-glass filters and to encase lanthanum borate glass rods, which then are drawn into fine fibers for fiberoptics applications. The element has been used as a coding agent for inks used in data handling systems. Small amounts also have been used for activating phosphate-type phosphors. The addition of samarium oxide produces a strong narrow emission in the near-infrared spectral region. The most significant use of samarium is in the permanent-magnet alloys $SmCo_5$ and Sm_2Co_{17}. The strength of these magnets Second to that of $Nd_2Fe_{14}B$ by only a small margin. But, because of costs the utilization of the $SmCo_5 - Sm_2Co_{17}$ alloys is much less than that of the Nd−based permanent magnets. The Sm−base permanent magnets have a much higher magnetic ordering temperature and are used in high temperature greater than 250 °F (or 120 °C) applications. See also **Cobalt**.

See references listed at ends of entries on Chemical Elements; and Rare-Earth Elements and Metals.

Note: This entry was revised and updated by K.A. Gschneidner, Jr, Director, and B. Evans, Assistant Chemist, Rare-earth Information Center, Institute for Physical Research and Technology, Iowa State University, Ames, Iowa.

SAMPLE-AND-HOLD AMPLIFIER.   Also known as a track-and-hold amplifier, this device has an output that is proportional to the input until a "hold" signal is received. Upon receipt of that signal, the amplifier output is maintained essentially constant even though there may be changes in the input signal. The input and output waveforms of a sample-and-hold amplifier are shown in Fig. 1.

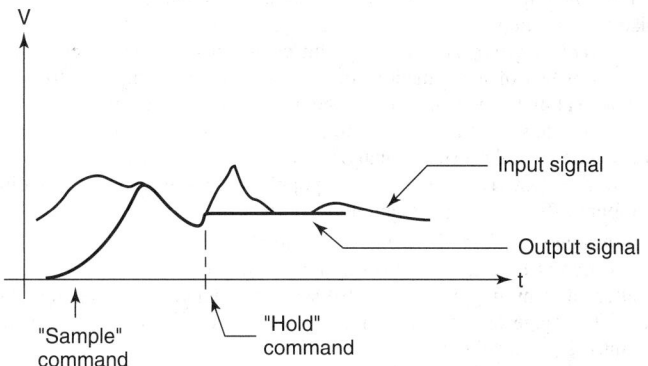

Fig. 1.   Action of sample-and-hold amplifier.

As shown in Fig. 2, the conceptual design of a sample-and-hold amplifier comprises two independent amplifiers connected by a switch. With sampling switch S_1 closed, holding capacitor C is charged when an input signal is applied to the first amplifier. Upon receipt of the "hold" command, switch S_1 is opened, thus leaving capacitor C charged at the instantaneous value of the input signal. Capacitor C is not discharged because the second amplifier has a high input impedance. The output of the second amplifier remains essentially steady for a period of time. The "hold" signal may be generated by an external circuit (coupled to a process or experiment) or by a computer or digital control unit under control of a stored program.

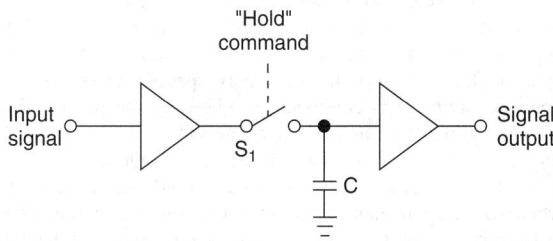

Fig. 2.   Schematic circuit of a sample-and-hold amplifier.

Sample-and-hold amplifiers meet certain specialized needs in digital-data acquisition systems. Usually one or both of the following requirements exist: (1) the value of a single signal must be determined at a precise instant of time, or (2) the values of two signals must be compared at a precise instant of time. These requirements cannot be met by a time-shared amplifier and analog-to-digital converter system because these systems require a finite settling and conversion time. Where a sample-and-hold amplifier is used, it is possible to retain the instantaneous value of one or more input signals over the time interval required to convert them to digital values.

THOMAS J. HARRISON, International Business Machines Corporation, Boca Raton, FL.

SAMPLE (Statistics).   A sample is a collection of individuals drawn from a population. Ordinarily, inferences are to be made from the sample to the population, and the one must be in some way representative of the other.

When sampling from an actual population, the simplest method is to draw a random sample. The members of the population are numbered off, and the sample selected with the aid of a table of random numbers or some similar device. To ensure more even coverage, stratified sampling may be adopted. The population is divided into a number of homogeneous groups (the strata) and a random sample is selected from each. If only a random selection of the strata are sampled, we have a two-stage sample; three or more stage samples are similarly constructed. In two-phase sampling one type of observation is taken only on a small calibrating sample, while another (which may be easier or cheaper to obtain) is taken on a larger sample; regression or ratio relationships derived from the small sample are used to predict values of the first measurement for individuals in the large sample.

The introduction of a random element can ensure the absence of bias from an estimate based on a sample, and also makes possible a valid estimate of the error to which such an estimate is subject. (For various types of samples, and various methods of sampling, see **Sampling (Statistics)**.

SAMPLING CONTROLLER.   An automatic controller using intermittently observed values of a signal, such as the setpoint signal, the actuating error signal, or the signal representing the controlled variable, to effect control action. Also termed *sampled data system*.

Applications of sampled data systems include those applications in which the actuating or error information is only available in sample form, as well as those arrangements in which deliberate conversion of continuous data to sampled data is made. Examples of the first situation include an automatic tracking radar system and a gas chromatograph. The radar system scans in two coordinates and thus can furnish information on a particular target only at the discrete time intervals when the antenna direction permits radio frequency every to intercept the target. In the chromatograph, information on a particular variable is available at the completion of each sample analysis. An illustration of the second possibility arises in direct digital control. In this application, the digital computer is used to operate, simultaneously, numerous feedback control systems. By use of sampled data techniques the computer may be used periodically to process data from one system, then from the second, and so on until it completes the operating cycle and accepts data from the first system again. Figure 1 is a typical schematic for this type of system.

The hold circuit shown in Fig. 1 is an integral part of any sampled data control system. Its function is to hold or remember the value of assigned variable at the sampling instant for a finite portion of the sampling period *T*. See Fig. 2. Its position in the loop will depend on the particular application. Wave form for a zero order hold following a sampler is shown in Fig. 3. Although higher order holds are possible, they are seldom used in practice due to their complex nature.

See also **Control Action**; and **Sample-and-Hold Amplifier**.

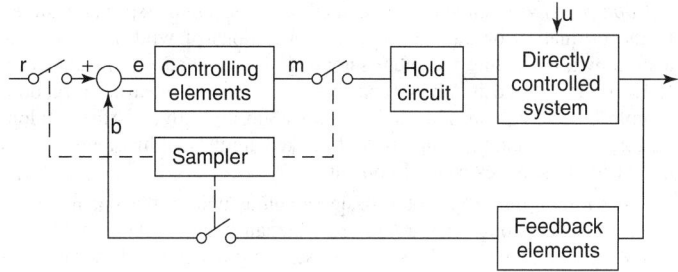

Fig. 1.   Block diagram of sampled data control system.

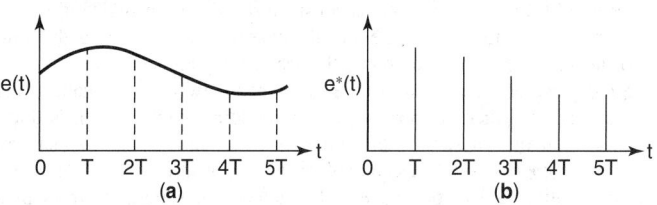

Fig. 2.   Actuating or error signal: (**a**) Continuous function of time; (**b**) sampled pulse train.

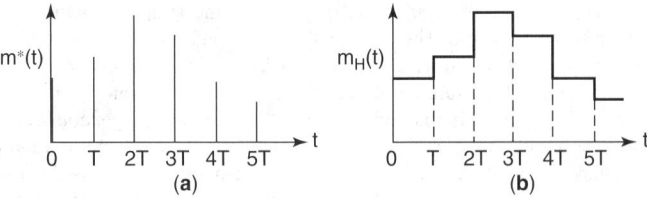

Fig. 3.   Output waveforms of sample-and-hold circuit.

SAMPLING (Statistics).   The process of taking a sample. If repeated samples are drawn from a probability distribution and the value of some statistic is calculated for each sample, the resulting set of values will define a new probability distribution known as the *sampling distribution* of the statistic. The *sampling fraction* is the proportion of the total number of sampling-units in the population, stratum, or higher-stage unit within which simple random sampling (with multiple counting of sample-units when sampled with replacement) is made. There are thus sampling fractions corresponding to different strata and different stages of sampling. Exactly the same definition is sometimes (loosely) applied to other sampling schemes, e.g., in sampling with variable probability or *multistage sampling* (ratio of total number of ultimate units included in the sample to total units in the population). However, for other general application it appears desirable to define it as the reciprocal of the raising-factor of the sample when it exists, i.e., when the sample is self-weighing. A *uniform sampling fraction* is obtained when a sample is selected from a population which has been grouped into strata, in such a way that the number of units selected from each stratum is proportional to the number of units in that stratum. If from a stratified population a simple random sample is selected from each stratum in such a way that the proportion of units sampled in each stratum varies from stratum to stratum, the sample is said to be selected with *variable sampling fraction*. Applicability of the term to other sampling schemes rests upon the general definition of sampling fraction. There are many methods of sampling depending upon the procedure to be followed and the distribution of samples, which are determined in turn by the character of the population to be sampled.

Area sampling is a method of sampling used when no complete frame of reference is available. The total area under investigation is divided into small sub-areas, which are sampled at random or by some restricted random process. Each of the chosen sub-areas is then fully inspected and enumerated, and may form a frame for further sampling if desired. The term may also be used (but it is not to be recommended) as meaning the sampling of a domain to determine area, e.g., under a crop.

Bulk sampling is the sampling of materials which are available in bulk form, that is to say, it is the population which is in bulk; the term does not mean the drawing of a sample in bulk. Examples of such sampling would be the sampling of a shipment of coal for ash-content, or tobacco for moisture content.

Capture release sampling is a method of sampling specially suited to the estimation of the size of total populations of wild animals. It is also known as capture-recapture sampling. The method was practiced by Lincoln (1930) and involves capturing, marking and releasing a random sample, say, of animals of a particular kind. Subsequently, a further random sample is taken and the proportion of marked animals in this sample forms the basis of estimates of total population.

Cluster sampling. When the basic sampling unit in the population is to be found in groups or clusters (e.g., human beings in households) the sampling is sometimes carried out by selecting a sample of clusters and observing all the members of each selected cluster.

Direct sampling is a term used when the sample units are the actual members of the population and not, for instance, some kind of record relating to such numbers, such as census form, ticket, or registration card. The term relates to the directness of the observation of the units that enter into the sample, not to the process by which they are selected.

Sampling error is that part of the difference between a population value and an estimate thereof, derived from a random sample, which is due to the fact that only a sample of values is observed. This is distinct from errors due to imperfect selection, bias in response or estimation, errors of observation and recording, etc. The totality of sampling errors in all possible samples of the same size generates the sampling distribution of the statistic used to estimate the parent value.

Extensive sampling is a term used to denote sampling where the subject matter, or geographical coverage, of a sample is diffuse or widespread as opposed to intensive, where it is narrowed to a small field. Extensive sampling may refer either to a case where a variety of topics is covered superficially (rather than a few topics in detail) or a large area is surveyed broadly (rather than a small area studied in detail). The term could also be used with reference to time, that is to say, of sampling covering a long period.

It would be convenient to distinguish the cases as space-extensive, item-extensive and time-extensive, respectively.

Grid sampling is a form of cluster sampling, the clusters being individual areas of a grid and hence consisting of groups of basic cells arranged in some standard geometrical pattern. The term *configurational sampling* is also used in the same sense.

Indirect sampling is sampling from documents, or some record of the characteristics of a population, rather than the recording of information obtained at first hand from units of the population themselves. For example, it is becoming customary to obtain preliminary information on the results of, say, a national census by analyzing a sample of the census forms before the full analysis is undertaken; the population is then subject to indirect sampling. (See also Direct sampling.)

Intensive sampling. Like extensive sampling this expression may mean two different things: either (a) sampling in a particular area with a dense scatter of sampling points, or (b) sampling wherein information on a restricted range of topics is sought by probing on them very deeply with an intricate schedule of questions. (See also Extensive sampling.)

Inverse sampling is a method of sampling that requires that drawings at random shall be continued until certain specified conditions dependent on the results of those drawings have been fulfilled, e.g., until a given number of individuals of specified type have emerged. In this sense, it is allied to sequential sampling. The term is not a good one.

Lattice sampling is a method of sampling in which substrata are selected (for the sampling of individuals) according to some pattern analogous to the allocation of treatments on a lattice experimental design. For example, if there are two criteria of stratification, each p-fold, so that there are p^2 substrata, it is possible to choose p sub-strata so that none occurs in more than one "row" or "column" of the array representing the p^2 possible substrata; in short, in the manner of a Latin square. Similar schemes are possible for three-way, or more, classification. Various schemes of the lattice type are known under the name of "deep stratification."

Line sampling is a method of sampling in a geographical area. Lines are drawn across the area and all members of the population falling on the line, or intersected by it, are included in the sample. If the lines are straight parallels equally spaced across the area concerned, then the sampling becomes one form of systematic sampling. If, instead of all intercepts on the lines, a series of evenly spaced points are chosen on each line, the sampling is equivalent to choosing the points on a lattice and may also be regarded as two-stage line sampling.

Lottery sampling is a method of drawing random samples from a population by constructing a miniature of the population (e.g., by inscribing the particulars of each member on to a card) and drawing members at random from it (e.g., by shuffling the cards and dealing a set haphazardly). It is the method usually employed at a lottery—hence its name—but suffers from the disadvantage that the preparation of the cards entails considerable labor and strict precautions must be taken in the shuffling process to guard against bias.

Mixed sampling is where a sampling plan envisages the use of two or more basic methods of sampling. For example, in a multistage sample, if the sampling units at one stage are drawn at random and those at another by a systematic method, the whole process is "mixed."

Usage is not uniform, but where samples at one stage were drawn at random with replacement and at another stage were drawn at random without replacement, it would seem better not to describe the whole process as "mixed," since the essential basic method of random selection being employed throughout.

Multiphase sampling. It is sometimes convenient and economical to collect certain items of information from the whole of the units of a sample and other items of (usually more detailed) information from a subsample of the units constituting the original sample. This may be termed *two-phase sampling*, e.g., if the collection of information concerning variate, y, is relatively expensive, and there exists some other variate, x, correlated with it, which is relatively cheap to investigate, it may be profitable to carry out sampling in two phases. At the first phase, x is investigated, and the information thus obtained is used either (a) to stratify the population at the second phase, when y is investigated, or (b) as supplementary information at the second phase, a ratio or regression estimate being used. Two-phase sampling is sometimes called *double sampling*. Further phases may be added if desired. It may be noted, however, that multiphase sampling does not necessarily imply the use of any relationships between the variates x and y. The expression is not to be confused with multistage sampling.

Probability sampling. Any method of selection of a sample based on the theory of probability; at any stage of the operation of selection the probability of any set of units being selected must be known. It is the only general method known which can provide a measure of precision of the estimate. Sometimes the term random sampling is used in the sense of probability sampling.

Proportional sampling is a method of selecting sample numbers from different strata so that the numbers chosen from the strata are proportional to the population numbers in those strata.

Quota sample is a sample (usually of human beings) in which each investigator is instructed to collect information from an assigned number of individuals (the quota) but the individuals are left to his personal choice. In practice, this choice is severely limited by "controls," e.g., he is instructed to secure certain numbers in assigned age-groups, equal numbers of the two sexes, certain numbers in particular social classes and so forth. Subject to these controls, which are designed to make the sample as representative as possible, he is not restricted to the contracting of assigned individuals as in most forms of probability sampling.

Representative sample. In the widest sense, a sample which is representative of a population. Some confusion arises according to whether "representative" is regarded as meaning "selected by some process which gives all samples an equal chance of appearing to represent the population"; or, alternatively, whether it means "typical in respect of certain characteristics, however chosen." On the whole, it seems best to confine the word "representative" to samples which turn out to be so, however chosen, rather than apply it to those chosen with the object of being representative.

Route sampling. A procedure similar to line sampling and used in surveys of crop acreage in districts that are well provided with roads. A route that adequately covers the area is chosen and the roadside lengths of the different crops recorded. Since the location of roads is unlikely to be random, estimates of acreage so obtained are likely to be biased, but changes in acreages may be estimated by using the same route for a number of years. The method of route sampling as a form of systematic sampling can also be applied to crop estimation.

Sampling with replacement. When a sampling unit is drawn from a finite population and is returned to that population, after its characteristic(s) have been recorded, before the next unit is drawn, the sampling is said to be "with replacement." In the contrary case, the sampling is "without replacement."

A different usage occurs in sample-surveys when samples are taken on successive occasions. If the same members are used for successive samples there is said to be no replacement; but if some members are retained and others are replaced by new individuals there is "partial replacement."

Unbiased sample. A sample drawn and recorded by a method which is free from bias. This implies not only freedom from bias in the method of selection (e.g., random sampling) but freedom from any bias of procedure, e.g., wrong definition, nonresponse, design of questions, interviewer bias, etc. An unbiased sample in these respects should be distinguished from unbiased estimating processes which may be employed upon the data.

See also **Quota Sample**; and **Random Selection**.

SAN ANDREAS FAULT. See **Earth Tectonics and Earthquakes**.

SANDBOX TREE. See **Euphorbiaceae**.

SAND CRICKET (*Insecta, Orthoptera*). *Stenopelmatus*. Thick-bodied clumsy insects with large heads and long slender antennae. They live in loose soil, usually under some protective object, in the western United States. They are not true crickets but are more closely related to the long-horned grasshoppers.

SAND DAB. See **Flatfishes**.

SAND DOLLAR (*Echinodermata, Echinoidea*). A sea urchin with a very thin body, almost circular in outline and with a diameter of less than 3 inches (7.6 centimeters). It is common on both coasts of North America.

SAND FLY FEVER. See **Phlebotomus Fever**.

SAND GROUSE. See **Pigeons and Doves**.

SANDPIPER. See **Waders, Shorebirds, and Gulls**.

SAND SHARKS. See **Sharks**.

SANDSTONE. Sand grains cemented by such substances as silica, carbonate of lime or iron oxide, so as to form a solid rock is called sandstone. It occurs usually in beds of varying thickness, depending upon the conditions under which the original sediments were laid down. Because it is normally well-jointed and easy to work, sandstone has been much used for building purposes. Unfortunately, however, as most sandstones are quite porous, the weathering action of the atmospheric agencies may have a very deleterious effect upon them.

SANDSTONE DIKE. Sandstone occurring in fissures that have been filled from above, or from beneath. The latter type are usually the result of earthquake fissures in great flood plain or delta deposits in which the sands have been injected from below.

SAPAJOUS. See **Monkeys and Baboons**.

SAPANWOOD. See **Brazilwood**.

SAPODILLA (*Achras zapota, Sapotaceae*). A large tree native to the forests of Central and tropical South America, the fruit of which is very desirable. The yellow-brown flesh is translucent and very sweet and wholesome. The greatest value of the tree is in its latex product, which yields chicle. The chicle-gathering industry is centered in Yucatán and Central America. The tapping is done in the rainy season. The tapper climbs to a height of 30 feet (9 meters), and with a machete cuts a series of connecting zig-zig diagonal gashes in the bark as he descends. At the bottom of this series of cuts he attaches a cup, into which the latex flows. The crude substance is collected, boiled down to eliminate much of its water and the coagulated product pressed into 20–25 pound (9–11.3 kilogram) blocks. This substance, chicle, varies in quality from the best grade, which is milk-white in color, to pinkish or darker grades, which have received less care in preparation. Each tree yields about $2\frac{1}{2}$ pounds (1.1 kilogram) of chicle during one season and may be tapped every 3–4 years. The blocks of chicle are shipped largely to the United States, where they are melted and cleaned, flavored and sweetened, and then marketed as the familiar chewing gum. This use of the latex of the Sapodilla is not new, since the Aztecs and their predecessors knew of it and used it. When first introduced into the United States it was tried as a rubber substitute, but proved unsuitable.

See Table 1 for the record sapodilla trees growing in the United States.

SAPPHIRE. See **Corundum**.

SAPROPHYTES. These are plants that obtain their food from nonliving organic material. Most of the saprophytes are fungi. Among the higher plants, a small number of flowering plants and perhaps a few mosses are also saprophytes. It is characteristic of these saprophytic plants that they have little or no chlorophyll, and so are not able to carry on photosynthesis. Their energy is derived from the complex organic substances they absorb. In many instances, the absorption of these substances is greatly advanced by the presence of mycorrhizae.

Especially is this the case with various species of saprophytic orchids which have mycorrhizae within the cells of the roots or rhizomes. The various species of Carol-roots (*Corallorrhiza*) are common saprophytic orchids of American woods. These orchids have no roots, absorption occurring in the much-branched fleshy rhizome which gives the plant its name. In this rhizome the mycorrhizae are found. These plants have erect stems, leaves reduced to scales, and no chlorophyll. Other saprophytic orchids occur in the continents of the Old World.

TABLE 1. RECORD SAPODILLA TREES IN THE UNITED STATES[1]

Specimen	Circumference[2]		Height		Spread		Location
	Inches	Centimeters	Feet	Meters	Feet	Meters	
Sapodilla (1992) (*Manilkara zapota*)	156	396	72	21.9	62	18.9	Florida
Sapodilla (1993) (*Manilkara zapota*)	174	442	56	17.1	53	16.2	Florida

[1]From the "National Register of Big Trees," American Forests (by permission).
[2]At 4.5 feet (1.4 meters).

SARAH (Search and Rescue and Homing). A radio homing device originally designed for personnel rescue and now used in spacecraft recovery operations at sea.

SARCOMA. A malignant tumor originating in connective tissue. These growths are composed of densely packed cells, diffusely imbedded in a homogeneous ground substance. Their degree of malignancy varies greatly. They spread by local infiltration and by blood stream invasion. The most frequent sites in which sarcomas develop are bone, lymph nodes, and subcutaneous tissue.

Sarcomas are much less common forms of malignant tumors than are carcinomas.

SARDINE. See **Herring**.

SARDONYX. See **Agate**.

SAROS. The fact that eclipses occur in periodic intervals was known to the ancient Chaldeans, and probably even in prehistoric times. This period of 18 years, $11\frac{1}{3}$ days ($10\frac{1}{3}$ days if there happen to be 5 leap years in the interval) is known as the Saros. If an eclipse should occur on January 1, 1977, at noon, another similar eclipse would occur on January 12, 1995, at eight o'clock in the evening. The eclipse would not occur at the same point on the earth but would be about 8 hours farther west in longitude.

During the course of a Saros there are about 29 lunar and 41 solar eclipses, each repeated during the next Saros, but not at the same portion of the earth. See also **Eclipse**.

SARSAPARILLA (*Smilax* sp.; *Liliaceae*). The genus *Smilax* contains some 200 species, most of which are tropical, though a few such as the carrion flower, *Smilax herbacea*, and the cat briar, *Smilax rotundifolia*, occur as far north as the New England states. The tropical species are mostly climbing shrubs or vines, usually with prickly stems. The leaves are entire and of oblong to ovate shape. At the base of the leaf is a pair of tendrils, which are perhaps to be interpreted as modified stipules, though such structures are not usually found in monocotyledons. The flowers are small, dioecious and borne in umbels. The fruit is a berry. Some of the South American species are the source of Sarsaparilla, which is obtained from the dried roots.

SASSABY. See **Antelope**.

SASSAFRAS. See **Laurel Family**.

SATELLITE (Astronomy). The term as used in astronomy usually refers to small, planet-like objects that are revolving about the individual planets in orbits. The moon is the satellite of the earth and has been known from remotest antiquity. Satellites over the years have served a useful purpose to astronomers since the mass of a planet can be determined accurately only if the planet has a satellite. By application of the rigorous expression for the harmonic Keplerian laws of planetary motion, the mass of any planet and satellite may be found in terms of the mass of the earth-moon system, after the distance of the planet from the satellite and its period of revolution are known. The problem of the determination of the masses of the satellites themselves is a more difficult problem.

Further information regarding the different satellites will be found in **Moon (Earth's)**; and **Planets and the Solar System**. See also entries on the specific planets.

SATELLITES (Communications and Navigation). Satellites are unmanned spacecraft orbiting the Earth above its atmosphere as components in information networks. *Communications satellites* serve as either broadcast distributors or switching/relaying stations for wireless communication circuits. These satellites receive and transmit microwave radio frequencies carrying voice, video, data, facsimile (fax), and/or paging signals generated by and addressed to subscribers in public and private networks; they also distribute coded television broadcast signals directly to residential and commercial receivers. *Communications satellites* have enabled the installation of telephones and television facilities in remote areas without investment in costly terrestrial networks. *Navigation satellites* provide global positioning services and extremely accurate time signals to receivers used on ships and boats, aircraft, many different types of land vehicles, and even persons on foot, enabling military and civilian users to accurately determine, at any time, the location on the planet of their vessel, aircraft, vehicle, or person. Such satellites are also used to guide military missiles to their targets.

Although satellites seem to be expensive, they cost far less than the terrestrial infrastructure that would be needed to perform the same functions. Until communications satellites were implemented, many areas of the world were isolated from the networks serving telephones, television, and data networking. In developing countries where funding simply is not available to construct microwave or wireline terrestrial networks in remote areas, the use of satellite communications has helped to unify their populations and spurred development. In times of catastrophe, satellites enable communication when terrestrial facilities are damaged or destroyed. Located far above the Earth's atmosphere, these versatile machines are practically immune to the destructive effects of storms, floods, fire and earthquakes. See Fig. 1.

A satellite requires a launch vehicle in order to reach its operating altitude. Most satellites have been placed in orbit by mounting them on expendable huge multi-stage rockets fired from terrestrial facilities. The chief sites for launching commercial satellites are located in the United States (Cape Canaveral in Florida, and Vandenberg Air Force Base, California), Kazakhstan, central Asia (a former state of the USSR), French Guiana (in South America), China, and Japan.

The expendable rocket serving as the launch vehicle commonly has three stages. The first raises the satellite to about 50 miles (80 kilometers) above the Earth, where that stage falls away. A second stage ignites to lift the satellite up to an altitude of about 100 miles (160 kilometers). The second stage then falls away and a third stage ignites, placing the satellite above the atmosphere in the *transfer orbit*, a temporary elliptical path around the Earth. The third stage then drops away, leaving the satellite in space, where an on-board *apogee kick motor* ignites, in the *orbit injection phase* of the launch, moving the satellite into its assigned orbital slot, at which point the motor is turned off. The entire process generates various discarded components, such as rocket casings, bolts, covers, etc., which remain floating in space as debris, usually in an orbit below an altitude of 1,200 miles (ca. 2,000 kilometers). Many thousands of such items are carefully tracked to minimize the danger of accidental collision between "space junk" and satellite launches or manned spacecraft. Such collisions would be catastrophic because of the speeds involved (thousands of miles per hour).

The total weight of a communications satellite at the time of launch may be between 1,000 pounds (453 kilograms) and as much as 9,000 pounds (4,080 kilograms). Some are not much larger than a washtub, while others are more than four stories high. See Fig. 2. Perhaps the largest unmanned satellite in orbit, launched in 1991, is the *Compton Gamma Ray Observatory (CGRO)*, with a mass of over 17 tons (17,273 kilograms) and measuring 70 feet across its two solar arrays. See also **Compton Gamma Ray Observatory**.

Commercial Communications Satellites
Geosynchronous Orbit

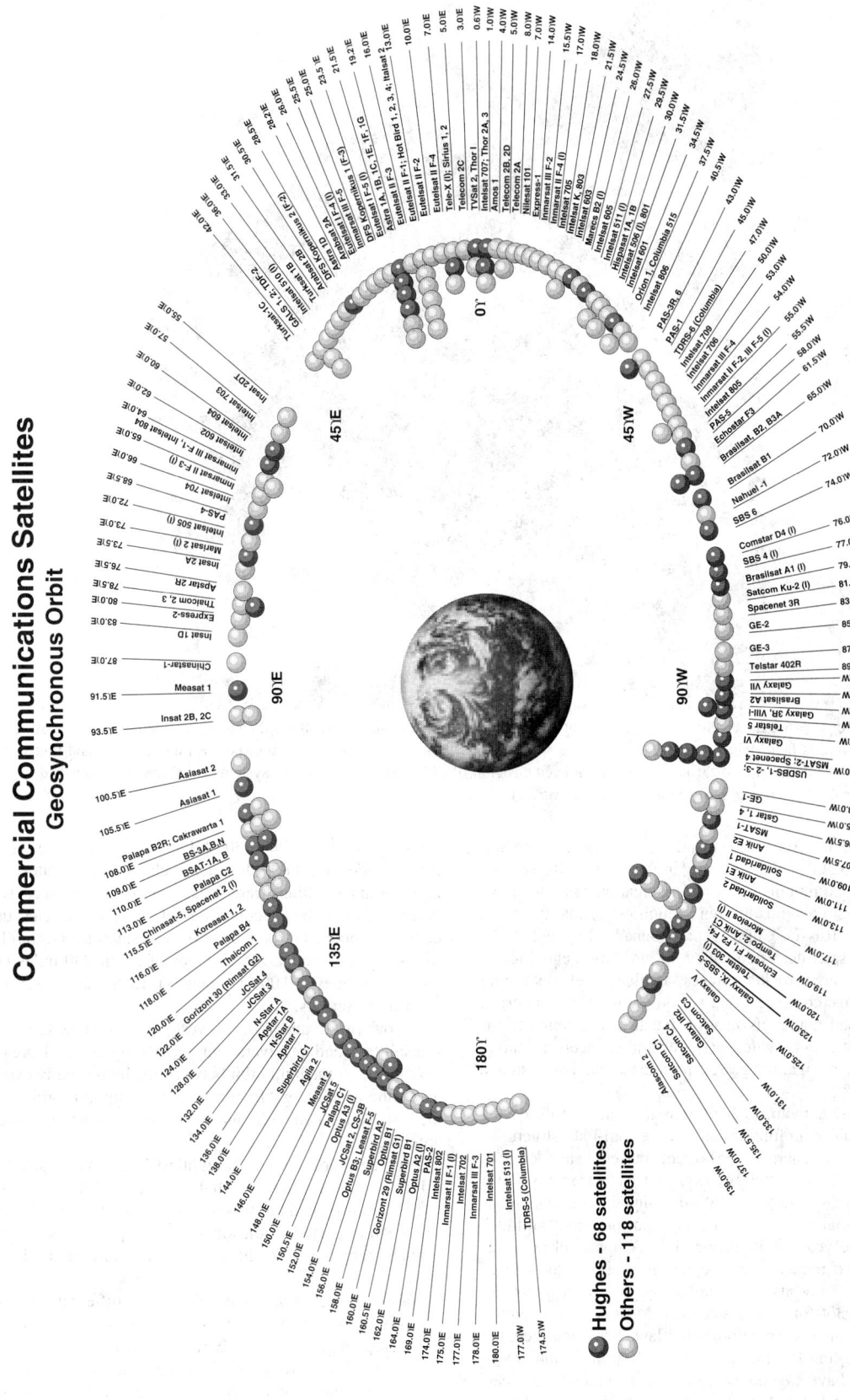

Fig. 1. As of late 1998, about 200 communications satellites were in geostationary orbit about 22,300 miles (35,800 kilometers) above the Earth's equator. This chart shows their positions and identifies each spacecraft. (*Courtesy of Hughes Space and Communications Company.*)

● Hughes - 68 satellites
○ Others - 118 satellites

3063

Fig. 2.   Even when folded together on its handling dolly in preparation for shipment, PanAmSat's PAS-8 communications satellite is an imposing bulk that represents the size of many spacecraft in this field. Weighing 8,360 pounds (3,800 kilograms), the craft has 24 C-band transponders and 24 Ku-band transponders. They provide video, telecommunications and Internet access throughout the Asia-Pacific region. PAS-8 was built by Space Systems/Loral and was launched into geostationary orbit from Khazakstan, November 4, 1998. (*Courtesy of PanAmSat Corporation.*)

According to the National Aeronautics and Space Agency (NASA) in the USA, more than 3,500 satellites were in orbit in 1998, serving many different purposes. Communications and navigation satellites have been and are being built in several different countries, primarily in the United States, but also in Russia, China, Japan, Canada, England, Italy, France, and Germany. A satellite, often called a "bird" by technologists, is a complex assembly of small rocket engines, auxiliary fuel, stabilizing and attitude control systems, computers and various subsystems, batteries, solar cells, and the payload that is its mission (such as communications amplifiers, frequency converters, switchgear, transmitter-receivers called "*transponders*," antennas, navigation gear, computers, imaging systems, etc.). See Figures 3, 4 and 5.

The solar power panels, sometimes built into the outer walls of the satellite housing but usually unfolding into large sail-like structures, convert the sun's photonic energy into direct current electricity to continuously charge the on-board batteries powering the electronic systems. The auxiliary propellant fuel–in multiple tanks or a single large tank–feeds "thruster" rocket engines that correct the satellite's position and "attitude" when necessary during the years of its service life. A final fuel reservoir is used to fire an on-board rocket when the satellite reaches the end of its useful life (the rocket "de-orbits" the satellite by pushing it into outer space, freeing its orbital slot for a new satellite). Most satellites require several years of assembly and testing before being launched into space, but the industry has been working toward speedier delivery of 18 months or less. Some manufacturers have created assembly-line techniques for fleets of related satellites, aiming for factory cycle times of 35 days including pre-assembly time for component testing.

Satellite Orbits

The orbit is the path of the satellite in relation to the planet as it moves through space above the Earth's atmosphere. The orbit must be at least 180 miles (300 kilometers) above sea level but is often 22,300 miles (35,680 kilometers) above the equator, where almost all commercial satellites were located as of 1998. Because of the two Van Allen radiation belts, consisting of ionized particles that damage solar cells and some electronic components, satellites are rarely orbited in "off-limits" spaces around the altitudes of about 930 miles (1,500 kilometers) and 6,213 miles (10,000 kilometers). Radiation caused by solar storms is even more serious.

A low orbit, under the lowest Van Allen belt, means the satellite's transmitters and receivers can serve only a small area of Earth at any given time. A medium orbit between the belts greatly expands the coverage area (the satellite's "footprint"). The highest orbit is far beyond the highest Van Allen belt and can serve almost half the planet with its antennas.

Although considered "weightless" in space, satellites are held in orbit by the Earth's gravitational grip. They do not fall down to Earth because of their forward speed or velocity. During the satellite's service life, periodic adjustments of its position are necessary to correct orbital changes due to the anisotropic nature of the Earth's gravitational field.

Various types of orbits are plotted for different altitudes, depending on the mission of the satellite. See Fig. 6. Most communications satellites are located high above the Earth's equator ("equatorial orbits") in a belt-like ring far out in space. Each position is identified by the longitude intersecting the equator. While such a satellite appears to hover above the same spot on Earth, it is in a *geostationary-Earth orbit* (GEO), 22,300 miles (35,800 kilometers) high, synchronized to keep pace with the Earth's rotation around its axis. The geostationary satellite is not actually orbiting *around* the Earth but rather moving *with* the Earth, which means it is actually moving through space at almost 7,000 miles/hour

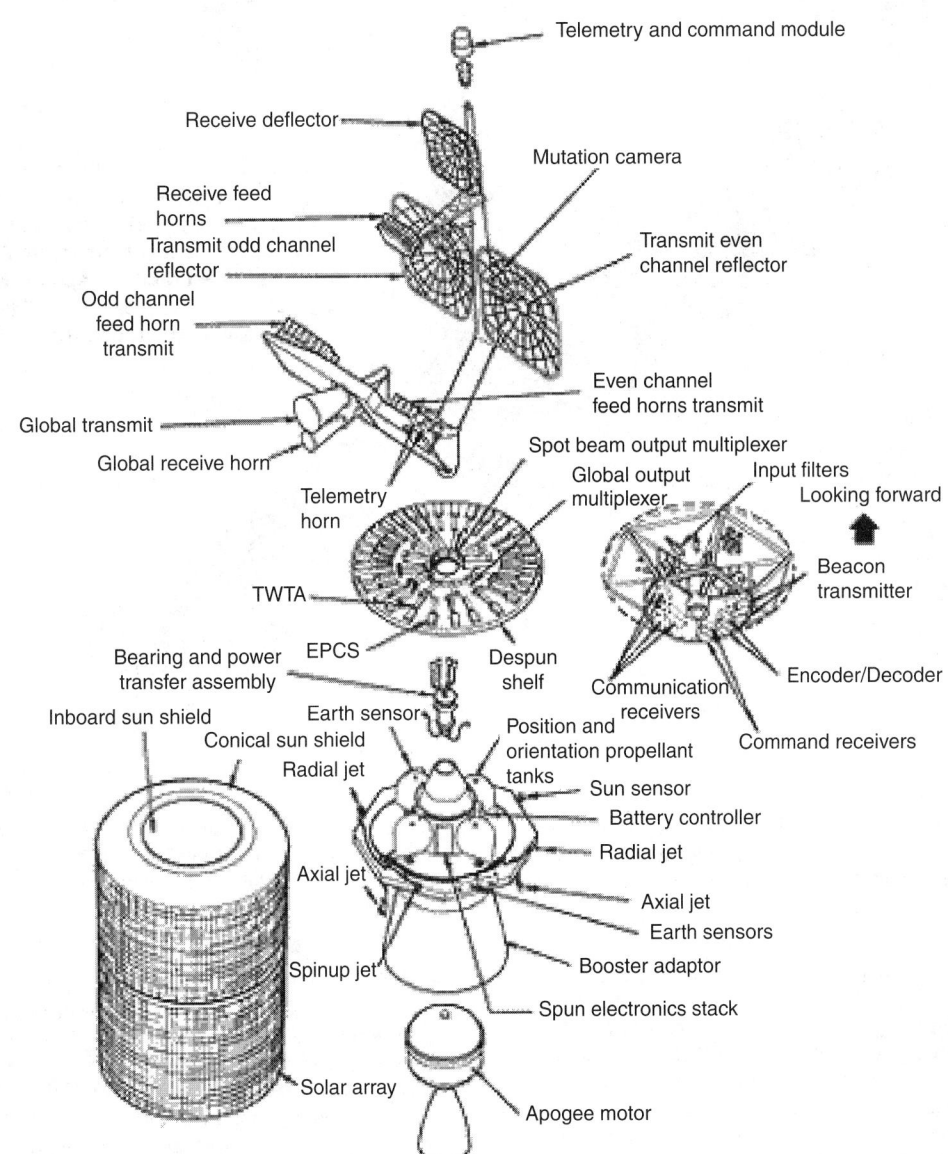

Fig. 3. This exploded view of and INTELSAT IVA communications satellite illustrates its chief components. Hughes Space and Communications Company built six INTELSAT IVAs (one was lost during its launch in 1977) to provide coverage of land masses on both sides of the Atlantic basin, using four spot beams so isolated from each other that the same frequencies can be used in the east and the west. This antenna design significantly increased satellite communications capability within the allocated frequency band. Each INTELSAT IVA had 20 transponders, compared to only 12 on the earlier INTELSAT IV satellites. The spacecraft weighed 3,335 pounds (1,515 kilograms) and had an overall height of almost 23 feet (7 meters) and a diameter of almost 8 feet (more than 2 meters). The first of the series was launched from Cape Canaveral on September 25, 1975, and the last two in 1978. Although their design life in orbit was specified as seven years, the INTELSAT IVA spacecraft operated an average of almost 11 years before being retired.

Cylindrical satellites are divided into two halves: the upper half spins to provide basic gyroscopic stability to the spacecraft; the bottom half remains pointed toward Earth. The spinning half of the INTELSAT IVA satellite was covered with nearly 17,000 solar cells, which absorbed energy from the sun and converted it to provide primary electrical power of 600 watts. The lower half contained the communications payload. The positioning and orientation (reaction control) subsystem was mounted on the rotor. Redundant jets were provided for orbital station-keeping, attitude control and spin-up with sufficient fuel (hydrazine) for the orbital life of the satellite.

The directional communication antennas were pointed to the roper point on Earth by the despin control system. Redundant sensor information (three earth sensors, two sun sensors, all rotor-mounted) was used by an on-board processor to establish the inertial attitude of the spacecraft and control the antenna platform. The apogee motor was carried in the aft half of the rotor section. The burn of that motor provided the needed impulse to place the satellite in its final synchronous orbit.

The despun section of the satellite contained the communications repeater electronics and telemetry and command electronics. Spacecraft commanding was performed through two cross-strapped command systems. Telemetry information from the spacecraft was provided by redundant pulse-code modulated systems with frequency modulated real-time mode available for transmission of certain data during spacecraft maneuvers. (*Courtesy of Hughes Space and Communications Company.*)

(11,265 kilometers/hour) as if it were attached to the spinning planet by an invisible tower. Navigation satellites and some communications satellites operate in a *medium-Earth orbit* (MEO), usually an altitude of about 12,500 miles (20,200 kilometers), and circle around the planet at an angle to the equator. Large fleets or "constellations" of communications satellites are scheduled for *low-Earth orbits* (LEO), circling the Earth at approximate altitudes as low as 485 miles (780 kilometers) to as high as 800 miles (almost 1,300 kilometers), with multiple satellites following each other in the same orbital path. (Various manned space flights and the *Mir* space station also have used or are using *low-Earth orbits*.)

A satellite in a "*polar orbit*" circles the Earth at right angles to the equator, passing near the planet's north and south poles. Still others circle the Earth in paths or "*planes*" at inclined angles to the equator. The altitude is usually scheduled to be about the same throughout the orbit. When an orbit is eccentric (not always the same distance from the planet), the most distant point from Earth is the *apogee*, while the closest point is the orbit's *perigee*. The difference between the *apogee* and the *perigee* is called the *orbit's degree of eccentricity*.

The first experimental communications satellites in 1962 and 1963 (AT&T's privately funded *Telstar I* and *Telstar II*, and RCA/NASA's *Relay*

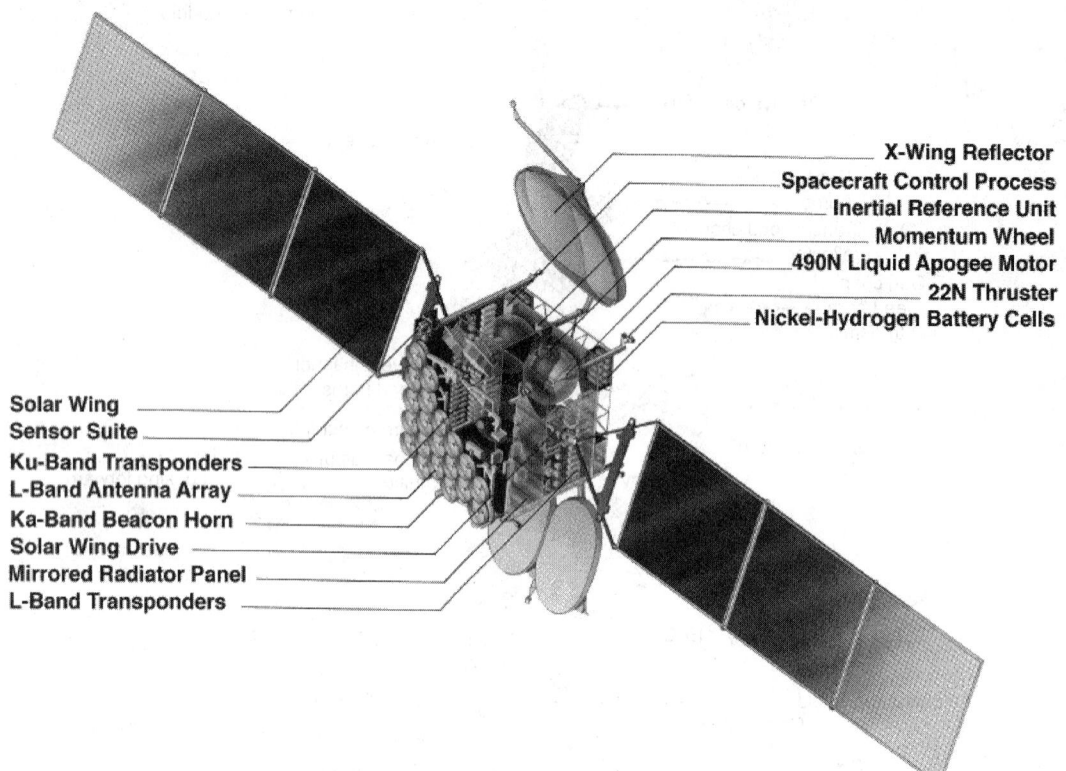

Fig. 4.   The Hughes 601 satellite is body-stabilized, with internal components providing three-axis stabilization. Introduced in 1987, it is the world's best-selling large spacecraft model. By the fourth quarter of 1998, the company had received orders for 72 HS 601 communications satellites. The basic configuration features as many as many as 48 transponders and offers up to 4,800 watts of power. A more powerful model, the HS 601 HP debuted in 1995 and features up to 60 transponders and provides up to 10,000 watts. (*Courtesy of Hughes Space and Communications Company.*)

I and *Relay II*) were designed for *low-orbit* operations. See Fig. 7. A Hughes-built geosynchronous satellite, *Syncom II*, followed them in 1963, with which numerous demonstrations were made to prove the feasibility of communications from a high orbit. See Fig. 8. (A geosynchronous orbit is inclined relative to the equator, so *Syncom II* described a figure-eight pattern rather than hovering over a single point as in a *geostationary orbit*; a geostationary orbit is always a *geosynchronous orbit*, but a geosynchronous orbit is not necessarily a geostationary orbit.) The first *geostationary satellite* was the Hughes-built *Syncom III*, launched in 1964.

Theoretically, only three geostationary satellites would be needed to provide total (except for the extreme Polar Regions) coverage of the Earth's surface, since each would have a gigantic "footprint." But three satellites would never have enough capacity to meet the world's demands for communications circuits. NASA reports more than 200 satellites in geostationary orbit, spaced some 900 miles (1,448 kilometers) apart along a circular path 165,000 miles (ca. 264,000 kilometers) long, far above the Earth's equator (which has a circumference of 24,901.6 miles or 40,075 kilometers).

There are thousands of fixed satellite earth stations, some with as many as 25 large and small antenna dishes, scattered around the globe. Most of these stations are linked by optical fiber cable and/or overland microwave radio circuits to either commercial or government television broadcasting facilities or electronic "gateway" switches serving public and private communication networks. In addition, there are many more mobile earth stations of varying sizes, such as those used by television reporting teams while on location for special events or catastrophes. During the 1990s, countless other satellite reception dishes–many as small as 12 inches (30 centimeters) in diameter–were installed throughout the world to receive television programming broadcast from satellites. Also in the 1990s, special handheld wireless telephones (sometimes called "*satphones*"), pagers, and text-messaging terminals were introduced, enabling a subscriber to link up with certain satellites using only a small built-in antenna on the handset; however, such systems still rely on the use of an earth station to complete the link to parties on terrestrial networks. In addition, a party on a conventional network phone can call a "satphone" user, whose terminal will be activated by a signal direct from the appropriate satellite.

Satellite Transmission Frequencies

Communications satellites are equipped with a variety of transponders (special transmitter-receivers). These operate within certain radio-frequency bands as specified by the Federal Communications Commission (FCC) and the International Telecommunication Union (ITU). The most common transponder settings employ C-band (*centimeter-band*, using wavelengths from 3.7 to 5.1 centimeters and operating within a frequency range of 4 to 6 gigahertz), and/or Ku-band (*K-under*, for a spectrum band under the K-band, operating within 11 to 14 gigahertz). Other settings include L-band (1 to 2 gigahertz) and S-band (2 to 3 gigahertz). The higher the frequency, the more susceptible the signal is to weakening or degrading when it passes through water vapor, rain or falling snow in which the size of the drops or crystals matches or exceeds the wavelength of the transmission frequency. Most geostationary communications satellites use the C-band, but that means television reception requires relatively large ground antennas (sometimes called "backyard antennas"). Since the introduction of direct-transmission of TV signals to businesses and residences, the Ku band has gained in popularity, since it can serve much smaller and less costly antenna dishes than the C-band.

A new frequency setting is the Ka-band (*K-above*, using a band above the K-band, for signals from 18 to 31 gigahertz). Introduced to low-orbit commercial use in 1998, this band has five times more spectrum than is available at lower frequency bands. The Ka band has been undergoing tests as part of NASA's *Advanced Communication Technology Satellite* (ACTS) project. ACTS was launched in 1993 and served in a geostationary slot before beginning inclined orbit operations in 1998. The craft also is testing new antenna systems for "multiple hopping beams," on-board switching to permit interconnectivity between users at the individual circuit level, and a microwave switch matrix to enable users to communicate at rates of gigabits per second.

Each of the thousands of satellites in space is assigned specific frequencies. The *uplink* (earth terminal to satellite) is a different frequency band than the *downlink* (satellite to earth terminal) so as to avoid signal interference. Because so many satellites are already in orbit, there is growing concern that the limited frequency spectrum may be fully assigned in the near future. Offsetting this fear is the ingenuity of communications

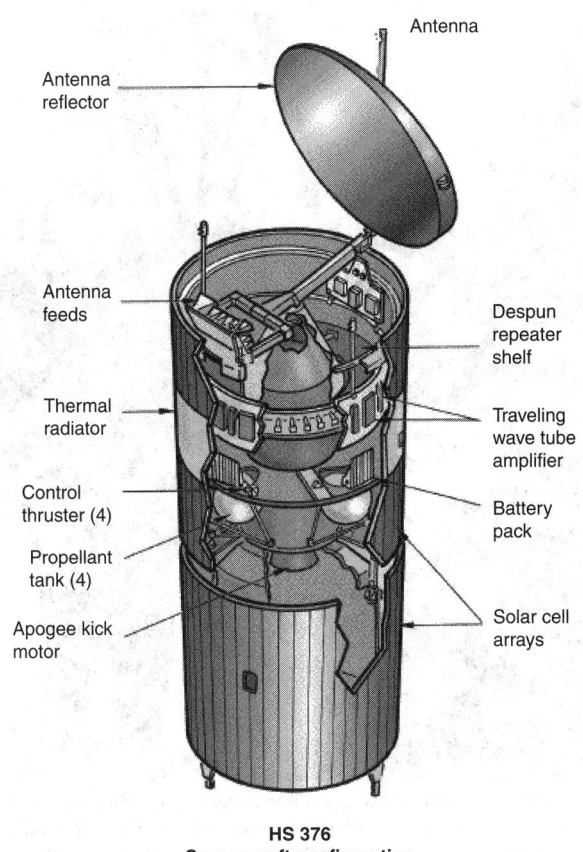

Antenna

Antenna reflector

Antenna feeds

Despun repeater shelf

Thermal radiator

Traveling wave tube amplifier

Control thruster (4)

Battery pack

Propellant tank (4)

Solar cell arrays

Apogee kick motor

HS 376
Speacecraft configuration

Fig. 5.   The Hughes 376 satellite was introduced in 1977 and is one of the most popular commercial communications satellite models. More than a dozen customers in 14 countries on five continents have installed or ordered 54 HS 376 spacecraft as of mid-1998. Cylindrical satellites like this are designed for spin-stabilization, in which the spinning upper half is the stabilizing element, with the communications equipment housed in the bottom half. (*Courtesy of Hughes Space and Communications Company.*)

engineers who have steadily improved their ability to compress the signals so they use less spectrum.

Geostationary Earth Orbit (GEO) Satellites

INTELSAT (INTernational TELecommunications SATellite), a consortium created August 24, 1964 by 11 nations, had 142 members by 1998. Since 1966, INTELSAT has covered all the continents with satellite communications. In 1998 INTELSAT diluted its holdings by spinning off a new company, New Skies Satellites, and transferring ownership of five in-orbit INTELSAT satellites, as well as the relevant assigned frequencies. The new company will compete in areas such as consumer-oriented video services and multimedia applications. Beginning the year 2000, INTELSAT had 17 satellites in operation. The United States is represented in INTELSAT by the Comsat Corporation, created by Congress as a quasi-private corporation via the Communications Satellite Act of 1962.

The first commercial geostationary satellite, the *INTELSAT I* or "*Early Bird*," was developed by Hughes Electronics for Comsat and launched in July, 1965. See Fig. 9. It weighed only 85 pounds and measured 23 inches high by 28 inches wide. *Early Bird* was equipped with two transponders capable of serving up to 240 telephone circuits, and was designed for a life expectancy of 18 months; the cost per circuit-year came to $30,000. By 1998, improved technology had raised geostationary satellite life expectancies to 15 years and more, while costs per circuit-year dropped to less than $1,000. Modern geostationary satellites usually have a launching mass of more than 3,000 pounds (1,360 kilograms) but can weigh as much as 9,000 pounds (over 4,000 kilograms). They can be either cylindrical or rectangular in shape, the housing measuring as much as four stories high plus large solar panels and antennas which are folded onto the satellite during the launch, then opened in space. Each such satellite can cost as much as $250 million or more and is insured against loss due to launching accidents and space failures.

Individual communications carriers, such as AT&T, British Telecom, MCI, GTE, etc., have long-term leases of transponders on the *INTELSAT* vehicles, and others can rent them for short-term individual projects, such as a conference. Although the original communication satellites were operated with analog circuits the modern transponders and associated circuitry usually employ digital signal processing to improve quality, efficiency, and capacity of transmission. Digital signals can be compressed and combined (or *multiplexed*) more easily than analog signals, resulting in more two-way telephone circuits on a given frequency band.

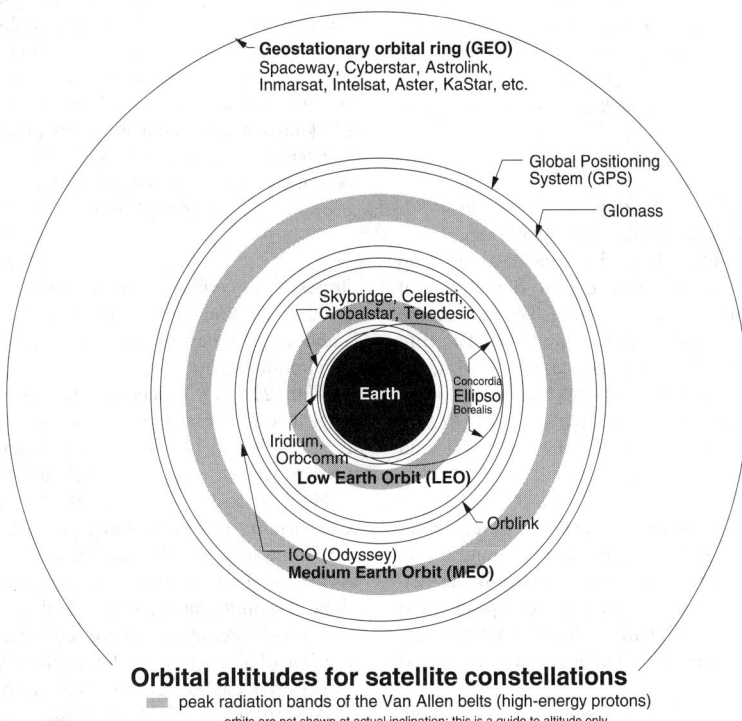

Geostationary orbital ring (GEO)
Spaceway, Cyberstar, Astrolink, Inmarsat, Intelsat, Aster, KaStar, etc.

Global Positioning System (GPS)

Glonass

Skybridge, Celestri, Globalstar, Teledesic

Earth

Concordia
Ellipso
Borealis

Iridium, Orbcomm
Low Earth Orbit (LEO)

Orblink

ICO (Odyssey)
Medium Earth Orbit (MEO)

Orbital altitudes for satellite constellations
peak radiation bands of the Van Allen belts (high-energy protons)
orbits are not shown at actual inclination; this is a guide to altitude only
from Lloyd's satellite constellations http://www.ee.surrey.ac.uk/Personal/L.Wood/constellations/

Fig. 6.   Satellite orbits are described in three general classes: low-Earth orbit (LEO); medium-Earth orbit (MEO); and, geostationary-Earth orbit (GEO). This chart shows the various altitudes above Earth to scale, as well as their relationship to the two Van Allen radiation belts. (*Illustration by Lloyd Wood, from a representation by Dr. Thor E. Wisloff.*)

Fig. 7. On July 10, 1962, the worlds first international communications satellite, Telstar I, was launched into a low-Earth orbit. Privately funded by AT&T and built by Bell Telephone Laboratories, the satellite weighed 170 pounds (77 kilograms) and was 34 inches (about 86 centimeters) in diameter. The surface of Telstar 1 carried 3,600 silicon solar cells that were the source of its primary power. More than 6,000 transistors and diodes were used as individual components, because the integrated-circuit era was barely dawning. The near-omnidirectional helical antenna on top of Telstar received commands and transmitted a continuous VHF signal for initial acquisition and tracking. The communication antennas for receiving and transmitting voice and video signals consisted of radiating boxes along the satellite's equator, interconnected by a series of printed-circuit hybrids to a single 6-gigahertz input and a single of 4-gigahertz output of the repeater. Telstar's service area included North America and Europe, using three major Earth stations: Andover, Maine; Goonhilly Downs, England; and Pleumeur-Bodou, France. No fees were charged to AT&T for any transmissions during the experimental satellite's availability, a period of three months (its circuits were damaged by excessive radioactivity caused by a nuclear test over the Pacific the day before Telstar I was launched). (*Courtesy of Bell Laboratories, Lucent Technologies.*)

For example, three *INTELSAT* series VIII satellites, orbited in 1997, each have 44 transponders (38 C-band, six Ku-band), capable of serving 22,500 two-way telephone circuits plus three TV channels, and up to 112,500 two-way telephone circuits with the use of digital circuit multiplication equipment (DCME) in the earth stations. The last in the series of the *INTELSAT* VIII/A satellites, orbited in 1998. The life expectancy of these satellites is 14 to 17 years in geostationary orbit. As of late 1998, seven *INTELSAT* series IX satellites were being developed. Spacecraft delivery and launch periods scheduled for the seven *INTELSAT* series IX satellites are as follows: the first four are to launch in 2001, and the remaining three in 2002.

PanAmSat Corporation, based in Greenwich, CT., is a leading provider of global video and data broadcasting services via satellite. The company builds, owns and operates networks that deliver entertainment and information to cable televisions, TV broadcast affiliates, direct-to-home TV operators, Internet service providers, telecommunications companies and corporations. PanAmSat merged in 1997 with part of Hughes Communications, a unit of Hughes Electronics Corporation, which is a wholly owned subsidiary of General Motors Corporation.

In September, 1998, PanAmSat launched a Hughes-built *PAS-7* satellite weighing 8,444 pounds (3,830 kilograms), for a geostationary slot above the Indian Ocean region. The spacecraft delivers satellite-based video and telecommunications services throughout that area. In 1999 PanAmSat launched the Galaxy XI satellite, the company's 20th spacecraft and the

Fig. 8. The first communications satellite to achieve geostationary orbit above the equator was NASA's Syncom III, launched in 1964. Syncom I was lost during its launch on Valentine's Day, 1963, but Syncom II achieved a synchronous orbit on July 26, 1963. Although Syncom II was in orbit 22,300 miles (35,680 kilometers) above the equator, it was not truly geostationary. However, the experimental satellite successfully relayed telephone conversations between President John F. Kennedy and Nigerian Prime Minister Abubaker Balewa in Africa in 1963 — the first live two-way call between heads of state by satellite relay. Built by Hughes Aircraft Company (now Hughes Space and Communications Company) for NASA, the satellite series was designed by Donald D. Williams and Thomas Hudspeth, under Dr. Harold Rosen. The photo shows Tom Hudspeth, left, with Dr. Rosen and a Syncom prototype during the Paris Air Show in 1961, where they displayed the actual-size satellite on the Eiffel Tower — and endured wags who declared, "This is as high as it will ever get." Syncom II measured two feet, four inches (71 centimeters) in diameter and was one foot, three inches (39 centimeters) high. Its mass in orbit was 78 pounds (35 kilograms). Its communications capability was either one-way TV, one two-way telephone channel or 16 teletype (text) channels. Syncom III had an improved design that provided more bandwidth for better TV transmission. (*Photo and drawing courtesy of Hughes Space and Communications Company.*)

largest commercial communications satellite ever deployed in space. In 2000 PanAmSat successfully launched the Galaxy XR spacecraft, the first satellite of the new millennium and the fifth in the company's Galaxy cable neighborhood.

With 21 *PAS* spacecraft in orbit today, PanAmSat has the world's largest commercial geostationary satellite network. See Fig. 10. PanAmSat is scheduled to launch four additional satellites by mid-2001, increasing total capacity to more than 900 usable transponders worldwide.

Hughes also is the largest investor in *American Mobile Satellite Corporation* (AMSC), which in 1995 flew the first satellite dedicated to providing land mobile services for individual users in North America. The *AMSC* system has three satellites in geostationary orbit, offering voice, data, facsimile, messaging and other services to travelers in trucks, autos, boats and aircraft across the continent. It uses frequencies in L-band, a portion of the spectrum lower than that assigned to satellites providing direct broadcast or serving fixed earth stations.

Another major family of geostationary satellites is the *Inmarsat* system, which had nine satellites in operation by late 1998. The international cooperative organization was created as the International Maritime Organization on July 16, 1979 and began operating its first satellite circuits in 1982, using leased transponders on three *Marisat* craft it inherited from *Comsat*,

Fig. 9. After the successful demonstrations of Syncom II and III, Hughes Aircraft built the world's first commercial communications satellite, popularly known as Early Bird, for the Communications Satellite Corporation (COMSAT). Launched into geostationary orbit on April 6, 1965, Early Bird began commercial service on June 28, 1965, providing line of sight communications between Europe and North America. Early Bird was capable of relaying 240 circuits for telephone, telegraph of facsimile transmission, or one television channel. Similar to Syncom in size and mass, Early Bird weighed 76 pounds (34 kilograms), a bit less than Syncom, but was a bit higher—one foot 11 inches (59 centimeters). Designed to last 18 months in orbit, Early Bird actually remained in full-time service for nearly four years before being placed on reserve duty. In June of 1969, it was recalled into service during the Apollo 11 mission, and was retired again two months later. (*Photo and drawing courtesy of Hughes Space and Communications Company.*)

plus two *Marecs* satellites from the *European Space Agency* (ESA), followed by Maritime communications subsystems on several *Intelsat V* satellites. *Inmarsat* launched the world's first dedicated mobile communications satellites–four *Inmarsat 2* series–in 1990. These were followed in 1996, 1997 and 1998 by five *Inmarsat 3* satellites (one a spare) designed and assembled by Lockheed Martin Telecommunications; Matra Marconi Space of the United Kingdom manufactured the communications payload, including the antennas, repeater and other communications electronics. The Series 3 satellites expanded the *Inmarsat* services to include the aviation community's ability to use both the American and Russian *global positioning system* (GPS) satellite navigation systems (see the **Navigation Satellites** section below). The nine *INMARSAT* geostationary spacecraft are connected to the terrestrial telecommunications networks via 39 marine and aeronautical ground earth stations serving as gateways, and in 1998 were being accessed by more than 100,000 customer terminals. (During the first 10 years, from 1982 to 1992, only 15,000 terminals were sold.) Service is global except for the extreme polar regions.

Originally a dependent of the United Nations, *Inmarsat* was founded to provide wireless voice and data links between maritime mobile equipment (i.e., ships and ocean drilling platforms) and ground networks, covering the four major ocean areas of the world. Its services now also include various governmental functions, such as backing up terrestrial communications for embassies around the world, and various customs and fisheries patrol craft. Peacekeeping task forces rely on *Inmarsat* systems to keep in touch with civil authorities, but the system cannot be employed for direct military use or acts of war, nor is it certified for use within the continental United States. By 1998, *Inmarsat* had 83 member countries in its cooperative. However, they plan to privatize the organization, dividing it into two entities: a public limited company seeking an initial public offering of stock by late 2000, and an intergovernmental structure ensuring that Inmarsat meets its public

service obligations. Those obligations include the Global Maritime Distress and Safety System (GMDSS).

A future satellite project involving Hughes Communications is the *Spaceway* system, which includes both geostationary and medium-orbit satellites. It currently plans to install eight satellites in geostationary orbit to provide high data rate transport services, and another 20 interconnected satellites in "nongeostationary" orbit to provide "advanced interactive broadband multimedia communications services in high traffic markets globally." The fleet of 20 *Spaceway* satellites will be divided into four different orbital paths or planes with five satellites in each plane, circling the planet at a medium-orbit altitude of about 6,400 miles (10,352 kilometers). The system is projected to be operating as early as 2002.

Lockheed Martin is developing another system, named *Astrolink*, as a major step forward in increasing data transmission rates. *Astrolink* will incorporate nine satellites in geostationary orbit, but the organization claims only five are needed for global connectivity. The users can select from a range of data transmission rates and pay only for what they use, as compared to paying the fixed cost of leased lines sized for the highest rate they might want. For example, a small office or home office could use a 35-inch (90-centimeter) diameter dish and be provided with data rates up to 416 kilobits per second. A medium-sized business could use data rates up to 2.1 megabits per second with a 39-inch (100-centimeter) dish, while a major enterprise could transmit data at up to 10.4 megabits per second with a six-foot (1.8-meter) dish. Up to 100 gateways will connect *Astrolink* to terrestrial networks worldwide. The first *Astrolink* satellite is tentatively scheduled for launching in 2001, at which time service could begin. The next four spacecraft will complete the global coverage, and the last four will be added as demand escalates.

Medium Earth Orbit (MEO) Satellites

A number of medium-orbit satellite systems involving multiple spacecraft were being developed in the late 1990s for mobile wireless telephony and data transmission. Several satellite constellations already existed in medium-orbit by the mid-1990s (see **Navigation Satellites**, later in this article).

The *ICONET* system, with Hughes Electronics as a major investor and satellite provider, plans to start operating services circa the summer of 2000. See Figs. 11, 12, and 13. The system will include 10 satellites (plus two spares) divided equally into two orbital planes 6,430 miles (about 10,350 kilometers) above the Earth and inclined by 45 degrees to the equator. Each satellite circles the planet once every six hours. The system will have 163 service-link transmit and receive beams, allowing frequency re-use by each satellite; the design supports at least 4,500 simultaneous telephone channels per satellite. Subscribers will use handheld mobile telephones and data terminals when the system is fully provisioned. Calls from a handset will be transmitted (*uplinked*) to an ICO satellite, processed and retransmitted (*downlinked*) on C-band frequencies to one of 12 earth stations called "*satellite access nodes*" (SANs) connected to gateway switches feeding into the world's terrestrial telephone networks. The ICO satellite, which handles the initial call, might pass out of the service area during the call, in which case the communication is handed off to the next satellite. Each satellite is designed to support up to 4,500 simultaneous telephone conversations. The *ICONET* system also carries a navigation and positioning payload to augment similar transmitting systems in the *global positioning system* (GPS) (see **Navigation Satellites**). ICO Global Communications is an outgrowth of *Inmarsat*, mentioned earlier under geostationary satellite systems.

In addition to the *ICONET* system, another medium-orbit system, known as the *Ellipso* project, is being planned for mobile phone and data customers, with operations starting in 2001. See Fig. 14. Design, development and deployment of the future system has been assigned to Boeing Corporation. *Ellipso* consists of a unique medium-orbit combination of two inclined orbital planes, each equipped with four active satellites and a spare, and an equatorial plane equipped initially with six satellites and a spare that can be augmented with four more satellites for future increased daytime capacity. The inclined orbits of the *Ellipso* constellation are truly elliptical: the apogees of the *Ellipso-Borealis* sub-constellation orbits, serving the northern temperate latitudes with 10 satellites in two planes, are 4,725 miles (7605 kilometers) and the perigees are 393 miles (633 kilometers) with a three-hour orbital period; the apogees are near the northern extremity of the orbits. The *Ellipso-Concordia* sub-constellation serves the tropical

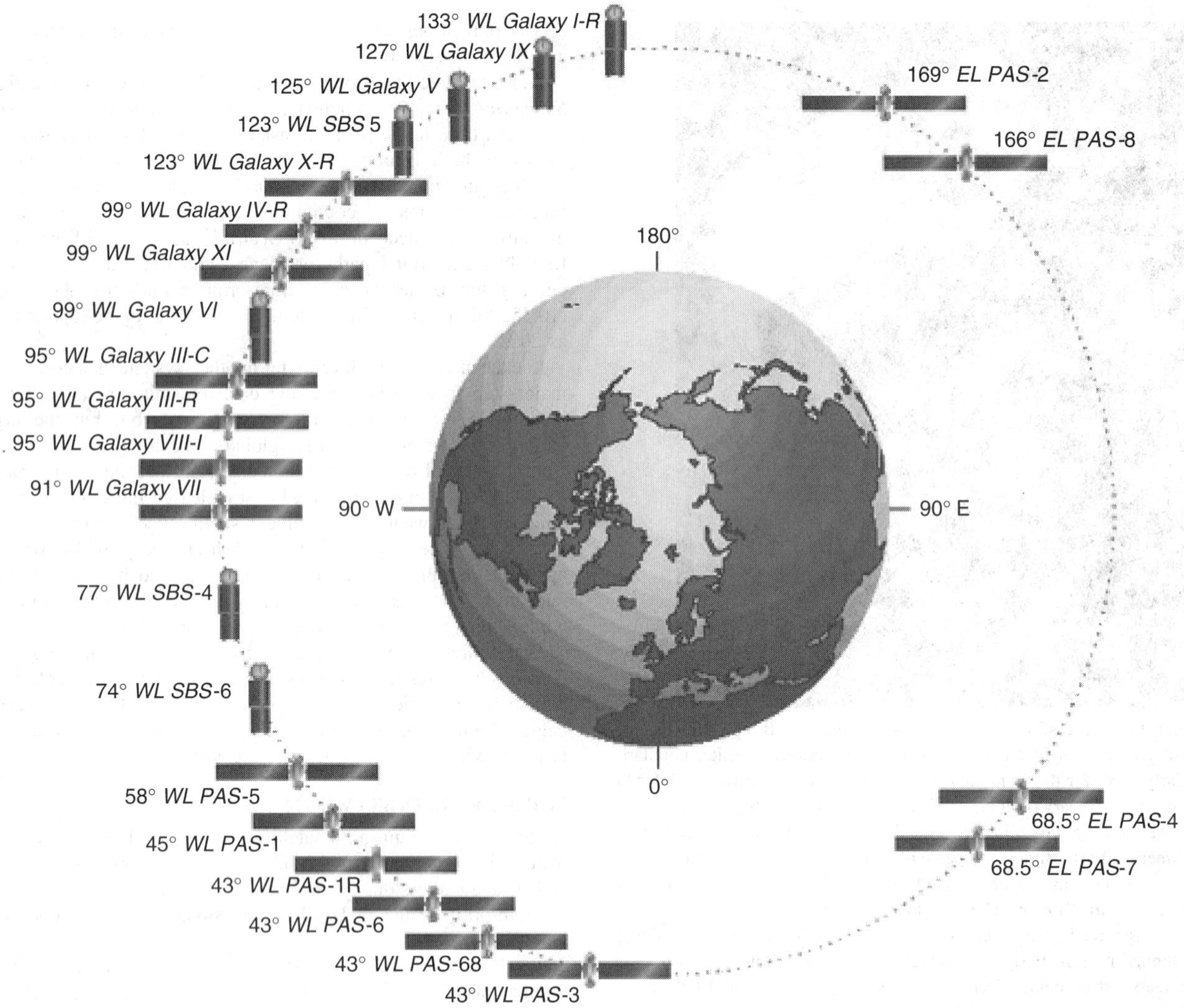

133° WL Galaxy I-R
127° WL Galaxy IX
125° WL Galaxy V
123° WL SBS 5
123° WL Galaxy X-R
99° WL Galaxy IV-R
99° WL Galaxy XI
99° WL Galaxy VI
95° WL Galaxy III-C
95° WL Galaxy III-R
95° WL Galaxy VIII-I
91° WL Galaxy VII
77° WL SBS-4
74° WL SBS-6
58° WL PAS-5
45° WL PAS-1
43° WL PAS-1R
43° WL PAS-6
43° WL PAS-68
43° WL PAS-3
169° EL PAS-2
166° EL PAS-8
180°
90° W
90° E
0°
68.5° EL PAS-4
68.5° EL PAS-7

Fig. 10. Looking down at the Earth's north pole, the equatorial ring of PanAmSat communications satellites in geostationary orbit displays their positions, which are 22,300 miles (35,800 kilometers) above the equator. More satellites are grouped above the American and European continents. (*Courtesy of PanAmSat Company.*)

SaVi

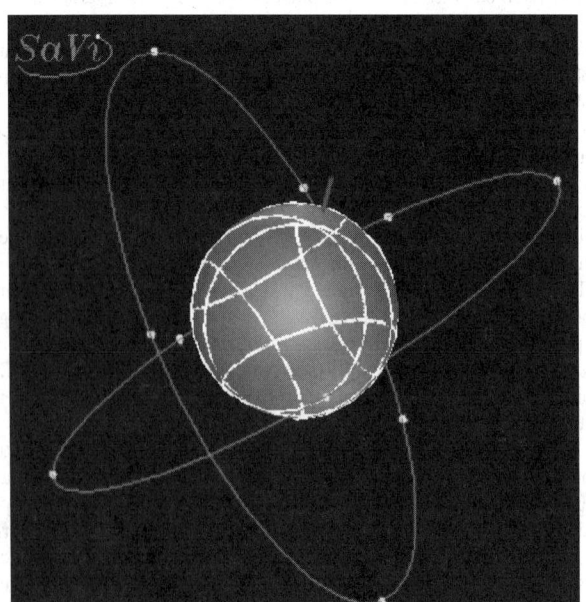

Fig. 11. This rendering shows the final design of the ICO constellation, consisting of two planes with five communications satellites in each. (*Constellation views rendered by Lloyd Wood, using SaVi, the Satellite Visualization software created by Patrick Worfolk and Robert Thurman at the Geometry Center, University of Minnesota.*)

ICO-05
ICO-12
ICO-06
ICO-04
ICO-11
ICO-03
ICO-08
ICO-00
ICO-07
ICO-02

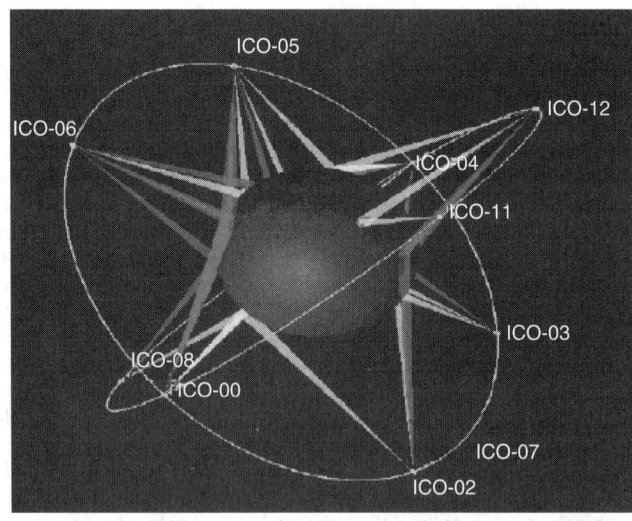

Fig. 12. The ICO constellation of 10 satellites (plus two spares) consists of two orbital planes 6,430 miles (about 10,350 kilometers) above the Earth, inclined by 45 degrees to the Equator. Each satellite contains spot beams designed to focus energy to a relatively small area so mobile telephones and data terminals can transmit and receive in direct links to and from the satellite passing overhead. As the satellite moves away from the user, the call is transferred to a spot beam of the next satellite. (*Animations by Raytheon Systems Company for ICO Global Communications.*)

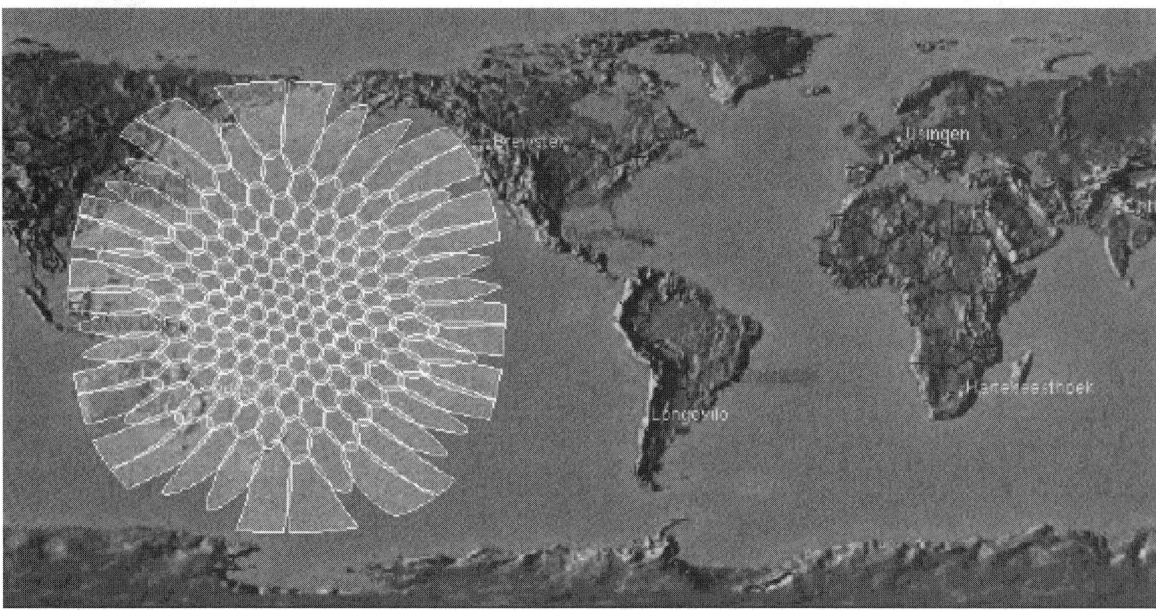

Fig. 13.   Each ICO satellite has 163 service-link transmit and receive beams capable of supporting at least 4,500 digital telephone channels using time-division multiple access (TDMA) multiplexing. This illustration shows the movement of an ICO satellite's footprint of spot beams as it moves across the world. (*Animations by Raytheon Systems Company for ICO Global Communications.*)

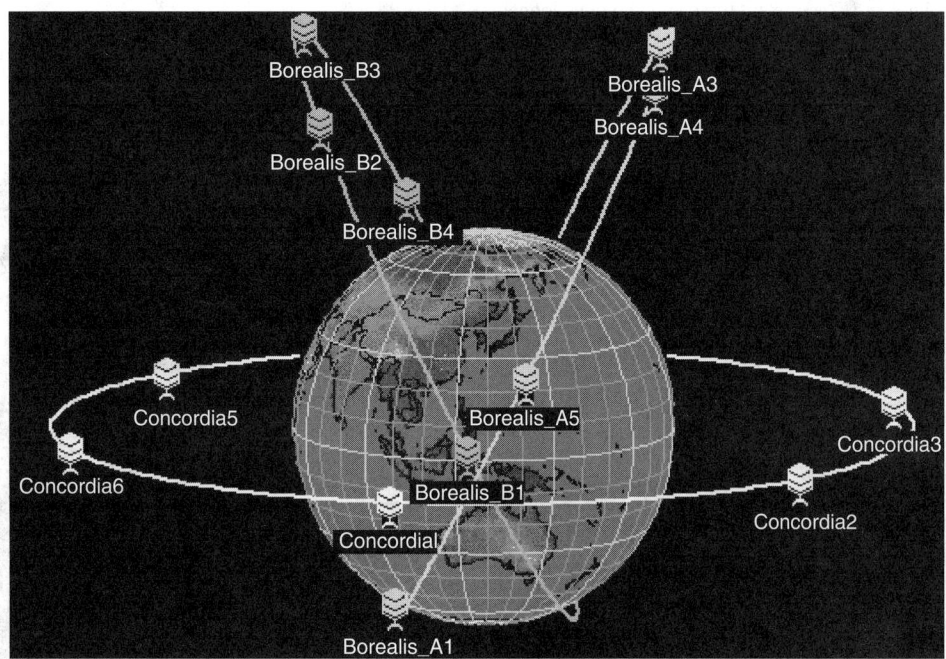

Fig. 14.   A unique, patented elliptical orbit configuration for the Ellipso project will use two complementary sub-constellations with a total of 17 satellites plus a spare when it is activated in 2001. Four active satellites (and a spare) in each of the two Borealis elliptical orbits will soar higher (up to 4,725 miles or 7,605 kilometers) in the northern latitudes so they will have a larger "footprint" in the great continental areas, and drop to an altitude of only 393 miles (633 kilometers) while passing "under" the Earth. The smaller land masses in the tropical and southern latitudes will be served by six active satellites (and a spare) in the equatorial Concordia orbit at an altitude of about 5,000 miles (8,050 kilometers). Ellipso is intended to bring affordable fixed, mobile and airborne voice and data telecommunications to new markets worldwide, according to Boeing. (*Courtesy of The Boeing Company.*)

and southern latitudes with seven satellites in a quasicircular equatorial orbit at an altitude of about 5,000 miles (8,050 kilometers). The system is designed to support voice telephony with digital voice encoding at several levels of quality depending on the customer's specification. *Ellipso* also will support various forms of data transmission, using *code division multiple access* (CDMA) or *"spread spectrum"* techniques to increase the number of voice and data channels carried within its assigned spectrum.

Low Earth Orbit (LEO) Satellites

Several organizations are developing low-orbit systems involving many satellites that trail each other along a given orbital plane. Most of these projects involve a number of orbital planes that are in contact with

numerous earth stations. An orbital plane with multiple satellites is called a "constellation." A combination of several orbital planes with multiple satellites in a single system is also called a constellation.

The concept resembles that of cellular telephony, in which a series of stationary antenna towers transfer a wireless telephone circuit from one antenna to the next as a caller moves through the system. However, the low-orbit satellite systems operate in reverse, with the caller being relatively stationary and the "antennas" moving across the sky as they transfer the call from one satellite to another.

Low-orbit systems for mobile communications have one major handicap: the satellites cross the sky rapidly and the signals can be blocked if buildings, trees, hills or other solid objects come between the subscriber's telephone and the satellite. Some satellites use "*spot beams*," narrowly focused radio transmission and reception beacons that "stick" with the terrestrial user until it is necessary to switch or transfer the user to the next satellite. New communications systems between the satellites will transfer signals from one spacecraft to the other in some systems, without downlinking to an earth station. Such signal transfers are via microwave radio at present, but research is underway on the feasibility of using laser beams.

The first such system to attain operating status with voice and data transmission was *Iridium*, which in 1998 had a constellation of 66 low-Earth orbit satellites, each weighing about 1,528 pounds (689 kilograms). See Fig. 15. Only 15 launches were needed over a one-year time frame to place the 66 satellites in orbit, with as many as seven spacecraft on a single launch. The *Iridium* satellites are grouped into six orbital planes inclined at 86.4 degrees to the equator and at an altitude of about 485 miles (780 kilometers). Each plane has 11 satellites plus one spare that will come to life if any of the 11 craft fails. Each satellite orbits the planet once every 100 minutes and 28 seconds and travels at a velocity of about 17,400 miles/hour (28,000 kilometers/hour) with a "*footprint*" diameter of 2,920 miles (4,700 kilometers). The system was conceived and built by the Motorola Corporation, which announced the concept in 1988. The estimated service life of each satellite is between five and eight years. Each satellite is equipped with 48 spot beams, permitting re-use of assigned frequencies in the L band (at 1616 to 1626.5 megahertz) for connection to subscriber phones, pay telephone booths, etc. See Figs. 16 and 17. Subscribers to the service will have to pay several thousand dollars for a special telephone, or hundreds of dollars for a pager, but the market is expected to consist of business and government travelers who have longed for a global communications system that provides them with a single number to use wherever they are on the planet. The speech quality is not as good as a wireline telephone, since the bit-rate is a mere 2.4 kilobits per second for full-duplex or two-way voice calls (network quality standards for a digital wireline voice circuit call for 64 kilobits, which can be acceptably reduced with special algorithms to lower digital rates). Data and fax transmission via *Iridium* is handled at 2400 baud, significantly slower than the setting of most computer modems. The satellites can communicate directly with each other, without relay by an earth station, using the Ka band at 23.18 to 23.38 gigahertz for intersatellite links.

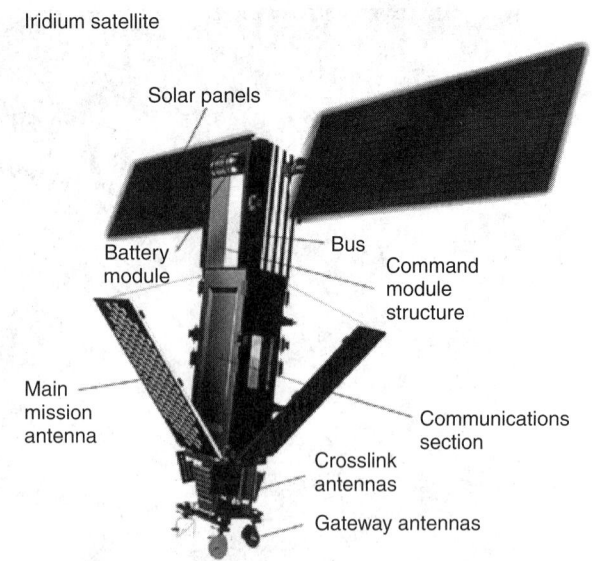

Fig. 16.   Assembled in production lines rather than as custom-built units, the 66 active communication satellites of the Iridium constellation can signal each other as well as the system's earth stations and the individual user's special mobile telephone. (*Rendering courtesy of Motorola Corporation.*)

Fig. 17.   The Iridium handheld telephone, costing more than $3,000, is capable of communicating with the Iridium satellites in low-Earth orbit as they pass overhead. (*Courtesy of Motorola Corporation.*)

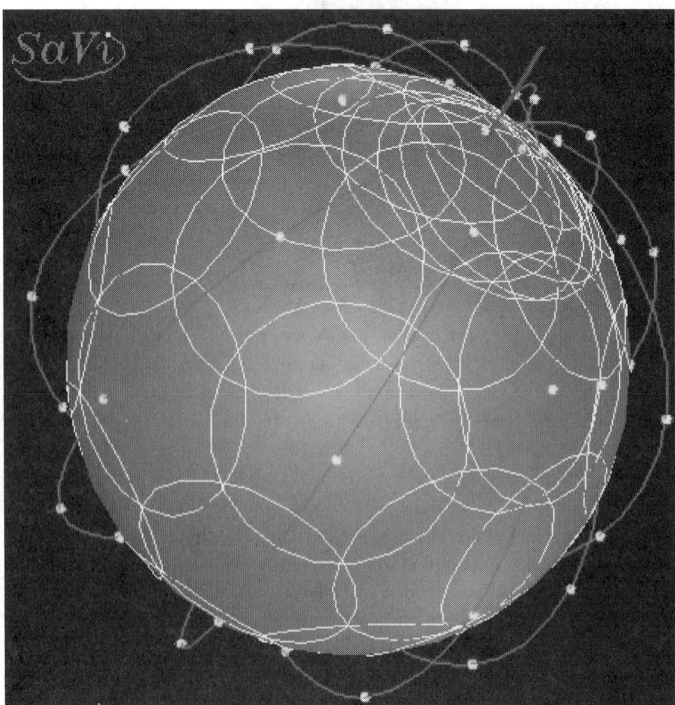

Fig. 15.   Activated in late 1998, the Iridium constellation has six orbital planes, each equipped with 11 satellites, or a total of 66 active satellites in low-Earth orbits. (*Constellation views rendered by Lloyd Wood, using SaVi, the Satellite Visualization software created by Patrick Worfolk and Robert Thurman at the Geometry Center, University of Minnesota.*)

When communicating with their gateways, their downlinks also use the Ka band, but at 19.4 to 19.6 gigahertz, while the uplinks employ 29.1 to 29.3 gigahertz. The system began operations with 12 terrestrial gateways.

Yet another low-orbit constellation project is *Globalstar*, founded in 1991 by Loral Space & Communications and QUALCOMM Inc. Strategic partners include Alenia Aerospazio (a Finmeccanica Company), China Telecom, DACOM, DaimlerChrysler Aerospace, Elasacom (a Finmeccanica Company), Hyundai, TE.SA.M (a joint venture between France Telecom and Alcatel), Space Systems/Loral and Vodafone AirTouch. This system became operational in the 3rd quarter of 1999. See Fig. 18. The

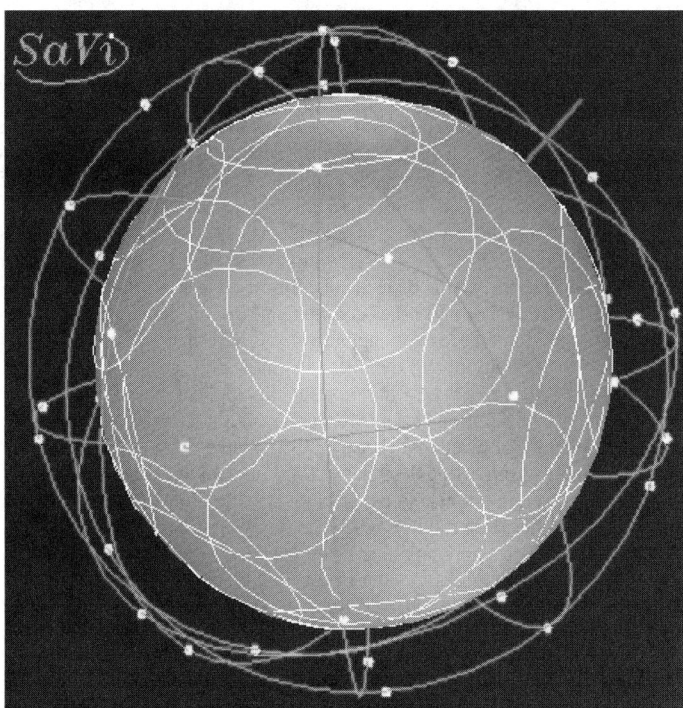

Fig. 18. The Globalstar constellation, scheduled for operation in late 1999, has eight orbital planes equipped with six active satellites in each plane, or a total of 48 active satellites (plus four on-orbit spares). (*Constellation views rendered by Lloyd Wood, using SaVi, the Satellite Visualization software created by Patrick Worfolk and Robert Thurman at the Geometry Center, University of Minnesota.*)

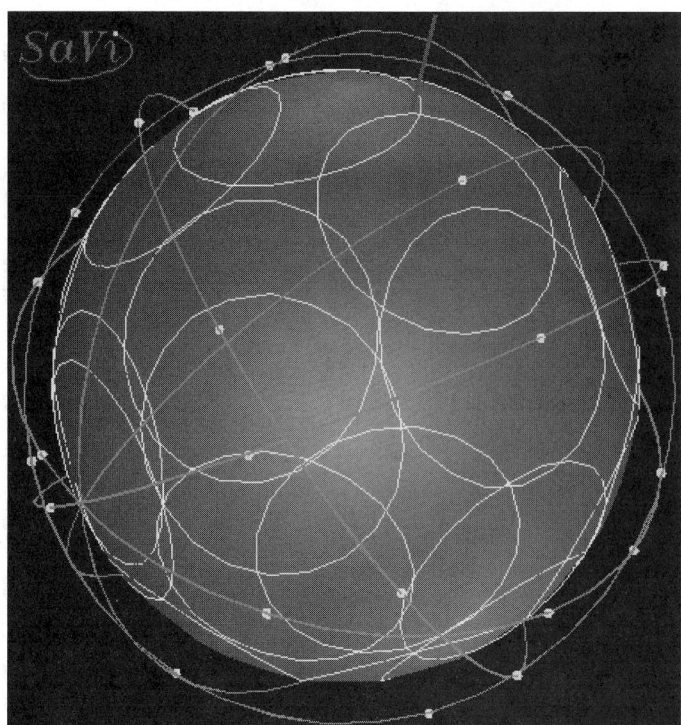

Fig. 19. The Orbomm satellite constellation has 36 active satellites in four planes of eight spacecraft each, and two higher-altitude planes of two spacecraft each. (*Constellation views rendered by Lloyd Wood, using SaVi, the Satellite Visualization software created by Patrick Worfolk and Robert Thurman at the Geometry Center, University of Minnesota.*)

Globalstar service is provided through a 48-satellite low-earth-orbiting (LEO) constellation. The system is divided into eight orbital planes having six satellites in each plane, plus four spare satellites remaining in orbit. The operating altitude is planned for 763 nautical miles (1,414 kilometers) and the orbital inclination is 52 degrees. Each satellite will circle the planet every 114 minutes. The participation by QUALCOMM means that the digital signals will be encoded using code-division multiple access (CDMA) technology rather than the frequency-division multiple access/time-division multiple access favored by many other satellite digital circuits. The project will support direct satellite links to pay phone stations equipped with *Globalstar* antenna, radio unit and optional digital telephone, but locally provided by an in-country service provider. In addition, the system includes vehicle-mounted car kits and handheld terminals. In use, the terminals will first try to connect with an existing cellular infrastructure; if that attempt fails, the units change to the satellite system. The call is then relayed via satellite down to a gateway antenna station for routing through the existing national terrestrial system to the end destination.

A unique, packets-only low-orbit system is operated by *ORBCOMM*, which placed two LEO satellites in orbit in April 1995 and reports having processed over 1.5 million messages that included personal communications and information on monitoring, positioning and tracking of commercial assets, environmental conditions and military resources on land and at sea. With 28 satellites in orbit (altitude: 508 miles or 818 kilometers) by September 1998, the company plans to have up to 36 small communication satellites launched by 1999, providing services in the USA and internationally. See Fig. 19. For launch, these compact spacecraft folded into a cylinder only 6-1/2 inches high and 41 inches in diameter and stacked eight high on a Pegasus rocket, which was launched from an aircraft at an altitude of 40,000 feet (12,200 meters). Each satellite has a mass of only 90 pounds (40 kilograms). An *ORBCOMM* satellite is designed to send and receive two-way alphanumeric packets, similar to two-way paging or e-mail, but does not handle voice. The craft is described as an "orbiting packet router" that "grabs" small data packets from sensors in vehicles, containers, vessels or remote fixed sites, and relays the packets through a tracking Earth station to a gateway control center, from which the information is sent to the addressee. Using a new hand-held, two-way text messenger, subscribers can communicate with any Internet e-mail address, sending or receiving text messages of up to 2,000 characters. The

device also integrates a *global positioning system* (GPS) receiver. This and other user devices, called "subscriber communicators," enable either monitoring of fixed, remote data assets, such as electric utility meters, oil or gas storage tanks, wells and pipelines and environmental projects, or mobile, two-way data and messaging communications. *ORBCOMM* operates in two frequency ranges: 137–138 megahertz and 400 megahertz, for transmissions down to mobile or fixed data communications devices, and 148–150 megahertz for uplinks.

Meanwhile, a much larger system involving more satellites than any other constellation system is being developed by a joint effort including Motorola, Lockheed Martin, The Boeing Company, cellular telephone pioneer Craig McCaw, Microsoft chairman Bill Gates, Saudi Prince Alwaleed Bin Talal, and Abu Dhabi Investment Company. Known as *Teledesic*, the system's initial design proposed a constellation of 840 satellites in 21 planes orbiting at an altitude of about 432 to 438 miles (695 to 705 kilometers), but this was eventually changed to 288 active satellites plus spares in 12 planes with 24 satellites per plane, and may be reconfigured again before it is implemented. See Fig. 20. The *Teledesic* network is promoted as a global, broadband "Internet-in-the-Sky™" system. Using advanced satellite technology, *Teledesic* and its partners are creating the world's first network to provide affordable, worldwide, "fiber-like" access to telecommunications services such as computer networking, broadband Internet access, interactive multimedia and high-quality voice. As of 2000, *Teledesic* service is targeted to begin in 2004.

Navigation Satellites

The various U.S. military forces have their own satellite operations offices within the Department of Defense. One of their most prominent installations is a navigation system–the *global positioning system* or *GPS*–completed in 1993. Developed and operated by the U.S. Department of Defense, it now serves thousands of civilian users as well as the armed forces. The fact that the U.S. military establishment controls *GPS* makes some European and Asian users nervous, so other systems are on the drawing boards.

The *GPS* satellites are operating in a medium orbit, about 12,000 miles or 20,200 kilometers high. At that altitude, at least four satellites are visible from any given spot on Earth at any time, day or night. *GPS* has a

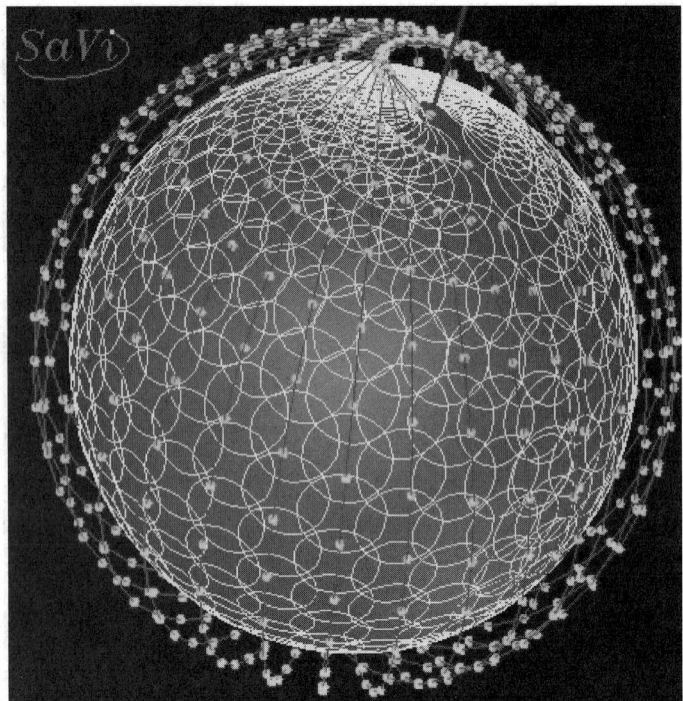

Fig. 20. The Teledesic project began with a mammoth proposal for 840 active satellites in 21 planes of 40 satellites each in low-Earth orbit. Boeing redesigned it to use 288 satellites in 12 planes of 24 satellites each. Target date for operations is 2003. (*Constellation views rendered by Lloyd Wood, using SaVi, the Satellite Visualization software created by Patrick Worfolk and Robert Thurman at the Geometry Center, University of Minnesota, and the Boeing Corporation.*)

constellation of 21 *Navstar* satellites, plus three operational spares in orbit. The first satellite was launched in 1978; it and the next nine spacecraft were developmental. They were followed by the present 24 satellites, launched from 1989 to 1994. These are distributed in six planes inclined at 55 degrees. Each satellite circles the globe in 12 hours. The average life expectancy of each *GPS* satellite is about 7 or 8 years. *GPS* employs the 1984 Worldwide Geodetic System of coordinates, similar to the latitude and longitude lines on global maps.

The Russian military establishment also has a navigation satellite system, known as *GLONASS* (Global Orbiting Navigation Satellite System). Like the American *GPS* constellation, *GLONASS* consists of 24 satellites (in 1998, only 15 were considered operational), but they are distributed in three orbital planes at an altitude of about 11,870 miles (ca. 19,100 kilometers). Although the basic concept is similar, there are important differences between *GPS* and *GLONASS*, especially in their radio signals, which are incompatible. There are dual-system receivers available for people who want to access both systems, and there are good reasons for doing so, because the 10 degree higher orbital inclination of the *GLONASS* satellites places them in portions of the sky where *GPS* satellites are never visible. Besides, the composite receiver can select from twice as many satellites, significantly improving the geometry for improved, real-time accuracy in determining location. Also, the radio frequency system used by *GLONASS* is compatible with planned future navigation aids sponsored by other countries, such as the European Geostationary Navigation Overlay System (EGNOS), which would be available to users of composite receivers. Life expectancy of *GLONASS* satellites is about half that of the *GPS* satellites, but the Russian administrators contend that by 2002 the design life will be about equal.

The *GPS* satellites are used by U.S. and allied military personnel to pinpoint the locations of troops, vehicles, and targets, using signals broadcast in an encrypted P code on two frequencies from each satellite, providing Precise Positioning Service (PPS). Commercial users employ only one frequency from each satellite's downlinks, in which a C/A code provides Standard Positioning Service (SPS) signals, which are processed in special receivers available for purchase. More than 100 different receiver models were in use by 1998. Essentially, the receiver uses the signals from three satellites to pinpoint by triangulation a two-dimensional fix (latitude and longitude), or from four satellites to determine the commercial user's

location and altitude. These are shown in a display on the receiver–but only within a space of about a city block or circa 300 feet (100 meters). The Department of Defense deliberately blurred the accuracy of the system for non-military users, except in special cases, so as to prevent abuse of the system by enemies of the U.S. Despite this, the commercial customers for the *GPS* spent $1.6 billion for its services in 1997.

By 1998, the U.S. government reported that military purchases of *GPS* equipment represented just a fraction of industry sales. Worldwide *GPS* equipment sales reached $4 billion in 1998 and are expected to double in 2000. See Fig. 21. The growing enthusiasm among commercial customers is due in part to the introduction of new services provided by *Inmarsat*, *ORBCOMM*, and other commercial satellite networks, in which their customers have access to supplementary information that reportedly improves the accuracy of the *GPS* signals, in some cases, to the diameter of a golf ball. Among some of the more public applications of the *GPS* by the military was Operation Desert Storm, when Allied troops attacked Iraqi forces to liberate Kuwait in February, 1991, using 9,000 *GPS* receivers to guide them through trackless deserts. *GPS* receivers also are used by submarines, aircraft, and warships. A major civilian project was the digging of the rail tunnel under the English Channel, in which *GPS* receivers (set to the high-precision capability with permission of the Department of Defense) helped engineers to steer excavators as they simultaneously dug from the English and French coastlines, so they met in the middle. See also **Tunnel Engineering**. Automobile manufacturers are offering moving-map displays (some with voice direction as well) guided by *GPS* receivers as an option, and many expect these systems will eventually be standard equipment.

Fig. 21. One of many GPS handheld terminals, the Magellan GPS ColorTrak receiver has nine graphic navigation screens (five can be custom-set), stores up to 500 locations, or 20 routes with 30 legs. The satellite status screen displays which satellites are being used and their relative signal strength, as well as their position overhead. The plotter screen tracks where the user has been (in red) and the user's destination (in blue). (*Courtesy GPS Store.*)

Fleet vehicles such as delivery trucks and buses often are *GPS*-equipped as a method of monitoring their locations at all times, especially in conjunction with the *ORBCOMM* system. Land surveyors, mining companies construction contractors, and motion-picture photographers rely on *GPS* for economic and precise positioning of critical activities. When combined with tiny implanted transmitters, *GPS* receivers can be used to track wildlife in the Arctic or in jungles. They may be a means of protecting helpless Alzheimer's disease patients who tend to wander away and become lost–or locating escaped criminals.

As might be expected, there are many more organizations plotting different installations of geostationary, geosynchronous, medium-orbit and low-orbit satellites. Some of these will be successfully implemented and others may never make it into space, depending on their finances and the expertness of their engineering. But there is no doubt that the satellite population will increase year by year.

The electronic interference of satellite transmission circuits with radio astronomy, as well as visual interference with optical astronomy, has raised concerns among researchers. Optical astronomers have already been irritated by this phenomenon, since the antennas of the low-orbit *Iridium* satellites have produced sun reflections (called "flares") while passing through the night sky, interrupting telescope viewing–and sometimes they are even visible as brief flashes in the sky in daylight. As more low-orbit satellites join those already in the sky, that problem may become more significant. For additional information on communication satellites, see **Telephony (Telecommunications)**. Weather (meteorological) satellites are described in the article on **Weather Technology**. Scientific and reconnaissance satellites are described in the following article.

Web References

For more current information on these various subjects, we suggest contacting the main website home pages of the different government agencies, manufacturers, and sponsors of satellite systems, as listed below in their worldwide web Universal Resource Locators (URLs):

Advanced Communication Technology Satellite project: *http://spacelink.nasa.gov/ Instructional.Materials/Curriculum.Support/Space.Science/Satellites/Advanced. Communication.Technology.ACTS/*
Astrolink project: *http://www.astrolink.com*
COMSAT Corporation: *http://www.comsat.com*
Globalstar project: *http://www.globalstar.com/*
GPS and GLONASS projects: *http://igscb.jpl.nasa.gov/projects/iglos/glonass.html*
Hughes Space & Communications Co.: *http://www.hughespace.com/flash/flash.html*
ICO Global Communications: *http://www.ico.com*
Inmarsat: *http://www.infosat.com/services/inmarsat/*
INTELSAT: *http://www.intelsat.com*
International Telecommunication Union (ITU): *http://www.itu.int*
Lloyd's Satellite Constellations (summaries of projects) "*http://www.ee.surrey.ac. uk/Personal/L.Wood*" *http://www.ee.surrey.ac.uk/Personal/L.Wood*
Motorola Corporation (for Iridium project): *http://www.mot.com*
National Aeronautics & Space Agency (NASA): *http://www.nasa.gov*
ORBCOMM project: *http://www.orbcomm.com*
PanAmSat Corporation: *http://www.panamsat.com*
Space Systems/Loral: *http://www.ssloral.com*
Teledesic project: *http://www.teledesic.com*
The Boeing Company (for Ellipso project): *http://www.boeing.com/flash.html*

RICHARD Q. HOFACKER, Jr., Bell Laboratories (retired)

SATELLITES (Scientific and Reconnaissance).

Satellites fall into three broad classifications: (1) commercially oriented missions, as represented by communication and navigation satellites described in the prior article; (2) scientifically dedicated satellites such as those found in space astronomy, cosmology, and earth sciences pertaining to the atmosphere, oceans, and terrain of the Earth; and (3) military-motivated missions. The latter are not described in this encyclopedia.

During the early years of satellite technology, rockets launched the orbiting vehicles. Many years later with the perfection of the space shuttle concept, orbiters within the handling capacity of the shuttle have been launched from that space vehicle. See also **Space Shuttle**. Power required for final orbit positioning, or when moving a given satellite to a new location, is accomplished by a power source (usually chemically propelled thrusters) that is built into the satellite by design. Control is managed by remote commands from ground stations on Earth.

Satellites generally orbit the Earth, although some satellites are research vehicles that have been designed to orbit a body other than Earth. Examples include the *Viking I* and *Viking II Orbiters*, which were designed to orbit the planet Earth, as opposed to *Viking I* and *Viking II Landers*, which actually set down on the surface of Mars. During the earlier Apollo lunar expedition, there was a lunar orbiter that was used to keep in touch with the crew on the surface of the moon. Pictures of the blind side of the moon were taken for the first time from the lunar orbiter.

More recently, satellites also refer to manned space platforms that orbit the Earth called space stations. Examples include the U.S. built *Skylab* of the 1970's, the Russian built *Salyut* series of the 1970's and 1980's, the Russian built *Mir Space Station* of the 1980's and 1990's and the multi-nationally owned and operated *International Space Station* which, as of late 1998, is under construction in space. See also **Space Stations**.

Satellite Sensing of Earth Resources

The earliest practical use of an Earth-orbiting satellite was simply that of viewing and studying Earth from the perspective that can only be achieved from a position in space far beyond that which can be attained by land-based aircraft. Weather satellites and land satellites, such as the *Landsat* program, share historical significance. The name *Landsat* was coined to differentiate the land surveillance satellite from a *Comsat* (communication satellite). Since its inception in 1965, the *Landsat* program has been highly successful in its scientific accomplishments and data collection. Remote sensing of numerous properties of Earth is central to the *Landsat* program, as well as to the Earth Science enterprise (formerly known as the *Mission to Planet Earth* program).

Early Remote Sensing Programs

Prior to the formal launching of the first earth resources technology satellites (initially called *ERTS* and later named *Landsat*), the results of earth surveys from high-altitude aircraft and balloons greatly aided scientists and conservationists in charting the features of the Earth. (It is interesting to note that, beginning in the 1990's, weather services in various countries and the National Center for Atmospheric Research, headquartered in Boulder, Colorado, USA, made extensive use of aircraft and balloons). Earth resources experiments were on the testing agenda of several of the earlier *Gemini* and *Apollo* manned space flights. Images were made of the Earth both in the visible light range and the infrared (IR) and ultraviolet (UV) portions of the electromagnetic spectrum. The results were promising and indicated that such data, properly interpreted, could assist in a wide array of programs–agricultural planning, charting the movement of sea life, monitoring concentrations of air and water pollutants, and furnishing more fundamental knowledge of the Earth in terms of its geography, cartography, forestry, geology, and anhydrology.

The History of the Landsat Program

The original concept of an earth resources monitoring satellite is generally attributed to the United States Geological Survey during the 1960's, although scientists at the United States Department of Agriculture were also thinking along similar lines during that period. In 1968, the United States Department of the Interior, which oversees the Geological Survey, requested funds to build a new satellite, which it called *EROS* (earth resources observation satellite). The Department of the Interior accepted data handling responsibility, and the National Aeronautics and Space Administration (NASA) agreed to develop and launch the satellite.

During the period that the first satellite was under development and construction, the name of the spacecraft was changed from *EROS* to *ERTS* (earth resources technology satellite). *The ERTS 1* was built into the same spacecraft frame as had been used for the *Nimbus* weather satellite of that period. As shown in Figs. 1 and 2, the spacecraft resembled an ocean

Fig. 1. The ERTS 1 (later named Landsat 1) earth resources technology satellite as it appeared in orbit after launching in 1972. The satellite operated in a polar orbit about 914 kilometers (about 570 miles) above the Earth. The orbiting observatory returned images from a multispectral scanner. A data collection system gathered environmental information from Earth-based platforms and relayed the information to a ground processing facility.

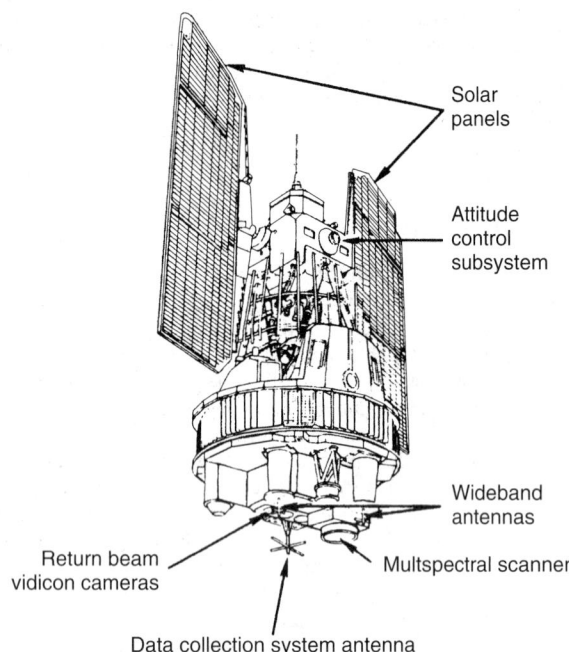

Fig. 2. Principal elements of the Landsat 1. Shortly after launching, the vidicon camera system failed. Thus, the satellite depended exclusively on the multispectral scanner (MSS), the results of which received wide scientific acclaim.

buoy, equipped with two rectangular photovoltaic "wings" to capture solar energy. *ERTS 1* was launched by NASA on July 23, 1972, on a Delta Rocket from Vandenberg Air Force Base, California, into polar orbit, which took the satellite over any given location on earth once every 18 days. Each day the satellite made 14 revolutions of the earth (about one every 103 minutes). Ground coverage proceeded westward until global coverage was completed. An active attitude control subsystem maintained the satellite's observing system within ±0.7 degree of the local vertical. The two solar panels provided 500 watts of electrical energy. Altitude of the satellite was maintained at 900 kilometers. The payload included:

- Three television cameras (return beam vidicons) for photographing the earth in color. This system failed shortly after launch.
- A multispectral scanner (MSS) was included as a secondary system. The MSS had previously been tested at high altitude in an airplane, but had not experienced orbital flight. The MSS became the primary system and performed well beyond expectations.

In addition to the viewing data, data was received from automatic data collection platforms located in the United States and Canada that concerned such local factors as stream flow, snow depth, and soil moisture. This information was relayed by the satellite to ground stations. Data from any platform was available to users within 24 hours from the time the sensor measurements were made. Data from the MSS were transmitted to tracking stations via dual wideband (2.2 giga-hertz) S-band data links for recording on magnetic tape. Narrowband telemetry provided for satellite housekeeping, the relay of data collection system data, and such payload-related data as attitude and timing. Telemetry, tracking, and command subsystems on board were compatible with tracking stations in NASA's space flight tracking and data network installed at that time.

As time permitted careful analysis of the data, users of the information showed considerable enthusiasm and the performance of *ERTS 1* was generally proclaimed a success. Thousands of images were collected and catalogued. The unusual colored views of features of the Earth were frequently seen in both the scientific and mainstream literature. Essentially, the first year of data gathering by the satellite was devoted by scientists in a variety of fields to learning how to use the data effectively, of literally coping with the immensity of the information, and in planning how to make the second satellite. Among the early accomplishments of the satellite were:

- Agriculture. Area coverage of Texas and Oklahoma categorized as to range and pasture land, forests and water; up to 11 types of crops identified for fields 20 acres (7 hectares) or larger for San Joaquin Valley and Imperial Valley, California.

- Cartography. Inaccuracies in map locations of Brazilian lakes and rivers discovered; unmapped lakes located in Iran. Geology. Nearly 50% additional unmapped faults found branching off the San Andreas Fault in California previously thought to be an accurately mapped seismologically active area.
- Hydrology. Shallow substrate water-bearing rocks detected in Nebraska, Illinois, and New York; polluted water drift charged off the coast of New Jersey; extent of Mississippi River flooding determined.
- Oceanography. Uncharted reefs detected off Bahaman Islands and Australian coast; navigation charts for Bahamas to depths of 4 to 8 meters (13 to 26 feet) produced more accurately than any prior maps.
- Urban Development. Dallas-Fort Worth images showed new roads, reservoirs, suburbs, and airports not on maps made three years prior.

A second satellite, *ERTS B*, was launched on January 22, 1975, and shortly thereafter was named *Landsat 2*. The original *ERTS 1* was renamed *Landsat 1*. A third satellite, *Landsat 3*, was launched on March 5, 1978. Design modifications were focused on improvements to the imaging system. By 1982, only one of the satellites, *Landsat 3*, remained active but the first of a second generation of surveillance satellites (*Landsat 4*, which was later renamed *Landsat D*) was launched that same year. The new satellite incorporated an improved multispectral scanner and a thematic mapper. See Figs. 3 and 4.

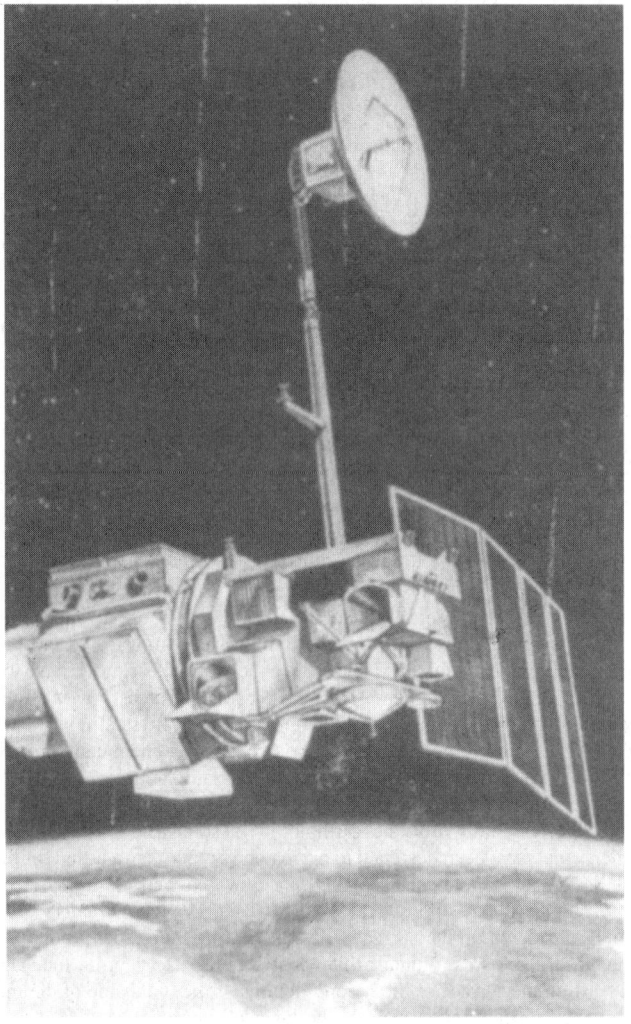

Fig. 3. View of Landsat D in orbit (artist's sketch). The satellite was launched into a sun-synchronous, polar orbit approximately 700 kilometers (435 miles) high in 1982. The satellite carried two remote sensing instruments as shown in Fig. 4. The satellite is sometimes referred to as Landsat 4. (*RCA Astro-Space Division, General Electric.*)

Landsat 1980–1998

Plans had been announced in the late 1970's for three additional *Landsat* satellites (*D'*, *''*, and *'''*) to round out the second generation of vehicles. In

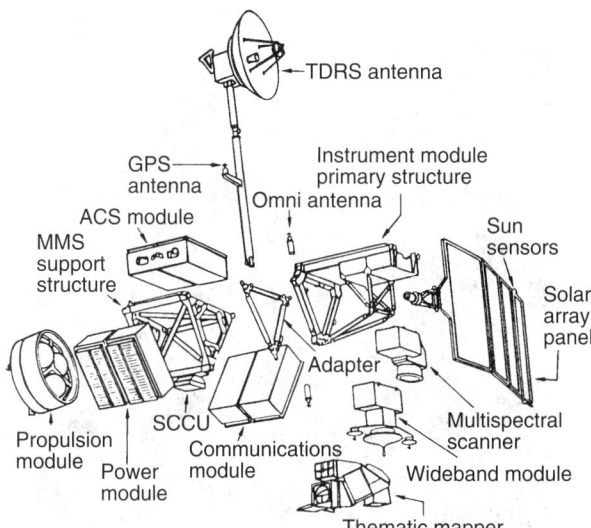

Fig. 4. Expanded view of Landsat D (or 4). The satellite carried a thematic mapper (TM), an experimental sensor of advanced design to provide scenes with 30-meter (about 100 feet) resolution. It is a 7-channel radiometer. Satellite also carried a multispectral scanner (MSS) with 4 channels, 80-meter (262.5-foot) resolution, which is identical to the sensors on Landsats 1 and 2. Landsat D had the capacity to generate 800 scenes per day (550 MMS; 250 TM) for all ground stations, as compared with 190 MSS scenes from Landsats 1, 2, and 3. (*National Aeronautics and Space Administration.*)

the very early 1980's and particularly after a change in Administration in 1981, interest in the program by Congress and the Administration waned, although a number of strong supporters remained. The use of data by local and state governments, other governments worldwide, and by private groups did not measure up to earlier expectations. Also, there had been a series of technical difficulties and disappointments. Much of the imagery of the world from space had been accomplished as witness the frequent use of images made by the earlier *Landsats*. Thus, with the exception of very specific projects, data already on hand served many needs. Financial support for the program as derived from the sale of data did not appear to be forthcoming. But the principal deterrent to the program was the Administration's plan to turn to the program over to the private sector. Nevertheless, *Landsat D'* symbol, which had been ready to launch for a considerable period, was indeed launched in March 1984.

In 1983, the Administration formally put the *Landsat* program out to bid to the private sector. A subsidy from the United States Government in the amount of $250 million to aid operating costs for a period of five years was an important part of the negotiations. In 1985, the Earth Observation Satellite Company (EOSAT), a partnership of Hughes Aircraft and RCA, was selected by the National Oceanic and Atmospheric Administration (NOAA) to operate the *Landsat* system under a ten year contract. Hughes announced in February 1986 that EOSAT would operate *Landsats 4* and *5*, (the latter formerly referred to as *D'*) and would be the primary source for marketing, ordering, and distributing the data from the satellites. EOSAT would commence construction of *Landsats 6* and *7* vehicles. The latter would be built by RCA and would carry imaging devices built by Hughes. See Fig. 5.

By mid-1990, Hughes Aircraft announced that testing had commenced on the *Landsat 6* mapper sensor. The spacecraft would orbit the Earth at the same altitude, inclination and equatorial crossing time as *Landsats 4* and *5*; however, the enhanced thematic mapper sensor would provide improved spatial resolution capable of discerning objects smaller than a tennis court.

In February 1992, the National Space Policy Directive No. 5, signed by President George Bush, formally recognized the importance of *Landsat* data. The Department of Defense (DoD) and NASA were instructed to develop and launch *Landsat 7* with performance capabilities at least equal to *Landsat 6*. In addition, DoD and NASA were instructed to define continuity requirements plus define a management plan that addressed future funding and operations responsibilities. The DoD accepted responsibility for the space segment, and NASA accepted responsibility for the ground segment. Later that year, the Land Remote Sensing Policy Act of 1992 recognized that commercialization of *Landsat* had not been successful. The *Landsat* Program Management team (LPM) was authorized to procure *Landsat 7* and to negotiate with contractors on data policy for Landsats 4–6.

The DoD signed a contract with General Electric (now Lockheed Martin Missiles and Space) for the construction and launch of *Landsat 7*. *Landsat 7* successfully launched onboard a Boeing Delta II on April 15, 1999. The satellite will carry an Enhanced Thematic Mapper-Plus sensor. This will enable the satellite to seasonally monitor small but significant processes on a global scale, such as the annual cycles of vegetation growth; deforestation; agricultural land use; erosion and other forms of land degradation; snow accumulation and melt and the associated fresh-water reservoir replenishment; and urbanization. NASA is responsible for the development and launch of the *Landsat 7* satellite and the development of the ground system. The *Landsat* Project at Goddard Space Flight Center manages these responsibilities. Hughes Santa Barbara Remote Sensing has contracted to build the sensor and Lockheed Martin Missiles and Space has contracted to develop the spacecraft. NOAA is responsible for operation and maintenance of the satellite, as well as the ground system for the life of the satellite. The United States Geological Survey (USGS), on behalf of NOAA, will capture, process, and distribute the data. The USGS is additionally responsible for maintaining an archive of *Landsat* data.

Landsat 7 is part of the global research program formerly known as NASA's *Mission to Planet Earth*, now called the *Earth Science enterprise*, which is a long-term program that is studying changes in Earth's global environment. The mission of the Earth Science enterprise is to study and

Fig. 5. Artist's concept of the Landsat satellite over Earth. (*NASA.*)

expand the knowledge of the total Earth system and the effects of change, both natural and human-induced, on the global environment.

Additional Remote Sensing Technologies

High-Altitude Aircraft. These are still used, using X-band, 3-centimeter, and L-band, 25 centimeter radar, as well as thermal infrared multispectral scanners. An 8-channel scanner has been used effectively to gather mineral signature information at wavelengths between 8 and 12 micrometers. Results indicate that igneous rock units can be identified from their free silica content and that carbonate as well as clay-bearing units are readily separable on digitally processed images. Considerable use of this tool has been made in the Death Valley region of California and the Cuprite area of Nevada.

Orbiting Laboratories. This type of craft known as space stations, for example the *Mir Space Station*, and the *International Space Station* (*ISS*), which is currently under construction, can be highly useful. See also **Space Stations**. Earth sciences research in land, ocean and atmospheric processes can be performed to document transient and long-term natural and human-induced changes in the global environment. The data collected on the *Mir*, and later on the *ISS*, will add to the documentation of atmospheric conditions, ecological, and seasonal changes over long time periods. During long-duration missions, astronauts and cosmonauts perform routine observations and photography of Earth. The data collected augments the 35-year database of Earth imagery gathered by the *Landsat* programs. Unlike the *Landsat*, which covers the globe completely, space station coverage is site specific. Site selection is based on joint US/Russian activity and long-term planning which has identified known regions of interest. Within those parameters, coverage is based on the trajectory and attitude of the space vehicle, current events, and weather patterns. Ad hoc photography of special, unpredictable events, such as tropical storms, floods, forest fires, volcanic eruptions and dust storms, can be requested of the crew real-time. A specific focus of current missions is to document the aftereffects of the 1998 El Niño event.

The Space Shuttle. An ocean color experiment (OCE), designed to map ocean features with an 8-channel scanning radiometer, was installed on one of the early Shuttle excursions. The primary aim of the OCE was to detect phytoplanktonic algae on a global basis and to determine the chlorophyll pigment concentrations. The OCE was set up to focus mainly on deep-water areas in contrast with coastal waters. Ocean images were obtained at three widely separated locations: Yellow Sea, Gulf of Cadiz, and Grand Bahama Bank. The researchers agreed that the method of using chlorophyll concentrations as a tracer may be applied in large areas. Plankton patches are natural drifters and can be tracked by satellites. Thus the color scanner proved its utility in studies of both ocean circulation and biological processes.

The database of the Space Shuttle Earth Observations Project (SSEOP) contains location records and descriptions of over 250,000 of the photographs that the astronauts took of Earth. These photographs show:

1. volcanic eruptions,
2. transatlantic duststorms,
3. continental-scale smoke palls,
4. deforestation grids in the rain forests of Brazil,
5. the bleeding of Madagascar's red soil out the Betsiboka Estuary,
6. plankton blooms tens of miles long,
7. the rise of the Great Salt Lake,
8. the fall of the Aral Sea, and
9. the effects of El Niño from droughts in Australia to floods in California.

For representative pictures of topographical conditions, see Figs. 6, 7, and 8.

Ocean-Atmosphere Specific Satellites. The Remote Ocean Sensing System (NROSS) was a United States Navy program of the early 1990's which was designed to provide information on ocean waves and eddies, as well as research data on surface winds. *TOPEX/Poseidon* was a joint United States-France mission that was launched during the same time period. See Fig. 9. The goal of *TOPEX/Poseidon* was to monitor global ocean circulation over a number of years in order to discover the relationships between the oceans and atmosphere, and improve climate predictions on a global scale. The *TOPEX/Poseidon* satellite remains in operation today. Every 10 days it measures global sea level

Fig. 6.   SSEOP—The Mir over Indian Ocean low pressure system. (*NASA*.)

Fig. 7.   SSEOP—Typhoon Violet, W. Pacific Ocean. (*NASA*.)

with unparalleled accuracy. This satellite was designed to carry a high-precision altimeter to measure the topography of the ocean surface, which results from the combined effects of wind, currents, and gravity. *TOPEX/Poseidon's* radar altimeter provided very precise measurements of sea surface heights, and the satellite data indicates that during 1994–95, sea levels increased. Although the data is not conclusive, continued increases in sea levels could worsen the effects of hurricanes and other storms, threatening coastal regions.

The combined data from *NROSS* and *TOPEX/Poseidon* yielded for the first time a comprehensive global description of oceanic circulation. A third and fourth satellite focusing on biology and geodesy, respectively, extended these studies. A later satellite, *Seasat*, equipped with synthetic aperture L-band radar, furnished important oceanographic information. See also **Ocean**.

Heat Capacity Mapping. A satellite of the same name (*HCMM*) performed heat capacity mapping. It provided thermal and visible images for mapping the thermal inertia of surface materials. See Figs. 10 and 11. Scientists from several countries including Australia, Canada, France, Germany, Italy, Morocco, Spain, Switzerland, and the United Kingdom, participated in investigations involving the *HCMM*. The satellite was used in connection with:

1. Urban heating patterns;
2. Freeze damage assessment with development of planting date advisories;
3. Evapotranspiration rates;

Fig. 8. SSEOP—North American and Pacific Tectonic Plates: An exceptional southeastward view of the boundary between the North American and Pacific tectonic plates in central California. On the west are the forested Sierra Nevada and the Central Valley, which formed during collision of the two plates about 90 million years ago. The collisional margin evolved to a transform one about 30 million years ago; the San Andreas fault is the best-known element of that transform fault system. Lake Tahoe and Mono Lake are visible in the eastern Sierra Nevada, beyond which extends the Basin and Range province of stretched continental crust. (*NASA.*)

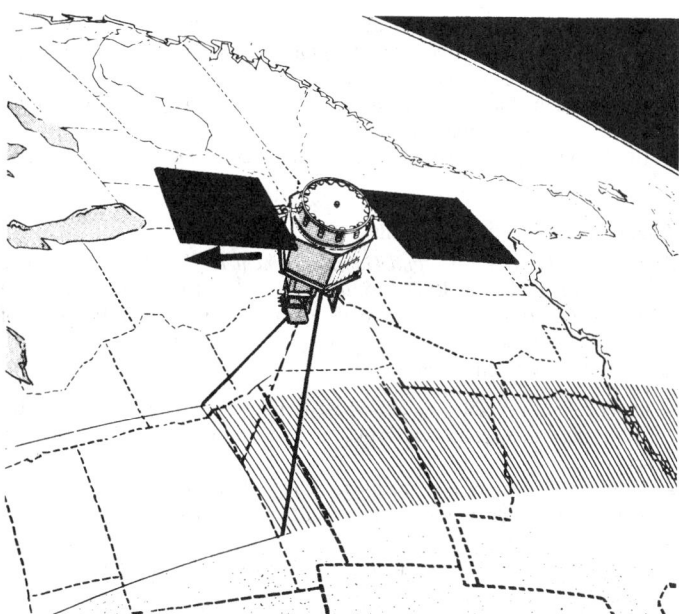

Fig. 10. *HCMM* (Heat Capacity Mapping Mission), the first spacecraft built to test the feasibility of measuring variations in the earth's temperature was launched on April 26, 1978. Purpose of mission–to produce thermal maps for discrimination of rock types, mineral resources, plant temperatures, soil moisture, snow fields, and water runoff. The experimental satellite traveled in a circular, sunsynchronous, 620-kilometer (385-mile) orbit that allowed for measuring mid latitude test areas of the earth's surface for their minimum temperatures and then measuring those same areas for maximum temperatures about 11 hours later. (*National Aeronautics and Space Administration.*)

Fig. 9. Mission to Planet Earth (1995) TOPEX/Poseidon is an altimeter-type satellite; the illustration shows how such satellites, working with tide-gauge networks and land-based and orbital geodetic location systems can enable accurate, long-term, continuous monitoring of climate and sea level. (*NASA.*)

4. Soil moisture assessment
5. Thermal mapping for discrimination of geologic units and energy or mineral resource areas, and evaluation of thermal modeling and satellite mapping techniques;

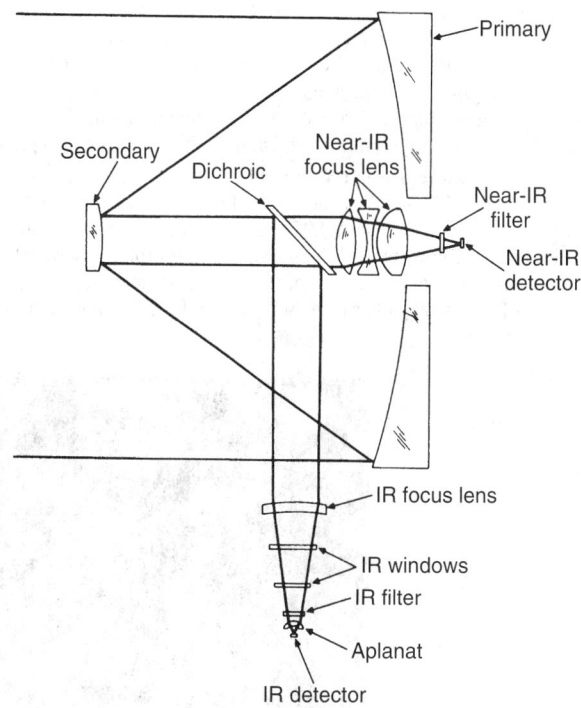

Fig. 11. Optical system of *HCMM* spacecraft. The instrument module was located on the spacecraft so that the Heat Capacity Mapping Radiometer (HCMR) was earth-pointing. The base radiometer instrument was a modified spare Surface Composition Mapping Radiometer similar to that flown on Nimbus 5. The HCMR had a small geometric field of view (less than 1 (1 milliradians), high radiometric accuracy, and a wide enough swath coverage on the ground so that selected areas could be covered within the 12-hour period corresponding to maximum and minimum temperatures.

The instrument operated in 2 channels, at approximately 0.5 to 1.1 micrometers (visible and near-infrared) and 10.5 to 12.5 micrometers (infrared), providing measurements of reflected solar energy and equivalent black-body temperatures.

6. Detection of high potential groundwater pollution;
7. Studies of snow hydrology programs;
8. Studies of estuarine currents;
9. Topoclimatological and snow-hydrological survey of various countries, such as Switzerland;
10. Monitoring of large-scale pollution effects in the North Sea among several others.

Earth Radiation Budget Satellite

The *Stratospheric Aerosol and Gas Experiment II* (*SAGE II*) is situated on the Earth Radiation Budget *Satellite* (*ERBS*). The mission objective of the SAGE II instrument is to determine the vertical distribution of stratospheric aerosols, ozone, nitrogen dioxide, and water vapor on a global scale. Specific objectives are:

- To develop a viable, satellite-based, remote sensing technique for measuring stratospheric aerosols, nitrogen dioxide, water vapor, and ozone;
- To utilize these measurements to study global circulation, transient stratospheric phenomena, sources and sinks, and long term trends of individual species;
- To investigate aerosol optical properties from the four wavelength aerosol extinction measurements.

The Mission to Planet Earth/Earth Science Enterprise

The title "Mission to Planet Earth" originated over a decade ago in a report on the U.S. civil space program by a commission led by former astronaut Dr. Sally Ride. The term summarized the concept of looking at Earth as NASA looks at other planets, with a focus on ecosystems. In 1998, the name *Mission to Planet Earth* (*MTPE*) was changed to *Earth Science enterprise* (*ESE*) to signify the long-term commitment by NASA to understanding the global implications of natural and human-induced changes to the environment. The *ESE* effort involves extensive interagency and international cooperation. Among the international participants are the U.S./France satellite *TOPEX/Poseidon*, discussed earlier, and the Russian satellite, *Meteor-3*, which carries the *ESE* instrument known as the *Total Ozone Mapping Spectrometer* (*TOMS*). The *ESE* uses satellites and other tools to generate data about the ecosystem of the Earth. *ESE* studies are expected to provide improved weather forecasts, along with tools for managing agriculture and forests (with particular attention to the rain forests), as well as information for the fishing industry and coastal planners.

Upper Atmosphere Research Satellite. The *Upper Atmosphere Research Satellite* (*UARS*) is considered the first major element in NASA's Earth Science enterprise. The *UARS* was deployed from the shuttle Discovery into a 585 kilometer, 57 degree inclination orbit on September 15, 1991. Its mission was to carry out the first systematic, comprehensive study of the stratosphere, as well as to furnish new data on the mesosphere and thermosphere. The Earth's upper atmosphere, beginning only 10–15 kilometers above the surface, remains a frontier largely unexplored from space.

The *UARS* provides simultaneous, coordinated measurements of the internal structure of the atmosphere including trace constituents, physical dynamics, radiative emission, thermal structure, and density. Additionally, the *UARS* provides measurements of the external influences that can act upon the upper atmosphere such as solar radiation, tropospheric conditions, and electric fields. The combination of orbit and instrument design will provide nearly global coverage. See Figs. 12 and 13.

The UARS mission objectives include the study of:

1. Upper atmosphere energy input and loss,
2. Upper atmosphere global photochemistry,
3. Upper atmosphere dynamics,
4. Coupling among these processes, and
5. The coupling between the upper and lower atmosphere.

A significant accomplishment of the *UARS* project was the mapping of an ozone-destroying radical, Chlorine Monoxide (ClO), within the Arctic region. The extent of ClO formation and its close association with polar stratospheric cloud formation was demonstrated. This was an important confirmation of earlier aircraft results, but it also showed the extent of the region with elevated ClO. Since these initial observations, *UARS* has continued to monitor both the late winter-spring ozone depletions in both the Arctic and Antarctic regions. *UARS* data showed that the Northern Hemisphere depletion in January-March 1996 was the largest ever recorded. Data from the *UARS* provided conclusive evidence that human-produced chemicals, including chlorofluorocarbons, are breaking down in the stratosphere and causing depletion of ozone over the Antarctic.

Shuttle-Based Tools. The Shuttle-based *Space Radar Laboratory* (*SRL*), which flew two 1994 missions, is another *ESE* tool that enabled an international team of scientists to observe shifting boundaries between temperate and boreal (northern) forests. These scientists analyzed the observations to study how such changes might affect changes in the climate. Results of the observations and analysis are also being used for map-making, study and interpretation. *SRL* data, along with aircraft and ground data, was also utilized by U.S. and Canadian scientists for the Boreal Ecosystem/Atmosphere Study of northern forests, a large scale investigation into how the forests and the atmosphere interchange energy, heat, water, carbon dioxide and other trace gases.

SeaStar Satellite. Orbital Imaging, a private U.S. company, launched the *SeaStar* satellite carrying the *Sea-viewing Wide Field-of-view Sensor* (*SeaWiFS*), in August 1997. See Fig. 14. The *SeaWiFS project* is an important part of NASA's *Earth Science enterprise*. Its goal is to obtain quantitative data on global ocean bio-optical properties over a 5-year period to aid researchers in the Earth science community. The

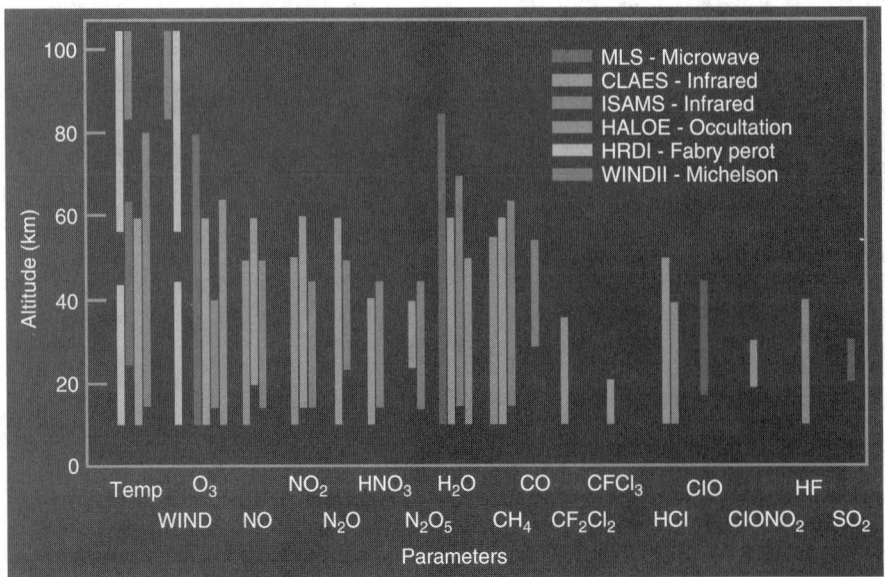

Fig. 12.   The Upper Atmosphere Research Satellite (UARS) contains chemistry and dynamics sensors that make measurements of temperature, pressure, wind velocity, and gas species concentrations in the altitude ranges shown in the figure above. (*NASA*.)

A program to study
global ozone change

Fig. 13.   The Upper Atmosphere Research Satellite (UARS) mission is to study global ozone change. (*NASA.*)

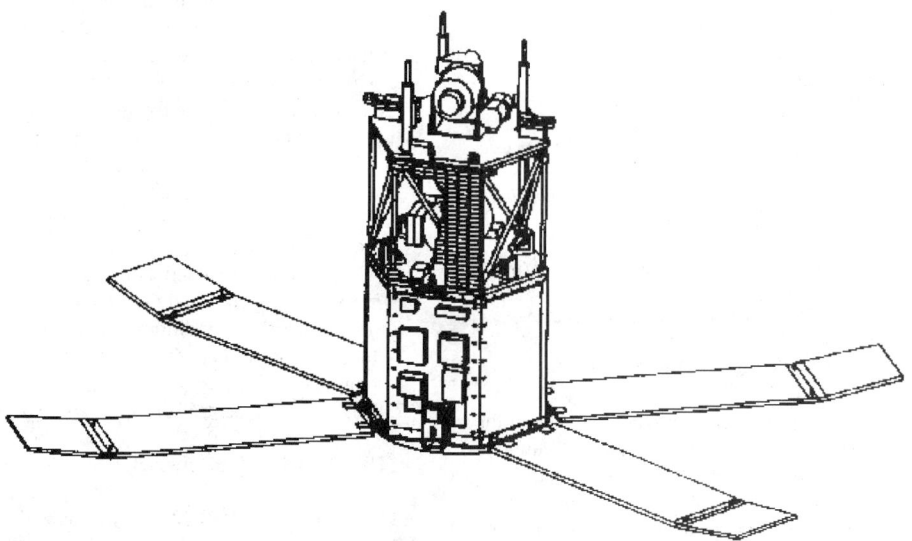

Fig. 14.   SeaStar Satellite, with its solar panels deployed, on which the Sea-viewing Wide Field-of-view
Sensor (SeaWiFS) ocean color sensor will be deployed. (*NASA.*)

SeaWiFS project will develop and operate a research data system that will process and distribute data received from an Earth-orbiting ocean color sensor. Subtle changes in ocean color are indicative of the presence of various types and quantities of marine phytoplankton, microscopic marine plants. The measurement of color, and its analysis, will enable researchers to determine the chlorophyll content of the ocean. Ultimately, scientists will be able to assess the productivity of the world's oceans. Goddard Space Flight Center (GSFC) will develop the data processing and archiving and has primary responsibility for the integrity of the final products.

The *Earth Science enterprise* will be an on-going effort for the foreseeable future as Phase II gets underway in 1999. The *ESE* strives to understand all earth systems, including its physical, chemical and biological components. To achieve this goal, the comprehensive *Earth Observing System* (*EOS*) is being developed. See Fig. 15. It consists of sensors being flown on a number of foreign and domestic satellites. *EOS* is the first integrated satellite/surface research system for observing how the Earth's parts (air, water, land and life) interact with each other. The *Earth Sciences enterprise* also includes the *EOS* Data and Information System (EOSDIS) and an active research and analysis program. The existing *Mir space station* complex affords several unique capabilities that will not be available from the *EOS* platforms. The moderate inclination orbit of 51.6 degrees of the *Mir* does not allow for the same global coverage as a polar orbit, but it does permit frequent revisits (approximately 3 days out of 4) to selected sites at mid-latitudes. It is anticipated that the *International Space Station* will provide the same excellent opportunities for earth science study.

Fig. 15.   Phase II of the Mission to Planet Earth, now called Earth Science enterprise, will begin in 1999 with the launch of the first Earth Observing System (EOS) spacecraft, EOS AM-1. (*NASA.*)

Web References

ACRIMSAT: *http://acrim.jpl.nasa.gov/*
AQUA: *http://eos-pm.gsfc.nasa.gov/*
Earth Science Enterprise: *http://www.esa.int/export/esaCP/index.html*
FLORIDA TODAY Space Online January 22, 1998, "Mission to Planet Earth name changed to Earth Science enterprise" *http://www.flatoday.com/space/explore/releases/1998/n98012.htm*
Goddard Space Flight Center–gateway to Landsat page: *http://landsat.gsfc.nasa.gov/*
JASON-1: *http://gaia.hq.nasa.gov/esemissions/launch.cfm?lau id =12*
Landsat Link page: *http://ls7pm3.gsfc.nasa.gov/links.html*
Landsat surveillance photos from the National Oceanic and Atmospheric Association: *http://www.nnic.noaa.gov/SOCC/gallery.htm   http://www.nnic.noaa.gov/SOCC/gallery.htm*
NASA Human Space Flight: *http://spaceflight.nasa.gov/*
Office of Earth Sciences–NASA Johnson Space Center: *http://www.jsc.nasa.gov/*
NASA Earth Observing System Project Science Office Homepage: *http://eospso.gsfc.nasa.gov/eospsohomepage.html*
http://eospso.gsfc.nasa.gov/eoshomepage/mischtml/nasafacts.html NASA Fact Sheets
SeaWiFS Project Homepage: *http://seawifs.gsfc.nasa.gov/SEAWIFS.html*
TERRA: *http://terra.nasa.gov/*
TOPEX/Poseidon Homepage: *http://topex-www.jpl.nasa.gov/* and *http://podaac.jpl.nasa.gov/toppos/indexold.html*
TROPICAL RAINFALL MEASURING MISSION: *http://trmm.gsfc.nasa.gov/*

SATURATED EDIBLE OILS.   See **Vegetable Oils (Edible).**

SATURATED VAPOR.   A vapor whose temperature corresponds to the boiling temperature at the pressure existing on it. A vapor is saturated when its temperature is a function of its pressure alone. A saturated vapor may be wet or dry; the term does not necessarily imply a wet vapor. A vapor of 100 percent quality, having no superheat, is said to be dry and saturated.

The physical attributes of saturated steam are pressure, temperature, volume, enthalpy, and entropy. These are always given for steam that is dry and saturated, leaving the reader to apply the quality factor when it occurs. The increase of volume on vaporization and the latent heat of evaporation are present in wet steam to the extent of the percent dryness of the steam. One of the important entries in the saturated steam table is that for atmospheric pressure. At 14.7 psi (1 atmosphere) absolute pressure, the saturation temperature of steam is 212 °F (100 °C). The heat contained in it as a boiling liquid is 180 Btu (45.4 kilogram-calories) (above 32 °F (0 °C)), and its latent heat of evaporation is 970.2 Btu/pound (539.1 kilogram-calories/kilogram).

SATURATION.   The state of being satisfied, or replete, or the action of bringing about that state. Following are specific uses of this term, some applying to single substance, entity, or region, and others to relations between more than one:

1. The condition in which the partial pressure (i.e., the pressure of a single component of a gaseous mixture, according to Dalton's Law) of any fluid constituent is equal to its maximum possible partial pressure under the existing environmental conditions, such that any increase in the amount of that constituent will initiate within it a change to a more condensed state. In molecular-kinetic terms, saturation is attained when the rate of return of molecules of a substance from the dissolved liquid or vapor phase to the more condensed parent phase is exactly equal to the rate of escape of molecules from the parent phase. In meteorology, the concept of saturation is applied, almost exclusively, to water vapor as a constituent of the atmosphere.
2. The term applied to the maximum current that will pass through a gas under definite conditions of ionization.
3. The attribute of any color perception possessing a hue, that determines the degree of its difference from the achromatic color perception most resembling it.
4. In a nuclear reactor, the maximum activity obtainable by activation in a definite flux.
5. The maximum magnetization (or the maximum permanent magnetization) of which a body or substance is capable.
6. The process or condition of dissolving in a solvent all of a solute that the solvent can absorb, under equilibrium conditions at a given temperature.
7. The complete neutralization of an acid or base.

SATURATION (Atmospheric).   See **Precipitation and Hydrometeors.**

SATURATION CURRENT.   The ionization current that results when the applied potential is sufficient to collect all ions; the maximum current that will pass through a gas under definite conditions of ionization. The saturation current is a measure of the charge carried by the ions produced in each second, and hence may be used as a measure of the radioactivity of a substance.

SATURN.   Sixth planet from the Sun, Saturn is unique in the solar system in that it is the only planet lighter than water, with a density of about 0.7 gram per cubic centimeter. Saturn is the second largest planet in the solar system, with a volume 815 times that of Earth, but with a mass only 95.2 times greater. Like Jupiter, Saturn's rapid rotation has caused the planet to be flattened at its poles. The equatorial radius is 60,330 kilometers (37,490 miles), while the polar radius is smaller — 54,000 kilometers (33,554 miles). The surface gravity of Saturn is 1.14 (Earth = 1.0). Saturn requires 29.46 Earth-years to complete one orbit around the Sun. Although a Saturnian year is long, its days are short, lasting only 10 hours, 39 minutes, 24 seconds, as first determined by

Voyager 1 and later reconfirmed by *Voyager 2*. See also **Voyager Missions to Jupiter and Saturn**.

In its slow orbit around the Sun, Saturn is perturbed by the other planets, notably Jupiter, so that its orbital path is not strictly elliptical. The planet wavers in its distance from the Sun in a region of between 9.0 AU and 10.1 AU. (One AU or astronomical unit equals distance from Sun to Earth, i.e., 149,597,860 kilometers; 92,955,806.8 miles.)

Saturn receives only about one-hundredth as much sunlight as is received by Earth. Like the three other gas giants (Jupiter, Neptune, and Uranus), Saturn does not have a solid surface comparable to that found on the terrestrial planets. Rather, it is a huge, multilayered globe of gases, notably a mixture of about 11% (mass) of helium, with nearly all the rest being hydrogen. There may be a small core predominantly of iron and rocky material. Gravity field analyses and temperature-profile measurements suggest that Saturn's core may extend out from the center by about 13,800 kilometers (8,575 miles), making it twice the size of Earth, but extremely small in relation to the huge size of the planet. It has been estimated that the core is probably so compressed that it may contain from 15 to 20 times the mass of Earth. It has been postulated that a layer of electrically conductive metallic hydrogen may surround the core. This form of hydrogen has not been observed on Earth because of the immense pressure required to produce it. The interaction between this inner atmosphere and the outer layers (hydrogen and helium) may explain Saturn's emission of heat. Both Saturn and Jupiter release about twice the amount of energy they receive from the Sun. But scientists observe that the two planets probably produce their energy in different ways. Whereas it is postulated that Jupiter emits energy left over from the gravitational contraction that occurred when the planet was formed (estimated about 4.6 billion years ago), it is suggested that Saturn's heat production may be the result of the separation of hydrogen and helium in the outer layer, with heavier helium sinking through the planet's liquid hydrogen interior. Other statistics are given in the entry on **Planets and the Solar System**. See Figs. 1, 2, and 3.

Missions to Saturn. To date, the environs of Saturn have been explored by three U.S. spacecraft, and by the Hubble Space Telescope, as described in the next several paragraphs. Most of the progress made during the past several years has been the analysis of massive amounts of data sent back to Earth from the Hubble Space Telescope and the former missions.

Saturn has been a frequent target of the Hubble Space Telescope, which has produced stunning views of long-lived hurricane-like storms in Saturn's atmosphere. The world's major telescopes, including Hubble, were recently trained on Saturn to observe the phenomenon known to astronomers as a Saturn ring plane crossing. The rings seen edge-on from the Earth's perspective on May 22, 1995, August 10, 1995, and February 11, 1996. Ring plane crossings provide astronomers with unique views of the Saturnian system. The Hubble Space Telescope more recently discovered "Great White Spot" that appears to be representative of periodic disturbances in the planet's atmosphere. This is described toward the end of this article. Considerable research from Earth also has gone forward with microwave radar imaging and is discussed later. See also **Hubble Space Telescope (HST)**.

The latest mission to Saturn and the most ambitious effort in planetary space exploration ever mounted. The *Cassini* spacecraft, was launched from Cape Canaveral, Florida, USA on October 15, 1997. *Cassini* is an international mission resulting from the joint efforts of NASA, the European Space Agency (ESA) and the Italian Space Agency, Agenzia Spaziale Italiana (ASI). *Cassini* is sending a sophisticated robotic spacecraft to orbit the ringed planet and study the Saturnian system. *Cassini* successfully performed a flyby of the planet Venus on April 26, 1998 coming about 284 kilometers (176 miles) from the Venusian surface. The flyby gave the *Cassini* spacecraft a boost in speed of about 7 kilometers per second (about 4 miles per second), which will help the spacecraft reach the planet Saturn as projected in July 2004. In its long trajectory to Saturn, *Cassini* will perform another flyby of Venus in June 1999, one of Earth in August 1999, and one of Jupiter in the year 2000. All of the flybys impart more speed to the spacecraft, which will enable it to reach its final destination of the Saturnian system. This flight pattern is referred to as a VVEJGA (Venus-Venus-Earth-Jupiter Gravity Assist) trajectory. It is anticipated that *Cassini* will enter orbit around Saturn in 2004 and remain in orbit for a minimum of four years. The goal of the *Cassini* mission primarily consists of delivering *Huygens*, a scientific probe, to Titan, Saturn's largest moon, and then remaining in orbit around Saturn to conduct detailed studies of the planet and its rings and satellites. The principal objectives are to:

1. determine the three-dimensional structure and dynamical behavior of the rings;
2. determine the composition of the satellite surfaces and the geological history of each object;

Fig. 1.   View of Saturn as seen with an Earth-based telescope (100 inch; 254 centimeters). (*Jet Propulsion Laboratory, National Aeronautics And Space Administration.*)

Fig. 2.   Image of Saturn showing three of its moons, Enceladus, Dione, and Tethys, that was taken by Voyager 1 on July 24, 1980 from a distance of 106 million kilometers (63.6 million miles). North is at 1 : 30. Bands can be seen in the Northern Hemisphere and structure can be identified in the rings, including the Cassini Division, toward the outer edge of the rings. The satellites, which can be seen in this image, are Enceladus, near the left edge of the rings, Dione, below Saturn, and Tethys, at the right edge of the frame. Saturn is 120,000 kilometers (72,000 miles) in diameter. (*Jet Propulsion Laboratory, National Aeronautics And Space Administration.*)

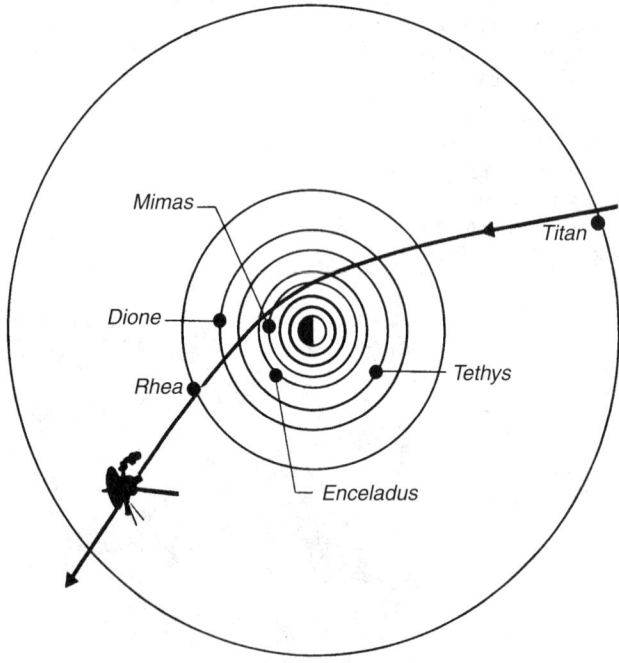

Fig. 4.   Trek of *Voyager 1* as it swept past Saturn in the fall of 1980. (*Jet Propulsion Laboratory.*)

Fig. 3.   Several dark spoke-like features can be seen across the broad B ring (left of planet). The moons Rhea and Dione appear as dots below and below left of Saturn, respectively. This photo was taken July 21, 1981, when *Voyager 2* was 33.9 million kilometers (21 million miles) from the planet. The spacecraft made its closest approach to Saturn on August 25, 1981. (*Jet Propulsion Laboratory, National Aeronautics And Space Administration.*)

3. determine the nature and origin of the dark material on Iapetus' leading hemisphere;
4. measure the three-dimensional structure and dynamical behavior of the magnetosphere;
5. study the dynamical behavior of Saturn's atmosphere at cloud level;

6. study the time variability of Titan's clouds and hazes; and,
7. characterize Titan's surface on a regional scale.

Cassini's instrumentation consists of: a composite infrared spectrometer, a CCD imaging system, a cosmic dust analyzer, a magnetospheric imaging instrument, an ion/neutral mass spectrometer, a magnetometer, a plasma spectrometer, a radar mapper, a radio and plasma wave experiment, an ultraviolet imaging spectrograph, and a visible/infrared mapping spectrometer. Telemetry from the communications antenna and other special transmitters will be used to make observations of the atmospheres of Titan and Saturn. They will also be used to measure the gravity fields of the planet Saturn and its satellites. See also **Cassini Mission to Saturn**.

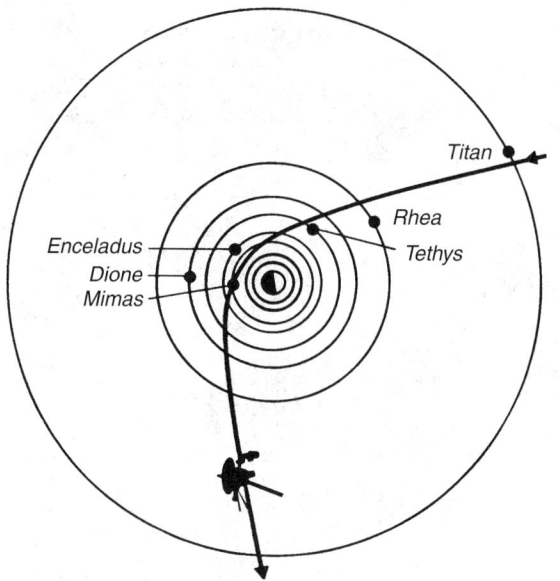

Fig. 5.   Trek of *Voyager 2* as it swept past Saturn in late-summer of 1981. (*Jet Propulsion Laboratory.*)

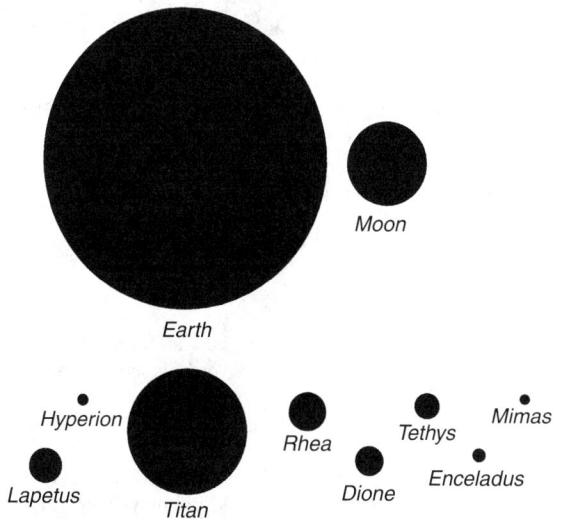

Fig. 6.   Size comparison (diameter) of Saturn's major satellites with Earth and its moon. (*Jet Propulsion Laboratory.*)

Past U.S. missions to Saturn include:

(1) *Pioneer 11*, in 1979, flew past the planet and on September 5 of that year, scientists reported the discovery of Saturn's 11th moon orbiting the planet and two more rings of debris encircling the planet. The space probe swept within 13,000 miles (20,917 kilometers) of the planet and then flew by Titan, Saturn's largest moon. Scientists had feared that the spacecraft would be damaged by some of the debris around the planet, but *Pioneer 11* passed through several potential trouble spots unharmed and beamed back a wealth of data to Earth. Much of the information supported the image of Saturn as had been previously determined by Earth-based observations.

(2) *Voyager 1*, in the fall of 1980, viewed the planet after the spacecraft had visited Jupiter early in the spring of 1979. Early pictures of the planet taken during the summer of 1980, as the spacecraft approached Saturn, were almost featureless with exception of the three classic rings that had been studied for years from Earth. As *Voyager 1* approached more closely, it appeared that there were not just three rings, but scores, then hundreds, and finally thousands of thin ringlets. It would turn out that they were not individual rings separated by gaps; some of the variations were caused by the gravitational attraction of nearby satellites pulling millions of particles into motion, spiraling outward across the rings like waves in an ocean. The photos from *Voyager 1* also revealed other puzzling phenomena—dark features that resembled spokes in the bright B-ring. As

Voyager 1 zeroed in on Saturn, more satellites were seen, until a total of six had been found—three from Earth observations and three by *Voyager*. Two that were discovered in *Voyager* images appeared to shepherd the narrow F-ring. Two more, discovered from Earth, had appeared to share the same orbit. Inspection of *Voyager* photos, however, showed that the satellites' orbits are about 31 miles (50 kilometers) apart. A little more calculation yielded the astonishing prediction that, as the two satellites approached each other in January 1982, they would trade orbits and continue on their way, to resume their game of "musical chairs" the next time they approached. A day before *Voyager 1* swept by Saturn, it flew within 2500 miles (4000 km) of the huge satellite Titan and passed directly behind it, making what scientists had predicted would be extremely important observations. Titan was shrouded by a thick, opaque haze that completely obscured its surface from the cameras. But the infrared instrument and the spacecraft radio probed the atmosphere to measure the diameter of the satellite and the thickness, temperature, and composition of its atmosphere. Once beyond Titan, *Voyager 1* flew past Saturn and briefly disappeared behind it. En route to the Earth, the radio signals penetrated Saturn's atmosphere and passed through the rings. Measurements of the way the atmosphere altered the signals, and the rings scattered them, would tell much about the atmosphere and help determine the sizes of particles that make up the rings. *Voyager 1* swept past its targets and took a new course upward from the plane in which the planets orbit the Sun, outward toward the edge of the solar system. Its cameras and its UV and IR instruments were turned off, but other instruments still probe for galactic cosmic rays, the edge of the solar system, and the beginning of interstellar space. The trek of *Voyager 1* is depicted in Fig. 4.

Fig. 7.   Sketch of Saturn by Galileo (1610). When Galileo first focused his telescope on Saturn, he realized that the appearance of the planet was unusual, but he did not ascertain its real character because the power of his telescope was far too low. He believed he was looking at three globes, one large and two small, which seemed to change slowly in appearance. In 1655, Huygens, after years of observing the planet, finally realized that these projections were actually a flat ring slightly separated from the main globe.

(3) *Voyager 2*, in late summer 1981, made it possible to view Saturn for a second time. Having had an opportunity to study the images taken by *Voyager 1*, scientists decided to arrange the second encounter for a closer study of the planet's rings and its satellites, other than Titan. As the summer of 1981 approached, scientists, having gained experience from *Voyager 1*, were able to set their camera exposures at more exact levels, to cope with the low light levels and the general blandness of Saturn. On this approach, Saturn presented alternating dark and bright bands of clouds and high-speed jet streams. Swirling cloud patterns, which were smaller versions of the large and intense storms seen on Jupiter, were also visible through Saturn's haze layer. *Voyager 2's* cameras zeroed in on the rings, and scientists searched for small satellites in the rings that might cause the multi-ringed appearance. Those moonlets, some scientists believed, might sweep up material in the rings, creating gaps. *Voyager 2* would soar closer to the rings, and the improved resolution of the pictures should show structures as small as 0.6 mile (1 km) in diameter. One of *Voyager's* most important experiments involved an instrument called a photopolarimeter, which measures light intensity. As *Voyager 2* passed above Saturn, the photopolarimeter detected changes in the starlight's intensity as it was altered by changes in the thickness of the rings. Quick analysis of data showed that the rings' structure was far different from what it appeared to be in the photos.

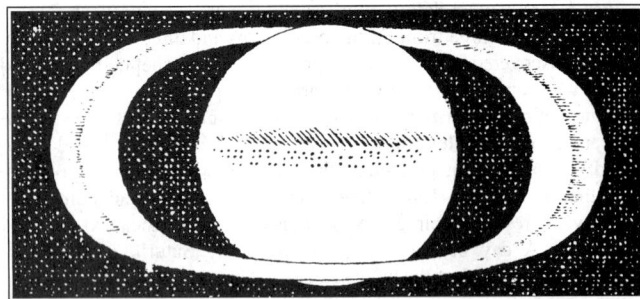

Fig. 8. Sketch of Saturn by Cassini (1675). Cassini found the first breach in the supposedly solid, rigid, and opaque ring when he discovered that it was divided into two parts by a dark line, now known as Cassini's division. In later years, Cassini also detected some of Saturn's moons. The earliest successful photograph of Saturn was taken in 1883 by Andrew Common. In 1895, James Keeler suggested that the rings are in fact a swarm of particles in near-independent orbits.

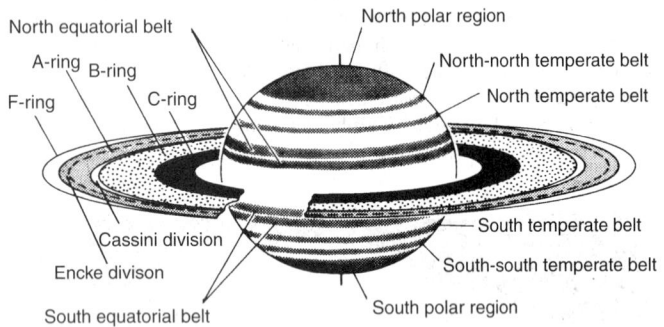

Fig. 9.   Principal visual features of Saturn.

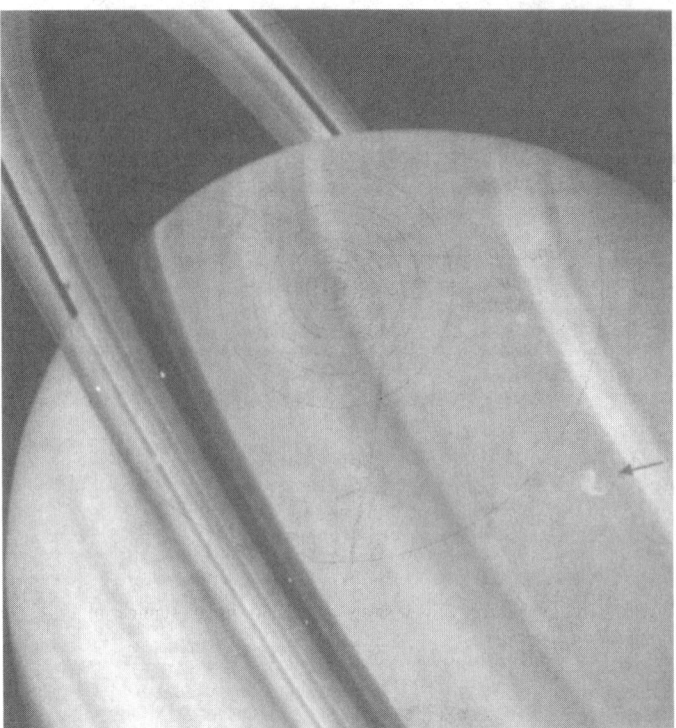

Fig. 10.   View of Saturn and its ring system returned by *Voyager 2* on August 11, 1981 when the spacecraft was 13.9 million kilometers (8.6 million miles) away from the planet and approaching it at about 1 million kilometers (620,000 miles) per day. The ring system's shadow is clearly cast in the equatorial region. Storm clouds and small-scale spots in the mid-latitudes are apparent. The so-called "ribbonlike" feature in the white cloud band marks a high-velocity jet at about 47°N. At this location, the westerly wind speeds are about 150 meters per second (330 miles per hour). The banding on Saturn extends toward both poles. (*Jet Propulsion Laboratory*.)

No region was totally empty of ring particles. The members of the photopolarimeter team have 800,000 samples, each one a 330-foot (100-meter) slice of the rings. Years will be required to complete this analysis. *Voyager 2* also photographed and measured all of the planet's satellites that were then known, a number that had expanded to 17. Upon completion of the sweep by Saturn, *Voyager 2* proceeded on toward Uranus and Neptune. See separate articles on these planets. The trek of *Voyager 2* past Saturn is depicted in Fig. 5.

Highlights of Voyager's Scientific Findings. After studying the volumes of data returned by the two *Voyager* spacecraft, although analyses are still in progress, scientists at NASA (National Aeronautics and Space Administration), and Jet Propulsion Laboratory (JPL), with over 30,000 photos of the Saturn system, have selected the following highlights:

1. There is less helium in the top of Saturn's atmosphere than in Jupiter's.
2. Subdued contrasts and color differences (Jupiter and Saturn) are primarily a result of either more horizontal mixing or less production of localized colors on Saturn than on Jupiter.
3. Winds blow at extremely high speeds on Saturn. Near the equator, the *Voyagers* measured winds of about 500 meters per second (1100 miles per hour). The winds blow primarily in an eastward direction.
4. The Voyagers found aurora like ultraviolet emissions at mid latitudes on Saturn, and auroras at higher latitudes.
5. The *Voyagers* discovered radio emissions from the planet with which they determined the length of Saturn's day to be 10 hours, 39 minutes, 24 seconds.
6. The complicated structure in Saturn's rings appears to be caused, at least in part, by density waves, which are created by gravitational interactions with several of the inner satellites. Few clear gaps exist anywhere in the rings.
7. Radial, spoke-like features were discovered in the rings and remain poorly understood.

8. Titan, Saturn's largest satellite, has an atmosphere composed of nitrogen, methane, and several organic compounds, including hydrogen cyanide.
9. Titan's surface atmospheric pressure is 1.6 bars (60% greater than the surface pressure on Earth).
10. The temperature at the surface of Titan is 95 K (−288°F; ~−172°C). Methane, therefore, possibly plays much the same role on Titan as water does on Earth.
11. Saturn's regular satellites appear to be composed primarily of ice. Phoebe, the outermost, is believed to be a captured asteroid.
12. Many new satellites have been discovered at Saturn, some from Earth-based observations, others by the two *Voyagers*. There may be over twenty. Scientists expect to find more. As indicated by Fig. 6, Saturn's satellites, with exception of the huge Titan, are smaller than Earth's moon.
13. The size of Saturn's magnetosphere, like Jupiter's, is controlled by external pressure of the solar wind.

More details of some of the foregoing observations are developed in this article.

Observations of Saturn by Early Astronomers. To the naked eye, Saturn appears like one of the brighter stars, but with the absence of twinkling. In a telescope, the planet itself has a belted appearance quite similar to that of Jupiter, but without as many distinctive surface features as are seen on the larger planet. Although Galileo realized that the planet had some unusual features, he did not live to know that what he saw was a ring system. Galileo's sketch, made in 1610, is shown in Fig. 7. Huygens first described the rings with reasonable accuracy for this period in 1650. Cassini, by 1675, identified a ring system that was divided into two parts by a dark line (now known as the *Cassini division*). Cassini also noted some belt or zone demarcations, as indicated by his sketch shown in Fig. 8. The first successful photo of Saturn, clearly showing the ring system, was not made until 1883. In 1895, Keller suggested that the rings were a swarm of particles in nearly independent orbits. Bond, in the late

Fig. 11. A computer-simulated view of particles ranging in size from marbles to beach balls, populating a 3 meter (10 foot) square section of Saturn's A Ring. This picture was constructed to illustrate the size distribution of ring particles based on Voyager Radio Science occultation data. In reality the particles are not smooth spheres, but their depiction here as spheres is representative of the properties often assumed (implicitly or explicitly) in mathematical models of the rings. (*NASA.*)

Fig. 12. This enhanced color image shows the outer half of Saturn's main ring system. The dark Cassini Division near the middle divides the outer A Ring from the B Ring. The Cassini Division was once thought to be empty, but this view shows that it contains several fainter ring structures. The most prominent structure in the A Ring is the Encke Gap, about one-third of the way from the ring's outer edge. This gap is "shepherded" open by the tiny embedded moon Pan. The B Ring is more densely packed with material than the A Ring (i.e., higher in optical depth) so it shows up as brighter in this image at a low phase angle. Several faint "spokes" are visible as dark streaks across the B Ring. At lower right, you can see the outer tip of the much fainter C Ring. (*NASA.*)

Fig. 13.   Saturn's outer A Ring in false color, constructed from Voyager 2 green, violet, and ultraviolet frames. The Cassini Division (lower right) is distinctly less red than the A Ring. Also visible, as a sequence of dark features in the outermost A Ring, is a series of orbital resonances with Saturn's inner moons. The 325 kilometer (195 mile) wide Encke Division (upper left center) contains the tiny moon Pan. However, the cause of the narrower Keeler Gap (closer to the outer edge of the A Ring) and the gaps in the Cassini Division is not known. (*NASA*.)

Fig. 14.   The outer part of Saturn's C Ring (blue) and the inner part of the B Ring (yellow) are shown here in false color. The color difference blurs over the boundary between the rings. The Maxwell Gap and its eccentric ringlet, which appears yellow due to image misalignment, are in the left center of this frame. (*NASA*.)

Fig. 15.   To illustrate the chemistry of the rings, this highly enhanced false color view of Saturn's rings was assembled from clear, orange and ultraviolet images. Possible variations in chemical composition from one part of Saturn's ring system to another are visible as subtle color variations. In addition to the previously known blue color of the C Ring and the Cassini Division, the picture shows additional color differences (red-green) within and between the A and B Rings. The distance to Saturn was 8.9 million kilometers (5.5 million miles). (*NASA.*)

1800s, found a second division of the rings and a third or "crepe" ring close to the planet itself.

Atmosphere of Saturn

Saturn has cloud bands similar to those of Jupiter, although they are more difficult to see and contrast less with the planetary disk. Images of Saturn confirm its rather bland appearance. The blandness may be a result of lower temperatures and reduced chemical and meteorological activity compared with Jupiter; or the presence of a relatively permanent and uniform high-altitude haze. As shown by Fig. 9, the principal features of Saturn's visible surface are stripes that parallel the equator. There are dark belts and light zones that have been seen continuously over two centuries of observation. Although *Voyager 1* saw only a few markings, *Voyager 2* saw many, including long-lived ovals, tilted features in east-west shear zones and others similar to, but smaller than, the features on Jupiter. One white oval on Saturn was noted to be 7000 by 5000 kilometers (4000 by 3000 miles), with a 100-meter-per-second (200-mile-per-hour) circumferential wind. See Fig. 10. Winds appear to blow at extremely high speeds on Saturn. Near the equator, the *Voyagers* measured winds about 500 meters per second (1100 miles per hour). The winds blow primarily in an easterly direction. The strongest winds are found near the equator, and velocity falls off uniformly at higher latitudes. At latitudes greater than 35°, the winds alternate eastward and westward as the latitude increases. The marked dominance of eastward jet streams indicates that winds are not confined to the cloud layer, but must extend inward at least 2000 kilometers (1240 miles). Furthermore, measurements by *Voyager 2* showing a striking north-south symmetry has led some scientists to suggest that the winds may extend from north to south clear through the interior of the planet.

When *Voyager 2* flew behind Saturn, its radio beam penetrated Saturn's atmosphere, measuring the upper-atmosphere temperature and density. Minimum temperatures of about 82 Kelvin (−312 °F; −191 °C) were measured at the 70-millibar level (surface pressure on Earth is about 1000 millibars). The temperature increased to 143 K (−202 °F; −130 °C) at the deepest levels probed — about 1200 millibars. Temperatures near the north pole were about 10 °C (18 °F) colder at the 100-millibar level than temperatures at mid-latitudes. Scientists believe the difference may be a seasonal effect.

The *Voyagers* found auroralike ultraviolet emissions of hydrogen at mid-latitudes (above 65°). Scientists have suggested the high-latitude auroral activity leads to formation of complex hydrocarbon molecules that are carried toward the equator. The mid-latitude auroras, which normally occur only in sunlit regions, remain a puzzle; bombardment by electrons and ions, known to cause auroras on Earth, occurs primarily at high latitudes.

Fig. 16.   Saturn's F Ring. The Voyager 2 PPS (Photopolarimeter) instrument performed a stellar occultation experiment at Saturn. Because of the instrument's high sensitivity and the small apparent size of the star used, delta Scorpii, structure was observed on a much finer scale—hundreds of meters—than would otherwise be possible. For comparison, the finest resolution obtained by the Voyager cameras was about 10 kilometers (6 miles). These synthetic images have been generated by "smearing out" sections of the one-dimensional occultation profile into two-dimensional pseudo-images. Regions of higher optical depth appear brighter in the images. These synthesized images show the very fine structure of Saturn's F Ring, including a dense outer core and a diffuse set of inner strands. Note that the uniformity and lack of kinks or knots here are only artifacts of the occultation experiment, which takes a single narrow slice of the rings. (*NASA.*)

Fig. 17.   This artist's rendering simulates Saturn's rings as seen from the surface of Pandora, one of the shepherds of the F Ring. The clumpy, wispy F Ring crosses the image from top to bottom, and the F Ring's other shepherd can be seen in the interior of the F Ring near the top. (*Courtesy of Dave Seal, JPL.*)

Saturn's Ring System

The alphabetical letter designations of Saturn's rings (Fig. 9) are based upon the chronology of discovery. The letters do not consistently relate to their positions relative to the planet. The A ring is the outermost ring visible with small telescopes. The B ring, the brightest ring, lies inside the A ring and is separated from it by Cassini's division. The C ring, barely visible with small telescopes, lies inside the B ring. The Cassini division between the B and A rings is the most prominent gap between the rings and is easily visible with Earth-based telescopes. A thin space within the outermost edge of the A ring is the Encke division. This was first identified by *Pioneer* spacecraft. The average thickness of the rings is estimated at no greater than 16 kilometers (10 miles).

Perhaps the greatest surprises and the most perplexing puzzles encountered by the *Voyagers* are the rings. *Voyager 1* found a great deal of unexpected structure in the classical A, B, and C rings. One suggestion was that the structure might be unresolved ringlets and gaps. Ring photos by *Voyager 1* were of lower resolution than those of *Voyager 2*, and scientists at first suggested that gaps might be created by tiny satellites orbiting within the rings and sweeping out bands of particles. One such large gap was detected at the inner edge of the Cassini division.

Cameras and instruments of much greater resolution than used in the past will be required on future missions to glean greater information pertaining to the variety and characteristics of particles actually making up what appears to be hundreds of thousands of rings. As some scientists believed, *Voyager* observations no longer support the concept that the gravity of small moonlets within the rings channel the trillions of ice chunks that compose the rings into so many narrow traffic lanes. The only moonlets of this nature detected were the "shepherds" discovered by *Voyager 1* which appear to maintain the thin outer F ring in place. As pointed out by Loudon (1981), there is more matter associated with the rings than would be expected from data that are converted into photographs. As the *Voyager 2* streaked through the ring system's thin 500-foot (150-meter) thick outer extension at a relative speed of 8 miles (13 kilometers) per second, the craft was bombarded by thousands of dust-sized particles not for 500 feet, but for about 1000 miles (1600 kilometers).

As observed by the Stanford University radio science team when *Voyager 1* passed behind Saturn and a radio signal was beamed to Earth through the rings, useful information was gained concerning the ringlets and the distribution of particle sizes within the ringlets. It was observed that in some ringlets, the fine particles are denser on the inner and outer edges than they are in the middle. The rings appear to consist of a tenuous background sheet of material, with empty gaps and denser ridges superimposed on it. It was noted that a typical ridge has a very sharp edge. The investigators, by comparing the attenuation of the microwave signal at two wavelengths, were able to estimate the size distribution of ring particles. It was found that there are about 120 to 200 particles per square kilometer in the size range of 9 to 11 meters. Very few larger particles were detected. The numbers as well as uncertainties increased rapidly with diminishing size. It was found in the 3- to 5-meter particle range, that there are from 1700 to 6000 particles per square kilometer. The investigators noted that there may be millions that are smaller than that; or there may be none; the radio data are consistent with either interpretation. See Fig. 11.

Voyager 2 measurements provided much of the data required for an understanding of structure. Higher-resolution photos of the inner edge of the Cassini division showed no sign of satellites down to about 5–9 kilometers (3–6 miles). No systematic searches were conducted in any of the other ring gaps. *Voyager 2's* photopolarimeter provided more surprises. The instrument measured changes in starlight from Delta Scorpii as the spacecraft flew above the rings when the starlight passed through the rings. The photopolarimeter was capable of resolving structure smaller than 300 meters (1000 feet). The star-occultation experiment showed that few clear gaps exist anywhere in the rings. The structure in the B ring instead appears to be variations in density of ring material, probably caused by traveling density waves or other, stationary forms of waves. The gravitational effects of Saturn's satellites form density waves. They propagate outward from positions where the ring particles orbit Saturn in harmony with the satellites. Resonant points are locations where a particle orbits Saturn in one-half or one-third the time required by a satellite, such as Mimas. For example, at the 2:1 resonant point with the satellite 1980 S1, a series of outward propagating density waves has characteristics that indicate there are about 60 grams of material per square centimeter of ring area and that the velocity of the particles relative to one another is about one millimeter per second. The small-scale structure of the rings may therefore be largely transitory, although larger-scale features, such as the Cassini and Encke divisions, appear to be more permanent. The edges of the rings where the few gaps exist are so sharp that the ring must be less than about 200 meters (650 feet) thick at such points. See Figs. 12–17.

In almost every case where clear gaps do appear in the rings, eccentric ringlets are found. These seem to show variations in brightness. In some cases, the differences are due to clumping or kinking; in others to nearly complete absence of ring material. Some scientists believe the only plausible explanation for the clear regions and kinky ringlets within them is the presence of nearby, yet undetected moonlets.

Two separate, discontinuous ringlets were found in the A-ring gap, about 73,000 kilometers (45,000 miles) from Saturn's cloud tops. At high resolution, at least one of these ringlets was noted to have multiple strands.

Pioneer 11 discovered Saturn's F ring in 1979. Photos of that ring taken by *Voyager 1* showed three separated strands that appeared twisted or braided. *Voyager 2* found five separate strands in a region that had

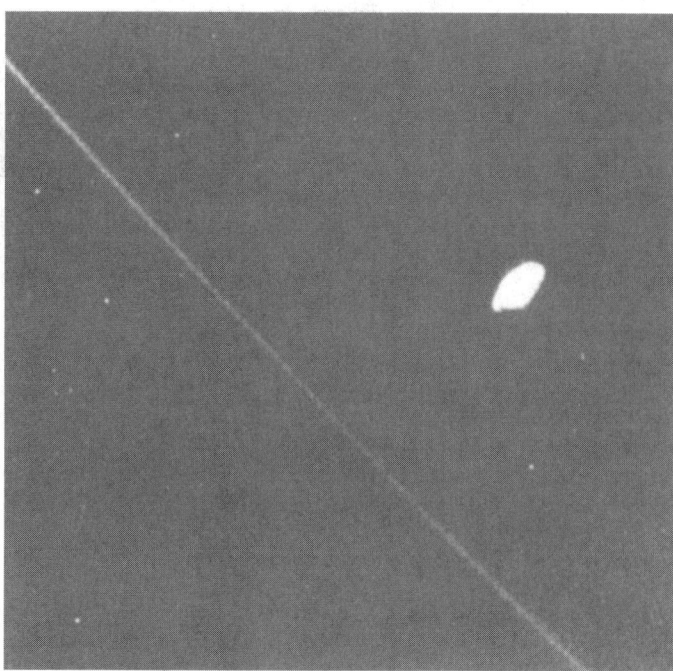

Fig. 18. Saturn's F ring and its inner shepherding satellite (1980S27) are pictured in this closeup *Voyager 2* image acquired on August 25, 1981 from a range of 365,000 kilometers (227,000 miles). Features as small as 6 kilometers (3.7 miles) are visible. The satellite is elongated and irregular, with its longest axis pointing toward the center of Saturn (toward upper right of view). As seen here, the F ring is thin and does not show the multiple, braided structure noted by Voyager 1 in 1980. There is no indication of a band or kink in the ring at its closest point to the shepherd; such a feature would be consistent with some of the theories advanced on the formation of the braids. (*Jet Propulsion Laboratory*.)

Fig. 19. *Voyager 2* obtained this high-resolution picture of Saturn's rings on August 22, 1981 when the spacecraft was 4 million kilometers (2.4 million miles) distant from the planet. Evident here are the numerous "spoke" features in the B ring; their very sharp, narrow appearance suggests short formation times. Some scientists believe that electromagnetic forces are responsible in some way for these features, but no detailed theory has been formulated. Pictures such as this and analyses of *Voyager 2's* spoke motion pictures may reveal additional clues about the origins of these complex structures. (*Jet Propulsion Laboratory*.)

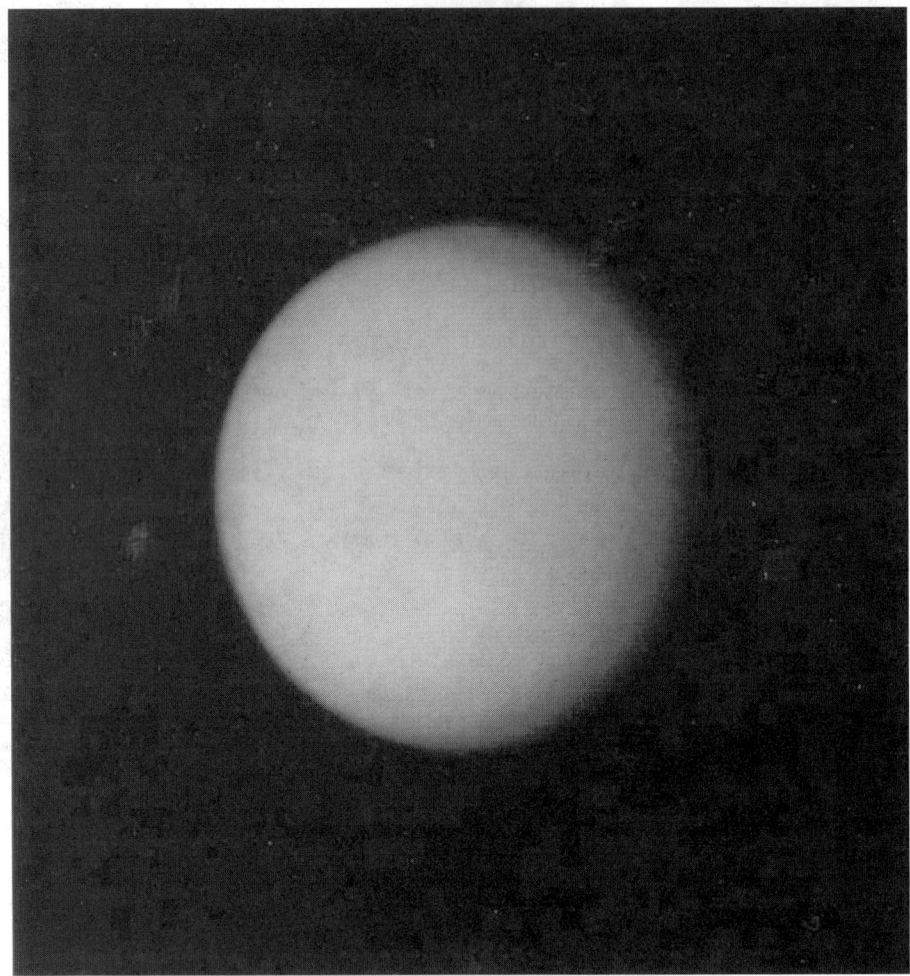

Fig. 20.   On August 25, 1981, *Voyager 2* obtained this view of Titan, Saturn's largest satellite, from a range of 4.5 million kilometers (2.7 million miles). Separate violet, blue and green frames were combined to make this photograph, in which cloud features of diameter greater than 85 kilometers (50 miles) are visible. Seen here are the bright Southern Hemisphere and the dark North Polar Region. At this resolution, Titan appears to be completely shrouded by a thick atmosphere; visible features consist of cloud bands parallel to the equator. There is the suggestion of a Northern Hemisphere feature diagonal to the equator, but this is the result of a loss of data in one of the images used to compose this photograph. The *Voyager* Project is managed for NASA by the Jet Propulsion Laboratory, Pasadena, Calif. (*Jet Propulsion Laboratory, California Institute Of Technology, National Aeronautics And Space Administration.*)

no apparent braiding, but did reveal apparent braiding in another region. The photopolarimeter found that the brightest of the F-ring strands was subdivided into at least ten strands. The twists in the F ring are believed to originate in gravitational perturbations caused by the two shepherding satellites, 1980 S26 and 1980 S27. See Fig. 18.

Clumps of material in the F ring appear fairly uniformly distributed around the ring every 9000 kilometers (5600 miles), a spacing that coincides with the relative motion of F-ring particles and the shepherding satellites in one orbital period. By analogy, scientists suggest that similar mechanisms may be operating for irregular ringlets that exist in gaps in the main ring system.

The spokes found in the B ring appear only at radial distances between 43,000 kilometers (27,000 miles) and 57,000 kilometers (35,000 miles) above Saturn's clouds. Some spokes, those that are narrow and have a radial alignment, may be recently formed. The broader, less radial spokes appear to have formed earlier than the narrow examples and seem to follow Keplerian orbits — individual areas rotate at speeds governed by distances from the center of the planet. In some cases, scientists suggest that they see evidence that new spokes are reprinted over older ones. Formation of the spokes is not restricted to regions near the planet's shadow. As both *Voyagers* approached Saturn, the spokes appeared dark against a bright ring background. As the spacecraft departed, the spokes appeared brighter than the surrounding ring areas, indicating they backscatter the reflected sunlight more efficiently. See Fig. 19.

Spokes are also visible at high phase angles in light reflected from Saturn on the unilluminated underside of the rings. This suggests, according to some scientists, that charging of the small particles by photoionization alone may not be responsible for levitating them above the bulk of the ring material.

Another challenge faced by investigators is an understanding of the observation that even general dimensions of the rings do not seem to remain true at all positions around the planet. The distance of the B ring's outer edge, near a 2:1 resonance with Mimas, varies by at least 140 kilometers (87 miles). Furthermore, the elliptical shape of the outer edge does not follow a Keplerian orbit, since Saturn is at the center of the ellipse, rather than at one focus. Although the gravitational effects of Mimas are most likely responsible for the elliptical shape, present theory predicts a somewhat smaller magnitude than that which was observed.

Voyager 1 measured radio waves that originate in sporadic electric discharges. The source of these discharges is still unknown. It is possible that they may originate in the rings. *Voyager 2* measured similar discharges, but at a rate only 10% that of the findings of *Voyager 1*, and with a different polarization.

Magnetosphere

The size of Saturn's magnetosphere is determined by external pressure of the solar wind. When *Voyager 2* entered the magnetosphere, the solar wind pressure was high and the magnetosphere extended only 19 Saturn radii (1.1 million kilometers; 712,000 miles) in the Sun's direction. Several hours later, however, the solar wind pressure dropped and Saturn's magnetosphere ballooned outward over a six-hour period. It apparently remained inflated for at least three days, since it was 70% larger than when *Voyager 2* crossed the magnetic boundary on the outbound leg.

Unlike all other planets whose magnetic fields have been measured, Saturn's field is tipped only about one degree relative to the rotation poles.

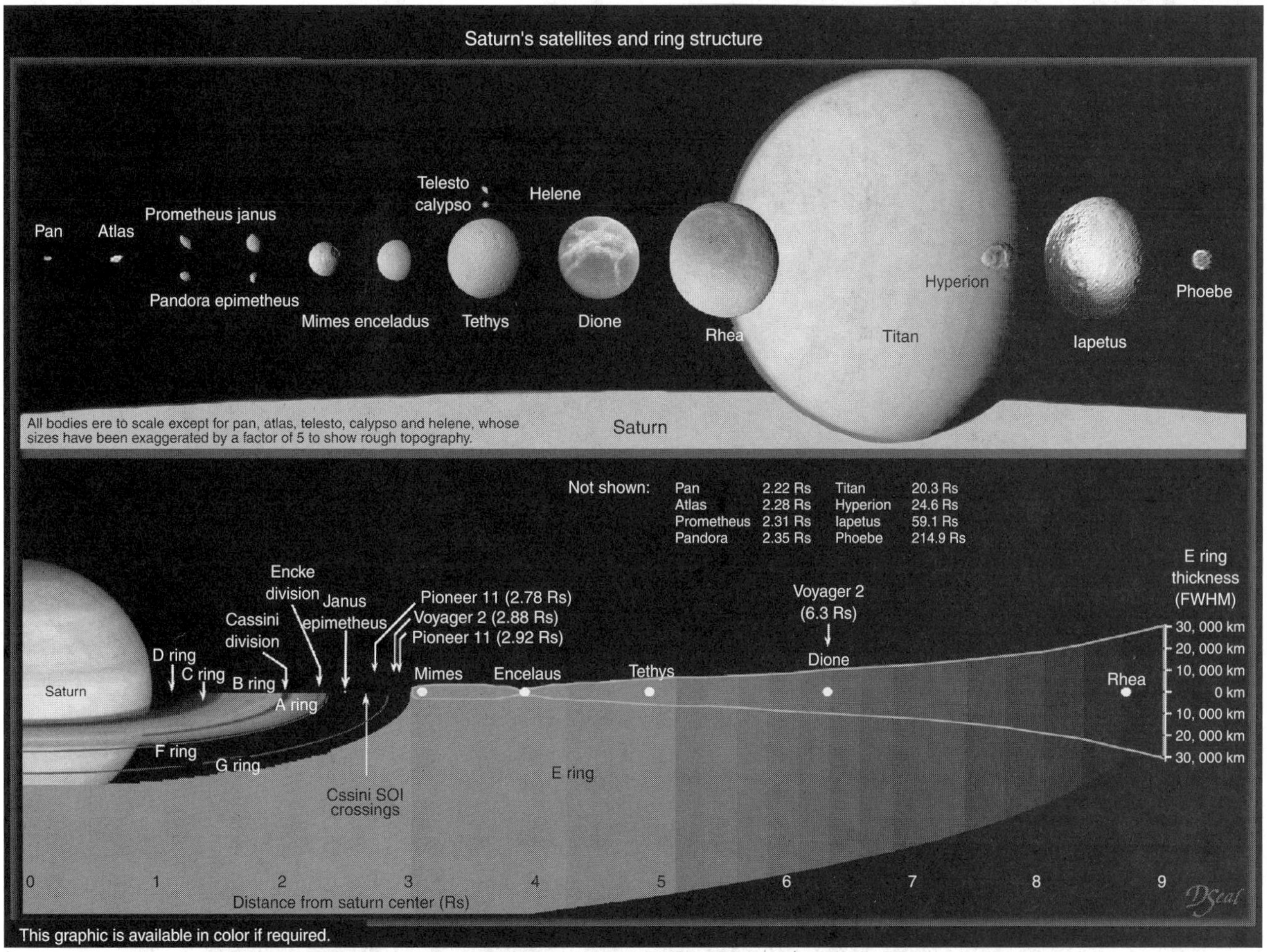

Saturn's satellites and ring structure

Pan   Atlas   Prometheus janus   Telesto calypso   Helene

Pandora epimetheus   Mimes enceladus   Tethys   Dione   Rhea   Hyperion   Titan   Iapetus   Phoebe

All bodies ere to scale except for pan, atlas, telesto, calypso and helene, whose sizes have been exaggerated by a factor of 5 to show rough topography.   Saturn

Not shown:
Pan	2.22 Rs	Titan	20.3 Rs
Atlas	2.28 Rs	Hyperion	24.6 Rs
Prometheus	2.31 Rs	Iapetus	59.1 Rs
Pandora	2.35 Rs	Phoebe	214.9 Rs

E ring thickness (FWHM)

Encke division   Janus   Pioneer 11 (2.78 Rs)   Voyager 2 (6.3 Rs)
Cassini division   epimetheus   Voyager 2 (2.88 Rs)   Pioneer 11 (2.92 Rs)

D ring   C ring   B ring   A ring   Mimes   Encelaus   Tethys   Dione   Rhea

Saturn

F ring   G ring   Cssini SOI crossings   E ring

30, 000 km
20, 000 km
10, 000 km
0 km
10, 000 km
20, 000 km
30, 000 km

Distance from saturn center (Rs)
0   1   2   3   4   5   6   7   8   9

This graphic is available in color if required.

Fig. 21.   Saturn has 18 named satellites, more than any other planet. There may very well also be several small ones yet to be discovered. Of those moons for which rotation rates are known, all but Phoebe and Hyperion rotate synchronously. The three pairs Mimas-Tethys, Enceladus-Dione and Titan-Hyperion interact gravitationally in such a way as to maintain stable relationships between their orbits: the period of Mimas' orbit is exactly half that of Tethys, they are thus said to be in a 1:2 resonance; Enceladus-Dione are also 1:2; Titan-Hyperion are in a 3:4 resonance. In addition to the 18 named satellites, at least a dozen more have been reported and given provisional designations. (*JPL, NASA*.)

Fig. 22.   This montage of images of the Saturnian system was prepared from an assemblage of images taken by the *Voyager 1* spacecraft during its Saturn encounter in November 1980. This artist's view shows Dione in the forefront, Saturn rising behind, Tethys and Mimas fading in the distance to the right, Enceladus and Rhea off Saturn's rings to the left, and Titan in its distant orbit at the top. The *Voyager* Project is managed for NASA by the Jet Propulsion Laboratory, Pasadena, California. (*Jet Propulsion Laboratory, California Institute Of Technology, National Aeronautics And Space Administration*.)

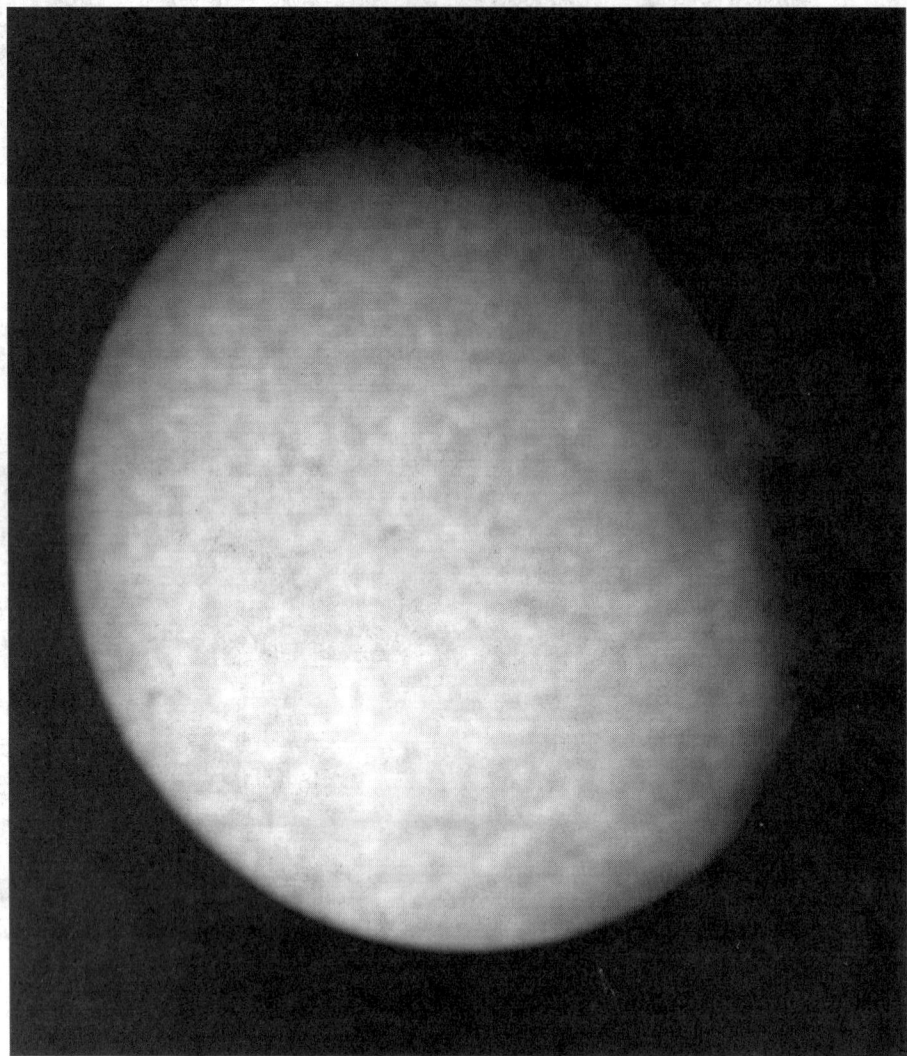

Fig. 23.   Little detail is visible on the surface of Saturn's satellite Enceladus in this color-enhanced image taken by NASA's *Voyager 1* on November 12, 1980 from a distance of 655,000 kilometers (393,000 miles). The lack of visible surface detail on the satellite indicates that Enceladus' surface is dramatically different from the surfaces of the other larger Saturnian moons. The *Voyager* Project is managed for NASA by the Jet Propulsion Laboratory, Pasadena, California. (*Jet Propulsion Laboratory, California Institute Of Technology, National Aeronautics And Space Administration*.)

That rare alignment was first measured by *Pioneer 11* in 1979 and was later confirmed by *Voyager 1*.

Several distinct regions have been identified within Saturn's magnetosphere. Inside about 400,000 kilometers (250,000 miles), there is a torus of H^+ and O^+ ions, probably originating from water ice sputtered from the surfaces of the satellites Dione and Tethys. These ions are positively charged atoms of hydrogen and oxygen that have lost one electron. Strong plasma-wave emissions appear to be associated with the inner torus.

At the outer regions of the inner torus, some ions have been accelerated to high velocities. In terms of temperatures, such velocities correspond to 400 to 500 million degrees Kelvin.

Outside the inner torus is a thick sheet of plasma that extends out to about 1 million kilometers (620,000 miles). The source for material in the outer plasma may be Saturn's ionosphere, Titan's atmosphere, and the neutral hydrogen torus that surrounds Titan between 500,000 kilometers (300,000 miles) and 1.5 million kilometers (1 million miles).

Radio emissions from Saturn had changed between the encounters of *Voyager 1* and *2*. *Voyager 2* detected Jupiter's magnetotail as it approached Saturn in the winter and early spring of 1981. Soon afterward, when Saturn was believed to be bathed in the Jovian magnetotail, the ringed planet's kilometric radio emissions were also undetectable.

During portions of *Voyager 2's* Saturn encounter, kilometric radio emissions again were not detected. The observations are consistent with effects caused by Jupiter's magnetotail, although *Voyager* scientists admit of no direct evidence that the shutdown of Saturn's natural radio signals was caused by Jupiter's magnetotail.

Satellites of Saturn

Titan. The largest of Saturn's satellites is Titan and, in fact, is the second largest satellite in the solar system. It is the only satellite known to have a dense atmosphere. Some scientists believe that Titan may prove to be the most interesting body in the solar system, i.e., from a terrestrial perspective. For almost two decades, space scientists have searched for clues to the primeval Earth. The chemistry taking place in Titan's atmosphere may be similar to that which occurred in the Earth's atmosphere several billion years ago.

Because of its thick, opaque atmosphere, astronomers for some years believed that Titan was the largest satellite in the solar system. Their measurements were necessarily limited to measurements at the cloud tops. *Voyager 1's* close approach and diametric radio occultation showed, however, that Titan's surface diameter is only 5050 kilometers (3200 miles), which is slightly smaller than Ganymede, Jupiter's largest satellite. Both these satellites are larger than the planet Mercury. Titan's density appears to be about twice that of water ice. It has been suggested that it may be composed of nearly equal amounts of rock and ice.

Titan's surface was not seen in any of the *Voyager* photos. The surface is hidden by a dense, optically thick photochemical haze whose main layer is about 50 kilometers (30 miles) thicker in the southern hemisphere than in the northern hemisphere. Several distinct, detached haze layers can be seen above the visibly opaque haze layer. See Fig. 20. These haze layers merge with the main layer over the north pole of Titan, forming what scientists first thought was a darkened hood. The hood was found, under the better viewing conditions of *Voyager 2*, to be a dark ring around the pole.

Fig. 24.   A heavily cratered surface is apparent in this photo of Saturn's satellite Tethys, taken November 12, 1980 by *Voyager 1* from a distance of 570,000 kilometers (354,000 miles). Other photos of Tethys showed the opposite hemisphere, dominated by what appears to be a large impact crater or hill. Tethys is about 1,000 kilometers (600 miles) in diameter and is composed largely of ice. *Voyager 1* made its closest approach to Saturn on November 12, 1980. (*Jet Propulsion Laboratory, California Institute Of Technology, National Aeronautics And Space Administration.*)

The southern hemisphere is slightly brighter than the northern hemisphere, possibly the result of seasonal effects. When the Voyagers flew past, the season on Titan was the equivalent of April on Earth; or early spring in the northern hemisphere and early autumn in the southern hemisphere.

The atmospheric pressure near Titan's surface is about 1.6 bars or 60% greater than that on Earth. The atmosphere is mostly nitrogen, also the major constituent of the Earth's atmosphere.

The surface temperature appears to be about 95 K ($-288\,°$F; $-180\,°$C), only 4 K above the triple-point temperature of methane. Methane, therefore, quite possibly plays the same role on Titan as water does on Earth—as rain, snow, and vapor. Rivers and lakes of methane may exist under a nitrogen sky. Clouds may drop liquid-methane precipitation. Titan's methane, through continuing photochemistry, may be converted to ethane, acetylene, an ethylene and, when combined with nitrogen, hydrogen cyanide. HCN is an especially important molecule, since it is a building block of amino acids. Titan's low temperature probably inhibits more complex organic chemistry.

Titan has no intrinsic magnetic field. Therefore, it has no electrically conducting and convecting liquid core. Its interaction with Saturn's magnetosphere creates a magnetic wake behind the satellite. Titan also serves as a source for both neutral and charged hydrogen atoms in Saturn's magnetosphere.

In later studies of Titan data (from *Voyager*), B.L. Lutz, C. de Bergh, and T. Owen reported discovery of carbon monoxide in the atmosphere of the satellite. The researchers reported that the 3-O rotation-vibration band of CO in the near-IR spectrum of Titan had been identified, and a reflecting layer model mixing ration of CO to molecular N of 6×10^{-5} had been determined. This result supported the earlier probable detection of CO_2

and strengthened possible analogies between the atmosphere of Titan and conditions on primitive Earth.

Eshelman, Lindal, and Tyler in post-*Voyager* data analysis, reported that Titan's dense and cold nitrogen atmosphere contains a small amount of methane under conditions at least approaching those at which one or both constituents would condense. The possibility of methane and nitrogen rain clouds and global methane oceans has been discussed widely. However, researchers report that, from specific features of radio occultation and other *Voyager* data, they have concluded that nitrogen does not condense on Titan, and that the satellite has neither global methane oceans nor a global cloud of liquid methane droplets. The investigators further observed that certain results indirectly support the conjecture that methane does not condense at any location. However, other considerations favor a methane ice haze in the troposphere, and liquid and solid methane may exist on the surface and as low clouds at polar latitudes.

F.M. Flaser (Goddard Space Flight Center), working on the same problem, observed that if global oceans of methane exist on Titan, the atmosphere above them must be within 2% of saturation. The two *Voyager* radio occultation soundings, made at low altitudes, probably occurred over land, since they imply a relative humidity & 70% near the surface. Oceans might exist at other low-altitude locations if the zonal wind velocities in the lowest 3 km are & 4 centimeters per second.

As a point of contrast, Lunine, Stevenson, and Yung (California Institute of Technology), from studies of Voyager data, suggested that Titan is covered by an ocean one to several kilometers deep consisting mainly of ethane. They report that if the ocean is in thermodynamic equilibrium with an atmosphere of 3% (mole fraction) methane, then its composition is roughly 70% ethane, 25% methane, and 5% nitrogen. Photochemical

Fig. 25.   The conspicuous crater on the surface of Saturn's moon Mimas is seen in this image taken by NASA's *Voyager 1* on November 12, 1980 when the spacecraft was 540,000 kilometers (324,000 miles) from the satellite. The massive crater, whose proportionate size (approximately 100 kilometers or 60 miles) is about one-quarter of the satellite's diameter (390 kilometers or 235 miles) is without precedent among the explored objects of the solar system. The impact that formed the crater was probably almost large enough to shatter Mimas into two or more fragments. (*Jet Propulsion Laboratory, California Institute Of Technology, National Aeronautics And Space Administration.*)

models predict that ethane is the dominant end product of methane photolysis so that the evolving ocean is both the source and the sink for continuing photolysis. The coexisting atmosphere is compatible with *Voyager* data.

As pointed out by D.O. Muhleman and colleagues, Division of Geological and Planetary Sciences, California Institute of Technology, "The present understanding of the atmosphere and surface conditions on Saturn's largest moon, Titan, including the stability of methane, and an application of thermodynamics leads to a strong prediction of liquid hydrocarbons in an ethane-methane mixture on the surface." These scientists utilized the Very Large Array radio telescope in New Mexico as the receiving instrument. Statistically significant echoes were obtained that show that Titan is not covered with a deep, global ocean of ethane, as previously thought. The researchers conclude their mid-1990 paper with, "The VLA/JPL (Very Large Array/Jet Propulsion Laboratory) radar measurements will be repeated as often as possible during the closest passages of Titan to the Earth. However, most of the questions concerning the Titan surface will not be answered until the *Cassini* spacecraft reaches the Saturn system. It is quite likely that such a spacecraft would have an imaging radar and altimeter that will reveal much of the surface structure and will

be very important for the study of Titan's geology, but will not be able to measure the surface reflectivity with an accuracy near to that of even current Earth-based radars."

Inner Satellites. Those Saturnian satellites which range in distance outward from the planet between 99,760 kilometers (62,000 miles) and 482,700 kilometers (300,000 miles) are known as *inner* satellites. These include Janus, Mimas, Enceladus, Tethys, Dione, and Rhea. These satellites are approximately spherical in shape and appear to be composed mostly of water ice. Enceladus reflects almost 100% of the sunlight that strikes it. These satellites represent a size of satellite not previously explored during pre-*Voyager* missions. Mimas, Tethys, Dione, and Rhea are all heavily cratered; Enceladus features fewer craters. See Figs. 21 and 22.

Enceladus. This satellite appears to have by far the most active surface of any satellite in the Saturnian system. At least five types of surface-terrain elements have been identified. Although craters can be seen across portions of its surface, the lack of craters in other areas implies an age less than a few hundred million years for the youngest region on the surface. See Fig. 23. It seems likely that portions of the surface of Enceladus are still undergoing change, since there are areas that are covered by ridged

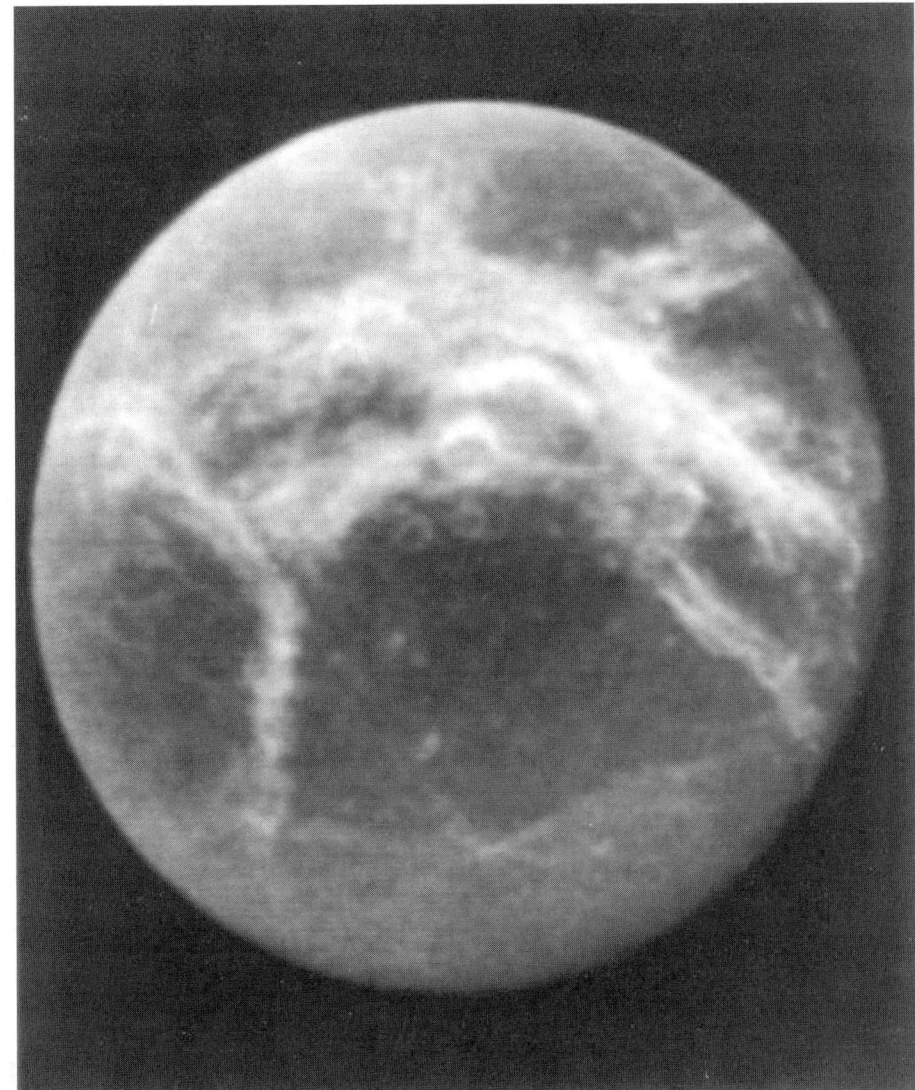

Fig. 26.   Large bright streaks are seen to cross the face of Saturn's moon Dione in this photograph taken by NASA's *Voyager 1* on November 12, 1980 from a distance of 695,000 kilometers (417,000 miles). Higher resolution views of Dione taken by *Voyager 1* show some of these streaks to be grooves that may be the result of fracturing in the satellite's surface. (*Jet Propulsion Laboratory, California Institute Of Technology, National Aeronautics And Space Administration.*)

plains with no evidence of cratering down to the limit of resolution of *Voyager 2* cameras (2 kilometers; 1.1 miles). Other areas are criss-crossed by a pattern of linear faults. It is unlikely that a body so small could contain enough radioactive material for the modification to have been produced internally. A more likely source of heating appears to be tidal interaction with Dione—similar to the action between Jupiter and its satellite, Io. For Enceladus' present orbit, however, current theories of tidal heating do not predict generation of sufficient energy to explain all the heating that must have occurred. Because the satellite reflects so much sunlight, Enceladus' current surface temperature is only 72 K (−330 °F; −201 °C).

Tethys. Photos of Tethys taken by *Voyager 2* show an enormous impact crater nearly one-third the diameter of the satellite and larger than Mimas. The crater appears to have formed when Tethys was relatively fluid because nearly the original shape of the satellite was restored after impact. A gigantic fracture covers three-fourths of Tethys' circumference. See Fig. 24. Scientists suggest that the fracture can be explained if Tethys were once fluid and its crust hardened before the interior. Freezing and consequent expansion of the interior could have caused a surface fracture about the size of that observed. However, expansion would not be expected to cause only one large crack. The canyon has been named Ithaca Chasma. Tethys' current surface temperature is about 86 K (−305 °F; −187 °C).

Mimas. Photos of Mimas show a huge crater. The crater is about one-third the diameter of the satellite. This crater has greater surface relief, implying that Mimas was much more rigid at the time the cratering event occurred. See Fig. 25.

Dione. The terrain of this satellite is described as "wispy," probably consisting of numerous cracks rimmed with ice. As shown in Fig. 26, a very large and wide fracture is noted in the southern hemisphere of the body. The wispy streaks stand out vividly against an already bright surface.

Rhea. This is a crater-saturated body, possibly impacted by many short-period comets. This is one of the better mapped satellites of Saturn. See Fig. 27.

Outer Satellites. In addition to Titan, previously described, the outer satellites of Saturn include Hyperion, Iapetus, and Phoebe. The outer satellites range in distance outward from the planet between about 1.15 million kilometers (720,000 miles) and 12.9 million kilometers (8 million miles).

Iapetus. This satellite has long been known to have large differences in surface brightness. Brightness of the surface material on the trailing side has been measured at 50%, while the leading-side material reflects only 5% of the sunlight that strikes it. Most of the dark material is distributed in a pattern directly centered on the leading surface, causing conjecture that the material was swept up as it spiralled inward, presumably from Phoebe. The trailing face of Iapetus, however, has several craters with dark floors. That implies that the dark material originated in the satellite's interior. It is possible that the dark material came both from Phoebe and from Iapetus' interior. See Fig. 28.

Hyperion. This satellite shows no evidence of internal activity. Its irregular shape and evidence of bombardment by meteoritic material makes it appear to be the oldest surface in the Saturnian system. Hyperion is shaped like a hamburger and is probably not in a gravitationally stable

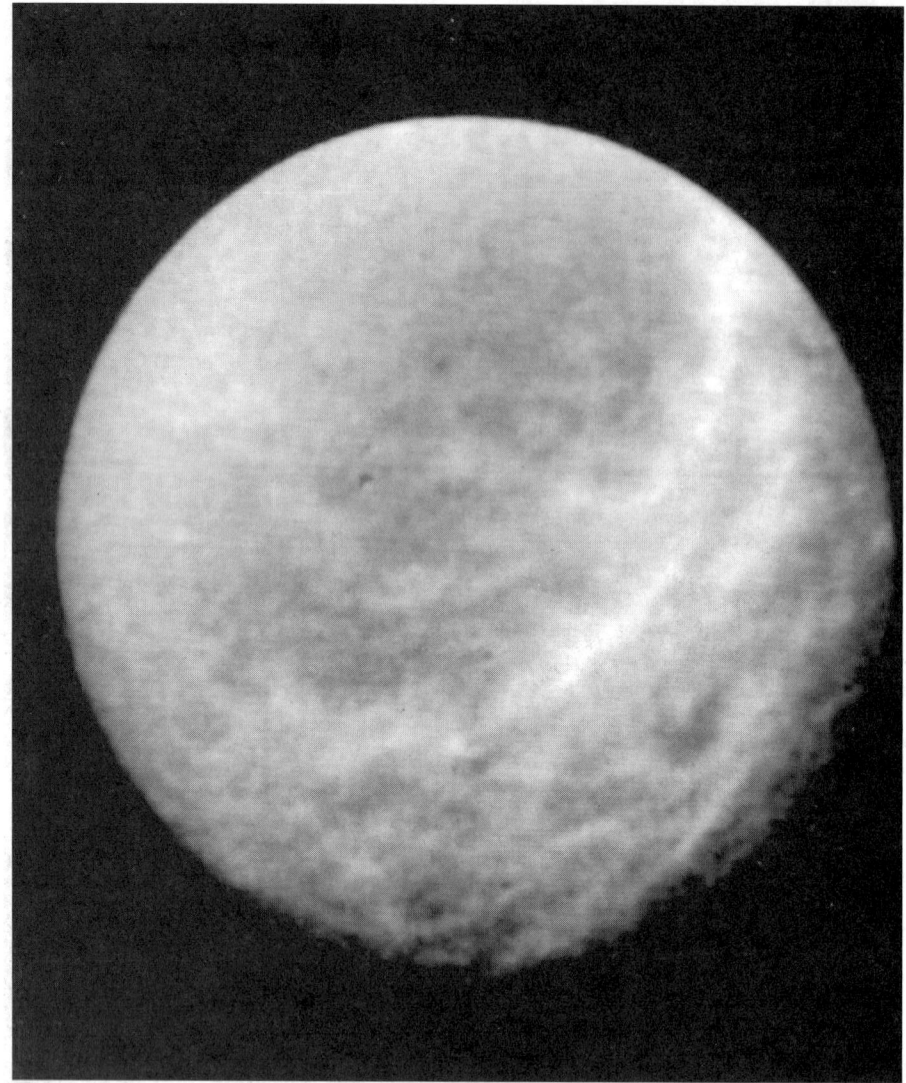

Fig. 27.   Special computer processing was used to enhance subtle color and brightness variations in this photo of Saturn's satellite, Rhea, taken by Voyager 1 on November 12, 1980, at a distance of 1.3 million kilometers (808,000 miles). Rhea's surface is composed mostly of ice, so is very reflective, and presents an almost uniformly white appearance. Of particular interest to Voyager scientists are the bright streaks that can be seen crossing Rhea's face. Scientists believe the streaks may be caused by fresh ice ejected from beneath the satellite's surface. (*Jet Propulsion Laboratory, California Institute Of Technology, National Aeronautics And Space Administration.*)

position. It is possible that one or more meteorites jostled Hyperion out of position and that the satellite will eventually swing back. See Fig. 29.

In studies of *Voyager* data, J. Wisdom and S. Peale (University of California, Santa Barbara) and F. Mignard (Research Center for the Study of Geodynamics and Astronomy, Grasse, France), reported that Hyperion tumbles chaotically, not just end over end, but first one way and then another, slowing down and then speeding up. These theorists believe that Saturn's gravity would synchronize Hyperion's rotation if the satellite were not of such an odd shape (115 × 145 × 190 km) and in an elongated, eccentric orbit. It is suggested that as Hyperion follows its eccentric orbit, Saturn tugs on different parts of the satellite with differing effects, sending it into chaotic tumbling. There is also the gravitational effect of Titan on the small satellite. Unfortunately, a difference of 18 months in the timing of observations is the basis of the current conclusions.

Further studies by Thomas (Cornell University) and colleagues of the brightness of Hyperion indicate a regular 13-day period of rotation during the 61 days of *Voyager 2's* encounter with Saturn. The case of Hyperion's rotation has not been solved conclusively.

Phoebe. This satellite was photographed by *Voyager 2* after the spacecraft passed Saturn. Phoebe orbits in a retrograde direction in a plane much closer to the ecliptic than to Saturn's equatorial plane. *Voyager 2* found that Phoebe is roughly circular in shape and reflects about 6% of the sunlight. It is also quite red in color. Phoebe rotates on its axis about once in nine hours. Thus, unlike the other Saturnian satellites, it does not always show the same face to the planet. If, as scientists suggest, Phoebe is a captured asteroid with its composition unmodified since its formation in the outer solar system, it is the first such object that has been photographed.

Newly Discovered, Very Small Satellites. A total of seven very small satellites was observed by *Voyagers 1* and *2*. See Fig. 30. These satellites are irregularly shaped bodies that have been highly cratered by the impact of cosmic debris. The irregularity is probably due to fracturing by large impacts and is sustained by the rigidity of the bodies. These objects range from about ten to several hundred kilometers across. Included are the two F-ring shepherding satellites, first seen by *Voyager 1* in 1980.

Post-Mission Analyses

Based upon the masses of information collected from prior Saturn missions, numerous scientists establish continuing programs of data analysis to refine prior concepts of the characteristics of the planet. As one example, J.E. Klepeis and colleagues at the University of California, Lawrence Livermore National Laboratory, reported on their experiences by establishing models of Saturn and making comparisons with the conditions on Jupiter. The group has observed, "Models of Jupiter and Saturn postulate a central rock core surrounded by a fluid mixture of hydrogen and helium. These models suggest that the mixture is undergoing phase separation in Saturn, but not Jupiter. State-of-the-art total energy calculations of the enthalpy of mixing for ordered alloys of hydrogen and helium confirm that at least partial phase separation has occurred on Saturn and predict that this process also has begun on Jupiter."

Fig. 28.   View made by *Voyager 2* at time of its closest approach to Iapetus, the outermost of Saturn's large satellites. Image was made on August 22, 1981 when the spacecraft was 1.1 million kilometers (680,000 miles) from the satellite. The camera resolution was about 21 kilometers (13 miles). This view, which is lit from above, primarily shows the heavily cratered northern hemisphere toward the bright trailing side of the satellite. The north pole itself is near the large central-peak crater seen partly in the shadow at the top of the image. Iapetus is noteworthy for the very dark material (lower and right-hand parts of this frame) that apparently covers the ice crust of the satellite primarily at its leading hemisphere. This dark material is red and reflects only about 5% of the incident sunlight; it may be of either external or internal origin. Study of its relationship with the underlying topography, for example, near the large crater at the border of the dark material, may help to resolve this mystery. (*Jet Propulsion Laboratory*.)

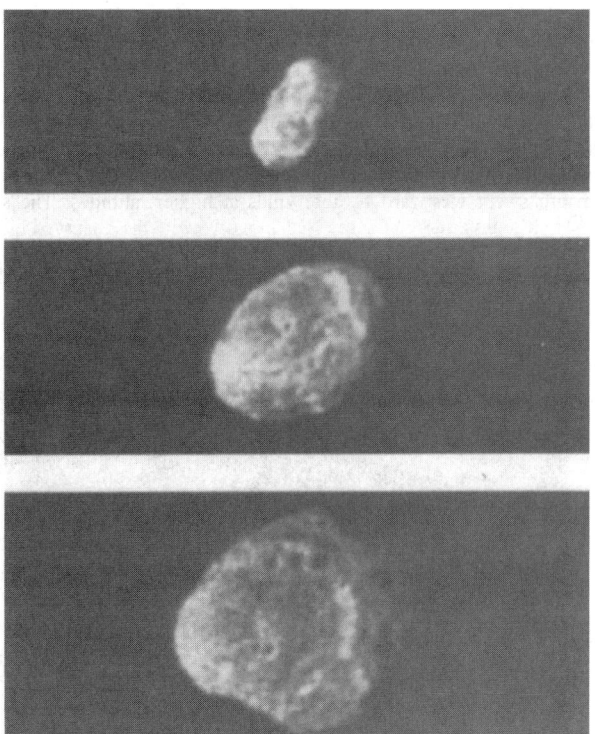

Fig. 29.   Three views of Hyperion obtained as *Voyager 2* flew by this satellite. The views were taken: top view, on the morning of August 23, 1981 from a range of 1.2 million kilometers (740,000 miles); middle view, on the morning of August 24 from 700,000 kilometers (430,000 miles); lower view, at noon on August 24 from 500,000 kilometers (310,000 miles). Together, these views show the changing aspect of the satellite as *Voyager 2* moved in for closer views. Hyperion, roughly 360 by 210 kilometers (220 by 130 miles) and shaped like a hamburger, is probably not in a gravitationally stable position. Its surface is pock-marked with many meteorite-impact craters. The large indentation at the bottom limb (lower view) is one such crater (about 100 kilometers; 60 miles across). The smallest visible crater pit is about 10–20 kilometers (6–12 miles.)

Radar Imaging from Earth. The *Voyager* missions to Saturn did not end all Earth-based astronomical imaging of the planet. These would include imaging by the Hubble space telescope mentioned later. However, much greater progress has occurred in connection with radar imaging. For example, A.W. Grossman and colleagues at the Division of Geological and Planetary Science, California Institute of Technology, describe their use of high-resolution microwave imaging of the planet. In a 1989 paper, the researchers used this technology to study the ring systems of Saturn. The radio interferometric observations were made at the Very Large Array telescope in New Mexico at wavelengths of 2 and 6 centimeters. With this information, the investigators prepared maps that show an increase in brightness temperature of about 3 K from equator to pole at both wavelengths, while a map made with 6-meter radiation indicated a bright band at northern mid-latitudes. "These data are consistent with a radiative transfer model of the atmosphere that constrains the well-mixed, fully saturated, NH_3 mixing ratio to be 1.2×10^{-4} in a region just below the NH_3 clouds, while the observed bright band indicates a 25 percent relative decrease of NH_3 in northern mid-latitudes. Brightness temperatures for the classical rings also are presented in the paper. Ring brightness shows a variation with azimuth and is linearly polarized at an average value of about 5 percent. The variations in ring polarization suggest that at least 20 percent of the ring brightness is the result of a single scattering process."

The scientists conclude their paper, "The observations represent the ultimate in resolution and sensitivity obtainable from Earth-based radio telescopes. A vast improvement in our knowledge of Saturn's deep atmosphere and rings can be obtained from a Saturn-orbiting spacecraft instrument observing at radio wavelengths."

Great White Spot of Saturn. When the Hubble space telescope was pointed toward Saturn in late August 1990, the planet's image, after computer corrections to compensate for Hubble's disability, exhibited no surprises. The planet appeared calm and much as expected. However, in late September, a number of amateur astronomers reported the sudden development of a white spot reminiscent of a gigantic storm system. Larger, professional telescopes were turned to view Saturn, and these better images confirmed some unusual activity on the planet spread across Saturn's equatorial region. Scientists then interrupted Hubble's planned schedule for viewing Saturn. The white spot was reconfirmed. See Fig. 31. Some scientists liken the spot to the Great Red Spot of Jupiter. Because magma in the interior of Saturn had not been detected or even suspected, a volcanic eruption was quickly ruled out. There was a press conference on November 20, 1990, at which time a color image was released. The spot was described as a brick-colored hurricane. One scientist has observed that apparently massive amounts of energy, if not magmatic, well up from its interior (this energy remains from the period when the planet was formed) and that this heat, when it reaches the planet's thin atmosphere, periodically causes considerable turbulence.

NASA Hubble Space Telescope image of the ringed planet Saturn shows a rare storm that appears as a white arrowhead-shaped feature near the planet's equator. See Fig. 32. An upwelling of warmer air, similar to a terrestrial thunderhead generates the storm. The east-west extent of this storm is equal to the diameter of the Earth (about 13,167 kilometers or 7,900 miles). Hubble provides new details about the effects of Saturn's prevailing winds on the storm. The new image shows that the storm's motion and size have changed little since its discovery in September, 1994. The storm was imaged with Hubble's Wide Field Planetary Camera 2 (WFPC2) in the wide field mode on December 1, 1994, when Saturn was 1,506.7 million kilometers (904 million miles) from the Earth. The picture is a composite of images taken through different color filters within a six minute interval to create a "true-color" rendition of the planet. The blue fringe on the right limb of the planet is an artifact of image processing used to compensate for the rotation of the planet between exposures.

The Hubble images are sharp enough to reveal that Saturn's prevailing winds shape a dark "wedge" that eats into the western (left) side of the bright central cloud. The planet's strongest eastward winds (clocked at 1,000 miles per hour from analysis of *Voyager* spacecraft images taken in 1980–81) are at the latitude of the wedge. To the north of this arrowhead-shaped feature, the winds decrease so that the storm center is moving eastward relative to the local flow. The clouds expanding north of the storm are swept westward by the winds at higher latitudes. The strong winds near the latitude of the dark wedge blow over the northern part of the storm, creating a secondary disturbance that generates the faint white clouds to the east (right) of the storm center. The storm's white clouds are ammonia ice crystals that form when an upward flow of warmer gases shoves its way through Saturn's frigid cloud tops. This current storm is larger than the white clouds associated with minor storms that have been reported more frequently as bright cloud features. Hubble observed a similar, though larger, storm in September 1990, which was one of three

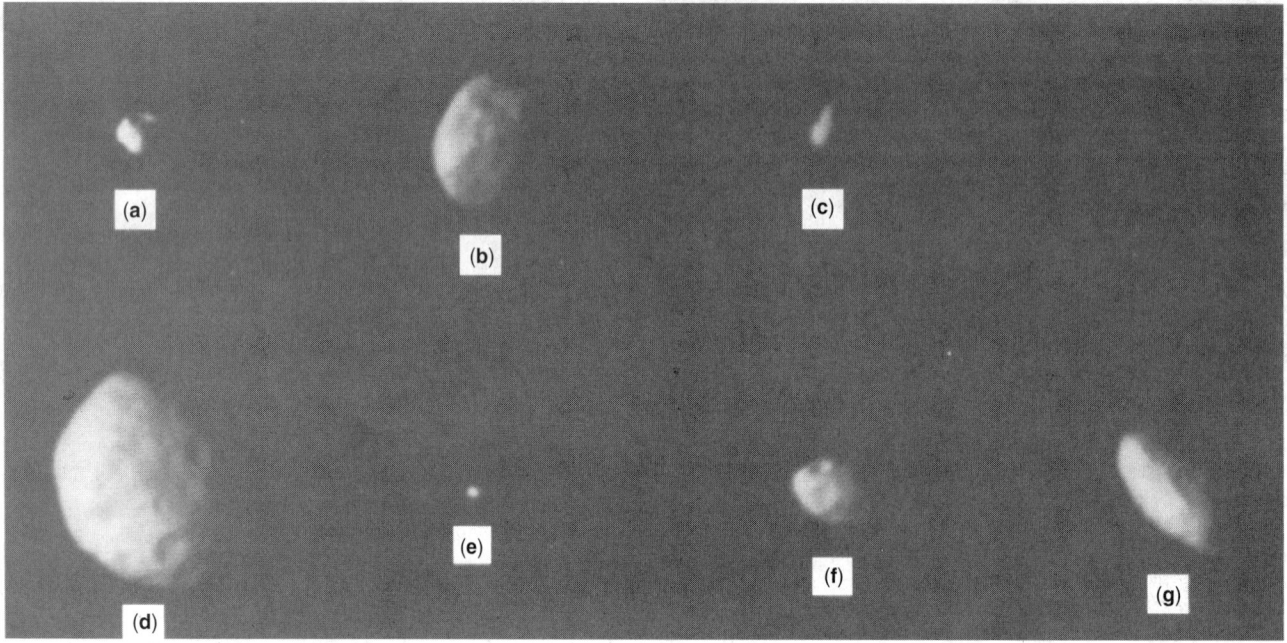

Fig. 30.   Composite showing seven of the very small satellites of Saturn as photographed on August 25, 1981 by *Voyager 2*. (**a**) 1980S6 (Dione trojan); (**b**) 1980S3 (Trailing co-orbital); (**c**) 1980S25 (Trailing Tethys trojan); (**d**) 1980S1 (Leading co-orbital); (**e**) 1980S13 (Leading Tethys trojan); (**f**) 1980S26 (Outer F-ring shepherd); (**g**) 1980S27 (Inner F-ring shepherd). It should be noted that these views taken at differing distances from the various satellites, ranging from 248,000 kilometers (154,000 miles) to 667,000 kilometers (414,000 miles). (*Jet Propulsion Laboratory*.)

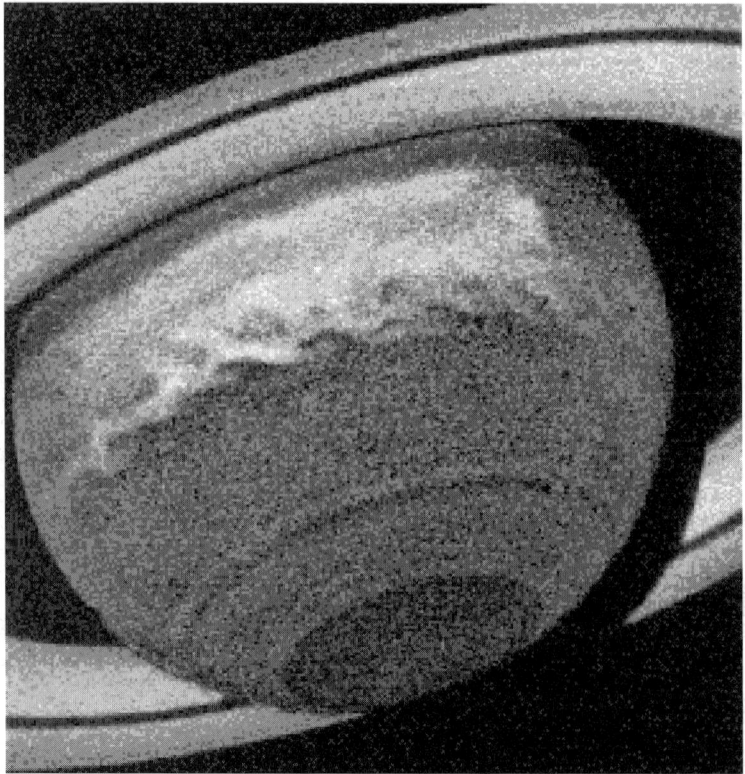

Fig. 31.   Saturn's Great White Spot. On September 24, 1990, Stuart Wilber discovered a giant white spot on Saturn using a 10-inch reflector at Las Cruces, New Mexico. The Hubble Space Telescope took this image on November 9, 1990, and the white spot had spread out around the planet. This was the most spectacular eruption on Saturn since 1933. (*NASA*).

Fig. 32.   This NASA Hubble Space Telescope image of the ringed planet Saturn shows a rare storm that appears as a white arrowhead-shaped feature near the planet's equator. The new image shows that the storm's motion and size have changed little since its discovery in September, 1994. The storm was imaged with Hubble's Wide Field Planetary Camera 2 (WFPC2) in the wide field mode on December 1, 1994, when Saturn was 1,506.7 million kilometers (904 million miles) from the Earth. The picture is a composite of images taken through different color filters within a six minute interval to create a "true-color" rendition of the planet. (*Credit: Reta Beebe (New Mexico State University), D. Gilmore, L. Bergeron (STScI), and NASA.*)

major Saturn storms seen over the past two centuries. Although these events were separated by about 57 years (approximately 2 Saturnian years) there is yet no explanation why they apparently follow a cycle — occurring when it is summer in Saturn's Northern Hemisphere. For additional photos of Saturn. See also **Hubble Space Telescope (HST)**.

Additional Reading

Allison, M., D.A. Godfrey, and R.F. Bebbe: "A Wave Dynamical Interpretation of Saturn's Polar Hexagon," *Science*, 1061 (March 2, 1990).

Bosh, A.S. and A.S. Rivkin: "Observations of Saturn's Inner Satellites During the May 1995 Ring-plane Crossing," *Science* 272, 518–521 1996.

Cherfas, J.: "Saturn Mission Backed, Europeans Relieved," *Science*, 628 (November 2, 1990).

Cowen, R.: "Spotting an Ephemeral Artifact on Saturn," *Science News*, 228 (October 13, 1990).

Dickson, D.: "Europeans Decide on a Trip to Saturn," *Science*, 1375 (December 9, 1988).

Eberhart, J.: "Saturn Ring Ripple Suggests 19th Moon," *Science News*, 31 (July 14, 1990).

Eberhart, J.: "Five-Year Hunt Locates Saturn's 18th Moon," *Science News*, 69 (August 4, 1990).

Godfrey, D.A.: "The Rotation Period of Saturn's Polar Hexagon," *Science*, 1206 (March 9, 1990).

Grossman, A.W., D.O. Muhelman, and G.L. Berge: "High-Resolution Microwave Images of Saturn," *Science*, 1211 (September 15, 1989).

Ingersoll, A.P.: "Atmospheric Dynamics of the Outer Planets," *Science*, 308 (April 20, 1990).

Kerr, R.A.: "A Passion for the Little Things Among the Planets," *Science*, 998 (November 24, 1989).

Klepeis, J.E. et al.: "Hydrogen-Helium Mixtures at Megabar Pressures: Implications for Jupiter and Saturn," *Science*, 986 (November 15, 1991).

Kunine, J.I.: "Origin and Evolution of Outer Solar System Atmospheres," *Science*, 141 (July 14, 1989).

McKay, C.P., J.B. Pollack, and R. Courtin: "The Greenhouse and Antigreenhouse Effects on Titan," *Science*, 1118 (September 6, 1991).

Muhleman, D.O. et al.: "Radar Reflectivity of Titan," *Science*, 975 (May 25, 1990).

Nicholson, P.D., M.R. Showalter, L. Dones, R.G. French, S.M. Larson, J.J. Lissauer, C.A. McGhee, P. Seitzer, B. Sicardy, and G.E. Danielson: "Observations of Saturn's Ring-plane Crossings in August and November 1995," *Science* 272, 509–515 1996.

Powell, C.S.: "Hubble Gags a Great White," *Sci. Amer.*, 26 (February 1991).

Rosen, P.A. and J.J. Lissauer: "The Titan Nodal Bending Wave in Saturn's Ring C," *Science*, 690 (August 5, 1988).

Waldrop, M.M.: "Titan: Continents in a Hydrocarbon Sea," *Science*, 129 (July 14, 1989).

Waldrop, M.M.: "Images of an Unquiet Planet (Saturn)," *Science*, 1201 (November 30, 1990).

Pre-1986 References

Bane, D.: "The Voyager 1 and 2 Saturn Science Results," Jet Propulsion Laboratory, Pasadena, CA, December 1981.

Davis, D.R. et al.: "Saturn Ring Particles as Dynamic Ephemeral Bodies," *Science*, **224**, 744–747 (1984).

Dyer, J.W.: "Pioneer Saturn" (Contains 14 papers relating to Pioneer 11 encounter with Saturn), *Science*, **207**, 400–453 (1980).

Eshleman, V.R., G.F. Lindal, and G.L. Tyler: "Is Titan Wet or Dry?" *Science*, **221**, 53–55 (1983).

Esposito, L.W. et al.: "Eccentric Ringlet in the Maxwell Gap at 1.45 Saturn Radii: Multi-Instrument Voyager Observations," *Science*, **222**, 57–59 (1983).

Flasar, F.M.: "Oceans on Titan?", *Science*, **221**, 55–57 (1983).

Gehrels, T. and M.S. Matthews, Eds.: "Saturn," University of Arizona Press, Tucson, AZ, 1984.

Ingersoll, A.P.: "Jupiter and Saturn," *Sci. Amer.*, 90–108 (December 1981).

Loudon, J.: "The Last Picture Show," *Technol. Rev. (MIT)*, **84**(2), 19 (1981).

Lunine, J.I., D.J. Stevenson, and Y.L. Yung: "Ethane Ocean on Titan," *Science*, **222**, 1229–1230 (1983).

Lutz, B.L., C. De Bergh, and T. Owens: "Titan: Discovery of Carbon Monoxide in Its Atmosphere," *Science*, **220**, 1374–1375 (1983).

Muhleman, D.O., G.L. Berge, and R.T. Clancy: "Microwave Measurements of Carbon Monoxide on Titan," *Science*, **223**, 393–396 (1984).

News: "Frigid Oceans for Triton and Titan: Chaotic Rotation Predicted for Hyperion; Could Saturn's Rings Have Melted Enceladus?" *Science*, **221**, 448–449 (1983).

News: "The Rotation of Saturn's Hyperion Looks Chaotic," *Science*, **230**, 1027 (1985).

Owen, T.: "Titan," *Sci. Amer.*, 98–109 (February 1982).

Soderblom, L.A. and T.V. Johnson: "The Moons of Saturn," *Sci. Amer.*, 100–116 (January 1982).

Stone, E.C. and E.D. Miner: "Voyager 1 Encounter with the Saturnian System," (Followed by a series of detailed articles). *Science*, **212**, 159–243 (1981).

Stone, E.C. et al.: "*Voyager 2* Encounter with the Saturnian System," *Science*, **215**, 499–594 (1982).

Zebker, H.A. and G.L. Tyler: "Thickness of Saturn's Rings Inferred from *Voyager 1* Observations of Microwave Scatter," *Science*, **223**, 396–398 (1984).

Web References

Cassini Project Home Page: *http://www.jpl.nasa.gov/cassini/*

Hubble Space Telescope Images of Saturn: *http://www.jpl.nasa.gov/saturn/hst_images.html*

NASA Space Link: *http://spacelink.msfc.nasa.gov/.index.html*

NSSDC Photo Gallery: Saturn: *http://nssdc.gsfc.nasa.gov/photo_gallery/photogallery-saturn.html*

Planetary Rings Node: *http://ringmaster.arc.nasa.gov/*

Solar System Exploration: Missions: Saturn: *Pioneer 11: http://solarsystem.nasa.gov/missions/sat_missns/sat-p11.html*

Solar System Exploration: Missions: Saturn: *Voyager 2: http://solarsystem.nasa.gov/missions/sat_missns/sat-voy2.html*

Solar System Exploration: Missions: Saturn: *Cassini & Huygens* Probe: *http://solar-system.nasa.gov/missions/sat_missns/sat-cassini.html*

Views of the Solar System: *http://www.hawastsoc.org/solar/eng/homepage.htm*

Voyager Images of Saturn: *http://www.jpl.nasa.gov/saturn/vgr_images.html*

Voyager Project Home Page: *http://vraptor.jpl.nasa.gov/*

Editor's Note: The staff of this encyclopedia is most appreciative of the cooperation extended by D. Bane and A.S. Woods and their associates at the Jet Propulsion Laboratory, Pasadena, California, for furnishing information for numerous portions of this article.

SAUERKRAUT. See **Brassica**.

SAUGER. See **Perches and Darters**.

SAVANNAH. A tropical or subtropical region of grassland and other drought-resistant (xerophilous) vegetation. This type of growth occurs in regions that have a long dry season but a heavy rainy season, and continuously warm temperatures. Africa has the most extensive areas of savannah, but they are also widespread in South America (the *campos*), and, to a lesser extent, in India and southeast Asia, Australia, and Central America. See also **Biome**.

SAWFISHES. See **Skates and Rays**.

SAWFLY (*Insecta, Hymenoptera*). A plant-feeding member of this order, whose more familiar species are the ants, bees, and wasps. The sawflies have four wings, somewhat like those of the wasps, but the abdomen is broadly connected with the thorax, in contrast with the thin-waisted bodies of the other forms. See Fig. 1. They are named from the saw-like ovipositor with which slits are cut in the tissues of plants to receive the eggs. Some species are of economic importance. Since the larvae eat leaves they can be destroyed by a number of control chemicals. Cultural practices are also extremely important.

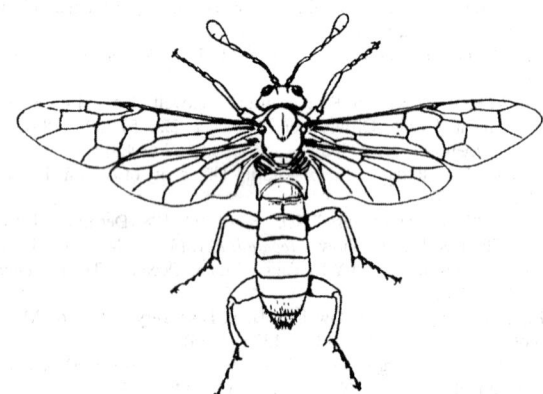

Fig. 1. Sawfly. (*USDA.*)

A native of North America, the *wheatstem sawfly* (*Cephus cinctus*, Norton) attacks wheat, as well as barley, spring rye, spelt, timothy, and a number of native grasses. The *European wheatstem sawfly* (*Cephus pygmacus*, Linne) is a closely related species. These insects are most damaging in the wheat-growing stages of the north and west of the Mississippi, although the European species is found mostly in the eastern states of New York and Pennsylvania. Without effective control measures, records indicate that in some years these insects have destroyed more than 50% of some crops. An infested field displays fallen straw (similar to the damage wrought by the hessian fly and jointworm). This is because the sawfly larvae have been feeding there. With the brown-headed, legless larva ($\frac{1}{3}$–$\frac{1}{2}$) inch; 8–12 millimeters long) will be straw cuttings that look like sawdust. This is the result of the cutting-type action characteristic of this insect. As the season progresses, the larva works their way down the stem and finally makes a V-shaped groove on the inside of the stem, causing the stem to break. The insect drops to the base of the plant and plugs up the opening, where it hibernates and ultimately pupates.

Control practices include turning the stubble shortly after harvest. Burning is not practical because the insects live close to ground level. Early cutting of grain, when possible, also reduces the damage. Rotation of crops, planting corn (maize), flax, alfalfa or sweet clover, is effective. Where the sawfly is a persistent problem, wheat varieties that are solid-stemmed should be selected.

The *raspberry sawfly* (*Monophadnoides geniculatus* or *Priophorus rubivorus*, Rowler) is sometimes abundant on raspberry leaves at a time in spring when the plant has reached its full foliage. The larva is a spiny, multilegged, pale-green worm that usually feeds along the edges of leaves. Unless controlled, the plant may be fully stripped of its leaves.

The *currant stem girdler* (*Janus integer*) is another economic species found in the United States.

Horntail flies are large sawflies (woodwasps) whose larvae bore in the trunks of trees. The adults have a cylindrical body; the female has a short, strong ovipositor which is the source of the name horn-tail. With this organ, holes are drilled into the wood of the tree for deposition of eggs. Among these is the species *Cimbex americana*, the *elm sawfly*.

The *pear slug* (*Caliroa cerasi*, Linne) is the olive to dark green or black larva of a shiny black sawfly. The larva resembles a snail or slug and ranges up to $\frac{1}{2}$ inch (12 millimeters) in length. The larva feeds on the upper surface of cherry, pear, and plum trees, retards growth and development of fruit, and generally weakens the tree. The pest is distributed throughout the United States. Chemical controls are the same as those for fruit curculios. See also **Curculio**. In small orchard operations, handpicking the larvae from the leaves and placing them in a pail containing kerosene is effective. Also, a few applications of lime diluted with water and applied to the leaves is effective. The slugs can be washed from the foliage with a strong stream of water and then destroyed at groundlevel.

Knerer and Atwood have made an extensive study of the polymorphism and speciation of *Diprionid* sawflies, and this is reported in *Science*, **179** (4078), 1090–1099 (1973). The *Diprionidae* represent only a small fraction of all the sawflies known, but they share the habits of most other leaf feeders. The family is interesting mainly because of the diversity of distinct races or physiological strains that are adapted to specific host plants, and because of the social behavior exhibited by the larvae in aggregations. Both phenomena illustrate various evolutionary mechanisms at work, simultaneously providing examples of newly emerging biologic units and of the origin of some of the most primitive social behavior found in insects. These factors are well summarized by Knerer and Atwood.

SAYBOLT VISCOSIMETER. See **Viscosity**.

S-BAND. A frequency band used in radar extending approximately from 1.55 to 5.2 kilomegacycles per second.

SCABIES. See **Dermatitis and Dermatosis**.

SCAD (*Osteichthyes*). Of the order *Percomorphi*, suborder *Percoidea*, family *Carangidae*, the scad is a species of fish common in the region of the West Indies. The scad is also called cigar fish or round robin. The big-eyed scad, goggler, or chicharro is related and although common in warmer waters, does range as far north as Cape Cod. The term *scad* also has been applied to the related horse mackerel, a European species occasionally taken on the Atlantic coast of North America.

SCALAR FIELD. Consider the space reflection $x' \to ; x; \mathbf{x}' = -\mathbf{x}$, $x'^0 = x^0$. Define the transformation of a spinless field operator under space reflection such that

$$\phi'(x') = P\phi(x)P^{-1}$$
$$= \eta\phi(x')$$

Where η_p is a complex constant and P the unitary operator which induces the transformation. If the operator of space reflection is performed twice, we must revert to the original field so that $\eta_{p^2} = 1$ or $\eta_p = \pm 1$. A field which transforms with $\eta_p = +1$ is called a *scalar field*, while one that transforms with $\eta_p = -1$ is called a *pseudoscalar field*. The particles described by scalar (pseudoscalar) fields have an intrinsic parity $+1(-1)$.

SCALAR PRODUCT. See **Vector Multiplication**.

SCALAR QUANTITY. A number, either in the sense of the ratio between two quantities of the same kind, or as distinguished from other mathematical constructs, e.g., vectors, matrices, etc.

SCALD FISHES.   See **Flatfishes**.

SCALE.   Six common uses of this term are:

1. A balance used for weighing.

2. A series of markings at regular intervals which are used for measurement or computation.

3. A defined set of values in terms of which quantities of the same nature as those defined may be measured, e.g., temperature scale.

4. A musical scale consisting of a series of notes (symbols, sensations or stimuli), arranged from high to low by a specified scheme of intervals, suitable for musical purposes.

5. In metallurgy, scale is the oxide layer that forms on the surface of metals upon heating in air or other oxidizing gases. The heavy scale that forms on steel ingots or billets upon heating for rolling or forging breaks away as the metal is deformed in the mill or under the hammer; however a lighter, often very adherent scale always remains after hot-working operations and after annealing or other heat treatments unless a nonoxidizing protective atmosphere is provided. Scale is removed from steel by pickling, generally in warm dilute sulfuric or hydrochloric acid. The scale which forms on stainless steel is much more resistant and its removal requires strong acids such as mixtures of hot nitric and hydrofluoric. A tight adherent scale is often left on steel for its protective value; for example, steam-blued steel sheets are used for stovepipe.

6. In animal anatomy, a scale is a flat structure developed as a covering. (A) The scales of fishes are in many cases arranged like shingles to form a complete armor at the surface of the body, and in other forms are small and scattered, merely adding to the resistant qualities of the skin. The elasmobranch fishes have placoid scales, whose form includes a broad base and a projecting enamel-covered point. This form of scale is much like the teeth of these fishes and is supposed to be ancestral to all vertebrate teeth. Ganoid scales, found in relatively few fishes, are regarded as a modification of the placoid type by the loss of the point and the addition of a hard outer layer known as ganoin. Some of the more primitive fishes of other groups have rounded scales, with smooth margins, known as cycloid, and others have ctenoid scales, with comb-like edges. The two last forms are found in the higher fishes. They bear neither enamel nor ganoin. (B) The scales found over the entire surface of the body in reptiles, on the legs of birds, and to a much more limited extent in mammals, as on the tails of rodents, are quite different from the scales of fishes. Each is a modified area of the skin, thickened and hardened by the development of the horny substance, keratin. Scales reach their highest development among the mammals in the pangolins and armadillos. In the latter they are underlaid by bony plates. (C) The butterflies and moths, a few beetles, and some other insects have the surface of the body and wings more or less covered by flattened scales. These structures are modified setae. They often contain pigments and in many species are so formed that they produce iridescent, metallic, or glossy physical colors by breaking up the light rays they reflect. (D) Scale insects, often called scales, are highly specialized sucking insects that live on plants. They belong to the order *Hemiptera* and are related to the plant lice and phylloxerans. Most species are minute. The young and the female adults are simplified and remain closely attached to the plant, secreting over themselves a protective covering or scale which gives them their name. This covering is usually characteristic of the species.

Scale insects include many species of economic importance. Among them are the useful cochineal insect, a source of dye, and the lac insect whose scale is the raw material from which shellac is made. China wax is also a scale insect product. Among the harmful species the purple scale of citrus fruits, the San José scale, and the oystershell scale are important. Because of the protective scale these species are not easily destroyed by the usual contact sprays. They are controlled by spraying but the concentration of poison must be high and spraying during the dormant stage of the plant is often necessary as a result. Fumigation of citrus trees with cyanide is practiced extensively. This method requires special equipment since the tree must be enclosed in a tent. See also **Scale Insects**.

SCALE FACTOR (Computer System).   In analog computing, a proportionality factor which relates the magnitude of a variable to its representation within a computer. In digital computing, the arbitrary factor that may be associated with numbers in a computer to adjust the position of the radix so that the significant digits occupy specified columns.

SCALE FISH.   See **Flatfishes**.

SCALE HEIGHT.   See **Atmospheric Pressure**.

SCALE INSECTS (*Insecta, Hemiptera*).   These insects include numerous species of the family *Coccidae*, of which the mealy bugs also are members. However, there is a marked contrast: Whereas the mealy bugs can move about freely, the scale insects, within a few hours after birth or hatching, remain fixed throughout the remainder of their life. The typical scale insect may be described as wingless and grub-like in appearance, but masked by a covering of a cottony, powdery, or waxy secretion. In some species, this coating hardens into a scale-like shell or skin — hence the term *scale insects*. See Fig. 1. Generally, the scale insects tend to become specialists, thriving on a few select trees and fruits as their source of food and habitat. For certain food crops, such as apple, citrus, and other deciduous fruits, the scale insects are regarded as very damaging pests. The insects are found throughout the world wherever these crops are cultivated.

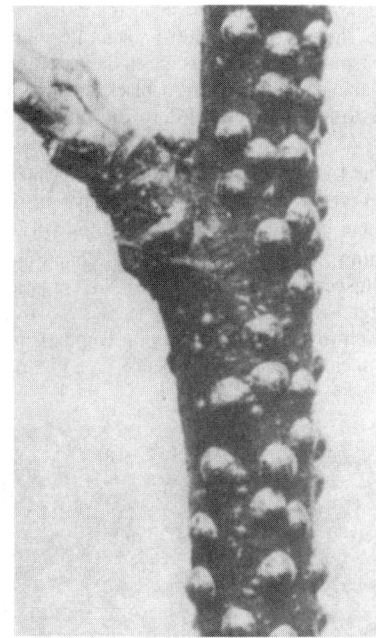

Fig. 1.   Terrapin scale attached to twig of peach tree.

The scale-like coverings make it difficult to kill these insects. They are very small and thus easy to transport as stock may be moved from a glasshouse or nursery for transplanting. They multiply very rapidly. For example, researchers have found that a single female of the *San Jose species* can rear from 100 to 600 young at a rate of 4 to 5 generations per year. It has been established that one *Lecanium scale insect* can lay over 2000 eggs. Fortunately, the scale insects have a number of natural enemies which most likely control the highest percentage of these pests, even though chemical intervention is often required to make up for any deficiencies of the natural enemies. A number of years ago, the *cottony cushion scale (Icerya purchasi)* was accidentally introduced into California. Originally an Australian insect, this scale multiplied very fast once free of its natural enemies and spread rapidly over the West Coast citrus crop. A very large measure of control was accomplished biologically by the purposeful introduction of lady beetles which have a great preference for this scale insect as a food source. Parasitic wasps and mites also are effective in naturally keeping the scale populations under control.

Because the scale insects have a considerable amount of built-in protection, however, including an ability to tightly attach themselves to the bark of trees, which makes them difficult to remove by washing, for years the most effective means of controlling really bad infestations has been the use of various fumigants, notably cyanogen gas. Gastight buildings for use in treating plants prior to transplanting have been used, as well as large tent-like enclosures over trees, glasshouses, and other buildings. For mild populations of the insects, however, various sprays, such as oil and kerosene emulsions, sometimes lime-sulfur formulations, and even soap solutions (under high pressure), are effective. In some cases, mechanical scraping of bark or twigs can be effective.

The principal damage caused by scale insects is their sucking out plant juices, causing discolored (sometimes red) spots on leaves, stems, and fruit. Wounds make excellent sites for invasion by fungus spores.

Following is an abridged list of various scale insects that are damaging to crops.

Barnacle Scale (*Ceroplastes cirripediformis*). Found on citrus and Eupatorium. Insect is about $\frac{1}{5}$ inch (5 millimeters) in length. Scales are dark brown and surface of scale sculptured to appear like miniature barnacles.

Black Scale (*Saissetia olex; Lecanium oleae*). Found on citrus and deciduous fruits; olive. Insect is oval-shape, about $\frac{1}{4}$ inch (6 millimeters) in diameter. Scale covering is black. On back of female, there is an H marking. Scales secrete honeydew, providing a locale for fungus to grow, thus smutting the fruit.

Black Parlatoria Scale (*Parlatoria zizyphus*). A pest to citrus in many parts of the world. The insect spends most of its life beneath a dull-black scale. It gives off a whitish or brownish secretion that sometimes extends beyond the scale. The female is about $\frac{1}{16}$ inch (1.5 millimeters) long; the male is somewhat smaller. The insects attach themselves to leaves and fruit of citrus. When abundant, they form a black crust. A heavy infestation causes the leaves to turn yellow and drop and prevents full development of the fruit.

Coconut Scale (*Aspidiotus destructor*). Found mainly on coconut. Insect is white to cream in color and has transparent scales. The insects frequently are found in large numbers on underside of leaves and fruit.

Cottony Cushion Scale (*Icerya purchasi*). Found on almond, apricot, citrus, and fig. Insect is red to yellowish in color and covered with comparatively large, white, fluted cottony masses for covering eggs. Insect is from $\frac{1}{4}$ to $\frac{1}{2}$ inch (6 to 12 millimeters) in length. Control is mainly by natural enemies, such as *Vedalia* lady beetle. See Fig. 2.

Fig. 2.   Cotton cushion scale insects on orange tree. (*USDA.*)

Date Palm Scale (*Parlatoria blanchardii*). Found mainly on date palm. A small, elongate insect having gray or black scales with white edges. The males are always white.

European Fruit Scale (*Lecanium corni*). Found mainly on plum. A rather large, circular-shaped insect which sometimes can be very destructive.

Euonymous Scale (*Chionaspis evonymi*). Found on small trees and shrubs of genus *Euonymous*. Insect is about $\frac{1}{12}$ inch (2 millimeters) in diameter with dark brown, convex scales on female and pure white scales on males.

Fig Scale (*Lepidosaphes ficus*, Signoret). This insect is a pest in fig orchards and traces of its presence may persist on the fruit after harvest. When infestation on the leaves becomes heavy, crawlers go to the developing fruit, settle, and feed on the green fig. Heavy infestations result in dried figs that are somewhat spotted and deformed.

Florida Red Scale (*Chrysomphalus aonidum*). Found on aspidistra (Asian herb and sometimes houseplant), *Betula* (alder and birch trees), citrus, coconut, and *Hedera* (woody vines; true ivy). Insect is circular with flat brown scales. Size ranges from $\frac{1}{16}$ to $\frac{1}{8}$ inch (1.5 to 3 millimeters) in diameter. Infestations require fumigation. See Fig. 3.

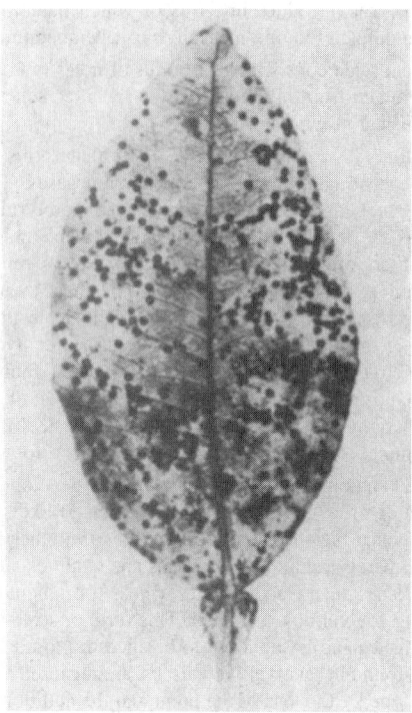

Fig. 3.   Infestation of Florida red scale insects on tree leaf. (*USDA.*)

Frosted Scale (*Eulecanium pruinosum*). Found on apricot, *Betula* (particularly birch in California), and *Laurus*. Insect is comparatively large (about $\frac{1}{2}$ inch; 12 millimeters long), is hemispherical in shape, and has a frosty-appearing wax covering.

Hemispherical Scale (*Saissetia hemisphaerica*). Found on *Betula*, citrus, coffee, *Cycas*, ferns, guava, *Hedera*. Insect is covered with a smooth, soft scale without markings. Very commonly found in glasshouses.

Japanese Wax Scale (*Ceroplaste ceriferus*). Found mainly on gardenia, but can spread to other plants. Insect is from $\frac{1}{4}$ to $\frac{1}{3}$ inch (6 to 8 millimeters) in diameter. It is covered with white-to-creamy waxy masses.

Juniper Scale (*Diaspis carueli*). Found mainly on juniper, but can spread to other plants. Insect is snow white with circular scales and with yellow central exuviae.

Oyster Scale (*Lepidosaphes ulmi*). Found on apple, *Betula*, *Ceanothus*, *Crataegus*, currant, and poplar. Sometimes also called the bark louse, this scale is about $\frac{1}{8}$ inch (3 millimeters) long and resembles an oyster shell in shape. The insect hibernates as small white eggs under old scales. The young hatch in late spring and appear as whitish lice crawling on the bark.

Pineapple Scale (*Diaspis bromeliae*). Found on *Billbergia*, olive, and pineapple. The insect is thin and circular, usually with very white scales and yellow exuviae infesting the leaves and fruit.

Pine Leaf Scale (*Chionaspis pinifoliae*). Found mainly on pine trees. The insect is quite small with white scales.

Pit-Making Oak Scale (*Asterolecanium variolosum*). Found mainly on oak. The insect is circular in shape and has glossy greenish-yellow scales. The scale is particularly damaging to the golden oak.

Purple Scale (*Lepidosaphes beckii*). Found on citrus, *Codiaeum*, fig, and olive. The insect ranges from a reddish-brown to a rich purple in color and has oyster-shell shaped scales. The insect is from $\frac{1}{16}$ to $\frac{1}{8}$ inch (1.5 to 3 millimeters) in length.

Rose Scale (*Aulacaspis rose*). Found mainly on rose. The insect is circular, with small whitish scales.

San Jose Scale (*Aspidiotus perniciosus*). One of the better known and more damaging of the scales. Found on almond, apple, cherry, *Cornus*, *Crataegus*, currant, pear, plum, *Sorbus*, etc. The insect is nearly circular and about as large as a pinhead. See Fig. 4. When in large numbers, the insects form a crust on branches, causing small red spots on the fruit. This scale multiplies with amazing rapidity. Three to four broods per year are possible. The young are born alive. Breeding goes on until quite late in the autumn, at which time all stages of the insect are killed, with exception of the small, half-grown black scales which often hibernate through the winter safely. The scale also sometimes attacks rose, gooseberry, elm, chestnut, oak, and walnut.

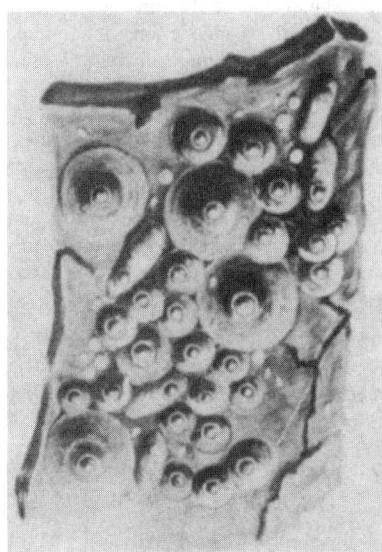

Fig. 4. Magnified view of San Jose scale insects fastened to branch of fruit tree. (*USDA.*)

Scurfy Scale (*Chionaspis furfurus*). Found on apple, *Crataegus*, and *Sorbus*. The insect is pearl-shaped, white to off-white in color, and about $\frac{1}{8}$ inch (3 millimeters) long. The insect frequently becomes intimately encrusted to the bark, with a resulting scurfy appearance of the bark. Hibernation is in the form of purple eggs located under old scales.

Soft Brown Scale (*Coccus hesperidum*). Found on citrus, *Cycas*, fig, *Hedera*, *Ipomoea*, jasmin, and *Laurus*. The insect is oval-shaped, flat, about $\frac{1}{4}$ inch (6 millimeters) long, and with flat, soft scales.

White Peach Scale (*Diaspis pentagonia*). Found on *Cycas*, *Diospyros*, *Laurus*, and peach. The insect is circular in shape, with gray scales and with exuviae at one side of center.

Several closely related scale insects have had economic significance outside the realm of agriculture over the years. Some have been used as the source of dyes, such as cochineal (*Coccus cacti*), a species very common in Mexico, and the scarlet grains (*Coccus poionicus*), which occur in Poland on the roots of the knawel plant and once were used as a source of dye for fabrics. One scale, known as the wax insect of China (*Coccus sinensis*) has been used for centuries as a source of wax used in candles, lacquers, sealers, and varnishes. The shell-lac insects (from which shellac is derived) are closely related.

The iridescent covering secreted by some of the scale insects, known as *ground pearl*, is useful in making ornaments.

More detail on scale insects can be found in *Foods and Food Production Encyclopedia* (D.M. and G.D. Considine, eds.), Van Nostrand Reinhold, New York, 1982.

SCALE (Weighing). See **Weighing**.

SCALES (Fish). See **Fishes**.

SCALING (Boiler). See **Feedwater (Boiler)**.

SCALING CIRCUIT. A circuit that produces an output pulse whenever a prescribed number of input pulses have been received. A binary scaler produces an output pulse whenever two input pulses have been received.

By putting binary scalers in sequences, scales of two, four, eight, sixteen, etc., are obtained. A decade scalar produces an output pulse whenever ten input pulses have been received. By putting decade scalers in sequence, scales of ten, one hundred, one thousand, etc., are obtained.

SCALLOP. See Mollusks.

SCANDIUM.   Chemical element, symbol Sc, at. no. 21, at. wt. 44.956, periodic table group 3, is a member of the rare earth group of elements along with yttrium and the lanthanides, mp 1540 °C, bp 2850 °C, density 2.989 g/cm^3 (alpha form), 3.73 g/cm^3 (beta form). The alpha form is close-packed hexagonal and beta is the face-centered cubic allotrope. Scandium is a relatively soft metal with a silvery luster. The metal is stable in air. Scandium combines readily with acids and at elevated temperatures with oxygen, halogens, and chalcogenides. ^{45}Sc occurs in nature and is not radioactive. Nine radioactive isotopes have been identified ^{40}Sc through ^{44}Sc and ^{46}Sc through ^{51}Sc, all with relatively short half-lives, ranging from a fraction of a second up to 84 days. Scandium occurs widely throughout nature, but in reasonably concentrated forms only in a few uncommon minerals. Abundance in the Earth's crust is estimated at approximately $5-6 \times 10^{-4}\%$, ranking it ahead of such elements as antimony, bismuth, silver, and gold. It is estimated that a cubic mile of seawater contains about 375 pounds of the element. Scandium was predicted by Mendeleev in 1869, at which time he called it *ekaboron* and foretold accurately a number of its properties. A small amount of scandium oxide was extracted from euxenite and gadolinite by Nilson in 1879, a material that Nilson called *scandia*. In the same year, Cleve isolated a greater quantity of the oxide, from which several compounds were prepared and favorably compared with Mendeleev's predictions for ekaboron. Ionic radius Sc^{3+} 0.745. First ionization potential, 6.54 eV; second, 12.80 eV; third, 24.76 eV. Other physical properties are given under **Chemical Elements**.

Scandium occurs in some ores with the Rare Earth Series elements. It is easily separated from the Lanthanides, as well as yttrium, by taking advantage of the greater solubility of its thiocyanate in ether. The three recognized scandium minerals are thortveilite, a silicate; and sterrettite and kolbeckite, both phosphates. Wiikite and bazzite, complex niobates and silicates, are known to contain more than 1% scandium. The recovery of scandium from uranium and tungsten from mining tailings, however is the major source of the element. Scandium has not been found without the other Rare earth elements. The element usually is separated from ore extracts and concentrates by precipitation as the oxalate. Scandium metal with a purity of 99.9% has been produced.

In water solutions, the scandium ion has a triple positive charge. Studies show, however, that the simple Sc^{3+} ion seldom exists. Rather, the form is highly polymerized and hydrolyzed—with hydroxy-bonded structures. In forming compounds, scandium parallels aluminum, yttrium, gallium, indium, and tellurium. Several carbides of scandium have been reported, the most stable carbide being ScC, including scandium clusters in fullerene cages.

Like the hydroxides of the Rare earth, scandium hydroxide, Sc(OH)$_3$, is precipitated by addition of alkalies to solutions of scandium salts; however, the latter is precipitated at pH 4.9, while the former require pH 6.3 or more, a property which is utilized in one method of separation. Upon heating the hydroxide (or certain oxyacid salts), scandium oxide, Sc$_2$O$_3$ is produced. Scandium hydroxide is less acidic than aluminum hydroxide, requiring boiling KOH solution to form the complex potassium compound, K$_2$[Sc(OH)$_5 \cdot$ H$_2$O] $\cdot$ 3H$_2$O.

All four trihalides of scandium are known. The trifluoride is very slightly soluble in H$_2$O, and is precipitated from scandium nitrate, Sc(NO$_3$)$_3$, solutions by hydrofluoric acid. It dissolves in alkali fluorides to yield the complex ion [ScF$_6$]$^{3-}$. The chloride is formed in solution by treating the hydroxide or oxide with HCl, yielding hydrated crystals on concentration, which give hydroxychlorides on heating. The bromide is also prepared from the oxide or hydroxide and hydrobromic acid, or in anhydrous form from the oxide, carbon, and bromine, on heating. The iodide is also prepared by the latter method.

The thiocyanate is prepared in solution by adding ammonium thiocyanate, NH$_4$SCN to HCl solutions of the chloride. Both basic and double carbonates are known. The former is precipitated from Sc^{3+} solutions by adding carbonate solutions, and is probably Sc(OH)CO$_3 \cdot$ H$_2$O. The latter are obtained by the use of an excess of the soluble carbonate. Normal, basic, and double sulfates are known. The first exists in several degrees of hydration; the second is obtained as $Sc(OH)SO_4 \cdot 2H_2O$, by treating the normal sulfate tetrahydrate with the hydroxide. The alkali double sulfates and alums are obtained by treating the sulfate solution with an excess of the alkali (or ammonium) sulfate solution.

The nitrate is readily obtained by action of dilute HNO_3 on the hydroxide. In aqueous solution, the anhydrous nitrate yields a monobasic nitrate on heating.

To date, the applications for scandium and its compounds have been limited, mainly because of its high cost. The main uses of scandium are in sporting goods equipment and in lighting, including automotive headlights. In former application scandium is added to improve the strength of aluminum, while in the latter ScI_3 is used in metal halide arc lamps to increase the intensity and to improve the color rendition to match that of natural sunlight. Minor applications include: scandium substitution for yttrium in yttrium garnets for electronic uses, analytical standards, semiconductor applications, welding wire, and metallurgical research.

Additional Reading

Carter, G.F. and D.E. Paul: "Materials Science and Engineering," ASM International, Materials Park, Ohio, 1991.

Gschneidner, K.A., Jr., B.J. Beaudry, and J. Capellen: "Rare Earth Metals," pp/720–732 in "Metals Handbook 10th Edition, Vol. 2, Properties and Selection: Nonferrous Alloys and Special Purpose Materials," ASM International, Metals Park, OH, 1990.

Gschenidner, K.A., Jr.: "Physical Properties of the Rare Earth Elements," pp. 4–112 to 4–121 in "CRC Handbook of Chemistry and Physics," 77th Edition, CRC Press, LLC., Boca Raton, FL, 1997.

Hammond, C.R.: "The Elements," pp. 4–1 to 4–34 in "CRC Handbook of Chemistry and Physics," 77th Edition, CRC Press, LLC., Boca Raton, FL, 1997.

Horovitz, C.T., K.A Gschneidner, Jr., G.A. Melson, D.H. Youngblood, and H.H. Schock: "Scandium. Its Occurrence, Chemistry, Physics, Metallurgy, Biology and Technology," Academic Press, London, UK, 1975.

Lewis, R.J. and N.I. Sax: "Dangerous Properties of Industrial Materials," 10th Edition, John Wiley & Sons, Inc., New York, NY, 1992.

NOTE: This entry was revised and updated by K.A. Gschneidner, Jr., Director, and B Evans, Assistant Chemist, Rare-earth Information Center, Institute for Physical Research and Technology, Iowa State University, Ames, IA.

SCANNING TUNNELING MICROSCOPE. This microscopic technique (STM), invented by Gerd K. Binnig and Heinrich Rohrer (IBM Research Laboratory, Zurich, Switzerland) in 1981, generates topographic images of surfaces with atomic resolution. With the STM, scientists have obtained previously unseen images of gold, silicon, nickel, oxygen, and carbon atoms. STM views can reveal flaws and contaminants in atomic surface structure. Detailed views of three-dimensional atomic "landscapes" are improving and will continue to improve the knowledge of surface physics and chemistry and thus be of exceptional value in technical fields as varied as metallurgy, magnetism, semiconductor technology, and biology. Before development of the STM technique, scientists had been puzzled about the exact surface structure of silicon, the basic material from which computer chips are made. Many models of the silicon surface have been constructed. The STM has enabled researchers to sort out prior assumptions and hypotheses. In connection with gold, the STM has revealed a surface structure created by the spontaneous formation of ribbon-like facets, features that previously had not been distinguished with such clarity. The renowned physicist, Wolfgang Pauli, many years ago observed, "The surface was invented by the devil," with reference to the fact that the surface of a solid is its boundary or interface with the environment outside and beyond the solid. Whereas atoms within a solid interact with other atoms within the solid, those atoms on the surface can interact only with those atoms directly underneath and those atoms in the environment beyond the surface. Thus, surface atoms behave with different rules.

Instrumental means for investigating the atomic pattern of surfaces have included electron microscopy and later (1950s), the field-ion microscope (invented by Edwin W. Müller). These techniques continue, but are generously abetted by the STM. In 1984, the inventors of the STM received the Hewlett-Packard Europhysics Prize for outstanding achievements in solid state physics and the King Faisal International Prize in Science. In 1986, Binnig and Rohrer shared in the Nobel Prize in Physics.

Expanded Applications of STM

Since its introduction for practical uses in the early 1980s, the scanning tunneling microscope has found scores of applications that were not contemplated at the outset of STM technology. These applications include the manufacture of optical grating masters, the manufacture of recording thin-film magnetic recording heads, the manufacture of compact disk stampers, and the repair of costly masks used for integrated circuit manufacture — just to mention a few examples of practical usage in the electronics field. STM can be operated under water and other fluids, permitting the examination of biological materials in a more natural setting. The versatility of the STM is demonstrated by the appearance of hundreds of technical papers in the literature over a 1-year period. The STM can achieve lateral resolution of 1 to 2 angstroms and, in the vertical dimension, better than 0.05 angstrom. To overcome the relatively few limitations of STM, the technology has spawned several other microscopic techniques, including the atomic force microscope, the friction force microscope, the magnetic force microscope, the electrostatic force microscope, the attractive mode force microscope, the scanning thermal microscope, the optical absorption microscope, the scanning ion-conductance microscope, the scanning near-field optical microscope, the scanning acoustic microscope, and the molecular dipstick microscope — all resulting from the character of innovative thinking that scientists have come to apply to the application of new forms of energy to microscopy, once wholly dependent upon materials interactions with visible light. STM has played an invaluable role, not only in the development of new materials, but in understanding how surface atoms differ from the permanently embedded atoms of a material.

STM Fundamentals

As put forth in their excellent 1985 paper (reference listed), Binnig and Rohrer credit *electron tunneling* as the phenomenon that underlies the

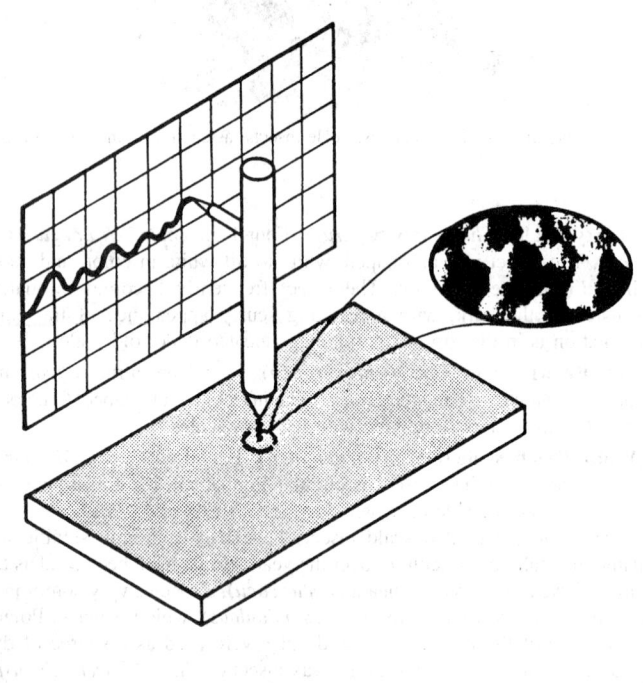

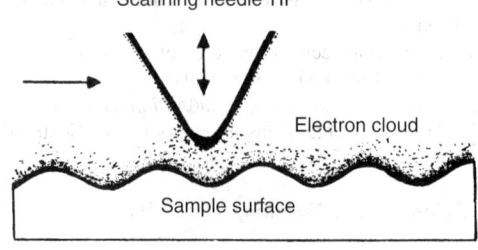

Fig. 1.   As the probe tip of the STM is scanned across the microscopic "hills and valleys" of a surface, the vertical position of the probe is precisely adjusted to maintain a constant tip-to-surface distance. This is accomplished by keeping the tunneling current constant. Consequently, the probe follows the surface contour as it moves, yielding a 2-dimensional, enlarged representation of the surface contour for each such scan. A full 3-dimensional image is obtained by assembling an entire sequence of scans. (*IBM Corporation.*)

operation of the STM. As indicated by Fig. 1, an electron cloud occupies the space between the surface of the sample and the needle tip used. The cloud is a consequence of the *indeterminacy*[1] of the electron's location (a result of its wavelike properties). Because the electron is "smeared out" so to speak, there is a probability that it can lie beyond the surface boundary of a conductor. The density of the electron cloud decreases exponentially with distance. A voltage-induced flow of electrons through the cloud thus is extremely sensitive to the distance between the surface and the tip. As the STM tip is swept across the surface, a feedback mechanism senses the flow (the *tunneling current*) and holds constant the height of the tip above the surface atoms. In this way, the tip follows the contours of the surface. The motion of the tip is read and processed by computer and displayed on a screen or plotter. By sweeping the tip through a pattern of parallel lines, a high-resolution, three-dimensional image of the surface is obtained.

Views of the STM are shown in Figs. 2 and 3. Binnig and Rohrer describe the STM as having two stages, suspended from springs, that operate within a cylindrical steel frame. The microscope mechanism is contained by the innermost stage. Vibration is a severe problem, requiring special preventive measures. To obtain images with high resolution (Fig. 4) of surface structures, obviously the microscope must be protected from even very small vibrations as may be caused by sound and footsteps and other disturbances within the laboratory. To subdue vibration, copper plates are attached to the bottom of the stainless-steel frame and magnets are attached to the bottom of the inner and outer stages. Any disturbance causes the copper plates to move up and down in the field generated by the magnets. The movement induces eddy currents in the plates. Interaction of the eddy currents with the magnetic field retards the motion of the plates and thus the motion of the microscope stages. Where investigations require vacuum conditions, a steel cover is placed over the outer frame of the microscope.

Fig. 3. Miniaturized version of the scanning tunneling microscope. The STM has become a very important tool for materials research and, in particular, to gaining new knowledge of computer chips. STMs are being used or are under construction by some fifty research groups worldwide in a broad range of physical, chemical, biological, and materials studies. (*IBM Corporation.*)

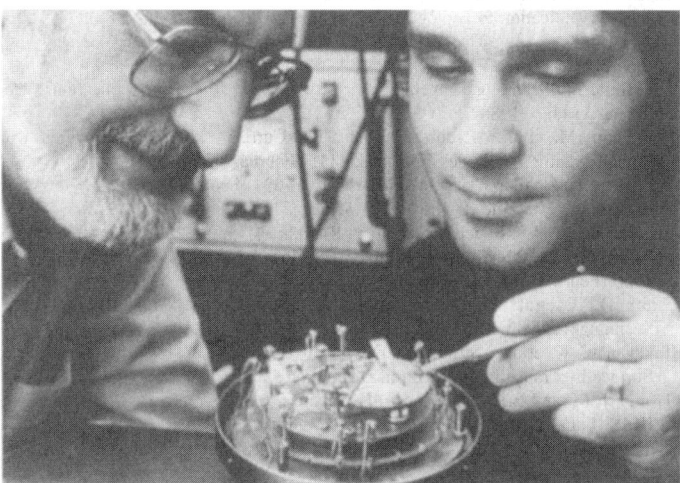

Fig. 2. Inventors Rohrer (left) and Binnig (right) shown adjusting the sample in the chamber of an early (1981) version of the scanning tunneling microscope. Later (1986), these scientists shared in the Nobel Prize in Physics for developing the STM instrument and technique. (*IBM Corporation.*)

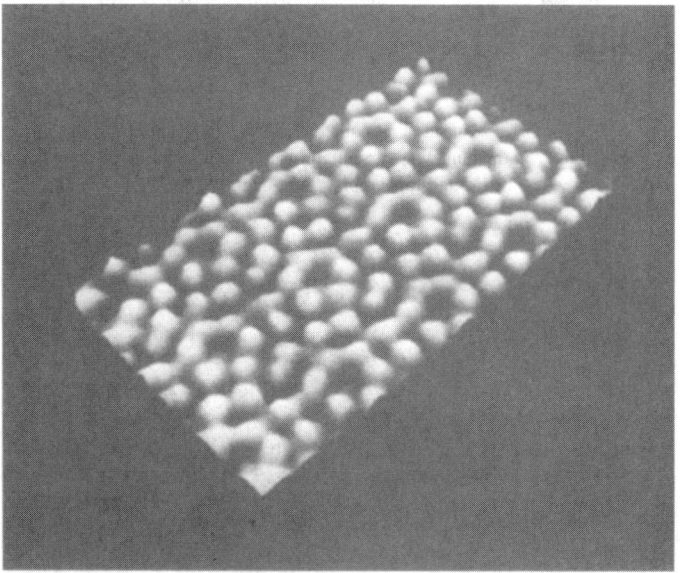

Fig. 4. Magnified millions of times are surface atoms of silicon. The image is computer-generated from data produced by a scanning tunneling microscope. (*IBM Corporation.*)

The microscopic device incorporates a sample and a scanning needle. Piezoelectric materials (expand or contract with applied voltage) make it possible for the device to resolve features that are about a hundredth the size of an atom. A piezoelectric drive positions the sample on a horizontal metal plate and a piezoelectric tripod then sweeps the scanning needle over the surface of the sample, simultaneously achieving high stability and precision. See also Figs. 5 through 8.

Not only does the STM portray atomic topography, but it also reveals atomic composition. The inventors observe that the tunneling current depends both on the tunnel distance and the electronic structure of the surface and on the fact that each atomic element has an electronic structure unique to itself. The ability of the STM to resolve both topography and electronic structure assures wide use of the technique in numerous fields. Unlike other imaging techniques, the STM does not alter or partially destroy the sample. The STM already has demonstrated its utility in biology even though lateral resolutions of only 10 angstroms can be achieved. In this instance, the relatively poor resolving power of the microscope is more than compensated for by its ability to provide a direct and nondestructive method of viewing biological samples. Researchers at the IBM Zurich Research Laboratory and the Swiss Federal Institute of Technology have scanned the surface of DNA and observed a series of zigzags which correspond to its helical structure. Researchers at the Autonomous University of Madrid have made detailed examinations of viruses, notably phi 29 which measures $40 \times 300 \times 200$ angstroms. It is currently visualized that the STM also will be very useful for testing electronic circuits, particularly as further reductions in size are achieved.

---

[1] The roots of atomic structure extend back over 50 years, to the development of the concept of quantum mechanics and experiments in 1927 by Davisson and Germer, the researchers who confirmed the wave nature of the electron. The first experimental verification of tunneling was made about three decades ago by Ivar Giaever.

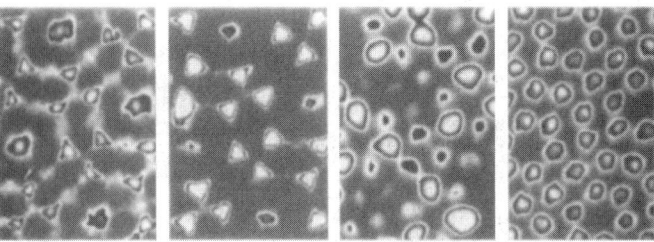

Fig. 5.   Looking successively deeper into the atomic structure of silicon. Sequence shows surface magnified some ten million times, to a depth of about nine angstroms (36 billionths of an inch). Shown from left to right are the geometric positions of the atoms and three different classes of electronic bonds. (**a**) shows the position of the top atoms; (**b**) the dangling bonds that reach up from those atoms; (**c**) the dangling bonds that reach up from other atoms in the second layer in the surface; and (**d**) bonds (called "back bonds") that reach out sideways from the atoms in the second layer in the surface. (*IBM Corporation.*)

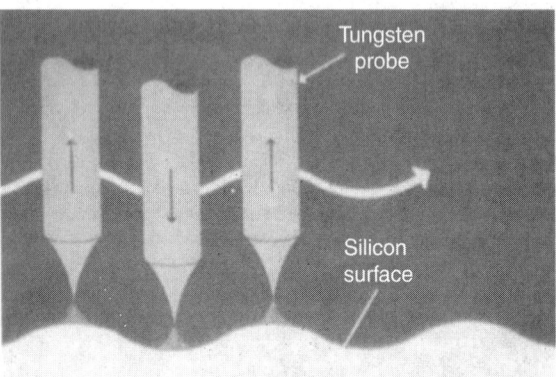

Fig. 6.   As probe is scanned across silicon surface, its height above the surface is adjusted to keep the tunneling current constant. The monitoring of those height changes provides the desired topographic information. (*IBM Corporation.*)

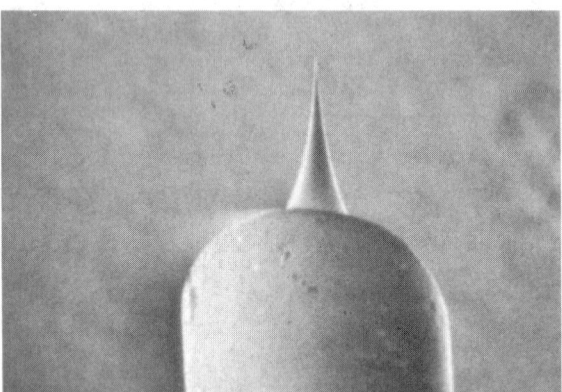

Fig. 7.   Tungsten probe tip for scanning tunneling microscope as seen by a scanning electron microscope. Such probes, produced by field-ion microscopy, can be constructed with tips only one atom in width. (*IBM Corporation.*)

Fig. 8.   Scanning tunneling microscope image of gallium arsenide (GaAs). Atoms to the left of each row are gallium; others are arsenic. (*IBM Corporation.*)

Additional Reading

Abelson, P.H.: "Phenomena at Interfaces," *Science*, 1357 (March 18, 1988).

Bai, C.: "Scanning Tunneling Microscopy and Its Application," 2nd Edition, Springer-Verlag Inc., New York, NY, 1999.

Bard, A.J. et al.: "Chemical Imaging of Surfaces with the Scanning Electrochemical Microscope," *Science*, 68 (October 4, 1991).

Baro, A.M., Binnig, G., Rohrer, H., Gerber, C., and et al.: "Real-Space Observation of 2 × 1 Structure of Chemisorbed Oxygen on NI(110) by Scanning Tunneling Microscopy," *Physical Review Letters*, **52**(15), 1304–1307 (April 9, 1984).

Beebe, T.P., Jr., et al.: "Direct Observation of Native DNA Structures with the Scanning Tunneling Microscope," *Science*, 370 (January 20, 1989).

Bindell, J.B.: "Elements of Scanning Electron Microscopy," *Advanced Materials & Processes*, 20 (March 1993).

Binnig, G., Rohrer, H., Gerber, C., and E. Weibel: "Facets as the Origin of Reconstructed Au(110) Surfaces," *Surface Science*, **131**, L379–L384 (1983).

Binnig, G. and H. Rohrer: "The Scanning Tunneling Microscope," *Sci. Amer.*, 50–56 (August 1985).

Binnig, G. and H. Rohrer: "The Scanning Tunneling Microscope," in The Laurates' Anthology, 72, Scientific American, Inc., New York, NY, 1990.

Birdi, K.S.: "Scanning Tunneling Microscopy (STM) and Atomic Force Microscopy (AFM): Applications in Surface and Colloid Chemistry," CRC Press, LLC., Boca Raton, FL, 2001.

Bonnell, D.A.: "Scanning Tunneling Microscopy and Spectroscopy: Theory, Techniques, and Applications," John Wiley & Sons, Inc., New York, NY, 1993.

Clemmer, C.R. and T.P. Beebe, Jr.: "Graphite: A Mimic for DNA and Other Biomolecules in Scanning Tunneling Microscope Studies," *Science*, 640 (February 8, 1991).

Chen, J.: "Introduction to Scanning Tunneling Microscopy," Oxford University Press, Inc., New York, NY, 1993.

Epstein, A.W.: "A Tunneling Microscope Can Cleave Molecules," *Sci. Amer.*, 26 (April 1988).

Flam, F.: "Scopes with a Light," *Science*, 30 (April 5, 1991).

Giaever, I.: "Energy Gap in Superconductors Measured by Electron Tunneling," *Physical Review Letters*, **5**(4), 147–148 (August 15, 1960).

Hansma, P.K. et al.: "Scanning Tunneling Microscopy and Atomic Force Microscopy: Application to Biology and Technology," *Science*, 209 (1988).

Hansma, P.K. et al.: "The Scanning Ion-Conductance Microscope," *Science*, 641 (February 3, 1989).

Kinoshita, J.: "Scanning Tunneling Microscope Spawns Diverse Applications," *Sci. Amer.*, 33 (July 1988).

Pomerantz, M. et al.: "Rectification of STM Current to Graphite Covered with Phthalocyanine Molecules," *Science*, 1115 (February 28, 1992).

Pool, R.: "The Children of the STM," *Science*, 634 (February 9, 1990).

Pool, R.: "A New Role for the STM," *Science*, 130 (December 7, 1990).

Schardt, B.C., Yau, Shueh-Lin, and F. Rinaldi: "Atomic Resolution Imaging of Adsorbates on Metal Surfaces in Air: Iodine Adsorption on Pt(III)," *Science*, 1050 (February 24, 1989).

Smith, D.P.E. et al.: "Smectic Liquid Crystal Monolayers on Graphite Observed by Scanning Tunneling Microscopy," *Science*, 43 (July 7, 1989).

Takayanagi, K. et al.: "Structure Analysis of the Silicon(111) 7 × 7 Reconstructed Surface by Transmission Electron Diffraction," *Surface Science*, 164, 367 (1985).

Tromp, R.M. and E.J. van Loenen: "Ion-Beam Crystallography on Silicon Surfaces III. Si(111)," *Surface Science*, 155, 441 (1985).

Whitman, L.J. et al.: "Manipulation of Adsorbed Atoms and Creation of New Structures on Room-Temperature Surfaces with the Scanning Tunneling Microscope," *Science*, 1206 (March 8, 1991).

Wiesendanger, R. and H.J. Guntherodt: "Scanning Tunneling Microscopy III: Theory of STM and Related Scanning Probe Methods," 2nd Edition, Springer-Verlag Inc., New York, NY, 1998.

SCAPIE.   See **Virus.**

SCAPOLITE.   The mineral scapolite is a silicate of calcium and aluminum that contains also some potassium, sodium, and chlorine. The name identifies all intermediate members of a series with the following end members: Marialite $3Na(AlSi^3)O8 \cdot NaCl$, Meinoite $3Ca(Al^2Si^2)$ $O^8 \cdot CaCO_3$. Its tetragonal crystals are coarse and thick, often very large. It occurs also in massive forms. It has a distinct prismatic cleavage; subconchoidal fracture; is brittle; hardness, 5.5–6; specific gravity, marialite, 2.5–2.62, meionite, 2.72–2.78; luster, vitreous to rather dull, color, white to gray, red, green, blue, or yellow; translucent to opaque, rarely transparent.

Scapolite is found in the metamorphic rocks, particularly those rich in calcium; also in contact metamorphic deposits in limestones. It has been found in basic igneous rocks, probably as a secondary mineral. Notable localities are Lake Baikal, Siberia; Arendal, Norway; and Madagascar. In the United States, it is found in Massachusetts, New York, and New Jersey. Greenville, in the Province of Quebec, Canada is an important locality. Superb transparent yellow gem crystals have recently been found in Brazil and Tanzania. Wernerite (scapolite) was named in honor of A.O. Werner, a famous German mineralogist (1749–1817).

SCARAB (*Insecta, Coleoptera*).   A species of beetle that was regarded as sacred by the ancient Egyptians. The term is applied especially to the sculptured likenesses of the beetle. From this name the family *Scarabeidae* has arisen, containing, among the many North American species, the June bugs or May beetles, the tumble bugs, and the rose chafer.

SCARLET FEVER.   Although staphylococci can induce a similar infection (scalded skin syndrome), Group A streptococci (*S. pyrogenes*) are essentially responsible for this disease, which is characteristically one of school children. There is no doubt that scarlet fever is caused by an erythrogenic toxin because its symptoms are reproduced by injection of filtrates of cultures of *S. pyrogenes* (the so-called Dick test). The disease is now regarded as a combination of the direct toxicity of the erythrogenic toxin together with a secondary toxicity due to hypersensitivity to a heat stable component of streptococcal toxin.

In the latter half of the nineteenth century, scarlet fever was the most common cause of death in children over one year old and was often simply referred to as "the fever." Although mild infections still are quite common, most instances escape attention and the disease has now virtually disappeared in the developed countries.

The disease is usually easy to recognize. An abrupt fever (100–103 °F; 37.8–39.5 °C) is accompanied by a disproportionate tachycardia. Coughing is notably absent. As well as an initial tonsillitis, an exanthem develops on the palate and the tongue is covered by a white fur through which red-tipped papillae project (*strawberry tongue*). The fur later peels off, leaving a raw red surface (*raspberry tongue*).

The rash, which is attributed to capillary damage, usually appears on the chest, neck, and arms within 24 hours of onset and lasts from 2 to 20 days before peeling begins with desquamation of fine flakes or large areas of skin.

All group A streptococci are still highly susceptible to benzyl penicillin and this, with closely related antibiotics, is the first choice for therapy. When penicillin allergy is suspected, erythromycin may be prescribed. In untreated patients, the possibility of subsequent development of rheumatic fever cannot be excluded. Otherwise, prognosis is excellent.

R. C. V.

SCARLET TANAGER.   See Tanager.

SCATTER DIAGRAM.   Also called scattergram. A plot representing corresponding observed values of two variables x and y as points in rectangular coordinates. If the two variables are functionally related, the points will be bunched, but if they are not functionally related, the points will be scattered uniformly over the plane.

Scatter diagrams are used to explore the influence of one variable upon another, strong relationships being revealed as a concentration around a definite curve. The Hertzsprung–Russel diagram showing individual stars is an example of a scatter diagram. See also **Coordinate System**.

SCATTERING.   In its general sense, this term refers to the redistribution of a group of entities, or bringing about a less orderly arrangement, either in position or direction. Thus when visible light enters a body of matter, however transparent, part of it is diffusely reflected or "scattered" in all directions. This is due to the interposition in the light stream of particles of varying size, from microscopic specks down to electrons, and the deflection of light quanta resulting from their encounters with these small obstacles. Similar effects are produced upon infrared, ultraviolet, x-rays, and other forms of electromagnetic radiation, and upon streams of particles such as cathode rays or alpha rays. More specifically, the term denotes the change in direction of particles or photons owing to collision with other particles or systems; it may also be regarded as the diffusion of a beam of sound or light (or other electromagnetic radiation) due to the anisotropy of the transmitting medium.

Coherent or elastic scattering, either of particles or photons, is scattering in which there are definite phase relations between incoming and scattered waves. Ordinary or Rayleigh scattering (defined below) is of this nature. In coherent scattering, interference occurs between the waves scattered by two or more centers. This type of scattering is exemplified by the Bragg

scattering of x-rays and the scattering of neutrons by crystals, which gives constructive interference only at certain angles, called Bragg angles.

Rayleigh scattering is coherent scattering in which the intensity of the light of wavelength λ, scattered in any direction making an angle θ with the incident direction, is directly proportional to $1 + \cos^2 \theta$ and inversely proportional to λ^4. The latter point is noteworthy in that it shows how much greater is the scattering of the short wavelengths. These relations apply when the scattering particles are much smaller than the wavelength of the radiation. Thus the sky is blue, and tobacco smoke appears blue, because blue light is scattered more than red. The unscattered light is of course complementary to blue, i.e., orange or yellow — which explains the "warm" hues of the sunset.

Mie scattering is any scattering produced by spherical particles without special regard to comparative size of radiation wavelength and particle diameter; to be contrasted, therefore, with Rayleigh scattering.

Rutherford scattering is a general term for the process in which moving charged particles are scattered at various angles by interaction with the nuclei of atoms of a solid material. In Rutherford's original work, high-speed α-particles from radon were focused in a narrow beam to strike a thin gold foil. Most of them pass through, but some are scattered.

On the other hand, *incoherent or inelastic scattering* is scattering in which the scattering elements act independently so that there are no definite phase relations between different parts of the scattered beam. The intensity of the scattered radiation at any point is determined by adding the intensities of the scattered radiation reaching the point from the independent scattering elements.

Thomson scattering is the scattering of electromagnetic radiation by free charged particles, computed either classically or according to nonrelativistic quantum theory. Scattering by electrons is interpreted classically as a process in which some of the energy of the primary radiation is reduced because electrons radiate when accelerated in the transverse electric field of the radiation. The scattering cross section is given by

$$\sigma_r = \frac{8}{3}\pi \left(\frac{e^2}{mc^2} \right)^2$$

which is 0.657 barn for an electron, and is called the *Thomson cross section*, or the *classical scattering cross section*.

Compton scattering is inelastic scattering of photons by electrons in the Compton effect. Because the total energy and total momentum are conserved in the collisions, the wavelength of the scattered radiation undergoes a change that depends in amount on the scattering angle. If the scattering electron is assumed to be at rest initially, the Compton shift is given by the equation:

$$\lambda' - \lambda = \lambda_0(1 - \cos\theta) = (h/m_e c)(1 - \cos\theta)$$

where λ' is the wavelength associated with the scattered photons, λ_0 is the wavelength of the incident photons, l0 is the (Compton) wavelength of the electron, and θ is the angle between the paths of incident and scattered photons.

Delbrück scattering is the scattering of light by a Coulomb field, a process which according to quantum electrodynamics occurs as a scattering of the light by the virtual electron-pairs produced by the Coulomb field. The total cross section is approximately 6 millibarns for uranium.

Single scattering is the deflection of a particle from its original path owing to one encounter with a single scattering center in the material traversed. This is to be distinguished from plural scattering and multiple scattering, which involve successive encounters with scattering centers.

Multiple scattering is any scattering of a particle or photon in which the final displacement is the sum of many displacements, usually small. A type of scattering intermediate between single and multiple scattering is called *plural scattering*.

Potential scattering is that portion of scattering by the nucleus of the atom that has its origin in reflection from the nuclear surface, thus leaving the interior of the nucleus undisturbed. The term usually is used in contradistinction to *resonance scattering*, which is the scattering arising from the part of the incident wave that penetrates the surface and interacts with the interior of the nucleus. In general, the scattered wave may have components arising from both kinds of scattering processes. The term potential scattering is also used to denote scattering of an incident wave by reflection at a change or discontinuity in the potential field.

Acoustic scattering is the irregular and diffuse reflection or diffraction of sound in many directions. Scattering frequently occurs when the reflecting surfaces or bodies are small compared with the wavelength of sound; in certain cases the reflecting bodies may be small inhomogeneities in the medium.

SCATTEROMETRY.   Scatterometry has its origin in early radar used in World War II. Early radar measurements over the oceans were corrupted by sea clutter (noise) and it was not known at the time the cause of this clutter. It was not until the 1960s that the noise in the radar signal was found to be the radar's response to the winds over the ocean.

The first scatterometer flew as part of the Skylab missions in 1973 and 1974, demonstrating that spaceborne scatterometers were indeed feasible. Then, from June to October 1978, the *Seasat-A Satellite Scatterometer* (SASS) proved that accurate wind velocity measurements could be made from space. In the 1990s, a single-swath (-beam) scatterometer was one of the instruments to fly on the European Space Agency's *ERS-1* (currently in stand-by mode) and *ERS-2* Remote Sensing Satellites, launched in July 1991 and April 1995 respectively.

The *NASA Scatterometer* (NSCAT) which launched aboard Japan's *ADEOS-Midori Satellite* in August 1996, was the first dual-swath, Ku-band scatterometer to fly since *Seasat*. From September 1996, when the instrument was first turned on, until premature termination of the mission due to satellite power loss in June 1997, NSCAT performed flawlessly and returned a continuous stream of global sea surface wind vector measurements. Unprecedented for coverage, resolution, and accuracy in the determination of ocean wind speed and direction, NSCAT data has already been applied to a wide variety of scientific and operational problems. These applications include such diverse areas as weather forecasting and the estimation of tropical rain forest reduction. Because of the success of the short-lived NSCAT mission, future Ku-band scatterometer instruments are now greatly anticipated by the ocean winds user community.

The Science of Scatterometry

Winds over the ocean modulate air–sea changes in heat, moisture, gases and particulates (matter in the form of small liquid or solid particles), regulating the crucial bond between atmosphere and ocean that establishes and maintains global and regional weather and climate. Data derived from ocean scatterometers is vital to researchers in their studies of air–sea interaction, ocean circulation, and their effects on weather patterns and global climate. In the past, weather data could be acquired over land, but our only knowledge of surface winds over oceans came from infrequent, and sometimes inaccurate, reports from ships and buoys. These data are also useful in the study of unusual weather phenomena such as El Niño, the long-term effects of deforestation on rain forests, and changes in the sea ice masses around the Polar Regions. These environmental changes all play a central role in regulating global climate. See also **El Niño**.

Winds over the Ocean.   Wind affects the full range of oceanic motion, from individual surface waves to complete current systems. The tropical Pacific Ocean and overlying atmosphere react to and influence each other. Easterly surface winds along the equator control the amount and temperature of the water that upwells (moves or flows upward) to the surface. This upwelling of cold water determines sea surface temperature distribution, which affects rainfall distribution. This in turn determines the strength of the easterly winds; a continuous cycle.

Applications: Observing Oceans from Space

As the human population grows and international commerce expands, more effective use of resources becomes more important than ever before. Efficient utilization of the sea is particularly vital to human survival. Today, satellite technology can be employed to observe Earth's oceans from space without much of the uncertainty encountered by mariners of times gone by. By measuring global sea–surface wind speed and direction, ocean scatterometer data can help meteorologists more accurately predict the marine phenomena that affect human life on a daily basis. Some examples follow:

Weather Forecasting.   Data from ocean scatterometers greatly enhance overall weather-forecasting capabilities. Most of the weather over the west coast of the United States, and some over the east coast, is generated over the oceans. The measurements derived from ocean scatterometers are assimilated into numerical models (computer programs that represent natural processes in terms of equations), which can be used to predict global and regional weather patterns. The data are delivered to the National

Oceanic and Atmospheric Administration (NOAA) within two hours, where they are used for timely, accurate weather forecasting.

Storm Detection.   The ocean scatterometer data can determine the location, direction, structure and strength of storms at sea. Severe marine storms; hurricanes near the Americas, typhoons in Asian waters, and mid-latitude cyclones worldwide are among the most destructive of all natural phenomena. In the United States alone hurricanes have been responsible for at least 17,000 deaths since 1900, and hundreds of millions of dollars in damage annually. If worldwide statistics are considered, the numbers are substantially higher. Although typically not as violent as hurricanes and typhoons, mid-latitude cyclones exact a heavy toll in casualties and material damage.

In recent years, our ability to detect and track severe storms has been dramatically enhanced by the advent of weather satellites. Cloud images from space are now routine on weather reports. Data from ocean scatterometers augment these familiar images by providing direct measurements of surface winds to compare with the observed cloud patterns. These wind data help meteorologists to more accurately identify the extent of gale force winds associated with a storm and provide inputs to numerical models that provide advanced warning of high waves and flooding.

Ship Routing.   Wind-observation data from ocean scatterometers are of particular significance in ship routing. Prior knowledge of wind behavior will enable shipmasters to choose routes that avoid heavy seas or high headwinds that may slow ships' progress, increase fuel consumption, or possibly cause damage to vessels and loss of life. In the past, ship captains relied on widely-spaced measurements from buoys and sporadic, sometimes unreliable reports from other ships. Data from satellite-based scatterometers are much more regular, extensive, and dependable.

Oil Production.   Earth's oceans are increasingly used as a source of fuel. As continental fossil-fuel supplies are depleted, the more challenging task of extracting oil and gas from the seabed becomes a necessity. Oil and gas production is already ongoing at numerous offshore sites around the world, including the Gulf of Mexico, the North Sea, the Persian Gulf. Thorough knowledge of the historical wind and wave conditions at any specific location is crucial to the design of drilling platforms. Safe, efficient drilling operations depend on an accurate understanding of the current state of the sea as well as warning of impending storms.

In the event of an oil spill, surface-wind information is key to determining how the oil will spread. Ocean scatterometer data could help clean-up and containment crews to minimize the environmental effects of such a disaster.

Food Production.   Perhaps the oldest use of the ocean is in the harvesting of food. Today, ocean fishing is a highly systematic activity that makes extensive use of advanced technology to reduce the cost and to increase the value of every catch. Detailed wind data from scatterometers can aid in the management of commercial seafood crops. The annual U.S. shrimp harvest in the Gulf of Mexico, for example, depends on favorable on-shore winds that help transport offshore plankton larvae to estuaries where the larvae can develop into adult shrimp. SeaWinds data can help fishermen monitor wind conditions during this critical phase in life of shrimp.

Missions

Seasat-A.   In the 1970s, JPL engineers and scientists realized that the sensors they were developing for interplanetary missions could be turned upon Earth itself to better understand our home planet. In 1978, JPL built an experimental satellite called *Seasat* to test a variety of oceanographic sensors including imaging radar, altimeters, radiometers and scatterometers.

The *Seasat* flight-tested four instruments that used radar to study Earth and its seas. Radars are useful tools because they can penetrate clouds, they operate in all weather conditions, and they provide their own illumination so they can function day and night. The radar instruments on *Seasat* measured the satellite's distance from the sea surface, measured near-surface ocean winds and took pictures using radar rather than light for illumination. *Seasat* also carried a visual and infrared radiometer that provided measurements that were used to judge the results of the radar instruments.

Seasat was launched on June 26, 1978, on an Atlas Agena rocket from Vandenberg Air Force Base, California (Mass: 2,300 kilograms (5,070 pounds) Science instruments: Radar altimeter, scanning multichannel microwave radiometer, microwave scatterometer, imaging radar, visual and infrared radiometer.) Unlike launches of interplanetary spacecraft from

Florida on eastward flight paths, launches from Vandenburg place the satellite in a north-south orbit that takes it close to Earth's poles. *Seasat* sent data to Earth for 106 days. See Figs. 1 and 2. Seasat-A was lost when a short circuit drained all power from its batteries

Fig. 1.   Painting of *Seasat-A* spacecraft as it studies oceans from Earth orbit. (*NASA/JPL.*)

Many later Earth-orbiting instruments developed at JPL owe their legacy to the *Seasat* mission. These include imaging radars flown on NASA's Space Shuttle as well as such Earth-orbiting satellites and instruments as *Topex/Poseidon*, the *NASA Scatterometer* (NSCAT), *QuikScat* and the planned *Jason 1*.

ERS-1 and ERS-2. The European Remote Sensing Satellites (ERS) were designed to study many processes of the Earth's oceans and land with a suite of instruments, one of which is a SAR on the AMI. *ERS-1* was launched in 1991 and *ERS-2* in 1995.

The Active Microwave Instrument (AMI) operates in three modes: (1) SAR Image Mode, for the acquisition of wide swath images over the oceans, polar ice caps and land areas. (2) SAR Wave Mode, yielding 5 km × 5 km images at regular intervals along track for the derivation of length and direction of ocean waves, and (3) Wind Scatterometer Mode, using three separate antennae for measurements of sea-surface, wind speed, and direction.

NASA Scatterometer (NSCAT).   NSCAT, a microwave radar scatterom-eter, measured near-surface wind vectors (both speed and direction) over the global oceans. This information is critical in determining regional weather patterns and global climate. NSCAT had two major systems - the spaceborne instrument system and the ground data processing system.

NSCAT was launched at 6:54 p.m. U.S. PDT, Friday, August 16, 1996, aboard the Advanced Earth Observing Satellite (ADEOS), a mission of the National Space Development Agency of Japan. ADEOS was launched into a near-polar Sun-synchronous orbit, by an H-II launch vehicle from Japan's Tanegashima Space Center. The largest satellite ever developed by Japan, ADEOS had a mass of approximately 3,500 kilograms (7,720 pounds) and a power-generation capability of approximately 4500 watts; its overall dimensions at launch were 4 × 4 × 5 meters (13 × 13 × 16 feet). When the NSCAT antenna and the solar array paddle were deployed, the satellite was an impressive 11 meters (36 feet) in height and the solar array extends outward 29 meters (95 feet).

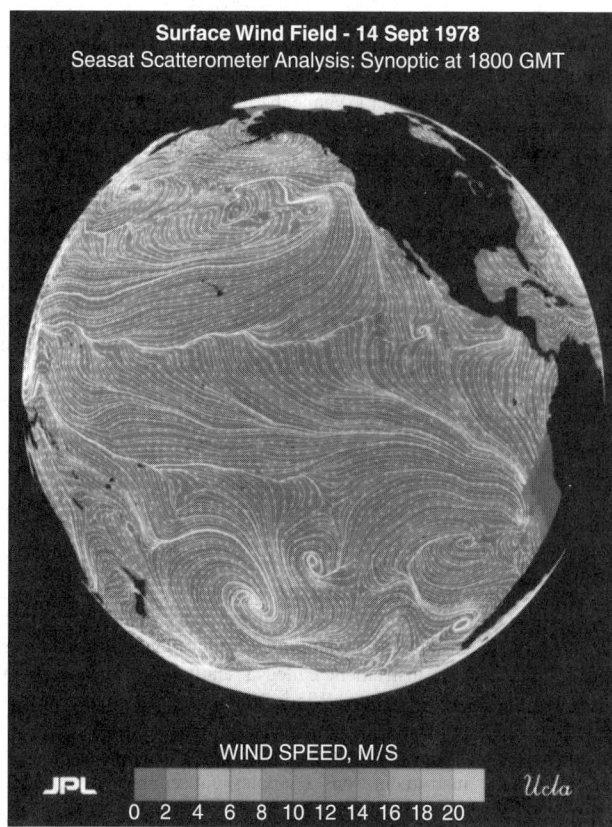

Fig. 2.   Global wind speed and direction are shown in this image using recently processed data acquired in 1978 by *Seasat-A* spacecraft.

Science Objectives:

- Acquire all-weather high-resolution measurements of near-surface winds over the global oceans.
- Determine atmospheric influences, ocean response and air-sea interactions on various spatial and temporal scales.
- Develop improved methods of assimilating wind data into numerical weather and wave prediction models.
- Combine wind data with measurements from various scientific disciplines to understand processes of global climatic change and weather.

Measurements:

- The instrument was operated continuously at a frequency of 13.995 GHz.
- Six dual-polarized, 3-meter long, stick-like antennas collected backscatter data with a resolution of 50 km for nine months before loss.
- Backscatter data was combined and processed to yield 268,000 globally distributed wind vectors per day.

Every two days, under all weather and cloud conditions, NSCAT measured wind speeds and directions over at least 90% of the Earth's ice-free oceans. Since oceans cover approximately 70% of Earth's surface, NSCAT played a key role in scientists' efforts to understand and predict complex global weather patterns and climate systems. NSCAT used eight antenna beams to scan two wide bands of ocean, one on each side of the instrument's orbital path. NSCAT transmitted short pulses of microwave energy to probe ocean surfaces and then measured the reflected or backscattered power. Variations in the magnitude of this backscattered power are caused by changes in small (centimeter-sized), wind-driven waves. Using a method called Doppler processing (a change in the observed frequency of the radio waves due to relative motion of source and observer), the measured backscattered power was separated into cells at specific locations on Earth's surface; these were then transmitted to the ground for processing. During ground processing, wind direction and speed was determined from these variations. Within two weeks of receiving the raw data, the ground system processed wind measurements.

Nine months after launch, the Midori satellite lost power, ending its and NSCAT's mission. NASA approved a rapid replacement mission called the *Quick Scatterometer* to take the place of NSCAT.

Quick Scatterometer. NASA's *Quick Scatterometer* (QuikSCAT) was lofted into space at 7:15 p.m. Pacific Daylight Time on Saturday, June 19, 1999 atop a U.S. Air Force Titan II launch vehicle from Space Launch Complex 4 West at California's Vandenberg Air Force Base. The satellite was launched in a south-southwesterly direction, soaring over the Pacific Ocean at sunset as it ascended into space to achieve an initial elliptical orbit with a maximum altitude of about 800 kilometers (500 miles) above Earth's surface. See Figs. 3 and 4.

Built in a record time of just 12 months, *QuikScat* will provide climatologists, meteorologists and oceanographers with daily, detailed snapshots of the winds swirling above the world's oceans. Labeled as NASA's next "El Niño watcher," *QuikSCAT* will be used to better understand global weather abnormalities and to generally improve weather forecasting.

QuikSCAT's predecessor, NSCAT (NASA Scatterometer), a microwave radar scatterometer, measured near-surface wind vectors (both speed and direction) over the global oceans starting in August 1996. The *QuikSCAT* mission is a "quick recovery" mission to fill the gap created by the loss of data from NSCAT, after the satellite it was flying on lost power in June 1997.

Winds play a major role in every aspect of weather on Earth. They directly affect the turbulent exchanges of heat, moisture and greenhouse gases between Earth's atmosphere and the ocean. To better understand their impact on oceans and improve weather forecasting, *QuikSCAT* carries a state-of-the-art radar instrument called a scatterometer for a two-year science mission. Known as "SeaWinds," this scatterometer operates by transmitting high-frequency microwave pulses to the ocean surface and measuring the "backscattered" or echoed radar pulses bounced back to the satellite. The instrument senses ripples caused by winds near the ocean's surface, from which scientists can compute the winds' speed and direction. The instruments can acquire hundreds of times more observations of surface wind velocity each day than can ships and buoys, and are the only remote-sensing systems able to provide continuous, accurate and high-resolution measurements of both wind speeds and direction regardless of weather conditions.

SeaWinds uses a rotating dish antenna with two spot beams that sweep in a circular pattern. The antenna radiates microwave pulses at a frequency of 13.4 gigahertz across broad regions on Earth's surface. The instrument is currently collecting data over ocean, land, and ice in a continuous, 1,800-kilometer-wide band (1,120 miles), making approximately 400,000 measurements and covering 90% of Earth's surface each day.

Science Objectives:

- Acquire all-weather, high-resolution measurements of near-surface winds over global oceans.
- Determine atmospheric forcing, ocean response, and air-sea interaction mechanisms on various spatial and temporal scales.
- Combine wind data with measurements from scientific instruments in other disciplines to help us better understand the mechanisms of global climate change and weather patterns.
- Study both annual and semi-annual rain forest vegetation changes.
- Study daily/seasonal sea ice edge movement and Arctic/Antarctic ice pack changes.

Operational Objectives:

- Improve weather forecasts near coastlines by using wind data in numerical weather- and wave-prediction models.
- Improve storm warning and monitoring.

Ground Systems:

- Tracking by Earth Polar Ground stations Svalbard, Norway; Poker Flats, Alaska; Wallops Island, Virginia; and McMurdo, Antarctica; Hatoyama, Japan (contingency station).
- High-quality research data products produced at JPL and distributed to science community within two weeks of receipt.

Fig. 3. NASA's *Quick Scatterometer* (*QuikScat*) lofted into space on June 19, 1999 from California's Vandenberg Air Force Base. *QuikScat* provides climatologists, meteorologists, and oceanographers with daily, detailed snapshots of ocean winds. The mission will greatly improve weather forecasting.

The satellite was launched on a U.S. Air Force Titan II launch vehicle soaring over the Pacific Ocean at sunset. Approximately two and a half minutes after launch, the Titan II first-stage engine shut down and the second stage ignited. A minute later, the nose cone separated in two halves and was jettisoned as planned. An hour into flight, *QuikScat* deployed its solar arrays.

Fig. 4.   This is an artist rendering of NASA's *Quick Scatterometer* (*QuikScat*). The large, dishlike feature at the bottom of the satellite is the scatterometer instrument that will measure winds over the ocean surface. (*NASA/JPL.*)

- Scatterometer science data products are distributed through the JPL Physical Oceanography Distributed Active Archive Center (PO.DAAC), a scientific data distribution site: *http://podaac.jpl.nasa.gov.*
- Operational data products produced at National Oceanic & Atmospheric Administration (NOAA) for international meteorological community within 3 hours of data collection. *http://www.noaa.gov/*

See also **Radar**; **Satellites (Scientific and Reconnaissance)**; and **Weather Technology**;

Web References

Air-Sea Interaction & Climate Team: *http://airsea-www.jpl.nasa.gov/*
Center for Ocean-Atmospheric Prediction Studies (COAPS). *http://www.coaps.fsu.edu/*
History of Scatterometry: *http://winds.jpl.nasa.gov/scatterometry/history.html*
Jet Propulsion Laboratory Homepage: *http://www.jpl.nasa.gov/*
Jet Propulsion Laboratory Winds Homepage: *http://winds.jpl.nasa.gov/*
Marine Observing Systems Team: *http://manati.wwb.noaa.gov/*
Oceanography Distributed Active Archive Center: *http://podaac.jpl.nasa.gov/*
Scatterometer Climate Record Pathfinder (SCP): *http://podaac.jpl.nasa.gov/scp/*
The ERS Missions: *http://earth.esa.int/ers/*
Note: For a higher resolution file for this artwork (A TIFF file is available) at *http://visibleearth.nasa.gov/cgi-bin/viewrecord?2941*

NASA's Jet Propulsion Laboratory, Pasadena, CA.

SCAUP.   See **Waterfowl**.

SCAVENGING.   1. The use of an unspecific precipitate to remove from solution by adsorption or coprecipitation a large fraction of one or more undesirable radionuclides. Voluminous gelatinous precipitates are usually used as scavengers, e.g., $Fe(OH)_3$. 2. The removal of impurities from molten metal by addition of substances to form slags, or other compounds that can readily be removed. 3. The removal of unwanted gases from systems, e.g., of products of combustion from an internal combustion engine, or residual gases from an evacuated tube.

SCHEELITE.   The mineral scheelite is calcium tungstate, $CaWO_4$, with molybdenum substituting for tungsten up to 25% in the molybdian scheelite variety. It is a tetragonal with an octahedral habit although also at times tabular, and may occur massive. It displays an octahedral cleavage; is brittle; hardness, 4.5–5; specific gravity, 6.1; luster, vitreous; color, white to yellowish, reddish, greenish and brownish; white streak; transparent to translucent. Scheelite is found in pegmatite and ore veins associated with granites, also as a contact metamorphic mineral. It is known from the Czech Republic and Slovakia, Saxony, Italy, Alsace, Finland; Cumberland and Cornwall in England; and Mexico. Crystals of exceptional length (6–10 inches; 15–25 centimeters) are found at various localities in Korea and Japan; and in the United States, in Connecticut, Colorado, South Dakota, Arizona, Nevada, and California. The Swedish chemist, Karl Wilhelm Scheele, discovered tungsten in this mineral, which later was named for him.

The mineral fluoresces vivid bluish white to white; or yellowish white with increasing molybdenum content under exposure to short-wave ultraviolet light.

SCHERING BRIDGE.   See **Bridge Circuits (Electrical/Electronic)**.

SCHILBEIDS.   See **Catfishes**.

SCHINUS TREE.   A shrubby tree capable of reaching a height of about 30 feet (9.1 meters) for some years has posed a wildlife and vegetation threat in the ecosystem of Everglades National Park, Florida. The schinus tree (*Schinus terebinthifolius*) is one of several exotic and tropical specimens introduced into Florida in the early 1890s as an ornamental shrub. It became known as Brazilian holly, Florida holly, or Brazilian pepper. In the early 1900s, the plant was studied in some detail at the U.S. Department of Agriculture's plant-introduction station in Miami and its remarkable growth rate was noted. It was found that seedlings would reach a height of 15 feet (4.6 meters) within a period of three years, producing fruit in the second year. Dwellers in south Florida, notably in the Homestead area, were pleased with the manner in which great numbers of birds were attracted to the red berries of the schinus. It was soon noted, however, that the tree was taking root in locations where it was not planted or wanted. The schinus was extremely difficult to control, and it spread rapidly and profusely. It was also found that the schinus irritated the skin much as poison ivy does (not surprising in retrospect, because the schinus is closely related to the poison ivy and poison sumac family). It was inevitable that the schinus would spread into parts of Everglades National Park, crowding out native species. The schinus also has spread into stands of native slash pine. It tolerates flooding and drought and sprouts prolifically after fires. Other imported trees (all with a positive intent on the part of the original introducers) which have caused somewhat similar problems include the *casuarina* and *melaleuca* trees/shrubs. The latter two plants are described in separate entries in this volume.

SCHIST.   The schists form a great group of metamorphic rocks chiefly notable for the preponderance of the lamellar minerals such as the micas, chlorite, talc, hornblende, graphite, etc. Quartz often occurs in drawn out grains to such an extent that a quartz schist is produced. Most schists have in all probability been derived from clays and muds which have passed through a series of metamorphic processes involving the production of shales, slates and phyllites as intermediate steps. Certain schists have been derived from fine-grained igneous rocks such as lavas and tuffs. Most schists are mica schists, but graphite and chlorite schists are common. Schists are named for the prominent or perhaps unusual mineral constituent, as garnet schist, tourmaline schist, glaucophane schist, etc. The word schist is derived from the Greek meaning to split, with reference to the easy separation of these rocks in a direction parallel to that in which the platy minerals lie.

SCHISTOSOMIASIS.   Once known as bilharziasis, this is an invasion of the body by a genus of trematode parasites or flukes, the *chistosomae*. The disease reaches far back into history and ancient Egyptians even had a hieroglyphic symbol for it — a penis dripping blood. The disease was indeed so common that blood in the urine was considered a puberty symbol for males. Approximately 5% of the world's population is affected, with most cases appearing in Africa, the Middle East, Central and South America, China, the Philippines, and Malaysia. Many cases are also found in the Caribbean countries.

Infected persons excrete the microscopic schistosome eggs in their feces and urine. If these reach fresh water, tiny embryos hatch and invade an intermediary host, usually a snail. Therein, they develop and multiply and 6 to 8 weeks later the snail releases cercariae, or schistosome larvae, in numbers amounting to some tens of thousands per day.

Humans are infected when swimming in the water, the infective larvae penetrating the skin very rapidly to the blood stream where they travel to various organs. Species of schistosomes select different target sites in which to reside. *S. mansoni* targets the mesenteric veins of the upper intestine; *S. japonicum*, the mesenteric veins of the lower intestine; *S. hematobium*, the bladder veins. Once settled, the schistosomes produce a steady output of eggs (1000 or more per day) and live for years in their selected sites.

About half the eggs produced migrate through the wall of the intestine or bladder and the remainder are swept to the liver by the blood stream where they cause a severe inflammatory reaction. The eggs that do exit to the intestine or bladder are excreted and able to enter a new host.

Schistosomiasis patients frequently suffer vomiting and diarrhea. An infection by *S. mansoni* or *S. japonicum* develops a hard, cirrhosis-like hepatomegaly; *S. hematobium* patients have intense bladder damage and heavy bleeding into the urine.

Diagnosis depends upon finding schistosome eggs in urine or feces or by serological tests, such as complement-fixation or ELISA (enzyme-linked immunosorbent assay).

Some progress has been made toward development of an anti-schistosome vaccine but, despite optimism, the body's immune system is continuously evaded by the organism's ability to camouflage itself with major histocompatibility antigens so that it resembles the host's own body. Meanwhile, praziquantel is the most effective drug against the parasite although metrifonate and niridazole are also of value.

R.C. VICKERY, M.D., D.Sc., Ph.D., Blanton/Dade City, FL.

SCHIZOPHRENIA.   A major medical illness. It has been estimated that a child living at the age of 15 years statistically has a 2 to 3% risk of being diagnosed as a schizophrenic sometime during its lifetime. Persons with the disease in the United States exceed the population of New York City. No real medical breakthroughs in diagnosing and curing the disease have occurred. Inasmuch as the disease is not fatal, lifetime costs of care are astonishing.

If one were to condense all that is known concerning the disease, three elements would predominate: (1) the disease appears to run in families; (2) neuroleptics (drugs that act in some manner with the brain's dopamine system) help to improve the symptoms of the disease; and (3) there may be something structurally awry in the schizophrenic brain.

Fundamentals still are argued among the experts: — Does schizophrenia represent a discrete mental disorder, or is it just one component of a spectrum of mental illness? Ming Tsuang (Harvard Medical School) has expressed the opinion that present data seem to favor the traditional view that schizophrenia is made up of two discrete, major psychoses: (1) schizophrenia that starts during adolescence and usually worsens; and (2) affective (mood) disorders that occur later in episodes and are less likely to incapacitate the patient. Researchers are now looking to molecular genetics for evidence to prove the two stages. Still other researchers, for example Timothy Crow (Northwick Park Hospital, Harrow, England), views affective illnesses and schizophrenia at opposite ends of a continuum of psychotic disorders. Depression is at one end of the spectrum, followed by manic depression, mixed schizophrenia and affective disorder, and schizophrenia sans affective disorder. Crow observes that there are more patients in the middle of the spectrum than at the extremes.

Numerous studies of identical twins have been made to get a handle on the familial or hereditary characteristics of the disease. The evidence of a concentration is strong. If one identical twin has the disease, the chances of the other identical twin having the disease is greater than 50%. Another research finding reveals the less obvious, i.e., the children of schizophrenics who are adopted by nonschizophrenic parents surprisingly show a higher incidence of schizophrenia than a control population. These and other similar studies have not convinced all specialists, however, that there is a strong hereditary factor. Researcher Kendler has observed that it is not fair to call schizophrenia a genetic disease; it is not a classic Mendelian trait inherited like the gene for Huntington's chorea. But, one cannot deny evidence that like coronary heart disease and early-onset hearing loss, schizophrenia tends to run in families. Presently, there are two principal hypotheses regarding the inheritance of schizophrenia: (1) a model based on a single major gene locus (supported by Philip Holzman, Harvard University); and (2) a model that suggests multiple genetic causes, of which Tsuang is an advocate. Some researchers consider that poor eye tracking may be a biological marker for schizophrenia. Others claim that poor eye tracking stems from other causes as well, including drug-induced effects, Parkinson's disease, multiple sclerosis, and some brain lesions.

Crow has suggested that a retrovirus may be involved. Perhaps the virus may incorporate into the human genome and thus be capable of passage from one generation to the next. There have been other viral postulates, supported by a survey which indicates over half of the schizophrenic population were born in winter or spring (virus season). Other researchers have associated the disease with the immune system.

Although very generalized, Sedvall (Karolinska Institute, Stockholm) observes that schizophrenia probably represents a complex interplay of genetic and environmental factors.

More recent research targets include direct studies of the brain—to observe blood flow, brain metabolism, and receptor mapping to monitor brain structure. The effects of various drugs can be observed with key instrumental tools in connection with human and experimental animals. These tools include computerized tomography (CT) scans, magnetic resonance imaging, and photon imaging.

Symptoms and Treatment

Several decades ago, schizophrenia was called *dementia praecox*. This terms means "a precocious demented state." In 1911, Blueler advocated the substitution of *schizophrenia — schizo* standing for splitting and *phrenia* for the mind, thus indicating a "breaking away" of the patient's mind from a normal evaluation of reality. There are several forms in which the illness manifests itself. However, denial of reality and inappropriate emotional responses are the most common symptoms. The distorted psyche is revealed by the patient's behavior. During some spans of time, the patient may be wild in behavior, as illustrated by the breaking up of furniture and the throwing of portable objects. During some periods, a patient may rip off clothing and go naked; or, during other periods, dress in a fantastic manner. The patient may laugh or cry without apparent cause. The patient may use a language consisting of jumbled fractions of words and phrases that are incomprehensible.

The patient may be confused as to identity and make fantastic claims as being someone of high repute. In all, the actions and mannerisms of the schizophrenic appear bizarre and unintelligible when viewed in the light of the real world. The actions are more easily understood when one realizes that this behavior is the product of a dream world, erected because the patient does not have the ability to perceive reality in a normal way.

When a schizophrenic becomes unable to find any solution that will enable the acceptance of a painful situation, the patient's normal defenses break down entirely and "reality" is imagined in terms of desires. Daydreams offset the poverty of true relationships and thus become more gratifying than reality.

Traditionally, patients who present the symptoms of acute schizophrenia are referred to a psychiatrist. A wide variety of antipsychotic drugs is available. Generally, these drugs can be used safely with most patients because of the wide difference between therapeutic and lethal doses. Further, most of these drugs are not addictive and, in most situations, the patient does not develop a tolerance (thus decreasing the effectiveness) to the antipsychotic effects of these drugs. Among the antipsychotic agents commonly used are the phenothiazines, thioxanthenes, dibenzoxazepines, butyrophenones, and indolones. Certain generalizations can be made concerning these drugs—rapid absorption $\frac{1}{2}$ to 1 hour with oral doses; 10 minutes with intramuscular administration); oxidative metabolism, usually in the liver, with byproducts excreted in the urinary tract. There are side effects of these drugs, particularly in elderly patients.

Often a considerable period of time will be required to find the most satisfactory antipsychotic agent for a given patient. With some of these drugs, however, a minimum period of three months is required to determine the full clinical response. Proper dosage of a given drug also varies considerably from one patient to the next. Acute schizophrenia may not require indefinite maintenance. In chronic schizophrenia, the drug maintenance program is long-term.

Additional Reading

Breier, A.: "Current Issues in the Psychopharmacology of Schizophrenia," Lippincott Williams & Wilkins, Philadelphia, PA, 2001.

Csernansky, J.C.: "Schizophrenia," Marcel Dekker, Inc., New York, NY, 2001.

Harrison, P.J. and G. Roberts: "The Neuropathology of Schizophrenia: Progress and Interpretation," Oxford University Press, Inc., New York, NY, 2000.

Heinrichs, R.W.: "In Search of Madness: Schizophrenia and Neuroscience," Oxford University Press, Inc., New York, NY, 2001.

Herz, M.I. and S. Marder: "Schizophrenia: A Comprehensive Text," Lippincott Williams & Wilkins, Philadelphia, PA, 2002.

Johnstone, E.C., M. Humphreys, F.H. Lang, S.M. Lawrie, and R. Sandler: "Schizophrenia: Concepts and Clinical Management," Cambridge University Press, New York, NY, 1999.

Lidow, M.S.: "Neurotransmitter Receptors in Actions of Antipsychotic Medications," CRC Press, LLC., Boca Raton, FL, 2000.

Wykes, T., S. Lewis, and N. Tarrier: "Outcome and Innovation in Psychological Treatment of Schizophrenia," John Wiley & Sons, Inc., New York, NY, 2000.

Web References

All About Schizophrenia - Mental Health Net: *http://mentalhelp.net/poc/center_index.php?id=7*

Merck Manual - Schizophrenia and Related Disorders: *http://www.merck.com/pubs/mmanual/section15/chapter193/193a.htm*

Mental Health Disorder Web Sites: *http://psychology2.semo.edu/websites/web12.htm*

Schizophrenia Fellowship of NSW: *http://www.sfnsw.webcentral.com.au/*

Schizophrenia Resources — NAMI: *http://council.nami.org/programs.html*

THE INSTITUTE OF LIVING, Hartford Hospital's Mental Health Network: *http://www.instituteofliving.org/schizophrenia/index.htm*

World Fellowship for Schizophrenia and Allied Disorders: *http://www.world-schizophrenia.org/*

SCHLIEREN. Refraction anomalies produced in transmitted light by differences in density or other anisotropy in parcels or strata of air (or other fluids). All the natural scintillation phenomena in the atmosphere result from density schlieren developed by turbulent processes. Schlieren optical systems are frequently used for observing and photographing colloidal refractive index gradients which arise in the study of electrophoresis, diffusion, sedimentation velocity, chromatography, and airflow in wind tunnels. Figure 1 shows a diagram of schlieren measurements in a wind tunnel, in which an interferometer is used to determine the interference effects produced by turbulence. Steady flow of air at constant density does not show on the interference pattern, but turbulent flow, shock waves and similar density variations do appear in the interference pattern. The word *schlieren* is used to describe the whole phenomenon. The interferometer plates are called *schlieren plates*, the interference pattern is called a *schlieren picture*, and the whole instrument is called a *schlieren interferometer* or simply *schlieren apparatus*.

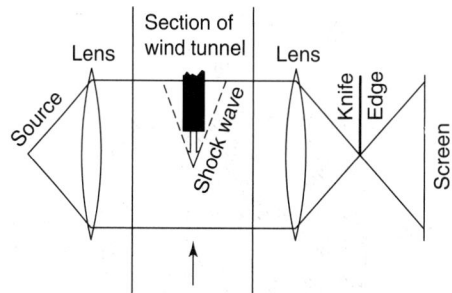

Fig. 1.   Idealized Schlieren apparatus for observing shock wave in supersonic tunnel.

SCHMIDT, MAARTEN (1929–   ). Schmidt was born in Groningen, the Netherlands. He received a Ph.D. in astronomy from the University of Leiden and worked at the Leiden Observatory. In 1959, he moved to American and joined the astronomy department at the California Institute of Technology where at first he continued working on the mass distribution and dynamics of the Galaxy. When Rudolph Minkowski retired, Schmidt took over his project of taking spectra of objects, which had been found to be radio emitters. In 1963 he discovered the first quasar, showing that these starlike objects exhibit ordinary hydrogen lines, but at redshifts far greater than those observed in stars. Since then he has investigated the evolution and distribution of quasars, discovering that they were more abundant when the universe was younger, one of the major reasons for the decline in favor of steady-state models of the universe. He is currently seeking to find the redshift above which there are no quasars. He also studies X-ray and gamma ray sources.

He won the Warner Prize of the American Astronomical Society in 1964 and in 1968, he was awarded the Rumford Award of the American Academy of Arts and Sciences. In 1992 he was awarded the Bruce Medal. See also **Quasars**.

J. M. I.

SCHMIDT OBJECTIVE. An objective for reflecting telescopes, designed to correct the aberration of the spherical mirror without introducing the coma (blurring) to which even a parabolic reflector is subject for wide fields. The results are obtained in somewhat the same way that spectacles correct for defects in vision. A slight chromatic aberration is introduced.

The Schmidt objective as originally designed consists of a concave spherical mirror, functioning in the same way as the objective of any reflecting telescope, but with a plate of glass interposed in front of it perpendicular to its axis at its center of curvature. This glass plate is not plane, but has one surface "figured" in such a way that, as the rays pass through it on their way to the mirror, it so modifies their course as to effect almost perfect correction for the spherical aberration and coma that the mirror would otherwise produce. In a later design, the objective consists of two coaxial cylinders of glass, in contact along a plane perpendicular to the axis. The rear surface of the rear piece is spherically convex and is silvered on the outside, thus presenting a concave spherical mirror to the interior of the glass cylinder. The front surface of the front piece, passing through the center of curvature of the mirror, is the correction surface, serving the same purpose as the glass plate in the older design. The reason for using two pieces is that the final real image F is of course produced between the mirror and the correction surface; and it is here that the plane of separation is located, so that the small photographic film used may be introduced. The general nature of these arrangements will be clear from Fig. 1. There are numerous other variations of this idea, all referred to as Schmidt telescopes.

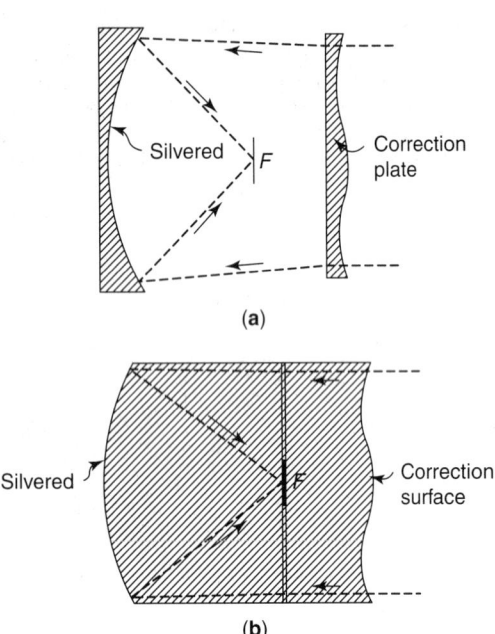

Fig. 1. Schmidt objective: (a) Original design; (b) the "solid" Schmidt (diagrammatic).

See also **Telescope (Astronomical-Optical)**.

SCHMIDT PROCESS. A method for converting a given set of vectors $\mathbf{u}_1, \mathbf{u}_2, \ldots, \mathbf{u}_n$ into an orthonormal set $\mathbf{v}_1, \mathbf{v}_2, \ldots, \mathbf{v}_n$. If the length of $\mathbf{v}_i$ is l_i and $\mathbf{e}_i = \mathbf{v}_i/l_i$ is a unit vector, the calculation may be performed with the recursion formula

$$\mathbf{v}_{i+1} = \mathbf{u}_{i+1} - \sum_{k=1}^{n} (\tilde{\mathbf{e}}_k \mathbf{u}_{i+1}) \mathbf{e}_k$$

where $\tilde{\mathbf{e}}_k$ is the transpose of $\mathbf{e}_k$. The procedure may also be used for functions with the appropriate integrations in place of summations. If, for example, the original functions are $1, x, x^2, \ldots$, defined over the range $x = \pm 1$, the orthonormal set obtained from them by the Schmidt process is a set of Legendre polynomials.

See also **Legendre Differential Equation**.

SCHMITT TRIGGER. A form of bistable multivibrator which furnishes a fast-acting on-off switch action, is capable of generating rectangular pulses from sinusoidal or other nonrectangular waveforms, and can determine when a signal reaches a specified D.C. level.

SCHOTTKY DEFECT. A lattice vacancy created by removing an ion from its site and placing it on the surface of the crystal. For electric

neutrality, the number of cation Schottky defects must equal the number of anion Schottky defects. The number, n, of Schottky defects is given by

$$\frac{n}{N-n} = C_s e^{-W/kT}$$

where there are N lattice points, and W is the energy required to remove an ion from a lattice point, and then add it to the surface. C_s is a numerical factor of the order of 10^3–10^4. This relation may be derived from thermodynamic arguments. It can be shown that the factor C_s includes a vibrational entropy term.

SCHRÖDINGER EQUATION. The basic equation of wave mechanics. It is developed by using the de Broglie wavelength in the description of a particle and then associating with the measurement of the energy E or of the x-component of momentum p_x of the particle a differential operator

$$E = i\hbar \frac{\partial}{\partial t} \text{ or } p_x = -i\hbar \frac{\partial}{\partial x}$$

where is the Dirac $\mathbf{h}$. The Hamiltonian function can be expressed either in terms of total energy or in terms of potential energy and momentum. Expressing it in both ways, one obtains:

$$-\frac{\hbar^2}{2m} \nabla^2 \psi + V(\mathbf{r})\psi = i\hbar \frac{\partial \psi}{\partial t} = E\psi$$

where ∇^2 is the Laplacian, m the mass of the particle, E, its total energy, and $V(\mathbf{r})$ its potential energy (usually a function of position). This is the time dependent Schrödinger equation for ψ.

In many instances, we are interested in the allowed values of E in stationary states of the system. Using the Planck law we may set $E = h\nu = 2pi\hbar\nu$ and write

$$\psi = \phi(\mathbf{r})e^{2\pi i \nu t} = \phi(\mathbf{r})e^{iEt/h}$$

where $\phi(\mathbf{r})$ is a function of position only.-We then obtain the time independent equation:

$$\left[\nabla^2 + \frac{2m}{\hbar}\{E - V(\mathbf{r})\} \right] \phi = 0$$

It is often found that solutions of the equation exist only for specific eigenvalues of E. To each eigenvalue E_n there corresponds an eigenfunction of the coordinates ϕ_n. The probability of finding the particle in a region of volume dV is

$$\int |\phi| dV = \int \phi\phi^* dV$$

assuming that ϕ has been normalized so that the integral over all space is unity.

See also **Quantum Mechanics**.

SCHRODINGER, ERWIN (1887–1961). Schrödinger was born in Vienna, Austria. He earned a Ph.D. in physics from the University of Vienna. He worked with Weyl and Debye researching a number of problems in theoretical physics. He wrote papers on problems of thermodynamics, he studied the theory of color vision and writing on color blindness, and more importantly, he studied the atom trying to find a wave equation for the motions executed by the electrons and their interactions. Today, Schrodinger is known for his discovery, Schrodinger's wave equation. He won the Nobel Prize in Physics in 1933, jointly with Paul Dirac for his wave theory of atomic phenomena.

Beginning in 1927, Schrodinger was professor of theoretical physics at the University of Berlin. When the Nazi regime came to power, he left for England. He later accepted a position at the University of Graz, Austria but when Germany annexed Austria, Schrodinger left and took a position in Ireland at the Institute for Advanced Studies. His work in physics continued here. He also became interested in philosophy and biology and published a book, *What Is Life?*, which was influential in the scientific world.

See also **Schrödinger Equation**.

J. M. I.

SCHUCHERT, CHARLES (1858–1942). Schuchert was an American paleontologist. He was a pioneer paleogeographer teaching historical geology at Yale from 1904–1923. He made significant contributions towards the geologic time scale. He is especially remembered for his book, *Historical Geology of North America*.

See also **Geologic Time Scale**.

J. M. I.

SCHWARZSCHILD, KARL (1873–1916). Schwarzschild was a Germany theoretical astrophysicist who computed solutions of Einstein's field equations in relativity. Knowledge of events is limited at the Schwarzschild radius and this has led to research on black holes. Shortly after this work, Schwarzschild died of a rare metabolic disorder.

See also **Gravitation**.

J. M. I.

SCHWARZSCHILD TELESCOPE. A reflecting telescope using two mirrors designed to be free of spherical aberration and coma. It requires a large secondary mirror, which causes a loss of light, and a long foretube to act as a light baffle. See also **Telescope (Astronomical-Optical)**.

SCINTILLATION (Astronomical). Any scintillation phenomena, such as irregular oscillatory motion, variation of intensity, and color fluctuation observed in the light emanating from an extraterrestrial source. To be distinguished from terrestrial scintillation primarily in that the light source for the latter lies somewhere within the Earth's atmosphere. Also called *stellar scintillation*. See also **Seeing (Astronomy)**.

SCINTILLATION (Atmospheric). Generic term for rapid variations in apparent position, brightness, or color of a distant luminous object viewed through the atmosphere. If the object lies outside the Earth's atmosphere, as in the case of stars and planets, the phenomenon is termed *astronomical scintillation*; if the luminous source lies within the earth's atmosphere, the phenomenon is termed *terrestrial scintillation*. As one of three principal factors governing astronomical "seeing" (i.e., the disturbing effects produced by the atmosphere upon the image quality of an observed astronomical body), scintillation is defined as variations in luminance, only.

It is clearly established that almost all scintillation effects are caused by anomalous refraction occurring in rather small parcels or strata of air (schlieren), whose temperatures and, hence, densities differ slightly from those of their surroundings. Normal wind motions transporting such schlieren across the observer's line of sight produce the irregular fluctuations characteristic of scintillation. Scintillation effects are always much more pronounced near the horizon than near the zenith. Parcels of the order of only centimeters to decimeters are believed to produce most of the scintillatory irregularities in the atmosphere.

See also **Schlieren**.

SCINTILLATION COUNTER. Radiation detectors that respond by emitting a flash of visible light are classified as scintillators or fluors. Typical scintillators used for counting gamma radiation are thallium doped crystals of sodium iodide, NaI(TI). Plastic (organic) scintillators are commonly used for counting beta radiation, and for counting alpha particles; zinc sulfide, ZnS, is a good detecting crystalline substance.

SCINTILLOMETER. A type of photoelectric photometer used in a method of determining high altitude winds on the assumption that stellar scintillation is caused by atmospheric inhomogeneities (schlieren) being carried along by the wind near tropopause level. Also called *scintillation meter*.

SCLERAL BUCKLE. A scleral buckle is the traditional surgical procedure used in repairing a detached retina. A detached retina occurs when a portion of the retina inside the back of the eye comes loose from the underlying epithelium. Because the retina is responsible for receiving and sending visual images to the brain, a detachment can seriously affect vision and should be repaired by a retina surgeon immediately.

There are two types of surgery available for retinal detachments: scleral buckling and pneumatic retinopexy. The scleral buckle is the more common treatment. During this surgery, the sclera (the outside covering or white of the eye) is indented or "buckled" inward, usually by attaching a piece of preserved sclera or silicone rubber to its surface.

The surgery is usually performed in an operating room under general anesthesia, although local anesthesia may be used in some cases. Because most retinal detachments are caused by a retinal tear, the first step is to repair the tear with either a laser or a freezing instrument known as a *cryoprobe*. This treatment creates a scar around the tear preventing further tears or detachment. Next, a piece of solid silicone or a silicone sponge is sewn onto the outside wall of the eye over the retinal tear site. The

silicone pushes in on the sclera until scarring from the cryotherapy (or laser treatment) seals the tear. The surgery is called *scleral buckling* because the sclera is buckled (pushed) in by the silicone. This buckle is left on the eye permanently and is not usually visible following surgery. On occasion, a vitrectomy may also be performed in conjunction with the scleral buckling surgery. During a vitrectomy, the vitreous fluid, which fills the eye, is removed and replaced with air or gas to push the retina back into place. These gases are gradually replaced by new vitreous fluid produced by the eye. The success of retina surgery depends on a number of factors including the length of time from when the retina detached until it was repaired, the size and location of the damage, and whether a fibrous growth has formed on the retina. If the macula (portion of the retina responsible for central or reading vision) was detached, vision rarely returns to normal.

In most cases, the retina can be successfully reattached, but reattachment does not necessarily mean restored vision. It may take many months after surgery for the best vision to occur and, if the first retinal detachment operation is not successful, a second operation may be performed. Treatment almost always prevents further loss of sight from this type of retinal problem.

Following scleral buckle surgery, the eye may be red and sore for a time and vision may be blurry. A scratchy sensation may also be felt as the stitches used to close the lining around the eye heal. The eye may be patched and eye drops given to prevent infection. Physical activity is usually restricted to ensure that the retina remains attached. Regular follow-up visits are necessary to ensure the retina is healing properly.

Serious complications from scleral buckling surgery are infrequent but can happen. The most common is redetachment of the retina. Occasionally, double vision can result from damage to the muscles that control eye movement. Other rare complications include cataract formation, bleeding beneath the retina, glaucoma, infection, and drooping of the eyelid.

Vision Rx, Inc., Elmsford, NY.

SCLERENCHYMA. A tissue composed of thick-walled cells of various forms. These cells have small pits and walls so thick that in many cases the cavity of the cell is nearly obliterated. Mature sclerenchyma cells are dead, containing no protoplasm. They serve to strengthen the part of the plant in which they are found or to protect the more delicate structures within. Sclerenchyma cells are of two kinds, stone cells and fibers. Stone cells are small, irregular in shape, and only slightly if at all elongated. They may be found in the cortex of the stem or elsewhere in the plant, but are particularly abundant in the endocarp of certain fruits and in seeds. The flesh of pears and blueberries contains many grit particles, which are groups of stone cells. Fibers are very much elongated cells, generally with long pointed ends and with simple pits in the walls. Hemp and flax are bundles of fibers of great value to people.

SCLERODERMA. This may be a localized disease (*scleroderma*) or a diffuse disease (*progressive systemic sclerosis*) which is caused by the deposition of fibrous connective tissue in the skin. Scleroderma is classified as one of the rheumatic diseases. The skin becomes thick and there is moderate swelling of affected tissue, followed by a deadening and wasting away of skin with accompanying loss of hair follicles, sebaceous and sweat glands. The skin tightens, loses pliability, and may be described as hide-like. In the course of the disease, there may be involvement of the lungs, heart, kidneys and other organs. Infrequently, systemic complications can be fatal. Where only the skin is affected, the disease may not be the primary determinant of life span. Sclerodermatous skin is seen most frequently in the hands and fingers and somewhat less frequently in the facial area, this latter involvement frequently reducing the size of the mouth. However, in some patients, the disease may be diffusive and involve all of the skin of the body.

The systemic aspects (sclerosis) of scleroderma are usually more serious than simple scleroderma. When the *musculoskeletal system* is involved, there may be a mild form of arthritis similar to rheumatoid arthritis. Sometimes *acral osteolysis* (shortening of fingers and toes) may develop. Also, there may be weakening of skeletal muscles. Scleroderma may involve the *gastrointestinal tract*, leading to diverticula. Where smooth muscle of the esophageal mucosa is replaced by fibrous tissue, *dysphagia* (difficulty in swallowing) may result — as well as gastric reflux and peptic esophagitis. *Lung involvement*, in which fibrosis occurs in the walls of the pulmonary arteries, may cause pulmonary hypertension, a condition that can lead to progressive respiratory failure. In *heart involvement*, fibrous tissue replaces extensive areas of cardiac muscle, precipitating arrhythmias, conduction disturbances, and congestive heart failure. In *kidney involvement*, scleroderma may cause chronic mild proteinuria and mild hypertension. The syndrome known as *malignant hypertension* may precipitate oliguric renal failure.

The etiology of scleroderma is unknown. The abnormal amount of collagen present in this disease appears to be normal rather than altered collagen.

SCLEROMETER. An apparatus for determining the hardness of a material by measuring the pressure on a standard point that is required to scratch the material. A scleroscope is a similar apparatus, which measures hardness by determining the rebound of a standard ball dropped on the subject material from a fixed height.

SCOLECITE. This mineral is a zeolite, a hydrous calcium-aluminum silicate, $CaAl_2Si_3O_{10} \cdot 3H_2O$. It occurs in slender monoclinic prisms and in fibrous and nodular masses. Hardness is 5; specific gravity, 2.27; luster vitreous to silky; transparent to translucent. When heated, some specimens of scolecite curl up like worms; hence its name, derived from the Greek meaning a worm. This mineral occurs with other zeolites, at Baden, Switzerland; Iceland; Greenland; the Deccan region of India; and in the United States at Golden, Colorado, and Paterson, New Jersey. Single crystals up to 12 inches (30 centimeters) in length have recently been found in a single large cavity in the basaltic trap rocks near Nasik, India.

SCOLIOSIS. A deformity, usually of the spine, in which there is abnormal lateral displacement of the spine (with at least one other compensatory curve in the opposite direction), which in extreme conditions causes what is commonly termed "hunchback." In *Friedreich's ataxia*, a hereditary spinocerebellar degradative disease, scoliosis is present. In some cases (uncommon), scoliosis can cause pulmonary or cardiac insufficiency, or both. Thoracic scoliosis is the result of a developmental anomaly and is precipitated by weakness in the paravertebral muscles.

SCOMBROID POISONING. See **Foodborne Diseases**.

SCORIA. The term applied to lava which is highly vesicular and slaggy in appearance, due to the escape of the volcanic gases while the lava is still viscous. Scoria may be considered as a very coarse variety of pumice, the vesicles occupying approximately the same amount of space as the solid material, and extremely variable in size and shape.

SCORODITE. This hydrated arsenate of ferric iron and aluminum $(Fe^{3+}, Mg^{3+})AsO_4 \cdot 2H_2O$, crystallizing in the orthorhombic system, is the iron-rich isomorphous end member of a complete series extending to the aluminum-rich mineral Mansfieldite. Crystals usually occur as drusy crusts. Also occurs as massive, compact, and earthy material. Hardness of 3.5–4, with specific gravity of 3.278. Vitreous to subadamantine luster, of pale green to liver-brown color.

Scorodite occurs as a secondary alteration mineral in the oxidized zone of metallic arsenic-containing veins. The mineral also may be a product of deposition from certain hot springs. World localities of note include Siberia; Laurium, Greece; Carinthia; Cornwall, England; and Nevada, Utah in the United States. Currently being deposited by hot springs at Yellowstone National Park in Wyoming.

SCORPION *(Scorpionida).* A terrestrial arthropod with a conspicuous pair of pincers and a slender terminal region of the body bearing a claw-like sting, and the order made up of these animals. See Fig. 1. They are grouped with the spiders, ticks and other forms in the class *Arachnida*.

The order is characterized by the conspicuous pincers and by the sting. The body is divided into cephalothorax and abdomen, and the latter consists of a broad anterior portion and a slender post-abdomen. The sting, a modified telson, bears the opening of the duct of a poison gland. The genital ducts open on the ventral surface of the first abdominal segment, just in front of a pair of comb-like pectines of the second segment which are regarded as accessory reproductive organs.

Scorpions are common in warm dry regions. In the United States, they occur as far north as Kentucky. Their poison is not virulent as a rule.

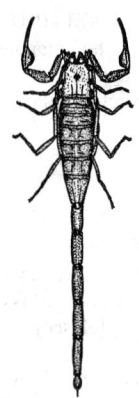

Fig. 1. Scorpion.

SCORPION FLY *(Insecta, Mecoptera).* Moderately large insects with four membranous wings. They are named from the peculiar modification of the terminal segments of the abdomen, which fairly resembles that of the scorpions. The apparent sting is, however, made up of the external genital organs. See also **Mecoptera**.

SCORPIUS (the scorpion). Scorpius is the eighth sign of the zodiac. The constellation is rather far south for observation in Europe and North America, but is a beautifully grouped constellation presenting more resemblance to the figure for which it is named than is the case with most of the others.

The brightest star in the group Antares (α Scorpii) is one of the most beautiful stars in the sky. It is distinctly reddish in color and gets its name from the fact that it opposes or rivals Mars, the red planet, in color. It is the largest star whose diameter has thus far been measured, having a diameter approximately 450 times that of the sun. The star is so large that if Antares were in the position of the sun it would extend out beyond the planet Mars. (See map accompanying entry on **Constellations**.)

SCOTOPIC VISION. See **Parafoveal Vision**

SCREAMER *(Aves, Ciconiiformes).* Peculiar South American birds with moderately long legs and large feet. The beak is like that of the domestic fowl and the wings are provided with two stout spurs on the front margin. The birds are as large as geese and swim readily, although the toes are not webbed. See also **Anseriformes**.

SCREW-WORM *(Insecta, Diptera).* The term *screw-worm* is usually reserved for those maggots of flies that attack a wounded or diseased animal, as contrasted with the egg-laying and larval activities of those flies (such as bot fly) that attack a healthy animal for the purpose of laying eggs within its body just below the skin. Screw-worms are of the family *Calliphoridae*, order *Diptera*. One species is *Callitroga* or *Cochliomyia hominivorax* (Coquerel), or *C. americana*. The adult screw-worm fly appears much like a house fly, but is about double its size. There are 3 prominent black stripes on the back behind the head. This fly lays its eggs on the edges of wounds of injured animals. The resulting maggots first feed on wound tissue and then proceed to sound tissue, a process aided by hooks in their mouth parts. Thus, old wounds do not heal, but spread and the affected animal becomes sullen and withdrawn. A chain reaction is activated because the odor from the spreading wound attracts more and more of the screw-worm flies to further the process. An untreated animal can die in relatively short order because, in this condition, the animal usually will not feed.

The species *Callitroga macellaria* (Fabricius), known as the *secondary screw-worm*, lays its eggs on the bodies of dead animals, but also attacks wounded animals. The species is responsible for a very high percentage of the animal screw-worm attacks in the southern states. Reactions from an infestation can include meningitis and peritonitis. Serious epidemics, involving tens of thousands of animals, have occurred in the Gulf States. Losses in some years run into the several millions of dollars.

Control is aided by continuing inspections of animals for any breaks in their skin that will attract the flies. Operations, such as dehorning, castrating, earmarking, docking of lamb tails, branding, etc., should be performed only during late fall and winter when flies are not active. Dogs should not be trained to bite or nip at livestock. Unavoidable injuries should be treated immediately. Infested wounds can be treated by applying smears of various formulations directly to the wound, using a small paint brush. The material should be thoroughly worked into any pockets where maggots may be hiding.

SCROPHULARIACEAE. This family contains some 2,500 species, most of which are herbs or shrubs. Its members are numerous in temperate regions, where many of them are common plants, as for example mullein, "butter-and-eggs," speedwell, and lousewort. Annuals, biennials and perennials are found in the family.

The flowers are zygomorphic, or bilaterally symmetrical, with the calyx and corolla both tubular and each composed of four or five lobes. In many plants of this family the corolla is distinctly 2-lipped, as in the Snapdragon. Usually there are four stamens, which are inserted on the corolla tube. The ovary, composed of two united carpels, becomes a dry capsule containing many small seeds. The flowers of this family are mostly pollinated by insects, such as bees, wasps, and flies, which seek the nectar secreted in a disk at the base of the ovary.

In this family are found many plants grown by man as ornamentals. Some, such as Foxglove (*Digitalis*) and *Veronica*, are hardy biennials, or perennials; others, like Snapdragon (*Antirrhinum*), are not hardy; while *Calceolaria*, a native of South America and Mexico, is a hothouse plant grown for its bizarre sac-like flowers of brilliant color. Drugs of medicinal value are also found in several plants of this family, the most important being digitalis, from species of foxglove. In early days many species were used as a source for homemade brews. Others are poisonous herbs.

SCUD. See **Clouds and Cloud Formation**.

SCULPINS (*Osteichthyes*). Of the order *Scleroparei*, family *Cottidae*, sculpins are peculiar fishes, usually small, with a broad depressed head and large pectoral fins. They are of no importance as food fishes. Many of the included species bear other names, including the little miller's thumbs of freshwaters and the sea raven, which ranges from Cape Cod to the Arctic. One species of the northern Atlantic coast is called the big sculpin, or daddy sculpin. The *Hemilepidotus hemilepidotus* (spotted irishlord) occurs in the American Pacific and reaches a length of about 20 inches (51 centimeters). *Enophrys bison* (buffalo sculpin) also occurs in the American Pacific. One of the larger sculpins is *Scorpaenichthys marmoratus* (crab-eating cabezon) which attains a length of about 30 inches (76 centimeters) and weighs up to 25 pounds (11 kilograms). It is found in Pacific waters from Lower California northward to British Columbia. Although the roe is poisonous, the flesh is considered good (even if a green color). The *Leptocottus armatus* (Pacific staghorn sculpin) is common and found in shallow waters of the Pacific coast, usually quite abundant in bays ranging from Lower California northward to Alaska. Their size usually does not exceed 6 inches (15 centimeters). *Hemitripterus americanus* (Atlantic sea raven) has qualities like a puffer in that it can inflate itself by swallowing air when taken from the water. Although rumored to be edible, its major use is for baiting lobster traps.

The grunt sculpin (*Rhamphocottus*) is of the family *Rhamphocottidae* and is named for the grunting noise it makes when taken from the water. Measuring only about 3 inches (7.5 centimeters), the grunt sculpin ranges in Pacific waters from northern California to Alaska. It is well known for its vertical temperature distribution. This sculpin does well in aquariums.

SCUP. See **Porgies**.

SCURVY. See **Ascorbic Acid (Vitamin C)**.

SCYPHOZOA. The jellyfishes. A class of the phylum *Coelenterata* made up entirely of marine animals which are, with very few exceptions, floating forms. The jellyfishes represent the highest development of the medusa form of coelenterates, and have lost the polyp stage with the exception of the reduced hydratuba larva.

Jellyfishes owe their name to the great development of the middle layer of the body (mesogloea), which is a bulky and jelly-like mass. They contain a high percentage of water, sometimes as great as 96%, and are consequently soft-bodied and without rigid support. In the water, however, they are delicate and beautiful. Many are filmy transparent creatures while

others are beautifully colored. They are found at various depths and in various seas, and in size they range from species less than 1 inch in diameter to the large *Cyanea* with a body 6–7 feet (1.8–2.1 meters) in diameter and tentacles 120 feet (36 meters) long. They are of no economic importance.

SEA ANEMONE. A complex polyp of the class Anthozoa (*Actinozoa*). Although closely related to the alcyonarians and corals the sea anemones are usually solitary and in some species are large and beautifully colored. They are without hard supporting structures such as the related forms possess.

SEA ARROW (*Mollusca, Cephalopoda*). Small slender squids, *Omnastrephes*, which swim very rapidly. Also called flying squids.

SEA BASS. See **Bass**.

SEA BEAR. See **Sea Lions and Seals**.

SEA BUTTERFLY (*Mollusca, Gasteropoda*). Mollusks with the foot formed into two wing-like lobes, which propel the animal through the sea by slow flapping movements. They make up the order *Pteropoda*.

SEA COWS (*Mammalia, Sirenia*). The manatee and dugong, collectively sea cows, constitute an order of mammals, *Sirenia*. The manatee is a fully aquatic animal with a horizontally flattened oval tail. There are no hind limbs, and the forelimbs are developed as flippers. See Fig. 1. They have a superficial similarity to whales but differ in many details of structure. They live only in shallow coastal waters and estuaries and eat aquatic plants. The several species of the genus *Manatus* are distributed on both shores of the Atlantic and in the Oriental and Australian regions. *M. latirostris* is found off Florida. The dugong is related to the American manatee and is found along the shores of the Oriental region. This animal (*Halicore*) has a blunt muzzle, a broad horizontal tail, and pectoral flippers. Principal diet is seaweed. The dugong is hunted for its flesh and oil.

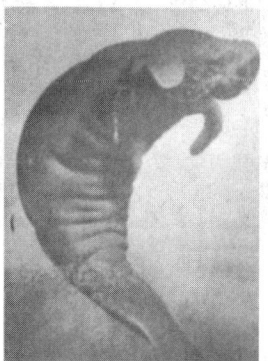

Fig. 1. Florida manatee.

SEA ELEPHANT. See **Sea Lions and Seals**.

SEA FAN (*Coelenterata, Anthozoa*). Marine polyps of the order *Alcyonaria* whose colonies are in the form of thin lacy fans.

SEA FEATHER (*Coelenterata, Anthozoa*). Marine polyps of the order *Alcyonaria*. The colony has a central stalk bearing lateral branches, the whole resembling a feather.

SEAFOOD POISONING. See **Foodborne Diseases**.

SEA GULL. See **Petrels and Albatrosses**; and **Waders, Shorebirds, and Gulls**.

SEA HARE (*Mollusca, Gasteropoda*). A marine mollusk of oval form with two ear-like tentacles near the anterior end which give it this name. The mantle almost conceals the shell and the foot forms two lobes by which the animal swims. Sea hares live on seaweed.

SEAHORSES (*Osteichthyes*).   Of the order *Solenichthys* (tube-mouthed fishes), family *Syngnathidae*, seahorses are of the same family as the pipefishes. See also **Pipefishes (Osteichthyes)**. Because of what might be termed their "cute" figures and their resemblance to equine mermaids, seahorses have been a favorite of commercial and scientific aquarium operators. They would also be favorites of fish hobbyists, were it not for the difficulties encountered in maintaining seahorses in tanks. See Fig. 1. There are about 25 species of seahorses, distributed well, but the majority occurring in the Indo-Australian area. Seahorses are exclusively marine in habitat and are quite small, ranging from a length of about $1\frac{1}{2}$ inches (3.8 centimeters) to a maximum of 8 inches (20 centimeters). There are numerous similarities between seahorses and pipefishes. For example, as with pipefishes, there is independent movement of the eyes. Seahorses are incapable of rapid movement and hence utilize an accurate suction mechanism which is operable within a range of about $1\frac{1}{2}$ inches (3.8 centimeters). It has been observed that juvenile seahorses may feed for a period of up to 10 hours and, during that time, consume from 3,000 to 4,000 brine shrimp. The large species (*Hippocampus hudsonius* or Atlantic seahorse) survives well in tanks but does not reproduce in captivity. In contrast, the *Hippocampus zosterae* (Gulf of Mexico) pigmy seahorse may survive through a number of generations. Seahorses are well known for the display of unusual courting habits. These include odd shaking movements, and in some species, the holding of tails, and, in the female, simply turning of the head stimulates the male. In the forward direction, the tail of the seahorse is used in a prehensile manner.

Fig. 1.   Seahorse.

SEA ICE.   See **Polar Research**.

SEAL.   See **Sea Lions and Seals**.

SEA LEMON (*Mollusca, Gasteropoda; Nudibranchiata*).   A flattened oval marine mollusk without shell, which feeds on sponges and other sessile animals. The roughened skin and the oval form suggest the fruit for which the animal is named.

SEA LEOPARD.   See **Sea Lions and Seals**.

SEA LEVEL.   See **Hydrology**; and **Ocean**.

SEA LIONS AND SEALS (*Mammalia, Pinnipedia*).   This order of mammals contains large swimming animals, including sea lions, seals, walruses, and sea elephants, among others. At one time, some authorities placed these animals under the *Carnivora*.

Sea lions are animals closely related to the seals. They belong to the same family as the fur seals and, with them, differ from the other seals in having external ears. The common sea lion, *Zalophus caliifornianus*, is found on the Pacific coast of Mexico and California, and Steller's sea lion,

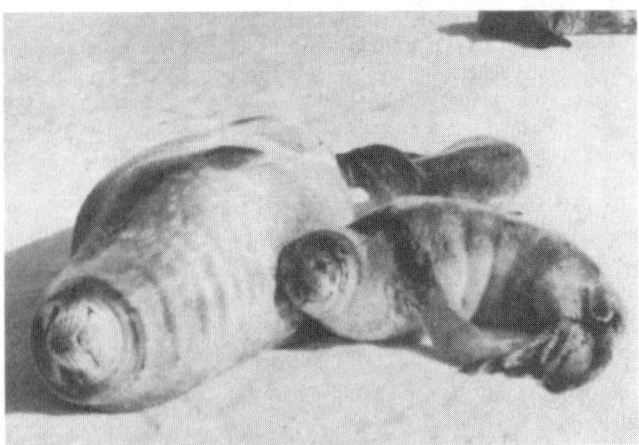

Fig. 1.   Weddel seal. (*A.M. Winchester.*)

Eumetopias stellari, ranges from the Bering Strait to southern California. See Fig. 1.

Seals are animals whose bodies are highly specialized for life in the ocean, although they are able to move about on land rather clumsily. The fore limbs are paddle-like flippers. The hind limbs are shifted so that they lie close together on opposite sides of the rudimentary tail and serve as a powerful propeller, increasing the effectiveness of the vertical undulating movements by which the animal swims. The body proper is formed to offer little resistance to the water in swimming. The fur seal represented on the Pacific coast by *Callorhinus alascanus* (*ursinus*) was once abundant in both hemispheres but was threatened with extermination by relentless hunting practices many years ago. The Pribilof Islands are an important center for these well-known seals. Fur seals are sometimes called sea bears. Seals are relatively long-lived, the average bull seal living up to 20 years of age. The bulls do not mate until they are 10 years old. An adult bull may mate many females within a few days. The cows live even longer, up to 30 years. Some species of seals are found throughout the North Atlantic and in waters around the British Isles. Reports indicate that seals were hunted for their meat as early as 500 B.C.

Although mature seals live to a good age, studies in which pups have been tagged and traced indicate that about 60% of the pups die before they reach one year of age. The pups are quite small and have only a relatively thin skin for protection. The adult seals develop thick layers (up to 2 inches); 5 centimeters of blubber immediately under the outer skin. The females identify their young by sense of smell. The pups are nursed, seal milk being among the richest of all mammal milks. Large bulls may attain a length of some 18 feet (5 meters) and a weight of about 3 tons (2.7 metric tons). The elephant seal may have a trunk some 2 feet (0.6 meter) in length. Also known as the sea elephant, the name refers to the proboscis of the male. They are found principally along the California coast. In arctic regions, polar bears are among the seals' worst enemies. The bears may pounce upon a seal sleeping on the ice, attack it in the water while swimming, or await a seal to come out of its breathing hole. Alaskan fur seals have been studied extensively. At the start of a mating season, a bull will stake out a specific area, to which females gradually arrive. Competing bulls herd as many females as possible into an area that can be protected effectively. The bulls are ferocious fighters and jealously guard their territories. See illustration that accompanies entry on **Sexual Selection**.

The sea leopard is a large seal, *Stenorhynchus leptonyx*, of the Southern Hemisphere. It is yellowish or tawny with gray spots.

The walrus is a giant marine animal related to the seals, but constituting a distinct family. Adults reach a length of more than 12 feet and a weight in excess of a ton. The feet of the walrus are adapted for swimming, but they are used mainly for clumsy locomotion on land, as in the case of seals. In early life, the body is covered with thick light brown fur that tends to disappear after middle age. The muzzle bears a number of very thick bristles. Walruses have the canine teeth of the upper jaw prolonged as tusks. The ivory of these tusks is valued by Eskimos. These animals are confined to the arctic seas and are commonly regarded as constituting an Atlantic, *Odobanus rosmarus*, and a Pacific, *O. obesus*, species, the latter with longer tusks. See Fig. 2.

Fig. 2.   Walruses.

Additional Reading

Bonner, W.N.: "Seals and Sea Lions of the World," Facts on File, Inc., New York, NY, 1994.

Rennick, P.: "Seals, Sea Lions and Sea Otters," Alaska Geographic Society, Anchorage, AK, 2000.

Ridgeway, S.H. and R.J. Harrison: "Handbook of Marine Mammals: The Walrus, Sea Lions, Fur Seals and Sea Otter," Vol. 1., Academic Press, Inc., San Diego, CA, 1998.

Riedman, M.: "The Pinnipeds: Seals, Sea Lions, and Walruses," University of California Press, Berkeley, CA, 1991.

SEA MOTHS *(Osteichthyes).* Of order the *Hypostomides,* family *Pegasidae,* sea moths are well named because, except upon close examination, they appear possibly more like an insect than a fish. In fact, biologists and zoologists have not fully agreed upon the exact fit for this creature. Doubtless, they belong in the great domain of fishes, but the exact place remains to be scientifically determined. Sea moths have been a curiosity for centuries as they were brought by travelers into Europe from the Orient and South Pacific. Because of similarities of the body armor, there is some resemblance to pipefishes, seahorses, and sea poachers. Some relationship with the flying gurnards is indicated by the expanded pectoral fins. Some five species of sea moths are known, all occurring in tropical waters from Hawaii to Africa. No specimens have been encountered in the Atlantic. The largest specimens measure about 5 inches (12.5 centimeters) in length.

SEAMOUNTS. Generally, isolated peaks rising 3,000 or more feet (915 meters) above the floor of the ocean basin. Also known as *guyots,* they are believed to be volcanic cones that once rose above sea level. Because they usually have flat tops, it is believed that sea erosion flattened them, after which the earth's crust in that location subsided, lowering the seamouts far below the surface to their present depths. See also **Ocean.**

SEA MOUSE *(Annelida, Polychaeta).* A marine worm, *Aphrodite* (*A. hastata* in American waters and *A. aculeata* in British waters), of compact oval form, covered above and on the sides with a felt-like material. It is recorded from Vineyard Sound on the Atlantic coast.

SEA OTTER. See **Mustelines.**

SEA RAVEN. See **Sculpins.**

SEA ROBINS *(Osteichthyes).* Of the suborder *Polynemoidea,* family *Triglidae,* sea robins occur in American Atlantic waters from Nova Scotia south to Venezuela. They attain a length of about 16 inches (40 centimeters), are highly colorful, with bony heads well protected with spines. Separate fanlike pectoral fins help the sea robin "walk" on the bottom. They are carnivorous, mollusks and small crustaceans comprising their main diet. Some of the larger members can attain a length of some 2 to 3 feet (0.6 to 0.9 meter). It is believed that all triglids are capable of noise-making and, as with other "acoustic" fish, such as croakers and grunts, their sounds are created by vibrating muscles with the assistance of a large air bladder which acts as a sounding box. The tubfish (*Trigla hirundo*),

also known as the yellow gurnard, is found in the waters off Europe and the African coasts. It is a very striking fish, featuring bright blue-edged pectorals. There is also a group of armored sea robins (*Peristediidae*) whose entire bodies are protected by bony projections. They prefer the deep tropical waters. Some of the larger triglids are considered good as food.

SEA SLUG *(Mollusca, Gasteropoda).* Marine mollusks with compact bodies and without shells. Some species have branching processes on the surface of the body by which they breathe. The creeping habits and general form are similar to those of the terrestrial slugs, although there is no closer relationship between the two. The sea slugs make up the order *Nudibranchiata.*

SEA SNAKES. See **Snakes.**

SEASON CRACKING. Spontaneous cracking of brass and other metals on standing. Intergranular cracks result from the action of residual internal stresses from cold-working operations aided by surface corrosion. Cold-worked high-zinc brasses sometimes fail during storage in ordinary atmosphere. Ammonia salts and other specific reagents greatly accelerate cracking in brasses subject to this defect. Many other metals, including stainless steels, are subject to cracking under certain conditions of stress and corrosion.

SEAWATER (Desalination). See **Desalination.**

SEAWEEDS. The production of edible seaweeds is a significant industry in the Orient. It is particularly well established in Japan where seaweed "farming" is practiced on a large scale. China and Korea are also notable producers of edible seaweeds, and improved hybrids have been reported from China. Increased demand for seaweeds has focused attention on methods of cultivating selected species. In the late 1950s, the Japanese *Gelidum* beds, which had proved adequate for centuries, declined rather mysteriously and caused a world shortage. This resulted in the large-scale exploitation of supplies of suitable seaweeds for the manufacture of agar in Chile and Portugal. By the early 1960s, the demand for carrageenin, and to a lesser extent for furcellaran, by the food processing industry caused a shortage of other red seaweeds. The main producers of these products constructed drying units in Nova Scotia and Prince Edward Island to encourage the collection of seaweed.

The economically important seaweeds fall into two main groups: (1) that which grows in the inter-tidal fringe; and (2) that which is permanently submerged. This apparently simple distinction is, however, complicated by many factors. The extent of the intertidal zone is partly controlled by the gradient of the beach, and it also increases with latitude and tidal range. There are considerable local variations in tidal range, varying from less than 3 meters (10 feet) in Shetland (British Isles) to nearly 10 meters (30 feet) on parts of the west coast of the isles. One prominent seaweed of the intertidal zone (*Chondrus crispus*) extends into the sublittoral; and a substantial part of the Canadian harvest grows sublittorally. (Littoral refers to of, on, or along the shore.) The important free-floating seaweeds are *Furcellaria fastigiata*, harvested in the central Kattegat; and *Gracilaria* spp., found off the coasts of Chile and Portugal. These forms are usually harvested by netting.

The cast seaweeds are important in some areas, such as the cast red seaweeds of Prince Edward Island and the cast stripes (*Laminaria hyperborea*) of the Scottish coasts.

Apart from edible uses, seaweed is essentially the raw material for the extraction of a range of carbohydrates, such as agar, carrageenin, and furcelleran from the red seaweeds; and sodium alginate from brown seaweeds—all products used in food processing. Edible seaweeds produced in Japan include: *Konbu* (*Laminaria* spp.); *wakeme* (*Undaria pinnatifida*); *amanori* (*Porphyra* spp.); and *aonori* (*Monostroma* spp. and *Enteromorpha* spp.). The production of amanori approximates upwards of 7000 sheets in Japan; and about one-third of that quantity in Korea, much of which is exported to Japan. Converted into weight, this production represents about 300,000 tons (270,000 metric tons) annually.

Additional Reading

Guiry, M.D. and G. Blunden: "Seaweed Resources in Europe: Uses and Potential," John Wiley & Sons, Inc., New York, NY, 1991.

Lobban, C.S. and P.J. Harrison: "Seaweed Ecology and Physiology," Cambridge University Press, New York, NY, 1996.

Luning, K., C. Yarish, and H. Kirkman: "Seaweeds: Their Environment, Biogeography and Ecophysiology," John Wiley & Sons, Inc., New York, NY, 1990.
Schneider, C.W. and R.B. Searles: "Seaweeds of the Southeastern United States: Cape Hatteras to Cape Canaveral," Duke University Press, Durham, NC, 1991.

SEBACEOUS CYST.   A cystic structure developing in the skin, due to plugging of a duct leading from a sebaceous gland. The cyst may increase in size, and some become infected. Sebaceous cysts are commonly called wens, and they are often seen on the scalp and face as well as other parts of the body. They always contain a white cheesy material.

Treatment is by surgical excision. If the entire cyst wall is not removed, recurrence is likely.

SEBACEOUS GLAND.   A gland of the skin of mammals which secretes an oily substance (sebum). These glands are usually associated with hair follicles and in man are especially abundant in the scalp, although they occur all over the body with the exception of the palms of the hands and soles of the feet. Their secretion keeps the skin pliable and anoints the hair.

In structure these glands are of the compound alveolar type. Their secretory parts are saccular, discharging to a common duct, which often opens in the hair follicle.

SEBORRHEA.   A variety of disorders of sebaceous glands are included in this clinical term. The dry, scaling form, "dandruff," is characterized by the presence of fine, branny, slightly greasy scales, which are readily shed. It is common on the scalp but may spread down over the face, neck and ears, and may even be a generalized dermatitis. The oily variety is associated with an excess secretion of oil by the sebaceous glands of the scalp and skin. The scalp is covered with a greasy layer which may mat the hair together; this type is associated with permanent baldness.

In some cases, psoriasis of the scalp may be involved. This disorder usually forms discrete plaques covered by silvery scales, whereas seborrheic dermatitis characteristically is diffuse and patchy and is associated with a greasy scale.

See also **Dermatitis and Dermatosis**.

SECOND ORDER REACTION.   See **Chemical Reaction Rate**.

SECOND ORDER SYSTEM.   A system whose performance characteristics are presented in the form of a second order differential equation

$$r(t) = k_1 \frac{f^2 C(t)}{dt^2} + k_2 \frac{dC(t)}{dt} + k_3 C(t) \ldots$$

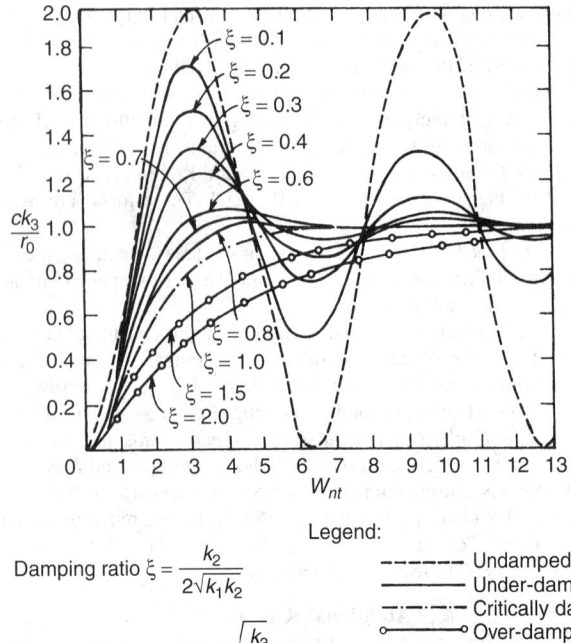

Fig. 1.   Time response curve of a second order system subjected to a step input.

where r = system input
C = system output
t = time

k_1, k_2 and k_3 are coefficients.

If all coefficients are constants, then the Laplace transform of foregoing equation also should be second order:

$$\frac{C(S)}{R(S)} = \frac{1}{k_1 S^2 + k_2 S + k_3} \ldots$$

For simplicity, all initial conditions were assumed zero. The time response of a second order system subjected to a step input, r_0, is shown graphically in Fig. 1.

SECONDARY EMISSION.   This term refers to the result of any of several different processes, in each of which some kind of "primary" emission, when it encounters some form of matter, gives rise to another emission of the same or of different character.

The most familiar example of a secondary emission is the x-rays, which have their origin in the impacts of high-speed electrons (cathode rays) upon atoms of matter. The resulting x-rays may themselves act, in turn, as the primary emission and, falling upon solid bodies, cause a secondary x-ray emission. Or they may fall upon a fluorescent substance (see also **Luminescence**) and give rise to a secondary radiation of visible light. X-rays, or any other photons, falling upon a photosensitive metal, may cause a secondary emission of photoelectrons. See also **Photoelectric Effect**. The "recoil" electrons from the Compton scattering of x-rays constitute one form of secondary emission. See also **Compton Effect**.

The most common use of the term denotes the emission of electrons from a solid as the result of the collision of higher energy electrons with the solid. Since the energy levels of the emitted electrons depend upon the atoms emitting them, as well as upon the initial energy, a method of surface layer analysis (i.e., analysis of atoms at the surface or several atomic layers below) has been based upon secondary electron emission. The method consists of placing the solid sample in a vacuum chamber and bombarding it with a beam of relatively low-energy electrons. The secondary electrons emitted (see also **Auger Effect**) are recorded according to their characteristic energies, which correspond to the atoms emitting them. The method is especially useful in detecting impurities of the lighter elements.

SECONDARY SEXUAL CHARACTERISTICS.   The characteristics of living things whose appearance is definitely associated with the sex of the individual, although the characteristics have no direct connection with the process of reproduction.

The different colors and patterns of the two sexes in many species of birds and insects are familiar examples of secondary sexual characteristics. Horns of some males, the manes of many male mammals, and the spurs of cocks are also in this category. In some species the differences resulting from such characteristics are so great that the two sexes can scarcely be associated by appearance. Cases are on record among the insects of the classification of males and females of the same species in different genera prior to the discovery of their relationship through other evidence. See also **Gonads**; and **Hormones**.

SECRETARY BIRD.   See **Eagle**.

SECRETION.   See **Kidney and Urinary Tract**; and **Urine**.

SECTILE.   Capable of being cut. In mineralogy, sectile refers to substances, such as talc, mica, and steatite, which can be cut smoothly by a knife.

SECTION MODULUS.   An inspection of flexure will reveal that the maximum stress in a member subjected to a transverse bending is directly proportional to the external bending moment, and inversely proportional to the ratio of

$$\frac{\text{moment of inertia}}{\text{Distance of the farthest stressed element from the neutral axis}}$$

It is apparent that this ratio is entirely a property of the shape and size of the cross section of the structural member. This ratio is known as the section modulus, and is an important property of rolled steel sections and other shapes used as structural members. When the bending moment to be withstood by a beam or column is divided by this section modulus, the quotient is the maximum bending stress which will exist in that member.

SECULAR DETERMINANT.

An equation of the form

$$K(\lambda) = |a_{ij} - b_{ij}\lambda| = 0$$

which becomes a polynomial in λ when the determinant is expanded. In matrix algebra it is the determinant of the characteristic matrix of A, $[\lambda E - A]$ where E is a unit matrix. In this case,

$$K(\lambda) = \lambda^n + a_1\lambda^{n-1} + \cdot + a_n = 0$$

is the characteristic equation or function, the coefficients a_i involve the elements of A, and the n roots of $K(\lambda)$ are the eigenvalues, latent or characteristic roots. All of them need not be different; if two or more of them are equal they are said to be degenerate. An important property of the latent roots is the following. Suppose two matrices A and B are related by a similarity transformation, $B = Q^{-1}AQ$, then the latent roots of A are identical with those of B.

See also **Determinant**.

SECULAR PARALLAX.

The parallactic motion of a star due to the sun's motion through space. See also **Stellar Parallax**.

SECULAR TERMS.

In the mathematical expression of an orbit, terms for very long period perturbations, in contrast to periodic terms, terms of short period.

SEDGE (Cyperaceae).

This family of monocotyledons is composed of grass-like plants which are found chiefly in marshy places. Most of its members are perennials with creeping rhizomes and grass-like leaves. The basal portion of the leaf is a sheath which completely surrounds the stem. The stem is generally solid and triangular in cross section. The inflorescence is a spike or a panicle, composed of one- to many-flowered spikelets. Each flower, borne in the axil of a bract, has three stamens and a single pistil; in some sedges there is also a perianth of six or many bristles. The fruit is an achene. All sedges are wind-pollinated.

Few of the sedges are of any importance. Some of them yield a coarse hay which may be fed to livestock, and is sometimes used for packing material. Papyrus, used as paper by the ancient peoples, was made from the stems of *Cyperus papyrus*. The erect stems and leaves of species of *Scirpus* are sometimes dried and woven to form chair seats; chairs finished with this material are called rush-bottomed chairs. The plants are sometimes called bulrushes. Several sedges form tubers, which are sometimes used for food.

SEEBECK EFFECT.

Production of electromotive force as a result of a circuit containing two conducting materials having two junctions between the materials at different temperatures. This effect is the basis of all thermocouples. The Seebeck effect is one of a number of related thermoelectric phenomena, is the inverse of the Peltier effect, and is closely related to the Thomson effect. See also **Thermocouple**.

SEED.

In the vegetable kingdom, seeds are a vital link to the past and to the future. For the majority of food plants, of all factors, including climate, water management, fertilizers, insecticides, etc., that are required to ensure a successful crop, healthy, vigorous seeds are, without qualification, the singular indispensable requirement. Seed requirements for planting are a significant cost factor and support a major seed growing, processing, and supply industry, as is evidenced by Table 1.

Seeds per se are consumed as food items and the oils and proteins they contain are expressed, extracted, and separated from them to yield edible fats and oils and vegetable protein. The grain industry is based upon seed processing. See also **Protein**; and **Vegetable Oils (Edible)**.

The collection and preservation of seeds as a source of germ plasm for improving crops in numerous ways (yields, resistance to disease, insects, and severe climates) are extremely important scientific undertakings.

Botanical Factors A seed consists of a dormant embryo, together with a quantity of stored food that may be absorbed in the embryo, or may surround it, and one or two seed coats or integuments. The seed develops from an ovule. In angiosperms, the seed is completely enclosed by the

TABLE 1. QUANTITY OF SEED REQUIRED FOR PLANTING SELECTED MAJOR CROPS[1]

Crop	Pounds/Acre	Kilograms/Hectare
Barley	80	90
Beans (dry edible)	54	60
Groundnuts (peanuts)	136	152
Maize (corn)	13.4	15
Oats	83	93
Peas (dry field)	167	187
Potatoes (Irish)	1830	1050
Rice	142	159
Rye	90	101
Sorghum	6.76	7.57
Soybeans	63	71
Sweet potatoes	600	672
Wheat (Durum)	89	100
Wheat (winter)	69	77

[1] Average of entire United States plantings.
Source: U.S. Department of Agriculture.

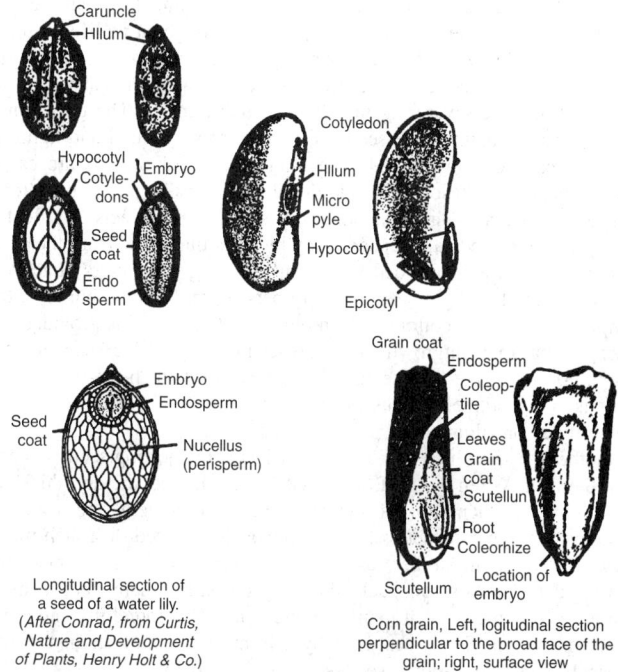

Longitudinal section of a seed of a water lily. (*After Conrad, from Curtis, Nature and Development of Plants, Henry Holt & Co.*)

Corn grain, Left, logitudinal section perpendicular to the broad face of the grain; right, surface view

Fig. 1.   Representative types of seeds.

ovary wall. In gymnosperms, the seed lies exposed on the surface of a scale of the cone. Representative seeds are shown in Fig. 1.

The fertilized egg develops into the embryo, which is a young plant contained in a seed. This embryo may be an undifferentiated mass of cells, as it is in the orchid family, but usually is more highly organized. It then consists of a short axis called the hypocotyl. At one end of the hypocotyl there is a primitive root called the radicle. At the other end is a terminal bud, called the plumule. This plumule may be nothing more than a small mass of undifferentiated cells, recognizable only as a small bulge at the apex of the hypocotyl, or it may be a well-developed shoot having a short internode and two distinct leaves. Borne laterally at the apex of the hypocotyl there are one or more seed leaves or cotyledons. In many seeds these cotyledons are thin and more or less leaf-like, while in others they are very fleshy, filled with stored food material, and form the greater part of the seed. The number of cotyledons varies. In monocotyledons there is usually only one cotyledon, in dicotyledons there are two, and in gymnosperms there are often many.

In angiosperms the endosperm is the tissue which results from the endosperm nucleus. It is a tissue rich in stored food. The food reserves stored in the seed are carbohydrates, especially starches and sugars, fats and proteins. The developing embryo gets its food from the endosperm. In many seeds, the embryo uses only a part of the endosperm during its development, so that the mature seed contains much endosperm surrounding the embryo. These are called albuminous seeds. In other plants, the food reserves of the endosperm have been entirely absorbed and restored in the embryo, especially in the cotyledons, which then becomes very fleshy. Such seeds are exalbuminous.

The embryo sac of the ovule is surrounded by a mass of tissue called the nucellus. In most plants, this is completely absorbed before the seed reaches maturity. In some seeds, it persists and it becomes much enlarged. It is then known as the perisperm, and serves as an additional source of stored food.

Surrounding all these are the seed coats, which develop from the integuments of the ovule. The outer coat may be variously modified to aid in the dissemination of the seeds. On the seeds of many plants there is a fleshy structure called the aril. This grows up around, and more or less covers, the outer integument. In some seeds, the outer coat produces an outgrowth called the caruncle, which seems to aid in absorbing water from the soil and passing it on to the seed. Passing through the integuments is a minute hole called the micropyle. It is through this that the pollen tube entered the young ovule; the radicle generally points directly toward it. The seed is attached to the ovary wall by a small stalk or funiculus which, when the seed falls off, leaves a scar called the hilum on the seed coats.

Natural Broadcasting of Seeds. Once started, without human intervention the plant must remain in a fixed position throughout its life. In higher plants, the vegetative parts are very rarely able to colonize new territories. The fruit, or less frequently the seed, is the part which is carried to new regions. There are several agents effecting this transfer. The most important is wind. Sometimes the seeds of a plant are very small and light, and so easily carried by the wind. The minute seeds of orchids are carried by this means; in these seeds, additional buoyancy is gained by a loose, thin case, which surrounds the embryo tissue within and acts as a float. In other plants the seeds are carried away by hairs that grow from the seeds. Milkweed seeds are provided with a tuft of long silky hairs attached at one end of the thin light seed. Cotton fibers serve the same purpose; they completely cover the cotton seed. In other plants the seed is provided with a wing, a flat thin outgrowth from the seed coats. Catalpa seeds are thus equipped, as also are pine seeds. The distances to which the wind carries seeds is considerable. By this means, plants are disseminated over many square miles or kilometers.

Another way in which new land is reached is by ejecting seeds violently from the fruit. When the fruit of the witch hazel is ripe, the dry, thick wall of the ovary suddenly snaps and hurls the seeds violently to a distance of many feet (several meters). The common Jewelweed or Touch-me-not scatters its seeds in similar fashion. At maturity the fruit abruptly splits open, and the valves roll back, throwing the seeds many feet away. In similar fashion the pods of many legumes split apart forcefully and scatter the seeds within. Seeds scattered by this means cannot attain the wide dispersal which wind-borne seeds do.

A few plants form seeds that float readily on water for some time without harm. Currents of water may carry the seeds to new shores. Other seeds are provided with hooks or barbs or have a sticky surface, which causes them to adhere to the bodies of passing animals that scatter them. In most plants, however, it is the fruit that is so carried. Such fruits, as beggar's lice, burdock, goldenrod and many others, are commonly mistaken for seeds.

Germination and Dormancy. Having reached an environment where suitable conditions exist, the seed germinates. The seeds of many plants must reach such a place in a very short time or perish, since they remain viable but a very short time. Those of the willow, for example, live only a few days after falling from the parent plant. The seeds of most garden vegetables grow best if planted within a year from the time they ripen, though they may retain their vitality for three or four years with decreasing vigor. On the other hand, some seeds lie dormant for a long time before germinating. Seeds of many weeds, including the common ragweed, the pollen of which causes hay-fever, may live for years before germinating, making it very difficult to eradicate the species by pulling up the plants for a single season. Tests show that the seeds of some plants may remain viable for 20–50 years. It is recorded that the seeds of the Asiatic lotus have germinated after lying dormant for 200 years. But records of viable seeds found in ancient vaults, such as contained the mummies of Egypt, are entirely unfounded.

Certain conditions favor the continued vitality of dormant seeds. Sometimes the seed has a very thick wall that is impervious to water or

to oxygen gas, and so excludes two things necessary to start germination. Until the wall has softened or rotted the seed does not germinate. In other seeds the thick wall resists the pressure of the developing embryo within. Many important crop plants have seeds that germinate slowly because of their thick coats. To hasten germination and to insure a uniform stand of plants the seeds are scarified before planting; that is, the seed coat is rubbed with abrasive substances that break down the impervious wall layers. Often the wall of the seed is sufficiently damaged during mechanical threshing to insure prompt germination when planted. Prolonged soaking sometimes hastens germination.

In other seeds, dormancy is inherent in the embryo itself. The embryo may be entirely undeveloped, requiring a long period of slow development before it can break the seed coats. Other seeds germinate only after a period of "after-ripening" which varies from a single winter to many years. The changes occurring during this period are as yet not well understood. The time of "after-ripening" may be considerably shortened by burying the seeds in sand or other suitable material and by keeping them cold and moist.

The external conditions necessary to cause germination are adequate water, suitable temperature and oxygen. With some seeds, light is an important factor. Most seeds contain very little water, which is one of the reasons why they can survive under adverse conditions such as cold and drought. To germinate they must receive additional water. This added water favors digestion, a process that makes available to the plant the stored food. Both water and oxygen are needed by the germinating embryo, because of the great increase in respiration, the process which frees to the plant the energy stored in the carbohydrates and other compounds.

The temperature at which a seed will germinate varies with different plants. For each there is a considerable range of temperature. The lowest temperature at which germination occurs is called the minimum temperature, and varies from $0-10\,°C$ or even higher. The maximum or highest temperature at which germination takes place is usually between 45 and $50\,°C$. The most favorable temperature, or optimum, is about $30\,°C$. Light favors the germination of many common plants, such as many grasses and troublesome weeds. Other plants, including many common crop plants, are unaffected by light.

Germination is the development of the embryo into a young plant. It becomes completed when the young plant is independent of the food stored in the seed. In most seeds the first visible change is swelling of the seed. This is a result of the increased water content. Often the seed coats are ruptured by the swelling of the contents of the seed. Respiration increases greatly, and much energy is made available in those regions where active growth occurs, that is, the hypocotyl and plumule. The radicle pushes out of the seed and attaches itself, by means of root hairs, to the soil particles. These then begin absorbing water from the soil. In some seeds, the hypocotyl elongates considerably, often forming an arch which subsequently straightens, lifting the cotyledons and plumule out of the soil and into the air. In other seeds, the cotyledons remain permanently underground, the plumule elongating and pushing out into the air. There the first leaves of the plant appear. With their formation the plant becomes independent.

Many foodstuffs are seeds, especially those of the cereal grains, rice, wheat, corn, barley, and oats. But seeds are used in many other ways. Many medicinal products are obtained from seeds. Linseed oil, soybean oil, and coconut oil are but a few of the many oils that come from seeds. Poppy seeds, caraway seeds, and mustard add flavor to other foods. Clothing is made from the hairy covering of the cotton seed. Beads, buttons, and ornaments of various kinds are also often made from seeds.

See also **Germ Plasm**; and **Plant Breeding**.

Additional Reading

Basra, A.S.: "Hybrid Seed Production in Vegetables: Rationale and Methods in Selected Crops," The Haworth Press, Inc., Binghamton, NY, 2000.

Copeland, L.O. and M.B. McDonald: "Principles of Seed Science and Technology," 4th Edition, Kluwer Academic Publishers, Norwell, MA, 2001.

Doijode, S.D.: "Seed Storage of Horticultural Crops," The Haworth Press, Inc., Binghamton, NY, 2001.

Fenner, M.: "Seeds: The Ecology of Regeneration in Plant Communities," 2nd Edition, Oxford University Press, Inc., New York, NY, 2001.

George, R.A.T.: "Vegetable Seed Production," 2nd Edition, Oxford University Press, Inc., New York, NY, 1999.

McDonald, M.B. and L.O. Copeland: "Seed Science and Technology Laboratory Manual," Iowa State Press, Ames, IA, 2000.

Sauer, J.D.: "Plant Migration: The Dynamics of Geographic Patterning in Seed Plant Species," University of California Press, Berkeley, CA, 2000.

Vazquez-Ramos, J., M. Black, and K.J. Bradford: "Seed Biology: Advances and Applications," Oxford University Press, Inc., New York, NY, 2000.

SEEDING (Gas Conductivity). Introduction of atoms, such as sodium, with a low ionization potential into a hot gas for the purpose of increasing the electrical conductivity.

SEEING (Astronomy). A term used by ground-based optical astronomers to describe image quality.

When describing stellar (point source) images, good seeing means the images are small and stationary; bad seeing means the images are large (sometimes changing in size in a rapid and irregular fashion) and unsteady. When describing images of extended objects (such as planets or nebulae), good seeing means the images are sharp and steady; bad seeing means the images are blurred and unsteady.

Seeing quality is determined by motions of air masses of differing index of refraction (due to differences in density and/or humidity) so that light rays are refracted or bent in a time-variable fashion. These motions can be both near the telescope or at some distance from it. A similar phenomenon is observed in viewing a distant object over a nearby fire or other source of heat. The heat causes rising currents of warmer air which produce time variable refraction so that the object's image is seen to "shimmer" or be blurred.

For most astronomical observations, the better the seeing, the fainter one can observe. Hence, seeing quality is an important criterion in selecting the site of a ground-based observatory. (Other important criteria are the number of clear hours, absence of anthropogenic night time illumination, and, for infrared observations, dryness of the atmosphere.) The sites with the best seeing are generally islands and mountains near oceans or in isolated flat areas. This is generally believed to be because the large-scale motions of the atmosphere in such places are largely laminar rather than turbulent. Local conditions, such as local topography, height of dome, temperature of dome floor, among other factors, can be very important and much effort is made to control these conditions.

See also **Light Pollution**.

PETER PESCH, Chairman, Astronomy Department, Case Western Reserve University, Cleveland, OH.

SEGER CONE. A series of substances having different fusion temperatures might serve roughly to measure the temperature of high-temperature regions such as furnaces, since, with a series of substances having progressively increasing fusing temperatures, the temperature naturally lies between the fusion temperature of the last substance fused, and that of the next not yet fused. A series of artificially prepared mixtures, mostly of the oxides such as clays, lime, feldspar, have been designed to form a series of "Seger cones." There are 60 mixtures covering a temperature range from 590 to 2,000 °C.

SEICHE. 1. A standing wave oscillation of an enclosed water body that continues, pendulum fashion, after the cessation of the originating force, which may have been either seismic or atmospheric.

2. An oscillation of a fluid body in response to a disturbing force having the same frequency as the natural frequency of the fluid system. Tides are now considered to be seiches induced primarily by the periodic forces caused by the sun and moon.

3. In the Great Lakes area, any sudden rise in the water of a harbor or a lake, whether or not it is oscillatory. Although inaccurate in a strict sense, this usage is well established in the Great Lakes area. See also **Estuary**.

SEISMIC SEA WAVE. See **Tsunami**.

SEISMOLOGY. See **Earth Tectonics and Earthquakes**.

SEIZURE (Neurological). Epilepsy is characterized by sudden, brief disturbances in brain function. A clinical seizure may be defined as one resulting from the excessive discharge of aggregates of neurons, which in some manner become depolarized in a synchronous fashion. Epilepsy is not a disease in the usual sense, but rather it is a disorder, the root causes of which remain poorly understood. Some researchers have suggested that the paroxysmal depolarization of neurons may derive from abnormalities

of neuronal membranes, disturbances in synaptic transmission, or defective functioning of glial cells. The causative factors will probably be revealed when there is a better understanding of the fundamental biochemistry and bioelectronics of the brain and the nervous system. Epilepsy may be divided into two broad categories: (1) *idiopathic epilepsy*, where there is no known organic injury to the brain prior to the first seizure; and (2) *symptomatic epilepsy*, where such damage has been confirmed. The genetics of seizure disorders (epilepsy) are poorly understood and difficult to research because of the probable large number of causative factors and, from a statistical standpoint, because of the prevalence of the disorder. It is estimated that a minimum of 0.5% of the population suffer from recurrent seizures. This, in a country of the size of the United States, translates into 1 million persons. It is generally concluded that the close relatives of persons with idiopathic epilepsy may experience a seizure incidence rate of three times that of the general population. Of course, where seizures are the result of injury (accidents to the brain and spine during and after birth — at any time of life), genetic factors do not enter.

Even though there is a poor understanding of the biochemistry and mechanics which precipitate seizures, over the years very effective treatment regimens have been developed empirically.

Types of Seizures

Some authorities have classified seizures on the basis of symptoms as follows:

> Generalized Seizures
> > Grand Mal (*tonic-clonic*)
> > Petit Mal
> > Drop Attack (*akinetic*)
> > Myoclonic
> Partial Seizures
> > Psychomotor
> > Focal Cortical
> Continual (continuing *status epilepticus*)

The treatment, including the drugs of choice, varies from one type to the next.

Grand Mal Epilepsy. In this type of seizure, the patient may feel unusually good or poor for a day or two prior to the attack. This vague state warns the patient of an impending attack. It is peculiar to the individual and identical in different attacks. There may be queer sensations in some part of the body. Flashes of light or color may be seen, strange sounds may be heard, and pleasant or disturbing emotions may be experienced. Consciousness is swiftly lost, often with a wild, harsh cry. Breathing stops; the legs and body stiffen; and the patient falls to the ground, the elbows bent at rigid right angles. During this part of the spasm, which lasts from 10 to 30 seconds, the bladder or bowel, or both, may empty. The face becomes blue as the features contort, and the patient seems about to die when the spasm breaks. Rhythmic muscular contractions, at first small and rapid, begin and then become slower and more powerful. Gasps for breath come through heavy froth, often blood-stained from a bitten tongue or cheek. The contractions become less and less frequent. This phase usually lasts 2 to 3 minutes, but may persist longer. At the end, the patient may sleep heavily for several hours or rouse with aimless, thrashing movements, dazed and forgetful and unable to understand what is said to him. There is often a severe headache for several hours. Occasionally, the patient may have one convulsion after another without regaining consciousness. The dangerous state is called *status epilepticus* and demands immediate attention by the physician, since it can result in death from exhaustion.

In summary, the clinical features of grand mal are generalized major motor convulsions with a loss of consciousness and a depression of cerebral function.

Petit Mal. In petit mal attacks, a milder form of epilepsy, the loss of consciousness is fleeting and variable — from 1 to 40 to 50 times per day. It may be so short that it goes unnoticed. The head may nod momentarily, the flow of speech may halt a second or two and then may be normally resumed; or perhaps only a vacant stare marks the attack. Sometimes there are one or two contractions of the arms flickering of the eyelids. In petit mal, an electroencephalogram a spike-and-wave pattern, which occurs at the rate of 3 per second.

Drop Attack. In this type of attack, there is a brief loss of consciousness and loss of muscle tone.

Myoclonic Type. The clinical features include isolated muscle jerks, which may be precipitated by various sensory stimuli. This condition is frequently associated with degenerative and metabolic brain disease.

Psychomotor Epilepsy. This manifestation may follow grand or petit mal seizures or occur independently. These attacks last from a few minutes to a day or two. The patient may remember nothing of the episode on recovery. The behavior is confused and unusual. There may be uncontrollable emotional outbursts, which may be violently destructive, or the patient may be dazed and apathetic. There is also evidence that some psychomotor cases may be the result of subtle injury to the brain at the time of birth. In this state, there is some impairment of consciousness.

Focal Seizures. Also known as Jacksonian epilepsy, these seizures assume two forms. There may be unusual sensations or uncontrolled movements remaining localized to one part of the body, while consciousness is undisturbed. Thus, the head and eyes may turn irresistibly to one side despite the patient's awareness. In the other form, the movements or sensations which begin in one part of the body may spread upward in a slow, orderly fashion, or they may cross to the other side. This form may develop into a typical grand mal attack with loss of consciousness. Many convulsive seizures resulting from brain injury are of the Jacksonian type.

Continual Seizures. There is incomplete recovery between attacks, which may be generalized or partial. Drugs of choice in treatment include phenytoin and phenobarbital. Patients are treated for *status epilepticus*. This condition is considered a medical emergency and invariably results in death when not treated. Factors which the physician and other emergency personnel must consider include: (1) measures for ensuring an unblocked airway and measures for ensuring adequate fluids, such as intravenous infusion with 5% dextrose; and (2) administration of drug of choice.

Electroencephalographic tracings can be of considerable value in the assessment of various seizure disorders. See Fig. 1. See also **Central and Peripheral Nervous Systems**.

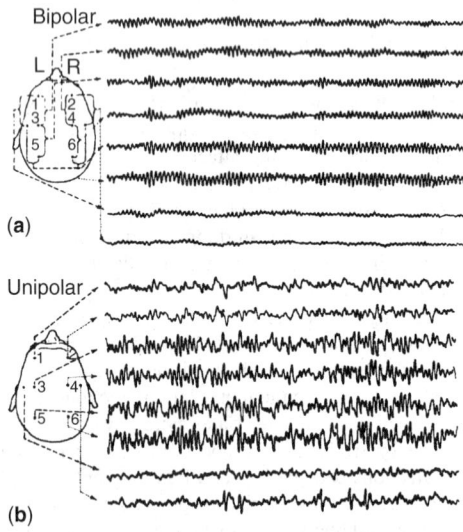

(a)

(b)

Fig. 1.   Electroencephalograms. (**a**) Normal adult, showing the low amplitude of the tracings that are obtained from electrodes variously placed on the head. (**b**) A 4-year-old child suffering from a convulsive disorder, showing high-amplitude waves, in marked contrast to those of the normal individual. (*Photography by F.W. Schmidth.*)

Additional Reading

Berkow, R. and M.H. Beers: "The Merck Manual," 17th Edition, Merck & Company, Inc., Ehitehouse Station, NJ, 1999.

Delanty, N.: "Seizures: Medical Causes and Management," Humana Press, Totowa, NJ, 2001.

Gates, J.R. and A.J. Rowan: "Non-Epileptic Seizures," 2nd Edition, Butterworth-Heinemann, Inc., Woburn, MA, 1999.

Holmes, G.L.: "Diagnosis and Management of Seizures in Children," W.B. Saunders Company, Philadelphia, PA, 1996.

Luders, H.C. and S. Noachtar: "Atlas of Epileptic Seizures and Syndromes," W.B. Saunders Company, Philadelphia, PA, 2000.

Luders, H.O. and S. Noachtar: "Epileptic Seizures: Pathophysiology and Clinical Semiology," W.B. Saunders Company, Philadelphia, PA, 2000.

Mizrahi, E.M. and P. Kellaway: "Diagnosis and Management of Neonatal Seizures," Lippincott-Raven Publishers, Philadelphia, PA, 1997.

Rowan, A.J. and R.E. Ramsay: "Seizures and Epilepsy in the Elderly," Butterworth-Heinemann, Inc., Woburn, MA, 1996.

Rowland, L.P.: "Merritt's Textbook of Neurology," 9th Edition, Lippincott, Williams & Wilkins, Philadelphia, PA, 1994.

Shinnar, S. and T.Z. Baram: "Febrile Seizures," Academic Press, Inc., San Diego, CA, 2001.

SELACHII.   See **Sharks**.

SELECTION RULES (Energy Levels).   It was found early in the study of atomic spectra that radiative transitions between certain pairs of energy levels seldom or never occur. A set of rules which are expressed in terms of the differences of the quantum numbers of the two states involved allow a prediction of allowed transitions and forbidden transitions. The conditions for allowed transitions are:

$$\Delta L \text{ (orbital angular momentum)} = \pm 1$$

$$\Delta J \text{ (total angular momentum)} = 0 \text{ or } \pm 1$$

$$\Delta M \text{ (magnetic orientation)} = 0 \text{ or } \pm 1$$

The selection rules are not rigorously obeyed. In atoms that do not exhibit Russell-Saunders coupling, the quantum numbers L and S are not defined. Even in atoms that do have this type of coupling, forbidden transitions are merely of lower probability than allowed ones, and they may occur from a state from which no transitions are allowed by the rules, if conditions are such that collisions of the second kind do not remove the atom from the initial state before it radiates (e.g., at extremely low pressures).

Similar selection rules hold for molecular spectra. In fact, let ψ_i and ψ_j be wave functions for two levels in any quantum mechanical system. Then if P is the appropriate operator, a transition between levels i and j is permitted if the matrix element

$$\int \psi_i^* P \psi_j \, d\tau$$

does not vanish. Here ψ_i^* is the complex conjugate of ψ_i, $d\tau$ is a volume element including all of the variables involved in the two wave functions, and the operator may refer to electric or magnetic dipole radiation, quadrupole radiation, polarizability, etc. If the integral vanishes, the transition is forbidden. Frequently, symmetry properties and group theory may be used to determine whether the matrix element does or does not vanish. This is very helpful since evaluation of the integral itself may be difficult or impossible.

SELECTION RULES (Nuclear).   A set of statements that serves to classify transitions of a given type (emission or absorption of radiation, beta decay, and so forth) in terms of the spin and parity (I and π) quantum numbers of the initial and final states of the systems involved in the transitions, in such a way that transitions of a given order of inherent probability (after making allowance for the influence of varying energy, charge and size of system, and so forth) are grouped together. The group having highest probability of taking place per unit time is said to consist of allowed transitions; all others are called forbidden transitions. Table 1 lists the selection rules for radiative transitions: each entry gives the character (E = electric, M = magnetic) and the multipole order (1 for dipole, 2 for quadrupole, 3 for octopole, ...) of the predominant radiation mechanism for the indicated spin change ΔI and parity change $\Delta \pi$; the entry "none" means that radiative transitions are strictly forbidden.

TABLE 1. SELECTIVE RULES FOR RADIATIVE TRANSITIONS

$\Delta \pi$	ΔI						
	0 $I = 0$	0 $I \neq 0$	1	2	3	4	5
No	None	$M1$	$M1$	$E2$	$M3$	$E4$	$M5$
Yes	None	$E1$	$E1$	$M2$	$E3$	$M4$	$E5$

Fermi selection rules and Gamow-Teller (GT) selection rules are alternative sets of rules for allowed beta transitions; both are currently believed to be valid, so that a transition allowed according to either set is actually allowed.

Table 2 lists the selection rules for beta decay: the entry A means that for the indicated spin and parity change the transition is allowed; I, means that it is first forbidden; II, second forbidden . . .

TABLE 2. SELECTION RULES FOR BETA DECAY

$\Delta\pi$	ΔI						
	0	1	2	3	4	5	6
No	A	A	II	II	IV	IV	VI
Yes	I	I	I	III	III	V	V

SELENIUM. Chemical element, symbol Se, at no. 34, at. wt. 78.96, periodic table group 16, mp 217 °C, bp 685 °C, density 4.82 g/cm³ (solid), 4.86 (single crystal). Selenium has a large number of allotropes, some of which have not been fully investigated. On heating selenium above its melting point and cooling it, a red vitreous mass is formed, probably a mixture of allotropes. A red, amorphous allotrope is precipitated by SO_2 from selenious acid solutions. On heating at above 150 °C, the red vitreous form changes to a gray hexagonal form, the stable form at ordinary temperatures, with metallic properties, one of which is photo-conductivity. By evaporation of a CS_2 solution of the red vitreous form below 72 °C, a red α-monoclinic form is obtained; evaporation above 72 °C gives β-monoclinic selenium. Black hexagonal selenium, believed to have a ring structure, is produced by heating amorphous selenium to near its melting point. Unlike sulfur, liquid selenium apparently has only one form.

There are six natural occurring isotopes: ^{74}Se, ^{76}Se through ^{78}Se, ^{80}Se, and ^{82}Se, and seven known radioactive isotopes ^{72}Se, ^{73}Se, ^{75}Se, ^{79}Se, ^{81}Se, ^{83}Se, and ^{84}Se. With exception of ^{79}Se which has a half-life of something less than 6×10^4 years, the half-lives of the other isotopes are comparatively short, measured in minutes, hours, or days. In terms of abundance, selenium ranks 34th among the elements occurring in the earth's crust. It is estimated that a cubic mile of seawater contains about 14 tons (3 metric tons per cubic kilometer) of selenium. First ionization potential 9.75 eV; second 21.3 eV; third, 33.9 eV; fourth, 42.72 eV; fifth, 72.8 eV. Oxidation potentials $H_2Se(aq) \rightarrow Se + 2H^+ + 2e^-$, 0.36 V; $Se + 3H_2O \rightarrow H_2SeO_3 + 4H^+ + 4e^-$, -0.740 V; $H_2SeO_3 + H_2O \rightarrow SeO_4^{2-} + 4H^+ + 2e^-$, -1.15 V; $Se^{2-} \rightarrow Se + 2e^-$, 0.78 V; $Se + 6OH^- \rightarrow SeO_3^{2-} + 3H_2O + 4e^-$, 0.36 V; $SeO_3^{2-} + 2OH^- \rightarrow SeO_4^{2-} + H_2O + 2e^-$, -0.03 V. Other important physical properties of selenium are given under **Chemical Elements**.

Selenium was first identified by Berzelius in 1817. The element is found associated with volcanic activity, as for example in cavities of Vesuvian lavas and in the volcanic tuff of Wyoming (about 150 parts per million).

Selenium occurs as selenide in many sulfide ores, especially those of copper, silver, lead, and iron, and is obtained as a by-product from the anode mud of copper refineries. The mud is (1) fused with sodium nitrate and silica, or (2) oxidized with HNO_3, and the H_2O extract is then treated with HCl and SO_2, whereupon free selenium is separated.

Uses: Selenium is widely used in photoelectric cells. The element alters its electrical resistance upon exposure to light. The response is proportional to the square root of incident energy. Selenium cells are most sensitive in the red portion of the spectrum. Although an external emf must be applied, the resistance is low and amplification is easy. In the selenium photovoltaic cell configuration, a thin film of vitreous or metallic selenium is coated onto a metal surface. Then, a transparent film of another metal, often platinum, is placed over the selenium. A cell of this type generates its own emf, with a decrease in internal resistance with increasing irradiation. The response essentially is proportional to incident energy. The cells are not importantly sensitive to small temperature changes.

Advantage of the unipolar conduction characteristic of selenium is taken in arc rectifiers. In a typical unit, a nickel or nickel-plated steel or an aluminum disk with a thin layer of selenium applied to one side is used. Selenium also is added to copper alloys and to stainless steel to increase machinability. Advantages claimed for selenium copper are high machinability, combined with hot-working properties and high electrical conductivity. As a decolorizer in glass, selenium counteracts green shades arising from ferrous ingredients. Sodium selenite is used in the production

of red enamels and in the manufacture of clear red glass. Addition of from 1 to 3% selenium to vulcanized rubber increases abrasion resistance. The element also is used in photographic and printing reproduction chemicals.

Selenium is also used as an additive to lead-antimony battery grid metal and as a vulcanizing agent to improve temperature and abrasion resistance of rubber.

Chemistry and Compounds: Due to its $4s^2 4p^4$ electron configuration, selenium, like sulfur, forms many divalent compounds with two covalent bonds and two lone pairs, and d hybridization is quite common, to form compounds with Se oxidation states of 4+ and 6+.

While selenium dioxide, SeO_2, can be produced by direct reaction of the element with oxygen activated by passage through HNO_3, the compound is easily made by heating selenious acid, H_2SeO_3. Selenium dioxide sublimes at 315–317 °C, and is readily reduced by SO_2 to elemental selenium. Selenium trioxide, SeO_3, is not prepared from the dioxide by oxidation, although selenium does react with oxygen to form SeO_3 and SeO_2 in an electric discharge. Preferred method of preparing SeO_3 is by refluxing potassium selenate with sulfur trioxide. The reverse reaction, hydration of SeO_3 to selenic acid, H_2SeO_4, occurs easily. Selenious acid, H_2SeO_3, produced by hydration of SeO_2, is a stronger oxidizing agent than sulfurous acid as judged by its quantitative oxidation of iodide ion in acid solution, but is a weaker acid (ionization constants 2.4×10^{-3} and 4.8×10^{-9} at 25 °C). It forms salts, the selenites, many of which, especially those of the heavy metals, are reduced to selenides by hydrazine. Many of the selenites, e.g., those of nickel, mercury, and ferric ion, are very slightly soluble in H_2O. Selenious acid is readily oxidized by halogens in the presence of silver ion or 30% H_2O_2 to selenic acid, H_2SeO_4. Selenic acid is as strong an acid as H_2SO_4, and it is more readily reduced, reacting with hydrobromic acid and hydriodic acid to form selenious acid or (at high concentration) elemental selenium. Like sulfate ion, SO_4^{2-}, SeO_4^{2-} is tetrahedral in crystals.

Hydrogen selenide, H_2Se, is a stronger acid than H_2S (ionization constants of H_2Se, 1.88×10^{-4} and about 10^{-10}) and is less readily obtained from selenides than H_2S from sulfides (the selenides of aluminum, iron and magnesium, Al_2Se_3, FeSe, and MgSe, require heating with H_2O or dilute acids). In general, the metal selenides are prepared by direct combination of the elements. Those of transition groups 3–8, 1 and 2 and main groups 3 and 4 exhibit many instances of well-defined compounds, berthollide compounds, and substitutional solid solutions. Thus, four intermediate phases are found in the palladium-selenium system, Pd_4Se, $Pd_{2.8}Se$, $Pd_{1.1}Se$, and $PdSe_2$.

Selenium hexafluoride, SeF_6, the only clearly defined hexahalide, is formed by reaction of fluorine with molten selenium. It is more reactive than the corresponding sulfur compound, SF_6, undergoing slow hydrolysis. Selenium forms tetrahalides with fluorine, chlorine, and bromine, and dihalides with chlorine and bromine. However, other halides can be found in complexes, e.g., treatment of the pyridine complex of SeF_4 in ether solution with HBr yields $(py)_2SeBr_6$. Selenium tetrafluoride also forms complexes with metal fluorides, giving $MSeF_5$ complexes with the alkali metals.

Selenium forms several oxyhalides, e.g., $SeOF_2$, $SeOCl_2$, and $SeOBr_2$, the first two being liquids and the last a crystalline solid, mp 41.6 °C. Selenium also forms tetraselenium tetranitride, Se_4N_4.

Selenocyanates, M^ISeCN, corresponding to the thiocyanates, are prepared by addition of selenium to soluble cyanides. They are similar to the thiocyanates except that HSeCN immediately decomposes in acid to selenium and hydrogen cyanide. The heavy metal selenocyanates are less soluble than the corresponding thiocyanates.

Selenium forms "thio"-type compounds, such as $SeSO_3$ by reaction of selenium and sulfur trioxide, $SeSO_3^{2-}$ (selenosulfates) by reaction of selenium and sulfites, SeS^{2-} (selenosulfides) by reaction of selenium with sulfides, as well as diselenides, Se_2^{2-}, and polyselenides, Se_x^{2-}.

Carbon diselenide is an evil-smelling liquid, and COSe and CSSe are also known.

Biological Role of Selenium: Some very interesting examples of the effect of soils on the nutritional quality of plants are associated with selenium. The element has not been found to be required by plants, but is required in very small amounts by warm-blooded animals and probably by humans. However, selenium in larger quantities can be very toxic to animals and humans.

In large areas of the world, the soils contain very little selenium in forms that can be taken up by plants. Crops produced in these areas are, therefore, very low in selenium. A selenium deficiency in livestock is a serious problem. A deficiency causes a form of muscular dystrophy in younger animals and poor reproductive qualities in the adult animals. For prevention, sodium selenate or sodium selenite, sometimes augmented with vitamin E, is added in proper proportions to feedstuffs. Some areas, including the Plains and Rocky Mountain states in the United States have soils that are rich in available selenium. In regions like these, selenium toxicity is a problem. The situation is particularly serious in Arizona, California, Montana, Nevada, New Mexico, and South Dakota.

An interesting feature of selenium is that it occurs naturally in several compounds and these vary greatly in their toxicity and in their value in preventing selenium-deficiency diseases. In its elemental form, selenium is essentially insoluble and biologically inactive. Inorganic selenates or selenites and some of the selenoamino acids in plants are very active biologically, whereas some of their metabolites that are excreted by animals are not biologically active. In well-drained alkaline soils, selenium tends to be oxidized to selenates and these are readily taken up by plants, even to levels that may be toxic to the animals that eat them. In acid and neutral soils, selenium tends to form selenites and these are insoluble and unavailable to plants. Selenium deficiency in livestock is most often found in areas with acid soils and especially soils formed from rocks low in selenium.

In 1934, the mysterious livestock maladies on certain farms and ranches of the Plains and Rocky Mountain states were discovered to be due to plants with so much selenium that they were poisoning grazing animals. Affected animals had sore feet and lost some of their hair; many died. Over the next 20 years, researchers found that the high levels of selenium occurred only in soils derived from certain geological formations of high selenium content. They also found that a group of plants, called *selenium accumulators*, had an extraordinary ability to extract selenium from the soil. These accumulators were mainly shrubs or weeds native to semiarid and desert range lands. They usually contained about 50 parts per million (ppm) or more of selenium, whereas range grasses and field crops growing nearby contained less than 5 ppm selenium. These findings helped ranchers to avoid the most dangerous areas when grazing livestock.

In 1957, selenium was found to be essential in preventing liver degeneration of laboratory rats. Since then, research workers have found that certain selenium compounds, either added to the diet or injected into the animal, would prevent some serious disease of lambs, calves, and chicks. That selenium is an essential nutrient element for birds and animals has been established.

In most diets used in livestock production, from 0.04 to 0.10 ppm of selenium protects the animal from deficiency diseases. If the diet is very high in vitamin E, the required level of selenium may be lower.

In terms of human dietary requirements, much of the wheat for breadmaking in the United States is produced in selenium-adequate sections of the country. Bread is generally a good source of dietary selenium.

Selenomethionine decomposes lipid peroxides and inhibits in vivo lipid peroxidation in tissues of vitamin-E-deficient chicks. Selenocystine catalyzes the decomposition of organic hydroperoxides. Selenoproteins show a high degree of inhibition of lipid peroxidation in livers of sheep, chickens, and rats. Thus, some forms of selenium exhibit in vivo antioxidant behavior.

Additional Reading

Carter, G.F. and D.E. Paul: "Materials Science and Engineering," ASM International, Materials Park, Ohio, 1991.
Considine, D.M. and G.D. Considine: "Van Nostrand Reinhold Encyclopedia of Chemistry, 4th Edition, Van Nostrand Reinhold Company, New York, NY, 1984.
Frankenberger, W.T. Jr. and R.A. Enberg: "Environmental Chemistry of Selenium," Marcel Dekker, Inc., New York, NY, 1998.
Greenwood, N.N. and A. Earnshaw: "Chemistry of the Elements," 2nd Edition, Butterworth-Heinemann, Inc., Wodurn, MA, 1997.
Hatfield, D.L.: "Selenium: Its Molecular Biology and Role in Human Health," Kluwer Academic Publishers, Norwell, MA, 2001.
Lewis, R.J. and N.I. Sax: "Sax's Dangerous Properties of Industrial Materials, 10th Edition, John Wiley & Sons, Inc., New York, NY, 1999.
Lide, D.R.: "CRC Handbook of Chemistry and Physics 2000–2001," 81st Edition, CRC Press, LLC., Boca Raton, FL, 2000.
Liotta, D. and R. Monahan III: "Selenium in Organic Synthesis," *Science*, **221** 356–361 (1986).
Marshall, E.: "High Selenium Levels Confirmed in Six States," *Science*, **231**, 111 (1986).
Meyers, R.A.: "Handbook of Chemicals Production Processes," McGraw-Hill Companies, Inc., New York, NY, 1986.
Reamer, D.C. and W.H. Zoller: "Selenium Biomethylation Products from Soil and Sewage Sludge," *Science*, **208**, 500–502 (1980).
Staff: "Plants Can Eat Toxic Soil," *National Food Review*, **42** (October–December, 1989).
Staff: "ASM Handbook — Properties and Selection: Nonferrous Alloys and Pure Metals," ASM International, Materials Park, Ohio, 1990.

SELENOGRAPHIC.   1.   Of or pertaining to the physical geography of the moon.

2.   Specifically, referring to positions on the moon measured in latitude from the moon's equator and in longitude from a reference meridian.

SELENOLOGY.   That branch of astronomy that treats of the moon, its magnitude, motion, constitution, and the like. *Selene* is Greek for *moon*.

SELF-DIAGNOSTICS.   A term commonly used in the sensors, instrumentation, process control, and computer field to designate separate circuitry and instructions (self-contained) to test and sometimes recalibrate an electronic measurement and supervisory piece of equipment.

SELF-ENERGY (Particle).   Classically, the energy of interaction between different parts of the particle (considered, for example, as a ball of charge). In quantized field theory (see also **Field Theory**) the contribution to the Hamiltonian arising from the virtual emission and absorption of other particles, especially photons or mesons. In terms of mass-energy equivalence, the energy equivalent of the rest mass of the particle.

SELF-INDUCTANCE.   The ratio of the magnetic flux linking a circuit to the flux-producing current in that circuit.

SELF-OPERATED CONTROLLER.   Often termed a *regulator*, a self-operated controller requires no external power for automatically controlling a variable, notably pressure, temperature, liquid level, or flow. Some of the styles of self-operated controllers are: (1) *directly actuated* (DA),

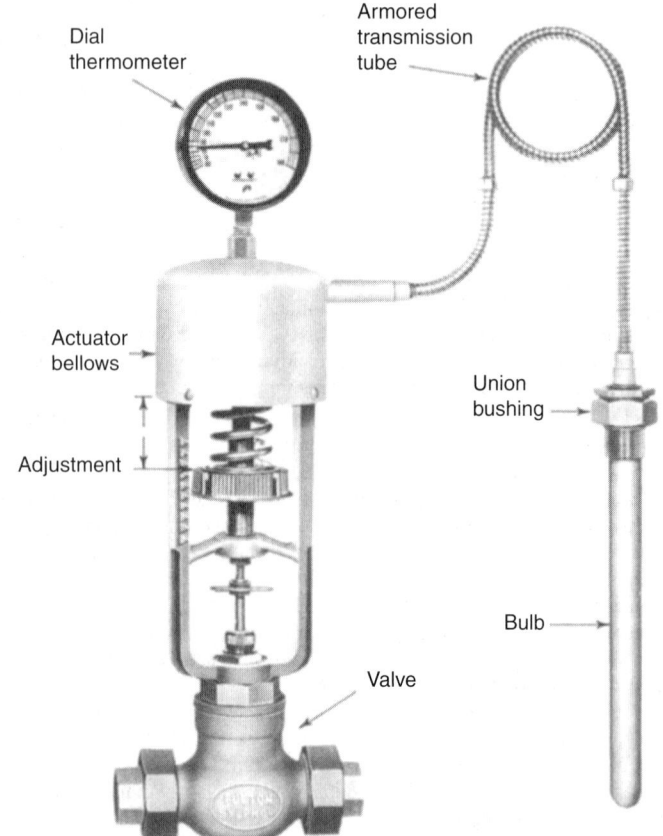

Fig. 1.   Direct-actuated remote-sensing self-operated temperature controller or regulator. (*Robertshaw.*)

where the valve is positioned directly by the measuring element; (2) *pilot-actuated* (PA), where the measuring element controls a pilot valve. The latter admits supply-line pressure to a piston or diaphragm to move the main valve; (3) *self-contained* (SC), where the measuring element is an intimate part of the valve structure. This type of device responds only to variations, such as pressure, temperature, or flow, that occur in the flowing medium; and (4) *remote-sensing* (RS), where the measuring element is separated from the valve.

Usually, regulators are furnished with only a proportional control action, with the proportional band established by the manufacturer. Control accuracy obtainable with self-operated controllers is on the order of: temperature regulators, $\pm 1\,^\circ$F; pressure regulators, ± 0.5 psi; level regulators, $\pm 1\%$. Self-operated controllers are advantageous where control specifications can be met because of their simplicity, ease of service, general ruggedness, comparatively low cost, no need for external power (hence extra wiring or piping), and normally long useful life. Limitations include the rather coarse accuracy performance previously mentioned, allowable pressure drop through the valve, limited availability of ranges, and the usually relatively large measuring element.

With reference to Fig. 1, the direct-actuated, remote-sensing temperature regulator employs a vapor-filled thermal system. When the temperature at the bulb rises, vapor pressure in the system is increased and this, in turn, is transmitted to the bellows. When the force produced equals the adjustable spring force, further pressure increase will cause a downward movement of the stem, thus causing the valve to commence closing. The reverse process occurs, of course, upon a drop in temperature.

SELLMEIER EQUATION. An equation often used for media that show anomalous dispersion

$$n^2 = 1 + \sum \frac{A\lambda^2}{\lambda^2 - \lambda_1^2} + \frac{B\lambda^2}{\lambda^2 - \lambda_2^2} + \cdots$$

where n is the refractive index, λ the wavelength of the light, λ_1 and λ_2 are the wavelengths of absorption lines, and $A, B, \ldots$ are constants to be determined from experimental data. If the equation is expanded in a series it becomes

$$n^2 = a + \frac{b}{\lambda^2} + \frac{c}{\lambda^4} \cdots$$

which has the same form as the Cauchy dispersion formula (but note that n occurs to the first power there).

SELSYN. (A trade name, from self-synchronous; often capitalized). An electrical remote indicating instrument operating on direct current, in which the angular position of the transmitter shaft, carrying a contact arm moving on a resistance strip, controls the pointer on the indicator dial.

SEMICARBAZONES. The products of the reaction between an aldehyde or a ketone with semicarbazide are termed *semicarbazones*.

$$\underset{\text{(acetaldehyde)}}{CH_3 \cdot CHO} + \underset{\text{(semicarbazide)}}{NH_2 \cdot CO \cdot NHNH_2} \longrightarrow$$

$$\underset{\text{(acetaldehyde semicarbazone)}}{CH_3 \cdot CH : N \cdot NH \cdot CONH_2} + H_2O$$

$$\underset{\text{(acetone)}}{(CH_3)_2CO} + NH_2 \cdot CO \cdot NH \cdot NH_2 \longrightarrow$$

$$\underset{\text{(acetone semicarbazone)}}{(CH_3)_2C : N \cdot NH \cdot CO \cdot NH_2} + H_2O$$

SEMICONDUCTORS. Materials and devices known as *semiconductors* have been the backbone of the electronics industry for many years. Semiconductors did not enter the industry in a major way, however, until several years after the vacuum tube (valve) had been well established. In terms of perspective, it is interesting to note that at least one semiconductor device predated the vacuum tube in the early days of radio communication. This was the then familiar galena crystal and accompanying whisker used in early crystal set radio receivers.

With the continuing attention to developing reliable and efficient vacuum tubes, which occurred over a long time span, interest in semiconductors stagnated—with the exception of the emerging radar technology of the World War II era. Massachusetts Institute of Technology's Radiation Laboratory became active in the investigation of crystal rectifiers and engaged in exploratory studies and in the development of very pure semiconducting materials, notably silicon and germanium. Several other

research institutions during this same period became interested in the theoretical aspects of solid state and semiconductors, including the energy levels of solids and the charged carrier transport in silicon and a few other materials. Out of this early phase of semiconductor technology came, in 1947, the invention of the transistor (contraction for transfer resistor). Inventors Shockley, Bardeen, and Brattain (Bell Laboratories) received the Nobel Prize in physics in 1956 for their accomplishment.

The transistor had a tremendous impact, constituting the birth of modern electronics. The transistor led to the development of almost innumerable semiconductor device configurations and ultimately to the phasing out of the vacuum tube, except for rather special and limited applications.

Innovations to fabricate semiconductors with improved performance and smaller size (microelectronics) for many years has been a continuing process without apparent end. As of the very late 1980s, the enhancement of semiconductors continues apace. Circuit integration has gone through several major phases—from IC (integrated circuit) to LCI (large-scale integration) to VLSI (very large-scale integration) to VHSIC (very high-speed integrated circuits) to ULSI (ultralarge-scale integration)—to the point where it is difficult to find appropriate adjectives to describe developments in ICs.

With few exceptions, it is only within recent years that serious concern has been expressed by experts in what the physical limits on size and performance may be. Fortunately, thus far the advancements in fabrication (chip-making processes), such as electron beam and molecular beam lithography, have allayed these concerns in the short term. Although some interest in nonsilicon materials has always been present, there has been a recent reawakening of interest in the use of gallium, indium, arsenic, phosphorus, and antimony—this interest intensifying because of the opportunities such elements offer over silicon. Although silicon which for years has been the basis for volume-produced devices, is not seriously threatened at this juncture, some authorities feel that, in the long term, the emphasis on silicon will be heavily shared with these other materials.

Even with the aforementioned materials, the microelectronics and optoelectronics industries have relied almost exclusively on *inorganic* materials. This is expected to change. Some authorities now forecast that in the early 2000s, there will be a major shift, namely, to *molecular electronics*. Even today, considerable attention for the long term is being paid to the *organic* solid state. The richness of the variety of organic molecular materials available offers enormous potential compared with the relative paucity of structures achievable with inorganic compounds, even when due allowance is made for the exciting developments in inorganic quantum well semiconductors. A hint of what may be achieved along these lines is given by the progress thus far made in liquid crystals, piezoelectric and photoconducting polymers. This implies, of course, that organics no longer will be confined to their traditional applications in electronics, such as for insulation, adhesion, or encapsulation.

Molecular electronics currently is defined as the use of organic molecular materials to perform an active function in the processing of information and its transmission and storage. An alternative definition has been suggested, namely, the achievement of *switching* on a molecular scale. As observed by G.G. Roberts (University of Oxford), "It is interesting to note that only a modest diminution in the size of electronic circuit components is required before the scale of individual molecules is reached; in fact many existing circuit elements could already be accommodated within the area occupied by a leukemia virus."

Some investigators forecast that during the first quarter of the 21st Century, molecular electronics will lead to *supermolecular electronics* where signal transport and control will be effected by nanometer-scale assemblies.

Numerous applications of contemporary semiconductors are described throughout this encyclopedia. See also **Molecular and Supermolecular Electronics**.

Nature of Semiconductors. From the standpoint of their use in electronics, semiconductors are distinguished from other classes of materials by their characteristic electrical conductivity σ. The electrical conductivities of materials vary by many orders of magnitude, and consequently can be classified as: (1) the perfectly conducting superconductors; (2) the highly conducting metals ($\sigma \approx 10^6$ mho/centimeter); (3) the somewhat less conducting semimetals ($\sigma \approx 10^4$ mho/centimeter); (4) the semiconductors covering a wide range of conductivities ($10^3 \gtrsim \sigma \gtrsim 10^{-7}$ mho/centimeter); and (5) the insulators, also covering a wide range ($10^{-10} \gtrsim \sigma \gtrsim 10^{-20}$ mho/centimeter).

These low-conductivity materials are characterized by the great sensitivity of their electrical conductivities to sample purity, crystal perfection, and external parameters, such as temperature, pressure, and frequency of the applied electric field. For example, the addition of less than 0.01% of a particular type of impurity can increase the electrical conductivity of a typical semiconductor like silicon or germanium by six or seven orders of magnitude. In contrast, the addition of impurities to typical metals and semimetals tends to decrease the electrical conductivity, but this decrease is usually small. Furthermore, the conductivity of semiconductors and insulators characteristically decreases by many orders of magnitude as the temperature is lowered from room temperature to 1 K. On the other hand, the conductivity of metals and semimetals characteristically increases in going to low temperatures, and the relative magnitude of this increase is much smaller than are the characteristic changes for semiconductors. The principal conduction mechanism in metals, semimetals, and semiconductors is electronic, whereas both electrons and the heavier charged ions may participate in the conduction processes of insulators.

Classification. It is customary to classify a semiconductor according to the sign of the majority of its charged carriers, so that a semiconductor with an excess of negatively charged carriers is termed *n*-type. A semiconductor with an excess of positively charged carriers is called *p*-type, while a material with no excess of charged carriers is considered to be perfectly compensated. Many of the important semiconductor devices depend upon fabricating a sharp discontinuity between the *n*- and *p*- type materials, the discontinuity being called a *p-n* junction.

Other Characteristics. Even though most semiconductors exhibit a metallic luster when inspected visually, this does not provide a reliable criterion for their classification, since the electrical conductivity of all materials is frequency dependent. Visual inspection tends to be sensitive to the conductivity properties at visible frequencies ($\sim 10^{15}$ Hz). Although materials with a high optical reflectivity tend also to exhibit high D.C. conductivity, these two properties are not necessarily correlated in semiconductors and metals. An example of a metal without metallic luster is ReO_3 (rhenium trioxide), a semitransparent reddish solid. On the other hand, most of the common semiconductors do exhibit metallic luster primarily because electronic excitation across their fundamental energy gaps can be achieved at infrared frequencies. At low frequencies, the principal conduction mechanism is free carrier conduction, which is important in metals and is present to some extent in semiconductors that contain impurities or are found at elevated temperatures. In contrast, interband transitions dominate the conduction process at very high frequencies. Interband transitions contribute to the conductivity by about the same order of magnitude in semiconductors, metals, and insulators.

Since the D.C. conductivity due to free carriers is characteristically low in semiconductors and insulators, the generation of free carriers by exposure to light at infrared, visible, and ultraviolet frequencies can lead to a large increase in the D.C. conductivity. This photoconductive effect, which is not observed in metals or semimetals, can be enormous in low-conductivity semiconductors (an increase in the D.C. conductivity of CdS (cadmium sulfide) by 8 orders of magnitude is observed). The effects of ultraviolet light, for example have been used successfully in reprogramming EPROM (electrically programmable read-only memory) devices, exposure to UV light causing trapped charges to leak off.

Because of the extreme sensitivity of semiconductors to impurities, temperature, pressure, light exposure, and certain other factors, these materials can be exploited in the fabrication of useful devices, such as the crystal diode, the transistor, integrated circuits, photodetectors, and light switches.

Flow of Current in Semiconductors. The flow of electric current depends upon the acceleration of charges by an externally applied electric field. Only those charges that resist collisions or scattering events are effective in the conduction process. Because of collisions, charged particles in a solid are not accelerated indefinitely by the applied field, but rather, after every scattering event, the velocity of a charged particle tends to be randomized. Thus the acceleration process must start anew after each scattering event and charged particles achieve only a finite velocity along the electric field **E**, the average value of the velocity being denoted by v_D, the drift velocity. The effectiveness of the charge transport by a particular charged particle is expressed by the mobility μ, which is defined as $\mu = v_D/E$. The mobility of a particle with charge e and mass m can be related directly to the mean time between scattering events (also called the relaxation time) by the expressed $\mu = e\tau/m$. The electrical conductivity

s depends upon the mobility of the charged carriers as well as on their concentration n, and is simply written as $\sigma = ne\mu$, where e is the charge of the carriers. The advantage of expressing the conductivity in this form is the explicit separation into a factor n that is highly sensitive to external parameters, such as temperature, pressure, optical excitation, irradiation, and into another factor μ, which depends characteristically on scattering mechanisms and on the electronic structure of the semiconductor.

The classical theory for electronic conduction in solids was developed by Drude in 1900. This theory has since been reinterpreted to explain why all contributions to the conductivity are made by electrons which can be excited into unoccupied states (Pauli principle) and why electrons moving through a perfectly periodic lattice are not scattered (wave-particle duality in quantum mechanics). Because of the wavelike character of an electron in quantum mechanics, the electron is subject to diffraction by the periodic array, yielding diffraction maxima in certain crystalline directions and diffraction minima in other directions. Although the periodic lattice does not scatter the electrons, it nevertheless modifies the mobility of the electrons. The cyclotron resonance technique is used in making detailed investigations in this field.

The origin of the energy barrier[1] for carrier generation is directly connected with the energy levels for electrons in a solid. Considering electrons in a solid from a tight-binding point of view, the discrete energy levels of the free atom broaden in the solid to form energy bands. For materials that are well described by the tight-binding approximation, the width of the energy bands is sufficiently small so that an energy gap is formed between the energy bands; in the forbidden energy gap there are no bound states. Of particular importance to the conduction properties of a solid is the fact that *all* of the available states in each band would be filled if each atom were to contribute exactly two electrons, thereby causing every solid with an odd number of electrons per atom to be metallic; while solids with an even number of electrons per atom would be insulating or semiconducting. The occurrence of energy bandgaps is also a consequence of the weak binding approximation, whereby the periodic potential itself is responsible for creating bandgaps through the mixing of states separated by a reciprocal lattice vector. See also **Solid-State Physics**.

For semiconductors, the excitation energy lies in the range 0.1 to about 2 eV. Thermal fluctuations are sufficient to excite a small, but significant, fraction of electrons from the occupied levels (the valence band) into the unoccupied levels (the conduction band). Both the excited electrons and the empty states in the valence band (aptly called *holes*) may move under the influence of an electric field, providing a means for conduction of current. (A hole acts like an electron with a positive charge.) Such electron-hole pairs may be produced not only by thermal energy, but also by incident light, providing photo-effects.

Crystallographic defects, in general, are also electronic defects. In metals, they provide scattering centers for electrons, increasing the resistance to charge flow. The resistance wire in many electric heaters, in fact, consists of an ordinary metal, such as iron, with additional alloying elements, such as nickel or chromium, providing scattering centers for electrons. In semiconductors and insulators, alloying elements and defects provide an even greater variety of effects, since they can change the electron-hole concentrations drastically in addition to providing scattering centers. The semiconductor industry has been built on the alloying of silicon, and a few other elements, including germanium, with *selected impurities* in carefully controlled concentration and geometry. Some of the other emerging semiconductor materials are described later.

Doping. This is a process for purposely adding impurities to a semiconductor (or production of a deviation from stoichiometric composition in order to achieve a desired characteristic). Doped material thus is no longer intrinsic but is impure and called extrinsic. If a trivalent impurity is introduced into silicon or germanium, *holes* are created and the material

---

[1] The sets of discrete but closely adjacent energy levels, equal in number to the number of atoms, that arise from each of the quantum states of the atoms of substance when the atoms condense to a solid from a nondegenerate gaseous condition, make up the energy band (also called the Bloch band). For a semiconductor, the highest energy level is the conduction band, containing only the excess electrons resulting from crystal impurities. The next highest level is the valence band, and it is usually completely filled with electrons. In between these bands is the forbidden band, which is wider for an insulating material than for a semiconductor and which vanishes in a conducting material.

is said to be *p*-type. Introduction of a pentavalent element into silicon or germanium, on the other hand, creates *free electrons* and the material is said to be *n*-type. Because of thermal effects, free electrons and holes are always being produced in silicon and germanium (intrinsic generation of electron-hole pairs). Consequently, there will be some electrons in *p*-type material and some holes in the *n*-type material. These carriers are referred to as *minority carriers*. Electrons in *n*-type material and holes in *p*-type material are termed *majority carriers*.

The process of placing impurities in the near-surface region of solids is accomplished by a procedure known as *implanting*. A commonly used implanting procedure is to accelerate impurity ions in an electrostatic field with the energy sufficient to impinge with the desired force on the solid target. Known as *ion implantation*, this carefully controlled and reproducible procedure has been widely used to dope semiconductors to create *p-n* junction formations. A certain amount of damage, however, occurs in the semiconductor material in some cases. The surface may become amorphous, or because the implanted dopants may not reach substitutional spots in the crystal lattice the ions may not become electrically active. Thus, it is necessary to anneal the solid for electrical activation of the implanted ions as well as to remove any damage.

Semiconductor Device Configurations

Representative of semiconductor devices is the diode, a two-terminal device which has the property of permitting current to flow with practically no resistance in one direction and offering nearly infinite resistance to current flow in the opposite direction. Applications of the diode are numerous, as in gating circuits used in digital computers.

A widely used semiconductor diode is the *p-n junction diode*. Imagine a crystal (single) of silicon doped so half the material is *p*-type and the other half is *n*-type. The internal boundary between the two extrinsic regions is a *p-n junction*, and the resulting device is a *diode*. See Fig. 1. Three possible configurations of the *p-n* junction are shown in Fig. 2. The energy diagrams for these three configurations are also shown in Fig. 2. Similar diagrams can be generated for holes. When the diode is unbiased, no net flow of electrons takes place across the junction. Assuming that some electrons on the *n*-side have sufficient energy to overcome the potential hill, electrons on the *p*-side (minority carriers) "slide down" the hill, making the net current flow zero. For the reverse biased example, the potential hill is raised and only the few minority carriers from the *p*-side slide down. This results in a minute reverse saturation current. When the diode is forward biased, the potential hill is lowered. This enables electrons to climb over the hill and current flow occurs. The same considerations apply to holes. In fact, the total diode current is equal to the sum of the electrons and holes flowing across the junction.

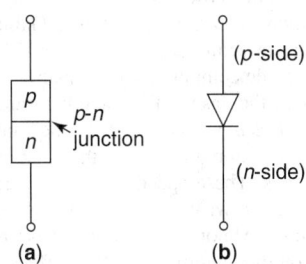

Fig. 1. (**a**) Configuration of *p-n* junction diode. (**b**) electrical symbol.

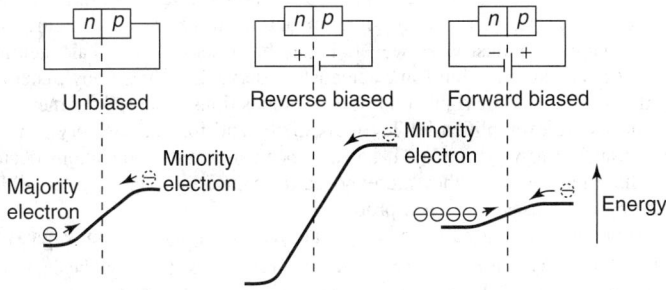

Fig. 2. Three possible configurations of a *p-n* junction diode.

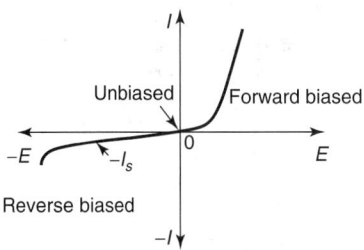

Fig. 3. Characteristic curve of a semiconductor diode.

The characteristic curve of a semiconductor diode is shown in Fig. 3. An equation for this curve, called the *rectifier equation*, is expressed as:

$$I = I_s(e^{-11600\,E/T} - 1)$$

where I = diode current, amperes

I_s = reverse saturated current (temperature dependent), amperes

E = diode biasing voltage ($+E$ for forward bias; $-E$ for reverse bias), volts

T = absolute temperature ($0\,°C + 273°$), degrees Kelvin

At room temperature (300 K) and $E > 0.1$ volt,

$$I \cong I_s^{39E}$$

Where E is more negative than 0.1 volt,

$$I \cong -I_s$$

An example of a simple rectifier employing a *p-n* junction diode is given in Fig. 4. During the positive half-cycle (0° to 180°) of the A.C. sinusoidal waveform v_s, the diode is forward-biased and conducts. The voltage v_L across load resistance R_L is, therefore, nearly identical to that of v_s for the positive half-cycle. For the negative half-cycle (180° to 360°), the diode is reverse biased and does not conduct. No current flows in R_L, and $v_L = 0$ during the negative half-cycle. Because the diode conducts for only one-half cycle, the circuit of Fig. 4 is called a *half-wave rectifier*. The waveform of v_L is only unidirectional. To obtain steady D.C., like that from a battery, a filter is required. An example of an elementary filter is a large-valued capacitor placed across the load resistor.

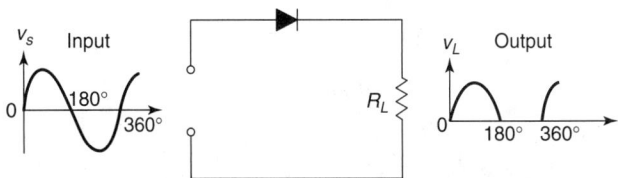

Fig. 4. Simple rectifier employing a *p-n* junction diode.

The circuit of Fig. 4 can also be used as a detector of amplitude-modulated (AM) radio waves. Fig. 5(a) illustrates the components of an AM wave. If this is applied to the input of Fig. 4, the wave is rectified and the output appears as shown in Fig. 5(b). Placing a small-valued capacitor across R_L filters out the carrier frequency and the desired modulating signal is obtained, as shown in Fig. 5(c).

A bipolar transistor on a single silicon crystal is diagrammed in Fig. 6.

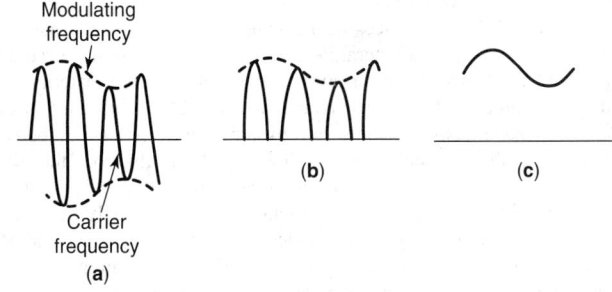

Fig. 5. Use of *p-n* junction diode as AM radio detector.

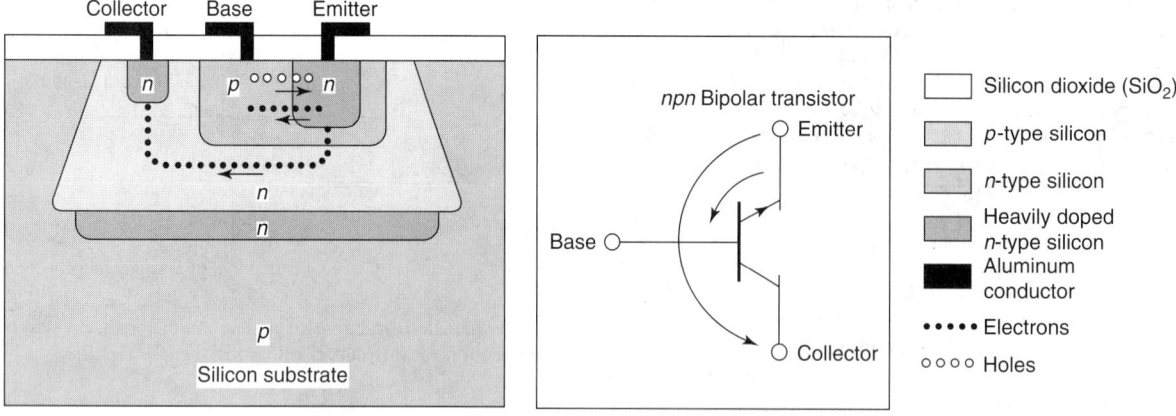

Fig. 6.   A bipolar transistor on a single silicon crystal. (*R.T. Kurnik, "Chemical vapor deposition in microelectronics," Chemical Engineering Progress, vol. 81, pp. 30–35, May, 1985.*)

Metal Oxide Semiconductors. The metal oxide semiconductor field effect transistor (MOSFET) is representative of another class of semiconductors. In *n*-MOS device, two islands of *n*-type silicon are created in a *p*-type silicon substrate. A thin layer of nonconducting SiO_2 lies on top of the silicon substrate. Direct connections on a *source* and *drain* are made to the two islands, while a metal *gate* is coupled to the silicon substrate by capacitance. Usually the source and substrate are electrically connected and held at a potential of zero volts. The drain is held at a positive voltage. In this condition, no current flows into the MOS device. When a positive potential is applied to the gate, the electric field attracts a majority of electrons to the thin layer at the surface of the crystal under the gate. Since this region is normally *p*-type, the surface becomes "inverted" creating a continuous *n*-type channel between source and drain, thus allowing large currents to flow. This creates a current amplification as in a bipolar transistor. An advantage of MOSFET over bipolar transistors is that they require no isolation islands and thus can be packed more closely on a silicon chip.

Complementary MOS devices (CMOS) have been widely used in recent years. See Fig. 7. The CMOS is made up of a *n*-MOS and a *p*-MOS. A main advantage of the CMOS is its low power consumption.

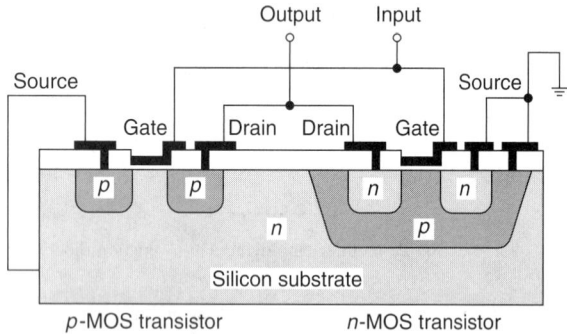

Fig. 7.   Complementary MOS device (CMOS) on a single silicon crystal. (*R.T. Kurnik, "Chemical vapor deposition in microelectronics," Chemical Engineering Progress, vol. 81, pp. 30–35, May, 1985.*)

Other Materials for Semiconductors. Although silicon (and germanium at one time) is the unquestioned principal semiconductor material to date, silicon does have limitations. For example, it is not easy to integrate electronic and photonic devices in the same microchip. Silicon has a relatively narrow range of temperature tolerance, is susceptible to radiation damage and has "slow" electrons compared with some other materials. Elements in Periodic Table Groups III and IV (now officially called Groups 13 and 15) have fast electrons. For example, the differences between electrons in gallium arsenide and silicon stem basically from the differing chemical characteristics. Electrons in gallium arsenide at low electric fields behave like very light particles which can move easily through the vibrating (and obstructing) crystal lattice of atoms. In contrast, the electrons in silicon behave like heavy particles that move sluggishly under the influence of an applied voltage. The result is significantly faster operating times in microchip operation. See Fig. 8. Fast electrons translate

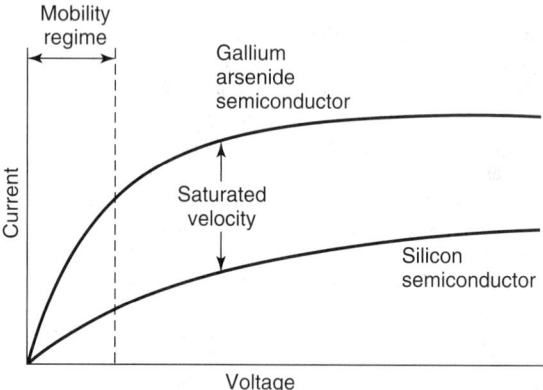

Fig. 8.   Current-voltage characteristics of two hypothetical devices of identical physical size. The gallium arsenide curve rises faster and reaches peak velocity faster than the silicon. This means that the group III–V (13–15) electrons produce significantly faster operating times in microchips. (*AT&T Technology.*)

into fast switches. Such switches, when multiplied by thousands or even hundreds of thousands, comprise the basic building blocks of a digital integrated circuit (commonly called a microchip). As pointed out by Allyn, Flahive, and Wemple (1986), there are two classes of speed: (1) Maximum speed achievable, no matter how much "push" is provided by the applied voltage. This is known as *saturated voltage*. Gallium arsenide materials have an advantage in saturated voltage over silicon of 1.5; and with indium — gallium arsenide compounds, the advantage reaches 2.5. (2) The second speed relates to the ease with which electrons can be brought up to full speed (*low-field electron mobility*). Higher mobility in the Groups 13–15 (III–V) semiconductors means that the electrons reach full speed at lower operating voltages. These speed advantages are particularly important in terms of interdevice wiring, which tends to dominate the speed of high-density microchips. Major emphasis on these newer semiconductor materials is directed on the Schottky gate field effect transistor. See Fig. 9.

In the 1950s and 1960s, considerable investigation was made of amorphous *chalcogenide glasses* for possible use in semiconductor devices. The glasses are named for the chalcogens (Group 16, formerly Group VI in the Periodic Table). Early in their consideration, these materials created a considerable controversy among solid-state physicists. Claims were made and challenged as regards their possible impact on further revolutionizing the semiconductor industry. However, it has been shown that chalcogenide glasses can "switch," but some scientists observe that almost any material will switch under the right conditions. Compositions proposed for memory switches are exemplified by $Te_{81}Ge_{15}Sb_2S_2$, and for non-memory switch materials, $Te_{40}As_{35}Si_{18}Ge_7$. It has also been shown that transitions occur in these glasses when they are exposed to intense light and thus possible photographic uses have been proposed.

Semiconductors used in solar cells are described under **Solar Energy**.

Gallium Arsenide Power Sources. GaAs was first synthesized in 1929 by V.M. Goldschmidt. Its semiconducting properties were not studied until 1952 by H. Welker. The first GaAs p-n junction used for

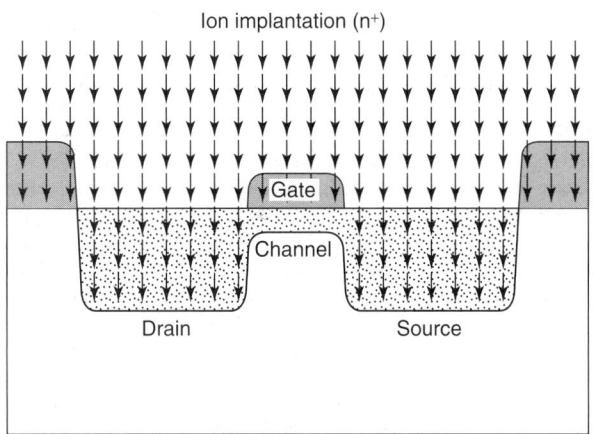

Fig. 9.   A Schottky barrier gate used in the metal-semiconductor field-effect transistor (MESFET) in AT&T gallium arsenide microchips. The tiny gate is only one micrometer wide (1/25,400 inch). The gate electrode is deposited before the ion-implantation process so that the gate material will "shade" the channel under it from the "ion rain" that doses the exposed material. (*AT&T Technology.*)

power generation at microwave frequencies was the tunnel diode. Later, GaAs varactor diodes were used in harmonic frequency multipliers and parametric amplifiers at microwave and mm-wave frequencies because of the inherent higher cut-off frequencies possible with gallium arsenide. In 1963, J.B. Gunn discovered the negative resistance property of GaAs, after which GaAs diodes and field effect transistors (FETs) were developed.

The idea for using diodes for generation and amplification of power at microwave frequencies was suggested by A. Uhlir, Jr. Frequency multipliers have been used for power generation since 1958. These devices depend on the nonlinear reactance or resistance characteristics of semiconductor diodes. Generally, there are three types of multiplier diodes — step recovery diodes, variable resistance multiplier diodes, and variable capacitance multiplier diodes.

T.B. Ramachandran (Microwave Device Technology Corporation) notes that there are two inherent major disadvantages for current GaAs FET power devices:

1. The devices are surface oriented. Since the active region is close to the surface, the surface effects tend to affect the device performance. This may be seen in the noise performance of GaAs FETS close to the carrier.
2. To increase the power output, the breakdown voltage must be increased. Active channel doping has to be decreased in order to increase the breakdown voltage. This reduction in doping density decreases the maximum current density, and this tends to reduce the total power output.

J.B. Gunn (International Business Machines) noticed in 1963 the current instabilities in GaAs at high electric fields. Known as the Gunn effect, or the transferred electron effect, Gunn diodes have been used as a low-cost source for microwaves since 1968. These components are comparatively easy to manufacture and hence the cost is low. GaAs Gunn diodes are used from C-band (4 GHz) through W band (100 GHz).

W.T. Read (AT&T Bell Laboratories) first reported microwave oscillations in silicon p-n junctions in 1965. During the interim, much research has gone into developing impact ionization avalanche transit time (IMPATT) diodes for a variety of applications.

Research and Development Trends

Quantum-Effect Devices. There is a limit on the components of ordinary integrated circuits because "smallness" of size can interfere with their functionality. Such problems may be overcome through the use of quantum-effect semiconductor devices.

It was predicted by a number of authorities that, before the year 2000, the physical laws that govern the behavior of circuit components would impede the ultimate shrinkage of the chip. As early as 1982, P.K. Chatterjee stressed how close the end point on downscaling components might be. Estimates of minimum feature size as of the early 1990s ranged between 100 and 500 billionths of a meter. As observed by R.T. Bate (Texas Instruments Incorporated), "The same solution that some of the very

phenomena that impose size limits on ordinary circuits could be exploited in a new generation of vastly more efficient devices. The functional bases for these devices are quantum-mechanical effects that carry semiconductor technology into a realm of physics where subatomic particles behave like waves and pass through formerly impenetrable barriers. With the so-called quantum semiconductor device, I believe it will be possible to put the circuitry of a supercomputer on a single chip."

Doped silicon, doped and undoped gallium arsenide, and aluminum gallium arsenide have been used as the basis for quantum devices. Of course, size reduction of these proportions pose difficult production tasks. In addition to shrinking size, quantum devices can be expected to be faster and more efficient. A prototype quantum chip, with features one-hundredth of the size of an ordinary chip may appear as shown in Fig. 10. An operational semiconductor device based upon the quantum effect should appear prior to the year 2000. Aggressive research currently is being carried out by AT&T Bell Laboratories, IBM Corporation, the Massachusetts Institute of Technology, Hughes Research Laboratories, Texas Instruments Corporation, the University of Cambridge, Philips Research Laboratory, and the University of Glasgow, among others. As stressed by R.T. Bate, "The commitment of so many research teams to a problematic technology attests to the tremendous potential of these devices and to the faith that they will take the lead in the next semiconductor revolution. The costs and risks involved must be borne in order to revitalize a rapidly maturing electronics industry; the results can only benefit a society that has learned to depend on integrated circuits in many ways."

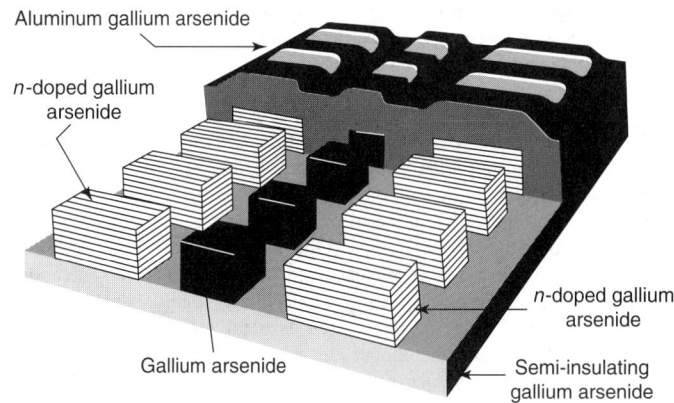

Fig. 10.   Quantum chip consisting of four materials. Final product is about $\frac{1}{100}$th size of conventional chip. Current flows from one negatively doped (n-doped) gallium arsenide block to another by way of a layer of aluminum gallium arsenide, a gallium arsenide cube, and thence to an other aluminum gallium arsenide layer. Current conductivity of a quantum device is extremely sensitivity and thus capable of exacting control. (*This idealized model is suggested by R.T. Bate in the scholarly reference cited.*)

Atom Switch. By employing the technique of the Scanning Tunneling Microscope, D.M. Eigler and a research team at the IBM Almaden Research Division, San Jose, California, have improvised an "atom switch." Through careful movement of a single xenon atom between the microscope's tip or a nickel surface, the researchers have altered the amount of tunneling current between tip and sample. When the xenon rests on the surface, this is tantamount to the switch's off position. The switch is turned on by applying a 64-millisecond (0.8 V pulse) to the tip. This causes the xenon to jump to the tip, thus increasing the tunneling current by a factor of about seven. As of the present, no practical applications of this switching action are planned because the apparatus involved is bulky and costly. Some scientists believe that the principle ultimately may be useful for information storage systems. Other scientists have observed that, if storing a bit in a cluster of 1000 atoms ever becomes practical, a machine could be developed that would store the contents of the U.S. Library of Congress on a silicon disk only 12 inches (30 cm) wide. For more detail, see Yam reference listed.

Dynamical Phenomena at Metal and Semiconductor Surfaces. This topic has been investigated in recent years through the use of ultrafast measuring techniques involving lasers and nonlinear optics. As reported by J. Bokor (AT&T Bell Laboratories), "Understanding of the rates and mechanisms for relaxation of optical excitation of the surface itself as well as those of adsorbates on the surface is providing new insight into surface chemistry, surface phase transitions, and surface recombination of charge carriers in semiconductors." The combination of lasers and nonlinear optical techniques is now being brought to bear on the next frontier in surface physics, namely surface dynamics. Ultrafast lasers allow for the study of picosecond and femtosecond processes directly in the time domain, circumventing the ambiguities attendant on linewidth measurements for the determination of lifetimes. One may anticipate continued growth in the diversity of applications of these techniques to the understanding of the complexities of surface dynamics. See Fig. 11.

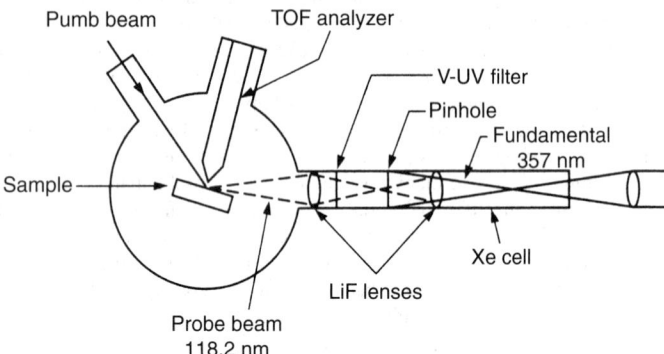

Fig. 11.   Experimental arrangement used for picosecond time- and angle-resolved photoemission spectroscopy. TOF = time of flight; V-UV = visible-ultraviolet; LiF = lithium fluoride Xe = xenon. (*Source: AT&T Bell Laboratories.*)

Amorphous Silicon. According to P.G. LeComber (University of Dundee), the most important difference between crystalline silicon and an amorphous semiconductor is that in the latter there is a continuous distribution of localized states within the forbidden energy gap. Another important difference concerns the mobility of the electrons or holes. LeComber observes, "In an amorphous material the periodicity of the lattice only extends over a few atomic spacings. Under these conditions, the electron transport may no longer be considered as band motion with occasional scattering, as in crystalline theory. In this case, the electron motion is essentially a diffusive process that can be considered to be similar to the Brownian motion of small particles in liquids. Properties of particular importance in the application of amorphous silicon films include:

- Thin films (about 1 micrometer thick).
- Low deposition temperature.
- Large area growth on many substrates, such as glass, metals, and flexible plastics.
- Mechanically very hard.
- Chemically very stable.
- Inert material.
- Extremely photoconductive.
- Room temperature electrical conductivity can be controlled over ten orders of magnitude by doping for both n-type and p-type material.
- Ease of sequentially producing p-type and n-type material by switching from one gas mixture to another.
- Easy to pattern arrays of devices using conventional photolithographic techniques developed for crystalline silicon.

Hybrid Ferromagnetic-Semiconductor Structures. G.A. Prinz and researchers at the Materials Science and Technology Division of the Naval Research Laboratory, Washington, DC, have been studying hybrid ferromagnetic-semiconductor materials through the use of modern thin-film techniques. Thus far, the team has researched and demonstrated combinations of Fe/Ge, Fe/GaAs, Fe/ZnSe, and Co/GaAs. The researchers observe, "Ultrahigh-vacuum growth techniques are being used to grow single-crystal films of magnetic materials. These growth procedures, carried out in the same molecular beam epitaxy systems commonly used for the growth of semiconductor films, have yielded a variety of new materials and structures that may prove useful for integrated electronics and integrated optical device applications."

Useful characteristics of hybrid ferromagnetic-semiconductor structures include:

1. Produce significant changes in the electrical and optical properties.
2. Coupling of devices to a radiation field, particularly in the microwave range.
3. Such devices provide a source of spin-polarized carriers.

Details of this research can be found in the Prinz reference listed.

Microclusters. These may be defined as small aggregates of atoms that make up a distinct phase of matter. The chemistry of clusters is highly reactive and selective. Their principle area of future application is catalysis. However, clusters also hold some promise for electronic applications. As observed by M.A. Duncan and D.H. Rouvray (University of Georgia), "Thin films of clusters possessing desirable electronic qualities could be of great interest in microelectronics. It is possible to envision applications in optical memories, image processing and superconductivity. Given the potential for construction of parts from networks of clusters, it may eventually be possible to make electronic devices on a molecular scale. Ultimately a machine might be designed that could serve as a link between solid-state electronics and biological systems, such as systems of neurons. Such a link might convey data from a television camera to the brain of a blind person."

See also separate article on **Molecular and Supermolecular Electronics**.

Additional Reading

Allison, J.: "Electronic Engineering Semiconductors and Devices," 2nd Edition, McGraw-Hill Companies, Inc., New York, NY, 1990.

Allyn, C.L., Flahive, P.G., and S.H. Wemple: "Choosing from Column III and Column IV," *Record (AT&T Bell Laboratories)*, 4–11 (January 1986).

Bate, R.T.: "The Quantum-Effect Device: Tomorrow's Transistor?" *Sci. Amer.*, 96 (March 1988).

Bierman, H.: "Material Advances Pave the Way for Device and System Improvements," *Microwave J.* 26 (October 1990).

Bokor, J.: "Ultrafast Dynamics at Semiconductor and Metal Surfaces," *Science*, 1130 (December 1, 1989).

Brennan, K.F.: "The Physics of Semiconductors: With Applications to Optoelectronic Devices," Cambridge University Press, New York, NY, 1999.

Brennan, K.F.: "Theory of Modern Electronic Semiconductor Devices," John Wiley & Sons, Inc., New York, NY, 2002.

Brodsky, M.H.: "Progress in Gallium Arsenide Semiconductors," *Sci. Amer.*, 68 (February 1990).

Brophy, J.J.: "Basic Electronics for Scientists," 5th Edition, McGraw-Hill Companies, Inc., New York, NY, 1990.

Dimitrijev, S.: "Understanding Semiconductor Devices," Oxford University Press, Inc., New York, NY, 2000.

DiSalvo, F.J.: "Solid-State Chemistry: A Rediscovered Chemical Frontier," *Science*, 649 (February 9, 1990).

Duncan, M.A. and D.H. Rouvray: "Microclusters," *Sci. Amer.*, 110 (December 1989).

Ellowitz, H.I.: "1991 U.S. GaAs Foundry Update," *Microwave J.*, 42 (August 1991).

Fink, D.G. and D. Christiansen: "Electronics Engineers' Handbook," 3rd Edition, McGraw-Hill Companies, Inc., New York, NY, 1989.

Fisk, Z. et al.: "Heavy-Electron Metals: New Highly Correlated States of Matter," *Science*, 33 (January 1, 1988).

Geinovatch, V.G.: "Prognostications from the Edge," *Microwave J.*, 26 (April 1991).

Geis, M.W. and J.C. Angus: "Diamond Film Semiconductors," *Sci. Amer.*, 84 (October 1992).

Goldstein, A.N., Echer, C.M., and A.P. Alivisatos: "Melting in Semiconductor Nanocrystals," *Science*, 1425 (June 5, 1992).

Kemerley, R.T. and D.F. Fayette: "Affordable MMICs for Air Force Systems," *Microwave J.*, 172 (May 1991).

LeComber, P.G.: "Amorphous Silicon — Electronics Into the 21st Century," University of Wales Review, 31 (Spring 1988).

Mouthaan, T.J.: "Semiconductor Devices Explained: Using Active Simulation," John Wiley & Sons, Inc., New York, NY, 1999.

Neamen, D.A.: "Semiconductor Physics and Devices: Basic Principles," 2nd Edition, McGraw-Hill Higher Education, New York, NY, 1997.

Pool, R.: "Clusters: Strange Morsels of Matter," *Science*, 1186 (June 8, 1990).

Prinz, G.A.: "Hybrid Ferromagnetic Semiconductor Structures," *Science*, 1092 (November 23, 1990).

Ramachandran, T.B.: "Gallium Arsenide Power Sources," *Microwave J.*, 91 (January 1990).

Schroder, D.K.: "Semiconductor Material and Device Characterization," 2nd Edition, John Wiley & Sons, Inc., New York, NY, 1998.

Singh, J.: "Semiconductor Devices: Basic Principles," John Wiley & Sons, Inc., New York, NY, 2000.

Soref, R.: "Silicon-Based Optical-Microwave Integrated Circuits," *Microwave J.*, 230 (May 1992).

Sze, S.M.M.: "Semiconductor Devices: Physics and Technology," 2nd Edition, John Wiley & Sons, Inc., New York, NY, 2001.

Van Zant, P.: "Microchip Fabrication," 4th Edition, McGraw-Hill Companies, Inc., New York, NY, 2000.

Whitaker, J.C.: "Semiconductor Devices and Circuits," CRC Press, LLC., Boca Raton, FL, 1999.

Wiley, J.B. and R.B. Kaner: "Rapid Solid-State Precursor Synthesis of Materials," *Science*, 1093 (February 28, 1992).

Yablonovitch, E.: "The Chemistry of Solid-State Electronics," *Science*, 347 (October 20, 1989).

Yam, P.: "Atomic Turn-On: First Atom Switch," *Sci. Amer.*, 20 (November 1991).

SEMIDIAMETER CORRECTION (Sextant).   When the altitude of a celestial object that is close enough to the earth to present a finite disk (e.g., moon, sun, planet) is measured with reference to the visible sea horizon or to an artificial horizon, other than that contained in the bubble sextant, it is more convenient to use the upper or lower limb (edge) of the disk than to estimate the center of the object. When solving the astronomical triangle using this measured altitude, as in determining position at sea, the position of the object given in the almanac is that of the center. Hence, to obtain the observed altitude of the center, a correction must be applied known as the correction for semidiameter. The correction is to be added or subtracted from that obtained with the sextant, depending upon whether the lower or upper limb of the object is observed. The value of the semidiameter is given in the almanac for the given date of observation. In using the moon for accurate determination of position, an additional factor, known as augmentation, must be considered due to the fact that the distance of the object from the observer varies with the altitude of the object. See also **Sextant**.

SEMIPERMEABLE MEMBRANE (or Semipermeable Diaphragm).
A membrane or septum through which one (or more) of the substances composing a mixture or solution may pass, but not all.

In osmotic pressure determinations, semipermeable membranes permit the passage of a solvent but not of certain colloidal or dissolved substances. Many natural membranes are semipermeable, e.g., cell walls; other membranes may be made artificially, e.g., by precipitating copper cyanoferrate(II) in the interstices of a porous cup, the cup serving as a frame to give the membrane stability.

Semipermeable membranes are also used in the separation of gases. See Fig. 1. When a semipermeable membrane is placed in a gas mixture, being impermeable to gas 2 and allowing gas 1 to pass, the force exerted on it will equal the area times the partial pressure of gas 2 only. While there are no ideal semipermeable membranes for gases, there exist in practice reasonable approximations to them, such as incandescent platinum or palladium sheets, which can be penetrated by hydrogen but not by other gases. A film of water also acts as a semipermeable membrane for gases, since it is pervious to NH_3 or SO_2 because of their solubility in water, but gases which are not easily soluble are held back.

See also **Desalination**.

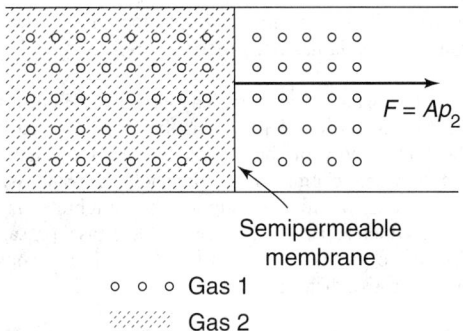

$$F = Ap_2$$

Semipermeable membrane

o o o Gas 1

Gas 2

Fig. 1.   Separation of gases by semipermeable membrane.

SEMITONE (Half-Step).   The interval between two sounds whose basic frequency ratio is approximately equal to the twelfth root of two. The interval, in equally tempered semitones, between any two frequencies, is 12 times the logarithm to the base 2 (or 39.86 times the logarithm to the base 10) of the frequency ratio.

SENSIBLE TEMPERATURE.   The temperature at which average indoor air of moderate humidity would induce, in a lightly clothed person, the same sensation of comfort as that induced by the actual environment. Sensible temperature depends on the air temperature; radiation from the sun, sky, and surrounding objects; relative humidity; and air motion. The wet-bulb temperature is often taken as an approximate measure.

SENSILLAE.   The sense organs of insects. The term is usually applied to the integumentary sense organs of the group but it is also extended to include the scolophores on which organs of hearing and chordotonal organs are based, and the ommatidia and retinulae of the eyes.

The sensillae of other kinds include some form of cuticular structure, often a projection, associated with a nerve ending and in some cases with gland cells. These organs include some of tactile function and chemoreceptors, both of taste and of smell. In form, their external parts are classed as six types. (1) Placoid sensillae end in a thin porous plate or membrane covering a canal. (2) Trichoid sensillae end with a slender seta. (3) Basiconic sensillae have a conical protuberance. (4) Styloconic sensillae have a fixed conical base bearing subordinate projections. (5) Coeloconic sensillae end with a depression containing a conical projection. (6) Ampulliform sensillae end with a slender projection in an expanded chamber at the inner end of a long tubule.

Tactile sensillae are distributed over the entire body but are often much more abundant on the legs and sensory appendages such as antennae and palpi. Organs of smell are often abundant on the antennae and in some species appear to be limited to these appendages. There is some possibility that they may occur on other parts of the body. Organs of taste are undoubtedly associated with the mouth parts, but they are supposed to be present in aquatic insects on the outer surface of the body as well.

SENSITIVITY (Instrument).   With reference to industrial and scientific instruments, the Instrument Society of America defines *sensitivity* as the ratio of a change in output magnitude to the change of input which causes it after the steady-state has been reached. Sensitivity is expressed as a ratio with the units of measurement of the two quantities stated. The ratio is constant over the range of a linear device. For a nonlinear device, the applicable input level must be stated.

Sensitivity has been used frequently to denote the *dead band*. However, its usage in this sense is deprecated since it is not in accord with accepted standard definitions of the term. See also **Dead Band (Instrument)**.

SENSITOMETRY.   The measurement of the light response characteristics of photographic film under specified conditions of exposure and development.

SENSOR (Measurement).   A device that detects or senses the value or change of value of a variable being measured. The first link in the measurement-system chain. In some cases, the sensor essentially comprises the total measurement system, as in the case of a liquid-in-glass thermometer where complete measurement is accomplished by calibrating a capillary column connected directly to the temperature-sensitive thermometer bulb. In other instances, the output of the sensor may be converted from one form of energy to another (for example, mechanical motion transduced to an electrical voltage or current) and amplified and conditioned one or more times before a useful output signal for display, recording, or control is obtained. There are hundreds of different kinds of sensors to detect changes, ranging over dozens of major variables. The term *sensor* is essentially synonymous with detector or primary element.

SENSORY ORGANS.   Structures in the animal body that are influenced by certain factors in the environment. Also known as receptors. The action of the environmental factor on the living substance is known as a stimulus. It results in the transmission of a nerve impulse to some nerve center and from this point may influence appropriate reactions of the animal or may be stored in memory. Special sense organs are found only in animals with nervous systems. Their development involves high specialization in some phase of the general property of living matter called irritability, and to some extent this property persists in all living tissue, whether nervous or sensory or not.

Stimuli arise from contacts with solid objects, from chemical compounds, either dissolved or in the gaseous state, from the incidence of light rays, and from factors which damage the body. It is known from various evidences that some animals perceive factors to which human organs are not sensitive, but as far as is known, the only stimulating factors are in the following groups.

Contact results in variable pressures to which organs of several kinds are sensitive. In vertebrates tactile corpuscles and other similar structures located in surface tissues may be classed as organs of touch. They are sensitive to simple pressure and give rise to images of form through the varying pressures due to uneven surfaces and gross contours. Since sound waves are due to rhythmic compression of the air, the ears and other auditory organs such as those of insects are also sensitive to pressures, but only to fluctuations of relatively high frequency (in man 30–30,000 per second). Between auditory organs and simple organs of touch are the lateral line organs of fishes and the chordotonal organs of insects, both related in some anatomical details to the auditory organs of the groups to which they belong. These organs are supposed to be sensitive to fluctuating pressures of lower than auditory frequency.

Tactile organs are also closely allied to sensory organs of insects, which apparently enable them to avoid obstacles when flying. Supposedly, these organs are sensitive to the changes of air pressure, often extremely delicate, resulting from approach to objects. They are located in the wings.

Dissolved substances stimulate organs of taste and gases or vapors act on organs of smell (olfactory organs). In the vertebrates the olfactory organs are associated with the nasal passages or occupy a similar position; nasal structures of fishes are limited to the olfactory function. Vertebrate organs of taste are known as taste buds and are located in the oral cavity principally, although aquatic forms may have them also in the skin. Sensory organs of this class in the invertebrates are extremely varied. Some are known as sensillae. In addition to organs of taste and smell, a general chemical sense is recognized. It is resident in various surface layers and is the least sensitive of the group. These sense organs are known collectively as chemoreceptors.

The varied integumentary sense organs of the human body are known to include some sensitive only to heat, cold, or pain. These organs may be the free nerve endings found in the skin. It has been suggested that pain may also result from overstimulation of other types of sense organs.

Sense organs that are stimulated by light are familiar to us in our own eyes. From this stage of complexity they range downward to simple light-sensitive cells. The transition includes organs capable of perceiving fluctuations of light, the direction from which it comes, and movements, as well as organs which form images in varying degrees of precision.

In contrast with sense organs of the kinds mentioned, which are classed as exteroreceptors, the body contains others called interoceptors. They are the source of sensations of hunger, thirst, nausea, and pain. Other interoceptors in the muscles, joints, and tendons are associated with the maintenance of equilibrium and are classed as proprioceptors. They are probably subject to varying pressure due to tension of muscles and shifting of the weight of the body. The semicircular canals of the inner ear of vertebrates are also organs of equilibrium.

Organs of many invertebrates, such as the tentaculocysts of jellyfishes, can be interpreted only by experimental evidence. By testing the animal with different stimuli definite conclusions can often be drawn from its reactions. In most cases there is evidence of functions like those of our own sense organs.

All sense organs consist of nerve endings associated with various specialized cells or tissues. The nerves are not limited to one type of stimulation but their response may be identical under various stimuli. Thus a mechanical shock to the eye produces a sensation of light. The nerve fibers leading from the sense organ toward the central system are sensory or afferent.

SENSORY RECEPTOR.   Specialized dendrites of certain neurons (sensory neurons; see also **Central and Peripheral Nervous Systems**), which are sensitive to some physical state such as stretch, temperature, or chemical environment, e.g., stretch receptor, cold receptor, etc. In the case of the visual, auditory, and olfactory systems the dendrites of the sensory nerve fiber form a synapse with a specialized sensory cell which is highly sensitive to the given physical agent, e.g., rods and cones of the retina are specifically sensitive to light.

Some receptors lie within the body (*interoceptors*) and some are on the surface (*exteroceptors*). It is the function of receptors to initiate sensory impulses.

SEPIOLITE.   The mineral sepiolite or meerschaum is soft, white, light in weight, and occurs in clay-like nodular masses. It is a complex, hydrous magnesium silicate corresponding to the formula $Mg_4Si_6O_{15}(OH)_2\cdot6H_2O$.

It crystallizes in the orthorhombic system; hardness, 2–2.5; specific gravity, 2; color, white, grayish white, sometimes a yellowish- or bluish-green; opaque. It is capable of floating on water, hence the name meerschaum or sea foam. It occurs in Asia Minor associated with serpentine and magnesite, and may be derived from the latter. Other deposits are in the Czech Republic and Slovakia, Morocco, and Spain; and in the United States in Pennsylvania and New Mexico. The name meerschaum is from the German. Sepiolite is from the Greek, meaning cuttlefish, referring to the similarity of the bone of that animal to the light, porous sepiolite. The material is used in the manufacture of smoking pipes.

SEPSIS.   A toxic condition of the human body as caused by the spread of bacteria from a focal infection to other portions of the body. See also **Septicemia**. *Sepsis neonatorum* is an invasive bacterial infection that occurs in the first week of life, accounting for 10–20% of neonatal deaths, particularly in premature, low-birth-weight infants.

SEPTARIAN STRUCTURE.   Mineralized irregular polyground joints or cracks in certain concretions. The structure resembles the pattern of cracks developed by desiccation of mud, and probably resulted from a similar cause—contraction due to desiccation of colloidal material.

SEPTICEMIA.   A condition wherein pathogenic organisms circulate in the blood, causing fever and other symptoms of their presence. A number of years ago, this condition was rather aptly called "blood poisoning," a term which purveyed the grave significance of septicemia prior to the appearance of antibiotics and other effectual therapies. Septicemia persists as a serious complication of numerous acute infections, as caused by hemolytic streptococci, staphylococci, pneumococci, meningococci, color bacilli, and other pathogens. Portals of entry of these pathogens into the blood include the lungs, middle ear, the mastoid process, the skin, and the genitourinary tract. Although there are no consistent signs of septicemia, the condition is suspected when, during the course of an infection, a patient develops unusual fever, hemorrhages into the skin (purpura) or joints, symptoms of endocarditis, jaundice, and widespread abscesses. In pneumonia, meningococcus meningitis, osteomyelitis, and puerperal or post-abortion infections, the physician is on the alert for the development of the signs of septicemia. Whenever the complication is suspected, a blood culture is made immediately.

Acute septicemia is a major manifestation of meningococcal disease caused by *Neisseria meningitidis*. In nongonococcal infection of the female genital tract, anaerobic bacteria of a type found in the intestinal tract are frequently the causative agents of septicemia. These infections may involve vulvovaginal abscesses, tubo-ovarian abscesses, and pelvic peritonitis. Similar microorganisms are present following gynecologic surgery, parturition, and abortion. A dreaded complication of septic abortion is clostridial myometritis with septicemia. Septicemia is sometimes a complication which may occur in patients on hemodialysis. See also **Kidney and Urinary Tract**.

Since the availability of antibiotics and other drugs, the former high mortality from septicemia has been markedly reduced.

SEPTUM.   A thin wall or partition. The term is applied to the radiating plates on the foot of the coral polyp, to the transverse partitions which subdivide the body cavity of the annelid worms into chambers, and to the partitions between chambers of the shell of Nautilus, among the invertebrates. Its most familiar use among the vertebrates is to designate the nasal septum which separates the right and left nasal passages, although it applies also to the partition between the right and left chambers of the heart and to numerous other structures.

SEQUENCE.   A set of quantities $s_1, s_2, \ldots, s_n, \ldots$, called elements, which can be arranged in an order so that when n is given, the nth member of the sequence s_n is completely specified. A relatively simple type of sequence is a progression. The elements are usually arranged by matching them up, one by one, with the positive integers $1, 2, 3, \ldots, n, \ldots$. A common symbol for a sequence is $\{S_n\}$. Let N be an arbitrary positive number. If it can be chosen so that $N \geq |s_n|$ for all absolute values of the members of a sequence, then the sequence is bounded. If there is at least one $|s_n| \geq N$ it is unbounded. This definition applies to an upper bound but a lower bound can be described in a similar way. A sequence is convergent if it has a limit; divergent otherwise. For example, the sequence of integers $1, 2, 3, \ldots, n, \ldots$ is divergent.

Examples of other sequences are series, finite or infinite; infinite products; continued fractions.

See also **Progression**.

SEQUENTIAL ANALYSIS. The analysis of material derived by a sequential method of sampling, that is to say, it is the data, not the analysis, which are sequential.

In sequential sampling the members are drawn one by one (or in groups) in order, and the results of the drawing at any stage decide whether sampling is to continue. The sample size is thus not fixed in advance but depends on the actual results and varies from one sample to another. The sampling terminates according to predetermined rules which are decided by the degree of precision required. See also **Sampling (Statistics)**.

SEQUESTERING AGENTS.   See **Chelates and Chelation**.

SEQUOIA.   See **Giant Sequoia**; and **Redwood (Coast)**.

SERANDITE.   The mineral serandite is a hydrated manganese-sodium silicate corresponding to the formula $Mn_2NaSi_3O_8(OH)$, crystallizing in the triclinic system, of pseudo-monoclinic character. Color, rose-red, pink; transparent; brittle, and uneven fracture. Prominent basal and prismatic cleavage; vitreous to pearly luster. Crystals thick tabular or prismatic, and as intergrown aggregates. Occurs as superb crystals in a carbonatite zone in a host body of nepheline-syenite in association with analcime, aegerine, and other rare minerals at Mt. St. Hilaire, Quebec, Canada. Its only other known world occurrence is on the Island of Rouma, Los Islands, Guinea.

SERIATE FABRIC.   A geological term proposed in 1906 by Cross, Iddings, Pirsson, and Washington, for the texture of an igneous rock whose granular crystals form a complete gradation in size.

SERIES.   An expression of the form $a_1 + a_2 + a_3 + \cdots + a_n + \cdots$ which may have a finite or an infinite number of terms. Its partial sums constitute a sequence $\{S_n\}$, where $s_1 = a_1, s_2 = a_1 + a_2, \ldots, s_n = \sum_{k=1}^{n} a_k$. If the number of terms in its partial sums is allowed to increase without limit, either: (a) s_n approaches a limit; (b) it does not approach a limit. In the first case, if $\lim_{n\to\infty} s_n = s$, then s is the sum of the convergent infinite series and one writes $\sum_{k=1}^{\infty} a_k = s$. If $\lim_{n\to\infty} s_n = \pm\infty$, the series is said to be definitely divergent to $\pm\infty$. It can also be indefinitely divergent, where M and m are upper and lower limits of the sequence and the series oscillates. A simple example is the alternating series, $S_n = 1 - 1 + 1 - 1 \pm \cdots + (-1)^{n-1}$, for its sum is either unity or zero, depending on whether n is even or odd.

Convergent series are generally the most useful type for practical applications (but see also **Asymptotic Series**), hence it is of great importance to test them for this property (see also **Convergence**).

Algebraic combination of series should not be made carelessly for it does not follow that the usual associative, distributive, and commutative laws of algebra will always hold. One important case is the Cauchy product. If

$$S_1 = \sum_{n=0}^{\infty} a_n \quad \text{and} \quad S_2 = \sum_{n=0}^{\infty} b_n$$

are absolutely convergent to the sums A and B, respectively, then the Cauchy product,

$$S_1 S_2 = \sum_{n=0}^{\infty} \sum_{i=0}^{\infty} a_i b_{n-1}$$

is absolutely convergent to the sum AB. This procedure is especially useful for power series, for if

$$y_1 = \sum_{n=0}^{\infty} a_n (x - x_0)^n \quad \text{and} \quad y_2 = \sum_{n=0}^{\infty} b_n (x - x_0)^n$$

then

$$z = y_1 y_2 = \sum_{n=0}^{\infty} \sum_{i=0}^{\infty} a_i b_{n-1} (x - x_0)^n$$

Similarly, a uniformly convergent series may be either differentiated or integrated, term by term. See also **Reversion of Series**.

SERIES MOTOR.   See **Motor (Electric)**.

SEROUS GLAND.   A gland that produces a watery secretion, in contrast with mucous glands, whose secretions are composed of or contain mucus. The term is used in connection with the salivary glands of vertebrates, which are partly serous and partly mixed. The serous gland cells are distinguished in part by their more granular cytoplasm and rounded nucleus, located near the middle of the cell.

SEROUS OTITIS MEDIA.   See **Hearing and the Ear**.

SEROWS.   See **Goats and Sheep**.

SERPENTINE.   This is a group name for minerals encompassing two principal polymorphic forms: *chrysotile* and *antigorite*. This monoclinic mineral of hydrous magnesium silicate composition $Mg_3Si_2O_5(OH)_4$ is essentially a product of metamorphic alteration of ultrabasic rocks rich in olivine, pyroxene, and amphibole. Serpentine crystals are unknown except as pseudomorphic replacements of other minerals, e.g., after clinochlore crystals at the Tilly Foster Mine, Brewster, New York, Antigorite occurs as platy masses; *chrysotile* as silky fibers. Most massive serpentine rocks are composed essentially of antigorite. The hardness is 2–5, specific gravity ranges from 2.2 (fibrous varieties) to 2.65 (massive varieties). Color usually mottled green. The name *serpentine* stems from the mottled character, somewhat resembling the skin of a serpent. There is a greasy to wax-like luster in massive material; silky in fibrous material. The minerals are translucent.

Chrysotile fibers are the source of commercial asbestos, although fibrous amphiboles also contribute to similar usage. Asbestos is economically valuable for its incombustibility and low conductivity of heat, thus as fireproofing and insulating material. See also **Asbestos**.

Chrysotile deposits of economic value are found in Quebec, Canada, in the former U.S.S.R., and in South Africa. Minor occurrences are found in the United States in Vermont, New York, New Jersey, and Arizona. *Verd antique* marble (serpentine marble) is quarried extensively near West Rutland, Vermont.

ELMER B. ROWLEY, F.M.S.A. Union College, Schenectady, NY.

SERVAL.   See **Cats**.

SERVICEBERRY.   See **Rose Family**.

SERVICE TREE.   See **Ash Trees**.

SERVOMECHANISM.   A closed-loop system which depends upon the feedback concept for operation. The terms *control system, regulator*, and *servomechanism* are often used interchangeably. There are some historical differences, but the distinctions frequently are fine. In terms of current usage, the terms servomechanism and regulator are gradually phasing out of the literature in deference to control system or controller.

Conventionally, for the servomechanism, it is assumed that the output or controlled variable is forced to be a preassigned function of the reference input, where the reference input is, in general, an arbitrary function of time. A typical example is a gun fire control system. Since the target is moving, the reference input (which could be a radar output signal of target position) must be variable. If the target is to be hit, gun position (controlled variable) must be forced to a preassigned function of the reference input. This characteristic of a servomechanism is contrasted with regulator operation in which the controlled variable is maintained substantially constant. Control of gas line pressure is an example of a regulator system. The reference input (desired value of line pressure) is a constant quantity. The system works to maintain the controlled variable (line pressure) at the desired value regardless of user load requirements. From an analytical standpoint, there is very little difference between the two systems.

A servomechanism also has been defined as a feedback control system in which one or more of the system signals represent mechanical motion. Thus, a servomechanism usually is used to control an output position mechanically in response to input signal changes.

Similarly, a regulator has been defined as a device that maintains a desired quantity at a predetermined value or varies it in accordance with a predetermined plan. A controller is aptly termed a regulator when it is relatively simple in design, easy to apply, relatively inexpensive, and has comparatively coarse performance characteristics. There are exceptions, of course, where manufacturers of sophisticated equipment elect to term their products regulators rather than controllers. Normally, a regulator does not incorporate means for transmitting or receiving signals from remote locations, indication, or recording. Usually a regulator is close-coupled to the equipment that it is controlling. Often, regulators are self-contained, self-actuating without requiring external energy sources. They depend upon energy derived from the equipment or system that they are regulating. A regulator in this category is logically termed a self-actuated regulator, or self-operated controller. See also **Self-Operated Controller**.

SERVOMOTORS. Electric motors are widely used in industry for effecting desired speeds and motions of machines, the positioning of parts or machines, and indirectly to control draw, thickness, stretch, and a number of other variables that must be controlled. Thus, servomotors are integral parts of automatic control and automation systems.

Servomotor and Servosystem Design Trends[1]

Whether it be *X-Y* or point-to-point positioning, or a constant or variable speed requirement, an electric motor provides precise motion control in a diverse group of products, ranging from simple conveyors to more complex machine tools and computer peripherals. The more complex systems utilize a 4-quadrant servo drive system in conjunction with the servomotor. With emphasis on increased industrial productivity and reliability, numerous advances in servotechnology have been made in the comparatively recent past. Advancements are leading to more effective use of the microprocessor in servo loop control. This article explores trends in both servomotors and servosystems, and how these trends relate to applications when a manufacturing firm upgrades its facilities to reflect the philosophy of modern control and automation technology.

Industrial Motors — A Perspective

The direct current (D.C.) motor was one of the first machines devised to convert electrical energy to mechanical power. Origin of the D.C. motor can be traced to machines conceived and tested by Michael Faraday.

Since D.C. motor speeds can easily be varied, they are utilized for applications where speed control, servo control, and/or positioning tasks exist. Most small motors used in industry are alternating current (A.C.) motors. These motors are relatively constant-speed devices and are applied where speed *control* is not required. The speed of an A.C. motor is determined by the frequency of the voltage applied (and the number of magnetic poles). See also **Motor (Electric)**.

Alternating Current (A.C.) Motors

There are basically two types of A.C. motors: (1) induction, and (2) synchronous. If the *induction motor* is viewed as a type of transformer, with the stator as the primary and the rotor as the secondary, it becomes easier to understand its operation. The currents that flow in the stator induce currents in the rotor, and two magnetic fields are set up. These two magnetic fields *interact* to produce *motion*. The speed of the magnetic field going around the stator will determine the speed of the rotor. The rotor will attempt to follow the stator's magnetic field, but will "slip" when a load is attached. Therefore, induction motors always rotate slower than the stator's rotating field. The *synchronous motor* is basically the same as the induction motor, but with slightly different rotor construction. The rotor is either (1) self-excited (same as induction), or (2) directly excited to set up the rotor field. The salient poles (or teeth) construction prevents slippage of the rotor field with respect to the stator field. Thus, this type of motor always rotates at the same speed (in synchronization) as the stator field. A single-phase A.C. motor is not self-starting. They employ a starting mechanism in order to start rotation — in the form of a start winding or a capacitor in a winding. Thus both motor types, induction and synchronous, utilize stators with rotating magnetic fields. They suffer from low starting torque, slow acceleration, and torque breakdown at overload.

Direct Current (D.C.) Motors

In a D.C. motor, the stator field can be set up by either permanent magnetics or a field winding. Thus, in contrast with the A.C. stator field which is rotating, the stator field is stationary. The second field, the rotor field, is set up by passing current through a commutator and into the rotor assembly.

[1] By John Mazurkiewicz, Manager, Application Engineering, Pacific Scientific, Rockford, Illinois.

The rotor field rotates in an effort to align itself with the stator field, but at the appropriate time (due to the commutator) the rotor field is switched. Thus, by this method the rotor field never catches up to the stator field. Rotational speed depends on the strength of the rotor field, i.e., the more voltage on the rotor, the faster the rotor will turn. Thus, the D.C. motor is straightforward — it has predictable speed-torque characteristics, and suffers none of the speed control problems associated with A.C. motors.

Brushless Motors

In recent years, there has been a trend to favor *brushless* motors. The brushless motor technology emerged in the 1930s, along with vacuum tube power technology, more sophisticated control systems, and the industrial needs for velocity control and position control of basic motors. The transistor became an efficient power-handling device in the 1950s when PWM (pulse width modulation) and PFM (pulse frequency modulation) techniques became practical. With subsequent developments of transistor circuits, analog operational amplifiers, low-cost logic components, memory arrays, and microprocessor chips, control systems became oriented toward the retention and processing of information and thus able to handle more complex calculations. In the 1970s, the development of new magnetic materials provided the opportunity to explore and to design innovation in terms of pulse-modulated D.C. motors. It was not until this rapid expansion of modern semiconductors and new magnetic materials, along with the requirements for upgrading existing products for the automation of industry, that brushless motor development and use began in earnest.

Motivations for considering the brushless motor include the desire for improved productivity, improved product requirements in terms of higher speed, greater acceleration/deceleration, reduced maintenance, reduced size, improved power-to-weight ratio, and increased reliability.

Operating Principle of Brushless Motors

A cross-sectional view of a brushless motor is shown in Fig. 1. Brushless motors are similar to A.C. motors in that a moving magnet field causes rotor movement or rotation. Both motor types use stator windings and have no brushes. Brushless motors are also similar to permanent magnet (PM) D.C. motors since they have linear characteristics. Also, both motor types use permanent magnets to generate one field. The brushless motor is, in essence, a *hybrid*, which combines the best attributes of both the A.C. and D.C. motors.

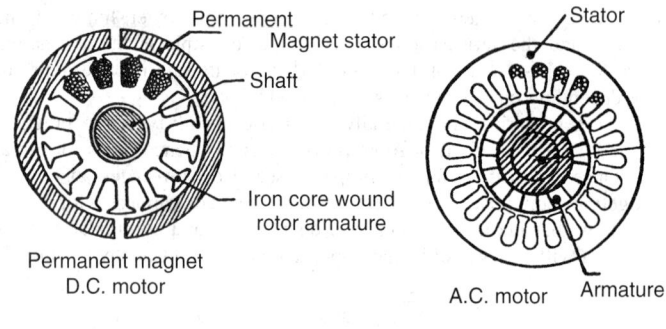

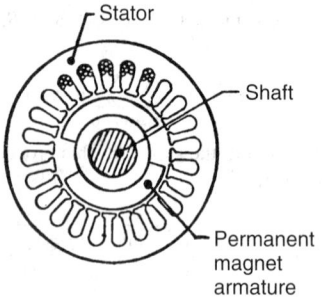

Fig. 1.   Cross section of various types of electric motors used in servosystems.

The configuration of the brushless motor most commonly used in contemporary systems is shown in Fig. 2(**a**). In this motor the rotor consists of permanent magnets and the stator consists of windings. These windings are termed "commutation" windings. By passing a current through a winding, a magnetic field is set up with which permanent magnets on the rotor interact. This results in rotation of the rotor. A representative family of modern brushless motors is shown in Fig. 2(**b**). Examples of brushless rotor assemblies are illustrated in Fig. 2(**c**).

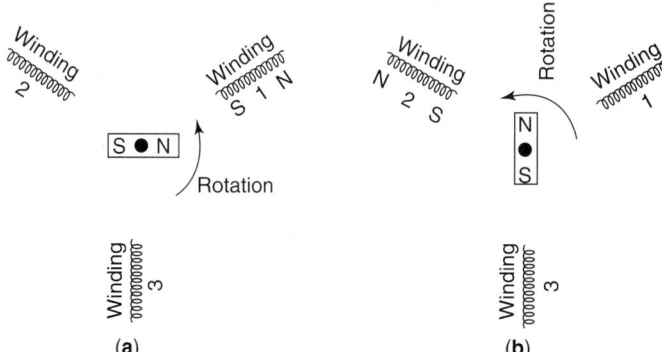

Fig. 3. Basic rotation of a brushless motor.

(**a**)

(**b**)

(**c**)

Fig. 2. Brushless motors: (**a**) cutaway view; (**b**) representative family of contemporary designs; (**c**) brushless rotor assemblies. (*Pacific Scientific, Rockford, Illinois.*)

Figure 3 illustrates, in simplified form, how rotation occurs. With a current passing through *winding 1* (Fig. 3(**a**)), a south pole is set up with which the permanent magnet will react and movement will begin. If, at the appropriate time, current is shut off in *winding 1* and turned on in *winding 2* (Fig. 3(**b**)), then the rotor will continue to move. By continuation of this timing sequence, complete rotation will occur as the rotor repeatedly tries to catch up to the stator magnetic field. In this example, the operation

is simplified for explanation by exciting only one winding at a time. In practical situations, two and sometimes three windings are energized at a time. This procedure permits the development of higher torques.

As indicated, if current is properly switched from winding to winding, the rotor will continue to rotate. Switching is accomplished in conjunction with a position sensor. Frequently, solid-state Hall-effect sensors are located on the shaft assembly. These extremely rapid sensors note the position of the shaft and provide an output signal. This output signal informs the motor's electronics when to switch current from winding to winding.

In comparing two motors that develop the same torque, the brushless type has advantages. Figure 4 illustrates this point by showing a locus of safe operating areas. The maximum speed is the maximum recommended top speed of the motor, determined by (1) commutator bar-to-bar breakdown voltage in a brush type motor, and (2) by mechanical centrifuge conditions in a brushless motor. The maximum temperature limit is determined by the motor's hot armature temperature. These are quite close inasmuch as the motors develop the same approximate stall torque. Operation above this line will result in the motor's armature temperature exceeding the recommended manufacturer's maximum limit.

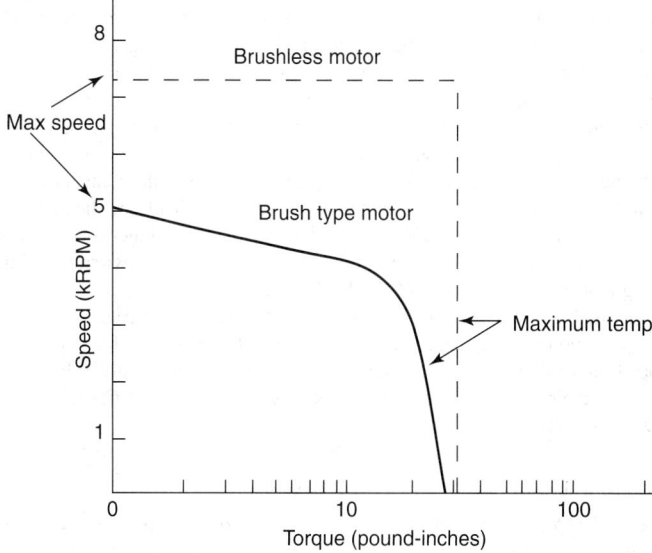

Fig. 4. Safe operating areas of two equivalent motors.

Control of Brushless Motors

Control of the brushless motor is accomplished by incorporating two additional output stages over the conventional servo control design. Figure 5 illustrates a simplified block diagram. *Basic operation*: Once the "run" command is given, the binary decoder compares outputs from the Hall sensors (which Halls are *on*). For example, if the shaft is sitting at "zero", input to the logic circuit (Hall sensor output) informs the logic that Hall sensors #1 and #3 are *on*. The binary decoder interprets these data and outputs a binary code. In this example, the output code from the switch

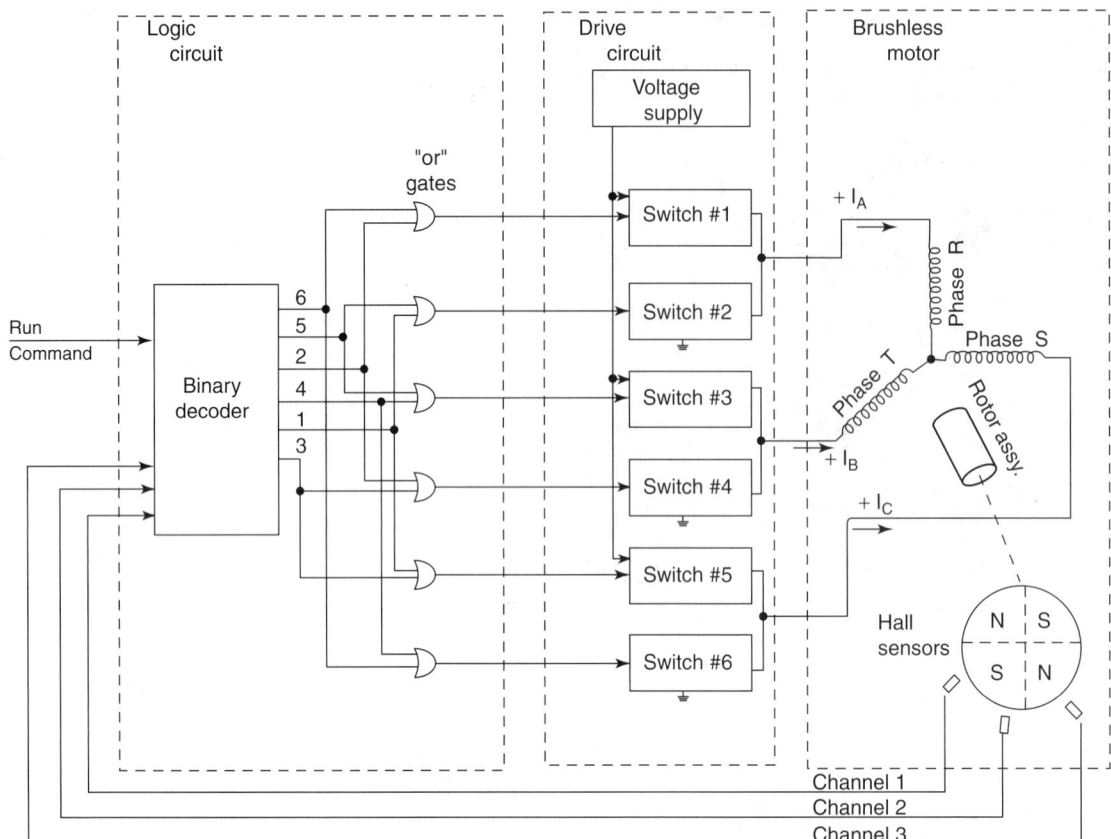

Fig. 5. Simplified block diagram of control system for brushless motor.

logic is a "5." This code will turn on appropriate OR gates, which turn on switches 3 and 2. The switches apply power from the voltage supply to motor windings (active legs T and R). As the motor rotates through 60°, the logic input will change state, as Hall #3 shuts off. The binary decoder output changes to a "1." This code will turn on appropriate OR gates, which turn on switches 2 and 5. (Note that during this transition, switch 2 has remained on, whereas switch 3 shuts off and switch 5 is turned on.) Again, the switches apply power to legs R and S. As the motor continues to rotate, the sequence continually progresses, changing current flow through the motor windings until complete rotation through 360° is attained, after which the sequence repeats. As current flow changes from winding to winding, the magnetic field also changes. In effect, the magnetic field is sequencing around the stator. The permanent magnet rotor will try to catch up to the field created, but never will—due to switching of the magnetic fields as a result of the Hall-effect sensor signals.

Thus, brushless rotation depends upon the stator field rotating (similar to A.C. motors). The significant difference is the presence of internal shaft position feedback in the brushless design. This element gives brushless motors their linear and predictable speed-torque characteristics (similar to D.C. motors).

Microprocessors in Servo Control Systems

Several types of controls have been developed over the years for D.C. motors, including SCR (silicon-controller rectifier), linear, pulse-width modulation, among others. These controls have served the needs of a diverse group of applications and generally they are basic and simple. The D.C. approach was chosen because no economically equivalent A.C. package existed. The A.C. motor manufacturers traditionally ignored this market. However, after the rapid escalation of energy costs in the 1970s, the motor industry perspective changed. Controls for A.C. motors quickly assumed a new level of sophistication. Two basic types of control emerged and are currently available: (1) the six step, and (2) pulse-width modulation. In development are more sophisticated controllers that employ SCR microprocessors and large scale integrated (LSI) dedicated electronics. In these cases, the control converts 60 Hz to direct current and then synthesizes a sine wave at a frequency to produce the desired speed.

The traditional cumbersome approach with discrete components would be unacceptable when using higher technology involving brushless designs. A search for a new approach began and, in light of the microprocessor, with its power and flexibility and its increased use in numerous applications, the concept of servosystem design was revolutionized. This approach has succeeded in providing greater flexibility in design of new systems and has potential to enhance the capabilities of existing systems. This could be termed the "intelligent" approach that utilizes microprocessor technology—an approach that impacts very favorably on design time, setup time, and implementation time.

Adjustable-Speed Brushless Control. A control of this type is self-contained, including power supply and heat sink, with the objective of driving a permanent-magnet brushless motor with Hall sensor feedback. The microprocessor-based control operates by energizing two of the motor's three windings, and switches power from winding to winding according to the feedback from the Hall sensors as previously described. This is coupled with pulse-width modulation drive techniques to make effective use of the output power transistors.

A user-friendly operating panel activates the control, while status lights provide easy readout. Also on the front panel is a speed adjustment pot, and a digital speed readout indicator. The tight speed regulation offered by the microprocessor approach improves system accuracy. Accuracy of ±5 RPM over speed ranges of 100 to 10,000 RPM is possible even with dynamic load variations of 50%. The digital readout also serves as a diagnostic indicator should any of the system's protection require activation to shut down the system. This provides simple, easy-to-use user interfacing.

The Servosystem (Servocontroller) Approach. This is used with brushless technology in the same basic manner as it is employed with other prime mover (motor) technologies. Traditionally, these closed-loop servosystem designs will involve determination of load conditions and velocity profile, then prime mover (motor) selection, and determination of amplifier requirements. Following is the tedious task of compensation for gain and bandwidth adjustments for accuracy and response. The latter may involve a paperwork analysis prior to breadboarding a prototype. Or, if the individual components are purchased from independent suppliers, the "tweeking" of potentiometers would begin—in an effort to set up and adjust the compensation values. This approach can be difficult. As an example, in some servo amplifiers, there are ten potentiometers for a variety of adjustments. Throughout the compensation process, the main consideration is to set up the compensation for a given load condition. If the load changes, then the entire process must be repeated. To alleviate this problem, the system is designed to accommodate worst case conditions. This results in a system design that will be overdamped for all conditions (except the worst case), thus severely compromising system performance (speed and accuracy).

A much less cumbersome and effective approach is to utilize microprocessor techniques—an approach that uses digital control and digital filtering. The "intelligent" system compensates by simply inputting or programming the controller with parameters that describe the servo loop parameters. The algorithms precisely control servo loop velocity. Execution is under microprocessor control and all servo loop compensation (motor, load, or environmental conditions) is monitored and controlled should changes occur. This simplifies design and gains flexibility for the system.

One approach is the use of a general-purpose microprocessor-based control as shown in Fig. 6. The advantages of this approach include fewer components needed, which in turn lowers cost, reduces circuitry, and improves reliability. Components must be added to complete the system design, such as a drive amplifier. However, there are microprocessor-based controls, which include an internal drive amplifier, available today in production quantities to drive a brushless motor. These devices are termed "intelligent servo drives" as shown in the block diagram of Fig. 7.

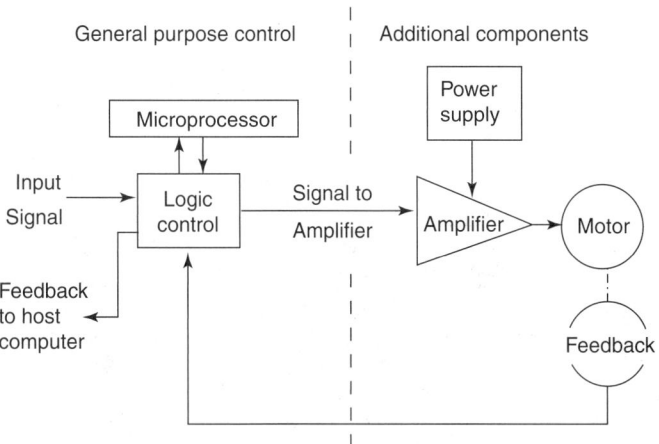

Fig. 6. A general-purpose microprocessor-based control system.

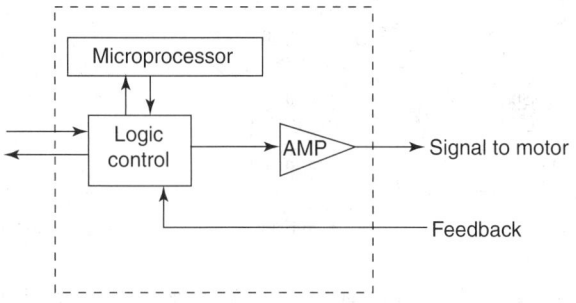

Fig. 7. Example of intelligent servo drive.

These brushless servocontrols are software compensated, thus eliminating the need for pot adjustments. The microprocessor-based control can accept a variety of inputs—either analog or digital (serial or parallel) velocity commands. Parameters are factory-loaded in nonvolatile EEPROM (electrically erasable/programmable read-only memory) so the engineer receives a stable system, thus reducing design, implementation, and setup time. Although most applications do not require further adjustments, the engineer may fine tune the system by way of a hand-held pendant, or any RS232C interface. The versatility of the microprocessor allows

incorporation of self-protection, which makes the overall system virtually indestructible. Protection features include peak, RMS, and short-circuit protection, as well as thermal protection, hardware and software protection, loss of feedback and velocity error. A brushless resolver serves as the feedback device and replaces the Hall sensor feedback. This assures operation at the optimum phase angle. The entire unit controls the brushless motor by way of PWM sinusoidal driving function, thus allowing smoother operation even at very low speeds.

The "intelligent control" closes the velocity loop and directly interfaces with most readily available programmable motion controllers for position control. Basic system operation is shown in Fig. 8. The first function is to receive instruction from the motion controller and generate the speed command for the velocity loop. The algorithm looks at the *present command* of the motor, the *previous command*, and the *next instantaneous position*. This is a periodic *sample* and *compare* to a desired reference value. The difference between the speeds at two time periods serves as an indication for velocity errors and is used for velocity corrections. The feedback signals from the brushless resolver are conditioned and routed to the CPU (central processing unit) through the interrupt control. All real-time inputs have interrupt-driven software. A control manipulation is calculated and subsequently used to command the drive system. The algorithm generates a command that is the difference of a term proportional to the velocity error and a term that is proportional to the integral of the velocity error. This velocity feedforward technique, included in the microprocessor servo reference generation circuitry and applied directly to the velocity loop, allows the controller system to operate with minimal error, even during hard acceleration and deceleration. Velocity feedforwarding is valuable for maintaining wide dynamic system response. This technique stabilizes the loop and allows the system to drive the brushless motor in a smooth manner regardless of the trajectory.

The microprocessor must also determine the sign and magnitude of the current/voltage for each of the brushless motor's three phases in order to drive the motor at the appropriate torque/speed. The function of the waveform is described by a sine wave. Since the three phases are shifted by 120°, and there are four poles in the motor, or two electrical cycles per revolution, the commands to the windings are:

Phase 1 = Sin $2\omega t$
Phase 2 = Sin $(2\omega t + 120°)$
Phase 3 = Sin $(2\omega t + 240°)$

Since this calculation is accomplished only once per sample period, the appropriate weighting factor can vary considerably from the first computation time to the next. The solution is to base the weighting factors on a period base at one-half sample later. This is a velocity lead on the commutation. To optimize this, the sine functions are stored in memory. For this scheme to work, the microprocessor must "know" the relationship between the brushless motor phases and the internal drive scheme. This is accomplished by the resolver feedback. The resolver interface consists of sine and cosine reference waveforms and a phase-shifted feedback which contains absolute position data. This signal is sent directly to the microprocessor through a resolver-to-digital converter, returning a velocity per sampling period. This feedback has proven superior for improvement of dynamics and straightforwarding for controllability.

The intelligent digital microprocessor-based control system, coupled with brushless motors, constitutes an excellent solution for most applications. This combination has the ability to produce higher torque at higher speeds. The numerous specific advantages have been previously described in detail.

Solid-State Variable-Speed Drives[2]

The past two decades have seen rapid growth in the availability and usage of solid-state variable speed drives. Today, there is a profusion of types which are suitable for virtually every type of electrical machine from the subfractional to the multithousand horsepower rating. See Fig. 9. Despite the diversity, there are two common properties of these drives. (1) All of them accept commonly available A.C. input power of fixed voltage and frequency and through switching power conversion, create an output of suitable characteristics to operate a particular type of electric

[2] By Richard H. Osman, Engineering Manager, A.C. Drives, Robicon Corporation (A Barber-Colman Company), Pittsburgh, Pennsylvania.

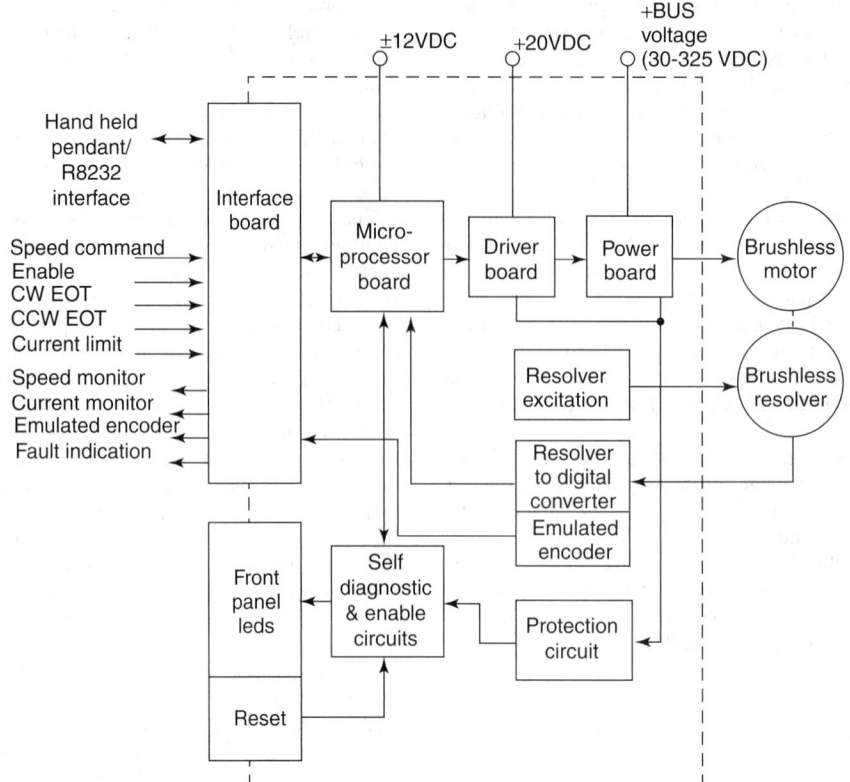

Fig. 8. Basic system operation of an intelligent brushless servo control. (CW = clockwise; CCW = counterclockwise; EDT = end of travel.) (*Pacific Scientific, Rockford, Illinois.*)

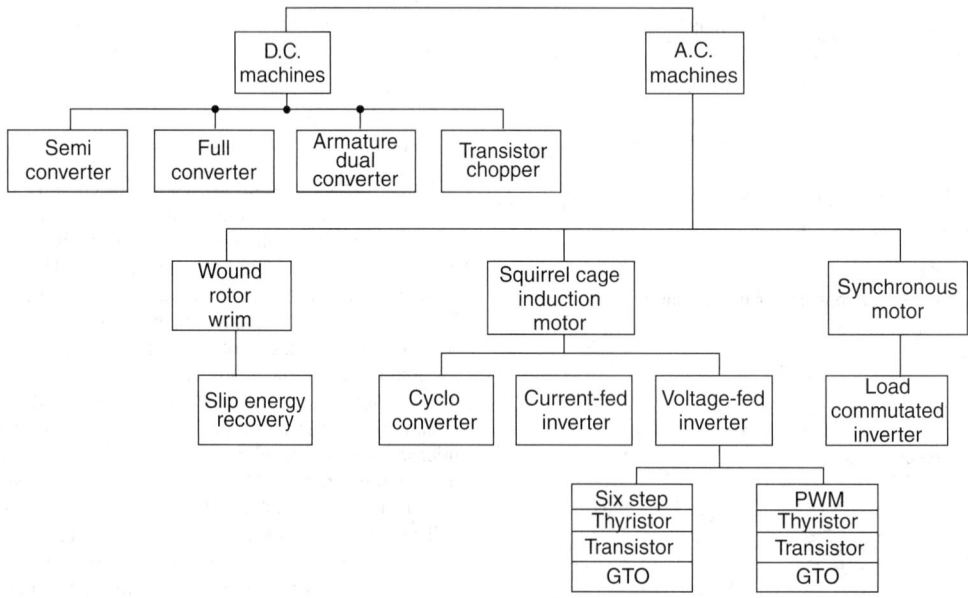

Fig. 9. General-purpose solid-state variable speed drive. (WRIM = wound rotor induction motor; GTO = gate-turn-off thyristors; PWM = pulse-width modulation.)

machine, i.e., they are *machine specific*. (2) All of them are based on solid-state switching devices. Even though many of the power conversion principles have been known as long as fifty years, when they were developed using mercury arc rectifiers, it was not until the invention of the thyristor in 1957 that variable speed drives became practical.

The thyristor (SCR) is a four-layer semiconductor device, which has many of the properties of an ideal switch. It has low leakage current in the off-state, a small voltage drop in the on-state, and takes only a small signal to initiate conduction (power gains of over 10^6 are common). When applied properly, the thyristor will last indefinitely. After its introduction, the current and voltage ratings increased rapidly. Today it has substantially higher power capability than any other solid-state device, and dominates power conversion in the medium and higher power ranges. The major

drawback of the thyristor is that it cannot be turned off by a gate signal, but the anode current must be interrupted in order for it to regain the blocking state. The inconvenience of having to commutate the thyristor in its anode circuit at a rather high energy level has encouraged the development of other related devices as power switches.

Development of Power-Switching Devices

Transistors predate thyristors, but their use as high-power switches was relatively restricted (compared to thyristors) until the ratings reached 50 A and 1,000 V in the same device, since the early 1980s. These devices are three-layer semiconductors which exhibit linear behavior, but are used only in saturation or cut off. In order to reduce the base drive requirements, most transistors used in variable speed drives are Darlington types. Even so, they have higher conduction losses and greater drive power requirements than thyristors. Nevertheless, because they can be turned on or off quickly via base signals, they are attractive candidates for drives within the scope of the transistor ratings, particularly for pulse-width-modulated inverters.

Gate-Turn-Off Thyristors. More recently, successful attempts to modify thyristors to permit them to be turned off by a gate signal have been made. These devices are four-layer types and are called *gate-turn-off thyristors*, or simply GTO's. Power GTO's have been around since at least 1965, but only relatively recently (1981) have devices rated more than a few tens of amps become available. Present GTO's have about the same forward drop as a Darlington transistor (twice that of a conventional thyristor). GTO's require a much more powerful gate drive, particularly for turn-off, but the lack of external commutation circuit requirements makes them desirable for inverter use. GTO's are available at higher voltage and current ratings than power transistors. Unlike transistors, once a GTO has been turned on or off with a gate pulse, it is not necessary to continue the gate signal due to the internal positive feedback mechanism inherent in four-layer devices.

Technological Base. The three aforementioned devices (thyristor, transistor, and GTO) form the technological base on which the solid-state variable-speed drive industry rests today. There are other devices in various stages of development which may or may not become significant depending on their cost and availability in large current (>50 A) and high voltage (1,000 V) ratings. These include: (1) the metal oxide semiconductor field effect transistor (MOSFET); (2) the insulated gate transistor; and (3) the static induction thyristor.

It has not yet been possible to construct *power MOSFET's* which have acceptably low ON resistance while still having the 1,000 V rating necessary for reliable power conversion at the 500 VAC level. Therefore, their use has been limited to small drives (under 10 HP). Power MOSFET's are the fastest power switching devices (100 ns) of the lot, and they also have very high gate impedance, thus greatly reducing the cost of drive circuits.

The *insulated gate transistor* is a combination of a power bipolar transistor and a MOSFET which combines the best properties of both devices. A most attractive feature is the very high input impedance, which permits them to be driven directly from lower power logic sources. Unfortunately, their ratings are not very impressive, being limited to a few tens of amperes and about 600 volts.

Static induction thyristors (SR's) are claimed to have the voltage and current ratings of GTO's, but with a much higher gate impedance so as to reduce driver requirements. The validity of these claims has not been proved in commercial use because SR's are just emerging from the development laboratory.

Variable-Speed Drive Hardware Development

Parallel to the development of power switching devices, there have been very significant advances in hardware for controlling variable speed drives. These controls are a mixture of analog and digital signal processing.

The advent of integrated circuit operational amplifiers and integrated circuit logic families made possible dramatic reductions in the size and cost of the drive control, while permitting more sophisticated and complex control algorithms without a reliability penalty. These developments occurred during the 1965–1975 period. Further consolidation of the control circuits occurred after that as large scale integrated circuits (LSI) became available. In fact, the pulse-width modulation (PMW) control technique was not practical until the appearance of LSI circuits because of the immense amount of combinational logic required. A significant trend has been the introduction of *microprocessors* into drive control circuits. While it is doubtful whether microprocessors will significantly reduce control circuit cost, there is general agreement that they are greatly expanding the capability of drive controls. The performance enhancements include: (1) more elaborate and detailed *diagnostics* due to the ability to store data relating to drive internal variables, such as current, speed, firing angle, and so on; (2) the ability to *communicate both ways* with user's central computers about drive status; (3) the ability to make *drive tuning adjustments* via keypads with parameters such as loop gains, ramp rates, current limit stored in memory rather than pot settings; (4) *self-tuning* drive controls; and (5) more adept techniques to overcome *power circuit nonlinearities*. The possibilities are large and are just beginning to become commercial realities.

D.C. Drives

The introduction of the thyristor had the most immediate impact in the D.C. drive area. Ward-Leonard (motor-generator) variable speed drives were quickly supplanted by thyristor D.C. drives of the type shown in Fig. 10 for reasons of lower cost, higher efficiency, and lower maintenance cost. This type of power circuit with three thyristors and three diodes in a three-phase bridge is generally referred to as a *semiconverter*. By phase control of the thyristors, it behaves as a programmable voltage source. Therefore, speed variation is obtained by adjustment of the armature voltage of the D.C. machine. Because the phase control is fast and precise, critical features like current limit are easily obtained. In fact, almost all thyristor DC drives today are configured as current regulators with a speed or voltage outer loop. The semiconverter is suitable for one-quadrant drives as it produces only one direction of current and output voltage. Input power factor is dependent on speed.

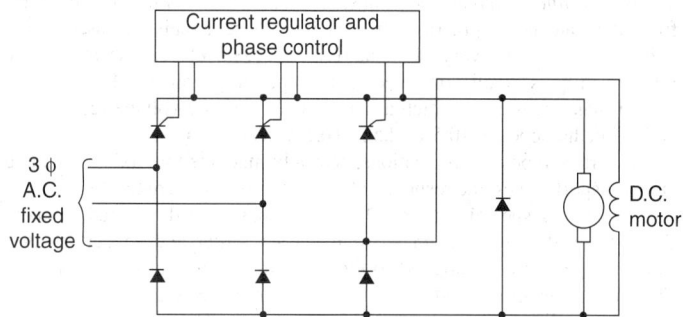

Fig. 10.   Thyristor D.C. drive-3-phase semiconverter.

Six-Thyristor Full Converter. As the cost differential between thyristors and diodes narrowed, the semiconverter was largely displaced by the six-thyristor full converter as shown in Fig. 11. This circuit arrangement (the Graetz circuit) has become the workhorse of the electrical variable speed drive industry as will be pointed out. The control techniques are very much the same as in the semiconverter. However, the full converter offers lower output ripple and the ability to regenerate, or return energy to the A.C. line. The system can be made into a four-quadrant drive by the addition of a bidirectional field controller. Torque direction is determined by field current direction. Due to the large field inductance, torque reversals are fairly slow (100–500 ms) but adequate for many applications.

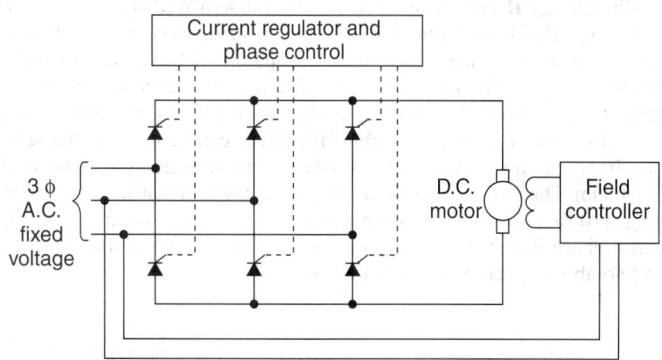

Fig. 11.   Thyristor D.C. drive-3 phase full converter.

Dual Armature Converter. For the best response of thyristor D.C. drives, the dual armature converter of Fig. 12 is preferred. This is simply two converters. See Fig. 11, connected back to back. Torque direction is determined by the direction of armature current, and since this is a low-inductance circuit, reversal can be accomplished in 10 ms (typically). Obviously, only one converter is conducting at one time with the other group of six thyristors not being gated. This is called "*bank selection.*"

Summary of Thyristor D.C. Drives. The three types of thyristor D.C. drives just described all share a common property in that the devices are turned off by the natural polarity reversal of the input line. This is called

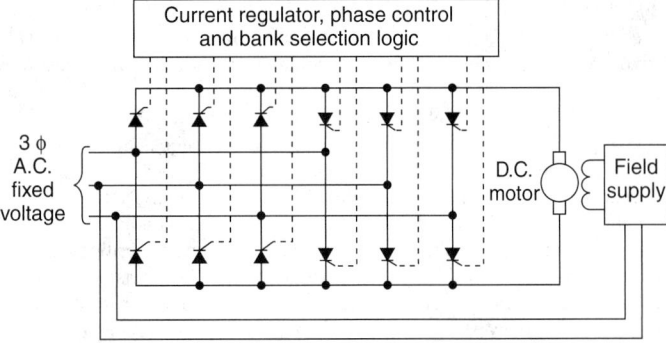

Fig. 12.   Thyristor D.C. drive-armature dual converter.

natural or *line commutation*. Thus, the inability to turn off a thyristor from the gate is no practical drawback in these circuits. Consequently, they are simple and very efficient (typically 98.5%) because the device forward drop is small compared to the operating voltage. These drives can be manufactured to match a D.C. machine of any voltage (commonly 500 V) or horsepower (0.5 to 2,500 HP, typically).

For certain types of applications, typically machine tool axis drives and tape transport drives, the response of phase controlled thyristor drives is not fast enough. A special class of D.C. drives has been developed. A fixed D.C. bus is developed from the line via a rectifier and capacitor filter. This voltage source is applied to the armature through power transistors. The voltage is modulated by duty cycle (or pulse width) control. The devices usually operate in the 1–5 kHz range. These specialty drives usually operate from 120 or 240 VAC and rarely exceed 10 HP. Frequently, they are applied with permanent magnet field D.C. machines.

A.C. Drives

The impact of the new solid-state switching devices was even larger on the A.C. variable speed drives, but it occurred somewhat later in time as compared to D.C. drives. A.C. drives are machine specific and more complex than D.C. drives. Solid-state variable-speed drives have been developed and marketed for wound-rotor induction motors (WRIM's), synchronous motors, and cage-type induction motors.

Historically, WRIM-based variable speed drives were commonly in use long before solid-state electronics. These drives operate on the principle of deliberately creating high-slip conditions in the machine and then disposing of the large rotor power which results. This is done by varying the resistance seen by the rotor windings.

Slip-Energy Recovery System. A more modern WRIM drive is shown in Fig. 13. This is called the *slip-energy recovery system* or *static Kramer drive*. The output of the rotor is rectified, and this D.C. voltage is coupled to the line via a thyristor converter. The line commutated converter is current regulated which effectively controls torque. Efficient, stepless speed control results. Very large (>1,000 HP) drives can be built, as the stator may be wound for medium voltage, while the rotor operates at 400–400 V maximum. The power conversion equipment may be downsized if a narrow range (100% to 70%) of control is adequate, for example in fan drives. The performance drawbacks are a poor system power factor if not corrected; and no above-synchronous-speed operation.

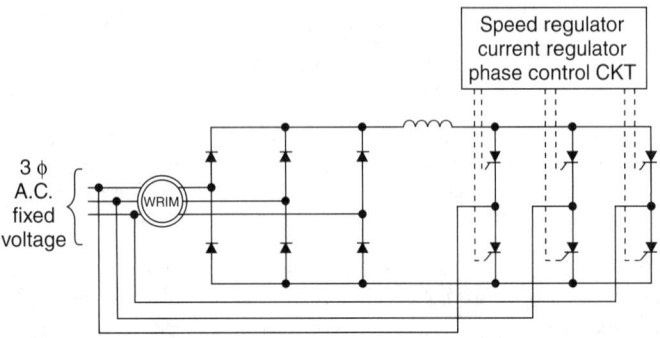

Fig. 13.   Slip energy recovery system-wound rotor inductor motor. (WRIM = wound rotor induction motor; CKT = circuit.)

The WRIM is the most expensive A.C. machine. This has made WRIM-based variable speed drives noncompetitive as compared to cage induction motor (IM) drives or load commutated inverters using synchronous machines. It appears that the WRIM will become a casualty of the tremendous progress in A.C. variable speed drives as applied to cage induction motors.

Load Commutated Inverter. As shown in Fig. 14, the load commutated inverter is based on a synchronous machine. It uses two thyristor bridges, one on the line side and one on the machine side. All devices are naturally commutated, because the back EMF of the machine commutates the load side converter. This requires the machine to operate with a leading power factor and, therefore, it requires substantially more field excitation and a special exciter compared to a normally applied synchronous motor. This also results in a reduction in the torque for a given current. The machine side devices are fired in exact synchronism with the rotation of the machine, so as to maintain constant torque angle and constant commutation margin. This is done either by rotor position feedback, or by phase control circuits driven by the machine terminal voltage. The line side converter is current regulated to control torque. A choke is used between converters to smooth the link current. Load commutated inverters (LCI's) came into commercial use about 1980, and are used mainly on very large drives (1,000–10,000 HP). At these power levels, multiple series devices are employed (typically 2.4 to 4 kV input) and conversion takes place directly at 2.4 or 4 kV or higher. The efficiency is excellent and reliability has been very good. Although they are capable of regeneration, LCI's are rarely used in 4-quadrant applications because of the difficulty in commutating at very low speeds where the machine voltage is negligible. Operation above line frequency is straightforward.

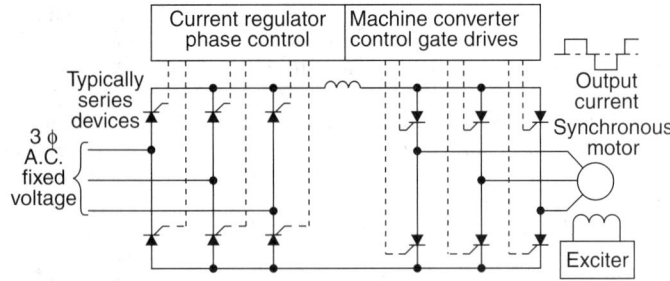

Fig. 14.   Load-commutated inverter (LCI).

Induction Motor Variable Speed Drives

Induction motor variable speed drives have the greatest diversity of power circuits. (See Fig. 9.) Because the squirrel cage induction motor is the least expensive, least complex, and most rugged electric machine, great effort has gone into drive development to exploit the machine's superior qualities. Due to its very simplicity, it is the least amenable to variable speed operation. Since it has only one electrical input port, the drive must control flux and torque simultaneously through this single input. As there is no access to the rotor, the power dissipation there raises its temperature—so very low-slip operation is essential.

Cycloconverter. One approach in an IM drive is to "synthesize" an A.C. voltage waveform from sections of the input voltage. This can be done with three dual converters and the circuit is called a *cycloconverter*. See Fig. 15. The output voltage is rich in harmonics, but of sufficient quality for IM drives as long as the output frequency does not exceed $\frac{1}{3}$ to $\frac{1}{2}$ of the input. The thyristors are line commutated, but there are 36 of them. (Sometimes half-wave circuits are used which need only 18 devices, but more harmonics result.) The cycloconverter is capable of heavy overloads and 4-quadrant operation, but it has a limited output frequency and poor input power factor. For special low-speed high horsepower (>1,000) applications, such as cement-kiln drives, the cycloconverter has been used.

Autosequentially Commutated Current-Fed Inverter. Still another approach to an IM drive is to generate a smooth D.C. current and feed that into different parts of windings of the machine so as to create a discretely rotating magnetomotive force (MMF). This type of inverter is called the *autosequentially commutated current fed-inverter* (ASCI). See Fig. 16. This circuit was invented later than other inverters and is much more popular in Europe and Japan than in the United States. Once

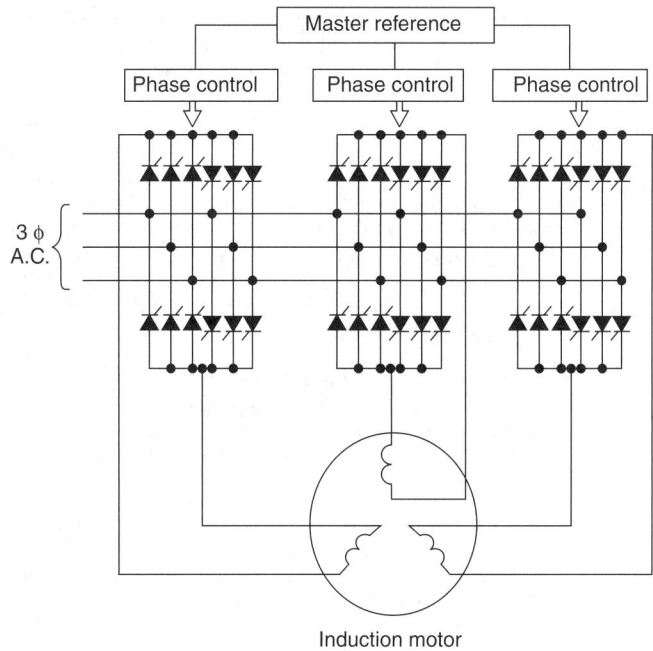

Fig. 15. Cycloconverter induction motor drive.

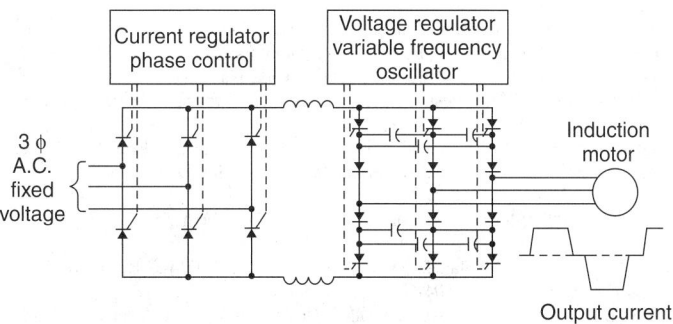

Fig. 16. Autosequentially commutated current-fed inverter (ASCI).

again, the input stage is a three-phase thyristor bridge which is current regulated. A link choke smooths the current going to the output stage. There, a thyristor bridge distributes the current into the motor windings with the same switching function as the input bridge, except at variable frequency. (Notice the similarity to the LCI.) The current waveform is a quasi-square wave whose frequency is set by the output switching rate and whose amplitude is controlled by the current regulator. The capacitors and rectifiers are used to store energy to commutate the thyristors, since the induction motor cannot provide this energy and remain magnetized, in contrast to the synchronous motor. This type of drive has simplicity, good efficiency (95%), excellent reliability, and four-quadrant operation up to about 120 Hz. Harmonics in the output current are reasonably low, giving a form factor of 1.05 (same as the LCI).

Moreover, harmonic currents are not machine dependent and decrease at light load. The input power factor is load and speed dependent, but much better than the cycloconverter. Above 100 HP, the ASCI is very cost effective. Because they are constructed with SCR's, they have recently (1984) become available at 2.4 and 4 kV direct conversion for very large (>1,000 HP) units. It is almost always possible to retrofit an existing motor with this type of drive. Due to the controlled current properties, this drive is virtually immune to damage from ground faults, load shorts and commutation failures.

Since MMF (current) is directly controlled and the drive is regenerative, ASCI's can readily be equipped with field oriented controls for the most demanding four-quadrant operation.

Field-Oriented Controls. Special mention should be made of A.C. induction motor drives with field oriented controls. They are the state of the art. The control technique keeps track of the flux and MMF vectors inside the machine in order to provide a fast and precise torque response

to an external reference. In addition, they are 4-quadrant drives, capable of producing either direction of torque in either direction of rotation. They are the A.C. drives of choice for the most demanding applications such as traction drives and machine tool axis drives.

Voltage-Source Inverter. The third approach to IM drives is to generate a smooth D.C. voltage and apply that to different combinations of the machine windings so as to create a rotating flux. This circuit is called a *voltage-source inverter* (VSI). An implementation using thyristors is shown in Fig. 17. This circuit was the first application of thyristors to IM drives. The input is a 3-phase thyristor bridge which feeds a capacitor filter bank forming a controlled low-impedance voltage source. The output stage consists of six main thyristors (1–6) in a bridge with antiparallel diodes. There are six auxiliary or commutating thyristors (11–16) which together with the L-C circuits, impulse commutate the main devices. The output waveform is a quasi-square wave of voltage whose amplitude is set by the D.C. link voltage. Here the output frequency is determined by the output switching rate, and the output voltage is set by the voltage regulator on the input converter. In order to reduce the size of the commutating L-C, special thyristors with fast turn-off times are required — in contrast to the ordinary phase control types used in D.C. drives, LCI's, and ASCI's. Despite the complexity, these drives have had a reasonably good reliability record. They have good efficiency (typically 95% at full speed, full load), speed dependent power factor, and can operate at very high frequencies (180 Hz and up). Regeneration to the line is not possible. Many of these units are in service, and they are still available from 50 to 55 HP, typically at 460 VAC. They are not available at over 600 V.

Transistors and GTO's in Voltage-Source Inverter. In order to reduce the cost and complexity of the VSI shown in Fig. 17, the thyristors and their commutation circuits have been replaced with transistors or with GTO's. The resulting circuit is shown in Fig. 18. The performance features are about the same as the thyristor version, but size and weight are substantially reduced. Although the conduction losses are higher due to the higher "on" voltages, commutation losses are reduced substantially — so efficiencies remain in the 95% range. The transistor version has been available since about 1982 at 460 VAC input; 230-volt units have been available since the mid-1970s. Presently, 100 HP transistor drives with single output devices are available; up to 300 HP with parallel output devices can be obtained.

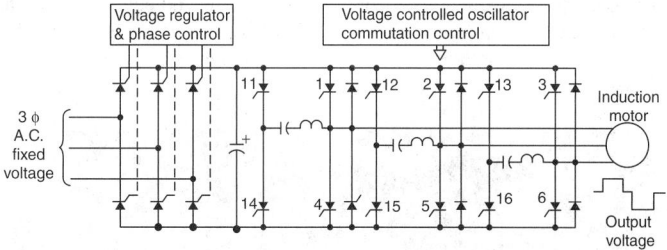

Fig. 17. Voltage-source inverter; six-step thyristor, impulse commutated.

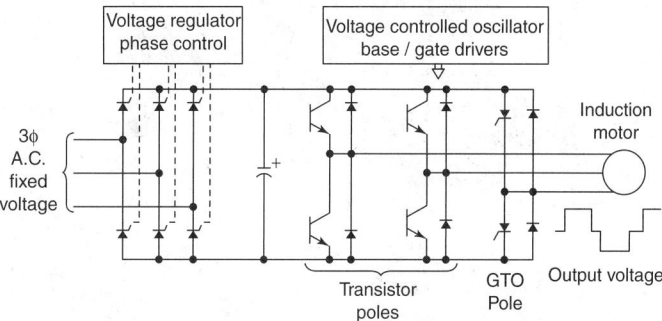

Fig. 18. Voltage-source inverter, six-step transistor or gage-turn-off thyristor (GTO).

Since GTO's have somewhat higher ratings, drives from 50 to 500 HP at 460 VAC are now available. It is difficult to forecast which device will be more successful, but transistors are much more widely used up to

100 HP than GTO's. There will be a three-way competition in the future among GTO, transistor, and thyristor drives in the 100–500 HP range to capture the bulk of the market. GTO and transistor costs will have to be substantially reduced to challenge thyristor designs at the upper end of this range.

Pulse-Width Modulated (PWM) Drives. The induction motor drives discussed thus far are all similar in that the amplitude of the output is controlled by the input converter. Another category of voltage source inverters controls both the frequency and amplitude by the output switches alone. A representative circuit based on transistors is shown in Fig. 19. Note that the input converter is replaced with a diode bridge so that the D.C. link operates at a fixed, unregulated voltage. The diode front end gives virtually unity power factor, independent of load and speed. This type of drive is called *pulse-width-modulated* (PWM).

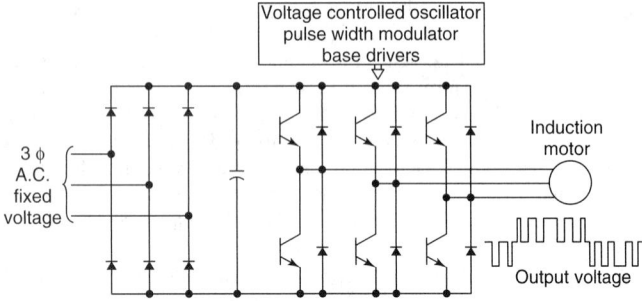

Fig. 19.   Pulse-width modulated (PWM) inverter-transistor implementation.

An output voltage waveform is synthesized from constant amplitude, variable-width pulses at a high (1–3 kHz) frequency so that a sinusoidal output is simulated; the lower harmonics (5,7,11,13,17,19, ...) in six-step waveforms are not present in sophisticated PWM's. One advantage is smooth torque, low harmonic currents, and no cogging. Some PWM designs merely encode the six-step square wave, but this results in having the worst features of both PWM and six-step designs. Although this approach eliminates the phase control requirements and cuts the front end losses somewhat, there are offsetting drawbacks. Since every switching causes an energy loss in the output devices and their suppressors, the total

Fig. 20.   Representative 2500HP, A.C. variable-frequency drive. (*Robicon Corporation, Pittsburgh, Pennsylvania.*)

losses at high speed go up considerably over six-step (six switches per cycle) if the same devices are used. To overcome this, many versions revert to six-step at 60 Hz. The output devices are stressed much more severely than in six-step. Many PWM designs do not have a voltage regulator; at any given output frequency, they deliver a preset fraction of the input voltage. If the input fluctuates, so does the output. Finally, the high-frequency switching may cause objectionable acoustic noises in the motor. There are both transistor and GTO PWM units on the market today in the range of 1–3,000 HP at 460 VAC. The transistor versions have a better reliability record. As with all voltage source inverters, regeneration to the line is not inherent.

A representative industrial solid-state variable speed drive is given in Figs. 20. See also **Stepper Motors**.

Additional Reading

Bailey, S.J.: "Servo Design Today: Hardware Elements Fade as Software Closes Feedback Loop," *Cont. Eng.*, 57–61, February 1985.

Bailey, S.J.: "Servomotor and Stepper: Key Elements of Motion Control," *Cont. Eng.*, 55–59 (February 1986).

Bailey, S.J.: "AC Motor Drives Selection," *Cont. Eng.*, 101–105 (April 1986).

Bailey, S.J.: "Lessening the Gap Between Incremental and Continuous Motion Control," *Cont. Eng.*, 72–76 (February 1987).

Bailey, S.J.: "AC Motor Drives Use Microprocessors to Set Top Specs for Motion Control," *Cont. Eng.*, 98–102 (April 1987).

Bailey, S.J.: "Servo Vs Stepper—Motion Control Design Decisions with Dynamic Overtones," *Cont. Eng.*, 68–72 (May 1987).

Bedford, B.D. and R.G. Hoft: "Principles of Inverter Circuits," Krieger Publishing Company, Melbourne, FL, 1985.

Bose, B.K.: "Adjustable Speed A.C. Drive Systems," John Wiley & Sons, Inc., New York, NY, 1981.

Brichant, F.: "Force-Commutated Inverters," Macmillan, New York, NY, 1984.

Dote, Y. and S. Kinoshita: "Brushless Servomotors: Fundamentals and Applications," Oxford University Press, Inc., New York, NY, 1991.

Mazurkiewicz, J.: "Brushless Motors Coming on Strong," *Electronic Products*, 61–75, September 1984.

Mazurkiewicz, J.: "Advances in Microprocessor Control for Brushless Motors," Electronic Motion and Control Association Conf., San Diego, Calif., November 1984.

Moreton, P.: "Brushless Servomotors," Butterworth-Heinemann, Inc., Woburn, MA, 1999.

Murphy, H.: "Star-Modulated, Variable Frequency A.C. Drives Bring Higher Performance," *Cont. Eng.*, 104–108 (April 1987).

Pelly, B.R.: "Thyristor Phase-Controlled Converters and Cycloconverters," John Wiley & Sons, Inc., New York, NY, 1971.

Scoles, G.J.: "Handbook of Rectifier Circuits," John Wiley & Sons, Inc., New York, NY, 1980.

Sen, P.C.: "Thyristor D.C. Drives," Krieger Publishing Company, Melbourne, FL, 1991.

SESAME SEED OIL.   See **Vegetable Oils (Edible)**.

SET (Mathematics).   A collection of numbers or symbols considered as a whole. For example, the set of all prime numbers or the set of all matrices with determinant equal to unity. Geometrically, the symbols in a set determine a *domain*.

SET (Permanent).   When a solid has been strained beyond the elastic limit and the deforming stress is completely removed, in general the strain does not decrease ultimately to zero but to some nonvanishing value, known as a permanent set.

SETTLING TIME.   The time required, following the initiation of a specified stimulus to a system, for the output to enter and remain within a specified narrow band centered on its steady-state value. See diagram that accompanies entry on **Response (Instrument)**. The stimulus may be a step impulse, ramp, parabola, or sinusoid. For a step or impulse, the band is often specified as ±2%. For nonlinear behavior, both magnitude and pattern of the stimulus should be specified.

SEWELLEL.   See **Rodentia**.

SEX.   The state of an individual as determined by its adaptations for a special part in biparental reproduction and modifications of the process. Also a category of individuals adapted for a special part in reproduction. The usual sexes are male and female. Neuter individuals exist among colonial invertebrates, including both sexless forms and abortive females whose limited reproductive powers are not exercised under normal conditions. Sex is also expressed in hermaphrodite animals which carry on the usual processes of sexual reproduction but have male and female organs in the same individual.

The differentiation of the sexes is associated with the development of two kinds of gametes in the process of sexual reproduction. Organs and ducts capable of producing the larger egg cells of the female and providing them with quantities of food material make up a reproductive system much different from that which produces the minute male spermatozoa and the seminal fluid in which they are discharged. The development of the external genitalia for internal insemination also results in conspicuous differences, since the male has a projecting penis or other intromittent organ, while the female has the terminal portion of the genital ducts specialized for the reception of this organ. In the mammals, the mammary glands of the female also constitute a conspicuous sexual distinction. These organs may be classed as essential and accessory organs of reproduction, the former category including the gonads and ducts and the latter such parts as the external genitalia and the mammary glands.

Sexes also differ in more or less conspicuous secondary sexual characters such as the beard of man, which are definitely associated with sex but have no direct part in reproduction.

The sexes of most animals are specialized in behavior as well as in structure for the performance of reproductive acts, for accessory functions such as the building of nests, and for subsequent duties of parental care. All of these phases of sexual differentiation are intricately variable among the many species of animals.

SEX (Flower).   See **Flower**.

SEX-INFLUENCED INHERITANCE.
Inheritance influenced by sex, but not restricted to one sex or the other. Baldness in man is often given as a possible case of sex-influenced inheritance. Baldness is supposed to be dominant in men and recessive in women. Hence, a man will be bald if he carries only one gene for baldness, but a woman must carry two such genes before she will be bald. The horns of sheep are also good examples. The horned condition is due to a gene that is dominant in the males and recessive in the females. The Dorset sheep are also homozygous for the gene for horns, so both sexes have horns. The Suffolk sheep are homozygous for the gene for hornless. When the two breeds are crossed, the male offspring all have horns and the females are all hornless. When these first generation offspring are crossed, the offspring show a ratio of 3 horned to 1 hornless in the males and 1 horned and 3 hornless in the females. This is to be expected on the basis of the method of inheritance of a sex-influenced gene.

SEX-LIMITED INHERITANCE.
A condition where certain genes are expressed in only one sex, although they may be carried by both sexes.

The human beard is a good example. It normally shows only in males, although a man inherits the characteristics of his beard just as much from his mother as from his father. Breast development in a woman and not in a man is another example. In general, the development of these traits in mammals is conditioned by the sex hormones. A woman who takes male sex hormones will develop a beard as a result of the action of the latent genes she has carried all her life. In cattle, the qualities of milk production are transmitted just as much from the bull as from the cow. The most outstanding cases of sexual dimorphism resulting from sex-limited genes is to be found in the birds. The brilliant plumage of the male pheasant or peacock stands out in great contrast to the drab pattern of feathers found in the female.

In invertebrate animals, sex-limited genes respond to the chromosomal make-up of the individual cells. Butterflies often show such great differences in the two sexes that they would never be taken for the same species had they not been seen mating. When a gynandromorph is found of the bilateral type, all of the tissue on one side of the body will show the typical male pattern, but that on the other side will show the female pattern. There is no mixing of traits as would be the case if sex hormones flowed to all parts of the body and influenced the expression of the genes.

SEX-LINKED INHERITANCE.
A pattern of inheritance resulting from genes on the nonhomologous portion of the X-chromosome of most animals, but it can also be the genes on the nonhomologous portion of the W-chromosome of the animals with WZ sex determination. In the former case, the female carries two of each of the sex-linked genes, whereas the male carries only one of each. The males, therefore, express all of their sex-linked genes, and the females express those genes that are dominant or homozygous recessive, or intermediate in their expression. A male receives all of his sex-linked genes from the female parent; a female receives a sex-linked gene of each kind from both parents. Hemophilia, bleeder's disease, is a typical sex-linked gene in man. It is recessive, but shows when present on a man's single X-chromosome. Hence, half of the sons of a heterozygous woman will have hemophilia, but all of the daughters will be normal because they will receive a gene for normal blood clotting from their father. The gene for hemophilia was prevalent in the royal families of Europe. See Fig. 1. Color blindness is another human trait resulting from a sex-linked gene.

The first experimental evidence of sex linkage was found in 1910 when Morgan discovered a white-eyed mutant of *Drosophila melanogaster*. When white-eyed males were mated with normal red-eyed females, the F_1 generation displayed all red eyes. However, the F_2 generation included both red and white eyes in the proportion of 3 red to 1 white. This ratio suggested that the mutant gene for white eyes was recessive. The important observation was that the only males of the F_2 generation had white eyes. Therefore, the recessive gene for white eyes expressed itself only in males.

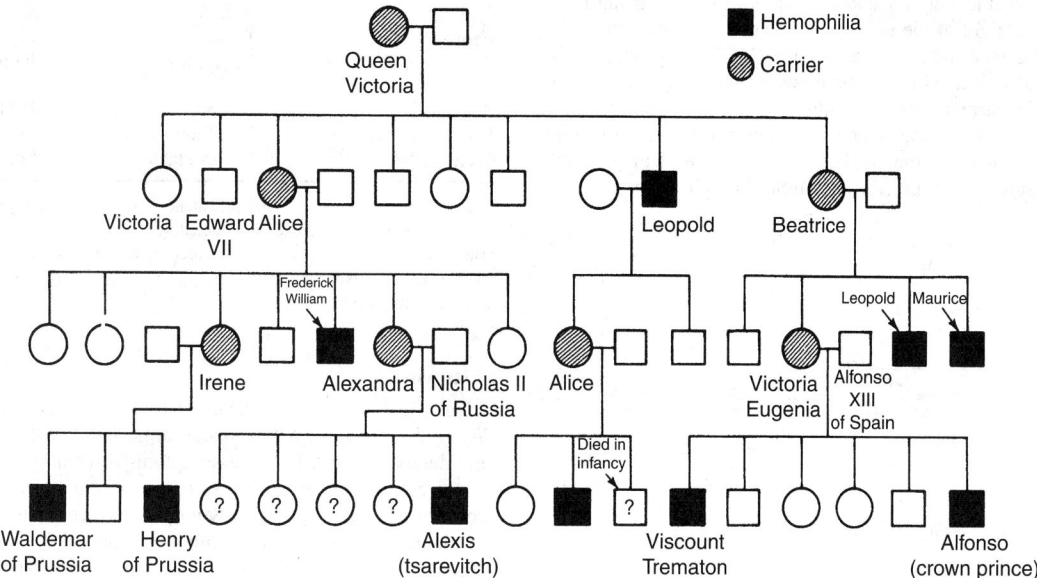

Fig. 1.   Pedigree of hemophilia in the royal families of Europe. The gene apparently arose as a mutation in the immediate ancestry of Queen Victoria of England and spread through the royal families of Europe, with a great impact on history.

Sex-linked genes in males are referred to as *hemizygous* since only one member of an allelic pair of genes is required for expression.

In birds the condition is reversed: the male has the two W-chromosomes and carries two of each sex-linked gene, whereas the female has a single W-chromosome and will express all the sex-linked genes. See also **Genetics and Gene Science (Classical)**; and **Heredity**.

Ann C. DeBaldo, Ph.D., Assoc. Prof., College of Public Health, University of South Florida, Tampa, FL.

SEXTANS.   A minor southern constellation located near the equator.

SEXTANT.   The sextant is a light, portable instrument designed for the purpose of measuring the angular distance between the two objects. It represents the most recent stage in a succession of portable devices for angular measurement advancing from the astrolabe through the cross-staff down to the modern instrument. The immediate predecessor of the sextant was the quadrant (the "hog-yoke" of the sailing ship era), but the design of this instrument is very similar to that of the sextant used today. Since the sextant is used by navigators for the purpose of measuring the apparent altitude of celestial objects, the opinion is prevalent that it is purely a navigational instrument. Such is far from the case, and the instrument is of value to explorers, surveyors, or any person who desires to measure angular distance.

The optical system of the sextant (and quadrant) is shown in the accompanying diagram. The mirror *A* (called the horizon glass) is divided into two sections by a line parallel to the plane of the instrument. The upper section is unsilvered, so that an observer looking through the telescope *T* can see directly through *A* to an object in the direction *H*. The mirror *B* (called the index mirror) may be rotated about an axis perpendicular to the plane of the instrument and coincident with the center of the graduated arc *CD*. Attached to the index mirror is a vernier arm, which sweeps along the arc.

With the mirror *B* strictly parallel to *A*, the observer will see two superimposed images of the object in the direction *H*. One of these is the direct image observed through the upper section of the horizon glass, and the other is a reflected image with light traveling along the path *HBAT*. Under these conditions, the index *C* should read zero; in case this is not so, an "index correction" must be applied to all observations with the instrument.

In case the angular distance between an object in the direction *H* and another in the direction *S* (angle *SBH*) is desired, the observer moves the index arm along the arc until an image of *S* (light path *SBAT*) is superimposed on the direct image of *H*. Application of the laws of reflection of light will show that the angle *CBD* through which the mirror is turned in one-half the angle *SBH*. To obviate the necessity of dividing each reading of the instrument by two, the arc is so graduated that when *CBD* is actually 60° the index will read 120°.

To obtain the altitude of a celestial object at sea, the instrument is directed toward the visible horizon and the index arm moved until the desired object is in the field of view. Now the sextant is rotated back and forth about the optic axis of the telescope (the line *HAT* in Fig. 1), and the image of the celestial object will apparently swing back and forth along a short arc. The index arm is now carefully adjusted until the celestial object is just tangent to the horizon at the lowest point of its apparent swing. Under these conditions the reading of the sextant will give the sextant altitude measured along the vertical circle through the object.

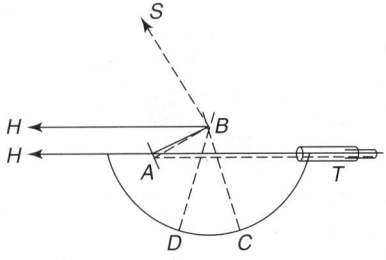

Fig. 1.   Optical system of sextant.

To overcome situations where it is difficult or impossible to see the horizon clearly, a small bubble made visible in the instrument's optical system is used as an indication of vertical. A gyro vertical also has been used as a horizon reference.

SEXUALLY TRANSMITTED DISEASES.   Once known universally as venereal diseases (VD), sexually transmitted diseases (STDs) are transmitted through close, intimate, and sexual contact. STDs afflict an estimated 200 to 400 million people worldwide, and the incidence of infection is rising. The costs of dealing with these diseases run into the billions of dollars annually.

In the United States, more than 65 million people are currently living with and incurable sexually transmitted disease (STD). An additional 15 million people become infected with one or more STDs each year, roughly half of whom contract lifelong infections (Cates). Yet, STDs are one of the most under-recognized health problems in the country today. Despite the fact that STDs are extremely widespread, have severe and sometimes deadly consequences, and add billions of dollars to the nation's healthcare costs each year, most people in the United States remain unaware of the risks and consequences of all but the most prominent STD — the human immunodeficiency virus or HIV.

While extremely common, STDs are difficult to track. Many people with these infections do not have symptoms and remain undiagnosed. Even diseases that are diagnosed are frequently not reported and counted. These "hidden" epidemics are magnified with each new infection that goes unrecognized and untreated.

More than 25 diseases are spread primarily through sexual activity, and the trends for each disease vary considerably, but together these infections comprise a significant public health problem.

The latest estimates indicate that there are 15 million new STD cases in the United States each year (Cates). Approximately one-fourth of these new infections are in teenagers. And while some STDs, such as syphilis, have been brought to all time lows, others, like genital herpes, gonorrhea, and chlamydia, continue to resurge and spread through the population.

Because there is no single STD epidemic, but rather multiple epidemics, discussions about trends over time and populations affected must focus on each specific STD. More is known about the frequency and trends of some STDs than others, since many of the diseases are difficult to track. Not including HIV, the most common STDs in the U.S. are chlamydia, gonorrhea, syphilis, genital herpes, human papillomavirus, hepatitis B, trichomoniasis, and bacterial vaginosis. The latest estimates of incidence and prevalence are provided in Table 1.

TABLE 1. INCIDENCE AND PREVALENCE OF STDs

	Incidence (Estimated number of new cases every year)	Prevalence[1] (Estimated number of people currently infected)
Chlamydia	3 million	2 million
Gonorrhea	650,000	Not available
Syphilis	70,000	Not available
Herpes	1 million	45 million
Human Papillomavirus (HPV)	5.5 million	20 million
Hepatitis B	120,000	417,000
Trichomoniasis	5 million	Not available
Bacterial Vaginosis[2]	Not available	Not available

[1]No recent surveys on national prevalence for gonorrhea, syphilis, trichomoniasis or bacterial vaginosis have been conducted.
[2]Bacterial vaginosis is a genital infection that is not sexually transmitted but is associated with sexual intercourse.
Source: Cates, 1999.

Chlamydia

Chlamydia is a common sexually transmitted disease (STD) caused by *Chlamydia trachomatis*, a bacterium, which can damage a woman's reproductive organs. Because symptoms of chlamydia are mild or absent, serious complications that cause irreversible damage, including infertility, can occur "silently" before a woman ever recognizes a problem.

Chlamydia is the most commonly reported infectious disease in the United States and may be one of the most dangerous sexually transmitted diseases among women today. While the disease can be easily cured with antibiotics, millions of cases go unrecognized. If left untreated, chlamydia can have severe consequences, particularly for women. Up to 40 percent of women with untreated chlamydia will develop pelvic inflammatory disease (PID), and one in five women with PID becomes infertile. Chlamydia also can cause prematurity, eye disease, and pneumonia in infants. Moreover, women infected with chlamydia are three to five times more likely to become infected with HIV, if exposed.

Seventy-five percent of women and 50 percent of men with chlamydia have no symptoms. The majority of cases therefore go undiagnosed and unreported. The number of reported cases about 660,000 cases in 1999, is merely the tip of the iceberg.

- An estimated three million people contact chlamydia each year (Cates).
- Chlamydia is believed to be declining overall in the United States, primarily because of increased efforts to screen and treat women for chlamydia. Chlamydia is estimated to have declined from well over four million annual infections in the early 1980s to the current level of three million annual infections (Cates).
- Reported chlamydia rates in women greatly exceed those in men largely because screening programs have been primarily directed at women. True rates are probably far more similar for women and men.
- Since 1994, there has been increased funding available for chlamydia screening in publicly funded family planning and STD clinics. The percentage of women testing positive for chlamydia (chlamydia positivity) in family planning clinics by state provides a good indication of where the disease remains most widespread. The highest level of infection tends to be seen in areas where screening and treatment
- have not been as widely implemented. The greatest declines have generally been in areas of the country with the most effective and prolonged screening programs.
- From 1988 to 1999, the Pacific Northwest — Washington, Oregon, Idaho, and Alaska — witnessed a 62 percent decline in infection among women tested for chlamydia in family planning clinics.
- In the Mid-Atlantic States — Delaware, Washington, DC, Maryland, Pennsylvania, Virginia, and West Virginia — similar trends are occurring, with a decline of 16 percent since 1994 (DSTDP, CDC).
- From 1997 to 1999, chlamydia test positivity in family planning clinics actually increased in eight out of 10 regions. However, these reported increases are most likely due to changes in newly available and better laboratory tests and expanded screening programs to populations with higher levels of disease (DSTDP, CDC).

Reducing the level of chlamydia will require continued expansion of screening and treatment among women and new efforts to reach men. While men experience symptoms and seek treatment on their own more often than women, half of men with chlamydia are asymptomatic. Reaching these men with treatment is critical to stem the spread of chlamydia and its severe consequences.

It is also critical to reach those at greatest risk, primarily young people, with programs to help reduce their risk of infection through safer sexual behaviors. See also **Chlamydia**.

For more information on Chlamydia, see the Division of Sexually Transmitted Diseases (CDC): *http://www.cdc.gov/nchstp/dstd/Fact_Sheets/FactsChlamydiaInfo.htm*

Gonorrhea

Gonorrhea is caused by *Neisseria gonorrhoeae*, a bacterium that can grow and multiply easily in mucous membranes of the body. Gonorrhea bacteria can grow in the warm, moist areas of the reproductive tract, including the cervix (opening to the womb), uterus (womb), and fallopian tubes (egg canals) in women, and in the urethra (urine canal) in women and men. The bacteria can also grow in the mouth, throat, and anus.

Gonorrhea is a sexually transmitted bacterial disease. Reported gonorrhea rates declined steadily until the late 1990s. From 1985 to 1996, rates of the disease decreased nearly 10 percent annually (Fox, CDC). However, rates stabilized between 1996 and 1997, and between 1997 and 1999, gonorrhea rates increased by nine percent (DSTPD, CDC). This increase, combined with signs of an increase in gonorrhea among gay and bisexual men, is reason for concern.

Rates of infection remain high among adolescents, young adults, and African Americans. Gonorrhea remains a major cause of pelvic inflammatory disease (PID) and subsequent infertility and tubal pregnancies in women. Additionally, studies have shown that gonorrhea can facilitate HIV transmission and may be contributing significantly to the spread of HIV in

the south. There is a critical need to reach populations that remain at high risk for gonorrhea with intensified prevention and treatment efforts. Gonorrhea can be easily cured, if detected early, and the long-term consequences can be prevented.

- An estimated 650,000 cases of gonorrhea occur each year in the United States (Cates).
- Cases reported to CDC are believed to represent about half of all annual infections. While an underestimate of actual cases, these reports provide a good indication of trends in the disease.
- The reported gonorrhea rate in the United States remains the highest of any industrialized country and is roughly 50 times that of Sweden and eight times that of Canada.
- Researchers have seen alarming indications that gonorrhea may be on the rise among gay and bisexual men — men who have sex with men (MSM). In the mid-1980s, reports of increased condom use and reduced risky sexual practices accompanied dramatic decreases in rectal gonorrhea among MSM in several cities (Fox, CDC). Yet, data from cities throughout the country suggest that this trend may be reversing, and that gonorrhea cases may be resurging with the prospect of facilitating the spread of HIV in the gay community.

The rise in gonorrhea rates should serve as a wake-up call to people at risk, including gay and bisexual men, that high-risk sexual behaviors continue to have very real consequences. See also **Gonorrhea and Gonococcemia**.

For more information on Gonorrhea, see the Division of Sexually Transmitted Diseases (CDC): *http://www.cdc.gov/nchstp/dstd/Fact_Sheets/FactsGonorrhea.htm*

Syphilis

Syphilis is a complex sexually transmitted disease (STD) caused by the bacterium *Treponema pallidum*. It has often been called "the great imitator" because so many of the signs and symptoms are indistinguishable from those of other diseases.

Syphilis is passed from person to person through direct contact with a syphilis sore. Sores occur mainly on the external genitals, vagina, anus, or in the rectum. Sores also can occur on the lips and in the mouth. Transmission of the organism occurs during vaginal, anal, or oral sex. Pregnant women with the disease can pass it to the babies they are carrying. Syphilis **cannot** be spread by toilet seats, door knobs, swimming pools, hot tubs, bath tubs, shared clothing, or eating utensils.

In the United States, over 35,600 cases of syphilis were reported by health officials in 1999, including 6,650 cases of primary and secondary syphilis (a decline of 5.4% from 1998) and 556 cases of congenital syphilis in newborns. More cases occur each year than come to the attention of health officials. Of the nine states with the highest 1999 syphilis rates (2–5 times higher than the national rate of 2.5 cases per 100,000), eight were in the South. Although syphilis rates remain higher in the South than in other regions, the South had a 32% decline in the primary and secondary syphilis rate from 1997 to 1999, illustrating that the greatest improvements in disease control have taken place where syphilis incidence has been the greatest. In 1999, 25 counties accounted for 50% of all primary and secondary syphilis cases. Two hundred sixty-five counties had syphilis rates above the U.S. Public Health Service's Healthy People 2000 objective of 4 cases per 100,000. These 265 counties (9% of the total number of counties in the U.S.) accounted for approximately 74% of the total primary and secondary syphilis cases reported in 1999.

In 1999, syphilis occurred primarily in persons aged 20 to 39, and the reported rate in men was 1.5 times greater than the rate in women. The incidence of syphilis was highest in women aged 20 to 29 years and in men 30 to 39. Some fundamental societal problems, such as poverty, inadequate access to health care, and lack of education are associated with disproportionately high levels of syphilis in certain populations. Cases of primary and secondary syphilis in 1999 had the following race or ethnicity distribution: African Americans 75%, whites 16%, Hispanics 8%, and others 1%. Syphilis reflects one of the most glaring examples of racial disparity in health status, with the rate for African Americans nearly 30 times the rate for whites.

Symptoms

Primary Stage. The time between infection with syphilis and the start of the first symptom can range from 10–90 days (average 21 days). The primary stage of syphilis is usually marked by the appearance of a single sore (called a chancre), but there may be multiple sores. The chancre is usually firm, round, small, and painless. It appears at the spot where syphilis entered the body. The chancre lasts 3–6 weeks, and it will heal on its own. If adequate treatment is not administered, the infection progresses to the secondary stage.

Secondary Stage. The second stage starts when one or more areas of the skin break into a rash that usually does not itch. Rashes can appear as the chancre is fading or can be delayed for weeks. The rash often appears as rough, red or reddish brown spots both on the palms of the hands and on the bottoms of the feet. The rash also may also appear on other parts of the body with different characteristics, some of which resemble other diseases. Sometimes the rashes are so faint that they are not noticed. Even without treatment, rashes clear up on their own. In addition to rashes, second-stage symptoms can include fever, swollen lymph glands, sore throat, patchy hair loss, headaches, weight loss, muscle aches, and tiredness. A person can easily pass the disease to sex partners when primary or secondary stage signs or symptoms are present.

Late Syphilis. The latent (hidden) stage of syphilis begins when the secondary symptoms disappear. Without treatment, the infected person still has syphilis even though there are no signs or symptoms. It remains in the body, and it may begin to damage the internal organs, including the brain, nerves, eyes, heart, blood vessels, liver, bones, and joints. This internal damage may show up many years later in the late or tertiary stage of syphilis. Late stage signs and symptoms include not being able to coordinate muscle movements, paralysis, numbness, gradual blindness and dementia. This damage may be serious enough to cause death.

Diagnosis and Treatment

A health care provider can diagnose syphilis by using dark field microscopy to examine material from infectious sores. If syphilis bacteria are present in the sore, they will show up with a characteristic appearance.

A blood test is another way to determine whether someone has syphilis. Shortly after infection occurs, the body produces syphilis antibodies that can be detected by an accurate, safe and inexpensive blood test. A low level of antibodies will stay in the blood for months or years even after the disease has been successfully treated. Because untreated syphilis in a pregnant woman can infect and possibly kill her developing baby, every pregnant woman should have a blood test for syphilis.

A single dose of penicillin, an antibiotic, will cure a person who has had syphilis for less than a year. Larger doses are needed to cure someone who has had it for longer than a year. For people who are allergic to penicillin, other antibiotics are available to treat syphilis. There are no home remedies or over-the-counter drugs that will cure syphilis. Penicillin treatment will kill the syphilis bacterium and prevent further damage, but it will not repair any damage already done. Persons who receive syphilis treatment must abstain from sexual contact with new partners until the syphilis sores are completely healed. Persons with syphilis must notify their sex partners so that they also can be tested, and, if necessary, receive treatment.

Having had syphilis does not protect a person from getting it again. Antibodies are produced as a person reacts to the disease, and, after treatment, these antibodies may offer partial protection from getting infected again, if exposed right away. Even though there may be a short period of protection, the antibody levels naturally decrease in the blood, and people become susceptible to syphilis infection again if they are sexually exposed to syphilis sores. See also **Syphilis**.

For more information on Syphilis, see the Division of Sexually Transmitted Diseases (CDC): *http://www.cdc.gov/std/*

Genital Warts (Human Papillomavirus HPV)

Genital HPV infection is a sexually transmitted disease (STD) that is caused by human papillomavirus (HPV). Human papillomavirus, or HPV, is the name of a group of viruses that includes more than 100 different strains or types. Over 30 of these are sexually transmitted, and they can infect the genital area, like the skin of the penis, vulva, labia, or anus, or the tissues covering the vagina and cervix.

Some of these viruses are considered "high-risk" types and may cause abnormal Pap smears and cancer of the cervix, anus, and penis. Others are "low-risk," and they may cause mild Pap smear abnormalities and genital warts. Genital warts are single or multiple growths or bumps that appear in the genital area, and sometimes form a cauliflower-like shape.

Approximately twenty million people are currently infected with HPV. Fifty to 75% of sexually active men and women acquire genital HPV infection at some point in their lives. About 5.5 million Americans get a new genital HPV infection each year.

The types of HPV that infect the genital area are spread primarily through sexual contact. Most HPV infections have no signs or symptoms; therefore, most infected persons are completely unaware they are infected, yet they can transmit the virus to a sex partner. Rarely, pregnant women can pass HPV to their baby during vaginal delivery. A newborn that is exposed to HPV during delivery can develop warts in the larynx (voice box).

Most people who have a genital HPV infection do not know they are infected. The virus lives in the skin or mucus membranes and usually causes no symptoms. Other people get visible genital warts. See also **Gonads**.

Genital warts usually appear as soft, moist, pink or red swellings. They can be raised or flat, single or multiple, small or large. Some cluster together forming a cauliflower-like shape. They can appear on the vulva, in or around the vagina or anus, on the cervix, and on the penis, scrotum, groin, or thigh. Warts can appear within several weeks after sexual contact with an infected person, or they can take months to appear.

Genital warts are diagnosed by inspection. Visible genital warts can be removed, but no treatment is better than another and no single treatment is ideal for all cases.

For more information on Human Papillomavirus (HPV), see the Division of Sexually Transmitted Diseases (CDC): *http://www.cdc.gov/std/*

Genital Herpes (Herpes Simplex Virus HSV)

Herpes is a sexually transmitted disease (STD) caused by the herpes simplex viruses type 1 (HSV-1) and type 2 (HSV-2). Most individuals have no or only minimal signs or symptoms from HSV-1 or HSV-2 infection. When signs do occur, they typically appear as one or more blisters on or around the genitals or rectum. The blisters break, leaving tender ulcers (sores) that may take two to four weeks to heal the first time they occur. Typically, another outbreak can appear weeks or months after the first, but it almost always is less severe and shorter than the first episode. Although the infection can stay in the body indefinitely, the number of outbreaks tends to go down over a period of years.

HSV-1 and HSV-2 can be found and released from the sores that the viruses cause, but they also are released between episodes from skin that does not appear to be broken or to have a sore. A person almost always gets HSV-2 infection during sexual contact with someone who has a genital HSV-2 infection. HSV-1 causes infections of the mouth and lips, so-called "fever blisters." A person can get HSV-1 by coming into contact with the saliva of an infected person. HSV-1 infection of the genitals almost always is caused by oral-genital sexual contact with a person who has the oral HSV-1 infection.

Results of a recent, nationally representative study show that genital herpes infection is common in the United States. Nationwide, 45 million people ages 12 and older, or one out of five of the total adolescent and adult population, are infected with HSV-2.

HSV-2 infection is more common in women (approximately one out of four women) than in men (almost one out of five). This may be due to male-to-female transmission being more efficient than female-to-male transmission. HSV-2 infection also is more common in blacks (45.9%) than in whites (17.6%). Race and ethnicity in the United States correlate with other, more fundamental determinants of health such as poverty, access to good quality health care, behavior for seeking health care, illicit drug use, and living in communities with a high prevalence of STDs.

Since the late 1970s, the number of Americans with genital herpes infection has increased 30%. The largest increase is currently occurring in young white teens. HSV-2 infection is now five times more common in 12- to 19-year-old whites, and it is twice as common in young adults ages 20 to 29 than it was 20 years ago.

HSV-2 usually produces only mild symptoms or signs or no symptoms at all. However, HSV-2 can cause recurrent painful genital sores in many adults, and HSV-2 infection can be severe in people with suppressed immune systems. Regardless of severity of symptoms, genital herpes frequently causes psychological distress in people who know they are infected.

In addition, HSV-2 can cause potentially fatal infections in infants if the mother is shedding virus at the time of delivery. It is important that women avoid contracting herpes during pregnancy because a first episode during pregnancy causes a greater risk of transmission to the newborn. If a woman has active genital herpes at delivery, a cesarean delivery is usually performed. Fortunately, infection of an infant from women with HSV-2 infection is rare.

In the United States, HSV-2 may play a major role in the heterosexual spread of HIV, the virus that causes AIDS. Herpes can make people more susceptible to HIV infection, and it can make HIV-infected individuals more infectious.

Most people infected with HSV-2 are not aware of their infection. However, if signs and symptoms occur during the first episode, they can be quite pronounced. The first episode usually occurs within two weeks after the virus is transmitted, and the sores typically heal within two to four weeks. Other signs and symptoms during the primary episode may include a second crop of sores, or flu-like symptoms, including fever and swollen glands. However, most individuals with HSV-2 infection may never have sores, or they may have very mild signs that they do not even notice or that they mistake for insect bites or a rash.

Most people diagnosed with a first episode of genital herpes can expect to have several symptomatic recurrences a year (typically four or five). These recurrences usually are most noticeable within the first year following the first episode.

The signs and symptoms associated with HSV-2 can vary greatly. Health care providers can diagnose genital herpes by visual inspection if the outbreak is typical, and by taking a sample from the sore(s). HSV infections can be difficult to diagnose between outbreaks. Blood tests which detect HSV-1 or HSV-2 infection may be helpful, although the results are not always clear cut.

There is no treatment that can cure herpes, but antiviral medications can shorten and prevent outbreaks during the period of time the person takes the medication. See also **Herpes Simplex Virus Diseases**.

Hepatitis B (HBV)

Hepatitis B (HBV) virus is a serious viral disease that attacks the liver, and can cause extreme illness and even death. An acute HBV infection is a newly acquired, symptomatic infection. In some people, the infection resolves and the virus is cleared. However, many will remain chronically infected with the virus after the symptoms associated with their new infection have subsided. People chronically infected with HBV face an increased risk of developing chronic liver disease, including cirrhosis (scarring) and liver cancer, and for transmitting HBV infection to others.

According to the Third National Health and Nutrition Examination Survey [NHANES III], about five percent of the U.S. population has ever been infected with hepatitis B, with an estimated 200,000 infections occurring each year (Coleman). Of these, it is believed that 120,000 infections are acquired through sexual transmission annually, mostly among young adults. An estimated 417,000 people are currently living with chronic sexually acquired HBV infection.

Infants and young children have the highest risk of chronic infection.

An estimated 5,000 to 6,000 deaths occur each year from chronic hepatitis B-related liver disease.

Hepatitis B vaccinations have been recommended for people with risk factors for HBV infection since the vaccine became available in 1981. However, many teens and young adults at risk through sexual or drug-related behavior have not been vaccinated for HBV. Intensified efforts to vaccinate high-risk groups are urgently needed.

In 1997, 10,416 cases of acute hepatitis B were reported to the CDC. However, reported cases dramatically underestimate the actual number of people who are infected with hepatitis B virus each year, an estimated 200,000.

A recent study demonstrates the high degree of under-vaccination among those at high risk. Among acute hepatitis B cases reported by sentinel counties in 1996, 70 percent had a missed opportunity for vaccination in the past. Of these, 42 percent had been treated for an STD in the past, 31 percent had been in prison or jail at some time in their lifetime, and 25 percent reported sexual or household contact with an HBV-infected person (Mast, CDC).

See also **Liver**.

Trichomoniasis

Trichomoniasis is a common sexually transmitted disease (STD) that is spread through penis-to-vagina intercourse or vulva-to-vulva contact with an infected partner. Women can acquire the disease from infected men or women, whereas men usually contract it only from infected women.

Trichomoniasis is caused by the single-celled protozoan parasite *Trichomonas vaginalis*. The vagina is the most common site of infection in women, and the urethra is the most common site of infection in men.

Trichomoniasis is the most common curable STD in young, sexually active women. An estimated 5 million new cases occur each year in women and men.

Most men with trichomoniasis do not have signs or symptoms. Men with symptoms may have an irritation inside the penis, mild discharge, or slight burning after urination or ejaculation.

Many women do have signs or symptoms of infection. In these women, trichomoniasis causes a frothy, yellow-green vaginal discharge with a strong odor. The infection may also cause discomfort during intercourse and urination. Irritation and itching of the female genital area and, in rare cases, lower abdominal pain can also occur.

Trichomoniasis in pregnant women may cause premature rupture of the membranes and preterm delivery. The genital inflammation caused by trichomoniasis might also increase a woman's risk of acquiring HIV infection if she is exposed to HIV. Trichomoniasis in a woman who is also infected with HIV can increase the chances of transmitting HIV infection to a sex partner.

To diagnose trichomoniasis, a health care provider must perform a physical examination and laboratory test. In women, a pelvic examination can reveal small red ulcerations on the vaginal wall or cervix. Laboratory tests are performed on a sample of vaginal fluid or urethral fluid to look for the disease-causing parasite. The parasite is harder to detect in men than in women.

Trichomoniasis can usually be cured with the prescription drug metronidazole given by mouth in a single dose. The symptoms of trichomoniasis in infected men may disappear within a few weeks without treatment. However, an infected man, even a man who has never had symptoms or whose symptoms have stopped, can continue to infect a female partner until he has been treated. Therefore, both partners should be treated at the same time to eliminate the parasite. Persons being treated for trichomoniasis should avoid sex until they and their sex partners complete treatment and have no symptoms. Metronidazole can be used by pregnant women.

See also **Gonads**.

Bacterial Vaginosis (BV)

Bacterial vaginosis (BV) is the most common vaginal infection in women of childbearing age. Women with BV often have an abnormal vaginal discharge with an unpleasant odor. Some women report a strong fish like odor, especially after intercourse. The discharge is usually white or gray; it can be thin. Women with BV may also have burning during urination or itching around the outside of the vagina, or both. Some women with BV report no signs or symptoms at all.

The cause of BV is not fully understood. BV is associated with an imbalance in the bacteria that are normally found in a woman's vagina. The vagina normally contains mostly "good" bacteria, and fewer "harmful" bacteria. BV develops when there is a change in the environment of the vagina that causes an increase in harmful bacteria.

Not much is known about how women get BV. Women who have a new sex partner or who have had multiple sex partners are more likely to develop BV. Women who have never had sexual intercourse are rarely affected. It is not clear what role sexual activity plays in the development of BV, and there are many unanswered questions about the role that harmful bacteria play in causing BV. Women do not get BV from toilet seats, bedding, swimming pools, or from touching objects around them.

Scientific studies suggest that BV is common in women of reproductive age. In the United States, as many as 16% of pregnant women have BV. This varies by race and ethnicity from 6% in Asians and 9% in whites to 16% in Hispanics and 23% in African Americans. BV is generally more commonly seen in women attending STD clinics than in those attending family planning or prenatal clinics.

A health care provider must examine the vagina for signs of BV (e.g., discharge) and perform laboratory tests on a sample of vaginal fluid to look for bacteria associated with BV.

Although BV will sometimes clear up without treatment, all women with symptoms of BV should be treated to avoid such complications as PID. Treatment is especially important for pregnant women. All pregnant women, regardless of symptoms, who have ever had a premature delivery or low birth weight baby should be considered for a BV examination and be treated when necessary. All pregnant women who have symptoms of BV should be checked and treated. Male partners generally do not need to be treated. However, BV may spread between female sex partners.

BV is treatable with antimicrobial medicines prescribed by a health care provider. Two different medicines are recommended as treatment for BV: metronidazole or clindamycin. Either can be used with non-pregnant or with pregnant women, but the recommended dosages differ. Women with BV who are HIV-positive should receive the same treatment as those who are HIV-negative. BV can recur after treatment.

The Role of STD Detection and Treatment in HIV Prevention

Testing and treatment of sexually transmitted diseases (STDs) can be an effective tool in preventing the spread of HIV, the virus that causes AIDS. An understanding of the relationship between STDs and HIV infection can help in the development of effective HIV prevention programs for persons with high-risk sexual behaviors.

Individuals who are infected with STDs are at least two to five times more likely than uninfected individuals to acquire HIV if they are exposed to the virus through sexual contact. In addition, if an HIV-infected individual is also infected with another STD, that person is more likely to transmit HIV through sexual contact than other HIV-infected persons (Wasserheit).

There is substantial biological evidence demonstrating that the presence of other STDs increases the likelihood of both transmitting and acquiring HIV (Fleming, Wasserheit).

Increased susceptibility. STDs probably increase susceptibility to HIV infection by two mechanisms. Genital ulcers (e.g., syphilis, herpes, or chancroid) result in breaks in the genital tract lining or skin. These breaks create a portal of entry for HIV. Nonulcerative STDs (e.g., chlamydia, gonorrhea, and trichomoniasis) increase the concentration of cells in genital secretions that can serve as targets for HIV (e.g., CD^{4+} cells).

Increased infectiousness. Studies have shown that when HIV-infected individuals are also infected with other STDs, they are more likely to have HIV in their genital secretions. For example, men who are infected with both gonorrhea and HIV are more than twice as likely to shed HIV in their genital secretions than are those who are infected only with HIV. Moreover, the median concentration of HIV in semen is as much as 10 times higher in men who are infected with both gonorrhea and HIV than in men infected only with HIV.

Evidence from intervention studies indicates that detecting and treating STDs can substantially reduce HIV transmission at the individual and community levels.

STD treatment reduces an individual's ability to transmit HIV. Studies have shown that treating STDs in HIV-infected individuals decreases both the amount of HIV they shed and how often they shed the virus (Fleming, Wasserheit).

STD treatment reduces the spread of HIV infection in communities. Two community-level, randomized trials have examined the role of STD treatment in HIV transmission. Together, their results have begun to clarify conditions under which STD treatment is likely to be most successful in reducing HIV transmission. First, continuous interventions to improve access to effective STD treatment services is likely to be more effective in reducing HIV transmission than intermittent interventions through strategies such as periodic mass treatment. Second, STD treatment is likely to be most effective in reducing HIV transmission where STD rates are high and the heterosexual HIV epidemic is young. Third, treatment of symptomatic STDs may be particularly important.

The first community trial, conducted in a rural area of Tanzania, demonstrated a decrease of about 40% in new, heterosexually transmitted HIV infections in communities with continuous access to improved treatment of symptomatic STDs, as compared to communities with minimal STD services, where incidence remained about the same (Grosskurth, Mosha, Todd, et al.). However, in the second trial conducted in Uganda, a reduction in HIV transmission was not demonstrated when the STD control approach was community-wide mass treatment administered to everyone every 10 months in the absence of regular access to improved STD services (Wawer, et al.).

What are the implications for HIV prevention programs?

Strong STD prevention, testing, and treatment can play a vital role in comprehensive programs to prevent sexual transmission of HIV.

Furthermore, STD trends can offer important insights into where the HIV epidemic may grow, making STD surveillance data helpful in forecasting where HIV rates are likely to increase. Better linkages are needed between HIV and STD prevention efforts nationwide in order to control both epidemics.

In the context of persistently high prevalence of STDs in many parts of the United States and with emerging evidence that the U.S. HIV epidemic increasingly is affecting populations with the highest rates of curable STDs, CDC's Advisory Committee on HIV and STD Prevention (ACHSP) has recommended the following:

Early detection and treatment of curable STDs should become a major, explicit component of comprehensive HIV prevention programs at national, state, and local levels.

In areas where STDs that facilitate HIV transmission are prevalent, screening and treatment programs should be expanded.

HIV and STD prevention programs in the United States, together with private and public sector partners, should take joint responsibility for implementing these strategies.

The ACHSP also notes that early detection and treatment of STDs should be only one component of a comprehensive HIV prevention program, which also must include a range of social, behavioral, and biomedical interventions.

See also **Acquired Immune Deficiency Syndrome (AIDS)**.

STDs and Pregnancy

Women who are pregnant can become infected with the same sexually transmitted diseases (STDs) as women who are not pregnant. Pregnancy does not provide women or their babies any protection against STDs. In fact, the consequences of an STD can be significantly more serious, even life threatening, for a woman and her baby if the woman becomes infected with an STD while she is pregnant. As the list of diseases known to be sexually transmitted continues to grow, it is increasingly important that women be aware of the harmful effects of these diseases and know how to protect themselves and their children against infection.

STDs can be transmitted from a pregnant woman to the fetus, newborn, or infant before, during, or after birth. Some STDs (like syphilis) cross the placenta and infect the fetus during its development. Other STDs (like gonorrhea, chlamydia, hepatitis B, and genital herpes) are transmitted from the mother to the infant as the infant passes through the birth canal. HIV infection can cross the placenta during pregnancy, infect the newborn during the birth process, and, unlike other STDs, infect an infant as a result of breast-feeding.

Harmful effects on the baby may include stillbirth, low birth weight, conjunctivitis (eye infection), pneumonia, neonatal sepsis (infection in the blood stream), neurologic damage (such as brain damage or motor disorder), congenital abnormalities (including blindness, deafness, or other organ damage), acute hepatitis, meningitis, chronic liver disease, and cirrhosis. Some of these consequences may be apparent at birth; others may not be detected until months or even years later.

STDs affect women of every socioeconomic and educational level, age, race, ethnicity, and religion. The CDC STD Treatment Guidelines (1997) recommend that pregnant women be screened for the following STDs:

- Chlamydia
- Gonorrhea
- Hepatitis B
- HIV
- Syphilis

Pregnant women should request these tests specifically because some doctors do not routinely perform them. New and increasingly accurate tests continue to become available. Even if a woman has been tested in the past, she should be tested again when she becomes pregnant.

Bacterial STDs (like chlamydia, gonorrhea, and syphilis) can be treated and cured with antibiotics during pregnancy. There is no cure for viral STDs such as genital herpes and HIV, but antiviral medication for herpes and HIV may reduce symptoms in the pregnant woman. In addition, the risk of passing HIV infection from mother to baby is dramatically reduced by treatment. For women who have active genital herpes lesions at the time of delivery, a cesarean section may be performed to protect the newborn against infection.

Pelvic Inflammatory Disease (PID)

Pelvic inflammatory disease (PID) is a general term that refers to infection of the fallopian tubes (tubes that carry eggs from the ovary to the womb) and of other internal reproductive organs in women. It is a common and serious complication of some sexually transmitted diseases (STDs). Inside the lower abdominal cavity, PID can damage the fallopian tubes and tissues in and near the uterus and ovaries. Untreated PID can lead to serious consequences including infertility, ectopic pregnancy, abscess formation, and chronic pelvic pain.

Each year in the United States, more than one million women experience an episode of acute PID. More than 100,000 women become infertile each year as a result of PID, and a large proportion of the ectopic pregnancies occurring every year are due to the consequences of PID. More than 150 women die from this infection every year.

PID occurs when bacteria move upward from a woman's vagina or cervix into the internal reproductive organs. Sexually active women in their childbearing years are most at risk. Many different organisms can cause PID, but most cases are associated with gonorrhea and chlamydia, two very common bacterial STDs. It is estimated that 10% to 80% of women with either of these STDs will develop symptomatic PID.

Symptoms of PID vary from none to severe. Particularly when it is caused by chlamydial infection, PID may produce only mild symptoms or no symptoms at all, even while it is seriously damaging the internal reproductive organs. Because of the vague symptoms, PID goes unrecognized both by women and by their health care providers about two thirds of the time. Women who do have symptoms of PID most commonly have lower abdominal pain. Other signs and symptoms include fever, unusual vaginal discharge that may have a foul odor, painful intercourse, painful urination, irregular menstrual bleeding, and pain in the right upper abdomen (rare).

Early and complete treatment can help prevent complications of PID. Without treatment, PID can cause permanent damage to the female internal reproductive organs. Infection-causing bacteria can silently invade the fallopian tubes, causing normal tissue to turn into scar tissue. Scar tissue blocks or interrupts the normal movement of eggs into the uterus. If the fallopian tubes are totally blocked by scar tissue, an egg will not be fertilized by sperm or move to the uterus to develop into a baby. Totally blocked fallopian tubes cause a woman to be infertile. Infertility can also occur if the fallopian tubes are partially blocked or even slightly damaged. About one in five women with PID becomes infertile. If a woman has multiple episodes of PID, her chances of becoming infertile are increased.

In addition, a partially blocked or slightly damaged fallopian tube may cause a fertilized egg to get stuck in the tube. This fertilized egg may begin to grow in the tube as if it were in the womb. This is an ectopic pregnancy, which is a pregnancy in the fallopian tube or elsewhere outside the uterus. As it grows, an ectopic pregnancy can rupture the fallopian tube and cause severe pain, internal bleeding, and even death. Scarring in the fallopian tubes and other pelvic structures can also cause chronic pelvic pain (pain that lasts for months or even years). Women with repeated episodes of PID are more likely than women with a single episode to suffer infertility, ectopic pregnancy, or chronic pelvic pain.

PID is difficult to diagnose because the symptoms are often subtle and mild. Many episodes of PID go undetected because the woman or her health care provider fails to recognize the implications of mild or nonspecific symptoms. Because there are no precise tests for PID, a diagnosis is usually based on clinical findings. If symptoms such as lower abdominal pain are present, a health care provider should perform a physical examination to determine the nature and location of the pain and check for fever, abnormal vaginal or cervical discharge, and for evidence of gonorrhea or chlamydia infection. If the findings suggest PID, treatment is necessary.

If more information is necessary, the health care provider may order other tests to identify the infection-causing organism or to distinguish between PID and other problems with similar symptoms. A pelvic ultrasound is a procedure that may be helpful in evaluating someone for PID. An ultrasound can view the pelvic area to see whether the fallopian tubes are enlarged or whether an abscess is present. In some cases, a laparoscopy may be necessary to confirm the diagnosis. A laparoscopy is a minor surgical procedure in which a thin, flexible tube with a lighted end (laparoscope) is inserted through a small incision in the lower abdomen. This procedure enables the doctor to view the internal pelvic organs and to take specimens for laboratory studies, if needed.

PID can be cured with antibiotics. If women have pelvic pain and other symptoms caused by PID, it is critical that they seek care immediately. Prompt antibiotic treatment can prevent severe damage to pelvic organs. The longer women delay treatment for PID, the more likely they are to be infertile or to have an ectopic pregnancy in the future because of damage to the tubes. However, antibiotic treatment does not reverse any damage that has already occurred to the reproductive organs.

Because of the difficulty in identifying organisms infecting the internal reproductive organs and because more than one organism may be responsible for an episode of PID, PID is usually treated with at least two antibiotics that are effective against a wide range of infectious agents. These antibiotics can be given by mouth or by vein. The symptoms may go away before the infection is cured. Even if symptoms do go away, women should finish taking all of the medicine. This will help prevent the infection from returning. Women on treatment for PID should be re-evaluated by their health care provider two to three days after starting treatment to be sure the antibiotics are working to cure the infection. In addition, women's sex partners should be treated to decrease the risk of re-infection, even if the partners have no symptoms. Many women with PID have sex partners who have no symptoms, although their sex partners may be infected with the organisms that can cause PID.

About one fourth of women with suspected PID must be hospitalized. Hospitalization may be recommended if the woman is severely ill (e.g., high fever) or pregnant; if she cannot take oral medication and needs intravenous antibiotics; if the diagnosis is uncertain; or in some cases, if she is infected with HIV (human immunodeficiency virus, the virus that causes AIDS). If symptoms continue or if an abscess does not resolve, surgery may be needed. Complications of PID, such as chronic pelvic pain and scarring are difficult to treat but are sometimes improved with surgery.

See also **Antibiotic**; and **Bacterial Diseases**.

Additional Reading

Anderson, J. and L. Dahlberg: "High-risk Sexual Behavior in the General Population. Results from a National Survey 1988-90," *Sex Transm Dis* **19**, 320–325 (1992).

Aral, S.O. and J.N. Wasserheit: "Interactions Among HIV, Other Sexually Transmitted Diseases, Socioeconomic Status, and Poverty in Women," In: O'Leary A. and L.S. Jemmott: *Women at Risk: Issues in the Primary Prevention of AIDS*, Plenum Press, New York, NY, 1995.

Baur, H.M., Y. Ting, C.E. Greer, et al.: "Genital Human Papillomavirus Infection in Female University Students as Determined by a PCR-Based Method," *jama* **265**(4), 472–477 (1991).

Bosch, F.X. et al.: "Prevalence of Human Papillomavirus in Cervical Cancer: A Worldwide Perspective," *J. Natl. Cancer Inst.* **87**, 796–802 (1995).

Bunnell, R., L. Dahlberg, K. Stone, et al.: "Misconceptions about STD Prevention and Associations with STD Prevalence and Incidence in Adolescent Females in a Southeastern City [abstract]," In: Program and Abstracts of the 1998 National STD Prevention Conference, December 1998.

Burk, R.D., G.Y. Ho, L. Beardsely, M. Lempa, et al.: "Sexual Behavior and Partner Characteristics are the Predominant Risk Factors for Genital Human Papillomavirus Infection in Young Women," *J. Infect. Dis.* **174**(4), 679–689 (1996).

Cates, W. et al.: "Estimates of the Incidence and Prevalence of Sexually Transmitted Diseases in the United States. *Sex. Trans. Dis.* **26**(suppl), S2–S7 (1999).

Coleman, P.J., G.M. McQuillan, L.A. Moyer, et al.: "Incidence of Hepatitis B Virus Infection in the United States, 1976-1994: Estimates from the National Health and Nutrition Examination Surveys," *J. Infect. Dis.* **178**, 954–960 (1998).

Cohen, D.A. et al.: "Repeated School-based Screening for Sexually Transmitted Diseases: A Feasible Strategy for Reaching Adolescents," *Pediatrics* **104**(6), 1281–1285 1999.

Eng, T.R. and W.T. Bulter: Institute of Medicine. "The Hidden Epidemic: Confronting Sexually Transmitted Diseases," National Academy Press, Washington, DC, 1997.

Fleming, D.T., G.M. McQuillian, R.E. Johnson, et al.: "Herpes Simplex Virus Type 2 in the United States, 1976 to 1994," *N. Engl. J. Med.* **337**(16), 1105–1111 (1997).

Fleming, D.T. and J.N. Wasserheit: "From Epidemiological Synergy to Public Health Policy and practice: The Contribution of Other Sexually Transmitted Diseases to Sexual Transmission of HIV Infection," *Sexually Transmitted Infections* **75**, 3–17 (1999).

Fox, K.K., W.L. Whittington, W.C. Levine, W.C., et al.: "Gonorrhea in the United States, 1981–1996. Demo-graphic and Geographic Trends," *Sex. Trans. Dis.* **26**(7), 386–393 (1998).

Goldenberg, R.L., W.W. Andrews, A.C. Yuan, H.T. MacKay, M.E. St. Louis: "Sexually Transmitted Diseases and Adverse Outcomes of Pregnancy," In: *Clinics in Perinatology: Infections in Perinatology* **24**(1): 23–41 1997.

Goldenberg, R.L., W.W. Andrews, A.C. Yuan, et al.: "Sexually Transmitted Diseases and Adverse Outcomes of Pregnancy," *Clin. Perinatal* **2491**, 23–41 (March 1997).

Grosskurth, H., F. Mosha, J. Todd, et al.: "Impact of Improved Treatment of Sexually Transmitted Diseases on HIV Infection in Rural Tanzania: Randomized Controlled Trial," *Lancet* **346**, 630–636 (1995).

Hillier, S. and K. Holmes,: "Bacterial vaginosis," In: Holmes, K., P. Mardh, P. Sparling, et al.: "Sexually Transmitted Diseases," 3rd Edition, The McGraw-Hill Companies, Inc., New York, NY, 1999, pp. 563–586.

Ho, G.Y.F., R. Bierman, L. Beardsley, et al.: "Natural History of Cervicovaginal Papillomavirus Infection in Young Women," *N. Engl. J. Med.* **338**(7), 423–428 (1998).

Hook, E.W. III and H.H. Handsfield: "Gonorrhea," In: Holmes, K., P. Markh, P. Sparling et al.: "Sexually Transmitted Diseases," 3rd Edition. The McGraw-Hill Companies, Inc., New York, NY, 1999, pp. 451–466.

Holmes, K.K., P.F. Sparling, P.A. Mardh," "Sexually Transmitted Diseases," 3rd Edition, The McGraw-Hill Companies, New York, NY, 1999.

Klausner J.D. et al.: "Tracing a Syphilis Outbreak Through Cyberspace," *JAMA* **284**, 447–449 (2000).

Koutsky, L.: "Epidemiology of Genital Human Papillomavirus Infection," *Am. J. Med.* **102**(suppl 5A), 3–8 (1997).

Koutsky, L.A. and N.B. Kiviat: In: Holmes, K., P. Mardh, P. Sparling, et al.: "Sexually Transmitted Diseases," 3rd Edition, The McGraw-Hill Companies, Inc., New York, NY, 1999, pp. 347–359.

Krieger, J.N. and J.F. Alderete: "Trichomonas vaginalis and Trichomoniasis," In: Holmes, K., P. Markh, P. Sparling, et al.: "Sexually Transmitted Diseases," 3rd Edition, The McGraw-Hill Companies, Inc., New York, NY, 1999, pp. 587–604.

Ku, L., F.L. Sonenstein, C.F. Turner, et al.: "The Promise of Integrated Representative Surveys About Sexually Transmitted Diseases and Behavior," *Sex Transm. Dis.*: 299–309 (May 1997).

Laumann, E.O., J.H. Gagnon, R.T. Michael, and S. Michaels: "The Number of Partners," In: The Social Organization of Sexuality: Sexual Practices in the United States," University of Chicago Press, Chicago, IL, 1994, 174–224.

Laumann, E.O., J.H. Gagnon, R.T. Michael, and S. Michaels: "Sexual Networks," In: "The Social Organization of Sexuality: Sexual Practices in the United States.," University of Chicago Press, Chicago, IL, 1994, pp. 225–268.

Martinez, J., R. Smith, M. Farmer, et al.: "High Prevalence of Genital Tract Papillomavirus Infection in Female Adolescents," *Pediatrics* **82**, 604–608 (1988).

Mast, E.E., F.J. Mahony, M.J. Alter, and H.S. Margolis: "Progress Toward Elimination of Hepatitis B Virus Transmission in the United States," *Vaccine* (suppl): S48–S51 (1998).

McQuillan, G.M.: "Implications of a National Survey for STDs: Results from the NHANES Survey. Presentation at 2000 Infectious Disease Society of America Conference. September 7–10, New Orleans," 2000.

Mertz, K.J., G.M. McQuillan, W.C. Levine, et al.: "A Pilot Study of the Prevalence of Chlamydial Infection in a National Household Survey," *Sex. Trans. Dis.* 225–228 (May 1998).

Mertz, K.J., D. Trees, W.C. Levine, et al.: "Etiology of Genital Ulcers and Prevalence of Human Immunodeficiency Virus Coinfection in 10 U.S. Cities," *J. Infect. Dis.* **178** (1998).

Moran, J.S., S.O. Aral, W.C. Jenkins, T.A. Peterman, and E.R. Alexander: "The Impact of Sexually Transmitted Diseases on Minority Populations in the United States," *Public Health Rep.* **104**, 560–565 (1989).

Palefsky, J.M.: "Human Papillomavirus Infection and Anogenital Neoplasia in Human Immunodeficiency Virus-Positive Men and Women," *Monogr. Natl. Cancer Inst.* **23**, 15–20 (1998).

Reitmeijer, C.A., K.J. Yamaguchi, C.G. Ortiz, et al.: "Feasibility and Yield of Screening Urine for Chlamydia trachomatis by Polymerase Chain Reaction Among High-Risk Male Youth in Field-Based and Other Nonclinic Settings. A New Strategy for Sexually Transmitted Disease Control," *Sex. Transm. Dis.* **24**(7), 429–435 (1997).

Seidman, S.N. and S.O. Aral: "Race Differentials in STD Transmission," *Am. J. Public Health (letter)* **82**, 1297 (1992).

Shah, K.V.: "Human Papillomaviruses and Genital Cancers," *N. Engl. J. Med.* **337**, 1386–1388 (1997).

Stamm, W.E., In: Holmes, K., P. Mardh, P. Sparling, et al.: "Sexually Transmitted Diseases," 3rd edition. The McGraw-Hill Companies, Inc., New York, NY, 1999, 407–422.

Staff: American Social Health Association: "Sexually Transmitted Diseases in America: How Many Cases and at What Cost?," Research Triangle Park, NC, 1998.

Staff CDC: "Gonorrhea — United States," *MMWR 2000*, **49**(24), 538–542, (1998).

Staff CDC: "1998 Guidelines for Treatment of SexuallyTransmitted Diseases." *MMWR* **47**(RR-1) (1998).

Staff: American College of Obstetricians and Gynecologists (ACOG). "Pelvic Inflammatory Disease. ACOG Patient Education Pamphlet, 1999.

Staff CDC: "Increases in Unsafe Sex and Rectal Gonorrhea Among Men Who Have Sex with Men — San Francisco, 1994–1997," *MMWR* **48**(35), 773–777 (1999).

Staff CDC: "Resurgent Bacterial Sexually Transmitted Disease Among Men Who Have Sex With Men — King County, Washington, 1997–1999," *MMWR* **48**(3), 773–777 (1999).

Staff CDC: "Fluoroquinolone-Resistance in Neisseria gonorrhoeae, Hawaii," 1999, and Decreased Susceptibility to Azithromycin in N. Gonorrhoeae Missouri, 1999," *MMWR* **49**(37), 833–837 (2000).

Staff Division of STD Prevention: "The National Plan to Eliminate Syphilis from the United States, National Center for HIV, STD, and TB Prevention," Centers for Disease Control and Prevention (CDC), Atlanta, GA, October 1999.

Staff Division of STD Prevention: "Prevention of Genital HPV Infection and Sequelae: Report of an External Consultants Meeting," Department of Health and Human Services, Centers for Disease Control and Prevention (CDC), Atlanta, GA, December 1999.

Staff Division of STD Prevention: "Sexually Transmitted Disease Surveillance, 1999," U.S. Department of Health and Human Services, Centers of Disease Control and Prevention (CDC), Atlanta, GA, September 2000.

Staff: National Committee for Quality Assurance (NCGA), HEDIS 2000: Technical Specifications, Washington, DC, 1999, pp. 68–70, 285–286.

St. Louis, M.E. and J.H. Wasserheit: "Elimination of Syphilis in the United States. *Science* **281**, 353–354 (1998).

Wasserheit, J.N.: "Epidemiologic Synergy: Interrelationships Between Human Immunodeficiency Virus Infection and Other Sexually Transmitted Diseases." *Sex. Trans. Dis.* **9**, 61–77 (1992).

Wawer, M.J., N.K. Sewankambo, D. Serwadda, et al.: "Control of Sexually Transmitted Diseases for AIDS Prevention in Uganda: A Randomized Community Trial," Rakai Project Study Group, *Lancet* **353**(9152), 525–535 (1999).

Web References

American Social Health Association: *http://www.ashastd.org/*
CDC NPIN: *http://www.cdcnpin.org/*
Division of Sexually Transmitted Diseases: *http://www.cdc.gov/std/*
National HPV and Cervical Cancer: *http://www.ashastd.org/hpvccrc/*

Centers for Disease Control and Prevention (CDC), Atlanta, GA.

SEXUAL SELECTION. A form of natural selection in which the sex of the individual plays an important part in the selection. In birds, we often find an extreme sexual dimorphism because of the effects of such selection. The male may have brilliant plumage, whereas the female may be much more drab in her plumage. This is supposed to have arisen because of the elaborate courtship of many birds in which the males display their plumage, and it is thought there is some degree of selection of mates by the females. In other species of animals, where the males have battles for possession of the females, as in the fur seals and the moose, selection favors strength and fighting ability in the males. See Fig. 1.

Fig. 1.   Sexual selection in the fur seals. The bull in the upper left has collected his harem of devoted females through countless victorious battles with other males. At the extreme right, an envious bachelor male looks on.

SEYCHELLES PALM.   See Palm Trees.

SEYFERT GALAXY.   Any galaxy having a very bright nucleus showing a high excitation spectrum with rather broad emission lines. Recently, some Seyfert Galaxies have been shown to be radio galaxies. See also **Galaxy**.

SFERICS.   (Also spelled spherics). The study of atmospherics, especially from a meteorological point of view. This involves techniques of locating and tracking atmospherics sources and evaluating received signals (waveform, frequency, etc.) in terms of source.

SFERICS FIX.   The estimated location of a source of atmospherics, presumably a lightning discharge.

SFERICS OBSERVATION.   An evaluation, from one or more sferics receivers, of the location of weather conditions with which lightning is associated. Such observations are more commonly obtained from networks of two or three widely spaced stations. Simultaneous observations of the azimuth of the discharge are made at all stations and the location of the storm is determined by triangulation.

SFERICS RECEIVER.   An instrument that measures, electronically, the direction of arrival, intensity, and rate of occurrence of atmospherics. In its simplest form the instrument consists of two orthogonally crossed antennas. Their output signals are connected to an oscillograph so that one loop measures the north-south component whereas the other measures the east-west component. These are combined vertically to give the azimuth. Also called *lightning recorder*.

SHACKLE.   A shackle is a piece used for connecting together two parts. The parts so connected can have some relative motion permitted by the shackle, but at the same time the extent of their freedom is limited by the restraining action of the shackle.

SHAD.   See **Gizzard Shad**; and **Herring**.

SHADBUSH.   See **Rose Family**.

SHALE.   A fine-grained sedimentary rock whose original constituents were clays or muds. It is characterized by thin laminae breaking with an irregular curving fracture, often splintery, and parallel to the often indistinguishable bedding planes.

SHANNON.   In information theory, a unit of logarithmic measures of information equal to the decision content of a set of two mutually exclusive events expressed by the logarithm to base two; e.g., the decision content of a character set of 8 characters equals 3 Shannons. Synonymous with information content binary unit. (*American National Dictionary for Information Processing*.) See also **Information Theory**.

SHANNON FORMULA.   A theorem in information theory that states that a method of coding exists whereby C binary digits per second may be transmitted with arbitrarily small frequency of error where C is given by

$$C = B \log_2 \left(1 + \frac{S}{N} \right)$$

and no higher rate can be transmitted; B is the bandwidth, and S/N is the signal-to-noise ratio. See also **Information Theory**.

SHAPLEY, HARLOW (1885–1972).   Shapley was an American astronomer who is especially remembered as the man who discovered that the Milky Way galaxy was much larger than originally thought and that Earth's solar system was at not the center of the galaxy.

Interestingly, Shapley had planned to study journalism in college but switched to astronomy. He did graduate work at the Princeton Observatory. His dissertation was on properties of stars known as eclipsing binary stars. Later he expanded his thesis and published it making a significant contribution to the field of astronomy.

Shapley made some of his most important discoveries while researching at the Mount Wilson Observatory in Pasadena, California. Shapley found that Cepheid variables were pulsating stars. He was able to determine a star's absolute brightness and then by comparing this to its observed brightness, he was able to calculate how far away the star was. From this measuring system, Shapley determined the placement of Earth's solar system in the Milky Way. He was also able to estimated that the center of the galaxy was located fifty thousand light-years from the sun.

In 1921, Shapley returned to Harvard Observatory and became its director. He was elected to the National Academy of Sciences in 1924 and in 1926 received the Draper Medal.

Shapley is credited with making Harvard Observatory the leading astronomy educational center in the United States during the 1920s.

Shapley recruited scientists worldwide and was instrumental in helping to bring many German Jewish scientists to America. After World War II, he helped form the United Nations Educational, Scientific and Cultural Organization. He is sometimes criticized for urging the United States to have scientific exchange and cooperation with the Soviets. Although his feelings on this issue were unpopular at the time, he was elected president of the American Association for the Advancement of Science. He spent his later years traveling and lecturing on astronomy and philosophy. He wrote the popular book, *Through Rugged Ways to the Stars*.

See also **Cosmology**; and **Galaxy**.

J. M. I.

SHARED TIME CONTROL.   In an automatic control system, control action in which one controller divides its computation or control time among several control loops rather than acting on all loops simultaneously.

SHARKS (*Chondrichthyes*).   Of the order *Selachii*, there are several families of sharks. A shark is a carnivorous fish with a cartilaginous skeleton. The mouth opens on the ventral surface of the head and is armed with many rows of sharp teeth attached to the skin and similar in structure to the placoid scales of the body. The tail is of the heterocercal form, having two lobes with the backbone extending into the upper lobe. The openings of the gill slits are separate. Of the principal families of sharks, there are: (1) the frilled shark (*Chlamydoselachidae*); (2) the sixgill and sevengill cowsharks (*Hexanchidae*); (3) sand sharks (*Carchariidae*); (4) goblin sharks (*Scapanorhynchidae*); (5) mackerel sharks (*Isuridae*); (6) thresher sharks (*Alopiidae*); (7) the basking shark (*Cetorhinidae*); (8) the whale shark (*Rhincodontidae*); (9) catsharks (*Scyliorhinidae*); (10) false catsharks (*Pseudotriakidae*); (11) smooth dogfishes (*Triakidae*); (12) the requiem sharks (*Carcharhinidae*); (13) hammerhead sharks (*Sphynidae*); (14) hornsharks (*Heterodontidae*); (15) saw sharks (*Pristiophoridae*); (16) spiny dogfishes (*Squalidae*); (17) spineless dogfishes (*Dalatiidae*); (18) the alligator dogfish (*Echinorhinidae*); (19) the angel sharks (*Squatinidae*); and (20) carpet and nurse sharks (*Orectolobidae*).

Because of their danger and threat of danger, sharks have received much attention — over the many hundreds of years that people have been swimming in the seas, whether for recreation or as the result of accidents at sea. However, the interest in the dangers of sharks increased markedly with the advent several years ago of diving (for military or constructional purposes), and particularly with the great expansion of skin diving as a hobby and sport. The lore pertaining to the true danger and the habits of sharks with relation to the presence of people obviously would consume many books. Sharks are well known for their curiosity. The problem, of course, is to differentiate between their curiosity and their possible more serious, aggressive intentions.

Many sharks are docile by nature and require certain stimuli to alter their temperament from one of curiosity to one of aggression and attack. An almost certain stimulus is the presence of blood or fish juices in the water. Thus, in shark-infested waters, a diver never should retain any speared or otherwise injured fishes in the vicinity of skin-diving operations. Records indicate that sharks may be interested only in the injured fishes and not in the persons nearby, but that, in seeking out the source of fish juices and blood, the sharks may make mistakes and thus cause injury to persons in the area. Also, sharks do not like to be disturbed — they do not like to have objects in their way and they do not like to be touched, as for example a skin diver attempting to catch a shark by the tail. A shark may suddenly reverse direction in such a situation and attack the diver. Because of the sensitive and unpredictable temperaments of many sharks, the diver never should consider a shark "friendly" or become careless as the result of past interesting and uneventful encounters with sharks. For example, it has been postulated by some authorities that just as among people, there will be found the occasional psychotic shark that does not conform to normal habits and patterns of behavior. In all situations, of course, the size of the shark as an index of potential danger should always be considered. It is a good rule to consider any shark that is over 10 feet (3 meters) in length as a potential source of trouble.

Sharks are also known to be rather indiscriminate when they are feeding. Experiments with sharks in large aquariums, for example, have indicated that during a feeding "frenzy," a shark may easily mistake an undesirable food item for the items that it really wants most. Feeding sharks in captivity always must be done with precaution. The sharks, obviously, are not aware of where their food ends and where the feeder's hand begins. Further, during a period of feeding excitement, a captive shark may attack and eat other creatures in the aquarium that it normally would not bother.

Some authorities have suggested that there may be as many persons injured by shark bumps as by shark bites. There is considerable postulation in attempts to describe the reasons behind such statistics. ("Shark Attack" by V.M. Coppleson, published by Angus and Robertson, Sydney, Australia, 1959).

There appears to be a correlation between the temperature of the water and the feeding habits of some sharks. Statistics tend to show that fewer shark attacks occur in waters below 60 °F (~16 °C) and that the number goes up in waters above 70 °F (~20 °C). This correlates with the fact that the sharks most dangerous to people are found in tropical waters. Also, shark attacks in temperate waters are usually limited to summer months. This fact, of course, could be easily explained because there are more people in the water during the summer than at other times. In the study by Dr. Coppleson, it is also pointed out that shark attacks in Australian waters tend to occur in mid-afternoon. Statistics also show that sharks prefer male to female human victims (a ratio of 20:1).

Much investigation has gone into the development of shark repellents. This topic has always been of interest to the military (abandonment of aircraft over shark-infested waters; activities of naval divers, etc.). One relatively successful repellent developed during World War II was copper acetate containing a nigrosene dye (ratio of 20–80%). This was found quite effective with Atlantic sharks, but quite ineffective with Pacific sharks. This difference still requires further investigation. It has been noted that captive sharks often will not feed if the aquarium water contains copper compounds introduced for controlling parasites. Apparently, these compounds cause an irritation of the sharks' nostrils, thus disturbing their sense of smell and consequent desire for food.

Because of its preference for deep waters, the data on the frilled shark are not extensive. The number of specimens caught has been limited. The adult attains a length of about $6\frac{1}{2}$ feet (2 meters), the young are born alive, the coloring is brown with no particular markings.

The more primitive sixgill and sevengill cowsharks are identified by counting the gill openings on each side of the head. The main groups of sharks always possess five gill openings per side. Coloring of cowsharks varies from gray to brown with no particular markings. A giant sixgill shark was caught in English waters over a hundred years ago. It measured over 26 feet (7.9 meters) in length. This is the largest sixgill shark on record. They are widely distributed in Atlantic and Pacific waters. The young are born alive, seldom exceeding 16 inches in length. The sixgills apparently prefer deep water during daytime hours, but frequent the surface after dark for feeding. The Pacific sevengill is the broad-headed *Notorhynchus maculatum*. It prefers offshore waters and is rarely seen in shallow water. The shark is dark gray with black spots. There is a nursery ground of these sharks in the southern end of San Francisco Bay, an exception to their preference for deeper water. Generally, the sevengills are not considered edible.

Although much remains to be learned about sharks, a number of past misconceptions have been exposed in recent years. For example, sharks are not ravenous eaters. Grey reef adults feed only once in six or even twelve days. Species living on the sea bottoms go without food for weeks, and the big basking sharks found in British waters appear to starve all winter. Nor, as once believed, do sharks depend entirely upon their sense of smell. This sense, of course, is extremely acute. It has been estimated that sharks can detect (by smell) dilutions of one part in a million. Their sense of hearing also registers sounds hundreds of meters away. Although the shark's vision is lacking in resolution of details, it does perform well under poor lighting conditions. Sharks also are able to sense electric fields in surrounding waters and may use this sense as one means of navigation. Although sharks must continue to swim to avoid sinking, only the very slightest movements are required by some species. It has been observed that the Australian gray nurse shark can virtually hover. It has been noted that the great white maneating shark can nearly pivot on its nose. The outstanding buoyancy of some shark species is attributed to very large livers containing light oils and fats, which act as swim bladders. At one time, it was believed that sharks had to keep moving in order to breathe, requiring a continuous flow of water past the gills, but it has been shown that many species, including some of the largest, such as the tiger and bull sharks, can achieve temporary breathing through muscular pumping.

The sand shark is a dangerous-looking creature with large mouth and wicked teeth. The sand tiger shark typifies one's visions of a dangerous,

wicked shark. There are no records of attack on people on the American side of the Atlantic, but it is highly feared in South African waters. The largest recorded specimen measured about $10\frac{1}{2}$ feet (3.2 meters) in length, with a weight of over 300 pounds (136 kilograms).

Of the mackerel sharks, the most famous is the great white shark (*Carcharodon carchiarias*), also known as the maneater or maneating-shark. According to records, one of the largest maneaters was caught off Port Fairey, Australia in 1870. The jaws of this shark are preserved and on display at the British Museum. The shark has been reported as measuring about $36\frac{1}{2}$ feet (11.1 meters) in length. However, the majority of maneaters caught have ranged between 20 and 25 feet (6.1 and 7.6 meters) in length. The bodies of these sharks are massive and hence a shark of only 17 feet (5.2 meters) in length may weigh up to nearly 3,000 pounds (1361 kilograms). The weight record may be held by a 21-foot (6.4 meters) shark caught in Cuban waters. It is estimated to have weighed about 7,000 pounds (3175 kilograms). J.E. Randall (*Science*, **181** (4095), 169–170 (1973)) casts some doubts over these previously reported dimensions.

The temperament of the maneater is considered very bad—with the vicious habit of considering just about anything it sees as edible. Some of the contents found in the stomachs of captured maneaters have included the remains (in some cases, the intact bodies) of large dogs, seals and sea lions themselves weighing in excess of 100 pounds (45.3 kilograms), and, interestingly, other sharks that may measure up to 7 feet (2.1 meters) in length.

The mako or sharp-nosed mackerel shark is known for its great activity once hooked, displaying marlin-like maneuvers and swimming much faster than its relative, the maneater. The *Isurus oxyrhinchus* is the Atlantic mako. The largest specimen on record attained a length of about 12 feet (3.6 meters) and weighed around 1,200 pounds (544 kilograms). The *Isurus glauca* is the Indo-Pacific mako. Both species of makos prefer tropical seas. The makos tend to gulp their food, as evidenced by finding large intact items in their stomachs. In one instance, a 120-pound (54.4-kilogram) swordfish was found in the stomach contents of a Bahaman mako which weighed over 700 pounds (317.5 kilograms). Other species of mako sharks include the *Lamna nasus* (common Atlantic mackerel shark, sometimes called porbeagle); and the *Lamna ditropis* (Pacific mackerel shark).

The thresher shark is noted for a very long whip-like tail. This may equal the length of the body, providing the fish with power and maneuverability. They are of offshore, tropical distribution. Large specimens run about 20 feet (6.1 meters) in length and weigh up to about 1,000 pounds (453.6 kilograms). The number of species of thresher sharks is quite limited.

The basking shark is well named because of its apparent preference to spend much time simply floating along the surface or cruising at very slow speeds. The second-largest known shark is *Cetorhinus maximus*, the giant basking shark, with recorded lengths up to 45 feet (13.7 meters). Very much like a mackerel shark, the basking shark differs in its preference for plankton rather than carnivorous food. It is believed that the young are born alive, probably up to 6 feet (1.8 meters) in length. Although found worldwide in temperate waters, there are significant concentrations reported in waters off southern California and Europe. Fishing for basking sharks is considered important commercially, notably because of the value placed upon their large livers (yielding oil). For example, a liver weighing over 850 pounds (386 kilograms) may be found in an 8,500-pound (3856-kilogram) shark about 30 feet (9.1 meters) in length. Unfortunately, the liver oil contains no vitamins, but it is used for various tanning processes.

Regarded as the largest of fish, the whale shark attains a length of 45 feet (13.7 meters) at minimum. There have been numerous reports and writings over the years pertaining to the docile nature of these large creatures. Although the whale shark feeds on very small substances for food, such as small fishes, squid, and crustaceans, it nevertheless possesses numerous small teeth. The sharks feed in a vertical position. In American Atlantic waters, the population center is around the Caribbean. There is a concentration in the Pacific off the Gulf of California. Whale sharks also have been seen in the Red Sea and off the Philippines.

Catsharks have beautiful stripes, bars, and mottling, prefer inshore waters, and reach a maximum length of about 3 feet (1 meter). They are elasmobranches (cartilaginous skeleton, plate-like scales, lack of air bladder). The common European spotted dogfishes (*Scyliorhinus caniculus* and *S. stellaris*) are catsharks (even though named dogfishes).

The *S. retifer*, also called the chain dogfish, is found in the waters off New Jersey. The South African "skaamoog" (*Holohalaelurus regani*) has markings suggestive of Egyptian hieroglyphics. The *Cephaloscyllium*, which attains a length of about 4 feet (1.2 meters), ranges the eastern Pacific, but is absent from Atlantic waters. The *Cephaloscyllium uter* is also known as a swell shark and occurs from Monterey Bay in California to Lower California waters. Its name is derived from the fact that, upon being pulled from the water, it imbibes large quantities of air, sometimes swelling to twice its normal size. If released to the water, the air is expelled and the fish returns to normal dimensions.

Only two species of the false catsharks are known. These sharks are quite large and prefer deep water. They are rarely seen. The identifying marking of these sharks is the long base of the dorsal fin. The largest specimen from the Atlantic measured somewhat less than 10 feet (3 meters). Of the specimens captured, all have been taken from waters ranging from 1,000 to 5,000 feet (305 to 1524 meters) in depth.

Small in size, usually less than 5 feet (1.5 meters) in length, there are relatively few species of the smooth dogfishes, which are more or less intermediate between nurse sharks and catsharks. Very abundant in American Atlantic coastal waters is the *Mustelus canis*, ranging from as far south as Uruguay northward to Cape Cod. They have been carefully studied and their biological traits and life history are well known. The smooth dogfish is known for its ability to change its coloration. It is not considered edible. *Rhinotriacis henlei* is the most abundant of sharks in American Pacific coastal waters. Other members of the family of smooth dogfishes (*Triakidae*) include the spotted shark (*Mustelus punctulatus*) found in the Mediterranean and the waters of South Africa. Achieving a length of about 6 feet (1.8 meters), this shark is covered with tiny black spots, hence its name. The *Triakis semifasciata* (leopard shark) is found in American Pacific coastal waters, ranging from Lower California northward as far as Oregon. San Francisco Bay is a nursery ground for this species.

There are well over 60 species of requiem sharks (*Carcharhinidae*). They enjoy the characteristics of typical sharks. The *Galeocerdo cuvieri* (tiger shark), shown in Fig. 1, is omnivorous, eating birds, fishes, animals, garbage, coal, turtles and so on. Its base color is a grayish-brown with a lighter-colored undersurface. There usually is some mottling along the upper surfaces. The tiger shark attains a length of about 18 feet (5.4 meters) in American waters; probably about 14 feet (4.2 meters) in Australian waters. The weight may be as much as 1,400 pounds (635 kilograms). Preferring deep water, the tiger shark on occasion will follow prey into shallow waters. Commonly found in the Caribbean and the waters around Florida, the species is not known along the northern European coasts or in the Mediterranean.

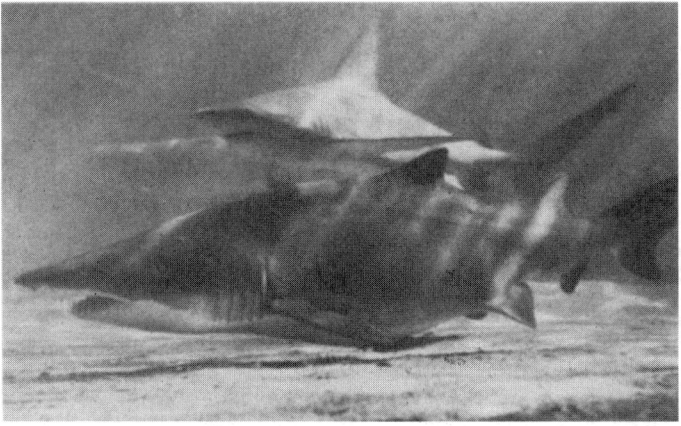

Fig. 1.   Tiger shark. (*A.M. Winchester.*)

During World War II, shark livers became an important source of vitamins as the result of the blockading of Norway, since cod-liver oil was previously obtained from Norway. Thus, so-called soupfin sharks were commercially sought, even on the California coast, until about 1950 when low-cost synthetic vitamins appeared. The best sharks for liver were the soupfin (*Galeorhinus zyopterus*) and the dogfish (*Squalus acanthias*). The *Galeorhinus australis* (Australian school sharks) bears resemblance to the American Pacific soupfin. However, the latter is not considered a food fish, whereas the school shark is important commercially.

The great blue shark is also a requiem shark, well named because of the indigo blue coloration of its upper surfaces. It attains a length of nearly 13 feet (3.9 meters) and is long, slender, and streamlined, with a sharp nose. It is considered a sporting fish, but not edible. Common along the American Atlantic coast is the lemon shark (*Negaprion brevirostiris*), frequenting the waters from Brazil as far northward as North Carolina. These sharks reach a length of from 7 to 11 feet (2.1 to 3.3 meters). They prefer inshore shallow waters. Biological studies of the lemon shark have been carried out at Florida's Cape Haze Marine Laboratory. The silky shark (*Carcharhinus floridanus*) is also a requiem shark and is abundant along both Atlantic and Pacific coasts. Of interest, despite its abundance, is the fact that it was not officially identified and catalogued until 1953. The white-tip shark (*Carcharhinus longimanus*) with a gray coloration is probably the most common of sharks in the Atlantic and Pacific coastal waters. It reaches a length of about 13 feet (3.9 meters).

A number of sharks can tolerate brackish water and may live for awhile in fresh water, but the only known species that lives permanently in fresh water is the maneating *Carcharhinus nicaraguenis* which inhabits Lake Nicaragua. This shark is gray with a very heavy body, reaching a maximum length of about 8 to 10 feet (2.4 to 3 meters).

Because of the most unusual and well-named head, the hammer-head shark (*Sphyrnidae*) is readily identified. It is interesting to note that in a large specimen of, say, 15 feet (4.5 meters) in length and weight up to 1,500 pounds (680 kilograms) the eyes will be separated from each other by as much as 3 feet (1 meter). Because of its peculiarly shaped head, another member of this family of sharks is called the "shovel-head." Hammerheads prefer tropical seas, but move north into temperate waters in the summer. In 1959, there was an epidemic of shark attacks on the West Coast of the United States, attributed in the main to hammerheads. The unusual shape of the hammer-type head has aroused much biological interest. Some experts postulate that the hammer may serve as a balancing mechanism. Because hammerheads do not survive long in captivity, studies have been difficult.

Additional Reading

Bright, M.: "Sharks," Smithsonian Institution Press, Washington, D.C., 2002.
Paxton, J.R. and W.N. Eschmeyer: "Encyclopedia of Fishes," Academic Press, Inc., San Diego, CA, 1998.
Perrine, D.: "Sharks and Rays of the World," Voyageur Press, Inc., Stillwater, MN, 1999.
Sharth, S.: "Sharks and Rays," Grolier Publishing, Danbury, CT, 2002.
Waller, G.: "SeaLife: A Complete Guide to the Marine Environment," Smithsonian Institution Press, Washington, DC, 1996.

SHASTA FIR.   See **Fir Trees.**

SHAULA (λ Scorpii).   Ranking twenty-third in apparent brightness among the stars, Shaula has a true brightness value of 1,700 as compared with unity for the sun. Shaula is a blue-white, spectral type B star and is located in the constellation Scorpius, a zodiacal constellation. Estimated distance from the earth is 300 light years. See also **Constellations.**

SHEAR.   A force that lies in the plane of an area or a parallel plane is called a shearing force. It is the force that tends to cause the plane of the area to slide on the adjacent planes.

The vertical shear for any section of a simple beam is the magnitude of the resultant of the transverse loads on either side of the section. Transverse loads are those which are at right angles to the length of the beam. If the loads are inclined, the vertical components, only, should be used in computing the vertical shear. The resisting shear at any section is the internal force that opposes the shearing action of the external loads. It is numerically equal to the external shear but in the opposite direction. Vertical shear, which is always accompanied by bending movement at a section of a beam, is numerically equal to the rate of change of this moment with respect to distance along the beam. This shear is arbitrarily assumed to be positive if the resultant of the vertical loads to the left of a section acts in an upward direction. In a symmetrically loaded simple beam the shear is equal to zero at the center of the beam. See also **Elasticity**.

In addition to vertical shear in a beam, there is always a horizontal shear which is a result of the difference in the flexural stresses (see also **Flexure**) between any two vertical planes. The tendency of adjacent horizontal planes to slide upon each other is caused by horizontal shear. The effect may be better understood by visualizing a beam composed of flat planks laid one on top of the other. As the beam bends, due to the applied loads, the bottom of one plank will slide upon the top surface of the one beneath it unless this effect is restrained by friction, nails, bolts, or other fastenings.

The unit stresses resulting from vertical shear are called vertical shearing stresses. At any point in a beam, these stresses are numerically equal to the horizontal shearing stresses. The variation of the unit shearing stresses over the cross section of a rectangular beam is parabolic, being equal to zero at the top and bottom surfaces and a maximum at the neutral axis. When horizontal and vertical shear, only, act at a point in a body, the body is said to be in a state of pure shear at the point.

Shear is not restricted to beams. It occurs wherever there is bending. Columns, which are subjected to bending caused by eccentric loads, or by inclined or lateral loads, must be designed to withstand the shearing stresses. Rivets and welds are also subjected to shear. If the riveted connection is made so that the shear occurs between two plates only, it is called a single shear. When the type of connection is such that the shearing force is opposed by resisting shears acting on two planes, as in the case of three plates riveted together, the condition is called double shear.

A shear diagram is a graphical representation of the variation of vertical shear on a beam. Figure 1 is an illustration of a shear diagram for an overhanging beam with a uniformly distributed load covering the entire length of the beam. The points where the shear changes sign are points of maximum bending moment. The area of the shear diagram between any two points is equal to the change of bending moment between these points.

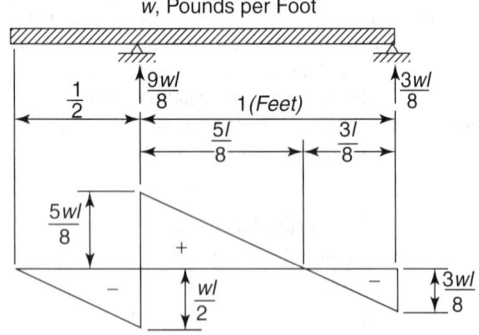

Fig. 1.   Shear diagram.

SHEAR CENTER.   The point through which the external shear must act at any cross section of a beam in order to eliminate torsional stresses is called the shear center of the particular cross section in question. The shear center coincides with the centroid of the internal shearing forces on the cross-section. If the beam is to bend without twisting, the loads must be applied in such a manner that the external shear at any section will pass through the shear center. The shear center has no meaning for sections where pure bending occurs because there is no shear and, therefore, no torsional stress can exist under these conditions.

SHEATHBILL.   See **Waders, Shorebirds, and Gulls**.

SHEEP.   See **Goats and Sheep**.

SHEEPSHEAD.   See **Porgies**.

SHEEP TICK *(Insecta, Diptera)*.   A wingless parasitic fly, *Melophagus ovinus* (Linne), that resembles a true tick only in its flattened body and leathery texture. Like the other flies, it has sucking mouth parts and hence draws blood from the skin of the host. Some scientists have described the insect as a degenerate, louse-like fly that has completely lost its wings. The insect is especially harmful to lambs. The tick is irritating to the sheep, causing it to rub and scratch, sometimes injuring the fleece and reducing its marketable value. It migrates from ewes to lambs at shearing time. It also attacks goats. Distribution is worldwide in all sheep-producing regions.

The adult sheep tick is about $\frac{1}{4}$ inch (6 millimeters) in length and has a covering of spiny hairs. The tick is also present on sheep in another stage of development, when it is somewhat like a brown seed, tenaciously fastened to the hair of the animal, especially inside the thighs, along the belly, and

around the neck. This is the pupal stage of the fly. The sheep tick does not lay eggs, but is live-borne. The young are about $\frac{1}{8}$ inch (3 millimeters) in length, oval in shape, and off-white in color.

Numerous commercial preparations are available as sprays and dips for controlling the sheep tick. Such formulations may include lindane, malathion, rotenone, or toxaphene. The natural pyrethrins or allethrin are also effective. Care must be taken in applying control chemicals and to make certain that such applications are not made just prior to shearing or marketing. Sheared wool should not be kept near flocks for fear of infestation traveling from one to the other and, after dipping or spraying, the animals should be turned into an area that is considered free of ticks. Previous areas of infestation should be vacated for up to 6 weeks.

SHELL.   A hard external covering secreted by folds of the body wall of many animals. The term applies properly to the shells of Brachipoda and Mollusca, although it is sometimes used in reference to the hard exoskeleton of crustaceans. Also a hard covering of eggs.

Brachiopod shells consist of two valves, one upper and one lower. They have an outer layer of organic matter known as the periostracum, under it a thin layer of calcium carbonate, and a thick inner layer of mixed organic and calcareous matter, deposited in prismatic form. The valves are opened and closed by a complex system of muscles.

Molluscan shells are usually spiral in form, like many common snail shells, or bivalve like those of mussels and the oyster. The valves of such shells are lateral in position. A third rarer form is the shell of nautilus, which is spirally coiled in one plane. Internal shells of slugs, chitons, and some cephalopods are in the form of plates of calcareous matter.

The external shells of mollusks have an external horny layer, the periostracum, a smooth lining of nacre or mother-of-pearl, and a thick calcareous middle layer.

Because of their permanence and beauty the shells of many marine mollusks have attracted the attention of collectors, and many have received common names. Among them the spiral staircase or wentletrap, periwinkle, conch, finger shell, cowry, coffee-bean, apple-seed, oyster drill, whelk, papal miter, cockle, gem, and others are to be found on the coasts of the United States.

The shells of eggs are calcareous or chitinous coverings secreted by a portion of the female reproductive ducts known as the shell gland. In the birds they are characteristically and often beautifully colored, and in the insects they may be beautifully sculptured. They are sometimes perforated by a minute opening or group of openings called the micropyle for the entry of the sperm. The egg shell of insects is also called the chorion, a term not to be confused with the chorion of vertebrate embryos. See also **Mollusca**.

SHELL (Atomic).   See **Chemical Elements**.

SHELLAC.   A secretion or excretion of the lac insect, *Coccus lacca*, found in the forests of Assam and Siam. Freed from wood it is called "seed lac." It is soluble in alkaline solutions such as ammonia, sodium borate, sodium carbonate and sodium hydroxide, and also in various organic chemicals. When dissolved in acetone or alcohol, shellac yields the familiar shellac varnish of superior gloss and hardness. Orange shellac is bleached with sodium hypochlorite solution to form white shellac. See also **Paints and Coatings**.

SHELLFISH.   Any sea animal that is protected by a hard shell, such as crabs, lobsters, crawfish and crayfish, and shrimp of the class *Crustacea*; and clams, quahogs, oysters, and snails of the phylum *Mollusca*. See also **Crustaceans (Edible)**; and **Mollusks**.

SHELLFISH POISONING.   See **Foodborne Diseases**.

SHELL MOLDING.   The forming of a mold cavity for casting metals by placing sand bonded with a thermosetting resin against a preheated metal pattern (150 to 300 °C).

SHEPPARD'S CORRECTIONS.   If a continuous frequency distribution is grouped, the moments derived by replacing each observation by the central value of the group into which it falls differ from those of the original distribution. Provided the distribution tapers smoothly to zero in both directions, average corrections to the grouped moments, known as Sheppard's corrections, may be applied as follows:

$$\mu_1 \text{ (corrected)} = \mu_1$$
$$\mu_2 \text{ (corrected)} = \mu_2 - h^2/12$$
$$\mu_3 \text{ (corrected)} = \mu_3$$
$$\mu_4 \text{ (corrected)} = \mu_4 - \tfrac{1}{2}h^2\mu_2 + 7h^4/240$$

SHERARDIZING. The process for applying an adherent protective coating of zinc to steel parts by heating at $700\,°F$ ($371\,°C$) in contact with zinc dust in a rotating-drum container.

SHIELD (Geology). From a tectonics viewpoint, a shield is a large area of exposed basement rocks in a craton commonly with a very gentle convex surface, surrounded by sediment-covered platforms; e.g., Canadian Shield, Baltic Shield. The rocks of virtually all shield areas are Precambrian. See also **Craton**. From a paleontological viewpoint, a shield is: (a) a protective cover or structure on an animal, likened to or resembling a shield; e.g., an ossicle of an ophiuroid, or the carapace of a crustacean, or a large scale on the head of a lizard; (b) a float or curved, lateral outgrowth at one or more levels of a tangential rod or needle in the skeleton of an acantharian radiolarian and forming by fusion the lattice shell; (c) one of the discoidal elements of the placolith coccolith.

SHIGELLOSIS. See **Foodborne Diseases**.

SHINGLES. See **Dermatitis and Dermatosis**.

SHIPWORM (*Mollusca, Lamellibranchiata*). A peculiar marine mollusk that bores into submerged wood and apparently is among the few animals that can digest cellulose and related materials. The shipworms belong to several genera of which *Teredo* is most often cited. All are slender worm-like creatures but they have the characteristic structures of the bivalves. The valves of the shell are small separate parts, located at the anterior end of the worm, and are used for excavating the burrow.

Shipworms do great damage to wooden hulls and marine piling, consequently they have been subject to detailed studies to determine methods of avoiding their destructive attacks.

SHITTIMWOOD. See **Acacia Trees**.

SHOCK SYNDROME. In the simplest of terms, shock is the inadequate delivery of blood to the major organs of the body. Unless immediately treated, deprivation of blood supply causes disturbance of the metabolism (sometimes a shift from aerobic to anaerobic metabolism at the cellular level) of the organs with resultant damage. Because of these profound consequences, the treatment of shock is considered an emergency procedure. Cellular damage may be reversed with very prompt treatment, but is otherwise irreversible, leading to the ultimate death of the patient. Recovery from shock depends upon promptness of treatment and the age and general underlying health of the patient. Authorities suggest five broad categories of shock, based upon causation.

(1) *Hypovolemic shock.* Several conditions cause a massive loss of blood, plasma, or extracellular fluid from the body or intravascular compartment. The latter may be lost from the gastrointestinal tract from vomiting or diarrhea, abusive use of diuretics, in extensive burns, as well as acute pancreatitis. The most common loss of blood and plasma is encountered in hemorrhagic shock, as that which may result from serious trauma or severe gastrointestinal bleeding. The arterial blood pressure is lowered, causing a deficient supply of blood to tissues, as the result of loss of fluid volume from the vascular space.

(2) *Septic shock.* This may occur as the result of septicemia caused by a gram-negative bacterial infection. Bacterial endotoxin (a complex lipopolysaccharide) is released into the bloodstream. Septic shock is only infrequently a consequence of infections by gram-positive organisms, viruses, fungi, or rickettsias. Sequestration and pooling of blood in various vascular compartments lowers the availability of blood for the perfusion of other vital organs.

(3) *Cardiogenic shock.* This may result from a massive myocardial infarction caused by extensive damage to the myocardium. This sometimes occurs in connection with cardiac surgery; less commonly by severe myocarditis. Cardiogenic shock also may be caused by an arrhythmia in a patient with serious heart disease. In essence, the heart fails to pump, causing a reduction in cardiac output and arterial pressure.

(4) *Obstruction to cardiac filling.* Where cardiac filling is prevented or lessened, as by a massive pulmonary embolism or tumors or other space-occupying lesions, this type of shock may be precipitated.

(5) *Neurogenic shock.* In this type of shock, there is loss of vasomotor tone. This may arise from general or spinal anesthesia, an injury to the spinal cord, or from the massive intake of depressant drugs, such as certain narcotics and barbiturates. Respiratory arrest causing sustained hypoxia also may be a factor in this syndrome.

Depending upon the type of shock, various drugs are administered, including the catecholamines, such as norepinephrine, epinephrine, metraminol, dopamine, and isoproterenol. These drugs increase the arterial perfusion pressure. Vasodilating agents also may be used, particularly in connection with the treatment of septic shock and in some cases of hemorrhagic shock. These drugs, which include sodium nitroprusside, nitroglycerin, isosorbide dinitrate, phentolamine, and adrenal corticosteroids, must be administered with extreme care while the patient is constantly monitored. In cases of cardiogenic shock, circulatory assistance may be provided by the use of intra-aortic balloon counterpulsation. The balloon is inserted through a femoral artery and positioned, usually with the assistance of fluoroscopy, in the descending thoracic aorta. The balloon is programmed from the electrocardiogram such that it will deflate just prior to systole and inflate in diastole. Even though the use of intra-aortic balloon counterpulsation has been successful in many cases, it is estimated that to date it has increased the survival rate from cardiogenic shock by only 5 to 10% of cases. Where possible, if the patient has survived for 2 days with balloon counterpulsation and if other conditions are favorable, angiography and coronary artery revascularization will be attempted. See also list of entries given in the entry on **Heart and Circulatory System (Human)**.

Additional Reading

Berger, P.B., D.R. Holmes, and A. Battler: "Cardiogenic Shock: Diagnosis and Treatment," Humana Press, Totowa, NJ, 2002.
Dhainaut, J.F., L.G. Thijs, and G. Park: "Septic Shock," W.B. Saunders Company, Philadelphia, PA, 1999.
Evans, T.J.: "Septic Shock: Methods and Protocols," Vol. 36, Humana Press, Totowa, NJ, 1999.
Neugebauer, E.A. and J.W. Holaday: "Handbook of Mediators in Septic Shock," CRC Press, LLC., Boca Raton, FL, 1993.
Rietschel, E.T. and H. Wagner: "Pathology of Septic Shock," Vol. 216, Springer-Verlag Inc., New York, NY, 1996.

SHOCK TUBE. A relatively long tube or pipe, in which very brief high-speed gas flows are produced by the sudden release of gas at very high pressure into a low-pressure portion of the tube; the high-speed flow moves into the region of low pressure behind a shock wave.

SHOCK WAVE. Infinitesimal disturbances in a fluid medium are propagated with a characteristic speed known as the sound speed. When the restriction on the amplitude of the disturbance is lifted, the linear approximation breaks down and the velocity of propagation becomes dependent on the amplitude of the disturbance. Another feature of this phenomenon is that the forward gradient of the disturbance rapidly steepens until it becomes a discontinuity and propagates as such. A *shock wave* is then a discontinuity in the physical properties of a fluid medium which propagates through the medium at supersonic velocity without further change. The strength of the shock is defined by the Mach number, the ratio of its velocity to the undisturbed sound speed. Such waves are generated by the detonation of explosive material, by high-speed aircraft and missiles, and by earthquakes. See also **Earth Tectonics and Earthquakes**.

Since all media are necessarily discrete, a true discontinuity is inconceivable, but, as the thickness of the shock transition corresponds to only a few mean free paths in a gas (or internuclear distances in a solid), the transition can be treated as a discontinuity to the same extent that the medium can be regarded as continuous. In comparison with an adiabatic or isentropic change, the shock wave is an irreversible process and hence leads to an increase in the entropy of the material. The pressure, density, and temperature of the medium are all raised on passage through the shock and the flow velocity is reduced. The latter is easily understood by observing that, with respect to the moving front, the molecules enter with an ordered flow motion at supersonic speed and the transport processes in the front transform a major fraction of this ordered flow into the random temperature or kinetic motions of the molecules.

The extent to which the various properties change through the transition depends on the magnitude or strength of the shock and on the thermodynamic properties of the fluid. For an essentially incompressible material, such as a liquid or solid, the major change normally occurs in the pressure variable, whereas for a gaseous medium the most significant change is in the temperature. Although shock waves in solid and liquid materials have been used to study physical properties at high pressures, the method is rather limited by the small test times available before the interaction of other wave phenomena that prevent the attainment of thermodynamic equilibrium, and it is in gases that shock waves have proved of most interest.

The detailed behavior of the shock transition is in itself a most important subject for study, since shock waves are associated with the flight of supersonic aircraft and with the reentry of ballistic missiles into the Earth's atmosphere. In addition to their own intrinsic interest, shock waves are important for other reasons. Since the transition involves the translational motions of the molecules, energy must eventually be transferred into other modes before the system reaches equilibrium. These subsequent relaxation processes involve the rotation, vibration, chemical reaction, electronic excitation, and even ionization of the molecules if the shock is sufficiently strong. The shock phenomenon thus provides an excellent method for studying energy transfer processes. For chemical reactions in particular, the shock wave provides a source of heat that is essentially instantaneous and is completely homogeneous. Also, provided the thermodynamic properties of the medium are known, the temperature is completely defined by a determination of the shock velocity.

Although shock waves can be created in many ways, including the detonation of high explosives and in wind tunnels, the simplest technique makes use of the *shock tube*, discovered in 1899 by Vieille. A long tube of uniform cross section is divided into two parts by a thin diaphragm and gas is admitted to these at different pressures. If the diaphragm is ruptured in some way, a shock wave is generated in the low-pressure gas and a corresponding rarefaction, or expansion, wave in the driver gas. Because the motion is restricted to a single dimension by the containing walls, the strength of the shock does not decrease with distance, as it would in a three-dimensional expansion, and the relaxation processes become simple functions of distances behind the front. This extremely simple piece of equipment can generate temperatures up to 20,000 K, since the strength (or velocity) of the shock depends only on the pressure ratio across the diaphragm immediately prior to rupture and on the thermodynamic properties of the gases in the two sections.

The disadvantage of all shock tube work is that the front moves so rapidly and subsequent wave interactions follow so soon afterwards that the available testing time is very short, often as little as 100 microseconds. Shock waves can also be created in highly ionized media where the forces are Coulombic in origin and the shocks are termed "collisionless." In this situation, shock waves lie more properly in the realm of plasma physics.

See also **Aerodynamics and Aerostatics**; and **Supersonic Aerodynamics**.

Additional Reading

Ben-Dor, G., T. Elperin, and O. Igra: "Handbook of Shock Waves," Vol. 5, Academic Press, Inc., San Diego, CA, 2000.

Harris, C.M. and A.G. Piersol: "Shock and Vibration Handbook," 5th Edition, McGraw-Hill Companies, Inc., New York, NY, 2001.

Kim, Y.W.: "Current Topics in Shock Waves," American Institute of Physics, College Pard, MD, 1997.

Shugaev, F.V. and L.S. Shtemenko: "Propagation and Reflection of Shock Waves," World Scientific Publishing Company, Inc., Riveredge, NJ, 1997.

Zel'dovich, YaB. and Y.P. Raizer: "Physics of Shock Waves and High-Temperature Hydrodynamic Phenomena," Dover Publications, Inc., Mineola, NY, 2002.

SHOOTING STAR.   The popular term used to designate meteors. These objects bear little if any relation to the stars, aside from their being seen as bright, rapidly moving objects against the dark sky.

SHORAN (Short-range Navigation).   A precision electronic position fixing system using a pulse transmitter and receiver and two transponder beacons at fixed points. High-precision shoran is called *hiran*.

SHORE EFFECT.   The change in the characteristics of an electromagnetic wave as it passes along a land-sea boundary, due to a difference in the propagation characteristics of the two regions. A source of error in radio direction-finders.

SHORT CIRCUIT.   An electrical circuit is considered to be shorted when the terminals are connected directly together, with only the impedance of the short connecting leads between them. Thus, for all practical purposes, there is no resistance between them, hence no voltage can exist between them. While shorting a circuit that does not contain, and is not connected to, any source of voltage will produce no harmful effects, shorting a set of terminals across which a voltage normally exists will produce, in many instances, disastrous current flows. In power circuits, protection is often provided by circuit breakers or fuses which open the circuit under the high values of current that will flow on short-circuits. Even then, the transient effects that result from short circuits may cause generators to "arc over." A short obviously puts the circuit out of use.

See also **Circuit Breaker**; and **Fuse (Electric)**.

SHORT-RANGE FORCE.   A force between two particles which is essentially ineffective when the interparticle separation exceeds a certain distance; usually applied to nuclear forces which have a range of several times 10^{-13} centimeter.

SHOVELLER.   See **Waterfowl**.

SHOWER.   See **Precipitation and Hydrometeors**.

SHREWS.   See **Moles and Shrews**.

SHRIKE (*Aves, Passeriformes*).   This bird (*Lanius*) is chiefly known for its habit of catching other birds and small animals and impaling uneaten remnants on thorns. It is also known as the butcher bird. The beak is notched and in some species hooked. The numerous species occur on all continents but South America. The minivet is a brightly colored shrike (cuckoo-shrike), several species of which inhabit eastern Asia and India. The bird is about 6 inches (15 centimeters) long, is black along the back, with bright orange underneath and on the tail and wings. The female tends to be dark gray with a dull yellow coloration. Nests are cup-shaped and made of roots, pine needles, spider webs, and twigs. There are three to four green eggs with pale pink spots.

SHUNT (Electrical).   An electrical bypath so arranged that an electric current divides and flows partially through a second path (termed *shunt*). Shunts are employed in instrument circuits and other electrical equipment. For example, in the shunt-wound generator, the field coils are shunted across the armature circuit. The shunt resistance is very high and consequently only a very small portion of the current flows through the shunt winding.

A current bypass often is used with permanent-magnet moving-coil electrical instruments because only currents up to about 50 milliamperes can be taken into a moving-coil through the springs. Direct current instruments in excess of this range require the use of a parallel resistance circuit formed by one or several shunts. Where these ranges are moderate, the shunts usually are self-contained. On higher ranges, the shunts become physically large and convert more than a few watts into heat. One kilowatt for 50 millivolts, 20,000 amperes. Thus, they are used as accessories external to the instrument. Small or large, these devices consist of one or several manganin conductors terminating in copper blocks, which are provided with separate terminals for connection to the instrument, to avoid errors.

The manganin sections are soldered into the copper blocks. Shunt construction is such that heat is carried off at a rate sufficient to keep the operating temperature below the softening point of the solder. Adequate conductors tightly fastened, clean contact surfaces, and free air circulation are important.

Ammeters for use with external shunts are provided with special leads for shunt connection. Shunts are usually made to produce a standard potential drop, such as 50 millivolts at rated current. The associated DC mechanism is then built to give full scale deflection on a slightly smaller potential so as to allow for lead resistance. As the leads form part of the mechanism circuit, shown in Fig. 1, their resistance must not be altered.

Usually the current through the mechanism is a negligible portion of the total and the potential drop across the shunt terminals is nearly the same with or without the mechanism in the circuit. In instruments of moderate precision and fairly high current range, this difference can be neglected and shunts and instruments made interchangeable. In instruments of high

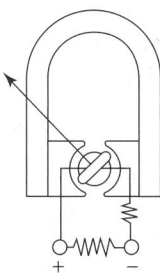

Fig. 1.   Use of a shunt with a permanent-magnet moving-coil ammeter.

precision and in instruments of low current range, the shunt adjustment must take into account the instrument current. Such combinations are usually not interchangeable.

Multirange shunts should be connected as shown in Fig. 2 to avoid the use of a switch with its variable contact resistance between shunts and mechanism.

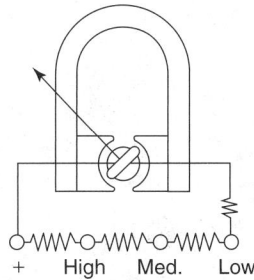

Fig. 2.   Preferable manner for connecting a multirange shunt to avoid use of switch.

Because of the relatively high current consumption of alternating current instruments, shunts are not used either for obtaining multiple ranges or for extending the base range. The division of current between mechanism and shunt would become unfavorable and would invite inaccuracies due to the difficulty of obtaining good pressure contacts, as well as the fact that the current division would be a function not merely of the relative resistances of the parallel circuits, but also of their relative impedances. Moreover, AC mechanisms are not suitable for low millivolt ranges so that shunts would have to have large potential drops and would generate excessive heat. The ranges of alternating current ammeters are, therefore, extended with the aid of current transformers.

SIAMANG.   See **Anthropoids**.

SIBERIAN HIGH.   See **Atmosphere (Earth)**.

SICKLE CELL ANEMIA.   See **Anemias**.

SIDEREAL DAY.   The duration of one rotation of the Earth on its axis, with respect to the vernal equinox. It is measured by successive transits of the vernal equinox over the upper branch of a meridian. Because of the precession of the equinoxes, the sidereal day thus defined is slightly less than the period of rotation with respect to the stars, but the difference is less than 0.01 second. The length of the mean sidereal day is 24 hours of sidereal time or 23 hours 56 minutes 4.09054 seconds of mean solar time. See also **Time**.

SIDEREAL HOUR ANGLE (SHA).   Angular distance west of the vernal equinox; the arc of the celestial equator, or the angle at the celestial pole, between the hour circle of the vernal equinox and the hour circle of a point on the celestial sphere, measured westward from the hour circle of the vernal equinox through 360 degrees. Angular distance east of the vernal equinox, through 24 hours, is right ascension.

SIDEREAL MONTH.   The average period of revolution of the Moon with respect to the stars, a period of 27 days 7 hours 43 minutes 11.5 seconds, or approximately $27\frac{1}{3}$ days.

SIDEREAL PERIOD.   The sidereal period of any object is its period of revolution around its primary. In general, sidereal period may be defined as the time required for an object to move from a particular position among the stars back to the same longitude again, as seen from the sun.

SIDEREAL TIME.   See **Time**.

SIDEREAL YEAR.   The period of one apparent revolution of the Earth around the Sun, with respect to the stars, averaging 365 days 6 hours 9 minutes 9.55 seconds in 1955, and increasing at the rate of 0.000095 second annually. Because of the precession of the equinoxes this is about 20 minutes longer than a tropical year. See also **Time**.

SIDERITE.   This mineral is a carbonate of iron, $FeCO_3$. It is hexagonal with rhombohedral crystals, and also occurs in various massive forms. It has a rhombohedral cleavage; uneven fracture; is brittle; hardness, 3.75–4.25; specific gravity, 3.96; luster, vitreous to pearly; color, gray, yellowish- or greenish-gray, green, reddish-brown and brown. Siderite is found as concretionary masses in the sedimentary rocks; as a replacement mineral from the action of iron solutions upon limestones; and in metalliferous veins as a gangue mineral. It is relatively common. Siderite is found in Austria, Saxony, the Czech Republic and Slovakia, France, England, Italy, Greenland, Australia, Brazil and Bolivia. In the United States important localities are in Connecticut, Pennsylvania, New Jersey, Ohio, and Washington. It is an iron ore. The mineral was at one time called chalybite.

SIDEWINDER.   See **Snakes**.

SIFAKA.   See **Lemur**.

SIGMA PARTICLE.   A hyperon with a rest-mass energy of about 1193.4 MeV, an isospin quantum number 1, an angular momentum spin quantum number $\frac{1}{2}$, and a strangeness quantum number 1. Symbol, Σ.

SIGMA$_T$ (symbol σ_t).   A conveniently abbreviated value of the density of a seawater sample of temperature t and salinity S:

$$\sigma_t = [\rho(S, t) - 1] \times 10^3$$

where $\rho(S, t)$ is the value of the seawater density in c.g.s. units at standard atmospheric pressure. If, for example, $\rho(S, t) = 1.02648$, then $\sigma t = 26.48$.

SIGMOIDOSCOPY.   An instrumental technique for examining the rectum and sigmoid colon. The patient is placed in the knee-chest position, preferably on a motorized table that allows support of the knees below the level of the abdomen. After introduction of the sigmoidoscope, any liquid colonic contents are aspirated. A cotton swab may be pressed against the mucosa and rotated 360 degrees to remove mucus and debris. The physician then searches for ulcerations, a granular mucosal surface, polyps, friability (bleeding), and other conditions, depending upon the exact purpose of the examination. Where information beyond the reach of the sigmoidoscope is required and when this cannot be fully obtained through a barium-contrast x-ray, colonoscopy may be used. This involves an instrument employing fiber optics that permits examination of the colonic surface from the anus to the ileocecal valve.

SIGNAL.   1.   An independent input variable.

2.   A visual, audible, or other indication used to convey information.

3.   The intelligence, message, or effect to be conveyed over a communication system.

4.   A signal wave.

SIGNAL CONDITIONING.   A process for modifying an input signal prior to introduction into an electronic system, such as a digital-data acquisition or instrumentation operation. The meaning of the term varies from one type of application to the next. Modification of an analog input signal prior to amplification may include: attenuation (scaling), filtering, conversion (current to voltage or voltage to current), impedance-level transformation, bridge or signal compensation, and in numerous instances specialized operations such as cold-junction compensation in the case of thermocouple inputs. Commercially available signal conditioning apparatus may include both amplification and conversion from analog to digital form. Because of the rather nebulous nature of this term, its use in procurement always should be accompanied by detailed specification of functions and operating parameters.

SIGNAL GENERATOR.   In the development, calibration, and testing of electronic hardware, instruments, and systems, frequently it is required to simulate certain electrical signals. A signal source provides the stimulus that creates the response to be measured. This source usually is an oscillator or a standard-signal generator of known characteristics which can be adjusted to establish a known set of conditions. These characteristics include: (1) the frequency; (2) the output voltage and impedance; (3) the carrier-signal waveform, which typically may be sine wave, square wave, pulse, or random noise; and (4) the modulation that carries the system information through variation of phase, frequency, amplitude, or timing of the carrier waveform.

Signal sources can be classified functionally as to whether the information that they yield is readily usable in *frequency domain*, or in *time-domain* analysis. Sine wave techniques form the basis of power generation and transmission systems and most communication systems, leading to ready frequency domain analysis. Many developments in information transmission and data handling, such as radar systems, digital computers, telemetry, and wire telegraphy, are based upon pulse techniques which yield most easily to time domain analysis.

Common to all of these systems are ultimate performance limitations, determined by system bandwidths and noise. Bandwidth and transient performance are closely related. They convert one into the other and can be measured as phenomena in either the frequency or the time domain at the convenience of the analyst. Noise is most easily measured by comparison with a noise source of known characteristics.

Types of Signal Sources.   These fall into five major categories:

1. *Oscillators*, or sine wave generators which embrace frequencies from 0.01 Hz to 7 GHz, with maximum output levels from a few milliwatts to 200 watts.

2. *Standard signal generators*, which are sine wave oscillators with accurately calibrated output voltage behind a standard impedance and with calibrated modulation capabilities. For wideband measurements, a sweep frequency instrument provides calibrated sweep bands as well as calibrated output. Mechanical sweep devices also are available for converting conventional signal generators and oscillators to sweep generators.

3. *Frequency synthesizers* generate output frequencies continuously adjustable over wide ranges, all coherently derived from a single quartz crystal oscillator. Commercial instruments are available with optional degrees of resolution from one part in 10^3 to one part in 10^9, with either manual or programmable control.

4. *Pulse generators* for time domain measurements. See also **Pulse Generator**.

5. *Random-noise generators* that produce wideband noise of known spectrum and energy distribution for noise and vibration testing in mechanical systems, noise measurements in communication circuits, and applications is psychological, probability, and information theory research. These devices are described under **Noise Generator**.

Oscillators

The variable frequency, sine wave oscillator is the basic general purpose signal source. With it one can make a series of measurements at specified frequencies which can be combined to specify performance in the frequency domain. These measurements may be made by manual settings, point-by-point, or by a frequency swept automatically over the desired range to display the system response on a chart recorder or a cathode ray oscilloscope.

Oscillators are of four basic types:

1. *LC Oscillators.* At radio frequencies where tuning can be accomplished with air capacitors, the LC circuit is the best and most economical frequency-determining system. They cover frequencies from 500 kHz to 1,050 MHz. At frequencies above 1,000 MHz, circuits with distributed constants are used.

2. *RC Oscillators.* The frequency is determined by resistive and capacitive elements in this type of instrument. These devices in various configurations cover a frequency range of 0.01 to over 1 MHz.

3. *Beat-frequency Oscillators.* In these devices, the output frequency is the difference between the frequencies of a variable-frequency and a fixed-frequency oscillator. Several decades of frequency can be covered in one band with a single control.

4. *Klystron Oscillators.* In these devices, the frequency is determined by a velocity-modulated electron stream, which excites a resonant cavity. In one form, a reflex klystron in a coaxial cavity with a noncontacting plunger is used to cover frequencies from 1.7 to 4.1 GHz. Internal square wave and frequency modulation are provided.

An automatic, audio frequency measuring system that combines a beat-frequency audio generator and a graphic level recorder for automatic plotting of frequency response data is shown in Fig. 1. This type of instrument is used widely for measuring the response of filters, attenuators, networks, loudspeakers, amplifiers, microphones, transducers, and complete acoustic systems.

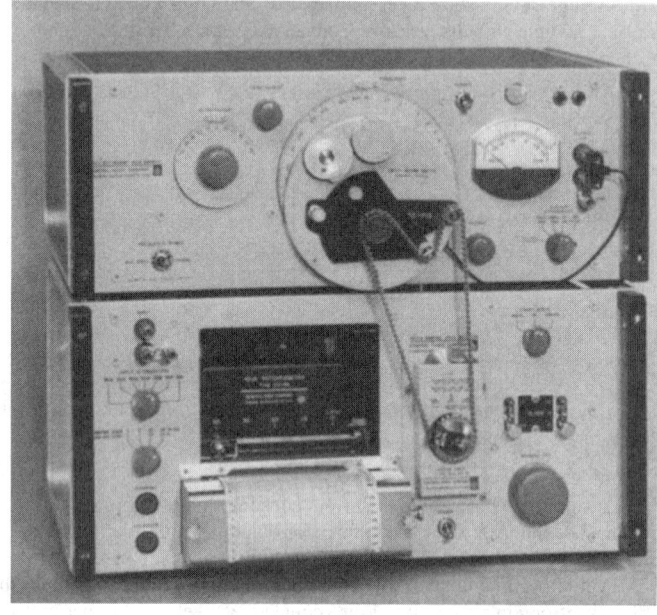

Fig. 1.   Automatic audio-frequency measuring system.

Standard Signal Generators

These instruments provide a source of alternating current energy of accurately known characteristics. The carrier, or center frequency, is indicated by a dial setting, the output voltage by a meter reading and associated attenuator setting, and the modulation by a meter reading that is set by appropriate knobs. See Fig. 2.

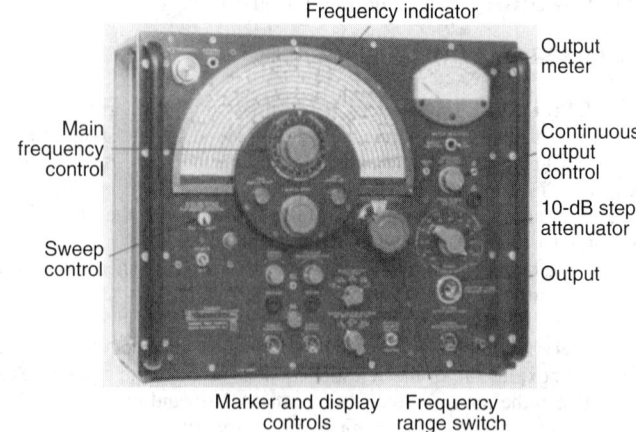

Fig. 2.   Standard sweep frequency generator.

Common types of modulation signals are sine wave, square wave, and pulse. The output signal either may be frequency or amplitude modulated by these signals. When the frequency modulation system produces a considerable excursion in frequency at a relatively low cyclical rate, the instrument is known as a sweep frequency generator and is particularly useful for automatic data display. Standard signal generators are used for testing radio receivers, as voltage standards over the range from a few microvolts to about one volt, and generally as power sources in measurement of gain, bandwidth, signal-to-noise ratio, standing wave ratio, and other circuit properties.

For use as a standard signal generator, the oscillator must be stable, have reasonably constant output over any one frequency range, have a good waveform, and have no appreciable hum or noise modulation. Careful overall shielding of the generator is essential to minimize stray fields.

The elements of an amplitude modulated standard signal generator are shown in Fig. 3. An amplifier may be added readily at lower frequencies, as shown in Fig. 4, to isolate the oscillator from the load and to minimize the incidental frequency modulation that usually results from amplitude modulation. The elements of a standard sweep frequency generator are shown in Fig. 5.

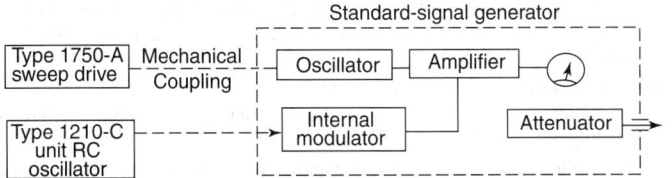

Fig. 4. Elements of an amplitude modulated standard signal generator with amplifier.

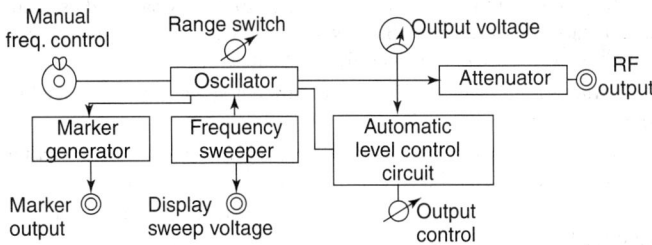

Fig. 5. Elements of a standard sweep frequency generator.

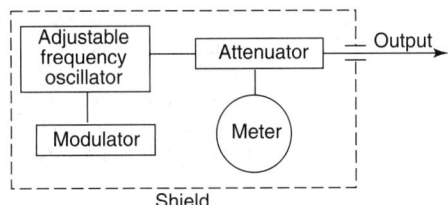

Fig. 3. Elements of a standard signal generator.

The specifications for three standard signal generators and for a standard sweep frequency generator are given in Table 1 to illustrate the overall performance parameters of these instruments.

Frequency Synthesizers

These devices are well suited for repeatable, high precision frequency response measurements on amplifiers, filters, transducers, and similar electronic devices. Either point-by-point or sweep techniques may be used. An instrument of this type is shown in Fig. 6 and the overall specifications of four configurations of it are given in Table 2 to illustrate the overall

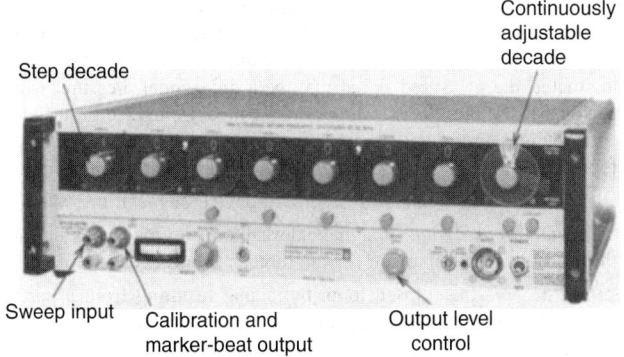

Fig. 6.   Coherent decade frequency synthesizer.

TABLE 1. PERFORMANCE PARAMETERS OF SIGNAL GENERATORS

Instrument Type	Frequency Range	Open-circuit Voltage	Impedance Ohms	Output Modulation %
Standard signal generators	5 kHz–50 MHz	0.1 µV–200 µV	10.50	0–80
	67 kHz–80 MHz	0.1 µV–6 V	50	95
	9.5 MHz–500 MHz	0.1 µV–10 V	50	95
Standard sweep frequency generators	0.7 MHz–230 MHz	0.3 µV–1 V	50	Sweep all bands
	0.45 MHz–10.7 MHz	0.3 µV–1 V	50	

TABLE 2. PERFORMANCE PARAMETERS OF FREQUENCY SYNTHESIZERS

Characteristic	Example #1	Example #2	Example #3	Example #4
Frequency range	0–100 KHz	0–1 MHz	30 Hz–12 MHz	10 kHz–70 MHz
Smallest digital step	0.01 Hz	0.1 Hz	1 Hz	10 Hz
Smallest direct-calibrated continuously-adjustable decade increments	0.0001 Hz	0.001 Hz	0.01 Hz	0.1 Hz
Maximum bandwidth controllable by continuously adjustable decade module	100 kHz	1 MHz	1 MHz	1.2 MHz
Spurious frequency outputs:				
Harmonic (a^+ maximum output)	< −40 dB	< −40 dB	< −30 dB	< −30 dB
Nonharmonic	< −80 dB	< −60 dB	< −60 dB	< −60 dB
Output	(Coupling switch at ac) Adjustable, 0 to 2 V, rms (Coupling switch at dc) Adjustable, 0 to 0.8 V, rms 0 to 2 V, rms		(Output impedance switch at 50 ohms) 0 to 2 V, rms (Output impedance switch at zero)	0.2 to 2 V, rms, metered and leveled behind 50 ohms ± 5%

performance parameters obtainable in these kinds of instruments. The output frequency is synthesized directly by repetitive arithmetic manipulations of frequency in a series of identical modules. The synthesizers are equipped for direct, front-panel, digit selection.

Frequency synthesizers combine the advantages of tunable oscillators, which are not usually highly stable, with those of frequency standards, which though very stable, are not tunable. In addition, frequency synthesizers can be tuned like a decade box, i.e., in discrete steps for precision and repeatability; or they can be swept over bands as wide as the instrument's full frequency range, or as narrow as 0.001 Hz.

SIGNAL (Instrument). With reference to industrial and scientific instruments, the Instrument Society of America defines *signal* as information about a variable that can be transmitted.

Actuating-Error Signal. The reference-input signal minus the feedback signal.

Error Signal. In a closed loop, the signal resulting from subtracting a particular return signal from its corresponding input signal.

Feedback Signal. The return signal that results from a measurement of the directly controlled variable.

Input Signal. A signal applied to a device, element, or system.

Measured Signal. The electrical, mechanical, pneumatic, or other variable applied to the input of a device. It is the analog of the measured variable produced by a transducer (when such is used).

In a thermocouple thermometer, the measured signal is an emf, which is the electrical analog of the temperature applied to the thermocouple. In a flowmeter, the measured signal may be a differential pressure, which is the analog of the rate of flow through an orifice. In an electric tachometer system, the measured signal may be a voltage, which is the electrical analog of the speed of rotation of the part coupled to the tachometer generator.

Output Signal. A signal delivered by a device element, or system.

Reference-Input Signal. A signal external to a control loop which serves as the standard of comparison for the directly controlled variable.

Return Signal. In a closed loop, the signal resulting from a particular input signal, and transmitted by the loop and to be subtracted from the input signal.

Signal Transducer. A transducer that converts one standardized transmission signal to another.

Signal-to-Noise Ratio. Ratio of signal amplitude to noise amplitude.

SIGNAL LEVEL. At any point in a transmission system, the difference of the measure of the signal at that point from the measure of an arbitrarily-specified signal chosen as a reference. In audio techniques, the measures of the signal are often expressed in decibels, thus their difference is conveniently expressed as a ratio.

SIGN CONVENTION (Lens and Mirror). Since every distance involved in lens computations must be measured from some origin, a convention of signs should be adopted to insure consistency in the derivation and use of formulas. Unfortunately, this has not been done by all authors. The following probability has the largest following: (1) Draw all figures with the light incident on the surface of reflection or refraction from the left. (2) Consider the object distance $p = PV$ positive when P is at the left of the vertex. (3) Consider the image distance $q = VP'$ positive when P' is at the right of the vertex. (4) Consider the radius of curvature $R = CV$ positive when the center of curvature lies to the right of the vertex. (5) Consider the slope angles positive when the axis must be rotated counterclockwise through less than $\pi/2$ to bring it into coincidence with the ray. (6) Consider angles of incidence and refraction positive when the

radius of curvature must be rotated counterclockwise through less than $\pi/2$ to bring it into coincidence with the ray. (7) Consider distances normal to the axis positive when measured upward.

In Fig. 1, only θ' is negative. When the convention is followed, the simple thin-lens formula results.

SIGNIFICANCE TESTS. Suppose that a sample provides an estimate θ of a parameter t, and that a certain hypothesis specifies a certain value for θ, $\theta = \theta_0$ say. t will differ from θ_0 by a discrepancy $d = \theta_0 - t$, and it may be possible to deduce from the sampling distribution of t the probability P that a discrepancy as large as d would have arisen if the hypothesis $\theta = \theta_0$ were true. If this probability is small, the sample may be taken to provide evidence against the truth of the hypothesis. This procedure is called a test of significance. If P is less than some value α (commonly chosen to be 0.05 or 0.01), we say that t is significantly different from θ_0 at the level α, or simply that d is significant at this level. The hypothesis $\theta = \theta_0$ (i.e., $d = 0$) is called the null hypothesis.

Alternative tests of the same hypothesis are often available, based on different statistics. To choose between them, we introduce the notion of an alternative hypothesis, $\theta = \theta_1$ say. We may now calculate, for a given α, the probability β of obtaining a significant result at this level when the alternative hypothesis is true. This probability (a function of α and θ_1) is called the power of the test. A few tests can be shown to be at least as powerful as any alternative test for all values of θ_1. Such tests are said to be uniformly most powerful and are clearly to be preferred in cases where they exist.

If we imagine some action being taken based on the result of a significance test, this action may be referred to as accepting or rejecting the null hypothesis. Sampling fluctuations may then lead to two types of incorrect action:

1. We may reject the null hypothesis when it is true.
2. We may accept the null hypothesis when it is false.

These are referred to as errors of the first and second kind; their probabilities are respectively α and $(1 + \beta)$. A test for which $(1 + \beta) \geq \alpha$ for all values of θ_1 is said to be unbiased.

Sir Maurice Kendall, International Statistical Institute, London.

SIGNIFICANT DIGITS. 1. The digits that determine the mantissa of the logarithm of the number beginning with the first digit on the left that is not zero and ending with the last digit on the right that is not zero.

2. The digits of a number that have a significance; the digits of a number beginning with the first nonzero digit on the left side of the decimal point, or with the first digit after the decimal point if there is no nonzero digit to the left of the decimal point, and ending with the last digit to the right. Note that the use of the final zero in the number 0.230 implies that the number is known to third place accuracy.

SIGNIFICANT WAVE. In ocean wave forecasting, a fictitious wave whose height and period are equal to the average height and period of the highest one-third of the actual waves that pass a fixed point.

SIKAS DEER. See **Deer**.

SIKES SCALE. See **Specific Gravity**.

SILAGE. A feedstuff resulting from the anaerobic preservation of moist forage crops or crop residues by acidification. This definition (M.E. McCullough, Georgia Experiment Station, University of Georgia) aptly confines silage to the production of a feedstuff and eliminates the confusion of nonagricultural uses. The definition confines the process to anaerobic conditions and excludes decomposition under aerobic fermentations in which the final result is the disposition of a waste product, such as sewage. The two important inclusions in the definition are the terms *forage* and *acidification*. This confines the definition to harvested farm products that are being stored for feed and limits the method of preservation to acidification. The latter limitation does permit the acidification process to include either or both the direct application of acids or the development of acids through fermentation within the silage mass. Silage must not be confused with simply wet grains in storage. McCullough points out that there are many terms in common use that have no scientific significance in silage terminology (such as *haylage, cornlege, oatledge*, etc.—all coined names used to designate a particular method for harvesting and storing

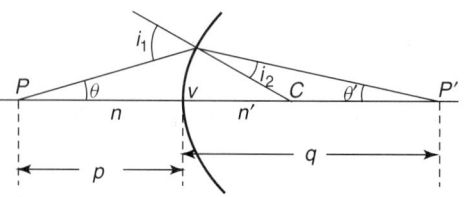

Fig. 1.   Demonstration of sign convention.

forage). They are like brand names in the feed industry, such as *dairy feed* or *hog supplement*, which have no meaning by themselves. Other terms used in connection with silage, such as *wilted, direct cut,* or *recut*, have value only as being descriptive of the methods used in harvesting or storing the crop.

Advantages and Limitations of Silage. In terms of the dairy farm, there are several advantages to making and feeding silage, including:

1. Silage may be fed at any time of the year or year-round.
2. Silage may be harvested in adverse weather as well as in fair weather.
3. There is very little waste in silage feeding processes.
4. The silage crop can be readily harvested at the most favorable stage of maturity.
5. Silage requires less storage space than hay.
6. Silage crops are removed from the land earlier, thus facilitating a double cropping system.
7. The use of silage increases the carrying capacity of cows per unit of land and per farm by saving more feed per unit of land.
8. The fire hazard in storing silage is minimal.
9. Forage stored as silage is the most economical method of preserving the greatest amounts of nutrients (TDN, total digestible nutrients) per unit of land.
10. Silage is an excellent substitute for pasture.
11. The making and feeding of silage can be completely mechanized.
12. A few weeds will not seriously affect the quality of silage, since most weeds seem to lose off-flavors in the ensiling process. Wild onion is an exception; its flavor is usually retained in the silage.
13. Silage may be kept for many years, almost indefinitely, if it is properly made and well stored.
14. A silo saves feed that might otherwise be wasted.
15. When forage is stored in the form of silage, it has some values that other stored roughages do not possess, including palatability, laxative effect, and the fact that it can be fed free-choice (self-fed) without danger to dairy cattle.

Limitations of silage include:

1. Initial costs of silo and silage-making machinery are high.
2. After a silo is opened, some silage should be fed daily to prevent spoilage.
3. Sale of silage is generally impractical; hence it must usually be fed on the farm where it is stored.
4. Silage requires some skill and knowledge to be handled successfully, thus keeping spoilage, other losses, and reduction in quality to a minimum.

SILANES. See **Silicon.**

SILICA GLASS. See **Glass.**

SILICATES (Soluble). The most common and commercially used soluble silicates are those of sodium and potassium. Soluble silicates are systems containing varying proportions of an alkali metal or quaternary ammonium ion and silica. The soluble silicates can be produced over a wide range of stoichiometric and nonstoichiometric composition and are distinguished by the ratio of *silica* to *alkali*. This ratio is generally expressed as the *weight percent ratio* of silica to alkali-metal oxide (SiO_2/M_2O). Particularly with lithium and quaternary ammonium silicates, the molar ratio is used.

Sodium silicates find wide application in many types of detergents and cleaning compounds and have been used for many years as adhesives and cements. Both sodium and potassium silicates are important bonding agents in a large variety of ceramic cement and refractory applications, notably because of their heat stability and resistance to chemicals. Alkali-metal silicate bonds are used in high-temperature ceramic products in the fabrication of electrical components. Soluble silicates find wide application for pelletizing, granulating, and briquetting finely divided particles, such as clays, fertilizers, and ores. Sodium silicates also are used as bonding materials for foundry mold and core compositions. Because of their adherence properties, soluble sodium and potassium silicates are widely used as coatings. Frequently, sodium silicates are used to protect against water-line corrosion in tanks. The ability to form sols and gels is an interesting and very useful characteristic of soluble silicates. Silica gels

are used in a major way as desiccants and as carriers for the production of petroleum-cracking catalysts, as well as raw materials in the manufacture of zeolites. Activated sols are used in water clarification.

Generally, sodium and potassium silicates are made by fusion of pure sand with alkali-metal carbonate or alkali-metal sulfate and carbon. This operation is carried out in large open-hearth furnaces heated to a temperature range of 1300–1500 °C. The resulting glasses may be used in this form, or dissolved in water to produce silicate solutions. Sodium and potassium silicate solutions also can be made by dissolving sand in sodium or potassium hydroxide solution at elevated temperatures and pressures. Lithium silicate glasses, although insoluble in water, can be made by dissolving silica gel in, or mixing silica sols with lithium hydroxide solutions. Anhydrous sodium metasilicate is made from the anhydrous melt. This salt crystallizes rapidly from its aqueous solution at temperatures in the range of 80–85 °C.

The most important property of sodium and potassium silicate glasses and hydrated amorphous powders is their solubility in water. The dissolution of vitreous alkali is a two-stage process. In an ion-exchange process between the alkali-metal ions in the glass and the hydrogen ions in the aqueous phase, the aqueous phase becomes alkaline, due to the excess of hydroxyl ions produced while a protective layer of silanol groups is formed in the surface of the glass. In the second phase, a nucleophilic depolymerization similar to the base-catalyzed depolymerization of silicate micelles in water takes place.

When sodium silicate solutions of intermediate ratios are concentrated to a thick gum, they become very sticky and tacky. This property is important to many of the adhesive applications. It is related to high cohesion and low surface tension rather than primarily to viscosity.

The stability of soluble silicate solutions depends strongly on pH and concentration. The addition of acids and acid-forming compounds gives rise to the formation of silica gels. Soluble alkali-metal silicate solutions are not compatible with most organic water-miscible solvents. The addition of alcohols and ketones causes phase separation into liquid layers. A few organic systems, however, particularly polyols, such as glycols, glycerins, sugars, and polyethylene glycols, are compatible and miscible with alkali-metal silicate solutions. See also **Adhesives**; and **Glass**.

SILICEOUS.   Containing silicon dioxide or one of its compounds.

SILICIC.   1. Containing or pertaining to silicon. 2. Containing silicic acid (ortho) H_4SiO_4; or silicic acid (meta) H_2SiO_3; or silicic acids of a higher degree of hydration (disilicic acids, trisilicic acids, etc.).

SILICIFICATION.   An important geochemical process by which certain sedimentary rocks such as limestones and dolomites, or calcareous fossils are partially or entirely replaced by silica, SiO_2. See also **Chert**; and **Flint**.

SILICON.   Chemical element, symbol Si, at. no. 14, at. wt. 28.086, periodic table group 14, mp 1408–1,412 °C, bp 2,355 °C, density 2.242 g/cm³ (solid crystalline, 20 °C), 2.32 g/cm³ (single crystal; 20 °C). Elemental silicon has a face-centered cubic crystal structure (diamond structure). The existence of a hexagonal form of silicon with a wurtzite-type structure and with lattice parameters $a = 3.80$ Å and $c = 6.28$ Å was established in 1963 (Wentorf-Kasper). Claims to different parameters were made by Jennings-Richman (1976). These differences are discussed by Kasper-Wentorf (1977). Much new knowledge concerning the crystalline structures and phase transitions of silicon has been gained during the mid-1980s, notably from research under immensely high pressures and investigations involving the tunneling microscope, as described shortly.

The common form of silicon is a dark-gray, hard solid. It can be obtained as a brown microcrystalline powder, which is not an allotrope of the gray form. Both forms are unaffected by air at ordinary temperatures, but when heated in air to high temperatures a protective layer of oxide is formed. Silicon reacts with nitrogen at high temperatures to form the nitride; with chlorine to form the chloride, with several metals to form silicides. Crystalline silicon is unattacked by HCl or HNO_3, or H_2SO_4, but is attacked by hydrofluoric acid to form silicon tetrafluoride gas. Silicon is soluble in NaOH solution forming sodium silicate and hydrogen gas. Silicon reacts with dry chlorine to form silicon tetrachloride.

There are three naturally occurring isotopes, ^{28}Si through ^{30}Si, and three radioactive isotopes have been identified, ^{27}Si, ^{31}Si, and ^{32}Si. The latter isotope has a half-life of approximately 700 years, while the half-lives of the other two are short, measured in terms of seconds and hours.

Lavoisier showed in 1787 that SiO_2 was not a single element and indicated that it was the oxide of a hitherto unknown element. In the early 1800s, Scheele, Davy, Gay-Lussac, and Thénard attempted to isolate the element, but were not successful. In 1871, Berzelius discovered silicon in a cast-iron melt and, in 1823, succeeded in isolating the element by reduction of potassium fluorosilicate with potassium. Small laboratory amounts were produced by H.E. Sainte-Claire Deville in 1854 and by C. Winkler in 1864. It was not until 1900 that the effective properties of silicon as a deoxidizing agent for steel production were observed. Shortly thereafter, ferrosilicon alloys, using quartzite, coke, and iron pellets, were produced in electric refining furnaces of the type already in use for making calcium carbide. With this technique, it was possible to produce silicon of about 98% purity. It remained for the rigid purity requirements of semiconductors many years later before silicon of higher purities was produced.

Silicon is ranked second in the order of chemical elements appearing in the earth's crust, an average of 27.72% occurring in igneous rocks. In terms of seawater, it is estimated that a cubic mile of seawater contains about 15,000 tons of silicon (3240 metric tons per cubic kilometer). In terms of abundance throughout the universe, silicon is ranked seventh. First ionization potential 8.149 eV; second, 16.27 eV; third, 33.30 eV; fourth, 44.95 eV. Oxidation potentials $Si + 2H_2O \rightarrow SiO_2 + 4H^+ + 4e^-$, 0.86 V; $Si + 6OH^- \rightarrow SiO_3^{2-} + 3H_2O + 4e^-$, 1.73 V.

Other important physical properties of silicon are given under **Chemical Elements**.

Because of its chemical reactivity, silicon does not occur in elemental form in nature. The element is present in igneous rocks and clays as alumino-silicate; as the oxide SiO_2 in quartz, sand. (Fig. 1), flint, and the gems amethyst, jasper, chalcedony, agate, onyx, tridymite, opal, crystobalite; as silicates in zircon (zirconium silicate, $ZrSiO_4$), in willemite (zinc silicate, Zn_2SiO_4), in wollastinite (calcium silicate, $CaSiO_3$), in serpentine (magnesium silicate, $Mg_3Si_2O_7$). Impure (up to 98% Si) silicon is obtained from the oxide (1) by igniting with aluminum powder, or (2) by reduction with carbon in an electric furnace. See also **Cancrinite**.

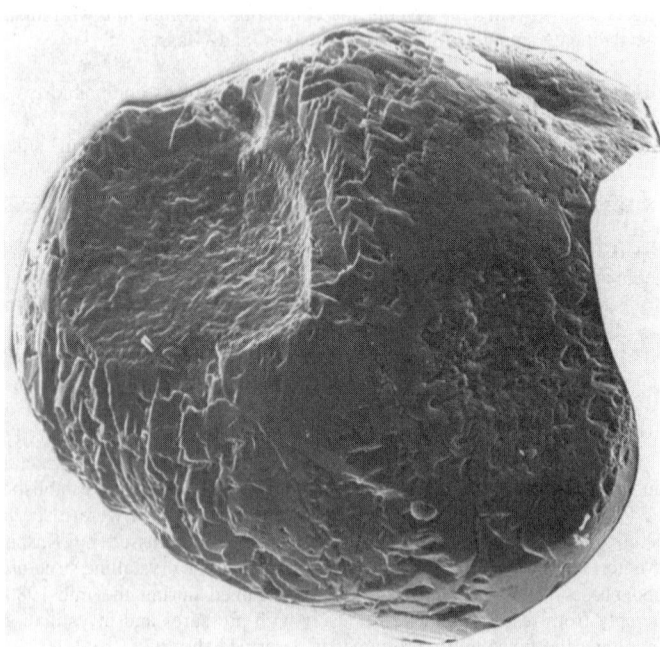

Fig. 1.   Grain of sand, originally magnified 100 ×.

Silicon Production for Alloys: production of raw steel requires about 1.6–1.7 kilograms of silicon per metric ton of steel. The silicon is used in the form of ferrosilicon, which contains about 20% silicon. It is estimated that about 3 million metric tons of ferrosilicon are consumed annually in steelmaking. The 20%-silicon-content ferrosilicon can be made in a conventional blast furnace. Ferrosilicons with higher silicon contents (45, 75, 90, and 98%) must be produced in electric furnaces. The raw materials are pure quartzites. The presence of impurities, such as Al_2O_3 and CaO, interfere with the melting process because of the formation of dross. The reducing agent used is chemical coke. For the very high concentrations of silicon (90–98%), ash-free petroleum coke or charcoal are used. Iron

is added in the form of small pellets or chips in the production of the 45–75%-silicon alloys.

For certain metal alloys, a calcium silicon alloy is required. This alloy also is used as a steel deoxidizer and is favored because it forms a low-melting-point calcium silicate product. A representative composition of the alloy is: 30–33% Ca, 60–64% Si, 3–5% Fe, 1–2% Al, 0.3–0.6% C, and less than 0.15% S and P.

Silicon is used in the primary and secondary aluminum industry. The purity of silicon for metallurgical purposes ranges from 96.7 to 98.5% silicon; 0.10 to 0.75% aluminum; 0.03 to 0.04% calcium; with the remainder being principally iron.

Silicon Carbide: This compound is an important industrial abrasive, having a hardness of 9.5 on the Mohs scale. In this compound, each silicon atom is surrounded tetrahedrally by four carbon atoms, and similarly, each carbon atom is surrounded by four silicon atoms. Silicon carbide is made by reducing pure quartz (glass-sand) with petroleum coke in an electric-resistance type furnace, known as the Acheson process. The product is hexagonal crystals ranging from light-green to black. It is used as a ceramic raw material for dross-repellent linings as well as for many abrasive applications.

Silicon carbide also has been recognized for many years because of its having a unique set of electronic material advantages over silicon and gallium arsenide. Not only can SiC withstand higher device operating temperatures (approximately 650 °C compared with silicon's 150 °C), but SiC devices can operate with ten times the voltage capability and three times the thermal conductance capability. And, they are mechanically much more robust than traditional semiconductors.

The foregoing characteristics enable the configuration of a whole new family of high-power microwave and high-temperature electronics that can withstand high radiation for military and commercial systems.

Only recently has it been possible to produce uniform, centimeter-sized crystals. A high-purity vapor transport growth process has been developed (Westinghouse) to produce 1.5-inch (3.8-cm) device-grade SiC crystals and wafers. These comprised the building blocks for a demonstration of microwave transistors in the early 1990s. See Fig. 2.

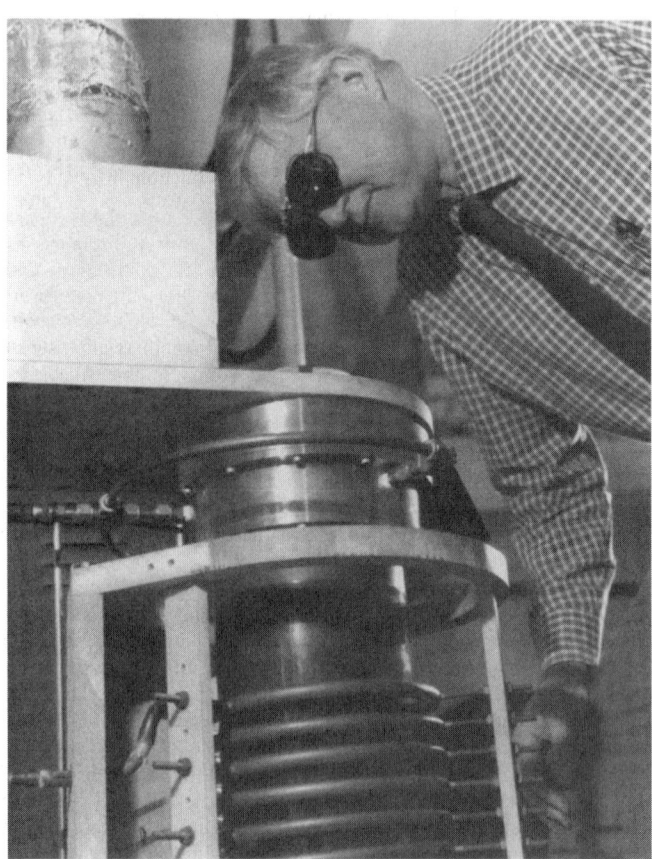

Fig. 2.   Researcher Dan Barrett (Westinghouse Science & Technology Center) checks the hot (2400 °C) crystal growth furnace that he designed for physical vapor transport growth of single crystals of silicon carbide.

Super- or Hyperpure Silicon: For semiconductor use, there can be only one atom of impurity for every 100,000 silicon atoms! The starting material for the manufacture of hyperpure silicon is silicon tetrachloride, $SiCl_4$, or trichlorosilane, $SiCl_3H$. Both of these materials can be reduced with hydrogen to yield a compact deposition of silicon on hot surfaces, ranging from $800-1,200°C$. The starting compounds are purified of boron and phosphorus by fractional distillation and absorption techniques. The process hydrogen is purified by passing it through molecular sieves under high pressure, followed by absorption techniques at a low temperature $(-190°C)$. With the highly purified starting ingredients an excess of hydrogen is circulated through heated quartz tubes. Or, the gas mixture may be blown into quartz bell jars, whereupon the silicon is deposited on filaments of tantalum or tungsten or on thin rods of hyperpure silicon, which may be heated by electrical resistance or radiofrequency energy. This process yields polycrystalline rods of silicon which range up to about one meter in length and 150 millimeters in diameter.

To be used in seimconductor devices, the polycrystalline silicon must be converted to single crystals of a defined, predetermined type of conductivity (*n* or *p* type). The crystals must be rigidly controlled as regards their resistivity, and possess the highest degree of crystallographic perfection. The two crystal-growing techniques used are: (1) crucible free vertical float zoning which removes all residual impurities, including phosphorus, arsenic, and oxygen, but boron is essentially irremovable by floating zoning, or (2) crucible pulling in which the crystals, particularly those of lower resistivity, are drawn out of a melt in a process known as the Czochralski technique. Both processes must be conducted under helium or argon, often under a vacuum of 10^{-5} torr.

Production of Ultrapure Silicon Crystal. In 1990, Westinghouse engineers reported the production of the purest crystal of silicon ever made—namely, four times purer than previously reported material. The crystal also is significantly larger, adding to its practicality in the manufacture of microelectronic circuits and devices.

The cylindrical structure, called a *boule*, weighs 22 pounds (10 kg) and is over a yard (meter) long, with a diameter of just over 3 inches (8 cm). Impurities are a few parts in 100 billion, compared with more than 10 parts in 100 billion previously reported for 1-inch (2.5-cm) diameter ultrapure crystals.

Crystal boules are sliced into wafers on which microelectronic circuits and power semiconductor devices are fabricated. An important use of the wafers is for infrared dectors for space, defense, and environmental applications.

Liquid-Solution Synthesis of Silicon Crystals. In late 1992, J.R. Heath (IBM Watson Research Laboratory) reported on a liquid-solution phase technique for preparing submicrometer-sized silicon single crystals. The synthesis is based on the reduction of $SiCl_4$ and $RSiCl_3$ (R = H, octyl) by sodium metal in a nonpolar organic solvent at high temperatures $(385°C)$ and high pressure (above 100 atmospheres). For R = H, the synthesis produces hexagonal silicon single crystals ranging from 5 to 2000 nanometers. For R = octyl, the synthesis also produces hexagonal-shaped silicon single crystals.

Light Emission from Silicon. Because of silicon's successes in the electronic components field, research has been going on to find a form of Si that will produce luminous radiation. Because of former failures, numbers of scientists have given up this research. However, independently in French and British laboratories during 1990, some success has been achieved. These researchers have found that, if one etches Si into structures so tiny that the electronic behavior of the material is transformed, full-color emission from what are termed "silicon quantum wires." A British researcher L. Canham (Royal Signals and Radar Establishment, Malvern, England) has observed that, to make silicon quantum wires, a process for sculpting silicon, known for some 30 years, is the basis. A silicon wafer is immersed in an acid electrochemical bath, which bores into the disk to produce extremely small so-called "wormholes." The latter are etched chemically, enlarging them until they meet one another. The result is a columnar structure of silicon. The latter are about a micron high, and 50 of them, stacked end to end, would span an area about equivalent to a cross-section of a human hair and are only a few nanometers thick $\left(\frac{1}{15,000}\right)$, smaller than a hair. The researchers have found that, when such a structure is bathed in ultraviolet light, light emission occurs. The emitted wavelength is determined by the porosity of the Si layer.

J.P. Harbison (Bellcore, Redbank, New Jersey) observes, "This is not the moment of the breakthrough for light-emitting silicon, but it is the moment when a lot of people are realizing its potential."

See also **Crystal**; and **Semiconductors**.

Research on Silicon Structure and Surface Properties. The rather unusual properties of silicon have intrigued scientists for many years. It possesses the physical properties of a *metalloid* (exhibits properties of both a metal and nonmetal). In several ways, silicon resembles germanium and, to a lesser extent, it resembles arsenic and boron. Silicon is a semiconductor of electricity, the conductivity rising with temperature. Silicon, in pure form, is intrinsically a semiconductor. The presence of impurities in very minute amounts markedly increases its conductivity. By introducing elements of group 13 (such as boron), which have a deficiency of electrons, the *p*-type silicon results. Therein, electricity is conducted by migration of electron vacancies or holes. On the other hand, introduction of elements of group 15, such as arsenic or phosphorus, in which there is no deficiency of electrons, the *n*-type silicon results, in which extra electrons carry an increased current because of their migration. Scientists have not been satisfied with oversimplified explanations such as that just given. As a key material in the microelectronics field, where the processing of silicon into chips and other configurations for electronic components is essentially effected at the surface of the silicon, particular interest concerns those crystalline structural details that play a role in the electronic nature of the element. However, prior to the emergence of solid-state technology, scientists were puzzled by what appeared to be crystal structure and surface anomalies and, consequently, research dates back many years, with progress largely determined by the instrumentation available to investigators. The *tunneling microscope*, the invention of which is accredited to G.K. Binnig and H. Rohrer and partially to E.W. Müller (who also invented the field-ion microscope in the 1950s), has contributed much toward an understanding of the surface of the silicon crystal, an understanding which is expected to be translated ultimately in manufacturing improvements and better final properties of silicon-based electronic components. See also **Scanning Tunneling Microscope**.

The relatively recent availability of means to create extremely high pressures (see also **Diamond Anvil High Pressure Cell**) has made it possible to gain further insights into the character of silicon. It has been learned from such experimentation that at a pressure of 110 kilobars (about 1.6 million pounds per square inch), silicon enters *truly metallic* phases. At the pressure stated, silicon abruptly assumes a structure similar to the beta form of tin. At this pressure and at a temperature of 6 degrees Kelvin (six Celsius degrees above absolute zero) the metal becomes superconducting, that is, it offers no resistance to the passage of electrons. At a pressure of 130 kilobars, the beta-tin form of silicon transforms into what has been designated as the *primitive hexagonal* phase, a phase first discovered in 1984. This research was conducted by Cohen, Chang, and Dacorogna (University of California, Berkeley). Prior to this experiment, it was not thought that such a phase would exist in the crystal of any chemical element. The researchers entered into theoretical calculations after the experiment to better understand the properties of the primitive hexagonal phase. The calculations were lengthy and required a CRAY/X-MP computer. A major finding was that the bonds linking the atoms in each of the planes defined by the hexagons should be weaker than the bonds linking the atoms in adjacent planes. This indicates that the electronic charge distribution should be inhomogeneous along one dimension. This inhomogeneity is an indication of a good superconductor because, in effect, it provides corridors through which electrons can move.

In their investigation, the researchers turned back to the much earlier hypotheses of quantum mechanics, including the properties of phonons (in quantum mechanics, a phonon can be treated as a particle, one that interacts with electrons). A strong interaction improves the opportunities for superconductivity. However, where the coupling is too strong, the integrity of the lattice may collapse and a structural phase transition may occur. Testing for superconductivity at such high pressures will be difficult. The Berkeley group predicts that the superconducting temperature will rise to a value greater than 10 degrees as the pressure nears the value required for the transition from the simple hexagonal phase to the hexagonal closed-packed phase. If expectations are proved, silicon could be the best superconductor of all chemical elements. Translating this to practical application, of course, may or may not be feasible at some future date. See also **Superconductivity**.

Silicon Chemistry and Compounds

Like carbon, silicon forms chiefly covalent bonds, but its greater atomic radius enables it to form positive ions more readily. Unlike carbon and tin, silicon is not allotropic, having only one elemental form, the diamond structure in which each atom is surrounded tetrahedrally by four others to which it is covalently bonded. An apparently amorphous brown powder, produced by combustion of silane, SiH_4, has been found to be a microcrystalline variety of this covalently-bonded structure. Much research has been conducted during the 1970s and 1980s pertaining to the more exotic silicon compounds, such as the *disilenes*, and to silicon-mediated organic synthesis. These topics are discussed later in this article. The following several paragraphs are devoted to the large number of traditional silicon compounds whose constitution and characteristics have been well established over the years.

Silicon dioxide, (SiO_2): This compound exists in at least eleven distinct crystalline forms. Several of them are obtained by heating α-quartz, which has a number of transition points, to produce β-quartz, and to give various forms of tridymite and crystobalite. The unit of structure is the tetrahedron in which each silicon atom is covalently bonded to four oxygen atoms, and the variation is in the ways these tetrahedra are interconnected (by oxygen atoms) to form a three-dimensional system.

Silicon dioxide is converted by hydrofluoric acid into silicon tetrafluoride, SiF_4, a gas. SiF_4 can also be produced directly from the elements, as can the other tetrahalides, silicon tetrachloride, $SiCl_4$ (a liquid), silicon tetrabromide, $SiBr_4$ (a liquid), and silicon tetraiodide, SiI_4 (a solid). The silicon halides hydrolyze much more readily than the carbon halides, because the unoccupied silicon 3_d orbitals are energetically not far above its 3_s and 3_p orbitals. This fact also permits the formation of the sp^3d^2 hybrid bonds of the fluorosilicate ion, SiF_6^{2-}, and additional compounds of the halides, e.g., $SiX_4 \cdot 2$ pyridine. Silicon also is intermediate between carbon and the higher members of main group 4 of the periodic table in forming a dichloride, $SiCl_3$, by strong heating of silicon with silicon tetrachloride.

Quartz and other forms of silica react very slightly with water to form monosilicic acid, $(SiO_2)_n + 2nH_2O \rightarrow nSi(OH)_4$. As shown, this reaction is a depolymerization followed by a hydrolysis, and proceeds rapidly with hot alkalis or fused alkali metal carbonates, yielding soluble silicates containing the SiO_4^{4-} and $(SiO_3^{2-})_n$ ions. The hydrolysis reaction is geologically important, because it is considered to be the starting point in the formation of the innumerable silica silicate minerals that occur so widely in nature, just as many of the silica minerals may have originated by the reverse reaction. Many of the more complex silicic acids are considered to form by polymerization of $Si(OH)_4$ molecules by sharing of $-OH$ ions between two silicon ions (octahedrally coordinated by six hydroxyl ions) followed by condensation with the loss of water to produce — $\underset{|}{\overset{|}{Si}}$ — O — $\underset{|}{\overset{|}{Si}}$ — linkages. The polymerization of silicic acid is carried out industrially to produce silica gel, a stable sol of colloidal particles. The various methods involve careful removal of H_2O, the catalytic effect of acid or alkali (or fluoride ion) and controlled pH. Many varieties of silica gel have been made, including the zerogels and aerogels, in which the aqueous phase is displaced by a gaseous one.

In 1992, Yeganeh-Haeri, Weidner, and Parise (Center for High Pressure Research, State University of New York, Stony Brook) used laser Brillouin spectroscopy to determine the adiabatic single-crystal elastic stiffness coefficients of silicon dioxide in the alpha-cristobalite structure. This SiO_2 polymorph, unlike other silicas and silicates, was found to exhibit a negative Poisson's ratio. Alpha-cristobalite contracts laterally when compressed and expands laterally when stretched. Tensorial analysis of the elastic coefficients showed that Poisson's ratio reached a maximum value of -0.5 in some directions, whereas averaged values for the single-phased aggregate yielded a Poisson's ratio of -0.16.

Silicon Dioxide as a Chemical Intermediate. In 1992, R.M. Laine (University of Michigan, Ann Arbor) announced the development of a process that transforms sand and other forms of silica into reactive silicates that can be used to synthesize unusual silicon-based chemicals, polymers, glasses, and ceramics. The Laine procedure produces pentacoordinate silicates directly from low-cost raw materials—silicon dioxide, ethylene glycol, and an alkali base. The mixture is approximately a 60:1 ratio of silica gel, fused silica (or sand) to metal hydroxide and ethylene glycol. Heating the mixture slowly, the ethylene glycol and water (used

to put the materials in solution) boils off. The resulting glycolatosilicates, unlike the hexa- and tetracoordinate forms, are reactive and offer potential for synthesizing a wide range of materials. Laine observes, "The new silicon chemistry could produce alternatives to many petrochemical-based products and could be competitive with or superior to present carbon-based materials." Thus far, a number of materials have been produced by the process:

1. A clear polymer capable of conducting electric current when spread in a thin layer across a flat surface. Potential includes applications in batteries, heated windshields, and electrochromic windshields.
2. A fire retardant polymer that is easily impregnated into wood to "petrify" the material, making it stronger and nonflammable.
3. Liquid-crystal polymers stable to about 425 °C (800 °F), with potential for uses in watch displays and aerospace instrumentation.
4. Silicate glasses capable of withstanding high temperatures.

Silicates: The great number of naturally occurring silicates result, as just indicated, from the polymerization and dehydration of monosilicic acid to form, ultimately, such groups and ions as $(Si_2O_7)^{6-}$, $(Si_3O_9)^{6-}$, $(Si_4O_{12})^{8-}$, and $(Si_6O_{18})^{12-}$. Various cations, such as those of boron, B^{3+}, aluminum, Al^{3+}, etc., in the structure lie at the centers of anionic polyhedra having as anions the O^{2-} ions of neighboring SiO_4 tetrahedra, in which each $Si-O$ bond has an electrostatic bond strength of 1. Cations of lower charge density, on the other hand, like sodium, Na^+, potassium, K^+, calcium, Ca^{2+}, etc., are located interstitially. The great variety of the silicates is due to the considerable degree of isomorphism, exhibited not only by elements of the same group, but by elements of different groups, whereby they partly replace each other in the complex silicates, and by no means necessarily in stoichiometric proportions. Thus troosite may be represented by the formula $(Zn, Mn)_2SiO_4$, chrysolite by $6Mg_2SiO_4 \cdot Fe_2SiO_4$, and vermiculite by $(Mg, Fe)_3(AlSi)_4O_{10} \cdot (OH)_2 \cdot 4H_2O$, even the silicon in vermiculite being partly replaced (by aluminum). One plane of classification of the silicates is upon the basis of the linking of the SiO_4 tetrahedra:

A. Discrete silicate radicals.
  1. Single tetrahedral (SiO_4^{4-}), e.g., phenacite, Be_2SiO_4.
  2. Two tetrahedra (Si_2O_7)$^{6-}$, e.g., hardystonite, $Ca_2ZnSi_2O_7$.
  3. Three tetrahedra (Si_3O_9)$^{6-}$, e.g., benitoite, $BaTiSi_3O_9$.
  4. Four tetrahedra (Si_4O_{12})$^{18-}$, e.g., axinite, (Fe, Mn) $Ca_2Al_2BO_3$ Si_4O_{12}.
  5. Six tetrahedra (Si_6O_{18})$^{12-}$, e.g., beryl, $Be_3Al_2Si_6O_{18}$.

B. Silicon-oxygen chains of indefinite length.
  1. Single chains with one silicon atom to three oxygen atoms, e.g., diopside, $CaMg(SiO_3)_2$.
  2. Double chains ($Si : O = 4 : 11$), e.g., tremolite, $Ca_2Mg_5(Si_4O_{11})_2$ $(OH)_2$.

C. Silicon-oxygen sheets. ($Si : O = 2 : 5$), e.g., talc, $Mg_3(SiO_5)_2(OH)_2$.

D. Silicon-oxygen spatial networks.
  1. Composition SiO_2 (composed of interlinked SiO_4 tetrahedral), e.g., quartz, SiO_2.
  2. Composition $M_n(Si, Al)_nO_{2n}$, e.g., feldspar, KSi_3AlO_8. These are probably based upon silicon and aluminum tetrahedra, variously linked.

Silanes: The increasingly large number of silicon compounds produced by industrial processes may be systematized about the silanes and their substitution products, just as the silicates are about the SiO_4 tetrahedron. Silicon, like carbon, forms a number of hydrides, though their number is much more limited. The silane series, analogous to the paraffin hydrocarbons, has at least six members, silane (SiH_4), disilane ($H_3Si-SiH_3$),... hexasilane ($H_3Si-SiH_2-SiH_2-SiH_2-SiH_2-SiH_3$). They are increasingly unstable, hexasilane dissociating at room temperature. They are halogenated with free halogens to form substituted silanes, and catalytically with the hydrogen halides. The halosilanes react with NH_3 to form silylamines or silazanes and are hydrolyzed by water to form siloxanes. Prosiloxane, H_2SiO, polymerizes readily but disiloxane, $H_3Si-O-SiO_3$, and the higher siloxanes, although they polymerize, can readily be studied. They have properties like the ethers and other analogous carbon compounds. Hydrogen-containing siloxanes, such as $HO_2Si-SiO_2H$ are also known and polymerize readily. There are also ring siloxanes, such as siloxen, which has a polymerized structure of epoxy form (a powerful reducing agent)

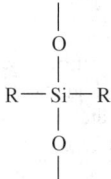

Silyl and polysilyl radicals also combine with nitrogen, arsenic, and other main group 5 elements, as with sulfur and selenium.

The silazanes are such compounds of silicon, nitrogen and hydrogen of the general formula $H_3Si(NHSiH_2)nNHSiH_3$, being called disilazane, trisilazane, etc., according to the number of silicon atoms present. (In disilazane, n in the above formula has a value of 0, in trisilazane it is 1, etc.)

The silthianes are sulfur compounds having the general formula $H_3Si(SSiH_2)_nSSiH_3$ which are called disilthiane, trisilthiane, etc., according to the number of silicon atoms present. They have the generic name *silthianes*. (In disilthiane, n in the above formula has a value of 0, in trisilthiane, a value of 1, etc.)

Silicones: These are semiorganic polymers with a quartz-like structure in which various organic groups are attached to the silicon atom. By varying the kind and number of organic groups, a variety of materials ranging from liquids through gels and elastomers to rigid solids (resins) can be produced.

The organosilicon compounds may be regarded as substituted silanes, although of course their preparation is not usually in this way. Thus, ethyl silicate, $Si(OC_2H_5)_4$, is prepared from silicon tetrachloride and ethyl alcohol, and tetraethyl silane, $Si(C_2H_5)_4$, is prepared from silicon tetrachloride and diethylzinc. The silicon-carbon bond, unlike the carbon-carbon bond, has about 12% of ionic character, varying somewhat with the atoms or groups attached to the two atoms. Other types of organosilicon compounds include the esters, the alkoxyhalosilanes, the higher tetra-alkylsilanes (prepared from silicon tetrachloride and Grignard reagents), the alkylsilanes (H partly replaced by R), the alkylhalosilanes, the alkylalkoxysilanes, the alkylsilylamines, some aryl compounds of the foregoing types, and many related derivatives of disilane and the polysilanes. Other types of compounds are those having silicon-carbon chains and the organosiloxane compounds, for which the name "silicones" is often used. These are essentially chains or networks of groups

$$R - \underset{\underset{O}{|}}{\overset{\overset{O}{|}}{Si}} - R$$

joined by oxygen atoms attached to the silicon atoms as shown. There are many other groups of silicon compounds, as well as individual ones.

Aluminates: Many complex silico-aluminates or aluminosilicates are formed in nature. Of these, clay in more or less pure form (pure clay, kaolinite; kaolin, china clay, $H_4Si_2Al_2O_9$ or $Al_2O_3 \cdot 2SiO_2 \cdot 2H_2O$) is of great importance. Clay is formed by the weathering of igneous rocks, and is used in the manufacture of bricks, pottery, porcelain, Portland cement. Sodium aluminosilicate is used in water purification to remove dissolved calcium compounds.

Fluosilicate: Sodium fluosilicate, Na_2SiF_6, white solid slightly soluble; magnesium fluosilicate, $MgSiF_6$, white solid, soluble.

Sulfides: Silicon monosulfide, SiS, yellow solid, somewhat volatile, formed by heating to redness crystalline silicon in sulfur vapor, reactive with water; silicon disulfide, SiS_2, white crystals, formed by heating amorphous silicon and sulfur, and then subliming, reactive with water.

Nitrides: Trisilicon tetranitride, Si_3N_4, by heating silicon oxide plus carbon to 1,500 °C in a current of nitrogen gas.

Silicates. See also **Adhesives**.

Silicon-Silicon Double Bond: Of the chemical elements, Si is closest to carbon in terms of its chemical properties. Multiple bonds pervade carbon chemistry and thus it is no surprise that investigators, over a period of many years, have been seeking evidence of multiple bonding in silicon. As early as 1911, Kipping reported compounds exhibiting this bonding, but these substances were later shown to be polymers or cyclic oligomers. It was not until the 1960s that good evidence was reported for the existence of Si=C (silene), Si=Si (disilene), and Si=O (silanone) compounds. The full reality of such compounds, however, was not reported until 1981. At that time, a silene and a disilene, each of which is stable at room temperature, were reported by two separate groups. Brook, et al. reported on a silene; West, et al. reported on a disilene. It has since been concluded that many disilenes can be prepared, including compounds that are unexpectedly stable. Molecules containing Si=Si bonds can be synthesized by several routes. The key to stabilization of these compounds is to provide large substituents bonded to the Si atoms so that polymerization is blocked. It has been determined that disilenes react chemically by addition across the double bond, as do alkenes. Tetramesityldisilene, as reported by West, also undergoes a wide variety of addition reactions previously unestablished in organic chemistry. The result is several "new" and unusual types of molecules, the details of which are reported in the West paper listed under references.

Silicon-Mediated Organic Synthesis: As reported by Paquette (reference listed), since the late 1960s, organic chemists have used the chemical properties of tetracovalent silicon to achieve a variety of new synthetic transformations. Paquette (Ohio State University) summarizes, "In carbon-functional silanes, exceptional stabilization is provided to a carbocation center in the beta position when the carbon-silicon bond lies in plane. This phenomenon directs electrophilic attack to the silicon-substituted carbon in aryl-, vinyl-, and alkynylsilanes and to carbon-3 in allylsilanes. For different reasons, silicon also stabilizes a carbon-metal bond in the alpha position. Consequently, access to many silicon-containing organometallics is readily available. The exceptional strength of silicon-oxygen and silicon-fluorine bonds is yet another factor that controls the chemical reactivity of silicon reagents. In recent developments, preparative chemists have taken advantage of these properties in imaginative and useful ways." In the Paquette paper, these observations are developed in exceptional and illustrated detail.

Reactions of Elemental Silicon. In the late 1980s, E.A. Pugar and P.E.D. Morgan and a team of researchers at the Rockwell International Science Center, Thousand Oaks, California, conducted a thorough effort to understand "Low Temperature Direct Reactions Between Elemental Silicon and Liquid Ammonia or Amines for Ceramics and Chemical Intermediates." Details are given in reference cited.

Because of the important potential applications of silicon nitride, the use of low-cost starting materials, such as elemental silicon and liquid ammonia or amines, may be more effective than the existing chloride method. In earlier work, this process was found to form silicon di-imide ($Si(NH)_2$), but required purification steps to remove chloride.

Pugar and Morgan elucidate their research and include a summary of the work of other researchers over the years. The report concludes: "Through the use of modern sensitive probes, direct elemental silicon reactions with liquid ammonia, silicon-hydrazine and silicon-organic amines have been discovered. The reaction of elemental silicon with nitrogen-containing reagents, under rather benign conditions, can produce ceramic precursors and with further chemical treatments can produce fibers, films, and other commercial and industrial products."

Nomenclature of Silicon Compounds: The name of the compound SiH_4 is *silane*. Compounds having the general formula $H_3Si \cdot [SiH_2]_n \cdot SiH_3$ are called disilane, trisilane, etc., according to the number of silicon atoms present. Compounds of the general formula Si_nH_{2n+2} have the generic name *silanes*. Example: Trisilane, $H_3Si \cdot SiH_2 \cdot SiH_3$.

Compounds having the formula $H_3Si \cdot [NH \cdot SiH_2]_n \cdot NH \cdot SiH_3$ are called disilazane, trisilazane, etc., according to the number of silicon atoms present; they have the generic name *silazanes*. Example: Trisilazaine, $H_3Si \cdot NH \cdot SiH_2 \cdot NH \cdot SiH_3$.

Compounds having the formula $H_3Si \cdot [S \cdot SiH_2]_n \cdot S \cdot SiH_3$ are called disilthiane, trisilthiane, etc., according to the number of silicon atoms present; they have the generic name *silthianes*. Example: Trisilthiane, $H_3Si \cdot S \cdot SiH_2 \cdot S \cdot SiH_3$.

Compounds having the formula $H_3Si \cdot [O \cdot SiH_2]_n \cdot O \cdot SiH_3$ are called disiloxane, trisiloxane, etc., according to the number of silicon atoms present; they have the generic name *siloxanes*. Example: Trisiloxane, $H_3Si \cdot O \cdot SiH_2 \cdot O \cdot SiH_3$.

For designating the positions of substituents on compounds named as silanes, silazanes, silthianes, and siloxanes, each member of the fundamental chain is numbered from one terminal silicon atom to the other. When two or more possibilities for numbering occur, the same principles are followed as for carbon compounds. Examples:

1-Butyl-2,3-dichloro-2pentyltrisilane
$Cl \cdot SiH_2 \cdot SiCl(C_5H_{11}) \cdot SiH_2 \cdot C_4H_9$
2-Methyl-3-pentyloxytrisilazane
$SiH_3 \cdot N(CH_3) \cdot SiH(OC_5H_{11}) \cdot$
$NH \cdot SiH_3$
1-Methoxytrisiloxane
$CH_3O \cdot SiH_2 \cdot O \cdot SiH_2 \cdot O \cdot SiH_3$

The names of representative radicals containing silicon are shown below. These illustrate the principles on which any further radical names should be formed.

Silicon, hydrogen

silyl	H_3Si-
silylene	$H_3Si=$
silylidyne	$HSi\equiv$
disilanyl	$H_3Si \cdot SiH_2-$
trisilanyl	$H_3Si \cdot SiH_2 SiH_2-$
disilanylene	$-SiH_2 \cdot SiH_2-$
trisilanylene	$-SiH_2 \cdot SiH_2 \cdot SiH_2-$
cyclohexasilanyl	$\begin{array}{l} SiH_2 \cdot SiH_2 \cdot SiH_2 \\ \mid \qquad\qquad \mid \\ SiH_2 \cdot SiH_2 \cdot SiH_2 \end{array}$

Silicon, hydrogen, oxygen

siloxy	$H_3Si \cdot O-$
disiloxanyl	$H_3Si \cdot O \cdot SiH_2-$
disilanoxy	$H_3Si \cdot SiH_2 \cdot O-$
disiloxanoxy	$H_3Si \cdot O \cdot SiH_2 \cdot O-$

Silicon, hydrogen, sulfur

silylthio	$H_3Si \cdot S-$
disilanylthio	$H_3Si \cdot SiH_2 \cdot S-$
disilthianyl	$H_3Si \cdot S \cdot SiH_2-$
disilthianylthio	$H_3Si \cdot S \cdot SiH_2 \cdot S-$

Silicon, hydrogen, sulfur, oxygen

disilthianoxy	$H_3Si \cdot S \cdot SiH_2 \cdot O-$
disiloxanylthio	$H_3Si \cdot O \cdot SiH_2 \cdot S-$

Silicon, hydrogen, nitrogen

silylamino	$H_3Si \cdot NH-$
disilanylamino	$H_3Si \cdot SiH_2 \cdot NH-$
disilazanyl	$H_3Si \cdot NH \cdot SiH_2-$
disilazanylamino	$H_3Si \cdot NH \cdot SiH_2 \cdot NH-$

Silicon, hydrogen, nitrogen, oxygen

disilazanoxy	$H_3Si \cdot NH \cdot SiH_2 \cdot O-$
disiloxanylamino	$H_3Si \cdot O \cdot SiH_2 \cdot NH-$

Compound radical names may be formed in the usual manner. Examples:

silyldisilanyl	$(H_2Si)_2SiH-$
disilyldisilanyl	$(H_3Si)_3Si-$
triphenylsilyl	$(C_6H_5)_3Si-$

Open-chain compounds which have the requirements for more than one of the structures already defined are named, if possible, in terms of silane, silazane, silthiane, or siloxane containing the largest number of silicon atoms. Examples:

3-Siloxytrisilthiane
$H_3Si \cdot S \cdot SiH \cdot SiH_3$
$\mid$
$O \cdot SiH_3$
1-Siloxy-3-(disilthianoxy)trisilthiane
$H_3Si \cdot S \cdot SiH \cdot S \cdot SiH_2 \cdot OSiH_3$
$\mid$
$O \cdot SiH_2 \cdot S \cdot SiH_3$

When there is a choice between two parent compounds possessing the same number of silicon atoms, the order of precedence is siloxanes,

silthianes, silazanes, and silanes. Examples:

1-Silylthiodisiloxane
$SiH_3 \cdot O \cdot SiH_2 \cdot S \cdot SiH_3$
1-Silylaminodisilthiane
$SiH_3 \cdot S \cdot SiH_2 \cdot NH \cdot SiH_3$
1-Phenyl-3-silyldisiloxane
$SiH_3 \cdot SiH_2 \cdot O \cdot SiH_2 \cdot C_6H_5$

Cyclic silicon compounds having the formula $[SiH_2]_n$ are called cyclotrisilane, cyclotetrasilane, etc., according to the number of members in the ring; they have the generic name *cyclosilanes*. Example:

Cyclotrisilane $SiH_2 \cdot SiH_2 \cdot SiH_2$

Cyclic compounds having the formula $[SiH_2 \cdot NH]_n$ are called cyclodisilazane, cyclotrisilazane, etc., according to the number of silicon atoms in the ring. They have the generic name *cyclosilazanes*. Example:

Cyclotrisilazane
$HN \cdot SiH_2 \cdot NH \cdot SiH_2 \cdot NH \cdot \qquad SiH_2$

Cyclic compounds having the formula $[SiH_2 \cdot S]_n$ have the generic name *cyclosilthianes* and are named similarly to the cyclosilazanes. Example:

Cyclotrisilthiane
$S \cdot SiH_2 \cdot S \cdot SiH_2 \cdot O \cdot \qquad SiH_2$

Cyclic compounds having the formula $[SiH_2 \cdot O]_n$ have the generic name *cyclosiloxanes* and are named similarly to the cyclosilazanes. Example:

Cyclotrisiloxane
$O \cdot SiH_2 \cdot O \cdot SiH_2 \cdot O \cdot \qquad SiH_2$

Cyclosilanes, cyclosilazanes, cyclosilthianes, and cyclosiloxanes are numbered in the same way as carbon compounds of similar nature. Examples:

2-Methoxycyclotrisilazane
$HN \cdot SiH_2 \cdot NH \cdot SiH_2 \cdot NH \cdot SiH \cdot OCH$

2-Methoxycyclorisilthiane
$S \cdot SiH_2 \cdot S \cdot SiH_2 \cdot S \cdot SiH \cdot OCH_3$

2-Methoxycyclotrisiloxane
$O \cdot SiH_2 \cdot O \cdot SiH_2 \cdot O \cdot SiH \cdot OCH_3$

Polycyclic siloxanes (polycyclic compounds whose members consist entirely of alternating silicon and oxygen atoms) are named as bicyclosiloxanes, tricyclosiloxanes, etc., or as spirosiloxanes, and are numbered according to methods in use for carbon compounds of similar nature. Polycyclic silthianes, silazanes, and silanes are treated similarly. Examples:

3,3,5,5,9,9-Hexamethyl-1,7-diphenylbicyclo[5,3,1]pentasiloxane

Tetramethyltricyclo[3,3,1,1[tetrasiloxane

The names of compounds containing silicon atoms as heteromembers (with or without other heteromembers) but not classifiable as (linear or cyclic) silanes, silazanes, silthianes or siloxanes are derived with the aid of the oxa-aza convention. Examples:

2,2,4,4,6,6-Hexamethyl-2,4,6-trisilaheptane
$(CH_3)_3Si \cdot CH_2 \cdot Si(CH_3)_2 \cdot CH_2 \cdot Si(CH_3)_3$
2,4,6,8,-Tetraoxa-5-carbonsoilane
$SiH_3 \cdot O \cdot SiH_2 \cdot O \cdot CH_2 \cdot O \cdot SiH_2 \cdot O \cdot SiH_3$

Octaphenyloxacyclopentasilane

$$
\begin{array}{c}
O \\
(C_6H_5)_2Si \qquad Si(C_6H_5)_2 \\
| \qquad\qquad | \\
(C_6H_5)_2Si —— Si(C_6H_5)_2
\end{array}
$$

Hydroxy-derivatives in which the hydroxyl groups are attached to a silicon atom are named by adding the suffixes *ol, diol, triol*, etc., to the name of the parent compound. Examples:

Silanol	$H_3Si \cdot OH$
Silanediol	$H_2Si(OH)_2$
Silanetriol	$HSi(OH)_3$
Disilanehexaol	$(HO)_3Si \cdot Si(OH)_3$
Disiloxanol	$H_3Si \cdot O \cdot SiH_2 \cdot OH$

Cyclohexasilanol
$$
\begin{array}{c}
SiH_2 \cdot SiH_2 \cdot SiH \cdot OH \\
| \qquad\qquad\qquad | \\
SiH_2 \cdot SiH_2 \cdot SiH_2
\end{array}
$$

Polyhydroxy-derivatives in which hydroxyl group is attached to a silicon atom are named wherever possible in accordance with the principle of treating like things alike. Example:

1,13,5,5-Pentamethyltrisiloxane-1,3,5-trio

$$
\begin{array}{c}
HO \quad HO \qquad CH_3 \quad HO \\
| \qquad\backslash \qquad / \qquad\quad | \\
(CH_3)_2Si — O — Si — O — Si(CH_3)_2
\end{array}
$$

Otherwise they are named in accordance with the principle of the largest parent compound. Example:

2-Hydroxysilyltetrasilane-1,4-dio

$$
\begin{array}{c}
SiH_2 \cdot OH \\
| \\
HO \cdot SiH_2 \cdot SiH_2 \cdot SiH \cdot SiH_2 \cdot OH
\end{array}
$$

Substituents other than hydroxyl groups (functional atoms or groups and hydrocarbon radicals) attached to silicon are expressed by appropriate prefixes or suffixes. Examples:

Ethyldisilane
$CH_3 \cdot CH_2 \cdot SiH_2 \cdot SiH_3$
Hexachlorodisiloxane
$Cl_3Si \cdot O \cdot SiCl_3$
Dibutyldichlorosilane
$(CH_3 \cdot CH_2 \cdot CH_2 \cdot CH_2)_2SiCl_2$
Silylamine
$H_3Si \cdot NH_2$
Silanediamine
$H_2Si(NH_2)_2$
Silanetriamine
$HSi(NH_2)_3$
N-Methylsilylamine
$H_3Si \cdot NH \cdot CH_3$
N,N-Dimethylsilylamine
$H_3Si \cdot N(CH_3)_2$
N,N'-Dimethylsilanediamine
$H_2Si(NH \cdot CH_3)_2$
N,N',N''-Trimethylsilanetriamine
$HSi(NH \cdot CH_3)_3$
Acetoxytrimethylsilane
$(CH_3)_3Si \cdot O \cdot OC \cdot CH_3$
Diacetoxydimethylsilane
$(CH_3)_2Si(O \cdot OC \cdot CH_3)_2$

Compounds containing carbon as well as silicon and in which there is a "reactive group" in the carbon-containing portion of the molecule

not shared by a silicon atom are named in terms of the organic parent compound wherever feasible. Examples:

α-Trimethylsilylacetanilide
$(CH_3)_3Si \cdot CH_2 \cdot NH \cdot C_6H_5$
1-Trichlorosilylethanol
$Cl_3Si \cdot CH(OH) \cdot CH_3$
2-Trimethylsilylethanol
$(CH_3)_3Si \cdot CH_2 \cdot CH_2OH$
(Hydroxydimethylsilyl)methanol
$$
\begin{array}{c}
(CH_3)_2Si \cdot CH_2OH \\
| \\
OH
\end{array}
$$
α-(Hydroxydimethylsilyl)acetanilide
$$
\begin{array}{c}
(CH_3)_2Si \cdot CH_2 \cdot CO \cdot NH \cdot C_6H_3 \\
| \\
OH
\end{array}
$$
(Silylmethyl)amine
$H_3Si \cdot CH_2 \cdot NH_2$
But by rules 70.16 and 70.17:
(Methoxymethyl)silanol
$CH_3O \cdot CH_2 \cdot SiH_2 \cdot OH$
N-Methylsilylamine
$H_3Si \cdot NH \cdot CH_3$

Compounds in which metals are combined directly with silicon are, in general, named as derivatives of the metal. Example:

(Triphenylsilyl)lithium
$(C_6H_5)_3SiLi$

However, in exceptional cases, the metal may be named as a substituent. Example:

Sodium *p*-(sodiosilyl)benzoate
$p\text{-}NaO_2C \cdot C_6H_4 \cdot SiH_2Na$

Metallic salts of hydroxy-derivatives may be named in the customary manner. Example:

Sodium salt of triphenylsilanol
$(C_6H_5)_3Si \cdot ONa$

Essentially Nonchemical Properties of Silicon

Were it not for the firm establishment of silicon as an indispensable material for modern electronics, the other exceptionally attractive properties of Si may have been overlooked for many years. Over the past few decades, electronics components manufacturers have mastered the skills required for manufacturing microminiature components, and this experience has given Si a head start for use in other subminiature structures. Si has been recognized as an outstanding material for making micromachined subminiature structures essentially just within the past decade, and it has become one of the key materials in the comparatively new field of *nanotechnology*.

As a mechanical material, silicon is stronger than steel, it does not show mechanical hysteresis, and it is highly sensitive to stress. This combination of properties qualifies Si as an excellent sensor for detecting acceleration, pressure, force, and other variables encountered in processing and manufacturing. One method of measuring fluid flow, for example, traditionally has depended upon sensing pressure differentials, as in the case of an orifice-type flowmeter. Thus, silicon sensors can be used. Silicon accelerometers can employ the same piezoresistive sensing technique used in pressure sensors.

In addition to sensors, Si can be used at the subminiature scale for the production of tiny pipe, nozzles, and valves required by automatic control systems. Thus, the weight and bulk of future control systems may be reduced by several orders of magnitude.

As of the early 1990s, engineers are working at the "edge" of a new kind of robotics—that is, subminiature handling devices that can master the handling requirements of the new nanomanufacturing technology. Such robots would be miniature, fully integrated silicon systems drawing heavily on the technologies of silicon-integrated electronics and micromachining. Semi-intelligent robots could be used in many manufacturing and control tasks. According to some researchers, such robots could have intelligence at the lowest possible system level, thus allowing them to function semiautonomously, with occasional input from a central control system.

Biological Applications of Silicon Technology. H.M. McConnell (Stanford University) and a team of researchers have developed a silicon-based device called a cytosometer (microphysiometer) that can be used to detect and monitor the response of cells to a variety of chemical substances, particularly ligands for specific plasma membrane receptors. As pointed out by McConnell, "The microphysiometer measures the rate of proton excretion from 10^4 to 10^6 cells. The instruments serves two distinct functions. In terms of detecting specific molecules, selected biological cells in this instrument serve as detectors and amplifiers. The microphysiometer can also investigate cell function and biochemistry. A major application of this instrument may prove to be screening for new receptor ligands. In this respect, the instrument appears to offer significant advantages over other techniques." More detail is given in the McConnell reference listed.

Additional Reading

Amato, I.: "Shine On, Holey Silicon," *Science*, 922 (May 17, 1991).

Aufderhaar, H.C.: "Silicon," in "Metals Handbook," 9th edition, Vol. 2, ASM International, Metals Park, OH, 1989.

Binnig, G. and H. Rohrer: "The Scanning Tunneling Microscope," *Sci. Amer.*, 253(2), 50–56 (August 1985).

Boland, J.J. and G.N. Parsons: "Bond Selectivity in Silicon Film Growth," *Science*, 1304 (May 29, 1992).

Bryzek, J., Mallon, J.R., Jr., and R.H. Grace: "Silicon's Synthesis: Sensors to Systems," *Instrumentation technology*, 40 (January 1989).

Carter, G.F. and D.E. Paul: "Materials Science and Engineering," ASM International, Materials Park, OH, 1991.

Chabal, Y.J.: "Fundamental Aspects of Silicon Oxidation," Springer-Verlag Inc., New York, NY, 2001.

Connally, J.A. and S.B. Brown: "Slow Crack Growth in Single-Crystal Silicon," *Science*, 1537 (June 12, 1992).

Corcoran, E.: "Holey Silicon," *Sci. Amer.*, 102 (March 1992).

Dunn, W.: "Micromachined Sensors for Automotive Applications," *Sensors*, 54 (September 1991).

Feng, Z.C. and R. Tsu: "Porous Silicon," World Scientific Publishing Company, Inc., Riveredge, NJ, 1994.

Golovchenko, J.A.: "The Tunneling Microscope: A New Look at the Atomic World," *Science*, 232, 48–53 (1986).

Greenwood, N.N. and A. Earnshaw: "Chemistry of the Elements," 2nd Edition, Butterworth-Heinemann, Woburn, MA, 1997.

Heath, J.R.: "A Liquid-Solution-Phase Synthesis of Crystalline Silicon," *Science*, 1131 (November 13, 1992).

Henkel, S.: "Silicon Microvalves Fabricated on Bimetallic Diaphragms," *Sensors*, 4 (December 1991).

Iyer, S.S. and Y-H Xie: "Light Emission from Silicon," *Science,* 40 (April 2, 1993).

Jackson, K.: "Silicon Devices: Structures and Processing," John Wiley & Sons, Inc., New York, NY, 1998.

Jennings, H.M. and M.H. Richman: *Science*, 193, 1242 (1976).

Kasper, J.S. and R.H. Wentorf, Jr.: "Hexagonal (Wurtzite) Silicon," *Science*, 197, 599 (1977).

Laine, R.M.: "Beach Sand: Material of the Future?" *Advanced Materials & Processes*, 6 (February 1992).

LeComber, P.G.: "Amorphous Silicon—Electronics Into the 21st Century," University of Wales Review, 31 (Spring 1988).

Lide, D.R.: "CRC Handbook of Chemistry and Physics 2000-2001," 81st Edition, CRC Press, LLC., Boca Raton, FL, 2000.

Link, B.: "Field-Qualified Silicon Accelerometers," *Sensors*, 28 (March 1993).

Maugh, T.H., II: "A New Route to Intermetallics (Metal Silicides)," *Science*, 225, 403 (1984).

McConnell et al.: "The Cytosensor Microphysiometer: Biological Applications of Silicon Technology," *Science*, 1906 (September 25, 1992).

Meyers, R.A.: "Handbook of Chemicals Production Processes," McGraw-Hill Companies, Inc., New York, NY, 1986.

Nalwa, H.S.: "Silicon-Based Materials and Devices," Academic Press, Inc., San Diego, CA, 2001.

Paquette, L.A.: "Silicon-Mediated Organic Synthesis," *Science*, 217, 793–800 (1982).

Pensl, G. and H. Matsunami: "Silicon Carbide: A Review of Fundamental Questions and Applications to Current Device Technology," John Wiley & Sons, Inc., New York, NY, 1997.

Pugar, E.A. and P.E.D. Morgan: "Low Temperature Direct Reactions Between Elemental Silicon and Liquid Ammonia or Amines for Ceramics and Chemical Intermediates," in Report issued by Rockwell International Science Center, Thousand Oaks, California (September 1988).

Rappoport, Z. and Y. Apeloig: "The Chemistry of Organic Silicon Compounds," John Wiley & Sons, Inc., New York, NY, 2001.

Robinson, A.L.: "Consensus on Silicon Surface Structure Near," *Science*, 232, 451–453 (1986).

Schubert, U.: "Silicon Chemistry," Springer-Verlag Inc., New York, NY, 1999.

Simpson, T.L. and B.E. Volcani: "Silicon and Siliceous Structures in Biological Systems," Springer-Verlag Inc., New York, NY, 1981.

Staff: "Grace with Pressure (Silicon)," *Sci. Amer.*, 253(2), 62–64 (August 1985).

Staff: "ASM Handbook—Properties and Selection: Nonferrous Alloys and Pure Metals," ASM International, Materials Park, OH, 1990.

Staff: "Silicon Atoms 'See the Light'," *Advanced Materials & Processes*, 6 (November 1990).

Staff: "Tough $MoSi_2$ Composites also Combat Oxidation," *Advanced Materials & Processes*, 26 (January 1991).

Strausser, Y.E.: "Characterization in Silicon Processing," Butterworth-Heinemann, Inc., Woburn, MA, 1993.

Street, R.A.: "Technology and Applications of Amorphous Silicon," Springer-Verlag Inc., New York, NY, 2001.

Tanaka, K. and H. Okamoto: "Amorphous Silicon," John Wiley & Sons, Inc., New York, NY, 1999.

Travis, J.: "Building a Silicon Surface, Atom by Atom," *Science*, 1354 (March 13, 1992).

Wentorf, R.H., Jr. and J.S. Kasper: *Science*, 139, 338 (1963).

West, R.: "Isolable Compounds Containing a Silicon-Silicon Double Bond," *Science*, 225, 1109–1114 (1984).

Yeganeh-Haeri, A., Weidner, D.J., and J.B. Parise: "Elasticity of Alpha-Cristobalite: A Silicon Dioxide with a Negative Poisson's Ratio," *Science*, 650 (July 31, 1992).

Yun, W. and R.T. Howe: "Recent Developments in Silicon Micro-accelerometers," *Sensors*, 31 (October 1992).

Yun, W. and R.T. Howe: "Sigma-Delta Modulator Interfacing with Silicon Micro-sensors," *Sensors*, 11 (May 1993).

Zdebick, M.: "A Revolutionary Actuator for Microstructures," *Sensors*, 26 (February 1993).

SILICON CONTROLLED RECTIFIER (or SCR).

A semiconductor device consisting of four alternate layers of n and p type silicon, which functions as a current controlled switch. The schematic symbol and its correspondence to the four layers is shown in Fig. 1. Two outstanding features are associated with the SCR, namely, the speed of switching and ratio of controlled to controlling currents. Load currents of tens to hundreds of amperes may be turned on in a few microseconds and turned off in about ten times as long a time. Typically, the controlling current is several orders of magnitude less than the controlled current.

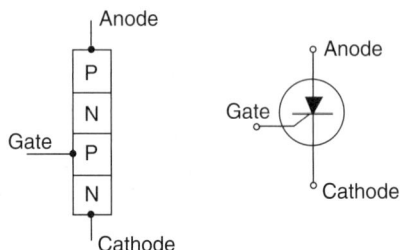

Fig. 1.   Diagrammatic representation of silicon-controlled rectifier and schematic symbol.

In operation, a comparatively small anode current flows when the gate current is zero and the anode-to-cathode voltage is held below a critical value known as the *forward breakover voltage*. A positive pulse of gate current of suitable magnitude triggers the device into a high conduction mode. At this time the anode current is determined almost entirely by the characteristics of the external circuit in series with the anode-cathode path, providing this current exceeds a small minimum value called the *holding current*. Once the high conduction mode is achieved, the gate current has no further control over the anode current. To turn off the anode current, the anode-to-cathode voltage must be reduced substantially (so that the anode current falls below the holding current). For turn-off in minimum time, it is necessary to reverse the polarity of the anode voltage.

SILICON CONTROLLED SWITCH (or SCS).

An electronic device consisting of four layers of semiconductor material alternately doped with p and n type impurities. It differs from the silicon controlled rectifier in that external connections are made to all four layers rather than to only three, as in the case of the latter. The schematic symbol and an indication of the equivalence of the device to a combination of two junction transistors are shown in Fig. 1. It is primarily used for switching and control applications.

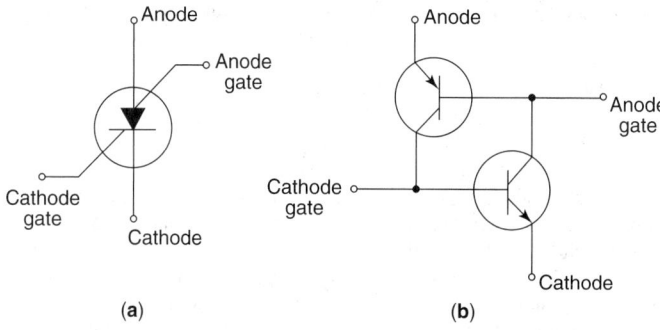

Fig. 1. Schematic symbol of silicon controlled switch (**a**) and equivalence to interconnection of two transistors (**b**).

SILK COTTON TREES. Of the family *Bombaceceae* (silk cotton family), there are several genera of these trees from which valuable commercial fibers and oils are obtained. The silk cotton tree (*Ceiba pentandra*), also called the ceiba or kapok tree is a tropical tree with a large trunk that may reach a height of about 100 feet (30.5 meters) and diameter of 10 feet (0.3 meter). The tree is found in Brazil, Java, India, Central America, and Ecuador. The trunk has large roots that stabilize the tree and taper from underground to buttress the natural trunk. The tree produces large pods, which yield silk cotton called *kapok*. This material is used for filling mattresses and in padding cushions and furniture. Kapok is very light and resilient and also can be used as an insulating material. Most of the kapok used in the United States comes from Java. The Javanese tree was introduced from Brazil where the tree is referred to locally as the samaúma. The long, silky, and white fibers appear much like cotton. However, the fibers cannot be spun because of their brittleness. Kapok is also sometimes referred to as Illiani silk and Java cotton. Chemically, kapok is similar to jute, but with different proportions of the constituents, kapok being lower in alpha cellulose and higher in pentosans and lignin.

Other important genera of the silk cotton family include: *Calotropis gigantea*, a shrub grown in the East Indies and India and which produces a silky cotton known as *madar*, of an inferior quality to kapok. *Calotropis procera*, found in the same region, produces *akund*. Sometimes these inferior fibers are mixed with kapok. The species *Bombax ceiba* and *B. malabaricum*, grown in India, yield a red silk cotton called *simal*. A fiber known as *shilo* is obtained from *B. ellipticum*. The Indian tree *Cochlospermum gossypium* yields a white silk cotton similar to kapok. Mexican kapok is produced from *Bombax palmeri* and from *Ceiba schottii* and *C. acuminata*. Ecuadoran kapok comes from the genus *Chorisia*. Balsa fiber from the genus *Ochroma*, produced in Colombia, Ecuador, and Central America, is of a darker color.

The seeds of the kapok tree yields a semidrying oil which finds use in the manufacture of margarine and soaps.

SILKWORM (*Insecta, Lepidoptera*). The caterpillar of the moth, *Bombyx mori*, which produces the silk of commerce. Caterpillars of the family *Saturniidae* also spin cocoons of silk, and are known as the giant silkworms. This family includes the common luna, cecropia, and polyphemus moths of North America, as well as many other species.

SILL. A tabular mass of igneous rock that has been intruded laterally between layers of sedimentary rock, beds of volcanic lava or ejectmenta, or even along the direction of the foliation in metamorphic rock. The term sill is synonymous with intrusive sheet.

SILLIMANITE. The mineral sillimanite is an aluminum silicate, the formula Al_2SiO_5 being like that of andalusite and kyanite. It is orthorhombic, usually in slender prisms, but may be fibrous or massive. Its hardness is 6.5–7.5; specific gravity, 3.23–3.27; luster, vitreous to silky; color, various shades of gray, grayish-green, and grayish-brown; transparent to translucent. It occurs in granites and gneisses as tiny prisms and aggregates, and is often associated with andalusite, cordierite and corundum. Sillimanite has been found in Bavaria, the Czech Republic and Slovakia, France, India, the Malagasy Republic, Myanmar and Ceylon, the latter two localities furnishing transparent sapphire-blue gem stones. In the United States sillimanite has been found in Connecticut, New York, Pennsylvania, Delaware, North Carolina, and in California, where in Inyo County is the largest deposit

in the world. This mineral was named in honor of Benjamin Silliman for many years professor of chemistry and natural science at Yale University. Sillimanite is used in the manufacture of spark plug "porcelains" and laboratory ware.

SILTSTONE. A term signifying a clastic sedimentary rock in which the particles are of silt grade.

SILURIAN PERIOD. A major subdivision of the Paleozoic Era. Type locality, Wales and Shropshire, England. The formations of this system were first studied and described by R.I. Murchison in 1835. The Silurian Period began 380 million years ago and lasted for 50 million years. In Murchison's time, and for some time afterward, the Paleozoic Era, below the Devonian, was divided into the Cambrian and Silurian. This practice still holds in parts of Europe. The Silurian formations are well exposed in Eastern North America, especially in New York State, and the length of the Appalachian geosyncline where the sediments are principally red sandstones and shales of delta origin. There was little or no volcanic activity except in southeastern Maine. The Silurian is also well exposed in nearly all of Western Europe, Northern Siberia, Myanmar, Central Asia, Himalayas, Morocco, Australia, Peru and Bolivia. The system is characterized by all types of sediments, with evidence in certain areas of an arid climate, including thick deposits of salt and gypsum. The maximum thickness of sediments is 15,000 feet (4,500 meters) in Britain.

SILVER. Chemical element, symbol Ag (from Latin *argentum*), at. no. 47, at. wt. 107.868 ± 0.003, periodic table group 11, mp 961.93 °C, bp approximately 2,212 °C, density 10.50 g/cm³ (20 °C). Elemental silver has a face-centered cubic crystal structure.

Silver is a white metal, softer than copper and harder than gold. When molten, silver is luminescent and occludes oxygen, but the oxygen is released upon solidification. As a conductor of heat and electricity, silver is superior to all other metals. Silver is soluble in HNO_3 containing a trace of nitrate; soluble in hot 80% H_2SO_4; insoluble in HCl or acetic acid; tarnished by H_2S, soluble sulfides and many sulfur-containing organic substances (e.g., proteins); not affected by air or H_2O at ordinary temperatures, but at 200 °C, a slight film of silver oxide is formed; not affected by alkalis, either in solution or fused. There are two stable, naturally occurring isotopes, ^{107}Ag and ^{109}Ag. In addition, there are reported to be 25 less stable isotopes, ranging in half-life from 5 seconds to 253 days.

In terms of cosmic abundance, the estimate of Harold C. Urey (1952), using silicon as a base with a figure of 10,000, silver was assigned an abundance figure of 0.023. In terms of abundance in sea water, silver is ranked number 43 among the elements, with an estimated content of 1.5 tons per cubic mile (0.324 metric ton per cubic kilometer) of sea water.

Electronic configuration is $1s^2 2s^2 2p^6 3s^2 3p^6 3d^{10} 4s^2 4p^6 4d^{10} 5s^1$. First ionization potential is 7.574 eV; second, 21.4 eV; third, 35.9 eV. Oxidation potentials: $Ag \rightarrow Ag^+ + e^-$, $E^0 = -0.7995$ V; $Ag^+ \rightarrow Ag^{2+} + e^-$, -1.98 V; $2Ag^+ \ OH^- \rightarrow Ag_2O + H_2O + 2e^-$, -0.344 V; $Ag_2O + 2OH^- \rightarrow 2AgO + H_2O + 2e^-$, -0.57 V; $2AgO + 2OH^- \rightarrow Ag_2O_3 + H_2O + 2e^-$, -0.74 V. Other important physical properties of silver are given under **Chemical Elements**.

Occurrence and Processing: Silver is widely distributed throughout the world. It rarely occurs in native form, but is found in ore bodies as silver chloride, or more frequently, as simple and complex sulfides. In former years, simple silver and gold-silver ores were processed by amalgamation or cyanidation processes. The availability of ores amenable to treatment by these means has declined. Most silver is now obtained as a byproduct or coproduct from base metal ores, particularly those of copper, lead, and zinc. Although these ores are different in mineral complexity and grade, processing is similar.

All the ores are concentrated in complex mills by selective froth flotation to produce individual copper, zinc, lead, and, infrequently, silver concentrates. The copper and lead concentrates are smelted to produce lead and copper bullions from which silver is recovered by electrolytic or fine refining. The silver bearing zinc concentrates are commonly processed by leaching and electrolytic methods. Silver is ultimately recovered as a byproduct from zinc plant residues. Canada is a leading silver mining country. Other important sources of silver are Mexico, the United States, Peru, the former U.S.S.R., and Australia. See also **Mineralogy**.

A substantial portion of the total world silver supply is obtained from recycled scrap. Much of this scrap comes from photographic film, jewelry and the electrical field. The high value of the scrap dictates accurate sampling and careful feed preparation. Efficient and fast processing is required to minimize metal losses and a tie-up of high-value materials. The highly complex nature of plant feed, with respect to physical form, chemical composition, and grade, requires use of complex and highly flexible processing procedures.

Uses of Silver: Silver in the twentieth century can be classified an industrial commodity. For most of the 19th century, silver was a monetary metal. Industrial consumption of silver is principally in photographic film, electrical contacts, batteries and brazing alloys. Sterling silver and silver plated copper alloys are used extensively for tableware and for jewelry and other decorative art. Recently, the field of commemorative and collector arts has become a substantial market for silver alloys, particularly sterling silver.

The predominant place of silver salts as photographic receptors is not the result of any unusual primary sensitivity to illumination, but is due to the fact that they undergo an unusual secondary amplification process called "development." Silver salts, like the salts of many metals, when immersed in solutions of many reducing agents, are changed to metallic silver. The photographic system depends upon the fact that when certain mild reducing agents (called "photographic developers") are chosen, the rate of reduction is increased many fold if the silver salt crystals carry very small amounts of metallic silver at the developer-crystal interface. The effect produced by the original light exposure is amplified in the development process by a factor of 100 billion. Whereas new photographic or recording devices are being developed not involving silver, none yet approach the packing density of a fine-grained image possible using silver. Thus, it appears that silver will be used in photographic recording for many years to come.

Among the electrical uses for silver are electrical contacts, printed circuits, and batteries. By far, the primary use is in electrical contacts where the high electrical and thermal conductivities, as well as corrosion and oxidation resistances, of silver are major reasons for its selection. Although silver has a strong tendency to weld under heavy currents, this is counteracted by alloying or by adding nonmetallic substances (such as cadmium oxide) to the silver matrix. The use of silver-cadmium oxide and silver-tungsten materials in electrical contact applications is widespread. The alloys used to improve the wear resistance and to reduce the sticking tendency of silver include silver-gold, silver-copper, silver-palladium, and silver-platinum. More complex alloys include silver-copper-nickel, silver-magnesium-nickel, silver-gold-cadmium-copper, and silver-cadmium-copper-nickel. Silver-cadmium oxide alloys are unique materials and are prepared either by combining silver and cadmium oxide by powder metal techniques or by the internal oxidation of a silver-cadmium alloy. Electrical alloys, which are impossible to combine by conventional melting, lend themselves to powder metal fabrication. Such composite structures as silver-graphite, silver-iron, and silver-tungsten are good examples of these types of materials.

In silver batteries, the silver oxide-zinc secondary battery has found its place in applications where energy delivered per unit of weight and space is of prime importance. The major disadvantages lie in their high cost and relatively short life. Consequently, a large part of the silver battery market is concerned with defense and space components. See also **Battery**.

Prior to World War II, consumption of silver in silverware and jewelry was the largest industrial use of silver. Competition from stainless steel in flatware and holloware has contributed to a decline in overall use. Most consumption of silver in silverware and jewelry is in the form of sterling silver, an alloy of silver with approximately 7.5 weight percent copper. Silver plate, which is silver electroplated on a base metal, varies widely in specification. The thickness, expressed for example in penny-weights of pure silver per gross of teaspoons can range from a low of 1 to as high as 200.

In the 1920s and 1930s, low-temperature silver-copper brazing alloys were found to be useful on copper and its alloys and iron and its alloys (including stainless steel). Silver and copper form a simple eutectic system with limited solid solubility. This system can absorb elements such as zinc, cadmium, tin, and indium. These additions lower its melting temperature. It also can absorb higher melting elements such as nickel or palladium. These raise its melting temperature, but may improve its wetting characteristics, corrosion resistance, and strength at elevated temperatures. Silver solders or brazing alloys have the ability of making joints far stronger and more durable than common soft-solder (such as lead-tin) alloys. They are used in most refrigeration systems to join copper tubing. Also, extensive use is found in the assembly of automotive parts, military components, aircraft assemblies, and other hard goods manufacture. The nominal composition of a popular brazing alloy, ASTM Classification BAg-1 is silver 45%, copper-15%, zinc-16%, cadmium-24%.

One silver alloy containing about 70% silver, 26% tin, 3% copper, and 1% zinc is unique in that it is used extensively by dentists in combination with mercury to fill cavities in teeth. The "amalgam" manufacturers supply dentists with the alloy in the form of powder (filed, or more recently, atomized). This is mixed with mercury, using from 8 to 5 parts of mercury to 5 of alloy, and the cavity is packed. In the cavity, a metallurgical reaction takes place in which the silver-tin compound in the alloy becomes a durable silver-tin-mercury compound.

Silver, its oxides, halides and other salts play important roles in chemistry. Silver is an excellent catalyst in oxidizing reactions such as in the production of formaldehyde from methanol and oxygen, ethylene oxide from ethylene and oxygen, and glyoxal from ethylene glycol and oxygen. Silver has oligodynamic properties, that is, the ability of minute amounts of silver in solution to kill bacteria. Modern technology has made use of this property in various ways, mainly as a means of purifying water.

Small amounts of silver are used annually in such diverse applications as a backing for mirrors, and in control rods for pressurized water nuclear reactors. Miscellaneous uses like this account for only a small fraction of total silver consumption.

Chemistry of Silver. Silver(I) oxide, Ag_2O, is made by action of oxygen under pressure on silver at 300 °C, or by precipitation of a silver salt with carbonate-free alkali metal hydroxide; it is covalent, each silver atom (in solid Ag_2O) having two collinear bonds and each oxygen atom four tetrahedral ones; two such interpenetrating lattices constitute the structure. Silver(I) oxide is the normal oxide of silver. Silver(II) oxide, AgO, is formed when ozone reacts with silver, and thus was once considered to be a peroxide. Silver(III) oxide, Ag_2O_3, has been obtained in impure state by anodic oxidation of silver.

All of the silver(I) halides of the four common halogens are well known. The fluoride may be prepared from the elements, the chloride by action of hydrogen chloride gas at 150 °C, upon silver, and the bromide and iodide by ionic reactions in solution. The chloride, bromide, and iodide are essentially insoluble in H_2O, but the fluoride is soluble. There is also a subfluoride, Ag_2F, which may be prepared as a cathodic deposit by electrolysis of silver(I) fluoride AgF, or by evaporation of finely divided silver with silver(I) fluoride in dilute hydrofluoric acid. It is an anisotropically conducting solid and is considered to be made up in the solid state of two silver layers, metallic-bonded to each other, and ionic-covalent bonded to a single fluorine layer. It has reverse cadmium iodide structure. Silver subchloride, Ag_2Cl is made by reaction of Ag_2F and phosphorus trichloride. Silver(II) fluoride, AgF_2, made by action of fluorine upon a silver(I) halide, is a fluorinating agent or catalyst for fluorinations. The silver(I) halides vary markedly in ionicity, the values given by Pauling being AgF 70%, AgCl 30%, AgBr 23% and AgI 11%. This is reflected in their crystal structures and in their solubility in water (or rather, their relative insolubility). The first three have sodium chloride structure, AgI has wurtzite structure; AgF has a molal solubility of 14, and the pK_{sp} values of the others are 9.75, 12. 27 and 16.08, respectively.

Silver differs markedly from copper in forming few oxy compounds. One of these is silver oxynitrate or silver(II, III) nitrate which has the empirical formula $AgO_{1.148}(NO_3)_{0.453}$, in which the average oxidation number of silver is 2.448. It is prepared by action of fluorine upon aqueous silver nitrate or is obtained as an anodic deposit by electrolysis of silver nitrate in dilute HNO_3.

Silver enters into complex formation with many ions and molecules. With halogens, the silver complexes are fewer than the copper ones. Silver chloride dissolves in HCl with the formation of such chloroargentate ions as $(AgCl_2)^-$, $(AgCl_3)^{2-}$, and possibly $(AgCl_4)^{3-}$. Complex ions with bromide, $(AgBr_2)^-$ and $(AgBr_3)^{2-}$ are more stable, as are those with iodide, than those with chloride. Complexes of the type Ag_2Cl^+, Ag_3Cl^{2+}, Ag_2Br^+, Ag_3Br^{2+}, Ag_4Br^{3+}, $Ag_2Br_6^{2-}$, Ag_2I^+, Ag_3I^{2+}, Ag_4I^{3+}, $Ag_2I_6^{4-}$, $Ag_2I_7^{5-}$ and $Ag_3I_8^{5-}$ are also known. With ammonia the ions $(Ag(NH_3)_2)^+$ and $(Ag(NH_3)_3)^+$ are definitely known and others may exist. Similar complexes are formed with amines and diamines. With cyanides, silver forms very stable complexes, the number of CN^- ions in the complex depending somewhat upon the excess of cyanide, so that $(Ag(CN)_2)^-$, $(Ag(CN)_3)^{2-}$, and $(Ag(CN)_4)^{3-}$ are definitely known. With thiosulfates, silver forms various complexes. In dilute solution, $(Ag_2(S_2O_3)_2)^{2-}$ exists, while in high concentration of $S_2O_3^{2-}$ ion, the complex $(Ag_2(S_2O_3)_6)^{10-}$ has been identified. In HNO_3 solution Ag^+ is easily oxidized to Ag^{2+} by peroxydisulfate. From this solution complex compounds of dipositive silver can be prepared, which are stable because coordination radically alters the oxidation potential of Ag(I) to Ag(II). They include pyridine complexes such as $(Ag(py)_4) \times (NO_3)_2$. 8-Hydroxyquinoline complexes containing the ions $(Ag(oxin)_2)^{2+}$, and o-phenanthroline complexes containing the ion $(Ag(o\text{-}phen)_2)^{2+}$. Silver(III) is known in the square, planar complex AgF_4^-, which has been prepared as $KAgF_4$ by direct fluorination of a mixture of potassium chloride and silver chloride. Silver(III), like Cu(III), also occurs in tellurate and periodate complexes.

Other silver compounds include: Silver chromate (Ag_2CrO_4), yellow to red to brown precipitate by reaction of silver nitrate solution and potassium chromate solution.

Silver dichromate $(Ag_2Cr_2O_7)$, red precipitate by reaction of silver nitrate solution and potassium dichromate solution, changing to silver chromate upon boiling with H_2O.

Silver phosphate (Ag_3PO_4), yellow precipitate, by reaction of silver nitrate solution and disodium hydrogen phosphate solution, soluble in HNO_3 and in NH_4OH, turns dark on exposure to light.

Silver sulfate (Ag_2SO_4), white precipitate, by the action of silver nitrate solution and potassium sodium or ammonium sulfate solution or H_2SO_4, mp of silver sulfate 652 °C.

Silver sulfide (Ag_2S), black precipitate, by the reaction of silver nitrate solution and hydrogen sulfide.

Silver forms several compounds or complexes with proteins by the action of silver oxide with gelatin in alkali solution, or with albumin, or by suspension in casein solution and by other methods. Such silver-protein complexes containing from 19 to 23% of silver are known as "mild silver protein" and are used as antiseptic solutions. They are readily soluble in H_2O.

Additional Reading

Carapella, S.C., Jr. and D.A. Corrigan: "Properties of Pure Silver," Metals Handbook, 9th Edition, Vol. 2, ASM International, Metals Park, OH, 1979.

Considine, D.M. and G.D. Considine: "Van Nostrand Reinhold Encyclopedia of Chemistry," 4th Edition, Van Nostrand Reinhold, New York, NY, 1984.

Coxe, C.D., McDonald, A.S., and G.H. Sistare, Jr.: "Properties of Silver and Silver Alloys," Metals Handbook, 9th Edition, Vol. 2, ASM International, Metals Park, OH, 1979.

Coxe, C.D., McDonald, A.S., and G.H. Sistare, Jr.: "Silver-Base Brazing Alloys," Metals Handbook, 9th Edition, Vol. 2, ASM International, Metals Park, OH, 1979.

Davis, J.R.: "Metals Handbook," 2nd Edition, ASM International, Metals Park, OH, 1998.

Friend, W.Z.: "Corrosion Resistance of Precious Metals," Metals Handbook, 9th Edition, Vol. 2, ASM International, Metals Park, OH, 1979.

Gale, N.H. and Z. Stos-Gale: "Lead and Silver in the Ancient Aegean," Sci. Amer., 176–192 (June 1981).

Greener, E.H.: "Dental Materials," Encyclopedia of Materials Science and Engineering, MIT Press, Cambridge, MA, 1986.

Greenwood, N.N. and A. Earnshaw: "Chemistry of the Elements," 2nd Edition, Butterworth-Heinemann, Inc., Woburn, MA, 1997.

Lechtman, H.: "Pre-Columbian Surface Metallurgy," Sci. Amer., 56–53 (June 1984).

Lide, D.R.: "CRC Handbook of Chemistry and Physics 2000-2001," 81st Edition, CRC Press, LLC., Boca Raton, FL, 2000.

Parker, P.: "McGraw-Hill Encyclopedia of Chemistry," 2nd Edition, The McGraw-Hill Companies, Inc., New York, NY, 1993.

Sinfelt, J.H.: "Bimetallic Catalysts," Sci. Amer., 90–98 (September 1985).

Stwertka, A. and E. Stwertka: "A Guide to the Elements," Oxford University Press, Inc., New York, NY, 1998.

Waterstrat, R.M. and G. Dickson: "Dental Amalgam (Hg, Ag, Sn, Cu, Zn)," Metals Handbook, 9th Edition, Vol. 2, ASM International, Metals Park, OH, 1979.

Zysk, E.D.: "Precious Metals and Their Use," Metals Handbook, 9th Edition, Vol. 2, ASM International, Metals Park, OH, 1979.

DONALD A. CORRIGAN, Handy & Harman, Fairfield, CT.

SILVER-CELL BATTERY.
A type of short-duration, high-power-density battery of light weight used for single-time, high-power applications in vehicles where weight is critical.

SILVER FIRS.
See **Fir Trees**.

SILVERFISH
(Insecta, Thysanura. Lepisma). One of the primitive wingless insects. It is about $\frac{1}{2}$ inch (12 millimeters) long, broad in front and tapering behind, with two long slender antennae and three similar processes at the caudal end of the body. It is covered with lustrous grayish scales, hence the common name. The insect eats starchy materials and sometimes defaces book bindings, wall paper, and laundry, but as a rule it is not sufficiently abundant to be a pest. It is often found in damp buildings. It is also called the fish moth.

SILVERSIDES
(Osteichthyes). Of the suborder *Mugiloidea*, family *Atherinidae*, the silversides are related to barracudas and mullet. They also are termed antherinid smelts, but should not be confused with the true smelts. See also **Smelts**. There are numerous species, ranging up to a length somewhat in excess of 20 inches (51 centimeters). The *Antherinopsis californiensis* (jack smelt) is a silvery fish found in the waters off the California coast. Fresh and brackish-water silversides are found in the Great Lakes and in inland waters as far south as Florida and as far west as Texas. Mexican freshwaters support the *Chirostoma*, a rather abundant fish reaching a length of about 20 inches (51 centimeters) and well regarded as a food fish. Several species occur in the Mediterranean, only a few of which are found in the more northern waters off the British Isles.

Occurring in the waters of Lower California and as far northward as southern California, the *Leuresthes tenuis* (grunion) attains a length of 5 to 7 inches (12.5 to 18 centimeters) and is well known for its spawning habits. The female grunion lays her eggs in a shallow indentation in the beach and always at night during a high tide. The male fertilizes the eggs immediately, whereupon the grunion "flop" back to sea, aided by the waves. Within just a few minutes after they are touched by the waters of the subsequent high tide, the eggs are hatched. Spawning normally occurs within one or two days after a full or new moon. The *Hubbsiella sardina* has habits similar to the grunion, but will spawn during daylight hours. Grunions mature at 1 year and live about 3 years.

SILVERY POUT.
See **Codfishes**.

SIMPLEX OPERATION (Communication System).
A method of operation in which communication between two stations takes place in one direction at a time. This includes ordinary transmit-receive operation, press-to-talk operation, voice-operated carrier, and other forms of manual or automatic switching from transmit to receive.

SIMULATOR (Computer System).
A device, data processing system, or computer program that represents a system or phenomenon, and that mirrors or maps the effects of various changes in the original, enabling the original to be studied, analyzed, and comprehended by means of the behavior of the model. The term also is used to designate a routine which is executed by one computer, but which imitates the operations of another computer.

SINGLE-ENDED AMPLIFIER.
A device designed to amplify signals between a single input terminal and the ground or signal reference point. Generally, the output signal is generated between a single output terminal and the same signal reference point.

A disadvantage of the design stems from installation difficulties caused by the common connection between the input and output circuits. Unfortunately, any potential that exists between the amplifier signal reference point (ground) and the reference point for the input signal appears in series with the input signal. Thus, this additional potential is amplified with the signal and cannot be distinguished from the true signal voltage. Since, by definition, a potential between two signal reference points is a common-mode voltage, the common-mode rejection of the single-ended amplifier is zero. See also **Common-Mode Rejection Ratio**.

Where amplifiers of this kind are used in digital-data acquisition and instrumentation systems, careful attention must be given to the grounding circuits. Obviously, a major requirement is for the input signal reference point to be at precisely the same potential as the signal reference point for the amplifier input. Thus, a low-impedance connection between these two reference points at all frequencies of concern is needed. This is a difficult condition to achieve where the signal source and the amplifier are separated by a significant distance. Thus, amplifiers of this type seldom are used where such separation exists. For small subsystems where the ground structure is well controlled, single-ended amplifiers are used.

SINGULAR. A term for unusual or peculiar behavior of a mathematical entity. Thus, a singular matrix is one for which the determinant vanishes and a transformation involving that matrix is also singular. An integral equation with infinite limits of integration or an infinite kernel is singular. An algebraic or transcendental equation or function often becomes singular at one or more special values of the variables defining it. This is called a singular point or a singularity. If one studies these equations as the description of a curve, one is interested in points where tangents to the curve become indeterminate or infinite (see also **Singular Point of a Curve**). On the other hand, the function itself might become infinite or show some peculiarity at a point. This is of special concern in the study of a differential equation (see also **Singular Point of a Function**). The solution of a differential equation can also be singular. Such a solution cannot be obtained from the general solution by specifying the value of its parameters. See also **Clairaut Equation**.

SINGULAR POINT OF A CURVE.

If the tangent to a curve becomes an indeterminate form at one or more points, either finite or infinite, that point is called a singular point. The methods of differential calculus can, however, be used to find the tangents at such points and to investigate the behavior of the curve there. Many possibilities exist but only a few of them can be treated here. Conic sections do not have singular points; hence, the results apply only to higher plane curves.

When a curve crosses itself, there will be two tangents at that point, which is called a *double point*. If the two tangents are different, it is a *node* (L. *nodus*, knot) or *crunode* (L. *crux*, cross). If the two tangents coincide, the point is a *cusp* (L. *cuspis, point*), a *parabolic point*, or a *spinode* (L. *spina*, thorn). A third possibility is that of imaginary tangents and the point is then a *conjugate* or *isolated point*, also sometimes called an *acnode* (L. *acus*, needle).

When k branches of a curve cross, the point is a multiple point of order k. When $k > 2$, several new kinds of behavior appear, the most common of which are: *point of osculation* (L. *osculare*, kiss) or *tacnode* (L. *tactus*, touch), the curve recedes from the point of tangency in opposite directions; *cusp of the second kind*, also called *rhamphoid* (Gr. $\pi\alpha\mu\phi\circ\varsigma$, beak), where the two branches of a curve lie on the same side of a common tangent. At a cusp of the first kind, also called *keratoid* (Gr. $\kappa\alpha\pi\alpha\varsigma$, horn), the two branches of the curve are on opposite sides of the tangent. This is the only kind of cusp which can occur at a double point.

The preceding cases apply generally to algebraic curves. There are further possibilities for transcendental curves: *end point, terminal point*, or *point d'arrêt*, the curve terminates abruptly; *salient point* or *point saillant*, two branches of the curve end without a continuous derivative, thus without a common tangent.

Typical examples of each of these types, where the singular point occurs at the origin, are: node, lemniscate of Bernoulli; point of osculation, $a^4 y^2 = x^4(a^2 - x^2)$; cusp of first kind, semi-cubical parabola or cissoid; cusp of second kind, $(y^2 - x^2)^2 = x^5$; conjugate point, $y^2 = x^2(x - 1)$; end point, $y = x \ln x$; salient point, $x = \cot y/x$.

See also **Cusp (Mathematics)**; and **Maximum and Minimum (Mathematics)**.

SINGULAR POINT OF A FUNCTION.

A value of the complex variable z, for which $f(z)$ is an analytic function, is called an ordinary point. Any point that is not an ordinary point is a singular point. According to a theorem of Liouville, if $f(z)$ has no singular point for z finite or infinite, it must be a constant. Another Liouville theorem, more familiar to chemists and physicists, occurs in statistical mechanics. See also **Liouville Equation**.

Singular points or singularities are classified as: (1) poles or unessential singularities; (2) essential singularities or poles of infinite order; (3) branch points caused by the fact that the function is not single-valued.

Let $w = u + iv$ be a single-valued function of $z = x + iy$, where u, v are real single-valued functions of x and y. Then $z = z_0$ is a pole of order k, provided that $(z - z_0)^k w(z)$ is analytic and not zero at $z = z_0$. The number k is an integer greater than unity and is the order of the pole. Singular points of this kind are nonessential because they are effectively removed if $w(z)$ is multiplied by $(z - z_0)^k$. They are called poles because a three-dimensional plot of w, x, y shows that w becomes infinite at the singular point and thus a pole of infinite height occurs there on the z-plane. Typical examples are $w = 1/z(z - a) \times (z - b)$, which has three simple poles at $z = 0, a, b$, respectively; $w = \csc z$, an infinite number of poles on the real axis.

A singularity that is not a pole or a branch point is called essential. It is actually a pole of infinite order. A simple example is $w(z) = \sin 1/z = 1/z - 1/3!z^3 + 1/5!z^5 - 1/7!z^7 \ldots$ and it is seen that no finite value of k in $z^k w(z)$ will remove the singular point of this function at $z = 0$.

The essential or nonessential character of a singularity can be investigated by expansion of the function in a Laurent series. In most cases, the result can be obtained more simply by inspection of the function. If the point $z = \infty$ is of interest, transfer to the new variable $z = \overline{1/\xi}$ and the function at $\xi = 0$.

When a function is multivalued its discontinuities are called branch points. Analytic continuation between two points will then give different values for $f(z)$ if the two paths include a branch point. Branch points always occur in pairs. The line joining them is a branch line and a contour crossing a branch line changes from one set of values of $f(z)$ to another. The two (or more) independent values of $f(z)$ are its branches. The values of the different branches of $f(z)$ are identical at a branch point. See also **Riemann Surface**.

Singular points are of considerable importance in the study of linear differential equations, especially when they are to be solved by expansions in series.

The point $z - z_0$ is a regular singular point of the second-order equation $(z - z_0)^2 y'' + P(z)(z - z_0)y' + Q(z)y = 0$, provided: (1) z_0 is not an ordinary point; (2) both $P(z)$ and $Q(z)$ are analytic functions at $z = z_0$. If these conditions are not met, the singularity is irregular. This classification of singular points may be extended to linear differential equations of any order.

At a regular singular point the difference between the two exponents of the indicial equation is arbitrary, except that it cannot be zero or integral. If this difference equals $\frac{1}{2}$, the singularity is elementary.

An irregular singular point arises from the confluence of two or more regular singular points. In the case of a second-order differential equation, the indicial equation at an irregular singular point is of first degree or less and there is only one series solution there or none.

See also **Complex Variable**.

SINKING SPEED (Aircraft).

The vertical component of the velocity of an airplane gliding without acceleration in power-off condition is the sinking speed. It is analogous to "rate of climb" of climbing flight. Methods exist for estimating sinking speed using other data of the airplane, to wit, air speed, altitude, weight, parasite area, wing span, and airplane efficiency factor. Gliding at minimum sinking speed should not be confused with the flattest glide.

Referring to Fig. 1, the following equations of statics may be obtained, using the lift-drag axes as positive axes:

$$L_w + L_t = W \cos\theta \tag{1}$$

$$D_T - W \sin\theta = 0 \tag{2}$$

where L_w = lift of the wing, L_t = lift of the horizontal tail surfaces small relative to L_w, and may therefore be neglected, D_T = total drag of the glider, W = weight of the glider.

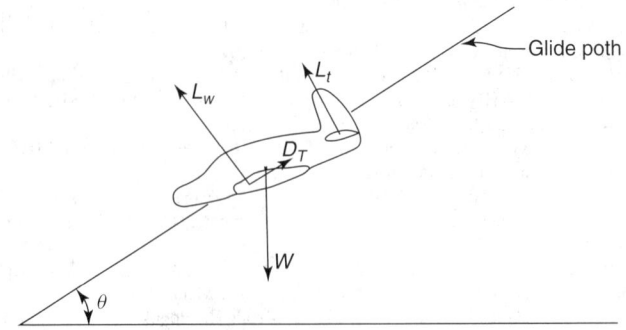

Fig. 1.   Forces acting on a glider in a glide.

With the assumption that the lift on the tail surfaces is negligible, the following relationships are obtainable from equations (1) and (2) and from the usual lift and drag equations:

$$L = \frac{\rho v_g^2 C_L S}{2} \text{ and } D = \frac{\rho v_g^2 C_D S}{2}$$

where v is the glide path speed.

$$L_w = W \cos\theta$$

$$\cot\theta = \frac{L_w}{D_T} = \frac{C_L}{C_D}$$

Also, the sinking speed v_s, equals the gliding speed v_g multiplied by $\sin\theta$, and the horizontal component equals the gliding speed v_g multiplied by $\cos\theta$.

SINTERING. The heating of an aggregate of fine metal particles at a temperature below their melting point so as to cause them to weld together and agglomerate. See also **Agglomeration**.

SINUS. A pouch or cavity in any organ or tissue, or an abnormal cavity or passage formed by the destruction of tissue. The term is applied to a very large number of such structures in the human body, such as: a dilated portion of a vein containing venous blood; a chronically infected tract such as a fistula; the air cavities within the cranial bones, especially those located near the nose and connecting with it. See illustration that accompanies entry on **Skeletal System**. They are called accessory sinuses of the nose or paranasal sinuses. They extend from the nasal passages into bones of the skull, and are named according to the bones in which they lie as the frontal, ethmoidal, sphenoidal, and maxillary sinuses. The maxillary sinuses are also called antra. The maxillary sinuses are found on either side of the nose. They are large in size, and lie between the floor of the eye socket above and the upper teeth below. The frontal sinuses lie in the forehead above the roof of the eye socket, one on either side of the midline of the forehead. The ethmoidal sinuses are three groups with numerous air cells, situated between the eyes in either side of the midline or septum of the nose. The sphenoidal sinuses, two in number, are situated above and behind the nose proper. They are all lined with a delicate mucous membrane continuous with the nasal mucous membrane. When the sinuses become infected, sinusitis is said to be present.

SINUSITIS. Infection of the mucous membrane lining of the paranasal sinuses. This is a common condition, but sometimes overdiagnosed. The disease may be acute or chronic. Rhinitis (viral or allergic) sometimes predisposes sinusitis. Also, the disease may follow blockage of the nasal septum by nasal polyps. Other less frequent causes include sudden changes in altitude, foreign bodies or tumors in the intranasal system, and certain processes of a systemic nature, such as cystic fibrosis. Any process which interferes with the drainage or ventilation of the sinuses may induce sinusitis. Common causative microorganisms include pneumococci, Group A and other streptococci, and *Haemophilus influenzae*. Less frequent causes are staphylococcus, *Mycobacterium tuberculosis*, *Aspergillus* fungi, *Mucor* fungi, viruses, and *Mycoplasma pneumoniae*. Involvement may be confined to one sinus, or more commonly to both sinuses in adults. Children are more inclined to suffer from *ethmoiditis*.

Symptoms of sinusitis vary with the site. See also **Sinus**. Infection of the *frontal sinus* (*frontal sinusitis*) causes pain and tenderness over the lower forehead. There may be purulent (pus-like) drainage. In *maxillary sinusitis*, there is pain and tenderness over the cheeks. The pain may be referred to the teeth; the hard palate may become inflamed. *Ethmoid sinusitis* (*ethmoiditis*), in the acute phase, is characterized by a dull headache, which may be moderate or intense. Ethmoid pertains to the bones at the front part of the base of the skull, forming part of the septum and walls of the nasal cavity. These are perforated bones through which the olfactory nerves pass. There may be swelling between the inner corner of the eye and nose. The nasal discharge may be profuse or entirely absent, depending upon whether the drainage passages are partially or fully blocked. In subacute ethmoiditis, headache may be bothersome, but pain is usually not severe. Chronic inflammation of the ethmoid sinuses is manifested by headache, cough, a general feeling of fatigue, and often a slight fever. Acute *sphenoid sinusitis* occurs relatively frequently, with accompanying headache, which has been described as excruciating and may be only partially relieved by drugs. Sphenoid pertains to the wedge-shaped compound bone at the base of the skull. Symptoms of sleeplessness and a fear of choking are sometimes present, caused by the thick discharge from the sinus. In the chronic form of the disease, the patient may experience pressure pains spreading in diverse directions, accompanied by a thick, sticky postnasal drip.

Therapy for acute sinusitis includes analgesics, topical heat, and decongestants. Pseudoephedrine is often used. Antihistamines may be used to aid in decongestion. There is no strong body of evidence to date that indicates the usefulness of antibiotics in the treatment of acute sinusitis per se, although these drugs are frequently administered as a preventive measure. When used, ampicillin, penicillin, cloxacilin, or erythromycin are frequently the drugs of choice. Antibiotics are always indicated in toxic patients and when treatment with decongestants is insufficient. In chronic sinusitis, irrigation and surgical drainage may be required.

Although relatively uncommon, sinusitis can predispose a number of conditions, such as osteomyelitis of the frontal bones (particularly in children), orbital infections, such as orbital cellulitis, septic cavernous sinus thrombophlebitis, among others. These are serious, sometimes life-threatening conditions, but uncommonly seen because of antibiotic therapy.

Additional Reading

Druce, H.M.: "Sinusitis: Pathophysiology and Treatment," Marcel Dekker, Inc., New York, NY, 1993.
Gershwin, M.E. and G. Incaudo: "Diseases of the Sinuses," Humana Press, Totowa, NJ, 1996.
Rosin, D.F.: "The Sinus SourceBook," Lowell House, Inc., Lowell, MA, 1997.
Williams, M.L.: "The Sinusitis Help Book: A Comprehensive Guide to a Common Problem," John Wiley & Sons, Inc., New York, NY, 1998.

SIPHON (Zoology). A passage between the mantle folds of bivalve mollusks through which water enters or leaves the mantle cavity. In some species these passages are developed into long muscular tubes and in others they are no more than poorly marked openings. They are two in number, a dorsal or excurrent siphon and a ventral or incurrent siphon. Water is taken in through the latter, passes through the gills to the chambers above them, and flows out through the dorsal siphon.

A part of the mantle border in some of the marine gasteropods also forms a tube through which water can be drawn into the mantle cavity. This tube is known as the siphon.

The term oral siphon is applied to the canal leading to the mouth of ascidians and the opening from the atrial cavity is sometimes known as the atrial siphon.

Unlike these organs, all of them associated with respiration, the siphon of sea urchins (*Echinoidea*) is a slender tube associated with the alimentary tract.

SIRENIA. See **Sea Cows**.

SIRIUS (α Canis Major). This is the brightest star in the sky and volumes have been written concerning its matchless brilliancy. Historically, it is undoubtedly the most interesting star in the heavens and references to it are found throughout all ancient literatures back to the earliest known writings. Aside from its surpassing brilliancy, the fact that it may be observed from every habitable portion of the earth has served to make it an object of veneration by all peoples. Sirius was worshiped in the valley of the Nile long before Rome was even heard of, and many ancient Egyptian temples were so arranged that the light from this star would penetrate to the inner altars.

Sirius has a true brightness value of 23 as compared with unity for the sun. Sirius is a white, spectral type A star and is relatively nearby ($\pi = 0.''375$), estimated distance from the earth 8.7 light years. The mass of Sirius is well known because it is a double star. The companion is a white dwarf. See also **Constellations**; and **Star**.

SISKIN. See **Finch**.

SKARN. A petrological term describing the process of contact metamorphism by which certain mineral silicates such as amphibole, pyroxene and garnet replace limestone and dolomite.

SKATES AND RAYS (*Chondrichthyes*). Like sharks, the skates and rays are cartilage fishes. Of the order *Batoidei*, skates and rays may be subdivided into: (1) Electric rays (*Torpedinidae*); (2) guitarfishes (*Rhinobatidae*); (3) sawfishes (*Pristidae*); (4) skates (*Rajidae*); (5) stingrays, whiprays, butterfly rays, and round rays (*Dasyatidae*); (6) eagle rays, bat rays, and cow-nosed rays (*Myliobatidae*); and (7) devil rays (*Mobulidae*). Certain features characterize the skates and rays as compared with other elasmobranches (cartilaginous skeleton, plate-like scales, lack of air bladder). Some of the differences include:—instead of to the head, the pectoral fins fasten to the sides of the head; the pectoral fins are much enlarged;

gills are located on the undersides of the fins; the method of respiration is altered, to accommodate for bottom dwelling in muddy environs; and there is no free upper eyelid, as found in sharks.

The electric ray carries a large electric organ under the wings and adjacent to the head. On the average, an electric ray can generate from 75 to 80 volts, but potentials as high as 200 volts have been recorded. Several days are needed by the ray to recharge the electric-producing apparatus. These fishes are of no food value and largely of interest because of their electrical characteristics. As with other fishes having electrical apparatus (see also *Catfishes*; and *Gymnotid Eels*); the electrical ability of the fish is used to stun potential or overactive food and for defense. However, it also has been postulated that if the fish can control the discharge to a low, possibly microvoltage level, the apparatus may be used for detection and navigation along the lines of sonar. This would be reasonable in the case of the rays, which live in muddy waters where visibility at best is poor. The electric ray holds a positive charge on its upper surface; negative on the undersurface. Approximately 40 species of electric rays are catalogued. The largest genus is *Torpedo*. The American Atlantic *Torpedo* attains a length of 5 feet (1.5 meters) or more and a weight of from 160 to 200 pounds (73 to 91 kilograms). The electric rays are ovoviviparous — eggs laid and hatched within the parent. Widely distributed in ocean waters of temperate or tropical regions, electric rays prefer small crustaceans for their diet and range from the surface down to depths of about 3,000 feet (900 meters).

In appearance, the guitarfish looks a bit like a shark and a bit like a ray, perhaps something like a flattened out shark. The guitarfishes are of nine genera with a total of over 150 species. The common Mediterranean guitarfish (*Rhinobatos rhinobatos*) frequents Mediterranean waters from Angola, Africa to Portugal; the *Rhinobatos horkelii* (Brazilian guitarfish) is found along the South American coast. The American Pacific guitarfish (*Rhinobatos productus*) is found in the waters of the Gulf of California and as far north as Monterey Bay. Guitarfishes are also ovoviviparous. They eat small crabs and fish. They survive quite well in captivity.

The sawfish looks like a guitarfish with a long, double-edged saw fastened to its nose. These fishes are classified as giant rays and six species are known. Records indicate that some species attain a length of up to 35 feet (10.5 meters) and a weight up to 5,000 pounds (2268 kilograms). Sawfishes prefer salt and brackish waters, but also migrate to fresh water. The landlocked Lake Nicaragua has a significant population of sawfishes. They also are found in the Indian River (Florida). The *Pristis pectinatus* is the common western Atlantic sawfish and possesses from 25 to 32 pair of rostral teeth. The *Pristis pristis* (eastern Atlantic sawfish) has fewer pairs of such teeth. The sawfish uses the saw for clubbing its prey and also for digging. In preparing for a meal, the sawfish sweeps through the water flailing the saw, thus wounding its victims, which then are consumed in due course. Even in captivity, the sawfish retains its habit of swinging the saw back and forth during feeding time.

Also an elasmobranch, the skate (*Rajidae*) has a very flat body and a pair of pectoral wings. Cartilage projected forward of the body creates a very elongated nose. Skates prefer tropical and temperate waters and are widely found with exception of the areas of Micronesia, Polynesia, the Hawaiian Islands and the waters off the northeastern coast of South America. They prefer sandy, muddy bottoms of shallow water, preferably less than 600 feet (180 meters) deep. The number of species exceeds 100. The hedgehog skate (*Raja erinacea*) is common and can be found in the western Atlantic from Nova Scotia south to South Carolina. Adults measure about 20 inches (51 centimeters) in length and weigh about a pound. The *Zearaja nasuta* that frequents New Zealand waters attains a length of over 6 feet and may weigh up to 70 pounds. Possibly the largest of these fishes is the *Raja binoculata*, occurring in the American Pacific, which attains a length of about 8 feet (2.4 meters). Some species of skates have electric organs, but the voltages are considered to be quite small. It is possible that the skate may have electrogenic capabilities. Skates found along European coasts are sold commercially, but they are not a major item.

Stingrays are characterized by venom spines. See Fig. 1. The spines can be replaced if lost. Death can result if the venom gland is driven into a swimmer. It has been recorded that large stingrays can thrust the venom spine through the planking of a small boat. There are approximately 80 species of dasyatid stingrays, ranging from fishes of about $1\frac{1}{2}$ pounds (0.7 kilogram) and about 1 foot (0.3 meter) across the wings to a weight of 750 pounds (340 kilograms) and wingspan of 6 to 7 feet (1.8 to 2.1 meters), as found in the Australian stingray. Stingray young are born alive.

Fig. 1.   Sting ray shown in the company of several lemon sharks. (*A.M. Winchester.*)

Although enjoying a very bad reputation, the devil rays (*Mobulidae*) under scientific investigation have proved to be quite docile. The *Mobula diabolis* (Australian devil ray) measures but 2 feet (0.6 meter) across the wingtips, whereas *Manta* may attain a wingspan of 22 feet and weigh up to 3,500 pounds (1588 kilograms). They usually travel singly or in pairs, seldom in schools. They are ovoviviparous.

Additional Reading

Demski, L.S. and J.P. Wourms: "The Reproduction and Development of Sharks, Skates, Rays, and Ratfishes," Kluwer Academic Publishers, Norwell, MA, 1993.

Hamlett, W.C.: "Sharks, Skates, and Rays: The Biology of Elasmobranch Fishes," Johns Hopkins University Press, Baltimore, MD, 1999.

McCormick, H.W., W.E. Young, and T. Allen: "Shadows in the Sea: The Sharks, Skates and Rays, Fully," The Lyons Press, New York, NY, 1996.

SKELETAL SYSTEM.   An aggregation of rigid or semirigid structures that provide mechanical support for the body and usually a lever system on which the muscles act.

The simplest type of skeletal system is found in the sponges (see also **Porifera**) and some of the coelenterates. Sponges have scattered spicules of calcareous or siliceous matter among their loosely integrated tissues or are held together by a meshwork of fibers composed of the material spongin, so familiar in the sponges of commerce. The commercial sponge is, in fact, the skeleton freed of organic matter. Spicules are variously formed bodies, some straight rods, some with radiating axes from three to six in number, and some expanded at the ends. Approximately similar scattered bodies are found also in alcyonarians, and in some of these animals they are united to form a continuous mass in which the polyps are imbedded or a solid core surrounded by softer material. Rock corals lay down a calcareous deposit beneath the base to which each polyp of the colony adds.

In many invertebrates the body wall is the sole support of the animal. Even though it is soft, the incorporation of muscular layers in it provides for movement without rigid skeletal structures.

The echinoderms differ in having calcareous plates (ossicles) throughout the integument. In the sea urchins they are closely joined to form a shell and in the sea cucumbers they are small and scattered. Other groups show an intermediate condition with closely associated but movable ossicles. These structures in the sea urchin also illustrate lever action in the use of the spines for locomotion and in the peculiar chewing organ known as Aristotle's lantern.

The arthropods and vertebrates differ in the presence of a complex skeletal system, which supports the body and forms the foundation of the jointed appendages. In no other phyla are lever-like appendages found.

The arthropods have an exoskeleton made up of hard plates (sclerites) of chitin and calcareous matter, formed in circumscribed areas of the cuticula. They are separated by flexible zones, which provide for freedom of movement, and are moved by muscles attached to their inner surfaces. The appendages are formed of rigid segments attached to each other and to the body by the flexible tissue of the joints. The skeleton forms a sheath enclosing the muscles in such appendages.

Among the half-million species of arthropods wide variation is inevitable. In general, the exoskeleton of the head forms a compact capsule. The skeleton of the thorax also tends to become compact, although its segments are distinctly recognizable in many species. This portion of the exoskeleton also forms a shield-like carapace in many species. The abdomen retains greater mobility, although its segments may be reduced in number and the exoskeleton of each may be consolidated to a ring surrounding the body. A moderately complex segment of the exoskeleton contains a dorsal, a ventral, and two lateral sclerites known respectively as the tergum or tergite, the sternum or sternite, and the pleura or pleurites. In many cases this plan is simplified by fusion or made more complex by subdivision of the sclerites.

An endoskeleton of limited extent in the arthropods is made up of internal projections from the exoskeleton known as apodemes. They serve as places of attachment for muscles.

One important result of exoskeletal support is the provision of wings in the insects as thin-walled sacs of the body wall. By the apposition of the upper and lower walls of these sacs they become thin membranous planes sufficiently strong and rigid to support the body in flight. Thus, the jointed appendages are freed from participation in adaptation for flight.

Exoskeletal structures also appear in the vertebrates but here they are merely hard parts without the usual supporting functions of the skeleton. They are derived from either or both layers of the skin. The category includes teeth and beaks, horns, claws, hoofs and nails, scales, feathers, and hair. Bony plates associated with scales, as in the alligator and armadillo, are also exoskeletal.

The supporting skeleton of the vertebrates (Fig. 1) is an endoskeleton composed of bones and cartilage. It is made up of two chief divisions, an axial skeleton consisting of a vertebral column, ribs, and skull, and an appendicular skeleton including pectoral and pelvic girdles, each bearing a pair of appendages. The pharyngeal wall of fishes is supported by the visceral skeleton, which persists to a limited extent in the more advanced classes. In the elasmobranch fishes now living, the skeleton is composed entirely of cartilage. In most vertebrates, it is made up very largely of bone.

the chondrocranium. In the bony fishes it is replaced by bones in the same position and is supplemented by superficial bony plates enclosing the remainder of the brain. The jaws of the elasmobranch are also supported by cartilages and in the bony fishes these cartilages are supplemented by bones. Above the fishes, a chondrocranium appears in the embryo and is replaced during development by the ethmoid bone and parts of the sphenoid, temporal, and occipital bones, taken in order from front to back. The remaining bones are not preformed in cartilage. They are more numerous in the lower form than in humans. In the human skull (Fig. 2), they are the pair of nasal bones in the bridge of the nose, the pair of lachrymals in the orbits, the vomer in the nasal septum, the large frontal bone of the forehead, the parietals in the top and sides of the skull, the upper part of the occipital forming the lower back wall of the skull, and the temporals above the ears. The upper jaw is based on the maxillary bone from which the palatine extends into the hard palate. A slender rod, Meckel's cartilage, supports the lower jaw of the human fetus but the permanent lower jaw is the mandible, a dermal bone formed independently.

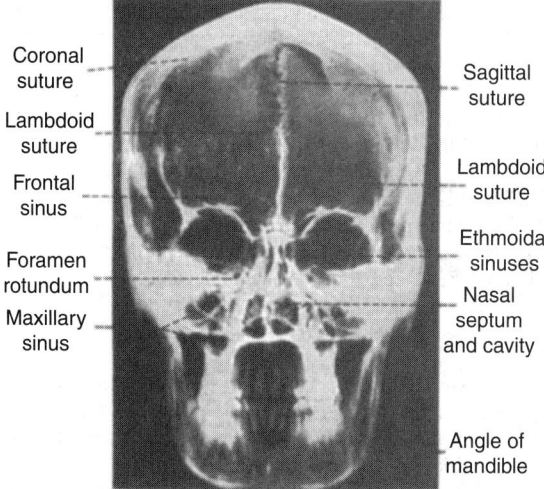

Fig. 2.   Antero-posterior radiography of a skull. (*Cunningham.*)

The vertebrae that make up the spinal column consist typically of a centrum bearing a neural arch above and a haemal arch below. The neural arch surrounds the spinal cord and is surmounted by a neural spine. In the fishes the haemal arch encloses the dorsal aorta and in the region of the body cavity open haemal arches form ribs known as fish ribs. In other classes, and in a few fishes, these ribs are replaced by similar slender bones known as true ribs. They develop between the segments of the body lateral to the fish ribs and above them, and articulate with processes of the vertebrae. These ribs join a median ventral structure, the sternum or breast bone, composed of bones and cartilages.

The vertebrae differ in various regions of the body. The first, with which the skull articulates, is the atlas. The second, the axis, is noteworthy for its odontoid process, a solid anterior extension of the centrum. This process is the centrum of the atlas. These two and the remainder of the series in the neck are called cervical vertebrae. The following series with which the ribs articulate are the thoracic vertebrae. The lumbar vertebrae extend to the articulation of the pelvic girdle, where one vertebra or a fused series constitute the sacrum. The following caudal vertebrae lie in the tail or, in the apes and man, are fused to form a mass called the os coccyx.

The pectoral and pelvic girdles are composed of three pairs of bones, two passing downward and toward the median line of the body from the articulations of the appendages and one upward. Of the more constant bones in the pectoral girdle, also called the shoulder girdle, the upper bone is the scapula (shoulder blade), the anterior of the lower bones is the clavicle (collar bone), and the posterior is the coracoid. The two pairs attach to the sternum. They are supplemented in amphibia and reptiles by other bones. The bones of each half of the pelvic girdle are the dorsal ilium, the anterior pubis, and the posterior ischium. This girdle in many forms is firmly attached to the sacrum.

The girdles are modified in various ways. The principal tendency of the pectoral girdle is toward simplification by the loss of bones, so that in

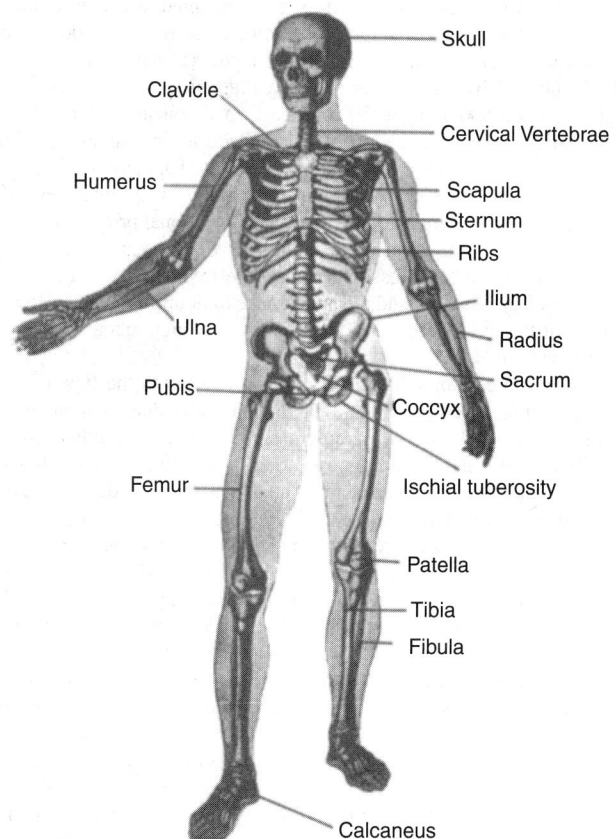

Fig. 1.   Skeletal system of human.

The skull in the elasmobranch fishes consists of a mass of cartilage below, behind, and partially enclosing the brain. This structure is called

many species only the scapula and clavicle remain, as in man, the coracoid being reduced to a process on the scapula. In the hoofed mammals only the scapula persists. In the moles, however, the girdle is large and strong. The pelvic girdle, in correlation with the stresses that it bears in locomotion, becomes compact. The bones of each side tend to fuse and the pubic symphysis, at the ventral union of the two halves, is very firm. The name pelvis or pelvic bone is often applied to this composite structure.

The primitive appendages are the paired fins on the fishes. These structures are precursors of the pentadactyl appendage of terrestrial vertebrates. From this basic plan the structure of appendages in various groups has come by modification of the proportions and relations of the bones, by loss of digits, and to a limited extent by the fusion of separate bones. Both loss and fusion have occurred in the wings of birds, and maximum loss is found in the legs of the horse and allied species, where a single functional digit remains.

The appendages also contain sesamoid bones, small rounded bones developed in the tendons spanning movable joints. The most constant of these bones is the patella or knee cap.

The visceral skeleton in its primitive condition consists of a series of small bones and cartilages supporting the branchial arches between the gill slits. In the existing vertebrates the embryonic primordia of the hinged jaws are derived from these arches but the visceral skeleton is otherwise greatly reduced except in the fishes. It persists in part as the small bones (ossicles) of the middle ear, the hyoid bones supporting the tongue, and the cartilages of the larynx and upper part of the trachea.

SKEWNESS (Mathematics).   That property of a frequency distribution involving its symmetry or asymmetry. Skewness is usually measured by the departure from zero of the third *moment* about the mean, standardized by division by σ^3 where σ is the standard deviation.

SKIMMER.   See **Waders, Shorebirds, and Gulls**.

SKIN.   The covering of the body of vertebrates. It consists of two parts, an outer epidermis and an inner dermis or corium. The former develops in the embryo from the outer germ layer, the ectoderm, and the latter from the middle germ layer, the mesoderm.

The epidermis is composed of many layers of cells in two principal strata, the stratum corneum and the stratum germinativum. The flattened cells next the surface are hardened by deposits of pareleidin, a substance related to keratin, and are said to be keratinized. They make up the stratum corneum. Below them are several layers of thicker cells whose active proliferation gives rise to the cells of the stratum corneum. These layers constitute the stratum germinativum. In the outer cells of this stratum granules of keratohyalin appear as forerunners of the pareleidin of the stratum corneum, forming the thin stratum granulosum. A thin clear zone just outside of the granular layer, known as the stratum lucidum, is regarded as the basal layer of the stratum corneum. In its cells, the granules of keratohyalin become a diffuse intermediate substance, eleidin.

The corium is a dense connective tissue layer extending from the fatty subcutaneous tissue. It is obscurely divided into an inner stratum reticulare and an outer stratum papillare, which rises in papillae beneath the epidermis. Epidermal derivatives including hair follicles, sweat glands, and sebaceous glands extend into the corium, and it contains nerve endings, tactile corpuscles, and blood vessels. Smooth muscle fibers attached to the hair follicles lie in it, and in some parts of the skin it contains other muscle fibers. The voluntary muscles of the face by which expression is controlled end in it.

Aging of the Skin.   The deep, felt-like layer (dermis) gives the skin its suppleness and substance. The connective tissue fibers of the dermis form a woven pattern that, in infants, gives the skin a soft, velvety texture. As the skin ages, this woven pattern becomes tighter and less resilient, tougher, and wrinkled. These essentially undesirable changes are hastened by certain deleterious factors, including overexposure to ultraviolet (UV) rays of the sun, exceptionally large doses of vitamin C (ascorbic acid), exposure to heat, wind, drying agents (such as alcohol), and frigid temperatures. These changes are noted as "wrinkling" and often include the formation of "crow's feet" at the outer corner of the eyes, as well as some withering at the joining of the lips.

To date, these aging processes are not subject to reversal. Moisturizing agents (wool fat and lanolin) as well as extra doses of vitamins A and E and retinoic acid (Retin-A) have been tried, but documentation to date has

been less than convincing. At best, these substances may slightly slow the natural skin aging processes.

Diseases of the skin are numerous and are described in the article on **Dermatitis and Dermatosis**. In the following article on **Skin Cancer**, the principle life-threatening skin diseases are described. The principal trauma to the skin includes burns, either from fire and heat or from chemical substances. See also **Burn**.

Progress has been made in using "artificial" skin, particularly in the treatment of burn cases. As reported by Erickson, some success has occurred in the development of so-called skin "stand-ins," or dermal substitutes, which may partially replace the need for testing skin medications and cosmetics on laboratory animals. These substitutes are comprised of complex cultures of cells, look like real skin, and are available in sheets. The possible use of such new dermal substitutes (*TestSkin*) ultimately may revolutionize skin-product testing and research and obsolete most needs for laboratory animals. The concept also is being investigated for possible use in burn therapy. (See *Sci. Amer.*, 168, September 1990).

Additional Reading

Berardesca, E., K.P. Wilhelm, and P. Elsner: "Bioengineering of the Skin: Skin Biomechanics," CRC Press, LLC., Boca Raton, FL, 2001.
Freinkel, R.K. and D.T. Woodley: "The Biology of the Skin," CRC Press, LLC, Boca Raton, FL, 2001.
Lebwohl, M.G.: "Atlas of the Skin and Systemic Disease," 2nd Edition, Churchill Livingstone, Inc., Philadelphia, PA, 2001.
Lookingbill, D.P. and J.G. Marks: "Principles of Dermatology," 3rd Edition, W.B. Saunders Company, Philadelphia, PA, 2000.
Yannas, I.V.: "Tissue and Organ Regeneration in Adults: Synthesis of Skin and Peripheral Nerves," Springer-Verlag, Inc., New York, NY, 2001.

SKIN CANCER.   Probably because the skin is the largest organ of the body and because it is so continuously exposed to the hazards of sunlight, ionizing radiation, abrasion, chemical attack, and other hostile conditions, it is more frequently the seat of cancer than any other site on the body. Although not responsible for all carcinomas of the skin, chronic exposure to sunlight, where the individual is not protected by intense melanin pigmentation, is the most important etiologic factor in the development of skin cancer. This itself may be further mediated genetically.

In the late 1990s, the number of newly diagnosed skin cancers per year ranged between 600,000 and 800,000 cases per year in the United States. These statistics are paralleled in other countries located in similar climate zones and with comparable life-styles. Part of the increase in the skin cancer treatment load may be attributed to the growing awareness of the general populace to public and institutional educational programs that have stressed the dangers of exposure to sunlight as well as a growing awareness of individuals to seek professional help more frequently when confronted with changes in their skin and the appearance of doubtful lesions. However, the bulk of the increase is due to new cancers brought about by failure to heed the warnings.

The death rate from skin cancer varies widely with the type of cancer and the timing of diagnosis and therapy. Deaths due to nonmelanoma cancers, especially basal-cell carcinomas, are comparatively low. In 1992, Weinstock et al. reported on a population-based study of mortality from nonmelanoma skin cancers that indicated that 59% of deaths were due to squamous-cell carcinoma and 20% were due to basal-cell carcinomas. Increased risk was associated with cancers on the ears and eyelids.

There were 32,000 deaths in the U.S. in 1991 attributed to cutaneous melanoma. This incidence has nearly tripled since the 1950s. Most increases were in young adult whites (ages 25 to 29 years). The annual incidence is higher than that of any other cancer. Sunlight is considered the most important environmental factor favoring the pathogenesis of melanoma. There appears to be some familial connection, with a risk factor of 2 to 8 times for individuals whose parents had melanoma. It was learned that dysplastic nevi are potential precursor lesions from which melanoma may develop.

The prognosis of malignant melanoma worsens with increasing thickness of the tumor. It also is related to site, age, and sex. Early diagnosis and removal of the lesion usually has been successful in warding off the usual incurable metastatic phase of the disease. In patients with tumors less than 0.75 mm thick, the 5-year survival rate is 96% dropping to 47% in cases where the tumor exceeds 4 mm. The 5-year survival rate falls to 36% in those persons with distant metastes.

In an excellent article by Phillips et al. (reference listed), the authors describe the latest surgical margins in the management of Stage I melanoma (i.e., those limited to the skin).

A number of benign skin lesions may develop into malignancies if left untreated. The most common of these is *actinic keratosis* (senile or *solar keratosis*) which will occur in areas frequently exposed to sunlight — the face, lower arms, neck, scalp (where not shielded by hair) and the dorsal surfaces of the hands. Varying in size up to one centimeter (diameter), the lesions may be pink to tan, smooth or scaly, and of almost any shape. *Leukoplakia* is much like actinic keratosis except that it will appear on mucous membranes (lips, oral mucosa, vulva, etc.) the lesions being somewhat elevated, irregularly shaped and sharply bordered white patches. Squamous cell carcinoma in situ (*Bowen's disease*) presents as single or multiple sharply defined plaques that are slightly thickened and brownish-red with varying amounts of scale. The lesions resemble eczema or psoriasis, but fail to respond to local therapy.

Although only about 5% of these lesions become invasive and less than 2% metastatize, more important is the recognition that the patient is at significant risk of developing carcinoma of the respiratory, gastrointestinal, and genitourinary systems. Intraepidermal metastases (*Paget's disease*) is an uncommon carcinoma presenting as a sharply defined, red, weeping, crusted or scaly lesion resembling atopic eczema which does not respond to topical corticosteroids. It may develop within the epithelium of the mammary ducts and extend upward into the epidermis or appear on the pubis, perineum, genitals and other sites related to underlying apocrine or eccrine sweat glands.

Basal Cell Carcinoma. This accounts for more than 75% of all skin cancers and these are among the most strongly antigenic of all human cancers. They arise from the epidermis, cytologically resemble the normal basal cells, and show little tendency to undergo the usual differentiation into squamous cells that produce keratin. Although these tumors rarely metastasize, they are locally invasive and may go deeply into nerves, bone, and brain. Like most cancers, they are remarkably painless in their course and this lack of symptoms leads to prolonged neglect of a lesion. The typical basal-cell carcinoma is an uninflamed smooth, waxy nodule which appears to be translucent and may show varying amounts of melanin pigment in the form of small dots. Such nodules often ulcerate and may re-epithelialize leading the patient to believe that the nodule is resolving. A patient with one basal-cell carcinoma is likely to have others at the same time or shortly thereafter. Because the tumors are malignant, they should be excised or otherwise destroyed after diagnosis has been confirmed. A border of normal tissue, adjacent to the tumor, must be removed to prevent recurrence from invasive strands of tumor cells.

Squamous Cell Carcinoma. This also arises from the epidermis, but shows significant squamous differentiation and usually keratin production. The lesions have a variable tendency to metastasize depending upon their size, extent of invasions, and their mode of origin. The typical squamous-cell carcinoma is a painless, firm, red nodule or plaque with visible scales on the surface. Ulceration may occur, in contrast with the basal-cell variety and, when this is so, they have the lowest frequency of metastasis. On the other hand, where they arise from mucous membranes, burn scars, sinus tracts, or from apparently normal skin, they have a much higher tendency to metastasize.

Primary Malignant Melanoma of the Skin. This is the most serious disease in dermatology. Pigmented moles are among the most common growths on the skin of humans, yet cancer involving pigment cells is relatively uncommon, constituting about 1.5% of all cancers. The diseases is, however, virtually incurable by chemotherapy or X-rays and so far has only responded to deep excision.

Even in its early stages, primary malignant melanoma is relatively easy to detect. The variegation of color and irregularities in surface pattern and configuration are characteristic even when the tumor has red, white, or blue surface color. Any of the several types of melanoma may exist for several years in the pre-invasive stage.

These tumors are usually seen in the fourth and sixth decades of life. White people with light complexions are more prone to melanomas than persons with dark pigmented skin and there is a significant relationship between incidence and exposure to sun. Three forms are recognized: (1) *Lentigo malignant melanoma* with a median onset of 70 years. The lesions are various shades of brown and black and the margin is flat. The lesions are usually found on the head and neck, being less common on hands and legs. They usually commence as freckles with irregular outlines

and may grow very slowly over 5 to 15 years. (2) *Superficial spreading melanoma* is the most common form, with a median onset of 55 to 60 years. The lesions may have several different colors from brown to gray and rose-pink. The margins are distinctly palpable and the lesions are found on all body surfaces, beginning as small irregular, brown pigmented areas with various shades of color. Growth into papules or nodules may require from 1 to 5 years. (3) *Nodular melanoma* has a median age of onset of about 50 years. The lesions are uniformly bluish-black with a depigmented halo and the margins are palpable. Usually commencing as a papule or nodule with a smooth, scaly, eroded surface, the lesions grow rapidly.

Survival rates are highest for the lentigo type and may be as high as 95%. In nodular melanoma where distant metastases frequently form, the prognosis is much poorer. Immunotherapy or chemotherapy when combined with surgery may improve the situation.

Sarcomas of the Skin. These may be primary or metastatic and single or multiple. They involve connective tissue, the vascular system, adipose tissue, muscle and/or the lympho-reticular system. Kaposi's idiopathic hemorrhagic sarcoma is manifested by firm, reddish-brown or bluish plaques or nodules on the hands and feet.

Mycosis Fungoides. This is the most common primary lymphoma (tumor of lymphoid tissue) of the skin and technically is a relatively indolent cutaneous T-cell lymphoma characterized by a neoplastic proliferation of mature helper T cells. Each year, about a thousand new cases are presented in the United States. However, the incidence is increasing. The lesion may present as a nonspecific erythematous, scaly, eczematous eruption that may go on for years before a diagnosis is made. It gradually develops thicker, scaly, annular plaques reminiscent of psoriasis. In the accelerated phase the disease is characterized by skin tumors, lymph-node involvement, and a spreading to the viscera. These conditions are associated with a survival of 2.5 years. Patients who present only superficial skin involvement have a better prognosis (some 12-plus years). Sézary syndrome, where there are plaques, tumors, erythrodermic, and lymph-node or blood involvement, but no visceral involvement, has a median survival of 5 years.

Protection from the Sun

Life-styles have changed a lot during the last few decades, and, unfortunately, the incidence of skin cancer has increased markedly. Cases of skin cancer in earlier years were almost always associated with outdoor manual laborers, and thus a tanned skin was shunned by many of the elite. In recent years, particularly with the younger generation, a tanned skin has become a status symbol by people who regard a nicely tanned body as a symbol of health and strength. Young males and females alike have taken to sun bathing, to using tanning parlors, and taking tanning pills. This is despite increasingly frequent warnings, by numerous public and private institutions, against excessive exposure to solar rays, and their relationship with skin cancer.

While tanning pills generally have been considered hepatotoxic (adverse effects on the liver), a large number of satisfactory and safe UV blocking and screening lotions has become available. The labels on these products should be read carefully to make certain that sufficient protection will be provided over what periods of exposure time. But, even with such safeguards, sunbathers must always act in moderation.

Professional dermatologists suggest:

1. Always wear sunglasses that absorb from 90 to 100% of UV light. Glasses with gray lenses are preferred to avoid undue alterations in color perception. Glasses that wrap around the side provide protection against reflected-light trauma. Sunglasses protect against *solar keratitis*, which is a sun-induced inflammation of the cornea of the eye. The glasses also provide long-term protection against later cataract development.

2. Apply lip balm to lips, zinc oxide paste to the nose, and wear a broad-brimmed hat. A weekly lanolin facial treatment is excellent for maintaining a good skin condition.

3. Use a sun block (screen) over all areas of the body that will be exposed to the UV light. Light-skinned people require a highly protective blocking agent (rated 12 to 20). Dark-skinned people can use a lighter blocking agent.

4. Under no circumstances should babies or older family members be permitted to bask in the sun. Exposure for just a few minutes can be damaging.

5. Do not relax precautions on cloudy days. Such days are a special hazard because clouds are radiolucent, offering no barrier to UV rays.

6. The foregoing also apply when in the water. A water film provides no protection and, in fact, may partially remove the blocking agent (usually PABA [para -amino benzoic acid]).

Additional Reading

Altmeyer, P., M. Freitag, K. Hoffmann, and M. Stucker: "Skin Cancer and UV-Radiation," Springer-Verlag, Inc., New York, NY, 2001.

Balch, C.M.: "Cutaneous Melanoma," 2nd Edition, Lippincott-Raven Publishers, Philadelphia, PA, 1991.

Burg, G. and J. Fletcher: "Atlas of Cancer of the Skin," Churchill Livingstone, Inc., Philadelphia, PA, 1999.

Chu, A.C. and R.L. Edelson: "Malignant Tumors of the Skin," Oxford University Press, Inc., New York, NY, 1999.

Garbe, C., C.E. Orfanos, and S. Schmitz: "Skin Cancer: Basic Science, Clinical Research, and Treatment," Vol. 139, Springer-Verlag, Inc., New York, NY, 1995.

Koh, H.K.: "Cutaneous Melanoma," *N. Eng. J. Med.*, 171–182 (July **18**, 1991).

Kruttman, J. and C.A. Elmets: "Photoimmunology," Blackwell Science, Inc., Malden, MA, 1995.

Miller, S.J. and M.E. Maloney: "Cutaneous Oncology," Blackwell Science, Inc., Malden, MA, 1997.

Moy, R.L., D. Taheri, and A. Ostad: "Practical Management of Skin Cancer," Lippincott-Raven Publishers, Philadelphia, PA, 1998.

Mukhtar, H.: "Skin Cancer: Mechanisms and Human Relevance," CRC Press, LLC., Boca Raton, FL, 1994.

Phillips, T.J. et al.: "Recent Advances in Dermatology," *N. Eng. J. Med.*, 167 (January 16, 1992).

van Vloten, W.A., R. Willemze, G.L. Vejlsgaard, and K. Thomsen: "Cutaneous Lymphoma." S. Karger Publishers, Inc., Farmington, CT, 1990.

Weinstock, M.A. et al.: "Non-Melanoma Skin Cancer Mortality," *Arch. Dermatol*, 1194 (July 1991).

Web References

American Academy of Dermatology Association: *http://www.aadassociation.org/*
American Academy of Dermatology: *http://www.aad.org/patient_intro.html*
American Cancer Society: *http://www.cancer.org/*
CancerNet: *http://www.cancernet.nci.nih.gov/*
Memorial Sloan-Kettering Cancer Center: *http://www.mskcc.org/mskcc/html/44.cfm*
National Cancer Institute: *http://www.nci.nih.gov/*

R.C. VICKERY, M.D., D.Sc., Ph.D., Blanton-Dade City, FL.

SKIN DISEASES. See **Dermatitis and Dermatosis**.

SKIN EFFECT. The alternating current flowing in a conductor is not uniformly distributed over the cross section of the conductor but becomes denser the farther the cross section being considered is from the center of the conductor. This effect, known as skin effect, is caused by the varying inductance of the different parts of the conductor. If a circular conductor is considered to be specific, and imagined to be composed of many elemental filaments in parallel it can be seen readily why the current flow is more dense as the distance from the center becomes greater. The current in each filament sets up a flux around it which links both it and any other filaments which may be within the lines of flux. It will be realized at once that the filaments in the center can be linked by many more lines than those on the outside, since the outer ones are beyond much of the flux of the inner ones. Thus the inductance of the inner filaments is greater than that of the outer ones and so the reactance of the inner ones is greater. This greater impedance of the inside causes more of the current to flow in the outer-layers and gives rise to the skin effect. As the effective cross-section of the conductor is decreased by this, the resistance to A.C. is greater than the resistance to D.C. In certain cases, such as the induction motor, where the conductor is largely surrounded by iron, the skin effect is very much greater than in a conductor in free space. In this motor the effect is utilized to obtain high starting torque without serious impairment of the running characteristics. In the layout of A.C. buses, various arrangements such as hollow squares, channels, etc., are used to save material since the inner part would be of very little use. At radio frequencies the effect is very pronounced so only a thin outer layer of the conductor is effective, thus greatly increasing the resistance. It is convenient to speak of a "skin depth" which is defined as the depth below the surface of a conductor at which the current density has been reduced to $1/e$ times its value at the surface of the conductor. The skin depth δ is given by the expression

$$\delta = \sqrt{\pi f \mu T}$$

where δ is depth in meters, f is frequency in cycles/second (Hz), μ is the permeability in henries/meter and T is the conductivity in mhos/meter.

SKIN FRICTION. The drag force on a body arises from tangential stresses at its surface, usually viscous, from the normal pressure distribution over it, and, for supersonic flow, from wave-drag. The component calculated from the tangential viscous stress is the skin friction and, for a bluff body, is the component most sensitive to change of Reynolds number. Roughly, the skin friction coefficient, the stress per unit area divided by the dynamic head of the free stream is proportional to the one-fifth-power of the Reynolds number.

SKINKS. Of the class *Reptilia*, subclass *Lepidosauria*, order *Squamata* (scaly reptiles), suborder *Sauria* (lizards), infraorder *Scincomorphs* (skinks and allies), these animals are found in three subfamilies: *Tiliquinae*, which includes the Giant, Cape Verde, Stumped-tail, Blue-tongued, and Spiny-tailed skinks; Scincinae, which includes the common skink and the Eastern, Hemprich's, Arabian, Sand, and Algerian skinks, among many others; and Lygosominae, which has many genera, including the East Indian Brown-sided, Emerald, Spotted, Casque-headed, Dart, and Blind skinks, among others. Closely related and part of *Scincomorpha* are two additional families, the tailed lizards and night lizards, which are described under **Lizards**.

Skinks are distinguished by the head, which is covered on top with large, symmetrically arranged, ossified plates, and by the round scales on the back and belly that overlap like roof shingles. The tongue is free and moderately long. It is slightly notched at the end and bears imbricating scale-like papillae. The body is usually cylindrical, and the head ends in a sharp snout. The feet in most species bear five toes; the tail is pointed at the tip. The skinks demonstrate a gradual transition from lizards that have four strong legs to the legless "snake-type" lizard. In this transition, it is hypothesized that, initially, the trunk became longer, the limbs became smaller and more delicate, and the number of fingers and toes was reduced until finally becoming stumpy vestiges, with the forelegs disappearing entirely and, lastly, the hindlegs vanishing as well. Thus, the manner of locomotion changed from walking to snakelike slithering. These changes reflected a move from the original mode of life above ground to the specialized life of the subterranean burrower. These changes are reflected in varying degree among the numerous genera of skinks. The form of the tail also was involved in the slowly evolving transition. Some genera of skinks, such as the Australian stumped-tailed skink (*Tiliqua rugosa*), has only a stump tail, whereas others have a spine-like tail. Ear openings also changed, wherein the external ear opening is found reduced and has become covered with scales. This may have occurred as the result of the animal's transition to a subterranean form of life — to protect or even close the openings of sense organs. In many skinks, according to I.E. Fuhn, the lower eyelid, customarily movable and scaly in lizards, evolved into a transparent window that allows the animal to see through when it is closed. If the lids fuse, a rigid, transparent "spectacle" is formed, much as in snakes. Where skinks colonized rocky regions or trees, it became advantageous to develop adhesive mechanisms on the fingers and toes. There are exceptions, as in the case of the giant skink, which evolved a prehensile tail.

Skinks are *thermophilous* (heat-loving) and are found mainly in the tropical and subtropical regions of the earth, such as southern Asia, Africa, and regions of Australia and Polynesia. Some, however, are found in the Americas and Europe. Because of their highly terrestrial preference, skinks are not usually found near bodies of water.

Formal classification of the skinks is based largely upon the structure of the skull, thus accounting for the three families previously mentioned.

Fuhn suggests that the *Tiliquinae* are relicts of the Tertiary Period. The giant skink (*Corucia zebrata*) is, on the average, about 65 centimeters long. It is presently found in the tropical forests on San Cristobal in the Solomon Islands. The largest of all modern skinks, this animal lives in trees, a characteristic that is rare among skinks. Especially adapted to arboreal life, its characteristics of long prehensile tail, strong limbs armed with claws, large ear and eye openings, broad head, and short, blunt nose are distinguishing features. The upper body is greenish-white, marked with irregular brown bands; the head is sometimes a reddish-brown. This animal has not been studied extensively in its habitat.

Also of *Tiliquinae* is the Cape Verde skink, which at one time was a favorite animal for terrariums. It is relatively easy to sustain in captivity, but, because of extensive hunting for its flesh by fishermen in its natural habitat, the species is threatened. A close relative of this creature became extinct as early as 1708.

Among the *Tiliqua*, the blue-tongued skink (*Tiliqua scincoides*) is well known and named because of its cobalt-blue tongue. Also studied is the stump-tailed skink (*T. rugosa*), which achieves a length of about 36 centimeters. The latter is found in Western Australia, South Australia, and New South Wales. Notably, this skink is quite short and wide, has a flat tail that is widened at the end and, because of its round end, that appears to have been chopped short (sometimes referred to as "bob-tail"). The threat display of the animal has been described by Robert Mertens: "It rolls itself into an arc and, hissing, opens its mouth to show the wide, long, cobalt-blue tongue in sharp contrast to the red mouth. Though its mouth, armed with strong teeth, is wide open, the animal does not bite, but rather tries to scare its enemy away simply by its threat display." The "bobtail" often lives on sandy dunes and can burrow into upper layers of sand. It moves slowly, but in an emergency can run quite fast. The animal's diet consists of vegetation, fruit, and small animals. The blue-tongued skink is ovoviviparous, the female bearing two or three young at a time. The animal has been featured in terrariums.

Of the subfamily *Scincinae*, the common skink is the best representative of this group. See Fig. 1. The species (*Scincus scincus*) generally is adapted to live in the desert or other sandy regions. The characteristic ability of these lizards to move about rapidly in the sand was noted in early Biblical times. At one period, in the late 1700s, it was believed that the lizard had aphrodisiac and medicinal properties. All members of this species have sturdy, cylindrical bodies. The tip of the snout has an elongated wedge-like shape, and the tail, conically shaped, is short. The legs each bear five toes with serrated fingers. The coat of scales is remarkably smooth and firm.

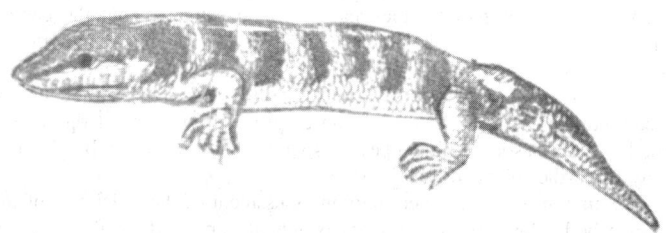

Fig. 1.   Common skink (*Scincus scincus*).

Scincinae also includes the genus *Chalcides*, with its well-known cylindrical skinks, which are distributed in the Mediterranean, from eastern Africa and Arabia into Iran, Pakistan, and India. The species *Chalcides chalcides* is a snake-like animal, with three toes on the limbs. This animal may reach a length of about 42 centimeters. During rapid locomotion, the legs are pressed against the body and not used. The upper side is olive to bronze in coloration, with up to 11 dark, longitudinal stripes. The tail is moderately long, and the head is not set off from the trunk. The limbs are short, with five toes. The smooth and shiny scales are brown. These animals are ovoviviparous, bearing about three young at a time. A popular terrarium specimen is *C. ocellatus*. Its natural habitat is Morocco and environs. It prefers valleys in which date palms grow or else groves of tamarisk on clay soil, where it uses the tracks formed during dry periods as hiding places. This skink is active all day until sunset, during which time it feeds on crickets. Because the animal has a snakelike appearance, there are numerous superstitions about it. For terrariums, Franz Werner observes: "It loves the sun and needs a warm, sunny container with a sandy floor and stones under which it can withdraw at twilight. One can feed it with small mealworms and flies. The skink can become quite tame and will come out to take food handed to it."

Additional Reading

Behler, J.L. and F.W. King: "The National Audubon Society Field Guide to North American Reptiles and Amphibians," Alfred A. Knopf, Inc., New York, NY, 1979.
Coborn, J.: "Prehensile-Tailed Skinks," T.F.H. Publications, Inc., Neptune, NJ, 1996.
Rogner, M.: "Lizards: Monitors, Skinks, and Other Lizards, Including Tuataras and Crocodilians," Vol. 2, Krieger Publishing Company, Melbourne, FL, 1997.
Walls, J.G.: "The Skinks Identification, Care & Breeding," T.F.H. Publications, Inc., Neptune, NJ, 1995.
Walls, J.G.: "Blue-Tongued Skinks," T.F.H. Publications, Inc., Neptune, NJ, 1996.

SKIP DISTANCE.   As the frequency of a radio wave is increased, the minimum angle of incidence at which the wave will be reflected from the ionosphere rather than pass on through becomes greater. This means that the higher the frequency the farther from the transmitter the reflected sky wave strikes the earth. This distance between the transmitter and the point closest to it at which the sky wave can be received is the skip distance. The ground wave is attenuated more rapidly the higher the frequency so that at high frequencies there may be a region in which the ground wave has become too weak for use and in which the sky wave cannot be received because of its skip. Above about 4 MHz this effect becomes very noticeable, the dead region for higher frequencies running to a distance of a few hundred miles from the transmitter.

SKIPJACK TUNA.   See **Tunas**.

SKIPPER (*Insecta, Lepidoptera*).   An insect much like the butterflies but in many ways intermediate between the butterflies and moths. Most species are distinguished by their knobbed and hooked antennae. The few that lack the hook are less easily distinguished from the butterflies except by the veins of the wings or by other details of structure. Most of the skippers of the temperate zones are of moderate size and modest colors but many tropical species are brilliant. They constitute the superfamily *Hesperioidea*.

The name skipper refers to the vigorous and often erratic flight of these insects. Their small wings and powerful muscles accompany rapid flight. Many species perch readily and make short darting flights from place to place.

SKUA.   See **Waders, Shorebirds, and Gulls**.

SKUNK.   See **Mustelines**.

SKUTTERUDITE.   This mineral includes an isomorphous series with *smaltite-chloanthite*, essentially cobalt/nickel arsenides, $(Co, Ni) As_{2-3}$, crystallizing in the isometric system. The usual habit is cubic, octahedral, or cubo-octahedral. The mineral also occurs in massive and granular forms. Skutterudite has a metallic luster; hardness of 5.5 to 6.0, a specific gravity of 6.5. The mineral is opaque with tin-white to silver-gray color. The nickel-rich material alters surficially to annabergite (green color); the cobalt-rich material to erythrite (rose color). The streak is black. The mineral is an essential ore of cobalt and nickel.

Skutterudite is found in moderate-temperature veins, commonly associated with other cobalt/nickel minerals, e.g., cobaltite and nickeline. The mineral was named for its occurrence at Skutterud, Norway. Important ore sources are Norway, Bohemia, Saxony, Spain, France, and New South Wales, Australia. Notable occurrences are in Ontario, Canada, mainly Sudbury, South Lorrain, and Gowganda.

SKY COVER.   In surface weather observations, *sky cover* is a term used to denote one or more of the following: (a) the amount of sky covered but not necessarily concealed by clouds or by obscuring phenomena, i.e., any atmospheric phenomenon that obscures a portion of the sky from the point of observation; (b) the amount of sky concealed by obscuring phenomena that reach the ground; or (c) the amount of sky covered or concealed by a combination of (a) and (b).

"Opaque" sky cover is the amount of sky cover that completely hides everything above it; *transparent sky cover* is that portion of sky cover through which higher clouds, blue sky, etc., may be observed; and *total sky cover* is the two taken together. Sky cover, for any level aloft, is described as "thin" if the ratio of transparent to total sky cover at and below that level is one-half or more.

Sky cover is reported in *eighths* under international observing and reporting practice. The "0" indicates $\frac{0}{8}$, or clear; and "8" indicates $\frac{8}{8}$, or total cover.

Sky cover is also reported in *tenths*, so that 0.0 indicates a clear sky; 1.0 ($\frac{10}{10}$) indicates a completely covered sky. According to the *summation principle* used in weather observing practice, the sky cover at any level is equal to the summation of the sky cover of the lowest layer plus the additional sky cover provided at all successively higher layers up to and including the layer in question. Thus, no layer can be assigned a sky cover that is less than a lower layer, and no sky cover can be greater than 1.0.

If, at any level, the ratio of transparent sky cover to total sky cover is one-half or more (excluding transparent portions of surface-based obscuring phenomena), the layer at that level is classified as thin.

The term *ceiling*, in aviation weather observing practice, refers to the ascribed height of the lowest layer of clouds or obscuring phenomena when that layer is reported as "broken," "overcast," or "obscuration," and not classified "thin" or "partial." When none of these conditions is satisfied, the ceiling is termed "unlimited." Whenever the height of a cirriform cloud layer is unknown, a slant "/" is reported in lieu of a height value; at all other items, the ceiling is expressed in feet above the surface.

The following classifications of sky cover are used in aviation weather observations:

a. Clear: sky cover less than 0.1 (no ceiling).
b. Scattered: sky cover 0.1 to 0.5 (clouds or obscuring phenomena aloft only; no ceiling).
c. Broken: sky cover 0.6 to 0.9 (minimum requirement for a ceiling; must be some clouds or obscuring phenomena aloft).
d. Overcast: sky cover 1.0 (must be some clouds or obscuring phenomena aloft).
e. Partial obscuration: sky cover 0.1 to 0.9 (surface-based obscuring phenomena only).
f. Obscuration: total sky cover (surface-based obscuring phenomena only).

See also **Clouds and Cloud Formation**; and **Weather Technology**.

PETER E. KRAGHT, Certified Consulting Meteorologist, Mabank, TX.

SKYHOOK BALLOON. (Originally a code name for a U.S. Navy Project.) A large free balloon having a plastic envelope, used especially for constant-level meteorological observations at very high altitudes.

SKYLARK. See **Lark**.

SLAG. Slag is a fused product occurring in connection with metallurgical and combustion processes. It is composed of the oxidized impurities in a metal, and of a fluxing substance, and of ash. In the steel industry, slag is the neutralized product of anhydrous compounds entering into the process. Slag is of great importance to the operator of a steel furnace or a cupola, in that, through the slag, impurities are separated and removed from the metal. By floating as a molten covering on the pool of metal, slag protects it from oxidation and serves to keep it clean. By controlling the character of slag, and continuous observation, the metallurgist insures that the metal is of the quality desired.

Molten ash is one of the products of combustion of coal in certain high-capacity boiler furnaces. It is also called slag. In some plants, the ash is removed from the furnace in this fluid form. Such furnaces are known as slag tap furnaces. Slag has some commercial value as ballast, coarse aggregate for concrete, road metal, etc.

SLATE. A fine-grained homogeneous sedimentary rock composed of clay or volcanic ash which has been metamorphosed (foliated) so as to develop a high degree of fissility or slaty cleavage which is usually at a high angle to the planes of stratification. This high degree of fissility makes the better grades of slates an extremely useful roofing material which, however, has been somewhat replaced in recent years by synthetic and manufactured substitutes. The finest slates in the world come from Wales, Britain.

SLAVE ANTENNA. A directional antenna that is positioned in azimuth and elevation by a servo system. The information controlling the servosystem is supplied by a tracking or positioning system.

SLEEP. Recurrent physiological loss of consciousness at regular intervals, predominantly in humans once in 24 hours and lasting for a more or less constant time of an average of about 8 hours for adults. The newborn infant sleeps 18 to 20 hours a day. The periods of sleep last from 2 to 3 hours, and usually are interrupted by hunger. As the baby's stomach grows and more food is consumed, the periods of hunger are farther apart and the periods of sleep are longer. When the child is 6 months old, it sleeps less—16 to 18 hours a day. At the age of about one year, the child sleeps about 12 hours at night, with usually a morning and an afternoon nap. Although it may be more difficult for older people to get the necessary sleep, there is little if any truth that older people do not require as much sleep as younger persons. Some of the problems of the aged can be derived because of a lack of sustained sleep. The amount of sleep for any person varies with each individual. The prime requisite is that the individual should sleep long enough to awaken rested and refreshed.

The human body cannot work around the clock. It must have a regular opportunity to catch up on its repair activities and to dispose of wastes that have accumulated during the day faster than they can be discarded. During sleep, the entire body slows down. The liver stores starches needed for the next day. The kidneys clean the blood. The rate of metabolism is at its lowest, being just sufficient to keep the vital parts of the body in operation. The blood pressure drops; the pulse rate is slower; and breathing is irregular and slowed down. The body is less sensitive to stimulation by pain, light, or sound. Even the temperature may drop by as much as a degree. Since the brain does not have muscles, it does not tire in the same way the body does. But there is evidence that the brain requires sleep for other reasons, particularly as sleep is related to memory.

The autonomic nervous system and the involuntary muscles, such as those of the heart and the respiratory mechanism, continue to work in sleep, at a slightly reduced rate, so that the whole body does not share equally in the recuperative process. A number of bodily changes are constantly observed in sleep; these include deep muscular relaxation, loss of some reflex activity, slowing of the heart rate and reduction in volume of many glandular secretions. Deep sleep brings no dreams; those which accompany light sleep are generally the result of external stimuli received at the time, and which, in the absence of the critical faculty of the cerebral cortex, are greatly misinterpreted.

Although research continues, there is no fully satisfactory explanation of the phenomenon of sleep. In particular, the biochemical aspects of sleep have received comparatively little detailed attention. There are many ancillary problems which have been investigated, but the central one—what is accomplished biochemically each day by long hours of sleep—has been attacked only a relatively few times and with no marked success. The symptomatic treatment of insomnia by the use of sleep-inducing drugs is largely on an empirical basis. Knowledge of the fundamental reasons behind a person's vital need to sleep can be obtained only by further biochemical investigation.

The amount of sleep enters into an assessment of the efficiency of the human body. Just how much work is actually obtained for the fuel that has been fed the body? When gasoline is burned in an engine, only about one-eighth of the energy released can be captured to drive the motor, and the remainder is lost in the form of the heat that is given off. Such an engine is said to have an efficiency of 12.5% because it wastes 87.5% of the fuel through heat loss. Some diesel engines have a 32% efficiency. If one were to calculate the work produced by a person performing strenuous work, it would be found that the efficiency of the human body is about 20%. Much of the fuel burned by the body is converted to heat to maintain the body temperature and is lost by the body to the surroundings without doing any real work. Since the human body cannot work regularly the full 24 hours of a day, its efficiency is in reality considerably lower—perhaps about 10 to 12%. Nevertheless, the human body is built to last a great many years as contrasted with most other energy-consuming mechanisms. Some of the basic accomplishments of the human body during a 24-hour period include:

—Kidneys filter over 150 quarts (142 liters) of fluids.
—Heart beats over 100,000 times.
—1,500 gallons (5678 liters) of blood are pumped through the body.
—Over 10 million red cells are destroyed every second, all of which must be replaced at an equally rapid pace.
—Nerves carry millions of messages.
—Eyelids blink thousands of times to cleanse and lubricate the eyes.

Electroencephalography (EEG) reveals a pattern of large slow waves which appear when a subject falls asleep. As shown in Fig. 1, this pattern can be made to appear experimentally in cats by partially cutting through the midbrain at a certain level. It is thought possible that such an incision cuts off the vital centers in the midbrain from certain postulated driving centers elsewhere in the brain, and that with the stream of awakening impulses removed, sleep occurs. But the means by which this process occurs almost unfailingly every 24 hours in the human being remains

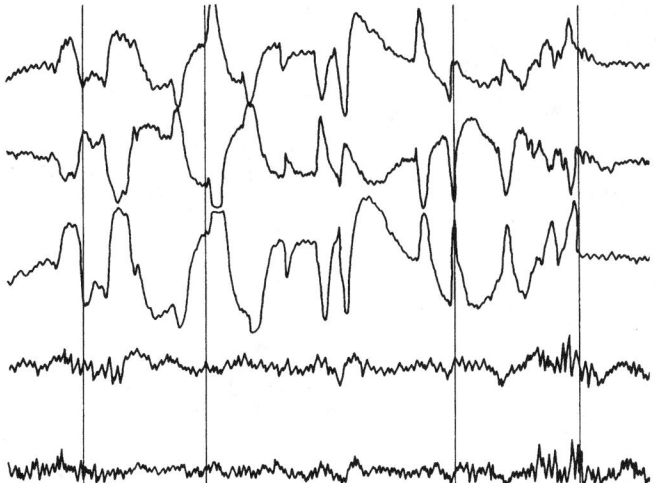

Fig. 1.   Alpha. The subject drowses, on the edges of sleep. Top three EEG channels show rolling movements of right eye, left eye, and both. Two bottom channels trace brain's electrical changes. The even pattern of alpha rhythm indicates relaxation.

obscure. EEG and eye-movement studies, however, are shedding some light on the process.

The alpha waves grow smaller as the subject passes through the gates of the unconscious. The eyes are very slowly rolling. For a moment, the subject may wake up during the early part of this descent, alerted by a sudden spasm that causes the body to jerk. This, like the brain waves, is a sign of neural changes within. It is known as the *myoclonic jerk*, resulting from a tiny burst of activity in the brain, somewhat related to that of an epileptic seizure. The myoclonic jerk, however, is normal in all human sleep. It is gone in a fraction of a second and descent continues. The subject does not note the peculiar transformation as true sleep is approached. The lower chart records of Fig. 1 indicate Stage 1 of sleep. The pattern of the sleeper's brain waves is small and pinched, low, irregular and rapidly changing. Occasionally the regular waves of the alpha rhythm break through. The sleeper may be enjoying a floating sensation or drifting with idle thoughts and dreams. The muscles are relaxing, the heart rate is slowing. The subject could be awakened easily and might insist that actually sleep had not occurred. But, after a few minutes at this port of entry, if not disturbed, the subject descends again to another level, another step removed from the real world.

As the sleeper passes into Stage II, the brain waves change again. Now they trace out quick bursts — a rapid crescendo and decrescendo, resembling a wire spindle, and unmistakable on the EEG chart. The eyes roll slowly from side to side, but if the experimenter gently opens a lid the sleeper will not see. If awakened, which can be accomplished easily with a modest sound, the subject may still think that actual sleep has not occurred. At this point, however, the subject has been soundly asleep for perhaps 10 minutes. Whatever happens now in the subject's imagination will be totally beyond conscious grasp.

Still the sleeper descends to Stage III. This is characterized by large slow waves that occur about one a second. Sometimes they are about five times the amplitude of the waking alpha rhythm. It will take a louder noise to awaken the subject, perhaps some repetition of the person's name. The muscles are very relaxed and the breathing is even. The heart rate has become slower; body temperature declines; blood pressure drops.

Some 20 to 30 minutes after the subject fell asleep, the sleeper reaches Stage IV, the deepest level. This is marked by large, slow brain waves, called delta waves (Fig. 2) that trace a pattern resembling jagged buttes. Stage IV is a relatively dreamless oblivion. The breathing is even, heart rate, blood pressure, and body temperature slowly falling. But, the sleeper does not remain in Stage IV. After 20 minutes or so in this depth, the subject begins to drift back up through the lighter levels toward the surface. By the time the subject has been asleep for about 90 minutes, brain waves of the lightest sleep, even resembling those of waking, will be indicated. Still the sleeper is not easy to awaken, lying limply, the eyes moving jerkily under closed lids as if watching something. This is a special variety of Stage I, known as REM (for rapid eye movement) sleep. See Fig. 3. If awakened during this period, the subject would almost certainly remember

dreaming and most often in vivid detail. After perhaps 10 minutes in the REM state, the sleeper will probably turn over in bed and begin shifting down the levels of sleep again to the depths, only to return in another hour or so for a longer REM dream. Each night, the entire cycle may be repeated 4 to 5 times.

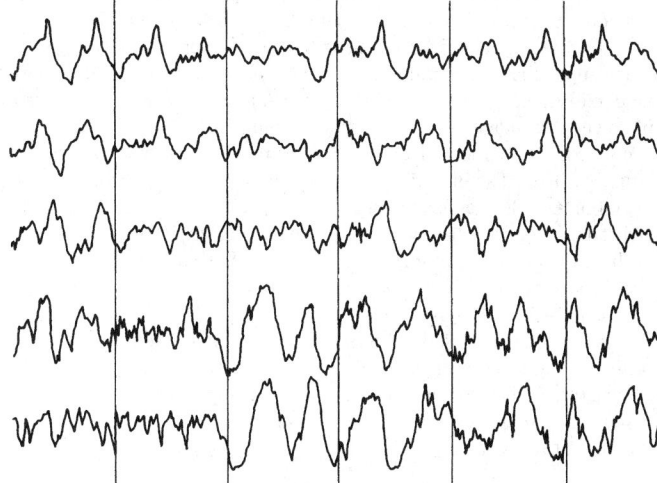

Fig. 2.   Delta. Marked by large, slow brain waves (bottom channels). This is the deepest stage of sleep, the first to be made up for after total deprivation.

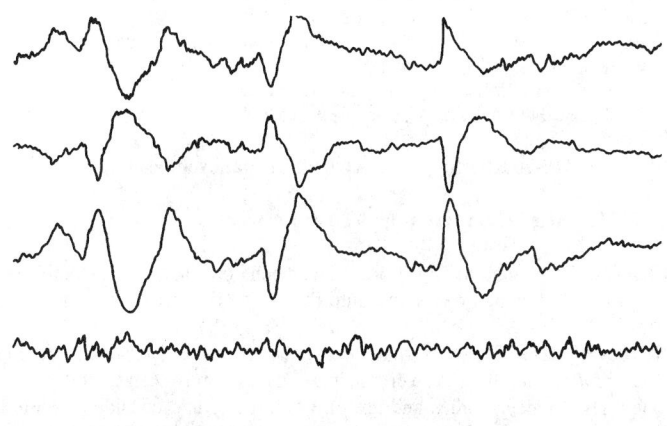

Fig. 3.   REM. Rapid eye movements, spaced in the top three channels, indicate dreaming. This is the most dramatic of the night's phases. Although brain waves are low, a psychological storm is raging in breathing, pulse, brain temperature, and hormone circulation.

In an excellent paper on sleep and memory (see references), Fowler, Sullivan, and Ekstrand describe experiments conducted to determine the relationship of sleep with ability to remember. The interference theory of forgetting essentially states that forgetting is due to interference from learning taking place during the retention interval. It then would possibly follow that by putting subjects to sleep, recall could be facilitated. The experiments described in the reference paper demonstrated that memory over an interval with relatively high amounts of rapid eye movement (REM) sleep was inferior to memory over an interval with relatively high amounts of Stage IV sleep. The findings to date suggest that, at least for humans, REM sleep does not facilitate memory consolidation and that Stage IV sleep may be beneficial to memory.

In studying patients complaining of insomnia, Guilleminault, Eldridge, and Dement uncovered a new clinical syndrome, sleep apnea (partial suspension of breathing) sometimes associated with insomnia. These findings came out of combining respiratory studies with sleep research. Quoting briefly from the aforementioned reference. "Sleep apneas have been reported in the cardiopulmonary syndrome of obesity (Pickwickian) and in other syndromes involving hypersomnia, such as narcolepsy. The apneas in hypersomniacs seem to be temporarily associated with sleep. Several distinct types of sleep apnea have been defined in these conditions. They include a "central" type, characterized first by cessation of breathing and then, after the apnea, by a simultaneous resumption of diaphragmatic movements and oral airflow; an "obstructive" or "peripheral" type characterized by the interruption of airflow secondary to upper airway obstruction, but with continuance of diaphragmatic and thoracic muscle contraction; and a "mixed" type, characterized by an initial central apnea followed by temporary upper airway obstruction at the subsequent resumption of diaphragmatic movements." It is concluded from these early studies that some unknown percentage of patients complaining of chronic insomnia may have profound disorders of respiratory control mechanisms, with resulting disruptive sleep. The investigators conclude with the suggestion that respiratory function during sleep should be evaluated in patients who complain of chronic insomnia characterized by several conscious arousals throughout the night and early morning, and who also have a short latency before onset of sleep and a history of heavy snoring.

Somnambulism (sleepwalking) and talking while asleep are symptoms most commonly observed in persons with conflicts. Some investigators ascribe the sleepwalking (and other actions performed by a person who is apparently otherwise asleep) to underlying unconscious distress. The phenomenon indicates that certain brain areas, more than others, are inhibited during sleep.

Additional Reading

Berkow, R. and M.H. Beers: "The Merck Manual," 17th Edition, Merck & Company, Inc., Whitthouse Station, NJ, 1999.

Carskadon, M.A.: "Adolescent Sleep Patterns," Cambridge University Press, New York, NY, 2002.

Chokroverty, S.: "100 Questions and Answers about Sleep and Sleep Disorders," Blackwell Science, Inc., Malden, MA, 2001.

Hauri, P., S. Linde, and M. Jarman: "No More Sleepless Nights Workbook," John Wiley & Sons, Inc., New York, NY, 2000.

Lugaresi, E. and P.L. Parmeggiani: "Somatic and Autonomic Regulations in Sleep: Physiological and Clinical Aspects," Springer-Verlag, Inc., New York, NY, 2001.

SLEEPING SICKNESS.   See **African Trypanosomiasis**.

SLEET.   See **Precipitation and Hydrometeors**.

SLEW.   To change the position of an antenna or range gear assembly by injecting a synthetic error signal into the positioning servo amplifier.

SLICKHEAD FISHES (*Osteichthyes*).   Of the order *Isospondyli*, family *Alepocephalidae*, these are very deep sea fishes, rarely seen in collections. They have slender bodies and are of a dark color and considered quite a small fish. The *Dolichopteryx* is equipped with "telescopic eyes."

SLIDER (*Reptilia, Chelonia*).   Turtles related to the painted turtles and land tortoises. The several species are brownish or greenish, marked with yellow and in some cases also with red. They constitute the genus *Pseudemys*. Most of the species are confined to the southern states but one, known as the red-bellied terrapin, is found near coastal rivers from Florida to Cape Cod and another lives in the Mississippi valley as far north as Iowa. Both of these species are edible. Also called cooters.

SLIDE RULE.   A hand-held, manually operated analog computer for multiplying, dividing, extracting roots, and obtaining powers of numbers mechanically by logarithmic means. In construction, the simple slide rule consists of two adjacent logarithmic scales, which may be so set that a reading on one is added to a reading on another. This represents the addition of logarithms when multiplying numbers. The slide rule for general use is found in several different forms, and there are many special slide rules such as stadia rules, electrical rules, hydraulic computing rules, etc.

Figure 1 illustrates the principle of multiplication by the slide rule. Two logarithmic scales, A and B, are arranged on a slide rule so that they may be mechanically added. In the illustration given, the process of multiplying three by two is shown. The end of the B scale is aligned over the 3 on the A scale, so that to 3 on the A scale may be added 2 on the B scale. Now if these scales were uniformly divided, the result, of course, would be five units on the A scale, but since they are logarithmically divided, the result

on the A scale is the product rather than the sum. Thus, under the 2 on the B scale, one reads the product of $3 \times 2 = 6$ on the A scale. A glass runner is provided so that the alignment of the numbers of the two scales may be facilitated. By halving the divisions, two complete scales could be placed on the rule. Using a full scale, such as A of the figure, and a double scale, one can extract square roots. A triple scale can be used for cube roots. Other scales contained on the ordinary slide rule are a uniform division scale for logarithms, and scales for reading the value of two trigonometric functions of an angle, the sine and tangent.

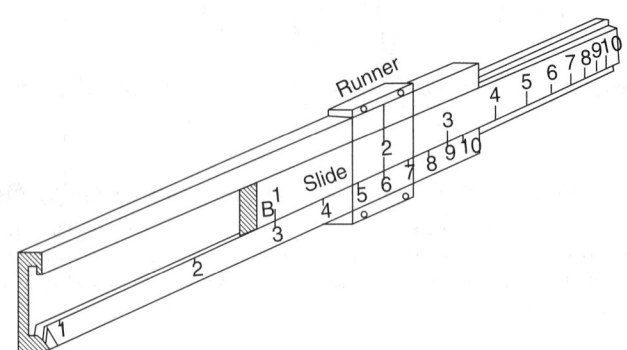

Fig. 1.   Simple slide rule.

With the advent of fast, low-cost, solid-state electronic pocket and portable calculators, the conventional engineer's slide rule has largely been replaced. But the slide rule is still in use, because the availability of plastic materials and manufacturing techniques has greatly lowered the cost. Customized slide rules directed to the solution of specific problems remain quite popular, particularly among manufacturers of equipment who frequently provide slide rules to users of their products as an aid to application, maintenance, and procurement decisions. The principal features of a simple slide rule are shown in Fig. 1.

SLIP RINGS.   These are conducting rings attached to a rotating part of an electrical machine to make connection through brushes with the stationary part of the circuit. They are used where it is not necessary to commutate the current being conducted.

SLIT.   The long narrow opening by which radiation enters or leaves certain diffraction instruments. Slits are often used as line sources of radiation or of particles, and combination of two or more slits are employed as a collimator.

SLIT LAMP.   The slit lamp, or biomicroscope, is a tabletop microscope that uses an adjustable beam of light to provide concentrated illumination and magnification of various parts of the eye. The light beam may be adjusted both as to width of the slit of light and the height of the slit, thus the term, *slit lamp*. This microscope magnifies the structure of the eye some 6 to 40 times, and the eye doctor uses it to detect and evaluate any abnormalities in the structures of the eye.

Comprehensive eye examinations comprise an evaluation of the physiologic function and anatomic status of the eye, visual system and its related structures, and always include a slit lamp examination. Before a slit lamp examination, the eyes are sometimes dilated with special eye drops to enlarge the pupil size. This provides the doctor with a larger "window" to examine the inside structures of the eye.

Vision Rx, Inc., Elmsford, NY.

SLOPE.   In rectangular coordinates, the ratio of the change of the ordinate to the corresponding change of the abscissa of a point moving along a line. If a straight line is determined by the points (x_1, y_1) and (x_2, y_2) its slope $m = (y_2 - y_1)/(x_2 - x_1)$. If the line is not straight its slope is given by

$$\lim_{x_2 \to x_1} m = (dy/dx)_x = {}_{x_1}$$

which is the tangent or derivative at the point.

See also **Coordinate System**.

SLOTH-BEAR. See **Bears**.

SLOTHS. See **Edentata**.

SLUDGE. When fresh sewage is admitted to settling tanks a certain amount of the solid matter in suspension will settle out, 50% more or less for sedimentation periods of an hour and a half or so. This collection of solids is known as fresh sludge. Such sludge will become actively putrescent in a short time and in modern treatment plants must be passed on from the sedimentation tank before this stage is reached. This may be done in two common ways. The fresh sludge may be passed through the slot in an Imhoff tank to the lower story or digestion chamber. Here, decomposition by anaerobic bacteria takes place with considerable liquefaction and reduction in volume. After the decomposition process has run its course (in 6–9 months) the resulting sludge is called "digested" sludge and is relatively inoffensive in character. It may be disposed of by drying on sludge drying beds and spreading on the land. It has little, if any, fertilizing value, being in the nature of humus. The sludge digestion chamber is operated on a periodic schedule of sludge withdrawals.

Alternatively, plain sedimentation basins with mechanical equipment for continuous collection of the fresh sludge may be used. The fresh sludge, so collected, is discharged into separate sludge digestion tanks which operate on the principle of the lower story of the Imhoff tank except that by means of higher and better temperature control the digestion cycle is much more rapid and efficient than for the Imhoff tank.

SLUG (*Mollusca, Gasteropoda*). Not an insect, but a mollusk, this pest is damaging to certain food crops and is a serious pest in glasshouse operations, notably on certain ornamental plants, such as coleus, geranium, marigold, and snapdragon. The slug is also a severe pest on lawns, particularly on dichondra. The pest eats very large holes, often entire leaves, always at night. The creatures are fleshy, worm-like, gray-to-brown, slimy. They are without body segmentation and are apodal. Slugs range up to 3 inches (7.5 centimeters) in length. See Fig. 1. During the day, the pests hide under leaves, debris, etc., but in a heavy infestation, they emerge at night in the scores. Some of the species include: the *spotted garden slug* (*Limax maximum*, Linne); the *gray garden slug* (*Deroceras reticulatum*, Müller); the *gray field slug* (*D. laeve*, Muller); and the *greenhouse slug* (*Milax gagates*, Linne).

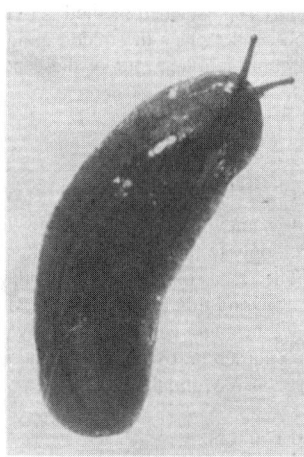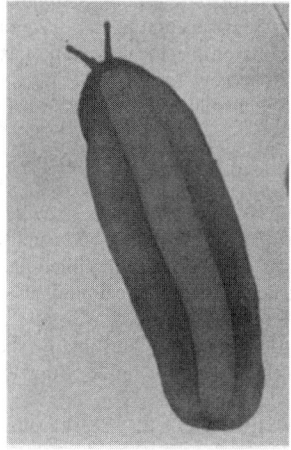

Fig. 1. Slugs. (*A.M. Winchester.*)

The control measure most frequently used is poisoned bait, placed in suspected locales at night. The quantity required varies with the strength of the mixture used. Ingredients frequently used include bran as the base, plus calcium arsenate, metaldehyde, blackstrap molasses, and water. The molasses is used as an attractant.

The *pear slug*, which damages cherry, pear, and plum trees, is not a true slug, but rather the larva of a species of sawfly. See also **Sawfly**.

SLUG TUNING. A means for varying the frequency of a resonant circuit by introducing a slug of material into either the electric or magnetic fields or both.

SLUICE. A channel through which water flows is, in some cases, called a sluice. A sluice may be a pressure conduit, or it may be an open flume. For instance, the word may be applied to an artificial channel to lead water from one point to another, especially a temporary wooden channel. Sluices are often incorporated in dam structures. A sluice through a dam is a conduit cast in the concrete and equipped with controls called sluice gates. The purpose of the sluice is to empty the reservoir if necessary, to control the water level, and to aid in passing floods.

SLURRY EXPLOSIVES. See **Explosive**.

SMALLPOX. This disease caused by a large, enveloped poxvirus featuring a complex helical nucleocapsid structure containing double-stranded DNA and several enzymes, was once described as the most devastating pestilence in human history. Worldwide eradication culminating in the last reported case in late 1977 (patient in Merka, Somalia recovered) indeed marked a major hallmark in the professions of public health and preventive medicine. In 1967, when the World Health Organization (WHO) instituted a campaign to eradicate smallpox from the last reservoirs on earth, the disease was still endemic in 33 countries and 11 other countries reported cases. The eradication process started many years ago, of course, with mandatory vaccination policies in a number of countries and with particularly tight control over travelers who could carry the disease from one country to the next. The thrust of the last decade by WHO concentrated on mass vaccinations of the populace of countries where the disease still persisted, as in Brazil, Indonesia, Bangladesh, India, Afghanistan, and Africa south of the Sahara. In the last few years of the WHO program, the emphasis also included strict isolation of smallpox patients. At a time when it appeared that the disease had finally been overcome, an outbreak occurred in 1977 among nomads in eastern Africa and in Mogadishu (capital of Somalia) during which time nearly 1400 cases were reported. One of the final steps was a last major mass vaccination program in Somalia.

Smallpox (also called variola), an acute contagious disease characterized by high fever, chills, headache, generalized pain, and vomiting and with an accompanying skin eruption that resulted in severe scarring, has been known since antiquity. Lesions have been seen on the skin of an Egyptian mummy of the twentieth dynasty. The classical scarring which labeled a person for life as a victim of smallpox was often accompanied by development of blindness. The disease originally spread from India and Central Asia to Europe and became especially prevalent at the time of the Crusades in the eleventh century. It was first introduced in America by a Negro slave of Cortex in 1520, producing an epidemic that killed several million people. One century later it appeared in New England. It was first called smallpox in Europe, to differentiate it from grand-pox (syphilis of the skin). The discovery of vaccination by Jenner in 1798 is one of the landmarks in the development of medical science.

Black or hemorrhagic smallpox, a virulent fulminating form of the disease, commonly caused death within two to six days. Fatalities from smallpox often arose from complications, including abscesses, pneumonia, septicemia, nephritis, and ulcers on the cornea of the eye. In recent decades, these complications were treated with chemotherapy. However, no specific drug to assure the cure of smallpox had been found by the time of its eradication.

Additional Reading

Basu, R.N., Jezek, Z., and N.A. Ward: "The Eradication of Smallpox from India," World Health Organization, Publications Centre, Albany, NY, 1980.

Bazin, H. and E. Jenner: "The Eradication of Smallpox," Academic Press, Inc., San Diego, CA, 2000.

Fenner, F., D.A. Henderson, I. Arita, Z. Jezek, and I.D. Ladnyi: "Smallpox and Its Eradication," World Health Organization, Geneva, Switzerland, 1988.

Frauenthal, J.C.: "Smallpox: When Should Routine Vaccination Be Discontinued," Birkhauser Verlag, Cambridge, MA, 1981.

Henderson, D.A.: "The Eradication of Smallpox," *Sci. Amer.* (October 1976).

Razzell, P.: "Edward Jenner's Cowpox Vaccine," Caliban Books, Firle, Lewes, Sussex, England, 1977.

Razzell, P.: "The Conquest of Smallpox," Caliban Books, Firle, Lewes, Sussex, England, 1977.

Web References

International World Health Organization: *http://www.who.int/home-page/*

Facts About Anthrax and Smallpox as Bioterrorism Weapons: *http://healthlink.mcw. edu/content/article/1004505206.html*

Smallpox: Clinical and Epidemiologic Features: *http://www.cdc.gov/ncidod/EID/vol5no4/henderson.htm*
Smallpox Information Center: *http://smallpox.phages.org/*

R.C. VICKERY, M.D.; D.Sc.; Ph.D., Blanton/Dade City, FL.

SMECTIC LIQUID CRYSTALS.   See **Liquid Crystals**.

SMELTING.   The process of heating ores to a high temperature in the presence of a reducing agent, such as carbon (coke), and of a fluxing agent to remove the accompanying rock gangue is termed smelting. Iron ore is the most abundantly smelted ore. It contains about 20% gangue (clay and sand). The ore is heated in an air blast furnace with coke and limestone (fluxing agent) at a temperature above the melting point of iron and slag (fusion mixture of impurities and flux). The molten iron (the more dense material) and molten slag (the less dense material) are removed separately from the furnace. See also **Arsenic**; **Cadmium**; **Cobalt**; **Copper**; **Indium**; **Iron Metals, Alloys, and Steels**; **Lead**; **Silver**; and **Tin**.

SMELTS (*Osteichthyes*).   Of the order *Isospondyli*, family *Osmeridae*, smelts (more properly termed true smelts) seldom exceed 14 inches in length, and thus classified as a small fish. They prefer the cold or temperate in-shore waters of the northern hemisphere. The osmerid smelts are frequently confused with the atherinid smelt (silverside family). See also **Silversides (Osteichthyes)**. The true smelts have a small adipose fin on the dorsal fin, which is absent in the silversides.

Most true smelts, numbering some 13 species, are found in the Pacific, with no occurrence in the Indian Ocean or the southern hemisphere. They can be extremely abundant. From time to time, for example, the *Spirinchus thaleichthys* (Sacramento smelt) has occurred in extremely large numbers in San Francisco Bay. Smelts are, of course, logical prey for larger fishes. Fisheries for adult smelts have long prospered.

The *Thaleichthys pacificus* (eulachon) is found in the waters off the American Pacific Northwest. Nicknamed "candlefish," the fish, when dried and because of its excessive oil content, can be burned as a candle or torch, a practice sometimes used by Indians. The *Hypomesus olidus*, which attains a length up to 12 inches (30 centimeters), is found in the Sacramento River delta (fresh water), whereas its Japanese counterpart is a marine form, frequenting fresh waters only for spawning. The *Osmerus mordax* (American Atlantic smelt) occurs from Virginia northward to the Gulf of St. Lawrence. In the early 1900s, this fish was introduced to the Great Lakes where it has done very well. On the Pacific coast, ranging from southern California northward to Alaska, *Hypomesus pretiosus* (surf smelt) prefers the surf. It is a popular bait. The *Osmerus eperlanus* (European smelt) is well regarded commercially and occurs in numerous European rivers, including the Seine.

SMITH DIAGRAM (or Smith Chart).   A diagram with polar coordinates, developed to aid in the solution of transmission line and waveguide problems. It is composed of the following sets of lines: (1) constant resistance circles; (2) constant reactance circles; (3) circles of constant standing-wave-ratio; (4) radius lines representing constant line-angle loci. The chart employs normalized quantities for maximum flexibility. See Fig. 1.

SMITHSONITE.   Smithsonite is zinc carbonate, $ZnCO_3$, a hexagonal mineral with a rhombohedral cleavage. It is a brittle mineral; hardness, 4–4.5; specific gravity, 4.3–4.5; luster, vitreous to dull; color, usually white, but may be colored yellowish or brownish or perhaps blue or green due to impurities. It is translucent to opaque. Smithsonite is a secondary mineral after sphalerite or may replace limestone or dolomite. It is sometimes called calamine (but true calamine is a zinc silicate) and often associated with it. Smithsonite occurs in Siberia, Greece, Rumania, Austria, Sardinia, Cumberland and Derbyshire, England; New South Wales, Australia; South West Africa, and Mexico. In the United States, it is found in Pennsylvania, Wisconsin, Missouri, Arkansas, and Utah. This mineral was named in honor of James Smithson, whose legacy founded the Smithsonian Institution at Washington, DC.

SMOG AND SMOKE POLLUTION.   See **Pollution (Air)**.

SMOKE TREE.   See **Cashew and Sumac Trees**.

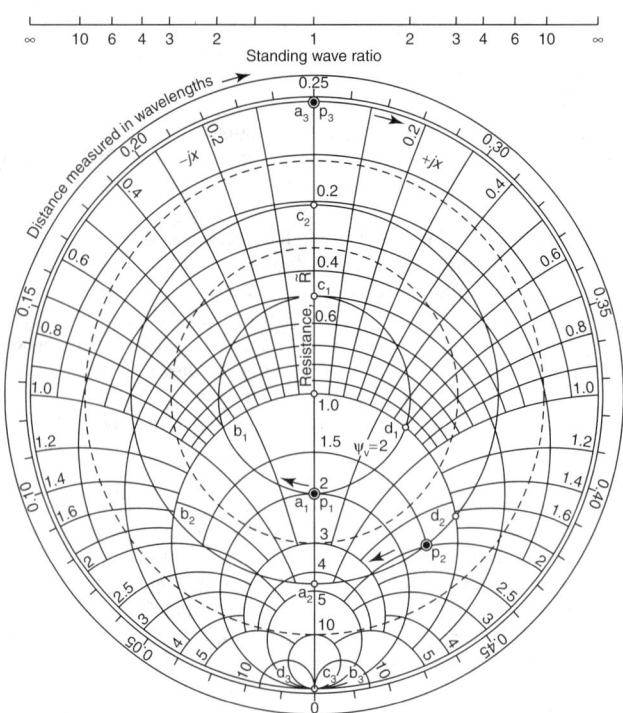

Fig. 1.   Smith diagram showing circles of constant standing wave ratio, each corresponding to a particular terminal impedance as follows: **(a)** The terminal impedance (p_1) is $\tilde{Z}(1) = 2 + j0$. **(b)** The terminal impedance (p_2) is $\tilde{Z}(1) = 1.5 + j2$. **(c)** The terminal impedance (p_3) is $\tilde{Z}(1) = 0 + j0$.

SMOOTHING (Of a Curve).   The replacement of a curve, or of a sequence of points by another that is in some sense more regular, and yet whose ordinates, for any abscissa, are changed as little as possible. The irregularities in a sequence of points may be due to errors in measurement. If theory requires the theoretically correct points to lie on a given curve, one may apply some method of curve fitting, possibly least squares. If not, one may select arbitrarily a simple function, possibly a polynomial, and fit it by least squares. If the purpose is merely to obtain a smooth graph, this may be drawn visually. A somewhat more sophisticated method is to take, say, 5 consecutive points, fit a parabola, and replace the middle point by the one on the parabola. The next parabola requires four of these points and one new one.

See also **Neighborhood of a Point**.

SMUT FUNGI.   See **Fungus**.

SNAIL (*Mollusca, Gasteropoda*).   A name properly applied to most members of this class, but usually given only to the small fresh-water and terrestrial species with coiled shells. The term does not correspond with the scientific classification of mollusks beyond this general application.

Snails are regarded as a food delicacy in many countries of the world. Closely related to the slug, the snail is a garden pest with tastes and habits much the same as those of the slug. See also **Mollusks**; and **Slug**.

SNAIL DARTER.   See **Perches and Darters**.

SNAKE FLY (*Insecta, Neuroptera*).   A peculiar insect found only in the western United States. It has four membranous wings and the body is prolonged at the anterior end. These insects are insect eaters and are found commonly on the bark and foliage of trees.

SNAKES.   According to one view, primitive snakes were specialized burrowers, whereas according to another view, they were surface-dwelling or aquatic boa-like snakes from which specialized burrowers evolved in one line and fast-moving terrestrial and arboreal types in another. Snakes are in the class *Reptilia* (reptiles), the order *Squamata* (scaly reptiles), and the suborder *Serpentes* (snakes). Although there are variations to this classification, this is the classification followed by Grzimek (1971).

About 3000 species and subspecies of snakes range from southern Canada, northern Sweden and the Kamchatka Peninsula southward throughout Africa and most of the Americas, Australia, and temperate and tropical islands except Polynesia (only one tiny species, for example, occurs in Hawaii). There are no snakes in Ireland or New Zealand. Islands have been populated by snakes from adjacent mainland or by subsequent chance. Ireland lacks them because it was separated from continental contact before the last glacier departed, and no introduction has since occurred.

As shown in Table 1, there are 3 infra-orders, 12 families, and nearly 20 subfamilies, but this number has been variously estimated in accordance with the degree of significance attached to specializations of diverse groups. Even with these complications, a few generalizations can be made.

The most primitive families of snakes possess internal vestiges of the skeletal girdle of the hind limbs. In some families, a vestige of the thigh (femur) remains, often visible externally as a small spur on either side of the anal opening. Snakes do not possess movable eyelids. The skin over the eye is in the form of a large scale flush with the rest of the cranial surface in the most primitive snakes. In more advanced forms, the scale fits the eye perfectly and is designated a *brille* or tertiary spectacle. Snakes sleep, but little outwardly visible evidence reveals when they are asleep.

The ears cannot directly receive airborne sounds, since the external ear and middle ear cavity are missing. The middle ear ossicle or columella is present, but is embedded in muscle except where it fits into the *fenestra ovalis* of the capsule of the inner ear. Snakes can detect some ground-borne vibrations, since the inner ear is largely intact. They are deaf to music of snake charmers and to their own noises. The tongue in all snakes is long, slender and forked at the tip. It functions primarily in picking up airborne particles that go into solution on its moist surface. By placing the tips of the tongue into the paired vomeronasal organs at the anterior tip of the roof of the mouth, the snake can detect odors.

Snakes periodically molt the outer epidermal layers of the entire skin, including eye covering and tips of tongue. Frequency varies with species and health, but is not known to be less than once per year. Molting

TABLE 1. CLASSIFICATION OF SNAKES

CLASS   EPTILIA (Reptiles)
ORDER   QUAMATA (Scaly Reptiles)
SUBORDER   ERPENTES (Snakes)

INFRAORDER: BLIND SNAKES (*Scolecophidia*)
FAMILY: Blind Snakes (*Typhlopidae*)
Examples: Peter's blind snake (*Typhlops dinga*); Common blind snake (*T. braminus*)
FAMILY: Slender Blind Snakes (*Leptoryphlopidae*)
Examples: Texas blind snake (*L. dulcis*); Western blind snake (*L. humilis*)

INFRAORDER: PRIMITIVE SNAKES (*Henophidia*)
FAMILY: PIPE SNAKES (*Aniliidae*)
Examples: False coral snake (*Anilus scytale*); Pipe snake (*Cylindrophis rufus*)
FAMILY: SHIELD-TAILED SNAKES (*Uropeltidae*)
FAMILY: SUNBEAM SNAKES (*Xenopeltidae*)
Example: Sunbeam snake (*Xenopeltis unicolor*)
FAMILY: ACROCHORDIDS (*Acrochordidae*)
Examples: Javan wart snakes (*Acrochordus javanicus*); Indian wart snake (*Chersydrus granulatus*)
FAMILY: PYTHONS AND BOAS (*Boidae*)
SUBFAMILY: *Loxoceminae*
Example: Mexican python (*Loxocemus bicolor*)
SUBFAMILY: *Pythoninae*
Examples: Reticulated python (*Python reticulatus*); Indian python (*P. molurus*); Rock python (*P. sebae*); Blood python (*P. curtus*); Timor python (*P. timorensis*); Ball or royal python (*P. regius*); Angola python (*P. anchietae*); Carpet python (*Morelia argus*); Brown python (*Liasis fuscus*); Boa (*Bothrochilus boa*); Black-headed python (*Aspidites* and *Chondropython*); Burrowing python (*Calabaria reinhardtii*)
SUBFAMILY: Boas (*Boinae*)
Examples: Wood snakes (*Tropidophis*); Cuban boa (*Epicrates angulifer*); Rainbow boa (*Epicrates cenchris*); Rubber boa (*Charina bottae*); California boa (*Lichanura roseofusca*); Madagascar boa constrictor (*Acrantophis madagascariensis*); Sand boas (*Eryx*); Brown sand boa (*Sanzinia johnii*); Emerald tree boa (*Corallus caninus*); Boa constrictor (*Boa constrictor*); Anaconda (*Bunectes murinus*); Yellow Anaconda (*Eunectes notaeus*)
SUBFAMILY: *Bolyeriinae* (Round Island Boas)

INFRAORDER: ADVANCED SNAKES (*Xenophidia* or *Caenophidia*)
FAMILY: COLUBRID SNAKES (*Colubridae*)
SUBFAMILY: *Xenoderminae* (Xenodermin Snakes)
SUBFAMILY: *Sibynophinae*
SUBFAMILY: *Xenodontinae*
Examples: Eastern hognosed snake (*Heterodon platyrhinos*); Western hognosed snake (*H. nasicus*)
SUBFAMILY: *Natricinae*
Examples: Ringed snake (*Natrix natrix*); Diced snake (*N. tessellata*); Common water snake (*N. sipedon*); Diamond-back water snake (*N. rhombifera*); Brown water snake (*N. taxispilota*); Glossy water snake (*Regina rigida*); Queen snake (*R. septemvittata*); Kirtland's water snake (*Clonophis kirtlandii*); Common garter snake (*Thamnophis sirtalis*); Mexican garter snake (*T. eques*)
SUBFAMILY: *Colubrinae*
Examples: King snakes (*Lampropeltis* spp.); Milk snake (*L. triangulum*); Common king snake (*L. getulus*); Green snakes (*Opheodrys* spp.); Racer snakes (*Coluber* spp.); Eastern coast coachwhip (*Masticophis flagellum*); Speckled racers (*Drymobius,* spp.); Patch-nosed snakes (*Salvadora* spp.); Leaf-nosed snakes (*Phyllorhynchus,* spp.); Rat snakes (*Elaphe* spp.); Rat snake (*E. obsoleta*); Fox snake (*E. vulpina*); False water cobra (*Hydrodynastes gigas*); Indian rat snake (*Ptyas korros*); Keeled rat snake (*P. carinatus*)
SUBFAMILY: *Calamarinae* (Dwarf Snakes)
Example: Linne's dwarf snake (*Calamaria linnaei*)
SUBFAMILY: *Lycodontinae* (Lycodontine Snakes)
Examples: Indian wolf snake (*Lycodon aulicus*); African wolf snake (*L. capense*); Cape file snake (*Mehelya capensis*); Brown house snake (*Boaedon fuliginosus*); Rainbow snake (*Farancia erythrogrammus*); Mud snake (*F. abacura*)
SUBFAMILY: *Dipsadinae*
SUBFAMILY: *Pareinae*
SUBFAMILY: *Dasypeltinae* (Egg-eating Snakes)
Examples: African egg-eating snake (*Dasypeltis scabra*); Indian egg-eating snake (*Elachistodon westermanni*)
SUBFAMILY: *Homalopsinae* (Homalopsine Colubrid Snakes)
Examples: White-bellied water snake (*Fordonia leucobalia*); Tentacled snake (*Erpeton tentaculatum*)
SUBFAMILY: *Boiginae* (Boigine Snakes or Boigine Vipers)
Examples: Tree snakes (*Boiga,* spp.); Mangrove snake (*B. dendrophila*); Twig snake (*Thelotornis kirtlandii*); Green vine snake (*Oxybelis fulgidus*); Flying snakes (*Chrysopelea* spp.)

(continued)

TABLE 1. (*Continued*)

FAMILY: COBRAS (*Elapidae*)
  Examples: King cobra (*Ophiophagus hannah*); True cobras (*Naja* spp.); Asian or Indian cobra (*N. naja*); Egyptian cobra (*N. haje*); Spitting cobra
    (*N. nigricollis*); Black-lipped cobra (*N. melanoleuca*); Cape cobra (*N. nivea*); Gold's tree cobra (*Pseudohaje goldii*); Desert black snakes (*Walterinnesia*
    spp.); Shield-nose snakes (*Aspidelaps* spp.); Water cobras or water snakes (*Boulengerina* spp.); Black mamba (*Dendroaspis polylepis*); Mamba (*D.
    angusticeps*); Kraits (*Bungarus* spp.); Oriental coral snakes (*Calliophis* spp.); Long-glanded coral snakes (*Maticora* spp.); American coral snakes
    (*Micurus* spp.); Southern coral snake (*M. frontalis*); Eastern coral snake or Harlequin snake (*M. fulvius*); Western coral snake (*Micruroides* spp.); Arizona
    coral snake (*M. euryxanthus*); Death adder (*Acanthophis antarcticus*); Australian tiger snakes (*Notechis* spp.); Australian copperhead (*Denisonia* spp.)
FAMILY: SEA SNAKES (*Hydrophiidae*)
  SUBFAMILY: *Laticaudinae*
    Examples: Black-banded sea krait (*Laticauda laticauda*); Olive-brown sea snake (*Aipysurus laevis*)
  SUBFAMILY: *Hydrophiinae*
    Examples: Yellow sea snake (*Hydrophis spiralis*); Beaked sea snake (*Enhydrina schistosa*)
FAMILY: VIPERS (*Viperidae*)
  Examples: True vipers (*Viptera* spp.): Adder or common European viper (*V. berus*); Asp viper or European asp (*V. aspis*); Sand viper (*V. ammodytes*);
    Mountain viper (*V. sznthia*); Saw-scaled viper (*Echis carinatus*); Horned viper (*Cerastes cerastes*); Puff adders (*Bitis* spp.); Rhinoceros viper
    (*B. nasicornis*); Dwarf puff adder (*B. peringueyi*); Bush vipers or tree vipers (*Atractaspis* spp.); Common burrowing viper (*A. irregularis*); Night adders
    (*Causus* spp.); Common night adder (*C. rhombeatus*)
FAMILY: PIT VIPERS (*Crotalidae*)
  Examples: Rattlesnakes (*Crotalus* spp.); Santa Catalina rattlesnake (*C. catalinensis*); Eastern diamond-back rattlesnake (*C. adamanteus*); Western
    diamond-back rattlesnake (*C. atrox*); Prairie rattlesnake (*C. viridis*); Timber rattlesnake (*C. horridus*); Green rattlesnake (*C. lepidus*); Sidewinder
    (*C. cerates*); Tropical rattlesnake (*C. durissus*); Pygmy rattlesnakes (*Sistrurus* spp.); Lance-head snakes (*Bothrops* spp.); Asian lance-head snakes
    (*Trimeresurus* spp.); Bushmaster (*Lachesis mutus*); Haly's viper (*Agkistrodon halys*); Himalayan viper (*A. himalayanus*); Chinese copperhead (*A. acutus*);
    Malaysian moccasin (*A. rhodostoma*); Cottonmouth or water moccasin (*A. piscivorus*); Copperhead (*A. contortrix*); Tropical moccasin (*A. bilineatus*)

Note: Some of the better known as well as other species selected at random are included in the various examples given. The examples represent only some of the species of snakes.

commonly occurs three times per year, but more frequently if the snake has been injured. The molt usually proceeds toward the posterior, starting at the snout and lower jaw where the skin is rubbed backward over the head, turning it wrongside out and leaving it usually in one piece. In preparation for the molt, the basal layer of epidermis rapidly proliferates, producing a completely new outer epidermis, with cuboidal and squamous cells. Breakdown of the basal cuboidal cells of the older layer forms a milky fluid, giving a dull cast to the snake as a whole and a milky appearance to the eye. At this time snakes can see but dimly, if at all, and tend to remain inactive and hidden from view. They are more likely during this period to be defensive if disturbed. The fluid disappears by evaporation or absorption after about three days, and about two days thereafter the slough is shed.

Many snakes lay eggs. Others bear their young live. Parental care is generally absent. The eggs may be abandoned after deposition in damp sand, soil, sawdust, ground litter, or rotting logs. Eggs hatch in about three days to three months. Some viviparous types possess a placenta. As pointed out later, some species, such as cobras, pay more attention to their eggs than other species.

Like other reptiles, snakes are ectothermic, their body temperatures being controlled largely by behavior and to only a slight degree by intrinsic mechanisms.

Males court by repeatedly rubbing chin and body along length of the female's body. In rare circumstances the two animals intertwine the bodies and rise straight upward until only tail and rear part of the body remain on the ground. Ordinarily the latter behavior is a "combat dance" between males. Snakes, like lizards, possess two copulatory organs (hemipenes). Copulation occurs with either and with both structures, but not with both at once, and persists for long periods, from one-half to several hours.

Snakes have numerous well-developed salivary glands, the secretions of which have been shown, experimentally, to be powerfully toxic in almost all species. Yet the bites of most snakes produce no signs or symptoms of poisoning because of the absence of venom-conducting fangs. So-called nonpoisonous snakes have only "solid" teeth, lacking a groove down one side. All "poisonous" species possess one to four pairs of fangs. Fangs are teeth specialized for injection of venom, having either a tube-like structure acting as a needle, or a groove down one side which conducts venom like a capillary tube when it is imbedded in a victim's flesh.

Snakes have teeth, usually all conical, slender, curved, of about the same size and shape. They are found on five paired jaw bones: the maxilla, palatine, pterygoid, dentary, and rarely the premaxilla. They vary in total number from about 20 to 200. Three sorts of fang modification have evolved. Snakes possess venoms containing numerous protein enzymes capable of promoting various types of tissue degeneration. Snakes feed upon animals, never plant material. They often scavenge. Snakes usually swallow the food head first. The trachea can be extended forward below the food far enough to enable the snake to breathe while swallowing.

Four modes of locomotion are concertina, sidewinding, rectilinear, and horizontal undulation.

Blind Snakes, Primitive Snakes, and Wart Snakes

The infraorder of blind snakes (*Scolecophidia*) contains two snake families whose members burrow in the ground: (1) Blind snakes (*Typhlopidae*), and (2) slender blind snakes (*Leptotyphlopidae*). These two families bear such a remarkable resemblance that it may appear as if they are very closely related and belong to the same family. However, differences in skull structure and skeletal elements cause them to be divided into distinct families. These are very primitive snakes. They have small, poorly differentiated scales that remain small on the belly and do not form the large scutes or ridges seen on more advanced snake species. And, they have vestiges of upper hind leg bones and the pelvic girdle. Some zoologists believe that blind snakes are actually closer to some lizards than to other snakes.

Both blind snake families are characterized by a subterranean mode of life. These snakes have a smooth, round body that makes them look like big worms, and a blunt head that merges smoothly with the rest of the body. The tail is short and often has a spiny structure whose function has not been clarified. Perhaps the spine helps anchor the snake in the ground so the body can move more easily. Some species use the spine as a weapon. The eyes are tiny and are covered by large head scales, but they gleam through these scales. The eyes are probably used to distinguish light and dark. The skull is used a great deal as blind snakes burrow beneath the ground, and it is shortened and forms a hard capsule.

The *common blind snake* (*Typhlops braminus*) has a habit of finding its way into flower pots that are left in the open for long periods of time. This species crawls about among the roots and feeds on the insect larvae and other creatures that may be dwelling there. This snake occurs in Madagascar, India, Sri Lanka, Southeast Asia, and the Indo-Australian islands, but is also found in southern Mexico. Other blind snake genera occur only in tropical South America.

The *Texas blind snake* (*Leptotyphlops dulcis*) and the western blind snake (*L. humilis*) occur as far north as the sandy, dry prairies of the southern United States, where they burrow in the ground, beneath rocks, or inside fallen trees. They often crawl about on the ground in the early evening hours, but they disappear again after dark. Their diet consists of insect larvae, ants, and termites. They usually lay a few elongated eggs.

The primitive snakes (*Henophidia*) comprise five primitive, nonpoisonous families, three of which have just a few species. Pipe snakes (*Aniliidae*) are three genera from South America and southern and southeastern Asia. They have vestiges of the pelvic girdle and the hind limbs. The hind limb remnants look like small claws on each side of the anal opening. In contrast to blind snakes, the head scales on pipe snakes have not fused much. The eyes are small, but do not lie beneath scales. The scales on the belly are clearly larger than those on the back, although they are not the size of those found in advanced snakes.

The *false coral snake (Anilius scytale)* has a length of some 75–85 centimeters (25–34 inches) and lives in South America. It is brilliant red and has black rings on its back, giving it an appearance not unlike poisonous coral snakes. It lacks the yellow bands. This snake bears live young. Diet consists of lizards and snakes.

Shield-tailed snakes (Uropeltidae) are called that because of the greatly enlarged, modified scale found near the tip of the tail. The function of this structure has not been clarified, but it possibly serves to anchor the animal to the ground. Shield-tailed snakes are thought by some to burrow by pushing the tail tip forward and turning it from side to side. They lack externally visible hind limb vestiges. Shield-tailed snakes give birth to 3 to 8 live young at a time. The diet consists of earthworms and insect larvae. Most species do not exceed 30 centimeters (12 inches) in length, but the *Rhinophis oxyrhynchus* achieve a length of about 60 centimeters (24 inches).

Sunbeam snakes (Xenopeltidae) are distributed across India and southeastern Asia. The toothed premaxillary bone, rudimentary pelvic girdle, and hind limb vestiges are all primitive characteristics. The sunbeam snake *(Xenopeltis unicolor)* achieves a length of about 1 meter (3.2 feet). The snake has a round body, and its dorsal scales shimmer with a rainbow effect. When a sunbeam snake is excited, it twitches its tail vigorously, with much the same action as in rattlesnakes. Sunbeam snakes lack rattles. Their prey consists of small snakes, frogs, small rodents, and birds.

Wart snakes (Arochordidae) are distributed from the coastal regions of India and Sri Lanka across the Indo-Australian islands as far as the Solomons. Unlike most other snakes, the scales of wart snakes do not overlap, but are set adjacent to each other (an arrangement often found in lizards) and bear a sharp ridge. The scale characteristics give the skin a rough, warty appearance. Wart snakes live in the brackish zone of rivers, but sometimes swim short distances in the sea. They are sluggish, slow-swimming snakes that can often stay underwater on the bottom for long periods without breathing. The nasal openings are on the upper side of the snout, enabling them to breathe by simply protruding this part out of the water. Wart snakes only rarely come onto land, where they crawl about clumsily.

Boids (*Boidea*)

These snakes arose toward the end of the Cretaceous period as species closely related to monitor-like lizards. Boids comprise not only the largest snakes alive today, but also small and medium-size species. The top of the head has small scales or large, symmetrical plates, and the back has small scales. The belly is covered with larger plates. The pupils are vertical. There are paired lungs. All three parts of the pelvic vestiges are present, and the skull bones are joined by flexible linkages. Prey is killed by constriction.

There are only four species in the entire boid family that pose a danger for people—the rock python, the Indian python, the reticulate python, and the anaconda. See Fig. 1. Boids are the only *large* animals on earth that are mute. At most, they can utter a hiss. They are also deaf and cannot perceive

Fig. 1.   Regal python. (*A.M. Winchester.*)

air vibrations. They can sharply perceive movements in the ground or on the substrate upon which they rest. In recent years, it has been shown that some snakes can perceive loud noises with the sensitive tips of the tongue.

As in all snakes, the eyes of boids lack lids and are covered by the epidermis. The skin is shed along with the rest at molting time, and a new epidermal covering develops. The membrane protects the eye. The eyes of snakes are less mobile and less elastic than avian and mammalian eyes, and therefore their accommodation capability is less. Boids have fairly keen vision, however. They can distinguish outlines well at short distances and can recognize immobile objects in the vicinity and climb about on them. Prey and conspecifics are distinguished primarily by the way they move and from their odor.

During the day, the boid pupils constrict and form vertical slits, while at dusk and at night they widen and look like the eyes of a cat. The firm covering offers effective protection for the eyes when the snake crawls through thickets and holes, and underwater, where they enable the snake to see. The protective membrane gives the impression that the snake is constantly staring. The eyes are only useful for perceiving prey at close range and for the perception of movement; boids have another organ that helps them detect prey nonvisually. Small rectangular openings in the scales on the lower and upper lips (the openings being analogous to the pit organ of pit vipers) enable the snake to perceive even faint warmth radiations. They can detect a human hand at a few dozen centimeters with this heat organ. The organ enables boids to find warm-blooded prey hiding in concealed sites. The finest of all senses, however, is olfaction (smelling), which arises from the Jacobson's organ in the roof of the mouth. The tongue leads small particles in the air to this organ. The heat organ and Jacobson's organ make snakes independent of light conditions, for regardless of bright or dark, they can pursue their prey and often overtake it.

Boids move not only by means of swinging movements in the horizontal plane, but also by stretching the front of the body forward and pulling the rest behind (earthworm movement). A boid on a smooth surface can leave a straight track rather than a winding one. Boids move even better in the water than on land. They swim with surprising agility, diving and surfacing. When swimming, they make use of their ability to pump air inside the body, preventing it from sinking. If the snake wants to dive it releases some of the air. Some boids (not just the purely aquatic South American anaconda) cover considerable distances in water, even undertaking extensive trips into the open sea, sometimes floating along with drifting tree stumps.

One boa constrictor is reported to have covered the distance from the South American mainland to Saint Vincent Island, a distance of 320 kilometers. When the volcano Krakatoa erupted in 1888, all life was destroyed on that Indonesian island. Biologists then recorded the return to the island of various plants and animals in the following decades. Reticulate pythons were one of the first reptiles to reappear there, arriving in 1908.

A great deal has been fancifully written about the strike of a python (for example, Kipling's descriptions in the *Jungle Book*). Snake breeders know that pythons never intentionally strike an object with their head. Nevertheless, they are very strong animals. The entire body, from head to tip of tail, is a single muscle package, which enables boids to gain a powerful hold on their prey. A strong man can probably defend himself against a 4-meter (13-foot) long python or anaconda, but he could not deal alone with one of these weighing 50 kilograms (110 pounds) or more. The boid seizes its prey or an enemy by darting quickly toward the animal and seizing it with its long, sharp teeth. Then the snake wraps around the prey two or more times, at the same time constricting its muscles. Boids do not inflict killing wounds with the teeth. They strangle their prey or kill it by causing vital blood vessels to burst.

Boids must first gain a hold with their teeth if they are to deal effectively with prey, so their teeth are of utmost importance. A reticulate python has about 100 very sharp teeth that point backward. If the snake grabs even a finger, the finger can not be pulled out. One would have to force open the snake's jaw and push the finger further in before it could be released from the teeth. Boids kept in zoos become fairly tame quickly. Even pythons caught in the wild defend themselves by biting and almost never by attempting to wrap themselves around the persons capturing them. In handling newly arrived boids at zoos, one man must be assigned for each meter (3.2 feet) of length of the snake. Each man must be able to grab the snake firmly and not let go.

Although boids become tame in a short time, snake charmers often literally cool their animal down before doing a performance. The dulled snake lets the charmer do about anything when it is cool and it does not regain its normal activity level until after the display.

It was formerly believed that boids cover their prey with saliva prior to swallowing them, but it is now known that this does not occur until the prey is in the jaws and inside the snake's body. A snake can regurgitate its prey if the snake is startled, but this is uncommon. However, a goat or antelope that is swallowed can cause so much discomfort with its horns that the snake may rid itself of the animal.

Although large snakes can swallow huge amounts at a single time, they actually eat modest amounts and require just a little more than their own weight in food per year. After swallowing large prey, a boid may not eat at all for many weeks and may fast for several months without losing any significant body weight.

Gases forming in the snake's stomach cause the body to become even more inflated several days after eating prey, and the gases sometimes cause the snake considerable discomfort. The swallowing capacity of boids has its limits, and reports about them swallowing horses, cows, and other large animals are considered fantasy.

Large boids reach sexual maturity in three years (in captivity). Males recognize females by the anal secretions of the female. Copulation may last for several hours. While all pythons lay eggs, all boa species are live-bearers. Boa ovoviviparity is to some extent believed to be the result of a more advanced stage of adaptation. Female pythons, which coil around their eggs and may incubate them two or more months, are in greater potential danger than female boas, which can move away right after they give birth. The young boas, which are able to crawl into all sorts of recesses immediately after birth, are also less vulnerable than eggs. The eggs that female boas develop within their bodies, are covered by a membrane instead of a shell.

Tree Boas (Corallus). The *emerald tree boa* (*C. caninus*), which attains a length up to 2 meters (6.4 feet) is from northern South America and Brazil and is one of the most beautiful boas of all. Its brilliant green coloration with whitish or yellowish bands offers superb camouflage when it crawls about in trees or shrubs as it hunts birds or lizards, or when simply resting in a tree. The body is compressed laterally, enabling the snake to press close to the tree branches. On the ground it is not a particularly agile snake. In the resting position, the prehensile tail often grasps a branch as the body is wound around the tail, with the coils of the body equally divided on both sides of the point where the tail is grasping. The long, powerful teeth enable tree boas to catch birds, which they snap at and then hold onto with these teeth.

The *garden tree boa* (*Corallus enhydris*) attains a length of up to 2.5 meters (8.2 feet) and is distributed in the Central and South American tropics, as well as the Antilles. The basic coloration may be brown, ochre, or gray, and sometimes it has a striking pattern. The garden tree boa uses an accordion-like climbing technique. The snake wraps the front of its body around a thin part of the tree trunk and then pulls itself higher, only to re-anchor the body again and climb a bit more.

Boa Constrictor (Boa constrictor). This snake usually will reach a maximum length of 4 to 4.5 meters (13.1 to 14.8 feet) and will weigh up to 60 kilograms (132 pounds). This snake is probably the most familiar of the boids, although by no means the largest. It is distributed from Mexico to northern Argentina. Three closely related species occur on the West Indies islands and the plains of western Argentina. Boa constrictors, whose pattern varies from one individual to the next, are among the most beautiful of snakes. They are ground dwellers and are particularly prevalent in mountain forests. These snakes live chiefly on small mammals, birds, tegus, and iguanas.

Anacondas (Eunectes). These boas are more aquatic than those just described. The *anaconda* (*E. murinus*) can attain a length of 9 to 9.6 meters (29.5 to 31.5 feet) and the large ones are the longest of reptiles in the world today. The *yellow anaconda* (*E. notaeus*) is considerably smaller.

Contrary to popular perceptions, most boids (excepting sand boas and a few others) cannot tolerate extremely severe heat. They prefer temperatures of 20–30 °C. Many species hibernate or estivate and can survive unfavorable seasons in a dormant state.

Colubrid Snakes

The colubrid family (*Colubridae*) contains more species than any other snake family. The relationship between the presence or absence of grooves and the individual species is evident when one inspects the venom gland, a structure that is necessarily found in every venomous snake. Venom glands probably developed through a process of adaptation from upper lip and salivary glands. The venom glands consist of the upper lip gland (which only produces mucus) and a poison-transporting venom gland. As early as 1894, investigators demonstrated that the secretion of a single venom gland from a ringed snake is potent enough to kill a guinea pig. The mucous gland lubricates the prey; the venom gland helps prepare it for digestion.

The number of species of colubrid snakes is extensive, as is obvious from the table, and only a few of the more interesting species can be highlighted here.

Ringed Snake (Natrix natrix). These snakes reach a length of about 1 meter (3.2 feet) on the average, but some species are longer. Ringed snakes are distributed across a large area, including most of Europe and a bit of northern Africa. The snake is named for the two fairly distinct orange, yellow, or whitish moonshaped spots on the rear of the head, which look something like a ring. It is difficult today to find a ringed snake more than 2 meters (6.5 feet) long. The longer ones are usually more cautious and therefore more difficult to catch. Due to their large size, they are forced to traverse a greater area and to visit many ponds in order to find sufficient nutrition. Females wander about seeking favorable egg-laying sites (often in dung heaps). Their enemies include cats, hedgehogs, raptorial birds, and humans. Many are killed on modern highways.

When ringed snakes are threatened, they usually respond with a series of stereotyped defensive behavior patterns — flicking the tongue and hissing; releasing the content of their stink gland; defecating; or regurgitating food. Only rarely do they actually bite, and a ringed snake bite has no harmful effects for humans.

Diced Snake (Natrix tessellata). These snakes are found in central Europe and the Near East. They can achieve a length of between 75 and 100 centimeters (29.5 and 39 inches), but eastern species may be as long as 150 centimeters (59 inches). The diced snake is rather slender with a narrow head. The eyes are medium in size, are situated slightly high, and have round pupils. The top of the neck often has a dark V-shaped spot. The underside is whitish or yellowish, or it may have reddish and black spots. Some authorities believe that the diced snake is a relic from the warmer postglacial period. It is even more dependent on water than the ringed snake. It consumes a larger amount of fishes and often lies for hours or days underwater near the shore on low rocks, under which it flees when people approach. The eyes of the diced snake are well adapted to a highly aquatic life. While ringed snakes can be found considerable distances from water, the diced snake occurs in the immediate vicinity of water.

Viperine Snake (Natrix maura). This species achieves a length of over 80 centimeters (31.5 inches). Distribution is similar to that of the ringed and diced snakes. The viperine name arises from the zigzag band extending along its back, giving it a similar appearance to the common viper. The snake is harmless. Viperine snakes can be distinguished from vipers by the round pupils of the viperines. The stripe on the back of this snake sometimes appears as two rows of spots, but usually forms the aforementioned zigzag pattern. Dark spots or rings run along both sides of the central stripe, and they enclose a white or yellowish spot. Viperine snakes feed on fishes and amphibians.

Water Snakes (Natrix spp.). There are many water snakes in North America, mainly in the United States. The *diamond-back water snake* (*N. rhombifera*) is about 1.4 meters (4.6 feet) long when fully grown. The back is brown or olive-brown, with a chain of rhomboid figures. The belly is yellow and has halfmoon-shaped brown spots. Distribution is from the midwestern United States southward into Mexico. The *plain-bellied water snake* (*N. erythrogaster*) also reaches a length of about 1.4 meters (4.6 feet). The upper side is uniform black to redbrown and the underside is reddish. This snake occurs in the southeastern United States. The *queen snake* (*Regina septemvittata*) attains a length of about 50 centimeters (19.6 inches) and is similar to the plain-bellied water snake except it is more slender. This species occurs from Pennsylvania to Wisconsin. The *brown water snake* (*Natrix taxispilota*) on the average has a length up to 1.3 to 1.5 meters (4.3 to 4.9 feet) and is the largest North American water snake. It is heavy and plump and has a clearly distinguishable head. It is brown to rust-brown, with large rectangular spots. This species occurs in the southern United States. The *Brazos water snake* (*N. harteri*), reaching a length of about 90 centimeters (35.4 inches) and is known only in the Palo Pinto region on the upper Brazos River in Texas. *Kirtland's water snake* (*Clonophis kirtlandii*) reaches a length of about 45 centimeters

(17.7 inches) and is a light-brown to gray snake with roundish black spots. Distribution is from the midwestern United States to Pennsylvania and New Jersey. *Graham's water snake (Regina grahami)* attains a length up to 60 centimeters (23.6 inches). It is dark brown with a broad yellow band on the first three rows of scales, bordered below by a narrow dark stripe. A pale stripe lined on both sides with black extends across the back, and the belly is yellowish with dark splotches. This snake occurs from Illinois to Louisiana and eastern Texas.

Water snakes generally live on fishes, anurans, and urodele amphibitans, crustaceans, and insects. Mating occurs in April and the young are born live between August and early October. Newborn common water snakes are about 22 centimeters (8.7 inches) long. With their plump body, dependence on water, and vigorous defensive behavior, the common water snake may be confused with water moccasins.

Garter Snake (Thamnophis). These snakes achieve a length of 50 to 150 centimeters (19.7 to 59 inches) and are the most prevalent colubrid snakes in North America. They are small, slender, and often considered "cute." There are several species. These ground-dwelling colubrid snakes are found throughout the 48 contiguous United States, as well as in southern Canada and northern Mexico. They are sometimes observed in urban areas. They differ from water snakes in having an undivided anal scute, and many species have three light stripes. All garter snakes are live-bearers. Some feed chiefly on earthworms, while others living near water concentrate on frogs.

Smooth Snakes (Coronella). These snakes have a small head, which is distinct from the neck. The medium-size eyes have round pupils. The scales are smooth and are arranged in rows. The group includes small to medium-size species of ground-dwelling, nonvenomous colubrid snakes distributed in Europe and Asia. Smooth snakes occur mostly in woods, inhabiting clearings, paths, forests, and particularly piles of wood and of rocks and shrub-covered ground, where they can sun themselves. That same habitat supports sand lizards, common lizards, and blind snakes and mice on which they feed. Like king snakes, smooth snakes also feed on other snakes.

Common King Snake (Lampropeltis getulus). These snakes attain a length up to 2 meters (6.6 feet). They prey upon other snakes, even poisonous ones. Their diet also includes small mammals, lizards, and fishes. When a king snake seizes another snake, it wraps its body firmly around the other animal and chokes it. Interestingly, rattlesnakes do not assume their typical position when they are brought face to face with a king snake. Instead, they press the front of the body against the ground and try to hit the king snake with another part of the body. The king snake has a light chain pattern on the back. Distribution is in the southern United States and northern Mexico.

Dwarf Snakes (subfamily Calamarinae). These small snakes are only from 25 to 30 centimeters (9.9 to 11.8 inches) long. The head is small and passes into the round, firm trunk without any noticeable neck section. There are some seven genera with approximately eight species, all from the Southeast Asian tropics, chiefly Indonesia and the Philippines. They feed primarily on earthworms and insects. They spend all their lives on the ground, concealing themselves in the soft earth beneath fallen tree stumps and rocks. Defenseless and slow, they are often eaten by other snakes.

Flying Snakes (Chrysopela). These snakes do not fly, but they drop from one branch to a lower one so quickly that it appears as though they are gliding. They attain a length up to 1.5 meters (4.9 feet). They have a longitudinal ridge on each side of the belly. The *C. ornata* has brilliant coloration with a complex pattern. These snakes are found in Indonesia and southeastern Asia.

Cobras and Sea Snakes

The cobra family contains some of the most greatly feared snakes, about which so much has been written: cobras, which perform before snake charmers; the legendary black mamba; and the colorful, red-yellow-and-black ringed American coral snakes. Two members of the family are usually thought of as being among the most venomous snakes in the world—the Southeast Asian king cobra and the Australian taipan. Cobra species occur in every continent except Europe. The family is particularly diverse in Australia, with about 75% of all Australian snakes being in the cobra family.

There are four dangerous venomous snake families known: cobras, sea snakes, vipers, and pit vipers. These four families form two major groups.

The fangs are always in front of the mouth, but in cobras and sea snakes, they have a groove, which indicates how the originally open venom canal developed over many centuries of adaptation.

The *Elapidae*, like other venomous snakes, arose from nonvenomous snakes. Their anatomy is much more like that of the colubrids than is that of the plump, short-tailed vipers. Also, in cobras the fangs are firm, relatively short structures, while in vipers the short, vertical upper jawbone and fangs are rotated back against the palate when the mouth is closed. Thirdly, cobra fangs have a closed groove, but it has a deep indentation. Finally, cobras as a rule produce neurotoxins in their venom, while viper venom generally has a hematoxic (blood poisoning) action.

Elapids are generally slender, highly agile snakes with a colubrid-like head that is not very distinct from the neck, and which bears large, colubrid-like scutes. The fangs are short, so the mouth does not have to open very wide for them to be useful. There are no other upper teeth in the highly developed species (mambas and coral snakes), while others have one or more small, ungrooved teeth behind the fangs in the upper jaw. Additional teeth are found on the palate, the pterygoid bone, and the lower jaw.

The length of these snakes varies from 30 centimeters (11.8 inches) in the bandy-bandy to over 5 meters (16.4 feet) in the king cobra. The body often has stripes that may be very colorful. Many cobras can flatten their body when they are excited, and *Naja* cobras can spread their neck ribs, forming the familiar hood. Cobras and sea snakes share a similar fang anatomy, but they differ in that cobras have a round tail. Their chief prey are small vertebrates. Several species prefer to feed on other snakes. They are ovoviviparous (live bearing, particularly in the Australian species) or oviparous (egglayers). Elapids do not constrict their prey.

The *Elapidae* contains more venomous snakes than does any other snake family. Most of them are ground-dwellers, often secreting themselves in rodent dens or burrowing in loose earth (coral snakes). A few are arboreal (mambas, cobras, and Gold's cobra), and one genus (winged water cobra) inhabits standing and slowly flowing water. Most elapid snakes are active at dusk and night and they avoid bright sunlight. There are 41 genera with approximately 180 species and 300 subspecies.

King Cobra (Ophiophagus hannah). This is the largest poisonous snake in the world. Specimens as long as 4 meters (13.1 feet) are not unusual, and the current record is 5.58 meters (18.3 feet). Whether the species is also the most dangerous poisonous snake is still being debated. The very large venom glands do produce a huge quantity of venom, and fatalities from king cobra bites have been recorded in Burma, India, and China. However, one should regard phrases, such as "the most dangerous venomous snake" with some skepticism. The effect of any individual bite on a human will vary with the conditions under which the incident occurs. The metabolic reactions of the person involved (death from shock has occurred, but not in every case) and the degree to which the snake is excited (usually exaggerated in reports) both play a role in the effect of a bite. It is not correct to measure potency of a snake's venom by using the number of people it has bitten. Actually, the most toxic snakes bite the least often (H.G. Petzold, 1972). Experimental studies have revealed that coral snake venom is much stronger than pit viper venom, but many more deaths are caused by pit vipers than by coral snakes because of the life habits and the prevalence of the species. See Fig. 2.

King cobras are much less aggressive than many other, smaller snakes, although one might not guess this when seeing a king cobra's hooded threat display. This display usually has a powerful effect on humans. The king cobra inhabits India, but not the nearby island of Sri Lanka. Distribution extends into southern China as far as Shanghai, Malaysia, the Sunda Islands to Bali, the Andaman Islands, and the Philippines. The head of the king cobra can grow to be near the size of a human head and coloration varies with habitat from olive-brown or gray to a deep, shining black. Forty to fifty narrow, irregular light stripes extend across the back. The eyes have a bronze hue. The two fangs are never more than 10 centimeters (4 inches) long, and there are usually three smaller upper teeth behind each of them.

The snake generally leads a secretive life. However, it feeds by day as well as by night. It prefers dense highland jungle, often near water, and in the Nilgiri Mountains of India it occurs as high as 2000 meters (6562 feet). King cobras often flee into water when they are pursued. Unlike most other venomous snakes, the king cobra is very agile and excitable. Its threat posture is almost universally known; the front of the body is raised and the loose neck skin is stretched wide as the movable neck ribs spread out. The king cobra's hood is not the disk-like rounded form of the smaller Asian cobra (also known as the Indian cobra), and the king cobra lacks

Fig. 2.   Cobra. (*A.M. Winchester.*)

the eyeglass markings. An upright king cobra can sway its forebody for several minutes, but it never does so in the swanlike manner of *Naja* cobras. However, only cobra species can move forward while it is in the threat posture. Hardly anyone would easily dismiss an encounter with a threatening king cobra, which not only assumes that characteristic posture, but also emits a high, penetrating hiss through the nasal openings.

King cobras are especially feared during the mating season, and in India, paths and streets may be closed off by officials if a cobra nest is discovered nearby. This leads to a fascinating aspect of king cobras. The eggs are incubated. The characteristic "two-story" nest, made of foliage, was first described early in the nineteenth century. The first "story" contains the 18 to 40 white eggs; the female coils her body around them, while the male is usually stationed near the nest. Until recent years, it was believed that the snakes randomly gathered twigs that were blown about by the wind. Definitive observations were made at the Bronx Zoo (New York) of a breeding pair of king cobras. The 4-meter (13.1-foot) female grabbed bamboo shoots, branches, and foliage by slinging her forebody around them, and in two days, she had constructed a nest nearly 1 meter (3.2 feet) wide. In their Indian habitat, newborn king cobras are about 0.5 meter (1.5 feet) long and are deep black with yellowish-white stripes. Later the coloration becomes lighter and the contrast between the various hues diminishes.

King cobra venom is extremely potent. A human can die within 15 minutes after being bitten unless treated with the proper antivenin. The symptoms for king cobra poisoning are typical for the neurotoxic action of elapid venom. Victims suffer from vision disabilities and severe dizziness. This is followed by a slowing down of the reflexes and respiratory difficulties. Work elephants are sometimes killed by king cobras in India and Thailand if the snake manages to bite a tender spot, such as the tip of the trunk or the area where the toenail meets the skin. An elephant bitten by a king cobra usually will die within 3 to 4 hours.

Not every king cobra strikes at humans without provocation. In terms of the "art" of snake charming, few king cobras are used in such performances, and those that are used are generally limited to religious performances, as in Burma. The Egyptian cobra and the Asian or Indian cobra are the more typical performing snakes. Snake charming is an ancient art; the snake that Aaron charmed before the pharaoh was an Egyptian cobra. Although many charmers work with artificially cooled-down snakes or fangless cobras, masters of the art use normal cobras. Snakes are deaf and do not dance in their baskets to the tune of their charmer's flute. With their eyes and forebody, they follow the movements of the tip of the flute and the swaying body of their master, who merely needs to move with the rhythm of his music to make it appear that the snake is dancing. The charmer knows the striking distance of the cobra as well as any professional herpetologist and adjusts himself to the snake accordingly.

True Cobras (Naja). These snakes are less surrounded by myths, but are more prevalent. The *Asian* or *Indian cobra* (*Naja naja*) is the most

prevalent and occurs over a large region from central Asia through India and southern China to the Sunda Islands and the Philippines. Coloration varies more within this species than in any other; there are light-brown, brown, olive-gray, and completely black common cobras, which may have lighter stripes. Asian cobras generally reach a length of 1.5 meters (4.9 feet), with the largest ones being found on Sri Lanka.

They inhabit a number of places — from the thickest jungle to open rice paddies, city parks and gardens, sheds, bazaars, and village streets. The proportion of the body that is raised when a cobra threatens depends upon the degree that the animal is excited and varies from one-fifth to one-third of the total body length. Raising of the hood is a sign of threat and defense and, together with raising the body, shows that the snake is prepared to attack. A threatening cobra does not bite upward; its strike is not directed higher than the mouth is at the time the snake is in the upright position.

Cobras are visually oriented animals; they show little response to mechanical manipulation.

The first attempts to develop immunology and serum therapy for snake bite occurred as early as 1891. Calmette was the first to collect cobra venom for injection into small animals. The first snake serum was available in 1895. When cobras are "milked," an expert holds the snake at the edge of a Petri dish and lightly massages its venom glands with thumb and middle finger. This releases about half the venom stored in the glands; only about one-fifth of the contents are released when a cobra bites. Cobra venom is extremely potent. As little as 0.00002 gram of dehydrated cobra venom can kill a guinea pig. Theoretically, one gram of the dehydrated cobra venom is sufficiently potent to kill 140 dogs, 167,000 mice, or 165 humans. The mongoose is reputed to be immune to snake bite, but this is not true. However, a mongoose can tolerate a dose of cobra venom that is eight times the quantity that would kill a rabbit.

Mambas (Dendroaspis). This snake is more feared and infamous in Africa than the resident cobras. The most familiar is the *black mamba* (*D. polylepis*), which attains a length of about 4 meters (13.1 feet) and is the largest venomous snake in Africa. It is characterized by lightning-quick, elegant movements in branches and by the vigor of its reactions. This large, slender snake has a narrow head, smooth scales arranged in diagonal rows, very large eyes with round pupils, and enlarged lower teeth, which, together with the poison fangs of the upper jaw, grab a firm hold onto prey, usually birds, arboreal lizards, and tree frogs.

Contrary to their reputation, mambas do not make wild attacks on harmless travelers. The snakes are shy and will flee into the foliage when people approach. Some mambas are highly territorial, and will stay in the same small area for months.

Banded Krait (Bungarus fasciatus). This is a lacquer-black snake with shiny yellow bands on a strangely ridged back, a coiled snake that lies in the shade. The slender head is hardly distinctly different from the rest of the body. The krait achieves a length of 1.5 to 2 meters (4.9 to 6.6 feet). Kraits are nocturnal and sensitive to light. They are specialists and eat almost nothing but other snakes, and they readily attack the large, powerful rat snakes, cobras, and even smaller conspecifics. The poison fangs are only 2 to 3 millimeters long and the amount of venom released (30–40 milligrams, based on measurements of dehydrated venom taken at East Berlin's zoo) is not great. However, the venom contains extremely potent neurotoxic material that paralyzes the respiratory center and can kill a human being in 30 minutes. However, it is extremely unusual for a human being to be bitten by a krait. Kraits are egg-layers.

Coral Snakes (Micrurus, Micruroides, and Leptomicrurus). These snakes usually range between 60 and 80 centimeters (23.6 and 31.5 inches) in length, seldom up to 1 meter (3.3 feet). They are characterized by their bright coloration and pattern. In the majority of American coral snakes (*Micrurus*), the only teeth in the upper jaw are the two small poison fangs, an indication that this genus is a phylogenetically recent one. Little venom is released with each bite; the maximum for the Brazilian species (*Micrurus corallinus*) is 50–200 milligrams of liquid (i.e., 60 milligrams dehydrated). Most coral snakes are in Mexico, which has about 30 species. Colombia has 28 and Central America, excluding Mexico, contains 26 species. The number of species is reduced as one moves further north or south; there are just three coral snake species in the United States; Argentina contains only four.

Coral snakes feed chiefly on lizards and small snakes and, less often, on young birds, frogs, and insects. When they bite, they "chew" several times in succession, a behavioral pattern that is rare in venomous snakes. Repeated biting increases the amount of venom injected. The swallowing of prey is a slow process, because the jaws are not very flexible. Coral snakes do not have a strong tendency to bite, and cases of bitten humans are unusual, but when they occur they are serious injuries.

The *eastern coral snake* (*Micrurus fulvius*) achieves a length up to 60 centimeters (23.6 inches) and occurs from southeastern United States to Mexico and is one of the most prevalent species. It often inhabits rodent dens, where its bright coloration makes it stand out. The snout is black. This species assumes a curious defensive position when it is irritated. The head is hidden; the short tail is raised several centimeters and moved back and forth.

The *Arizona coral snakes* (*Micruroides euryxanthus*) also attain a length of about 60 centimeters (23.6 inches). These snakes inhabit arid terrain. Arizona coral snakes have a small, smooth tooth behind each poison fang. This species has red, yellow, and black bands. It is interesting to note that nonpoisonous colubrid snakes in North, Central, and South America bear the same red, yellow, and black ring markings as the poisonous coral snakes. However, the distinction between the harmless and dangerous snakes is rather easy to make in the case of North American snakes. In the coral (dangerous) snakes, a yellow (or white) ring is always bordered by a red one, while in the harmless colubrid snakes, such as the brilliantly colored milk snake (*Lampropeltis doliata syspila*), red and yellow are separated by a black ring. The problem of identification is more difficult in South America because of the presence of "true" and "false" coral snakes, which have nearly identical markings.

In Australia, where elapids are particularly prevalent and diverse, there are numerous small, primitive, slightly potent species. There are as well some species that are among the most dangerous members in the entire family. Their bite can be fatal to humans. They include the *taipan* (*Oxyuranus scutulatus*), which attains a length of 3 to 4 meters (9.8 to 13.1 feet). This is the most poisonous of Australian snakes. It inhabits remote coastal stretches of northern Australia and New Guinea, as well as the islands of the Torres Strait. The snake is brown to brown-black, and melanistic individuals have been found in New Guinea. About 80% of taipan bites are fatal unless antivenin is administered in time. Even with serum injections, many people die. It has been reported that a riding horse in northern Queensland died within five minutes after being bitten by a taipan. The species has long poison fangs. The venom not only paralyzes the central nervous system, but also destroys red blood corpuscles.

When preying (chiefly on rats), the taipan makes several bites in rapid succession. The snake often remains in rat holes and is active early in the morning and late in the evening.

The gray-brown, dark-banded *death adder* (*Acanthopis antarcticus*) is a large Australian elapid, which attains a length of about 80 centimeters (31 inches). It is one of the more prevalent elapids in Australia and New Guinea. The appearance is much like that of a viper. It has keeled scales and, unlike nearly every other cobra species, has a broad, clearly distinct, nearly triangular head (usually a characteristic exclusive to vipers). When excited, the death adder can inflate itself into a sausage shape. Its venom (up to 236 milligrams of liquid per milking) is more potent than cobra venom. Although the bite of a death adder leaves practically no mark, about half of all untreated bites in humans result in death due to respiratory arrest. Death adders are chiefly active at dusk. Australian farmers intensely dislike the species, because their sheep are in areas where the snake is most prevalent (central and western Australia), and a sheep that steps on a death adder will probably die, because the snake reacts by biting the animal. Female death adders bear from 10 to 12 live young.

Sea Snakes (Hydrophiidae). These snakes have the same basic characteristics as the *Elapidae*. However, they have some unique features that reflect their marine existence; the tail is laterally compressed and forms a rudder-like structure. The trunk, particularly the rear portion, also is often laterally compressed. The scale covering is nearly uniform. These highly specialized marine inhabitants have lost the broad ventral scutes seen in terrestrial snakes, so the scales are nearly the same all over the body. The external nasal openings are on the top of the head and can be closed by a valve. Sea snakes have salt glands in the head for eliminating excess salt.

Most sea snakes belong to the subfamily *Hydrophiinae*, which contains 13 genera. Members of this subfamily share the paddle-shaped tail with all other sea snakes. These snakes have lost all dependence upon terrestrial life, and never go on land. Body musculature is degenerate. The strong current in the water takes over much of the work that the muscles would otherwise have to provide. Heavy specimens are rather helpless on land and can suffocate since their vestigial supporting muscles cannot even fill the lungs with air. One of the biological puzzles is the striking similarity in body shape, coloration, and patterning between several moray-like fishes and various sea snakes of the Indian and Pacific Oceans.

Vipers

Two snake families, whose venom apparatus has reached an impressive degree of effectiveness, are at the peak of phylogenetic development of the large suborder of snakes. Their poison fangs have no sign of grooves; they are solenoglyph teeth, which means that they actually have enclosed canals within the fangs that transmit the venom out of the body, very much like a hypodermic needle. Furthermore, members of these two families, excepting just a few species, have lost the large head scutes found in colubrid snakes and cobras and instead have many small scales on the head. These families are the vipers or adders (*Viperidae*) and the pit vipers (*Crotalidae*). Pit vipers are considered to be a distinct family because of their unique pit organs.

Vipers are found only in Europe, Asia, and Africa. Australia lacks both vipers and pit vipers, an indication that both families are phylogenetically recent, developed after the Australian continent became a separate land mass. Whether vipers developed from cobra-like ancestors or directly from nonvenomous colubrid snakes remains obscure. It has been well established that vipers arose somewhere in western Asia.

The most important distinguishing characteristic of the venom apparatus in vipers is that the two upper jaws, which bear the fangs, are greatly shortened. Each upper jaw has a special joint that permits the jaw, along with the fang anchored firmly within it, to rotate 90 degrees. If the viper closes its mouth, the fangs lie back, tips inward, and are covered by a fold in the mucous membrane. When the mouth is opened, a lifting mechanism is activated, putting the fangs into a vertical position. The fangs (or more precisely, the upper jaw bones) are laid back with the same action as when a pocket knife is snapped together. The adaptation of folding back the fangs permits them to be extremely long, far exceeding the length of those in cobras. The fangs of the giant king cobra are not much longer than those of the rather small adder. The long fangs enable vipers to bite deeply into the tissues and cause the victim to suffer severe necrosis. The fangs snap back into the mouth after they are withdrawn from the victim.

Viper venom contains mainly hematoxic materials (substances that injure the blood and associated vessels). Thus, a viper bite has a very different effect from a cobra or mamba bite (their venoms being neurotoxic). Viper bites are accompanied by prominent local irritation and symptoms of severe blood poisoning, with burning pain, inflammation, pronounced discoloration, sudden drop in blood pressure, internal bleeding, degeneration of the tissues, and the formation of an abscess. Death ensues because the heart stops, not as the result of respiratory arrest as in the case of cobra bites.

Vipers or adders are generally compact, sturdy snakes. The head is triangular and is distinct from the rest of the body. Their length ranges from 30 centimeters (11.8 inches), as in the case of the dwarf puff adder, to about 180 centimeters (5.8 feet), as in the Gaboon viper. The pupils are usually vertical and elliptical. All face bones are movable. Each of the two shortened, retractile upper jaw bones bears only the tubular venom fang (which can only be activated for a short period) and often one to several significantly smaller reserve teeth of various sizes, none being a firmly positioned poison fang. The tail is short, and coloration is usually drab. In the genus *Vipera*, it is often of a dark zigzag pattern or a rhomboid band along the back. Desert species are sand-yellow, while jungle vipers often have a colorful carpet marking. There are 10 genera with 60 species and 110 subspecies.

Some vipers inflate their bodies into sausage shapes when excited. Almost all vipers assume a plate-shaped coiled position as a threat gesture, in which they lift up the neck and hold it in an S-shape. Other threat behaviors include loud hissing and rapid, forward jerks of the head. Some sand dwellers, such as the saw-scaled viper, create a particularly impressive sound by rubbing their scales together. Vipers feed chiefly on small vertebrates, particularly rats, mice, and lizards and less often on frogs and birds, paralyzing or killing their prey by biting it. Some of the smallest vipers prefer locusts. Many vipers are useful for controlling rodent pests.

The true vipers (*Vipera*) are distributed in Europe and Asia, with 8 species and, in Africa, with 4 species. The *adder* (*Vipera berus*) attains a length of 50 to 60 centimeters (19.7 to 23.6 inches), seldom over 80 centimeters (31.5 inches). The adder can tolerate the coldest climates of any viper species. Its distribution essentially follows that of deciduous forests, mixed forests, northern pine forests, and high-altitude forests. In Europe, the species occurs as far north as the Arctic Circle and also inhabits the Carpathian, Balkan, and Caucasus Mountains. The *asp viper* (*V. aspis*), about the same size as the adder, occurs chiefly in the Mediterranean region, extending northward as far as the southernmost part of the Black Forest of Germany and moving from there into central France and western and southern Switzerland.

Adders do not have the compact appearance of many other vipers. The head is only slightly broadened toward the rear and it is covered with rather large scales. Basic coloration in males is gray; in females, brown. Adders hibernate underground at a site which is moist, but protected from flooding. Live-bearing (ovoviviparity) permits the adder to penetrate far north and high up on mountains. The number of young varies between 6 and 20. Two vipers are seldom found together outside the mating season, although their home territories overlap and they may live within several meters of each other. Vipers hunt by lying in wait and grabbing appropriate prey that passes within striking distance. The chief enemies of European vipers are serpent eagles and the hedgehog.

Attempts to find snake antivenins were being made in antiquity, but truly effective venom antidotes were not developed until the end of the 19th century with the work of Calmette, as previously mentioned in connection with cobra antivenom, and the work of Phisalix and Bertrand, who produced asp viper antivenin at about the same time. Asp viper antivenin became available in 1896. The *sand viper* (*Vipera ammodytes*), which attains a length up to 90 centimeters (35.4 inches), is found in southeastern Europe and is the most dangerous (venomous) snake in Europe. Sand viper bites cause fatalities in remote parts of the Balkans. However, its venom does not compare with the potency of that in cobras or rattlesnakes.

The Gaboon viper, from the jungles of western, central, and eastern Africa is the largest of all vipers. It is the most prevalent of venomous snake species in the mountains of Cameroon. Its patterning is like that on ornate Oriental carpets; the deep purple-brown background has geometrical patterns in yellow, light brown, and blue, usually in an hourglass arrangement along the sides with long, rectangular, deep-purple bordered rings. The pale-brown head with its dark medial line has a chocolate-colored wedge-shaped spot whose tip extends to the eyes. With most exceptions, Gaboon vipers are docile and do not readily bite.

Pit Vipers

Pit vipers and vipers share a common extinct ancestor. Pit vipers probably arose once the egg-laying vipers had evolved and were developing along that line. Pit vipers survived because their new organ, the pit organ, was biologically successful, and this recent snake group is still evolving. More than two-thirds of all pit vipers inhabit the Americas, but their origin is probably in tropical Asia, where vipers and pit vipers occur together. There are no pit vipers in Africa or Australia, but the species has reached the southeastern edge of Europe, on the Caspian Sea.

In pit vipers, the pupils are vertical and elliptical. The fangs are long and solenoglyph (tubular). They often remain inside the prey, and may be excreted in the pit viper's feces. Fangs are replaced two to four times annually by reserve teeth. Contractions of the muscles surrounding the venom glands control the amount of venom released and direct the venom via a mucous membrane covering to the teeth. Their length varies from 40 centimeters (15.7 inches) to 3.75 meters (12.3 feet). Coloration is typically gray, brown, or olive with a distinctive light-dark diamond pattern or dark spots. There is often a dark band extending from each eye to the corner of the mouth. Arboreal pit vipers (with prehensile tails) are green or yellowish. Behaviorally, pit vipers resemble vipers, but the threat posture consists of slightly flattening the body and rolling it together into a spiral, and then lifting the forebody from the ground and bending it into an S-shape. The end of the tail is also raised, and tail shaking occurs even in those pit vipers lacking rattles. The rattling is a warning reaction. The diet is the same as that of the vipers.

Venom in most pit vipers contains hematoxins like that of vipers. The tropical rattlesnake is an exception. Since pit vipers play a more significant role in North and South America than vipers do in Europe, organized medical aid for bitten people is more extensive in the Americas. The Instituto Butantan, near São Paulo, Brazil, was the pioneer establishment for dealing with the treatment of venomous snake bites. In 1901, Mineiro, who had gained fame by combating typhoid and pestilence epidemics in Brazilian ports, was assigned the task of establishing a serum therapeutic institute for the State of São Paulo.

All pit vipers have a deep pit between the nostril and the eye; its opening is larger than the nostril but smaller than the eye. The pit, which has been depicted in old Indian drawings, lies in an indentation of the upper jaw bone. Scientists puzzled for decades over the meaning of these pits. As early as 1824, it was known that tiny nerves lead to them, and that the pits were therefore sensory organs. But there were arguments over just what sense modality they served. It was not until the late 1930s that most of the mysteries were cleared up. It is now established that the pit organs are heat-detecting sensors. They do not function like a thermometer, as once thought, but respond to temperature differences between the inside of the snake and the outside. Sensitivity has been suggested as being a small fraction of one degree Celsius. With this organ, the pit viper can sense the presence of warm-blooded prey at a distance up to 50 centimeters (19.6 inches). The pit organ is particularly helpful at night when the difference between the temperature of the prey and the ambient temperature is the greatest.

Rattlesnakes (Crotalus). These snakes are among the best-researched and most impressive snakes known. All rattlesnakes have the rattle at the tip of the tail. The biological function of the rattle was discussed for many years. The hard, dry, chain-like, loosely jointed rattle elements at the tail tip are the remains of previous moltings. See Fig. 3. In newborn rattlesnakes, the tail ends in a spherical scale. While all other snakes cast off this terminal scale at every molt, rattlesnakes do so only at their first molt, which occurs 7 to 10 days after birth. This last scale hardens into a piston-like hollow structure before the onset of the second molting; it has a ring-like construction. During subsequent molts, this last scale loosens like all the others, but it is not cast off. In the meantime, a new terminal scale has developed from inside this first one, and the new one has a greater diameter and holds the first one. Another terminal scale develops at the next molt, and so it goes, adding more and more rattles to the chain. Each of the hollow, pod-like scales in the chain of rattles was, at one time, the closest to the snake's body. Thus, contrary to widespread popular concepts, the number of rattles does not indicate the rattlesnake's age in years. They show how many times the snake has molted and rattlesnakes molt at an average of three times each year. Rattles tend to break off and with an irregular molting schedule, there is no direct connection between number of rattles and a snake's age. Rattlers can shake their tails much faster than a human eye can perceive—up to forty or 60 cycles per second. The sound is more like a hissing buzz or whir than a rattle. The rattlesnake does not hear its own rattling because it is deaf. The most likely meaning of rattling is that it is a warning sound used to intimidate enemies—a tool for upsetting the enemy. One exception to the foregoing is the Santa Catalina rattlesnake (*Crotalus catalinensis*), which does not have rattles.

Fig. 3.   Rattlesnake. (*A.M. Winchester.*)

Most species of rattlesnakes prefer dry, rocky, shrub-covered terrain, where they can conceal themselves inside crevices in the rocks or in mouse holes. Forest dwellers are more active during bright daylight hours than are species that inhabit more open country. Northern species often hibernate in large groups. A single mass winter quarter can support rattlesnakes from a large region and usually the same site will be used year after year. Northern populations of prairie rattlesnakes have been known to form groups numbering 200 to 300 on the average, but sometimes up to a thousand snakes. Rattlesnakes often share their quarters with nonvenomous snakes. Symbiotic relationships also occur among rattlesnakes, burrowing owls, and prairie dogs.

The diet of American rattlesnakes consists of mammals, primarily white-footed mice, pocket mice, voles, wood rats, and chipmunks. Large rattlers also may feed on ground squirrels and wild rabbits. Birds, such as domestic chicks and quail, are eaten less frequently.

Mating occurs in the spring in the southern United States. All rattlesnakes bear live young, born between August and October. Further north, especially in Canada, litters are born every other year in June and July. Most species bear 8 to 15 young per clutch.

Eastern Diamond-Back Rattlesnake (Crotalus adamanteus). This is the largest of all rattlesnakes and achieves a length up to 2.5 meters (8.2 feet) and a weight of 10 kilograms (22 pounds). This massive snake with its light-edged diamond figures is one of the most dangerous venomous snakes, chiefly because of the large amount of venom it produces. In a single 'milking,' this snake can yield 1050 milligrams of liquid venom (equal to 300–500 milligrams of dehydrated venom).

Western Diamond-Back Rattlesnake (Crotalus atrox). This snake attains a length of 2.2 meters (7.2 feet), but usually is somewhat shorter. The snake is aggressive and easily excitable and causes the majority of fatalities from snake bite in the United States. The species is useful in controlling rodents. The species is highly fertile, bearing 10 to 20 young at a time.

Prairie Rattlesnake (Crotalus viridis). This is a greenish-olive snake with longitudinal rows of round, brown spots. It is one of the most widely distributed of the rattlesnakes and exists in many subspecies. It is found in the Sierra Nevada mountains in California up to 4,000 meters (13,124 feet) and extends as far north as Canada. These snakes live primarily on mice.

Sidewinder (Crotalus cerastes). This small rattlesnake achieves a length of 60 to 70 centimeters (23.6 to 27.6 inches). Its sidewinding movement is an adaptation to desert life. Sidewinders can move at speeds of 3–4 kilometers (1.8–2.5 miles) per hour, making it one of the speediest of all rattlesnakes. It is almost purely nocturnal. Sidewinders are not particularly venomous; their prey includes other reptiles, especially smooth-throated lizards and whiptail lizards.

Tropical Rattlesnake (Crotalus durissus). Also called the cascaval or South American rattlesnake, this is probably the most dangerous of the rattlesnake species. Its venom contains substantial amounts of neurotoxic material. It is estimated that a cascaval bite is about as potent as a puff adder and cobra bite combined. A very special antivenin was developed in order to lower the death rate from bites in South America, particularly in Brazil. Approximately 75% of all untreated adults bitten die from a cascaval bite. A peculiar aspect of a cascaval bite is paralysis of the neck muscles of the victim. This causes the head to suddenly drop to one side. Severe neural disturbances later appear, including auditory and visual impediments that lead to blindness and unconsciousness.

The Agkistrodon Snakes. Much more is known about American *Agkistrodon* snakes than about the Asian species. They are as familiar to Americans living in the southern and eastern United States as are rattlesnakes. The most important of the American *Agkistrodons* include the cottonmouth or water moccasin, and the copperheads.

Cottonmouth or Water Moccasin (Agkistrodon piscivorus). This is a compact, powerful snake and is treated with great respect when it is encountered in rice fields in the southern United States. A cottonmouth bite is seldom fatal, but it has extremely unpleasant consequences. The popular name is due to the white inside of the jaw, which is visible when the snake is irritated, the open mouth being a threat gesture. Large cottonmouths are a deep blue-black, quite unlike the juveniles, which have jagged, red-brown, light-edged bands on a flesh-colored background. Five to 15 young are born after a long gestation period. Females reproduce every other year.

Cottonmouths often stay in swamps and on low branches overhanging the water. They swim well and flee into the water when alarmed. The species is not a true aquatic snake, since it does not catch most of its prey (frogs and fishes) while swimming, but rather by lying in wait on the shore. Cottonmouths apparently form pairs frequently, as shown by studies on several islands off Florida. Like rattlesnakes, they often winter in groups.

Copperhead (Agkistrodon contortrix). Although the fatalities from snake bites in the United States are mainly attributed to the western diamond-back rattlesnake, the number of incidents is greater from copperhead bites. The copperhead attains a length up to 1 meter (3.2 feet) and is distributed in four subspecies across many eastern and southern states. Copperheads are strikingly colored, beautiful snakes whose bands, arranged in an irregular zigzag, may be narrow or broad and red-brown or cinnamon on a gray or pinkish background. The top of the head is a light reddish-brown. This may be an adaptation to their favorite habitat, which is foliage-covered forest floors. They adapt their coloration to their background. Copperheads are extremely adaptable snakes. In some heavily populated areas where rattlesnakes disappeared many years ago, such as along the Hudson River in New York State, copperheads are still prevalent. Copperheads winter in groups, frequently with rattlesnakes. A female bears 8 to 12 live young, which have sulfur-yellow tail tips.

Additional Reading

Beardsley, T.M.: "Snakes in the Grass," *Sci. Amer.*, 22 (February 1990).

Behler, J.L. and F.W. King: "The National Audubon Society Field Guide to North American Reptiles and Amphibians," Alfred A. Knopf, Inc., New York, NY, 1979.

Coborn, J.: "The Atlas of Snakes of the World," T.F.H. Publications, Inc., Neptune, FL, 1991.

Crews, D. and W.R. Garstka: "The Ecological Physiology of a Garter Snake," *Sci. Amer.*, **247**(5), 158–168 (November 1982).

Culotta, W.A. and G.V. Pickwell: "Venomous Sea Snakes: A Comprehensive Bibliography," Krieger Publishing Company, Melbourne, FL, 1993.

Ernst, C.H.: "Venomous Reptiles of North America," Smithsonian Institution Press, Washington, DC, 1999.

Goldstein, E.J.C. et al.: Bacteriology of Rattlesnake Venom and Implications for Therapy," *J. Infect. Dis.*, **140**, 818 (1979).

Greene, H.W. and R.W. McDiarmid: "Coral Snake Mimicry," *Science*, **213**, 1207–1212 (1981).

Greene, H.W.: "Snakes: The Evolution of Mystery in Nature," University of California Press, Berkeley, CA, 2000.

Gruber, U.: "Blind Snakes, Primitive Snakes, and Wart Snakes," in "Grzimek's Animal Life Encyclopedia," Vol. 6, pp. 359–380, Van Nostrand Reinhold, New York, NY, 1972.

Harvey, A.L.: "Snake Toxins," Elsevier Science, New York, NY, 1991.

Hediger, H.: "Snakes," in "Grzimek's Animal Life Encyclopedia," Vol. 6, pp. 345–358, Van Nostrand Reinhold, New York, NY, 1972.

Holman, J.A.: "Fossil Snakes of North America: Origin, Evolution, Distribution, Paleoecology," Indiana University Press, Bloomington, IN, 2000.

Lillywhite, H.B.: "Snakes, Blood Circulation, and Gravity," *Sci. Amer.*, 92 (December 1988).

Mara, W.P.: "Water Snakes of North America," T.F.H. Publications, Inc., Neptune, NJ, 1995.

Mara, W.P.: "Desert Snakes of North America," T.F.H. Publications, Inc., Neptune, NJ, 1997.

Marais, J.: "Complete Guide to Snakes of South Africa," Krieger Publishing Company, Melbourne, FL, 1992.

Mattison, C.: "The Encyclopedia of Snakes," Facts on File, Inc., New York, NY, 1999.

Murphy, J.C. and R.W. Henderson: "Tales of Giant Snakes: A Historical Natural History of Anacondas and Pythons," Krieger Publishing Company, Melbourne, FL, 1997.

Newman, E.A. and P.H. Hartline: "The Infrared 'Vision' of Snakes," *Sci. Amer.*, **246**, 3, 116–126 (1982).

Petzold, H.G.: "Cobras and Sea Snakes," pp. 415–438; and "Vipers and Pit Vipers," pp. 439–485 in "Grzimek's Animal Life Encyclopedia," Vol. 6, Van Nostrand Reinhold, New York, NY, 1972.

Reinhard, W. and Z. Vogel: "Colubrid Snakes," in "Grzimek's Animal Life Encyclopedia," Vol. 6, pp. 381–414, Van Nostrand Reinhold, New York, NY, 1972.

Ricciuti, E.R.: "The Snake Almanac," The Globe Pequot Press, Inc., Guilford, CT, 2000.

Rodda, G.H., D. Chiszar, and H. Tanaka: "Problem Snake Management: The Habu and the Brown Treesnake," Cornell University Press, Ithaca, NY, 1998.

Rossman, D.A., R.A. Seigel, and N.B. Ford: "The Garter Snakes: Evolution and Ecology," Vol. 2, University of Oklahoma Press, Norman, OK, 1996.

Savitzky, A.H.: "Hinged Teeth in Snakes," *Science*, **212**, 346–349 (1981).

Seigel, R.A. and J.T. Collins: "Snakes: Ecology and Behavior," The McGraw-Hill Companies, Inc., New York, NY, 1993.

Shine, R.: "Australian Snakes: A Natural History," Cornell University Press, Ithaca, NY, 1995.

Staff: "Treatment of Snakebite in the USA," *Med. Lett. Drugs Ther.*, **24**, 87 (1982).

Sutherland, S.K. et al.: "Rationalisation of First Aid Measures for Elapid Snakebite," *Lancet*, **1**, 183 (1979).

Thorpe, R.S., W. Wuster, and A. Malhotra: "Venomous Snakes: Ecology, Evolution and Snakebite," Oxford University Press, Inc., New York, NY, 1996.

Watt, C.H., Jr.: "Poisonous Snakebite Treatment in the United States, *JAMA*, **240**, 654 (1978).

Weiss, R.: "Snakebite Succor: Researchers Foresee Antivenom Improvements," *Science News*, 360 (December 8, 1990).

Werler, J.E. and J.R. Dixon: "Texas Snakes: Identification, Distribution, and Natural History," University of Texas Press, Austin, TX, 2000.

Wright, A.H. and A.A. Wright: "Handbook of Snakes of the United States and Canada," Cornell University Press, Ithaca, NY, 1994.

SNAPPERS *(Osteichthyes).*   Of the order *Percomorphi*, family *Lutianidae*, snappers prefer tropical shallow inshore waters and usually occur in large schools. They possess bright, iridescent coloration, usually of blue, red, and yellow shades. They are known to migrate over great distances in search of food. They are carnivorous. They are an abundant fish and particularly in the Indo-Pacific region. The *Ocyurus chrysurus* (yellowtail snapper) is found in the tropical Atlantic and attains a length of about 2 feet (0.6 meter). The *L. synagris* (spot snapper) and *L. analis* (muttonfish) are found in American Atlantic waters. The established reputation of snappers as good food fishes has been tarnished from time to time by outbreaks of tropical fish poisoning, largely attributed to certain species of snappers. The poisonous qualities, however, appear to vary with particular region and time of year. Snappers considered in the poisonous category include: *Lutjanus bohar* (onespot snapper); *L. gibbus* and *L. vaigiensis* (red snappers); and *Aprion virescens* (blue-gray snapper). See also **Foodborne Diseases**.

SNELLEN EYE CHART.   The Snellen Eye Chart is the standard used in measuring the eye's ability to distinguish detail and shapes (visual acuity). The chart is based on the work of a Dutch ophthalmologist, Dr. Hermann Snellen, who in 1862 designed this system for describing human vision.

The Snellen Chart is made up of a series of letters, numbers or symbols of progressively smaller size, the largest at the top. "Normal" vision is 20/20. A person who can clearly read a one-inch letter at a distance of 20 feet is considered to have normal vision. All measurements obtained from use of the Snellen Chart are a comparison to that standard. Persons who have 20/40 vision, for example, can read at 20 feet what people with normal vision can read at 40 feet.

Visual acuity is normally checked before proceeding with any other part of an eye examination, including administering any medications to the patient's eyes. The patient is allowed to keep on any eyeglasses or contact lens and is positioned 20 feet in front of the Snellen Eye Chart. Covering one eye at a time, the patient is asked to read progressively smaller letters until it is no longer possible to read the small letters. Then the test is repeated with the other eye.

The Snellen Eye Chart measures only visual acuity. It does not prove the absence of an eye disease or any other eye problem. See also **Visual Acuity**.

Vision Rx, Inc., Elmsford, NY.

SNIPE.   See **Waders, Shorebirds, and Gulls**.

SNIPE FLY *(Insecta, Diptera).*   A two-winged fly with long legs and a conical abdomen. It belongs to the family *Rhagionidae*, sometimes called *Leptidae*. Some of the species suck blood, and in the western part of the United States are very annoying to human beings. This insect receives the name deer fly in the west, a term applied to a small horsefly in the east.

SNOW.   See **Precipitation and Hydrometeors**.

SNOW LEOPARD.   See **Cats**.

SOAP.   See **Colloid System**.

SOAPS.   Chemically, a soap is defined as any salt of a fatty acid containing 8 or more carbon atoms. Structurally a soap consists of a hydrophilic (water compatible) carboxylic acid which is attached to a hydrophobic (water repellent) hydrocarbon. Soap molecules thus combine two types of behavior in one structure; part of the molecule is attracted to water and the other part is attracted to oil. This feature underlies the function of these materials as surface active agents, or surfactants. Soaps are one class of surfactants. The other classes generally are called detergents. See also **Detergents**.

All surfactants, including soaps, demonstrate a common physical property — when they dissolve they preferentially concentrate at solution surfaces. These surfaces are known as the *interfacial regions* or regions where one continuous phase, such as water, stops and another, such as oil, begins. By their presence at the interface, surfactants lower the total energy associated with maintaining that boundary and thereby stabilize it. Without surfactants, a mixture of oil and water will soon separate into two distinct phases where the total surface area across which water and oil contact each other will be minimal. Adding soap to the water reduces its surface tension — the energy needed to maintain contact between the oil and the water. The oil then can be broken into microscopic droplets, which are dispersed in the water. Creation of these droplets, however, is accompanied by a huge increase in the interfacial contact area between oil and water. The dispersion of the oil in water is only possible, and only can be maintained over a period of time, because the surfactant reduces the energy associated with the large surface over which oil and water are in contact with each other. This phenomenon is the basis for the cleansing action of soaps and other detergents. Stabilization of the interface between the water used to cleanse and oils and other water-insoluble soils facilitates the dispersion of these materials into the water.

Although soaps and synthetic detergents have similar physical properties, several factors distinguish between them. Soap is generally made from natural fats and oils (oleochemicals). Some important synthetic detergents are also derived from oleochemicals, but almost no ordinary soaps are produced from petrochemicals. Fats and oils are triglycerides which contain three fatty acids, the basic structural unit of soaps, chemically linked to a glycerine backbone. As the "soap" chemical structure basically exists in natural triglycerides, with relatively straightforward processing operations, soap can be obtained from fats and oils.

Another important distinguishing feature of soaps is that they form a curdy, insoluble compound in hard water due to interaction between the carboxylate soap structure and calcium and magnesium ions in the water. Synthetic detergents, which generally are based on sulfate or sulfonate chemical structures for the water-attracting portion of the molecule, have less affinity for these metals and thus work well in all types of water. In addition, since these synthetics maintain their surfactancy, they also function to disperse objectionable curd. For these reasons, the synthetic detergents have generally replaced soaps in heavy-duty cleaning (laundry, floors, woodwork). Soaps, however, remain popular for mild cleaning and particularly for personal cleansing.

Personal Cleansing Soap Products

The major soap-based products which one commonly encounters are soap bars. Two broad categories of bar soaps may be defined: *basic cleaning bars*, which are natural soaps without extra ingredients and comprise about 20% of the market; and bars with special ingredients to provide a benefit beyond fundamental cleansing. The latter category may be further subdivided into *deodorant soaps* and *skin care bars*. Generally most of these bars command a higher retail price than basic cleaning bars, with skin care bars priced above deodorant soaps.

Deodorant soaps add fragrances that are partially substantive to the skin and that mask body odors, and antimicrobial agents. The antimicrobials, such as *Triclocarban*®, are deposited on the skin and inhibit bacterial growth and associated malodors.

Skin care bars are formulated with ingredients for which specific skin benefits are claimed. Consumers generally recognize and are concerned that personal cleansing products can dry the skin, leaving it feeling rough, itchy, and tight, and looking powdery and scaly. To counter these effects, particularly during the dry winter months, they may elect to use a cleansing bar containing a moisturizer, as well as increasing their use of body oils and hand and body lotions. Skin care claims for these products are based on the inclusion of moisturizers such as glycerin, cocoa butter, lanolin, cold cream and vitamins to the soap.

The mildness of soap bars toward the skin can also be enhanced by the process of *superfatting*. In superfatting, excess fatty acid is added to the soap during processing. This water-insoluble material functions as an emollient, significantly improving the mildness and the lathering of the bar.

Manufacture of Soap

Ingredients.   The primary materials used in the manufacture of bar soaps are natural fats and oils. The performance and physical properties of soap bars can be varied by altering the blend of fats and oils used to make the neat soap. The most common materials used are top-quality animal tallows and coconut oil with blends ranging from 50% to 85% tallow. Generally it is found that bars containing higher proportions of coconut soap are physically harder, more brittle, lather more, and are more expensive to produce due to the higher cost of coconut oil. It is therefore common practice to vary the blend of tallows and coconut oil to meet the desired properties and price of each product.

These basic materials eventually are converted to their neutral salts by use of some alkaline material, such as sodium hydroxide. Additional, minor ingredients are added, e.g. sodium silicate or magnesium sulfate, to control alkalinity, odor, and aging stability.

The basic process is that of reacting fat stocks with alkali to form soap (direct saponification) and glycerin, followed by washing to remove the glycerin. Two methods of direct saponification are in common use (*kettle method* and *continuous saponification*). An alternative method is splitting fat stocks with water (hydrolysis) to form fatty acids and glycerine, followed by neutralization of the fatty acids with alkali.

Kettle Method. The pioneers used a simplified kettle process when they boiled animal fat and wood ashes (for alkalinity) for several hours in a large pot. The modern soap kettle has a capacity of 60,000–300,000 pounds (27,216–136,080 kg) and is equipped for heating, settling, and blending the fats, alkali, salt, and water.

The kettle first is charged with fat and a sodium hydroxide solution. Then follows a sequence of heating, separating, and washing to convert the raw materials to *finished base soap* and to separate the impurities and byproducts. The process normally takes several days for any single kettle. Although there have been improvements in handling and purification such as continuous centrifugation, the basic kettle process of saponifying fats directly with caustic remains unchanged.

Continuous Saponification. Fat stocks, plus caustic and salt solutions, are fed continuously into an autoclave operating under pressure at typically about 250 °F (120 °C). A recycle stream provides sufficient soap concentration to solubilize the fat stream for good contacting with the caustic. The soap-lye-glycerin mix moves to a mixer/cooler to complete saponification. The cooler temperature reduces the solubility of soap in the lye and aids separation.

Glycerin and excess caustic are removed by several stages of countercurrent washing with fresh washing solution. The washing and separation stages usually take the form of a series of mixers and centrifugal separators or a continuous countercurrent contactor, such as a rotating disk contractor (RDC) in a vertical column. The mix from the saponifier is fed near the bottom of the RDC and washing solution near the top. The lower-density soap rises through the falling wash solution. Washed soap exits at the top while spent lye (glycerin plus lye solution) exits out the bottom of the RDC column. Spent lye is processed to recover the glycerin.

The washed soap is converted to finished base soap (neat soap) by a final composition adjustment called *fitting*. Fitting is accomplished by adding water (plus salt as needed), which causes a phase separation. Depending on the salt concentration the separated phase is either a lye or niger phase. A centrifuge or kettles can be used to separate the two phases.

Hydrolyzer Process. The development of continuous hydrolysis provides basic improvements in the processing of fats into soap. There are several advantages over the kettle process: (1) better quality soaps can be made from darker fats; (2) glycerin recovery is simplified, because no salt is needed and the resulting finished glycerin is of higher quality; (3) a single hydrolyzer unit produces about the same quantity of soap as 10 kettles, thus effecting savings in manufacturing space and a reduction of in-process inventory; and (4) greater flexibility is possible in controlling the chemical and physical properties of the finished soap. The hydrolyzing process consists essentially of (1) hydrolysis, (2) fatty acid distillation, (3) post-hardening (optional), (4) neutralization, and (5) glycerin recovery. The basic hydrolyzing process is shown in Fig. 1.

Hydrolysis. Development of continuous hydrolyzing was the key step toward this continuous soap making process. In this reaction, fat and water react to form fatty acid and glycerin:

$$(RCOO)_3C_3H_5O + H_2O \rightleftharpoons 3\ RCOOH + C_3H_5(OH)_3$$

where R is an alkyl of C_8 or larger. This equation represents the complete hydrolysis. Actually, the reaction takes place in a stepwise fashion, forming intermediate diglyceride and monoglyceride.

The reaction can be accomplished only through intimate contact between water and fat molecules. High temperature makes it possible to dissolve an

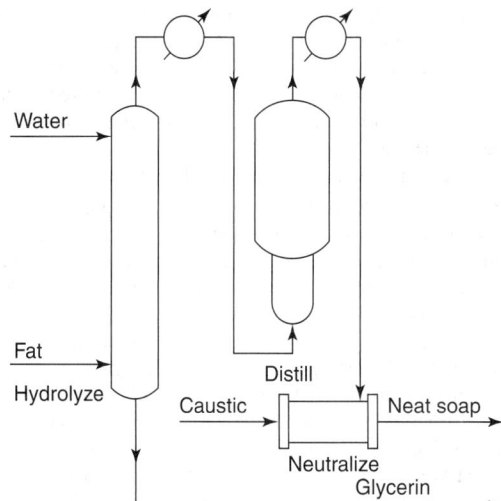

Fig. 1.   Basic hydrolyzer process used in soap manufacture.

appreciable quantity of water in the fat phase and to obtain this intimate contact. At room temperature, water and fat are essentially insoluble. At elevated temperature, the solubility of water increases to 12–25%, depending upon the type of fat. At the higher temperatures, high pressures also are necessary to keep the water from flashing into steam.

The reaction is reversible. In order to make it proceed to the right, the proportion of water to fat can be increased or the glycerin can be removed. Removal of glycerin is used as the reaction-forcing method. The required combination of high temperature, high pressure, and continuous glycerin removal is accomplished in a countercurrent hydrolyzer column. Fat stocks, blended in the proper formula, are mixed with dry zinc oxide catalyst. The mixture is maintained at about 212 °F (100°) to ensure dryness and to keep the catalyst in solution. Hot water for the hydrolysis reactions is put under high pressure by piston-type feed pumps with adjustable drives so that the rates and proportions of fat to water can be accurately controlled. The fat and water are heated to the hydrolyzing temperature by direct steam injection or by heat exchangers. The fats are pumped into the column near the bottom, and the water enters near the top. Thus, a countercurrent flow of water downward through rising fatty material is obtained.

The hydrolysis occurs in a two-phase reaction system. The fats and fatty acids flow continuously with droplets of water falling through them. Glycerin from the hydrolysis is dissolved in the excess water falling through the column. The rate-limiting factor is the transfer of glycerin into the water droplets. Zinc oxide catalyzes the reaction of forming zinc soap, which increases the glycerin transfer across the oil-water interface. Fresh water entering the column at the top reduces the glycerin to the lowest possible point, while a glycerin-water seat maintained at the bottom of the column (where the glycerin content is highest) prevents fat from washing out.

The fatty material passes upward through the column with about 99% completeness in splitting. The fatty acids, saturated with water, are discharged through an orifice into a flash tank. The dissolved water vaporizes, cooling the fatty acids and blanketing them with steam. The fatty acid contains the zinc soap catalyst and the remaining unsplit fat.

The column, pumps, and piping in contact with the hot fatty acid are made from corrosion-resistant stainless steel. The column is a hollow vessel, containing no baffles, trays, or packing material of any kind. The quality of the hydrolyzing operation is determined by the degree of split obtained on the fat. The fatty acid stream contains very little free glycerin, if any.

Distillation. The second key step in continuous soapmaking is distillation. Originally, fatty acids made in hydrolyzers were acid washed to split out the zinc soap and then bleached to improve color, but continuous distillation of the hydrolyzer fatty acids results in lighter soap from darker stocks at lower cost.

The fatty acids from the hydrolyzer are collected in the still feed tank and vacuum-dried to reduce moisture to low levels. Then they are flash-distilled at an absolute pressure of 2–5 mm Hg. The still bottoms are recirculated through heat exchangers back to the still to carry the heat necessary for vaporizing the fatty acids. The still bottoms, which contain the zinc soap catalyst and unsplit fat, are removed from the system, acidulated to remove the zinc, and frequently used in animal feeds. The fatty acid vapors from the still pass to several water condensers in series. The condensed fatty acids drop to a surge tank for posthardening or directly for neutralization.

The two prime objectives of this process, maintenance of good odor and color in the distillate and proper bottoms yield, are achieved by effective control over vacuum, temperature, and distillation rate.

Posthardening. Not shown in the figure is an optional further treatment of the fatty acids known as *posthardening*. This operation involves hydrogenation of some of the unsaturated carbon-carbon bonds on the fatty acid molecules. Originally, the purpose of this step was to improve color and odor. As such, the hardening was intended only to eliminate polyunsaturates, leaving the majority of the monounsaturates unaffected. A greater amount of hardening can be performed, however, to tailor some of the physical properties of the finished bar characteristics. The fatty acids from distillation are heated and passed with a metered hydrogen supply through hardening tubes which contain a fixed bed of granular nickel catalyst where the hydrogenation takes place. The hardened fatty acids flow through a filter to remove traces of catalyst. The filtered stock drops to a flash tank, where excess hydrogen is removed. Hardening is controlled by temperature, pressure, hydrogen flow, residence time, and catalyst age. The fatty acids then are cooled for neutralization.

Neutralization. The saponification reaction between alkaline solutions and fatty acids is almost instantaneous:

$$RCOOH + NaOH \longrightarrow RCOONa + H_2O$$

Each reactant is metered accurately into the neutralizer, where intimate mixing occurs and the reaction takes place. Soap from the neutralizer is discharged at about 200 °F (93 °C) to a blend tank equipped with agitation and recirculation to ensure uniform composition of the soap. This base soap (or *neat soap*) is stored until required for subsequent processing into finished bars.

The characteristics of the neat soap are controlled easily by accurately governing the composition of the alkaline solution used. Normal hydrolyzer neat soap contains about 69% actual soap, 30% water, and less than 1% NaCl, plus other stabilizers. Neat soap is a uniform, translucent, white, viscous fluid at 180–200 °F (82–93 °C).

Glycerin Recovery. The glycerin water stream from the hydrolyzer is concentrated by evaporation, purified, and subsequently sold or used in other processes.

Milled Bar Soap Manufacture. Milled soap is a high-grade soap in which critical crystal-phase changes have been brought about through the use of mixers, milling rolls, and plodders. The milled soap is made by drying a good grade of neat soap to about 15% moisture content, breaking up the crystalline structure that develops during drying and cooling, plasticizing and converting a sufficient portion of the soap to a desirable phase condition, de-aerating and compacting the resulting mass, and forming it into bars. Perfume, coloring matter, preservatives, and special additives are incorporated prior to the milling operation. A milled bar is particularly hard, dense, and smooth, and it lathers freely without forming excessive soft soap on the surface of the bar.

Drying. Liquid base soap is dried from a 30% water liquid form to a solid of about 15% water content. If desired, some minor ingredients may be blended into the soap stream prior to drying. Methods of drying used in common practice are (1) chip drying, (2) atmospheric flash drying, and (3) vacuum flash drying.

Chip Drying. Sometimes called *ribbon drying*, this process involves spreading a thin layer of hot base soap on a large chilled drum, which cools and firms up the soap. Drying is promoted primarily by the difference in water vapor pressure between the soap chips and the air surrounding them. No attempt is made to increase drying rate by heating the soap itself.

Atmospheric Flash Drying. A tower similar to a synthetic-granules spray-drying tower is used. The heat for drying, however, is put into the soap by heating it under high pressure before flashing it into the tower. During flashing, the pressure on the soap is abruptly relieved and soap moisture flashes to steam. Air to the tower is used for cooling. The soap temperature as it enters the flashing nozzles determines the final moisture of the dried soap.

An alternative method involves flashing the soap from the nozzle onto the surface of a chilled drum. The resultant solid soap is scraped off in flake form. This process called *chill flake drying*, is the method of choice for drying sticky soap/synthetic combination formulas.

Vacuum Flash Drying. In this most recent technique, drying takes place in a vacuum vessel similar to an atmospheric tower but smaller. The soap is similarly heated before flashing but under less pressure, so that boiling (actually drying of the soap) occurs in the heat exchangers. Since there is boiling in the heaters, the moisture of the dried soap depends primarily upon soap flow rate, soap pressure, and steam pressure to the heater and to a minor extent on the absolute pressure in the vacuum chamber. The final temperature of the soap depends entirely upon the absolute pressure in the vacuum chamber.

Mixing. After drying, the soap noodles or flakes are mixed with all additional ingredients required by the final product formula. Mixing is done in batch processes or continuously. These ingredients include dye, perfume, preservatives, deodorants, opacifier, and special purpose items. The type and proportion of these materials is largely what makes one brand of milled soap different from another.

In batch mixing, dried soap and additives are measured and dumped into a dry blender, where macro-mixing occurs. The batch process is cumbersome and slow, and it is difficult to maintain uniform quality. Continuous mixing operations for improving economy and efficiency of mixing include precision metering devices to measure the additives into the soap noodles as they are pulverized and conveyed through the mixer. Although these ingredients constitute but a small portion of the total product, their effect on the physical properties, e.g., softness, resistance to cracking, lathering, and resistance to dissolving, are considerable.

Milling. The three objectives of milling are: (1) thorough and intimate final mixing of the soap, perfume, and other ingredients without overheating; (2) crushing lumps of overdried soap and pulverizing them into pieces too small to appear as lumps or hard specks in the finished bar; and (3) conversion of a sufficient portion of the soap into the waxy, plastic phase of cold working. Soap is milled by forcing it through a series of rolls, thus subjecting it to a strong shearing action. This cold working at the proper moisture content changes the crystalline structure or phase of the soap. Temperature control during milling is important. If the temperature is too low, the wrong crystal structure will be formed, resulting in soft soap or a hard, brittle structure prone to cracking. If the temperature is too high, the soap will become sticky and difficult to process further.

Another method for complete mixing and working uses multiple plodding and screening, in which the soap and additives are pushed together through finer and finer mesh screens.

Plodding. After milling, it is necessary to form the soap into a shape for making the final bar. This usually is accomplished with a plodder, which essentially is a large-size meat grinder with a barrel that terminates in a cone. The plodder functions to compact the pellets or flakes of soap into a solid mass, squeeze out any pockets of entrapped air, and extrude it as a firm, uniform, and continuous strip.

Operations that follow include cutting, stamping, wrapping, and packing.

Transparent Bar Manufacture. Most milled bars are opaque and contain a whitening agent (titanium dioxide) to create a uniform appearance. By eliminating this whitener and carefully controlling processing conditions, a bar that is transparent can be produced. This transparency results when the soap crystals are reduced to microscopic size which then allows light to pass through the structure. It is also important to achieve the correct soap phase. The control of soap phase is a function of the ratio of tallow to coconut soaps, the milling temperature, and soap moisture which must be maintained within very rigid limits.

Floating-Bar Manufacture. Base soap made from the desired blends of fat and oil first is flash-dried to a moisture content of about 22%. It then enters a mechanical mixer called a *crutcher*, where it is thoroughly mixed with perfume, preservatives, and air. The amount of air controls the density of the final product, giving the bar a density of less than one and making it floatable.

From the crutcher, the mix goes to a freezer to reduce the temperature of the soap to the point where it will hold its shape when extruded. In the earlier steps, the soap mix is in liquid form. Rapid chilling is required to put it into a solid state. The machine is similar to a commercial ice-cream freezer, consisting of a horizontal cylinder surrounded by a jacket and housing a rotating shaft (mutator) on which scraping blades are mounted. The liquid soap mix from the crutcher is pumped into one end of the cylinder. A refrigerated brine solution is circulated through the jacket to chill the soap. The scraping blades on the mutator remove the chilled soap from the cylinder walls and maintain uniformity of the mix. The nose of the freezer is equipped with an oblong orifice through which the soap is extruded, after chilling, in the form of a continuous ribbon, which has the same cross section as the final bar. There follows a series of cooling, storing, stamping, and packaging operations.

Additional Reading

Bailey, A.E. and Y.H. Hui: "Bailey's Industrial Oil and Fat Products," 5th Edition, John Wiley & Sons, Inc., New York, NY, 1996.

Basta, N.: "Shreve's Chemical Process Industries Handbook," 6th Edition, The McGraw-Hill Companies, Inc., New York, NY, 1993.

Woolatt, E.: "The Manufacture of Soaps, Other Detergents, and Glycerin," John Wiley & Sons, Inc., New York, NY, 1985.

Web References

Detergent Chemistry: History: *http://www.chemistry.co.nz/deterghistory.htm*

The Procter & Gamble Company: *http://www.pg.com/sitesearch/google.jhtml?_DARGS=%2Fsitesearch%2Fgoogle.jhtml*

R. MARC DAHLGREN and JOHN N. KALBERG,
Ivory Technical Center, The Procter & Gamble Company,
Cincinnati, OH.

SOCIETY (Ecology).   A group of individuals of the same species living together for mutual benefit, with some division of labor. The society is a high expression of colonial organization in the animal kingdom, and is not sharply separated from simpler forms of colonies.

In the simplest type of colony, as found among the 1-celled animals, the associated individuals are similar and each is capable of complete existence in itself. In the same group a slight division of labor appears, accompanied by structural differentiation of the individuals for different tasks. This form of organization persists in the coelenterates, bryozoans, and ascidians, with varying degrees of structural continuity in the colony and varying degrees of differentiation and division of labor.

Among the more highly organized animals the possibility of association of individuals in a complex society involving division of labor is expressed only among the social insects and man. The insect society, or colony as it is often called, continues the associated principle of structural specialization of the individual for its particular duties, with the resulting castes exemplified in a simple form by the queen, drone, and worker honeybees. Among the termites and ants the differentiation is much more extreme and complex. To a moderate degree the honeybee colony also shows specialization of behavior among the workers, which may engage in various activities within their powers according to the requirements of the colony at different times.

The human society differs from that of insects in the restriction of inherent fitness for special duties to less evident details of organization. While inherent fitness undoubtedly exists among men, they are structurally of approximately the same form and their specialization is largely a result of training. In other words, specialization of the individual in human society is conspicuously a specialization of behavior. Lack of structural specialization is compensated by the use of tools.

In all cases, the society is an extension of the prevailing biological principle that biological units of any degree of complexity can be associated together as component parts of a larger coordinated unit.

SODALITE.   An isometric mineral, a sodium aluminum silicate containing sodium chloride, with the chemical composition $Na_4Al_3 (SiO_4)_3Cl$, potassium sometimes replacing a small amount of sodium. It is commonly found as dodecahedrons or simply massive. When observed sodalite has a dodecahedral cleavage; conchoidal to uneven fracture; brittle; hardness, 5.5–6; specific gravity, 2.14–2.30; luster, vitreous to greasy; color grayish to greenish or yellowish, may be white. It is often a beautiful blue and may sometimes be red. It is transparent to translucent; streak, white. Sodalite is found in igneous rocks of nephelite-syenite type which have been produced from soda rich magmas. Sodalite also has been found in the lavas of Vesuvius. Common minerals associated with it are nephelite and cancrinite. It occurs in the Ilmen Mountains of the former U.S.S.R.; at Vesuvius and Monte Somma, Italy; in Norway and Greenland. In Canada, in British Columbia and in Ontario, beautiful blue sodalite is found; and in the United States similar material comes from Kennebec County, Maine. The mineral derives its name from the fact of its soda content.

SODA NITRE.   The mineral soda nitre or Chile saltpeter is naturally occurring sodium nitrate, $NaNO_3$. Its hexagonal crystals are rare, this mineral usually being found in crystalline aggregates, crusts or masses. It is soft; hardness, 1.5–2; specific gravity, 2.266; vitreous luster; colorless or white to yellow or gray; transparent to opaque. Soda nitre is a most important mineral commercially, being used in the manufacture of nitric acid, other nitrates and fertilizers. The chief soda nitre deposits of the world are those found in the Atacama and Tarapaca deserts of northern Chile, although others exist in the Argentine and Bolivia. Some small deposits have been found in California, New Mexico and Nevada. The origin of these nitrate deposits is far from being well understood. They have been regarded as nitrates formed originally by oxidation of organic matter and subsequently leached out. Guano, the excrement of birds, might be the original source of the nitrates. Ground water and ancient marine deposits have been suggested as well as the possibility of derivation from nitric acid produced in the atmosphere during electrical storms. Some investigators consider that the nitrates may have come from volcanic sources.

SODIUM.   Chemical element, symbol Na, at. no. 11, at. wt. 22.9898, periodic table group 1 (alkali metals), mp 97.82 °C, bp 882.9 °C, density 0.971 g/cm^3 (solid at 0 °C), 0.9268 g/cm^3 (liquid at mp). Elemental sodium has a face-centered cubic crystal structure.

Sodium is a silvery-white metal. It can be readily molded and cut by knife. It oxidizes instantly on exposure to air, and reacts with water violently, yielding sodium hydroxide and hydrogen gas, consequently is preserved under kerosene, and burns in air at a red heat with yellow flame. Discovered by Davy in 1807.

There is only one naturally occurring isotope, ^{23}Na. There are five known radioactive isotopes, ^{20}Na through ^{22}Na, and ^{25}Na, all with short half-lives except ^{22}Na with a half-life of 2.6 years. See also **Radioactivity**. In terms of abundance, sodium ranks sixth among the elements occurring in the earth's crust, with an average of 2.9% sodium in igneous rocks. In terms of content in seawater, the element ranks fourth (due mainly to excellent solubility of its compounds), with an estimated 50,000,000 tons of sodium per cubic mile of seawater. First ionization potential 5.138 eV. Oxidation potential Na → Na$^+$ + e$^-$, 2.712 V.

Other important physical properties of sodium are given under **Chemical Elements**.

Sodium does not occur in nature in the free state because of its great chemical reactivity. Sodium occurs as sodium chloride in the ocean (1.14% Na); in salt deposits (salt, halite, NaCl), e.g., in Michigan, New York, Louisiana, in Great Britain, and in Germany; in salt lakes, e.g., the Dead Sea (3% Na), Great Salt Lake; in common rocks (average of the solid shell of the earth 2.75% Na) as sodium nitrate (Chile saltpeter, $NaNO_3$) in Chile; as sodium borate (rasorite, kernite, $Na_2B_3O_7 \cdot 4H_2O$, in California; tinkal, $Na_2B_4O_7 \cdot 10H_2O$, in Tibet); and as sodium carbonate Na_2CO_2 and sulfate Na_2SO_4 in certain salt lake areas. See also **Sodium Chloride**.

Although sodium metal was isolated in 1807, it remained a laboratory curiosity until Oersted discovered in 1824 that sodium metal will reduce aluminum chloride to produce pure aluminum metal. This discovery led to the development of a commercial process for the manufacture of sodium. The first cell was designed by Castner in 1886 and a plant was built in Niagara Falls, N.Y., because of availability of low-cost electric power, for the electrolysis of fused NaOH. This process was made obsolete in 1921 by introduction of the Downs process in which a mixture of fused sodium chloride and calcium chloride is electrolyzed to produce metallic sodium. The modern cells have four anodes (graphite) surrounded by a steel cathode. Wire mesh diaphragms extend down into the electrolysis zone to prevent recombination of product sodium and chlorine. The use of calcium chloride in the cell significantly lowers the melting point of the mix. Sodium chloride has a mp 800 °C, calcium chloride, mp 772 °C, the two-salt eutectic, mp 505 °C. Calcium has limited solubility in sodium. The excess calcium reacts with the sodium chloride present, Ca + 2NaCl → 2Na + CaCl$_2$, and thus does not contaminate the sodium metal to a large degree. The sodium, which is saturated with calcium, is cooled in a riser pipe. This reduces the solubility of Ca in Na, precipitating Ca, which falls back into the cell, where it reacts to form more Na. The Na that overflows at the top of the riser pipe contains 1% or less of Ca. The Na is further purified by filtration at a temperature near its melting point, reducing the Ca content to about 0.05%. The cells operate at about 8 V, with groups of 25 to 40 cells connected in series.

Uses. Like so many of the chemical elements, the compounds of sodium are far more important than elemental sodium—by several orders of magnitude.

Among the attractions of molten sodium metal as a heat-transfer medium are: (1) low density compared with other metals and combinations of salts, contributing to low cost per unit volume and thus relative ease of pumping, sodium being about one-half that of the more commonly applied nitrate-nitrite heat-transfer salts; (2) relatively low vapor pressure even at temperatures as high as $550\,°C$; (3) greater heat capacity than most common metals in liquid form, the thermal conductivity being 5 to $10×$ greater than the conductivities of lead or mercury and $50×$ higher than for most organic heat-transfer media; and (4) the viscosity of molten sodium is quite low. Despite these fine qualifications, however, the use of sodium as a heat-transfer medium has enjoyed a mixed reception over the years, partially attributable to a lack of marketing thrust in its behalf. Sodium is fifth among the metals in terms of electrical conductivity—hence bus bars are constructed from steel pipe filled with sodium. The characteristic yellow sodium light, created by the passage of an electric current through sodium vapor, is used for commercial and industrial lighting. Sodium is used to modify aluminum-silicon alloys. Normally coarse and brittle, such alloys can be transformed into fine-grained alloys with good casting properties through the addition of a fraction of 1% of sodium. Sodium also has been used as a hardening agent in bearing metals. When added with an alkaline-earth metal, such as calcium, sodium increases the hardness of lead. The German alloy "Bahnmetal" is an alloy of this type.

Generally, plain carbon steel containers are sufficient for handling metallic sodium at temperatures not in excess of the metal's boiling point. All-welded pipeline construction and bellows-sealed packless valves are usually used. Because of the metal's violent reactivity with H_2O, conventional fire extinguishers, including CO_2 and chlorinated hydrocarbons, should not be used. The preferred fire-retarding agents are salt, graphite, and soda ash, but they must be dry. Sand usually is not recommended because it is difficult to obtain perfectly dry sand in an emergency. In manufacturing operations involving sodium, particularly at reasonably high temperatures, an apron, leggings, and a complete face covering should be used. At normal temperatures, or where only small quantities of the metal are required, as may be the case in a research laboratory, conventional protective gear and goggles and gloves usually suffice.

Chemistry and Compounds: Sodium metal is obtained by electrolysis of fused sodium chloride or hydroxide out of contact with air. Its uses are limited in extent, but important in particular cases, as in the liberation of a metal from its chloride by reaction of sodium to form sodium chloride, and in certain reactions of organic chemistry.

The ionization potential of sodium (5.138 eV) is second to that of lithium and higher than those of the other alkali metals. However, the measured value of its oxidation potential against a normal aqueous solution of its ion is 2.712 V, the lowest of the group. Potassium is more electropositive in many of its reactions, even with water; though both react vigorously to produce the hydroxide and hydrogen, the reaction of potassium is more vigorous. With bromine, sodium reacts only slowly without heating and with iodine scarcely at all even on heating; potassium reacts violently with bromine, and with iodine on heating.

Because of the ease of removal of its single $3s$ electron (5.138 eV) and the great difficulty of removing a second electron (47.29 eV), sodium is exclusively monovalent in its compounds, which are electrovalent. Some experimental work indicates that the sodium alkyls may be covalent, but even they form conducting solutions in other metal alkyls.

Sodium Atoms Confined. In an interesting experiment conducted at the National Bureau of Standards in 1985, Migdall and colleagues trapped slow-moving neutral Na atoms in a magnetic field that created an energy well for the atoms. Robinson (1985) reported that approximately 10^5 sodium atoms in a trap volume of 20 cubic centimeters were stopped for a brief instant of time. For many years, spectroscopists have visualized the ideal sample where a collection of atoms or molecules would reside motionless in space for a period of time. Because of this experiment, researchers are closer to this goal. In the experiment, it was found that the particles gradually leak out, with time constants ranging from 0.1 to 1 second. Similar experiments have been conducted at AT&T Bell Laboratories (Holmdel, New Jersey). The theoretical trapping time in a perfect vacuum has been estimated as greater than 1000 seconds. The importance of these experiments is explored in considerable detail by Robinson in the reference listed.

Like lithium, sodium and its compounds have been studied extensively in solution in liquid NH_3. Sodium metal in such solutions slowly or with catalysis forms the amide, $NaNH_2$. The solution of the metal is a powerful reducing agent, reacting with metallic salts to free the metal, with which it may form an intermetallic compound

$$Na + AgCl \longrightarrow NaCl + Ag$$

$$9\,Na + 4\,Zn(CN)_2 \longrightarrow 8\,NaCN + NaZn_4$$

Sodium chloride also forms the amide, or at low temperatures the pentammoniate, $NaCl\cdot 5NH_3$.

Like the other alkali metals, sodium forms compounds with virtually all the anions, organic as well as inorganic. These compounds are remarkable for their great variety and for the fact that the reactivity of sodium bicarbonate with many metallic oxides permits preparation of many compounds that are unstable in aqueous solution. While other alkali bicarbonates react similarly, the general discussion of these compounds, and of the inorganic alkali salts generally, is appropriately given in this book under this entry for sodium, from which such a great number of inorganic (as well as organic) salts has been prepared.

Thus, normal (ortho) sodium arsenates $Na_3AsO_4 \cdot xH_2O$ and acid arsenates exist both in solution and in the solid state, whereas the meta- and pyroarsenates exist only as solids, but are readily prepared by heating arsenic pentoxide, As_2O_5, and sodium bicarbonate in correct proportions to produce the primary and secondary sodium arsenates, whence the meta- and pyroarsenates are obtained by heating

$$NaH_2AsO_4 \xrightarrow{\text{Heat}} NaAsO_3 + H_2O$$

$$2\,Na_2HAsO_4 \xrightarrow{\text{Heat}} Na_4AsO_2O_7 + H_2O$$

Similarly, the boron salts include metaborates, $NaBO_2 \cdot xH_2O$, tetraborates, $Na_2B_4O_7 \cdot xH_2O$, other polyborates, $Na_2B_{10}O_{16} \cdot xH_2O$, at least one orthoborate, Na_3BO_3, and peroxyborates, such as $NaBO_3\cdot H_2O$. See also **Boron**. Other important sodium salts include the carbonates, cyanides, cyanates, hexacyanoferrates, $Na_4Fe(CN)_4$ and $Na_3Fe(CN)_6$, halides, polyhalides, hypohalites, halites, halates, perhalates, permanganates, ortho-, pyro-, meta-, fluoro-, and peroxyphosphates, hyposulfites, sulfites, sulfates, thiosulfates, peroxysulfates, polythionates, tungstates, vanadates, uranates, etc.

In addition to the simple compounds, sodium forms double salts of various types, although because of the relatively small size of the Na^+ ion, the number of sodium alums (see also **Alum**) is relatively small.

In addition to the inorganic salts, sodium forms such binary compounds as a phosphide, Na_3P, by direct union with phosphorus, a nitride, Na_3N, by direct union with nitrogen when activated electrically (which decomposes partly to give sodium amide, NaN_3, also obtained by heating sodium nitrate with sodium amide) and the oxides. Sodium monoxide, Na_2O, is obtained by heating the nitrite with the metal, displacing the nitrogen. Sodium peroxide, Na_2O_2, is the most stable oxide, obtained by reaction of the elements. Sodium superoxide is known, NaO_2, and one other oxide, Na_2O_3, has been reported. Sodium hydroxide, $NaOH$, is very soluble in H_2O and soluble in alcohol. It is almost completely ionized in water at ordinary concentrations, although its basic character is less than those of the higher elements in the group ($pK_B = -0.70$).

The detailed chemistry and applications of some of the more important compounds, other than those already discussed, follow.

Aluminate: Sodium aluminate, $NaAlO_2$, white solid, (1) by reaction of aluminum hydroxide and NaOH solution, (2) by fusion of aluminum oxide and sodium carbonate, the solution reacts with CO_2 to form aluminum hydroxide. Used as a mordant, and in water purification. See also **Aluminum**.

Aluminosilicate: Sodium aluminosilicate is used as a water softener for the removal of dissolved calcium compounds.

Amide: Sodamide, sodamine, $NaNH_2$, white solid, formed by reaction of sodium metal and dry NH_3 gas at $350\,°C$, or by solution of the metal in liquid ammonia. Reacts with carbon upon heating, to form sodium cyanide, and with nitrous oxide to form sodium azide, NaN_3.

Bromide: Sodium bromide, $NaBr$, white solid, soluble, mp $755\,°C$. Used in photography and in medicine. See also **Bromine**.

Carbonates: Sodium carbonate (anhydrous), soda ash, Na_2CO_3, sodium carbonate decahydrate, washing soda, sal soda, $Na_2CO_3\cdot 10H_2O$, white solid, soluble, mp 851 °C, formed by heating sodium hydrogen carbonate, either dry or in solution. Commonly bought and sold in quantity on the basis of oxide Na_2O determined by analysis (58.5% Na_2O equivalent to 100.0% Na_2CO_3).

Soda ash is a very-high-tonnage chemical raw material and approaches a production rate of 10 million tons/year in the United States. About 40% of soda ash is used in glassmaking; approximately 35% goes into the production of sodium chemicals, such as sodium chromates, phosphates, and silicates; nearly 10% is used by the pulp and paper industry; the remainder going into the production of soaps and detergents and in nonferrous metals refining. The first process for preparing soda ash was developed by Leblanc during the first French Revolution. In the Leblanc process, sodium chloride first is converted to sodium sulfate and subsequently the sulfate is heated with limestone and coke: (1) $Na_2SO_4 + 2C \rightarrow Na_2S + 2CO_2$; (2) $Na_2S + CaCO_3 \rightarrow Na_2CO_3 + CaS$. During the mid-1800s, the Solvay process was introduced. In this process, CO_2 is passed through an NH_3-saturated sodium chloride solution to form sodium bicarbonate, then followed by calcination of the bicarbonate: (1) $NH_3 + CO_2 + NaCl + H_2O \rightarrow HNaCO_3 + NH_4Cl$; (2) $2HNaCO_3 + heat + Na_2CO_3 + CO_2 + H_2O$. A large proportion of soda ash now is derived from the natural mineral trona, which occurs in great abundance near Green River, Wyoming. Chemically trona is sodium sesquicarbonate, $Na_2CO_3 \cdot NaHCO_3 \cdot 2H_2O$. After crushing, the natural ore is dissolved in agitated tanks to form a concentrated solution. Most of the impurities (boron oxides, calcium carbonate silica, sodium silicate, and shale rock) are insoluble in hot H_2O and separate out upon settling. Upon cooling, the filtered sesquicarbonate solution forms fine needle-like crystals in a vacuum crystallizer. After centrifuging, the sesquicarbonate crystals are heated to about 240 °C in rotary calciners whereupon CO_2 and bound H_2O are released to form natural soda ash. The crystals have a purity of 99.88% or more and handle easily without abrading or forming dust and thus assisting glassmakers and other users in obtaining uniform and homogeneous mixes.

Chlorate: Sodium chlorate, chlorate of soda, $NaClO_3$, white solid, soluble, mp 260 °C, powerful oxidizing agent and consequently a fire hazard with dry organic materials, such as clothes, and with sulfur; upon heating oxygen is liberated and the residue is sodium chloride; formed by electrolysis of sodium chloride solution under proper conditions. Used (1) as a weedkiller (above hazard), (2) in matches, and explosives, (3) in the textile and leather industries.

Chloride: Sodium chloride, common salt, rock salt, halite, NaCl, white solid, soluble, mp 804 °C. See also **Sodium Chloride**.

Chromate: Sodium chromate, $Na_2CrO_4 \cdot 10H_2O$, yellow solid, soluble, formed by reaction of sodium carbonate and chromite at high temperatures in a current of air, and then extracting with water and evaporating the solution. Used (1) as a source of chromate, (2) in leather tanning, (3) in textile dyeing, (4) in inks.

Citrate: Sodium citrate, $Na_3C_6H_5O_7 \cdot 5\frac{1}{2}H_2O$ white solid, soluble, formed (1) by reaction of sodium carbonate or hydroxide and citric acid, (2) by reaction of calcium citrate and sodium sulfate or carbonate solution, and then filtering and evaporating the filtrate. Used in soft drinks and in medicine.

Cyanide: Sodium cyanide, NaCN, white solid, soluble, very poisonous, formed (1) by reaction of sodamide and carbon at high temperature, (2) by reaction of calcium cyanamide and sodium chloride at high temperature, reacts in dilute solution in air with gold or silver to form soluble sodium gold or silver cyanide, and used for this purpose in the cyanide process for recovery of gold. The percentage of available cyanide is greater than in potassium cyanide previously used. Used as a source of cyanide, and for hydrocyanic acid.

Dichromate: Sodium dichromate, $Na_2Cr_2O_7 \cdot 2H_2O$, red solid, soluble, powerful oxidizing agent, and consequently a fire hazard with dry carbonaceous materials. Formed by acidifying sodium chromate solution, and then evaporating. Used (1) in matches and pyrotechnics, (2) in leather tanning and in the textile industry, (3) as a source of chromate, cheaper than potassium dichromate.

Dithionate: Sodium dithionate, "sodium hyposulfate," $Na_2S_2O_6 \cdot 2H_2O$, white solid, soluble, formed from manganese dithionate solution and sodium carbonate solution, and then filtering and evaporating the filtrate.

Fluorides: Sodium fluoride NaF, white solid, soluble, formed by reaction of sodium carbonate and hydrofluoric acid, and then evaporating. Used (1) as an antiseptic and antifermentative in alcohol distilleries,

(2) as a food preservative, (3) as a poison for rats and roaches, (4) as a constituent of ceramic enamels and fluxes; sodium hydrogen fluoride, sodium difluoride, sodium acid fluoride, $NaHF_2$, white solid, soluble, formed by reaction of sodium carbonate and excess hydrofluoric acid, and then evaporating. Used (1) as an antiseptic, (2) for etching glass, (3) as a food preservative, (4) for preserving zoological specimens.

Fluosilicate: Sodium fluosilicate, Na_2SiF_6, white solid, very slightly soluble in cold H_2O, formed by reaction of sodium carbonate and hydrofluosilicic acid. Used (1) in ceramic glazes and opal glass, (2) in laundering, (3) as an antiseptic.

Formate: Sodium formate, $NaCHO_2$, white solid, soluble, formed by reaction of NaOH and carbon monoxide under pressure at about 200 °C. Used (1) as a source of formate and formic acid, (2) as a reducing agent in organic chemistry, (3) as a mordant in dyeing, (4) in medicine.

Hydride: Sodium hydride, NaH, white solid, reactive with water yielding hydrogen gas and NaOH solution, formed by reaction of sodium and hydrogen at about 360 °C. Used as a powerful reducing agent.

Hydroxide: Sodium hydroxide, caustic soda, sodium hydrate, "lye," NaOH, white solid, soluble, mp 318 °C, an important strong alkali, not as cheap as calcium oxide (a strong alkali) nor sodium carbonate (a mild alkali), but of wide use. Formed (1) by reaction of sodium carbonate and calcium hydroxide in H_2O, and then separation of the solution and evaporation, (2) by electrolysis of sodium chloride solution under the proper conditions, and evaporation. Commonly bought and sold in quantity on the basis of oxide Na_2O determined by analysis (77.5% Na_2O equivalent to 100.0% NaOH). Used (1) in the manufacture of soap, rayon, paper ("soda process"), (2) in petroleum and vegetable oil refining, (3) in the rubber industry, in the textile and tanning industries, (4) in the preparation of sodium salts, (a) in solution, (b) upon fusion. See Fig. 1.

Fig. 1. Triple-effect evaporator used in concentrating soda solutions and preparation of solid sodium hydroxide.

Hypochlorite: Sodium hypochlorite, NaOCl, commonly in solution by (1) electrolysis of sodium chloride solution under proper conditions, (2) reaction of calcium hypochlorite suspension in water and sodium carbonate solution, and then filtering. Used (1) as a bleaching agent for textiles and paper pulp, (2) as a disinfectant, especially for water, (3) as an oxidizing reagent.

Hypophosphite: Sodium hypophosphite, $NaH_2PO_2 \cdot H_2O$, white solid, soluble, formed (1) by reaction of hypophosphorous acid and sodium carbonate solution, and then evaporating, (2) by reaction of NaOH solution and phosphorus on heating (poisonous phosphine gas evolved).

Hyposulfite: Sodium hyposulfite, sodium hydrosulfite (not sodium thiosulfate), $Na_2S_2O_4$, white solid, soluble, formed by reaction of sodium hydrogen sulfite and zinc metal powder, and then precipitating sodium hyposulfite by sodium chloride in concentrated solution. Used as an important reducing agent in the textile industry, e.g., bleaching, color discharge.

Iodide: Sodium iodide, NaI, white solid, soluble, mp 651 °C, formed by reaction of sodium carbonate or hydroxide and hydriodic acid, and then evaporating. Used in photography, in medicine and as a source of iodide.

Manganate: Sodium manganate, Na_2MnO_4, green solid, soluble, permanent in alkali, formed by heating to high temperature manganese dioxide and sodium carbonate, and then extracting with water and evaporating the solution. The first step in the preparation of sodium permanganate from pyrolusite.

Nitrate: Sodium nitrate, nitrate of soda, Chile saltpeter, "caliche," $NaNO3$, white solid, soluble, mp $308\,°C$, source in nature is Chile, in the fixation of atmospheric nitrogen HNO_3 is frequently transformed by sodium carbonate into sodium nitrate, and the solution evaporated. Used (1) as an important nitrogenous fertilizer, (2) as a source of nitrate and HNO_3, (3) in pyrotechnics, (4) in fluxes.

Nitrite: Sodium nitrite, $NaNO_2$, yellowish-white solid, soluble, formed (1) by reaction of nitric oxide plus nitrogen dioxide and sodium carbonate or hydroxide, and then evaporating, (2) by heating sodium nitrate and lead to a high temperature, and then extracting the soluble portion (lead monoxide insoluble) with H_2O and evaporating. Used as an important reagent (diazotizing) in organic chemistry.

Oleate: Sodium oleate, $NaC_{18}H_{33}O_2$, white solid, soluble, froth or foam upon shaking the H_2O solution (soap), formed by reaction of $NaOH$ and oleic acid (in alcoholic solution) and evaporating. Used as a source of oleate.

Oxalates: Sodium oxalate, $Na_2C_2O_4$, white solid, moderately soluble, formed (1) by reaction of sodium carbonate or hydroxide and oxalic acid, and then evaporating, (2) by heating sodium formate rapidly, with loss of hydrogen. Used as a source of oxalate; sodium hydrogen oxalate, sodium binoxalate, sodium acid oxalate, $NaHC_2O_4·H_2O$, white solid, moderately soluble.

Palmitate: Sodium palmitate, $NaC_{16}H_{31}O_2$, white solid, soluble, froth or foam upon shaking the H_2O solution (soap), formed by reaction of $NaOH$ and palmitic acid (in alcoholic solution) and evaporating. Used as a source of palmitate.

Permanganate: Sodium permanganate, permanganate of soda, $NaMnO_4$, purple solid, soluble, formed by oxidation of acidified sodium manganate solution with chlorine, and then evaporating. Used (1) as disinfectant and bactericide, (2) in medicine.

Phenate: Sodium phenate, sodium phenoxide, sodium phenolate, $NaOC_6H_5$, white solid, soluble, formed by reaction of sodium hydroxide (not carbonate) solution and phenol, and then evaporating. Used in the preparation of sodium salicylate.

Phosphates: Trisodium phosphate, tribasic sodium phosphate, $Na_3PO_4·12H_2O$, white solid, soluble, formed (1) by reaction of sodium hydroxide and the requisite amount of phosphoric acid, and then evaporating, (2) by reaction of disodium hydrogen phosphate plus sodium hydroxide, and then evaporating. Used (1) as a cleansing and laundering agent, (2) as a water softener, (3) in photography, (4) in tanning, (5) in the purification of sugar solutions; disodium hydrogen phosphate, dibasic sodium phosphate, $Na_2HPO_4·12H_2O$, white solid, soluble, formed (1) by reaction of dicalcium hydrogen phosphate and sodium carbonate solution, and then evaporating the solution, (2) by reaction of sodium carbonate and the requisite amount of phosphoric acid, and then evaporating. Used (1) in weighting silk, (2) in dyeing and printing textiles, (3) in fireproofing wood, paper, fabrics, (4) in ceramic glazes, (5) in baking powders, (6) to prepare sodium pyrophosphate; sodium dihydrogen phosphate, monobasic sodium phosphate, $NaH_2PO_4·H_2O$, white solid, soluble, formed (1) by reaction of sodium carbonate and the requisite amount of phosphoric acid, and then evaporating, (2) by reaction of calcium monohydrogen phosphate and sodium carbonate solution, and then evaporating the solution. Used (1) in baking powders, (2) in medicine, (3) to prepare sodium metaphosphate; sodium pyrophosphate, $Na_4P_2O_7·10H_2O$, white solid, soluble, mp about $900\,°C$, formed by heating disodium hydrogen phosphate to complete loss of water, followed by crystallization from water solution. Used in electroanalysis; sodium metaphosphate, $NaPO_3$, white solid, soluble, mp $617\,°C$, formed by heating sodium dihydrogen phosphate or sodium ammonium phosphate to complete loss of water, is an easily fusible phosphate forming colored phosphates with many metallic oxides, e.g., cobalt oxide. The hexametaphosphate, $(NaPO_3)_6$, is an important water-conditioning agent forming soluble complex compounds with many cations, e.g., Ca^{2+}, Mg^{2+}. Many polyphosphate compounds are known; their various uses include water softening and ion exchange. They are widely formulated in detergents, as are several of the simpler phosphates.

Phosphites: Disodium hydrogen phosphite, $Na_2HPO_3·5H_2O$, white solid, soluble, formed by reaction of phosphorous acid and sodium carbonate, and then evaporating at a low temperature, mp of anhydrous salt is $53\,°C$, at higher temperatures yields sodium phosphate and phosphine gas; sodium dihydrogen phosphite, $NaH_2PO_3 · 2\frac{1}{2}H_2O$, white solid, soluble, formed by reaction of phosphorous acid and $NaOH$ cooled to $-23\,°C$ when the crystalline salt separates.

Salicylate: Sodium salicylate, $NaC_7H_5O_3$, white solid, soluble, formed by reaction of sodium phenate and CO_2 under pressure. Used as a source of salicylate and for salicylic acid.

Silicate: Sodium silicate, sodium metasilicate, "water glass," Na_2SiO_3, colorless (when pure) glass, soluble, mp $1,088\,°C$, formed by reaction of silicon oxide and sodium carbonate at high temperature; solution reacts with CO_2 of the air, or with sodium carbonate solution or ammonium chloride solution, yielding silicic acid, gelatinous precipitate. Sodium silicate solution is used (1) in soaps, (2) for preserving eggs, (3) for treating wood against decay, (4) for rendering cloth, paper, wood noninflammable, (5) in dyeing and printing textiles, (6) as an adhesive (e.g., for paper boxes) and cement. Sold as granular, crystals, or $40°$ Baumé solution.

Silicoaluminate: (See aluminosilicate, above.)

Silicofluoride: (See fluosilicate, above.)

Stearate: Sodium stearate, $NaC_{18}H_{35}O_2$, white solid, soluble, froth or foam upon shaking the water solution (soap), formed by reaction of $NaOH$ and stearic acid (in alcoholic solution) and evaporating. Used as a source of stearate.

Sulfates: Sodium sulfate (anhydrous), "salt cake," Na_2SO_4, sodium sulfate, decahydrate, "Glauber's salt," $Na_2SO_4·10H_2O$, white solid, soluble, formed by reaction of sodium chloride and H_2SO_4 upon heating with evolution of hydrogen chloride gas. Used (1) in dyeing, (2) along with carbon in the manufacture of glass, (3) as a source of sulfate, (4) to prepare sodium sulfide; sodium hydrogen sulfate, sodium bisulfate, sodium acid sulfate, "nitre cake," $NaHSO_4$, white solid, soluble, formed by reaction of sodium nitrate and H_2SO_4, upon heating, with evolution of HNO_3. Used (1) as a cheap substitute for H_2SO_4. (2) in dyeing, (3) as a flux in metallurgy; sodium pyrosulfate, $Na_2S_2O_7$, white solid, soluble, formed by heating sodium hydrogen sulfate to complete loss of H_2O.

Sulfides: Sodium sulfide, Na_2S, yellowish to reddish solid, soluble, formed (1) by heating sodium sulfate and carbon to a high temperature. Used (1) as the cooking liquor reagent (along with sodium hydroxide) in the "sulfate" or "kraft" process of converting wood into paper pulp, (2) as a depilatory, (3) in sheep dips, (4) in photography, engraving and lithography, (5) in organic reactions, (6) as a source of sulfide, (7) as a reducing agent; sodium hydrogen sulfide, sodium bisulfide, sodium acid sulfide, $NaHS$, formed in solution by reaction of $NaOH$ or carbonate solution and excess H_2S.

Sulfites: Sodium sulfite, Na_2SO_3, white solid, soluble, dilute solution readily oxidized in air, but retarded by mannitol (carbohydrates), formed by reaction of sodium carbonate or hydroxide solution and the requisite amount of SO_2, at high temperature yields sodium sulfate and sodium sulfide. Used (1) as a source of sulfite, (2) as a reducing agent, (3) to prepare sodium thiosulfate, (4) as a food preservative, (5) as a photographic developer, (6) as a bleaching agent and antichlor in the textile industry; sodium hydrogen sulfite, sodium bisulfite, sodium acid sulfite, $NaHSO_3$, white solid, soluble, formed by reaction of sodium carbonate solution and excess sulfurous acid. Uses similar to those of sodium sulfite.

Tartrate: Sodium tartrate, $Na_2C_4H_4O_6·2H_2O$, white solid, soluble, formed by reaction of sodium carbonate solution and tartaric acid. Used in medicine; sodium potassium tartrate, Rochelle salt, $NaKC_4H_4O_6·4H_2O$, white solid, soluble. Used (1) in medicine, (2) as a source of tartrate.

Thiosulfate: Sodium thiosulfate, "Hypo" $Na_2S_2O_3·5H_2O$, white solid, soluble, formed by reaction of sodium sulfite and sulfur upon boiling, and then evaporating. Used (1) in photography as fixing agent to dissolve unchanged silver salt, (2) as a reducing agent and antichlor. See also **Sodium Thiosulfate**.

Tungstate: Sodium tungstate, $Na_2WO_4·2H_2O$, white solid, soluble, by reaction of $NaOH$ solution and tungsten trioxide upon boiling, and then evaporating. Used (1) in fireproofing fabrics, (2) as a source of tungsten for chemical reactions.

Uranate: Sodium uranate, uranium yellow, Na_2UO_4, yellow solid, insoluble, formed by reaction of soluble uranyl salt solution and excess sodium carbonate solution. Used (1) in the manufacture of yellowish-green fluorescent glass, (2) in ceramic enamels, (3) as a source of uranium for chemical reactions.

Vanadate: Sodium vanadate, sodium orthovanadate, Na_3VO_4, white solid, soluble, formed by fusion of vanadium pentoxide and sodium carbonate. Used (1) in inks, (2) in photography, (3) in dyeing of furs, (4) in inoculation of plant life.

The larger number of organic compounds of sodium are for great part derivatives of oxygen-containing compounds such as salts of organic acids (several of which are discussed above), alcoholic and phenolic compounds (carboxylates, alkoxides, phenoxides, etc.). However, in some cases, sodium derivatives of nitrogen-containing compounds, as sodium benzamide, $C_6H_5C(O)NHNa$, and sodium anilide, C_6H_5NHNa, contain sodium-nitrogen bonds, while even sodium-boron bonds exist in certain boron-containing compounds, as sodium triphenylborene, $NaB(C_6H_5)_3$, and others; and in a number of compounds sodium is carbon-connected, as in methylsodium, CH_3Na, ethylsodium, C_2H_5Na, cyclopentadienylsodium, C_5H_5Na, and sodium triphenylmethane, $NaC(C_6H_5)_3$.

The organometallic compounds of sodium may be divided into two groups, differing in properties. One group, e.g., ethylsodium, consists of compounds that are colorless, insoluble in organic solvents, and that electrolyze readily in diethylzinc solution. Another group, e.g., benzylsodium, $C_6H_5CH_2Na$, are colored, and soluble in organic solvents.

Like all the alkali metals, sodium coordinates with salicylaldehyde. Its tetracovalent compounds, with those of potassium, are the more stable of the group, for the following reasons: (1) Increasing ionic size carries with it increasing electropositiveness and ease of ionization, which diminishes the tendency to coordinate. (2) The increasing distance of the nucleus from the coordinating electrons with increasing atomic volume makes it less likely that additional electrons will be held with ease. (3) On the other hand, there is an increase in the maximum coordination number with the elements of higher atomic number. These factors are in keeping with a maximum stability for the tetracovalent compounds occurring with sodium.

Sodium in Biological Systems: Sodium is essential to higher animals which regulate the composition of their body fluids and to some marine organisms. The several important roles played by the sodium cation in biological systems, frequently in concert with the potassium cation are described in the entry on **Potassium and Sodium (In Biological System)**.

Additional Reading

Considine, D.M. and G.D. Considine: "Van Nostrand Encyclopedia of Chemistry," 4th Edition, Van Nostrand Reinhold Company, Inc., New York, NY, 1984.

Emsley, J. and J. Neruda: "The Elements," Oxford University Press, Inc., New York, NY, 1996.

Garrett, D.E.: "Sodium Sulfate: Handbook of Deposits, Processing and Use," Academic Press, Inc., San Diego, CA, 2001.

Greenwood, N.N. and A. Earnshaw: "Chemistry of the Elements," 2nd Edition, Butterworth-Heinemann, Inc., Woburn, MA, 1997.

Kent, J.A.: "Riegel's Handbook of Industrial Chemistry," 9th Edition, Chapman & Hall, New York, NY, 1992.

Krebs, R.E.: "The History and Use of Our Earth's Chemical Elements: A Reference Guide," Greenwood Publishing Group, Inc., Westport, CT, 1998.

Lagowski, J.J.: "MacMillan Encyclopedia of Chemistry," Vol. 1, MacMillan Library Reference, New York, NY, 1997.

Lewis, R.J. and N.I. Sax: "Sax's Dangerous Properties of Industrial Materials," 10th Edition, John Wiley & Sons, Inc., New York, NY, 2000.

Lide, D.R.: "CRC Handbook of Chemistry and Physics 2000-2001," 81st Edition, CRC Press, LLC., Boca Raton, FL, 2000.

Meyers, R.A.: "Handbook of Chemicals Production Processes," The McGraw-Hill, Companies, Inc., New York, NY, 1986.

Parker, S.P.: "McGraw-Hill Encyclopedia of Chemistry," 2nd Edition, The McGraw-Hill Companies, Inc., New York, NY, 1993.

Robinson, A.L.: "Sodium Atoms Stopped and Confined," *Science*, 229, 39–41 (1985).

Staff: "ASM Handbook—Properties and Selection: Nonferrous Alloys and Pure Metals," ASM International, Materials Park, OH, 1990.

Stwertka, A. and E. Stwertka: "A Guide to the Elements," Oxford University Press, Inc., New York, NY, 1998.

SODIUM CHLORIDE. NaCl, formula weight 58.44, white solid, cubic crystal structure, mp 800.6 °C, bp 1,413 °C, sp gr 2.165. Commonly called "salt," the mineral name for rock salt is *halite*. See also **Halite (Rock Salt)**. The compound is soluble in H_2O (35.7 g/100 g H_2O at 0 °C; 39.8 g/100 g H_2O at 100 °C), only slightly soluble in alcohol, and insoluble in HCl. Sodium chloride is produced in nearly all nations of the world, but

some only have a sufficient supply for local needs. The leading salt-producing nations include the United States, China, the former U.S.S.R., West Germany, France, the United Kingdom, India, Italy, Canada, and Mexico. In 28 states of the United States and in several provinces of Canada, salt occurs as bedded or domed deposits. Most of the rock salt produced in the United States comes from Michigan, New York, Texas, Ohio, Louisiana, and Kansas. Purity ranges from 97% NaCl for Kansas salt to 99% purity and higher for Louisiana salt. The main impurities are calcium sulfate (0.5–2%), dolomite, quartz, calcite, and traces of iron oxides. Natural rock salt is mined much as coal and usually marketed without purification, after crushing and screening. For most industrial and consumer requirements, the impurities are harmless. There is no evidence that bacteria exist in rock salt. Additionally, there is some solar salt production in the Great Salt Lake area of Utah and on the west coast. Salt deposits date back to past geologic ages and are believed to be the results of evaporated impounded sea water.

Purified salt for table and industrial processing requirements of a special nature is made by dissolving raw sodium chloride in H_2O and then evaporating the H_2O to form a final product. There are several types of evaporated salt, including *granulated salt* in which each crystal is a tiny cube, and *grainer* or *flake salt*, made up of irregularly shaped crystals, often thin and flaky and unusually soft. A process for producing evaporated salt is shown in Fig. 1. Holes are drilled into the salt deposits, after which H_2O is pumped into the beds to create a brine which then is brought to the surface for refining. In this method, all insolubles are left in the bed. After some pretreatment to remove hardness and dissolved gases, the semipure brine is evaporated in multiple-effect vacuum pans. The salt crystallizes as perfect cubes of NaCl. In the system shown, each vacuum pan performs not only as an evaporator, but also as a boiler. The vapors from a preceding pan are used to heat the contents of the following pan. This system of heat economizing is possible because each succeeding pan in the series is under less pressure—hence the contents boil at a lower temperature. See Fig. 2. The lower pressure in succeeding pans results from condensation of the vapors as well as assistance from vacuum pumps. Crystal size is controlled by evaporation rate, the latter depending on the degree of vacuum, temperature, and agitation maintained. When grown to proper size, the crystals drop to the bottom of the pans and fall into the salt legs, from which they are drawn continuously in the form of a slurry. After washing, filtering, cooling, and screening, they are packaged. See also **Evaporation**.

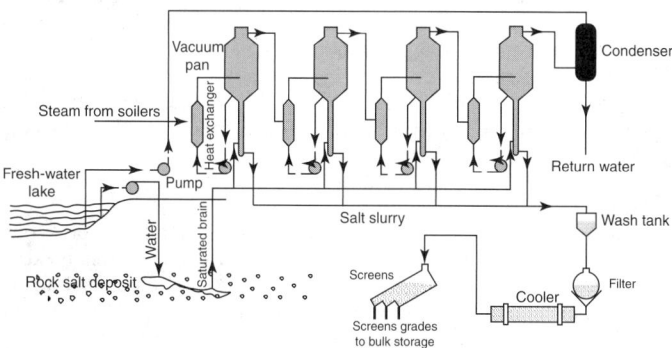

Fig. 1. Multiple-effect vacuum pans used in production of sodium chloride from brine. The saturated brine is formed by pumping fresh water directly into the rock salt deposit, leaving insoluble materials in the deposit.

Grainer salt is made by surface evaporation of brine in flat pans open to the atmosphere. Heat usually is furnished by steam pipes located a few inches below the tank bottom. Crystals form at the surface of the brine and are held there temporarily by surface tension. Thus, they grow laterally for awhile and form thin flakes. But, as they grow, they tend to sink and this process imparts a peculiar, hollow pyramid-like structure to them. Such crystals are called *hopper crystals*. Ultimately, the crystals sink to the bottom where they are scraped to one end of the pan. The crystals are fragile and during handling they break up, finally assuming a flake-like shape. Thus, the term *flake salt*.

In the *recrystallizer process* for making salt, advantage is taken of the fact that the solubility of NaCl increases with temperature whereas the solubility of the principal impurity, $CaSO_4$, decreases with temperature. In *solar* facilities, the raw brine is pumped into concentrating ponds where

Fig. 2. Portion of train of evaporator bodies in a multiple-effect vacuum evaporation system used in production of sodium chloride.

most of the H_2O is evaporated. Some of the impurities are precipitated out in this stage, after which the saturated brine is transferred to crystallization ponds where the salt crystallizes out at a high degree of purity. Since evaporation occurs at the surface of the ponds, hopper crystals are formed as in the grainer process, with flake salt being the final product.

Uses. Sodium chloride is a very high-tonnage material. In addition to its familiar use in the diets of man and animals, representing a small part of total production, large quantities are used by highway departments to control icy road conditions, in agriculture, and as a basic chemical raw material. The chemical industry consumes about two-thirds of the salt produced, the majority of it going to electrolytic plants. Some of the basic inorganic chemicals that require salt as a starting material include soda ash, calcium chloride, caustic soda, sodium sulfate, sodium bisulfate, HCl, sodium cyanide, sodium hypochlorite, and chlorine. See also **Chlorine**; and **Sodium**.

Salt and the Diet

In food processing, the preservative and organoleptic qualities of salt are well established and it is fully appreciated why use of salt even to excess is attractive to food processors. Excessive usage is also habitual among people who "salt first and taste later." Reports show over 1 million tons (900,000 metric tons) of salt are used in foods and in connection with eating in the United States in a given year. Nearly an additional 2 million tons (1.8 million metric tons) are used in the agricultural field, much of which is consumed by livestock. The total daily intake of the North American consumer as of 1980 is estimated to be in the range of 10–12 grams of salt, which reduces to a range of 3900–4700 milligrams of sodium. Highly salted snack foods, the consumption of which has increased markedly in many parts of the world during recent years, accounts for a significant consumption of salt. In addition, certain other food ingredients, such as monosodium glutamate and soy sauce, sometimes used in excess, also contribute to the average intake of sodium.

Sodium and chloride are not normally retained in the body even when there is a high intake. See also **Chloride (Biological Aspects)**. Amounts consumed in excess of need are excreted, so that the level in the body is maintained within very narrow limits, as is also the chloride, regardless of intake. The primary route of excretion is via the urine, with substantial amounts also lost in sweat and feces. About 50% of the sodium in the human body is located in the extracellular body fluids; 10% inside the cells; and 40% in the bones. Chloride is found mainly in the gastric juice and other body fluids.

Essential though sodium is to the normal functioning of the human body, there has been considerable concern over the last few years, about the amount of salt in the diet. This concern centers mainly on possible relationship between salt and hypertension (high blood pressure).

Hypertension afflicts more than 20% of the world population, with an estimated 24 million cases in the United States as of 1980. In 1976, Marx reported that, in about 90% of these cases, the actual causes of hypertension cannot be pinpointed.—This was in face of the fact that research on the possible role of sodium in essential hypertension had been underway for 60 years or longer. Tests of unmedicated persons with essential hypertension have been found to indicate a lowering of blood pressure when sodium intake is restricted below one gram per day—and that the blood pressure rises again if additional sodium is taken. However, in other studies, some persons have retained a normal blood pressure level even when fed substantially increased amounts of salt (or other sodium-ion-furnishing substances). In 1976, Freis reported positive correlations between estimated average salt consumption of various ethnic populations and their incidence of hypertension. But such studies are complicated by many factors, including the inability to control or eliminate other possible causes of hypertension, such as obesity, genetic predisposition, general nutritional status, and potassium intake. It also has been generally proved extremely difficult to determine differences between individuals within these cultures.

Nevertheless, the concern remains on the part of a large number of professional people who feel that someday a definitive correlation will be made. And, with considerable awareness of the lay public in this regard, very definite pressures are being exerted on food processors to reduce salt usage and to more accurately label their merchandise in this regard.

The physiology of the sodium-potassium relationship is explained in some detail in the entry on **Potassium and Sodium (In Biological Systems)**.

Concerning the sodium content (much of which is derived from salt), the following composition data may be of interest. The figures in parentheses are milligrams of sodium per 100 grams of food.

Meats: Canadian bacon (2555), bacon (1077), cured ham (860), beef liver (136), pork chops (60), ground beef (48).
Cheeses: Parmesan (1848), process (1421), blue (1396), brick (557), cream cheese (294).
Other dairy products: Ice cream (83), whole milk (50), sherbet (45).
Miscellaneous foods: Pretzels (7800), soy sauce (regular) (6082), dill pickles (4000-5000), soy sauce (mild) (3569), green olives (2400), soda crackers (1100), salted peanuts (groundnuts) (418), eggs (122).
Vegetables: Beet greens (130), celery (126), dandelion greens (76), kale (75), spinach (60), beets (60), watercress (52), turnips (49), carrots (47), artichokes (43), collards (43), mustard greens (32), Chinese cabbage (20). Other common vegetables range between (10) and (18).

Sodium Chloride and Energy

As pointed out by Wick (*Oceanus*, **22**, 4, 28, 1980), most of the energy in the oceans is bound in thermal and chemical forms. Although thermal energy is presently commanding the most attention, within the past few years another, rather unusual, form has received notice. Where rivers flow into the oceans a completely untapped source of energy exists—represented by a large osmotic pressure difference between fresh and salt water. If economical ways to tap these salinity gradients could be developed, large quantities of energy would be available. See also **Solar Energy**.

SODIUM THIOSULFATE. $Na_2S_2O_3 \cdot 5H_2O$, formula weight 248.19, white crystalline solid, decomposes above 48 °C, sp gr 1.685. Also known as "hypo" and sometimes misnamed "hyposulfite," sodium thiosulfate is very soluble in H_2O (301.8 parts in 100 parts H_2O at 60 °C), soluble in ammonia solutions, and very slightly soluble in alcohol. When sodium thiosulfate is added to an acid, thiosulfuric acid $H_2S_2O_3$ may be formed, but only for an instant, immediately decomposing into sulfur and SO_2.

Sodium thiosulfate is used: (1) to dissolve silver chloride, bromide, iodide in the photographic "fixing" bath, soluble sodium silver thiosulfate being formed plus sodium chloride, bromide, iodide; (2) in reaction with iodine in solution, sodium tetrathionate and sodium iodide being simultaneously formed, or with ferric salt solution, sodium tetrathionate and ferrous being simultaneously formed; and, (3) in reaction with chlorine as an "antichlor" forming sulfate and chloride. Sodium thiosulfate reacts with silver nitrate solutions yielding silver sulfide, brown precipitate, and with permanganate yielding manganous. Sodium amalgam changes sodium thiosulfate to sodium sulfide plus sodium sulfite.

Sodium thiosulfate is formed: (1) by reaction of sodium sulfite solution and sulfur upon warming; (2) by reaction of sodium sulfite solid and sulfur upon heating; and, (3) by complex reaction of sulfur and sodium hydroxide solution upon warming. Sulfur yields sodium sulfide plus sodium sulfite, and the latter reacts with excess sulfur, forming sodium thiosulfate. The sodium sulfide present may be converted into sodium thiosulfate by passing in SO_2 until the solution changes from yellow to colorless.

There are numerous other thiosulfates, including potassium, magnesium, calcium, barium, mercury, lead, and silver. All are soluble in H_2O except Ba, Pb, and Ag thiosulfates.

Thiosulfates are commonly identified as follows:

1. Dilute acids precipitate sulfur from thiosulfates (difference from sulfides and sulfites).

2. Zinc sulfate and sodium hexacyanoferrate(II) give no color (difference from sulfites).

SOFAR (Sound Fixing and Ranging).   A system of navigation providing hyperbolic lines of position determined by shore listening stations which receive sound signals produced by depth charges dropped at sea and exploding in a sound channel which is at a considerable depth in most areas. This system was used in Project Mercury for locating spacecraft down at sea.

SOFTWARE (Computer System).   The totality of programs, procedures, rules, and (possibly) documentation used in conjunction with computers, such as compilers, assemblers, narrators, routines, and subroutines. References are made to software and hardware parts of a system where the hardware comprises the physical (mechanical and electronic) components of the system. In some machines, the instructions are microprogrammed in a special control storage section of the machine using a more basic code, which is actually wired into the machine. This is contrasted with the situation where the instructions are wired into the control unit. The microprogram technique permits the economic construction of various size machines that appear to have identical instruction sets. However, microprograms are not generally considered as software. They are sometimes called firmware. See also **Digital Computer Systems**; and **Microprogram**.

SOFTWARE ENGINEERING.   Engineering is the disciplined application of established principles for design and development of entities and systems. Software engineering deals with software systems. Software engineering is increasingly important because many, newer devices and systems include or are even based upon software. For example, communication systems (such as telephones and television devices and the systems that distribute phone and television data) are being changed from analog to digital information and from analog processing to software-controlled signal processing and switching. The very nature of modern business is being transformed by software tools for office automation and for electronic commerce (e.g., the network connections of businesses to businesses and organizations to individuals.)

A key point about software engineering is that engineers do different work than other information technology workers. Many people are able to write small programs, but designing larger systems requires additional skills beyond just programming. As in other fields of engineering, design encompasses issues of correctness, reliability, usability, economy, etc. Development includes issues of documentation, distribution, and maintenance, as well as management of teams of engineers and technical workers. Technicians, typically called programmers, may do may do actual coding, testing, maintenance, and customer communication and support.

The Status of Software Engineering

Software development is yet an emerging discipline. The first programming languages (FORTRAN, COBOL) date to 1960. The first use of the concept of "software engineering" dates to 1968. There is not yet a common B.S. curriculum for software engineering. Instead, professional education for information technology workers is in fields of computer science, information systems, and even business systems management. Computer science programs are reviewed and accredited by Computing Science Accreditation Board (CSAB), which is moving towards merging with ABET, which accredits other engineering programs. Within CSAB

guidelines, there is little requirement for software engineering courses. Yet, within CS curricula, there is a steady shift towards more software engineering content. For example, a decade ago, core C.S. courses were programming, data structures, and analysis of algorithms. Today the beginning programming course typically includes simple design structures, the data structures course includes software engineering patterns and use of standard library packages, the third core course focuses on software architecture and design, and lastly, the curriculum concludes with a capstone one-year course on software engineering. That course emphasizes project management, software design and testing, and a group project design and implementation. Students often report that the hardest part of the project experience is the group dynamics of working together and managing a team.

At the M.S. level, several schools offer a Master of Software Engineering degree (or some similar degree) (Graduate Programs in Software Engineering). These degrees are often taught as distance education programs for persons working in the software field. There are several reasons for this pattern. First, because of good opportunities and salaries for B.S. holders, few American students continue directly to graduate work in computing. Second, students in many different disciplines of study include core courses in computer science. Later when some of these persons become involved in developing software for their discipline, the need arises for specific training in software engineering. Many master's programs are tailored for persons who have individual programming experience and who then desire to manage software development projects.

The Ph.D. degree is typically required for persons who teach or who work in research. Most Ph.D.-level work in software engineering is done within computer science departments. One exception is the Ph.D. in Software Engineering offered at Carnegie Mellon University.

There are two professional societies for computing and software professionals, the IEEE Computer Society and the Association for Computing Machinery (ACM). Both have names not associated with software, which reflects that historically, the computer itself was deemed more significant and challenging than the software.

There is yet no professional engineering certification for software engineers. Indeed, there is no official designation of who can be called a software engineer. Should it require a degree in the discipline, a degree in computer science, demonstration of specific skills and knowledge, or just experience in programming? This is significant because there have been several cases reported wherein project failure or even catastrophe has been attributed to software errors. There is effort to establish national certification for software engineers (IEEE); however, the ACM recently withdrew from support of that effort. There is a professional code of ethics for software engineers.

Because there is no well-established base of trained software engineers, it was observed that there is a need to provide assistance to companies adopting software engineering concepts. Towards that end, the Software Engineering Institute was established with grants from the U.S. Department of Defense. The SEI is a nonprofit organization that promotes software engineering by hosting conferences and by providing guidelines for education and for the software development process. One development of the SEI is the Capability Maturity Model for software development, which defines and ranks levels of industrial processes for software development.

Foundation Knowledge

What distinguishes software engineering from computer programming or information systems management? In the second case, software engineering deals with the management of the development of new software rather than the management of the operations of information systems. In the first case, the difference lies in the process and scale of software development and in the issue of control of quality, particularly correctness.

Many people today learn to write programs in some programming language. Many can develop complex functions using Rapid-Application-Development tools such as Visual Basic wizards and controls for graphic user interface and for data base access. Many can develop simple web pages using HTML and even applets using Java. These skills are not the core of software engineering. Experience has shown the pitfalls in developing large software systems. For large systems, errors are introduced in the interactions between people and between software components. One person can easily develop a system with ten software components, whereas managing and working within a team of 200 people developing a system with 2000 components requires a different process and skills.

The number of potential points of communication and hence the number of possible places for errors in communication increases with both the square of the number of persons and the square of the number of components.

From an academic point of view, the foundations of software engineering are shown in the core courses for the MSE degree. Even though there is no formal accreditation for MSE education, there is general conformity among schools that offer an MSE degree. From an industrial point of view, the foundations of software engineering are reflected in common practices, tools, and standards (Software Engineering Standard Committee). The following items correspond to core course in a MSE degree program:

- Breadth of knowledge of software development and management: For example, the text and web site by Pressman covers topics including: development process and management, measurements and planning, quality assurance, requirements analysis, architecture design, coding practices, testing, and software tools for all of these items. The web site for his book includes details of these topics, outlines, and sample assessment quizzes. Another view of the breadth of software engineering is the Software Engineering Body of Knowledge report, which is intended to identify common knowledge of the field. Professional competency in software engineering should also include the ability to read applied papers in software engineering journals, such as IEEE Transactions on Software Engineering.
- Object oriented software design and architecture: Software architecture consists of the organization of components. Design consists of defining models for both architectures and dynamic behaviors. The common notation for designs is the Unified Modeling Language.
- Formal methods in software specification: Mathematical models are used to describe properties of software. In Europe, the most accepted modeling languages are Z and VDM (Virtual Library), but these have not gained widespread use in the United States. The Object Constraint Language (OCL) component of the Unified Modeling Language seems likely to be used for such formal specification of software. Checking of properties of finite state models of software is now recognized to be effective for software verification (Formal Methods), but it is not yet widely used.
- Software project management: Managers direct and evaluate documents that define a project and its scheduling.
- Software project and portfolio: The crucial aspect of software engineering preparation is project experience with one-on-one review of all project work and documents. Such close review and feedback is not duplicated in most undergraduate classes. Major items of the portfolio are: Personal log of activities, requirements statement and use-case model; size and cost estimates and project work schedule; a software quality assurance plan (including documents to be reviewed); software design using UML; a formal model using Z or OCL; project inspection reports; test plan, including test cases and expected results; program source code; test results and analysis; project demonstration report; user documentation, including guidance for installation and use; and project evaluation, including evaluation of the size and schedule estimate.

Summary

Software engineering is a discipline distinct from just computer programming. It is concerned with software design and development and with management of teams of programmers and engineers. Most undergraduate computer science programs introduce software engineering concepts, but most training in software engineering is done through industrial training or through graduate studies.

See also **Software Metrics**; and **Artificial Intelligence**.

Additional Reading

"A Summary of the ACM Position on Software Engineering as a Licensed Engineering Profession," July 17, 2000, *http://www.acm.org/serving/se_policy/selep_main.html#executive_summary*

Graduate Programs in Software Engineering, FASE, **8**(9), *http://www.cs.ttu.edu/fase/v8n09.txt*

"Guide to the Software Engineering Body of Knowledge", November 1999, *http://www.swebok.org/*

Pressman R.: "Software Engineering: A Practitioner's Approach", McGraw-Hill, 2000, ISBN: 0073655783; Supporting materials at *http://www.mhhe.com/engcs/compsci/pressman/*

"Software Engineering Code of Ethics", IEEE Computer Society, October 1999, *http://www.computer.org/computer/Code-of-Ethics.pdf*

Web References

Association for Computing Machinery, *http://www.acm.org/*

Carnegie Mellon University, Ph.D. Program in Software Engineering, *http://www.isri.cs.cmu.edu/phd.html*

Computing Sciences Accreditation Board, *http://csab.org/*

FASE: Forum for Advancing Software Engineering Education, *http://www.cs.ttu.edu/fase/*

Formal Methods — Model Checking, *http://www.cs.cmu.edu/~modelcheck/*

I.E.E.E. Computer Society, *http://www.computer.org/*

IEEE Transactions on Software Engineering, *http://www.computer.org/tse/*

Object Constraint Language Specification, September 1997, *http://www.rational.com/media/uml/resources/media/ad970808 UML11 OCL.pdf*

Software Engineering Coordinating Committee, IEEE Computer Society, *http://computer.org/tab/swecc.htm*

Software Engineering Institute, Carnegie Mellon University, *http://www.sei.cmu.edu/*

Software Engineering Standards Committee, *http://www.computer.org/standards/sesc/*

UML: Unified Modeling Language, Rational Corp., *http://www.rational.com/uml/index.jsp*

Virtual Library: Formal Methods, *http://archive.comlab.ox.ac.uk/formal-methods.html*

WILLIAM HANKLEY, Kansas State University, http://www.cis.ksu.edu/

SOFTWARE METRICS.   A software metric is a rule for quantifying a characteristic or attribute of a computer software entity. That is, software metrics is an attempt to attach to software entities meaningful numbers that represent the extent that the entity has the attribute of interest.

For example, a simple metric is the FileSize metric. It is the total number of characters in the source file(s) of a program. Using the FileSize metric, one can determine the measure of a particular program, such as 3K bytes. The File Size metric provides a concrete measure of the abstract attribute of program size.

Another example of a software metric is a complexity metric. Complexity metrics are intended to provide a quantitative measure of the abstract attribute of complexity. Intuitively, if one software entity is more complex than another software entity, then it follows that the more complex entity may have more errors, may be harder to understand, or may be more difficult to maintain. Unfortunately, complexity metrics provide only an imperfect measurement of the attribute of complexity.

In addition to metrics for source programs, other metrics can be used for software entities such as requirements documents, design object models, or data base structure models.

Metrics for requirements and design documents can be used to guide decisions about development and as a basis for predictions, such as for cost and effort. Metrics for programs can be used to support decisions about testing and maintenance and as a basis for comparing different versions of programs. Ideally, metrics would be desired for the development cost of software and for the quality of the resultant program; but, lacking those, metrics for size and complexity are commonly used.

If one could attach a number (quantification) to a software entity that matches the complexity (attribute) of the software, then one could make better management decisions about the need for additional testing or the need to rewrite the software, etc. Unfortunately, it is not possible to meaningfully quantify an attribute as broad or as ill-specified as complexity.

Maurice Halstead introduced several metrics that better define program size. He found that the number of unique operators, η_1 (operators are symbols such as "+" and function and procedure names), and unique operands, η_2 (generally operands are the variables and numbers in the program) could be used to predict software size as: $N = \eta_1 \log_2 \eta_1 + \eta_2 \log_2 \eta_2$. He also defined metrics that relate to complexity, but such metrics have not been widely used.

Thomas McCabe introduced a complexity metric for programs. It was based on the cyclomatic number used in graph theory. McCabe measured the cyclomatic number on the control flow graph of a program or design pseudo-code. He showed that the cyclomatic number was equal to one plus the number of decisions in the control graph. The cyclomatic number is commonly used as a measure of complexity. A good "rule of thumb" is that program modules with cyclomatic number greater than ten should be avoided. Whereas the cyclomatic number is intended to measure complexity of single program units (a function, procedure method, etc.), a number of metrics have been introduced to measure the complexity of object models and programs.

The earliest size metric for programs was Lines of Code (LOC). Unfortunately, LOC is not a good predictor of cost of development, likelihood of errors, etc. Additional, there is no standard way to measure LOC; variations include counting or not counting blank lines, comments, declarations, and extra lines used for formatting. Still, LOC is also commonly used.

Albrecht and Gaffney developed the Function Points (FP) metric. It is used to predict the size and cost of development of a proposed software. The developer identifies and counts all inputs, outputs, inquiries, files, and interfaces and rates each as simple, average, or complex. Additional characteristics of the project are ranked on a scale of 1 to 5.

More recent work (Baker et al.) has applied measurement theory to software metrics. Measurement theory has been used for many years in other sciences. Important ideas include the representation condition and the notion of measurement scales.

The representation condition requires that if one entity is less than another entity in terms of a selected attribute, then any metric for that attribute must associate a smaller number to the first entity than it associates with the second entity. For example, the FileSize "metric" does not meet the representation condition. If a "small" program is taken and all variables with very long names are replaced, the measure of FileSize of the "small" program can be larger than that of some "larger" program which uses short variable names.

There are five measurement scales: nominal, ordinal, interval, rational, and absolute. Every software metric must be one of the five types of scales (Gustafson and Parsed). A nominal scale just attaches numbers to entities, such as numbers on sport uniforms. Ordinal scales rank entities, like being first or second. Interval scales have a fixed difference between values. Temperature is an example of an interval scale. Rational scales have a universally accepted zero and a fixed difference. Length is an example of a rational scale. An absolute scale counts the number of items.

The type of scale determines the types of statistics that can be applied to the metric. A metric must be at least an interval scale to take averages and must be at least a rational scale to consider ratios between values. For example, the cyclomatic number is an interval scale and thus it would be correct to calculate the average cyclomatic number of a set of modules. However, one could not correctly say that one module has twice the cyclomatic number as another module. Some recent research has directed toward formal approaches to defining and identifying properties of software metrics (Gustafson and Parsed).

Dieter Rombach and Victor Basili have presented the Goal-Question-Metric (GQM). The first step in this method is to select goals that should be maximized or minimized. For example, a goal might be customer satisfaction. Next, questions are developed that are related to that goal. Such as, "Is the customer finding too many errors?" Finally, metrics are proposed that quantify answers to the questions. For example, a metric could be the number of customer reported errors. Since the metrics are closely related to the goals, use of the metrics should contribute to the achievement of the goal.

See also **Software Engineering**.

Additional Reading

Albrecht, A.J.: "Measuring application development productivity," in Proc. IBM Applications Development Joint SHARE/GUIDE Symposium, 1979 pp. 83–92.

Baker, A. et al.: "A Philosophy for Software Measurement," *Journal of Systems and Software* **12**, 277–281.

Basili, V. and D. Rombach: "The Tame Project: Towards improvement-oriented software environments," *IEEE Trans on Software Engineering* **14**(6), 758–773 (1988).

Fenton, N E.: "Software Metrics," PWS Publishing Company, Boston, MA, 1998.

Gustafson, D. and B. Parsed: "Properties of Software Measures," in *Proc. of FACS* Workshop on Software Metrics, 1991.

Halstead, M.: "Elements of Software Science," Elsevier, North-Holland, 1977.

Kan, S.H.: "Metrics and Models in Software Quality Engineering," Addison Wesley Longman, Inc., Reading, MA, 1994.

McCabe, T.: "A Complexity Measure," *IEEE Trans on Software Engineering*, **2**(4), 308–320, (1976).

Melton, A.: "Software Measurement," International Thomson Computer Press, Boston, MA, 1995.

Menasce, D.A. and A.F. Virgilio: "Capacity Planning for Web Performance: metrics, models, and methods," Prentice-Hall Inc., Upper Saddle River, N.J, 1998.

Moody, J.A., W.L. Chapman, and F.D. Voorhees: "Metrics and Case Studies for Evaluating Engineering Designs," Prentice-Hall Inc., Upper Saddle River, N.J, 1997.

Poulin, J.S.: "Measuring Software Reuse," Addison Wesley Longman, Inc., Reading, MA., 1996.

Zuse, H.: "Software Complexity Measures and Methods," W. de Gruyter, Berlin, 1991.

DAVID A. GUSTAFSON, James Madison University, Harrisonburg, VA.

SOFTWOODS. See **Wood (or Timber; Lumber)**.

SOIL. All consolidated earth material over bedrock. Soil is approximately equivalent to regolith.[1] Agriculturally, soil is any one of many varied natural media that support or can support land plant growth outdoors; or, when in containers, indoors. The lower limit of *topsoil* is normally the lower limit of biologic activity, which usually coincides with the common rooting of native perennial land plants. The word *soil* is derived from the Latin *solum* for "ground."

The upper part of the regolith is divided into topsoil and subsoil. The topsoil is usually a relatively thin layer or zone of the more highly decomposed mineral constituents of the regolith and contains a varying proportion of organic material called *humus*. This soil zone is the habitat of the shallow-rooted plants, such as most grasses. The topsoil usually passes gradationally into the subsoil, which supplies some of the moisture and food for the deeply rooted plants and trees. The subsoil may or may not pass gradationally into the underlying bedrock. Topsoil is easily destroyed by erosion when not protected by a mantle of vegetation.

Soil serves: (1) As a foundation for holding plants in place, whether tiny grasses or huge trees; (2) as a protective covering for the root structures of plants; (3) as a source and/or medium of exchange for supplying plants with nutrients; (4) and as a reservoir for moisture upon which growing plants can draw. Soil also must be capable of allowing excessive moisture to pass through its pores and drain to a lower level so that the soil will not remain excessively wet. Properties such as permeability and strength are not only of importance to the agriculturist, but to civil engineers and construction people who excavate, drill through, and handle soil in connection with buildings, roads, tunnels, etc.

Soil is a subsystem that interacts as part of a four-element system: (1) The *climatic or environmental subsystem* prevails immediately above the ground level and thus is the microclimate for a particular location. The principal variables of this subsystem are temperature, humidity, precipitation, and solar radiation—all of which interact constantly with soil. (2) The characteristics and patterns of the *hydrologic subsystem* determine essentially how water reaches the soil, both from above and below, and how water is carried away or drained from the soil. See also **Hydrology**. (3) The *plant subsystem* reduces the nutrient and moisture content of the soil, depending upon the particular uptake characteristics of a given plant. The plant subsystem also contributes in a major way to hold the soil in place and to protect it from disintegration and destruction by water and wind erosion. The plant subsystem also distributes moisture over the top of the soil so that all porous paths of the soil can be used to transport water rather than overloading and hence enlarging only some of the pores. The plant subsystem also protects the top of the soil against drying into a hard crust during periods of drought. Once disturbed, the characteristics of soil are difficult to replicate—a problem that arises when large projects, such as strip-mining, remove vast amounts of soil. See also **Revegetation**. (4) The *soil subsystem*, the properties of which are described briefly in this entry.

Lack of sufficient attention to the long-term protection of soil has caused innumerable problems and losses over the years. Although warnings of gross problems arising from soil destruction and land mismanagement were given in North America as early as the latter part of the 1600s, it required the rudest of awakenings to precipitate national interest and action in soil conservation. This came in the early 1930s in the form of the Great Dust Bowl, a national disaster that affected some 96 million acres (38.4 million hectares) of farmland in the southern part of the Great Plains region, involving parts of Kansas, Oklahoma, Texas, New Mexico, and Colorado. During just a few years of severe droughts, accompanied

---

[1] A general term for the entire layer or mantle of fragmental and loose, incoherent, or unconsolidated rock material, of whatever origin (residual or transported) and of a much varied character, that nearly everywhere forms the surface of the land and overlies or covers the more coherent bedrock. It includes rock debris (weathered in place) of all kinds, volcanic ash, glacial drift, alluvium, loess and aeolian deposits, vegetal accumulations, and soils.

by frequent high winds, literally billions of tons of soil were lost. Organic matter, clay, and silt were lifted and carried for great distances. There were times when the heavily laden skies as far east as the Atlantic coast were darkened. Sand and silt dunes from 4 to 10 feet (1.2 to 3 meters) in height were formed in many locations. In some parts of the Dust Bowl, as much as 80% of the land suffered from wind erosion. Parallel situations have occurred in several other areas of the world.

The Soil Conservation Service of the U.S. Department of Agriculture was established in 1935. Concentrated and participative programs with land users in the Great Plains region from the mid-1950s to the present time have provided impressive improvements: (1) 2.4 million acres (1 million hectares) of permanent vegetative cover have been established; (2) 1.0 million acres (0.4 million hectares) of field and wind stripcropping have been introduced; (3) 169 thousand (68 thousand hectares) of grasslands have been reestablished; (4) 41 thousand acres (16.8 thousand hectares) of trees or shrubs have been placed as windbreaks; (5) 81 thousand miles (150 thousand kilometers) of terraces have been constructed; (6) 5.4 million acres (2.2 million hectares) of brush control have been provided; and (7) 9 thousand miles (16.7 thousand kilometers) of pipelines to provide water for livestock grazing lands have been installed. See also **Erosion (Geology)**.

Soil Characteristics and Classification

A soil is a naturally occurring three-dimensional body with morphology and properties resulting from effects of climate, flora and fauna, parent rock materials, topography, and time. A soil occupies a portion of the land surface, is mappable and is composed of horizons that parallel the land surface. A vertical section downward through all the horizons of the soil is called a *soil profile*. See Fig. 1.

Characterization of a soil requires selection of a representative profile that is described as quantitatively as possible, utilizing comparative charts for color, structure, and other properties, and accurately measuring soil horizons. Soils are collected from horizons and analyzed for particle-size distribution, pH, organic carbon, nitrogen, free iron oxide, calcium carbonate equivalent, moisture tension, cation-exchange capacity, extractable cations (calcium, magnesium, hydrogen, sodium, potassium), base saturation, and bulk density, among other factors.

Soil classification has been oriented to soil properties in recent years, but still is tempered with concepts of soil genesis, with external associations, and with the use of the soil. The first systematic classification was by Dokuchaiev in Russia in 1882. Based upon field and laboratory characteristics, soils were grouped into three categories — *normal soils* of the dry-land vegetative zones and moors, *transitional soils* of washed or dry land sediments; and *abnormal soils*. The system involved properties of the soil with external associations of climate and vegetation. Later, an associate (Sibirtsev) renamed the highest classes *zonal*, *intrazonal*, and *azonal*.

A traditional classification of soils includes three categories: (1) *Young soils*. These usually show their relationship to the parent material and are typical flood plain and hilly land deposits, when the soil surfaces are constantly being replenished or disturbed. (2) *Mature Soils*. These usually cover relatively flat lands where there are good drainage conditions but relatively little erosion. The development of these soils has gone so far in some cases, particularly in semi-arid regions, that little relation is shown to the parent material and their nature has therefore been principally determined by climatic and organic factors. (3) *Old Soils*. These usually cover old flat surfaces that have not been disturbed by erosion or sedimentation for a long time. Such soils, due to the dominance of climatic factors in their formation, have lost many of their original characteristics and have, therefore, developed abnormal features. When soils are intensively cultivated their mineral and organic constituents are rapidly depleted and must be replenished by rotation of crops and the application of natural fertilizers. The method of allowing the land to remain fallow is now known to be inefficient. The complete removal of vegetable cover, such as may result from overgrazing, deforestation, or dry farming, exposes the soil to rapid erosion and destruction.

A more scientific classification of soils, adapted by the U.S. Department of Agriculture, is given in Table 1. Systematic classification of soils in the United States began with the work of Coffey in 1912 and resulted in the first comprehensive system by Marbut in 1936. Considering the size of the earth and the large number of soils represented, the detailed cataloging of soils for any country is a tremendous task. To simplify the task to some extent and to make findings more meaningful from a practical viewpoint, the European Commission on Agriculture (Working Party on Soil Classification and Survey) in 1966 correlated types of soils

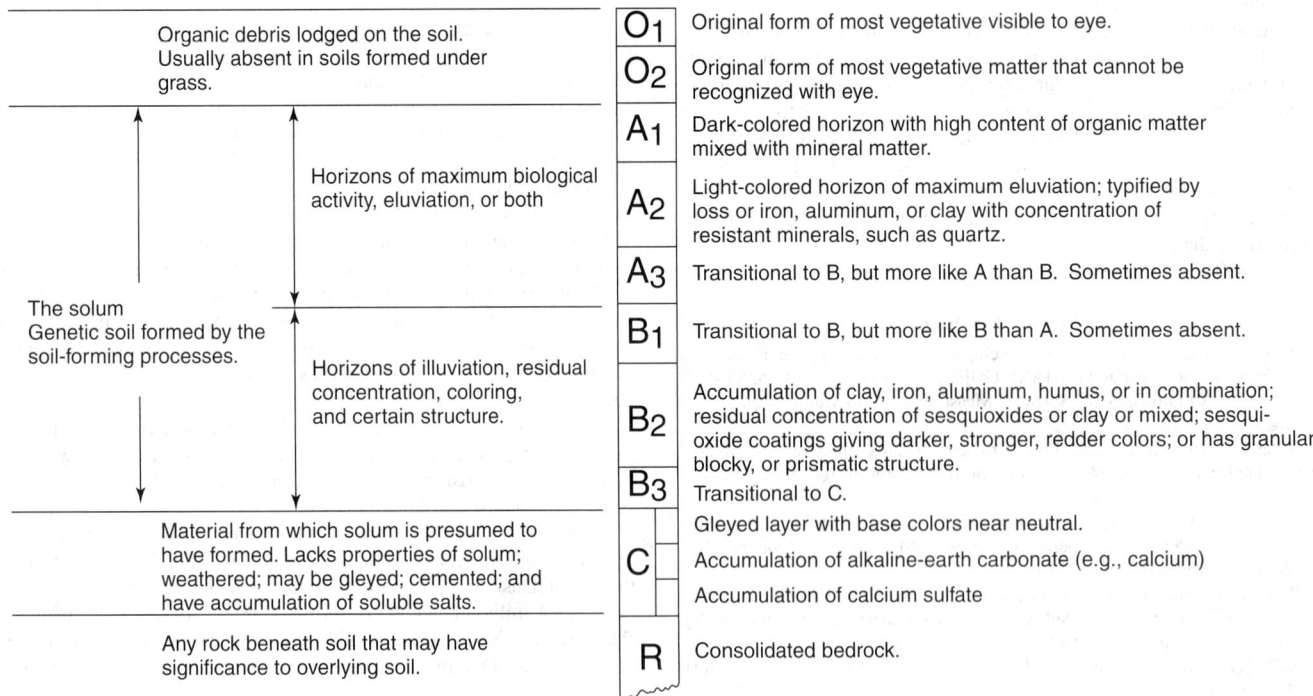

Fig. 1.   Hypothetical soil profile that has all principal horizons. Not all horizons shown are present in any given profile, but every profile has some of them. Terms used in diagram: *Eluviation* is the downward movement of soluble or suspended material in a soil from the A horizon to the B horizon by groundwater percolation. The term refers especially, but not exclusively, to the movement of colloids, whereas the term *leaching* refers to the complete removal of soluble materials. *Illuviation* is the accumulation of soluble or suspended material in a lower soil horizon that was transported from an upper horizon by the process of eluviation. *Gleying* is soil mottling, caused by partial oxidation and reduction of its constituent ferric iron compounds, due to conditions of intermittent water saturation. Process is also called *gleization*. (*Adapted from USDA diagram.*)

TABLE 1. SOIL CLASSIFICATION SYSTEM (U.S.D.A.)

Order and Suborders	Definitions and Properties
Entisols	Weakly developed soils on freshly exposed rock or recent alluvium without genetic horizons. While alluvium may be rich in plant nutrients, entisols often are too shallow, too wet, or too dry for agricultural purposes.
E1	*Aquents.* Seasonally or perennially wet.
E2	*Orthents.* Loam or clay texture, often shallow to bedrock.
E3	*Psamments.* Sand or loamy sand texture.
Vertisols	Clay soils that have deep wide cracks during periods of moisture deficiency. During rainfall, vertisols swell, slide and produce warping.
V1	*Uderts.* Usually moist, with cracks open less than 90 days/year.
V2	*Usterts.* Dry and cracked more than 90 days/year.
Inceptisols	Soils that are beginning to show development of genetic horizons. Inceptisols lack evidence of weathering and usually are found in humid climates where leaching is active.
I1	*Andepts.* Soils containing amorphous or allophanic clay, often associated with volcanic ash and/or pumice.
I2	*Aquepts.* Seasonally or perennially wet.
I3	*Ochrepts.* Soil with thin, light colored surface horizons.
14	*Tropepts.* Continuously warm or hot.
15	*Umbrepts.* Dark surface horizons; medium to low base supply.
Aridisols	Soils which contain little organic matter or nitrogen. They are usually dry for more than 6 months/year. In numerous areas, salts accumulate on or near the soil surface. Since the nutrient content, except nitrogen, of aridisols is often high, these soils can be productive with irrigation and nitrogen application. Salt accumulation can be a problem with some crops.
D1	*Undifferentiated aridisols.*
D2	*Argids.* Soils with horizons of clay accumulation.
Mollisols	Soils with dark, thick, organic-rich surface horizon, high base supply. Mollisols are highly fertile and can support a variety of crops.
M1	*Albolls.* Soils with seasonally high water tables.
M2	*Borolls.* Cool or cold soils.
M3	*Rendolls.* Soils with subsurface accumulations of calcium carbonate, but no clay.
M4	*Udolls.* Temperate or warm, usually moist.
M5	*Ustolls.* Temperate or hot. Dry more than 90 days/year.
M6	*Xerolls.* Cool to warm. Moist in winter and continuously dry more than 60 days/year.
Spodosols	Soils found primarily in cool and humid forested regions. Spodosols have subsurface accumulations of amorphous materials, mainly iron and aluminum oxides. These soils are usually strongly leached, but can be used for crop support with addition of lime and fertilizer.
S1	*Undifferentiated spodosols.*
S2	*Aquods.* Seasonally wet.
S3	*Humods.* Soils with subsurface accumulations of organic matter.
S4	*Orthods.* Soils with subsurface accumulations of organic matter, iron, and aluminum.
Alfisols	Soils of middle latitudes and degraded grasslands soils. Alfisols are strongly weathered, with gray to brown surface horizons, a subsurface clay accumulation, and a medium-to-high base supply. With adequate lime and fertilizer, the alfisols will continue to produce a variety of crops.
A1	*Boralfs.* Cool soils.
A2	*Udalfs.* Temperate to hot. Usually moist.
A3	*Ustalfs.* Temperate to hot. Dry more than 90 days/year.
A4	*Xeralfs.* Temperate to warm. Moist in winter and continuously dry more than 60 days in summer.
Ultisols	Strongly weathered soils of the middle and low latitudes. Ultisols are usually moist and low in organic matter. These soils have experienced a high degree of mineral alteration and extensive leaching. With fertilizer additions and good management, ultisols can support crops.
U1	*Aquults.* Seasonally wet.
U2	*Humults.* Temperate or warm. Moist all year. High content of organic matter.
U3	*Udults.* Temperate to hot. Usually moist.
U4	*Ustults.* Warm or hot. Dry more than 90 days/year.
Oxisols	The predominant soils of the Tropics. Oxisols have experienced the greatest degree of mineral alteration and horizon development of any soil. The humus breakdown is rapid and the soils are usually deep and porous. Oxisols require fertilization to support continued crop production.
O1	*Orthox.* Hot and nearly always moist.
O2	*Ustox.* Warm or hot. Dry for long periods, but moist for at least 90 days/year.
Histosols	Bog or peat soils composed primarily of vegetative debris in various stages of decomposition.
Mountain	Soils with various temperature and moisture parameters. Altitude, aspect, steepness of slope, and relief cause these soils to vary greatly within short distances. In many places, soil will be entirely absent.
Soil-absent	Rugged mountains and icefields.

Note: Further details can be obtained from "Soil Classification, A Comprehensive System, 7th Approximation," Soil Conservation Service, U.S. Department of Agriculture, Washington, DC. (Published periodically).

(soil units) with several regions designated by geography, geology, and climatology, and, to some extent, by the traditional use of the soils. Mixed criteria enter into soil classification schemes simply because the various physical or chemical characteristics, considered separately or together, do not fully identify a soil. The principal categories adopted by the European Commission include: Lowlands, Mountainous Areas and Highlands, Volcanic Areas, Zones of the Tundras and Fields, Zones of the Boreal Forests of Conifers and Birch, Zones of Mixed Forests of Conifers and Broadleaved Trees, Zones of the Central European Beech Forests and Oak Forests, Zones of the Oak Forests and of the Atlantic Heaths, Zones of the Continental Oak Forest, Zones of the European Grassland, Zones of the Mediterranean Sclerophyll Forests, Zones of the Juniper Forests and the Mediterranean Steppes, Zones of the Montane Mediterranean Forests, Zones of the Subalpine Mediterranean Forests, and Zones of the Arabo-Caspian Steppes.

Soil Genesis. The origin and processes of soil formation usually are inferred, by relating measured morphological, physical, and chemical properties of a part to other parts of a given soil. And, during the last several decades, laboratory experimentation has revealed a better understanding of many of these processes. A factor to be stressed is that, in general, these processes occur over very long periods of time and frequently under multivariate conditions—conditions that are extremely difficult to duplicate and speed up in the laboratory. In the late 1950s, an interesting group of experiments revealed information concerning the formation of something similar to podzolic soil. Organic and distilled water leachates from tree leaves were passed through columns of different soil materials. Bleached surface layers and subjacent layers of stronger color formed in the columns. Effluent solutions from the base of the columns contained detectable amounts of calcium, magnesium, iron, manganese, phosphorus, potassium, and sodium. Very fine silicate clays, e.g., illite, montomorillonite, vermiculite, and chlorite, also were suspended in the effluent. Removal, transfer, and transformation were demonstrable experimentally. Examination of the columns showed that clay was partially removed from the bleached layers and was deposited in voids in the lower layers. The experiments showed removal (*eluviation*) and addition (*illuviation*) actually occurring and at a much accelerated pace.

Organic matter probably best illustrates additions to a soil and is formed in the biological decomposition of plant and animal residues by soil microorganisms. Plants supply most of the organic matter as a dry material added to the soil surface and as roots in the subsurface. It has been estimated that short grass prairie in semiarid regions may annually add 0.7 ton/acre (1.6 metric tons/hectare) of dry matter; tall grass prairie in subhumid regions, 0.8 to 1.7 tons/acre (1.8 to 3.8 tons/hectare); pine forest in more humid areas, 2.1 tons/acre (4.7 metric tons/hectare); and tropical rain forest, from 45 to 90 tons/acre (101 to 202 metric tons/hectare). Under bluegrass roots, additions may amount to 2.4 tons/ acre (5.4 metric tons/hectare) in the top 4 inches (10 centimeters) of soil.

During decomposition, plant materials are converted to carbon dioxide, water, mineral elements, and other chemically altered substances. Less-resistant materials are consumed first by soil microbes—so that more resistant plant materials remain with the new organic compounds that are synthesized by the organisms. At any time, the organic matter at a place in the soil reflects an equilibrium state of the addition of new material to the system, removal of more readily decomposable materials, and transformation to other forms by microorganisms and other agents. Organic matter also may be transferred within the soil by physical and physicochemical processes. Burrowing animals, worms, and insects turn over the soil and physically mix adjacent portions. Freezing and thawing and wetting and drying also assist in the process. Colloidal organic matter may be flushed downward or laterally and coagulate as coatings on structural aggregates in the soil.

The more unstable organic compounds are rapidly oxidized to carbon dioxide and water by various biochemical processes, while the more stable fractions accumulate. Conjugated ring compounds containing carbon, hydrogen, oxygen, nitrogen, phosphorus, and sulfur and other elements in small quantities accumulate in relatively stable organic and organomineral colloidal complexes. Lignin-like, phytin-like, and nucleo-protein-like compounds are included. Sorption of the organic matter on mineral colloid surfaces, particularly layer silicates, such as montmorillonite, helps to stabilize the organic matter against biochemical oxidation. In tropical soils, high stability of soil organic matter is imparted by coatings of aluminum hydroxide and red ferric oxide. Organic and iron oxide colloids, when fairly abundant, stabilize the soil into porous aggregates through which ample air and water can circulate.

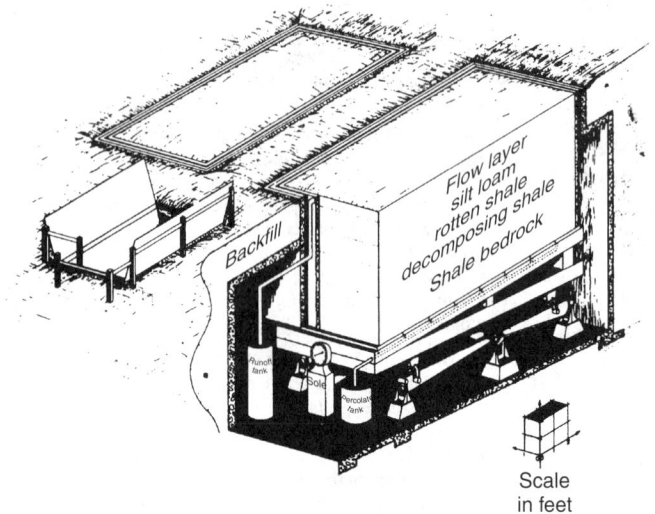

Fig. 3. A lysimeter, which provides a means for isolating soil masses and recording weight changes and water percolation. Such instruments provide accurate assessments of moisture behavior in soil. The lysimeter shown here represents $\frac{1}{500}$ acre (0.0008 hectare) and is 8 feet (2.4 meters) deep. The soil weighs 65 tons (58.5 metric tons), yet can be weighed to a precision of 5 pounds (2.3 kilograms). Soil scientists use lysimeters to study evapotranspiration, moisture consumption by crops, precipitation, water movement, and pollution. (*USDA diagram.*)

Fig. 2. Interaction of raindrops with the soil surface is an important component of the erosion process. Frames shown here were made by a 16-millimeter movie camera capable of speeds from 150 to 8000 frames per second. In the experiment, water drops are released from a tower 40 feet (12 meters) high and strike a plate glass target table. Drops from 5 to 6 millimeters in diameter are produced. Target plate is covered with water approximately 0.5 millimeter deep. (*North Central Soil Conservation Research Center, U.S. Department of Agriculture, Morris, Minnesota.*)

Localized spots of decomposing organic matter are important in reducing small but important quantities of iron to ferrous form and manganese to divalent form so that they become available to plants. Moderately to highly alkaline soils sometimes have inadequate activity of the reduced forms of iron and manganese, particularly in the absence of sufficient organic matter.

Radiocarbon dates of organic matter from the surface horizons of soils not only reflect the equilibrium status, but point out the turnover of the

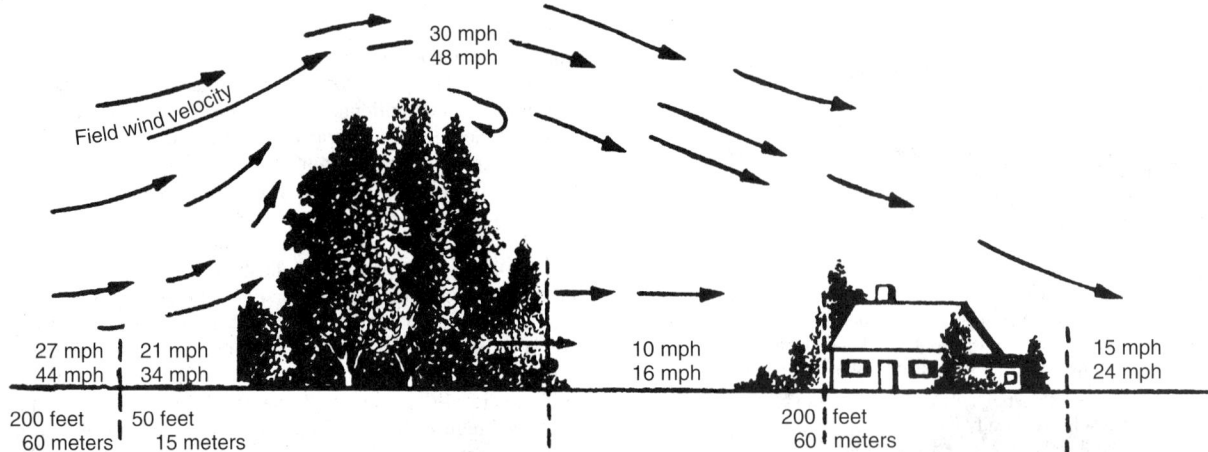

Fig. 4. Windbreaks reduce wind currents. Part of the air current is diverted over the top of the trees and part of it filters through the trees. Breaks like this reduce wind erosion of soil. Farmstead, livestock, and wildlife windbreaks should be relatively dense and wide to permit maximum protection close to the trees. Field, orchard, and garden-type windbreaks need not be so wide and dense. (*USDA diagram.*)

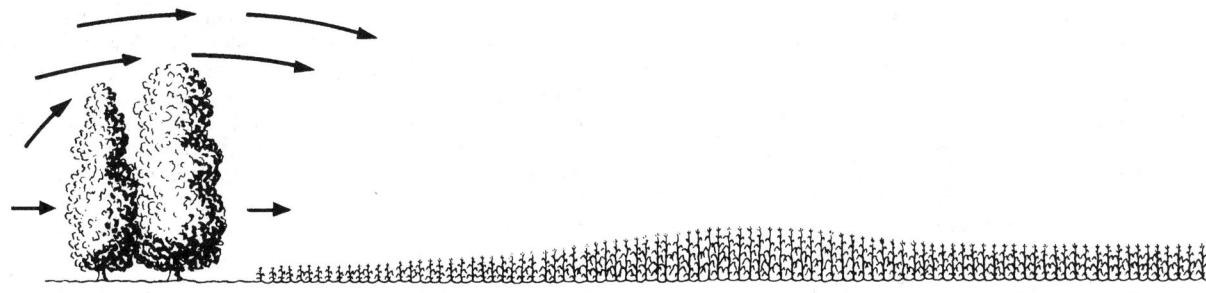

Fig. 5. Sometimes crop yields are lowest next to a tree windbreak. A common error made by some growers is to observe only the immediately adjacent area. The greatest gains are from a distance of 30 to 45 feet (9 to 13.5 meters) from the windbreak. (*USDA diagram.*)

system. One example of research, for example, has shown that in the Edina soil in southern Iowa, organic matter is 410 ± 110 years old in the 6-inch (15-centimeter) top layer. In the next subjacent layer, the age is 840 ± 220 years. At depths of 23 to 25 inches (58.4 to 63.5 centimeters), the organic carbon is 1545 ± 110 years old. The entire soil has been estimated as 14,000 years old.

The four kinds of changes that develop soil horizons are dependent upon many basic processes, such as hydration, oxidation, reduction, solution, precipitation, freezing, thawing, wetting, drying, among others. These processes, in turn, are dependent upon the four fundamental factors of soil formation: (1) nature of the parent material; (2) topography; (3) climate; and (4) biological activity that occurs in the upper strata of the soil. To these factors must be added time and imposed manual and mechanical manipulation (as by tilling, planting, etc.) and chemical manipulation (as by fertilizing and use of various control chemicals that seep into the soil).

Soil is destroyed by two principal processes—*water erosion* and *wind erosion*. The word *erosion* is the physical removal of all or part of established soil by washing or blowing away. Erosion, in some instances, also brings soil to convenient locations, but usually in so doing, unless carried out over long periods as in the development of bottomlands and deltas where crops can be grown, the new muddy, fine, highly unconsolidated and disintegrated soil causes more problems than immediate benefits. The bringing in or transfer of soil by water is commonly referred to as *sedimentation*. Much research has been carried out in connection with water and wind erosion. Typical of fundamental research is the study of splash patterns as shown in Fig. 2. The lysimeter, as shown in Fig. 3, also has been effectively used. The effects of wind erosion have been extensively studied, as exemplified by Figs. 4 and 5.

Additional Reading

Angers, D.A., E.G. Gregorich, L.W. Turchenek, et al.: "Soil and Environmental Science Dictionary," CRC Press, LLC., Boca Raton, F, 2001.

Bohn, H.L., B.L. McNeal, and G.A. O'Connor: "Soil Chemistry," 3rd Edition, John Wiley & Sons, Inc., New York, NY, 2001.

Bowles, J.E.: "Foundation Analysis and Design," 5th Edition, The McGraw-Hill Companies, Inc., New York, NY, 1995.

Brady, N.C. and R.R. Weil: "The Nature and Properties of Soils," 13th Edition, Prentice Hall, Inc., Upper Saddle River, NJ, 2001.

Carroll, R.C. and J.H. Vandermeer: "Agroecology," The McGraw-Hill Companies, Inc., New York, NY, 1990.

Das, B.M.: "Soil Mechanics," Oxford University Press, Inc., New York, NY, 2001.

FAO: "Soil Maps of the World," Food and Agriculture Organization of the United Nations, Rome, Italy. (Revised periodically.)

Fisher, R.F. and W.L. Prtichett: "Ecology and Management of Forest Soils," 3rd Edition, John Wiley & Sons, Inc., New York, NY, 1999.

Franklin, J.A. and M.B. Dusseaultz: "Rock Engineering Applications," The McGraw-Hill Companies, Inc., New York, NY, 1991.

Frenkel, H. and A. Meiri: "Soil Salinity," John Wiley & Sons, Inc., New York, NY, 1985,

Goldman, S. et al.: "Erosion and Sediment Control Handbook," The McGraw-Hill Companies, Inc., New York, NY,1986.

Hausmann, M.R.: "Engineering Principles of Ground Modification," The McGraw-Hill Companies, Inc., New York, NY, 1990.

Lal, R.: "Soil Erosion in the Tropics," The McGraw-Hill Companies, Inc., New York, NY, 1990.

Levy, R.: "Chemistry of Irrigated Soils," John Wiley & Sons, Inc., New York, NY, 1984.

Pierzynski, G.M., G.F. Vance, and J.T. Sims: "Soils and Environmental Quality," 2nd Edition, CRC Press, LLC., Boca Raton, FL 2000.

Rendig, V.V. and H.M. Taylor: "Principles of Soil-Plant Interrelationships," The McGraw-Hill Companies, Inc., New York, NY, 1989.

Singer, M.J. and D.N. Munns: "Soils: An Introduction," 5th Edition, Prentice Hall, Inc., Upper Saddle River, NJ, 2001.

USDA: "Soil Conservation Reports," National Soil Survey Laboratory, U.S. Department of Agriculture, Washington, DC, (Published periodically.)

Warrick, A.W.: "Soil Physics Companion," CRC Press, LLC., Boca Raton, FL, 2001.

SOIL BEARING VALUE. See Foundations.

SOIL PERMEABILITY. See Hydrology.

SOL. A word sometimes used to describe the solar day on the planets and their satellites. A sol or solar day is the interval between two successive transits of the sun past a given meridian. The length of the solar day ranges widely from one planetary body to the next.

SOLANACEAE (Potato Family). A relatively small plant family of some 1,700 species, in about 80 genera. The members include herbs, shrubs, and a few trees, most of the last being found in the tropics. The family has many important food and medicinal plants. All of its members contain poisonous alkaloids, as solanin, capsaicin, nicotine, and atropine.

Members of this family have leaves that are usually alternate and variously shaped, often lobed or dissected or pinnately compound. The variability frequently occurs in the leaves of a single plant. The flowers occur either singly or in cymes, and are regular with an inferior five-lobed calyx, a five-lobed corolla of various shapes. These are often large and conspicuously colored, with five stamens and a pistil composed of two united carpels, a long style, and a single terminal stigma. The ovary contains many ovules. The fruit is either a berry (potato and tomato) or a capsule (tobacco).

The most important member of this family is the Potato, *Solanum tuberosum*, commonly called the Irish potato, common potato, White Potato, or English potato, to distinguish it from the sweet potato. The plant is a branched herb, the branches tending to spread out more or less, and growing 2–4 feet in height. The green stems are annual, but the tubers, which are modified stems, give to the plant a perennial nature. The leaves are pinnately compound and rather irregular. The flowers are $1-1\frac{1}{2}$ inches in diameter, white, often with blue or purple tones or stripes, and with a tubular corolla. The fruit is a globular berry containing many small seeds embedded in a green pulp. Common names given these potato berries are potato balls, potato apples or seed balls. When mature, they are either green or brown; often they contain few seeds, but since they are rarely used in propagating the potato, this is of little consequence. They are used experimentally in the production of new varieties. Once a desirable variety is found, it is perpetuated vegetatively by means of the tubers. These tubers are formed at the tips of modified underground branches, or rhizomes, which radiate outward from the basal portion of the stem and to the casual observer resemble roots. The length of these stems varies from a few inches to a foot or more, depending somewhat on the variety. Rhizome formation and tuber development start soon after the tops appear above ground, and are advanced by darkness and low temperatures. The tubers continue to grow throughout the growing life of the plant. A mature tuber has the structure of a stem, but very much modified. Externally, one may recognize nodes and internodes, the nodes being determined by the eyes, depressions in the surface of the potato, each depression or eye containing a tiny bud. Internally the tuber is largely parenchymatous tissue with the cells filled with starch-grains. Near the surface of the potato, cross sections show a faint dark line which is the vascular tissue, very much reduced. Potato tubers vary greatly in shape, as well as in size; the better varieties are oblong or oval, with smooth skin and rather shallow eyes. In color, tubers range from brown through yellow to red, according to variety. The length of time required to mature a marketable tuber also varies greatly, some early varieties reaching a desirable size in about two months, while late varieties require five or six months.

The tubers are the principal means of propagation. For this purpose a tuber is cut into irregular pieces, each of which contains two or three eyes. To prevent disease the cut pieces are treated with fungicides and allowed to dry slightly, after which they are planted. Sprouting starts at once, indeed frequently occurs while the potato is still in the storage bin. If the bin is dark the sprouts formed will be long, slender and white. Sprouts formed in light are short, stout and dark green.

The potato is a native plant of cool upland regions of South America, where it has been long cultivated by the natives, who use the tubers for food. From America, it was carried to the southern part of Europe, where at first it was grown largely as a curiosity. From Europe the plant was brought back to America and introduced into what is now the United States, thus explaining its name of English or Irish Potato. It is now extensively grown in regions having a cool climate, Maine and Idaho being particularly noted for their potato crops.

In all green parts of the potato a poisonous alkaloid, solanin, occurs. This substance may also occur in the tubers, particularly if the latter are exposed to light long enough to become green in color.

Another important food plant of this family is the tomato, *Lycopersicum esculentum*, and related species. This is a native of South America, in which continent the tomato is still found wild. The tomato is a coarse branching perennial herb which is usually grown as an annual. It is an important food which has become immensely popular. Large quantities are consumed raw or canned. Tomato juice is important in many diets. Greater greenhouse production and improved rapid transportation from southern states makes it possible to enjoy fresh tomatoes in any season.

The tomato was carried to Europe by the early Spanish explorers and became an important food plant in southern European countries. In England, and also in the United States, it was widely grown before 1830 as a garden ornamental known as "Love Apple," which however was not eaten, but was held by many to be rankly poisonous. The Italians seem to be the first to have braved the danger of tasting the delicious fruit of this "poisonous" plant.

A third member of the family providing food to people is the egg plant, *Solanum melongena*, a coarse, somewhat woody, branching herb native to India. The plant has a rough stem, 2 or 3 feet tall, large sinuate-lobed ovate leaves and purplish flowers. See Fig. 1. The fruit is a berry, very large in some varieties. It is eaten either baked or sliced and fried.

Fig. 1. Large-fruited Black Beauty variety eggplant, the most popular of eggplant varieties in the United States. (*USDA photo.*)

Serving rather as a condiment than a food, but still used as a food stuff in some of its varieties, is the pepper, *Capsicum annuum*, another native

plant of tropical America. Peppers are either annual or biennial plants, with branching stems 1–3 feet tall, smooth shining leaves and white flowers. There are many varieties, which bear fruits of a variety of sizes and shapes (Fig. 2), as well as degrees of pungency. Some varieties are known as sweet peppers, and are used while yet green in salads or stuffed and baked; others are hot peppers. The pungent taste is due to an acrid compound, capsaicin, which in hot peppers occurs throughout the fruit, but in sweet peppers is largely restricted to the immediate region of the seeds; since only the fleshy pericarp is eaten, the seed being removed, this pungent substance is lost.

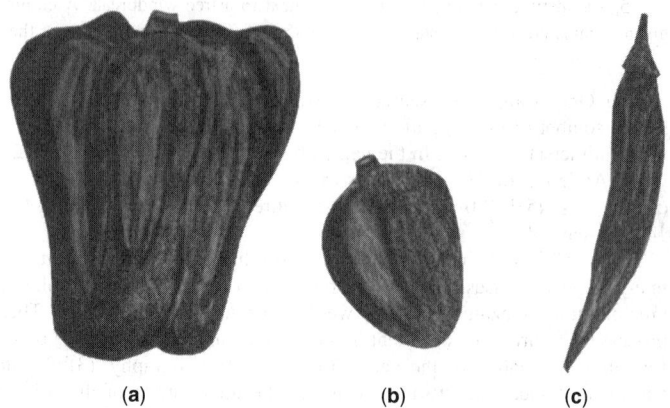

Fig. 2.   Comparative sizes and shapes of common peppers: (**a**) bell or bullnose pepper; (**b**) pimiento; (**c**) Cayenne or chili pepper.

Peppers are used in many ways other than as food. Small hot peppers are a frequent component of mixed pickles, and also are used in salads. The whole fruit of some varieties is ground up to a powder, and becomes Cayenne pepper, an extremely pungent condiment. Small smooth fruits of var. *conoides* are preserved whole in brine or vinegar, and known as Tabasco sauce or Tabasco peppers. Red peppers are used medicinally. Many varieties of pepper are grown as ornamental plants, the brilliant fruit offering a startling contrast to the dark green leaves.

More widely grown for its ornamental properties is the *Petunia*, introduced from South America.

Many members of the Potato Family are important drug plants. Here belong *Atropa belladonna*, yielding atropine, and *Datura stramonium*, the Thorn Apple, called also Jimson Weed or Jamestown Weed. The latter is a tall, rather coarse, branching annual having broad leaves with sinuate margins and large trumpet-shaped white flowers, borne singly in the axils of the leaves (a related species, *D. Tatula*, has purple flowers). The fruit is a prickly capsule. In all parts of the plant, but especially in the seed, are found several drugs. The most abundant is hyoscyamine, with small amounts of atropine and scopolamin present. The powdered leaves and seeds are used medicinally, chiefly in treating asthma.

Less important medicinally are several plants such as *Hyoscyamus niger*, *Solanum dulcamara*, the Bittersweet, and *Solanum niger*, the garden Nightshade, all held to be poisonous plants if eaten in sufficient quantity by man or domestic animals.

Another member of this family, ranking along with the potato in importance, is the tobacco plant, *Nicotiana tabacum*, also a native plant of tropical America. It is an annual plant growing 3–6 feet or more in height, and stout. The leaves are alternate, simple and rather large and, like the stem, covered with sticky hairs. The rather large flowers are borne in terminal racemes and have a funnel-shaped corolla which is yellow, white, pink or purple in different species and varieties. The fruit is a capsule containing many small seeds.

SOL AND SOLATION.   See **Colloid System**.

SOLAR APEX.   The point on the celestial sphere toward which the sun is traveling. Also called *apex of the sun's way*. The solar apex is at approximately right ascension 270 degrees declination 34 degrees N. The point diametrically opposite the solar apex on the celestial sphere is the solar antapex, right ascension 90 degrees declination 34 degrees S.

SOLAR CELLS.   See **Solar Energy**.

SOLAR ENERGY. The vast quantity of energy received by Earth from the sun and the potential for converting that energy into more useful forms for society has intrigued scientists, engineers, and social planners for decades. This interest was sharpened by the oil embargo of the 1970s and, for about a decade after that, tremendous interest was displayed, by the scientific and lay community alike, in alternative energy sources, including a turn to solar energy. Energy from the sun was considered by many people as a relatively low-cost and essentially pollution-free source, particularly in contrast with polluting, nonrenewable, so-called fossil fuels and with nuclear fuels, which many people consider in a negative light. During the 1970s, but tapering off in the 1980s, many, many millions of dollars were invested by governments worldwide and by private institutions, architectural and solar equipment firms, and energy supply firms toward the development of practical, economically competitive solar energy systems.

As a result of that activity, progress in designing passive solar energy systems into office and factory buildings has been impressive, but not nearly so extensive as once estimated. Active solar energy systems, in which solar radiation is converted into another energy form (usually electrical) has also progressed, but the number of outstandingly successful installations is relatively limited, and essentially these are presently regarded as still in an experimental phase. In contrast, considerable research continues to be directed toward solar cells (solid-state devices that convert solar radiation into electric power), but it should be immediately stressed that solar cells for communication and other satellites and space vehicles are vitally needed, because they provide an energy source difficult to obtain in other ways. Cost, in this instance, is not supercritical, but one finds that solar cells for building, etc. heating and power still are essentially noncompetitive. Some relatively low-cost, small solar-powered devices designed mainly for public consumption have appeared in recent years.

To put solar energy into perspective, one should review other energy resources as well. Fortunately, throughout this encyclopedia, such energy information is available. See list of articles at end of this description and also consult alphabetical index.

Availability of Solar Energy

Not to be confused with insulation, the word *insolation* (acronym for "incoming solar radiation") defines the rate at which direct solar radiation is incident upon a unit horizontal surface at any point on or above the surface of the earth. The unit of insolation is the Langley, named after Samuel Pierpoint Langley (1834–1906), an American astronomer, physicist and pioneer in the utilization of solar energy:

$$1 \text{ langley} = 1 \text{ gram calorie per square centimeter}$$

$$= 3.687 \text{ Btu per square foot}$$

Fundamental to the practical application of real-time solar energy systems is the amount of energy received from the sun at any given location at any particular time. The energy received varies with the geometry of the sun-earth system—and thus varies with latitude, season of the year, time of day, as well as with local weather conditions. On a typical day in June, anywhere from 500 to 700 langleys of solar energy can be expected in most parts of the United States, whereas in December, with shorter days and more inclement weather, only from 100 to 300 langleys can be expected. In June, for example, both Saskatoon, Canada and Tampa, Florida receive about 600 langleys of solar energy daily, but in December, the amount of such energy received in Saskatoon drops to 75 langleys per day (only 12% of the June value at that location), whereas, in Tampa, more than 50% of the June amount is received in December. In terms of equipment costs, this translates into needing four times the solar-energy collector surface in Saskatoon as that required in Tampa in order to supply the same amount of power year-round. Maps of the type shown in Fig. 1 can be helpful in this regard.

The solar constant (intensity of solar radiation outside the Earth's atmosphere at the mean distance between the earth and the sun) has been determined by measurements from satellites and high-altitude aircraft and is 1.353 kilowatts per square meter. This extraterrestrial radiation, which corresponds closely to that of a blackbody at 5762 °K, is 7% in the ultraviolet range (wavelength less than 0.39 micrometer) and 47% in the visible range (wavelengths from 0.38 to 0.78 micrometer), with the balance in the near-infrared (largely with wavelengths of less than 3 micrometers). Radiation is depleted as it passes through the atmosphere by a combination of scattering and absorption; the radiation that reaches the ground—the

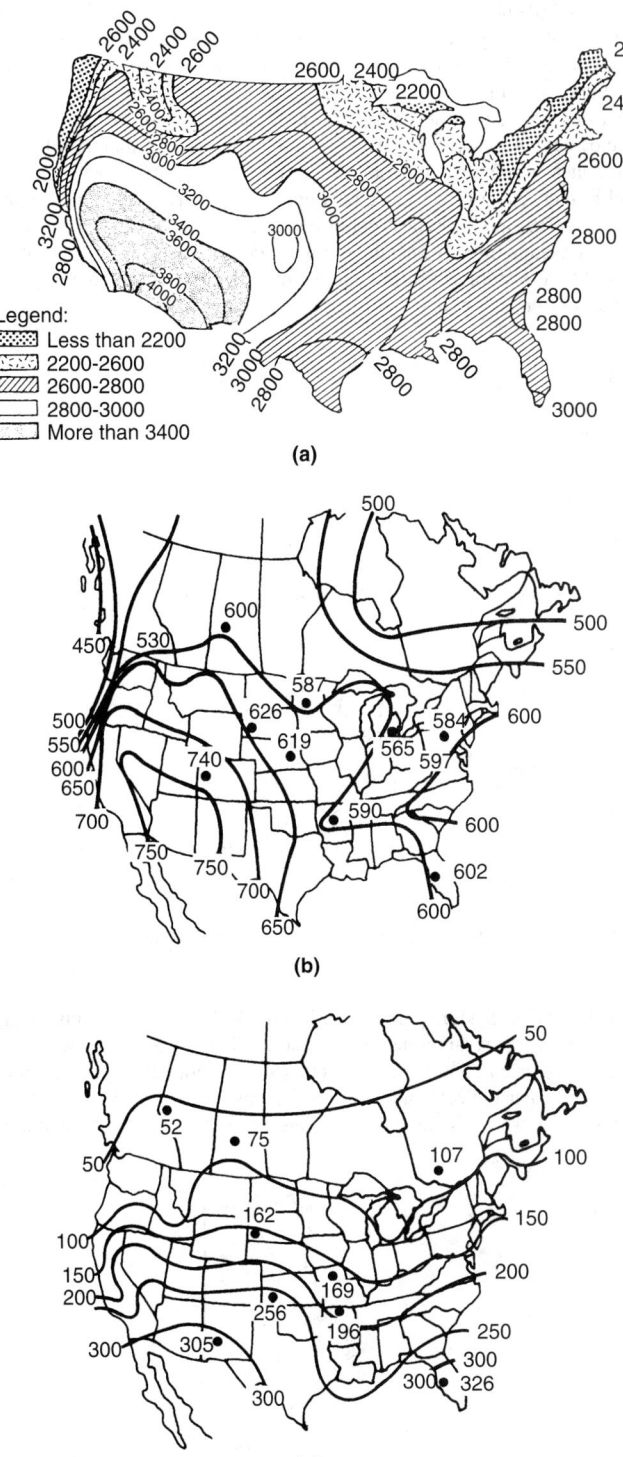

Fig. 1. Availability of solar energy (insolation): (**a**) Average number of hours of sunshine per year (United States); (**b**) median daily insolation in langleys (North America in June); (**c**) in December. (*National Oceanic and Atmospheric Administration.*)

raw material of this energy source—can vary from almost none, under heavy cloud cover, to 85–90% of the solar constant under *very clear* skies.

Solar Energy for Building and Residence Comfort

The application of solar energy for residences and commercial and public buildings tends to fall into three categories of increasing complexity: (1) heating only; (2) cooling only; and (3) combined heating and cooling.

Simple Heating System with Hot Storage. As shown in Fig. 2, the basic elements of this system, exclusive of pumps, valves, and controls, are: (a) a solar collector; (b) an auxiliary heating device; (c) hot storage system; and (d) heater element fan and air duct system. The solar collector absorbs heat energy from the sun and transfers it to a heat-transfer fluid, which conveys the heat to a hot storage system. From the hot storage, heat is withdrawn from storage through a heater coil, where an air-circulating fan carries heat from the coils into an air duct system. When the solar collector cannot provide an adequate amount of heat to maintain the hot storage at a minimum temperature, the auxiliary heating device, such as a fuel burner, electric resistance heating, or an electric driven heat pump, comes on. This auxiliary heat could be added to hot storage as shown in Fig. 2, or used to directly heat the room air. For the case of electric heating, heat addition to storage will provide the opportunity to limit auxiliary heat addition to nonpeak hours.

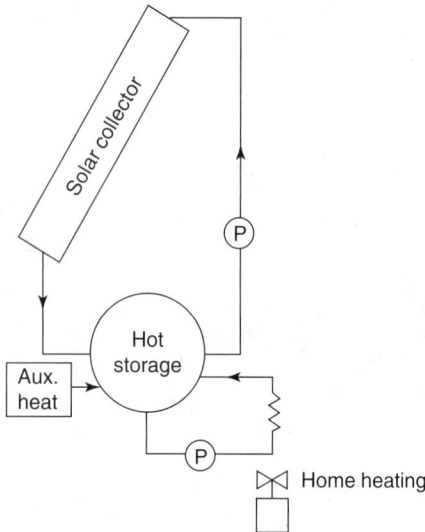

Fig. 2.   Basic elements of solar heating system.

Solar Cooling System. As shown in Fig. 3, the basic elements of this system are: (a) a solar collector; (b) an auxiliary heating device; (c) a cold storage system; and (d) a heat-actuated refrigeration loop (absorption cycle). The solar collector absorbs heat energy and transfers it to a heat-transfer fluid, which in turn conveys the solar heat to the generator or

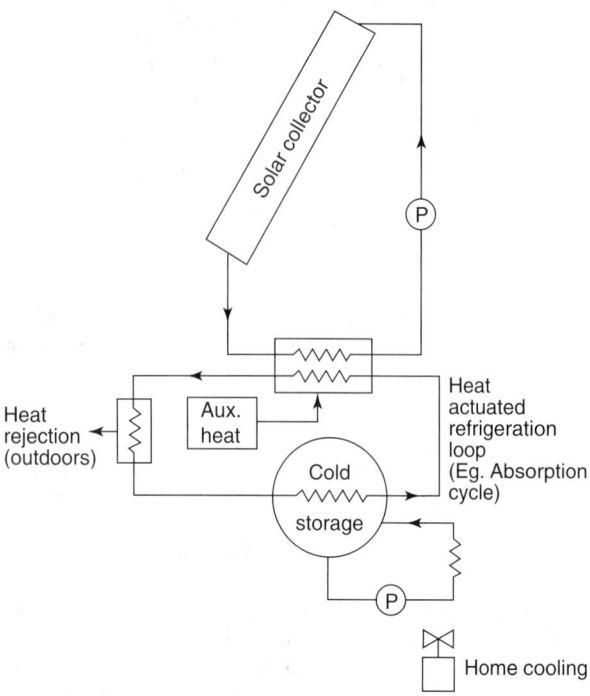

Fig. 3.   Basic elements of solar cooling system. Cold storage only.

boiler of the heat-actuated refrigeration loop. This loop can also be driven by auxiliary heating when the solar heat input is not adequate. If the heat-actuated refrigeration loop were of the Rankine-cycle type, rather than an absorption cycle, it might also be possible to drive the refrigeration loop with auxiliary power rather than auxiliary heating — a more favorable situation if the auxiliary is electric rather than fuel. The refrigeration loop cools the cold storage reservoir from which home cooling is supplied upon demand.

Combined Solar Heating and Cooling System. A system of this type is shown in Fig. 4. This is only one of a variety of possible systems. The major elements of this system are: (a) a solar collector; (b) an auxiliary furnace with heating coils; (c) a storage system (hot in winter; cold in summer); (d) absorption refrigeration cycle; and (e) necessary valving and controls. The system is designed to provide both heating and cooling upon demand. The heat energy generated by the solar collector is directed to either the hot storage tank or the refrigeration cycle generator, according to the seasonal mode of operation. When solar energy (either direct or stored) cannot supply the required heating or cooling load, auxiliary heating or cooling can be used.

In the winter mode of operation, solar energy is gathered at the collector and is pumped directly to the storage system. From the storage system, heat is extracted according to household needs through the heating/cooling coil in the main air duct. This coil is controlled by 3-way valves which are open to heating and shut to cooling. Heat is then extracted from the coil by air fans and carried into the house. At certain times, the solar collector and storage system will not be able to provide enough heat to maintain the heat needs of the household. In these cases, the auxiliary furnace will assume the heating load until the solar system is able to provide heating. This furnace will also provide auxiliary heat for domestic hot water when solar energy cannot provide this function. At times, while on winter mode operation, there will be days when cooling may be required. When this condition exists, the auxiliary furnace will drive the absorption cycle. The two 3-way valves on the heating/cooling coil will be actuated to permit chilled water from the absorption machine to circulate through this coil.

In the summer mode of operation, solar energy is gathered at the collector and pumped in the form of heat directly to the absorption refrigeration machine and to the domestic hot water heater. The collected energy serves as the main driving force for the cooling system. When the collector is unable to provide the necessary energy for the cooling system, the auxiliary furnace is activated to supplement the energy load. Once the cooling cycle is activated, the cooling produced is directed to the same storage system used for storing heat in the winter. Cooling is extracted per household needs from the storage system through the heating/cooling coil. In this mode, the 3-way valves of the coil are open to storage system and shut to direct cooling from the cooling cycle. Should the storage system not be able to provide the cooling, these valves would be closed to the storage system and open to direct cooling. While on the summer mode, the auxiliary furnace can be used to heat the house on occasional cold nights.

The foregoing systems are representative of general concepts and not necessarily of final designs or optimum arrangements. The final detail system design will, in particular, be dependent upon whether fuel or electricity is used for auxiliary heating. For the near term, natural gas or fuel oil may be preferable to electric heating. However, in the longer term, as developments in solar collection and heat storage reduce the amount of auxiliary heat required, and as heat pump technology is improved, electricity may become more attractive.

Architectural and Building Factors. Solar climate control systems will have to be integrated with different building designs. New buildings can be designed to fit the requirement of a solar climate control system while applications to existing buildings will have to be determined individually. Collectors can be installed on flat roofs of buildings, or designed to fit the sloping roofs of a wide variety of buildings. Collectors could serve a single building or a cluster of buildings. As previously pointed out, the geometry of the collector installation varies with location latitude. Economics to a large degree will be determined by availability of reasonably sustained sunshine.

Flat-Plate Collectors. Essentially since the outset of solar energy technology for environmental heating and cooling of living and working enclosures, the flat-plate collector has dominated the field. Within recent years, however, some of the initial needs for collectors of this nature have been obviated by more attention being given to the design of passive solar collectors, described a bit later. Also, for very large commercial

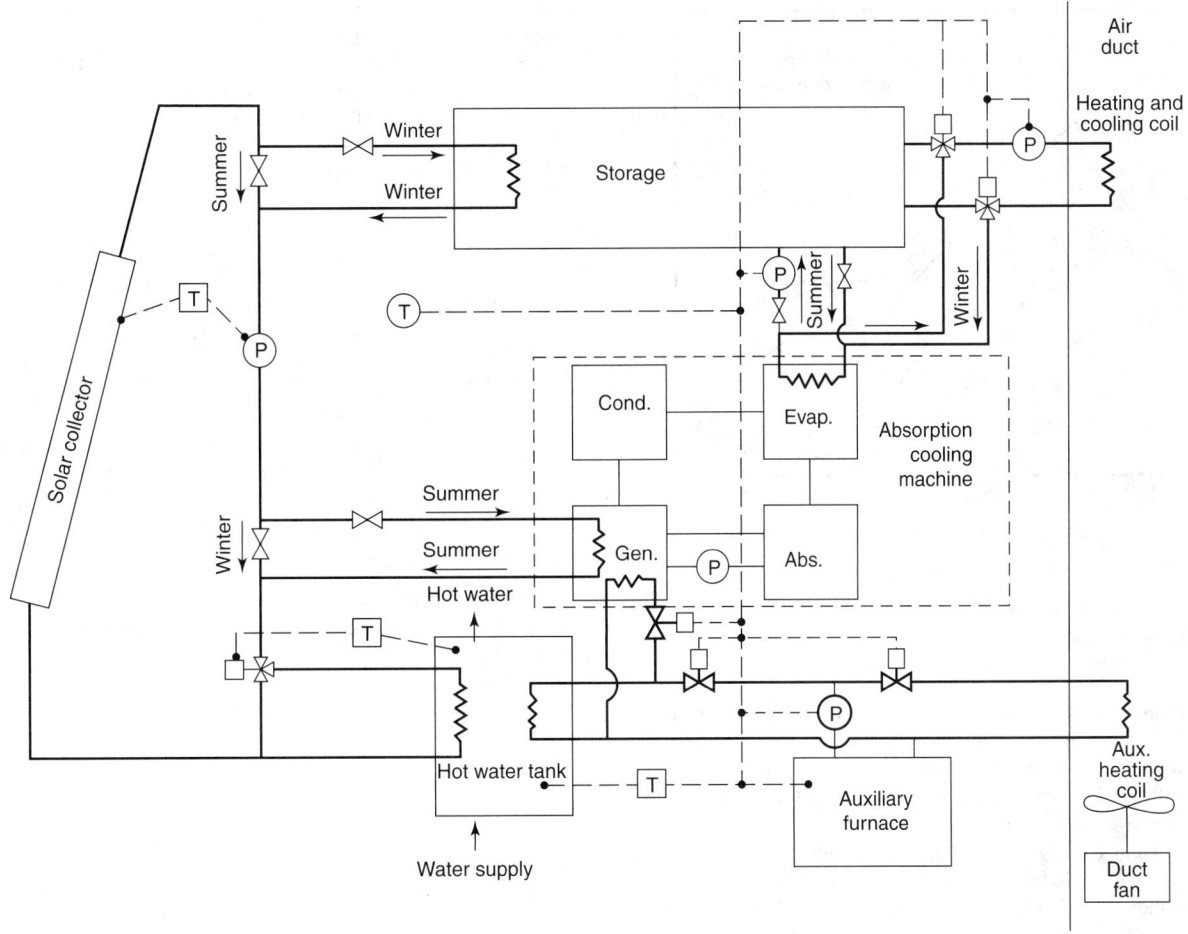

Fig. 4.   Combined solar climate heating and cooling system.

installations, there has been a trend toward the use of nonfocusing or trough or line-focusing concentrators, also described later.

The essential features of a flat-plate solar collector are shown in Fig. 5. A blackened receiver surface covered by one or more special glass plates is used. Since the glass is transparent to the incident solar radiation, but opaque to the reradiated energy, the solar collector, like a greenhouse, serves to trap solar energy. The working fluid used to remove the heat from the collector can be either air flowing between the blackened surface and the glass plate or water (or some other liquid) flowing in tubes attached to the blackened plate. Solar collection efficiency is defined as the ratio of usable energy collected per unit time to the incident solar flux. Efficiency, η, may be calculated as:

$$\eta = \alpha\tau - \frac{q_L}{q_{in}}$$

where: α is the absorptivity for sunlight; τ is the transmittance of the glass plate; q_L is the heat loss from the collector; and q_{in} is the incident solar flux. The heat loss q_L is, in turn, dependent upon the emissivity, ε, for low-temperature radiation. Typical performance characteristics for a solar energy collector are shown in Fig. 6, which is a plot of collector efficiency versus temperature of the absorber plate for an incident radiation of 300 Btu/(hour) (square feet); 814 kcal/(hour) (square meter). Note that the efficiency falls off as absorber temperature rises. Efficiency, of course, also drops off rapidly as the incident radiation is reduced, since the heat loss term is a function of absorber temperature only.

A problem that must be faced by architects and engineers is the need to integrate collectors into building and residence design in a way to maximize thermal performance and, at the same time, provide an esthetically satisfactory structure. A major variable, depending upon energy requirements and the average solar insolation over a year, is the amount of collector area needed. Obviously, the larger the energy requirements and the less favorable the insolation, the greater the problem. Because collectors must be oriented within rather narrow limits if they are to maximize their capture of solar radiation, the problem of retrofitting many existing structures sometimes renders a project impractical. Thus, the general pattern has been one of concentrating on new construction, particularly for low-rise, flat-roofed buildings, such as schools and shopping centers, where cooling may be more important than heating. Where surrounding area is available, as, for example, an adjacent parking area or facility, the

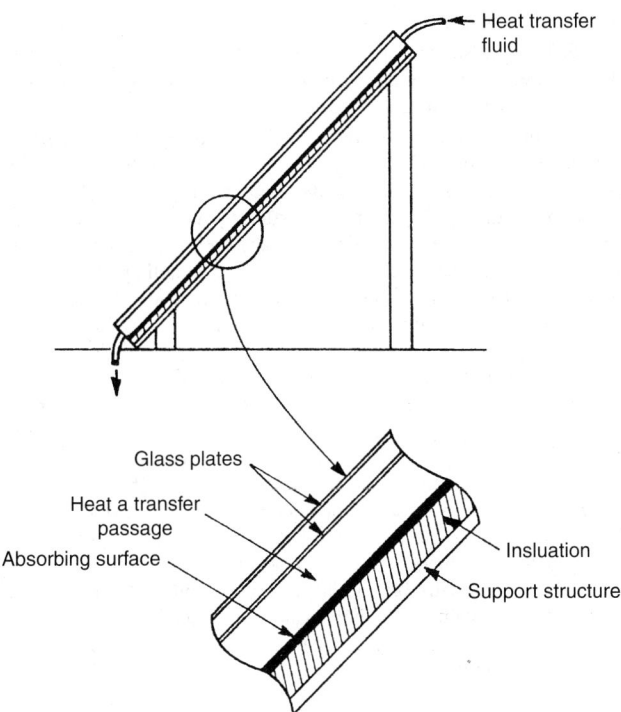

Fig. 5.   Basic elements of a flat-plate solar collector.

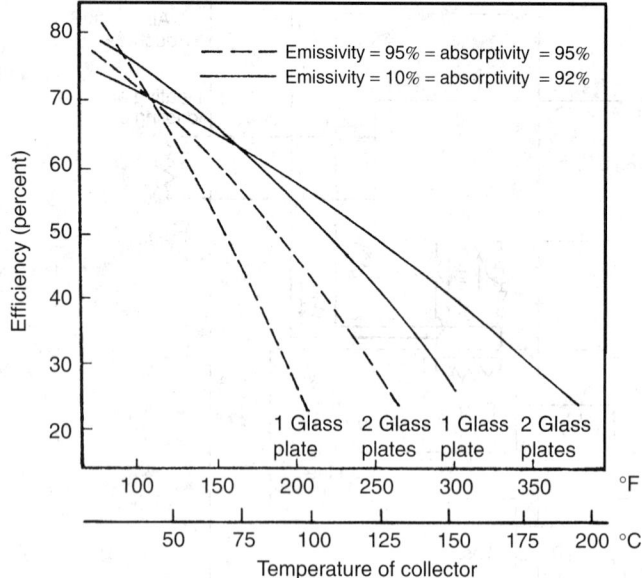

Fig. 6.   Relative solar collection efficiency of plate-type solar collectors with different emissivities.

collectors can be installed apart from the structure to be heated and/or cooled. Also, by turning to advanced collector designs, which provide greater efficiency, even at greater initial cost, architects and builders can better cope with the problem of collector area required. Passive heating of buildings is also a possibility in many cases.

Evacuated-Tube Collectors. In this type of collector, an inner glass cylinder, blackened to absorb solar radiation, is enclosed within an outer protective cylinder. The space between the two cylinders is evacuated. The inner cylinder is usually coated with material that reduces energy loss through reradiation. Transfer of heat is accomplished by a fluid (air or liquid) that flows through the inner cylinder. These collectors are similar to flat-plate collectors in that they can use both direct and diffuse light. However, the evacuated-tube collectors operate better during the early and late parts of the day. The vacuum provides such excellent insulation that they are less affected by high winds and cold weather than the flat-plate collectors. The output of the evacuated-tube collectors is essentially independent of ambient temperature and their efficiency is generally 40–50%. Ordinarily these collectors operate at about 180 °F (82 °C) for space-heating applications and certain process heating uses. Equipped with reflectors, they can operate up to 240 °F (116 °C), which is sufficient to drive absorption air conditioners. Cylindrical evacuated tube collectors can absorb radiation coming from any direction (360° aperture). Usually, they are mounted in arrays with a spacing of about one cylinder diameter between tubes and with a reflective material behind them.

Large-scale collectors, with or without focusing (radiation concentrators) are described a bit later.

Heat Storage. A comparison of heat-storage capacity on a volumetric basis between various storage media shows that water can store 62.5 Btu per cubic foot per degree Fahrenheit (311.5 kcal per cubic meter per degree Celsius. Rocks, bricks and gravel can store about 36 Btu/cu. ft/degree F (179.4 kcal/cu. meter/degree C). In addition to fluid media and solid media, advantage can be taken of the latent heat of a phase transition. Some salts, which melt in the desired temperature range, can store about 60 Btu/cu. ft/degree F (299 kcal/cu. meter/degree C) as sensible heat, and 9500 Btu/cu. ft./degree F (47,348 kcal/cu. meter/degree C) at the melting point as heat of fusion. However, these salts tend to undercool rather than crystallize during the cooling cycle. While undercooling can be prevented by use of nucleating agents, the fixed rate of crystal growth is very slow in most salt hydrates. Heat cannot be withdrawn more rapidly than it can be supplied by the growth of crystals. This is a serious problem, which can only be partially overcome by the design of the storage container to provide a large heat-transfer area. Based upon these alternatives, an insulated tank of water to store heat is one of the most efficient solutions. Early work concentrated on $Na_2SO_4 \cdot 10H_2O$ which undergoes a phase transition when heated at 32 °C (89.6 °F). Because phase separation of this hydrate occurs

on cycling, other chemical systems are being sought which can undergo thousands of cycles without loss of storage capacity.

Typically optimum storage capacity varies from 1 to 3 typical winter days' solar energy supply, depending upon the site, and for a medium-size residence or small commercial building would be in the range of 1000–2000 gallons (38–76 hectoliters). A section of a comparatively small solar-heated building is shown in Fig. 7.

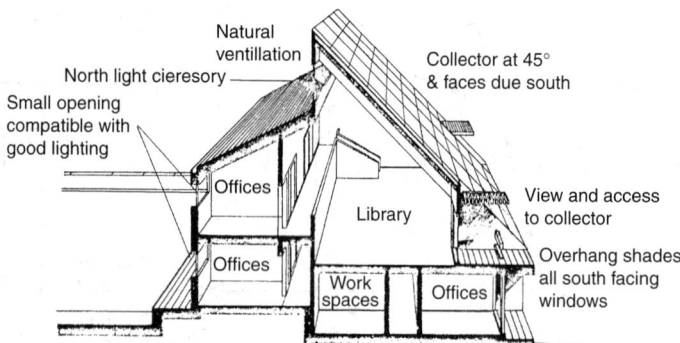

Fig. 7.   Section of solar-heated building. Solar collector has an area of 3500 square feet (325 square meters) facing south at an angle of 45 degrees. There are about 8000 square feet (743 square meters) of working space. Estimates of heat loss indicate heat demand is in range of 40,000–70,000 Btu (10,080–17,640 kcal) per day. Located in the northeastern United States, the building was designed to furnish between 65 and 75% of total seasonal heating load.

Collectors also can be used as energy dissipaters by designing them to lose heat by convection and radiation to the clear night sky. The role of the collector is thus fully reversed for the cooling cycle. This requires a system for moving insulation, unless design compromises in collector design are made for both the heating and cooling cycles.

Solar collectors also can be combined with heat pumps. The latter can serve as an independent (auxiliary) source of heating energy, or the collector-storage system can serve as the energy source for the evaporator of the heat pump. The latter system has apparent advantages of lowering mean collector temperature and raising the mean evaporator temperature of the heat pump, thus improving the performance of each. See Fig. 8; see also the entry on **Heat Pump**.

In addition to close cycle absorption cooling, open cycles are of potential interest. Desiccants can be used to absorb water vapor from room air, which then can be evaporatively cooled. The desiccant is regenerated and recycled. Löf has suggested the use of triethylene glycol as a desiccant, with solar-heated air for regeneration. Lithium chloride also has been proposed as a desiccant.

Heat-Actuated Refrigeration. A variety of heat-actuated refrigeration cycles has been proposed for solar air conditioning. These can be divided into heat engine types, such as the Rankine and Stirling cycles, and the absorption machines. Most successful to date have been the lithium bromide-water and the ammonia-water absorption cycles. Regardless of type, operating temperature is a tradeoff when coupling a solar collector to a heat-actuated refrigeration machine. The efficiency of solar collectors decreases with temperature. On the other hand, the coefficient of heat-actuated refrigerators increases with generator temperature. Figure 9 shows Carnot coefficient of performance as a function of generator temperature for an evaporator temperature of 45 °F (7 °C) and for heat rejection temperatures of 120 °F (49 °C) (typical of air cooling) and 100 °F (38 °C) (warm water cooling, such as might be achieved with well water). For the case of absorption machines, it is assumed that absorber and condenser operate at a common heat-rejection temperature. Using these plots, the collection efficiency–Carnot coefficiency of performance product for the simple single-pane black collector reaches a maximum rate at about 175 °F (80 °C) and falls off rapidly at higher temperatures. For a more refined collector, such as a single pane with selective surface collector, the efficiency coefficient of performance-product is maximum and nearly constant in the range of 200 to 300 °F (93 to 149 °C).

To achieve temperatures above the boiling point of water, concentrators are required as well as collectors. One of the largest high-temperature solar energy systems for building heating and cooling commenced operation in

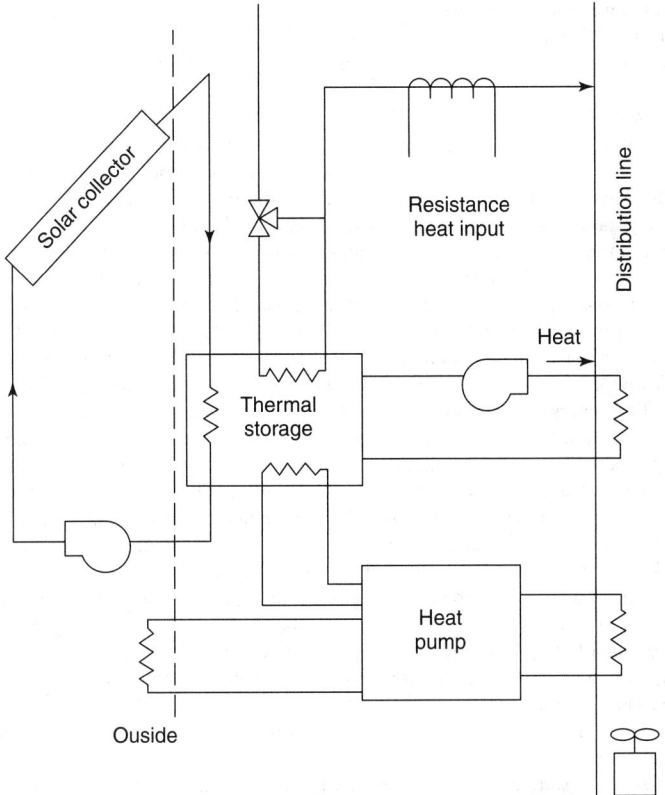

Fig. 8. Solar heating system with heat pump auxiliary.

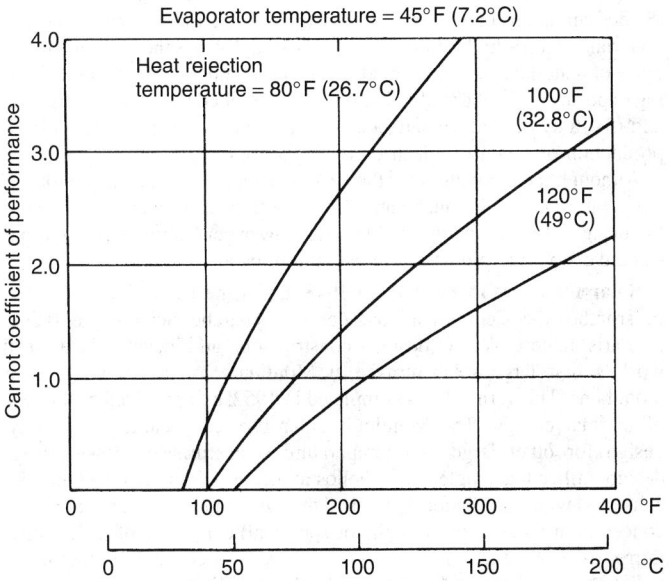

Evaporator temperature = 45°F (7.2°C)

Heat rejection
temperature = 80°F (26.7°C)

100°F
(32.8°C)

120°F
(49°C)

Fig. 9. Carnot coefficient of performance of heat-actuated refrigerators.

safety control circuits are continuously monitored to protect the collector field from damage by weather excesses.

Passive Solar Heating and Cooling. Although not always the practical solution, one of the most sensible approaches to the utilization of solar energy does not require pumps, fans, etc., but utilizes the building or residence structure per se as the solar radiation absorber (or insulator). An example of the modern approach to passive approaches is, ironically, incorporated in Montezuma's Castle, built around 700 A.D. by the cliffdwelling Indians of Arizona. The basic philosophy is to design the structure to capture and retain heat during winter and to remain cool during summer. For example, the windows that face north are made small or largely eliminated. These windows do not contribute much to heat collection during summer or winter and thus, if small, diminish heat leakage in either direction during all seasons. South-facing windows are made large because they are required as radiation collectors during winter. However, they are protected during the summer season by an overhanging roof. These features alone, of course, are common-sense ideas that have been used by some designers for many years. Passive systems also take advantage of heavy masonry walls (or other sources of thermal mass) which can absorb solar radiation during the day and reradiate some of it at night. Such an arrangement could be called a "concrete collector." The designer can improve the effectiveness of extensive south-facing windows by constructing an interior masonry wall adjacent to the windows (lighting becomes a problem for special design). In an experimental building (Wallasey School, Liverpool, England), the south wall of the two-story concrete structure is made up essentially of double-glass windows with a heat-storage (or insulating) wall. The only supplementary heat required is derived from body heat of the students and heat radiated by electric lights. A structure of this type, of course, is subject to wider interior temperature variations than those to which much of society has become accustomed. Temperature swings can be reduced by partially decoupling thermal storage from the living and working space. In this concept, solar radiation entering the south-facing windows is absorbed by a masonry wall (sometimes called a Trombe wall) or by a water wall in which water-filled drums are placed. This wall insulates the building from high temperatures during daytime and transmits stored energy to the structure for warming during nighttime. There is an office building and warehouse in northern New Mexico which incorporates a water wall passive system and which provides 95% of the energy required for heating.

Some designers use a "roof pond," in which plastic bags filled with liquid are exposed to the sun during the day. They are covered with an insulating panel at night so that they can radiate stored heat downward to the structure. The cycle can be reversed to provide cooling during summer.

Investigations indicate that the optimum thickness for concrete thermal storage walls is about 30–40 centimeters (1–1.3 feet). Innovations in passive systems are appearing at a rapid rate. Some of these include movable insulation for shielding glass areas at night, and more compact thermal storage systems, such as ceiling tiles which have been developed by the Massachusetts Institute of Technology. These tiles contain a material that undergoes a phase change at 75°F (24°C), storing heat as the material melts and later releasing the heat to the room as the material solidifies. It is expected that, as passive systems improve, there will be a considerable impact on the traditional flat-plate collector.

Large-Scale, High-Temperature Solar Energy Systems

Systems in this category require concentration of solar radiation prior to its collection and utilization. Concentrators fall into three categories:

(1) *Nonfocusing concentrators* have the advantage that they do not have to continuously track the sun and thus do not require optical precision. Also, they can utilize both diffuse and direct radiation and thus are partially operable on cloudy days. They do not, however, operate as efficiently as focusing types, particularly during the early and late periods of the day. A simply designed nonfocusing concentrator essentially will consist of a stationary mirror or reflector located next to a flat-plate collector. In another approach, placing reflectors behind evacuated-tube collectors also accomplishes a modest degree of concentration. An advanced nonfocusing concentrator, known as the *compound parabolic concentrator* (CPC) was developed by the Argonne National Laboratory as the result of experience gained from designing light-concentrating devices for use in high-energy physics experiments. The device incorporates a parabolic surface designed to provide the maximum amount of radiation to an absorber for a given concentration ratio. In one configuration, the radiation is concentrated 1.8

late 1978, at an eight-story office building (Minneapolis), which houses about 500 employees, and has over 100,000 square feet (929 square meters) of working space. During an average year, the system was designed to generate about 50% of heating needs, 80% of cooling energy needs, and 100% of heat for hot water needs. On the roof of a parking ramp adjacent to the building, 252 trough-like collectors track the sun to focus its rays on liquid-filled pipes. The liquid, which may reach 177°C (350°F), is pumped to an isolation heat exchanger that allows different fluids, pressures, and flow rates to be used in the two-loop system, dictated because of the cold winter temperatures. Excess heat is stored underground in two 18,000-gallon (681 hectoliters) tanks until required. Each row of solar collectors is under the control of a local system that uses a photosensitive sun tracker and bidirectional electric drive A.C. motor. Wind, solar isolation and other

times and, when operated in conjunction with a stationary collector, will reach temperatures as high as 250 °F (121 °C). Units with even higher concentration ratios (3× up to 6×) have been developed for use with evacuated-tube collectors and can achieve temperatures between 300 and 450 °F (150 and 232 °C). At a concentration ratio of 6, it is necessary to reorient the collector once each month. Currently, manufacturing costs are relatively high, but the CPC holds promise for a number of future applications.

(2) *Trough or line-focusing concentrators* track the sun by focusing in one direction only. Concentrations in this category of device range from 10× to 100× and they are capable of achieving temperatures between 200 and 600 °F (93 and 316 °C). On average, these devices deliver a minimum of 50% of the solar energy available to the heat-transfer medium in the absorber. In one configuration, mirrors form a parabolic trough that focuses radiation onto a linear absorber. Usually the mirrors are constructed of polished metal or coated plastic; the absorbers are blackened metallic pipe or evacuated-glass tubes. See Fig. 10. The entire assembly or array tracks the sun. Although normally considered in terms of relatively large thermal capacities, versions have been offered for residence and small building applications. Other, more sophisticated versions, operating at the high-temperature range, are used to drive Rankine-cycle heat engines for pumping irrigation water in a number of locations in the southwestern United States. Other installations include water heating for industrial processes. In another type of line-focusing concentrator, the optics are altered to utilize plastic Fresnel lenses for focusing the radiation onto the absorbers. Apparently a similar thermal result can be accomplished, as compared with the parabolic mirror approach, with a smaller optical surface.

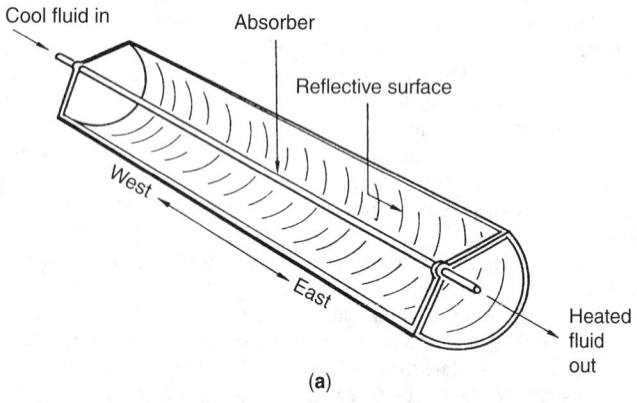

(a)

(b)

Fig. 10.   (a) One configuration of an optical-type concentrator with axial absorber; (b) parabolic troughs under test for various uses, including irrigation pumping for croplands.

(3) In another one-axis tracking concentrator, a fixed trough is made up of flat mirrors and tracking is accomplished by moving the absorber.

Known as the *Russell collector*, this unit apparently can reach temperatures up to 900 °F (482 °C).

Two-axis focusing systems will be described a bit later.

High-Temperature Solar Energy

Prior to the serious consideration of high-temperature solar energy as a source of electric power, much was learned from the design of solar furnaces, the objective of which was the production of extremely high temperatures for materials testing, a very useful research function that continues. Much knowledge was and is continuing to be learned from the operation of solar furnaces — knowledge that is helpful in the design of solar power plants. It is fitting here, as a backdrop to describing the solar power tower concept, to present information on solar furnaces. Historically, solar furnaces have been selected for high-temperature research and development activities where a highly concentrated source of nonpolluting radiant energy is required. Generally such activities can be categorized as: (1) high-temperature chemistry involving the formation of very pure or otherwise unique materials; (2) high-temperature processing by which a material is fused, purified or otherwise improved; (3) high temperature property measurements involving the determination of the behavior of a material under conditions which require a noncontaminating environment; (4) determination of the thermal shock resistance or other behavior of materials in a high-temperature, high-heat flux radiant energy environment; and (5) study of high temperature solar-thermal conversion systems. Certain of these applications may be refined further by conducting the operation in an optically transparent vessel or one containing a transparent window such as fused quartz through which the radiant energy may pass and in which the composition and pressure of the atmosphere can be controlled.

A few examples of the types of high temperature studies which have been conducted in the previously described categories are: (1) gas phase reactions to form pyrolytic graphite; (2) production of very high purity fused aluminum oxide and fused silica, the production of stabilized zirconia and the purification of reactive metals in a controlled atmosphere; (3) determination of microwave transmission characteristics of dielectric materials at very high temperatures; (4) study of the thermal shock resistance of materials under high heat flux thermal radiation conditions simulating exposure to the thermal radiation pulse provided by a nuclear explosion; and (5) study of heat exchangers, such as boilers and superheaters for the production of steam for electric power generation.

Although the motivation for the design of such furnaces may be for high-temperature research, much can be learned from them that is applicable to the design of solar energy facilities for power generation. Up to the point of conversion, the problems are essentially parallel.

Solar Furnaces in France. In 1948, under the leadership of Professor F. Trombe, the Centre National de la Recherche Scientifique (CNRS) in Paris undertook the design, construction, and development of the world's first large solar furnace at Montlouis in the French Pyrenees mountains. This furnace was completed in 1952, and provided 50 kilowatts of thermal energy. The Montlouis solar furnace became the prototype design for other large high-temperature solar furnaces. Basically, this design utilized a single large heliostat (array of numerous flat mirror elements) which continuously tracked the sun to direct the sun's rays onto a concentrating reflector (parabolic or spherical) consisting of many smaller mirror elements each of which was contoured to concentrate the incident radiation at a common focal point. In the case of the Montlouis furnace, the heliostat was 43 feet (13.1 meters) wide and 34 feet (10.4 meters) tall and contained 540 flat mirrors each 50 × 50 centimeters. The concentrating reflector was made up of 3,500 mirrors 16 × 16 centimeters arranged in a parabolic configuration 36 feet (11 meters) wide and 30 feet (9.1 meters) high with a focal length of 6 meters. Each of the 3,500 flat mirror elements in the parabolic concentrator was mechanically contoured and aligned to focus the radiation received from the heliostat onto the focal point of the parabola.

The successful performance of Montlouis solar furnace led to the use of its design as the prototype for the next three large single heliostat-concentrator solar furnaces which were to be built during the next twenty years. All three of these furnaces were similar to the Montlouis furnace in size, operation and thermal power level and were constructed by: (1) U.S. Army Quartermaster Corps, Natick, Massachusetts; (2) Tohoku University, Sendai, Japan; and (3) the French Army's Laboratoire, Central de L'Armement, Odeillo, Font-Romeu, France. In 1973 the U.S. Army's solar furnace was moved to the Nuclear Weapon Effects Laboratory, White Sands Missile Range, New Mexico, where it became operational in 1974.

Although the Montlouis solar furnace played a major role in developing applications for high-temperature solar energy, and in providing design information for the three other large solar furnaces, its most valuable contribution to the field of high-temperature solar energy was the experience and background it provided the CNRS Solar Energy Laboratory. This led them to design and construct the CNRS 1,000-kilowatt solar furnace.

The CNRS 1,000 kilowatt solar furnace is located at Odeillo, Font-Romeu, altitude of 5,900 feet (1798 meters) about 25 miles (40 kilometers) east of Andorra and 5 miles (8 kilometers) west of Montlouis. At this location, the sun shines as many as 180 days a year and solar intensities as high as 1,000 watts per square meter are common. The solar furnace was completed on October 1, 1970, after more than 10 years of construction.

Figure 11 is a schematic of the CNRS 1,000-kilowatt solar furnace. This furnace utilizes 63 heliostats to direct the sun's rays onto the surface of the giant parabolic concentrator.

Fig. 12. Large parabolic reflector and focal building in foreground. Concentrated energy is directed at the solar furnace located within the focal building. Installation is at Odeillo, Font Romeu, France. (*Photo by Glenn D. Considine.*)

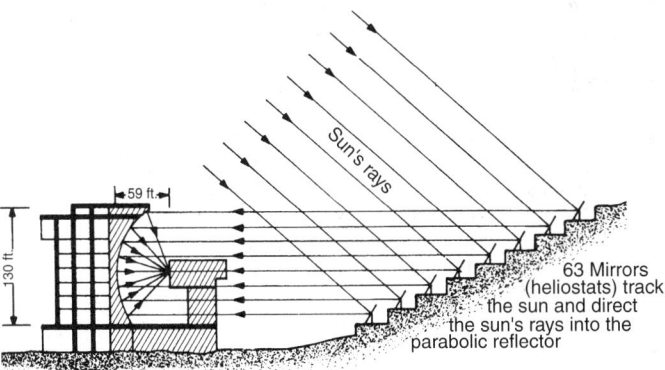

Fig. 11. Schematic representation of the 1000-kilowatt solar furnace at Odeillo, Font-Romeau, France. (*Centre National de la Recherche Scientifique.*)

Fig. 13. Field of heliostat-controlled collector-reflector mirrors, which direct their energy to the parabolic reflector at the solar furnace installation at Odeillo, Font-Romeu, France. (*Photo by Glenn D. Considine.*)

The 63 heliostats are each 7.5 meters wide by 6 meters high and contain 180 single flat mirror elements 50×50 centimeters. The total area of mirror surface in the 63 heliostats is 2,835 square meters or over one-half the playing area of a football field.

The heliostats are located directly north of the parabola and are arranged on eight terraces. Each terrace corresponds in elevation to one of the floors of the building supporting the concentrating parabola. A solar beam of constant energy is thus directed horizontally and southward from the heliostats to the mirrors that make up the concentrating parabola.

Each heliostat is designed to illuminate a specific area on the parabola and is equipped with a dual optical control system, which maintains the proper orientation for each heliostat by means of a dual hydraulic system. This dual system permits each heliostat to be operated in either a "search" or "track" mode. In both cases, the optical guidance system uses an optical tube, which contains four photodiodes that control the heliostat motion in east-west and up-down direction.

When operating in the "search" mode a short (10-centimeter) optical tube with a 40 degree acceptance angle is used to activate the "fast" hydraulic system, which operates in an on-off mode to quickly bring the heliostat to within the operating range of the "track" system. In the "track" mode a 100-centimeter optical tube is used to control a slower acting hydraulic system which operates in a proportional control mode. The size of the sun's image at the base of the 100-centimeter tube is $\frac{1}{2}$-inch (13 mm) in diameter and the accuracy of the control is one minute of arc.

The concentrating parabola has a focal length of 18 meters, is 40 meters high and 54 meters wide, and the focal axis is 13 meters above the first floor. The parabola consists of 9,500 initially flat glass mirrors that were mechanically curved and adjusted to provide a solar image of minimum diameter at the focal point. Almost two years were required to accomplish these two precise adjustments which were completed on 1 October 1970. Figure 12 shows the parabola and the focal building into which the concentrated solar energy is directed. See also Fig. 13.

The solar energy incident on an area of about 2,000 square meters is concentrated by the parabolic reflector onto an area of less than 0.3 square

meters. Sixty percent of the total thermal energy (about 600 kilowatts) is concentrated in an area of less than 0.08 square meter at the center of the focal plane of the parabola.

The diameter of the image of the sun at the focal point is 17 centimeters and 27% of the thermal energy (about 270 kilowatts) is concentrated in this area. Heat flux data in watts per square centimeter in the focal area are presented graphically in Fig. 14. Curve 0 represents the heat flux at the focal plane. Curve d/2 shows the heat flux and temperature at a plane

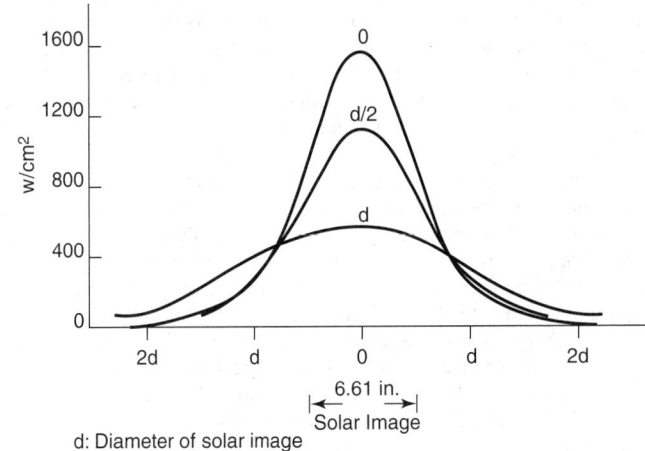

Fig. 14. Solar energy versus distance from focal point in Odeillo solar furnace. (*Georgia Institute of Technology.*)

removed one-half the diameter of the solar image (8.5 centimeters) behind the focal plane. Curve d presents the same data on a vertical plane removed one diameter of the solar image (17 centimeters) behind the focal plane.

Solar Tower Energy Collector

Authorities have observed that solar energy can be usefully collected optically from one square mile (2.6 square kilometers) of surface area, or even larger, and concentrated onto a central receiver by an array of heliostats, i.e., independently steered mirrors. By judiciously spacing mirrors over 35% of the area, such a system in the desert southwest of the United States, for example, could collect 2800 megawatt-hours thermal per day in midwinter and almost twice that amount of energy in midsummer. In order that the reflected radiation from this field be efficiently intercepted, the central receiver would have to be several hundred meters high.

Unlike the Odeillo installation, previously described, where a field of heliostats finally focus their energy to a small aperture by way of a huge parabolic reflector, in the solar tower approach, the energy from each mirror is directed to a central receiving tower, located high above the field, as shown schematically in Fig. 15. Shown is a large array of heliostats by which essentially flat mirrors are automatically steered to reflect or redirect the incident solar radiation to a high tower. It is assumed that the terrain of the heliostat field is flat. However, a gentle slope southward would be advantageous. After reflection from the mirrors, the redirected solar energy can be absorbed and converted to heat by a black body receiver placed in the focal region. The heat can be transported down the supporting tower by way of liquid metal and/or steam lines and can be stored or used to operate a conventional turbine generating station. Alternate uses would be direct conversion to electricity by way of high-power-density solar cells placed in the focal region, or use of the heat to produce a fuel thermodynamically. Two-axis control can be obtained by either hydraulically or electrically operated servo-mechanisms that derive a signal from a simple position-sensing element.

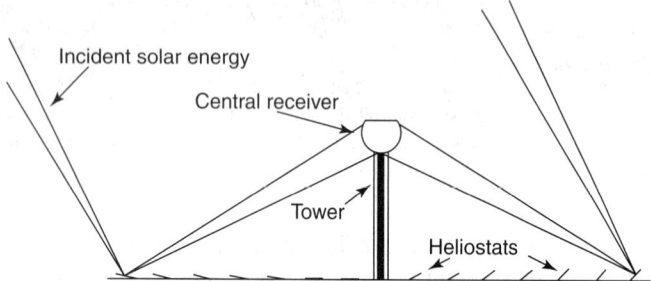

Fig. 15.   Schematic diagram of solar tower energy concept.

It requires energy collected from an area like a square mile to be of interest to power utilities. Energy collection from hundreds of such installations would be required to make a significant impact on the energy supply. If one replaces the 300-meter tower with geometrically identical systems using 100-meter towers, it is found that 9 such towers would be required. Although the cost of nine smaller towers would compare with the cost of a single, large tower, additional costs and losses would be incurred in connecting heat-transfer lines to a central generator to handle 9 collectors. Also, heliostats smaller than about 20 square meters are not economical because the cost per heliostat of the support, actuator, and steering systems have a substantial fixed component.

The first choice of heat-transfer fluid would be steam because it would appear not to require any new technology. However, because the flux density that must be absorbed and transferred to the fluid can be appreciably higher than in conventional steam plants, efficient operation may require some new technology. Also, because of the large daily and seasonal heat flux variations, the design of the receiver is not trivial and may ultimately be best accomplished by utilization of liquid metal, such as sodium, for heat transfer from the receiver surface to a steam line. Liquid sodium technology has been developed for nuclear reactors and operating temperatures of 550 to 650 °C appear reasonable. Sodium also presents a promising high-temperature thermal storage medium. In general, the thermal cycling due to the intermittent nature of solar energy of this high-temperature system will have to be considered in any detailed design. The black body boiler

surface should not deteriorate at high temperature in the presence of air. If deterioration is a problem and if convective losses are appreciable an inverted cavity design can be used in order to avoid the use of vacuum jacketing.

Solar energy may best be utilized at first by a steam generator operating in a solar-only mode with short-term storage of an hour or less to provide operational stability. Because utilities require very high reliability, little new plant capacity credit would be given such a plant because of possible cloudiness. New plant credit could be given if there were possibilities of using liquid fuels, such as liquid petroleum gas or oil for backup. This could be accomplished by adding a simple low-cost, but possibly inefficient burner to add heat. Although liquid fuels may be in short supply, they are easily stored and afford an ideal way of giving a solar plant high reliability as an intermediate plant. Depending on actual operating conditions, it may be that very little liquid fuel is burned. The comparison of such a hybrid plant should be made with a solar plant that has a gas turbine as a backup. An alternate approach might be to store solar energy as sensible heat, perhaps in underground cavities, or as latent heat.

Some authorities believe that close ties between solar power and conventional electric power plants — so-called solar thermal electric designs — represent an ideal approach to the large-scale use of solar energy. In this concept, instead of a solar-powered facility being linked to traditional power-generating facilities by way of the electric grid, a fossil fuel backup for a solar system would be located at the same site as the solar electric plant. Savings could be realized through the common usage of certain equipment items. In one design, the oil-fired backup would use the same turbine as a power tower system. Some designers have estimated that this additional capability would add only about 0.26% of the total cost of the solar electric facility.

There are other authorities who believe that adapting solar energy to electric utilities will limit the economic potential of solar energy. The basic problem, is that both technologies are very capital intensive and that the electric utility, because of the high fixed costs of generation, transmission, and distribution capacity, represents a poor backup for solar energy systems. On the other hand, the solar collection system, because it represents pure, high-cost capital and because of its outage problems, cannot be considered as a part-load source of auxiliary energy for the electric utility system.

Solar Energy Plant at Electric Utility Level

Construction commenced in 1975 on an experimental 10 MWe central receiver pilot plant in a combined effort by two electric utilities in the southwestern United States, Southern California Edison and the Los Angeles Department of Water & Power, who worked in cooperation with the U.S. Department of Energy and the California Energy Commission. The start of continuous electric power production commenced in August 1984 and the plant is now up to its design capacity of 10 MWe. The plant, known as *Solar One*, is located in Daggett, California just off Highway 40 and east of Barstow. A panoramic view of the facility, clearly showing nearly 2000 heliostat-controlled mirrors focusing their collected energy on a boiler atop a 300-foot (91-meter) tower, is given in Fig. 16.

Although large and very impressive, *Solar One* is regarded as an experimental pilot plant for proving and testing technological improvements that can be incorporated in future commercial-size plants. The Daggett plant is a scale model of a 100 MWe generating plant. On its own, *Solar One* is currently furnishing the electricity requirements for a community of about 6000 people. *Solar One* relies on a combination of both old and new solar technology. Certain features not found in typical commercial generating plants allow great flexibility in plant operation. Several different types of solar central receiver plants can be simulated within this one project.

The Basic Concept of *Solar One*. Computer-controlled mirrors (heliostats) totaling 1818 in number form a circular array around a central tower. Within the receiver, the solar energy is transformed into high-temperature thermal energy in a water-steam heat transport fluid. The thermal energy can be converted to electric power immediately or stored to extend plant operation. See Fig. 17. The collected solar energy is most efficiently put to work as receiver steam to power a turbine-generator (Path A). If the energy is to be stored, receiver steam follows path B and heats oil that is routed to and from the thermal storage tank. Energy is discharged from storage by using hot oil from the tank (path C) to generate steam, which is then sent to the turbine along path D.

The thermal storage system uses oil as both a thermal storage medium and a heat transport fluid. The maximum operating temperature of the

Fig. 16.   Panoramic view of the world's largest solar thermal electric power plant, located on the Mojave Desert, near Daggett, California. The 10 megawatt (electric) facility commenced operation in 1982 and achieved design capacity in 1984. The collector tower (receiver upon which solar energy is reflected) is located atop a 300-foot (91-meter) tower. The north field of heliostats (mirrors kept in synchronism with the movement of the sun), 1240 in number, is shown in background; the south field (578 heliostats) is shown in foreground. Operating facilities, turbine generators, and storage tank are shown in circular middle section under the tower. The facility is operated by Southern California Edison and the Los Angeles Department of Water and Power, in conjunction with the U.S. Department of Energy and the California Energy Commission.

storage system is 575 °F (300 °C). As a result, electricity is generated less efficiently than when 960 °F (515 °C) receiver-supplied steam is used directly in the turbine.

The operating temperature of the storage system simulates steam generation conditions in industrial plants and the chemical processing industry. Furthermore, because storage-supplied heat can supplement solar-supplied energy, *Solar One* can simulate a plant that uses both conventional fuels and solar energy.

Heliostats. The facility receives 3600 to 4000 hours of sunlight/year (9.8 to 10.9 hours/day). Construction of the 1818 heliostats for the pilot plant demonstrated that prototype designs can be successfully produced in volume quantities with conventional manufacturing techniques. Each heliostat has a reflective area of 430 square feet (39.3 square meters). The heliostat glass is specially formulated to contain a minimum amount of impurities. As a result, 91% of the incident sunlight can be reflected when the mirror surface is clean. A close-up of a set of mirrors (in a vertical position for demonstration purposes) is given in Fig. 18.

The vertical and horizontal movement of the heliostats is directed by a control system—a microprocessor in each heliostat, a controller to regulate groups of up to 32 heliostats, and a central computer. Over 97% of the heliostats are available more than 98% of the time. Operation of the heliostats has suggested areas for further research and development; for example, rain water may be sufficient to maintain the cleanliness of the mirrors, and mechanical rinsing may be required only in dry months.

The heliostats, as shown in Fig. 19, are distributed in a south field (578) and a north field (1240). The mirrors are slightly concave (approximately 1000 foot focal length with $\frac{1}{6}$-inch curvature in a 10-foot length). The total weight of a heliostat, as previously shown in Fig. 18, is 4312 pounds (1956 kg). The heliostats are normally stowed in a vertical position except

when high wind conditions exist. During daylight hours, of course, the mirrors are rotated by a drive mechanism to follow the direct solar rays as closely as possible. It is interesting to note that the sun's position is calculated rather than sensed—so that even when clouds briefly cover the sun, maximum energy is reflected.

Receiver. On top of the steel tower rests the cylindrical receiver, a superheated steam boiler that is 14 meters (46 feet) tall and 7 meters (23 feet) in diameter. The receiver weighs almost 50 tons and is positioned over 20 stories above the ground. Feedwater is pumped to the bottom of the receiver, where it is vaporized to superheated steam in a single pass to the receiver's top. The steam is then piped to the turbine-generator at the foot of the tower. This steam can also provide heat to the thermal storage system.

Thermal Storage System. On a clear day, the receiver can generate sufficient steam to simultaneously operate the turbine and also deliver heat to the storage system. The thermal storage can generate power in the evening or during periods of cloud cover. It also provides steam for start-up in the morning and for keeping selected portions of the plant warm when the plant is not operating. The steel-walled insulated storage tank has a capacity of 3.5 million liters and sits on lightweight insulating concrete. The tank is filled with sand, rocks, and a high-temperature thermal oil. Steam from the receiver is routed through heat exchangers to heat the thermal oil, which is then pumped into the tank to heat the rock and sand. This stored thermal energy can then be transferred to the turbine generator for electrical power production.

Power Generation Station. The turbine-generator is rated at 12.5 megawatts and is sized to handle the full plant system output plus all internal plantloads. The dual-admission turbine has a high-pressure steam inlet for steam produced by the receiver and a low-pressure inlet for

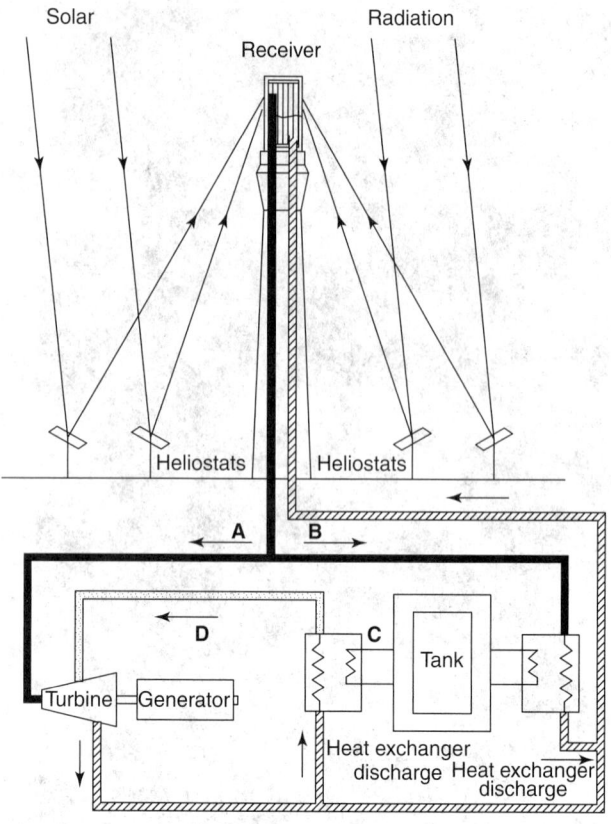

Fig. 17.   General concept of *Solar One*. Within the receiver, the solar energy is transformed into high-temperature thermal energy in a water-steam heat transport fluid. The thermal energy can be converted to electric power immediately or stored to extend plant operation. The collected solar energy is most efficiently used as *receiver steam* to power a turbine-generator (Path A). If the energy is to be stored, receiver steam follows Path B and heats oil that is routed to and from the thermal storage tank (Path C) to generate steam, which is then sent to the turbine along Path D. The thermal storage system uses oil as both a thermal storage medium and as a heat-transport fluid.

Fig. 18.   Close-up of a heliostat rack assembly shown in vertical position for demonstration purposes. Note reflection of tower in mirrors. *Solar One* has a total of 1818 heliostats.

steam produced from thermal storage. The rated turbine thermal-to-electric efficiency from receiver to steam is 35%. The efficiency is 25% from the lower quality thermal storage system.

Master Control System. In the morning the operator, through keyboard commands, positions the heliostats at standby operating points, begins water circulation in the receiver, and then issues a command to the system to start up the plant. At this point, a computer takes over and automatically directs heliostats to track the receiver. When receiver steam conditions are correct, steam is routed to the turbine. The operator then synchronizes the turbine to the electric grid, after which the minimal manual attention is needed. If conditions change, such as a cloud passing over, the control system automatically makes adjustments to keep the plant in the best operating state. If some abnormal event occurs, alarm messages tell the operator which parameters are out of normal operating range. The operator can, at any time, make changes in any plant operating condition. While the pilot plant control system was designed for controlling a water-steam central receiver solar plant, the basic functions and operating philosophy are readily adaptable to other power plants.

Performance. The requirement for production of 10 MWe was exceeded by a peak production of 12.1 MWe. Similarly, the required 7 MWe net generation from storage was exceeded by an output of 7.3 MWe. The plant also has successfully operated down to 0.5 MWe, which is considerably lower than the designed minimum operating production level of 2 MWe. The minimum sunlight threshold for operation was designed as 450 W per square meter, yet the plant has operated in direct solar radiation levels as low as 300 W per square meter. In an endurance test, the receiver and storage system kept the turbine continuously on-line for 33.6 hours and generated 127 MWe net.

Solar One was designed to have 95% of the heliostats available at any one time. Between April 1982 and April 1983, 98% of the heliostats

were available for operation. This percentage later increased to 99%. The establishment of the sharp thermal gradient (thermocline) needed for the storage system has been verified. Gradients of 49°C/meter have been measured. Equally important is the very low rate of heat loss from the storage tank. The tank heat loss has been measured at 1.3% day.

Central Receiver Test Facility

The largest facility specifically designed for testing central receiver components and subsystems was built in Albuquerque, New Mexico in the late 1970s. At this facility, a 15-meter (49-foot) diameter concrete-and-steel receiver tower rises some 60 meters (197 feet) above the ground. Within the tower, three test bays at different levels are used for experiments. A huge elevator can transport equipment weighing as much as 100 tons to these test bays. There is a total of 222 heliostats; when all of them are focused on a receiver in one of the test bays, temperatures in excess of 2000°C (3632°F) can be generated. In practice, lower temperatures are used for receiver testing. The test facility has been used to try out innovative receivers that use gas, liquid sodium, or molten nitrate salts for thermal transport. In one system, molten salt has been used as the heat transport fluid and storage medium in an integrated central receiver system to produce an electrical power output of 750 kWe.

Solid-Particle Central Receiver. A new type of receiver has been under investigation. A novel concept for a central receiver uses sandsize refractory particles that free-fall in a cavity receiver. A conceptual design is shown in Fig. 20. Scientists observe that the advantages of a solid particle receiver over traditional fluid in-tube receivers are: (1) the particles can directly absorb solar radiation, and (2) the particles maintain their integrity at high temperatures. These advantages, coupled with the possibility that the particles can serve as the storage medium, could provide a cost-effective means of high-temperature solar energy utilization. High temperatures are attractive for fuels and chemical production, industrial process heat

Collector field layout

North field (1240 heliostats)

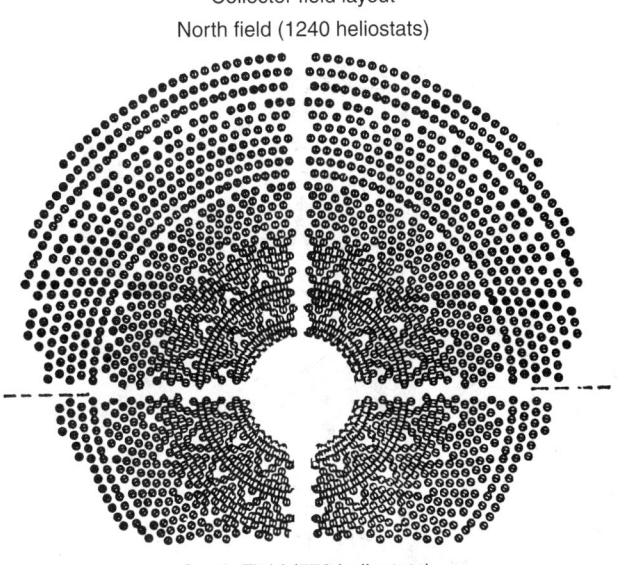

South Field (578 heliostats)

(a)

Collector field Segmentation

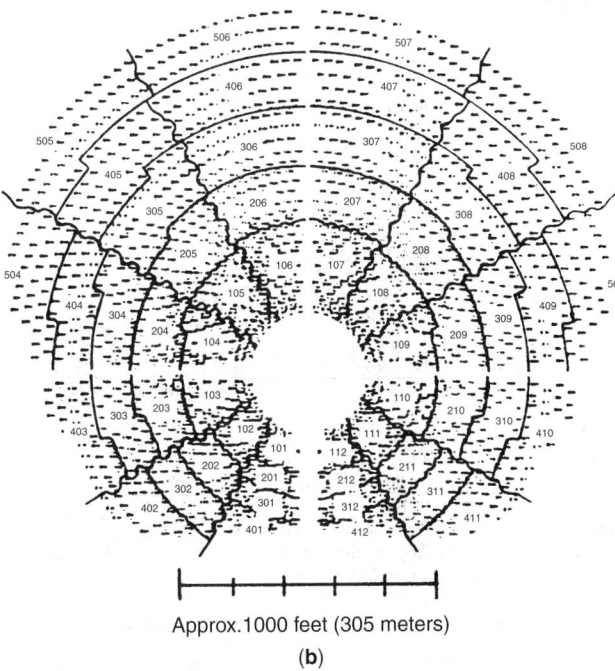

Approx.1000 feet (305 meters)

(b)

Fig. 19. (a) The heliostats are distributed in two fields—North and South; (b) for control purposes, the heliostats are segmented.

applications, or Brayton cycle electricity generation. The concept is in an early experimental stage.

Heat Engine Cycles for Solar Power

Heat engines for conversion of solar energy to electric power ideally should have the following attributes: (1) low cost per kilowatt output capacity; (2) long life and reliable operation with minimal maintenance; (3) safe and environmentally acceptable operation; (4) characteristics compatible with cycle top temperatures up to 1,000 K; and (5) efficiency approaching Carnot values.

Heat engines that are potential candidates for coupling a solar heat source include thermoelectric, thermionic, thermochemical, magnetohydro-dynamic, Rankine, Brayton (simple or recuperated), and cascaded cycles.

Rankine Cycle. The steam-Rankine cycle employing steam turbines has been the mainstay of utility thermal electric power generation for many years. The cycle, as developed over the years, is sophisticated and efficient. The equipment is dependable and readily available. A typical

cycle (Fig. 21) uses superheat, reheat, and regeneration. Heat exchange between flue gas and inlet air adds several percentage points to boiler efficiency in fossil-fueled plants. Modern steam Rankine systems operate at a cycle top temperature of about 800 K with efficiencies of about 40%. All characteristics of this cycle are well suited to use in solar plants.

Brayton Cycle. In recent years, attention has been drawn to the Brayton cycle as a potential and practical alternative to the steam Rankine cycle for solar power and for high-temperature gas-cooled nuclear reactors. The Brayton cycle is most familiar in its open form as used in aircraft gas turbines. The open Brayton cycle cannot compete with steam-Rankine in efficiency. In a power-generation application, cycle efficiencies on the order of 20% would be expected. However, the Brayton cycle can achieve higher efficiency through recuperation, sometimes called regeneration. A representative cycle diagram is given in Fig. 22. The working fluid is an inert gas, typically helium. Inert gas mixtures, such as helium-xenon, have been studied and have potential advantages.

The recuperated Brayton cycle approaches Carnot efficiency in the ideal limit. As compressor and turbine work are reduced, the average temperatures for heat addition and rejection approach the cycle limit temperature. The limit is reached as compressor and turbine work (and cycle pressure ratio) approach zero and fluid mass flow per unit power output approaches infinity. It can be expected from this that practical recuperated Brayton cycles would operate at relatively low pressure ratios, but be very sensitive to pressure drop. With the assumption of constant gas specific heat over the cycle temperature range, a good assumption for helium, the cycle efficiency of a recuperated Brayton cycle may be expressed:

$$\eta e = 1 - \left[\frac{\dfrac{r_{pc}\zeta - 1}{\eta_b} + \dfrac{\Delta T_r}{T_0}}{\dfrac{T_3}{T_0}\eta_T \left[1 - \left(\dfrac{G}{r_{pc}} \right)^{\zeta} \right] + \dfrac{\Delta T_r}{T_0}} \right]$$

where r_{pc} is compressor pressure ratio (>1)
  η_b is compressor efficiency
  ΔT_r is temperature difference across recuperator ($T_4 - T_2$ in Fig. 21)
  T_0 is cycle lower limit temperature
  T_3 is cycle top temperature
  η_T is turbine efficiency
  ζ is specific heat factor, $(\gamma - 1)/\gamma = 0.4$ for $\gamma = 1.67$ as for helium
  G is the product of the four pressure drop factors,

$$\left(\frac{P_1}{P_2} \right) \quad \left(\frac{P_2}{P_3} \right) \quad \left(\frac{P_4}{P_5} \right) \quad \left(\frac{P_5}{P_0} \right)$$

See also **Gas and Expansion Turbines.**

Solar Energy for Industrial/Metallurgical Processes

It is interesting to note that experiments on melting large masses of metals were conducted at the French Odeillo solar furnace as early as the mid-1970s. Since the mid-1980s, solar furnaces in other locations have researched what promises to be an important technology in the near future—namely, Solar Induced Surface Transformation of Materials (SISTM). In the past, numerous heat-treating methodologies have been practiced widely. Less traditional methods have been developed in recent years, including cladding/coating, self-propagating high-temperature synthesis, thin-film deposition, and ion-, laser-, and electron-beam processes.

Researchers have found that, for delivering large fluxes on target, the solar furnace is less capital intensive than competing methods that require an intermediate energy-conversion step. For example, a 1-square-meter (11 ft²) solar dish can deliver about 1 kW of optical power to a target. To deliver the same amount of radiant energy, an arc lamp would have to be powered by the electrical output of an 11-square-meter (120 ft²) dish, and a carbon dioxide laser would require the energy supplied by 26-square-meter (280 ft²) disk.

The automotive industry, which turned to laser transformation hardening of engine and other drivetrain components, is now seriously investigating the "solar hardening" process.

Solar Furnaces No Longer Uncommon. The past decade has brought increased interest in using solar energy for industrial processes, not just

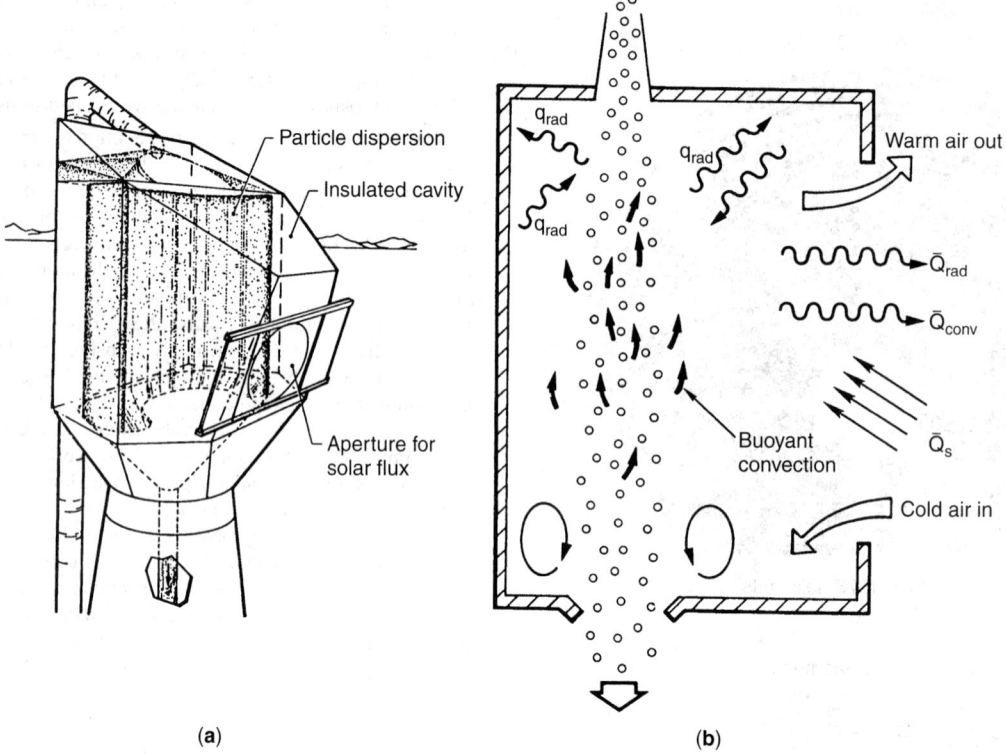

(a)   **(b)**

Fig. 20.   Solid-particle central solar energy receiver: (**a**) conceptual design; (**b**) thermal phenomena in a solid particle receiver. (*Sandia National Laboratories.*)

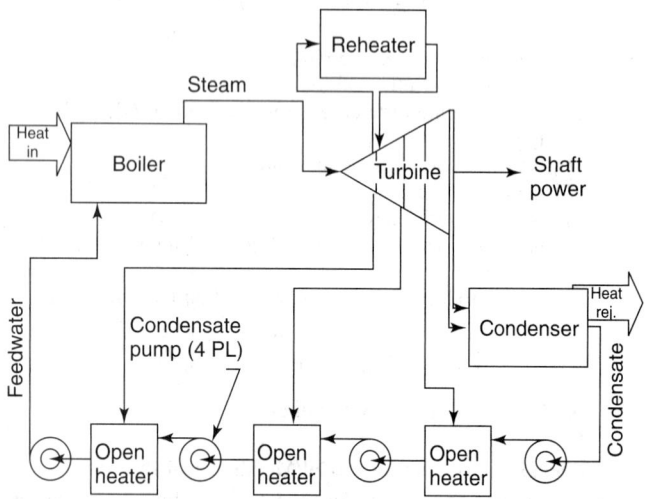

Fig. 21.   Schematic diagram of regenerative-reheat steam Rankine cycle.

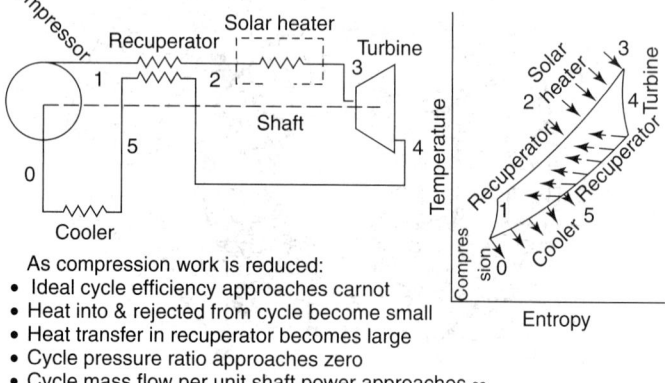

As compression work is reduced:
- Ideal cycle efficiency approaches carnot
- Heat into & rejected from cycle become small
- Heat transfer in recuperator becomes large
- Cycle pressure ratio approaches zero
- Cycle mass flow per unit shaft power approaches ∞

Fig. 22.   Recuperated Brayton cycle diagram.

to conserve energy or avoid the use of fossil fuels. Some of the important installations, as of the early 1990s, are listed in Table 1.

The time required to bring a part to treating temperature also is an important consideration. Solar furnaces have no competition on this point. See Table 2.

Chemicals Directly from Solar Energy. Whereas the early solar furnace located in Odeillo, France, was constructed for solar energy studies in general and later used for melting huge masses of metals and materials and the solar plant in the Mojave Desert in California is being used directly to generate electric power, scientists in the early 1990s were taking a somewhat different approach.

Researchers have recognized that the regions where ample sunshine is available for harvesting seldom coincide with the population and industrial centers of the world. Further, solar energy is of an intermittent nature because it depends upon clear skies with no cloud cover. I. Dostrovsky (Weizmann Institute of Science) observes that most of the limited amount of research on solar energy focuses on converting sunlight to electricity, mainly by photovoltaic and thermal methods, or, in the case of the

Daggett, California, utility installation, to furnish heat to supply steam turbines.

An alternative approach is that of using solar energy to produce chemicals that can be stored and transported much as present fossil fuels. One installation along these lines is now undergoing demonstration operation in Saudi Arabia. This project, known as *HYSOLAR*, is a joint venture of Saudi Arabia and Germany. In essence, the facility consists of a plant that produces hydrogen. See also **Hydrogen (Fuel)**.

One process involves the high-temperature decomposition of sulfuric acid, using recoverable iodine as an intermediate reactant. In a first stage, sulfuric acid yields water and sulfur dioxide. In a second stage, the sulfur dioxide plus water and iodine yield hydrogen iodide and sulfuric acid. In the third stage, the hydrogen iodide yields hydrogen (the desired product) and recoverable iodine.

In an electrolytic process, hydrogen is produced by electrolyzing a mixture of sulfur dioxide and water to produce sulfuric acid and hydrogen. In still another electrolytic process, bromine, sulfur dioxide, and water react to form hydrogen bromide and sulfuric acid. By applying 0.62 V to the hydrogen bromide molecules, hydrogen is yielded and the original bromine is recovered.

TABLE 1. COMPARISON OF HIGH-FLUX SOLAR FACILITIES

Location	Total Power kW	Peak Flux MW/m^2
Albuquerque, New Mexico		
Central receiver test facility	5000	2.4
Furnace	22	3.0
Atlanta, Georgia		
Furnace	1.3	9.5
Golden, Colorado	10.0	2.5
(measured using nonimaging secondary concentrator)		20.0
White Sands, New Mexico	30	3.6
Odeillo, France		
Horizontal furnace	1000	16.0
Vertical furnace	6.5	15.0
Rehovot, Israel		
Central receiver test facility	2900	
Furnace	16	11.0
Uzbek, Russia	1000	17.0

Source: Solar Energy Research Institute.

TABLE 2. SOLAR FURNACE TIME TO REACH MELTING POINT OF MATERIALS
(When exposed to absorbed solar flux of 20 MW/m^2)

Material	Melting point, T_m, °C	Time to reach T_m, ^{1}sec
Carbides		
TiC	3,200	9.14
NbC	3,500	0.86
ZrC	3,540	0.112
SiC	3,830	0.56
Metals		
Al	660	0.42
Cu	1,083	3.10
Ni	1,453	1.03
Steel	1,535	0.79
Ti	1,670	0.23
Cr	1,857	1.46
Mo	2,617	4.70
W	3,407	9.80
Nitrides		
Si$_3$N$_4$	1,900	0.059
AlN	2,200	1.320
BN	3,000	0.545
TiN	3,200	0.611
Oxides		
SiO$_2$	1,720	0.014
TiO$_2$	1,870	0.044
Al$_2$O$_3$	2,050	1.00
V$_2$O$_3$	2,410	0.107
CaO	2,580	1.66
HfO$_2$	2,780	0.188
MgO	2,800	2.59
ZrO$_2$	2,900	0.089

Source: Solar Energy Research Institute.

This type of approach is somewhat reminiscent of the chemistry of coal gasification. More detail is given in the Dostrovsky reference listed.

Solar Energy for Detoxifying Hazardous Chemicals. The Solar Energy Research Institute, Golden, Colorado, has developed a system for detoxifying hazardous chemicals in polluted groundwater. In essence, a photocatalyst is added to the polluted water and then pumped through long, narrow glass tubes that are exposed to sunlight. High-energy photons activate the catalyst, which in turn breaks the pollutants down into nontoxic components. The tubes are mounted in reflecting glass troughs to improve the efficiency of the process. The system has proved particularly effective against trichloroethylene, once a common industrial cleaner. In this application, the polluted water is mixed with titanium dioxide catalyst. Hydroxyl radicals are created that break the offending solvent into water, carbon dioxide, and very dilute hydrochloric acid. The next step is that of determining how effective the process may be in removing other chlorinated hydrocarbons, as well as such substances as benzene, various pesticides, and textile dyes.

Photovoltaic Conversion (Solar Cells)

Photovoltaic devices made of selenium have been known since the 19th Century. Pioneering research in semiconductors, which led to the invention of the transistor in 1947, formed the basis of the modern theory of photovoltaic performance. From this research, the silicon solar cell was the first known photovoltaic device that could convert a sufficient amount of the sun's energy to power complex electronic circuits. The conventional silicon cell is a solid-state device in which a junction is formed between single crystals of silicon separately doped with impurity atoms in order to create *n* (negative) regions and *p* (positive) regions which respectively are receptors to electrons and to "holes" (absence of electrons). See also **Semiconductors**. The first solar cell to be demonstrated occurred at Bell Laboratories (now AT&T Bell Laboratories) in Murray Hill, New Jersey in 1954.

In a photovoltaic device, the energy in light is transferred to electrons in the semiconductor when a photon collides with an atom in the material with enough energy to dislodge an electron from a fixed position in the material. A common technique for producing a voltage is by creating an abrupt discontinuity in the conductivity of the cell material (typically silicon) through the addition of dopants. A basic limit on the performance of these devices stems from the fact that light photons lacking the energy needed to lift electrons from the valence to the conduction bands ("band gap" energy) cannot contribute to photovoltaic current, and from the fact that the energy given to electrons which exceeds the minimum excitation threshold cannot be recovered as useful electric current. Most of the photon energy not recovered as electricity is converted to thermal energy in the cell.

Photon energies in the visible light spectrum vary from 1.8 eV (deep red) to 3 eV (violet). In silicon, about 1.1 eV is needed to produce a photovoltaic electron; in gallium arsenide (GaAs), this is about 1.4 eV. Silicon is a comparatively poor absorber of light and consequently silicon cells must be from 100 to 200 micrometers thick to capture an acceptable fraction of the incident light. This places limitations on crystal grain size and thus, with present technology, single crystals must be used. Polycrystalline materials may alter this problem favorably and much research is being directed toward developing polycrystalline materials and, in general, for finding methods to minimize the impact of grain boundaries.

Thin films of gallium arsenide (GaAs) and cadmium sulfide/copper sulfide (CdS/Cu$_2$S) show potential because they are better absorbers of light and can be made thinner than crystalline silicon. Smaller crystal grains can be tolerated better than with crystalline silicon. See also **Thin Films**. These can be spray- or vapor-deposited, thus simplifying manufacturing. One possible drawback of the CdS and GaAs materials is their toxicity, particularly hazardous during manufacturing operations.

Where solar cells are used in *concentrated* sunlight, efficiency becomes of particular importance because of its effect upon total collector area needed, this being a major cost component of a solar energy system. A number of ingenious collector configurations have been developed. Further, there is the concept of the *thermophotovoltaic cell*, which may be able to achieve efficiencies as high as 30–50% through shifting the spectrum of light reaching the cell to a range where most of the photons are close to the minimum excitation threshold for silicon cells. High efficiencies in intense radiation can be achieved, for example, with GaAs cells by covering them with a layer of Ga$_x$Al$_{1-x}$As, a material that reduces surface and contact losses. Clearly the interface between cell and solar radiation is of as great importance as development of new cell materials per se.

So-called *wet solar cells* show promise, particularly because of their relative ease of fabrication. In this type of photovoltaic cell, the junction is formed between a semiconductor and a liquid electrolyte. No doping is required because a junction forms spontaneously when a suitable semiconductor, such as GaAs, is contacted with a suitable electrolyte. Three knotty problems (accelerated oxidation of surface of semiconductor; exchange of ions between semiconductor and electrolyte forming a blocking layer; and deposition of ions of impurities on the surface of the semiconductor) all have been solved and thus the concept now appears technically viable.

Over a number of years, the photovoltaic cell developers received large financial incentives from the U.S. government. For example, the National Photovoltaics Act of 1978 was passed by the U.S. Congress, which authorized an expenditure of $1.5 billion for research, development, and demonstration of solar cell systems for converting sunlight into electric power. Also, in connection with the Federal Non-Nuclear Energy Research and Development Act of 1974, which established the concept of "net energy"—that is, the effect of new devices and systems on the overall energy balance. Projects were evaluated on the basis of their "potential for production of net energy." Although there have been some breakthroughs of particular significance to scientists and a gradually expanding market for photovoltaics in addition to use in space, particularly in various consumer products, the long awaited and ultimate application (generation of electric power in impressive amounts at competitive prices) has remained elusive. Scores of analyses have been made and forecasts range from very pessimistic to quite optimistic. The era of practically achieving this goal on the part of the photovoltaic cell community tends to be progressively shifted outward into the future. Forecasts usually are based upon numerous assumptions that are subject to periodic change and their reporting is best left to the periodicals and thus are not detailed here.

It is in order, however, to sum up the observations of the Electric Power Research Institute (EPRI): Photovoltaics need significant additional research to reduce cost and increase efficiency of the cells as well as their support systems (tracking and D.C.-to-A.C. power conversion) before they can be competitive with conventional electricity supply technologies. Current manufacturing costs for flat-panel arrays of interconnected, encapsulated cells are approximately $5000 per peak kW, and balance-of-plant costs double the effective system cost to $10,000 per peak kW. This compares roughly with about $300 per kW for combustion turbines; $1400/kW for pulverized coal plants; and $2500/kW for nuclear plants. Two classes of photovoltaic converters that appear to show the most promise for producing large amounts of power are (1) the inexpensive, flat-plate, thin-film devices with target prices of less than $1500 per peak kW and efficiencies of 15%, compared with their current costs of $5000 per peak kW and efficiencies of about 10%; and (2) very high-efficiency, high-concentration devices with target prices less than $1500 per peak kW and efficiencies of 25%, compared with their current costs of $7000 per peak kW and efficiencies to utilities (largely subsidized or experimental programs).

Some authorities estimate that photovoltaic utility capacity could range from 0.6 to 16 GW by the year 2010, provided that needed technical performance is achieved.

Satellite Energy Collectors

Having proven their value in connection with relatively small space satellites, probes, etc., a huge satellite energy collector was first proposed in the late 1960s and, largely on the basis of national concerns with energy supplies precipitated by the oil embargo of the 1970s, considerable attention was given at the design level and in the literature to a solar power satellite (SPS). One proposal called for a space-based array requiring about 90 square kilometers (55 square miles)! That is about the size of Manhattan Island. The satellite would be in a geosynchronous orbit some 36,000 kilometers (22,000 miles) above Earth. Because nearly all authorities now consider such a project very "futuristic," no further details are reported here.

NOTE: To construct a perspective on the energy situation for the 1990s, some readers may wish to refer to the following articles in this encyclopedia: **Energy; Coal; Coal Conversion Processes; Electric Power Production and Distribution; Fuel Cells; Geothermal Energy; Hydroelectric Power; Hydrogen (Fuel); Natural Gas; Nuclear Power; Petroleum; Semiconductors; Tar Sands;** and **Tidal Energy.** See also alphabetical index.

Additional Reading

Asbury, J.G. and R.O. Mueller: "Solar Energy and Electric Utilities,: *Science,* **195**, 445–450 (1977).

Asbury, J.G., Maslowski, C., and R.O. Mueller: "Solar Availability for Winter Space Heating," *Science,* **206**, 679–681 (1979).

Beattie, D.A.: "History and Overview of Solar Heat Technologies," MIT Press, Cambridge, MA, 1997.

Becker, M.: "Solar Thermal Central Receiver Systems," Springer-Verlag, Inc., New York, NY, 1987.

Considine, D.M.: "Solar Absorption Coating and Heat-Pipe System," in "Energy Technology Handbook: (D.M. Considine, editor), The McGraw-Hill Companies, Inc., New York, NY, 1977.

Dostrovsky, I.: "Energy and the Missing Resource," Cambridge University Press, New York, NY, 1988.

Dostrovsky, I.: "Chemical Fuels from the Sun," *Sci. Amer.,* 102 (December 1991).

Flood, D.J.: "Space Solar Cell Research," *Chem. Eng. Progress,* 62 (April 1989).

Goswami, D.Y., F. Kreith, and J.F. Kreider: "Principles of Solar Engineering," 2nd Edition, Taylor & Francis, Inc., Philadelphia, PA, 1999.

Gupta, B.P.: "Solar Thermal Technology: Research and Development and Applications," Proceedings of the Fourth International Symposium, Albuquerque, NM, 1990.

Holden, C.: "Sunlight Breaks Down Hazardous Chemicals," *Science,* 1215 (September 13, 1991).

Hubbard, H.M.: "Photovoltaics Today and Tomorrow," *Science,* 297 (April 21, 1989).

Laird, F.N.: "Solar Energy, Technology Policy, and Institutional Values," Cambridge University Press, New York, NY, 2001.

Stanley, J.T., Fields, C.L., and J.R. Pitts: "Surface Treating with Sunbeams," *Advanced Materials & Processes,* 16 (December 1990).

Waterbury, R.C.: "Solar Pump Delivers Remote Power," *InTech,* 74 (January 1990).

Wieder, S.: "An Introduction to Solar Energy for Scientists and Engineers," Krieger Publishing Company, Melbourne, FL, 1992.

Wilson, H.G., MacCready, P.B., and C.R. Kyle: "Lessons of Sunnyracer," *Sci. Amer.,* 90 (March 1989).

Winter, Carl-Jochen and J. Nitsch: "Hydrogen as an Energy Carrier: Technologies, Systems, Economy," Springer-Verlag, Inc., New York, NY, 1988.

NOTE: For earlier references on solar energy, see prior edition of this encyclopedia.

Web References

American Solar Energy Society: *http://www.ases.org/*

Electric Power Research Institute: *http://www.epri.com/*

Georgia Institute of Technology University Center of Excellence for Photovoltaics Research and Education: *http://www.ece.gatech.edu/research/UCEP/*

National Renewable Energy Laboratory: *http://www.nrel.gov/*

National Solar Thermal Test Facility: *http://www.sandia.gov/Renewable_Energy/solarthermal/nsttf.html*

Office of Power Technologies: *http://www.eren.doe.gov/power/*

University of Florida Solar Energy and Energy Conversion Laboratory: *http://www.me.ufl.edu/SOLAR/*

University of Massachusetts Renewable Energy Research Laboratory: *http://www.ecs.umass.edu/mie/labs/rerl/*

US Environmental Protection Agency Clean Energy: *http://www.epa.gov/globalwarming/actions/cleanenergy/index.html*

100 Top Energy Sites: *http://www.100topenergysites.com/*

SOLAR PARALLAX. The angle subtended by the equatorial radius of the earth at the distance of the sun is called the solar parallax. It is related directly to the astronomical unit which scales the solar system and is the baseline for the determination of stellar parallax.

One method for obtaining the solar parallax is to determine accurately the distance between the earth and another planet. Since their orbits are extremely well observed (on unit scale), knowing the separation at any given moment allows one to place an absolute value to the scale. The minor planet Eros has been used because of its close approach to Earth of only 26,000,000 kilometers and its point-like appearance on the photographic plate. Such observations have yielded a value for the solar parallax of

$$\pi = 8''.7984 \pm 0.0004$$

Radar observations of Venus accomplish the same purpose and yield a value for the astronomical unit of

$$\text{A.U.} = 149,598,640 \pm 250 \text{ kilometers}$$

which results in a solar parallax of

$$\pi = 8''.79414 \pm 0''.00002$$

and a lunar parallactic term in longitude of

$$P = 124''.987 \pm 0''.001$$

The radar observations are believed to be the better values by an order of magnitude. It is interesting to note that the results from lunar occultations lead to the value of

$$\pi = 8''.793 \pm 0''.003$$

$$P = 124''.97 \pm 0''.04$$

SOLAR TIME. Time based upon the rotation of the Earth relative to the Sun. Solar time may be designated as mean or astronomical if the mean Sun is the reference, or apparent if the apparent Sun is the reference. The difference between mean and apparent time is called equation of time. Solar time may be further designated according to the reference meridian, either the local or Greenwich meridian or additionally in the case of mean time, a designated zone meridian. Standard or daylight-saving are variations of zone time. Time may also be designated according to the timepiece, as chronometer time or watch time, the time indicated by these instruments. See also **Time**.

SOLENOIDAL.

Applied to vector field having zero divergence, hence, one that may be expressed as the curl of another vector:

$$\mathbf{a} = \nabla \times \mathbf{b}$$

where **a** is the solenoidal vector and **b** is a vector field (sometimes called the vector potential of **a**), which can be determined from the differential equation. An equivalent definition of a solenoidal vector is one of which the integral over every reducible surface in its field is zero.

See also **Divergence (Mathematics)**.

SOLENOID (Electrical).

An electrically energized coil that may consist of one or more layers of windings. It is the basis of all forms of the electromagnet and is thus part of the operating mechanism of many electrically operated devices. One of the simplest forms and at the same time a widely used one is the plunger type solenoid. This is a coil wound on a non-magnetic form in which a magnetic plunger may move. Energizing the coil pulls the plunger up into the coil and thus operates the associated mechanism. The iron-clad solenoid is similar except for an iron case surrounding the coil. This increases the magnetic pull on the plunger. Other types use a fixed core and various types of external armatures. Solenoids are widely used for operating circuit breakers, track switches, valves, and many other electromechanical devices.

SOLFATARA.

A type of fumarole (volcanic), the gases of which are characteristically sulfurous. The *solfateric stage* is a late or decadent type of volcanic activity, characterized by sulfurous gases emitted from the vent. Also may refer to Etymol, the Solfatara volcano in Italy.

SOLID ANGLE.

Consider a small cone with a base of area dS and a vertex at a fixed point P. This cone will cut out an area $d\sigma$ on a sphere of radius r with center at P. The solid angle subtended by dS at P is defined as $d\omega = d\sigma/r^2$. It is numerically equal to the area cut out by the same cone on a sphere of unit radius at the same point P. The unit used for measuring a solid angle is the steradian.

See also **Angle (Mathematics)**.

SOLID (Geometry).

A solid is a limited portion of space, bounded by a surface. A distinction should be made between the surface and the solid, thus a conical surface and a cone are not identical. A surface of revolution is a solid generated by the revolution of a plane area about a line, the axis of revolution. The study of solid properties is the main concern of solid geometry.

Among the figures considered are: anchor ring, cone, cylinder, ellipsoid, hyperboloid, paraboloid, parallelepiped, polyhedron, prism, prismatoid, pyramid, sphere.

See also **Conic Section**; and **Geometry**.

SOLID ROCKETS.

See **Rocket Propellants**.

SOLID-STATE PHYSICS.

The study of the physical properties (crystallographic, electrical and electronic, magnetic, acoustic, optical, thermal, mechanical, etc.) of substances in the solid phase.

In years past, much emphasis has been given to crystalline solids and this continues, but there has been a growing shift of interest to polymeric and amorphous substances as well. Much attention in the past has been given to metals and this also continues apace, but other substances are now under very serious investigation, including the ceramics, glasses, and organics. Interest in the solid state, of course, was given a tremendous boost by the discovery of semiconductors in the 1940s. During the intervening years, this interest has been spurred by other electronic and electrical materials, including dielectrics, piezoelectrics, ferroelectrics, conductors and superconductors, electrodes, insulators, contacts, and polymers and macromolecular materials, notably those that are electroactive. Interest outside the electronics field, notably in the science of ceramics, glasses, and entirely new materials, such as composites, also has been adding to the body of knowledge of the solid state. However, because of the great

need for solid materials with special properties for a host of applications, solid-state theory has tended to lag practice.

Nevertheless, solid-state theory has made excellent progress during the past decade. Just a few examples would include:

Excitonic Matter. The interaction of light with solid matter is a phenomenon of fundamental importance for exploring the quantum mechanics of materials. This field dates back to Einstein's finding that light energy is carried by quantized packets of radiation (photons). More recently, it has been found that a conduction electron can combine with a positively charged "hole" in a semiconductor to create an *exciton*, which, in turn, can form molecules and liquids. Some authorities consider the exciton as a new phase of matter. It was learned several years ago that the energy of incident photons can be converted inside a crystal into what might be termed short-lived neutral entities, i.e., excitons. As reported in an excellent paper by Wolfe and Mysyrowicz (1984), the exciton resembles the hydrogen atom. It consists of two oppositely charged carriers bound together by electrostatic attraction. In the hydrogen atom, the positive charge is a proton, which is surrounded by the negatively charged electron. In the exciton, the positive charge has a mass of an estimated $\frac{1}{1000}$th that of the proton. In the Wolfe/Mysyrowicz paper (details far beyond the scope of this encyclopedia), the investigators address several interesting questions. Can the exciton propagate freely through the crystal like a free hydrogen atom in a gas? Can two or more excitons combine to form a molecule? Can the excitonic "atoms" or the molecules made up of them form liquid or solid phases? Can more exotic phases of condensed excitonic matter come into being? How are excitons created by light in a crystal? Why does a crystal absorb light at all?

Electron Transport in Solids. It is well established (elucidated in several articles in this encyclopedia) that the production of integrated circuits (ICs) requires manufacturing techniques of extreme precision and sophistication. The purity of materials used is also far higher than experienced by most other materials-processing industries. It has been observed by Howard, Jackel, Mankiewich, and Skocpol (AT&T Bell Laboratories), in a 1986 paper, that a single-crystal silicon wafer 15 cm or more in diameter can be obtained with concentrations of undesired dopants at less than 1 part in 10 billion and with only about one defect per square centimeter. Accuracy in recent years is in terms of a few nanometers, and feature sizes in commercial circuits are down to 1 micrometer (micron) and getting smaller. Thus, it is no surprise that the silicon transistor can serve as a model for investigating numerous areas of the solid state. Using new patterning techniques, devices almost $\frac{1}{100}$th the size of commercial ICs can be made, making it possible to study transport physics in microstructures only a few hundred atoms across.

In 1985, two research institutions (IBM and AT&T Bell Laboratories) reported that electrons can travel through a semiconductor without being slowed by collisions (*ballistically*). The report was based upon experimental data showing a ballistic peak in the electron energy spectra of gallium arsenide (GaAs) test devices. This is reported in more detail in article on **Arsenic**.

Electroactive Polymers and Macromolecular Electronics. Electroactive polymers are of particular interest in connection with their use in fabricating improved electronic microstructures. Scientists at AT&T Bell Laboratories have been active in the investigation and development of electroactive polymers notably for electrodes. As reported by Chidsey and Murray (1986), electrodes can be coated with electrochemically reactive polymers in several microstructural formats called sandwich, array, bilayer, micro-, and ion-gate electrodes. These microstructures can be used to study the transport of electrons and ions through the polymers as a function of the polymer oxidation state, which is essential for understanding the conductivity properties of these new chemical materials. The microstructures also exhibit potentially useful electrical and optical responses, including current rectification, charge storage and amplification, electron-hole pair separation, and gates for ion flow. In their well-illustrated paper, the investigators explore the three broad categories of electroactive polymers: (1) pi-conjugated, electronically conducting polymers; (2) polymers with covalently linked redox groups (redox polymers); and (3) ion-exchange polymers. In summary, the authors observe that although macromolecular electronics is still at a rudimentary level, the concepts involved are quite novel and with continued development may lead to practical applications. See also entry in this encyclopedia, **Molecular and Supermolecular Electronics**.

Quantized Hall Effect. In 1980, at the Max Planck Institute, Klaus von Klitzing discovered the quantized Hall effect, a phenomenon that occurs in certain semiconductor devices at low temperatures in very strong magnetic fields. As pointed out by Halperin (1986), the quantized Hall effect is observed in artificial structures known as two-dimensional electron systems. The conduction electrons in these systems are trapped in a very thin layer, such that the electronic motion perpendicular to the layer is frozen into its lowest quantum mechanical stage and thus plays no role in the conductivity of the device. In his experiment, Klitzing worked with a silicon field effect transistor (MOSFET). Electrons are trapped in what is called an inversion layer near the surface of a silicon crystal that is covered with a film of insulating silicon oxide, on top of which is deposited a metal gate electrode, used to control the density of conduction electrons in the inversion layer. This effect had been predicted as early as 1975 by Japanese investigators. Considerable detail pertaining to von Klitzing's experimental apparatus is given in the Halperin paper.

Surface Physics. Closely allied with solid state physics is the discipline of surface science. Investigations in this area have been quite intense during the past decade, notably in connection with catalysts. A catalyst is a species that changes the rate of a reaction and yet is regenerated by that reaction so that it seems to be unchanged in the net reaction. Although there are enzyme catalysts, for example, the majority of industrially interesting catalysts are found among the metals, the surfaces of which serve to catalyze reactions. The first catalytic phenomena were observed as early as 1835 by Berzelius and later better quantified by Ostwald in 1894. Aided by the great volume of catalysts used industrially ($ billions/year), the incentive for research is large. See articles on **Catalysis; Scanning Tunneling Microscope**; and **Silicon**.

Extremely High-Pressure Research. The invention and refinement of the modern diamond anvil cell (Carnegie Institution) occurred in the mid-1980s. This is a tool par excellence for optical, infrared and Raman spectroscopy and enables the researcher to study the changes in the electronic structure and chemical binding caused by the application of high pressure. Phase transitions, which involve changes in the atomic architecture can be determined with the diamond cell using the x-ray diffraction technique. Studies with the diamond anvil cell have been particularly valuable for obtaining geophysical information—for example, the state of silicate minerals and oxides in the mantle region right up to the core-mantle boundary to provide a view of the earth's interior, where high pressure and high temperature conditions exist. In solid-state physics, there is the fascinating challenge of making metallic hydrogen under ultrahigh pressure. This extraordinary change from a very good insulating to a metallic state in hydrogen is predicted to occur near 3 to 4 million atmospheres. See article on **Diamond Anvil High Pressure Cell**.

The foregoing examples are but a few to indicate the continuing vigorous research into the nature of the solid state. See also **Superconductivity**.

Concepts of Solids Simplified

The atoms that comprise a solid can be considered for many purposes to be hard balls which rest against each other in a regular repetitive pattern called the crystal structure. Most elements have relatively simple crystal structures of high symmetry, but many compounds have complex crystal structures of low symmetry. The determination of crystal structures, of atom location in the crystal, and of the dependence of many physical properties upon the inherent characteristics of the perfect solid is an absorbing study, one that has occupied the lives of numerous geologists, mineralogists, physicists, and other scientists for many years.

The rigid, hard-ball model is not adequate to explain many properties of solids. To begin with, solids can be deformed by finite forces, thus solids must not be completely rigid. Furthermore, atoms in a solid possess vibrational energy, so the atoms must not be precisely fixed to mathematically defined lattice points. This deformability of solids is built into the model by the assignment of deformable bonds (springs) between nearest atom neighbors. This ball-and-spring model has many successes; one important early use was that of Einstein to devise a reasonably successful theory of specific heats. Later incorporation by Debye of coupled motion of groups of atoms led to an even more successful theory.

Several measures exist of the strength of these bonds. One is the size of the elastic constants—for most solids, Young's modulus is about 10^{11} newtons per square meter. The other is the frequency of vibration of the atoms—values around 10^{13} to 10^{14} Hz are found.

The lack of perfection occasioned by elastic deformation of solids is but one of many kinds of crystalline imperfections. Defects are frequently found in crystals, produced in nature and in the laboratory. These defects may be characterized by three principal parameters—their geometry, size, and energy of formation.

All real crystals have atoms which occupy external surface sites and which do not possess the correct number of nearest neighbors as a consequence. Thus, a surface is a seat of energy and is characterized by surface tension. Furthermore, internal surfaces exist, grain boundaries and twin boundaries across which atoms are incorrectly positioned. In a crystal of reasonable size—say 1 cubic centimeter, these two-dimensional defects, called *surface defects*, contain only about 1 atom in 106, a rather small fraction. Even so, surfaces are important attributes of solids.

Some defects have extent in only one dimension—*line defects*. The most prominent of these, the dislocation, is a line in the crystal along which atoms have either an incorrect number of neighbors or neighbors which have not the correct distance or angle. In 1 cubic centimeter of a real crystal, one might find a wide variation of length of dislocations present—from near zero to perhaps 10^{11} centimeters.

Defects which have extent of only about an atomic diameter also exist in crystals—the *point defects*. Vacant lattice sites may occur—*vacancies*. Extra atoms—*interstitials*—may be inserted between regular crystal atoms. Atoms of the wrong chemical species—*impurities*—also may be present.

The properties of defects are intimately related to their energy of formation. A standard against which this energy can be compared is provided by the energy of sublimation—the energy necessary to separate the ions of a solid into neutral, noninteracting atoms. This energy is about 81,000 calories per mole for a typical metal, copper, at room temperature, about 3.5 eV per atom. Energies of surfaces, both free surfaces and grain boundaries, are about 1000 ergs per square centimeter, about 1 eV per surface atom. Dislocation energies are of similar size per atom length of dislocation, about 1 to 5 eV, so the energy of a dislocation is about 10^{-4} erg per centimeter of length. Point defects, too, possess an inherent energy of about 1 eV each. Vacancies in copper have an energy of about 1 eV; self-interstitials, 2 or 3 eV.

The energies per atom of these various defects, surface, line, and point, are all much larger than the average thermal energy per atom in a solid at reasonable temperatures. This thermal energy kT is only about $\frac{1}{40}$ eV at room temperature. Thus, defects can be produced only by conditions which exist during manufacturing (artificial and natural) by external means, such as plastic deformation or particle bombardment; or by large local fluctuations in thermal energy away from the average.

The total amount of energy bound up in ordinary concentrations of these defects is not large as compared to the total thermal energy of a solid at normal temperatures. All the vacancies in equilibrium in copper, even at the melting point, comprise less than 10 calories of energy per mole, much less than the enthalpy at 1357 K (the melting point) of more than 7000 calories per mole. In a material with very heavy dislocation density, 10^{12} centimeters per cubic centimeter, the total dislocation energy is only a few calories per mole. And the total energy of a free surface of a compact block of 1 mole of copper is even less: about 10^{-3} calorie. Thus, the inherent energy of these defects is not large; even so they are immensely important in controlling many phenomena in crystals—as in the case of semiconductor devices.

Crystallographic defects need not remain stationary in the crystal; they may move about with time. Some of these movements may reduce the overall free energy of the solid; others (these are chiefly movement of the point defects) may simply be the wandering of random walk. Since these movements require larger than kT, the motion of defects depends upon rather large local fluctuations in energy. Consequently, their rate of motion depends upon temperature through a Boltzmann factor $\exp(-\Delta H/RT)$, where ΔH is the enthalpy increase necessary to move the defect from the lowest-energy site to the top of the barrier.

A convenient description of the crystalline structure of solids is thus seen to consist of successive stages of approximation. First, the mathematically perfect geometrical model is described; then departures from this perfect regularity are permitted. The deformability of solids is allowed for by letting the force constants between adjacent atoms be finite, not infinite. Then, misplacement of atoms is permitted and a variety of crystalline irregularities, called defects, is described. Some of these defects have intrinsic features which affect properties of the crystal; other affect the properties by their motion from site to site in the crystal. In spite of their relatively small number, defects are of immense importance.

Electronic Structure of Solids. In principle, the electronic structure of solids is determined by the electronic structure of the *free atoms* of which the solid is composed. Since the free atom structure is known rather well, especially for atoms of lower atomic number, the electronic structure of solids should be subject to determination by calculation. This is not the case. A wide variety of interactions occur between the electrons on adjacent atoms as they approach the equilibrium distance characteristic of solids. These interactions are of such complex nature that they tend to defy concise definition and involve such a host of charged particles, electrons, and ion cores, that only approximate calculations can usually be made. Nevertheless, the use of approximate models allows many general features of the electronic structure to be deduced, especially when close interplay between theory and experiment is established. As for the crystalline structure of solids, two stages are useful in understanding the electronic structure. First, the perfect electronic structure is defined. Then, irregularities in this structure, again termed defects, are described. Although both the geometry and energy of crystalline defects are defined, description of the geometry of the charge distribution of many of the electron defects is difficult, and one must generally be content with description of the formation energy of the defect.

The nuclei of the atoms in a solid and the inner electrons form ion cores with energy levels little different from corresponding levels in free atoms. The characteristics of the valence electrons are modified greatly, however. The state functions of these outer electrons greatly overlap those of neighboring atoms. Restrictions of the Pauli Exclusion Principle and the Uncertainty Principle force modification of the state functions, and the development of a set of split energy levels becomes a quasi-continuous band of levels of width, which are several electron volts for most solids. Importantly, unoccupied levels of the atoms are also split into bands. The electronic characteristics of solids are determined by the relative position in energy of the occupied and unoccupied levels as well as by the characteristics of the electrons within a band.

Metals. The solid is called a *metal* if excitation of electrons from the highest filled levels to the lowest unoccupied levels can occur with infinitesimal expenditure of energy. Thus, excitation can occur by means of many external forces, such as electric fields, heat, light, radio waves. Metals are, therefore, good conductors of electricity and of heat; they are opaque to light and they reflect radio waves.

Insulators. Some solids have wide spacing between the occupied and the unoccupied energy states—2 eV or more. Such solids are called *insulators* since normal electric fields cannot cause extensive motion of the electrons. Examples are diamond, sodium chloride, sulfur, quartz, mica. They are poor conductors of electricity and heat and are usually transparent to light (when not filled with impurities or defects).

Semiconductors. Solids with conductivity properties intermediate between those of metals and insulators are called *semiconductors*. For them, the excitation energy lies in the range 0.1 to about 2 eV. Thermal fluctuations are sufficient to excite a small, but significant, fraction of electrons from the occupied levels (the valence band) into the unoccupied levels (the conductance band). Both the excited electrons and the empty states in the valence band (aptly called *holes*) may move under the influence of an electric field, providing a means for conduction of current. Such electron-hole pairs may be produced not only by thermal energy, but also by incident light, providing photo-effects. The inverse process, emission of light by annihilation of electrons and holes in suitably prepared materials, provides a highly efficient light source (example, light-emitting diodes).

Crystallographic defects, in general, are also electronic defects. In metals, they provide scattering centers for electrons, increasing the resistance to charge flow. The resistance wire in many electric heaters consists of an ordinary metal, such as iron with additional alloying elements such as nickel or chromium providing scattering centers for electrons. In semiconductors and insulators, however, alloying elements and defects provide an even greater variety of effects, since they can change the electron-hole concentrations drastically in addition to providing scattering centers. This is the basis of semiconductor technology. See also **Semiconductors**.

Interactions of Solids with Light. Solids are useful because of their interaction with external forces or stimuli, such as electric and magnetic fields, heat, and mechanical forces. Yet among these interactions, probably the most important is the interrelation between matter and light. This interaction, important to all photosynthetic phenomena and the production of food; to the artificial generation of light; to the use of phosphors in cathode-ray tubes—is also the basis of spectroscopy and its use in the study of solids. In this field, first came investigation of emission and absorption of radiation from free atoms. Later investigations included emission and absorption of radiation by atoms in solids—giving rise to maser and laser phenomena, Mossbauer spectroscopy, nuclear magnetic resonance, X-ray diffraction, infrared spectroscopy, fluorescence, the Raman effect, microwave emission and absorption, among many other useful effects.

Band Theory of Solids

The success of the simple free electron theory of metals was so striking that it was natural to ask how the same ideas could be applied to other types of solids, such as semiconductors and insulators. The basic assumption of the free electron theory is that the atoms may be stripped of their outer electrons, the resulting ions arranged in the crystalline lattice, and the electrons then poured into the space between.

The free electron model results from the neglect of the interaction of the various atoms and of the periodic variation of the potential in which the electrons move, i.e., as their distance from the nearest metallic ion changes. When the former is taken into account, it is found that each energy eigenstate of an isolated atom is split into N non-degenerate states, where N is the number of atoms in the crystal. The group of levels that result from a single atomic state form an *allowed band*. If we start from the free electron picture and consider the effect of the periodic variations of potential, the Bloch theorem leads to the conclusion that there will be discontinuities in the plot of energy vs. momentum whenever the wave vector $\mathbf{k}$ has magnitude and direction such that it satisfies the Bragg law for reflection, in which λ may be set equal to $1/\mathbf{k}$ to give $\mathbf{k} \cdot \mathbf{d} = n$. Here $\mathbf{k}$ is the wave vector, $\mathbf{d}$ is the vector separation of two atomic planes in the crystal, and n is an integer equal to the scalar product. As with the atomic interaction model, the number of eigenstates between two energy breaks is equal to the number of atoms in the crystal. Thus, either approach leads to the existence of a manifold of energy levels occurring in groups of N closely spaced levels, the groups being separated by energies that are often very large compared with the spacing of levels within a group, somewhat as shown in Fig. 1. Each group of levels is known as an *allowed band*; the energies between groups are said to be in a *forbidden band*. Because these levels depend on the properties of the body as a whole, the entire

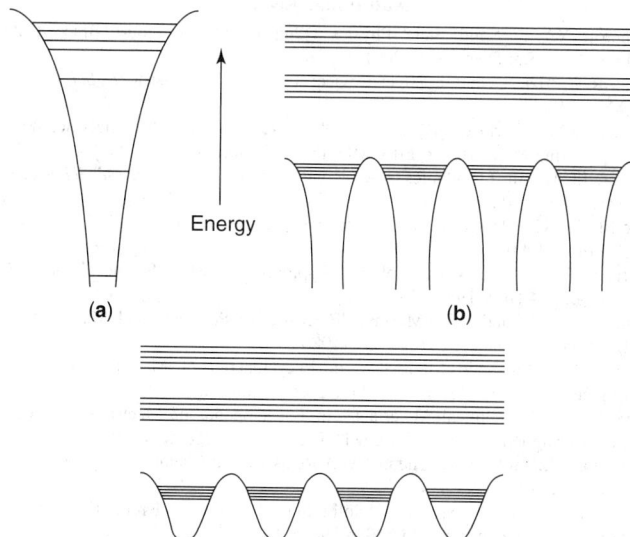

Fig. 1. Origin of the energy levels in a crystalline solid. The curves represent potential energy versus distance. At (**a**), the potential energy is that of an isolated ion; the energy levels, represented by the horizontal lines, are sharp. At (**b**), the overlap of the fields of the ions lowers the potential energy curve between the atomic positions and results in a splitting of each atomic level into a band of allowed levels. At (**c**), the model is derived from one in which the electrons are free, subject only to a periodic potential resulting from the ionic fields.

macroscopic crystal may be considered to be a single giant molecule. The electrical, mechanical, and thermal properties of the crystal are then largely determined by the electrons in the energy levels within the highest occupied bands.

Because electrons obey the Pauli Exclusion Principle, not more than two of them (with oppositely directed spin) can exist in any single energy level. In thermal equilibrium at the zero of absolute temperature, than, all of the levels up to some particular energy, determined by the number of electrons present, will be occupied and all above this energy will be vacant. This highest level is known as the *Fermi level*. At higher temperatures there will not be a sharp discontinuity in occupancy—some of the levels below the Fermi level will be vacant and some above it will be occupied. The *Fermi level* is then defined as the energy of the state that has a 50% chance of being occupied.

The Fermi level is determined by the number of electrons present, and the properties of the material are therefore dependent on whether this energy falls near the bottom, top, or middle of an allowed band. If the number of electrons is such as to exactly fill certain bands, with a wide gap above them, the material will be an insulator (m). If the gap is very narrow, or if there are impurities present to create extra levels, the substance may be semiconducting (o). See Fig. 2. In these cases, it is difficult to supply sufficient thermal energy to an electron to promote it into the conduction band above the gap, where alone it is free to carry an electric current. In a metal (n), however, there is always a partially filled band, in which the electrons behave in many respects as if they were free. The existence of the partially filled band may be due either to the fact that each atom contains an odd number of electrons or to the overlapping of two allowed bands, each of which will be partly filled. Direct evidence for the existence of bands is provided by the soft X-ray emission spectra, but the importance of the theory is not so much its correctness in detail as the simplicity of the band scheme by which the energy relations between various phenomena may be shown on a single diagram.

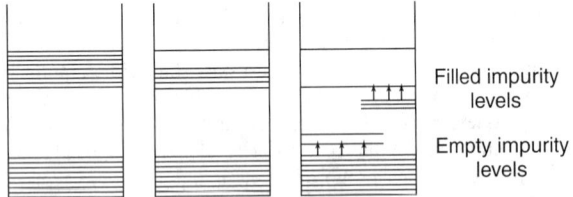

Fig. 2.   Band diagrams of *m*, insulator; *n*, metal; and *o*, semiconductor.

Additional Reading

Ashcroft, N.W.: "Solid State Physics," 2nd Edition, Harcourt Brace College Publishers, San Diego, CA, 2001.

Bate, R.T.: "The Quantum-Effect Device: Tomorrow's Transistor?" *Sci. Amer.*, 96 (March 1988).

Blakely, J.M.: "Surfaces and Interfaces," in Encyclopedia of Materials Science and Engineering (M.B. Bever, Ed.), MIT Press, Cambridge, MA, 1988.

Bokor, J.: "Ultrafast Dynamics at Semiconductor and Metal Surfaces," *Science*, 1130 (December 1, 1989).

Brodsky, M.H.: "Progress in Gallium Arsenide Semiconductors," *Sci. Amer.*, 68 (February 1990).

Caruana, C.M.: "The Interdisciplinary Approach to Surface Science," *Chem. Eng. Progress*, 64 (July 1987).

Chidsey, C.E.D. and R.W. Murray: "Electroactive Polymers and Macromolecular Electronics," *Science*, **231**, 25–31 (1986).

Chin, G.Y.: "Magnetic Materials," in Encyclopedia of Materials Science and Engineering (M.B. Bever, Ed.), MIT Press, Cambridge, MA, 1988.

DeShazer, L.G.: "Optical Materials," in Encyclopedia of Materials Science and Engineering (M.B. Bever, Ed.), MIT Press, Cambridge, MA, 1988.

DiSalvo, F.J.: "Solid-State Chemistry: A Rediscovered Chemical Frontier," *Science*, 649 (February 9, 1990).

Ehrenreich, H. and F. Spaepen: "Solid State Physics: Fullerene Fundamentals," Vol. 48, Academic Press, Inc., San Diego, CA, 1997.

Ehrenreich, H. and F. Saepen: "Solid State Physics: Advances in Research and Applications," Vol. 55, Academic Press, Inc., San Diego, CA, 2000.

Fisk, Z. et al.: "Heavy-Electron Metals: New Highly Correlated States of Matter," *Science*, 33 (January 1, 1988).

Halperin, B.I.: "The 1985 Noble Prize in Physics (Quantized Hall Effect)," *Science*, **231**, 820–822 (1986).

Heiblum, M. and L.F. Eastman: "Ballistic Electrons in Semiconductors," *Sci. Amer.*, 102–111 (February 1987).

Howard, R.E. et al.: "Electrons in Silicon Microstructures," *Science*, **231**, 346–349 (1986).

Karasz, F.E. and T.S. Ellis: "Polymers: Structure, Properties, and Structure-Property Relations," in Encyclopedia of Materials Science and Engineering (M.B. Bever, Ed.), MIT Press, Cambridge, MA, 1988.

Kittel, C.: "Introduction to Solid State Physics," 7th Edition, John Wiley & Sons, Inc., New York, NY, 1995.

Kramer, B.: "Advances in Solid State Physics 41," Vol. **41**, Springer-Verlag Inc., New York, NY, 2001.

Landman, U. et al.: "Atomistic Mechanisms and Dynamics of Adhesion, Nanoindentation, and Fracture," *Science*, 454 (April 27, 1990).

LeComber, P.G.: "Amorphous Silicon—Electronics into the 21st Century," University of Wales Review, 31 (Spring 1988).

Lovinger, A.J.: "Ferroelectric Polymers," *Science*, **220**, 1116–1121 (1983).

Mott, N.F. and E.A. Davis: "Electronic Processes in Non-Crystalline Materials," Oxford University Press, Inc., New York, NY, 1979.

Pool R.: "Clusters: Strange Morsels of Matter," *Science*, 1184 (June 8, 1990).

Pool, R.: "A Transistor That Works Electron by Electron," *Science*, 629 (August 10, 1990).

Prinz, G.A.: "Hybrid Ferromagnetic Semiconductor Structures," *Science*, 1092 (November 23, 1990).

Williams, E.D. and N.C. Bartelt: "Thermodynamics of Surface Morphology," *Science*, 393 (January 25, 1991).

Wolfe, J.P.: "Thermodynamics of Excitons," *Physics Today*, **35**(12), 46–54 (March 1982).

Wolfe, J.P. and A. Mysyrowicz: "Excitonic Matter," *Sci. Amer.*, 98–107 (March 1984).

Yablonovitch, E.: "The Chemistry of Solid-State Electronics," *Science*, 347 (October 20, 1989).

SOLIDUS CURVE.   A curve representing the equilibrium between the solid phase and the liquid phase in a condensed system of two components. The relationship is reduced to a two-dimensional curve by disregarding the influence of the vapor phase. The points on the solidus curve are obtained by plotting the temperature at which the last of the liquid phase solidifies, against the composition, usually in terms of the percentage composition of one of the two components.

SOLION.   A small electrochemical oxidation-reduction cell consisting of a small cylinder containing a solution and divided into sections by platinum gauze, porous ceramics, or other materials. A type of solion for detecting sound waves consists of a potassium iodide-iodine solution in which the iodide ions are oxidized to triiodide ions at the anode, and the reverse process occurs at the cathode. The cell is constructed so that the sound waves cause agitation of the solution between the electrodes, and thus change the current. In addition to detection of sound, solions can be designed to detect changes in other conditions, such as temperature, pressure, and acceleration.

SOLSTICE.   Either of the two points on the sun's apparent annual path, where it is displaced farthest, north or south, from the earth's equator, i.e., a point of greatest deviation of the ecliptic from the celestial equator.

The point north of the celestial equator is termed the summer solstice, and the point south of the equator is called the winter solstice, inasmuch as the sun is at these respective points at the commencement of summer and winter in the Northern Hemisphere.

SOLUBILITY.   A property of a substance by virtue of which it forms mixtures with other substances which are chemically and physically homogeneous throughout. The degree of solubility is the concentration of a solute in a saturated solution at any given temperature. The degree of solubility of most substances increases with a rise in temperature, but there are cases (notably the organic salts of calcium) where a substance is more soluble in cold than in hot solvents.

SOLUBILITY CURVE.   The graph showing the variation with temperature of the concentration by a substance in its saturated solution in a solvent.

SOLUBILITY PRODUCT.   A numerical quantity dependent upon the temperature and the solvent, characteristic of electrolytes. It is the product of the concentrations of ions in a saturated solution and defines the degree of solubility of the substance. When the product of the ion concentrations exceeds the solubility product, precipitation commonly results. Strictly speaking, the product of the activities of the ions should be used to determine the solubility product, but in many cases the results obtained using concentrations, as suggested by Nernst, are correct.

SOLVENT. The term solvent generally denotes a liquid that dissolves another compound to form a homogeneous liquid mixture in one phase. More broadly, the term is used to mean that component of a liquid, gaseous, or solid mixture which is present in excess over all other components of the system. A *chemical solvent* is the term used for solvents in those instances where the process of solution is attended by a chemical reaction between the solvent and the solute. In contrast, a *physical solvent* is one that does not react with the solute. A *dissociating solvent* is one in which solutes that associate in many other solvents enter into solution as single molecules. For instance, various carboxylic acids associate and thus give abnormal elevations of the boiling point, abnormal depressions of the freezing point, etc., in many organic solvents; but in water, however, they do not associate. For this reason water is called a dissociating solvent for such solutes. A liquid that dissolves or extracts a substance from solution in another solvent without itself being very soluble in that other solvent is termed an *immiscible solvent*. A solvent whose constituent molecules do not possess permanent dipole moments and do not form ionized solutions, is termed a *nonpolar solvent*. *Polar solvents*, on the other hand, consist of polar molecules, that is, molecules that exert local electrical forces. In such solvents, acids, bases, and salts, that is, electrolytes, in general, dissociate into ions and form electrically conducting solutions. Water, ammonia, and sulfur dioxide are typical polar solvents. A *normal solvent* is one that does not undergo chemical association, namely, the formation of complexes between its molecules.

A *leveling solvent* is a solvent in which the acidity or basicity of a solute is limited (or leveled) by the acidity or basicity of the solvent itself. For example, the strongest acid that can exist in water is oxonium ion, H_3O^+. Consequently, even though HCl (for example) is intrinsically a much stronger acid than H_3O^+, its acidity in aqueous solution is "leveled" to that of H_3O^+ through the reaction $HCl + H_2O \rightleftharpoons H_3O^+ + Cl^-$. Likewise the very strong base KNH_2 is leveled in water to the basicity of OH^-

$$KNH_2 + H_2O \rightleftharpoons K^+ + OH^- + NH_3$$

The solvents that are leveling to both acids and bases are self-ionized solvents, e.g., water, ammonia, alcohols, carboxylic acids, nitric acid, etc. Basic non-protonic solvents are leveling to acids, but not to bases (i.e., they are differentiating toward bases), e.g., pyridine, ethers, ketones, etc., since the strongest acid attainable is the protonated solvent molecule (e.g., $C_5H_5N + HCl \rightleftharpoons C_5H_5NH^+ + Cl^-$), whereas there is no corresponding basic species derived from the solvent. Though solvents leveling to bases but not to acids are in principle much more difficult to find, in practice, very strong acids like H_2SO_4 and $HClO_4^-$ are limiting to bases because the species HSO_4^- and ClO_4^-, which will be formed by almost any basic substance, are the strongest bases attainable in these solvents — $B^- + HClO_4 \rightleftharpoons HB + ClO_4^-$ — whereas practically no other acid is capable of producing the cations $H_3SO_4^+$ and $H_2ClO_4^+$ in these solvents (i.e., they are differentiating toward acids).

Differentiating solvents are solvents in which neither the acidity of acids nor the basicity of bases is limited by the nature of the solvent. These solvents are not self-ionized. The aliphatic hydrocarbons and the halogenated hydrocarbons are such solvents.

In industry it is generally understood that solvents are simple or complex, pure or impure, compounds or mixtures of compounds (either natural or synthetic), which dissolve many water-insoluble products like fats, waxes, resins, etc., forming homogeneous solutions; that such organic solvents dissolve these water-insoluble products in various proportions depending on the solvent power of the solvent, the degree of solubility of the solute, and the temperature; and that the solute can be recovered with its original properties by the removal of the solvent from the solution. It is also understood in industry that there is a much more limited number of solvents which do not have the properties given above but which nevertheless are of considerable importance; they are the inorganic solvents like water, liquid ammonia, liquid metals, and the like.

Solvents have been classified on various arbitrary bases: (1) boiling point, (2) evaporation rate, (3) polarity, (4) industrial applications, (5) chemical composition, (6) proton donor and proton acceptor relationships, and (7) behavior toward a dye, Magdala Red. Thus on the basis of industrial application one can classify solvents as those for (1) acetyl-cellulose, (2) pyroxylin, (3) resins and rubber, (4) cellulose

ether, (5) chlorinated rubber, (6) synthetic resins, and (7) solvents and blending agents for cellulose ester lacquers. Solvents classified according to chemical composition are noted below.

The term *solvent action* is understood to mean any process of making substances water-soluble; but in a broader interpretation the term is understood to be the phenomenon of making a substance soluble in a solvent. Solvent power, diluting power, solvency and similar expressions indicate the property of solvents to disperse the molecules of a solute or vehicle thereby causing a decrease in viscosity.

The most common solvent is water. Water dissolves a great many gases, liquids, and solids, and is much used for this purpose. Other liquids similarly dissolve many substances without reacting chemically with them. Important considerations in connection with the choice of solvent for a given case are (1) vapor pressure and boiling point, (2) solvent power under stated conditions of temperature, (3) ease and completeness of recoverability by evaporation and condensation, and completeness of separation from dissolved material by evaporation, (4) heat of vaporization, (5) miscibility with water or other liquid, if present, (6) inertness to chemical reaction with the materials present, and with the apparatus, (7) inflammability and explosiveness, (8) odor and toxicity; (9) cost of solvent, loss in process, cost of recovering.

See also **Pollution (Air)**.

Colligative Properties of Solutions. When solute is added to a pure solvent, thus forming a solution, properties of the solvent are altered, including (1) osmotic pressure; (2) vapor pressure (lowered); (3) melting point (lowered); and (4) boiling point (elevated). These properties bear a relationship to the number of solute molecules in solution and not to the nature of the molecules. These phenomena are explained by enhanced tension in the solvent. Complete explanation of these changes is beyond the scope of this book, but reference is suggested to H.T. Hammel's article on "Colligative Properties of a Solution" (*Science*, **192**, 748–756, 1976).

SOLVOLYSIS. A generalized conception of the relation between a solvent and a solute (i.e., a relation between two components of a single-phase homogeneous system) whereby new compounds are produced. In most instances, the solvent molecule donates a proton to, or accepts a proton from a molecule of solute, or both, forming one or more different molecules. A particular case of special interest occurs when water is used as solvent, in which case the interaction between solute and solvent is called *hydrolysis*.

SOMATOPLASM. The tissue of an organism exclusive of the reproductive cells (germ plasm). The somatoplasm of animals typically has the diploid chromosome number, and the reproductive cells have the haploid chromosome number.

SOMITE. One of the longitudinal series of segments into which the bodies of many animals are divided. These segments are clearly shown in a simple form in the earthworms. In humans, they are made evident by the structure of the spinal column and the series of spinal nerves, but they are overshadowed externally by the high development of the appendages. The term is synonymous with metamere.

The segmental masses of mesoderm in the vertebrate embryo are also called somites. They are the primordia of the axial skeleton, voluntary muscles of the body and appendages, and the inner layer of the skin.

SONAR. A coined word derived from the phrase, "sound navigation and ranging." The term generally refers to the principles employed in the design and operation of systems that utilize acoustic energy transmitted in an ocean medium; while the systems themselves are referred to as sonar systems. Thus, sonar may be defined as a branch of applied acoustics concerned with the utilization of the ocean as the transmitting medium.

The problem of sonar is threefold: (a) understanding the transmission of acoustic energy through the transmitting medium, (b) developing sources which convert mechanical or electrical energy into acoustic energy, and (c) developing receivers which convert the acoustic energy back into mechanical or electrical energy. See Fig. 1.

Whenever a body vibrates in a fluid, longitudinal waves are formed, which propagate outward from the vibrating body. The particles of the fluid are set in motion, and temporary stresses are produced which increase and decrease during each vibration. The motion of the particles gives the

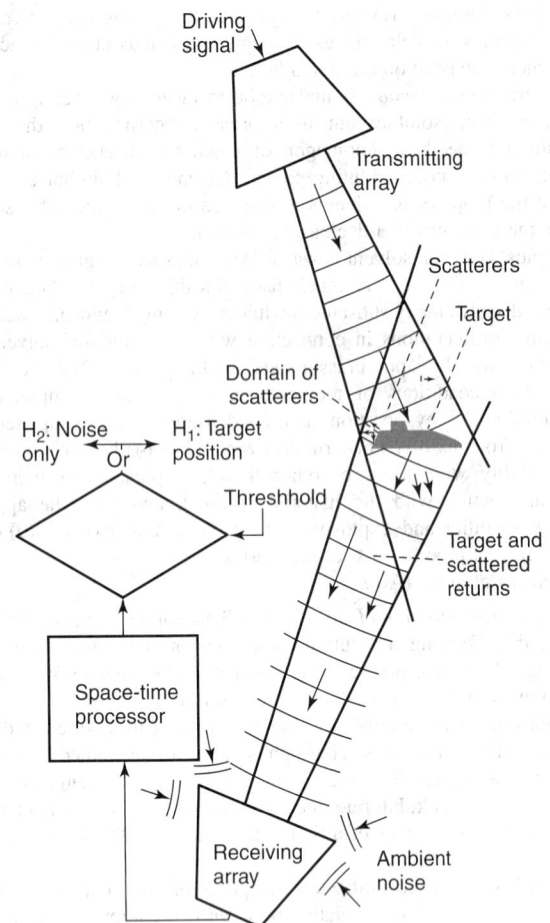

Fig. 1. General operating principle of sonar detection system. (*U.S. Naval Underwater Systems Center, Newport, Rhode Island.*)

fluid kinetic energy while the stresses induce potential energy. The sum of the two energies is called acoustic energy.

Traditionally, the starting point for a discussion of the transmission of acoustic energy in a fluid is to assume a point source radiating acoustic energy in an ideal homogeneous nonabsorptive medium of infinite extent. Under these assumptions, the energy from the source will radiate outwards with the wave front forming a spherical shell. As the radius of the shell increases, the sound intensity decreases. In practice, it is customary to express the sound intensity by means of a logarithmic scale. The most generally used logarithmic scale is the decibel. The intensity level in decibels of a sound of intensity I is defined as $10\log(I/I_0)$ where I_0 is a reference intensity. The intensity level can also be expressed as $20\log(P/P_0)$ where P is the pressure and P_0 the reference pressure, usually 1 dyne/cm^2 in underwater acoustics. In this discussion, the terms in all equations are expressed in decibels. The decrease in intensity as the shell increases in radius is called the spreading loss. The spreading loss from a unit range of R_0 to a range of R is $10\log[(\text{intensity at } R_0)/(\text{intensity at } R)] = 10\log(R^2/I^2) = 20\log R$. In most applications the use of such a simple model has been inadequate.

A more realistic model considers the following factors: the water-earth interface (bottom), the water-atmosphere interface (surface), the absorption of acoustic energy in the medium, the presence of foreign material in the medium, and the distribution of sound velocity. Considered as an acoustic medium, the waters of the ocean form a thin layer on the Earth's surface. Some of the acoustic energy radiated into this layer by a source will reach either the surface or the bottom. At either of these surfaces abrupt discontinuities in acoustic properties occur. Because of these discontinuities part of the intercepted energy is reflected, part may be transmitted across the interface, and part may be scattered within the medium. Since the transmission of an acoustic wave in water is accompanied by a compression and expansion of the medium, friction will occur between water molecules. This friction results in the conversion of some of the acoustic energy into thermal energy. In addition to this

frictional, or viscous, loss there is another loss of energy in seawater related to the salts that continuously undergo chemical changes because of pressure fluctuations. Energy losses associated with both of these phenomena are called absorption losses. Due to the presence of foreign bodies in the volume of water, reflection and scattering is not limited to the surface and bottom boundaries. Foreign matter and biological content vary widely in size and acoustic characteristics. All ocean waters contain such bodies that modify the direction in which the acoustic energy is transmitted. In sufficient number they may also modify and increase the total absorption loss. The effect of variations in sound velocity is to bend the wave front in the direction of the lower velocity. This bending of the wave front is referred to as refraction. Both refraction and reflection can result in the guiding of acoustic energy in certain directions.

The factors above affect the propagation of acoustic energy in seawater in two different ways. The first results in a spreading loss already mentioned, and the second results in a loss referred to as attenuation. Attenuation consists of both the scattering and absorption losses. The spreading and attenuation are related to the distance the acoustic energy travels in different ways. An important difference is that the spreading loss frequently is relatively independent of frequency while the attenuation is a function of frequency.

There are three basic types of sonar systems: direct listening systems, echo-ranging systems, and communication systems.

In direct listening the acoustic energy is radiated by the target, which is the primary source. The acoustic transmission is a one-way process. In their more elementary forms direct listening sonar systems may be nondirectional and only give a warning that a primary source is in the vicinity of the searching vehicle; or directional, and permit determination of the bearing of individual primary sources relative to the listening platform. They generally do not give range. Direct listening is limited by the magnitude of the signal when it reaches the receiving point and the magnitude of the interfering noise that tends to obscure its reception.

In echo ranging, the sonar system projects acoustic energy into the water with the expectation that this energy will strike a target and enough of the energy will be reflected back to the searching platform so that it can be recognized as a target echo. The primary source of acoustic energy is in the searching platform, with the target, upon reflection of the energy, becoming a secondary acoustic source. The transmission of the energy is a two-way process. Echo-ranging sonar systems permit a determination of the bearing of a silent target, and by timing the echo-signal transmission and by knowing the velocity of sound in seawater, a range may also be obtained. Echo ranging is limited by the relative magnitudes of the signal and of the locally generated interference. In some cases the sonar performance is limited by reverberation, which is the acoustic energy returning by reflectors other than the target of interest.

Acoustically, sonar communications systems are similar to direct listening systems in that they utilize a one-way transmission path. Instrumentally, they are similar to echo-ranging systems, one located at each of the two points between which communications is to be established. In these systems coded pulses or voice modulated signals are transmitted by one system and received by the other.

To hear a target by direct listening, it is necessary that the acoustic level of the target less the transmission loss along the acoustic path from the target to the listening equipment be equal to or greater than the level of the background noise. This may be expressed as $L - H \geq N$ where L is the source level of the target, H is the one-way transmission loss, and N is the noise level. The size of this inequality depends upon operator skill, signal processing, and method of presentation. It is called the signal excess, E. This inequality can be written as an equation where $E = L - H - N$. This equation is called the direct-listening sonar equation for an omnidirectional listening hydrophone. When using directional hydrophones a factor called the directivity index must be added to the right-hand member of the equation. The source level, L, is a measure of the amount of acoustic energy put into the water by the target vehicle and is equal to 10 log (sound intensity at unit distance from the source). The transmission loss, H, is the sum of the losses related to refraction, surface and bottom reflection, absorption, and scattering. The noise, N, results from unwanted acoustic energies arriving from many different sources and normally consists of thermal, ambient, and self noises.

To see a target by echo ranging, it is necessary that the acoustic level of the primary source, less twice the transmission loss along the acoustic path from source to target plus the target strength, be equal to or greater than the noise. This may be expressed, for a nondirectional receiver against a noise background, as $L - 2H + T \geq N$ where T is the target strength, a function of the reflecting characteristics of the target. Against a reverberation background the inequality becomes $L - 2H + T \geq R$ where R is the reverberation level. As in the case of direct listening, the inequalities can be expressed in terms of the signal excess as $E_N = L - 2H + T - N$ or $E_R = L - 2H + T - R$ where E_N and E_R are the signal excesses for noise and reverberation. The noise, N, comes from own-ship's noise and target noise. Reverberation, R, is the energy that is returned from the outgoing acoustic energy to the receiving equipment after having been reflected from reflectors in the medium other than the target. Reverberation sources usually are backscattering from the surface, bottom, and foreign particles in the water.

Echolocating bats (*Eptesicus fuscus*) are capable of detecting changes as short as 500 nanoseconds in the arrival time of sonar echoes when these changes appear as jitter or alternations in arrival time from one echo to the next. The psychophysical function relating the bat's performance to the magnitude of the jitter corresponds to the half-wave rectified cross-correlation function between the emitted sonar signals and the echoes. The bat perceives the phase or period structure of the sounds, which cover the 25- to 100-kilohertz frequency range. The acoustic image of a sonar target is apparently derived from time-domain or periodicity information processing by the nervous system. The biological sonar of bats in the suborder Microchiroptera is of much interest because bat sonar represents a rather well-defined example of a biological communications system.

For use of sonar in medicine, see **Electrocardiography**; in oceanography, see **Ocean**; and **Ocean Research Vessels**; in petroleum exploration, **Petroleum**; in photography, **Photography and Imagery**.

Additional Reading

Minkoff, J.R.: "Signals, Noise, and Active Sensors: Radar, Sonar, Laser Radar," John Wiley & Sons, Inc., New York, NY, 1991.

Murton, B.J. and P. Blondel: "Handbook of Seafloor Sonar Imagery," John Wiley & Sons, Inc., New York, NY, 1997.

Stergiopoulos, S.: "Advanced Signal Processing Handbook: Theory and Implementation for Radar, Sonar, and Medical Imaging Real-Time Systems," CRC Press, LLC., Boca Raton, FL, 2000.

Urick, R.J.: "Principles of Underwater Sound," 3rd Edition, The McGraw-Hill Companies, Inc., New York, NY, 1983.

SONE.   A unit of loudness. A simple tone of frequency 1000 cycles per second, 40 decibels above a listener's threshold, produces a loudness of 1 sone. The loudness of any sound that is judged by the listener to be n times that of the 1-sone tone is n sones. A millisone is equal to 0.001 sone. The loudness scale is a relation between loudness and level above threshold for a particular listener. In presenting data relating loudness in sones to sound pressure level, or in averaging the loudness scales of several listeners, the thresholds (measured or assumed) should be specified.

SONIC BARRIER.   A popular term for the large increase in drag that acts upon an aircraft approaching acoustic velocity; the point at which the speed of sound is attained and existing subsonic and supersonic flow theories are rather indefinite. Also called *sound barrier*.

SONIC BOOM.   A noise caused by a shock wave that emanates from an aircraft or other object traveling at or above sonic velocity. A shock wave is a pressure disturbance and is received by the ear as a noise or clap.

SONICS.   The technology of sound in processing and analysis. Sonics includes the use of sound in any noncommunication process.

SONIC SPEED.   Acoustic velocity; by extension, the speed of a body traveling at a Mach number of 1.

SORGHUM (*Andropogon Sorghum; Gramineae*).   Sorghums are annual grasses of tropical origin. They have an extensive system of wiry roots and solid stems 3–15 feet (0.9–45 meters) tall. The leaves are smaller than those of corn, and capable of rolling up tightly during periods of drought, and quickly unrolling and starting to function when favorable moisture conditions return. Because of this habit sorghums are often grown in regions subject to frequent drought. The inflorescence is a panicle usually of very compact habit. Ordinarily the spikelets are paired, one of the pair being sessile or stemless, the other having a short stem or pedicel. The former is fertile, the latter staminate. The grains are enclosed in the glumes and vary considerably in shape in different varieties. There are two main groups of sorghums; the sweet or saccharine sorghums, the juicy pitch of which is a source of syrup; and the grain sorghums, which yield grain, stock food and ensilage. Kaffir is one of the latter group. In Asia, grain sorghums are employed in a countless variety of ways, as for fuel, brooms, mats, fences, windbreaks, roof thatch, and in making a fermented drink. In the United States, grain sorghums are finding increasing popularity in several industrial processes, as a source of starch, wax, and other products. See Figs. 1 and 2. See also **Broomcorn**.

Fig. 1.   Head of Schrock grain sorghum. (*USDA diagram.*)

Fig. 2.   Sorghum residue being shredded for return to the soil to add organic matter and increase fertility. (*USDA photo.*)

SORPTION.   A generalized term for the many phenomena commonly included under the terms adsorption and absorption when the nature of the phenomenon involved in a particular case is unknown or indefinite.

SOUTH ATLANTIC CENTRAL WATER.   An oceanic water mass, extending at the surface roughly from southern Africa to Patagonia, where the Subantarctic Water is on the surface. This region of junction is south of the subtropical convergence in the South Atlantic. Temperature range 6–18°C (42.8–64.4°F), salinity range, 34.5%–36.0%.

SOUTH ATLANTIC CURRENT.   An eastward flowing current of the South Atlantic Ocean that is continuous with the northern edge of the Antarctic Circumpolar Water.

SOUTH EQUATORIAL CURRENT.   In the Atlantic and Pacific Oceans, the westward counterclockwise drift of surface water in the Equatorial latitudes south of the Equator that is separated from the North Equatorial Current by the Equatorial countercurrent. Driven westward by the trade winds, these great currents are deflected by the land masses and also by the Coriolis effect. In the Southern hemisphere they are deflected to their left. Upon turning poleward, the current becomes a part of the west-wind drift, a product of the prevailing westerlies.

SOUTH PACIFIC CENTRAL WATER.   Due to the great size of the Pacific Ocean, it contains more well-developed oceanic water masses than the Atlantic. Thus there are both an Eastern and Western South Pacific Central Water mass. Extending from the Pacific Equatorial Water on the north to the Subantarctic Water on the south, and covering the width of the Ocean, except for a transition zone on the eastern side, they differ from each other in properties to a greater degree than the Eastern and Western masses in the North Pacific because the Western South Central Water is closely similar to the Central Water of the Indian Ocean, having a temperature range of 8–15°C (46.4–59°F) and the salinity of 34.6–35.5%. The Eastern South Pacific Water has a temperature range of 10–18°C (50–64.4°F) and the salinity of 34.0–35.0%.

SOUTH PACIFIC CURRENT.   An eastward flowing current of the South Pacific Ocean that is continuous with the northern edge of the Antarctic Circumpolar Water.

SOWBUG (*Crustacea, Isopoda*).   Related to crayfish, the sowbug is a crustacean and not an insect. The sowbug is a common but not severely damaging pest on food crops. The pest is found most often in connection with glasshouse or home gardening operations. The sowbug does feed on the tender portions of some plants and also attacks mushrooms. The sowbug (*Porcellio leavis*, Koch) reproduces by way of eggs, but the eggs remain in a marsupium for approximately 2 months. The pest is slow in maturing, requiring nearly a year to become a fully developed adult. The sowbug is a notable pest of watercress. See Fig. 1.

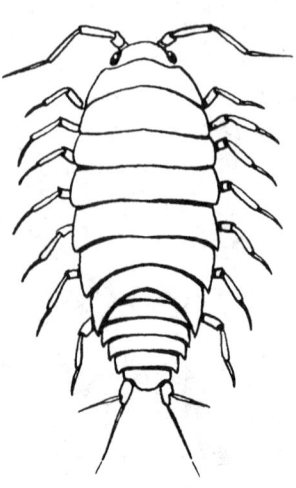

Fig. 1.   Sowbug. (*USDA*.)

The *pillbug* (*Armadillidium vulgate*, Latrielle), is closely related to the sowbug and is of similar habit. The pillbug is distinguished by its ability to curl up into nearly a perfect spheroidal form when disturbed.

SOY PROTEIN.   See **Protein**.

SOYBEAN.   Of the family *Leguminosae*, subfamily *Papilionaceae*, and the genus *Glycine max*, the soybean is a typical legume seed differing in color, size, and shape, depending upon variety. The common field varieties grown in the United States are nearly spherical and are yellow in color, (e.g., the Lincoln soybean). See Fig. 1. The food reserves of the seed are stored in the cotyledons (90% seed), the interior of which is filled with elongated palisade-like cells, themselves filled with proteins and oil. The bulk of the protein is stored in protein bodies varying between 2 μm (microns or micrometers) and 20 μm in diameter. The oil is located in smaller structures, the spherosomes, 0.2 μm to 0.5 μm, interspersed between the protein bodies. Isolated protein bodies may contain as much as 90% protein and together account for 60 to 70% total protein of defatted soy flour. Proximate analysis of the whole bean shows that protein (40%) and oil (21%) make up about 60% of the bean, the remaining third consisting of nonstarch carbohydrates, including polysaccharides, stachyose (3.8%), raffinose (1.1%), and sucrose. Nucleic acids are present only as minor constituents of the soybean, unlike some of the new single-cell protein sources, where nucleic acids can account for between 5 and 20% cell weight and may be included within the total nitrogen value of the isolate. Other minor constituents, such as the antitrypsin factor and so-called "beany flavor" components, have caused some adverse comment against the use of soybean products in the human dietary. Moreover, every legume (bean, pea, field bean) also has the flatus problem, but in processing this problem is not insurmountable.

Fig. 1.   Recently pulled soybean plant lying on table for examination. Root structure is at right. Mature pods drop from vines. (*USDA photo*.)

Importance of the Soybean.   During the past couple of decades, a major source of protein, namely that derived from the soybean, has developed. In times when there is concern over the ability of the world to feed its growing population, soybean protein technology is one of several scientific tools for disproving the Malthusian prophecy.

   In terms of present knowledge, soybeans are capable of producing the greatest amount of protein per unit of land of any major plant or animal source used as food by people today. See Table 1.

TABLE 1. COMPARISON OF PROTEIN-YIELDING COMMODITIES

Protein Source	Kilograms from 1 Hectare	Pounds from 1 Acre
Beef cattle	65	58
Wheat	202	180
Maize (corn)	362	323
Soybean	560	500

In considering the present importance and future potential of the soybean, there are at least two factors of merit: (1) Large numbers of people in the lesser developed areas of the world suffer from a protein deficiency in their diets, leading to serious health problems; and (2) in the more developed areas, many people are accustomed to meat, milk, and eggs and desire to expand consumption of these products at lower costs. Many countries have the capacity to expand livestock and poultry production and, in the process, to lower prices of these products. This potential accounts for much of the market growth opportunity for soybeans and other feedstuffs. But, even in areas where livestock and poultry development may come more slowly, there is the opportunity to expand protein consumption in other ways. Soy protein provides a means of both extending the animal proteins and of replacing them with high-quality protein that is relatively inexpensive to produce–thus, the opportunity to further utilize soybeans directly as human food.

A technical factor of large significance is the general excellence of soybean protein as a protein. Soybean protein has a high content of essential amino acids, particularly lysine, leucine, and isoleucine. Detailed analysis is given later. Only in the sulfur-containing amino acids, cystine and methionine, is the soybean low, indicating that methionine is the first limiting amino acid to be considered when using soy products in a diet. Full fat soy flour also offers a valuable nutritional contribution from the high proportion of essential fatty acids, linoleic (51%) and linolenic (9%), that its oil contains. Since the oil represents some 20% of the total flour, the use of soy flour as a supplement in high-protein breads also elevates the essential fatty acid content of the product.

The soybean originated in the northern provinces of China and was first described in about 2000 B.C. as one of the most important cultivated legumes and one of the five sacred grains essential to Chinese civilization. The soybean was used as a basic food and a source of medicinals from the Middle Ages until the early 1700s, when greater interest was shown in the legume. Two major developments occurred at that time: (1) The first extraction of oil from the soybean, which led to the development of a new industry in Manchuria; and (2) introduction of the soybean to Europe in 1712 and the United States in 1804.

SOYBEAN OIL. See Vegetable Oils (Edible).

SPACE CHARGE.

The electric charge on a conductor is to be regarded as confined to an infinitely thin layer at the surface and thus, in a sense, as geometrically two-dimensional. Even when a current is flowing through the conductor, the quantities of positive and negative electricity within any element of volume at any instant are equal, thus neutralizing each other's electrostatic effect. For this reason, Laplace's equation applies to the interior of a current-bearing conductor as well as to a complete vacuum. This is true whether the conductor be of the metallic or the electrolytic type.

Quite different are the conditions in a vacuum tube, where streams of electrons or of positive ions (e.g., canal rays) may occupy sizable regions to the virtual exclusion of carriers of the opposite sign. See also **Ionized Gases**. The electricity thus monopolizing such a region is called a "space charge," and may exert marked influence on the performance of the tube.

A typical instance of this occurs in thermionic rectifiers. If the cathode is emitting no electrons, the potential gradient across from cathode to anode is nearly uniform; that is, the potential increases by the same number of volts for each centimeter from the one electrode to the other. If now a small emission begins from the cathode (because of its being heated), the electron swarm is more dense near the cathode, due to their accelerated motion, much as the stream of water descending from a faucet is broader near the faucet than farther down. The potential gradient is no longer uniform, and the total increase of potential, that is, the "plate voltage," is less than before. Further increase in emission may result in an actual

drop of potential near the cathode (followed by a rise farther on), with still further decrease of plate potential. Such a potential minimum is contrary to the condition expressed by Laplace's equation, which therefore does not hold in a region occupied by a space charge. (Another formula, called Poisson's equation, here applies instead. It is $\nabla^2 V = 4\pi\rho$, where V is the potential and r is the electric volume density at the point in question.)

In gas-filled discharge tubes the effect of electronic space charge is largely offset by the presence of great numbers of less mobile positive ions. As a result, such tubes not only have much lower resistance but have a more uniform potential gradient and give a better approximation to Ohm's law.

In atmospheric electricity, space charge refers to a preponderance of either negative or positive ions within any given portion of the atmosphere.

SPACE CHARGE LIMITATION OF CURRENTS.

It has been shown by Child that the current between a plane cathode and a parallel plane anode at a distance d from it, when the anode potential is V, cannot exceed a certain maximum value, determined by the modification of the electric field near the cathode as a result of the space charge of electrons in that region. If the electrons leave the cathode with zero speed, the maximum current per unit area of the cathode is

$$i = \frac{4\varepsilon_0}{9}\sqrt{\frac{2e}{m}}\frac{v^{3/2}}{d^2}$$

where e and m are the electronic charge and mass, respectively, and ε_0 is the electric permittivity constant in any self-consistent system of units. Langmuir has extended the equation to include the case of a cylindrical cathode of radius a, surrounded by a coaxial cylindrical anode of radius b. The maximum current per unit length is then

$$i = \frac{8\pi\varepsilon_0}{9}\sqrt{\frac{2e}{m}}\frac{v^{3/2}}{b(\ln b/a)^2}$$

The dependence of the current on the $\frac{3}{2}$ power of the potential difference is general, and is the basis of the definition of perveance.

SPACECRAFT POWER SYSTEMS.

Spacecraft travel millions of miles away from Earth, often encountering hostile space environments along the way. Power for these spacecraft must be efficient and reliable, providing power to the spacecraft's instruments and other subsystems (communication, navigation, etc). Roughly 300 Watts to 2500 Watts of electricity is required to power a spacecraft like *Voyager*, *Galileo*, or *Magellan*. The power supply must provide a large percentage of its power over a lifetime measured in years or decades. Choices of technology to meet these requirements are constrained largely to two: photovoltaics and radioisotope power systems.

A spacecraft's electrical power system generally consists of the primary power generating unit (solar or radioisotope) and a power management and distribution system. The electrical power system may also contain an energy storage unit (battery) for other spacecraft power needs.

This article discusses the various ways NASA provides power for spacecraft that are orbiting Earth as well as traveling to distant planets.

Battery Power

A battery, which is actually an electric cell, is a device that produces electricity from a chemical reaction. Strictly speaking, a battery consists of two or more cells connected in series or parallel, but the term is generally used for a single cell. A cell consists of a negative electrode; an electrolyte, which conducts ions; a separator, also an ion conductor; and a positive electrode. The electrolyte may be aqueous (composed of water) or nonaqueous (not composed of water), in liquid, paste, or solid form. When the cell is connected to an external load, or device to be powered, the negative electrode supplies a current of electrons that flow through the load and are accepted by the positive electrode. When the external load is removed the reaction ceases.

A primary battery is one that can convert its chemicals into electricity only once and then must be discarded. A secondary battery has electrodes that can be reconstituted by passing electricity back through it; also called a storage or *rechargeable battery*, it can be reused many times.

One of the first space batteries was the silver–zinc battery, which dominated the industry in the 1960s. This is a premium system with very high specific power and energy, but is quite expensive due to the use of silver. They are still used in selected applications, such as launch vehicles (rockets) and torpedoes. The Mars *Pathfinder* also used a silver–zinc battery, but it was designed to be rechargeable. They have a relatively short cycle life, and are not used for multiyear missions. This type of battery is used commonly in the commercial market for hearing aids.

Batteries come in several styles; the most familiar are single-use alkaline batteries. NASA spacecraft usually use rechargeable nickel–cadmium or nickel–hydride batteries like those found in laptop computers or cellular phones. Engineers think of batteries as a place to store electricity in a chemical form.

Battery technology is part of the power system, storing and discharging energy on each orbit of the spacecraft. The batteries help provide a constant source of power to the spacecraft by storing energy when excess is provided by the solar cells and discharging stored energy when the solar cells are not providing any during periods of eclipse.

Nickel-Cadmium Batteries. Nickel–cadmium has been the most common space battery since the 1970s. They were used in all commercial communications satellites, in most earth orbiters, and in some space probes. They are generally a prismatic (resembling, or being a prism) design, and packaged very efficiently. This means that the batteries can be stored on the spacecraft in a very compact form, eliminating the need for extraneous space. They have been known to last for ten to twenty years in space. They are still in use in selected space applications, including small satellites and for missions that encounter very severe radiation environments.

This battery uses nickel oxide in its positive electrode (cathode), a cadmium compound in its negative electrode (anode), and potassium hydroxide solution as its electrolyte. The nickel–cadmium battery is rechargeable, so it can cycle repeatedly. A nickel–cadmium battery converts chemical energy to electrical energy upon discharge and converts electrical energy back to chemical energy upon recharge. In a fully discharged NiCd battery, the cathode contains nickel hydroxide, $Ni(OH)_2$, and cadmium hydroxide, $Cd(OH)_2$ in the anode. When the battery is charged, the chemical composition of the cathode is transformed and the nickel hydroxide changes to nickel oxyhydroxide, $NiOOH$. In the anode, cadmium hydroxide is transformed to cadmium. As the battery is discharged, the process is reversed, as shown in the following formula.

$$Cd + 2\,H_2O + 2\,NiOOH \longrightarrow 2\,Ni(OH)_2 + Cd(OH)_2$$

Nickel–cadmium is the most commonly used battery for Low Earth Orbit (LEO) missions. A spacecraft battery consists of series-connected cells, the number of which depends upon bus voltage requirements and output voltage of the individual cells. See also **Battery**.

Nickel–Hydrogen Batteries. The nickel–hydrogen battery is currently the most popular space battery. It can be considered a hybrid between the nickel–cadmium battery and the fuel cell. The cadmium electrode was replaced with a hydrogen gas electrode. This battery is visually much different from the nickel–cadmium battery, because the cell is a pressure vessel, which must contain over one thousand pounds per square inch (psi) of hydrogen gas. It is significantly lighter than nickel–cadmium, but is more difficult to package, much like a crate of eggs. It is the longest-lived space battery yet built, with 10 to 20 year lifetimes being common. This battery is too expensive for commercial applications, and few terrestrial examples have been built.

Nickel–hydrogen batteries are sometimes confused with nickel–metal hydride batteries, the batteries commonly found in cell phones and laptops. The nickel–metal hydride system is rarely used in space due to its limited life. Nickel–hydrogen, as well as nickel–cadmium batteries use the same electrolyte, a solution of potassium hydroxide, which is commonly called *lye*.

Incentives for developing nickel–metal hydride (Ni-MH) batteries comes from pressing health and environmental concerns to find replacements for the nickel–cadmium rechargeable batteries. Due to worker's safety requirements, processing of cadmium for batteries in the U.S. is already in the process of being phased out. Furthermore, environmental legislation for the 1990s and the 21st century will most likely make it imperative to curtail the use of cadmium in batteries for consumer use. In spite of these pressures, next to the lead–acid battery, the nickel–cadmium battery still has the largest share of the rechargeable battery market. Further incentives for researching hydrogen-based batteries comes from the general belief that hydrogen and electricity will displace and eventually replace a significant fraction of the energy-carrying contributions of fossil-fuel resources, becoming the foundation for a sustainable energy system

based on renewable sources. Finally, there is considerable interest in the development of Ni–MH batteries for electric vehicles and hybrid vehicles.

The nickel–metal hydride battery operates in concentrated KOH (potassium hydroxide) electrolyte. The electrode reactions in a nickel–metal hydride battery are as follows:

$$\text{Cathode } (+) : NiOOH + H_2O + e^- \; Ni(OH)_2 + OH^- \qquad (1)$$

$$\text{Anode } (-) : (1/x)MH_x + OH^- (1/x)M + H_2O + e^- \qquad (2)$$

$$\text{Overall} : (1/x)MH_x + NiOOH(1/x)M + Ni(OH)_2 \qquad (3)$$

The KOH electrolyte can only transport the OH-ions and, to balance the charge transport, electrons must circulate through the external load. The nickel oxy–hydroxide electrode (equation 1) has been extensively researched and characterized, and its application has been widely demonstrated for both terrestrial and aerospace applications. Most of the current research in Ni–metal hydride batteries has involved improving the performance of the metal hydride anode. Specifically, this requires the development of a hydride electrode with the following characteristics: (1) long cycle life, (2) high capacity, (3) high rate of charge and discharge at constant voltage, and (4) retention capacity.

Lithium Batteries. These systems are different from all of the previously mentioned batteries, in that no water is used in the electrolyte. They use a nonaqueous electrolyte instead, which is composed of organic liquids and salts of lithium to provide ionic conductivity. This system has much higher cell voltages than the aqueous electrolyte systems. Without water, the evolution of hydrogen and oxygen gases is eliminated and cells can operate with much wider potentials. They also require a more complex assembly, as it must be done in a nearly perfectly dry atmosphere.

A number of nonrechargeable batteries were first developed with lithium metal as the anode. Commercial coin cells used for today's watch batteries are mostly a lithium chemistry. These systems use a variety of cathode systems that are safe enough for consumer use. The cathodes are made of various materials, such as carbon monofluoride, copper oxide, or vanadium pentoxide. All solid cathode systems are limited in the discharge rate they will support.

To obtain a higher discharge rate, liquid cathode systems were developed. The electrolyte is reactive in these designs and reacts at the porous cathode, which provides catalytic sites and electrical current collection. Several examples of these systems include lithium–thionyl chloride and lithium–sulfur dioxide. These batteries are used in space and for military applications, as well as for emergency beacons on the ground. They are generally not available to the public because they are less safe than the solid cathode systems.

The rechargeable lithium battery field is a growing area for space, and the latest technology for space is the lithium ion battery. The high voltage lithium-ion cells are moving into space for short to moderate length missions. They are easy to package, and very light. In this system, the lithium metal anode is replaced with a carbon electrode, which inserts lithium ions from the electrolyte, storing them in a solid solution phase. This configuration has improved safety over previous lithium rechargeable, and yet gives good rate capability.

The next step in lithium ion battery technology is believed to be the lithium polymer battery. This battery replaces the liquid electrolyte with either a gelled electrolyte or a true solid electrolyte. These batteries are supposed to be even lighter than lithium ion batteries, but there are currently no plans to fly this technology in space. It is also not commonly available in the commercial market, although it may be just around the corner.

In retrospect, we have come a long way since the leaky flashlight batteries of the sixties, when space flight was born. There is a wide range of solutions available to meet the many demands of space flight, 80 below zero to the high temperatures of a solar fly by. It is possible to handle massive radiation, decades of service, and loads reaching tens of kilowatts. There will be a continued evolution of this technology and a constant striving toward improved batteries.

Radioisotope Power

Radioisotope power systems convert the heat (thermal energy) generated from the decay of radioisotopes to electricity. The heat may be used to produce electrical power either by a static or a dynamic conversion process. See **Radioisotope Power System Technology (RPS)**. Static processes use nonmoving energy conversion devices such as a thermocouple or a thermionic converter to convert the heat into useful electrical energy.

Radioisotope Heater Units (RHUs) provide proven and reliable continuous heat to sensitive spacecraft components and instruments, enabling their successful operation throughout a mission. RHUs generate heat from the radioactive decay of a small pellet of plutonium. This heat is transferred to spacecraft structures, systems, and instruments directly without moving parts or intervening electronic components.

Radioisotope Power System Technology (RPS). The traditional means to satisfy electrical power requirements for planetary space probes that cannot rely on solar or battery power alone is through Radioisotope Power Systems (RPS). Advanced RPSs, using higher efficiency thermal conversion, are under study to reduce the quantity of radioisotope material necessary to accomplish a mission and to provide a smaller, more efficient radioisotope power system.

The basic method of radioisotope power energy conversion is based upon thermoelectrics. The thermoelectric conversion concept is based on the principle that when two dissimilar metals are joined in a closed circuit and two junctions are kept at different temperatures, an electrical voltage, and subsequent electrical current, is produced. An RPS is based on the thermoelectric concept consisting of two parts: a source of heat and a system for converting the heat to electricity. The heat source contains a radioisotope, such as plutonium dioxide, that becomes physically hot from its own radioactive decay. This heat is converted to electricity by a thermoelectric converter, which uses the Seebeck effect. All of the U.S. radioisotopic power generators launched to date have employed thermoelectric converters.

Radioisotope Heater Unit (RHU). Most spacecraft can use solar energy to provide heat to keep their structure, systems, and instruments warm enough to operate effectively. However, when solar or other heat source technologies are not feasible, an alternative heat source is required for the spacecraft.

By using Radioisotope Heater Units (RHUs), the spacecraft designer can allocate scarce spacecraft electrical power to operate the spacecraft systems and instruments. RHUs also provide the added benefit of reducing the potential for electromagnetic interference generated by electrical heating systems.

Characteristics of RHUs include:

- Highly reliable, continuous and predictable output of heat.
- No moving parts.
- Compact structure.
- Resistance to radiation and meteorite damage.
- Heat produced is independent of distance from the sun.

RHUs provide proven and reliable continuous heat to sensitive spacecraft instruments and scientific experiments enabling their successful operation throughout the mission.

RHUs generate heat from the natural radioactive decay of a small pellet of plutonium dioxide (mostly plutonium-238). This heat is transferred to spacecraft structures, systems, and instruments directly without moving parts or intervening electronic components.

RHUs are very compact, 3.2 centimeters (1.3 inches) long and 2.6 centimeters (1 inch) in diameter. The fuel pellet is about the size and shape of a pencil eraser weighing approximately 2.7 grams (0.1 ounces). All together each RHU weighs about 40 grams (1.4 ounces).

RHUs have a very rugged containment system to prevent or minimize the release of plutonium dioxide fuel even when subjected to severe accident conditions. Containment is achieved through multiple layers, which are resistant to the heat and impact that might be encountered during a spacecraft accident. An external graphite aeroshell (a reentry shield) and a graphite insulator protect the fuel from impacts, fires and atmospheric reentry conditions. Internally, the fuel is encapsulated in a high-strength, platinum–rhodium metal shell (or "clad") which further contains and protects the fuel during potential accidents. See Fig. 1.

In addition to this containment, the plutonium dioxide fuel is used in a ceramic form of which tends to break into large pieces rather than dispersing as fine particles. This minimizes interaction of the fuel with the environment and the potential for human exposure in the extremely unlikely event the multiple fuel containment barriers are breached. Since each RHU fuel pellet is individually encapsulated in its own aeroshell and fuel clad, the potential for a single event to affect more than one pellet is reduced.

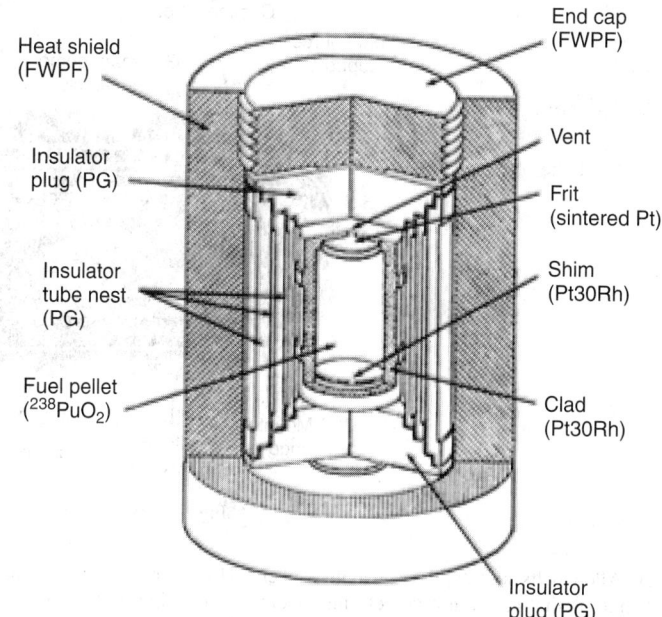

Lightweight radioisotope heater unit

Heat shield (FWPF)
End cap (FWPF)
Insulator plug (PG)
Vent
Frit (sintered Pt)
Insulator tube nest (PG)
Shim (Pt30Rh)
Fuel pellet (²³⁸PuO₂)
Clad (Pt30Rh)
Insulator plug (PG)

Fig. 1.   Lightweight radioisotope heater unit. (*NASA/JPL.*)

RHUs have been subjected to a rigorous series of laboratory and field tests. This testing has demonstrated that the RHUs are extremely rugged and reliable devices that safely provide fuel containment in a wide range of accident scenarios. No releases occurred for those events, which were within the limits of anticipated accident scenarios. In summary, the RHUs are extremely dependable devices that have been designed and tested to contain their fuel in a wide range of mission accidents.

Radioisotope Thermoelectric Generators (RTGs). RTGs are a compact space power system, which convert thermal energy from the decay of radioisotope materials to electrical current through the use of solid-state thermoelectric converters. These power systems are reliable even when used in the harsh environment of outer space. RTGs are not nuclear reactors and have no moving parts. They use neither fission nor fusion processes to produce energy. Instead, they provide power through the radioactive decay of plutonium. The heat generated by this process is changed into electricity by solid-state thermoelectric converters.

For about three decades, the United States has used RTGs to meet the thermal and electrical energy requirements for some spacecraft. These power requirements include operation of the on-board scientific experiments, spacecraft maintenance and monitoring, temperature control, and data transmission.

Radioisotope power systems convert the heat (thermal energy) generated from the decay of radioisotopes to electricity. The heat may be used to produce electrical power either by a static or a dynamic conversion process. See **Radioisotope Power System Technology (RPS)**. Static processes use nonmoving energy conversion devices such as a thermocouple or a thermionic converter to convert the heat into useful electrical energy. The thermocouple (thermoelectric conversion) is based on the principle that when two dissimilar metals are joined in a closed circuit and the two junctions are kept at different temperatures, an electrical voltage, and subsequent electrical current, is produced. This conversion technology is used in RTGs. Thermionic converters produce electricity by boiling off electrons from a heated surface or cathode. A collector or anode then captures the electrons. The cathode and the anode are enclosed in a tube with conductive vapors that enhances current flows created by the directional movement of the electrons. All of the U.S. radioisotopic power generators launched to date have employed thermoelectric converters, i.e., have been RTGs.

One beneficial contribution of RTGs is their simplicity: because there are no moving parts and no moving fluids to contain, they are highly reliable. RTGs are long-lasting and relatively insensitive to the chilling cold of space and virtually invulnerable to high radiation fields, such as Earth's

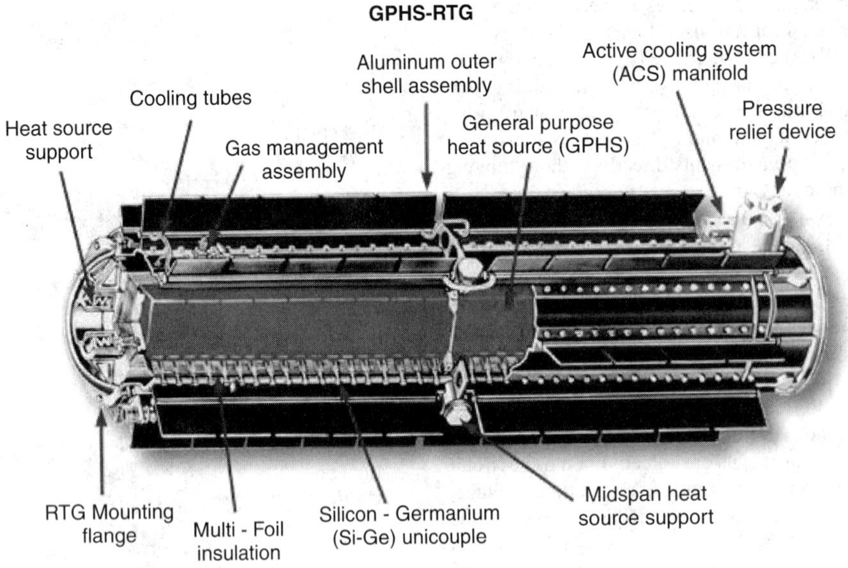

GPHS-RTG

Fig. 2.   Radioisotope thermoelectric generator. (*NASA/JPL.*)

Van Allen belts and Jupiter's magnetosphere. However, one unfortunate attribute is that they are relatively inefficient in producing electricity.

RTGs as currently designed for space missions contain several kilograms of an isotopic mixture of the radioactive element plutonium in the form of an oxide, (i.e., plutonium dioxide) pressed into a ceramic pellet. The primary constituent of these fuel pellets is the isotope 238 (Pu-238). The pellets are arranged in a converter housing and function as a heat source. The radioactive decay of the plutonium dioxide produces heat, some of which is converted into electricity by an array of thermocouples made of silicon-germanium. Waste heat is radiated into space from the external housing surface and the attached array of metal fins. Each 56-kg (123.3-pound) General Purpose Heat Source (GPHS) RTG contains approximately 11 kg (24 pounds) of plutonium dioxide fuel. See Fig. 2.

The Department of Energy (DOE) has demonstrated that RTGs will minimize the possibility of fuel release during the generators' lifetime, particularly in the event of an accident. The very low probability of a plutonium release results from the protective-layering design of a spacecraft's RTGs. The radioisotope energy source for the GPHS-RTG is a stacked column of 18 individual GPHS modules. Each module consists of a graphite aeroshell, two-carbon-bonded carbon fiber (CBCF) insulator sleeves, two graphite impact shells (GISs), and four fueled clads. The graphite (carbon-carbon composite) aeroshell serves as the module's primary heat shield to protect the internal components from direct exposure to a reentry's thermal and aerodynamic environment. The two GISs contained in each GPHS module provide the primary resistance to impact or mechanical loads. Each GIS assembly is thermally insulated from the aeroshell by a low thermal-conducting CBCF insulator sleeve. Each fueled clad, separated by a graphite floating membrane, consists of one fuel pellet of ceramic (or solid) plutonium dioxide encased in an iridium shell. The iridium shell protects and immobilizes the fuel. The iridium alloy is compatible with the plutonium dioxide fuel material, resists oxidation to air and has a high melting temperature. Each clad also contains a vent designed to release the helium generated by the alpha particle decay of the fuel. The protective layers are specifically designed to safeguard the plutonium from fires, explosions, fragment impacts and the heat of atmospheric re-entry.

The DOE has designed the GPHS to assure that the plutonium dioxide fuel is contained or immobilized to the maximum extent practical during all mission phases, including ground handling, transportation, launch, and unplanned events, such as atmospheric reentry from Earth orbit, Earth impact, and post-impact situations. The design features for the GPHS-RTG assembly incorporate many safety-related considerations. The graphitic (carbon–carbon composite) materials (i.e., the aeroshell, CBCFs, and GISs) contained in the GPHS modules perform several safety functions. As stated previously, the primary function of the aeroshell is to protect the fueled clads against the hostile environment of atmospheric heating. The GISs protect the fueled clads from ground or debris impact in the event of an accident. Each GIS also serves as a redundant heat shield

in the event of a GPHS failure. In addition, the GIS also serves as a redundant reentry aeroshell. Assembly of the graphitic materials used for the aeroshell is started with a three dimensional weave of high-strength graphite fibers. The fibers upon impregnation with a carbonaceous resin and graphitization of the matrix, form a material which has outstanding high temperature strength capabilities required to accommodate heat shield mechanical and thermal stresses that occur during reentry. This material is one of the best available for reentry applications.

More than 30 years have been invested in the engineering, safety analysis and testing of RTGs. The Safety features listed below are incorporated into the RTG design and extensive testing has demonstrated that they are expected to contain or limit the releases during most launch or reentry accident scenarios.

First, the plutonium dioxide fuel is used in its heat-resistant, ceramic form, which reduces its chance of vaporizing in fire or reentry environments. This ceramic-form fuel is also highly insoluble, has a low chemical reactivity, and if fractured, tends to break into large, non-respirable particles and chunks. These characteristics help to mitigate the potential health effects from accidents involving the release of this fuel.

Second, the fuel is divided among 18 small, independent modular units, each with its own heat shield and impact shell. This design reduces the chances of fuel release in an accident because all modules would not be equally impacted in an accident.

Third, multiple layers of protective materials, are used to protect the fuel and prevent its accidental release. Iridium is a metal that has a very high melting point and is strong, corrosion-resistant and chemically compatible with plutonium dioxide. These characteristics make iridium useful for protecting and containing each fuel pellet. Graphite is used because it is lightweight and highly heat-resistant.

In the United States, the current RTG design philosophy is fuel containment; that is, if a mission must be aborted the RTGs are designed to retain the plutonium dioxide when the GPHS modules re-enter the Earth's atmosphere. The success of this design philosophy has been tested twice during the course of the U.S. space program. In both cases, the RTGs performed as designed and no fuel was released.

Overall, the DOE has spent more than twelve years in engineering, fabrication, safety testing, and evaluation of the GPHS modules. Testing of the present design has demonstrated the ability of the GPHS modules to retain the fuel in a range of accident conditions. In addition, DOE is constantly striving to improve upon their research and improve the safety of the GPHS design, including the use of alternative materials.

Stirling Power Systems Technology. NASA Glenn Research Center (GRC) and the Department of Energy (DOE) are researching a Stirling convertor for an advanced radioisotope power system that could provide more efficient spacecraft power for NASA missions. NASA GRC is also analyzing systems using Stirling convertors for other space applications including solar dynamic power systems for space-based radar and as a potential alternative to radioisotope thermoelectric generators (RTGs) for space applications. The Stirling engine converts the heat from radioisotope decay to a reciprocating mechanical motion and in turn, that mechanical motion is converted to an electrical current for use by the spacecraft. These systems use the heat generated from the radioactive decay of radioisotopes to warm a gas or a liquid to rotate a turbine or move a piston in a cylinder. The turbine or piston in turn drives a generator or linear alternator that produces electricity. The free-piston design was introduced to eliminate seals and lubrication in the power linkage, thus making it useful for space applications.

Stirling efforts that have focused on a 55 watt free-piston Stirling convertor, are known as the *Technology Demonstration Convertor* (TDC). The TDC has successfully tested the free-piston design in a configuration relevant to space applications. Further work would be required to demonstrate a feasible Stirling system design.

In a Stirling Conversion system, heat from some source (i.e., radioisotopes) is used to power a Stirling engine. The engine has a working gas contained within it that absorbs the source heat through conduction, convection, or radiation. Attaching an alternator to the drive shaft can then produce electricity. The alternator produces single phase, alternating current (AC) power typically between 50 and 70 Vrms (volts root mean square). Most Stirling power systems would include power electronics to convert the AC into the usable direct current (DC) for the spacecraft.

The GRC technology effort is complemented by a DOE Stirling Radioisotope Power System (SRPS) integration activity. It is expected that many of the configuration, operational and reliability issues associated with Stirling will be addressed through these efforts.

Solar Power

Solar panels are devices that convert light into electricity. They are called solar after the sun or "Sol" because the sun is the most powerful source of the light to use. They are sometimes called *photovoltaics*, which means "light-electricity". Solar cells or PV cells rely on the photovoltaic effect to absorb the energy of the sun and cause current to flow between two oppositely charge layers. Most NASA spacecraft are powered by solar panels (collections of solar cells), including all NASA spacecraft that orbit the Earth and Mars and spacecraft that are going to nearby comets or asteroids.

Crystalline silicon and gallium arsenide are typical choices of materials for solar panels for deep-space missions. Gallium arsenide crystals are grown especially for photovoltaic use, but silicon crystals are available in less-expensive standard ingots, which are produced mainly for consumption in the microelectronics industry.

When exposed to direct sunlight at 1 AU, a 6-centimeter (2.4-inch) diameter silicon cell can produce a current of about 0.5 ampere at 0.5 volt. Gallium arsenide is more efficient. Crystalline ingots are sliced into wafer-thin disks, polished to remove slicing damage, dopants are introduced into the wafers, and metallic conductors are deposited onto each surface: a thin grid on the sun-facing side and usually a flat sheet on the other. Spacecraft solar panels are constructed of these cells cut into appropriate shapes, protected from radiation and handling damage on the front surface by bonding on a cover glass, and cemented onto a substrate (either a rigid panel or a flexible blanket), and electrical connections are made in series-parallel to determine total output voltage. The cement and the substrate must be thermally conductive, because in flight the cells tend to heat up from absorbing infrared energy that is not converted to electricity. Since cell heating reduces the operating efficiency it is desirable to minimize the heating. The substrate is supported on a deployable structural framework. The resulting assemblies are called *solar panels* or *solar arrays*.

A solar panel is a collection of solar cells. Although each solar cell provides a relatively small amount of power, many solar cells spread over a large area can provide enough power to be useful. To get the most power, solar panels have to be pointed directly at the Sun. Spacecraft are built so that the solar panels can be pivoted as the spacecraft moves. Thus, they can always stay in the direct path of the light rays no matter how the spacecraft is pointed. Spacecraft are usually designed with solar panels that can always be pointed at the Sun, even as the rest of the body of the spacecraft moves around, much as a tank turret can be aimed independently of where the tank is going. A tracking mechanism is often incorporated into the solar arrays to keep the array pointed towards the sun.

Solar panels need to have a lot of surface area that can be pointed towards the Sun as the spacecraft moves. More exposed surface area means more electricity can be converted from light energy from the Sun. Sometimes, satellite scientists purposefully orient the solar panels to "off point," or out of direct alignment from the Sun. This happens if the batteries are completely charged and the amount of electricity needed is lower than the amount of electricity made. The extra power will just be vented by a shunt into space as heat.

Solar panels are very hardy. Compared to alternative power sources, they wear out very slowly. The principal factor affecting the loss in power with time is the Space radiation environment. For low radiation environments, such as low Earth orbiting, their effectiveness decreases around 1 to 2 percent a year. This means after a five year mission the solar panels will still be making more than 90% of what they made at the beginning of the mission (as long as they have not gotten farther away from the Sun). In contrast, for missions in higher radiation environments, such as mid altitude Earth orbit 1,240 to 6,210 miles (2,000 to 10,000 kilometers), arrays can lose half their power within 1 year. That is one reason few missions fly in this orbital range.

To date, solar power has been practical for spacecraft operating no farther from the sun than the orbit of Mars. For example, *Magellan, Mars Global Surveyor*, and *Mars Observer* used solar power as did the Earth-orbiting, Hubble Space Telescope. For future missions, it is desirable to reduce solar array mass, and to increase the power generated per unit area. This will reduce overall spacecraft mass, and may make the operation of solar-powered spacecraft feasible at larger distances from the sun.

Solar array mass could be reduced with thin-film photovoltaic cells, flexible blanket substrates, and composite support structures. Solar array efficiency could be improved by using new photovoltaic cell materials and solar concentrators that intensify the incident sunlight.

Photovoltaic concentrator solar arrays for primary spacecraft power are devices, which intensify the sunlight on the photovoltaics. This design uses lenses, called *Fresnel lenses*, which take a large area of sunlight and direct it towards a specific spot by bending the rays of light and focusing them. Some people use the same principle when they use a magnifying lens to focus the Sun's rays on a pile of kindling or paper to start fires.

Solar concentrators put one of these lenses over every solar cell. This focuses light from the large concentrator area down to the smaller cell area. This allows the quantity of expensive solar cells to be reduced by the amount of concentration. Concentrators work best when there is a single source of light and the concentrator can be pointed right at it. This is ideal in space, where the Sun is a single light source. Solar cells are the most expensive part of solar arrays, and arrays are often a very expensive part of the spacecraft. This technology allows costs to be cut significantly due to the utilization of less material.

Fresnel lenses have been around since Augustin Jean Fresnel invented them in 1822. Theaters use them for spotlights and lighthouses use them to make their lights visible at greater distances. See also **Fresnel Mirror**.

Degradation of Solar Arrays

- Solar flares, which are unpredictable severe bouts of radiation that can degrade solar cell output.
- Micrometeorites, which are tiny, gravel-sized bits of rock and other space junk floating in space can pit or penetrate solar cells. The solar concentrator as well as a thick layer of glass can provide some protection to the solar panels from these.
- If a satellite's mission path takes it away from the Sun (farther out into the solar system) solar panels will become less and less useful.
- Conversely, when used nearer to the sun, the solar array mechanical integrity and solar cell output can be reduced by severe heating. Most long lifetime solar arrays have generally been limited to the orbit of Venus.

Development Projects for Solar Arrays

- AEC-Able, manufacturer of the SCARLET array is developing a higher efficiency, lower mass, solar array with linear solar concentrators for use in the New Millennium program.
- The Mir Cooperative Solar Array is a joint U.S. / Russian flight demonstration of the solar array technology to be used on the International Space Station.
- The AstroEdge Concentrator Solar Array is a flight demonstration of reflective solar concentrators and lightweight rigid panel array structures on the SSTI

Clark spacecraft.

- The Hughes (now Boeing) HS702 commercial communications satellite solar array uses a low ratio reflective concentrator and has been operational since 1999.
- AEC-Able Engineering challenged the Solar Array industry by lowering the achievable limits on weight and stowage volume. Approximately 1/4 of the weight of conventional designs, UltraFlex enables spacecraft to maximize payload or reduce launch costs especially for interplanetary missions.
- Sharp is developing higher efficiency silicon solar cells for space applications.

See also **Plutonium**; **Solar Energy**; and **Space Stations**.

Additional Reading

Bube, R.H.: "Photovoltaic Materials," World Scientific Publishing Company, Inc., Riveredge, NJ, 2000.

Goetzberger, A., J. Knobloch, and B. Voss: "Crystalline Silicon Solar Cells," John Wiley & Sons, Inc., New York, NY, 1998.

Hyder, A.K. and R.L. Wiley: "Spacecraft Power Technologies," Imperial College Press, London, UK, 1998.

Markvart, T.: "Solar Electricity," 2nd Edition, John Wiley & Sons, Inc., New York, NY, 2000.

Yamamoto, O., M. Wakihara: "Lithium Ion Batteries: Fundamentals and Performance," John Wiley & Sons, Inc., New York, NY, 1998.

Web References

Office of Nuclear Energy, Science & Technology — U.S. Department of Energy: *http://www.ne.doe.gov/*

U.S. Department of Energy Photovoltaics Program: *http://www.eren.doe.gov/pv/*

U.S. Department of Energy National Center for Photovoltaics: *http://www.nrel.gov/ncpv/*

Jet Propulsion Laboratory, Pasadena, CA.

SPACE FRAME (or Space Structure). A three-dimensional framed structure. It may be composed of triangles, rectangles or a combination of these forms. Space frames are statically determinate or indeterminate depending on the number of members, support conditions and the rigidity of the joints. Bridges, framed domes, transmission towers, radio towers and building frames are all space structures.

If the space frame is statically determinate it may be analyzed by means of the equations for three-dimensional statics. These equations state that the summation of all forces parallel to three axes, usually taken as mutually perpendicular, must equal zero and the moments of all forces about these axes must also equal zero.

Indeterminate space frames may be analyzed by methods which are applicable to indeterminate planar structures. In recent years numerical methods have been developed for the analysis of rigid space frames. Bridges and building frames are highly indeterminate when considered as space frames because the joints have a certain amount of rigidity even though they may not have been designed to resist moment. Thus, all parts do participate to some extent in distributing the loads to the foundation. Due to this action some members may be subjected to torsional stress as well as direct and bending stresses.

SPACE INFRARED TELESCOPE FACILITY (SIRTF). The Space InfraRed Telescope Facility (SIRTF) is a space-borne, cryogenically-cooled infrared observatory capable of studying objects ranging from our Solar System to the distant reaches of the Universe. SIRTF is the final element in **NASA's Great Observatories Program**, and an important scientific and technical cornerstone of the new **Astronomical Search for Origins Program**. See also **Origins Program**.

To grasp the wonders of the cosmos, and understand its infinite variety and splendor, we must collect and analyze radiation emitted by phenomena throughout the entire electromagnetic (EM) spectrum. Towards that end, NASA proposed the concept of Great Observatories, a series of four space-borne observatories designed to conduct astronomical studies over many different wavelengths. An important aspect of the Great Observatory program was to overlap the operations phases of the missions to enable astronomers to make contemporaneous observations of an object at different spectral wavelengths.

The first element of the program, and the best known, is the Hubble Space Telescope (HST). The Hubble telescope was deployed by the Space Shuttle *Discovery* on April 24, 1990. A subsequent Shuttle mission in 1993 serviced HST and recovered its full capability. A second successful servicing mission took place in 1997. Subsequent servicing missions have added additional capabilities to HST, which observes the Universe at ultraviolet, visual, and near-infrared wavelengths. See also **Hubble Space Telescope (HST)**.

The second Great Observatory was launched and deployed by the Space Shuttle *Atlantis* on April 5, 1991: the Compton Gamma-Ray Observatory (CGRO). This mission collected data on some of the most violent physical processes in the Universe, characterized by their extremely high energies. The CGRO was safely de-orbited and re-entered the Earth's atmosphere on June 4, 2000 after nine years in service. See also **Gamma-Ray Astronomy**; and **Compton Gamma-Ray Observatory (CGRO)**.

The third member of the Great Observatory family, the Chandra X-Ray Observatory (CXO), was deployed from the Space Shuttle *Columbia* and boosted into a high-Earth orbit, on July 23, 1999. This observatory is observing such objects as black holes, quasars, and high-temperature gases throughout the x-ray portion of the EM spectrum. See also **X-Ray Astronomy**; and **Chandra X-Ray Observatory**.

The Space InfraRed Telescope Facility (SIRTF) represents the fourth and final element in NASA's Great Observatory program. The observatory is scheduled for launched July 2002 and will have a cryogenic lifetime of at least 2.5 years. SIRTF will fill in an important gap in wavelength coverage not available from the ground—the thermal infrared.

Mission Overview

SIRTF will explore the birth and evolution of the universe and the objects within it with unprecedented sensitivity, viewing phenomena not seen through other astronomical techniques. The SIRTF mission is managed for NASA by the Jet Propulsion Laboratory (JPL), a division of the California Institute of Technology. Infrared wavelengths, redder than the human eye can see, can be thought of as heat radiation. The colder infrared detectors are, the better they can "see" objects in the sky that glow in the infrared. In space, infrared telescopes are above the warmth of Earth's atmosphere, and can be cooled to very low temperatures, enabling studies of objects that range from far below room temperature to many thousands of degrees.

SIRTF will be the first space observatory to combine such a highly sensitive cryogenically cooled telescope with the imaging and spectroscopic power of a new generation of infrared detector arrays. SIRTF will see through clouds of dust and gas that obscure much of the universe from view. It will sense the infrared radiation, or heat, emitted by objects from the oldest galaxies to forming stars and emerging planetary systems. Scientists expect the observatory to help them discover the answers to important questions about the origin and evolution of the universe, its galaxies and stars, and to seek out stars like the Sun that may hold planets like Earth orbiting about them.

SIRTF's specially designed orbit will completely remove the spacecraft from the heat of the Earth and from the obscuring atmosphere that blocks our view of most of the infrared light from celestial sources. The observatory will see wavelengths of light critical to NASA's quest for an understanding of our origins, allowing astronomers to discover the basic processes underlying the formation of planets and the birth of stars and galaxies.

Data received through NASA's Deep Space Network from SIRTF will be processed and disseminated to the science community at the JPL/Caltech Infared Processing and Analysis Center located at the Caltech campus.

The National Academy of Sciences designated SIRTF the highest priority major mission for all of U.S. astronomy in the 1990s. With U.S. technology developments, innovative engineering and streamlined mission planning, SIRTF has undergone radical redesign to be consistent with the new NASA paradigm of faster/better/cheaper. The result is an affordable yet enormously powerful observatory.

SIRTF's Telescope and Spacecraft Overview

Cryogenic Telescope Assembly. SIRTF's Cryogenic Telescope Assembly (CTA) consists of four main parts: a superfluid helium cryostat, a light-weight 85centimeter (33.5 inch) Ritchey-Chretien telescope, an Outer Shell Group, and a Multiple Instrument Chamber which houses the Science Instruments. Everything that is cold is part of the CTA. The CTA is mechanically mounted to, but thermally isolated from, the spacecraft by means of low conductivity struts and thermal radiation shields. The solar array and spacecraft shields block the CTA from the sun and spacecraft components at all times, helping to keep the CTA as cold as possible.

Telescope. The SIRTF telescope fabricated by Hughes Danbury Optical Systems is a lightweight reflector of Ritchey-Chrétien design. It weighs less than 50 kilograms (110 pounds) and is designed to operate at an extremely low temperature. The telescope has an 85 centimeter (33.5 inch) diameter aperture. All of its parts, except for the mirror supports, are made of light-weight beryllium. Beryllium is a very strong material, which works well in the construction of infrared space telescopes because it has a low heat capacity at very low temperatures. The telescope is attached to the top of the vapor-cooled cryostat vacuum shell, which keeps the science instruments very cold.

The design philosophy of the telescope assembly is based on the following guidelines:

- Maximize the use of materials with a very high stiffness/density ratio, high thermal conductivity, and low cryogenic specific heat.
- Build the entire telescope of the same material to prevent thermal expansion mismatch complications, and to make the telescope assembly, as dimensionally stable as possible.
- Select a configuration that minimizes the size of the major elements of the telescope assembly.
- Strive for the simplest possible design to minimize the number of parts, thereby reducing the time and cost for design, fabrication, and integration.

Multiple Instrument Chamber. SIRTF's Multiple Instrument Chamber contains the cold parts of SIRTF's three science instruments as well as the pointing calibration reference sensor which is part of the pointing control system. The Multiple Instrument Chamber is built to be so tight that no light can get through it except that which is allowed to be detected directly by the instruments. The chamber is 84 centimeters (33.1 inches) in diameter by 20 centimeters (7.9 inches) high. It has an aluminum baseplate and cover and is mounted directly to the helium tank.

Infrared Array Camera (IRAC). The Infrared Array Camera (IRAC) is one of SIRTF's three science instruments, and provides imaging capabilities at near- and mid-infrared wavelengths. It is a general-purpose camera that will be used by observers on SIRTF for a wide variety of astronomical research programs.

IRAC is a four-channel camera that provides simultaneous 5.12×5.12 arcmin images at 3.6, 4.5, 5.8, and 8 microns. Each of the four detector arrays in the camera are 256×256 pixels in size. IRAC uses two sets of detector arrays. The two short-wavelength channels are imaged by composite detectors made from indium and antimony. The long-wavelength channels use silicon detectors that have been specially treated with arsenic. The only moving part in IRAC is the camera shutter.

Infrared Spectrograph (IRS). The Infrared Spectrograph(IRS) is one of the three instruments onboard SIRTF and provides both high- and low-resolution spectroscopy at mid-infrared wavelengths. Spectrometers are instruments, which spread light out into its constituent wavelengths creating a spectra. Within this spectra astronomers can study emission and absorption lines: which are the fingerprints of atoms and molecules.

The IRS has four separate modules: a low-resolution, short-wavelength mode covering the 5.3-14 micron interval; a high-resolution, short-wavelength mode covering 10-19.5 microns; a low-resolution, long-wavelength mode for observations at 14-40 microns; and a high resolution, long-wavelength mode for 19-37 microns. Each module has its own entrance slit to let infrared light in. The detectors are 128×128 arrays. The shorter-wavelength silicon detectors are treated with arsenic; the longer-wavelength silicon detectors are treated with antimony.

The IRS instrument consists of two physically separated parts, the cold assemblies which are located within the SIRTF multiple instrument chamber and the warm electronics, which are located in the SIRTF spacecraft bus. The IRS has no moving parts!

Multiband Imaging Photometer (MIPS). The Multiband Imaging Photometer for SIRTF (MIPS) is one of the three science instruments that will fly on the Observatory and will provide imaging and limited spectroscopic data at far-infrared wavelengths. It has three detector arrays. A 128×128 array for imaging at 24 microns is composed of silicon, specially treated with arsenic. A 32×32 array for imaging at 70 microns, and a 2×20 array for imaging at 160 microns both use germanium, treated with gallium. The 32×32 array will also take spectra from 50–100 microns. The MIPS field of view varies from about 5×5 arcmin at the shortest wavelength to about 0.5×5 arcmin at the longest wavelength.

The three arrays, calibrators, scan mirror, and optics compose the cryogenic part of the MIPS. This assembly is mounted in the SIRTF

cold instrument chamber. In addition, the MIPS and the IRS share warm electronics that controls their operation. The only moving part in MIPS is a scan mirror used to efficiently map large areas of the sky.

Cryostat. In general, SIRTF's science payload needs to be kept very cold to work properly. This is because it uses infrared detectors that are very sensitive to heat. If the instruments are not very cold, the heat from the instruments themselves will interfere with the study of the faint infrared radiation from objects in space. SIRTF's cryostat will keep the science instruments at temperatures as low as 1.4 degrees Kelvin ($-272\,^{\circ}$C or $-457\,^{\circ}$F) for up to 5 years.

SIRTF's cryostat will keep the instruments cold by venting helium vapor from a liquid helium tank. The cryostat consists of the vacuum shell, inner and middle vapor cooled shields, helium tank, and the fluid management system. The tank holds about 360 liters of superfluid helium.

The telescope is attached to the top of the vapor-cooled cryostat. The telescope and cryostat shell is launched warm, and cool down to about 35 degrees Kelvin ($-238\,^{\circ}$C or $397\,^{\circ}$F) once in orbit. The cryostat vacuum shell must be sealed during ground operations and launch. Once the cryostat is cooled and in the vacuum of space, a cryostat door on the top of the shell is opened to allow light into the Multiple Instrument Chamber.

Outer Shell. SIRTF's outer shell is made up of a dust cover, outer shield (cooled by helium vapor), thermal shields (which block radiation from space), and the solar panels. The telescope and cryostat are surrounded by the outer shell. The shell keeps exterior heat from reaching the telescope and instuments by radiating it out into cold space.

The outer shell is made of lightweight, aluminum honeycomb sandwich construction. It is painted black on the anti-sun side, for efficient thermal radiation to cold space, and is shiny on the sun-facing side, for efficient rejection of heat from the sun and solar panels. To maximize the helium lifetime, the outer shell temperature must be as low as possible. The sources of heat to the outer shell are the spacecraft bus, solar panel, and star trackers.

The outer shell is closed at the top by an ejectable cover, which lets out unwanted dry nitrogen which, if left in, could contaminate the telescope's mirrors and other surfaces. The thermal shields are sized and shaped such that the observatory is shaded at all times when pointed properly.

Spacecraft. SIRTF's spacecraft refers to the warm portion of the Observatory, including the solar panel, the bus structure, and the components mounted in the bus that provide the Observatory engineering functions. These components include: the solar arrays, the command and data handling unit, the reaction control subsystem, the telecommunications subsystem, the power generation and distribution subsystem, the pointing control subsystem, and the flight software.

The spacecraft consists of an octagonal bus structure in which the avionics and the science instrument's warm electronics are housed, and a solar panel that provides electrical power to the vehicle and serves to shade the cryo-telescope assembly from direct exposure to the sun. The spacecraft provides electrical power to the science instruments, orients and stabilizes the boresight of the telescope, collects and compressed data from the science instruments for later transmission to the ground, executes stored commands to direct science instrument activities, and communicates with the ground system. All communications with SIRTF will be conducted through NASA's Deep Space Network. The spacecraft is responsible for the safe operation of the Observatory.

Solar Panel Assembly. SIRTF's Solar Panel Assembly is designed to provide the electrical power needed to operate SIRTF for up to 5 years. The Solar Panel Assembly is made up of two solar panels each of which has 392 solar cells. Each solar cell is 5.5 centimeters (2.2 inches) by 6.5 centimeters (2.6 inches) in area. Together, these cells will convert radiation from the Sun into a total of 427 Watts of electrical energy. SIRTF cannot point more than 120 degrees away from the sun. If it does, sunlight will not hit the solar panels properly.

The solar panel has 50 percent of its area covered with solar cells and 50 percent covered with optical solar reflectors. These reflectors reduce the solar panel temperature to about 330 K. The wedge shaped solar panel shield is angled away from the outer shell in order to improve its view factor to space and to improve the thermal isolation between it and the outer shell.

Spacecraft Bus. SIRTF's Spacecraft Bus is an octagonal structure in which the avionics and the science instrument warm electronics are housed. The Spacecraft Bus provides electrical power to the science instruments, orients and stabilizes the boresight of the telescope, collects and compresses data from the science instruments for later transmission to the ground, executes stored commands to direct science instrument activities, and communicates with the ground system.

The bus structure makes up the nine bays within which the bus equipment is housed. Four side modules, four corner modules, two heatpipe equipment panels, and six equipment panels make up the nine bays. The equipment panels are sandwiched between the modules and are bolted and bonded together for maximum stiffness and structural integrity. On the outside of the bays, panels are bolted to the modules. These panels are removable and provide access to the outer bays during integration and test. The center bay is accessible from the top.

Science Overview

SIRTF will study a wide variety of astronomical phenomena, extending from our Solar System to the distant reaches of the early Universe. Providing coverage at infrared wavelength between 3 and 180 microns, SIRTF provides an important scientific complement to the Hubble Space Telescope and the Chandra X-Ray Observatory. The shortest IR wavelengths will pierce through the heavy obscuration of dust and allow astronomers to study newly formed stars. The longer IR wavelengths are well suited to studying the distribution of dust, an important ingredient of future planet and star formation, throughout the Milky Way Galaxy.

Nearly 80 percent of the observing time on SIRTF will be available to the scientific community-at-large through competitive proposals solicited by the SIRTF Science Center. To date, about one-fifth of the SIRTF mission (assuming a 5-year mission) has been defined through the Legacy Science Program, the First-Look Survey and the Guaranteed Time Observations. Solicitations for the balance of the science program will be issued on an annual basis, starting shortly before launch.

One consequence of the re-designs of SIRTF in the early 1990s was the decision to design SIRTF such that it would make important and lasting contributions in these four science themes:

- **The Search for Brown Dwarfs and Super-Planets:** These objects have too little mass to ignite the fusion reactions which power stars, but are larger and warmer than planets found in our Solar System. Astronomers are now starting to detect these long-sought objects and want to know the extent to which they might account for the elusive dark matter that is thought to permeate the Universe. SIRTF will provide invaluable information about their population and physical characteristics.
- **The Study of Ultraluminous Infrared Galaxies and Active Galactic Nuclei:** Many galaxies emit more radiation at infrared wavelengths than in all other regions of the electromagnetic spectrum combined. These ultraluminous IR galaxies could be triggered by intense bursts of star formation stimulated by colliding galaxies, or by dust-enshrouded active galactic nuclei (including quasars) powered by black holes. SIRTF will trace the origins and evolution of these objects to cosmological distances.
- **The Study of the Early Universe:** Cosmological redshifts result from the expansion of the Universe, shifting the observed light from astronomical phenomena to longer wavelengths. Objects at high redshifts are seen as they existed long ago. Most of the optical and ultraviolet radiation emitted from stars and galaxies since the beginning of time is now shifted into the infrared. SIRTF will provide important insights into when and how the first stars and galaxies formed.

Apart from being scientifically interesting in their own right, these themes are directly relevant to NASA's Origins Program, which seeks to understand the origins of the Universe, galaxies, stars, and planets.

While these themes drove the budget-mandated re-design of the Observatory, it should be emphasized that SIRTF's powerful capabilities will be applied to a wide range of other astronomical topics. Moreover, SIRTF offers unparalleled capabilities in a space-borne infrared telescope. History has repeatedly demonstrated that such huge advances lead to serendipitous discoveries of unanticipated phenomena. With SIRTF, astronomers expect the unexpected!

More information on the SIRTF mission is available from the project's home page at *http://sirtf.caltech.edu/index.html*

Jet Propulsion Laboratory, California Institute of Technology, Pasadena, CA.

SPACE MEDICINE. A branch of aerospace medicine concerned specifically with the health of persons who make, or expect to make, flights into space beyond the sensible atmosphere.

SPACE MICROELECTRONICS TECHNOLOGY. Future space travel means small spacecraft. In today's constrained budget environment, large space missions can no longer be afforded. At the Jet Propulsion Laboratory (JPL), planners have adjusted the size of missions and spacecraft to be more moderate, primarily by taking more risks and asking the spacecraft to do less. In the future, they will be using very advanced technology in order to reduce the spacecraft size even more, while retaining the functionality of today's spacecraft. The spacecraft of the future will provide world-class science in focused areas for an affordable price, making it possible for many of these microspacecraft to be launched, perhaps as many as one per month.

The Jet Propulsion Laboratory, Center for Space Microelectronics Technology (CSMT) concentrates on innovative, high-risk, high-payoff concepts and devices that hold the potential to enable new space missions or to enhance current and planned space missions. The center conducts research and development in microsensors and microinstruments, advanced detectors and high performance computing. Once the concepts are proved through demonstrations, the successes are transferred to mission applications or to industry for commercialization.

The center focuses on areas of microelectronics and advanced computing that are unique to space applications. This includes sensors studying objects in space in portions of the electromagnetic spectrum that can not be readily studied from Earth because of atmospheric interference, as well as high-performance ground computing for mission data analysis and visualization.

In 1987, NASA and several Department of Defense agencies with space responsibilities established CSMT in order to create a program in space microelectronics with world-class facilities, equipment and staff.

Microdevices Laboratory

Much of CSMT's research and development work takes place in the Microdevices Laboratory, a unique 38,000-square-footfacility at JPL. It includes clean rooms for thin-film material deposition, lithography, device processing, and optical characterization.

Work at the Microdevices Laboratory encompasses a wide range of sensors designed for use on future spacecraft, including passive optical instruments, dust detectors and a variety of spectrometer instruments. A mass spectrometer, for instance, is used to identify and measure gases. An X-ray fluorescence spectrometer would be used to measure the absolute abundance of elements at the surface of an airless body, such as an asteroid. The center also is developing new micro instruments for Earth remote sensing.

Other sensors measure infrared radiation. The intensity of infrared radiation in our environment is second only to that of visible light. All objects produce radiation, most of which is emitted in the infrared waveband of the electromagnetic spectrum. Infrared extends from just beyond red light to the beginning of microwaves. NASA has a significant interest in space instruments that map in the infrared. The agency is developing instruments that look upward into the universe as well as downward to observe Earth. The center has developed a gallium arsenide-based quantum well infrared photo detector (QWIP) that has been made into large arrays and packaged into camcorder sized portable cameras. QWIP detectors have a narrow bandwidth ($\sim$1 μm) and have been made in various wavelength responses from 6 μm to 22 μm. Because gallium arsenide is a mature and producible technology, 256 by 256 and 640 by 480 element arrays can be routinely made.

In addition, scientists are currently working to improve spacecraft "eyes." Wireless digital imaging cameras the size of two sugar cubes are being developed which can operate in the ultraviolet, visible, and near-infrared realms. The active pixel sensor, a compact, solid-state image sensor technology, makes possible a veritable "camera on a chip." In many imaging applications, such sensors may ultimately replace charge-coupled devices (CCDs), which measure light digitally and are used in many still and video cameras. An active pixel sensor camera uses 100 times less power than the standard CCD camera, has superior resolution, is less susceptible to radiation damage in space, and can be made in a standard semiconductor factory. The camera features automatic exposure control and electronic pan and zoom. In addition, it comes with a built-in transmitter that can send images more than a mile. These "cameras-on-a chip" will be used in orbiters, landers, and rovers.

Still, CCDs will be used on many future space missions, and a new process has been developed at the center to build CCDs to image at ultraviolet wavelengths. That enhancement will allow a new set of NASA space instruments for ultraviolet astronomy and other remote sensing projects.

In the area of microinstruments, scientists are developing highly sensitive, light, compact, robust, low-power microaccelerometers and seismometers for planetary, cometary, microgravity, and terrestrial applications. Conventional seismometers are ill-suited for space applications; despite good sensitivity, they require careful deployment and are delicate, heavy and power-hungry. The miniature seismometer weights less than 200 grams (about 7 ounces); normal seismometers are about the size of an overnight bag and weigh about 50 times more. To understand if Mars has a molten core, scientists want to measure Mars quakes with seismometers capable of measuring one billionth the acceleration of gravity, all within a package the size of a message pager. A prototype of a microseismometer has successfully measured earthquakes in laboratory demonstrations.

Several microdevices needed for future space missions have been developed by the center, including instruments for a microweather station that may go to Mars. Imagine fleets of nearly autonomous microlanders dispersed by one small spacecraft, each with a weather station aboard. A network of these stations could be established to supply information about the humidity and wind and the composition and temperature of the planetary atmosphere. Because of the importance of water to the atmospheric science of both Earth and Mars, the microhygrometer is the most scientifically important component of the micro weather station. A microhygrometer for direct dew-point measurements has been developed and successfully tested on NASA's DC8 for upper troposphere measurements of humidity. It has demonstrated superior performance compared to conventional, large, chilled-mirror hygrometers. Other instruments being developed in the program include micromachined silicon sensors for wind, pressure, and air temperature, a radiation densitometer to measure radiation and a micro laser Doppler anemometer to measure wind speed and direction. Surface micro weather stations also have applications in military tactical situations as tools for gathering critical information on surface conditions on land or sea. This effort has resulted in the world's first demonstration of single mode lasers suitable for spectroscopy applications that operate at ambient temperatures.

To enable scientists to learn more about the role water played in Mars' past and its impact on the planet today, the Microdevices Laboratory has developed and delivered space-qualified tunable diode lasers to detect water, as well as a variety of other gases. Measurement of water in the Martian atmosphere will be done by near-infrared diode lasers the size of pencil points. These lasers can be assembled with their electronics into instruments the size of a roll of pennies. Any data attained could help answer questions regarding the possibility of life on the planet.

One of the most innovative projects at the Microdevices Laboratory has been the development of the ballistic electron emission microscope (BEEM). Scientists working in solid-state physics must know the conditions existing at the interface of two separate materials. The BEEM method uses a scanning tunneling microscope to inject an electron tunnel current into a structure with one or more buried interfaces between different materials. See also **Scanning Tunneling Microscope**.

Electrons injected into the structure are sent ballistically–that is, without scattering or loss of energy–for distances as small as tens of billionths of a meter (nanometers). The stabilized microscope tip is scanned over the surface as the electron current is detected crossing the buried interface. The transmission of ballistic electrons gives information on the material and interface quality.

High-Performance Computing

High-performance computers are needed onboard spacecraft to enable spacecraft autonomy and to analyze and compress scientific data prior to transmission to Earth. High-performance computers are also needed on the ground to analyze and visualize data as part of the process of turning data into knowledge. Ground-based computers are also used for space mission and spacecraft design and simulation, and for theoretical studies and modeling of physical phenomena.

The center's activities in on-board computing have focused on developing a miniature flight computer using advanced technologies, including multi-chip modules and three-dimensional chip and module stacking.

Although the initial emphasis has been on system miniaturization, the limited power onboard deep space missions, has made low-power consumption a major new R&D direction. Under NASA's High Performance Computing and Communications program, the center is developing a scalable, low-power flight computer architecture that relies on commercial

microprocessors and implements the fault tolerance needed for space applications in software. A single mode of the computer could be used for a simple microlander application, whereas a 50-processor parallel machine can do onboard processing of hyper-spectral science imagery.

Other scientists at the center are developing innovative magnetic and optical data storage techniques and optical processing. A major thrust is directed toward electronic neural networks modelled on the human brain, capable of pattern recognition and vehicle control in real time. See also **Artificial Intelligence: Neural Networks**.

JPL and the California Institute of Technology have been pioneers in developing technologies for massively parallel computing. A dozen years ago, the Center was building parallel supercomputers because there was no industry. Caltech/JPL partnerships with Intel, Cray Research and, most recently, Hewlett Packard have been instrumental in turning high performance parallel supercomputing into the industry known today.

The center concentrates on software and applications of high performance parallel supercomputers for NASA and Defense Department applications. These include ocean modeling, data visualization, mission design, and radar processing. The new 256-processor Hewlett Packard Exemplar system has a peak performance of 184 billion operations per second, with a memory of 64 gigabytes. This is 700 times faster than the JPL's CRAY X-MP supercomputer of a decade ago. The memory is 4,000 times larger than that of a typical desktop system. The large memory allows huge problems to be tackled. For example, all the data NASA has from previous missions can fit into the machine's memory and can be processed and visualized in real time.

Microspacecraft

In the future, we will be using more and more very advanced technology in order to reduce spacecraft size, all the while retaining the functionality of today's spacecraft.

By the year 2010, second-generation microspacecraft the size of toaster ovens that weigh 5-1/2 kilo-grams (about 12 pounds) and use 5 watts of power will travel a billion miles away and send data back to Earth. They will be able to figure out their location and navigate autonomously, all by the position of the stars.

These microspacecraft will be enabled by advances in space technology. Many of the key technologies will be derived from those in such commercial products as cell phones, low-power palm top computers, and pagers. Others are being developed specifically for space. These include a micromachined gyro, accelerometer and other micro-electro-mechanical systems, microthrusters, neural networks and spacecraft autonomy software.

The center is developing ground-based micro-spacecraft prototypes that include many of the above components to begin to investigate the systems issues that will arise when building these new miniature spacecraft.

Management

Dr. Barbara Wilson is director of the Center for Space Microelectronics Technology. Policy guidance and program oversight are provided by a board of governors. Board members include the major sponsors of the center, together with the JPL director, Caltech president and Caltech provost.

The center's scientific advisory board, composed of seven world-renowned scientists, reviews the technical program and provides advice to the board of governors and the center's director.

Programs

Many of the center's technologies have commercial as well as government mission applications. The U.S. Department of Commerce joined the center in 1991 and urged the center and its sponsors to emphasize technology applications for business. As a result, the center has initiated programs with a strong emphasis on dual-use technologies and partnerships with industry.

Currently there are 39 cooperative agreements with U.S. industry in the areas of electronics, computing, communications, automotive, and health care. Those collaborations are with companies both large and small, as well as minority- and women-owned businesses.

Key programs under the center include:

- Low- and high-temperature superconductors.
- Semiconductor lasers.
- Microsensors and microinstruments.
- Microelectro-mechanical systems.

- Infrared, visible and ultraviolet detectors.
- Submillimeter receivers.
- Advanced flight computer.
- High-performance computing.
- Advanced networking.
- Neural networks.
- Vertical Bloch line memory.

For more information, visit the CSMT web site at *http://csmt.jpl.nasa.gov*

Jet Propulsion Laboratory, California Institute of Technology, Pasadena, CA.

SPACE (Minkowski).   A flat space of four dimensions, of which three specify the position (x, y, z) of a point in space, and the fourth represents the time t at which an event occurs at that point. Usually, the coordinates in the space are denoted by x_μ ($\mu = 1, 2, 3, 4$) with $x_1 = x$, $x_2 = y$, $x_3 = z$, $x_4 = ict$, where $i^2 = -1$. It is also possible to write $x_4 = ct$, but then it is necessary to associate with the space the metric g^{ij}, where $g^{11} = g^{22} = g^{33} = 1$, $g^{44} = -1$.

SPACE SHUTTLE.   The space shuttle is a two-stage spacecraft designed to transport personnel and cargo between Earth and orbiting satellites or space stations in outer space. The space shuttle program is part of the National Aeronautics and Space Administration (NASA) Space Transportation System (STS).

History

In September 1969, two months after the first manned lunar landing, a Space Task Group appointed by the President of the United States to study the future course of U.S. space research and exploration made the recommendation that... "The United States accept the basic goal of a balanced manned and unmanned space program. To achieve this goal, the United States should ... develop new systems of technology for space operation...through a program directed initially toward development of a new space transportation capability..."

In early 1970, NASA initiated extensive engineering, design, and cost studies of a space shuttle. These studies covered a wide variety of concepts ranging from a fully reusable manned booster and orbiter to dual strap-on solid propellant rocket motors and an expendable liquid propellant tank. Each concept evaluated development risks and costs in relation to the suitability and the overall economics of the entire system.

On January 5, 1972, President Richard M. Nixon announced that NASA would proceed with the development of a reusable low cost space shuttle system. NASA and its aerospace industry contractors continued engineering studies through January and February of 1972; finally on March 15, 1972, NASA announced that the shuttle would use two solid propellant rocket motors. The decision was based on information developed by studies that showed that the solid rocket system offered lower development cost and lower technical risk.

On September 17, 1976, the first suborbital spacecraft, *Enterprise*, was rolled out. A total of thirteen test flights were performed. The *Enterprise* was built as a test vehicle and not equipped for space flight.

Five captive flights, with the *Enterprise* perched atop a 747 jumbo jet with no crew and unpowered, were conducted to test the structural integrity of the craft. Three crewed captive flights followed with the crew operating the flight control systems in preparation for the first orbiter free flight. Finally, five free flights occurred with an astronaut crew separating the orbiter from the 747 shuttle carrier and maneuvering to a landing at Edwards Air Force Base. See Fig. 1.

For all of the captive flights and the first three free flights, the orbiter was outfitted with a tail cone covering its aft section to reduce aerodynamic drag and turbulence. The final two free flights were made without the tail cone, and the three simulated space shuttle main engines and two orbital maneuvering system engines were exposed aerodynamically.

After numerous tests across the United States, the *Enterprise* was ferried across the Atlantic for several air shows across Europe. Finally, on November 18, 1985, the *Enterprise* was ferried from Kennedy Space Center to Washington, DC and became the property of the Smithsonian Institution.

Space Shuttle Overview

The space shuttle is the world's first reusable spacecraft, and the first spacecraft in history that can carry large satellites both to and from orbit. The shuttle launches like a rocket, maneuvers in Earth orbit like a spacecraft and lands like an airplane. Each of the four space shuttle orbiters now in operation, *Columbia, Discovery, Atlantis* and *Endeavour*, is designed to fly at least 100 missions. So far, altogether they have flown a combined total of less than one-fourth of that.

The space shuttle consists of three major components: the orbiter which houses the crew; a large external fuel tank that holds fuel for the main engines; and two solid rocket boosters which provide most of the shuttle's

Fig. 1.   The Shuttle *Enterprise* separates from the 747 shuttle carrier aircraft during the approach and landing tests. (*NASA.*)

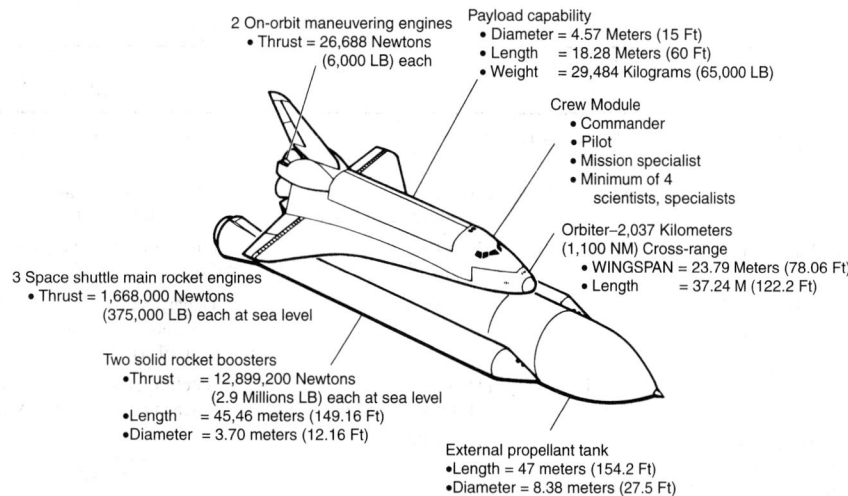

2 On-orbit maneuvering engines
• Thrust = 26,688 Newtons
  (6,000 LB) each

Payload capability
• Diameter = 4.57 Meters (15 Ft)
• Length   = 18.28 Meters (60 Ft)
• Weight   = 29,484 Kilograms (65,000 LB)

Crew Module
• Commander
• Pilot
• Mission specialist
• Minimum of 4
  scientists, specialists

Orbiter–2,037 Kilometers
(1,100 NM) Cross-range
• WINGSPAN = 23.79 Meters (78.06 Ft)
• Length   = 37.24 M (122.2 Ft)

3 Space shuttle main rocket engines
• Thrust = 1,668,000 Newtons
  (375,000 LB) each at sea level

Two solid rocket boosters
•Thrust   = 12,899,200 Newtons
  (2.9 Millions LB) each at sea level
•Length   = 45,46 meters (149.16 Ft)
•Diameter = 3.70 meters (12.16 Ft)

External propellant tank
•Length = 47 meters (154.2 Ft)
•Diameter = 8.38 meters (27.5 Ft)

Fig. 2.   Principal features of the space shuttle (*NASA.*)

lift during the first two minutes of flight. All of the components are reused except for the external fuel tank, which burns up in the atmosphere after each launch. See Fig. 2.

The empty weight at rollout is 158,289 pounds (71,800 kilograms) and 178,000 pounds (80,700 kilograms) with main engines installed. About the size of a commercial DC-9 jetliner, the Orbiter can deliver to orbit single or multiunit payloads up to 65,000 pounds (29,484 kilograms) in a cargo bay that measures 15 × 60 feet (4.5 × 18 meters), and bring back payloads weighing up to 32,000 pounds (14,515 kilograms).

The longest the shuttle has stayed in orbit on any single mission is 17.5 days on mission STS-80 in November 1996. Normally, missions may be planned for anywhere from five to 16 days in duration. The smallest crew ever to fly on the shuttle numbered two people on the first few missions. The largest crew numbered eight people. Normally, crews may range in size from five to seven people. The shuttle is designed to reach orbits ranging from about 185 kilometers to 643 kilometers (115 statute miles to 400 statute miles) high.

The shuttle has the most reliable launch record of any rocket now in operation. Since 1981, it has boosted more than 1.36 million kilograms (three million pounds) of cargo into orbit. More than 600 crewmembers have flown on its missions. Although it has been in operation for over 20 years, the shuttle has continually evolved and is significantly different today than when it first was launched. NASA has made literally thousands of major and minor modifications to the original design that have made it safer, more reliable and more capable today than ever before.

Since 1992 alone, NASA has made engine and system improvements that are estimated to have tripled the safety of flying the space shuttle, and the number of problems experienced while a space shuttle is in flight has decreased by 70 percent. During the same period, the cost of operating the shuttle has decreased by one and a quarter billion dollars annually, a reduction of more than 40 percent. At the same time, because of weight reductions and other improvements, the cargo the shuttle can carry has increased by 7.3 metric tons (8 tons.)

NASA plans to continue to improve the shuttle during the next five years, with goals of increasing its safety by improving the highest risk components. In managing and operating the space shuttle, NASA holds the safety of the crew as the highest priority. NASA is prepared to continue flying the shuttle for at least the next decade.

NASA's Orbiter Fleet

In the day-to-day world of Shuttle operations and processing, Space Shuttle orbiters go by a more prosaic designation. *Columbia* is commonly refereed to as OV-102, for Orbiter Vehicle-102. *Challenger, Discovery, Atlantis,* and *Endeavour* are, respectively, OV-99, OV-103, OV-104, and OV-105.

America's fleet of Space Shuttle orbiters is named after pioneering vessels that established new frontiers in research and exploration. NASA delved through the history books to find ships, which achieved historical significance through discoveries about the world's oceans or the Earth itself. Another important criterion in the selection process was consideration for the international nature of the Space Shuttle program.

Columbia (OV-102). Columbia, the oldest orbiter in the Shuttle fleet and the first to fly in space, is named after the Boston, Massachussetts-based sloop captained by American Robert Gray. On May 11, 1792, Gray and his crew maneuvered the "Columbia" past the dangerous sandbar at the mouth of a river extending more than 1,000 miles (1,610 meters) through what is today southeastern British Columbia, Canada, and the Washington-Oregon border. The river was later named after the ship. Gray also led Columbia and its crew on the first American circumnavigation of the globe, carrying a cargo of otter skins to Canton, China, and then returning to Boston.

Other sailing ships have further enhanced the luster of the name "Columbia." The first U.S. Navy ship to circle the globe bore that title, as did the command module for *Apollo* 11, the first lunar landing mission.

On a more directly patriotic note, "Columbia" is considered to be the feminine personification of the United States. The name is derived from that of another famous explorer, Christopher Columbus.

The first of NASA's orbiter fleet was delivered to Kennedy Space Center in March 1979. After several years of design, construction, and modification, *Columbia* initiated the Space Shuttle flight program when it lifted off Pad A at the Kennedy Space Center's Launch Complex 39 located on Merritt Island, Florida, just north of Cape Canaveral on April 12, 1981 at 7:00 a.m. (EST). Touchdown occurred at Edwards Air Force Base, California, at 10:21 a.m. (PST) on April 14, for a mission span of 54 hours, 21 minutes in an orbit 198 miles (319 kilometers) from Earth. The orbital speed of the ship was 17,000 miles (27,353 kilometers) per hour. Astronauts John W. Young and Robert L. Crippen flew the space shuttle *Columbia* on the first flight of the Space Transportation System (STS-1). *Columbia* became the first airplanelike craft to land from orbit for refurbishment and use on future missions. The STS-1 mission also was the first to employ both liquid and solid propellant rocket engines for the launch of a spacecraft carrying humans.

The second mission occurred with liftoff from the Kennedy Space Center's Launch Complex on November 12, 1981. Scheduled for a total time in space of about five days, the mission had to be shortened because of a malfunction in one of the three fuel cells that generated electrical power for the *Columbia*. During the shortened mission, the abilities of the craft's manipulating arm to place satellites and other objects into space orbit, as well as to retrieve them, were satisfactorily demonstrated, as was the feasibility of a number of scientific experiments placed aboard the craft. For a chronology of *Columbia* (OV-102) flights with primary payloads; see Table 1.

Upgrades and Features

Columbia was the first on-line orbiter to undergo the scheduled inspection and retrofit program. It was transported August 10, 1991, after its completion of mission STS-40, to prime Shuttle contractor Rockwell International's Palmdale, California assembly plant. The oldest orbiter in the fleet underwent approximately 50 modifications, including the addition of carbon brakes, drag chute, improved nose wheel steering, removal of development flight instrumentation and an enhancement of its thermal

TABLE 1. FLIGHTS OF *COLUMBIA* (OV-102)

(1981 to date)

Mission	Crew	Launch Date	Landing Date and Site	Primary Payload
STS-1[1]	Young, Crippen	04/12/1981	04/14/1981 Edwards Air Force Base	Development Flight Instrumentation package
STS-2	Engle, Truly	11/12/1981	11/14/1981 Edwards Air Force Base	Office of Space and Terrestrial Applications-1 (OSTA-1)
STS-3	Lousma, Fullerton	03/22/1982	03/30/1982 White Sands	Office of Space Science- 1 (OSS-1)
STS-4	Mattingly, Hartsfield	06/27/1982	07/04/1982 Edwards Air Force Base	Department of Defense (DOD 82-1) and Continuous Flow Electrophoresis System (CFES)
STS-5[2]	Brand, Overmyer, Lenoir, Allen	11/11/1982	11/16/1982 Edwards Air Force Base	Satellite Business System (CSBS-C) and Canadian Satellite (ANIK C-3)
STS-9[3]	Young, Shaw, Parker, Garriott, Merbold, Lichtenberg	11/28/1983	11/28/1983 Edwards Air Force Base	Spacelab-1
STS-61-C	Gibson, Bolden, Chang-Diaz, Hawley, Nelson, Cenker, Congressman Nelson	01/12/1986	01/18/1986 Edwards Air Force Base	SATCOM Ku-1 (RCA Americom) Satellite
STS-28	Shaw, Richards, Leestma, Adamson, Brown	08/08/1989	08/13/1989 Edwards Air Force Base	Department of Defense (DOD)
STS-32	Brandenstein, Wetherbee, Dunbar, Ivins, Low	01/09/1990	01/20/1990 Edwards Air Force Base	Synchronous Communication Satellite IV-F5 (SYNCOM IV-5) and Long Duration Exposure Facility Retrieval (LDEF)
STS-35	Brand, Gardner, Lounge, Hoffman, Parker, Parise, Durrance	12/02/1990	12/10/1990 Edwards Air Force Base	Ultraviolet and X-ray astronomy (ASTRO-1)
STS-40	O,Connor, Gutierrez, Jernigan, Seddon, Bagian, Gaffney, Hughes-Fulford	06/05/1991	06/14/1991 Edwards Air Force Base	Spacelab Life Sciences-1 (SLS-1)
STS-50	Richards, Bowersox, Dunbar, Baker, Meade, DeLucas, Trinh	06/25/1992	07/09/1992 Kennedy Space Center	United States Microgravity Laboratory-1 (USML-1)
STS-52	Wetherbee, Baker, Veach, Jernigan, Shepard, MacLean	10/22/1992	11/01/1992 Kennedy Space Center	Laser Geodynamic Satellite-II (LAGEOS II) and U.S. Microgravity Payload-1(USMP-1)
STS-55	Nagel, Hendricks, Ross, Precourt, Harris, Walter, Schlegel	04/26/1993	05/06/1993 Kennedy Space Center	Spacelab-D2 (second German-dedicated Spacelab)
STS-58	Blaha, Searfoss, Seddon, McArthur, Wolf, Lucid, Fettman	10/18/1993	11/1/1993 Edwards Air Force Base	Spacelab (SLS-2)
STS-62	Casper, Allen, Thuot, Gemar, Ivins	03/04/1994	03/18/1994 Kennedy Space Center	Office of Aeronautics and Space Technology-2 (OAST-2) and United States Microgravity Payload-2 (USMP-2)
STS-65[4]	Cabana, Halsell, Haib, Thomas, Walz, Chiao, Naito-Mukai	07/08/1994	07/23/1994 Kennedy Space Center	International Microgravity Laboratory-2 (IML-2)
STS-73[5]	Bowersox, Rominger, Thornton, Lopez-Alegria, Coleman, Sacco, Leslie	10/20/1995	11/05/1995 Kennedy Space Center	United States Microgravity Laboratory-2 (USML-2)
STS-75	Allen, Horowitz, Chang-Diaz, Cheli, Hoffman, Nicollier, Guidoni	02/22/1996	03/09/1996 Kennedy Space Center	Tethered Satellite System Reflight (TSS-1R) United States and Microgravity Payload-3 (USMP-3)
STS-78	Henricks, Kregel, Helms, Linnehan, Brady, Favier, Thirsk	06/20/1996	07/07/1996 Kennedy Space Center	Life and Microgravity Spacelab (LMS)
STS-80	Cockrell, Rominger, Jernigan, Jones, Musgrave	11/19/1996	12/07/1996 Kennedy Space Center	Orbiting and Retrievable Far and Extreme Ultraviolet Spectrograph-Shuttle Pallet Satellite II (ORFEUS-SPAS II) and Wake Shield Facility-3 (WSF-3)
STS-83[6]	Halsell, Still, Voss, Thomas, Gernhardt, Crouch, Linteris	04/04/1997	04/08/1997 Kennedy Space Center	Microgravity Science Laboratory-1 (MSL-1)

TABLE 1. (*Continued*)

Mission	Crew	Launch Date	Landing Date Date	Primary Payload
STS-94[7]	Halsell, Still, Voss, Thomas, Gernhardt, Crouch, Linteris	07/01/1997	07/17/1997 Kennedy Space Center	Microgravity Science Laboratory-1 (MSL-1)
STS-87	Lindsey, Kregel, Scott, Chawla, Doi, Kadenyuk	11/19/1997	12/05/1997 Kennedy Space Center	United States Microgravity Payload-4 (USMP- 4) and SPARTAN 201-04 rescue
STS-90	Searfoss, Altman, Linnehan, Williams, Hire, Buckey, Pawelczyk	04/17/1998	05/03/1998 Kennedy Space Center	Final Spacelab Mission
STS-91[8]	Precourt, Pudwill Gorie, Lawrence, Chang-Diaz, Kavandi, Ryumin,	06/02/1998	06/12/1998 Kennedy Space Center	9th and Final Shuttle-*Mir* Docking
STS-93[9]	Collins, Ashby, Hawley, Coleman, Tognini	07/23/1999	07/27/1999 Kennedy Space Center	Chandra X-ray Observatory

[1] The primary mission objectives for STS-1 were to check out all the systems on the space shuttle, accomplish a safe ascent into orbit and to return to Earth for a safe landing. All of these objectives were met successfully.

The main payload carried on STS-1 was a Development Flight Instrumentation package, which contained sensors and measuring devices to record orbiter performance and the stresses that occurred during launch, ascent, orbital flight, descent and landing.

[2] First Shuttle operational mission deployed two commercial communications satellites, ANIK C-3 for TELESAT Canada and SBS-C for Satellite Business Systems. Each equipped with Payload Assist Module-D (PAM-D) solid rocket motor, which fired about 45 minutes after deployment, placing each satellite into highly elliptical orbit.

First scheduled space walk in Shuttle program canceled due to malfunction of space suit.

[3] Flight carried first Spacelab mission and first astronaut to represent European Space Agency (ESA), Ulf Merbold of Germany. ESA and NASA jointly sponsored Spacelab-1 and conducted investigations, which demonstrated capability for advanced research in space. Spacelab is an orbital laboratory and observations platform composed of cylindrical pressurized modules and U-shaped unpressurized pallets, which remain in orbiter's cargo bay during flight.

[4] Payload Specialist Chiaki Mukai became first Japanese woman to fly in space; she also set record for longest flight to date by female astronaut.

[5] Crew took time out from Spacelab work to tape ceremonial first pitch for Game Five of baseball World Series, marking first time the thrower was not actually in the ballpark for the pitch.

[6] STS-83. First flight of the Microgravity Science Laboratory-1 (MSL-1) cut short due to concerns about one of three fuel cells, marking only the third time in the Shuttle program history a mission ended early.

[7] STS-94. Marked the first reflight of same vehicle, crew and payloads, following shortened STS-83 mission in April due to indications of a fuel cell problem.

[8] STS-91. return of 7th and last U.S. astronaut (Andrew S.W. Thomas) to live and work aboard *Mir*. Ninth and Final Shuttle-*Mir* Docking.

[9] STS-93. On the first day of the scheduled five-day mission, the Chandra X-ray Observatory was deployed from *Columbia's* payload bay. Chandra's two-stage Inertial Upper Stage (IUS) propelled the observatory into a transfer orbit of 205 miles by 44,759 miles in altitude. See also **Chandra X-ray Observatory.**

For more detailed information on *Columbia* missions see: *http://spaceflight.nasa.gov/shuttle/archives/year1981.html*

protection system. The orbiter returned to KSC February 9, 1992 to begin processing for mission STS-50 in June of that year.

On October 8, 1994, *Columbia* was transported to Palmdale California for its first orbiter maintenance down time (ODMP.) This orbiter modification and refurbishment time is expected to take approximately six months. About 90 modifications will be made to *Columbia* during its stay in California. Upgrades will be made to the main landing gear thermal barrier, tire pressure monitoring system and radiator drive circuitry. Repairs will be made to the radiators in areas where micrometeorites have made impacts. Other work planned includes intensive structural inspections and the application of an upgraded corrosion control coating on the wings and rudder.

On September 24, 1999, *Columbia* was transported to Palmdale California for its second ODMP. While in California, workers will perform more than 100 modifications on the vehicle. *Columbia* will be the second orbiter outfitted with the multifunctional electronic display system (MEDS) or "glass cockpit". See Fig. 3. During the past year, Shuttle *Atlantis* had the full-color, flat-panel displays installed on its flight deck during an OMDP. The new system improves crew interaction with the orbiter during flight and reduces the high cost of maintaining the outdated electromechanical cockpit displays currently onboard.

While her sister ships are being outfitted with external airlocks in support of the *International Space Station* assembly, *Columbia's* internal airlock will not be removed during this OMDP. Thus, *Columbia* will continue to be able to accommodate payloads requiring the orbiter's 60-foot long cargo bay. Though not currently slated to dock with the *International Space Station*, *Columbia* will be given additional wire harnesses and connectors while at Palmdale to allow installation of the Orbiter Docking System at Kennedy Space Center. This prepares *Columbia* for docking operations with the space station if plans change.

While at Palmdale, *Columbia's* 100 miles of wiring will be given a thorough inspection. This is part of NASA's fleet wide wiring inspection. The wiring problem was first identified on *Columbia* as a result of the STS-93 mission.

Preparation work for an enhanced Global Positioning Satellite system capability will also be performed on *Columbia*. When installed, the new system will more accurately pinpoint the orbiter's location in flight. A space-to-space orbiter radio and wireless video modification will increase communication capabilities for *Columbia's* future crewmembers and space walkers. In addition to scheduled weight saving modifications, *Columbia's* radiators or coolant lines will be enhanced for protection from orbital debris.

On February 21, 2002 *Columbia* resumed operation. Astronauts Scott D. Altman and Duane G. Carey piloted space shuttle *Columbia* (STS-109) on a third servicing mission of the Hubble Space Telescope. Mission Specialist installed new science instruments, the Advanced Camera for Surveys (ACS), new rigid Solar Arrays (SA3), new Power Control Unit (PCU) and a new Cryocooler for the Near Infrared Camera and Multi-Object Spectrometer (NICMOS). STS-109 will also reboost HST to a higher orbit.

Challenger (OV-99). The second orbiter to become operational at Kennedy Space Center was named after the British Naval research vessel *HMS Challenger* that sailed the Atlantic and Pacific oceans during the 1870's. The Apollo 17 lunar module also carried the name of *Challenger*. Like her historic predecessors, Space Shuttle *Challenger* and her crews made significant contributions to America's scientific growth. For a chronology of *Challenger* (OV-99) flights with primary payloads; see Table 2 on p. 3251.

Challenger joined NASA fleet of reusable winged spaceships in July 1982. It flew nine successful Space Shuttle missions. On January 28, 1986,

Fig. 3. This "fish-eye" view illustrates NASA's Multifunction Electronic Display Subsystem (MEDS), otherwise known as "glass cockpit." It represents a number of important modifications that have been accomplished on the Orbiter's flight deck. This photo is actually a recent one of the fixed base Space Shuttle mission simulator in the Johnson Space Center's (JSC) Mission Simulation and Training Facility. The fixed base simulator has been outfitted with MEDS to be used by flight crews for training. (*NASA*.)

the *Challenger* and its seven-member crew were lost 73 seconds after launch when a booster failure resulted in the breakup of the vehicle.

Discovery (OV-103). The third orbiter to become operational at Kennedy Space Center, was named after one of two ships that were used by the British explorer James Cook in the 1770s during voyages in the South Pacific that led to the discovery of the Hawaiian Islands. Another of his ships was the *Endeavour*, the namesake of NASA's newest orbiter.

Cook also used *Discovery* to explore the coasts of southern Alaska and northwestern Canada. During the American Revolutionary War, Benjamin Franklin made a safe conduct with request for the British vessel because of the scientific importance of its research.

Other famous ships have carried the name *Discovery*, including one used to explore Hudson Bay in Canada as well as search for what was hoped to be the northwest passage from the Atlantic to the Pacific in 1610 and 1611. Another, based on whaling ship design, was used by the British Royal Geographical Society for an expedition to the North Pole in 1875. The organization then built another *Discovery* in 1901 to conduct its Antarctic expedition that concluded in 1904. This ship still exists and is being preserved by the Society.

Discovery (OV-103) arrived at Kennedy Space Center in November 1983. It was launched on its first mission, flight STS-41-D, on August 30, 1984. It carried aloft three communications satellites for deployment by its astronaut crew.

Other *Discovery* milestones include the deployment of the Hubble Space Telescope on mission STS-31 in April 1990, the launching of the *Ulysses* spacecraft to explore the Sun's polar regions on mission STS-41 in October of that year and the deployment of the Upper Atmosphere Research Satellite (UARS) in September 1991. For a chronology of *Discovery* (OV-103) flights with primary payloads (see Table 3 on pp. 3251 and 3252).

Upgrades and Features

Discovery benefited from lessons learned in the construction and testing of *Enterprise, Columbia* and *Challenger*. At rollout, its weight was some 6,870 pounds (3,120 kilograms) less than *Columbia*. Two orbiters, *Challenger* and *Discovery*, were modified at KSC to enable them to carry the Centaur upper stage in the payload bay. These modifications included extra plumbing to load and vent Centaur's cryogenic (L02/LH2) propellants (other IUS/PAM upper stages use solid propellants), and controls on the aft flight deck for loading and monitoring the Centaur stage. No Centaur flight was ever flown and after the loss of *Challenger* it was decided that the risk was too great to launch a shuttle with a fueled Centaur upper stage in the payload bay.

On August 25, 1995, *Discovery* underwent a nine-month Orbiter Maintenance Down Period (OMDP) in Palmdale California. The vehicle was outfitted with a 5th set of cryogenic tanks and an external airlock to support missions to the International Space Station. *Discovery* departed Palmdale, CA, riding piggy-back on a modified Boeing 747 at 10:01am EDT June 28, 1996 and arrived at KSC on June 29, 1996. See Fig. 4 on p. 3254.

On February 11, 1997 *Discovery* resumed operations. Astronauts Kenneth D. Bowersox and Scott J. Horowitz piloted space shuttle *Discovery* (STS-82) on a second in a series of planned servicing missions to the orbiting Hubble Space Telescope (HST). Work performed on the telescope will significantly upgrade the scientific capabilities of the HST and keep the telescope functioning smoothly until the next scheduled servicing missions in 2002.

STS-82 demonstrated anew the capability of the Space Shuttle to service orbiting spacecraft as well as the benefits of human spaceflight. The seven-member crew completed servicing and upgrading of the Hubble Space

TABLE 2. FLIGHTS OF *CHALLENGER* (OV-99)
(1983 to 1986)

Mission	Crew	Launch Date	Landing Date Date	Primary Payload
STS-6[1]	Weitz, Bobko, Peterson, Musgrave	04/04/1983	04/09/1983 Edwards Air Force Base	Tracking and Data Relay Satellite-1 (TDRS-1)
STS-7[2]	Crippen, Hauck, Fabian, Ride, Thagard	06/18/1983	96/24/1983 Edwards Air Force Base	Two Communications Satellites ANIK C-2 for TELESAT Canada PALAPA B1 for Indonesia
STS-8[2]	Truly, Brandenstein, Gardner, Bluford, Thornton	08/30/1983	09/05/1983 Edwards Air Force Base	Multipurpose satellite for India (INSAT-1B)
STS-41-B[4]	Brand, Gibson, McCandless, McNair, Stewart	02/03/1984	02/11/1984 Kennedy Space Center	Two satellites deployed WESTAR-VI and PALAPA-B2
STS-41-C[5]	Crippen, Scobee, Nelson, van Hoften, Hart	04/06/1984	04/13/1984 Edwards Air Force Base	Long Duration Exposure Facility (LDEF)
STS-41-G[6]	Crippen, McBride, Sullivan, Ride, Leestma, Garneau, Scully-Power	10/05/1984	10/13/1094 Kennedy Space Center	Earth Radiation Budget Satellite (ERBS) Office of Space and Terrestrial Applications-3 (OSTA-3)
STS-51-B	Overmyer, Gregory, Lind, Thagard, Thornton, Wang, van den Berg	04/29/1985	05/06/1985 Edwards Air Force Base	Spacelab-3
STS-51-F	Fullerton, Bridges, Musgrave, England, Henize, Acton, Bartoe	07/29/1985	08/06/1985 Edwards Air Force Base	Spacelab-2 Life Sciences, Plasma Physics, Astronomy, High-Energy Astrophysics, Solar Physics, Atmospheric Physics and Technology Research
STS-61-A	Hartsfield, Nagel, Buchli, Bluford, Jr., Dunbar, Furrer, Messerschmid, Ockels	10/30/1985	11/06/1985 Edwards Air Force Base	D-1 Spacelab Mission (First German-Dedicated Spacelab)
STS-51-L[7]	Scobee, Smith, Resnik, Onizuka, McNair, Jarvis, McAuliffe	01/28/1986	N/A	Tracking Data Relay Satellite-2 (TDRS-2) and Shuttle-Pointed Tool for Astronomy (SPARTAN-203)

[1] First flight of *Challenger* (OV-99) and first space walk of Shuttle program performed by Peterson and Musgrave, lasting about four hours, 17 minutes.
[2] Sally K. Ride became the first American woman to fly in space.
[3] Guion S. Bluford became the first African-American to fly in space.
First night launch and night landing of a space shuttle.
[4] First untethered space walks by McCandless and Stewart, using manned maneuvering unit, and first Kennedy Space Center landing.
[5] First on-orbit spacecraft repair, and first direct ascent trajectory for Space Shuttle.
[6] First flight to include two women, Sally K. Ride and Kathryn D. Sullivan.
Kathryn D. Sullivan first American woman to walk in space.
[7] Explosion 73 seconds after liftoff claimed crew and vehicle. Shuttle flights halted while extensive investigation into accident and assessment of Shuttle program conducted.

TABLE 3. FLIGHTS OF *DISCOVERY* (OV-103)
(1984 to date)

Mission	Crew	Launch Date	Landing Date Date	Primary Payload
STS-41-D	Hartsfield, Coats, Resnik, Hawley, Mullane, Walker	08/30/1984	09/05/1984 Edwards Air Force Base	Satellite Business System (SBS-D) SYNCOM IV-2 (also known as LEASAT 2) and TELSTAR
STS-51-A[1]	Hauck, Walker, Fisher, Gardner, Allen	11/08/1984	11/16/1984 Kennedy Space Center	Canadian communications satellite (TELESAT-H) and Defense communications satellite (SYNCOM IV-1 also known as LEASAT-1)
STS-51-C[2]	Mattingly, Shriver, Onizuka, Buchli, Payton	01/24/1985	01/27/1985 Kennedy Space Center	Department of Defense (DOD)
STS-51-D	Bobko, Williams, Seddon, Hoffman, Griggs, Walker, Garn	04/12/1985	04/19/1985 Kennedy Space Center	Communications Satellite deployed TELESAT-I (ANIK C-1) and SYNCOM IV-3 failed to deploy
STS-51-G	Brandenstein, Lucid Creighton, Nagel, Thornton, Baudry, Al-Saud	06/17/1985	06/24/1985 Edwards Air Force Base	Three communications satellites deployed: MORELOS-A, for Mexico; ARABSAT-A, for Arab Satellite Communications Organization; and TELSTAR-3D, for AT&T

(continued)

TABLE 3. (*Continued*)

Mission	Crew	Launch Date	Landing Date Date	Primary Payload
STS-51-I[3]	Engle, Covey, van Hoften, Lounge, Fisher	08/27/1985	09/03/1985 Edwards Air Force Base	Three communications satellites deployed: ASC-1, for American Satellite Company; AUSSAT-1, an Australian Communications Satellite; and SYNCOM IV-4
STS-26[4]	Hauck, Covey, Lounge, Nelson, Hilmers	09/29/1988	10/03/1988 Edwards Air Force Base	NASA Tracking and Data Relay Satellite-3 (TDRS-3)
STS-29	Coats, Blaha, Bagian, Buchli, Springer	03/13/1989	03/18/1989 Edwards Air Force Base	Tracking and Data Relay Satellite-4 (TDRS-4)
STS-33[5]	Gregory, Blaha, Musgrave, Carter Jr., Thornton	11/22/1989	11/27/1989 Edwards Air Force Base	Department of Defense (DOD)
STS-31[6]	Shriver, Bolden, Jr., Hawley, McCandless II, Sullivan	04/24/1990	04/29/1990 Edwards Air Force Base	Hubble Space Telescope (HST) deploy
STS-41[7]	Richards, Cabana, Shepherd, Melnick, Akers	10/06/1990	10/10/1990 Edwards Air Force Base	ESA-built Ulysses spacecraft
STS-39	Coats, Hammond, Jr., Bluford Jr., Hieb, Harbaugh, Veach, McMonagle	04/28/1991	05/06/1991 Kennedy Space Center	Dedicated Department of Defense Mission
STS-48[8]	Creighton, Reightler, Jr., Buchli, Gernar, Brown	09/12/1991	09/18/1991 Edwards Air Force Base	Upper Atmosphere Research Satellite (UARS)
STS-42	Grabe, Oswald, Thagard, Hilmers, Readdy, Bondar, Merbold	01/22/1992	01/30/1992 Edwards Air Force Base	International Microgravity Laboratory-1 (IML-1)
STS-53[9]	Walker, Cabana, Bluford Jr., Voss, Clifford	12/02/1992	12/9/1992 Edwards Air Force Base	Department of Defense (DOD)
STS-56	Cameron, Oswald, Foale, Cockrell, Ochoa	04/08/1993	04/17/1993 Kennedy Space Center	Atmospheric Laboratory for Applications and Science-2 (ATLAS-2)
STS-51[10]	Culbertson Jr., Readdy, Newman, Bursch, Walz	09/12/1993	09/22/1993 Kennedy Space Center	Advanced Communications Technology Satellite (ACTS) and Orbiting and Retrievable Far and Extreme Ultraviolet Spectrograph-Shuttle Pallet Satellite (OERFEUS- SPAS)
STS-60[11]	Bolden, Reightler Jr., Davis, Sega, Chang-Diaz, Sergei K. Krikalev (Russia)	02/03/1994	02/11/1994 Kennedy Space Center	Wake Shield Facility-1 (WSF-1) and SPACEHAB pressurized module
STS-64[12]	Richards, Hammond, Jr., Linenger, Helms, Meade, Lee	09/09/1994	09/20/1994 Edwards Air Force Base	Lidar In-space Technology Experiment LITE
STS-63[13]	Wetherbee, Collins, Foale, Voss, Harris, Vladimar G. Titov	02/03/1995	02/11/1995 Kennedy Space Center	SPACEHAB-3 and Mir Rendezvous
STS-70	Henricks, Kregel, Currie, Thomas, Weber	07/13/1995	07/22/1995 Kennedy Space Center	Tracking and Data Relay Satellite-G deployed (TDRS-G)
STS-82[14]	Bowersox, Horowitz, Lee, Hawley, Harbaugh, Smith, Tanner	02/11/1997	02/21/1997 Kennedy Space Center	2nd Hubble Space Telescope (HST) servicing
STS-85[15]	Brown, Rominger, Davis, Curbeam, Robinson, Tryggvason,	08/07/1997	08/19/1997 Kennedy Space Center	Cryogenic Infrared Spectrometers and Telescopes for the Atmosphere-Shuttle Pallet Satellite-2 (CRISTA-SPAS02)
STS-91[16]	Precourt, Pudwill Gorie, Lawrence, Kavandi, Chang-Diaz, Valery Victorovitch Ryumin	06/02/1998	06/12/1998 Kennedy Space Center	9th and Final Shuttle-*Mir* docking
STS-95[17]	Brown, Lindsey, Parazynski, Duque, Robinson, Mukai, John H. Glenn	10/29/1998	11/07/1998 Kennedy Space Center	John Glenn's Flight and SPACEHAB module
STS-96[18]	Rominger, Husband, Ochoa, Jernigan, Barry, Payette, Valery Ivanovich Tokarev	05/27/1999	06/06/1999 Kennedy Space Center,	2nd Space Station Flight (ISS)
STS-103[19]	Brown, Kelly, Smith, Foale, Grunsfeld, Nicollier, Clervoy	12/19/1999	12/27/1999 Kennedy Space Center	Hubble Space Telescope repair-3A
STS-92	Duffy, Melroy, Wakata, Chiao, Wisoff, McArthur, Lopez-Alegria	10/11/2000	10/24/2000 Edwards Air Force Base	*International Space Station* Flight 3A Integrated Truss Structure (ITS) Z1, Pressurized Mating Adapter 3 (PMA 3), Ku-band Communications System, Control Moment Gyros (CMGs)

TABLE 3. (*Continued*)

Mission	Crew	Launch Date	Landing Date Date	Primary Payload
STS-102[20]	Wetherbee, Kelly, Thomas, Richards, Voss, Helms, Usachev	03/08/2001	03/21/2001 Kennedy Space Center	*International Space Station* Flight 5A.1
STS-105[20]	Horowitz, Sturckow, Barry, Forrester	08/10/2001	08/22/2001 Kennedy Space Center	*International Space Station* Flight 7A.1

[1] Allen and Gardner, wearing jet-propelled manned maneuvering units, retrieved two malfunctioning satellites: PALAPA-B2 and WESTAR-VI, both deployed on Mission STS-41-B.

[2] First mission dedicated to Department of Defense. U.S. Air Force Inertial Upper Stage booster deployed and met mission objectives.

[3] Fisher and van Hoften performed two extravehicular activities (EVAs) totaling 11 hours, 51 minutes. Part of time spent retrieving, repairing and redeploying LEASAT-3, deployed on Mission STS-51-D.

[4] Mission marked resumption of Shuttle flights after 1986 STS-51-L *Challenger* accident.

[5] Fifth mission dedicated to Department of Defense.

[6] Primary payload, Hubble Space Telescope, deployed in a 380-statute-mile orbit. See also **Hubble Space Telescope (HST)**.

[7] Primary payload, ESA-built Ulysses spacecraft to explore polar regions of Sun, deployed. See also **Ulysses Mission**.

[8] Primary payload, the Upper Atmosphere Research Satellite (UARS), deployed on the third day of the mission. During its planned 18-month mission, the 14,500-pound observatory will make the most extensive study ever conducted of the Earth's troposphere, the upper level of the planet's envelope of life-sustaining gases which also include the protective ozone layer. UARS has ten sensing and measuring devices: Cryogenic Limb Array Etalon Spectrometer (CLAES); Improved Stratospheric and Mesospheric Sounder (ISAMS); Microwave Limb Sounder (MLS); Halogen Occultation Experiment (HALOE); High Resolution Doppler Imager (HRDI); Wind Imaging Interferometer (WINDII); Solar Ultraviolet Spectral Irradiance Monitor (SUSIM); Solar/Stellar Irradiance Comparison Experiment (SOLSTICE); Particle Environment Monitor (PEM) and Active Cavity Radiometer Irradiance Monitor (ACRIM II).

[9] Final Shuttle flight for Department of Defense (DOD). Classified DOD payload deployed on flight day one, after which flight activities became unclassified.

[10] On Sept. 16, Mission Specialists Newman and Walz performed extravehicular activity (EVA) lasting seven hours, five minutes and 28 seconds. Final in series of generic space walks begun earlier in year. Astronauts also evaluated tools, tethers and foot restraint platform intended for upcoming Hubble Space Telescope servicing mission.

[11] First Shuttle flight of 1994 marked the first flight of Russian cosmonaut Sergei K. Krikalev on U.S. Space Shuttle as first element in implementing Agreement on NASA/Russian Space Agency Cooperation in Human Space Flight.

[12] First untethered U.S. extravehicular activity (EVA) in 10 years.

[13] First Shuttle flight of 1995 included several history- making achievements: First flight of a female Shuttle pilot and, as part of Phase I of international space station program, second flight of Russian cosmonaut on Shuttle and first approach and flyaround by Shuttle with Russian space station *Mir*.

[14] STS-82 demonstrated anew the capability of the Space Shuttle to service orbiting spacecraft as well as the benefits of human spaceflight. Six-member crew completed servicing and upgrading of the Hubble Space Telescope during four planned extravehicular activities (EVAs) and then performed a fifth unscheduled space walk to repair insulation on the telescope.

HST deployed in April 1990 during STS-31. It was designed to undergo periodic servicing and upgrading over its 15-year lifespan, with first servicing performed during STS-61 in December 1993. Hawley, who originally deployed the telescope, operated the orbiter Remote Manipulator System arm on STS-82 to retrieve HST for second servicing at 3:34 a.m. EST, Feb. 13, and positioned it in payload bay less than half an hour later.

[15] STS-85 carried a complement of payloads in the cargo bay that focused on Mission to Planet Earth objectives as well as preparations for International Space Station assembly.

[16] At hatch opening, Andy Thomas officially became a member of Discovery's crew, completing 130 days of living and working on *Mir*. The transfer wrapped up a total of 907 days spent by seven U.S. astronauts aboard the Russian space station as long-duration crewmembers.

[17] The scientific research mission also returned space pioneer John Glenn to orbit — 36 years, eight months and nine days after he became the first American to orbit the Earth. Medical research during the mission included a battery of tests on Payload Specialist Glenn and Mission Specialist Pedro Duque to further research how the absence of gravity affects balance and perception, immune system response, bone and muscle density, metabolism and blood flow, and sleep.

[18] All major objectives were accomplished during the mission. On May 29, Discovery made the first docking to the *International Space Station* (ISS). Rominger eased the Shuttle to a textbook linkup with Unity's Pressurized Mating Adapter #2 as the orbiter and the ISS flew over the Russian-Kazakh border. The 45th space walk in Space Shuttle history and the fourth of the ISS era lasted 7 hours and 55 minutes, making it the second-longest ever conducted. The crew transferred 3,567 pounds of material, including clothing, sleeping bags, spare parts, medical equipment, supplies, hardware and about 84 gallons of water, to the interior of the station. The astronauts also installed parts of a wireless strain gauge system that will help engineers track the effects of adding modules to the station throughout its assembly, cleaning filters and checking smoke detectors. Eighteen items weighing 197 pounds were moved from the station to Discovery for a return to Earth.

[19] The Hubble Space Telescope is alive and well and back on duty after a successful December, 1999 Servicing Mission (SM3A). To prove it, NASA released two stunning images taken by Hubble just two weeks after *Discovery's* Christmas-time service call. *Discovery's* seven-member crew included two Hubble Servicing Mission veterans. What was originally conceived as a mission of preventive maintenance turned more urgent on November 13, 1999 when the fourth of six gyros failed and Hubble temporarily closed its eyes on the Universe. Unable to conduct science without three working gyros, Hubble entered a state of dormancy called safe mode. Essentially, Hubble "went to sleep" while it waited for help. NASA decided to split the Third Servicing Mission (SM3) into two parts, SM3A and SM3B, after the third of Hubble's six gyroscopes failed. In accordance with NASA's flight rules, a "call-up" mission was quickly conceived and developed and executed in a record seven months!

[20] STS-102 Swaps International Space Station Crews: Space Shuttle *Discovery* spent 13 days in orbit, with nearly nine of those days docked to the ISS. While at the orbital outpost, the STS-102 crew attached the Leonardo Multi-Purpose Logistics Module, transferred supplies and equipment to the station. Space walkers spent a total of 15 hours and 26 minutes during two excursions outside the docked complex. The first space walk was the longest in space history.

Discovery delivered the Expedition Two Crew Commander Yury Usachev and Flight Engineers Jim Voss and Susan Helms for its extended stay aboard the space station. It returned to Earth with Expedition One Commander Bill Shepherd, Flight Engineer Sergei Krikalev and Soyuz Commander Yuri Gidzenko, who had spent 4.5 months living on the station.

[21] Space Shuttle *Discovery* spent 12 days in orbit, with eight of those days docked to the ISS. While at the orbital outpost, the STS-105 crew attached the Leonardo Multi-Purpose Logistics Module, transferred supplies and equipment to the station, completed two space walks and deployed a small spacecraft called Simplesat.

Discovery delivered the Expedition Three crew Commander Frank Culbertson, Pilot Vladimir Dezhurov and Flight Engineer Mikhail Tyurin for its extended stay aboard the space station. It returned to Earth with Expedition Two crewmembers Commander Yury Usachev and Flight Engineers Jim Voss and Susan Helms who had spent about five months living on the station.

Fig. 4.   Shuttle returning Piggybacked to a Boeing 747. (*NASA.*)

Telescope during four planned extravehicular activities (EVAs) and then performed a fifth unscheduled space walk to repair insulation on the telescope. See also **Hubble Space Telescope (HST)**.

Atlantis (OV-104). Atlantis, the fourth orbiter to become operational in the fleet launched from KSC, is named after the primary research vessel for the Woods Hole Oceanographic Institute in Massachusetts from 1930 to 1966. The two-masted, 460-ton (417 ton) ketch was the first U.S. vessel to be used for oceanographic research. Such research was considered to be one of the last bastions of the sailing vessel as steam-and-diesel powered vessels dominated the waterways.

The steel-hulled ocean research ship was approximately 140 feet long (42.7 meters) and 29 feet wide (8.84 meters) to add to her stability. She featured a crew of 17 and room for five scientists. The research personnel worked in two onboard laboratories, examining water samples and marine life brought to the surface by two large winches from thousands of feet below the surface. The water samples taken at different depths varied in temperature, providing clues to the flow of ocean currents. The crew also used the first electronic sounding devices to map the ocean floor. See also **Ocean Research Vessels**.

Atlantis (OV-104) was delivered to Kennedy Space Center in April 1985. It lifted off on its maiden voyage on October 3, 1985, on mission STS-51-J, the second dedicated Department of Defense flight. *Atlantis* has carried on the spirit of the sailing vessel with several important voyages of its own, including the Galileo interplanetary probe to Jupiter on STS-34 in October 1989, and STS-37 the deployment of the Arthur Holley Compton Gamma Ray Observatory (CGRO) as its primary payload, in April 1991. For a chronology of *Atlantis* (OV-104) flights with primary payloads, see Table 4. See also **Compton Gamma-Ray Observatory (CGRO)**; and **Galileo Mission to Jupiter**.

Upgrades and Features

Atlantis benefited from lessons learned in the construction and testing of *Enterprise, Columbia,* and *Challenger*. At rollout, its weight was some 6,974 pounds (3,160 kilograms) less than *Columbia*. The Experience gained during the Orbiter assembly process also enabled *Atlantis* to be completed with a 49.5 percent reduction in man-hours (compared to *Columbia*). Much of this decrease can be attributed to the greater use of thermal protection blankets on the upper orbiter body instead of tiles. During the construction of *Discovery* and *Atlantis*, NASA opted to have the various contractors manufacture a set of "structural spares" to facilitate

the repair of an Orbiter if one was damaged during an accident. This contract was valued at $389 million and consisted of a spare aft-fuselage, mid-fuselage, forward fuselage halves, vertical tail and rudder, wings, elevons and a body flap. These spares were later assembled into the orbiter *Endeavour. Atlantis* was shipped to California November 5, 1997 to undergo upgrades and modifications. These modifications included a drag chute, new plumbing lines that configure the orbiter for extended duration, more than 800 new heat protection tiles and blankets and new insulation for the main landing gear doors, structural mods to the *Atlantis* airframe. Altogether, 165 modifications were made to *Atlantis* over the 20 months it spent in Palmdale, California.

On May 19, 2000 *Atlantis* resumed operations. Astronauts Terrence W. Wilcutt and Scott D. Altman piloted space shuttle *Atlantis* (STS-106) on a fourth International Space Station assembly flight (ISS-2A.2b.)

Veteran Astronaut Terry Wilcutt (Col., USMC) leads the seven-man crew, commanding his second Shuttle flight and making his fourth trip into space. During the 11-day mission, Wilcutt and his crew mates spent a week inside the ISS unloading supplies from both a double SPACEHAB cargo module in the rear of *Atlantis's* cargo bay and from a Russian Progress M-1 resupply craft docked to the aft end of the Zvezda Service Module. Zvezda, which linked up to the ISS on July 26, 1999 will serve as the early living quarters for the station and is the cornerstone of the Russian contribution to the ISS.

The goal of the flight is to prepare Zvezda for the arrival of the first resident, or Expedition, and the start of a permanent human presence on the new outpost. Expedition One Commander Bill Shepherd, Soyuz Commander Yuri Gidzenko and Flight Engineer Sergei Krikalev were launched in a Soyuz capsule from the Baikonur Cosmodrome in Kazakhstan on October 31, 2000. On November 2, 2000 the crew docked with the ISS and spent 136 days as the initial or "shake down" crew for the continuous human presence on the International Space Station. During this time, Shepherd kept a log of the activities aboard the orbiting outpost. See also *http://spaceflight.nasa.gov/station/crew/exp1/ex1logs.html*.

See also **Space Stations**.

In addition, Dr. Ed Lu and Yuri Malenchenko conducted a $6\frac{1}{2}$-hour space walk to hook up electrical communications and telemetry cables between Zvezda and the Zarya Control Module, whose computers handed over commanding functions to the Service Module's computers in a smooth transition in late July 2000. Lu and Malenchenko also installed a magnetometer to the exterior of Zvezda. The magnetometer will serve

TABLE 4. FLIGHTS OF *ATLANTIS* (OV-104)

(1985 to date)

Mission	Crew	Launch Date	Landing Date / Date	Primary Payload
STS-51-J[1]	Bobko, Grabe, Hilmers, Stewart, Pailes	10/03/1985	10/07/1985 Edwards Air Force Base	Second mission dedicated to Department of Defense (DOD)
STS-61-B	Shaw, O'Connor, Cleave, Spring, Ross, Vela, Walker	11/26/1985	12/03/1985 Edwards Air Force Base	Three communications satellites deployed: MORELOS-B (Mexico) AUSSAT-2 (Australia) SATCOM KU-2 (RCA Americom)
STS-27	Gibson, Gardner, Mullane, Ross, Shepherd	12/02/1988	12/06/1988 Edwards Air Force Base	Third mission dedicated to Department of Defense (DOD)
STS-30[2]	Walker, Grabe, Thagard, Cleave, Lee	05/04/1989	05/08/1989 Edwards Air Force Base	Magellan/Venus Radar Mapper Spacecraft.
STS-34[3]	Williams, McCulley, Chang-Diaz, Lucid, Baker	10/18/1989	10/23/1989 Edwards Air Force Base	Galileo/Jupiter spacecraft
STS-36	Creighton, Casper, Mullane, Hilmers, Thuot	02/28/1990	03/04/1990 Edwards Air Force Base	Sixth mission dedicated to Department of Defense (DOD)
STS-38	Covey, Culbertson, Springer, Meade, Gemar	11/15/1990	11/20/1990 Kennedy Space Center	Seventh mission dedicated to Department of Defense (DOD)
STS-37[4]	Nagel, Cameron, Ross, Apt, Godwin	04/05/1991	04/11/1991 Kennedy Space Center	Compton Gamma Ray Observatory (CGRO.)
STS-43	Blaha, Baker, Lucid, Adamson, Low	08/02/1991	08/11/1991 Kennedy Space Center	Tracking and Data Relay Satellite-5 (TDRS-5)
STS-44	Gregory, Henricks, Musgrave, Runco, Voss, Hennen	11/24/1991	12/01/1991 Edwards Air Force Base	Dedicated Department of Defense Mission (DOD) and Defense Support Program (DSP) satellite
STS-45[5]	Bolden, Duffy, Sullivan, Leestma, Foale, Lichtenberg, Frimout	03/24/1992	04/02/1992 Kennedy Space Center	Atmospheric Laboratory for Applications and Science-1 (ATLAS-1)
STS-46[6]	Shriver, Allen, Hoffman, Chang-Diaz, Nicollier, Ivins, Malerba	07/31/1992	08/08/1992 Kennedy Space Center	European Space Agency's European Retrievable Carrier (EURECA)
STS-66	McMonagle, Brown, Ochoa, Parazynski, Tanner, Clervoy	11/03/1994	11/14/1994 Edwards Air Force Base	Atmospheric Laboratory for Applications and Science-3 (ATLAS-3)
STS-71[7]	Gibson, Precourt, Baker, Dunbar, Harbaugh	06/27/1995	07/07/1995	1st Shuttle-*Mir* Docking
STS-74[8]	Cameron, Halsell, Ross, McArthur, Hadfield	12/12/1995	12/20/1995 Kennedy Space Center	2nd Shuttle-*Mir* Docking
STS-76[9]	Chilton, Searfoss, Lucid, Godwin, Clifford, Sega	03/22/1996	03/31/1996 Edwards Air Force Base	3rd Shuttle-*Mir* Docking
STS-79[10]	Readdy, Wilcutt, Akers, Blaha, Apt, Walz	09/16/1996	09/26/1996 Kennedy Space Center	4th Shuttle-*Mir* Docking
STS-81[11]	Baker, Jett, Grunsfeld, Ivins, Wisoff, Linenger	01/12/1997	01/22/1997 Kennedy Space Center	5th Shuttle-*Mir* Docking
STS-84[12]	Precourt, Collins, Foale, Noriega, Clervoy, Kondakova	05/15/1997	05/24/1997 Kennedy Space Center	6th Shuttle-*Mir* Docking
STS-86[13]	Wetherbee, Bloomfield, Titov, Parazynski, Chretien, Lawrence, Wolf	09/25/1997	10/06/1997 Kennedy Space Center	7th Shuttle-*Mir* Docking
STS-101[14]	Halsell, Jr., Horowitz, Helms, Usachev, Voss, Weber, Williams	05/19/2000	05/26/2000 Kennedy Space Center	International Space Station (ISS) Flight 2A.2a
STS-106[15]	Wilcutt, Altman, Burbank, Lu, Malenchenko, Mastracchio, Morukov	09/08/2000	09/20/2000 Kennedy Space Center	International Space Station (ISS) Flight 2A.2b
STS-98[16]	Cockrell, Polansky, Curbeam, Jones, Ivins	02/07/2001	02/20/2001 Kennedy Space Center	International Space Station (ISS) Flight 5A

(continued)

TABLE 4. (*Continued*)

Mission	Crew	Launch Date	Landing Date Date	Primary Payload
STS-104[17]	Lindsey, Hobaugh, Gernhardt, Kavandi, Reilly	07/12/2001	07/24/2001 Kennedy Space Center	International Space Station (ISS) Flight 7A

[1]First flight of Space Shuttle *Atlantis*.

[2]First U.S. planetary mission in 11 years; and first on Shuttle. Primary payload, Magellan/Venus radar mapper spacecraft and attached Inertial Upper Stage (IUS), deployed six hours, 14 minutes into flight. IUS first and second stage fired as planned, boosting Magellan spacecraft on proper trajectory for 15-month journey to Venus. See also **Magellan Mission to Venus**; and **Venus**.

[3]Primary payload, Galileo/Jupiter spacecraft and attached Inertial Upper Stage (IUS), deployed six hours, 30 minutes into flight. IUS stages fired, placing Galileo on trajectory for six-year trip to Jupiter via gravitational boosts from Venus and Earth and possible observational brushes with asteroids Gaspra and Ida. See also **Galileo Mission to Jupiter**; and **Jupiter**.

[4]Primary payload, Compton Gamma Ray Observatory (CGRO), deployed on flight day three. CGRO high-gain antenna failed to deploy on command; finally freed and manually deployed by Ross and Apt during unscheduled contingency space walk, first since April 1985. Following day, two astronauts performed first scheduled space walk since November 1985 to test means for astronauts to move themselves and equipment about while maintaining planned Space Station Freedom. See also **Compton Gamma-Ray Observatory (CGRO)**.

[5]Mission marked first flight of Atmospheric Laboratory for Applications and Science-1 (ATLAS-1), mounted on nondeployable Spacelab pallets in orbiter cargo bay. U.S., France, Germany, Belgium, United Kingdom, Switzerland, The Netherlands, and Japan provided 12 instruments designed to perform 14 investigations in four fields.

[6]Primary objective was deployment of the European Space Agency's European Retrievable Carrier (EURECA) and operation of the joint NASA/Italian Space Agency Tethered Satellite System (TSS).

[7]STS-71 marked a number of historic firsts in human spaceflight history: 100th U.S. human space launch conducted from Cape; first U.S. Space Shuttle-Russian Space Station *Mir* docking and joint on-orbit operations; largest spacecraft ever in orbit; and first on-orbit changeout of Shuttle crew.

When linked, *Atlantis* and *Mir* formed largest spacecraft ever in orbit, with a total mass of almost one-half million pounds (about 225 tons) orbiting some 218 nautical miles above the Earth. After hatches on each side opened, STS-71 crew passed into Mir for welcoming ceremony. On same day, *Mir* 18 crew officially transferred responsibility for station to *Mir* 19 crew, and two crews switched spacecraft.

[8]STS-74 marked second docking of U.S. Space Shuttle to Russian Space Station *Mir*, continuing Phase I activities leading to construction of international space station later this decade. Unlike first docking flight during which crew exchange took place, second docking focused on delivery of equipment to Mir. Primary payload of mission was Russian-built Docking Module (DM), designed to become permanent extension on *Mir* to afford better clearances for Shuttle-*Mir* linkups. Two solar arrays were stowed on DM for later transfer to Mir by spacewalking cosmonauts.

[9]Third linkup between U.S. Space Shuttle and Russian Space Station *Mir* highlighted by transfer of veteran Astronaut Shannon Lucid to *Mir* to become first American woman to live on station. Her approximately four-and-a-half month stay also will eclipse long-duration U.S. spaceflight record set by first American to live on *Mir*, Norm Thagard. Lucid will be succeeded by astronaut John Blaha during STS-79 in August, giving her distinction of membership in four different flight crews, two U.S. and two Russian, and her stay on *Mir* kicked off continuous U.S. presence in space for next two years.

[10]STS-79 highlighted by return to Earth of U.S. Astronaut Lucid after 188 days in space, first U.S. crew exchange aboard Russian Space Station *Mir*, and fourth Shuttle-*Mir* docking. Lucid's long-duration spaceflight set new U.S. record as well as world record for a woman. She embarked to *Mir* March 22 with STS-76 mission. Succeeding her on *Mir* for an approximately four-month stay is Blaha, who will return in January 1997 with STS-81 crew; U.S. Astronaut Jerry Linenger will replace him.

[11]First Shuttle flight of 1997 highlighted by return of U.S. Astronaut John Blaha to Earth after 118-day stay aboard Russian Space Station *Mir* and the largest transfer to date of logistics between the two spacecraft. *Atlantis* also returned carrying the first plants to complete a life cycle in space, a crop of wheat grown from seed to seed. This fifth of nine planned dockings continued Phase 1B of the NASA/Russian Space Agency cooperative effort, with Linenger becoming the third U.S. astronaut in succession to live on *Mir*. Same payload configuration flown on previous docking flight featuring SPACEHAB Double module–flown again.

[12]Sixth Shuttle-*Mir* docking highlighted by transfer of fourth successive U.S. crew member to the Russian Space Station. U.S. Astronaut Mike Foale exchanged places with Jerry Linenger, who arrived at *Mir* Jan. 15 with the crew of Shuttle Mission STS-81. Linenger spent 123 days on *Mir* and just over 132 days in space from launch to landing, placing him second behind U.S. Astronaut Shannon Lucid for most time spent on-orbit by an American. Another milestone reached during his stay was one-year anniversary of continuous U.S. presence in space that began with Lucid's arrival at *Mir* March 22, 1996.

Other significant events during Linenger's stay included first U.S.-Russian space walk. On April 29, Linenger participated in five-hour extravehicular activity (EVA) with *Mir* 23 Commander Vasily Tsibliev to attach a monitor to the outside of the station. The Optical Properties Monitor (OPM) was to remain on *Mir* for nine months to allow study of the effect of the space environment on optical properties, such as mirrors used in telescopes.

[13]The seventh *Mir* docking mission continued the presence of a U.S. astronaut on the Russian space station with the transfer of physician David A. Wolf to *Mir*. Wolf became the sixth U.S. Astronaut in succession to live on Mir to continue Phase 1B of the NASA/Russian Space agency cooperative effort.

Foale returned to Earth after spending 145 days in space, 134 of them aboard Mir. His estimated mileage logged was 58 million miles (93 million kilometers), making his the second longest U.S. space flight, behind Shannon Lucid's record of 188 days.

First joint U.S.-Russian extravehicular activity during a Shuttle mission, which was also the 39th in the Space Shuttle program, was conducted by Titov and Parazynski. During the five-hour, one-minute space walk on Oct. 1, the pair affixed a 121-pound Solar Array Cap to the docking module for future use by Mir crewmembers to seal off the suspected leak in Spektr's hull.

[14]Space Shuttle *Atlantis* spent nearly 10 days in space six of which were spent docked with the ISS. While docked with the space station, the crew refurbished and replaced components in both the Zarya and Unity Modules. Voss and Williams performed a 6.5-hour space walk the day after docking to install a Russian Strela cargo boom on the outside of Zarya. They also replaced a faulty radio antenna and performed several other tasks in advance of space walks on future station assembly missions.

[15]The STS-106 crew spent five days, 9 hours and 21 minutes inside the International Space Station. The seven crewmembers completed a long checklist aimed at making the station a home for its first residents. Lu and Malenchenko performed a space walk to connect power, data and communications cables to the newly arrived Zvezda Service Module and the station.

[16]The primary objective of STS-98, International Space Station Assembly Mission 5A, was to deliver and install the U.S. Destiny Laboratory onto the ISS. The centerpiece of research on this world-class scientific orbiting outpost, this workshop in space will support experiments and studies in cancer, diabetes and materials, just to name a few. The arrival of the Destiny Lab brought the space stations mass to about 101.6 metric tons (112 tons), surpassing that of the Russian Mir space station for the first time. See also **Space Stations**.

[17]Top priority of the STS-104 mission of *Atlantis* was installation on the *International Space Station* of the Joint Airlock. This gave the station crewmembers the capability of conducting spacewalks from the orbiting laboratory using either the Russian Orlan spacesuits or U.S. spacesuits. The airlock was attached to the station's Unity Node, on the module's starboard Common Berthing Mechanism, and its "survival heaters" activated. The installation used the station's new Canadarm2 robotic arm during the mission's first spacewalk.

as a three-dimensional compass designed to minimize Zvezda propellent usage by relaying information to the module's computers regarding its orientation relative to the Earth.

Endeavour (OV-105). *Endeavour*, the newest addition to the orbiter fleet, is named after the first ship commanded by James Cook, the 18th century British explorer, navigator and astronomer. On *Endeavour's* maiden voyage in August 1768, Cook sailed to the South Pacific to observe and record the infrequent event of the planet Venus passing between the Earth and the sun. Determining the transit of Venus enabled early astronomers to find the distance of the sun from the Earth, which then could be used as a unit of measurement in calculating the parameters of the universe. Cook also discovered and charted New Zealand, surveyed the eastern coast of Australia, and navigated the Great Barrier Reef there.

Cook's voyage on the *Endeavour* also established the usefulness of sending scientists on voyages of exploration. While sailing with Cook, naturalists Joseph Banks and Carl Solander collected many new families and species of plants, and encountered numerous new species of animals.

Endeavour and her crew reportedly made the first long-distance voyage on which no crewman died from scurvy, the dietary disease caused by lack of ascorbic acids. Cook is credited with being the first captain to use diet as a cure for scurvy, when he made his crew eat cress, sauerkraut, and an orange extract.

The *Endeavour* was small at about 368 tons (334 tons), 100 feet (30.5 meters) in length and 20 feet (6.1 meters) in width. See had a round bluff bow and a flat bottom. The ship's career ended on a reef along Rhode Island.

Authorization to construct the fifth Space Shuttle orbiter as a replacement for Challenger was granted by Congress on August 1, 1987. For the first time, a national competition involving students in elementary and secondary schools produced the name of the new orbiter; it was announced by President George Bush in 1989.

The space shuttle *Endeavour* (OV-105) was delivered to the Kennedy Space Center's (KSC's) Shuttle Landing Facility May 7, 1991, atop NASA's new Shuttle Carrier Aircraft (NASA 911). The space agency's newest orbiter began flight operations on May 7, 1992; STS-49 highlighted by the dramatic rescue of a stranded communications satellite. For a chronology of *Endeavour* (OV-105) flights with primary payloads, see Table 5.

Upgrades and Features

Endeavour features new hardware designed to improve and expand orbiter capabilities. Most of this equipment was later incorporated into the other three orbiters during out-of-service major inspection and modification programs. *Endeavour's* upgrades include:

- A 40-foot diameter (12.2-meter) drag chute that is expected to reduce the orbiter's rollout distance by 1,000 to 2,000 feet (305 to 610 meters).
- The plumbing and electrical connections needed for Extended Duration Orbiter (EDO) modifications to allow up to 28-day missions.
- Updated avionics systems that include advanced general purpose computers, improved inertial measurement units and tactical air navigation systems, enhanced master events controllers and multiplexer-demultiplexers, a solid-state star tracker and improved nose wheel steering mechanisms.
- An improved version of the Auxiliary Power Units (APU's) that provide power to operate the Shuttle's hydraulic systems.

On July 30, 1996 the orbiter *Endeavour* riding atop the modified 747 Shuttle Carrier Aircraft (SCA) arrived at Palmdale, CA, at 3:45 p.m. EDT. *Endeavour* remained in California eight months undergoing numerous modifications and structural inspections as part of its first Orbiter Maintenance Down Period (OMDP). The most significant modification will be in the installation of an external air lock making *Endeavour* capable of docking with the International Space Station.

On January 22, 1998, *Endeavour* returned to space after completing its first (OMDP), becoming the first orbiter other than *Atlantis* to dock with *Mir*. Astronauts Terrence W. Wilcutt and Joe F. Edwards, Jr. piloted NASA's first Shuttle mission of 1998 (STS-89). The continuing cooperative effort in space exploration between the United States and Russia and a joint spacewalk was the focus of the mission. During the mission, more than 7,000 pounds of experiments, supplies and hardware were transferred between the two spacecraft.

Space Transportation System (STS)

Launch Sites. There are two launch sites for the Space Shuttle. Kennedy Space Center (KSC) in Florida is used for launches to place the orbiter in equatorial orbits (around the equator), and Vandenberg Air Force Base launch site in California will be used for launches that place the orbiter in polar orbit missions.

Landing sites are located at the KSC and Vandenberg. Additional landing sites are provided at Edwards Air Force Base in California and White Sands, N.M. Contingency landing sites are also provided in the event the orbiter must return to Earth in an emergency.

Since the late 1960s, Pads A and B at Kennedy Space Center's Launch Complex 39 have served as backdrops for America's most significant manned space flight endeavors; *Apollo*, *Skylab*, *Apollo-Soyuz*, and the *Space Shuttle*.

Located on Merritt Island, Florida, just north of Cape Canaveral, the pads were originally built for the huge *Apollo/Saturn V* rockets that launched American astronauts on their historic journeys to the Moon and back.

TABLE 5. FLIGHTS OF *ENDEAVOUR* (OV-105)

(1992 to date)

Mission	Crew	Launch Date	Landing Date and Site	Primary Payload
STS-49[1]	Brandenstein, Chilton, Thuot, Thornton, Hieb, Akers, Melnick	05/07/1992	05/16/1992 Edwards Air Force Base	Captured and redeployed INTELSAT VI (F-3) satellite
STS-47[2]	Gibson, Brown, Lee, Davis, Apt, Jemison, Mohri	09/12/1992	09/20/1992 Kennedy Space Center	Spacelab- J (SL-J), utilized pressurized Spacelab module
STS-54	Casper, McMonagle, Runco, Harbaugh, Helms	01/13/1993	01/19/1993 Kennedy Space Center	Fifth Tracking and Data Relay Satellite (TDRS-6)
STS-57[3]	Grabe, Duffy, Low, Sherlock, Wisoff, Voss	06/21/1993	01/01/1993 Kennedy Space Center	SPACEHAB-1 and (EURECA) retrieval European Retrievable Carrier
STS-61[4]	Covey, Bowersox, Musgrave, Thornton, Nicollier, Hoffman, Akers	12/02/1993	12/13/1993	Hubble Space Telescope 1st servicing mission
STS-59[5]	Gutierrez, Chilton, Godwin, Apt, Clifford, Jones	04/09/1994	04/20/1994 Edwards Air Force Base	Space Radar Laboratory (SRL-1)
STS-68[6]	Baker, Wilcutt, Jones, Smith, Bursch, Wisoff	09/30/1994	10/11/1994 Edwards Air Force Base	Space Radar Laboratory (SLR-2)
STS-67[7]	Oswald, Gregory, Jernigan, Grunsfeld, Lawrence, Parise, Durrance	03/02/1995	03/18/1995 Edwards Air Force Base	ASTRO-2

(continued)

TABLE 5. (*Continued*)

Mission	Crew	Launch Date	Landing Date Date	Primary Payload
STS-69[8]	Walker, Cockrell, Voss, Newman, Gernhardt	09/07/1995	09/18/1995 Kennedy Space Center	Spartan 201-03, deployed and Wake Shield Facility-2 (WSF-2) deployed
STS-72[9]	Duffy, Jett, Chiao, Barry, Scott, Wakata	01/11/1996	01/20/1996 Kennedy Space Center	Space Flyer Unit (SFU) retrieval and Office of Aeronautics and Space Technology-Flyer (OAST-Flyer)
STS-77	Casper, Brown, Bursch, Runco, Garneau, Thomas	05/19/1996	05/29/1996 Kennedy Space Center	SPACEHAB-4 pressurized research module and Spartan free-flyer deployed
STS-89[10]	Wilcutt, Edwards, Dunbar, Anderson, Reilly, Sharipov, Thomas	01/22/1998	01/31/1998 Kennedy Space Center	8th Shuttle-*MIR* Docking
STS-88[11]	Cabana, Sturckow, Ross, Currie, Newman, Krikalev	12/04/1998	12/15/1998 Kennedy Space Center	First International Space Station Flight UNITY Connecting Module
STS-99[12]	Kregel, Gorie, Thiele, Kavandi, Voss, Mohri	02/11/2000	02/22/2000 Kennedy Space Center	Shuttle Radar Topography Mission
STS-97[13]	Jett, Bloomfield, Tanner, Noriega, Garneau	11/30/2000	12/11/2000 Kennedy Space Center	ISS Assembly Flight 4A P6 Integrated Truss Structure
STS-100[14]	Rominger, Ashby, Hadfield, Phillips, Parazynski, Guidoni, Rosaviakosmos	04/19/2001	05/01/2001 Kennedy Space Center	ISS Assembly Flight 6A. Canadarm2 robotic arm
STS-108[15]	Gorie, Kelly, Godwin, Tani,	12/05/2001	12/17/2001	International Space Station Flight UF-1 Raffaello Multi-Purpose Logistics Module and STARSHINE 2

[1]First flight of Space Shuttle *Endeavour*.

First U.S. orbital flight to feature four extravehicular activities (EVAs), two of these longest in U.S. space flight history to date (eight hours, 29 minutes and seven hours, 45 minutes), and longest to date by a female Astronaut. First space flight ever to involve three crewmembers simultaneously working outside spacecraft; first time astronauts attached live rocket motor to orbiting satellite.

[2]Mae C. Jemison first African-American woman to fly in space, and Mamoru Mohri the first Japanese to fly on Shuttle.

[3]STS-57 marked first flight of commercially developed SPACEHAB, pressurized laboratory designed to more than double pressurized workspace for crew-tended experiments. Altogether 22 experiments were flown, covering materials and life sciences, and wastewater recycling experiment for space station.

[4]Final Shuttle flight of 1993 was one of most challenging and complex manned missions ever attempted. During record five back-to-back space walks totaling 35 hours and 28 minutes, two teams of astronauts completed first servicing of Hubble Space Telescope (HST). In many instances, tasks completed sooner than expected and few contingencies that did arise handled smoothly. See also **Hubble Space Telescope (HST)**.

[5]First flight of Toughened Uni-Piece Fibrous Insulation, known as TUFI, an improved thermal protection tile. Several test tiles were placed on orbiter's base heat shield between three main engines.

[6]STS-68 marked second flight in 1994 of Space Radar Laboratory (first flight was STS-59 in April), part of NASA's Mission to Planet Earth.

[7]STS-67 became first advertised Shuttle mission connected to Internet. Users of more than 200,000 computers from 59 countries logged on to Astro-2 home page at Marshall Space Flight Center; more than 2.4 million requests were recorded during mission, many answered by crew on-orbit.

Endeavour logged 6.9 million miles (11 million kilometers) in completing longest Shuttle flight to date, allowing sustained examination of "hidden universe" of ultraviolet light. Primary payload, Astro Observatory, flown once before (on STS-35 in December 1990) but second flight had almost twice the duration. Planned Astro-2 observations built on discoveries made by Astro-1, as well as seeking answers to other questions.

[8]STS-69 marked first time two different payloads were retrieved and deployed during same mission. Also featured an extravehicular activity to practice for International Space Station activities and to evaluate space suit design modifications.

[9]First Shuttle flight of 1996 highlighted by retrieval of a Japanese satellite, deployment and retrieval of a NASA science payload, and two spacewalks.

[10]Docking of *Endeavour* to *Mir* occurred at 3:14 p.m., Jan. 24, at an altitude of 214 nautical miles. Hatches opened at 5:25 p.m. the same day. Transfer of Andy Thomas to *Mir* and return of David Wolf to the U.S. orbiter occurred at 6:35 p.m., Jan. 25. Initially, Thomas thought his Sokol pressure suit did not fit, and the crew exchange was allowed to proceed only after Wolf's suit was adjusted to fit Thomas. Once on Mir, Thomas was able to make adequate adjustments to his own suit (which would be worn should the crew need to return to Earth in the Soyuz capsule) and this remained on *Mir* with him. Wolf spent a total of 119 days aboard *Mir*, and after landing his total on-orbit time was 128 days.

[11]The STS-88 "Unity" mission is the first manned International Space Station assembly flight. The primary mission objective is to rendezvous with the already launched Zarya control module and successfully attach the Unity connecting module, providing the foundation for future ISS components.

[12]The primary objective of the Shuttle Radar Topography Mission is to acquire a high-resolution topographic map of the Earth's land mass (between 60 °N and 56 °S) and to test new technologies for deployment of large rigid structures and measurement of their distortions to extremely high precision.

The Shuttle Radar Topography Mission represents a breakthrough in the science of remote-sensing and will produce topographic maps of Earth 30 times as precise as the best global maps in use today. The information will be used to attempt to produce one of the most comprehensive and accurate maps of Earth ever assembled.

In addition, this mission offers a number of applications for data products and science, including: geology, geophysics, earthquake research, volcano monitoring; hydrologic modeling; ecology; co-registration and terrain correction of remotely-acquired image data; atmospheric modeling; flood inundation modeling; urban planning; natural hazard consequence assessments; fire spread models; and transportation/infrastructure planning.

[13]STS-97 delivers giant solar arrays to International Space Station. During its 11-day mission, the crew of Space Shuttle *Endeavour* saw the ISS spread its wings, giant solar arrays that quintupled the station's electrical power.

The 73-meter (240-foot) long solar array structure attached and unfolded by *Endeavour's* international crew of five, is the longest human-made object ever to fly in space. *Endeavour* carried aloft the U.S.-developed solar arrays, associated electronics, batteries, cooling radiator and support structure. The entire 15.4-metric ton (17-ton) package is called the P6 Integrated Truss Segment, and is the heaviest and largest element yet delivered to the station aboard a space shuttle.

The addition of the huge solar arrays clearly distinguishes the ISS from any predecessor spacecraft. The arrays provide the station with more electrical power, a key to successful modern research.

[14]*Endeavour* and its crew spent almost 12 days on orbit, eight of which were spent in joint operations with the ISS crew. *Endeavour's* crew delivered and installed a new robotic arm (Candarm2) and helped to transfer equipment and supplies between vehicles.

Mission Specialists Chris Hadfield of the Canadian Space Agency and Scott Parazynski of NASA performed two spacewalks to install the new 17.6-meter (57.7-foot) robotic arm onto the ISS. Canadarm2, a beefier second-generation version of the shuttle's robot arm, is essential to the continued assembly of the space station at the outpost grows beyond the reach of the shuttle's arm.

[15]Top priorities for the STS-108 (UF-1) mission of *Endeavour* are rotation of the International Space Station Expedition Three and Expedition Four crews, bringing water, equipment and supplies to the station and completion of spacewalk and robotics tasks.

Following the joint U.S.-Soviet *Apollo-Soyuz* Test Project mission of July 1975, the pads were modified to support Space Shuttle operations.

Both pads were designed to support the concept of mobile launch operations, in which space vehicles are assembled and checked out in the protected environment of the Vehicle Assembly Building, then transported by large tracked vehicles to the launch pad for final processing and launch. During the *Apollo* era, key pad service structures were mobile. For the Space Shuttle, two permanent service towers were installed at each pad for the first time, the Fixed Service Structure and the Rotating Service Structure.

On April 12, 1981, Shuttle operations commenced at Pad A with the launch of Columbia on STS-1. After 23 more successful launches from A, the first Space Shuttle to lift off from Pad B was the ill-fated *Challenger* in January 1986. Pad B was designated for the resumption of Shuttle flights in September 1988, followed by the reactivation of Pad A in January 1990.

Major Features. Both pads are octagonally shaped and share identical features. Pad A is located 18,159 feet (5,530 meters) from the Vehicle Assembly Building via the crawlerway, Pad B 22,400 feet (6,830 meters). The pads are 8,716 (2,660 meters) feet apart. Each pad covers about a quarter-square mile of land. Launches are conducted from atop a concrete hardstand 390 by 325 feet (119 by 99 meters), located at the center of the pad area. The Pad A and Pad B hardstands are 48 feet (14.6 meters) and 55 feet (16.8 meters) above sea level, respectively.

Space Shuttle Requirements. The Shuttle will transport cargo into near Earth orbit 100 to 217 nautical miles (115 to 250 statute miles) above the Earth. This cargo, or payload, is carried in a bay 15 feet (4.57 meters) in diameter and 60 feet (18.3 meters) long.

Major system requirements are that the orbiter and the two solid rocket boosters be reusable.

Other features of the Shuttle:

- The orbiter has carried a flight crew of up to eight persons. A total of 10 persons could be carried under emergency conditions.
- The basic mission is seven days in space.
- The crew compartment has a shirtsleeve environment, and the acceleration load is never greater than 3 Gs.
- In its return to Earth, the orbiter has a cross-range maneuvering capability of 1,100 nautical miles (1,265 statute miles).

The Space Shuttle is launched in an upright position, with thrust provided by the three Space Shuttle engines and the two SRBs. See Fig. 5. After about two minutes, the two solid rocket boosters are spent and are separated from the external tank. See Fig. 6. They fall into the ocean at predetermined points and are recovered. A homing device on each booster guides recovery craft to it for towing the equipment back to shore. The boosters are then refurbished, refueled, and made ready for another flight.

Fig. 5. NASA's re-useable space vehicle, the Space Shuttle, takes off like a rocket, operates in orbit like a spacecraft, then lands like an aircraft. Here it lifts off the launchpad at Kennedy Space Center near Orlando, Florida. At this moment five rockets are firing: the Shuttle's three main engines and two solid rocket boosters. Together they are generating almost seven million pounds of thrust at the moment of lift-off. (*JPL/NASA.*)

Fig. 6.   After lift-off, the solid rocket boosters attached to the sides of the large, reddish, external fuel tank continue to burn for approximately two minutes. When their fuel is spent, they separate from the tank, fall back into the sea, and are retrieved for re-use on subsequent Shuttle missions. The Shuttle's three main engines continue to fire, using half a million gallons of liquid hydrogen and liquid oxygen. Between 8 and 9 minutes into the flight Earth orbit is achieved, the external tank is cast away, and the astronauts use small maneuvering engines to ease the Shuttle into its final orbit. (*JPL/NASA.*)

The Space Shuttle main engines continue firing for about eight minutes. They shut down just before the craft is inserted into orbit. The external tank is then separated from the orbiter. It follows a ballistic trajectory into a remote area of the ocean but is not recovered.

There are 38 primary Reaction Control System (RCS) engines and six vernier RCS engines located on the orbiter. The first use of selected primary reaction control system engines occurs at orbiter/external tank separation. The selected primary reaction control system engines are used in the separation sequence to provide an attitude hold for separation. Then they move the orbiter away from the external tank to ensure orbiter clearance from the arc of the rotating external tank. Finally, they return to an attitude hold prior to the initiation of the firing of the Orbital Maneuvering System (OMS) engines to place the orbiter into orbit.

The primary and/or vernier RCS engines are used normally on orbit to provide attitude pitch, roll and yaw maneuvers as well as translation maneuvers.

The two OMS engines are used to place the orbiter on orbit, for major velocity maneuvers on orbit and to slow the orbiter for reentry, called the deorbit maneuver. Normally, two OMS engine thrusting sequences are used to place the orbiter on orbit, and only one thrusting sequence is used for deorbit.

The orbiter's velocity on orbit is approximately 25,405 feet per second (17,322 statute miles per hour). The deorbit maneuver decreases this velocity approximately 300 fps (205 mph) for reentry.

In some missions, only one OMS thrusting sequence is used to place the orbiter on orbit. This is referred to as direct insertion. Direct insertion is a technique used in some missions where there are high-performance requirements, such as a heavy payload or a high orbital altitude. This technique uses the Space Shuttle main engines to achieve the desired apogee (high point in an orbit) altitude, thus conserving orbital maneuvering system propellants. Following jettison of the external tank, only one OMS thrusting sequence is required to establish the desired orbit altitude.

The crew then begins their payload operations performing numerous assigned tasks, depending upon the purpose of the mission, such as retrieving satellites from Earth's orbit, servicing orbiting satellites, conducting experiments in space, and studying the Earth and deep space. See Fig. 7.

For deorbit, the orbiter is rotated tail first in the direction of the velocity by the primary reaction control system engines. Then the OMS engines are used to decrease the orbiter's velocity.

During the initial entry sequence, selected primary RCS engines are used to control the orbiter's attitude (pitch, roll, and yaw). As aerodynamic

Fig. 7.   In this image, a space-suited astronaut prepares a satellite for its release from the Shuttle's cargo bay. The bay, 18.3 meters long and 4.6 meters wide (60 feet by 15 feet), has payload attachment points along its full length, and is adaptable enough to accommodate as many as five unmanned spacecraft of various sizes and shapes in one mission. The cargo bay is equipped also with a 15 meter (50 foot) robot arm for lifting and releasing, or grasping and retrieving satellites in space. (*JPL/NASA.*)

pressure builds up, the orbiter flight control surfaces become active and the primary reaction control system engines are inhibited.

During entry, the thermal protection system covering the entire orbiter provides the protection for the orbiter to survive the extremely high temperatures encountered during entry. Portions of the orbiter's exterior will reach temperatures of up to 2300 °F (1260 °C). The thermal protection system is reusable (it does not burn off or ablate during entry).

The unpowered orbiter glides to Earth and lands on a runway like an airplane. Nominal touchdown speed varies from 184 to 196 knots (213 to 225 miles per hour). See Fig. 8.

The main landing gear wheels have a braking system for stopping the orbiter on the runway, and the nose wheel is steerable, again similar to a conventional airplane.

Main Engine. The shuttle's main engine, along with two solid rocket boosters, is the most advanced liquid-fueled rocket engine built to date. With variable thrust permitting the engine thrust to be tailored to the mission needs, it can operate effectively at both high and low altitudes. This system can operate up to $7\frac{1}{2}$ hours of accumulated firing time before major maintenance or overhaul is required. It is expected that the main engine will be reusable for up to 55 separate shuttle missions.

Three main engines are mounted on the orbiter aft fuselage in a triangular pattern. The engines are so spaced that they can be gimbaled during flight and, in conjunction with the two solid rocket-boosters, are used to steer the shuttle during flight as well as to provide thrust for launch. Fuel for the engines, liquid hydrogen and liquid oxygen, is carried to be supplied from the tank at a rate of about 45,283 gallons (63,588 liters) per minute of hydrogen and about 16,800 gallons (63,588 liters) per minute of oxygen.

Numerous features have been designed into the engine to satisfy the performance, life, reliability, and maintainability requirements of the shuttle. One feature includes the use of a stage combustion power cycle coupled with high combustion chamber pressures. In the staged combustion

cycle, the liquid hydrogen is partially burned at high pressure and relatively low temperature in preburners and then completely combusted at high temperature and pressure in the main combustion chamber before expanding through the high-area ratio nozzle. The rapid mixing of propellants under these conditions is so complete that a combustion efficiency of about 99% is attained. The engine also uses hydrogen fuel to cool all combustion devices in direct contact with high-temperature combustion products, thereby contributing to long engine life.

Each engine has three primary levels of thrust or power: minimum, rated, and full power. Engine thrust, however, can be varied throughout the range from minimum to full power level, depending on mission needs. On some shuttle flights where heavy payloads dictate an extra measure of power, up to full power level thrust can be commanded. This level equals 109% of rated power. During the latter part of ascent, engine thrust is reduced to ensure that an acceleration force of no more than three times that of Earth's gravity is reached. This acceleration level, permitted by the throttleable shuttle engines, is about one-third the acceleration experienced in prior manned space flights and is well under the physical stress limits of non-astronaut scientists who will fly aboard the space shuttle. The lowest throttle setting, the minimum power level, equals 65% of rated power.

Solid Rocket Boosters. Two solid rocket boosters are used for each shuttle flight to provide, along with the orbiter main engines, the initial ascent thrust to lift the shuttle with its payload from the launch pad to an altitude of about 27.5 miles (44 kilometers). Prior to launch, the entire shuttle weight is supported by the two boosters. Each solid rocket booster is made up of six subsystems: the solid rocket motor, structures, thrust vector control, separation, recovery, and electrical and instrumentation subsystems. The overall length of the solid rocket booster is 149.1 feet (45.5 meters). The diameter is 12.2 feet (3.7 meters).

The heart of each booster is the motor, the largest solid rocket motor ever flown to date and the first designed for reuse. The motor is made up of

Fig. 8.   The return to Earth at the end of a mission is a critical and demanding time for the Shuttle crew. Astronauts slow the craft and ease it into Earth's atmosphere by using small rockets in the craft's nose and tail. Then after re-entry, they steer it like an airplane using rudders and flaps as the craft glides Earthward. Here, after having slowed down from 1,000 to 100 meters per second (3,280 to 328 feet per second), and withstood temperatures of up to 1000 °C (1830 °F), the Space Shuttle *Columbia* touches down at Edwards Air Force Base, thirty busy minutes after having left Earth orbit. (*JPL/ NASA.*)

eleven segments joined together to make four loading segments, which are filled with propellant at the manufacturer's site. The segmented design permits ease of fabrication, handling and transportation. The segments are shipped from the manufacturer's site to the launch sites by rail in specially built canisters carried on a flat rail car. At the launch site, they are assembled to make up a complete motor.

Propellant loading of the motor segments is completed in pairs from batches of propellant ingredients to minimize any thrust imbalances between boosters used for a single shuttle flight. Propellant loading, using different internal propellant shapes (cores), is done in such a way that will cause a regressive thrust 55 seconds into the shuttle flight. This prevents overstressing the shuttle vehicle during the critical phase of flight, which is called the period of maximum dynamic pressure. Each motor, when assembled, contains about 1.1 million pounds (.5 million kilograms) of propellant, and at launch develops a thrust of 2.9 million pounds (1.3 million kilograms).

The exhaust nozzle in the aft segment of each motor, in conjunction with the orbiter engines, steers the shuttle during flight. The nozzle can be moved up to 6.65 degrees by the booster's hydraulically operated thrust vector control subsystem. The latter is controlled by the orbiter's guidance and control computer.

Throughout the flight, measurements are taken to verify proper booster performance. The signals are routed to the orbiter for data recording and transmission to the ground. Electrical power for the solid rocket booster subsystems is supplied from the orbiter fuel cells through interconnect cabling from the external tank.

At burnout, the two solid rocket boosters are separated from the external tank by pyrotechnic (explosive) devices, and moved away from the shuttle vehicle by eight separation motors, four housed in the forward nose frustum and four on the aft skirt. Each of the eight separation motors, fired at solid rocket motor burnout, develops a thrust of 22,000 pounds (9,900 kilograms) for a duration of a little more than one-half second, just sufficient to move the boosters away from the still-accelerating orbiter and external tank. Also a part of the system is a device for separating electrical interconnection with the external tank.

The facility for providing the method to control the boosters' final descent velocity and altitude after separation is the recovery subsystem. This system in the forward section of each booster and within the nose cap consists of parachutes and location aids to help in the search and retrieval operations for each expended booster and its parachute.

Following separation and entry into the lower atmosphere at about 15,420 feet (4,700 meters), each booster is slowed by a pilot and drogue parachute, and finally by three main parachutes, each 104 feet (31.7 meters) in diameter. Impact with the water is aft end first about 160 miles (257.5 kilometers) downrange, at a speed of about 60 miles (95.5 kilometers) per hour. By entering the water in this way, the air in the hollow boosters is trapped and compressed, causing the boosters to float with the forward end out of the water. At booster impact, the main parachutes are disconnected and the direction finding beacons and lights are actuated to guide recovery craft to the boosters and parachutes. The parachutes are picked up by the recovery craft, and the boosters are towed to shore, where they are disassembled and refurbished. The motor segments are shipped to the manufacturer by rail for refurbishment and reloading for a subsequent shuttle flight. The other systems are refurbished either at the launch site or at the respective manufacturers' locations. The thrust vector control subsystem, structural subsystem, and the electrical subsystems are planned for 20 flights and the recovery subsystem for 10 flights.

The separation system is not planned for reuse. At the launch site, the two boosters are assembled vertically on the launch platform. Following assembly of the boosters, the external tank is attached to the boosters, and finally the orbiter and payload to the external tank, thus making the shuttle ready for checkout and another flight.

External Tank. This tank has two principal roles in the space shuttle program: (1) to contain and deliver quality propellants, liquid hydrogen and liquid oxygen, to the engines; and (2) to serve as the structural backbone of the shuttle during launch operations.

The external tank is composed of essentially two tanks (a large hydrogen tank and a smaller oxygen tank) joined together to form one large propellant storage container that is 154.2 feet (47 meters) in length and 27.5 feet (8.4 meters) in diameter. See Fig. 7. The oxygen tank is the forward portion of the external tank and, when loaded, contains 1,331,783 pounds (603,983 kilograms) of liquid oxygen. By comparison, the oxygen tank has considerably more volume than that of a house with a floor square area of 2,000 square feet (186 square meters). The forward end of the oxygen tank curves to a point to reduce aerodynamic drag and to provide lightning protection for the shuttle vehicle once the shuttle has cleared the launch pad. Prior to launch, the launch tower provides lightning protection.

The liquid hydrogen tank, aft of the oxygen tank, is about 2.5 times larger than the oxygen tank. In this tank is stored approximately 223,814 pounds (101,503 kilograms) of cold liquid hydrogen (approximately $-240\,°F$ or $-251\,°C$).

The intertank joins the two tanks and provides a protective compartment to house some of the instrumentation components in the space between the two propellant tanks. For launch, the external tank supports the orbiter and solid rocket boosters at attach points on the tank. Since the thrust is generated by the orbiter main engines and the solid rocket boosters, the external tank must absorb the thrust loads for the shuttle during launch. The intertank takes the major thrust loads from the solid rocket boosters; the orbiter main engine thrust loads are transferred through other attach fittings on the tank.

Much of the outer surface of the tank is protected thermally. Spray-on foam insulation is applied over the forward portion of the oxygen tank, the intertank, and the sides of the hydrogen tank. The foam insulation is required for reducing ice or frost formation on the tank during launch preparation, thus protecting the orbiter insulation from free-falling ice during flight. Additionally, it serves to minimize heat leaks into the tank, which would cause excessive boiling of the liquid propellants, and to prevent condensation and solidification of the air next to the tank.

An ablating material (a substance that *chars away*) is applied to the external tank bulges and projections to protect them from aerodynamic heating during flight through the atmosphere. Sometimes a combination foam and *ablator* is used where both heating and insulation protection are needed. Protection is also required in areas where exhausts of launch engines provide high radiant energy to the tank and where separation motor exhaust plumes may strike the tank.

The external tank, having no electrical power, obtains required electrical energy from the orbiter fuel cells. It does, however, provide the needed cabling to carry power and signals to the external tank electronics and instrumentation components and to the two solid rocket boosters. Fluid controls and valves, except the vent valves, for operation of the engines are located in the orbiter. This is done to minimize throwaway costs inasmuch as the external tank is not reused.

During flight, gases supplied from the three engines pressurize the two tanks. Pressurization is required for structural support of the tank and for operating pressure requirements of the engine pumps. Near the end of the launch phase of the shuttle mission, when the orbiter is just short of orbital velocity, the main engines are cut off. After about ten seconds, the external tank is severed from its attachment to the orbiter, playing a totally passive role in the separation sequence. Just prior to separation, the external tank tumbling system is activated, opening a valve and venting the oxygen tank through the nose cap. This causes the external tank to pitch away from the orbiter and begin to tumble at a rate that will assure that the tank will break up upon reentry and fall within the designated ocean impact area.

Orbiter Structure. The orbiter structure is divided into nine major sections: the forward fuselage, which consists of upper and lower sections fit around a pressurized crew compartment; wings; midfuselage; payload bay doors; aft fuselage; forward reaction control system; vertical tail; orbital maneuvering system/reaction control system pods; and body flap. The majority of the sections are constructed of conventional aluminum and protected by reusable surface insulation.

The forward fuselage structure is composed of 2024 aluminum alloy skin-stringer panels, frames, and bulkheads. The crew compartment is supported within the forward fuselage at four attachment points, which minimizes thermal conductivity, and is welded to create a pressure-tight vessel. The two major attach points are located at the aft end of the crew compartment at the flight deck floor level. The three-level compartment has a wide hatch for normal passage and hatches in the airlock to permit extravehicular and intravehicular activities. The side hatch can be jettisoned.

Redundant pressure windowpanes are provided in the six forward windshields, the two overhead viewing windows, the two aft viewing windows and the side hatch windows. Approximately 300 penetrations in the pressure shell are sealed with plates and fittings. A large removable panel in the aft bulkhead provides access to the interior of the crew compartment during initial fabrication and assembly and provides for airlock installation and removal.

The crew compartment is configured to accommodate a crew of four on the flight deck and three to four in the middeck. The crew compartment is pressurized to 14.7 psia, plus or minus 0.2 psia, and is maintained at an 80-percent nitrogen and 20-percent oxygen composition that provides a shirtsleeve environment for the flight crew. The crew compartment is designed for 16 psia. The crew compartment's volume with the airlock in the middeck is 2,325 cubic feet (69.7 cubic meters). If the airlock is in the payload bay, the crew compartment's cabin volume is 2,625 cubic feet (78.7 cubic meters). The crew cabin arrangement consists of a flight deck, middeck and lower level equipment bay in a combination of living, working, and storage areas.

The flight deck is the uppermost compartment of the cabin. The commander's and pilot's workstations are positioned side by side in the forward portion of the flight deck. These stations have controls and displays for maintaining autonomous control of the vehicle throughout all mission phases. Directly behind and to the sides of the commander and pilot centerline are the mission specialist seats. The aft flight deck has two overhead and aft viewing windows for viewing orbital operations.

The aft flight deck station also contains displays and controls for executing attitude or translational maneuvers for rendezvous, station keeping, docking, payload deployment and retrieval, payload monitoring, remote manipulator system controls and displays, payload bay door operations and closed-circuit television operations. The aft flight deck is approximately 40 square feet (3.6 square meters). The forward flight deck, which includes the center console and seats, is approximately 24 square feet (2.2 square meters). However, the side console controls and displays add approximately 3.5 square feet more (0.3 square meters).

Directly beneath the flight deck is the mid-deck. Access to the mid-deck is through two interdeck openings, which measure 26 by 28 inches (650 by 700 millimeters). Normally, the right interdeck opening is closed and the left is open. A ladder attached to the left interdeck access allows easy passage in 1-g conditions. The mid-deck provides crew accommodations and contains three avionics equipment bays. The two forward avionics bays utilize the complete width of the cabin and extend into the mid-deck 39 inches (975 millimeters) from the forward bulkhead. The aft bay extends into the mid-deck 39 inches (975 millimeters) from the aft bulkhead on the right side of the airlock. Just forward of the waste management system is the side hatch. The completely stripped mid-deck is approximately 160 square feet (14.4 square meters); the gross mobility area is approximately 100 square feet (9 square meters).

The side hatch in the mid-deck is used for normal crew entrance/exit and may be operated from within the crew cabin mid-deck or externally. It can be jettisoned for emergencies. It is attached to the crew cabin tunnel by hinges, a torque tube and support fittings. The hatch opens outwardly 90 degrees down with the orbiter horizontal or 90 degrees sideways with the orbiter vertical. It is 40 inches (1,000 millimeters) in diameter and has a 10-inch (250 millimeters) clear-view window in the center of the hatch. The window consists of three panes of glass.

Depending on the mission requirements, bunk sleep stations and a galley can be installed in the mid-deck. In addition, three or four seats of the same type as the mission specialists' seats on the flight deck can be installed in the mid-deck. Three seats over the normal three could be installed in the mid-deck for rescue missions if the bunk sleep stations were removed. The waste management system, located in the mid-deck, can also accommodate payloads in the pressurized crew compartment environment. The mid-deck also provides a stowage volume of 140 cubic feet (4 cubic meters). Accommodations are included for dining, sleeping, maintenance, exercising, and data management. On the orbiter centerline, just aft of the forward avionics equipment bay, an opening in the ceiling provides access to the inertial measurement units.

The mid-deck equipment bay below the mid-deck floor houses the major components of the waste management and air revitalization systems, such as pumps, fans, lithium hydroxide, absorbers, heat exchangers, and ducting. This compartment has space for stowing lithium hydroxide canisters and five separate spaces for crew equipment stowage with a volume of 29.9 cubic feet (0.8 cubic meters). Modular stowage lockers are used to store the flight crew's personal gear, mission-necessary equipment, personal hygiene equipment, and experiments. The modular lockers are made of sandwich panels of Kevlar/epoxy and a nonmetallic core. This reduced the lockers' weight by 83 percent compared to all-aluminum lockers, a reduction of approximately 150 pounds (67 kilograms).

Seating for three passengers/researchers, along with the habitability provisions, are located on the lower level or deck. The habitability provisions consist of a galley for food preparation, which includes an oven and hot and cold water dispensers for the preparation of rehydratable freeze-dried foods, an eating area, personal hygiene facilities, and sleeping accommodations.

The wing is constructed of a conventional aluminum alloy, using a corrugated spar web, truss-type ribs and riveted skin-stringer and honeycomb covers. The elevons are constructed of aluminum honeycomb and are split into two segments to minimize hinge binding and interaction with the wing.

The midfuselage is a 60-foot (18-meter) section of primary load-carrying structure between the forward and aft fuselages. It includes the wing carry-through structure and the payload bay doors. The skins consist of integral-machined aluminum panels and aluminum honeycomb sandwich panels. The frames are constructed from a combination of aluminum panels with riveted or machined integral stiffeners and a truss structure center section.

Payload bay doors make up the upper half of the midfuselage and are hinged along the side and split at the top centerline. The doors are graphite epoxy frames and honeycomb panel construction.

The aft fuselage includes a truss-type internal structure of diffusion-bonded elements that transfer the main engine thrust loads to the midfuselage and external tank. The aft fuselage's external surface is of standard construction except for the removable orbital maneuvering system/reaction control system pods, which are constructed of graphite epoxy skins and frames. An aluminum bulkhead shield with reusable insulation at the rear of the orbiter protects the rear portion of the aft fuselage.

The vertical tail, a conventional aluminum alloy structure, is a two-spar, multirib, integrally machined skin assembly. The tail is attached to the aft fuselage by bolted fittings at the two main spars. The rudder/speed brake assembly is divided into upper and lower sections, which are split longitudinally and actuated individually to serve as both rudder and speed brake.

The body flap, which has aluminum honeycomb covers, is attached to the lower aft fuselage by four rotary actuators.

These major structural assemblies are mated and held together by rivets and bolts. The midfuselage is joined to the forward and aft fuselage primarily by shear ties. The wing is attached to the midfuselage and aft fuselage primarily by shear ties, except in the area of the wing carry-through, where the upper panels are attached with tension bolts. The vertical tail is attached to the aft fuselage with bolts that work in both shear and tension.

The orbiter's exterior is covered with thermal protective materials to protect the spacecraft from solar radiation and the extreme heat of atmospheric reentry. Two types of reusable surface insulation, coated silica tiles and coated flexible sheets, cover the top and sides of the orbiter. The tiles protect the surfaces up to 1200°F (649°C); the flexible insulation provides protection up to 700°F (371°C). The coating on both types of insulation gives the orbiter a nearly white color and has optical properties that reflect solar radiation. On the bottom of the orbiter and on the leading edge of the tail, a high-temperature reusable surface insulation, made of coated silica tiles, protects the aluminum structure up to 2300°F (1260°C). The high-temperature coating is of a glossy black appearance. A reinforced carbon–carbon material is used for the nose cap and the wing leading edges, where the temperatures exceed 2300°F (1260°C).

Shuttle Landing Facility (SLF). Orbiter landings at the Kennedy Space Center are made on one of the largest runways in the world. The runway is located 3.2 kilometers (2 miles) northwest of the Vehicle Assembly Building and is 4,572 meters (15,000 feet) long and 91.4 meters (300 feet) wide, about as wide as the length of a football field. It has 305 meters (1000 feet) of paved overruns at each end and the paving thickness is 40.6 cm (15 inches) at the center.

The facility includes a 150 × 168 meters (490 × 550 feet) parking apron and a 3.2 kilometers (2 miles) tow-way connecting it with the Orbiter Processing Facility. Located adjacent to the parking apron is a Landing Aids Control Building (LACB) which supports landing operations and houses operations personnel.

Located at the northeast corner of the parking apron is the Mate/Demate device (MDD) used to raise and lower the orbiter from its 747 carrier aircraft during ferry operations. The open-truss steel structure is equipped with hoists, adapters and movable platforms for access to certain orbiter components and equipment. It also is equipped with lightning protection

devices. The MDD is 45.7 meters (150 feet) long, 28.3 meters (93 feet) wide and 32 meters (105 feet) high.

The Shuttle Landing Facility is equipped with a number of navigation and landing aids to assist Shuttle pilots in landing. There are four sophisticated Microwave Scanning Beam Landing System (MSBLS) ground stations, two located at each end of the runway, that provide elevation and directional/distance measurement for landing approaches from the northwest (runway 15) or southeast (runway 33). Equipment onboard the orbiter receives the data from the MSBLS stations and automatically makes any needed adjustments to the glide slope.

A Tactical Air Navigation (TACAN) system, located at mid-field off the east side of the runway, is used by pilots to execute an instrument landing approach to the runway. The TACAN has a range of 483 kilometers (300 miles), and is received by the orbiter when it emerges from the reentry blackout period. The final approach is guided by the MSBLS system.

Visual aids are provided by Precision Approach Path Indicators, known as the PAPI system. They utilize arrays of red and white lights that, when lined up properly by the pilot, will indicate the proper glide slope. A ball/bar light system is used for inner glide slope information on final approach to inform the pilot whether he is on, above or below the glide slope for an orbiter touchdown point marked on the runway.

A Recovery Convoy Staging Area, located just east of the runway about midway along its length, houses trailers, mobile units and specially designed vehicles that are used to "safe" the orbiter immediately after landing for crew egress and transfer of the orbiter to the Orbiter Processing Facility. For information on the Kennedy Space Center facilities see *http://science.ksc.nasa.gov/facilities/tour.html.*

Orbiter Ground Turnaround. Approximately 160 Space Shuttle launch operations team members support spacecraft recovery operations at the nominal end-of-mission landing site. Ground team members wearing self-contained atmospheric protective ensemble suits that protect them from toxic chemicals approach the spacecraft as soon as it stops rolling. The ground team members take sensor measurements to ensure the atmosphere in the vicinity of the spacecraft is not explosive. In the event of propellant leaks, a wind machine truck carrying a large fan will be moved into the area to create a turbulent airflow that will break up gas concentrations and reduce the potential for an explosion.

A ground support equipment air-conditioning purge unit is attached to the right-hand orbiter T-0 umbilical so cool air can be directed through the orbiter's aft fuselage, payload bay, forward fuselage, wings, vertical stabilizer, and orbital maneuvering system/reaction control system pods to dissipate the heat of entry.

A second ground support equipment ground cooling unit is connected to the left-hand orbiter T-0 umbilical spacecraft Freon Coolant loops to provide cooling for the flight crew and avionics during the postlanding and system checks. The spacecraft fuel cells remain powered up at this time. The flight crew will then exit the spacecraft, and a ground crew will power down the spacecraft.

AT KSC, the orbiter and ground support equipment convoy move from the runway to the Orbiter Processing Facility.

If the spacecraft lands at Edwards Air Force Base, the same procedures and ground support equipment are used as at the KSC after the orbiter has stopped on the runway. The orbiter and ground support equipment convoy move from the runway to the orbiter mate and demate facility at Edwards. After detailed inspection, the spacecraft is prepared to be ferried atop the Shuttle carrier aircraft from Edwards to KSC. For ferrying, a tail cone is installed over the aft section of the orbiter.

In the event of a landing at an alternate site, a crew of about eight team members will move to the landing site to assist the astronaut crew in preparing the orbiter for loading aboard the Shuttle carrier aircraft for transport back to the KSC. For landings outside the United States, personnel at the contingency landing sites will be provided minimum training on safe handling of the orbiter with emphasis on crash rescue training, how tow the orbiter to a safe area, and prevention of propellant conflagration.

Upon its return to the Orbiter Processing Facility (OPF) at KSC, the orbiter is safed (ordnance devices safed), the payload (if any) is removed, and the orbiter payload bay is reconfigured from the previous mission for the next mission. Any required maintenance and inspections are also performed while the orbiter is in the OPF. A payload for the orbiter's next mission may be installed in the orbiter's payload bay in the OPF or may be installed in the payload bay when the orbiter is at the launch pad.

The spacecraft is then towed to the Vehicle Assembly Building and mated to the external tank. The external tank and solid rocket boosters are stacked and mated on the mobile launcher platform while the orbiter is being refurbished. Space Shuttle orbiter connections are made and the integrated vehicle is checked and ordnance is installed.

The mobile launcher platform moves the entire space shuttle system on four crawlers to the launch pad, where connections are made and servicing and checkout activities begin. If the payload was not installed in the OPF, it will be installed at the launch pad followed by prelaunch activities.

Kennedy Space Center Launch Operations has responsibility for all mating, prelaunch testing and launch control ground activities until the Space Shuttle vehicle clears the launch pad tower. Responsibility is then turned over to Mission Control Center-Houston. The Mission Control Center's responsibility includes ascent, on-orbit operations, entry, approach and landing until landing runout completion, at which time the orbiter is handed over to the postlanding operations at the landing site for turnaround and re-launch. At the launch site the SRBs and external tank are processed for launch and the SRBs are recycled for reuse.

Editor's Note. The staff of this encyclopedia is most appreciative of the cooperation extended by National Aeronautics and Space Administration (NASA) for furnishing information for numerous portions of this article.

Web References

A complete listing of shuttle flights to date. *http://www.hq.nasa.gov/osf/shuttle/*
Fact Sheet Library: *http://spaceflight.nasa.gov/spacenews/factsheets/index.html#firstflights*
NASA Human Spaceflight: *http://spaceflight.nasa.gov/*
NASA History Office: *http://www.hq.nasa.gov/office/pao/History/shuttlehistory.html*
NASA History of the Space Shuttle: An Annotated Bibliography: *http://www.hq.nasa.gov/office/pao/History/Shuttlebib/contents.html*
NASA Space Transportation System: *http://science.ksc.nasa.gov/shuttle/technology/sts-newsref/*
NASA Shuttle Archives: *http://spaceflight.nasa.gov/shuttle/archives/year1990.html*
NASA Kennedy Space Center Fact Sheets: *http://www-pao.ksc.nasa.gov/kscpao/educate/docs.htm*

SPACE STATIONS. Space stations have long been seen as laboratories for learning about the effects of space conditions and as a springboard to the Moon and Mars. These space stations are man-made satellites designed to maintain a fixed orbit around the Earth while providing a platform for research activities in outer space. Crews live on board the space stations and travel back and forth to Earth via a space transportation system (STS). In the United States, the STS is known as the *Space Shuttle*. See also **Space Shuttle**. In Russia, the primary transport system is the *Soyez-TM*.

The first satellites designed as space stations were Russia's *Salyut* and the United States *Skylab*. These were programs of the early 1970s that served as the foundation of today's space station design and development. Long-duration microgravity research began on *Skylab*, and on *Salyut*, has continued on the *Mir Space Station* and will be further expanded and refined on the *International Space Station* (ISS). See also **Microgravity and Materials Processing**; and **Satellites (Scientific and Reconnaissance)**.

The Beginning (1869–1957)

The concept of a staffed outpost in Earth orbit dates from just after the Civil War. In 1869, American writer Edward Everett Hale published a science fiction tale called "The Brick Moon" in the *Atlantic Monthly*. Hale's manned satellite was a navigational aid for ships at sea. Hale proved prophetic. The fictional designers of the *Brick Moon* encountered many of the same problems with redesigns and funding that NASA would with its station more than a century later.

In 1903, Russian schoolteacher Konstantin Tsiolkovsky wrote "Beyond the Planet Earth," a work of fiction based on sound science. In it, he described orbiting space stations where humans would learn to live in space. Tsiolkovsky believed these would lead to self-contained space settlements and expeditions to the moon, Mars, and the asteroids. Tsiolkovsky wrote about rocketry and space travel until his death in 1935, inspiring generations of Russian space engineers.

In 1923, Hermann Oberth, a Romanian, coined the term "space station." Oberth's station was the starting point for flights to the Moon and Mars. Herman Noordung, an Austrian published the first space station blueprint in 1928. Like today's *International Space Station* (ISS), it had modules with different functions. Both men wrote that space station parts would be launched into space by rockets.

In 1926, American Robert H. Goddard made a major breakthrough by launching the first liquid-fueled rocket, setting the stage for the large, powerful rockets needed to launch space station parts into orbit. Rocketry advanced rapidly during World War II, especially in Germany, where the ideas of Oberth and Noordung had great influence. The German V-2 rocket, a missile with a range of about 300 miles (483 kilometers), became a prototype for both U.S. and Russian rockets after the war. See also **Goddard, Robert H.**

In 1945, renowned German rocket engineer Wernher Von Braun came to the U.S. to build rockets for the U.S. Army. In the 1950s, he worked with *Collier's* magazine and Walt Disney Studios to produce articles and documentaries on spaceflight. In them, he described a wheel-shaped space station reached by reusable winged spacecraft. Von Braun saw the station as an Earth-observation post, a laboratory, an observatory, and a springboard for Moon and Mars flights. See also **Von Braun, Werner (1912–1977)**.

Soviet engineers began work on large rockets in the 1930s. In May 1995, work began on the Baikonur launch site in central Asia. In August 1957, the world's first intercontinental ballistic missile lifted off from Baikonur on a test flight, followed by the launch of *Sputnik 1*, the world's first artificial satellite, on October 4, 1957. This triggered the Cold War competition between the U.S. and Soviet Union in space, which characterized the early years of the Space Age—competition replaced today by cooperation in the International Space Station Program. In response to *Sputnik*, the U.S. established the National Aeronautics and Space Administration (NASA) in 1958 and started its first man-in-space program, Project Mercury, in 1959.

Apollo and Space Stations (1958–1973)

Project Mercury had hardly begun when NASA and the Congress looked beyond it, to space stations and a permanent human presence in space. Space stations were seen as the next step after humans reached orbit. In 1959, a NASA committee recommended that a space station be established before a trip to the Moon, and the U.S. House of Representatives Space Committee declared a space station a logical follow-on to Project Mercury.

In April 1961, the Soviet Union launched the first human, Yuri Gagarin, into space in the *Vostok 1* spacecraft. President John F. Kennedy reviewed many options for a response to prove that the U.S. would not yield space to the Soviet Union, including a space station, but a man on the Moon won out. Getting to the Moon required so much work that the U.S. and Soviet Union were starting the race about even. In addition, the Moon landing was an unequivocal achievement, while a space station could take many different forms.

Space station studies continued within NASA and the aerospace industry, aided by the heightened interest in spaceflight attending Apollo. In 1964, seeds were planted for *Skylab*, a post-Apollo first-generation space station. Wernher von Braun, who became the first director of NASA's Marshall Space Flight Center, was instrumental in *Skylab's* development.

By 1968, a space station was NASA's leading candidate for a post-Apollo goal. In 1969, the year *Apollo 11* landed on the Moon, the agency proposed a 100-person permanent space station, with assembly completion scheduled for 1975. The station, called Space Base, was to be a laboratory for scientific and industrial experiments. Space Base was envisioned as home port for nuclear-powered tugs designed to carry people and supplies to an outpost on the Moon.

NASA realized that the cost of shipping supplies to a space station using expendable rockets would quickly exceed the station's construction cost. The agency also foresaw the need to be able to return things from a space station. A reusable spacecraft was the obvious solution. In 1968, NASA first called such a spacecraft a space shuttle.

First-Generation Stations (1964–1977)

First-generation space stations had one docking port and could not be resupplied or refueled. The stations were launched unmanned and later occupied by crews. There were two types: *Almaz* military stations and *Salyut* civilian stations. To confuse Western observers the Soviets called both kinds *Salyut*.

The *Almaz* military station program was the first approved. When proposed in 1964, it had three parts: the *Almaz* military surveillance space station, Transport Logistics Spacecraft for delivering soldier-cosmonauts and cargo, and Proton rockets for launching both. All of these spacecraft were built, but none was used as originally planned.

Soviet engineers completed several *Almaz* station hulls by 1970. The Soviet leadership ordered *Almaz* hulls transferred to a crash program to launch a civilian space station. Work on the Transport Logistics Spacecraft was deferred, and the *Soyuz* spacecraft originally built for the Soviet manned Moon program was reapplied to ferry crews to space stations. *Salyut 1*, the first space station in history, reached orbit unmanned atop a *Proton* rocket on April 19, 1971. The *Salyut 1* weighed 18,600 kilograms (41,000 pounds) and was approximately 14.5 meters (47.4 feet) in length. It had an interior space of 99 cubic meters (3,500 cubic feet) with 7 experimental workstations.

The early first-generation stations were plagued by failures. The crew of *Soyuz 10*, the first spacecraft sent to *Salyut 1*, was unable to enter the station because of a docking mechanism problem. The *Soyuz 11* crew lived aboard *Salyut 1* for three weeks, but died during return to Earth because the air escaped from their *Soyuz* spacecraft. Then, three first-generation stations failed to reach orbit or broke up in orbit before crews could reach them. The second failed station was *Salyut 2*, the first *Almaz* military station to fly.

The Soviets recovered rapidly from these failures. *Salyut 3* (1974–1975), *Salyut 4* 1974–1977), see Fig. 1, and *Salyut 5* (1976–1977) supported a total of five crews. In addition to military surveillance and scientific and industrial experiments, the cosmonauts performed engineering tests to help develop the second-generation space stations. See **Second-Generation Space Stations (1977–1985)** discussed later in this article.

Skylab (1973–1974)

The *Skylab* space station was launched May 14, 1973, from the NASA Kennedy Space Center by a huge *Saturn V* launch vehicle, the moon rocket of the Apollo Space Program. See Figs. 2 and 3. Sixty-three seconds after liftoff, the meteoroid shield—designed also to shade *Skylab's* workshop—deployed inadvertently. It was torn from the space station by atmospheric drag. This event and its effects started a ten-day period in which *Skylab* was beset with problems that had to be conquered before the space station would be safe and habitable for the three manned periods of its planned eight-month mission.

When the meteoroid shield ripped loose, it disturbed the mounting of workshop solar array "wing" two and caused it to partially deploy. The exhaust plume of the second stage retro-rockets impacted the partially deployed solar array and literally blew it into space. Also, a strap of debris from the meteoroid shield overlapped solar array "wing" number one such that when the programmed deployment signal occurred, wing number one was held in a slightly opened position where it was able to generate virtually no power.

In the meantime, the space station had achieved a near-circular orbit at the desired altitude of 435 kilometers (270 miles). All other major functions including payload shroud jettison, deployment of the Apollo Telescope Mount (*Skylab's* solar observatory) and its solar arrays, and pressurization of the space station occurred as planned.

Scientists, engineers, astronauts, and management personnel at the NASA Marshall Space Flight Center and elsewhere worked throughout the first ten-day period of *Skylab's* flight to devise the means for its rescue. Simultaneously, *Skylab*, seriously overheating, was maneuvered through varying nose-up attitudes that would best maintain an acceptable "holding" condition. During that ten-day period and for some time thereafter, the space station operated on less than half of its designed electrical system, in the partially nose-up attitudes, was generating power at reduced efficiency. The optimum condition that maintained the most favorable balance between *Skylab* temperatures and its power generation capability occurred at approximately 50 degrees nose-up.

Skylab's achievements are a summary of the accomplishments of many ground-based persons as well as its three separate crews who were launched in *Apollo*-type command modules by Saturn IB vehicles on May 25, July 28, and November 16, 1973. In *Skylab*, both the man-hours in space and the man-hours spent in performance of extravehicular activities (EVA) under microgravity conditions exceeded the combined totals of all of the world's previous space flights up to that time.

By deploying the parasol-type sun shield through *Skylab's* solar scientific airlock and later releasing workshop solar array wing number one

Salyut 4 civilian orbital station

1 Soyuz manned transport craft
2 EVA/occess hot can side hidden in this
  illustration
3 Rendeavour antenna
4 One of three starable solar panel
5 Gas storage for life support system
6 Food and storage lockers
7 Attitude control jets
8 KTDU-35 main propulsion system
9 Rendeavour transponder
10 Propellant links for main propulsion system
11 DST-1 apparatus
12 Chibis tower body negative pressure suit
13 Treadmill
14 Table
15 Main control console
16 Forward bulkhead of work compartment
17 Spheres of pressurant gas
18 Television camera system

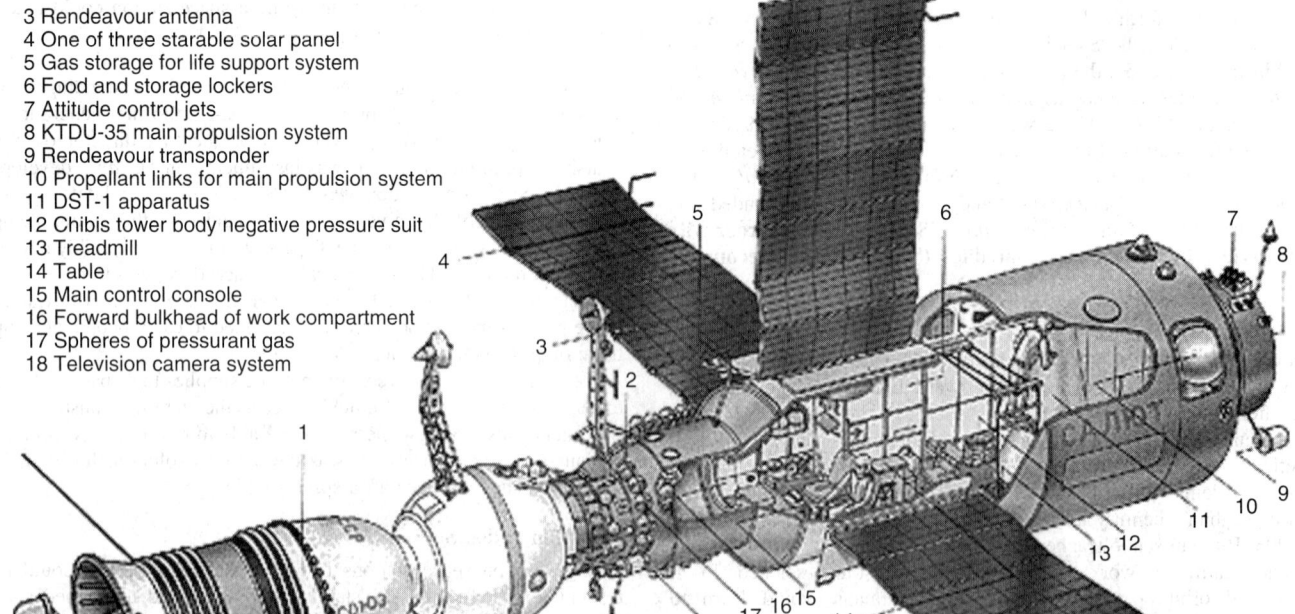

Fig. 1.   Diagram of *Salyut 4*, December 1974–February 1977, in the *Salyut* series of early space stations. The instrumentation included optical sensors, which were mounted on the outside of the station together with the X-ray detectors, and power supply and measurement units, which were inside the station. (*NASA*.)

Fig. 2.   An overhead view of the *Skylab* space station cluster in Earth orbit as photographed from the *Skylab 4* Command and Service Modules (CSM) during the final fly-around by the CSM before returning home. The space station is contrasted against a cloud-covered Earth. Note the solar shield, which was deployed by the second crew of *Skylab* and from which a micrometeoroid shield has been missing since the cluster was launched on May 14, 1973. The OWS solar panel on the left side was also lost on workshop launch day. (*NASA.*)

Fig. 3.   A Martin Marietta artist's concept of the *Skylab* space station cluster in Earth's orbit. The cluster is composed of the Apollo Command/Service Module, Orbital Workshop, Apollo Telescope Mount (ATM), Multiple Docking Adapter, and Airlock Module. In this concept, a member of the three-man astronaut crew is working atop the ATM in the zero gravity of space. Note the arrays of solar cell panels that turn sunlight into electric power for the space station. (*NASA.*)

during EVA, the first crew made the remainder of the mission possible. The second crew, also during EVA, erected another sun shield, a twin-pole device.

The effectiveness of *Skylab* crews exceeded expectations, especially in their ability to perform complex repair tasks. They demonstrated excellent mobility, both internal and external to the space station, showing humans to be a positive asset in conducting research from space. By selecting and photographing targets of opportunity on the Sun, and by evaluating weather conditions on Earth and recommending Earth Resources opportunities,

crewmen were instrumental in attaining extremely high quality solar and Earth oriented data.

All three crews demonstrated technical skills for scientific, operational, and maintenance functions. Their manual control of the space station, their fine pointing of experiments, and their reasoning and judgment throughout the manned periods were highly effective.

The capability to conduct longer manned missions was conclusively demonstrated in *Skylab*, first by the crew returning from the 28 day mission and, more forcefully, by the good health and physical condition of the second and third *Skylab* crews who stayed in weightless space for 59 and 84 days respectively. Also, resupply of space vehicles was attempted for the first time in *Skylab* and was proven to be effective.

During their time in space, all three crews exceeded the operational and experimental requirements placed upon them by the premission flight plan and schedule. In addition, the third crew performed a number of sightings of Comet Kohoutek which were not initially scheduled.

Following the final manned phase of the *Skylab* mission in February 1974, ground controllers performed some engineering tests of certain *Skylab* systems — tests that ground personnel were reluctant to do while men were aboard. Results from these tests helped to determine causes of failures during the mission and to obtain data on long term degradation of space systems.

Upon completion of the engineering tests, *Skylab* was positioned into a stable attitude and systems were shut down. It was expected that *Skylab* would remain in orbit eight to ten years. However, in the fall of 1977, it was determined that *Skylab* was no longer in a stable attitude as a result of greater than predicted solar activity.

On July 11, 1979, *Skylab* impacted the Earth surface. The debris dispersion area stretched from the Southeastern Indian Ocean across a sparsely populated section of Western Australia.

Second-Generation Stations (1977–1985)

With the second-generation stations, the Soviet space station program evolved from short-duration to long-duration stays. Like the first-generation stations, they were launched unmanned and their crews arrived later in *Soyuz* spacecraft. Second-generation stations had two docking ports. This permitted refueling and resupply by automated *Progress* freighters derived from *Soyuz*. *Progress* docked automatically at the aft port, and was then opened and unloaded by cosmonauts on the station. Transfer of fuel to the station took place automatically under supervision from the ground.

A second docking port also meant long-duration resident crews could receive visitors. Visiting crews often included cosmonaut-researchers from Soviet bloc countries or countries sympathetic to the Soviet Union. Vladimir Remek of Czechoslovakia, the first space traveler not from the U.S. or the Soviet Union, visited *Salyut 6* in 1978.

Visiting crews relieved the monotony of a long stay in space. They often traded their *Soyuz* spacecraft for the one already docked at the station because *Soyuz* had only a limited lifetime in orbit. Lifetime was gradually extended from 60–90 days for the *Soyuz Ferry* to more than 180 days for the *Soyuz*-TM.

Salyut 6 Key Facts:

- Civilian space station from 1977–1982.
- The station received 16 cosmonaut crews, including six long-duration crews. The longest stay time for a *Salyut 6* crew was 185 days. The first *Salyut 6* long-duration crew stayed in orbit for 96 days, beating the 84-day world record for space endurance established in 1974 by the last *Skylab* crew.
- The station hosted cosmonauts from Hungary, Poland, Romania, Cuba, Mongolia, Vietnam, and East Germany.
- Twelve *Progress* freighters delivered more than 20 tons of equipment, supplies, and fuel.
- An experimental transport logistics spacecraft called *Cosmos 1267* docked with *Salyut 6* in 1982. The transport logistics spacecraft was originally designed for the *Almaz* program. *Cosmos 1267* proved that large modules could dock automatically with space stations, a major step toward the multimodular *Mir* station and the International Space Station.

Salyut 7 Key Facts:

- Civilian space station from 1982–1991.
- *Salyut 7*, a near twin of *Salyut 6*, was home to 10 cosmonaut crews, including six long-duration crews. The longest stay time was 237 days.
- Cosmonauts from France and India worked aboard the station, as did the first female space traveler since 1963.
- Thirteen *Progress* freighters delivered more than 25 tons of equipment, supplies, and fuel to *Salyut 7*.
- Two experimental transport logistics spacecraft, *Cosmos 1443* and *Cosmos 1686*, docked with *Salyut 7*. *Cosmos 1686* was a transitional vehicle, a transport logistics spacecraft redesigned to serve as an experimental space station module.

- *Salyut 7* was abandoned in 1986 and reentered Earth's atmosphere over Argentina in 1991.

Third-Generation Station: *Mir* (1986–2001)

The *Mir* Space Station was a manned space station of Russian design that had been in orbit for 15 years. The *Mir* Space Station was designed as a successor to the *Salyut* series and was highly successful. It was comprised of a docking module and 6 research modules, each with a specific design and purpose, which were launched independently and assembled together in space. See Figs. 4 and 5 for illustrations of the configuration. Figure 6 is a close-up view of the *Mir* Space Station as seen from the Space Shuttle.

The core module, *Mir*, which means "Peace" or "Village" in Russian, was the first element of the complex launched from Baikonur Cosmodrome in Kazakhstan, central Asia, on February 20, 1986 at an inclination of 51.6 degrees to the equator. The *Mir* complex developed over the next 10 years with the additions of the component modules: *Kvant-1*, *Kvant-2*, *Kristall*, *Spektr*, the *Docking Module*, and *Piroda*. *Mir* had two axial docking ports, fore and aft, for *Soyuz-TM* manned transports and automated, unmanned *Progress-M* supply ships, plus four radial berthing ports for expansion modules. Manned operations began on March 15, 1986 and there had been near continuous manned operations on *Mir* until 2001.

The *Mir* Space Station, a civilian, third-generation permanent space station, was owned by the Russian Space Agency. The Russian Space Agency (RKA) was formed after the breakup of the former Soviet Union when the existing Soviet space program was eliminated. The RKA uses the technology and launch sites that formerly belonged to the Soviet space program. Centralized control of Russia's civilian space program, including all manned and unmanned nonmilitary space flights falls under the jurisdiction of the RKA. The RKA employs about 300 people, and for much of the work, hires outside contractors. Energia Rocket and Space Complex is the prime contractor used by the RKA. Energia operated the *Mir* Space Station and also owns and operates the Mission Control Center in Kaliningrad. The military counterpart of the RKA is the Military Space Forces (VKS). The RKA and VKS share control of facilities such as the Baikonur Cosmodrome and the Gagarin Cosmonaut Training Center.

Mir Modules

The *Mir* core module contained the space station control center, the crew's living quarters, and power and life support equipment. The core module was 13.1 meters (43.2 feet) long with a maximum diameter of 4.1 meters (13.5 feet). It weighed 20,900 kilograms (46,000 pounds) and had a volume of 90 cubic meters (3,200 cubic feet). See Fig. 7. The module was comprised of two distinct zones, one for living and one for operations. Familiar spatial orientation was used in the living area for the comfort of the crew. The floors were carpeted in dark green, the walls were painted a light green, and the ceilings were painted white and fitted with fluorescent lighting. Although up and down have no meaning in the near total absence of gravity, the design served to provide a semblance of normalcy for the crew.

The crew's living area had a galley with cooking elements, refrigerator, dining table and trash receptacle; a bathroom for personal hygiene with sink, shower, and toilet; and the crew's cabins. Each crewmember had his/her own cabin with a sleeping bag, chair, mirror, and porthole. Also inside this module were medical monitoring equipment and a bicycle ergometer for exercise. In addition, the living area was equipped with a ham radio operation and a video tape recorder with a library of tapes for entertainment and instruction.

The operations area provided the control center for the entire *Mir* complex. The crew was able to monitor and command core systems, the piloting station, and science equipment and facilities from this central operations area. The multiple docking ports of the transfer compartment allowed for the addition of secondary station modules or transport craft. The intermediate compartment was a pressurized tunnel 1.8 meters (6 feet) in diameter that connects the working module to the aft docking port. Main engine and fuel tanks were located in the non-pressurized assembly compartment.

Kvant-1 ("Quantum-1"), the astrophysics module, was launched on March 31, 1987. See Fig. 8. *Kvant-1* was 5.8 meters (19.1 feet) long with a maximum diameter of 4.1 meters (13.5 feet). *Kvant-1* weighed 11,050 kilograms (24,300 pounds) and had a volume of 40 cubic meters (1,440 cubic feet). *Kvant-1* was successfully docked after several attempts when the cosmonauts on board *Mir*, Yuri Romanenko and Alexander

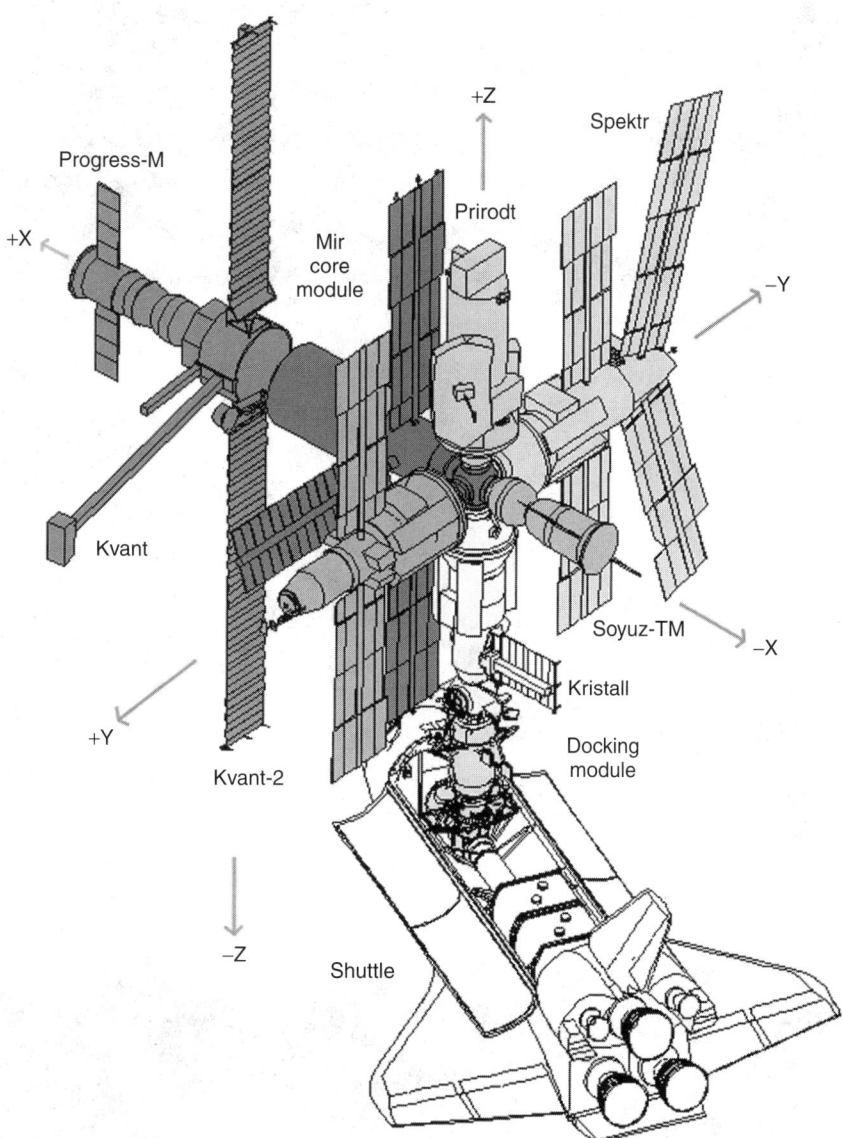

Fig. 4. *Mir* Space Station configuration with shuttle docked and module components labeled. (*NASA.*)

Laveikin, engaged in a spacewalk to remove a plastic bag from the docking port that was preventing a solid connection. *Kvant-1* was located on the aft docking port of the *Mir* core and had its own passive-docking unit that is used by *Progress-M* and *Soyuz-TM* spacecraft. It contained the equipment necessary to gather information and conduct research into the physics of active galaxies, quasars, neutron stars, and other astrophysical phenomena by measuring electromagnetic spectra and x-ray emissions. *Kvant-1* also contained control and life support equipment for the station. Part of the module was a pressurized area for working and living, while the rest was nonpressurized space for equipment. It had gyrostabilizers that could change the station attitude without propulsive fuel.

Kvant-2, ("Quantum-2"), the second module that was added to the *Mir* core, was the Scientific and Airlock module, which was launched on December 6, 1989. See Fig. 9. *Kvant-2* was 13.7 meters (45.2 feet) long with a maximum diameter of 4.3 meters (14.2 feet). *Kvant-2* weighed 18,500 kilograms (40,700 pounds) and had a volume of 61.3 cubic meters (2,200 cubic feet). This module was originally developed as a logistical transport spacecraft for the Soviet Union's *Almaz* military space station program. *Kvant-2* included Earth observation photographic equipment, scientific equipment for biotechnology research, life support equipment, as well as shower and washing facilities. Two solar panels were attached to the exterior of the module. It was also equipped with an EVA airlock. The airlock allowed access to the outside of *Mir*, which enabled the crewmembers to conduct experiments on the effects of space exposure on electronics and construction materials. Although the airlock included a manned maneuvering unit similar to one tested on American shuttle flights

in the 1980s, it had not been utilized. It was tested twice in February 1990 and never used again.

Kristall ("Crystal") was the third module added to the *Mir* core. See Fig. 10. Launched on May 31, 1990, *Kristall* was 13.7 meters (45.2 feet) long and had a maximum diameter of 4.3 meters (14.2 feet). *Kristall* weighed 19,650 kilograms (43,230 pounds) and had a volume of 60.8 cubic meters (2,180 cubic feet). It contained technological research equipment that produced semiconductors and other high-tech materials that benefit from the low gravity environment of space. It was used primarily to develop biological and materials processing technology in the space environment, but it also contained a greenhouse to cultivate plants in zero gravity. *Kristall* was equipped with a *Shuttle* docking port.

Spektr ("Spectrum") was the fourth module to join *Mir* and was launched in May 1995. See Fig. 11. *Spektr* was 13 meters (42.9 feet) long with a maximum diameter of 4.3 meters (14.2 feet). It weighed 19,640 kilograms (43,200 pounds) and had a volume of 61.9 cubic meters (2,200 cubic feet). This module was designed for scientific research specifically related to Earth observation. The module included equipment for atmospheric and surface research, as well as detectors to study the x-ray and gamma-ray background outside the station. *Spektr* served as the living quarters for American astronauts staying on *Mir* as part of the joint Russian-American program. The module also had 4 solar panels, which before June 1997 generated nearly half of the station's electrical power. The collision of a *Progress* cargo vessel with the *Spektr* module on June 25, 1997 resulted in the loss of that module when the hull was

Fig. 5. The *Mir* Space Station, once completed, had a mass of more than 100 tons (90.7 tonnes). It consists of seven modules launched separately and brought together in space over a period of 10 years. *Mir* measures more than 107 feet (32.6 meters) long with docked *Progress-M* and *Soyuz-TM* spacecraft, and is about 90 feet (27.4 meters) wide across its modules. (*NASA*.)

Fig. 6. The *Mir* Space Station in a close-up view as seen from the Space Shuttle *Atlantis*. (*NASA*.)

Fig. 7.   The 20.4-ton (18.5 tonnes) *Mir* Core Module was launched in February 1986. (*NASA.*)

Fig. 8.   *Kvant-1*, a small 11-ton (10 tonnes) module, was added to the *Mir* core's aft port in 1987. It contains astrophysics instruments, life support and attitude control equipment. (*NASA.*)

punctured, depressurizing the module, and the solar panels were damaged. To view the damaged solar array panels, see Figs. 12 and 13.

The Docking Module, the next component module added to *Mir*, was delivered and installed by the shuttle *Atlantis* mission STS-74 in November 1995. The *Docking* module, attached to the aft end of the *Kristall* module, made it possible for the space shuttle to dock more easily with *Mir*. On a prior mission, STS-71 in June 1995, the shuttle docked with the *Kristall* module on *Mir*. However, to make that docking possible, the *Kristall* configuration had to be changed to give the shuttle enough clearance to dock without bumping into any of the solar arrays in the vicinity. Russian cosmonauts had to perform a spacewalk to move the *Kristall* module from a radial axis to a longitudinal axis, relative to *Mir*. After the shuttle departed, *Kristall* was moved back to its original location. Modules for *Mir's* radial berthing ports first dock at the front port. Each module carried a manipulator arm that locks into a socket on *Mir*. The arm pivots the module into place at the proper radial port.

Priroda (**"Nature"**) was the final component module, which was launched in April 1996, completing the assembly of research components in the *Mir* space station complex. See Fig. 14. *Priroda* was 13 meters (42.9 feet) long and has a maximum diameter of 4.3 meters (14.2 feet). It weighed 19,700 kilograms (43,300 pounds) and had a volume of 66 cubic meters (2,370 cubic feet). The module carried Earth observing equipment as well as experiments and other equipment for the joint American-Russian missions on *Mir* (including more than a ton of materials for American astronaut Shannon Lucid, who was on the station when *Priroda* arrived.) It contained active, passive, and infra-red radiometers, a synthetic aperture radar, and several types of spectrometers used for measuring ozone and aerosol concentrations in the atmosphere. The Earth remote-sensing mission of *Priroda* was multifold. It was designed to study the atmosphere and oceans, and in particular, the environmental impact of human activities and pollution. It was also designed to conduct geological surveys that can be used to locate mineral resources and water reserves and study the effects of erosion on crops and forests. It was enabled to receive and relay information from "emergency buoys" located in seismically active areas, around nuclear power plants, and other zones, as part of an emergency warning system.

Soyuz-TM module was used to transport crews and cargo to and from the *Mir* Space Station. In addition, it served as a lifeboat and shelter should an emergency situation occur that would require the crew to abandon *Mir*. The *Soyuz-TM* was 7 meters (23.1 feet) in length and had a maximum diameter of 2.7 meters (8.9 feet). It weighed 7,100 kilograms (15,600 pounds) and had a volume of 10 cubic meters (360 cubic feet). It was made up of three compartments: the orbital module, the descent module and the instrumentation module. *Soyuz* was first launched in 1963, and was a two-stage rocket that could deliver a payload of over 15,000 pounds into Low-Earth Orbit at a 51.6-degree inclination. This rocket is the primary launcher used for manned Russian space flights and nearly half of all Russian space missions (manned and unmanned) are launched using the *Soyuz* vehicle. The *Soyuz* rocket is also used to deploy low-altitude reconnaissance satellites.

Progress-M Cargo Transport was an unmanned cargo and resupply vehicle used to send science equipment and data, food supplies and crew mail to and from *Mir*. The *Progress-M Transport* is 7 meters (23.1 feet) in length with a maximum diameter of 2.7 meters (8.9 feet). It weighed 7,200 kilograms (15,800 pounds) and had a volume of 7.6 cubic meters (270 cubic feet). After the *Progress* spacecraft has docked and been unloaded, it remained in place for use as a trash storage site. Once the spacecraft was full and shortly before the arrival of the new *Progress*

Fig. 9.   *Kvant-2*, added in 1989, carries an EVA airlock, solar arrays, and life support equipment as well as the provisions for personal hygiene care, water and oxygen. The 19.6-ton (17.8 tonnes) module is based on the transport logistics spacecraft originally intended for Russia's Almaz military space station program of the early 1970s. (*NASA.*)

Fig. 10.   *Kristall* added in 1990, carries scientific equipment, retractable solar arrays, and a docking node equipped with a special androgynous docking mechanism designed to receive spacecraft weighing up to 100 tons (90.7 tonnes). Originally, the Russian Buran shuttle, which made one unmanned orbital test flight in 1988, would have docked with *Mir* using the androgynous unit. Space Shuttle *Atlantis* used the androgynous unit to dock with *Mir* for the first time on the STS-71 mission in July 1995. On STS-74, in November 1995, *Atlantis* permanently attached a Docking Module to *Kristall's* androgynous docking unit. The Docking Module improved clearance between *Atlantis* and *Mir*'s solar arrays on subsequent docking flights. The 19.6-ton (17.8 tonnes) *Kristall* module is based on the transport logistics spacecraft originally designed to carry Soviet soldier-cosmonauts to the Almaz military space stations. (*NASA.*)

spacecraft, it undocked. Early trash-filled *Progress* ships were deorbited and allowed to burn up in the Earth's atmosphere. Later *Progress-M* freighters were improved and returned to Earth carrying primarily trash, but occasionally with materials for analysis as well. See Fig. 15 for a view of the *Mir* configuration with both *Soyuz* and *Progress* transports docked.

Mir Key Facts

- An important goal of the *Mir* program was to maintain a permanent human presence in space. Except for two brief periods (July

1986–February 1987; April–September 1989), Russian cosmonauts had lived aboard *Mir* continuously for the past 12 years, demonstrating proven experience in space station operations.
- Dr. Valeri Polyakov arrived on *Mir* on *Soyuz-TM 18* in January 1994 and returned to Earth on *Soyuz-TM* 20 on March 21, 1995. He lived in orbit for more than 438 days, a world record.
- Cosmonaut-researchers from Afghanistan, Austria, Britain, Bulgaria, the European Space Agency, France, Germany, Japan, Kazakhstan, and Syria visited *Mir*. European and French cosmonauts lived on *Mir* for

Fig. 11. *Spektr* was launched on a Russian Proton rocket from the Baikonur launch center in central Asia on May 20, 1995. The module was berthed at the radial port opposite *Kvant 2* after *Kristall* was moved out of the way. *Spektr* carries four solar arrays and scientific equipment, including more than 1600 pounds (726 kilograms) of U.S. equipment. The focus of scientific study for this module is Earth observation, specifically natural resources and atmosphere. The equipment onboard is supplied by both Russia and the United States. (*NASA.*)

Fig. 12. View of the damage to the solar array panel on *Spektr*, and the module body, the result of a collision with a *Progress* cargo ship. (*NASA.*)

as long as a month. U.S. astronauts have spent up to 5 months on the station.

- More than 40 *Progress* and *Progress-M* freighters have delivered more than 100 tons of supplies and fuel to *Mir*. The improved *Progress-M* occasionally carried a capsule for returning to Earth a small quantity of experiment results and industrial products from the station. Occasionally cargo came back to Earth with cosmonauts in *Soyuz-TM* capsules.

- Beginning with STS-71, the shuttle had returned to Earth more industrial products and experiment samples than would be possible using the *Progress-M* capsules or *Soyuz-TM*. In addition, the shuttle could be

- *Mir* had docked with 31 spacecraft.
- *Mir* had docked with 64 cargo vessels.
- *Mir* docked with the space shuttle 9 times.
- There were 17 space expeditions to *Mir*.
- There were 28 long-term crews on board *Mir*.
- 125 cosmonauts/astronauts from 12 different countries visited *Mir*.

For a complete list of *Mir* highlights, see the following website: *http://www.satobs.org/mir.html*

Born in the Cold War, *Mir's* ultimate legacy is that it taught the two-superpower enemies, the USSR and the USA, to co-operate in space. *Mir* showed man how to live and work in space and how to overcome unforeseen problems and survive potentially fatal situations. In terms of science, *Mir* returned very little but in no way does that diminish its achievements. It was mankind's first true long-term home in space and many will lament its passing.

Mir celebrated 15 years in space in 2001–10 years longer than the original projected lifespan of the station. But on 23 March the final command was sent to the pride of the Russian space program. The final stages of reentry all went according to plan and were a triumph for Russian ground controllers.

Since 1978-present spacecraft, including several previous space stations, have been safely brought to Earth in the South Pacific and *Mir* was no exception. At 0530GMT the disintegrating remains, which had split into approximately nine to twelve pieces, passed over Japan. The 25 tonnes (27.6 tons) of wreckage finally streaked across the sky and hit the intended target area in the South Pacific, 5,800 kilometers (3,600 miles) off the eastern coast of Australia, at about 0600GMT.

Mir/Shuttle Program (1994–1998)

On June 17, 1992 in Washington DC, George Bush, the President of the United States, and Boris Yeltsin, President of the Russian Federation, signed the "Agreement between the United States of America and the Russian Federation Concerning Cooperation in the Exploration and Use of Outer Space for Peaceful Purposes." This agreement states that one of the areas of cooperation will include a "Space Shuttle and *Mir* Space Station mission involving the participation of U.S. astronauts and Russian Cosmonauts." At this Washington meeting the leaders further agreed to flight's of Russian cosmonauts on the Shuttle in 1993, flight of an U.S. astronaut on a long-duration mission on *Mir* in 1994, and a docking mission between the Shuttle and the *Mir* in 1995. This was the beginning of the Phase 1 (*Mir*/Shuttle) Program).

Fig. 13. Close-up view of the damaged solar array panel on *Spektr* as seen from the *Soyuz* transport. (*NASA.*)

used to return components from *Mir's* exterior, such as solar arrays, for studying the effects of long exposure to space conditions — a capability not available with *Progress-M* and *Soyuz-TM*.

- Important lessons from *Mir* operations and *Shuttle–Mir* operations and research are being incorporated into the *International Space Station* design and planning.

Fig. 14. *Piroda*, the last science module to be added to the *Mir*, launched from Baikonur on April 23, 1996, and docked to the space station as scheduled on April 26. Its primary purpose is to add Earth remote sensing capability to *Mir*. It also contains the hardware and supplies for several joint U.S.-Russian science experiments. (*NASA.*)

Fig. 15.   Russia's *Mir* space station during rendezvous operations with the Space Shuttle *Discovery*. Docked at the bottom (nearest portion where longest solar array panel is visible) is a *Soyuz* vehicle. On the opposite end is a Progress spacecraft. (*NASA*.)

On October 5, 1992, in Moscow, Daniel Goldin, Administrator of NASA, and Yuri Koptev, Director General of RSA, signed the "Implementing Agreement between the National Aeronautics and Space Administration of the United States of America and the Russian Space Agency of the Russian Federation on Human Space Flight Cooperation." This agreement further outlined details of cooperation that included:

- A Russian cosmonaut flying on the Shuttle mission STS-60 as a mission specialist.
- A U.S. astronaut launching on a *Soyuz*, flying more than 90 days on the *Mir*, and returning on a Shuttle.
- Russian cosmonauts on *Mir* being "changed out" via the Shuttle on the same flight that would return the U.S. astronaut.
- The evaluation of and possible contract for the Russian Androgynous Peripheral Docking Assembly developed by NPO Energia for use on the Shuttle.

This program was called the *Mir*-Shuttle Program.

Later, the American side proposed expansion of the joint program: It would include up to 10 dockings of the Shuttle with *Mir*; and would increase the presence of American astronauts on *Mir* to up to two years; and deliver up to two tons of hardware on board the Russian *Spektr* and *Priroda modules*. Separate flights of up to six months were proposed for American astronauts on board *Mir*. In June, a contract was concluded for work between RSA and NASA. This program was called *Mir*-NASA. The work performed for the *Mir*-Shuttle and *Mir*-NASA program are considered as Phase 1 of the preparation for the creation of the *International Space Station*.

The Phase 1 Program became a formal stand-alone program on the NASA side on October 6, 1994, which included 11 space shuttle flights over a four-year period. See Fig. 16. The United States and Russia combined their existing assets, primarily U.S. space shuttle orbiters and the Russian Space Station *Mir*, in a joint operation of the program, which lasted from 1994–1998. The goal of the Shuttle–*Mir*, a complicated interlocking program, was to build joint space experience, which would be necessary to complete the International Space Station (ISS), and to begin joint scientific research. A significant challenge of the program was to incorporate the diverse and unique working styles and philosophies of the U.S. and Russian space agencies and their international partners.

Four major science objectives were established at the beginning of the Shuttle-*Mir* Program, International Space Station, Phase I. They were:

- To get engineering and operational experience in a research program on a long-duration space platform. (The most recent U.S. experience had been aboard *Skylab* in the 1970s. Tremendous knowledge and experience was gained on *Mir*, and many of the lessons are applicable to the *International Space Station* program.)
- To learn more about *Mir* and the microgravity environment in general for conducting research programs in various disciplines.
- To use *Mir* as a test bed for space station technologies.
- To learn how to do space walks with Russian counterparts and systems. (One Phase 1 EVA was planned initially. Three were performed.)

NASA astronauts underwent specialized training before living aboard *Mir*. As a prerequisite for the assignment, they had to acquire cosmonaut certification training. They also learned to speak Russian and attained proficiency with the experiments they would perform and the work they would have to do. The first American, Norman Thagard, to live aboard *Mir* reported feelings of loneliness and isolation, and steps were taken to prevent that happening to his successors.

Shuttle-*Mir*, International Space Station, Phase 1 served as a four-year prologue to station assembly in Phases 2 and 3. NASA and its international partners found that Phase 1 was invaluable in learning about all aspects of living and working in low-Earth orbit. For the list of long-duration flight missions that took place during the Shuttle-*Mir* Program, see Table 1.

International Space Station

The International Space Station Program has been in the planning stages since 1984, when President Ronald Reagan called for a space station in his State of the Union address. He said that the space station program was to include participation by U.S. allies.

With the presidential mandate in place, NASA set up the Space Station Program Office in April 1984, and issued a Request for Proposal to U.S. industry in September 1984. In April 1985, NASA let contracts on four work packages, each involving a different mix of contractors and managed by a separate NASA field center. (This was consolidated into three work packages in 1991.)

This marked the start of Space Station Phase B development, which aimed at defining the station's shape. By March 1986, the baseline design was the dual keel, a rectangular framework with a truss across the middle for holding the station's living and working modules and solar arrays.

By the spring of 1985, Japan, Canada, and the European Space Agency each signed a bilateral memorandum of understanding with the U.S. for participation in the space station project. In May 1985, NASA held the first space station international user workshop in Copenhagen, Denmark. By mid-1986, the partners reached agreement on their respective hardware

Fig. 16. Shuttle docked with *Mir*. (*NASA*.)

contributions. Canada would build a remote manipulator system similar to the one it had built for the space shuttle, while Japan and Europe would each contribute laboratory modules. Formal agreements were signed in September 1988. These partners' contributions remain generally unchanged for the *International Space Station*.

In 1987, the dual keel configuration was revised to take into account a reduced space shuttle flight rate in the wake of the *Challenger* accident. The revised baseline had a single truss with the built-in option to upgrade to the dual keel design. The need for a space station lifeboat, called the assured crew return vehicle, was also identified.

In 1988, Reagan gave the station a name–*Freedom*. Space Station *Freedom's* design underwent modifications with each annual budget cycle as Congress called for its cost to be reduced. The truss was shortened and the U.S. Habitation and Laboratory modules reduced in size. The truss was to be launched in sections with subsystems already in place. Despite the redesigns, NASA and contractors produced a substantial amount of hardware. In 1992, in moves presaging the current increased cooperation between the U.S. and Russia, the U.S. agreed to buy Russian *Soyuz* vehicles to serve as Freedom's lifeboats (these are now known as *Soyuz* crew transfer vehicles) and the Shuttle-*Mir* Program got its start. See **Shuttle-*Mir* Program** described early in this article.

In 1993, President William Clinton called for the station to be redesigned once again to reduce costs and include more international involvement. To stimulate innovation, teams from different NASA centers competed to develop three distinct station redesign options. The White House selected the option dubbed Alpha. In its new form, the station uses 75 percent of the hardware designs originally intended for the *Freedom* program. After the Russians agreed to supply major hardware elements, many originally intended for their *Mir 2* space station program, the station became known as the *International Space Station*. Russian participation reduces the station's cost to the U.S. while permitting expansion to basic operational capability much earlier than *Freedom*. This provides new opportunities to all the station partners by permitting early scientific research.

The program's management was also redesigned. Johnson Space Center became lead center for the space station program, and Boeing became prime contractor. NASA and Boeing teams are housed together at JSC to increase efficiency through improved communications.

The first phase of the International Space Station Program, the Shuttle-*Mir* Program, kicked off in February 1994 with STS-60, when Sergei Krikalev became the first Russian astronaut to fly on a shuttle. See **Shuttle-Mir Program** and Table 1.

The *International Space Station* (ISS) will be a permanent research platform in space, a laboratory orbiting the Earth. The *International Space Station* is the single largest international aerospace project ever undertaken by humankind. Assembly of the *International Space Station* involves complex on-orbit operations and hardware/software integration tasks that are unprecedented in space history. Drawing upon the scientific and technological expertise of 16 nations: the United States, Canada, Japan, Russia, 11 member nations of the European Space Agency (ESA), and

TABLE 1. TIMELINE OF SHUTTLE-*MIR* (1994–1998)

Mission	Duration	Highlights
STS-60	February 3–11, 1994	The first shuttle flight of 1994 marked the first flight of a Russian cosmonaut, Sergei Krikalev, onboard the U.S. space shuttle. Part of an international agreement on human space flight, the mission also was the second flight of the Spacehab pressurized module and marked the 100th "Get Away Special" payload to fly in space.
STS-63	February 3–11, 1995	Referred to as the "near *Mir*" mission, STS-63 had special importance as a dress rehearsal for the missions that would rendezvous and dock with *Mir*. With Cosmonaut Vladimir Titov aboard, *Discovery* rendezvoused with Mir, closed to within 37 feet (11.3 meters) then backed off to 400 feet (122 meters) and performed a fly-around, but did not dock. Docking equipment was tested for future missions, and communications procedures between the two mission control centers were validated.
Mir 18	March 14–July 7, 1995	Norman Thagard was the first American astronaut to train in Russia, the first to launch into space aboard a *Soyuz TM-21* (with cosmonauts Vladimir Dezhurov and Gennady Strekalov), and the first to complete a residency aboard *Mir*, setting an American space endurance record of 115 days in orbit. During that time, the *Spektr* science module was launched to *Mir*, with more than 1,500 pounds of research equipment from the U.S. and other countries. Thagard returned to Earth with the space shuttle mission STS-71, the first shuttle flight to dock with *Mir*.
STS-71	June 27–July 7, 1995	STS-71 marked the 100th U.S. human space launch; the first docking of the space shuttle with *Mir*; the largest combined spacecraft ever in orbit; and the first shuttle changeout of a *Mir* crew, Cosmonauts Anatoly Soloyev and Nikolai Budarin, and returned Dezhurov, Strekalov and Thagard to Earth.
STS-74	November 12–20, 1994	STS-74 marked the second docking of the U.S. space shuttle to Russian Space Station *Mir* and illustrated the international flavor of the space station effort. The mission's primary payload was the Russian-built Docking Module (DM), designed to become a permanent extension on *Mir* to afford better clearances for Shuttle-*Mir* linkups. Unlike the STS-71 crew exchange flight, this second docking focused on delivery of equipment to *Mir*.
STS-76	March 22–31, 1996	STS-76 was the third docking between a U.S. space shuttle and the Russian Space Station *Mir* and was highlighted by the transfer to *Mir* of astronaut Shannon Lucid. Lucid became the first American woman to live on the station and commenced a continuous U.S. presence in space for the next two years. Her 188-day stay eclipsed the long-duration U.S. spaceflight record set by the first American *Mir* astronaut, Norman Thagard. STS-76 also marked the first flight of the Spacehab pressurized module to support Shuttle-Mir dockings. Spacehab primarily served as a stowage area for supplies and equipment.
STS-79	August 16–26, 1996	STS-79 brought back to Earth U.S. astronaut Shannon Lucid after her 188 days in space, a new U.S. spaceflight record and a world record for a woman. Astronaut John Blaha succeeded Lucid on *Mir*, joining *Mir*-22 commander Valeri Korzun and flight engineer Alexander Kaleri. The mission conducted the first U.S. Shuttle-Mir crew exchange, the fourth Shuttle-Mir docking, and the first flight of the Spacehab Double Module configuration.
STS-81	January 12–22, 1997	STS-81 replaced U.S. astronaut John Blaha with Jerry Linenger, after Blaha's 118-day stay aboard Russian Space Station *Mir*. It transferred the most materials to date, during this fifth of nine dockings. *Atlantis* also returned the first plants to complete a life cycle in space — a crop of wheat planted by Shannon Lucid and grown from seed to seed. During Linenger's stay aboard, the fire of February 1997 occurred, offering new challenges and new information. Linenger conducted the first spacewalk by an U.S. astronaut outside *Mir* wearing a Russian spacesuit.
STS-84	May 15–24, 1997	STS-84 delivered U.S. Astronaut Michael Foale to *Mir* and brought back to Earth astronaut Jerry Linenger, who had spent 123 days on *Mir* and who was, at that time, second behind Shannon Lucid for most time spent on-orbit by an American. One of the first items transferred to *Mir* was an Elektron oxygen-generating unit, especially important after the Feb. 23, 1997 fire that had occurred on *Mir*.
STS-86	September 25–October 6, 1997	STS-86 performed the seventh Shuttle-*Mir* docking and continued the presence of a U.S. astronaut onboard *Mir*, by transferring the sixth U.S. *Mir* astronaut David Wolf, in exchange for Michael Foale, who returned to Earth after 145 days in space. Crewmembers Titov and Parazynski conducted the first joint U.S.-Russian extravehicular activity during a shuttle mission, and the first in which a Russian wore a U.S. spacesuit. During the 5-hour spacewalk on October 1, 1997, the pair affixed a 121-pound (54.9 kilogram) Solar Array Cap to the Docking Module, for a future attempt by *Mir* crew members to seal off the suspected leak in *Spektr's* hull. Parazynski and Titov also retrieved four Mir Environmental Effects Payloads (MEEPs) from the outside of Mir and tested several components of the Simplified Aid for EVA Rescue (SAFER) jetpacks.
STS-89	January 22–31, 1998	STS-89 delivered Cosmonaut Salizhan Sharipov to *Mir* and replaced U.S. Mir astronaut David Wolf with Andy Thomas, after Wolf had spent 119 days onboard Mir and a total of 128 days on-orbit. *Endeavour* was the first orbiter after *Atlantis* to dock with *Mir*.
STS-91	June 2–12, 1998	STS-91 closed out the Shuttle-*Mir* Program, when it brought back to Earth U.S. *Mir* astronaut Andy Thomas, after he had spent 130 days on *Mir*. The transfer wrapped up a total of 907 days spent by seven U.S. astronauts aboard the Russian space station. When the hatches closed for undocking at 9:07 a.m. EDT June 8 and the spacecraft separated at 12:01 p.m. EDT that day, the final Shuttle-*Mir* docking mission was concluded and Phase 1 of the International Space Station (ISS) program came to an end.

For more information on the Shuttle-*Mir* Program timeline see *http://spaceflight.nasa.gov/history/shuttle-mir/history/h-timeline.htm.*

Fig. 17. An early conceptualization of the *International Space Station*, which will be a permanent research platform in space, a laboratory orbiting the Earth. (*NASA.*)

Brazil, enabling scientists from around the globe to perform long-duration research in the unique environment of microgravity. See Fig. 17.

More than four times as large as the Russian *Mir* space station, the completed *International Space Station* will have a mass (weight) of about 1 million pounds (453,592 kilograms). The ISS inclination will be approximately 51.6 degrees to the equator. The operating altitude is 220 nautical miles average (407 kilometers) because of orbital decay and the reboost schedule. There are 19 scheduled reboost flights during the assembly phase. The wingspan width is 356 feet (108.5 meters) and the length is 290 feet (88.4 meters). Power generated by the completed station will be about 110 kilowatts with 46 kilowatts dedicated to scientific research. This power will be generated by 4 photovoltaic modules, each with 2 arrays measuring 34.1 meters by 11.2 meters (112 feet by 39 feet). The atmosphere inside will be 14.7 psi (101.36 kilopascals) the same as Earth. Total pressurized volume of the completed station will be approximately 43,000 cubic feet (1,218 m^3) in 6 laboratories.

The pressurized living and working space aboard the completed ISS will be about the size of three average American homes. It giant solar arrays will generate the electricity needed to power about 50 average American homes. Inside the ISS its weightless environment will be maintained at "shirt sleeve" temperatures.

Six main laboratories will house research facilities, which include:

- Two U.S.—a laboratory module called "Destiny" and a Centrifuge Accommodations Module (CAM).
- One European Space Agency (ESA) laboratory named "Columbus."
- One Japanese Experiment Module named "Kibo."
- Two Russian Research Modules.

Researchers in government agencies, commercial enterprises, and scientific institutions will utilize the space station state-of-the-art laboratory facilities. The highly advanced data computing capabilities of the *International Space Station* will enable researchers to continuously operate multiple experiments. Scientists will be able to monitor and conduct their experiments from computer terminals in laboratories all over the world. With the exception of the centrifuge, all laboratory facilities will be designed to fit into a standard payload rack, called an "International Standard Payload Rack" (ISPR). See Fig. 18.

The central girder, called the truss, will connect the modules and four giant solar arrays making the ISS larger than a football field. The Canadian-built Remote Manipulator System, a 55-foot robot arm and a grappling

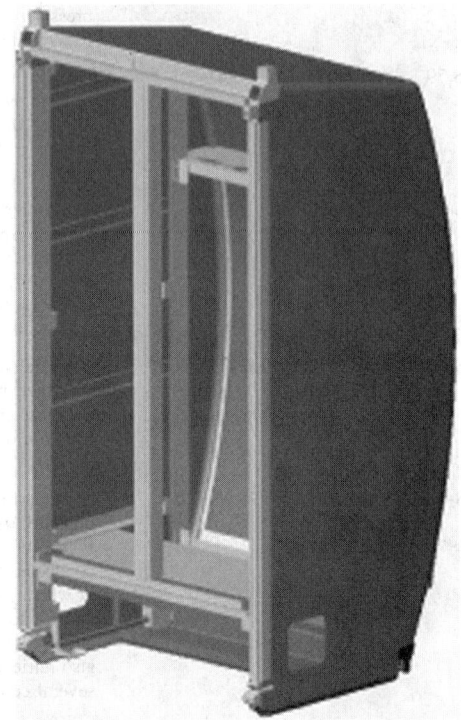

Fig. 18. The standard payload rack, called an International Standard Payload Rack (ISPR) will allow payloads to be easily changed out as needed and will allow facilities and payloads developed by one partner to fit into rack space supplied by another. (*NASA.*)

mechanism called the Special Purpose Dexterous Manipulator (SPDM) will move along the truss on a mobile base transporter to perform assembly and maintenance work.

External sites for mounting experiments intended for looking down to Earth and out into space or for direct exposure to space are provided at four locations on the truss structure, along with 10 on the Japanese *Kibo* Module's back porch and 4 on the ESA *Columbus* Module exposed facility.

These external experiment sites vary as to the number of payloads that can be accommodated.

A three-person Russian Soyuz capsule provides emergency crew return.

A variety of vehicles will be visiting the ISS to ferry crew and supplies to and from Earth. Crew exchanges will be accomplished with the Space Shuttle and *Soyuz*. Russian Progress spacecraft, Japanese H-II Transfer Vehicle, and Europe's Autonomous Transfer Vehicle (ATV) will provide resupply and reboost.

Responsibility for the planning and operations of the *International Space Station* is divided among the partners. NASA has the overall integration role for systems and payload operations. The Space Station Control Center (SSCC) in Houston is responsible for crew safety, rendezvous and docking, and systems such as life support, propulsion, communications, and power. The Payload Operations Integration Center (POIC) in Huntsville is responsible for integrated payload operations and is the focal point for the science planning and operations that will occur around the world.

The other partners' control centers work with the SSCC and POIC to keep the ISS running smoothly and to ensure that specific goals and program objectives are being met. The Space Station Information and Promotion Center (SSIPC, Japan), Attached Pressurized Module Control Center (APM-CC, Europe), and Mission Control Center-Moscow (MCC-M, Russia) are all responsible for planning and operating the payloads and systems that reside within their respective modules.

Progress to Date

The *International Space Station* and its crew of three are orbiting the Earth at an altitude of over 230 miles (373 kilometers) and at a speed exceeding 17,000 mph (27,358 km/h).

Operational Progress. A new chapter in human space flight history unfolded with the launch of the first permanent occupants of the new *International Space Station* (ISS), the Expedition One crew. Liftoff occurred October 31, 2000 aboard a Russian *Soyuz* rocket from the Baikonur Cosmodrome in Kazakhstan. U.S. astronaut, Bill Shepherd, Russian cosomonauts, Yuri Gidzenko as *Soyuz* Commander and Sergei Krikalev as Flight Engineer, docked with the ISS on November 2, 2000 to begin a pioneering four-month mission to prepare the facility for future scientific research. The Expedition One crewmembers returned to Earth March 21, 2001 aboard the Space Shuttle *Discovery*.

On March 8, 2001 *Discovery* STS-102 delivered the second resident crew Expedition Two. Relieving a crew that transformed the *International Space Station* into a home, the second station crew opened the research outpost for business, beginning scientific work while continuing the growth in size and sophistication.

The mission of the station's Expedition Two crew, Commander Yury Usachev and Flight Engineers Jim Voss and Susan Helms, was highlighted by the arrival of the complex's first major experiments; the installation and checkout of a Canadian-developed station robotic arm; and the installation and initial space walks from a new station airlock designed for both U.S. and Russian spacesuits.

The Expedition Two crew worked with 18 different scientific experiments. The research included studies of the human body in space, space radiation, observations of the Earth, crystal growth in weightlessness and plant growth in space.

When Usachev, Voss, and Helms handed off their assignments to the Expedition Three crew, during space shuttle mission STS-105 on August 12, 2001 the station had reached a significant plateau of self-reliance: able to sustain and maintain itself for long periods and capable of ongoing research activities.

On August 22, 2001 the Expedition Three crew, Commander Frank Culbertson, Pilot Vladimir Dezhurov, and Flight Engineer Mikhail Tyurin, assumed command of a four-month expedition, which was highlighted by the extension of scientific research and the addition of a new Russian docking port to the expanding complex.

Expedition Three science focused on the effects of space flight on bone and muscle mass during extended stays in space and how such phenomena may relate to similar conditions on Earth. Additional experiments in the fields of human life sciences, biotechnology, education, and video technology were conducted.

The Expedition Three crewmembers returned to Earth December 17, 2001 aboard the Space Shuttle *Endeavour*.

On December 5, 2001 the Expedition Four crew, Commander Yury I. Onufrienko, Flight Engineers Daniel W. Bursch and Carl E. Walz, assumed command of a four- month expedition. The number of science investigations aboard the *International Space Station* will almost double during Expedition Four, and new equipment will make the research outpost even more productive.

With the successful installation of Quest the Airlock, in July 2001, the ISS has completed it Phase II Assembly milestone and is now a fully functioning operational facility. Key operational systems activated and confirmed include the following:

- Life-support systems needed to maintain a "shirt-sleeve" environment for the crew.
- Power systems needed to generate, store, and distribute renewable electricity from the solar arrays.
- Flight control systems needed to maintain a stable orbiting platform.
- Voice systems needed to communicate among the orbiting modules and with ground control data systems for commanding and monitoring ISS health and status.
- Airlock and space suit systems needed to conduct spacewalks without the Space Shuttle present.

Mission Control Center-Houston (MCC-H) has assumed responsibility from MCC-Moscow as the lead ISS Control Center.

Development and Assembly Progress

Each International Partner has made contributions to the ISS configuration on orbit:

- *Destiny*, the U.S. Laboratory, the most complex and capable piece of the ISS.
- Russia's Zvezda Service Module.
- NASA-developed Neutron Detector in the U.S. Lab, *Kibo*, Japan's Experiment Module and principal contribution, continues on schedule for a 2004 launch.
- European Space Agency's (ESA's) data management equipment on the Service Module. The Columbus Module, ESA's principal contribution, is scheduled for a 2005 launch.
- Canada's robotic arm was delivered to orbit in April of 2001.
- In addition, two of the three U.S.-owned, Italian-built Multi-Purpose Logistics Modules have been delivered and are making logistics flights to and from the ISS.

Approximately 75 percent of all U.S. flight hardware has been delivered to Kennedy Space Center for deployed to orbit.

More than 46 space flights over five years and at least three space vehicles — the Space Shuttle, the Russian *Soyuz* rocket and the Russian Proton rocket — will deliver the various space station components to Earth orbit. Assembly of the more than 100 components, will require a combination of human space walks and robot technologies.

ISS Elements Launched to Date

Sixteen flights, which includes 12 space shuttle missions, have already occurred in the *International Space Station* era:

- November 20, 1998, ISS Mission 1A/R, the *Zarya* Control Module (Functional Cargo Block—FGB) was launched into orbit aboard a Russian Proton rocket.
- December 4, 1998, ISS Mission 2A Space Shuttle *Endeavour* (STS-88) attached the *Unity module* and two Pressurized Mating Adapters to *Zarya*.
- May 27, 1999, ISS Mission 2A.1 Space Shuttle *Discovery* (STS-96). For flight 2A.1 the double SPACEHAB module carried internal and resupply cargo for station outfitting. The STS-96 Shuttle Discovery flight will go into the books as the first docking of a shuttle with the *International Space Station*.
- May 19, 2000, ISS Mission 2A.2a Space Shuttle *Atlantis* (STS-101) prepared the station for the arrival of the *Zvezda* Service Module. Installed four new batteries, 10 new smoke detectors and four new cooling fans on the *Zarya* module; Installed the final parts of the Strela crane on Pressurized Mating Adapter 1; Removed and replaced the early communications system antenna; Installed handrails on Node 1 (*Unity module*).
- July 12, 2000, ISS Mission 1R attached the *Zvezda* Service Module. The Service Module is the primary Russian station contribution and an early station living quarters. It provides life support system functions to all early elements.

- September 8, 2000, ISS Mission 2A.2a Space Shuttle *Atlantis* (STS-106). The crew spent five days, 9 hours and 21 minutes inside the *International Space Station*. The seven crewmembers completed a long checklist aimed at making the station a home for its first residents.
- October 11, 2000, ISS Mission 3A, Space Shuttle *Discovery* (STS-92) attached the Integrated Truss Structure (ITS) Z1, Pressurized Mating Adapter 3 (PMA 3), Ku-band Communications System and the Control Moment Gyros (CMGs).
- October 31, 2000, ISS Mission 2R (*Soyuz*) Test Flight and Assembly. Provides Russian assured crew return capability without the space shuttle present. The initial three-person crew (Expedition One) establishes first human presence on the space station.
- Nov. 30, 2000, ISS Mission 4A, Space Shuttle *Endeavour* (STS-97) attached the first set of U.S. solar arrays and batteries, called the "Photovoltaic (PV) Module," radiators, and the P6 Truss.
- February 7, 2001, ISS Mission 5A, Space Shuttle *Atlantis* (STS-98) attached the U.S. *Destiny* Laboratory Module.
- March 8, 2001, ISS Mission 5A.1, Space Shuttle *Discovery* (STS-102) delivered the second resident crew and attached *Leonardo*, the Multi-Purpose Logistics Module (MPLM).
- April 19, 2001, ISS Mission 6A, Space Shuttle *Endeavour* (STS-100) delivers the Canada's Robot Arm (Canadarm2), the station's mechanical arm, which is needed to perform assembly operations on later flights. Attached the *Raffaello* the Italian-built Multi-Purpose Logistics Module (MPLM) which carries six system racks and two storage racks for the U.S. Lab.
- July 12, 2001, ISS Mission 7A, Space Shuttle *Atlantis* (STS-104) delivered the Joint Airlock, enabling spacewalks without the Space Shuttle present.
- August 10, 2001, ISS Mission 7A.1 Space Shuttle *Discovery* (STS-105) delivered the Expedition Three crew and attached the *Leonardo* Multi-Purpose Logistics Module, transferred supplies and equipment to the station.
- September 14, 2001, ISS Mission 4R (*Soyuz*) attached the Docking Compartment 1 (DC-1) and Strela Boom. These provide additional egress and ingress location for Russian-based space walks and a *Soyuz* docking port.
- December 5, 2001, ISS Mission UF1, Space Shuttle *Endeavour* (STS-108) attached the *Raffaello* Multi-Purpose Logistics Module to the station so that about 2.7 metric tons (3 tons) of equipment and supplies could be unloaded. The crew later returned *Raffaello* to *Endeavour's* payload bay for the trip home. Expedition Four crew delivered.

International Space Station elements scheduled for launched over the next 5 years are shown in Table 2.

For a view of the *International Space Station*, assembled complete, in its most current configuration design, along with the components labeled by country of origin, see Fig. 19.

International Space Station Modules and Components

Zarya Control Module: The *Zarya Control Module*, also known by the technical term Functional Cargo Block and the Russian acronym (FGB), was the first component launched for the *International Space Station*. This module was designed to provide the station's initial propulsion and power. The 19,323-kilogram (42,600-pound) pressurized module was launched on a Russian Proton rocket November 20, 1998.

The U.S.-funded and Russian-built *Zarya*, which means "Sunrise" when translated to English, is a U.S. component of the station although it was built and launched by Russia. The module was built by the Khrunichev State Research and Production Space Center, which is also known as KhSC, in Moscow under a subcontract to The Boeing Company for NASA. Only weeks after *Zarya* reached orbit, Space Shuttle *Endeavour* made a rendezvous and attached a U.S.-built connecting module called *Node 1*, or *Unity*. The *Zarya* module provided orientation control, communications and electrical power attached to the passive *Node 1* while the station awaited launch of the third component, a Russian-provided crew living quarters and early station core known as the *Zvezda Service Module*. The Service Module enhanced or replaced many functions of *Zarya*. Later in the station assembly sequence, the *Zarya module* will be used primarily for its storage capacity and external fuel tanks.

The *Zarya module* is 12.6 meters (41.2 feet) long and 4.1 meters (13.5 feet wide) at its widest point. See Fig. 20. It has an operational lifetime of at least 15 years. Its solar arrays and six nickel–cadmium batteries can provide an average of 3 kilowatts of electrical power. Using the Russian Kurs system, the *Zarya* performed an automated and remotely piloted rendezvous and docking with the Service Module in orbit. Its side docking ports will accommodate Russian *Soyuz* piloted spacecraft and unpiloted *Progress* resupply spacecraft. See Figs. 21 and 22. Each of the two solar arrays is 10.7 meters (35 feet) long and 3.4 meters (11 feet) wide. The module's 16 fuel tanks combined can hold more than 5.4 metric tons (6 tons) of propellant. The attitude control system for the module includes 24 large steering jets and 12 small steering jets. Two large engines are available for reboosting the spacecraft and making major orbital changes.

Construction of the *Zarya* module began at KhSC in December 1994. It was shipped to the Baikonur Cosmodrome, Kazakhstan, launch site to begin launch preparations in January 1998. The three-stage proton rocket launched the module into a 220.4- by 339.6-kilometer (137- by 211-statute mile) orbit. See Fig. 23. During launch, the module's systems were in an idle mode to conserve battery power. After reaching the initial elliptical orbit and separating from the Proton's third stage, a set of preprogrammed commands automatically activated the module's systems and deployed the solar arrays and communications antennas. After several days of operational tests, the module was commanded to fire its engines and circularize its orbit at an altitude of about 386.2 kilometers (240 statute miles), the orbit at which *Endeavour* made rendezvous and captured the spacecraft to attach it to the U.S.-built *Unity Connecting Module*.

TABLE 2. *INTERNATIONAL SPACE STATION* PROPOSED SCHEDULE FOR UPCOMING MISSIONS

Mission	Launch Date	Launch Vehicle	Elements
8A	04/04/2002	Shuttle *Atlantis*	Central Truss Segment (ITS S0) and Mobile Transporter (MT)
UF-2	05/02/2002	Shuttle *Endeavour*	*Multipurpose Logistics Module* (MPLM) with payload racks. The Mobile Base System is installed on the Mobile Transporter to complete the *Canadian Mobile Servicing System* (MSS).
9A	08/01/2002	Space Shuttle	First right-side Truss segment (ITSS1) with radiators; Crew and Equipment Translation Aid (CETA) Cart A.
11A	09/06/2002	Space Shuttle	First left-side truss segment (ITSP1); Crew and Equipment Translation Aid (CETA) Cart B.
12A	01/2003	Space Shuttle	Second left-side truss segment (ITS P3/P4); Second set of Solar arrays and batteries (Photovoltaic Module).
12A.1	04/2003	Space Shuttle	Third left-side truss segment (ITSP5), logistics and supplies
13A	05/2003	Space Shuttle	Second right-side truss segment (ITS S3/S4); The Mobile Base System is installed on the Mobile Transporter to complete the Canadian Mobile Servicing System, or MSS. Third set of U.S. Solar arrays and batteries (Photovoltaic Module).

TABLE 2. (*Continued*)

Mission	Launch Date	Launch Vehicle	Elements
15A	11/2003	Space Shuttle	Fourth and final set of U.S. solar arrays delivered along with fourth starboard truss segment, the S6 Truss.
3R	TBD	Russian Proton Rocket	Universal Docking Module.
5R	TBD	Russian *Soyuz* Rocket	Docking Compartment 2 (DC2).
10A	02/2004	Space Shuttle	The second of three station connecting modules, *Node 2*, attaches to end of U.S. Lab and provides attach locations for the Japanese laboratory, European laboratory, the *Centrifuge Accomodation Module* and later *Multipurpose Logistics Modules*.
9A.1	04/2004	Space Shuttle	Russian provided Science Power Platform (SPP) with four solar arrays and European Robotic Arm, or ERA, which is a second station mechanical arm that will be used to maintain the SPP.
1J/A	08/2004	Space Shuttle	Japanese Experiment Module Experiment Logistics Module (JEMELM PS) and two additional solar arrays for the Russian Science Power Platform, or SPP.
1J	09/2004	Space Shuttle	The primary Japanese contribution, the Japanese Experiment Module, or JEM, laboratory (*Kibo*), is delivered and begins use and a Japanese robotic arm attached to the Japanese Experiment Module is delivered.
UF-3	01/2005	Space Shuttle	Multipurpose Logistics Module (MPLM) and *EXPRESS Pallet*.
1E	02/2005	Space Shuttle	The European Space Agency's primary contribution to the station, the Columbus Orbital Facility laboratory (*Columbus Module*).
UF-4	05/2005	Space Shuttle	Spacelab Pallet carrying "Canada Hand" (Special Purpose Dexterous Manipulator) and Alpha Magnetic Spectrometer experiment to be attached to station truss site.
2J/A	06/2005	Space Shuttle	The Japanese "back porch," the Exposed Facility, for Japanese laboratory along with external experiments carried in a Japanese exterior logistics carrier.
UF-5	09/2005	Space Shuttle	*Multipurpose Logistics Module* and *EXPRESS Pallet*.
UF-6	TBD	Space Shuttle	*Multipurpose Logistics Module* (MPLM) and Batteries; *EXPRESS Pallet*.
14A	TBD	Space Shuttle	*Cupola* with eight windows provides station crew with direct viewing capability for some robotics operations, space walks and experiments and Science Power Platform (SPP) Solar Arrays; Zvezda Micrometeroid and Orbital Debris Shields (MMOD); MT/CETA Rails.
16A	TBD	Space Shuttle	U.S. *Habitation Module* to enhance crew accommodations and provide for a station crew with as many as seven members.
17A	TBD	Space Shuttle	*Multipurpose Logistics Module*; Destiny racks; Common Berthing Mechanism—an interface between the Crew Return Vehicle and U.S. Node 3.
18A	TBD	Space Shuttle	X-38 Crew Return Vehicle is attached to the station provides additional seven-person crew return capability, added to already existing three-person *Soyuz* crew return capability.
20A	TBD	Space Shuttle	U.S. Node 3. Node 3 provides attachment points for the U.S. Habitation Module, Crew Return Vehicle, pressurized mating adapter and any future station additions.
8R	TBD	Russian *Soyuz* Rocket	First of two Russian laboratories providing experiment and research facilities (Research Module 1).
9R	TBD	Russian Proton Rocket	*Docking and Stowage Module* (DSM). Provides additional on-orbit stowage and *Soyuz* docking location
10R	TBD	Russian *Soyuz* Rocket	Second Russian laboratory to house experiments and research facilities (Research Module 2).
19A	TBD	Space Shuttle	*Multipurpose Logistics Module* (MPLM)
UF-7	TBD	Space Shuttle	*Centrifuge Accommodation Module* (CAM), completes the complement of station laboratory facilities providing a facility to control gravity for research activities. The CAM attaches to *Node 2*.

International space station

- Science power platform
- Service module
- Docking compartment
- Universal docking module
- Research module
- Soyuz
- Research module
- Zarya (Sunrise) Contol Module
- Pressurized mating adaptor 1
- Docking and stowage module
- Soyuz
- S3 truss segment
- Express pallet
- Thermal control panels
- S0 truss segment
- Mobile servicing system
- Unity (Node 1)
- CSA remote manipulator system
- Z1 truss segment
- P3 truss segment
- P1 truss segment
- Solar alpha rotary joint
- P5 truss segment
- P4 Truss segment
- P6 Truss segment
- Port photovoltaic arrays
- Centrifuge accomodation module
- JEM experiment logistics module
- JEM remote manipulator system
- JEM exposed facility
- Japanese experiment module (JEM)
- Pressurized mating adapter 2
- Multi-purpose logistics module
- European lab - Columbus orbital facility
- Node 2
- U.S. Lab
- Cupola
- Airlock
- Habitation module
- Node 3
- Pressurized mating adaptor 3
- Crew return vehicle
- Solar alpha rotary joint
- S4 truss segment
- S5 truss segment
- S6 truss segment
- Starboard photovoltaic arrays
- S1 truss segment

Legend:
- United States
- Russia
- Japan
- Europe
- Canada
- Italy
- Brazil

ISS-ov 7.24.98
http://station.nasa.gov/gallery

Fig. 19.   Source for a component view of the International Space Station with each element identified by its name and ownership at assembly complete. (Rev. D) (*NASA.*)

3282

Fig. 20.   The *Zarya* module, also known as the Functional Cargo Block, of the *International Space Station* in Khrunichev's manufacturing facility in Moscow. (*NASA.*)

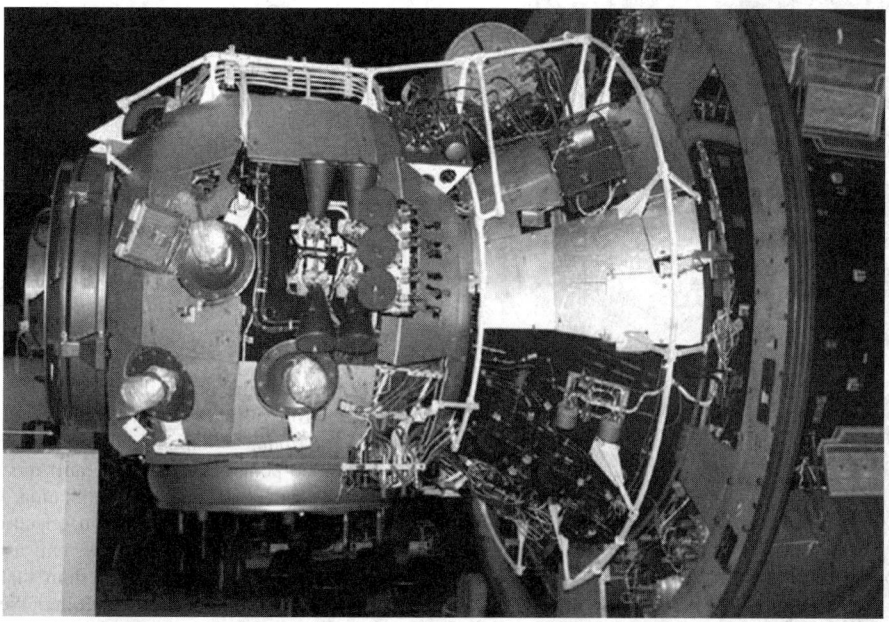

Fig. 21.   Two docking ports on the control module, or Functional Cargo Block (Russian acronym FGB), are shown in this view as the module is prepared for shipment from its Moscow factory to the launch site in Kazakstan in January 1998. (*NASA.*)

Unity Connecting Module: The first U.S.-built component of the *International Space Station*, a six-sided connecting module and passageway, or node, named *Unity*, was the primary cargo of Space Shuttle mission STS-88, launched on December 4, 1998 as the first mission dedicated to assembly of the station.

Now permanently attached to the *Zarya control module* in orbit, the *Unity connecting module* lays a foundation for all future U.S. *International Space Station* modules. *Unity* has six berthing ports, one on each side, one of which already is attached to *Zarya*. Future U.S. station modules and station components will attach to the remaining five ports. Built by the Boeing Company at a manufacturing facility at the Marshall Space Flight Center in Huntsville, Alabama, *Unity* is the first of three such connecting modules that will be built for the station. Sometimes referred to as *Node 1*, the *Unity module* measures 15 feet (4.6 meters) in diameter and 18 feet (5.5 meters) long. Including mating adapters attached at each end, as is currently the case in orbit, the overall component measures about 34 feet (10.4 meters) long. See Fig. 24.

Essential space station resources such as fluids, environmental control and life support systems, electrical and data systems are routed through *Unity* to supply work and living areas of the station. More than 50,000 mechanical items, 216 lines to carry fluids and gases, and 121 internal and external electrical cables using six miles of wire were installed in the *Unity node*. *Unity* is made of aluminum. See Fig. 25.

The remote power control modules (RPCM's) that are contained in the Unity node were also designed and developed by NASA Glenn Research Center. Each of the 12 small metal containers is filled with "circuit breakers" that provide switching and protection in case of a short circuit during construction of the ISS or during operation of the completed station. They can be controlled by astronauts via laptop computers or by ground personnel.

Two conical docking adapters were attached to each end of *Unity* prior to its launch aboard *Endeavour*. The adapters, called *pressurized mating adapters* (PMAs), allow the docking systems used by the Space Shuttle and by Russian modules to attach to the node's hatches and

Fig. 22.   The aft docking port on the control module, or Functional Cargo Block (Russian acronym FGB), is seen in this view as the module is prepared for shipment from its Moscow factory to the launch site in Kazakstan in January 1998. (*NASA.*)

Fig. 23.   *Zarya*, the first component of the *International Space Station* (ISS) launched flawlessly at 1:40 a.m. (EST) on November 20, 1998, from Kazahkstan. Atop a Russian Proton rocket, the *Zarya Module* was lifted into orbit from the Baikonur Cosmodrome in Kazahkstan. (*NASA.*)

berthing mechanisms. One of the conical adapters now permanently attaches *Unity* to *Zarya*, while the other provides a Shuttle docking port. *Unity* and the two mating adapters weigh about 25,600 pounds (11,600 kilograms). Attached to the exterior of the mating adapter that permanently attaches *Unity* to *Zarya* are computers, or multiplexer-demultiplexers (MDMs), which provide early command and control of *Unity*. *Unity* also is outfitted with an early communications system that

allows data, voice and low data rate video with Mission Control, Houston, to supplement Russian communications systems during the early station assembly activities.

The two remaining station connecting modules, or nodes, are being built by the European Space Agency (ESA) for NASA in Italy by Alenia Aerospazio. Nodes 2 and 3 will be slightly longer than the *Unity node*, measuring almost 21 feet (6.4 meters) long, and each will hold eight standard space station equipment racks in addition to six berthing ports. ESA is building the two additional nodes as partial payment for the launch of the ESA *Columbus laboratory module* and other equipment on the Space Shuttle. *Unity* holds four equipment racks.

Zvezda Service Module: The *Zvezda Service Module* was the first fully Russian contribution to the *International Space Station* and will serve as the early cornerstone for the first human habitation of the station. The *Zvezda Service Module* a unpiloted component was launched on July 12, 2000 as the third station component to reach orbit, docking by remote control with the already orbiting *Zarya Control Module* and *Unity Connecting Module* at an altitude of about 250 statute miles.

The 42,000-pound (19,100-kilogram) module, similar in layout to the core module of Russia's *Mir* Space Station, will provide the early station living quarters; life support system; electrical power distribution; data processing system; flight control system; and propulsion system. It also will provide a communications system that includes remote command capabilities from ground flight controllers. See Fig. 26.

Although many of these systems will be supplemented or replaced by later U.S. station components, the Service Module will always remain the structural and functional center of the Russian segment of the International Space Station.

The module has a wingspan of 97.5 feet (29.9 meters) from tip to tip of the solar arrays, and 43 feet (13.1 meters) long from end to end. The Service Module contains three pressurized compartments: a small, spherical Transfer Compartment at the forward end; the long, cylindrical main Work Compartment; and the small, cylindrical Transfer Chamber at the aft end. An unpressurized Assembly Compartment is wrapped around the exterior of the Transfer Chamber at the aft of the module. The Assembly Compartment holds external equipment such as propellant tanks, thrusters and communications antennas.

The Service Module includes four docking ports, one in the aft Transfer Chamber and three in the spherical forward Transfer Compartment—one facing forward, one facing up and one facing down. See Fig. 27. The aft docking port has a probe and cone docking mechanism to allow dockings by the *Progress* resupply spacecraft and *Soyuz* piloted spacecraft. It is also outfitted with an automated rendezvous and docking system. The forward docking ports have hybrid docking mechanisms to allow docking with the *Zarya* using the forward-facing port; with a Russian Science Power

Fig. 24.   A close-up view of the Node 1 in its work stand in the Space Station Processing Facility at Kennedy Space Center shows two of its six hatches that will serve as docking ports. (*NASA*.)

Fig. 25.   The interior of the first U.S. module to be launched to the *International Space Station*, a connecting module called *Node 1*, is shown as it neared the completion of manufacturing in the spring of 1997. The node was shipped from the Boeing manufacturing facility at the Marshall Space Flight Center in Huntsville, Alabama, to the Kennedy Space Center, Florida. (*NASA*.)

Platform to be delivered later in the station's assembly using the up-facing port; and with a Russian Universal Docking Module to be delivered later in assembly using the down-facing port.

Living accommodations on the Service Module include personal sleeping quarters for the crew; a toilet and hygiene facilities; a galley with a refrigerator-freezer; and a table for securing meals while eating. The module has a total of 14 windows, including three 9-inch (22.9-centimeter) diameter windows in the forward Transfer Compartment for viewing docking activities; one large 16-inch (40.4-centimeter) diameter window in the Working Compartment; an individual window in each crew compartment; and additional windows positioned for Earth and intramodule observations. Exercise equipment will include a NASA-provided treadmill and a

stationary bicycle. The crew's wastewater and condensation water will be recycled for use in oxygen-generating devices on the module, but it is not planned to be recycled for use as drinking water. Spacewalks using Russian Orlan-M spacesuits can be performed from the Service Module by using the Transfer Compartment as an airlock. The module also provides data, voice and television communications with Mission Control Centers in Moscow and in Houston.

Z1 Integrated Truss Structure: The *Z1 Truss* was the first permanent latticework structure for the space station, very much like a girder, setting the stage for the future addition of the station's major trusses or backbones. The Z1 fixture also serves as the platform on which the huge U.S. solar arrays were mounted on the following shuttle assembly flight, STS-97. It

Fig. 26.   Starboard side of the *Zvezda Service Module* exterior. The aft end is to the left. (*NASA*.)

Fig. 27.   Transfer compartment on the forward section of the *Zvezda Service Module*. (*NASA*.)

includes power distribution components, four flat discs that will be used to control the station's attitude, communications equipment, temperature control system hardware, space walk/extravehicular aids and power, data and coolant connections. See Fig. 28.

The crew of shuttle mission STS-92 attached the *Z1 Truss* to the station on October 14, 2000.

P6 Integrated Truss Structure: During shuttle mission STS-97, Space Shuttle *Endeavour* delivered the first set of U.S.-provided solar arrays and batteries, called the *P6 Photovoltaic Module*. They were temporarily installed on the *P6 Integrated Truss Structure* on the *Z1 Truss* until it can be relocated to its permanent location on the *P5 Truss* during a later assembly mission. See **Photovoltaic Cell**.

Fig. 28.   Not long after separation of the Space Shuttle *Discovery* from the *International Space Station* (ISS), a crewmember onboard the shuttle was able to use a 70 mm handheld camera to grab this "edge-on" image of the station, featuring its newest additions. Backdropped against the blackness of space, the *Z1 Truss* structure and its antenna, as well as the new pressurized mating adapter (PMA-3), are visible in the foreground. (*NASA.*)

The *P6 Integrated Truss Structure* contains three discrete elements: the Photovoltaic Array Assembly, the Integrated Equipment Assembly, and the Long Spacer.

The *P6* has four primary functions: the conversion or generation, storage, regulation and distribution of electrical power for the space station. The station derives its power from the conversion of solar energy into electrical power. The *Photovoltaic Power Module* performs this energy conversion.

The *P6 Integrated Truss Structure* was installed on Sunday, December 3, 2000.

Photovoltaic Module: Electrical power is the most critical resource for the *International Space Station* (ISS) because it allows the crew to live comfortably, to safely operate the station, and to perform scientific experiments. So, whether it is used to power the life support system, run a furnace that makes crystals, manage a computerized data network, or operate a centrifuge, electricity is essential. Since the only readily available source of energy for spacecraft is sunlight, NASA Glenn Research Center has pioneered, and continues to develop, technologies to efficiently convert solar energy to electrical power. One method of harnessing this energy, called *photovoltaics*, uses purified silicon solar cells to directly convert light to electricity. Large numbers of cells are assembled in arrays to produce high power levels.

However, a spacecraft orbiting the Earth is not always in direct sunlight. Therefore, the ISS relies on nickel-hydrogen rechargeable batteries to provide continuous power during the "eclipse" part of the orbit. The batteries ensure that the station is never without power to sustain life-support systems and experiments. During the sunlit part of the orbit, the batteries are recharged. The process of collecting sunlight, converting it to electricity, and managing and distributing this electricity builds up excess heat that can damage spacecraft equipment. This heat must be eliminated for reliable operation of the Space Station in orbit. The ISS power system uses radiators to dissipate the heat away from the spacecraft. The radiators are shaded from sunlight and aligned toward the cold void of deep space.

The power management and distribution subsystem disburses power at 160 volts of direct current (abbreviated as "dc") around the station through a series of switches. These switches have built-in microprocessors that are controlled by software and are connected to a computer network running throughout the station. To meet operational requirements, dc-to-dc converter units step down and condition the voltage from 160 to 120 volts dc to form a secondary power system to service the loads. The converters also isolate the secondary system from the primary system and maintain uniform power quality throughout the station.

The NASA Glenn Research Center has decades of experience in designing, building, and testing space power systems. Therefore, it became the NASA Center initially responsible for designing the Space Station power system—the largest power system ever constructed in space. Engineers at Glenn combined state-of-the-art electrical designs with complex computer-aided analyses.

The *International Space Station's* electrical power system (EPS) will use eight "photovoltaic solar arrays" to convert sunlight to electricity. Each of the eight solar arrays will be 112 feet (34.1 meters) long by 39 feet (11.9 meters) wide. With all eight arrays installed, the complete Space Station is large enough to cover a football field. See Fig. 29.

Because the Space Station needs very high power levels, the solar arrays will require more than 250,000 silicon solar cells. NASA has developed a method of mounting the solar arrays on a "blanket" that can be folded like an accordion for delivery to space. Once in orbit, astronauts will deploy the blankets to their full size. Gimbals will be used to rotate the arrays so that they face the Sun to provide maximum power to the Space Station.

The complete power system, consisting of U.S. and Russian hardware, will generate 110 kW (kilowatts) total power, about as much as 55 houses would typically use. Approximately 46 kW will be available for research activities.

Radiator System: The ISS radiator system maintains the temperatures of systems and components. It was tested at the NASA Glenn Space

Fig. 29.   The *International Space Station* is shown with its eight "photovoltaic solar arrays" installed. These solar arrays will convert sunlight to electricity. (*NASA.*)

Power Facility (SPF) at Glenn's Plum Brook Station in Sandusky, Ohio. This facility is the world's largest space environment simulation chamber 100 feet (30.5 meters) in diameter by 122 (37.2 meters) in height and is used to ground-test large space-bound hardware. It can simulate the severe conditions of space such as vacuum, low temperatures, and unfiltered sunlight.

The radiator system on the ISS consists of seven panels each about 6 by 12 feet (1.8 by 3.7 meters) designed to deploy in orbit from a 2-foot (0.6 meter) high stowed position to a 50-foot (15.2 meter) long extended position. The first round of tests confirmed that the deployment mechanism would operate properly in the cold void of space. The next phase of tests used a NASA Glenn-designed and assembled ammonia flow system to evaluate radiator performance. The ammonia collects heat from the Space Station's electronic equipment and module cooling components and transfers it to the radiator panels to be dissipated into space. Ammonia was selected because it was found to be the best heat transport fluid that meets all of NASA's thermal performance and safety requirements (i.e., toxicity, flammability, freeze temperature, stability, cost, and successful commercial and industrial use).

Destiny Laboratory Module: The *Destiny Laboratory Module* is the centerpiece of the *International Space Station*, where unprecedented science experiments will be performed in the near-zero gravity of space.

The 28-foot (8.5-meter) long, 14-foot (4.3-meter) diameter laboratory weighs 31,000 pounds (14, 100 kilograms) and cost approximately $1.4 billion. It is the most sophisticated and versatile space laboratory ever built and eventually will house an additional 18 racks for crew support and scientific research that can be removed and replaced periodically as experiment operations warrant. This versatility will allow researchers from around the world to conduct experiments in the unique microgravity environment of space as never before.

Destiny will provide many more functions of the station in addition to serving as the platform for experiment operations. It was launched with five system racks already installed. They include two avionics racks, two thermal control system racks and an atmosphere revitalization system rack. Each is capable of being tilted downward to provide access to the area behind.

The two avionics racks house equipment controlling the Communications and Tracking, Environmental Control and Life Support System, Thermal Control System, Command and Data Handling, and the Electrical Power System. These racks manage the audio equipment, video switching, and the computer switching boxes, called *Multiplexer/Demultiplexers*, which provide computer control of lab systems.

There are two thermal control system racks that circulate chilled water to cool other racks and the cabin air. One is a "low temperature" system, with water chilled to about 4 °C (39.2 °F). The low temperature system provides cooling to selected racks. The 'moderate temperature' system, with water chilled to about 17 °C (62.6 °F), cools selected racks and the cabin air.

The atmosphere revitalization (AR) rack provides for carbon dioxide removal, trace contaminant control, and monitoring of the cabin air. On orbit, the AR rack was moved from its launch location to its operational location during lab outfitting.

Additionally, the lab controls the function of the gyroscopes, called *Control Moment Gyros* (CMGs). These were preintegrated into the *Z1 Truss* launched on STS-92 in the fall of 2000 and were activated during STS-98. The four CMGs are electrically powered and provide attitude control of the station. At least two are required to provide attitude control without the need for supplementary control using the thrusters on *Zvezda* or *Zarya*.

The lab has three cylindrical sections and two endcones with hatches that will be mated to other station components. See Fig. 30. A 50.9-centimeter (20-inch) diameter window is located on one side of the center module segment. This pressurized module is designed to accommodate pressurized payloads. It has a capacity of 24 rack locations. Payload racks will occupy 13 locations specially designed to support experiments.

An exterior waffle pattern strengthens the hull of the lab. The exterior is covered by a debris shield blanket made of a material similar to that used in bulletproof vests on Earth. A thin aluminum debris shield is placed over the blanket for additional protection.

Destiny was launched into space aboard shuttle mission STS-98, station assembly flight 5A, on February 7, 2001.

Multi-Purpose Logistics Modules: The three *Multi-Purpose Logistics Modules*, which were built by the Italian Space Agency (ASI), are pressurized modules that serve as the *International Space Station's* "moving vans," carrying laboratory racks filled with equipment, experiments and supplies to and from the station aboard the space shuttle.

The unpiloted, reusable Logistics Modules function as both a cargo carrier and a space station module when they are flown. Mounted in the space shuttle's cargo bay for launch and landing they are berthed to the station using the shuttle's robotic arm after the shuttle has docked. While berthed to the station, racks of equipment are unloaded from the module and then old racks and equipment may be reloaded to be taken back to Earth. The Logistics Module is then detached from the station and positioned back into the shuttle's cargo bay for the trip home. When in the cargo bay, the cargo module is independent of the shuttle cabin, and there is no passageway for shuttle crewmembers to travel from the shuttle cabin to the module.

In order to function as an attached station module as well as a cargo transport, the Logistics Modules also include components that provide some life support, fire detection and suppression, electrical distribution and computer functions. Eventually, the modules also will carry refrigerator freezers for transporting experiment samples and food to and from the station. Although built in Italy, the logistics modules, technically known as *Multi-Purpose Logistics Modules* or MPLMs, are owned by the U.S. and provided in exchange for Italian access to U.S. research time on the station.

Fig. 30. The U.S. *Destiny* Laboratory module for the *International Space Station* is shown under construction in the fall of 1997 at the Marshall Space Flight Center station manufacturing facility in Huntsville, Al. (*NASA*.)

Fig. 31. The *Leonardo Multi Purpose Logistics Module* rests in *Discovery's* payload bay in this view taken from the station by a crewmember using a digital still camera. (*NASA*.)

The first MPLM, named *Leonardo*, was launched on the space shuttle *Discovery* mission STS-102 on March 8, 2001. See Fig. 31. On that flight, *Leonardo* was filled with equipment and supplies to outfit the U.S. *Destiny Laboratory Module*, which was carried to the station on STS-98. The second module, named *Raffaello*, was launched on space shuttle *Endeavour* (STS-100) on April 19, 2001. *Leonardo* flew a second time on the shuttle *Discovery* mission (STS-105) and will fly for a third time on STS-111 scheduled for April 2002. *Raffaello* flew for the second time on the space shuttle *Endeavour* mission (STS-108) December 5, 2001.

Construction of ASI's *Leonardo* module began in April 1996 at the Alenia Aerospazio factory in Turin, Italy. *Leonardo* was delivered to Kennedy from Italy in August 1998 by a special Beluga cargo aircraft. The cylindrical module is approximately 6.4 meters (21 feet) long and 4.6 meters (15 feet) in diameter, weighing almost 4.1 metric tons (4.5 tons). It can carry up to 9.1 metric tons (10 tons) of cargo packed into 16 standard space station equipment racks. Of the 16 racks the module can carry, five can be furnished with power, data and fluid to support a refrigerator freezer. *Raffaello* arrived at Kennedy in August 1999. The third module, named *Donatello*, was delivered to Kennedy on February 1, 2001.

The Italian Space Agency chose the names of the modules because they denote some of the great talents in Italian history. Leonardo da Vinci, an extraordinary inventor-scientist, civil engineer, architect, artist and military planner and weapons designer. Donato di Niccolo di Betto Bardi, one of the greatest sculptors of all time and one of the founders of modern sculpture; and Raffaello Sanzio, an artist whose work stands alone for its visual achievement of human grandeur, both in clarity of form and ease of composition.

Space Station *Remote Manipulator System* (SSRMS): Canada is contributing an essential component of the *International Space Station*, the *Mobile Servicing System*. This robotic system will play a key role in space station assembly and maintenance: moving equipment and supplies around the station, supporting astronauts working in space, and servicing instruments and other payloads attached to the space station. Astronauts will receive robotics training to enable them to perform these functions with the arm.

The Mobile Servicing System, which is also known as the MSS, has three parts:

- *Canadarm 2*: Launched on the space shuttle *Edneavour* mission STS-100 on April 19, 2001, the next generation Canadarm is a bigger, better, smarter version of the space shuttle's robotic arm. The robotic arm is essential to the continued assembly or the space station as the outpost grows beyond the reach of the shuttle's arm. It is 17.6 meters (57.7 feet) long when fully extended and has seven motorized joints. This arm is capable of handling large payloads and assisting with docking the space shuttle. The Space Station *Remote Manipulator System*, or SSRMS, is self-relocatable with a Latching End Effector, so it can be attached to complementary ports spread throughout the station's exterior surfaces. See Fig. 32.
- *Mobile Base System*: A work platform that moves along rails covering the length of the space station, the *Mobile Base System*, or MBS, will provide lateral mobility for the Canadarm as it traverses the main trusses. The *Mobile Base System* is scheduled for ISS Mission UF-2 (STS-111) April 2002.
- *Special Purpose Dexterous Manipulator*: The Special Purpose Dexterous Manipulator, or Canada Hand, is a smaller two-armed robot capable of handling the delicate assembly tasks currently handled by astronauts during space walks. The *Special Purpose Dexterous Manipulator* is scheduled for ISS Mission UF-4, May 2005.

Joint Airlock: The space shuttle Atlantis delivered the *Joint Airlock* to the *International Space Station* on the STS-104 Mission July 12, 2001, giving station crewmembers the ability to conduct spacewalks using either U.S. or Russian suits. The airlock also will vent less precious air into space than the shuttle airlock.

The airlock is a critical space station element because of design differences between American and Russian spacesuits. American suits will not fit through Russian-designed airlocks. During a series of integration tests, the Russian suits were connected to the airlock to assure that they worked together. The airlock is specially designed to accommodate both suits, providing a chamber where astronauts from every nation can suit up for spacewalks to conduct science experiments and perform maintenance outside the station. See Fig. 33.

The airlock serves two key purposes: to keep air from escaping when the hatch to space is opened and to regulate the air pressure before an astronaut enters or leaves the ISS. It has two compartments: the crew lock, from which astronauts will enter and leave the station; and the equipment lock, where the spacewalkers will change into and out of their suits and stow all necessary gear.

Built at a cost of $164 million (including associated tanks) the airlock was designed and built by Boeing at NASA's Marshall Space Flight Center (MSFC) in Huntsville, AL. The Airlock is constructed of aluminum and weighs 6,064 kilograms (13,368 pounds). The overall length is 5.5 meters (18 feet) and 4 meters (13 feet) in diameter with a volume of 34 cubic meters (1,200 cubic feet). Boeing-MSFC manufactured the equipment lock pressure shell, mated the equipment lock with the crew lock, installed all subsystems, and successfully performed airlock qualification testing.

Japanese Experiment Module (Kibo): Japan is obliged to develop an experiment module called *Kibo*. Kibo is the first manned facility of Japan in which a maximum of four astronauts can perform experimental activities for a long duration of time. Kibo consists of four components. Two experimental facilities, the *Pressurized Module* and *Exposed Facility*, logistics modules attached to each of them and, a Manipulator to be used for experiments or for ORU changeout tasks.

Pressurized Module (PM) is a facility where astronauts conduct experiments or control the total JEM facility. Inside PM, the air composition and pressure are kept the same as on Earth, and temperature and humidity are controlled so as to be comfortable for astronauts' activity all the time. Therefore, astronauts can work in the PM in a shirt-sleeve environment as we do on the ground. See Fig. 34.

Devices on board PM can be categorized into two types, system devices, which are vital to maintain the JEM facility itself, and experiment devices.

System devices are required to maintain JEM's function or affect astronauts' activities. Power supply, communications, air conditioning, device cooling water control or devices with space experiment support functions are included. Also, the manipulator console used to exchange

Fig. 32.   Astronaut Chris A. Hadfield, STS-100 mission specialist representing the Canadian Space Agency (CSA), stands on one Canadian-built robot arm to work with another one. Called *Canadarm2*, the newest addition to the *International Space Station* (ISS) was ferried up to the orbital outpost by the STS-100 crew. Hadfield's feet are secured on a special foot restraint attached to the end of the Remote Manipulator System (RMS) arm, which represents one of the standard shuttle components for the majority of the 100-plus STS missions thus far. The picture was recorded with a digital still camera. (*NASA.*)

Fig. 33. The airlock for the *International Space Station* is lifted by crane in the fall of 1997, during manufacturing in the Space Station Manufacturing Building at the Marshall Space Flight Center in Huntsville, AL. The airlock includes two sections: the larger "equipment lock" on the left that will store spacesuits and associated gear and the narrower "crewlock" on the right from which astronauts will exit into space. (*NASA.*)

Fig. 34. The (*Kibo*) *Pressurized Module* (PM) during assembly. (*NASA.*)

devices on the Exposed Facility, intersatellite communication equipment, or airlock through which devices can be exchanged between the EF and PM are very important, as such, if any of them were to be lost, the operation of JEM will face a serious situation.

Experiment devices are dedicated to conducting experiments invited from the public. They can perform their functions only in cooperation with the System Devices stated above. The PM contains a total of ten experiment racks equipped with mainly biological and material experiment functions. Researchers will attempt new material development or will try to solve scientific problems.

Experiment Logistics Module Pressurized Section (ELM-PS): When the *Experiment Logistics Module Pressurized Section* (ELM-PS) is launched for the first time, it will be used as a container to carry experiment racks and system racks. Once on orbit, it will be mainly used as a storage facility. System devices, tools to be used for maintenance of experiment devices, materials for experiments or spare items for trouble are stored inside the ELM-PS. The ELM-PS volume is less than that of the PM. See Fig. 35.

Although ELM-PS has not received much attention among the JEM components, it is going to be the first manned spacecraft of Japan. JEM

Fig. 35.   The (*Kibo*) Experiment Logistics Module Pressurized Section (ELM-PS) during assembly. (*NASA*.)

will be launched by three separate Space Shuttle missions, and the ELM-PS will be the first of JEM's components to be launched to orbit.

When JEM assembly is completed on orbit, ELM-PS will be attached to the top of the PM. However, since it will be launched prior to the PM, it will first be attached to a US module with a common berthing mechanism. When PM arrives on orbit, the ELM-PS will be moved to its normal position. The ELM-PS will thus be the one most awaiting the arrival of PM.

Exposed Facility (EF): The *Exposed Facility* is a multipurpose experiment space where various activities such as scientific experiments, Earth observation, communication experiments, scientific and engineering experiments or material experiments can be conducted by utilizing environment which is exposed to space characterized with microgravity, high level vacuum and vast area.

On ISS, JEM and the Truss are the only facilities that allow experiments those being exposed to space environment, which many researchers are interested in.

EF will be utilized not only for activities peculiar to Japan, but will be utilized also as a facility for international cooperation.

As stated above, JEM EF is expected to contribute to the evolution of science and technology toward next generation, and also to maintain and to promote world wide cooperation.

The EF with its size of 6-meter × 5-meter × 4-meter (20-feet × 16.7-feet × 13.3-feet) weighs approximately 4,000 kilograms (8,890 pounds) at time of launch. Experiment payloads of EF can be exchanged, which allows various experiments to be conducted. The EF will be operated for ten years on orbit supporting exposed experiments. It will supply electric power, circulates coolant for cooling the experiment devices or collects experiment data.

The standard payload envelope is assumed to be of 1.85-meter × 1-meter × 0.8-meter (6.2-feet × 3.3-feet × 2.7-feet) and weighs 500 kilograms (1,110 pounds).

Remote Manipulator System: The Remote Manipulator System (JEM-RMS) serves as an arm to support experiments conducted on the Exposed Facility. The main arm handles large items, for delicate tasks, the small fine arm can attached at the end of the main arm. The main arm is equipped with a TV camera, which allows astronauts to monitor the operation from inside the pressurized module.

Express Pallet: EXPRESS is the acronym for EXpedite The PRocessing of Experiments to Space Station. The EXPRESS program consists of two separate systems: the EXPRESS rack for pressurized payloads and the EXPRESS pallet for attached payloads. See Fig. 36.

The EXPRESS pallet for attached payloads is provided by Brazil as a participant through a bilateral arrangement with the United States and the Brazilian Space Agency, which is also known as AEB.

Attached Payloads are located outside of the pressurized volume of the Space Station on the truss or the Japanese *Experiment Module Exposed Facility* (JEM EF). The EXPRESS Pallet can be located at any site on the truss segment. The EXPRESS Pallet has six robotically replaceable adapters for payloads or payload complements.

Attached payloads may use the Station power, command and data handling system and video. The crew will interface using robotics for installation and removal of the attached payloads, with no nominal EVA operations anticipated.

Columbus Laboratory: The *Columbus* Laboratory is a pressurized, habitable module, which will be attached to *Node 2* of the Station. The *Columbus* Laboratory's structure is derived from the Italian *Mini-Pressurised Logistics Module* (MPLM).

The internal facilities of the Columbus Laboratory consist of the Microgravity Facilities for *Columbus* (MFC) and the European Drawer Rack. The Microgravity Facilities for Columbus cover the development of five multi-user laboratories in the fields of Biology, Human Physiology, Materials and Fluid Science.

It is the goal of the MFC program to maintain the four disciplines constantly present and generating scientific data throughout the lifetime of the station. This will provide European and international scientists with a wide envelope of research opportunities offered from the unique vantage point of a world class facility in space.

The first phase of the MFC program covers the initial period between 1997 and 2003. It involves the design and development of the following facilities:

- Biolab.
- Fluid Science Laboratory (FSL).
- European Physiology Modules (EPM's).
- Material Science Laboratory (MSL).

During this phase, experiment specific hardware (e.g., experiment containers for fluid science and biology, cartridges for material science) will be developed to support the multi-user facilities. This phase will also see the development of second generation modules and facilities (different furnaces, a bioreactor, new physiology equipment, upgraded diagnostics) designed to enhance the capabilities of the current laboratories and provide research opportunities for the widest possible range of European researchers.

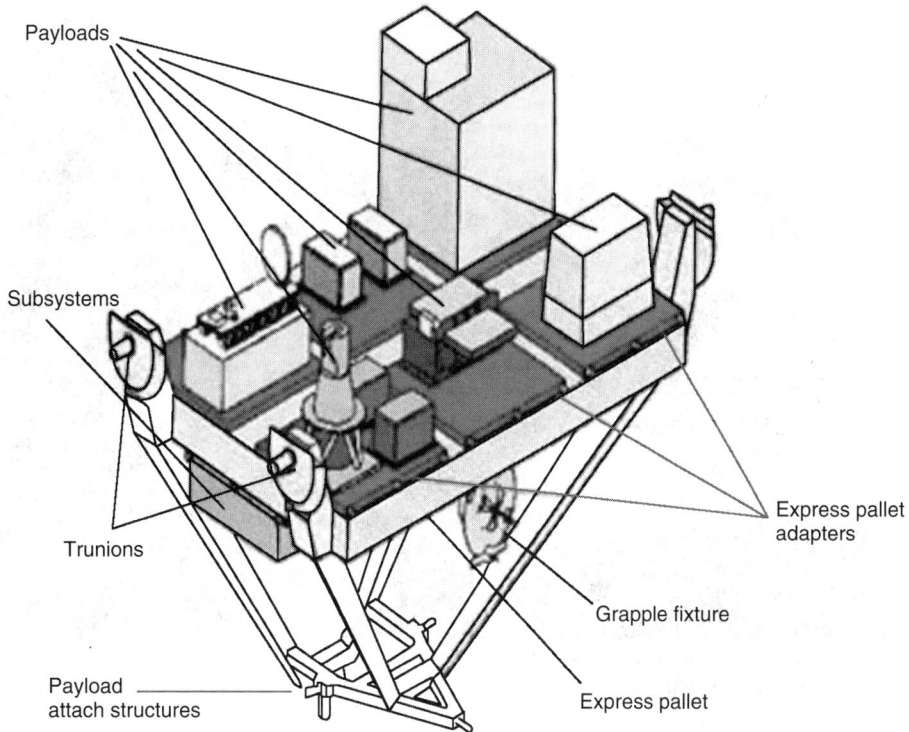

Fig. 36.   Example of an EXPRESS pallet with payloads. (*NASA*.)

These multi-user facilities will be modular in design to allow for upgrading, easy refurbishment and repair. They will also feature advanced telescience capabilities designed to alleviate crew involvement and facilitate user access. Facility Operations will be handled through dedicated Facility Responsible Centres (FRC's); User Home Bases (UHB's), located at universities and research centers, will be the direct link between the user and the space laboratory.

The facilities are presently in the last stages of their Definition and Preliminary Design phase (Phase B), or early into their Design and Development phase (Phase C/D), with the exception of the EPMs, which is currently undergoing its Preliminary Analysis phase (phase A). Each facility is supported by a dedicated science team, which is following its development and advises the Agency on its best scientific use. Scientists are actively participating in the project reviews in order to assure that ESA receives proper feedback and attends to their needs as future users.

The *Columbus* Laboratory is scheduled for launch in April 2004.

X-38: Crew Return Vehicles: With technologies that blaze a trail for future human spacecraft, NASA's X-38 project is developing, at an unprecedented low cost, a prototype rescue vehicle that could provide astronauts on the *International Space Station* an immediate return home in an emergency.

The X-38 couples a proven shape, taken largely from Air Force's X-24A project from the 1970s, with dozens of new technologies—the world's largest parafoil parachute; the first all-electric spacecraft controls; flight software developed in a quarter of the time required for past spacecraft; laser-initiated explosive mechanisms for deploying parachutes; and global positioning system-based navigation.

An innovative combination of a shape first tested in the 1970s and today's latest aerospace technology, the X-38 already is flying in the actual conditions in which it must perform. Since 1997, increasingly complex, unpiloted atmospheric test flights of the X-38 have been under way at the Dryden Flight Research Center in California. An unpiloted X-38 space test vehicle, now under construction at the Johnson Space Center in Houston, Texas, will fly aboard the space shuttle in 2003 and descend to a landing independently. The X-38 is designed to fit the unique needs of a space station "lifeboat"—long-term, maintenance-free reliability that is always in "turnkey" condition, ready to provide the crew a quick, safe trip home under any circumstance. See Fig. 37.

In addition to contributions from companies and NASA centers coast-to-coast, international space agencies are participating with the United States in the X-38's development. Contributions to the X-38 are being made by Germany, Belgium, Italy, Netherlands, France, Spain, Sweden and Switzerland and 22 companies throughout Europe.

For more information on the X-38 see *http://spaceflight.nasa.gov/ spacenews/factsheets/pdfs/fs_x-38.pdf*.

Research on the *International Space Station*

The *International Space Station* will establish an unprecedented state-of-the-art laboratory complex in orbit, more than four times the size and with almost 60 times the electrical power for experiments, critical for research capability, of Russia's *Mir*. Research in the station's six laboratories will lead to discoveries in medicine, materials and fundamental science that will benefit people all over the world. Through its research and technology, the station also will serve as an indispensable step in preparation for future human space exploration. Examples of the types of U.S. research that will be performed aboard the station include:

- **Protein crystal studies:** More pure protein crystals may be grown in space than on Earth. Analysis of these crystals helps scientists better understand the nature of proteins, enzymes and viruses, perhaps leading to the development of new drugs and a better understanding of the fundamental building blocks of life. Similar experiments have been conducted on the Space Shuttle, although they are limited by the short duration of Shuttle flights. This type of research could lead to the study of possible treatments for cancer, diabetes, emphysema and immune system disorders, among other research.

- **Tissue culture:** Living cells can be grown in a laboratory environment in space where they are not distorted by gravity. NASA already has developed a Bioreactor device that is used on Earth to simulate, for such cultures, the effect of reduced gravity. Still, these devices are limited by gravity. Growing cultures for long periods aboard the station will further advance this research. Such cultures can be used to test new treatments for cancer without risking harm to patients, among other uses.

- **Life in low gravity:** The effects of long-term exposure to reduced gravity on humans weakening muscles; changes in how the heart, arteries and veins work; and the loss of bone density, among others, will be studied aboard the station. Studies of these effects may lead to a better understanding of the body's systems and similar ailments on Earth. A thorough understanding of such effects and possible methods of counteracting them is needed to prepare for future long-term human exploration of the solar system. In addition, studies of the gravitational effects on plants, animals and the function of living cells will be conducted aboard the station. A centrifuge, located in the *Centrifuge*

Fig. 37.   The first atmospheric test vehicle for the X-38 prototype International Space Station crew return vehicle goes through a "captive carry" flight test in fall 1997 under the wing of a B-52 aircraft at NASA's Dryden Flight Research Center, California. (*NASA.*)

Accommodation Module, will use centrifugal force to generate simulated gravity ranging from almost zero to twice that of Earth. This facility will imitate Earth's gravity for comparison purposes; eliminate variables in experiments; and simulate the gravity on the Moon or Mars for experiments that can provide information useful for future space travels.

- **Flames, fluids and metal in space:** Fluids, flames, molten metal and other materials will be the subject of basic research on the station. Even flames burn differently without gravity. Reduced gravity reduces convection currents, the currents that cause warm air or fluid to rise and cool air or fluid to sink on Earth. This absence of convection alters the flame shape in orbit and allows studies of the combustion process that are impossible on Earth, a research field called *Combustion Science*. The absence of convection allows molten metals or other materials to be mixed more thoroughly in orbit than on Earth. Scientists plan to study this field, called *Materials Science*, to create better metal alloys and more perfect materials for applications such as computer chips. The study of all of these areas may lead to developments that can enhance many industries on Earth.

- **The nature of space:** Some experiments aboard the station will take place on the exterior of the station modules. Such exterior experiments can study the space environment and how long-term exposure to space, the vacuum and the debris, affects materials. This research can provide future spacecraft designers and scientists a better understanding of the nature of space and enhance spacecraft design. Some experiments will study the basic forces of nature, a field called Fundamental Physics, where experiments take advantage of weightlessness to study forces that are weak and difficult to study when subject to gravity on Earth. Experiments in this field may help explain how the universe developed. Investigations that use lasers to cool atoms to near absolute zero may help us understand gravity itself. In addition to investigating basic questions about nature, this research could lead to down-to-Earth developments that may include clocks a thousand times more accurate than today's atomic clocks; better weather forecasting; and stronger materials.

- **Watching the Earth:** Observations of the Earth from orbit help the study of large-scale, long-term changes in the environment. Studies in this field can increase understanding of the forests, oceans and mountains. The effects of volcanoes, ancient meteorite impacts, hurricanes and typhoons can be studied. In addition, changes to the Earth that are caused by the human race can be observed. The effects of air pollution, such as smog over cities; of deforestation, the cutting and burning of forests; and of water pollution, such as oil spills, are visible from space and can be captured in images that provide a global perspective unavailable from the ground.

- **Commercialization:** As part of the Commercialization of space research on the station, industries will participate in research by conducting experiments and studies aimed at developing new products and services. The results may benefit those on Earth not only by providing innovative new products as a result, but also by creating new jobs to make the products.

Editor's Note: The staff of this encyclopedia is most appreciative of the cooperation extended by National Aeronautics and Space Administration (NASA) for furnishing information for numerous portions of this article.

For further information on space stations, and the *International Space Station* in particular, please see the list of additional sources below.

Web References

Apollo *Soyuz*: *http://www-pao.ksc.nasa.gov/kscpao/history/astp/astp.html*
Brazilian International Space Station Program: *http://www.inpe.br/programas/iss/ingles/default.htm*
Canadian Space Agency: *http://www.space.gc.ca/*
European Space Agency: *http://www.space.gc.ca/*
History of Shuttle-*Mir*: *http://spaceflight.nasa.gov/history/shuttle-mir/*
ISS Assembly Line Drawings: *http://spaceflight.nasa.gov/gallery/images/station/lineart/ndxpage1.html*
ISS Fact Sheet Library: *http://spaceflight.nasa.gov/spacenews/factsheets/index.html #firstflights*
ISS Assembly: *http://spaceflight.nasa.gov/station/assembly/index.html*
Jonathan's Space Report: *http://hea-www.harvard.edu/QEDT/jcm/space/jsr/jsr.html*
Jonathan's Space Home Page: *http://hea-www.harvard.edu/QEDT/jcm/space/space.html*
Kennedy Space Center: *http://www.ksc.nasa.gov/*
Marshall Space Flight Center Fact Sheets: *http://www1.msfc.nasa.gov/NEWSROOM/background/facts.htm*
NASA Fact Sheet Library: *http://spaceflight.nasa.gov/spacenews/factsheets/index.html*
NASA Human Spaceflight: *http://spaceflight.nasa.gov/*
NASA Shuttle-*Mir*: *http://spaceflight.nasa.gov/*
NASA Space Station Science: *http://spaceflight.nasa.gov/station/science/index.html*
NASA Space Station Crew: *http://spaceflight.nasa.gov/station/crew/index.html*
NASA Space Station Extravehicular Activity: *http://spaceflight.nasa.gov/station/eva/index.html*
NASA International Space Station Reference: *http://spaceflight.nasa.gov/station/reference/index.html*
NASA Space Station Gallery: *http://spaceflight.nasa.gov/gallery/images/station/*
NASA Space Shuttle Missions: *http://spaceflight.nasa.gov/shuttle/archives/*
Russian Space Agency: *http://liftoff.msfc.nasa.gov/rsa*
Russian Aviation page: *http://aeroweb.lucia.it/~agretch/RAP.html*
The Skylab Project: *http://www-pao.ksc.nasa.gov/kscpao/history/skylab/skylab.htm*
The Salyut 4 Space Station: *http://astroe.gsfc.nasa.gov/docs/heasarc/missions/salyut4.html*

SPACE-TIME.   A space of four dimensions which specify the space and time coordinates of an event. In the absence of a gravitational field, space-time reduces to Minkowski space.

See also **Space (Minkowski)**; **Gravitation**; and **Relativity and Relativity Theory**.

SPACE VEHICLE GUIDANCE AND CONTROL.   The term guidance means the sum total of the orders and instructions a space vehicle must be given to keep it on the chosen path to its objective. They may be transmitted from Earth upon the basis of the course and position of the craft as signaled or observed, they may be given by instruments on board, or, in the case of manned space vehicles, they may originate, partly at least, from a human pilot. Then, on the basis of this information, the control system operates devices which change a course as necessary.

Obviously, the guidance program for a particular mission begins with the path through space which the space vehicle is to follow. That path may be suborbital, such as that taken by sounding rockets, which return to Earth relatively promptly. It may be orbital, as in the case of satellites, which go into a closed orbit about the Earth. Or it may leave the Earth's gravitational field to extend into space, as in the case of space probes and corresponding manned spacecraft. Of the paths followed by these various craft, the ideal, circular orbit is the most simple to calculate.

Consider a satellite launched from Earth which, when it reaches the height of its projected orbit, is turned into a horizontal position and given a horizontal velocity (*injected*) into its orbit. What must this velocity be for a stable orbit, i.e., a satellite that is to continue circling the Earth at that distance?

If we ignore certain complications, such as the fact that the Earth is not perfectly spherical and its mass is not uniformly distributed, and also the effect of the drag upon the satellite of the atmosphere, corresponding to its density at that altitude, and so consider only the height and injection velocity of the satellite, the calculation becomes relatively simple. From Newton's First Law (see also **Newton's Laws of Dynamics**) we know that if no force acted upon the satellite it would continue its horizontal motion (i.e., motion tangent to its orbit) and travel away into space. On the other hand, if it were not moving, it would fall straight to Earth due to the acceleration of gravity. Since both effects are present, its velocity is the resultant of the injection (horizontal) velocity and that due to gravity, and there is obviously a value of the former at which the orbit is stable. This value is given by the expression

$$v_H = \sqrt{\frac{g_0 R_0^2}{R_0 + h}} \qquad (1)$$

where g_0 is the acceleration of gravity at the Earth's surface (32.16 ft-sec^{-2}), R_0 is the radius of the Earth, and h is the height of the satellite above the surface of the Earth. Since g_0 and R_0 are almost constant, this equation shows that the greater the height of a satellite above the surface of the Earth, the smaller the injection velocity necessary to stabilize it there.

If the injection velocity does not have the value calculated for that height by Equation (1), then the orbit will not be circular. If the injection velocity is greater than the calculated value of v_H but not greater than $\sqrt{2}v_H$, the orbit will be an ellipse. If the injection velocity equals or exceeds $\sqrt{2}v_H$, then the satellite will cease to orbit about the Earth, and move away into space. If the injection velocity is less than the calculated value of v_H, the satellite will fall toward Earth along a curved path, the closer the velocity to v_H, the longer the path. In fact, if the injection point is at a sufficient height and the injection velocity is not too far below the value of v_H calculated by Equation (1), this path will not intersect the Earth, but will take the form of an ellipse. It should also be noted that even if the injection velocity should have the exact value calculated by Equation (1), and if the injection is in a direction above or below the horizontal, the orbit will be an ellipse.

In view of these circumstances, it is not surprising that the satellites that have been launched have elliptical orbits. Further reasons are the facts that the mass distribution of the Earth is not uniformly spherical and that the atmosphere exerts a drag upon satellites, especially those that approach the Earth closely. Neither of these effects were taken into account in the derivation of Equation (1).

The path to be followed by a probe or manned spacecraft that is to leave the gravitational field of the Earth to travel to such objects as the moon or one of the planets presents a more complicated problem than an ideal satellite. The difference is due partly to the fact that the destination is also a moving object, and its motion must be included in the calculations.

The first step, the escape velocity from the earth's gravitational field, as already noted in the discussion of satellites, is $\sqrt{2}v_H$, or by substituting into Equation (1),

$$v_E = \sqrt{2}\sqrt{\frac{g_0 R_0^2}{R_0 + h}} = \frac{2g_0 R_0^2}{R_0 + h} \qquad (2)$$

If the starting point is at the Earth's surface, the term h drops out of the equation, and the value so calculated is about 7 miles per second (neglecting the drag of the atmosphere). The first requirement for space journeys from Earth is that the craft attain a velocity exceeding v_E. Note that if the point of departure is at a height h above the Earth's surface, such as would be provided by a space platform or by a spacecraft already in orbit, the value of v_E is reduced accordingly.

After the problem of escaping Earth's effective gravitational field is solved, there are several ways of calculating the best path to another planet. We can calculate the path that minimizes the time, which obviously requires a greater expenditure of energy than a path that minimizes the energy. The latter calculation was made by W. Hohmann on the simplifying assumption of circular planetary orbits and the results, which are ellipses, are called Hohmann transfer ellipses. They show that for minimum energy, the spacecraft should leave the one planet tangentially to its orbit and land tangentially to the orbit of the second planet. Of course, this simple picture must be considerably modified in plotting actual interplanetary paths, and the result is usually a compromise between minimum-energy and minimum-time paths.

Having determined the path the space vehicle is to follow, the next step is to determine the guidance necessary to keep it on that path. For convenience in planning, the guidance program is often divided into three stages: initial guidance (prior to orbital injection), mid-course guidance, and terminal guidance. Note that initial guidance applies not only to satellites, but also to probes and manned spacecraft that are to leave the Earth's gravitational field, since they are commonly injected into an orbit from which they then leave for the moon or another planet.

Methods of guidance include: (1) *Command Guidance*, in which the space vehicle is tracked from Earth, and commands are sent to it as necessary. Since radio (or radar) is used to send these instructions, as well as to receive positional data from the space vehicle, the method is also called *radio guidance*. (2) *Inertial guidance*, in which all guidance operations are carried out by instruments aboard the space vehicle. In a combination of (1) and (2), compensating adjustments to the inertial guidance system are sent from Earth, the method being called *radio-inertial guidance*. (3) *Celestial guidance*, operated by celestial navigation methods applied automatically by instruments aboard the spacecraft.

As to the first method, it should be noted that, although it is called radio guidance, the frequencies actually used are those in the radar range. These signals are used not only for control systems, but also for determining the distance of the vehicle and its velocity component away from the observer. Then the trajectory of the vehicle is adjusted to change its orientation, which is done by controls that turn it about its three axes. The essential elements in these controls are gyroscopes, which are controlled by radar signals when used in a command guidance system.

In the second method, inertial guidance gyroscopes play an even more important part. They are used in the stable platform system, which is set before the vehicle is launched and which provides a reference for the attitude of the vehicle throughout its flight. They are used as integrating devices to convert to velocity readings the acceleration measurements made by a number of devices, including spring-mass accelerometers, vibrating-element accelerometers, and pendulum accelerometers, which measure acceleration by its effect upon, respectively, a spring-held mass, an element in vibration, or the motion of the pendulum. Similarly, the velocity indications so obtained are integrated again to obtain the distance traveled. See also **Acceleration**; **Acceleration (Due to Gravity)**; and **Accelerometer**.

Since gyroscopes have a tendency to drift, the combined radio-inertial guidance system sends command instructions to correct these errors. This is done by a radar communication hookup in which the space vehicle receives continuously its range data (as observed from Earth) from an Earth-based station. These observed range data signals are then fed into the vehicle's guidance system to correct instrumental errors that may have developed by such means as gyroscope drift.

Obviously some of the guidance functions described up to this point, such as changes in the altitude of the space vehicle, or even the close adjustment of gyroscopes, are characteristic of the early part of a mission, i.e., initial guidance. In fact, mid-course guidance is rarely used on satellites or other relatively short missions, but is important chiefly on missions to the moon or farther. Due to the far greater distances, the radio command method must be modified. If doppler shift measurements are used, a two-way system is necessary, in which the radar signals from Earth actuate a transmitter on the spacecraft that amplifies and transmits the signals back.

In addition to doppler measurements, the time for signal transmission is determined and the phase difference between the Earth signal and the spacecraft signal is found by interferometer methods using radio telescopes. With these various data, the distance, angular position, and radial velocity of the spacecraft can be calculated. In view of the amount of calculation necessary in mid-course guidance, it is not surprising that computers are a prominent feature of these operations. See Fig. 1.

Astronomical observations from Earth can be plotted to establish the path of the spacecraft. Whatever method of path measurement is used, corrections to it are made by first actuating gyroscopes to turn the craft to the correct attitude, and then signaling a rocket aboard it to fire.

Terminal guidance begins with the correction of the trajectory of the spacecraft, satellite, or vehicle to completion of its mission. The vehicle continues as necessary up to the end, whether that be a landing on return to Earth, or a landing on or orbiting of a planet or the moon. For landing on Earth, the path to be followed is easier to calculate in advance and to control, since the properties of the Earth's atmosphere are known, and the attitude and position of the vehicle can be observed from Earth. The velocity of the vehicle in its path depends upon its position and attitude

at the commencement of the terminal stage, the acceleration of gravity, and the drag of the atmosphere, which also has a heating effect that must be held within limits. The function of guidance is to modify the effect of these variables by adjusting the attitude by means of gyroscopes and small rockets, as discussed earlier in this entry for initial guidance, and then to decrease the velocity by means of the larger retrorockets.

The space vehicle control system must act in concert with the guidance system. The chief controlled variables are attitude and velocity, the latter including both speed and direction of motion. The extent to which these variables are controlled varies with the kind of space vehicle and the stage of its mission, i.e., it differs for the initial, midcourse, and terminal stages.

Attitude control can be effected during the initial (or launch) phase by changing the direction of the thrust of the rocket vehicle. The thrust is the force accelerating the vehicle in a forward direction, due to reaction to the opposite force exerted by the stream of gases leaving the rear of the vehicle. Under stable conditions of motion, the direction of the thrust is along the central axis of the vehicle. In one method of turning the vehicle, jet vanes, which are placed in the exhaust gas stream, can be moved so as to deflect the stream in various directions, thus exerting a turning moment which continues until the direction of the central axis of the vehicle corresponds to the new direction of thrust. These vanes may be moved by electric motors, or hydraulic or pneumatic systems actuated by the guidance system. Similar in its action to the jet vanes is the jetavator, which may be mounted in the exhaust gas stream, but consists of a collar-shaped member, which is free to rotate when moved by an actuator. Since it rotates about only one axis, three or more of these devices may be used. Other methods turn the space vehicle by turning the nozzle itself, which is mounted on a swivel. In smaller vehicles, the entire motor may be attached

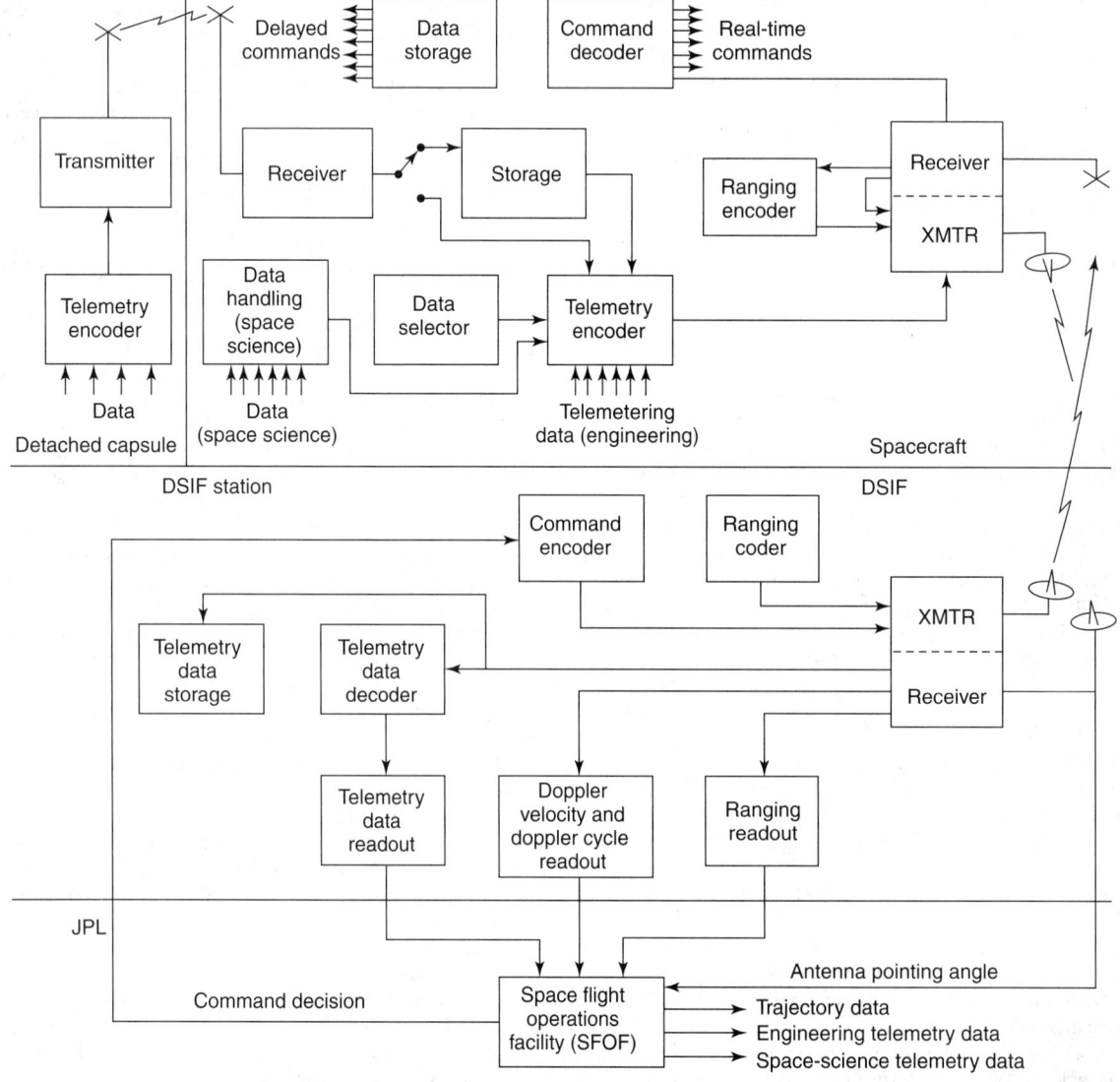

Fig. 1.   Fundamentals of space-probe communication subsystem. SFOF = space flight operations facility; DSIF = deep space instrumentation facility.

to a hinge, so that it is free to turn. Such methods are not usually applicable to large rockets; they are usually turned by small (vernier) rocket motors. Since the control system must be capable of turning the space vehicle as much as is required about all three of the space axes, there may be six such motors mounted on swivels. Often these motors will operate on a medium different from that of the main rocket engine. Compressed gases or liquid monopropellants, such as hydrogen peroxide and hydrazine can be used. Another method of rocket attitude control is by means of a jet of noncombustible liquid that is discharged into the exhaust gas stream to change its direction.

Since velocity has direction as well as magnitude (speed), attitude control also controls the direction of the velocity. Control of the magnitude of velocity is of particular importance during the terminal stage of some missions, chiefly for landings, but also for rendezvous missions, or for missions which have orbits about a body as their destinations. Retrorockets may be used to effect slow-downs for such maneuvers. See also **Rocket Propellants**.

See also **Astronautics**.

SPACE VELOCITY (Stellar).   The space velocity of a star is its actual velocity in space relative to the sun. The term *fixed star* is one which has been handed down to us from the era when philosophers believed that the stars were fixed on a sphere, rotating about the earth, which they referred to as the celestial sphere. At present, we believe that practically all of the stars are actually in motion in space with velocities comparable in magnitude to the velocities with which the planets are moving about the sun in their orbits. In measuring any velocity it is necessary to have a definite reference point and the space velocity of a star is its velocity referred to the sun.

In Fig. 1 we have the space velocity of a star represented by the vector (directed straight line proportional to the velocity in direction and magnitude) AB. The angle β, which this vector makes with the direction of the star from the sun at any particular instant, gives the direction of the space velocity at that instant.

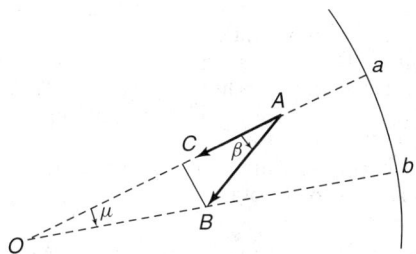

Fig. 1.   Vector representation of space velocity.

As observed from the solar system this space velocity may be resolved into two components: AC the radial velocity, and m the proper motion. The radial velocity is determined directly in terms of linear velocity (i.e., in miles or kilometers per second), but the proper motion may only be determined in angular units, usually expressed in seconds of arc per year. In order that the space velocity may be known, the proper motion must be converted into a linear velocity, commonly known as the transverse velocity of the star. This may be accomplished only if the distance of the star is known. Expressing the distance in terms of the stellar parallax, π, and calling the transverse velocity T (the line CB in the figure), the following relations may be derived:

$$T = 4.74 \; \mu/\pi \; \text{kilometers/second}$$

or

$$T = 2.94 \; \mu/\pi \; \text{miles/second}$$

With both the transverse velocity, T, and the radial velocity, R, known in the same units, the problem of determining the space velocity S in the same units is merely that of solving the plane right-triangle ACB. This solution yields

$$S^2 = T^2 + R^2 \quad \cos\beta = R/S \quad \sin\beta = T/S$$

SPAR.   In marine parlance, a spar is a round timber used to extend a sail. Used with this meaning, it could be a mast, a boom, or a yard.

A structural member used similarly to extend a surface to obtain an air reaction is found in the wing of an airplane. There, the spar is a principal structural member running the length of the wing. Usually there are two, parallel to one another, but wings have been built having only one spar. In such a design the spar is required to take torsion and bending as well as compression if the wing is externally braced. The spars support the ribs upon which the wing covering is stretched. The reaction of the air upon the latter is transmitted to the spars, which in turn are attached to the fuselage. The air reaction may be carried by the spars entirely by bending as in the case of a truly cantilever wing, or it may be carried by spars which are braced outboard of the fuselage by wires or struts.

SPARK CHAMBER.   A device used for the study of high-energy nuclear reactions and related particle physics. The device consists of a chamber filled with an inert gas, often helium or helium and neon, in which there is a stack of parallel conducting plates. These plates are connected alternately to the positive and negative terminals of a high-voltage source (10,000 volts upward). A chamber may have from 25 to 100 plates, each about 1 millimeter thick and 1 millimeter square, spaced about 0.6 centimeter apart. The chamber is operated in conjunction with a counter used as a detector for ionizing particles entering the chamber. The counter is used to connect the high-voltage circuit momentarily to the plates, and to trigger the shutter of a camera when a particle enters the chamber. As a particle traverses the chamber, it produces many ion-pairs along its path, so that the gas becomes conducting, and sparks occur between the plates in the region through which the ionizing particles pass. The light from them is focused by the lens system so that stereoscopic photographs may be obtained of the events. See also **Bubble Chamber**; **Cloud Chamber**; and **Particles (Sub-atomic)**.

SPARROW   *(Aves, Passeriformes).*   A common form of small seedeating bird *(Aves)* of the family *Fringillidae*. The many species are for the greater part rather quietly colored in browns and grays with streaked and spotted plumage, but some bear conspicuous black or white marks and the browns of some species are very bright. Some of the sparrows have beautiful songs, although none rival such outstanding singers as the brown thrasher and the mockingbird.

The field sparrow, *Spizella pusilla*, chipping sparrow, *S. passerina*, white-throated, *Zonotrichia albicollis*, and white-crowned, *Z. leucophrys*, sparrows, song sparrow, *Melospiza melodia*, and in the west the lark sparrow, *Chondestes grammacus*, are among the well known North American species.

SPAWNING.   Reproduction and the bringing forth of young in fishes. In the cases of many species of fishes, the reproductive process is quite simple in that the male sheds his sperm and the female sheds her eggs in adjacent water wherein the mingling of these two materials causes fertilization. But, in numerous species, the process is varied and often reasonably complex. Most fishes have particular preferences for what might be termed idealized spawning conditions for their particular species. These conditions include water temperature, water salinity (fresh versus marine waters), and geographic location. The relative stillness of the water also plays a role. Because temperature, in particular, varies with season, the spawning habits usually have a very marked seasonal influence. This factor alone can cause long migrations of fishes to their desired water conditions, but salinity essentially does not change with seasons and, therefore, migrations from marine waters to fresh waters (or vice versa) is a major factor. These water conditions not only affect spawning, but more importantly they affect the survival of the fertilized eggs and the early development of the young. The eggs, which usually float on the surface of the water with many species, are subject to all manner of danger. In compensation for the great degree of chance involved in survival of the early life-giving processes, most species produce eggs in astronomical numbers. The eggs sometimes number into the several millions. For this reason, in many species, the ovaries are extensive, for example comprising about one-fifth of the body weight of a female salmon. At a single spawning, the cod routinely deposits from 4 to 6 million eggs. Most fish eggs have a yolk, upon which the embryo feeds during incubation and often for a period after hatching. Then the very tiny mouth of the tiny fish commences to consume plankton. In the very early stages of life, the tiny fish frequently is referred to as a *larva* or *fry*. It is interesting to note that the larva forms of many species appear quite different in form from the adult fish, although a larval cod, for example, leaves no question but that it is a very small cod.

Another protective means occurs in the species of some fishes by way of the phenomenon of changes in sex and hermaphroditism. Some species are always self-fertilizing.

The oviparous (egg-laying) species often exhibit unusual ways of depositing and protecting their eggs: burrowing in a sandy bottom; attaching eggs by way of sticky substances to vegetation; or depositing eggs by drilling into the bodies or shells of other creatures, where the eggs are left for safekeeping and development. An unusual adaptation is found in the male Australian kurtus where eggs are incubated in a pouch on the male's forehead. Some species deposit their eggs in moist places on land, where they may remain for a few minutes or few days. The grunion is very precise about time of year, phase of moon, and tide, depositing its eggs in wet sand, whereupon they are immediately fertilized by the male. Then, within a few minutes, they are hatched and carried back to sea by the waves. See also **Silversides (Osteichthyes)**.

The spawning habits of several species are described briefly in this volume. Refer to list of entries under **Fishes**.

There are numerous species where the young are born alive. These are termed ovoviviparous, the incubation and hatching of the eggs occurring within the body of the female. The young in these instances usually require careful attention and monitoring of one or both parents. In the discus, for example, the young "nurse" upon fluids and mucus present on the sides of both male and female parent. The parents take turns in this procedure.

SPECIFIC ACTIVITY.   Three common uses of this term are: 1. The activity of a radio element per unit weight of element present in the sample. 2. The activity per unit mass of a pure radionuclide. 3. The activity per unit weight of any sample of radioactive material. Specific activity is commonly given in a wide variety of units (e.g., millicuries per gram, disintegrations per second per milligram, counts per minute per milligram, etc.). See also **Radioactivity**.

SPECIFIC GRAVITY.   For a given liquid, the specific gravity may be defined as the ratio of the density of the liquid to the density of water. Because the density of water varies, particularly with changes in temperature, the temperature of the water to which a specific gravity measurement is referred should be stated. In exacting, scientific observations, the reference may be to pure (double-distilled) water at $4\,°C$ ($39.2\,°F$). In engineering practice, the reference frequently is to pure water at $15.6\,°C$ ($60\,°F$). A value of unity is established for water. Thus, liquids with a specific gravity less than 1 are lighter than water; those with a specific gravity greater than 1 are heavier than water. From a practical standpoint, it usually is more meaningful to express the specific gravity of gases with reference to pure air rather than to pure water. Thus, for a given gas, the specific gravity may be defined as the ratio of the density of the gas to the density of air. Since the density of air varies markedly with both temperature and pressure, exacting observations should reflect both conditions. Common reference conditions are $0\,°C$ and 1 atmosphere pressure (760 torr; 760 millimeters Hg; 29.92 inches Hg).

Specific Gravity Scales.   Arising essentially from a lack of communication between various scientific and industrial communities, a number of different specific gravity scales were formulated in earlier times. Because so much data and experience have been accumulated in terms of these scales, several methods of expressing specific gravity persist in common use. The most important of these scales are defined here.

API Scale — This scale was selected in 1921 by the American Petroleum Institute, the U.S. Bureau of Mines, and the National Bureau of Standards (Washington, DC) as the standard for petroleum products in the United States.

$$\text{Degrees hydrometer scale (at } 15\,°C; 60\,°F) = \frac{141.5}{\text{sp gr}} - 131.5$$

Balling Scale — This scale is used mainly in the brewing industry to estimate percent wort but also is used to indicate percent by weight of either dissolved solids or sugar liquors. Hydrometers are graduated in percent weight at $60\,°F$ or $17.5\,°C$.

Barkometer Scale — This scale is used essentially in the tanning and tanning-extract industry. Water equals zero. Each scale degree equals a change of 0.001 in specific gravity. The following formula applies:

$$\text{Sp gr} = 1.000 \pm 0.001 \times (\text{degrees Barkometer})$$

Baumé Scale — This scale is used widely in connection with the measurement of acids and light and heavy liquids, such as syrups. The

scale originally was proposed by Antoine Baumé, a French chemist, in 1768. The scale has been widely accepted because of the simplicity of the numbers which represent liquid specific gravity. Two scales are in use:

$$\text{For light liquids, } °\text{Bé} = \frac{140}{\text{sp gr}} - 130$$

$$\text{For heavy liquids, } °\text{Bé} = 145 - \frac{145}{\text{sp gr}}$$

The standard temperature for these formulas is $15.6\,°C$ ($60\,°F$).

To calibrate his instrument for heavy liquids, Baumé prepared a solution of 15 parts by weight of sodium chloride in water. On his hydrometer, Baumé marked zero at the point to which the float submerged in pure water; and he marked the scale 15 at the point to which the float submerged in the salt solution. He then divided the distance between the two marks into 15 equal spaces (or degrees as he termed them). In connection with liquids lighter than water, Baumé prepared a 10% sodium chloride solution. In this case, he marked the scale zero at the point to which the float submerged in the salt solution; and he marked the scale 10 at the point to which the float submerged in pure water. Thus, he created a scale which provided increasing numbers with decreases in density.

Users of the Baumé method found that the scale generally read 66 when the float was submerged in oil of vitriol. Thus, early manufacturers of hydrometers calibrated the instruments by this method. There were variations in the Baumé scale, however, because of lack of standardization in hydrometer calibration. Consequently, in 1904, the National Bureau of Standards made a careful survey and finally adopted the scales previously given for light and for heavy liquids.

Brix Scale — This scale is used almost exclusively by the sugar industry. Degrees on the scale represent percent pure sucrose by weight at $17.5\,°C$ ($63.5\,°F$).

Quevenne Scale — This scale is used for milk testing and essentially represents an abbreviation of specific gravity. For example, $20°$ Quevenne indicates a specific gravity of 1.020; $40°$ Quevenne, a specific gravity of 1.040, and so on. One lactometer unit approximates $0.29°$ Quevenne.

Richter, Sikes, and Tralles Scales — These are alcoholometer scales, which indicate directly in percent ethyl alcohol by weight in water.

Twaddle Scale — This scale is the result of attempting to simplify the measurement of industrial liquids heavier than water. The range of specific gravity from 1.000 to 2.000 is divided into 200 equal parts. Thus, $1°$ Twaddle equals 0.005 sp gr.

An abridged compilation of specific gravity conversions is given in Table 1. The specific gravities of numerous materials are given throughout this volume.

Determination of Specific Gravity. The principal means for measuring specific gravity (and density) of liquids and gases are listed in Table 2.

Hand Hydrometer. This instrument consists essentially of a long, slender glass float weighted at the lower end and provided with a scale so graduated that the depth to which the instrument sinks in the liquid indicates the specific gravity by direct reading of the scale. See Fig. 1. The numbering of the scale increases from the top downward. The instrument sometimes is proportioned so that the numbering begins with unity at the top, being applicable only to liquids heavier than water; in others, it increases from the top to unity at the lower end and is for use with liquids lighter than water; in still other designs, unity is marked at the middle of the scale and thus the instrument may be used for both light and heavy liquids. To be sensitive, the stem carrying the scale must be slender. It may be observed that the scale intervals corresponding to equal increments of density cannot be equal where the stem is of uniform diameter. These intervals, in fact, are inversely proportional to the square of the density, being much smaller at the lower than at the upper end of the scale. To avoid this, some hydrometers are graduated with an arbitrary scale having uniform spacing, as in the case of the Baumé scale, the readings of which may be converted into density by reference to tables. Nicholson devised a hydrometer for measuring the densities of small solids, the specimen being placed on the hydrometer, first above and then below the surface of the water in which the instrument floats, and its volume deduced from the resulting alteration in buoyant force. See Fig. 2. Hand hydrometers are used extensively where automatic, continuous, and remote readings of specific gravity or density are not required. However, a simple hydrometer can be used in a standpipe, equipped with an overflow at reading level, thus allowing for visual observations of a continuously flowing liquid.

TABLE 1. SPECIFIC GRAVITY SCALE EQUIVALENTS

Specific Gravity 60°/60°F	°Baume	°API	Specific Gravity 60°/60°F	°Baume	°API
0.600	103.33	104.33	0.800	45.00	45.38
0.620	95.81	96.73	0.820	40.73	41.06
0.640	88.75	89.59	0.840	36.67	36.95
0.660	82.12	82.89	0.860	32.79	33.03
0.680	75.88	76.59	0.880	29.09	29.30
0.700	70.00	70.64	0.900	25.56	25.72
0.720	64.44	65.03	0.920	22.17	22.30
0.740	59.19	59.72	0.940	18.94	19.03
0.760	54.21	54.68	0.960	15.83	15.90
0.780	49.49	49.91	0.980	12.86	12.89
			1.000	10.00	10.00

Specific Gravity 60°/60°F	°Baume	°Twaddle	Specific Gravity 60°/60°F	°Baume	°Twaddle
1.020	2.84	4	1.500	48.33	100
1.040	5.58	8	1.520	49.61	104
1.060	8.21	12	1.540	50.84	108
1.080	10.74	16	1.560	52.05	112
1.100	13.18	20	1.580	53.23	116
1.120	15.54	24	1.600	54.38	120
1.140	17.81	28	1.620	55.49	124
1.160	20.00	32	1.640	56.59	128
1.180	22.12	36	1.660	57.65	132
1.200	24.17	40	1.680	58.69	136
1.220	26.14	44	1.700	59.71	140
1.240	28.06	48	1.720	60.70	144
1.260	29.92	52	1.740	61.67	148
1.280	31.72	56	1.760	62.61	152
1.300	33.46	60	1.780	63.54	156
1.320	35.15	64	1.800	64.66	160
1.340	36.79	68	1.820	65.33	164
1.360	38.38	72	1.840	66.20	168
1.380	39.93	76	1.860	67.04	172
1.400	41.43	80	1.880	67.87	176
1.420	42.89	84	1.900	68.68	180
1.440	44.31	88	1.920	64.98	184
1.460	45.68	92	1.940	70.26	188
1.480	47.03	96	1.960	71.02	192
			1.980	71.77	196
			2.000	72.50	200

Note: 60°F = 15.6°C

TABLE 2. SPECIFIC GRAVITY AND DENSITY INSTRUMENTATION

Instrument	Liquids	Gases	Solids
Hydrometers			
Nicholson's hydrometer			x
Hand type	x		
Photoelectric type	x		
Inductance-bridge type	x		
Specific-gravity balance		x	
Fixed-volume methods			
Balanced-flow vessel	x		
Displacement meter	x		
Chain-balanced plummet	x		
Buoyancy gas balance		x	
Differential-pressure methods			
Liquid-purge systems	x		
Air-bubbler systems	x		
Viscous-drag method		x	
Boiling-point rise system	x		
Radiation gages	x		
Pycnometer[a]	x		x

[a]See separate editorial entry under **Pycnometer**.

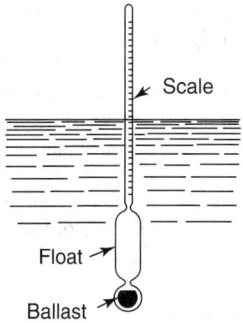

Fig. 1. Hydrometer for liquids.

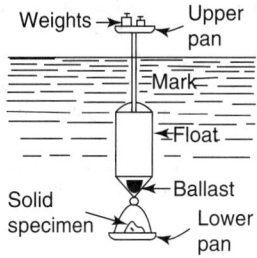

Fig. 2. Nicholson's hydrometer for small solids.

Automated Hydrometers. One form of hydrometer utilizes an opaque stem which, as the stem rises and falls, effects the amount of light which passes through a slit to a photocell. In this way, the photocell output is proportional to specific gravity and may be recorded by an electric instrument. In another industrial version, designed for remote transmission, the hydrometer is contained within a metal cylinder. A rod connects the bottom portion of the float to an armature, which moves vertically between inductance coils. Changes in inductance are transmitted by cable to a central instrument panel receiving instrument.

Chain-balanced Float. In this device, a submerged plummet, which is self-centering and operates essentially without friction, moves up or down with changes in specific gravity. A section of chain is fastened to the bottom of the plummet to provide a counter-buoyancy force. The effective chain weight varies as the plummet moves up and down. For each value of density or specific gravity, the plummet assumes a specific point of equilibrium. By means of a differential transformer, readings may be transmitted to a receiving instrument. A resistance thermometer bridge may be used where compensation for density changes with temperature is required. See Fig. 3.

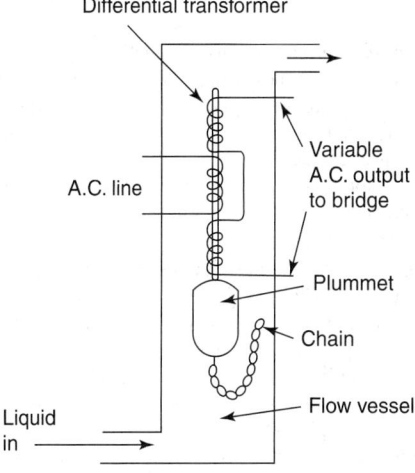

Fig. 3. Chain-balanced float or plummet-type specific gravity or density meter.

Balanced-flow Vessel. In this method of specific gravity measurement, a vessel with a fixed-volume is weighed automatically by means of a scale, spring, or use of a pneumatic force-balance system. The liquid being measured flows continuously into and out of the vessel by way of flexible connections. Accuracy of the system is very good.

Displacement-type Meter. In this instrument, the liquid being measured flows continuously through a chamber. A displacer element, usually a hollow metal sphere or cylinder containing air, is submerged fully in the liquid. The buoyant force on the displacer is measured, often by a pneumatic force-balance system, and any variations in this force reflect changes in the specific gravity or density of the liquid. The system can be compensated for changes in liquid temperature by thermostatically heating the chamber.

Differential-pressure Method. As shown in Fig. 4, two bubbler tubes, the exit of one being lower than the other, are installed in a vessel containing the liquid being measured. Air is bubbled into the liquid through these tubes. The difference in pressure required by each tube represents the weight of a constant-height column of the liquid equal to the difference in level of the ends of the two tubes. Thus, the instrument can be calibrated directly in terms of specific gravity or density. The accuracy ranges from 0.3 to 1% and provides a good approach for liquids that do not tend to crystallize in the measuring pipes. The system is used extensively in the pulp and paper industry for the measurement of white liquor, light black liquor, and bleach. There are several variations of the air-bubbler method, including the use of a reference column and a system with range suppression.

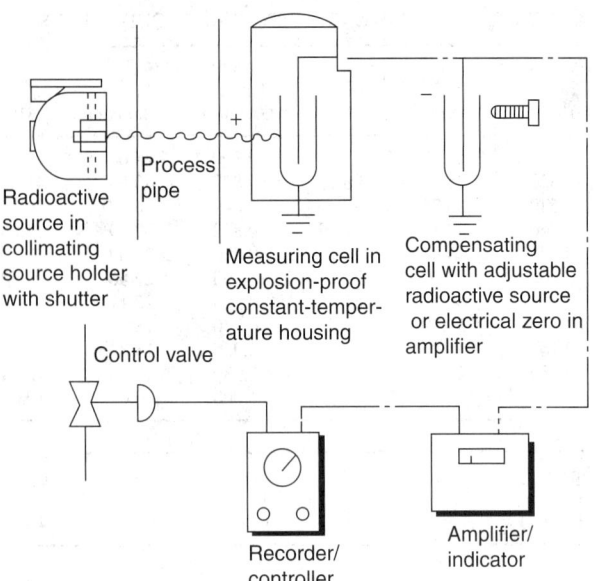

Fig. 5.   Nuclear radiation-type density meter.

weight measurement is converted to the motion of an indicating pointer or recording pen over a graduated scale, calibrated in units of density or specific gravity. In a viscous-drag type of instrument, one set of impellers is driven in a chamber containing a standard reference gas, and the power required to achieve rotation is measured. A matching set of nonrotating impellers is located in a chamber containing the gas under test. The sets of impellers are connected by a linkage and measure the relative drag shown by the tendency of the impellers to rotate in the test gas. The balance point is a function of relative gas density. In the buoyancy gas balance, a displacer is mounted on a balance beam in the vessel containing the test gas. A reading of the air pressure required to maintain the displacer at a perfectly horizontal position (observed through a window in the chamber) is read from a manometer. Then, the air is displaced in the chamber by the gas under test. The foregoing procedure is repeated. The ratio of pressure with air to pressure with the gas provides a measure of the density of the gas relative to air. This is primarily a laboratory type measurement.

SPECIFIC HEAT.   Sometimes called specific heat capacity. The quantity of heat required to raise the temperature of unit mass of a substance by one degree of temperature. The units commonly used for its expression are the unit mass of one gram, the unit quantity of heat in terms of the calorie. See also **Heat**.

Specific Heat at Constant Pressure. The amount of heat required to raise unit mass of a substance through one degree of temperature without change of pressure. Usually denoted by C_p, when the mole is the unit of mass, and c_p when the gram is the unit of mass.

Specific Heat at Constant Volume. The amount of heat required to raise unit mass of a substance through one degree of temperature without change of volume. Usually denoted by C_v, when the mole is the unit of mass, and c_v when the gram is the unit of mass.

SPECIFIC HEAT (Electronic).   In the original formulation of the Drude free electron theory of metals, it was assumed that the electrons formed a classical gas whose specific heat is just $3Nk$ (N being the number of particles, k, Boltzmann's constant). No such specific heat was observed, but it was pointed out by Sommerfedd that the electrons should be treated as a Fermi-Dirac gas, for which the heat capacity at constant volume per mole of electrons is given by

$$C_v = \tfrac{1}{2}\pi^2 NkT/T_F$$

where T_F is the Fermi temperature. This formula suggests that only the fraction $(\pi^2/6)(T/T_F)$ of the electrons can actually contribute to the specific heat. Since this fraction is of the order of 10^{-3} at room temperatures, the electronic specific heat is negligible compared with the lattice specific heat except at temperatures of a few degrees absolute. In the band theory of solids, C_v is roughly proportional to the effective mass of the electrons.

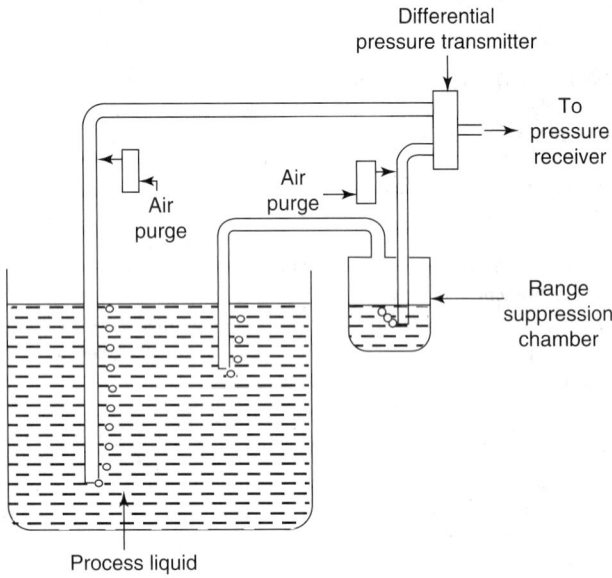

Fig. 4.   Differential-pressure method for specific gravity measurement.

Boiling-point Rise System. For certain liquids, the temperature of a boiling solution of the unknown may be compared with that of boiling water at the same pressure. For a given solution, the boiling-point elevation may be calibrated in terms of specific gravity at standard temperature. Usually two resistance thermometers are used. The system finds use in the control of evaporators to determine the endpoint of evaporation. Good accuracy is achieved in the determination of one dissolved component, or of mixtures of fixed composition.

Nuclear-type Density Meters. As illustrated in Fig. 5, a radioisotope source is placed on one side of a pipeline while a detector is placed on the opposite side. Transmitted radiation is in proportion to the density of the material within the pipeline. Standard radiation detectors and associated electrical instrumentation can be used. An A.C. type amplifier is used for accurate determinations when the measurement spans are narrow. For wider spans, D.C. amplifiers may be used. A compensating cell may be used for zero suppression. This essentially eliminates zero drift rate that results from radioisotope source decay. The method is particularly attractive for the density and specific gravity determination of slurries. See also **Radioactivity**.

Gas Density Measurements. In the gas specific gravity balance, a tall column of gas is weighed by the floating bottom of the vessel. This

SPECIFIC HEAT (Humphries Equation).

An expression for the ratio of specific heats of moist air, useful in the calculation of the velocity of sound in the atmosphere:

$$C_p/C_v = \gamma = 1.40 - 0.1e/p$$

where γ is the ratio of specific heats, p is the total atmospheric pressure, e is the water vapor pressure.

SPECIFIC HUMIDITY.

In a system of moist air, the (dimensionless) ratio of the mass of water vapor to the total mass of the system. The specific humidity may be approximated by the mixing ratio for many purposes: $q = w/(1 + w)$ where q is the specific humidity and w is the mixing ratio.

SPECIFIC SURFACE.

The surface, or area, of a substance or entity per unit volume; obtained by dividing the area by the volume, and expressed in reciprocal units of length.

SPECIFIC VOLUME.

The volume of a substance or entity per unit mass, obtained by dividing the volume by the mass; and expressed in units of length to the third power and reciprocal units of mass. Reciprocal of the density.

See also **Density**.

SPECTRAL ANALYSIS.

A method of analyzing stationary time-series into a series of harmonic terms. In effect, the series is correlated with trigonometrical functions of type $\cos(\omega t + \beta)$ for a range of values of ω. A high correlation indicates the possibility of a harmonic of corresponding period in the original series; and in any case the pattern of variation of the correlation over a range of ω is typical of the constitution of the series. The process is analogous to the splitting of a ray of polychromatic light into a spectrum exhibiting its monochromatic constituents.

The correlation, or some simple function of it called the intensity, is usually graphed as ordinate either against the period $2\pi/\omega$, giving the *periodogram*, or against the frequency ω, giving the *power spectrum*.

The word periodogram is also used for other methods that attempt to detect periodicity in the series.

Similar methods can be applied to examine the relationship between two or more series. For two series this leads to two components, real and imaginary, of the spectral density. The real component is called the co-spectrum and the imaginary component the quadrature spectrum. The two are amalgamated to form a measure known as the coherence.

The spectrum can be regarded as a Fourier transform of the autocor-relation fraction. See also **Spurious Correlation**. A more sophisticated generalization concerns the Fourier transform of third-order moments and is known as the bi-spectrum, not a term to be recommended.

SIR MAURICE KENDALL, International Statistical Institute, London.

SPECTRAL CENTROID.

An average wavelength, computed especially for light filters and other light-transmitting devices, by taking a weighted average, for each wavelength, of the spectral energy distribution, of the incident light, the transmittance of the device, and the luminosity data of the eye.

SPECTRAL CHARACTERISTIC.

A relation, usually shown by a graph, between wavelength and some other variable. 1. In the case of a luminescent screen, the spectral characteristic is the relation between wavelength and emitted radiant power per unit wavelength interval. 2. In a photoelectric device, it is the relation between wavelength and sensitivity per unit wavelength interval.

SPECTRAL CLASS.

A casual examination of the stars with the eye or small telescope reveals that stars have different colors. If stars radiate approximately as black bodies, then it will be expected that they can be graded according to their temperature, i.e., their color. The first attempt to grade stars into groups or classes was at Harvard in the famous Henry Draper catalogue. This system has been refined into the present powerful system called the Yerkes or MKK classification and is contained in the Yerkes Atlas of Stellar Spectra.

In the modern classification, the Harvard classes are still retained; thus, there is the sequence of stars W, O, B, A, F, G, K, M, R, N, and S. The Yerkes classification is essentially one of ratios of various ionized and excited levels of atoms. For example, a B0-type star is determined primarily by the fact that the ratio of the 4552 Si III line to that of the 4089 Si IV line is less than one; whereas, in a B1-type star, this ratio is greater than one. For the later types of stars (in particular, the M, R, N, and S stars), the intensity of the TiO bands is read for classification. The classes from B through M are subdivided into ten units; thus, there is B0, B1, B2, etc., to B9 and G0, G1, G2, etc., to G9.

At the time when this spectral classification was developed, it was apparent that the stars were of different luminosity; i.e., that there were giant and supergiant stars. A criteria was developed for increasing luminosity based essentially on ratios of certain lines. The ratios will be slightly different due to the sharpening of lines in a giant atmosphere. This is due to the reduced pressure in these atmospheres.

The luminosity types are Ia, the most luminous supergiants; Ib, the less luminous supergiants; II, the luminous giants; III, the giants; IV, the subgiants; V, the main sequence stars or dwarfs; and VI, the subdwarfs. It is apparent, then, that there is a two-dimensional classification involving luminosity and spectral type. This is called the luminosity spectral diagram or the Hertzsprung-Russell diagram, and is exhibited in the article on the **Giant and Dwarf Stars**.

In order to use the MKK classifications, one merely obtains spectrograms at the dispersion available to his spectrograph and compares them with a series of MKK standards. To do this, he selects an unknown star, takes its spectrum at the same dispersion, and compares it with his catalogue of MKK standards. In this way, he can easily classify within one subdivision of a spectral type many, many stars in any given period of time.

A second and perhaps much more powerful method of classification has been developed by Barbier, Chalonge, and Divan. The spectrum of a star is taken at very low dispersion, traced out by means of a microphotometer, and the Balmer decrement is measured. The three coordinates are obtained — Φ, D, and λ. Thus there is a three-dimensional classification, which differentiates very neatly between Population I and Population II. However, the resulting three-dimensional surface applies only to spectral types O through approximately F6 and in the various luminosity classes, and so has the drawback that the fainter or redder end of the spectral sequence is not considered.

SPECTRAL COLORS.

1. Colors present in the spectrum of white light. 2. Colors that are represented by points on the chromaticity diagram that lie on straight lines between the achromatic point and the spectrum locus.

SPECTRAL ENERGY DISTRIBUTION.

When radiation exhibiting a continuous spectrum, as that from a hot stove or the light from an incandescent lamp, is quantitatively analyzed, it is found that quite different amounts of power are represented by the radiation within equal ranges of wavelength or of frequency having different limits. The proportion in any such range depends upon the character of the source. Thus, in the radiation from a candle, the ratio of the energy output between 6,500 and 6,600 angstroms (red) to that between 4,500 and 4,600 angstroms (blue-violet) is greater than the corresponding ratio for the radiation from an arc lamp. If we divide the spectrum into small intervals of wavelength, say 10 angstroms, and plot the power output for each range as ordinate, with the mean wavelength of the interval as abscissa, the result is a curve showing the distribution of power through the spectrum. When the radiation is due to high temperature, as in the above examples, there is always a wavelength interval having maximum power, that is, the curve has a "peak," from which the ordinates fall off in both directions. Wien pointed out that the higher the temperature of the source, the farther toward the short-wavelength end of the spectrum does this peak lie. See Fig. 1.

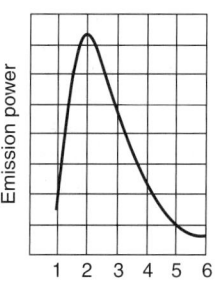

Fig. 1. Spectral energy distribution for black body at 1170° (1443° absolute) with peak at 2000 angstroms (0.2 micrometer).

An instrument utilizing a prism for dispersing the radiation, together with a thermocouple or similar device for measuring its flux density in different ranges, may be used to analyze infrared thermal radiation, and is called a spectroradiometer. The spectrophotometer performs a similar service for visible light, except that the results in this case are usually tabulated in terms of the visibility rather than the actual power of the emission. See also **Wien Laws**.

SPECTRAL FUNCTION.   A necessary and sufficient condition for $\rho(\tau)$, where $\tau = 0, 1, 2, \ldots$, to be an autocorrelation function of a discrete stationary stochastic process is that it is expressible in the form

$$\rho(\tau) = \frac{1}{\pi} \int_0^\pi \cos \tau\omega \, dF(\omega)$$

where $F(\omega)$ is a nondecreasing function with $F(0) = 0$, $F(\pi) = \pi$. For a continuous process the corresponding condition is that:

$$\rho(\tau) = \frac{1}{\pi} \int_0^\infty \cos \tau\omega \, dF(\omega)$$

With $F(0) = 0$, $F(\infty) = 1$. Conversely, we have

$$F(\omega) = \omega + 2 \sum_{j=i}^\infty \frac{\rho_j}{j} \sin j\omega, 0 \le \omega \le \pi$$

For the discontinuous process, and

$$F(\omega) = \frac{2}{\pi} \int_0^\infty \rho(x) \frac{\sin x\omega}{x} dx, 0 \le \omega \le \infty$$

for the continuous case. The function $F(\omega)$ is variously called the spectral function, *integrated function*, power spectrum or integrated power spectrum; the first appearing to be the simplest usage. Similarly, $dF(\omega)/d\omega$ is called the *spectral density*.

Both spectral function and spectral density can be defined directly without invoking the concept of autocorrelation.

SPECTRAL SENSITIVITY.   Three uses of this term are: 1. The sensitivity of a detector measured for narrow spectral bands throughout the spectrum. 2. The emitted radiant-power wavelength distribution of a luminescent screen under a given condition of excitation. 3. The sensitivity of a photoelectric device in relation to the wavelength of the incident radiant energy.

Spectral sensitivity is usually displayed on a spectral characteristic.

SPECTROCHEMICAL ANALYSIS (Visible).   Chemical systems that exhibit a selective light absorptive capacity are colored. Hence, the terms *colorimetric analysis* and *colorimetry* often are used to designate the measurement of such systems when the objective is to determine the concentration of the constituent responsible for the color. The use of the term *colorimetry* in this respect is not to be confused with the use of the same term in physics where the term refers strictly to the measurement of color. See also **Colorimetry**.

Visible spectrometry may be used to determine a constituent constituting the major part of a sample, but it also may be applied to the determination of trace quantities. Some methods are applicable to amounts of a few parts per million, with some tests sensitive to 0.01 ppm or less.

Like other areas of spectroscopy, visible spectrometry has a wide range of applications. Included are most of the elements, many anions, functional groups, and innumerable compounds.

The main practical problems in the methodology of visible spectrometry are: (1) to prepare a suitable colored solution; and (2) to measure the light absorptive capacity of this solution, or to compare it with that of a colored solution of known concentration.

Visual spectrochemical analysis has largely been displaced by other, more automated instrumental methods.

SPECTRO INSTRUMENTS.   *Spectro* is used as a prefix for a wide assortment of analytical instruments. *Spectro* is derived from *spectrum*, which originally referred to the component colors that make up visible light, the so-called rainbow colors of violet, indigo, blue, green, yellow, orange, and red. A very simple device made up of a glass prism to break up sunlight into color bands is referred to as a *spectroscope*. Much more sophisticated instruments are available for manual manipulation and

observation, which still rely on this basic, simple principle; these are termed visual spectroscopes, and the field is called visual spectroscopy.

Over the years, the term *spectrum* has increased in application and meaning and now embraces the total electromagnetic radiation span — no longer being confined to the visible portion. Concurrently, the term *spectroscope* has broadened in meaning so that mention of spectroscope no longer signifies an instrument which operates in the visible region. This situation gave rise to the need for modifying words for use with the term *spectroscope* to signify the portion of the electromagnetic spectrum with which the instrument is concerned. Thus, there are infrared spectroscopes, x-ray spectroscopes, ultraviolet spectroscopes, microwave spectroscopes, and so on. Further, the term *spectrum* has widened to include practically any orderly array of phenomena. For example, a mass spectrometer sorts atoms or radicals by their atomic weight — over a spectrum of values. Other spectrometers analyze the decay "spectra" of radioactive isotopes.

In terms of what is measured or observed, there are (1) portions of the electromagnetic spectrum: gamma-ray, cosmic ray, x-ray, ultraviolet, infrared, far-infrared, microwave, and radiowave instruments; (2) regions pertaining to the energies of particles: beta ray (electrons), protons, neutrons, and mass associated instruments; and (3) instruments dealing with other "spectra" such as radioactive decay and Mossbauer effects.

The suffix *scope* was ample when spectro instruments were essentially manual and confined to visual observations with the unaided eye. As the functions of these instruments increased, new suffixes were required to completely describe the instruments. Thus, if the instrument provides a record of the measurement, either by means of photographic film or by pen recording on a chart, the suffix *graph* is used. Where a meter is utilized to display the information detected by a device such as a bolometer, thermocouple, or thermistor, and so on, the suffix *meter* may be used. In cases where an instrument is designed to measure the intensity of various portions of spectra (again not confined to visible light), the suffix *photometer* may be used. Thus, there are spectrographs (spectrography), spectrometers (spectrometry), and spectrophotometers (spectrophotometry).

But, with increasing complications of instrument design and flexibility of use, the foregoing terms still are not sufficient to fully describe many instruments. Although some instruments display a continuous spectrum of what is being measured, other instruments filter out, for example, certain incident radiation. In other words, certain radiation may be absorbed, giving rise to the terms *absorption spectrometer* or *absorption spectrograph*. In some cases, the energy level of the specimen under analysis must be raised, as by means of flame heating or a spark discharge. Instruments in this category are of the emission type — and hence such terms as *flame photometer* and *optical emission spectrometer*. The names of spectro instruments can be complicated further where some special function may be incorporated in the title. A few examples are:

Spectroheliograph: a spectrograph designed for use with a telescope and for making photographs of the sun, in which the radiation of a particular element in the radiation of the sun may be recorded.

Ultraviolet-visible Spectropolarimeter: an instrument for measuring the rotation of the plane of polarization in accordance with wavelength and light intensity.

Spectrophotofluorometer: an instrument which provides means for both controlling the exciting wavelength and for identifying and measuring light output of a fluorescing sample. If the term *fluorometer* alone were used, it could indicate an instrument with a filtered light input that will measure the fluorescent light from a sample, usually by wavelength.

SPECTROSCOPE.   Several types of instruments for producing and viewing spectra are included under this term. Variations in form are due, not only to differences in principle, but also to the type of radiation or phenomena to be examined. Terminology employed in this field of instrumentation is described under **Spectro Instruments**.

The earlier spectroscopes, developed by Fraunhofer, Ångström, and others in the early part of the last century, adapted Newton's discovery of the dispersion of light by a prism. The essential features shown in Fig. 1 are a slit, S, a collimator, C, for rendering the light from the slit parallel before entering the prism, one or more dispersing prisms, P, and a telescope, T (or a camera), for forming images of the slit in the various wavelengths and thus providing a method for viewing or photographing the spectrum. The light passes through these in the order

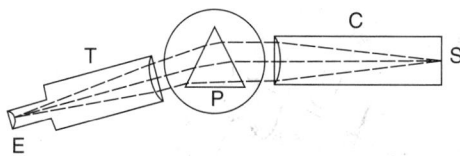

Fig. 1.   Simple prism spectroscope: Collimator, C; slit, S; prism, P; telescope, T; and eyepiece for viewing spectrum, E.

named, being deviated by the prism through various angles according to the wavelength. When a spectroscope is provided with a graduated circle for measuring deviations, it is called a "spectrometer." The "direct-vision" or nondeviation spectroscope, employing an Amici prism, is a compact instrument for qualitative purposes. The photographic spectroscope, known as a "spectrograph," is now almost universally used in spectral research.

Many modern spectroscopes employ the diffraction grating instead of the prism. In the concave grating spectroscope, developed by Rowland, the collimating lens and telescope or camera objective are unnecessary because of the focusing effect of the grating itself. See Fig. 2.

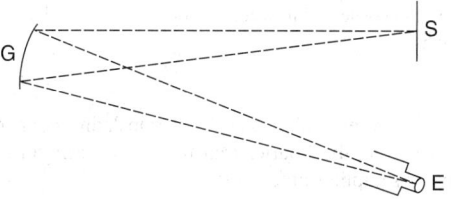

Fig. 2.   Concave grating spectroscope: Slit, S; grating, G, eyepiece or plate-holder, E.

The growth of the importance of infrared spectrography and spectrophotometry in determining the structure of compounds and the composition of substances has led to the development of many infrared spectroscopes and other instruments. Most infrared spectroscopes and spectrophotometers employ front-surface mirrors instead of lenses. This eliminates the necessity for energy to pass through glass, quartz, or similar material. Furthermore, it would be difficult or impossible to make lenses that would bring rays of the widely varying wavelengths in the infrared region to focus at one point. Occasionally a rock salt lens will be found in an infrared spectrophotometer, but such lenses are usually not essential optical components. Parabolic mirrors bring energy of all wavelengths to focus at one point. Reflection from most metallic surfaces is generally very efficient in the infrared region. Both gratings and prisms can be used for dispersing the energy, but prisms seem to be more common, perhaps because energy in the infrared region is at a premium and none can be wasted in higher order spectra. The materials most suitable in the infrared region are quartz, calcium fluoride, sodium chloride, and sodium bromide, all in the form of single crystals.

The energy source for an infrared spectroscope or spectrophotometer is a Nernst glower or a Globar. The receiving element for the infrared radiation must be a bolometer, Golay cell or thermistor since photocells are not sensitive in this region of the spectrum. Indeed, the thermoelectric element chosen must be extremely sensitive, since the average energy in the dispersed beam is very small.

Most available infrared instruments use the Littrow mount for the prism, the beam being reflected from a plane mirror behind the prism and thus returning it through the prism a second time. This doubles the dispersion produced. Actually, a double-pass system is also used so that the beam goes through the prism four times. Other design modifications include those with single beam and double monochromator, double beam and double monochromator, and related combinations. See also **Infrared Radiation**.

Check alphabetical index for a wide range of spectroscopic instruments that are based upon numerous materials-energy relations. Also see **Analysis (Chemical)**.

SPECTROSCOPIC BINARIES.   Discovery of the brighter component of Mizar in 1889 introduced a class of binary stars, which are not double stars in the ordinary sense of the term. The components are too close together for them to be observed separately even with optical telescopes

of high resolving power. In such binaries the period is usually short and the orbital velocities high. Unless the orbit plane happens to be perpendicular to the line of sight, the orbital velocities will have components in the line of sight and the observed radial velocity of the system will vary periodically. Since radial velocity is measured with the spectroscope, employing the Doppler-Fizeau principle, the binaries so observed are known as spectroscopic binaries. In some spectroscopic binaries the spectra of both stars are visible and the lines are alternately double and single. Such stars are known as double-line binaries. In others the spectrum of only one component is seen and the lines in this spectrum move periodically from violet toward the red and back again. See Fig. 1.

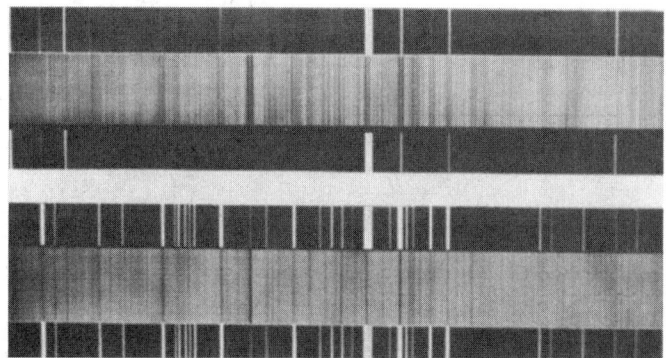

Fig. 1.   Spectrogram of *Mizar*. The lines of the two components are separated in the upper photo and superposed in the lower. This was a classical observation made several decades ago at Yerkes observatory.

The determination of the orbit of a spectroscopic binary is made from a long series of observations of the radial velocity of one or more components of the system. The observations are first plotted against time and from the resulting curve the period may be obtained. With this period determined observations are then reduced to a single epoch and the best possible curve drawn through the points, obtaining what is known as the velocity curve of the system. If the orbit is circular, the velocity curve will be a sine curve. If the orbit is elliptical, the shape of the curve will depend upon the eccentricity of the ellipse and the orientation of the major axis with reference to the line of sight. From the shape of the velocity curve the orbit of the system in space may be determined. In the solution of the spectroscopic orbit it is impossible to determine individually the semimajor axis, a, and the inclination of the orbit plane, i. However, the product of the semimajor axis by the sine of the inclination (i.e., $a \sin i$) may be determined directly in linear units (i.e., in either miles or kilometers). If either a or i can be obtained from other types of observations, as in the case of eclipsing binaries, a complete solution for the orbit can be made.

See also **Binary Stars**; **Eclipsing Binary**; and **Visual Binaries**.

SPECTROSCOPIC PARALLAX.   The term *spectroscopic parallax* of a star is applied to a determination of the distance of a star in which the stellar parallax is determined from observations of spectral peculiarities of the star, together with determinations of the apparent brightness of the object.

The apparent brightness of a star depends upon two fundamental factors: the intrinsic brightness of the star and its distance from the observer. Expressed on the stellar magnitude scale, we find the apparent magnitude, m, the absolute magnitude, M, and the stellar parallax, p, to be connected by the analytical expressions: $M = m + 5 + 5 \log_{10} \pi$. The apparent magnitude of a star may be determined by a variety of methods of stellar photometry, and if a method is available for the determination of the absolute magnitude the value of the stellar parallax may be determined.

The relative intensities of certain spectral lines are different in giant and dwarf stars of the same spectral type. The relative intensities of selected pairs of lines may be compared in stars of the same spectral type and known absolute magnitudes and a "calibration curve" obtained. The relative intensities of the same pairs may then be found in stars of unknown absolute magnitudes and the calibration curves used to determine the absolute magnitude. Thus, the parallax may be determined from a study of spectra. The accuracy of the determinations of spectroscopic parallax compares very favorably with parallaxes obtained from the relative trigonometric methods.

SPECTRUM ANALYSIS. An electronic measurement technique for quantifying the level of electrical signals over a specific frequency range. The spectrum analyzer, an electronic instrument for making spectrum analysis measurements, conveys its signal information usually by displaying a rectangular plot of signal amplitudes (level) versus signal frequency (spectrum) on a cathode ray tube (CRT). Figure 1 shows a diagrammatical plot of a frequency spectrum and the comparable CRT photo of a spectrum analyzer output.

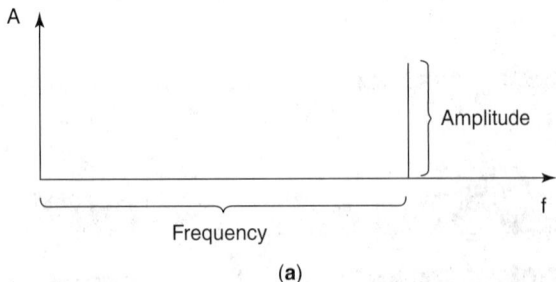

(a)

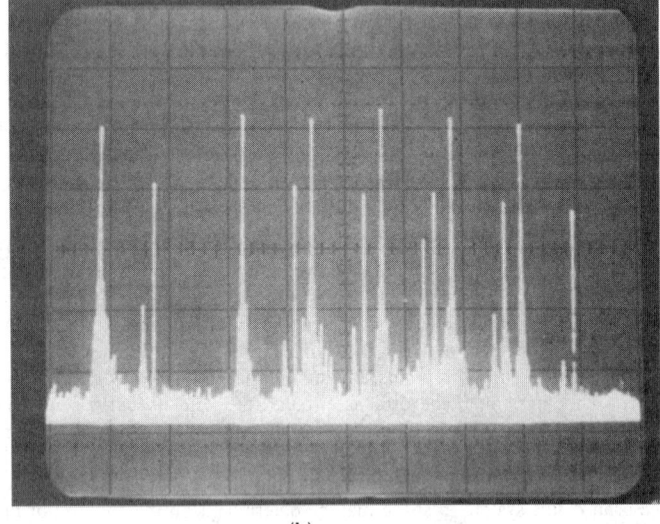

(b)

Fig. 1. (**a**) Information content of spectrum analyzer display; (**b**) Six-channel spectrum analyzer CRT display of six television broadcasts.

Signal analysis. An electrical *signal* transfers energy from one point to another. A signal may be strictly for power transfer, such as in high voltage power distribution, or it may be for information transfer, such as a telephone voice channel. It is necessary to quantify electronic signals for the effective design and maintenance of all types of equipment and machinery.

An electrical signal is composed of one or more frequency components channeled into a single transmission path, such as a cable, waveguide, or antenna. (A dc signal is no exception, it has one signal component at zero frequency.) An instrument that is capable of identifying and quantifying each frequency component is a signal analyzer. A spectrum analyzer is a type of *signal analyzer*.

The number of different types of signal analyzers are differentiated mainly by the electronic measurement techniques used. Each has a significant contribution to make to spectrum analysis measurements. These techniques are: (1) Fourier transformation of an input time domain signal by mathematical computation (Fourier analyzer); (2) filtering an input signal through a number of piecewise tuned bandpass filters (real-time spectrum analyzer); (3) scanning an input signal with a tunable filter (spectrum viewer); and (4) heterodyne receiver (spectrum analyzer, tuned voltmeter, and wave analyzer).

Fourier analyzer. Any nonsinusoidal periodic signal has sinusoidal components predicted by the mathematical Fourier series equations. The Fourier analyzer processes a given input voltage signal with a digital computer, transforming the signal into the frequency domain with both phase and amplitude information. See Fig. 2.

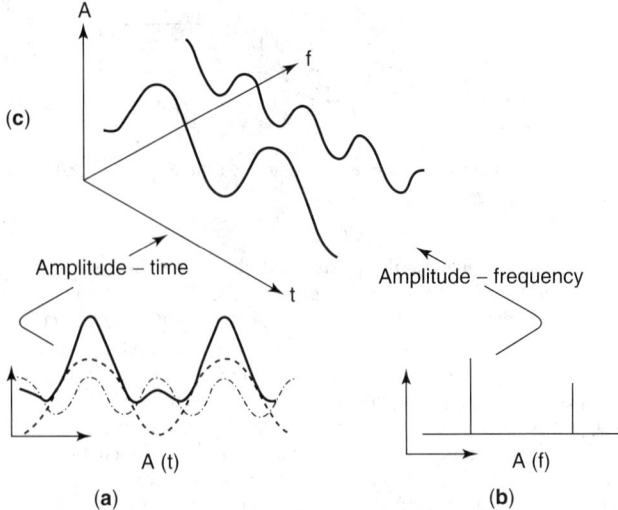

Fig. 2. (**a**) A time-varying input signal; (**b**) the Fourier analyzer transforms the signal into the frequency domain (with phase information also available, but not shown; (**c**) composite 3-dimensional plot showing relationship of time and frequency graphs.

The Fourier analyzer is also capable of translating nonperiodic voltage signals by calculating the Fourier transform in a form convenient for the digital computer sampling processes:

$$S_x''(t) = \Delta t \sum_{n=-\infty}^{n=+\infty} x(n\,\Delta t)e^{-2\pi f n\,\Delta t}$$

where $x(n\,\Delta t)$ are the analyzer-measured values of the input voltage signal.

Any time-varying voltage can be transformed. When a signal is not periodic, the total elapsed input time, T, is taken as the period. The analyzer develops a frequency spectrum (with phase information) of the signal as if it were being repeated at a $1/T$ rate. See Fig. 3. The CRT display is capable of showing either the real or imaginary frequency components on a rectangular grid or the magnitude/phase diagram on a polar plot.

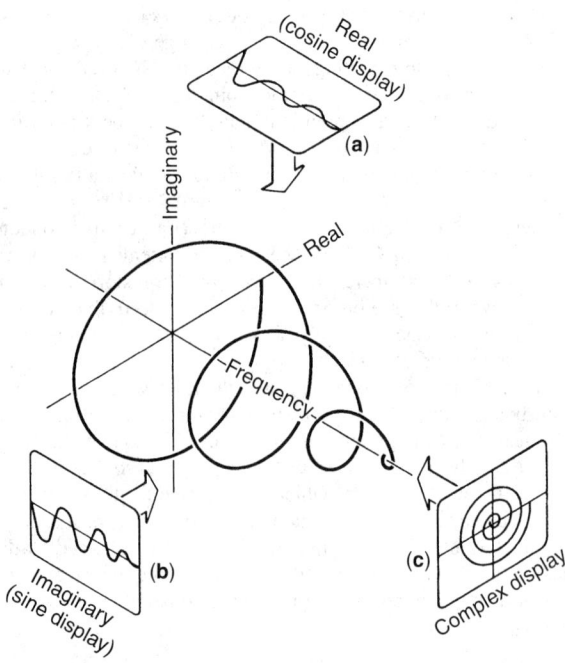

Fig. 3. Fourier analyzer processing a nonperiodic input voltage signal with the digital Fourier transform. (**a**) The real frequency component envelope (individual frequency spikes are not shown); (**b**) imaginary frequency component envelope; and (**c**) complex amplitude and phase display.

The fundamental parameters that govern the Fourier analyzer's performance capability can be summarized by these equations:

$$T = n \times \Delta t \text{ (for the time domain)}$$

where
T = total time of the sample, in seconds
n = number of samples taken, usually a power of 2, reflecting the amount of computer storage available
Δt = time between samples, in seconds
$F_{max} = n/2 \times \Delta f$ (for the frequency domain)

where F_{max} = maximum frequency of display, bandwidth, in Hz
$n/2$ = the number of frequency points, $\frac{1}{2}$ of n points are dedicated to either the real or imaginary frequency planes
Δf = the number of Hz between displayed frequency points, frequency resolution

Solving these to eliminate n:

$$\Delta t = \tfrac{1}{2} F_{max} = \text{sample spacing in time}$$

$$\Delta f = 1/T = \text{frequency resolution}$$

F_{max} is typically 100 kHz or below. For a given n, either F_{max} or Δf can be optimized only at the expense of the other. However, the selection of sample times and bandwidths can be tailored to the individual application requirement.

Because of the Fourier analyzer's measurement and computational strength, it can be used to simulate the transfer function for processes that have measurable input/output voltage signals. The analyzer can then be used to simulate the processed frequency spectrum response to stimuli too difficult, dangerous, or costly to input to the real process.

Real-time spectrum analyzer. This instrument, shown in Fig. 4, displays the frequency spectral components on a CRT as they occur. The display response is directly proportional to the input signal at each instant of time. (The Fourier analyzer presents a computed display that simulates this real-time response, and thus it is sometimes referred to as a real-time analyzer.)

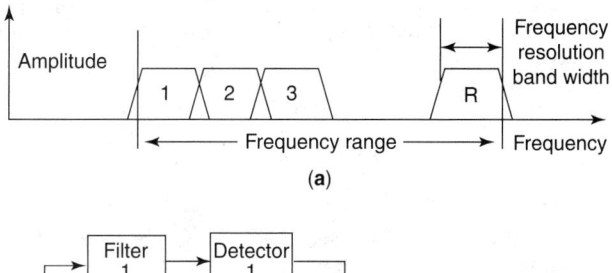

(a)

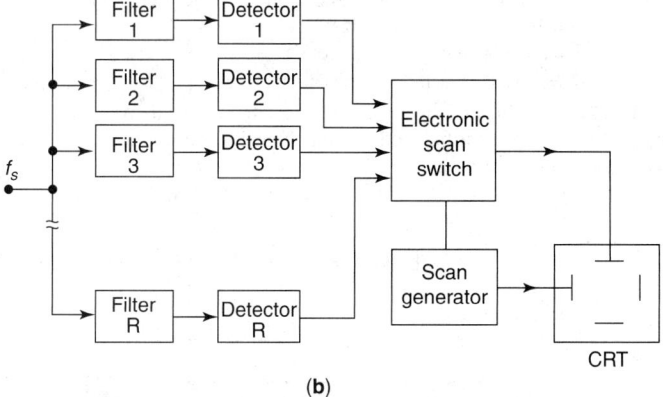

(b)

Fig. 4. (**a**) The real-time spectrum analyzer consists of a number of passband filters aligned such that their passbands continuously cover the spectrum of interest; (**b**) the electronic scan switch samples the filters fast enough to display instantaneous response of the input signals.

A real-time spectrum analyzer is able to respond to input stimulus essentially instantaneously because it passes the input directly to the display through a set of filters and detectors which quantify the spectral energy. See Fig. 5.

The real-time spectrum analyzer shows accurate stimulus-response characteristics up to 10 kHz. With as many as 500 high-resolution bandpass

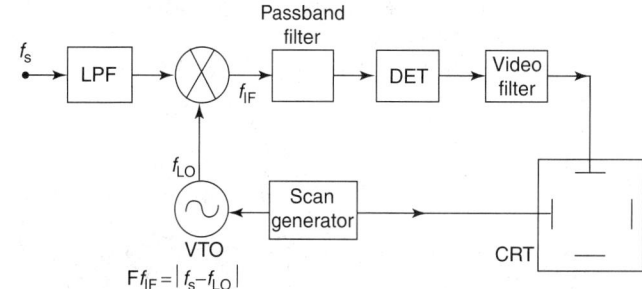

Fig. 5. Application of superheterodyne principle in real-time spectrum analyzer.

filters, the analyzer can display signals close together in frequency even when they differ in amplitude. Filter rolloff is typically a 15:1 voltage decrease per octave.

Spectrum viewer. The *spectrum viewer* uses the same input filter-detector as the real-time spectrum analyzer except the viewer uses only a single filter which is swept-tuned through the frequency band of interest. See Fig. 6.

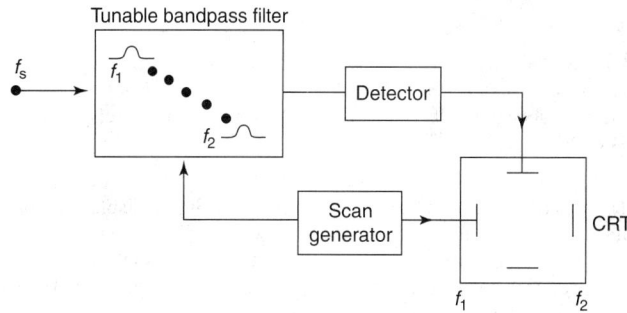

Fig. 6. Spectrum viewer operation. Synchronizing the tunable filter center frequency with the cathode ray tube horizontal deflection results in a spectral display for each sweep.

The advantage is cost effective octave or greater band spectral coverage from 2 GHz to over 18 GHz made possible by a tunable yttrium iron garnet (YIG) filter. As with all swept-tuned analyzers, the spectrum viewer is not a real-time analyzer. Since it views only one small segment of frequency spectrum in any one instant, the filter must charge and discharge as it is tuned past various spectral signals. (The parallel filters of the real-time analyzer always remain responsive to input spectra, even though they are sampled for display by sweeping process.)

The spectrum viewer is the least expensive technique for displaying frequencies above 1 GHz; however, it generally does not have absolute amplitude calibration, high frequency resolution, or high sensitivity.

Superheterodyne receiver spectrum analyzer. The final category of signal analyzers utilizes the same technique used in almost every commercial radio receiver, the *superheterodyne* principle. The technique uses a bandpass filter as in the other analyzers, but in the heterodyne process the filter is not centered at the input frequency. Rather the filter is at a fixed frequency called the intermediate frequency (IF). Since most of the instrument's signal processes are done at the single IF, more sophisticated filtering and scaling is possible. This results in accuracy and high performance.

The local oscillator (LO) frequency is varied from a control on the instrument's front panel until the following condition is met:

$$f_{LO} - f_S = f_{IF}$$

where f_{LO} = local oscillator frequency
f_S = input signal frequency
f_{IF} = intermediate frequency

The resulting IF signal is processed with the familiar bandpass filter and displayed. This technique is operable from dc to over 40 GHz.

Tuning the superheterodyne signal analyzer is done either manually, reading only one point in the spectrum at a time, or automatically, sweeping and reading over a frequency range called a *frequency span*. Traditionally only the swept-tuned analyzer is called a *spectrum analyzer*.

See Fig. 7. The manually tuned analyzer, or *wave analyzer*, is capable of spectrum analyzer measurements. Many of the operational characteristics are common to both the wave and spectrum analyzers.

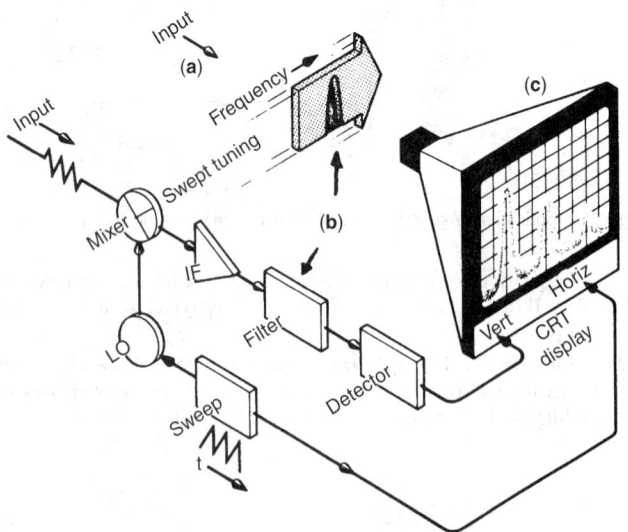

Fig. 7. Swept-tuned spectrum analyzer. Input spectrum (**a**) is mixed with sweeping local oscillator to produce three successive responses at f_{IF}. These are ushered through the IF filter (**b**) to the display (**c**) Note the displayed signal takes on the filter shape.

The output for the wave analyzer is a meter or digital display, calibrated in units convenient for the industry to be served. Most spectrum analyzer detection methods are proportional to power; thus, units of power and voltage referred to a transmission characteristic impedance (Z_o) are the most commonly used.

Here are their definitions:

V = voltas (rms) into Z_0

mV = millivolts into Z_0

μV = microvolts into Z_0

dBm = decibel above a milliwatt

$= 10 \log \dfrac{P}{10^{-3}}$

dBmV = decibel above a milliwatt referred to Z_0

$= 20 \log \dfrac{v}{10^{-3}}$

dBmV = decibel above a milliwat referred to Z_0

$= 20 \dfrac{v}{10^{-6}}$

where P = power in watts

v = voltage in rms volts

Z_0 = transmission line characteristic impedance, such as 50 Ω, 75 Ω, 600 Ω

Absolute amplitude calibration means each point on the meter or CRT display represents a specific voltage or power level. *Dynamic range* is the amplitude ratio of the largest to the smallest signal that can be displayed

simultaneously with no analyzer distortion products present. (Distortion products are harmonic and intermodulation signals generated inside the spectrum analyzer, which are displayed as legitimate signals.)

The lowest level signal that can be detected by the analyzer is specified by *sensitivity*. A spectrum or wave analyzer's sensitivity is limited by its inherent average noise. An unknown signal can be detected when the signal power is equal to the inherent average noise power. On the CRT this signal will appear 3 dB above the noise level. This relationship is given by

$$\frac{P + n}{N} = 2$$

where P = signal power, in watts

N = inherent analyzer noise power, in watts.

The ability to distinguish between two input signals that are close together in frequency is *resolution*. In spectrum analyzers this capability is primarily dependent upon IF resolution bandwidth, the 3 dB down points (half power) on the filter response curve. See Fig. 8. The narrower the resolution bandwidth, the better the resolution. However, since the IF filter must respond fully to the input signal ($f_{\text{IF}} = f_{\text{s}} - f_{\text{LO}}$) on each sweep, the sweep speed (Hz/second) must be slow enough to accommodate the filter. This relationship is shown by

$$T \propto \frac{S}{(BW)^2}$$

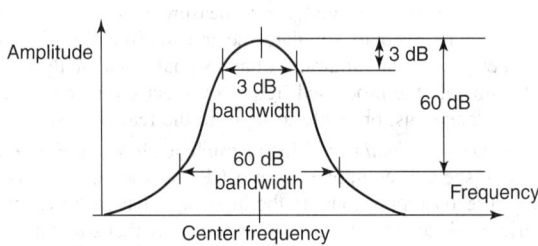

Fig. 8.   Typical Gaussian filter. Signals of equal amplitude can just be resolved when they are separated by the 3 dB bandwidth. Unequal signals can be resolved if they are separated by greater than half the bandwidth at the amplitude difference between them. Shape factor is defined as the ratio of the 60 dB bandwidth to the 3 dB bandwidth.

or

$$BW \simeq \frac{S}{T} \times K$$

for Gaussian shaped filter,

where BW = 3 dB bandwidth, or resolution bandwidth of filter, in Hz

S = frequency span, in Hz

T = time to sweep S, in seconds

$K = 0.94 (\text{Hz sec})^{1/2}$

Applications. Spectral response from electrical signals provides insight in a number of scientific fields. With the wide selection of transducers available, almost every industry and technology has areas where spectrum analysis can make a useful contribution. Some of these are summarized in Table 1.

JEFFREY L. THOMAS, Hewlett-Packard, Santa Rosa, CA.

TABLE 1. APPLICATIONS OF SPECTRUM ANALYSIS

Type of Instrument	Frequency Range	Application	Examples
Fourier analyzer	dc-100 kHz	Sound and vibration	Acoustic imprints for voice, noise pollution, and sonar; structural analysis for rotating machinery; stimulus-response testing
		Time-to-frequency transformation	Mechanical structure models; real-time transfer function determination; improving signal-to-noise ratios (noise rejection)
Real-time analyzer	dc-10 kHz	Sound and vibration	Real-time acoustic and structural analysis
		Electronic circuit analysis (100 kHz)	Filter design
Swept-tuned analyzer (Spectrum analyzer)	10 Hz-40 GHz	Electronic circuit analysis	General circuit design measurements; modulation, noise, distortion, power, frequency, and stability

SPEECH CLIPPING. The clipping of peak speech signals (peak clipping) or the reduction of weaker speech signals to zero (center clipping) in intelligibility tests.

SPEED. The magnitude of the vector velocity. Speed is a scalar quantity and is expressed in units of length divided by time. See also **Velocity**.

SPERRYLITE. A mineral diarsenide of platinum, $PtAs_2$. Crystallizes in the isometric system. Hardness, 6–7; specific gravity, 10.58; color, white; opaque. Named after Francis L. Sperry, Sudbury, Ontario.

SPHALERITE BLENDE. Also known as zinc blende, this mineral is zinc sulfide, $(Zn, Fe)S$, practically always containing some iron, crystallizing in the isometric system frequently as tetrahedrons, sometimes as cubes or dodecahedrons, but usually massive with easy cleavage, which is dodecahedral. It is a brittle mineral with a conchoidal fracture; hardness, 2.5–4; specific gravity, 3.9–4.1; luster, adamantine to resinous, commonly the latter. It is usually some shade of yellow brown or brownish-black, less often red, green, whitish, or colorless; streak, yellowish or brownish, sometimes white; transparent to translucent. Certain varieties are phosphorescent or fluorescent. Sphalerite is the commonest of the zinc-bearing minerals, and is found associated with galena, chalcopyrite, tetrahedrite, barite, and fluorite, as a result of contact metamorphism, and as replacements and vein deposits.

There are very many European localities, including Saxony; Bohemia; Switzerland; Cornwall, in England; Spain; Sweden; Japan; and elsewhere. In the United States, sphalerite is found in Arkansas, Iowa, Wisconsin, Illinois, Colorado, New Jersey, Pennsylvania, Ohio, and especially in the area that includes parts of Kansas, Missouri, and Oklahoma. The word sphalerite is derived from the Greek, meaning treacherous, and its older name, blende, meaning blind or deceiving, refers to the fact that it was often mistaken for lead ore.

SPHENE. This mineral occurs as a yellow, green, gray, or brown calcium titanosilicate, corresponding to formula $CaTiSiO_2$, crystallizing in the monoclinic system. Fracture conchoidal to uneven; brittle; habit usually wedge-shaped and flattened crystals, also massive and lamellar; luster, resinous to adamantine; transparent to opaque; hardness, 5–5.5; specific gravity, 3.45–3.55.

Sphene is an accessory mineral of widespread occurrence in igneous rocks, and calcium-rich schists and gneisses of metamorphic origin, and very common in nepheline-syenites. In the United States sphene is found in Arkansas, California, New Jersey, New York; in Ontario and Quebec in Canada; and from Greenland, Brazil, Norway, France, Austria, Finland, Russia, Madagascar and New Zealand, as well as many other world localities.

SPHENISCIFORMES *(Aves, Spheniscidae).* A group of flightless sea-birds with very distinguishing characteristics. Their relationship to other orders of birds is uncertain, and therefore some ornithologists separate the penguins as a superorder or even a subclass which is distinct from all other living birds. Their closest relatives appear to be the Tube-Nosed Swimmers *(Procellariiformes)*.

The morphological characteristics that distinguish these birds from other birds are a result of their adaptation to life in the water. The length is from 40 to 115 centimeters (16 to 45 inches), and the weight ranges from 1 to 30 kilograms (2 to 66 pounds). The long, spindle-shaped body has legs which are inserted far back, so they are most effective as oars and steering organs. The tail, as a steering rudder, is streamlined and triangular; the wings are transformed into flippers, but contain all the bones of a wing suitable for flying. The bones, however, are shortened, flattened, and tightly connected by ligaments, thus forming a rigid surface. The breast muscles (wing muscles) are large, taking up the whole front from the neck on down to the lower abdomen. The trachea is, as in *Procellariiformes*, divided lengthwise. The body is uniformly covered with feathers except for the brood patch; they have thick subcutaneous fat-pads. There are 6 genera with 18 species, which are confined to the Southern Hemisphere. See also **Penguin**.

SPHERE. A solid bounded by a spherical surface. It may be generated by revolving a semicircle about its diameter as an axis. A *radius* of a sphere is a straight line from the center to the surface of the solid. A *diameter* is twice as long as a radius for it passes through the center of the sphere and ends on the surface. All radii of a sphere are equal and all diameters are equal. A great circle of a sphere is determined by a plane passing through the center; if the plane does not pass through the center, the section is a small circle. An *axis* of a sphere is any diameter and its ends are called poles, often a north or a south pole.

Two arcs of great circles intersecting on a sphere determine a spherical angle. A spherical polygon is a part of the surface of a sphere bounded by three or more arcs of great circles. When it has three sides it is a spherical triangle, which may be right, obtuse, acute, equilateral, isosceles, or scalene. See also **Angle (Mathematics)**.

The volume of a sphere, $V = 4\pi r^3/3$, where r is its radius and its area, $A = 4\pi r^2$, which also equals the lateral area of a cylinder circumscribed about it. If the sphere is hollow, that is a closed spherical surface, so that its radius to the inside wall is r_1 and to the outside wall r_2, then $V = 4p(r_2^3 - r_1^3)/3$.

For area and volume of torus, see also **Anchor Ring (or Torus)**.

For other properties of a sphere, see also **Spherical Surface**.

SPHERE OF INFLUENCE (Planet). The surface in space about a planet where the ratio of the force with which the sun perturbs the motion of a particle about the planet, to the force of attraction of the planet equals the ratio of the force with which the planet perturbs the motion of a particle about the sun, to the force of attraction of the sun on the particle. The volume inside this surface defines the region where the attracting body exerts the primary influence on a particle. See also **Planets and the Solar System**.

SPHERICAL ABERRATION. If the surfaces of a lens or the reflecting surface of a mirror are spherical, the rays refracted through or reflected from the outer portions will be brought to a focus in a different plane than those from the center, thus producing a blurring of the resultant image, known as spherical aberration. This effect is more pronounced in short-focus lenses or mirrors than in long-focus instruments, for the curvature of the surfaces of the short-focus instruments is greater.

The decrease of spherical aberration with increase in focal length was discovered very early in the history of optical instruments, and during the seventeenth century we find telescope builders increasing the focal lengths of their instruments to tremendous proportions. Telescopes with focal lengths between 100 and 200 feet were not uncommon during this period, and the problem of supporting the long thin tubes so that they could be used for astronomical purposes and remain straight was one calling for great ingenuity. Descartes in 1637 published the theory of spherical aberration and showed that theoretically it could be corrected by grinding the surfaces of lenses and mirrors in curves other than spheres. The difficulties of grinding the required lens curves were apparently greater than those of operating the long-focus telescopes. About the middle of the eighteenth century John Dolland published the fact that spherical aberration could be corrected by the same method employed for the treatment of chromatic aberration, i.e., by using two lenses, one convergent and the other divergent. All modern telescopic lenses of good quality now employ the double object glass with the figures of the lenses and the separation between the components depending upon the ideas of the makers. For wide-angle, short-focus lenses, such as are used in modern hand cameras and in astrographic cameras, the simple pair of lenses does not provide sufficient correction for all of the aberrations, and three or more lenses are used in combination. Perhaps the most common is the so-called doublet, in which two pairs of lenses are used with a considerable separation between the pairs.

Spherical aberration may be corrected in the case of a mirror by grinding the concave surface in the form of a paraboloid of revolution instead of in the form of a sphere. This provides almost complete correction for rays entering the mirror parallel to the axis of revolution of the paraboloid, i.e., along the principal axis of the mirror; but for rays entering at a moderately large angle, spherical and various other aberrations make their appearance in the resultant image. Hence, while the reflecting type of telescope can be more easily corrected for the aberrations along the axis, nevertheless it cannot be used to obtain photographs of large areas with good definition throughout.

SPHERICAL HARMONICS. In analogy to harmonic functions in the plane, the solutions of the Laplace equation in spherical coordinates.

Spherical surface harmonics are special sets taken over the surface of a sphere; therefore, the harmonic components are restricted to an integral number of waves over the sphere. Spherical harmonics have been applied in the study of the large-scale oscillations of the atmosphere.

SPHERICAL POLAR COORDINATES.   A curvilinear coordinate system. Its parameters are: (1) the radius vector r from an origin or pole to the point; (2) the colatitude q, an angle made by r and a fixed axis, the polar axis; (3) the longitude f made by the plane of q with a fixed plane through the polar axis, called the meridian plane. The coordinate surfaces are: (1) concentric spherical surfaces about the origin, $r = $ const.; (2) right circular conical surfaces with apex at the origin and axis along the Z-axis, $\phi = $ const.; (3) planes from the Z-axis, $\phi = $ const. The range of the variables is $0 \le r \le \infty$; $0 \le \pi$; $0 \le \phi\pi$. In terms of a right-handed rectangular system with the same origin:

$$x = r \sin\theta \, \cos\phi \quad r^2 = x^2 + y^2 + z^2$$
$$y = r \sin\theta \, \cos\phi \quad \theta = \cot^{-1} z / \sqrt{x^2 + y^2}$$
$$z = r \cos\theta; \quad\quad \phi = \tan^{-1} y/k$$

Synonymous terms are spherical coordinates or polar coordinates in space. If $\theta = \pi/2$, the point lies in the XY-plane and if the longitude ϕ is then called θ, as is customary, the system becomes that of polar coordinates in a plane.

See also **Conical Surface; Coordinate System and Polar Coordinates**.

SPHERICAL SURFACE.   A surface all points of which are at a fixed distance, the *radius*, from a fixed point, the *center*. The term sphere is frequently used for this surface but it more properly means a solid bounded by a spherical surface.

In rectangular coordinates its general equation is

$$x^2 + y^2 + z^2 + Gx + Hy + Kz + L = 0$$

but if the center is taken at the origin of the coordinate system, the equation becomes $x^2 + y^2 + z^2 = r^2$, where r is the radius of the spherical surface.

The surface is measured in terms of the following parts: (1) *Zone*, a portion of the surface included between parallel planes. The bases of the zones are circumferences made by the planes but if one of the bounding planes is tangent to the surface, it is a zone of one base. The distance between the planes is the *altitude* of the zone. (2) *Lune*, a part of the surface bounded by the circumferences of two great circles. (3) Spherical *pyramid*, part of a sphere bounded by a spherical polygon and the planes of its sides. (4) Spherical *sector*, part of a surface generated by the revolution of a circular sector about any diameter of which the sector is a part. (5) Spherical *segment*, a portion of a spherical surface between two parallel planes. (6) Spherical *wedge*, a portion of the surface bounded by a line and two great semicircles.

If A is surface area, the various cases give for it: sphere, $4\pi r^2$; zone, $2\pi r h$; lune, $\pi r^2 a/90$; triangle or polygon, $\pi r^2 E/180$, where r is the radius of the surface, h is the altitude of the zone, a is the number of degrees in the angle, and E is the spherical excess, defined by $E = T - 180\,(n-2)$, where T is the sum of the angles and there are n sides to the polygon.

See also **Sphere**.

SPHERICAL TRIGONOMETRY.   A generalization of plane trigonometry, spherical trigonometry is primarily concerned with the solution of spherical triangles. It is used in many navigation and astronomical problems, as well as in the construction of certain kinds of maps.

Solution of a spherical triangle means the finding of unknown sides and angles from given values for other sides and angles. The right spherical triangle is the simplest case, but it, unlike the plane right triangle, can have two or even three right angles. If, however, it has two right angles, the sides opposite them are quadrants and the third angle has the same measure as its opposite side. If all three angles are right angles, the measure of each side is 90°. These cases are all relatively simple; hence, we will consider a triangle that has only one right angle, its opposite side in general not being 90°.

Let a, b, c be the sides of a right spherical triangle, measured by the angle subtended at the center of a sphere, and with opposite angles A, B,

C, where $A = 90°$. Then 10 relations exist, as follows:

$$\sin a = \sin A \, \sin c \quad\quad \sin b = \sin B \, \sin c$$
$$\sin a = \tan b \, \cot B \quad\quad \sin b = \tan a \, \cot A$$
$$\cos A = \cos a \, \sin B \quad\quad \cos B = \cos b \, \sin A$$
$$\cos A = \tan b \, \cot c \quad\quad \sin B = \tan a \, \cot c$$
$$\cos c = \cot A \, \cot B \quad\quad \cos c = \cos a \, \cos b$$

Since these equations are rather awkward. Napier's rules are convenient in actual use. They can be stated as follows: let co-A, co-c, co-B mean the complements of the angles (co-A = 90° − A) and arrange the parts as shown in the diagram. Then, calling any angle a middle part, it will have two parts adjacent to it and two parts opposite to it. Napier's rules are then: (i) the sine of a middle part equals the product of the tangents of the adjacent parts; (ii) the sine of part equals the product of the cosines of the opposite parts. See Fig. 1.

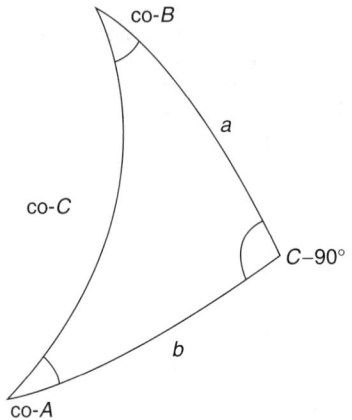

Fig. 1.   Designation of angles in Napier's rules.

In the case of an oblique spherical triangle, the following relations may be obtained. (1) Law of sines:

$$\frac{\sin A}{\sin a} = \frac{\sin B}{\sin b} = \frac{\sin C}{\sin c}$$

(2) Law of cosines: $\cos a = \cos b \cos c + \sin b \sin c \cos A$ and $\cos A = -\cos B \cos C + \sin B \sin C \cos a$. (3) Haversine law: have $a = \text{hav}(b - c) + \sin b \sin c \, \text{hav} \, A$. (4) Half-angle formulas:

$$\sin A/2 = \sqrt{\frac{\sin(s-b)\sin(s-c)}{\sin b \sin c}}$$

and

$$\tan A/2 = \frac{r}{\sin(s-a)},$$

where $2s = (a+b+c)$ and $r^2 \sin s = \sin(s-a)\sin(s-b)\sin(s-c)$. (5) Napier's analogies:

$$\frac{\tan(a+b)/2}{\tan c/2} = \frac{\cos(A-B)/2}{\cos(A+B)/2}.$$

(6) Gauss formulas:

$$\sin(A-B)/2 = \frac{\sin(a-b)/2 \cos C/2}{\sin c/2}.$$

(7) Rule of quadrants: in any spherical triangle, one-half the sum of two angles is in the same quadrant as one-half the sum of the sides opposite. (8) Spherical excess: $\tan^2 E/4 = \tan s/2 \tan(s-a)/2 \tan(s-b)/2 \tan(s-c)/2$. There are additional relations in several of these cases (for example, 3, 4, 5), obtainable by cyclic permutation of A, B, C and a, b, c.

In solving a spherical triangle there are six possible cases, depending on the parts given, as follows: (I) a, b, C; (II) A, B, c; (III) b, c, C; (IV) A, B, a; (V) a, b, c; (VI) A, B, C. Case (III) is an ambiguous case, since there may be two solutions or only one. Each of the six cases may be solved by several combinations of the relations given in the preceding paragraph.

See also **Trigonometry**.

SPHEROID.   An ellipsoid. Also called ellipsoid of revolution, from the fact that it can be formed by revolving an ellipse about one of its axes.

If the shorter axis is used as the axis of revolution, an *oblate spheroid* results, and if the longer axis is used, a *prolate spheroid* results. The earth is approximately an oblate spheroid.

See also **Ellipsoid**; and **Surface (Of Revolution)**.

SPHEROIDAL COORDINATE.   A degenerate system of curvilinear coordinates obtained from ellipsoidal coordinates when two axes of the quadric surface are equal in length. There are two special cases: *oblate* and *prolate* spheroidal coordinates. In the oblate case, the coordinate surfaces are families of oblate ellipsoids of revolution around the Z-axis with semi-axes $a = c\sqrt{1 + \xi^2}$, $b = c\xi(\xi = \text{const.})$; hyperboloids of revolution of one sheet with $a = c\sqrt{1 + \eta^2}$, $b = c\eta(\eta = \text{const.})$ and planes from the Z-axis ($\phi = \text{const.}$). The following additional relations hold:

$$0 \leq \xi \leq \infty; \quad -1 \leq \eta \leq 1; \quad 0 \leq \phi \leq 2\pi$$

$$x = c\sqrt{(1 + \xi^2)(1 - \eta^2)}\cos\phi$$

$$y = c\sqrt{(1 + \xi^2)(1 - \eta^2)}\sin\phi$$

$$z = c\xi\eta$$

Alternative variables often used are $\xi = \sinh u$; $\eta = \cos v$; $0 \leq u \leq \infty$; $0 \leq v \leq \pi$.

For prolate spheroidal coordinates, the coordinate surfaces are families of prolate ellipsoids of revolution around the Z-axis with semi-axes $a = c\xi$, $b = v\sqrt{\xi^2 - 1})$, $\xi = \text{const.}$; hyperboloids of revolution of two sheets with $a = c\eta$, $b = c\sqrt{1 - \eta^2}$, $\eta = \text{const.}$ and planes from the Z-axis, $\phi = \text{const.}$ The coordinates are limited in range: $-1 \leq \eta \leq 1$; $1 \leq \xi \leq \infty$; $0 \leq \phi \leq 2\pi$ and in terms of rectangular coordinates

$$x = c\sqrt{(1 - \eta^2)(\xi^2 - 1)}\cos\phi$$

$y = c\sqrt{(1 - \eta^2)(\xi^2 - 1)}\sin\phi$; $z = c\xi\eta$. The variables $\xi = \cosh u$; $\eta = \cos v$; $0 \leq u \leq \infty$; $0 \leq v \leq \pi$ are often used.

This system is particularly useful in quantum mechanical two-center problems, for if the centers are taken at the foci of the system, the focal radii to a point where the surfaces of revolution intersect satisfy the relations $(r_1 + r_2) = 2c\xi$; $(r_1 - r_2) = 2c\eta$.

See also **Coordinate System**; **Curvilinear Orthogonal Coordinates**; **Ellipsoidal Coordinate**; and **Hyperboloid**.

SPHEROMETER.   An instrument for measuring the curvature of solid spherical surfaces, either convex or concave, such as those of lenses; a measurement in which high precision is not easily attained. See Fig. 1. The most familiar mechanical device for this purpose is a form of micrometer. It resembles a small three-legged stool, the sharp steel points of whose legs form an equilateral triangle. The micrometer screw, also with a sharp point, is mounted at the center of this supporting trivet, and is adjusted to read zero when all four points are in one plane, as determined by standing the instrument on a flat plate of glass and screwing the micrometer point down to it. The distance from each of the legs to the central axis must be accurately known. It is called k, and if the elevation (or depression) of the

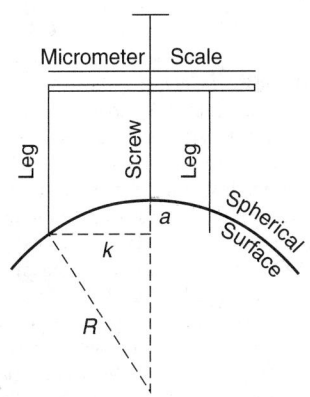

Fig. 1.   Diagrammatic section of spherometer.

micrometer point to fit a given spherical surface is a, then the radius of curvature of the surface is readily calculated as

$$R = \frac{k^2 + a^2}{2a}$$

The chief source of error are in determining just when contact takes place between micrometer point and surface, and in measuring k.

See also **Circular Curves**; and **Sphere**.

SPHYGMOMANOMETER.   An apparatus for measuring blood pressure. It consists of a rubber-bag cuff, which is wrapped around the upper arm. This is inflated by a hand bulb. The cuff is connected by rubber tubing to a measuring device that is either a sealed column of mercury or a spring scale. Sufficient pressure is pumped into the rubber cuff to compress the brachial artery in the upper arm. A stethoscope is applied over the artery below the cuff and air is gradually allowed to escape from the cuff until the pulse can be heard. The reading on the scale or column of mercury at this point indicates the systolic pressure or the highest pressure in the arteries during contraction of the heart. The deflation of the cuff is continued, and that point on the scale when the last sound of the disappearing pulse is heard is the distolic pressure, or lowest pressure in the artery during diastole, or relaxation of the heart muscle between beats. The normal systolic reading of an adult varies from 110 to 130 or 140 millimeters of mercury. Normal diastolic readings vary from 60 to 90 millimeters of mercury. See also **Hypertension (High Blood Pressure)**.

SPICA (α Virginis).   Ranking fifteenth in apparent brightness among the stars, Spica has a true brightness value of 2,800 as compared with unity for the sun. Spica is a blue-white, spectral type B star and is located in the constellation Virgo, a zodiacal constellation. Estimated distance from the earth is 260 light years.

Spica is particularly interesting in that it is believed to be the star that provided Hipparchus with the data that enabled him to discover the precession of the equinoxes. The temple at Thebes was oriented with reference to Spica in about 3200 B.C. Later temples which were oriented to this same star indicated the motion of the star due to precession and provided the necessary data.

See also **Constellations**.

SPICULES.   Bright spikes extending into the chromosphere of the Sun from below. They are several hundred miles in diameter and extend outward 5000 to 10,000 miles. Spicules have a lifetime of several minutes and may be related to granules.

SPIDER (*Arachnida, Araneida*).   Arthropods of almost exclusively terrestrial habits, known commonly for their ability to spin silken webs. They differ from other arthropods in one or more of the following characters: The body is divided into cephalothorax and abdomen, and the latter is unsegmented. The head bears a group of simple eyes. Four pairs of legs are present. The jaws are perforated by the ducts of poison glands. The ventral surface of the abdomen bears the openings of the lung books, respiratory organs of peculiar form, and a group of spinnerettes through which the ducts of the silk glands open.

Spiders are among the most interesting of all animals, and with few exceptions, e.g., the black widow spider, are harmless. Even though they secrete poison most of them are too small to bite a human being unless on a very thin fold of tissue, and most seem to have no inclination to bite. Even the large hairy species commonly called banana spiders or tarantulas are mild-mannered creatures.

To what extent the bad reputation of the black widow, *Latrodectus mactans*, is deserved seems difficult to establish. See Fig. 1. Apparently its bite is severely poisonous and occasionally fatal, and apparently it is vicious in its habits. There is contradictory evidence, however, so the case is not wholly settled.

Spiders vary greatly in habits. Some spin funnel-like webs in which they hide to await their prey, others form irregular webs, and still others make the orb webs so beautifully demonstrated in our gardens on dewy mornings. Other forms spin no web but capture their prey by pouncing on it from concealment or by open chase. Among these forms are the crab spiders, named from their short broad form, which lie in wait on plants and are sometimes almost perfectly hidden in flowers by their concealing coloration. The wolf spiders are stout hairy species, often black in color. They hunt like the predators for which they are named.

Fig. 1.   Black widow spider. (*A.M. Winchester.*)

The spiders that capture prey without the use of webs have other uses for silk, such as the formation of cocoons or egg-sacs in which the eggs are deposited, and the construction of a smooth lining for their hiding places. The most remarkable example of the latter use is the nest of the trapdoor spiders of warm regions. These nests consist of a silk-lined burrow with a beveled margin at the surface of the ground. A lid hinged with tough silk fits perfectly into this beveled depression and can be held shut by the spider, which provides in its lining two depressions to be gripped by the claws.

The mating habits of spiders are also remarkable. The male, in many species much smaller than the female, goes through a courting procedure as complex as that of the birds, and is often killed by his consort. Reproductive adaptations are also peculiar in the males, in that the palpi are modified to convey the seminal fluid to the genital passages of the female. When sexually mature the male spins a web in which the contents of the reproductive organs are discharged, to be taken up into the cavities in the palpi. When the individual is successful in securing a mate he thrusts the palpi one at a time into her genital aperture.

See also **Arachnida**.

Additional Reading

Chapman, R.F.: "The Insects: Structure and Function," 4th Edition, Cambridge University Press, New York, NY, 1998.

Choe, J.C., B.J. Crespi: "The Evolution of Mating Systems in Insects and Arachnids," Cambridge University Press, New York, NY, 1997.

Eisner, T., S. Nowicki: "Spider Web Protection through Visual Advertisement: Role of the Stabilimentum," *Science*, **219**, 185–187 (1983).

Foelix, R.F.: "Biology of Spiders," 2nd Edition, Oxford University Press, Inc., New York, NY, 1996.

Gullan, P.J., P.S. Cranston: "he Insects: An Outline of Entomology," Blackwell Science, Inc., Malden, MA, 2000.

Hadley, N.F.: "The Arthropod Cuticle," *Sci. Amer.*, 104–114 (July 1986).

Jackson, R.R.: "A Web-Building Jumping Spider," *Sci. Amer.*, 102–115 (September 1985).

Kaston, B.J., E.T. Cawley, J. Bamrick, et al.: "How to Know the Spiders," 3rd Edition, The McGraw-Hill Companies Inc., New York, NY, 1998.

Lubin, Y.D., M.H. Robinson: "Dispersal by Swarming in a Social Spider," *Science*, **216**, 319–321 (1982).

Masters, W.M., H. Markl: "Vibration Signal Transmission in Spider Orb Webs," *Science*, **213**, 363–365 (1981).

Punzo, F.: "The Biology of Camel-Spiders: (*Arachnida, Solifugae*)," Kluwer Academic Publishers, Norwell, MA, 1998.

Schowalter, T.D.: "Insect Ecology: An Ecosystem Approach," Academic Press, Inc., San Diego, CA, 2000.

Speight, M.R., M.D. Hunter, and A.D. Watt: "Ecology of Insects: Concepts and Applications," Blackwell Science, Inc., Malden, MA, 1999.

Witt, P.N. and J.S. Rovner: "Spider-Communication," Princeton University Press, Princeton, NJ, 1982.

SPIDER MONKEY.   See Monkeys and Baboons.

SPILLWAY.   One of the important adjuncts of a dam of the overflow type is a spillway, which is simply an opening through or over which excess water may flow when the reservoir is full. The spillway may be a certain overflow section of the dam, or it may be located at one side of the dam.

SPINACH.   Of the family *Chenopodiacea* (goosefoot family), spinach (*Spinacia oleracea*) is a pot herb, cooked and eaten much as other greens, such as chard, turnip greens, and mustard greens. Not all greens, of course, are members of the same botanical family. Many are brassicas, members of the *Cruciferae* (mustard) family. Spinach is a hardy cool-weather plant that withstands winter conditions in the southern United States. In most of the northern states, spinach is primarily an early-spring or late-fall crop, but in some areas, where summer temperatures are mild, spinach may be grown continuously from early spring until late fall. Winter culture of spinach is possible only where moderate temperatures, as found in California, prevail.

It is believed that the cultivation of spinach commenced in Persia (Iran) about A.D. 300 or 400. There is no mention of spinach in very ancient records. One writing in China, dating back to A.D. 647, refers to spinach as the "herb of Persia." The vegetable was introduced into Spain by the Moors in about A.D. 1100. By the late 1200s, spinach was known and consumed throughout most of Europe. An English writer referred to the vegetable as "spynoches" in 1390. There are no records, but it is assumed that spinach was brought to the United States by the early colonists. By 1806, three varieties were listed in seed catalogs. The first savoyed-leaf variety was introduced in 1828.

There are two principal variations of spinach: (1) the *savoy* or *wrinkled-leaf* kind; and (2) the *semisavoy* or *flat-leafed* kind. For the fresh market, the savoy type is usually preferred. The semisavoy kind is used for processing. Some botanists also classify spinach in terms of whether a variety is *prickly seeded* or *smooth seeded*. Modern growers greatly prefer the smooth-seeded varieties for ease and precision of planting. A savoy-type plant is shown in Fig. 1. One further classification distinguishes the *long-standing* varieties. Long-standing refers to the fact that the plant is slow in bolting—that is, it does not go to seed early, a desirable characteristic for growers.

Fig. 1.   A healthy spinach plant of *Long Standing Bloomsdale variety. (Spinacia oleracea). (Ferry Morse Seed Co.)*

SPINACH LEAF MINER (*Insecta, Diptera*).   One of several leaf-mining flies and maggots that attack a variety of plants. The spinach leaf miner (*Pegomya hyoscyami*, Panzer) attacks beet, chard, mango, spinach, and sugarbeet, as well as a number of weeds, such as chickweed and lamb's quarters. The insect is of much greater economic importance in Europe than in North America. It is believed that the insect was introduced into North America in the early 1880s. The small maggots produce blister-like blotches on the leaves of the aforementioned plants, rendering them unfit for market. Even where the leaves of such plants may not be used, the roots and seeds do not develop properly if the plant is vigorously attacked by this insect. The adult is a slender, grayish-black, two-winged fly, about $\frac{1}{4}$ inch (6 millimeters) long. The adult emerges in mid-spring and mating and egg-laying occur almost immediately. The eggs are laid on the underside of leaves. The very small maggots immediately commence feeding on leaves when hatched. They migrate from one leaf to the next and thus destruction is not localized. There are from 3 to 4 generations of the insect per year. Effective control can be achieved by destroying nearby host weeds. Screening of the plant beds with cheese-cloth also can be an effective control measure. Spinach planted in early spring or late fall usually is not affected. Where cultural methods are not sufficient, chemical controls, such as spraying or dusting parathion or diazinon in the affected areas will bring about control. Care must be exercised to avoid using chemicals near the time of harvest because of the possible presence of poisonous residuals on the leaves.

SPIN-DEPENDENT FORCE. The force between two particles which depends on their relative spin orientations and possibly on their spin directions relative to the line joining the particles. Physical basis could be the interaction between the magnetic moments of the particles, or in the case of nuclear forces, to the exchange of π-mesons between the nucleons.

SPINE. The vertebral column or backbone, in humans composed of 33 vertebrae which, grouped according to regions, are: 7 cervical; 12 thoracic; 5 lumbar, 5 sacral, and 4 coccygeal vertebrae. See also **Skeletal System**.

SPINEL. The mineral spinel is one of a group of minerals which crystallize in the isometric system with an octahedral habit, and whose chemical compositions are analogous. These minerals are combinations of bivalent and trivalent oxides of magnesium, zinc, iron, manganese, aluminum, and chromium, the general formula being represented as $R''O \cdot R_2''O_3$. The bivalent oxides may be MgO, ZnO, FeO, and MnO, and the trivalent oxides Al_2O_3, Fe_2O_3, Mn_2O_3, and Cr_2O_3. The more important members of the spinel group are spinel, $MgAl_2O_4$; gahnite, zinc spinel, $ZnAl_2O_4$, franklinite $(Zn, Mn^{2+}, Fe^{2+})(Fe^{3+}, Mn^{3+})_2O_4$, and chromite, $FeCr_2O_4$. True spinel has long been found in the gem-bearing gravels of Sri Lanka and in limestones of Burma and Thailand.

Spinel usually occurs in isometric crystals, octahedrons, often twinned. It has an imperfect octahedral cleavage; conchoidal fracture; is brittle; hardness, 7.5–8; specific gravity, 3.58; luster, vitreous to dull; transparent to opaque; streak white; may be colorless, rarely through various shades of red, blue, green, yellow, brown, or black. These colors are doubtless due to small amounts of impurities. The clear red spinels are called spinel-rubies or balas-rubies and were often confused with genuine rubies in times past. Rubicelle is a yellow spinel. A violet-colored manganese-bearing spinel is called almandine spinel.

Spinel is found as a metamorphic mineral, and also as a primary mineral in basic rocks, because in such magmas the absence of alkalies prevents the formation of feldspars, and any aluminum oxide present will form corundum or combine with magnesia to form spinel. This fact accounts for the finding of both ruby and spinel together. In addition to the localities mentioned above, which yield beautiful specimens, spinel is found in Italy and Sweden and in Madagascar. Also in the United States in Orange County, New York, and in Sussex County, New Jersey, are many well-known spinel localities. Spinel is found also in Macon County, North Carolina, and in Canada in Quebec and Ontario.

The name spinel is derived from the Greek, meaning a spark, in reference to the fire-red color of the sort much used for gems. Balas ruby is derived from Balascia, the ancient name for Badakhshan, a region of central Asia situated in the upper valley of the Kokcha River, one of the principal tributaries of the Oxus.

ELMER B. ROWLEY, F.M.S.A., formerly Mineral Curator, Department of Civil Engineering, Union College, Schenectady, NY.

SPINNERET. A spinning organ of spiders. The spinnerets are located on the ventral surface of the abdomen, near or at its tip, and vary from one to three pairs. They are conical to cylindrical in form. Each has a membranous terminal portion called the spinning field, through which run many minute spinning tubes from the silk glands. The nature of the spinning tubes varies, different tubes producing different kinds of silk.

In the act of spinning the liquid silk is forced through the spinning tubes, to harden on exposure to the air as silk. The spinnerets bear spines, some of them apparently tactile, and are moved by muscles, so that the spider is able to form and place its threads with precision.

Glass fibers and various synthetic fibers, such as polyesters, are also spun in a process known as *melt spinning*. In modern spinning plants, the polymer is heated and conveyed to the spinning head by means of a melt extruder. If the polymer is initially in the form of quenched chips, it must be thoroughly dried before melting or the molten resin will degrade by ester hydrolysis during the spinning process. The molten polymer in its passage from extruder to jet has a viscosity of the order of 1,000 poises. Pressures in the area range from 1,000 to 5,000 psi (6.9 to 34.5 mPa). The spinneret is made of stainless steel with a number of holes ranging from 14 to several hundred, depending upon the filament count of the yarn to be spun. The holes have diameters ranging from 0.008 to about 0.025 inch (0.2 to 0.6 millimeter), in accordance with the denier being spun. The depth of the holes is 1.3 to 3 times the diameter. See also **Fibers**.

SPIN (Nuclear). The total angular momentum of an atomic nucleus, when it is considered as a single particle.

SPIN STATE. A system is said to be in a definite spin state when the quantum mechanical wave function describing the system is an eigenfunction of the various spin operators corresponding to the square of the total spin angular momentum and the component(s) of the spin angular momentum being used to designate the system. For example, the spin state of a single particle might be designated by specifying the particle's spin projection on its direction of motion. The operator corresponding to this dynamical variable is $\mathbf{s} \cdot \mathbf{p}$ where $\mathbf{s}$ is the vector operator corresponding to the spin angular momentum of the particle and $\mathbf{p}$ is the vector operator corresponding to the linear momentum. For a particle to be in a definite spin state according to this method of designation, it must be described by a wave function which is an eigenfunction of $\mathbf{s} \cdot \mathbf{p}$. As another example, the spin state of a two particle system might be specified in one of two ways — either by specifying S^2 and S_z, the square of the total spin angular momentum and its z-component respectively, or by specifying S_{1z} and S_{2z}, the z-components of the spin angular momentum of each particle. If the case for the two particle system are considered in detail, using nonrelativistic quantum mechanics and the Pauli spin operators, one has

$$\mathbf{S}_1^2 = \tfrac{1}{4}\hbar^2\boldsymbol{\sigma}_1^2$$

$$\mathbf{S}_2^2 = \tfrac{1}{4}\hbar^2\boldsymbol{\sigma}_2^2$$

$$\mathbf{S}^2 = (\mathbf{S}_1 + \mathbf{S}_2)^2 = \mathbf{S}_1^2 + \mathbf{S}_2^2 + \mathbf{S}_1 \cdot \mathbf{S}_2$$

$$= \tfrac{1}{2}\hbar^2(\boldsymbol{\sigma}_1^2 + \boldsymbol{\sigma}_2^2 + \boldsymbol{\sigma}_1 \cdot \boldsymbol{\sigma}_2)$$

and

$$S_z = S_{1z} + S_{2a}$$

with

$$S_{1z} = \tfrac{1}{2}\sigma_{1z} \text{ and } S_{2z} = \tfrac{1}{2}\sigma_{2z}$$

Then letting α represent the column matrix $\binom{1}{0}$ and β represent the column matrix $\binom{0}{1}$, one has

$$S_1^2\alpha_1 = \tfrac{1}{2}\left(\tfrac{1}{2} + 1\right)\hbar^2\alpha_1$$

$$S_{1z}\alpha_1 = \tfrac{1}{2}\hbar\alpha_1$$

$$S_1^2\beta_1 = \tfrac{1}{2}(\tfrac{1}{2} + 1)\hbar^2\beta_1$$

$$S_{1z}\beta_1 = -\tfrac{1}{2}\hbar\beta_1$$

Thus $\alpha_1\alpha_2$, $\alpha_1\beta_2$, $\beta_1\alpha_2$, and $\beta_1\beta_2$ represent spin states with, respectively,

$$S_{1z} = \tfrac{1}{2}\hbar, \quad S_{2z} = \tfrac{1}{2}\hbar$$

$$S_{1z} = \tfrac{1}{2}\hbar, \quad S_{2z} = -\tfrac{1}{2}\hbar$$

$$S_{1z} = -\tfrac{1}{2}\hbar, \quad S_{2z} = \tfrac{1}{2}\hbar$$

$$S_{1z} = -\tfrac{1}{2}\hbar, \quad S_{2z} = -\tfrac{1}{2}\hbar$$

Also $\alpha_1\alpha_2$; $\tfrac{1}{2}(\alpha_1\beta_2 + \beta_1\alpha_2)$; and $\beta_1\beta_2$ all represent spin states with $S^2 = 2^{-2}$ but $S_z = \check{}$; $S_z = 0$, and $S_z = -\check{}$ respectively. The state $\tfrac{1}{2}(\alpha_1\beta_2 - \beta_1\alpha_2)$ is one with $S^2 = 0$ and $S_z = 0$.

SPINY ANTEATER. See **Monotremeta**.

SPIRACLE. 1. A small opening on the surface of the insect body, leading into an air tube of the respiratory system. Insect spiracles are simple openings in some species, but usually they are guarded by some structure that prevents the entrance of foreign particles. This guard varies from a fringe of hairs to an elaborate sieve-like plate. Some insects depend on a closing apparatus of the associated trachea for the exclusion of particles.

In the primitive condition, spiracles apparently were paired and metameric, one opening on each side of each segment. This arrangement is only slightly modified in some species, which are said to be peripneustic. In other cases the number of openings is greatly restricted, only one or two pairs providing the sole entrance to the respiratory apparatus. Insects with a single pair at the anterior end of the body are called propneustic, those with a pair at the posterior end metapneustic, and those with a pair at each end amphipneustic. The spiracles may also be associated with tubes, either extending the tracheae beyond the general surface of the body or forming external conduits for air, by which the insect may reach the air from within the water or decaying matter in which it lives. 2. The greatly reduced external opening representing the first of the series of gill slits in elasmobranch fishes.

SPIRAL CURVE. In railway or highway alignments a spiral curve, sometimes called an easement or transition curve, is one which provides a gradual change of curvature when passing from a tangent (straight line) to a circular curve. See Fig. 1. The cubic parabola and cubic spiral are curves well suited for this purpose. The former is used when spirals are to be laid out by offsets from the tangent, since the offset to any point on the curve varies as the cube of the distance along the tangent from the point where the curve begins. When the curve is to be laid out by deflection angles the cubic spiral is used. In this curve the offset from the tangent is proportional to the cube of the distance along the curve measured from the point of tangency. Spiral curves are particularly well adapted for the use of superelevation since they furnish a means for gradually increasing this quantity from zero to the amount required on the circular portion of the curve. Railroads use a combination of spiral and circular curves to connect the tangents on their main lines. The value of the spiral is now recognized in modern highway design for high-speed traffic.

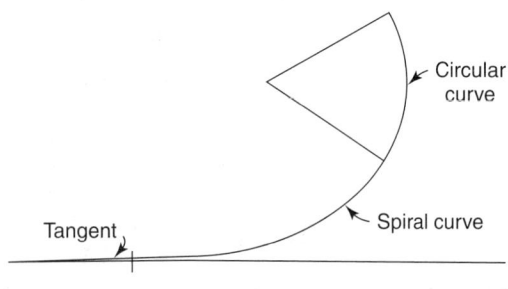

Fig. 1.   Spiral curve.

See also **Superelevation**.

SPIRAL (Mathematics). A transcendental plane curve, for which the equation in many cases can be written in polar coordinates and in general form as $r = a_0\theta^n + a_1\theta^{n-1} + \cdots + a_n$. A spiral can also be defined as the locus of a point that moves about a fixed pole, while its radius vector **r** and its vectorial angle θ continually increases or decreases according to some law. The best-known cases have special names as follows: Archimedes spiral, Cornu spiral, Sici spiral, hyperbolic spiral, lituus, logarithmic spiral, parabolic spiral.

See also **Curve (Plane)**.

SPIRAL VALVE. A thin ridge projecting from the lining of the intestine of elasmobranch fishes to the center of the cavity and following a spiral course through the length of the tube. The structure provides greatly increased surface for the digestion and absorption of food. It is so arranged that the food must follow a spiral course, but beyond this regulating effect is not a valve in the usual sense. It merely slows the passage of food.

SPITTLE BUG (*Insecta, Homoptera*). Of the family *Cercopidae*, the spittle insects, also commonly called frog hoppers because of their remote resemblance to frogs, are quite damaging to certain crops. The immature spittle bug sucks the sap and juices of plants and secretes a protective frothy mass about itself. Control chemicals must be applied at just the right time—about one week after the first appearance of the nymphs. There are no apparent natural enemies to these insects. When infestations are severe, reaching the proportion of several hundred nymphs per plant, as much as 30–40% of a leguminous hay crop and nearly all of an alfalfa seed crop can be lost. Infested strawberry plants also show great reductions in yield as the result of the spittle bug.

The insect winters over as an egg, only about 1 millimeter long. Masses of up to 30 eggs will be deposited in grain stubble several inches above ground level. The eggs hatch from early spring to early summer and the first instar nymphs migrate to host plants and attempt to find locations where the humidity will be relatively high. Immediately, the nymph covers itself with the aforementioned secretion, a procedure that is highly effective against desiccation. There are 5 nymphal instars in the cycle, which ranges from 1 to 3 months, largely depending upon temperature.

SPLEEN. An organ of vertebrates derived from mesenchyme and lying in the mesentery. It is closely associated with the circulatory system. The organ consists of masses of tissue of granular appearance, known as lymphoid tissue, located around fine terminal branches of veins and arteries. According to one interpretation, these vessels are connected through the spleen pulp by modified capillaries called splenic sinuses. The pulp is supported by a reticular tissue foundation and contains blood cells of all kinds in addition to the characteristic mesenchymal cells. The functions of the organ are the formation of blood cells, the destruction of old red cells, the removal of other debris from the blood stream, and as a reservoir for blood. It is estimated that the spleen can store from one-fifth to one-third of the total volume of blood. By undergoing periodic contractions during severe exercises, lowering of barometric pressure, and in cases of carbon monoxide poisoning, asphyxia, and hemorrhage, the organ can accommodate emergency calls for blood to be added to the circulatory system.

In humans, the spleen is situated in the left upper part of the abdomen, behind the stomach, just below the diaphragm. The organ, in an adult human, measures about $5 \times 3 \times 2$ inches ($12.5 \times 7.5 \times 5$ centimeters) in size. In some diseases, it enlarges (*splenomegaly*) and may even fill a large portion of the left side of the abdomen. Enlargement occurs in certain pathological conditions, such as Banti's disease, Gaucher's disease, and certain anemias. *Splenectomy* (surgical removal of the spleen) frequently gives favorable results in these cases. Enlargement also occurs in malaria, bacterial endocarditis, leukemia, and Hodgkin's disease, among others. In these latter cases, removal of the spleen is medically contraindicated. In persons who have undergone a splenectomy, or who are functionally asplenic (such as in sickle cell disease), there are attendant risks of overwhelming bacteremias with certain microorganisms, such as *Streptococcus pneumoniae* and *Haemophilas influenzae* Type B. This is particularly true of children.

The spleen is classified as a ductless gland and is also regarded as one of the centers of activity of the reticuloendothelial system. Its presence is not necessary for life. It may be removed surgically and often is, following abdominal injuries with rupture and hemorrhage from the spleen, or in the treatment of certain blood diseases (hemorrhagic purpura, familial jaundice, etc.), or for the removal of splenic tumors or cysts. Congenital anomalies such as accessory spleens occur, and rarely has the spleen been found to be completely absent.

Additional Reading

Bowdler, A.J.: "The Complete Spleen: Structure, Function, and Clinical Disorders," 2nd Edition, Humana Press, Totowa, NJ, 2001.

Cuschieri, A. and C.D. Forbes: "Disorders of the Spleen," Blackwell Science, Inc., Malden, MA, 1994.

Neiman, R.S., A. Orazi, and M. Strauss: "Disorders of the Spleen, Vol. 38," 2nd Edition, W.B. Saunders Company, Philadelphia, PA, 1999.

Phillips, E.H., L. Morgenstern, and J.R. Hiatt: "Surgical Diseases of the Spleen," Springer-Verlag, Inc., New York, NY, 2000.

Wilkins, B.S. and D.H. Wright: "Illustrated Pathology of the Spleen," Cambridge University Press, New York, NY, 2000.

SPODUMENE. The mineral spodumene is a lithium aluminum silicate corresponding to the formula $LiAlSi_2O_6$ and occurs in monoclinic prismatic crystals, occasionally of very large size. It also occurs massive.

Spodumene has a perfect prismatic cleavage often very noticeable; uneven to splintery fracture; brittle; hardness, 6.5–7.5; specific gravity, 3–3.2; luster, vitreous to dull; color, grayish- to greenish-white, green, yellow and purple. Its streak is white; it is transparent to translucent. Spodumene is characteristically a mineral of the pegmatites, and it is found in Sweden, Ireland, Madagascar and Brazil. In the United States it is found especially in the pegmatites of Oxford County, Maine; in the towns of Goshen, Huntington and Chesterfield in western Massachusetts; at Branchville, Connecticut; in North Carolina; in South Dakota in huge crystals and in San Diego and Riverside Counties in California.

The name spodumene is derived from the Greek meaning ash-colored, particularly appropriate for the slightly weathered varieties. Hiddenite, the beautiful emerald-green or yellow-green spodumene that is used as a gem, was named for W.E. Hidden. Kunzite, named in honor of George F. Kunz, is a transparent lilac to rose-colored spodumene from Madagascar and California, and recently as magnificent, large gem crystals of both purple and yellow color from the Hindu-Kush Mountains, Nuristan Province in Afghanistan. Beautiful gem stones are cut from such crystals, but its easy cleavage discourages its use as a wearable gem. Spodumene alters rather readily to a mass of albite and muscovite. The commercial use of spodumene is chiefly as a source of lithium compounds.

See also **Lithium**.

SPONDYLARTHROPATHIES.

During the early part of the twentieth century, all inflammatory diseases of or in the region of the joints were called "rheumatoid arthritis." Differences, such as the variations in sites involved, were simply assumed to be a reflection of the rather random character of a single disease. Dissatisfaction with this oversimplified view developed over the years and new names for specific manifestations of such conditions appeared. Rheumatoid spondylitis, for example, came into vogue to identify rheumatoid arthritis of the spine. Today there is an entire group of rheumatic diseases that are not directly related to rheumatoid arthritis. These diseases are called the *spondylarthropathies*. Ankylosing spondylitis is the most common disease of this group, which also includes Reiter's syndrome. Findings during and since the early 1970s have shown an immunogenetic connection between the spondylarthropathies and indeed a relationship with several other diseases. As one authority has observed, the group typified by ankylosing spondylitis may be distinguished from rheumatoid arthritis on historic, geographic, epidemologic, genetic, immunogenetic, clinical, immunologic, pathologic, radiologic, and therapeutic grounds.

In research in this area, human leukocyte antigen (HLA-B27) has been the key to developing the relationship of ankylosing spondylitis, for example, with certain other diseases. Statistics indicate that 90–100% of patients with ankylosing spondylitis are B27-positive; 70–90% of patients with endemic Reiter's syndrome (combination of urethritis, conjunctivitis, and arthritis, first discovered by Reiter in 1916); 80–90% with *Salmonella*-reactive arthropathy; 80% with epidemic Reiter's syndrome; 80% with *Yersinia*-reactive arthropathy; 50–70% with inflammatory bowel disease with sacroiliitis; 50–60% with psoriatic arthropy with sacroiliitis; 40–60% with juvenile chronic polyarthropathy; among other striking comparisons.

Ankylosing spondylitis, one time regarded as an uncommon disease, is now believed to occur at about the same rate as rheumatoid arthritis. Although the B27 gene occurs in about 7% of white people, only about 1.5% develop the disease. The disease is rare among black people. It is believed that available statistics may be low because of earlier misdiagnosis and inaccurate reporting. At one time, the disease was considered predominant among white males under 40 years of age, but it is now suggested that there may be a reasonable balance of occurrence in males and females.

If left untreated, ankylosing spondylitis ultimately renders the spine rigid. Fortunately, based upon the recently collected knowledge pertaining to the nature of the disease and, in combination with early treatment, the prognosis is quite good. Medication is directed toward relieving pain and decreasing inflammation. As of the early 1980s, because of the relative unavailability of detailed controlled studies defining the treatment of the disease, current therapy may be altered as new information is gained.

Psoriatic Arthropathy. The association of psoriasis with psoriatic arthropathy ("psoriatic arthritis") has been a subject of debate for many years. The disease is frequently seen and it is estimated that possibly 20% of individuals with psoriasis (principally with psoriatic nail disease) may develop psoriatic arthropathy. The occurrence in men significantly exceeds that in women. The disease may take the form of asymmetric involvement of both large and small joints. The so-called sausage-shaped digit is considered typical. Although a connection between psoriasis and psoriatic arthropathy has been made, much remains to be learned. It is known, for example, that some forms of psoriatic arthropathy are distinguished only with difficulty from rheumatoid arthritis and that the latter can be present in a patient with psoriasis coincidentally and with no connection between the two diseases.

SPONGES. See **Porifera**.

SPORE.

A special type of reproductive cell that develops directly into a new plant. Spores are of many kinds. In the Thallophytes they may be asexual cells or zygotes. Thick-walled spores formed after the union of isogametes are called zygospores. Oöspores are fertilized eggs. In the higher plants spores are always produced by diploid plants and develop into haploid gametophytes. In Selaginella and the seed plants small spores, microspores, produce male gametophytes; large spores, megaspores, produce female gametophytes. Pollen grains in the seed plants develop from microspores.

Some bacterial cells become thick-walled spores. This allows them to survive unfavorable conditions. Fungi also develop spores. See also **Fungus**.

SPOROPHYLL.

A spore-bearing leaf. It actually bears the sporangia that contain the spores. It may be greatly modified in structure and appearance.

SPOROTRICHOSIS.

This disease of worldwide distribution is caused by the dimorphic fungus *Sporothrix schenckii*, a fungus that thrives well in a variety of soils and decaying vegetation. It habituates areas of temperate and tropical climates. Exceptionally rich sources of the fungus are found in sphagnum moss, barberry or rose thorns, some soils, and splinters from rotting wood. Infection usually results from subcutaneous inoculation of the infectious spores as the result of contacting a sharp object that contains the spores. The disease is only secondarily contracted by inhalation of spores. Incidence of the disease is higher among males than females. The disease usually is manifested by subcutaneous nodules with an overlying purplish or pinkish coloration. This is followed by involvement of the draining lymph nodes. The disease progresses slowly. The cutaneous-lymphatic form of sporotrichosis usually responds well to treatment with a saturated solution of potassium iodide administered 3 or 4 times daily. Iodide therapy may be required for approximately one month. There are some side effects and the treatment should be supervised by a physician. Where there is an intolerance to iodide therapy, amphotericin B, a microbial agent used in treating other fungus infections, may be administered. A small percentage of patients display pulmonary manifestations of the disease. The clinical signs are almost identical with those of tuberculosis.

R. C. V.

SPRAIN.

A variety of injuries in or about a join, that occur when the movement of the joint is carried beyond its normal range, or carried forcibly in a direction where its range is limited. Sprains occur in the ankle, knee, wrist, elbow, and spine, in that order of frequency. An injury in which a sudden wrenching or twisting produces tearing of the ligaments or tendons is a common form of sprain. Displacement of cartilages between joints, tearing of muscles around the joint, and tearing of the synovial membrane are others. The symptoms of a sprain are those of inflammation with pain, swelling, and limitation of motion of the part.

Treatment consists of immobilization of the joint with an adhesive strapping or an elastic bandage, elevation of the extremity, and sometimes aspiration of the joint cavity if an effusion or outpouring of inflammatory fluid occurs.

SPRAY DRYING.

A process used in the production of numerous chemical and food products. It is widely used in connection with the production of powdered milk and instant coffee preparations. The spray drying is unique among dryers in that it dries a finely divided droplet by direct contact with the drying medium (usually air) in an extremely short retention time (3 to 30 seconds). This short contact time results in minimum heat degradation of the dried product, a feature that led to the popularity of the spray dryer in the food and dairy industries during its early development. In the case of coffee extract, water in the feed will range from 50 to 70%.

Atomization. In as much as the spray dryer operates by drying a finely divided droplet, the feed to the dryer must be capable of being atomized sufficiently to ensure that the largest droplet produced will be dried within the retention time provided. There are different requirements on the degree of atomization needed to result in the desired product. These factors including minimizing the fine and/or coarse fractions, controlling particle dryness, and controlling bulk density. All commercial atomizers, whether of the centrifugal-wheel, pressure-nozzle, or other types, will produce a particle-size distribution that follows a probability curve. As the total energy input increases, the average particle size will decrease and the particle-size distribution will improve, i.e., the spread between the largest and smallest particles will be less.

Centrifugal-wheel Atomizers. The wheel consists of a disk, which is rotated at very high speed (1700–50,000 revolutions per minute). See Fig. 1. Feed generally is introduced to the center, with centrifugal force

dispersing the feed and throwing out a thin film to the periphery. As the film leaves the disk, it breaks up into a thread, which in turn forms droplets. The disk is located in the hot-air stream so that even though droplets are thrown toward the wall of the dryer, the hot air travels cocurrently and dries the particle sufficiently to prevent wall build-up upon contact. A spray dryer with a wheel atomizer must be relatively large in diameter and shorter than a dryer with pressure nozzles.

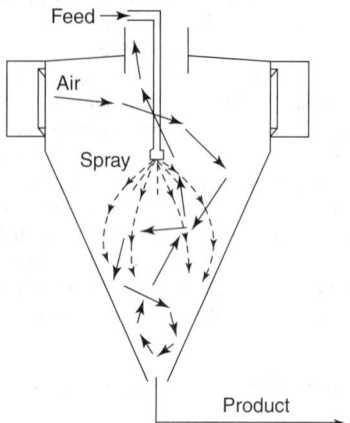

Fig. 3. Mixed-flow spray dryer.

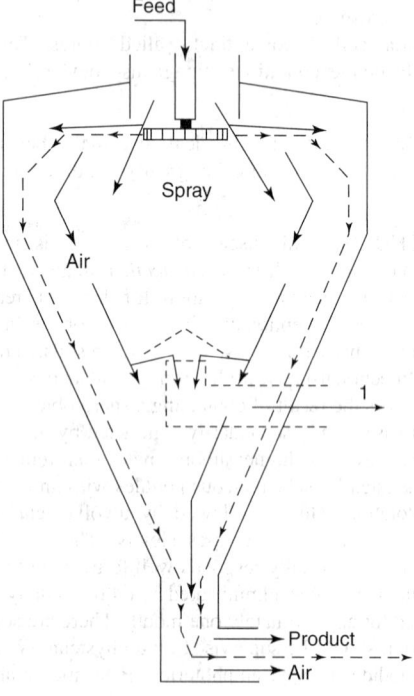

Fig. 1. Spray dryer with wheel atomizer: (1) Air outlet when drying chamber is used for initial separation.

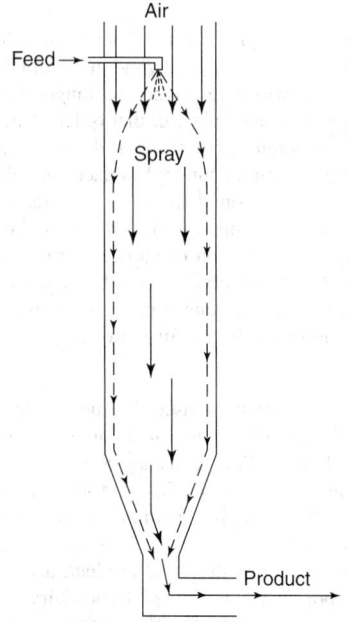

Fig. 2. Spray dryer featuring parallel flow.

Fig. 4. Tall-form chamber employing nozzle atomization. System is particularly suited for dense particles requiring high-pressure atomization. (*Stork-Bowen.*)

Pressure-Nozzle Atomizers. This system consists of an orifice placed after a fixed mechanism, called a core, swirl chamber, or whizzer, depending upon the manufacturer. A high-pressure pump moves the feed to the nozzle body at a pressure of from 250 to 8000 pounds per square inch (17 to 544 atmospheres). The feed slurry is pumped through high-pressure piping to the whizzer, where a spin is imparted to the fluid before it enters the nozzle orifices. This results in a hollow-cone spray which throws droplets either cocurrent or countercurrent to the air flow.

The flow pattern is such that a cocurrent spray dryer must be relatively long and small in diameter (Fig. 2), whereas a countercurrent dryer is shorter and larger in diameter. A third type, sometimes referred to as a mixed-flow dryer (Fig. 3), uses an air pattern similar to a cyclone collector, i.e., the spray is introduced at the upcoming air stream (countercurrent) and the particles transfer to the air sweeping the wall (cocurrent).

A multistory, tall-form dryer chamber using nozzle atomization is shown in Fig. 4.

Additional Reading

Considine, D.M. and G.D. Considine: "Foods and Food Production Encyclopedia," Van Nostrand Reinhold Company, Inc., New York, NY, 1982.
Farrall, A.W.: "Engineering for Dairy and Food Products," 2nd Edition, Krieger Publishing Company, Melbourne, FL, 1979.
Flink, M.M.: "Energy Analysis in Dehydration Processes," *Food Technology*, **31**(3), 77–84 (1977).
Masters, K.: "Spray Drying Handbook," 5th Edition, Halsted Press, Washington, DC, 1991.
Toledo, R.T.: "Fundamentals of Food Process Engineering," 3rd Edition, Aspen Publishers, Inc., Gaithersburg, MD, 1999.

Web Reference

Pulse Combustion Spray Drying Systems: *http://www.pulsedry.com/*

SPREADING COEFFICIENT. A thermodynamic expression for the work done in the spreading of one liquid on another. It is the difference between the work of adhesion between the two liquids and the work of cohesion of the liquid spreading, which may be expressed by the equation

$$F_s = \gamma_B - \gamma_A - \gamma_{AB}$$

where F_s is the spreading coefficient, γ_B is the surface tension of the stationary liquid, γ_A is the surface tension of the spreading liquid, and γ_{AB} is the interfacial tension between the liquids.

SPRING. Devices used to absorb energy or shock as in automobile springs; to serve as a source of power as in clocks or watches; and to provide a force to maintain pressure between contacting surfaces as in friction clutches. Springs with ground ends are generally more satisfactory than those with plain ends; compression springs are more desirable for heavy loads than extension springs, because of the possibility of stress concentration in the loop of the extension spring. Compression springs can, however, be employed for tensile loading. Conical coil springs, if properly designed, may be compressed flat under load. Disk springs represent a recent development that is being extensively employed for heavy loads. Laminated or leaf springs are used in vehicles of various types, although coil springs are now being used in automotive applications. Coil springs may be made of square, rectangular, or round wire. See Fig. 1.

SPRING CLOCK. A clock that uses a mainspring to power its balance wheel. Until the fifteenth century, all European clocks were weight-driven. Early spring clocks, developed before the seventeenth century invention of the balance wheel and hairspring by Huygens, had no effective control over the mainspring's progressive loss of power as it unwound. Consequently, first the *fusee* and later the *stackfreed* were developed for equalizing the spring's declining force.

The term fusee is derived from the Latin *fusata*—a cone-shaped device spirally wound with a thread, attached to the spring, designed to compensate through leverage for the declining power of the unwinding clock spring. The term stackfreed is probably derived from the German—a device for compensating through leverage for the declining force of an unwinding clock or watch spring, employing a pinion fixed to the arbor around which the spring is coiled and a second spring with an eccentric cam.

Although not accurate, the resulting clocks were useful because, unlike weight-driven clocks, the spring clocks worked continuously while and after being moved; that is, they were portable. Huygens used the hairspring to control the periodic oscillation of the balance wheel which, in turn, determined the periodic freeing of the pallet from the escape wheel. The latter rotated to work the clock's wheel train which operated the hands on the dial.

The eighteenth-century English horologist John Harrison designed and built the first accurate spring clock for use as a ship's chronometer. Harrison used a mainspring-powered balance wheel movement (incorporating a fusee) set in a watch-like case about 5 inches (12.5 centimeters) in diameter. In 1770, Thomas Mudge of England invented the detached lever escapement that freed the balance wheel from direct contact with the escape wheel, thus completing the basic mechanical clock movement. This basic movement (without a fusee), versions of which still are used in clocks, was miniaturized in the twentieth century for use in small jeweled-lever travel clocks and wristwatches. The mainspring-powered balance wheel normally oscillates two and one-half times a second, causing the jeweled lever escapement to tick five times a second.

Spring-operated timing devices cover a range of intervals, which may vary from a fractional second to a week, or a month or more. The typical lever-type escape movement used today is dependent for its accuracy on precision of manufacture, temperature compensation, and the effect of external loads. Spring movements are limited to driving light loads because of low torque and lack of constant torque. Accuracy may be within 0.2 seconds absolute in a good stopwatch and 5 seconds per day in a good clock movement.

SPRING INDEX AND SPRING RATE. The spring index is the ratio between the mean diameter of the coil of a helical compression spring and the diameter or radial thickness of the spring wire or strip. The spring index is used as a modifier in computing the safe load a spring may carry. For the majority of spring types, the deflection of the spring is proportional to the load on it. The ratio of load to deflection, which may have the units of pounds per inch or inch-pounds per radian, is known as the spring rate, spring constant, or spring scale.

SPRUCE TREES. Members of the family *Pinaceae* (pine family), these trees are of several species. Spruces are of the genus *Picea*. In terms of tree population and regional areas covered, the spruces exceed the fir trees. See also **Fir Trees.** Some of the species of spruce are extensively used for pulp wood in paper production. The fiber length of the Sitka spruce is a little over 3 millimeters ($\frac{1}{8}$ inch) in length, putting it just behind the Douglas fir and the longleaf pine in this respect; and considerably ahead of the jack pine and lodgepole pine, all used as pulp woods.

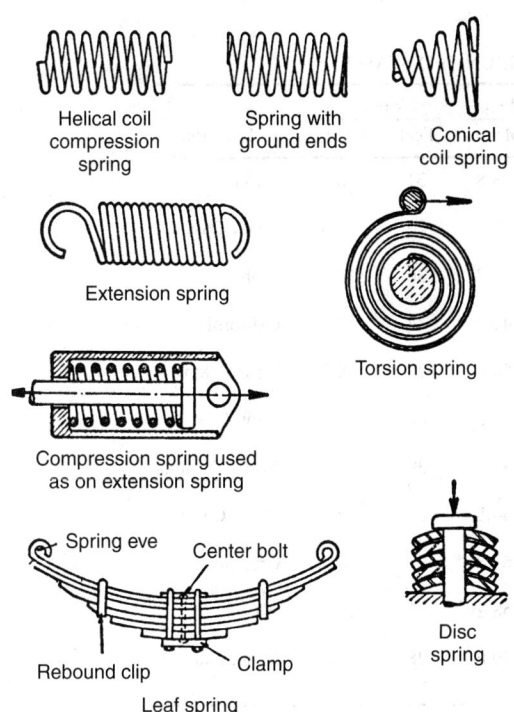

Helical coil compression spring

Spring with ground ends

Conical coil spring

Extension spring

Torsion spring

Compression spring used as on extension spring

Spring eve   Center bolt

Rebound clip   Clamp

Leaf spring

Disc spring

Fig. 1. Types of springs.

Important species of spruce trees not listed on accompanying Tables include:

Chinese spruce	*P. asperata*
Colorado spruce	See Blue spruce
Dragon spruce	See Chinese spruce
Eastern Himalayan weeping spruce	*P. Spinulosa*
Himalayan weeping spruce	*P. smithiana*
Japanese tiger-tail spruce	*P. polita*
Likiang spruce	*P. likiangensis*
Oriental spruce	*P. orientalis*
Sargent spruce	*P. brachytyla*
Schrenk's spruce	*P. schrenkiana*
Serbian spruce	*P. omorika*
Weeping spruce	See Brewer's spruce
Yezo spruce	*P. jezoensis*

Spruce trees are considered very hardy. They are evergreen trees, often conical. The record spruce trees in the United States are detailed in Table 1. See also **Conifers**.

The white spruce (sometimes called cat spruce or skunk spruce) is a tall tree, usually attaining a height of 70 to 100 feet (21 to 30 meters) in favorable circumstances, somewhat resembles the balsam fir in color and contour. The numerous branches tend to be pendulous. When the needles are bruised, they exude a pungent and unpleasant odor. In addition to pulp wood for paper production, the light, soft, straight-grained and pale yellow wood is used for interiors, flooring, and general construction. The tree ranges from Labrador, Newfoundland, and Nova Scotia all the way west across Canada to British Columbia and is found in the northern parts of the northeastern United States, extending westward at that latitude as far as the Rocky Mountains. It is also found in the mountainous regions of Massachusetts and Connecticut.

The timber of the red spruce is quite strong. The tree sometimes attains a height of about 100 feet (30 meters). Near the center of the tree, the branches are essentially spread horizontally, but the upper branches ascend at an angle of about 45°. This tree ranges from New foundland westward through Pennsylvania and on to Minnesota. It is found in the White Mountains of New England as high as 5000 feet (1500 meters) and in the Adirondacks and Allegheny Mountains to about 4000 feet (1220 meters). The tree ranges along the Allegheny Mountains as far south as Georgia. Some of the better timber from this tree is used for piano sounding boards, but the majority is used for pulp wood, for sheathing and flooring in building construction. Early in this century, the red spruce was threatened by over-harvesting, averted by the passage of the Weeks Act in 1911 for the purchase of national forest at the headwaters of navigable streams. This provided government supervision and control over about one-hundred thousand acres in the White Mountain region.

Closely related, but smaller than the red spruce, the black spruce (or bog spruce) is a smaller tree found in the mountains of the southern states. This tree ranges widely from Labrador, Newfoundland, and Nova Scotia westward past Hudson Bay and on across Canada into southern Alaska. Occurrence in the south is essentially dictated by the presence of swampy regions where it thrives. Black spruce is also a pulp wood and is also used for heavy construction, such as piles, posts, and ship construction.

The Sitka spruce is a western tree, ranging from northern California into Alaska and at elevations ranging from sea level to about 3000 feet (900 meters). Normally, under favorable conditions, the tree may reach a height of some 150 to 180 feet (45 to 54 meters). A Sitka spruce is illustrated in Fig. 1. Parts of fallen Sitka spruces have been found, indicating the trees once approached 300 feet (90 meters) in height. The rugged characteristics of this tree have made it a favorite for planting forests in northern Europe.

The Englemann spruce is also a western tree and is found at high elevations from northwestern Canada (Alberta and British Columbia) southward in a wide swath through Montana, Idaho, Washington, and Oregon to Arizona and New Mexico. The timber from the tree finds general construction use, but is not as strong as most timbers. The blue spruce (sometimes called Colorado blue spruce or silver spruce) is of light coloration, with a distinctive light bluish-gray color, appearing silverish in some light. The tree is a favorite for gardens and landscaping. Normally, the tree does not attain a height much in excess of 40 feet (12 meters) under favorable conditions, but there are exceptions as will be noted from Table 1. The natural region of the tree includes the Colorado and eastern Utah Rocky Mountains and northward into Wyoming. However, the tree has been planted widely throughout the United States. The color tends to vary with location, sometimes becoming more of a bluish-green or light green.

Brewer's or the weeping spruce is also a western tree, found from northwestern California northward into Oregon. Some trees also are found in the Coastal range on the northern slopes. The tree normally attains a height of less than 75 feet (22.5 meters), but can well exceed this figure as shown in Table 1. The tree branches downward to the ground, giving it the weeping appearance. The tree prefers altitudes between 4000 and 8000 feet (1200 and 2440 meters).

The tallest native tree of Europe is the Norway spruce, attaining a height of about 200 feet (60 meters) under favorable conditions. It is the common Christmas Tree of Europe. After the Ice Ages had eliminated the tree from

TABLE 1. RECORD SPRUCE TREES IN THE UNITED STATES[1]

Specimen	Circumference[2]		Height		Spread		Location
	Inches	Centimeters	Feet	Meters	Feet	Meters	
Black spruce (1989) (*Picea mariana*)	62	157	78	23.8	21	6.4	Wisconsin
Black Hills spruce (1996) (*Picea glauca* var. *densata*)	104	264	122	37.2	25	7.6	South Dakota
Blue spruce (1980) (*Picea pungens*)	186	472	122	37.2	36	11	Utah
Brewer spruce (1999) (*Picea brewerana*)	272	691	137	41.8	44	13.4	California
Engelmann spruce (1995) (*Picea engelmannii*)	283	719	179	54.6	27	8.2	Washington
Norway spruce (1994) (*Picea abies*)	180	457	120	36.6	66	20.1	New York
Red spruce (1986) (*Picea rubens*)	169	429	123	37.5	39	11.9	North Carolina
Red spruce (1987) (*Picea rubens*)	144	366	146	44.5	34	10.4	North Carolina
Sitka spruce (1987) (*Picea sitchensis*)	673	1709	206	62.8	93	28.3	Oregon
Sitka spruce (1987) (*Picea sitchensis*)	707	1796	191	58.2	96	29.3	Washington
White spruce (typ.) (1995) (*Picea glauca* var. *glauca*)	125	318	130	39.6	28	8.5	Minnesota

[1] From the "National Register of Big Trees," American Forests (by permission).
[2] At 4.5 feet (1.4 meters).

Fig. 1. Sitka spruce located at Seaside, Oregon. (*W. Gucterian, Portland, Oregon.*)

Britain, it was absent until about the year 1500 when it was reintroduced to the British Isles. The tree ranges from Norway southward and eastward, reaching the Italian Alps. The tree was introduced into North America many years ago and can be found in parks and private grounds of northern cities. Another rugged European spruce is the Serbian spruce, which is mainly found in the mountainous regions of eastern Yugoslavia. It is a very fast growing tree. The oriental spruce actually is not from the orient, but is native to the Caucasus Mountains. It is a popular garden species in Europe.

Asian spruces include the Sargent spruce, which is found in western China. Some authorities compare it favorably with the Norway spruce. Because of its attractive flowers, the Likiang spruce, found in China's southwest Yunnan province, is planted mainly for reasons of decor. The Himalayan weeping spruce is often compared with the North American weeping spruce (Brewer's spruce). Schrenk's spruce is found in central Asia.

The engineering characteristics of some spruce timber are given in Table 2.

Spruce Budworm For many years, the spruce budworm (*Choristoneura fumiferana*) has been recognized as one of nature's most destructive forest insects. The first documented epidemic occurred in 1704, and since then, there have been a number of epidemics, each lasting from 5 to 11 years. Authorities believe that the insect has been infecting the spruce-fir forest at intervals of 50 to 100 years as far back as the Ice Age. A major epidemic occurred over the years 1910 to 1920 that spread through Quebec, New Brunswick, Maine, and northern Minnesota, with an estimated destruction of 225 million cords of pulpwood. In some areas, 90% of the fir trees were reported killed during that epidemic. During relatively recent decades, widespread aerial spraying of chemical insecticides was the remedial measure of choice. But, despite extremely large expenditures for materials and personnel for undertaking such massive spray projects, chemical pesticides proved only partially successful in suppressing the budworm. In the early 1980s, forest managers turned to integrated pest management, by which all available necessary control techniques are consolidated into a concerted program to manage insect populations in ways that avoid or reduce economic damage and minimize adverse environmental side effects. An important part of integrated pest management is a study of the dynamics of the spruce budworm itself. What is its role and how does it interact in the forest ecosystem?

As pointed out by M.J. Jones (*American Forests*, 18–23, June 1980), "The budworm itself is part of an integrated system, a natural cycle that assures the continual regeneration of the spruce-fir forest. Balsam fir and red spruce enjoy a shared role in the coniferous stands north of the pine belt and south of the fir belt in the east. These two species differ in character, yet cooperate in returning land to a coniferous state after a major disturbance, such as windthrow, logging and insect attack. Balsam fir begins to produce seed early (at 15 years) and often (every 2 years). Fir seedlings usually are abundant and grow rapidly when competing trees are removed. Red spruce begins to produce seed later (at 25 years) and less frequently (every 3 to 8 years), and does not grow up as rapidly. Thus, a disturbance in a spruce-fir stand tends initially to shift the composition balance toward fir. And balsam fir is the preferred food of the spruce budworm, despite its name. To a lesser extent, the budworm also attacks red, white, and black spruce, and occasionally tamarack and hemlock. In the very act of eating, the budworm prepares for future meals. In stands where it reaches epidemic levels, the bug destroys the host species in such a way that ensures the development of a new stand of that same species — for future infestation. The factors that contribute to an outbreak are not completely understood,

TABLE 2. ENGINEERING DATA ON SPRUCE TREES

| Common Name for Species | Green Condition | | | Air Dried (12% Moisture) | | Maximum Crushing Strength (Parallel to Grain) | | Maximum Tensile Strength (Perpendicular to Grain) | |
	Moisture Content (Percent)	Weight/ Cu. Foot (Pounds)	Weight/ Cu. Meter (Kilograms)	Weight/ Cu. Foot (Pounds)	Weight/ Cu. Meter (Kilograms)	(Psi)	(MPa)	(Psi)	(MPa)
Eastern spruce	45	34	545	28	449	2600	17.9	200	1.4
Englemann spruce	80	39	625	23	368	2190	15.1	240	1.7
Sitka spruce	42	33	529	28	449	2670	18.4	250	1.7

Source: U.S. Forest Products Laboratory.

but it is believed that for budworm populations to explode, mature and overmature fir must be combined with a series of warm, dry summers. Where these conditions coincide, budworms thrive and multiply tenfold each year. At the same time, the fir produces prolific seed and establishes a new stand."

In addition to other biological approaches used in recent years, pheromones (sex attractants) have been synthesized for use as bait in trapping amorous males. Pheromones and other biological controls are discussed in entry on **Insecticide and Pesticide Technology**.

SPRUE (Tropical).   A disease that mainly occurs in India, Puerto Rico, and Vietnam. The disease features a malabsorption of two or more substances—fats, proteins, carbohydrates, minerals, vitamins, and even water. These losses occur in the feces in an acute diarrhea with pale bulky stools. Steatorrhea is present in 95% of cases. There is abdominal distention with colic, weakness and wasting, megaloblastic anemia and edema.

The disease is now considered to be caused by colonization of the upper intestinal tract by coliform bacilli. This, in turn, produces lesions based on the villi of the jejunum and ileum, which broaden and fuse.

Normally self-limiting, sprue has caused mortality up to 35% in some villages in India. Treatment involves antibiotics—tetracycline and ampicillin—together with folic acid and Vitamin B_{12} therapy.

R. C. V.

SPUR-FOWL.   See **Partridge**.

SPUR GEARING.   This form of toothed gearing is used for transmitting power between shafts whose axes are parallel. The velocity ratio of a spur gear set is the ratio of the number of revolutions of one gear to the number of revolutions of the other. The pitch circles of a pair of spur gears are those imaginary circles that are equivalent to the peripheries of a pair of friction wheels that would operate without slipping at the same velocity ratio and center distance as the gears themselves. The point of tangency P of the pitch circles on the line of centers is the pitch point of the gearing. The smaller of the two gears is usually referred to as the pinion.

SPURIOUS CORRELATION.   Correlation that is misleading. Take two variables x and y, which are independent. Consider a third variable z, which is correlated with x and with y. Form $v = f(x, z)$ and $w = f(z, y)$, then frequently v and w will show considerable dependence which may be traced to the correlation of x with z and y with z. In general, then, spurious correlations may be defined as correlation which is introduced by other variables rather than the ones under study.

The real question at issue in correlation is actually this: Are the variables in which we are interested x and y or v and w? If the variables are x and y then the correlation of y with w is spurious. If the casual variables, however, are v and w and not x and y, then the correlation of v and w is valid and not spurious. The term "spurious" is perhaps itself misleading. It does not imply that the correlation does not exist; only that it may be due to a rather circuitous train of casual influences.

Sir Maurice Kendall, International Statistical Institute, London.

SPURIOUS RADIATION.   Any radiation from a transmitter other than that produced by the carrier and its normal sidebands. A radiated harmonic of the carrier is one example of a spurious radiation.

SPUTNIK.   The first artificial satellite, one of a series of Russian earth-orbiting satellites, launched on October 4, 1957.

SPUTTERING.   1. In a gas discharge, material is removed, as though by evaporation, from the electrodes, even though they remain cold. This phenomenon is known as sputtering. 2. The term is also used for the corresponding phenomenon when the discharge is through a liquid. In the first case, sputtering is a nuisance that limits the life of a device; in the second case, it is put to work to make colloidal solutions of metals. 3. A result of the disintegration of the metal cathode in a vacuum tube due to bombardment by positive ions. Atoms of the metal are ejected in various directions, leaving the cathode surface in an abraded and roughened condition. The ejected atoms alight upon and cling firmly to the tube walls and other adjacent surfaces, forming a blackish or lustrous metallic film. This effect is often utilized to form very fine-grained coatings of metal

upon surfaces of glass, quartz, etc., purposely exposed to the sputtering. Films of different metals can be obtained by using cathodes made of these metals. Glass plates may be thus silvered, or suspension fibers of spun quartz rendered conducting for use in electrometers, etc.

SQUALL.   See **Fronts and Storms**.

SQUARE AND SQUARE ROOT.   The square of a number or quantity is the product of that number or quantity when multiplied by itself. Hence, 4×4 yields 16, the latter being the square of 4. Similarly, the number 4 is termed the square root of the number 16. The process of raising any number to any integral *power*, as in squaring (a power of 2) or in cubing (a power of 3) is known as *involution*. The process of extracting roots, that is, of finding that 4 is the square root of 16, or of finding that 3 is the cube root of 27, is known as *evolution*. See also **Exponent**; **Number Theory**; **Radical (Mathematics)**; and **Root (Mathematics)**.

SQUARE LAW MODULATOR.   A device whose output is proportional to the square of its input. The carrier and modulating signal are added in the input to produce a modulated carrier in the output.

SQUARE WAVE.   A square wave, as the name indicates, has the wave form shown in Fig. 1.

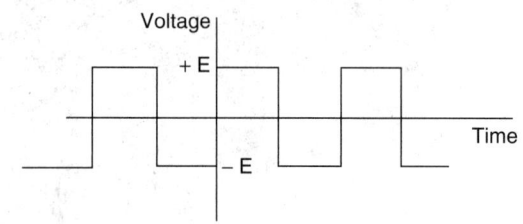

Fig. 1.   Square wave.

Any periodic wave, regardless of its shape, may be analyzed into a series of sine and cosine components whose frequencies are harmonically related. The number of these components will be determined by the shape of the wave, but in general the sharper the corners of the original wave the more component terms. Thus a square wave will require a wide range of frequencies to express it. These components are not mere mathematical fictions but are true electrical components in the case of an electric wave. They may be separated and examined by means of proper filter circuits. Since a square wave will contain a long series of frequencies, it may be used for rapidly determining the frequency response of a piece of equipment by applying the wave to the input and noting the distortion of the output wave. The distortion is due to certain frequencies of the original wave being attenuated or amplified out of proportion in passing through the circuit. Thus the necessity of making a laborious series of tests at various frequencies using sine waves is avoided. When an operator is properly trained in interpreting the results of such testing it offers a rapid means of checking amplifiers, networks, etc. These square waves may be generated by a variety of electronic circuits.

SQUASH.   Of the family *Cucurbitaceae* (gourd family), squash plants are of three major species and one minor species of the genus *Cucurbita*. These several species, plus the designations of summer and winter squash, tend to complicate a classification of these plants. There is not a direct relationship between species or growth pattern and whether a plant is a summer or a winter variety. Although a majority of squash plants assume an indeterminate growth pattern as vine-like, tendril-bearing herbs, some take the more determinate form of bushes or semibushes. In terms of food value, the winter squash is rated very high among all vegetables. Winter squash usually is baked or used in pies. Varieties of winter squash more closely resemble the pumpkin, and canning and freezing processes are very similar. Many, but not all varieties of winter squashes are members of the species *Cucurbita maxima*. Some also are members of *C. moschata* and *C. mixta*. Varieties of *C. maxima* and *C. moschata* can be crossed artificially, but such crossing does not occur naturally in the field.

Specific Varieties.   The genus *Cucurbita* is indigenous to the Americas. The largest concentration of wild species is in Mexico, in a vast area from just south of Mexico to the border between Mexico and Guatemala. The cultivated species *C. pepo*, *C. mixta*, and *C. moschata* are North American, whereas *C. maxima* is from South America. Because of the uncharted movement of pre-Columbian peoples and their crops in the Americas, it is difficult to pinpoint the exact area of origin of the cultivated species. One authority suggests that *C. pepo* is native to the southwestern United States and northern Mexico. *C. moschata* and *C. mixta* are lowland species ranging from Vera Cruz, Mexico, southward through Central America. In this area, cultivars of *C. mixta* are much used for their tasty, edible seeds.

Through years of cultivation, research, and experimentation, numerous varieties of squash have been developed. See Fig. 1.

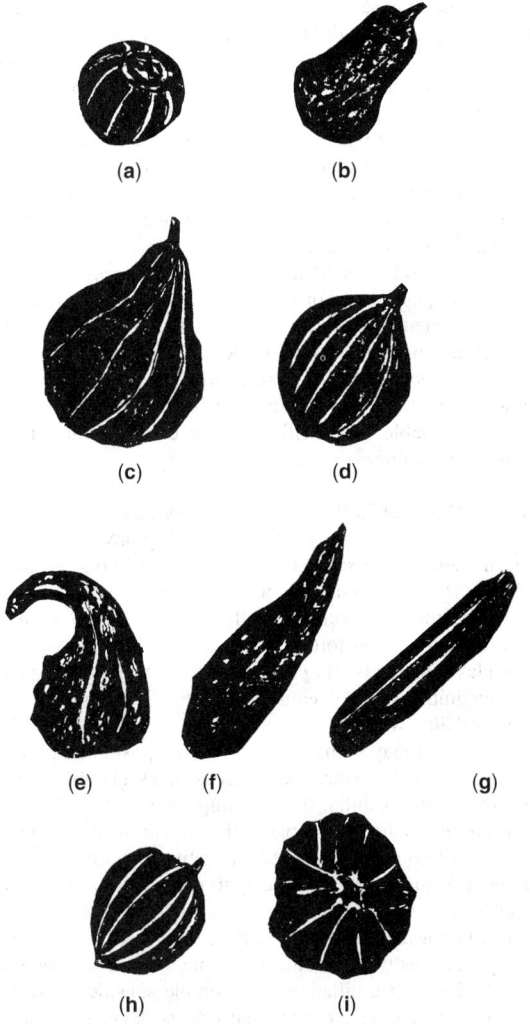

Fig. 1.   Various types of squash. Winter squash: (**a**) Buttercup; (**b**) Butternut; (**c**) Hubbard; (**d**) Gold Nugget. Summer squash: (**e**) Crookneck; (**f**) Straightneck; (**g**) Zucchini; (**h**) Acorn; (**i**) Scallop.

Pollination. Squash plants have unisexual flowers. In order to produce fruit, squashes require cross-pollination—that is, the transfer of pollen from the anthers of the male flower to the stigma of the female flower. Inadequate pollination results in reduced yield and misshapen fruit. Bees provide the primary means of pollen transfer, and often beehives are used. Experienced growers will have one hive for every 3 to 5 acres (1.2 to 2 hectares) of squash plants. Beehives are placed in the center of the field so that bees do not have to travel more than a few hundred feet to feed. Obviously, there should be no insecticide applications during the flowering period.

Nutritional Aspects. Winter squash is high in vitamin A. Summer squash is a good source of vitamin C. Squash has fair amounts of iron. It is low in sodium and protein. Squashes, particularly the summer varieties, are considered low-calorie foods.

Prepared squash is one of the more popular baby foods.

SQUASH BORER (*Insecta, Lepidoptera*).   The larva of a moth, *Melittia satyriniformis*, which bores in the root and stem of squash vines, often killing the plant. The moth is a beautiful species with olive fore wings, transparent hind wings, and legs tufted with orange and black.

The larva thrusts waste material out of holes in the stem. When detected by this means it can be cut out of the plant and killed. In large fields where hand control is impossible, deep plowing when the vines are dead kills many of the insects. Other methods depend on an accurate knowledge of the time of deposition of the eggs. Spraying is effective during that period.

SQUASH BUG (*Insecta, Hemiptera*).   A true bug, *Anasa tristis*, about $\frac{5}{8}$ inch (16 millimeters) long, gray-brown above and mottled with yellow below, showing red in flight from areas beneath the wings. It sucks the juices of pumpkins, squash, and related vines and spreads a bacterial wilt.

The adults hibernate beneath debris on the ground and may be trapped under pieces of board and destroyed.

SQUID.   See Mollusks.

SQUIRRELFISHES (*Osteichthyes*).   Of the order *Berycomorphi*, family *Holocentridae*, squirrelfishes, also known as soldierfishes, are usually of some shade of bright red and prefer tropical reefs. There are about 70 species, well distributed throughout world tropical waters. The largest (*Holocentrus spinifer*) attains a length of about 2 feet (0.6 meter) and is found in the eastern Pacific. *Ostichthys japonicus* (deep-water squirrelfish) differs from other fishes of the family by preferring deeper water. The fish is marketed commercially in Hawaii under the name *menpachi* and is considered a premium food fish. It has been reported that squirrel-fishes can produce sounds.

SQUIRRELS AND OTHER SCIUROMORPHS (*Mammalia, Rodentia*).   This large family of rodents also includes the chipmunks, gophers, spermophiles, prairie dogs, woodchucks, marmots, and whistlers, among others.

The squirrel is an arboreal or terrestrial rodent with a long bushy tail. Representatives of the true squirrels are found on all continents but Australia. North America has ten species of true squirrels, some widely distributed and extremely variable, and two species of flying squirrels. The little red squirrel, *Sciurus hudsonicus*, is the most widely known, with the gray, *S. carolinensis*, and fox, *S. niger*, also well known and widely distributed. The latter two animals are valued as small game, many people considering the flesh to be excellent. These squirrels may gather as much as 20 pounds (9 kilograms) of nuts per season for storage to be used during the winter months. Some species also accumulate thick layers of fat in preparation for winter.

Flying squirrels belong to the genus *Sciuropterus* and are distinguished from true squirrels and related forms by the presence of folds of skin stretching from the front to the hind legs along the sides of the body. These folds, held extended by the legs, support the animal in the air during long gliding leaps. See Fig. 1. The African flying squirrels belong to a different family, the *Anomaluridae*. The sciuroptera are found from Scandinavia eastward to Japan and from northern Canada southward to Honduras. Large numbers are found in India and the Malayan regions. Size ranges widely, but some species approach 3 feet (.9 meter) in length. General coloration is a fox-like red, or red spotted with white or black. In the larger species, the fold of the gliding membrane connects the hind limbs with the tail. The smaller species have the membrane along the side from front to back feet. All species are omnivorous and have large, dark eyes. It has been reported that glides of up to 200 feet (60 meters) from high to low tree branches occur.

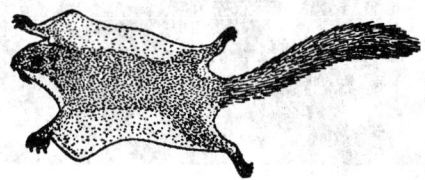

Fig. 1.   Flying squirrel.

Chipmunks are small burrowing rodents of squirrel-like appearance, but with the tail shorter and not bushy. They are brown or grayish with longitudinal stripes on the back or sides. They are omnivorous and, while not usually troublesome, sometimes do destroy flowering bulbs during the winter. Chipmunks belong to several genera and numerous species and subspecies. Some of the western species are called golden chipmunks or rock squirrels and others antelope chipmunks or ground squirrels. They are related to the ground squirrels and gophers. While predominantly North American, some species are found in Siberia.

Pocket gophers are stout-bodied burrowing animals of several species, found throughout the United States. They have fur-lined cheek pouches opening at the sides of the mouth. So-called ground squirrels are slender burrowing species of the gopher and are found in the central and western states. Gophers are injurious to crops. The pocket gophers eat roots, and the ground squirrels are more injurious to grain. The term *gopher* is from the French, meaning "to tunnel." Gophers live in tunnels. The entrance to a tunnel is hidden, but the presence of a large mound of dirt nearby indicates the proximity of a tunnel. The tunnels have numerous entries and ports of escape and can be long and complex. Gophers are particularly destructive in dry seasons, when they gnaw roots for their moisture content. In size, the gopher is comparable to a large rat, but with larger proportioned claws. Because of the grinding sound made when cutting roots, the animal sometimes can be heard while working.

The spermophile is a small slender animal with short legs, small external ears, a short hairy tail, and large cheek pouches. They are also called ground squirrels or gophers. Most of the several species of spermophiles are confined to limited areas in the arid lands of the western United States. The striped species, commonly called gopher, ranges from central Ohio to the Rocky Mountains and from Canada to Texas. This species is brown with alternating clay-yellow stripes and rows of spots on the back and sides. The animal is destructive to crops and known for damaging lawns.

The woodchuck is a large heavy-bodied animal of the Northern Hemisphere. See Fig. 2. Woodchucks have short stout legs and are powerful burrowing animals, penetrating many feet into the ground. They also climb readily, although somewhat clumsily. They eat vegetation of many kinds and sometimes become troublesome in fields and gardens. North America has three species of woodchucks, the common woodchuck, *Marmota monax*, the yellow-bellied woodchuck, *M. flaviventris*, and the whistler, *M. caligata*. The first occurs from Kansas to Georgia, northward to Alaska and Hudson Bay. The second ranges from the Rocky Mountains to the Pacific, and the third is also western, ranging from Montana and Washington to Alaska. The names *marmot* and *groundhog* are also generally applied to them. The Old World species are widely distributed in Europe and Asia, where they are more commonly called marmots. Some of these include the bobac, *M. bobac*, and the alpine marmot, *M. marmota*. The fur is sparse and rather coarse, but it is used to a limited extent. The flesh of woodchucks is eaten, but it is inferior to that of other common rodents. While its flavor has been described as good, the flesh is coarse in texture, as compared with that of squirrels and rabbits.

Fig. 2.   Eastern woodchuck. (*W. Goodpaster.*)

The prairie dog is a burrowing animal found chiefly west of the Mississippi River but has been introduced into a few eastern states. The

animal is small and stout-bodied, with shallow cheek pouches. The prairie dog is a plant feeder and in settled regions can damage crops severely. The animal is also called the prairie marmot. The prairie dog ranges from 10 to 12 inches (25 to 30 centimeters) in length, plus a tail of 2 to 4 inches (5 to 10 centimeters) in length. Because of their damaging ways, the population is small due to extensive extermination.

STABILITY.   In general, the tendency to remain in a given state or condition, without spontaneous change; and thus that attribute of a system which enables it to develop restoring forces between its elements, equal to or greater than the disturbing forces, so as to restore a state of equilibrium between the elements. Thus, a body of air is in a stable state if, when displaced somewhat from its original position, it tends to return thereto. A chemical compound is said to be stable if it is not readily decomposed.

In meteorology, *static stability* (also called *hydrostatic stability*, *vertical stability*, *convectional stability*) is the stability of an atmosphere in hydrostatic equilibrium with respect to vertical displacements, usually considered by the parcel method. The criterion for stability is that the displaced parcel be subjected to a buoyant force opposite to its displacement, e.g., that a parcel displaced upward be colder than its new environment. This is the case if $\gamma < \Gamma$, where γ is the environmental lapse rate and Γ the process lapse rate, dry-adiabatic for unsaturated air and saturation-adiabatic for saturated air.

Neutral stability (also called *indifferent stability*, *indifferent equilibrium*) is the state of an unsaturated or saturated column of air in the atmosphere when its environmental lapse rate is equal to the dry-adiabatic lapse rate or the saturation-adiabatic lapse rate, respectively. Under such conditions, a parcel of air displaced vertically will experience no buoyant acceleration.

Conditional stability is a condition of a negative feedback system which causes it to be stable (nonoscillating) for certain values of gain, and unstable for other values.

STABILITY (Mechanical).   Mechanical stability is that property of a body that causes it to develop forces opposing any position or motion disturbing influence. The subject may be divided into *static stability* and *dynamic stability*. The former is concerned with the production of the restoring forces, the latter with the oscillations that are set up in the system as a result of the restoring forces.

Another classification is: (1) *positive stability*, when the displaced object returns to an initial state of equilibrium after a temporary disturbance; (2) *neutral stability*, when the object tends to remain in a definite position but when disturbed may come to rest in a new position; and (3) *negative stability* (i.e., instability), when the object assumes an entirely new position when disturbed from its initial state. A simple damped pendulum illustrates the first; a sphere on a horizontal plane, the second; while a slender cylinder standing vertically on end is a case of negative stability.

Let it be assumed that an object at rest or in a state of uniform motion receives a disturbing force. Depending on the kind of stability possessed, it might react with one of the motions shown in Fig. 1. If it is dynamically stable as well as statically stable, its motion-time history may be one of diminishing oscillation or of simple subsidence, depending on the magnitude of damping, and inertial effects. Dynamic instability may occur with either static stability or static instability. These lead to divergent oscillation, or to complete divergence.

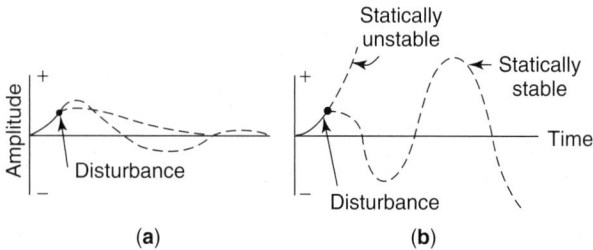

Fig. 1.   (**a**) Positive stability (both statically and dynamically stable); (**b**) negative stability (both dynamically unstable).

STABILITY (System).   For control and feedback system analysis, a system is stable only if all roots of its characteristics equation lie to the left of the imaginary axis of the *s*-plane. Thus, if the roots of the characteristic equation can be determined, the question of stability is answered. If, however, the characteristic equation is of high order, the determination of roots may involve a great deal of work. In cases such as this the existence of roots in the right half of the s-plane may be determined using either the Routh or Hurwitz criterion. The methods are as follows:

Routh Criterion. If, as is the case with all lumped parameter systems, the characteristic equation can be written in polynomial form

$$a_0 + a_1 s + a_2 s^2 + \cdots + a_n s^n = 0$$

the array

$$
\begin{array}{cccc}
a_0 & a_2 & a_4 & \cdots \\
a_1 & a_3 & a_5 & \cdots \\
A_1 & A_2 & A_3 & \cdots \\
B_1 & B_2 & \cdots & \cdots \\
C_1 & \cdots & \cdots & \cdots
\end{array}
$$

is set up, in which $A_1 = (a_1 a_2 - a_0 a_3)/a_1$, $A_2 = (a_1 a_4 - a_0 a_5)/a_1$, $A_3 = (a_1 a_6 - a_0 a_7)/a_1$, etc., and succeeding rows are obtained from the preceding two rows in the same manner as the third is obtained from the first two. The rows will be found to get shorter by one element every two rows and to be $(n + 1)$ in number.

The system is stable if and only if all the elements of the first column have the same sign. (In fact, the number of changes of sign on going down the first column is the number of roots of the characteristic equation in the right half-plane.)

If a zero is formed in the first column before the array is complete (which prohibits the calculation of further rows), it may indicate either instability or critical stability. If the zero is preceded by two rows consisting of the same number of elements, such that the ratio of corresponding elements in the two rows is constant, roots on the imaginary axis are indicated. Moreover, in this case, the elements of either of these rows, read from right to left, successively multiplied by s^0, s, s^2, etc., added and equated to zero will give the equation whose roots are the roots of the characteristic equation lying (in conjugate pairs) on the imaginary axis. If, on the other hand, the zero is not preceded by two rows with proportional elements, then there is at least one root in the right half-plane and the system is unstable.

A necessary but, except when $n = 1$ or 2, an insufficient condition for stability is that all the polynomial coefficients shall be nonzero and all have the same sign. This condition is always implicitly contained in the Routh criterion.

Hurwitz Criterion. Using the same polynomial form of the characteristic equation as before, the array

$$
\begin{array}{ccccccc}
a_1 & a_0 & 0 & 0 & 0 & \cdots & \cdots \\
a_3 & a_2 & a_1 & a_0 & 0 & 0 & \cdots \\
a_5 & a_4 & a_3 & a_2 & a_1 & a_0 & \cdots \\
a_7 & a_6 & \cdots & \cdots & \cdots & \cdots & \cdots
\end{array}
$$

is drawn up, extending for $(n - 1)$ rows and columns, absent coefficients being replaced by zeros.

The system is stable if and only if:

1. All the coefficients of the polynomial are of the same sign, which must be assumed positive; and
2. The values of the determinants of order 2 to $(n - 1)$ and having the top left array element as their top left element are also positive.

The application of either the Routh or Hurwitz criteria is a valuable tool in the analysis of system stability. However, there are limitations that must be recognized. To use either of these techniques the system equation must be known. In many cases, and this is especially true in process control applications, the system equation is not available. In applying these techniques little information can be obtained concerning the relative stability of the system. Furthermore, it is difficult to determine the effect on stability of individual parameters for design considerations. Due to these limitations, the stability of a system is normally determined from frequency response curves using the Nyquist stability criterion. The frequency response curves may be obtained from the system equations or determined experimentally. A general description of the Nyquist method is as follows.

Nyquist Criterion; 1 (for a control system). It is assumed that if $G(s)$ is the output to error transfer function, then as $s \to \infty$, $G(s) \to 0$. This is normally the case in any practical system.

If the system is stable on closed loop, the vector locus of $G(i\omega)$ drawn from $\omega \to -\infty$ to $\omega \to +\infty$ in the sense of increasing ω (i.e., the vector locus of $G(s)$ corresponding to following the imaginary axis in the s-plane) encircles the point $(-1, 0)$ P times counterclockwise, where P is the number of poles of $G(s)$ to the right of the imaginary axis, multiple poles counting according to their order.

The following points should be noted:

1. If $P \neq 0$ the system is unstable on open loop, i.e., with the (-1) feedback from output to error broken.
2. The value of P may be found, in lumped parameter systems, by writing $G(s) = N(s)/D(s)$, where N and D are polynomials; P is then the number of zeros of $D(s)$ to the right of the imaginary axis, which may be found most simply by the Routh criterion.
3. If $G(s)$ has poles *on* the imaginary axis, these must be circumvented in the s-plane by infinitesimal semicircles in the right half-plane having these poles as centers. These indentations will correspond in the plane of $G(s)$ to a clockwise rotation through $180°$ at indefinitely large radius for every simple pole so encountered, multiple poles again counting according to their order. These infinite semicircles must be taken into account in assessing the number of encirclements of the point $(-1, 0)$.
4. Since $G(-i\omega)$ is the complex conjugate of $G(i\omega)$, the locus of $G(s)$ for negative real frequencies is the mirror image, in the real axis of the $G(s)$ plane, of the locus for corresponding positive real frequencies.
5. If and only if $P = 0$, an adequate simpler criterion is that the locus of $G(i\omega)$ drawn in the sense of increasing ω, shall leave the point $(-1, 0)$ to its left. See also **Feedback**. This same concept may be applied in using the Bode plot. In this context the system gain must be less than 1.0 when the phase angle is equal to minus $180°$. It is easily seen that this corresponds to the point $(-1, 0)$ on the Nyquist plot.

Nyquist Criterion; 2 (for a general feedback system). It is, of course, always possible to reduce the basic equations of the system

$$a_{1j}(s)Q_1(s) + a_{2j}(s)Q_2(s) + \cdots + a_{nj}(s)Q_n(s) = f_j(s), \quad j = 1, 2, \ldots n$$

by eliminating all but two of the Q's, to a pair of equations relating these two quantities only, which can be put in the form

$$Q_1(s) - \beta s)Q_2(s) = F_1(s)$$

$$Q_2(s) - \mu(s)Q_1(s) = F_2(s)$$

This, in effect, reduces the system to a single-loop system of loop transfer function $\mu(s)\beta s)$. Since the characteristic equation is

$$\Delta(s) = 1 - \mu(s)\beta s) = 0$$

the above formulation of the Nyquist criterion may be applied, replacing $G(s)$ in the preceding section by $-\mu(s)\beta s$. Thus the system will be stable provided that:

1. The locus of $-\mu(i\omega)\beta i\omega)$ encircles the point $(-1, 0)$ P times counterclockwise, or
2. The locus of $+\mu(i\omega)\beta i\omega)$ encircles the point $(+1, 0)$ P times counterclockwise, or
3. The locus of $1 - \mu(i\omega)\beta i\omega)$ encircles the origin P times counterclockwise, P being in each case the number of simple poles of $\mu(s)\beta s)$ lying in the right-hand half-plane, any poles on the imaginary axis being circumvented as explained above.

In a multiloop system, however, μ or β or both may comprise subsidiary, possibly unstable loops, so that in general $P \neq 0$. To avoid the reduction of the equations to the above form and the subsequent finding of the value of P by the Routh criterion, the Nyquist method may be extended as follows.

We start from the premise that a passive system is necessarily stable. In other words, if, in the given system, the active elements are made inactive, we have a stable system. We then imagine each of the active elements to be activized one at a time until the system is back to its normal state, investigating at each stage the behavior of the return difference for the reactivized element. (The order in which the elements are reactivized is entirely a matter of convenience.)

Rendering an element inactive has the effect, in the set of general system equations, of making some parameter of that element zero. In the case of an electron tube, the amplification factor or the mutual conductance vanish;

in the case of a rotary machine amplifier, the armature voltage per ampere-turn vanishes, and analogously for other types of active elements. Further, it can be shown that the return difference for any parameter is the ratio of the values assumed by the system determinant when the parameter has its normal value, to when this value is made zero, provided only that the determinant is a linear function of the parameter.

We suppose then that there are m active elements and, having placed these in some arbitrary order, we denote by Δ_r the value of the system determinant when the first r of these elements are activized, the remainder being inactive. Starting from the completely inactive system, the return difference for the first element is Δ_1/Δ_0. The return difference for the second element (the first one remaining active) is Δ_2/Δ_1. The return difference for the third (the first two remaining active) is Δ_3/Δ_2 and so on, until finally the return difference for the mth element (all other remaining active) is Δ_m/Δ_{m-1} in which, furthermore, $\Delta_m \equiv \Delta$, the normal system determinant. Now the number of counterclockwise encirclement of the origin made by the Nyquist locus of the return difference of the rth element Δ_r/Δ_{r-1} is $(z_{r-1} - z_r)$, where z_r denotes the number of zeros of Δ_r in the right half-plane. Hence the sum of such encirclements made by all the successive return difference loci is $(z_0 - z_1) + (z_1 - z_2) + \cdots + (z_{m-1} - z_m) = z_0 - z_m$. But z_0 is zero since the system is then passive. For stability, moreover, z_m must be zero. Hence, *for stability of the normal system, the sum of the encirclements of the origin made by the various return difference Nyquist loci must be zero.*

Additional Reading

Higgins, S.P. Jr., and J.M. Nelson: "Process Control Techniques," in Process Instruments and Controls Handbook, 3rd Ed. (D.M. and G.D. Considine, Eds.), The McGraw-Hill Companies, Inc., New York, NY, 1985.

Matley, J. et al.: "Practical Process Instrumentation and Control," Vol. 2, The McGraw-Hill Companies, Inc., New York, NY, 1986.

Osborne, R.L.: "Fundamentals of Automatic Process Control," in Process Instruments and Controls Handbook, 3rd Ed. (D.M. and G.D. Considine, Eds.), The McGraw-Hill Companies, Inc., New York, NY, 1985.

Padiyar, K.R.: "Power System Dynamics: Stability and Control," John Wiley & Sons, Inc., New York, NY, 1999.

Shinskey, F.G.: "Process Control Systems," 4th Edition, The McGraw-Hill Companies, Inc., New York, NY, 1996.

STABILIZATION (Ship). In addition to the obvious method of suitable design of the hull, a number of specific devices have been developed for damping the motions of a ship among the waves, especially the rolling motion. Ships have been built carrying large gyroscopes, but in general this method proved too costly in money and added weight for the effect obtained. A more widely used method is that of *anti-rolling tanks*, also called *water chambers*, which are usually placed in stabilizing bulges on each side of the hull. By controlling the flow of water from the tanks on one side to those on the other, so that the period of the contained water is proportionate (ideally about 70%) to the ship's period of roll, the maximum stabilization can be obtained. (Its amount is much less than that of a gyroscope, but far less costly.) Control of the water flow may be effected by a valve-controlled air line connecting the upper parts of the U-tubes connecting the tanks. This control can also be used to shut off the system in wave systems of very irregular pattern, which cause erratic behavior of the ship when the water is shifted among the anti-rolling tanks.

STABLE FLY (*Insecta, Diptera*). Similar to the house fly in appearance, the small stable fly, *Stomoxys calcitrans* (Linne), is a severe pest against horses and mules, usually biting the legs of the animals. It also attacks cattle, hogs, goats, sheep, rabbits, dogs, rats, and humans. The insect is widely distributed throughout the United States. In the northern states, the insect winters over as larvae and pupae, usually in wet areas near straw and manure. In the southern states, the fly can be found in all stages all year. Where stable flies are abundant, animals can lose a substantial amount of blood during the course of a day because each stable fly will require from 1 to 2 drops of blood each time it bites. The bites are painful and irritating and generally contribute to deterioration of the health of the animals attacked. Dairy cows yield less milk; cattle lose weight; horses can become unmanageable. The fly is also suspected of carrying a number of diseases, such as anthrax, leprosy, surra, and swamp fever (infectious anemia of horses), although no solid proof of this has been available.

Prevention through cleanliness of animal quarters cannot be over-stressed as an effective measure. Highly valued animals can be protected with blankets and coverings. Fly traps, mechanical or electrical, are

suggested for dairy barns and other areas frequented by the flies. Numerous commercial repellent formulations are available — for use in disinfecting animal quarters as well as for spraying on affected parts of the animals once or twice per week.

STACK. In geology, a rock pillar or monument that occurs relatively close to marine cliffs which have usually been developed in hard, horizontally bedded, but jointed, sedimentary rocks. In some cases, a stack is the remaining pillar of an arch. Both sea arches and stacks are well developed on the coast lines of the Gaspé Peninsula and the North East Highlands of Scotland.

STADIMETER. An instrument for determining the distance to an object of know dimension by measuring the angle subtended at the observer by the object.

STAGE. 1. A self-propelled separable element of a rocket vehicle. 2. A step or process through which a fluid passes, especially in compression or expansion. 3. A set of stator blades and a set of rotor blades in an axial-flow compressor or a turbine (see also **Turbine (Steam)**); an impeller wheel in a radial-flow compressor.

STALACTITE AND STALAGMITE. A *stalactite* is a deposit of a mineralized solution, commonly calcium carbonate, which hangs like an icicle from the roof or wall of a limestone cavern. See Fig. 1. The formation of the stalactite usually is quite a slow process. Corresponding columnar structures built upward from the floors of caves beneath the stalactites are called *stalagmites*. Stalactite is derived from the Greek, meaning to fall in drops. Stalagmite, also derived from the Greek, means that which drops. When a stalactite from the top of a cave and a stalagmite from the floor of the cave join, the resulting singular structure is called a *column*.

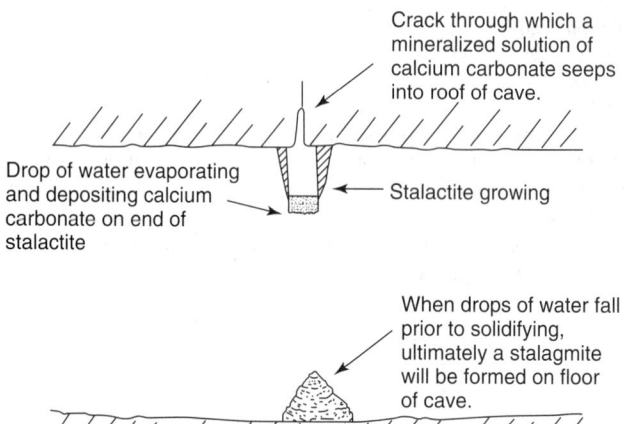

Fig. 1.   Formation of stalactite and stalagmite.

STALAGMOMETER. An apparatus for the measurement of surface tension by the drop-weight method.

STALING (Bakery and Food Products). Most authorities agree that staling of bakery products commences immediately after the product leaves the oven and that it is a reasonably progressive process, not involving a delayed start or of a cyclic nature. Post-processing conditions obviously can favor or be unfavorable to the process of staling. In bread, the moisture content of the crumb averages about 45% (weight) when the loaf leaves the oven. The crust, on the other hand, has a moisture content of 5–7%. Under these conditions (lack of equilibrium), the crust loses moisture on the outside to the atmosphere, while it picks up moisture from the crumb on the inside of the loaf. The result is a toughening of the crust, accompanied and aided by formation of a soggy undersurface. The crumb, of course, becomes drier and drier with time. As moisture is expelled to the atmosphere, along with other volatiles, flavor decreases. Flavor loss also is the result of essential deactivation of certain flavoring components when they are absorbed by the inner solids. It is also likely that unexplained chemical reactions occur (oxidation, etc.) that alter the chemical structure and hence aroma and flavor of the flavoring components. With some products, the staling processes are not necessarily fully irreversible. Some breads, such as Italian breads, can be moistened and placed in an oven and returned to a reasonable degree of organoleptic acceptability. Some bakers purposely underbake their product to slow up the staling process.

Many authorities believe that starch is in some way the key to most of the staling phenomena. A tie-in with starch is evident in a practical way by observing that bread put in the refrigerator tends to become stale sooner than when left at a reasonable room temperature; but if the bread is placed in a freezer, it may be retained for long periods, but with some evidence of staleness once the bread is thawed. It is known that starch retrogrades at a faster rate as the temperature is lowered, but beyond a certain point the starch degradation is greatly slowed at very low temperatures and hence the staling rate is slowed. Various additives have alleviated the staling process to some degree.

STALL.   A term in common use to describe the condition of an aerodynamic burble upon a wing. An airplane that has "stalled" has had the streamline flow over the upper wing surface partly destroyed. The lift is partially lost so weight and lift are out of equilibrium. Hence, a stall is followed by a nosing down motion accompanied by rapid loss of altitude. The term "stall" here applies to an aerodynamic phenomenon. The term is also used when an aircraft engine ceases to operate, when it is said to stall.

See also **Aerodynamics and Aerostatics**; and **Airplane**.

STANDARD CONDITIONS.   Many physical and chemical phenomena and substances are defined in terms of *standard* conditions. In some instances, a temperature commonly prevailing in a chemical laboratory may be selected. Thus, when comparing a number of substances, such as their index of refraction, one may find lists in handbooks that give these values as measured by a given temperature and pressure. The Smithsonian Tables, for example, include such data for scores of substances. The researcher then can seek formulas for converting such values to other temperatures/pressures. Interpolations in some instances can be linear over a wide range of values, or the relationships may be nonlinear.

In the case of gases, properties may be tabulated in terms of their existence at 0 °C and 760 mm pressure. To determine the volume of a gas at some different temperature and pressure, corrections derived from known relationships (Charles', Amonton's, Gay-Lussac's, and other laws) must be applied as appropriate. In the case of pH values given at some measured value (standard for comparison), the same situation applies. Commonly, lists of pH values are based upon measurements taken at 25 °C. The pH of pure water at 22 °C is 7.00; at 25 °C, 6.998; and at 100 °C, 6.13. Modern pH instruments compensate for temperature differences through application of the Nernst equation.

Standard conditions are not necessarily consistent with standards definitions. The careful researcher will always take note of the conditions stated for determining values in a tabulated list.

STANDARD DEVIATION.   The standard deviation of a probability distribution is the square root of its variance. It is the most useful measure of dispersion.

STANDARD ERROR.   A name often given to the standard deviation of a sampling distribution.

STANDARD FREE ENERGY INCREASE.   Often referred to as standard free energy. The increase in Gibbs free energy (see also **Free Energy**) when the reactants in a chemical change, all in their standard states (e.g., unit concentration, or at 1 atmosphere pressure) are converted into the products in their standard states. Given by

$$\Delta G^{\circ} = -RT \ln K$$

where R is the gas constant, T, the absolute temperature, K, the equilibrium constant. See also **Free Energy Change**.

STANDARD STATE.   The stable form of a substance at unit activity. The stable state for each substance of a gaseous system is the ideal gas at 1 atmosphere pressure; for a solution it is taken at unit mole fraction; and for a solid or liquid element it is taken at 1 atmosphere pressure and ordinary temperature.

STANDARD UNIT (Statistics).   A variate may be changed to standard units by the transformation

$$t = \frac{x - \bar{x}}{\sigma_x}$$

where t is in standard units, x is the variate, $\bar{x}$ is the mean, and σ_x is the standard deviation of the distribution. In such cases t has a mean of zero and a standard deviation of 1 and is said to be in standard measure.

STANDING-WAVE RATIO.   Any transmission line such as a waveguide or an acoustic transmission system, unless terminated by its characteristic impedance, will exhibit a superposition of standing and progressive waves. The standing-wave ratio is a measure of the relative amplitudes of the two types of wave. It is defined as the ratio of the maximum amplitude of pressure (or voltage) to the minimum amplitude of pressure (or voltage) measured along the path of the waves. Thus, at a given frequency in a uniform waveguide, the standing-wave ratio is the ratio of the maximum to the minimum amplitudes of corresponding components of the field (or the voltage or current) along the waveguide in the direction of propagation. Alternatively, the standing-wave ratio may be expressed as the reciprocal of the ratio defined above.

STANNITE (Mineral).   This mineral is a sulfo-stannate of copper and iron, sometimes with some zinc, corresponding to the formula, Cu_2FeSnS_4. It is tetragonal; brittle with uneven fracture; hardness, 4; specific gravity, 4.3–4.5; metallic luster; color, gray to black, sometimes tarnished by chalcopyrite; streak, black; opaque. The mineral occurs associated with cassiterite, chalcopyrite, tetrahedrite, and pyrite, probably the result of deposition by hot alkaline solution. Stannite occurs in Bohemia; Cornwall, England; Tasmania; Bolivia; and in the United States in South Dakota. It derives its name from the Latin word for "tin," *stannum*.

STAR.   There is probably no one class of physical objects that has attracted more popular and scientific attention throughout the ages than have the stars. A small portion of the mass of mythological material that is associated with these bodies will be found in the various articles dealing with the individual constellations and a few of the brighter stars.

A star, as distinguished from a *planet*, is a self-gravitating object capable of generating its own energy by nuclear processes, or the remnant (white dwarf or neutron star) of such an object. The *protostellar phase* would be seen as the formative part of the life of a star. The contraction of the gas cloud leads to core temperatures high enough to ignite the central engine. A lower mass limit of 0.08 solar masses appears to be required in order to initiate the simplest nuclear reactions, while the upper stable mass may be as great as several hundred solar masses.

The methods for determining the physical characteristics of the stars will be found under such titles as: stellar parallax, stellar magnitude, spectral class, binary stars, variable stars, cepheids, giant and dwarf stars, etc. The characteristics of a typical main-sequence G-type star will be found in the articles dealing with the sun and various solar characteristics. The source of the energy radiated by the stars is discussed in the article on the sun, and under carbon cycle and proton-proton chain. See Tables 1, 2, and 3.

Numerous classes and specific stars are described in various entries throughout this encyclopedia. Consult alphabetical index.

STEVEN N. SHORE, University of Indiana South Bend, South Bend, IN.

STAR CATALOGUES.   Any listing of stars, usually arranged in order of increasing right ascension. Originally, star catalogues were intended merely for the purpose of providing accurate positions of the stars for use by navigators, but many modern catalogues are designed to provide particular characteristics of the stars.

The oldest existing star catalogue is contained in the Almagest of Ptolemy issued about 137 A.D. The Almagest catalogue was the standard one used throughout the Middle East and Latin West to the end of the 15th Century. Tycho Brahe's catalogue of 1580 marks the dawn of the modern era of star catalogues, and since that time, many others have appeared. Probably the most comprehensive catalogue issued is the Bonner Durchmusterung, which first appeared about 1850, together with the various extensions that have since been published. During the last half of the nineteenth century the Astronomische Gesellschaft sponsored

TABLE 1. PHYSICAL CHARACTERISTICS OF SOME TYPICAL STARS[a]

Star	Spectral Class	Temperature in °K	Density in Terms of Water	Referred to Sun as Unity: Luminosity	Mass	Diameter
			Giants			
Antares	M1 Ib	3,100	0.0000003	3,500	30	480
Aldeberan	K5 III	3,300	0.00002	90	4	60
Arcturus	$K2_p$ III	4,100	0.0003	100	8	30
Capella	G8 III?+F	5,500	0.002	150	4.2	12
β Centauri	B III	21,000	0.02	3,100	25	11
			Main Sequence			
Vega	A0 V	11,200	0.1	50	3	2.4
Sirius A	A1 V	11,200	0.4	26	2.4	1.8
Altair	A7 V	8,600	0.6	9.2	2	1.4
Procyon	F5 V	6,500	1.2	5.4	1.75	1.7
α Centauri A	G2 V	6,000	1.1	1.12	1.1	1.2
The Sun	G2 V	6,000	1.4	1	1	1.0
70 Ophiuchi A	K0 V	5,100	0.9	0.42	0.9	1.0
61 Cygni A	K7 V	3,800	1.3	0.21	0.6	0.7
Krueger 60A	M3 V	3,300	9	0.002	0.3	0.3
			White Dwarfs			
Sirius B	F	7,500	27,000	0.10	0.96	0.034
O_2 Eridani B	A0	11,000	64,000	0.003	0.44	0.019

[a]For a general discussion of Population I and Population II stars, see **Giant and Dwarf Stars**.

TABLE 2. THE TEN BRIGHTEST STARS
(*Courtesy* van de Kamp *Publications of the Astronomical Society of the Pacific*, 1953, vol. 65)

Number	Name	Right Ascension 1900	Declination 1900	Visual Magnitude and Spectrum A	B	C	Annual Proper Motion	Parallax		Distance in Light Years	Visual Absolute Magnitude A	B	C	Visual Luminosity ⊙ A	B	C
1	Sirius	$6^h40^m.7$	−16° 35′	−1.6 A0	7.1 A5	—	1″.32	0″.375	±″.004	8.7	+1.3	+10.0	—	23	0.008	—
2	Canopus	6 21 .7	−52 38	−0.9 F0	—	—	0 .02	0 .018	6	180:	−4.6:	—	—	5,200:	—	—
3	α Centauri	14 32 .8	−60 25	+0.3 G0	1.7 K5	11 M	3 .68	0 .760	5	4.29	+4.7	+6.1	+15.4	1.0	0.28	0.000052
4	Vega	18 33 .6	+38 41	0.1 A0	—	—	0 .35	0 .123	5	26.5	+0.5	—	—	48	—	—
5	Capella	5 9 .3	+45 54	0.2 G0	10.0 M1	13.7 M5	0 .44	0 .073	4	45	−0.5	+9.3	+13.0	120	0.014	0.0005
6	Arcturus	14 11 .1	+19 42	0.2 K0	—	—	2 .29	0 .090	5	36	0.0	—	—	76	—	—
7	Rigel	5 9 .7	− 8 19	0.3 B8p	—	—	0 .01	0 .005:		650:	−6.2:	—	—	23,000:	—	—
8	Procyon	7 34 .1	+ 5 29	0.5 F5	10.8	—	1 .25	0 .288	4	11.3	+2.8	+13.1	—	5.8	0.00044	—
9	Archernar	1 34 .0	−57 45	0.6 B5	—	—	0 .09	0 .023	42	140:	−2.6:	—	—	800:	—	—
10	β Centauri	13 56 .8	−59 53	0.9 B1	—	—	0 .04	0 .016	11	200:	−3.1:	—	—	1,300:	—	—

TABLE 3. STARS NEARER THAN FIVE PARSECS
(*Courtesy* van de Kamp *Publications of the Astronomical Society of the Pacific*, 1953, vol. 65)

Number	Name	Right Ascension 1950	Declination 1950	Parallax	Distance in Light Years	Cross Proper Motion	Motion KM/sec	Position Angle	Radial Velocity KM/sec	Visual Magnitude and Spectrum A	B	C	Visual Absolute Magnitude A	B	C
1	Sun	—	—	—	—	—	—	—	0	−26.9 G0	—	—	4.7	—	—
2	α Centauri	$14^h36^m.2$	−60°38	0″.745	4.3	3″.68	23	281°	−25	0.3 G2	1.7 K5	11 M5e	4.7	6.1	15.4
3	Barnard's star	17 55.4	+4 33	0.552	6.0	10.30	90	356	−108	9.5 M5	*	—	13.2	*	—
4	Wolf 359	10 54.2	+7 20	0.429	7.7	4.71	54	235	+13	13.5 M6e	—	—	16.6	—	—
5	Luyten 726–8	1 36.4	−18 13	0.367	7.9	3.35	38	80	+29	12.5 M6e	13.0 M6e	—	15.6	16.1	—
6	Lalande 21185	11 0.6	+36 18	0.398	8.2	4.78	57	187	−86	7.5 M2	*	—	10.5	*	—
7	Sirius	6 42.9	−16 39	0.375	8.7	2.32	16	204	−8	−1.6 A1	7.1 wd	—	1.3	10.0	—
8	Ross 154	18 46.7	−23 53	0.345	9.3	0.72	9	106	−4	10.6 M5e	—	—	13.3	—	—
9	Ross 248	23 39.4	+43 55	0.316	10.3	2.58	23	176	−81	12.2 M6e	—	—	14.7	—	—
10	ε Eridani	3 30.6	−9 38	0.303	10.8	0.97	15	271	+15	3.8 K2	—	—	6.2	—	—
11	Ross 128	11 45.1	+17	0.301	10.9	1.37	22	151	−13	11.1 M5	—	—	13.5	—	—
12	61 Cygni	21 4.7	+38 30	0.293	11.1	5.22	84	52	−64	5.6 K6	6.3 M0	*	7.9	8.6	*
13	Luyten 789–6	22 35.7	−15 37	0.305	11.2	3.27	53	46	−60	12.2 M6	—	—	14.5	—	—
14	Procyon	7 36.7	+5 21	0.288	11.3	1.25	20	214	−3	0.5 F5	10.8 wd	—	2.8	13.1	
15	ε Indi	21 59.6	−57 0	0.285	11.4	4.67	77	123	−40	4.7 K5	—	—	7.0	—	—
16	Σ 2398	18 42.2	+59 33	0.280	11.6	2.29	38	324	+1	8.9 M4	9.7 M4	—	11.1	11.9	—
17	Groombridge 34	0 15.5	+43 44	0.280	11.7	2.91	49	82	+14	8.1 M2e	10.9 M4e	—	10.3	13.1	—
18	τ Ceti	1 41.7	−16 12	0.275	11.8	1.92	33	297	−16	3.6 G4	—	—	5.8	—	—
19	Lacaille 9352	23 2.6	−36 9	0.279	11.9	6.90	118	79	+10	7.2 M2	—	—	9.4	—	—
20	BD + 50° 1668	7 24.7	+5 29	0.268	12.4	3.73	67	171	+26	10.1 M4	—	—	12.2	—	—

*The stars nearest the sun are often referred to as nearby stars. They are generally main sequence dwarf stars.

a catalogue of accurate positions of the majority of the stars contained in the Bonner Durchmusterung.

With application of photography to astronomy, several projects have been launched for obtaining comprehensive star catalogues. By far the most ambitious of all of the photographic catalogues is the so-called Astrographic Catalogue, which was started as a cooperative effort of 18 observatories in 1887. It is not entirely completed. However, there has been completed a photographic survey of the sky known as the Sky Atlas, the most recent such atlas being the Palomar Sky Atlas, and European Southern Observatory atlas.

As part of the Space Telescope project, a general catalogue of potential guide stars is being prepared. This will also serve as a permanent archival data set for future magnitude and positional studies. The European Space Agency is building an astrometric satellite (*HIPPARCHOS*) designed to observe positions and motions of several hundred thousand stars over the next decade.

Not all catalogues are restricted to positional and magnitude data. The largest spectroscopic catalogue is the Henry Draper project, first completed in the 1920s and currently being revised. The document contains information on the spectral characteristics of over 200,000 stars.

See also **Bonner Durchmusterung**.

STEVEN N. SHORE, University of Indiana South Bend, South Bend, IN.

STARCH.   Chemically, starch is a homopolymer of α-D-glucopyranoside of two distinct types. The linear polysaccharide, amylose, has a degree of polymerization on the order of several hundred glucose residues connected by alpha-D-(1 → 4)-glucosidic linkages. The branched polymer, amylopectin, has a DP (degree of polymerization) on the order of several hundred thousand glucose residues. The segments between the branched points average about 25 glucose residues linked by alpha-D-(1 → 4)-glucosidic bonds, while the branched points are linked by alpha-D-(1 → 6)-bonds. See Fig. 1.

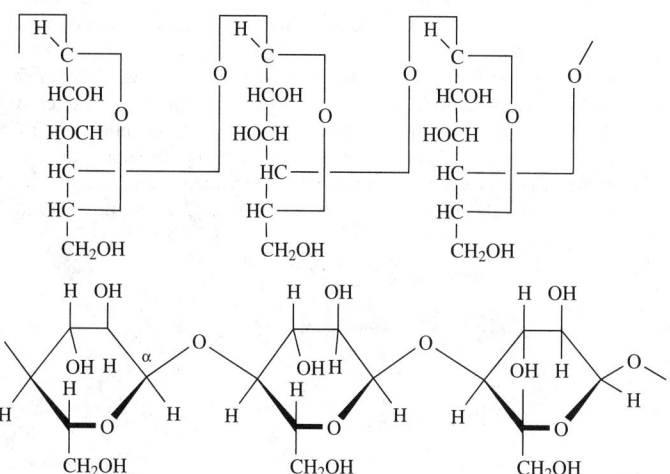

Fig. 1.   A segment of the starch molecule.

Most cereal starches are made up of about 75% amylopectin and 25% amylose molecules. However, root starches are slightly higher in amylopectin, while waxy corn* and waxy milo starch contain almost 100% amylopectin. At the other extreme, high amylose corn starch and wrinkled pea starches contain 60–80% amylose. The molecules of amylose and amylopectin are synthesized by enzymes inside the living cell in plastids known as amyloplasts and are deposited as starch granules. These granules are microscopic in size, ranging from 3–8 micrometers in diameter for rice starch up to 100 micrometers for the larger potato starch granules. Corn starch usually falls in a range of 5–25 micrometers. An experienced observer usually can identify the genetic origin of a sample of starch by

* With exception of North America, where the plant is called *corn*, other English-speaking people call it *maize*. French = *mais*; Spanish = *maiz*.

the size and shape of the granules. The granules are insoluble in cold water, but swell rapidly when heated to the gelatinization temperature range for the particular starch involved. As the granules swell, they lose their characteristic cross under polarized light and imbibe water rapidly until they are many times their original size. Upon continued heating or mechanical shear, the swollen granules begin to disintegrate and the viscosity, having reached a maximum, begins to decrease. However, there usually are some granules and some segments of granules that do not completely disperse in aqueous systems even under the most stringent conditions.

As the partially dissolved paste is cooled, the hydrated molecules and segments of granules begin to precipitate. In a dilute system (approximately 1%), the segments and molecules retrograde or precipitate. At higher concentrations, sufficient intermolecular and intersegment bonds form to fix the entire system into three-dimensional gel. The rigidity of this gel is affected by many factors, but the amylose content is perhaps the most significant. High amylose starches, when thoroughly cooked, form very rigid gels. Waxy corn or waxy milo starch paste form little, if any, gel structure when cooled.

While some wheat and potatoes are processed in the United States, over 90% of all starch is produced from corn in what is called the *corn wet milling industry*. Close to one-quarter of a billion bushels of corn, representing about 5% of the total corn crop, is converted into wet-process products. The corn refining process is illustrated in Fig. 2. Shelled corn is delivered to the wet-milling plant in boxcars containing an average of 2,000 bushels (50.8 metric tons) per car, and unloaded into a grated pit. The corn is elevated to temporary storage bins, and then to scale hoppers for weighing and sampling. The corn passes through mechanical cleaners designed to separate unwanted substances, such as pieces of cobs, sticks, and husks, as well as metal and stones. The cleaners agitate the kernels over a series of perforated metal sheets; the smaller foreign materials drop through the perforations, while a blast of air blows away chaff and dust, and electromagnets draw out nails and bits of metal. Coming out of the storage bins, the corn is given a second cleaning before going into very large "steep" tanks.

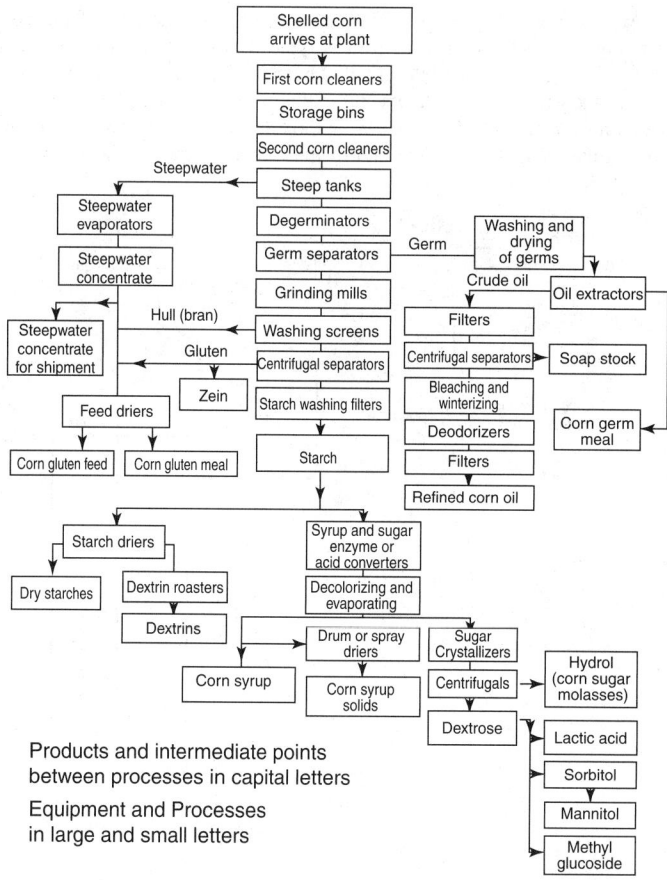

Fig. 2.   The corn (maize) refining process. (*Corn Refiners Association, Inc.*)

Fig. 3. Triple-effect steepwater evaporator. The third effect (forced circulation) is shown in background; second effect (falling-film, recirculating) is middle unit. The first effect vapor head is shown in foreground. (*Swenson, Whiting Corp.*)

contains much of the soluble protein, carbohydrates, and minerals of the corn kernel, and is drawn off as the first by-product of the process. Steepwater, unmodified and modified, is an essential nutrient for production of antibiotic drugs, vitamins, amino acids, and fermentation chemicals. See Fig. 3. It is also an effective growth supplement for animal feeds.

From the steeps, the softened kernels go through degerminating mills, which are designed not for fine grinding, but rather for tearing the soft kernels apart into coarse particles, freeing the rubbery oil-bearing germ without crushing it, and loosening the bran. The wet, macerated kernels then are sluiced into flotation tanks, called germ separators, or centrifugal hydrocyclones. The germs, lighter than the other components of the kernel, float to the surface, and are skimmed off. By oil expellers or extractors (heat and pressure) and by means of solvents, practically all of the oil is removed as another byproduct to be settled, filtered, refined, and otherwise processed into clear, edible oil for salad dressing and frying, and "corn oil foods" or "soap stock" for soap manufacture. The residue of the germ, after oil-extraction, is ground and marketed as corn germ meal, or may become a part of corn gluten feed or meal.

The remaining mixture of starch, gluten, and bran (hull), which is finely ground, is washed through a series of screens to sieve the bran from the starch and gluten. The hull becomes part of corn gluten feed.

The remaining mixture of gluten and starch is pumped from the shakers to high speed centrifugal machines, which, because of the difference in specific gravity, separate the relatively heavier starch from the lighter gluten. After further processing, the protein-rich gluten is marketed as such, or becomes corn gluten meal, or may be mixed with steepwater, corn oil meal, and hulls to become corn gluten feed. Gluten may also be made to yield a highly versatile protein, *zein*; amino acids, such as glumatic acid, leucine, and tyrosine; and xanthophyll oil, for poultry rations.

Having been separated from the kernels, the starch is now ready for washing, drying, or further processing into numerous dry-starch products, or into dextrin, or for conversion into syrup and sugar. From a 56-pound (25 kilograms) bushel of corn, approximately 32 pounds (14.5 kilograms) of starch result, about 14.5 pounds (6.6 kilograms) of feed and feed products, about 2 pounds (0.9 kilogram) of oil, the remainder being water.

Starch Conversion. More than half of the total production of starch is converted into syrup dextrins or dextrose by acid hydrolysis and/or enzyme action or heat treatment.

Starch, mixed with water, and heated in the presence of weak hydrochloric acid, breaks down chemically by hydrolysis. If the hydrolysis or conversion of corn starch is interrupted before final conversion, a noncrystallizing corn syrup is obtained. Many varieties may be made by supplemental use of enzymes to meet specific functional requirements. The solids content is varied to suit the requirements of the users. Corn syrup is used in

At this point, the use of water becomes an essential part of the corn refining process. The cleaned corn is typically moved into large wooden or metal tanks holding 2,000 to 6,000 bushels (50.8 to 152.4 metric tons), and soaked for 36 to 48 hours in circulating warm water 49 °C (120 °F) containing a small amount of sulfur dioxide to control fermentation and to facilitate softening. At the end of the steeping process, the steepwater

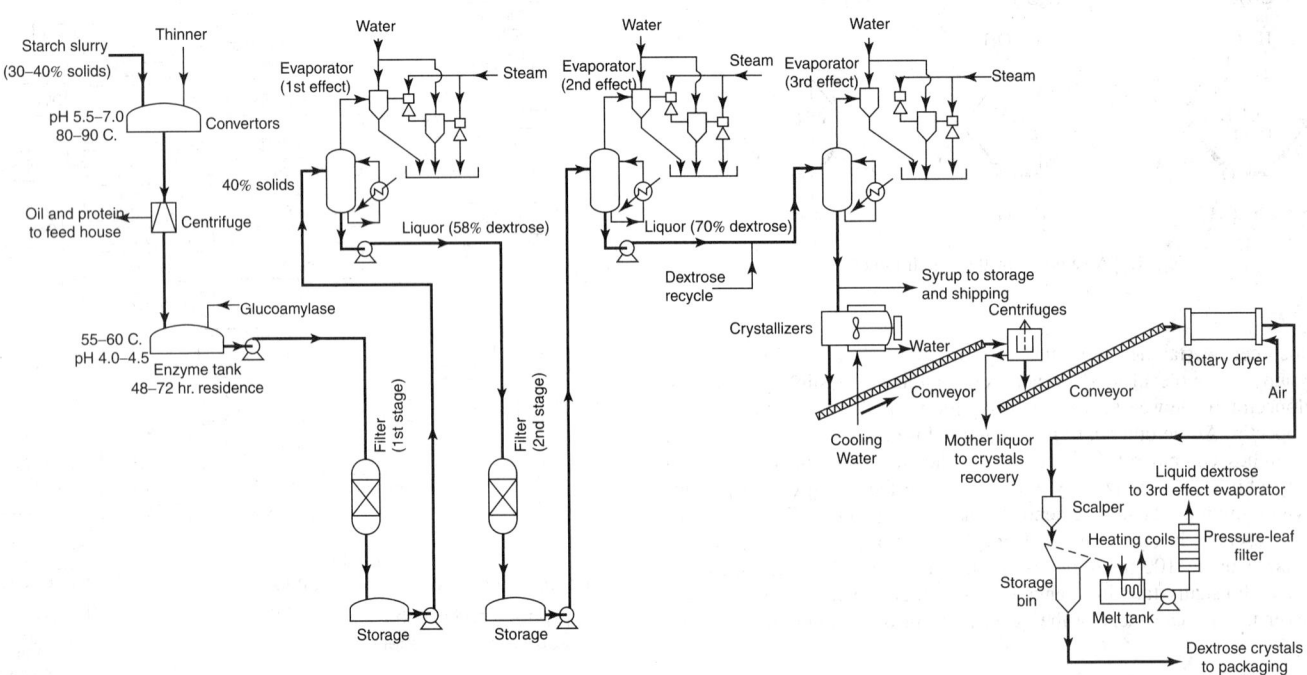

Fig. 4. Enzyme process for converting starch into dextrose. (*A.E. Staley Mfg. Co.*)

a wide variety of food products, including baby foods, breakfast foods, cheese spreads, chewing gum, chocolate products, confectionary, cordials, frostings and icings, peanut butter, sausage, and for numerous industrial products, including adhesives, dyes and inks, explosives, metal plating, plasticizers, polishes, textile finishes, and in leather tanning.

A process for converting starch to dextrose is shown in Fig. 4. The enzyme process shown overcomes flavor and color difficulties of the hydrochloric acid method. The enzyme is obtained by growing a mold (*Aspergillus phoenicis*, a member of the *Aspergillus niger* group). The mold yields the key glucoamylase as well as transglucosylase. The latter must be eliminated because it catalyzes the formation of undesirable glucosidic linkages. Through a special process, almost pure glucoamylase is obtained.

Purified starch slurry (30–40% solids), made from dent corn, is received in the converters from basic processing at the corn plant. A preliminary conversion using alpha-amylase enzyme or acid is carried out at 80–90 °C (176–194 °F), during which 15–25% of the starch is converted into dextrose. This thins the starch slurry, allowing easier addition of the glucoamylase enzyme. It also prevents formation of unhydrolyzable gelatinous material during the main conversion, and results in increased dextrose yields of from 3–4%. Thinning also reduces evaporation costs because starch concentrations of 30–40% can be handled compared with the 12–20% limit for the acid process. Before the main conversion, the starch-dextrose slurry is centrifuged to remove oil and protein by-products, which are processed for animal feed.

The slurry then goes to a 25,000-gallon (946-hectoliter) enzyme tank where, at pH 4.0–4.5 and 60 °C, the major reaction with the glucoamylase takes place. It is a batch operation requiring about 72 hours. When conversion is complete, the batch (97–98.5% dextrose on a dry basis) is passed through a preliminary decolorizing filter of powdered carbon and then pumped on to the first of three evaporators. The remaining operations are evaporation and crystallization, followed by centrifuging, and rotary drying for dextrose crystals; and by a remelting and filtering process for handling of outsize crystals, the resulting liquid being returned to the third effect evaporator for reprocessing.

Additional Reading

Bourne, G.H.: "Nutritional Value of Cereal Products, Beans and Starches," S. Karger Publishers, Inc., Farmington, CT, 1989.
Galliard, T.: "Starch: Properties and Potential," John Wiley & Sons, Inc., New York, NY, 1987.
Preiss, J., M.N. Sival, and S.L. Taylor: "Starch: Basic Science to Biotechnology," Vol. 41, Academic Press, Inc., San Diego, CA, 1998.
Schenck, F.W., R.E. Hebeda: "Starch Hydrolysis Products: Worldwide Technology, Production, and Applications," John Wiley & Sons, Inc., New York, NY, 1992.
Whistler, R.L.: "Starch," 3rd Edition, Academic Press, Inc., San Diego, CA, 2001.

STAR CONNECTION. The connection of the various phases of an A.C. machine or circuit in which one end of all phases is connected to a common point, the other end of each phase going to a line. The *Y* is the three-phase star and is the most common example of this type connection.

STARDUST MISSION. NASA's *Stardust* mission launched on February 7, 1999, from Cape Canaveral Florida, will send a spacecraft flying through the cloud of dust that surrounds the nucleus of a comet, and for the first time ever, bring cometary material back to Earth.

Comets, which periodically grace our sky like celestial bottle rockets, are thought to hold many of the original ingredients of the recipe that created the planets and brought plentiful water to Earth. They are also rich in organic material, which provided our planet with many of the ready-to-mix molecules that could give rise to life. They may be the oldest, most primitive bodies in the solar system, a preserved record of the original nebula that formed the Sun and the planets.

"Scientists have long sought a sample directly from a known comet because of the unique chemical and physical information these bodies contain about the earliest history of the solar system," said Dr. Edward Weiler, NASA's associate administrator for space science. "Locked within comet molecules and atoms could be the record of the formation of the planets and the materials from which they were made."

Stardust is the first U.S. space mission dedicated solely to the exploration of a comet, and will be the first to return extraterrestrial material from outside the orbit of the Moon. The primary goal of *Stardust* is to collect dust and carbon-based samples from a well-preserved comet called *Wild 2* -pronounced "Vilt 2" after the name of its Swiss discoverer.

The spacecraft will also collect interstellar dust from a recently discovered flow of particles that passes through our solar system from interstellar space. As in the proverbial "from dust to dust," this interstellar dust represents the ultimate in recycled material; it is the stuff from which all solid objects in the universe are made, and the state to which everything eventually returns. Scientists want to discover the composition of this "stardust" to determine the history, chemistry, physics and mineralogy of nature's most fundamental building blocks.

Because it would be virtually impossible to equip a spacecraft with the most sophisticated lab instrumentation needed to analyze such material in space, the *Stardust* spacecraft is more of a robotic lab assistant whose job it is pick up and deliver a sample to scientists back on Earth. The spacecraft will, however, radio some on-the-spot analytical observations of the comet and interstellar dust.

The *Stardust* Mission is a collaborative effort between NASA, university and industry partners:

- The Principal Investigator is Dr. Donald E. Brownlee of the University of Washington, well known for his discovery of cosmic particles in the stratosphere known as Brownlee Particles. He also co-authored the bestseller "Rare Earth: Why Complex Life Is Uncommon," which puts forward a hypothesis predicting that simple, microbial life will be widespread in the universe, while complex animal or plant life will be extremely rare.
- Dr. Peter Tsou of the Jet Propulsion Lab (JPL), innovator in aerogel technology serves as Deputy Investigator.
- The contractor for the *Stardust* spacecraft is Lockheed Martin Astronautics, Denver, CO.
- The Jet Propulsion Laboratory has an experienced project management team, led by Thomas C. Duxbury. In addition, JPL provided the optical navigation camera.
- The Max Planck Institute (MPI) of Germany provided the real-time dust composition analyzer for the spacecraft.
- Ames Research Center provided the heat shield.
- Johnson Space Center will provide the planetary materials curatorial facility where the samples can be preserved and tests conducted.
- University of Chicago provided the Navigation Camera.

NASA's Discovery Program

Stardust is a mission under NASA's Discovery Program, which sponsors low-cost solar system exploration projects with highly focused science goals. Created in 1992, the Discovery Program competitively selects proposals submitted by teams led by scientists called principal investigators and supported by organizations, which provide project management and build and fly the spacecraft. In recent years, NASA has identified several finalists form dozens of mission proposals submitted. These finalists receive funding to conduct feasibility studies for and additional period of time before a final selection is made. *Stardust* was competitively selected in the fall of 1995 under NASA's Discovery Program of low-cost, highly focused science missions. As the fourth Discovery mission chosen, *Stardust* has met a fast development schedule, which uses a small Delta launch vehicle, is cost-capped at less than $200 million, and is the product of a partnership involving NASA, academia and industry.

Other missions in the Discovery Program are:

- *Near Earth Asteroid Rendezvous* (NEAR) was launched in February 1996 and became the first spacecraft to orbit an asteroid when it reached Eros in February 2000. NEAR is managed by Johns Hopkins University's Applied Physics Laboratory.
- *Mars Pathfinder* was launched in December 1996 and reached Mars on July 4, 1997, demonstrating a unique way of landing with airbags to deliver a small robotic rover. The Jet Propulsion Laboratory, Pasadena, CA, managed the Mars Pathfinder. See also **Pathfinder Mission to Mars**.
- Launched in January 1998, *Lunar Prospector* entered orbit around Earth's Moon five days later, circling at an altitude of about 60 miles (100 kilometers). Principal investigator is Dr. Alan Binder of the Lunar Research Institute, Gilroy, CA, with project management by NASA's Ames Research Center. See also **Lunar Prospector Mission**.
- *Genesis* is a mission that will return samples of solar wind particles following launch in January 2001. Principal investigator is Dr. Donald Burnett of the California Institute of Technology, with project management by the Jet Propulsion Laboratory.

- *Comet Nucleus Tour (Contour)* will be launched in July 2002 to execute close flybys of three comets: Encke, Schwassmann-Wachmann-3 and d'Arrest. Principal investigator is Dr. Joseph Veverka of Cornell University, with project management by Johns Hopkins University's Applied Physics Laboratory.
- In November 1998, NASA selected five new Discovery proposals for feasibility studies before making a final selection of one or two missions in June 1999. The five are *Aladdin*, a mission to gather samples from Mars' moons Phobos and Deimos. *Deep Impact*, which would fire a projectile into Comet P/Tempel 1 to expose its pristine interior ice and rock. *Inside Jupiter*, an orbiter that would study the giant planet's interior, measuring its gravitational and magnetic fields. *Messenger*, an orbiter that would globally image and study Mercury; and *Vesper*, an orbiter that would study the atmosphere of Venus. In this round of selections, NASA also decided to fund part of an instrument to fly on the European Space Agency's *Mars Express* spacecraft in 2003 to study the interaction between the solar wind and Mars' atmosphere.

See also **Jet Propulsion Laboratory (JPL)**.

Why Stardust?

Far beyond the orbits of the planets on the outer fringes of the solar system, a vast swarm of perhaps a trillion dormant comets circles the Sun. Frozen balls of ice, rocks, and dust, they are the undercooked leftovers that remained after a sprawling cloud of gas and dust condensed to form the Sun and planets about 4.6 billion years ago. From time to time, the gravitational pull of a passing star will nudge some of them out of their orbits, plunging them into the inner solar system, where they erupt with glowing tails as they loop around the Sun.

Closer to home, a stream of interstellar dust flows continuously through the solar system. Each perhaps 1/50th the width of a human hair, these tiny particles are the pulverized flotsam of the galaxy, bits of ancient stars that exploded as they died. This "stardust" is literally the stuff of, which life on earth all made. It is the source of nearly all of the elements on Earth heavier than oxygen.

These two niches bearing clues of the dawn of the solar system are the target for NASA's *Stardust* mission. The spacecraft will use a collector mechanism that employs a unique substance called aerogel to snag comet particles as well as interstellar dust flowing through the solar system, returning them to Earth for detailed study in laboratories.

Data returned from the *Stardust* spacecraft and the precious samples it returns to Earth will provide opportunities for significant breakthroughs in areas of key interest to astrophysics, planetary science and astrobiology. The samples will provide scientists with direct information on the solid particles that permeate our galaxy.

Stardust's cometary dust and interstellar dust samples will help provide answers to fundamental questions about the origin of solar systems, planets and life: How and when did the elements that led to life enter the solar system? How were these materials transformed within the solar system by forces such as heating and exposure to ultraviolet light? How were they distributed among planetary bodies, and in what molecular and mineral-based forms? These questions are of major importance for astrobiology and the search for life-generating processes and environments elsewhere in the universe. See also **Astrobiology**.

Comets

Though frequently beautiful, comets traditionally have stricken terror as often as they have generated excitement as they wheel across the sky during their passage around the Sun. Astrologers interpreted the sudden appearances of the glowing visitors as ill omens presaging famine, flood or the death of kings. Even as recently as the 1910 appearance of Halley's Comet, entrepreneurs did a brisk business selling gas masks to people who feared Earth's passage through the comet's tail.

In the 4th century B.C., the Greek philosopher Aristotle concluded that comets were some kind of emission from Earth that rose into the sky. The heavens, he maintained, were perfect and orderly; a phenomenon as unexpected and erratic as a comet surely could not be part of the celestial vault. In 1577, Danish astronomer Tycho Brahe carefully examined the positions of a comet and the Moon against the stars during the evening and predawn morning. Due to parallax, a close object will appear to change its position against the stars more than a distant object will, similar to holding up a finger and looking at it while closing one eye and then the other. The Moon appeared to move more against the stars from evening to morning

than the comet did, leading Tycho to conclude that the comet was at least four times farther away.

A hundred years later, the English physicist, Isaac Newton, established that a comet appearing in 1680 followed a nearly parabolic orbit. The English astronomer Edmund Halley used Newton's method to study the orbits of two dozen documented cometary visits. Three comet passages in 1531, 1607, and 1682 were so similar that he concluded they in fact were appearances of a single comet wheeling around the Sun in a closed ellipse every 75 years. He successfully predicted another visit in 1758–1759, and the comet thereafter bore his name.

Since then, astronomers have concluded that some comets return relatively frequently, in intervals ranging from 3 to 200 years; these are the so-called "short-period" comets. Others have enormous orbits that bring them back only once in many centuries.

In the mid-1800s, scientists also began to turn their attention to the question of comets' composition. Astronomers noted that several major meteor showers took place when Earth passed through the known orbits of comets, leading them to conclude that the objects are clumps of dust or sand. By the early 20th century, scientists studied comets using the technique of spectroscopy, breaking down the color spectrum of light given off by an object to reveal the chemical makeup of the object. They concluded that comets also emitted gases as well as molecular ions.

In 1950, the American astronomer, Fred L. Whipple, authored a major paper proposing the "dirty snowball" model of the cometary nucleus. This model, which has since been widely adopted, pictures the nucleus as a mixture of dark organic material, rocky grains and water ice. ("Organic" means that the compound is carbon-based, but not necessarily biological in origin.) Most comets range in size from about 2 to 7 kilometers (1 to 5 miles) in diameter. Shields that protect the *Stardust* spacecraft from dust impacts were named for Whipple in honor of his role in cometary science.

If comets contain icy material, they must originate somewhere much colder than the relatively warm inner solar system. Also in 1950, the Dutch astronomer Jan Hendrick Oort (1900–1992) used indirect reasoning from observations to establish the existence of a vast cloud of comets orbiting many billions of miles from the Sun, perhaps 50,000 astronomical units (AU) away (one AU is the distance from Earth to the Sun), or nearly halfway to the nearest star. This region has since become known as the Oort Cloud.

A year later, the Dutch-born American astronomer, Gerard Kuiper (1905–1973), made the point that the Oort Cloud is too distant to act as the nursery for short-period comets. He suggested the existence of a belt of dormant comets lying just outside the orbits of the planets at perhaps 30 to 100 AU from the Sun; this has become known as the Kuiper Belt. Jupiter's gravity periodically influences one of these bodies to take up a new orbit around the Sun. The Oort Cloud, by contrast, would be the home of long-period comets. They are periodically nudged from their orbits by any one of several influences, perhaps the gravitational pull of a passing star or giant molecular cloud, or tidal forces of the Milky Way Galaxy.

In addition to the length of time between their visits, another feature distinguishes short- and long-period comets. The orbits of short-period comets are all fairly close to the ecliptic, the plane in which Earth and most other planets orbit the sun. Long-period comets, by contrast, dive inwards toward the Sun from virtually any part of the sky. This suggests that the Kuiper Belt is a relatively flat belt, whereas the Oort Cloud is a three-dimensional sphere surrounding the solar system.

Where did the Oort Cloud and Kuiper Belt come from? Most astronomers now believe that the material that became comets condensed in the outer solar system around the orbits of Uranus and Neptune. Gravitational effects from those giant planets flung the comets outward.

Residing at the farthest reaches of the Sun's influence, comets did not undergo the same heating as the rest of the objects in the solar system, so they retain, largely unchanged, the original composition of solar system materials. As the preserved building blocks of the outer solar system, comets offer clues to the chemical mixture from which the planets formed some 4.6 billion years ago.

The geologic record of the planets shows that, about 3.9 billion years ago, a period of heavy comet and asteroid bombardment tapered off. The earliest evidence of life on Earth dates from just after the end of this heavy bombardment. The constant barrage of debris had vaporized Earth's oceans, leaving the planet too hot for the survival of fragile carbon-based molecules upon which life is based. Scientists therefore wonder: How could life form so quickly when there was so little liquid water or carbon-based molecules on Earth's surface? The answer may be that comets, which are abundant in both water and carbon-based molecules, delivered essential ingredients for life to begin.

Comets are also at least partially responsible for the replenishment of Earth's ocean after the vaporization of an early ocean during the late heavy bombardment. While Earth has long been regarded as the "water planet," it and the other terrestrial planets (Mercury, Venus and Mars) are actually poor in the percentage of water and in carbon-based molecules they contain when compared to objects that reside in the outer solar system at Jupiter's orbit or beyond. Comets are about 50 percent water by weight and about 10 to 20 percent carbon by weight. It has long been suspected that what little carbon and water there is on Earth was delivered here by objects such as comets that came from a more water-rich part of the solar system.

While comets are a likely source for life's building blocks, they have also played a devastating role in altering life on our planet. A comet or asteroid is credited as the likely source of the impact that changed Earth's climate, wiped out the dinosaurs and gave rise to the age of mammals 65 million years ago. A catastrophic collision between a comet or asteroid and Earth is estimated to happen at intervals of several tens of millions of years. See also **Comet**.

Other Comet Missions

Several other spacecraft have studied comets, not all of which were originally designed for that purpose. Several new missions to comets are being developed for launch in coming years.

Past comet missions include:

- In 1985, NASA modified the orbit of the International Sun-Earth Explorer (ISEE) spacecraft to execute a flyby of Comet Giacobini-Zinner. The spacecraft was renamed *International Comet Explorer* (ICE).
- An international armada of robotic spacecraft flew out to greet Halley's Comet during its return in 1986. The fleet included the European Space Agency's *Giotto*, the Soviet Union's *Vega 1* and *Vega 2*, and Japan's *Sakigake* and *Suisei* spacecraft.
- Comet Shoemaker-Levy 9's spectacular collision with Jupiter in 1994 was observed by NASA's Hubble Space Telescope, the Jupiter-bound *Galileo* spacecraft and the Sun-orbiting *Ulysses* spacecraft. See also **Hubble Space Telescope (HST)**; **Galileo Mission to Jupiter**; and **Ulysses Mission**.

Future comet missions are:

- *Comet Nucleus Tour* (Contour) will take images and comparative spectral maps of three comets — Encke, Schwassmann-Wachmann-3 and d'Arrest — following launch in July 2002.
- *Deep Space 4/Champollion* will perform the first landing on a cometary nucleus and demonstrate technologies for collecting samples when it reaches Comet Tempel 1 following launch in 2003.
- A European Space Agency mission, *Rosetta* will be launched in 2003 to orbit Comet Wirtanen and deliver a surface science package to the comet's surface. NASA is providing science instruments for the comet orbiter.

Interstellar Dust

In 1990, NASA launched the *Ulysses* spacecraft on a flight path that would take it close to Jupiter, flinging it into an orbit around the Sun far above and below the ecliptic plane. While en route from Earth to Jupiter, the spacecraft's dust detector measured a constant flow of particles, each about a micron in size, or 1/50th the diameter of a human hair, entering the solar system from interstellar space. This observation was corroborated by the similar dust detector on the *Galileo* spacecraft, which reached Jupiter in 1995.

Scientists believe that in fact interstellar dust is ubiquitous in the space between the stars of the Milky Way Galaxy. The dust curtains huge areas of the sky; the broad, dark line across the length of the Milky Way that can be seen with the naked eye is a blanket of interstellar dust. As the Sun orbits the galactic center, it cuts through the dust like a ship passing through waves. From our perspective within the solar system, the dust seems to be flowing from approximately the direction that the Sun is moving toward, a point called the ihsolar apexla in the constellation Hercules. Outward pressure from the solar wind sweeps the inner solar system near Earth clean, but interstellar dust is more easily detected beyond the orbit of Mars.

Interstellar dust provided the building blocks for solid materials on Earth and other planets. In the cycle of star formation and death, light elements such as hydrogen coalesce to form a star, which in time develops enough mass to burst into an ongoing nuclear reaction. Many stars die in a spectacular explosion, converting light elements into heavier elements. The resulting interstellar particles contain a record of the processes at work in their parent stars as well as the environments they have passed through in the galaxy. This information is retained in particles at a scale smaller than a micron.

Interstellar dust forms by condensation in circumstellar regions around evolved stars of many different types, including red giants, carbon stars, novas and supernovas. The process gives rise to silicate grains when there is more oxygen than carbon in the star, and carbon-based grains when the carbon content exceeds that of oxygen. Pristine grains will retain the radioactive signatures of the environment they formed in.

In the past decade, scientists have gained new understanding about the formation and early evolution of the solar system and the role of interstellar dust and comets in that process. Studies of interstellar dust have been conducted with Earth and space-based telescopes; in addition, scientists have collected and studied dust in the stratosphere and in Earth orbit. Since the late 1960s, collections of ocean sediment have brought up microscopic glass and metallic spherules, space particles that melted during atmospheric entry. The sample of interstellar dust returned by *Stardust* will be compared with these and others to help define how dust evolves from its interstellar state to help create stars, planets and life in the universe.

Infrared observations have also provided new knowledge of star-formation and the role that dust plays in that process. Scientists have found many similarities between interstellar dust and cometary composition. The same gases, ice particles and silicates believed to be in comets also are found in interstellar clouds.

Even though the interstellar dust samples will be small and partly eroded, they will open a significant new window of information on galactic and nebular processes, materials and environments. Having actual samples in hand provides many unique advantages. Just as the return of lunar samples by the *Apollo* missions of the 1960s and 1970s revolutionized our understanding of the Moon, scientists expect that the *Stardust* mission's sample return will also have a pro- found impact on our knowledge of comets and stars.

Mission Overview

Stardust was launched (February 7, 1999) from Space Launch Complex 17A at Cape Canaveral Air Station, FL, on a variant of the Delta II launch vehicle known as a Delta 7426, one of the new series of launch vehicles procured under NASA's Med-Lite program.

Launch Vehicle: The first stage of the Delta II is augmented by four strap-on solid rocket motors. The solid rocket motors are designed to be jettisoned from the vehicle within 66 seconds after launch, after they have exhausted all of their solid propellant. Each of the four solid rocket motors is 1 meter (3.28 feet) in diameter and 13 meters (42.6 feet) long; each contains 11,765 kilograms (25,937 pounds) of hydroxyl-terminated polybutadiene (HTPB) propellant and provides an average thrust of 446,023 newtons (100,270 pounds) at sea level. The casings of the solid rocket motors are made of lightweight graphite epoxy.

The main body of the first stage is 2.4 meters (8 feet) in diameter and 26.1 meters (85.6 feet) long. It is powered by an RS-27A engine, which uses 96,160 kilograms (212,000 pounds) of RP-1 (rocket propellant 1, a highly refined kerosene) and liquid oxygen as its fuel and oxidizer.

The Delta II's second stage is 2.4 meters (8 feet) in diameter and 6 meters (19.7 feet) long, and is powered by an AJ10-118 K engine. The propellant is 5,900 kilograms (13,000 pounds) of Aerozine 50 (A-50), a mixture of hydrazine and unsymmetrical dimethyl hydrazine (UDMH), with nitrogen tetroxide as the oxidizer. This engine is restartable, and will perform two separate burns during *Stardust's* launch.

The launch vehicle's third and final stage is a Thiokol Star 37FM booster, measuring 1.7 meter (5.5 feet) long and 0.9 meter (3 feet) wide. Its motor carries 1,090 kilograms (2,400 pounds) of solid propellant, composed of a mixture of aluminum, ammonium perchlorate and hydroxyl-terminated polybutadiene (HTPB). The third stage includes a spin table supporting small rockets, which are used to spin up the third stage itself and the attached Stardust spacecraft. After *Stardust* separates from the third stage, a weight on a cable unwinds to tumble the third stage to avoid recontact. See Fig. 1.

Launch Timing: The timing of Stardust's launch is based upon the trajectory the spacecraft must fly to rendezvous with Comet Wild-2 in 2004.

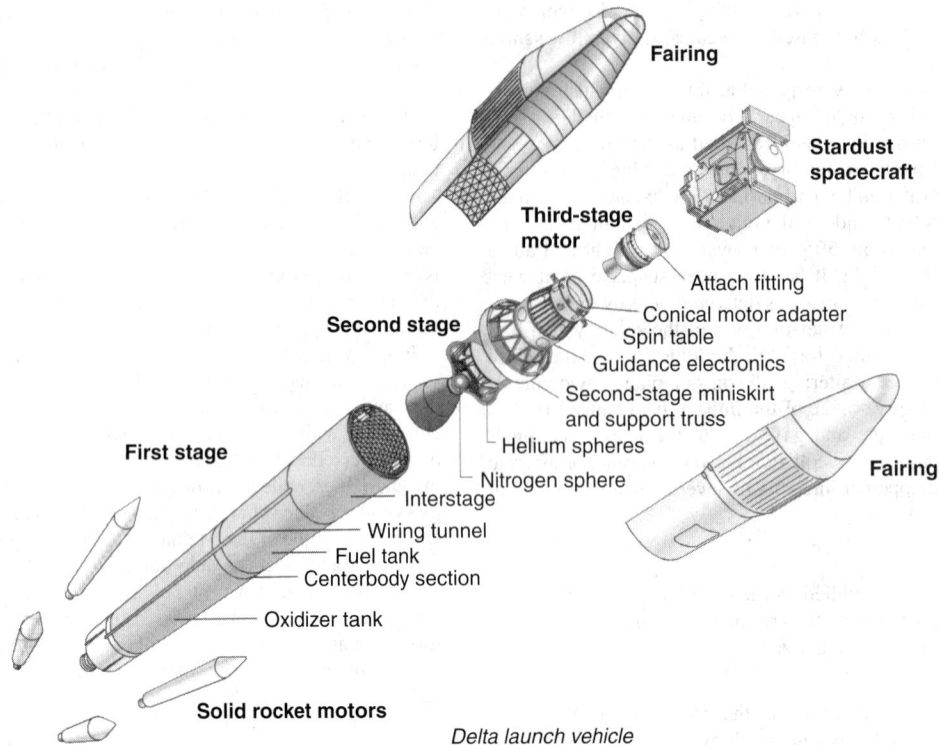

Fig. 1.   Delta launch vehicle. (*Jet Propulsion Laboratory, Pasadena, CA.*)

The launch period opened February 6 and would have continued through February 25, 1999. One near-instantaneous launch opportunity exists each day during the period. On February 6, the launch opportunity is at 4:07 p.m. Eastern Standard Time. The time of each daily opportunity varies (sometimes later or earlier) on later days of the launch period. *Stardust* was launched at 4:04 EST. February 7, 1999.

Launch Events: Launch occurred in three phases, consisting of liftoff and insertion into a 189-kilometer (102-mile) parking orbit; a coast of about a half hour until the vehicle position was properly aligned relative to the direction it must leave Earth; and final injection to an escape trajectory. The total time needed to complete the process was a little under an hour.

Sixty-six seconds after liftoff, the four solid rocket motors were discarded while the first stage continued to burn. About 4 minutes, 24 seconds after liftoff, the first stage stopped firing and was discarded eight seconds later. About five seconds later, the second stage engine ignited. The fairing or nose-cone enclosure of the launch vehicle was discarded 4 minutes, 42 seconds after liftoff. The first burn of the rocket's second stage ended about 11 minutes, 22 seconds after liftoff.

About 21 minutes after launch, the second stage was restarted. At about 24 minutes into the flight, the third stage separated and burned for about two minutes, after which the *Stardust* spacecraft separated from the third stage and began its first orbit of the Sun.

Immediately after separation from the Delta's third stage, *Stardust* stopped its own spinning by firing its thrusters. About 4 minutes after separation, the spacecraft's solar array unfolded and was pointed toward the Sun. Shortly thereafter, the 34-meter-diameter (112-foot) antenna at the Deep Space Network complex in Canberra, Australia acquired *Stardust's* signal. See Figs. 2 and 3.

Cruise: *Stardust's* first two years of flight included one orbital loop of the Sun. In March of 2000, when *Stardust* was between the orbits of Mars and Jupiter, the greatest distance that the spacecraft will be from the Sun, *Stardust's* onboard engines fired to put the spacecraft on course for a later gravity assist swingby of Earth.

As *Stardust* traveled back inward toward the Sun on the latter part of its first orbit, the spacecraft passed through a region where interstellar particles flow through the solar system. From March through May 2000, the spacecraft opened its collector mechanism and captured such interstellar particles. The "B side" of the collector mechanism was used, reserving the "A side" for the spacecraft's later dust collection mission at Comet Wild-2.

At the conclusion of its first orbit of the Sun, the spacecraft flew past Earth on January 15, 2001 at an altitude of 5,964 kilometers (3,706 miles). This increased the size of *Stardust's* flight path to a 2-1/2-year loop of the Sun. The spacecraft will stay this course through two more orbits.

As the spacecraft travels back inward toward the Sun on the latter part of its second orbit, it again will collect interstellar particles flowing through the solar system from July through December 2002. See Fig. 4.

Comet Flyby: One hundred and sixty days before encountering Wild-2 or about July 25, 2003 the spacecraft is scheduled to fire its thrusters to fine-tune its flight path through the comet's coma, based on updated information on the position of the comet's nucleus provided by the navigation camera. The encounter of Wild-2 will occur on January 2, 2004, when the comet is 1.86 astronomical units from the Sun (almost twice Earth's distance from the Sun) and 97 days after the comet has rounded the Sun.

At its closest approach, *Stardust* will be traveling at a speed of 6.1 kilometers per second (about 13,650 miles per hour) relative to the comet. The velocity has been chosen to optimize the capture of particles by the aerogel collector; at this speed, it can "soft-catch" the comet samples without changing them greatly. The passage through the most intensive rain of debris within the coma will last about eight minutes.

Wild-2 will be far from its peak period of activity and be relatively safe for a close flyby. The spacecraft will approach Wild-2 from above the comet's orbital plane, then dip slightly below it. In effect, the comet will "run over" the spacecraft. Approaching from the sunlit side and northern portion of the comet, the spacecraft's flight path through the coma will take *Stardust* within about 150 kilometers (100 miles) of the comet's nucleus. This "miss distance" was selected to balance between the need to protect the spacecraft and the objective of sampling the freshest possible material off the comet's nucleus.

Stardust's navigation camera, which is fixed to the spacecraft body, will take images of the comet nucleus. The camera will be protected from direct hits because it faces away from the direction of the particle onslaught. It will record images of the comet through the reflection in a movable mirror. The mirror will provide image-motion compensation, that is, it will move to keep the reflection of the comet in the camera's field-of-view and minimize image smearing during the flyby.

Wild-2 is an ideal target in part because it has only recently been deflected from a distant orbit into its current orbit, which brings it into the inner solar system. Its drastic orbit change resulted from a very close approach that the comet made to Jupiter in September 1974. Before that, the comet was in a much longer orbit and had made fewer passages of

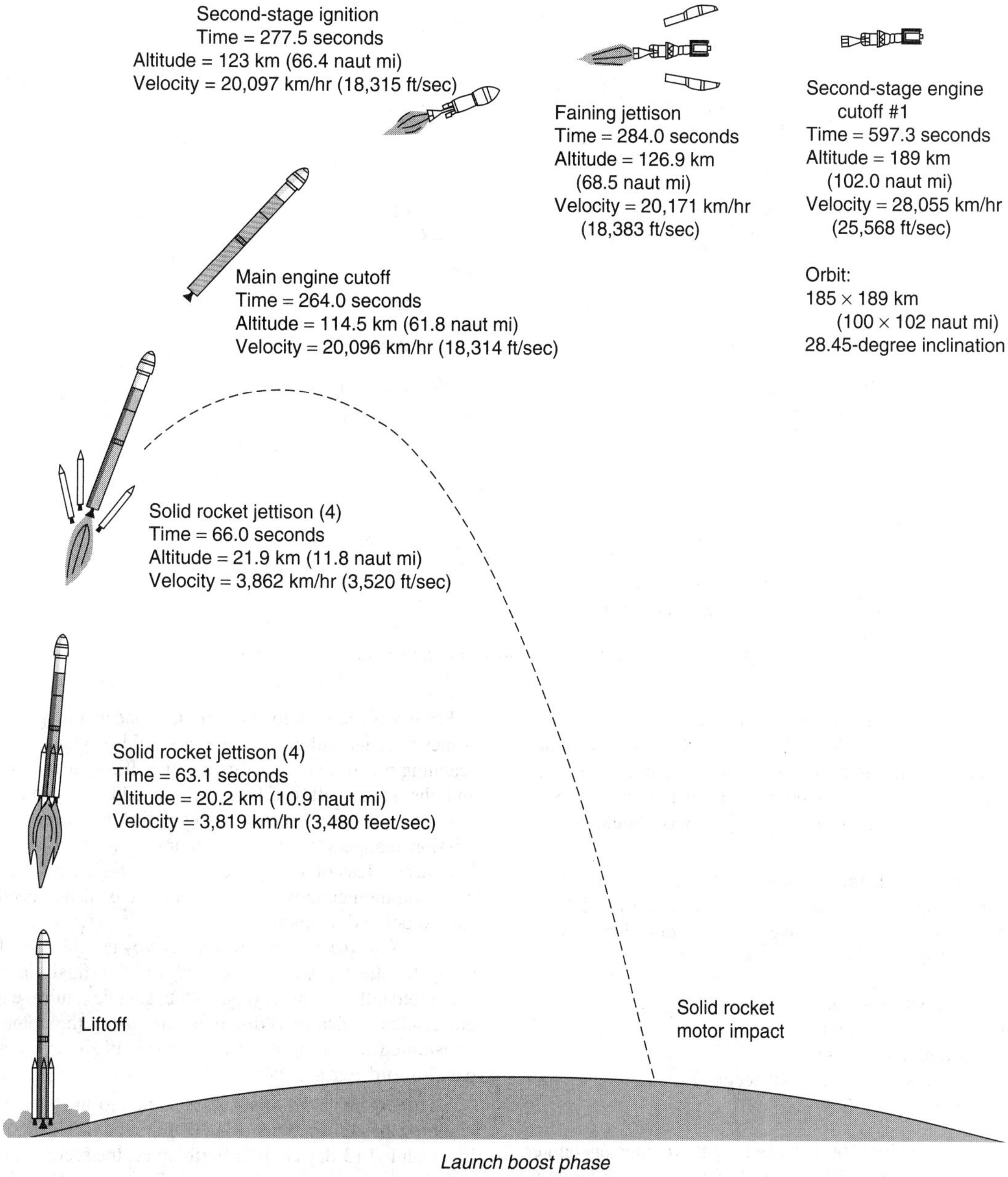

Second-stage ignition
Time = 277.5 seconds
Altitude = 123 km (66.4 naut mi)
Velocity = 20,097 km/hr (18,315 ft/sec)

Faining jettison
Time = 284.0 seconds
Altitude = 126.9 km
(68.5 naut mi)
Velocity = 20,171 km/hr
(18,383 ft/sec)

Second-stage engine
cutoff #1
Time = 597.3 seconds
Altitude = 189 km
(102.0 naut mi)
Velocity = 28,055 km/hr
(25,568 ft/sec)

Orbit:
185 × 189 km
(100 × 102 naut mi)
28.45-degree inclination

Main engine cutoff
Time = 264.0 seconds
Altitude = 114.5 km (61.8 naut mi)
Velocity = 20,096 km/hr (18,314 ft/sec)

Solid rocket jettison (4)
Time = 66.0 seconds
Altitude = 21.9 km (11.8 naut mi)
Velocity = 3,862 km/hr (3,520 ft/sec)

Solid rocket jettison (4)
Time = 63.1 seconds
Altitude = 20.2 km (10.9 naut mi)
Velocity = 3,819 km/hr (3,480 feet/sec)

Liftoff

Solid rocket
motor impact

Launch boost phase

Fig. 2.   Launch boost phase. (*Jet Propulsion Laboratory, Pasadena, CA.*)

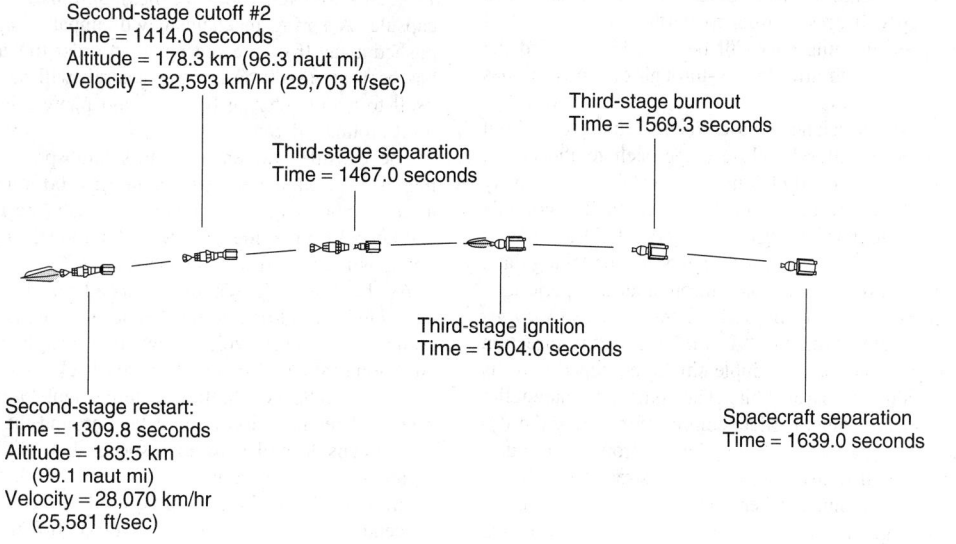

Second-stage cutoff #2
Time = 1414.0 seconds
Altitude = 178.3 km (96.3 naut mi)
Velocity = 32,593 km/hr (29,703 ft/sec)

Third-stage burnout
Time = 1569.3 seconds

Third-stage separation
Time = 1467.0 seconds

Third-stage ignition
Time = 1504.0 seconds

Second-stage restart:
Time = 1309.8 seconds
Altitude = 183.5 km
(99.1 naut mi)
Velocity = 28,070 km/hr
(25,581 ft/sec)

Spacecraft separation
Time = 1639.0 seconds

Fig. 3.   Launch injection phase. (*Jet Propulsion Laboratory, Pasadena, CA.*)

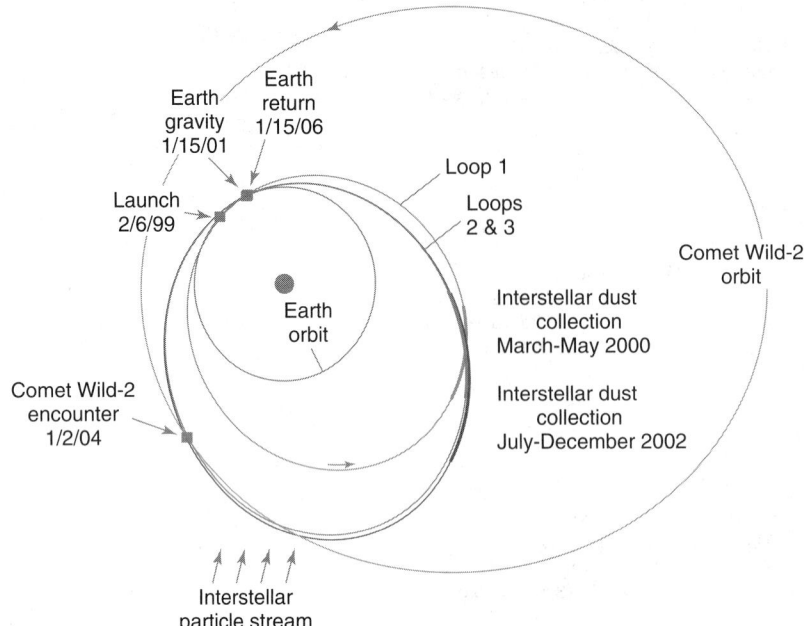

Fig. 4.   Mission trajectory. (*Jet Propulsion Laboratory, Pasadena, CA.*)

the Sun, so it is more pristine than most short-period comets. Since Comet Wild-2's orbit change in 1974, it has looped in toward the Sun three times. Comet scientists anticipate that 1 to 5 percent of Wild-2's nucleus surface could be active with gas and dust jets erupting from the surface. With a relatively slow flyby, the existence and activity of jets should be well observed.

Encounter Timeline: For planning purposes, the mission team has defined an encounter period spanning from 100 days before to 150 days after the comet flyby. This is divided into five time segments during which various mission activities are planned:

- **Far Encounter** −100 days to −1 day
- **Near Encounter** −1 day to −5 hours
- **Close Encounter** −5 hours to +5 hours
- **Closest Encounter** −240 seconds to +240 seconds
- **Post Encounter** −0 days to +150 days

During the segment beginning 100 days before the comet encounter, the navigation camera will image Wild-2 weekly to assist in targeting the spacecraft. *Stardust* will repeatedly image the comet through the camera's eight different filters, acquiring as many images of the comet as can be practically returned until 12 hours before the closest approach to the nucleus. One closeup will be sent back to Earth near the point of closest approach. Subsequent imaging data will be recorded onboard the spacecraft for transmission to Earth after the fly-through of the coma has been completed.

The mission's core period for science data collection runs from about five hours before to five hours after its closest approach to the comet nucleus. Five hours and about 100,000 kilometers (60,000 miles) away from the nucleus, the spacecraft will begin to enter the coma. The comet's nucleus should begin to emerge in the navigation camera's field-of-view as an extended dark body. All comet science experiments will be taking data, and the spacecraft will be tracked continuously throughout this period.

Just before the spacecraft's closest approach to the nucleus, *Stardust* will deploy its aerogel collector, with the "A side" facing the direction of incoming comet particles. The main Whipple shield, equipped with its dust flux monitor, will be counting particle hits. The comet and interstellar dust analyzer will be measuring comet particle composition during the fly-through. From −5 hours through −4 minutes, and again from +4 minutes to +5 hours, the spacecraft will transmit a continuous stream of imaging and other observations. At −4 minutes, when the comet nucleus occupies 60 by 60 picture elements (pixels) in the camera's field-of-view, a final black-and-white picture of the nucleus will be transmitted directly to Earth. Any images taken from −4 minutes to +4 minutes will be stored onboard for later transmission.

From −3 minutes to +4 minutes, one major goal will be to keep the comet nucleus within the camera's field-of-view. To accomplish this, the scanning mirror can adjust its position from viewing forward to backward, and the spacecraft itself can tilt to add a second axis to the mirror's position. This is expected to result in significant spacecraft motion, so mission planners expect to temporarily lose radio contact with *Stardust*. Should this loss of signal occur, *Stardust's* medium-gain antenna will take over communications functions until the dish-shaped high-gain antenna can be pointed again at Earth.

Once *Stardust* has completed its voyage through the coma, Stardust's aerogel collector will be stowed for the final time, sealing its comet and interstellar samples inside the sample return capsule. Imaging and other science data recorded onboard during the coma fly-through will be transmitted to Earth, and *Stardust* will begin the final leg of its journey back toward Earth.

Earth Return: *Stardust* is scheduled to fire its thrusters three times as it approaches Earth to fine-tune its flight path. The first trajectory maneuver is scheduled 13 days before Earth entry, the second three days before entry and the final maneuver 3 hours before entry. Earth entry will take place on January 15, 2006.

Soon after the final trajectory maneuver at an altitude of 110,728 kilometers (68,805 miles), *Stardust* will release the sample return capsule. A spring mechanism will impart a spin to the capsule as it is pushed away from the spacecraft in order to stabilize it. After the capsule has been released, the main spacecraft will perform a maneuver to divert itself to avoid entering Earth's atmosphere. The spacecraft will remain in orbit around the Sun.

The capsule will enter Earth's atmosphere at a velocity of approximately 12.8 kilometers per second (28,600 miles per hour). The capsule's aerodynamic shape and center of gravity are designed like a badminton shuttlecock so that the capsule will automatically orient itself with its nose down as it enters the atmosphere.

As the capsule descends, its speed will be reduced by friction on its heat shield, a 60-degree half-angle blunt cone made of a graphite-epoxy composite covered with a new, light-weight thermal protection system. Additional ablative material on the back shell that is similar to material used on the Space Shuttle's external tank protects the capsule from the effects of recirculation flow of heat around the capsule.

The capsule will slow to a speed about 1.4 times the speed of sound at an altitude of about 30 kilometers (100,000 feet), at which time a small pyrotechnic charge will be fired, releasing a drogue parachute. After descending to about 3 kilometers (10,000 feet), a line holding the drogue chute will be cut, allowing the drogue to pull out a larger parachute that will carry the capsule to its soft landing. At touchdown, the capsule will be traveling at approximately 4.5 meters per second (14.8 feet per second), or about 16 kilometers per hour (10 miles per hour). In all, about 10 minutes will elapse between the beginning of the entry into Earth's atmosphere until the parachute is deployed.

The landing site at the Utah Test and Training Range near Salt Lake City was chosen because the area is a vast, desolate and unoccupied salt flat controlled by the U.S. Air Force with the Army. The landing footprint for the sample return capsule will be about 30 by 84 kilometers (18 by 52 miles), an ample space to allow for aerodynamic uncertainties and winds that might affect the direction the capsule travels in the atmosphere. To land within the footprint, the capsule's trajectory must achieve an entry accuracy of 0.08 degree. The sample return capsule will approach the landing zone on a heading of approximately 122 degrees on a northwest to southeast trajectory. Landing time will take place at 3 a.m. Mountain Standard Time on January 15, 2006.

The actual landing footprint will be predicted by tracking the spacecraft just before the capsule's release. Roughly six hours before entry, an updated footprint will be provided to the capsule recovery team.

Ground Recovery: A UHF radio beacon on the capsule will transmit a signal as the capsule descends to Earth, while the parachute and capsule will be tracked by radar. A helicopter will be used to fly the retrieval crew to the landing site. Given the small size and mass of the capsule, mission planners do not expect that its recovery and transportation will require extraordinary handling measures or hardware other than a specialized handling fixture to cradle the capsule during transport. The capsule will be transported to a staging area at the Utah Test and Training Range where the sample canister will be extracted. The sample canister then will be transported to its final destination, the planetary material curatorial facility at NASA's Johnson Space Center, Houston, TX.

Planetary Protection: The U.S. is a signatory to the United Nations' 1966 Treaty of Principles Governing the Activities of States in the Exploration and Use of Outer Space, Including the Moon and Other Celestial Bodies. Known as the "Outer Space Treaty," this document states in part that exploration of the Moon and other celestial bodies shall be conducted "so as to avoid their harmful contamination and also adverse changes in the environment of the Earth resulting from the introduction of extraterrestrial matter."

Comets are believed to be primordial bodies made up of material that is virtually unchanged since their creation when the solar system formed 4.6 billion years ago. This means that any evolutionary processes leading to the emergence of life have not occurred. There is no scientific reason to believe that bacteria or viruses or any other life exist on comets. One of the objectives of the *Stardust* mission is to investigate whether the chemical building blocks of life exist on comets. But even if such building blocks do reside there, comets have not provided the hospitable environment required over millions of years to accommodate the complex processes that could result in the emergence of even single-celled organic life.

On *Stardust*, all comet particles that are collected will be heated to extremely high temperature due to their impact speed on the aerogel collector. The temperature caused by the compression interaction between aerogel and any given particle is calculated to be at least 10,000 °C (more than 18,000 °F). In fact, the collector material literally melts to encapsulate the captured particles. Such high temperatures are naturally sterilizing. As a particle hits the aerogel sample collector, it will come to a dead stop within a microsecond, having traveled about 3 centimeters (1.2 inches) into the aerogel. By that point, the aerogel, which is silica-based, will have melted around the particle, trapping it in glass.

It should be noted that particles from space, including material from comets, fall onto Earth's surface at a rate of approximately 40,000 tons per year, and some of this material is believed to survive atmospheric entry without severe heating.

Outreach: The *Stardust* project has forged partnerships with several educational enterprises to increase public awareness of the mission's goals and strategy, and to broaden the distribution of new knowledge the mission will produce:

- **Challenger Center**, based in Alexandria, VA, has partnered with the project to manage the Stardust Educator Fellowship program. Twenty-five educators selected from across the country were trained in the science and engineering behind the mission. They also received a workshop package and educational materials, and will conduct their own teacher training on *Stardust* throughout the mission. More information about Challenger Center is available at *http://www.challenger.org/*

- The **Jason Project**, founded by oceanographer and underwater explorer Dr. Robert Ballard after he found the wreckage of the Titanic on the sea floor, has created the "Jason IX Curriculum" in partnership with the *Stardust* project. The curriculum focuses on a study of life on Earth and in its oceans. The tie that links *Stardust* to oceanography is the presence of particles from space that lie on the ocean floor. Students will work with representatives of the *Stardust* mission to help find answers to the three key questions of the Jason Project: What are Earth's physical systems? How do these systems affect life on Earth? What technologies do we use to study these systems, and why? More information about The Jason Project is available at *http://www.jasonproject.org*

- The **Kirkpatrick Science and Air Space Museum**, also known as **Omniplex**, is a hands-on science center located in Oklahoma City, OK. In partnership with the *Stardust* project, Omniplex has produced a planetarium show and a traveling "Stardust Café" exhibit dedicated to the science topics *Stardust* will investigate. More information about Omniplex is available at *http://www.omniplex.org*

The Stardust project sponsored a "Send Your Name to a Comet" campaign that invited people from around the world to submit their names via the Internet to fly onboard the *Stardust* spacecraft. Two microchips bearing the names of more than 1.136 million people are onboard the *Stardust* spacecraft.

The names were electronically etched onto fingernail-size silicon chips at JPL's Microdevices Lab. Writing on the microchips is so small that about 80 letters would equal the width of a human hair. The names can be read only with the aid of an electron microscope.

The first *Stardust* microchip contains 136,000 names collected from October through November 1997 from persons all over the world. That microchip has been placed inside the sample return capsule. The second microchip contains more than a million names from members of the public, and was placed on the back of the arm that holds the dust collector. In addition to holding names from the public-at-large, the second microchip contains all 58,214 names inscribed on the Vietnam Veterans Memorial in Washington, DC, as a tribute to those who died in that war.

Spacecraft Profile

The *Stardust* spacecraft incorporates innovative, state-of-the-art technologies pioneered by other recent missions with off-the-shelf spacecraft components and, in some cases, spare parts and instrumentation left over from previous missions.

The *Stardust* spacecraft is derived from a rectangular deep-space bus called SpaceProbe developed by Lockheed Martin Astronautics, Denver, CO. The total weight of the spacecraft, including the sample return capsule and propellant carried onboard for trajectory adjustments, is 385 kilograms (848 pounds). The main bus is 1.7 meters (5.6 feet) high, 0.66 meter (2.16 feet) wide and 0.66 meter (2.16 feet) deep, about the size of an average office desk. Panels are made of a core of aluminum honeycomb, with outer layers of graphite fibers and polycyanate face sheets. When its two parallel solar panels are deployed in space, the spacecraft takes on the shape of a letter H. See Fig. 5.

There are three dedicated science packages on *Stardust*; the two-sided dust collector, the comet and interstellar dust analyzer, and the dust flux monitor. Science data are also obtained without dedicated hardware. The navigation camera, for example, will provide images of the comet both for targeting accuracy and scientific analysis.

Aerogel Dust Collectors

To collect particles without damaging them, Stardust will use an extraordinary substance called *aerogel*, a silicon-based solid with a porous, sponge-like structure in which 99 percent of the volume is empty space. Originally invented in 1933 by a researcher at the College of the Pacific in Northern California, aerogel is made from fine silica mixed with a solvent. The mixture is set in molds of the desired shape and thickness, and then pressure-cooked at high temperature. Over the past several years, aerogel has been made and flight-qualified at the Jet Propulsion Laboratory for space missions. See Fig. 6.

A cube of aerogel looks like solid, pale-blue smoke. It is the lightest-weight, lowest-mass solid known, and has been found to be ideal for capturing tiny particles in space. There is extensive experience, both in laboratory and space flight experiments, in using aerogel to collect hypervelocity particles. Eight Space Shuttle flights have been equipped with aerogel collectors.

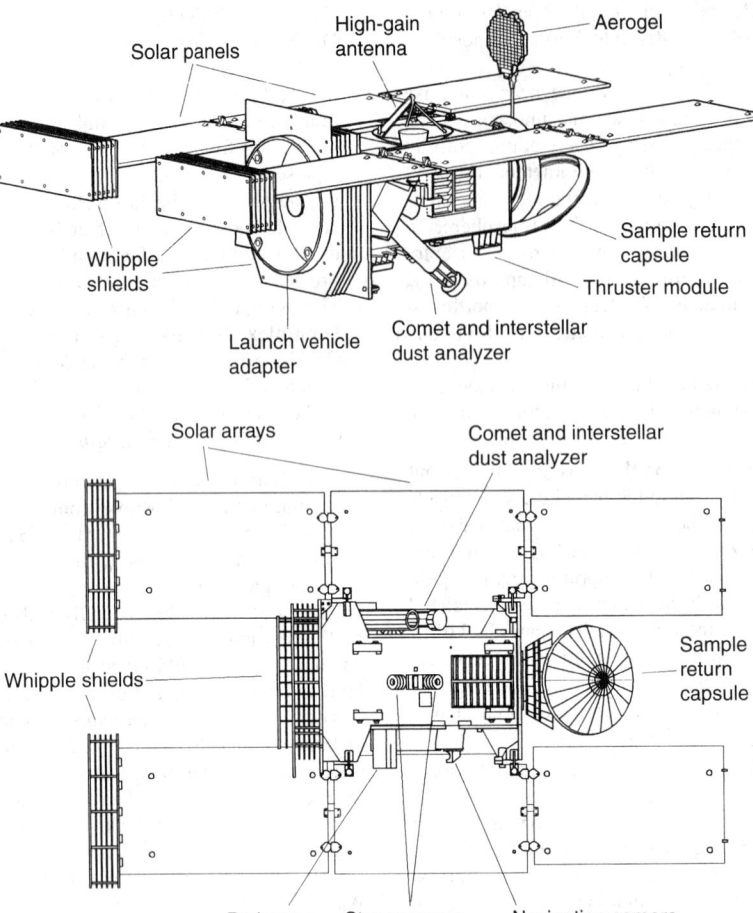

Fig. 5.   *Stardust* spacecraft. (*Jet Propulsion Laboratory, Pasadena, CA.*)

Fig. 6.   This photograph illustrates the excellent insulating properties of aerogel. The crayons on top of the aerogel are protected from the flame underneath, and are not melting. (*Jet Propulsion Laboratory, Pasadena, CA.*)

The exotic material has many unusual properties, such as uniquely low thermal and sound conductivity, in addition to its exceptional ability to capture hypervelocity dust. Aerogel was also used as a lightweight thermal insulator on Mars *Pathfinder's Sojourner* rover.

When *Stardust* flies through the comet's coma, the impact velocity of particles as they are captured will be up to six times the speed of a bullet fired from a high-powered rifle. The Whipple shields can protect the spacecraft from impacts of particles the size of a pea, but larger particles present a more severe hazard.

Although the particles captured in aerogel will each be smaller than a grain of sand, high-speed capture in most substances would alter their shape and chemical composition, or vaporize them entirely. With aerogel, however, particles are softly caught in the material and slowed to a stop. When a particle hits the aerogel, it will bury itself, creating a carrot-shaped track in the aerogel up to 200 times its own length as it slows down and comes to a stop. The aerogel made for the *Stardust* mission has extraordinary, water-like clarity that will allow scientists to locate a particle at the end of each track etched in the substance. Each narrow, hollow cone leading to a particle will easily be seen in the aerogel with a stereo microscope.

The sizes of the particles collected in the aerogel are expected to range mostly from about a micron (a millionth of a meter, or 1/25,000th of an inch, or about 1/50th of the width of a human hair) to 100 microns (a tenth of a millimeter, or 1/250th of an inch, or about twice the width of a human hair). *Stardust* scientists anticipate that the aerogel will return a few particles at the upper end of this size range, and a million particles in the submicron range.

Most of the scientific analysis will be devoted to particles that are 15 microns (about 1/1,700th of an inch, or about one-third the width of a human hair) in size. The Stardust science team expects that the samples returned will be profoundly complex, and each particle will be probed for years in research labs.

One side of the dust collection module, called the "A side," will be used for the comet encounter, while the opposite side ("B side") will

Fig. 7.   Dust collector with aerogel. (*Jet Propulsion Laboratory, Pasadena, CA.*)

be used for interstellar collection. More than 1,000 square centimeters (160 square inches) of collection area is provided on each side. Stardust's cometary and interstellar dust collectors each have 130 rectangular blocks of aerogel measuring 2 by 4 centimeters (0.8 by 1.6 inches), plus two slightly smaller rhomboidal blocks. The thickness of the aerogel on the cometary particle collection side is 3 centimeters (1.2 inches), while the thickness of the aerogel on the interstellar dust particle collection side is 1 centimeter (0.4 inch). The density of the aerogel is graded—less dense at the point of particle entry, and progressively denser deeper in the material. Each block of aerogel is held in a frame with thin aluminum sheeting. See Fig. 7.

Overall, the collection unit resembles a metal ice tray set in an oversize tennis racket. It is similar to previous systems used to collect particles in Earth orbit on SpaceHab and other Space Shuttle-borne experiments. The sample return capsule is a little less than a meter (or yard) in diameter, and opens like a clamshell to extend the dust collector into the dust stream. After collecting samples, the cell assembly will fold down for stowage into the sample return capsule.

Sample Return Capsule

The sample return capsule is a blunt-nosed cone with a diameter of 81 centimeters (32 inches). It has five major components: a heat shield, back shell, sample canister, parachute system and avionics. The total mass of the capsule, including the parachute system, is 45.7 kilograms (101 pounds).

A hinged clamshell mechanism opens and closes the capsule. The dust collector fits inside, extending on hinges to collect samples and retracting to fold down back inside the capsule. The capsule is encased in ablative materials to protect the samples stowed in its interior from the heat of reentry.

The heat shield is made of a graphite-epoxy composite covered with a thermal protection system. The thermal protection system is made of a phenolic-impregnated carbon ablator developed by NASA's Ames Research Center for use on high-speed reentry vehicles. The capsule's heat shield will remain attached to the capsule throughout descent and serves as a protective cover for the sample canister at touchdown.

The back-shell structure is also made of a graphite-epoxy composite covered with a thermal protection system that is made of a cork-based material called *SLA 561V*. This material was developed by Lockheed Martin for use on the *Viking* missions to Mars in the 1970s, and is currently used on the Space Shuttle's external tank. The back shell provides the attach points for the parachute system.

The sample canister is an aluminum enclosure that holds the aerogel and the mechanism used to deploy and stow the aerogel collector during the mission. The canister is mounted on an equipment deck suspended between the backshell and heat shield.

The parachute system incorporates a drogue and main parachute inside a single canister. As the capsule descends toward Earth, a gravity-switch sensor and timer will trigger a pyrotechnic gas cartridge that will pressurize a mortar tube and expel the drogue chute. The drogue chute will be deployed at an altitude of approximately 30 kilometers (100,000 feet) at a speed of about mach 1.4 to provide stability to the capsule. Based on information from timer and backup pressure transducers, a small pyrotechnic device will cut the drogue chute from the capsule at an altitude of approximately 3 kilometers (10,000 feet). As the drogue chute moves away, it will extract the 8.2-meter-diameter (27-foot) main chute from the canister. Upon touchdown, cutters will fire to cut the main chute cables so that winds do not drag the capsule across the terrain. See Fig. 8.

The capsule carries an ultra-high-frequency (UHF) radio locator beacon to be used in conjunction with locator equipment on the recovery helicopters. The beacon will be turned on at main parachute deployment and will remain on until turned off by recovery personnel. The beacon is powered by redundant sets of lithium sulfur dioxide batteries, which have long shelf life and tolerance to wide temperature extremes, and are safe to handle. The capsule carries sufficient battery capacity to operate the UHF beacon for at least 40 hours. See Fig. 9.

Comet and Interstellar Dust Analyzer

The comet and interstellar dust analyzer is derived from the design of an instrument that flew on the European Space Agency's *Giotto* spacecraft and the Soviet Union's *Vega* spacecraft when they encountered Comet Halley in 1986. The instrument obtained unique data on the chemical composition of individual particulates in Halley's coma. *Stardust's* version of the instrument will study the chemical composition of particulates in the coma of Comet Wild-2.

The purpose of the analyzer instrument is to intercept and perform instantaneous compositional analysis of dust as it is encountered by the spacecraft. Data will be transmitted to Earth as soon as a communication link is available.

The instrument is a mass spectrometer, which separates the masses of ions by comparing differences in their flight times. When a dust particle hits the instrument's target, the impact creates ions, which are extracted from the particle by an electrostatic grid. Depending on the polarity of the

Fig. 8. Sample return capsule parachuting down to earth. (*Jet Propulsion Laboratory, Pasadena, CA.*)

Fig. 9. Sample return capsule landing. (*Jet Propulsion Laboratory, Pasadena, CA.*)

target, positive or negative ions can be extracted. As extracted ions move through the instrument, they are reflected and then detected. Heavier ions take more time to travel through the instrument than lighter ones, so the flight times of the ions are then used to calculate their masses. From this information, the ion's chemical identification can be made. See Fig. 10.

In all, the instrument consists of a particle inlet, a target, an ion extractor, a mass spectrometer and an ion detector.

Co-investigator in charge of the comet and interstellar dust analyzer is Dr. Jochen Kissel of the Max-Planck-Institut für Extraterrestrische Physik, Garching, Germany. The instrument was developed and fabricated by von Hoerner & Sulger GmbH, Schwetzingen, Germany, under contract to the German Space Agency and the Max-Plank-Institut. Software for the instrument was developed by the Finnish Meteorological Institute, Helsinki, Finland, under subcontract to von Hoerner & Sulger.

Dust Flux Monitor

The dust flux monitor measures the size and frequency of dust particles in the comet's coma. The instrument consists of two film sensors and two vibration sensors. The film material responds to particle impacts by generating a small electrical signal when penetrated by dust particles. The mass of the particle is determined by measuring the size of the electrical signals. The number of particles is determined by counting the number of signals. By using two film sensors with different diameters and thicknesses, the instrument will provide data on what particle sizes were encountered and what the size distribution of the particles is.

The two vibration sensors are designed to provide similar data for larger particles, and are installed on the Whipple shield that protect the spacecraft's main bus. These sensors will detect the impact of large comet dust particles that penetrate the outer layers of the shield. This system, essentially a particle impact counter, will give mission engineers information about the potential dust hazard when the spacecraft flies through the coma environment. Co-investigator in charge of the dust flux monitor is Dr. Anthony Tuzzolino of the University of Chicago, where the monitor was developed and constructed.

Navigation Camera

The navigation camera is capable of acquiring images with a resolution of 6 meters (21.6 feet) per pixel at a distance of 100 kilometers (62 miles).

The camera is a combination of adapted spare components left over from previous space missions. The main camera is a spare wide-angle unit left over from the two *Voyager* spacecraft missions launched to the outer planets in 1977. The camera uses a single *Voyager* eight-position filter wheel, thermal housing, and spare optics and mechanisms. Designers added a thermal radiator to the *Voyager* camera. Combined with it is a modernized sensor head left over from the *Galileo* mission to Jupiter launched in 1989. The sensor head uses the existing *Galileo* design updated with a 1024-by-1024-pixel array charge-coupled device (CCD) from the *Cassini* mission to Saturn, but has been modified to use new miniature electronics.

The eight-position filter wheel is fitted with three gas filters, three dust filters, a high-resolution filter and a clear filter. The gas filters will permit study of the comet's dust jets as possible sources of gas, as well as a comparison of the sources of gas and dust on the surface of the comet nucleus. The three dust filters will allow study of the color and scattering properties of the dust. The high-resolution filter will be used for the highest-resolution imaging of the comet nucleus. The clear filter will be used for the study of structure and brightness differences on the comet nucleus. See Fig. 11.

During distant imaging of the comet's coma, the camera will take pictures through a periscope in order to protect the camera's primary optics as the spacecraft enters the coma. In the periscope, light is reflected off mirrors made of highly polished metals designed to minimize image degradation while withstanding particle impacts. During close approach, the nucleus is tracked and several images taken with a rotating mirror that no longer views through the periscope.

Propulsion System

The *Stardust* spacecraft needs only a relatively modest propulsion system because it is on a low-energy trajectory for its flyby of Comet Wild-2 and subsequent return to Earth, and because it is aided by a gravity-assisted boost maneuver when it flies by the Earth for the first time.

correction maneuvers or turning the spacecraft. Eight smaller thrusters producing 0.9 newton (0.2 pound) of thrust each will be used for attitude control. The thrusters are in four clusters located on the opposite side of the spacecraft from the deployed aerogel. In all, the spacecraft carries 85 kilograms (187 pounds) of hydrazine propellant.

Attitude Control

The attitude control system manages the spacecraft's orientation in space. Like most planetary spacecraft, *Stardust* is three-axis stabilized, meaning that its orientation is held fixed in relation to space, as opposed to spacecraft that stabilize themselves by spinning.

Stardust determines its orientation at any given time using a star camera or one of two inertial measurement units, each of which consists of three ring-laser gyroscopes and three accelerometers. The spacecraft's orientation is changed by firing thrusters. The inertial measurement units are needed only during trajectory correction maneuvers and during the fly-through of the cometary coma when stars may be difficult to detect. Otherwise, the vehicle can be operated in a mode using only stellar guidance for spacecraft positioning. Two Sun sensors will serve as backup units, coming into play if needed to augment or replace the information provided by the rest of the attitude control system's elements.

Command and Data Handling

The spacecraft's computer is embedded in the spacecraft's command and data-handling subsystem, and provides computing capability for all spacecraft subsystems. At its heart is a RAD6000 processor, a radiation-hardened version of the PowerPC chip used on some models of Macintosh computers. It can be switched between clock speeds of 5, 10 or 20 MHz. The computer includes 128 megabytes of random-access memory (RAM); unlike many other spacecraft, *Stardust* does not have an onboard tape recorder, but instead stores data in its RAM for transmission to Earth. The computer also has 3 megabytes of programmable memory that can store data even when the computer is powered off.

The spacecraft uses about 20 percent of the 128 megabytes of data storage for its own internal housekeeping. The rest of the memory is used to store science data and for computer programs, which control science observations. Memory allocated to specific instruments includes about 75 megabytes for images taken by the navigation camera, 13 megabytes for data from the comet and interstellar dust analyzer, and 2 megabytes for data from the dust flux monitor.

Power

Two solar array panels affixed to the spacecraft are deployed shortly after launch. Together they provide 6.6 square meters (7.9 square yards) of solar collecting area using high-efficiency silicon solar cells. One 16-amp-hour

Fig. 10.   Comet and Interstellar Dust Analyzer (CIDA). (*Jet Propulsion Laboratory, Pasadena, CA.*)

The spacecraft is equipped with two sets of thrusters, which use hydrazine as a monopropellant. Eight larger thrusters, each of, which puts out 4.4 newtons (1 pound) of thrust, will be used for trajectory

Fig. 11.   *Stardust* navigation camera. (*Jet Propulsion Laboratory, Pasadena, CA.*)

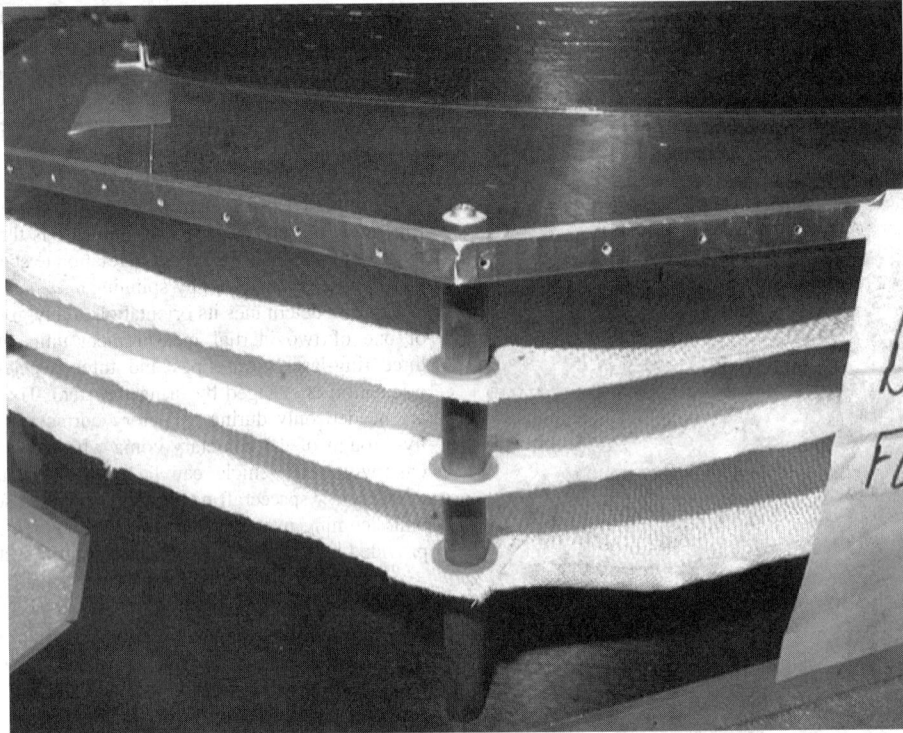

Fig. 12.   Whipple Shields. (*Jet Propulsion Laboratory, Pasadena, CA.*)

nickel-hydrogen battery provides power when the solar arrays are pointed away from the Sun and during peak power operations. See also **Spacecraft Power Systems**.

Thermal Control

Stardust's thermal control subsystem uses louvers to control the temperature of the inertial measurement units and the telecommunications system's solid-state power amplifiers. Thermal coatings and multi-layer insulation blankets and heaters are used to control the temperature of other parts of the spacecraft.

Telecommunications

Stardust is equipped with a transponder (radio transmitter/receiver) originally developed for the *Cassini* mission to Saturn, as well as a 15-watt radio frequency solid-state amplifier. Communications will be mainly conducted through the spacecraft's medium-gain antenna. Three low-gain antennas are used for initial communications near Earth and to receive commands when the spacecraft is in nearly any orientation. Data rates will range from 40 to 4,000 bits per second.

A 0.6-meter-diameter (2-foot) high-gain dish antenna is used primarily for communication during the comet encounter. *Stardust* will use it to transmit images of the comet nucleus, as well as data from the comet and interstellar dust analyzer and the dust flux monitor, at a high data rate to minimize the transmission time and the risk of losing data during the extended time that would be required to transmit the data through the medium-gain antenna.

Most data from the spacecraft will be received through the Deep Space Network's 34- meter-diameter (112-foot) ground antennas, but 70-meter (230-foot) antennas will be used during some critical telecommunications phases, such as when *Stardust* transmits science data during and after the comet encounter. See also **Deep Space Network**.

Redundancy

Virtually all spacecraft components are redundant, with critical items "cross-strapped," or interconnected so that they can be switched in or out most efficiently. The battery includes an extra pair of cells. Fault protection software is designed so that the spacecraft is protected from reasonable, credible faults without unnecessarily putting the spacecraft into a safe mode due to unanticipated but probably benign glitches.

Whipple Shields

The shields that will protect *Stardust* from the blast of cometary particles is named for American astronomer, Dr. Fred L. Whipple, who in 1950

accurately predicted the "dirty snow-ball" model of the cometary nucleus as a mixture of dark organic material, rocky grains and water ice. The system includes two bumpers at the front of the spacecraft, which protect the solar panels, and another shield protecting the main spacecraft body. Each of the shields is built around composite panels designed to disperse particles as they impact, augmented by blankets of Nextel ceramic cloth that further dissipate and spread particle debris. See Fig. 12.

Web References

Boeing designed and built the Delta rocket, that blasted Stardust into space in February 1999: *http://www.boeing.com/defense-space/space/*

Comet Definitions: *http://encke.jpl.nasa.gov/define.html*

Comet Fact Sheet, NSSDC: *http://nssdc.gsfc.nasa.gov/planetary/factsheet/cometfact. html*

Comet Hale-Bopp Home Page: *http://www.jpl.nasa.gov/comet/*

Frequently Asked Questions: Comets, Asteroids, Planets, NSSDC *http://nssdc.gsfc.nasa.gov/planetary/Comets*

http://nssdc.gsfc.nasa.gov/planetary/Asteroids

http://nssdc.gsfc.nasa.gov/planetary/planetaryfaq.html

For additional technical information about aerogel see, Berekley National Laboratory at *http://eande.lbl.gov/ECS/aerogels/satoc.htm*

History of Comet Wild 2 and How to Observe It: *http://stardust.jpl.nasa.gov/ comets/wild2.html*

Jet Propulsion Laboratory *Stardust* Mission: *http://stardust.jpl.nasa.gov/*

Lockheed Martin Astronautics; Lockheed Martin Astronautics brings years of experience in space missions to the project. The company built the lightweight, low-cost Stardust spacecraft and Sample-Return Capsule: *http://www.ast.lmco.com/*

NASA Ames Research Center developed a new carbon—based heat shield that will protect the return capsule, when it returns to Earth. *http://www.arc.nasa.gov/*

NASA Johnson Space Center; The *Stardust* Curation Team, located at JSC in Houston, Texas, handles the Sample Return Capsule and its precious cargo of interstellar dust grains and cometary dust particles: *http://www-curator.jsc.nasa.gov/curator/stardust/*

The Max-Planck-Institut fur extraterrestrische Physik provided the Cometary and Interstellar Dust Analyzer Instrument. *http://www.mpe.mpg.de/*

University of Chicago; The University of Chicago provided the Dust Flux Monitor Instrument. *http://astro.uchicago.edu/*

University of Washington; The principal investigator, Dr. Donald Brownlee, is from the University of Washington. Dr. Brownlee is well known for his work on cosmic particles in the stratosphere, known as Brownlee particles. *http://www.astro.washington.edu/brownlee/*

Jet Propulsion Laboratory, Pasadena, CA.

STARK EFFECT. In 1913, Stark showed that every line in the Balmer series of hydrogen, when excited in a strong electric field of 100 kilovolts per square centimeter or more, is split into several components. If the spectrum is observed perpendicular to the field, some members of the line pattern are plane-polarized with the electric vector parallel to the field (*p*-components) and the others are polarized with the electric vector normal to the field (*s*-components). When the spectrum is observed parallel to the field, only unpolarized *s*-components are observed. A similar splitting of lines is noted in the cathode dark space of a discharge tube. The Stark effect is similar, in many respects, to the Zeeman effect but it is generally more difficult to study experimentally because of the high potential gradients needed in the light source. Its theory is quite different, and the observed spectral pattern varies markedly in character and in number of components as the field intensity increases.

STARLING (Aves, Passeriformes).

A bird of any of several species native to Europe, Asia, and northern Africa. The African glossy starlings belong to a family containing also the Asiatic grackles of hill mynas, while the true starlings are placed in a closely related family.

The one species common in North America is the common European starling, *Sturnus vulgaris*, which was introduced into New York in 1890 and has since spread more than a third of the way across the continent. The males are black with green and blue iridescence and light tips on many of the smaller feathers, and the females are brownish-gray. The beak is rather large. In Europe the starlings are valued as destroyers of insects and for their song. They are able mimics. The same good qualities are worthy of consideration in America, but the birds are also destructive of native species, and in the fall they become a nuisance by gathering in great flocks to roost in countless thousands in the trees of residential districts. The net verdict is against them, and they are commonly regarded as pests.

STARLING, ERNEST H. (1866–1927).

Starling was an English physiologist who is remembered for his work on the cardiovascular system. He is most famous for his work with his collaborator, William Maddock Bayliss, for which he is credited with the discovery of secretin, a chemical that releases digestive juices by the pancreas. He suggested "hormone" be given to any chemical, such as secretin, that transmits a message from one part of the body and causes an effect in another part of the body.

Also Starling is credited in overseeing the construction and equipping of a new physiology building at Guy's Hospital Medical College in London in 1899 and making it one of the best research facilities for its time in physiology.

See also **Heart and Circulatory System (Physiology)**.

J. M. I.

STARVATION.

A state of existence without food or with inadequate food. When animals are completely deprived of food their only source of energy for the essential processes of metabolism is the material already present in the body. The carbohydrate stored as glycogen is quickly used up, leaving only the stored fat, the circulating protein (see also **Amino Acids**), and the protein of the tissues. Studies of the progress of starvation in mammals have shown that the fat is almost completely used up and that the greatest loss of tissue proteins is from the muscles, due to their great bulk. In percentages, however, the liver, spleen, and gonads lose more of their bulk than other parts of the body, and the heart and central nervous system are maintained at the expense of the other parts with very little loss of substance. Extensive observations of the details of metabolism and bodily changes during starvation allowed to continue to the death of the animal have been recorded.

A remarkable result of starvation in some of the lower invertebrates is a progressive shrinkage of the body as a whole, in contrast with the emaciation that results in vertebrates. Observations on the flatworm, *Planaria*, have shown that ultimately it even retraces its development to assume an embryonic form.

STATE.

1. In its fundamental connotation, this term refers to the condition of a substance, as its state of aggregation, which may be solid, liquid, or gaseous—compact or dispersed.

2. As extended to a particle, the state may denote its condition of oxidation, as the state of oxidation of an atom, or the energy level, as the orbital of an electron, or in fact, the energy level of any particle.

3. In quantum mechanics, the word state is used in its most general context to refer to the condition of a system described by a wave function satisfying the Schrödinger equation for the system, when this wave function is simultaneously an eigenfunction of one or more quantum mechanical operators corresponding to one or more dynamical variables. If this set of operators includes all those which will commute with the ones in the set, then the state of the system is as completely specified as the Heisenberg uncertainty principle (see also **Uncertainty Principle**) allows and is characterized by the eigenvalues of these operators. These eigenvalues are the results which will always be found if measurements of the corresponding dynamical variables are made. In its more limited sense, the word state is used to refer to the condition of the system when its wave function is simultaneously an eigenfunction of the Hamiltonian operator. In this case the system is characterized by a definite value of the energy, i.e., the eigenvalue of the Hamiltonian operator, and is said to be an *energy eigenstate*, a *stationary state*, or a definite *energy state*.

STATIC AND NOISE.

Names commonly applied to all the various random electrical disturbances picked up by a radio receiver. These can be divided into two general classes, natural and artificially generated (as from machines) static. The first is caused by various types of natural electrical discharges, the most pronounced being those of lightning. However, a static-producing discharge is not necessarily, or even usually, a visible lightning discharge. Various static charges are often continually building up and discharging in the atmosphere and hence inducing disturbances in the receiver. Cosmic radiations are also responsible for static. These types of natural static are often called atmospherics. The types of artificial static are almost as numerous as the electrical machines which people have developed. Any sparking contact or poor electrical connection will produce static which will be picked up by nearby receivers. Unfortunately many types of this interference may be fed back along the power lines and directly into the receiver. X-ray and diathermy machines are also sources of interference but cannot be properly classed as static.

The elimination of static presents a particularly difficult problem since the frequencies in the static pulse cover a wide band, certain types being more prevalent in some frequency ranges than others. Artificial static is best eliminated by correcting the fault at the source although a filter in the power line often helps if the disturbance is coming into the set through the line. Natural static can be minimized, but not eliminated. For amplitude modulation systems limiting the frequency band to which the receiver responds will reduce the noise. Various types of limiters will also reduce the effect since the static signal is frequently greater than the desired one. Frequency modulation is inherently less susceptible to interfering noise and offers almost noise-free reception.

Noise in Industrial Measurement Systems.[1] Signals entering a data acquisition and control system include unwanted noise. Whether this noise is troublesome depends on the signal-to-noise ratio and the specific application. In general it is desirable to minimize noise to achieve high accuracy. Digital signals are relatively immune to noise because of their discrete (and high-level) nature. In contrast, analog signals are directly influenced by relatively low-level disturbances. The major noise-transfer mechanisms include conductive, inductive (magnetic), and capacitive coupling. Examples include the following:

- Switching of high-current loads in nearby wiring can induce noise signals by magnetic coupling (transformer action).
- Signal wires running close to ac power cables can pick up 50- or 60-Hz noise by capacitive coupling.
- Allowing more than one power or signal return path can produce ground loops that inject errors by conduction

Conductance involves current flowing through ohmic paths (direct contact), as opposed to inductance or capacitance.

Interference via capacitive or magnetic mechanisms usually requires that the disturbing source be close to the affected circuit. At high frequencies, however, radiated emissions (electromagnetic signals) can be propagated over long distances.

In all cases, the induced noise level will depend on several user-influenced factors:

- Signal source output impedance.
- Signal source load impedance (input impedance to the data-acquisition system).

[1] Information furnished by H.L. Skolnik, *Intelligent Instrumentation, Inc.*, Tucson, Arizona.

- Lead-wire length, shielding, and grounding.
- Proximity to noise source or sources.
- Signal and noise amplitude

Transducers that can be modeled by a current source are inherently less sensitive to magnetically induced noise pickup than are voltage-driven devices. An error voltage coupled magnetically into the connecting wires appears in series with the signal source. This has the effect of modulating the voltage across the transducer. However, if the transducer approaches ideal current-source characteristics, no significant change in the signal current will result. When the transducer appears as a voltage source (regardless of impedance), the magnetically induced errors add directly to the signal source without attenuation.

Errors also are caused by capacitive coupling in both current and voltage transducer circuits. With capacitive coupling, a voltage divider is formed by the coupling capacitor and the load impedance. The error signal induced is proportional to $2\pi f RC$, where R is the load resistor, C is the coupling capacitance, and f is the interfering frequency. Clearly, the smaller the capacitance (or frequency), the smaller the induced error voltage. However, reducing the resistance only improves voltage-type transducer circuits.

Example. Assume that the interfering signal is a 110-volt ac 60-Hz power line, the equivalent coupling capacitance is 100 pF, and the terminating resistance is 250 ohms (typical for a 4- to 20-mA current loop). The resulting induced error voltage will be about 1 mV, which is less than 1 LSB in a 12-bit 10-volt system.

If the load impedance were 100 kΩ, as it could be in a voltage input application, the induced error could be much larger. The equivalent R seen by the interfering source depends on not only the load impedance, but also the source impedance and the distributed nature of the connecting wires. Under worst-case conditions, where the wire inductance separates the load and source impedances, the induced error could be as large as 0.4 volt. This represents about an 8% full-scale error.

Even though current-type signals are usually converted to a voltage at the input to the data-acquisition system, with a low-value resistor this does not improve noise performance. This is because both the noise and the transducer signals are proportional to the same load impedance.

It should be pointed out that this example does not take advantage of or benefit from shielding, grounding, and filtering techniques.

STATICS.   Statics is the branch of mechanics that deals with particles or bodies in equilibrium under the action of forces or of torques. It embraces the composition and resolution of forces, the equilibrium of bodies under balanced forces, and such properties of bodies as center of gravity and moment of inertia.

Statics is the oldest branch of mechanics, some of its principles having been used by the ancient Egyptians and Babylonians in their constructions of temples and pyramids. As a science it was established by Archytas of Taras (ca. 380 B.C.) and primarily by Archimedes (287–212 B.C.). It was further developed by medieval writers on the "science of weights," such as Jordanus de Nemore (thirteenth century) and Blasius of Parma (fourteenth century). In the sixteenth century, it was revived by Leonardo da Vinci, Guido Ubaldi, and particularly by Simon Stevin (1548–1620) who laid the foundations of modern statics (inclined plane, equilibrium of pulleys, parallelogram of forces, etc.).

Although the laws of statics can in principle be derived from those of dynamics as a limiting case for vanished velocities or accelerations, statics has been developed, since the end of the eighteenth century, independently of dynamics. Its fundamental theme, like that of dynamics, is the concept of force, representing the action of one body on another and characterized by its point of application, its magnitude, and its direction (line of action), or briefly by a vector.

The following four principles may serve as the basic postulates for statics: (1) The principle of composition (addition) of forces; (2) the principle of transmissibility of force; (3) the principle of equilibrium; and (4) the principle of action and reaction.

Statistical analysis of framed structures or trusses, collections of straight members pinned or jointed together at the ends, is based on these principles.

STATICS (Graphical).   The equilibrium of forces is often treated graphically in such practical problems as the stresses in the members of a framed structure. If three concurrent forces are in equilibrium, the three vectors drawn to a common scale to represent them may be made to form a closed triangle (Fig. 1); or if more than three, a closed polygon (Fig. 2). The principle may be extended and is much used in the calculation of the forces in a truss by means of the so-called stress diagram. A simple example is shown in Fig. 3, which represents a small roof-truss with equal loads resting on it at the joints A, B, C, D, E, and supported by the upward reactions of the walls at A and E. The several compartments of the figure are numbered, and both the external forces and the forces acting along the members between these compartments are represented, both in magnitude and direction, by the lines joining the corresponding numbers in the stress diagram. For example, the compressive force in the strut BF is represented by the line 8–9, while the tension in the vertical rod CF is given by 9–10. The closed figure 5–4–10–9–5 in the stress diagram indicates the equilibrium of the forces acting at the joint C. This method of analysis is attributed to Maxwell. See also **Bow's Notation**.

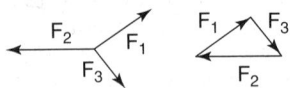

Fig. 1.   Three forces in equilibrium.

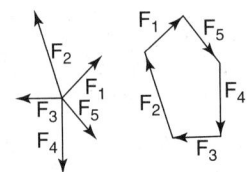

Fig. 2.   Five forces in equilibrium.

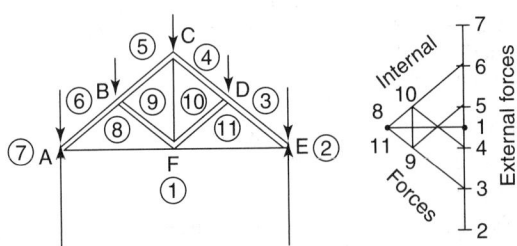

Fig. 3.   Elevation of truss with corresponding stress diagram.

In the graphical solution of some types of trusses, it is found, on reaching a particular joint, that the arrangement of members is such that there are more than two unknowns. It is then necessary to replace the unknowns by a substitute system, which reduces the number of unknowns at the joint to two. The substitute system consists of a single member inserted in such a way that the truss remains stable and determinate. See also **Determinate Structure**. This member is called a substitute, fictitious or phantom member. When the solution reaches a joint where the stress in the members is unaffected by the substitution, the substitute arrangement is replaced by the original arrangement. The graphic procedure is reversed in direction to find the stress in the original members.

STATIONARY STATE.   1.   In a quantum-mechanical system, a state corresponding to a definite value of the energy.

2.   A thermodynamic system is in a stationary state when its intensive variables are time-independent. The definition includes two classes of situations;

(1) *Equilibrium states*: the system is in thermodynamic equilibrium; no irreversible processes occur in it and there are no exchanges with the outside world. The entropy production vanishes.

(2) *Stationary nonequilibrium states*: in this case irreversible processes proceed in the system. The time-independence of the intensive properties is due to a compensation of the effect of the irreversible processes by exchanges with the outside world.

An isolated piece of metal initially at a nonuniform temperature will, of course, reach a state of thermodynamic equilibrium. But if we cool it at one end and heat it at the other, it will reach a stationary nonequilibrium state. Another example is afforded by a system that receives a component M from the outside environment and transforms it through a certain number of intermediate states into a final product F which is returned to the external environment. After some time a stationary state arises when the concentration of the intermediate components no longer vary with time.

One may say that stationary nonequilibrium states are reached when there exist some constraints that prevent the system from reaching true thermodynamic equilibrium. In the first example it is the difference of temperature at both ends, in the second, it is the value of the concentrations of the initial and final products in the external environment.

STATISTIC.   A statistic is a function of the observations in a sample designed to estimate a parameter of the population from which the sample was drawn, or to carry out a test of significance of a hypothesis.

STATISTICAL MECHANICS.   One major problem of physics involves the prediction of the macroscopic properties of matter in terms of the properties of the molecules of which it is composed. According to the ideas of classical physics, this could have been accomplished by a determination of the detailed motion of each molecule and by a subsequent superposition or summation of their effects. The Heisenberg indeterminacy principle now indicates that this process is impossible, since we cannot acquire sufficient information about the initial state of the molecules. Even if this were not so, the problem would be practically insoluble because of the extremely large numbers of molecules involved in nearly all observations. Many successful predictions can be made, however, by considering only the average, or most probable, behavior of the molecules, rather than the behavior of individuals. This is the method used in statistical mechanics.

In the general approach to classical statistical mechanics, each particle is considered to occupy a point in phase space, i.e., to have a definite position and momentum, at a given instant. The probability that the point corresponding to a particle will fall in any small volume of the phase space is taken proportional to the volume. The probability of a specific arrangement of points is proportional to the number of ways that the total ensemble of molecules could be permuted to achieve the arrangement. When this is done, and it is further required that the number of molecules and their total energy remain constant, one can obtain a description of the most probable distribution of the molecules in phase space. The Maxwell-Boltzmann distribution law results.

When the ideas of symmetry and of microscopic reversibility are combined with those of probability, statistical mechanics can deal with many stationary state nonequilibrium problems as well as with equilibrium distributions. Equations for such properties as viscosity, thermal conductivity, diffusion, and others are derived in this way.

The development of quantum theory, particularly of quantum mechanics, forced certain changes in statistical mechanics. In the development of the resulting quantum statistics, the phase space is divided into cells of volume h^f, where h is the Planck constant and f is the number of degrees of freedom. In considering the permutations of the molecules, it is recognized that the interchange of two identical particles does not lead to a new state. With these two new ideas, one arrives at the Bose-Einstein statistics. These statistics must be further modified for particles, such as electrons, to which the Pauli exclusion principle applies, and the Fermi-Dirac statistics follow.

It is often possible to obtain similar or identical results from statistical mechanics and from thermodynamics, and the assumption that a system will be in a state of maximal probability in equilibrium is equivalent to the law of entropy. The major difference between the two approaches is that thermodynamics starts with macroscopic laws of great generality and its results are independent of any particular molecular model of the system, while statistical methods always depend on some such model.

STATOR.   In machinery, a part or assembly that remains stationary with respect to a rotating or moving part or assembly, such as the field frame of an electric motor or generator, or the stationary casing and blades surrounding an axial-flow compressor rotor or turbine wheel; a stator blade.

STAUROLITE.   The mineral staurolite is a complex silicate of iron and aluminum corresponding to the formula $(Fe,Mg,Zn)_2Al_9Si_4O_{23}$ (OH) but somewhat varying and may carry magnesium or zinc. It is orthorhombic,

prismatic, twins common, often producing cruciform crystals. It is a brittle mineral; fracture, subconchoidal; hardness, 7–7.5; specific gravity, 3.65–3.83; luster, subvitreous to resinous; color, dark brown, sometimes reddish to nearly black; grayish streak; translucent to opaque. Staurolite is a metamorphic mineral usually the result of regional rather than contact metamorphism, and is common in schists, phyllites and gneisses together with garnet, kyanite, and tourmaline.

Well-known European localities are in Switzerland and Brittany; and in the United States this mineral is common in the schists of New England, and those of the southern Alleghenies. Frequently the crystals are found loose in the soil after the disintegration of the country rock. The name staurolite is derived from the Greek meaning a cross, in reference to the twin crystals, the more nearly perfect crosses being somewhat in demand as curios.

STAYBOLT.   The surfaces of pressure vessels, such as boiler drums and tanks, which are not of a natural bulged shape, such as the cylinder or the sphere, must be stayed against bulging by special tension rods called staybolts.

STEADY STATE.   A characteristic of a condition, such as value, rate, periodicity, or amplitude, exhibiting only negligible change over an arbitrary long period of time. It may describe a condition in which some characteristics are static; others dynamic. The *steady-state deviation* of a system may be defined as the system deviation after transients have expired.

In medical parlance, a relatively steady condition of processes within the body sometimes is referred to as *homeostasis*.

STEAM CYCLES (Diagram Factor).   This is a factor relating particularly to piston and cylinder engines. Although it has a meaning applied to internal combustion engines, diagram factor is principally a dimension of the steam engine. Certain analyses of the steam engine, especially those concerned with predicting the performance of a given unit under stated steam conditions, are most easily made with the use of the diagram factor. This factor is defined as the ratio of the actual mean effective pressure to the theoretical effective pressure; also as the actual work to the theoretical work. The theoretical case is that of a steam engine having no compression, no wire drawing, and no clearance.

The ratio of the area of this cycle to that of the actual engine operating between the same pressure limits, is typical for any one class of engine. The following table gives some values of diagram factor for steam engines. Knowing the steam conditions, and the type of engine, the probable pressure and horsepower realizable can be closely estimated by calculating the theoretical quantities and multiplying them by the diagram factor. In this light, diagram factor is a means of modifying theoretical calculations to bring them in line with actual experience.

DIAGRAM FACTORS

High-speed, simple automatic	0.70–0.85
Low-speed, releasing gear	0.80–0.90
Uniflows	
Full compression, condensing	0.75–0.85
Full compression, noncondensing	0.70–0.80
Controlled compression, condensing	0.85–0.90
Controlled compression, noncondensing	0.80–0.85

$$P_a = P_t \times f$$

$$IHP = \frac{2P_aLAN}{33,000}$$

P_a = actual mean effective pressure, in pounds per square inch
P_t = theoretical mean effective pressure, in pounds per square inch
f = diagram factor
IHP = indicated horsepower (steam engine)
L = stroke in feet
A = piston area, in square inches
N = rotative speed, in revolutions per minute

STEAM ENGINE.   A positive displacement piston and cylinder machine, which, when supplied with steam at a pressure above its exhaust pressure, uses that steam expansively for the production of power that it makes available as a rotating torque at a crankshaft or flywheel. The steam engine is, with few exceptions, double acting. The steam engine is characterized by moderate or low speeds (100–500 rpm), the use of atmospheric exhaust (or a moderate vacuum), high starting torque, and ease of conversion to reversible operation if desired. The excellence of the steam turbine when a large amount of power is to be generated, especially where the exhaust is at a high vacuum, coupled with the high efficiency of the diesel engine as a prime mover, has restricted the steam engine as a prime mover. However, where a boiler must be supplied anyway, as for the generation of heating steam, the steam engine is usually superior to the diesel engine as a source of power. And, where exhaust pressures are high (often the case in industry, where the exhaust is process steam), the steam engine offers advantages which are not seriously challenged by the other types of prime movers.

The principal parts of a steam engine are:

1. The frame or bedplate. In a multicylindered engine, this takes the form of a crankcase to which the cylinders are attached, but in a single-cylindered engine, the cylinder is often integral with the frame.
2. Cylinder, with valve chest.
3. Piston, piston rod, crosshead, connecting rod. This mechanical linkage receives a push from steam pressure at one end, and delivers it as a torque force on the crank.
4. Crankshaft, bearings, and flywheel. This part of the engine accomplishes the conversion of reciprocating to rotary motion, supports the shaft for power offtake, and steadies the speed.
5. Valves and valve gear. The device for admission and release of the steam to and from the cylinder, together with the means for actuating it from the crankshaft.
6. Governor. Stationary engines are automatically regulated for constant speed by means of a governor.
7. Lubrication. The piston and cylinder are lubricated by oil mixed with the steam.

A steam engine converts from 5 to 15% of the heat supplied to it into work, depending on the state of the steam supplied, and on the exhaust pressure. The heat unconverted is composed: first, of heat remaining in the exhaust steam; second, initial condensation; third, incomplete expansion; fourth, wire drawing; fifth, friction; and sixth, radiation (negligible). The first of these is the largest, and is reducible only within certain limits. Incomplete expansion results from the release of the steam at the end of the stroke at a pressure higher than the exhaust. By using a longer stroke, this could be eliminated, but there is a point beyond which the increased cost of the engine more than offsets the gain derived by eliminating this loss.

The cycle upon which the engine operates is briefly described as follows (Fig. 1): slightly before the piston reaches the dead-center position corresponding to minimum cylinder volume, the valve connects the cylinder with the steam line so that as the piston starts on its outward travel, the full steam pressure is acting on it. The beginning of this action is known as the event of *admission*. When some 20 to 30% of the stroke has been completed, the valve closes the port on the event known as *cutoff*, and during the remainder of the stroke, the steam is expanded adiabatically

to the accompaniment of decreasing pressure. Near the end of its stroke, the valve again opens the port, this time connecting the cylinder with the exhaust line. This event is known as *release*. The cylinder remains connected with the exhaust during the return stroke of the piston, and the steam is expelled until approximately $\frac{2}{3}$ of the return stroke has been completed. The valve then closes the port, and the remaining steam is trapped in the cylinder and compressed. The beginning of this process is known as the event of *compression*. The four events just described govern the form of the steam engine cycle. The engine using it will be able to develop a horsepower hour using from 10 to 25 pounds of steam, depending upon the expansion permitted by the terminal conditions of the steam.

The steam engine may be mechanically controlled to give variable output so that when it is connected to a load that varies, it maintains nearly constant speed. There are two methods of accomplishing this result. In one, called cutoff governing, the percent of the stroke during which the valve connects the cylinder with the boiler is varied, and in this way, different amounts of steam are admitted to the cylinder at one pressure. The mechanism to effect this type of control is incorporated in the valve drive. The other method, called throttling governing, consists of interposing an artificial resistance to create a pressure drop between the boiler and the engine, so that although the same volume is admitted on each stroke (the cutoff being constant), the weight of steam admitted will vary because of the variation in density created by throttling. The governor, in this case, operates on a throttle valve located at the steam inlet.

Additional Reading

Avallone, E.A. and T. Baumeister: "Marks' Standard Handbook for Mechanical Engineers," 10th Edition, The McGraw-Hill Companies Inc., New York, NY, 1996.

Boulton, M. and J. Tann: "The Selected Papers of Boulton and Watt: Vol. I: The Engine Partnership, 1775–1825," MIT Press, Cambridge, MA, 1981.

Hills, R.L.: "Power from Steam: A History of the Stationary Steam Engine," Cambridge University Press, New York, NY, 1993.

Jones, F.D. and H.H. Ryffel: "Machinery's Handbook Guide,' 26th Edition, Industrial Press, New York, NY, 2001.

Sproule, A. and M. Pollard: "James Watt: Master of the Steam Engine," Gale Group, Farmington Hills, MI, 2001.

STEARIC ACID AND STEARATES. Stearic acid $H \cdot C_{18}H_{35}O_2$ or $C_{17}H_{35} \cdot COOH$ or $CH_3(CH_2)_{16} \cdot COOH$ is a white solid, melting point 69 °C, boiling point 383 °C, insoluble in water, slightly soluble in alcohol, soluble in ether. Stearic acid may be obtained from glyceryl tristearate, present in many solid fats such as tallow, and in smaller percentage in semisolid fats (lard) and liquid vegetable oils (cottonseed oil, corn oil), by hydrolysis. The crude stearic acid, after separation of the water solution of glycerol, is cooled to fractionally crystallize the stearic and palmitic acids, which are then separated by filtration (oleic acid in the liquid), and fractional distillation under diminished pressure. With sodium hydroxide, stearic acid forms sodium stearate, a soap. Most soaps are mixtures of sodium stearate, palmitate and oleate.

The following are representative esters of stearic acid: Methyl stearate $C_{17}H_{35}COOCH_3$, melting point 38 °C, boiling point 215 °C at 15 millimeters pressure; ethyl stearate $C_{17}H_{35}COOC_2H_5$, melting point 35 °C, boiling point 200 °C at 10 millimeters pressure; glyceryl tristearate [tristearin $C_3H_5(COOC_{17}H_{35})_3$], melting point 70 °C approximately.

Stearic acid is used (1) in the preparation of metallic stearates, such as aluminum stearate for thickening lubricating oils, for waterproofing materials, and for varnish driers, (2) in the manufacture of "stearin" candles, and is added in small amounts to paraffin wax candles. As the glyceryl ester, stearic acid is one of the constituents of many vegetable and animal oils and fats.

See also **Rubber (Natural)**.

STEELYARD. An early type of scale that employs an unequal first class lever arm constructed so that the fulcrum point can be suspended from above, such as by a hook on the end of a chain, the other end of the chain being fastened to a ceiling support. The load is suspended from the end of the short lever arm. A sliding poise on the long end of the lever arm is positioned to balance the load. Although largely replaced by spring scales, the steelyard, because of its portability and low cost, still finds limited use, particularly in underdeveloped nations. See Fig. 1.

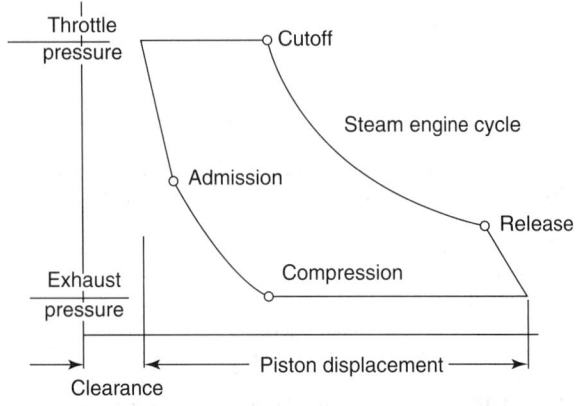

Fig. 1.   Steam engine cycle.

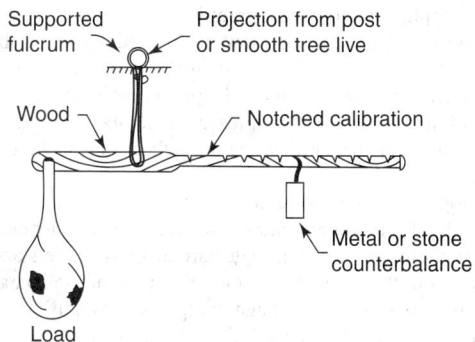

Fig. 1.   Steelyard used for weighing since ancient times.

STELE.   The vascular tissue of the axis of a plant is called the stele or central cylinder. The principal tissues composing it are the xylem and the phloem. Others are pith, pith rays, cambium, and pericycle. The most primitive type is the protostele, which consists of a solid central mass of xylem surrounded by a cylinder of phloem. There is no pith in a protostele. Protosteles are found in the roots of all plants and in the stems of some of the ferns. Many would set off the stele of the root as a separate type known as a radial stele, since the central mass of xylem is not a cylinder, but has several arms projecting outward from its surface, with the phloem concentrated between these arms. In stems, only those of certain Lycopsida like *Lycopodium*, have radial steles.

STELLAR HEAT INDEX.   As applied to a star, the difference between the visual and radiometric magnitudes: i.e.,

$$\text{heat index} = m_v - m_r$$

where m_r is the radiometric magnitude and m_v is the visual magnitude. A radiometric magnitude is obtained by measuring the overall energy brightness of a star by means of a thermocouple or bolometer, and by reducing the readings to outside atmosphere after correcting for losses in the optical system.

STELLAR LUMINOSITY.   The intrinsic brightness of a star may be expressed either in terms of its absolute magnitude or in terms of the sun's brightness as unity. The luminosity of a star is defined as its intrinsic brightness in terms of the brightness of the sun as unity. That is, if our sun were replaced by a star of luminosity 100, the light received by the earth would be 100 times as great.

STELLAR MAGNITUDE.   In the first star catalogues issued by Hipparchus and Ptolemy, the relative apparent brightness of the stars was designated by a system of six numbers referred to as the magnitudes of the stars. Twenty of the brightest stars were referred to as first magnitude, and those at the limit of visibility were called sixth magnitude. The stars with brightness intermediate between the two extremes were assigned to a magnitude number with the numbers increasing with faintness of the stars. With the application of the telescope to astronomy, many faint stars were discovered and the need for additional magnitude numbers became evident. Unfortunately for modern astronomers, the attempt was made to amplify the ancient magnitude system not only to include the fainter stars, but also to indicate finer gradations of brightness by a decimal system. The result is that astronomers are now using a system that was started about 2,000 years ago and has all the clumsiness and inconvenience for modern observers that is characteristic of so many of the ancient scientific instruments.

There is no definite evidence that Hipparchus or Ptolemy had any idea in mind at the time they first used the magnitude system other than to provide a rough descriptive term for the stars. In the early part of the nineteenth century, Sir John Herschel found that the apparent brightness of a first magnitude star is about 100 times that of a sixth magnitude star. In 1850 Pogson proposed a fixed scale of stellar magnitudes based upon the original scale of Hipparchus and Ptolemy, but so adjusted that it would agree at the sixth magnitude with the system employed by Argelander in his famous Bonner Durchmusterung. Adopting the announcement of Herschel that the ratio of brightness of a first and sixth magnitude star is approximately 100, Pogson proposed that the ratio between successive magnitudes should be

$\sqrt[5]{100}$ or approximately 2.512. This leads to an analytical expression for the magnitude scale as follows:

Let I_1 be the intensity of a star of apparent magnitude m_1, and $I2$ the intensity of a star of apparent magnitude m_2; then

$$\frac{I_1}{I_2} = 2.512^{(m_2 - m_1)}$$

or

$$\log I_1 - \log I_2 = 0.4(m_2 - m_1)$$

Since the magnitude scale is a scale of relative brightness, it is necessary to establish a system of standards. For this purpose, a group of stars in the immediate vicinity of the north celestial pole has been selected. The magnitudes of the stars in this "north polar sequence" have been very carefully determined and agreed upon by the International Astronomical Union. All magnitudes determined at the present time should be referred, either directly or indirectly, to this standard sequence. More recently, Johnson and Morgan have set up a series of "photometric standard stars" whose magnitudes are strictly given on a specified system.

The magnitude scale as originally established referred to the relative apparent visual brightnesses of the stars. With the application of photography to astronomy, difficulty with the magnitude scale immediately became evident. If we have two stars of the same visual magnitude, one of them blue and the other red, the photographic image of the blue star will be much stronger than the photographic image of the red star. The colors of the stars in the sky vary with the different spectral types, and the visual magnitude differences between a number of stars of different spectral types will differ considerably from the magnitude differences obtained by photographic means. Furthermore, the photographic magnitudes, so-called, will be different, depending upon the type of plate used and the characteristics of different telescopes, and it becomes necessary to be very explicit in defining the particular range of wavelengths of spectral energy that are to be used in any magnitude scale. The difference between the photographic and visual magnitudes of a star is known as the color index of the star, the term arising from the fact that the color is the determining factor in the magnitude scale difference.

With the application of various other types of radiation measuring instruments, such as bolometers and radiometers, to the measurement of the apparent brightness of the stars, the necessity has arisen for various different magnitude scales, such as bolometric magnitude, radiometric magnitude, etc. The problem of the intercorrelation of the different systems is at present in a very confused state and much research is being carried on in this important field. It is hoped that in the future some system of expressing the apparent brightnesses of the stars may be devised that will replace the present complicated inverse logarithmic scale of magnitudes.

For the purpose of expressing the intrinsic brightness of a star, independent of the distance of the star from the earth, a system of absolute magnitudes has been devised.

The apparent brightness of a star, or any other luminous object, depends both upon the intrinsic brightness of the object and also upon its distance from the observer. In the case of the stars, the apparent brightness, expressed as stellar magnitude, may be determined by any one of the standard methods of stellar photometry. In case the distance of the star is known, the intrinsic brightness may be immediately calculated. Conversely, if there is any method available for determining the intrinsic brightness of a star independently of a knowledge of the distance, this distance may be computed from the ratio between the apparent and intrinsic brightness.

The absolute magnitude of a star is the apparent brightness, expressed on the magnitude scale, that a star would have if it were situated at a distance of ten parsecs from the sun or, in other words, if the stellar parallax of the star were $\frac{1}{10}$ of a second. Analytically, the absolute magnitude, M, of a star is connected with the apparent magnitude, m, and the stellar parallax, π, by:

$$M = m + 5 + 5 \log \pi$$

On this scale, we find the sun, with apparent magnitude -26.72 and parallax $206265''$, to have an absolute magnitude of 4.85. Antares with parallax $0''.009$ and apparent magnitude 1.22 is found to have an absolute magnitude of -4.0. On the basis of these absolute magnitudes and the defining relation of the magnitude scale, we find that the brightness ratio of Antares to the sun is 3470, or that the star Antares is actually 3470 times brighter than the sun.

The zero point of the absolute magnitude scale is arbitrary. By convention, a normal A0 main sequence star has an absolute magnitude of zero and a color index of zero, as well.

STELLAR PARALLAX. A term used by astronomers as a means for expressing the distance of a given star. Technically defined, stellar parallax is the angle that would be subtended by the mean distance of the Earth from the sun (one astronomical unit) at the distance of the star from the sun. The symbol π is used, and as the angle is always small, it is always given in seconds of arc; this measure is understood unless explicitly stated otherwise.

From the earliest days of the Pythagoreans, any theory of the structure of the universe that postulated that the Earth might move about the sun was objected to, on the ground that such motion should produce an apparent motion of the stars. Copernicus, in proposing his heliocentric theory, met this objection by postulating that the distances of the stars were so incomparably greater than the distance of the Earth from the sun that no instrumental methods would be capable of detecting the motion even if it did exist. The attempts to test the Copernican doctrine by searching for this so-called stellar parallax gave a tremendous impetus to the design of accurate instruments, but even with the improvement in instrumental equipment, the effect was not observed, and the Copernican theory lost ground. It was not until 1838 that Bessel was able definitely to prove that the effect is present.

The type of effect to be looked for is illustrated in Fig. 1. The type of curve that the stars should apparently follow due to the Earth's motion about the sun varies from an eclipse with eccentricity equal to that of the Earth's orbit for stars at the pole of the ecliptic, to oscillations back and forth along a straight line for stars in the plane of the ecliptic.

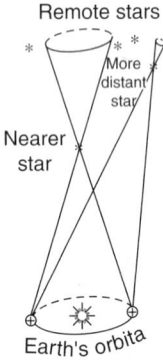

Fig. 1. Parallaxes of the stars. Owing to Earth's revolution, the nearer stars describe parallax orbits annually with respect to the remote stars.

The problem of determination of stellar parallax is theoretically very simple. All that is necessary is to make a series of observations of the positions of a star on any system of spherical coordinates (e.g., right ascension and declination) and from the observed changes in position throughout the year determine the stellar parallax. This so-called absolute method was attempted many times but failed to reveal any definite value because of the fact that the instrumental errors were larger than the effect sought. This is not surprising when we consider that the largest stellar parallax ever found (for the star Proxima Centauri) has a value of $0''.783$, equivalent to the angle subtended by a ten-cent piece at a distance of approximately 3 miles. (See also **Spherical Polar Coordinates**.)

With the failure of the absolute method to yield values for the parallaxes of the stars, Bessel and Struve decided upon an indirect or relative method for determining the desired quantity. This method is based upon the assumption that certain stars are at such a great distance that their parallaxes are too small for detection, but that there are other stars closer to the sun which should show motion relative to the distant background. Bessel selected the star 61 Cygni, which was assumed to be relatively close to the Earth from a large proper motion, while Struve selected the star Vega, which has an appreciable proper motion and is also so bright as to imply closeness to the Earth. Proceeding by different methods, both Bessel and Struve, in 1838, were able to show that the stars they had selected showed parallactic motion relative to the background of stars.

Until the application of photography to astronomy, the problem of determination of stellar parallaxes was very tedious and laborious, and up to 1880, the distances of fewer than 25 stars had been determined. With the application of photography, the progress of parallax determination became much more rapid, and at present, a number of long programs of observations in both the northern and southern hemispheres have been completed.

The photographic method consists in first selecting stars that are suspected, either from proper motion, spectral type, or other characteristics, to be relatively close to the sun. Plates are taken of these stars, great care being exercised in the guiding, and the brightness of the "parallax star" is reduced until its photographic image compares favorably with the images of the fainter "background stars." The plates are all taken at the same hour angle, either east or west, to eliminate so far as possible atmospheric effects, and the dates on which the plates are taken are separated as much as they can be, to make the effect of the Earth's motion as large as possible. Twenty or thirty plates, extending over several years, are taken, and the position of the parallax star is carefully measured with reference to half-a-dozen background stars. A least squares solution will yield the motion of the star relative to the background. This motion will consist both of the proper motion and the parallactic shift. The former may be separated from the latter, because proper motion is linear in character, whereas the parallactic shift is periodic. For stars within 5 million times the sun's distance from the earth (parallax $0''.04$), the mean of two or three determinations will be correct within 20 percent. For twice this distance, the results are only accurate enough for statistical purposes, and beyond this distance, the trigonometric method is practically valueless. Occasionally, due to an unfortunate choice of parallax star or comparison stars, the value of the stellar parallax comes out to be a negative quantity. Such a "negative parallax" simply means that the star under observation is more distant than those selected for comparison purposes. For the more distant stars beyond the range of the trigonometric method, certain other methods are available, such as: parallaxes of members of moving clusters; mean parallaxes; dynamical parallaxes of double stars; and spectroscopic parallaxes.

See also **Dynamical Parallax**; and **Star**.

STEM (Plant). The stem of a plant is that part that bears the leaves and flowers and later fruits. Commonly it grows erect, lifting these various organs up above the ground. It is readily distinguished from the root by being separated into joints or nodes and internodes, and by bearing leaves or by having on its surface scars left by the falling of leaves.

Stems may be classified in several ways. If one considers the duration of the stem, it becomes an annual lasting but one year, a biennial living 2 years, or a perennial stem, which grows for several years. If one considers the internal structure of the stem, he finds it to be herbaceous or woody. An herbaceous stem is one which is largely made up of parenchymatous cells, without a great mass of woody tissue. In temperate regions such stems last but a single year, at the end of which they die. In annuals, the entire plant dies, while in herbaceous perennials the top dies, but the basal portion including the root and the lower stem lives on. Woody plants are those in which the stem is predominantly composed of vascular tissues. Such plants are either trees or shrubs or vines. Trees are commonly distinguished by the existence of a single stem or trunk, which does not branch at its base, whereas in shrubs no single trunk exists, but several of equal size result from basal branchings. Branching is either excurrent or deliquescent. When excurrent, the trunk is distinctly recognizable throughout its length, the branches coming from it being much smaller, as in many conifers such as spruce or fir. Usually such trees have a conical shape, due to the progressively smaller branches from bottom to top of the tree. In deliquescent branching, the main trunk branches into several large branches, which in turn divide, as in the elm. Vines are distinguished by their long relatively slender stems, which usually require external support. Another classification separates erect stems from those which are procumbent or trailing on the ground, from stems which clamber over supports, and from twining stems, which wind tightly around any supporting object.

In many plants the stem is very much reduced in size, appearing as a small often flattened ball, as in the common Cyclamen or in certain Cactus plants. Other plants are said to be stemless, the stem existing only as a small object at the top of the root, the leaves arising from it seeming to come directly from the top of the root. A familiar example is the dandelion.

The first year of growth in many biennials results in a similarly stemless plant; carrots, beets, and parsnips are common examples.

As previously noted, one of the functions of the stem is to elevate the leaves into a position where they may function most efficiently, and to elevate the flowers to a position where they may become more conspicuous, and where the resulting fruits may be better scattered. Not only does the stem perform this function, but it also permits a great increase in the number of leaves and flowers which may be borne. The stem is the organ through which sap ascends from the root to the leaves, and through which organic materials elaborated in the leaves pass to the place of storage. The stem itself may be the place in which materials are stored.

The stem develops from a bud. In the seeds, the embryo has a terminal rudimentary bud, the plumule, from which the first stem develops. The tip of this stem bears a terminal bud, from which further increase in the stem is developed. In the axils of the leaves lateral buds are formed which develop into branches. Elongation of the stem takes place only in the tip, extending downward therefrom a few inches, and is caused by the cellular changes like those that occur also in the elongating root. In some plants, as in Grasses, intercalary growth also occurs. This is growth in the region of the older nodes of the stem. Increase in diameter of the stem results from the divisions of special cells called cambium cells.

When the extreme variations of stem are considered, with their range from tiny plants less than half an inch in height to forest giants towering 300 feet (91 meters) and over, and from vines with slender wiry stem less than $\frac{1}{16}$ inch in diameter through succulent herbs to sturdy trunks 40 feet or more through, it is not surprising that stem structure should be very variable and often complex. Yet they are all composed of the same types of cells and are all arranged on two fundamental patterns, one found in dicotyledonous plants, the other characterizing the monocotyledons.

In dicotyledons, the growing tip of the young stem has cells that are all alike, having a dense cytoplasm, and large nuclei, which, if the stem is growing, will be dividing frequently. These cells comprise the promeristem, the region where active cell division occurs, but little change in cell form.

As these cells increase in number, some are carried ahead, while others remain unchanged in position. The latter gradually show very evident changes in size and shape, and in the nature of their walls. The outermost layer, called the dermatogen or protoderm, is made up of somewhat flattened cells which will become the epidermis, a protective covering against entrance of disease-producing organisms and against excessive loss of water. Within the body of the stem tip certain strands of cells become distinct by their elongate shape and dense protoplasmic content. These procambium strands are the beginnings of the vascular tissue presently to appear. In cross sections of the stem the procambium appears as a ring of separate masses of cells. The remaining cells of the growing tip are parenchymatous cells, called the ground meristem, changed but little from the promeristem condition.

As the procambial cells grow older, they gradually change in form. Those cells which are nearest the center of the stem become xylem cells, those toward the circumference of the stem become phloem, with cambium cells separating the two types. In some stems the differentiation of cells continues until the procambial strands have united to form a continuous cylinder which gives place to concentric rings of xylem and phloem cells separated by a band of cambium cells. In other stems, the strands remain distinct, forming separate vascular bundles. The ground meristem or parenchyma in the center of the stem becomes the pith, that surrounding the strands becomes the cortex and pericycle, while the radiating masses of parenchyma cells between the separate strands make up the pith rays. All these tissues, derived indirectly from the differentiation of the promeristem cells, form the primary body, composed of primary tissues.

The epidermal cells are often somewhat elongated in the direction of the length of the stem. The outer wall of an epidermal cell is frequently much thickened and cutinized, so that it becomes impervious to water. Many stomata are found in the epidermis. As the diameter of the stem increases, the epidermal cells gradually become stretched until they finally break apart and are lost.

The cortex inside the epidermis is comprised of several kinds of cells. See Fig. 1. Those nearest the surface are the collenchyma cells. These are modified parenchyma cells, the walls of which are thickened in their angles. These cells usually contain chloroplasts. Because of their thickened walls, collenchyma cells give support to the stem, while at the same time they manufacture food. The parenchyma cells of the cortex are thin-walled, and

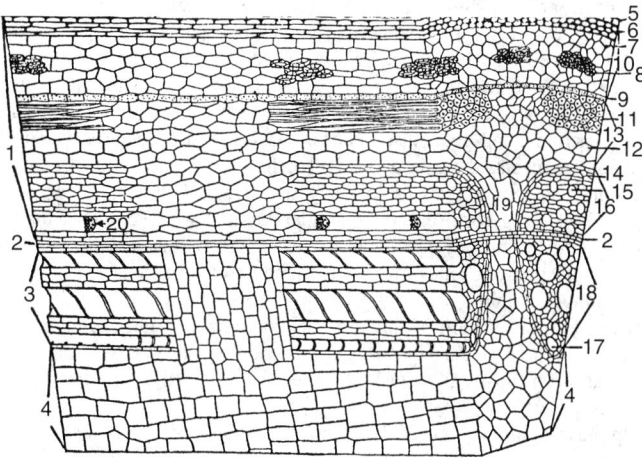

Fig. 1. Diagram showing the tissues derived from the primary meristems: (1) bark; (2) cambium; (3) wood; (4) pith; (5) epidermis; (6) collenchyma; (7) parenchyma of cortex; (8) stone cells; (9) starch sheath; (10) primary cortex; (11) pericyclic fibers; (12) pericyclic parenchyma; (13) pericycle; (14) phloem parenchyma; (15) sieve tube; (16) phloem; (17) vessel; (18) xylem; (19) pith ray; (20) sieve plate. (*After Stevens.*)

either rounded in shape or, through mutual pressure, more or less angular. Those near the surface of the stem often contain chloroplasts, and carry on photosynthesis. They give rigidity to the stem because of their turgor pressure, and also serve as storage tissue. In some stems there are also found in the cortex thick-walled sclerenchyma cells. These may be either long slender fibers or short stone cells. In a few plants the innermost cells of the cortex form a definite endodermis, the walls of each cell being much thickened.

The tissues inside the cortex include the pericycle, the vascular bundles, pith rays, and the pith. This is the stele.

The cells of the pericycle are very similar to those of the cortex, so much so that it is often very difficult to distinguish one from the other. In the pericycle of stems, which usually is much thicker than that in roots, sclerenchyma cells, both fibers and stone cells, may be found along with the parenchyma, just as in the cortex.

The cells comprising the vascular system — xylem and phloem — are very much modified. The procambial cells toward the center of the stem first elongate greatly, without appreciably increasing in diameter. Soon changes appear in the wall, secondary deposits of cellulose being laid down against the primary wall. The manner in which the wall is thickened varies in different cells. In some the thick deposits are in rings: these are formed while the cell is still elongating and so are gradually separated. Such cells are called annular cells. In other cells of the first-formed xylem, which is called protoxylem, the wall thickenings are in the form of spirals, producing spiral cells, which allow a certain amount of growth even when the thick wall is formed. In cells that differentiate later, when elongation has been completed, the thickening of the wall will be much more extensive, only irregularly distributed, narrow slits being left unthickened. These slits extend transversely in the wall of the cell. Such cells are called scalariform or reticulate cells. Differentiation of the xylem cells continues until most of the inner part of the procambial strand has been changed to xylem cells. Those that form after the narrow protoxylem cells are called metaxylem cells. Protoxylem and metaxylem together make up the primary xylem. It should be understood that, as the xylem cells develop, the end walls are cut away forming long tubes, called vessels or tracheae. Water moves rapidly through such tubes.

The cells on the outside of the procambial strand become the primary phloem cells. These are the sieve tubes and companion cells. Early stages in the formation of these cells are much like those of xylem cells, elongation first occurring and then changes in the cell wall. In the formation of phloem cells, a single cell divides into two, which become very unequal; the larger one continues to increase in size and loses its nucleus; the smaller one frequently divides again. The walls of these cells remain comparatively thin, with characteristic perforated thin places, called sieve plates, in the end walls of the larger cells, which form the sieve tubes. The smaller cells, called companion cells, are characterized by a dense cytoplasmic content, small vacuoles and prominent nucleus. They are connected with the sieve tubes by numerous small thin places, called simple pits, in their walls. Phloem cells are channels in which food passes through the stem. In addition to these various cells, the phloem contains parenchyma cells, which are used for storage of materials and in some plants long thick-walled fibers, called phloem or bast fibers.

Parenchyma and fibers also occur in the xylem. Between the xylem and phloem elements, there is a band of cells that remains unmodified and becomes an important tissue in many stems. This is the cambium, which by its divisions gives rise to the secondary tissues, which compose the bulk of the stems of woody plants. These secondary tissues are the secondary xylem and phloem, and differ only slightly from the primary xylem and phloem, being unlike mainly in their origin. Secondary rays, which provide for lateral translocation of food, are also produced by the cambium.

In the center of the stem is the pith, composed of large, thin-walled cells arranged in irregular fashion. They function principally as places of storage of food.

In nearly all monocotyledons, no cambium is formed, therefore the monocot stem is composed entirely of primary tissues. The arrangement of these tissues is vastly different from that of the stems of dicotyledons. The vascular tissues occur in the form of separate small bundles scattered throughout the stem. It is impossible to distinguish any limit that separates cortex from pericycle and pith. In many monocots the central portion of the stem is entirely free from bundles and recognized as a pith. Often the pith breaks up, forming a hollow stem.

All gymnosperms have woody stems. The development and structure of these are quite similar to that of dicotyledons, but, except in a few uncommon species, the xylem is composed entirely of tracheids, distinguished from the tracheae of the angiosperms by the fact that they are elongated cells instead of long tubes formed from many cells, and no companion cells are formed in the phloem.

In most plants, the function of the stem is to display the leaves and reproductive organs in the most favorable position and to carry materials from one part of the plant to another. In many plants, the stem is a highly specialized structure with different functions. Often these specialized stems take over the function of one of the other organs of the plant.

The outer tissues of the young stems of nearly all plants are green. Therefore some photosynthesis takes place in these tissues. There are many plants in which the stem is the principal, if not the only, place where photosynthesis occurs. In many cases, as in some of the cacti, the appearance of the stem differs very little from that of any other plant; but leaves are very much reduced or entirely lacking, all photosynthesis occurring in the stem. In other species of cactus, such as the Prickly Pear and the widely cultivated crab or Christmas cactus, the stem is very much flattened, but still distinctly recognizable as a stem. In some plants, however, the modification has become extreme. The ultimate branches of the stem have become very much flattened and have a shape which gives them every appearance of a leaf. Only their position in the axil of a tiny scale, the real leaf, betrays their true nature. The dainty Smilax, *Asparagus asparagoides*, of the florist, has branches of this kind. So also does the Butcher's Broom, a marsh plant of Europe, which is widely cultivated and appears during the Christmas season, stained a brilliant scarlet. The inconspicuous greenish-white flowers of this plant occur in the center of that part which is commonly assumed to be a leaf. The tiny needle-like "leaves" of the garden asparagus are really branches.

In a few plants, the stem becomes a very important reproductive part. See Fig. 2. Runners, long slender branches from the base of the stem, grow out horizontally over the surface of the ground, and root at their tip. There a new plant is formed. With death and disintegration of the connecting stem, the young plant becomes separate from its parent. This is a form of vegetative reproduction. A stolon differs very little from a runner; it is a prostrate branch, which regularly roots at its nodes, and sends up new plants not only at its tip but also from the nodes. Rootstocks or rhizomes are spreading underground branches, which produce adventitious roots at their nodes and often spread widely.

Many rhizomes become very fleshy because of an accumulation in them of food materials. Their principal function then is storage. Storage occurs in the stems of many different plants. If only a portion of the rhizome becomes enlarged, it is called a tuber. The common white potato is a very familiar tuber, which is formed at the tip of a slender rhizome. The "eyes" of the potato are really the nodes of the stem; from them buds will give rise to branches when the potato grows. More modified than the tuber is the corm. This is a short thick erect rootstock. Often it is much broader

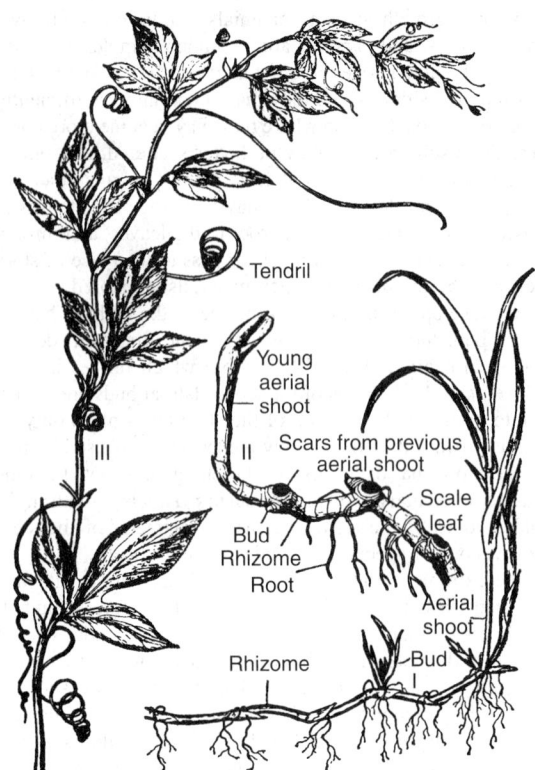

Fig. 2.   Types of modified stems: (1) rhizome or quack grass (*Agropyron repens*); (II) rhizome of Solomon's Seal (*Polygonatum commutatum*); (III) stem of passion flower (*Passiflora incarnata*) with tendrils that are modified stems.

than it is long. Buds are formed on the upper surface of the corm. Each of these buds grows into a new plant, exhausting the substance of the old corm. New corms form at the base of the old one. Corms known to all are those of *gladiolus* and *crocus*. These are usually incorrectly called bulbs. See Fig. 3. A bulb is a fleshy bud, composed of a short thick stem and many fleshy or scaly leaves or leaf bases. Onions form true bulbs, as do tulips, hyacinths and many lilies.

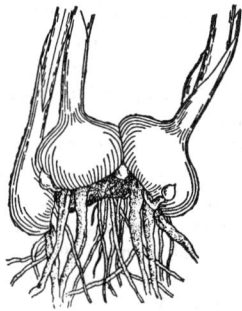

Fig. 3.   Corms of Gladiolus, consisting chiefly of fleshy stems.

The materials stored in the stems so far considered are mainly reserve food. Other stems become swollen with stored water. Many cacti and Euphorbias have stems of this sort.

Thorns are usually stems or branches that have become stiff and pointed and serve to protect the plant. Tendrils are organs that support a plant as it grows up through other plants. Not all tendrils are stems. Some, like those of the grape, are definitely so; others are leaves or parts of leaves. Usually plants which have tendrils have slender stems which lack sufficient strength to support themselves. Plants of this type are vines. There are several types of vines, or climbing plants. One of them supports itself by twining tightly around any available support. The direction of twining is very constant for any species, some invariably turning in a clockwise direction and others counterclockwise.

Other climbers support themselves by adventitious roots, which form in abundance and cling tightly to any support. Other climbing plants are supported solely by the presence of many spines or prickles, often hooked, or pointed backwards so that the stem does not easily slide off any object on which it rests. Climbing roses illustrate this type of climber. But many tropical vines are much better illustrations. Often these grow to great lengths, hanging in long festoons from the tops of tall trees, or growing in tangled masses over low shrubby plants. In these tropical climbers, which are commonly called lianas, the stems often assume curious flattened or fluted or irregular shapes. In diameter they vary from a fraction of an inch to many inches; they may attain a length of 400 or 500 feet (122 or 152 meters). They are one of the most characteristic and annoying features of the tropical rain forest.

Additional Reading

Behnke, H.D., K. Esser, J.W. Kadereit, and U. Luttge: "Progress in Botany: Genetics, Cell Biology, Physiology Ecology and Vegetation Science," Springer-Verlag, Inc., New York, NY, 1998.

Fosket, D.E.: "Plant Growth and Development: A Molecular Approach," Academic Press, Inc., San Diego, CA, 1994.

Gartner,B.L.: "Plant Stems: Physiology and Functional Morphology," Academic Press, Inc., San Diego, CA, 1995.

Steeves, T.A. and I.M. Sussex: "Patterns in Plant Development," 2nd Edition, Cambridge University Press, New York, NY, 1989.

Stern, K.R.: "Introduction to Plant Biology," 8th Edition, The McGraw-Hill Companies, Inc., New York, NY, 2000.

Taiz, L. and E. Zeiger: "Plant Physiology;" Singular Publishing Group, Inc., San Diego, CA, 1998.

STEPHANITE.   The mineral stephanite, silver antimony sulfide, Ag_5SbS_4, is found in short prismatic or tabular orthorhombic crystals. It is a brittle mineral; hardness, 2–2.5; specific gravity, 6.25; metallic luster; color, black; streak, black; opaque.

Stephanite occurs associated with other silver minerals and is believed to be primary in character. Localities are in the Czech Republic and Slovakia, Saxony, the Harz Mountains, Sardinia; Cornwall, England; Chile and Mexico. In the United States it is found in Nevada, where it is an important silver ore. It was named for the Archduke Stephan of Austria, mining director of that country at the time this mineral was first described.

STEPPE.   An extensive, treeless grassland area in southeastern Europe and Asia developing in the semiarid mid-latitudes of that region. They are generally considered drier than the prairie that develops in the subhumid mid-latitudes of the United States.

STEPPER MOTORS.   The importance and usage of the electric stepper motor has increased immensely during the past two decades. Stepper motors, for example, are widely used in modern electronic typewriters, word processors, and other computerized products. They are important components of industrial robotic systems, large and small, and in other machines where repeatable positioning control is required. Stepper motors for most applications must be designed to work reliably over hundreds of thousands to millions of cycles.

The advantages of stepper motors are low cost, ruggedness, construction simplicity, high reliability, no maintenance, wide acceptance, no "tweaking" to stabilize, and no feedback components needed; they are inherently fail-safe and tolerate most environments. Steppers are simple to drive and control in an open-loop configuration. They provide excellent torque at low speed, up to 5 times the continuous torque of a brush motor of the same frame size or double the torque of the equivalent brushless motor. Frequently, a gearbox can be eliminated. A stepper-driven system is inherently stiff, with known limits to the dynamic position error.

There are three main types of stepper motors.

Permanent-Magnet Motors. The tin-can, or "canstack," motor shown in Fig. 1 is perhaps the most widely used type in commercial, nonindustrial applications. It is essentially a low-cost, low-torque, low-speed device ideally suited for use in computer peripherals, for example. The motor construction results in relatively large step angles, but the overall simplicity favors high-volume, low-cost production. The axial-air gap or disk motor is a variant of the permanent-magnet design, which achieves higher performance mainly because of its very low rotor inertia. This does restrict the applications of the motor to those situations involving little inertia, such as positioning the print wheel in a daisy-wheel printer.

Disadvantages of the stepper motor include resonance effects, relatively long settling times, and rough performance unless a microstepper is

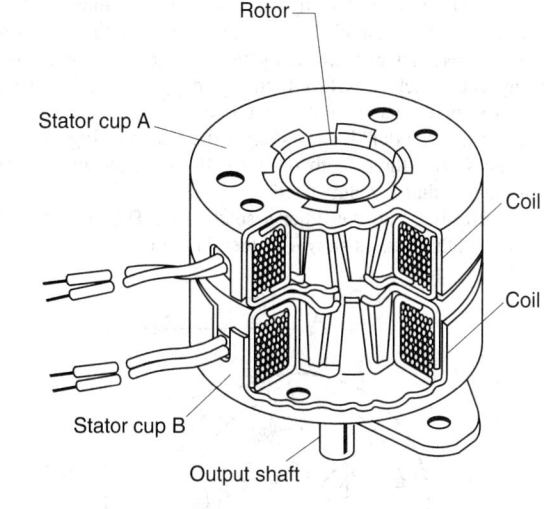

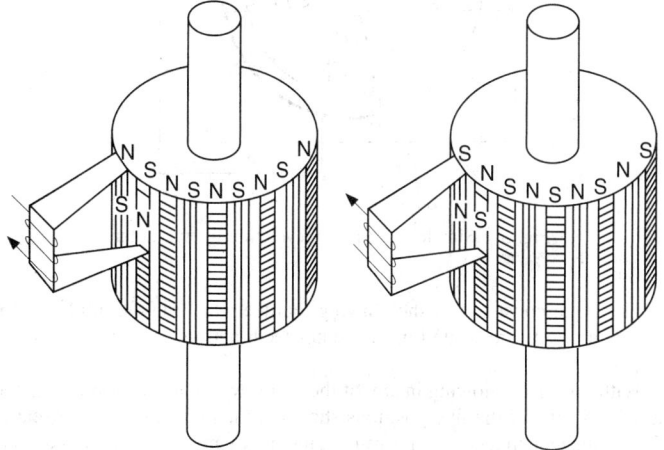

Fig. 1. Canstack, or permanent-magnet, stepper motor. (*Top*) sectional view; (*bottom*) as used for positioning print wheel in a daisy-wheel printer. (*Airpax Corp., U.S.*)

used. Undetected position loss may result in open-loop systems. Steppers consume current regardless of load conditions, and they tend to run hot. Steppers tend to be noisy, especially when operated at high speeds. Some of the foregoing limitations can be overcome by use of a closed-loop system.

Variable-Reluctance Motors. There is no permanent magnet in a variable-reluctance motor. Thus the rotor spins freely without detent torque. Torque output for a given frame size is restricted, although the torque-to-inertia ratio is good. This type of motor is frequently used in small sizes for applications such as micropositioning tables. Variable-reluctance motors are seldom used in industrial applications. Having no permanent magnet, these motors are not sensitive to current polarity and thus require a different driving arrangement compared to other types. See Fig. 2.

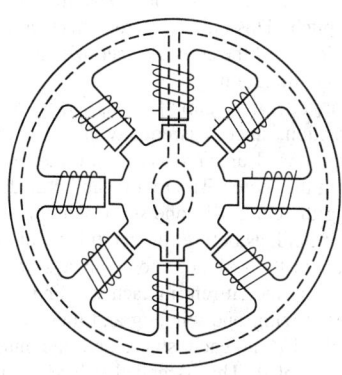

Fig. 2. Variable-reluctance motor.

Hybrid Stepper Motors. The hybrid motor is the most widely used stepper motor in industrial applications. Most hybrid motors are two-phase, although five-phase designs are available. A recent development is the enhanced hybrid, which uses flux-focusing magnets to give a significant improvement in performance, but at extra cost.

The rotor of the "model" hybrid stepper illustrated in Fig. 3 consists of two pole pieces with three teeth on each. Between the pole pieces is a permanent magnet that is magnetized along the axis of the motor, making one end a north pole, and the other a south pole. The teeth are offset at the north and south ends, as shown in the diagram.

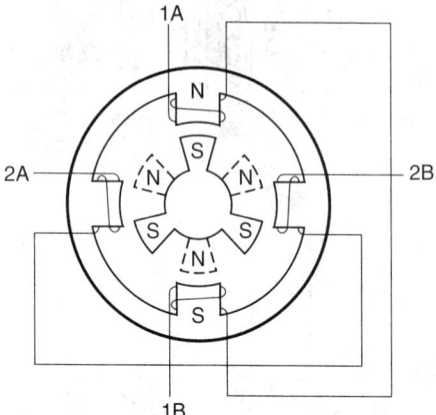

Fig. 3.   Simple 12-step-per-revolution hybrid motor.

The stator consists of a shell having four teeth that run the full length of the rotor. Coils are wound on the stator teeth and are connected together in pairs.

With no current flowing in any of the motor windings, the rotor will tend to take up one of the five positions shown in the diagram. This is because the permanent magnet in the rotor attempts to minimize the reluctance, or magnetic resistance, of the flux path from one end to the other. This occurs when a pair of north- and south-pole rotor teeth are aligned with two of the stator poles. The torque, tending to hold the motor in one of these positions, is usually small and called the detent torque. The motor shown has 12 possible detent positions.

If current is passed through one pair of stator windings, as shown in Fig. 4(a), the resulting north and south stator poles will attract teeth of the opposite polarity on each end of the rotor. Thus there are only three stable positions for the rotor, the same as the number of rotor teeth. The torque required to deflect the rotor from its stable position is thus much greater and is referred to as the holding torque.

By changing the current flow from the first to the second set of stator windings [Fig. 4(b)], the stator field rotates through 90° and attracts a new pair of rotor poles. This results in the rotor turning through 30°, corresponding to one full step. Reverting to the first set of stator windings, but energizing them in the opposite direction, the stator field will be rotated through another 90° and the rotor will take another 30° step, as shown in Fig. 4(c). Finally, the second set of windings is energized in the opposite direction [Fig. 4(d)] to give a third step position. Returning to the first condition [Fig. 4(a)] and after these four steps, the rotor will have moved through one tooth pitch. This simple motor, therefore, performs 12 steps per revolution. Obviously, if the coils are energized in the reverse sequence, the motor will change direction.

If the two coils are energized simultaneously (Fig. 5), the rotor takes up an intermediate position, since it is equally attracted to two stator poles. Greater torque is produced under these conditions because all the stator poles are influencing the motor. The motor can be made to take a full step simply by reversing the current in one set of windings. This causes a 90° rotation of the stator field, as before. In fact, this would be the normal way of driving the motor in the full-step mode, always keeping two windings energized and reversing the current in each winding alternately.

By alternately energizing one winding and then two (Fig. 6), the rotor moves through only 15° at each stage, and the number of steps per revolution will be doubled. This is called half-stepping. Most industrial applications make use of this stepping mode. Although sometimes there is

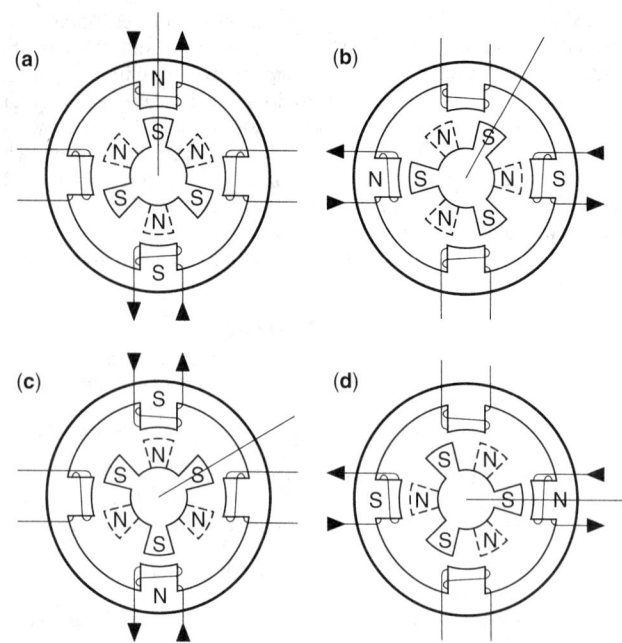

Fig. 4.   Full stepping, one phase on. (*Parker-Hannifin Corp., Compumotor Div.*)

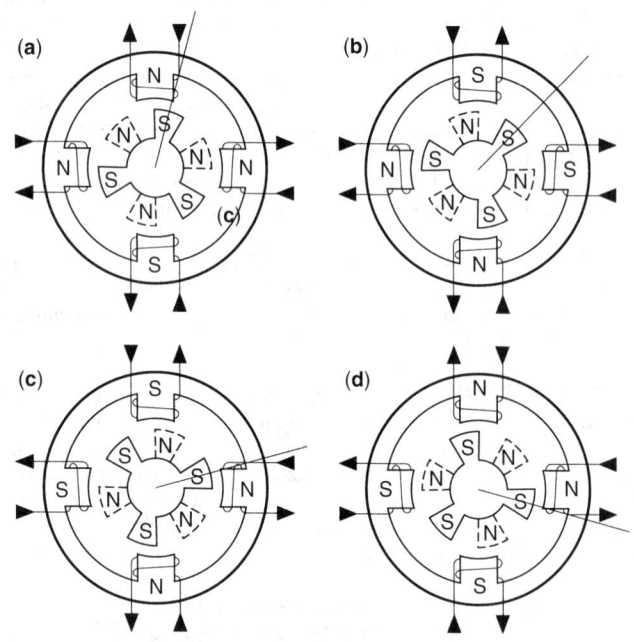

Fig. 5.   Full stepping, two phases on. (*Parker-Hannifin Corp., Compumotor Div.*)

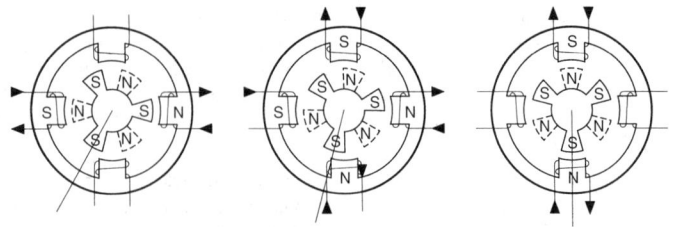

Fig. 6.   Half stepping. (*Parker Hannifin Corp., Compumotor Div.*)

a slight loss of torque, this mode results in much better smoothness at low speeds, and less overshoot and ringing occur at the end of each step.

Current Patterns in Motor Winding. When the motor is driven in its full-step mode, energizing two windings, or phases, at a time (Fig. 7), the torque available on each step will be the same (subject to very small variations in the motor and drive characteristics). In the half-step mode,

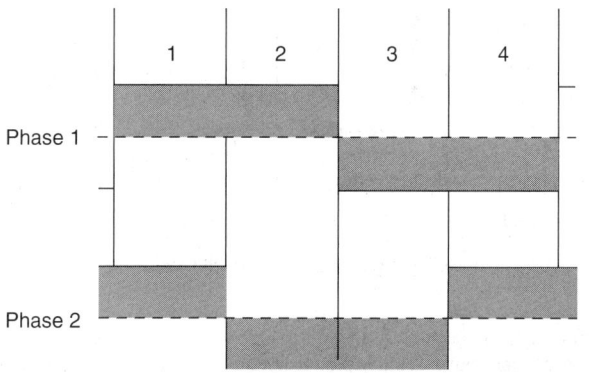

Fig. 7. Full step current, two phases on. (*Parker-Hannifin Corp., Compumotor Div.*)

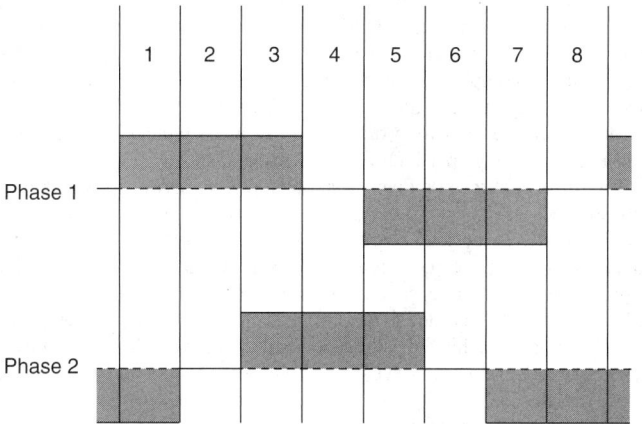

Fig. 8. Half-step current (*Parker-Hannifan Corp., Compumotor Div.*)

two phases are alternately energized and then only one, as shown in Fig. 8. Assuming the drive delivers the same winding current in each case, this will cause greater torque to be produced when there are two windings energized—that is, alternate steps will be strong and weak. Although the available torque obviously is limited by the weaker step, there is a improvement in low-speed smoothness over the full-step mode.

The motor designer would like to produce approximately equal torque on every step and to have this torque be at the level of the stronger step. This goal can be achieved by using a higher current level when there is only one winding energized. This does not overly dissipate the motor because the manufacturer's current rating will assume two phases to be energized. (The current rating is based on the allowable case temperature.) With only one phase energized, the same total power will be dissipated if the current is increased by 40%. Using the higher current in the one-phase-on state produces approximately equal torque on alternate steps, as indicated in Fig. 9.

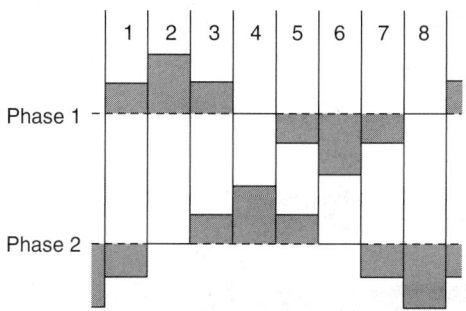

Fig. 9. Half-step current, profiled. (*Parker-Hannifin Corp., Compumotor Div.*)

Microstepping. It will be noted from the prior discussion that energizing both phases with equal currents produces an intermediate step position halfway between the one-phase-on positions. If the two phase currents

are unequal, the rotor position will be shifted toward the stronger pole. This effect is utilized in the microstepping drive, which subdivides the basic motor step by proportioning the current in the two windings. In this way the step size is reduced and the low-speed smoothness is improved dramatically. High-resolution microstep drives divide the full motor step into as many as 500 microsteps, giving 100,000 steps per revolution. In this situation the current pattern in the windings closely resembles two sine waves with a 90° phase shift between them (Fig. 10). Thus the motor is driven very much as though it were a conventional ac synchronous motor. In fact, the stepper motor can be driven in this manner from a 60-Hz (U.S.) or 50-Hz (Europe) sine-wave source by including a capacitor in series with one phase. It will rotate at 60 r/min.

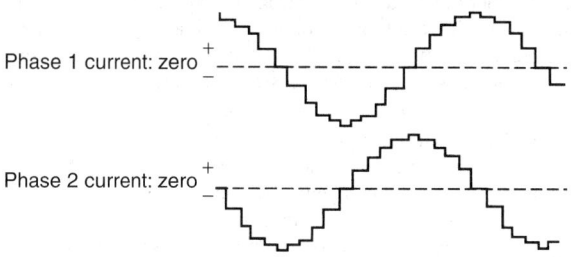

Fig. 10. Phase currents in microstep mode. (*Parker-Hannifin Corp., Compumotor Div.*)

Standard 200-Step Hybrid Motor. The standard stepper motor operates in the manner just described as a model, but it has a greater number of teeth on the rotor and stator, giving a smaller basic step size. The rotor is in two sections, as described previously, but has 50 teeth on each section. The half-tooth displacement between the two sections is retained. The stator has eight poles, each with five teeth, making a total of 40 teeth (Fig. 11).

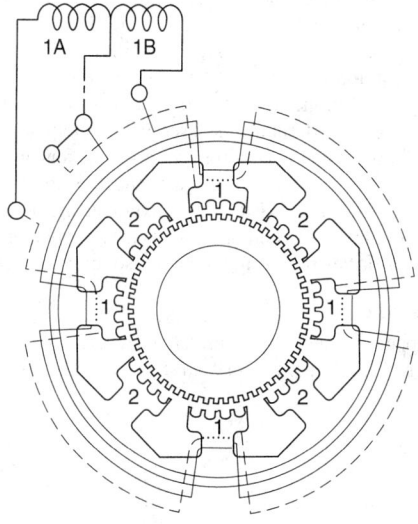

Phase 1 wind shown
Phase 2 windings on intermediate poles

Fig. 11. 200-step hybrid motor. (*Parker-Hannifin, Compumotor Div.*)

Visualize that a tooth is placed in each of the gaps between the stator poles, in which case there would be a total of 48 teeth, two less than the number of rotor teeth. If rotor and stator teeth were aligned at 12 o'clock, they would also be aligned at 6 o'clock. But at 3 and 9 o'clock the teeth would be misaligned. However, due to the displacement between the sets of rotor teeth, alignment will occur at 3 o'clock and 9 o'clock at the other end of the rotor.

In practice, the windings are arranged in sets of four and wound such that diametrically opposite poles are the same. Thus, referring to Fig. 11, the north poles at 12 and 6 o'clock attract the south-pole teeth at the front of the rotor and the south poles at 3 and 9 o'clock attract the north-pole teeth at the back. By switching current to the second set of coils, the stator field pattern rotates through 45°, but to align with this new field, the rotor only has to turn through 1.8°. This is equivalent to one-quarter of a tooth pitch on the rotor, giving 200 full steps per revolution.

Note that there are as many detent positions as there are full steps per revolution, namely, 200. The detent positions correspond with rotor teeth being fully aligned with stator teeth. When power is applied to a stepper drive, it is usual for it to energize in the zero-phase state in which there is current in both sets of windings. The resulting rotor position does not correspond with a natural detent position, so an unloaded motor will always move by at least one-half step at power on. Of course, if the system were turned off other than in the zero-phase state, or the motor is moved in the meantime, a greater movement may be seen at power-up.

For a given current pattern in the windings there are as many stable positions as there are rotor teeth (50 for a 200-step motor). If a motor is desynchronized, the resulting position error will always be a whole number of rotor teeth, or a multiple of 7.2°. A motor cannot "miss" individual steps. Position errors of one or two steps may be due to noise, spurious step pulses, or a controller fault.

Bifilar Windings. Most motors are described as being bifilar wound, which means there are two identical sets of windings on each pole. Two lengths of wire are wound together as though they were a single coil. This produces two windings that are electrically and magnetically almost identical. If one coil were wound on top of the other, even with the same number of turns, the magnetic characteristics would be different.

The origin of the bifilar winding goes back to the unipolar drive. Rather than reversing the current in one winding, the field may be reversed by transferring current to a second coil wound in the opposite direction. (Although the two coils are wound the same way, interchanging the ends has the same effect.) Thus, with a bifilar-wound motor, the drive can be kept simple. However, this requirement has now largely disappeared with the widespread availability of the more efficient bipolar drive. Nevertheless, the two sets of windings do provide additional flexibility.

If all the coils in a bifilar-wound motor are brought out separately, there will be a total of eight leads (Fig. 12). This is becoming the most common configuration, since it gives the greatest flexibility. However, there are still a number of motors produced with only six leads, one lead serving as a common connection to each winding in a bifilar pair. This arrangement limits the range of applications of the motor, since the windings cannot be connected in parallel. Some motors are made with only four leads. These are not bifilar-wound and cannot be used with a unipolar drive. There is obviously no alternative connection method with a four-lead motor, but in many applications this is not a drawback and the problem of insulating unused leads is avoided. Occasionally a five-lead motor may be encountered. These should be avoided inasmuch as they cannot be used with conventional bipolar drives requiring electrical isolation between the phases.

Linear Stepper Motors

The linear stepper is essentially a conventional rotary stepper that has been "unwrapped" so that it operates in a straight line. The moving component

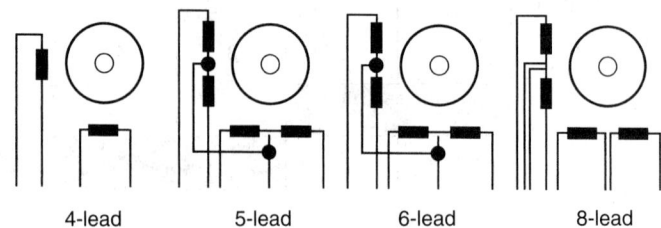

Fig. 12. Motor lead configurations.

is referred to as the forcer, and it travels along a fixed element, or platen. For operational purposes the platen is equivalent to the rotor in a normal stepper, although it is an entirely passive device and has no permanent magnet. The magnet is incorporated in the moving forcer together with the coils (Fig. 13).

The forcer is equipped with four pole pieces, each having three teeth. The teeth are staggered in pitch with respect to those on the platens so that switching the current in the coils will bring the next set of teeth into alignment. A complete switching cycle (four full steps) is equivalent to one tooth pitch on the platen. Like the rotary stepper, the linear motor can be driven from a microstep drive. In this case a typical linear resolution will be 12,500 steps per inch (4921 steps/cm).

The linear motor finds favor in applications involving a low mass to be moved at very high speed. In a lead-screw-driven system the predominant inertia usually is the lead screw rather than the load to be moved. Hence most of the motor torque goes to accelerate the lead screw, and this problem becomes more severe the longer the travel required. In using a linear motor, all the developed force is applied directly to the load and the performance achieved is independent of the length of the move. A screw-driven system can develop greater linear force and better stiffness. However, the maximum speed may be as much as 10 times higher with the equivalent linear motor.

With further reference to Fig. 13, the forcer consists of two electromagnets A and B and a strong rare-earth permanent magnet. The two pole faces of each electromagnet are toothed to concentrate the magnetic flux. Four sets of teeth on the forcer are spaced in quadrature so that only one set at a time can be aligned with the platen teeth.

The magnetic flux passing between the forcer and the platen gives rise to a very strong force of attraction between the two pieces. The attractive force can be up to 10 times the peak holding force of the motor, requiring a bearing arrangement to maintain precise clearance between the pole faces and the platen teeth. Either mechanical roller bearings or air bearings are used to maintain the required clearance.

When current is established in a field winding, the resulting magnetic field tends to reinforce permanent magnetic flux at one pole face and cancel it at the other. By reversing the current, the reinforcement and cancellation are exchanged. Removing current divides the permanent magnetic flux

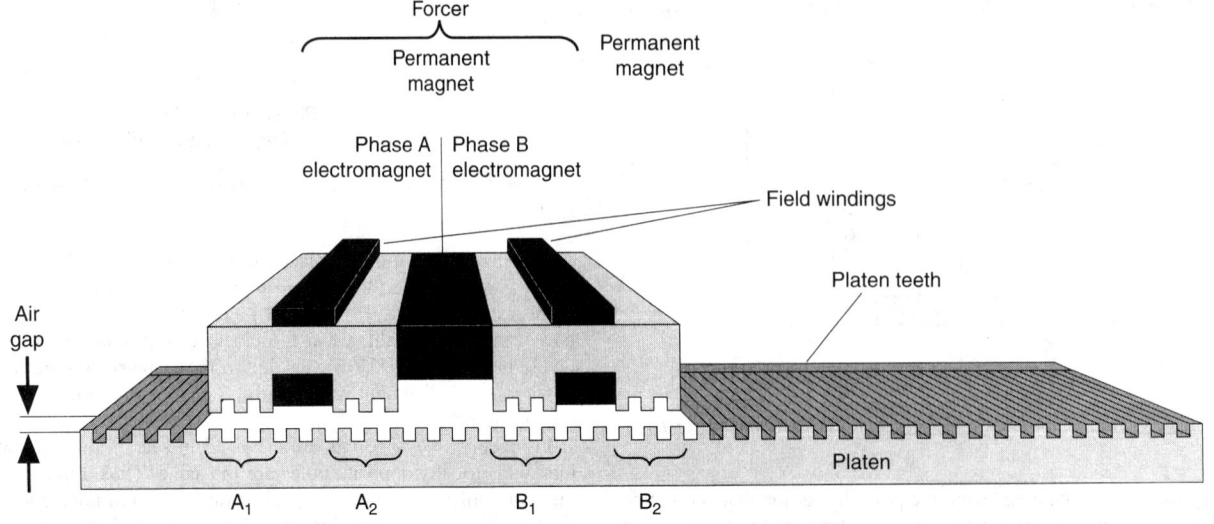

Fig. 13. Principle of linear stepping motor.

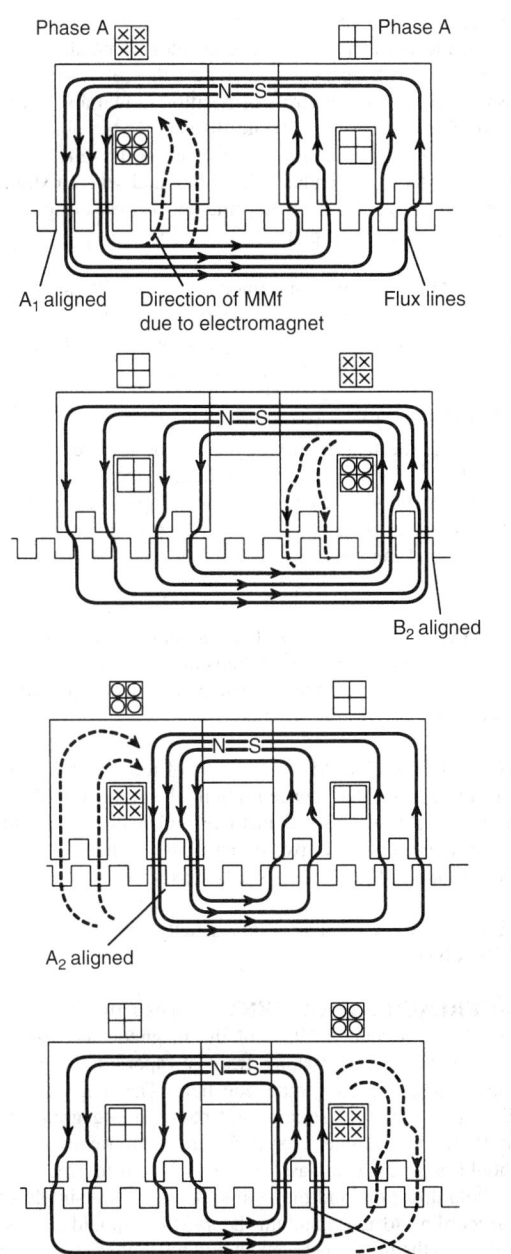

Fig. 14.   Four cardinal states of full steps of the force. (*Parker-Hannifin Corp., Compumotor Div.*)

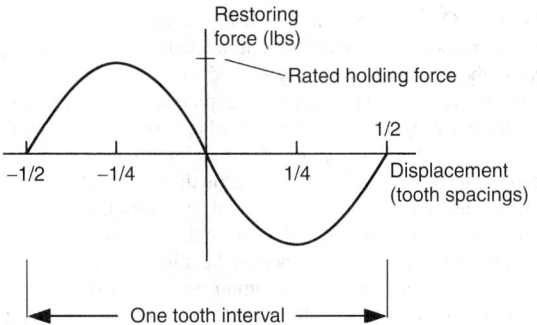

Fig. 15.   Force versus displacement (linear stepper). (*Parker-Hannifin Corp., Compumotor Div.*)

equally between the pole faces. By selectively applying current to phases *A* and *B* it is possible to concentrate the flux at any of the forcer's four pole faces. The face receiving the highest flux concentration will attempt to align its teeth with the platen. Fig. 14 shows the four primary states

or full steps of the forcer. The four steps result in motion of one tooth interval to the right. Reversing the sequence moves the forcer to the left.

Repeating the sequence in the example will cause the forcer to continue its movement. When the sequence is stopped, the forcer stops, with the appropriate tooth set aligned. At rest, the forcer develops a holding force that opposes any attempt to displace it. As the resting motor is displaced from equilibrium, the restoring force increases until the displacement reaches one-quarter of a tooth interval (Fig. 15). Beyond this point the restoring force drops. If the motor is pushed over the crest of its holding force, it slips or jumps rather sharply and comes to rest at an integral number of tooth intervals away from its original location. If this occurs while the forcer is traveling along the platen, it is referred to as a stall condition.

Linear Step Motor Characteristics. These include velocity ripple, platen mounting, environment, life expectancy, yaw (plus pitch and roll), and accuracy. To summarize, the worstcase accuracy of a linear step motor can be given by

$$\text{Accuracy} = A + B + C + D + E + F$$

where A = cyclic error due to motor magnetics, which recurs once every pole pitch as measured on motor body
B = unidirectional repeatability — error measured by repeated moves to the same point from different distances in the same direction
C = hysteresis — backlash of motor when changing direction due to magnetic nonlinearity and mechanical friction
D = cumulative platen error — linear error of platen as measured on motor body
E = random platen error — nonlinear errors remaining in platen after linear error is disregarded
F = thermal expansion error — error caused by change in temperature, expanding or contracting the platen

Additional Reading

Gieras, J.F. and M. Wing: "Permanent Magnet Motor Technology: Design and Applications," Marcel Dekker, Inc., New York, NY, 1996.
Gottlieb, I.M.: "Electric Motors and Control Techniques," 2nd Edition, The McGraw-Hill Companies, Inc., New York, NY, 1994.
Hand, A.: "Electric Motor Maintenance and Troubleshooting," The McGraw-Hill Companies, Inc., New York, NY, 2001.
Hanselman, D.C.: "Brushless Permanent-Magnet Motor Design," The McGraw-Hill Companies, Inc., New York, NY, 1994.
Hendershot, J.R. and T.J. Miller: "Design of Brushless Permanent-Magnet Motors," Vol. 37, Oxford University Press, Inc., New York, NY, 1994.
Yeadon, W.H. and A.W. Yeadon: "Handbook of Small Electric Motors," The McGraw-Hill Companies, Inc., New York, NY, 2001.

STEP RESPONSE TIME. Of a system or an element, the time required for an output to make the change from an initial value to a large specified percentage of the final steady-state value, either before or in the absence of overshoot, as a result of a step change to the input. See figure that accompanies entry on **Response (Instrument)**. This time is usually stated for 90, 95 or 99% change.

Time Constant. The value T in an exponential response term $A \exp(-t/T)$ or in one of the transform factors $1 + sT$, $1 + j\omega T$, $1/(1 + sT)$, $1/(1 + j\omega T)$, where s = complex variable; t = time, seconds; T = time constant; $j = \sqrt{-1}$; ω = frequency, radian/second.

For the output of a first-order (lag or lead) system forced by a step or an impulse, T is the time required to complete 63.2% of the total rise or decay; at any instant during the process, T is the quotient of the instantaneous rate of change divided into the change to be completed. In higher-order systems, there is a time constant for each of the first-order components of the process. In a Bode diagram, breakpoints occur at $\omega = 1/T$.

STEPTOE. A hill or mountain whose top projects above a lava flow which has surrounded its lower flanks.

STERADIAN. The unit solid angle that cuts unit area from the surface of a sphere of unit radius centered at the vertex of the solid angle. There are 4π steradians in a sphere.

STEREO BROADCASTING. A system of radio broadcasting using frequency modulation in which the modulating signal contains information obtained from a stereo microphone which following its modulation, transmission, and reception can subsequently be reproduced by a stereo loudspeaker. The complete communication process involves generating electric signals that correspond to variations in sound intensity at two different points in the physical location where the program originates, transmitting a combination of these two signals by means of a frequency modulated radio transmitter, separating the signals in the listener's receiver and supplying the components to the loudspeaker system.

In the United States, transmission is in accordance with standards established by the Federal Communications Commission. The electrical signals obtained from the two microphones are arbitrarily designated Left (L) and Right (R) signals. They are combined to produce a sum signal (L + R) and difference signal (L − R). According to the standards, each of the audio signals may contain frequency components in the range of 50–15,000 Hz. A frequency translation of the L − R signal spectrum is effected by amplitude modulation on a subcarrier of 38 kHz which is then suppressed. To the combination of the L + R spectrum and the translated L − R spectrum is added a 19 kHz pilot signal, which is to be used to aid in the restoration of the L and R signals after demodulation of the transmitted signal in the listener's receiver. The distribution of the composite modulating signal, which is used for the frequency modulation of the transmitter, is depicted in Fig. 1. Radio transmission takes place in the frequency range 88.1 to 107.9 Hz.

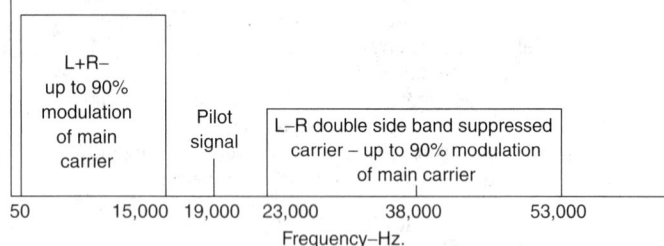

Fig. 1.   Frequency distribution of composite modulating signal for stereo broadcasting.

Upon reception, after demodulation, which recovers the composite modulating signal, the signal requires processing to recover the L and R audio signals originally developed by the two microphones. The inverse of the frequency translation which occurred prior to the frequency modulation is now performed on the L − R signal. The composite signal is passed through a bandpass filter, which passes only frequencies in the range from 23 to 53 kHz. The 19 kHz pilot signal is also extracted using a filter and is then passed through a doubler-amplifier circuit from which a signal of 38 kHz is obtained. This signal and the output of the bandpass filter are applied to a detector in which *heterodyne* action occurs. The resultant output from the detector has the waveform of the L − R audio signal at the transmitter, together with some extraneous signals that were introduced by the modulation and demodulation processes. The L + R and L − R signals are now fed to a circuit which separates them into the L and R signals in a manner inverse to that in which they were obtained at the transmitter. After further amplification, the L and R signals are furnished to an appropriate stereophonic loudspeaker system. It should be noted that the L + R signal not only provides one portion of the stereo signal but also serves as the compatible monophonic signal for those listeners whose receivers are not equipped for stereo reception and reproduction.

See also **Radio Communication**.

STEREOGRAPHIC PROJECTION.   1.   The stereographic projection of a sphere on a plane is defined as follows: For a given point P, called the pole, on the surface of a sphere S, and for a given plane M not passing through P, and perpendicular to a diameter through P, the line joining P with a variable point Q and M intersects S in a second point R. This mapping of the points R of the sphere S on the points Q of M is called a stereographic projection of S on M.

2.   This type of map projection is used to some extent by navigators, but more commonly for maps of the celestial sphere, particularly in constructing the basic map for star finders. A stereographic projection has a valuable property in that circles on the sphere appear as circles on the projected map. Because of this property, several attempts have been made to use the projection for drawing lines of position obtained from celestial objects. In the proposed methods, the subastral point would be plotted on the chart and the circle of position drawn about this point with a radius equal to the observed zenith distance of the object. This method is proposed for use by aviators, where speed in drawing lines of position and obtaining a fix is particularly desirable, and where extreme accuracy is never possible because of errors inherent in measuring altitude with a bubble sextant in a moving plane.

STEREO-POWER.   For prism binoculars or similar stereo systems, the ratio of the distance between the objective axes to the distance between eyepiece axes multiplied by the magnifying power. A measure of the stereoscopic radius.

STEREOSCOPE.   The sensation of depth in an object is due to binocular vision; that is, to the fact that two eyes do not each see exactly the same view. By taking two pictures with a camera moved a few inches — or with a double stereoscopic camera — two slightly different pictures are obtained. A stereoscope is a device by which each eye sees only one of these pictures, and the same sensation of depth is obtained as with direct binocular vision.

STEREO SYSTEM.   An acoustical system in which a plurality of microphones (or other transducers), transmission channels and reproducers are arranged so as to provide a sensation of spatial distribution of the original sound sources to the listener.

STEREOSPECTROGRAM.   A method of representing spectral data in which the three variables, concentration of solute, optical density, and wavelength of light, are plotted in three dimensions to produce a three-dimensional figure; or else in two dimensions by choosing an oblique axis in addition to the customary x-axis and y-axis.

STERLING SILVER.   Silver alloy, usually with copper, containing at least 92.5% silver.

STERN-GERLACH EXPERIMENT.   An experimental test by O. Stern and W. Gerlach (Germany, 1924) of the magnetic moment of atoms. A stream of metallic atoms, issuing from a vaporizing furnace through a narrow slit, entered a strong magnetic field. The magnetic intensity was perpendicular to the atom stream, and had a strong gradient in its own direction. If magnetic moments of atoms are due to revolving electrons, the atoms should, according to classical theory, begin to precess at all angles about the field direction, and the atomic beam should simply broaden into a band. According to the quantum theory, they should precess at certain angles only, and the original stream should be divided into distinct streams. Since the beam was split into $2J + 1$ different beams, the experiment showed that in a magnetic field not all orientations to the field, but only $2J + 1$ discrete directions, are possible. See also **Magnetism**; and **Precession**.

STEROIDS.   Organic compounds characterized from a structural standpoint by the cyclopentanophenanthrene nucleus as shown in Fig. 1. Biochemically, the steroids are closely related to the terpenes. Steroids occur widely in nature, both in animals and plants. Many steroids are hormones, such as estrogens and cortisone, which are produced by the body's endocrine system and which are of great importance in the regulation of numerous body processes, such as growth and metabolism. Similarly, steroid hormones are important to several physiological processes within plants. Auxin, for example, is a plant growth hormone that regulates longitudinal cell structure so as to permit bending of the stalk or stem in phototropic response. The most common animal steroid, a steroid alcohol (or *sterol*) is *cholesterol*, which is the precursor of bile acids, steroid hormones, and provitamin D3. It should be stressed, however, that not all steroids are hormones; and not all hormones are steroids.

Over the last 35 to 40 years, steroid therapy, that is, the medical augmentation of steroid hormones that are insufficient in the body (as well as treating diseases and elements which result from an overabundance of certain steroid hormones in the body), is one of the major chapters in the history of medical progress. Similarly, the understanding of steroid chemistry has contributed markedly to plant biology, notably to plant

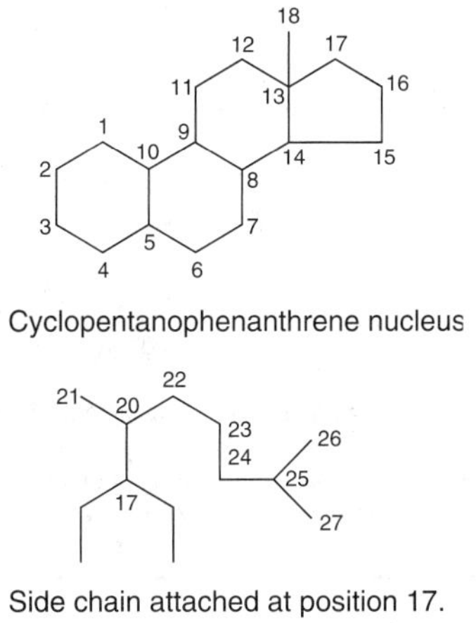

Cyclopentanophenanthrene nucleus

Side chain attached at position 17.

Fig. 1. Steroid molecule nucleus and side chain.

of synthesis, but for which there are no known counterparts in nature. Some of the steroid hormones, including steroidal synthetics, of importance medically and to scientific investigations, are listed and described briefly in Table 1.

The early history of steroid chemistry commenced with the observations of Mauthner, Windlaus, Wieland, Jacobs, Diels, and other organic chemists and biologists in their early observations on the products of oxidation, aromatization, and other reactions of cholesterol, bile acids, and plant glycosides. The interrelationship between sterols and bile acids was recognized early during these investigations. Shortly after the steroid character of the female and male sex hormones had been established, x-ray crystallography demonstrated errors in early theories concerning molecular structure. Rosenheim and King, working with monomolecular layers and x-ray evidence gained by Bernal, formulated the concept of the cyclopentanophenanthrene nucleus.

Once the skeleton of the steroids was established, there remained the task of understanding the steric relationships of the molecules. With reference to Fig. 1, it will be noted that there are nine asymmetric carbon atoms in the steroid skeleton—C_5, C_{10}, C_9, C_8, C_{14}, and C_{13} in the ring system. There are also two asymmetric carbons in the side chain attached at C_{17}. These are C_{20} and C_{25}. With reference to Figs. 1 and 2, the relative configuration of C_5 and C_{10}, of C_9 and C_8, and of C_{14} and C_{13}, determines whether the junctions between rings a/b, b/c, and c/d, respectively, are *trans* or *cis*. According to an arbitrary convention, one designates the substituent

breeding and the development of plant growth regulators. Commencing with the isolation of steroid hormones from natural sources, techniques were later developed to synthesize a number of hormones. Then, a further step involved the application of steroid hormones, developed by way

---

TABLE 1. REPRESENTATIVE STEROID HORMONES AND STEROIDAL SYNTHETICS[1]

Antiinflammatory, Antiallergic, and Antirheumatic Agents (Adrenal Corticosteroids)

Betamethasone (9-fluoro-16β-methylprednisolone; 16β-methyl-11b,17α,21-trihydroxy-9α-fluoro-1,4-pregnadiene-3,20-dione). $C_{22}H_{29}O_5F$, mw = 329.5. Also, the *betamethasone acetate*, $C_{24}H_{31}O_6F$, mw = 434.5; and *betamethasone disodium phosphate*, $C_{22}H_{28}O_8FNa_2P$, mw = 516.4. Both of the latter compounds are used for treating carpal tunnel syndrome, the most common of the entrapment neuropathies. The median nerve is subjected to compression and possibly ischemia in the confined space between the carpal bones and the flexor retinaculum of the wrist.

Chloroprednisone acetate (6α-chloroprednisone acetate; 6α-chloro-$\Delta^{1,4}$-pregnidien-17β,21-diol-3,11,20-trione-21-acetate). Mw = 436.6. Multiple uses.

Corticosterone (11,21-dihydroxyprogesterone; Δ^4-pregnene-11β,21-diol-3,20-dione; 11β,21-dihydroxy-4-pregnene-3,20-dione). $C_{21}H_{30}O_4$, mw = 346.4. Multiple uses.

Cortisone (17-hydroxy-11-dehydrocorticosterone; 17α,21-dihydroxy-4-pregnene-3,11,20-trione; Δ^4-pregnene-17α,21-diol-3,11,20- trione; Kendall compound; Wintersteiner compound F). $C_{21}H_{28}O_5$, mw = 360.4. Multiple uses.

Desoxycorticosterone (deoxycorticosterone; 11-desoxycorticosterone; 21-hydroxyprogesterone; 4-pregnen-21-ol-3,20-dione; Kendall desoxy compound B; Reichstein substance Q). $C_{21}H_{30}O_3$, mw = 330.2. Also, the *desoxycorticosterone acetate* (DCA). $C_{23}H_{32}O_4$, mw = 372.4; and *desoxycorticosterone pivalate*, $C_{26}H_{38}O_4$, mw = 414.6. Multiple uses.

Dexamethasone (hexadecadrol; 9α-fluoro-16α-21-trihydroxy-16α-methyl-1,4-pregnadiene-3,20-dione). $C_{22}H_{29}FO_5$, mw = 392.4. Widely used in the treatment of benign intracranial hypertension, brain abscess, brain metastases, brain tumor, cerebral thrombosis, Cushing's syndrome, encephalitis, hypertensive encephalopathy, lumbar disk disease, meningococcal cerebral edema, shock, superior vena cava obstruction in cancer patients, ulcerative colitis.

Dichlorisone acetate (9α,11β-dichloro-1,4-pregnadiene-17α,21-diol-3,20-dione-21-acetate). $C_{23}H_{28}O_5Cl$, mw = 455.3.

Fluocinolone acetonide (6α,9α-difluoro-16α hydroxyprednisolone-16,17-acetonide). $C_{24}H_{30}O_6F_2$, mw = 452.50.

Fluorohydrocortisone (fluorocortisone; 9α-fluoro-11β,17α,21-trihydroxy-4-pregnene-3,20-dione). $C_{21}H_{29}O_5$, mw = 380.4. Used in treating Shy-Drager syndrome (parenchymatous degeneration of the central nervous system); also in treating orthostatic hypotension (a cause of temporary loss of consciousness when a person rises to an erect position). Also *fluorometholone* (9α-fluoro-11β,17α-dihydroxy-6α-methyl-1,4-pregnadiene-3,20-dione). $C_{22}H_{24}FO_4$, mw = 376.4; and *fluprednisolone* (6α-fluoroprednisolone), $C_{21}H_{27}FO_3$, mw = 378.4; and flurandrenolone (6-fluoro-16α-hydroxyhydrocortisone-16,17-acetonide), $C_{24}H_{33}O_6F$, mw = 436.5.

Hydrocortisone (cortisol; 11β,17α,21-trihydroxy-4-pregnene-3,20-dione), $C_{21}H_{30}O_5$, mw = 362.5. Used in treating adrenal insufficiency, notably in cancer patients, contact dermatitis, panhypopituitarism, psoriasis, shock, and urticaria. Also *hydrocortisone acetate* (cortisol acetate), $C_{23}H_{32}O_6$, mw = 404.5, used in treating rheumatoid arthritis; and *hydrocortisone sodium succinate* (11β,17α,21-trihydroxy-4-pregnene-3,20-dione-21-hydrogen succinate, sodium salt), $C_{25}H_{33}O_8Na$, mw = 484.5. The latter compound is used in treating ulcerative colitis.

Methylprednisolone (D1-6a-methylhydrocortisone). $C_{22}H_{30}O_5$, mw = 374.5. Used in treating thrombocytopenia with intracranial hemorrhage, Gram-negative bacteremia, posttransfusion purpura, and shock. Also *methylprednisolone sodium succinate*, $C_{26}H_{33}O_8Na$, mw = 496.5.

Paramethasone (6α-fluoro-16α-methylprednisolone). $C_{22}H_{30}O_5$, mw = 392.45. Also *paramethasone acetate*, $C_{24}H_{31}O_6F$, mw = 434.5.

Prednisolone (methacortandralone; 1,4-pregnadiene-3,20-dione-11β,17α-triol). $C_{21}H_{28}O_5$, mw = 360.4. Also *prednisolone phosphate sodium* (disodium prednisolone-21-phosphate), $C_{21}H_{27}Na_2O_8P$, mw = 484.4, used in treating ulcerative colitis. Also *prednisolone pivalate* (prednisolone trimethylacetate), $C_{26}H_{36}O_6$, mw = 444.6.

Prednisone (metacortandricin; 17α,21-dihydroxy-1,4-pregnadiene-3,11,20-trione). $C_{21}H_{26}O_5$, mw = 358.4. Used in the treatment of scores of ailments and diseases. To mention a few: acute erythroleukemia, acute gouty arthritis, acute pericarditis, aspiration pneumonitis, autoimmune hemolytic anemia, breast cancer, bronchial asthma, chronic hepatitis, dermatomyositis, desquamative interstitial pneumonia, Hodgkin's disease, hypercalcemia, immune neutropenia, lymphocytic leukemia, osteoporosis, hemoglobinuria, prostate cancer, psoriasis, radiation enteritis, rheumatoid arthritis, trichinosis, ulcerative colitis, usual interstitial pneumonia, viral anthropathies.

Triamcinolone (9α-fluoro-16α-hydroxyprednisolone). $C_{21}H_{27}FO_6$, mw = 394.4. Used in treating acute gouty arthritis and uremic pericarditis. Also *triamcinolone acetonide* (9α-fluoro-11β,21-dihydroxy-16α,17α-isopropylidenedioxy-1,4-pregnadiene-3,20-dione), $C_{24}H_{31}FO_6$, mw = 434.4. Used in treating acne vulgaris. Also *triamcinolone diacetate* (9α-fluoro-16α-hydroxyprednisolone-16,21-diacetate), $C_{25}H_{31}FO_8$, mw = 478.49.

Androgens and Anabolic Agents

Androsterone (3α-hydroxy-17-androstenone). $C_{19}H_{30}2$, mw = 290.4. Also *fluoxymesterone* (9α-fluoro-11β, 17β-dihydroxy-17α-methyl-4-androsten-3-one), $C_{20}H_{29}FO_3$, mw = 336.4. Used in treating paroxysmal nocturnal hemoglobinuria. Also *aldosterone* (electrocortin; 18-formyl-11β,21-dihydroxy-4-pregnene-3,20-dione) $C_{21}H_{28}O_5$, mw = 360.4.

(continued)

TABLE 1. (*Continued*)

Androgens and Anabolic Agents

Hydroxydione sodium (21-hydroxypregnane-3,20-dione-21-sodium hemisuccinate). $C_{25}H_{35}O_6Na$, mw = 454.5.
Spironolactone (3-(30-oxo-7α-acethylthio-17β-hydroxy-4-androsten-17α-yl)-propionic acid γ-lactone), $C_{24}H_{32}O_4S$, mw = 416.5. Used in treating congestive heart failure, hypertension, hypokalemia.
Methandrostenolone (17α-methyl-17β-hydroxy-1,4-androstadien-3-one). $C_{20}H_{28}O_2$, mw = 300.4.
Methylandrostenediol (MAD; methandriol; 17α-methyl-5-androsten-3β,17β-diol). $C_{20}H_{32}O_2$, mw = 304.4.
Methyl testosterone (17α-methyl-Δ^4-androsten-17-β-0 1-3-one). $C_{20}H_{30}O_2$, mw = 302.4.
Norethandrolone (17α-ethyl-19-nortestosterone). $C_{20}H_{30}O2$, mw = 302.4. Also *oxandroline* (17β-hydroxy-17α-methyl-2-oxa-5α-androstane-3-one), $C_{19}H_{30}O_3$, mw = 306.4.
Oxymetholone (2-hydroxymethylene-17-α-methyldihydrotestosterone). $C_{21}H_{32}O_3$, mw = 332.4. Used in treating agnogenic myeloid metaplasia and hereditary angioedema. Also *prometholone* (2α-methyl-dihydro-testosterone propionate), mw = 360.5.
Testosterone (trans-testosterone; 17β-hydroxy-4-androsten-3-one). $C_{19}H_{28}O_2$, mw = 288.4. Used in treating acne vulgaris, impotence, polycystic ovary syndrome, male hypogonadism. Also *testosterone cypionate*, $C_{27}H_{40}O_3$, mw = 412.6; *testosterone enanthate*, $C_{26}H_{40}O_3$, mw = 400.6; *testosterone phenylacetate*, $C_{27}H_{34}O_3$, mw = 406.5; *testosterone propionate*, $C_{22}H_{32}O_3$, mw = 344.4.

Estrogens

Equilenin (1,3,5-10,6,8-estrapentaen-3-ol-17-one). $C_{18}H_{18}O_2$, mw = 266.3. Also *equilin* (1,3,5,7-estratetraen-3-ol-17-one), $C_{18}H_{20}O_2$, mw = 268.3.
Estradiol (β-estradiol; dihydrofolliculin, dihydroxyestrin; 3,17-ephidhydroxyestratriene). $C_{18}H_{24}O_2$, mw = 272.3. Also *estradiol benzoate*, $C_{25}H_{28}O_3$, mw = 376.4; *estradiol cypionate*, $C_{26}H_{36}O_2$, mw = 396.6; *estradiol diprionate*, $C_{24}H_{32}O_4$, mw = 384.5.
Estriol (trihydroxyestrin; 1,3,5-estratriene-3,16α,17β-triol). $C_{18}H_{24}O_3$, mw = 288.3.
Estrone (folliculin; ketohydroxyestrin; 1,3,5-estratriene-3-ol-17-one). $C_{18}H_{22}O_2$, mw = 270.3. Also *estrone benzoate*, $C_{25}H_{26}O_3$, mw = 374.4.
Ethynyl estradiol (17-ethinyl estradiol; 17α-ethynyl-1,3,5-estratriene-3,17β-diol). $C_{20}H_{24}O_2$, mw = 296.4. Used to treat acne vulgaris, osteoporosis.
Mestranol (ethylestradiol-3-methylether; 3-methoxy-19-nor-17α-pregnα-1,3,5-trien-20-yn-17-ol). $C_{21}H_{26}O_2$, mw = 310.4. Used in treating acne vulgaris.

Progestogens and Progestins

Acetoxypregnenolone (21-acetoxypregnenolone; 3-hydroxy-21-acetoxy-5-pregnen-20-one). $C_{23}H_{34}O_4$, mw = 374.5.
Anagestone acetate (6α-methyl-4-pregnen-17α-ol-20-one). $C_{24}H_{36}O_3$, mw = 372.6.
Chlormadinone acetate (6-chloro-D4,6-pregnadiene-17α-ol-3,20-dione acetate). $C_{23}H_{29}ClO_4$, mw = 414.9.
Dimethisterone (17β-hydroxy-6α-methyl-17α-(prop-1-nyl)-androst-4-ene-3-one). $C_{23}H_{32}O_2 \cdot H_2O$, mw = 358.5. Also *ethisterone*, $C_{21}H_{28}O_2$, mw = 312.4.
Ethynodiol diacetate (19-nor-17α-pregn-4-en-20-yne-3β,17-diol diacetate). $C_{24}H_{32}O_4$, mw = 384.5. Used in treating acne vulgaris.
Flurogestone acetate (17α-acetoxy-9α-fluoro-11β-hydroxy-4-pregnene-3,20-dione). $C_{23}H_{31}O_5F$, mw = 406.5.
Hydroxymethylprogesterone (medroxyprogesterone; 17α-hydroxy-6α-methyl-4-pregnene-3,20-dione). $C_{22}H_{23}O_3$, mw = 344.5. Used for treating menopausal symptoms, secondary amenorrhea. Also *hydroxymethylprogesterone acetate*, $C_{24}H_{24}O_3$, mw = 386.5. Used in treating hypogonadal females. Also *hydroxyprogesterone* (4-pregen-17α-ol-3,20-dione), $C_{21}H_{30}O_3$, mw = 330.4. Also *hydroxyprogesterone caproate* (17α-hydroxy-4-pregnene-3,20-dione caproate), $C_{27}H_{40}O_4$, mw = 428.6.
Melengestrol acetate (MGA; 6-dehydro-17-hydroxy-6-methyl-16-methylene-progesterone acetate), $C_{25}H_{32}O_4$, mw = 396.51.
Norethindrone (norethisterone; 17α-ethynyl-17-hydroxy-19-nor-17α-4-en-20-yn-3-one). $C_{20}H_{26}O_2$, mw = 298.4. Used in treating acne vulgaris. Also *norethindrone acetate*, $C_{22}H_{28}O_3$, mw = 340.4. Also *norethynodrel*, $C_{20}H_{26}O_2$, mw = 298.4. Also *normethisterone*, $C_{19}H_{28}O_2$, mw = 288.4.
Pregnenolone (D5-pregnen-3β-ol-20-one). $C_{21}H_{32}O_2$, mw = 308.4. Important in the synthesis of adrenal hormones.
Progesterone (progestin; progestone; Δ^4-pregnene-3,20-dione). $C_{21}H_{30}O_2$, mw = 314.4. Used in treating excessive uterine bleeding, hypogonadal females, menopausal symptoms, polycystic ovary syndrome, secondary amenorrhea.

Diuretic, Antidiuretic and Anesthetic Agents

Aldosterone and spironolactone are described earlier in this list.
Hydroxydione sodium (21-hydroxypregnane-3,20-dione-21-sodium hemisuccinate). $C_{25}H_{35}O_6Na$, mw = 454.5

---

[1]This is an abridged list of steroid hormones. Some are much more important and widely used than others. Some are relatively recent to steroid therapy; others have been used more widely in the past than presently. See also **Hormones**.

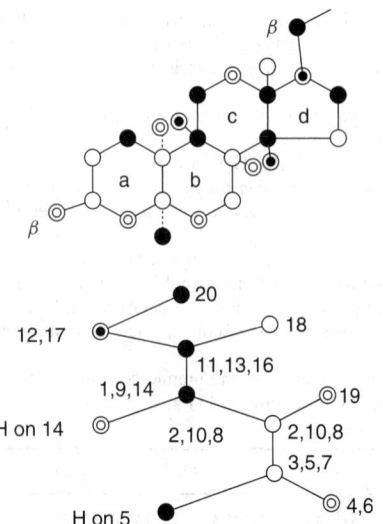

Fig. 2.   Constellation of a saturated beta-sterol; ● = atoms in bottom plane; ◉ = atoms in second plane; ○ = atoms in third plane; ◎ = atoms in top plane. Lateral view of molecule is shown at right.

groups α or β depending upon whether they are situated below the plane of the molecule, when depicted in a certain way. Usually, in a structural diagram of a steroid molecule, a dotted line connection will be used between atoms to designate an alpha position; a regular solid line for a beta position.

Figure 2 illustrates one of the two most important configurations of steroid skeletons, among 63 other possibilities, as they are found in nature. The side chain is usually attached in β-position to C_{17}. The configurations on C_{20} and C_{24} have likewise been determined and are known to produce steric isomerisms.

The sterols, from which the name of the entire group is derived, are monovalent alcohols with a secondary hydroxyl group on C_3 usually in β-position. The best known representative is *cholesterol*. See Fig. 3. This compound forms esters with a great variety of acids. Both the free and esterified sterols accompany the neutral fat and the phosphatides in most animal and plant fat. Upon alkaline hydrolysis, the other lipid constituents form fatty acid soaps; the fraction which remains insoluble in aqueous alkaline solution is called "unsaponifiable" and consists primarily of sterols. The variations in the cholesterol content of blood of animals, particularly in humans, are of significance for the diagnosis of various diseases. See also **Cholesterol**. Cholesterol is the principal sterol of all vertebrate animals. It is also found in some mollusks and in crustaceans, where it may be of alimentary origin.

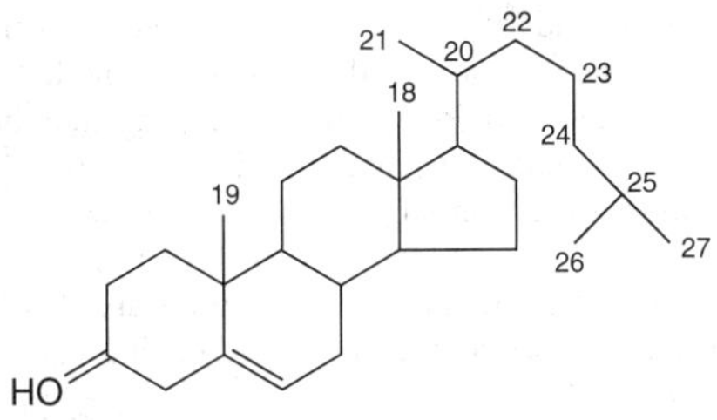

Fig. 3.   Cholesterol.

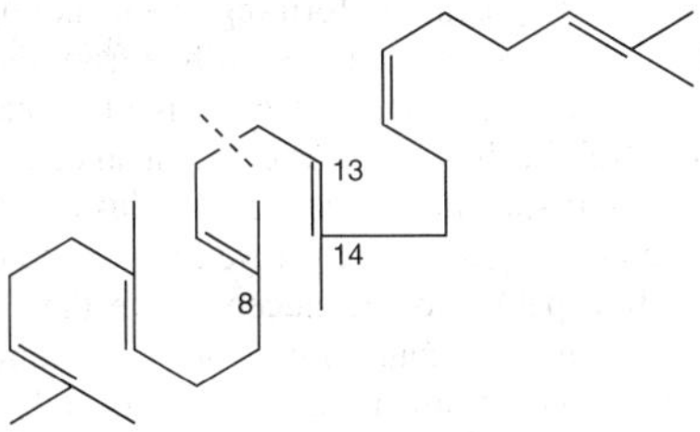

Fig. 4.   Squalene.

Sterol Biogenesis. The finding that ingestion of radioactively labeled acetic acid leads to the synthesis of radioactive cholesterol was the first step in the elucidation of sterol biosynthesis. A growth factor for *Lactobacilli*, replaceable by acetic acid, was found to have the structure $HOCH_2 \cdot CH_2 \cdot C(CH_3)(OH) \cdot CH_2 \cdot CO_2H$. This compound was termed *mevalonic acid*. Its close relationship to a trimer of acetic acid is evident. Six molecules of the C_6 acid polymerize, losing their carboxyl groups, to the linear isoprenol *squalene*, a hydrocarbon occurring in nature. Twelve of the carbon atoms shown as circles in Fig. 4, originate from the carboxy groups of the original acetic acid, the remaining 18 from the methyl groups. Squalene folds in the manner indicated in the formula and yields (with a two-step rearrangement of the methyl group from C_8 and C_{13}) *lanosterol*, a "protosterol" found in wool and fat. This protosterol loses three methyl groups in positions 4, 4, and 14 in the course of biosynthesis, yielding *zymosterol*, found in yeast, which is convertible to cholesterol. The gradual oxidative degradation of the side chain in cholesterol to bile acids and subsequently to the various steroid hormones in animals is well established and has been confirmed by C-14 tracer studies. Many of the enzymes operative during these hormone syntheses in the insertion of hydroxyl groups on individual carbon atoms have been separated and localized in various cell constituents. Major steps in sterol biogenesis are shown in Fig. 5.

Classification of Medically Important and Useful Steroid Hormones

In addition to the medical uses of steroid hormones for alleviating conditions brought about by insufficiencies or overabundance of any particular hormone of this class within the body, and thus returning the desirable hormone balance, numerous therapies do not fall directly into these two categories. Rather, steroid hormones are used in connection with some ailments and diseases because of positive clinical results even though much remains to be learned concerning the details of their function.

Steroid hormones are difficult to classify because some of them serve large numbers of uses. The conventional approach places them into four categories.

Androgens and Anabolic Agents. Androgen is the male sex hormone. The androgenic hormones are synthesized in the body by the testis, the cortex of the adrenal gland, and, to a slight extent, by the ovary. The androgens have a number of sexually related functions. Androgens also serve as anabolic agents, i.e., in nutrition of muscle and bone in both male and female persons. A number of androgens have been synthesized. Some of these reduce or eliminate the production of male secondary sex characteristics when administered to females (growth of facial hair, lowering of the voice, etc.)

Estrogens. Estrogen is a general term for female sex hormones. They are responsible for the development of the female secondary sex characteristics, such as the deposition of fat and the development of the breasts. Estrogens are produced by the ovary, and, to a lesser degree, by the adrenal cortex and testis. Some synthetic *nonsteroid* compounds, such as diethylstilbestrol and hexestrol, have estrogenic activity.

Progestogens and Progestins. Progesterone (Δ^4-pregnene-3,20-dione), $C_{21}H_{30}O_2$, is the female sex hormone secreted in the body by the corpus luteum, by the adrenal cortex, or by the placenta during pregnancy. It is important in the preparation of the uterus for pregnancy, and for the maintenance of pregnancy. Progesterone is believed to be the precursor of the adrenal steroid hormones.

Adrenal Corticosteroids. Among these hormones are compounds which have been found to be antiinflammatory, antiallergic, and antirheumatic

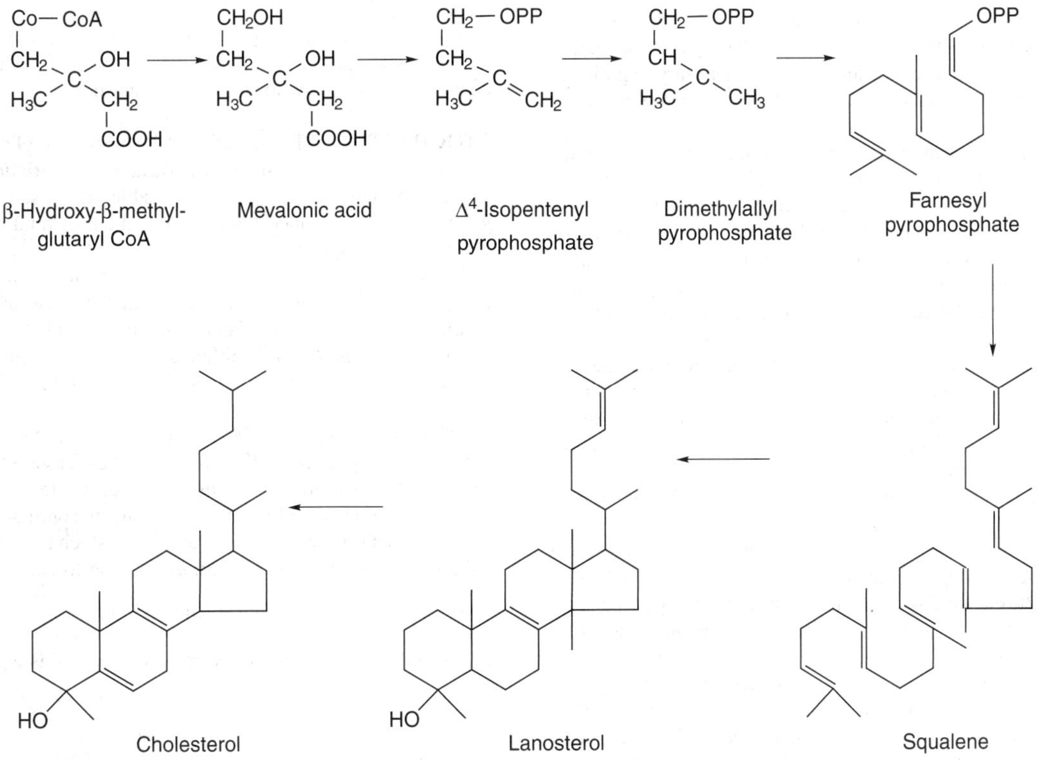

Fig. 5.   Structures of key compounds involved in biogenesis of sterols.

agents and consequently are very important in steroid therapy. Cortisone was first applied in 1949 and became a major drug for treating rheumatoid arthritis, among other ailments, soon thereafter. Hydrocortisone followed and a bit later several synthetic analogues (not found in the body) were developed. Well known among these is prednisone, found particularly effective in cases of diseases of collagen tissue. But its use is much more widespread. Several examples are given in the accompanying table. See also **Hormones**. Steroid therapy throughout the years has had to cope with production of numerous side effects. Problems like this provide an incentive to continued vigorous research for new compounds.

In addition to the foregoing four major classifications (by application), there are also steroid hormones which are effective diuretic, antidiuretic, and local anesthetic agents. Among these are aldosterone, spironolactone, and hydroxydione sodium.

Bile Acids

The bile acids are monocarboxylic acids of the steroid group with 24 carbon atoms and 1–3 secondary hydroxyl groups. They occur in the bile of all vertebrates from the teleosts upward, mostly in peptidic conjugation with glycine and taurine. The bile acids are described in the entry on **Bile**.

Additional Reading

Connolly, S.: "Steroids," Heinemann Library, Woburn, MA, 2000.

Genazzani, A.R., F. Petraglia, and R.H. Purdy: "The Brain: Source and Target for Sex Steroid Hormones," CRC Press, LLC, Boca Raton, FL, 1996.

Handa, R.J., S. Hayashi, E. Terasawa, and M. Kawata: "Neuroplasticity, Development, and Steroid Hormone Action," CRC Press, LLC, Boca Raton, FL, 2001.

Karch, S.B.: "The Pathology of Drug Abuse," 2nd Edition, CRC Press, LLC., Boca Raton, FL, 1996.

Kirk, D.N., B. Hill, H.L. Makin, and G.M. Murphy: "Dictionary of Steroids: Chemical Data Structure," CRC Press, LLC, Boca Raton, FL, 1999.

Lukas, S.E.: "Steroids," Enslow Publishers, Inc., Berkeley Heights, NJ, 2001.

Milne G.W.A.: "Ashgate Handbook of Endocrine Agents and Steroids," Ashgate Publishing Company, Brookfield, VT, 2000.

Monroe, J.: "Steroid Drug Dangers," Enslow Publishers, Inc., Berkeley Heights, NJ, 1999.

Veldhuis, J.D. and A. Giustina: "Sex-Steroid Interactions with Growth Hormone," Springer-Verlag, Inc., New York, NY, 1999.

Web References

AboutSteroids.com: *http://www.aboutsteroids.com/*

American Academy of Pediatrics: Steroids: *http://www.aap.org/family/steroids.htm*

Anabolic Steroid Abuse: *http://www.steroidabuse.org/*

SteroidsInfo.com: *http://www.steroidsinfo.com/*

STETHOSCOPE.   An instrument for listening to sounds originating within the body, especially of the heart and lungs. The stethoscope is a low-pass filter the exact acoustical characteristics of which vary significantly from one manufacturer to another. Electronic stethoscopes are available.

STIBNITE.   The mineral stibnite, antimony sulfide, Sb_2S_3, is found in radiated groups of acicular orthorhombic crystals or in other sorts of aggregates, as well as blades, also as columnar or granular masses. It shows a highly perfect pinacoidal cleavage; conchoidal fracture; hardness, 2; specific gravity, 4.63–4.66; luster, metallic and very brilliant on cleavage faces or freshly fractured surfaces. Its color is a steely gray; the streak very similar in color, may be covered with a black, sometimes iridescent tarnish.

Stibnite is the most common antimony mineral known and is the chief ore of that metal. It is a primary ore mineral and occurs with other antimony minerals and galena, sphalerite, and silver ores. It is found in Germany, Rumania, the Balkans, Italy, Borneo, Peru, Japan, China, Mexico; and in the United States in California and Nevada.

The name stibnite is derived from the Latin word for antimony, *stibium*.

STIFFNESS.   In general, the ability of a system to resist a prescribed deviation. In the case of a deformable elastic medium, stiffness is the ratio of a steady force to the elastic displacement produced by it, e.g., for a spring the force required to produce unit stretch. The term is applied most often to an elastic system vibrating about a position of equilibrium. Acoustic stiffness is the quantity which, when divided by 2π times the frequency, gives the acoustic reactance associated with the potential energy of the medium or its boundaries. The unit commonly used is dyne/centimeter. Mechanical stiffness is expressed in terms of the various elastic constants and moduli.

STIGMA.   1. A secondary sexual mark of insects. In many species of butterflies, it consists of a patch on the wing of the male bearing modified scales on a more or less modified area of the wing membrane. 2. A term used by some entomologists in place of spiracle. 3. A pigmented spot or "eye spot" sensitive to light in some flagellate protozoa.

STIGMATIC.   Two uses of this term in optics are: (1) For a bundle of rays, homocentric. (2) For an optical system, having equal focal power in all meridians.

STILBITE.   The mineral stilbite, $NaCa_2(Al_5Si_{13})O_{36}\cdot14H_2O$, is a zeolite, the compound monoclinic crystals of which are usually grouped in approximately parallel positions, forming sheaf-like aggregates, which have a soft pearly luster, whence the name stilbite from the Greek, meaning luster. The less commonly used term desmine is likewise from the Greek, meaning a bundle. Stilbite has one perfect cleavage; uneven fracture; is brittle; hardness, 3.5–4; specific gravity, 2–2.2; luster, vitreous to pearly; color, usually white but may be brownish, yellowish, red or pink. Its streak is white, and it is transparent to translucent. Like the other zeolites stilbite occurs in cavities in basalts and traps, rarely in granites and gneisses. Of the many localities may be mentioned Trentino, Italy; the Harz Mountains; Valais, Switzerland; Arendal, Norway; the Ghats Mountains of India; and Mexico. The Triassic traps of New Jersey and Pennsylvania furnish specimens as do also rocks of the same age in Nova Scotia. This mineral sometimes is called *desmine*.

STIMULUS (Nerve).   Physical agent used to set up a nerve impulse. Since the nerve fiber is excitable by electric currents, a common stimulus is a brief current pulse whose amplitude and duration and rate of delivery can be controlled with precision. Other physical agents can be used, depending upon the input under examination, e.g., sudden stretch in the case of stretch receptor, sound wave for auditory input, and light in the case of the eye.

STINK BUG (*Insecta, Hemiptera*).   A flattened bug of generally ovate form, in many species with an angular outline. Most species are moderately large, reaching a length of about half an inch. The many species, constituting the family *Pentatomidae*, are also characterized by the fetid odor of the secretion discharged from glands opening on the lower surface of the body.

One species, the harlequin cabbage bug or calico-back, is a troublesome pest. It is best controlled by clean cultivation of fields, hand picking of bugs and their eggs, and the use of trap crops, planted early to attract the insects.

Some members of the group eat other insects and are probably beneficial in destroying pests, but unfortunately even the harmless species may contaminate berries with their unpleasant odors.

STOCHASTIC.   The adjective "stochastic" implies the presence of a random variable; e.g., stochastic variation is variation in which at least one of the elements is a random variable, and a stochastic process is one wherein the system incorporates an element of randomness as opposed to a deterministic system.

The word derives from Greek στόχος, a target, and a stochastiches was a person who forecast a future event in the sense of aiming at the truth. In this sense it occurs in sixteenth-century English writings. Bernoulli in the *Ars Conjectandi* (1719) refers to the "ars conjectandi sive stochastice." The word passed out of usage until revived in the twentieth century.

STOCHASTIC PROCESS.   A family of variates (x_t) where t assumes values in a certain range T. In most practical cases x_t is the observation at time t and T is a time-range, but t may also refer to distribution in space and may be considered for discontinuous or continuous values.

A stochastic process (x_t) is said to be stochastically continuous if, for values t, $t + h_1$, $t = h_2$, ... with h_n tending to zero as n tends to infinity

$$\lim_{n \to \infty} x_t + h_n$$

exists in the sense of stochastic convergence and is equal to x. Likewise, if

$$\lim \frac{x_{t+h_n} - x_t}{h_n}$$

exists in the sense of stochastic convergence, the process is said to be stochastically differentiable. And if the process exists in $a \le t \le b$ and the

Riemann integral

$$\int_b^a x_t \, dt$$

exists in the sense of stochastic convergence, the process is said to be stochastically integrable.

STOKES FLOW. Flow of a viscous fluid at a very small Reynolds number when inertial, acceleration forces are negligible and the Navier-Stokes equations reduce to

$$\mu \nabla^2 \mathbf{v} - \text{grad } p = 0$$

The approximation may not be possible in all parts of the flow even at very low speeds.

STOKES LAW FOR VISCOSITY. A solid sphere moving with velocity V through a fluid of viscosity μ experiences a resistance to motion

$$F = 6\pi\mu a \mathbf{V}$$

where a is the radius of the sphere. The law is accurate only if the flow Reynolds number $\rho a V / \mu$ is less than 0.1.

STOKES THEOREM. The surface integral of the curl of a vector function equals the line-integral of that function around a closed curve bounding the surface.

$$\int_S \nabla \times \nabla \cdot d\mathbf{S} = \mathbf{V} \cdot d\mathbf{S}$$

If the components of $\mathbf{V}$ in rectangular Cartesian coordinates are u, v, w and the direction cosines of the normal to $d\mathbf{S}$ are λ, μ, v the theorem may also be given as

$$\int_S \left[\lambda \left(\frac{\partial w}{\partial y} - \frac{\partial v}{\partial z} \right) + \mu \left(\frac{\partial u}{\partial z} - \frac{\partial w}{\partial x} \right) + v \left(\frac{\partial v}{\partial x} - \frac{\partial u}{\partial y} \right) \right] dS$$

$$= \int_C (u \, dx + v \, dy + w \, dz)$$

See also **Curl**; and **Vector**.

STOLON. 1. In botany, a branch that grows out horizontally from the base of the stem, takes root, and gives rise to a new plant at the nodes or at the tip.

2. In zoology, a shoot growing from the base of an animal or a colony, from which other individuals arise by budding.

STOMATE (or Stoma). The name applied to the minute pores that occur abundantly in the epidermis of leaves and, less abundantly, in the epidermis of young stems, flower parts and fruits. Each stomate is located between two distinctive epidermal cells called *guard cells*. The size and shape of the guard cells varies considerably from one species to another. Unlike other epidermal cells the *guard cells* contain chloroplasts. In many species of plants stomates occur in both the upper and lower epidermis of the leaf. In many other species, especially of woody plants, stomates occur only in the lower epidermis. Even in those species in which the stomates are present in both epidermises there are commonly, although not invariably, more per unit area in the lower epidermis. In floating leaves, such as those of the water lily, stomates are present only in the upper epidermis. In many kinds of plants, the stomates are restricted to grooves or furrows in the leaf.

The number of stomates per unit area varies with the kind of plant and also, within limits, with the conditions under which the plant has developed. The range is from a few thousand to about a hundred thousand per square centimeter of leaf surface. A single corn plant has been estimated to bear from 140 to 240 million stomates and the number on a large tree could be represented only by a figure of astronomical dimensions. The size of the individual stomates also varies greatly from species to species, their dimensions being expressed in microns. In some species the fully open stomates may be as large as 8 to 10 × 30 to 40 micromillimeters, as measured along the two axes of the elliptical pore, but in most species they are smaller. Species in which the stomates are relatively small usually have more per unit area than those which have relatively large stomates.

The structure of the cell wall of a guard cell is quite complex, varying considerably in thickness and elasticity from one part of the cell to another. This wall structure is such that when the guard cells increase in turgidity

their inner walls—those bounding the pore—bow away from each other causing a widening of the stomate. In general, therefore, when the guard cells are turgid the stomate is open; when the guard cells are flaccid the stomach is closed.

Shifts in the turgidity of the guard cells causing opening and closing of stomates are conditioned by a number of factors among which the most important are light, temperature, and the internal water supply of the leaf. In general stomates are open in the light and closed in the dark although there are many exceptions to this statement. In general, low temperatures are unfavorable to stomatal opening, and when the temperature falls below the optimum for stomatal opening for a given species, the stomates will remain closed or open only incompletely, even if light conditions and water supply are favorable. Similarly, drought conditions, by resulting in a reduction in the water content of the leaves, are usually unfavorable to stomatal opening even if light and temperature conditions are favorable. During prolonged droughts, the stomates of many species remain nearly or completely closed most of the time. Night opening of the stomates is of regular occurrence in some species such as certain cacti, in which they are not usually open in the daytime, and may occur in some other species under certain conditions.

When the stomates are open they serve as the principal pathways through which gases diffuse into or out of the leaf; when the stomates are closed all gaseous exchanges between a leaf and its environment are greatly retarded. The gases of greatest physiological importance which enter or depart from a leaf principally through the stomates are oxygen, carbon dioxide, and water vapor. Loss of water vapor in the process of transpiration occurs principally through the stomates. Similarly the inward diffusion of carbon dioxide and outward diffusion of oxygen, the gaseous exchanges accompanying photosynthesis, occur principally through the stomates.

STONE AGE. An archeological term to designate a cultural level that is characterized by the use of stone implements. Classically it is divided into the Eolithic, Paleolithic, and Neolithic Periods. The Stone Age is the first of the so-called three-age system (Bronze Age and Iron Age following).

STONE ROLLER *(Osteichthyes)*. Of the group *Cypriniformes*, family *Cyprinidae*, the stone roller is a bottom-feeding fish of moderate size found in small streams from Wyoming to New York and south to the Gulf. It is herbivorous.

STOPPING POWER. A measure of the effect of a substance upon the kinetic energy of a charged particle passing through it. The linear stopping power S_l is the energy loss per unit distance and is given by $S_l = -dE/dx$, where E is the kinetic energy of the particle and x is the distance traversed in the medium. The *mass stopping power* S_m is the energy loss per unit surface density traversed, and is given by $S_m = S_l/\rho$, where ρ is the density of the substance. The *atomic stopping power* S_a of an element is the energy loss per atom, per unit area normal to the particle's motion, and is given by $S_a = S_l/n = S_m A/N$, where n is the number of atoms per unit volume, N is the Avogadro number, and A is the atomic weight. The *molecular stopping power* of a compound is similarly defined in terms of molecules; it is very nearly if not exactly equal to the sum of the atomic stopping powers of the constituent atoms. The *relative stopping power* is the ratio of the stopping power of a given substance to that of a standard substance, commonly aluminum, oxygen or air. The *stopping equivalent* for a given thickness of a substance is that thickness of a standard substance capable of producing the same energy loss. The *air equivalent* is the stopping equivalent in terms of air at 15 °C and 1 atmosphere as the standard substance. The term equivalent stopping power is not clearly defined, but sometimes is used synonymously with relative stopping power and sometimes with stopping equivalent.

STORAGE BATTERY. See **Battery**.

STORAGE (Computer). Any medium capable of storing information. As generally defined, however, a storage unit is a device on or in which data can be stored, read, and erased. The major classifications of storage devices associated with computer systems are: (1) *immediate-access*; (2) *random-access*; and (3) *sequential-access*. As a general rule, the cost per bit of information is greater for immediate-access storage devices, but the access time is considerably faster than for the other two types.

The various physical means to effect storage are described under **Memory (Electronic)**.

Immediate-Access Storage. In these devices, information can be read in a microsecond or less. Usually an array of storage elements can be directly addressed and thus all information in the array requires the same amount of time to be read. Specific storage configurations in this class include core storage, and monolithic storage.

Random-Access Storage. Storage devices in which the time required to obtain information is independent of the location of the information most recently obtained. This strict definition must be qualified by the observation that what is meant is *relatively* random. Thus, magnetic drums are relatively nonrandom access when compared with monolithic storage, but are relatively random access when compared with magnetic tapes for file storage. Disk-storage and drum-storage units usually are referred to as random-access storage devices. The time required to read or write information on these units generally is in the 10- to 200-millisecond range. This time is dependent upon where the information is recorded, with respect to the read/write head at the time the data are addressed.

Sequential-Access Storage. Storage devices in which the items of information stored become available only in a one-after-the-other sequence, whether or not all the information or only some of it is desired. Storage on magnetic tape is an example.

Some other computer storage configurations are defined by:

Auxiliary Storage. A storage device in addition to the main storage of a computer; e.g., magnetic tape, disk, diskette, or magnetic drum. Auxiliary storage usually holds much larger amounts of information than the main storage, and the information is accessible less rapidly.

Buffer Storage. (1) A synchronizing element between two different forms of storage, usually between internal and external. (2) An input device in which information is assembled from external or secondary storage and stored ready for transfer to internal storage. (3) An output device into which information is copied from internal storage and held for transfer to secondary or external storage. Computation continues while transfers between buffer storage and secondary or internal storage or vice versa take place. (4) Any device that stores information temporarily during data transfers.

Circulating Storage. A device or unit which stores information in a train or pattern of pulses, where the pattern of pulses issuing at the final end are sensed, amplified, reshaped and reinserted into the device at the beginning end.

External Storage. (1) The storage of data on a device that is not an integral part of a computer, but in a form prescribed for use by the computer. (2) A facility or device, not an integral part of a computer, on which data usable by a computer is stored such as off-line magnetic tape units, or punch card devices.

Internal Storage. (1) The storage of data on a device which is an integral part of a computer. (2) The storage facilities forming an integral physical part of the computer and directly controlled by the computer. In such facilities, all data are automatically accessible to the computer; e.g., magnetic core, and magnetic tape on-line.

Main Storage. Usually the fastest storage device of a computer and the one from which instructions are executed.

Program Storage. A portion of the internal storage reserved for the storage of programs, routines, and subroutines. In many systems protection devices are used to prevent inadvertent alteration of the contents of the program storage.

Serial Storage. A storage technique in which time is one of the factors used to locate any given bit, character, word, or groups of words appearing one after the other in time sequence, and in which access time includes a variable latency or waiting time of from zero to many word times. A storage is said to be serial by word when the individual bits comprising a word appear serially in time; or a storage is serial by character when the characters representing coded decimal or other nonbinary numbers appear serially in time; e.g., magnetic drums are usually serial by word but may be serial by bit, or parallel by bit, or serial by character and parallel by bit.

Working Storage. A portion of the internal storage reserved for the data upon which operations are being performed. Synonymous with working space and temporary storage and contrasted with program storage.

THOMAS J. HARRISON, International Business Machines corporation, Boca Raton, FL.

STORKS *(Aves, Ciconniformes).* Large birds of the Old World and South America. They range from Argentina northward to Mexico, but not into the United States. They are found widely in Europe; Asia, Africa, and in parts of Australia. Their size ranges from about 42 to 60 inches (107 to 152 centimeters) tall when standing and ranges from 4 to 6.5 feet (1.2 to 2 meters) in length. The color varies with species, ranging from white through dull gray, sometimes having a somewhat greenish sheen. There are numerous species, of which at least 28 tropical species are known. Most storks are migrating birds, flying with their neck stretched and legs slanted backward in a V-shape. The migration pattern is from colder to warmer areas in the fall and returning to the colder climates for spring and summer. Some species prefer chimney tops as nesting locations, while other species like marshes near water and in tall grasses. There are usually 3 to 5 blunt, oval, white eggs, the eggs being laid at intervals of about 2 days. Both parents incubate the eggs for 30 to 38 days. Some species fly in large flocks, particularly those migrating from Africa to Europe and Asia in early spring. The return flight is usually made some time in August. Numerous bird-watching groups have banded the legs of the baby storks to keep track of these migrations. After banding, a baby will often "play dead" for quite a period. The young storks are called siblings.

The common white stork of Europe, *Ciconia alba,* which nests on the tops of chimneys is probably the best-known species. See Fig. 1. The jabiru (*Jabiru mycteria*) is a white stork with green head and green coloring on the partly-naked neck. The upper portion of the neck also may have dark blue and orange coloring. This bird may achieve a length of some 55 inches (140 centimeters) and is one of the larger flying birds. Nesting preference is pine trees. Eggs are light-green in color. This stork is known to prey on grass fires, catching small animals as they are flushed out by the heat. Other dietary items include small fish and a variety of insects. Some varieties of jabiru have a bronze coloration on their wings and a bit of green color in the tail — with coral color legs.

Fig. 1.   Storks. (*A.M. Winchester.*)

Adjutant is a name applied to stork-like birds of several species. These birds occur in Africa and the Oriental region where they are valuable as scavengers. From at least one species (*Leptoptilus crumenifer*) the soft downy feathers known as marabou are secured. Birds as tall as 6–7 feet (1.8–2.1 meters) have been reported in India. These birds feed on frogs, carrion, and fish. They are known for their success in chasing vultures away from carrion.

The name stork sometimes is confused with that of the related herons, as in the case of the whale-headed or shoe-billed species of the White Nile. This species, called both heron and stork, has an enormous and powerful beak, both broad and deep and provided with a strong hook at the tip. See also **Ciconiiformes.**

STRABISMUS (Cross-Eyes). Strabismus is a functional defect where the eyes are misaligned and point in different directions. The brain's ability to see three-dimensional objects depends on proper alignment of the eyes. When both eyes are properly aligned and aimed at the same target, the visual portion of the brain fuses the forms into a single image. When one eye turns inward, outward, upward, or downward, two different pictures are sent to the brain. This causes loss of depth perception and binocular vision. The turned eye may be straight at times, and the misalignment may come and go. Strabismus occurs in about 4% of all children in the United States, equally in males and females, and is sometimes hereditary. The condition can also develop later in life.

There are two forms of strabismus, esotropia and exotropia. In esotropia, one eye deviates inward toward the nose. In exotropia, one eye deviates outward away from the nose. Esotropia is the most common type of strabismus in infants. Exotropia often begins between the ages of 2 and 4.

In young children with any form of strabismus, the brain may learn to ignore the misaligned eye's image and see only the image from the best-seeing eye. This is called *amblyopia*, or lazy eye, and results in a loss of depth perception. When an adult develops strabismus, double vision sometimes occurs because the brain has already been trained to receive images from both eyes and cannot ignore the image from the turned eye.

The causes of strabismus are not fully understood. There are six muscles that control eye movement, four that move it up and down and two that move it side to side. All these muscles must be coordinated and working properly in order for the brain to see a single image. When one or more of these muscles does not work properly, some form of strabismus may occur. Strabismus is more common in children with disorders that affect the brain such as cerebral palsy, Down syndrome, hydrocephalus, and brain tumors.

The earliest sign of strabismus is usually a noticeable deviation of one eye. This symptom may at first be intermittent, occurring when a child is daydreaming, not feeling well, or tired. The deviation may also be more noticeable when the child looks at something in the distance. Frequent rubbing of the eyes is also common with strabismus. In bright light, children may squint or tilt their heads in order to use both eyes together. Few children ever complain about double vision, although they may close one eye to compensate for the problem.

The eyes of infants and small children can sometimes seem to be crossed when they have a wide, flat nose and a fold of skin at the inner eyelid. Called *false strabismus*, it usually disappears as the child grows. True strabismus will not be outgrown. An eye doctor can tell the difference between true and false strabismus.

To diagnose strabismus, an eye doctor will perform a comprehensive eye examination including an ocular motility (eye movement) evaluation and an evaluation of the internal ocular structures. The most reliable treatment for strabismus is usually eye muscle surgery, but this is often preceded by eye patching and/or eyeglass therapy, especially if amblyopia (lazy eye) is present. This therapy is designed to maximize the existing vision in the "bad" eye. In some children, this therapy may eliminate the need for surgery.

To correct strabismus, the eye surgeon makes a small incision in the tissue covering the eye and accesses the eye muscles. The appropriate muscles are then repositioned to allow the eye to move properly. The procedure is usually performed under general anesthesia. Recovery time is rapid, and most people are able to resume normal activities within a few days. Following surgery, corrective eyeglasses may be needed, and, in many cases, further surgery is required later to keep the eyes straight.

When a child requires surgery for strabismus, the procedure is usually performed before school age. This is easier for the child and gives the eyes a better chance to work together. As with all surgery, there are some risks. However, strabismus surgery is usually a safe and effective treatment. See also **Amblyopia**; **Estropia**; **Extropia**; and **Vision and the Eye**.

Vision Rx, Inc., Elmsford, NY.

STRAIGHT EDGE. One of three surfaces, any two of which when placed together, coincide throughout their length. A straight-edge also is the name of a hand device in the form of a piece of material whose edge is "perfectly" straight and which is used for testing plane surfaces and drafting straight lines. A ruler is a straight-edge.

STRAIN ENERGY. A term that usually denotes the elastic energy stored in a stressed body that can be recovered as work upon unloading.

STRAIN GAGE. An instrumental device used to measure the dimensional change within or on the surface of a specimen. The electrical-type strain gage may operate on the measurement of a capacitance, inductance, or a resistance change that is proportional to strain. The bonded resistance-type strain gage is the most widely used in many fields. The principle of operation was discovered in 1856 by Lord Kelvin. The fundamental relationships between resistance change and strain are shown by Fig. 1. When

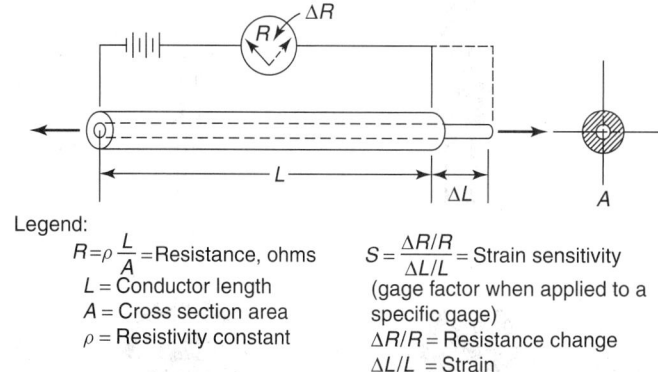

Legend:
$R = \rho \dfrac{L}{A}$ = Resistance, ohms $S = \dfrac{\Delta R/R}{\Delta L/L}$ = Strain sensitivity
L = Conductor length   (gage factor when applied to a
A = Cross section area   specific gage)
ρ = Resistivity constant $\Delta R/R$ = Resistance change
    $\Delta L/L$ = Strain

Fig. 1. Basic relationships between resistance change and strain in resistance-type strain gage.

a conductor of length L and cross-sectional area A is elongated, the length increases and the area decreases by Poisson's effect to produce an increase in resistance. The resistance change $\Delta R/R$ then is related to the length change $\Delta L/L$, or strain ε by the strain sensitivity or gage factor. If the strain sensitivity were dependent upon dimensional change only, resulting from the usual Poisson ratio of 0.3, then all metallic conductors would have a theoretical value of 1.6 in the elastic range and 2.0 in the plastic range of the alloy used where the Poisson ratio becomes 0.5. However, the resistivity alters with strain in order to produce the gage-factor range of 2.0 to 4.5 experienced by the alloys most often used for metallic strain gages.

Early bonded strain gages were developed in the 1930s by Dr. Arthur C. Ruge at the Massachusetts Institute of Technology. The gage was made from fine strain-sensitive wire attached to a thin paper carrier with nitrocellulose cement for dimensional stability and to provide electrical isolation from a metal specimen. The carrier was bonded to the specimen with the same cement so that the specimen surface strain would be reliably transmitted into the fine-wire grid. The grid shape was designed to provide maximum gage resistance in the smallest possible gage length and width. Foil also can be used as the strain element. In addition to material variations, a large number of grid shapes is available in modern strain gages, particularly in the foil construction to meet the needs of specific applications. Gage lengths range from 6 to $\frac{1}{64}$th inch and resistance values from 1,000 to 60 ohms.

Strain gage sensing element materials most commonly used are: (1) *Constantan* (copper-nickel alloy) used mainly for static strain measurement because of low and controllable temperature coefficient; (2) *Nichrome V* (nickel-chrome alloy) frequently used for high-temperature static and dynamic strain measurements; (3) *Dynaloy* and *IsoElastic* (nickel-iron alloy plus other proprietary ingredients), used for dynamic tests where the larger temperature coefficients of these materials are of no consequence; (4) *Stabiloy* and *Karma* (nickel-chrome alloys) containing other ingredients which provide wider temperature compensation range; and (5) platinum alloy (usually tungsten) which shows unusual stability and fatigue life at elevated temperatures. Additionally, semiconductor strain gages are used. These are similar to conventional metallic gages, the principal difference being the greater response of semiconductor gages to both strain and temperature. They have large and nonlinear resistance versus strain, arising primarily to the piezoresistive effect. They have found their main use in development of high-output transducers, such as load cells and pressure cells.

The application of strain gages breaks down into two main areas: (1) applications where the gage measures strain as the primary objective of measurement (stress analysis of various structures—bridges, boat hulls, etc.); and (2) uses where the measurement of strain, in turn, is a measure of another variable, such as pressure, impact, acceleration, and other force-associated variables, i.e., the strain gage becomes a transducer. Because strain gages, when properly compensated for temperature effects, can achieve overall accuracies of plus-or-minus 0.10% or better and because of their great flexibility, relatively low cost, and availability of numerous configurations, they are widely used in numerous kinds of transducers. A few representative strain gage configurations are illustrated in Fig. 2.

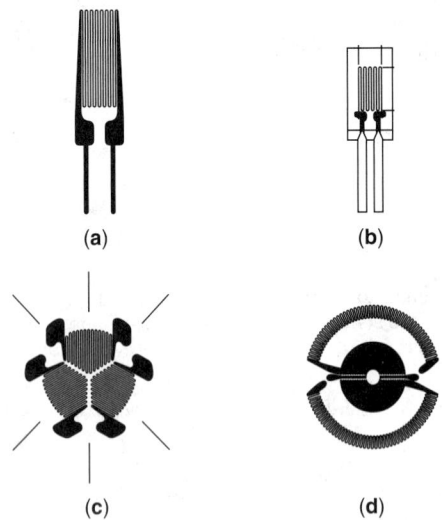

Fig. 2.   Representative configurations of strain gages: (**a**) Uniaxial strain gage, wire type; (**b**) uniaxial strain gage, foil type; (**c**) rosette gage (3-element foil) rosette, 60 degrees planar; and (**d**) 4-element "diaphragm" gage which takes advantage of the tangential tensile strains developed at the center of the diaphragm and the compressive radial strains present at the edge.

STRAIN HARDENING EXPONENT.   A measure of the rate of hardening with strain as in a tensile test, as expressed by the constant "n" in the equation $\sigma = \sigma_0 \delta^n$, where σ is the true stress, σ_0 the true stress at unit strain, and δ the true strain.

STRAIN THEORY.   A theory first proposed by von Baeyer to explain the relative stability of various carbon compounds. It may be stated in the form: The regular tetrahedral-symmetric position is the most stable of all possible positions of neighboring carbon compounds; variations from this position produces increased energy content, and hence strain. Since the angle at the vertex of a regular tetrahedron is $109°28$; this theory ascribes minimum strain to cyclopentane, of the polymethylenes. The theory is borne out by the lesser stability of cyclobutane and cyclopropane, but not to the degree that might be expected by the stability of some of the higher-membered rings. In that case, the lesser strain is often due to a spatial or three-dimensional structure.

Extensions of the strain theory have been made, with varying success, to other hydrocarbon ring structures, saturated and unsaturated, to ring compounds in which the hydrogen atoms have been variously substituted, and to rings containing atoms other than carbon, as well as to bicyclic and polycyclic systems.

STRATH.   1. In geography, a broad alluviated valley. 2. In geomorphology, a valley or confluent valleys which represent a local base level of erosion or a local and incipient peneplain.

STRATIFICATION (Ecology).   As described under **Ecology**, the primary classification of ecological communities is by the broad nature of the environment—fresh-water, salt-water, terrestrial, etc. communities. Vertical stratification also occurs in many of these communities as, for example, certain fishes that prefer shallow water, others that inhabit deep water; or in a tropical forest, animals and birds that prefer the tree tops, others that prefer any one of several intervening levels between the surface and the top canopy. Further, some organisms prefer an underground habitat.

STRATIGRAPHY.   The study of the origin and chronological successions of the observable rocks of the lithosphere, in which each lithologic unit is considered to be a formation. The term formation is usually confined to bedded or stratified rocks, including lava flows and volcanic ashes. The major principles involved in the correlation (dating) of formations are: (1) The law of superposition, or that the chronological sequence of any stratigraphic section depends upon the original order in which the formations were laid down; thus the fundamental basis of stratigraphy is structural geology. (2) Index fossils, or those species of fossils whose stratigraphic age are already known. (3) lithology. Igneous rocks may be dated by the age of the sedimentary rocks which they intrude, or overlie;

or by radioactive minerals. See also **Paleontology**; **Structural Geology**; and **Radioactivity**.

STREAM FUNCTION.   A parameter of two-dimensional, nondivergent flow, the value of which is constant along each streamline. For flow in the (x, y)-plane, the stream function ψ is related to the respective coordinate velocities u and v by the equations

$$u = -\frac{\partial \psi}{\partial y}, \quad v = \frac{\partial \psi}{\partial x}$$

Stokes's stream function. (also called *current function*). If the flow is three-dimensional but is axisymmetric (i.e., the same in every plane containing the axis of symmetry), a Stokes's stream function ψ will exist such that

$$v_s = -\frac{1}{r} \frac{\partial \psi}{\partial n}$$

where v_s is the speed in an arbitrary direction s, r the distance from the axis of symmetry, and n is normal to the direction s, increasing to the left. Note that Stokes's stream function has dimensions $L^3 T^{-1}$.

Stream functions can also be defined for more complex three-dimensional flows.

STREAMING (Molecular).   Application of kinetic theory to the flow of gas through a tube at low pressures, such that the mean free path is large compared with the diameter of the tube. In this case, the streaming of the gas is due to the random motion of the molecules, and to the density gradient down the tube, so that the numbers of molecules traversing a given cross section in opposite directions is different. For a tube of circular cross section, the mass flowing per second is proportional to the pressure difference and the cube of the radius.

STRESS RELIEVING.   A heat treatment to relieve residual stresses.

STRESS-RUPTURE TEST.   A form of short time creep test in which a tensile specimen is deformed to rupture (fracture) under constant load and temperature.

STRESS-STRAIN CURVE.   A graphical representation of the relation between unit stress and unit deformation in a stressed body as a gradually increasing load is applied.

STRESS (Structural).   A quantitative expression of a condition within an elastic material due to deformation, or strain, brought about by external forces, inequalities of temperature, or otherwise. Its measure is always the ratio of a force to an area. By some, stress is interpreted as a force distributed over an area, and the above ratio is called the "unit stress." Central, torsional and bending loads cause stress. The total resisting force acting at any section of the body divided by the area of the section is the average stress, commonly expressed in pounds per square inch. The unit stress is the resisting force on a unit of area. The component of a stress that acts at right angles to a surface is known as the normal stress. If this stress is produced by a load whose resultant passes through the center of gravity of the area, it is called an axial or direct stress. A normal resisting force that causes the fibers to increase in length is a tensile stress, while one which shortens the fibers is a compressive stress. The latter is often called a bearing stress. The component of any stress that lies in the plane of the area is a shearing stress. See also **Elasticity**.

The "conventional stress" as applied to tension or compression tests is the instantaneous value of the load divided by the original cross-section area. This type of stress is used by engineers for design purposes since their primary concern is how large a load or stress a structure is capable of carrying, as measured in terms of the structures' original dimensions. Frequently, however, for scientific purposes, it is important to know the true stress determined by dividing the force at a given instant by the area existing at that time.

Direct tensile or compressive stresses are known as primary stresses. The bending stress, resulting from deflection, is called a secondary stress. The stresses developed in a column due to the lateral deflections are of a secondary nature. The rigidity of the riveted or welded joints of a truss that has deflected due to the axial deformation of its members causes bending stresses in the members which are classified as secondary stresses. The resistance offered by a body to a combination of direct and bending loads is frequently called a combined stress. A normal stress which occurs at a point in a plane on which the shearing stress is zero is known as a principal stress. If this normal stress is tensile, it is often called a diagonal tension stress; if compressive it is known as a diagonal compression stress.

The internal resisting force that arises in a restrained body due to temperature changes is a thermal stress. The adhesive resistance developed in the concrete surrounding the steel reinforcing rods when a reinforced concrete member is subjected to load is known as bond stress. Safe unit resisting forces used in design are called working stresses. These are usually taken as a percentage of the ultimate stress or the elastic limit of the material.

The stress developed in bridge members as a result of traction between the wheels of the live load and the supporting surface is called a traction stress. The effect of these stresses is usually neglected in highway bridge design but must be considered in railway bridges.

STRIATION.   Three uses of this term are: 1. A striped appearance of the positive column of a Crookes tube. 2. A defect of optical materials, such as optical glass, having the appearance of streaks through the material, and seriously affecting the material for use as lenses or windows. 3. The scratches on bedrock or on pebbles and boulders which are the result of glaciation. These are called glacial striae to distinguish them from the striae which occur on the surfaces of fault planes.

STRIGIFORMES *(Aves)*.   This order of birds is distinctly set off from other bird kinds. In physical structure they are remarkably homogeneous. They are recognized immediately by their large heads, their forward-directed eyes, the seeming absence of a neck, and their soft plumage. Their length is 15–80 centimeters (6–31 inches) and the weight is 55–4200 grams (2 ounces to 9 pounds). They have 11 primaries and 12 (rarely 10) tail feathers. The fourth toe is abducted and reversed diagonally. The beak is drawn down steeply, making it appear small. The upper mandible is hooked. The plumage is particularly well suited for nocturnal activity: gray, brown, black, and white shades arranged in patches, stripes, bands, and streaks effect a muted coloration. See accompanying illustration. They possess calls, mostly sounding an "oo" or an "ee," that serve to find and to keep contact with their conspecifics during their nocturnal activities. Their clutch consists of 1–12 white, roundish eggs. The 2 families are: 1. Barn Owls (*Tytonidae*); and 2. Owls in the narrower sense (*Strigidae*). Altogether there are 28 genera and 144 species.

Owls are to be distinguished from raptors not only by both external and internal morphology, but quite notably by their behavior. Owl young, unlike those of raptors, are born blind. Most owls are active at dusk or during the night, and some during the day. They feed chiefly on smaller vertebrates, especially rodents, but also on insects and worms. Some, such as the fishing owls, are adapted for a specialized diet. Some smaller species prefer to consume insects. Indigestible parts of the prey (fur, feathers, bones, and chitin) are regurgitated.

The notion that owls cannot see well in daylight is still widespread, but untrue. Owls are indeed far-sighted, yet their vision is excellent in daylight; in complete darkness, they are just as helpless as humans. The particular structure of their eyes, however, enables them to orient themselves adequately in very dim light. The number of light-sensitive cells in the retina (rods) is greatly multiplied in those owls which are active at night, such as the tawny owl (*Strix aluco*). See Fig. 1. On the other hand, such species have very limited color discrimination, while owls that are active at twilight or during the day do recognize colors. The retina possesses not only the rods, but also more cones than that of nocturnal species. The owl's eyes, which are usually very large, are immovable, forcing the bird to turn its head for a change of visual field.

The bony ring supporting the sclera (the dense, fibrous, opaque white outer coat of the eyeball), which is peculiar to a bird's eye, has developed into a regular cylinder in owls; it unites the dioptric apparatus (lens, cornea, and iris) with the relatively restricted retina. This results in an eye whose shape is strongly reminiscent of the "telescopic eyes" of some deep sea fish and some nocturnal mammals. Such "telescopic eyes" are adjusted to seeing in the dimmest of lights. The angle of the corneal fenestra is very wide (approximately 160 degrees), and the light gathered from a large field of vision is reduced to a small image on the retina.

The sense of hearing is superbly developed in owls. The margins of the auditory meatus are remodeled into feathered flaps, which can close over the ear completely and so protect the sensitive inner parts. When erected, the flaps become wide, movable sound-funnels, which enable the owl to

Fig. 1.   Face of tawny owl (*Strix aluco.*)

pick up the faintest sounds from various directions. The feathery tufts on the head of various owl species are not ears, but decoration.

All owls lay pure white, more or less round eggs, which, as a rule, are incubated only by the female. During that period, it is usually the male alone that takes care of the feeding; they often pile up a store of food near the brooding ground. Both parents share in the rearing of the young.

No owl builds a genuine nest. The closest to it is a more or less thick layer of usually dry plant material put together by the short-eared owl, which is a ground-brooder. Other species dig a hollow at their brooding place and tear up boluses or food leftovers to use as a kind of nest bedding. In most cases, however, the eggs are deposited directly on the ground of the chosen brood site, sometimes in an abandoned crow's nest or in the aerie of a predator.

When young owls hatch, they are blind and covered with a whitish down, and their eyes are closed. Often the several young of one brood hatch at intervals of several days, for the eggs are laid in intervals and most species begin incubation after the first egg is laid. After one week, on the average, the ears and eyes open and gradually the young begin to get a second dress which in some species is downy and in others resembles the dress of the adult birds.

The family of the barn owls (*Tytonidae*) stands well removed based on recent studies, phylogenetically, from the true owls. The facial veil is more or less heart-shaped; the talon of the middle toe has a "comb;" and the posterior edge of the sternum has either two notches or none at all. There are 2 genera with a total of 11 species: (1) Barn Owls (*Tyto*), with 9 species; and (2) Bay Owls (*Pholidus*), with 2 species.

The 9 species of barn owls (*Tyto*) are as follows: 1. The Barn Owl (*Tyto alba*) reaches a length of 34 centimeters (13 inches), with a wingspread of 95 centimeters (37 inches), and weighs about 300 grams ($10\frac{1}{2}$ ounces). The facial veil is heart-shaped. The eyes are relatively small and black-brown; the wings are long and rather pointed. Since the intertarsal joints are close to each other when the bird is perching, it gives a knock-kneed impression. 2. The Cape Grass Owl (*Tyto capensis*) is found in southern Africa; it is a ground-breeding bird. 3. The Grass Owl (*Tyto longimembris*) occurs in the grasslands of India and southern China, some of the Sunda Islands, and Australia. 4. The Celebes Barn Owl (*Tyto rosenbergii*) inhabits the rain forests of Celebes. 5. The Minnahassa Barn Owl (*Tyto inexpectata*) inhabits northern Celebes. 6. The Madagascar Grass Owl (*Tyto soumagnei*) is relatively small, has long wings, and inhabits woodland glades in the rain forests of Madagascar; it feeds mainly on amphibia. 7. The Masked Owl (*Tyto novaehollandiae*) is found in the jungles of Australia and some Sunda Islands. 8. The Sooty Owl (*Tyto tenebricosa*) is the only species without yellow-brown or yellow-orange shades in its plumage. Its habitat is humid parts of the jungle and rain forests in New Guinea and Australia. 9. The New Britain Barn Owl (*Tyto aurantia*) occurs only in New Pomerani (New Britain).

The 2 species of bay owls (*Pholidus*) are as follows: 1. The Bay Owl (*Pholidus badius*) reaches a length of 30 centimeters (12 inches). 2. The Congo Bay Owl (*Pholidus prigoginei*) was discovered only in 1951, in the highlands northwest of Lake Tanganyika.

Bays owls are strictly nocturnally active birds, inhabiting the jungle. Wide, natural cavities serve as brood sites; the 3–5 purely white, roundish eggs are deposited directly on the floor, which is usually covered with a layer of decayed wood. The exact breeding season is not known. Bay owls feed on small vertebrates, such as small mammals, birds, reptiles, and amphibia, and possibly also on fish, for they like to hunt near the water. They are known by a hollow-sounding "hoo."

The family Owls (*Strigidae*) comprises 24 genera with 133 species. The claw of the middle toe has no pecten (comb). The 2 subfamilies are: (1) True Owls (*Buboninae*) and (2) Long-Eared Owls and Disk-Eyed Owls (*Striginae*).

The 18 genera which comprise the subfamily *Buboninae* are as follows: 1. The Scops Owls (*Otus*); 2. *Jubula* (Maned Owl, *Jubula lettii*); 3. *Lophostrix* (Crested Owl, *Lophostrix cristata*); 4. the Eagle Owls (*Bubo*); 5. the Fish Owls (*Ketupa*); 6. the Fishing Owls (*Scotopelia*); 7. the Spectacled Owls (*Pulsatrix*); 8. *Nyctea* (Snowy Owl, *Nyctea scandiaca*); 9. *Surnia* (Hawk Owl, *Surnia ulula*); 10. the Pygmy Owl (*Glaucidium*); 11. *Micrathene* (Elf Owl, *Micrathene whitneyi*); 12. *Uroglaux* (New Guinea Hawk Owl, *Uroglaux dimorpha*); 13. the Hawk Owls (*Ninox*); 14. *Gymnoglaux* (Bare-legged Owl, *Gymnoglaux lawrencii*); 15. *Sceloglaux* (Laughing Owl, *Sceloglaux albifacies*); 16. *Athene* (Little Owl, *Athene noctua*); 17. *Speotyto* (Burrowing Owl, *Speotyto cunicularia*); 18. the Tropical Tawny Owls (*Ciccaba*).

The 6 genera which comprise the subfamily of Long-eared Owls and Disk-eyed Owls (*Striginae*) are as follows: 1. The Disk-eyed Owls (*Strix*); 2. *Rhinoptynx* (*Rhinoptynx clamator*); 3. the Long-eared Owls (*Asio*); 4. *Pseudoscops* (Jamaican Owl, *Pseudoscops grammicus*); 5. *Nesasio* (Fearful Owl, *Nesasio solomonensis*); 6. the Tengmalm's Owls (*Aegolius*). See also **Owls**.

STRINGER. 1. A configuration of nonmetallic inclusions in wrought metals that is elongated in the direction of working. In steels, stringers are usually formed of oxides or sulfides. 2. One of the longitudinal beams in the floor system of a bridge.

STRIPPING. In nuclear reactions, an effect observed primarily in bombardment with deuterons, whereby only part of the incident particle merges with the target nucleus, and the remainder proceeds with most of its original momentum in a direction determined solely by its electromagnetic interactions with the target nucleus. The effect is strongly marked in deuteron bombardment, and in this case typically leads to directional neutron and proton beams that emerge from the target (if the latter is sufficiently thin) predominantly in the forward direction. The angular divergence of the beams decreases with increasing energy of the incident deuterons. Subsidiary peaks are sometimes observed at rather small angles relative to the forward direction. When the (*d, p*) type of stripping occurs with deuterons having energies smaller than or comparable with the Coulomb barrier of the target nucleus, it is often called the Oppenheimer-Phillips process.

For the use of the term stripping in chemical technology, see also **Distillation**.

STROBOSCOPE. An instrument that permits intermittent observations of a cyclically moving object in such a way as to produce an optical illusion of stopped or slowed motion. This phenomenon is readily apparent, for example, when rewinding a tape at many revolutions per minute when the tape deck is located under a 60-Hz incandescent lamp. Patterns on the reel tend to slow and then appear to stop before reversing their direction. Of course, stroboscopic effects have been known for decades,[1] one of the first scientific applications being found in very high-speed

photography. Intermittency of observation can be provided by mechanical interruption of the line of sight (as with a motion picture camera) or by intermittent illumination of the object being viewed. The industrial stroboscope basically is a lamp plus the electronic circuits required to turn the lamp on and off very rapidly, at rates as high as 150,000 flashes per minute and higher.

The schematic diagram of an electronic stroboscope is shown in Fig. 1. The device includes a strobotron tube with its associated discharge capacitors, a triggering tube to fire the strobotron, an oscillator to determine the flashing rate, and a power supply. With the use of harmonic techniques, speeds up to 1 million r/min can be measured. Accuracy is nominally ±1% of the dial reading after calibration.

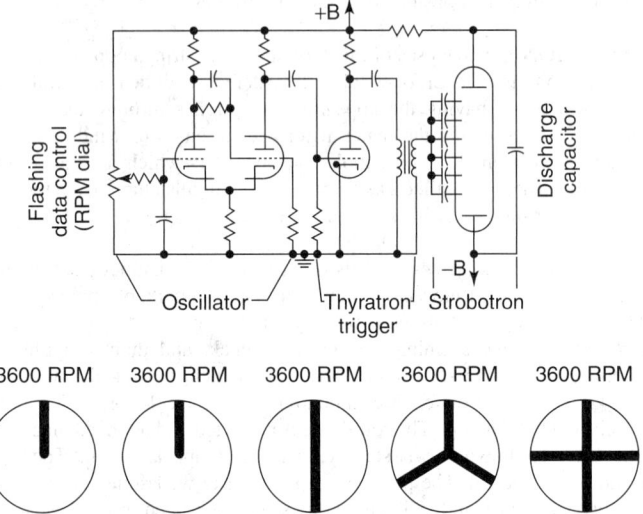

Fig. 1. Electronic stroboscope, (*a*) Schematic representation of circuit. (*b*) Images obtained at harmonic and subharmonic flashing rates of a stroboscope. Even with an asymmetrical object, the correct fundamental image is repeated when the stroboscope is flashing at one-half, one-third, and so on, the speed of the object. The proper setting for a fundamental speed measurement is the highest setting at which a single stationary image can be achieved. This does not hold, however, if the fundamental is beyond the flashing rate of the stroboscope. There are several ways to distinguish fundamental from submultiple images. The flashing rate can be decreased until another single image appears. If this occurs at half the first reading, the first reading was the actual speed of the device. If it occurs at some other value, then the first reading was a submultiple. Or the user can double the flashing rate and check for a double image. Or the user can flip the range switch to the next higher range. Because of the 6:1 relationship between ranges, a 6:1 pattern should appear. The 6:1 relationship between ranges also makes it convenient to convert speed readings from revolutions per minute into cycles per second. One simply flips to the next lower range and divides the new reading by 10.

Because of their portability and easy setup, stroboscopes find a variety of applications, principally in machine and vehicle research, development, and testing. (*GenRad.*)

To serve as a tachometer, a stroboscope must have its own flashing-rate control circuits and calibrated dial. Stroboscope tachometer test disks are available. These disks can be cut out and mounted on light cardboard or metal. The center must be carefully located and fitted onto the drive shaft. Although more automatic means are available to measure belt slippage, this was commonly accomplished by stroboscopes in earlier times.

STROMATOLITE. A term that has been generally applied to variously shaped (often domal), laminated, calcareous sedimentary structures formed in a shallow-water environment under the influence of a mat or assemblage of sediment-binding blue-green algae that trap fine (silty) detritus and precipitate calcium carbonate and that commonly develop colonies or irregular accumulations of a constant shape, but with little or no microstructure. It has a variety of gross forms, from near-horizontal to markedly convex, columnar, and subspherical. Stromatolites were originally considered animal fossils, and although they are still regarded as fossils because they are the products of organic growth, they are not fossils of any specific organism, but rather consist of associations of different genera and species of

---

[1] Invented independently by Stampfer of Vienna and Plateau of Ghent in 1832. Stampfer chose the name "stroboscope," which is derived from the Greek words meaning "whirling watcher."

organisms that can no longer be recognized and named or that are without organic structures. An excellent treatise on stromatolites is *Stromatolites* (M.R. Walter, editor), Elsevier, New York, 1976.

STROMATOLITH.   A term proposed by Foye, in 1916, for banded gneisses composed of alternate layers of igneous and metamorphic (schistose) rocks.

STRONTIANITE.   The mineral strontianite is strontium carbonate, $SrCO_3$, usually occurring in whitish-yellow or whitish-green masses of radiated acicular crystals, or in fibrous or granular form. When distinctly crystallized it is obviously orthorhombic, but such crystals are rare. It has a nearly perfect prismatic cleavage; uneven fracture; brittle; hardness, 3.5; specific gravity, 3.785; luster, vitreous; color, as above, also green, gray and colorless; streak, white; transparent to translucent. Strontianite occurs in veins, chiefly in limestones, occasionally in the crystalline rocks, and it is usually associated with calcite and celestite. It is found in the metalliferous veins in the Harz Mountains and Saxony. It is commercially important in Westphalia where it is mined for use in the beet sugar industry. In the United States, crystalline masses and gorges of strontianite are found in Schoharie County, New York, long a famous locality for this mineral.

STRONTIUM.   Chemical element, symbol Sr, at. no. 38, at. wt. 87.62, periodic table group 2, mp 769 °C, bp 1384 °C, density 2.54 g/cm^3 (20 °C). Below 215 °C, elemental strontium has a face-centered cubic crystal structure; between 215–605 °C, a hexagonal close-packed crystal structure; and above 605 °C, a body-centered cubic crystal structure.

Strontium is a silver-white metal, soft as lead, malleable, ductile, oxidizes rapidly on exposure to air, burns when heated in air emitting a brilliant light and forming oxide and nitride, reacts with H_2O yielding strontium hydroxide and hydrogen gas. Discovered by Hope and by Klaproth in 1793, and isolated by Davy in 1808.

There are four stable isotopes, ^{84}Sr and ^{86}Sr through ^{88}Sr, and seven known radioactive isotopes, ^{82}Sr, ^{83}Sr, ^{85}Sr, and ^{89}Sr through ^{92}Sr, all with relatively short half-lives measurable in hours or days except 90Sr which has a half-life of about 26 years. The latter isotope represents a hazard from nuclear blasting activities because of its long half-life, tendency to contaminate food products, such as milk, and retention in the body. See also **Radioactivity**. In terms of abundance, strontium is 21st among the elements occurring in the rocks of the earth's crust. In terms of the content of sea water, the element ranks 11th, with an estimated 38,000 tons of strontium per cubic mile (9120 tons/cubic kilometer) of seawater. First ionization potential 5.692 eV; second, 10.98 eV. Oxidation potentials $Sr \rightarrow Sr^{2+} + 2e^-$, 2.89 V; $Sr + 2OH^- + 8H_2O \rightarrow Sr(OH)_2 \cdot 8H_2O + 2e^-$, 2.99 V. Other important physical properties of strontium are given under **Chemical Elements**.

Occurrence and Characteristics. Strontium occurs chiefly as sulfate (celestite, $SrSO_4$) and carbonate (strontianite, $SrCO_3$) although widely distributed in small concentration. The commercially exploited deposits are mainly in England. The sulfate or carbonate is transformed into chloride, and the electrolysis of the fused chloride yields strontium metal.

As is to be expected from its high oxidation potential (2.89 V) strontium, like calcium and barium, reacts readily with all halogens, oxygen and sulfur to form halides, oxide and sulfide. See also **Celestite**; and **Strontianite**. In all its compounds it is divalent. It reacts vigorously with H_2O to form the hydroxide, displacing hydrogen and it forms a hydride with hydrogen. Strontium hydroxide forms a peroxide on treatment with H_2O_2 in the cold. Strontium exhibits little tendency to form complexes; the ammines formed with NH_3 are unstable, the β-diketones and alcoholates are not well characterized, and the chelates formed with ethylenediamine and related compounds are the only representatives of the type. Common compounds of strontium are the following:

Strontium acetate, $Sr(C_2H_3O_2)_2$, white crystals, soluble, formed by reaction of strontium carbonate or hydroxide and acetic acid.

Strontium carbide (acetylide), SrC_2, black solid, formed by reaction of strontium oxide and carbon at electric furnace temperature; the carbide reacts with water yielding acetylene gas and strontium hydroxide.

Strontium carbonate, $SrCO_3$, white solid, insoluble ($K_{sp} = 9.4 \times 10^{-10}$), formed (1) by reaction of strontium salt solution and sodium carbonate or bicarbonate solution, (2) by reaction of strontium hydroxide solution and CO_2. Strontium carbonate decomposes at 1,200 °C to form strontium oxide and CO_2, and is dissolved by excess CO_2, forming strontium bicarbonate, $Sr(HCO_3)_2$, solution.

Strontium chloride, $SrCl_2 \cdot 6H_2O$, white crystals, soluble, formed by reaction of strontium carbonate or hydroxide and HCl. Anhydrous strontium chloride, $SrCl_2$, absorbs dry NH_3 gas.

Strontium chromate, $SrCrO_4$, yellow precipitate ($K_{sp} = 3.75 \times 10^{-5}$) formed by reaction of strontium salt solution and potassium chromate solution.

Strontium cyanamide, $SrCN_2$, formed with the cyanide, $Sr(CN)_2$, by heating strontium carbide at 1,200 °C with nitrogen.

Strontium hydride, SrH_2, white solid, formed by heating strontium metal or amalgam in hydrogen gas at 250 °C. Is reactive with H_2O, yielding strontium hydroxide and hydrogen gas.

Strontium nitrate, $Sr(NO_3)_2$, white crystals, soluble, formed by reaction of strontium carbonate or hydroxide and HNO_3.

Strontium oxide, SrO, white solid, mp about 2,400 °C, reactive with H_2O to form strontium hydroxide ($K_{sp} = 3.2 \times 10^{-4}$); strontium peroxide, $SrO_2 \cdot 8H_2O$, white precipitate, by reaction of strontium salt solution and hydrogen or sodium peroxide, yields anhydrous strontium peroxide SrO_2, upon heating at 130 °C in a current of dry air.

Strontium oxalate, SrC_2O_4, white precipitate ($K_{sp} = 5.6 \times 10^{-8}$) formed by reaction of strontium salt solution and ammonium oxalate solution.

Strontium sulfate, $SrSO_4$, white precipitate ($K_{sp} = 3.2 \times 10^{-7}$), formed by reaction of strontium salt solution and H_2SO_4 or sodium sulfate solution, insoluble in acids. On heating with carbon strontium sulfate yields strontium sulfide, SrS, while on boiling with sodium carbonate solution, $SrSO_4$ yields strontium carbonate.

Strontium sulfide, SrS, grayish-white solid (thermodynamic K_{sp} 500) reactive with water to form strontium hydrosulfide, $Sr(SH)_2$, solution. Strontium hydrosulfide is formed (1) by reaction of strontium sulfide and H_2O, (2) by saturation of strontium hydroxide solution with H_2S. Strontium polysulfides are formed by boiling strontium hydrosulfide with sulfur.

Editor's Note re Strontium Isotope Research. At any given time, the Sr isotope composition in seawater is uniform throughout the ocean because the oceanic residence time of Sr (5 million years) exceeds the mixing time of the oceans ($\sim$1000 years). However, over geologic time, the $^{87}Sr/^{86}Sr$ ratio in seawater has varied as the result of fluxes of Sr to the oceans from various sources. These would include submarine hydrothermal activity, fluxes from rivers, and submarine recycling, the latter occurring by limestone recrystallization and erosion of ancient sedimentary carbonate. J. Hess and colleagues (University of Rhode Island) reported in 1986 that the seawater Sr isotope composition appears to be a smoothly varying function of time and can be useful for high-precision correlations of oceanic sediments for certain periods of time. These researchers prepared a detailed record of the Sr isotope ratio during the last 100 million years by measuring this ratio in well over a hundred foraminifera samples. Sample preservation was evaluated from scanning electron microscopy studies, measured Sr/Ca ratios, and pore water Sr isotope ratios. Results show that the marine Sr isotope composition can be used for correlating and dating well-preserved authigenic marine sediments throughout much of the Cenozoic to a precision of $\pm$1 mil years. See also **Condrite**.

In 1990, R.C. Capo and D.J. DePaolo (University of California, Los Angeles and Berkeley, respectively) reported that "marine carbonate samples indicate that during the past 2.5 million years the $^{87}Sr/^{86}Sr$ ratio of seawater has increased by 14×10^{-7}. The high average rate of increase of this ratio indicates that continental weathering rates were exceptionally high. Nonuniformity in the rate of increase suggests that weathering rates fluctuated by as much as $\pm$30 percent of present-day values. Some of the observed shifts in weathering rate are contemporaneous with climatic changes inferred from records of oxygen isotopes and carbonate preservation in deep sea sediments."

Studies of Metamorphism. As reported by J.N. Christensen, J.L. Rosenfield, and D.J. DePaolo (University of California, Berkeley), "Measurement of the radial variation of the $^{87}Sr/^{86}Sr$ ratio in a single crystal from a metamorphic rock can be used to determine the crystal's growth rate. Such variation records the accumulation of ^{87}Sr from radioactive decay of ^{87}Rb (rubidium) in the rock matrix from which the crystal grew. This method can be used to study the rates of petrological processes associated with mountain building." This methodology has been applied by the researchers mentioned to the study of the rates of tectonometamorphic processes from rubidium and strontium isotopes in garnet."

Isotopic Tests for Upwelling Water. In studies of the Yucca Mountain, Nevada, area as a potential site for a high-level nuclear waste repository, the area has been aggressively scrutinized geologically for possible upwelling of deep-seated waters. Strontium and uranium isotopic compositions of hydrogenic materials were used by scientists J.S. Stuckless, Z.E. Peterman, and D.R. Muhs (U.S. Geological Survey, Denver, Colorado) to assist in confirming other geological methods. Their findings indicated in 1991 that the vein deposits are isotopically distinct from groundwater in the two aquifers that underline Yucca Mountain, thus indicating that the calcite could not have precipitated from groundwater and thus providing evidence against upwelling water at the site.

Additional Reading

Capo, R.C. and D.J. DePaolo: "Seawater Strontium Isotopic Variations from 2.5 Million Years Ago to the Present," *Science*, **51** (July 6, 1990).

Christensen, J.N., J.L. Rosenfeld, and D.J. DePaolo: "Rates of Tectonometamorphic Processes from Rubidium and Strontium Isotopes in Garnet," *Science*, 1405 (June 21, 1989).

Greenwood, N.N. and A. Earnshaw: "Chemistry of the Elements," 2nd Edition, Butterworth-Heinemann, Inc., Woburn, MA, 1997.

Hess, J., M.L. Bender, and J.-G. Schilling: "Evolution of the Ratio of Strontium-87 to Strontium-86 in Seawater from Cretaceous to Present," *Science*, **231**, 979–983 (1986).

Krebs, R.E.: "The History and Use of Our Earth's Chemical Elements: A Reference Guide," Greenwood Publishing Group, Inc., Westport, CT, 1998.

Lewis, R.J. and N.I. Sax: "Sax's Dangerous Properties of Industrial Materials," 10th Edition, John Wiley & Sons, Inc., New York, NY, 1999.

Lide, D.R.: "CRC Handbook of Chemistry and Physics 2000–2001," 81st Edition, CRC Press, LLC, Boca Raton, FL, 2000.

Macdougall, J.D.: "Seawater Strontium Isotopes, Acid Rain, and the Cretaceous-Tertiary Boundary," *Science*, 485 (January 29, 1988).

Meyers, R.A.: "Handbook of Chemicals Production," The McGraw-Hill Companies, Inc., New York, NY, 1986.

Parker, S.P.: "McGraw-Hill Encyclopedia of Chemistry," 2nd Edition, The McGraw-Hill Companies, Inc., New York, NY, 1993.

Staff: "ASM Handbook — Properties and Selection: Nonferrous Alloys and Pure Metals," ASM International, Materials Park, OH, 1990.

Stuckless, J.S., Z.E. Peterman, and D.R. Muhs: "U and Sr Isotopes in Ground Water and Calcite, Yucca Mountain, Nevada: Evidence Against Upwelling Water," *Science*, 551 (October 25, 1991).

Stwertka, A. and E. Stwertka: "A Guide to the Elements," Oxford University Press, Inc., New York, NY, 1998.

STEPHEN E. HLUCHAN, Business Manager, Calcium Metal Products, Minerals, Pigments & Metals Division, Pfizer Inc., Wallingford, CT.

STRUCTURAL GEOLOGY.   That branch of geology which deals with the form, arrangement, and internal structure of the rocks, and especially with the description, representation, and analysis of structures, chiefly on a moderate to small scale. The subject is similar to tectonics, but the latter is generally used for the broader regional or historical phases. Structural petrology is the study of the internal structure or fabric of a rock, commonly with the aim of clarifying the rock's deformational history. (*American Geological Institute.*)

STRUT.   A structural member subjected to compression. Its conditions of loading and analysis are the same as for a column. If there is any difference between strut and column, it rests on the following points. A column is usually thought of as being a fairly large compression member, vertical in position. Small columns are frequently called struts; also, struts are compression members which are incorporated into structures in many positions besides the vertical.

STUFFING BOX.   A device for preventing leakage or transfer of fluid between moving parts, usually consisting of a relatively soft packing compressed or confined by an adjustable member called a gland. Stuffing box packings differ from gaskets in that they are used in confined spaces, and do not of themselves withstand stresses due to fluid pressure. In the usual form, the stuffing box consists of a hollow cylinder surrounding the moving (reciprocating or rotating rod or shaft) member; the space between the hole and the rod or shaft is filled with packing compressed by the gland. Packings may consist of relatively plastic material, bonding

such substances as cotton fabric, rope, or asbestos, but packings of rubber, leather, pressed graphite, or molded plastics are also used.

In high-speed machinery, such as turbines and rotary compressors, where no adequate cooling is available, and in high-pressure equipment where small clearances are required, and for which packings are inadequate, devices known as labyrinths may be used. A labyrinth consists of a series of projections on the rotating element, running in close contact with grooves on the stationary element. To enable a labyrinth to function as a seal, there must be some fluid flow; the fluid first passes a restriction and then expands into a chamber, which consumes a certain amount of energy. After a series of such expansions, a considerable pressure drop will exist between the initial and terminal points of the labyrinth. In order to render the device operative, there must always be some leakage at the terminal point. The effective operation of a labyrinth is less marked for liquids than for vapors.

STURGEONS (*Osteichthyes*).   Of the order *Chondrostei*, family *Acipenseridae*, sturgeons are fishes of moderate to large size, found in the oceans and fresh waters of the Northern Hemisphere. Their chief external characteristic is the series of bony plates arranged in rows along the back and sides, separated by wide spaces containing only small hard elements. See Fig. 1. The sturgeons are considered excellent food fishes and are the source of caviar. There are about fifteen European species and nine North American species.

Fig. 1.   Sturgeon.

Sturgeons have poor vision, but this is partly compensated by fleshy whiskers, which trail in the sand and assist in locating food at the bottom. Rare among the fishes, the sturgeons have taste buds external of the mouth. It is believed that these also assist in locating a food supply. Also quite rare among fishes is the fact that sturgeons eat quite slowly. The diet includes crawfish, insect larvae, snails and some small fish.

The largest of the sturgeons is the *Huso huso* (giant beluga) found in the Black and Caspian seas and the Volga River. Records indicate the largest of these to weigh 2860 pounds (1297 kilograms) with a length of 28 feet (8.4 meters). The age was unknown. A 75-year old sturgeon was weighed at 2200 pounds (998 kilograms) length 13 feet (3.9 meters). The Eurasian species frequently weigh in excess of a half-ton. The marine sturgeon *Acipenser sturo* is found in the temperate waters of the Atlantic, usually weighs about 500 pounds (227 kilograms) as an adult and attains a length of about 10 feet (3 meters). Marine sturgeons also prefer the muddy bottom and diet on shrimp, clams, worms, crustaceans, and small fishes.

The largest of American fresh water fishes is the *Acipenser transmontanus* (Pacific coast white sturgeon). They usually weigh about 300 pounds (136 kilograms) but records indicate weights up to 1300 pounds (590 kilograms) and more. The *Acipenser medirostris* is also a sturgeon of the American Pacific and averages about 350 pounds (159 kilograms) length of about 7 feet (2.1 meters).

Sturgeons are probably best known for the caviar processed from the roe. Caviar production is no longer limited to Eastern Europe, but has been undertaken in other areas, including the United States. In the interest of conservation, eggs can be stripped from the living fish (somewhat similar to practices in trout hatcheries). Various means are used to free the eggs from the egg membranes. The eggs are then placed in a brine which extracts the liquid from the eggs. After draining the brine, the eggs can be packed and sold commercially.

STYE (Hordeolum).   A stye is a lump or pimple on the edge of the eyelid caused by an infection or inflammation involving the hair follicles of the eyelashes. It most often occurs on the edge of the upper lid, but may also occur on the lower lid. An internal hordeolum is the same as a stye, but involves an internal eyelid gland located farther back on the eyelid. A stye can be painful and affect the entire eyelid.

A chalazion is similar in appearance to a stye and is caused by the blockage of one or more of the small oil-producing glands in the eyelid. The secretions within the gland harden and form a bump. A chalazion is usually painless unless it becomes infected. Styes and chalazions are quite common and easily treated.

Eyelids contain many mucus-producing glands called *Meibomian glands*, and when one of these glands becomes plugged, blocked, or infected, a stye forms. Styes are usually caused by a bacterial infection, often the staphylococcus germ, and are common among children, those with chronic eyelid infections, diabetics, and sometimes patients with poor hygiene. Styes are unrelated to any disease and are usually not harmful to the eye or vision if properly treated.

When a stye is in its early stages, the site looks red and swollen, and there is tenderness or pain on the edge of the top or bottom eyelid. Light sensitivity as well as increased tear production may be experienced. Sometimes, an abscess forms.

The best treatment for a stye is to apply hot, moist compresses to the eyelid on a frequent basis. This relieves pain and inflammation and helps speed up the formation of a white-head. Once the stye comes to a head, it should drain on its own. However, if it does not empty on its own, surgical draining may be required. The head of the stye is usually on the outside, but it can occur on the underside of the lid. Antibiotic ointment or cortisone drops may be prescribed to keep down the bacteria count of the eyelid. If the tissues surrounding the stye appear swollen and infected, oral antibiotics may also be prescribed.

Although styes are infectious, they are not contagious. It is important not to spread the infection and to practice careful personal hygiene. Do not share washcloths, hand towels and avoid close personal contact during the acute phase of the disease. If styes occur frequently, it is necessary to see eye doctor for an examination.

Vision Rx, Inc., Elmsford, NY.

STYLET.   Small sharp structures used for piercing. The term applies to the calcareous spines of the proboscis of some nemertine worms and to the slender piercing organs of some insect mouths. The latter are modified mandibles and maxillae.

STYLOLITE.   A columnal-like structure, at right angles, or highly inclined to the bedding planes of certain limestones, believed to be produced by differential vertical movements induced under great pressure. The term is derived from the Greek words meaning column and a stone.

STYRENE-MALEIC ANHYDRIDE.   A thermoplastic copolymer made by the copolymerization of styrene and maleic anhydride. Two types of polymers are available — impact-modified SMA terpolymer alloys (*Cadon*®) and SMA copolymers, with and without rubber impact modifiers (*Dylark*®). These products are distinguished by higher heat resistance than the parent styrenic and ABS families. The MA functionality also provides improved adhesion to glass fiber reinforcement systems. Recent developments include terpolymer alloy systems with high-speed impact performance and low-temperature ductile fail characteristics required by automotive instrument panel usage.

Copolymers show chemical resistance generally similar to that of polystyrene and terpolymers similar to that of ABS (acrylonitrile-butadiene-styrene). Neither type is recommended for use in strongly alkaline environments. All impact versions have good natural color and products are available in a wide range of colors. Copolymer crystal grades have good clarity and gloss.

Glass-reinforced SMA polymers are used as electrical connectors, consoles, top pads, and as supports for urethane-padded instrument panels. There are several additional automotive uses. SMA are also found in coffee makers, steam curlers, power tools, audio cassette components, business machines, vacuum cleaners, solar heat collectors, electrical housing, and fan blades, among others.

SUBACUTE SCLEROSING PANENCEPHALITIS.   This is a form of subacute encephalitis affecting children and young adults; it is due to persistent infection with the measles virus. SSPE and inclusion-body encephalitis are now recognized as the same disease. It most frequently follows by about six years measles contracted at an earlier age and is twice as common in boys as girls. The same vaccination programs that have controlled measles in the United States have also reduced the incidence of SSPE. In the United Kingdom, some twenty cases are notified annually.

It is not known whether persistence of the virus in the brain is the result of an abnormal incomplete form of the virus or of immunodeficiency on the part of the host.

Onset of the syndrome is usually insidious with loss of energy and interest. After some time increasing clumsiness draws attention to the organic nature of the disease and at this stage involuntary movements or myoclonus usually appear. Visual signs may be prominent with chloroidoretinal scarring present in about 30% of cases. Further progression is marked by intellectual deterioration, rigidity, and spacticity, and increasing helplessness with death in a year or two. The disease may be arrested in a state of complete incapacity or rarely some improvement may occur, but not full recovery. No treatment is known to be effective.

R. C. V.

SUBARCTIC PACIFIC WATER.   An oceanic water mass extending from the Aleutians on the north to the North Pacific Central Water masses on the south, and far out into the Western Pacific. It is a major surface water mass of low temperature (2–4 °C (35.6–39.2 °F) at 50° north latitude) and low salinity, 32.2–34.1%, the last figure existing at its lower depths (400–500 meters) (1320–1650 feet). Its eastern part, which extends to the coast of America is higher in temperature and salinity, due to effects of warm air from the land.

SUBGRAPH.   A graph containing a subset of the edges of the original graph.

The *complement of a subgraph* is the set of all elements of the graph not in the subgraph.

A *connected graph* is essentially one that is in "one piece." An unconnected graph must therefore be decomposable into several *connected* "pieces." This notion can be made precise by using the notion of a maximal connected subgraph.

Let two vertices β_1 and β_2 of a graph G be defined to be equivalent if there exists a path between them and denote this relation symbolically by $\beta_1 \doteq \beta_2$. Clearly, if

(a) $\beta_1 \doteq \beta_2$ and $\beta_2 \doteq \beta_3$, then $\beta_1 \doteq \beta_3$
(b) and for the same vertex, $\beta_1 \doteq \beta_1$
(c) If $\beta_1 \doteq \beta_2$, then $\beta_2 \doteq \beta_1$

Thus the relation is an equivalence relation in the strict mathematical sense because it is (a) transitive, (b) reflexive and (c) symmetric. The collection of all vertices of G is partitioned into disjoint equivalence classes such that two vertices belong to the same class if and only if they are equivalent. If G is finite, the number of these classes is also finite and can be enumerated $S_1, S_2, \ldots, S_P$. A little thought reveals that the set of all edges whose two endpoints are vertices in $S_r(r = 1, 2, \ldots, P)$ is a *connected subgraph* G_r of G. Moreover $G_r(r = 1, 2, \ldots, P)$ is a *maximal connected subgraph* of G in the sense that the addition of any more vertices to G_r renders it unconnected. Evidently, $P = 1$ if and only if G is connected.

A *proper subgraph* is a subgraph which does not contain all the edges of the graph.

Disjoint subgraphs are subgraphs of a graph which have no vertices in common.

See also **Graph (Mathematics)**.

SUBLIMATION.   The direct transition, under suitable conditions, between the vapor and the solid state of a substance. If solid iodine is placed in a tube and slightly warmed, it vaporizes and the vapor reforms into crystals on the cooler parts of the tube. Many crystalline substances, both metallic and nonmetallic, may be similarly sublimated in a vacuum; fairly large crystals of selenium have been thus prepared. The most familiar sublimates are frost and snow. As in the case of other changes of state, sublimation is accompanied by the absorption or evolution of heat, the quantity of which per unit mass is called the heat of sublimation of the substance. At pressures near the triple point the heat of sublimation is approximately equal to the sum of the heats of fusion and vaporization. In physical and chemical literature, it is customary to regard as sublimation only the transition from solid to vapor, not from vapor to solid; but meteorologists do not make this distinction.

Sublimation plays a major role in the freeze-drying of foods. See also **Freeze-Drying**.

SUBLIMATION (Heat of).   The quantity of heat required at constant temperature (and pressure) to evaporate unit mass of a solid. In sublimation, the change is directly from solid to vapor, without appearance of the liquid phase.

SUBROUTINE (Computer System).   1. The set of instructions necessary to direct a computer to carry out a well-defined mathematical or logical operation. 2. A subunit of a routine. A subroutine is often written in relative or symbolic coding even when the routine to which it belongs is not. 3. A portion of a routine that causes a computer to carry out a well-defined mathematical or logical operation. 4. A routine arranged so that control may be transferred to it from a master routine and so that, at the conclusion of the subroutine, control reverts to the master routine. Such a subroutine is usually a closed subroutine. 5. A single routine may simultaneously be both a subroutine with respect to another routine and a master routine with respect to a third. Usually control is transferred to a single subroutine from more than one place in the master routine; the reason for using the subroutine is to avoid having to repeat the same sequence of instructions in different places in the master routine.

Closed Subroutine. A subroutine not stored in the main path of the routine. Such a subroutine is entered by a jump operation and provision is made to return control to the main routine at the end of the operation. The instructions related to the entry and reentry function constitute a linkage.

Open Subroutine. A subroutine, of which a replica must be inserted at each place in the computer program where the subroutine is used. Although requiring more storage, this approach avoids linkage and housekeeping overhead.

T. J. H.

SUBSIDENCE.   Subsiding air is sinking air and is associated with lateral divergence. Subsidence in the atmosphere is a stabilizing influence; it also decreases relative humidity within the sinking air as it warms the air. Atmospheric pressure usually rises under the influence of subsidence, which is normally associated with anticyclones. Clear or partially clouded skies are the usual weather in a region of subsidence.

SUBSOLAR POINT.   The geographical position of the Sun; that point on the Earth at which the Sun is in the zenith at a specified time.

SUBTRACTION.   An operation that is the inverse of addition. The − sign is used to denote the operation of subtraction. In the equation $M - S = R$, M is the *minuend*, S the *subtrahend*, and R is the *remainder*. To perform the operation, calculate $M + (-S) = R$, i.e., change the sign of the subtrahend and add.

SUBTRACTIVE COLOR PROCESS.   A method of photographic color synthesis using two or more superimposed colorants, which selectively absorb their complementary colors from white light.

Most modern processes of color photography make use of a subtractive synthesis to yield prints or transparencies. In a three-color process, the colorants cyan, magenta, and yellow are used to control the amounts of red, green and blue in a beam of white light. See Fig. 1. This beam of

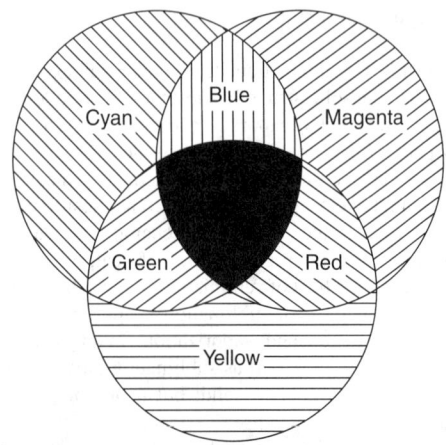

Fig. 1.   Superimposed color filters.

white light may be either that of a projector with its color transparency, or the light reflected from a white support, such as paper, on which the color reproduction is printed. In the first case, the light passes through the colorants once, while in the print viewed by reflection the light must traverse the colorants twice.

The colorants are positive or negative images. A cyan positive image (a cyan colorant controls red light), for example, may be prepared from the negative that recorded the red present in the subject. The magenta and yellow colorant images are likewise made from green and blue record negatives. These colorant images are superimposed in register to yield the final reproduction. The three colorants may be in separate removable layers or they may be physically inseparable as in the modern integral tripacks. The contrast of the colorant images must be approximately double for a picture to be viewed by transmitted light as compared to one to be viewed by reflection. The accuracy of color reproduction by a subtractive synthesis as compared to an additive is chiefly dependent on how satisfactorily the three colorants cyan, magenta, and yellow fulfill their role, as red, green and blue absorbers respectively. Color correction is often adopted to improve the accuracy of reproduction when using the colorants generally available.

See also **Photography and Imagery**.

SUCCESSION (Plant).   The gradual replacement of one plant association by another. Succession is caused by slow changes in the environmental factors which influence the establishment, development and survival of plants. Many of these changes, such as shading, increase in the humus content or porosity of the soil, decrease in soil temperature, etc., are brought about by plants themselves.

A good example of a plant succession occurs in the filling of a shallow pond. At first the deeper parts of such a pond are occupied only by submersed aquatic species of plants. As the pond becomes shallower, as a result of the accumulation of plant remains and the deposition of the silt caught in such remains, floating-leaf species such as water lilies invade the area and largely shade out the underwater species. With further decrease in the depth of the water the floating leaf aquatics are replaced by cattails, reeds, bulrushes and other typical marsh plants. At a still later stage in succession the former pond will be occupied by a wet meadow composed largely of sedges, rushes, and spike rushes. Still later swamp shrubs such as willows, alders, and buttombush invade the area only to be replaced in turn by trees of species that can tolerate a poorly drained substratum. Still further successional stages may ultimately result in the development of a climax association on the area.

Another example of plant succession that can be observed in many parts of the country is the natural reforestation of abandoned farm land. The pioneer invaders of such an area are mostly herbaceous plants, the very first to appear being mostly annuals. Soon shrubs invade the area to be followed in turn by certain species of trees. The first tree species to occupy the area, however, seldom represent a stable plant association. In large parts of the Eastern United States, for example, pines of one species or another are usually the first kind of tree to occupy abandoned fields, but they are, in time, replaced by other species, most commonly by oaks or hickories, or both. In many regions still further successional stages will be passed through before a stable climax association is attained.

Other examples of kinds of areas on which plant succession is initiated are sand dunes, burned over forest lands, cliffs or raw talus slopes, and the bottoms of lakes which have been exposed by drainage. Important consideration should be given to the principles of plant succession in forest and range management and in other land utilization problems.

SUCCESSIVE-APPROXIMATION A/D CONVERTER.   This type of analog-to-digital converter makes a direct comparison between an unknown input signal and a reference signal. The converter, as shown schematically in Fig. 1, includes a digital-to-analog (D/A) Converter. This provides an output voltage V_R, a precise fraction of the reference source voltage. The setting of analog switches, which are controlled by the digital outputs of the output register, determine the attenuation factor of the converter. Digital-output information of the A/D converter also is stored in the output register. Two input AND logic gates are used as conditional-reset gates. The shift register also functions as a step counter.

The A/D converter shown in the diagram is a binary n-bit converter. A "start convert" pulse causes all registers to clear and sets a 1 into the first position of the shift register. In turn, this action also turns on the first switch in the A/D converter and sets the first stage of the output register

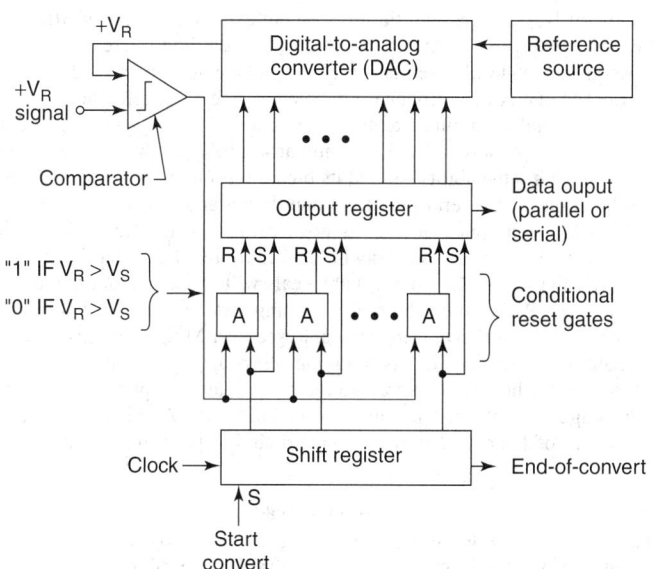

Fig. 1.   Successive-approximation analog-to-digital converter.

to 1. The output of the D/A converter is equal to one-half the full-scale D/A converter output voltage inasmuch as the D/A converter is binary. The D/A output voltage is compared directly with signal voltage V_S by the comparator. The output of the latter is a binary 1 if V_R is greater than V_S and a binary 0 if V_R is less than V_S.

The first bit in the output register may be reset to 0 or may remain as a 1, depending upon the comparator output. Upon appearance of the next clock pulse, the 1 in the shift register moves to the second bit at which instant the comparison process is repeated — with the exception that the output of the D/A converter is either 0.25 or 0.75 of the full-scale D/A converted output. This is dependent upon whether the first bit was reset to 0 or remained as a 1, respectively.

Codes other than binary can be used in successive-approximation A/D converters. What is required is the production of an attenuation that is proportional to the weighting value of the code digit. A binary-coded decimal converter can be constructed by using a binary-coded decimal digital-to-analog converter. The D/A converter will provide attenuation ratios of 0.8, 0.4, 0.2, 0.1, 0.08 and so on. Of course, appropriate coding of the output register is necessary so that the information can be stored in coded form.

Successive-approximation A/D converters can be used at conversion speeds up to approximately 100,000 samples/second at resolutions up to 16 bits (not including sign). At lower resolution, speeds over 250,000 samples/second are practical. Factors to be considered in the design and application of these A/D converters include: (1) stability and regulation of the reference source, (2) overload and recovery characteristics of the comparator, (3) analog-switch characteristics, and (4) speed and response of the ladder network.

THOMAS J. HARRISON, International Business Machines Corporation, Boca Raton, FL.

SUCCULENTS.   Succulent plants are those that are fleshy, that is, have their stems or leaves greatly enlarged to serve as water-storage organs. Succulents occur wild in regions in which there is a very limited supply of available water, as in the desert lands of western America, central Asia, and many parts of Africa. Among stem succulents the outstanding representatives are the members of the Cactus family, with the milkweed and spurge families also ranking high. Leaf succulents include many plants such as *Sedums, Crassulas, Bryophyllums,* and *Gasterias.* Many of the leaf succulents are plants of extremely curious habit. In some, the leaves are so swollen with stored water that they form a nearly spherical mass growing almost buried in the ground; in others the masses of leaves form compact rosettes at the tips of the branches; while in others the swollen leaves appear like small green beads along a slender stem.

SUCKERS *(Osteichthyes).*   Of the group *Cypriniformes,* family *Catostomidae,* suckers live near the bottom of streams, feeding on vegetation and small animals. The mouth is usually provided with fleshy lips and in some species opens at a downward angle. They are not highly valued as food fishes, but the common sucker (*Catostomus commersoni*) often abundant in small streams and lakes is considered of excellent taste although bony.

Buffalo fishes are in this family and have the appearance of a goldfish or carp. The largemouth buffalo (*Ictiobus cyprinellus*) which can attain a length up to 3 feet (0.9 meters) is common throughout the central portion of the United States. *Carpiodes cyprinus* (quillback carp-sucker) will often measure up to 26 inches (66 centimeters) in length and weigh up to about 12 pounds (5.4 kilograms). This fish is considered a commercial item in Lake Erie and is also found in the eastern and central regions of the United States.

Suckers are essentially an American fish. Other species include: *Catostomus latipinnis* (flannelmouth sucker) found in the Colorado River; *Minytrema melanops* (spotted sucker) found in the eastern states; *Xyaruchen texanus* (humpback sucker) found in the Colorado River Basin and attains a length up to 2 feet (0.6 meter); and *Myxocyprinus asiaticus,* similar to the *Xyaruchen,* which is one of the few suckers found outside of America, in Asia.

Suckers are sometimes difficult to differentiate from minnows. A method is described in a book by Hubbs and Lagler, "Fishes of the Great Lakes Region," Cranbrook Institute of Science, 1958.

SUDDEN INFANT DEATH SYNDROME (SIDS).   Known as SIDS (sometimes crib or cot death), this syndrome may be defined as the precipitous (sudden), unexpected death of an apparently healthy infant from whom an autopsy fails to identify the cause of death. In other words, these infants die of mysterious causes, which most experts as of the late 1900s considered a puzzle that, to date, has defied statistical research. There may be preventive measures, but physicians question — on the basis of what?

Of each one thousand live births in modern industrialized nations, 990 infants can be expected to live to their first birthday (and normally beyond for many years). Of the ten babies who die before they are one year old, six do not survive the first month of life — succumbing to gross errors in growth and development, usually the result of a very low birth weight, prematurity, or some congenital defect, i.e., causes of death that are explainable. Two or three of the remaining four infants will die in hospital some time during the first year, usually as the result of a serious treatment-resisting infection or because of a late-detected birth abnormality. The remaining two or three infants will die *unexpectedly,* often in their own cribs, usually discovered too late to take to hospital, but even then, death will usually occur and a thorough necropsy examination will not yield a plausible cause of death. Even though the number two or three out of a sample of 1000 infants is small statistically, nevertheless in a large nation, there will be several thousand cases of SIDS. (Estimated to be 10,000 cases/year in the United States and the United Kingdom, collectively.)

Golding, Limerick, and Macfarlane in a 1986 book on SIDS (see references) describe a typical SIDS situation: "Three-month-old David's mother is a nurse in her early 30s. He is her third child; happily married to a physician, she does not smoke. Through normal birth, she has just mothered a normal, contented baby. At two months, he was screened at the health clinic; no abnormality at all. At three months, not long before his first immunization, she put him in his crib for his afternoon sleep. Warm but not encumbered, he went right to sleep. She heard no sound, but in 30 minutes she went up to check on him. He was very pale; when she picked him up, he was floppy and around his mouth was a blood-tinged froth. She cleared the tiny mouth, phoned for an ambulance and tried resuscitation. Within 10 minutes David was in hospital, still warm but without breath or heartbeat. The emergency staff gave up after an hour's effort. The postmortem was entirely normal except for some vomit in the infant's upper mouth and airways. Certificate of Death — Sudden Infant Death (SID)."

Studies on infant mortality in efforts to relate SIDS statistically with one or more causative factors have been conducted in Britain, Canada, the United States, Australia, New Zealand and in some European countries as well as in Japan and Israel. Statistics thus far have shown that, in contrast with other causes of infant fatality (notable decreases in most countries in recent years), the incidence of SIDS has held steady. (Only exceptions reported are in two large Swedish cities and these have not been explained satisfactorily.) Over the past 15 years, there have been at least five international conferences on the subject.

Although no convincing cause of SIDS is now under present consideration, many hypotheses have been offered over the years, only to be disproved by statistics. In a recent study in the United Kingdom in a medically well-tended region with about 14,000 births per year, some very general criteria have been offered, but these do not contain much help in terms of SIDS prevention.

- Age of mother—SIDS is relatively higher for young mothers, notably those in the lower economic sector of society and who smoke or use drugs. (Yet for one-half of cases, the mothers were not under economic stress and one-third did not smoke.)
- Season of year—SIDS occurs more frequently during winter months.
- Male babies are at higher risk of SIDS than females.
- Babies of exceptionally low birth weight are more susceptible to SIDS. (But, 80% of babies in the sample weighed a normal amount at birth.)
- Siblings of dead infants do not appear to be at much greater risk.
- Infanticide is seldom masked as SIDS.

Some of the past proposed hypotheses that have essentially been ruled out by most authorities today include: (1) sudden arrest of breathing—failure in bioelectronic development; (2) use of the birth control pill; (3) vaccines and medications normally administered to infants; (4) selenium and other trace element deficiencies in the diet; (5) overheating, perhaps augmented by fever, causing death by heat stroke (febrile convulsions are common in young children, but occur at ages far advanced of the peak of SIDS); (6) vague muscular effects. A point that has added some confusion to the diagnosis of SIDS (as defined here) is respiratory syncytial virus (common virus infection leading to bronchitis and pneumonia)—because this condition matches that of the usual period for SIDS and also has the same ratio of more male than female infants. Similar viruses may show similar patterns. Some authorities believe that anaphylactic shock to the sensitized infant in reaction to respiratory virus infection may precipitate SIDS, but they cautiously suggest that this could account for only some SIDS cases as currently understood. The strongest statistical connection for the virus hypothesis is that SIDS occurs more frequently in winter than in summer.

The real puzzler of SIDS is the usual lack of evidence at post mortem examination.

Updated Appraisals of SIDS

In an 1991 study by J.S. Kemp and B.T. Thach (Washington University of Medicine, St. Louis), data on 25 cases of SIDS were analyzed carefully. Efforts were made to distinguish the difference on postmortem examination of (a) accidental suffocation and (b) SIDS. The researchers point out that three basic assumptions generally can be used to separate the two causes of death. (1) Healthy infants will not suffocate on ordinary bedding. (2) Infants will suffocate only if their heads are firmly restrained when the airway is occluded. (3) Infants lying with their faces straight down and their noses and mouths pressed into the bedding can turn their heads to obtain access to fresh air. In the 25% or more of infants with SIDS who are found with their faces straight down, the posture is thus considered coincidental and unrelated to the cause of death.

The Washington University study produced one very important finding—namely, that the deaths of several of the previously diagnosed SIDS cases actually were deaths from suffocation that occurred in a manner not previously reported in infants. The deaths all occurred on a particular type of cushion filled with polystyrene beads and marketed for infants.

To support their study, the researchers experimented by measuring the mechanics (simulated by machine) involved in breathing with head down on a pliable material. Thus it was found that an infant lying face down on certain materials may, in fact, rebreathe expired gases. In the tests, gas concentrations were measured. Carbon dioxide gas was used in the simulations. Thus, simulated infants with faces pressed down on certain materials resulted not in a sudden cessation of breathing, but in breathing high concentrations of carbon dioxide. Not only is air flow through such materials impeded, but the carbon dioxide content rose.

In mid-1991, the researchers concluded: "Accelerated suffocation by rebreathing was the most likely cause of death in most of the 25 infant cases studied. Consequently, there is a need to reassess the cause of death in the 28 to 52 percent of the victims of SIDS who are found with their faces straight down. Safety regulations setting standards for softness, malleability, and the potential for rebreathing are needed for infant bedding."

Obviously, the foregoing findings cannot explain all cases of SIDS, and the findings have not gone unchallenged. See Balding reference listed.

Another avenue of research on SIDS includes that of a medium-chain acyl coenzyme A dehydrogenase deficiency. See Ding reference listed.

The unusual breathing habits of seals also is being investigated for possible leads to causes of SIDS in humans. It has been known for decades that the seal has the ability to hold its breath for long periods of time. Seals halt breathing for several minutes when diving and also when they sleep. A form of sleep apnea in seals is positive in that this slows down their metabolism and conserves energy in the form of blubber. It has been found, however, that breath-holding in baby seals will cause random fluctuations in heart rate. Part of the baby seal's learning process is that of finding how to stabilize its heartbeat during apnea. Researcher M. Castellini (University of California, Santa Cruz) is studying sleep apnea in baby seals in an effort to learn how they overcome the early heart-rate problem; also, this knowledge may provide insights on human SIDS. Another group at the University of Pennsylvania is studying a similar phenomenon that occurs in bulldogs.

Additional Reading

Balding, L.E.: "SIDS and Suffocation," *N. Eng. J. Med.*, 1806 (December 19, 1991).

Bergman, A.B.: "The Discovery of Sudden Infant Death Syndrome: Lessons in the Practice of Political Medicine," Greenwood Publishers Group, Inc., Westport, CT, 1986.

Berkow, R., M.H. Beers: "The Merck Manual," 17th Edition, Merck & Company, Inc., Whitehouse Station, NJ, 1999.

Byard, R.W., S.D. Cohle: "Sudden Death in Infancy, Childhood and Adolescence," Cambridge University Press, New York, NY, 1994.

Byard, R.W., H.F. Krous: "Sudden Infant Death Syndrome: Problems, Progress and Possibilities," Oxford University Press, Inc., New York, NY, 2001.

Ding, J.H., C.R. Roe, and A.K. Iafolla: "Medium-Chain Acyl-Coenzyme A Dehydrogenase Deficiency and Sudden Infant Death," *N. Eng. J. Med.*, 61 (July 4, 1991).

Flam, F.: "The SIDS-Seal Connection," *Science*, 1613 (June 21, 1991).

Gilbert, E.F., K. Kenison: "Fetal Hemoglobin Levels in SIDS," *N. Eng. J. Med.*, 1281 (November 1, 1990).

Golding, J., S. Limerick, and A. Macfarlane: "Sudden Infant Death: Patterns, Puzzles, and Problems," University of Washington Press, Seattle, WA, 1986.

Kemp, J.S., B.T. Thach: "Sudden Death in Infants Sleeping on Polystyrene-Filled Cushions," *N. Eng. J. Med.*, 1858 (June 27, 1991).

Lagercrantz, H., T.A. Slotkin: "The 'Stress' of Being Born," *Sci. Amer.*, 100–107 (April 1986).

Long, W. et al.: "A Controlled Trial of Synthetic Surfactant in Infants Weighing 1250 g or more with Respiratory Distress Syndrome," *N. Eng. J. Med.*, 1696 (December 12, 1991).

McCormick, M.C.: "The Contribution of Low Birth Weight to Infant Mortality and Childhood Morbidity," *N. Engl. J. Med.*, **312**(2), 82–90 (January 10, 1985).

Perlman, J.M., F. Moya: "Synthetic Surfactants in Infants with Respiratory Distress Syndrome," *N. Eng. J. Med.*, 1703 (June 18, 1992).

Sears, W.M.D.: "SIDS: A Parents Guide to Understanding and Preventing Sudden Infant Death Syndrome," Vol. 1, Little, Brown & Company, Boston, MA, 1996.

Ziai, M. Ed.: "Pediatrics," 3rd Edition, Little, Brown & Company, Boston, MA, 1984.

Web References

Centers for Disease Control and Prevention: *http://search.cdc.gov/search97cgi/s97-cgi.exe?Action=Search&Collection=CDCALL1&ResultTemplate=cdcnormal.hts&queryText=Sudden+Infant+Death+Syndrome*

Sudden Infant Death Alliance: *http://sidsalliance.org/index/default.asp*

Sudden Infant Death Syndrome Network: *http://sids-network.org/*

Sudden Infant Death Syndrome (SIDS): *http://www.cdc.gov/ncidod/hip/abc/facts40.htm*

SUFFICIENT STATISTIC. A sufficient statistic is one which summarizes all the information in the sample concerning the relevant parameter. Thus if t_0 is a sufficient estimator of a parameter θ, and t_1 is an alternative estimator calculated from the same sample, a combination of t_0 and t_1 (such as their mean) is no better as an estimate of θ than t_0 alone. A sufficient statistic is necessarily efficient.

SUGARCANE. A tall tropical perennial grass (*Saccharum officinarium*) and the source of over half of the world's supply of sucrose (table sugar, saccharose), the major sweetener for foods and beverages. The remaining sucrose comes from the sugar beet (*Beta vulgaris*) and the grain sorghum. Honey and maple trees furnish a small amount of sugar.

Sugarcane grows from 5 to 15 feet (1.5 to 4.5 meters) in height, with stalks that range from 1 to 2 inches (2.5 to 5 centimeters) in diameter. The sugarcane stalk is made up of a series of nodes and internodal sections, ranging from 1 to 10 inches (1.5 to 25 centimeters) in length. At each node there is a vegetative bud and root primordia that will "germinate" or sprout to produce a new shoot and nodal roots when placed in the proper environment. The saccharine content of the nodes is very low, whereas the internodes contain most of the sugar. As the plant commences to grow, leaves are formed on alternate sides of the nodes, reaching a length of some 3 feet (0.9 meter). At the base, the leaves are about 2 inches (5 centimeters) wide and taper gradually throughout their length to a sharp point at the end. The plant begins to display nodes and internodes when only a few weeks old. Early in the life of the plant, these parts are green, but after a while some of these parts may change color, depending upon variety. Although some may continue green, others will take on a striped pattern or a yellow-green or purple coloration. If disease-free, maturity of the plant is first indicated by yellowing and dropping of the lower leaves. Although the bottom part of the plant may be quite barren, the upper stalk will be green and vigorous. In some regions, frost may halt maturity, but if permitted to mature this will take the form of development of flowers and seeds. From 12 to 15 months are required to achieve full maturity by some varieties and in some regions, notably in the more tropical areas.

Sugarcane is not an agricultural crop that is likely to diminish over the years ahead, even though the role of sucrose in food products has been the subject of penetrating analysis and criticism during the past few years—critiques that go far beyond the more traditional negative roles of sucrose in terms of dental caries. The emergence of corn (maize) and other sugar sources also has posed a serious threat. Sugar is frequently the first substance to be attacked when obesity is mentioned, and dieting (even though usually not of a long-term nature or a permanent accomplishment) is almost a status symbol in some countries. Sugar has been identified as an ugly "empty calorie." Nevertheless, despite many of the very well justified critiques of sugar excesses in various diets, population increases in some countries, coupled with increased consumption of snacks and convenience foods in other countries, has enabled sugar to maintain a relatively high volume of production.

But, even if improvements in nutrition ultimately pose a relatively serious threat to the continued high production of sugarcane and sugar beets, it is now believed by many authorities that sugarcane production, in particular, will increase several fold during the next decade or two—not as a food product, but rather as a raw material for the chemical industry, replacing to some extent the present declining raw materials for the petrochemical industry.

Efforts of the Brazilian government to produce alcohol from sugarcane and blend that with hydrocarbons for automotive fuel are exemplary of the important role that sugarcane can play as an industrial chemical raw material.

Additional Reading

Bakker, H.: "Sugar Cane Cultivation and Management," Kluwer Academic Publishers, Norwell, MA, 1999.

James, G.: "Sugarcane," Iowa State University Press, Ames, IA, 2002.

Naik, G.R.: "Sugarcane Biotechnology," Science Publishers, Inc., Enfield, NH, 2001.

Rao, G.P., A.B. Filho, J.C. Autrey, and R.C. Megarey: "Sugarcane Pathology Vol. 1: Fungal Diseases," Science Publishers, Inc., Enfield, NH, 2000.

Rao, G.P., R.E. Ford, M. Tosic, and D.S. Teakle: "Sugarcane Pathology Vol. 2: Virus and Phytoplasma Diseases," Science Publishers, Inc., Enfield, NH, 2001.

Rao, G.P., R.E. Ford, M. Tosic, and D.S. Teakle: "Sugarcane Pathology Vol. 3: Bacterial and Nematode Diseases," Science Publishers, Inc., Enfield, NH, 2001.

SULFONAMIDE DRUGS. In 1935, Domagk, a German researcher, was the first to observe the clinical value of *prontosil*, a red compound derived from azo dyes. Paraaminobenzenesulfonamide was shown to be the effective portion of the prontosil molecule. This substance was given the name *sulfanilamide*. This was the first of a group of related drugs to receive wide clinical trial. It was found to be effective in the treatment of hemolytic streptococcal and staphylococcal infections. Within a short span of years, related drugs were synthesized and given clinical trials. These included *sulfapyridine, sulfathiazole, sulfaguanidine, sulfadiazine,* and *sulfamerazine*. These drugs acted by inhibiting the growth of bacteria rather than by killing organisms.

Even though numerous adverse side effects were observed over a period of time, the sulfonamides played an important role in medicine prior to the advent of the antibiotics. In recent years, the importance of the so-called *sulfa drugs* has diminished considerably, but for certain situations they

are still considered important antimicrobials. Presently the sulfonamides are mainly used to treat uncomplicated urinary tract infections, including prostatitis, due to *E. coli*. They are also used to treat a number of noncardial infections. At one time the sulfa drugs were widely used in the treatment of meningococcal meningitis and bacillary dysentery. Unfortunately, the bacilli responsible for these diseases developed, over the years, a resistance to the drugs, severely reducing their efficacy.

Within the last few years, some new sulfa drugs have been introduced, including trimethoprim-sulfamethoxazole. This drug has broadened the scope in treatment of urinary tract infections derived from species in addition to *E. coli*, namely, *Klebsiella, Enterobacter,* and *Porteus* species. This drug also is used for the treatment of acute otitis media in children, particularly those instances where strains of *H. influenzae* and *streptococcus pneumoniae* may be suspected. The drug is also used to treat systemic infections that may arise from chloramphenicol- and ampicillin-resistant *Salmonella*; as well as infections attributed to *Pneumocystis carinii*.

Also, the nature of sulfonamide compounds (relatively short duration of action, capability of entering into synergism with other drugs, poor absorption, and topical effectiveness, not to mention relatively low cost) is taken advantage of in what is sometimes called short-acting sulfonamides. Short-acting sulfonamides include sulfisoxazole, sulfadiazine, and trisulfapyrimidines. An intermediate-acting sulfonamide in current use is sulfamethoxazole. This drug does tend to cause renal damage arising from sulfonamide crystalluria.

Sulfacetamide eyedrops continue to be used for treatment of superficial ocular infections. Sometimes silver-sulfadiazine cream is applied to burn surfaces to minimize or prevent bacterial growth, as well as preventing invasive infection.

The adverse effects of sulfonamides include hypersensitivity reactions, as manifested by rashes, photodermatitis (allergic reaction to light), so-called drug fever, nausea, and vomiting. These reactions occur with some frequency when sulfonamides are administered. Less frequently encountered is crystalluria, previously mentioned, but with the risk lessened in the case of sulfisoxazole. Sulfa drugs also occasionally cause hemolytic anemia, agranulocytosis, and kernicterus (in infants) when the drugs are given to nursing mothers. In rare instances, sulfa drugs may precipitate hepatitis, aplastic anemia, renal tubular necrosis, and certain blood disorders.

SULFONE POLYMERS. Polysulfone is a transparent, heat-resistant, ultrastable and high-performance engineering thermoplastic. It is amorphous and has low flammability and smoke emission. Electrical properties are good; the material remains essentially unchanged up to near its glass transition temperature, 190 °C (374 °F). The molecular structure of polysulfone features the diaryl sulfone group, a group that tends to attract electrons from the phenyl rings. Oxygen atoms para to the sulfone group enhance resonance and produce oxidation resistance. High resonance also strengthens the bonds spatially, fixing the grouping into a planar configuration. Thus, the polymer has good thermal stability and rigidity at high temperatures. Ether linkages provide chain flexibility, thus imparting good impact strength.

The resistance to acids, alkalies, and salt solutions is high and also good in terms of detergents, oils, and alcohols even at elevated temperatures under moderate stress. Polysulfones, however, are attacked by polar organic solvents, such as ketones, chlorinated hydrocarbons, and aromatic hydrocarbons. The material can be used continuously in steam up to temperatures of 93 °C (300 °F). Maximum stress in water at about 82 °C (180 °F) is 2000 psi (steady loads) and 2500 psi (intermittent loads). In long-term performance at 150 °C (300 °F), polysulfone increases about 10% in strength and modulus values, retaining 90% of its dielectric strength and 70% of its impact strength.

Polysulfone is widely used in medical instrumentation and trays for holding instruments during sterilization. Food processing applications, such as piping, scraper blades, steam tables, microwave oven cookware, and beverage dispensing tanks, are numerous. Electrical/electronic applications include connectors, automotive fuses and switch housings, soil bobbins and cores, television components, capacitor film, and structural circuit boards. In chemical processing equipment, uses include corrosion-resistant piping, tower packing, pump parts, filter modules, and membranes. Polysulfone is available in both molding and extrusion grades. A special medical grade is available. Also available are polysulfone compounds with glass fiber or beads, as well as fillers, such as *Teflon*®.

SULFUR. Chemical element, symbol S, at. no. 16, at. wt. 32.064, periodic table group 16, mp 112.8 °C (rhombic), 119.0 °C (monoclinic), 120.0 °C (amorphous), bp 444.7 °C (all forms), sp gr 2.07 (rhombic), 1.96 (monoclinic), 2.046 (amorphous). Atomic weight varies slightly because of naturally occurring isotopes 32, 33, 34, and 36, the total possible variation amounting to ±0.003.

The stable isotopes of sulfur are ^{32}S, ^{33}S, ^{34}S, and ^{36}S. There are three known radioactive isotopes, ^{31}S, ^{35}S, and ^{37}S, with ^{35}S having the longest half-life (87.1 days). See also **Radioactivity.** Electronic configuration $1s^2 2s^2 2p^6 3s^2 3p^4$. Ionic radius S^{2-} 1.855 Å, S^{6+} 0.29 Å (Pauling). Covalent radius 1.07 Å. In terms of abundance, sulfur ranks fourteenth among the elements occurring in the earth's crust, with an estimated 520 grams per metric ton. In seawater, the element ranks fifth, with an estimated 894 grams per metric ton.

First ionization potential 10.357 eV; second, 23.3 eV; third, 34.9 eV; fourth, 47.08 eV; fifth, 63.0 eV; sixth 87.67 eV. Oxidation potentials $H_2S(aq) \rightarrow S + 2H^+ + 2e^-$, −0.141 V; $H_2SO_3 + H_2O \rightarrow SO_4^{2-} + 4H^+ + 2e^-$, −0.20 V; $S + 3H_2O \rightarrow H_2SO_3 + 4H^+ + 4e^-$, −0.45 V; $SO_3^{2-} + 2OH^- \rightarrow SO_4^{2-} + H_2O + 2e^-$, 0.90 V; $S^{2-} \rightarrow S + 2e^-$, 0.508 V; $HS^- + OH^- \rightarrow S + H_2O + 2e^-$, 0.478 V. Other important physical properties of sulfur are given under **Chemical Elements.**

Sulfur has a large number of allotropes. The ordinary form, α-sulfur, is rhombic having a crystal unit cell composed of sixteen S_8 molecules. At 95.5 °C it undergoes transition to β-sulfur, which is monoclinic and also has a molecular weight (in solution in carbon disulfide) corresponding to S_8. Four other monoclinic forms have been identified microscopically: γ-sulfur, prepared by heating α-sulfur to 150 °C, cooling to 90 °C, and inducing crystallization by friction, ρ-sulfur, S_6, prepared by extracting an acidulated sodium thiosulfate solution with toluene, as well as δ-sulfur, and λ-sulfur. There is also a tetrahedral form, θ-sulfur, crystallized from a carbon disulfide solution of rhombic sulfur treated with balsam. The first liquid form to appear is λ-sulfur, a pale yellow liquid, obtained on heating sulfur to 120 °C. Above 160 °C, this form changes to a viscous, dark-brown liquid consisting mainly of μ-sulfur. A third liquid allotrope, π-sulfur is considered to exist in molten sulfur, in equilibrium with the other two forms, having its greatest concentration at about 180 °C. Sulfur vapor has been shown to contain S_8, S_6, S_4, and S_2 molecules. Several other allotropes of sulfur have been produced, including two paramagnetic forms, purple and green in color, by low-temperature processing.

Sulfur occurs as free sulfur in many volcanic districts, and may have been formed in part by sublimation, by decomposition of hydrogen sulfide, or metallic sulfides, or by organic agencies. It is often associated with limestones and gypsum. Sulfur is found in Spain, Iceland, Japan, Mexico, and Italy. It occurs especially in Sicily, which was the producer for the world until about the beginning of the twentieth century, when Herman Frasch, by inventing the superheated-water method of mining sulfur, made available the great Louisiana and Texas deposits. This method of mining is at the same time a method of purifying sulfur, because in the process of heating, accompanying materials remain unmelted at the temperature at which sulfur melts and is drawn off. In the Louisiana and Texas deposits the sulfur is associated with gypsum, occurring in the caprock overlying the salt plugs that have pierced the strata underlying the Gulf coastal plain. In the United States, sulfur is also found in California, Colorado, Nevada, and Wyoming. Sulfur also occurs as (1) sulfides, e.g. cobaltite, iron disulfide, pyrite, FeS_2, lead sulfide, galenite, PbS, copper iron sulfide, copper pyrite, $CuFeS_2$, zinc sulfide, zinc blende, ZnS, mercury sulfide, cinnabar, HgS; and (2) as sulfates, e.g., calcium sulfate, gypsum, $CaSO_4 \cdot 2H_2O$, barium sulfate, barite, $BaSO_4$. Several of these minerals are described under separate alphabetical entries.

Sulfur Production and Use: The manufacture of H_2SO_4 accounts for nearly 90% of all sulfur consumed. Of this, about 50% of the H_2SO_4 goes into fertilizer production, nearly 20% into chemical manufacture, 5% into pigments, about 3% each for iron and steel production and the manufacture of rayon and synthetic films, and about 2% for various petroleum processes. The balance of over 15% of H_2SO_4 is consumed by a large number of other industries, this all giving credence to the use of H_2SO_4 production figures as an overall economic index. The 10% of the sulfur not going into H_2SO_4 is converted into numerous chemicals that are consumed by a variety of industries, the largest among these being pulp and paper production and the manufacture of carbon disulfide.

Sulfur Compounds: In addition to the compounds described in the following paragraphs, see also **Hydrogen Sulfide; Mercaptans;**

Sodium Thiosulfate; Sulfuric Acid; Sulfurous Acid; Thiocyanic Acid; Thioethers; Thiophene; and **Thiourea.**

Sulfur-Oxygen Compounds: Due to its $3s^2 3p^4$ electron configuration sulfur, like oxygen, forms many divalent compounds with two covalent bonds and two lone electron pairs, but d-hybridization is quite common, to form compounds with oxidation of 4+ and 6+.

A number of suboxides of sulfur have been reported, but in general their composition has not been clearly established. Polysulfur oxides of formula $S_{8-16}O_2$ are formed by reaction of hydrogen sulfide and sulfur dioxide. Also, when sulfur is burned with oxygen in very limited supply disulfur monoxide, S_2O is formed. This has the structure

A mixture of sulfur dioxide, SO_2, and sulfur vapor, at low pressure and with an electric discharge, forms sulfur monoxide, SO. Its presence is shown from its absorption spectrum, but upon separation it disproportionates at once to sulfur and SO_2. Sulfur sesquioxide, S_2O_3, is formed by reaction of powdered sulfur with anhydrous SO_3; S_2O_3 also disproportionates (at 20 °C in nitrogen) to sulfur and SO_2. Sulfur dioxide, SO_2, is formed by the combustion in air or oxygen of sulfur and sulfur compounds generally, except those in which sulfur is in a higher state of oxidation. Sulfur dioxide has an O—S—O bond angle of 119.5°. The sigma bonds utilize essentially sulfur p orbitals, with d p hybridization for the pi bonds. Its oxidation to sulfur trioxide, SO_3, by atmospheric oxygen attains a significant rate only at higher temperatures, but can be materially increased by catalysts. Sulfur trioxide is also evolved from oleum on heating. It exists in the vapor state chiefly as the planar monomer, in which the oxygen atoms are spaced symmetrically (120° angles) about the sulfur atom, and it has S—O bond lengths of 1.43 Å. Liquid SO_3 is partly trimerized, and exists in three physical forms.

Sulfur tetroxide is formed by reaction of pure oxygen and sulfur dioxide under the silent electric discharge. It is not obtained pure, but in a variable SO_3/SO_4 ratio, and as a polymerized white solid. Another peroxide, $(SO_2OOSO_2O)_x$, which is written as S_2O_7, is known.

Of the 16 oxyacids of sulfur that are recognized, only four have been isolated. The more important oxyacids of sulfur are: (1) Thiosulfurous acid, $H_2S_2O_2$, structure not established, existing only in compounds, an oxidizing agent for Fe^{2+}, H_2S and HI; (2) Sulfoxylic acid, H_2SO_2, existing only in salts and other compounds, e.g., $ZnSO_2$, SCl_2, $S(OR)_2$, structure probably

$$H : \overset{..}{O} : \overset{..}{S} : \overset{..}{O} : H$$

(3) Dithionous acid (or hydrosulfurous acid), $H_2S_2O_4$, existing only in compounds, widely used reducing agent, chiefly as the sodium salt, for organic substances, also reduces Sb^{3+}, Ag^+, Pb^{2+}, Cu^{2+} to the elements, structure

$$H : \overset{..}{\underset{..}{O}} : \overset{..}{\underset{}{S}} : \overset{..}{\underset{}{S}} : \overset{..}{\underset{..}{O}} : H$$

(4) Sulfurous acid, H_2SO_3, produced by hydration of SO_2, not isolated but existing in many salts, the sulfites and acid sulfites, and many organic compounds, including the dialkyl or diaryl sulfites and the alkyl or aryl sulfonic acid esters, which suggests two possible structures, $(HO)_2SO$ and $H(HO)SO_2$, although the acid dissociation constants (first, 1.25×10^{-2}, and second, 5.6×10^{-8}) suggest the structure with only one unhydrogenated oxygen atom. Sulfurous acid and sulfites are fairly strong reducing agents, but the HSO_3^- ion may act as an oxidizing agent, as for formates and related compounds. Other compounds of SO_2 are the metabisulfites or pyrosulfites, containing the ion

$$^- : \overset{..}{\underset{..}{O}} : \overset{..}{\underset{}{S}} : \overset{..}{\underset{..}{O}} : \overset{..}{\underset{}{S}} : \overset{..}{\underset{..}{O}} : ^-$$

which enters into equilibrium with water to form acid sulfite. (5) Thiosulfuric acid, $H_2S_2O_3$, existing only in compounds, the anion having the structure

$$-:\overset{\overset{\displaystyle :\ddot{O}:}{|}}{\underset{\underset{\displaystyle :\ddot{O}:}{|}}{\overset{\displaystyle }{O}:\overset{\displaystyle }{S}:\ddot{O}:}}-$$

and widely used as a coordinating ion for forming complexes with metals; it also is an oxidizing agent, and is used in iodometric titrations. (6) Dithionic acid, $H_2S_2O_6$, existing only in compounds but stable in dilute solution at room temperature, and differing in its stability to hydrolysis and oxidation from the polythionates,

$$-:\ddot{O}:\overset{\overset{\displaystyle :\ddot{O}:}{|}}{\underset{\underset{\displaystyle :\ddot{O}:}{|}}{S}}\; : \;\overset{\overset{\displaystyle :\ddot{O}:}{|}}{\underset{\underset{\displaystyle :\ddot{O}:}{|}}{S}}:\ddot{O}:-$$

(7) Polythionic acids, $H_2S_nO_6$, in which n has values of 3, 4, 5, 6 and others, some of which have been reported to have values indefinitely high (20–80), structure not established, though there is evidence that they consist of two sulfonic acid groups connected by a linear chain of sulfur atoms. An interesting property of the polythionates that are very rich in sulfur ($n > 20$) is their slight tendency to decompose to give free S. (8) Sulfuric acid, H_2SO_4, structure

$$H:\overset{\overset{\displaystyle :\ddot{O}:}{|}}{\underset{\underset{\displaystyle :\ddot{O}:}{|}}{\ddot{O}:S:\ddot{O}:}}H$$

strong acid, formed by hydration of sulfur trioxide, completely dissociated (first ionization) in aqueous solutions up to 40%; above that concentration dissociation decreases and hydrate formation occurs. Both normal and acid sulfates are formed by metallic elements, though the products of their direct reaction with the acid vary with temperature. (9) Sulfuric acid dissolves SO_3, the product of a 1:1 ratio being pyrosulfuric or disulfuric acid. $H_2S_2O_7$, which forms the pyrosulfates, also obtainable by heating acid sulfates, structure $HO(O)(O)SOS(O)(O)OH$. Two series of alkali metal pyrosulfates are known: those formed from SO_3 and the metal sulfates and those formed from H_2SO_4 and the metal sulfates, which have the pyrosulfuric acid structure. (10) Peroxymonosulfuric acid is produced by addition of SO_3 to concentrated H_2O_2, its salts are fairly stable, and it has the structure $HOS(O)(O)OOH$. (11) Peroxydisulfuric acid is produced by reaction of concentrated H_2O_2 on H_2SO_4 or by electrolysis of acid sulfate solutions; its salts are fairly stable and it has the structure $HOS(O)(O)OOS(O)(O)OH$.

Hydrogen Sulfide: H_2S is a weak acid ($pK_{A1} = 7.00$), ($pK_{A2} = 12.92$) stronger than water but weaker than H_2Se, as expected from its position in the periodic system; its reducing strength exhibits the same relation. Its long use in analytical chemistry is due to the differential solubility of many sulfides with variation of the pH of an aqueous solution. Hydrogen persulfide, H_2S_2, structure HSSH, with an S—S bond distance of 2.05 Å, formed from an alkali metal polysulfide solution and HCl at low temperatures, is the first of a group of hydrogen polysulfides of the general formula H_2S_x.

Sulfur Halides: Many are known. Those that have been identified and whose properties have been determined include the fluorine compounds, S_2F_2, SF_4, SF_6, S_2F_{10}, the chlorine compounds, S_2Cl_2, SCl_2, SCl_4 and the bromine compound, S_2Br_2. Sulfur chlorides of general formula S_nCl_2 are known up to $n \approx 20$. A similar series of cyanides, $S_n(CN)_2$, is known. Derivatives of SCl_4, e.g., SCl_3CN, have been prepared and the list of derivatives of SF_6 is rapidly growing, including S_2F_{10}, $(SF_5)_2O$, $(SF_5)_2O_2$, SF_5Cl, SF_5Cl_3, $SF_4(CF_3)_2$, $(SF_5)_2CF_2$, SF_5OF, SF_5OSO_2F, etc. Derivatives of SF_4 include $C_6H_5SF_3$, $(SF_3)_2CF_2$, etc. All of them except the higher fluorides hydrolyze readily, and they are essentially covalent in character. The simple compounds can be prepared directly from the elements, the activity of the halogen determining the product obtained: fluorine yielding SF_6 and S_2F_{10} and the other fluorides being prepared from those; chlorine and bromine yielding the monohalides, from which the others are obtained by continued halogenation.

Sulfur Oxyhalides: Four general compositions of oxyhalides of sulfur have been known for many years. In one of these, sulfur has a 4+ oxidation state, the thionyl halides, SOX_2, and in three of which it has a 6+ oxidation state, thionyl tetrafluoride, the sulfuryl and pyrosulfuryl halides, SOF_4,

SO_3X_2 and $S_2O_5X_2$, respectively. As is the case for the simple halides, no iodine compounds are known, but polyhalogen ones, such as SOFCl and SO_2FCl exist.

Isolable Oxysulfuranes: Sulfuranes, as described by Musher (1969), are compounds of sulfur(IV) in which four ligands are attached to sulfur and have in common with rare-gas compounds such as XeF_2 an electronic structure involving a formal expansion of the valence shell of the central atom from 8 to 10 electrons. Martin and Perozzi (1976) pointed out that the incorporation of oxygen ligands makes possible a wide range of new structural types that illustrate structure-reactivity relationships in a particularly illuminating way.

For many years, it was postulated that most types of sulfuranes were intermediate (not isolable) compounds. However, the isolable halosulfuranes have been well established for many years. The first known of these, SCl_4, was prepared by Michaelis and Schifferdecker in 1873. In 1911, it was found that SF_4, while highly reactive, was thermally stable. However, the compound was not fully described until 1929. Development of SF_4 led to the creation of a family of stable fluorosulfuranes and their derivatives. It was found that the fluorines in these compounds can be replaced by aryl or perfluoroalkyl groups (Tyczkowski, 1953). Kimura and Bauer (1963) described the geometry of SF_4 as a distorted trigonal bipyramid with two fluorines and a lone pair of electrons occupying equatorial positions, with the other two fluorines in apical positions. The postulated structures of SF_4 (a), and of a derivative (b) are shown below:

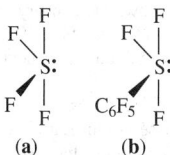

(a)   (b)

In the early 1970s, Sheppard, by reacting SF_4 with pentafluorophenyllithium, prepared an isolated sulfurane with four carbon-centered ligands, namely, *tetrakis*-(pentafluorophenyl)sulfurane, $(C_6F_5)_4S$. Martin and Perozzi (1976) prepared the first isolable diaryldialkyloxysulfurane. If it is protected against moisture, the researchers found the compound to be stable over an indefinite period at room temperature. The research in this interesting area continues, some of the details of which are well described by Martin–Perozzi (1976). Summarizing the situation, the researchers observe that the development of synthetic methods for oxysulfuranes has made a wide range of isolable compounds of hypervalent sulfur available for study. Structure-reactivity correlations are now becoming evident as a result of such study. The fact that oxygen is dicoordinate makes it possible to sythesize cyclic oxysulfuranes and to use the pronounced changes of reactivity which accompany cyclization to design new, potentially useful sulfurane reagents stable enough to allow isolation.

Sulfur-Nitrogen Compounds: Many of the sulfur-nitrogen compounds are sulfuric acid derivatives. Three of these compounds correspond to replacement of the hydrogen atoms of ammonia with one, two and three —SO_3H radicals, the monosubstituted compound being aminesulfonic (sulfamic) acid, and being readily separated, the others known only in their salts, the aminedisulfonates (imidodisulfonates) and aminetrisulfonates (nitrilotrisulfonates). Other amines, such as hydroxylamine and hydrazine have similarly related compounds. See also **Hydrazine**; and **Hydroxylamine**. Diamino derivatives of the sulfoxy acids are also known, such as sulfamide, $H_2NSO_2NH_2$. Imidosulfinamide, HN (SONH$_2$)$_2$, has been prepared by reaction of $SOCl_2$ and ammonia (also directly from SO_2 and ammonia), and a trimer of sulfimide, $(O_2SNH)_3$, by ammoniation of SO_2Cl_2. It is cyclic in structure, composed of alternate >NH and >SO_2 groups. Nitrosulfonates, containing the ion SO_3 NO^- and dinitrososulfonates, containing $SO_3N_2O_2^{2-}$, are also known.

The most important sulfur-nitrogen compound is tetrasulfur tetranitride, S_4N_4, prepared in many ways, including the direct reaction of ammonia and sulfur. All data on its structure are in accord with a puckered eight-member ring, or a cage with N—S connections. The question as to whether there are also transannular N—N or S—S bonds has not been clearly settled. On hydrogenation it adds 4 H atoms, on fluorination it forms $S_4N_4F_4$, structure

$$\underset{\underline{\hspace{4cm}}}{\overset{\overset{\displaystyle F}{|}}{S}-N-\overset{\overset{\displaystyle F}{|}}{S}-N-\overset{\overset{\displaystyle F}{|}}{S}-N-\overset{\overset{\displaystyle F}{|}}{S}-N}$$

and SN_2F_2 the latter reacting with SNF to form SNF_3, structure F_2SNF. Other thiazyl compounds, prepared from S_4N_4 and the halogens or sulfur halides, include $(ClSN)_3$, S_4N_3Cl, S_4N_3Br, S_4N_3I. These last are salts, i.e., $[N_4S_3]X$, and salts of other anions can also be prepared. Other sulfur-nitrogen compounds known are SN_2, S_4N_2, S_5N_2, and S_2N_2, the last being formed by heating S_4N_4.

Thiocyanogen, $(SCN)_2$, is formed by treatment of a metal thiocyanate with bromine in an organic solvent. It reacts with organic compounds in a manner completely analogous to the free halogens, lying between bromine and iodine in oxidizing power. The alkali metal and alkaline earth metal thiocyanates are prepared by fusing the cyanides with sulfur, and the other metal thiocyanates, as well as the organic ones, are usually prepared from the alkali metal thiocyanates. Many selenium analogs of thio compounds can be made, including $SeSO_3$, SO_3Se^{2-}, SSe^{2-}, etc.

In addition to carbon disulfide (odorless when pure), carbon subsulfide, $S=C=C=C=S$, an evil-smelling red oil and carbon monosulfide, (CS_x), are known as well as COS, CSSe and CSTe. Because of its similarity to oxygen, and the reactivity of its acids, sulfur enters widely into organic compounds.

For biological aspects of sulfur: See also **Sulfur (In Biological Systems)**.

Additional Reading

Dalrymple, D.A. and T.W.Trofe: "An Overview of Liquid Redox Sulfur Recovery," *Chem. Eng. Progress*, 43 (March 1989).

Greenwood, N.N. and A. Earnshaw: "Chemistry of the Elements," 2nd Edition, Butterworth-Heinemann, Inc., Woburn, MA, 1997.

Kent, J.A.: "Riegel's Handbook of Industrial Chemistry," 9th Edition, Chapman & Hall, New York, NY, 1992.

Kimura, K. and S.H. Bauer: *J. Chem. Phys.*, **39**, 3172 (1963).

Krebs, R.E.: "The History and Use of Our Earth's Chemical Elements: A Reference Guide," Greenwood Publishing Group, Inc., Westport, CT, 1998.

Lewis, R.J. and N.I. Sax: "Sax's Dangerous Properties of Industrial Materials," 10th Edition, John Wiley & Sons, Inc., New York, NY, 1999.

Lide, D.R.: "CRC Handbook of Chemistry and Physics 2000–2001," 81st Edition, CRC Press, LLC, Boca Raton, FL, 2000.

Loretta, J. and P.W. Atkins: "Chemistry: Molecules, Matter and Change," W.H. Freeman and Company, New York, NY, 1999.

Martin, J.C. and E.F. Perozzi: "Isolable Oxysulfuranes in Organic Chemistry," *Science*, **191**, 154–159 (1976).

Meyers, R.A.: "Handbook of Chemicals Production Processes," The McGraw-Hill Companies, Inc., New York, NY, 1986.

Mollare, P.D.: "From Calcasieu to Caminada: A Brief History of the Louisiana Sulfur Industry," *Chem. Eng. Progress*, 73 (March 1989).

Parker, S.P.: "McGraw-Hill Encyclopedia of Chemistry," 2nd Edition, The McGraw-Hill Companies, Inc., New York, NY, 1993.

Stwertka, A. and E. Stwertka: "A Guide to the Elements," Oxford University Press, Inc., New York, NY, 1998.

Trofe, T.W., D.A. Dalrymple, and F.A. Scheffel: "Stetford Process Status and R&D Needs," Topical Report GRI-87/0021, Gas Research Institute, Chicago, IL, 1987.

Tyczkowski, E.A. and L.A. Bigelow: *J. Amer. Chem. Soc.*, **75**, 3523 (1953).

SULFUR (In Biological Systems). Sulfur, in some form, is required by all living organisms. It is utilized in various oxidation states, including sulfide, elemental sulfur, sulfite, sulfate, and thiosulfate by lower forms and in organic combinations by all. The more important sulfur-containing organic compounds include the amino acids (cysteine, cystine, and methionine, which are components of proteins); the vitamins thiamine and biotin; the cofactors lipoic acid and coenzyme A; certain complex lipids of nerve tissues, the sulfatides; components of mucopolysaccharides, the sulfated polysaccharides; various low-molecular-weight compounds, such as glutathione and the hormones vasopressin and oxytocin; and many therapeutic agents, such as the sulfonamides and penicillins, as well as oral hypoglycemic agents sometimes used in treatment of diabetes mellitus. Sulfhydryl groups of the cysteine residues in enzyme proteins and related compounds, such as hemoglobin, play a key role in many biocatalytic processes; sulfhydryl-disulfide interchange reactions involving the cysteine residues of proteins are critical events in the immune processes, in transport across cell membranes, and in blood clotting. The S—S bridges between these residues are important in the maintenance of the tertiary structure of most proteins.

The electronic structure of sulfur is such that a variety of oxidation states are readily obtainable. It can be said that a sulfur cycle exists in nature, as noted in Fig. 1.

The oxidation and reduction of elemental sulfur and sulfide occur in different species of bacteria, e.g., the oxidation of sulfides via elemental

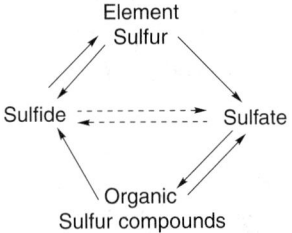

Fig. 1. Sulfur cycle.

sulfur to sulfate takes place in *Chromatia*, the alternative oxidation to sulfate in *Thiobacillus*. The reduction of sulfate to sulfide occurs in *Desulfovibrio*. The biosynthesis of organic sulfur compounds from sulfate takes place mainly in plants and bacteria, and the oxidation of these compounds to sulfate is characteristic of animal species and of heterotrophic bacteria.

The amino acids cysteine and cystine are interconverted by oxidation-reduction reactions, as shown by

$$
\begin{array}{ccccc}
S & S & & S \\
| & | & & | \\
CH_2 & CH_2 & & CH_2 \\
| & | & & | \\
CNNH_2 & CNNH_2 + 2\,H \rightleftharpoons & & 2\,CNNH_2 \\
| & | & & | \\
COOH & COOH & & COOH \\
\text{(Cysteine)} & & & \text{(Cystine)}
\end{array}
$$

Cystine was first isolated from a urinary calculus by Wollaston in 1805. It was shown to be a component of protein by Morner in 1899 and independently by Embden in 1900. Proof of its structure was given by Friedman in 1902. See also **Amino Acids**; **Coenzymes**; **Proteins**; and **Vitamins**.

In the chain from soils to plants to humans, inorganic sulfur, or more accurately, the sulfate ion (SO_4^{2-}), is taken up by plants and converted within the plant to organic compounds (the sulfur amino acids). These amino acids combine with other amino acids to make up plant protein. When the plant is eaten by a human or by livestock animals, the protein is broken down and the amino acids are absorbed from the digestive tract and recombined in the proteins of the animal body. The most important feature of sulfur in the food chain is that plants use inorganic sulfur compounds to make sulfur amino acids, whereas animals and humans use the sulfur amino acids for their own processes and excrete inorganic sulfur compounds resulting from the metabolism of the sulfur amino acids.

Ruminants, such as cattle, sheep, and goats, can use inorganic sulfur in their diets because the microorganisms in the rumen convert the inorganic sulfur into sulfur amino acids and these are then absorbed farther along in the digestive tract.

Soils very low in available sulfur are common in a number of regions of the world. In the United States, low-sulfur soils are frequently found in the Pacific Northwest and in some parts of the Great Lakes states. For many years, sulfur in the form of calcium sulfate was an accessory part of most commercial phosphate fertilizers, and this probably helped to prevent development of widespread sulfur deficiency in crops grown where these fertilizers were used. Volatile sulfur compounds from smoke, particularly before tight pollution controls, were an important source of sulfur for plants growing near industrial centers. In some cases, excessive sulfur in the air can cause injury to the plants. The trend toward high-analysis fertilizers without sulfur and air pollution abatement diminishes some of the inadvertent sources of sulfur for plants and crops and creates a need for more deliberate use of sulfur-containing fertilizers.

The extent to which any plant will convert inorganic sulfur taken up from the soil into amino acids and incorporate these into protein is controlled by the genetics of the plant. Increasing the available sulfur in soils to levels in excess of those needed for optimum plant growth will not increase the concentration of sulfur amino acids in plant tissues. To meet the requirements for sulfur amino acids in human diets, the use of food plant species with the inherited ability to build proteins with high levels of sulfur amino acids is required in addition to that supplied by way of the soil.

Since animals tend to concentrate in their own proteins the sulfur amino acids contained in the plants they eat, such animal products (meat, eggs, and cheese) are valuable sources of the essential sulfur amino acids in human diets. In regions where the diet is composed almost entirely of foods of plant origin, deficiencies of sulfur amino acids may be critical in human nutrition. Frequently, persons in such areas (also voluntary vegetarians) are also likely to suffer from a number of other dietary insufficiencies unless supplemental sources are used.

Diets of corn (maize) and soybean meal are usually fortified with sulfur amino acids for pigs and chickens. Sometimes fishmeal, a good source of sulfur amino acids, is added to the diets, or sulfur amino acids synthesized by organic chemical processes may be used.

Since ruminants can utilize a wide variety of sulfur compounds, any practice to increase the sulfur in plants may help to meet the requirements of these animals. Sheep appear to have a higher requirement for sulfur than most other animals, perhaps because wool contains a fairly high level of sulfur. Adding sulfur fertilizers to soils used to produce forage for sheep may improve growth and wool production, even though no increased yield of the forage crop per se may be noted.

Sulfate and Organic Sulfates. Inorganic sulfate ion (SO_4^{2-}) occurs widely in nature. Thus, it is not surprising that this ion can be used in a number of ways in biological systems. These uses can be divided primarily into two categories: (1) formation of sulfate esters and the reduction of sulfate to a form that will serve as a precursor of the amino acids cysteine and methionine; and (2) certain specialized bacteria use sulfate to oxidize carbon compounds and thus reduce sulfate to sulfide, while other specialized bacterial species derive energy from the oxidation of inorganic sulfur compounds to sulfate.

Among the variety of sulfate esters formed by living cells are the sulfate esters of phenolic and steroid compounds excreted by animals, sulfate polysaccharides, and simple esters, such as choline sulfate. The key intermediate in the formation of all of these compounds has been shown to be 3'-phosphoadenosine-5'-phosphosulfate (PAPS). This nucleotide also serves as an intermediate in sulfate reduction.

In organisms that utilize sulfate as a source of sulfur for synthesis of cysteine and methionine, the first step in the reduction process is the formation of PAPS. This is not surprising since the direct reduction of sulfate ion itself is an extremely difficult chemical process. It is known that the reduction of esters and anhydrides occurs much more readily than the reduction of corresponding anions. Following activation, the sulfuryl group of PAPS is reduced to sulfite ion (SO_3^{2-}) by reduced triphosphopyridine nucleotide (TPNH) and a complex enzyme system. Following the reduction of PAPS to sulfite, additional reduction steps readily produce hydrogen sulfide, which appears to be a direct precursor of the amino acid cysteine.

Sulfur Compounds in Onion and Garlic. Dating back to antiquity, there have claims made for the curative and preventative physiological powers of onion and garlic. Dr. Eric Block (State University of New York at Albany), a specialist in the organic chemistry of sulfur, and colleagues, have investigated the chemistry of onion and garlic over a period of years. Some of the results were reported in the Block (1985) reference listed. As pointed out by Block, the cutting of an onion or a garlic bulb releases a number of low-molecular-weight organic molecules that incorporate sulfur atoms in bonding forms rarely encountered in nature. These molecules are highly reactive and they change spontaneously into other organic sulfur compounds, which in turn participate in further transformations. Researchers have cataloged a number of biological effects of the extracts from these bulbs, including antibacterial and antifungal properties. Other extracts act as antithrombotic agents (inhibit blood platelets). As early as 1721, a drink consisting of wine and macerated garlic (*vinaigre des quatre voleurs*) was used as an antibiotic in France and is still available today! Pasteur (1858) reported on the antibacterial properties of garlic. Albert Schweitzer is reported to have used garlic in the treatment of amoebic dysentery in Africa. As reported by Block, laboratory investigations have shown that garlic juice diluted in one part in 125,000 inhibits the growth of bacteria of the genera *Staphylococcus, Streptococcus, Vibrio* (including *V. cholerae*) and *Bacillus* (including *B. typhosus, B. dysenteriae,* and *B. Enteritidis*). Lacrimatory factors contained in these bulbs are well known.

Serious research commenced in 1844 by Theodor Wertheim, a German chemist. He attributed some of the properties of garlic, "*mainly to the presence of a sulfur-containing, liquid body, the so-called garlic oil. All that is known about the material is limited to some meager facts about the pure product, which is obtained by steam distillation of bulbs of Allium sativum. Since sulfur bonding has been little investigated so far, a study of this material promises to supply useful results for science.*" Wertheim

suggested the name *allyl* for the oil. Today, allyl is used for chemicals in the C_3H_5 series (CH_2=$CHCH_2$).

Another German investigator (Semmler, 1892) also produced garlic oil via steam distillation. The oil yielded diallyl disulfide, CH_2=$CHCH_2$ $SSCH_2CH$≡CH_2, with minor amounts of diallyl trisulfide and diallyl tetrasulfide present. The oil yielded by similar experimentation with onions was different, containing essentially propionaldehyde, $C2H5CHO$, plus a number of sulfur compounds, of which dipropyl disulfide, $C_6H_{12}S_2$, was one.

Using less harsh methods, Cavallito (1944) produced an oil, $C_6H_{10}S_2O$, by extracting the garlic with ethyl alcohol (room temperature). This oil was found to be more powerful than penicillin or sulfaguanidine against *B. typhosus*. The exact formula of Cavallito's oil was found to be allyl-2-propenethiosulfinate, CH_2=$CHCH_2S(O)SCH_2CH$=CH_2. Cavallito gave this substance the common name, *allicin*. Precursor molecules for allicin have been identified and it has been established that allicin is not developed in garlic until it is initiated by an enzyme, termed *allinase*.

As pointed out by Block, allinin is the "first natural substance found to display optical isomerism due to mirror-image forms at sulfur as well as at carbon." The research by Block and others is well delineated in the Block (1985) reference. Although much remains to be learned, there is now a long line of hard scientific evidence that garlic and onion have beneficial physiological properties. Most scientists to date suggest that the properties of these bulbs are best exploited by consuming the fresh products, rather than from extracts, particularly when the latter are derived from harsh methodologies, such as steam distillation.

Preservatives. Sulfur compounds, such as sulfur dioxide and sodium bisulfite, are used commercially to preserve the color of various food products, such as orange juice, dehydrated fruits and vegetables, such as apricots, carrots, peaches, pears, potatoes, and many others. Concentrated sulfur dioxide is used in wine-making to destroy certain bacteria. The color preservation of canned green beans and peas is enhanced by dipping the produce in a sulfite solution prior to canning. In 1986, some of these compounds and uses were put under closer regulation in the United States.

The sulfatases are a widely distributed group of enzymes that hydrolyze simple sulfate esters to inorganic sulfate.

Sulfur-Based Pesticides. Sulfur (elemental) has been used as an effective acaricide, fungicide, and insecticide. For ease of use, a number of special formulations are available, ranging from sulfur dusts (up to 95% sulfur); a wettable powder (30 to 90%); and paste-like solutions in which the sulfur is ground to a fine colloidal form. Such formulations may contain up to 50% sulfur. Target plant diseases of sulfur when used as a fungicide include: apple scab, brown rot, downy and powdery mildew, and peach scab. Against insects, sulfur is effective for mite, scale, and thrip. Most formulations are not injurious to honeybees.

Specific sulfur control chemicals include: (1) Calcium polysulfide (lime-sulfur), dating back to the 1850s and available as a solution (up to 31% sulfur) or as a dry powder (up to 70% sulfur). The compound is effective against anthracnose, apple scab, brown rot, powdery mildew, and peach leaf curl—and against mite and scale insects. (2) Sodium polysulfide and sodium thisulfate mixtures—used for spraying and dipping fruit, adding color to the product, and prolonging the period during which the fruit can be picked. (3) Sodium thiosulfate pentahydrate, which prevents discoloration of some green vegetables (use is regulated).

Role of Sulfur in Tidal Wetlands. As pointed out by Luther, et al. (1985 reference listed), the biogeochemical role of sulfur in tidal wetlands presently is subject to considerable research. Sulfur is an important redox element under natural aquatic conditions and is responsible for several important biogeochemical processes, including (1) sulfate reduction, (2) pyrite formation, (3) metal cycling, (4) salt-marsh ecosystem energetics, and (5) atmospheric sulfur emissions. These processes depend upon the formation of one or more sulfur intermediates, which may have any oxidation state between +6 and −2. The intermediate oxidation states may be organic or inorganic. In their study, Luther and colleagues analyzed sulfur species in pore waters of the Great Marsh, Delaware. Anticipated findings reported were bisulfide increases with depth due to sulfate reduction and subsurface sulfate excesses and pH minima, the result of a seasonal redox cycle. Not expected was the pervasive presence of thiols, such as glutathione, particularly during periods of biological production. It appears that salt marshes may be unique among marine systems in producing high concentrations of thiols. Polysulfides, thiosulfate, and tetrathionate also showed seasonal subsurface maxima. The findings suggest a dynamic seasonal cycling of sulfur in salt marshes involving abiological and biological reactions and dissolved and solid sulfur species. The researchers suggest that the chemosynthetic turnover of pyrite to organic sulfur is the likely pathway for this sulfur cycling. It follows that the material, chemical, and energy cycles in wetlands appear to be optimally synergistic.

Additional Reading

Block, E.: "The Chemistry (Sulfur Compounds) of Garlic and Onions," *Sci. American*, **252**(3), 114–119 (1985).

Considine, D.M. and G.D. Considine: "Food and Food Production Encyclopedia," Van Nostrand Reinhold Company, Inc., New York, NY, 1982.

Greyson, J.C.: "Carbon, Nitrogen and Sulfur Pollutants and Their Determination in Air and Water," Marcel Dekker, Inc., New York, NY, 1990.

Lide, D.R.: "CRC Handbook of Chemistry & Physics 2000-2001," 81st Edition, CRC Press, LLC., Boca Raton, FL, 2000.

Luther, G.W., III et al.: "Inorganic and Organic Sulfur Cycling in Salt-Marsh Pore Waters," *Science*, **232**, 746–749 (1986).

Maynard, D.G.: "Sulfur in the Environment," Marcel Dekker, Inc., New York, NY, 1998.

Mitchell, S.C.: "Biological Interactions of Sulfur Compounds," Taylor & Francis, Inc., Philadelphia, PA, 1996.

Mudahar, M.S., J.S. Kanwar: "Fertilizer Sulfur and Food Production," Kluwer Academic Publishers, Norwell, MA, 1986.

Stevenson, F.J., M.A. Cole: "Cycles of Soils: Carbon, Nitrogen, Phosphorus, Sulfur, Micronutrients," 2nd Edition, John Wiley & Sons, Inc., New York, NY, 1999.

SULFURIC ACID. Infrequently termed "oil of vitriol," sulfuric acid, H_2SO_4, is a colorless, oily liquid, dense, highly reactive, and miscible with water in all proportions. Much heat is evolved when concentrated sulfuric acid is mixed with water and, as a safety precaution to prevent spluttering, the acid is poured into the water rather than vice versa. Sulfuric acid will dissolve most metals. The concentrated acid oxidizes, dehydrates, or sulfonates most organic compounds, sometimes causing charring. There are numerous commercial and industrial uses for H_2SO_4 and these include the manufacture of fertilizers, chemicals, inorganic pigments, petroleum refining, etching, as a catalyst in alkylation processes, in electroplating baths, for pickling and other operations in iron and steel production, in rayon and film manufacture, in the making of explosives, and in nonferrous metallurgy, to mention only some of its numerous uses. Because of its wide use industrially, some economists over the years have included sulfuric acid consumption among their economic indicators.

Most countries with significant industrial activity and particularly in chemicals production will have significant capacities for making sulfuric acid. In some countries, H_2SO_4 is the leading chemical in terms of tonnage production. Depending upon suppliers, H_2SO_4 is commercially available in a number of strengths, ranging from 77.7% H_2SO_4 (60° Baumé, sp. gr. 1.71) through 93.2% H_2SO_4 (66° Baumé), 98% H_2SO_4, 99% H_2SO_4, and 100% H_2SO_4 (sp. gr. 1.84).

Fundamentally, there are two kinds of sulfuric acid plants: (1) those that use the dry gas (sulfur burning) process; and (2) those that use the wet gas process. In the first type, the raw materials are elemental sulfur and water. In the second type, the sulfur dioxide feed may come from a variety of sources, including metallurgical smelters (copper, zinc, lead, etc.), pyrite roasters, waste acid decomposition furnaces, and hydrogen sulfide burners. In these plants, the SO_2 gas stream enters the acid plant containing a large amount of water vapor. The gas is usually hot (260–430 °C) and dusty, and also may contain a number of impurities, such as fluorides, that could harm the catalyst in the contact section of the plant. These incoming gases thus require cooling and purification in the series of scrubbers and electrostatic precipitators, followed by drying prior to entering the contact section of the plant.

In either type of plant, sulfur dioxide is converted to sulfur trioxide in the contact portion of the plant. The reaction $SO_2 + \frac{1}{2}O_2 \rightarrow SO_3$ is effected by passing the SO_2 over a catalyst, usually vanadium pentoxide (V_2O_5). The catalyst in the converter vessel is usually in the form of small pellets and typically arranged in four layers. Provision is made for removal of the heat of reaction after each layer or stage. The catalyst may be used for a number of years with only a very moderate decrease in activity.

From this fundamental point, the sulfuric acid plant designer has a number of alternatives and options to consider. Two factors are of major import in sulfuric acid plant design today, namely, recovery and conservation of energy, and minimizing environmental impact. For example, in the relatively simple plants of a few years age, the SO_2 need contact the catalyst but once and the absorption of the resulting SO_3 in water (a solution of sulfuric acid) could be handled in a single absorption tower. Recycling could be kept to a minimum. In the modern sulfuric acid plant, double contact (DC) of the gases with catalyst and double absorption (DA) of the gases is commonly practiced. Designs are available in numerous configurations, each offering various advantages in terms of energy conservation, pollution minimization, initial and operating costs. A typical sulfur-burning DC/DA sulfuric acid plant is shown in Fig. 1.

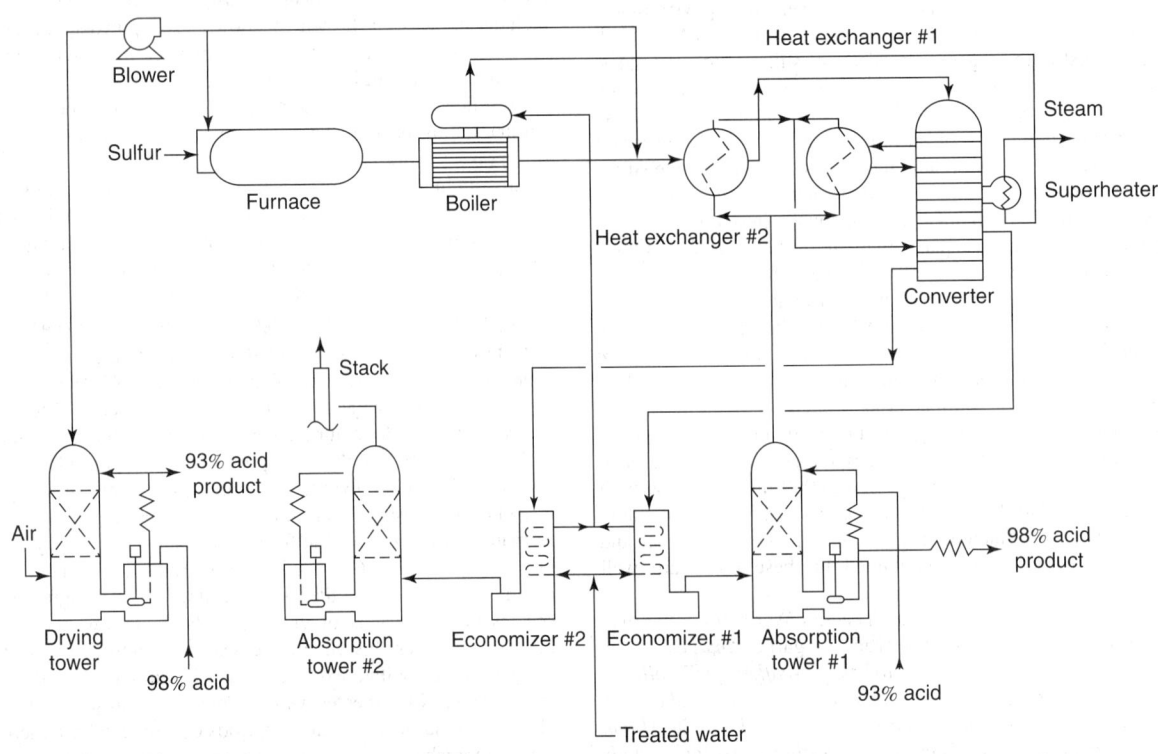

Fig. 1.   Representative sulfuric acid plant of the sulfur-burning, double contact (DC), double absorption (DA) type.

Of the approximately 40 million tons (36 million metric tons) of sulfuric acid manufactured in the United States per year, about 90% is used in the production of fertilizers and other inorganic chemicals. Much of the remaining 10% of H_2SO_4 is used by the petroleum, petrochemical, and organic chemicals industries. Much of this latter acid is involved in recycling kinds of processes. As pollution regulations in various countries become more restrictive, spent acid may become a much more attractive raw material than has been the case in the past.

As pointed out by Sander and Daradimos (1978), a regeneration of sulfuric acid of high quality can only be attained by thermal decomposition back to sulfur dioxide at high temperatures, where all organic impurities are completely burned—followed by reprocessing the SO_2 gases by the contact process to concentrated acid or oleum.

Reactivity of Sulfuric Acid. Dilute sulfuric acid reacts: (1) with many hydroxides, e.g., sodium hydroxide, to yield two series of sulfates (the acid is dibasic), e.g., sodium sulfate or sodium hydrogen sulfate, depending upon the ratio of acid to base reacting, (2) with many ordinary oxides, e.g., magnesium oxide, to yield the corresponding sulfate, e.g., magnesium sulfate solution, (3) with some carbonates, e.g., zinc carbonate, to yield the corresponding sulfate, e.g., zinc sulfate solution plus carbon dioxide gas (calcium carbonate is soon coated by a layer of calcium sulfate, which prevents further reaction), (4) with some sulfides, e.g., ferrous sulfide, to yield the corresponding sulfate, e.g., ferrous sulfate plus hydrogen sulfide gas, (5) with many metals, e.g., zinc, if not too pure (but not copper), to yield the corresponding sulfate, e.g., zinc sulfate solution plus hydrogen gas, (6) with solutions of some salts to yield the corresponding sulfate, e.g., barium chloride, changed to barium sulfate precipitate, calcium citrate, malate, tartrate to calcium sulfate precipitate and the free organic acid in solution.

Higher strengths of sulfuric acid react similarly in kind to the cases of (1), (2), (3), (6) above, but not, in general, as in cases (4) and (5) above. Copper and concentrated sulfuric acid yield copper sulfate and sulfur dioxide gas. Iron reacts similarly, yielding ferric sulfate in the place of copper sulfate.

A number of other reactions of sulfuric acid are characteristic of its higher strengths. Concentrated sulfuric acid is thus (7) an oxidizing agent, and a further example is the oxidation of sulfur to sulfur dioxide (the reacting sulfuric acid is reduced to sulfur dioxide), (8) a sulfonating agent, e.g., naphthalene sulfonated to naphthalene-sulfonic acids (mono-, alpha or beta, di- several), (9) an esterification agent, e.g., methyl alcohol esterified to dimethyl sulfate $(CH_3O)_2SO_2$, melting point $-32\,°C$, boiling point $189\,°C$, or methyl hydrogen sulfate $CH_3O \cdot SO_2OH$, ethyl alcohol esterified to diethyl sulfate $(C_2H_5O)_2SO_2$, melting point $-26\,°C$, boiling point $208\,°C$, or ethyl hydrogen sulfate $C_2H_5O \cdot SO_2OH$, (10) a dehydration agent, e.g., formic acid into carbon monoxide, sugar blackened with separation of carbon, (11) an addition agent, e.g., ethylene into ethyl hydrogen sulfate, (12) a nonvolatile acid upon heating, e.g., with sodium chlorite or nitrate, hydrogen chloride or nitric acid, respectively, is volatilized and sodium sulfate or sodium hydrogen sulfate remains as a residue.

Additional Reading

Behrens, D.: "DECHEMA Corrosion Handbook: Corrosive Agents and Their Interaction with Materials, Sulfuric Acid," Vol. 8, John Wiley & Sons, Inc., New York, NY, 1991.

Kent, J.A.: "Riegel's Handbook of Industrial Chemistry," 9th Edition, Chapman & Hall, New York, NY, 1992.

Lewis, R.J., N.I. Sax: "Sax's Dangerous Properties of Industrial Materials," 10th Edition, John Wiley & Sons, Inc., New York, NY, 1999.

Lide, D.R.: "CRC Handbook of Chemistry and Physics 2000-2001," 81st Edition, CRC Press, LLC, Boca Raton, FL, 2000.

Parker, S.P.: "McGraw-Hill Encyclopedia of Chemistry," 2nd Edition, The McGraw-Hill Companies, Inc., New York, NY, 1993.

Sander, U., G. Daradimos: "Regenerating Spent Acid," *Chem. Eng. Progress,* **74,** 57–67 (1978).

SULFUROUS ACID. H_2SO_3, formula weight 82.08, colorless liquid, prepared by dissolving SO_2 in H_2O. Reagent grade H_2SO_3 contains approximately 6% SO_2 in solution. As a bleaching agent, sulfurous acid is used for whitening wool, silk, feathers, sponge, straw, wood, and other natural products. In some areas, its use is permitted for bleaching and preserving dried fruits. The salts of sulfurous acid are sulfites.

Sulfurous acid is a strong reducing agent, being oxidized to H_2SO_4: (1) on standing in contact with air; (2) by chlorine, bromine, iodine, yielding HCl, HBr, or HI, respectively; (3) by HNO_3 or nitrous acid yielding nitric oxide; and (4) by permanganate. Sulfurous acid is itself reduced by zinc and dilute H_2SO_4 to H_2S. Sulfurous acid also may be formed by the reaction of a sulfite or bisulfite solution and an acid.

Sodium sulfite Na_2SO_3 and sodium hydrogen sulfite $NaHSO_3$ are formed by the reaction of sulfurous acid and NaOH or sodium carbonate in the proper proportions and concentrations. Sodium sulfite, when dry and upon heating, yields sodium sulfate and sodium sulfide. Sodium pyrosulfite (sodium metabisulfite) $Na_2S_2O_5$ is a common sulfite. Crystalline sulfites are obtained by warming the corresponding bisulfite solutions. Calcium hydrogen sulfite $Ca(HSO_3)_2$ is used in conjunction with excess sulfurous acid in converting wood to paper pulp. Sodium sulfite and silver nitrate solutions react to yield silver sulfite, a white precipitate, which upon boiling decomposes forming silver sulfide, a brown precipitate.

An esterification agent, sulfurous acid forms dimethyl sulfite $(CH_3O)_2SO$, bp $126\,°C$ and diethyl sulfite $(C_2H_5O)_2SO$, bp $161\,°C$. Sulfites give a white precipitate with barium chloride, soluble in HCl with evolution of SO_2. Sulfites decolorize iodine in acid solution.

SUM. The answer when two or more quantities are combined by addition. In group theory, the term direct sum has a special meaning. See also **Representation of Groups**.

The usual symbol for the operation of summation is the Greek capital letter *sigma* (Σ). Thus, Insert Equation

$$\sum_{n=0}^{\infty} a_n x^n = f(x)$$

means to add the terms $a_n x^n$, assigning every integral value from zero to infinity to the letter n. When two or more sums are to be taken with respect to several subscripts or indices only one summation sign may be used and the summation limits are often omitted, if they are understood from the accompanying text. Even the summation sign can be omitted and this is common, especially in tensor analysis. Thus, an expression like

$$A^m = \frac{\partial x^{-m}}{\partial x^i} A^i$$

by convention means that summation is to be made over the repeated index, i. Still further condensation is customary in matrix algebra, where $\mathbf{Ab} = \mathbf{C}$ means that the matrix $\mathbf{C}$ has elements C_{ij} calculated by the equation

$$c_{ij} = \sum A_{ik} B_{kj}$$

with A_{ik} and B_{kj} as elements of $\mathbf{A}$ and $\mathbf{B}$. The summation over k is to be extended from unity to n, where n is the number of columns in $\mathbf{A}$ and the number of rows in $\mathbf{B}$.

SUMMER SOLSTICE. 1. That point on the ecliptic occupied by the Sun at maximum northerly declination. Sometimes called *June solstice, first point of Cancer.*

2. That instant at which the Sun reaches the point of maximum northerly declination, about June 21.

SUMNER LINE. A line of position obtained from the observation of altitude of some celestial object. This method for obtaining a line of position was discovered by Captain Thomas H. Sumner in 1837, and circumstances leading to the discovery are described in "The American Practical Navigator," Bowditch.

The method employed by Sumner for obtaining a celestial line of position was standard procedure for American ship masters until the early part of the twentieth century. For this reason, it is worthy of consideration here, in spite of the fact that it has been almost completely superseded by the methods described in the article on celestial navigation. Captain Sumner used the old-fashioned method for determining latitude and longitude at sea from an observation of the altitude of a celestial object. These methods require the solution of the astronomical triangle. To solve this triangle, at least three parts must be known. At sea, the altitude of the object, as obtained from sextant observations, will give one part, and the declination of the object will give another. To obtain the third part, either the latitude or the longitude of the observer must be known. Since latitude and longitude are the coordinates that the navigator is seeking, it would seem at first glance that a solution of the problem is impossible. However, if the object observed is approximately due east or west, a change in latitude of several miles will produce but slight effect on the computed longitude. On the other hand, if the object is nearly due north or south, a change in longitude will produce but slight effect in the computed latitude. An approximate value of latitude and longitude can be obtained by methods of dead reckoning (DR). If the observed object is within 45° of east or west, the DR latitude is used and the longitude is computed. When the object is within 45° of the meridian, the DR longitude is used in the computation of the latitude. If the DR position is known to within 20 miles (32 kilometers), the computed latitude or longitude will be accurate to within $\frac{1}{4}$ of a mile, unless the observed object is close to the zenith.

In Captain Sumner's case, his ship had experienced gales and fog for a number of days, and the DR position was very uncertain. When the clouds broke away, he obtained an altitude of the sun in the forenoon. Since his DR position was uncertain, he assumed several values for latitude separated by about 20 miles (32 kilometers), and computed the corresponding longitudes. On plotting these positions, he found that they lay along a straight line on a mercator chart. He made the assumption, which has since been established as sound, that his ship must be on the line.

After the publication of this discovery, the use of Sumner lines became the standard procedure for most navigators. If two objects, differing in bearing by at least 45°, are available for observation, two Sumner lines can be obtained and a fix determined. If, as is frequently the case during daylight hours, only one object is available, this one object is observed twice, with an interval of time between the two observations sufficient to produce a change in bearing of at least 45°. The fix of these two lines is obtained by moving one line to the time of the other by the method of running fix.

A Sumner line is, in reality, a small circle on the earth, with a point on the earth directly under the observed object as center. Unless the object is within 10° of the zenith, the curvature is so slight as to be negligible in drawing the line on a mercator chart.

See also **Course**; and **Navigation**.

SUMPTNER PRINCIPLE. When a source of light is placed at any point inside a sphere with perfectly diffusing walls, every part of the interior is equally illuminated.

SUN (The). The sun is a self-luminous body; it is our nearest star, 93,000,000 miles (~150 million kilometers) from the Earth, from which we receive light and heat to generate and maintain all life processes on the Earth. The sun is 109 times the Earth's diameter and 300,000 times its mass, but when we compare it to other stars for size, mass, and brightness we find that it falls in the middle of their range. It is a typical dwarf star of spectral class G2. Its energy output in the form of light and heat, generated by nuclear processes, appears to be remarkably constant and to have been so over hundreds of millions of years, and is expected to be so for many billions of years into the future. Being the closest star, we can examine its surface in great detail. As such it serves as a testing ground for astrophysical theories. The sun's atmosphere consists of the visible surface or photosphere and two higher-temperature envelopes, the chromosphere and corona. Telescopic views show the surface of the sun to be in great turmoil, covered by a granular pattern of convection cells and punctuated

with sunspots, great magnetic storms in the sun's atmosphere. Magnetism plays the dominant role in the sun's variability; in addition to sunspots it is fundamental to flares, with their X-ray and radio bursts, it controls the motions and behavior of prominences, governs the solar wind streaming past the Earth, and indirectly produces a wide variety of solar-terrestrial events: auroras, magnetic storms, radio fadeouts and possible changes in the climate. Physical data for the sun are given in Table 1.

Modern solar astronomy can be roughly divided into three major subdivisions:

1. *Velocity fields*: Solar rotation, granulation and supergranulation, solar oscillations.
2. *Magnetic fields*: General field of the sun, sunspots, flares, prominences.
3. *Composition*: Interior structure, chemical composition, energy output.

These will be discussed in the following sections of this entry.

The Photosphere

The visible surface of the sun is called the photosphere — literally, light sphere. If the surface of the sun is examined carefully through a telescope of 4 inches (~10.2 centimeters) aperture or larger, at the times of best viewing the surface is seen to be mottled, a structure referred to as *granulation*. Figure 1 shows granules in a photograph taken on very high contrast film. Also clearly visible, even in a small telescope, and sometimes with the naked eye, are *sunspots*, dark only in contrast to their brilliant surroundings.

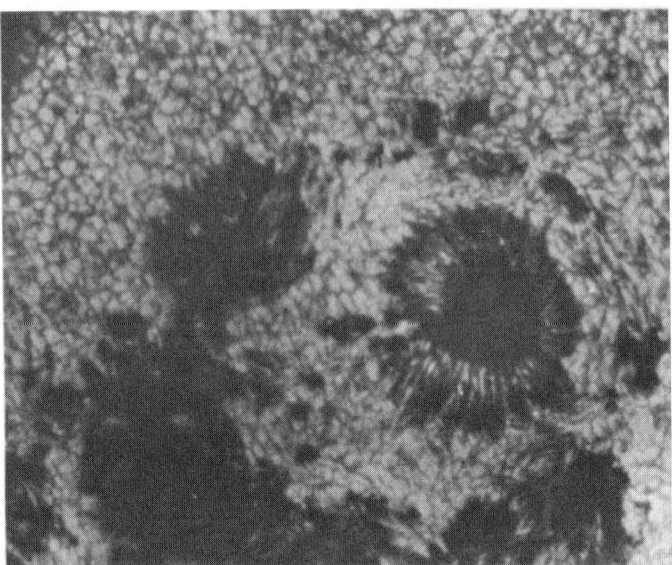

Fig. 1.   Portion of sun's disk photographed from a balloon at an altitude of about 24 kilometers (80,000 feet). (*Princeton University*.)

TABLE 1. DIMENSIONS AND PHYSICAL DATA FOR THE SUN

Diameter	864,000 miles (1,390,180 kilometers); 110 × diameter of the earth
Volume	1.412×10^{33} cm^3 (1.3 million times the volume of the Earth)
Mass	1.989×10^{33} gm (330,000 times mass of the Earth)
Mean density	1.41 gm cm^{-3} (1/4 the earth's mean density)
Central density	148 gm cm^{-3}
Force of gravity of surface	Nearly 28 times that at earth's surface (A mass weighing 100 pounds on Earth would weigh nearly 1.4 tons on the sun)
Mean distance from Earth	92,955,807 miles (149,597,870 kilometers = 1.0 astronomical unit)
Time for light to travel from sun to Earth	499 seconds (slightly over 8 minutes)
Velocity in space	About 12.4 miles per second (20 kilometers per second) toward a direction in space not far from the star Vega
Solar constant (rate at which solar radiation is received outside the Earth's atmosphere on a surface normal to the incident radiation and at the Earth's mean distance from the sun)	1.94 calories per square centimeter per minute
Candle power of sun	2.4×10^{27} candles
Average illumination of sun at zenith	100,000 meter-candles

Three to four million granules, each about the size of an American state (200–1000 miles; 322–1609 kilometers across), cover the surface of the sun. Similar to the convection in a boiling pot of cereal, granulation is one of the ways energy is brought to the surface from the deeper, hotter layers. Time lapse photography reveals the 5–10 minute life history of a granule: each starts out as a small bright area that grows in several minutes to about 1000 miles (1610 kilometers) diameter, divides into several smaller units, which may coalesce with other granules or they fade and then are replaced by a new granule. A spectrum of the photosphere shows Doppler shifts of the spectrum lines indicating that each granule rises in the atmosphere with an average velocity of $\frac{1}{3}$ mile (~0.5 kilometer) per second. As the granule cools and dissipates the gas returns to the sun in the darker intergranular spaces. Groups of granules appear to be in a constant up and down oscillation with a 5-minute period.

The photospheric layer is about 500 miles (805 kilometers) thick — hotter (10,000 °K), denser, and more opaque at the bottom, cooler (4200 °K), much less dense, and quite transparent at the top. The energy radiated from the photosphere can be characterized by a continuous spectrum with a temperature of 5740 °K — the effective temperature of the sun. In appearance the surface is brilliant white, brighter at the center of the disk, where we see to the deeper hotter layers, and darker at the limb of the sun, where we see only the outermost, cooler layers of emitting gas. The variation in brightness, center to limb, is known as the solar *limb darkening* (visible also in spectrum lines, i.e., Hα of Fig. 2).

The amount of limb darkening varies with wavelength. In the red and infrared there is very little variation between center and limb, whereas in the violet the intensity falls rapidly away from disk center to about 10% of the central value. From the observed limb darkening we can derive from an integral equation the temperature variation with depth in the sun's atmosphere.

The Chromosphere

A layer of the sun's atmosphere about 6200 miles (10,000 kilometers) thick just above the photosphere was named the chromosphere by eclipse observers. At eclipse time, when the moon has covered the bright photosphere, it is revealed for a few seconds before and after totality, as a thin crescent colored bright pink (from the hydrogen Hα spectrum line). Examined through a telescope the top of the layer is resolved into a great multitude of small, short-lived (5–15 minutes), geyser-like jets called *spicules*, each about 310 miles (500 kilometers) diameter and 3100–6200 miles (5000–10,000 kilometers) tall, projecting outward at all angles to the sun's surface.

Spicules are not randomly distributed over the sun's surface. High-resolution spectroheliograms show that they are arranged in a "network" pattern that covers the entire sun. Piled up at the network boundaries by slow, outward directed motion, each element of the net is a giant convection cell, typically 18,600 miles (30,000 kilometers) across, in which material flows up and outward from the center carrying with it the chromospheric magnetic fields concentrated in the spicules, to the cell boundary, and then descends. The motion is slow, horizontally about $\frac{1}{3}$ mile ($\frac{1}{2}$ kilometer) per second. The lifetime of a cell is about a day.

The disk of the sun when viewed in the light of Hα through a spectrohelioscope or monochromatic filter presents a totally different view from its white light photospheric image. Prominences, bright at the limb show very black on the disk. See Fig. 2. Bright areas appearing around sunspots, given the French name for beach, are called *plages*. In the central regions of the disk we see a composite of very small, overlapping, worm-like prominence structures referred to as chromosphericmottling. Near sunspots these structures are drawn out into long threads and filaments which often exhibit a spiral structure tracing out magnetic fields emerging from the spot.

The gases of the chromosphere are transparent. When viewed at the extreme limb, and not projected onto the bright underlying photosphere, the chromospheric spectrum consists of emission lines which in the last crescent phase of an eclipse momentarily flash into view, i.e., the *flash spectrum*. See Fig. 3. From the temperature minimum (4200 °K) at the top of the photosphere the temperature rises to a plateau of 6000 °K for the layers 620–1240 miles (1000–2000 kilometers) high then rises very rapidly to 30,000 °K at 1550 miles (2500 kilometers) and then to coronal values. The lower pressure and higher temperature of the chromosphere results in greater ionization and excitation than in the photosphere, thus the spectrum consists of many lines from singly ionized elements.

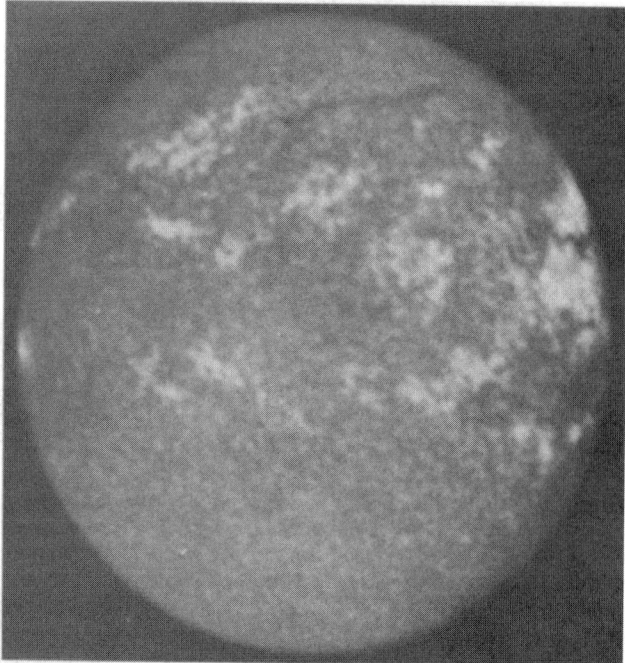

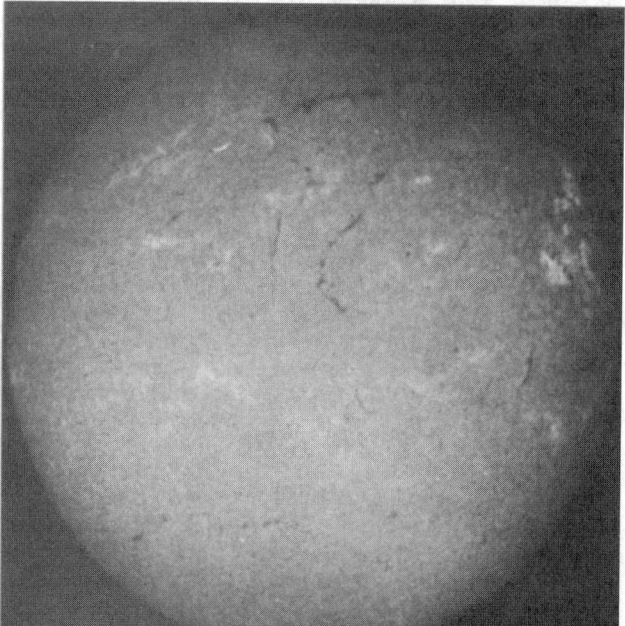

Fig. 2.   Photographs of sun taken in the K line of Ca II. (*Top*) Calcium plages are indicated. (*Bottom*) Simultaneous photograph in the Balmer H alpha line. (*U.S. Navy.*)

Corona

At the time of a total solar eclipse a pearly white radiance, called the corona, can be seen around the sun. See Fig. 4. The corona is composed of three parts, labeled L, K and F. The L-corona refers to a low level portion, the light from which is emitted by highly ionized atoms. The light of the K-corona, one hundred times brighter than L, is sunlight scattered by fast moving free electrons. The F component is produced by sunlight scattered by interplanetary dust particles; it is an extension of zodical light inward to the sun. Only the L and K components represent the true coronal atmosphere of the sun. Though the inner corona can be studied with a very specialized telescope called a Coronagraph or from space, its beauty can only be seen and appreciated at the time of a total eclipse, when one can discern outward streaming rays, arches, filaments, and brilliant red prominences, apparently projected on the corona. At sunspot maximum short streamers extending outward in all directions give the corona a globular shape; at sunspot minimum there may be only a few long equatorial streamers together with a polar plume or crown composed

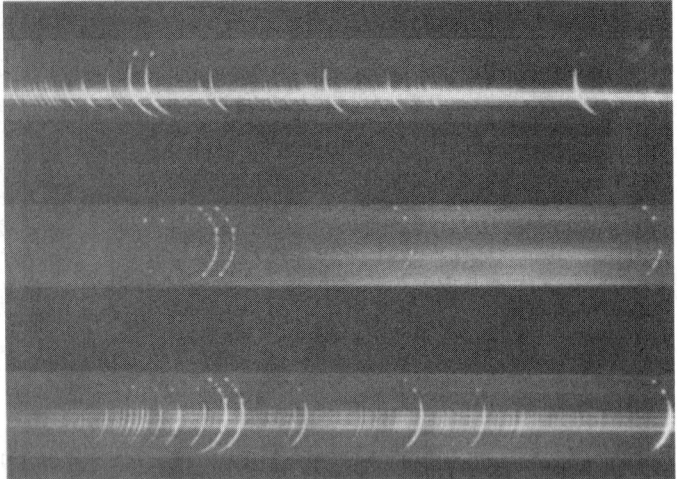

Fig. 3. Solar flash spectrum. (*Top*) Just before second contact; (*Center*) Bailey's — beads light passing through valleys in the moon's limb form two rings that are the violet lines of ionized calcium; (*Bottom*) Just after third contact. (*Mount Wilson and Palomar Observatories.*)

Fig. 4. The corona shown at time of total eclipse of the sun.

of fine rays curving outward from the north and south poles like iron filings at the poles of a magnet.

Though normally visible only at the limb of the sun, when the bright surface is occulted, the corona can also be seen projected on the disk, if photographed from space in X-ray wavelengths. The Japanese YOHKOH satellite, launched at the peak of the 1991 sunspot cycle, aimed in X-rays at the million-degree corona and yielded spectacular motion pictures of intense, active regions in the sun's corona, visible in these wavelengths both on the sun's disk and off the sun's limb. The hope is to solve the puzzle of the flare mechanism. Photographs from Skylab revealed a complex structure of bright points surrounded by rays and arch structures. Between them large dark irregular areas appear, *coronal holes*. These are low-density regions from which magnetic lines of force lead out into interplanetary space, allowing the plasma to escape. Moving with a velocity of about 400 km/sec near the earth, this plasma is called the solar wind. This represents a steady mass flow from the sun of 10^{-14} solar masses/year. Coronal transients, outwardly expanding loops or clouds, contribute another 5%. Parenthetically this is a very small fraction of the loss shown by many early-type hot stars and cool giants and supergiants.

The spectrum of the corona, first observed in 1869, exhibited many unidentified spectrum lines. Not duplicated in the laboratory, the composition of the corona remained a mystery for many years, leading to the suggestion of a light hypothetical gas, "coronium," as their source. W. Grotrian and B. Edlen (1939–1942) found the clue and showed that the spectrum was that of common elements — iron, calcium, nickel, etc. — which had

lost 10 to 15 electrons, thus indicating a temperature of several million degrees for the corona. As yet no satisfactory theory accounts for the temperature rise from about 4200 °K at the top of the photosphere to 2,000,000 °K in the corona.

Spectrum Analysis

The latter half of the nineteenth century saw not only great advances in the observations of the solar spectrum, but also the beginnings of the theory by which those observations could be understood. The first breakthrough occurred when Gustav Kirchoff (1859) noted that incandescent solids or liquids gave off continuous spectra of all colors, while hot gases emitted light in bright lines, their color or wavelength being characteristic of the chemical composition of the gas. If the continuous spectrum of an incandescent solid was passed through a cooler gas, dark lines would appear in the spectrum at exactly the same positions as the bright lines displayed by the incandescent gas when viewed alone. This fact formed the basis for the analysis of the solar spectrum. Kirchoff applied his results to derive a model of the sun as a hot liquid sphere covered by an atmosphere of gases, which produced the dark lines in the spectrum. We now know that a dense gas may also produce a continuous spectrum and that the sun is totally gaseous, but for the 1860s Kirchoff's ideas were quite reasonable.

Since each gaseous element produces a characteristic pattern of spectral lines it was possible to analyze the chemical composition of the sun, a task first undertaken by Kirchoff together with a Heidelberg Professor of Chemistry, R. Bunsen. The solar spectrum was set beside the spectrum produced by a laboratory sample (usually heated to an incandescent vapor in a spark discharge or an arc source) and visually noting the coincidences of the lines; in the later years of the nineteenth century a photographic comparison was made with far greater precision. Kirchoff and Bunsen identified solar spectrum lines from the following elements: sodium, calcium, barium, strontium, magnesium, copper, iron, chromium, nickel, cobalt, zinc, and gold. Later workers such as Angstrom extended this list to include hydrogen, which we now know to be the major constituent of the sun. To date most (about 65) of the naturally occurring nonradioactive elements have been found to be present in the solar atmosphere. At the time of the total solar eclipse of 1869, Lockyer and Frankland observed in the flash spectrum a bright yellow line which had never before been seen in laboratory spectra. The element producing this line was not known on the earth and was named helium (after *helios* — the Greek word for sun). It was not until 1895 that helium was first detected on the earth. Despite its rarity on Earth, helium is the second most abundant gas in the sun (after hydrogen).

Cosmology deals with the chemical formation and evolution of the universe. The chemical composition of the sun, Earth, and meteorites serve as standards of comparison; hence a great deal of effort has been expended in their determination. Solar abundances can only be determined for the sun's atmosphere, but with few exceptions, the sun is considered to have the same composition throughout.

Solar (and stellar) atomic and molecular abundances can be determined from spectrum analysis, i.e., by knowing the atomic parameters from laboratory determinations and by comparing the observed intensity of Fraunhofer lines with theory. As the abundance increases a Fraunhofer line becomes darker in the center until black, i.e., saturated, then slowly broadens, finally developing resonance wings — the so-called *curve of growth*. On the theoretical side the strength of a line is determined by: (1) the abundance of the element, (2) the number of atoms of the element in a state able to produce the line — determined by the pressure and temperature of the gas — and (3) the atomic transition probability between the atomic levels connected with the line. A plot of the theoretical intensity vs. observed intensity yields the atomic abundance of the element. Hydrogen and helium make up 99.9% by volume of the sun. For every 1,000,000 atoms of hydrogen there are 63,000 of He, 690 of O, 32 of Fe, 3 of Al, 2 of Na, etc.

Sunspots

The discovery and first scientific study of sunspots was carried out telescopically by Galileo Galilei in 1610. Many of his contemporaries to whom he showed them refused to believe their eyes for they considered the sun to be immaculate and hence they had to be objects in front of the telescope. However, Galileo noted their daily westward movement, their motion in parallel lines, and their slower apparent motion near the limb, from which he concluded that the sun was a sphere rotating in about 27 days and that the spots were on its surface.

When examined through a telescope a large sunspot is seen to consist of two distinct parts: a dark central *umbra* surrounded by a less dark *penumbra*. On birth, one sees several small round pores each about 2000 miles (3220 kilometers) in diameter separated by tens of thousands of miles. If their development continues—most small sunspots last less than a day—the leader and follower grow rapidly in size and separate, resulting in two large spots having umbra and penumbra together with a number of smaller spots forming a complex group. Old age sets in after a week or two; one of the large spots disintegrates, as it becomes covered by light bridges, leaving a single spot which over several days diminishes in size until gone. On occasion exceptionally large, complex groups develop, exhibiting great variety of shape, often covered by brilliant "bridges" of light, giving the impression of great turmoil. Such spots, which may extend over 100,000 miles (160,000 kilometers) of the sun's surface, are easily visible to the naked eye when the sun's light is reduced by fog or by absorption of the earth's atmosphere at sunset. Russian records of naked eye sunspots describe them as looking like nail heads. Chinese annals, some before Christ, record many objects (sunspots) on the sun and describe them as like a flight of birds or as having the appearance of pigeon eggs.

Sunspots are dark only in contrast to the photosphere. Brighter than an electric arc, examined spectroscopically one finds atomic and molecular lines from which a temperature of about 3800 °K is obtained. George Ellery Hale (1914) found that some of the atomic lines were divided and polarized in the sunspot spectrum (the Zeeman effect), demonstrating the strong magnetic field associated with sunspots. Further work showed that pairs of spots were of opposite polarity, that the leader spot in the Northern Hemisphere had a magnetic polarity opposite that of the leader in the Southern Hemisphere, and that the polarity in each hemisphere reversed every 11 years.

Telescopic records for 400 years and scattered naked eye observations for 2000 years show that the number of sunspots varies with an average period of 11.2 years. Wolf, of Zurich, introduced the term *sunspot number*, $R = f + 10g$, as a measure of solar activity, where f is the total number of individual sunspots and g is the number of spot regions, either groups or spots. This number is quite arbitrary, but in practice serves very well to define solar activity. See Fig. 5. The last maximum occurred in 1979, at which time as many as 100 spots were visible on the sun's surface. At minimum, months may pass without a single record of a sunspot. The cause of this periodicity is not known. It has been suggested that Jupiter, which revolves around the sun with a period of 11.8 years and has a mass $\frac{1}{1000}$ that of the sun, could introduce tidal action and sunspots. Other suggestions, such as swarms of meteors striking the surface, have been made—all without success.

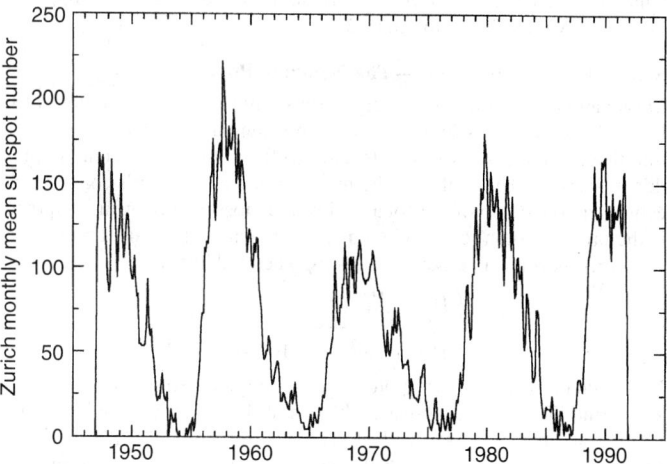

Fig. 5.   Sunspot cycle.

We should here remark that the sun rotates not like the earth, as a rigid body, but rather nonuniformly. Being gaseous throughout, it rotates once in 25 days in the equatorial regions, whereas the polar regions require 35 days. This nonuniform rotation results in deep-seated stirring of the interior. The sun is a magnet like the earth. According to a recent theory due to Babcock, its nonuniform rotation causes the magnetic lines of force to wind up like string on a top. As the magnetic lines deep in the interior become drawn out parallel to the equator and get tighter and closer together, magnetic buoyancy and turbulent motion carries a kink in the lines to the surface. The visible manifestation at the surface then is a pair of spots of opposite magnetic polarity with a field strength of 1000 to 3000 gauss, a field equal to that in a modern power-plant generator but distinguished from it by the fact that it extends over an area larger than the size of the Earth. The time required for the windup of the magnetic lines of force is 3–4 years. The time required for them to appear at the surface is 6–7 years; hence, the 11-year cyclic period. Each solar cycle begins with appearance of spots in two zones north and south about 30° from the equator. During the cycle the zone of spots drifts equatorward, ending at ±5°.

Strangely, the cycle of sunspot activity died out for about 75 years in the late seventeenth century. Indirect evidence of solar activity before the invention of the telescope in 1610, from isotopic carbon 14 found in tree rings, suggests that even longer periods of quiescence have occurred. Eddy has noted that the periods of low solar activity coincides with climatic lows in European temperatures and thus a connection between solar activity and climate.

Prominences

Prominences first seen at solar total eclipses stand as cloud-like forms above the sun's surface. See Fig. 6. They are classified by appearance and formation into the following types: quiescent, coronal, eruptive, sunspot, and tornado. A typical quiescent prominence is a large, thin, caterpillar-like form with several of its feet attached to the sun's surface; some may extend over 125,000 miles (201,125 kilometers) in length; 30,000 miles (48,270 kilometers) in height by 5,000 miles (8,045 kilometers) thick. They generally develop in the sunspot zone, then drift north or south to the polar regions; drawn out by differential solar rotation they may last for several years as a polar crown. Coronal forms appear to condense from the hot coronal gas and stand free of the sun's surface. Active prominences are generally associated with underlying or nearby sunspots.

Fig. 6.   Hedge-row prominence. (*Big Bear Solar Observatory.*)

Though formerly viewed at an eclipse or spectroscopically, today prominences are more easily seen and studied through a narrow passband interference filter (5 angstroms or less) or through a Lyot-Ohman polarizing filter (0.2–1.0 Angstrom passband) centered on Hα—the red line of hydrogen, $\lambda = 6563$ angstroms. Time-lapse motion pictures of active prominences show the strong control exerted by magnetic fields on their support and behavior. Often appearing as great inverted funnels, condensing from the corona, there is a general downward flow of prominence material along curved arcs into neighboring sunspots. Over a sunspot region, particularly after a flare, multiple arches may appear, condensing at several apices, the material flowing downward along curved lines of magnetic force with velocities of 12–24 miles (20–40 kilometers) per second. Greatly enhanced activity is often triggered by a nearby solar flare that destroys the equilibrium of the magnetic field structure; the whole or a portion of a prominence may erupt from the sun with great acceleration reaching velocities of several hundreds of miles/second.

Spectroscopic studies of prominences show that they may be divided into cool, quiescent (8000–10,000 °K) forms and hot, active arch-prominences with temperatures up to 100,000 °K. The same elements that appear in the sun's atmosphere exist in the prominence, their appearance modified only by its temperature and pressure.

Flares

A flare is a short-lived brightening of a small area of the sun comprising an evolving bipolar magnetic structure. Optically, except for the very rare and outstanding flares visible in white light, most flares are seen through the spectrohelioscope or filter as intense bursts in the Hα light of hydrogen; they are one of the most spectacular sights connected with the sun. Flares are chromospheric phenomena. They are by no means uncommon; at the peak of the sunspot cycle there are about 20–30 flares per year of class 3, having a duration 2–4 hours, and a thousand of class 1 lasting on the average 25 minutes. Preflare activity is often noted in quiescent prominences which takes the form of enhanced internal motions, a slow rise into the corona, and when the flare starts sudden acceleration and complete disappearance, sometimes to reappear with hours or days in nearly the same shape and position. The flare phenomenon often includes the ejection of material: surges from small flares take place along lines of magnetic force, generally at an angle to the sun's surface, the material returning to the sun along the same path. Great flares are accompanied by sprays of explosively ejected material with velocities as high as 930 miles (1500 kilometers)/second.

The flare phenomenon is exceedingly complex. We will describe some of the events in a great flare. Typically, individual brightenings merge along the sides of a disappearing quiescent prominence to form two ribbons, or starting as a bright single ribbon in a young active region it divides into two which initially separate with speeds up to 62 miles (100 kilometers)/second, but soon stop. See Fig. 7. The flare starts with a very rapid increase of temperature to 5–10 million degrees Kelvin high in the corona, accompanied by a soft X-ray burst, followed by a slow exponential decrease. Next the flare energy moves down into the chromosphere, heating a thin layer to 10,000 °K and becoming visible as

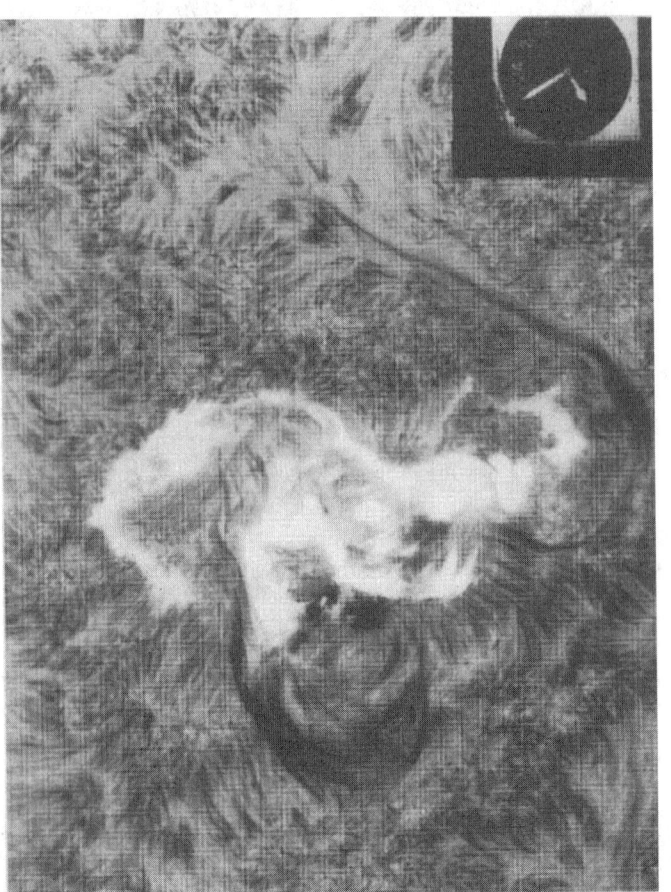

Fig. 7.   An intense two-ribbon flare with associated magnetic arches. (*Big Bear Solar Observatory*.)

ribbons in Hα. In the flash phase, within the first 10 minutes, electrons accelerated to 10–100 keV produce hard X-rays, impulsive radio, and far-ultraviolet bursts. Streams of electrons moving outward at 100,000 km/sec produce intense radio frequency bursts and can be tracked at low radio frequencies to the vicinity of the earth and recorded by spacecraft. In some large flares a blast wave originates, accelerating electrons and protons to hundreds of millions of electron volts which, in turn, when bombarding the photosphere, produce nuclear reactions and associated gamma rays.

Among the many terrestrial effects of a flare is the short-wave radio fadeout (SWF). Hard X-rays emitted at the onset of large flare illuminate the sunlight side of the earth, causing an SWF for all transmission paths over this hemispherical cap. The X-rays ionize the lower layers of the earth's atmosphere (60 km), freeing electrons which, excited to oscillation by radio waves, quickly lose their energy of oscillation by collisions with abundant air molecules. The loss of signal by absorption of the radio waves may last for several hours. Very large flares have been known to produce marked changes in the earth's geomagnetic field, which in turn induce electric currents in long-distance transmission lines strong enough to trip circuit breakers and to stop telegraph communication.

On March 10, 1989, a great flare occurred in a large sunspot region on the sun. Protons and electrons arriving at the earth two days later produced brilliant auroras over Canada and the United States. Transformers tied to the Hydro-Quebec power grid and to the United States saturated, overheated, and were destroyed from induced, nearly DC, earth-surface potentials of 1–10 V per kilometer of transmission line. The cost of outages and loss of transformers exceeded tens of millions of dollars from this one event.

The Solar Constant

The solar constant is defined as the quantity of solar energy received at normal incidence outside the earth's atmosphere at the earth's mean distance from the sun. It is expressed in calories per square centimeter per minute or kilowatts per square meter. The latest measurements give 1.968 cal cm^{-2} min^{-1} or 1.373 kW m^{-2}.

Variations in the solar output in X-ray, far infrared, and radio wavelengths are well known, but very little energy comes from these wavelengths (less than 1.0%), hence the larger question of any appreciable variation in the total luminosity of the sun over long periods of time goes begging. The evidence from paleontology suggests that it has been nearly constant (less than 10–15% variation) over several billion years. Today a variation of 0.5% or less is of interest to climatologists. For decades, C.G. Abbot with instruments of his own devising made a long series of measurements from mountains located in arid lands, particularly the Atacama desert of Northern Chile. He observed variations of up to 1.5%, which he attempted to correlate with rainfall. Recent measures from satellites show variations up to ±0.04%.

Solar Energy Generation — The Neutrino Puzzle

Mass can be converted into energy. Substituting the energy output of the sun (3.86 × 10^{33} ergs/sec) into Einstein's famous formula $E = mc^2$, we find that m, the mass loss, amounts to 4,700,000 tons per second. Large as this is, it is only 10^{-11} of the solar mass per year. In the sun's core, between center and one-tenth of the solar radius, the most important fusion process is the proton-proton reaction in which hydrogen nuclei, called protons, are converted into a stable helium nucleus. The reaction is:

$$2 \times \begin{cases} H^1 + H^1 \longrightarrow e^+ + \nu \\ H^2 + H^1 \longrightarrow He^3 + \gamma \\ He^3 + He^3 \longrightarrow He^4 + H^1 + H^1 \end{cases}$$

In this chain reaction, two protons fuse giving deuterium, a positron, and a neutrino. The neutrino escapes the sun at the velocity of light, carrying a small amount of energy with it. The positron collides with an electron and both are annihilated, with the release of gamma rays. The fusion of the deuterium nucleus with another abundant proton results in the light isotope of helium, with two protons and one neutron, and the emission of more gamma rays. Finally, two helium-3 atoms combine to make a stable helium nucleus plus two nuclei of hydrogen. With a loss of six hydrogens and two electrons we gain one helium and two hydrogens, additionally two neutrinos and 5 gamma rays are released in the process. From the known weights of He^4 and H^1 one calculates that 4.3 × 10^{-5} ergs is released for every He^4 nucleus formed.

Ninety percent of the sun's energy is produced through this reaction. Another reaction chain of the proton-proton cycle gives beryllium and boron with the emission of gamma rays, a positron, and a high energy (14 MeV) neutrino. It is this neutrino that can be detected by its absorption in a chlorine 37 nucleus, giving a radioactive argon 37 atom. The experiment performed by Davis uses 100,000 gallons (3785 hectoliters) of perchloroethylene enclosed in a sealed tank situated in a deep gold mine in South Dakota. One atom per day of radioactive argon is observed. This rate falls short by a factor of 3 from the rate calculated from the standard model of the sun, that is, its temperature, pressure, energy output, and chemical composition. The discrepancy can be equated either to a lack of knowledge of the sun's interior, or to a lack of knowledge of the neutrino's properties. This experiment has stimulated a great deal of theoretical work in solar and neutrino physics.

Low-energy neutrinos from the proton-proton fusion reaction, which contributes most of the neutrinos, can be detected through a nuclear transmutation of gallium, by neutrino capture, giving radioactive germanium. At the earth's distance from the sun, there are 1000 billion billion of these neutrinos passing through our bodies every second. A neutrino, moving with the velocity of light and having passed from the sun's core to its surface (and through the Earth if the sun is on the other side and we are in darkness), spends very little time in the neighborhood of a gallium atom. The chance of capture is very, very small. Thirty tons of gallium in solution sits in a laboratory under a high peak in the Italian Apennines, and in another experiment 57 tons of gallium metal is deep under a mountain in the Caucasus. Thirty tons of gallium should yield 1.2 atoms of germanium per day. From 300 days of collection, the number of captures was 63% of that predicted. Further tests are under way.

Solar Seismology

Like the earth, the sun also oscillates with a variety of shapes and frequencies. Though such oscillation is extremely difficult to detect from 93 million miles (~150 million kilometers) away, from precise measurements of its shape and from Doppler measurements of the up and down motion of its surface we can now probe the sun's interior. The 5-minute oscillation, discovered in the 1960s, of large groups of solar granules covering areas 2480–9300 miles (4000–15,000 kilometers) across, represents a standing pattern of sound waves trapped within the convection zone. Globally the entire convection zone (124,000 miles; 200,000 kilometers deep) is oscillating, with several nodes around the sun's circumference and with a number of nodes between the surface and the bottom of the zone. See Fig. 8. The observations are interpreted in terms of spherical harmonics of low-angular (l) and high-radial ($n \sim 15$) overtones with an amplitude of 10–100 cm/sec. l-values of 0, 1, 2, 3, 4 have been observed in a 5-day Doppler-shift measurement recorded in Antarctica at the South Pole during its summer, where the sun may be seen continuously for long periods of time. Because of the global

coherence in space and time, it has been possible to derive the depth of the convection zone and the solar rotation rate down to the bottom of the zone, where it is found that the sun rotates 5% faster than at the surface. In addition to the well established 5-minute oscillation, work in Crimea, the United Kingdom, and the United States has yielded a 160-minute period of oscillation.

The Sun's Motion in Space

Our sun, together with the stars in its immediate neighborhood, slowly orbits around the galactic center, located 33,000 light years away in the direction of the Milky Way constellation of Sagittarius, moving with a velocity of 155 miles (250 kilometers) per second. The "cosmic year" or period of rotation is 2400 million years. Additionally the sun has its own motion within the local group of naked eye stars. A study of radial velocities and proper motions places the direction in which the sun is moving near the summer constellation of Lyra. From the radial velocities its speed toward the solar apex is found to be approximately 12.1 miles (19.5 kilometers) per second.

Solar Telescopes and Instrumentation

Because of the enormous light and heat of the sun, and because the full surface of the star can be examined, unusual telescopes and instruments have evolved. The size of the sun's image in a camera or telescope is $\frac{1}{108}$th of the focal length. Thus, in the common 35-millimeter camera of 50 millimeters focal length, the sun's image on the film is only $\frac{1}{2}$ millimeter $\frac{1}{50}$th inch). The McMath-Pierce Solar Telescope Facility at Kitt Peak, Arizona, with a 300-foot (91.4-meter) focal length, yields an image 3 feet (0.9 meter) in diameter. Like most solar telescopes, the McMath-Pierce instrument is not pointed to the sun, but is fixed in direction. Sunlight goes 500 feet (152 meters) down the south polar axis at an angle of 32°. The heliostat mounting is driven to follow the sun during the course of the day. The beam, focused by a 60-inch (152-centimeter) concave mirror, is returned slightly below the incoming light to a 48-inch (122-centimeter) flat that directs the image vertically downward to spectrographs and other instrumentation located in the observing room. Great care has been taken to eliminate thermal disturbances, which distort the image. Thus, the 80-inch (203-centimeter) mirror is located 100 feet (30.5 meters) above the local terrain to eliminate ground effects. Additionally, the whole telescope, two-thirds of it underground, is cooled to ambient air temperature. The problem of internal seeing in solar telescopes is so great that the bold step of evacuating the whole telescope has recently been taken. The newer 30-inch (76.2-centimeter), 180-foot (54.9-meter) focal length tower at Sacramento Peak in New Mexico, and the 24-inch (61-centimeter) Cassegrain instrument located at San Fernando, California are good examples of vacuum telescopes.

The McMath-Pierce solar telescope is illustrated in the diagram of Fig. 9. An overall external view of the telescope is shown in Fig. 10. An image is shown in Fig. 11.

Coronagraph. A telescope for observing the corona surrounding the sun, at times other than at a solar eclipse, is called a *coronagraph*. An artificial eclipse of the sun is produced by placing an opaque circular disk slightly larger than the image of the sun at the focus of the telescope. Light from the corona passing the edge of the occulting disk is re-imaged by a lens onto a photographic plate or other detector. Suitable diaphragms capture the diffracted light from the objective. The brightness of the corona is one millionth that of the sun, hence, the success of the instrument depends upon the great care in selecting the glass for the singlet objective, free of bubbles and stria, and in polishing the surfaces free of all scratches. Furthermore, great pains must be taken to avoid dust on the objective. To minimize the atmospheric scattering, coronagraphs are mounted at high elevation sites (10,000 feet; 3050 meters).

The appearance of strong emission lines in the spectrum of the corona permits the instrument to be used monochromatically with narrow band polarizing filters or with a spectrograph.

Spectroheliograph. Essentially the instrument consists of a high-dispersion spectrograph with a second slit placed directly in front of the photographic plate so that the radiation from only one spectral line is received on the plate. If the instrument is so placed that the first slit is in the principal focus of a telescope directed toward the sun, a narrow strip of the sun's image will be admitted by the first slit and an image of that narrow section of the sun will be formed on the photographic plate in the

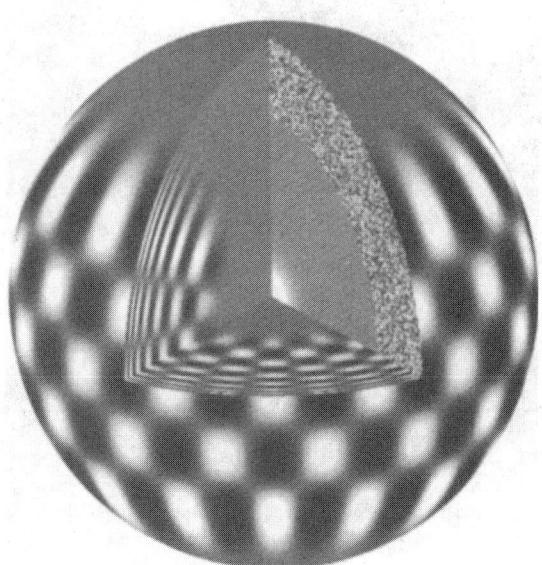

Fig. 8.   The up-down motions (solar oscillations) of one of the 10 million modes of the solar surface. This mode has an angular degree $l = 20$, an azimuthal order $m = 16$, a radial order $n = 15$, and a frequency corresponding to a period of 5 minutes. (*Global Oscillation Network Group, National Solar Observatories.*)

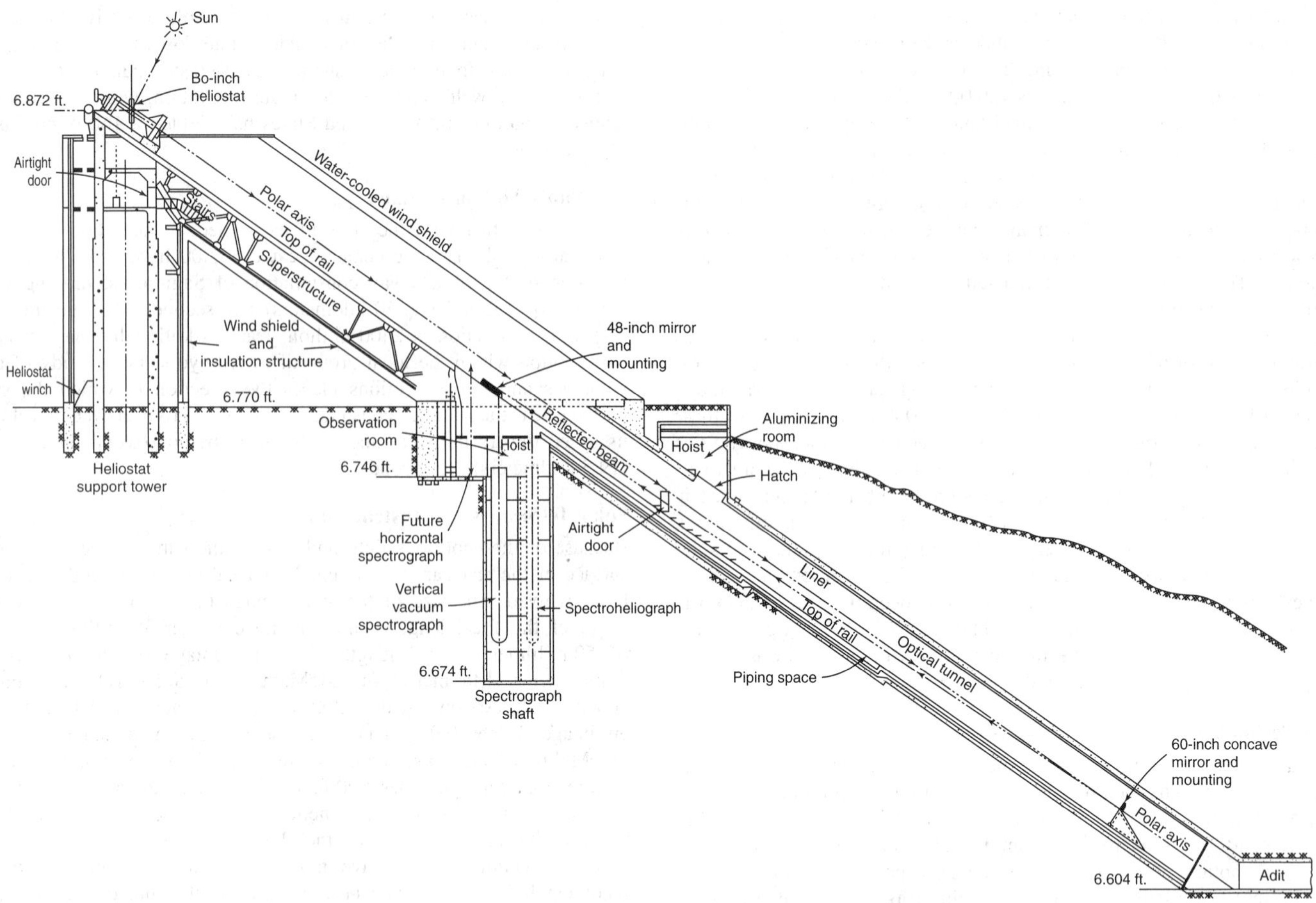

Fig. 9.   Sectional view of the McMath-Pierce solar telescope located at Kitt Peak, Arizona. (*Kitt Peak National Observatory.*)

Fig. 10.   Heliostat support tower and part of water-cooled wind shield of the McMath-Pierce solar telescope. (*Kitt Peak National Observatory.*)

Fig. 11.   Inspection of solar image. (*Kitt Peak National Observatory.*)

particular radiation for which the second slit is adjusted. The instrument is so constructed that the first slit may be moved across the image and at the same time the second slit will move across the photographic plate at the same rate. Hence, it is possible to obtain a photograph of the sun in the almost monochromatic radiation of any particular element, say calcium.

Birefringent Filters. B. Lyot (1933) investigated the transmission of a series of crystalline plates in polarized light. Independently in 1937,

Y. Ohman built such a filter capable of isolating the Hα line of hydrogen and was able to photograph prominences through it. The filter consists of a series of quartz plates in which each plate has a thickness of one-half the preceding plate. Polaroid is sandwiched between each plate.

By suitable design, a narrow passband (0.25–5 Å) can be obtained anywhere within the spectrum. With their use, the whole sun can be examined in the light of one element, usually hydrogen. These filters have been widely used particularly in cinematography of solar phenomena.

Solar Experiments in Space

The selective absorption of ozone, carbon dioxide, and water vapor in the Earth's atmosphere effectively blocks the ultraviolet solar spectrum short of 2900 angstroms and in bands beyond 7000 angstroms in the infrared. The earliest (1946) rockets carrying small spectrographs above 70 kilometers (43 miles) extended the solar spectrum from the ultraviolet ozone cut off at 2900 angstroms to 2100 angstroms. These unstabilized instruments were soon superseded by high resolution spectrographs on stabilized platforms capable of guiding to a fraction of a second of arc. The *Skylab* manned mission of 1973 mounted a complex of instruments designed to obtain high resolution direct photographs of the sun, prominences and flares in X-ray and far UV wavelengths. The disk of the sun was observed to be covered with thousands of X-ray bright points bringing magnetic fields to the surface. *Skylab*'s white light coronagraph observed for the first time huge blast waves—coronal transients—moving outward through the corona at speeds of 600 km (375 mi) per second, set in motion by an underlying flare.

The solar maximum mission of 1980 carried seven instruments into space: a gamma ray spectrometer, several X-ray spectrometers, a coronagraph, and an active cavity radiometer for measuring the solar irradiance. This instrument is still (1987) providing data. Future programs look to high resolution imaging of the solar granulation.

The Six Basic Problems of Solar Physics

Giovanelli has listed five problems of solar physics to which we add a sixth, i.e., the neutrino puzzle: (1) the cause of the sunspot cycle, (2) the structure of the convection zone comprising the outer 200,000 km (125,000 mi) of the sun's radius, (3) differential rotation with latitude and depth in the sun's atmosphere, (4) the cause of flares, and (5) the heating mechanism of the chromosphere and corona.

Additional Reading

Bahcall, J.N.: "Neutrino Astrophysics," Cambridge University Press, New York, NY, 1990.
Bahcall, J.N.: "The Solar Neutrino Problem," *Sci. American*, **262**, May, 54 (1990).
Billings, D.E.: "A Guide to the Solar Corona," Academic Press, Inc., San Diego, CA, 1966.
Christensen-Dalsgaard, J.D. Gough, and J. Toomre: "Seismology of the Sun," *Science*, **229**, 923 (1985).
Cox, A.N., W.C. Livingston, and M.S. Matthews: "Solar Interior and Atmosphere," University of Arizona Press, Tucson, AZ, 1998.
Cox, A.N. and C.W. Allen: "Astrophysical Quantities," 4th Edition, Springer-Verlag, Inc., New York, NY, 1999.
Culhane, J.L.: "The Physics of Solar Flares," Cambridge University Press, New York, NY, 1992.
Culhane, J.L. and E. Hiei: "Solar Flare, Coronal and Heliospheric Dynamics," Elsevier Science, New York, NY, 1995.
Eddy, J.A.: "The New Sun, the Solar Results from Skylab," National Aeronautics and Space Administration, Washington, DC, 1979.
Giovanelli, R.G.: "Secrets of the Sun," Cambridge University Press, New York, NY, 1984.
Golub, L. and J.M. Pasachoff: "Nearest Star: The Surprising Science of Our Sun," Harvard University Press, Cambridge, MA, 2001.
Harvey, J.W., J.R. Kennedy, and J.W. Leibacher: "GONG: To See Inside Our Sun," *Sky and Telescope*, 470 (1987).
Kappenman, J.G. and V.D. Albertson: "Bracing for the Geomagnetic Storms," *IEEE, Spectrum*, **27**, March, 27 (1990).
Lang, K.R.: "The Cambridge Encyclopedia of the Sun," Cambridge University Press, New York, NY, 2001.
Noyes, R.W.: "The Sun, Our Star," Harvard University Press, Cambridge, MA, 1982.
Pepin, R.A., J.A. Eddy, and R.B. Merrill: "Ancient Sun: Proceedings of the Conference on the Ancient Sun: Fossil Record in the Earth, Moon and Meteorites, Boulder Colorado, October 16–19, 1979," Elsevier Science, New York, NY, 1980.
Pierce, A.K.: "The McMath Solar Telescope of the Kitt Peak National Observatory," *Applied Optics*, 3, 12, 1337–1346 (1964).
Sonett, C.P., M.S. Giampapa, and M.S. Matthews: "The Sun in Time," University of Arizona Press, Tucson, AZ, 1991.
Thomas, J.H., N.O. Weiss: "Sunspots: Theory and Observations," Kluwer Academic Publishers, Norwell, MA, 1992.
Zirin, H.: "Astrophysics of the Sun," Cambridge University Press, Cambridge, MA, 1988.
Zirker, J.B.: "Total Eclipses of the Sun," Princeton University Press, Princeton, NJ, 1995.

SUN COMPASS. A device utilizing the direction of the sun for direction or orientation purposes. The sun compass operates on much the same principle as that of the sun dial. In the sun dial, the gnomon for casting the shadow is set accurately parallel to the earth's axis of rotation, and the direction of the shadow indicates local apparent time. In the sun compass, the dial is set for local apparent time, and the direction of the shadow is used in connection with a compass card. The instrument is quite complicated, for it must be set for terrestrial latitude, longitude, and local apparent time. It has been of great service in connection with flights in the polar regions of the earth, where the weakness and uncertainty of the horizontal component of the earth's magnetic field render the use of the magnetic compass most unreliable.

SUNDIAL. It is logical to suppose that from the earliest times, mankind has used the apparently moving sun as a means for reckoning time. As the sun appears to move across the heavens during the day, the position and length of the shadow cast by an opaque rod will continually change. The positions of lengths of this shadow may be used for the purpose of subdividing the period between sunrise and sunset. Any device that utilizes the shadow cast by the sun for the purpose of subdividing the day into equal parts is known as a sun dial.

It is difficult to say just when the first sun dial was constructed. The earliest written record is found in Isaiah 38:8, which was written approximately 700 years before the Christian era. The earliest instrument still existing is one built in Egypt, for which the exact date of construction is unknown. Sun dials came into general use during the thirteenth century, and the different types were then rapidly developed. By the time that mechanical clocks and watches made their appearance in the fifteenth century, a multitude of types of sun dials had been constructed, and many volumes had been written regarding the theory of the various devices.

There are two fundamental types of sun dials: fixed and portable. The most common fixed type marks the divisions of the day by means of the shadow thrown by the sun. The dial itself may be set at any desired angle, but generally, the plate is horizontal, and the style, which casts the shadow, is so placed as to be parallel to the axis of rotation of the earth. The portable dial makes use of the fact that the length of the shadow of the sun varies throughout the day, being the shortest at noon and the longest at sunrise and sunset. The great difficulty with this type is that because of the change in declination of the sun with season, it is necessary to have different scales of time for different periods of the year.

A multitude of ingenious and beautiful types of sun dials have been used in the past for the purpose of keeping time and are in use at present as ornaments or items of curiosity. In adjusting the horizontal type of sun dial, such as may be purchased from a number of dealers in garden supplies or curios, it is important to remember that the style should be parallel to the axis of the earth. That is, it should lie exactly in the true north-south plane, and the north end should be so elevated that the angle which the style makes with the horizontal plate is equal to the latitude of the observer. When properly adjusted, the sun dial will read local apparent time. This time will differ from ordinary clock time by the longitude difference between the position of the dial and the standard time meridian, and also by the equation of time.

See also **Equation of Time**.

SUNFISHES (*Osteichthyes*). Of the suborder *Percoidea*, family *Centrarchidae*, various species of sunfishes are among the favorite food and sporting fishes, particularly in North American fresh waters. The largemouth black bass, crappie, and bluegill are members of this family. Although initially American fishes, a number of species have been introduced elsewhere, notably Europe. The largest of the species is *Micropterus salmoides* (largemouth black bass) which measures up to about 30 inches (76 centimeters) and weighs up to 25 pounds (11 kilograms). The *Micropterus dolomieui* (smallmouth) is somewhat smaller, ranging up to 27 inches (69 centimeters) and weighing up to 12 pounds (5.4 kilograms). The *Lepomis macrochirus* (common bluegill) is frequently used as a small forage fish when stocking the larger species. Capable of attaining a length of about 15 inches (38 centimeters) and weight of nearly 5 pounds (2.3 kilograms), the average adult bluegill usually does not exceed about

4 inches (10 centimeters). Crappies are popular with sportsmen and have been distributed widely throughout North America. The *Pomoxis annularis* (white crappie) prefers muddy and turbid waters. In contrast, the *Pomoxis nigromaculatus* (black crappie) normally is found in clear waters. Records indicate that crappies up to 21 inches (53 centimeters) in length and 5 pounds (2.3 kilograms) have been caught. Archoplites interruptus (Sacramento perch), attaining a length of about 12 inches (30 centimeters), is found in the San Joaquin and Sacramento basins of California.

SUPERCHARGER.

The performance of an internal combustion engine is indicated, among other things, by the brake horsepower output. A review of the factors affecting power indicates that atmospheric conditions have a significant effect. A naturally aspirated (unsupercharged) engine is able to draw into the cylinders on suction strokes only from 70–85% of the fuel charge which it is theoretically capable of inducing. Consequently, the mean effective pressures are not as large as they might be, and power output per cubic inch of piston displacement does not reach its maximum possible value. Compression of the incoming air, or air-fuel mixture, somewhat above ambient pressure is a natural way of increasing output at sea level or of regaining it at altitudes. A compressor used for this purpose is designated a *supercharger*.

In recent years in connection with automotive engines, there has been considerable emphasis on the use of a turbocharger boost device to increase downsized engine power while increasing fuel economy and, in essence, retaining packaging advantages.

The Society of Automotive Engineers observed that recently there has been a high level of activity involving the use of the mechanically driven supercharger as a boost device. The supercharger can provide improved engine torque response at low engine speeds as compared to the turbocharger. The latest production turbochargers using variable geometry housings and ceramic turbines still take four times as long as a positive displacement, mechanically driven supercharger to produce maximum boost. In addition, the turbocompressor does not reach its maximum efficiency range until high speeds and airflows are achieved later in the vehicle acceleration event. This contrasts with the almost immediate boost response of the supercharger, which takes approximately 0.4 second to produce 50 kPa boost. The supercharger is continuously driven at full boost speed for the given rpm, and as soon as the bypass valve can be closed, the intake system is pressurized. There is no need to accelerate a mechanical device to high rotational speeds prior to production of boost pressure.

In testing commercially available vehicles, some equipped with turbochargers and others with superchargers, the researchers found that in attempting to accelerate from a stop in a turbocharged car resulted in a sluggish feel until 2 seconds after pressing the accelerator pedal. Then, the power rapidly climbed making it necessary to readjust the pedal position. Usually, the lack of turbocharger response at low engine speeds resulted in a large throttle opening and rich fuel during the first 3 seconds of acceleration then a gradual throttle closing. Fuel economy suffered and a high level of driver interaction was required to maintain vehicle control.

When driving a supercharged car, much less throttle modulation was needed to set the desired level of vehicle acceleration. The engine torque level increased almost immediately upon application of the throttle and did not suddenly change seconds later as with the turbocharged vehicles. This reduced the amount of driver interaction needed to maintain control of the vehicle and gave the same feel as a much larger, naturally aspirated engine under the same conditions.

SUPERCONDUCTIVITY.

A property of a material that is characterized by zero electric resistivity and, ideally, zero permeability. The phenomenon of superconductivity was discovered in 1911 by Heike Kamerlingh Onnes (University of Leiden) as the outcome of a remarkable achievement in those years—the liquefaction of helium for the first time. Helium condenses at atmospheric pressure at 4.2 K. Onnes, using the newly available very low temperature substance, proceeded to investigate the electrical resistance of various metals at low temperatures. Even prior to quantum mechanics, it had been predicted that if absolute zero could be achieved in a metal having a perfectly regular interatomic structure, the electrical resistance of the metal would be zero. Onnes found that the resistance of a mercury wire suddenly dropped to zero at 4.2 K, which indeed is a very low temperature, but still well above absolute zero. Onnes and other investigators at Leiden researched superconductivity in other metals,

such as lead, which was found to become superconductive at 7.2 K. It should be mentioned at this point that scientists of that period were biased in their thinking of conductivity and superconductivity in terms of metals. The first stable conducting organic material was not synthesized until 1960 and it was not until 1979 that a superconducting organic material was isolated. The so-called "hottest" superconductor was announced in mid-1993 by a research team at Eidgenossische Technische Hochschule in Zurich. The new material, considered toxic, is made of two distinct compounds that commence to be superconductive at 133 K. Prior record was claimed for $Tl_2Ca_2Ba_2Cu_3O_{10}$ at 127 K.

Rediscovery. The topic and potential of superconductivity essentially was rediscovered in the late 1970s and early 1980s, during which period the research activity safely can be described as zealous. This was fueled by the scores of probable applications of superconductors, but which to date largely remain as promises. Research continues at a good pace, and the topic is much better understood as compared with a decade ago. As pointed out later in this article, much excellent technological fallout has occurred from many millions of dollars invested in research. The search for the ultimately practical superconductor, although still elusive, has reinforced the multidisciplinary sciences involved, notably physics and chemistry.

Application Targets. Among the ultimately practical applications for superconductivity, especially those of significant future commercial values, are: (1) magnetic shielding, (2) magnetic resonance imaging magnets for medicine and research, (3) electric utility transmission lines and load-leveling storage coils, (4) magnetic separators for materials processing, (5) higher-speed (10×) switching and signal transmission for computers, (6) more compact electronics with finer interconnect lines, (7) extremely compact electric motors and actuators, (8) no-loss portable electrical storage "batteries," and (9) noncontact bearings and magnetically levitated vehicles. Special areas of interest by the military in superconductors include (1) infrared optical detector elements, (2) high-speed millimeter and submillimeter-wave electronics for advanced radar countermeasures, (3) magnetic anomaly detection of submarines, and (4) free-electron laser components.

Considerable research of a contemplative or assumptive nature continues to go forward—that is, studying how the "ideal" superconductor can be applied, once developed. Examples of progress are shown in Figs. 1, 2, and 3.

Fundamental Research Findings

For several years following the previously mentioned work of the Louden scholars, investigators concentrated principally on metallic elements and alloys, including indium, tin, vanadium, molybdenum, niobium-zirconium, and niobium-tin, among many others. A number of basic discoveries were made. For example, finding that the property of superconductivity could be destroyed by the application of a magnetic field equal to or greater than a critical field H_c. This H_c, for a given superconductor, is a function of the temperature given approximately by

$$H_c = H_0(1 - T^2/T_c^2) \qquad (1)$$

where H_0, the critical field at 0K, is in general different for different superconductors and has values from a few gauss to a couple of thousand gauss. For applied magnetic fields less than H_c, the flux is excluded from the bulk of the superconducting sample, penetrating only to a small depth λ into the surface. The value of λ (called the penetration depth) is in the range 10^{-5} to 10^{-6} centimeter. Thus the magnetization curve for a superconductor is

$$B \text{ (inside)} = 0 \qquad \text{for } H < H_c$$

$$B \text{ (inside)} = B \text{ (outside)} \qquad \text{for } H > H_c$$

This magnetization behavior is reversible and cannot therefore be explained entirely on the basis of the zero resistance. The reversible magnetization behavior is called the Meissner effect.

The existence of the penetration depth λ suggests that a sample having at least one dimension less than λ should have unusual superconducting properties, and such is indeed the case. Thin superconducting films, of thickness d less than λ, have critical fields higher than the bulk critical field, approximately in the ratio of λ to d. This result follows qualitatively from the thermodynamics of the Meissner effect: the metal in the superconducting state has a lower free energy than in the normal state, and the transition to the normal state occurs when the energy needed to

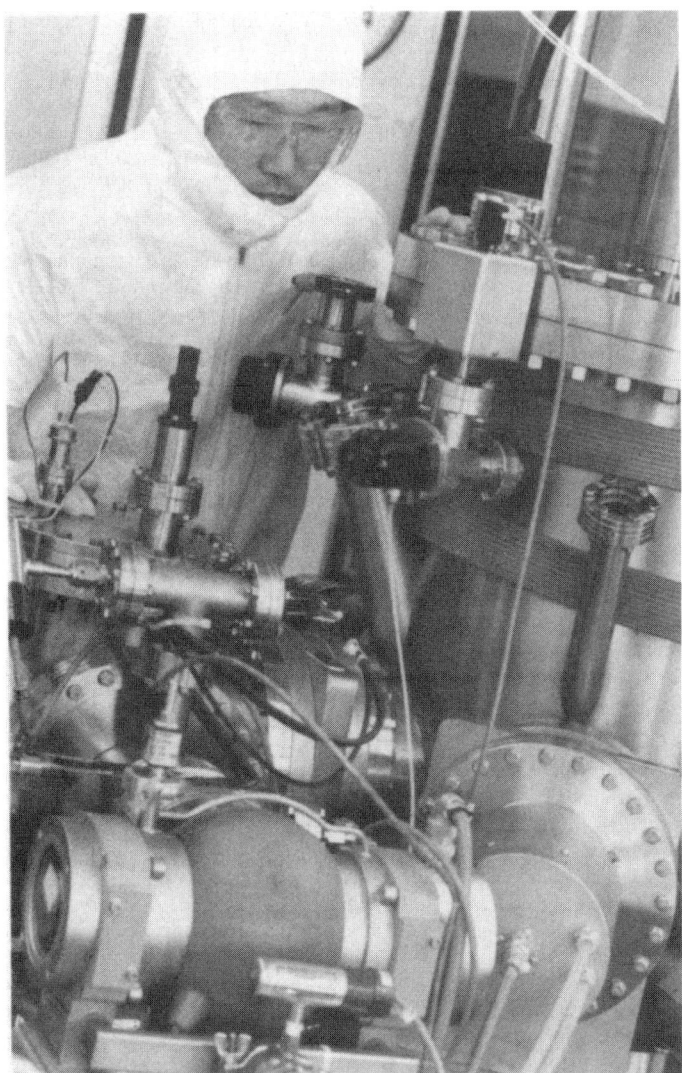

Fig. 1.   Special equipment required to fabricate low-temperature superconducting junctions. Josephson junctions are comprised of aluminum oxide sandwiched between layers of niobium. These trilayer devices are considered vital to the very-high-speed signal processing demands of next-generation computers, radar, and communication systems. Shown in illustration is scientist Dr. Joonhee Kang. (*Westinghouse Electric Corporation.*)

Fig. 2.   Electrical lead comprised of a high-temperature superconductor can carry a current of 2000 amperes. A variety of uses include magnetic resonance imaging and superconducting magnetic energy storage. (*Westinghouse Electric Corporation.*)

keep the flux out becomes equal to this free energy difference. But in the case of a thin film with $d < \lambda$, there is partial penetration of the flux into the film, and thus one must go to a higher applied field before the free energy difference is compensated by the magnetic energy.

It is clear that the existence of the critical field also implies the existence of a critical transport electrical current in a superconducting wire, i.e., that current I_c, which produces the critical field H_c at the surface of the wire. For example, in a cylindrical wire of radius r, $I_c = \frac{1}{2} r H_c$. This result is called the Silsbee rule.

All of the above properties distinguish superconductors from "normal" metals. There is another very important distinction, which contains a clue to understanding some of the properties of superconductors. In a normal metal at 0 K, the electrons, which obey Fermi statistics, occupy all available states of energy below a certain maximum energy called the Fermi energy ζ. Raising the temperature of the metal causes electrons to be singly excited to states just above the Fermi energy. There is for all practical purposes a continuum of such excited energy states available above the Fermi energy. The situation is quite different in a superconductor; it turns out that in a superconductor, the lowest excited state for an electron is separated by an energy gap ε from the ground state. The existence of this gap in the excitation spectrum has been confirmed by a wide range of measurements: electronic heat capacity, thermal conductivity, ultrasonic attenuation, far infrared and microwave absorption, and tunneling. The energy gap is a monotonically decreasing function of temperature, having

a value ~ 3.5 kT_c at 0 K (where k is the Boltzmann constant) and vanishing at T_c.

The superconducting state has a lower entropy than the normal state, and therefore one concludes that superconducting electrons are in a more ordered state. Without, for the present, inquiring more deeply into the nature of this ordering, one can state that a spatial change in this order produced say by a magnetic field will occur, not discontinuously, but over a finite distance ζ, which is called the *coherence length*. The coherence length represents the range of order in the superconducting state and is typically about 10^{-4} centimeter, though we shall see later that it can in some superconductors take much lower values and lead to some remarkable properties.

Measurements of the transition temperature on different isotopes of the same superconductor showed that T_c is proportional to $M^{-1/2}$, where M is the isotopic mass. This isotope effect suggests that the mechanism underlying superconductivity must involve the properties of the lattice, in addition to those of the electrons. Another indication of this is given by the behavior of allotropic modifications of the same element: white tin is superconducting, while grey tin is not, and the hexagonal and face-centered cubic phases of lanthanum have different transition temperatures. A third, and most striking, indication is that the current vs voltage characteristic of a superconducting tunneling junction shows a structure which is intimately related to the phonon spectrum of the superconductor.

The superconducting properties of alloys present a bewildering variety of phenomena. They show a great deal of magnetic hysteresis, with little indication of a perfect Meissner effect. The Silsbee rule is inapplicable, and the resistive transition occurs at fields generally very much higher than in pure superconductors. For example, a wire of Nb_3Sn can carry a current of 10^5 amperes/cubic centimeter in an applied field of 100 kilogauss, while a similar wire of lead would carry about 10^3 amperes/cubic centimeter in a field of only 100 gauss. When experiments are done using well-annealed (preferably single-crystal) alloys, it is found that the critical currents drop considerably, and the magnetic behavior becomes reversible but still quite unlike that of pure superconductors. The flux is excluded

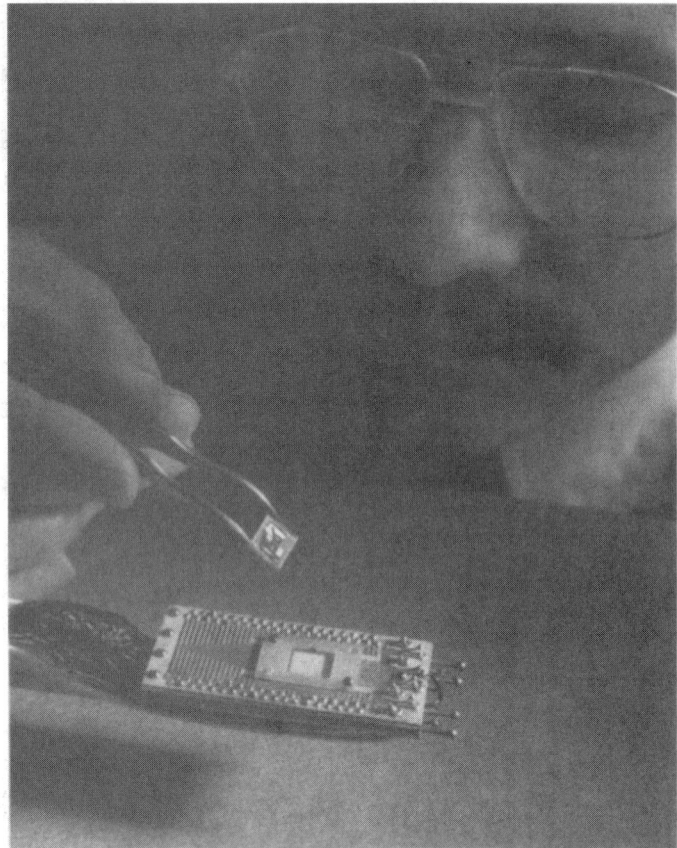

Fig. 3.   Scientist Donald L. Miller holds an integrated circuit chip comprising a high-resolution superconducting analog-to-digital converter. The one-square-centimeter chip, known as a counting converter, holds promise as an unprecedented combination of high resolution and low power consumption, as needed in future air traffic control radar and infrared space-tracking applications. The 12-bit circuit (Josephson junction) has a resolution of 1 part in 4000. (*Westinghouse Electric Corporation.*)

from the interior of the sample up to a well-defined field H_{c1}. When the applied field is raised further, flux begins to penetrate, even though the resistance remains zero, until a second critical field H_{c2} is reached, at which the flux penetration is complete, and normal resistance is abruptly restored.

Superconductivity Theory

The theory of superconductivity has developed along two lines, the phenomenological and the microscopic. The phenomenological treatment was initiated by F. London, who modified the Maxwell electromagnetic equations so as to allow for the Meissner effect. His theory explained the existence and order of magnitude of the penetration depth, and gave a qualitative account of some of the electrodynamic properties. The treatment was extended by V.L. Ginzburg and L.D. Landau, and by A.B. Pippard, who in particular emphasized the concept of the range of coherence. A.A. Abrikosov used these ideas to develop a model for alloy superconductors. He showed that if the electronic structure of the superconductor were such that the coherence length ζ becomes smaller than the penetration depth λ, one would get magnetic behavior similar to that observed in alloys, with two critical fields H_{c1} and H_{c2}. The problem of high critical currents in unannealed (or otherwise metallurgically imperfect) alloys and compounds is more complicated because it involves the interaction between the microscopic metallurgical structure and the superconducting properties. This is an area of great research activity because of the technological implication to be mentioned later.

The microscopic theory of superconductivity was initiated by H. Fröhlich, who first recognized the importance of the interactions of electrons with lattice vibrations and in fact predicted the isotope effect before its experimental observation. The detailed microscopic theory was developed by J. Bardeen, L.N. Cooper and J.R. Schrieffer in 1957, and represents one of the outstanding landmarks in the modern theory of solids. The BCS theory, as it is called, considers a system of electrons interacting with

the phonons, which are the quantized vibrations of the lattice. There is a screened coulomb repulsion between pairs of electrons, but in addition there is also an attraction between them via the electron-phonon interaction. If the net effect of these two interactions is attractive, then the lowest energy state of the electron system has a strong correlation between pairs of electrons with equal and opposite momenta and opposite spin and having energies within the range $k\theta$ (where θ is the Debye temperature) about the Fermi energy. This correlation causes a lowering of the energy of each of these Cooper pairs (named after L.N. Cooper who first pointed out their existence on the basis of some general arguments) by an amount ε relative to the Fermi energy. The energy ε may be regarded as the binding energy of the pair, and is therefore the minimum energy which must be supplied in order to raise an electron to an excited state. We see thus that the experimentally observed energy gap follows from the theory. The magnitude ε_0 of the gap at 0K is

$$\varepsilon_0 \approx 4k\theta \exp\left(-\frac{1}{NV}\right)$$

where N is the density of electronic states at the Fermi energy and V is the net electron-electron interaction energy. The superconducting transition temperature T_c is given by

$$3.5kT_c \approx \varepsilon_0$$

It has been shown that the BCS theory does lead to the phenomenological equations of London, Pippard and Ginzburg and Landau, and one may therefore state that the basic phenomena of superconductivity are now understood from a microscopic point of view, i.e., in terms of the atomic and electronic structure of solids. It is true, however, that we cannot yet, *ab initio*, calculate V for a given metal and therefore predict whether it will be superconducting or not. The difficulty here is our ignorance of the exact wave functions to be used in describing the electrons and phonons in a specific metal, and their interactions. However, we believe that the problem is soluble in principle at least.

The range of coherence follows naturally from the BCS theory, and we see now why it becomes short in alloys. The electron mean free path is much shorter in an alloy than in a pure metal, and electron scattering tends to break up the correlated pairs, so that for very short mean free paths one would expect the coherence length to become comparable to the mean free path. Then the ratio $\kappa \approx \lambda/\zeta$ (called the Ginzburg-Landau order parameter) becomes greater than unity, and the observed magnetic properties of alloy superconductors can be derived. The two kinds of superconductors, namely those with $\kappa < 1/\sqrt(2)$ and those with $\kappa > 1/\sqrt(2)$ (the inequalities follow from the detailed theory) are called respectively type I and type II superconductors.

Challenges to Established Theories. It is interesting to note that some theoreticians struggle with describing how superconductivity occurs at high temperatures in the newer, ceramic superconductors. This is understandable because the classic theory of superconductivity is tied to metals. Most ceramic superconductors discovered to date incorporate distinctive layers of copper and oxygen atoms. One question posed by some researchers, "Is the mechanism of high-temperature superconductivity the same in hole superconductors as it is in electron superconductors?"

Researchers at the Brookhaven National Laboratory, in applying X-ray techniques to a cerium-doped electron superconductor developed at the University of Tokyo, found that the holes of a hole superconductor are linked to oxygen atoms in the copper-oxygen layers, whereas in an electron superconductor the electrons are associated with copper atoms. This is exemplary of how easy it is for former theories to become outdated when new material combinations are tested for their superconductivity.

Superconductivity Research

In 1962, B. Josephson recognized the implications of the complex order parameter for the dynamics of the superconductor, and in particular when one considers a system consisting of two bulk conductors connected by a "weak link." This research led to the development of a series of weak link devices commonly called Josephson junctions. See also **Josephson Junctions**. These devices hold much promise for achieving ultra high-speed computers where switching time is of the order of 10^{-11} second.

Good success also has been achieved in the use of certain type II superconductors, such as Nb-Zr and Nb-Ti alloys, and Nb_3Sn, in making electromagnets. In a conventional electromagnet employing normal conductors, the entire electric power applied to the magnet is consumed as

Joule heating. For a magnet to produce 100 kilogauss in a reasonable volume, the power requirement can run into megawatts. In striking contrast, a superconducting magnet develops no Joule heat because its resistance is zero. Indeed, if such a magnet has a superconducting shunt placed across it after it is energized, the external power supply can be removed, and the current continues to flow indefinitely through the magnet and shunt, maintaining the field constant. Superconducting magnets have been constructed producing very strong fields in usable volumes. There is a natural upper limit to the critical field possible in such superconductors, given by the paramagnetic energy of the electrons (due to their spin moment) in the normal state becoming equal to the condensation energy of the Cooper pairs in the superconducting state. This leads to a limit of about 360 kilogauss for a superconductor with a T_c of 20 K.

As investigators accumulated data upon data, many emphasized the practical as well as theoretical aspects of superconductors. The ultimate superconductor, of course, would be one that operated at room temperature or above. The materials must be manufacturable in a useful form, such as strong ductile wires for high-field magnets, electrical machinery, and power transmission lines, situations which could be even more important in commercial and industrial application than their value to science per se. (Traditionally, superconducting materials have been hard, brittle, and difficult to process.) Although superconductors that would operate at room temperature and above present a long-range target, lesser targets, including practical ways to cool them with liquid nitrogen instead of liquid helium and possibly, even better, operate them within a closed-cycle refrigeration system is the goal in the shortrange. Useful superconductors in large-scale applications must retain their properties not only at high temperatures, but also in the presence of high magnetic fields and while carrying large electrical currents. Praveen Chaudhari (IBM) has observed that new superconductors will enter the marketplace rapidly when intensive materials engineering produces easily cooled, mechanically robust conductor configurations that can handle high current densities ($100{,}000+$ A/cm^2) under powerful magnetic fields ($10+$ T), while maintaining stable superconductivity.

Johannes Georg Bednorz and Karl Alexander Mueller (IBM Zurich Research Laboratory) after several years devoted to a study of oxide compounds (not in terms of superconductivity) proceeded with the working hypothesis that an increase in the density of charge carriers in a material (either as electrons or as positively charged "holes") possibly would lead to a rise in transition temperature. They commenced a search for nickel- and copper-containing oxides. Early in 1986, they found a certain form of barium lanthanum copper oxide that evidenced the onset of superconductivity at temperatures as high as 35 K (12 degrees over the previous record). They encountered skepticism, because the facts did not square with accepted theory that limits the phenomenon to well below 35 K. Shortly thereafter, however, researchers at the University of Tokyo, the University of Houston, and AT&T Bell Laboratories confirmed the Bednorz-Mueller findings. On October 14, 1987, the Nobel prize in physics was awarded to these two researchers and a speaker for the Royal Swedish Academy of Sciences observed that their work inspired "the explosive development in which hundreds of laboratories the world over commenced work on similar material." [It should be observed that Ching-Wu Chu and colleagues (University of Houston) did announce in February 1987 that a related class of ceramics (a certain form of yttrium barium copper oxide) remained superconducting up to 94 K, proclaiming that to be the first superconductor which could be cooled by liquid nitrogen (bp = 77 K) instead of requiring helium.] That announcement in itself also precipitated a "rush" of researchers to the ceramics.

Technological Fallout of Superconductivity Research

While in the course of finding viable superconductors for commercial applications, researchers have produced valuable ancillary information.

Quantization of Energy. In a scholarly paper, D.G. McDonald (U.S. National Institute of Standards and Technology, Boulder, Colorado) observes, "Ideas about quantized energy levels originated in atomic physics, but research in superconductivity has led to unparalleled precision in the measurement of energy levels. Microscopic things can be identical; macroscopic things cannot. This proposition is so imbued in the minds of physicists that it is interesting to see that it is false in the following sense. In the past, physicists believed that only atoms and molecules could have identical states of energy, but recent experiments have shown that much larger bodies, superconductors in macroscopic quantum states, have

equally well-defined energies." In the paper, McDonald uses the novelty of the Josephson effect to illustrate the primary point of the technical paper. See McDonald reference listed.

Structural Chemistry. In an enlightening paper, R.J. Cava (AT&T Bell Laboratories) asserts, "The discovery of high-temperature superconductivity in oxides based on copper and rare and alkaline earths at first caught the solid-state physics and materials science communities completely by surprise. Since the earliest 30 to 40 K superconductors based on $La_{2-x}(Ca,Sr,Ba)_x CuO_4$, many new superconducting copper oxides have been discovered, with ever-increasing chemical and structural complexity. The current record transition temperature is held[1] by $Tl^2 Ba^2 Ca^2 Cu^3 O^{10}$, a material whose processing requires the stoichiometric control of five elements, each with considerably different chemical characteristics." In the Cava paper, the crystal structures of the known copper oxide superconductors are described, with particular emphasis on the manner in which they fall into structural families. The local charge picture — a framework for understanding the influence of chemical composition, stoichiometry, and doping on the electrical properties of complex structures — is also described.

This probing of complex and previously unattended solid materials typifies technical fallout from superconductor research.

Impact on Materials Processing and Chemical Engineering. In an interesting paper, R. Kumar (Indian Institute of Science, Bangalore) points out how processing considerations for achieving high-temperature superconductors has introduced new process engineering problems not contemplated heretofore. In a paper (reference listed), Kumar observes that processes involve multicomponent solid-solid reactions; mixing of fine powders; simultaneous precipitation of many ions from solutions, emulsions, microemulsions, and liquid membranes; the flow of cohesive powders with and without binders; the flow of thin films over partially wetted particles; grain boundary growth and composition; quick evaporation using pulsed lasers; mixing of molecules during their flight paths and the influence of oxygen jets; deposition of particles on substrates; and other relatively unfamiliar processing techniques.

Additional Reading

Amato, I.: "Finally, a Hotter Superconductor," *Science*, 755 (May 7, 1993).

Beardsley, T.M.: "Unsuperconductivity," *Sci. Amer.*, 22D (April 1989).

Bishop, D.J., P.L. Gammel, and D.A. Huse: "Resistance in High-Temperature Superconductors," *Sci. Amer.*, 48 (February 1993).

Brosha, E.L. et al.: "Metastability of Superconducting Compounds in the Y-Ba-Cu-O System," *Science*, 196 (April 9, 1993).

Caruana, C.M.: "Superconductivity: The Near and Long Term Outlook," *Chem. Eng. Progress*, 72 (May 1988).

Cava, R.J.: "Structural Chemistry and the Local Charge Picture of Copper Oxide Superconductors," *Science*, 656 (February 9, 1990).

Conradson, S.D., I.D. Raistrick, and A.R. Bishop: "Axial-Oxygen-Centered Lattice Instabilities and High-Temperature Superconductivity," *Science*, 1394 (June 15, 1990).

Erwin, S.C. and W.E. Pickett: "Theoretical Fermi-Surface Properties and Superconducting Parameters for K3C60," *Science*, 842 (November 8, 1991).

Fisk, Z. et al.: "Heavy-Electron Metals: New Highly Correlated States of Matter," *Science*, 33 (January 1, 1988).

Fisk, Z., G. Aeppli: "Superstructures and Superconductivity," *Science*, 38 (April 2, 1993).

Foner, S., T.P. Orlando: "Superconductors: The Long Road Ahead," *Technology Review (MIT)*, 36 (February 1988).

Gabelle, T.H., J.K. Hulm: "Superconductivity — The State that Came in from the Cold," *Science*, 367 (January 22, 1988).

Haroche, S., J.-M. Raimond: "Cavity Quantum Electrodynamics," *Sci. Amer.*, 54 (April 1993).

Hazen, R.M.: "Perovskites," *Sci. Amer.*, 74 (June 1988).

Iqbal, Z. et al.: "Superconductivity at 45 K in Rb/Tl Codoped C60 and C60/C70 Mixtures," *Science*, 826 (November 8, 1991).

Ishiguiro, T., K. Yamaji, and G. Saito: "Organic Superconductors," 2nd Edition, Springer-Verlag, Inc., New York, NY, 1997.

Ketterson, J.B. and S. Song: "Superconductivity," Cambridge University Press, New York, NY, 1999.

Kumar, R.: "Chemical Engineering and the Development of Hot Superconductors," *Chem. Eng. Progress*, 17 (April 1990).

Laughlin, R.B.: "The Relationship Between High-Temperature Superconductivity and the Fractional Quantum Hall Effect," *Science*, 525 (October 28, 1988).

Lee, P.J.: "Engineering Superconductivity," Wiley-IEEE Press, New York, NY, 2001.

[1] As of 1990.

Little, W.A.: "Experimental Constraints on Theories of High-Transition Temperature Superconductors," *Science*, 1390 (December 9, 1988).

Luss, D. et al.: "Processing High-Temperature Superconductors," *Chem. Eng. Progress*, 40 (September 1989).

McDonald, D.G.: "Superconductivity and the Quantization of Energy," *Science*, 177 (January 12, 1990).

Murphy, D.W. et al.: "Processing Techniques for the 93 K Superconductor $Ba_2YCu_3O_7$," *Science*, 922 (August 19, 1988).

Pool, R.: "Superconductor Patents: Four Groups Duke It Out," *Science*, 931 (September 1, 1989).

Pool, R.: "Superconductivity Stars React to the Market," *Science*, 373 (January 25, 1991).

Poole, C.P., H.A. Farach, and R.J. Creswick: "Superconductivity," Academic Press, Inc., San Diego, CA, 1996.

Poole, C.P., H.A. Farach, and R.J. Creswick: "Handbook of Superconductivity," Academic Press, Inc., San Diego, CA, 1999.

Ross, P. and R. Ruthen: "Squeezed Hydrogen Forms Metal with Superconducting Potential," *Sci. Amer.*, **26** (November 1989).

Schrieffer, J.R.: "The Theory of Superconductivity," Perseus Publishing, Boulder, CO, 1999.

Shrivastava, K.N.: "Superconductivity," World Scientific Publishing Company, Inc., Riveredge, NJ, 2000.

Shumay, W.C. Jr.: "Superconductor Materials Engineering," Advanced Materials & Processes, 49 (November 1988).

Sleight, A.W.: "Chemistry of High-Temperature Superconductors," *Science*, 1519 (December 16, 1988).

Staff: "Trying to Cooperate in Order to Compete," *Technology Review (MIT)*, 13 (February/March 1991).

Stix, G.: "Superconducting SQUIDS," *Sci. Amer.*, 112 (March 1991).

Sun, J.Z. et al.: "Elimination of Current Dissipation in High Transition Temperature Superconductors," *Science*, 307 (January 19, 1990).

Tinkham, M.: "Introduction to Superconductivity," 2nd Edition, The McGraw-Hill Companies, Inc., New York, 1995.

Wolsky, A.M., R.F. Giese, and E.J. Daniels: "The New Superconductors: Prospects for Applications," *Sci. Amer.*, 60 (February 1989).

SUPERCOOLING. The cooling of a liquid below its freezing point without the separation of the solid phase. This is a condition of metastable equilibrium, as is shown by solidification of the supercooled liquid upon the addition of the solid phase, or the application of certain stresses, or simply upon prolonged standing.

SUPERELEVATION. When the plane of a roadway is tilted on a curve (commonly known as banked), it is said to be superelevated. The purpose of superelevation is to permit a vehicle to round a curve on a roadway at high speed without danger of overturning or skidding. See Fig. 1. The superelevation can be made so that the resultant of dead weight and centrifugal force passes through the vertical plane of symmetry of the vehicle. In this condition, no side sway would be felt by the occupants. However, the superelevation necessary to accomplish this is different for each vehicle speed, so that it is apparent that the superelevation of a highway presupposes an average vehicle speed. The same is true of railways, although the variation of speeds with which the trains round curves is less than in the case of highway traffic.

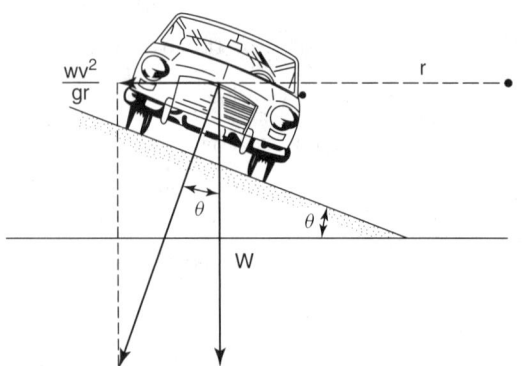

Fig. 1. Demonstration of superelevation principle applied to highway.

To illustrate how the superelevation depends upon vehicle speed, let it be assumed that an automobile approaches a curve on a highway at a speed of V (feet per second). If the radius of the turn is r, the centrifugal acceleration is V^2/r. Furthermore, assume that the weight is W pounds. While negotiating the curve, the car is subject to two forces, one, the weight vertically downward, the other, centrifugal force acting horizontally away from the center of curvature, and having a magnitude WV^2/gr. The surface of the road must be perpendicular to the resultant for no "side sway." If superelevation is given as the angle of bank (see figure), the angle of superelevation θ has a tangent equal to centrifugal force divided by weight. This tangent is V^2/gr, demonstrating that the superelevation must be made with respect to the radius of curvature and the velocity of the vehicle. It is independent of the dimensions and weight of the vehicle.

See also **Spiral Curve**.

SUPERFINISHING. An abrasive process for removing smear metal, scratches and ridges produced by machining and grinding operations, and other surface irregularities from parts that are to have a highly finished surface. The process resembles lapping in that a lubricated abrasive stone is applied to the surface at comparatively low speeds and light pressures. A superfinishing head whose base is attached to the cross-slide of an engine lathe may be used. The base supports two vertical cylindrical guides on which the head proper may be manually adjusted to the work by a hand-operated lever. The abrasive stone is carried in a vertical slide, which is subjected to the action of a spring for applying pressure to the stone. The stone pressure may be regulated to suit the requirements of the work by turning the screw at the top of the slideway.

SUPERFLUIDITY. The term used to describe a property of condensed matter in which a resistance-less flow of current occurs. The mass-four isotope of helium in the liquid state, plus over 20 metallic elements, are known to exhibit this phenomenon. In the case of liquid helium, these currents are hydrodynamic. For the metallic elements, they consist of electron streams. The effect occurs only at very low temperatures in the vicinity of the absolute zero ($-273.16\,°C$ or 0 K). In the case of helium, the maximum temperature at which the effect occurs is about 2.2 K. For metals, the highest temperature is in the vicinity of 20 K.

If one of the metals (commonly referred to as superconductors) is cast in the form of a ring and an external magnetic field is applied perpendicularly to its plane and then removed, a current will flow round the ring induced by Faraday induction. This current will produce a magnetic field, proportional to the current, and the size of the current may be observed by measuring this field. Were the ring (e.g., one made of lead) at a temperature above 7.2 K, this current and field would decay to zero in a fraction of a second. But with the metal at a temperature below 7.2 K before the external field is removed, this current shows no sign of decay even when observations extend over a period of a year. As a result of such measurements, it has been estimated that it would require 10^{99} years for the supercurrent to decay. Such persistent or "frictionless" currents in superconductors were observed in the early 1900s—hence they are not a recent discovery.

In the case of liquid helium, these currents are hydrodynamic, i.e., they consist of streams of neutral (uncharged) helium atoms flowing in rings. Since, unlike electrons, the helium atoms carry no charge, there is no resulting magnetic field. This makes such currents much more difficult to create and detect. Nevertheless, as a result of research carried out in England and the United States during the late 1950s and early 1960s, the existence of supercurrents in liquid helium has been established.

SUPERIMPOSED RIVER VALLEY. A river valley that is independent of present structural control may be described as either superimposed or antecedent. In the former case it is implied that the river has been able to maintain its course across resistant structures such as ridges, because it started as a consequent stream and has been "let down" on the underlying or nonconformable structure.

SUPERNOVA 1987A AND 1993J. What is there about an event that occurred 160,000 years ago that so fires the imagination and excites scientists into a veritable frenzy of observation? Supernovae are the results of violent explosions and are perhaps the most spectacular outbursts of energy we can ever see. A nova is an explosive variable star that brightens by a factor of thousands within hours or days and then fades to its previous brightness within months. The name "nova" is shortened from "stella nova," the Latin for "new star." It was once believed that these were really new stars, but we now know that they are actually old stars which suddenly explode and become much brighter. If the star was too faint to see

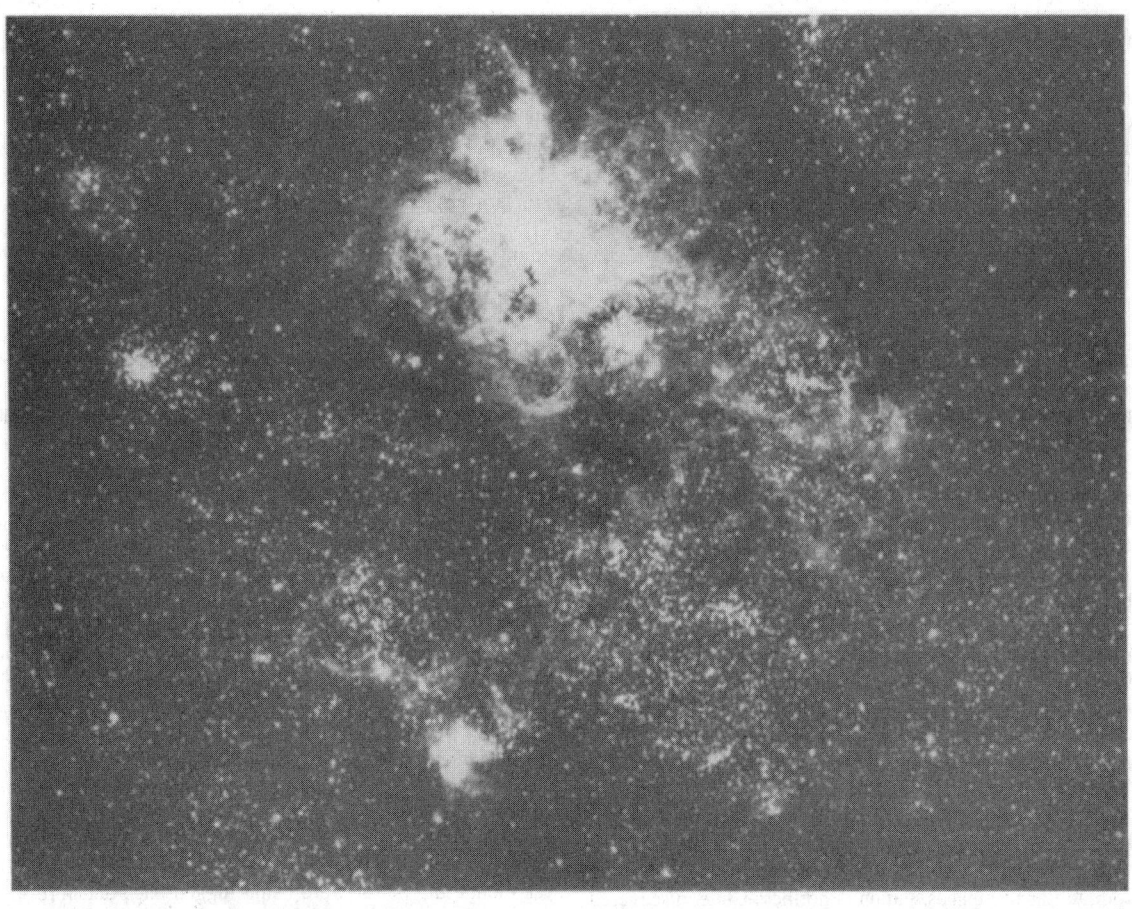

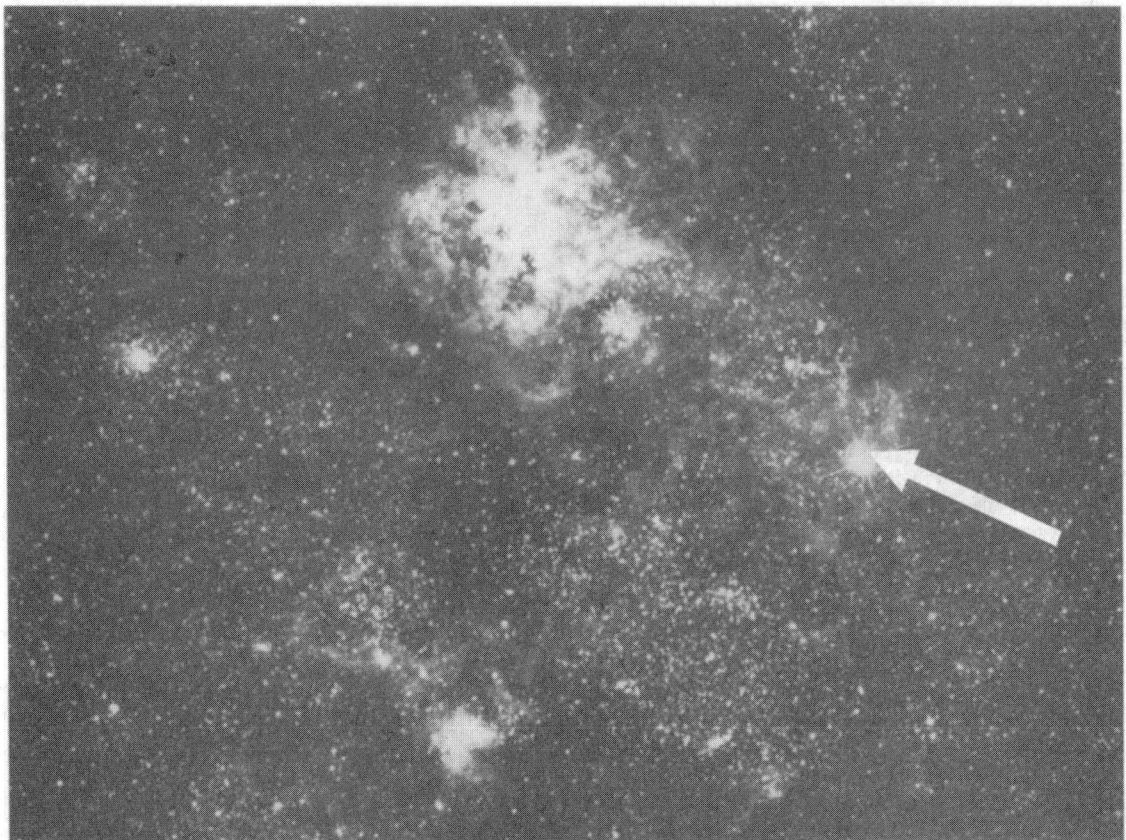

Fig. 1. Supernova 1987A: (*Top*) Large Magellanic Cloud taken with a 1.5 meter telescope at NOAO's Cerro Tololo (Chile) Inter-American Observatory in 1969 by Dr. Victor Blanco, former observatory director. (*Bottom*) Image of same area of sky made with the Cerro Tololo's Schmidt telescope on February 26, 1987 by Dr. Wendy Borbers of the Harvard-Smithsonian Center for Astrophysics. Bright image at right-center of photo shows the supernova detected in this nearby galaxy. (*National Optical Astronomy Observatories.*)

before the explosive brightening, then it appeared as if it were a "new" star. Observations show that novae occur in close binary systems, one of whose components is a white dwarf star. White dwarf stars represent a final stage in the life history of a star, when it has consumed its nuclear fuel and has shrunk to the dimensions of the earth. In such a close binary system, mass transfer onto the white dwarf can occur when the companion star begins to expand as a result of its evolution. Detailed numerical calculations (in lieu of experiments!) show that when a critical amount of mass has been dumped onto the white dwarf star, explosive nuclear fusion occurs and the outer layer of accumulated matter is blown off.

A supernova is superficially similar to a nova in that it suddenly increases in brightness. A supernova, however, increases in brightness by factors of many millions. Hence the name *super nova*. Careful searches through historical records of visual observations from around the world suggest that supernovae were observed in AD 185, 393, 1006, 1054, 1181, 1572 and 1604.

The 1054 AD event left a spectacular remnant, widely known today as the Crab Nebula, and an associated pulsar. The events of 1572 AD and 1604 AD were observed by the famous astronomers Tycho Brahe and Johannes Kepler respectively and occurred before the invention of the telescope. No new supernova has been observed in our Milky Way since 1604 AD and thus our knowledge is almost entirely based on observations of supernovae in external galaxies. Because most external galaxies are so distant, supernovae in them appear quite faint to us, in spite of their intrinsic brightness, which at maximum light can rival that of billions of stars. Astronomers have long hoped for a bright, nearby supernova that would permit more detailed observations.

Additional information on novas and supernovas can be found in article on **Nova and Supernova**.

Supernova 1987A

On the night of 23/24 February 1987, Ian Shelton, working at the University of Toronto Las Campanas Station in northern Chile, discovered the brightest supernova seen since 1604. See Fig. 1. Not surprisingly, several other people, in Australia, Chile and New Zealand, saw it that same night, when it had a visual magnitude of about five. The official designation is Supernova 1987A (the letter A indicating that it is the first supernova discovered in 1987) in the Large Magellanic Cloud, because it appears to belong to this nearby companion galaxy to our own Milky Way. The Large and Small Magellanic Clouds are the nearest external galaxies and are named after the famous Portuguese navigator Ferdinand Magellen who traveled around the world in the early 1500s. The Magellanic Clouds are visible only from the Earth's Southern Hemisphere and appear to the unaided eyes as illuminated clouds. "Nearby" is a relative concept, of course. The Large Magellanic Cloud is approximately 160,000 light years away, which means that what Shelton saw that night actually happened some 160,000 years ago.

Within hours, astronomers all over the world were notified of Shelton's discovery. Astronomy is not an experimental science, and so astronomers must make the most of opportunities offered by unscheduled "experiments" which occur in the universe around us. For this purpose, the International Astronomical Union operates a Central Bureau for Astronomical Telegrams which sends out telegrams and circulars (which are distributed via postal and electronic mail services). All the suitably located telescopes were thus able to join quickly in gathering data about this serendipitous discovery, thus ensuring that Shelton's supernova already had become the best observed supernova ever. See accompanying illustration.

EDITOR'S NOTES: At the end of 1987, Supernova 1987A continued to attract scientific attention. By December 1987, the object had faded at a steady 0.01 magnitude rate per day and was by then observable only instrumentally. It is regarded as the closest (160,000 light-years from Earth) and the most studied supernova in nearly 400 years (and, in fact, in history because of the unavailability of adequate instrumentation centuries ago). For quite some time after the event, observations, including the detection of neutrinos and the long-term falloff in luminosity aided in authenticating some theoretical predictions. During the investigation, a few intriguing problems surfaced, one of these being the blue color of the progenitor star, which was dimmer by a factor of 10 than what had been predicted. As one scientist observed, "We started out looking at the most dramatic event in the universe, and now we're arguing about what seemed like the most pedestrian thing—hydrogen burning and stellar evolution." Indeed, the event triggered a detailed reexamination of the former knowledge of stars.

Traditional theory provides the following approximate scenario for Sandu-leak—the star, losing sufficient thermonuclear fusion to prevent collapse, could no longer support its own weight and lost to "gravity." Hydrogen was converted to helium; the temperature and pressure in the core of the star progressively increased. The most tightly bound elements, oxygen, silicon, and ultimately iron, were involved, in a situation about to turn critical. The steadily growing core passed a threshold of about 1.5 solar masses; a multibillion-degree iron plasma underwent a phase transition. Thence, the iron nuclei commenced to "boil," disintegrating into helium nuclei, thus eliminating the supporting core. The core gave away and a supernova event occurred.

The gravitational potential energy of the collapsing core radiated neutrinos. The gravitational potential energy given up by the collapse is estimated at 3×10^{53} ergs (equivalent to about that required to convert about 10% of the mass of the sun into energy). Astronomers detected a neutrino pulse from the object at the large proton-decay detector near Cleveland, Ohio, as well as by a similar detector in Japan. (A prior burst of neutrino events observed at the Mount Blanc proton decay detector in Europe about 4 hours earlier is now believed to be spurious.)

Stanford Woosley (University of California at Santa Cruz) has observed that the emergence of cobalt-56 is yet another confirmation of standard theory. The resemblance to radioactive decay is no accident. Since June 1987, the supernova has been shining by the light of cobalt-56.

In February 1989, astronomers (University of California, Berkeley) suggested that they caught a glimpse of the ultra dense object at its heart, namely, a pulsar spinning furiously, almost 2000 times per second—so fast that it actually had broken apart. See Waldrop reference. However, also as reported by Waldrop, in late March 1989, the pulsar no longer was in view. As pointed out by one investigator, "Extensive computer analysis of the data showed that, on the night of 18 January (1989), the supernova was flickering ever so faintly at 1968.629 times per second, or more than twice as fast as any other pulsar ever seen. The rate gently rose and fell as though the pulsar were being tugged back and forth by the gravity of some kind of companion object." The foregoing frenzy of observations was disproved by a team at the Cerro Tololo Inter-American Observatory in Chile when it was announced that "the pulsar-like signals actually came from a television camera used to transmit images from the telescope to a monitor."

Supernova 1993J

An amateur astronomer, Francisco Garcia (Lugo, Spain), first reported this supernova on the night of March 28, 1993. On the following night, astronomers at the University of California, Berkeley, confirmed that Garcia had discovered a new supernova and, in fact, the brightest to shine in the Northern Hemisphere since 1937. Although supernova 1987A was far brighter, it could be seen only from the Southern Hemisphere. Professional astronomers consider 1993J to be in the same class as 1987A. The suspect object is believed to be a red supergiant, a larger and cooler star than the blue supergiant that exploded in 1987A. The object now is the target of numerous astronomical instruments.

Additional Reading

Danziger, J.: "Supernovae Remnants," Kluwer Academic Publishers, Norwell, MA, 1983.

Horgan, J.: "Supernova 1987 Confirms and Contradicts Theories," *Sci. Amer.*, 26 (March 1988).

Horgan, J.: "Astronomers May Have Detected Supernova 1987A's Core," *Sci. Amer.*, 22D (April 1989).

Horgan, J.: "A Remarkable 'Pulsar' Was Just a Flash in the Pan," *Sci. Amer.*, 30 (May 1990).

Imshennik, V.S., D.K. Nadezhin, and R.A. Syunyaev: "Astrophysics and Space Physics Reviews: Supernova 1987a," Vol. 8, Gordon & Breach Science Publishers, Newark, NJ, 1989.

Jayawardhana, R.: "A New Supernova in the Northern Sky," *Science*, 163 (April 9, 1993).

Mann, A.K.: "Shadow of a Star: The Neutrino Story of Supernova 1987a," W.H. Freeman Company, New York, NY, 1997.

Marschall, L.A.: "The Supernova Story," Princeton University Press, Princeton, NJ, 1994.

McCray, R. and Z. Wang: "Supernovae and Supernova Remnants: Proceedings International Astronomical Union Colloquium 145, Held in Xian, China, May 24–29, 1993," Cambridge University Press, New York, NY, 1994.

News: "Sighting of a Supernova," *Science*, **235**, 1143 (1987).

News: "The Supernova 1987A Shows a Mind of Its Own—and a Burst of Neutrinos," *Science*, **235**, 1322–1323 (1987).

News: "Supernova Neutrinos," *Sci. Amer.*, 18–19 (June 1987).

Pasachoff, J.M.: "A Field Guide to the Stars and Planets," 4th Edition, Houghton Mifflin Company, New York, NY, 1999.

Thompson, G.D. and J.T. Bryan, Jr.: "Supernova Search Charts and Handbook," Cambridge University Press, New York, NY, 1990.

Waldrop, M.M.: "Supernova 1987A: Facts and Fancies," *Science*, 460 (January 29, 1988).

Waldrop, M.M.: "The Supernova 1987A Pulsar: Found?" *Science*, 892 (February 17, 1989).

Waldrop, M.M.: "Pulsar, Pulsar, Where Art Thou, Pulsar?" *Science*, 1553 (March 24, 1989).

PETER PESCH, Chairman, Astronomy Department, Case Western Reserve University, Cleveland, OH.

SUPERPOSITION (Law of).

The fundamental law in stratigraphy and historical geology stating that underlying strata are older than overlying strata unless the formations have been inverted by folding or by low-angle thrusts.

SUPERPOSITION (Nernst Principle of).

The potential difference between junctions in similar pairs of solutions which have the same ratio of concentrations are the same even if the absolute concentrations are different, e.g., the same potential difference exists between normal solutions of HCl and KCl as exists between tenth-normal solutions of HCl and KCl.

SUPERPOSITION (Principle of).

If a physical system is acted on by a number of independent influences, the resultant influence is the sum (vector or algebraic as circumstances dictate) of the individual influences. The principle takes on many specific forms depending on the nature of the system and the influence in question. For example, when two forces act simultaneously on a particle the resultant force is the vector sum of the two. Another example is provided by the small oscillations of a system about a state of equilibrium. Thus the total displacement of a vibrating string is the algebraic sum of all its various harmonic modes of oscillation which add without interfering with each other. The principle is validated in this case by the fact that the wave equation governing the oscillations is linear. Superposition does not apply to nonlinear systems.

The principle can also be applied to quantum mechanics. Here it is exemplified by the postulate that any state function of a given quantum mechanical system corresponding to a given observable (e.g., the energy) can be expressed as a linear expansion of the eigenstates of the system for the same observable.

SUPERSATURATED VAPOR.

A vapor that remains dry, although its heat content is less than that of dry and saturated vapor at the pressure. Supersaturation is an unstable condition, and is found in the steam emerging from the nozzles of a steam turbine. The abnormality of the phenomenon is similar to that of supercooling. Supersaturation of the steam probably results from the very rapid expansion of steam in the nozzle, permitting the traverse of a short distance before the condensation of moisture is completed. At a certain definite point, however, known as the Williams limit, the supersaturation vanishes, and the steam regains the wet state which would be normal in view of the pressure and the heat content. Supersaturation of vapor is impossible in the presence of numerous charged ions or dust particles.

SUPERSATURATION (Chemical).

The condition existing in a solution when it contains more solute than is needed to cause saturation. Thermodynamically, this type of supersaturation is closely allied to supersaturation of a vapor, since the solute cannot crystallize out in solutions free from impurities or seed crystals of the solute. See also **Supersaturated Vapor**.

SUPERSONIC AERODYNAMICS.

In the entry on **Aerodynamics and Aerostatics**, the topic is generally covered from the standpoint of subsonic flows. World War II fighters were the first aircraft to attain speeds which produced, on critical points of the airplane, local flow velocities of Mach 1.0 and higher, although the airplane speed was less than the speed of sound. Only in the last couple of decades has aerodynamic experience passed beyond a limited theoretical knowledge. But since the early 1950s, a considerable bank of knowledge on supersonic flight has been collected.

For flight at low speeds, below about 300 miles per hour, air acts as an incompressible fluid. However, as velocity increases, air density changes about the airplane, and this effect becomes increasingly important as speeds are increased. When flow velocities reach sonic speeds at some point on an airplane, the airplane's drag begins to increase at a rate much greater than that indicated by subsonic aerodynamic theory. Subsonic flow principles are invalid at all speeds above this point.

Certain new definitions and concepts are necessary in dealing with air as a compressible fluid and with supersonic speeds:

Mach number is the ratio of the speed of motion to the speed of sound in air. The term comes from an Austrian physicist, Ernst Mach. An airplane traveling at the speed of sound is traveling at Mach 1.0.

A *shock wave* (compression wave) occurs in supersonic flow when the air must enter a volume smaller than the volume it has been occupying.

An *expansion wave* occurs in supersonic flow when the air is entering a larger volume than it has been occupying.

Supersonic Flow Characteristics. When an airplane flies at subsonic speeds, the air ahead is "warned" of the airplane's approach by a pressure change transmitted ahead of the airplane at the speed of sound. Because of this warning, the air begins to move aside before the airplane arrives and is prepared to let it pass easily. If the airplane travels at supersonic speeds, the air ahead receives no advance warning of the airplane's approach because the airplane is outspeeding its own pressure waves. Sound pressure changes are felt only within a cone-shaped region behind the nose of the airplane. Since the air is unprepared for the airplane's arrival, it must move to one side abruptly to let the airplane pass. This sudden displacement of the air is accomplished through a "shock wave." See Fig. 1.

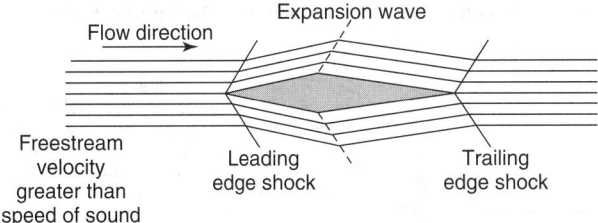

Fig. 1.   Typical supersonic flow pattern.

The water-wave analogy furnishes a good physical picture of the subsonic "warning" system and supersonic "shock" formation. If one drops pebbles into a smooth pond of water, one each second, from the same point, each pebble will produce a water-wave moving outward with constantly increasing radius, as shown in Fig. 2. This is similar to the pattern of sound waves produced by an airplane sitting on the runway before take-off. Even though one cannot see the airplane, its presence is signaled by these outward rolling waves of engine noise.

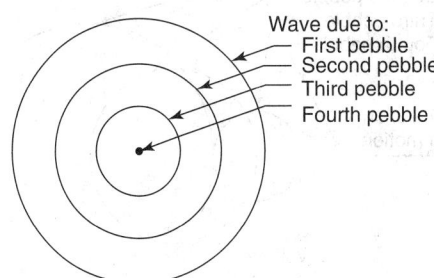

Fig. 2.   Supersonic aerodynamics.

Now suppose we move slowly over the pond dropping pebbles at regular intervals. The picture of the waves is changed to that shown in Fig. 3. Each pebble still produces a circular wave, but the circles are crowded together on the side toward which we are moving; the center of each succeeding circle is displaced from the preceding one by a distance proportional to the speed at which we are traveling over the water. The wave pattern is identical to the pattern of sound waves around an airplane flying at subsonic speeds. The air ahead of the airplane is warned of the imminent arrival of the airplane and the warning time decreases with increasing airplane speed. The warning time is zero when the airplane is flying at exactly

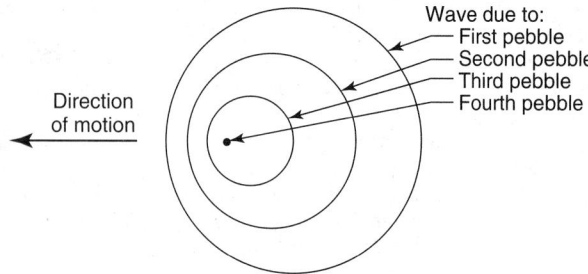

Fig. 3.   Waves for motion at subsonic speed.

sonic speed. The corresponding water wave pattern is shown in Fig. 4. If we move across the water more rapidly than the water-wave speed, the wave pattern is markedly different from the previously illustrated patterns. The smaller circles are no longer completely inside the next larger ones. Now all the circles are included within a wedge-shaped region as shown in Fig. 5. This is similar to the sound wave pattern for an airplane flying at supersonic speed. The airplane is, in fact, a continuous disturbance in the air rather than an intermittent one, such as the pebbles falling regularly into the pond. Therefore, instead of several circles, there is an envelope surface of countless circles. The wedge on the surface of the pond looks like a section through the cone formed by an airplane in the air. Figure 5 indicates that there is considerable overrunning and interference between the wave circles; we might suspect from this that such interference will change the envelope shape for an actual airplane. This is, in fact, the case.

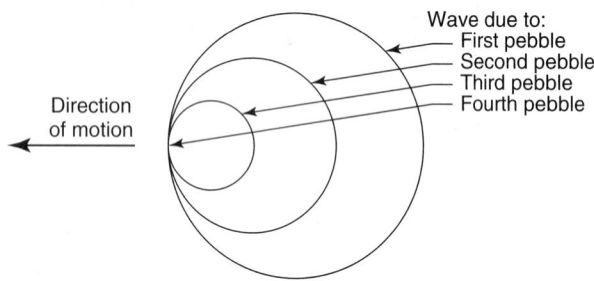

Fig. 4.   Waves simulating Mach 1.0 motion.

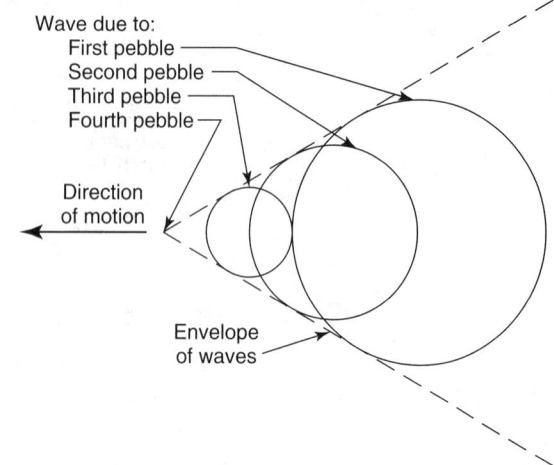

Fig. 5.   Waves simulating supersonic motion.

If the airplane is very streamlined and has a long, sharply pointed nose, then the air is not required to move a great distance suddenly in allowing the airplane to pass. In this case, the interference between sound waves is slight; the envelope is defined by the Mach angle, μ. Figure 6 shows that the Mach angle is the angle whose sine is the speed of sound divided by body velocity, or $1/V$. Thus, the Mach angle is $90°$ at a Mach number of

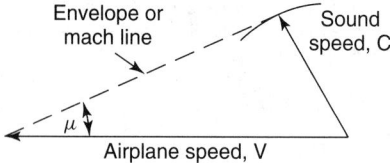

Fig. 6.   Mach angle.

1.0; $30°$ at a Mach number of 2.0; and $10°$ at a Mach number of 5.75, for example. This envelope is called a Mach line in two dimensions or a Mach cone in three dimensions.

Types of Supersonic Waves. It is now apparent that waves are formed about any disturbance in a supersonic stream of air. The type of wave formed depends upon the nature of the disturbing influence. In our case, an airplane is the disturbing influence; its shape determines the location and characteristics of the waves formed. A wave of some type will exist whenever the air is required to change direction. The wave caused by a slight disturbance was defined as a Mach line. Air passing through a Mach line undergoes an infinitesimally small amount of temperature increase, pressure increase, and decrease in velocity. The Mach line envelope due to a long, slender, sharply-pointed body is conical in shape and is called a Mach cone. See Fig. 7. The envelope for a very thin wing is a wedge at center span bounded by a Mach cone at each tip, as shown in Fig. 8. The apex angle of these wedges and cones is the Mach angle, μ.

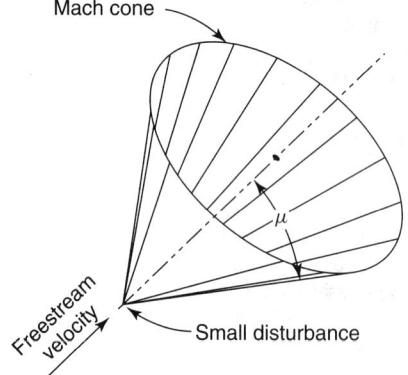

Fig. 7.   Conical shock wave.

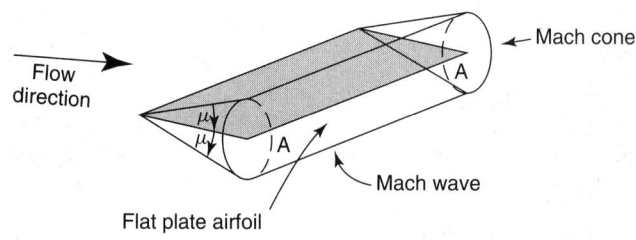

Fig. 8.   Shock wave and Mach cones.

Mach waves are associated only with very small or gradual changes in the flow direction of the passing air. The bodies small enough to produce Mach waves are too slender to be incorporated on an actual airplane. To be practical, an airplane must be too blunt and thick for Mach waves to form; instead, a blunt body causes shock waves, called shocks.

These shocks are formed by the interference of sound waves mentioned earlier in the water wave discussion. Shocks are like Mach lines in that the pressure and temperature of the air passing through are abruptly increased and the air velocity is decreased. However, the magnitude of these changes through a shock is many times greater than the magnitude of these changes through a Mach line.

The difference in the magnitude of these changes is the essential difference between a Mach line and a shock wave. Since the drag of an object is dependent upon the pressure on its surface, the drag caused by a shock is very high compared to that caused by a Mach wave on the same body. Fundamentally, a Mach wave may be thought of as a shock of negligible strength, a shock through which air undergoes the smallest pressure, temperature, and velocity changes. The magnitude of the changes in these properties is used to measure the strength of a shock. The strength of a shock is dependent upon its angle with the freestream and the freestream Mach number. Strong shocks are associated with high drag. The strongest shocks are normal shocks, so-called because they stand at right angles to the freestream. All shocks standing at an angle of less than 90° to the freestream are called oblique shocks. Figure 9 shows examples of these two general cases.

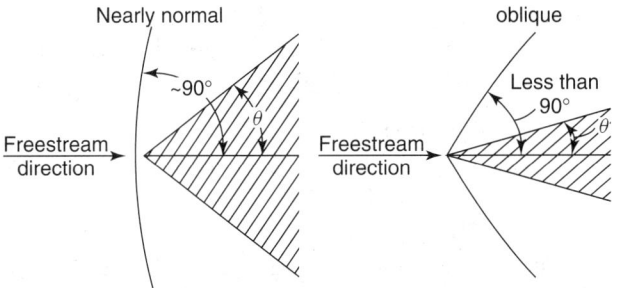

Fig. 9. Types of shock waves.

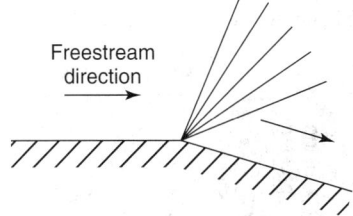

Fig. 10. Expansion wave.

The Mach line and shock wave are compression waves. There is also the expansion wave or fan, which has characteristics opposite to the compression wave. In passing through an expansion wave, air velocity increases, while temperature and pressure are reduced. Expansion waves occur where bodies begin to narrow, making more space available for the passing air to occupy. Figure 10 illustrates a typical expansion wave. Since compression and expansion waves are opposite in nature, they tend to cancel each other when they intersect, and the shock's strength is reduced accordingly. Figure 11 shows a complete wave pattern on a double-wedge airfoil. The airfoil is at zero angle of attack. Shocks are formed at the leading and trailing edges while expansion fans occur at the surface angles or discontinuities. To obtain lift, the airfoil would have to have some finite angle of attack; the resultant wave pattern is shown in Fig. 12.

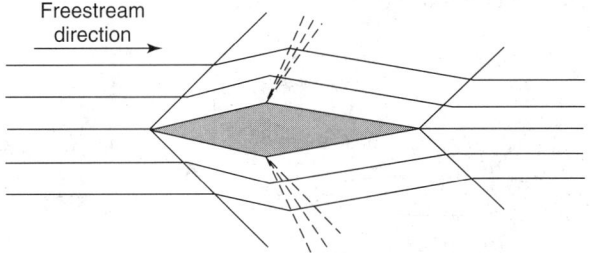

Fig. 11. Wave pattern on double-wedge airfoil where angle of attack is zero degrees.

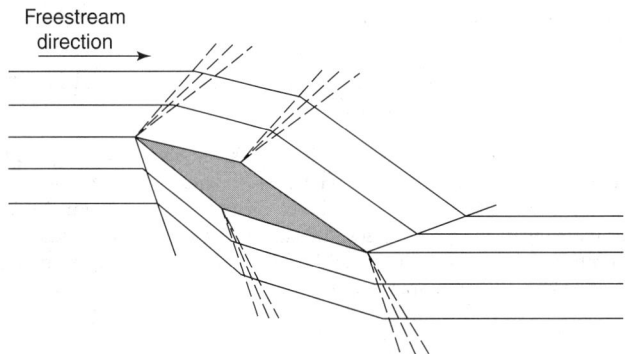

Fig. 12. Wave pattern on a double-wedge airfoil at angle of attack.

The oblique shocks at the upper surface of the nose and lower surface of the rail are replaced by expansion fans in Fig. 12. This is a graphic illustration of the effect of slope change on wave formation. In one case, the air is pushed away to allow the airfoil to pass; in the other, the air has room to expand after passing over the airfoil nose, but undergoes high compression at the lower surface nose.

Airfoil Characteristics. The double-wedge airfoil used to illustrate wave patterns is convenient to study because the flow changes in direction only at four definite points. Another typical airfoil section might be as shown in Fig. 13. Here there are leading- and trailing-edge shocks, but the expansion is continuous over the entire surface between. The expansion waves intersect the leading-edge shock and progressively weaken it, thus making it a curved shock. The local pressure coefficient along the chord is plotted as a part of Fig. 13. The pressure coefficient is proportional to the local slope. The local slope is influenced by airfoil shape and angle of attack as well as by freestream Mach number. The upper and lower surfaces produce about equal amounts of lifting pressure. This represents a departure from the subsonic case where most of the lifting pressure comes from the upper surface.

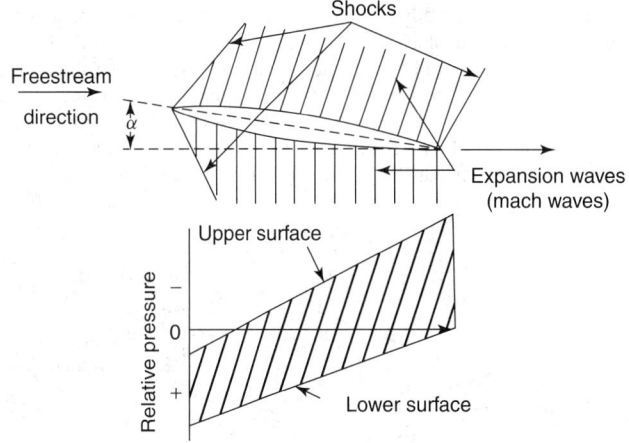

Fig. 13. Pressure distribution of an airfoil.

For supersonic airfoils, the center of pressure is very close to the half-chord point. This is a marked departure from the quarter-chord location of the center of pressure in subsonic flight.

The lift coefficient for supersonic airfoils of infinite span at moderate angle of attack and Mach number is

$$C_L = \frac{4\alpha}{\sqrt{M^2 - 1}}$$

where C_L = lift coefficient
   α = wing angle of attack, radians
   M = freestream Mach number

The expression for lift ($C_L(\rho V^2/2)S$) is still applicable, and

$$L = \left(\frac{4\alpha}{\sqrt{M^2 - 1}}\right)\left(\frac{\rho V^2}{2}S\right) = \frac{2\alpha \rho V^2 S}{\sqrt{M^2 - 1}}$$

where V = velocity, feet per second.
 S = wing area, square feet.
 $\dfrac{\rho V^2}{2}$ = dynamic pressure, pounds per square foot (ρ = air density in slugs per cubic foot).

The drag coefficient of airfoils in supersonic flight is composed of several equally important parts. These are pressure drag, friction drag, and drag due to lift. Pressure drag is the drag due to pressure distribution over the wing at the angle of zero lift and does not vary appreciably. It is also referred to as wave drag, and is similar to subsonic form drag. The approximately pressure drag coefficient is

$$C_D = \frac{5.33(t/c)}{\sqrt{M^2 - 1}}$$

where C_D = drag coefficient
 t/c = thickness ratio

The pressure drag coefficient increases with increasing wing thickness and decreases with increasing Mach number. Consequently, supersonic airfoils are thin sections.

Friction drag results from the viscosity of the air. This tendency of the air to cling to a surface and to itself is felt with supersonic flow just as with subsonic.

Drag due to lift is the component of the resultant pressure force, which acts in the drag direction. It too has an exact counterpart in subsonic flow. Drag due to lift is

$$C_D = \frac{4\alpha^2}{\sqrt{M^2 - 1}}$$

The term induced drag used for subsonic flow is not applied to supersonic drags. Induced drag in the subsonic case is due to tip losses. Tip losses occur in supersonic flow, but not in the same proportions as in subsonic flow.

Wing (Finite Span) Characteristics. It should be recalled that in subsonic flow, a wing of finite span experiences a three-dimensional flow which includes tip vortices, a downwash field, and induced velocities locally along the wing surface. See also **Aerodynamics and Aerostatics**. This is not true in supersonic flow. In Fig. 14, note that the pressure along the wing between the tip Mach cones is the same as for an airfoil of infinite length. Vortices produced within the tip Mach cones reduce the pressure from the airfoil value to zero at the tip, with the average lifting pressure in the tip region one-half the airfoil value. With a low aspect ratio, the absence of any airfoil circulation, as in subsonic flow, results in a marked decrease in wing average lifting pressure, and lift coefficient.

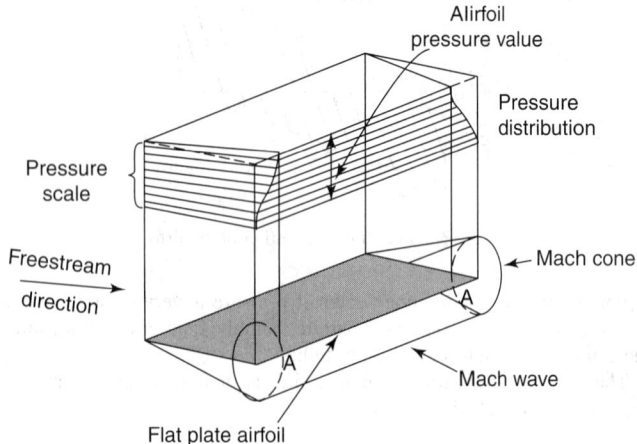

Fig. 14.   Supersonic pressure distribution on a rectangular wing.

The supersonic drag due to lift depends upon the airfoil and angle of attack, while the subsonic induced drag is a function of lift coefficient and aspect ratio.

If a wing with a planform other than rectangular is used, tip losses can be eliminated. The delta, or triangular, wing planform accomplishes this and can be illustrated by the two possible pressure patterns over a delta

wing, depending on the relationship between freestream Mach number and wing leading-edge sweep.

In this case (Fig. 15), the components of velocity perpendicular to the leading edge are subsonic, even though the freestream flow is supersonic. The lifting pressure is maximum along the leading edge and decreases rapidly toward the center of the wing. The average lift coefficient is less than would be obtained by a similar airfoil in subsonic flow.

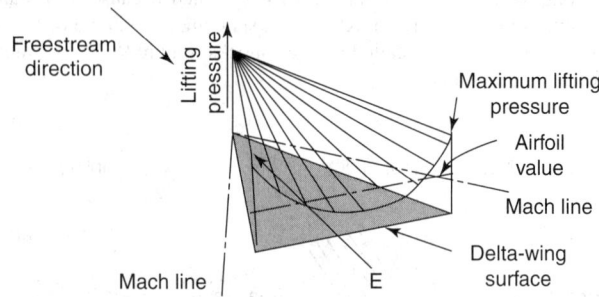

Fig. 15.   Pressure distribution on a delta wing (leading edge inside the Mach line).

In the other pattern, the wing leading edge lies ahead of the tip Mach cone. Figure 16 illustrates that the highest lifting pressure still exists along the wing leading edge. In this case, however, the lifting pressure remains constant at this peak value in the region between the leading edge and the Mach cone. Inside the Mach cone, the lifting pressure again decreases, but the wing's average lift coefficient is as high as can be obtained with an airfoil of a similar cross section. When the leading edge is outside, the pressure drag coefficient is lower than airfoil pressure drag. Wing pressure drag reaches a maximum when the Mach cone lies along the leading edge.

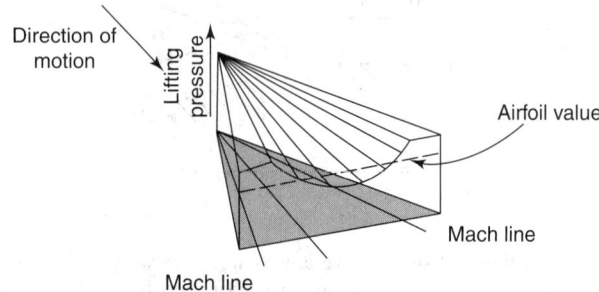

Fig. 16.   Pressure distribution on a delta wing (leading edge outside the Mach line).

Consider the effects of modifying the delta planform. If area is added at the trailing edge to make a diamond planform, it is being added where the local lift coefficient is low. In this case, the average wing lift coefficient is less than that obtainable with a delta planform. Conversely, cutting out area to given an arrow planform will increase the average lift coefficient.

Body Characteristics. This discussion pertains only to bodies of revolution. A body of revolution is one whose cross section perpendicular to its longitudinal axis is always circular (Fig. 17). Airplane fuselages are generally as near circular in cross section as volume requirements will permit. Studies and tests indicate that a parabolic longitudinal shape is desirable. The pressure distribution on a body of revolution at zero angle of attack

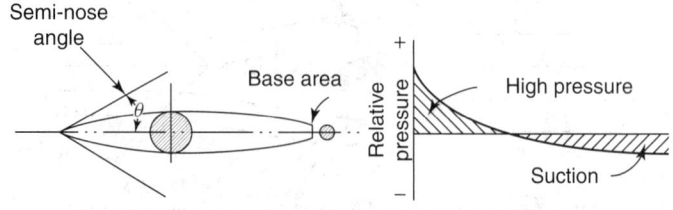

Fig. 17.   Pressure distribution on a body of revolution.

shows a positive value at the nose lower than for an airfoil of the same "seminose angle." See Fig. 17. Following this, the air finds more room in which to expand than in the case of the airfoil; it can fill the "ring" all around the body. The pressure drops so rapidly that the pressure coefficient returns to zero before the body slope has returned to zero, i.e., before the body contour becomes parallel to the longitudinal axis of the body.

The expansion continues over the aft part of the body and the pressure coefficient becomes more negative. However, the largest negative value of the pressure coefficient is limited by the occurrences of a complete expansion to a vacuum. Usually, the positive pressure coefficient at the nose is larger than this maximum negative value.

The pressure drag of bodies of revolution depends upon their shape, angle of attack, and flight Mach number. Angles of attack other than zero usually do not cause a large drag increase, nor will they cause the body to produce much lift. Drag force increases with increasing Mach number just as does wing lift force. Body shape has a strong influence on body drag. The longitudinal lines should be parabolic; the exact equation of the lines is a function of Mach number. A good fineness ratio for low drag is a length to maximum diameter ratio of 8 to 12. Many bodies are designed so that the pressure drag coefficient reaches its peak at very low supersonic Mach numbers, then drops rapidly until it begins to approach a minimum value at about Mach 2. The center of pressure of such bodies will be close to the half-length point if the tail is pointed. As base area increases, the center of pressure moves forward.

Wing-Body Combinations. In consideration of subsonic drag, it is apparent that a factor is interference drag, resulting from the effect of angles between surfaces, such as wing and fuselage. This drag is reduced by the use of smooth fairings, thus avoiding sharp corners — in effect it is corrected "locally" at the point it occurs.

In supersonic airplanes, the interference problem is much more critical and cannot be solved locally. It has been stated that the ideal streamline shape is a body of revolution having a longitudinal parabolic curve, that is, if the cross-sectional areas of the ideal body, taken at even increments along its axis, were plotted, the result would be a parabolic curve. Studies in the supersonic area indicate that in supersonic aircraft, the parabolic cross-sectional area distribution from nose to tail must be based on the complete airplane cross section, not just the fuselage cross section. The so-called "Coke-bottle" shape of the fuselage of some supersonic airplanes is the result of the application of this principle.

Supersonic Stability and Control. The analysis of the stability and control characteristics of an airplane capable of flight at supersonic speeds is essentially the same as that used for subsonic aircraft. A supersonic airplane must be statically and dynamically stable; it must have a control system that gives the pilot accurate, safe control of the airplane throughout its flight regime. However, the magnitude of control forces and moments and the changes in these forces and moments resulting from a displacement of the airplane about one or more of the stability axes are much greater at supersonic speeds than at subsonic speeds. Consequently, supersonic airplanes have power control systems to enable the pilot to move the control surfaces. The power control systems used on these aircraft can be designed so that the pilot has the same stick feel for a maneuver of given severity at all speeds despite the variation of control force with speed changes. In general, these airplanes also need 3-axis stability-augmentation systems to provide satisfactory dynamic stability.

Aerodynamic Heating. This results from the conversion of kinetic energy to thermal energy. At the surface of an aircraft, the air is slowed to zero velocity. This means that freestream kinetic energy has been converted to thermal energy. The resultant stagnation temperature is a function of freestream temperature and Mach number. Most of the aircraft surface does not reach this stagnation temperature because the conversion of kinetic energy to thermal energy is not 100% efficient. The maximum resultant skin temperature, known as adiabatic wall temperature, is 85 to 95% as high as the stagnation temperature. Radiation reduces skin temperature to less than adiabatic wall temperature; this effect increases as altitude and Mach number increase. Adiabatic wall temperature minus heat loss to radiation gives equilibrium temperature; this is the temperature to which an aircraft is designed. Typical stagnation temperatures are about $260\,°F$ ($127\,°C$) at Mach 2.0; and $1,550\,°F$ ($843\,°C$) at Mach 5. These temperatures serve to explain the structural and airconditioning problems involved in the design of supersonic airplanes.

Noise and Supersonic "Boom." The high-thrust jet engines used to power modern aircraft produce a sound level that tends to exceed human tolerance. Users of these aircraft equip their ground crews and mechanics with ear plugs to prevent physical and mental damage due to this noise. In addition to their effect on people, these sounds, or pressure waves, must be considered in the design of any part of the aircraft on which they impinge.

Supersonic flight is sufficiently commonplace to the public, but an associated phenomenon, known as the sonic boom or supersonic bang, has not met with the same public acceptance. A boom will occur whenever a shock wave (pressure wave) emanating from an aircraft flying supersonically reaches an observer. According to the water-wave analogy given earlier, a supersonic boom is similar to a water wave moving past a floating leaf and causing the leaf to bob up and down.

There are many shocks emanating from an aircraft flying supersonically, but these usually interact and coalesce to form two main shocks — one from the nose and one from the aft end of the aircraft. For this reason, the shape of an aircraft has little or no influence on the number of shocks reaching an observer. The bow and tail shock waves gradually diverge as they extend outward from the aircraft. This divergence is due to a slight difference in propagation velocities of the two waves. The observer will hear two bangs or booms if the time interval between the passing of the two waves is of the order of 0.10 second or greater. If this time interval is much less than this figure, as would be experienced during a low-level pass of a small aircraft, the ear could not discriminate between the two shock waves and would hear only one.

The loudness of the boom is a function of the distance between airplane flight path and observer, Mach number, aircraft size and shape, atmospheric pressure, temperature, and winds. The factor having the strongest influence on loudness is the distance between observer and airplane flight path. The loudness, as an increase in pressure above atmospheric, is inversely proportional to the three-quarter power of this distance.

Atmospheric temperature and wind gradients cause a bending or refraction of shock waves. Other atmospheric conditions, such as clouds, cause a diffusion of the waves. Hence in some cases the shockwave pattern from the airplane may become so distorted and attenuated that the bang is not even heard on the ground. Prediction of these effects under actual flight conditions is extremely difficult, though by precise control of airplane speed and flight path, shock waves may be focused at a point in space or on the ground. Such conditions can produce a supersonic boom on the order of ten times the normal intensity.

Several means of avoiding sonic boom from all causes have been tested. One, based on extremely accurate prediction of temperature and wind at the time of take-off, would adjust the angle of climb for subsonic speed. A two-degree error in temperature prediction, however, could negate a good plan and possibly the resultant noise would be worse than were no effort made to control it. Changing the climb angle, of course, also could adversely affect engine performance and fuel consumption. Another method advocates deceleration to subsonic speed before descent from cruise level, solving only one phase of the problem. Another plan embraces routing supersonic aircraft over unpopulated areas. Flying at greater altitudes is an alternative. However, the result of higher altitudes might decrease the intensity of the shock wave at the ground, but it would also spread the area affected. As of the early 1980s, the best approach appears to be a combination of very careful operational control at low altitudes and cruise altitudes sufficiently high to prevent or reduce the intensity effect of shock waves reaching the ground.

See also **Aerodynamics and Aerostatics**; and **Airplane**.

Additional Reading

Champion, M.: "IUTAM Symposium on Combustion in Supersonic Flows: Proceedings of the IUTAM Symposium Held in Poitiers, France, 2–6 Oct. 1995," Kluwer Academic Publishers, Norwell, MA, 1996.

Courant, R., K.O. Friedrichs: "Supersonic Flow and Shock Waves," Springer-Verlag, Inc., New York, NY, 1992.

Mikohailov, G.K. and V.Z. Parton: "Super-Hypersonic Aerodynamic and Heat Transfer," CRC Press, LLC, Boca Raton, FL, 1992.

Smits, A.J. and Jan-Paul Dussauge: "Turbulent Shear Layers in Supersonic Flow," Springer-Verlag, Inc., New York, NY, 1996.

SUPERVISORY CONTROL. In control system terminology, a control action in which the control loops operate independently, subject to intermittent corrective action, e.g., setpoint changes from an external source. Generally, a supervisory control system is assumed to incorporate a computer that provides broad, possibly relatively infrequent commands to a control system. The latter continues to perform its usual control functions during those periods between commands provided by the supervisory computer. In this sense, supervisory control is superimposed over the fundamental, second-to-second control system. A supervisory control computer thus replaces or assists the human operator in making periodic adjustments to a control system in an effort to better realize process objectives. The computer in a supervisory control system decides according to the information it receives, assisted by certain programmed calculations or logic procedures, what adjustments in the control system should be made. Where the computer is completely *off-time*, the human operator will make the control adjustments manually based upon the computer inputs he receives. In other configurations, the computer may be interconnected with the process control system on a periodic or continuous basis. There are several shades of difference between a purely off-line and a purely on-line situation.

SUPPRESSOR (Electrical Noise). An element or device used in electric or electronic components or circuits to prevent or reduce undesired actions or currents. See also **Static and Noise**.

SUPPURATION. The formation and/or discharge of pus (leukocytes, serum, microorganisms, and necrotic debris) — as from a pustule (abscess, bubo, pimple, or other type of sore).

SURFACE. The locus of points satisfying an equation in three variables $f(x, y, z) = 0$. If u, v are parameters, the equation $x = \phi_1(u, v)$, $y = \phi_2(u, v)$, $z = \phi_3(u, v)$ are often useful. Figures which lie on the surface of a plane are studied in plane geometry. They include triangles, quadrilaterals, polygons, and circles. The differential properties of surfaces are a generalization of those that occur for plane and space curves but they become fairly complicated. They can be studied by means of differential calculus and differential geometry but tensor methods are especially suitable. A ruled surface is one that can be generated by the motion of a straight line called the generatrix. Any quadric surface can be produced in this way but the generatrix is real only for the cone, cylinder, hyperboloid of one sheet, and the hyperbolic paraboloid. In the other possible cases, the generatrix is imaginary. A surface of revolution results when a plane curve is revolved about a line lying in its plane. Special cases of each of the quadric surfaces may also be produced in this way.

Properly one should distinguish between a surface and a solid but often, for example, a sphere and a spherical surface are considered to be the same mathematical concept.

See also **Area**.

SURFACE (of Revolution). A surface which may be generated by rotating a plane curve about an axis in its plane. Sections of the surface perpendicular to the axis are circles called *parallel circles*, or *parallels*, or *lines of latitude*. Sections of the surface by planes containing the axis are called *meridian sections*, or *meridians*, or *lines of longitude*.

See also **Conic Section**.

SURFACE TENSION. Fluid surfaces exhibit certain features resembling the properties of a stretched elastic membrane; hence the term surface tension. Thus, one may lay a needle or a safety-razor blade upon the surface of water, and it will lie at rest in a shallow depression caused by its weight, much as if it were on a rubber air-cushion. A soap bubble, likewise, tends to contract, and actually creates a pressure inside, somewhat after the manner of a rubber balloon. The analogy is imperfect, however, since the tension in the rubber increases with the radius of the balloon, and the pressure inside, which would otherwise decrease, remains approximately constant; while the liquid "film tension" remains constant and the pressure in the bubble falls off as the bubble is blown.

Surface tension results from the tendency of a liquid surface to contract. It is given by the tension σ across a unit length of a line on the surface of the liquid. The surface tension of a liquid depends on the temperature; it diminishes as temperature increases and becomes 0 at the critical temperature. For water σ is 0.073 newtons/meter at $20\,^{\circ}$C, and for mercury, it is 0.47 newtons/meter at $18\,^{\circ}$C.

Surface tension is intimately connected with capillarity, that is, rise or depression of liquid inside a tube of small bore when the tube is dipped into the liquid. Another factor related to this phenomenon is the angle of contact. If a liquid is in contact with a solid and with air along a line, the angle θ between the solid-liquid interface and the liquid-air interface is

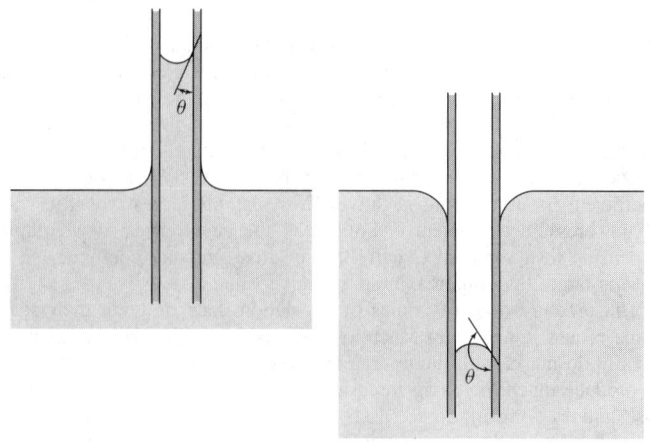

Fig. 1.   Interrelationship between surface tension and capillarity: (*Left*) Case where angle theta is less than 90° (water); (*Right*) case where angle theta is greater than 90° (mercury).

called the angle of contact. See Fig. 1. If $\theta = 0$, the liquid is said to wet the tube thoroughly. If θ is less than 90°, the liquid rises in the capillary; and if more than 90°, the liquid does not wet the solid, but is depressed in the tube. For mercury on glass, the angle of contact is 140°, so that mercury is depressed when a glass capillary is dipped into mercury. The rise h of the liquid in the capillary is given by $h = 2\sigma \cos \theta / r \rho g$, where r is the radius of the tube, ρ the density of the liquid, and g is the acceleration due to gravity.

Surface tension can be explained on the basis of molecular theory. If the surface area of liquid is expanded, some of the molecules inside the liquid rise to the surface. Because a molecule inside a mass of liquid is under the forces of the surrounding molecules, while a molecule on the surface is only partly surrounded by other molecules, work is necessary to bring molecules from the inside to the surface. This indicates that force must be applied along the surface in order to increase the area of the surface. This force appears as tension on the surface and when expressed as tension per unit length of a line lying on the surface, it is called the surface tension of the liquid.

The molecular theory of surface tension was dealt with by Laplace (1749–1827). But, as a result of the clarification of the nature of intermolecular forces by quantum mechanics and of the more recent developments in the study of molecular distribution in liquids, the nature and value of surface tension have been better understood from a molecular viewpoint. Surface tension is closely associated with a sudden, but continuous change in the density from the value for bulk liquid to the value for the gaseous state in traversing the surface. See Fig. 2. As a result of this inhomogeneity, the stress across a strip parallel to the boundary — ρ_N per unit area — is different from that across a strip perpendicular to the boundary — ρ_T per unit area. This is in contrast with the case of homogeneous fluid in which the stress across any elementary plane has the same value regardless of the direction of the plane.

The stress ρ_T is a function of the coordinate z, the z-axis being taken normal to the surface and directed from liquid to vapor. The stress ρ_N is

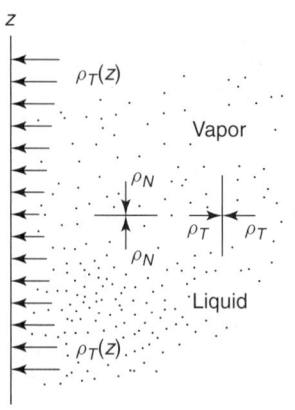

Fig. 2.   Stress relationships in surface tension.

constant throughout the liquid and the vapor. Figure 2 shows the stress ρ_N and ρ_T. The stress $\rho_T(z)$ as a function of z is also shown on the left side of the figure.

SURFACE TENSION (Gibbs Formula). The total differential of the surface tension γ in the variables temperature T, and chemical potentials μ, is

$$d\gamma = -s^a \, dT - \sum_i \Gamma_i \, d\mu_i \qquad (1)$$

where s^a is the entropy per unit area of the surface phase and Γ_i the adsorption of component i.

This formula is the analog for a surface phase, of the Gibbs-Duhem equation for a bulk phase.

Another basic formula also due to Gibbs, that relates the surface tension to the thermodynamic functions of the surface phase, is

$$\gamma = A^a - \sum_i \Gamma_i \mu_i \qquad (2)$$

where A^a is the Helmholtz free energy (see also **Free Energy (2)**) per unit area of the surface phase. This expression is the analog of the relation between the Gibbs free energy and the chemical potentials

$$G = \sum_i n_i \mu_i \qquad (3)$$

valid for a bulk phase.

SURGE. In electric terminology, a surge is an oscillation of relatively great magnitude set up by an electric discharge in a line or system. In meteorology, a surge is a general change in barometric pressure apparently superposed upon cyclonic and normal diurnal changes.

SURGEONFISHES *(Osteichthyes)*. Of the suborder *Acanthuroidea*, family *Acanthuridae*, the surgeonfishes are well named because of the presence of a sharp, knifelike structure on the sides of the caudal peduncle. This is located just ahead of the tail. The fisherman unless careful can receive a serious cut from this structure. There are some 100 species of herbivorous surgeonfishes, the majority of which do not exceed 20 inches (51 centimeters) in length. They are considered food fishes, but the *Acanthurus triostegus* (Indo-Pacific convict fish) has been implicated in tropical fish poisoning instances. The general food supply of the surgeonfish is algae gathered from rocks and coral. Few of the most commonly seen species (*Ctenochaetus strigosus* and *C. striatus*) do well in aquariums.

SURGE TANK OR VESSEL. A liquid-holding chamber used to adsorb irregularities in flow. A surge tank may be used in a process where the total amount of fluid flowing around the closed cycle is constant, but where the volume passing one point in the cycle varies from that at another.

SUTURE. 1. The line of union of the adjacent flat bones making up the skull. 2. The surgical sewing-up of a wound or incision. Suture material is generally classified either as absorbable or nonabsorbable. The absorbable type is often made of either plain or chromic catgut. Nonabsorbable material is silk, linen, fine wire, or strands of synthetic composition such as nylon.

A fascial suture is one fashioned of a strip of fascia which is usually removed from the thigh where it covers the external muscles. Such sutures are often described as living sutures as they act similarly to a graft. They are used principally in the repair of large herniae where the tissues are weak.

SWALLOW *(Aves, Passeriformes)*. An insect-eating bird with a short wide beak, long and relatively narrow wings, weak feet and legs, and usually a forked tail. The distribution of the swallows is worldwide. They constitute the well-marked family *Hirundinidae*.

The swallows' nests are built in burrows, holes in trees, and about human dwellings, hence some of the species are familiar friends. In North America the purple martin, *Progne subis*, is widely known from the large colonies that nest in bird houses year after year. Among the true swallows of this continent the barn swallow, *Hirundo erythrogaster*, is among the most beautiful and is undoubtedly the most widely known. The bank, *Riparia riparia*, cliff, *Petrochelidon albifrons*, and rough-winged, *Stelgidopteryx ruficollis*, swallows are also widely distributed and locally common, though less beautiful than the tree swallow and the western violet-green swallow,

Tachycineta thalassina. The glossy blue and green shades of the upper parts of the last two species are very striking, but both species are found in wild areas, and are thus less familiar than those mentioned above.

SWALLOW FLOAT (or Swallow Plinger). An electronic device that can be set to float in the seas at a predetermined depth, from which it emits sounds that can be detected by appropriate devices at distances up to several miles.

SWALLOWTAIL *(Insecta, Lepidoptera)*. A large butterfly, usually with slender tails extending from the hinder angles of the hind wings. The many species of swallowtails belong to the family *Papilionidae*. They are widely distributed in the tropical and temperate zones, especially in the former where some are very beautiful and brilliantly colored. Twenty-one species occur in North America. Most of them are yellow with black markings or vice versa but the common pawpaw swallowtail of the eastern and southern states is greenish-white with black bands and some red marks. Some of our species have metallic blue or green scales on the hind wings. Although some species lack the tails, they are also called swallowtails by association with the typical forms.

SWAMP. Where the flatness of the land, the presence of impervious soils or bedrock, or abnormal amounts of plant material obstruct or entirely prevent the normal drainage of an area, an excess of moisture will accumulate to the point of saturation, and a swamp will come into existence. While most swamps are level, this is not a necessary condition, for hillside swamps are by no means uncommon, due to a constant supply of percolating groundwater which maintains the swampy condition.

Lake basins are occasionally filled with vegetation and sediment, thus becoming swamps; these are frequently referred to as muskegs, a word of American Indian origin. Swamps may be formed on the flood plains of rivers as well as upon their deltas; they are characteristic of the flat ill-drained areas of the Atlantic Coastal Plain, examples of which are the Great Dismal Swamp which covers about 2,000 square miles (5,180 square kilometers) in the states of Virginia and North Carolina, and the Everglades of Florida, covering about 4,000 square miles (10,360 square kilometers).

Coastal saltwater swamps may develop in the zone between high and low tides or extend up river estuaries; examples of these are common along the Atlantic and Gulf coasts of the United States. In certain northern latitudes swamps develop into peat bogs. Peat bogs are an important source of fuel in Northern Europe, and also serve as an interesting illustration of the origin of coal, as exemplified in the peat, lignite, bituminous coal series. Figure 1 illustrates the formation of a peat bog.

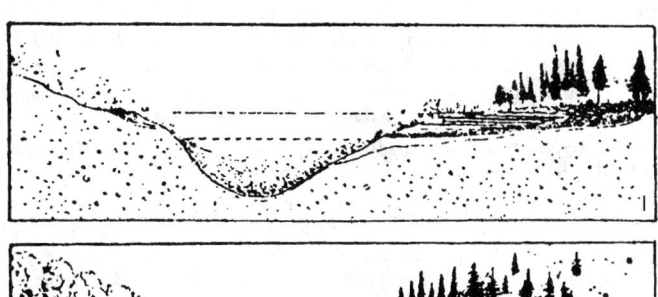

Fig. 1. Origin and development of a peat bog, illustrating the successive stages in filling of a pond by the growth of the peat bog. (*After Dachnowski.*)

SWEEPBACK (Aircraft Wing). The acute angle between a line perpendicular to the plane of symmetry and the locus spanwise of the aerodynamic centers of the airfoils comprising the wing. For airplanes operating at subsonic speeds, sweepback may be employed to obtain the proper relationship between the center of gravity of the airplane and the aerodynamic center of the wing. For designs operating at or above sonic velocities, sweepback is employed to reduce the drag of the wing and the sweepback angle is equal to or greater than the Mach angle.

Sweepback has the same effect as incorporating a positive dihedral angle in the wing design for producing lateral stability. Since the sweepback for very high-speed aircraft is high (30 to 60 degrees) the excess lateral stability due to sweepback has to be counteracted by a negative dihedral, thereby producing a drooping or "tired" look to the wings. See also **Aerodynamics and Aerostatics**; and **Airplane**.

SWEEP CIRCUIT. A circuit utilizing the transient voltage produced in a resistor-capacitor combination which furnishes the voltage or current for deflecting the electron beam in a cathode ray tube (see also **Cathode-Ray Tube**). Figure 1a shows a simple but widely used sweep circuit. The capacitor is charged through the resistance R until the plate to cathode voltage of the thyratron reaches the breakdown value, at which time it begins to conduct and permits the capacitor to discharge very rapidly.

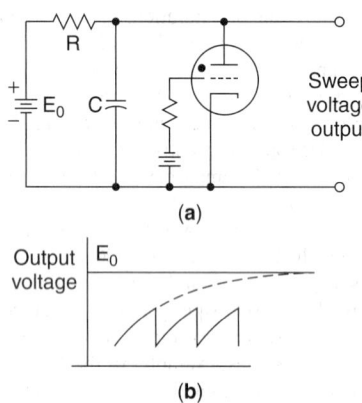

(a)

(b)

Fig. 1. (a) Sweep circuit using thyratron; (b) voltage waveform at output of a sweep circuit.

The voltage across the capacitor thus varies between the breakdown and extinction voltages of the thyratron. The transition from the lower to the higher value is along an exponential path determined by R, C, and E_0. The process is then repeated and the resultant voltage output is as shown. Although this circuit is free-running, i.e., does not require an external signal to produce the sweep voltage, the output voltage may be synchronized with another signal by the introduction of a synchronizing signal at the grid. Virtually the same properties are obtained from the circuit of Fig. 2 in which a unijunction transistor is employed. Capacitor C is charged through resistor R until the emitter voltage of the unijunction transistor reaches the peak point value, which results in the emitter junction becoming forward biased, with a corresponding sharp reduction in the dynamic resistance between the emitter and the base connected to the negative side of the battery. Capacitor C then discharges through this low impedance emitter-to-base path until the emitter voltage becomes so low that the emitter junction fails to conduct. At this time, charging of capacitor C resumes

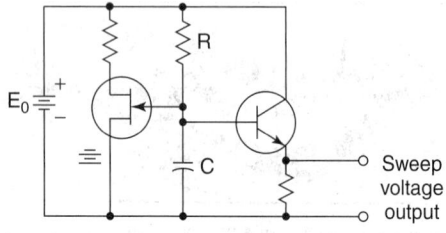

Fig. 2. Sweep circuit using unijunction transistor.

and the cycle is repeated. The voltage waveforms are the same as shown in Fig. 1a. The second transistor is an emitter follower used to minimize the effect of the circuit to which the sweep voltage is supplied on the operation of the unijunction transistor circuit.

A circuit that provides a single sweep output for each input pulse is shown in Fig. 3a; the associated voltage waveforms are given in Fig. 3b. Prior to time t_1, the triode conducts at zero bias, and the voltage across capacitor C is only a small fraction of the battery voltage because of the voltage drop in the large resistor R. At time t_1, the grid to cathode voltage is made negative enough to cut off the flow of plate current. This condition is maintained until time t_2. During the interval of time that the tube is cut off, the capacitor C is charged from the battery through the resistor R. The voltage across the capacitor increases along the exponential curve associated with the transient in a resistor-capacitor circuit. Ultimately, e_b would become equal to the battery voltage. The exponential rise is not permitted to continue after time t_2, however, for when the tube is returned to zero bias at that time, the capacitor discharges through the tube and its voltage rapidly assumes the value which it had at the start of the charging process. Many refinements, including feedback, may be added to the circuit to increase the degree of linearity of the voltage change between t_1 and t_2.

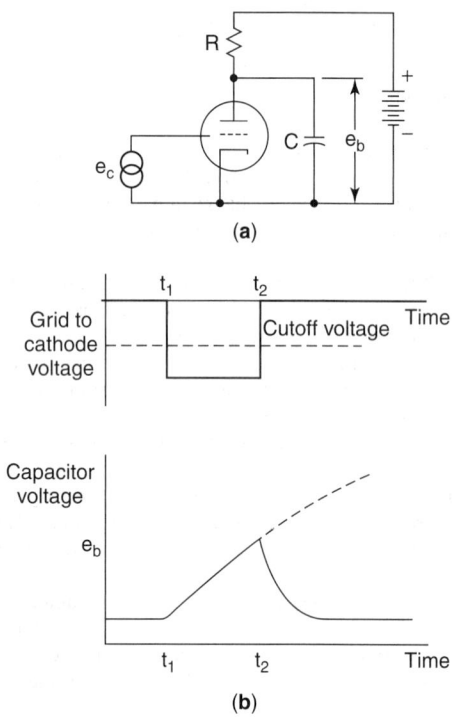

(a)

(b)

Fig. 3. (a) Typical sweep generator circuit; (b) input and output voltage waveforms for sweep circuits.

SWEETENERS. Drawings in Egyptian tombs depicting beekeeping practices and honey production attest that the demand for sweet-tasting substances dates back to 2600 B.C. Sugar consumption varies considerably from one country to the next as shown in Table 1.

In terms of sugar consumption in the United States, until the early 1940s, sucrose from sugarcane and sugar beets accounted for a very high volume of the fundamental sweeteners. Since that time, there has been continuously increasing consumption of corn sweeteners and other caloric sweeteners, notably high-fructose corn syrup (HFCS). Of course, a marked impact on sucrose consumption occurred with the introduction of artificial sweeteners, particularly of saccharin and aspartame. Sweeteners fall into two general categories — *nutritive* and *nonnutritive*.

Nutritive Sweeteners

In addition to their sweetening power, nutritive sweeteners are effective preservatives in numerous foods. Sweeteners tie up water, essential for microorganism growth, thus preventing or inhibiting spoilage. Nutritive sweeteners also serve as food for yeasts and other fermenting agents, so important in many processes, including baking. The principal functional properties of sucrose are (1) browning reactions,

TABLE 1. SUGAR CONSUMPTION PER CAPITA PER YEAR

Country	Refined Sugar	
	Pounds	Kilograms
Israel	150	68.0
Bulgaria	130	59.0
Australia	119	54.0
New Zealand	110	49.9
Costa Rica	108	49.0
Cuba	107	48.5
Switzerland	106	48.1
United States	102	46.3
Hungary	99	44.9
Iceland	98	44.9
Poland	95	43.1
Sweden	94	43.1
Austria	92	41.7
Czechoslovakia	92	41.7
European Economic Community	89	40.4
Norway	87	39.5

Source: International Sugar Organization.

(2) fermentability, (3) flavor enhancement, (4) freezing-point depression, (5) nutritive solids source, (6) osmotic pressure, (7) sweetness, (8) texture tenderizer, and (9) viscosity/bodying agent.

Among the principal *natural sugars* are fructose, glucose (also called dextrose), honey, invert sugar, lactose, maltose, raffinose and stachyose, sucrose, sugar alcohols, and xylitol.

Dextrose Equivalent. A means for comparing one sugar with another. The total amount of reducing sugars, expressed as dextrose (glucose), that is present in a given sugar syrup is calculated as a percentage of the total dry substance. More technically, the dextrose equivalent (DE) is the number of reducing ends of sugar that will react with copper. The DE can be measured in several ways.

Fructose. Also called *levulose* or *fruit sugar*, $C_6H_{12}O_6$. It is the sweetest of the common sugars, being from 1.1 to 2.0 times as sweet as sucrose. Fructose is generally found in fruits and honey. An apple is 4% sucrose, 6% fructose, and 1% glucose (by weight). A grape (*Vitis labrusca*) is about 2% sucrose, 8% fructose, 7% glucose, and 2% maltose (by weight) (Shallenberger). See also **Apple**; and **Grapes and Wines**. Commercially processed fructose is available as white crystals, soluble in water, alcohol, and ether, with a melting point between 103 and 105 °C (217.4 and 221 °F) (decomposition). Fructose can be derived by the hydrolysis of inulin; by the hydrolysis of beet sugar followed by lime separation; and from cornstarch by enzymic or microbial action.

Dry crystalline fructose is reported to have a sweetness level of 180 on a scale in which sucrose is represented at 100 (Andres, 1977). In cool, weak solutions and at lower pH, sweetness value is reported to be 140–150. At neutral pH or higher temperatures, the sweetness level drops, and at 50 °C (122 °F) sweetness equals that of a corresponding sucrose solution. A synergistic sweetness effect is reported between sucrose and fructose. A 40–60% fructose/sucrose mixture in a 100% water solution is sweeter than either component under comparable conditions (Unpublished report, University of Helsinki, 1972).

Glucose. Also known as *grape sugar* or *dextrose*, this is the main compound into which other sugars and carbohydrates are converted in the human body and thus is the major sugar found in blood. Glucose is naturally present in many fruits and is the basic "repeating" unit of the starches found in many vegetables, such as potato. Purified glucose takes the form of colorless crystals or white granular powder, odorless, with a sweet taste. Soluble in water, slightly soluble in alcohol. Melting point is 146 °C (294.8 °F). Glucose finds many uses—confectionery, infant foods, brewing and winemaking, caramel coloring, baking, and canning. Glucose is derived from the hydrolysis of corn starch with acids or enzymes. Glucose is a component of invert sugar and glucose syrup. Glucose was first obtained (1974) from cellulose by enzyme hydrolysis.

Corn (maize) syrup is a sweetener derived from corn starch by a process that was first commercialized in the 1920s. Corn syrup is composed of glucose and a variety of sugars described as the "maltose series of oligosaccharides." These syrups are not as sweet as sucrose, but are very often used in conjunction with sugar in confections and other food products.

Five types of corn sweeteners are commercially available: (1) *Corn syrup* (glucose syrup), with a DE of 20 or more, is a purified and concentrated aqueous solution of mono-, di-, and oligosaccharides. High fructose corn syrup (HFCS) is prepared by enzymatically converting glucose to fructose with glucose isomerase. (2) *Maltodextrin*, concentrated solutions or dried powders of disaccharides, characterized with a DE of less than 20. The manufacturing process is similar to that of corn syrup except that the conversion process is stopped at an earlier stage. (3) *Dried corn syrup* is a granular, crystalline, or powder product, from which a portion of the water has been removed. (4) *Dextrose monohydrate* is a purified and crystallized form of D-glucose, and contains one molecule of water of crystallization per molecule of D-glucose. (5) *Dextrose anhydrous* is primarily D-glucose with no water of crystallization.

Galactose. A monosaccharide commonly occurring in milk sugar or lactose. Formula, $C_6H_{12}O_6$.

Honey. A natural syrup which varies in composition and flavor, depending upon the plant source from which the nectar was collected by the honeybee, the amount of processing, and the duration of storage. The principal sugars contained in honey are fructose and glucose, the same components as in table sugar. There are minute amounts of vitamins and minerals in honey, but these are not usually considered in terms of calculating minimum requirements. See also the entry on **Honey**.

Invert Sugar. A mixture of 50% glucose and 50% fructose obtained by the hydrolysis of sucrose. Invert sugar absorbs water readily, and is usually only handled as a syrup. Because of its fructose content, invert sugar is levorotatory in solution, and sweeter than sucrose. Invert sugar is often incorporated in products where loss of water must be minimized. Commercially, invert sugar is obtained from the inversion of a 96% cane sugar solution. This sugar is used in various foods, in the brewing industry, confectionery field, and in tobacco curing.

Lactose. *Milk sugar* or *saccharum lactis*, $C_{12}H_{22}O_{11} \cdot H_2O$. Purified lactose is a white, hard, crystalline mass or white powder with a sweet taste, odorless. It is stable in air, soluble in water, insoluble in ether and chloroform, very slightly soluble in alcohol. The compound decomposes at 203.5 °C (398.3 °F). Lactose is derived from whey, by concentration and crystallization. Cow's milk contains about 5% lactose. Because of its relative lack of sweetening power, lactose is not considered a sweetener in the usual sense. It is used as a bulking agent in numerous food products. Lactose can be used effectively as a carrier for artificial sweeteners to give a free-flowing powder that is easily handled. There has been interest in the hydrolysis of lactose into glucose and galactose, both enzymatically and chemically. It has been reported that glucose and galactose are known to be sweeter than lactose itself. The relative sweetness of sugars is not a constant relationship, but depends upon many factors, including pH, temperature, and presence of other constituents. Mixtures of sugars can make a different sweetness impression than that of individual sugars alone. Synergistic sweetness often results from a combination of sugars.

Maltose. Also known as *malt sugar*, maltose is a product of the fermentation of starches by enzymes or yeast. Barley malt, which is used as an adjunct in brewing, enhances the flavor and color of beer because of its maltose content. Maltose also is formed by yeast during breadmaking. Maltose is the most common reducing disaccharide, $C_{12}H_{22}O_{11} \cdot H_2O$, composed of two molecules of glucose. It is found in starch and glycogen. Purified maltose takes the form of colorless crystals, melting point, 102–103 °C (215.6–217.4 °F). Soluble in alcohol; insoluble in ether. Combustible. Maltose is used as a nutrient, sweetener, and culture medium.

Raffinose and Stachyose. These are sugars found in significant amounts in some foods, such as beans. These sugars are not digested in the stomach and upper intestine as are other disaccharides. They are fermented by bacteria in the lower digestive tract, producing gases and sometimes causing discomfort from flatulence. Raffinose is a trisaccharide composed of one molecule each of D(+)-galactose, D(+)-glucose, and D(−)-fructose, $C_{18}H_{32}O_{16} \cdot 5H_2O$. Raffinose is sometimes used in the preparation of other saccharides.

Sucrose. Table sugar, also known as *saccharose*. Sucrose is a disaccharide, composed of two simple sugars, glucose and fructose, chemically bound together, $C_{12}H_{22}O_{11}$. Hard, white, dry crystals, lumps, or powder. Sweet taste, odorless. Soluble in water; slightly soluble in alcohol. Solutions are neutral to litmus. Decomposes in range of 160–186 °C

(320–366.8 °F). Combustible. Optical rotation = +33.6°. Derived from sugarcane or sugar beets and also obtainable from sorghum. Sucrose is the most abundant free sugar in the plant kingdom and has been used since antiquity (Mead and Chem, 1977).

Raw sugar, centrifugal sugar, refined sugar, and molasses are described in the entry on **Sugarcane**. *Turbinado sugar* is raw sugar that has been refined to remove impurities and most of the molasses. It is edible when produced under sanitary conditions and has a molasses flavor. *Brown sugar* consists of sucrose crystals covered with a film of molasses syrup that give the characteristic color and flavor. The sucrose content varies from 91 to 96%. *Confectioner's* or *powdered sugar* is another form of sucrose made by grinding the sugar crystals. It is usually mixed with about 3% starch to prevent clumping. It is used for household baking, canning, and table use, or industrially where rapid solution in cold liquids is desirable.

Sugar Alcohols. These are polyols, chemically reduced carbohydrates. Important in this group are *sorbitol, mannitol, maltitol,* and *xylitol.* Xylitol is described later.

Polyols are frequently used sugar substitutes and are particularly suited to situations where their different sensory and functional properties are attractive. In addition to sweetness, some of the polyols have other useful properties. For example, although it contains the same number of calories/gram as other sweeteners, sorbitol is absorbed more slowly from the digestive tract than is sucrose. It is, therefore, useful in making foods intended for special diets. When consumed in large quantities (1–2 oz; 25059 g)/day, sorbitol can have a laxative effect, apparently because of its comparatively slow intestinal absorption.

When sugar alcohols are ingested, the body converts them first to fructose, which does not require insulin to facilitate its entry into the cells. For this reason, ingesting these sweeteners (including fructose itself) does not cause the immediate increase in blood sugar level which occurs upon eating glucose or sucrose. Within the body, however, the fructose is rapidly converted to other compounds, which *do* require insulin in their metabolism. One effect of this stepwise metabolism is to "damp out" the peaks in blood sugar levels which occur immediately after ingesting sucrose, but which are absent after ingesting fructose, even if the eventual insulin requirements are the same. Thus, individuals with metabolic problems should not make the assumption that fruit sugars are perfectly all right to consume, but first should consult their physicians. In fact, some health scientists are dubious about pursuing the apparent claims for substituting fructose and sugar alcohols for sucrose as a major sweetener, particularly for diabetics, until more research is done on their long-range nutritional and biophysiological consequences. Research interest has also focused on these sweeteners because of their relatively low potential for causing dental caries. Studies have shown about a 30% reduction in dental caries in laboratory animals on sorbitol and mannitol diets, and virtually complete elimination of caries in the animals when on xylitol diets.

Xylitol. This is a 5-carbon sugar alcohol that occurs widely in nature—raspberries, strawberries, yellow plums, cauliflower, spinach, and many other plants. Although widely distributed in nature, it is present in low concentrations and this makes it uneconomic to extract the substance directly from plants. Thus, commercial xylitol must be produced from xylan or xylose-rich precursors through the use of chemical, enzymatic, and other bioprocessing conversions. A frequently used source has been birch tree chips. Other appropriate starting materials include beech and other hardwood chips, almond and pecan shells, cottonseed hulls, straw, cornstalks (maize), and corn cobs. The base source in the aforementioned agricultural waste materials is hemicellulose xylan. The hemicellulose is acid hydrolyzed to yield xylose which, followed by hydrogenation and chromatographic separation, yields xylitol.

Xylitol is equally as sweet as sucrose. This property is of advantage to food processors because in reformulating a product from sucrose to xylitol, approximately the same amounts of xylitol can be used. Because xylitol has a negative heat of solution, the substance cools the saliva, producing a perceived sensation of coolness, quite desirable in some food products, notably beverages. Recently, this property has been used in an iced-tea-flavored candy distributed in the European market. As of the late 1980s, 28 countries have ruled positively in terms of xylitol for use in commercial products. Xylitol has been found particularly attractive for use in chewing gum, mint and hard candies, and as a coating for pharmaceutical products. Xylitol has the structural formula shown below, with a molecular weight of 152.1. It is a crystalline, white, sweet, odorless powder, soluble

in water and slightly soluble in ethanol and methanol. It has no optical activity.

$$HOCH_2 - \overset{\overset{\displaystyle H}{|}}{C} - \overset{\overset{\displaystyle OH}{|}}{C} - \overset{\overset{\displaystyle H}{|}}{C} CH_2OH$$
$$\overset{|}{OH} \quad \overset{|}{H} \quad \overset{|}{OH}$$

Isomalt. Developed in Germany, isomalt is described as an energy-reduced bulk sweetener and marketed in Europe under the tradename *Palatinit*™. The compound is produced from sucrose in a two-step process, as shown in Fig. 1.

In the first step, the easily hydrolyzable 1-2 glucoside linkage between the glucose and fructose moieties of sucrose are catalyzed by immobilized enzymes to produce isomaltulose, *Palatinos.*™ After crystallization, the isomaltulose is hydrogenated in a neutral aqueous solution using a nickel catalyst.

It is claimed that isomalt is odorless, white, crystalline, and sweet tasting without the accompanying taste or aftertaste. Sweetening power is from 0.45 to 0.6 that of sucrose. A synergistic effect is achieved when isomalt is combined with other artificial sweeteners and sugar substitutes. Principal applications are in confections, pan-coated goods, and chewing gum. The substance was approved for use in most European countries in 1985. Classification of isomalt as a GRAS substance was petitioned in the United States. (GRAS = generally regarded as safe.)

Aspartame. This synthetic sweetener is included with the nutritive sweeteners because it does have some caloric value (when metabolized as a protein, it releases 4 kcal/g). The relationship between sweetness of aspartame and sucrose is almost linear when plotted on a log-log scale. Aspartame is 182 times sweeter than a 2% sucrose solution, but only 43 times sweeter than a 30% solution. The clean, full sweetness of aspartame is similar to that of sucrose and complements other flavors.

The full name of aspartame is *aspartylphenylalanine*, a dipeptide that degrades to a simple amino acid. It has been reported as easily metabolized by humans. Aspartame was accidentally discovered in 1965 with the synthesis of a product for ulcer therapy. Aspartame is metabolized by the same biochemical pathway as proteins, yielding phenylalanine, aspartic acid, and methanol. Because of the byproduct phenylalanine, which some individuals are unable to metabolize, appropriate labeling is required. This is a concern for individuals with phenylketonuria (PKU). Aspartame was first approved in the United States in 1974, then banned in 1975. In July of 1981, it was approved for use in various foods, dry beverage mixes, and in tabletop sweeteners. Approval for use in carbonated beverages was granted in July 1983.

Currently, aspartame is used in tabletop sweeteners (*Equal* in the U.S.; *Egal* in Quebec, Canada; and *Canderal* in Europe and the U.K.). Aspartame currently is incorporated as the exclusive sweetening ingredient in nearly all diet soft drinks in the United States. In other countries, it may be blended with saccharin at a level close to 50% of the saccharin level. Soft-drink manufacturers have taken some measures to enhance stability by raising pH slightly and by more closely controlling the inventory for carbonated soft drinks. Notable differences in sweetness are perceived at a 40% loss in aspartame level.

Crystalline Maltitol. Classified as a bulk sweetener with taste and mouthfeel similar to sucrose, crystalline maltitol contains maltitol as the major component (88+%), with small amounts of sorbitol, maltotriitrol, and hydrogenated oligosaccharides. Its use is in tabletop sweeteners, chocolate, candy, and baked goods. Maltitol has been a major component of hydrogenated glucose syrup in the United States since 1977 and has been used in Japan since 1963. The product was introduced in Europe in 1984. Classification of crystalline maltitol as a GRAS substance was petitioned in the United States in 1986.

Nonnutritive Sweeteners

There are several currently used and a number of potential noncaloric sweeteners, including saccharin, cyclamate (banned in the U.S., but permitted in approximately 40 other countries), acesulfame K, monellin (from the serendipity berry), stevioside, glycyrrhizin, hernandulcin, neo-sugar, miraculin (from miracle fruit), and a sweetener-enhancer, thaumatin, are being investigated.

Saccharin. A noncaloric sweetener that is about 300 times as sweet as sugar. The compound is manufactured on a large scale in several countries.

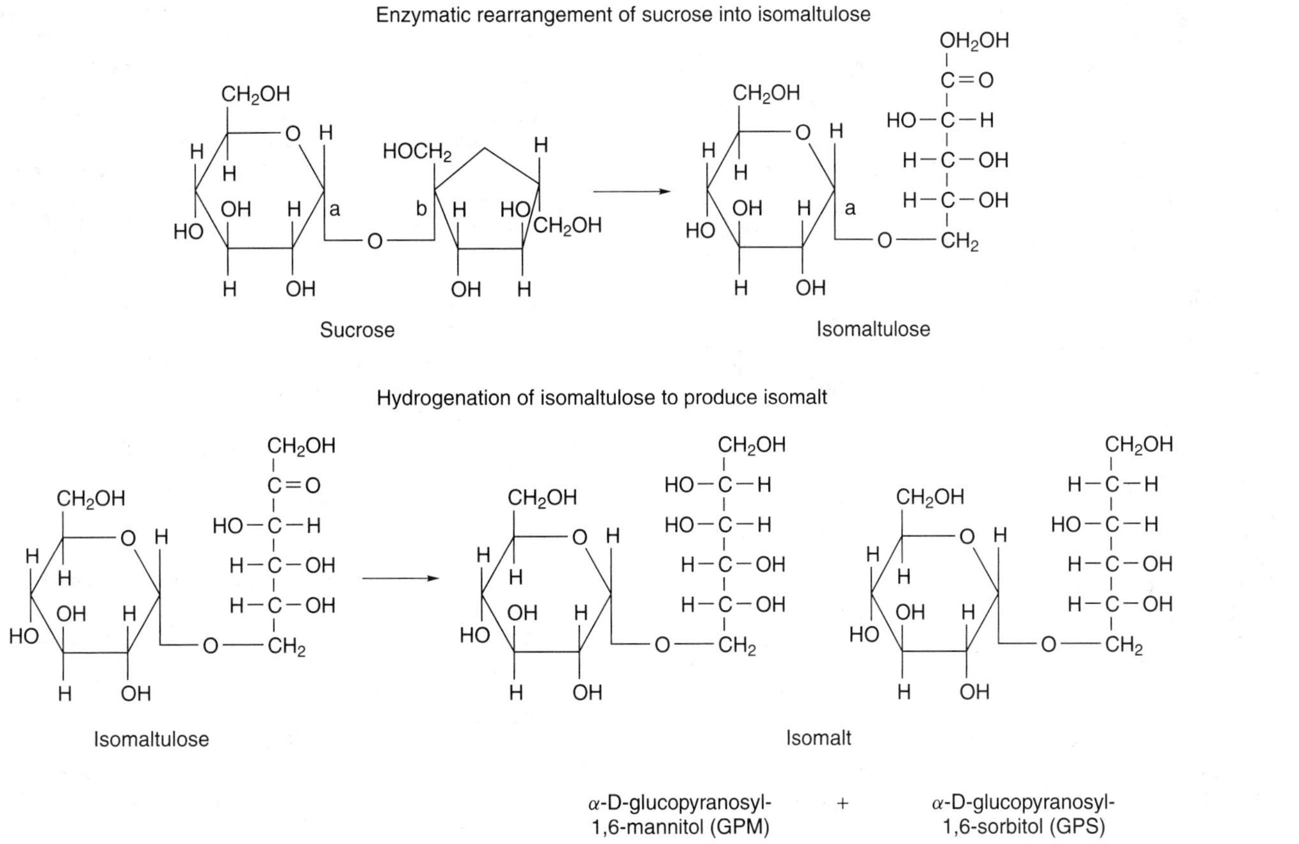

Enzymatic rearrangement of sucrose into isomaltulose

Hydrogenation of isomaltulose to produce isomalt

Fig. 1.   Isomalt.

It is made as saccharin, sodium saccharin, and calcium saccharin, as shown by formulas below.

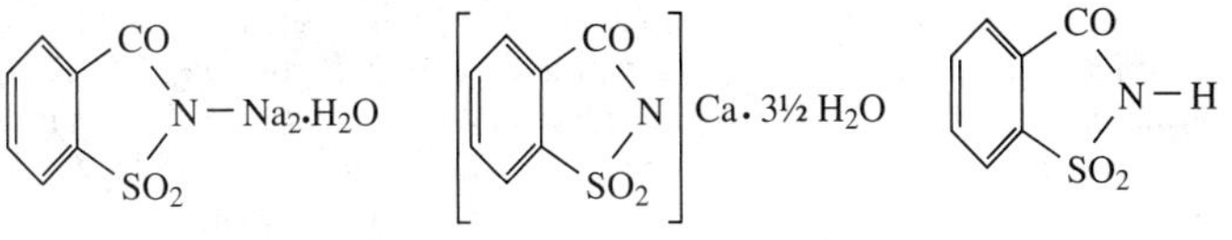

Sodium saccharin   Calcium saccharin   Saccharin

Saccharin (ortho-benzosulfimide) was discovered in 1879 by I. Remsen and C. Fahlberg when they were researching the oxidation products of toluene sulfone amide. The most common forms of saccharin are sodium and calcium saccharin, although ammonium and other salts have been prepared and used to a very limited extent. The saccharins are white, crystalline powders, with melting points between 226 and 230 °C (438.8 and 446 °F). Soluble in amyl acetate, ethyl acetate, benzene, and alcohol; slightly soluble in water, chloroform, and ether. Saccharin is derived from a mixture of toluenesulfonic acids. They are converted into the sodium salts, then distilled with phosphorus trichloride and chlorine to obtain the ortho-toluene sulfonyl chloride, which by means of ammonia is converted into ortho-toluenesulfamide. This is oxidized with permanganate, then treated with acid, and saccharin is crystallized out. In food formulations, saccharin is used mainly in the form of its sodium and calcium salts. Sodium bicarbonate may be added to provide improved water solubility.

Saccharin is used in conjunction with aspartame in carbonated beverages. Other uses include tabletop sweeteners, dry beverage blends, canned fruits, gelatin desserts, cooked and instant puddings, salad dressings, jams, jellies, preserves, and baked goods.

For many years, saccharin has been under investigation by a number of countries. As of the late 1900s, some questions remained unresolved.

Cyclamate. Group name for synthetic, nonnutritive sweeteners derived from cyclohexylamine or cyclamic acid. The series includes sodium, potassium, and calcium cyclamates. Cyclamates occur as white crystals, or as white crystalline powders. They are odorless and in dilute solution are about 30 times as sweet as sucrose. The purity of commercially available compounds is approximately 98%.

Discovered in 1937 and patented in 1940, cyclamate is a derivative of cyclohexylamine, specifically, cyclohexane sulfonic acid. The sodium salt form is normally used, but the calcium salt may be substituted in low-sodium diets. See structural formulas below.

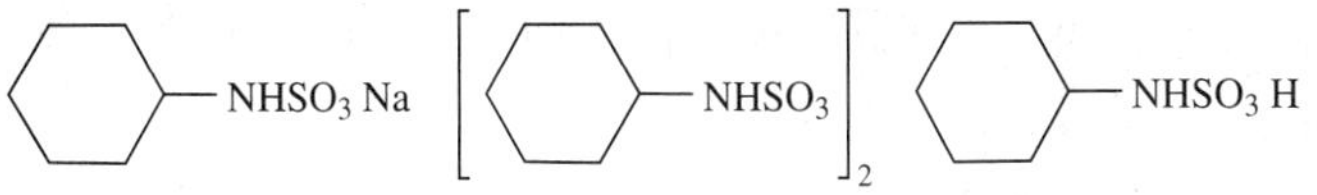

Sodium cyclamate   Calcium cyclamate   Cyclamic acid

Once widely used, cyclamate was prohibited in the United States in 1970. Although used in many other countries, reapproval in the United States has not yet been established. An independent review of the possible carcinogenicity of cyclamate was conducted in April 1985 by the National Academy of Sciences/National Research Council at the request of the Food and Drug Administration. The review concluded that cyclamate itself is not a carcinogen, although it may serve as a promotor or cocarcinogen in the presence of other substances.

Acesulfame-K. This substance (potassium salt of the cyclic sulfano-mide), 6-methyl-1,2,3-oxathiazine-4(3H)-1,2,2-dioxide, shown below, was developed by Karl Clauss (Hoechst Celanese Corporation, Somerville, New Jersey) in 1967. The compound is a white, odorless, crystalline substance with a sweetening power 200 times that of sucrose. A synergistic effect is produced when the substance is combined with a number of other sweeteners. The substance is calorie-free and not metabolized in the human body. Approval of the use of Acesulfame-K was given by the Food and Drug Administration (FDA) in the United States in 1983 and it is found in scores of popular retail products, including yogurt, rice pudding, and soft drinks.

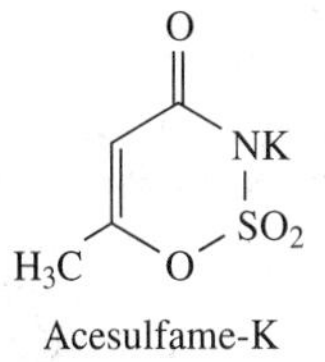

Acesulfame-K

Sucralose. Developed in England during the mid-1980s, testing and evaluation commenced in 1988. The structural formula of the compound (a chlorinated disaccharide derived from sucrose) is shown below.

Sucralose

Sucralose is absorbed poorly in humans and other mammalian species. The small portion that is absorbed is not broken down and is quickly excreted. It has been reported that an extensive array of studies has demonstrated that sucralose is nontoxic — not carcinogenic, teratogenic, mutagenic, or caloric.

Monellin. The sweetness of this compound is claimed to be 1500 to 3000 times that of sucrose, but a different flavor profile prevails. The detection of a sweet taste is slow, commencing after a few seconds in contact with the taste buds, then gradually increasing to its peak intensity. The sweet taste can persist for up to an hour. The source is the relatively rare serendipity berry, the fruit of a noncultivated West African vine. Extraction of the sweet component is effected by treating the berry with a series of enzymes (pectinase and bromelain), followed by dialysis and chromatographic separation. The compound resulting contains the protein *monoellin* with a molecular weight of about 10,700 and composed of two nonidentical polypeptide chains of 50 and 43 amino acids. Neither of the individual chains imparts sweetness. Regulatory measures have not been instituted because of the compound's apparent instability and limited raw resources for processing. However, to date, tests with mice have shown no evidence of toxicity.

Stevioside. Derived from the roots of the herb *Stevia rebaudiana*, this compound has found limited use in Japan and a few other countries as a low-calorie sweetener having about 300 times the sweetening power of sucrose. The compound has not been investigated thoroughly by a number of countries with strong regulatory agencies and, therefore, is not on the immediate horizon for wide consideration as a sweetener.

The dried leaves of *S. rebaudiana* have been used in Paraguay for many years to sweeten bitter drinks. From 3 to 8% of the dried leaves is stevioside, which is a diterpene glycoside as shown by the formula below.

Stevioside

Glycyrrhizins. These are noncaloric sweeteners approximately 50 times as sweet as sugar and used as a flavor enhancer under the GRAS classification in the United States. Glycyrrhizins, which have a pronounced licorice taste, are used in tobacco, pharmaceuticals, and some confectionary products. They are available in powder or liquid form and with color, or as odorless, colorless products. These compounds are stable at high temperatures (132°C; 270°F) for a short time and thus can be used in

bakery products. In some chocolate-based products, the sweetener has been used to replace up to 20% of the cocoa. The sweetener also has excellent foaming and emulsifying action in aqueous solutions. Typical products in which these sweeteners may have application include cake mixes, ice creams, candies, cookies, desserts, beverages, meat products, sauces, and seasonings, as well as some fruit and vegetable products. Generally available as malted and ammoniated glycyrrhizin.

The basic compound is a triterpene glycoside. It is extracted from the licorice root, of which the principal sources are China, Russia, Spain, Italy, France, Iran, Iraq, and Turkey. The roots, containing 10% moisture, are dried and shredded, after which they are extracted with aqueous ammonia, concentrated in vacuum evaporators, precipitated with sulfuric acid, and crystallized with 95% ethyl alcohol.

Hernandulcin. Tasting panels have estimated that this substance is 1000 times sweeter than sucrose, but the flavor profile is described as somewhat less pleasant than that of sucrose. Hernandulcin is derived from a plant, *Lippia dulcis* Trev, commonly known as "sweet herb" by the Aztecs as early as the 1570s. It has been categorized as noncarcinogenic, based upon standard bacterial mutagenicity tests. The economic potential is being studied.

Neosugar. This is another substance in early stages of development and testing. The compound is composed of sucrose attached in a beta(2-1) linkage to 2, 3, or 4 fructose units.

Miraculin. Rather than a sweet-tasting substance, miraculin is described as a taste-modifying substance that elicits a sweet taste to tart foods. The product has been reported as used by African cultures for over a century. The compound is derived from a shrub (*Synsepalum dulcificum*) which grows in West Africa. Miraculin is a glycoprotein with a molecular weight ranging from 42,000 to 44,000. Approval of the Food and Drug Administration has thus far been denied, awaiting further tests. A GRAS category was denied in 1974.

Thaumatin. This is a protein extracted and purified from *Thaumatococcus danielli*, a plant that is found in West Africa. The leaves of the plant have been used for many years in Africa for wrapping food during cooking. Claims have been made that thaumatin is from 2000 to 2500 times sweeter than 8–10% solution of sucrose. The final product is odorless, cream-colored and imparts a lingering licorice-like aftertaste. The substance synergizes well with monosodium glutamate (MSG) and is used in typical Japanese seasonings as well as in chewing gum, pet foods, and certain pharmaceuticals (to mask unpleasant flavor notes). Use in Japan has been approved since 1979. It is considered a GRAS substance in the United States for use in chewing gum. In this application, thaumatin extends the flavor and boosts the perceived duration of flavor. The compound is normally applied as a dust to the surface of gum. Some authorities believe that the use of thaumatin in pet foods has high potential.

Sweeteners in Formulating and Processing

In using low-calorie sweeteners in various food products, the problems are not limited to flavor, but often much more importantly involve texture, acidity, storage stability, and preservability, among others. Acceptable nonnutritively sweetened products cannot be developed by the simple substitution of artificial sweeteners for sugars. Rather, the new product must be completely reformulated from the beginning. Three examples follow.

Jams, Jellies, and Preserves. Traditional products in this category contain 65% or more soluble solids. In low-calorie analogs, soluble solids range from 15% to 20%. Under these circumstances, commonly used pectins (high methoxyl content) do not suffice. Thus, special LM (low methoxyl) pectins must be used, along with additional gelling agents, such as locust bean gum, guar gum, and other gums and mucilagenous substances, some of which may require some masking. In the absence of sugar, a preservative, such as ascorbic acid, sorbic acid, sorbate salts, propionate salts, and benzoates, usually is required to the extent of about 0.1% (weight).

Soft Drinks. In addition to providing sweetness, sugar also functions to provide mouthfeel and to stabilize the carbon dioxide of soft drinks. To contribute to mouthfeel, the use of hydrocolloids and sorbitol has been attempted with limited success. Hydrocolloids also help to some degree with the problem of carbonation retention, but the principal solutions to this problem involve avoiding all factors which contribute to carbonation loss. Thus, the requirement for very well filtered water to eliminate particulates as possible nucleation points; any substances that promote foaming must be avoided; any emulsifying agents used in connection with flavoring agents must be handled carefully to avoid foaming; carbonation should be carried out at low temperature (34 °F; 1.1 °C); and trace quantities of metals must be absent from the water.

Bakery Products. These foods are among the most difficult as regards the use of artificial sweeteners. A listing of the functions of sugar in baked goods beyond that of providing sweetness is indicative of these problems. Sugar contributes to texture in forming structures, in providing moist and tender crumbs by counteracting the toughening characteristics of flour, milk, and egg solids. In the emulsification process required to retain gas during leavening, sugar is an effective accessory agent. Ingredients frequently used in bakery products to compensate for the absence of sugar include carboxymethylcellulose, mannitol, sorbitol, and dextrins, but, generally, these have not been very satisfactory—either to processor or consumer. This remains a large area of challenge for the food processors and ingredient manufacturers.

Evaluating Synthetic Sweeteners. Evaluation of new sweeteners, unlike that of most functional food ingredients, is not possible using totally objective means. There are no general rules leading to structure/function relationships for all classes of sweeteners. The principal judgments must rely on human sensory panel tests. Matters of this type are described from a general standpoint in the entry on **Sensory Evaluation** The training and administration of sensory panels for sweeteners are beyond the scope of this volume.

Additional Reading

Andres, C.: "Alternate Sweeteners," *Food Processing*, **38**(5), 50–52 (1977).

Barndt, R.L. and G. Jackson: "Stability of Sucralose in Baked Goods," *Food Technology*, 62 (January 1990).

Bartoshuk, L.M.: "Sweetness: History, Preference, and Genetic Variability," *Food Technology*, 108 (November 1991).

Birch, G.G.: "Chemical and Biochemical Mechanisms of Sweetness," *Food Technology*, 121 (November 1991).

Chen, J.C.P. and Chung-Chi Chou: "Chen-Chou Cane Sugar Handbook: A Manual for Cane Sugar Manufacturers and Their Chemists," 12th Edition, John Wiley & Sons, Inc., New York, NY, 1993.

Corti, A.: "Low-Calorie Sweeteners: Present and Future," S. Karger Publishers, Inc., Farmington, CT, 1999.

DeMan, J.M.: "Principles of Food Chemistry," 3rd Edition, Aspen Publishers, Inc., Gaithersburg, MD, 1999.

Farber, S.A.: "The Price of Sweetness," *Technology Review (MIT)*, 46 (January 1990).

Fennema, O.R.: "Food Chemistry," 3rd Edition, Marcel Dekker, Inc., New York, NY, 1998.

Grenby, T.H.: "Advances in Sweeteners," Blackie Academic & Professional, New York, NY, 1999.

Igoe, R.S. and Y.H. Hui: "Dictionary of Food Ingredients," 4th Edition, Aspen Publishers, Inc., Gaithersburg, MD, 2001.

Keller, W.E. et al.: "Formulation of Aspartame-Sweetened Frozen Dairy Dessert without Bulking Agents," *Food Technology*, 102 (February 1991).

Kretchmer, N. and C. Hollenbeck: "Sugars and Sweeteners," CRC Press, LLC., Boca Raton, FL, 1991.

Lindley, M.G.: "From Basic Research on Sweetness to the Development of Sweeteners," *Food Technology*, 134 (November 1991).

Nabors, L. O'Brien: "Alternative Sweeteners," 3rd Edition, Marcel Dekker, Inc., New York, NY, 2001.

Noble, A.C., N.L. Matysiak, and S. Bonnans: "Factors Affecting the Time- Intensity Parameters of Sweetness," *Food Technology*, 128 (November 1991).

O'Mahony, M.: "Techniques and Problems in Measuring Sweet Taste," *Food Technology*, 128 (November 1991).

Pepper, T. and P.M. Olinger: "Xylitol in Sugar-Free Confections," *Food Technology*, 98 (October 1988).

Read, N.W. and J. Donelly: "Food and Nutritional Supplements: Their Role in Health and Disease," Springer-Verlag, Inc., New York, NY, 2001.

Shallenberger, R.S.: "Predicting Sweetness from Chemical Structure and Knowledge of the Chemoreception Mechanism of Sweetness, *Institute of Food Technologists Symposium*, Saint Louis, MO, 1979.

Staff: "Applications of Aspartame in Baking," *Food Technology*, 56 (January 1988).

Staff: "Evaluation of Advanced Sweeteners," *Food Technology*, 60 (January 1988).

Staff: "FDA Clears Hoechst's Non-Caloric Sweetener for Use in Dry Foods," *Food Technology*, 108 (October 1988).

Welti-Chanes, J. and G.V. Barbosa-Canovas: "Engineering and Food for the 21st Century," CRC Press, LLC, Boca Raton, FL, 2002.

Wnnia, S.M.: "Modeling the Sweet Taste of Mixtures," *Food Technology*, 140 (November 1991).

Wong, D.W.S.: "Mechanism and Theory in Food Chemistry," Chapman & Hall, New York, NY, 1999.

SWEETGUM TREE. A relatively large American tree (*Liquidambar styraciflua*). The tree achieves its largest stature in the south Atlantic states. The mature tree rises straight up on a tall trunk that is usually free of branches for up to 70 feet (21.3 meters) above ground. In the immature tree, the head is narrow and pyramidal. As the tree matures, the head becomes irregular when not pruned. The leaves are star-shaped (5 lobes), glossy, dark, and rather thick. They turn into many brilliant colors in autumn. The flowers are quite inconspicuous. Brown and twisted seed capsules tend to hang by long stalks through most of winter. The wood has a tendency to warp and is difficult to season; otherwise it would rival black walnut for working. Sweetgum trees are known in Asia (*L. orientale*).

The record sweetgum tree growing in the United States is located in North Carolina. As compiled by the American Forests, this specimen has a circumference (at 4.5 feet; 1.4 meters above ground level) of 178 inches (452 cm), a height of 136 feet (41.5 meters), and a spread of 66 feet (20.1 meters).

SWEET POTATO (*Ipomoea Batatas; Convolvulaceae*). The plant is a trailing perennial, the stems of which twine in counter-clockwise direction around supporting objects. These stems arise from much-thickened roots, which are rich in starch. In cultivation many varieties have been developed, with many different leaf shapes. Dark green, heart-shaped leaves with shining surface occur in several varieties, while in others the leaves are variously lobed and dissected. The flowers, seldom produced in plants grown in northern latitudes, are about 2 inches (5 centimeters) across, purple, and borne either singly or in small axillary cymes. The fruit is a capsule.

Various methods of propagation are employed. Small roots may be planted whole, adventitious buds soon forming and giving rise to shoots which appear above the ground in about four weeks. Root cuttings from growing plants may also be used, especially in regions where the growing season is long. On occasion stem cuttings may be used, but necessarily demand a long growing season and reach maturity very late in the season. Seed may be grown, but germination is slow and uneven, and the product not uniform.

The sweet potato is an American plant, a native of the West Indies and Central America. In cultivation it has gradually spread out of tropical lands, new varieties being developed to suit new localities. At present the crop is grown as far north as Cape Cod.

The principal use of the sweet potato is for human consumption, although some are fed to swine. The vines are frequently used as stock food. From the roots, starch, flour, glucose, and alcohol are extracted, to a limited extent.

Sweet potatoes are frequently called yams. This application of the name yam to the sweet potato is confusing, since the true yam is an entirely different plant, *Dioscorea alata* (*Dioscoreaceae*), widely grown in tropical lands for its edible tubers, which are rich in sugar, watery, and soft when cooked. The flowers are white. Propagation is mainly by cuttings of tubers, each containing one or more eyes, or small buds, such as are found in the white potato tuber. Yams are widely used as food.

Another species of *Ipomoea*, *I. purpurea*, is the Morning Glory, frequently cultivated for its showy flowers.

SWIFT (Lizard). (*Reptilia, Squamata*). A lizard. The name is applied without scientific accuracy to some of the iguanas, including small species of two different genera. Most of these lizards occur in the southwestern United States and Mexico but one species, the pine or fence lizard, *Sceloporus undulatus*, is found as far north as Michigan, New Jersey, and Oregon.

SWIFTS AND HUMMINGBIRDS (*Aves, Apodiformes*). The swift is a small bird with a short wide beak and long slender wings. The swift is superficially like the swallows, but is more closely related to the hummingbirds. The numerous species of swifts are widely distributed in both hemispheres, five species occurring in North America. The common chimney swift, *Chaetura pelagica*, which has abandoned its original habit of nesting in hollow trees to occupy chimneys, is both widely distributed and abundant. The remaining species are found only in the far West and Southwest. The swift makes its nest of various materials cemented together and fastened to their support with saliva. One species of the Oriental region uses the secretion alone, without foreign materials. The nests of this species are attached to the walls of caves and are the famous edible bird nests of Chinese epicures.

The hummingbird is a small bird of the New World whose wings are moved so rapidly in flight that they produce a low-pitched sound. The hummingbird is capable of hovering in one spot in the air, and habitually poises before flowers which are visited for nectar and insects. The wings may beat up to 90 times per second. There are about 320 species of hummingbirds, many of these occurring in the northern Andes of South America. However, hummingbirds range from Alaska southward into much of the South American continent. They are found up to the snow line, 16,000 feet (4,800 meters), in Ecuador. The hummingbird has short legs, small feet, narrow wings, and long feathers. The bill may be from $\frac{1}{3}$ to 5 inches (0.8–13 centimeters) in length, depending upon the species. It usually is straight. The tongue is specialized and tube-like. The voice is comprised of a weak chippering sound. The nest is very small, cup-like, built of plant fibers and spiderwebs, and lined with down. The nest opening is only about 1 inch (2.5 centimeters) in diameter. There are usually two white eggs.

The hummingbird exhibits rather brave behavior around homes and people, wherever food is put out for them, or where flowers are in bloom. The bird can appear and disappear from the scene with great rapidity. The male hummingbird shows no concern for family affairs, leaving the nest building care and feeding of the young to the female. The female is often quite pugnacious when watching over her eggs and young, and can be quite dominating in terms of other hummingbirds when they encroach on a favorite feeding cup or flower bed. A few species of hummingbirds hibernate in winter. The ruby-throated hummingbird is one of the few species that inhabits the eastern United States. See Fig. 1. The coquette is one of a group of species found from southern Mexico to southern Brazil. These birds are distinguished by the crested head and conspicuous frills at the sides of the neck. See also **Apodiformes**.

Fig. 1.   Ruby-throated hummingbird. (*Harrison.*)

SWINE (*Mammalia, Artiodactyla*).   Pigs, boars, hogs, and peccaries comprise one of the more primitive groups of the order *Artiodactyla* (even-toed hoofed animals). The group is not large in terms of identifiable species in the wild, as contrasted, for example, with the antelopines and even the bovines, but because they are so highly valued in terms of the domestic breeds, they do comprise an extremely large population group of mammals.

Pigs are native to the warmer parts of Europe and Asia, the Oriental region, and Africa. None are attributed to originating in North or South America, although the closely related Peccaries have inhabited these continents for many, many centuries. Pigs are not ruminants, i.e., they do not chew their cuds. They all dig with their muzzles and have a preference in nature for vegetable matter. They are characterized by having large litters. There are certain misconceptions concerning pigs that should be clarified. In general, these animals are not at all stupid as sometimes portrayed, but are very intelligent and rate only second to the Great Apes among the mammals, except humans, for their mental abilities. By nature, they are not dirty. When in the wild and left to their own habits, these animals are exceptionally orderly and clean. Thus, likening a person to a pig really should not be considered uncomplimentary. It is true that most species of *Swine* enjoy mud wallows where they go to cool off, to remove external parasites from their bodies, and to cleanse themselves.

Mud, as some beauticians acclaim, is an excellent cleaner, containing helpful antibiotics. The pig in captivity that wallows in mud that is littered with excrement, rotting garbage, etc., does not do this out of choice. In the wild, these animals never excrete in their mud wallows.

In connection with the domesticated pigs, certain terminology often requires clarification. The terms *pig* and *swine* are generally considered synonymous in most parts of the world, but in the United States, swine usually refers to pigs that are under three months of age. In England, Canada, and Australia, pigs are swine of any age or weight. Informally, untamed, wild pigs are sometimes referred to as *wild swine*. Adult animals are often referred to as hogs, particularly the marketable and commercial animals. Hog also refers to a castrated boar. A *gilt* is a young female pig, often referring to an animal with its first litter. A *sow* is an adult female pig. A *boar* is a well-developed male used for breeding service. A *boar pig* is a male animal under breeding age (usually less than six months old). A *stag* is a male animal that has been castrated in maturity—after the tusks, shields, enlarged sheath, crest, and other characteristics have developed. A *barrow* is a male animal that has been castrated before sexual characteristics have developed. A *shote* is an immature animal of either sex.

There are over 400 breeds of domestic pigs. Some of the more important are listed here. *Berkshire* (Fig. 1), originated in England and Japan. It is of medium size, black, with white on face, legs, and tail tip. *Chester White* (Fig. 2) originated in Chester County, Pennsylvania. The animal has a pink skin and is highly regarded for its lard. *Duroc* originated in New York State, is of red coloration and considered an excellent pig for large-scale production. *Essex* is black with white saddle on shoulder, legs, nose, and tail. *Gloucestershire* originated in England, is lop-eared, white with black spots, and is highly considered for bacon and cross breeding. *Hampshire* is derived from the Saddleback and appears much like it; it is highly regarded for bacon. *Hereford* is red with white head, a good producer. *Iberian* is a red-colored animal with short ears; it resembles pigs of medieval times. It is of no commercial importance outside Spain and Portugal. *Landrace* is a white animal with lop ears, popular in Denmark, Germany, Norway, and Sweden. This breed was exported from Sweden to England in 1953. It is a highly regarded breed. The bacon from these

Fig. 1.   Berkshire boar. (*USDA.*)

Fig. 2.   Chester White gilt. (*USDA.*)

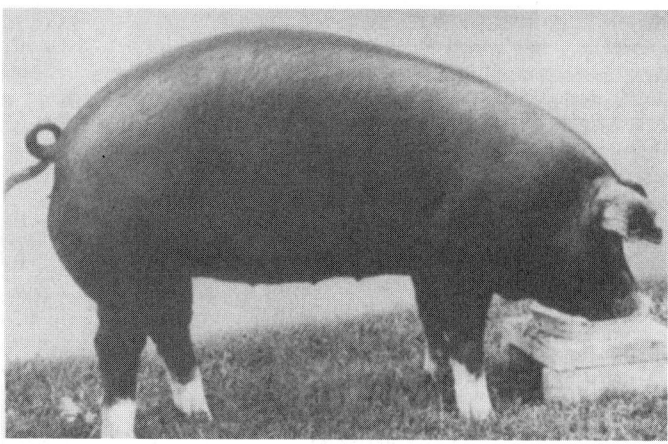

Fig. 3.   Poland china gilt. (*USDA*.)

animals is often called Wiltshire bacon. *Cornwall* is an all-black animal with lop ears; it originated in England and is mainly raised for bacon and cross-breeding. *Mangalitza* is a white pig, is much like the Chester White, but smaller and dishfaced. *Pietrain* originated in Belgium, is lop-eared and off-white color. This breed has attracted much attention since 1960 and essentially has been confined to Belgium, France, England, Denmark, and the Netherlands. *Poland China* (Fig. 3) originated in Ohio (Warren and Butler counties). It is a very large animal, black with white face, feet, and tail tip, and drooping ears. *Spotted Poland China* originated in Indiana. It is much like the Poland China except with white spots over body. It is an excellent lard producer. *Tanworth* originated in England and is found in Canada, Australia, and New Zealand. It is golden red with long snout and erect ears. Its major use is for bacon and cross breeding. *Welch* is white with lop ears; excellent for bacon; found in Africa, France, Poland, and the former U.S.S.R. *Wessex Saddleback* is black with white saddle. It originated in England; is extensively produced in Australia. Major uses are for bacon and crossing with white boars. It was introduced into the United States several years ago as a special breed. *West French White* is lop-eared, white; a major pork producer. *Yorkshire* is one of the most widely distributed of all breeds and a good producer. It is a white animal with erect ears. Other important breeds include the *Cheshire, Duroc-Jersey, Large Black,* and *Razorback.* Breeders look for: (1) prolific sows that farrow and raise large litters; (2) animals that grow rapidly and that show economical gain during feeding periods; and (3) animals that are resistant to infection and parasites and that have carcasses that withstand handling, transporting, and processing well. In particular, boars are sought that are prolific, and have a high degree of masculinity, good disposition, and an ability to sire strong, healthy pigs. In selecting sows, the length of the body and femininity are important, particularly a well developed udder and two rows of teats with a minimum of six teats in a row.

Young boars may breed at about 8 months of age. They usually will be mated to 20 to 30 sows during their first breeding season. Mature boars will be mated with from 40 to 60 sows per season. The sow's period of estrus (heat) is about 3 days, occurring every 3 weeks. Mating usually occurs on the second or third day. Gestation period is 114 days. A gilt will have her first litter at about one year of age.

Pigs also can be classified by the main type of product for which they are best suited. Lard producers are large of frame. Animals with longer frames are more suitable for bacon. Because of increased demands for lean meat, an animal with a lean carcass is desired. Usually white pigs require more intensive care than the colored varieties. Most of the white breeds originated in Canada and the United States. They usually are excellent lard producers.

Of the Eurasian Pigs found in the wild state, most species are found in the Oriental Region. However, the Wild Boar, of which there are many races, occurs in Western Europe, North Africa, across India and to southeast and central Asia. However, the animal is not found north of the Caspian Sea, in the U.S.S.R., or the great Asian mountain chains. They are characterized by long, pointed muzzles. Coloration is red-brown-chocolate-black. The hair is stiff and bristly. The amount of hair varies with climate. Some authorities consider the bite of the wild boar the worst of any mammal with the exception of the Killer Whale. The bite is more of a ripping than a slicing nature and is considerably worse than that inflicted

by the Great Cats. The smaller Pigmy Pig is found in the Himalayas. The mature boars stand only about 1 foot (0.3 meter) high at the shoulders. On the other hand, the Giant Bornean Pig will attain a length of some 6 feet (1.8 meters), with very large heads. They are characterized by an upturned moustache.

The African Bush-Pigs and Forest-Hogs are found widely throughout that continent. The Bush-Pigs are known for their rooting abilities. These animals like to turn over fallen logs for the snakes, rats, snails, and fungi found thereunder. It is interesting to note that the Giant Forest-Hog was not discovered by naturalists until the present century. It is a giant, low-slung animal with big head and widespread ears. Warts are located below the eyes. The body is covered sparsely with long, stiff bristles which, as the animal matures, could almost be described as spines.

The Wart-Hog is well known for its ugly appearance, characterized by a large flattened head out of which grow most unattractive warts. They are known for their erratic behavior. The Babirusa of the East Indies is a nocturnal, forest-loving beast. The males have pronounced teeth. They are known for their excellent flesh.

It is believed that the peccaries migrated to North America when there was a land bridge with Asia. The collared peccary or muskhog ranges from the southwestern United States to southern South America, and the white-lipped peccary is found only from Belize to Paraguay. The latter species is gregarious, living in large bands. Its vicious nature makes it dangerous to encounter, although a single animal is too small to trouble a human being. The collared peccary lives singly or in small groups and is inoffensive. The collared peccary attains a length of about 3 feet (1 meter) and a height of close to 16 inches (40 centimeters).

The early Chinese in America are credited with bringing the breed of domesticated true pig to the New World. This animal readily takes to the wilds and breeds with wild species. This gave rise to the feral or so-called razorback hogs of the American south.

Additional Reading

Bollen, P.J., A.K. Hansen, and H.J. Rasmussen: "The Laboratory Swine," CRC Press, LLC, Boca Raton, FL, 1999.
Considine, D.M. and G.D. Considine: "Foods and Food Production Encyclopedia," Van Nostrand Reinhold Company, Inc., New York, NY, 1982.
Cowart, R.P. and S.W. Casteel: "An Outline of Swine Diseases," 2nd Edition, Iowa State Press, Ames, IA, 2001.
Gordon, I.R.: "Controlled Reproduction in Pigs," CAB International, New York, NY, 1997.
Harris, D.L.: "Multi-Site Pig Production," Iowa State Press, Ames, IA, 1999.
Kyriazakis, I.: "Quantitative Biology of the Pig," CAB International, New York, NY, 1999.
Lewis, A.J. and L.L. Southern: "Swine Nutrition," 2nd Edition, CRC Press, LLC., Boca Raton, FL, 2001.
Rothschild, M.R. and A. Ruvinsky: "The Genetics of the Pig," CAB International, New York, NY, 1998.
Taverner, M.R. and A.C. Dunkin: "Pig Production," Elsevier Science, New York, NY, 1996.
Whittemoore, C.T.: "The Science and Practice of Pig Production," Blackwell Science, Inc., Malden, MA, 1999.

Web References

Directories of Agriculture-related Internet Information Resources: *http://www.agnic.org/diragis/*
The PigSite.com - Global Swine Industry Portal: *http://www.thepigsite.com/*

SWORD-BEARER (*Insecta, Orthoptera*).   A long-horned grasshopper of the group known as cone-headed grasshoppers, from the conical prolongation of the head. The sword-bearer is named from the long ovipositor of the female. This organ is a slender slightly curved blade longer than the entire body. The species is found in the northern states of the United States, east of the Rockies. See also **Grasshopper**.

SYCAMORE AND PLANE TREES.   Of the family *Platanaceae* (plane family), these trees are known as plane trees in Europe, but in America are more commonly referred to as sycamores or buttonwoods. To add to the confusion of nomenclature, the maples of Europe (family *Aceraceae*) of particular species, such as *A. pseudoplatanus*, are known as sycamore-maples. Plane trees are very common in Europe. At one time, it was estimated that over 60% of the trees in London were plane trees. The London plane is considered to be an early cross between the eastern plane (*Platanus orientalis*), which is native to Turkey, and the western plane (*P. occidentalis*), known as the sycamore or button wood in America. This

TABLE 1. RECORD SYCAMORE TREES IN THE UNITED STATES[1]

Specimen	Circumference[2]		Height		Spread		Location
	Inches	Centimeters	Feet	Meters	Feet	Meters	
Western sycamore (1999) (*Platanus occidentalis*)	334	848	95	29	105	32	Virginia
Arizona sycamore (1981) (*Platanus wrightii*)	283	719	114	34.7	116	35.4	New Mexico
Arizona sycamore (1999) (*Platanus wrightii*)	335	851	69	21	88	26.8	Arizona
California sycamore (1998) (*Platanus racemosa*)	344	874	94	28.7	104	31.7	California

[1]From the "National Register of Big Trees," American Forests (by permission).
[2]At 4.5 feet (1.4 meters).

tree is found extensively in the eastern United States, as well as in Turkey, Greece, southeastern Europe, and Asia Minor. The tree ranges from 75 to 100 feet (22.5 to 30 meters) in height, with a large trunk and broad spread. See Table 1. The bark is dark gray, peels easily, with a lighter gray color underneath. The leaf is large, smooth, coarse, and permanently lobed. The flower is unisexual, about 1 inch across. The wood of the sycamore is light yellow, with a red-brown heartwood. It is tough, close-grained, and firm. Commercial sycamore wood, when cut and quartered, appears much like oak. The surface is lustrous and takes readily to a fine polish. The wood is used for tool handles, rollers, flooring, cooperage, furniture, and veneers. Other species include *P. silver*, native to the eastern United States; *P. racemosa*, found mainly in California; and *P. wrightii*, found in the southwest United States, notably Arizona.

SYENITE. A coarse-grained, granular, therefore intrusive, igneous rock of the general composition of granite except that quartz is either absent or present in a relatively small amount. The feldspars are alkaline in character and the dark mineral is usually hornblende. Soda-lime feldspars may be present in small quantities. The term syenite was originally applied to hornblende granite like that of Syene in Egypt from whence the name is derived. Syenite is not a common rock, some of the more important occurrences being, in the United States, in New England, Arkansas, Montana, and New York State (syenite gneisses), and elsewhere, in Switzerland, Germany, and Norway.

SYLVANITE. A mineral, a telluride of gold and silver approximating the formula AgAuTe$_4$. Sylvanite is monoclinic, occurring in bladed, columnar, and granular forms as well as arborescent and branching. It is a brittle mineral; hardness, 1.5–2; specific gravity, 8.16; luster, metallic; color and streak, steel gray to yellowish-gray. This mineral is found associated with gold and tellurides of gold and silver or with sulfides such as pyrite. It is found in Rumania, Australia, Colorado and California. It was named for Rumanian Transylvania where it was first found.

Krennerite is another telluride of gold and silver with a similar composition to sylvanite, but crystallizing in the orthorhombic system. Calaverite is a gold telluride with only a small silver content.

SYLVITEA. mineral, potassium chloride, KCl, occurring in cubes, or as cubes modified by octahedra. Sylvite is therefore isometric. It has a perfect cubic cleavage; uneven fracture; is brittle; hardness, 2; specific gravity, 1.9; luster, vitreous; colorless when pure but may be white, bluish, yellowish or reddish due to impurities. It is soluble in water. It is much rarer than halite and has been found as sublimates at Mt. Vesuvius and as bedded deposits at Stassfurt, Germany. Extensive deposits occur in sedimentary deposits in the Permian basin of southwestern New Mexico, near Carlsbad, in the United States.

It is used as a source of potash salts. Potassium chloride was called by the early chemists *sal digestivus Sylvii*, whence the name of the mineral.

SYMBIOSIS (Ecology). A close association between two organisms of different species in which at least one of the two benefits. The two organisms may be both plants, one may be a plant and the other an animal, or both may be animals. Three different kinds of symbiosis are recognized by ecologists:

1. *Mutualism*, the condition where both of the symbionts benefit from the association. A lichen, for instance, is a composite plant formed in actuality from two different plants. One, a fungus, gains nutrients from the alga, and the other, an alga, gains protection and an increased supply of water from the fungus. The result is a highly efficient plant structure, which can grow in places where either type of plant alone could not exist. The nitrogen-fixing bacterium, *Rhizobus leguminosum*, lives in the roots of the legumes. The bacteria are protected in the nodules which form on the roots and the legumes benefit by an increased supply of nitrates fixed by the bacteria. The pollination of flowers by insects is an example of a more loosely knit mutualistic association, involving plants and animals. The rhinoceros and the tick bird provide a good example of mutualism among two higher animals. See Fig. 1. The tick bird picks the ticks off the rhinoceros and gets food, and the rhinoceros gets rid of its ticks. Also, the tick bird has much keener eyesight than the nearsighted rhinoceros and gives warnings in times of danger by jumping up and down and uttering shrill cries.

Fig. 1. An example of mutualism: a tick bird on the back of a rhinoceros. (*A.M. Winchester.*)

In some cases, the mutualism is *facultative* — either organism can exist without the other; but in other cases, the association is *obligatory* — neither organism can live without the other. An example of the latter is the association of termites and certain protozoa that live in their intestines. The termites eat wood, but have no enzymes for digesting wood. The protozoa have such enzymes, but cannot live on wood unless it is first chewed to a pulp by the termites. Termites die if the protozoa are killed by antibiotics and the protozoa die when they leave the body of the termites. The Smyrna fig and a tiny wasp are also entirely dependent upon one another. The fig can produce neither fruit nor seeds unless this species of wasp enters the young flower and accomplishes cross-fertilization. The wasps die without the fig because the eggs are laid within some of the figs.

2. *Commensalism*. When one symbiont benefits from the association and the other is neither harmed nor benefited. The shark sucker, *Remora*, which attaches itself to sharks and gets a free ride as well as some share of the food killed by the shark, benefits from the association, but the shark

is not affected one way or another. Many of the bacteria living in the human mouth and intestine are also classed as commensals. The green hydra, *Chlorohydra viridisima*, is green in color because there are small one-celled algae in its gastrodermal cells. The hydra neither benefits nor is harmed by the presence of these plants; it gets along just as well without them. Hence, this is a case of commensalism.

3. *Parasitism.* When one symbiont benefits and the other is harmed. A tapeworm living within the intestine of a vertebrate animal is a good example of a parasite, since the host animal is definitely harmed. The tapeworm is an endoparasite, since it lives within the body of another animal; ectoparasites, such as fleas, leeches, and ticks, live outside the body and have a more temporary association.

SYMMETRIC. Arranged in accordance with a certain similarity with reference to a certain geometrical entity or position, which may be a point (center or point of symmetry), a line (axis of symmetry), or a plane (plane of symmetry), etc.

SYMMETRIC FUNCTION. A function whose value remains unchanged under any permutation of its independent variables. That is, any function $f(x_1, x_2, \ldots, x_n)$ not affected by an interchange of any x_i and x_j. If these are roots of an algebraic equation

$$x^n - c_1 x^{n-1} + c_2 x^{n-2} - \cdot \pm c_n = 0$$

then

$$c_1 = \sum x_i, \quad c_2 = \sum x_i x_j, \quad c_3 = \sum x_i x_j x_k \ldots$$

are the elementary symmetric functions, where the summations are extended over all distinct products of distinct factors, the x_i being themselves considered independently varying. Any rational symmetric function of the x_i is a rational function of the c_i.

See also **Permutation**.

SYMMETRY (Axis of). A line drawn within a body or within a set of points in such a location and direction that a rotation of the body through an angle $(2\pi/n)$ radians about the line as an axis, n being an integer, greater than unity, results in a configuration indistinguishable from the original configuration. A body or set having such an axis is said to have n-fold symmetry, and the line is said to be an n-fold axis. Thus a line through the center of a cube and parallel to a face is a fourfold axis of symmetry, while a body diagonal of the cube is a twofold axis.

SYMMETRY (Center of). A symmetry element such that any line through it will intersect the crystal at equal distances on either side. Schoenflies symbol, subscript i.

SYMMETRY (Plane of). A plane passed through a body or through a set of points in such a location and direction that the reflection of all points in the plane results in a configuration indistinguishable from the original configuration. Thus, a cube has many planes of symmetry through its center, including those parallel to the faces and those passing through face diagonals.

SYMMETRY (Zoology). The arrangements of the parts of animal bodies in relation to centralized axes. The bodies of some one-celled animals are asymmetrical and of others, notably the Heliozoa, spherically symmetrical with the hard parts of the skeleton radiating in various directions from a common center. By far the most common forms of symmetry, however, are those known as radial and bilateral.

Radial symmetry is especially common among the sessile animals such as sea anemones and the related jellyfishes whose movements are weak. These animals have a principal axis passing through the mouth from which similar structures extend on several radii. The same form of symmetry appears in the echinoderms although these animals begin life as bilaterally symmetrical larvae. The radial symmetry of the adult accompanies sluggish movement and in some forms food-securing habits like those of sessile animals.

Bilaterally symmetrical animals have similar halves flanking a median plane in the principal axis of the body. Sense organs are concentrated near the end that goes first in locomotion, forming a head in which the mouth opens as a rule. This end of the body is the cephalic end, in contrast with the opposite caudal end where the tail is attached in the vertebrates.

The originally upper and lower or dorsal and ventral surfaces are also differentiated, since the animal rests on the latter while the former is exposed to surrounding influences, and the sides of the body are known as right and left. This type of symmetry prevails in all actively moving animals.

The value of each type of symmetry is clearly correlated with the mode of life in which it is found. Sessile animals receive food and are subjected to dangers only when the responsible factors approach under their own powers of locomotion or on currents in the water. It is an advantage to the animal to be able to perceive such factors as easily in one direction as another. Bilaterally symmetrical animals move about in search of food. Hence, the end of the body that normally goes first has the chief need of powers of perception, while the upper and lower surfaces are exposed to different environmental conditions and the sides are similar in their contacts.

SYMPHYLA. Small and rare animals living in moist debris at the surface of the ground. They are related to primitive insects and centipedes and are usually regarded as a class of the phylum *Arthropoda*. They have a pair of antennae but no eyes. The segments of the body are well marked, bearing 11 or 12 pairs of legs. The animals breathe by means of tracheae.

SYNAPSE. The association between nerve cells of animals above the level of coelenterates. In the latter, the nerve network is composed of cells whose processes are structurally connected, but in higher nervous systems, the fine terminal branches of nerve processes merely come into close contact with those of adjacent cells.

Synapses can be defined as regions of structural specialization between two or more neurons. Since, as a rule, synapses conduct impulses only unidirectionally, some type of asymmetry might be expected in their structure — and there being regions of apposition between adjacent neurons, they may involve almost any parts of the two neuronal surfaces. The most common type of synaptic junction is that between an axon and a dendrite or a soma, the efferent fiber being expanded at its end to form a small bulb. When a nervous impulse reaches the terminal bulb of the axon, acetylcholine is liberated and depolarizes the secondary neuron, thus creating a nervous impulse throughout its length. Cholinesterase at the synapse, however, quickly inactivates the acetylcholine and prevents it from stimulating the dendrites of the secondary neuron more than once.

In the intact nervous system, the impulses normally travel only from the dendrites to cell body or soma, to the axon and finally to the end bulb.

How neurons form synapses in the development nervous system and distinguish appropriate from inappropriate synapses remains one of the central, unsolved problems in neurobiology. However, it has recently been found that cyclic AMP (adenosine $3'$: 5-cyclic phosphate) effects synaptogenesis by regulating the expression of voltage sensitive Ca^{2+} channels suggesting that cyclic AMF affects post-translational modification of some glycoproteins and the cellular levels of certain proteins.

The nervous system mediates adaptive response of an organism to environmental changes or to changes in the organism itself. To this end, the nervous system is uniquely modifiable or plastic. Neuronal plasticity is largely the capability of synapses to modify their function, to be replaced, and to increase or decrease their number when required. Neuronal plasticity is maximal during development and is expressed after maturity in response to external or internal perturbations, such as changes in hormonal levels, environmental modification, or injury.

See also **Central and Peripheral Nervous Systems**.

Additional Reading

Cowan, W.M., C.F. Stevens, and T.C. Sudhof: "Synapses," Johns Hopkins University Press, Baltimore, MD, 2000.

Geiser, C.: "From Sound to Synapse: Physiology of the Mammalian Ear," Oxford University Press, Inc., New York, NY, 1998.

Kuno, M.: "The Synapse: Function, Plasticity, and Neurotrophism," Oxford University Press, Inc., New York, NY, 1995.

Llinas, R.R.: "The Squid Giant Synapse as a Model for Chemical Transmission," Oxford University Press, Inc., New York, NY, 1999.

Nicholls, D.G.: "Proteins, Transmitters, and Synapses," Blackwell Science, Inc., Malden MA, 1994.

Shepherd, G.M.: "The Synaptic Organization of the Brain," 4th Edition, Oxford University Press, Inc., New York, NY, 1997.

Stanford, S.C., J.A. Gray, and P. Salmon: "Stress: From Synapse to Syndrome," Academic Press, Inc., San Diego, CA, 1993.

R.C. VICKERY, M.D., D.Sc., Ph.D., Blanton/Dade City, FL.

SYNCHRONOUS. A synchronous operation takes place in a fixed time relationship to another operation or event, such as a clock pulse. See also **Asynchronous.** When a set of contacts is sampled at a fixed time interval, the operation is termed synchronous. This situation is to be contrasted with that where the contacts may be sampled randomly under the control of an external signal. Generally, the read operation of a main storage unit is synchronous. The turning on of the X and Y selection drivers and the sampling of the storage output on the sense line are controlled by a fixed-frequency clock.

SYNCHRONOUS COMPUTER. A computer in which the starting time of every ordinary operational cycle is controlled by signals which occur at regular intervals.

SYNCLASTIC SURFACE. A surface, or portion of a surface, on which the two principal radii of curvature at each point have the same sign. Also called *surface of positive total curvature.*

SYNCLINE. The syncline is a structure in which the strata are bent downward in an inverted arch, the sides of which are designated the limbs. The syncline may be a broad open fold or tightly compressed with steep dips, and pitch either upward or downward.

SYNCOPE. A fainting spell, in which the unconsciousness is due to a temporary cerebral anemia, i.e., insufficient circulation of blood in the brain. Differing only in degree from *true syncope* is dizziness (light-headedness). Syncope also may occur from noncardiovascular causes, such as head injury, hypoglycemia, seizures, and hysteria.

SYNDROME. A group of symptoms characterizing or occurring in any abnormal state or disease.

SYNERESIS. The contraction of a gel with accompanying pressing out of the interstitial solution or serum. Observed in the clotting of blood, with silicic acid gels, etc. See also **Colloid System.**

SYNERGIC CURVE. A curve plotted for the ascent of a rocket vehicle calculated to give the vehicle an optimum economy in fuel with an optimum velocity. This curve, plotted to minimize air resistance, starts off vertically, but bends toward the horizontal between 20 and 60 miles (32 and 96 kilometers) altitude to minimize the thrust required for vertical ascent.

SYNODIC PERIOD. The synodic period of any member of the solar system is the time required for the object to go from some particular position relative to the sun as seen from the earth back to the same position. In the case of the moon, the synodic period is the time required to go from conjunction, or new moon, back to conjunction. This period of approximately 29.5 days is the original month as used by ancient astronomers in the construction of the calendar.

Since a planet is best observed at opposition, the synodic period of the planet gives the interval of time between successive positions of favorable observation. The synodic period is related to the sidereal period, i.e., the actual period of revolution of an object about the sun, by a simple relationship:

let P be the sidereal period of the object;
   S the synodic period of the same object;
   E the sidereal period of the earth (approximately 365.25 days);

then, for planets with orbits inside that of the earth, $1/S = 1/P - 1/E$; and for planets with orbits outside that of the earth, $1/S = 1/E - 1/P$.

SYNTHESIS (Chemical). The process of building chemical compounds through a planned series of steps (reactions, separations, etc.). Synthesis usually is the method of choice: (1) when the desired compound is not present in natural materials from which it can be isolated: (2) when the compound cannot be easily obtained from reacting readily available materials in a few simple steps; and (3) although a compound may be available within a natural complex, the economic separation and purification are prohibitive, or often in the case of biochemicals, too little natural raw material is available to meet the demand.

Even more important, synthesis plays a key role in developing new, untried chemical structures which, on paper, appear to have properties that may be of great value, e.g., a new synthetic material, a new drug, or a new fuel. Chemicals by design from prior knowledge of related materials generally are created via the route of synthesis. Further, synthesis is fundamental to broadening the base of chemical knowledge. Sometimes unexpected results occur, i.e., compounds with unusual, unexpected, and often desirable practical chemical and/or physical properties.

Because of the hundreds of thousands of organic substances already established, but many yet remaining to be "built," organic synthesis predominates. Most of the synthetics (elastomers, fibers, and other polymers, coatings, films, adhesives, and numerous other products) that have appeared during the last 30 to 40 years resulted from research involving organic synthesis. Some of the early work in organic synthesis dealt with the creation of certain fatty acids and ketones. A few examples are given to provide an insight into the workings of synthesis.

In the following examples, only the main starting ingredients and products are shown. No attempt is made to indicate byproducts or the conditions of the reactions involved:

(a) Target compound: Ethylpropylacetic acid, $(C_2H_5)(C_3H_7)CH:COOH$
   (1) Acetic anhydride $\rightarrow$ ethyl acetate
      (+ alcohol)
   (2) Ethyl acetate $\rightarrow$ ethyl acetoacetate
      (sodium + dilute acids)
   (3) Ethyl acetoacetate $\rightarrow$ sodium derivative of ethyl acetoacetate
      (+ sodium ethoxide)
   (4) Sodium derivative of ethyl acetoacetate $\rightarrow$ ethyl ethylpropyl acetoacetate
      (+ propyl iodide)
   (5) Ethyl ethylpropyl acetoacetate $\rightarrow$ ethylpropylacetic acid
      (concentrated alcohol and potash)

(b) Target compound: Butyl acetone, $CH_3 \cdot CO \cdot CH_2 \cdot C_2H_4$
   (1) through (3), same as given in example (a)
   (4) Sodium derivative of ethyl acetoacetate $\rightarrow$ ethylbutylpropyl acetoacetate
      (+ butyl iodide)
   (5) Ethylbutylpropyl acetoacetate $\rightarrow$ butyl acetone
      (+ dilute alcohol and potash)

(c) Target compound: n-valeric acid, $CH3 \cdot CH_2 \cdot CH_2 CH_2 COOH$
   (1) Potassium chloroacetate $\rightarrow$ potassium cyanoacetate
      (+ potassium cyanide)
   (2) Potassium cyanoacetate $\rightarrow$ ethyl malonate
      (+ alcohol and hydrogen chloride)
   (3) Ethyl malonate $\rightarrow$ sodium derivative of ethyl malonate
      (+ sodium ethoxide)
   (4) Sodium derivative of ethyl malonate $\rightarrow$ ethylpropyl malonate
      (+ propyl iodide)

The compounds on the right-hand side of intermediate reactions are often called *intermediates.* See also **Intermediate (Chemical).**

Some of the notable syntheses from the early history of the technique include:

Inorganic Syntheses
   1746   Sulfuric acid (chamber process)
   1800   Soda ash (Le Blanc process)
   1861   Soda ash (Solvay process)
   1890   Sulfuric acid (contact process)
   1912   Ammonia (Haber-Bosch process)
Organic Syntheses
   1828   Urea (Wohler)
   1857   Mauveine (Perkin)
   1869   Celluloid (Hyatt)
   1877   Ethylbenzene (Friedel-Crafts)
   1884   Rayon (Chardonnet)
   1910   Phenolic resins (Baekeland)
   1910   Neoarsphenamine (Ehrlich)
   1920   Aldehydes, alcohols (Oxo synthesis)
   1925   Insulin (Banting)
   1927   Methanol
   1930   Neoprene (Nieuwland)
   1935   Nylon (Carothers)
   1940   Styrene-butadiene rubber
   1950   Polyisoprene

SYNTHESIS GAS. For a number of industrial organic syntheses that proceed in the gaseous phase, it is advantageous to prepare a chargestock to specification. When a mixture of gases is so prepared, the term *synthesis gas* is often used. Thus, there are several mixtures which qualify under this definition: (1) a mixture of H_2 and N_2 used for NH_3 synthesis; (2) a mixture of CO and H_2 for methyl alcohol synthesis; and (3) a mixture of CO, H_2, and olefins for the synthesis of oxo-alcohols. Ammonia synthesis gas is described briefly here.

The hydrogen required for NH_3 synthesis gas may be obtained in commercial quantities from coke oven water gas; from steam reforming of hydrocarbons; from the partial oxidation of hydrocarbon chargestocks; or from the electrolysis of H_2O. The nitrogen required may come from the introduction of air to the process, or where specifically required, pure nitrogen may be obtained from an air separation plant. Since NH_3 synthesis occurs under high pressure, it is advantageous to generate the synthesis gas at high pressure and thus avoid additional high compression costs. For this and other economic situations, coke oven gas and hydrogen from electrolysis are eliminated. This leaves hydrocarbons as the logical choice.

In the steam-hydrocarbon reforming process, steam at temperatures up to $850\,^{\circ}C$ and pressures up to 30 atmospheres reacts with the desulfurized hydrocarbon feed, in the presence of a nickel catalyst, to produce H_2, CO, CO_2, CH_4, and some undecomposed steam. In a second process stage, these product gases are further reformed. Air also is added at this stage to introduce nitrogen into the gas mixture. The exit gases from this stage are further purified to provide the desired 3 parts H_2 to 1 part N_2 which is the correct empirical ratio for NH_3 synthesis. See also **Ammonia**.

SYNTHETIC DIVISION. An abbreviated process, using detached coefficients, for finding the quotient of a polynomial in one variable x by a divisor of the form $x - r$, where r is a constant. The procedure may be illustrated by the polynomial $a_0x^4 + a_1x^3 + a_3x + a_4$. The results can be obtained in the following form, which should be self-explanatory:

$$
\begin{array}{ccc}
a_0 & a_1 & a_2 \\
 & a_0r & A_1r \\
a_0 \quad A_1 = a_0r + a_1, & A_2 = A_1r + a_2 \\
\end{array}
$$

$$
\begin{array}{cc}
a_3 & a_4 \quad r \\
A_2r & A_3r \\
A_3 = A_2r + a_3 & R = A_3r + a_4
\end{array}
$$

The quotient is $a_0x^3 + A_1x^2 + A_2x + A_3$ and the remainder is R. A similar process applies to a polynomial of any other degree. In the general case, if the coefficient of any power of x is missing, a zero should be supplied in the appropriate place in the first line of the scheme. If the remainder in the synthetic division is zero, then the divisor is a factor of the given polynomial.

See also **Polynomial**.

SYPHILIS. Syphilis is a complex sexually transmitted disease (STD), congenital or acquired, caused by the bacterium, *Treponema pallidum*. It has often been called "the great imitator" because so many of the signs and symptoms are indistinguishable from those of other diseases. Syphilis is characterized by primary and secondary stages, during which it is highly contagious, and a noncontagious, late, or tertiary stage, marked by involvement of many organs and tissues. In the congenital form, there is no primary lesion, and late manifestations predominate.

Epidemiology. The number of cases of syphilis reported and the reporting requirements range widely from one country to the next. In most industrialized nations, the disease is reasonably well documented, but even then most health officials believe that the disease is underreported, mainly in the underdeveloped countries, as exemplified by parts of Africa. There were alarming trends of case numbers in the United States since the mid-1950s. As observed by E.W. Hook (University of Alabama, Birmingham) and C.M. Marra (American Social Health Association) in April 1992, "In the early 20th century, syphilis was a major public health problem. In the 1920s more than 20 percent of the patients in U.S. mental institutions had tertiary syphilis (general paresis). Even before the therapeutic efficacy of penicillin was described, public health measures such as serologic-screening programs and treatment through government-funded rapid-treatment centers had begun to reduce the rate of syphilis in the United States. By 1941, only 10 percent of patients in mental institutions had tertiary syphilis. These changes accelerated after World War II as penicillin became widely available. In 1979 only 162 syphilis-related admissions to mental institutions were reported, 98 percent fewer than 60 years earlier. Similarly, the number of syphilis-related deaths declined more than 50-fold between the early 1940s and the 1980s."

By the mid-1950s only about 6500 cases of primary and secondary syphilis were reported per year in the United States. In the late 1950s and early 1960s, the number of cases increased, fluctuating between 19,000 and 26,000 cases per year until 1978. The rates increased slightly in the late 1970s and early 1980s, and a disproportionate number of cases occurred in homosexual men. In the mid-1980s, primarily because of behavioral changes adopted in response to the AIDS epidemic, syphilis declined among homosexual men, as did the male-to-female ratio of cases.

In 1985, the incidence of syphilis among heterosexual men and women began to increase rapidly, a trend that also led to dramatic increases in congenital syphilis. Between 1983 and 1990, when 50,225 cases of primary and secondary syphilis were reported, there was a 75% increase in the incidence of the disease. Although widespread, the changes were not geographically or racially uniform. The rates among black men increased 126% (from 69 to 156 per 100,000) and among black women 231% (from 35 to 116 per 100,000), whereas the rates for other subgroups of the population remained stable or even declined. Rates for non-Hispanic white men declined 50%, from 6 to 3 per 100,000 during the same period. Although rates of syphilis remain highest in urban areas of the southeastern United States and Texas, changes have been reported nationwide in both rural and urban areas. From 1989 to 1990 some of the greatest increases occurred in Midwestern cities where syphilis had previously been uncommon. For example, in three Ohio cities the rates increased between 235% and 293%, and similar increases occurred in St. Louis (172%) and Milwaukee (153%).

Several factors have contributed to the changing epidemiology of syphilis: (1) limited access to health care has delayed the diagnosis of syphilis; (2) a linkage with concomitant epidemics of illegal drug use, particularly of "crack" cocaine because cocaine is associated with prolific engagement in sexual practices as well as the "exchange" of sex in payment of drugs; and (3) the difficulty in tracing sexual partners among drug users. To address these problems, public health officials in some areas have modified traditional intervention activities, performing screening at crack houses and considering mass treatment of well-defined, identifiable risk groups.

By 1999 over 35,600 cases of syphilis in the United States, were reported by health officials in, including 6,650 cases of primary and secondary syphilis (a decline of 5.4% from 1998) and 556 cases of congenital syphilis in newborns. More cases occur each year than come to the attention of health officials. Of the nine states with the highest 1999 syphilis rates (2–5 times higher than the national rate of 2.5 cases per 100,000), eight were in the South. Although syphilis rates remain higher in the South than in other regions, the South had a 32% decline in the primary and secondary syphilis rate from 1997 to 1999, illustrating that the greatest improvements in disease control have taken place where syphilis incidence has been the greatest. In 1999, 25 counties accounted for 50% of all primary and secondary syphilis cases. Two hundred sixty-five counties had syphilis rates above the U.S. Public Health Service's Healthy People 2000 objective of 4 cases per 100,000. These 265 counties (9% of the total number of counties in the U.S.) accounted for approximately 74% of the total primary and secondary syphilis cases reported in 1999.

In 1999, syphilis occurred primarily in persons aged 20 to 39, and the reported rate in men was 1.5 times greater than the rate in women. The incidence of syphilis was highest in women aged 20 to 29 years and in men 30 to 39. Some fundamental societal problems, such as poverty, inadequate access to health care, and lack of education are associated with disproportionately high levels of syphilis in certain populations. Cases of primary and secondary syphilis in 1999 had the following race or ethnicity distribution: African Americans 75%, whites 16%, Hispanics 8%, and others 1%. Syphilis reflects one of the most glaring examples of racial disparity in health status, with the rate for African Americans nearly 30 times the rate for whites.

Social Vectors of the Disease. A.M. Brandt (Harvard Medical School) has reviewed how past experience in connection with the sociological aspects of syphilis may be helpful in relation to the current crisis concerning AIDS (acquired immunodeficiency syndrome). Brandt observes that AIDS, like syphilis in the past, engenders powerful social conflicts about the meaning, nature, and risks of sexuality; the nature and role of the state in protecting and promoting public health; the significance of individual rights in regard to communal good; and the nature of the doctor-patient relationship and social responsibility. The analogs that AIDS poses to syphilis are striking: the pervasive fear of contagion, concerns about casual transmission, the stigmatization of victims, and the conflicts between public health and civil liberties. The importance of the history of syphilis (reviewed by Brandt in *Science*, **239**, 375–369, January 22, 1988) is that it reminds the public of that range of forces that influence disease, health, and social policy.

Chronology of Syphilis. The early history of syphilis is not entirely clear. It is thought by some historians that the disease was first introduced into Europe by Columbus' returning sailors, and subsequently spread through Italy where it became a great scourge, by the soldiers of Charles V. Another view is that the disease has existed in civilized man since antiquity. Early Egyptian and Assyrian inscriptions as well as bony changes found in mummies are interpreted as supporting this theory. In the Middle Ages, syphilis occurred in severe, widespread epidemics with enormous mortality rates. Later it became a milder disease, and its venereal nature was recognized. It was not until 1905, when Schaudinn discovered the *Treponema pallidum* to be the causative organism, that syphilis and gonorrhea were recognized as two distinct diseases. Development of the dark field microscope made it possible for the first time to detect the causative spirochete.

Acquired Syphilis

The disease is almost always transmitted during sexual intercourse. The causative agent gains entry through an abraded surface, usually on the genital organs. In the male, the primary sore is commonly on the penis, where it is easily seen. In the female, it is found either on the mucous surface of the vulva or within the vagina, where it may remain unnoticed.

The primary stage is marked by the appearance of the chancre, which develops at the site of invasion, usually 2–4 weeks after exposure. Before the chancre develops, the organisms often have invaded the body tissues by way of the lymph channels and blood stream. A fully developed chancre appears as a clean, slightly raised, hard circumscribed ulcer, which exudes a thin, highly infectious secretion. By darkfield microscopic examination, this secretion can be seen to teem with spirochetes. The lesion is painless, but tender lymph nodes may develop in the groin.

The secondary stage begins 6 weeks to 6 months after the primary. By the time of onset of this period, the organisms have invaded the tissues, and the body's defense reactions are active. Antibodies are being produced, and the Wassermann reaction is positive. The second stage may be so mild as to pass unnoticed, but it is usually ushered in with a rash over the skin and mucous membranes. The eruption is extremely variable and may imitate any skin disease. The commonest type is the macular, in which the lesions consist of flat, or slightly raised, rose-colored spots, most prominent over the abdomen and chest. Characteristic features of the secondary eruption are its symmetrical distribution, painless, non-itching nature, and its tendency to appear on the palms and soles. Its duration is variable from weeks to months, and it may fade and leave a faint pigmentation for a time. With treatment the rash disappears promptly. Spirochetes are present in skin and mucous membrane lesions, and are particularly easy to demonstrate in the soft moist sores on the genitalia, the condylomata.

Constitutional symptoms in the secondary stage are usually mild. They consist of sore throat, slight fever, headache, and some enlargement of the superficial lymph nodes.

The latent period is the interval between the secondary and tertiary phases of the disease. During this time, the patient may enjoy excellent health, and be unaware that he has syphilis. The duration is variable — from a few weeks to many years — even 25 to 30. During this time, the various organs and tissues are harboring spirochetes, but these are inactive, hibernating, as it were.

The tertiary stage is the late picture, which is usually seen only in untreated syphilis. It is marked by the development of gumma, tumor-like masses in the skin and visceral organs, and inflammatory changes in the cardiovascular and central nervous systems. Since any organ or tissue may be attacked in tertiary syphilis, the signs and symptoms are extremely varied.

The most common types of tertiary syphilis are those involving the cardiovascular and central nervous systems. Cardiovascular involvement usually appears earlier — 10 to 15 years after infection. It occurs as aortitis,

aortic aneurysm and sometimes coronary occlusion. In syphilitic heart disease, aortitis is commonly followed by insufficiency of the aortic valve. This mechanical defect puts a tremendous strain on the heart, which compensates by hypertrophy of the muscle. Eventually the strain is too great for the reserve, and heart failure and finally death are the result.

Central nervous system syphilis may not develop for 30 years after the primary infection. It occurs as a meningitis, as an arteritis or inflammation of the small cerebral vessels with secondary changes in the brain tissue, and as involvement of the cells of the brain and cord in tabes dorsalis and paresis.

Congenital Syphilis

It is important that no pregnant woman have syphilis, for the causative organism of this disease is one of the few that can pass through the placental barrier to the unborn child. Once the disease establishes itself in the baby, abortion or stillbirth may follow. If syphilis is diagnosed early in pregnancy, treatment can be started which may curb the disease and allow the baby to develop normally. In the most severe forms of infection, the child may show at birth an extensive skin eruption, fissures about the angles of the mouth and nose, a characteristic nasal discharge, bone lesions, and enlargement of the liver and spleen. Underdevelopment and poor nutrition are conspicuous. Such a child rarely lives longer than a few days. In other instances, the infected infant may appear normal at birth, yet after a few months develop the typical signs and symptoms of the disease.

Children who survive the active congenital disease, or those in whom the infection remains latent, often show certain permanent stigmata which make the diagnosis apparent at a glance. "Saddle nose" and deformed (Hutchinson's) teeth — peg-shaped notched incisors which are widely spaced — are characteristic. Bony lesions, inflammation of the cornea of the eye (interstitial keratitis), deafness, and central nervous system syphilis, such as occur in the acquired form, are also seen in children with congenital syphilis.

A test for syphilis should be performed on all pregnant women at the initial visit to the obstetrician, and repeated during the third trimester in all cases where there may be a risk of acquiring syphilis. Where diagnosis is positive, the patient should be treated. Early syphilis in pregnancy is treated with the same dosage of antibiotics as for any other person in whom the disease has been diagnosed. Tetracycline is not recommended for syphilitic infections in pregnancy because of potential toxicity for mother and fetus. Monthly tests for syphilis should be made during the full period of pregnancy. Where effective treatment is given, the risk of congenital syphilis in the newborn is minimized.

Diagnosis of Syphilis

The first blood test for the disease was developed by August von Wassermann, a German bacteriologist, in 1907. The Wassermann test was used for many years, but during the interim a number of other serologic tests have been developed. These tests fall into two main categories: (1) treponemal tests which are designed to detect the treponemal antibody produced in response to syphilitic infection; and (2) nontreponemal or reagin tests, which detect an antibody-like substance (reagin), the latter found in the serum of an infected patient. Reagin is assumed to result from the interaction of *T. pallidum* with body tissue. Compared with the nontreponemal tests, the treponemal tests are relatively time-consuming and costly.

To guide therapeutic decisions and disease-intervention activities, syphilis is divided into a series of clinical stages. Despite its usefulness, clinical staging is imprecise. Patients with late stages of disease may have no recollection of signs of earlier stages, possibly because most syphilitic lesions are painless or because some patients may not have clinically apparent primary or secondary lesions. Also, there is considerable overlap between stages. For greater detail pertaining to diagnostic procedures, refer to the Hook-Marra paper.

Therapy for Syphilis. The primary goals of therapy are to prevent transmission and avoid the late complications in affected patients. *Treponema pallidum* cannot readily be propagated in vitro. Consequently, much research has been conducted using rabbit models.

Penicillin remains the best-studied and the preferred therapy for syphilis. A single dose of 2.4 million units of penicillin G benzathine results in a serum penicillin concentration of more than 0.018 micrograms per milliliter, which remains effective for about 3 weeks. Studies have shown that this dose for early syphilis is not complete and that retreatment is required in up to 10% of patients. There have been reports that penicillin G benzathine does not relieve neurological and ocular syphilis. However, prior to the era of the HIV epidemic, penicillin G benzathine therapy for early syphilis had a success rate of 95% or higher. This high rate was attributed to the fact that the therapeutic regimen and an intact immune system act together to clear peripheral organisms as well as those sequestered in the central nervous system and the eye. When symptoms of neurosyphilis are evidenced, the preferred treatment may be the use of high-dose intravenous penicillin (sterile) or a combination of intramuscular penicillin G procaine with oral probenecid.

Treatment of persons who are allergic to penicillin, repeated dosage of tetracyclines (although not for pregnant women), and possibly erythromycin may be considered. Information is still being collected on the efficacy of other antibiotics. Persons with the HIV infection present a considerably more complex situation. Both syphilis and HIV are protean diseases, and they interact on a number of levels. Syphilitic genital ulcers may enhance the acquisition and transmission of HIV. The natural history of syphilis may be modified in patients co-infected with HIV (human immunodeficiency virus). The results of laboratory tests for syphilis may be different in HIV-infected persons, thus misleading the proper therapy to be used. Numerous studies have shown that syphilis, as well as other genital-ulcer diseases, are disproportionately common in HIV-infected patients and vice versa, suggesting that syphilis increases the risk of the acquisition and transmission of HIV. Much more detail on this subject is given in the Hook-Marra reference listed.

Untreated Syphilis

As a cause of death, untreated syphilis has received extensive attention of medical statisticians and long and complex studies of this topic continue. Questions for which specific answers have not been fully provided include: To what degree does syphilis masquerade as another disease? If left untreated, what is the ultimate pathway of destruction caused by syphilis? To what degree is syphilis indirectly associated with death from other causes?

In an extensive study (Gjestland) of Figures gathered by Boeck and Brusgaard pertaining to 2,000 syphilitic patients who lived in Oslo, Norway during the period 1891–1910 and who had received no specific therapy for the disease, the following findings were revealed: (1) untreated syphilitic patients exceeded their expected mortality rates by 53% (male) and 63% (female); (2) the male patients developed cardiovascular syphilis in 13.0% of the cases—7.6% (female); (3) neurosyphilis developed in 9.4% (male) and 5.0% (female); (4) the mortality rate from syphilis among males was double that of females; (5) clinical or autopsy evidence of syphilitic pathology of a serious nature was noted in 23% of all patients.

A similar study was made of males 25 years or older in the United States. Known as the Tuskegee Study, these findings have been summarized as follows: (1) Only one-fourth of the untreated syphilitics were normal after an infection of 15 years' duration. Most of the abnormal findings were associated with the cardiovascular system; (2) During the first 12 years of observation, 25% of the syphilitics and 14% of the control group (no syphilis) of comparable ages had died. At age 25, untreated male syphilitics would have a reduction of life expectancy of about 20%; (3) After 25 years of follow-up to this study, the life expectancy of individuals with syphilis (ages 25 to 50) was found to be 17% of the control group. A male with untreated syphilis of more than 10 years' duration and a sustained reactive serology would have approximately a 1:1 chance of having demonstrable cardiovascular involvement (autopsy findings).

Another interesting study concerned with the outcome of untreated syphilitic infection was done by Rosahn (Yale) by analyzing autopsy records from 1917 to 1941, consisting of 3,907 cases over age 20. Major observations included the following. (1) About 9.7% of the population studied had clinical laboratory or anatomic evidence of syphilis. Earlier, in 1938, Vonderlehr and Usilton estimated that one person in ten in the United States would have syphilis before dying. (2) Males with lesions at autopsy comprised 4.7% of the male population; 2.7% of the females, indicating a sexual bearing upon the resistance to tissue changes of late syphilis. (3) Syphilis significantly reduced longevity, whether or not there were tissue lesions. (4) No greater frequency of anatomic lesions was apparent among blacks than whites, but whites with such lesions died at a greater rate than blacks, indicating that as previously believed, syphilis does not run a more fatal course among blacks than among whites. (5) About 39% of untreated syphilitics indicated anatomic evidence of syphilis at autopsy.

(6) About 23% of the untreated syphilitics died primarily as the result of the disease. There were several parallels between the Yale study and the Brusgaard study previously described.

Additional Reading

Barton, S.E. and P.E.Hay: "Handbook of Genitourinary Medicine," Oxford University Press, Inc., New York, NY, 1999.
Borchart, K.A. and M.A. Noble: "Sexually Transmitted Diseases: Epidemiology, Pathology, Diagnosis," CRC Press, LLC, Boca Raton, FL, 1997.
Brandt, A.M.: "The Syphilis Epidemic and Its Relation to AIDS," *Science*, 375 (January 22, 1988).
Brandt, A.M.: "No Magic Bullet: A Social History of Venereal Disease in the United States Since 1880," Oxford University Press, Inc., New York, NY, 1990.
Elsner, P. and A. Eichmann: "Sexually Transmitted Diseases: Advances in Diagnosis and Treatment," S. Karger Publishers, Inc., Farmington, CT, 1996.
Grapes, B.J. and B. Scott: "Sexually Transmitted Diseases," Gale Group, Farmington Hills, MI, 2001.
Handsfield, H.H.: "Color Atlas and Synopsis of Sexually Transmitted Diseases," 2nd Edition, The McGraw-Hill Companies, Inc., New York, NY, 2000.
Holmes, K.K., Per-Anders Mardh, and J. Wasserheit: "Sexually Transmitted Diseases," 3rd Edition, The McGraw-Hill Companies, Inc., New York, NY, 1999.
Hook, E.W. and C.M. Marra: "Acquired Syphilis in Adults," *N. Eng. J. Med.*, 1060 (April 16, 1992).
Hutchinson, C.M. and E.W. Hook: "Syphilis in Adults," *Medical Clinician North America*, **74**, 1389 (1990).
Lynch, P.J. and L. Edwards: "Genital Dermatology," Churchill Livingstone, Inc., Philadelphia, PA, 1994.
Mandell, G.L., R. Dolin, J.E. Bennett, and R.G. Douglas: "Mandell, Douglas, and Bennett's Principles and Practice of Infectious Diseases," 5th Edition, Churchill Livingstone, Inc., Philadelphia, PA, 1998.
Marr, L.: "Sexually Transmitted Diseases: A Physician Tells You What You Need to Know," Johns Hopkins University Press, Baltimore, MD, 1998.
Morse, S.A., K.K. Holmes, and A.A. Moreland: "Atlas of Sexually Transmitted Diseases and AIDS," 2nd Edition, Mosby-Year Book, Inc., St Louis, MO, 1995.
Peeling, R.W. and P.F. Sparling: "Sexually Transmitted Diseases: Methods and Protocols," Vol. 20, Humana Press, Totowa, NJ, 1998.
Quetel, C. and J. Braddock: "The History of Syphilis," Johns Hopkins University Press, Baltimore, MD, 1990.
Reverby, S.M.: "Tuskegee's Truths: Rethinking the Tuskegee Syphilis Study," University of North Carolina Press, Chapel Hill, NC, 2000.
Sparling, P.F.: "Natural History of Syphilis," in Sexually Transmitted Diseases (K.K. Holmes and P.A. Mandh, Editors), The McGraw-Hill Companies, Inc., New York, NY, 1990.
Stanberry, L.R. and D. Bernstein: "Sexually Transmitted Diseases: Vaccines, Prevention and Control," Academic Press, Inc., San Diego, CA, 2000.
White, D.O. and F.J. Fenner: "Medical Virology," 4th Edition, Academic Press, Inc., San Diego, CA, 1994.

Web References

Centers for Disease Control and Prevention: *http://www.cdc.gov/std/*
Mayo Clinic: Reproductive Diseases and Disorders: *http://www.mayoclinic.com/invoke.cfm?id = HQ01378*
Sexually Transmitted Diseases: *http://www.stdjournal.com/*
STD: *http://www.i-std.com/*
Syphilis: *http://www.niaid.nih.gov/factsheets/stdsyph.htm*
The CDC National Prevention Information Network: *http://www.cdcnpin.org/std/start.htm*

SYSTEMIC LUPUS ERYTHEMATOSUS.   Generally grouped with the rheumatic diseases, *systemic lupus erythematosus* (frequently shortened to *lupus* and abbreviated, SLE) was, until a few years ago, considered to be a rather rare condition. With an improved understanding of the disease and the use of more definitive diagnosis and precise nomenclature, the disease has become better identified. It is now variously estimated to affect 5 cases per 100,000 population—to as high as 50,000 new cases per year in the United States alone. Records to date indicate that SLE occurs in women at a rate nearly ten times that of men and that it usually appears during the second and third decade of life. Thus, the generalization can be made that SLE is predominantly a disease of young women.

Systemic lupus erythematosus is a chronic multisystem inflammatory disease with currently unknown origin. Clinical features are quite variable. SLE ranges from a moderate, benign condition to one with serious consequences that can lead to death. As with rheumatoid arthritis, SLE is an autoimmune disease, in which the body's natural defense mechanisms behave erratically. See **Immune System and Immunology**.

The diagnosis of SLE is frequently complex, but the physician will suspect the condition when a young female is seen with a combination of fever, skin rash, and arthritis. The fever is usually low-grade, but in a so-called "lupus crisis" may reach 104.9 °F (40.5 °C). The typical skin lesion appears in the facial region, particularly involving the nose and cheeks, producing what is sometimes called a "butterfly rash." Alopecia (spotty loss of hair) also may occur. There also may be ulcerations of the mouth and lips. It is believed that the skin conditions arise from a superabundance of keratin, which plugs the hair follicles and sweat and sebaceous glands. The arthritis most often involves the hands, wrists, elbows, knees, and ankles. Cartilage or bone degradation is not usually found. Tendonitis is not uncommon.

Further examination may show involvement of SLE in other organs. causing, for example, acute hemolytic anemia, thrombocytopenic purpura, or pericarditis. Most frequently, after a flare-up, there will be characteristic remissions, which may extend from months to years. The risk of an acute attack depends to a large degree upon the number of organs that have been affected. Prognosis of recovery and remission diminishes in cases where the kidneys and central nervous system are involved. Situations which appear to precipitate the occurrence of SLE and which aggravate symptoms appear to include infections, stress (emotional and physical), surgery, pregnancy, some drugs, and undue exposure to sunlight.

Moderate involvement of the kidneys may occur in nearly all cases. Less commonly, renal malfunction may be serious and pose a life-threatening situation. Nervous system involvements occur in about half the cases, these varying considerably in magnitude. These may be manifested in psychoses and, in particular, organic psychoses, which include deterioration of intellectual capacity, memory, and disorientation. These manifestations are usually of a transient nature, but they may occur any time during the course of the disease. Cardiac involvement is usually confined to pericarditis, but endocarditis occurs in some cases. Involvement of the respiratory system may cause pleuritic chest pain. Pleural fluid, in such cases, will almost always show high concentrations of protein. In some cases, bacterial pneumonia, pulmonary infarcts, uremic pneumonitis, and congestive heart failure may develop.

The treatment of SLE, as of the early 1990's remains somewhat controversial and consequently nonuniform. This will change with time as both fundamental scientific information, better clinical statistical data, and empirical observations are collected. Treatment is complicated by the variety of symptoms presented by SLE in various patients. Generally, therapy is directed to greater periods of rest to reduce the effects of emotional and physical stress. Aspirin is commonly suggested as an anti-inflammatory drug. When skin rashes are difficult to control, antimalarials, such as chloroquine and hydroxychloroquine, have been used.

Glucocorticoids will demonstrate marked suppression of SLE manifestations in many patients, but these drugs are usually reserved for patients with serious life-threatening conditions.

Currently, there is much research directed toward a better understanding of the origin and course of SLE. This research is motivated not only in the interest of developing improved therapy for SLE patients, but also as a probable rewarding avenue to a better understanding of many autoimmune phenomena. Histocompatibility antigens located on cell surfaces are antigens that elicit rejection of transplanted organs by the immune system. Researchers are interested in these antigens as regards SLE to determine if they are directly involved in provoking autoimmunity or that they may be indirectly involved because of their genetic association with immune response genes. Faulty control of immune responses has been strongly implicated in the etiology of lupus.

Additional Reading

Hughes, G.R.: "Lupus," Oxford University Press, Inc., New York, NY, 2000.

Kammer, G.M. and G.C. Tsokos: "Lupus: Molecular and Cellular Pathogenesis," Vol. 7, Humana Press, Totowa, NJ, 1999.

Lahita, R.G.: "Systemic Lupus Erythematosus," 3rd Edition, Academic Press, Inc., San Diego, CA, 1998.

Maraux, A.: "Everything You Need to Know about Lupus," Rosen Publishing Group, Inc," New York, NY, 2000.

Miescher, P.A.: "Systemic Lupus Erythematosus," Springer-Verlag, Inc., New York, NY, 1995.

Schur, P.H.: "The Clinical Management of Systemic Lupus Erythematosus," 2nd Edition, Lippincott-Raven Publishers, Philadelphia, PA, 1996.

Wallace, D.J.: "The Lupus Book: A Guide for Patients and Their Families," 2nd Edition, Oxford University Press, Inc., New York, NY, 2000.

Wallace, D.J. and B.H. Hahn: "DuBois' Lupus Erythematosus," 6th Edition, Lippincott Williams & Wilkins, Philadelphia, PA, 2001.

Web References

Lupus Foundation of America: *http://www.lupus.org/*

Lupus Information Center: *http://www.geocities.com/HotSprings/Spa/9534/facts.html*

SYSTEMS ENGINEERING.   A term widely used in engineering and industrial planning, defined somewhat differently in various organizations. Probably there is general agreement that it is the application of the scientific method to a communications system, a data processing system, or an aircraft, a ship or an entire business. Moreover, most of its practitioners would extend the definition to groups of businesses, that is, to industries or even entire societies.

SYZYGY.   The points in the moon's orbit about the earth at which the moon is new or full.

T

TABER ICE. A seam, lens, or layer of ground ice, usually pure, formed by the drawing in of water to a growing ice crystal as the ground becomes frozen. Sometimes referred to as *segretation ice* or *sirlin-type ice*.

TABES. (1) A progressive wasting of the body or part of it. (2) *Tabes dorsalis* or locomotor ataxia is a syphilitic infection of the spinal cord. This form of syphilis occurs in about 5% of untreated syphilitics around the ages of 25 to 40 years, that is from 8 to 15 years after the original syphilitic infection. Cure is extremely difficult and often impossible. The longer the infection has been present the less chance there is of a cure or even of arresting the disease. See also **Syphilis**.

The symptoms of spinal cord syphilis are numerous and do not make their appearance in any set order. The first group of symptoms usually noticed are: lightning pains, that is, severe shooting pains of a few seconds' duration, usually of the leg (or prolonged pains resembling "rheumatism"; (2) Absence of knee jerks; (3) Fixed and irregular pupils; (4) Difficulty in urination.

The second stage symptoms merge gradually with the first so that in addition to the foregoing symptom difficulty in walking and in-coordination of all forms of muscular movement gradually increase. The walk is often characteristic. The foot is raised high, thrown forward forcibly, and slapped to the ground.

In the last or third stage paralysis often occurs. There is increased mobility about the joints producing a "flail" joint. Eye symptoms sometimes lead to blindness. Impotence is usual. Paroxysms of acute pain simulating surgical lesions occur in various abdominal organs. These may last several days and may be accompanied by vomiting. Perforating ulcers appear on the feet, muscular wasting, weakness, and loss of control of muscle, progresses and death occurs finally from some terminal infection such as pneumonia.

The diagnosis is made from the symptoms, physical signs, and serologic tests of the blood and spinal fluid.

Treatment is carried out in courses lasting several years. The various anti-syphilitic drugs are used intramuscularly, intravenously and intra-spinally. In cases not responding to this form of treatment, the patient is inoculated with malaria. The high fever produced by this disease sometimes checks the syphilis of the nervous system. It is more often used with paresis, which may also be present with *tabes dorsalis*. The malaria is then cured with quinine. Other forms of fever therapy may also be used.

TABLELAND. (a) A general term for a broad, elevated region of land with a nearly level or undulating surface of considerable extent, as found in much of South Africa. Sometimes referred to as a *continental plateau*. (b) A plateau bordered by abrupt cliff-like edges rising sharply from the surrounding lowland, sometimes called a *mesa*.

TABULAR ICEBERG. A flat-topped iceberg that may be very large (up to 160 km long and more than 500 m thick), with cliff-like sides. Tabular icebergs are usually detached from an ice shelf and are numerous in the Antarctic Sea. Sometimes called *ice lands*.

TACHYCARDIAS. Tachycardia means "fast heart rate" and refers to heart rates greater than 100 beats per minute. Tachycardias can be an appropriate response to exercise or other physiologic causes such fever. Other tachycardias can be inappropriate manifestations due to heart disease or problems with the intrinsic electrical system of the heart. Three mechanisms are thought to explain how tachycardias occur; *abnormal automaticity*, *reentry*, and *triggered activity*.

Automaticity refers to the spontaneous generation of impulses within a tissue. Automaticity is a property of all the heart tissues and is responsible for the normal pacemaker function of the heart. Metabolic abnormalities of cells in the atria, AV junction, or ventricles may cause an increase in the rate of spontaneous impulse formation. When this occurs, tachycardia occurs. A variety of illnesses and drugs can affect automaticity, but abnormal automaticity is thought to account for less than 10% of tachycardias. The most common automatic tachycardias are seen in acutely ill patients with underlying metabolic disease.

Reentry refers to the repetitive cycling of an impulse over a fixed circuit. In reentry, two roughly parallel conducting pathways with common origins connect to form a potential electrical circuit. Normally, impulses travel down the pathways side by side and therefore do not interfere with the rhythm. When a premature beat enters the circuit, however, the difference in conduction properties between the two pathways allows the impulse to travel down only one of the pathways. When the impulse reaches the intersection with the second pathway at the opposite end of the circuit where the pathways again intersect, the impulse travels backward up the second pathway to the intersection with the first pathway. The impulse then "reenters" the first pathway and travels back down it to restart the cycle. The continuously circulating impulse spins around and around the reentrant loop and overtakes the heart's rhythm when it exits the circuit during each lap. Reentrant circuits occur rather frequently both in normal and abnormal hearts, and some may be present at birth. Reentrant circuits may also be acquired as a result of cardiac damage, as is thought to be the case in many patients when ventricular tachycardia appears following a myocardial infarction. See also **Arrhythmias (Cardiac)**.

TACHYLYTE (or Tachylite). Pure tachylite is a natural, basic black glass, which may form along the chilled contacts of dikes or sills. It also occurs as a rind on basic pillow lavas that have been suddenly chilled by plunging into water. Occasionally it forms entire flows from certain Hawaiian volcanoes.

TACONITE. Term for the ferruginous cherts associated with the iron ores of Lake Superior district.

TACTILE ORGANS. Organs of touch. Sensory organs located at the surface of the body that are stimulated by pressure. The chief tactile organs of the human body are known as tactile corpuscles. They consist of an elliptical bulb about $\frac{1}{300}$ inch (0.008 centimeter) or less in length, enclosed in a connective tissue sheath and divided incompletely by plates of the same tissue. From one to several coarse nerve fibers enter the corpuscle and follow a winding course inside, ending between or on the connective tissue cells. These corpuscles are especially abundant in the skin of the finger tips and in other parts where the sense of touch is highly developed, hence there can be no reasonable doubt that they serve this sense. They are related to other corpuscles of more limited occurrence which may also be tactile, and are supplemented in the skin by free nerve endings, some of them expanded into disk-like nets, which are regarded as tactile.

In many invertebrates hairlike projections at the surface of the body serve for ready reception of tactile stimuli and in the vertebrates true hairs, often bristlelike, transmit such stimuli to the nerve endings at their bases. From the sensitiveness of the human scalp to light contacts transmitted through erect hairs, it is easy to judge the value of these tactile hairs. They are well-illustrated by the whiskers or vibrissae of the cat and other mammals. See also **Sensory Organs**.

Tactile Feedback. It is common practice in electronic equipment today for designers to rely on coding and switching devices that provide the operator no tactile confirmation that a switch, for example, has been actuated. Shielded, surface-type switches are widely used in cash registers, microwave ranges, and industrial controls, where a switching action is not "felt" by the operator. For maximum assurance to the operator and thus to safety, many designers prefer to use mechanical-type switches that "click" or in some other tactile manner confirm the operation to the operator. Several types of switches offering tactile feedback are shown in Fig. 1.

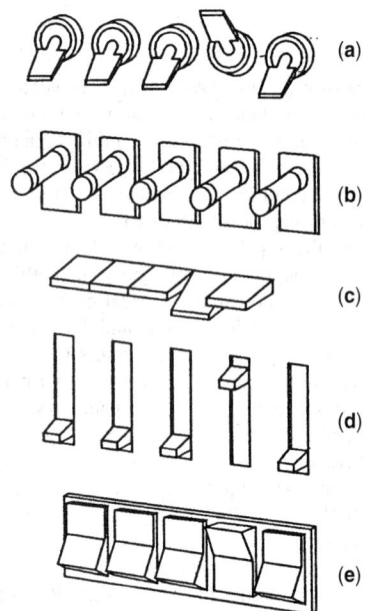

(a)

(b)

(c)

(d)

(e)

Fig. 1.   Types of industrial switches that provide tactile feedback to the operator: (**a**) toggle, (**b**) push-pull, (**c**) paddle, (**d**) slide, and (**e**) rocker switches. (*Microswitch.*)

TADPOLE.   See Frogs and Toads.

TAHR.   See Goats and Sheep.

TAILWIND.   See Jet Streams.

TAIPAN.   See Snakes.

TAKIN.   See Goats and Sheep.

TALBOT LAW.   The apparent brightness of an object viewed through a slotted-disk, rotating above a critical frequency, is proportional to the ratio, of the angular aperture of the opening to the opaque sectors.

TALC.   The mineral talc is a magnesium silicate corresponding to the formula $Mg_3Si_4O_{10}(OH)_2$ which occurs as foliated to fibrous masses, its monoclinic crystals being so rare as to be almost unknown. It has a perfect basal cleavage, the folia nonelastic although slightly flexible; it is sectile and very soft; hardness, 1; specific gravity, 2.5–2.8; luster, waxlike or pearly; color, white to gray or green; translucent to opaque. It has a distinctly greasy feel. Talc is a metamorphic mineral resulting from the alteration of silicates of magnesium like pyroxenes, amphiboles, olivine and similar minerals.

It is found chiefly in the metamorphic rocks, often those of a more basic type due to the alteration of the minerals above mentioned. Some localities are the Austrian Tyrol, the St. Gotthard district of Switzerland, Bavaria and Cornwall, England. In Canada, talc is found in Brome County, Quebec and Hastings County, Ontario. In the United States, well-known localities are to be found in Vermont, New Hampshire, Massachusetts, Rhode Island, New York, Pennsylvania, Maryland, and North Carolina.

A coarse grayish-green talc rock has been called soapstone or steatite and was formerly much used for stoves, sinks, electrical switchboards, etc. Talc finds use as a cosmetic, for lubricants and as a filler in paper manufacturing. Most tailor's "chalk" consists of talc. The origin of the word talc is not definitely known.

See also terms listed under **Mineralogy**.

TALUS.   The mass of coarse rock fragments which accumulate at the foot of a cliff as a result of the processes of weathering and gravity. In Great Britain, the term scree is used for such material.

TAMARACK.   See Larch Trees.

TAMM LEVELS.   Surface states; the extra electron energy levels found at crystal surfaces.

TANAGER (*Aves, Passeriformes*).   An American bird of brilliant coloration, related to the finches. The tanagers are chiefly birds of the Central and South American tropics, where over 200 species are found. Of the few species that enter the United States, only the scarlet tanager, *Piranga erythromelas*, is widely known. The brilliant red of its body, contrasting with the black wings and tail, makes it one of our most striking species. The dull olive plumage of the female exemplifies a common contrast between the sexes in the entire group.

TANDEM (Cascade).   Two-terminal pair networks are in tandem when the output terminals of one network are directly connected to the input terminals of the other network.

TANGENT (Geometry).   Take a point P on a curve, a neighboring point P', also on the curve, and draw the secant line PP'. As P' approaches P, the secant approaches a limiting position which is the tangent to the curve at P. The slope of the tangent at P is frequently called the slope of the curve at that point. See Fig 1.

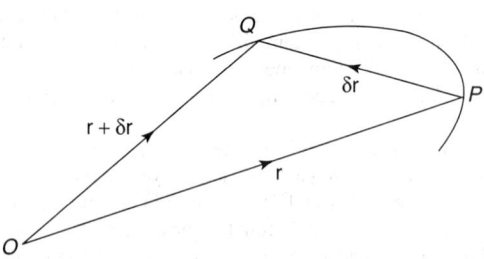

Fig. 1.   Tangent to a curve at a point P. Let $\mathbf{r}$ and $\mathbf{r} + \delta\mathbf{r}$ be the vectors drawn from an origin O to the point P and a neighboring point Q on the curve. Let δs be the distance measured along the curve from P to Q. Denote the limit $\lim_{\delta s \to 0} \delta\mathbf{r}/\delta s$, if it exists, by $\mathbf{t}$. Then $\mathbf{t}$ is the unit tangent to the curve at P. A line through P in the direction of $\mathbf{t}$ is called the tangent to the curve at P.

The equation of a plane curve may be given, in rectangular coordinates, in either one of the equivalent forms: (a) $y = f(x)$; (b) $x = f_1(t)$, $y = f_2(t)$, where t is a parameter; (c) $F(x, y) = 0$. If ϕ is the angle between the tangent to the curve and the X-axis, then $\tan \phi$ for the three cases is given by: (a) y'; (b) f_2'/f_1'; (c) $-F_x/F_y$. The tangent to the curve is $y \csc \phi$. Additional quantities of interest are: the subtangent, $y \cot \phi = y/y'$; the normal, $y \sec \phi = y\sqrt{1 + y'^2}$; the subnormal, $y \tan \phi = yy'$. The intercept of the tangent on the X-axis is $x - y/y'$ and on the Y-axis, $y - xy'$. Corresponding quantities for the normal are $x + yy'$ and $y + x/y'$, respectively.

Equations of the tangent at (x_0, y_0) for the three forms of the curve are: (a) $y - y_0 = f'(x_0)(x - x_0)$; (b) $(y - y_0)f_1'(t) = (xx_0)f_2'(t)$; (c) $(x - x_0)F_x + (y - y_0)F_y = 0$, where the derivatives are to be evaluated at (x_0, y_0). The corresponding equations for the normal are:

$$\text{(a)} \quad f'(x_0)(y - y_0) + (x - x_0) = 0$$

$$\text{(b)} \quad (y - y_0)f_2'(t) + (x - x_0)f_1'(t) = 0$$

$$\text{(c)} \quad (x - x_0)F_y - (y - y_0)F_x = 0$$

All tangents to a surface at a given point lie in a tangent plane to the surface. If the surface is given by $z = f(x, y)$, the equation of the tangent plane is $(x - x_0)F_x + (y - y_0)F_y = (z - z_0)$, where the partial derivatives are to be evaluated at the given point.

See also **Circle (Geometry)**; and **Circular Curves**.

TANGENTIAL ACCELERATION.   1. In meteorology, the component of the acceleration directed along the velocity vector (streamline), with magnitude equal to the rate of change of speed of the parcel dV/dt, where V is the speed. See also **Atmosphere (Earth)**.

2. In horizontal, frictionless atmospheric flow, the tangential acceleration is balanced by the tangential pressure force

$$\frac{dV}{dt} = -\alpha \frac{\partial p}{\partial s}$$

where α is the specific volume, p the pressure, and s a coordinate along the streamline. Thus, flow without tangential acceleration is along the isobars, and the wind is the gradient wind.

TANGERINE. See **Citrus Trees**.

TANK CIRCUIT. In a signal transmission system, a circuit consisting of inductance and capacitance, capable of storing electric energy over a band of frequencies continuously distributed about a single frequency at which the circuit is said to be resonant (tuned). The selectivity of the circuit is proportional to the ratio of the energy stored in the circuit to the energy dissipated. This ratio is sometimes termed the Q of the circuit.

TANNIN. Substances found in many plants; generally related to one of the phenols, pyrogallol or catechol. By their action on animal skins, they cause changes that make the skins resistant to decomposition and at the same time leave them flexible and very strong, greatly improved in wearing qualities. Skins so treated are said to be tanned, and are called leather. Tanning is a very old art, having been practiced in China since long before the Christian era. It was also known to the American Indians before the arrival of the Europeans.

Tannins are found in various parts of the plant, appearing frequently in leaves, and in the cortical tissues of stems. Tannins may be found in the walls of cells or in the vacuoles; often their presence causes the cell to appear dark-colored. Many fruits, such as the persimmon, contain large amounts of tannin, especially before they are ripe. Wound tissues, and especially the hypertrophied tissues known as galls, which result from the bits of certain insects, are particularly rich in tannins. Tannins appear to be by-products of the metabolism of the plant. When present in the epidermal cells, tannins are seen as a deterrent to snails, which might injure the leaf by feeding on it, to parasitic fungi, which might otherwise enter the leaf tissue, and as a protection against desiccation, since they form substances impervious to water.

An important source of tannin is the bark of various trees, especially that of the hemlock and several species of oaks. The bark is removed from the tree in sheets approximately 4 feet long. Stripping from the tree is usually done in the spring, when the cambial cells are most active and the bark separates easily. To remove the bark, two rings are cut completely through the bark and around the tree. A longitudinal slit is made through the bark from one ring to the other. With the use of a blunt, long-handled implement, the bark is then pried loose from the tree and allowed to dry. By felling the tree, the entire trunk may be stripped of its bark in this way. The dried bark is shipped to mills, where the tannin is extracted. Tannins from these barks are used to tan leather for shoe-soles and other heavy leathers. The wood of the chestnut tree yields a tannin similarly used.

Trees of the genus *Schinopsis*, native to the southern part of South America, including southern Brazil, Bolivia and other southern countries are very important source of tannin. These trees are known by the name "quebracho," which means "ax-breaker," because of their very hard, dense, heavy, dark-red wood, which is cut with difficulty. The heartwood of the tree contains 20–27% tannin, which is obtained by cutting the wood into small chips and extracting with water. This tannin is often used in combination with tannins from other plants.

The bark of many other trees yields large amounts of tannins. Among these are the mangrove, and several species of Acacia, known as wattles, natives of Australia. Fruits also may be a source of tannin. The fruits of *Terminalia chebula*, called *myrobalans*, are an important tannin source. The tree is a native of tropical Asia. Another fruit rich in tannin is *divi-divi*, the pods of a legume, *Caesalpinia coriaria*, which is native in tropical America and the West Indies. Sumac leaves, especially those of *Rhus coriaria*, a shrub or small tree native in Mediterranean Europe, are rich in tannins. To obtain the tannin, the plants are cut down and spread out to dry. The leaves are then removed from the stems and packed into bags, which are shipped to the mills. There the leaves are first cleaned and then ground up. The tannins from this source are used in manufacturing fine leathers, like glove

leathers. Leaves of other species of sumac, including the various American sumacs, also contain tannins which, however, are not so valuable and are little used.

Tannins are solids, soluble in water or alcohol, usually extracted by hot water, insoluble in ether, chloroform, carbon disulfide, benzene, soluble in alcohol-ether mixture, and in ethyl acetate, possessing a bitter astringent taste. Tannins (1) yield precipitates with gelatin, proteins (connected with the property of making leather from hides), alkaline salt solutions of many heavy metals, e.g., lead acetate, copper acetate (precipitate brown), antimonyl tartrate, concentrated dichromate solution, also by chromic acid (1% CrO_3); (2) yield dark blue or green coloration with ferric salt solutions; (3) in alkaline solution, absorb oxygen and yield dark colored solution; (4) with iodine in potassium iodide plus small proportion of ammonium hydroxide, yield red color (5) with dilute solution potassium hexacyanoferrate(II) in ammonium hydroxide, yield a red to brown coloration (care not to use excess reagent).

While tannins probably vary in composition, the type generally termed tannic acid is a pentadigalloylglucose for hydrolysis yields diagallic acid and glucose.

Additional Reading

Hemingway, R.W. and J.J. Karchesy: "Chemistry and Significance of Condensed Tannins," Perseus Books, Boulder, CO, 1989.
Hemingway, R.W. and P.E. Laks: "Plant Polyphenols: Synthesis, Properties and Significance," Kluwer Academic Publishers, Norwell, MA, 1992.
Lemmens, R.H. and N. Wulijarni-Soetjipto: "Dye and Tannin-Producing Plants," Balogh Scientific Books, Champaign, IL, 1991.
Salunkhe, D.K., J.K. Chavan, and S.S. Kadam: "Dietary Tannins: Consequences and Remedies," CRC Press, LLC., Boca Raton, FL, 1990.

TANTALITE. This black mineral, (Fe, Mn)(Ta, Nb)$_2$O$_6$, is isomorphous with columbite and dimorphous with tapiolite. Tantalite occurs in pegmatites and is a principal ore of tantalum.

TANTALUM. A major portion of this article was furnished by M. Schussler, Fansteel, North Chicago, Illinois. Chemical element symbol Ta, at. no. 73, at. wt. 180.948, periodic table group 5, mp 2,996 °C, bp 5,427 °C, density 16.65 g/cm³ (solid at 20 °C), 17.1 (single crystal). Elemental tantalum has a body-centered cubic crystal structure. Because of high mp, it is considered a refractory metal.

Tantalum is a slightly bluish metal; ductile, malleable, and when polished resembles platinum; burns upon being heated in air; insoluble in HCl or HNO$_3$, but soluble in hydrofluoric acid or a mixture of hydrofluoric and HNO$_3$. The tough, impermeable oxide film formed on the metal when exposed to air makes tantalum the most resistant of all metals to atmospheric corrosion. Tantalum was first identified by Ekeberg as a new element in yttrium minerals in 1802 and was first obtained in pure form by Berzelius in 1820 by heating potassium tantalofluoride with potassium. There is one, naturally occurring stable isotope ^{181}Ta. ^{180}Ta also occurs naturally (isotopic abundance 0.012%), with a half-life of something greater than 10^7 years. At least nine other radioactive isotopes have been identified ^{176}Ta through ^{179}Ta and ^{182}Ta through ^{186}Ta. With exception of ^{179}Ta (half-life of about 600 days), the remaining half-lives are expressed in minutes, hours, or days. ^{182}Ta has been used as a source of gamma rays. See also **Radioactivity**. In terms of abundance, tantalum does not appear on the list of the first 36 elements that occur in the earth's crust and hence is relatively scarce. Also, tantalum does not appear on the list of the first 65 elements that are found in seawater. First ionization potential, 7.7 eV. Oxidation potential 2Ta + 5H$_2$O ← Ta$_2$O$_5$ + 10H$^+$ + 10e$^-$, 0.71 V. Other important physical properties of tantalum are given under **Chemical Elements**.

Tantalum is found in a number of oxide minerals, which almost invariably also contain niobium (columbium). The most important tantalum-bearing minerals are tantalite and columbite, which are variations of the same natural compound (Fe, Mn)(Ta, Nb)$_2$O$_6$. Much of the tantalum concentrates has been obtained as a byproduct from tin mining; in recent years, tin slags, which are a byproduct of the smelting of cassiterite ores, such as those found in the Republic of Congo. Nigeria, Portugal, Malaya, and Thailand have been an important raw material source for tantalum.

The first successful industrial process used to extract tantalum and niobium from the tantalite-columbite-containing minerals employed alkali fusion to decompose the ore, acid treatment to remove most of the impurities, and the historic Marignac fractional-crystallization method to separate the tantalum from the niobium and to purify the resulting K_2TaF_7. Most tantalum production now employs recovery of the tantalum and niobium values by dissolution of the ore or ore concentrate in hydrofluoric acid. Then the dissolved tantalum and niobium values are selectively stripped from the appropriately acidified aqueous solution and separated from each other in a liquid-liquid extraction process using methyl isobutyl ketone (MIBK) or other suitable organic solvent. The resulting purified tantalum-bearing solution is generally treated with potassium fluoride or hydroxide to recover the tantalum in the form of potassium tantalum fluoride, K_2TaF_7, or with ammonium hydroxide to precipitate tantalum hydroxide, which is subsequently calcined to obtain tantalum pentoxide, Ta_2O_5. Tantalum metal is generally obtained by sodium reduction of K_2TaF_7, although electrolysis of K_2TaF_7 and carbon reduction of Ta_2O_5 in an electric furnace have also been used. Tantalum metal can absorb large volumes of hydrogen during heating in a hydrogen-bearing atmosphere at an intermediate temperature range (450–700 °C). The hydrogen is readily removed by heating in vacuum at higher temperatures.

Uses: Tantalum is used widely, although in small quantities in the electronics industry in electrolytic capacitors, emitters, and getters. The corrosion resistance of tantalum has been compared with that of glass. Additionally, the metal has a high heat-transfer coefficient and is easy to fabricate. Consequently, it finds use in equipment that must resist strong corrosive attack, as in the manufacture of HCl, hydrogen peroxide, in chromium plating baths, in bromine heaters and stills, and in the preparation of corrosive fine chemicals, such as ethyl bromide. The metal also has been used in resistance heaters in very high-temperature furnaces and for some nuclear reactor parts.

Alloys: Tantalum is added to nickel and nickel-cobalt superalloys for gas-turbine and jet-engine parts. Several surgical applications for tantalum have developed because of the inertness of the metal to body fluids and the tolerance of the body for the metal. Tantalum may be placed in the skull or other body parts without rejection. Strips and screws made of tantalum are used for holding broken pieces of bone and tantalum wire mesh is used for surgical staples, braid for sutures, and reinforcements. Tantalum-base alloys are used for aerospace structures and space power systems, principally because of the high-temperature stability and strength of these alloys. They operate satisfactorily at temperatures in excess of 1,600 °C. Tantalum alloys are used in heat exchangers. See Fig. 1. Small additions of zirconium to tantalum increases its tensile strength at normal temperatures and up to approximately 1,200 °C. Also, when added in amounts of about 5%, hafnium, molybdenum, rhenium, tungsten, and vanadium also increase the strength of tantalum. The tensile strength of ternary alloys of tantalum (Ta with 30% Nb and 5% Zr of V) at room temperature is about 3X that of tantalum alone. A tantalum-tungsten alloy is used for fabricating springs for high temperature and high vacuum applications.

The trend in the chemical industry is to use increasingly high processing temperatures and pressures, requiring stronger materials with better corrosion resistance. With these objectives in mind, researchers have been studying the influence of the alloying elements tungsten, molybdenum, niobium (columbium), hafnium, zirconium, and rhenium on both the mechanical properties and corrosion behavior of pure tantalum. They have found that additions of only 1% to 3% molybdenum to tantalum, for example, has a marked effect in decreasing the susceptibility of pure tantalum to hydrogen embrittlement in severely corrosive conditions. Not only is the corrosion rate of tantalum decreased, but mechanical

Fig. 1.   High-heat-transfer bayonet-style exchangers employing tantalum alloy tubes. Each exchanger uses 104 tubes. (*Fansteel.*)

properties, such as strength and room temperature workability, are also improved.

Tantalum-tungsten alloys have been successfully developed, which exhibit at least four key advantages: (1) tungsten causes a considerable solid solution hardening (SSH) effect in tantalum; (2) tungsten shows almost no evaporation during electron beam melting; (3) tungsten is less costly than tantalum; and (4) the corrosion rate of tantalum is but slightly influenced by the addition of tungsten up to about 10% (wt).

Tantalum-hafnium and tantalum-zirconium alloys are less suitable for aggressive acid environments. Compared with tantalum-tungsten alloys, tantalum rhenium alloys are superior in corrosion resistance and to hydrogen embrittlement, but the major disadvantage is the high cost of rhenium.

Some of the properties of tantalum and its alloys are given in Table 1.

Chemistry and Compounds. As might be expected from its 5d 36s 2 electron configuration, tantalum forms pentavalent compounds. In fact, they constitute the great majority of tantalum compounds, although the valences 2, 3, and 4 are known. However, the existence of the Ta^{5+} ion is very brief, since it readily coordinates with H_2O, OH^-, and other anions or molecules. Tantalum is extremely resistant to chemical action, not being attacked by acids other than hydrofluoric acid, and by alkalies only upon fusion. Even fluorine and oxygen react only on heating.

Tantalum pentoxide, formed by heating the metal with oxygen, reacts with hydrofluoric acid, alkali bisulfates or alkali hydroxides, forming tantalates with the latter. It reacts with a number of halogen compounds to give tantalum pentafluoride, pentachloride and pentabromide, TaF_5, $TaCl_5$, and $TaBr_5$. (Carbon tetrachloride is often used in this preparation of $TaCl_5$.) These compounds readily undergo hydrolysis, and may form oxyhalides, such as TaO_2F and $TaOBr_3$. They may be reduced, but with difficulty, $TaCl_5$ when heated with aluminum yielding the tetrachloride, $TaCl_4$. The trihalides, $TaCl_3$ and $TaBr_3$ have also been prepared. Tantalum(V) fluoride

TABLE 1. REPRESENTATIVE PROPERTIES OF TANTALUM ALLOYS

Alloy Additions (Weight, %)	Forms Commercially Available	Code Name	Typical High-temperature Strength			
			Temperature, °C	Tensile, MPa	Temperature, °C	10-h Rupture, MPa
None	All	Unalloyed Ta	1315	59	1315	7
7.5 W (P/M alloy)	Wire, strip	FS61	25	1140	—	—
2.5 W, 0.15 Nb	All	FS63	95	315	—	—
25 W	All	KBI 6	95	315	—	—
0 W	All	Ta-10W	1315	345	1315	140
8 W, 2 Hf	All	T-111	1315	255	—	—
8 W, 1 Re, 1 Hf, 0.025 C	All	Astar 811C	1315	275	—	—
40 Nb	All	KBI 40	260	290	—	—
37.5 Nb, 2.5 W, 2 Mo	All	KBI 41	260	515	—	—

(*After Advanced Materials & Processes*).

combines with other fluorides, notably the alkali metal fluorides, to yield complexes, such as K_2TaF_7 and Na_3TaF_8.

Other complexes of tantalum(V) are formed with oxygen-function compounds, such as o-dihydroxybenzene and acetylacetone.

In addition to Ta_2O_5, another oxide is known, TaO_2, which may be formed by active-metal reduction (as is the tetrachloride), except that the pentoxide is heated with magnesium rather than aluminum. It forms with alkali metals the metatantalates, $MTaO_3$, the orthotantalates, M_3TaO_4, and pyrotantalates, $M_4Ta_2O_7$, as well as such polytantalates as $M_8Ta_6O_{19}$, the latter requiring fusion with the alkali hydroxides.

The only known sulfide, which is produced by heating with carbon disulfide, is TaS_2, but at least two nitrides are known, TaN and Ta_3N_5, the latter being unstable.

The organometallic compounds of tantalum all involve oxygen bonding, with the exception of a dicyclopentadienyl compound, $(C_5H_5)_2TaBr_3$. The others are alkoxy compounds, such as $(C_2H_5O)_3TaCl_2$, $Ta(OC_2H_5)_5$, $Ta(OCH(C_2H_5)CH_3)_5$, etc., with the exception of bis(fluorosulfonyloxy) trichlorotantalane, $Cl_3Ta(OS(O_2)F)_2$.

Additional Reading

Cardonne, S.M. et al.: "Tantalum and Its Alloys," *Advanced Materials & Processing,* 16 (September 1992).
Carter, G.F. and D.E. Paul: "Materials Science and Engineering," ASM International, Materials Park, OH, 1991.
Davis, J.R.: "Metals Handbook," 2nd Edition, ASM International, Materials Park, OH, 1998.
Greenwood, N.N. and A. Earnshaw: "Chemistry of the Elements," 2nd Edition, Butterworth-Heinemann, Inc., Woburn, MA, 1997.
Gypen, L.A. and A. Deruyttere: "New Tantalum Base Alloys for Chemical Industry Applications," *Metal Progress,* 127(2), 27–34 (February 1985).
Hala, J.: "Halides, Oxyhalides and Salts of Halogen Complexes of Titanium, Zirconium, Hafnium, Vanadium, Niobium and Tantalum," Vol. 40, Elsevier Science, New York, NY, 1989.
Hawley, G.G. and R.J. Lewis: "Hawley's Condensed Chemical Dictionary," 13th Edition, John Wiley & Sons, Inc., New York, NY, 1999.
Krebs, R.E.: "The History and Use of Our Earth's Chemical Elements: A Reference Guide," Greenwood Publishing Group, Inc., Westport, CT, 1998.
Lide, D.R.: "CRC Handbook of Chemistry and Physics 2000–2001," 81st Edition, CRC Press, LLC., Boca Raton, FL, 2000.
Staff: "Properties and Selection: Nonferrous Alloys and Pure Metals," ASM International, Materials Park, OH, 1990.
Yau, Te-Lin and K.W. Bird: "Know Which Reactive and Refractory Metals Work for You," *Chem. Eng. Progress,* 65 (February 1992).

M. SCHUSSLER, North Chicago, IL.

TANTOCHRONE. A curve in a vertical plane such that a particle sliding frictionless along it under the sole influence of gravity will vibrate with the same period for all amplitudes. This curve is a cycloid. The period of vibration is

$$T = 4\pi\sqrt{\frac{r}{g}}$$

Such a pendulum is also known as Huygen's pendulum.

TAPEWORM. Tapeworms (cestodes) are parasitic flat worms, segmented in form, and having a head portion (scolex) which can continually give rise to new segments. As long as this head remains in the human intestine the worm has not been destroyed, for from this an entire new body can grow. These worms do not possess a digestive canal and are usually hermaphrodites. Nourishment is absorbed from the host through the surface of the worm's body. Fertilization is accomplished between segments of different worms or between different segments of the same worm.

Certain tapeworms apparently do not greatly harm the host while others may produce marked symptoms such as severe anemia. Larval forms of the worm may invade the various human tissues in certain forms of cestode infestation and cause serious trouble.

The kinds of tapeworms commonly found in humans are as follows: (1) The fish tapeworm (*Diphyllobothrium latum*) is found throughout the world and infests humans when infected fish, which has not been thoroughly cooked, is eaten. The growing popularity of raw fish dishes has increased the risk of infestation. Serious symptoms may not develop for years, but often a condition resembling pernicious anemia develops. Recovery follows the expulsion of the worm. (2) The dwarf tapeworm (*Hymenolopsis nana*) is also found throughout the world and is the most common of tapeworms occurring in the United States. It is of small size,

being only a few centimeters long, but when present in large numbers the worms cause diarrhea, loss of weight and appetite, and nervous manifestations. The symptoms disappear when the parasites are expelled. (3) The pork tapeworm (*Taenia solium*) is large, measuring 9 to 12 feet (2.7 to 3.6 meters) in length. Most of the symptoms and dangers of this worm are due to the invasion of the tissue by its larval forms. This form is rare in the United States. (4) The beef tapeworm (*Taenia saginata*) is common in the world wherever beef is eaten. Cold storage will kill the parasite. The adult forms are extremely large worms; some measure 30 to 50 feet (9 to 15 meters) in length. Colic and pain may be present in patients infected with this worm. Spontaneous discharge of a long piece of worm may inform the patient of his guest.

The diagnosis of worm infestation is made by finding either segments of worms or eggs in the feces. The white blood count often reveals an eosinophilia.

Treatment with various antihelminths is usually successful if correctly employed. The drugs most frequently used are niclosamide or praziquantel.

R. C. V.

TAPHONOMY. The study of fossils in their geological contexts with a view to sorting out the factors intervening between death and definitive burial that bias the fossil record in certain ways and render paleoecological reconstruction difficult. This topic of science was proposed by the Russian paleontologist I.A. Efremov as a branch of paleontology in 1940. Activity in the field has not been extensive. In 1976, the Wenner-Gren Foundation for Anthropological Research sponsored a symposium, "Taphonomy and Paleoecology, with Special Reference to Sub-Saharan Africa." The papers were published in a book entitled "Vertebrate Taphonomy and Paleoecology," K. Behrensmeyer and A.P. Hill, editors, Univ. Chicago Press, Chicago, Illinois, 1988.

TAPIOCA. See **Euphorbiaceae**.

TAPIR *(Mammalia, Perissodactyla; Tapirus)*. Moderately large animals found in the neighborhood of water in the forests of Central and South America and the Malayan region. See Fig. 1. They reach a length of 8 feet (2.4 meters) and a height of a little over 3 feet (0.9 meter) and are stoutly built. Their most conspicuous characteristic is the short trunk into which the snout is prolonged. Only one of the five species occurs in the Old World.

Fig. 1.   American tapir. (*A.M. Winchester.*)

The natives of the American tropics hunt tapirs for their flesh and hides, although the value of the latter for leather is limited.

TARANTULA. 1. *Arachnida, Araneida.* A name applied indiscriminately to many of the large hairy spiders of the warmer parts of the Americas, perhaps most commonly to the forms so often imported into temperate latitudes in bunches of bananas. The name comes from the generic name *Tarantula*, which has been applied to an entirely different group of spiders. 2. *Arachnida, Pedipalpi.* A genus of whip scorpions. The application of this name is dependent upon rules of nomenclature, which have apparently received no authoritative interpretation. It will undoubtedly remain in popular usage as a name for many large spiders, whatever its scientific disposition. See Fig. 1.

Fig. 1.   Tarantula. (*A.M. Winchester.*)

TARO.   Several perennial herbaceous plants of the genus *Colocasia* of the *Arum* family are commonly referred to as *taro* and, in some parts of the world, by the name *cocco* or *cocoyam*. These large and leafy plants are cultivated for their fleshy roots, which comprise a main foodstuff in parts of Polynesia and in southeast Asian countries. In some areas, the roots are called *eddoes*. The tubers contain a high percentage of starch, but considerably more protein than another common starchy vegetable, the potato. The leaves of taro plants are large and heart-shaped, resulting in one variety being called the *elephant's ear*. The plant is characterized by greenish flowers that resemble a calla, without petals, a floral envelope, stamens, or pistils. The fruit of the plant is a berry. Many varieties and forms of taro are known to the natives of the Pacific islands.

Colocasia esculenta is the principal and most commonly known form of taro. This is the main form of taro, cultivated in Hawaii, for example, and is the basis of *poi*. This pastelike material is prepared by steaming or boiling the tubers, after which they are peeled and pounded to form a mash. This mash is then permitted to ferment for several days. The taro is also eaten much as the potato, usually parboiled, after which it is baked. The tender, unopen leaves of the taro plant, sometimes termed *luau*, are used as greens.

The dasheen is a variety of taro grown in the southern United States. It is a long-season crop, adapted for culture only in southern regions, where there is normally a very warm frostless season of at least 7 months. Dasheen requires a rich loamy soil, an abundance of moisture with good drainage, and a fairly moist atmosphere. The plant was originally brought to the United States in about 1905 from Trinidad and Puerto Rico.

Small tubers from 2 to 5 ounces (57 to 142 grams) in weight are used for planting in much the same way as potatoes. Planting may be done 2 or 3 weeks before frosts are over, and the season may be lengthened by starting the plants indoors and setting them out after the last frost. The plants usually are set in 3.5- to 4-foot (about 1 to 1.3 meters) rows, about 2 feet (0.6 meter) apart in the rows. Dasheen tubers may be dug and dried on the ground in much the same way as sweet potatoes, and stored at 50 °F (10 °C) with ventilation.

Colocasia antiquorum, a plant of great antiquity, still is grown in Egypt in limited quantities. The plant is inferior to most other forms of taro, but is edible. This plant is sometimes called the Egyptian or Indian taro and in Egypt is known as *golgas*.

Cyrtosperma chamissonis, actually the *mwang* plant, is called taro in Micronesia. This particular plant is very much larger than the taro, having a rootstock weighing up to 50 pounds (22.7 kilograms) as compared with the ordinary taro rootstock weighing usually less than 5 pounds (2.3 kilograms).

Xanthosma sagittifolium, also known as *yautia*, is grown in the West Indies. This plant resembles, and is sometimes called, taro, but the plant is much larger. The root is high in starch and has a greater caloric value per unit of weight than the potato.

See also **Aroids**.

TARPAN.   See **Horses, Asses, and Zebras**.

TARPON (*Osteichthyes*).   Of the order *Isospondyli*, family *Elopidae*, the tarpon is probably the most primitive of the suborder of clupeids. It is a large marine fish of silvery color, found in the waters of the West Indies, along the Gulf Coast, and northward to a limited extent along the Atlantic coast. It reaches a length of about 6 feet (1.8 meters) and is known for its very large scales, which may measure 3 inches (7.5 centimeters) across. Also known as the silver fish, silver king, savanilla, and sabalo. The tarpon is among the great game fishes, but is not commonly regarded as desirable for food. *Megalops atlanticus* (the Atlantic tarpon) has been recorded at 200 pounds (90.9 kilograms), length of 8 feet (2.4 meters) on several catches; the largest recorded has an 8-foot (2.4-meter), 340 pounder (154 kilograms). The *Megalops cyprinoides* (Pacific tarpon) is smaller, rarely exceeding 40 inches (1 meter) in length.

TARRAGON.   See **Artemisia**; and **Flavorings**.

TAR SANDS.   Also called *bituminous sands* and *oil sands*, tar sands represent a vast potential of petroleumlike energy reserves and a reservoir of materials for the preparation of syncrudes. In 1988, the processing of tar sands is steadily approaching a state of economic viability. Although there are major technological problems remaining in the recovery and processing of tar sands into practical fuels, the overriding factor affecting progress in this field is a combination of economics and technology.

The heavy, viscous petroleum substances impregnating the tar sands are called *asphaltic oils*. Other names used to describe these oils include *maltha*, *brea*, and *chapapote*. Asphaltic petroleums are most commonly confused with, but are *not* related to *asphaltites* (gilsonite, glance pitch, and grahamite); the *asphaltic pyrobitumens* (elaterite, wurtzilite, albertite, and impsonite); the native *mineral wax* (ozokerite); and the *pyrogenous distillates* of bituminous substances (tar and pitch).

Tar sands are composed of a mixture of 84–88% sand and mineral-rich clays, 4% water, and 8–12% bitumen. Bitumen is a dense, sticky, semisolid that is about 83% carbon. The substance does not flow at room temperature and is heavier than water. At higher temperatures, it flows freely and floats on water. Characteristics of tar sands important to mining, recovery, and processing include grain size, composition, sortability, porosity, permeability, and microscopic habitat.

Tar Sand Resources

The presence of tar sands in North America was noted by American Indians several centuries ago. Pitch recovered from surface deposits was used for waterproofing canoes. It is reported that Columbus observed asphalt from Pitch Lake in Trinidad and used the material for repairing his ships on his third voyage to the West Indies in 1498. The same bitumen deposit was reported by Sir Walter Raleigh in 1595. For several centuries the material was used for repairing vessels.

In 1962, reports were received of tar sands in the Bjorne Formation of Triassic age on Melville Island in the Canadian Arctic Archipelago near the southern margin of the Sverdrup sedimentary basin. Subsequent investigations of these deposits by officers of the Geological Survey of Canada revealed them to be a seepage derived from the oxidation and polymerization of 19–31° API gravity oil, with a total in-place reserves of only 30 million barrels. The largest find of tar sands in Canada (and possibly in the world) was located in the subsurface of northern Alberta in the valley of the lower Athabasca River, along a distance of 160 kilometers (100 miles). This is now known as the Athabasca deposit. As indicated by Mossop (1980), McMurray Formation deposition began in the Athabasca region in Early Cretaceous time. The surface on which the initial sediments were laid down was an exposed landscape of Devonian limestone. It is envisioned that, during the McMurray period, the region underwent gradual subsidence, with the Boreal Sea[1] slowly transgressing across it from the north. The McMurray sand deposition stopped when the sea

[1] Boreal is a term that pertains to the north, or to things located in northern regions. The Boreal region is characterized by tundra and taiga—a climatic zone having a definite winter that experiences snow and a short summer that is generally

eventually transgressed the entire area, giving subsequent rise to deposition of Clearwater Formation marine shales. Not all geologists are agreed upon the details of the formation and possible later biodegradation of tar sand deposits. Geographically close to the Athabasca deposit are the Cold Lake, Wabasca, and Peace River deposits. Underlying these Cretaceous tar sands are Devonian carbonate rocks (limestone and dolomite) impregnated with bitumen of essentially the same composition as the bitumen in the Cretaceous sands. See Fig. 1.

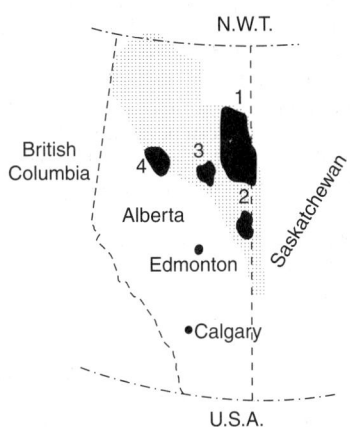

Fig. 1.   Location of major tar (oil) sand deposits in Alberta, Canada: (1) Athabasca, (2) Cold Lake, (3) Wabasca, and (4) Peace River, N.W.T. (Northwest Territories).

Along with other alternative sources of energy during the energy crisis of the 1970s, considerable attention was devoted to the exploitation of tar sands. Once, it was predicted that deposits in Canada could yield a light synthetic crude oil to the extent of a million barrels per day, or about one-third of Canada's petroleum requirements. Later, when serious environmental concern over fossil fuels was indicated, research turned essentially elsewhere. It was estimated in the late 1970s that tar sands reserves in the United States, mainly in Utah, would have the petroleum equivalent of 90 billion barrels.

It is interesting to note that tar sands worldwide contain the largest accumulations of liquid hydrocarbons in the earth's crust.

The seriousness of the Canadian tar sands effort is demonstrated by a view of a plant in Alberta as of about 1980. See Fig. 2.

Fig. 2.   General view of upgrading plant for processing bitumen mined from Athabasca deposit. In the foreground is the sulfur stockpile; in the background is a tailings pond. Right background shows the extraction plant; left is the coke pile. (*Alberta Government Photographic Service.*)

hot, characterized by a large annual range of temperature. This region includes parts of North America, central Europe, and Asia, generally between latitudes 60°N and 40°N.

TARSIOIDS (*Mammalia, Primates*).   Comprised of small creatures known as Tarsiers, this is the smallest of the divisions of the *Primates*. The tarsier is a peculiar animal found on some islands of the Oriental Region, notably in Borneo, Celebes, Java, Sumatra, and the Philippines. Tarsiers are forest dwellers and they are about as large as a medium-size rat, with very short muzzles and large eyes, circular, close together, and directed forward. The hind legs, particularly the ankles, are long, and the tips of the digits are expanded into fleshy disks. The tarsier moves about on its hind legs by spring leaps. They are of considerable scientific interest because they appear to be a leftover from some ancient line of animals. It is reasoned that the ancient ancestry may trace to the *Tupaioids* (tree-shrews). However, some authorities identify the Tarsier more with the *Lorisoids*. Tarsiers also are of interest because they are insectivorous, feeding at night, and literally catching their prey with their hands and conveying it to their mouths for eating. They are covered with short, thick, soft, and wooly fur, with a long furry tail at the base, but with essentially a naked terminal. With exception of their extra-long legs, due to elongated ankle bones, they have somewhat the appearance of a squirrel with very large eyes.

TARSUS.   1. The terminal division of the leg of an insect. It consists of five segments in the typical form, the terminal segment bearing a pair of claws. Tarsi are modified in various species by the fusion or loss of segments, and in one genus, *Bittacus*, the terminal joint folds back on the next to form a grasping organ. 2. The shank of the leg of a bird (*Aves*). 3. The proximal portion of the foot of vertebrates, containing several tarsal bones. 4. The framework of connective tissue which gives shape to the eyelid.

TARTARIC ACID. $(CHOHCO_2H)_2$, formula weight 150.09, white crystalline solid with four physical isomers, three of which are optically active: (1) dextro- and (2) levotartaric acid, both with same mp 168–170°C and sp gr 1.760, (3) racemic acid (dextrolevo), mp 205–206°C, sp gr 1.697, and (4) mesotartaric acid (inactive), mp 159–160°C, sp gr 1.737. Racemic acid crystallizes with one molecule of H_2O. All forms decompose before reaching the boiling point at atmospheric pressure. All forms are soluble in H_2O, slightly soluble in alcohol, and essentially insoluble in ether. Tartaric acid is a primary example of optical isomerism and one of the earliest compounds studied in this regard. Tartaric acid is a dibasic acid with two series of salts and esters.

Tartrates (like citrates) in solution change silver of ammonio-silver nitrate into metallic silver. Potassium hydrogen tartrate and calcium tartrate, on account of their solubility characteristics, are of importance in the separation and recovery of tartaric acid. The former salt is readily converted into the latter, and the resulting calcium tartrate plus dilute sulfuric acid yields tartaric acid plus calcium sulfate, and the latter may be separated by filtration. Tartaric acid may be obtained by evaporation of the filtrate. Ester: Diethyl tartrate $COOC_2H_5(CHOH)_2COOC_2H_5$, melting point 17°C, boiling point 280°C. Tartaric acid may be obtained (1) from some natural products, e.g., in the juice of grapes and acid fruits, often in conjunction with citric or malic acid; potassium hydrogen tartrate, "argol," in the residue of wine vats, (2) by synthesis.

Tartaric acid is used: (1) in baking powders as potassium hydrogen tartrate ("cream of tartar") with sodium bicarbonate; (2) in medicine, e.g., potassium antimonyl tartrate ("Tartar emetic"); (3) in effervescent medicinal salts; (4) in blue printing as ferric tartrate; and (5) in silvering mirrors—ammonio-silver nitrate yielding a smooth deposit of silver. Sodium potassium tartrate ("Rochelle salt," $NaKC_4H_4O_6 \cdot 4H_2O$) is used in medicine, and in the preparation of Fehling's solution, which is an alkaline cupric solution made by mixing copper sulfate solution, sodium potassium tartrate solution and sodium hydroxide solution, and is used as an oxidizing reagent in the case of many organic compounds, such as glucose and reducing sugars, and aldehydes, with which cuprous oxide, red to yellow precipitate, is formed.

See also **Isomerism.**

TASMANIAN DEVIL.   See **Marsupialia.**

TASMANIAN GUM TREE.   See **Eucalyptus Trees.**

TASTE BUD.   A sensory organ of the vertebrates, sensitive to contact with substances in solution and those in a liquid state. A taste bud consists of a spindle-shaped group of cells in the epithelium of the vertebrate tongue and in some species in the lining of the mouth and pharynx. Taste buds have been reported in the skin of some aquatic animals. The bud includes supporting cells of thick spindle shape and slender taste cells ending with a short taste hair which projects into a minute pit at the free end of the bud. The interpretation of these two types of cells is a subject of disagreement; it is possible that they represent stages in the development of a single form of cell.

The action of taste buds results in four fundamental taste sensations: sweet, sour or acid, salty, and bitter. Perception of these chemical properties is localized in the tissues containing organs of taste, but differentiation of the organs accompanying this localization has not been demonstrated.

More detail on taste and odor receptors is given in the entry on **Flavorings**.

TATLER.   See Waders, Shorebirds, and Gulls.

TAU PARTICLE.   Discovered in 1975, the tau particle is a lepton with a mass of 1.8 GeV, almost twice that of the proton. Like other leptons, the tau particle is considered as pointlike. See also **Particles (Subatomic)**.

TAURUS (the bull).   A constellation of great antiquity; the second sign of the zodiac. Two of the open clusters of Taurus, the Pleiades and the Hyades, are frequently referred to in the Bible, and Aldebaran, its brightest star, is mentioned by both Homer and Hesiod. See also **Aldebaran**.

The two asterisms, the Pleiades and the Hyades, are both open clusters, i.e., groups of stars moving together through space. The Pleiades group is also noteworthy in that it is filled with diffuse nebulous material.

There are a number of double stars within Taurus available for observers with small telescopes. With a wide-field instrument, such as an opera glass, the two doubles, Sigma and Theta, can both be seen at once and, with the surrounding stars, make a very interesting spectable.

See map accompanying the entry on **Constellations**.

TAUTOMERISM.   See Isomerism.

TAXODIACEAE.   See Cypress Trees; Giant Sequoia; and Redwood (Coast).

TAXONOMY.   The science of classification. In dealing with the many details involved in classifying the half-million known species of animals, it has been found necessary to adopt rules of procedure in order to secure approximate uniformity and stability of results.

For many years, zoologists and botanists tried to classify animals and plants into a system that represented a survey of the abundance of forms in fauna and flora. Scientists found early in their work toward a perfect classification that new knowledge constantly required alterations in the system and, of course, to a degree this continues today. Classification always has been difficult, too, because of fundamental differences of opinion among the persons who were constructing the system. Often these differences are insignificant — yet they can cause difficulty for persons who attempt to locate information in the literature.

The animal kingdom, for example, has been split into several subkingdoms and these have been divided into further sections, subsections, etc. The scale of the most important systematic categories follows in a descending rank order, as shown below:

Kingdom
Subkingdom
Phylum
Subphylum
Class
Subclass
Superorder
Infraorder
Family
Subfamily
Tribe
Genus
Subgenus
Species
Subspecies

The scientific names of animals and their spelling are determined by the international rules for the zoological nomenclature as agreed by the

XV International Congress for Zoology and are obligatory for zoological publications. The name of the genus, which is a Latin or Latinized noun, is singular and capitalized or otherwise typeset (such as italics) to distinguish it from the remainder of words in a sentence or listing. The names of species and subspecies may be nouns or adjectives and they are spelled in the lower case, preferably italicized. The name of a subgenus, which is formed in the same manner as a genus, may be added in brackets following the name of the genus. The names of the tribes, subfamilies, families, and superfamilies are usually plural nouns, sometimes fully capitalized or with an initial capital. They are formed from the name of a given genus by adding to the principal word the endings -ini for the tribe, -inae for the subfamily, -idae for the family, and -oidea for the superfamily. The names of the authors who were the first to describe and to name a species, subspecies, or group of animals can be cited with the year of this naming. In a description involving several mentions of a species, subspecies, etc., the author acknowledgment need be made only once — when the term is first mentioned in a text.

The principal objectives of taxonomy are: (1) to discriminate among organisms and to provide means for the subsequent recognition (identification) of the discriminated entities (taxa); (2) to develop a suitable procedure (nomenclature) for designating taxa for reference purposes; and (3) to devise and perfect a scheme of classification in which the named taxa can be arranged. The functions of the classification are: (1) to provide a means for the communication and retrieval of information concerning each of the approximately 1.5 million different kinds of organisms; (2) to facilitate the gathering of new information by permitting the prediction of characters in unfamiliar organisms; and (3) to demonstrate at once the unit and diversity of organic life by expressing past adaptive patterns of the various taxa.

TAYLOR, BROOK (1685–1731).   Taylor was a British mathematician most remembered for his work that helped in the development of calculus. He grew up to be an accomplished musician and painter besides mathematician. In 1708 even before graduating from St. John's College Cambridge, Taylor had written his first important paper producing a solution to the problem of the center of oscillation. It was not published until 1714 and resulted in a priority dispute with the work of Johann Bernoulli.

Also in the year 1714, Taylor was elected Secretary to the Royal Society and during his years there (until 1718) he produced most his mathematics. His "Direct and Indirect Methods of Incrementation" added a new branch to mathematics called the calculus of finite differences. He also invented integration by parts, and discovered the celebrated series known as Taylor's expansion. His work *Methodus* contains the famous formula known as Taylor's theorem. Not until 1772, however, was its importance really understood and proclaimed by the French mathematician, Lagrange, as the basic principle of differential calculus.

A gifted artist, Taylor devised the basic principles of perspective in *Linear Perspective*. His work gives the first general treatment of vanishing points.

See also **Taylor Series**.

J. M. I.

TAYLOR SERIES.   A convergent power series is generally useful as a representation of a function. Let us assume that $f(x)$ can be so given as an infinite series in powers of $(x - a)$, expecting the series to be valid for values of x near a. Then,

$$f(x) = b_0 + b_1(x - a) + b_2(x - a)^2 + \cdots + b_k(x - a)^k + \cdots$$

and the coefficients can be obtained by successive differentiations to give

$$b_0 = f(a); b_1 < f'(a);   b_2 = f''(a)/2!; \cdots$$

The result, which is known as Taylor's series (Brook Taylor, 1685–1731, was an English mathematician), is

$$f(x) = f(a) + f'(a)(x - a) + f''(a)\frac{(x - a)^2}{2!} + \cdots$$

$$+ f^{(k)}(a)\frac{(x - a)^k}{k!} + \cdots$$

The procedure, so far, has been purely formal for it has only been assumed that such an expansion is possible. One can, however, apply the extended mean value theorem of differential calculus which shows that the Taylor series expansion and the extended mean value theorem are identical, if $x = b$ and $a < z < x$, provided we stop after n terms. Let the first n terms in the Taylor series (the nth term involves $f^{(n-1)}$ or $(n-1) = k$) be called S_n and the last term in the extended mean value theorem be called the remainder.

$$R = f^{(n)}(z)(x-a)^n/n!$$

so that $f(x) - S_n = R$. Now if $x = x_0$ and R approaches a limit, $\lim_{n \to \infty} S_n = f(x_0)$ so that the Taylor series does indeed converge to give $f(x_0)$ at $x = x_0$.

Another useful form of the Taylor series is obtained if we let $h = (x - a)$ and replace a by x_0. Finally dropping the subscript on x_0, the result is

$$f(x+h) = f(x) + f'(x)h + f''(x)\frac{h^2}{2!} + \cdots + f^{(k)}(x)\frac{h^k}{k!} + \cdots$$

The remainder after n terms can be given in several equivalent forms. Some of them are

$$R_n(x) = \frac{(x-a)^n f^{(n)}}{n!}\{a + \phi(x-a)\}. \ 0 < \phi < 1$$

$$R_n(x) = \frac{1}{(n-1)!}\int_0^{x-a} f^{(n)}(x-t)t^{n-1}dt$$

In the second form of the series

$$R_n(x) = \frac{h^n f^{(n)}}{n!}(x + \phi h)$$

$$R_n(x) = \frac{1}{(n-1)!}\int_0^{h} f^{(n)}(x+h-t)t^{n-1}dt$$

Use of the Taylor series requires that the function and its derivatives of all orders must exist at $x = a$. The binomial series is a special case of it. See also **Maclaurin Series**.

The Taylor series may be extensively generalized. Thus, in several variables

$$f(x+h, y+k, \ldots) = f(x, y, \ldots) + df(x, y, \ldots)$$
$$+ \frac{1}{2!}d^2 f(x, y, \ldots) + \cdots + \frac{1}{k!}d^k f(x, y, \ldots)$$
$$+ \cdots + R_n$$

It can also be used for the complex variable. In that case the boundary of the region of the complex plane in which a Taylor series converges is called the circle of convergence. For a given function, the radius of this circle depends on the point (center of the circle) about which the series is developed. On the boundary the power series diverges because there are one or more singular points of the function. Frequently it may be found outside of the circle of convergence by analytic continuation.

In starting the numerical solution of a differential equation, suppose $y' = f(x, y)$ is given with initial values (x_0, y_0). Then the solution in a Taylor series is

$$y = y_0 + (x-x_0)y_0' + \frac{(x-x_0)^2}{2!}y_0'' + \cdots + \frac{(x-x_0)^k}{k!}y_0^{(k)} + \cdots$$

If the derivatives can be evaluated readily, the solution can be computed numerically. Other methods (see also **Differential Equation, Numerical Solution of**) are preferred for continuing the solution. See also **Algebraic Equations**.

TAYRA. See **Mustelines**.

t DISTRIBUTION. If $\bar{x}$ is the mean of n independent observations, from a normal population with mean zero, and if s^2 is an estimate of the population variance based on the sample variance with $n - 1$ degrees of freedom, the quantity $t = \bar{x}/s$ has the distribution

$$\frac{\Gamma\left(\frac{1}{2}n\right)}{(n\pi)^{1/2}\Gamma\left(\frac{1}{2}n - 1\right)}\left(1 + \frac{t^2}{n}\right)^{-1/2(n+1)}, -\infty < t < \infty$$

The t distribution is the basis for significance tests and confidence of fiducial statements referring to means, regression coefficients, and so on. It does not depend on the variances of the normal population.

T DISTRIBUTION. A generalization, due to H. Hotelling, of the t-distribution to multivariate statistics. Given n values of the p variables $x_1, x_2, \ldots, x_p$ let D_{jk} be the inverse of the dispersion matrix. Define

$$T^2 = n\sum_{j,k=1}^{p} D_{jk}\bar{x}_j\bar{x}_k$$

Then T^2 has, in samples from a multivariate normal distribution, the distribution

$$dF = \frac{1}{B\left\{\frac{1}{2}(n-p), \frac{1}{2}p\right\}}\frac{\{T^2/(n-1)\}\frac{1}{2}(p-2)}{\{1 + T^2/(n-1)\}\frac{1}{2}(p-2)}d\{T^2/(n-1)\}$$

TEA *(Camellia sinensis; Ternstroemiaceae).* The tea plant is an evergreen shrub probably indigenous to China, where it has been cultivated since early times. The plant possesses alternate elliptical leaves which when mature are tough, and vary in length from 2 to 5 inches (5 to 12.5 centimeters). The flowers are axillary and appear singly or in small groups. They are white, slightly fragrant, and about an inch in diameter. Each flower has numerous stamens and a single pistil composed of three carpels. The fruit is a woody capsule containing three large seeds.

For successful growth, tea must be planted in regions having abundant rainfall. In China, most of the tea plants are grown on small farms. The plants are grown from seed or from nursery stock and begin to yield crops when 3 or 4 years old, continuing to do so thereafter for 50 years. The young shoots appear in flushes, growing rapidly for a time. From these flushes are picked the young leaves used for tea. Several flushes occur each year.

TEAK TREE *(Tectona grandis; Verbenaceae).* The teak tree is tall, with very rough-surfaced oblong leaves from 10–20 inches (25–51 centimeters) long and 8–15 inches (20–38 centimeters) broad, and small white or blue-tinted flowers borne in large panicles. The tree is native in the tropics of Asia and is frequently grown in plantations in India, Java, and other Asian countries for its hard wood. This wood is very durable and much used in shipbuilding and in the making of fine furniture. The wood is very heavy and dries slowly. Drying is hastened somewhat by girdling the tree at the base and leaving it standing for a year or more. During this time it dies and dries out, after which it is felled and floated to the shipping port.

An adult teak tree may attain a height of from 75 to 100 feet (23 to 30.5 meters). The tree grows quite slowly. The wood is similar in coloration to the black walnut, being yellowish-golden brown. The durability of the wood is dependent largely upon the locale of the tree. The weight per cubic foot is approximately 40 pounds (641 kilograms per cubic meter). While the wood is similar to walnut in many respects, the tree per se resembles an oak. African teak is a term used for the wood from the *Chlorophora excelsa*, also called *iroko*, which grows well in parts of western Africa. Surinam teak is from *Hymenea courbaril*, a tree found in the West Indies and northern coast of South America. Australian teak is from the *Flindersia australis*, which grows in New South Wales. Most of the wood from the southeastern Asian countries is from *Dipterocarpus tuberculatus*.

TEAL. See **Waterfowl**.

TEARDROP BALLOON. A sounding balloon which, when operationally inflated, resembles an inverted teardrop. This shape was determined primarily by aerodynamic considerations of the problem obtaining maximum stable rates of balloon ascension.

TEARS. Tears are the salty fluid that lubricates and helps protect the cornea, the membrane that covers the front of the eye. Tears flow into the eye through ducts from tiny glands located under the upper eyelids and are spread over the eye each time a person blinks, which is about every six seconds. Tears keep the eye moist and free of dust and other eye irritants.

Tears drain into tiny openings on the edge of the upper and lower eyelids that lead to the nasolacrimal tear ducts near the bridge of the nose. From there, they are channeled into the nasal cavity where they are either swallowed or drain through the nostrils. That is why the nose runs when one cry. When eyes are irritated, they produce additional tears called reflex or irritant tears. Emotional tears are produced when a person is are either happy or sad. Excess tears often overflow the lower lid and run down the cheeks.

Tears provide five functions.

1. Keeping the epithelium moist, thereby protecting the outer covering of the eye from damage because of dryness.
2. Creating a smooth optical surface on the front of the cornea.
3. Acting as the main supplier of oxygen and other nutrients to the cornea.
4. Carrying waste products away from the cornea.
5. Providing enzymes that destroy bacteria that can harm the eye.

Normal tears that cover the corneal surface comprise three layers:

The *lipid*, or oil layer, which is the outer layer of the tear film and helps prevent the lacrimal layer beneath it from evaporating or overflowing the lower eyelid.

The *lacrimal*, or watery layer, which is the middle layer and contains salts, proteins, and an enzyme called lysozyme that actually protect and nourish the eye.

The *mucoid*, or mucus layer, which is the bottom layer of tears. It contains cells called goblet cells that cause the tears to adhere to the eye.

All three layers of tears are necessary for proper lubrication.

When an eye produces too few tears or tears of faulty composition, a condition called dry eye results. Ironically, one of the symptoms of dry eye can be excessive watering of the eyes. The watering is a natural reflex caused by irritation to the eye because the composition of the tears is wrong. In addition, if one has dry eye, the eye can feel scratchy, dry, irritated, or generally uncomfortable.

An eye doctor diagnoses dry eye by measuring production, thickness, and chemical makeup of the tear film. To treat dry eye, use a tear substitute, also called artificial tears. Administered several times a day, artificial tears bring relief in most cases. If left untreated, dry eye can damage eye tissue and possibly scar the cornea.

When artificial tears do not sufficiently lubricate your eye, an eye care professional may insert a small plug in the corner of the eye to slow drainage and loss of tears. In rare cases, the eye doctor may recommend surgery.

Vision Rx, Inc., Elmsford, NY.

TECHNETIUM. Chemical element symbol Tc, at. no. 43, at. wt. 98.906, periodic table group 7, mp 2172 °C, bp 4877 °C, does not occur in nature. The present location of technetium in the periodic table was vacant for many years, during which time several claims to having found the element were made, but never confirmed. One such claimant termed the element masurium. Technetium has been detected in certain stars and this discovery must be resolved with current theories of stellar evolution and element synthesis.

^{97}Tc, the first isotope to be isolated, was extracted by Perrier and Segré from molybdenum which had been bombarded with deuterons in the Berkeley cyclotron. The reaction was ^{96}Mo$(d, n)^{97}$Tc. The isotope with the longest half-life, ^{99}Tc (half-life $= 2.12 \times 10^5$ years), is found in relatively large amounts among the fission products of uranium. It is also produced by neutron irradiation of ^{98}Mo, by the reaction

$$^{98}\text{Mo}(n, \gamma)^{99}\text{Mo}(\beta\text{-decay})^{99m}\text{Tc(isomeric transition)}^{99}\text{Tc}$$

Significant quantities have been isolated and considerably larger quantities could be made available if applications for it were developed. A U.S. government-owned invention available for licensing concerns a method for recovering technetium from nuclear fuel reprocessing waste solutions. ^{99}Tc has found some application in diagnostic medicine. Ingested soluble technetium compounds tend to concentrate in the liver and are valuable in labeling and in radiological examination of that organ, and this was the basis of the early medical uses. However, the ideal nuclear properties of ^{99m}Tc have led to expanded usage in medical diagnostics. By technetium labeling of suitable compounds (or blood serum components), diseases involving the circulatory system and organs other than the liver can be diagnosed.

In all, sixteen isotopes of technetium have been reported of mass numbers 92–105, 107, and 108.

Superconductivity has been observed in technetium metal and in alloys based on technetium with additions of Pd, Os, Rh, Ru, Sn, V, Ti, Re, W, or C.

A study of the chemistry of technetium shows it to have, as expected, properties intermediate between those of its homologues manganese and rhenium, the resemblance to the latter being perhaps greater than to the former. Like rhenium, technetium apparently exists in (IV), (VI), and (VII) oxidation states. Pure technetium metal has been prepared by passing hydrogen gas at 1,000 °C over the sulfide obtained by precipitation with H_2S from HCl solution. The metal has been shown to have the same crystal structure as rhenium and the adjacent elements osmium and ruthenium. Among its compounds are the ditechnetium heptasulfide, Te_2S_7, readily precipitated by H_2S from oxidized solutions, the corresponding oxide, Tc_2O_7 produced directly from the elements at higher temperatures, which reacts with NH_3 to form ammonium pertechnate, NH_4TcO_4, and the hexachloro complex ion, $TcCl_6{}^{2-}$, which like the corresponding rhenium ion, has a magnetic moment corresponding to three unpaired electron spins.

In 1991, cardiologists (University of California, Los Angeles) reported the use of a new combination of mixtures for yielding images of healthy and damaged areas of the heart. This enables physicians to assess the effectiveness of clot-busting drugs and other cardiac therapies. Use of the technique as a preventive measure for determining persons at risk of sudden blood-flow blockages and thus sudden heart attacks also has been suggested. The product (DuPont-Merck) is called *technetium-99 m sestamibi*. Technetium-99 m is a tracer, and sestamibi is an effective "heart-seeking" compound. This technique is superior to use of the thallium radioisotope because of the requirements of thallium to process images within 30 minutes of injection. Technetium is not that time-sensitive.

Additional Reading

Considine, D.M. and G.D. Considine: "Van Nostrand Reinhold Encyclopedia of Chemistry," 4th Edition, Van Nostrand Reinhold, New York, NY, 1984.

Fackelmann, K.A.: "Diagnostic Duo Highlights Heart Damage," *Science News*, 4 (January 5, 1991).

Greenwood, N.N. and A. Earnshaw: "Chemistry of the Elements," 2nd Edition, Butterworth-Heinemann, Inc., Woburn, MA, 1997.

Krebs, R.E.: "The History and Use of Our Earth's Chemical Elements: A Reference Guide," Greenwood Publishing Group, Inc., Westport, CT, 1998.

Lewis, R.J. and N.I. Sax: "Sax's Dangerous Properties of Industrial Materials," 10th Edition, John Wiley & Sons, Inc., New York, NY, 1999.

Lide, D.R.: "CRC Handbook of Chemistry and Physics 2000–2001," 81st Edition, CRC Press, LLC., Boca Raton, FL, 2000.

Sinflet, J.H. "Bimetallic Catalysts," *Sci. Amer.*, **90** (March 1985).

ROBERT Q. BARR, Director, Technical Information, Climax Molybdenum Company, Greenwich, CT.

TECTONICS. See **Earth Tectonics and Earthquakes**.

TEJU (*Reptilia, Sauria; Tupinambis*). A large lizard of the West Indies and South America. Lives in the forests near water but is not aquatic. It reaches a length of a yard, with long slender tail and heavy forequarters. It is chiefly olive and black.

TEKTITE. A small (usually walnut-size), rounded, pitted, jet-black to olive-greenish or yellowish body of silicate glass of nonvolcanic origin, found usually in groups in several widely separated areas of the earth's surface and apparently bearing no relation to the associated geologic formations. Most tektites have uniformly high silica (68–82%) and very low water contents (average, 0.005%). Their composition is unlike that of obsidian and more like that of shale. They have various shapes, strongly suggesting modeling by aerodynamic forces and they average a few grams in weight. The largest found weighs 3.2 kilograms. Some authorities believe that tektites are of extraterrestrial origin, or alternatively the product of large hypervelocity meteorite impacts on terrestrial rocks. The term was proposed by Suess in 1900 who believed they were meteorites which at one time had undergone melting.

TELECENTRIC SYSTEM. A telescopic system having the aperture stop at one of the foci of the objective lens. If the aperture stop is placed at the focus on the side of the image, the system is "telecentric" on the side of the image. If the aperture stop is placed on the side of the object, the system is "telecentric" on the side of the object. This is useful in measuring telescopes, since a slight change from exact focus will not greatly change the apparent size of the object.

With a telecentric system, either the entrance or the exit pupil is at infinity. See also **Telescope (Astronomical-Optical)**.

TELEOPERATION. As defined by Bejczy, "Mechanical activities performed by mechanical devices at a remote site under remote control." A rather comprehensive review of the history and current state of the art is given in Bejczy's article in *Science*, **208**, 1327–1334 (1980).

TELEOSTEICA. Suborder of the *Osteichthyes* (bony fishes). Includes all species with exception of *Chondrosteica* (sturgeons and lobedfinned fishes); *Holosteica* (alligator gars and bowfins); and *Choanichthyes* (coelacanths and lungfishes).

TELEPHONY (Telecommunications). Telephony is the technology of communicating speech and other sounds between telephones or other terminals, using analog or digital signals transmitted via electrical, wireless, or lightwave circuits. When the circuits also carry data, facsimile, or video signals, the combination is known as *telecommunications*. See also the **Satellites (Communication and Navigation)** entry and numerous other articles in this encyclopedia which relate to communications.

Early History of Telephony

As early as 1667, Robert Hooke (England) transmitted sounds mechanically over a distance of several meters by way of an extended wire. Charles Bourseul (France), in 1854, introduced the concept of a diaphragm for making and breaking electrical contacts when the diaphragm was vibrated by sound waves, but was unsuccessful in converting the electrical pulses back into sound. However, his work introduced the concept of a microphone, one of the key components of a telephone. Johann Philipp Reis (Germany), working on the basis of Bourseul's concept a few years later, developed a collodion membrane which, when vibrated by sound waves, interrupted a battery circuit. Because he lacked financial backing, and since Germany at that time lacked a credible patent protection system, Reis was frustrated in his efforts to achieve a marketable device that replicated the human voice.

It was not until the research of Alexander Graham Bell during the 1870s that a practical system capable of conducting the human voice over a distance with reasonable quality and timbre was achieved. Bell was granted a United States patent[1] in 1876 and the first commercial telephone exchange was installed in New Haven, Connecticut. See Figs. 1, 2, 3, and 4. It is of interest to note that the initial purpose of Bell's experiments was to develop a 'Harmonic Telegraph, "with which six or eight Morse messages could be sent on a single wire at the same time without interference. The machines did not work as well as Bell had hoped, but further experiments by the team of Bell and Thomas Watson accidentally discovered how to make a current of electricity vary in intensity to match the variations of sounds in air.

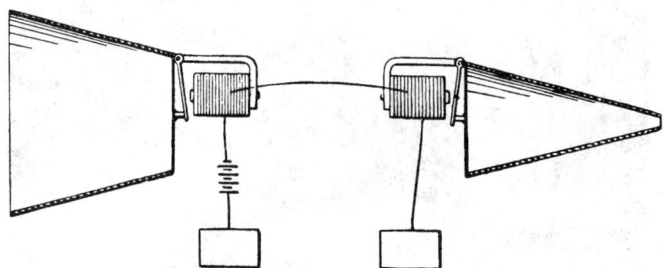

Fig. 1.   Illustration from U.S. Patent 174,465, issued to Alexander Graham Bell on March 7, 1876. The transmitter, a receiver, and a connection between them illustrates the basic principle of the "Electric-Speaking Telephone." (*AT&T.*)

Bell's first telephone was called a "*gallows*" because of its resemblance to that structure. The commercial version was the "*iron box*" (possibly the first *black box* of technology), of which some 6,000 units were built for customer use. The device did not have a ringer or a dial; it was connected by iron wire to another black box. A user simply shouted into the telephone to attract the attention of someone at the other phone. An electromagnetic converter alternately served as a transmitter and receiver. The next model was the "*butterstamp*" phone, in which a corded wire extending from the

[1] It is interesting to note that later on the same day in 1876 (February 14), the U.S. Patent Office also received an application for patent from Elisha Gray, who in 1867 had received his first patent on a self-adjusting telegraph relay. This is an example of that rare human phenomenon, simultaneous invention. This is well-detailed by Hounshell (1981).

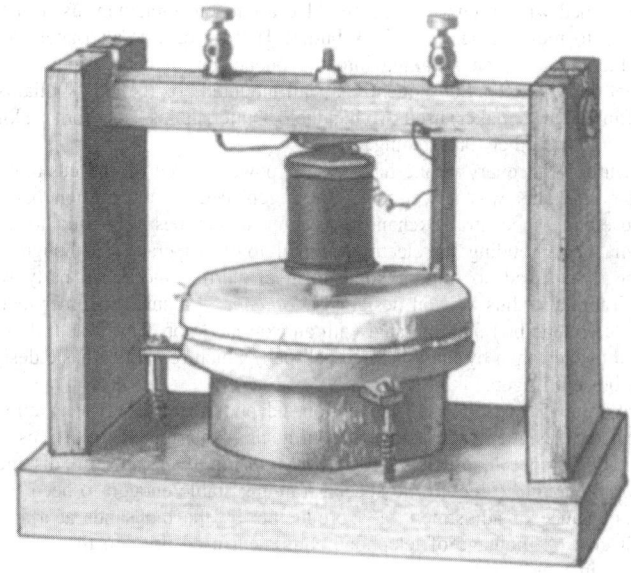

Fig. 2.   First attempted telephone transmitter with its parchment diaphragm attached to the magnetized metallic reed. This instrument, sometimes called a "gallows telephone" because of its resemblance to that device, was used to transmit the first speech sounds electrically. While Bell was experimenting with vibrating reeds, he was shaping the concept that a current might be given the form of a sound wave (analog). When one of his transmitting reeds by accident generated an undulating current—a sound-shaped current—instead of an intermittent current, and he heard the corresponding sound as he listened to the receiving reed, he realized that he was on the track. This experiment was conducted on June 2, 1875. (*AT&T.*)

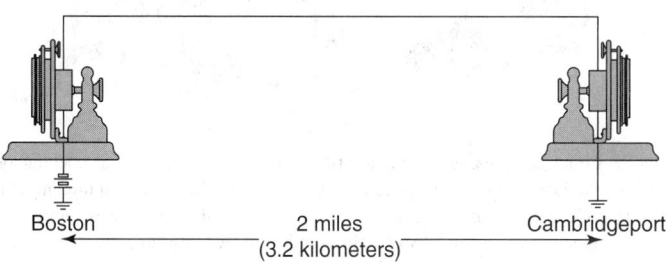

Fig. 3.   The circuit and instruments used by Bell on October 9, 1876 in a trial of two-way telephoning over a private telegraph wire between an office in Boston and a factory in Cambridgeport, Massachusetts. This was the first time that satisfactory and sustained conversation was carried on by electrical means between persons a considerable distance apart. (*AT&T.*)

Fig. 4.   Sketch of first telephone switchboard installed in New Haven, Connecticut in January of 1878. This board served 21 customers over 8 lines. The primitive switchboard used multiple-contact switches and a magnetic annunciator or "drop" (shown in upper right) to indicate that a connection was wanted. (*AT&T.*)

box ended with a device that resembled a kitchen implement used in the 1800s to mold patties from bulk butter. The device was alternately held to the mouth when speaking into the phone, or to the ear to hear the other party's voice. Subsequently, a wall-mounted box (called "Williams's coffin" after its maker) used two handheld "butterstamps," one for speaking into and the other for listening.

Until 1894, every phone had its own power source on the customer's premises. This was a wet-cell battery, replaced by a common battery housed in the central exchange. A crank sometimes was used to turn a magneto, sending an electrical signal to the operator. Although the first phones had no ringers, some form of alerting the called party was desirable and thus Watson designed a *thumper*, a small hammer inside the telephone box and operated with an external knob. That was followed by a buzzer, then by the polarized call bell, which remains a basic design for present ringers.

Within 15 years of Bell's invention, telephone technology was able to meet the designer's long-range design goals—instruments with a usable quality of speech, efficient enough to transmit over a useful distance, simple enough to require no special training, stable enough to need little maintenance or adjustment, and producible by the thousands at a small unit cost. A montage of telephone equipment used prior to the 1930s is given in Figs. 5–11.

Fig. 7.   Magneto wall-mounted telephone set (1882). This handsome instrument, encased in oak and using the Blake Transmitter and Bell's hand receiver, was the first telephone built for the Bell System by Western Electric. It was in service for many years and was one of the first side-winder models on which a crank was turned to signal the operator. (*AT&T.*)

Fig. 5.   First commercial telephone (1877). The round, cameralike opening on the boxlike instrument served as transmitter and receiver and thus required mouth-to-ear shifts when conversing. Developed in the fall of 1876, it was introduced to customer service when a Boston banker leased two instruments. These were attached to a line between the banker's office and his home in Somerville, Massachusetts. (*AT&T.*)

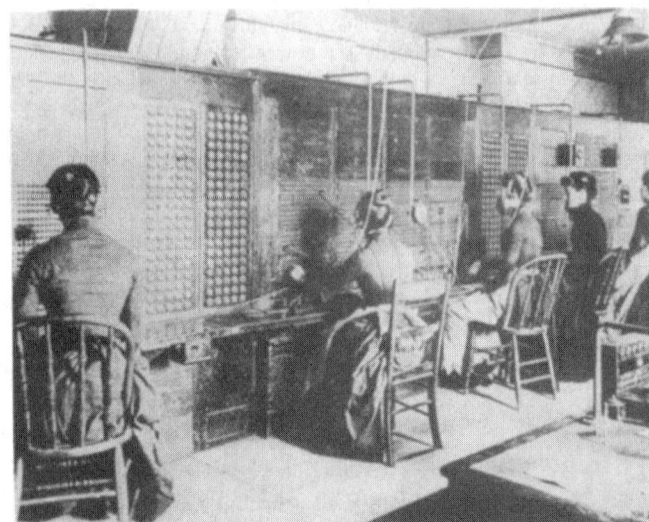

Fig. 8.   Switchboard (1883) with operators seated on wooden chairs and separated by panels known as "annunciator drops" between operator positions. The annunciator drops indicated telephone lines requesting service. (*AT&T.*)

Fig. 6.   An early telephone system receiving considerable use was that of the Gold and Stock Company, originally formed to broadcast stock market and banking reports by means of telegraph printers. A large switchboard used by the firm is depicted here. The operators were recruited from the ranks of telegraph messengers. (*AT&T.*)

While the development of the telephone instrument was vital to the new field of telephony, other factors also were important. The telephone sets had to be connected by wires and switchboards in cities and towns, and eventually the cities were interconnected as well. Long distance lines were constructed up and down the eastern coast of North America, and finally, in 1915, across the continent. (During the ceremonies inaugurating the first transcontinental telephone line on January 15, 1915, Alexander Graham Bell in New York City connected a replica of his 1875 gallows-type telephone to the line and repeated the first sentence ever spoken over the telephone, "Mr. Watson, come here, I want you." Thomas A. Watson, Bell's former assistant, was in San Francisco for the occasion and replied that it would take him a week to get there.)

The man chiefly responsible for driving the expansion of telephone service during the years 1907 through 1920 (the year of his death) was

Fig. 9. Desk telephone set (1897). In the early 1890s, the telephone began to assume the shape in which it remained familiar to users in North America for the following three to four decades. The ancestor of the upright desk set, shown here, represented a refinement of earlier models. The set was made of cast brass. (*AT&T.*)

Fig. 10. Dial telephone (1919). The first dial telephone exchange is credited to Almon B. Strowger, a mortician, who introduced it in LaPorte, Indiana in 1892. It was many years, however, before switching equipment was developed to permit dial installations in larger cities. Installations commenced in New York City in 1922. (*AT&T.*)

Theodore N. Vail, AT&T's president, who in 1908 created the term "Bell System" and originated the theme of "One Policy, One System, Universal Service." Vail's goal was to eliminate dual services by either buying competitive companies or abandoning the market to the competing service. Dual services referred to two competing companies in the same city, a situation which raised costs to telephone users, since they often had to subscribe to both services because the competing networks were not interlinked as they are today. Although the concept of assigning telephone service to a monopoly in each market disturbed some people, it remained a fact until the introduction of cellular wireless services in 1984, when the Federal Communications Commission (FCC) designated two competing entities to serve each cellular system market, and 10 years later

Fig. 11. Toll system operating room (1929) showing 800-line switchboard installed at the Illinois Bell Telephone Company. Note billing messengers on roller skates. (*AT&T.*)

added more wireless competitors in the field of personal communications services (PCS). Today, there are many competing long distance telephone services, and numerous wireless phone companies—but the wired local networks are still basically a monopoly in each market. There are legal and commercial efforts underway to change that monopoly.

Perhaps the most intensive research sponsored by the telephone companies and others was devoted to automating the call-routing system. Until the dial telephone was invented and automatic switches were introduced, all telephone calls—even those across town—had to be switched by operators plugging cords into sockets after asking what number was desired. The first operators were boys recruited from the ranks of telegraph messengers, but their manners proved to be too crude, so refined young ladies replaced them. As the telephone network expanded and thousands of switchboards were installed, the network planners could foresee the day when the use of human operators to connect long distance calls would become too unwieldy. For example, in the 1920s, the banks of operators in some large cities became so long that agile young ladies on roller skates were used to pick up call charge slips from individual operators to bring them to central desks so they could be sent by pneumatic tube to the billing department. By the 1950s, customers were provided with direct dialing across the North American continent, followed 20 years later by international direct dialing. Billing for toll calls became automated. Operators are now the exception rather than the rule in placing long distance calls, domestic or international.

Modern Telephony

Three basic groupings of equipment are used in modern telephony:

1. Customer premises equipment (CPE) such as corded and cordless telephones, facsimile or "fax" machines, office switchboards or PBXs, and desktop computer terminals or personal computers. Wireless cellular mobile telephones and personal communications services (PCS) pocket phones are included in this group.
2. Transmission equipment, such as metallic and fiber optic cables, wireless (microwave) radio systems and satellites in space.
3. Switching equipment for both local (or central office) exchanges and toll (or long distance) facilities.

This equipment is linked together as a network by an ingenious assortment of computers and software programs called *operations systems*[2]. Operations systems are software programs designed to help communications companies perform specific network maintenance and administration functions, such as management of customer records, billing, inventory updates, and generating usage forecasts. They also enable the introduction and use of automated customer services. The operations systems employed

[2] Operations systems are not to be confused with *operating systems* such as the UNIX ® system or Microsoft's MS DOS system. *Operating systems* are software programs that manage a computer's hardware and software components, determining when and how to run programs, and are not pertinent to the operation of a communications carrier's network.

by telephone companies and other carriers help to improve the reliability of today's telecommunications equipment and services, balancing the ever-increasing complexity behind that equipment.

Because of rapid advances in the technology of telephony, these equipment groups are capable of far more than the traditional basic service, sometimes referred to in the communications industry as *"plain old telephone service"* or POTS. In addition to coping with a rapidly changing technology, the U.S. communications system user must now adjust to organizational changes as well[3]. Similar changes are occurring on other continents. Many countries throughout the world are restructuring their telecommunications organizations in order to match changing technologies to the needs and applications of customers for information services.

In the United States, the responsibility for determining communications standards is assigned to the American National Standards Institute (ANSI), while the authorization and monitoring of standards is assigned to the Federal Communications Commission (FCC). Although each local telephone company and long distance company is responsible for its own equipment maintenance and operation, the American telecommunications network still functions as a single technical entity for most users. A long-distance telephone call may use the facilities of several companies, but their involvement is usually "transparent" to the user, i.e., the call crosses various corporate, state and even international boundaries automatically. In fact, the combined facilities of the American network (which is usually referred to as the *Public Switched Telephone Network* or *PSTN*) have been called "the world's largest computer system," in which each telephone is a terminal that can access any other telephone (terminal), or be accessed from it, via the central system.

As of 1998, the United States telephone network served more than 160-million telephone lines (the actual number of telephones is unknown since a single line can have numerous extensions), more than any other nation in the world. The FCC reports that telephone subscribership nationwide amounts to 94 percent of US households, and has a reliability rating of 99.99 percent. If you, the reader, look around, chances are excellent you will see one or more telephones near you—in many cases, even if you are outdoors.

At the same time, the ratio of fixed telephones to mobile telephones has been changing rapidly. Researchers report that by 1998 there were 50 million *wireless* telephone subscribers in North America, a figure which continues to grow at a startling rate. Part of this domestic growth seems due to the American desire to be mobile rather than tied to a telephone cord.

However, there appears to be a different reason for the growth of wireless systems in other countries. Despite strong efforts in many nations to enlarge their wired phone networks (in 1994 some 38-million new wirelines were added to the world's networks, plus another 45 million new lines in 1995), in 1995 more than 43 million applicants in emerging markets were still on waiting lists for a telephone wireline connection, and the average waiting time was more than a year. In 1994 19 million *new wireless* subscribers were added throughout the world, and this escalated by another 33 million *new wireless* subscribers in 1995. Part of the international growth is

---

[3] The industry's basic structure in the United States was dismantled on January 1, 1984, when the Bell System's operating telephone companies were divested by court order from their parent firm, the American Telephone and Telegraph Company (AT&T). The telephone system serving the United States then included seven separate regional holding companies (RHCs)–Ameritech, Bell Atlantic, BellSouth, NYNEX, Pacific Telesis, Southwestern Bell Corporation, and US West–containing 21 of the 23 former Bell operating companies (BOCs). In recent years a number of mergers have reduced that number to four RHCs–Bell Atlantic, Bell South, Southwestern Bell Corporation (now SBC Communications), and US West. The largest single telephone operating company is GTE, which in 1991 merged with Contel Corporation to form the only nationwide local Telephone Company (since the breakup of the Bell System. (In 1998, GTE agreed to merge with Bell Atlantic.) In addition, some 1200 other phone companies of various sizes (formerly called *independents*), serve local customers. Assorted inter-exchange carriers, sometimes called *long-distance companies*, are led by AT&T, MCI, and Sprint. The technology base for AT&T's systems and services is provided by AT&T Labs, since the Bell Telephone Laboratories organization in 1996 was spun off from AT&T with the former Network Systems Group (known for many years as Western Electric), which is now known as Lucent Technologies. Applied research and technical support for the regional holding companies of the former Bell System is now contained in each organization, since the former Bell Communications Research (Bellcore) organization was sold by them in 1998.

caused by the difficulty of financing and constructing expansion of existing terrestrial wirelines.

Wireless systems based on cellular and/or communication satellite technologies provide virtually *"instant"* networks at relatively low cost, especially in difficult terrain. Experts predict that before 2000, there will be 200 million wireless phone users worldwide. This estimate may well be too conservative, depending on the success of new satellite telephone systems.

The growth of international telephone-call traffic averaged 15 percent per year between 1975 (4 billion minutes) and 1995 (over 60 billion minutes), according to the International Telecommunication Union (ITU). Because the world's telecommunications systems are vital to multinational businesses and governmental operations on a global scale, the United Nations is becoming increasingly important as a means to coordinate differing national systems standards. The ITU is a specialized agency of the United Nations, operating with five permanent organs: the ITU General Secretariat (sponsor of the World Administrative Radio Conferences that help write the International Radio Regulations); the Bureau of Telecommunication Development (providing technical support to developing nations); the International Frequency Registration Board (coordinating frequency usage by radio stations within different nations); the Telecommunication Standardization Bureau (TSB), created in 1992 as the replacement for the former Consultative Committee for International Telegraphy and Telephony (CCITT); and the Radio Communications Bureau (RCB), formerly the Consultative Committee for International Radio (CCIR).

Membership in these ITU organizations includes 181 governments, 88 "recognized operating agencies" or licensed network operators and service providers (such as AT&T, MCI, Sprint, British Telecom, France Telecom, Deutsche Telekom, etc.), 146 equipment manufacturers and scientific and industrial organizations, and 38 international organizations whose operations rely heavily on communications technology (i.e., airline companies, banks, stock exchanges, etc.). The task of developing global standards is assigned to study groups, which can have up to 400 or more participants. These are divided into working parties for each specific area of work, and then subdivided into expert teams. The process is not conducive to expedited approvals, but given the global scale involved, that is to be expected. The reviews extend all the way from individual telephone sets to transoceanic cables and communications satellites. Some governments regard these "standards" as mandatory, while the United States and some other countries view them as "advisory."

An example of an ITU activity is the international numbering plan, developed in 1963 and based on the level of national telephone development forecast by each country for the year 2000. The plan allows a maximum number of 12 digits in the world number, including a single-digit World Zone (there are nine Zones), a two-digit country code if needed, a three-digit area code or two-digit city code, and up to seven digits for the exchange and telephone line number. The actual number of digits varies from country to country. Because of its size and the number of telephone access lines—almost 210 million or more than 41 percent of those in the world—North America was assigned a single digit Zone code (the numeral "1") shared by Canada and the United States, followed by a three-digit area code and the combined seven-digit exchange-and-line number.

In the 1983 Federal Court proceedings that divested the Bell Telephone companies from AT&T, a geographic patchwork of about 160 service areas (each known as a Local Access and Transport Area or LATA) was established across the United States. Wireline telephone service within each LATA is provided only by an authorized Local Exchange Company or LEC, which also is one of the two authorized cellular telephone providers (the other is a separate company) in each major market. The LEC may be either a Bell Operating Company (BOC) or one of some 1,200 "independent" companies. In the mid-1990s, a new wireless service known as Personal Communications Services (PCS) was permitted by the FCC in each city, but these service companies were established by lottery sales of frequency bands, conducted by the FCC. Service between LATAs or across several LATAs is provided by Inter-Exchange Carriers (IECs), also known as long-distance companies. LEC services are priced according to tariffs authorized by state public utility regulators, while IEC tariffs are partially regulated by the FCC.

There is a dark horse in the race for telephone subscribers: the Internet/World Wide Web. Although this amorphous network of personal computers, "servers" and data bases has been focused on linking computers for transmission of text, data, video and recorded sound, the temptation to include two-way voice telephony has been strong. The reason for the attraction is the absence of long distance charges based on the use of a circuit for a specific period of time. On the Internet/World Wide Web, access charges (by many different service providers) are often a fixed monthly fee, regardless of how long a connection may last or how far apart the participants are, even if they are on different continents. The drawback has been the technology—packet-switched transmission—which shreds the message into many small parts and scatters them through many different circuits on their way to its final destination where all the packets are reassembled into the message before delivery to the end customer. This works fine for text, video and recorded sound, but two-way speech is more sensitive to time anomalies caused by the translation from analog sound to digital packets and then from packets back to analog signals. As the technology continues to improve, the Internet/World Wide Web may cause significant drops in long distance telephone revenues—unless changes occur in the billing procedures by either the telephone companies or the Internet access providers. Ironically, many of the same physical transmission and switching facilities are used by both sides.

Customer Premises Equipment (CPE)

A telephone, or *station set*, is actually a *compact transmitter* and *receiver* of electrical signals, either from a wireline or through an antenna. As a transmitter, it uses either a carbon-granule button or an electret microphone that converts sound waves to an electrical signal, produced with analog techniques. See Fig. 12. This means that the electrical signal continuously matches a limited range of speech sounds with changes in amplitude and frequency. As a receiver, the set contains a miniature loudspeaker that converts the incoming electrical signal to analog sound waves, reproducing a limited range of audible voice frequencies that can instantly be identified as a "telephone call." Digital telephones operate differently, using a microchip called a "*codec*"—for coder/decoder—to convert the analog speech frequencies into a series of digital number sets called "samples," each transforming the analog signal into an eight-bit digital code. Sampling is performed at the rate of 8,000 times a second, and the digital bits are transmitted and received at the rate of 64,000 bits per second in a network-quality system.

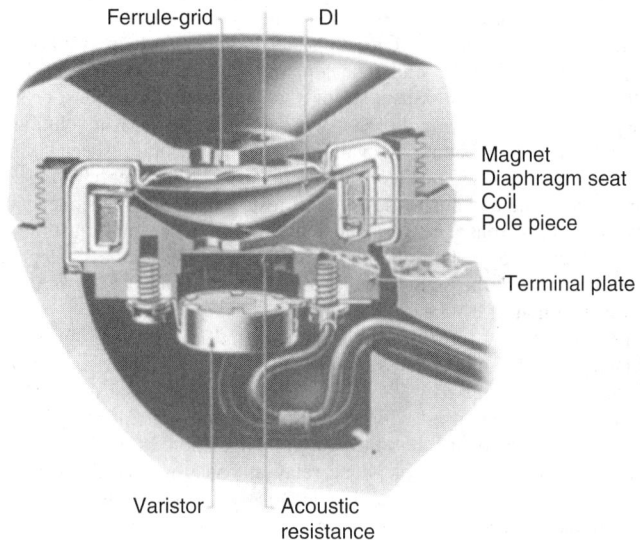

Fig. 12. Sectional view of handset of a touch-tone telephone. (*AT&T.*)

As of 1992, the approved digital voice bit rate could be as slow as 16,000 bits (16 kilobits) per second on wire or 8,000 bits (8 kilobits) per second for cellular wireless transmission/reception. Although a digital telephone may be connected to an analog line, using a device called a "*modem*," most digital phones are connected to digital transmission lines, either within a private network or provided by a telephone company as an extra-cost option.

In addition, the basic telephone apparatus includes a *switchhook* (which lets the power from the telephone line or the handset battery flow through the transmitter/receiver), an electromagnetic or electronic *ringer*

Fig. 13. Closeups of rotary dial configuration and touch-tone dial for desktop telephones. The pushbuttons on the touch-tone dial include special signals (*) and (#), which are used to activate special services provided in digital switches. (*AT&T.*)

(to announce incoming calls), and a *dial* for entering the customer identification number of the telephone being called. See Fig. 13.

In North America, most customer identification numbers consist of ten digits, comprised of a three-digit area code, a three-digit central office (local exchange) code, and a four-digit station number. Each digit is selected from a dial with ten digits, "1" through "0"; eight of these digits are accompanied by a three-letter group ("2" has "**ABC**," "3" has "**DEF**," "4" has "**GHI**," etc.). The letters originally were proposed as a coding method for dialing letters that suggested the exchange name, such as Murray Hill or "**PE**nnsylvania" (as in Glenn Miller's famous big-band tune, "**PE**nnsylvania 6-5000"). Today, the letters often are used as a mnemonic code for commercial names instead of numbers, i.e., **AIR-LINE**.

The digits "0" and "1" have other functions as well. "0" can be used to dial the telephone operator, while the digit "1" can be used to request extra service in long-distance and wide-area dialing, as described under **Switching Equipment** later in this article.

The dialing mechanism on the telephone is either *rotary* or *pushbutton*—often called *touch-tone*. The touch-tone dial is preferred when the user wants to access services associated with the digital telephone network already installed in the United States but still evolving around the world. Some telephones automatically dial a number at the touch of one button (so-called "speed dialing"). The repertory of numbers is stored in electronic memory within the telephone set. Many of these telephones automatically store the last manually dialed number and redial it when the user pushes a button.

A rotary dial usually has ten finger-holes in a rotating finger wheel mounted on a number plate. The user dials by selecting each numeral, turning the dial until the finger stop is met, and then releasing the finger wheel. The rotary dial signals the number being called by means of dial pulses, i.e., controlled interruptions (nominally ten pulses per second) of the 48-volt direct-current power supplied from the central battery serving that customer's area.

The touch-tone pad has 12 pushbuttons—digits "1" through "0", plus an asterisk or star (*) and a pound (#) sign for special services available from electronic switches in modern exchanges. Associated oscillator circuitry in the touch-tone dial phone generates tones in the *voiceband* range (a nominal bandwidth of 4000 Hertz, extending from a low of 300 Hz to a high tone of about 3800 Hz) used in telephony. Each touch-tone digit and symbol is uniquely represented by two tones, one from each of two mutually exclusive groups of four frequencies (a signaling type often called dual-tone multifrequency). The switch acts instantly on each numeral rather than waiting until all the numerals have been dialed.

The telephone switchhook is used to signal the status of the phone, i.e., *idle* or *busy*. When the telephone is idle, or *on-hook*, the switchhook contacts are open, and when the set is busy, or *off-hook*, the switchhook contacts are closed. These basic signals allow other equipment to recognize a call's origination, answer, and termination. The customer is alerted to an incoming call by a bell or tone ringer (or a light in some cases). If the customer picks up the phone to place a call, a *dial tone* must be heard that indicates that a connection exists to the telephone company's nearest exchange. Lack of a dial tone commonly means there is no connection. If a number is dialed, the caller can hear a signal indicating that the called phone is ringing, or if the phone is being used already, the caller hears a *line busy tone* consisting of an on-off buzzing sound with 60 interruptions per minute. A fast busy tone (120 interruptions per minute) indicates the network circuits are unavailable for the call being attempted.

Although the electrical signals representing the voice usually are transmitted over 48 volts of dc power (in wireline systems), the ringer is driven by a nominal voltage of 86 volts root-mean-square (rms) at 20 Hz superimposed temporarily on the telephone line at the switch in the central office or exchange. The power systems serving public telephone network circuits are usually independent of the electric utility lines in a given area, so most telephones can still be used in if the electric utility power is cut off, provided the telephone lines were not damaged. However, some types of customer premises equipment, such as electronic switchboards or answering machines, may depend on utility power for operation, so they may become inoperative if a power failure occurs.

Many shapes and styles of telephone sets are available to telephone customers. Some sets are adapted to the special needs of people with disabilities, such as extra-large dial buttons for the visually impaired, and handset volume controls for the hearing impaired, while others include speakerphones for hands-free conversations. Multikey telephones enable users to connect a single telephone set to a number of different telephone lines, or link a single telephone to various other extensions. *Call director* telephones function as a miniature office switchboard, allowing a receptionist or secretary, for example, to answer an incoming call and direct it to an appropriate desk. Electronic telephones often include a liquid crystal display (LCD) panel or light-emitting diodes (LEDs) that can show the user a number as it is being dialed or, in some systems, display an incoming caller's number before the telephone is picked up. So-called "smart phones" also have appeared in the 1990s. These are usually a combination of a telephone and a small computer with a miniature keyboard and LCD screen, capable of e-mail access and faxing as well as conventional voice calls.

Even the public coin telephone has been upgraded and now includes phones activated by inserting a special credit card that either identifies the user's billing account or automatically deducts the call charges from the card's built-in prepaid deposit. Some of the most sophisticated public phone facilities are in developing countries, where special booths are equipped with satellite telephone facilities capable of direct links to new multi-satellite "constellations."

In 1964, AT&T introduced the first public video-telephones, called the "*Picturephone*." They required a special telephone line between the Picturephone and the local switch, and each party on the circuit had to have a Picturephone in order to "enjoy" the black-and-white image, which was about equal in quality to that of commercial TV. For various reasons (the relatively high monthly charges, the need for a special phone line between the switch and each phone, and an individual reluctance to appear "on camera" whenever the telephone rang, among others) the market rejected the product. In 1992, new video telephones were introduced, allowing the user to simply plug the set into an ordinary telephone outlet; of course, the video transmission functions only when the other party also has a video telephone. Again, the product failed to inspire much public enthusiasm, for various reasons (the image quality was mediocre, with poor color and resolution, as well as limited motion, and the public was still reluctant to be seen whenever the phone rang). However, many companies have found it useful to equip or lease special meeting rooms with video cameras that enable groups of people in various facilities to conduct meetings by television via the telephone network, with the signals carried over temporarily leased terrestrial and satellite circuits, a procedure that saves travel time and expense for many participants.

In recent years, the *cordless* telephone has been introduced to business and residential users. This device is actually a small short-range radio set, in the form of a wireless battery-powered handset (or headset) with a short antenna stub, that sends to and receives from a nearby base unit plugged into a telephone wall outlet and a power outlet in the office or home. Depending on the design, the cordless phone can transmit over distances of a city block or more. This design allows the user freedom of movement while still being served by a wireline connection. However, the ringer does not operate if the handset's battery switch is turned off. The handset's wireless mode can operate with either analog or digital transmission, depending on the model selected. Some models employ a variety of optional security techniques to prevent or minimize unauthorized access by other nearby cordless phone sets.

Early cordless analog phones (ca. 1980s) and some modern, inexpensive models operate within a radio frequency range of 46 to 49 megahertz, which is susceptible to interference from various other electrical devices (baby monitors, computers, electric motors, etc.) as well as signal reflections or blockage due to building materials. Newer cordless telephones are designed to operate at 900 megahertz, which is much less prone to such problems. In addition, the newer telephones are available with various accessories such as headsets, answering machines, spare battery packs, speakerphones, caller identification, and even a flip-up LCD screen and computer keyboard on the base component, enabling sending and receiving of e-mail via the Internet. Most users are aware that the voice signal quality of cordless telephones is not as good as that of a wire line, due to signal compression techniques.

The battery of a cordless phone is in the transmitter-receiver handset, and is charged while the handset is resting on the base unit. When the battery power in the cordless handset drops too low, due to extended use, the phone will "go dead," so a conventional line telephone is recommended as a backup.

Another version of cordless or "mobile" telephones is designed to be used in vehicles such as cars, vans, and trucks. The communications industry reported that, by early 1991, more than five million customers were using such mobile telephone equipment, called "cellular" telephones. More information on this technology, called "*cellular*" telephones, is provided in the "**Transmission**" section of this article.

The wireless mobile telephone sets themselves have shrunk in size since the technology was introduced for the general public in 1946. Those early mobile telephones, accessing a single large antenna serving an entire city, required the handset to be wired to a large, bulky transmitter/receiver, usually installed in the trunk of an automobile. Today a cellular telephone unit need be no larger than a cordless handset, and includes the entire necessary transmitter and receiver components as well as the power supply. In fact, many users carry their cellular telephone with them in a briefcase or purse and use it while walking, shopping, riding in mass transit facilities, sitting in a car (although it is strongly recommended that the user avoid telephoning while driving the car), etc. This technology has revolutionized the communication habits and availability of many people, especially in the business community. Travelers with "*cellphones*" no longer have to search for a pay phone and businesses such as restaurants and luncheonettes find that offering patrons a cellular phone as a convenience can offset the cost of installing a pay phone booth. However, like the cordless telephones described above, the sound quality of a cellular transmission is usually lower than that of a wired line.

By 1990 a new form of wireless telephony began taking shape—the Personal Communications Network (PCN), also known as *Personal Communications Services* (*PCS*). The PCS customer uses a small, pocket-sized telephone equipped with a tiny stub antenna, similar in appearance to the cellular telephone, but different in its operation. In the PCS concept, a network of small radiotelephone relay antennas is spaced much more closely than conventional mobile-telephone cell sites (possibly mounted on alternate street lampposts, for example). The relatively weak signal from the PCS telephone is strong enough to reach the nearest cell antenna, which emits microwave radio signals at far lower energy levels than are needed for vehicular mobile telephony.

The battery of a PCS phone provides enough power for several hours of talk-time and 50 to 200 hours of standby time. By 1998, numerous models with many different features were available for purchase, and PCS networks were installed in many cities. Operating in the radio frequency range of about 2 gigahertz for customer connections, the PCS caller often is linked to cellular networks as well. PCS telephones are all digital, using either of two different signal coding methods, depending on the type of antenna used by the service provider. Like the cordless and cellular telephones, the PCS phones do not equal the sound quality of a wire line.

During the past few decades, customer equipment has been expanded to include signals other than acoustical sounds. These signals are generally classified as "*data*" because they consist of text or images coded as digital pulses. Generally, the data signals are received or transmitted by facsimile (fax) machines or personal computers. An electronic peripheral device called a *modem* (for MOdulator-DEModulator) is used to couple a digital[4] computer or terminal to an analog telephone line so it can exchange data with another computer or terminal. The transmission capacity of a modem is usually stated in "*baud*" or "*kilobits-per-second.*" In 1998, the maximum transmission rate for consumer-market modems was about 56 kilobits/second, based on the technical limitations of a conventional four-wire telephone line, but scientists are pushing that barrier and aiming at transmission of multi-megabit signal streams over ordinary phone lines equipped with special terminals at each end. Two-way signal flow is known as *duplex* transmission (one-way voice transmission is called a *simplex* circuit, in which each speaker, when ready to cede the channel to the other person, says "over," and when the transmission is ended, says "Roger and out") and occurs on either a *dial-up* telephone line, accessed by manual or automatic dialing, or on a *dedicated* or *reserved* private line. When a telephone and a computer/modem combination share the same line, they cannot be used simultaneously except in special systems. In the late 1990s, some special arrangements were devised to enable voice transmission over the digital Internet and World Wide Web, which employ packet signal transmission rather than conventional dedicated circuits. Wireless modems are now available that permit portable computers to interface with wireless voice systems.

Computers generally operate with digital technology, so the modem converts the digital data signals into a form of analog signal, discrete in time value, that is suitable for transmission over analog voiceband circuits, and vice-versa for incoming data signals. Various speeds of operation, measured in bit-rate-per-second or "*baud*," are available, with the cost of the equipment rising as the rate increases. The fastest practical speed of a modem over normal telephone lines is about 56,000 bits (56 kilobits) per second. But some telephone lines are incapable of handling such data speed without causing transmission errors, so bit rates as slow as 300 baud are used, depending on economics as well as system considerations. The modem may provide an acoustical coupler, a cradle into which the telephone receiver is placed for transmission, or the modem may be wired directly to the telephone outlet. In addition to computer terminals and personal computers, modems also may be linked to facsimile machines used to transmit images of printed, typed, or illustrated documents. Many modern fax machines already have a modem built into their circuitry.

Fax machines are an example of the handicaps imposed by proprietary, incompatible standards. The effects of a change in such standards can be seen in what happened after the issuance in 1980 of the Group 3 facsimile standard, which suddenly enabled different brands of fax machines to communicate with each other. Prior to this, a Brand X machine could not send or receive a fax involving a Brand Y or a Brand Z machine. Fax machines that previously were installed only in corporate mailrooms suddenly appeared in individual offices and in small businesses, even pizza parlors, and in homes as well. All that had been needed was a standard to overcome the same-brand hurdle that blocked universal application of the technology.

Earlier fax machines required six minutes to transmit a single page at a nominal resolution of 100×100 dots per inch. With the Group 3 standard, transmission time for a typical business letter is 20 to 30 seconds per page, and resolution is adjustable to as fine as 200 dots per inch vertically and horizontally. (A Group 4 facsimile standard was set in 1984, designed for use with digital networks, in particular the *Integrated Services Digital Network* or *ISDN*, at resolutions of up to 400 dots per inch and transmission speeds of 56 or 64 kilobits per second.)

Other CPE items linked to telephone lines include: teletypewriters; private branch exchanges (PBXs, both manual and automatic) that resemble small exchange switches and link private telephone extensions, such as in an office, hotel, or hospital, to each other and to the public switched network; key system sets (often found on receptionists' desks); and various sensors, monitors, and alarms. Until recently, all these devices used analog signal techniques to link to an analog telephone line. However, modern PBX systems, capable of handling up to 100,000 lines or extensions, can link directly into a digital telephone trunk. The analog-to-digital (A/D) or digital-to-analog (D/A) conversion takes place inside the PBX instead of in a modem at the telephone.

Business customers and other organizations that do not want to install a PBX on their premises can use a telephone company service called *Centrex*, in which their telephones are connected to a reserved portion of the telephone company's local exchange switch. Incoming and outgoing calls are handled by the Centrex system as if it was an on-premises PBX.

Digital communications lines on customer premises are usually part of private networks called LANs (*Local Area Networks*) with circuits that are not switched by public network equipment. However, digital switched lines are becoming more prevalent on customer premises as a concept of telephony, called *Integrated Services Digital Network* or *ISDN*, is accepted by businesses and organizations—and residences as well. Special new telephones, known as ISDN sets and often paired with personal computers or other terminals, permit direct connection to digital lines via universal wall outlets, and can be moved from one location to another without rewiring telephone lines or making other changes in the system. Modems are not necessary when digital equipment is connected to a digital line. Elimination of the analog/digital conversions improves the quality of end-to-end digital data transmission.

A very special telephone set, now available only to authorized agencies of the U.S. federal government, is certified by the National Security Agency as a secure telephone terminal. Its genesis was the innovation of radio-telephony between Los Angeles and Santa Catalina Island in 1920. At that time, persons with radio receivers could listen in to the wireless telephone conversations, so a transmitter/receiver system was devised that "*scrambled*" frequency bands to prevent unauthorized eavesdropping. Scrambler phones were commonplace in military operations in World War II, but enemy engineers quickly found ways to decode these radio voice transmissions. A unique radio scrambling system known as SIGSALY (and often called "*The Green Hornet*" because the signal sounded like the buzzing theme of a popular radio program) eventually was developed by Bell Laboratories for use by President Franklin Delano Roosevelt and Prime Minister Winston Churchill; the system's voice-coding technique remained unbroken for the duration of the war and many years after.

Today's secure telephone units, available from several manufacturers, are known as the STU-III model and use digital voice-compression techniques as well as controlled access systems. STU-III terminals can make ordinary calls if so desired, but with the use of a crypto ignition key (equipped with a memory chip that unlocks the phone's security features) they can encrypt/decrypt voice, data, and fax transmissions as well as scanned photographs and other graphics. Variations on the secure telephone units are available for state and local governments, as well as for commercial applications in the U.S. and abroad.

Transmission Equipment

Signals used in telephony and telecommunications are carried on a variety of media. These include metallic wires, terrestrial and satellite microwave channels, and optical fiber or lightwave systems.

At the customer premises, the customary link is called *telephone station wire*, which is a four-conductor color-coded copper wire in a plastic jacket, suitable for installing inside walls or structures. Between the handheld receiver-transmitter and the body of the telephone set is a coiled cord that usually is modular (i.e., it can be unplugged at both ends and replaced by another cord of similar design or different length). Today, the station wire between the body and the telephone line in the wall also is modular, allowing easy replacement of the cord with a selection of lengths. Premises wiring is of the twisted pair cable type when it carries signals from several sets on separate circuits; the wires are twisted to minimize the accidental mixing by induction of two separate signals, which would produce "*cross-talk*."

Local Telephone Loop. The wiring between the customer premises and the switch in a local exchange office is called the loop. See Fig. 14. In its simplest arrangement, this wiring is connected directly from each premises to the switch, establishing a single-party line. (Before World War II, many residential telephones were on multiparty circuits or so-called *party lines*, in which up to ten different telephones shared the same line. Distinctive

---

[4] Digital signals are a method of coding a transmission with binary digits, i.e., 1 and 0, in which each components of the signal is structured as a "*bit*" and assembled with other bits into a "*byte*," which in turn is assembled into a "*word*." Analog signals that make up the voice signal for telephony are limited to a bandwidth of 4000 Hertz, including control signaling; when coded as a digital signal, the voice transmission frat for network quality on a wireline is 64 kilobits per second. Because human speech is analog in nature, the digital signal must be decoded before it can be heard by human ear.

The telephone loop

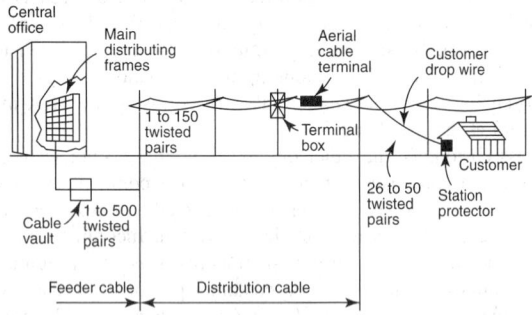

Fig. 14. Local telephone loop connects customer's premises with the local switching machine. The loop is actually an insulated pair of wires protected from the elements by a cable sheath. Telephone cables contain many such "pairs" and these cables may be buried in underground conduit, strung on poles, or buried directly in the ground. Telephone personnel usually refer to "*feeder*" and "*distribution*" cable. Feeder cable runs from the central office through major population centers to a given geographical area. At various points, the distribution cable branches off from the feeder cable, bringing cable pairs closer to the customers. Distribution cables themselves may branch into smaller entities. Each of these cables is separately identified in telecommunications company records and in the predictive system's database. The system attempts to locate a cable failure by correlating trouble messages with different cables and combinations of pairs. For example, it may point to a failure in either the feeder cable or in a segment of the distribution cable, or both. This information helps identify where the outside craft personnel should begin to look for the problem (*AT&T*).

rings enabled subscribers to identify who was being called—but privacy depended on the cooperation of the party-line customers, who could listen in on other conversations if they so chose. By 1981, multiparty lines were rare; about 97 percent of the former Bell System was single-party in 1984 when the Bell companies were divested from AT&T.

Most loops today are arranged in a tree-like structure. A paired cable is the primary transmission facility in the loop network. Individual drop cables connect to a distribution cable that combines a number of customer wire pairs. The distribution cables in turn are connected to feeder cables. Near the central office serving the loop, the analog feeder cables are generally quite large—1200 to 3600 wire pairs.

To reduce the number of cables required, digital loop carrier systems use digital multiplexing and transmission techniques, thereby enabling a number of customer loops to share a smaller number of wire pairs and electronics. This is done by using analog/digital conversion and time-division multiplexing at a collection point in the loop, between the customer premises and the switch. This equipment is small enough to fit individual systems onto a pedestal but is often installed underground. Customer signals are transmitted digitally from the loop carrier remote terminal to the exchange office, called the *central office* or *CO*.

The communication link is usually a T1 digital carrier system, with a transmission capacity of 1.544 million bits (megabits) per second (Mb/s). An individual T1 carrier, composed of two wire pairs and regenerators, can handle 24 voiceband channels, each rated at 64,000 bits (64 kilobits) per second, or 64 kb/s. The first T1 digital carrier, a short-haul (metropolitan area) system using paired cable, was introduced into the network in 1962 near Chicago, IL. At that time, there were no digital switches in the network, so digital-to-analog converters were required at the T1 system's end points to make it compatible with the analog network connections. Today, converters are unnecessary when the T1 connects to a digital switch or other digital terminal.

When equipped with digital repeaters[5] to regenerate the signal at critical points (about a mile apart) in the transmission path, the T1 digital line can

[5] In trunk systems between offices, the telephone circuit may pass through a series of repeaters. These are devices that equalize or amplify the analog voice signal to correct losses of strength so the signal does not fade away before it reaches its destination. Earlier analog repeater designs often amplified electrical "noise" as well, so the signal was sometimes masked by crackling or other disturbances after transmission through many repeaters. Repeaters are now increasingly rare in telephone circuits. Modern repeaters, when their use is necessary, normally filter out unwanted "noise," and digital circuits

extend up to about 50 miles between the remote loop carrier site and the central office. The maximum distances between consumer premises and the remote loop carrier sites can be 26,000 feet (7,900 meters) with 22-gauge loaded cable and almost 50,000 feet (15,240 meters) on 19-gauge cable. Microwave radio systems that handle T1-type transmission also can be used between the remote and central office terminals, with relay stations every 20 miles (32 kilometers) or so at horizon points.

In some rural areas, telephone customers are still served by open-wire lines, which consist of uninsulated pairs of wires supported on poles spaced about 125 feet (about 38 meters) apart. The wires range in diameter from 0.08-inch (2.03 millimeters) with 12 gauge to 0.165-inch (4.19 millimeters) with 6 gauge and may be copper, copper-clad steel, or galvanized steel. Such systems are, of course, becoming increasingly rare as the nation's telecommunications network is upgraded.

Coaxial Cable. In 1936, a new type of cable, called coaxial, was experimentally demonstrated as the first broadband transmission long distance system. *Broadband* systems transmit a band of electrical frequencies several million cycles (megahertz) in width. Terminal equipment at each end subdivides the bandwidth into frequency bands (in the case of analog transmission) or time slots (in digital transmission). These subdivisions keep the various voice and data signals distinct as they follow their assigned paths.

Today, a coaxial conductor is usually a flexible 3/8th-inch copper tube in which a No. 10 gauge (0.102-inch or 2.59 millimeter) copper wire is held precisely centered by small plastic insulators spaced about 1 inch (25.4 millimeters) apart. A coaxial cable containing eight tubes is about 2 inches (5 centimeters) wide and weighs almost 3 pounds (1.3 kilograms) per linear foot (30 centimeters), while a 20-tube cable is about 3 inches (7.5 centimeters) thick and weighs 9 pounds (4 kilograms) per foot (30 centimeters).

Most cable television networks are equipped with coaxial cable networks, which may be utilized for voice telephone calls in the future if proposed mergers between long distance companies and cable TV operators take place. However, satellite TV distribution systems are not designed to handle two-way signal feeds through small earth antennas. See also **Satellites (Communication and Navigation).**

The availability of coaxial cable in the 1930s and 1940s made it possible to provide commercial radio networks with an efficient method of distributing their programs from a central source to various affiliate stations around the nation, over telephone circuits. Actually, the first ideas of using telephone wires for one-way transmission of radio broadcast signals developed after World War I, when radio was still in its infancy. Coaxial cable improved the quality of sound transmission to match the improved quality of the transmitters and receivers as the years progressed. Although still widely in use, especially by the cable television industry, coaxial cable is no longer favored in specifications for new broadband communications systems in the public switched telephone network, having been replaced in preference by microwave radio and optical fiber lightwave systems.

Microwave Links. After World War II, television broadcasters set up similar arrangements with the telephone companies. Although their first broadcasts were sent over coaxial cable lines, hundreds of voice circuits were displaced to carry a single TV channel, so more facilities were needed. Fortunately, A new broadband transmission technology evolved during the early 1940s and became a major component of the national network for a time. Known as *microwave radio* systems, these lines did not need copper wire or other physical media to link points in a network. Instead, they employed antenna towers and radio transmitters that sent extremely high-frequency signals from relay tower to relay tower along a line of sight. The distance between relay towers usually ranged from 15 to 30 miles (24 to 48 km). One of the electromagnetic spectrum ranges assigned by the Federal Communications Commission (FCC) for this service extends from 3.7 billion to 4.2 billion hertz (gigahertz) to carry voice signals over 12 channels in each direction. Each channel can carry one television transmission or 1800 voice circuits; however, one channel is reserved as a spare or protection channel. Other bands in use for commercial terrestrial microwave radio transmission are in the vicinity of 2, 6, 11, 18, and 23 gigahertz (Ghz).

virtually eliminate the problem. When digital signal weakens during transit, it is *regenerated* rather than simply amplified. Regeneration eliminates electronic "noise" by retiming and restructuring the square-wave pulses that characterize pulse-code modulated (PCM) digital signals.

The first commercial system using microwave radio in telephony was placed in service between New York and Chicago in 1950. Before the end of the following year, a transcontinental system was in operation. From then on, these systems grew in several ways: they covered an increasing number of routes, they also increased in capacity, and they eventually expanded their versatility through the introduction of digital transmission.

Eventually, microwave radio filled the majority of the former Bell System's needs for high-capacity long-haul transmission. In 1986, about 70 percent of long-distance traffic in the United States domestic network was handled by microwave carrier. However, this dominance was challenged by the rapid development of fiber optic lightwave systems, which were impervious to atmospheric disruptions. By 1990, terrestrial microwave transmission systems had been replaced by high-capacity fiber optic cables in AT&T's long-distance network, except for certain leased routes serving private networks.

Satellite Systems. A special class of microwave radio systems involves satellite systems. See also **Satellites (Communication and Navigation)**. A communications satellite system of the 1990s is an extremely complex combination of fixed and mobile earth stations (transmitter/receivers equipped with dish-shaped earth antennas ranging in diameter from about one yard or meter to more than ten yards or meters) and massive cylindrical or rectangular assemblies orbiting the planet above the atmosphere. Each satellite, often weighing two to four tons or more at launch, is equipped with multiple transponders (receiver-amplifier-transmitter devices which relay voice, video, fax and data signals from one earth station to another over precisely tuned microwave frequencies) as well as power supplies and appropriate electronic control equipment. Depending on their design, each of these huge structures can handle tens of thousands of two-way telephone calls simultaneously and has an orbital life expectancy as long as 15 or more years. Shortly before 2000, new low-Earth orbit communications satellite fleets were introduced, providing direct signal links between pocket-size telephones and the satellites used to relay the calls to other terrestrial telephones.

The world's first broadband (high-capacity) international communications satellite was AT&T's *Telstar I*, which was placed in a low nonsynchronous orbit around Earth on July 10, 1962. *Telstar II* was placed in orbit on May 7, 1963. They were both experimental in nature, providing the first domestic and international TV transmission by active satellite, using an earth station in Andover, Maine, USA, another in Goonhilly Downs, Cornwall, England, and a third in Pleumeur-bodou, Brittany, France. Two other experimental low-orbit satellites, known as *Relay I* and *Relay II*, were placed in orbit in the same time frame. But the experimental satellites most resembling today's communication satellites were *Syncom II* and *Syncom III*, which were positioned in high orbits 22,300 miles (35,881 kilometers) above the Earth's equator in 1963. The first commercial communications satellite, *Intelsat I*, was launched in 1965; known as "Early Bird," the 85-pound (38 kilograms) satellite transmitted the equivalent of 120 simultaneous analog telephone conversations between the United States and Europe. *Early Bird* also was the first commercial satellite to be placed in a geostationary or synchronous orbit, in which its orbital velocity exactly matched the Earth's rotation around its axis, so the satellite appeared to be fixed in space above an assigned spot along the equator, from which its antennas could reach almost half the planet. Succeeding generations of *Intelsat* satellites orbit above the Atlantic, Pacific, and Indian Oceans, handling telephone messages and color television signals among scores of nations around the globe.

A new class of satellite communications equipment designed primarily for private networks operating outside the public switched telephone network is called the *Very Small Aperture Terminal* (*VSAT*, pronounced vee-sat). An example of a VSAT network would be an automotive manufacturer with a computer center in Michigan which is linked by satellite to several thousand automobile dealers across the U.S. The network center in Michigan would be equipped with a satellite transmitter-receiver capable of broadcasting the same one-way program (voice and video, data, fax) to all the dealers simultaneously, or setting up a multiparty videoconference or voice conference call, or connecting the center to a single dealer location for voice, video, data or fax transmission. VSAT systems operate with digital packet-switching technology over leased transponders. U.S. businesses were the first VSAT users, starting in the 1980s.

Another type of telephony equipment intended for land-based digital private networks is increasingly being adapted for satellite use by remote communities, especially in Alaska. Known as the *integrated access and cross-connect system* (*IACS*), this wideband packet product can be used for digital compression of voice, video, fax and voiceband data. A remote community can link its local telephone lines (usually about 100 or so) into a digital PBX that feeds its multiplexed circuit signals into the IACS for transmission up to a satellite for relay down to a land station connected to the global terrestrial telephone network. The system is capable of compressing 120 digital voice channels into a North American satellite transmission standard link for 24 digital channels, using bit compression, statistical multiplexing, and packetized transmission. This technology thus replaces a public radio broadcast system with a public telephone system offering communications privacy. During Operation Desert Storm in 1990/91, more than a dozen such systems were installed in Saudi Arabia, enabling U.S. troops to telephone home from remote locations in the Middle Eastern desert.

Telephony between continents actually began as radiotelephony in 1927. Although crude by modern standards, the wireless system worked well enough so that callers were willing to pay $75 (the equivalent of more than $800 in 1998 currency) for three minutes or less. During the first year of service, only six to eight calls per day were made, but the public warmed up to the concept and by mid-1928, the volume reached more than 30 calls per day. On October 16, 1928, one call between New York City and Paris lasted one hour and 37 minutes, costing the caller $1,527.50 in 1928 money.

However, the wireless technology was susceptible to interruption and fading caused by atmospheric disturbances, especially sunspot activity. Sometimes it was out of service for days, even weeks.

Submarine Cables. In 1956, the first transatlantic undersea telephone cable was installed between North America and Europe, using two cables, each carrying 36 analog voice circuits in opposite directions. The cable landing sites were in Oban, Scotland, and Clarenville, Newfoundland, Canada—the shortest undersea distance between North America and the networks linking the United Kingdom and European nations: 1950 nautical miles (or about 2200 statute miles or 3550 kilometers). Each cable required 51 repeaters to amplify the signal along the route, but since solid state electronics were still evolving, those repeaters were equipped with vacuum tubes. (Transatlantic telegraph cables were introduced during the middle to late 1800s, but could not carry speech.) By 1985, seven submarine coaxial copper cables had been laid by special cable ships across the floor of the Atlantic Ocean, See Fig. 15, and other cables criss-crossed the Pacific between the U.S., Australia, Japan, Hong Kong, Taiwan, and Southeast Asia, See Fig. 16. For the map of submarine cable service to Europe, see Fig. 17.

Undersea telecommunications cables are susceptible to damage from fishing trawlers. In the relatively shallow waters of continental shelves, the cables have to be buried a few feet under the ocean bottom. This is done by a remote controlled vehicle which uses water jets to plow a furrow and then cover the cable as the laying procedure moves along the planned route. The route is carefully laid out according to the shortest distance between the two landing points, but also to avoid deep canyons and jagged ocean bottom features that could damage the cable. The cable plow is not needed after the cable ship reaches the end of the continental shelf, where the cable slowly drops to the ocean floor.

All the copper submarine coaxial cables used analog signal technology, so any computer transmissions were limited to modem signals carried within the voiceband circuit.

Satellite circuits could be used for high-speed digital transmission of computer signals, but the long distances involved [22,300 miles (35,881 kilometers) of uplink, 22,300 miles (35,881 kilometers) of downlink, for a total of 44,600 miles (77,761 kilometers)] caused problems for some systems due to the time lag incurred. But, as often happens, a new technology—replacing copper with fiber or lightguide cables—emerged from the laboratories and development facilities to meet the challenging growth of both data and voice in global telecommunications traffic. See Fig. 18.

Optical Fiber Cables. The medium now favored for broadband (sometimes called "wideband") digital transmission is called *lightguide* or *optical fiber cable*. Introduced during the late 1970s for terrestrial systems, lightguide cables consist of hair-thin glass fibers specially formulated and manufactured for communications use. The fibers are solid, not hollow. Instead of transmitting electrons (as is done with metallic conductors), the lightguide carries pulses of light from the infrared end of

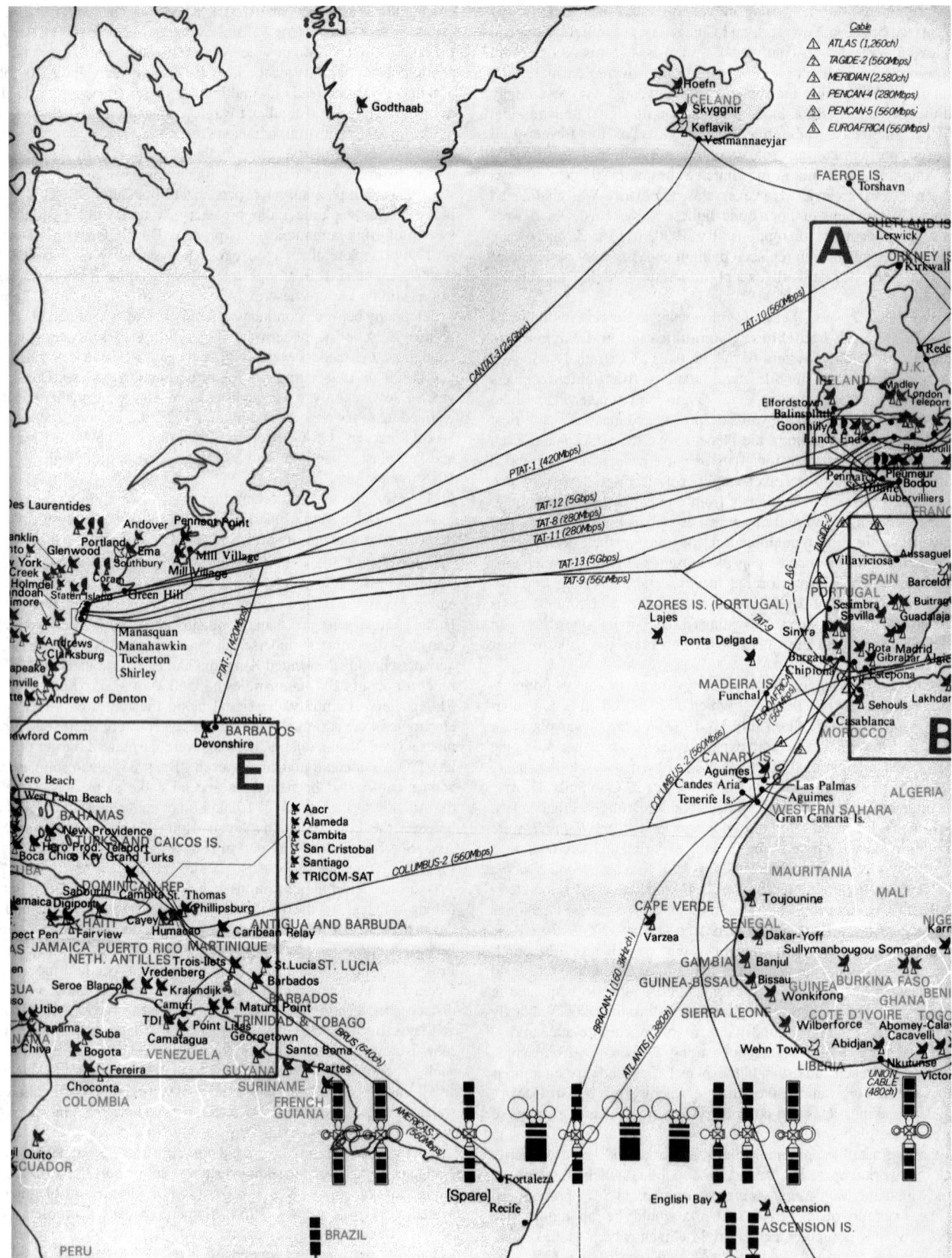

Fig. 15.   Map of submarine telecommunications cables across the Atlantic Ocean (1996). (*KDD Co.*)

the electromagnetic spectrum. These pulses are generated by either light-emitting diodes (LEDs) or semiconductor lasers at a rate of many millions of bits (megabits) or even billions of bits (gigabits) per second, using pulse code modulation (PCM) techniques. At the receiving end, a special semiconductor detector diode converts the incoming light pulses into coded electrical pulses. Copper wires run parallel to the fibers, carrying control

pulses. When the submarine fiber cable is buried in the continental shelves, and emerges to drop to the ocean floor, special cable armor is used in that critical point to sheath the electrical and optical fiber components from damage by sharks. Tests have shown that the sharks are attracted to the cable by the electrical fields emanating from the control circuits. When the sharks bite the cable, they can puncture the insulation and allow sea water

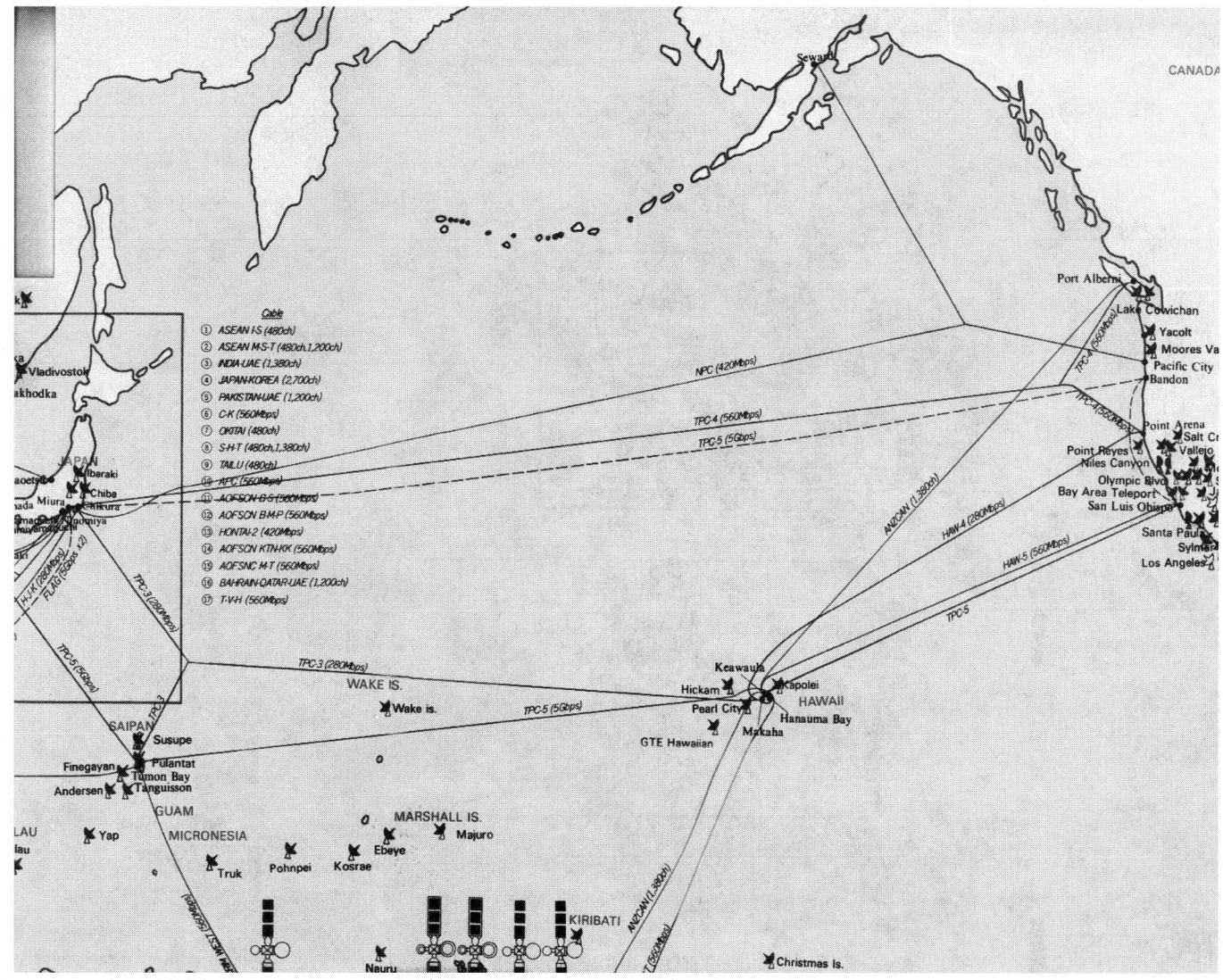

Fig. 16.   Map of submarine telecommunications cables across the Pacific Ocean (1996). (*KDD Co.*)

to short out the electrical components. (Some people seriously think that the optical fibers in new cables are hollow, and that shark bites compress the fiber and block the light signals from passing through. This is not true.)

Lightguides are not yet used to feed signals directly out of or into telephone sets, but are now specified as the transmission medium of choice on most new trunk routes because they are much lighter and less expensive than metallic cable of comparable capacity.

The communications fibers are supplied as either multimode or single-mode designs. The fibers are made up of layers of various types of glass, so a cross-section resembles that of a solid tree trunk. A multimode fiber has a relatively wide core and can use either lasers or light-emitting diodes (LEDs) as light sources. Single-mode fibers require better sources such as lasers, but have a much greater capacity for carrying signals. All long-distance lightwave lines now use single-mode fiber, while multimode fiber is often used in short links, especially if the light source is to be an LED. The highest bit rate for multimode fiber, without incurring significant modal dispersion problems, is about 180 megabits per second; single-mode fibers are already handling more than 3.4 gigabits per fiber in commercial trunks between selected U.S. cities.

In fact, demonstrations of experimental lightwave systems often are followed within a year or two by the commercial introduction of systems matching the new super transmission speeds. Multi-gigabit submarine cables already are on the drawing boards, and ultra-long-distance cables without digital signal regenerators have been demonstrated, using novel, high-speed all-optical logic gates and transmitting solitons (light pulses that retain their shape over long distances without blurring).

In September 1991 a nine-member Bell Labs team transmitted solitons at 32 gigabits per second over 55 miles (90 km) of optical fiber. Another researcher has transmitted solitons more than 6,000 miles (9,666 km) through optical fiber without electronic regeneration. The elimination of regeneration repeaters in transoceanic submarine cables would significantly lower their cost, improving the economics of global communications.

Major communications routes throughout the United States now use lightwave systems that replaced metallic or wireless radio systems. New transatlantic and transpacific submarine cable systems also are designed to use single-mode lightguide fiber. High-speed digital transmission on the first fiber-optic submarine cables, deployed in 1988, uses six single-mode optical fibers carrying laser signals at the rate of 296 megabits (million bits) per second. Only four of the fibers are operating at any given time to provide the equivalent of 40,000 digital voice circuits; the remaining pair of fibers are standby. The first Atlantic fiber cable has fewer than 125 regenerating repeaters along its 3145-nautical-mile length, while the first Pacific fiber cable has fewer than 250 repeaters along its more than 10,000 nautical miles (18,520 km). Both systems featured the first underwater branching of submarine cables, so the single landing point at the American end of each cable is linked to two landing points at the other end, i.e., the United Kingdom and France, or Japan and Guam. The projected life of both lightwave cables is 25 years of operation (to about 2013).

The public's growing enthusiasm for international telephony required the enormous increase in traffic-handling capacity provided by fiber optic cables. The first transatlantic copper cable, known as "*TAT-1*," carried 10 million telephone calls during its 22-year service life (from 1956 to 1978). The first new fiber optic cable, known as "*TAT-8*," carried 10 million calls in only two days.

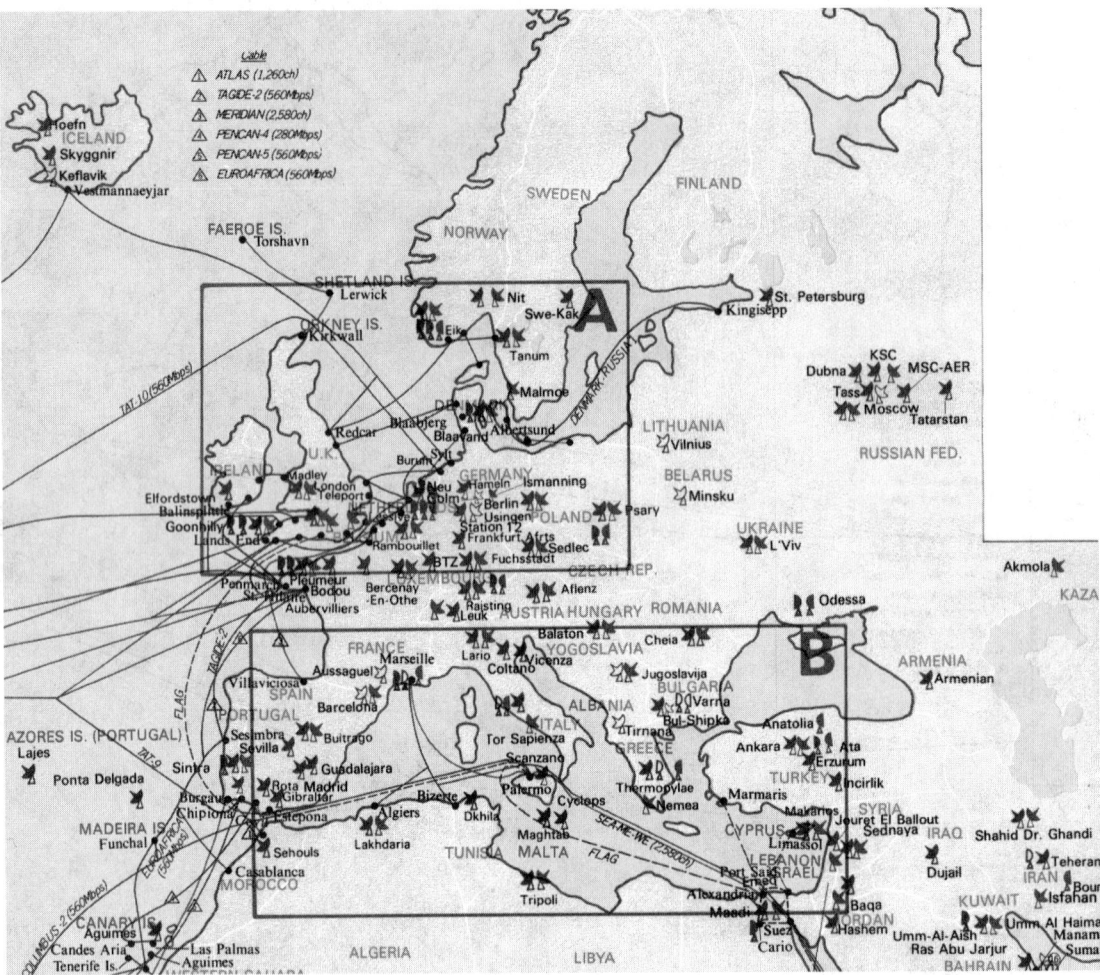

Fig. 17.   Map of submarine telecommunications cables serving Europe (1996). (*KDD Co.*)

In 1992 a second-generation fiber-optic system was installed, operating at 560 megabits (million bits) per second and adding the equivalent of 80,000 digital voice circuits. The *TAT-9* Atlantic cable has three branching units, one to enable two North American landing points, the others to provide separate landing points in England, France and Spain. The *TAT-10* Atlantic cable, activated in 1993, was the first direct cable between North America and Germany, and added the equivalent of 80,000 digital voice circuits. Three more cables quickly followed these to European landing points, including in 1995 the *TAT-12* and *TAT-13* fiber-optic cables, each of which transmits at the astonishing rate of five *gigabits* (five *billion* bits) per second.

Yet even a 5-gigabit per second transmission rate seems slow compared to the upcoming fiber-optic cables. In the Pacific, where telecommunications traffic between Japan, Hawaii and the United States seems to fill the cables almost as soon as they are installed, a new design has been created for the combined transpacific TPC-6/WorldCom Japan-U.S. fiber-optic cable, scheduled to begin operations in mid-2000. It will operate at 80 gigabits per second at the start (equivalent to 967,680 simultaneous phone calls), and has the potential to expand to 640 gigabits (equivalent to 7 million, 741 thousand, 440 simultaneous calls). And scientists at Bell Laboratories have demonstrated new cables transmitting in the *terabits* (*trillion bits*) per second.

Trunks. A major method of keeping telephony economical is to concentrate high volumes of traffic onto relatively few transmission media linking two major points. The high-volume media are known as trunks, and usually serve as links between central exchange offices or toll offices (which handle only long distance calls). In 1983, the communications network then operated by the former Bell System contained over 11,000 switching systems connected by over 7.5 million trunks arranged in 400,000 separate trunk groups.

Multiplexers. Calls are concentrated or consolidated by means of equipment called a *multiplexer*. For analog signals, the multiplexer

uses a technique called *frequency division*, in which voice channels are sandwiched into broadband frequency layers by means of sophisticated filtering systems. For example, 12 analog voice channels are stacked to form a *basic group* within a bandwidth of 60 to 108 kilohertz (kHz). Then 60 circuits are multiplexed in layers to form a *supergroup* signal (312 kHz to 552 kHz). The next stage is assembling of 600 circuits into a *mastergroup* signal (564 kHz to 3084 kHz), followed by the highest analog level, 3600 circuits in a *jumbogroup* signal (0.546 to 17.548 megahertz, or MHz) to be fed to a microwave radio or cable transmission system. The four multiplexing steps are called the *transmission hierarchy* by telephony engineers.

Digital signals also have a hierarchy, but the multiplexer uses another technique: time division. The digital signals use thousands of coded bits per second that are interleaved with bits from other channels into high-speed signal streams of millions or billions of bits per second. *Time-division multiplexing* (*TDM*) relies on increases of bit-rates per second at each of five levels of the *digital hierarchy*. The basic rate for a digital voice channel in North America is 64 kb/s (known as the DS0 signal). Of course, a computer terminal equipped with keyboard input operates at a much slower rate, so most such lines function at 9.6 kb/s or less; these slower speeds are known as *subrate data* and are usually multiplexed onto a DS0 signal before entering the network.

The lowest digital network transmission rate, equivalent to 24 voice channels combined, is DS1 (1.544 million bits or megabits, per second). The next step up isDS1C (3.152 megabits per second), followed by DS2 (6.312 megabits per second), and DS3 (44.736 megabits per second), which is the highest digital signal rate provided to customer premises by the telephone network. The DS4 rate (274.176 megabits per second) is the highest level of hierarchical transmission and is used exclusively for international communications links. Internally in the AT&T network, even higher digital signal levels are used with lightwave (fiber optic) technology. These super-links operate at 417 megabits per second, 1.7 gigabits per

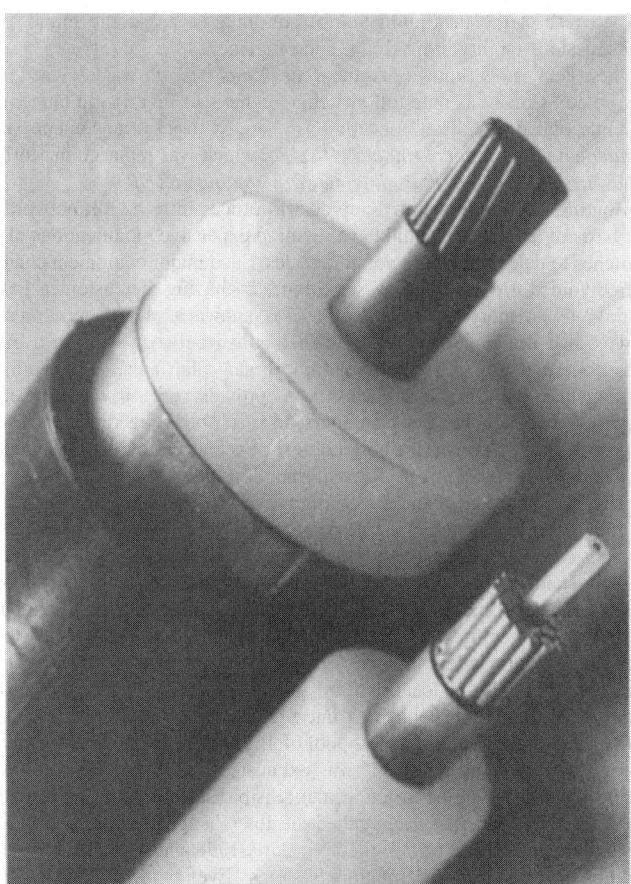

Fig. 18.   The single submarine lightwave cable shown here (bottom) more than doubled the number of available transatlantic cable circuits when the TAT-8 system was completed in 1988. The fiber cable is half the size (and one-third the weight) of the TAT-7 coaxial copper cable shown in photo (top). New submarine lightwave cables use fiber optic technology that far exceeds the capacity of the first such system. (*AT&T*.)

second, and 3.4 gigabits per second between special terminals but are not assigned a DS hierarchy code.

In the 1990s, plans were introduced for implementing a new optical transport system in the world's public switched networks. The worldwide designation for this eight-level hierarchy is *SDH* (*Synchronous Digital Hierarchy*), but the slightly different U.S. version is known as *SONET* (*synchronous Optical NETwork*). The basic optical signal of SONET is OC-1, operating at a line rate of 51.840 megabits (million bits) per second. This is multiplexed in seven different stages to the highest SONET level, OC-48 operating at 2.488 gigabits (billion bits) per second. Higher transmission rates are currently used in major transmission cables, but these are not rated for switching.

Digital multiplexing circuitry is much simpler than that of analog equipment, so digital terminal costs are considerably lower. In addition, the digital multiplexer provides more versatile means of monitoring system performance and isolating faults. All lightwave systems in the North American telecommunications network use digital coding and multiplexing.

Cellular Mobile Transmission. Another important technology in transmission media is known as *cellular mobile* transmission. This wireless design dramatically expands the number of mobile telephone circuits previously available with conventional mobile telephone systems. (Mobile telephones differ from other mobile radio service in that the conversations are individually connected to the telephone network, while other radio services are usually of the broadcast type, i.e., police or radio dispatchers, and can be overheard on any receiver tuned to the frequency being used.)

The previously allocated frequency spectrum limited the number of mobile telephone circuits to fewer than 25 in major metropolitan areas, such as New York, Chicago, or Los Angeles. Cellular technology expands that to thousands of available circuits for each market by reusing a given frequency many times without interference among the callers.

In a cellular mobile communications system for an urban area, the region is divided into relatively small neighboring *cells*, each containing a base station [a small radio tower with a service radius of 1 to 8 miles (1.6 to 13 kilometers)]. The cells are generally depicted on plan layouts as octagon-shaped areas butted against each other. The mobile telephone set transmits and receives an analog or digital voiceband signal, depending on the cell type. As the vehicle moves from one cell to another, the call signal is "*handed off*" from one radio tower to the other, an intricate process controlled by a special digital switch and software at the heart of the cellular system. Landlines connect the towers to the central switch, which in turn is connected to the rest of the telephone network by an access line feeding into the nearest exchange or central office of the local Telephone Company.

The first mobile telephone system for the public switched network was authorized by the FCC in 1946 and was based on a single radio tower for an entire metropolitan area. Because of the limited frequency spectrum allocated for the new service, and the relatively primitive available technology of the time, the network could serve only a few hundred customers in any major city, so long waiting lists quickly developed. Bell Laboratories proposed a cellular subdivision of the service area in 1946, which would re-use the same frequencies in some of the cells, thus enabling the system to handle more calls simultaneously. But the technology of the day was not capable of providing such facilities.

Work continued until the early 1970s, when installation of cellular systems became feasible. However, the FCC repeatedly denied permission for introducing the new approach while arguments raged in the industry as to whether the technology was developed enough. Finally, cellular mobile phone technology was test-marketed in 1979 in Chicago, servicing 2000 customers in an area of 2100 square miles (ca. 5460 sq. km.) divided into ten cells. About 100 cars could be served simultaneously by a single analog cell, depending on the system's design. The trial system operated between 825 and 890 megahertz. Each radio channel connecting the mobile phones with the cell towers consisted of a pair of frequency-modulated (FM) one-way channels, each channel 30 kilohertz wide, separated by 45 megahertz. The radio spectrum assigned to cellular technology in the United States started with the 800 megahertz band, but now includes 1200 megahertz as well because of the influx of European transmission equipment.

In 1991, the first digital cellular system was tested in the U.S., using new terminal sets provided by six different manufacturers. The system was based on a technology known as *TDMA* (*Time Division Multiplex Access*) and tripled the service capacity of each cell, compared to analog technology. In the U.S., the terminal must be a dual digital/analog design, capable of automatically switching back and forth to accept signals coming from either digital or analog antennas. The digital bit-rate approved for this service is eight bits per second, accomplished through innovative voice signal compression techniques, but the acoustical quality of speech is not as clean as that of wireline systems.

Some cellular telephones are designed to be portable and can be removed from the vehicle so that they can be used by a pedestrian if an antenna is within transmission range. Other mobile telephone systems include fax and small ("laptop" or "notebook") portable computer equipment operated in a wireless mode. In the early 1990s, such data transmissions to and from analog towers were made via modems from a stationary point, with the vehicle parked.

As cellular mobile phone systems continue to be installed throughout the world, it may become possible to transmit and receive speech, data, fax and video from a moving vehicle anywhere to a location anywhere else in the world. And because digital data transmission to and from portable computers would be possible with these systems, the communications potential for the future is staggering.

Switching Equipment

Obviously, it would be impractical to extend separate wires from every telephone to all other telephones. For example, four interconnected telephones would require six lines, while 40 phone would need 780 lines and 400 telephones would require 79,800 direct lines.

To conserve transmission media and dollars, each telephone is instead connected by a pair of wires to a nearby central point, called an exchange or central office, where individual connections to other telephones can be made as requested. The device that makes those connections is called a *switch*. The telephone network consists of many thousands of switches connected to each other as well as to millions of telephones.

In 1951, when engineers at Bell Labs prepared to introduce automatic long-distance dialing with the Nationwide Numbering Plan, each dial office was designed to serve up to 10,000 numbers with a limitation of four digits, plus the three digits that identified the central office switch. Smaller communities often used only four digits for cross-town calling if only one switch was involved, but bigger cities with multiple switches had to assign extra digits to identify which switch had the destination number.

Switching apparatus evolved in five major successive phases:

1. *step-by-step* switches, controlled directly by power pulses fed from the telephone dial;
2. *panel* switches, controlled indirectly by relays operating the switch from the dial;
3. *crossbar* switches, introducing a common-control concept that greatly increased the efficiency of switching operations;
4. *analog electronic* switches, which replaced electromechanical switching with solid-state electronic switching and introduced stored-program control (SPC) or software concepts using switching programs called "generics"; and,
5. *digital electronic* switches, which added time-division concepts to the technology (all previous switches were based on space-division concepts). By combining digital switching and digital transmission, synergistic benefits were obtained. Although most customer premises connections were still analog in 1998, the goal of a total digital network capable of extremely sophisticated features is now in sight.

The earliest switches were electromechanical machines, activated either by the pulses from the telephone dial or by relays. Stored-program control replaced the hard-wired logic of common control with software consisting of detailed instructions for a program guiding a computer that operated the actual switch hardware. Changes in services, features, and other switch functions could now be done by program changes instead of hardware changes. The first electronic switching system using stored-program control for a commercial installation was placed in use ("cut over" in telephone parlance) in 1963 at Succasunna, New Jersey. The 4000-line office was served by an analog system known as a No. 1 ESS switch, using space-division switching. Many of these switches were still in use by local exchange companies in the early 1990s.

The concept of analog space-division electronic switching dominated the field until challenged in the mid-1970s by the first digital time-division electronic switching system, the AT&T No. 4ESS toll switch. The new digital switch was first used in the AT&T Long Lines Trunk Operations Center in Chicago. The current era of switching technology is now based on digital signals and time-division multiplexing in which the *pulse code modulated (PCM)* signals are separated and switched by a time-slot interchange. Time-division switching lowers the cost of voice communications, is compatible with high-speed transmission of digital data, and can be modified to fit the packet switching systems used in the Internet, World Wide Web, and Integrated Services Digital Network (ISDN) applications. The goal is end-to-end digital connectivity, automatic switching of every type of voice, data, and image signal without any analog carriers.

In the first half of the 20th century, switching was automatic only for local telephone calls. Long-distance, or toll, calls were placed by human operators. In 1951, nationwide customer-dialed calling began when *direct-distance dialing (DDD)* was introduced in several areas of the country. A substantial telephone company investment in new equipment was required nationwide, including the "independent" telephone companies that were not part of the former Bell System. Further modifications resulted in international direct distance dialing (IDDD), starting in 1970 with calls to the United Kingdom and later into some 140 other countries throughout the world. Americans can reach about 250 countries by long distance, more than 100 of those do not qualify for IDDD service for economic reasons or because their governments restrict direct dialing to their telephone users from outside their national borders, so operators serve as intermediaries in those nations.

The need for direct dialing was clearly evident in the data compiling the escalation of long distance calling. In 1987, AT&T alone handled about 85 million long-distance calls during a typical 24-hour business-week day. Three years later, that number had risen to circa 115 to 120 million long-distance calls for the same period, by 1995, the average volume of long-distance calls was up to 185 million calls per day.

For nationwide DDD toll dialing, complete end-to-end connectivity is required for universal service through the network. The switching plan

had to restrict the maximum number of links to a specific number for transmission stability, uniformity, and efficiency.

A system, now known as the *hierarchical plan*, was devised that permitted as many as nine toll switching centers to function in tandem on a single dialed call. That concept was changed to an arrangement called *dynamic nonhierarchical routing (DNHR)*, which was replaced in 1991 by a newer system called *real-time network routing (RTNR)*.

Routing calls from step-by-step central offices into the toll network on the basis of area codes would have required expensive modifications of the switches to detect the dialing of area codes. Such major equipment changes were avoided for most step-by-step offices by having the customer dial an initial "1" on most toll calls. The "1" is known as a toll barrier prefix and is also used in some states to distinguish between calls in a given area code and toll-calling within a numbering plan area. Also, with the introduction of DDD it was necessary to provide codes in the numbering plan for operators, but not customers, to reach other operators in distant centers for assistance in reaching points not yet accessible by DDD, or for other forms of routing or special call-handling procedures.

The interpretation of the dialed digits was the most important element in the design of switching systems for DDD. The changes in the step-by-step offices were relatively simple. For common-control local offices, register-sender capacity for the additional three digits of the area code was needed. The crossbar toll offices were to become the backbone of the DDD network; features for implementing DDD service were assigned to those offices. The nationwide number, transmission, and routing plans not only required the ability of local and toll offices to receive 10 digits, but also required the common control to interpret the three area code and three office code digits. The area code pool of 152 three-digit numbers, with the second digit limited to 0 or 1, was exhausted before the year 2000. By lifting the second-digit restriction, the feasible selection was expanded to 792 area codes. The program, called the Interchangeable Numbering Plan Areas, was transferred from Bellcore in early 1998 and is now administered by Lockheed Martin Information Systems. Even with 792 area codes, experts predict that North America will run out of area codes by 2025, so various proposals are under consideration to resolve the problem.

Network planners assigned the first area codes based on the ease of dialing (on a rotary phone) and the volume of calls to a particular city. That's why the area code for New York City is 212, Chicago is 312, and Los Angeles is 213, and so on. The combinations 211 to 911 were disqualified as area codes because they were reserved as service codes. The combinations of 200 to 900 were also set aside for special uses: 700 is the prefix for special services such as AT&T's switched digital services and teleconferencing calls, while 800 is for *WATS (Wide Area Telecommunications Service)* or *Freephone*, and 900 is for "dial-it" calls in which the caller is charged a fee by the party (usually a business) being called.

The first digital switching system was the AT&T 4ESS switch, introduced for long-distance (toll) service in 1976. Today, there are about 100 4ESS electronic switching systems serving the United States toll network, plus others in the Middle East and Far East. A 4ESS digital switch can serve up to 670,000 peak busy hour calls.

Digital switching also is performed by the AT&T 5ESS system introduced in Seneca, Illinois, in 1982. Designed for use in central offices serving from less than 1000 to more than 100,000 lines, the 5ESS switch is now also performing as an international gateway switch in several countries, functioning as the entry and exit point for international calls.

Remote and sparsely populated U.S. areas are provided with sophisticated telephone services by remote switching modules connected to modern central offices as far as 100 miles distant. The main switch serves as a host to the remote module, although the remote switch usually is capable of still connecting its local subscribers to each other even if it is temporarily cut off from its host.

The 5ESS switch was the first such system designed to allow connection to lightwave (fiber optic) links. However, the lightwave, or *optical*, signals must be converted to electronic signals before they can be switched. Switching of optical signals is still experimental and is not expected to become commercially available before the year 2000.

Traditionally, telephone calls across the long-distance network were routed through a hierarchy of switching centers, under the supervision of a technical staff headquartered in New Jersey. During the 1980s, a new concept was introduced, using computers in various locations to coordinate long-distance calling paths while monitoring the status of switches. At first, the new approach reset the predetermined routing paths of calls up to 10 times a day, providing up to 21 paths between each switch. In 1991 an improved system, known as *Real-Time Network Routing* (*RTNR*), was introduced, changing the routing path of a call automatically in "real time" (during the actual call setup) to transit around congested traffic areas or equipment failures. Human experts can still intercede for special purposes, such as selectively blocking calls into an area affected by a catastrophe like an earthquake. This prevents the area's switches from becoming overloaded by incoming calls.

RTNR will not, however, protect dedicated lines that are not switched through the 4ESS toll switches. A different protection system, known as *fast automatic restoration* (*FASTAR*) uses a routing algorithm to monitor the status of dedicated circuit equipment known as digital access and cross-connect systems (DACS), providing remote-controlled cross connections that make possible automated restoration of dedicated circuits.

A key factor in implementing computer-controlled routing was *common channel interoffice signaling* (*CCIS*), a high-speed, high-capacity signaling system for linking electronic switching systems on a network separate from the channels that carry voice signals. This is described as an *out-of-band* packet-switched network, out-of-band meaning that the control signals are outside the voiceband signals. (Typical control signals inside the voiceband are the dial tone, rotary dialing signals, touch-tone signals, busy signals, etc.) See the section titled **Global Telephone Networks**.

Supplementing the American public switched network is the growth of private voice-and-data switched networks, including government, business, and institutional customers. The federal government, for example, has the world's largest private switched network, known as the Federal Telecommunications System (FTS). It is a largely analog system, activated in 1964, which offers voice and low-speed data over 11 million circuit transmission miles with close to 35,000 access lines and 15,000 trunks serving some 3500 government locations in the continental United States, Puerto Rico, and the U.S. Virgin Islands. In size, it equals the former Bell System interstate network of 1957 and accounts for more than 1.5 billion minutes of use each year. A new FTS system was introduced in the 1990s as an option for government agencies, providing a digital voice/data network. Meanwhile, the Department of Defense (DoD) has installed the *Defense Commercial Telecommunications Network* (*DCTN*), a digital system designed to serve users in the DoD and other federal agencies at about 150 locations across the continental United States. Another DoD system is the *Automated Voice Network* (*AUTOVON*), which was the first worldwide switched network for private telephone and data transmission. AUTOVON is a global analog network, divided into two parts, the United States (CONUS) portion and the overseas portion. AUTOVON service began in April 1964.

Many businesses have installed private voice-and-data switched networks that link regional or national offices and plants. Such networks normally operate outside the facilities of the local or regional telephone companies, although certain trunks or other transmission media may occasionally be leased on an exclusive basis as interfaces with public networks.

In addition, the interexchange (long distance) carriers have created "*overlay networks*" for leased private circuits that partially use the public switched network. Such overlay networks provide reserved or *dedicated* analog or digital circuits for large communications customers, thus minimizing their equipment investments. Sophisticated software packages, plus advanced microelectronics and fiber optics technology, have eliminated much of the *hard wiring* that limited the flexibility of earlier private networks.

Today, private computer networks are often classified as a LAN (local area network) or a WAN (wide area network). A LAN connects computers in a specific area, such as an office building or campus, an industrial park, or a factory complex. A WAN connects computers and terminals spread out over a region, usually communicating over phone lines which are part of the Digital Data System (DDS), a nationwide overlay network operated by AT&T.

Global Telephone Networks

The North American telephone network (continental United States and Canada) has more telephone lines than any other network in the world. However, many other networks are still huge in terms of the concentration of lines within their territories. North America is listed as having 41 percent of the world 500-million-plus telephones, while Europe reports 37.4 percent. The expanding global economy of the 1990s, fostered by many multinational companies, required efficient and reliable linking of the telecommunications networks on all continents.

Until the 1980s, only the U.S. assigned the responsibility for its national telephone networks to business enterprises operating as monopolies. However, these enterprises were—and still are, in the case of local exchange carriers and AT&T—regulated by state public utility commissions in terms of their rates and authorizations for services. Other nations preferred to control the operation of their networks through government ministries, usually referred to as a Post, Telephone and Telegraph (PTT) agency.

Despite the coordinating role of the International Telecommunications Union of the United Nations, various nations established differing technical standards for their national telephone system, often as a means to protect local manufacturers who supplied those systems with switches, transmission media and customer premises equipment. These differing standards are gradually being eliminated in favor of a universal global set of standards. However, certain significant differences still remain by 2000 and must be accommodated for global digital networking.

For example, at the heart of digital signal coding is the process called *pulse code modulation* (*PCM*). The American and Japanese systems use a process named mu-law PCM, while virtually the entire remaining world adopted the A-law PCM process. The two approaches are incompatible, so mu-law PCM/A-law PCM transcoders had to be created. A decision reached within the ITU assigned the responsibility for operating these transcoders to the mu-law PCM countries (the U.S., Canada and Japan). For this reason, the digital signals transmitted over the transatlantic TAT fiber-optic cables are A-law PCM, transcoded in North America from the mu-law PCM signals otherwise used throughout the U.S. and Canadian networks.

The new and expanded telecommunications systems now evolving throughout the world are expected to generate many innovative applications as they become available to wider publics. Although voice transmission is still the dominant traffic component in world communications, the volume of fax, data and video traffic is increasing rapidly. For example, as of 1991 fax transmissions made up more than 25 percent of the communications traffic originating in the United States and connected to Japan. High speed computer data exchange over global digital telephone circuits is vital to many businesses (American Airlines, for example, owns and operates a computer system called *Sabre*, or Semi-Automated Business Research Environment, which provides fares and schedules for 665 airlines, as well as information on prices and availability for more than 20,000 hotels and 52 rental car companies via 85,500 Sabre terminals used in travel agencies in 47 countries). In addition, global videoconferencing is gaining in popularity as connection set-up arrangements are simplified. Compared to time and expense costs for international traveling, multi-point videoconferencing is far more economical, efficient and productive.

Video and data communications require more transmission capacity or "*bandwidth*" than ordinary voice communications, even though video signals are being compressed during transmission with the help of algorithms. A new optical transport system for the global public networks, called the *Synchronous Digital Hierarchy* (SDH) by the ITU's international communities and the *Synchronous Optical NETwork* (SONET) by ANSI in the U.S. is gradually being introduced as a means for providing a greater range of potential services. The eight transport levels for SONET start at 51.84 megabits per second but only 49.536 megabits are traffic—the remaining 2.034 megabits are assigned to the control overhead. Each level is multiplexed into the next level until the highest rate is reached (2.488 gigabits per second). The expanded control overhead enables network operators to develop future services using data bases housed in computers called "servers," which are strategically located throughout the networks.

The most significant program on an international scale is the *Integrated Services Digital Network* (ISDN). ISDN is expected to serve the world's business community initially, replacing the confusing mixture of incompatible computer communications networks with an economical, flexible, and reliable method based on universal access standards. Although the American common carriers are offering ISDN capabilities to limited regions at first, the various "ISDN islands" were joined in late 1992 by the first transcontinental U.S. transmission of ISDN messages. This connectivity was made possible by the widespread implementation of the ITU international protocol using a new out-of-band signaling system which ITU calls CCS7 (common channel signaling system 7) and which North America calls SS7. The individual digital network switches are upgraded to ISDN internetworking by new software packages, known as "*generics*," that reprogram digital switches to enable ISDN transmission between switches.

The ISDN concept is being implemented in two stages, known as "*narrow-band*" ISDN (ranging in capacity from 144,000 bits or 144 kilobits per second to as high as 45 million bits or 45 megabits per second) and "broadband" ISDN (viewed as based on fiber optic technology and starting at 45 megabits per second). Narrow-band ISDN was implemented in the 1980s, but broadband ISDN is still evolving.

Customer premises equipment (CPE) hardware for narrow-band ISDN is usually based on the availability of two bearer or "B" channels operating at 64 kilobits per second, or 64 kb/s of transmission capacity on a telephone line, plus a 16 kb/s data or "D" signaling channel. This capability is expected to enable simultaneous transmission of voice, data, and image signals on the "bearer" channels. In ISDN terminology, this line is known as a digital subscriber line (DSL). The DSL can be connected from the customer premises directly to the switch in the central office, via a basic rate interface (BRI), or multiplexed via a primary rate interface (PRI) that contains up to 23 64 kb/s channels and one 64 kb/s signaling and data channel, carried on a T-carrier link operated at 1.544 million bits (megabits) per second. The PRI functions will usually be served by a digital PBX, linking the premises to the central office.

Voice and data transmissions are integrated in ISDN, so there is no need to run separate wire pairs for each type of signal. Because the terminals in an ISDN system can identify themselves to the system, computer and telephone users can move from one location to another without having to physically rearrange wiring and other hardware connections. In the first application of ISDN in the United States, installed in 1986 in Oak Brook, Illinois at the facilities of the McDonald's Corporation by Illinois Bell Telephone together with various vendors, the ISDN network eventually replaced a variety of different voice and data networks.

A key technological factor in ISDN is the use of packet switching with the "data" or "D" channel. A packet system divides each signal, whether voice or data, into small blocks of information (typically 16 bytes in size, including coded address data identifying the destination). These packets are transmitted on circuits shared by other signals, much like people riding a bus; as each packet arrives at its destination, it leaves the "bus" and enters the module to which it was addressed. The original signal, of course, constitutes many packets that are separated into numerous channels or "buses," so the packets do not all arrive simultaneously; rather, they must be reorganized at the destination point before being displayed or printed. All this takes place at extreme speed—an average transmission bus can handle 42,000 packets a second. The information in a packet may be voice, data, or image signals.

The chief advantages of packet switching are flexibility and economy. A voice and data network that uses packet switches needs fewer lines to interconnect large numbers of endpoints, because each line can carry simultaneous two-way transmissions of both voice and data. Unlike conventional transmissions, the many parts of a single message can be sent along many different routes to the final address, where they will be assembled in proper order for the receiving equipment.

If the ISDN customer wants to set up a dedicated circuit to a specific number, instead of letting the network ship the messages via the usual packet switching methods, the ISDN packet switching concept allows the user to set up a *temporary* or *virtual* circuit, employing software instead of hard-wired connections. The virtual circuit can remain in place for seconds, days, or years, depending on the customer's needs. Unlike a dedicated circuit, the virtual circuit is paid for on the basis of actual time used rather than for the time in place, since others can use its capacity when no information is being transmitted.

Other applications of packet switching (however, not in the ISDN format) include the *Common Channel Interoffice Signaling* (CCIS) mentioned under the "**Switching Equipment**" heading. Previously used only by the common carriers, an advanced version of CCIS is playing an important role in an optional innovation called *Local Area Signaling Services* (LASS), in which telephone customers are able to use CCIS functions to trace obscene or harassing calls without contacting the telephone company in advance. In addition, LASS enables people—with the proper type of telephone or accessory equipment—to see a display of a calling number and name (or note that the number and name were blocked by the caller) before picking up the phone. They also can block specified numbers from being able to call into the line, or automatically recall someone who hung up just before the phone was answered. Telemarketers, medical services and others with computerized client files use a version of this technology to instantly display customer data on a terminal when a call is received from that customer's telephone.

Another future aspect in telecommunications is the use of lightwave systems in the loop, especially to homes. The high capacity of these fiber optics systems greatly exceeds the amount of telephone traffic currently generated by residences, but video image transmission—especially *digital high-definition television* (digital HDTV)—would quickly fill the available capacity. The fiber optic lines are not susceptible to interference by lightning, induction, or other electrical disturbances. However, most telephone companies want to depreciate their copper wire installations over 25 years before they invest in new fiber optic loop equipment.

The growing sophistication of the global telecommunications system is matched by a growing sophistication of the world's telephone users, large numbers of whom now use computers in their workplace and at home. Communications carriers in many nations are installing or expanding *intelligent networks*, using strategically located computers to provide on-demand services that were impossible only a few years ago. The CCS7 or SS7 signaling protocol enables the interworking of these global networks. By 1995, about three-quarters of the lines served by Bell operating companies had access to SS7 and its potential for intelligent networking.

The SS7 protocol is based on earlier versions of common channel signaling systems (CCSS), transmitted out-of-band between the telephones or customer terminals and the switches, or between the switches along a calling path. A simple example of the "intelligence" provided by the new systems is a national pizza calling line, in which the customer dials an 800 or 888 number from anywhere in the nation and is automatically connected to the nearest pizza outlet operated by the 800 sponsor. That sponsor also can insert customized messages for various outlets, announcing specials, business hours, etc.

Perhaps the most radical change in telephone network traffic in the 1990s was the sudden emergence—especially at the residential user level—of two-way data transmission produced by the Internet and the World Wide Web. Although the explosive growth of this new international "*virtual community*" with its "*chat rooms*" (which use text messages or e-mail) and direct marketing aspects was not expected by the telephone companies, they were quick to meet the demands on their networks. However, the local and long distance carriers are unable to control the quality of speech transmission over the Internet or World Wide Web, since those channels are designed for packet-switched data signals (using the Transmission Control Protocol/Internet Protocol, or TCP/IP rule sets) and are not optimized for the requirements of two-way voice conversations. Various vendors are developing new equipment aimed at improving the quality of speech transmission over the Internet/World Wide Web facilities.

Meanwhile, new approaches to network facilities are emerging as challenges to the existing frameworks. Satellites in geostationary orbit are being challenged by fleets of low-Earth orbit satellites that would permit direct calling from pocket-size telephones. See the **Satellites (Communication and Navigation)** entry. Cellular towers are being challenged by fleets of various types of aircraft, both winged and balloon, that will carry switching and relay equipment while flying in the stratosphere. Among these proposals are "*Halo*" (High Altitude Long Operation, in which the communications gear are located in a pod suspended below the aircraft fuselage, and each of three manned aircraft flies an eight-hour shift circling above a given metropolitan area), "*Pathfinder Plus*" (a propeller-driven unmanned aircraft powered by solar panels in its wings), and "*Sky Station International*" (a fleet of 250 blimps offering video, Internet access, and other types of high-speed data services around the world, as of 2001). Wireline systems are being challenged by such wireless technology as cellular and PCS systems. And the Internet is emerging as a challenge to the long distance telephone companies' billing structure.

A major fiber-optic scheme is a worldwide network called "*Project Oxygen*," a privately funded undersea facility with about 17,000 miles (27,000 kilometers) of special lightwave cable. Lucent Technologies announced in 1998 that its scientists had devised the ability to transmit as many as 10 million calls over a single fiber, dividing the strand into 80 separate wavelengths of light instead of 16. Using a technique called "*erbium doping*," the system will avoid the need for repeaters to regenerate the light signals at various points in the long cable runs. Connecting all the continents, Project Oxygen is seen as a global "*backbone*" network for video (it could transmit up to 10,000 video channels), Internet traffic, images and voice traffic. The promoters claim that costs for international voice calls will become as low as local calls. The entire network is expected to be completed by 2002.

Many other applications are evolving as the telecommunications networks of the United States and other nations apply new technologies.

Scientists and engineers engaged in software development as well as micro-electronics and photonic technologies are revolutionizing every aspect of communications. Current research into such intriguing concepts as "*virtual reality*" (actually an enhanced approach to techniques used in the aviation industry's advanced flight simulators) could introduce telephone users to a broader range of services featuring computer-driven three-dimensional vision and touch as well as audio.

The International Telecommunication Union compiled a "World Tele-communication Development Report 1996/1997," issued in February 1997 (an executive summary can be viewed on URL *http://www.itu.int/ti/ publications/world/summary/wtdr96.htm*). The "Overview" section concluded: "The convergence of the telecommunications sector with the computer and broadcasting worlds is creating new synergy, most evident in the growth of the Internet, which continues to double in size every year. At the start of 1997, there were more than 16 million host computers connected to the Internet and more than 50 million users. The significance of the Internet lies not so much in what it is, but what it could become. It can be regarded as the prototype of a global information infrastructure which will lay the platform for the electronic commerce of the 21st century."

Note: For more current information on these subjects, we suggest contacting the main website home pages of the different government agencies, manufacturers and sponsors of telephony and telecommunications systems, as listed below in their worldwide web Universal Resource Locators (URLs):

Alcatel: *http://www.alcatel.com*
American Telephone & Telegraph (AT&T): *http://www.att.com*
Bell Atlantic Corporation: *http://www.bellatlantic.com*
BellSouth: *http://www.bellsouth.com*
Cellular Telecommunications Industry Association (CTIA): *http://www. wow-com.com*
Ericsson: *http://www.ericsson.com*
Federal Communications Commission (FCC): *http://www.fcc.gov*
GTE: *http://www.gte.com*
International Telecommunications Union (ITU): *http://www.itu.int*
Lucent Technologies: *http://www.lucent.com*
MCI Worldcom, Inc.: *http://www.mciworldcom.com*
Motorola Corporation: *http://www.mot.com*
Nokia: *http://www.nokia.com*
SBC Communications: *http://www.sbc.com*
Siemens: *http://www.siemens.de*
Sprint Communications Co., L.P.: *http://www.sprint.com*
Submarine Systems Owners and Manufacturers: *http://www.teleport. com/~iscw*
United States Telephone Association (USTA): *http://www.usta.org*
US West: *http://www.uswest.com*

RICHARD Q. HOFACKER, Jr., Retired, AT&T Bell Laboratories.

TELESCOPE (Astronomical-Optical). Since the time of Galileo (early 1600s), optical telescopes have been used to study the planets of the solar system and the stars and other objects and phenomena that comprise the cosmos. Essentially prior to World War II, optical telescopes were confined to the visible light portion of the electromagnetic spectrum. Radar and the beginnings of sophisticated electronics for purely scientific purposes were fallouts of military usage during the war, as also were the application of infrared, ultraviolet, x-ray, and other portions of the spectrum for probing the cosmos. *Sputnik* (1957) introduced the space age as a concept of reality rather than a notion out of science fiction.

These and other factors affected, slowly but steadily, the future role of ground-based optical telescopes, which over prior centuries had enjoyed exclusive rights to the disciplines of astronomy, astrophysics, and other so-called astro-related subdisciplines. Over a relatively short time span, a few authorities seriously questioned the long-term role of the optical telescope, but they also appreciated the continuing need in the short and mid term. Later it was established that improved optical telescopes would continue to play an indispensable, but not an exclusive, role in searching for the truth of the cosmos.

By the 1970s and continuing into the present, a true renaissance in optical telescope design and performance is going forward at a rapid pace. These improvements have included not only the telescope's ability to collect more photons from cosmic objects, by way of larger and better mirrors and other aspects of the optical system, but also in vastly

improved detectors (sensors) as represented (for example, by charge-couple devices [CCDs]), that have markedly increased the instrument's resolution [image sharpness]). These advantages have become possible through the application of various forms of *adaptive optics* and other means that rely on computer power for their success.

Even with the dedication of several new optical observatories during the past decade, researchers still must wait in queues for instrument availability. Although funding problems worldwide often retard the construction of a new facility, progress has been impressive. Progress has been abetted by much innovative thinking in the astrophysics community. One example of this would be that of increasing the performance of optical telescopes by using them in arrays, as has been done with radio telescopes for a number of years. Serious thinking is now taking place concerning the location of optical telescopes in Antarctica and, at some future time, on the lunar surface. The verve existent at the time of dedication of the 200-inch (5-meter) reflector atop Mt. Palomar (California) in 1949 has returned to the community of research astronomers who use optical telescopes.

Chronology and Principles of Optical Telescopes

Over the centuries, optical telescopes have fallen into two major classifications: (1) refractors and (2) reflectors. Several design modes of each class have been developed.

Refracting Telescopes. Galileo, accredited with the development of the refractor (circa 1609), used the instrument to explore the Milky Way, which he referred to as a "vast crowd of stars." Over a brief span of time, Galileo discovered the craters on the moon, noted the phases of Venus, and found some of the moons of Jupiter. His simple refracting telescope was well suited for observing the coarse features of nearby astronomical bodies. The Galilean telescope operated along the lines of what was later described as an *opera glass*. See Fig. 1.

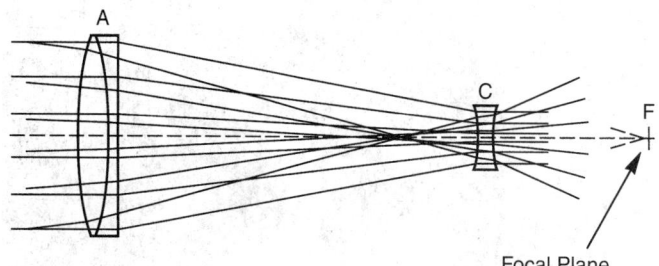

Fig. 1. Galilean telescope similar to opera glass. The objective A condenses the light from the object observed, making an image at F. However, before the rays reach their focus, they are intercepted by double concave eyepiece C.

Early scientists found that for more precise observations, the refractor had numerous disadvantages. For example, in a refractor, the light of each color is focused at a different point and thus images are blurred. When extra lenses are used to correct for chromatic aberration, a significant portion of the light received is lost to absorption by the lenses. Also, there are fundamental and practical size limitations of the refractor. Consequently, the largest refractor ever to be constructed was the 40-inch (1-meter) telescope at Yerkes Observatory (Williams Bay, Wisconsin), installed in 1897. However, in terms of large size, the "Great Refractor" was installed at Harvard University in 1843. As pointed out by P.J. Pauly (Rutgers University), "Many average Americans considered the great comet that appeared in early 1843 a confirmation of the well-known evangelist William Miller's prophecy that the world would end that year. Responsible citizens funded the Great Refractor in large part to combat such ignorant apocalyptic beliefs. In succeeding decades, Harvard astronomy prospered as part of international science. See Fig. 2.

Essentially, small, portable refractors find favor among many amateur astronomers. See Fig. 3.

The size and location of several famous refracting telescopes constructed over past years are given in Table 1. The current status of these instruments is indeterminate as of the early 1990s, but several have been retained for educational and experimentation purposes for which they are capable of handling.

Reflecting Telescopes. Sir Isaac Newton appreciated the limitations of the refractor at an early date and, in response, invented in 1668 the reflecting telescope that uses mirrors instead of lenses. In the basic reflector

TABLE 1. LARGE, WELL-KNOWN REFRACTING TELESCOPES
(Current Status Not Known in All Case)

Diameter		Geographical Location	Operator
Approx. Meters	Inches		
1	40	Williams Bay, Wisconsin	Yerkes Observatory
0.9	36	Mt. Hamilton, California	Lick Observatory
0.9	32	Potsdam, East Germany	Astrophysics Observatory
0.9	32	Meuden, France	Paris Observatory
0.8	30	Pittsburgh, Pennsylvania	Allegheny Observatory
0.8	30	Nice, France	University of Paris
0.7	28	Herstmonceux, England	Royal Greenwich Observatory
0.7	26.5	Johannesburg, South Africa	Union Observatory
0.7	26.5	Vienna, Austria	Universitats-Sternwarte
0.7	26	Charlottesville, Virginia	University of Virginia
0.7	26	Pulkova, Russia	Academy of Sciences Observatory
0.7	26	Belgrade, Eastern Europe	Astronomical Observatory
0.7	26	Tokyo, Japan	Mitaka Observatory
0.7	26	Charlottesville, Virginia	Leander McCormick Observatory
0.7	26	Washington, DC.	U.S. Naval Observatory
0.7	26	Canberra, Australia	Mt. Stromlo Observatory

Fig. 2.   Reasonable facsimile of woodcut of the "Great Refractor" installed at Harvard University (College) in 1843. (*Source: Adler Planetarium.*)

Fig. 3.   A wide-field refractor popular with amateur astronomers. The instrument provides a wide range of astronomical observations. (*Orion.*)

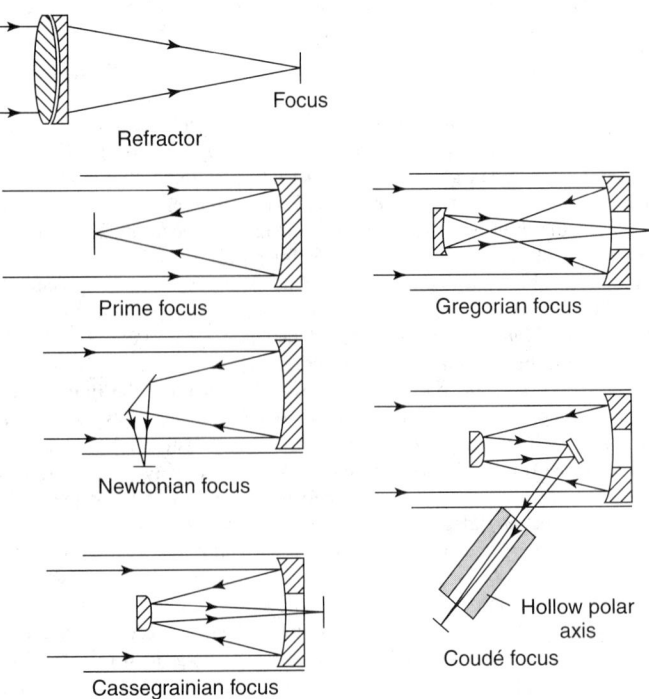

Fig. 4.   Optical paths of representative traditional telescopes.

design, a mirror collects light and focuses all colors at the same point, while using a secondary mirror to guide the received light to a point for viewing, photographing, or otherwise processing the image. Reflectors are well suited for research in the infrared (IT) as well. See also **Infrared Astronomy**.

Herschel, in the 1780s, built reflectors up to 40 feet (12 meters) in length with primaries as wide as 48 inches (1.2 meter). A diagram of one of Herschel's telescopes is given in the article on **Astronomy**. Herschel discovered Uranus with one of these instruments.

The optical paths, as compared with the refractor and as used in representative telescopes over the past several years, are shown in Fig. 4.

TABLE 2. REFLECTING TELESCOPES INSTALLED AS OF 1988[1]
(Does Not Include Several of the New-Generation Instruments)

Diameter		Geographical Location	Operator
Approx. Meters	Inches		
6	236	Zelenchukskaya, Russia	Special Astrophysical Observatory
5	200	Palomar Mountain, California	Hale Observatories
4.5	176	Mt. Hopkins, Arizona	Sacramento Peak Observatory
4	158	Kitt Peak, Arizona	Kitt Peak National Observatory
4	158	Cerro Toledo, Chile	Cerro Tololo Inter-American Observatory
4	153	Siding Spring, Australia	—
3.6	141	La Sila, Chile	—
3	120	Mt. Hamilton, California	Lick Observatory
2.7	107	Fort Davis, Texas	McDonald Observatory
2.6	104	Nauchny, Crimea, Russia	Crimean Astrophysics Observatory
2.6	102	Armenia	Byurakan Observatory
2.6	100	Mt. Wilson, California	Hale Observatories
2.5	98	Heistmonceaux, England	Royal Greenwich Observatory
2.3	90	Tucson, Arizona	Steward Observatory
2.2	88	Manoa, Hawaii	Mauna Kea Observatory
2.1	84	Manoa, Hawaii	Mauna Kea Observatory
2.1	84	Kitt Peak, Arizona	Kitt Peak National Observatory
2.1	82	Fort Davis, Texas	McDonald Observatory
2	79	Azerbaijan	Shemakha Astrophysics Observatory
2	77	Basses Alpes, France	Observatoire Saint Michel
2	76	Haute Provence, France	—
1.9	74	Pretoria, South Africa	Radcliffe Observatory
1.9	74	Tokyo, Japan	Tokyo Observatory
1.9	74	Okayama-ken, Japan	Astrophysics Observatory
1.9	74	Ontario, Canada	David Dunlap Observatory
1.9	74	Helwan, Egypt	Helwan Observatory
1.9	74	Mt. Stromlo, Australia	Mt. Stromlo Observatory
1.9	73	Victoria, B.C. Canada	Dominion Astrophysics Observatory
1.8	72	Flagstaff, Arizona	Perkins Observatory
1.8	72	Asiago, Italy	Padua University Observatory
1.6	61	Bosque Alegre Station, Argentina	National Observatory
1.6	61	Cambridge, Massachusetts	Agassiz Station Harvard Observatory
1.6	61	Flagstaff, Arizona	U.S. Naval Observatory
1.6	61	Tucson, Arizona	University of Arizona Observatory
1.5	60	Vienna, Austria	Figl Astrophysics Observatory
1.5	60	Pasadena, California	Mt. Wilson Observatory
1.5	60	Bloemfontein, South Africa	Boyden Observatory
1.5	60	Tuscon, Arizona	University of Arizona Observatory
1.4–2*	54–80*	Tautenberg, Germany	Observatorium der Deutschen
1.3	50	Crimea, Russia	Sternberg Astronomical Institute
1.3	50	Merate, Como, Italy	Observatorio Astronomica
1.3	50	Canberra, Australia	Mt. Stromlo Observatory
1.2	49	Berlin, Germany	Berlin-Babelsberg Observatory
1.2	48	Basses Alpes, France	Observatorie Saint Michel
1.2	48	Asiago, Italy	Padua University Observatory
1.2	48	Nauchny, Crimea, Russia	Astrophysics Observatory
1.2	48	Victoria, B.C. Canada	Dominion Astrophysics Observatory
1.2	48	Hyderabad, India	Nizamiah Observatory
1.2–1	48–72*	Mt. Palomar, California	Palomar Observatory
1.2	47	St. Michel, France	Paris Observatory
0.8–1	33–46*	Belgium	Uccle Observatory
1	40	Saltsjobaden, Sweden	Stockholm Observatory
1	40	Flagstaff, Arizona	U.S. Naval Observatory
1	40	Cape of Good Hope, South Africa	Royal Observatory

[1] Status of some of these instruments currently is unknown.
*Indicates a Schmidt telescope. First figure is diameter of correcting lens; second figure is diameter of mirror.

The pathways of the more recent, new-generation instruments are shown later in the article.

The major reflecting telescopes located worldwide prior to 1988, a time just prior to phasing in the new-generation instruments, is given in Table 2. The latter instruments are addressed separately later in this article. See Figures 5, 6, 7, 8, and 9.

Although the larger instruments are given the most attention in the literature, there are scores of smaller instruments that serve very useful purposes, not only for education, but also frequently in connection with special surveys and projects where their light-gathering power suffices. It is interesting to note from the table that relatively few observatories are located in the Southern Hemisphere. Astronomers have been aware of this shortcoming for a number of years and this fact demonstrates the significance of the Cerro Tololo (Chile) and the Las Campanas (Chile) installations. Some of the *new-generation* instruments are scheduled for the Southern Hemisphere.

Optical Fundamentals of Telescopes

The larger telescopes offer advantages of light-gathering power and resolution, but there are limitations on the latter. If a wave front of photons falls on Earth from a distant star, the number of photons per unit area is constant; thus a large telescope of area A_1 collects A_1/A_2 more photons than a telescope of area A_2. Since the intensity is proportional to area, the difference in limiting magnitude will be

$$\Delta m = 2.5 \log \frac{A_1}{A_2}$$

Fig. 5.   The Hale telescope and dome located atop Mt. Palomar, California. The primary mirror is 200 inches (approximately 5 meters) in diameter.

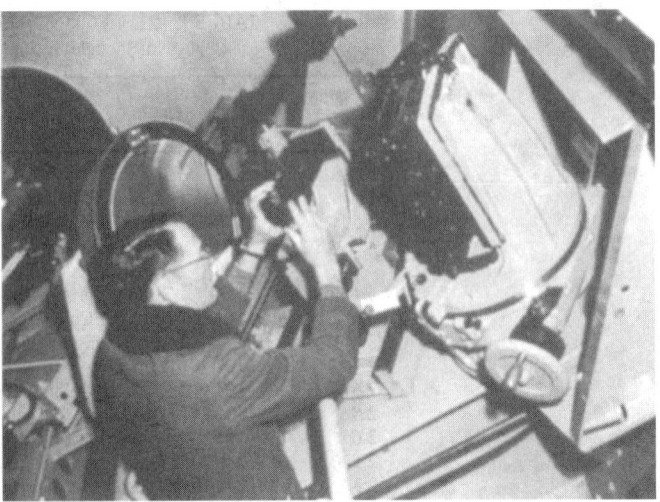

Fig. 6.   Interior of the Tode grating spectrographs used with the Hale telescope.

Hence, a telescope with ten times the area of another will see 2.5 magnitudes fainter.

Since the resolving power of a telescope is given by

$$2.52 \times 10^5 \lambda/a$$

where λ and a (the aperture) are in the same units, the larger the telescope, the finer the separation observable. This is certainly so for telescopes up

Fig. 7.   Aerial view of Kitt Peak, Arizona, a site which includes the telescopes of six observatories: (1) Kitt Peak National Observatory; (2) National Solar Observatory; (3) Steward Observatory (University of Arizona); (4) Warner and Swasey Observatory (Case Western Reserve University); (5) McGraw-Hill Observatory (Massachusetts Institute of Technology, Dartmouth, and University of Michigan consortium); and (6) the National Radio Astronomy Observatory (NRAO). The large dome in the lower left foreground of view is the 4-meter (158 in.) Mayall telescope and beside it is the Steward Observatory instrument. In the left background is the McMath Solar Telescope. The white dome at the upper center of view is the 2.1-meter (83-in.) instrument. The NRAO and McGraw-Hill domes are not shown. (*National Optical Astronomy Observatories.*)

Fig. 8. Aerial view of the Cerro Tololo Inter-American Observatory (Chile), which is dominated by the 4-meter (158-in.) telescope dome in the background. (*National Optical Astronomy Observatories.*)

Fig. 9. The Curtis-Schmidt telescope of the University of Michigan/Cerro Tololo Inter-American Observatory. This is a twin of the Case Western Reserve University's instrument on Kitt Peak, Arizona. The Schmidt type telescope is designed to take wide-angle photographs and spectrographs of the sky. (*National Optical Astronomy Observatories.*)

to a certain size; but at some point, resolution is not increased, because the resolving power meets the limitation on seeing. The seeing disk is seldom better than 1′ of arc. Hence, the smallest resolution possible is 0′.5 of arc, which is obtainable with a 30-centimeter (12-inch) telescope. For larger telescopes, the only gain is in light-gathering power.

The magnification of a telescope is given by $m = f_t/f_e$, where f_t is the focal length of the telescope and f_e is the focal length of the eyepiece. It is important to note that there is usually no magnification when a telescope is used photographically. Since observatory telescopes are used photographically rather than visually, it is more customary to speak of a telescope's F number and scale factor. The F number is simply the ratio of the focal length to the diameter of the objective: $F = f_t/D$. It may be seen that for a given diameter D, the optical speed of a system increases as the F number increases. This is because the seeing disk is finite, and hence the telescope is putting all the photos into a smaller spot. An F 6 telescope is faster than an F 18 telescope of the same aperture.

The scale factor s is usually given in seconds of arc per millimeter on the plate and is $s, = 206.265/f_t$, where f_t is in meters.

Refracting telescopes are divided into two classes: (1) astrographs, and (2) long-focus refractors. The astrographs, having small F numbers (~6) and large fields are used for statistical and survey work. For large field surveys, the Schmidt telescope has superseded the astrograph. The long-focus refractor has large F numbers and large plate scales; hence, it is ideally suited for astronomical position measurements (astrometry). The aperture of a refractor is limited by the plastic nature of glass, and all refractors are subject to chromatic aberration. See also **Schmidt Objective**.

Reflectors are of various types: the paraboloid, the Schwarzschild, the Ritchey-Chrétien, etc. Each type has variations that are named according to the position of the secondary or focus. Thus, a paraboloidal instrument may be said to be prime focus, Newtonian, Cassegrainian, Georgian, Coudé, etc. The various positions or secondaries and/or usable foci serve distinctive purposes. See Fig. 4. The prime focus and the Newtonian focus (which diverts the prime focus to the side of the tube by means of a flat diagonal mirror) supply a fast system, i.e., low F number. The Cassegrainian position uses a hyperboloidal secondary to reflect the beam back to an aperture in the primary, and is thus at a position of convenience. By using different hyperboloids, different effective focal lengths can be achieved, and hence, different F numbers and plate scales. The Coudé always has a large F number and is advantageous because its fixed position allows the use of large cumbersome equipment. The catadioptric telescopes are the Schmidt and the Maksutov, and the correction-plate and two mirror systems. See also **Coudé**; **Maksutov-Bouwers Telescope**.

Low Light-Level Detectors and Image Sharpness

Methods for improving telescopic images date back about 70 years. In 1921, Michelson and Pease introduced interferometric techniques to telescopes. Speckle interferometry was pioneered by Labeyrie many years later (1972) and an intensity interferometer was proposed by Brown and Tiss in 1974. Some of the early methods required the use of a bright nearby reference star (a difficult requirement for most observations).

The principles of active and adaptive optics have been known for several years, but it required the technological fallout from the Strategic Defense Initiative (SDI) and the availability of appropriate computer power to enable their present rapidly advancing applications to astronomy.

Traditional Photographic Plates. For over a century, astronomers have used photography for capturing the images of celestial objects. The photographic plate has had many advantages. Over the years, emulsions have improved immensely.

The photographic plate can read huge amounts of information—as many as 10^8 pixels on a single plate. The information is furnished in a way that is easy to comprehend, as contrasted, for example, with digitized presentations. The photographic plate also permits the integration of information over long exposure spans, making it possible to detect very faint objects. Plates are easy to study and to store permanently. As astronomical research advanced, however, the limitations of photographic plates became increasingly important. These limitations include: (1) photographic plates are nonlinear in their response; (2) they have a limited dynamic range; (3) there is a reduction in sensitivity with increased exposure time (reciprocity factor); (4) there are so-called adjacency effects, e.e., adjacent pixels may not be individually resolvable; (5) there is a low quantum efficiency; and (6) there are limitations as the result of finite granularity. Some of these limitations have been reduced or alleviated through advances in image processing, notably the use of digital technology. See also **Photography and Imagery**.

Even with the success of photography over many years, in the 1950s researchers began to seek photodetectors with comparatively high quantum efficiency and an efficiency that would extend over a wide range of wavelengths. This search was spurred, of course, by the increasingly available devices stemming from solid-state and semiconductor technology. During the 1970s, a number of approaches were taken, one being that of using a silicon intensified target (SIT) vidicon. Experiments were conducted at Mt. Palomar and Kitt Peak with considerable success.

Traditionally, astronomers analyzed photographic sky surveys by inspecting photographic plates on a light box, using powerful jeweler's-type magnifiers, a slow and tedious task. Among the modern analytical systems developed during the past few years is the Automatic Plate Measuring System (APM), developed at the University of Cambridge. In this particular system, each photograph is laser scanned in connection with a microdensitometer, which digitizes the information. A plate may contain as many as 4 gigabytes of information.

Charge-Coupled Device. The CCD, invented in 1969 (AT&T Bell Laboratories) consists of a three-layered semiconductor device, one layer of metallic electrodes and another of silicon crystal, separated by an insulating layer of silicon dioxide. The CCD uses the familiar phenomenon found in many microelectronic devices, namely, the ability to permit negatively charged electrons (or positively charged "holes") to move about controllably in semiconductor material. Most devices use this characteristic to change the electrical current flowing through them—as amplifiers or switches. The CCD stores and transfers information in the form of packets of electrical charge analogous to the tiny magnetic domains in bubble devices.

After wide acceptance of the CCD for use in television cameras, they were incorporated into a number of astronomical imaging systems. The topic is described in considerable detail in the Kristian-Blouke article (reference listed). See also **Charge-Coupled Device (CCD)**; **Microelectronics**; and **Semiconductors**.

Adaptive Optics. This expanding technique for use in optical telescopes largely is based upon technological fallout of the Strategic Defense Initiative (SDI) program in the United States during the 1980s. The original objective was that of eliminating atmospheric distortion of images of foreign satellites in orbits so that laser beams fired into space could disable such missiles at great distances. In early 1982, astronomers at the Mount Wilson Observatory (Pasadena, California) commenced applying the principles of adaptive optics to the 60-inch (1.5-meter) telescope at that location. The driving force for this was the researchers' estimate that the technique may allow telescopes to see images of objects 10 to 20 times sharper that may be from 100 to 400 times fainter than as captured by traditional optical telescopes. In addition to objects in far outer space, the technique also can provide more detail of the planets in the solar system, including the study of dust storms on Mars, weather conditions on Jupiter, and volcanic eruptions on Jupiter's moon, Io. It is predicted that adaptive optics ultimately will resolve, at least partially, the tens of thousands of stars that make up the globular cluster that occurs near the center of our galaxy. As one observer states, "The effect is that of a zoom lens focusing in on previously unattainable details."

Transfer of the previously exclusive military technology is proceeding at a good pace. Prior research in the military phase was conducted at the Kirtland Air Force Base, New Mexico, the University of Chicago, MIT's Lincoln Laboratory, and several independent military contractors. There is a consensus among researchers that upgrading land-based optical telescopes may be less costly than operating huge telescopes, such as the Hubble instrument. As of late 1993, much of the SDI information has not been declassified. Thus, the technique may be considered as still being in an early phase of practical development.

Central to a contemporary adaptive optics system is a small deformable mirror (in the range of 4 inches (10.2 cm)) that is capable of deforming and thus reshaping its surface at a rate of hundreds of times per second in the interest of neutralizing atmospheric effects. In one design, a laser beam is reflected from the upper atmosphere onto sensors within the telescope. The reflected light is analyzed by computer through the use of light sensors. Based upon this, very small mechanical fingers can "tweak" the surface on the order of. 001 millimeter. Currently, there are a number of detailed design options. For example, some researchers prefer to rely on a guide star, wherein the computer would reference that as a sharp point for compensating any blurring. Johns Hopkins University, the Georgia Institute of Technology, and Georgia State University, among others, are engaged in the development of new designs.

Adaptive optics research requires a thorough understanding of light pathway optics and mathematics.

Adaptive optics does have some limitations. The corrected field of view astronomers see in the sky covers only a partial area of the field of view of a large telescope (i.e., smaller than some galaxies appear).

The atmosphere over Mauna Kea (Hawaii) is one of the cleanest in the world, which is the reason, of course, why that particular mountain at an elevation of 13,780 feet (4200 meters) above sea level was selected for the Keck telescopes and other important instruments. Some astronomers have suggested that instruments at that location, when equipped with some form of adaptive optics, should be as close as one can come to an orbiting or a lunar-based optical telescope. Adaptive optics is particularly effective in the infrared range, where atmospheric distortion tends to be less than that for visible light.

Even with the power of telescopes located on Mauna Kea, stars do not appear as tiny bright points, but as jittery blurs approximately one-half of 1 arc second across, because of fine-grain atmospheric turbulence. Some mechanisms of adaptive optics can correct the image.

The use of a laser as a measure of atmospheric turbulence was first suggested by the French astronomers Renaud Foy and Antoine Labeyrie. The procedure suggested by these researchers, as reported by A. Finkbeiner: "Create an artificial star by shooting a laser into the atmosphere, where it is reflected by dust or creates a spot of fluorescence in the thin layer of sodium ions 900 kilometers up: Place the artificial star next to the dim object; Do the adaptive corrections on the artificial star; Apply the same corrections to the dim object you really want to see."

An ultrasimplistic diagram of this form of adaptive optics is given in Fig. 10.

In another related technique, a reimaging lens is used. After light enters the telescope, the light will strike a mirror of flexible *Mylar* stretched over a bundle of piezoelectric actuators. The latter deform the *Mylar* lens in response to computer commands. A second rigid mirror tilts to perform larger-scale corrections.

Active Optics. H.W. Babcock (The Observatories of the Carnegie Institution of Washington, Pasadena, California) has made serious reference to the use of *active optics* in optical telescope design. Babcock observes, "Closely related to adaptive optics there is a much slower type of system now known as *active optics*. It may be used for occasional adjustment of the several supports of the primary mirror; through flexure, the adjustments improve the effective optical figure. This procedure which I proposed in 1953, may be conducted with any reasonably bright reference star; it does not require rapid response."

It is interesting to note that active optical systems were demonstrated by researchers at the European Southern Observatory in the late 1980s and have stated that active optics provides improvements in performance that warrant its serious consideration for all large telescopes in the future.

Interferometry in Optical Telescopes

Interferometry at optical wavelengths first was demonstrated by Albert Michelson (Case Western Reserve University, Cleveland, Ohio) in 1887, for which he was awarded the 1907 Nobel Prize (physics). While Michelson's work figured prominently in connection with the Michelson-Morley experiment, which disproved the presence of an "ether" and served as an initiating point in Einstein's development of the theory of relativity, the serious application of interferometric principles in astrophysics occurred in the late 1940s and early 1950s when researchers, applying early electronics technology developed during World War II, found that the interferometric techniques were much more easily applicable to energy at radio wavelengths, culminating in such impressive installations as the Very Large Array radio telescope in New Mexico.

A simple Michelson interferometer is shown in Fig. 11.

Closure Phase. The principles of the closure phase, a number measured by triplets of Michelson interferometers, is fully independent of certain types of otherwise severe instrumental errors. Since the 1950s, when the principle was first noted, it has been applied to several problems involved with astronomical imaging. As pointed out by T.J. Cornwell (National Radio Astronomy Observatory, Socorro, New Mexico), "Methods based on the closure phase now allow imaging of complex objects in the presence of severe aberrations and are vital to the success of modern, high-resolution astronomical imaging both at radio and at optical wavelengths. Over the past 10 to 15 years, the concept of closure phase has been extended and generalized. One of the most important advances has been the development of automatic or self-calibration techniques."

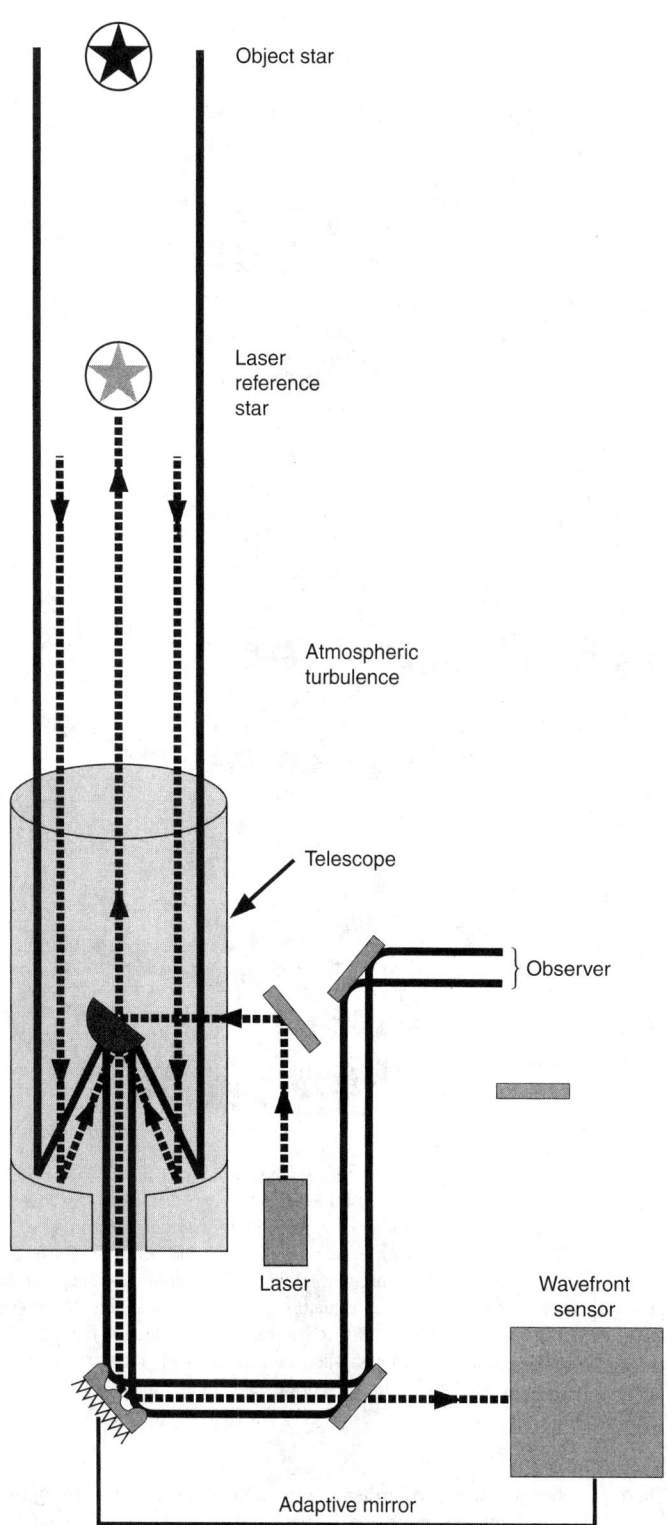

Fig. 10. Highly schematic diagram of one technique for achieving adaptive optics.

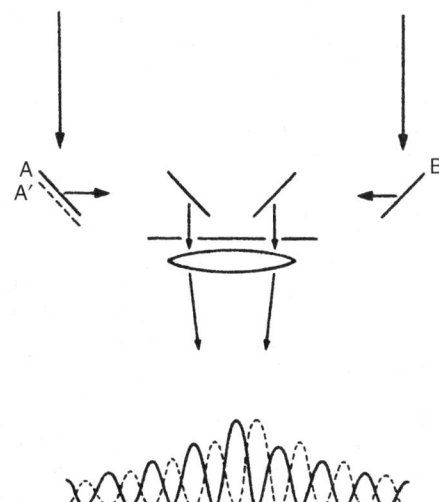

Fig. 11. Elements of a Michelson interferometer. Light collected at the outlying mirrors A and B is brought to a common focus where fringes are observed. Shifting the mirror at left from A to A′ in the one arm shifts the fringes as shown below. (*After Cornwell, National Radio Astronomy Observatory.*)

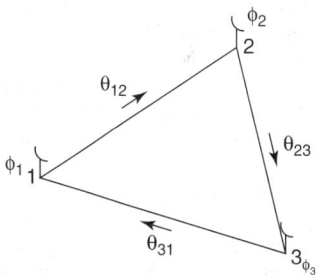

Fig. 12. Demonstration of closure phases. The element-based errors 01 vanish in the sum of the observed phases around a loop, a phenomenon that is known as Jennison's closure phase. (*After Cornwell, National Radio Astronomy Observatory.*)

A primary disadvantage of the masking approach is the loss of most incident photons. This problem, however, can be overcome by using the so-called "speckles" technique. Observers with years of experience have noted that fine detail (speckles) can be seen in the seeing disk for very brief periods of time. The use of this technique, plus other applications of the closure phase, are well explored by Cornwell, who summarizes his paper (reference listed) as follows: "Closure phase methods are applicable to any instrument that measures correlations between the quantities measured by error-prone sensors. The closure phase principle underlies the imaging in many different astronomical contexts: radio interferometry, high-resolution imaging with single optical telescopes, aberration correction in single radio telescopes, and scintillation phenomena."

Fully-Filled Aperture Technique. A computerized imaging technique first was used in connection with the 200-inch (5-meter) Hale telescope on Mount Palomar, California, as early as the late 1980s. Astronomers (California Institute of Technology) have reported that this technique and another technique known as Non-Redundant Masking have virtually eliminated the distorting effects of the atmosphere, making it possible for this (and any other optical telescope) to approach its theoretical maximum resolution. Improvements at the Hale installation have been estimated as a factor of 20, from roughly 1 arc second to about 50 milliarc seconds.

The fully-filled aperture technique requires no direct telescopic hardware. The blurred image simply utilizes the *speckle interferometry technique*. The computational demands have been described as horrendous, requiring the full parallel processing power of Caltech's 512-node "hypercube" supercomputer.

In the non-redundant masking technique, an opaque screen pierced with from five to seven very small holes is placed at the prime focus of the telescope. The separate rays of light that pass through the screen are recombined into an *undistorted* image through the application of

As applied to optical imaging, Cornwell observes that closure phase can be applied to optical imaging by "forcing the optical aperture to mimic the sparse filling found in radio interferometry." In so doing, the aperture of a conventional optical telescope can be considered as being made up of many interferometers formed between constant atmospheric phase. See Fig. 12. Thus, an image at the telescope's focal plan can be considered as being made from the combination of all these many interferometers. It is essentially impossible to sort out the contributions of the individual interferometers. However, a mask in the pupil can be used to force nonredundancy of measurement, thus making it "possible to derive closure phases from the Fourier transform of a single short snapshot."

mathematical algorithms. The latter are available from developments for use with radio telescope observations.

In either case, a satisfactory, clear image results, but at a high cost of computer power.

Mirrors and Lenses. From the time of Galileo, lens grinding for telescopes has been an arduous, painstaking task. The measurements required to control the process and the final results of the process traditionally have been inadequate for assuring perfection. New and revolutionary optical component facilities that utilize laser measurement devices and computer controls have appeared during the past decade. For example, at the Optical Science Laboratory (University College, London) a 30-ton grinding and polishing machine is capable of making mirrors and lenses with diameters up to 2.5 meters for cutting telescope blanks to within one micrometer or better.

During the period of the renaissance in telescope design commencing in the early 1980s, much attention has been given to mirror design and production. In addition to monolithic main mirrors, new honeycomb, segmented, and meniscus mirrors have been produced. With the emergence of various forms of adaptive adoptics, mirrors must be engineered as components of the total telescope system. Innovative designs are shown in Fig. 13.

In connection with the new WIYN (University of Wisconsin, Indiana University, Yale University, and National Optical Astronomy Observatories) telescope, under construction as of early 1994 and which will be the second largest telescope on Kitt Peak, unprecedented attention has been given to mirror design and production. The 3.5-meter (137.8-in) mirror was spun-cast and then polished at the University of Arizona Mirror Laboratory. A polishing technique known as stressed lap polishing, developed at the U of A laboratory, involves the use of a computer-controlled shaping tool that makes it possible to control the final figure of the mirror over the entire surface to a high degree of accuracy. Just over 150 hours of machine grinding and polishing, plus ten sessions of hand polishing, led to the final, extremely smooth surface.

Smoothness of a mirror's surface is one criterion of mirror quality, and it usually is expressed as a measure of the average amount of "bumpiness" on the surface. This "root mean square" measurement for the WIYN mirror is 16 nanometers, or less than one millionth of an inch. As one researcher stated, "One can visualize this degree of smoothness by picturing a mirror of a size stretching between Los Angeles and New York. In that case the rms measurement would show the average bump height of the mirror to be about an inch."

An important advance made during the polishing of the mirror was the application of pressurization polishing the cavities that exist within the mirror. Because of the honeycomb structure of the mirror, the top surface of the mirror is supported by honeycomb-shaped ribs around open cavities to minimize the weight of glass in the mirror. A problem called "print-through" can occur during polishing. As the polishing tool, called a lap, pushes on the mirror, portions directly over a supporting rib tend to be stiffer, leading to unevenness near the positions of the ribs. This problem was solved by putting the cavities between the ribs under air pressure equal to the downward pressure of the polishing lap, thus stiffening the glass over the cavities so that the surface was polished evenly.

Gravitational Lensing — Gratis Magnification!

The phenomenon of gravitational lensing may become an important strategy for astronomers in probing the currently uncharted regions of dark matter that permeate regions of the universe. Dark matter remains poorly understood. Some authorities believe that it may be made up of dead stars, planets, mini black holes, *or exotic particles* (?).

Although gravitational lenses are not part of a researcher's costly optical equipment, but exist naturally in certain situations in space, these "free" lenses can be put to use. The existence of a gravitational lens is consistent with earlier observations made by Einstein in connection with the general theory of relativity. As pointed out by the article on **Quasars**, a discovery of a "pair" of quasars noted in 1979 by researchers ultimately was concluded to be a pair of images of a single quasar, as the consequence of a nearby gravitational lens. The interesting and illusory effects of a gravitational lens were discovered by the Hubble Space Telescope some time after it was launched. At first considered to be an artifact caused by the telescope's "fuzzy" imaging, it later was diagnosed as "the most spectacular example of gravitational lensing yet," according to R. Ellis (Durham University) and D. Weedman (Pennsylvania State University).

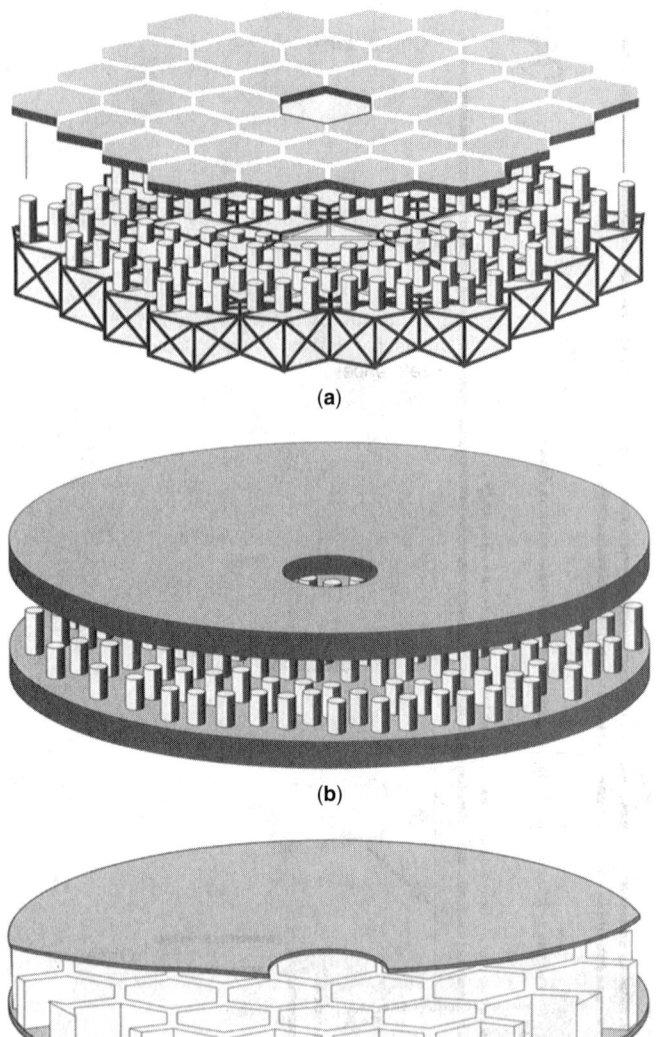

(a)

(b)

(c)

Fig. 13.   Types of telescope mirrors: (**a**) Segmented mirror. Rather than using a single (monolithic) large mirror, small, hexagonally shaped mirrors can be used. The segments are aligned by motorized controllers to maintain the correct shape for creating a single image. (**b**) Meniscus mirror. A solid reflecting surface is used. The reflector surface is so thin that it cannot maintain a rigid shape unless supported from below. Mechanical actuators are used to maintain the desired shape. (**c**) Honeycomb mirror. A thin reflective surface rides on the top of a glass honeycomb structure. Although such mirrors are stiff, they are relatively lightweight because of honeycomb-supporting structure. (*Source: C.F. Powell reference listed.*)

The researchers commented further, "Gravitational lenses are nothing new for astronomers, but normally the image of the galaxy that is stretched or multiplied by the intervening mass turns into *just a blur of light*. This latest example — mirror images of a galaxy about 10 billion light years away — is the first to preserve the detail in the distant galaxy. And that makes this image a powerful probe of the object that is doing the lensing, i.e., a cluster of 799 galaxies perhaps 4 billion light-years away called AC 114."

For some years, a Laser Interferometer Gravitational Wave Observator (LIGO) experiment has been underway. The project, largely a development of the California Institute of Technology and Massachusetts Institute of Technology, would be directed toward locating the presence of gravity waves in the cosmos. In essence, the objectives of the LIGO experiment have been debated within the science community. One researcher on the project observes, "We're talking about the collapsing cores of supernovas, or the collisions of black holes and neutron stars — events you have no hope of seeing with light. A prime example of such an event would be the death spiral of a binary pulsar." Our own galaxy is already known to contain at least one binary pulsar. Other scientists comment that LIGO is a costly project just to verify a prediction of general relativity, particularly because a majority of physicists believe in the theory without further proof.

Renaissance in Telescope Design

When the 5-meter (200-in.) Pyrex glass disk was cast for the huge reflector on Mt. Palomar. California in 1948 that instrument represented the acme of telescope making technology. A number of similar, but somewhat smaller instruments were put in place at various locations in the years to follow, but there were no real breakthroughs in telescope design until the early 1980s. Arthur F. Davidson (Center for Astrophysical Sciences, Johns Hopkins University) has observed. "Historically, apertures have doubled every 40 or 50 years, which means that we're due for the next phase in the 1990s. And, in fact, the technologies are falling into place."

Announcements made in later 1986 provided evidence of a renaissance in ground-based astronomy, emphasizing the continuing and expanding role of the reflecting telescope. Some of the programs planned already have reached fruition; others are in a late stage of development. Some more recently proposed programs are in the planning stage.

Keck I and II Telescopes on Mauna Kea, Hawaii

Keck I features a 10-meter (394-in) segmented primary mirror, composed of 36 separate hexagonal pieces. Construction commenced in the late 1980s. Keck I is a joint project of the University of California and the California Institute of Technology. In 1985, the W.M. Keck Foundation made a $70 million gift toward funding the project. Because the huge mirror is built up of segments, the individual segments are thin and lightweight. Each mirror segment is precisely controlled by computer commands for maintaining the mirror surface as optically perfect as possible. A complex of sensors and mechanical actuators is required. The optical path of Keck I is shown schematically in Fig. 14.

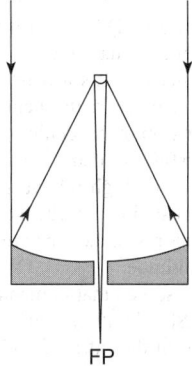

FP

Fig. 14. Optical pathway of the W.M. Keck Observatory reflector. Main mirror is made up of 36 hexagon segments FP = focal point.

As a preliminary test of the optics, when only 9 of its total of 36 mirror segments were in place, the instrument was turned skyward to gather in its first light. This image was that of the galaxy NGC 1232. Results of the pre-trial were very satisfactory. As of 1993, the entire system essentially is operational.

The gigantic mirror rests on complex machinery. Springs and flex discs keep each segment stress-free as the telescope angles toward different parts of the sky. Hydraulic actuators align the segments into a single 32-foot (3.8-meter) reflector.

Keck I incorporates a 256-by-256–element indium antimonide infrared array. The $94 million Keck I, the first of a new generation of superscopes, is among four infrared telescopes atop the 14,000-foot extinct Mauna Kea volcano in Hawaii with optical and infrared capability. The Keck telescope is expected to complement spaceborne optical telescopes, such as Hubble, by reading the chemical "signatures" provided by the space telescopes.

As of 1993, a companion (side-by-side) Keck II instrument is in the late planning phases. It will be jointly by the Keck Foundation, Cal Tech, and MIT. The two Keck telescopes essentially will be identical, but Keck II will incorporate any improvements as noted in the early operations of Keck I. When completed, the two Keck telescopes will have eight times the

light-collecting power of the 200-inch (5-meter) Hale telescope on Mount Palomar (California).

The National New Technology Telescope (NNTT)

This project of the National Optical Astronomy Observatories, as designed, will include four 8-meter (315-in) telescopes mounted, as shown by the optical pathway diagram of Fig. 15. When operated together, the combined telescopes will equal the power of a 16-meter (630-in) single telescope. The project throughout its history (first conceived in 1978) has been the subject of some controversy and budgetary problems. Early on, it was envisioned that the NNTT essentially could serve the interests of astronomers across the board, whereas priorities for working time on large, privately funded instruments could be somewhat biased. The original charter for funding the project emanated from the National Science Foundation (NSF).

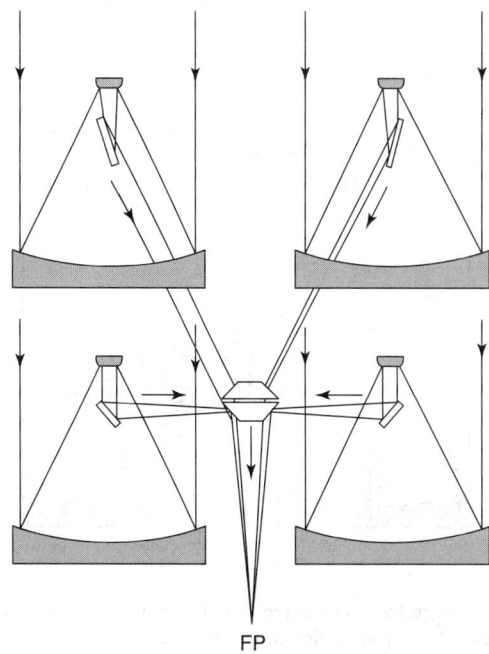

FP

Fig. 15. Optical pathway of proposed National New Technology Telescope (NNTT). The system includes four 8-meter (315-in.) telescopes, which when operated together will be the equivalent of a 16-meter (630-in.) instrument. FP = focal point.

Very Large Telescope (VLT) at the European Southern Observatory (ESO)

This design consists of four 8-meter (315 in.) reflectors mounted in a linear array. The telescopes may be used independently or operated together to provide the equivalent of a 16-meter (630-in) reflector. The light pathway is shown schematically in Fig. 16. The VLT now under construction will be located on a mountaintop site in Cerro Paranai, Chile, nearby the location of ESO's New Technology Telescope (NTT), which was completed in 1990 and located at La Silla, Chile. The latter is a 3.5-meter (138-in) reflector.

Columbus Telescope (Mount Graham, Arizona)

In October 1986, the University of Arizona, Ohio State University, and the University of Chicago, with the participation of the Osservatoril Astrofiscio di Arceri (Italy), announced a joint effort to construct a "binocular"-type telescope on Mount Graham, Arizona, with the intent of completion during the early 1990s. Each of the two telescopes will be 60% larger than the veteran Mount Palomar instrument in California, and the light-gathering power will be 2.5 times greater. The proposed light pathway is indicated in Fig. 17.

The project was slowed, partially due the pullout of Ohio State from the venture. Until the needed funds are fully available, the consortium plans to commence construction with only one of the twin 8.4-meter (33-in) mirrors called for in the original design, but with the intent for ultimate completion of the initial plan. Very early construction procedures commenced in mid-1993, after authorities gave approval, thus denying the objections of a limited number of environmentalists.

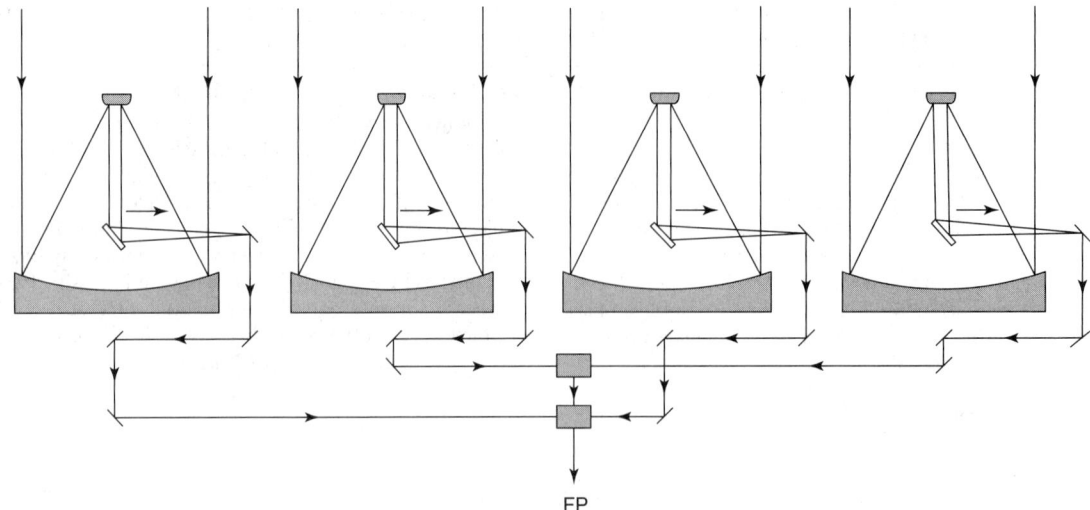

Fig. 16.   Optical pathway of proposed Very Large Telescope (VLT) intended for installation at the European Southern Observatory. The system will comprise four 8-meter (315-in.) reflectors mounted in a linear array. Each of the instruments may be used alone or in varying combinations. FP = focal point.

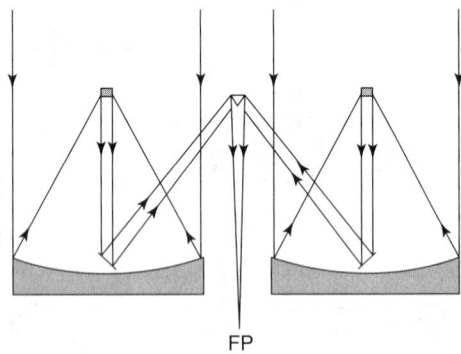

Fig. 17.   Optical pathway of the proposed Columbus (Two-Shooter) telescope that is designed like a pair of binoculars. FP = focal point.

Summary of Expansion Plans

There is insufficient space available in this article to cover numerous other examples of the new-generation telescopes, completed or in the planning phase. A few brief mentions are in order.

In the fall of 1986, the Carnegie Institution of Washington, the Johns Hopkins University, and the University of Arizona announced a collaborative effort for the construction of a new 8-meter (315 in.) telescope for installation at Carnegie's Las Campanas Observatory in Chile. Although the instrument will have only one mirror, the telescope will become the largest in the Southern Hemisphere, with an excellent view of the Magellanic Clouds and the center of the galaxy.

Other projects include the Multiple Mirror Telescope Upgrade (Smithsonian), Mount Hopkins, Arizona; Magellan Carnegie Institution and Johns Hopkins University, Las Campana, Chile; and the Japanese National Large Telescope, proposed for Mauna Kea, Hawaii.

In 1988, the University of Texas and Pennsylvania State University proposed a new Spectroscopic Survey Telescope (SST) for location at McDonald Observatory in Texas. The proposal promised twice the light-gathering power of any existing instrument at a strikingly low cost. The basic concept is that of constructing an instrument that does nothing but spectroscopy. The essence of the concept is the mounting. The SST would be permanently offset from the vertical direction by 30° and could only pivot around the vertical axis. Unlike conventional telescope designs, the instrument would not track stars across the sky, but rather, as it is positioned and pointed toward a given object in space, it would remain fixed in place while the earth's rotation would carry the object across the instrument's field of view.

Large Arrays of Smaller Telescopes

Some authorities have voiced the opinion that at some future date large arrays of smaller optical telescopes (along the lines of large-array radio telescopes) may play a major role in astronomy some years ahead. It is interesting to note that since 1986 researchers at the Mount Wilson (California) Observatory have experimented with two mobile light collectors that can be placed up to 31 meters (102 feet) apart, and, using interferometric measurements, surprisingly good results have been obtained.

Optical Telescopes in Antarctica

Although not a new topic, increasing interest is being shown by researchers in the location of telescopes in the Antarctic. Martin Pomerantz (Bartol Research Institute, University of Delaware) has been pointing out to the scientific community the benefits that may derive from telescopes installed in that region. He does concede: "It's a hard place to do science, unless you compare it to space." However, the benefits of thin, clear, and dry atmospheres of the region cannot be denied. Cosmic ray research from that area has been successful for many years. It is interesting to note that much of Antarctica is quite high (above sea level). For example, the Amundsen-Scott experiment station is located atop nearly 1.9 miles (3 km) of ice. Low winter temperatures reduce air pressure, adding an extra half kilometer to the effective altitude.

Some researchers have observed that unmanned balloons launched from Antarctica compete with Shuttle flights for observing hard x-rays and gamma rays. For observing in the infrared and submillimeter wavebands, the Antarctic is particularly attractive because of the very low concentration of water vapor, which absorbs radiation at this end of the spectrum.

As pointed out by J. Bally (AT&T Bell Laboratories), "Carbon provides a way to study how newborn stars literally destroy their placental cloud environments by radiation and winds." But sub-millimeter radiation from atomic carbon is absorbed by water vapor in most locations on Earth, causing interested astrophysicists to reserve time on NASA's Kuiper Airborne Observatory (a converted jet transport).

In early 1991, the National Science Foundation (NSF) established the Center for Astrophysical Research in Antarctica (CARA), with a plan for building a series of long-wavelength telescopes near the south pole.

Lunar-Based Telescopes

Concurrent with plans for visiting the Moon early in the age of space exploration, the concept of using the moon as a base for astronomical observations has been debated. Most of the arguments, pro and con, of lunar observatories are well covered in the literature, and particular reference to the J.O. Burns and B.F. Burke references listed is suggested.

The Hubble Space Telescope

This instrument named after Edwin P. Hubble, a renowned astronomer (1889–1953), is an imaging, plarimetric, astrometric, and spectroscopic 2.4 meter (94.5 inch) mirror observatory designed to operate in low-earth orbit. It is a combined ultraviolet (UV) and optical telescope. It is the first of several space telescopes; later generations are to be designed for more specialized service. See also **Hubble Space Telescope (HST)**.

Additional Reading

Anderson, P.H.: "Astrophysics Goes South," *Science*, 1494 (June 14, 1991).

Ashbrook, J.: "The Astronomical Scrapbook," Cambridge University Press, New York, NY, 1985.

Babcock, H.W.: "Adaptive Optics Revisited," *Science*, 253 (July 20, 1990).

Bahcal, J.N.: "Prioritizing Scientific Initiatives," *Science*, 1412 (March 22, 1991).

Blanco, V.M. and M.M. Phillips: "Progress and Opportunities In Southern Hemisphere Optical Astronomy," Brigham Young University Press, Provo, UT, 1988.

Burke, B.F.: "Astrophysics from the Moon," *Science*, 1365 (December 7, 1990).

Burns, J.O., et al.: "Observatories on the Moon," *Sci. Amer.*, 42 (March 1990).

Chaisson, E.J.: "Early Results from the Hubble Space Telescope," *Sci. Amer.*, 44 (June 1992).

Cornwell, T.J.: "The Application of Closure Phase to Astronomical Imaging," *Science*, 263 (July 21, 1989).

Cowen, R.: "Dawn of a Big Telescope," *Science News*, 348 (December 1, 1990).

Dickson, D.: "Britain Reveals Astronomy Plan," *Science*, 471 (January 27, 1989).

Dickman, S.: "New Telescopes Bring Europe Closer to the United States," *Science*, 465 (April 24, 1992).

Elliott, C.A. and M.W. Rossiter: "Science at Harvard University," Lehigh University Press, Bethlehem, PA, 1992.

Espinosa, J.M., F. Sanchez, and A. Herrero: "Instrumentation for Large Telescopes," Cambridge University Press, New York, NY, 1997.

Finkbeiner, A.: "Untwinkling the Stars," *Science*, 1786 (June 28, 1991).

Flam, F.: "Putting a Cosmic Illusion to Work," *Science*, 30 (April 3, 1992).

Flam, F.: "Peering Through a Lens, Sharply," *Science*, 393 (October 16, 1992).

Flam, F.: "NASA Stakes Its Reputation on Fix for Hubble Telescope," *Science*, 887 (February 12, 1993).

Florence, R.: "The Perfect Machine: Building the Palomar Telescope," Harper Trade, New York, NY, 1995.

Gibbons, A.: "Astronomers Want New Optical Telescopes, but...," *Science*, 806 (May 18, 1990).

Graham, D.: "A Sharper Image of the Cosmos," *Technology Review (MIT)*, 12 (October 1992).

Hamilton, D.P.: "Mount Graham Telescope Endangered by Ohio State Pullout," *Science*, 1199 (September 13, 1991).

Hamilton, D.P.: "Columbus Telescope Project Back on Track," *Science*, 1507 (June 12, 1992).

Hardy, J.W.: "Adaptive Optics for Astronomical Telescopes," Oxford University Press, Inc., New York, NY, 1998.

Hewitt, A.: "Optical and Infrared Telescopes for the 1990s," Kitt Peak National Observatory, Tucson, AZ, 1981.

Horgan, J.: "Quantum Philosophy," *Sci. Amer.*, 94 (July 1992).

Kitchin, C.R.: "Telescopes and Techniques: An Introduction to Practical Astronomy," Springer-Verlag, Inc., New York, NY, 1997.

Korsch, D.G.: "Reflective Optics," Academic Press, Inc., San Diego, CA, 2000.

Lena, P., F. Mignard, and F. Lebrun: "Observational Astrophysics," 2nd Edition Springer-Verlag, Inc., New York, NY, 1998.

Manly, P.L.: "Unusual Telescopes," Cambridge University Press, New York, NY, 1995.

Manly, P.L.: "The 20-CM Schmidt-Cassegrain Telescope: A Practical Observing Guide," Cambridge University Press, New York, NY, 1994.

Maran, S.P.: "A New Generation of Giant Eyes Gets Ready to Probe the Universe," *Smithsonian*, 40–53 (June 1987).

Marx, S.: "Astrophotography with the Schmidt," Cambridge University Press, New York, NY, 1992.

Mazzoldi, P.: "From Galileo's "Occhialino" to Optoelectronics," World Scientific Publishing Company, Inc., River Edge, NJ, 1993.

Moore, P.: "Eyes on the Universe: the Story of the Telescope," Springer-Verlag, Inc., New York, NY, 1997.

Powell, C.S.: "Gone in a Flash: Rapidly Moving Telescope," *Sci. Amer.*, 22 (April 1990).

Powell, C.S.: "Mirroring the Cosmos," *Sci. Amer.*, 112 (November 1991).

Powell, D.S.: "Astronomers Beat a Path to High Resolution," *Sci. Amer.*, 21 (July 1993).

Smith, R.C.: "Observational Astrophysics," Cambridge University Press, New York, NY, 1995.

Staff: "Keck 2 May See the Light," *Science*, 1301 (March 15, 1991).

Waldrom, M.M.: "Keck Telescope Ushers in a New Era," *Science*, 1244 (September 14, 1990).

Waldrop, M.M.: "Computer-Age Stargazing," *Science*, 1191 (September 15, 1989).

Waldrop, M.M.: "New Technology Telescope of the European Southern Observatory (ESO)," *Science*, 917 (February 23, 1990).

Waldrop, M.M.: "Keck's First Light," *Science*, 1511 (December 14, 1990).

Weiss, P.L.: "Reflections on Refraction," *Science News*, 236 (October 13, 1990).

Williams, D.: "'Optician' to the Stars," *Case Alumnus*, 2 (Winter 1993).

Wilson, R.N.: "Reflecting Telescope Optics l: Basic Design Theory and Its Historical Development," Vol. 1, Springer-Verlag, Inc., New York, NY, 1996.

Web References

Capilla Peak Observatory: *http://www.phys.unm.edu/~cpo/*

Carnegie Observatories: *http://www.ociw.edu/*

David Dunlap Observatory: *http://ddo.astro.utoronto.ca/*

Gemini Observatory: *http://www.gemini.edu/public/*

Isaac Newton Group of Telescopes, William Herschel Telescope (WHT) *http://www.ing.iac.es/PR/wht_info/*

Kitt Peak, The National Optical Astronomy Observatories: *http://www.noao.edu/kpno/pubpamph/pub.html*

Large Binocular Telescope: *http://lbtwww.arcetri.astro.it/*

Las Campanas Observatory: *http://www.ociw.edu/magellan/*

McDonald Observatory, Hobby-Eberly Telescope: *http://www.as.utexas.edu/mcdonald/het/het.html*

Mt. Wilson Observatory: *http://www.astrophys-assist.com/wilobs/*

National Solar Observatory, Sacramento Peak: *http://www.sunspot.noao.edu/telescopes.html*

Palomar Observatory: *http://www.astro.caltech.edu/observatories/palomar/*

Perkins Observatory: *http://www.perkins-observatory.org/*

South African Large Telescope: *http://www.salt.ac.za/*

Steward Observatory: *http://www.as.arizona.edu/steward/*

The Cerro Tololo Astronomical Observatory: *http://www.ctio.noao.edu/site/index.html*

The Very Large Telescope Project: *http://www.eso.org/projects/vlt/*

UCO/Lick Observatory: *http://www.ucolick.org/*

University of London Observatory: *http://www.ulo.ucl.ac.uk/telescopes/*

U.S. Naval Observatory: *http://www.usno.navy.mil/*

W.M. Keck Observatory: *http://www2.keck.hawaii.edu:3636/*

Yerkes Observatory University of Chicago: *http://astro.uchicago.edu/yerkes/*

TELESCOPIUM.   A southern constellation located near Scorpius.

TELEVISION (TV).   Once simply called "seeing at a distance," television is defined by the Institute of Electrical and Electronics Engineers (IEEE) as "the electric transmission and reception of transient visual images." Kiver and Kaufman stipulate TV as "the science of transmitting rapidly changing pictures from one place to another."

In the modern interpretation of TV, the term *television system* more appropriately connotes not only the visual (video) transmission of information, but the accompanying transmission of sound (audio) information, as well as means for storing and retrieving both video and audio data to extend the utility of the system beyond the simultaneous transmission and reception of information in real time. Video recording in the form of video cassette recorders (VCRs), for example, has significantly enhanced the value of TV in that the user not only can record "live" programs for later and innumerable "playbacks," but a large variety of prerecorded (video and audio) programs, ranging from individually and commercially made movies, educational, etc. presentations, may become "alive" on the TV set. Television is rapidly becoming a *multimedia* communications tool for education and entertainment.

In terms of the video elements of a television system, the principal elements required include: (1) a camera-type device for picking up the scene, (2) a transducer to convert the light impulses of the scene to corresponding electrical signals, (3) a transmitter to convert the electrical signals into suitable form for transmission to a distant receiver, (4) a receiver to pick up the transmitted signals and convert them to the appropriate form to apply to a further transducer, which (5) converts the electrical signals back into light and thus reproduces the original scene.

Chronology of Television

The concept of television, unlike some inventions that can be attributed to a single individual or group of individuals, represents a melding together of the findings of numerous scientists and inventive-type persons in a number of countries. Most of the early research commenced in the 1880s, during which period several rather crude systems for transmitting pictures over electrical wire were introduced. The concept of scanning was first suggested by Paul Nipkow in 1884 when he invented the *scanning disc*.

During the course of TV development, a breakthrough occurred with the realization that a "picture" should be broken up prior to transmitting it. Introduction of the cathode-ray tube (CRT) in 1900 represented a major accomplishment, and, of course, this basic technology remains today as a key element of a TV system. It was learned, however, that the early CRT, although well qualified for *reception*, was not useful for *transmission*. It was not until 1933 that the *iconoscope* was designed by Vladimir Zworykin and was used in early systems (TV camera). The iconoscope essentially was the inverse of the receiving CRT, in that it was capable of translating into electrical currents the light images that were focused upon a plate inside the tube. Further innovations were made, including the design of a scanning tube (image dissector) by Philo T. Farnsworth just a few years later.

Television on a very limited basis commenced in Britain in 1936. The system employed 405 lines operating at 50 frames per second. The range of the initial service was only 35 miles (56.3 km), and thus the number of viewers was too small for the transmission service to have any impact as a social force. In Britain, the 405-line system persisted for about 40 years, until adoption of the 625-line system in 1964. Television broadcasts of any major proportion did not commence in the United States until after World War II.

Color TV was introduced in the late 1960s. With color TV came the need for the application of digital techniques. Cable television, a system in which signals from distant stations are picked up by a master antenna and sent by cable to paying individual subscribers, was introduced in the United States in the early 1960s. During approximately the same time, closed-circuit television, in which the signal is transmitted by cable rather than by radio waves, was introduced for applications in business for staging remote conferences, in hospitals for observation medical procedures, and by industry for remotely operating heavy equipment. Satellite-direct-to-home television became possible, in lieu of community antenna television service (CATV) in the early 1970s and continues. In the early 1980s, the so-called "Sky Cable" concept was proposed. The system also is called *DirectTv*. The *Hughes HS601* satellite is scheduled for very early launching. The satellite will deliver news, sports, movies, and specialty TV programs to households equipped with low-cost, 18-inch (45.7-cm) antennas that, for example, will be attachable to any flat surface, such as a windowsill. The system is described further in the article on **Satellites (Communication and Navigation)**.

High-definition television (HDTV) and flatter, thinner TV screens are addressed later in this article.

Fundamental Television Principles

Scanning. In television, it is necessary to break up the scene into minute elements and utilize these elements in an orderly sequence. This process is called *scanning*[1] and is somewhat similar to the process of reading a text line-by-line from a printed page. In reading, the eye progresses along a line word by word and then returns to the beginning of the next line until, in essence, the entire page has been scanned. The basic principle involved in scanning a picture is shown in Fig. 1.

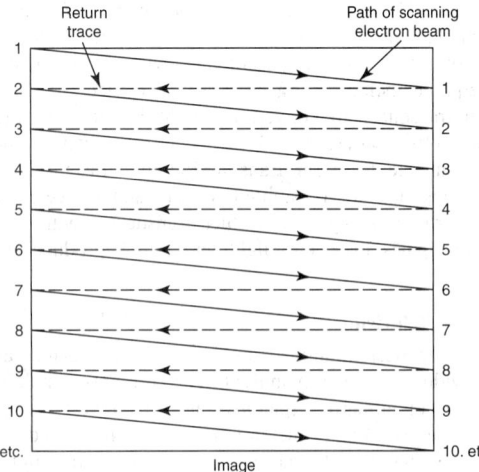

Fig. 1. Sampled representation of procedure for scanning a television picture. Solid lines indicate actual scanning; dashed lines indicate beam retraces.

Starting at the top left side, an electron beam moves from left to right across the screen along line 1–1. After the beam reaches the right side of the screen at point 1, it rapidly moves to the left side again along dotted line 1–2. During the motion from right to left, no picture information is transmitted and thus there is no "mark" made on the screen. Marked line 1–1 is called a *trace*; unmarked path 1–2 is called a return trace, or *retrace*. When the standards for television were established, a rectangular picture (or rectangular *raster*) was selected and the relationship between the height and width of the rectangle was established. Width of the rectangle divided by the height is termed *aspect ratio*. This ratio is 4/3 in the United States and United Kingdom. The aspect ratio holds regardless of the screen size. By adjusting voltages to the picture-tube circuitry, it is possible to produce a rectangle with different aspect ratios. However, the picture is transmitted in an aspect ratio of 4/3. For minimum distortion, the receiver picture should have an identical aspect ratio.

Because of the physical principles involved, there is a relationship between brightness and the rate of scanning, brightness going down with increased scanning rate. As with motion pictures, to provide a true and realistic feeling of motion to all actions displayed, it is necessary to present a number of pictures or *frames* per second. Early designers in compromising on brightness, utilizing a reasonable transmission bandwidth, and frame frequency, selected a rate of display of 30 complete pictures per second. A frame frequency of only 30 pictures/second would result in a flicker discernible to the eye, so each picture is divided into two parts called *fields*. Two fields must be produced in order to make one complete picture or frame. The field frequency, then, is 60 fields/second and the frame frequency is 30 frames/second. Each field contains one-half of the total picture elements. The system is fundamentally similar to that used in motion picture projectors wherein each individual picture is moved into the projector and flashed onto the screen—then a shutter comes down in front of the picture momentarily and the same picture is shown again. In motion pictures, each frame is shown 24 times per second and each frame is divided into two parts, or fields, so that the actual repetition rate is 48 fields/second. No flicker is observed by the eye. The choice of 60 fields/second and 30 frames/second was made so that the television picture could be synchronized to the standard power-line frequency of 60 Hz in the United States. With each frame divided into two parts, each field has one-half of 525 lines[2] from its beginning to the start of the next field. With *interlaced* scanning, each frame is broken up into an even-line field and an odd-line field. At the transmitter, all scanning frequencies are derived from the 60-Hz line frequency, and the receivers are automatically synchronized to the transmitted synchronizing pulses. As a result of this power-line frequency synchronization, the effects of hum on the picture, which might be caused by imperfect power supply filtering, will be stationary on the television receiver screen. If this synchronization did not occur, the hum effects would cause vertically moving patterns to pass through the picture. Adjacent cities, having television transmitters, sometimes use synchronized power-line frequencies to prevent this unwanted type of interference. When color television programs are broadcast, the field frequency is not 60 Hz, but is reduced slightly to 59.94 Hz. Thus, while hum interference patterns are not perfectly synchronized for color broadcasts, they move vertically at the rate of only 0.06 Hz, which is a slow rate and not readily detectable.

Basic Circuitry. A simplified complete television system is shown in Fig. 2. Note that there are two transmitters—one for transmitting the picture (video) signal; and another for transmitting the sound (aural) signal. The television camera converts the scene being televised into electrical impulses which are amplified and transmitted. Scanning and synchronizing voltages must be delivered to the camera. A *synchronizing signal* must also be transmitted so that the scene in the receiver can keep in step with the scene at the transmitter. The required scanning and synchronizing signals are also delivered to the video amplifiers and mixers in the transmitter. The output signals from the video amplifier and mixers are delivered to an amplitude-modulation (AM) system. Instead of both sidebands being transmitted, as in an AM broadcast transmitter, only one full sideband

---

[1] Scanning (television)—the process of analyzing or synthesizing successively, according to a predetermined method, the light values of picture elements constituting a picture area (ANSI-IEEE). *Interlaced scanning (television)*—a scanning process in which the distance from center to center of successively scanned lines is two or more times the nominal line width, and in which adjacent lines belong to different fields (ANSI-IEEE).

[2] The number of 525 lines was selected because of: (1) the frequency bandwidth available to the transmission of TV signals (required bandwidth increases with the number of lines), (2) the amount of detail required for a well-produced image, and (3) the ease with which the synchronizing (and blanking) signals can be generated. These figures will be altered when the final standards for HDTV are worked out.

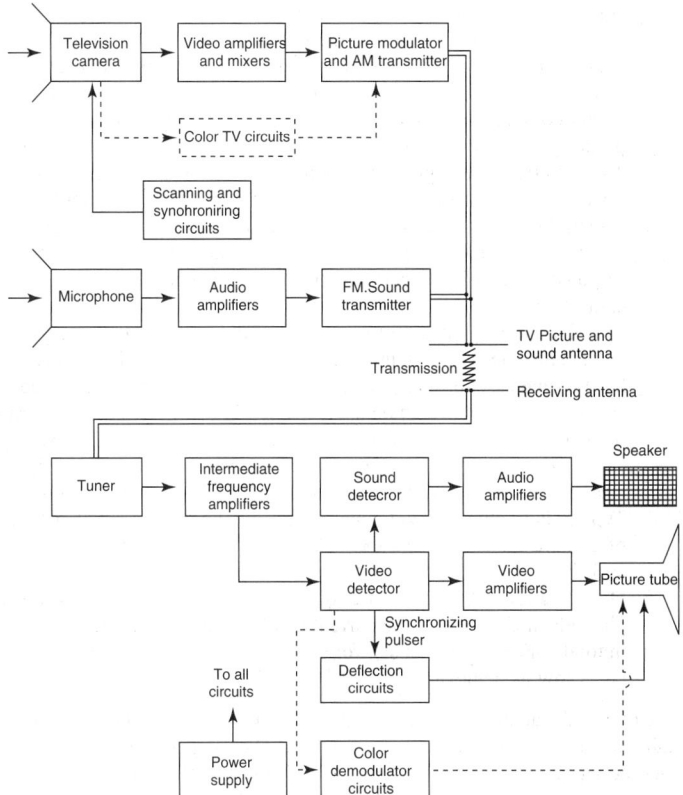

Fig. 2. Simplified block diagram of television system—transmitting and receiving.

and a part of the other is contained in the output signal. Elimination of most of one sideband conserves the electromagnetic spectrum so that more stations can transmit in a given range of frequencies. Transmitting with one full sideband and one partial sideband is termed *vestigial sideband transmission*, not to be confused with single sideband transmission in which one of the sidebands is fully removed.

If a color television scene is being transmitted, then the output from the color television camera goes through special color-processing circuits. These are shown by the dashed lines on Fig. 2. In the United States, when a color signal is being transmitted, an equivalent monochrome signal also is being transmitted, necessary so that color and black-and-white systems will be compatible. Basically, color transmission involves the use of special color television circuits to superimpose the color signal onto the already-existing monochrome signal. The audio system in a television transmitter is fully separate from the video section. It consists of a microphone, audio amplifiers, and a standard frequency-modulation (FM) transmitter. Both signals are transmitted from the same antenna by use of a special coupling device known as a *diplexer*.

In Fig. 2, the television receiver is depicted in the bottom portion of the diagram. There are four sections common to all receivers—an *antenna* system, a method of *selection*, a method of *detection*, and a device for *reproducing the intelligence*. For television receivers, the antenna is considerably more elaborate than those for radio receivers. This is the case because the signal is affected much more by reflections from large objects and also because the signal has a more limited distance over which it can travel. However, instead of transmitting the television signal by means of electromagnetic waves, it may be transmitted along cables in what is known as *closed-circuit television*. The signal is transmitted from the camera to one or many receivers along a coaxial cable. Such signals cannot be received by the general public. However, in *cable television* (cablevision), television signals can be transmitted over coaxial cables from stations that are normally out of range of the viewer's receiver. Cablevision permits the viewer to select from a larger number of programs. Subscribers usually pay for the service on a monthly basis. In areas where television may be difficult to receive because of interference by hills and mountains, for example, a *master antenna system* may be used. The master receiving antenna is located on top of the hill, for example, where reception is excellent. From this point, the signal is distributed by coaxial cable to

subscribers. A system of this type usually is referred to as *community antenna television*.

The selection portion of the television receiver is accomplished in the tuner, as shown in Fig. 2. It makes possible the selection of one channel and rejection of all others. Contemporary television receivers use superheterodyne types, so the tuner section includes a local oscillator for converting the RF signal to an intermediate frequency. The output of the tuner is delivered to the intermediate-frequency (IF) amplifiers and then to a detector. Detection involves two steps. The video signal is amplitude-modulated and, therefore, an AM detector is needed for that part of the signal. The sound portion is frequency-modulated, so obviously an FM detector is needed for that portion of the signal. The picture and sound signals are amplified by the video and audio amplifiers, respectively. Video signals are reproduced in the picture tube; sound signals in the receiver speaker.

The previously mentioned synchronizing pulses (sync pulses) required to synchronize the received picture with the one that is transmitted, are used to control the frequency of the deflection circuit currents. These currents cause the electron beam to scan the picture-tube screen.

If a color signal is present, special color circuit demodulators are present as shown by the dashed lines of Fig. 2. There is an automatic circuit in a color television receiver that prevents the color demodulators from operating when there is no color signal present. This is accomplished automatically. The transmitter may switch back and forth between color pictures and monochrome pictures, and the receiver control circuits will automatically determine which circuitry will operate to produce the monochrome or color picture.

Television signals are transmitted in the range of frequencies between 54 and 890 MHz. This range includes frequencies in the very high frequencies (VHF) and in the ultra-high frequencies (UHF). Unlike the lower frequencies used for AM broadcast radio, these frequencies are not reflected from the ionosphere. Instead, their transmission is limited to line-of-sight distances, which means a distance of about 45 miles from the transmitter. The value of 45 miles is an average because the terrain between the transmitter and the receiver antennas, and the heights of these antennas, must be taken into consideration. Transmission across wide spaces, such as the oceans, can be achieved by using active satellites. For transmitting signals coast-to-coast, microwave relay stations are used. These are simply repeater stations that receive the signal, amplify it, and the retransmit it. The relay stations are located at intervals of 50 to 100 miles (80 to 160 km). The use of fiber optics is also increasing. See also **Optical Fiber Systems**.

The television signal as transmitted must comply with strict standards established by governmental regulatory agencies (FCC in the United States). Each VHF and UHF television station is assigned a channel that is 6 MHz wide. The complete signal, called the composite signal, must be fitted into this bandwidth. The composite signal includes the video carrier, one vestigial sideband for the video signal, one complete sideband for the video signal, the color signals, the synchronizing pulses, and the FM sound signal.

It will be recalled that the video part of the composite signal is amplitude-modulated. In order to obtain the same number of frequencies for the video signal, using FM, the required bandwidth would be prohibitive. The synchronizing pulses are also transmitted by AM, and these pulses are superimposed onto the video signal. The color information is transmitted by slipping the color signals in between spaces in the monochrome video signals—a process called *interleaving*. There are 525 scanning lines generated for each frame, but only about 480 of these are actually used to make up the picture. Those lines that are used for producing the picture are called active lines. The lines that are not used for making the picture are generated during the time that it takes to get the beam from the bottom of the picture (at the end of the field) back to the top of the picture (at the start of a new field). The sound signal is frequency-modulated, and as with FM broadcast systems, it may also be produced by the indirect FM method. This actually produces phase-modulation (PM), but any FM detector that can demodulate FM signals can also demodulate a PM signal. The maximum deviation for television is plus or minus 25 kHz.

It should be mentioned once again that, with the advent of HDTV, channel widths assigned by the FCC and other communications regulatory agencies will have to be altered—because HDTV requires the transmission of considerably more information than is required by current TV transmission.

Development of TV Cameras. In order to assure high-quality images at the TV receiver, the camera system must resolve the scene into as many picture elements as may be practical. The quality of the reproduced picture increases with the number of picture elements. The scanning beam in the pickup tube must produce electrical signals that accurately represent each of the picture elements. Thus, the optical-electrical conversion must have a signal-to-noise ratio sufficiently high to assure effective pickup sensitivity at the lowest-light-level scenes that may be telecast. When no incident light is present, there must be very little or, ideally, no output signal.

Even though color TV cameras are the most widely used, for simplicity the diagram of a typical monochrome TV camera is given in Fig. 3. Color cameras are described later. In the early years of television, vacuum tubes were used in the electronic circuits of the camera. A major problem was the production of heat by the vacuum tubes, this requiring that the electronic circuits had to be realigned several times each day during regular program production. Tubes also contributed to the bulk and weight of camera handling. Transistors, integrated circuits, and modular construction have made remarkable improvements in the design and use of TV cameras. Temperature-compensated deflection circuits, the use of feedback, and improved regulated power supplies have resulted in at least a hundredfold improvement in camera stability. Because of reduction in bulk, it no longer is unusual for one person to operate several TV cameras.

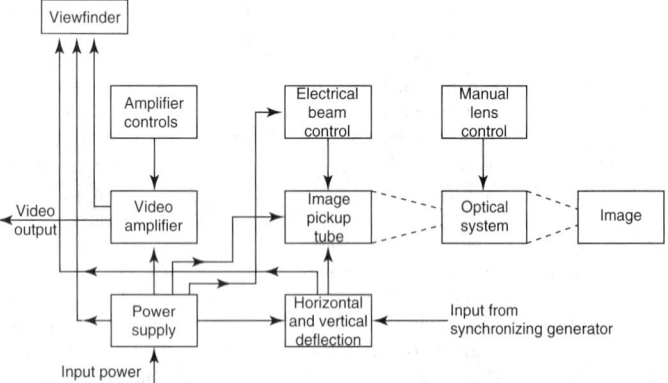

Fig. 3. Monochrome television camera. Light reflected by the scene is focused by an optical system onto the faceplate of the camera tube. By way of a photoelectric process, the light image is transformed into a virtual electronic replica in which each picture element is represented by a voltage. A scanning beam in the pickup tube next converts the picture (element by element) into electrical impulses. At the output, an electrical sequence develops, representing the original scene. Camera tube output is then amplified to provide the video signal for the transmitter. A sample of the video signal also is provided for observation in a cathode-ray tube (CRT) viewfinder mounted on the camera housing. Electronic circuits that provide the necessary control, synchronization, and power supply voltages operate the TV camera tube. A deflection system in the TV camera tube controls the movement of the camera tube scanning beam. In some situations, the cameras receive their synchronizing pulses from a studio-controlled unit. Such a unit also provides the synchronizing pulses that keep the receiver in step with the camera. Some TC cameras generate their own control signals. Manual controls are provided on the camera for setting the optical lens and for zooming.

When color television was introduced, a number of far reaching changes in TV camera technology were required. Some of the color-associated specifications which the color TV camera must meet include:

1. *Light-Transfer Characteristic.* This may be defined as the ratio of the faceplate illumination (in footcandles) to the output signal current (nanoamperes, nA). This is sometimes regarded as a measure of the "efficiency" of a camera tube. Typical values of output current range from 200 to 400 nA.
2. *Gamma Value.* This characteristic applies both to camera tubes and picture tubes and is a number that expresses the compression or expansion of original light values. Variations of gamma are inherent in the operation of some camera or picture tubes. With most camera tubes, the gamma value is unity — a linear characteristic that does not change the light values from the original scene when they are translated into electronic impulses. (Because of the desire for better

contrast in a picture tube, a gamma value up to 3, depending upon tube manufacturer, may be in the specifications

3. *Spectral Response.* As nearly as possible, the spectral response of the camera tube should match that of the human eye. This is necessary to render colors in their proper tones and also important in producing the proper gray scale in connection with black-and-white pictures. Color TV cameras have a greater response to each of the primary colors. Spectral response distribution has made it possible to manufacture camera tubes that are also sensitive to infrared (IR), ultraviolet (UV), and even X-radiation for special requirements.[3] Variations in spectral response have little effect on the other operating characteristics of the camera tube
4. *Lag.* This term applies to the time lag during which the image on a camera tube decays to an unnoticeable value. Camera tubes inherently have a tendency to retain images for short periods after the image is removed. Lag is detected on a television picture in the form of smears (comet trails), which appear following rapidly moving objects. Lag can be expressed as a percentage of the initial value of the signal current remaining 1/20 second after the illumination is removed. Typical lag values for vidicons range from 1.5% to 5%. Other types of camera tubes, such as the "Plumbicon," have lag values as low as 1.5%.
5. *Dark Current.* Characteristic of camera tubes as well as other photoelectric devices is the current that flows through the device even in total darkness. Obviously, for a camera tube, the dark current must be as low as achievable.

Color TV cameras are of two principal types: (1) a single (special) camera tube, and (2) three camera tubes. The single-tube configuration incorporates a vertical stripe filter that separates the colors prior to reaching the scanning beam. In the three-tube arrangement, separate tubes are used for each of the three primary colors (red, green, blue). A color filter system separates the incoming light from the image into the three colors, after which the light is focused on the faceplate of the appropriate camera tube. The three tubes are identical with exception of the photosensitive material used in each tube. The outputs from each tube are then amplified to provide three channels of video signals, one channel for each color.

Evolution of the Vidicon. The vidicon is a comparatively simple and compact device. A typical tube may be about 1/2 in. (~11/4 cm) in diameter and some 31/2 in. (~9 cm) long. Tubes of larger dimensions are not uncommon. Vidicons are widely used for closed-circuit television (CCTV) and for TV studio and film cameras. It is interesting to note that some vidicons can produce acceptable pictures when operating in near-total darkness or, in contrast, in near-direct sunlight.

The operating principle of the vidicon camera tube is shown in Fig. 4. The target consists of a transparent conducting film (the signal electrode) on the inner surface of the faceplate and a thin photoconductive layer

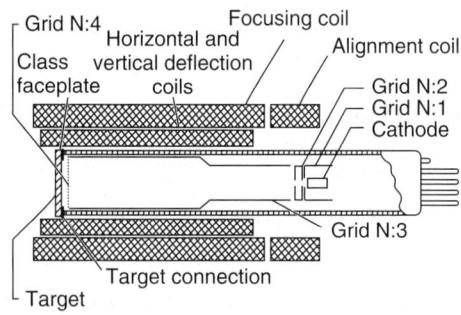

Fig. 4. Sectional view of internal construction of a vidicon.

[3] If the photosensitive material in a camera tube was able to emit an electron for each photon of light focused upon the material, the quantum efficiency of the material would be 100%, as shown by Q_{eff} = Electrons/Photons. Although not achievable, Q_{eff} is a practical way to compare photosensitive surfaces in the camera tube. In this comparison, photocurrent per lumen is measured using a standard light source. The source usually used for the measurement is a tungsten-filament light operating at a color temperature of 2870 K. Inasmuch as the lumen is a measure of brightness stimulation to the human eye, Q_{eff} is a useful way to express the sensitivity of the image-pickup tube.

deposited on the film. Each cross-sectional element of the photoconductive layer is an insulator in the dark, but becomes slightly conductive when it is illuminated. Such an element acts like a leaky capacitor, having one plate at the positive potential of the signal electrode and the other one floating. When light from the scene being televised is focused onto the surface of the photoconductive layer next to the faceplate, each illuminated element conducts slightly, the current depending upon the amount of light reaching the element. This causes the potential of its opposite surface (the gun side) to rise toward the signal electrode potential. Hence, there appears on the gun side of the entire layer surface a positive-potential replica of the scene composed of various element potentials corresponding to the pattern of light which is focused onto the photoconductive layer. When the gun side of the photoconductive layer, with its positive-potential replica, is scanned by the electron beam, electrons are deposited from the beam until the surface potential is reduced to that of the cathode in the gun. This action produces a change in the difference of potential between the two surfaces of the element being scanned. When the two surfaces of the element, which in effect form a charged capacitor, are connected through the external target (signal electrode) circuit and a scanning beam, a current is produced which constitutes the video signal. The amount of current flow is proportional to the surface potential of the element being scanned and to the rate of the scan. The video signal current is then used to develop a signal-output voltage across the load resistor. The signal polarity is such that for highlights in the image, the input to the first video amplifier swings in the negative direction. In the interval between scans, wherever the photoconductive layer is exposed to light, the migration of the charge through the layer causes its surface potential to rise toward that of the signal plate. On the next scan, sufficient electrons are deposited by the beam to return the surface to the cathode potential.

The electron gun contains a cathode, a control grid, and an accelerating grid. The beam is focused on the surface of the photoconductive layer by the combined action of the uniform magnetic field of an external coil and the electrostatic field of a third grid. A fourth grid serves to provide a uniform decelerating field between itself and the photoconductive layer, so that the electron beam will tend to approach the layer in a direction perpendicular to it, a condition that is necessary for driving the surface to the cathode potential. The beam-electrons approach the layer at a low velocity because of the low operating voltage of the signal electrode. Deflection of the beam across the photoconductive substance is obtained by external coils placed within the focusing field.

Evolution of the Plumbicon. In this tube, developed by Philips of Holland, the principles of the vidicon were generally applied. The main difference is in the target.

The inner surface of the glass faceplate is coated with a thin transparent conductive layer of tin oxide. This layer forms the signal plate of the target. On the scanning side of the signal plate, a photoconductive layer of lead monoxide is deposited. These layers are specially prepared to function as three sublayers, each with a different conduction mode. On the inner side of the faceplate, the tin oxide layer is a strong n-type semiconductor, commonly found in transistors. Next to this n-type region is a layer consisting of almost pure lead monoxide, which is an intrinsic semiconductor. The scanning side of the lead monoxide has been doped to form a p-type semiconductor. Together, these three layers form a p-i-n junction. The spectral response of the plumbicon can be varied during its manufacture to suit almost any particular application. The tube has gained wide acceptance in color television cameras.

Solid-State TV Cameras. These devices utilize charge-coupled devices (CCDs), which were discovered by scientists at the AT&T Bell Laboratories and first found keen usage in military applications and later in detectors for large telescopes. The devices were incorporated in solid-state TV cameras as probably the most forward step taken in recent years toward upgrading the cameras. Operating principles of CCDs are described in the article Charge-Coupled Devices.

When light from an image or scene is focused on a CCD, a pattern of electrical charges is created. The charges vary in proportion to the amount of light and serve as an accurate electrical representation of picture elements. These charges can be stored, transmitted out of the CCD chip sequentially, and later reassembled on a conventional TV screen or facsimile readout. Thus, the CCD is the basis for an all-solid-state TV camera, greatly simplifying camera design and lessening camera bulk and weight. Other advantages of CCD cameras include long life, low power consumption, and very significant size reduction. No warm-up time is

required, and, importantly, the CCD imaging device is not susceptible to lag (smearing caused by bright moving areas) or to burn-in (damage that can be caused by intense light or electron-beam bombardment). The CCD camera is more sensitive, an important advantage for acquiring pictures at low light levels. At an early stage of development, the first black-and-white camera for commercial use was demonstrated as early as 1971.

Development of TV Picture Tubes. The many decades required to develop a useful technique for converting electrical signals into light were described earlier in this article. In the interest of simplicity, a monochrome picture tube will be described first. This is a specialized form of cathode-ray tube.

An electron gun in the tube directs a beam of electrons toward a fluorescent material on the screen, which glows when struck by the electrons. Between the gun and the screen are deflection coils which deflect the beam horizontally and vertically to form a raster. The brightness of the screen at any point depends upon the number and velocity of electrons striking that point. Therefore, the brightness of the picture may be controlled by varying the grid-bias voltage with respect to the cathode voltage. See Fig. 5. A color-picture tube operates on the same basis principle, except that the screen is coated with different types of phosphors which produce colors when struck by electron beams. Combinations of red, green, and blue produce all colors. In a three-gun color picture tube there is a separate gun for each of the color phosphors. Color phosphor dots or strips are located so close together on the screen of the picture tube that the eye cannot distinguish between them. When all three adjacent colors are stimulated, the screen will radiate a white light. The various mixture colors are obtained by controlling the strength of the individual electron beams striking each color phosphor. In order to insure that each electron beam strikes only one particular color phosphor dot or strip, a shadow mask, or aperture mask, is located near the screen. It is very important, of course, that each electron beam be carefully directed to pass through the desired hole in the mask and thus strike only the desired color phosphor. See Fig. 6.

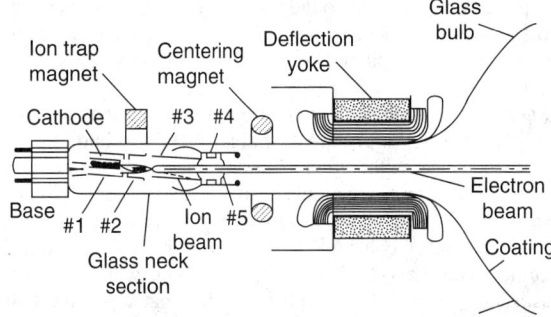

Fig. 5. Internal structure of an electrostatic focus television tube. The deflection is achieved magnetically.

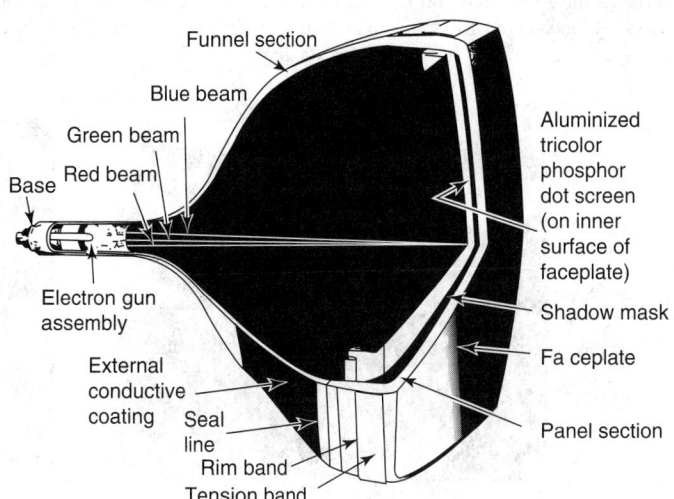

Fig. 6. Cutaway view of delta-gun color picture tube that uses color phosphor dots.

High-Definition Television (HDTV)

Efforts to improve the clarity of television presentations have been continuous since the first experimental transmissions over telephone wires and through the atmosphere. Black and white television gradually became color television, but the process was slowed by bitter competition between incompatible technologies and resistance by broadcasters, who had to invest in expensive new equipment. In the United States, commercial ownership of radio and TV stations as well as their programming is customary, although the Federal Communications Commission (FCC) has certain regulatory powers over them. In addition, in most nations the government (until recently) owned and operated the radio and television stations, providing the programming as well. Individual national television broadcasting facilities were often based on proprietary technology that prevented citizens from receiving broadcasts emanating from neighboring countries. These disparate national arrangements affected the introduction of new technologies.

The resolution of a television image is based on the number of horizontal lines and the number of picture elements (*pixels*) across each line, as well as the method of scanning the lines across the face of the picture tube. The conventional television broadcasting standards, known as *NTSC* (*National Television Systems Committee*) in the U.S., were established in 1941 and tailored to the technological capabilities of those days, which were based on vacuum tubes. In the U.S., the scanning rate was set at 525 horizontal lines, of which 480 are intended for display on the face of the "*picture tube*." Actually, most TV sets for decades displayed 300 lines or less, depending on the accuracy of the manufacturer's settings, which meant part of the image was cut off and part of the resolution was lost.

The technology still uses *interlaced scanning*, in which every second line is bypassed during the first scan, which traces 240 lines, after which the scanner returns to the top of the tube face and traces the next 240 lines between the previous tracing, thus forming a complete image with two scanning passes. This results in some disparity between some of the image's elements, creating "*jaggies*" and other artifacts in the image. Interlacing is performed on each of 30 picture frames per second (the standard projection speed for sound motion pictures). The shape of the TV image (its aspect ratio or ratio of width versus height) also was patterned after that of the motion picture industry's projection screens: 4 by 3. See Fig. 7.

The *interlaced scanning* used in television was rejected by the computer industry, which uses progressive scanning. In this technology, each line is traced in sequence, without skipping any lines, until the entire frame is scanned. The effect is a sharper image, so it is favored for the improvement of television clarity. However, it is more expensive to produce.

The introduction of color TV broadcasting in 1953 was based on its compatibility with the NTSC standards. England, which had set its TV broadcast standards in 1941, accepted a quality that was somewhat poorer than that of the U.S. (405 lines). However, such European nations as France and Germany waited until a few years after World War II, and thus established TV systems—now known as *PAL/SECAM*—with superior resolution (i.e., more scan lines per inch or per centimeter) compared to those of the Americans and British. However, all these standards remained virtually unchanged for more than four decades (except that the British changed to the NTSC standard color standards in 1973).

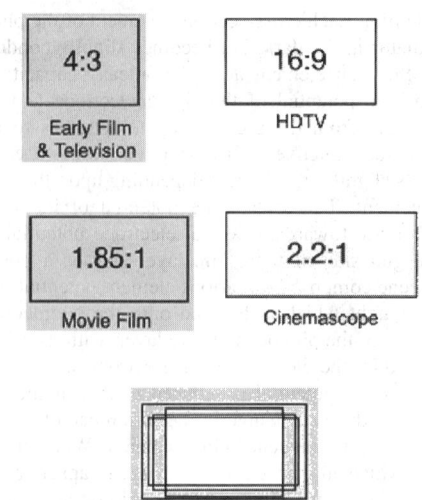

Fig. 7.   This figure represents a series of concentric rectangles, each representing a different set of screen dimensions widely in use.

Again, it was the motion picture industry that triggered the moves toward a new screen format and better clarity. The introduction of various wide-screen formats in the 1950s changed the traditional projection screen aspect ratio from the virtually-square 4:3 format to the clearly horizontal 16:9 and even wider formats. See Fig. 8. In addition, the motion picture studios converted from optical sound tracks to electronic sound tracks and added stereo sound to their films, improving the audio quality. The television industry, in turn, improved their image quality by replacing vacuum tube circuits with transistors and other solid-state electronic components, and developing the "simulcast," in which stereophonic audio was broadcast through an FM radio station in synchronization with the TV image broadcast; the viewer listened to the audio from the stereo radio while watching the image on the TV set. This was necessary because the NTSC standards were for monaural audio only.

In the early 1980s, Japanese television specialists introduced a new type of TV standard, which the industry called *high-definition television* (*HDTV*). The first true *HDTV* receivers and broadcasts were made by NHK in Japan in the early 1980s, using analog signals that required enormous bandwidth for each channel. The system was known as "*MUSE*" and operated with 1,125 horizontal scan lines. To the consternation of American television interests, the Japanese proposed worldwide adoption of the *MUSE* technology. However, their *HDTV* programming was limited to experimental broadcasts for demonstration purposes and no sets were available for purchase.

In 1987, the FCC undertook an Inquiry, i.e., a study to determine the effect of the development of *HDTV* on the existing broadcasting service. Most broadcasters were not enthusiastic about the new technology, since it would require expensive retooling of their broadcast facilities and could result in the loss of some of their assigned frequency bandwidth to meet the high capacity required for *HDTV* transmission.

Fig. 8.   Compared to standard television (NTSC), the HDTV image has twice the luminance definition—vertically and horizontally—and is twenty-five percent wider. Standard television aspect ratio is 4:3 (four units wide, three units high)—the HD aspect ratio is 16:9.

Meanwhile, the American manufacturers and broadcasters, as well as the FCC, were exploring many different options, all of which used analog signal transmission. But in 1990, the General Instrument Company—functioning as a research and development arm of the cable-television industry—developed a digital encoding method. This was preferable to analog because it was much less susceptible to distortions caused by atmospheric conditions and electrical interference. Later, new all-digital systems were announced by Zenith and AT&T, the Philips-Thomson-Sarnoff consortium, and the Massachusetts Institute of Technology (MIT).

General Instrument's method triggered a revolution in the American planning of *HDTV* systems, resulting in the discarding of analog transmission but development of a number of disparate digital approaches that culminated in a so-called "*Grand Alliance*" between the various interested—but often warring—parties [General Instrument, Zenith Electronics, Lucent Technologies (Bell Laboratories), MIT, Philips Electronics North America Corporation, Sarnoff Corporation and Thomson Consumer Electronics].

Defenders of the NTSC standards redesigned some analog receivers and called them "*Improved Definition Television* " or *IDT*, which presented significantly sharper images without requiring changes in broadcasting. Those sets cost several thousand dollars each, significantly more than the standard TV sets of the mid-1980s and did not become a serious factor in the marketplace, perhaps because of the growth of cable TV systems, in which subscribers usually saw an improvement in image quality compared to roof-antenna reception.

Promoters of the *HDTV* concept point out that the public can appreciate the new clarity of this technology only after seeing it on a large screen (*HDTV* sets are now often twice as large as modern *NTSC* receivers) or viewing it at closer range. Early viewing tests in 1987 compared the images of *HDTV* and *NTSC* on side-by-side similar-size receivers (a 26-inch high-end *NTSC* receiver and a 28-inch *HDTV* receiver) but produced discouraging reports: beyond a distance of five picture heights, viewers were unable to tell the difference (except for the different shapes of the picture tubes). See Fig. 9. The most favorable responses came from viewers closer to the *HDTV* receiver (three picture heights), and the most negative from those people seven picture heights away. The first integrated *digital HDTV* receiver offered in late 1998 by Zenith Electronics is a 64-inch diagonal widescreen rear-projection model, providing what the company calls "five times more picture than today's analog TV." It also includes digital audio comparable to the quality of compact-disc audio. It is expected to sell for more than $10,000, plus ca. $6,000 for an *HDTV* receiver/decoder. The Zenith decoder is designed to receive all 18 scanning formats included in the new *Digital TV* (*DTV*) standards accepted by the FCC, although 12 of those are not designated as *HDTV* quality.

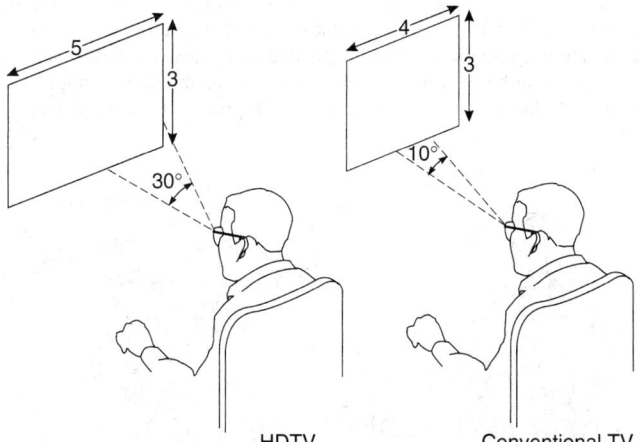

Fig. 9.   Approximate viewing difference between conventional TV and HDTV. The HDTV screen is about 25% wider than that of a traditional set and offers double the resolution. The result is a three-fold increase in the viewing angle from 10° to 30°.

As of 1998, *high-definition television* (*HDTV*) was one of several formats that have been proposed for the improvement of television signals and recordings. An international group, the *Advanced Television Systems Committee* (*ATSC*), was formed in 1982 to establish voluntary technical standards for *HDTV* and *SDTV* (standard-definition television), which were released in February 1998. That report states:

"There are six video formats in the *ATSC DTV* (digital television) standard which are '*High Definition Television*.' They are the 1080-line by 1920-pixel formats at all picture rates (24, 30 and 60 pictures per second) and the 720-line by 1280-pixel formats at these same picture rates. All of these formats have a 16:9 [displayed image] aspect ratio.

The remaining twelve video formats, while representing some significant improvements over analog *NTSC* (*National Television Systems Committee* standards established in 1941 for black and white and the 1950s for color TV) are not *High Definition Television*. They are referred to as '*Standard Definition Television*.' These are the 480-line by 704-pixel formats in 16:9 widescreen and 4:3 aspect ratios, at the picture rates listed above, and the 480-line by 640-pixel format at a 4:3 aspect ratio, at the same picture rates.

These definitions are restatements of terms clearly established and supported by the written record of the ten-year process of the FCC, the FCC Advisory Committee on Advanced Television Service (ACATS), and the ATSC. They also support the industry definitions for digital television receivers established in January, 1998 by the *Consumer Electronics Manufacturers Association* (*CEMA*). These definitions are also fully supported by the technical specifications for the various formats as measured against the internationally accepted definition of *HDTV* established in 1989 by the International Telecommunications Union, and the definitions cited by the FCC during the DTV standards development process."

The Advanced Television Systems Committee is composed of 136 member corporations, industry associations, standards bodies, research laboratories, and educational institutions. Its URL on the World Wide Web is: *http://www.atsc.org*. The Consumer Electronics Manufacturers Association provides simplified explanations of the technologies and advantages of *HDTV* and *SDTV* on their URL: *http://www.cemacity.org/digital*.

Every digital TV set will be able to receive any of four main formats: 1080-I (1080 horizontal lines, interlaced scan); 1080-P (1080 horizontal lines, progressive scan—the highest resolution format); 720-P (720 horizontal lines, progressive scan, which some experts contend looks at least as good as 1080-I); 480-P (480 horizontal lines, which look clearer and sharper than today's analog programs that are broadcast using 480-I, a format with the same number of lines, but interlaced).

However, according to a participant, the Grand Alliance system is flawed because it resulted in numerous formats, no provision for inexpensive receivers or set-top converters, and includes image interlace scanning technology as well as progressive scanning technology. Part of the problem results from the broadcasters wanting to maintain their NTSC standards as long as possible by requiring that HDTV be "*backward-compatible* " with NTSC. The FCC was caught in the middle between the warring factions, and—unable to persuade them to reconcile their differences—decided to adopt the DTV standard of the ATSC, which was based on the Grand Alliance recommendations. The broadcasters promised that each of the top ten TV markets will have at least one digital station by the end of 1998. They are required to transition to digital in the top ten markets (reaching half of the U.S. TV households) by November 1999. Some networks say they will broadcast only the *SDTV* signals, while others plan to transmit *SDTV* programs part of the time, and *HDTV* the rest of the time. Since no one knows what the public acceptance of *HDTV* will be in the future, the networks may change their minds later.

When *digital HDTV* was first developed, the signal was extremely bandwidth-hungry. Each color pixel requires 16 bits (binary digits) of information, and there are 1920 pixels in each scan line, times 1,080 scan lines per frame, and 30 frames per second. The result is 995,328,000 bits (995 megabits) per second of video information. In addition, the signal contains 48,000 audio samples per second, with 16 bits per sample—and eight audio channels for surround plus a second language, or a total of 6,144,000 bits (six megabits) per second of audio information. Adding in the ancillary services (closed captioning, program guides, system timing and synchronization, etc., totaling 256,000 bits per second), the entire combined video/audio source signal comes to more than a billion bits (a gigabit) per second—requiring XX megahertz of broadcast frequency spectrum bandwidth. The only way to squeeze all that information into the six-megahertz bandwidth assigned to each channel by the FCC was to use a process called *data compression*.

The video compression technique used in *HDTV* employs the MPEG-2 method. (MPEG refers to the Motion Picture Experts Group of the

International Standards Organization.) The 995 megabits of video is compressed to 18 megabits (55:1 compression). The audio information is compressed using a method proposed by Dolby Labs. Six of the source channels are compressed together as a "surround-sound" package, and the other two are compressed as a stereo pair for second-language use. The six megabits per second of audio information is compressed to 512 kilobits per second (12:1 compression). The ancillary services are typically not compressed. The total payload is thus reduced from more than a billion bits (a gigabit) per second to less than 19 megabits per second. The new six-megahertz digital signal will reach slightly more consumers from a typical transmitter site, and do so with less transmitted power. The picture will be digitally perfect throughout the coverage area, regardless of how far away the receiver is from the broadcast transmitter tower.

A standard NTSC analog broadcast TV channel requires a bandwidth of six megahertz. Each television network was assigned a second six-megahertz channel by the FCC in 1997 for the transition to digital broadcasting, but that same channel can be used either as an *HDTV* channel or to transmit four separate digital channels with ordinary picture quality (*SDTV*). In addition to broadcaster reluctance to invest in *HDTV* equipment, the broadcasters also realize that they can make extra profits in advertising over four program channels in digital technology instead of one *HDTV* channel. However, the entire digital feed of the network will be assigned a single set of call letters and channel positions followed by the letters "DT" (for Digital Television), i.e., WCBS-DT, Channel 2 in New York, even though the actual channel number for the digital broadcasts may be much higher, say Channel 65 or Channel 90. The viewers will not have to know that, since the receivers (under the ATSC scheme) will be programmed to automatically find the correct digital frequency assignment for the selected network. If a network elects to broadcast four different programs within the single six megahertz channel, a system of sub-numbers will have to be devised—but the FCC and the ATSC did not specify how that would work, thus leaving it up to the manufacturers of the *HDTV/digital receivers*.

The proliferation of digital programming channels could create problems among viewers trying to plan a viewing schedule, since the eventual program listings could be too long to print on paper. The plan in new digital sets is to display an on-screen electronic program grid covering nine hours for each channel, encoding each digital transmission with the program's name, length and start time.

Meanwhile, however, the current analog NTSC broadcasting system will continue for a transition period of at least eight years, ending in 2006. During this interim period, *digital SDTV* and *HDTV* will share the broadcast spectrum with the NTSC programming, but the latter will eventually be eliminated. Manufacturers of TV sets were understandably concerned about losing sales of standard NTSC receivers in the interim years, since consumers are confused about the introduction of *HDTV* and *digital TV*.

The availability of *HDTV* programming depends on many factors. Like the introduction of color programming in the 1950s and 1960s, programs in the *HDTV* formats will be sparse at the start, due to a lack of suitable material and the need by local TV stations to install appropriate equipment. Live broadcasts of sporting events and similar programming are the most likely candidates for the first year or two. As of 1999, *HDTV* programming will be available only through atmospheric transmission received by an antenna. If the signal is too weak, the receiver will simply not display any picture; with *HDTV*, it is all or nothing at all. (Experiments in Washington, DC. showed that indoor antennas received *HDTV* broadcasts for less than 60 percent of the test participants, but the percentage was much higher for those using rooftop antennas.)

Satellite owners also are interested in proving their ability to transmit *HDTV*, so in July 1998, the *PanAmSat* Corporation, owners and operators of more than 17 geostationary satellites [see also **Satellites (Communication and Navigation)** entry], sponsored a series of tests during a two-day demonstration involving vendors of *HDTV* equipment Using the same high-definition source material (delivered at 1.5 gigabits per second), each vendor used its own system to encode the *HDTV* signal into a post-production quality video signal at 45 megabits per second or distribution-quality video signal at 19.3 megabits per second. The encoded material was then uplinked to a geostationary satellite (*PAS-2*) in the 1080-I high-definition format in compliance with MPEG-2/DVB, and the satellite returned each signal to the teleport at Napa, California, where it was demodulated and decoded into *HDTV* format by the respective vendor's

integrated receiver/decoder for display on an *HDTV* monitor and spectrum analyzer.

A key question is how the cable television companies will cope with the new technology. The digital set-top converter boxes now offered by some cable TV companies are used with analog TV receivers and were not designed for connection to digital TV sets, so a special digital cable will be required—and nobody knew in 1998 what the specifications would be for that cable. As a result, none of the first generation *HDTVs* have the proper jacks to accept *HDTV* from a cable box.

But these problems are not expected to block the introduction of digital television and *HDTV* broadcasting. The new television sets are now being characterized as "*information appliances*" instead of simply being an entertainment medium.

Flat Display Panels

As of the early 1990s, most authorities in the field sensed that the traditional CRT, which requires considerable depth and hence an awkward dimension for electronic display devices, such as television sets, ultimately will be replaced by flat (picture-frame) displays that can be wall-mounted and carried in a briefcase. Technological development of the flat panel follows an all-too-familiar pattern of early technical leadership, as exemplified in this instance by such firms as Westinghouse and RCA, only to go essentially unattended for a number of years and later to be acted on by progressive firms in other countries.

Fundamentally, in any electronic device, signals are converted to points of light that form patterns and images. For example, in some laboratory and desk-top computers that are finding increasing use, rows and columns of electrodes are used to convey information. Thus, a grid of pixels (picture elements) is formed. At a given row-column intersection, the display (light-producing element) will be activated only when there is a voltage present in the given row and given column. This *multiplexing* technique is inferior by comparison with a CRT display. The problem of "fuzziness" is exacerbated as the number of pixels increases. The voltage that passes through the numerous display elements tends to permeate the element such that pixel displays are not fully turned off. In a busy display, black and white take on indistinguishable shades of gray. Also, multiplexing is unfit for displaying object movement on the screen. By their fundamental nature, a liquid crystal display (LCD) element responds in proportion to the voltage applied. In a technique invented by T. Peter Brody in the early 1980s, the foregoing problems can be overcome by way of what is called an *active matrix*.

In this technique, a thin-film transistor (TFT) acts as a switch at each pixel. As pointed out by Herb Brody, "The TFT can transfer and store enough voltage to quickly switch a liquid-crystal pixel from light to dark, resulting in a sharp image with no blurring. With the TFT active matrix, the electric charge used to switch one pixel no longer spills over into neighboring ones. As a result, the contrast between an on and off pixel can proach 50 or 100 to one—versus six to one in a typical multiplexed LCD. And because the TFTs are deposited on a transparent glass substrate, the display can be lit from behind, further enhancing its viewability. Red, green, and blue filters can be placed at each pixel to form a color display.

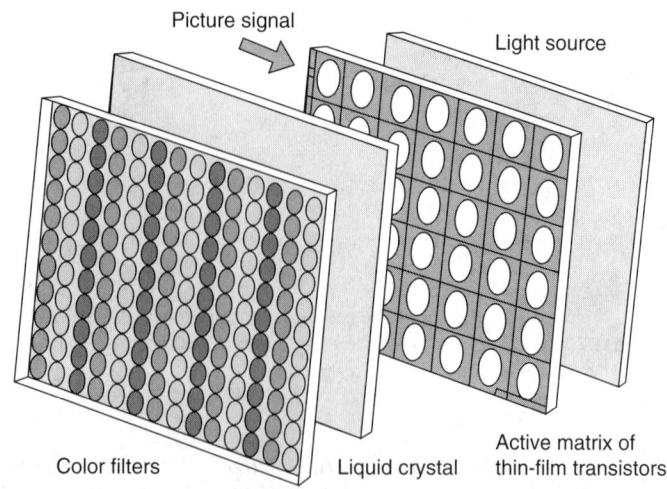

Fig. 10.   Principle of active matrix display technology. (*After Florida and Brody*.)

Because of its high contrast capability, an active matrix display can produce full color images that look as good as or better than those produced by CRTs." See Fig. 10. See also the family of articles catalogued under **Flat Panel Display**.

Additional Reading

Brinkley, J.: "Defining Vision: The Battle for the Future of Television," Harcourt, San Diego, CA, 1997.

Brody, H.: "What Is an Active Matrix Design," *Technology Review (MIT)*, 45 (August/September 1991).

Browne, N.: "American Television New Directions in History and Theory: New Directions in History and Theory," Vol. 2, Gordon & Breach Publishing Group, Newark, NJ, 1994.

Buscombe, C.G.: "Television and Video Systems: Operation, Maintenance, Troubleshooting and Repair," 2nd Edition, Prentice-Hall, Inc., Upper Saddle River, NJ, 1998.

Corcoran, E.: "A Technological Fix," *Sci. Amer.*, 60 (August 1989).

Corcoran, E.: "Picture Perfect," *Sci. Amer.*, 94 (February 1992).

Donahue, H.C.: "Choosing TV of the Future," *Technology Review (MIT)*, 30 (April 1989).

Florida, R. and D. Browdy: "The Invention that Got Away," *Technology Review (MIT)*, 41 (August/September 1991).

Grob, B., E. Herndon, and C. Herndon: "Basic Television and Video Systems," 6th Edition, The McGraw-Hill Companies, Inc., New York, NY, 1998.

Hartwig, R.L.: "Basic TV Technology: Digital and Analog," 3rd Edition, Butterworth-Heinemann, Inc., Woburn, MA, 2000.

Norman, C.: "HDTV: The Technology du Jour," *Science*, 761 (May 19, 1989).

Norman, C.: "Without Standards, U.S. HDTV Lacks Definition," *Science*, 762 (May 19, 1989).

Pool, R.: "Setting a New Standard (HDTV)," *Science*, 29 (October 7, 1988).

Pool, R.: "A Chance to Retake TV Markets?" *Science*, 30 (October 7, 1988).

Roberts, G.G.: "The Bridge of Technology," University of Wales J., 57 (Spring 1989).

Roush, W.: "Television and Industrial Policy," *Technology Review (MIT)*, 62 (November/December 1990).

Schreiber, W.F.: "Television and the Economy," *Technology Review (MIT)*, 35 (April 1989).

Stix, G.: "Don't Change the Channel, Rearrange that Face (Multimedia Systems)," *Sci. Amer.*, 120 (October 1991.)

Stoddard, A.G. and M.D. Dibner: "Europe's HDTV: Tuning Out Japan," *Technology Review (MIT)*, 39 (April 1989).

Watkinson, J.: "Television Fundamentals," Butterworth-Heinemann, Inc., Woburn, MA, 1996.

Weiss, S.M.: "Issues in Advanced Television Technology," Butterworth-Heinemann, Inc., Woburn, MA, 1996.

Winner, L.: "Who Needs HDTV?" *Technology Review (MIT)*, 20 (May/June 1989).

Web References

HDTV Industry Links: *http://ourworld.compuserve.com/homepages/calcote/hdlinks.htm http://www.zenith.com/main/about/aspect_ratio.html*

TELLURIC LINES. Absorption lines in a solar spectrum produced by constituents of the atmosphere of the Earth itself rather than by gases in the outer solar atmosphere such as those responsible for the Fraunhofer lines. The terrestrial nature of the absorption processes responsible for telluric lines is revealed by their intensity variation with solar zenith angle and by their freedom from any Doppler broadening due to their solar rotation. Water vapor produces the strongest of the telluric lines in the visible spectrum. See also **Adsorption (Process)**.

TELLURIUM. Chemical element, symbol Te, at. no. 52, at. wt. 127.60, periodic table group 6, mp 450 °C, bp 690 °C, density 6.24 g/cm^3 (crystalline form at 25 °C), 6.00 (amorphous form at 25 °C). Elemental tellurium has a hexagonal crystal structure with trigonal symmetry. Tellurium is a silver-white brittle semi-metal, stable in air, and in boiling H_2O, insoluble in HCl, but dissolved by HNO_3 or aqua regia to form telluric acid. The element is dissolved by NaOH solution and combines with chlorine upon heating to form tellurium tetrachloride.

In observing a peculiar phase in gold ores of the Transylvania region, Franz Müller von Reichenstein first identified the element in 1782. There are several natural occurring isotopes ^{120}Te, ^{122}Te through ^{126}Te, ^{128}Te, and ^{130}Te. Nine radioactive isotopes have been identified ^{118}Te, ^{119}Te, ^{121}Te, ^{123}Te, ^{127}Te, ^{129}Te, and ^{131}Te through ^{133}Te. With exception of ^{123}Te which has a half-life something greater than 10^{13} years, all of the other radioactive isotopes have half-lives measurable in terms of minutes, hours, or days. In terms of abundance, tellurium does not appear on the list of the first 36 elements that occur in the Earth's crust and hence is relatively scarce. Terrestrial abundance is estimated on the order of 0.002 ppm. Tellurium is found in seawater to the estimated extent of about 95 pounds per cubic mile of seawater. First ionization potential 9.01 eV; second, 18.6 eV; third, 30.5 eV; fourth 37.7 eV; fifth, 59.95 eV. Oxidation potentials $H_2Te(aq) \rightarrow Te + 2H^+ + 2e^-$, 0.69 V; $Te + 2H_2O \rightarrow ReO_2(s) + 4H^+ + 4e^-$, -0.529 V; $TeO_2(s) + 4H_2O \rightarrow H_6TeO_6(s) + 2H^+ + 2e^-$, -1.02 V; $Te^{2-} \rightarrow Te + 2e^-$, 0.92 V; $Te + 6OH^- \rightarrow TeO_3{}^{2-} + 3H_2O + 4e^-$, 0.02 V; $Te \rightarrow Te^{4+} + 4e^-$, -0.564 V. Electronic configuration $1s^2 2s^2 2p^6 3s^2 e p^6 3d^{10} - 4s^2 4p^6 4d^{10} 5s^2 5p^4$. Other important physical properties of tellurium are given under **Chemical Elements**.

Tellurium occurs chiefly as telluride in gold, silver, copper, lead, and nickel ores in Colorado, California, Ontario, Mexico, and Peru, and infrequently as free tellurium and tellurite (tellurium dioxide, TeO_2). The anode mud from copper and lead refineries, or the flue dust from roasting telluride gold ores is treated by fusion with sodium nitrate and carbonate and the melt extracted with water. The resulting solution is acidified carefully with H_2SO_4, whereupon tellurium dioxide is precipitated, and the dioxide reduced to free tellurium by heating with carbon.

Uses: On the scale of most other commercial metals, the production of elemental tellurium is relatively limited—approximately $\frac{1}{2}$ million pounds annually. Commercial tellurium is marketed at a purity of about 99.7%, although much purer forms are obtainable—up to 99.999%. The application of tellurium and tellurium compounds as catalysts is expanding. Small quantities are used in various electronic components, including solar cells, infrared detectors, emitters, and thermoelectric generators. Tellurium also is sometimes used as a dopant for semiconductor devices. The metal has been used in primer fuses for explosives. The main applications have been in metallurgy. Small additions of tellurium improve the machinability of low-carbon steels, stainless steels, and copper. The metal stabilizes the carbide in cast irons. Tellurium also helps to control pinhole porosity in steel castings. The very small addition (0.05%) of tellurium to lead improves a number of the physical properties of lead sheet, foil, and other shapes. To some extent, tellurium has been used as a curing agent and accelerator in rubber compounds.

Chemistry and Compounds. Tellurium occurs in the same periodic classification as sulfur, selenium, and polonium. Tellurium, unlike sulfur and selenium, has only two allotropic forms. Due to its $5s^2 5p^4$ electron configuration, tellurium, like sulfur and selenium, forms many divalent compounds with covalent bonds and two lone pairs, and d-hybridization is quite common, to form compounds with tellurium oxidation states of +4 and +6.

Tellurium dioxide, TeO_2, made directly from the element or by heating tellurous acid, H_2TeO_3, is a solid, subliming at 450 °C, insoluble in H_2O, which dissolves in acids and alkalis, exemplifying the increasing metallic character with atomic weight of the main group 6 elements. Tellurium dioxide accepts a proton from strong acids to form the ion $TeOOH^+$. Dehydration of telluric acid at 400 °C produces TeO_3, tellurium trioxide. It is not nearly as reactive as sulfur trioxide and selenium trioxide, but reacts with alkali hydroxides to form tellurates.

Tellurous acid, H_2TeO_3, can exist only in very dilute aqueous solutions (due to insolubility of TeO_2). It is a weak acid (ionization constants 2×10^{-3} and 2×10^{-8}). The salts of tellurous acid, the tellurites, may often be formed by reaction of TeO_2 with metal salts. Telluric acid, H_6TeO_6, is prepared by oxidation of tellurium with strong oxidizing agents, such as 30% hydrogen peroxide or boiling HNO_3 and catalyst. Various values of the ionization constants of telluric acid have been reported, on the order of 10^{-7} and 10^{-11}, but the best values would appear to be $pK_{A1} = 7.7$, $pK_{A2} = 11.0$, $pK_{A3} = 14.5$. Telluric acid is a quite strong oxidizing agent, forming halogens from hydrohalides in solution (except hydrogen fluoride). The alkali metal tellurates have the composition $M_2H_4TeO_6$, although metal tellurates with all H's replaced exist, such as Hg_3TeO_6 and Zn_3TeO_6.

Hydrogen telluride is a stronger acid than H_2S (ionization constants 2.27×10^{-3} and 10^{-11} (?) at 18 °C) and is less readily obtained from tellurides than hydrogen sulfide from sulfides. Aluminum telluride, Al_2Te_3, requires heating with H_2O or dilute acids. In general the metal tellurides are prepared by direct combination of the elements. Those of the transition metals and the zinc, gallium and germanium families exhibit many instances of both well defined compounds and non-daltonide compositions, as well as substitutional solid solutions. Also six intermediate phases are found in the palladium-tellurium system, Pd_4Te, Pd_3Te, $Pd_{2.5}Te$, Pd_2Te, $PdTe$, and $PdTe_2$.

Tellurium hexafluoride, TeF_6, the only clearly defined tellurium hexahalide, is formed directly from the elements at 150 °C, while at 0 °C the product is mainly the decafluoride, Te_2F_{10}. TeF_6 is a relatively weak Lewis acid, forming complexes with pyridine and other nitrogen bases. Tellurium tetrahalides of all four halogens exist, the TeF_4 formed from $TeCl_2$ and fluorine, the $TeCl_4$ from $TeCl_2$ and chlorine, the $TeBr_4$ from bromotrifluoromethane, CF_3Br, and molten tellurium, and the TeI_4 directly from the elements. Tellurium dichloride, $TeCl_2$, is prepared by passing dichlorodifluoromethane, CF_2Cl_2, over molten tellurium. $TeCl_2$ is quite reactive, disproportionating to tellurium and $TeCl_4$, and useful in preparing other tellurium compounds. $TeBr_2$ is obtained by distillation of the mixture of tellurium and $TeBr_4$ obtained in the reaction between bromotrifluoromethane and tellurium. The tetrahalides of tellurium form many addition compounds with other halides.

Organotellurium compounds corresponding generally to those of sulfur and selenium are known. Although carbon ditelluride has not been prepared, COTe and CSTe have been.

Toxicity. Tellurium and compounds are toxic. Acceptable concentration limit for an 8-hour daily exposure to dust and fumes in air is 0.1 milligrams of tellurium per cubic meter of air. Even exposure at this level may cause what is termed "garlic breath." Proper ventilation, appropriate hygienic practices, and good housekeeping should be observed in handling tellurium. Although elemental tellurium causes no apparent problems in contact handling, skin contact with soluble tellurium compounds must be avoided.

Additional Reading

Chizhikov, D.M. and V.P. Schastsivity: "Tellurium and Tellurides," (Translated from the Russian by E.M. Elms), Collet's Wellingbourough, Northants, England, 1970. (A Classic Reference.)

Considine, D.M. and G.D. Considine: "Van Nostrand Reinhold Encyclopedia of Chemistry," 4th Edition, Van Nostrand Reinhold, New York, NY, 1984. (A Classic Reference).

Greenwood, N.N. and A. Earnshaw: "Chemistry of the Elements," 2nd Edition, Butterworth-Heinemann, Inc., Woburn, MA, 1997.

Krebs, R.E.: "The History and Use of Our Earth's Chemical Elements: A Reference Guide," Greenwood Publishing Group, Inc., Westport, CT, 1998.

Lewis, R.J., N.I. Sax: "Sax's Dangerous Properties of Industrial Materials," 10th Edition, John Wiley & Sons, Inc., New York, NY, 1999.

Lide, D.R.: "CRC Handbook of Chemistry and Physics 2000-2001," 81st Edition, CRC Press, LLC., Boca Raton, FL, 2000.

Staff: "ASM Handbook — Properties and Selection: Nonferrous Alloys and Pure Metals," ASM International, Materials Park, OH, 1990.

S.C. CARAPELLA, Jr., ASARCO Incorporated, South Plainfield, NJ.

TELLUROMETER. See **Geodimeter.**

TEMBLOR. See **Earth Tectonics and Earthquakes.**

TEMPER. A heat-treating process for increasing the toughness and eliminating warping and cracking of hardened steels and other metals. Temper also refers to the degree of hardness induced by a coldworking process, such as sheet rolling or wire drawing.

TEMPERATURE. That property of systems which determines whether they are in thermodynamic equilibrium. Two systems are in equilibrium when their temperatures (measured on the same temperature scale) are equal. The existence of the property defined as temperature is a consequence of the zeroth law of thermodynamics. The zeroth law of thermodynamics leads to the conclusion that in the case of all systems there exist functions of their independent properties x_i such that at equilibrium

$$\phi_a(x_{ia}) = \phi_b(x_{ib}) = \theta \tag{1}$$

where subscripts a and b refer to two systems a and b each described by n_a properties x_{ia}, and n_b properties x_{ib}, respectively. The hypersurface

$$\theta = \phi(x_i) = \text{constant}$$

is called an *isotherm*, and the pairs of hypersurfaces, Equation (1), at equilibrium are called *corresponding isotherms*.

In order to establish a *temperature scale* it is necessary to assign numerical values θ to these corresponding isotherms in an arbitrary manner,

subject only to the condition that the resulting function shall be single-valued. A temperature scale is established by taking the following steps:

1. An arbitrary system is chosen (*thermometer*).

2. It is agreed to maintain $n - 1$ properties of the system constant, and to use the nth property (*thermometric property* $x_n = X$) as a measure of temperature θ.

3. A single-valued *thermometric function* is assumed. Usually the function is simple, for example

$$\theta = aX \tag{2}$$

or

$$\theta = AX + B. \tag{3}$$

The function usually contains one or several constants (a, A, B, etc.).

4. The values of the constants in the thermometric function are determined with reference to fixed thermometric points whose temperatures are arbitrarily assumed. The *fixed thermometric points* most frequently employed are: the *ice point*, *steam point* and *triple point* of water.

It is not surprising that there exist many different scales of temperature, so-called *empirical temperature scales*, because of the large amount of arbitrariness inherent in the choice.

A *Centigrade* (or *Celsius*) *temperature scale* is obtained by choosing the thermometric function, Equation (3), and assigning the following arbitrary values of temperature, θ, to the ice point (θ_i) and steam point (θ_s) respectively

$$\theta_i = 0\,^\circ C.$$

$$\theta_s = 100\,^\circ C.$$

Hence on a Centigrade scale

$$\theta = 100 \frac{X - X_i}{X_s - X_i} \tag{4}$$

(X_s measured at θ_s, X_i measured at θ_i).

A Fahrenheit temperature scale is obtained by using Equation (3) but with

$$\theta_i = 32\,^\circ F$$

$$\theta_s = 212\,^\circ F.$$

Hence

$$\theta = 32^\circ + 180^\circ \frac{X - X_i}{X_s - X_i}. \tag{5}$$

An *absolute scale* is obtained by choosing Equation (2). Depending on the value assigned to a, we obtain the *Kelvin* or *Rankine scale*. The *Kelvin absolute temperature scale* assigns to the triple point of water the value

$$\theta_3 = 273.16 \text{ K} \tag{6}$$

which thus becomes a *universal constant of physics*. Hence for the Kelvin scale

$$\theta = 273.16 \frac{X}{X_3} \tag{7}$$

where X_3 is measured at θ_3. For the *Rankine absolute temperature* scale we assume

$$\theta_3 = 273.16 \times 1.8\,^\circ R = 491.69\,^\circ R. \tag{8}$$

It is clear that by definition

$$1 \text{ K} = 1.8\,^\circ R, \tag{9}$$

and that on the Rankine scale

$$\theta = 491.69^\circ \frac{X}{X_3}. \tag{10}$$

Different empirical temperature scales will naturally differ from each other except at the respective fixed thermometric points. Even different scales of the same type (say different Centigrade scales) will differ at all temperatures, except the steam point and ice point, depending on the fortuitous properties of the system chosen as a thermometer. It is, therefore, necessary to remove these differences and to obtain a more universal scale. This has been achieved in two ways. The practical way of achieving uniformity is to lay down detailed rules concerning the thermometer (actually different thermometers depending on the range of temperatures to be measured). Such rules have been agreed on internationally and constitute the *international temperature scale*. Another way is to derive a universal scale from the principles of thermodynamics. The latter is called a *thermodynamic temperature scale*. Some authors refer to it as the *absolute temperature scale* which may be a source of confusion with the Kelvin and Rankine scales described earlier.

The *thermodynamic temperature scale T* is defined by the second law of thermodynamics. It can be shown that the *thermodynamic temperature scale* is identical with the *perfect-gas temperature scale* defined as follows:

1. The system is a gas thermometer (filled with a *real* gas).

2. The thermometric property is the product *pV extrapolated to* zero pressure, i.e.,

$$r = \lim_{p \to 0} (pV) \tag{11}$$

Hence the *thermodynamic Kelvin scale* is given by

$$T = 273.16 \frac{r}{r_3} \text{(K abs.)} \tag{12}$$

the *thermodynamic Centigrade scale* is given by

$$t - 100 \frac{r - r_i}{r_s - r_i} \text{(degrees C),} \tag{13}$$

the *thermodynamic Rankine scale* is given by

$$T = 491.69 \frac{r}{r_3} \text{(degrees R abs.),} \tag{14}$$

and the *thermodynamic Fahrenheit scale* is given by

$$t = 32 + 180 \frac{r - r_i}{r_s - r_i} \text{(degrees F).} \tag{15}$$

The zeros on the Kelvin and Rankine scales coincide and are termed the *absolute zero of temperature*. The absolute zero of temperature cannot be achieved by any finite process, as stated in the third law of thermodynamics.

The relation between the Centigrade and Kelvin thermodynamic scales is determined by

$$T_i = 273.16 \text{ K} = 0 \,^{\circ}\text{C} \tag{16}$$

and that between the Fahrenheit and Rankine scales is determined by

$$T_i = 491.69 \,^{\circ}\text{R} = 32 \,^{\circ}\text{F.} \tag{17}$$

Hence the absolute zero of temperature is at

$$-273.16 \,^{\circ}\text{C or} \; -459.69 \,^{\circ}\text{F.} \tag{18}$$

The Comité Consultatif of the International Committee of Weights and Measures selected 273.16 K as the value for the triple point of water. This set the ice-point at 273.15 K.

The relation between the *international temperature scale* and the thermodynamic temperature scale must be determined empirically with the aid of careful measurements involving gas thermometers.

See also **Units and Standards**.

TEMPERATURE (Human). The normal temperature of the interior of the human body is usually said to be 98.6 °F (37 °C). Actually, this normally varies in the course of 24 hours, so that variation between maximal and minimal temperatures during this period is about 1.8 °F (1 °C). The above Figures are those obtained when the temperature is taken by mouth. The rectal temperature is 1 °F (0.6 °C) higher.

TEMPERATURE SCALES. See **Heat**.

TEMPERATURE TRANSFER STANDARD. A device for the transfer of a temperature scale from one standardizing laboratory to another. One form consists of a sample of a purified material, the freezing point of which (when realized by a prescribed technique) is reproducible within narrow limits. Materials commonly employed are metals, such as zinc and tin, and organic compounds, such as benzoic acid, phenol, naphthalene, and phthalic anhydride. Another form is a tungsten ribbon-filament lamp, characterized by a stable lamp current-brightness temperature relation. This device is particularly useful for temperatures above 1,050 °C.

TEMPORAL. A bone of the vertebrate skull. In the human skull it lies at the side, centering about the ear. It forms the posterior part of the cheek bone with the articulation of the lower jaw, and bears the mastoid process behind the ear. It included the squamosal bone and the bones of the ear.

TEMPORAL LOBE. See **Central and Peripheral Nervous Systems**.

TENDON. Connective tissue structures connecting muscles with their skeletal supports. They are composed of parallel white fibers, closely bound into bundles between which the cells of the tendon are compressed. The entire tendon is surrounded by a fibrous sheath known as the vagina fibrosa, which is split in some cases to form a cavity containing a mucoid liquid. Such a sheath is called a vaginal mucosa. It develops where a wide range of movement is necessary.

Aponeuroses, ligaments, and fasciae are fibrous structures resembling tendons in structure. The first are broadly expanded and usually bind down muscles. The second connect bones. The third are fibrous coverings of muscles and other organs.

TENORITE. A mineral oxide and ore of copper, CuO. Crystallizes in the monoclinic system. Hardness, 3.5; specific gravity, 6.45; color, gray to black with metallic luster. Named after M. Tenore (1780–1861), Naples.

TENRECIDS. See **Moles and Shrews**.

TENSILE STRENGTH. The resistance offered by a material to tensile stresses, as measured by the tensile force per unit cross-sectional area required to break it.

TENSILE STRESS. See **Stress (Structural)**.

TENSIMETER. An apparatus for determining transition points (phase change) of materials in an indirect manner by measuring small changes in vapor pressure.

TENSIOMETER. An apparatus for measuring the surface tension of a liquid by registering the force necessary to detach a metal ring from the surface.

TENSION. In structural engineering, tension is used to denote the longitudinal force which causes the fibers of a member to elongate, thus giving rise to tensile stress.

TENSION (Interfacial). See **Interfacial Tension**.

TENSION MEMBER. Any member of a structure subjected to a primary tensile stress is called a tension member. The analysis or design depends on that part of the cross-sectional area that is effective in resisting the tensile stress. The effective area at any section is called the net section. It is the gross area minus the area of any holes including those filled by rivets or bolts. The net section may be based on a right section or a zig-zag section. The net section of a threaded member such as a bolt or tie rod is the area at the root of the thread. The strength of a tension member is a function of the minimum net section.

If the member is composed of two or more independent parts other than eye bars, the parts may be connected by the tie plates or by tie plates and lacing bars to equalize the stress and prevent excessive distortion during fabrication, shipping or erection.

TENSION TEST. Next to hardness tests, tension tests are the most frequently used to determine the mechanical properties of metals. Tension test specimens necessarily vary in form with the product to be tested. A machined cylindrical specimen with threaded or shouldered ends for gripping is used when the material is sufficiently thick. Standard flat specimens are used for flatrolled products. While both types of specimens have a reduced central section to ensure breaking within a measured gage length, wires and certain special shapes such as steel reinforcing bars for concrete are tested in full section without preparation. Special cast test specimens are often attached to castings, or cast separately, and these are generally tested without machining.

The significant loads determined in the test are reported as unit stresses based on the area of the original section (Stress equals load divided by area). The elongation is expressed as percent increase in length of the gage-marked section. The initial gage length is generally 2 inches

(5 centimeters), although an 8-inch (20.3-centimeter) gage length is used for certain flat specimens and other gage lengths are used for special specimens. The gage length should be specified when elongation is reported since percent elongation values are higher for short than for long gage lengths.

The elongation measured over a fixed gage length, and the reduction of area of the section at the fracture are measures of ductility. In cylindrical specimens, the area is readily determined from the final diameter at the fracture. The percent reduction of area is then determined as: original area minus final area, divided by original area.

Autographic load-deformation curves are often drawn during the test. From such a curve, the modulus of elasticity, proportional limit, and yield strength can be determined.

A typical curve has an essentially linear portion (OA) in which the deformation is proportional to the applied load. See Fig. 1. It follows that the unit stress (load divided by original area) is proportional to the unit strain (deformation divided by original gage length) in accordance with Hooke's Law. The numerical value of this ratio (e.g., in psi) is known as Young's Modulus or Modulus of Elasticity.

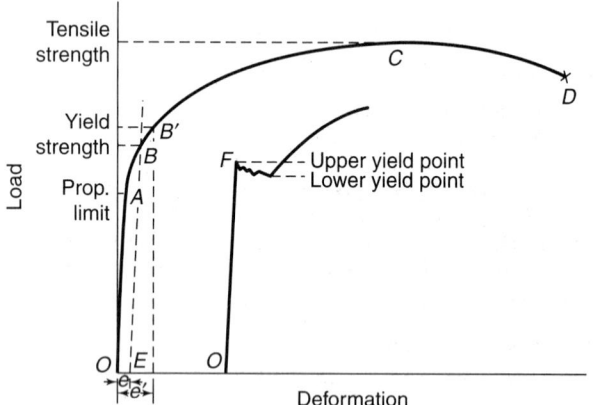

Fig. 1.   Stress-strain diagram.

The maximum stress that is developed without deviation from proportionality of stress to strain is the proportional limit (the stress corresponding to load A). The maximum stress that can be applied without causing permanent deformation upon release of the load is the elastic limit. Usually, there is little difference between the proportional limit and the elastic limit. Both are dependent on the sensitivity of the measuring devices used and certain details of testing technique. For this reason, the yield strength is generally used as a practical measure of the elastic properties of metals.

The yield strength is the stress at which the stress-strain curve deviates from the initial straight line by a specified increment of strain. The yield strength corresponding to the load at B is based on the specified strain deviation or offset e. The value of e may be as low as 0.0001 inch (0.0025 millimeter) of gage length but the most commonly used value is 0.002 inch (0.05 millimeter), or 0.2% strain.

If the load should be released after reaching B, the load deformation relationship will follow the line BE, or a curve line terminating between O and E. Thus the permanent strain will be e or a somewhat smaller value. When the final or permanent strain is specified, the stress is known as the proof stress.

An alternate type of yield strength is based on a specified total extension under load, such as 0.5%. If the specified extension is e′ the load B′ determines the "extension-underload" yield strength. Load B′ may be greater or less than B.

The tensile strength, or ultimate tensile strength, is the maximum stress developed in the tension test (load C divided by original area).

The breaking stress, corresponding to load D, is seldom determined or reported.

In loading tension specimens of many soft irons and steels, a point is reached where stretching continues without increase in load. The unit stress obtained by dividing this load, F, by the original area of the section is called the yield point. The elongation of the specimen at the yield point may reach 8% in some instances, after which the load will again increase to a maximum in the normal manner. Upper and lower yield points are

indicated at F; both are used, but the upper yield point is influenced by variations in testing technique such as alignment of the specimen in the testing machine and speed of test. Yield points occur only rarely in the nonferrous metals.

Conventional stress-strain curves are necessarily similar to the load-deformation curves from which they are derived. True stress-strain curves can also be derived in which the stress is based on the actual or instantaneous area of the cross-section. Such curves do not have a maximum corresponding to C, but increase continuously to the breaking load.

TENSOR.   A set of n^r components, which are functions of the coordinates of any point in n-dimensional space. They transform linearly and homogeneously, according to certain rules, when a transformation of coordinates is made. Tensors are called covariant, contravariant, or mixed, according to the law of transformation. The number r is called the rank or order of the tensor.

Suppose $(x^1, x^2, \ldots, x^n)$ are the coordinates of a point referred to a given coordinate system and $(\bar{x}^1, \bar{x}^2, \ldots, \bar{x}^n)$ are the coordinates of the same point referred to another coordinate system. Then n quantities $(A^1 A^2, \ldots A^n)$, functions of these coordinates are the contravariant components of a tensor of rank one if they transform as follows:

$$A^i = \frac{\partial \bar{x}^i}{\partial x^r} A^r, i = 1, 2, \ldots, n$$

The summation for the summy index r is understood and the upper suffix is not an exponent but a means of designating the different components. Contravariant tensors of higher rank are defined in a similar way. Thus, for a tensor of rank two,

$$A^{ij} = \frac{\partial \bar{x}^i}{\partial x^r} \frac{\partial \bar{x}^j}{\partial x^s} A^{rs}$$

A tensor of rank zero is a scalar; of rank one, a vector.

The transformation laws for covariant tensors of rank one and two, respectively are

$$A_i = \frac{\partial x^r}{\partial \bar{x}^i} A_r; \quad A_{ij} = \frac{\partial x^r}{\partial \bar{x}^i} \frac{\partial x^s}{\partial \bar{x}^j} A_{rs}$$

The transformation law for a mixed tensor of second rank is

$$A^i_j = \frac{\partial \bar{x}^i}{\partial x^r} \frac{\partial x^s}{\partial \bar{x}^j} A^r_s$$

Mixed tensors of higher rank may be defined by analogy. An important mixed tensor is the Kronecker delta,

$$\delta^m_n = 1, \quad m = n$$

$$\delta^m_n = 0, \quad m \neq v$$

If the components of a tensor satisfy the relation $A^{ij} = A^{ji}$, the tensor is symmetric, if $A^{ij} = -A^{ji}$, skew-symmetric. If the components are neither symmetric nor skew-symmetric, they may always be expressed as $A^{ij} = S^{ij} + T^{ij}$, where $S^{ij} = (A^{ij} + A^{ji})/2$, which is symmetric and $T^{ij} = (A^{ij} - A^{ji})/2$, which is skew-symmetric.

For further properties of tensors see also **Christoffel Symbol**; **Field Theory**; **Tensor Contraction**; and **Tensor Differentiation**. See also **Dyadic**.

TENSOR CONTRACTION.   Suppose that two indexes in a tensor, one upper and one lower, are identical, A^m_{npm}. Then summation over the repeated index shows that the tensor is actually of lower rank, for its transformation law is

$$A^m_{npm} = \frac{\partial \bar{x}^m}{\partial x^i} \frac{\partial x^j}{\partial \bar{x}^n} \frac{\partial x^k}{\partial \bar{x}^p} \frac{\partial x^h}{\partial \bar{x}^m} A^j_{jkh}$$

$$= \frac{\partial x^j}{\partial \bar{x}^n} \frac{\partial x^j}{\partial \bar{x}^p} \delta^h_i A^i_{jkh} = \frac{\partial x^j}{\partial \bar{x}^n} \frac{\partial x^k}{\partial \bar{x}^p} A^i_{jki}$$

where δ^h_i is a mixed tensor. The result could properly be written A_{jk}, since the tensor transforms like a covariant tensor of rank two. The process of contraction, which can be used for tensors of any rank, thus consists of deleting a repeated upper and lower index. It always reduces the rank of a mixed tensor by two.

An outer product results when A^m is multiplied by B_n to give a mixed tensor. Such a product of any rank or type can be formed. Multiplication, followed by contraction gives an inner product. This process is similar to the scalar product in vector analysis for $A^m B_m$ is a scalar.

These properties, together with addition, show how new tensors can be formed from two or more given tensors. Still another such operation is possible and it also shows the relation between a covariant and a contravariant tensor. Consider a symmetric tensor of the former type, g_{mn} with determinant g and cofactor, G^{mn}. (Note that G^{mn} is not a tensor.) Then, if we define $g^{mn} = G^{mn}/g$, the rule for the expansion of a determinant gives $g_{mn}g^{in} = \delta_m^i$. As suggested by the notation g^{mn} is a contravariant tensor, for if A^n is a vector, then $B_m = g_{mn}A^n$ is also a vector and $g^{mn}B_m = g^{mn}g_{mi}A^i = \delta_i^n A^i = A^n$. Thus, g^{mn} changes a covariant into a contravariant vector. Two vectors related by either g_{mn} or g^{mn} are called associated. In fact, $A^m = g^{mn}A_n$, and $A_m = g_{mn}A^n$ are often regarded as the same vector with covariant components, A_m and contravariant components, A^m.

The quantity g_{mn} is usually called the metric tensor. Consider a rectangular coordinate system, then the square of the line element ds is $ds^2 = dx^2 + dy^2 + dz^2$. In generalized coordinates, $q^{''t}$, one has $ds^2 = g_{mn}dq^m dq^n$ and in still another coordinate system $\bar{q}^i$, $ds^2 = g_{ij}d\bar{q}^i d\bar{q}^j$. (Note that the superscripts on ds, dx, dy, and dz are squares, not tensor indices.)

TENSOR DIFFERENTIATION.
Differentiation of a covariant tensor does not yield a tensor for the derivative and does not transform properly. The covariant derivative of a covariant tensor of rank one is then defined as

$$A_{i,j} = \frac{\partial A \cdot}{\partial x^j} - \{ij, h\}A_h$$

where $\{ij, h\}$ is a Christoffel three-index symbol (not a tensor) and this function transforms like a covariant tensor of rank two. Derivatives of contravariant and mixed tensors of all ranks may be defined in a similar way.

See also **Christoffel Symbol**.

TENSOR FIELD.
In each n-dimensional curvilinear coordinate system n^{M+N} functions of position are specified. Using the indicial notation let us denote those defined in the coordinate system x by $t_{k_1 k_2 \cdots k_N}^{j_1 j_2 \cdots j_M}$ and those defined in the coordinate system $\bar{x}$ by $\bar{t}_{k_1 k_2 \cdots k_N}^{j_1 j_2 \cdots j_M}$. In each case the M superscripts and N subscript take values ranging over $1, 2, \ldots, n$ independently. The functions are related by

$$t_{k_1 k_2 \cdots k_N}^{-j_1 j_2 \cdots j_M} \left| \frac{\partial x}{\partial \bar{x}} \right|^W \frac{\partial \bar{x}^{j_1}}{\partial x^{p_1}} \frac{\partial \bar{x}^{j_2}}{\partial x^{p_2}}$$

$$\cdots \frac{\partial \bar{x}^{j^M}}{\partial x^{p_M}} \frac{\partial x^{q_1}}{\partial \bar{x}^{k_1}} \frac{\partial x^{q_2}}{\partial \bar{x}^{k_2}} \cdots \frac{\partial x^{q_N}}{\partial \bar{x}^{k_N}} t_{q_1 q_2 \cdots q_N}^{p_1 p_2 \cdots p_M}$$

in which the summation convention is employed, $|\partial x/\partial \bar{x}|$ *denotes the Jacobian determinant* $\partial(x^1, x^2, \ldots, x^n)/\partial(\bar{x}^1, \bar{x}^2, \ldots, \bar{x}^n)$, W is an integer (positive or negative) and the summation convention is employed. The aggregate of sets of functions defined in this manner in every curvi-linear coordinate system is called an n-dimensional *tensor field of weight W*. The tensor field is said to be *contravariant* of order M and *covariant* of order N. It is said to have *total order* or *order* $M + N$. If $M = 0$ the tensor field is said to be a *covariant tensor field*. If $N = 0$, it is said to be a *contravariant tensor field*. If neither M nor N is zero, it is said to be a *mixed tensor field*. For summation convention, see also **Tensor**.

The set of functions $t_{k_1 k_2 \cdots k_N}^{j_1 j_2 \cdots j_M}$ are called the *components of the tensor field* in the coordinate system x; similarly the set of functions $\bar{t}_{k_1 k_2 \cdots k_N}^{j_1 j_2 \cdots j_M}$ are called the components of the tensor field in the coordinate system $\bar{x}$. The superscripts are called *contravariant indices* and the subscripts *covariant indices*.

At each point of space at which the tensor field is defined, it determines a set of n^{M+N} quantities in each curvilinear coordinate system. The aggregate of these sets of quantities is called an n-dimensional tensor at the point considered. The set of n^{M+N} quantities determined in the coordinate system x are the components of the tensor in the coordinate system x. The definitions of weight, contravariance, and order of a tensor are analogous to those for a tensor field.

A tensor field is often referred to as a *tensor*, when no confusion is likely to arise.

TENSOR FIELD (Cartesian).
In each n-dimensional rectangular Cartesian coordinate system n^M functions of position are specified. Let us denote the functions defined in the rectangular Cartesian coordinate systems x and $\bar{x}$ by $t_{i_1 i_2 \cdots i_M}$ and $\bar{t}_{i_1 i_2 \cdots i_M}$ respectively. In each case the M subscripts $i_1, i_2, \ldots, i_M$ take values ranging over $1, 2, \ldots, M$ independently. The functions are related by

$$\bar{t}_{i_1 i_2 \cdots i_M} = \frac{\partial \bar{x}_{i_1}}{\partial x_{j_1}} \frac{\partial \bar{x}_{i_2}}{\partial x_{j_2}} \cdots \frac{\partial \bar{x}_{i_M}}{\partial x_{j_M}} t_{j_1 i_1 \cdots j_M}$$

where the summation convention is used and the relation is valid for all choices of the rectangular Cartesian coordinate systems x and $\bar{x}$. The aggregate of sets of n^M functions so defined is called an n-dimensional Cartesian tensor field. M is called the *order* of the Cartesian tensor field. The set of functions $t_{i_1 i_2 \cdots i_M}$ are called the components of the Cartesian tensor field in the coordinate system x; similarly the set of functions $\bar{t}_{i_1 i_2 \cdots i_M}$ are called the components of the Cartesian tensor field in the coordinate system $\bar{x}$.

At each point of space at which the Cartesian tensor field is defined, it determines a set of n^M quantities in each rectangular Cartesian coordinate system. The aggregate of these sets of quantities is called an n-dimensional *Cartesian tensor* at the point considered. The set of n^M quantities determined in the coordinate system x are called the *components of the tensor* in the coordinate system x.

A Cartesian tensor field is often referred to as a *Cartesian tensor* and, particularly in applied mathematical contexts, as a *tensor*.

TEPHRA.
From the Greek word meaning "ash," *tephra* collectively refers to airborne fragments from an erupting volcano. Thus, tephra is a general term for all pyroclastics of a volcano. See also **Pyroclastic**.

TERBIUM.
Chemical element symbol Tb, at. no. 65, at. wt. 158.92, eighth in the Lanthanide Series in the periodic table, mp 1365 °C, bp 3230 °C, density 8.230 g/cm^3 (20 °C). Elemental terbium has a close-packed hexagonal crystal structure at 25 °C. The pure metallic terbium is silver-gray in color, and is stable in ambient air conditions. When pure, the metal is malleable. There is one natural isotope of terbium ^{159}Tb. The isotope is not radioactive and has a low acute-toxicity rating. Seventeen artificial isotopes have been identified. Little is known concerning the characteristics of terbium alloys and intermetallic compounds. Average content of the earth's crust is estimated at 0.9 ppm terbium, making this element the second least abundant of the rare-earth elements. Even at this level, however, terbium is potentially more available than antimony, bismuth, cadmium, or mercury. The element was first identified by C.G. Mosander in 1843. Electronic configuration of the ground state is mixed:

$$1s^2 2s^2 2p^6 3s^2 3p^6 3d^{10} 4s^2 4p^6 4d^{10} 4f^8 5s^2 5p^6 5d^1 6s^2.$$

Ionic radius Tb^{3+} 0.93 Å, Tb^{4+} 0.76 Å. Metallic radius 1.783 Å. First ionization potential 5.84 eV; second 11.52 eV. Other physical properties of terbium are given under **Rare-Earth Elements and Metals**. See also **Chemical Elements**.

Terbium occurs in apatite and xenotime and is derived from these minerals as a minor coproduct in the processing of yttrium. Processing involves organic ion-exchange or solvent extraction operations. Elemental terbium is produced by calcium reduction of anhydrous TbF$_3$ in a reactor under an inert atmosphere. Both the oxides and the metal are available at 99.9% purity.

To date, the uses for terbium have been quite limited. Terbium-activated lanthanum oxysulfide Tb:La$_2$O$_2$S is a phosphor finding use as an image intensifier for x-ray screens. Terbium-activated indium borate Tb:InBO$_3$ phosphor emits an intense narrow green light (5,450–5,500 Å) and has found use in information display systems where there are high ambient-light conditions. Future color television tubes may use terbium-activated yttrium silicate Tb:Y$_2$SiO$_5$ or yttrium phosphate Tb:YPO$_4$ green phosphors. They appear to be highly efficient post-deflection focused phosphors for elimination of the need for a shadow mask in a television tube. Although terbium oxide may be used as a stain for ceramics, the compound does not color glass. In soda-lime glass, a small quantity of terbium provides a strong green-blue fluorescence under ultraviolet radiation. Another important use of terbium is in amorphous Tb-Co(fe) alloys for magnetic recording and information storage.

NOTE: This entry was revised and updated by K.A. Gschneidner, Jr., Director, and B. Evans, Assistant Chemist, Rare-earth Information Center, Institute for Physical Research and Technology, Iowa State University, Ames, IA.

TEREPHTHALIC ACID. $C_6H_4(COOH)_2$, formula weight 166.13, crystalline solid sublimes upon heating, sp gr 1.510. The compound is almost insoluble in H_2O, only slightly soluble in warm alcohol, and insoluble in ether. Terephthalic acid (TPA) is a high-tonnage chemical, widely used in the production of synthetic materials, notably polyester fibers (poly-(ethylene terephthalate)).

There are several processes for making terephthalic acid on a large scale: (1) Benzoic acid, phthalic acid and other benzene-carboxylic acids in the form of alkali-metal salts, comprise the chargestock. In a first step, the alkali-metal salts (usually potassium) are converted to terephthalates when heated to a temperature exceeding 350°C. The dried potassium salts (of benzoic acid or o- or isophthalic acid) are heated in anhydrous form to approximately 420°C in an inert atmosphere (CO_2) and in the presence of a catalyst (usually cadmium benzoate, phthalate, oxide, or carbonate). The corresponding zinc compounds also have been used as catalysts. In a following step, the reaction products are dissolved in H_2O and the terephthalic acid precipitated out with dilute H_2SO_4. The yield of terephthalic acid ranges from 95 to 98%.

(2) Toluene, formaldehyde, HCl, calcium hydroxide, and HNO_3 comprise the chargestock. In step 1 of this process, the toluene is reacted with concentrated HCl at about 70°C along with paraformaldehyde. This accomplishes chloromethylation of approximately 98% of the toluene. In step 2, saponification of the chloromethyltoluene is effected with lime and H_2O under pressure and at about 125°C. The product is methylbenzyl alcohol. In step 3, the methylbenzyl alcohol is oxidized with HNO_3 (dilute) under a pressure of about 20 atmospheres and at a temperature of about 170°C. The main products are o-phthalic acid in HNO_3 solution and insoluble terephthalic acid.

(3) Paraxylene and air comprise the chargestock. These materials, along with a proprietary catalyst and solvent, are fed to a liquid-phase oxidation reactor, operated at moderate pressure and temperature. The reaction is:

$$C_6H_4(CH_3)_2 + 3\ O_2 \longrightarrow C_6H_4(COOH)_2 + 2\ H_2O.$$

The design details of these processes are proprietary. There are several other processes which essentially are variations of the foregoing descriptions. See also **Intermediate (Chemical)**; **Phthalic Acid**; and **Phthalic Anhydride**.

TERMINAL MORAINE (or End Moraine). When balance is maintained between the melting of a glacier and its forward advance, the debris carried on (superglacial); within (englacial); and dragged along the bottom (subglacial); is dumped at that point and builds up a heterogeneous mass of the transported material called the terminal moraine. If a glacier is slowly retreating and makes successive halts farther and farther up the valley, a series of terminal moraines are formed which are spoken of as recessional moraines. See also **Drumlin**; and **Glacial Deposits**.

TERMITE *(Insecta, Isoptera)*. Also incorrectly known as the white ant, the termite is of large economic importance. Because of their wood-eating habits, termites sometimes do great damage to farm structures and storage buildings used in connection with food production. While highly destructive in this way, the termite is not injurious to crops or livestock. See Fig. 1. See also **Isoptera (Termites)**; and **Termitophile**.

In recent years, some entomologists have given increased attention to the biology and physiology of the termite and, in particular, the presence of microorganisms in the termite, the sophisticated chemical defense mechanisms displayed by the termite, and the gaseous byproducts, including carbon dioxide, methane, and molecular hydrogen, generated by termites as a result of their metabolization of large amounts of biomass. These generous amounts of natural atmospheric pollutants are currently estimated to exceed the emissions, particularly of CO_2, from fossil-fired industrial plants.

Over a century ago, entomologists attributed the termite's ability to metabolize cellulose to enzymes synthesized by microorganisms in the insect's gut. This early observation has led to more recent investigations of the microorganisms present and to date at least one hundred species of protozoa and bacteria have been identified in the termite gut. The microscopic

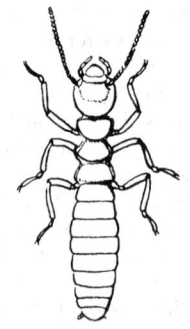

Fig. 1.   Termite. (*USDA.*)

ecosystem of the termite has been found to be astonishingly complex. One group of investigators has used the desert termite (*Pterotermes occidentis*) in laboratory studies. This is a rather rare species of termite that lives by consuming the dead interior wood of several species of desert plant, including the saguaro cactus and the palo verde tree. The majority of microorganisms that inhabit the gut of this termite are found in the paunch of the insect (the third segment of its digestive tract). The population density of the protozoa found is between a thousand and a million individuals per milliliter of the gut fluid. In this same fluid are found millions to billions of bacteria that swim around freely in the chamber of the paunch. As yet, it has not been determined if the protozoa, the bacilli, or both participate in generating the termite's wood-digesting enzymes. Many of the microorganisms are anaerobic and cannot live outside the termite gut. It also has been suggested that the anaerobes may depend on bacteria that serve as oxygen scavengers and thus protect them from toxic concentrations. It has been found that several of the microorganisms are sensitive to the pH of the gut fluid.

As pointed out by Prestwich (1983), it is estimated that the insect world has used chemicals both to attract prey and to repel predators for over 100 million years. Insect predators attempt to subdue their prey with toxic venoms or to lure them with seductive perfumes. Some insect prey species take counter actions by ejecting irritating, sticky, hot, or poisonous substances or by themselves becoming inedible. The termite is considered exceptional in terms of its arsenal of chemical defenses. It has been suggested that the termite even outperforms the army ant in this respect. Like the cockroach, the termite has evolved over millions of years from an ancestral stock that is believed to resemble modern roaches. There are over 2000 species of termite in the world today, mostly in tropical regions. Much of this successful survival can be attributed to chemical defenses.

Belgian researchers have suggested three main methods of chemical defense used by the termite: (1) *biting* (introduction of oily or toxic substances into the wound); (2) *daubing* (as with a paintbrush) a predator with contact poison applied to the cuticle—some termite species have an enlarged labrum (upper lip) for this purpose; (3) *glue squirting*—spraying an irritating, viscous entangling agent. Like ants, bees, and many other insects, termite colonies are extremely well organized. For example, so-called solider termites carry over 500 times as much defensive chemicals as do other members of a colony who are devoted to food gathering and mound repair tasks. The most common raiders of termite colonies are ants. In one defense against the ant, the biting and daubing mechanisms are quite successful. When the ant's cuticle is punctured, a hydrocarbon (normally harmless) softens the wound, interfering with the coagulation of the ant's hemolymph and preventing the resclerotization (natural repair) of cuticle damage.

There is considerable variation in the types of chemicals synthesized by various termite species. Predominant, however, are diterpenes, quinones, and modified fatty acids. In research on numerous species of termite, investigators have encountered some surprises, particularly in terms of the concept of parallel evolution. Too detailed for coverage here, an excellent summary is given by Prestwich. An essentially unanswered question: How do termites protect themselves from the actions of their own toxic chemicals?

As reported by Zimmerman et al. (1982), termites may be a potentially large source of atmospheric methane, carbon dioxide, and molecular hydrogen. Termites occur on about two-thirds of the earth's land surface. They process large amounts of biomass. Their digestion is mainly dependent upon the anaerobic decomposition by symbiotic bacteria (higher termite family, *Termitidae*), and by protozoa (lower termites). Their digestion efficiency (carbon ingested less carbon in feces divided by carbon ingested) is usually greater than 60%. The clearing of tropical forests and conversion of forests to grazing and agriculture tend to increase the density of termites.

Early studies had indicated that termites could contribute more CO_2 to the atmosphere than the combustion of fossil fuels. Follow-up research in the late 1980s revised the estimate downward to a probable 2% contribution of CO_2 to the atmosphere by termites worldwide.

Additional Reading

Abe, T., M. Higashi, D.E. Bignell, and M. Higashi: "Termites: Evolution, Sociality, Symbioses, Ecology," Kluwer Academic Publishers, Norwell, MA, 2000.

Pearce, M.J.: "Termites: Biology and Pest Management," CAB International, New York, NY, 1997.

Prestwich, G.D.: "The Chemical Defenses of Termites," *Sci. Amer.*, **249**(2), 78–87 (August 1983).

Sands, W.A.: "The Identification of Worker Castes of Termite Genera from Soils of Africa and the Middle East," CAB International, New York, NY, 1998.

Staff: "Microorganisms Present in Insects: Termites," *Sci. Amer.*, **246**(2), 78 (February 1982).

Staff: "Termites Generate Very Large Amounts of CO_2," *Sci. Amer.*, **248**(1), 73 (January 1983).

Staff: "Termites Not to Blame for Methane," *Science News*, 268 (April 28, 1990).

Zimmerman, P.R., et al.: "Termites: A Potentially Large Source of Atmospheric Methane, Carbon Dioxide, and Molecular Hydrogen," *Science*, **218**, 563–565 (1982).

TERMITOPHILE.   An insect of another form living in the nest of termites. Both ant and termite colonies are inhabited by other insects, some apparently living as scavengers and profiting by the supply of food and by the protection afforded by the colony while others produce secretions used by their hosts and so live in a commensal relationship. Even in the latter cases the termitophiles may eat the young of the termites, but since they render some return they are not to be regarded as parasitic on the colony. Some, however, are nourished by food supplied by the termites in return for the desired secretion. Insects of such habits are often very strangely formed, differing conspicuously from other members of the orders to which they belong.

Among the known termitophiles are flies, larvae of moths, one species of *Homoptera*, and many beetles.

TERN.   See **Waders, Shorebirds, and Gulls**.

TERNE.   See **Tin**.

TERRAPIN (*Reptilia, Chelonia*).   Certain of the pond and land turtles, especially those of the genera *Malaclemmys* and *Pseudemys*. The related genus *Terrapene* includes the box turtles, and the common painted turtles are also closely connected forms. Sale of painted turtles in the United States has been outlawed for reasons of health—the turtles being considered as carriers of dangerous microorganisms. The species whose flesh is so highly esteemed is the diamond-back terrapin or salt-marsh turtle found in salt marshes along the Atlantic coast from Massachusetts to Florida. This and a related species of the Gulf coast make up the genus *Malaclemmys*. The species of *Pseudemys* are commonly called sliders or cooters, one species bearing the name red-bellied terrapin. They are edible but are less highly valued than the diamond-back. Terrapins are widely distributed in the northern hemisphere. Elsewhere they are limited to a few species in Central and South America.

See also **Testudinata**; and **Turtles**.

TERRESTRIAL SCINTILLATION.   Generic term for scintillation phenomena observed in light that reaches the eye from sources lying within the Earth's atmosphere; to be differentiated from astronomical scintillation, which is observed in light from extraterrestrial sources such as stars. Also called *atmospheric boil*, *atmospheric shimmer*, *shimmer*, or *optical haze*. Terrestrial scintillation is produced by irregular refraction effects due to passage, across the line of sight, of air parcels (schlieren) whose densities differ slightly from that of their surroundings. Density irregularities with dimensions of the order of centimeters, or at most decimeters, are responsible for most such scintillatory effects.

TERTIARY.   The first period of the Cenozoic era (after the Cretaceous of the Mesozoic era and before the Quaternary), believed to have covered the span of time between 65 million and 3 to 2 million years ago. The Tertiary is divided into five epochs; the Paleocene, Eocene, Oligocene, Miocene, and Pliocene. The name was originally assigned as an era rather than a period designation. In this latter sense, it may be considered to have either four periods (Eocene, Oligocene, Miocene, Pliocene), or two (Paleogene and Neogene), with the Pleistocene and Holocene included in the Neogene.

TESLA COIL.   Essentially an air-core transformer used in connection with a spark gap and capacitor to produce a high voltage at a high frequency.

TESTING (Blood).   See **Blood**.

TESTING (Fuels).   See **Calorimetry**; **Coal**; and **Petroleum**.

TESTING (Impact).   See **Impact Testing**.

TESTOSTERONE.   See **Gonads**; **Hormones**; and **Steroids**.

TEST PATTERN.   In television, a special chart on which lines and other detail are arranged to indicate certain characteristics of the system through which the television signal obtained from this chart passes.

TESTUDINATA.   The turtle and tortoises, constituting an order of the class *Reptilia*, also known as the order *Chelonia*. These animals have a short broad body enclosed in a shell composed, in most species, of closely joined bony plates covered with thin plates of horn or tortoise shell. The bony plates are flattened ribs and vertebrae, with supplementary dermal plates. The upper part of the shell is the carapace and the plate below the body is the plastron; the two are firmly united along the sides in most species. The skin is covered with scales and the jaws are without teeth, forming a horn-sheathed cutting beak.

Most species of the order are partially aquatic but some are strictly terrestrial. The marine turtles are the most thoroughly aquatic species, but even they come to shore to lay their eggs. These species have all four limbs developed as paddlelike flippers while the other members of the group merely have webbed toes.

TETANUS.   A serious, often life-threatening disease caused by neurotoxin (*tetanospasmin*) generated by the anaerobic, Gram-negative bacillus *Clostridium tetani* which occurs in soil and in the intestines of domestic animals. Spores of the organism can also be recovered from clothing, house dust, and even the air of occupied buildings, such as hospitals. These and/or the *Clostridia* are introduced into the human body by way of penetrating wounds in which deep layers of tissue are punctured and exposed. When the spores multiply and spread, they generate two toxins, *tetanospasmin* and *tetanolysin*. Only the former has clinical effects, reaching the spinal cord and brain by blood spread and retrograde axon transport along peripheral nerves. The toxin interferes with the normal control of reflexes and muscle spasms result. These spasms vary in accordance with the relative strengths of opposing muscles. Thus, a spasm may produce an extension of hips and knees, flexion of the biceps, and/or *trismus* (lockjaw). Early symptoms of the disease are pain associated with muscle spasms, restlessness, stiffness (back, neck, thighs, abdomen), difficulty in opening the mouth (this being the first symptom in over half of cases). There also may be difficulty in swallowing. The early phase of the disease, with all or some of the foregoing symptoms, has a duration of about 72 hours, after which a phase known as general tetanus takes place. In persons who recover from the disease, this general phase will persist for about one week, during which time there are violent muscular spasms, although the patient retains consciousness. A sneering facial expression (*risus sardonicus*) may develop. Usually any sudden stimulus (bright light, loud noise) will stimulate what is known as a *tonic seizure*. The fever may rise to 101–102 °F (38.3–38.9 °C). Laryngeal spasm is the most dramatic and perhaps the most dangerous event in tetanus. Airway obstruction compromises gaseous exchange or renders it impossible; cyanosis occurs and death follows if the spasm does not pass off or is relieved by treatment. After a peaking of the general tetanus phase, several weeks are required for full recovery, during which time respiratory assistance is usually needed for 2 to 4 weeks. The patient usually is hospitalized for a total of 5–8 weeks.

Diagnosis of tetanus is usually not difficult because of the rapid progress of the disease and its specific early symptoms. Acute strychnine poisoning is the only disease with very similar symptoms. Trismus alone, however, is not fully confirmative of tetanus because this also occurs in encephalitis. There is an immediate differentiation observed, however, in that the person with tetanus usually remains conscious even during violent spasms, whereas in encephalitic trismus, most patients will be comatose. Trismus also can be produced by certain drugs, to which some individuals are particularly sensitive. These include phenothiazine. Trismus in such instances is called the *grimacing syndrome*.

Respiratory and cardiovascular complications may arise during the course of the disease and fractures may occur through frequency and severity of muscular spasms.

In the treatment of tetanus, the physician has three major objectives: (1) the prevention of further violent muscle spasms and the dangerous consequences which these pose, particularly as regards breathing; (2) the prevention of respiratory complications and preparing to handle such when they arise (tracheotomy, for example); and (3) slowing the source of the toxin. Muscle relaxants are used in conjunction with sedatives to control spasms and tonic seizures. The usual drug of choice is diazepam (Valium®) because of its rapid action. When spasms do not respond fully to this therapy, very strong neuromuscular blocking agents, such as *d*-Turbocurarine or pancuronium, may be required. This procedure of inducing muscle relaxation and of inducing, depending upon dosage, partial, temporary paralysis, is referred to as *curarization* (curare = a toxic mixture of about 40 alkaloids occurring in several species of South American trees with powerful action on the nervous system). Where facilities are available, the total paralysis regimen is now the method of choice for treating patients with severe tetanus. Hypertension, tachycardia, and profuse sweating brought on by overactivity induced by the disease are sometimes treated with a beta-adrenergic blocking drug, such as propranolol. Rarely is the intravenous use of morphine required. To provide an adequate airway, nasotracheal intubation will be used. In some cases, tracheostomy is required. Morbid anatomical lesions do not explain all deaths in tetanus, but the respiratory system is involved in most cases treated conservatively. Overwhelming tetanus intoxication of the brainstem is another important factor. Although antibiotics have not been proved to assist in the treatment of tetanus in a direct way, they are sometimes administered to reduce possible infection by other microorganisms as the result of tissue damaged by the *C. tetant* toxins. Although tetanus antitoxin apparently does not neutralize toxin already present in the nervous system of the patient, antitoxin can reduce the quantity of circulating toxin. Available in the United States and some other countries is human tetanus immune globulin, which is better tolerated than equine antiserum and has a longer effective life. Active immunization with toxoid is the surest way of preventing tetanus. If proof is required, the high death rate from tetanus in World War I when active immunization was not practiced should be compared with the virtual absence of the disease in World War II when immunization with toxoid was routine. After undertaking emergency medical treatment in the early phase of the disease, the physician will, of course, take prompt action in debriding and prophylaxis of the wound.

Additional Reading

DasGupta, B.R.: "Botulinum and Tetanus Neurotoxins: Neurotransmission and Biomedical Aspects," Kluwer Academic Publishers, Norwell, MA, 1993.

Evans, A.S. and P.S. Brachman: "Bacterial Infections of Humans. Epidemiology and Control," 3rd Edition, Plenum Medical Book Company, New York, NY, 1998.

Montecucco, C.: "Clostridial Neurotoxins: The Molecular Pathogenesis of Tetanus and Botulism," Vol. 195, Springer-Verlag, Inc., New York, NY, 1995.

Udwadia, F.E.E.: "Tetanus," Oxford University Press, Inc., New York, NY, 1994.

Web Reference

Centers for Disease Control and Prevention: *http://www.cdc.gov/health/diseases.htm*

R.C. VICKERY, M.D., D.Sc., Ph.D., Blanton/Dade City, FL.

TETANY. See **Parathyroid Glands**.

TETRACHORIC CORRELATION. If a bivariate normal distribution is dichotomized according to each variable, the value of ρ, the correlation coefficient, is uniquely determined by the proportions falling into the four resulting categories. If an observed 2×2 contingency table is regarded as having arisen in this way, an estimate of ρ can be obtained, using Tables of the bivariate normal distribution. Such an estimate is called a *tetrachoric correlation*.

TETRADIC. A dyadic converts a vector into another vector and a tetradic converts a dyadic into another dyadic. Thus, if ϕ and ψ are dyadics, the tetradic with 81 components can be represented by the Hebrew character (gimel) and $\psi =: \phi$. The components of the tetradic require four subscripts, while only two are needed for the dyadics. The relation between various components is

$$\psi_{ij} = \sum_{k,l} G_{ijkl}\phi_{kl}$$

An indemfactor, a unit tetradic, a conjugate, etc., can be defined as for a dyadic. These operators can be applied to the study of stress and strain in nonisotropic solids. See also **Dyadic**.

TETRADYMITE. A mineral, bismuth tellurium sulfide, corresponding to the formula Bi_2Te_2S. It is rhombohedral. Tetradymite occurs usually in gold quartz veins. It is found in Norway, Sweden, England, Bolivia, British Columbia; and in the United States, in Virginia, North Carolina, Georgia, Montana, Colorado, and elsewhere. It derives its name from the Greek word meaning fourfold, in reference to the double twin crystals occasionally developed.

TETRAHEDRITE. A mineral of the composition, $(Cu, Fe)_{12}As_4S_{13}$, isomorphous with tennantite. The color ranges from steel-gray to iron-black. The mineral frequently contains cobalt, lead, mercury, nickel, silver, or zinc in replacement of the copper. Tetrahedrite usually occurs in tetrahedral crystals associated with copper ores. The mineral is considered an important copper ore and sometimes is a valuable ore for silver. The mineral sometimes is referred to as *fahlore, gray copper ore*, and *stylotypite*.

TETRAHEDRON. See **Polyhedron**.

THALAMUS. See **Central and Peripheral Nervous Systems**.

THALLIUM. Chemical element symbol Tl, at. no. 81, at. wt. 204.38, periodic table group 3, mp 303.5 °C, bp 1447–1467 °C, density 11.85 g/cm³ (20 °C). Elemental thallium has a hexagonal close-packed crystal structure normally, but also exhibits a face-centered cubic crystal structure.

Thallium metal is bluish-gray upon fresh exposure, changing to dark gray on standing, this oxidation increased with temperature above 25 °C; soft, and may be easily cut with a knife. It is malleable but of low tenacity, so that it must be extruded to form wire; HNO_3 is the best solvent; forms alloys with many metals, e.g., mercury, cadmium, zinc, silver, copper, magnesium.

The element was first identified by Sir William Crookes spectrographically in 1861. While seeking tellurium, Crookes observed the characteristic bright green lines in the emission spectrum of thallium. At just about the same time, A. Lamy identified the element. Thallium occurs naturally as ^{203}Tl and ^{205}Tl. Eleven radioactive isotopes have been identified ^{198}Tl through ^{202}Tl, ^{204}Tl, and ^{206}Tl through ^{210}Tl. With exception of ^{204}Tl which has a half-life of 4.07 years, the other isotopes have relatively short half-lives expressed in minutes, hours, and days. See also **Radioactivity**. Thallium is not considered an abundant element, estimates of occurrence in the earth's crust ranging from 0.3 to 3.0 ppm. In a list of 65 chemicals found in seawater, thallium does not appear. First ionization potential 6.106 eV; second, 20.32 eV; third, 29.7 eV. Oxidation potentials $Tl \rightarrow Tl^+ + e^-$, 0.336 V; $Tl^+ \rightarrow Tl^{3+} + 2e^-$, −1.25 V; $Tl + OH^- \rightarrow TlOH + e^-$, 0.3445 V; $TlOH + 2OH^- \rightarrow Tl(OH)_3 + 2e^-$, 0.05 V. Electron configuration $1s^2 2s^2 2p^6 3s^2 3p^6 3d^{10} 4s^2 4p^6 4d^{10} 4f^{14} 5d^2 5p^6 5d^{10} 6s^2 6p^1$. Other important physical properties of thallium are given under **Chemical Elements**.

Thallium occurs in small amounts in pyrite, zinc blende, and hematite of certain localities, and in a few rare minerals in Sweden and Macedonia. For the recovery of thallium from the flue dust of pyrite burners, the dust is boiled with H_2O, allowed to stand some time, filtered, and HCl added to the filtrate, whereupon crude thallous chloride is precipitated. This is purified by further treatment, and thallium metal obtained (1) by electrolysis of the sulfate solution or (2) by fusion of the chloride with sodium cyanide and carbonate.

Uses: Because thallium is recovered from smelting lead and zinc concentrates, it is available to fulfill any new uses up to several thousand pounds per year. To date, practical applications have been relatively limited. Thallium-activated sodium iodide crystals find use in photomultiplier tubes. It has been learned that thallium bromoiodide crystals transmit infrared radiation and that crystals of thallium oxysulfide detect infrared radiation. A combination of these crystals has been used in military communication systems. Because of their density, both thallous formate and thallous malonate have been used in the preparation of heavy-liquid sink-float solutions used in the gravity separation of minerals. Mixtures of thallium, arsenic, sulfur, and selenium form low-melting-point glasses for encapsulation of semiconductors has been under investigation. It has been found that the addition of small amounts of thallium to the counterelectrode alloy used in selenium rectifiers will improve the performance of the rectifiers. Claims have been made that the addition of a thallium salt to absorb traces of oxygen in tungsten-filament incandescent lamps will increase lamp life. Also, it has been shown that the addition of thallium to various glass formulations will improve optical properties and increase the refractive index.

For a number of years, thallium sulfate had been used in rodenticides. Some use of thallium has been made in connection with alloys for low-temperature applications, particularly for switches, seals, and thermometers. The ternary eutectic mercury-thallium-indium alloy has a freezing point of $-63.3\,°C$, while the binary eutectic mercury-thallium alloy has a freezing point of $-60\,°C$. These freezing points are considerably lower than that of mercury usually used for similar applications at higher temperatures. Mercury freezes at $-38.87\,°C$.

Toxicity: Thallium and thallium compounds are toxic and skin contact must be avoided. Impervious gloves and aprons should be worn and excellent ventilation and masks should be provided where dusts and fumes may be present.

Chemistry and Compounds: Oxidation states: thallous, Tl^+; thallic, Tl^{3+}.

Because of low oxidation potential of thallium to form Tl^+, thallium is quite reactive, dissolving slowly in most dilute mineral acids to form thallium(I) solutions. The thallium(I) halides are insoluble in water, but thallium trihalides are soluble; the latter are formed by treatment of the thallium(I) halide in solution with the corresponding halogen. Thallium(III) iodide, however, does not exist, TlI_3 being $[Tl^+][I_3^-]$.

Thallium(III) compounds are readily reduced to the thallium(I) state (see difference in oxidation potentials above) and are thus fairly strong oxidizing agents. Thallium(I) compounds resemble those of the alkali metals in many respects, including a soluble, strong basic hydroxide (TlOH) ($K_B = 0.14$), a soluble carbonate (Tl_2CO_3), the formation of well crystallized salts, including those with complex anions, the formation of polysulfides (Tl_2S_5), and polyiodides (thus TlI_3 contains the monovalent metal ion, like rubidium iodide, RbI_3 and cesium iodide, CsI_3). Thallium(I) ion resembles silver in forming insoluble halides, sulfide and chromate. The thallium(I) ion forms only weak complexes (probably because of its larger size and low charge) but the thallium(III) ion forms strong ones. There are four complex chloro ions $[TlCl_4]^-$, $[TlCl_5]^{2-}$, $[TlCl_6]^{3-}$, and $[Tl_2Cl_9]^{3-}$, the last having the six-coordinated structure

$$\begin{bmatrix} Cl & & Cl & & Cl \\ Cl\!-\!Tl\cdots & Cl\cdots & Tl\cdots & Cl \\ Cl & & Cl & & Cl \end{bmatrix}^{3-}$$

The complex compounds include the chelates, such as the oxine chelate, and also such compounds as $Tl(TlCl_4)$, $Tl_3(TlBr_6)$, $TlCl_3 \cdot 3NH_3$, $14Rb_3TlBr_6 \cdot 16H_2O$ (here the presence of the ion $[Tl(H_2O)_8]^+$ has been shown), oxalates, such as $H[Tl(C_2O_4)_2]$ (dioxalatothallic acid), $K_2[Tl(C_2O_4)_2(NO_2)_2] \cdot H_2O$ and a number of complex hydrides, as well as the unstable binary hydride TlH_3.

The most readily prepared organometallic compounds are the dialkyl ones of the type R_2TlX, where X is an acid radical accompanying the ion $[R_2Tl]^+$. The trialkyl compounds of the type TlR_3 are immediately decomposed by H_2O, giving RH and R_2TlOH, in which the thallium atom is isoelectronic with the mercury atom in R_2Hg, and which is a strong base: $(C_2H_5)_2TlOH$, $K_B = 0.90$.

Additional Reading

Carter, G.F. and D.E. Paul: "Materials Science and Engineering," ASM International, Materials Park, OH, 1991.
Considine, D.M. and G.D. Considine: "Van Nostrand Reinhold Encyclopedia of Chemistry," 4th Edition, Van Nostrand Reinhold, New York, NY, 1984. (A Classic Reference).
Greenwood, N.N. and A. Earnshaw: "Chemistry of the Elements," 2nd Edition, Butterworth-Heinemann, Inc., Woburn, MA, 1997.
Hermann, A.M. and J.V. Yakhmi: "Thallium-Based High-Temperature Super Conductors," Marcel Dekker, Inc., New York, NY, 1994.
Krebs, R.E.: "The History and Use of Our Earth's Chemical Elements: A Reference Guide," Greenwood Publishing Group, Inc., Westport, CT, 1998.
Lewis, R.J. and N.I. Sax: "Sax's Dangerous Properties of Industrial Materials," 10th Edition, John Wiley & Sons, Inc., New York, NY, 1999.
Lide, D.R.: "CRC Handbook of Chemistry and Physics 2000–2001," 81st Edition, CRC Press, LLC., Boca Raton, FL, 2000.
Nriagu, J.O.: "Thallium in the Environment," Vol. 30, John Wiley & Sons, Inc., New York, NY, 1998.
Staff: "ASM Handbook—Properties and Selection: Nonferrous Alloys and Pure Metals," ASM International, Materials Park, OH, 1990.

S.C. CARAPELLA, Jr., ASARCO Incorporated,
South Plainfield, NJ.

THALLOPHYTA.   See **Fungus**.

THAMIN DEER.   See **Deer**.

THEODOLITE.   An optical instrument that consists of a sighting telescope so mounted that it is free to rotate around horizontal and vertical axes. The instrument incorporates graduated scales so that angles of rotation may be measured. Theodolites normally are calibrated for (1) collimation value, i.e., the centering of the cross hair to the telescope, (2) trunnion error, i.e., the truth of rotation of the telescope in a true vertical plane, and (3) the centering of the axis of rotation. Accuracy of the circle, position of the plate bubble, and accuracy of the micrometers also are measurable and correctable features. Values usually may be read in fractions of a second of arc.

Theodolites are used for alignment of large jigs and fixtures in manufacturing operations. Through triangulation, the instruments also can be used for length measurements. With addition of an autocollimation system, the theodolite can be used to establish incremental spacing of divided circles or any similar object, such as the tooth spacing of a large gear. A theodolite with a two-sided reading system eliminates errors of centering of circles and, therefore, when using the instrument with an autocollimating eyepiece, it is not necessary that the theodolite be precisely centered to the part under measurement. Since the instrument is rotatable in both the horizontal and vertical axis, the device essentially combines for the metrologist the value of two optical dividing units.

In meteorology, the theodolite is used to observe the motion of a pilot balloon. The telescope is usually fitted with a right-angle prism so that the observer continues to look horizontally into the eyepiece whatever the variation of the elevation.

THEOREM.   An assertion of which there is a proof. With few exceptions, notably Fermat's last theorem, any assertion referred to as a theorem has been rigorously proved. Examples from algebra are the Fundamental Theorem of Algebra, the Remainder Theorem, and the Factor Theorem.

THEORY OF EQUATIONS.   The branch of mathematics concerned with determining the solvability of algebraic equations and with developing methods for solving them. An equation of the form

$$a_0x^n + a_1x^{n-1} + \cdots + a_n = 0$$

where n is a positive integer and $a_0, a_1, \ldots, a_n$ are constants contained in the field of complex numbers of which $a_0 \neq 0$, is called an integral rational equation of the nth degree in x. (See also **Algebraic Equations**.) The left hand member is called a polynomial of the nth degree in x.

The remainder theorem states that the remainder obtained in dividing a polynomial $f(x)$ by a monomial $x - c$ is equal to $f(c)$. The factor theorem states that if r is a zero of a polynomial $f(x)$, then $(x - r)$ is a factor of $f(x)$. This indicates that if r is a root of an integral rational equation $f(x) = 0$, then $(x - r)$ is a factor of $f(x)$. The fundamental theorem of algebra states that every integral rational equation in a single variable x has at least one root and a corollary states that every polynomial $f(x)$ of degree n can be decomposed into exactly n linear factors. A repeated factor indicates a multiple root of the equation $f(x) = 0$. If a factor $x - r$ occurs k times with $k > 1$, r is called a root of multiplicity k. Every integral rational equation of degree n has exactly n roots, where a root of multiplicity k is counted as k roots. If the coefficients of the polynomial $(a_0, a_1, \ldots, a_n)$ are limited to the real number field, or any other field contained in the complex number field, the above theorems still hold; however, complete factorization will, in general, require complex roots.

By expressing a general polynomial $f(x)$ as a product of n linear factors and equating coefficients of like powers of x one can determine relationships between the coefficients of an integral rational equation

$$a_0 x^n + a_1 x^{n-1} + \cdots + a_n = 0$$

and its n roots $r_1, r_2, \ldots, r_n$. The relationships between $a_0, a_1, \ldots, a_n$ and $r_1, r_2, \ldots, r_n$ are

$$\frac{a_1}{a_0} = -(r_1 + r_2 + \cdots + r_n),$$

$$\frac{a_2}{a_0} = (r_1 r_2 + r_1 r_3 + \cdots r_1 r_n + \cdots + r_{n-1} r_n),$$

$$\frac{a_3}{a_0} = -(r_1 r_2 r_3 + r_1 r_2 r_4 + \cdots + r_1 r_2 r_n + \cdots + r_{n-2} r_{n-1} r_n),$$

$$\cdots\cdots\cdots\cdots\cdots\cdots\cdots\cdots\cdots$$

$$\frac{a_n}{a_0} = (-1)^n (r_1 r_2 r_3 r_4 \cdots r_n).$$

For an equation with real coefficients, it is known that if $a + ib$ is a root where a and b are real then $a - ib$ is also a root. For an equation with rational coefficients, if $a + \sqrt{b}$, where a and b are rational but $\sqrt{b}$ is not, is a root, then $a - \sqrt{b}$ is also a root. For an equation with integral coefficients, all rational roots may be found by determining whether the fractions whose numerators are factors of the constant term, and whose denominators are factors of the coefficient of the highest power of x, satisfy the equation.

Various methods can be used for determining bounds for the real roots of an equation. For an equation with real coefficients, it can be shown that the equation cannot have a real root greater than $|a_k/a_0| + 1$ or less than $-|a_k/a_0| - 1$, where a_0 is the coefficient of the highest power of x and a_k is the coefficient of greatest numerical value. Also, it can be shown that there are no positive roots less than $|a_n|(|a_n| + |a_k|)$, where a_k is the coefficient of greatest numerical value as before and a_n is the constant term.

An equation is said to be solved in terms of radicals if the roots can be expressed in terms of the coefficients using only the operations of addition, subtraction, multiplication, division, and extraction of roots. By the middle of the sixteenth century, general equations through degree 4 had been solved in terms of radicals. See also **Quadratic Equation** ($n = 2$), **Cubic Equation** ($n = 3$), and **Biquadratic Equation** ($n = 4$). Attempts to solve higher degree equations led nowhere for over 200 years. Ruffini and Abel are credited with proving that equations of degree higher than 4 cannot be solved by radicals. Those who credit Ruffini admit that his proof was incomplete. Those who credit Abel mention that Ruffini's proof was earlier but incomplete. Other authorities state that Abel's proof contained two oversights and credit Galois with the first definitive proof. Galois's theory proves the impossibility of solving equations of degree higher than 4 by establishing a correspondence between the permutations of roots of equations with rational coefficients in the algebraic relations holding among the roots and groups. If the permutation of roots preserves the validity of all the algebraic relations among the coefficients it is called a permissible permutation. The set of permissible permutations is called the Galois group of the equation. A theorem states that an algebraic equation is solvable if and only if its Galois group is solvable. It has been shown that the group of all permutations on n symbols is not solvable when $n \geq 5$, and that there exist equations with rational coefficients which have these groups as Galois groups. Therefore, it follows that solution by radicals is impossible for $n > 4$. Hermite succeeded in solving the general equation of degree 5 in terms of elliptic functions. Recent work in this direction solves the general equation of degree n in terms of Fuchsian functions.

For equations of degree higher than 2, it is usually more practical to determine the roots by an approximate method. The first step is to isolate the roots, that is, identify intervals which contain only one (possibly a multiple) root. See also **Rolle Theorem**; and **Budan Theorem**.

Once the roots have been isolated the roots can be obtained to the desired degree of accuracy using Horner's method of Newton's method for real roots. The Graeffe method can be used for real or complex roots. Newton's method is not limited to algebraic functions. Let $f(x)$ be a function with a root in the interval $a \leq x \leq b$. Let $f'(x)$ denote the derivative of $f(x)$. Newton's method consists of evaluating $f(x)$ and $f'(x)$ for a value $x = x_0$ where $a \leq x_0 \leq b$ and solving the equation

$$f(x_0) + f'(x_0)(x_1 - x_0) = 0$$

for x_1. The solution for x_1 is then used as a new value for x_0 and the process is repeated until the desired accuracy is achieved. As with any method which may diverge or oscillate some care must be taken during its use.

A set of simultaneous equations in several variables can be reduced to a set of equations each of which contains only one variable. This is easily accomplished for linear equations by the use of determinants. See also **Determinant**. For higher degree equations, it may be useful to calculate the resultant. Let

$$f(x) = a_0 x^n + a_1 x^{n-1} + \cdots + a_n = 0,$$

$$g(x) = b_0 x^m + b_1 x^{m-1} + \cdots + b_m = 0$$

be two equations whose coefficients may involve another unknown y. Suppose $a_0 \neq 0$ and let $\alpha_1, \alpha_2, \ldots, \alpha_n$ denote the roots of $f(x) = 0$. If one of these is also a root of $g(x) = 0$ then $a_0^m g(\alpha_1) g(\alpha_2) \cdots g(\alpha_n) = 0$. The left hand side of this equation is a homogeneous polynomial of dimension m in the coefficients $a_0, a_1, \ldots, a_n$, and a homogeneous polynomial of dimension n in the coefficients $b_0, b_1, \ldots, b_m$. This polynomial is called the resultant of $f(x)$ and $g(x)$ and is denoted by $R(f, g)$. The vanishing of the resultant is a necessary and sufficient condition for the two equations $f(x) = 0$ and $g(x) = 0$ to have a common root, provide $a_0 \neq 0$. If $b_0 \neq 0$, a necessary and sufficient condition is $R(g, f) = 0$.

It can be shown that

$$R(g, f) = (-1)^{mn} R(f, g).$$

For the two polynomials

$$f(x) = a_0 x^n + a_1 x^{n-1} + \cdots + a_n \text{ and}$$

$$g(x) = b_0 x^m + b_1 x^{m-1} + \cdots + b_m,$$

the determinant $D(f, g)$ given by

$$D(f, g) = \begin{vmatrix} a_0 & a_1 & a_2 \cdots a_n & & & \\ & a_0 & a_1 \cdots a_n & & & \\ & & \cdots\cdots & & & \\ & & & a_0 & a_1 \cdots a_n & \\ b_0 & b_1 \cdots b_m & & & & \\ & b_0 & b_1 \cdots b_m & & & \\ & & \cdots\cdots & & & \\ & & & b_0 & b_1 \cdots b_m & \end{vmatrix} \begin{matrix} \\ \left.\right\} m \text{ rows} \\ \\ \\ \left.\right\} m \text{ rows} \\ \\ \end{matrix}$$

is known as Sylvester's determinant. The vanishing of Sylvester's determinant is a necessary and sufficient condition that f and g have a common divisor of degree > 0, provided that $a_0 \neq 0$. It can be shown that Sylvester's determinant and the resultant are identical polynomials in the coefficients of f and g. Therefore, the resultant can be calculated by using Sylvester's determinant.

Let $f(x)$ represent nth degree polynomial as before. Let D represent the discriminant. See also **Discriminant**. The discriminant vanishes if and only if the equation $f(x) = 0$ has equal roots. Let $f^1(x)$ be the derivative of $f(x)$. It can be shown that the discriminant is related to the resultant of f and f^1 as

$$(-1)^{n(n-1)/2} a_0 D = R(f, f^1).$$

DONALD R. HODGE, The BDM Corporation, Vienna, VA.

THEORY OF SAMPLING.   See **Sampling**.

THERMAL CONDUCTIVITY.   An important property of materials that determines the rate at which thermal energy may flow through the material. Materials are selected for various purposes based upon their ability to conduct heat well (as in heat exchangers) or upon their resistance to conduct heat (as in insulation products).

In engineering terms, thermal conductivity is the time rate of transfer of heat by conduction through a unit thickness across a unit area for a unit difference of temperature. The units usually used are *calories per second* per *square centimeter* for a thickness of 1 centimeter and for a *temperature difference* of 1 °C.

THERMAL CONVECTION.   See **Convection**.

THERMAL DIFFUSION.   See **Diffusion**.

THERMAL DIFFUSIVITY.   The quantity $K\rho/C_v$, where K is the thermal conductivity, ρ is the density, and C_v is the specific heat per unit mass. The magnitude of this quantity determines the rate at which a body with a nonuniform temperature approaches equilibrium.

THERMAL EFFICIENCY.   Thermal efficiency is output, in heat units, divided by the heat supplied or chargeable. Thermal efficiency may partially define the operating condition of both a machine or a static piece of equipment. In the case of static equipment, if it is well insulated it may have very high thermal efficiencies. A machine which converts heat supplied into work output (the steam engine, Diesel engine, etc.) is always characterized by low thermal efficiencies because the conversion of the low-grade type of heat energy into high-grade energy of mechanical work is accomplished with considerable difficulty. The thermal efficiencies of the best prime movers today rarely exceed 35%, and are not found higher than 40% even when optimum conditions of loading, maintenance, and fuel employed are present. Thermal efficiency of a prime mover may be based on the output at the shaft per unit of heat supplied, or upon the electrical output in the case of a generator drive. It may also be based on cylinder horsepower in piston and cylinder prime movers. Heat supplied and chargeable to internal combustion engines is the heat in the fuel. Heat supplied to steam prime movers is the heat in the steam at the throttle; the heat chargeable is the heat supplied after deducting the heat of condensate in the exhaust.

THERMAL EQUATOR.   The belt of maximum temperature surrounding the earth, which moves north and south with (but lagging) the sun's motion. It is also spoken of as the center of the area bounded by the yearly mean isotherms of 80 °F (26.7 °C). See also **Climate**.

THERMAL GRADIENT (or Geothermal Gradient).   1. According to Smithsonian Physical Tables, the rate of variation of temperature in soil and rock from the surface of the earth down to depths of the order of kilometers. The thermal gradient varies greatly from place to place, depending on the geological history of the region, the radioactivity of the underlying rocks, and the conductivity of the upper rocks. An average figure is about +10 °C per kilometer.

2. In the atmosphere, the maximum rate of decrease of temperature with distance. Temperature gradients directed vertically are known as lapse rates.

THERMAL INSTABILITY.   See **Atmosphere (Earth)**.

THERMAL JET ENGINE.   A jet engine, that utilizes heat to expand gases for rearward ejection. This is the usual form of aircraft jet engine.

THERMAL NOISE.   See **Noise**.

THERMAL RADIATION.   All bodies that are not at absolute zero emit radiation excited by the thermal agitation of their molecules or atoms, whether there are other causes of excitation or not. This thermal radiation ranges in wavelength from the longest infrared to the shortest ultraviolet rays, its spectral energy distribution, however, depending upon the nature of the body and upon its temperature. The total emissive power of a surface at any temperature is the rate at which it emits energy of all wavelengths and in all directions, per unit area of radiating surface. The flux density (per unit solid angle) in various directions obeys the cosine emission law approximately; but strictly only in the case of a black body. See also **Wien Laws**.

THERMAL STABILITY.   A phase must satisfy certain conditions if it is to be stable (or metastable). For systems consisting of a single component the necessary and sufficient conditions are the conditions of thermal stability and of mechanical stability. The condition of thermal stability is

$$C_v > 0$$

The heat capacity at constant volume of all stable (or metastable) phases is positive.

THERMAL WIND.   See **Winds and Air Movement**.

THERMIONIC CONVERSION.   The process whereby electrons released by thermionic emission are collected and utilized as electric current. The simplest example of this is provided by a vacuum tube, in which the electrons released from a heated anode are collected at the cathode or plate. Used as a method of producing electrical power for spacecraft.

THERMIONIC EMISSION.   Direct ejection of electrons as the result of heating and material, which raises electron energy beyond the binding energy that holds the electron in the material.

THERMISTOR.   The name "thermistor" is derived from *therm*ally sensitive re*sistor*, since the resistance of a thermistor varies as a function of temperature. Although semiconductors have been known as a class of materials for many years, only since the 1950s have sufficient theory and manufacturing techniques been developed to spur their use as primary temperature elements.

A thermistor is an electrical device made of a solid semiconductor with a high-temperature coefficient of resistivity, which exhibits a linear voltage-current characteristic when its temperature is held constant. When a thermistor is used as a temperature-sensing element, the relationship between resistance and temperature is of primary concern. The approximate relationship applying to most thermistors is:

$$R_T = R_0 \exp B \left(\frac{1}{T} - \frac{1}{T_0} \right) \quad (1)$$

where R_0 = resistance value at reference temperature T_0 °K, ohms.
    R_T = resistance at temperature T °K, ohms.
    B = constant over temperature range, dependent upon manufacturing process and construction characteristics (specified by supplier)

$$B \cong \frac{E}{K} \quad (2)$$

where E = electron volt energy level
    K = Boltzmann's constant (8.625×10^{-5} eV/°K)

A second form of the approximate resistance-temperature relationship is written in the form

$$R_T = R_\infty e^{B/T} \quad (3)$$

where R_∞ = thermistor resistance as temperature approaches infinity, ohms.

Both equations (1) and (2) are only best approximations and, therefore, are of limited use in making highly accurate temperature measurements. However, they do serve to compare thermistor characteristics and thermistor types. The temperature coefficient usually is expressed as a percentage change in resistance per degree of temperature change and is approximately related to B by

$$a = \frac{dR}{dT} \left(\frac{1}{R} \right) = \frac{-B}{T_0^2} \quad (4)$$

where T_0 is in degrees Kelvin. It should be noted that the resistance of the thermometer is solely a function of its absolute temperature. Furthermore, it is apparent that the thermistor's resistance-temperature function has a characteristic high negative coefficient as well as a high degree of nonlinearity. The value of the coefficient for a common commercial thermistor is on the order of 2–6% per °K at room temperature. This value is approximately ten times that of metals used in the manufacture of resistance thermometers.

Resultant considerations due to the high coefficient characteristic of thermistors include inherent high sensitivity and high level of output, eliminating the need for extremely sensitive readout devices and leadwire matching techniques. However, limitations on interchangeability (particularly over wide temperature ranges), calibration, and stability — also inherent to thermistors — are quite restrictive. The high degree of nonlinearity in the resistance-temperature function usually limits the range of the readout instrumentation. In many applications, special prelinearization circuits must be used before interfacing with related system instrumentation. The negative temperature coefficient also may require an inversion (to positive form) when interfacing with some analog and/or digital instrumentation.

A number of metal oxides and their mixtures, including the oxides of cobalt, copper, iron, magnesium, manganese, nickel, tin, titanium, uranium, and zinc are among the most common semiconducting materials used in the construction of thermistors. Usually compressed into the desired shape from specially formulated powder, the oxides are then recrystallized by heat treatment, resulting in a dense ceramic body. The leadwires are then attached while electric contact is maintained, and the finished thermistor is then encapsulated. Some common configurations of thermistors are shown in Fig. 1. A comparison of the resistance-temperature characteristics of thermistors with platinum metal used in resistance thermometers is given in Fig. 2.

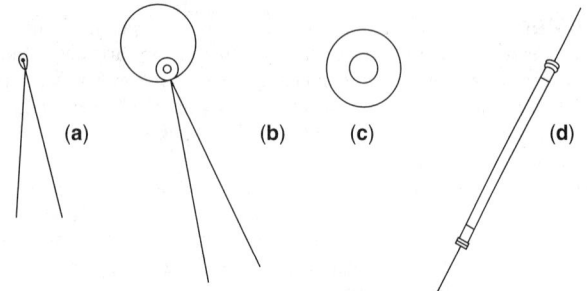

Fig. 1.   Thermistor configurations: (**a**) bead, (**b**) disc, (**c**) washer, (**d**) rod.

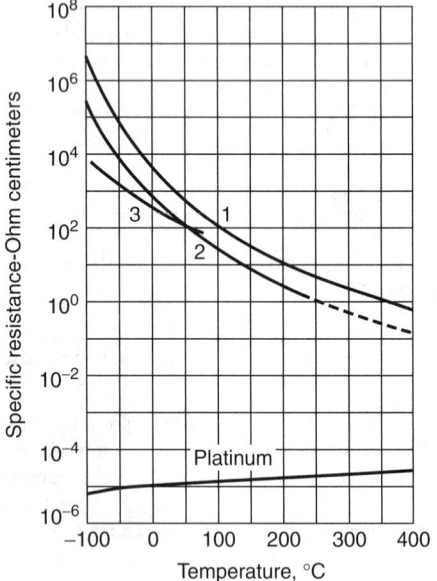

Fig. 2.   Resistance-temperature characteristics of thermistors as compared with platinum metal used in resistance thermometers.

THERMOCHEMISTRY.   That aspect of chemistry which deals with the heat changes which accompany chemical reactions and processes, the heat produced by them, and the influence of temperature and other thermal quantities upon them.

THERMOCLINE.   That layer of water in the oceans situated between the relatively warm surface water and the much colder main mass of water at the bottom. The temperature falls rather sharply in this layer, or zone, in contrast to the others that are relatively uniform. See also **Ocean**; and **Solar Energy**.

THERMOCOUPLE.   In 1821, Seebeck discovered that an electric current flows in a continuous circuit of two metals if the two junctions are at different temperatures, as shown in Fig. 1. A and B are two metals, T_1

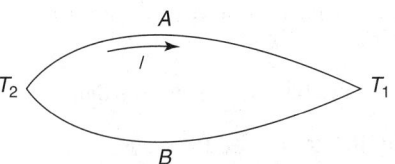

Fig. 1.   Simple thermocouple circuit.

and T_2 are the temperatures of the junctions. I is the thermoelectric current. A is thermoelectrically positive to B if T_1 is the colder junction. In 1834, Peltier found that current flowing across a junction of dissimilar metals causes heat to be absorbed or liberated. The direction of heat flow reverses if current flow is reversed. Rate of heat flow is proportional to current but depends upon both temperature and the materials at the junction. Heat transfer rate is given by PI, where P is the Peltier coefficient in watts per ampere, or the Peltier emf in volts. Many studies of the characteristics of thermocouples have led to the formulation of three fundamental laws:

1. *Law of the homogeneous circuit.* An electric current cannot be sustained in a circuit of a single homogeneous metal; however it may vary in section, by the application of heat alone.

2. *Law of intermediate metals.* If in any circuit of solid conductors the temperature is uniform from any point P through all the conducting matter to a point Q, the algebraic sum of the thermoelectromotive forces in the entire circuit is totally independent of this intermediate matter and is the same as if P and Q were put in contact.

3. *Law of successive or intermediate temperatures.* The thermal emf developed by any thermocouple of homogeneous metals with its junctions at any two temperatures T_1 and T_3 is the algebraic sum of the emf of the thermocouple with one junction at T_1 and the other at any other temperature T_2 and the emf of the same thermocouple with its junctions at T_2 and T_3. See Fig. 2.

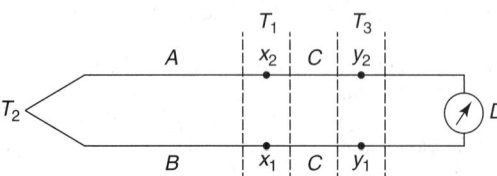

Fig. 2.   Law of Intermediate Metals makes it possible to use "foreign" wires to connect thermocouple to measuring instrument. Thermocouple materials A and B can be connected to the instrument by use of connecting materials C and D. If the temperatures at X_1 and X_2 are both at T_1 and if temperatures at Y_1 and Y_2 are both at T_3, the emf of the circuit will be independent of materials C and D.

Common thermocouple wire combinations used in industry are listed in Table 1. A choice of different metals is needed to fulfill a broad range of temperatures as well as for oxidizing or reducing conditions in use. The temperature-thermal emf curves for common types of thermocouples are given in Fig. 3. The hot junction of a thermocouple may be joined by any means that will ensure good electrical continuity when in use. Commonly,

TABLE 1. COMMONLY USED THERMOCOUPLES AND TEMPERATURE RANGES

ANSI Type	Positive Element	Negative Element	Normal Temperature Range	
			°F	°C
B	Platinum 20% Rhodium	Platinum 6% Rhodium	1,600–3,100	870–1,700
E	Chromel	Constantan	32–1,600	0–870
J	Iron	Constantan	32–1,400	0–760
K	Chromel	Alumel	32–2,300	0–1,260
R	Platinum 13% Rhodium	Platinum	32–2,700	0–1,480
S	Platinum 10% Rhodium	Platinum	32–2,700	0–1,480
T	Copper	Constantan	−300–+700	−180–+370

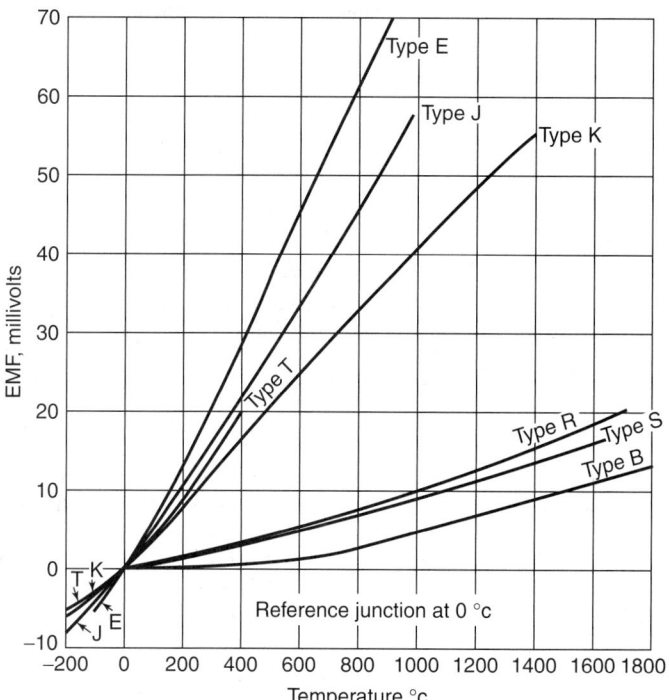

Fig. 3.   Temperature-thermal emf curves for common types of thermocouples. (*Honeywell.*)

Additional Reading

Jackson, D.A. and A.E. Mushin: "Thermocouples," in Process Instruments and Controls Handbook, 3rd Ed (D.M. and G.D. Considine, Eds.), The McGraw-Hill Companies, New York, NY, 1985.

Kerlin, T.W.: "Practical Thermocouple Thermometry," ISA, Research Triangle Park, NC, 1998.

Pollock, D.D.: "Thermocouples: Theory and Properties," CRC Press, LLC., Boca Raton, FL, 1991.

Staff: "Temperature Measurement Handbook," Omega Press, Stamford, CT, 1982.

Staff: "Manual on the Use of Thermocouples in Temperature Measurements," STP 470B, American Society for Testing and Materials, Philadelphia, PA, 1993.

THERMODYNAMIC EQUILIBRIUM.   See Equilibrium.

THERMODYNAMICS.   Classical thermodynamics is a theory which on the basis of four main laws and some ancillary assumptions deals with general limitations exhibited by the behavior of macroscopic systems. Phenomenologically it takes no cognizance of the atomic constitution of matter. All *mechanical* concepts such as kinetic energy or work are presupposed. Thermodynamics is motivated by the existence of dissipative mechanical systems. A *thermodynamic system* K may be thought of as a collection of bodies in bulk; when its condition is found to be unchanging in time (on a reasonable time scale) it is *in equilibrium*. It is then characterized by the values of a finite set of, say, n physical quantities, it being supposed that none of these is redundant. Such a set of quantities constitutes the *coordinates* of K, denoted by $x(= x_1, \ldots, x_n)$. Any set of values of these is a *state* $\mathfrak{S}$ of K. In virtue of these definitions, K is in a state only when it is in equilibrium. The passage of K from a state $\mathfrak{S}$ 1 to a state $\mathfrak{S}'$ is a *transition* of K. A transition is *quasi-static* if in its course it goes through a continuous sequence of states, and if the forces that do work on the system are just those which hold it in equilibrium. A transition is *reversible* if there exists a second transition that restores the initial state, the final condition of the surroundings of K being the same as the initial condition. Reversible transitions are assumed to be quasi-static.

An enclosure, such that the equilibrium of a system contained within it can only be disturbed by mechanical means, is *adiabatic*, otherwise it is *diathermic*. For instance, stirring, or the passage of an electric current, constitute "mechanical means." A system K_0 in an adiabatic enclosure is *adiabatically isolated*, but this does not preclude mechanical interactions with the surroundings. Its transitions are then called adiabatic.

For the time being, the masses of all substances present will be supposed fixed, and to achieve simplicity it will be given that (1) there are no substances present whose properties depend on their previous histories, and (2) capillary forces as well as long-range interactions are absent. Further, it will be supposed that of the n coordinates of K, just $n - 1$ have geometrical character (*deformation coordinates*, e.g., volumes of enclosures), so that the work done by K in a quasi-state transition is

$$\int dW = \int \sum_{k=1}^{n-1} P_k(x)\, dx_k \qquad (1)$$

Such a system will be called a *standard system* ($n - 1$ enclosures in diathermic contact, each containing a simple fluid, may serve as example, x_n being any one of the pressures).

The Zeroth Law.   Suppose two systems $K_A(x)$ and $K_B(y)$ to be in mutual diathermic contact. Experience shows that the states $\mathfrak{S}_A$ and $\mathfrak{S}_B$ cannot be assigned arbitrarily, but that there exists a necessary relation of the form

$$f(x; y) \equiv f(x_1, \ldots, x_n; y_1, \ldots, y_m) = 0 \qquad (2)$$

between them. If K_C is a third system, its diathermic equilibrium with K_B on the one hand, or with K_A on the other, is governed by conditions

$$g(y; z) = 0 \qquad (3)$$

and

$$h(z; x) = 0 \qquad (4)$$

respectively. That these three functions are not independent is expressed by the *Zeroth Law: If each of two systems is in equilibrium with a third system then they are in equilibrium with each other.* It follows that any two of Equations (2) through (4) imply the third, i.e., they must be equivalent to equations of the form

$$\xi(x) = \eta(y) = \zeta(z) \qquad (5)$$

the two wires are twisted together and either welded or silver-soldered. Simple clamping of wires together provides adequate connection for short-term use in clean atmospheres at lower temperatures. For industrial applications, the thermocouple is usually placed within a protecting tube. A typical assembly is shown in Fig. 4. For lower temperatures, carbon steel may be used. As the temperature goes up, wrought iron, stainless steel, nickel, nickel-chromium-iron, fused silica, silica-alumina, silicon carbide, alumina, and beryllia may be used. Beryllia protecting tubes will withstand operating temperatures of up to 4,000 °F (~2,200 °C). For some applications, disposable-tip thermocouples have been developed. These are particularly effective for high-temperature molten-metal temperature measurements.

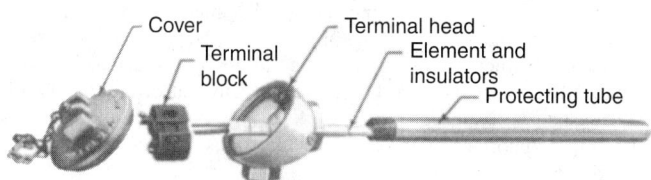

Fig. 4.   Assembly of industrial thermocouple, terminal block, and protecting tube. (*Honeywell.*)

The emf developed by a thermocouple depends upon the temperature of both the measuring and reference junctions. Thus, to determine temperature, the following data must be known: (1) the calibration data for the particular thermocouple; (2) the measured emf; and (3) the temperature of the reference junction. In laboratory cases, the reference junction can be maintained at the freezing temperature of water. However, in most modern instruments, the ambient temperature of the reference junction is sensed, and the correction is incorporated in the measurement circuitry.

Multiple thermocouples may be used in parallel and connected to a single instrument. A typical application is a fire-warning system. Multiple thermocouples also may be connected in series. This is a means for obtaining the average temperature of an object. Also two or more thermocouples may be connected in series so that the emf outputs of the couples are additive. In application installations, thermopiles sometimes are used to detect the presence or absence of the pilot flame and cause a relay to turn off the main gas supply valve.

Thus, with each system there is now associated a function, its *empirical temperature function*, such that two systems can be in equilibrium if and only if their *empirical temperatures* (i.e., the values of their empirical temperature functions) are equal. Write $t = \xi(x)$; so that one has the *equation of state* of K_A. Also, t may be introduced in place of any one of the x_k. Note that the empirical temperature is not uniquely determined since $t_A = t_B$ may be replaced by $\phi(t_A) = \phi(t_B)$ where the function ϕ is monotonic but otherwise arbitrary: one has a choice of *temperature scales*. For a system not in equilibrium, temperature is not defined.

The First Law. It is obvious that one can do mechanical work upon a system (say by stirring) while its initial and final states are the same. (Nothing is being said about the surroundings!) In this sense mechanical energy is not conserved. One might, however, hope that it is conserved at least in a restricted class of transitions. That this is so is asserted by the *First Law: The work W_0 done by a system K_0 in an adiabatic transition depends on the terminal states alone.* Thus, if $\mathfrak{S}'(x')$, $\mathfrak{S}''(x'')$ are the terminal states

$$W_0 = F(x'; x'')$$

If $\mathfrak{S}'''(x''')$ is a third state, and the previous transition proceeds via $\mathfrak{S}'''$, W_0 must not depend on x''', i.e.,

$$F(x'; x''') + F(x'''; x'') \equiv F(x'; x'')$$

It follows that there must exist a function $U(x)$, defined to within an arbitrary additive constant, such that

$$F(x'; x'') = U(x') - U(x'')(= -\Delta U, \text{ say})$$

$U(x)$ is the *internal energy function* of K. (To make sure that U is in fact defined for all states, one assumes that *some* adiabatic transition always exists between any pair of given states.) The energy of a compound standard system is the sum of the energies of its constituent standard systems. Further, U must be a monotonic function of t, and it is convenient to choose the scale of t such that $\partial U / \partial t > 0$.

When the transition from $\mathfrak{S}'$ to $\mathfrak{S}''$ is adiabatic, $W_0 + \Delta U$ vanishes by definition of U. If the transition is not adiabatic and W is the work done by K, the quantity

$$\Delta U + W(= Q, \text{ say}) \tag{6}$$

will in general fail to vanish. Q is then called the *heat absorbed* by K. Every element of a quasistatic adiabatic transition is subject to $dQ = 0$, i.e., by Equations (1) and (6), to the differential equation

$$\sum_{k=1}^{n-1} \left(P_k(x) + \frac{\partial U(x)}{\partial x_k} \right) dx_k + \frac{\partial U}{\partial t} dt = 0 \tag{7}$$

The Second Law. Experiment shows that if $\mathfrak{S}'$ and $\mathfrak{S}''$ are arbitrarily prescribed states, then it may be that no adiabatic transition from $\mathfrak{S}'$ to $\mathfrak{S}''$ exists. When this is the case one says that $\mathfrak{S}''$ is *inaccessible* from $\mathfrak{S}'$, but $\mathfrak{S}'$ is then accessible from $\mathfrak{S}''$, as has been already assumed. The states may of course happen to be mutually accessible. The existence of states adiabatically inaccessible from a given state is asserted precisely by the *Second Law: In every neighborhood of any state $\mathfrak{S}$ of an adiabatically isolated system there are states inaccessible from $\mathfrak{S}$.* (This formulation of the Second Law is known as the *Principle of Carathéodory.*) A fortiori this law applies to quasistatic transitions, i.e., those which satisfy Equation (7). It asserts there are states $\mathfrak{S}''$ near $\mathfrak{S}'$ such that no functions $x_k(t)$ exist which satisfy Equation (7) and whose values when $t = t''$ are just x_k'', $(k = 1, \ldots, n-1)$. It is merely a mathematical problem (the Theorem of Carathéodory) to prove that this is the case if and only if there exist functions $\lambda(x)$ and $s(x)$, $(x_n \equiv t)$ such that the left-hand member is identically equal to λds, where ds is the total differential of s. Thus, the Second Law entails that

$$dQ = dU + dW = \lambda ds \tag{8}$$

(dQ is of course not a total differential), s is called the *empirical entropy function* of K. It is not uniquely determined, since it may be replaced by any monotonic function of s. If two standard systems K_A and K_B in diathermic contact make up a compound system K_C; $dQ_C = dQ_A + dQ_B$, i.e., because of Equation (8),

$$\lambda_A ds_A + \lambda_B ds_B = \lambda_C ds_C$$

By including s_A, s_B and the common empirical temperature t among the coordinates of K_C, one infers that

$$\lambda_A = T(t)\theta_A(s_A), \quad \lambda_B = T(t)\theta_B(s_B),$$
$$\lambda_C = T(t)\theta(s_A, s_B)$$

The common function $T(t)$ is called the *absolute temperature function*, while

$$S_A(s_A) = \int \theta_A(s_A) ds_A$$

is the *metrical entropy* of K_A. The "element of heat" dQ of any standard system thus splits up into the product of a universal function of the empirical entropy and the total differential $dS(x)$ of the metrical entropy function:

$$T dS = dU + dW \tag{9}$$

By multiplying T by a constant and dividing S by the same constant, T can be arranged to be positive.

If one now chooses $x_n = S$ and recalls that the $x_k (k < n)$ are freely adjustable, the Second Law would be violated if S were also adjustable at will (by means of non-static adiabatic transitions). Taking continuity requirements into account, it follows that S can either never decrease or never increase. The single example of the sudden expansion of a real gas shows that it can never decrease. One has the *Principle of Increase of Entropy: The entropy of an adiabatically isolated system can never decrease.*

The Third Law. It is known from experiment that for given values of the deformation coordinates, the energy function has a lower bound U_0. The question rises whether the entropy S has an analogous property. It is found in practice that the specific heats $\partial U / \partial T$ of all substances appear to go to zero at least linearly with T as $T \to 0$. This ensures that the function S goes to a finite limit S_0 as $T \to 0$. Experiment shows, however, further that as $T \to 0$, the derivatives of S with respect to the deformation coordinates also go to zero. In contrast with U_0, S_0 has therefore the remarkable property that it is independent of the deformation coordinates. One thus arrives at the *Third Law: The entropy of any given system attains the same finite least value for every state of least energy.* One immediate consequence of this is that the so-called *classical ideal gas* (the product of whose volume V and pressure P is proportional to T, and whose energy is a function of T only) cannot exist in nature. Further, no system can have its absolute temperature reduced to zero. The Third Law is therefore a statement about the properties of functions, not of systems, at $T = 0$.

The practical applications of the theory just outlined divide themselves into two broad classes: (1) Those which are based on the existence and properties of the functions U and S and some others related to them—all "thermodynamic identities" being merely the integrability condition for the total differentials of these functions; and (2) those which are based on the Principle of Increase of Entropy: the entropy of the actual state of an adiabatically enclosed system being greater than that of any neighboring "virtual" state.

The most important of the auxiliary functions just mentioned are the *Helmholtz Function*:

$$F = U - TS \tag{10}$$

the *Gibbs Functions*:

$$G = U - TS + \sum_{k=1}^{n-1} P_k x_k \tag{11}$$

the *Enthalpy*:

$$H = U + \sum_{k=1}^{n-1} P_k x_k \tag{12}$$

sometimes called *thermodynamic potentials*. Then, e.g.,

$$dF = -S dT - dW$$

F therefore contains all available quantitative information about K, since

$$S = -\frac{\partial F}{\partial G}, \quad \text{and} \quad P_k = -\frac{\partial F}{\partial x_k} \tag{13}$$

The same is true of G for instance, since

$$S = -\frac{\partial G}{\partial T}, \quad \text{and} \quad x_k = -\frac{\partial G}{\partial P_k}$$

F and G are naturally taken as functions of $x_1, \ldots, x_{n-1}, T$ and of $P_1, \ldots, P_{n-1}, T$, respectively. At times one speaks of F as the "Helmholtz free energy" and of G as the "Gibbs free energy." In an *isothermal* reversible transition, the amount W of work done by a system is equal not to the decrease of its energy U but to the decrease $-\Delta F$ of its (Helmholtz) free energy. In the presence of internal sources of irreversibility

$$W < -\Delta F$$

In considering physicochemical equilibria, that is to say, if one is interested in the internal constitution of a system in equilibrium when changes of phase and chemical reactions are admitted, one introduces the *constitutive coordinates* n_i^α; this being the number of moles of the ith constituent C_i in the αth phase. The definitions of Equations (10) through (12) remain unaltered, for the n_i^α do not enter into the description of the interaction of the system with its surroundings. Let an amount dn_i^α of C_i be introduced quasistatically into the αth phase of the system. The work done on K shall be $\mu_i^\alpha \, dn_i$. The quantity μ_i^α so defined is the *chemical potential* of C_i in the αth phase. It is in general a function of all the coordinates of K. Then, identically.

$$dG = \sum_{k=1}^{n-1} x_k \, dP_k - S \, dT + \sum_i \sum_a \mu_i^\alpha \, dn_i^\alpha$$

Integrability conditions such as

$$\partial \mu_i^\alpha / \partial T = -\partial S / \partial n_i^\alpha$$

are applications of the first kind. On the other hand, the minimal property of G, derived from the maximal property of S, requires that

$$\sum_i \sum_a \mu_i^\alpha \, dn_i^\alpha = 0$$

when all virtual states differ only in the values of the constitutive coordinates. If the system is chemically inert, the dn_i^α are subject only to the requirements of the conservation of matter. One then concludes that if there are c constituents and p phases, i.e., $n + pc$ coordinates in all, then the number f of these to which arbitrary values may be assigned is

$$f = c - p + n$$

This typical application of the second kind is the Gibbs Phase Rule (for inert systems). This rule is often stated merely for systems with only two external coordinates ($n = 2$, e.g., $x_1 = P$, $x_2 = T$). There must then be no internal partitions within the system, nor may it, for instance, contain magnetic substances in the presence of external magnetic fields.

The beauty and power of phenomenological thermodynamics lies just in the generality and paucity of its basic laws which hold independently of any assumptions concerning the microscopic structure of the systems which they govern. Its quantitative content is limited to conditions of equilibrium. Its conceptual framework is too narrow to permit the description of the temporal behavior of systems, except to the extent that it makes it possible to decide which, of any pair of states of an adiabatically enclosed system, must have been the earlier state.

Statistical thermodynamics seeks to remedy these deficiencies by making specific assumptions about the microscopic structure of the system K, and relating its macroscopic behavior to that of its atomic constituents. K is then to be regarded as an *assembly* of a very large number of particles, which, on a non-quantal level, is a mechanical system with, say, N degrees of freedom. A *microstate* of K is a set of values of its N coordinates and its N conjugate momenta. It is out of the question to measure all these at a given time. One therefore constructs a *representative ensemble* $\mathcal{E}_K$ of K, which is an abstract collection of a very large number of identical copies of K. At any time t, the members of $\mathcal{E}_K$ will be in different microstates. Let the fractional number of members of the ensemble whose microstates lie in the range dp, dq about p, be $\phi \, dp \, dq$. Then ϕ is the *probability-in-phase*, and with $d\Gamma = dp \, dq$

$$\int \phi \, d\Gamma = 1 \tag{14}$$

The reason for this terminology is implicit in the *Postulate: The probability that a given assembly K will, at time t, be in a microstate lying in the range $d\Gamma$ about p,q, is equal to the probability $\phi \, d\Gamma$ that the microstate of a member of $\mathcal{E}_k$, selected at random at time t, lies in the same range.*

The mean value $\langle f \rangle$ of a dynamical quantity f is defined to be

$$\langle f \rangle = \int f \phi \, d\Gamma$$

If N is sufficiently large, fluctuations about the mean will usually be negligible.

When K is in equilibrium, ϕ must be constant in time, and this will be the case if it is a function of the (time-independent) Hamiltonian H of K. Ensemble averages are now assumed to coincide with temporal averages. When, in particular, K is in diathermic equilibrium with its surroundings one can show that ϕ must have the form

$$\phi = \exp[(\phi - H)/\phi] \tag{15}$$

where ϕ and θ are independent of p,q. Then

$$\theta \langle \ln \theta \rangle = \phi - \langle H \rangle \tag{16}$$

and, because of Equation (14)

$$d \int \exp[(\phi - H)/\theta] \, d\Gamma = 0 \langle d[(\phi - H)/\theta]$$

where d refers to a variation of the macroscopic coordinates of K. Using Equation (16) and its variation, the relation

$$-\theta d\langle \ln \phi \rangle = d\langle H \rangle - \langle dH \rangle \tag{17}$$

follows. Now $\langle H \rangle$ ($= \overline{U}$, say) is the total energy of the assembly, while $\langle dH \rangle$ is the average of the change of the potential energy, i.e., the work $-dW$ done by the external forces on K. If one writes

$$\overline{S} = -k \langle \ln \phi \rangle$$

where k is a constant, Equation (17) becomes

$$k^{-1} \theta \, d\overline{S} = d\overline{U} + dW$$

This is identical with the phenomenological relation of Equation (9) if one formally identifies S with $\overline{S}$, U with $\overline{U}$ and θ with kT. In this way, contact with the phenomenological theory has been established, and the quantities characteristic of the one theory have been *correlated* with that of the other. With this correlation, or interpretation, ϕ becomes F. However, because of Equations (14) and (15),

$$F = -kT \ln \int \exp(-H/kT) \, d\Gamma$$

so that if only H is known, the integral on the right (the *partition function*), and thus F, can be calculated. The equation of state of a real gas can thus in principle be obtained from a knowledge of the forces operating within the assembly. This illustrates how the additional information put into the theory yields a correspondingly greater output. Phenomenologically, such an equation of state might be written as

$$PV = \sum_{n=1}^{\infty} B_n(T) V^{1-n}$$

but here each of the *virial coefficients* $B_1, B_2, \ldots$ must be measured separately.

See also **Heat**; and **Heat Transfer**.

Additional Reading

Carter, A.H.: "Classical and Statistical Thermodynamics," Prentice-Hall, Inc., Upper Saddle River, NJ, 2000.

Cengel, Y.A. and M.A. Boles: "Thermodynamics: An Engineering Approach," 4th Edition, The McGraw-Hill Companies, Inc., New York, NY, 2001.

Mansoori, G.A.: "Thermodynamics: The Application of Classical and Statistical Thermodynamics to the Prediction of Equilibrium Properties," Taylor & Francis, Inc., Philadelphia, PA, 1991.

Russell, L.D.: "Classical Thermodynamics," Oxford University Press, Inc., New York, NY, 1995.

Sandler, S.L.: "Chemical and Engineering Thermodynamics," 3rd Edition, John Wiley & Sons, Inc., New York, NY, 1998.

Sonntag, R.E. and G.J. Van Wylen: "Introduction to Thermodynamics: Classical and Statistical," 3rd Edition, John Wiley & Sons, Inc., New York, NY, 1991.

THERMOELECTRIC COOLING. Like conventional refrigeration systems, thermoelectric systems obey the same basic laws of thermodynamics. Both in principle and result, thermoelectric cooling has much in common with conventional refrigeration methods. In a conventional refrigeration system, the main working parts are the freezer, condenser, and compressor. The freezer surface is where the liquid refrigerant boils, changes to vapor, and absorbs heat energy. The compressor circulates the refrigerant above ambient level. The condenser helps to discharge the absorbed heat into surrounding ambient. In thermoelectric refrigeration, the refrigerant in both liquid and vapor forms is replaced by two dissimilar conductors. The freezer surface becomes cold through absorption of energy by electrons as they pass from one semiconductor to another, instead of energy absorption by the refrigerant as it changes from liquid to vapor. The compressor is replaced by a direct current power source which pumps the electrons from one semiconductor to another. A heat sink replaces the conventional condenser fins, discharging the accumulated heat energy from the system.

The components of a thermoelectric cooler are indicated by the cross section of a typical unit shown in Fig. 1. Thermoelectric coolers such as this are actually small *heat pumps* that operate on the physical principles well established over a century ago. Semiconductor materials with dissimilar characteristics are connected electrically in series and thermally in parallel, so that two junctions are created. The semiconductor materials are *n*- and *p*-type and are so named because either they have more electrons than necessary to complete a perfect molecular lattice structure (*n*-type), or not enough electrons to complete a lattice structure (*p*-type). The extra electrons in the *n*-type material and the holes left in the *p*-type material are called *carriers* and they are the agents that move the heat energy from the cold to the hot junction.

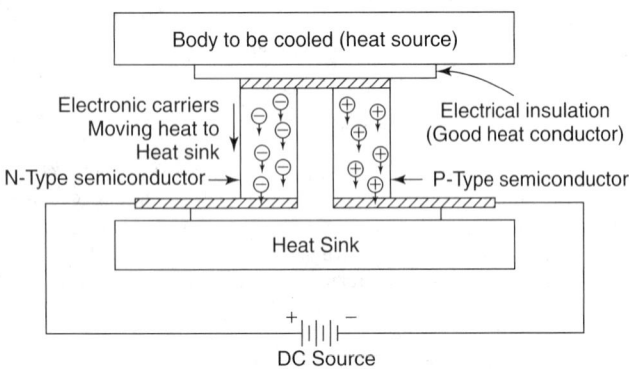

Fig. 1.   Cross section of thermoelectric cooler.

Heat absorbed at the cold junction is pumped to the hot junction at a rate proportional to carrier current passing through the circuit and the number of couples. Good thermoelectric semiconductor materials, such as bismuth telluride, greatly impede conventional heat conduction from hot to cold areas, yet provide an easy flow for the carriers. In addition, these materials have carriers with a capacity for carrying more heat. Only since the refinement of semiconductor materials in the early 1950s has thermoelectric refrigeration been considered practical for many applications.

In practical use, couples are combined in a module where they are connected in series electrically and in parallel thermally. See Fig. 2. Normally, a module is the smallest component available. The user can tailor quantity, size, or capacity of the module to fit exact requirements

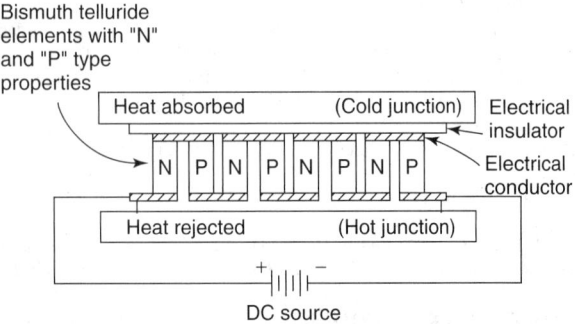

Fig. 2.   Thermoelectric module assembly. Elements are electrically in series, thermally in parallel.

without procuring more total capacity than is actually required. Modules are available in a variety of sizes, shapes, operating currents, operating voltages, number of couples, and ranges of heat-pumping levels. The present trend is toward a larger number of couples operating at a low current.

Thermoelectric coolers find three basic categories of applications. (1) Use in electronic components; (2) in temperature control units; and (3) in medical and laboratory instruments.

Modules normally contain from 2 to 71 couples with ceramic-metal laminate plates. If modules are to be used in cooling chambers of large components, a total surface area of virtually any size can be made by placing the appropriate number of modules side by side.

The interfaces at the cold junction and the hot junction must be constructed to transfer heat in and out of the module with little difference in temperature. This is accomplished with metal-ceramic laminate plates that give strength and permit good thermal bonding between the two interfaces. The outer plate surface is usually tinned to facilitate soldering to heat sinks. Where soldering is not practical, as in the case of thermal expansion differences, heat transfer grease is recommended. Epoxy bonding agents are available where a more permanent solderless bond is required.

The single-stage module is capable of pumping heat where the difference in temperature of the cold junction and hot junction is 70°C or less; however, in those applications requiring higher delta T_s, the modules can be cascaded. Cascading is a mechanical stacking of the modules so that the cold junction of one module becomes the heat sink for a smaller module placed on top. In addition to the heat pumped by any given stage, the next lower stage must also pump the heat resulting from the input power to that upper stage. Consequently, each succeeding stage must be larger and larger from the top of the cascade downward.

With any given set of heat sink and cold spot temperatures, there exists an optimum heat-pumping capacity or "size" ratio between each adjacent pair of stages. The optimum size ratio increases as the overall delta T increases, but decreases as the number of stages increases. It is not necessarily a constant from stage to stage, even with delta T and number of stages fixed. True optimization of a cascade design requires accurate temperature-dependent data on the thermoelectric materials in combination with a computerized numerical design theory.

Applications requiring low-temperature thermoelectric coolers usually have strict limitations on available power. Therefore, it is not practical to fabricate and stock numerous different cascades that can be optimized for only one set of conditions. On the other hand, fully optimized prototypes involve engineering and manufacturing costs that may prove uneconomical for some applications. Therefore, a low-cost alternative approach has been developed for responding to such requirements. Standard cascades are fabricated by assembling *partials* of standard modules. The number of different standard cascades is virtually unlimited due to "free" variables, such as number of stages, couple distribution, and the basic building block module. In order to determine the best standard cascade for a given application, the desired hot-side temperature, cold-side temperature, and thermal load are entered into a computer system. The result is a list of numerous standard coolers, which meet these specifications with various combinations of input power and cost. Generally, the lowest input power devices are of higher cost and vice versa.

The heat sink design is very important. The heat sink must carry heat away with minimum rise of temperature. It should be stressed that all the thermoelectric cooler does is to move energy from the load to the heat sink where it must be dissipated to another medium, the latter required to be cooler than the hot junction.

Power supply capabilities for thermoelectric coolers range from the simple open-loop direct current supply with a switch to sophisticated feedback systems with close temperature regulation and fast response. The only limitation on the supply is that ripple be maintained at a point lower than 10 to 15%. Open-loop systems will generally contain a transformer, rectifiers, choke, and chassis with heat sink for the rectifiers. In feedback systems, a thermistor is used to sense temperature at the cold junction. This signal is compared with the desired temperature setting to obtain an error signal.

THERMOGRAPHY.   The various types of detectors available for measuring infrared radiation are described in entry on **Infrared Radiation**. For coverage of an extended field of view, mosaics of conventional detectors have been used, as well as electronic imaging tubes, such as infrared vidicons and orthicons. Whether by such detectors or by rasterlike scanning of a small elemental field of view, it is possible to study the temperature pattern of an extended source of thermal radiation. Thermography is such a process, in which the temperature gradients are displayed in visible form, e.g., on photographic film, or in real time, on a cathode ray tube. When properly interpreted, the thermograph can be a valuable diagnostic aid to the physician. Industrial thermography, as applied, for example, to the non-destructive testing of integrated electronic circuit boards is useful for design purposes or for quality assurance. In interpreting the observed patterns, one must bear in mind that radiance variations may result from differences in either or both surface emitting and surface temperature from point to point. This fact is less troublesome in medical thermography because the emittance of the body is close to unity.

THERMOGRAVIMETRIC ANALYSIS. This analytical technique (TGA), also sometimes referred to as thermogravimetry, is a method whereby the weight and temperature of a sample under test are continuously recorded as the sample is heated at a constant and linear rate. The heating, usually within the range from ambient up to 1,100 or 1,500 °C, is achieved by a furnace, which surrounds, but does not touch, the sample. Thus, the sample remains freely suspended from the balancing mechanism, which is actuated as the sample mass alters in response to chemical reactions produced as its temperature is progressively increased. Weight-loss curves are preferably produced by means of an X-Y recorder. Of course, the term *weight-loss* must be used with reservation because in some atmospheres a sample may gain weight. Indirectly, TGA information can be applied to studies of rates of reaction and energies of activation for the reaction, sublimation, or vaporization of chemical compounds and minerals. Weight gains are caused by the adsorption and absorption of gases by solid samples, direct reaction as in oxidation, corrosion, recarbonation, and hydration, or the reaction of gases to produce solids. Conversely, weight losses are produced by the desorption of gases, dehydration, vaporization, sublimation, gaseous desolvation, and gas-liberating reactions of both organic and inorganic substances. Specific fields that have widely used TGA include: studies involving the thermal stability of minerals and mineral mixtures; the pyrolysis of coals, petroleum, and cellulose; the analysis of soils; roasting and calcining; thermochemical reactions of ceramics and cements; dehydration hygroscopicity studies; solid-state reactions; effect of radiation on various materials; corrosion studies; and the detection of short-lived unstable intermediate compounds.

THERMOGRAVITATIONAL COLUMN.   See **Diffusion**.

THERMOMETER.   An instrument that measures temperature. A thermometer may take advantage of one of several physical properties of materials that change when they are subjected to a change in temperature. Liquids, gases, and solids expand with increasing temperature; decrease in volume with decreasing temperature. Thus, there are liquid-in-glass thermometers, which depend upon the volumetric relation of mercury or other liquids with temperature. The difference in the energy radiating from materials at various temperatures is the basis of optical and radiation pyrometers. Bimetallic thermometers depend upon the differing expansiveness of different metals. Changes in electrical resistance with temperature are utilized in resistance thermometers and thermistors. Thermocouples depend upon the Seebeck, Peltier, and Thomson effects, wherein the emf in an electrical circuit comprising dissimilar metals bears a relationship to the temperature difference between a cold junction and the temperature being measured.

Several kinds of thermometers are described in this volume. See alphabetical index.

THERMOMETER (Filled-System).   A representative filled-system thermometer is shown in Fig. 1. The temperature-sensing element (bulb) contains a fluid that changes its volume or pressure with temperature. The pressure-sensitive element (bourdon) responds to these changes by delivering a motion or force to a device that transduces the signal to a usable form. This is commonly a mechanical linkage which drives a pointer or pen, but may be a pneumatic or electric device which transmits the temperature signal over long distances. These signals frequently are used for process control purposes.

Filled-system thermometers may be placed into one of two fundamental categories: (1) Those in which the bourdon responds to a volume change; and (2) those in which the bourdon responds to a *pressure change*. Those

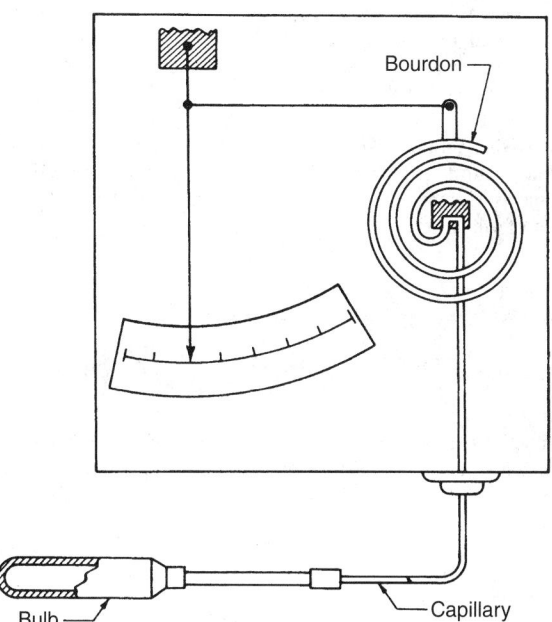

Fig. 1.   Filled-system thermometer.

that respond to volume changes are completely filled with a liquid. The variation in liquid expansiveness with temperature is greater than that of the bulb metal, the net volume change being communicated to the bourdon. In this design, an internal-system pressure change is not of primary importance. In those systems that respond to pressure changes, the bulb is either filled with a gas, or partially filled with a volatile liquid. Changes in gas or vapor pressure with changes in bulb temperature are communicated to the bourdon. The bourdon will increase in volume with increase in pressure, but in this design, this effect is not of primary importance.

In liquid-filled thermal systems, mercury or various organic fills are used. In vapor-filled systems, a number of hydrocarbons, including ethane, ethyl chloride, ethyl ether, chlorobenzene, and propane, among others, are used. Nitrogen or other fully-dried and purified gases may be used in a gas thermometer.

THERMOPHONE.   An electroacoustic transducer in which sound waves of calculable magnitude result from the expansion and contraction of the air adjacent to a conductor whose temperature varies in response to a current input. When used for the calibration of pressure microphones, a thermophone is generally used in a cavity, the dimensions of which are small compared to a wavelength.

THERMOPILE.   See **Thermocouple**.

THERMOPLASTIC.   See **Plastics**.

THERMOSTATIC BIMETAL.   See **Bimetal Thermometer**.

THERMOTROPIC ATMOSPHERE.   See **Atmosphere (Earth)**.

THETA FUNCTION.   Four functions of infinite series, written in terms of $q = e^{\pi/\tau}$, where τ is a complex number having its imaginary part positive. A representative theta function is

$$\theta(z, q) = 1 + 2\sum_{n=1}^{\infty} qn^2 \cos 2nz$$

They are related to elliptic functions and are especially useful in making numerical calculations with such functions. See also **Elliptic Integral**; **Jacobi Elliptic Function**; and **Weierstrass Function**.

THETA PINCH.   See **Nuclear Reactor**.

THEVENIN THEOREM.   The current in any terminating impedance Z_T connected to any network is the same as if Z_T were connected to a generator whose voltage is the open circuit voltage of the network, and whose internal impedance Z_R is the impedance looking back from the terminals of Z_T, with all generators replaced by impedance equal to the internal impedance of these generators.

THIAMINE (Vitamin B$_1$). Some earlier designations for this substance included aneurin, antineuritic factor, antiberiberi factor, and oryzamin. Thiamine is metabolically active as thiamine pyrophosphate (TPP), the formula of which is:

TPP functions as a coenzyme which participates in decarboxylation of α-keto acids. Dehydrogenation and decarboxylation must precede the formation of "active acetate" in the initial reaction of the TCA cycle (citric acid cycle):

This reaction is a good example of the interrelationship of vitamin B coenzymes. Four vitamin coenzymes are necessary for this one reaction: (1) thiamine (in TPP) for decarboxylation; (2) nicotinic acid in nicotinamide adenine dinucleotide (NAD); (3) riboflavin in flavin adenine dinucleotide (FAD); and (4) pantothenic acid in coenzyme A (CoA) for activation of the acetate fragment.

TPP also mediates the oxidative decarboxylation of α-ketoglutaric acid, another intermediate of carboxydrate metabolism in the citric acid cycle. The nutritional requirement for thiamine increases as dietary carbohydrate increases because of a greater demand for TPP.

The structure of thiamine hydrochloride is:

In this form and as other salts, such as thiamine mononitrate, the vitamin is available as a dietary supplement.

Diseases and disorders resulting from a deficiency of thiamine include beriberi, opisthotonos (in birds), polyneuritis, hyperesthesia, bradycardia, and edema. Rather than a specific disease, beriberi may be described as a clinical state resulting from a thiamine deficiency. In body cells, thiamine pyrophosphate is required for removing carbon dioxide from various substances, including pyruvic acid. Actually, this is accomplished by a decarboxylase of which thiamine pyrophosphate is a part. Where thiamine is deficient, the process of oxidation necessary for converting food into energy is impeded, causing a variety of manifestations throughout the body.

In so-called *dry* beriberi, pathologic alterations in neurons and nerve fibers occurs, leading in some instances to degeneration of peripheral nerves. This condition is termed peripheral neuritis, generally affecting the nerves in the arms and legs. There often is altered skin sensitivity to touch in the extremities and pain on pressure over large nerves. There is a gradual loss of muscle strength, which may lead to paralysis of a limb. In *wet* beriberi, there is a lessening of strength of the heart muscles. There is enlargement of the heart, dyspnoea, increased pulse rate, palpitation, and edema. Pathologically, degenerative changes are found in the nervous tissue, heart muscle, and gastrointestinal tract. In later stages, marked enlargement of the heart and liver may be noted.

In one form of thiamine deficiency, Wernicke's syndrome may be noted. Therein is paralysis, or weakness of the muscles that causes motion of the eyeball. Closely associated with thiamine deficiency are dietary problems of alcoholism. The psychotic disturbances of alcoholism, including delirium tremens, frequently respond to thiamine and other B complex vitamins. Injections of thiamine often produce dramatic improvements in persons suffering from beriberi. Beriberi sometimes occurs in infants who are breast-fed by mothers who suffer a thiamine deficiency. Beriberi remains of concern in the Orient where polished rice is a dietary staple.

In cattle, a thiamine deficiency causes podioencephalomalcia (PEM), characterized by blindness, decreased feed intake, incoordination, failure of rumen to contract, spasms, and paralysis. In swine, a deficiency retards growth and sometimes causes cyanosis (insufficient oxygen in blood), enlarged heart, accompanied by fatty degeneration of heart muscles. Chicks suffer from paralysis of peripheral nerves, causing polyneuritis (head drawn back).

Distribution and Sources. Relatively few natural foods are considered high in thiamine content.

High thiamine content (1,000–10,000 micrograms/100 grams). Ham, rice bran, soybean flour, wheat germ, yeast.

Medium thiamine content (100–1000 micrograms/100 grams). Almond, asparagus, barley, brazil nut, bean (kidney, lima, snap, soy, wax), beef, beet greens, broccoli, Brussels sprouts, carp, cashew, cauliflower, chicory, chestnut, chicken, clam, cod, corn (maize), dandelion greens, eggs, endive, gooseberry, groundnut (peanut), hazelnut, kale, kohlrabi, leek, lentil (dry), lobster, mackerel, milk, mushroom, oats, oyster, parsley, pea, pecan, plum, pork, potato, prune (dry), raisin (dry), rice (brown), salmon, turkey, veal, walnut, watercress.

Low thiamine content (10–100 micrograms/100 grams). Apple, apricot, artichoke, avocado, banana, beet, berry (black-, blue-, cran-, rasp-, straw-), cabbage, carrot, celery, cheeses, cherry, coconut, cucumber, currant, date (dry), eggplant, fig, flounder, grape, grapefruit, haddock, halibut, herring, lemon, lettuce, melons, orange, parsnip, peach, pear, pepper (sweet), pike, pineapple, prune, sardine, scallop, shrimp, tangerine, trout, tuna, turnip.

Commercial thiamine dietary supplements are prepared by synthesis: Pyrimidine + thiazole nuclei synthesized separately and then condensed; also build on pyrimidine with acetamidine. Precursors in the biosynthesis of thiamine include thiazole and pyrimidine pyrophosphate, with thiamine phosphate as an intermediate. In plants, production sites are found in grain and cereal germ.

Bioavailability of Thiamine. Factors which contribute to a lessening of thiamine bioavailability include: (1) cooking, inasmuch as the vitamin is heat labile and water soluble; (2) presence of certain enzymes in food, such as thiaminase for vitamin breakdown; (3) destruction by calcium carbonate, dibasic potassium phosphate, and manganous sulfate; (4) destruction by nitrites and sulfites; (5) diuresis and gastrointestinal diseases; and (6) presence of live yeasts and alkalis. An increase in availability can result from: (1) presence of cellulose in diet, which increases intestinal synthesis; (2) storage capacity in heart, liver, and kidney; and (3) stimulation of bacterial synthesis in intestine (normally none).

Antagonists of thiamine include pyrithiamine, oxythiamine, and 2-*n*-butyl homologue. Synergists include vitamins B$_2$, B$_6$, B$_{12}$, and niacin, pantothenic acid, and somatotrophin (growth hormone).

Unusual features of thiamine as observed by some researchers include: (1) it exerts a hormonal function in plants, controlling root growth; (2) it aids phosphorylation in liver, dephosphorylation in kidney; (3) it easily poisoned by heavy metals, acetyl iodide; (4) plant and animal cocarboxylases are identical; (5) it exerts a diuretic effect and it is constipating; (6) it can be allergenic on injection; (7) it is not available from intestinal bacteria; and (8) blood contains most cocarboxylase in leukocytes.

Thiamine is soluble in water and easily destroyed by heat. These two properties account for appreciable losses of thiamine from processed and stored foods. An acid medium favors the retention of thiamine, whereas an alkaline medium is detrimental to retention.

Determination of Thiamine. Bioassay methods include yeast fermentation; polyneuritic rate of cure in rat; bacterial metabolism. Physicochemical methods include thiochrome fluorescence; polarography; chromatography; absorption in neutral and acid solutions.

THICKENER (Food).   See **Additives (Food).**

THICKENER (Process).   See **Classifying (Process).**

THICKNESS MEASUREMENT AND GAGING SYSTEMS.   The requirement to measure and control material thickness appears in numerous industries, particularly those situations involving sheets, webbed materials, and precisely made parts. Thickness and related dimensions may be measured in several ways, including: (1) pneumatic gaging, (2) electric and electronic gaging, (3) nuclear-radiation gaging, (4) ultrasonic thickness gaging, and (5) x-ray thickness gaging, among others.

Pneumatic Gaging. A pneumatic gaging system consists essentially of components that provide a constant-pressure air supply, an indicating means, and a metering orifice. The principle of operation is based on the effects of varying the flow of air from the metering orifice. See also **Pneumatic Controller**; and **Transmission (Pneumatic).** For example, as flow from the orifice is obstructed, pressure in the system will build up to the regulated valve, and flow through the system will drop. Over a significant range of values in such a gaging system, there exists a linear relationship between flow or pressure and the size of the escapement orifice. In many gaging applications, the linear relationship is equated to the clearance that separates a sized metering orifice and an obstruction. The indicating component shows the change in flow or pressure in terms of a linear measurement. The orifice obstruction can be either the workpiece being gaged or an integral part of the gage spindle. An example of the first

is found in the gaging of a bore by inserting a gaging spindle containing two diametrically opposed orifices through which air is flowing. Here, there is no contact between the gaging spindle and the part being measured; the closeness of fit serves as the obstruction. In the second kind of gage, the gage spindle includes a component that mechanically contacts the work being measured. Movement of this component relative to a sized orifice affects air flow and establishes the basis for measurement. The kinds of measurements that can be made by different orifice arrangements are illustrated in Fig. 1.

Electric and Electronic Gaging. The use of air jets in pneumatic systems has many advantages because, to some degree, the concept does approach universality of application, not being severely affected by temperature, vibration, and other ambient conditions. In one widely used electrical system, the gage head contains two coils with a sintered iron core centered between them. This core is attached to the gage spindle and moves axially, as shown in Fig. 2. Its position relative to the coils affects the impedance of the coils. The coils and a symmetrical transformer in the oscillator form a bridge. When the core rests equally between the two coils, the bridge is balanced and the output signal from the gage head is zero. When the core is displaced, the impedance of the coils is changed. The impedance of one coil is reduced while that of the other is increased, which generates a signal in proportion to the amount of displacement of the core. The signal then is amplified, rectified, and indicated on a direct current meter, which is calibrated in units of length. Typical applications include

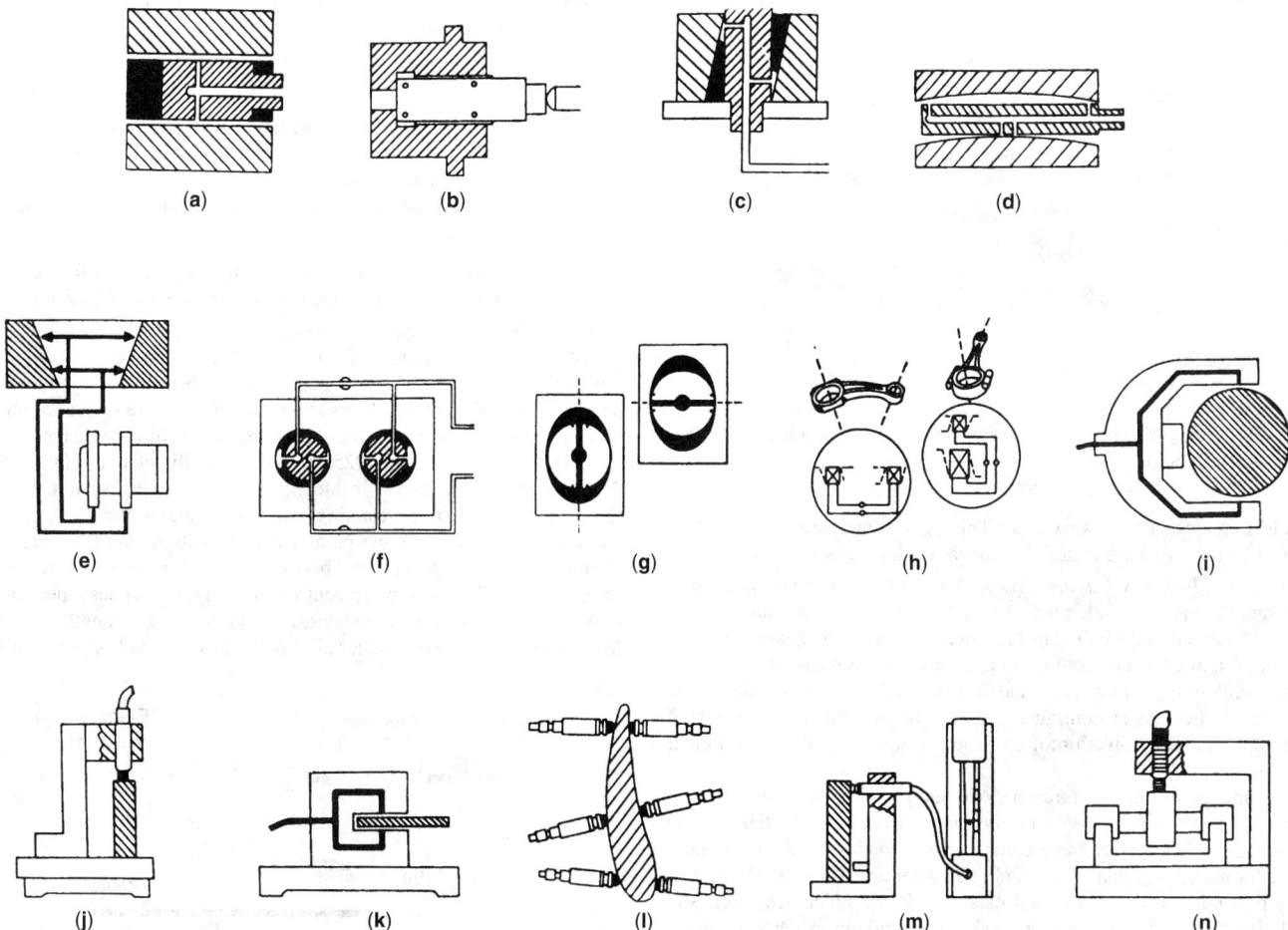

Fig. 1.   Basic applications of pneumatic gages: (**a**) two diametrically opposed open jets check true diameter of holes having tolerances of 0.005-inch (< 0.1 millimeter or less); (**b**) contact-type gaging head for measuring diameters of interrupted bores and bores with keyways; (**c**) spindle with opposed open orifices spaced longitudinally to check squareness of a bore axis with a face; (**d**) camber or straightness of hole is checked by rotating through 180) a spindle having four jets; (**e**) taper is indicated as any spindle is passed through a bore; (**f**) center distance between holes is checked by a fixture with two spindles, each having two opposed jets; (**g**) out-of-round is indicated when any spindle is rotated through 90); (**h**) parallelism of holes is indicated by combining two spindles of type used in squareness checking; (**i**) outside diameter is checked by two opposed standard jets; tolerance usually 0.002 inch (0.05 millimeter) or less; (**j**) height, width, or depth can be measured by a suitably fixtured contact-type gaging head; (**k**) thickness gage for thin parts and items having close tolerance incorporates two opposing standard jets; (**l**) multiple-contact-type gage heads mounted in a suitable fixture for checking contour; (**m**) squareness of surfaces for close-tolerance parts having good surface finish can be checked by a fixtured spindle; and (**n**) concentricity can be indicated by one or more contact-type spindles mounted in a suitable gage fixture.

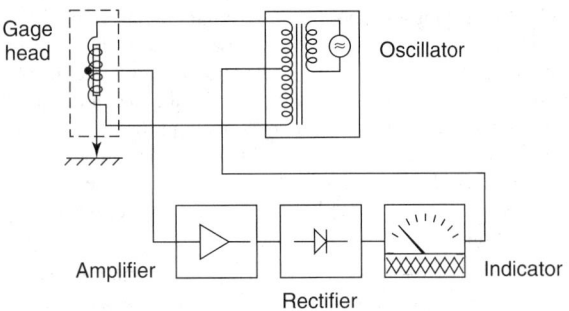

Fig. 2.   Impedance-type gaging system.

checking roundness, concentricity, parallelism, thickness, cam contours, tapers, flatness, and squareness—without requiring precision fixturing.

Nuclear-Radiation Thickness Gaging. Noncontact gaging techniques are required for several applications—for example, thickness control of paper and plastic sheeting. As shown by Fig. 3, the basic single-head gage holds a beta radioisotope source beneath the sheet to be measured and a detector cell above. The gage can measure a single point or scan the sheet automatically. Transmitted radiation is converted into an electric current and amplified for readout or automatic control. Calibration is usually in terms of basis weight (weight per unit area).

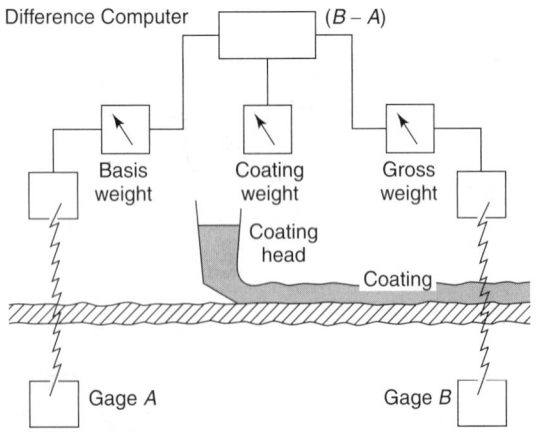

Fig. 3.   Double-gage beta-radiation system for control of coatings on paper, textiles, and plastics. (*Ohmart.*)

Nuclear fluorescence is also used. This system measures only coating weight regardless of base material hardness or thickness.

Ultrasonic Thickness Gaging. These gages find extensive application in the measurement of wall thickness and are useful in measuring most structural materials that will transmit sound. Almost all metals, plastics, and ceramics, and various composite materials can be measured this way. Measurements are instantaneous and can be made from one side of the part. Suitable transducer configurations also permit readings at elevated temperatures for determination of corrosion losses in chemical equipment while "on stream."

Ultrasonic gages utilize ultrasonic vibrations, most commonly in the portion of the frequency spectrum between 1 MHz and 15 MHz. Sound waves in these frequencies have certain characteristics which make them suitable for making measurements and finding defects. The sound beam can readily pass through most structural materials. It will reflect from acoustic boundaries, either internal flaws or geometric boundaries, such as the back surface of the part. Sound travels at characteristic velocity in the material, and this permits timing of the wave propagation through the material.

The basic element of an ultrasonic gaging system consists of a *generator* to produce the high-frequency electronic vibrations that, in turn, activate the crystal *transducer*. The transducer converts the electric signals to mechanical vibrations, or ultrasound; and these are introduced into the part. Echoes from various boundaries in the part are picked up by the same transducer and are reversibly converted from mechanical vibrations to an electric signal. This is further amplified and processed, and the information is displayed on one of several types of data presentations.

There are two categories of ultrasonic gages: (1) *resonance*, and (2) *pulse-echo.* Each has distinct electronic characteristics and readout methods. Both are dependent on transmission of the acoustic waves, reflections, and velocity in the material.

The *resonance* ultrasonic gages produce a frequency-modulated continuous-wave signal. This provides a corresponding swept frequency of sound waves introduced into the part. When the thickness of the part equals one-half wavelength, or multiples thereof, standing-wave conditions or mechanical resonances occur. The frequency of the fundamental resonance, or the difference frequency between two harmonic resonances is determined in the instrumentation. The thickness is determined by the formula

$$Th = \frac{Vel}{2F}$$

where Th = thickness of part under transducer
       Vel = speed of sound in the material
       F = frequency, Hz.

A popular readout for ultrasonic resonance gages is a large cathode-ray tube. Frequency indications on the tube are compared to an overlaid scale calibrated in inches, centimeters, etc., and thus direct thickness readings can be obtained. These gages can produce accuracies of 1% of nominal thickness.

The *pulse-echo* technique operates somewhat like a refined sonar system. Very short electric pulses, usually at discrete frequencies, are generated. These produce short acoustic pulses from the transducer which, in turn, pass through the material and reflect from the boundaries, just as in the resonance approach. The *transit time* of the pulses through the material is measured. The formula is

$$Th = \frac{Vel \times T}{2}$$

where Th = actual thickness of part under transducer or within sound
            beams
       Vel = speed of sound in material
       T = transit time of sound pulse through one round trip in the
            material.

One limitation of ultrasonic gaging is the requirement for coupling of the sound beam between transducer and part. See also **Ultrasonics**.

X-Ray Thickness Gaging. Noncontact X-ray gages measure thickness or density of hot or cold materials while in motion or when stationary. Steel, aluminum, brass, copper, glass, paper, and rubber—in continuous strip or sheets—as well as plastic films and material coats are typical of materials that can be gaged by this method. Different styles of X-ray gages will span a thickness range from 0.00025 inch (0.006 millimeter) in plastic through 2 inches (5.1 centimeters) in steel. An X-ray gaging system comprises three basic units: (a) a scanning unit that contains the X-ray generator and a detecting unit; (2) an operator control station; and (3) a power unit. Continuous X-ray gaging has become part of many computer-operated steel rolling mills. Automatic control based upon thickness measurements permits rolling to close tolerances, maximum on-gage length per ton, and less scrap or out-of-gage material at both ends of a coil or run. See Fig. 4.

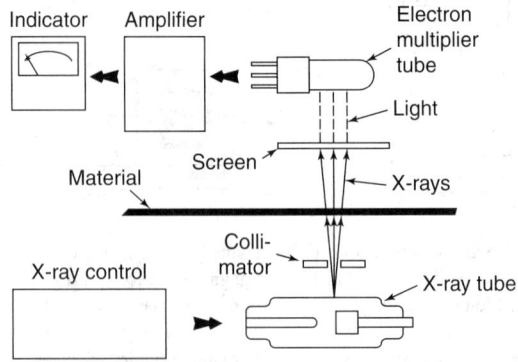

Fig. 4.   Principal elements of an x-ray thickness gage.

THIN FILMS.   The term *thin films* is used for a wide variety of physical structures. Self-supporting solid sheets usually are called foils when thinned from thicker material by such methods as rolling, beating, or etching, and films when obtained by stripping a deposited layer from its substrate. Supported thin films are deposited on planar or (in special cases) curved substrates by such methods as vacuum evaporation, cathode sputtering, electroplating, electroless plating, spraying, and various chemical surface reactions in a controlled atmosphere or electrolyte. Thicknesses of such supported films range from less than an atomic monolayer to a few micrometers (1 μm = 10^{-4} centimeter). Thin films not forming a continuous sheet are called "island films." Particularly noble metals may condense as islands of considerable thickness (up to ∼10^2 micrometer).

In scientific studies and technical applications, the use of well-controllable deposition methods such as vacuum evaporation and cathode sputtering are generally preferred. The film structure is markedly influenced by such deposition parameters as substrate composition and surface structure, source and substrate temperatures, deposition rate, and composition and pressure of the ambient atmosphere (where applicable). In general, the structure of films is more disordered than the corresponding bulk material. Smaller grains, higher dislocation concentrations, and deviations from stoichiometry are typical, and films approach bulk structure only as a limiting case. Under certain growth conditions, films exhibit preferential crystal orientations or even epitaxy. (Epitaxy means that the film structure is determined by the crystal structure and orientation of the underlying substrate.)

Solid thin films are common study objects in most phases of solid state physics. They supply the samples for the study of general structural and physical properties of solid matter where special beam methods require small quantities of material or extremely thin layers. For example, thin films are used in transmission electron microscopy and diffraction, neutron diffraction, ultraviolet spectroscopy, and X-ray diffraction and spectroscopy. Thin films represent the best means for studying physical effects, where these effects are caused by the extreme thinness of the material itself. Examples are the rotational switching of ferromagnetic films, electron tunneling phenomena, electromagnetic skin effects of various kinds, and certain optical interference phenomena. Films also are convenient vehicles for the investigation of nucleation and crystal growth, and for states of extremely disturbed thermodynamic equilibrium.

Presently, films find three major industrial uses: the decorative finishing of plastics, optical coatings of various kinds (mainly antireflection coatings, reflection increasing films, multilayer interference filters, and fluorescent coatings), and in electronic components.

Nucleation, Growth and Mechanical Properties of Films. In vacuum evaporation, molecules or atoms of thermal energy are deposited at a uniform angle of incidence and under well-defined environmental conditions. Most nucleation and growth studies, therefore, have been made on evaporated films. A particle approaching the substrate enters close to its surface a field of attracting short-range London forces with an exchange energy proportional to $-1/r^6$. At a still shorter distance r, repulsive forces proportional to $e^{-r/\text{constant}}$ resist the penetration of the electron clouds of the surface atoms. Due to the atomic or crystalline structure of the substrate, this potential field exhibits periodicity or quasiperiodicity in the substrate plane. The freshly condensed particles migrate over the surface with a jump frequency $i_D \propto \exp(-Q_D/kT)$, or desorb with a frequency $i_{ad} \propto \exp(-Q_{ad}/kT)$, where the activation energy Q_D is often approximately one-fourth of Q_{ad}. Permanent condensation occurs in most cases at distinct nucleation centers which may consist of deep potential wells of the substrate, clusters of condensed particles, or previously deposited "seed" particles of a different material. The number of nuclei formed in the second case is strongly temperature and rate dependent.

Most metals always condense in crystalline form, but the grain size is extremely small at low temperatures (in the order of a few micrometers) and increases markedly with increasing substrate temperatures. Grain size decreases with increasing deposition rates. The condensation of amorphous or quasi-liquid phases at low temperatures has been observed for such metals as antimony and bismuth and a few dielectrics. Some of these materials, on annealing, pass through otherwise unobserved and probably metastable phases.

Stresses of considerable magnitude are often observed in deposited films. The main causes of these stresses are a mismatch of expansion coefficients between substrate and film, enclosed impurity atoms, a high concentration of lattice defects and in very thin films, a variety of surface effects. Often, the stresses resulting from lattice defects can be minimized by the choice of a higher substrate temperature during deposition, or they can be reduced

by a post-deposition anneal. Metal films frequently exhibit tensile strengths that are considerably larger than those of the corresponding bulk materials.

Thin-film Optics. Deposited metal mirrors probably represent the oldest optical application of films. High-quality mirrors usually are produced by the vacuum evaporation of aluminum on an appropriately shaped glass substrate. Often, a glow-discharge cleaning of the substrate or a chromium undercoat is first applied to increase the adhesion of the aluminum. After deposition, the aluminum is protected by anodic oxidation or an evaporated overcoat of SiO, SiO_2, or Al_2O_3.

For SiO, maximum reflectance in the visible spectral region is achieved at a thickness of about 1400 micrometers. Rapid SiO evaporation reduces the reflectance at shorter wavelengths.

Single or multilayer coatings find increasing use as optical interference filters. These film stacks may consist solely of transparent films of different refractive indices n_f, or a combination of absorbing and nonabsorbing layers. Common low-index materials for glass coatings in the visible region of the spectrum are MgF_2 (n_f = 1.32 to 1.37), and cryolite Na_3AlF_6 (n_f = 1.28 to 1.34); high-index materials are SiO (n_f = 1.97), ZnS ($n_f \approx 2.34$), TiO_2 (n_f = 2.66 to 2.69) and CeO_2 (n_f = 2.2 to 2.4). The indices are given for the sodium D-line. Various semiconductors are used for infrared coatings.

At each air-film, film-film, or film-substrate interface, the incident light amplitude is split into a reflected and a transmitted fraction according to the Fresnel coefficients

$$f_{j-1} = (\hat{\mathcal{N}}_{j-1} - \hat{\mathcal{N}}_j)/(\hat{\mathcal{N}}_{j-1} + \hat{\mathcal{N}}_j) \text{ and } g_{j-1} = 2\hat{\mathcal{N}}_{j-1}/(\hat{\mathcal{N}}_{j-1} + \hat{\mathcal{N}}_j)$$

where j and $j-1$ denote the number of the optical layer counted from the side of the incident beam, $\hat{\mathcal{N}}_j = \mathcal{N}/\cos\Theta_j$ for p polarization or $\hat{\mathcal{N}}_j = \mathcal{N}_j \cos\Theta_j$ for s polarization is the effective refractive index, and $\mathcal{N}_j = n_j - ik_j$ the refractive index of the j layer.

$$\cos\Theta_j = \sqrt{(\sqrt{p_j^2 + q_j^2} + p_j)/2} - i\sqrt{(\sqrt{p_j^2 + q_j^2} + p_j)/2}$$

$$p_j = 1 + (k_j^2 - n_j^2)[n_0 \sin\theta_0/(n_j^2 + k_j^2)]^2$$

$$q_j = -2n_j k_j[n_0 \sin\theta_0/(n_j^2 + k_j^2)]^2$$

The symbol θ_0 is the angle of incidence in the incident medium.

For nonabsorbing film stacks ($k_i = 0$; $i = 1, 2, \dots, m+1$), the overall reflectance and transmittance may be obtained by summing the multiple coherent reflections between the film boundaries. A more general treatment based on electromagnetic theory yields for amplitude reflectance and transmittance the recursions formulas

$$r_{(j-1)-} = (f_{j-1} + r_{j-} \exp(-2i\hat{\Phi}_j))/(1 + f_{j-1}r_{j-} \exp(-2i\hat{\Phi}_j))$$

and

$$t_{(j-1)-} = (g_{j-1}t_{j-} \exp(-i\hat{\Phi}_j))/(1 + f_{j-1}r_{j-} \exp(-2i\hat{\Phi}_j))$$

$\hat{\Phi}_j = \Phi_j \cos\Theta_j$ is the effective phase thickness. $\Phi_j = (2\pi/\lambda) . \mathcal{N}_j l_j$ where λ is the wavelength in vacuo, and l_j is the geometrical film thickness. The recursion is started on the side of emergence, using the initial conditions $r_{m-} = f_m$ and $t_{m-} = g_m$. Intensities are given by $R = |r_{0-}|^2$ and $T = (\mathcal{R}\mathcal{N}_{m+1}/n_0)|t_{0-}|^2$ where $\mathcal{R}$ denotes "real part of." If A_j is the absorption in the layer j, $R + T + \Sigma_j A_j = 1$.

A single antireflection coating of $\lambda/4$ optical thickness $n_f l_f$ yields zero reflectance at $n_f = \sqrt{n_{\text{glass}}}$. A double layer coating of $\lambda/4$ films requires $n_2/n_1 = \sqrt{n_g}$. The transmission of a Fabry-Perot interference filter consisting of a dielectric spacer layer between two partially reflecting metal films is given by $I/I_0 = [(1 + A/T)^2 + (4R/T^2)\sin^2(\delta - 1)]^{-1}$ where $\Phi = 2\pi nl \cos\theta/\lambda$. R, T, and A are the reflection, transmission, and absorption coefficients of the reflecting layers. The refractive index and thickness of the spacer film are n and l. θ is the angle of refraction in the spacer, and δ the phase change for reflections at the spacer-metal film interfaces. $(I/I_0)_{\text{max}} = (T/(1-R))^2$ and $(I/I_0)_{\text{min}} = (T/(1+R))^2$. The band pass half-width is $\Delta\lambda_{1/2} \simeq \lambda(1-R)/m\pi R^{1/2}$ for the interference order $m(m\pi = \Phi)$. More complex coatings and filters, and their various applications, cannot be discussed here. It should be mentioned, however, that films play a very important role today in the accurate determination of the optical constants of many materials, but particularly of metals.

Thin-Film Electronics. Deposited dielectric film materials are SiO, MgF_2, ZnS, and various organic compounds. Thin capacitive layers in the 100 to 500 micrometers thickness region are often produced by the anodization of tantalum and aluminum to Ta_2O_5 or Al_2O_3, respectively. The breakdown strength and dielectric constant of films approach bulk values, but might be reduced by surface roughness, structural faults, and lower density. According to the Lorentz-Lorenz formula, the dielectric constant D changes with reduced density ρ as $dD/d\rho = 3C/(1 - C\rho)^2$, where C is a constant depending on the material. On metal-dielectric-metal films, quantum mechanical tunneling through the dielectric film becomes observable below a dielectric thickness of about 100 micrometers. For applied voltages less than the metal-insulator work function ϕ, the tunneling current density J is proportional to the applied voltage V, demonstrating that the low-voltage tunneling resistance is ohmic. $J = (qV/h^2s) \times (2m^*\phi)^{1/2} \exp[-(4\pi s/h)(2m^*\phi)^{1/2}]$. At high applied voltages ($qV > \phi$), the current increases very rapidly:

$$J = (q^2V^2/8\pi h\phi_s^2) \exp[-(8\pi s/3hqV)(2m^*)^{1/2}\phi^{3/2}].$$

s is the insulator thickness, m^* the electronic effective mass, and q the electron charge. Thicker dielectric films may exhibit in high fields appreciable Schottky or avalanche currents when they are greatly disordered.

Polycrystalline metal films generally show, due to their low structural order, a larger resistivity than the bulk material. According to Matthiessen's rule, the total resistivity can be expressed as $\rho = \rho(t) + \rho(i)$ where $\rho(t)$ is the temperature-dependent resistivity associated with scattering by lattice vibrations, and $\rho(i)$ is a temperature-independent resistivity caused by impurity or imperfection scattering. Very thin specimens with a thickness comparable to the electron mean free path show a $\rho(i)$ rapidly increasing with decreasing thickness. This increase is caused by an increasing contribution of non-specular electron scattering at the film surfaces. By annealing a metal film, $\rho(i)$ might be reduced permanently. A large $\rho(i)$ results in a small temperature coefficient α.

Many known superconductors can be deposited as superconductive films.

Through thin-film experiments, the energy gap in semiconductors can be measured, and material parameters, such as the penetration depth of magnetic fields, can be studied at dimensions less than the coherence range.

Studies of semiconductor films have shown many facets. The properties of epitaxial films have mainly been investigated on Ge and Si, and to a lesser degree on III–V compounds. Much work has been done on polycrystalline II–VI films, particularly with regard to the stoichiometry of the deposits, doping and post-deposition treatments, conductivity and carrier mobility, photo-conductance, fluorescence, electroluminescence, and metal-semiconductor junction properties. Among other semiconductors, selenium, tellurium, and a few transition metal oxides have found some interest.

Film resistors, capacitors, and interconnected R-C networks on planar glass or ceramic substrates are finding widespread industrial use. Common resistor materials are carbon, nichrome, and tin oxide in individual components; and nichrome, tantalum, tantalum nitride, SiO-chromium cermet, and cermet glazes in planar networks. Gold, copper, aluminum, or tantalum is used for termination lands, connection leads, and capacitor plates. SiO, MgF_2, and Ta_2O_5 serve as film capacitor dielectrics and crossover insulation. The geometrical configuration of the desired component or circuit pattern is obtained either by deposition through mechanical masks or by removing from a continuous sheet the undesired portions after the deposition process is completed. This removal is frequently accomplished by a combination of photolitho-graphic and etch processes.

The minimum length l and width w of a resistor are calculated from the given resistance R, the sheet resistance $\mathscr{R}$ in ohms per square, dissipated power P, and permissible power dissipation per square inch $\mathscr{P}$ by use of the formulas $w = \sqrt{(P \cdot \mathscr{R})/\mathscr{P} \cdot R}$ and $l = wR/\mathscr{R}$. The capacitance of film capacitors is given by $C = 0.225\, D(N - 1)\, A/t$, where C is the capacitance in picofarads, D the dielectric constant, N the number of plates, A the area in square inches, and t the dielectric thickness in inches.

In retrospect, it is gratifying to note how much progress in thin-film electronics was made prior to a more penetrating and fundamental understanding of surface phenomena. It is only relatively recently that such experimental tools as angle-resolved photoelectron spectroscopy, synchrotron far-ultraviolet and X-ray spectroscopies, and back-scattering and channeling of energetic ions, have been used. New computational methods are now available for calculating detailed maps of the electron distribution at a surface. It has been found that the surface geometry is either a relaxed version of the bulk structure, or a reconstructed arrangement with symmetry wholly different from the bulk. A comparison of the electron spectroscopy results with theoretical predictions of surface state energy spectra based on a particular surface model allows confirmation of the assumed atomic arrangement. Together, the theoretical and experimental approaches have provided the first complete description of surfaces, including identification of the atomic species present, their atomic arrangement, and the distribution of valence electrons in space and energy. It is quite possible that these approaches will lead to a detailed picture at the atomic level of the interface structure, electron states, charges, reconstruction, and the related junction electronic properties of semiconductor-semiconductor and metal-semiconductor interfaces.

Various barrier layer diodes have exhibited impressive rectification ratios, but limited breakdown strength and low speed due to their large specific capacitance. Of the many film transistor concepts studied, the insulated gate field effect device has been promising. Its structure consists of a minute metal-dielectric-semiconductor capacitor. The semiconductor strip carries current between two terminals called source and drain. A field applied between metal "gate" and source modulates the semiconductor conductance and consequently the source-drain current. Usable semiconductor materials with a sufficiently low concentration of interface states are CdS, CdSe, and tellurium. These devices exhibit pentodelike characteristics with voltage gains ranging from 2.5 at 60 megahertz to 8.5 at 2.5 megahertz. The gain-bandwidth product GB, which is equal to the transconductance divided by 2π times the gate capacitance, reaches values of about 20 megahertz. It is determined by $GB = \mu_d V_D/2\pi L^2$, where μ_d is the effective drift mobility of the electrons, V_D the source-drain potential, and L the source-drain spacing, which is usually selected between 5 and 50 micrometers.

An outgrowth of prior thin-film technology and of basic surface science research has been molecular beam epitaxy — the MBE formation of compound semiconductor films. The MBE technology involves the use of separate atomic and molecular beams from multiple thermal sources in high vacuum which irradiate a substrate at intensities selected to grow films having the desired composition and doping. The ability to achieve slow growth rates, together with independent control of the separate beam sources, permits the fabrication of semiconductor junction profiles, both in doping and in composition, with a precision approaching that of a single atomic layer. To date MBE has been used to prepare films and layer structures involving a number of GaAs and $Ga_xAl_{1-x}As$ devices. Included in such devices are varactor diodes having highly controlled hyperabrupt capacitance-voltage characteristics, IMPATT diodes, microwave mixer diodes, Schottky barrier field-effect transistors (FETs), injection lasers, optical waveguides, and integrated optical structures. Some authorities believe that the potential for MBE in solid-state electronics may be greatest for microwave and optical solid-state devices as well as for circuits where submicrometer layer structures are required. A recently demonstrated MBE GaAs Schottky barrier diode cryogenic mixer with a noise temperature of 315 K at 102 gigahertz is exemplary of the potential of MBE technology for millimeter wave electronics. See also **Schottky Defect**.

MBE superlattice structures also are very promising. These superlattice structures, with periodicities of 50–100 micrometers, show negative resistance characteristics attributed to resonant tunneling into the quantized energy states associated with the narrow potential wells formed by the layers. Detailed studies have shown that the potential well distributions may be controlled and positioned to a precision of a few atomic layers.

Thin-film technology has also played an important role in developing Josephson superconducting devices, which offer outstanding advantages in constructing ultrahigh-speed computers. See also **Josephson Superconducting Devices**. These are tunnel-junction type devices.

Thin-film and surface phenomena are fundamental to the successful development, production, and use of solid-state devices. The research in this area is extensive. See also **Molecular and Supermolecular Electronics**.

Magnetic Films. Magnetic thin films of nickel-iron (usually deposited at an 80:20 composition by weight) exhibit a number of unusual properties, which have led to many experimental and theoretical studies, as well as to important applications in binary storage and switching, magnetic amplifiers, and magneto-optical Kerr-effect displays.

Such "Permalloy" films have two stable states of magnetization, corresponding to positive and negative remanence. When deposited in a magnetic field or at an oblique angle, they exhibit uniaxial anisotropy. In practice, this anisotropy shows some dispersion, since it results from the alignment of local lattice disturbances. The stable states result from the minimization of the free energy $E = MH_L \cos\theta - MH_T \sin\theta + K\sin^2\theta$, where the last term represents the anisotropy energy, and θ is the angle between the magnetization M and the easy axis. From an inspection of the derivatives of this equation follows the hard-direction straight-line and the easy-direction square hysteresis loops of aniso-tropic films. In the latter case, the magnetization is always either $+M$ or $-M$, and the change occurs at $H_L = \pm H_K$. The transitions from unstable to stable states occur at $\partial^2 E/\partial\theta^2 = 0$, resulting in a critical curve $H_L^{2/3} + H_T^{2/3} = H_K^{2/3}$ which has the form of an asteroid enclosing the origin (see also **Magnetism**).

An important feature of magnetic films is the high speed with which the state of magnetization can be reversed. Dependent on film properties and magnetic fields, three modes of magnetization reversal occur: Domain wall motion, incoherent rotations, and the extremely fast coherent rotation of the magnetization. Wall motion switching is expected when the driving fields are smaller than the critical values. During the past decade, magnetic garnet films have gained prominence in a number of research and industrial applications.

Additional Reading

Brundle, C.R. and S. Wilson: "Encyclopedia of Materials Characterization: Surfaces, Interfaces, Thin Films," Butterworth-Heinemann, Inc., Woburn, MA, 1992.

Cohen, E.D.: "Coatings: Going Below the Surface," *Chem. Eng. Progress*, 19 (September 1990).

Elshabini-Riad, A.R. and F.D. Barlow: "Thin Film Technology Handbook," The McGraw-Hill Companies, Inc., New York, NY, 1996.

Ferendeci, A.M.: "Physical Foundations of Solid State and Electron Devices," The McGraw-Hill Companies, Inc., New York, NY, 1991.

Fink, D.G. and H.W. Beaty: "Standard Handbook for Electrical Engineers," 14th Edition, The McGraw-Hill Companies, Inc., New York, NY, 1999.

Feldman, L.C. and J.W. Mayer: "Fundamentals of Surface Thin Film," Prentice-Hall, Inc., Upper Saddle River, NJ, 1998.

Karim, A. and S. Kumar: "Polymer Surfaces, Interfaces and Thin Films," World Scientific Publishing Company, Inc., River Edge, NJ, 1999.

Lisenskey, G.C., et al.: "Electro-Optical Evidence for the Chelate Effect at Semiconductor Surfaces," *Science*, 840 (May 18, 1990).

Matacotta, F.C. and G. Ottaviani: "Science and Technology of Thin Films," World Scientific Publishing Company, Inc., River Edge, NJ, 1995.

Mittal, K.L. and P. Kumar: "Emulsions, Foams, and Thin Films," Marcel Dekker, Inc., New York, NY, 2000.

Ohring, M.: "The Materials Science of Thin Films," Harcourt Brace & Company, San Diego, CA, 1991.

Sayer, M. and K. Sreenivas: "Ceramic Thin Films: Fabrication and Applications," *Science*, 1056 (March 2, 1990).

Scriven, L.E. and W.J. Suszynski: "Take a Closer Look at Coating Problems," *Chem. Eng. Progress*, 24 (September 1990).

Staff: "Future Directions in Thin Film," World Scientific Publishing Company Inc., River Edge, NJ, 1997.

Staff: "Range of Critical Temperatures Observed for Superconductive Elements in Thin Films Condensed Usually at Low Temperatures," in Handbook of Chemistry and Physics, CRC Press, Boca Raton, Florida, 73rd Edition (1992–1993).

Stuart, R.V.: "Vacuum Technology, Thin Films and Sputtering: An Introduction," Academic Press, Inc., San Diego, CA, 1983.

Venables, J.A.: "Introduction to Surface and Thin Film Processes," Cambridge University Press, New York, NY, 2000.

Vossen, J.L. and W. Kern: "Thin Film Processes II," Academic Press, Inc., San Diego, CA, 1991.

Williams, E.D. and N.C. Bartelt: "Thermodynamics of Surface Morphology," *Science*, 393 (January 25, 1991).

THIN-LENS RELATIONSHIPS. Formulas relating image distance, object distance, focal length, index of refraction, curvature of surfaces, etc., of a lens which is sufficiently thin that, for the purpose of the calculation, the thickness may be neglected. The formula

$$\frac{1}{f} = \frac{1}{p} + \frac{1}{q} = (n-1)\left(\frac{1}{r_1} - \frac{1}{r_2}\right)$$

is a typical thin-lens formula.

THIOCYANIC ACID. Aqueous solution of hydrogen thiocyanate, HSCN, formula weight 59.08, yellow solid below mp 5 °C, unstable gas at room temperature. The acid is moderately stable only when dilute and cold. The salts of this acid are known as thiocyanates.

Thiocyanic acid is formed by reaction of barium thiocyanate solution and dilute sulfuric acid, and filtering off barium sulfate, or by the action of hydrogen sulfide on silver thiocyanate, filtering off silver sulfide.

Sodium, potassium, barium, or calcium thiocyanate may be made by reaction of sulfur and the corresponding cyanide by heating to fusion. Ammonium thiocyanate (plus ammonium sulfide) may be made by reaction of ammonia and carbon disulfide, a reaction which probably accounts for the presence of ammonium thiocyanate in the products of the destructive distillation of coal. This reaction corresponds to the formation of ammonium cyanate from ammonia and carbon dioxide.

Silver, lead, copper(I), and thallium(I) thiocyanates are insoluble and mercury(II), bismuth, and tin(II) thiocyanates slightly soluble. All of these are soluble in excess of soluble (e.g., ammonium) thiocyanate, forming complexes. Iron(III) thiocyanate gives a blood-red solution, used in detecting either Fe(III) or thiocyanate in solution, and is extracted from water by amyl alcohol. It is not formed in the presence of fluoride, phosphate and other strongly complexing ions.

When thiocyanic acid is treated with certain oxidizing agents, e.g., nitric acid, sulfuric acid and hydrocyanic acid are formed, but the action of lead tetraacetate on the acid, or of bromine in ether on lead(II) thiocyanate, gives thiocyanogen ("Rhodan") NCSSCN, a yellow, volatile oil, mp about -3 °C, which polymerizes irreversibly at room temperature to insoluble, brick-red parathiocyanogen $(NCS)_x$. Thiocyanogen reacts with organic compounds like a free halogen. It liberates iodine from iodides. In water it is rapidly hydrolyzed to sulfuric and hydrocyanic acids. When thiocyanic acid is treated with reducing agents, e.g., aluminum and dilute hydrochloric acid, hydrogen sulfide plus carbon plus ammonium chloride are formed.

Esters. Ethyl thiocyanate $C_2H_5 \cdot SCN$, colorless liquid, bp 142 °C. Formed by reaction (1) of potassium thiocyanate and potassium ethyl sulfate, (2) of cyanogen chloride and ethane-thiol. Oxidizable with fuming nitric acid to ethyl sulfonic acid $C_2H_5 \cdot SO_2OH$, and reducible with zinc and dilute sulfuric acid to ethane thiol C_2H_5SH. Ethyl isothiocyanate $C_2H_5 \cdot NCS$, colorless, odorous liquid, bp 132 °C. Formed by reaction of ethyl amine and carbon disulfide (cf. the formation of ammonium thiocyanate from ammonia and carbon disulfide). Reducible to ethyl amine $C_2H_5NH_2$ plus methylene sulfide CH_2S. Allyl isothiocyanate ("mustard oil") $C_3H_5 \cdot NCS$ liquid, bp 151 °C, odor of mustard, and causes blisters in contact with the skin.

THIOETHERS. Hydrogen sulfide yields two classes of organic compounds: (1) hydrosulfides, and (2) sulfides. The sulfides are termed thioethers. A more general term, *thiols*, also is used. This term not only embraces thioethers, but also covers thioalcohols, sulfhydrates, and thiophenols.

Ethyl sulfide $(C_2H_5)_2S$, one of the better known thioethers, is an odorous, inflammable liquid, mp -102.1 °C, bp 91.6 °C, sp gr 0.837. The compound is insoluble in H_2O and soluble in alcohol and ether. It is prepared by distilling ethyl potassium sulfate with potassium sulfide. Chemically, ethyl sulfide behaves much like the ethers. For example, none of the hydrogen atoms can be displaced by metals and generally the compound is very inert. Additional thioethers can be prepared in a similar manner with the corresponding proper ingredients. Upon oxidation with HNO_3, thioethers are converted to sulfones. The latter are stable crystalline substances. An example is ethyl sulfone $(C_2H_5)_2SO_2$.

THIOKOL RUBBERS. See **Elastomers**.

THIOPHENE. $\langle(CH:CH)_2\rangle S$, formula weight 84.13, colorless liquid resembling benzene in odor, mp -30 °C, bp 84 °C, sp gr 1.070. Thiophene and its derivatives closely resemble benzene and its derivatives in physical and chemical properties. Thiophene is present in coal tar and is recovered in the benzene distillation fraction (up to about 0.5% of the benzene present). Its removal from benzene is accomplished by mixing with concentrated sulfuric acid, soluble thiophene sulfonic acid being formed. Thiophene gives a characteristic blue coloration with isatin in concentrated sulfuric acid.

Thiophene may be formed (1) by passing ethyl sulfide (diethyl sulfide) through a red-hot tube, (2) by reduction of sodium succinate and phosphorus trisulfide. Chlorine and bromine yield chloro- and bromo-substitution products, respectively, cold fuming nitric acid yields thiophene sulfonic acid. Thiophene aldehyde $C_4H_3S \cdot CHO$, liquid, bp 198 °C, resembles benzaldehyde chemically rather than furfural. The corresponding primary alcohol and carboxylic acid are known. By comparison, where the sulfur atom of thiophene is occupied by oxygen, furane is the resulting compound.

Where the sulfur atom of thiophene is occupied by a nitrogen group (NH), pyrrole is the resulting compound.

Benzothiophene $C_6H_4 \cdot (CH)_2S$ is a solid, mp 31 °C, bp 221 °C, with physical and chemical properties that resemble naphthalene. By comparison, where the sulfur atom of benzothiophene is occupied by oxygen, the resulting compound is benzofurane (coumarone). Where the sulfur atom of benzothiophene is occupied by a nitrogen group (NH), indole is the resulting compound.

THIOUREA. $(NH_2)_2CS$, formula weight 76.12, white crystalline solid, mp 180–182 °C, decomposes before boiling at atmospheric pressure, sp gr 1.405. Thiourea is moderately soluble in H_2O, soluble in alcohol, and slightly soluble in ether. Sometimes referred to as thiocarbamide, sulfurea, and sulfocarbamide, thiourea may be considered chemically analogous to urea and is oxidized to urea by cold potassium permanganate solution. The compound is easily hydrolyzed to NH_3, CO_2, and H_2S. Upon long heating below the melting point, thiourea is transformed to ammonium thiocyanate. Thiourea is attractive for plastics manufacture because of the greater ease with which substitution can be made on the sulfur atom of thiourea than on the oxygen atom of urea.

Thiourea is formed by heating ammonium thiocyanate at 170 °C. After about an hour, 25% conversion is achieved. With HCl, thiourea forms thiourea hydrochloride; with mercuric oxide, thiourea forms a salt; and with silver chloride, it forms a complex salt.

Symmetrical diphenyl thiourea (thiocarbanilide) $(C_6H_5NH)_2CS$ is a solid, mp 154 °C. When heated with concentrated HCl, the compound yields aniline plus phenylisocyanate. Formed by the reaction of aniline and CS_2, symmetrical diethylthiourea $(C_2H_5NH)_2CS$ is a solid, mp 77 °C.

In addition to its use in plastics manufacture, thiourea is used in some photographic processes and photocopying papers; in organic synthesis as an intermediate (drugs, dyes, cosmetics); in rubber accelerators; and as a mold inhibitor.

THISTLE FAMILY.   See **Artemisia**.

THIXOTROPIC SUBSTANCE.   See **Colloid System**; and **Fluid**.

THOMAS-FERMI DIFFERENTIAL EQUATION.   An equation which occurs in studying the electron distribution in an atom

$$y'' \sqrt{x} = y^{3/2}$$

Although a special solution can be given as $x^3 y = 144$, it is required for physical reasons that $y(0) = 1$, $y(\infty) = 0$. Solutions satisfying these boundary conditions must be found by numerical or graphical methods.

THOMSON, J.J. (1856–1940).   Joseph John Thomson was an English physicist. At age fourteen, his father sent him to Owens College for preparatory scientific training. His attendance here was important to his career because this college had an outstanding science faculty and it also offered many experimental physics courses.

Thomson earned his engineering degree from Owens. Then he attended Trinity College of Cambridge University and studied mathematics and theoretical physics. When he graduated he began working in the Cavendish Laboratory at Cambridge and by age twenty-seven became its director. Thomson's main research was on the conduction of electricity through gases. After, Roentgen's discovery of X-rays in 1895, Thomson started working with Rutherford and found that passing X-rays through gases greatly increased their ability to conduct electricity. Much of Thomson's further research dealt with the composition of cathode rays. He believed that cathode rays were streams of tiny charged particles. His work concluded that the atom was not the fundamental unit of matter. He devised a model of the atom incorporating his theory of *corpuscles*. Later, the name "electron" was adopted.

In 1906, Thomson received the Nobel Prize for Physics "in recognition of the great merits of his theoretical and experimental investigations on the conduction of electricity by gases." Between the years of 1906 and 1914, Thomson studied canal rays and worked on separating the different kinds of atoms and atomic groupings present in them. In 1903, Thomson proposed a discontinuous theory of light, with light rays being composed of separate particles rather than continuous streams, and later Einstein developed the photon theory of light.

Thomson's work revolutionized scientific understanding of the atom and ushered in a new era in physical science. He is also remembered for his excellent teaching at Cavendish Laboratory.

See also **Electricity**; and **Electron Theory**.

J. M. I.

THOMSON (Law of).   (Or *Law of Helmholtz; Thomson's Rule*.) In an electric cell, the heat of reaction is a direct measure of the electromotive force, i.e., the chemical energy is simply converted into electrical energy. This is only approximately true, for part of the energy of an electric cell appears as heat, either absorbed or evolved, as the case may be.

THOMSON PARABOLA METHOD.   The method of investigating the charge-to-mass ratio of positive ions in which the ions are acted upon by electric and magnetic fields applied in the same direction normal to the path of the ions. It can be shown that ions of a given charge-to-mass ratio but different velocities will be deflected so as to form a parabola.

THOMSON PRINCIPLE.   The hypothesis that, if thermodynamically reversible and irreversible processes take place simultaneously in a system, the laws of thermodynamics may be applied to the reversible process while ignoring for this purpose the creation of entropy due to the irreversible process. Applied originally by Thomson to the case of thermoelectric effects. Also used in the treatment of electrochemical cells, thermal diffusion.

THOMSON SCATTERING.   See **Scattering**.

THORAX.   1.   The division of the arthropod body between the head and abdomen. It consists of three or four metameric segments of the body, often closely united so that their boundaries are difficult to determine. It bears a pair of jointed appendages on each segment, developed for walking or swimming, and in some species for grasping, and in the insects the two posterior segments may bear a pair of wings each. The three segments of the insect thorax are known as the pro-, meso-, and metathorax.

2.   A division of the trunk of vertebrates, just behind the head and neck. This region is supported by the ribs and contains the thoracic cavity, a division of the coelom. In the mammals it is separated from the abdominal cavity by the diaphragm and is further subdivided into the pleural cavities containing the lungs and the pericardial cavity containing the heart. It is a bony cage made of the sternum in front, the vertebral column behind, and the ribs connecting the two.

THORIANITE.   This mineral of thorium oxide, ThO_2, is isomorphous with uraninite and occurs in black, nearly opaque cubic crystals in Ceylon and in Madagascar. Often containing rare-earth metals and uranium, the ore is strongly radioactive. Because of its radioactivity, it is valuable in helping to date the relative ages of rocks in which it occurs.

THORITE.   The mineral thorite is a silicate of the rare element thorium and corresponds to the formula $ThSiO_4$. It is tetragonal and exhibits a prismatic cleavage. The original thorite was black in color with a specific gravity of 4.4–4.8. A variety orangite, so called from its orange-yellow color, has a specific gravity of 5.19–5.40. It has been found partly altered to thorite. Uranothorite contains uranium oxide. Thorite occurs in Norway in augite syenites. Thorite and orangite occur in Sweden, and orangite and uranothorite are found in Madagascar. Uranothorite is found in Ontario.

THORIUM.   Chemical element symbol Th, at. no. 90, at. wt. 232.038, radioactive metal of the Actinide Series, mp 1750 °C, bp 4790 °C, density 11.5–11.9 g/cm³ (17 °C). Thorium metal is dark gray, dissolves in HCl, is made passive in HNO_3, and is not affected by fusion with alkalis. The element combines with chlorine or sulfur at 450 °C; with hydrogen or nitrogen at 650 °C. All thorium-containing substances are radioactive. The element was discovered by J.J. Berzelius in 1829. The electronic configuration

$$1s^2 2s^2 2p^6 3s^2 3p^6 3d^{10} 4s^2 4p^6 4d^{10} 4f^{14} 5s^2 5p^6 5d^{10} 6s^2 6p^6 6d^2 7s^2.$$

Ionic radii Th^{3+} 1.08 Å, Th^{4+} 0.95 Å (Zachariasen). Metallic radius 1.797_5 Å. First ionization potential 5.7 eV; second, 16.2 eV; third, 29.4 eV. Oxidation potentials $Th \rightarrow Th^{4+} + 4e^-$, 1.90 V; $Th + 4OH^- \rightarrow ThO_2 + 2H_2O + 4e^-$, 2.48 V. See also **Chemical Elements**.

The isotopes of thorium include mass numbers 223–234. ^{232}Th has a half-life of 1.39×10^{10} years. See also **Radioactivity**. It emits an alpha-particle and forms meso-thorium 1 (radium-228), which is also radioactive, having a half-life of 6.7 years, emitting a beta-particle. Since ^{232}Th captures slow neutrons to form, by a series of nuclear reactions, ^{233}U which is fissionable, thorium can be used as a fuel for nuclear reactors of the breeder type. Thorium occurs in earth minerals, an average content estimated at about 12 ppm. Findings of the *Apollo 11* space flight indicated that thorium concentrations in some lunar rocks are about the same as the concentrations in terrestrial basalts.

Thorium occurs in monazite sand in Brazil, India, North and South Carolina; this ore contains 3–9% thorium oxide, and is the chief source; thorium is also found in thorite containing about 60% oxide and in thorianite, about 80% oxide. When heated with concentrated H_2SO_4 the minerals form thorium sulfate, from which, by a series of reactions, thorium nitrate, the chief commercial compound, is obtained.

Thorium has the oxidation state of (IV) in all of its important compounds. Its oxide, ThO_2, and its hydroxide are entirely basic. The nature of the ions present in a number of solutions of the soluble compounds is not known with certainty. Complex ions involving sulfate are suggested by the increased solubility of the sulfate in solutions of the acid sulfates. Similarly, other complex ions are suggested by the solubility of the carbonate in excess alkali carbonate and of the oxalate in ammonium oxalate. Such ready complex ion formation is consistent with the high positive charge of the thorium-(IV) ion.

Although the exact extent is not known accurately, hydrolysis of various salts is known to occur. Since the hydroxide is not precipitated it is assumed that the hydrolysis product is some ion on the form $Th(OH)_2^{++}$ or $ThOH^{3+}$. The solution chemistry of thorium is made more complicated because of the hydrolytic phenomena observed and the polynuclear complex ions that are formed at low acidities and higher thorium concentrations.

Studies of the complex ions formed by Th^{4+} with various complexing anions have given much information. For example, the equilibria and ionic species involved in the chloride complexing of aqueous thorium have been studied through the method of measuring the distribution between H_2O and benzene containing thenoyltrifluoroacetone. The conclusion: that there is successive complexing involving the species $ThCl^{3+}$, $ThCl_2^{2+}$, $ThCl_3^+$ and $ThCl_4$. Similarly, all the intermediate chelate complex ions between thorium and acetylacetone exist in aqueous solution of proper acidity.

Thorium dioxide (face-centered cubic structure) is very insoluble in H_2O, but dissolves in acids to yield salts.

Thorium forms one series of halides, another one of oxyhalides, and also a series of double or complex halides. In general, stability of these compounds toward heat decreases as the atomic weight of the halogen increases. These compounds are often isostructural with the corresponding compounds of other actinide elements in the (IV) oxidation state.

Thorium metal reacts with hydrogen at moderately elevated temperatures to yield two hydrides: (1) ThH_2, which has a pseudotetragonal body-centered unit containing two metal atoms, isomorphous or pseudoisomorphous with thorium carbide, zirconium hydride, and zirconium carbide, ThC_2, ZrH_2, and ZrC_2; and (2) a hydride of approximate composition $ThH_{3.75}$ or ThH_4, possessing a unique cubic structure unrelated to that of the parent metal.

Thorium sulfide, ThS_2, is obtained by the action of H_2S or sulfur on thorium metal. The oxysulfide, $ThOS$, has been obtained in several ways, one of which is by the action of CS_2 on thorium dioxide at elevated temperatures. At 800 °C and under pressure, sulfur combines with thorium to yield compounds with approximately the formulas ThS, Th_2S_3, and Th_3S_7. The first two have semimetallic properties and may be employed as ceramics for use with highly electropositive metals, whereas the last appears to be a polysulfide.

Anhydrous thorium sulfate, $Th(SO_4)_2$, is obtained by the action of concentrated H_2SO_4 on thoria (ThO_2). A solution of this salt deposits crystals of $Th(SO_4)_2 \cdot 9H_2O$ at about 15 °C, $Th(SO_4)_2 \cdot 8H_2O$ near 24 °C, and $Th(SO_4)_2 \cdot 4H_2O$ around 45 °C. At 100 °C other hydrates change to $Th(SO_4)_2 \cdot 2H_2O$. In aqueous solution, the salt is considerably hydrolyzed to an oxysulfate—for instance, $ThOSO_4 \cdot H_2O$.

Thorium nitrate, $Th(NO_3)_4 \cdot 12H_2O$, is obtained by dissolving thorium hydroxide in HNO_3.

Thorium orthophosphate, $Th_3(PO_4)_4 \cdot 4H_2O$, is precipitated by adding a solution of sodium phosphate to an acidic solution of a thorium salt. Thorium pyrophosphate, $ThP_2O_7 \cdot 2H_2O$, precipitates when an acidic solution of thorium nitrate is treated with one of tetrasodium pyrophosphate.

Thorium has been used as a fuel for nuclear reactors since it is a fertile material for the generation of fissionable uranium-233. Some experts have estimated that the energy available from the world's reserves of thorium is greater than all of the remaining fossil fuels (coal and petroleum) and of all of the remaining uranium, combined. Thorium oxide is used for gas mantles. The oxide also helps to control grain size in tungsten filaments and strengthens nickel alloys (TD nickel). Thorium is also used as an alloying addition in magnesium technology and as a deoxidant for molybdenum, iron, and other metals. Several applications for thorium are found in electronic technology.

Thorium oxide has a high refractive index and low dispersion and thus finds use in high-quality camera and scientific instrument lenses. Thorium oxide also is used as a catalyst in the conversion of ammonia to nitric acid, in petroleum cracking, and in sulfuric acid production.

Handling. ^{232}Th is sufficiently reactive to expose a photographic plate within a few hours. Thorium disintegrates with the production of thoron (^{220}radon). The latter is an alpha emitter and a radiation hazard. Areas where thorium is stored should be well ventilated and all precautions in the handling of thorium materials must be taken.

Additional Reading

Elvers, B., S. Hawkins, and W.E. Russey: "Ullmann's Encyclopedia of Industrial Chemistry: Thorium and Thorium Compounds to Vitamins," 5th Edition, John Wiley & Sons, Inc., New York, NY, 1997.
Finlayson-Dutton, G.: "Tinkering with Glass and Ceramic Structures," *Science*, 627 (August 10, 1990).
Greenwood, N.N. and A. Earnshaw: "Chemistry of the Elements," 2nd Edition, Butterworth-Heinemann, Inc., Woburn, MA, 1997.
Krebs, R.E.: "The History and Use of Our Earth's Chemical Elements: A Reference Guide," Greenwood Publishing Group, Inc., Westport, CT, 1998.
LaTourrette, T.Z., A.K. Kennedy, and G.J. Wasserburg: "Thorium-Uranium Fractionation by Garnet: Evidence for a Deep Source and Rapid Rise of Oceanic Basalts," *Science*, 739 (August 6, 1993).
Lewis, R.J. and N.I. Sax: "Sax's Dangerous Properties of Industrial Materials," 10th Edition, John Wiley & Sons, Inc., New York, NY, 1999.
Lide, D.R.: "CRC Handbook of Chemistry and Physics 2000–2001," 81st Edition, CRC Press, LLC., Boca Raton, FL, 2000.
Smith, J.F. et al.: "Thorium Preparation and Properties," Iowa State University Press, Ames, Iowa, 1975.
Staff: International Atomic Energy Agency, "Utilization of Thorium in Power Reactors," Bernan Associates, Lanham, MD, 1996.

THORIUM OXIDE.   See **Thorianite**.

THORIUM SERIES.   See **Radioactivity**.

THRASHER *(Aves, Passeriformes)*.   Moderately large North American birds related to the wrens, the mockingbird, and the catbird. They have a long curved beak. Several of the species are among our finest singers.

The brown thrasher, *Toxostoma rufum*, is a familiar bird east of the Rockies. Our six other species are western and southwestern.

THREADWORMS.   See **Nematodes**.

THREE-BODY PROBLEM (Astronomy).   Assuming that three or more objects exist in the universe, and that each of them attracts the others in accordance with the law of gravitation, there is the problem of predicting subsequent positions and motions. This problem is commonly referred to as the three-body, or *n*-body, problem.

The two-body problem has been completely solved, and the full solution may be expressed in comparatively few words and symbols. However, if one or more bodies is added, the solution becomes an exceedingly complex one that has never been reached in any form at all suitable for computational purposes. In fact, only one complete solution has ever been arrived at, in spite of the labors of practically all the great mathematicians of the past three centuries.

Despite the fact that no general solution of the problem is available, there are several practical computational methods for determining the positions of planets and other members of the solar system, taking into account the gravitational attraction of all effective members. All such solutions are made by successive approximations and various methods of computing perturbations rather than by the application of any general solution. In addition, a number of particular solutions of the three-body problem have been made by mathematicians. Lagrange, notably, showed that it is possible for an asteroid to be stable in such a position that it is equidistant from both the sun and Jupiter. In such a case, the three objects would be on the vertices of an equilateral triangle, and the asteroid orbit would have the same period as Jupiter. This case is illustrated in nature by the members of the so-called Trojan group.

THREE-PHASE EQUILIBRIUM.   For every pure, chemically stable substance there is a certain temperature and pressure at which it can exist in all three states or phases, solid, liquid, and vapor, each phase being in equilibrium with each of the others. At higher temperatures and pressures than those at this so-called "triple point," the liquid and vapor states may attain equilibrium; solid-vapor equilibrium is possible at lower temperatures and pressures; and solid liquid equilibrium can be obtained at higher pressures and at lower or higher temperatures according as the substance contracts or expands upon melting. These three equilibria may be represented by three temperature-pressure graphs, which converge at the triple point. Figure 1 illustrates the case of water.

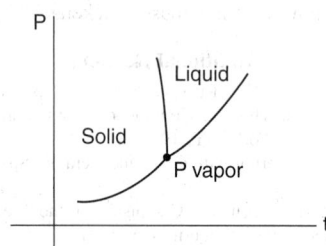

Fig. 1.   Triple-point (*P*) on temperature-pressure diagram.

In 1954, the thermodynamic temperature scale (i.e., the absolute Kelvin scale was redefined by setting the triple point temperature for water equal to exactly 273.16 K).

THRESHER SHARK.   See **Sharks**.

THRESHOLD DECODING.   See **Information Theory**.

THRESHOLD DETECTOR.   A circuit that provides an indication (digital) that a signal input is in excess of a predetermined magnitude. Assuming a predetermined threshold value were 6 V, an input of 6.5 V will result in an output representing a binary 1 and an input of 5.5 V causes a digital output representing a binary 0. Thus, a threshold detector is a form of comparator. The terms sometimes are used interchangeably. See also **Comparator Amplifier**. However, the term *comparator* may refer only to a circuit that detects whether or not an input signal has changed polarity.

From a practical standpoint, a band of uncertainty surrounds the predetermined threshold value. Thus, an input signal with a value that falls within the deadband will produce an indeterminant output, that is, one that cannot be interpreted as either a 1 or a 0. Increasing the gain of the threshold circuit will reduce the magnitude of the deadband. However, stability and overload considerations place limitations upon gain.

Threshold circuits are extensively used in data-acquisition and instrumentation systems. The terms *voltage sense* and *level sense* sometimes are used to describe a threshold circuit in digital systems. The devices are commonly used in instrumentation alarm systems for monitoring circuits, as for example, the output of a thermocouple to initiate a visual or audible alarm should a temperature rise too high or fall too low in a process.

THRESHOLD ILLUMINANCE.   The lowest value of illuminance that the eye is capable of detecting under specified conditions of background luminance and degree of dark adaptation of the eye. Also called *flux-density threshold*. This threshold, which controls the visibility of point light sources, especially at night, cannot be assigned any universal value, but nonflashing lights can generally be seen by a fully dark-adapted eye when the lights yield an illuminance of the order of 0.1 lumen per square kilometer at the eye.

THRIP (*Insecta, Thysanoptera*).   An order of minute insects, with or without wings. Their mouths are formed for sucking, the tarsi (feet) have an expanded tip, and the metamorphosis is gradual. When wings are present, there are two pairs, both formed of slender membranes with very long fringes. The insects are quite small, seldom over $\frac{1}{8}$ inch (3 millimeters) long. The thrips generally like to live in the floral portion of plants, but are found elsewhere, always feeding upon the sap. Some species are very serious pests on such crops as vegetables, fruits, field crops, and decorative flowers, notably those cultivated in glasshouses. The mouth portion of a thrip can be described as having an action that lies between that of the chewing insect and the piercing-sucking insect. Some thrips live on other small insect eggs and mites and thus, in certain cases, have an economic value. The thrips tend to specialize. Some of the major species injurious to crops are described in the remainder of this entry.

Bean Thrip (*Hercothrips fasciatus*, Pergande).   These insects feed on the foliage of bean, pea, and several other vegetable crops. They leave the plant with a silvered-bleached appearance and covered with very small black spots of excrement of the nymphs. The latter covering is sometimes called soot. Left uncontrolled, an infestation of thrips causes wilting and ultimately kills the plant.

Citrus Thrips.   There are three principal species that seriously damage citrus and many deciduous fruits. The species we describe here is *Scirtothrips citri* (Moulton). This insect can be quite damaging to citrus crops in Arizona and California, particularly in very dry, warm weather. The insect is less active in Florida. This insect winters in the egg stage. The eggs are deposited on the leaves and stems of infested trees. A female may deposit up to 250 eggs.

Dusting with very fine sulfur powder can be effective. Sulfur sprays also can be used.

Damage rendered by the thrip is by way of attacking buds and new growth, thus retarding the general development of an entire tree. The thrip also injures the fruit, causing a scarring of the skin and deformation of the fruit. Sometimes there is a telltale round mark or ring around the stem end of the fruit, indicating where thrips have fed beneath the sepals at the time the fruit was quite small. In very heavy infestations, crop losses approaching 90% have been recorded. See Fig 1.

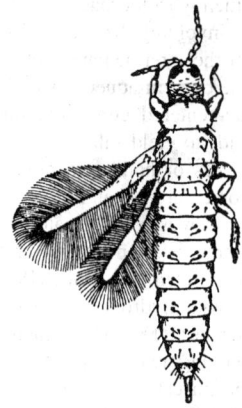

Fig. 1.   Thrip.

Onion Thrip (*Thrips tabaci*, Lindeman).   This insect damages nearly all garden crops, some weeds and field crops. This thrip is particularly injurious to bean, beet, cabbage, cauliflower, cucumber, melon, onion, squash, sweet clover, tomato, and turnip. It is distributed in all regions where onions are grown. On these plants, the thrip punctures leaves and sucks the exuding sap, leaving whitish areas on the leaves. In a severe infestation, most of the leaves are destroyed and the plant winters and dies. Unless controlled, the pest can destroy entire crops, notably in very dry seasons.

Pear Thrip (*Taeniothrips inconsequens*, Uzel).   In addition to pear, which this insect attacks vigorously, it also damages apple, apricot, cherry, grape, peach, plum, prune, and other fruit trees and bushes. The pear thrip was observed in the United States for the first time in the early 1900s in California. It has spread throughout the Pacific coastal area to British Columbia. It is also now found in the eastern states, principally Maryland, New York, and Pennsylvania.

THROAT. Internally, the pharynx: the cavity behind the mouth into which the nasal passages open and from which the esophagus and trachea lead. Externally, the ventral part of the neck.

THROMBIN. See **Anticoagulants**; and **Blood**.

THROMBOCYTES. See **Blood**.

THROMBOPLASTIN. See **Blood**.

THROTTLE. To throttle means to choke. Throttling a fluid flow is to decrease the cross-sectional area of flow at some point, as by partly closing a valve, damper, or gate in the conduit.

This word is in common usage to denote the adjustable control on the quantity of the working medium entering prime movers and hence has become synonymous, in common usage, with the control of prime mover power.

THROUGHPUT. In vacuum technology, the quantity of gas in pressure-volume units at a specified temperature flowing per unit time across a specified open section of a pump or pipeline. The specified temperature may be the actual temperature of the gas or a standard reference temperature. It is recommended that throughput be referred to standard room temperature. The recommended unit of throughput is the torr liter per second at 20 °C. Other units of throughput in common use are micron liters per second at 25 °C and micron cubic feet per minute at 68 °F. Under conditions of steady-state conservative flow the throughput across the entrance to a pipe is equal to the throughput at the exit. In this case throughput can be defined as the quantity of gas flowing through a pipe in pressure-volume units per unit time at room temperature.

THROWING POWER. A term used to denote the relative effectiveness of various electrolytic cells for the deposition of metal at the cathode. It is commonly determined by means of two half-cathodes which give the values for the relation between the distance from the anode and the weight of metal deposited.

THRUSH (*Aves, Passeriformes*). In the broad sense, a bird of the family *Turdidae*, but the term is most widely applied to the members of this family that retain a spotted breast as adults, while other species which lose the spots as their adult plumage develops receive other names. Among the latter are the robin and bluebirds of North America. The North American thrushes are of moderate or small size, brown, gray, or olive above and white below, with spots similar to the back. The wood thrush, *Hylocichla mustelina*, is widely distributed in the eastern half of the United States (Fig. 1), and the hermit thrush, *H. guttata*, is even more widely distributed, though less commonly known. The veery, also called Wilson's thrush (*Hylocichla muscescens*) also is found in eastern North America.

Fig. 1. Wood thrush. Bright brown above; white below, with large round spots on the breast and sides.

The family includes the European fieldfare, redwing, and blackbird, the ring-ouzel, and the Old World chats. In North America it is also represented by the Townsend solitaire and the wheatear, in addition to the robin and the several species of bluebirds.

The song-thrush of Europe, *Turdus philomelus*, is also called the mavis, and in America the term *thrush* is misapplied to the Louisiana water thrush, which is a warbler, because of its similar color and pattern.

See also **Passeriformes**.

THRUST. 1. The pushing or pulling force developed by an aircraft engine or a rocket engine. 2. The force exerted in any direction by a fluid jet or by a powered screw, as the thrust of an antitorque rotor. 3. Specifically, in rocketry, $F = mv$, where m is propellant mass flow and v is exhaust velocity relative to the vehicle. Also called *momentum thrust*.

THRUST AND DRAG. See **Aerodynamics**; **Airplane**; **Helicopters and V/STOL Craft**; and **Supersonic Aerodynamics**.

THUJA. See **Arborvitae**; and **Cedar Trees**.

THULIUM. Chemical element, symbol Tm, at. no. 69, at. wt. 168.934, twelfth in the Lanthanide Series in the periodic table, mp 1545 °C, bp 1950 °C, density 9.321 g/cm³ (20 °C). Elemental thulium has a close-packed hexagonal crystal structure at 25 °C. The pure metallic thulium is gray in color, with no evidence of tarnishing up to a temperature of 200 °C. Above 200 °C, the element combines with oxygen, sulfur, nitrogen, carbon, and hydrogen and will form intermetallic compounds with most metals. At higher temperatures, halogen gases react vigorously with the element to form trihalides. There is one natural isotope of thulium ^{169}Tm. Seventeen artificial isotopes have been produced. Average content of the earth's crust is estimated at 0.48 ppm thulium, making this element the least abundant of the rare-earth elements. Even at this level, however, thulium is potentially more available than antimony, bismuth, cadmium, or mercury. The element was first identified by P.T. Cleve in 1879. Electronic configuration

$$1s^2 2s^2 2p^6 3s^2 3p^6 3d^{10} 4s^2 4p^6 4d^{10} 4f^{12} 5s^2 5p^6 5d^1 6s^2.$$

Ionic radius Tm³⁺ 0.880 Å. Metallic radius 1.746 Å. First ionization potential 6.18 V; second 12.05 V. Other physical properties of thulium are given under **Rare-Earth Elements and Metals**.

Thulium occurs in apatite and xenotime and is derived from these minerals as a minor coproduct in the processing of yttrium. Processing involves organic ion-exchange, liquid-liquid, or solid-liquid, techniques. Prior to the development of cation exchange resins capable of separating the chemically similar rare earths, thulium was practically unavailable in pure form. Thulium metal is made by the direct reduction of thulium oxide by lanthanum metal at high temperature in a vacuum.

Important scientific and industrial applications for thulium and its compounds remain to be developed. In particular, the photoelectric, semiconductor, and thermoelectric properties of the element and compounds, particularly behavior in the near-infrared region of the spectrum, are being studied. Thulium has been used in phosphors, ferrite bubble devices, and catalysts. Irradiated thulium (^{169}Tm) is used in a portable x-ray unit.

NOTE: This entry was revised and updated by K.A. Gschneidner, Jr., Director, and B. Evans, Assistant Chemist, Rare-earth Information Center, Institute for Physical Research and Technology, Iowa State University, Ames, IA.

Additional Reading

Lewis, R.J. and N.I. Sax: "Sax's Dangerous Properties of Industrial Materials," 10th Edition, John Wiley & Sons, Inc., New York, NY, 1999.
Lide, D.R.: "CRC Handbook of Chemistry and Physics 2000–2001," 81st Edition, CRC Press, LLC., Boca Raton, FL, 2000.

THUNDERHEAD. See **Clouds and Cloud Formation**.

THUNDERSTORM. See **Fronts and Storms**.

THYME. See **Flavorings**.

THYMUS GLAND. Situated within the chest cavity close to the heart, the thymus gland has been known for centuries and was believed to be the seat of courage and affection in ancient times because it is situated close to the heart. The word *thymus* is derived from a Greek word having two meanings, "courage" and "thyme." The shape of the organ resembles the leaves of the thyme plant. The thymus is not always the same size, and it does not grow progressively as most organs do. From the day of birth to the age of approximately 12 years, the thymus grows gradually until it reaches a maximum size and weight. At this time, the organ is the size of a small egg — with about the same weight. At about age 15, the thymus begins to diminish in size and weight. In old age, the gland is practically nonexistent. During pregnancy, the thymus usually becomes smaller than normal but recovers its original size after delivery. Infectious diseases, x-rays, and hormones also can act on the thymus to decrease its size.

The thymus has a central role in establishing the immunological capacities of the body. The thymus elaborates hormones responsible for the production of cells with the capability of making antibodies and rejecting foreign elements in the body. During the first weeks of life, the thymus produces the *basic cells* that are then distributed throughout the body to other lymphocyte production centers, namely, the lymph nodes and spleen. At short notice, these organs can mass-produce lymphocytes and carry on the production of antibodies, which protect the body against invading microbes or foreign tissues.

During the 1960s, it was found that the thymus processes bone marrow stem cells into T-cells, the latter cells known to provide the body with cellular immunity. It was also learned that the thymus produces protein hormones. Initially, two thymic hormones were identified — *thymosin* and *thymopoietin*. Others have since been identified. These hormones participate in varying ways in the formation of different types of T cells. In the mid-1970s, thymic hormones were administered to children with immunodeficiency diseases resulting from an underdeveloped thymus and immature T cells; and to children whose T cells had been depressed by drugs, notably immunosuppresant drugs used in cancer treatment. Good results generally were obtained.

When the role of the thymus was further delineated, a distinction was made between thymus-*independent* (B) lymphocytes and thymus-*dependent* (T) lymphocytes. Although the specificity of B lymphocytes in their immune roles is well understood, information on the specificity of T lymphocyte receptors is still limited. It is known that the T cells display a restricted range of specificity as compared with the B cells. As pointed out by Paul and Benacerraf (1977), thymus-dependent lymphocytes react particularly to thymus-dependent protein and cell surface antigens and to products of genes encoded in the major histocompatibility complex (MHC). Additionally, T lymphocytes have important functions in regulating the immune response and in discriminating of self from nonself. Recent work indicates that individual T lymphocytes possess receptors that interact with both thymus-dependent antigens and MHC gene products, either independently or as associated structures. See **Immune System and Immunology**.

Some authorities have suggested that thymosin, which has been administered on a limited basis to some cancer patients has increased survival by nearly 200 days in cases of small cell or "oat cell" lung cancer. This is probably the result of increasing the number of T cells. No direct action of thymosin on the tumors per se has been demonstrated. Research also suggests that thymosin hormone may provide better protection for the elderly who possess an aging immune system.

The hormones from the adrenal cortex and the sex hormones act on the thymus and inhibit its function. But the hormone from the thyroid gland has a stimulating effect on the organ. The pituitary gland, by acting on the adrenals and gonads, inhibits the thymus; by acting on the thyroid, it stimulates the thymus.

The thymus on occasion may become larger than normal. The condition usually is associated with disturbances in other endocrine glands. In cases of adrenal insufficiency, the thymus and other lymphoid organs increase in size; similar effects are observed when the thyroid gland becomes overactive. This enlargement is thought to be in rare instances associated in some way with the cause of sudden death in apparently healthy children or young adults exposed to some stress, such as surgery.

The thymus, like all organs of the body, may be the site of tumor formation. There are several types of tumors that may develop, some rare, others more common. Whatever type of tumor, the symptoms manifest themselves by local pressure, difficulty in breathing, and bluish coloration of the skin. The symptoms are probably the result of pressure by the tumor on the windpipe and nerves in the chest. Tumors of the thymus are removed surgically; or if surgery is not possible, they may be destroyed by X-rays. When surgery is done early, it often effects permanent cure.

The thymus has been found to be enlarged in many cases of *myasthenia gravis*, a disease of the muscles. Some patients suffering from this disease have benefited by a removal of the enlarged thymus. Recent studies indicate that thymosin may also alleviate this condition.

Additional Reading

Dabrowski, M.P. and B.K. Dabrowska-Bernstein: "Immunoregulatory Role of Thymus," CRC Press, LLC., Boca Raton, FL, 1990.
Henry, K. and W.S. Symmers: "Thymus, Lymph Nodes, Spleen and Lymphatics," 3rd Edition, Churchill Livingstone, Inc., Philadelphia, PA, 1992.
Kornstein, M.J. and G.G. DeBlois: "Pathology of the Thymus and Mediastinum," Vol. 33, W.B. Saunders Company, Philadelphia, PA, 1995.
Paul and Benacerraf: *Science*, **195**, 1293–1301, 1977.
Willich, E. and W.R. Webb: "Thymus: Diagnostic Imaging, Functions, and Pathologic Anatomy," Springer-Verlag, Inc., New York, NY, 1992.

THYRISTOR. A generic title referring to a class of electronic devices consisting of four layers of semiconductor material alternately doped with *p* and *n* type impurities (see also **Transistor**) to form a *p-n-p-n* structure. Included in this category are the silicon-controlled rectifier (SCR), silicon-controlled switch (SCS), gate-turn-off switch (GTO), various forms of the *p-n-p-n* diode, light-activated SCR (LASCR), and three-element static ac switch (TRIAC).

THYROID GLAND. An endocrine gland located in the neck, the thyroid is shaped something like a butterfly. It is one of the largest endocrine glands in the body, averaging 0.7 ounce ($\sim$20 grams) in weight. It is the first gland to develop both in the individual and in the evolvement of the species. The gland is situated in the front of the neck where the latter joins the chest. The "wings" of the butterfly lie on either side of the windpipe and are referred to as lobes. The body of the butterfly is represented by a small bridge (isthmus) which passes in front of the windpipe and connects the "wings." The normal thyroid is packed so neatly between other adjoining structures in the neck that it can neither be seen as a bulge, nor felt as a distinct organ.

The gland has a brownish red, pulpy appearance and has one of the most copious blood supplies of any organ in the body. The average adult possesses approximately 5 quarts (4.7 liters) of blood, an amount that passes through the thyroid gland once per hour. The rapid flow, which brings about thyrotrophic hormone (a thyroid-stimulating hormone of the anterior pituitary gland) and iodide (for building thyroid hormones), exerts a considerable influence on thyroid activity.

Viewed through a microscope, the thyroid is seen to be composed of a great many small, hollow, ball-like units known as *follicles*. The average diameter of each follicle is about 1/100th inch (25 millimeters), although there is considerable variation in size in the healthy gland, and even variations when diseases are present. The functional cells of the thyroid (*acinar cells*) make up these follicles, the walls of which are one cell in thickness. Each follicle has its own small artery that supplies blood to a network of capillaries which themselves subsequently reunite to form a small vein.

Inside each follicle is a jelly-like material composed of thyroglobulin and called *colloid*. The raw material for hormone production is brought to the thyroid cells by the blood, and the finished products, known as thyroxine and triiodothyronine, are necessary for growth, development, and metabolism.

The metabolism of iodine is centrally involved in thyroid physiology. The daily ration of iodine is absorbed into the blood from the gastrointestinal tract as iodide. Iodide circulating in the blood enters the thyroid by a mechanism known as the "iodide trap," which concentrates iodide to a level 25 times that in the blood. The thyroid then builds the hormones *triiodothyronine* and *thyroxine* which are stored in the thyroid colloid for release as required.

In response to the stimulus of a center in the brain, thyroxine and triiodothyronine are released into the blood stream. A small portion of the thyroxine is "free" in the blood stream (only free thyroxine can enter the cells of the body), but most of it is chemically bound to a protein. Triiodothyronine is less closely bound to protein. Probably more than 80% of the iodine in the blood is in the form of thyroxine; the rest in triiodothyronine. Protein-bound thyroid hormone normally is greatly increased during pregnancy and during treatment with estrogen. See also **Endocrine System**.

The two thyroid hormones have the same effects on the body qualitatively, but quantitatively there are differences. Their function is to stimulate metabolism of nearly all cells except those of the brain, thyroid, testis, spleen, and uterus. The effect of thyroxine is not observed until one or more days after administration, but the effect is maintained and is lost only after several days. The response to triiodothyronine may occur within 6 hours after administration, and by 36 to 48 hours the response has disappeared.

The growth and function of the thyroid are under the control of the *thyrotrophic hormone* secreted by the anterior pituitary. These cells, in turn, vary their secretion rate in response to the amount of thyroid hormone in the blood. Thus, a fall in the amount of thyroid hormone in the blood is countered by an increased secretion of thyrotrophin, which stimulates the thyroid to increase the production of hormones. This feedback system maintains a balance in thyroid activity. The role of pituitary in this system is at least partially controlled by stimulation from the hypothalamus.

The exact means by which the thyroid hormone influences the cells of the body are not fully understood. One of the most striking effects is upon the consumption of oxygen by living cells. One of the main tasks of the thyroid hormone appears to be the maintenance of the "burning" process (*metabolism*) at an optimal level. In this connection, the quantity of thyroid hormone is small, but powerful. Even in the severest form of hyperthyroidism, the amount is increased to only about 0.000007 unit per unit of blood.

Tests of Thyroid Function. The principal objectives of thyroid function tests are to assess the rate of production of thyroid hormone and to explain findings obtained by physical examination of the gland. The total serum thyroxine (T_4) level may be determined by radioimmunoassay and most closely approximates the functional state of the thyroid gland in most patients. The radioactive iodine uptake test measures the percent of a tracer dose that enters the thyroid in any given period of time. Overactive glands generally show an increased avidity for iodine; the reverse holds for underactive glands. The test is used less often than in the past because of a number of shortcomings. Other radioimmunoassay determinations may include serum thyroglobulin, thyroid-stimulating hormone (TSH), and thyrotropin-releasing hormone (TRH) testing. The probable pathologic basis of thyroid enlargement may be studied by using a thyroid scan. Scintiscanning with technetium radioisotope demonstrates iodide-concentrating capacity in the thyroid. Scanning with isotopes of iodine reflects both concentration and binding. Scanning by ultrasound also provides a valuable approach in exploring nodular lesions of the thyroid. Excellent results have been obtained in discriminating between cystic and solid thyroid nodules.

Thyroid Disorders

The thyroid gland is subject to a variety of diseases. In addition to hypo- or hyperactivity, the gland may become enlarged for a number of reasons.

Goiter. The most common disease of the thyroid. The term simply means enlargement of the thyroid. The degree of enlargement is not always related to the degree of function. The largest goiters seldom produce excessive thyroid hormone. *Endemic goiter* is the most common form. This occurs most frequently in those areas of the world where the supply of iodine has been washed out of the soil (common in mountain areas). Enlargement of the thyroid seemingly results as a means of compensating for a lack of iodine. Sometimes this compensation is successful; often not. Of course, no degree of enlargement can compensate fully if there is an absolute deficiency of iodine. Upon gaining a supply of iodine, an enlarged gland will shrink to some extent but will not return to normal size. Individuals affected with this kind of goiter are likely to have low basal metabolic rates. If the condition exists during infancy, there is a serious interference with normal growth and development, and the resulting individual is known as a *cretin*. See also **Iodine.**

Lack of iodine in the diet is not the only circumstance that may lead to development of goiter. A large number of vegetables have goiter-producing qualities. Cabbage, Brussels sprouts, cauliflower, turnips, rutabaga, and soybeans, if taken daily into the body over long periods, may lead to the development of goiters. This does not mean that these foods are unsuitable or dangerous; quite the contrary is true. Only great amounts and high frequency of use need be avoided. There are few authentic cases on record of patients who ingested sufficiently large amounts of any of these foods to cause serious difficulty.

Fortunately, the systematic consumption of iodized table salt has largely eliminated the endemic goiter problem in the United States and to a considerable degree in other traditional goiter belts of the world. In Switzerland, at one time, the problem was very serious.

Simple goiters (growths) are not related to the presence of iodine. The cause of simple goiter is largely unknown. Since some foods are known to inhibit the production of thyroid hormone, it has been postulated that dietary factors may be responsible for simple goiter. Simple goiter is seen most often in adolescent girls; thus endocrine factors may play a role.

Nodular goiter, not related to the presence of iodine, may develop in an otherwise normal thyroid gland. Single or multiple nodules most often occur in older individuals. Many of the nodules, if examined microscopically, resemble the embryonic thyroid gland. Some produce no thyroid hormone; others produce no more per gram of tissue than the surrounding normal thyroid gland; still others produce abnormally large quantities of hormone. In the latter event, the cells in the normal thyroid tissue, in which the adenoma is growing, go into a resting phase, but even this shutdown is not sufficient to prevent the blood level of thyroid hormone form becoming too high. The incidence of cancer is much lower if more than one nodule, or adenoma, exists in the thyroid gland, or if the thyroid is overactive. The possibility that a nodule in this organ is a cancer is even greater if the patient is in the younger age groups.

Hyperthyroidism (Graves' Disease). There are several forms of this disease. The most spectacular type is sometimes called *exophthalmic goiter* or *Graves' disease*. In the fully developed case, patients begin to notice increased nervousness and irritability, often crying over trivial incidents. There is a great increase in appetite; food may be taken almost constantly through the day and part of the night. There may be a tendency toward insomnia. The heart beats rapidly and forcibly. Sometimes each beat visibly shakes the whole body. Despite increased intake of food, there is constant weight loss. Perspiration increases and the affected individual feels uncomfortably warm under circumstances that are comfortable to others. The skin becomes hot, moist, smooth, almost velvety. The nails are thin, concave, and may actually separate from the tissues of the finger or toe for some distance from their free margin. The hair becomes silky. If the hands are extended, there is an uncontrollable trembling. As weight loss increases, weakness becomes pronounced, and the patient may require assistance to rise from a chair. One or both eyes may protrude from their sockets (*exophthalmos*), in serious cases to such a degree that the eyelids cannot be closed, and even in sleep the eyes may remain open. The lids and "white" of the eyes become puffy. Because of the protrusion of the eyes, the face has the appearance of fright or anger.

The eye symptoms in patients with hyperthyroidism vary in degree from minimum to very severe exophthalmos. Fortunately, in most patients, the eye symptoms are mild; they are characterized by a retraction of the lids, resulting in a stare or a wide-eyed look. The patient also may blink frequently.

In patients with this disease, the thyroid gland enlarges, and the rate of blood flow through the organ is considerably increased. Typically, the basal metabolic rate is elevated and the degree of elevation correlates with the severity of the syndrome of hyperthyroidism. The thyroid gland may be from $2\frac{1}{2}$ to 4 times its normal size. In extreme situations, it may be 10 or more times normal size. The colloid is usually completely lost, and the thyroid acinar cells lose their normal appearance.

Most of the difficulty seen in Graves' disease can be traced directly to the abnormally high level of hormone circulating in the blood, or to some abnormality in the secretion itself. Everything is speeded up; more food is burned; more heat is produced; the sweat glands secrete faster; the heart beats faster; the muscles move faster; and the brain thinks faster. The human body, while capable of responding to emergencies, is not able to withstand continuous operation at a maximum rate. Thus, unchecked hyperthyroidism materially shortens life.

Much remains to be learned concerning the cause of hyperthyroidism. There appears to be a genetic relationship involved. In one study group of patients suffering from Graves' disease, it was found that 50% of the relatives had a change in thyroid function. Patients with this disorder also often relate the beginning of their symptoms to major emotional or traumatic crises in their lives. It has not been proved, however, that these are precipitating factors. Some endocrinologists believe that Graves' disease is basically a familial condition in which the full-blown affliction is simply triggered into existence by emotional or physical stress.

In many patients, the overactivity of the thyroid gland can be controlled by treatment with antithyroid drugs or iodide. For others, removal of the thyroid gland is practical and a fairly simple operation. However, patients with hyperthyroidism are poor surgical risks unless certain steps are taken to control their disease before the gland is removed. Most of the symptoms, except those involving the eyes, may be relieved either partially or fully by the use of iodine. If the disease has not run a long course, the surgical removal of most, but not all, of the diseased thyroid gland usually results in restoration of good health and relief of symptoms. The eyes, however, may not return to normal, but there often is improvement.

Certain foodstuffs are known to produce goiter. These chemical substances may be useful in the treatment of patients with hyperthyroidism. The substances, known as *thiouracils*, have made possible improvement in the medical treatment of patients with Graves' disease during recent years.

The specific treatment that best serves a given patient with hyperthyroidism is not always easy to select. Antithyroid drugs, such as propylthiouracil or methimazole, are reversible, effective in a large number of cases, and usually safe. The principal side effects are allergic reactions, which occur in a minority of cases. Treatment may be required for several weeks and sometimes up to 4 or 6 months. Some physicians maintain the treatment for about a year, carefully observing the patient after the drug is withdrawn. Other physicians withdraw treatment as soon as the patient becomes euthyroid (symptomless). If the treatment is too rapid or maintained too long, the patient may show signs of hypothyroidism. It is estimated that only about 30% of patients remain euthyroid after antithyroid drugs are discontinued. The rate of permanent remission is considerably lower than once believed. Some physicians use propanolol as an adjunctive therapy, taking advantage of the betablocking effect of this drug to ameliorate several of the symptoms of the disease, such as palpitations, excess sweating, tachycardia, nervousness, and tremor. Because of possible side effects of propanolol, the drug must be given judiciously. When applicable, the adjunctive therapy provides the patient with relief from several of the disturbing symptoms while the basic condition is being treated.

Subtotal thyroidectomy is an approach that has been used for many years, particularly prior to the availability of effective antithyroid drugs. The operation is considered difficult. Many specialists reserve this approach as one of the last elective measures.

Since the mid-1940s, therapy with radioactive iodine has been used widely and it estimated that nearly three-fourths of a million patients have been treated this way in the United States during the interim. After years of experience with the therapy, only two principal risks appear to be involved — that of developing hypothyroidism and, difficult to measure, possible genetic damage to gametes (of particular concern to women of childbearing age).

In a comparatively few patients, a so-called thyroid storm may develop. The manifestations are fever, severe tachycardia, extreme sweating, marked restlessness, leading, if left untreated, to dehydration and shock. The episode rarely occurs as the initial symptom of hyperthyroidism. It may in some instances, however, be the precipitating cause for the patient first consulting a physician. Regimens have been developed for the prompt and effective relief of such an episode.

Hypothyroidism and Myxedema. These conditions arise from the inadequate production of thyroid hormone. Patients may be classified as having (1) adult myxedema, (2) juvenile myxedema, or (3) cretinism. A majority of the body tissues become infiltrated with a mucuslike fluid which causes puffiness of the skin and may even cause interference with the normal function of certain of the internal organs. The skin becomes excessively dry; the hair is brittle, falls out, and is difficult to comb or curl; the fingernails are brittle; and the loose tissues of the face become characteristically puffy. The result is that the affected individual appears to be wearing a mask. The skin may become slightly tinged with yellow as a result of inability of the body to convert carotene, a yellow material present in vegetables, to vitamin A. Subsequently, vitamin A deficiency leads to roughening of the skin on certain areas of the body, notably around the elbows. The tongue enlarges. The vocal cords are affected such as to produce a peculiar huskiness or hoarseness and deepening of the voice. The speed of thinking is slowed. Patients with myxedema tire easily and usually feel sleepy most of the time. They often feel cold, even in warm weather. If left untreated, the heart may become enlarged and mental weakness may develop.

The treatment of hypothyroidism has improved markedly during the last several years. Fortunately, thyroid hormone replacement is reliable, nontoxic, nonallergenic, and inexpensive, but reliable diagnosis is a required prelude. Shortly after L-thyroxine is administered, the patient senses much improvement. However, several months of therapy may be required before the patient is fully restored to a normal state. Older patients and younger patients (with cardiac disease) must commence the

therapy on small doses. Careful monitoring of thyroid-stimulating hormone (TSH) is required during therapy. In years past, hypothyroidism was considered irreversible and usually progressive. Long-term therapy, of course, is required. A suggested correlation between thyroid therapy for hypothyroidism and increased incidence of breast cancer has not been proved and thus most physicians do not consider this a deterrent to the use of the therapy.

Thyroiditis. The thyroid gland is no exception to susceptibility to inflammatory infection. The majority of patients with this disease are thought to suffer from infection by small viruslike particles, which are not harmed by such agents as penicillin, aureomycin, and so on. The common type of thyroiditis occasionally gives evidence of being contagious. A satisfactory treatment appears to be the administration of *prednisone*, a compound related to the hormones of the adrenal cortex. The swelling of the gland is often quickly relieved, and with it, the discomfort. Hydrocortisone, ACTH, and a small amount of X-ray therapy also may be used. Some of the noninfectious diseases of the thyroid, not fully understood, include *Hashimoto's struma*, where the gland becomes moderately enlarged and has been compared in consistency to the granular structure of an apple. The disorder may lead to the destruction of the gland and resulting hypothyroidism. In *Riedel's struma*, which usually occurs after middle-life in women, the effect on the thyroid gland is one of great disorganization and destruction of the functional architecture. The gland may become enlarged but usually manages to produce sufficient hormone to prevent the average person from developing serious hypothyroidism or myxedema.

Tumors. There is a wide spectrum of activity or aggressiveness on the part of cancers arising in the thyroid gland. Cancer of the thyroid gland can occur at any age. There is a sudden increase in the rate at about the time of puberty. The usual treatment of thyroid cancer is early, thorough removal by surgery. If untreated, the spread of the tumor is limited for a time to the lymph nodes in the neck; secondary tumors later may appear in all tissues of the body, the lungs and bones being frequent sites. It is estimated that about 60% of cases treated by adequate surgical methods have no further trouble. Since 1940, radioactive iodine has been successful in partially and temporarily controlling certain secondary tumors from the thyroid. See also **Hormones**.

Additional Reading

Bayliss, R.I.S. and W.M.G. Tunbridge: "Thyroid Disease," 3rd Edition, Oxford University Press, Inc., New York, NY, 1998.

Burrow, G.N., R. Volpe, and J.H. Oppenheimer: "Thyroid Function and Disease," W.B. Saunders Company, Philadelphia, PA, 1997.

Cooper, D.S.: "Medical Management of Thyroid Disease," Marcel Dekker, Inc., New York, NY, 2001.

DeGroot, L.J., G. Hennemann, and P.R. Larsen: "The Thyroid and Its Diseases," 6th Edition, Churchill Livingstone, Inc., Philadelphia, PA, 1996.

Falk, S.A.: "Thyroid Disease: Endocrinology, Surgery, Nuclear Medicine, and Radiotherapy," 2nd Edition, Lippincott-Raven Publishers, Philadelphia, PA, 1997.

Grinspoon, S.K. and J.P. Bilezikian: "HIV Disease and the Endocrine System," *N. Eng. J. Med.*, 1360 (November 5, 1992).

Korenman, S.G. and M.I. Surks: "Atlas of Clinical Endocrinology: Thyroid Diseases," Vol. 1, Blackwell Science, Inc., Malden, MA, 1999.

Mandel, S.J., et al.: "Increased Need for Thyroxine During Pregnancy in Women with Primary Hypothyroidism," *N. Eng. J. Med.*, 91 (July 12, 1990).

McDougall, I.R.: "Thyroid Disease in Clinical Practice," Oxford University Press, Inc., New York, NY, 1992.

McGregor, A.M.: "Immunoendocrine Interactions and Autoimmunity," *N. Eng. J. Med.*, 1739 (June 14, 1990).

Medeiros-Neto, G.: "Administration of Thyroxine in Treated Graves Disease," *N. Eng. J. Med.*, 660 (August 29, 1991).

Pop, V.J.M., et al.: "Postpartum Thyroid Dysfunction and Depression in an Unselected Population," *N. Eng. J. Med.*, 1815 (June 20, 1991).

Stanbury, J.B.: "Inherited Disorders of the Thyroid System," CRC Press, LLC., Boca Raton, FL, 1994.

Takasu, N., et al.: "Exacerbation of Autoimmune Thyroid Dysfunction After Unilateral Adrenalectomy in Patients with Cushing's Syndrome," *N. Eng. J. Med.*, 1708 (June 14, 1990).

Takasu, N., et al.: "Disappearance of Thyrotropin-Blocking Antibodies and Spontaneous Recovery from Hypothyroidism in Autoimmune Thyroiditis," *N. Eng. J. Med.*, 513 (February 20, 1992).

Thorpe-Beeston, J.G.: "Maturation of the Secretion of Thyroid Hormone and Thyroid-Stimulating Hormone in the Fetus," *N. Eng. J. Med.*, 532 (February 21, 1991).

Utiger, R.D.: "Therapy of Hypothyroidism," *N. Eng. J. Med.*, 125 (July 12, 1990).

Utiger, R.D.: "Recognition of Thyroid Disease in the Fetus," *N. Eng. J. Med.*, 559 (February 21, 1991).

Utiger, R.D.: "The Pathogenesis of Autoimmune Thyroid Disease," *N. Eng. J. Med.*, 278 (July 25, 1991).

Utiger, R.D.: "Vanishing Hypothyroidism," *N. Eng. J. Med.*, 562 (February 20, 1992).

TICK *(Arachnida, Acarina).* A blood-sucking parasite related to the spiders and mites. The body of the tick is leathery and sack-like, with piercing mouth parts forming a small protuberance called the capitulum and a shield or scutum marking the dorsal surface in most species. See Fig. 1. The ticks are larger than their relatives, the mites, ranging from $\frac{1}{8}$ to $\frac{1}{4}$ inch (3 to 6 millimeters) in length. When filled with blood, a female tick may reach a length of about $\frac{1}{2}$ inch (12 millimeters). Of major importance to food-production operations, the cattle tick, the fowl tick, and the wood tick are widely distributed. The sheep tick is not a true tick, but is a member of *Diptera* and described separately. See also **Sheep Tick**. The Rocky Mountain wood tick, also called castor-bean tick or dog tick, is a dangerous pest to human beings and rodents, causing Rocky Mountain spotted fever. See also **Rickettsial Diseases**.

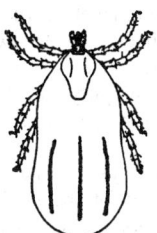

Fig. 1. Tick.

Cattle tick *(Boophilus* or *Margaropus annulatus,* Say). Of the family *Ixodidae*, this tick is considered the most destructive of the parasites that attack cattle, particularly in the southern United States. The tick not only attacks cattle, but also is a pest on sheep, mules, horses, and deer. One of the most injurious factors ascribed to this pest is that it is a vector for transmitting *Texas fever* (also called cattle fever, splenetic fever, and tick fever). Symptoms of this disease in cattle include: (1) pink to red coloration of the urine; (2) a dry muzzle; (3) an arched back; (4) a high fever; (5) drooping ears; and, upon a closer examination, (6) an enlarged spleen; and (7) congested liver. The cattle tick carries the cause of the disease, a minute parasite or protozoan (*Babesia bigemina*, Smith and Killbourne). The parasite destroys red blood corpuscles. Ticks do not transfer from one host to the next, but transmission occurs by way of the tick eggs and the young ticks resulting from their hatching. In some years, the loss of cattle to Texas fever has run into the many millions of dollars. The damage caused by the cattle tick is further magnified by the fact that the punctures made by the tick in the animal's flesh also attract the screw-worm fly which causes a condition known as *myiasis*, another severely damaging disease. See also **Screw-Worm**.

Losses arise not only from the death of cattle. (About 10% of those cattle attacked during summer months die; about 90% die if attacked during the fall and winter. The high Figures are notably true in the case of cattle that may have been imported to the south from the northern states.) In the case of dairy cows, milk production is often severely reduced even though the cow may recover from the fever. Hides also are badly damaged and their market price is similarly lowered.

In the United States, the cattle tick is found mainly in those states south of 36 °N. Although other species of tick may be involved, the same disease is found in Central and South America and portions of Africa, Europe, and southeast Asia, including the Philippines. The tick cannot survive winters in northern climes.

In affected areas, the cattle tick winters over as eggs or very small, 6-legged ticks, which are about $\frac{1}{32}$ inch (0.8 millimeter) in diameter (they are essentially round), and which are referred to as "seed ticks." These small ticks are found in different places, including grass, weeds, debris, almost wherever infested cattle have been located. Adult and nymph ticks also winter over in the skin of the affected cattle. Mating occurs within the host animal. After mating and when the female tick is satiated with the host's blood, she drops to the ground. At this time, the female is close to $\frac{1}{2}$

inch (12 millimeters) in length and may be of an olive-green or blush-gray hue and with a bean-shaped body. The skin is hard and shiny, but a bit wrinkled. In a period ranging from 2 to 6 weeks, the female then lays many hundreds of eggs (up to 5000), which are ejected from the front portion of her head. The mass of eggs is placed in the soil. Hatching requires from 2 to 6 weeks, although the eggs may be viable for several months. These eggs are also referred to as seed ticks. The newly hatched young crawl about on the ground, in grass, etc. and soon climb aboard a likely host animal. The larvae feed and then transform into nymphs and hence to adults, all the time clinging to and feeding on the host animal. The full life cycle is about 2 months.

Control in the United States and a number of regions of the world has been accomplished mainly by quarantine procedures. Control chemicals include arsenic trioxide, lindane, and toxaphene. Dipping of all cattle and farm animals once every 2 weeks in the proper formulation is an excellent procedure for eradicating the pest. Rotation of pasture land is also a good practice, leaving infested parcels of land free of cattle for a time sufficient to permit all ticks to die of starvation. In recent measures, these controls and quarantining procedures have proved very effective.

Ear Tick *(Otobius* or *Ornithodoros megnini*, Duges). This pest also prevails in the southwestern United States. The tick is named because it inhabits the outer ears of cattle and other farm animals, including dogs, for a period of months. The tick ranges up to $\frac{1}{3}$ inch (8 millimeters) in length. The invasion is by the nymph stage; adult ear ticks do not live on the host. The procedure of producing seed ticks and reinvasion of a host is similar to that described for the cattle tick.

Gulf Coast Tick *(Amblyomma maculatum*, Koch). This is another pest of the southern states and is extremely irritating to cattle. The tick is also found on certain of the native birds, including quail and meadow larks. The tick is similar in habit to the ear tick.

Control of the ear and Gulf Coast ticks is by use of a formulation containing lindane, xylene, and pine oil.

Wood Tick *(Dermacentor variabilis*, Say). Also known as the American dog tick, a serious interest in this pest, which attacks cattle, hogs, sheep, dogs, cats, rabbits, various wild animals, as well as humans, was established when it was found that this tick is the carrier of Rocky Mountain spotted fever (caused by *Rickettsia rickettsii*), in the midwestern and eastern United States. The tick also is a carrier of bovine anaplasmosis and tularaemia. Control chemicals effective against this tick include chlordane, dieldrin, toxaphene and DDT (or substitutes where DDT is not available).

Fowl Tick *(Argas persicus*, Oken). Also known as the *adobe tick*, or *bluebug*, the effects of an attack by the fowl tick is similar to that of the poultry mite, described in the entry on **Mite**. Most fowls are attacked, but the parasite prefers chickens. The pest is distributed in warm, dry regions throughout the world. In the United States, it is found in the southwestern states as well as in Florida. The presence of the fowl tick is complicated by the fact that it is the carrier of the dangerous poultry disease known as *fowl spirochaetosis*. For control measures, see previously mentioned coverage of poultry mite in entry on **Mite**. See also **Lyme Disease**.

Additional Reading

Camus, E., G. Uilenberg, and J.A. House: "Vector-Borne Pathogens: International Trade and Tropical Animal Diseases," New York Academy of Sciences, New York, NY, 1996.

Calisher, C.H.: "Hemorrhagic Fever with Renal Syndrome, Tick- and Mosquito-Borne Viruses, Supplementum 1: Archives of Virology," Springer-Verlag, Inc., New York, NY, 1991.

Hoffman, R.L.: "Lyme Disease: How to Avoid, Detect and Treat this Dangerous Tick-Borne Plague," Keats Publishing, Inc., Chicago, IL, 1994.

Jerace, C.K.: "Are Tick-Borne Diseases Also Horse-Borne?" *N. Eng. J. Med.*, 72 (January 2, 1992).

Mather, T.M. and D.E. Sonenshine: "Ecological Dynamics of Tick-Borne Zoonoses," Oxford University Press, Inc., New York, NY, 1997.

Relman, D.A., et al.: "The Agent of Bacillary Angiomatosis," *N. Eng. J. Med.*, 1573 (December 6, 1990).

Shapiro, E.D., et al.: "A Controlled Trial of Antimicrobial Prophylaxis for Lyme Disease After Deer-Tick Bites," *N. Eng. J. Med.*, 1769 (December 17, 1992).

Sonenshine, D.E.: "Biology of Ticks," Vol. 1, Oxford University Press, Inc., New York, NY, 1991.

Steere, A.C., E. Dwyer, and R. Winchester: "Association of Chronic Lyme Arthritis with HLA-DR4 and HLA-DR2 Alleles," *N. Eng. J. Med.*, 219 (July 26, 1990).

Szer, I.S., Taylor, E., and A.C. Steere: "The Long-Term Course of Lyme Arthritis in Children," *N. Eng. J. Med.*, 159 (July 18, 1991).

Web References

Centers for Disease Control and Prevention: *http://www.cdc.gov/health/diseases.htm#T*

FAO Tick Web Site: *http://www.fao.org/WAICENT/faoInfo/Agricult/AGA/AGAH/PD/pages/DEFAULT.HTM*

Insects on WWW: *http://www.isis.vt.edu/~fanjun/text/Links.html*

Tick Biology: *http://entomology.ucdavis.edu/faculty/rbkimsey/tickbio.html*

TIDAL BORE. See **Bore (Tidal)**.

TIDAL DEFORMATION (Earth). See **Earth**.

TIDAL ENERGY.   Although Sir Isaac Newton explained the underlying causes of the tides, i.e., the attraction exerted by the moon and the sun on the molecules of the oceans, and thus provided an immediate explanation for the variation of the time of high tide with the moon, for the modification of the tidal range during the half lunar month, and for the spring tides at suitable conjunctions of the moon and the sun, Newton nevertheless did not offer explanations of why tidal variations exist from one place on the Earth to the next. For example, Newton's explanations did not clarify such situations as: (1) Why do the high and low tides in Tahiti occur at the same time every day regardless of the phases of the moon? (2) Why is the tidal range at Mont Saint-Michel in the Rance Estuary (France) more than 13 meters, whereas it is measured in decimeters in the Mediterranean? (3) Why, although a double impulse causes the tide to ebb and flow along the coasts each day, a simultaneous reversal of flow and the rise in water level takes place in the Atlantic but is not observed in the English Channel? or (4) Why is there only one high tide every 24 hours at Do-Son in Tonkin? See Figures 1 and 2. Over the years, the principal explanations of these puzzles have stemmed from studies of tides as resonance phenomena.

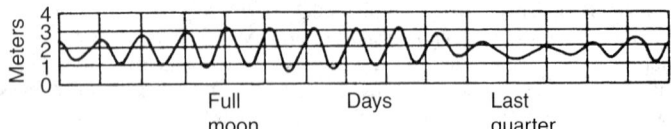

Fig. 1.   One full tide daily at Do-Son. (Tonkin.)

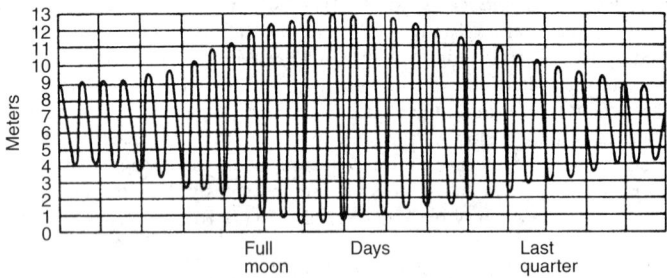

Fig. 2.   Two full tides daily at Mont Saint-Michel, France.

Laplace, in 1774, introduced hydrodynamics into tidal theory with the development of the concept of resonance between the periods of the moon and sun and the oscillatory periods of various water basins, such as the English Channel. While Newton earlier concentrated on basic cause, scientists who followed had to direct their attention to the response characteristics of various bodies of water to the prime cause. Notably, during the first half of the twentieth century, the problem was attacked by investigators including Allar, Belidor, Bonnefile, Caquot, Gibrat, Jeffreys, Lacombe, and many others.

The greatest variation the moon produces in the attraction exerted by the Earth on a molecule of water is six million times smaller than gravity, and the sun has a still smaller effect than the moon. The following observations are a summary of the observations made by Robert Gibrat, Consulting Engineer to Electricité de France for tidal power stations, as interpreted by reference to the literature. Altogether, if the free surface of the sea were to remain in equilibrium, and that is Newton's basic assumption, there would be a tide amounting to but a few decimeters

as a result of the variation in the horizontal component of gravity and one amounting to 0.9 meter (2.95 feet) as a result of variation in the vertical component. This obviously would leave the 13-meter tide at Mont-Saint-Michel unexplained. Newton's theory thus is inadequate for a full understanding. As the sea's surface cannot remain in equilibrium, the water flows back and forth and never comes to rest. As programs of capturing tidal energy, such as utilization of the enormous store of energy in the English Channel, come to the fore, considerably more sophisticated investigations of tidal phenomena must continue. The ratio of the energy taken up to the energy actually made available by the sea can no longer be neglected. The tides will certainly be modified and possibly in an unexpected manner since one is concerned with resonance, which, by definition, is very sensitive to variations of the main parameters. So, in considering a large tidal energy installation, it is necessary to know how and to what extent it may modify prior tidal conditions.

A few calculations of an elementary nature can clarify the essential concepts and provide an introduction to the methods of developing tidal energy. Utilizing tidal energy is very simple. A dam is built in a bay or estuary separating a basin from the sea. Suitable generating sets, which operate in both directions, make it possible to use the water to produce electrical energy as it flows back and forth. If it is assumed that the tide has a range A, the difference in level between high and low tide, and that a volume V can be stored between these two levels below elevation A, what then is the upper limit, if it exists, of the energy that can be used during each tide? To further simplify calculations, let it be assumed that the machines have an efficiency of unity and that there is no limitation on the discharges that can be used.

Under head H, a discharge q (using suitable units) will produce a power output qH. If $S(z)$ is the area of the pool at elevation Z, the energy generated by emptying a section dz is $S(z)z\,dz$. The energy produced per cycle will be:

$$\int_0^A S(z)\,dz \qquad \text{(during emptying)}$$

$$\int_0^A S(z)(A-z)\,dz \qquad \text{(during Filling)}$$

on the assumption that horizontal sections can be emptied or filled at the turn of the tide, the sea being at elevation O or A. With the simplifying assumptions made, the total energy per cycle with double working will be:

$$A\int_0^A S(z)\,dz = AV$$

or the production of the tidal range and the volume used, a formula which, it may be observed, is valid for a basin of any shape. Thus, one can call this quantity the "natural" energy of the basin for tide A. But, this name can be deceptive.

Each cubic meter of water entering the basin is thus used under a head A. If all tides had the same range A, at the end of the year, the 705 tides[1] would have provided a cumulative head of 705 A for each cubic meter of basin storage capacity. However, in fact, A varies throughout the year, together with V which is directly dependent upon it, since the basin does not fill or empty completely except at the maximum spring tides. The question may be asked, "Is the 'natural' energy the maximum amount of energy that can be obtained from a single tide?" Surge-wave propagation phenomena will quickly show that this is not so. See also **Tides**.

As early as 1942, a scale-model test made by a number of scientists in France had demonstrated how to exceed AV. At high tide (level A), the gates are suddenly opened without using the energy of the water as it passes through. A wave is propagated up the estuary and, increasing in amplitude, is reflected at the other end. The gates are closed at the moment when the wave returns and when the direction of flow at the dam is about to change. Finally, the water surface settles down to a level higher than A. On a scale model, 1.5A was attained without any difficulty. In a rectangular canal with a constant cross section, it can easily be seen that the level theoretically attains 2A in the absence of losses. In this process, the energy transformed while filling is zero; during emptying at low-tide slack water, it can be as much as:

$$\tfrac{1}{2}S(2A)^2 = 2SA^2 = 2AV$$

[1] The lunar day is $\sim 24^{\text{h}}50^{\text{m}}28^{\text{s}}$ in length versus a calendar day of 24^{h}. Thus, $2 \times 365 \times 86{,}400\text{s}/89{,}428\text{s} = \sim 705$ tides.

The energy obtained thus is twice the "natural" energy, and one enters into the realm of paradoxes. It might be hoped that this is a result of the surge wave and that all is worked out by nature. The answer is negative—because with a single basin, one could consider using power from the grid, using a double working cycle including pumping, a forerunner of the most complex real cycles.

Consider four phases as follows:

1. At the end of a cycle when the basin has been emptied to low-tide level and the gates are closed, power from the grid is used to pump water to the sea, lowering the level in the basin to $-B$. The energy used is:

$$E_p = \int_{-B}^{0} S(z)(-z)\,dz \qquad (z \leq 0)$$

2. The tide rises and at high tide the basin is filled during slack water, producing filling energy as follows:

$$E_r = \int_{-B}^{A} S(z)(A - z)\,dz \qquad (-B \leq z \leq A)$$

3. Still at the turn of the tide, water is pumped from the sea to the basin, using power from the grid a second time, and the water level rises to C. The energy thus used is:

$$E_{p^1} = \int_{A}^{C} S(z)(z - A)\,dz \qquad (A \leq z \leq C)$$

4. At low tide slack, the basin is emptied from C to zero, producing the following energy:

$$E_v = \int_{C}^{0} S(z)\,dz \qquad (0 \leq z \leq C)$$

After having accounted for the energy E_p and E_p^1 used for pumping, there remains a total of:

$$E_r + E_v - E_{p^1} = A\int_{-B}^{C} S(dz) = A(V + V_p + V_{p^1})$$

The transformed energy is equal to the sum of the "natural" energy plus *A times the total volume pumped and can thus be as large as desired if the area of the basin for $z > A$ can be made large enough.*

A double working cycle including pumping would seem capable of using energy without any restrictions other than those resulting from plant limitations.

This cannot be so, and it is essential to properly understand why. If not, one would make the same mistake as an electrician who knows the voltage at a power point and wants to work out the available wattage without taking into account the voltage drop and phase change resulting from plugging in an appliance. To forget the connection impedance would be disastrous. It becomes necessary at this point to analyze the physical process of energy transfer by the tides and to understand that the actual propagation of the energy drawn to the power station produces modifications to the range of the tide and the tidal currents which can appreciably alter the movement of energy. Certainly there is a "natural" limit for a given site, but basin capacity and tidal range are not enough to determine it.

The analogy of the energy of rivers may be helpful. In developing a site on a river, the energy produced is taken from the energy which was formerly dissipated naturally in whirlpools, eddies, and various forms of friction before the dam was built. The transformation energy limit is obviously the total of these losses. In similar fashion should the energy dissipated by the tide be taken as the "natural" limit that is being sought? The answer is negative—because a calculation shows that the mean power dissipated by friction in the Rance Estuary (France), for example, is at most 60,000 kilowatts, while the installed power plant facility there has a capacity of 240,000 kilowatts and is far from taking up all the utilizable energy. It is estimated that the additional energy imparted by the heavenly bodies is less than 1,000 kilowatts for the whole estuary. Thus, a further paradox. On the average, the total power dissipated in the entire English Channel is estimated at 157 million kilowatts.

Tidal energy is of a very special nature, however, and one cannot draw an analogy with the energy of a river without risking grave errors. The tidal phenomenon giving rise to this energy is essentially a resonance effect, which is very exceptional. Further, using such energy often causes the modifications thus made to the tidal regime to spread over a great distance and thereby considerably distort the initial phenomenon.

According to calculations made by Jeffreys, the mean power dissipated by the tide over the entire world is about 1,100 million kilowatts, most being supplied by shallow narrow seas like the English Channel, the Irish Sea, the North Sea, and mainly the Bering Straits, which account for 70% of the total.

At the time of Jeffrey's calculations (1952), a comparison between observations of the moon's movements and the numerical consequences of the theory led astronomers to assume the existence of a residual secular acceleration not arising from celestial mechanics. It was thus shown that the Earth's rotational speed was decreasing and releasing about 1,400 million kilowatts, which agreed quite satisfactorily with the order of magnitude of Jeffrey's calculations. It consequently seemed plausible to assume that the energy of the tide was derived from that of the Earth's rotation by friction between the water and the sea bed, thus slowing up the Earth.

During the early 1960s, however, the picture became much *less* clear. Observations made in the nuclear submarine *Nautilus* indicated that Jeffrey's Figures should be reduced by three-fourths. Also, some leading astronomers were of the opinion that the rotational speed of the Earth might be constant. The basic question remains without full proof, although there are numerous scientific opinions, "Is the Earth's kinetic energy, or the sun's thermal energy the source of tidal energy?"

Obviously, this complex topic is well beyond the scope of this volume. Reference to sources listed at the end of this article is suggested.

Tidal Energy Installations. The tides as a source of energy were exploited as early as the Middle Ages. For example, tidal power was used in the coastal areas of Brittany in France and in the British Isles to drive watermills from the 12th century onwards. See Fig. 3. In continuous operation for eight centuries, a mill on the Deben River at Woodstock in Suffolk, England was not retired until 1956, at which time it became an historic monument. In Brittany, as recently as 1959, a few small tidal mills were in operation—at the time Electricité de France decided to proceed with the construction of what is today the world's largest tidal generation station, installed in the maritime estuary of the Rance River (between Saint Malo and Dinard in Brittany). The French people have made the greatest advances in understanding and utilizing tidal energy to date.

Fig. 3.   Early tidal mill at Saint Suliac, Brittany.

In 1967, a tidal power plant was installed at Kislaya Bay in the U.S.S.R. Power engineers and scientists in Canada and the United States have considered the exploitation of locations in the Northeast, notably the Bay of Fundy which lies between the Canadian provinces of New Brunswick and Nova Scotia, and, in the United States, Passamaquoddy and Cobscook Bays in Maine. In Canada several decades ago, the "Little Kodiak" project in the Bay of Fundy studied the potential of tidal power. In the United States, an abrupt start was made with the Passamaquoddy project, also on the Bay of Fundy, in the early 1930s. At that time, National Recovery Administration (NRA) funds were made available to hire about 5000 persons for the project, but a subsequent and serious study showed that the scheme would not be economical and the work was halted. In North America, interest in tidal energy waned for several decades and it was not until the 1970s that serious consideration of tidal energy potential was shown. Current activities are described a bit later.

Potential Tidal Energy Plant Locations. For a viable tidal energy plant, a mean tidal range of 5 meters (16.4 feet) is considered mandatory.

TABLE 1. RANGES OF SPRING TIDES IN EXCESS OF 5 METERS

Location	Tidal Range	
	Meters	Feet
Bruntcoat Head, Nova Scotia (Bay of Fundy)	14.49	47.5
Rance Estuary, France	13.5	44.3
Anchorage, Alaska	9.03	29.6
Liverpool, England	8.27	27.1
St. John, New Brunswick, Canada	7.20	23.6
Dover, England	5.67	18.6
Cherbourg, France	5.49	18.0
Antwerp, Belgium	5.43	17.8
Rangoon, Burma	5.19	17.0
Juneau, Alaska	5.06	16.6
Panama (Pacific Side)	5.01	16.4

Nonconventional approaches have been proposed within the last few years and these are described later. Known ranges of spring tides in excess of 5 meters are shown in Table 1. However, these areas do not necessarily coincide with the present selections of some scientists. For example, as observed by Isaacs and Schmitt (1980), nearshore areas with a great tidal range and potential for tide power development are widely distributed and include the coasts of Alaska and British Columbia, the Gulf of California, the Bay of Biscay, the White Sea, the central Indian Ocean, and the coasts of Maine and eastern Canada. Or, as observed by Ryan (1980), internationally, the areas that appear to hold the most immediate promise are the upper Bay of Fundy, Canada; Chausey in the Bay of Mont St. Michel in France; the Gulf of Mezen in the Soviet Union; the Severn River Estuary in England; the Walcott Inlet in Australia; San José, Argentina; and Asan Bay in South Korea. Of these areas, the most active projects are in South Korea, France, and Canada. A 450-megawatt plant is in the planning stage in the inner basin and an 810-megawatt plant in the outer zone of Asan Bay (South Korea). Studies also have been made for a 330-megawatt project in Incheon and at Garorim.

Duff (1978) suggests that, in addition to the Bay of Fundy and currently operating locations, appropriate locations for extraction of tidal power include Cook Inlet, Alaska, the coastline of the English Channel and the Irish Sea, the Sea of Okhotsk off the coast of the former U.S.S.R., the coast of the Yellow Sea off Korea, and the inlets of the Kimberly Coast in Australia, indicating these as convenient sites for constructing tidal power conversion machinery.

Rance Estuary Facility. Study of methods of equipping the estuary of the Rance (Brittany) for production of electrical energy on a large scale (240 megawatts) was commenced quite early in the twentieth century, although detailed design and construction did not commence until the early 1960s. Final connection of the 24 turboalternator sets to the electrical grid occurred in early December 1967.

Successful operation of a plant of this magnitude had to await the development of three major factors: (1) a far greater, if not full, understanding of tidal phenomena; (2) development of a new type of turboalternator set, known as the bulb unit, which is able to use low heads to drive a turbine in either direction of flow, and also to function as a pump; and (3) cutting off the estuary of the Rance, which meant damming a river with a maximum flow of 18,000 cubic meters per second, both at flood and at ebb. The first factor, of course, involved highly analytical mathematical and theoretical investigations. The latter two factors involved solutions to difficult problems in mechanical, electrical, civil, hydraulic, and materials engineering.

The installation extends from Pointe de la Brebis on the left bank of the estuary to Point de la Briantais on the right bank, a distance of some 750 meters (~2,460 feet). See Fig. 4. The tidal generating facilities consist basically of the tidal basin constituted by the estuary of the river Rance, separated from the sea by a dam. Sluice gates built into the dam can be operated to adjust the natural rate of drainage from the basin to the sea with outgoing tides, or the rate at which the basin is filled with incoming tides. The difference in levels between the basin and the sea thus can be controlled and, when it reaches an appropriate figure, the basin is filled (or emptied) through the dam. The flow of water is used to move the wheels of hydraulic turbines which, in turn, drive alternators. These turboalternator sets, of which there are 24, produce electrical energy.

Fig. 4.   View of Rance Estuary tidal power plant in France. (*Photo by Glenn D. Considine.*)

Apart from the three functional elements (dam, sluices, turboalternators), additional requirements had to be met, including: (a) the dam was not to be an obstacle to navigation between the sea and the estuary and so had to incorporate a lock; (b) the general appearance of the installation had to respect the natural features of the site; and (c) the construction had to serve as the basis for a road across the Rance. Geologically, the Rance estuary was formed by subsidence, and its banks are sharply outlined. The useful volume of the tidal basin is 184 million cubic meters (~6.5 billion cubic feet) between the extreme of zero elevation at ebb tide and the 13.5 meters reached in exceptionally heavy spring tides. The corresponding areas of the basin are 4.3 square kilometers (~1.7 square miles) at zero elevation and 22 square kilometers (8.5 square miles) at 13.5 meters.

In the early 1940s, designers were well familiar with low-head river site harnessing problems as the result of river installations in France and other parts of Europe, but this knowledge could not be applied directly to tidal plant design. One of the particular features of a tidal plant is its range of turbine operation in either direction of flow, with both head and discharge varying considerably. Power plant operational studies soon pointed to the advantage of pumping at around high tide slack water so as to overfill the basin; and so as to overempty it at low tide. This required machines capable of running as low-delivery head pumps. Design studies by Robert Gibrat clearly showed the tremendous potential scope of tidal power plant operating cycles. In order to derive maximum benefit from these cycles, it was essential to develop units capable of operating as turbines, pumps, or orifices with flow in either direction. Cross section of a bulb unit is shown in Fig. 5.

As of the early 1980s, the French were reevaluating the feasibility of establishing a giant multi-basin tidal power station (6,000 to 12,000 megawatts) at Chausey in the Bay of Mont. St. Michel, a location not far from the present Rance Estuary facility.

Canadian Plans. It was not until 1977 that the Federal/Provincial Tidal Power Review Board (TPRB) of Canada concluded that tidal power from the Bay of Fundy can compete on economic terms with electricity from fossil fuels. In 1969, a study had suggested three Bay of Fundy sites. See Fig. 6. One of these is Economy Point on the Minas Basin. Some of the world's highest diurnal tides are experienced at this location. Other locations with potential are the Cumberland Basin and Shepody Bay near the border between New Brunswick and Nova Scotia. It has been envisioned that a plant located at Economy Point would have a capacity of 3800 megawatts, about equivalent to four nuclear facilities, with an electric power output considerably greater than the requirements of Canada's Maritime Provinces.

The importance of resonance in the exploitation of tidal power resources, previously mentioned in this entry, has been a major consideration of the Canadian planners in selecting Bay of Fundy sites. Even though maximum exploitation may not be envisioned for many years, an early site based upon short-term needs could adversely affect the ultimate tidal power available at later sites. For example, a reduction in the length of the Bay, resulting from construction of a barrier, will result in some disturbance of the natural resonance. It has been estimated that resonant high tides in the Bay are likely to occur with a natural period of 12 hours, 25 minutes.

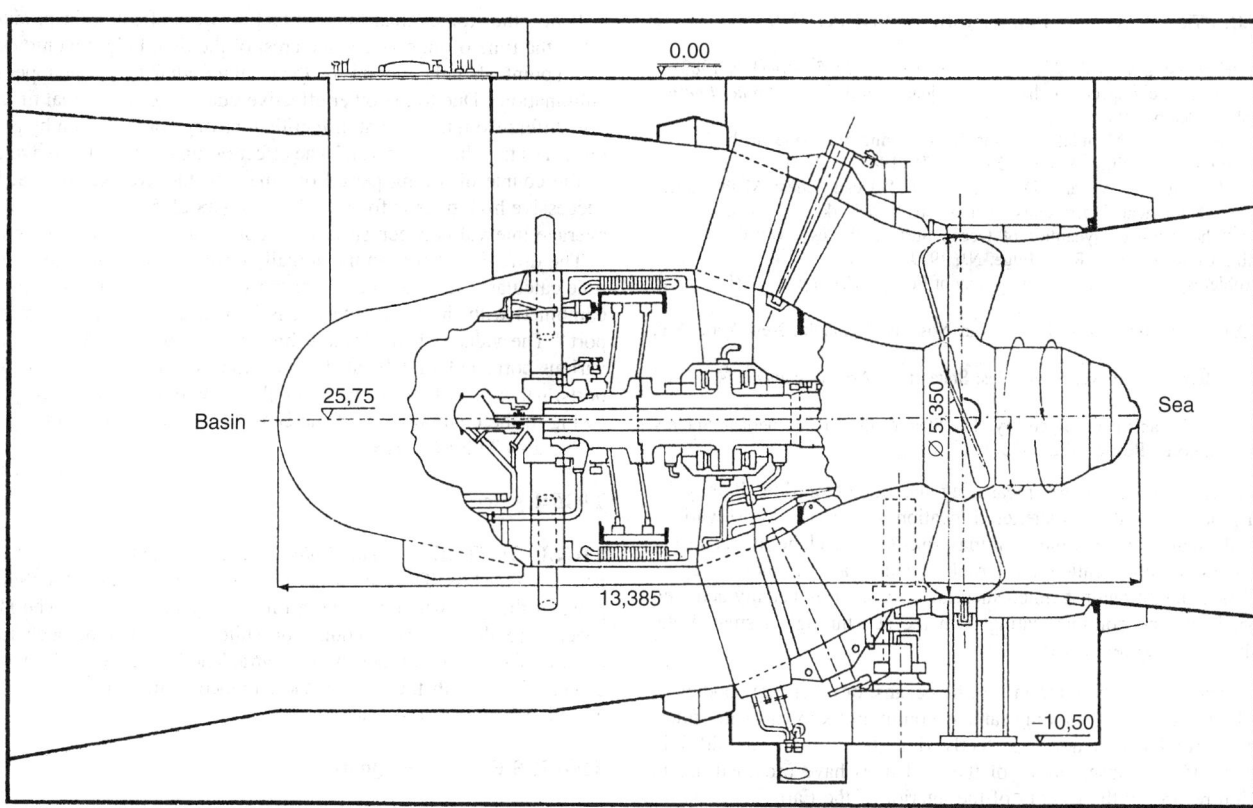

Fig. 5.   Sectional view of reversing bulb-type unit used at the Rance estuary tidal power plant. A total of 24 units of this type is used. (*Electricité de France.*)

Canadian engineers have estimated that a barrier at a wide site, such as Chignetco Bay (Fig. 6), a point where the Bay divides, might reduce the tidal amplitude by as much as one-third and thus greatly reduce the effectiveness of plants at Shepody Bay and Cumberland Basin. The TPRB reached a general conclusion that sites of tidal barriers should be located as near to the heads of bays as possible.

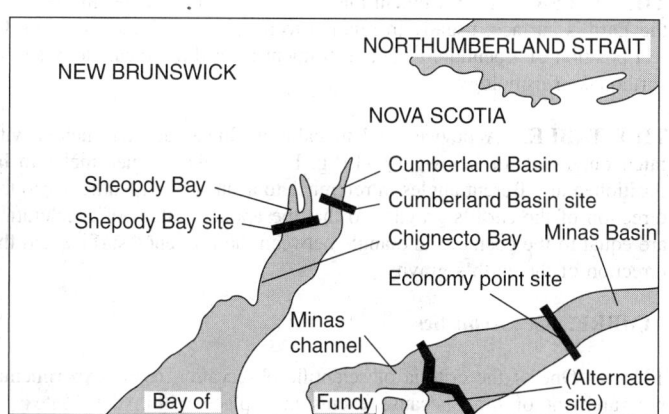

Fig. 6.   Selection of tidal power plants to take advantage of the large tides in the Bay of Fundy.

Calculations of tidal resonance and behavior do not always match actual observed behavior. For example, early calculations indicated that the natural period of the Bay of Fundy is 9 hours, much lower than observed values. Models and calculations must take adjacent large bays into consideration, even though at some distance, because the resonance appears to be an integration of a whole system of bays, not an isolated region. Thus, the Gulf of Maine must be considered in making plans for the Bay of Fundy. Further, it is now believed that, because of the shallowness of the Georges Bank, much of the tidal wave energy enters the Bay of Fundy by way of a submarine canyon some 150 fathoms deep. The future progress of sedimentation as the result of barrier construction also must be taken into consideration.

Tidal Power Potential in the United States. In the United States, tidal power facilities appear to be limited principally to Cook Inlet, Alaska (7.5 meters; 24.6 feet); and Passamaquoddy and Cobscook Bays in Maine, each with a tide of about 5.5 meters (18 feet).

In recent years, there has been more interest shown by energy planners in the United States to newer, unconventional methods for extracting tidal energy as contrasted with the kinds of installations previously described in this entry. As recently as March 1979, the U.S. Army Corps of Engineers concluded a preliminary economic study of the Cobscook Bay area in Maine and observed that although tidal power from this site is more competitive than in decades past, a program still is not justified.

In a new concept proposed and patented by A.M. Gorlov (Northeastern University), a very thin plastic barrier (called a water sail) would replace the conventional dam. The reinforced plastic membrane would be hermetically anchored to the bottom and sides of the bay. It is proposed that the membrane would be constructed in sections so that, if necessary, it could be pulled aside. As explained by Ryan (1980), "The top of the barrier would be supported by a cable spanning the entrance to the bay. The cable would be fixed to several specially designed floats that would keep the barrier above the surface of the water, maintaining the desired differential level—approximately 2 meters (6.5 feet) of head between the ocean and basin side—during rising and receding tides. The underwater part of the barrier would be exposed to a net water pressure equal to the difference in water levels across the dams. It would therefore be called upon to withstand pressure of 2 meters of water (0.2 atmosphere)—well within the strength limits of today's reinforced plastic materials."

Conversion of tidal energy would be through the use of compressed air. It is envisioned that two chambers connected to an air motor, essentially a large piston, would be used. As of the early 1980s, the concept is undergoing early feasibility tests.

In the United Kingdom, as of the early 1980s, there was considerable competition in technical circles between using conventional approaches to exploiting tidal power as, for example, the Severn estuary between England and Wales, and the use of wave power.

Additional Reading

Brosche, P. and J. Sundermann: "Tidal Friction and the Earth's Rotation," Springer-Verlag, Inc., New York, NY, 1983.

Duff, G.F.D.: "Tidal Power in the Bay of Fundy," *Technol. Rev. (MIT)*, **812**, 34–42 (1978).

Fay, J.A.: "Harnessing the Tides," *Technology Review (MIT)*, 51–61 (July 1983).

Gibrat, R.: "Scientific Aspects of the Use of Tidal Energy," *Rev. Franc. Energie* (September–October 1966).

Ku, L.F., et al.: "Nodal Modulation of the Lunar Semidurnal Tide in the Bay of Fundy and Gulf of Maine," *Science*, **230**, 69–71 (1985).

Marchuk, G.I., Kagan, B.A., and D.E. Cartwright: "Ocean Tides: Mathematical Models and Numerical Experiments," Pergamon, New York, NY, 1984.

Mei, C.C.: "The Applied Dynamics of Ocean Surface Waves," World Scientific Publishing Company, Inc., River Edge, NJ, 1989.

News: "Problems with Tidal Power in the Bay of Fundy," *Sci. Amer.*, 70 (December 1984).

Redfield, A.C.: "Introduction to Tides," Van Nostrand Reinhold, New York, NY, 1982.

Ryan, P.R.: "Harnessing Power from Tides: State of the Art," *Oceanus*, **224**, 64–67 (1980).

Seymour, R.J.: "Ocean Energy Recovery: The State of the Art," American Society of Civil Engineers, Reston, VA, 1992.

TIDE GAGE. A device for measuring the height of tide. It may be simply a graduated staff in a sheltered location where visual observations can be made at any desired time; or it may consist of an elaborate recording instrument (sometimes called marigraph) making a continuous graphic record of tide height against time. Such an instrument is usually actuated by a float in a pipe communicating with the sea through a small hole, which filters out shorter waves.

TIDES. The periodic rise and fall of the oceans, or other large bodies of water, relative to the surrounding land is commonly known as the tides. Since the Earth itself is not a perfectly rigid body, there are tides in the Earth itself, and observations of the land tides have provided useful information regarding the rigidity of the interior of the Earth.

Tides are produced by the combination of a number of external forces, with the principal force being the gravitational attraction of the moon. Owing to the differences of distance of the moon from various portions of the Earth, the amount of the attractive force will be different in different places and tend to produce a deformation. Figure 1 indicates (on a very much exaggerated scale) the deformation forces, and the resultant tides in a very deep ocean surrounding a rigid Earth.

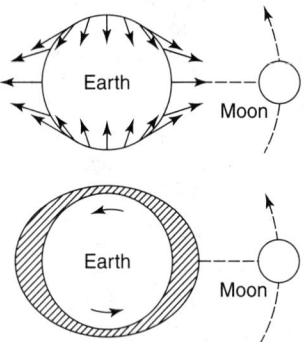

Fig. 1. Tides in a very deep ocean. If the whole earth were covered by very deep water and if the Earth's rotation were slower, high tides would occur under and opposite the moon. The upper figure shows the tide-raising force of the moon at different places.

The tidal force due to the gravitational attraction of the sun ranks second in importance to the tide-raising forces from the moon. Twice during each month, at times of full and new moon, the forces from the sun and moon are acting in parallel directions and produce the maximum tide range known as "spring tides." At times of quarter-moon the forces are acting at right angles to each other and the resultant minimum range of tides is known as "neap tides." Other tidal forces due to the rotation of the Earth, the revolution of the Earth-moon system, and the revolution of the sun-Earth system, make the problem of tide prediction one of great complexity. Detailed harmonic analysis of the observed tides has indicated the presence of tidal forces due to the attractions of several of the major planets. In the open ocean, from observations taken on isolated islands, the average range of ocean tides relative to the land is about $2\frac{1}{2}$ feet (0.75 meters). Configuration of shore line and contour of the ocean floor greatly increase this range for most stations along the coasts of the continents. The maximum range of tides in the land is about 9 inches.

If no other forces than lunar attraction were effective in producing the tides, the time of high water (the crest of the tidal bulge) would occur at a given point when the moon is on the local meridian (either at upper or lower culmination). Due to the other effective tidal forces the actual time of high tide differs from the instant of meridian passage of the moon by an amount known as the "lunar interval." The effect of lunar interval will average zero in the course of a long period of time, and the average interval between successive high tides is found to be 12 hours 25.5 minutes, which is $\frac{1}{2}$ the average interval between successive upper culminations of the moon.

The difference between the actually observed time of high tide and the time calculated from the instant of transit of the moon across the meridian, combined with the lunar interval, is known as the "establishment of the port." The values of the "establishment" are obtained observationally for various ports and are tabulated in tide Tables. The values obtained for two ports, which may be separated by only a few miles, may be very different owing to the configuration of the coast line between the ports.

See also **Tidal Energy**.

TIGER. See **Cats**.

TIGER BEETLE (*Insecta, Coleoptera*). A small longlegged beetle of predacious habits. They are usually found on exposed earth or sand in the glare of the sun, where they both run and fly very rapidly. The numerous species are blue or green, reddish, or white, with or without a characteristic pattern of spots and dashes on the elytra. The larvae live in burrows in the ground, lying with the head at the entrance to the burrow ready to seize any victim that comes near.

TIGER'S EYE. See **Quartz**.

TIGER SHARK. See **Sharks**.

TILL. A general term for coarsely graded and extremely heterogeneous sediments of glacial origin. Till is generally classified as unstratified drift which may vary from clays to mixtures of clay, sand, and boulders. A particularly sticky form of clay till is called gumbo.

See also **Glacial Deposits (or Drift)**.

TILL (Blacialk). See **Glacial Deposits (or Drift)**.

TILTMETER. An instrument that measures slight changes in the tilt of the Earth's surface, usually in relation to a liquid-level surface or to the rest position of a pendulum. The instrument is used in volcanology and in earthquake seismology.

TILT TABLE. A device used to calibrate linear accelerometers with rated ranges of, or below, +/−1.0 g. It allows the accelerometer to be positioned at different angles in reference to a surface perpendicular to the direction of the earth's gravity, so that the applied values of acceleration are equal to the cosine of the angle between the reference surface and the direction of the earth's gravity.

TIMBRE. See **Acoustics**.

TIME. One of the criteria of scientific observation is the experimental measurement of time because events take place within the context of time reference systems. These systems, or clocks, include the frequency rates, or cycles, of the Earth—its spin on its axis or orbital revolution about the sun; the action of a pendulum or of a watch's balance wheel; controlled alternating electric current; the vibrations of a tuning fork or quartz crystals; radioactive decay; and the oscillations of the atom of cesium-133. Consequently, *time is what the clock reads*. There is, in short, no absolute time standard. Generally speaking, the more stable the cyclical system, the more precise and accurate will be the clock reading.

Apparent Solar Time, or so-called *sundial time*, is measured by the diurnal motion of the sun as observed from a point on the Earth's surface as the Earth (which is not a perfect sphere) rotates (not uniformly) on its axis, in an elliptical solar orbit in an orbital plane that does not coincide with the plane of its rotation at the equator. As a result, apparent solar days vary in length.

Mean Solar Time is apparent solar time averaged out to eliminate rotational and orbital variables, so that the mean solar day is the average of all apparent solar days in the orbital year. This is divided by 86,400 to determine the *mean solar second*. The *equation of time*, with a numerical value of from 0 to about 16 minutes, describes the difference between the apparent and mean solar time; it is at maximum in February and November.

The Gregorian Calendar, instituted in 1582 by Pope Gregory XIII, reformed the Julian calendar, which had a mean year of 365.25 days, equivalent to 11 minutes and 14 seconds longer than the mean solar year, so that the Gregorian mean year has 365.2425 days. This reform involved (1) omitting a 10-day (October 5–14, 1582) period without interrupting the sequence of the days of the week to restore the date of the actual vernal equinox to March 21 — so as to (2) regulate Easter coincident with new rules for ecclesiastical feast days; and (3) omitting the Julian intercalary day in centurial years not divisible by 400 (such as 1800), although otherwise retaining the intercalary day inserted in every fourth, or Leap year.

The Gregorian calendar, adopted in 1752 by England and its American colonies, later came into almost universal civil use. Exceptions include Russia, which adopted the calendar in 1918 but has since developed a variant arrangement for intercalary days.

The calendar system based on the Julian Day Number, sometimes called the Astronomical Day System takes January 1, 4713 B.C. of the Julian calendar, starting at 12 noon, as Day 1, and counts all subsequent Days consecutively, as measured by the spin of the earth on its axis, without reference to the solar, or tropical year. Julian Days are divided decimally, rather than into hours, minutes, and seconds. See also **Calendar**.

> Editor's Note: In the paragraphs that follow, the NBS (National Bureau of Standards) is mentioned several times. Just a few years ago, the NBS was renamed and now is known as the National Institute of Technology and Standards (NIST).

Universal Time (UT), also called *Greenwich Mean Time (GMT)* because it is referenced to the Prime (or zero) Meridian at Greenwich, England, is also based on the earth's rotation; if uncorrected, therefore, its units are equal to the mean solar second. As a time scale, these units are designated as UT_0. The UT_1 scale is UT_0 corrected to allow for the polar motion, or nutation, of the earth. The UT_2 scale is UT_1 corrected for seasonal variations in the earth's rotation. UT is announced in International Morse Code each 5 minutes by radio stations WWV (Fort Collins, Colorado) and WWVH (Puuene, Maui, Hawaii) of the U.S. National Bureau of Standards. Other time signals broadcast by the stations are kept in close agreement with UT_2.

UT is also expressed in terms of the 24 *Standard Time Zones* that serially cover the Earth's surface at coincident intervals of 15 degrees longitude and 60 minutes UT, as agreed at the Washington Meridian Conference of 1884, thus accounting respectively for each of the 24 hours of the calendar day. These local *civil time zones*, with irregular statutory boundaries, were chosen so that noon occurs when the sun appears approximately overhead at the zone's central meridian. *Standard Time zones* were established in the United States in 1883. *Daylight Saving Time* was proposed by Benjamin Franklin and first adopted in Europe during World War I to cut electric power consumption. Since 1967, federal legislation in the United States has required the states to observe it uniformly. However, cities and other communities are not bound by this legislation. Daylight Saving Time was extended during the winter of 1973–1974 because of the energy crisis, was changed again in the autumn of 1974, and possibly may undergo additional shifts if this simple arrangement can be used effectively to save energy. An interesting and somewhat detailed explanation of the relationship between Standard and Daylight Saving Time will be found in "Standard and Daylight-Saving Time," by I.R. Bartky and E. Harrison, *Scientific American*, **240**, 5, 46–53 (1979).

The one-hour shift in standard time each 15 degrees longitude requires a one-day jump forward in date when moving from America to Asia and a one-day jump backward in date when moving from Asia to America. This data boundary running irregularly by treaty from the poles through the Pacific is the *International Date Line*.

Apparent Sidereal Time, the time scale generally used by astronomers, is measured by the apparent diurnal motion of a distant star (not the sun) as observed from a point on the earth's surface. This scale is more easily observed than apparent solar time.

Ephemeris Time. A definition of time based on idealized motions of the Sun and Moon that was adopted and introduced in the astronomical ephemerides on January 1, 1960, to free astronomical computations from the effect of the irregularities of the Earth's rotation. One second of Ephemeris Time is a specified fraction of the tropical year 1900 (equals 31,556,925,974.7 seconds of Ephemeris Time); the difference between Ephemeris Time and Greenwich Standard Time (Universal Time) is derived for each year from a comparison of the observed and computed position of the Moon.

Atomic Time is based on the atomic second, officially defined in 1967 by the 13th General Conference of Weights and Measures as 9,192,631,770 oscillations of the atom of cesium-133 (although this value was provisionally defined in 1964 as 9,192,631,770 plus or minus 20). This value expresses the ET second as closely as possible in terms of an atomic standard and was derived by a joint experiment of the U.S. Naval Observatory and Great Britain's National Physical Laboratory, using a dual-rate moon-position camera, developed by Dr. William Markowitz of the U.S. Naval Observatory, and a cesium-beam clock. The U.S. National Bureau of Standards used a cesium-beam clock as its NBS atomic frequency standard and, as a service, the Bureau provides UT_2 in cooperation with the U.S. Naval Observatory.

Corrections to the frequencies transmitted by NBS radio stations WWV and WWVH are determined with regard to the NBS atomic standard and are published monthly by the "Proceedings of the Institute of Electrical and Electronics Engineers." The stations' carrier and modulation frequencies are intentionally offset from the atomic frequency standard by a precisely known amount so that, on the average, the time signals broadcast remain as close as possible to UT_2. In 1966, 1967, and 1968, the offset was minus 300 parts in 10^{10} parts. This offset compensates, in effect, for the progressive changes in the speed of rotation of the earth so as to permit the radio service to offer UT_2 while retaining the atomic time standard to provide the unit of time, the second. The carrier frequency of NBS radio station WWVL (Fort Collins, Colorado) is also offset by the same amount, but NBS very-low frequency radio station WWVB (also in Fort Collins) transmits without offset to permit absolute frequency comparisons by user.

The cesium-beam clock of the National Bureau of Standards is referred to as a primary clock because independently of any other reference it provides highly precise and accurate time to scientific, research, industrial, defense, and public needs. Secondary clocks must, unlike primary clocks, be referenced to more accurate clocks for calibration. Today, quartz crystal clocks are the most frequently used secondary time standards. Master clocks, in contrast, are systems used to directly control subsidiary clocks. For example, the 60-cycle alternating current plug-in electric clock employs a synchronous motor controlled by the alternating current acting as a master clock system. A wrist timepiece, instead, has its own frequency standard, either a 2.5 to Hz balance wheel or a 360 Hz tuning fork.

The International Astronomical Union and the General Conference of Weights and Measures officially define international time standards, while the Bureau Internationale de l'Heure, at the Paris Observatory, Paris, France, provides and receives data to assist in the uniform international observation of time standards.

Continued Quest for Better Clocks and Frequency Standards. As observed by Wineland (1984), there is reason to believe that the inaccuracy of a time standard based on stored ions can eventually be much smaller than that of the cesium clock, which can have an accuracy of about one part in 10^{13} or less. In an experiment at the National Bureau of Standards (Boulder, Colorado) in 1983, the frequency of a particular hyperfine transition in the ground state of beryllium atomic ions was measured with an inaccuracy of only about one part in 10^{13}. In this experiment, the ions were confined or "trapped" in a small region of space by using static electromagnetic fields and their kinetic temperature was lowered to less than 1 K by a process sometimes called "*laser cooling.*" In all of physics, only a few measurements can achieve a higher accuracy. Those experiments measure similar transitions in neutral cesium atoms. Potential accuracies of frequency standards and clocks based on such experiments are anticipated to be better than one part in 10^{15}.

Time and Philosophy. Over the centuries, naturalists and poets alike have commented on the characteristics of time with relation to earthly and human events. Lord Byron in the first act of his play, *Cain, a Mystery*, quotes the fallen angel Lucifer as saying: "With us acts are exempt from time, and we can crowd eternity into an hour, or stretch an hour into eternity. We breathe not by a mortal measurement — but that's a mystery." Alfred North Whitehead said in The *Concept of Nature*, "it is impossible to meditate on time and the mystery of the creative passage of nature without an overwhelming emotion at the limitations of human intelligence." Martin Gardner, writing for "Mathematical Games" in *Scientific American*, **240**, 3, 21–27 (1979) explores two bizarre questions about time, namely: (1) Is it meaningful to speak of time stopping? (2) Is it meaningful to speak of altering the past?

An excellent book pertaining to the development of the mechanical clock in the last half of the 13th Century and of the clock's impact on society, industry, and commerce is "Revolution in Time" by David S. Landes, Harvard University Press, Cambridge, Massachusetts, 1983.

TIME (Biological).   See **Circadian Clock**.

TIME CONSTANT.   All physical systems require a certain amount of time to transmit signals from input to output. This time is simple resistance-capacitance and inductance-resistance systems is characterized by the system time constant. In the resistance-capacitance circuit (most common example of this concept), the time constant is equal to the product of the resistance and capacitance. In this case the resistance is measured in ohms and the capacitance in farads. The resultant time constant value is in seconds and is a measure of the time required for the circuit variable (voltage) to reach $1-\frac{1}{e}$ or 63% of the final value. In this notation, e is the base of the natural logarithms (2.71828...).

The differential equation for this simple circuit, using Fig. 1 as a reference, is as follows:

$$i = \frac{E_i - E_0}{R} \qquad (1)$$

$$i = C\frac{dE_0}{dt} \qquad (2)$$

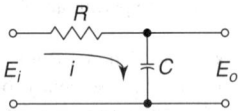

Fig. 1.   Simple RC circuit.

Combining Equations (1) and (2) gives

$$\frac{E_i - E_0}{RC} = \frac{dE_0}{dt} \qquad (3)$$

The solution of the first-order differential equation for a unit step change in E_i is

$$E_0 = 1 - e^{-t/T} \qquad (4)$$

where $T = RC$. The time response for E_0 is shown in Fig. 2. Similar equations may be developed for the inductance-resistance circuit. In this circuit the time constant is equal to the inductance in henries divided by the resistance in ohms.

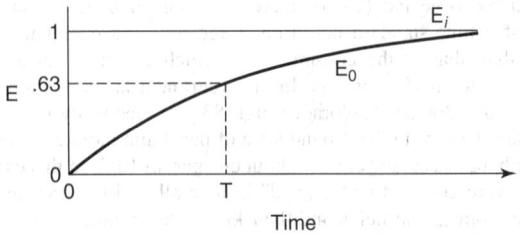

Fig. 2.   Simple RC circuit time response.

The use of electrical examples in defining the time constant is in no way intended to limit this concept to electrical systems. The same concept may be applied to mechanical, thermal, pneumatic, hydraulic, and any other type of system having resistive and capacitive elements. This similarity between systems is the basis for all analog computer work. Due to the widespread application of this concept in systems analysis it has been assigned many names other than the time constant. Some of the terms in use, which have the same meaning, are exponential lag, first-order lag, and single capacity lag.

TIME DELAY.   The time required by a specific voltage or current to travel through a circuit.

TIME DEMODULATION.   The process by which information is obtained from a time-modulated wave about the signal imparted to the wave in time modulation.

TIME DISCRIMINATOR.   A circuit that indicates the time equality of two events, or the sense and approximate magnitude of the inequality.

TIME-DIVISION MULTIPLEXING.   The process or device in which each modulating wave modulates a separate pulse subcarrier, the pulse subcarriers being spaced in time so that no two pulses occupy the same time interval. Time division permits the transmission of two or more signals over a common path by using different time intervals for the transmission of the intelligence of each message signal.

TIME (Equation).   See **Equation of Time**.

TIME GATE.   A transducer that gives output only during chosen time intervals.

TIME MODULATION.   Modulation in which the time of appearance of a definite portion of a waveform, measured with respect to a reference time, is varied in accordance with a signal.

TIME PATTERN.   A picture-tube presentation of horizontal and vertical lines or dot-rows generated by two stable frequency sources operating at multiples of the line and field frequencies.

TIME REVERSAL INVARIANCE.   See **Conservation Laws and Symmetry**.

TIME (Rise).   See **Rise Time**.

TIME SERIES.   The values of a variable generated successively in time. A continuous barograph trace is an example of a *continuous time series*, whereas a sequence of hourly pressures is an example of a *discrete time series*. Graphically, a time series is usually plotted with time as the abscissa and the values of the function as the ordinate.

TIME SHARING.   The use of a device, particularly a computer or data-processing machine, for two or more purposes during the same overall time interval, accomplished by interspersing component and subsystem actions in time. In the case of a digital computer, time sharing generally connotes the process of using main storage for the concurrent execution of more than one job by temporarily storing all jobs, but the one in control, on auxiliary storage. This technique allows a computer memory to be used by several independent programs (users). The method most often is associated with a computer-controlled terminal system used in an interactive mode.

The prime objective in using an interactive time-sharing system is to permit more efficient use of the computer system in proceeding from initial specification of a program to a final checked-out operating revision and also the more efficient use of people in the solution of problems by providing answers and services in a short time on demand. The traditional batch-processing type of operation necessitates the coding and preparation of a program as a separate first step. Submission of the job to a computer control center for inclusion in a batch run can result in several hours of delay. The problem of writing and checking out a large, complex program in this fashion does not efficiently use the programmer's time because of loss of continuity in the problem. Time-sharing systems are designed to allow a user to operate the computer in an interactive manner, by way of a variety of console devices, and thus obtain immediate responses to requests for answers, while also receiving immediate indications of errors in programming.

Modern systems have the capability of handling many terminal users concurrently and still use available time to process batch jobs in the normal fashion. The magnitude of the capability for this type of operation depends upon computer speed, storage size, auxiliary storage size and speed, the operating system used to manage and control these facilities, and the type and amount of terminal service to be offered. Time slicing is a common approach to allocating time for each user in the terminal system. See also **Time Slicing**.

Time sharing also is used in nonterminal systems for the purpose of sharing a portion of the system memory between jobs of different priority. This usually is called multiprogramming. In some control and data-acquisition systems, a portion of the main storage (a partition) is allocated to run nonprocess-related jobs (background jobs) in the conventional batch-processing manner. However, when the system is notified that a process-related program requires the partition, the background job is stored on auxiliary storage until the higher-priority program has been executed. The background job then is restored and resumed, if in the meantime, no other higher-priority job is waiting for service.

TIME SIGNAL. 1. An accurate signal marking a specified time or time interval. It is used primarily for determining errors of timepieces. Such signals are usually sent from an observatory by radio or telegraph.

2. In photography, a time indication registered on the film to serve as a time reference for interpretation of the date recorded on the film.

TIME SLICING. A technique that allows several users to employ a computer facility as though each had complete control of the machine. Several users can be serviced one at a time, unaware of each other, because of the relative speed between computer operation and human response. Essentially, time slicing is used by a software control system to control the allocation of facilities of a computer to tasks requesting service. The allocation basis is a fixed time interval. Each job is executed for the time period used in the time slice. The job is then temporarily stored on an auxiliary storage device (time sharing) or suspended (multiprogramming), while another job is being run. Each job, therefore, is run in an incremental fashion until complete. See also **Multiprogramming (Computer System)**.

TIME-STRATIGRAPHIC UNIT. This has been described as "a subdivision of rocks considered solely as the record of a specific interval of geologic time" (ACSN, 1961, art. 26). It is a material unit or body of strata based upon actual sections or sequences of rock, thus representing the preserved record of rocks formed during an arbitrary interval of finite geologic time, the time extending from the commencement to the ending of its deposition or intrusion. As stated by Hedberg (1976), a time-stratigraphic unit "provides us with the only available technique of determining approximate isochronous surfaces and thereby geographically extending approximate working boundaries for time-stratigraphic units." The magnitude of the unit is measured by the length of the time interval to which its rocks correspond, not by their thickness. The term *chronolith* is synonymous with time-stratigraphic unit and is not to be confused with *chonolith*.

Additional Reading

Eicher, D.L.: "Geologic Time," Prentice-Hall, Inc., Upper Saddle River, NJ, 1976.
Hedberg, D.: "International Stratigraphic Guide," John Wiley & Sons, Inc., New York, NY, 1976.

TIME ZONE SYSTEM. See **Zone Time**.

TIMING BELT. A specially designed belt that provides a nonslip positive speed ratio between the driving and driven pulleys of an apparatus or machine. Timing belts are commonly used on numerically controlled machine tools and similar applications where light-weight, zero backlash, and positive drive are important to the servo operation. The belts also are used to drive the camshafts in some engines and for motor driven pumps, fans, and similar applications where quiet operation, long life, and minimum maintenance are required. Special materials for operation in hot oil or at very low temperatures are available.

A timing belt has an integral number of teeth (similar to gear teeth) molded on the inner surface. The pulleys carry grooves milled across the face to receive the belt teeth. Like a gear or chain drive, the belt provides an exact speed ratio corresponding to the ratio of the numbers of teeth on the pulleys.

The belts are molded of neoprene or similar materials in an endless loop having a flat backing strip with the teeth molded to shape on the inner belt surface. Molded within the backing strip are steel cables helically wound around the periphery of the belt for strength and stiffness in tension. The faces of the teeth are covered with a woven nylon fabric molded integrally with the belt.

Belt and pulley combinations are provided in a range of pitches varying from $\frac{1}{5}$ inch to $1\frac{1}{4}$ inch (spacing between teeth) and widths from $\frac{1}{4}$ inch to 5 inches. The corresponding range of power is from fractional horsepower to over 185 horsepower. Belts are available in a range of lengths permitting a wide range of shaft center distances. The advantages found in timing belt drives, as compared with gears, V-belts, and chains, are relatively low cost, light weight, low noise, zero backlash, and no requirement for lubrication. Timing belts provide a positive drive without high tension and have a long life as compared with V-belts.

TIMOTHY. See **Grasses**.

TIN. Chemical element, symbol Sn, at. no. 50, at. wt. 118.69, periodic table group 4, mp 231.97 °C, bp 2,270 °C, density 7.29 g/cm^3 (white tin at 15 °C), 5.77 (gray tin at 13 °C), 6.97 (liquid at mp). There are two allotropic forms of tin: (1) the more common soft white beta tin has a body-centered tetragonal crystal form, (2) the brittle gray alpha tin has a diamond-type cubic crystal form. The cubic form, α-tin, stable below 18 °C, is an intrinsic semiconductor, and is gray. At 161 °C, white tin undergoes a transition to rhombic or γ-tin.

Tin is a silver-white metal with a bluish tinge, softer than zinc and harder than lead. It is malleable, ductile at 100 °C; can be powdered at 200 °C, and upon exposure to temperatures below 18 °C, it crumbles to a grayish powder due to the "tin pest," which is caused by the transformation of white to gray tin (the reverse transformation may be brought about by heating gray tin to about 100 °C). When a bar of tin is bent a marked creaking sound is emitted due to the friction of the crystals. Tin: is not oxidized on exposure to air at ordinary temperatures; burns to stannic oxide when heated to high temperatures in air or oxygen; is soluble in HCl to form stannous chloride; is converted by concentrated HNO$_3$ into soluble beta-stannic acid; is soluble in aqua regia to form stannic chloride; is soluble in NaOH solution slowly to form sodium stannite and hydrogen gas; and reacts with chlorine to form volatile stannic chloride. Discovery, prehistoric.

Tin has the largest number of naturally occurring isotopes ^{112}Sn, ^{114}Sn through ^{120}Sn, ^{122}Sn, and ^{124}Sn. Five radioactive isotopes have been identified ^{111}Sn, ^{113}Sn, ^{121}Sn, ^{123}Sn, and ^{125}Sn. With exception of ^{121}Sn, which has a half-life of about 5 years, the half-lives of the other isotopes are comparatively short, expressed in minutes and days. Tin occurs in the Earth's crust to the extent of about 40 grams/ton. It is estimated that a cubic mile of seawater contains about 15 tons of tin. First ionization potential 7.332 eV; second, 14.52 eV; third, 30.49 eV; fourth, 40.57 eV. Oxidation potential Sn → Sn^{2+} + 2e$^-$, 0.406 V; Sn^{2+} → Sn^{4+} + 2e$^-$, −0.14 V; HSnO$_2^-$ + 3OH$^-$ + H$_2$O → Sn(OH)$_6^{2-}$ + 2e$^+$, 0.96 V; Sn + 3OH$^-$ → HSnO$_2^-$ + H$_2$O + 2e$^-$, 0.79 V. Electronic configuration $1s^2 2s^2 2p^6 3s^2 3p^6 3d^{10} 4s^2 4p^6 4d^{10} 5s^2 5p^2$.

Other important physical properties of tin are given under **Chemical Elements**.

Tin occurs as oxide (cassiterite, tin stone, stannic oxide, SnO$_2$), obtained commercially in Malaysia, Indonesia, Thailand, and Bolivia. The ore is concentrated and then roasted to oxide (83–88% stannic oxide). The product is treated in a blast furnace and crude tin recovered. Refining is conducted by electrolysis, or by fractional fusion.

Tin also occurs as complex sulfidic ores. The economic working of these ores is essentially confined to Bolivia. The ores include SnS$_2$ · Cu$_2$S · FeS (stannite), SnS (herzenbergite, SnS · PbS (teallite), 2SnS$_2$ · Sb$_2$S$_3$ · 5PbS (franckeite), Sn$_6$Pb$_6$Sb$_2$S$_{11}$ (cylindrite), 2SnS$_2$ · 2PbS · 2(FeZn)S · Sb$_2$S$_3$ (plumbostannite), and 4Ag$_2$S · SnS$_2$ (canfieldite).

Secondary tin is an important source of the metal. Tinplate scrap may be detinned electrolytically or chemically. The alkaline chemical process is the most widely used and involves a caustic solution, which contains an oxidizing agent to remove both tin and the underlying iron-tin alloy from the steel. The solution formed then is either (1) crystallized to form sodium stannate, (2) electrolyzed to recover tin metal, or (3) acidified with CO$_2$, H$_2$SO$_4$, or acidic gases to precipitate hydrated tin oxide. There are several secondary tin smelters in the United States, but only one primary smelter. The main primary tin smelters are located in the United Kingdom, Malaysia, and Thailand.

Uses: Not including former Soviet Bloc nations, world consumption of tin is in excess of 187,000 metric tons annually. Principal uses are: (1) tinplate, 35%; (2) solder, 24%; (3) bronze, 9%; (4) other alloys, 8%; (5) tinning, 4%; (6) chemicals, 5%; (7) other uses, 15%. In the United States, tin consumption in 1979 was: (1) tinplate, 29%; (2) solder, 29%;

(3) bronze and brass, 14%; (4) chemicals, 8%; (5) other alloys, 9%; (6) tinning, 4%; (7) other uses, 7%. In the United States, the major portion of tinplate is used in the making of cans. The advantages of tin for cans and food-processing equipment include its nontoxic nature, resistance to corrosive attack by acids and other aqueous solutions, and when combined with other metals, strength.

Tin plate. Tin coating may be applied to steel by (1) electroplating, usually as part of a high-speed, continuous process, or (2) by dipping cut sheets in a bath of molten tin. Electrolytic tin plate essentially is a sandwich in which the central core is strip steel. This core is thoroughly cleaned in a pickling solution prior to electroplating. The actual plating occurs as the strip moves through horizontal or vertical tanks containing electrolyte. The moving strip then is heated as it passes between high-frequency electric induction coils, whereupon the tin coating melts and flows to form a lustrous coat. The average thickness of tin on the end-product sheet is 0.00003 inch (0.0008 millimeter) on each side. A complex system of instrumentation is used to control process conditions and to inspect the moving sheet for any perforations in the plate. In hot-dip tinning, individual steel sheets are pickled and washed. A layer of hot palm oil is maintained on top of the molten tin bath to prevent oxidation of the molten tin by air and to prevent the molten tin from freezing too rapidly on the plate, thus providing a more even coating with a high luster.

Terne plate. This is a sheet-steel product that is coated with an alloy of tin and lead. The coatings range from 50–50 mixtures of lead and tin to as low as 12% tin and 88% lead. Plate used for roofing normally is about 25% tin and 75% lead. In addition to roofing, terne plate is used in the manufacture of gasoline tanks for automotive vehicles, oil cans, and containers for solvents, resins, etc.

Alloys. Tin is widely used as both a major and minor ingredient of alloy metals. These applications are summarized in Tables 1, 2, and 3. Phosphor bronzes (Table 3) actually contain very little phosphorus, ranging from 0.03 to 0.50%, and hence the alloys are poorly designated. Tin bronzes is the better term. High-silicon bronzes contain about 2.8% tin; low-silicon bronzes about 2.0% tin. Gun metals are tin bronze casting alloys with a 5–10% zinc content. Some wrought copper-base alloys contain tin: (1) Inhibited Admiralty metal, 1% tin; (2) manganese bronze, 1% tin; (3) naval brass, 0.75% tin, (4) leaded naval brass, 0.75% tin. See also **Copper**.

Copper-nickel-tin alloys (UNS C72500, 2700, and 2900) are *spinodal*[1] materials that can be age hardened after forming. They combine tensile

TABLE 1. REPRESENTATIVE SOFT SOLDERS

Composition, %				Temperature		Working Temperature Freezing Range
Tin	Lead	Antimony	Silver	Solidus	Liquidus	
80	20	0	0	183 °C	203 °C	20 °C
70	30	0	0	183	192	9
60	40	0	0	183	189	6
50	50	0	0	183	216	33
49	50	1	0	186	210	24
40	60	0	0	183	234	51
39	60	1	0	186	230	44
30	70	0	0	183	252	69
29	70	1	0	186	252	66
20	80	0	0	183	273	90
20	78.75	0	1.25	180	270	90
19	80	1	0	185	273	88
10	88.50	0	1.50	178	290	112
10	90	0	0	183	297	114
5	95	0	0	270	311	41
2.5	97.5	0	0	301	319	18

Tin-bearing solders are considered soft solders as contrasted with the hard solders which contain substantial quantities of silver. However, small quantities of silver are added to some tin solders to increase strength of the resulting joint and to adjust working temperature range. Antimony also is added in some cases for the latter purpose. Wiping solders usually have a tin content ranging from 35 to 40%. Solders used in automotive-body work require a wide plastic (working temperature) range. Solders with a low tin content (below 25%) are generally used. For very low-temperature melting, bismuth and cadmium also may be added to tin-lead solders.

[1] Spinodal structure is a fine homogeneous mixture of two phases that form by the growth of composition waves in a solid solution during suitable heat treatment.

strengths as high as 1380 MPa (200×10^3 psi) with resistance to oxidation, stress relaxation, fatigue, and stress-corrosion cracking. For use in round-wire form, these alloys are challenging the traditional position held by phosphor bronzes for electronic leads, contact pins, and sockets. In flatwire form, they compete with beryllium copper for eyeglass frames, circuit boards, and electronic-contact clips. The alloys also have been used for rivets, self-threading screws, and a variety of cold-headed parts.

Chemistry and Compounds: Tin forms two series of compounds: tin(II) or stannous compounds and tin(IV) or stannic compounds. Tin(II) oxide, SnO, insoluble in water, is formed by precipitation of an SnO hydrate from an $SnCl_2$ solution with alkali and later treatment in water (near the boiling point and at constant pH). It is amphiprotic, but only slightly acid, forming stannites slowly with strong alkalis. Sodium stannite is conveniently prepared from tin(II) chloride: $SnCl_2 + 3NaOH \rightarrow Na[Sn(OH)_3] + 2NaCl$. Tin(IV) oxide, SnO_2, is much more acidic; it readily reacts with NaOH to form stannate ions, $Sn(OH)_6{}^{2-}$. In fact, no hydroxide of the formula $Sn(OH)_4$ has ever been obtained. The metal metastannates, e.g., $M^{II}SnO_3$, are generally made by fusion methods and have three-dimensional polymeric anions in which each tin atom is surrounded octahedrally by six oxygen atoms. There are, however, two forms of stannic acid, H_2SnO_3. The α-stannic acid is a white, gelatinous precipitate obtained by treating $SnCl_4$ with NH_4OH. The α-stannic acid (also called metastannic acid) is a white powder obtained by action of concentrated HNO_3 on tin; unlike the α-form it is insoluble in concentrated acids and alkali metal hydroxides.

Tin forms dihalides and tetrahalides with all of the common halogens. These compounds may be prepared by direct combination of the elements, the tetrahalides being favored. Like the halides of the lower main group 4 elements, all are essentially covalent. Their hydrolysis requires, therefore, an initial step consisting of the coordinative addition of two molecules of water, followed by the loss of one molecule of HX, the process being repeated until the end product $H_2Sn(OH)_6$ is obtained. The most significant commercial tin halides are stannous chloride, stannic chloride, and stannous fluoride.

The increasingly electropositive character of main group 4 as tin is reached is evident from the fact that its hydrides are much less stable than those of silicon and germanium. Known are Sn_2H_6 and SnH_4, which is obtained by hydrolysis of magnesium stannide, Mg_2Sn, or by electrolysis of a phosphoric acid solution with a tin cathode.

Among the other inorganic compounds of tin are the:

Nitrates. Stannous nitrate $Sn(NO_3)_2$, white solid, by reaction of tin metal and dilute HNO_3 and crystallization, soluble in water with slight excess of HNO_3.

Sulfates: Stannous sulfate, a white powder soluble in water and H_2SO_4, is obtained commercially by action of H_2SO_4 on $SnCl_4$ or Sn. Stannic sulfate may be formed by the solution of stannic hydroxide in dilute H_2SO_4, or by action of oxidizing agents on stannous salts.

Sulfides. Stannous sulfide SnS, dark brown precipitate, by reaction of stannous salt solution and H_2S, insoluble in sodium sulfide solution but soluble in sodium polysulfide solution, forming sodium thiostannate; stannic sulfide SnS_2, yellow precipitate, by reaction of stannic salt solution and H_2S, soluble in sodium sulfide solution, forming sodium thiostannate.

Organometallic Compounds: In common with the other elements of main group 4, tin forms many organometallic compounds; the range of possible combinations is virtually limitless. They include:

(1) *Tetraorganotins*, R_4Sn, prepared either by alkylation of tin halides with Grignard Reagents or alkyl lithium; by reaction of an organic halide with a tin-sodium alloy; by direct reaction of tin with an organic halide; or by reaction of stannic chloride with alkyl aluminum compounds.

(2) *Organotin halides*, $RSnX_3$, R_2SnX_2, and R_3SnX, prepared by disproportionation of the tetraorganotin with stannic halide or by direct alkylation of stannic halide.

(3) *Organotin oxides*, R_2SnO or $(R_3Sn)_2O$, prepared by treatment of the organotin halides with alkali.

(4) *Stannoic acids*, RSnOOH.

The organotin halides and oxides are usually the intermediates used in the synthesis of other organotin derivatives, such as the organotin

The phases of a spinodal differ in composition from each other and from the parent phase, but have the same crystal structure as the parent phase.

TABLE 2. REPRESENTATIVE BABBITT METALS

Ingredients	[a]SAE 10 %	SAE 11 %	SAE 12 %	SAE 13 %	SAE 14 %	SAE 15 %
Tin	90	86	88.25	4.5–5.5	9.25–10.75	0.9–1.25
Antimony	4–5	6–7.5	7–8.5	9.25–10.75	14–16	14.5–15.5
Lead	0.35	0.35	0.35	86	76	Remainder
Copper	4–5	5–6.5	2.25–3.75	0.50	0.50	0.6
Iron	0.08	0.08	0.08	—	—	—
Arsenic	0.10	0.10	0.10	0.60	0.60	0.8–1.10
Bismuth	0.08	0.08	0.08	—	—	—

[a]Society of Automotive Engineers.

TABLE 3. REPRESENTATIVE TIN-BEARING BRONZES
Phosphor Bronzes

	1.25% Tin	4% Tin[c]	5% Tin	8% Tin	10% Tin
Copper[a]	98.75%	88.00%	95.00%	92.00%	90.00%
Melting point, °C	1,077	1,000	1,050	1,027	1,000
Tensile strength, 1,000 psi					
Hard sheet	65	58	81	93	100
Soft sheet	40	44	47	55	66
Rockwell hardness					
Hard sheet	75B	68B	87B	93B	97B
Soft sheet	60F	65F	73F	75F	55B
Electrical conductivity					
% IACS[b]	48	19	18	13	11
Thermal conductivity					
Btu(ft^2)(ft)(°F) at 68°F	120	50	47	36	29
Major Uses	Electrical contact wire Messenger cable Flexible metal hose Pole-line hardware	Bearings Bushings Gears Pinions Shafts Screw-machine products Washers Valve parts	Bearings Bellows Bourdons Gears Rivets Springs Wire cloth Truss wire	Bearings Bellows Bourdons Fasteners Washers Springs Switch parts Chemical hardware	Heavy bars, plates Bridge and expansion plates Heavy springs

[a]Small amounts of zinc, lead, iron, antimony, and phosphorus also present.
[b]International Annealed Copper Standard.
[c]Free-cutting phosphor bronze.

carboxylates, organotin sulfur-derivatives, organotin hydroxides, etc. The most significant commercial organotins include dibutyltin and dioctyltin carboxylates and sulfur derivatives, used as polyvinyl chloride (PVC) stabilizers and as catalysts in polymer systems; bis(tributyltin)oxide, triphenyltin fluoride, and tributyltin fluoride, used as antifoulants for marine paints, fungicides, bactericides, sanitizing agents, and wood preservatives; tricyclohexyltin hydroxide as an insecticide; triphenyltin hydroxide and triphenyltin acetate as agricultural fungicides; and dibutyltin dilaurate as a poultry anthelmintic.

Biochemical Aspects of Tin. Scientists at the University of Maryland have found that sediment microflora (from Chesapeake Bay sediments) can produce dimethyltin and trimethyltin species from inorganic Sn(IV) compounds. The results were consistent with a geocycle of tin proposed by Ridley et al. in 1977. The findings support the hypothesis that tin can be biotransformed in an estuarine environment.

Researchers J. Versieck and L. Vanballenberghe (University Hospital, Ghent, Belgium) have observed, "Tin has chemical properties offering potentials for a biological function. The element has a tendency to form truly covalent linkages as well as coordination complexes; hence, it was hypothesized that it could well contribute to the tertiary structure of proteins or other biologically important macromolecules, such as nucleic acids. The oxidation-reduction potential of $Sn^{2+} \rightleftharpoons Sn^{4+}$ being at 0.13 V, well within the range of physiological oxidation-reduction reactions, it was also speculated that the element could function as the active site of metalloenzymes."

During the late 1960s, several experiments led to the conclusion that tin was indispensable for the growth of rats fed purified amino acid diets in trace-element-controlled isolators. However, no additional evidence for biological essentiality has been added since that time. Information on the tin content of foods is meager. Estimates have shown that the intake is less than 1 milligram per day. However, because of contamination from packaging, it has been estimated that this figure occasionally could rise to 50 mg/day. The aforementioned researchers have developed what appears to be a reliable method for determining tin in human blood serum by radiochemical neutron activation analysis.

Additional Reading

Considine, D.M. and G.D. Considine: "Van Nostrand Reinhold Encyclopedia of Chemistry," 4th Edition, Van Nostrand Reinhold, New York, NY, 1984. (A Classic Reference.)

Davis, J.R.: "Metals Handbook," 2nd Edition, ASM International, Materials Park, OH, 1998.

Franklin, A.D., J.S. Olin, and T.A. Wertime (editors): "The Search for Ancient Tin," Smithsonian Institution, Washington, DC, 1978.

Greenwood, N.N. and A. Earnshaw: "Chemistry of the Elements," 2nd Edition, Butterworth-Heinemann, Inc., Woburn, MA, 1997.

Hallas, L.E., Means, J.C., and J.J. Cooney: "Methylation of Tin by Estuarine Microorganisms," Science, 215, 1505–1507 (1982).

Krebs, R.E.: "The History and Use of Our Earth's Chemical Elements: A Reference Guide," Greenwood Publishing Group, Inc., Westport, CT, 1998.

Lewis, R.J. and N.I. Sax: "Sax's Dangerous Properties of Industrial Materials," 10th Edition, John Wiley & Sons, Inc., New York, NY, 1999.

Lide, D.R.: "CRC Handbook of Chemistry and Physics 2000–2001," 81st Edition, CRC Press, LLC., Boca Raton, FL, 1999.

Meyer, C.: "Ore Metals Through Geologic History," Science, 227, 1421–1428 (1985).

Patai, S.E.: "The Chemistry of Organic Germanium, Tin and Lead Compounds," John Wiley & Sons, Inc., New York, NY, 1995.

Staff: "Forecast '91 Metals," Advanced Materials & Processes, 24 (January 1991).

Staff: Various publications on tin and its compounds, including Tin and Its Uses (quarterly): Tin International (monthly); and statistical publications on tin (periodically), International Tin Research Institute, Middlesex, England.

Staff: "Annual Review of the World Tin Industry," Rayner-Harwill Ltd., London, published yearly.

Staff: "Tin Chemicals for Industry," Tin Research Institute, Greenford, Middlesex, UB6 7AQ, England (published periodically).

Versieck, J. and L. Vanballenberghe: "Determination of Tin in Human Blood Serum by Radiochemical Neutron Activation Analysis," *Analytical Chemistry*, 1143 (June 1, 1991).

Web References

International Tin Research Institute (ITRI) Ltd: *http://www.itri.co.uk/index.htm*

Tin: *http://me.mit.edu/2.01/Taxonomy/Characteristics/Tin.html*

USGS Minerals Information, Tin: *http://minerals.usgs.gov/minerals/pubs/commodity/tin/*

TINAMIFORMES. An order of ground birds (*Aves*) with a compact form, a slender neck, a relatively narrow head, and a rather short beak which is slender and curves slightly downward. Their wings are short and hence their capacity for flight is poor. Their feet are strong, with only three well developed forward toes; the hind toe is located in a high position, being either retrogressed or absent. The tail is very short, and in a few species it is hidden under the tail coverts; this and the abundant rump feathering cause the rounded body shape. The length is 20–53 centimeters (7.9–21 inches), and the weight is 450–2300 grams (1–5 pounds).

They live in tropical parts of South America; in the north they live only slightly beyond the Tropic of Cancer (Zimttao in northwest Mexico), but in the south they are widely distributed in the temperate zone. They inhabit varied habitats: rainforest, thickets, bushland, tree steppes, and treeless grassland up to an altitude of 5,000 meters (16,405 feet).

With their beautiful, moving songs, their magnificent eggs, and their unusual family life it is unfortunate that they are so shy, which is partly because they are constantly hunted for their tasty meat. See also **Tinamous (Tinamidae)**.

TINAMOUS (*Tinamidae*). In older natural history books, the description of the birds generally began with the ratites, including the rheas, ostriches, cassowaries, kiwis, and the extinct moas and elephant birds of Madagascar. Since these large, flightless birds have no keel on the sternum, which is otherwise characteristic of birds, they are known as flat-chested birds (ratites) in contrast to all the other modern birds, which are keel-breasted (*Carinatae*). Today we know that the ratites descended from flying, keel-breasted ancestors. The tinamous (*Tinamidae*) are a living primitive bird family which are close to the ancestral group of the ratites and particularly to the nandus (rheas). The tinamous used to be classified with gallinaceous birds, because externally they resemble guinea fowl. Because of their relationship to the ratites, they are today considered a separate order, *Tinamiformes*. See also **Tinamiformes**.

There are 9 genera with 43 species, divided into 2 subfamilies: (a) the Woodland Tinamous (*Tinaminae*) and (b) the Steppe-Tinamous (*Rhynchotinae*).

The Woodland Tinamous have connected nostrils in the middle of the beak or before it. There are 3 genera, which are distinguished by the back of their tarsometatarsus: (1) The genus Rough Taos (*Tinamus*) contains the following species: the Great Tinamou (*Tinamu major*) and Gray Tinamou (*Tinamus tao*); (2) Scaly Taos (*Nothocercus*): Bonaparte's Tinamou (*Nothocercus bonapartei*) and the Black-Capped Tinamou (*Nothocercus nigracapillus*); (3) the genus Smooth Tinamous (*Crypturellus*) has many species, among them the Little Tinamou (*Crypturellus soui*), the Variegated Tinamou (*Crypturellus variegatus*), the Brushland Tinamou (*Crypturellus cinnamomeus*), the Slaty-Breasted Tinamou (*Crypturellus boucardi*), and Tatupa (*Crypturellus tataupa*).

The Steppe-Tinamous have nostrils which are connected at the base of the beak. There are 6 genera: (1) the Red-Winged Tinamou with one species, the Red-Winged Tinamou (*Rhynchotus rufescens*); this bird is hunted in the pampas as a "partridge substitute;" (2) the Crested Tinamou (*Eudromia*); among them is the martineta Tinamou (*Eudromia elegans*; Fig. 1), which has no hind toe; (3) the Puna Tinamou (*Tinamotis pentlandii*; Fig. 1); (4) the Partridge Tinamous (*Nothoprocta*) comprise many species, among them the Ornate Tinamou (*Nothoprocta ornata*) and *Nothoprocta cinerascens*; (5) the Quail Tinamous (*Nothura*) comprise many species, among them the Spotted Tinamou (*Nothura maculosa*); (6) the Peacock Tinamous, with only one species (*Taoniscus nanus*), found in southeast Brazil and northeast Argentina; the upper tail coverts of this species are extended into a train.

Fig. 1.   Tinamous: (A) Puna (*Tinamotis pentlandii*), (B) Martineta (*Eudromia elegans*). (*Sketch by Glenn D. Considine.*)

Walking or running, the tinamous move almost exclusively on the ground. When approached by man, they hide in thick ground cover or steal away unobserved. When they are hard-pressed in open country, they often crawl into holes which another animal has dug. Some species are very reluctant to fly. But when surprised by a larger animal or when followed too closely, they rise suddenly with loud frightening wing beats, and often they call. They disappear swiftly and alight in thick vegetation. The effectiveness of this escape behavior is evident from the fact that one of the largest species, the great tao, can still be found in forests in which men and dogs have exterminated all other large birds.

Tinamous eat mainly small fruits and seeds. They pick them from the ground or from the plants they can reach from the ground; sometimes they jump up about 10 centimeters (4 inches) to reach a particularly tempting fruit. Seeds with winglike appendages, which make swallowing difficult, are beaten against the ground or vigorously shaken in order to remove them. The tinamous also eat opening buds, tender leaves, blossoms, and even roots. For variety they catch insects and their larvae, worms, and, in moist places, mollusks. They swallow small animals whole; they first peck at larger ones, then shake them or beat them against the ground. When searching for food, they scatter fallen leaves and other ground cover aside with their beaks, but they do not scratch with their feet. When searching for worms or larvae in moist places, brushland tinamou and probably other species fling the earth aside with their beaks, digging 2–3 centimeters ($\frac{3}{4}$ – 1 inch) deep. Like many other birds, they swallow small stones and grains of sand. Those species living in deserts drink either very rarely or not at all.

Hearing the calls made by tinamous is one of the most unforgettable experiences. Although the calls or songs of many species are simple, the purity and softness of their tones are attained only by a few birds. In several species the songs sound like organ tones, while others are flutelike whistles or trills that have a melancholy quality. The tones of the song sometimes have the same pitch, but in some species they may rise, and in others they rise and fall again. In contrast to most tinamous, the male Bonaparte's tinamou utters a rough crowing or barking call, which can be heard for several kilometers through the mountain forests. In some species, males and females can be readily distinguished by their calls. Tinamous sing mainly during the breeding season; they are noisiest in the early morning and late evening. When startled tinamous fly off or follow one another, they shriek or crow hoarsely.

Most adult tinamous live singly except during the reproductive season. The Martineta tinamous form strings of 6 to 30 or more birds. Reproductive behavior differs from that of most other birds. Only the males care for the eggs and young. So far no exception to this rule is known. In the few species that have been adequately studied, the males live in polygyny and the females in polyandry. A male in breeding condition attracts 2, 3 or even more females by continuously calling. They all lay their eggs in the same nest and then leave their incubation to the male. When the male has raised his young or has lost the eggs, he begins to call again and attracts new hens, which then supply him with another nest full of eggs. This breeding behavior has been observed in such different species as Bonaparte's tinamou, the brushland tinamou and the slaty-breasted tinamou. The variegated tinamou, however, cares for only a single egg.

The shiny tinamou eggs are among the most beautiful natural products known. They may be green, turquoise-blue, purple, wine-red, slate gray, or a chocolate color, and they often have a purple or violet lustre. They are always uniformly colored without spots or blotches. Their shape is oval or elliptical.

Incubating male tinamous sit continuously on the eggs for many hours. In most species, they leave the clutch only once a day to look for food. These excursions are undertaken during the mornings or in the afternoon, depending upon the weather; they last from 45 minutes up to 4 or 5 hours.

Newly hatched tinamous are densely covered with long soft down which in some species is marked in dulled colors. On the first day after hatching, the male leads the young out of the nest, moving slowly and calling the young with repeated soft whistles or whining tones. On leaving the nest the chicks of the smaller tinamous are very delicate but move so skillfully and well-hidden through the dense vegetation that little is known about their lives after they leave the nest. They develop rapidly and soon differ little in color and size from the adults.

TINCTURE. An extract containing the active principles of a drug or flavor in solution in alcohol or a mixture of alcohol and water. See also **Flavors and Essences.**

TINEA. See **Dermatitis and Dermatosis.**

TINNITUS. See **Hearing and the Ear.**

TIN STONE. See **Cassiterite.**

TINTOMETER. A type of colorimeter, in which the intensity of color, and hence the concentration of the colored substance in a colored solution, is determined by comparison with colored glass slides or standard solutions.

TISSUE. An aggregation of cells of characteristic form together with their intercellular matrix, specialized for the performance of some limited function or functions. All cells of the multicellular animal body take part in the formation of tissues, and tissues in turn are to a great extent incorporated in organs. The cells in a tissue may be alike, or several kinds may be present, and in some tissues the cells are supplemented by a conspicuous bulk of intercellular materials.

Animal tissues are divided into five classes: epithelial, muscular, nervous, connective, and vascular.

Plant tissues are: conducting, supporting, protective, food producing, food storage, and embryonic or meristematic.

TIT (*Aves, Passeriformes*). A small insect-eating bird of agile habits and friendly nature, exemplified by the chickadees, *Penthestes*, and titmice, *Baeolophus*, of North America. They are chiefly birds of the northern hemisphere, but the group is represented in the Australian region. The colors of most species are quiet, ranging from white to black through bluish grays, relieved by a limited amount of buff or chestnut. The blue tit, *Parus caeruleus*, and the azure tit, *P. cyanus*, of the Old World, however, are much more brightly colored, as their names imply.

In North America some species of chickadee is to be found in almost every locality, while titmice occur in the far west and in the states east of the Mississippi.

These species are usually fearless, and although they have no reputation as singers their cheery calls are always welcome. In the winter, they visit feed boxes readily, and they soon become accustomed to close observation if the observer avoids sudden movements. See also **Chickadee.**

TITANITE. A yellow or brown calcium silicotitanite, $CaTiSiO_5$, having a waxy luster, and often containing niobium (columbium), chromium, fluorine, and other elements. Titanite occurs in wedge-shaped monoclinic crystals, usually as an accessory mineral in granitic rocks and in calcium-rich metamorphic rocks. See also **Sphene.**

TITANIUM. Chemical element, symbol Ti, at. no. 22, at. wt. 47.9, periodic table group 4, mp 1650–1670°C, bp 3,287°C, density 4.507 g/cm³ (20°C). Below 885°C, elemental titanium has a hexagonal closepacked crystal structure; above this temperature, it has a body-centered cubic crystal structure. Compact titanium is a white metal, when cold it is brittle and may be powdered, but at red heat may be forged and drawn into wire. Titanium exhibits some passivity in air due to formation of coatings of oxide or nitride. At 610°C, titanium reacts with oxygen to form titanium dioxide; at 800°C, it reacts with nitrogen to form titanium nitride. Upon heating with chlorine, the metal

forms titanium tetrachloride. Cold, dilute H_2SO_4 readily dissolves the metal to form titanous sulfate. Hot, concentrated H_2SO_4 yields titanic sulfate. The element was first identified by Gregor in 1789 and later named titanium by Klaproth (1795). A metal of 95% purity was not produced until 1887 when it was made by the reduction of titanium tetrachloride with sodium. The first commercial uses date back to 1860 when ferrotitanium was used as an alloying element in steel and a bit later as a deoxidizer in the production of steel. There are five natural isotopes of the metal ^{46}Ti through ^{50}Ti, and three radioactive isotopes have been identified ^{44}Ti, ^{45}Ti, and ^{51}Ti, the latter two with relatively short half-lives measured in minutes and hours. ^{44}Ti has a half-life of approximately 10^3 years. Titanium is relatively abundant, ranking 8th in the list of chemical elements occurring in the Earth's crust. Titanium ranks 35th among the elements in terms of content in seawater, with an estimated 5 tons of titanium per cubic mile (1.1 kilograms per cubic kilometer) of seawater. First ionization potential 6.83 eV; second, 13.60 eV; third, 27.6 eV; fourth, 44.66 eV. Oxidation potential $Ti + 2H_2O \rightarrow TiO_2 + 4H^+ + 4e^-$, 0.95 V; $Ti \rightarrow Ti^2 + 2e^-$, 1.75 V; $Ti^{2+} \rightarrow Ti^{3+} + e^-$, 0.37 V. Electronic configuration $1s^2 2s^2 2p^6 3s^2 3p^6 3d^2 4s^2$.

Other physical properties of titanium are given under **Chemical Elements.**

Titanium occurs in practically all rocks and is an important constituent of many minerals. Only rutile TiO_2, however, is of commercial importance. The most important sources of this mineral are the sand dunes of Australia and Florida. Presently, Australia furnishes over 80% of the rutile requirements. Projects are underway to beneficiate (reduce) the other major potential titanium source, e.g., ilmenite. The known reserves of ilmenite $FeTiO_3$ are estimated 50 × greater than those of rutile. For mining the sand deposits for rutile, large floating dredge concentrators are used. Gravity concentration, followed by magnetic and electrostatic separation, yield a raw rutile of about 95% TiO_2 content. See also **Ilmenite**; and **Rutile.**

Production of titanium metal first involves the preparation of $TiCl_4$, a colorless liquid. Rutile and coke are charged into a continuous chlorinator. Upon the addition of chlorine gas, $TiCl_4$ is yielded in an exothermic reaction. To separate the metal, the $TiCl_4$, in a separate process, is reacted with molten magnesium metal pigs at about 50°C. The products are magnesium chloride $MgCl_2$ and titanium metal sponge. The by-product $MgCl_2$ is electrolyzed and the resulting magnesium and chlorine are recycled in the process. In another process, sodium metal is used instead of magnesium. And in still another process, the $TiCl_4$ may be electrolyzed.

Uses: The major uses for titanium are in various alloys, although unalloyed titanium finds some application. Titanium alloys are classified as alpha, alpha-beta, or beta, determined by the phases present in the alloy at room temperature. The alpha alloys usually result when the main elements present are the alpha stabilizers, e.g., oxygen, nitrogen, hydrogen, and carbon. Alpha-beta alloys and beta alloys contain increasing amounts of beta stabilizers, mainly vanadium, molybdenum, iron, chromium, manganese, tantalum, and niobium (columbium). The alpha-beta class of alloys normally has great room-temperature strength and may be heat treated. The annealed beta alloys show poor thermal stability over about 230°C, but do have good formability and weldability. The beta alloys may be age heat treated wherein some alpha phase is precipitated and this results in a very high room-temperature strength. The complexity of titanium alloys is brought about by the fact that the element is allotropic and undergoes a phase transformation at about 885°C, changing from one crystalline form to another as mentioned at the start of this entry. The variations in strength and percent elongation for the three major types of alloys and for pure titanium are given in Table 1.

Many diversified applications have been found for titanium and its alloys. Moreover, the number of these applications tends to increase steadily as greater production and improved processes reduce costs. At the present time, titanium is still an expensive material and is only used where its light weight, high strength, and corrosion resistance justify its cost. Aeronautical and missile design engineers find titanium and its alloys to be materials whose light weight and high strength, particularly at elevated temperatures (600°C), give them many applications in aircraft and missile construction. About 99% of all titanium materials are used in these fields.

Titanium and its alloys are widely used in compressor blades, turbine disks and many other forged parts of the jet engine. Here they offer resistance to high temperature, as well as weight-saving. The latter quality is increasing their use in the structural airplane parts, ranging from engines and air frames to skin and fastenings. Titanium sheet finds application in

TABLE 1. TITANIUM ALLOYS

Alloy	Tensile Strength		Yield Strength		Percent Elongation
	psi	mpa	psi	mPa	
Pure titanium					
High purity (99.9%)					
Annealed	34,000	237	20,000	138	54
Commercial purity					
(99.0%)	79,000	545	63,000	435	27
Alpha alloy					
Ti-5Al-2.5 Sn					
Annealed	125,000	863	120,000	828	18
Alpha-beta alloy					
Ti-6Al-4V					
Annealed	135,000	932	120,000	828	11
Heat treated	170,000	1173	150,000	1035	7
Beta alloy					
Ti-3Al-13V-11Cr					
Heat treated	180,000	1242	170,000	1173	6

Classification of alloys by application:
Airframe Alloys
  Ti-75A, Ti-5Al-2.5Sn, Ti-6Al-6V-2Sn, Ti-6Al-4 V, Ti-7Al-4Mo,
    Ti-4Al-3Mo-IV, Ti-8Mn, Ti-13V-11Cr-3Al
Engine Alloys
  Ti-8Al-1Mo-IV, Ti-5Al-2.5Sn, Ti-6Al-4V, Ti-6Al-2Sn-4Zr-2Mo,
    Ti-6Al-2Sn-4Zr-6Mo
Corrosion-resistant Alloys
  Ti-35A, Ti-50A, Ti-65A, Ti-0.2Pd, Ti-2Ni

shroud assemblies, cable shrouds and ammunition tracks. Titanium alloy sheet is formed into ribs for use as stiffeners, as well as fuselage frames and bulk heads. Other uses of titanium in aircraft include channel sections, flat rubbing strips, landing gear doors, hydraulic lines, baffles, tail cones, longerons, etc. Other uses of titanium alloys include bulk heads, ducts, fire walls, etc.

The light weight of titanium and its alloys, coupled with their corrosion resistance, has brought them into use in ships, especially naval ships. Here many investigations show the important advantages of the metal and its alloys as wet exhaust muffles for submarine diesel engines, and as meter disks, and heat exchanger tubes which offer improved service for widespread use in salt water. Military applications of titanium extend from cannon and guided missiles to light-weight armor-plate for tanks. These materials offer other weight savings in other parts of military vehicles, such as piston rods and transmissions, which may extend to the transportation industries generally.

Throughout the chemical industry, titanium is used extensively both in plant and in laboratory. Among important present-day applications are heat exchangers, autoclave heads, autoclave coils for cooling and heating, chemical processing racks, and valves and tanks where corrosion resistance is necessary.

Advancements in Titanium Technology. Increasingly, titanium alloys are competing with nickel-base alloys on the basis of cost, strength, and corrosion resistance. The alloy Ti-3Al-2.5 V, for example, is finding expanded use in the process industries because of its resistance to mildly reducing chloride environments.

Authorities in the field do not believe that the demand for titanium in the aerospace industry will be adversely affected by the increasing use of polymer-matrix composites in airframes and engines. Titanium has been replaced outright by composite materials in only a few aerospace applications. In those instances, the primary considerations have been weight reduction and "stealth" characteristics. By contrast, titanium has been selected instead of composites for several applications where the primary considerations have been titanium's superior stiffness and toughness as well as its multidirectional strength characteristics. Titanium also has gained favor because of its compatibility with composites. An example of this compatibility is found in titanium-aluminide foils. These are used to fabricate honeycomb structures. Fiber-reinforced alpha 2 titanium aluminide-matrix composites are fabricated by consolidating a foil/fiber/foil layup, using hot isostatic processing. The reinforcements are carbon-coated silicon-carbide continuous fibers.

Near-Net-Shape Processing. Even greater use of titanium in aircraft has been limited because of its relatively high cost (as compared with aluminum, by a factor of $10\times$ to $20\times$) and also because of its lower machinability. Gains are being made by titanium on both of these counts, however, by initially making titanium parts very close to the shape and dimensions of the desired end product (near-net-shape or NNS). These gains have been made through advanced titanium processing techniques and include precision casting, hot isostatic pressing, blended elemental (BE) powder techniques, and precision forging.

Increasingly, titanium alloys are competing with nickel-based alloys on the basis of cost, strength, and corrosion resistance.

Titanium Diboride. TiB_2 cermet is an extremely hard material. For several years, there was no successful process for depositing the material on parts for achieving wear resistance. A process was introduced by Montreal Carbide Co. in 1990, however, that deposits microspheres of TiB_2 in a metallic matrix. Using a conventional plasma-spray technique, a series of cermet coatings, in which the ceramic phase is finely and uniformly dispersed, yields a wear-resistant surface. TiB_2 is synthesized during spraying by a reaction between powders of titanium-bearing alloys and boron-bearing alloys. Due to the rapid solidification, the TiB_2 crystals that form are finely dispersed in a metallic matrix, which originated from the alloys used in the process.

Chemistry and Compounds: Due to its $3d^2 4s^2$ electron configuration, titanium forms tetravalent compounds readily, although the Ti^{4+} ion does not exist as such in aqueous solution, except at very low or high pH values, the common cation being hydrated TiO^{2+} (or more probably $Ti(OH)_2^{2+}$). Many of the tetravalent compounds are largely covalent. There are also Ti(III) and a few Ti(II) compounds, the latter being very easily oxidized. Titanium dioxide, TiO_2, is well known both as a mineral, of which three structural forms exist, and as an industrial product obtained from ferrous titanate, $FeTiO_3$ ores or by oxidation of tin(IV) chloride, $TiCl_4$. See also **Titanium Dioxide.** Moreover, the precipitate obtained by action of alkali metal hydroxides upon solutions of tetravalent titanium is a hydrated oxide. The latter is readily soluble in acids to form oxysalts, which are usually formulated in terms of the TiO^{2+} ion, without including its water of hydration, e.g., as $NaTiOPO_4$. The hydrated TiO^{2+} ion is not amphiprotic, in that it does not dissolve in alkali hydroxides. However, it does react on fusion with alkali carbonates to form such compounds as M_2TiO_3 and $M_2Ti_2O_5$, these compounds having been shown to be mixed oxides rather than titanates. The alkaline earth titanates have the face-centered perovskite structure, and barium titanate, widely used for its electrical properties, has been produced in other crystalline forms.

Lower oxides of titanium, Ti_2O_3 and TiO, have been produced by reduction of TiO_2.

All four of the common halogens form tetrahalides of titanium, $TiCl_4$ being a liquid at ordinary temperatures, while TiF_4, $TiBr_4$, and TiI_4 are solids. They are readily hydrolyzed, yielding as end products TiO_2 and the hydrogen halide, in the case of $TiCl_4$ an intermediate addition product of the type $H_2O \cdot TiCl_4$ is considered to be formed. This is in accordance with the behavior of $TiCl_4$ and $TiBr_4$ as Lewis acids to form such unstable adducts, not only with water, but with oxygen-function organic compounds. Likewise, titanium chelates are formed with oxygen donor compounds such as acetylacetone.

The dihalides of titanium, formed by reduction of the tetrahalides, are vigorous reducing agents and unstable; $TiCl_2$ is inflammable in air. The trihalides, though more stable than the dihalides, are effective reducing agents. Ti(III) occurs in aqueous solutions as $Ti(H_2O)_6^{3+}$.

Normal oxyacid salts of titanium are unknown, but many basic salts, formulated as stated above, in terms of TiO^{2+}, though more or less hydrated, have been prepared.

Like the oxide, halides and sulfide, the nitride, boride, and carbide of titanium(IV) can be made by heating the elements together at high temperatures. The last three compounds are alloy-like in character, they can vary in composition without becoming unstable and they are extremely hard.

The halogen complexes are the most stable complex ions of titanium. The hexafluorotitanate ion, TiF_6^{2-} is very stable, as are the peroxo-complexes, containing $-Ti-O-O-$. The $TiCl_6^{2-}$ and $TiBr_6^{2-}$ complexes are less stable, except in concentrated solutions of the hydrogen halides. A number of compounds of the $TiCl_5^{2-}$ ion are known, especially of the higher alkali metals, e.g., $M_2TiCl_5 \cdot H_2O$.

Additional Reading

Bauccio, M.L.: "ASM Engineered Materials Reference Book," 2nd Edition, ASM International, Materials Park, OH, 1994.

Brady, G.S., H.R. Clauser, and J.A. Vaccari: "Materials Handbook," 14th Edition, McGraw-Hill Professional Book Group, New York, NY, 1996.

Carter, G.F. and D.E. Paul: "Materials Science and Engineering," ASM International, Materials Park, OH, 1991.

Chiles, J.R.: "Titanium" *Smithsonian*, 86 (May 1987).

Collings, E.W. and G. Welsch: "Materials Properties Handbook: Titanium Alloys," ASM International, Materials Park, OH, 1995.

Copley, S.M.: "Applied General and Nonferrous Physical Metallurgy," Encyclopedia of Materials Science and Engineering, MIT Press, Cambridge, MA, 1986.

Davis, J.R.: "ASM Materials Engineering Dictionary," ASM International, Materials Park, OH, 1992.

Donachie, M.J.: "Titanium: A Technical Guide," 2nd Edition, ASM International, Materials Park, OH, 2000.

Gauthier, M.M.: "Engineered Materials Handbook," ASM International, Materials Park, OH, 1995.

Greenwood, N.N. and A. Earnshaw: "Chemistry of the Elements," 2nd Edition, Butterworth-Heinemann, Inc., Woburn, MA, 1997.

Jha, S.C., et al.: "Titanium-Aluminide Foils," *Advanced Materials & Processes*, 87 (April 1991).

Krebs, R.E.: "The History and Use of Our Earth's Chemical Elements: A Reference Guide," Greenwood Publishing Group, Inc., Westport, CT, 1998.

Kubel, E.J., Jr.: "Titanium Near-Net-Shape Technology Shaping Up," *Advanced Materials & Processes*, 46 (February 1987).

Lide, D.R.: "CRC Handbook of Chemistry and Physics 2000–2001," 81st Edition, CRC Press, LLC., Boca Raton, FL, 2000.

Nelson, O.E.: "Titanium Staves Off Composites," *Advanced Materials & Processes*, 18 (June 1991).

Square, M.: "Titanium Boride Cermet: New Wear-Resistant Coating," *Advanced Materials & Processes*, 117 (April 1990).

Staff: "Navy Studies Titanium Applications," *Advanced Materials & Processes*, 14 (September 1989).

Staff: "ASM Handbook — Properties and Selection: Nonferrous Alloys and Pure Metals," ASM International, Materials Park, OH, 1990.

Staff: "Titanium and Titanium Aluminides," *Advanced Materials & Processes*, 71 (April 1990).

Staff: "Titanium Forecast," *Advanced Materials & Processes*, 18 (January 1991).

Staff: "Profile of Titanium Manufacturers," *Advanced Materials & Processes*, 52 (June 1991).

TITANIUM DIOXIDE. TiO_2, formula weight 79.90, variously colored, depending upon source, but white when purified and sold in commerce. Decomposes at about $1,640\,°C$ before melting, density 4.26 g/cm^3, insoluble in H_2O, soluble in H_2SO_4 or alkalis. Titanium dioxide is a very high-tonnage material and is the principal white pigment of commerce. The compound has an exceptionally high refractive index, great inertness, and a negligible color, all qualities that make it close to an ideal white pigment. Annual production approximates two million metric tons, of which nearly one-half of this amount is produced in the United States. Major uses of TiO_2 pigments are: (1) paint, 60%, (2) paper, 14%, (3) plastics and floor coverings, 12%, (4) printing inks, 3%, and (5) various applications including rubber, ceramics, roofing granules, and textiles, 11%.

Two major processes are used for producing raw titanium dioxide pigment: (1) the sulfate process, a batch process accounting for over half of current production, introduced by European makers in the early 1930s; and (2) the chloride process, a continuous process, introduced in the late 1950s and accounting for most of the new plant construction since the mid-1960s. The sulfate process can handle both rutile and anatase, but the chloride process is limited to rutile.

In the sulfate process, ilmenite (45–60% TiO_2) or a slag rich in titanium (70% TiO_2) obtained from electric smelting of ilmenite, is the feedstock. The raw materials first are digested: $FeTiO_3 + 2H_2SO_4 \rightarrow FeSO_4 + TiO \cdot SO_4 + 2H_2O$. In a second step, the concentrated liquor is nucleated, diluted with H_2O, and boiled until nearly all of the titanium has precipitated out in the form of flocculated titanium dioxide (anatase) hydrate: $TiO \cdot SO_4 + 2H_2O \rightarrow TiO_2 \cdot H_2O + H_2SO_4$. After filtering, the cake is leached under reducing conditions to remove residual iron. Conditioning agents are added, after which the hydrate is dried and calcined in a rotary kiln at approximately $900\,°C$: $TiO_2 \cdot H_2O \rightarrow TiO_2 + H_2O$. The conditioning agents usually consist of a phosphate and a potassium salt, as well as zinc, antimony, and aluminum compounds. The purpose of these additions is to improve the final properties of the pigment, including color, photochemical stability, and dispersibility, as well as to catalyze the formation of rutile from the anatase hydrate.

In the chloride process, the feedstock must be high in titanium and low in iron. Mineral rutile (95% TiO_2) is best suited, but leucoxene (65% TiO_2) can be used. See also **Brookite**. An economical conversion of ilmenite for use as a chloride process feedstock has not been developed to date. The ore is mixed with coke and chlorinated at about $900\,°C$ in a fluidized bed. The principal product is titanium tetrachloride, but other impurities including iron also are chlorinated and thus must be removed by selective condensation and distillation. Up to this point, the process is similar to that of producing titanium metal as described under **Titanium**. By selective reduction prior to distillation, vanadium present is removed as $VOCl_3$. In the next step, the purified $TiCl_4$ reacts with oxygen at a temperature of about $1,000\,°C$. The presence of $AlCl_3$ in this reaction promotes the formation of rutile instead of anatase. The two major steps are: (1) chlorination: $3TiO_2 + 4C + 6Cl_2 \rightarrow 3TiCl_4 + 2CO + 2CO_2$; and (2) oxidation: $TiCl_4 + O_2 \rightarrow TiO_2 + 2Cl_2$. The chlorine is recycled. The raw titanium dioxide product generally is neutralized by washing in an aqueous solution of proper pH.

Many grades of titanium dioxide pigments are offered commercially. They range in crystal structure (anatase or rutile), particle shape and size, the type of hydrous oxide coating applied, and the type and quantity of additives applied. Generally, the commercial pigments contain 80–99% TiO_2, the remainder of the formulation comprised of alumina and silica hydrates. Nonpigmentary grades of titanium dioxide for the glass, welding-rod, electroceramic, and vitreous-enamel industries contain 99% TiO_2.

TITEL. See **Antelope**.

TITRATION COULOMETER. See **Coulometer**.

TITRATION (Potentiometric). This analytical method is based in principle upon the Nernst equation, which may be written in the form

$$E = E^0 - \frac{0.059}{n} \log \frac{A_{\text{ox}}}{A_{\text{red}}}$$

where E is the measured electromotive force, E^0 is the standard value of the electromotive force (electrode potential) when the substances of the electrochemical reaction are in their standard states, n is the valence change (change in number of electrons per mole of the reactants) and the A-terms are the activities of the oxidized and reduced forms of the reactants. See also **Activity Coefficient**. Activities are proportional to concentrations, so that the concentration of one of the reactants may be determined if that of the other is known. Thus the concentration of Cu^{2+} ions in a solution could be found by use of an electrode of metallic copper (unstrained metals are assumed to be at unit activity). Or the concentration of Cl^- ions in a solution could be found by use of an electrode of (insoluble) silver chloride $AgCl$ deposited on an electrode of metallic silver. Or the concentrations in a solution containing two ions in different states of oxidation, such as Fe^{2+} and Fe^{3+}, could be found by the use of an inert electrode, such as one of platinum. The three reactions and corresponding forms of Nernst's equation (using the approximation of substituting concentrations for activities) are:

$$Cu^0 \rightleftharpoons Cu^{2+} + 2e^-$$

$$E = E^0 - \frac{0.0591}{2} \log c_{Cu^{2+}}$$

$$Ag^+ + Cl^- \rightleftharpoons AgCl \downarrow$$

$$E = E^0 - (0.0591) \log K_{sp} + (0.0591) \log c_{Cl^-}$$

(where K_{sp} is the solubility product of $AgCl$)

$$Fe^{2+} \rightleftharpoons Fe^{3+} + e^-$$

$$E = E^0 - (0.0591) \log \frac{c_{Fe^{3+}}}{c_{Fe^{2+}}}$$

In constructing an electric cell for potentiometric titrations it is necessary, of course, to use a second electrode to complete the circuit, in addition to the measuring electrodes (commonly called *indicator electrodes*) described above. Ideally the second electrode would be a hydrogen electrode which (as explained in the entry on electrode potential) is the standard reference electrode for which the potential, in equilibrium with its ions, is defined as zero. Since it is awkward to use, other electrodes of known potential, such as the calomel electrode or the glass electrode, are commonly used

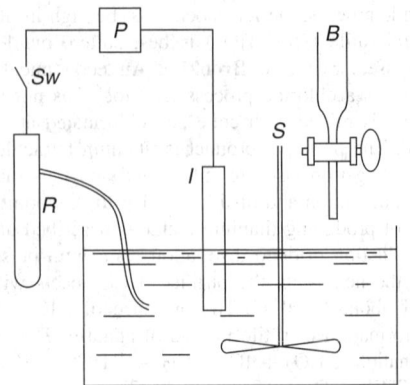

Fig. 1. Apparatus for potentiometric titration: (*R*) reference electrode; (*I*) indicating electrode; (*P*) potentiometer; (*B*) burette; (*S*) stirrer; (*Sw*) switch.

as reference electrodes. The arrangement of the apparatus is as shown in Fig. 1.

The procedure in a potentiometric titration is to determine the potential of the indicator electrode after each addition of the titrating solution. This is done by closing the switch in the circuit shown in Fig. 1 just long enough to read the potential on the sensitive measuring device used, such as a potentiometer or vacuum tube voltmeter. Of course, very small additions of titrant are made as the expected endpoint is approached. Then the readings of potential are plotted against volume of titrant added, as shown in Fig. 2. Reactions suitable for determination by this method show sharply defined endpoints as pictured in the figure. To correspond to the true stoichiometric endpoint, certain conditions should be met, including reversibility of the reaction and allowing time for the electrodes to reach equilibrium before closing the switch to read the potential difference. While the description given was on the basis of titration with zero current drawn from the electrodes between readings, there is also a method in which a small constant current is drawn, not large enough to affect their potentials, but serving to eliminate some of the sources of error in the null method.

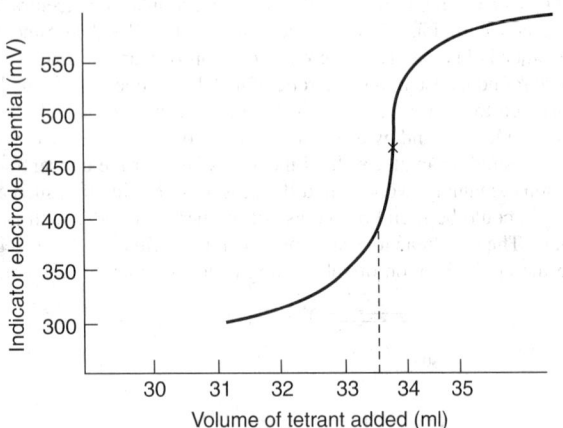

Fig. 2. Potentiometric titration curve.

Types of titrations for which potentiometric methods of determining endpoints are particularly useful including titrations of halide mixtures, of various metals, of alkaloids, in non-aqueous solvents, and various titrations with oxidizing agents, such as permanganate, dichromate, iodate, and ceric sulfate.

TITRATION (Thermometric). This technique consists of the detection and measurement of the change in temperature of a solution as the titrant is added to it, under as near adiabatic conditions as possible. Experimentally, the titrant is added from a constant-delivery burette into the titrate (solution to be titrated) which is contained in an insulated container such as a Dewar flask. The resultant temperature-volume (or time) curve thus obtained is similar to other titration curves, e.g., acid-base, in that the end point of the reaction can be readily ascertained. Since all reactions involve a detectable endothermic or exothermic enthalpy change, the technique has potentially wide application in analytical chemistry, especially in those cases where other more common methods are not applicable.

Several idealized thermometric titration curves for an exothermic reaction are given in Fig. 1. A titration curve in which the titrant and the titrate are at the same initial temperature is illustrated in (a). The actual titration is preceded by a blank run, *CD*, in which no titrant is added to the titrate. At *D*, titrant is added, causing the temperature of the titrate to rise rapidly, reaching a maximum value at *E*. Beyond *E*, additional titrant causes no further change in the temperature of the titrate, hence, the horizontal excess reagent line, *EF*. The temperature rise of the titrate (and titration vessel), ΔT, is obtained by determining the vertical distance between the excess reagent line, *ED'*, and *CD*. In curve (b), the conditions of the titration were identical to those of (a) except that the titrant was at a higher temperature than the titrate, hence, the sloping excess reagent line, *EF*. For curve (c), conditions were also identical to the above except that the titrant was at a lower temperature than the titrate. The excess reagent line, *EF*, thus slopes in an opposite direction to that of curve (b). For an endothermic reaction, the curves would be identical to the above except that the temperature changes would be in an opposite direction.

TOAD. See **Frogs and Toads**.

TOAD BUG (*Insecta, Hemiptera*). A small broad bug found on the muddy shores of ponds and streams. It is peculiarly like a toad in appearance and habits, capturing insects as prey and burrowing at times. Only a few species are known.

TOADSTOOL. See **Agarics**; **Basidiomycetes**; and **Fungus**.

TOBACCO WORM (*Insecta, Lepidoptera*). The caterpillar of a large sphinx moth, *Protoparce sexta*, which eats the leaves of tobacco, tomato, and other plants. The moth is gray with a row of six orange spots on each side of the abdomen and the larva is green, about 3 inches (7.5 centimeters) long, with a stout horn near the caudal end of the body. Its sides are marked with oblique whitish lines.

Crop rotation as prescribed for the region involved, and dusting with lead arsenate (see also **Lead**), are effective methods of control on large fields. The larvae are so easily seen when they become large enough to cause severe damage that hand picking is effective on truck crops of smaller extent.

Tobacco is attacked by other caterpillars, hence the term tobacco worm is not always restricted to this form.

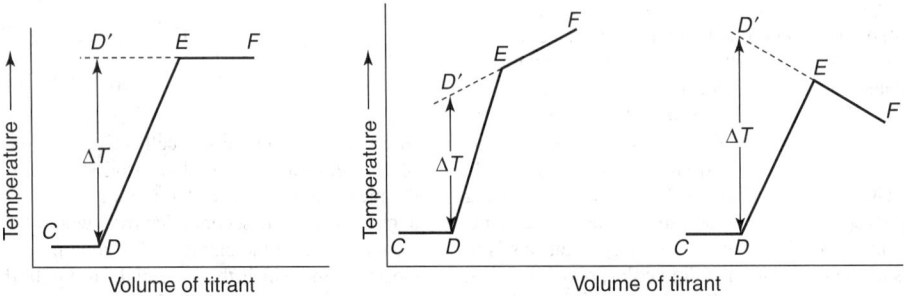

Fig. 1.   Thermometric curves for an exothermic reaction (ideal).

TOCOPHEROLS. See **Vitamin E**.

TODOROKITES. Calcium-bearing manganese oxides found in terrestrial manganese ore depositions, in weathering products of manganese-bearing rocks, and in some manganese nodules. Todorokites are, in many instances, principal constituents of manganese nodules. Host copper and nickel within the modules is potentially of economic importance. Knowledge of todorokites is important for understanding how nodules form and how they concentrate transition elements from ocean waters. See **Ocean Resources (Mineral)**.

Formerly believed to be a single phase, todorokite was discredited as a mineral by the International Mineralogical Association Commission on New Minerals and Mineral Names in 1970 when evidence was submitted showing it to be a complex mixture of several compounds. As reported by Turner and Buseck (1981), many mineralogists felt the decision was incorrect since recognizable x-ray diffraction patterns could be produced from todorokite collected from widespread deposits. X-ray patterns have not been adequate for structure determination. Further confusion is caused by the variable morphology of todorokite, which appears fibrous in some samples and platy in others. Todorokite material appears to have a range of related structures. High-resolution transmission electron microscopy reveals that terrestrial todorokites consist of tunnel structures of previously unreported dimensions and that these tunnel structures are intergrown coherently on a unit cell scale. As shown in Fig. 1, many types and degrees of disorder are evident. The widespread presence of disorder explains prior confusion with x-ray diffraction patterns and hence the difficulty to relate x-rays studies with specific structures. Turner and Buseck suggested a revised nomenclature of tunnel manganese oxides. This is described in *Science*, **212**, 1024–1027 (1981).

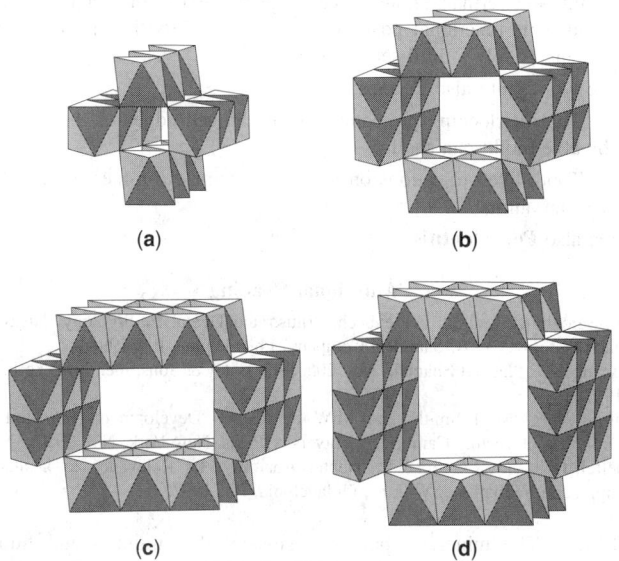

(a)   **(b)**

(c)   **(d)**

Fig. 1. Diagram of various manganese oxide tunnel structures. The common smallest unit of the structures is a $[Mn^{+3}, Mn^{+4}]o_6$ octahedron. The octahedra are edge linked to form long chains that are corner linked to form the framework of tunnel minerals: (**a**) 1 × 1, pyrolusite, (**b**) 2 × 2, hollandite, (**c**) 2 × 3, romanechite and (**d**) 3 × 3, todorokite. The larger tunnel structures (hollandite, romanechite, todorokite) can accommodate large cations and water. There are many more possible tunnel structures and those with a common dimension (for instance the double chain sides of hollandite and romanechite) can intergrow on a unit cell level. As far as is now known, todorokite occurs only as an intergrowth of differently sized tunnel structures that have a triple chain in common (3 × 2, 3 × 3, 3 × 5, 3 × 6, etc.) The 3 × 3 structure is the most prevalent. (*Source:* S. Turner.)

TODY. See **Kingfishers and Other Coraciiformes**.

TOLUENE. $C_6H_5CH_3$, formula weight 92.13, colorless, odorous liquid, mp $-95\,°C$, bp $110.8\,°C$, sp gr 0.866. Toluene, a homologue of benzene (C_nH_{2n-6} series of aromatic hydrocarbons), essentially is insoluble in H_2O, but is fully miscible with alcohol, ether, chloroform, and many

other organic liquids. Toluene dissolves iodine, sulfur, oils, fats, resins, and phosgene. When ignited, toluene burns with a smoky flame. Unlike benzene, toluene cannot be easily purified by crystallization.

Industrial-grade toluene distills between 108.6 and 112 °C, is water-white and has a flash point of 2 to 5 °C. The specific gravity ranges from 0.854 to 0.874. Toluene is a high-tonnage industrial chemical with United States production approximating 3 million tons/year. Petroleum sources account for about 95% of toluene production, the remainder coming from coal gas and coal tar. The dealkylation of toluene is a prime source of benzene, accounting for about one-half of toluene consumption. The production of diisocyanates from toluene is increasing. As a component of fuels, the use of toluene is lessening. Toluene takes part in several industrially-important syntheses. The hydrogenation of toluene yields methyl cyclohexane $C_6H_{11}CH_3$, a solvent for fats, oils, rubbers, and waves. Trinitrotoluene (TNT) $C_6H_2(CH_3)(NO_2)_3$ is a major component of several explosives. When reacted with H_2SO_4, toluene yields o- and p-toluene sulfonic acids $CH_3C_6H_4SO_3H$. Saccharin is a derivative of the ortho acid; chloramine T (an antiseptic) is a derivative of the para acid. A widely used solvent for synthetic resins and rubber, monochlorotoluene $CH_3C_6H_4Cl$ is a derivative of toluene. Toluene also is used in the manufacture of benzoic acid, the latter an important ingredient for phenol production.

One modern toluene production process commences with mixed hydrocarbon stocks and can be used for making both toluene and benzene, separately or simultaneously. The process essentially is a combination of extraction and distillation. An aqueous dimethyl sulfoxide (DMSO) solution is passed countercurrently against the mixed hydrocarbon feed. A mixture of aromatic and paraffinic hydrocarbons serves as reflux.

In other industrially-important processes, toluene is a source of benzyl chloride $C_6H_5CH_2Cl$, benzal chloride $C_6H_5CHCl_2$, benzotrichloride $C_6H_5CCl_3$, benzyl alcohol $C_6H_5CH_2OH$, benzaldehyde, C_6H_5CHO, and sodium benzoate C_6H_5COONa.

TOMATO. See **Solanaceae (Potato Family)**.

TOMATO WORM (*Insecta, Lepidoptera*). A large caterpillar, *Protoparce quinquemaculata*, similar to the tobacco worm in appearance and habits and belonging to a closely related species. The moths of the two species are similar but are easily distinguished by comparison.

TOMB-BAT. See **Bats**.

TOMBOLO. A type of sand bar that connects one island with another, or an island to the mainland.

TONE. See **Musical Sound**.

TONE CONTROL. The control for regulating the frequency response of an audio amplifier. In its most common form as found in most radio receivers it consists of a capacitor in series with a variable resistance shunted across the circuit at some point. Since such a combination passes high frequencies more easily than low values, the high frequencies will be attenuated, the degree being determined by the setting of the variable resistance and the impedance of the circuit to which the tone control is connected. Thus by varying the resistance more or less attenuation may be given the high frequencies, and the effect is as if the bass response were being varied. In more elaborate tone controls, separate bass and treble controls are provided, thus allowing a balanced control of the response.

TONGUE (Anatomy). An organ associated with the floor of the oral cavity, usually projecting or protrusile. The true tongue is a vertebrate structure, occurring in all classes above the fishes. It is made up largely of voluntary muscle fibers, so distributed that it can be protruded and withdrawn and swung in every possible direction. It arises embryonically from the floor of the anterior part of the pharynx, extending forward into the oral cavity.

The tongue is used in man and other species for the manipulation of foods in chewing, and in some of the amphibians, reptiles and birds it aids in capturing food, usually by adhesion. In man it is an important organ of speech, aiding in the modulation of sounds produced by the larynx.

The word is sometimes applied to the radula of the snails and to parts of the insect mouth, but only from superficial similarity with the true tongue of the vertebrates.

TONGUE SOLE.   See **Flatfishes**.

TONKA BEAN.   See **Bean**.

TONOMETER.   A tonometer is an instrument used by eye care professionals to measure the intraocular pressure of a patient's eye. This pressure measurement is an important part of every eye examination because an increase in pressure may signal the onset of glaucoma, a sight-threatening disease. Pressure is measured in millimeters of mercury, and, as a general guideline, pressure above 21 millimeters is considered to be elevated pressure, though not all persons with that reading have glaucoma.

The most popular tonometry test is with the noncontact tonometer, often referred to as the "air-puff" test. The patient looks through a machine as it blows a gentle puff of air at the eye. This puff of air flattens the cornea slightly in order to give the pressure reading. The machine is gas pressurized and does not require direct contact with the eye.

Another form of tonometry, and one of the most accurate, is the contact tonometer, an instrument that looks like a pen. When using the contact tonometer, numbing eye-drops are administered, and then the tip of the tonometer touches the eye and measures the pressure.

An applanation tonometer, measures the force required to flatten a small area of central cornea. A topical anesthetic and fluorescein dye, are instilled before the measurement is taken. One type of applanation tonometer attaches to a slit lamp, another is hand-held.

The Schiotz tonometer measures the amount the cornea is indented by a fixed weight that artificially raises the pressure. It is a simple, portable instrument, but is considered to be somewhat less accurate than other types of measuring devices. See also **Glaucoma**.

Vision Rx, Inc., Elmsford, NY.

TOOTH.   1.   A hard structure projecting from the wall of the mouth or the anterior part of the alimentary tract (digestive system) and used for grasping and breaking up food. Teeth vary from the chitinous projections on the radula of mollusks to the complex structures of the vertebrates. Chitinous teeth of annelid worms, located on the walls of the pharynx, are also known as jaws.

Vertebrate teeth are formed of two layers of hard materials over a living papilla, the pulp, which contains blood vessels and nerves. They form in the embryo from ingrowths of the outermost layer of the body, the ectoderm, associated with masses differentiated in the middle layer, the mesoderm. The mesoderm forms the dental papilla and around its outer surface, except at the end that is to remain connected with the body, lays down a layer of dentine. This material is similar to bone in composition but not in minute structure. It is harder, contains less organic matter than bone, and is the ivory of commerce. The ectodermal cap over the dental papilla develops into an enamel organ and deposits a layer of enamel on the dentine. Enamel is made up of minute prisms and is the hardest substance produced by the animal body, containing only 2–4% organic matter. When the tooth takes its place in the jaw in the process of eruption the enamel organ is destroyed, hence no more of this substance can be produced. Dentine may be deposited later in life, however, encroaching on the pulp cavity.

The teeth of sharks are the most primitive vertebrate teeth. They are formed like the placoid scales of the body, with a principal point and sometimes smaller supplementary points, and are superficially attached in rows on the jaws. Both scales and teeth contain dentine covered with enamel.

In other fishes and amphibians teeth are distributed in various parts of the oral cavity and are associated with the bones of the jaws and skull. In reptiles and mammals they are limited to the jaws and are seated in sockets (alveoli) in the bones by means of a third hard substance, cementum, deposited between the dentine of the roots and the bone of the jaw.

Primitive teeth are simple conical structures. In mammals this type persists only in the canines. The other teeth are sharp-edged incisors, broad molars with projecting cusps on the apposed surfaces, and premolars of intermediate form. The premolars of man are also called bicuspids. An assemblage of teeth of these kinds, variously specialized for cutting, tearing and holding, and grinding and crushing, is known as a heterodont dentition, and is the fundamental plan of dentition in the mammals. The various forms of teeth are the basis of specialization in different groups of mammals for the use of limited types of food. Thus species that depend on plant tissues have greater need of crushing teeth while carnivorous forms need cutting teeth for their tougher food. The molars of the former are broad and are so folded that the worn apposed (occlusal) surfaces show alternating bands of dentine, enamel, and cementum in patterns characteristic of the kind of animal. In the elephants these materials form transverse ridges. The incisors of grazing animals are also flattened, forming clipping, rather than cutting, teeth. The canines are reduced or lacking. In carnivores the incisors are greatly reduced, the canines are highly developed, and the molars are sharp-edged. Molars of the two jaws work together like the blades of shears. Some mammals, notably certain anteaters, have lost all trace of teeth.

The tusks of various animals are greatly elongated teeth projecting from the jaws for use in fighting and digging. They are usually devoid of enamel. Those of the elephants are upper incisors and in early growth are provided with enamel tips. Those of walruses and swine are canines. Single tusks of elephants weighing almost 200 pounds (90.7 kilograms) have been recorded but the average is less than 100 pounds (45.4 kilograms). These tusks provide the finest ivory.

Some teeth grow constantly as they are worn away, while others attain a fixed form within a short period and still others undergo a partial compensation for wear. The incisors of rodents are of the first type. These chisel-like teeth wear away in gnawing but are constantly renewed by growth at their bases to preserve a uniform length. The molars and premolars of horses are elevated in the jaws by lengthening of the roots, but the amount available for use during the life of the animal is regulated by the height of the crown at the initial formation of the tooth. Human teeth assume a fixed form. The only compensation for wear is the partial filling in of the pulp cavity by the deposition of dentine.

A compensation for breakage and wear is also found in the replacements that occur in lower vertebrates. Mammals normally have no more than two sets, milk and permanent, and once the permanent teeth of adult life have assumed their functional positions in the jaws, loss through accident is permanent. Rarely a third set of teeth develops.

The word tooth also denotes:

2.   The interlocking projections at the hinged edges of the valves of the bivalve molluscan shell.

3.   Tooth-like projections on a hard structure or on the surface of the body of an animal.

See also **Periodontitis**.

Additional Reading

Bath-Balogh, M. and M. Fehrenbach: "Illustrated Dental Embryology, Histology, and Anatomy," W.B. Saunders Company, Philadelphia, PA, 1997.
Slavkin, H.C.: "Dental Enamel," Vol. 205, John Wiley & Sons, Inc., New York, NY, 1997.
Teaford, M.F., M.M. Smith, and M.W. Ferguson: "Development, Function and Evolution of Teeth," Cambridge University Press, New York, NY, 2000.
Woelfel, J.B. and R.C. Scheid: "Dental Anatomy: Its Relevance to Dentistry," Lippincott Williams & Wilkins, Philadelphia, PA, 1997.

TOPAZ.   The mineral topaz is a silicate of aluminum and fluorine corresponding to the formula $Al_2SiO_4(F, OH)_2$. It is orthorhombic and its crystals are mostly prismatic terminated by pyramidal and other faces, the basal pinacoid being often present. Massive varieties are known. It has an easy and perfect basal cleavage, hence for this reason gems or fine specimens should be handled with care to avoid developing cleavage flaws. The fracture is conchoidal to uneven; hardness, 8; specific gravity, 3.4–3.6; luster, vitreous; color, of typical topaz, wine or straw-yellow but may be colorless, white, gray, green, blue or reddish-yellow; transparent to translucent. When heated, yellow topaz often becomes a reddish pink.

Topaz is found associated with the more acid rocks of the granite and rhyolite type and may occur with fluorite and cassiterite. Topaz comes from many localities, a few of which are: the former U.S.S.R. in the Urals and the Ilmen Mountains; the Czech Republic and Slovakia, Saxony, Norway, Sweden, Japan, Brazil and Mexico. In the United States, topaz has been found in Oxford County, Maine; Carroll County, New Hampshire; Fairfield County, Connecticut; El Paso and Chaffee Counties, Colorado; and in Texas, Utah and California.

The name *topaz* is derived from the Greek meaning to seek. It was the name of an island in the Red Sea that was difficult to find, and from which a yellow stone, now believed to be a yellowish-olivine, was obtained in ancient times. In the Middle Ages, any yellow stone was called topaz, but now the name is properly applied only to the species here described.

TOPOCENTRIC. Of measurements or coordinates, referred to the position of the observer on the Earth as the origin.

TOPOGRAPHICAL MAPPING. Topographical mapping consists of representing on a map the physical features of a given section of land, by showing thereon contours or hachures representing the elevation, and noting various other physical features by conventionalized signs. Thus trees, streams, marshes, and roads may be part of a topographic survey. The method of topographic surveying and mapping of large areas consists of establishing points for horizontal and vertical control, and surveying the adjacent area from these points by the use of leveling, stadia, or the plane table.

TOPOLOGICAL GROUP (or Continuous Group). A group for whose elements a topology is defined in such a way that the group operations (formation of product and of inverse) are continuous. Thus, the set of real numbers forms a topological group if the group operation is given by addition and the distance between two numbers by the absolute value of their difference. Consider, e.g., the (one-parameter) group of rotations of the Euclidean plane given by the elementary formulas

$$X = x \cos\theta - y \sin\theta, \quad Y = x \sin\theta + y \cos\theta$$

corresponding to the group of matrices

$$\begin{pmatrix} \cos\theta & -\sin\theta \\ \sin\theta & \cos\theta \end{pmatrix}$$

depending on the parameter θ. If we now define two matrices as being close together if every element of one is close to the corresponding element of the other (see under **Topology**), then it is clear that multiplication of matrices is continuous, i.e., that if $A = A(\theta_1)$ is a fixed matrix and $B = B(\theta_1)$ is a variable one, then small changes in B will produce only small changes in AB.

Most groups which occur in applications are either finite or are topological groups, and often express the symmetry of a given system. Most useful is the full linear group of all $n \times n$ real or complex matrices and certain distinguished subgroups such as the special linear group, the orthogonal group, etc.

TOPOLOGICAL SPACE. An abstract set with a structure based on the notion of nearness sufficient for the study of the limit concept.

A topological space is a set X along with a collection $\mathcal{J}$ of distinguished subsets of X called *open* sets which satisfies: (1) each x in X is contained in some open set, (2) the intersection of two open sets is open, and (3) the union of a subcollection of open sets is open.

A subset K of X is said to be *closed* if the complement $X - K$ is open. The intersection of a collection of closed sets is closed and the smallest closed set containing a subset M of X is called the *closure* of M and is denoted by $\overline{M}$. A point x is said to be an *accumulation* or *limit point* of a set M if x is in the closure of the set obtained by deleting x from M.

It is usually assumed that a topological space has additional separation properties. In particular, a topological space is said to be a *Hausdorff space* if distinct points are contained in disjoint open sets.

Most of the familiar examples of topological spaces are metrizable. See also **Metric Space**. Not all topological spaces are metrizable, however.

A sequence $x_1, x_2, x_3, \ldots$ of points in a topological space is said to converge to a point x if for every open set U containing x, there exists an integer N so that for every $n > N$ the point x_n is in U. The notion of sequential convergence can be used to provide an alternate definition of metrizable topological spaces in terms of accumulation points but a more general notion of convergence is necessary for general topological spaces.

A mapping f between topological spaces X and Y is said to be continuous if for each point x in X and open set U in Y containing $f(x)$, there is an open subset V of X so that the image of V is contained in U. A continuous mapping can bend, can twist, and can change lengths, angles and shapes but cannot tear or break. If the mapping f is also one-to-one, and onto itself, and the inverse of f is continuous, then f is said to be a *homeomorphism* or a *topological transformation*. The mapping $f(x) = \tan x$ is a homeomorphism between the open interval $(-\pi/2, \pi/2)$ and the real line. Topological spaces are said to be homeomorphic if there

exists a homeomorphism between them. A property of a space is said to be a *topological property* if every homeomorphic copy also possesses it. Being a Hausdorff space is a topological property, while being a triangle is not.

A topological space is said to have a property *locally* if some open set about each point has this property. For example, a space is said to be *locally Euclidean* if each point is contained in an open set which is homeomorphic to an open subset of Euclidean space. If in a locally Euclidean space overlapping open sets match up properly, the space is said to be a *manifold*.

Usually, the collection of possible states of a physical system possess a natural topology, that is, there is a notion of two systems being "close." Further, the collection of all states having a given energy forms a submanifold.

R.G. DOUGLAS, State University of New York at Stony Brook, Stony Brook, NY.

TOPOLOGY. A branch of mathematics which studies the *topological properties* of figures. See Fig. 1 on p. 3506. Although certain topological problems had been studied previously, the subject reached maturity in this century and has become a separate field of study.

Two basic problems in topology can be stated very generally as follows: (1) Given topological spaces X and Y, when are they homeomorphic? (2) Given subsets A of X and B of Y and a continuous mapping f from A to B, when does there exist a mapping F from X to Y which agrees with f on A? Questions as general as these have complete answers only in very special cases.

An example of the first problem is obtained by considering a circle, a square, a triangle, and a figure-eight; any two of the first three are homeomorphic, while the latter is homeomorphic to none of the others. An example of the second problem is obtained by considering two simple loops (non-self-intersecting) in the plane; the loops are homeomorphic and the *Jordan curve theorem* states that there exists a homeomorphism on the plane which extends this mapping. This implies, in particular, that a simple loop separates the plane into an inside region and an outside. On the other hand, if the similar problem is considered in three-space with one loop being knotted and the other with no knot, then the loops are homeomorphic, but the homeomorphism between them can not be extended to a homeomorphism on three-space.

Topology is composed of several branches. *General topology* is the study of the most abstract topological spaces and is related to certain parts of functional analysis. *Point-set topology* is the study of arbitrary subsets of Euclidean spaces, mainly of dimensions two, three and four. *Differentiable topology* is the study of manifolds with sufficient additional structure to define the notion of differentiable mapping between such manifolds. The most important examples of *differentiable manifolds* are Euclidean n-space and the Euclidean n-sphere, which consists of all points in $n + 1$-space at distance one from the origin.

Combinatorial topology is the study of geometric objects by decomposing them into simpler objects (simplexes) which are adjoined to each other in a regular way. A triangle is decomposed into three line segments or 1-simplexes. *Algebraic topology* is the study of topology using algebraic methods. There are many natural ways of associating with a topological space some algebraic object such as a group so that homeomorphic spaces have isomorphic groups. The first problem mentioned has a negative answer in case the groups associated with the spaces X and Y are not isomorphic. See also **Four-Color Map Theorem**.

R.G. DOUGLAS, State University of New York at Stony Brook, Stony Brook, NY.

TOPSOIL. See **Soil**.

TORBERNITE. An ore of uranium with the composition, $Cu(UO_2)_2$ $(PO_4)_2 \cdot 8-12H_2O$, green, radioactive, tetragonal, and isomorphous with autunite. Occurring in tabular crystals or in foliated form, the mineral is commonly a secondary mineral.

TOROIDAL COORDINATE. A curvilinear coordinate system closely related to bipolar coordinates. If the traces of the surfaces in that system are taken in the XY-plane as two families of mutually orthogonal circles, then rotation of the circles about the Z-axis forms a family of spherical surfaces and a family of anchor rings. The toroidal coordinate surfaces are

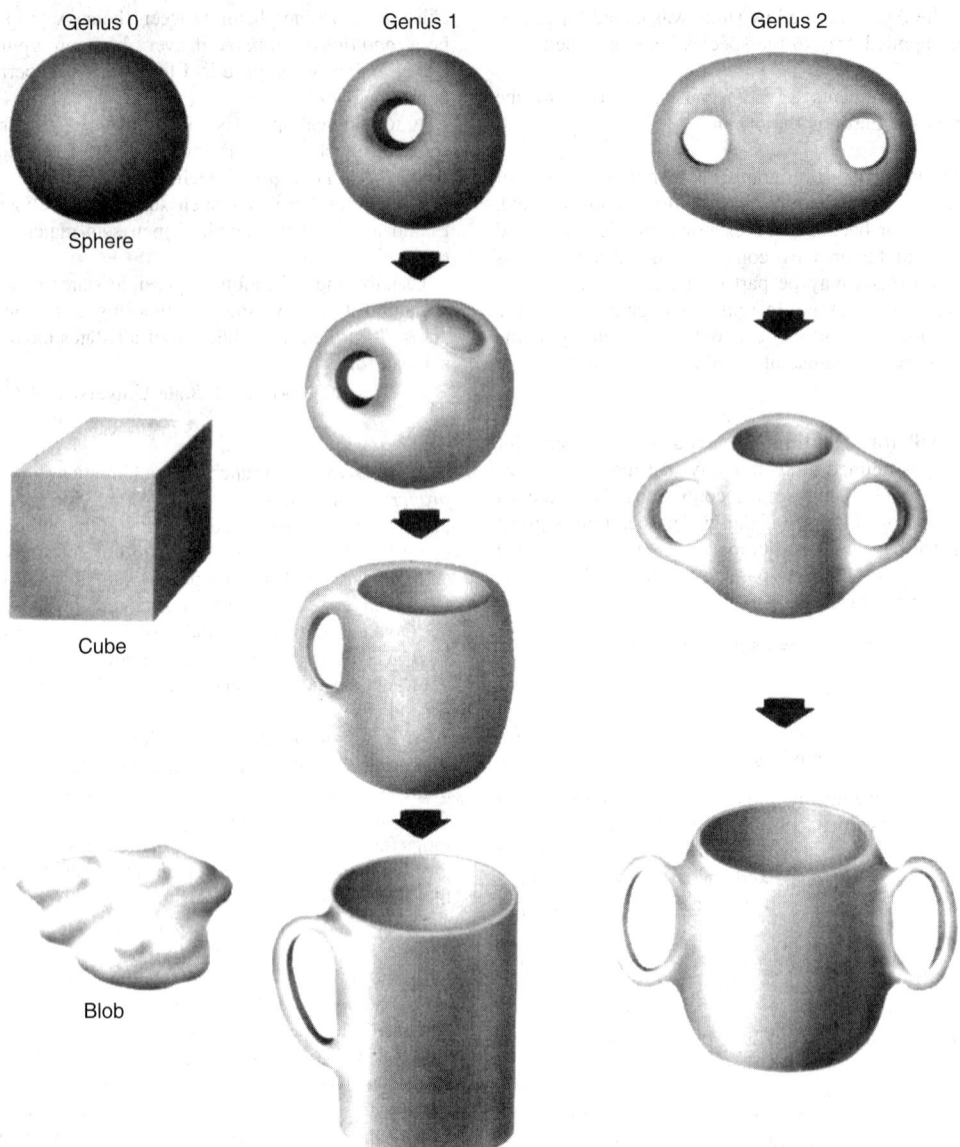

Fig. 1. Topological transformation. The sphere, cube, and blob are basically the same topologically and all of Genus 0. The doughnut typifies Genus 1 and can be transformed into innumerable shapes and still retain the topological basic features of Genus 1. Similarly a Genus 2 surface can be transformed into many different shapes. One may envision all of the foregoing from a plastic substance, such as clay. The principal differences between the classes are (0) no hole; (1) one hole; (2) two holes.

then taken as: (1) spherical, with centers on the Z-axis at a distance of $\pm a \cot \xi$ from the origin and with radii of $a \csc \xi$, $\xi =$ const.; (2) anchor rings with radii $a \coth \eta$ for the axial circles and circular cross sections of radii $a \csch \eta$, $\eta =$ const.; (3) planes through the Z-axis, $\psi =$ const., where $\psi = \tan^{-1} y/x$. These coordinates are related to rectangular coordinates by the equations

$$x = r \cos \psi, \quad y = r \sin \psi$$

$$r = \frac{a \sinh \eta}{\cosh \eta - \cos \xi}; \quad z = \frac{a \sinh \xi}{\cosh \eta - \cos \xi}$$

$0 \le \xi \le 2\pi$; $0 \le \psi \le 2\pi$; $0 \le \eta \le \infty$.

See also **Coordinate System**; and **Spheroidal Coordinate**.

TORQUE. A torque is a concept that may be simply expressed as the effectiveness of a force in setting a body into rotation. 1. For a single particle the torque is the moment of the resultant force on the particle with respect to a particular origin. This is expressed by the vector relation $\mathbf{L} = \mathbf{r} \times \mathbf{F}$, where $\mathbf{L}$ is the torque, $\mathbf{r}$ is the position vector with respect to the origin, and $\mathbf{F}$ is the resultant force. The torque is equal to the time rate of change of the moment of momentum. 2. For a rigid body, the torque

with respect to a set of axes is expressed by the relation

$$\mathbf{L} = \int_v \mathbf{r} \times \mathbf{F}_v \, dv$$

where $\mathbf{F}_v$ is the resultant force per unit volume due to external forces on the element dv, and $\mathbf{r}$ is the position vector of the volume element.

For a rigid body undergoing free rotation about a single axis, the torque $\mathbf{L} = I\alpha$, where I is the moment of inertia and α is the angular acceleration. 3. In engineering mechanics usage, a torque often refers to the torsional or twisting moment or couple which tends to twist a rigidly fixed object such as a shaft about an axis of rotation. In Fig. 1, a shaft of diameter d is connected to a rigid support. The forces $\mathbf{P}$ form a couple with a torque $\mathbf{L} = \mathbf{P}d$ and tend to rotate the shaft in a counterclockwise direction. See also **Moment**.

TORQUE AMPLIFIER. A device possessing input and output shafts and supplying work to rotate the output shaft, without imposing any significant torque on the input shaft. The speed of the output shaft is equal or proportional to that of the input shaft, regardless of the load on the former.

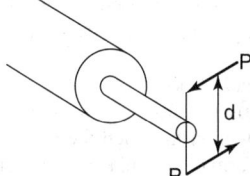

Fig. 1.   Parameters of torque.

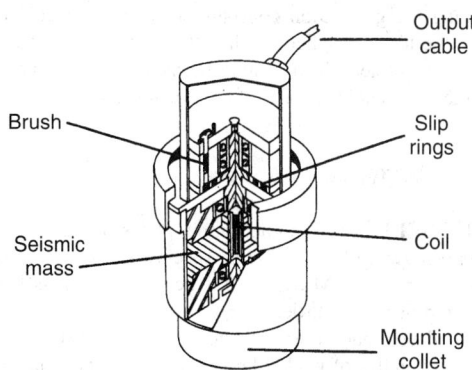

Fig. 1.   Sectional view of seismic mass-type transducer for measurement of torsional vibration.

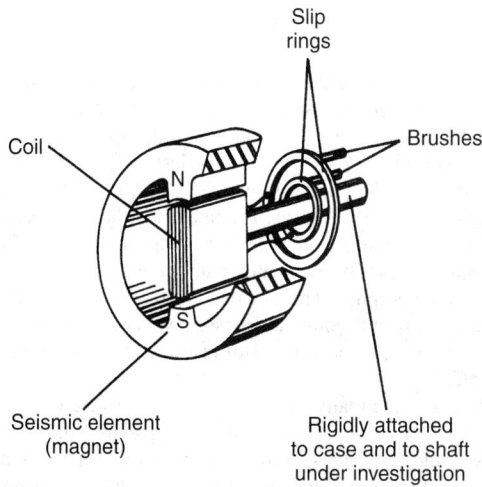

Fig. 2.   Mounting of seismic mass-type transducers for torsional vibration measurement.

TORQUE CONVERTER.   A device for changing the ratio of torque to speed in a system transmitting power by exerting torque. The most general case is that of two shafts connected by a gear system or a clutch. In the case of the former, or of the latter if it is of positive or of friction type, the torque in the driven shaft, T_d, is equal to $T_i(N_d/N_i)$, where T_i is the torque of the input shaft, N_i the rotational speed of the input shaft, and N_d is that of the driven shaft, assuming 100% efficiency. If the efficiency is less, T_d is decreased proportionately.

Among the other types of torque converters, the hydraulic type is most widely used. It consists essentially of an impeller mounted on the input shaft which pumps fluid so that it impinges on the blades of a turbine mounted on the output shaft. On leaving the turbine the fluid meets the blades of a reactor that is attached to the outer casing, and the fluid is thus returned into the impeller.

A number of modifications of this simple arrangement are found in practice. The torque converter may consist of more than one stage; the reactor may be mounted so that it is free to rotate with the fluid at high speeds, since no power at all is transmitted when the shaft speed equals the engine speed (efficiency zero); and the reactor may be divided into sections to obtain more uniform efficiency.

TORR.   Provisional international standard term to replace the English term *millimeter of mercury* and its abbreviation *mm of Hg* (or the French *mm de Hg*). The torr is defined as 1/760 of a standard atmosphere or 1,013,250/760 dynes per square centimeter. This is equivalent to defining the torr as 1333.22 microbars and differs by only one part in seven million form the International Standard millimeter of mercury. The prefixes *milli* and *micro* are attached without hyphenation.

TORREY PINE.   See Pine Trees.

TORSIONAL STRESS.   The shearing stress which occurs at any point in a body as the result of an applied torque or torsional load is called a torsional stress. If the body is a circular shaft, the stress is found from the formula

$$s_s = \frac{Tc}{J}$$

in which

  s_s = required unit stress
  T = torque
  c = radial distance from the center of the shaft to the point
  J = polar moment of inertia of the cross-sectional area

This formula is based on the assumption that a plane section before twisting remains a plane after the torque is applied; also that the radii remain straight. The theory of elasticity shows that these assumptions are true for all sections except those adjacent to the applied torque and the supports.

The cross section of a noncircular body becomes decidedly warped after twisting, so the stress must be obtained by formulae developed from the theory of elasticity, by experimental means or by approximate formulae which are given in textbooks on advanced strength of materials.

TORSIONAL VIBRATION MEASUREMENT.   Variations from uniform shaft speed reduce the efficiency of engines and machinery. Thus, it is important that reciprocating engine performance be evaluated and that machine tool vibration, as two examples, be analyzed and corrected. A seismic mass type transducer is commonly used for such analyses. Sectional and mounting views of such a device are shown in Figs. 1 and 2.

Internal construction of the device features an *Alnico V* magnet seismic mass suspended on ball bearings with no end play. Built to withstand transverse and longitudinal vibratory movements, the seismic mass is restricted to torsional vibration. Springs provide an elastic coupling to the case. The coil, rigidly affixed to the case, is oriented in the field produced by the magnet (seismic mass) case. Any relative motion between the two generates a voltage which is a function of the instantaneous velocity of this angular vibration. Solid silver slip rings and dual sets of silver-graphic brushes assure conduction of transducer coil output. Precision bearings and a silicone damping fluid acting as a lubricant and combine to give the device excellent amplitude linearity and no tendency to stick at low frequencies and amplitudes.

The instrument is designed to measure vibration precisely even when superimposed on modes other than angular. Output voltages resulting from shaft speed variations can be fed to a vibration meter for direct indications of vibrational angular velocity and amplitude. Or, the output can be fed to a recording oscillograph or to an oscilloscope. The device shown is suitable for applications up to a maximum of 6,000 revolutions per minute, and has a frequency response of 10 to 1,000 Hz, a natural frequency of about 3 Hz, a coil resistance of 700 ohms, and a temperature range of 0–150 °F (−17.8–66 °C).

TORSION BALANCE.   The torsion balance is an instrument for measuring very feeble forces of attraction or repulsion. It has played an important role in the demonstration of Coulomb's laws and in measuring the gravitation constant and radiation pressure. It consists essentially of a light horizontal rod suspended by a slender elastic wire or fiber and carrying at each end a small ball, vane, or other object upon which the unknown force is to act tangentially. The resulting torsional deflection is measured by an optical lever, the small mirror of which is attached at the base of the suspension. To calculate the corresponding torque (and hence the force), the "torsion coefficient" of the suspension must have been previously determined. This is the torque required to twist the suspension through an angle of 1 radian, and is conveniently obtained by using the same suspension as the support for a torsion pendulum of known moment of inertia, and measuring the period. In the Cavendish method for determining the gravitation constant, small balls of gold or platinum are used and the torque applied by the attractions of much more massive balls or cylinders of lead in the same horizontal plane. The Eötvös balance is a highly specialized form of the torsion balance, used in gravity measurements.

TORTOISE. See **Turtles**.

TORTOISE SHELL. The mottled horny plates of the shell of a marine turtle, *Eretmochelys imbricata*. This species is found in warm seas, extending as far north as Massachusetts occasionally. It is known as the hawksbill or tortoise-shell turtle.

Tortoise shell was once highly valued for ornamental use, as in toilet articles and the handles of pocket knives, but it has been largely replaced by synthetic materials.

TOUCAN. See **Woodpeckers and Toucans**.

TOUCH (Sense). See **Tactile Organs**.

TOURMALINE. The mineral tourmaline is a complex silicate of aluminum and boron, but because of isomorphous replacements this mineral varies widely in chemical composition, iron, magnesium, and lithium entering into combination to a greater or less extent with the aluminum and boron. Its general formula is $(Na, Ca)(Mg, Fe^{2+}, Fe^{3+}, Al, Li)_3Al_6(BO_3)_3(Si_6O_{18})(OH, F)_4$. Tourmaline belongs to the hexagonal system, its crystals are usually prismatic, tending to be long and slender, often acicular. The crystals are ordinarily terminated with three faces of a rhombohedron and usually hemimorphic. The smaller crystals are frequently found in radial arrangement, and columnar masses are common. The prisms are usually three-, six-, or nine-sided with heavy vertical striations producing a rounded effect.

Tourmaline is essentially without cleavage; fracture, conchoidal to uneven; brittle; hardness, 7–7.5; specific gravity, 3.03–3.25; luster, vitreous inclining to resinous; color, in common tourmaline black, bluish-black, brown, blue, green, red or pink, and in the transparent varieties colorless (rare), various shades of rose and pink, greens, blues and browns. The color arrangement in tourmaline is of considerable interest; bicolored crystals are common and may be green at one end and pink at the other, or green on the outside, and pink within, which, in the case of transparent or translucent crystals, is very attractive.

The opaque black tourmaline is called schorl, a term which was applied to all tourmaline until 1703 when the word *tourmaline* was introduced, it being a corruption of the Ceylonese word, *turamali*. The origin of the word *schorl* is not known, but is perhaps Scandinavian, and is used to identify the iron-bearing black tourmalines; elbaites and liddicoatites tend to light shades of blue, red, green, and their bicolored combinations; the brown colored tourmalines of varying shades of dark brown to yellow to nearly colorless are called dravites and uvites (with the exception of the black tourmalines found at Pierrepont, New York, which have been identified as uvites); the completely colorless variety, achroite, falls within the elbaite group. Small tourmalines are found in granites and some gneisses.

Due to the mineralizing action of magmatic vapors, tourmaline is found particularly well developed in pegmatites, and as a contact metamorphic mineral. A few of the important localities are: the Ural Mountains; Bohemia; Saxony; the Island of Elba; Norway; Devonshire and Cornwall, England; Greenland; Madagascar. Magnificent elbaite crystals are obtained from Madagascar, Brazil, and Afghanistan; liddicoatite crystals from Madagascar. In the United States in Oxford and Androscoggin Counties, Maine; Grafton and Sullivan Counties, New Hampshire; Hampshire County, Massachusetts; Haddam and Fairfield Counties, Connecticut; St. Lawrence County, New York; Sussex County, New Jersey; Delaware County, Pennsylvania; and San Diego County, California.

See also terms listed under **Mineralogy**.

ELMER B. ROWLEY, Union College, Schenectady, NY.

TOWNSEND AVALANCHE. A term used in gas-filled counter technology to describe a process which is essentially a cascade multiplication of ions. In this process an ion produces another ion by collision, and the new and original ions produce still others by further collisions, resulting finally in an "avalanche" of ions (or electrons). The terms "cumulative

ionization" and "cascade" are also used to describe this process. It occurs in a nonself-maintained gas discharge, where ions have sufficient energy.

TOXEMIA. The spread of poisonous substances, either those produced by the body cells or those resulting from the growth of microorganisms. There is generalized infection in which the blood contains toxins, but not necessarily bacteria. A form of toxemia arises in certain pregnancies. A rise in blood pressure is often the earliest symptom of toxemia resulting from a kidney disturbance. Analysis of the urine demonstrates the presence or absence of albumen which also may be indicative of toxemia. The cause of toxemia during pregnancy is not fully understood, but it is usually accompanied by accumulations of salt and water with a marked tendency to retain salt.

TOXICOLOGY. The technology of poisonous substances—their detection, and counteractions. Basic to this branch of science is the realization that chemical compounds vary in their danger to humans and their environment. Poisons can be simple or complex, inorganic or organic chemical compounds, bacterial or viral byproducts (toxins), animal-produced substances, such as venom—all of which produce ill effects on humans ranging from a low level of debilitation to almost instant death. Drugs used in countering diseases or physiological deficiencies can, in some doses, act as poisons. Many metals or elements are essential for life, but their body concentration for optimum health varies from element to element and depends somewhat on bodily weight. Once these optimum concentrations are exceeded, the metals or elements become contaminating, polluting, and often harmful.

Levels of toxicity ratings range from unknown, through low or light toxicity to moderate and severe. Exposure may also be acute, sub-acute, or chronic. Dusts, fumes, mists, vapors, gases or liquids may be absorbed through the skin, orally, or through the lungs.

Sources of information pertaining to toxic substances include local and national health organizations in many countries. Several treatises on the subject have been prepared, including the broad spectrum "Dangerous Properties of Industrial Materials," compiled by Sax (Van Nostrand Reinhold, New York).

R. C. V.

TOXOID. A toxin treated to destroy its toxicity, but leaving it capable of inducing formation of antibodies.

TOXOPLASMOSIS. An infection with *Toxoplasma gondii*, a coccidian protozoan parasite of cosmopolitan distribution. Asexual reproduction of the parasite occurs in many species of mammals and birds, but sexual reproduction seems to occur only in cats. The mode of transmission to humans is unclear, but may be transplacental, by consumption of undercooked meat containing *T. gondii* cysts, or by food contamination by *T. gondii* oocysts from cat feces. The trophozoites are liberated in the human intestine, penetrate the intestinal mucosa and invade the blood stream. Subsequently, they infect other organs where they multiply and may eventually encyst. In human cases, the incubation appears to be 10–23 days. The organisms are obligate intracellular parasites, which may destroy the host cell. Muscle and liver are especially affected. Serologic evidence indicates that most cases are asymptomatic but a few develop a systematic, mononucleosislike illness. Intrauterine infection is most dangerous and carries the risk of head deformity, mental retardation, and blindness. In the immunosuppressed patient, *T. gondii* can be an opportunistic pathogen, causing devastating neurological or disseminated disease. In many instances, this syndrome appears to result from defective cell-mediated immunity, which allows the reactivation of latent infection. Neurological abnormalities predominate in at least 50% of these patients. Diagnosis is by inoculation of biopsy material into susceptible animals, embryonated eggs, or tissue culture. The indirect fluorescent antibody test is the most widely used serologic test. The Sabin-Feldman dye test is also useful but requires the use of living toxoplasmas. Treatment consists of a 1 month combined regimen of pyrimethamine and trisulfapyrimidines. Prevention consists of careful handling and cooking of meat. A temperature of 150 °F (66 °C) should be reached throughout the meat. Pregnant women should be especially careful to avoid exposure to cat litter pans or contaminated soil.

R.C. VICKERY, M.D., D.Sc., Ph.D., Blanton/Date City, Florida.

TRACE (Mathematics). The sum of the diagonal elements of a matrix, indicated by $Tr\ \mathbf{A} = \Sigma A_{ii}$. Its properties include $Tr\ \mathbf{AB} = Tr\ \mathbf{BA}$; $Tr\ \mathbf{C} = Tr\ \mathbf{A} \cdot Tr\ \mathbf{B}$, where $\mathbf{C} = \mathbf{A} \times \mathbf{B}$, the direct product. It is also called *Spur* from the German word. When a matrix is a representation of a group, its trace is called the *character* of the representation. If $\mathbf{A}$ and $\mathbf{B}$ are two matrices related by a similarity transformation, so that $\mathbf{B} = \mathbf{Q}^{-1}\mathbf{AQ}$, then $Tr\ \mathbf{A} = Tr\ \mathbf{B}$.

TRACER (Radioactive). see **Radioactivity**.

TRACHEA. 1. An air tube of the arthropod respiratory system. These tubes open at the surface of the body through spiracles and lead inward, branching extensively and in some species expanding to form air sacs. At their inner extremities they connect with very fine tubules called tracheoles, of independent origin. Tracheae are ingrowths of the outer layer of the body and are lined with a continuation of the cuticula, which forms spiral rods in their walls. Near the spiracle they are often provided with a muscular closing device by which air can be excluded if it contains harmful materials. The tracheoles form within cells of the lining of tracheae, later breaking out and assuming their connection with the larger tubes. They have no spiral supporting rods (taenidia).

2. The principal air tube of the vertebrate respiratory system, also called the wind pipe. It leads from the pharynx to the major branches (bronchi) connecting it with the lungs, and its wall is supported by rings of cartilage. At the pharyngeal end in amphibians and mammals it forms an expanded larynx containing vocal cords, and at the point where it forks to form the bronchi in birds the remarkable vocal organ called the syrinx is developed.

TRACHEOSTOMY (Tracheotomy). For relieving certain types of air-way obstruction, a surgical opening is made in the forward wall of the windpipe (trachea). A tube is inserted through this small aperture to provide access to an air supply for the patient. The procedure is used in life-saving situations, particularly in the case of children where complete closure of the larynx may occur. Other conditions which sometimes require the procedure include choking in connection with eating, injuries to the chest, poliomyelitis, extensive burns, and severe respiratory infections. Once the underlying cause of the injury or disease is ascertained and normal breathing can be restored, the tube is removed, and the aperture is closed. Usually the wound will close itself in a relatively short period without any after-effects.

TRACHOMA. Trachoma is a chronic infection of the conjunctiva, the thin membrane that covers the outer surface of the eye and the inside lining of the eyelids. It is caused by a microorganism that grows only within eyelid cells, and it is one of the oldest infectious diseases, dating back several thousand years. Because the disease is not fatal and causes progressive rather than sudden damage to the eyes, it is often not taken seriously and may even be accepted as a fact of life in some countries. The trachoma infection is the leading cause of blindness in the world.

The disease is rare in the United States and is most often seen in third-world and developing nations where poverty, overcrowding, personal hygiene, and access to clean water and health care are problems. It is prevalent in many African countries, parts of Central and South America, and some countries in the Eastern Mediterranean and in Asia.

Trachoma is caused by infection with the microorganism *Chlamydia trachomatis* and is highly infectious, usually spread by contact with eye or nose discharge from an infected person. It is also transmitted by flies and gnats, which are attracted to eyes and runny noses. Because it is easily transmitted from person to person, the disease often strikes entire communities.

In the early stages, trachoma resembles conjunctivitis. It starts with an inflamed outer lining of the eye (conjunctiva) and is accompanied with eye discharge and sticky, red eyes. Tearing, light sensitivity, swollen eyelids, pain, and corneal inflammation are some of the first symptoms. As the disease progresses over time, the recurring infections scarring on the inside of the eyelid. Eventually the inside of the eyelids cause are scarred so badly that they turn inward with the eyelashes rubbing the eyeballs, scratching the lenses of the eyes with each blink and causing corneal ulcerations that become infected. This scarring leads to blindness. Although trachoma begins in childhood, blindness usually does not occur until the age of 40 or 50 after repeated infections cause inflammation and scarring.

In the early stages of the infection, antibiotics can effectively treat trachoma. However, if the disease progresses, as is often the case in underdeveloped countries, treatment becomes difficult. After scarring begins, a simple surgical procedure known as tarsal rotation is required to reverse the inturned eyelashes. Surgeries to repair lid deformities and corneal transplants are also options, but are not available to most people who are afflicted by the disease. Trachoma is preventable with proper diet, sanitation and education. By simply washing their faces and hands, children will not spread the infection.

Although blindness from trachoma has been eliminated in several countries around the world, the disease continues to be a serious public health problem in many underdeveloped areas. See also **Conjunctiva**; and **Vision and the Eye**.

Vision Rx, Inc., Elmsford, NY.

TRACHYTE. The name of an extrusive, igneous, fine-grained or porphyritic rock, the surface equivalent of syenite. Trachyte is predominant in alkali feldspar and usually contains biotite and augite. Trachyte is an old name, proposed by Brongiart in 1813, and has never been altered or supplanted. It is derived from the Greek, meaning rough.

TRACKING. In radio circuits, this term is used to designate the maintenance of a relation between the resonant frequencies of two or more circuits tuned from the same shaft. Thus in the superheterodyne receiver the oscillator tuning must track with the tuned radio frequency circuits so the difference in frequency will always be the constant intermediate frequency.

In gunnery the term is used to designate the following of the target by the director or gun.

With radars used for following a target, tracking may be manual, where the operator turns azimuth and elevation angle controls in such a manner as to realize a maximum returned signal from the target; or it may be automatic, where a narrow conical or pencil-shaped beam is electronically "locked" on the target.

In correlation tracking and ranging, a nonambiguous trajectory-measuring system for aircraft and space vehicles, using short-baseline, single-station, continuous-wave phase-comparison measuring two direction cosines and a slant range, is used. In correlation tracking and triangulation, a trajectory measuring system for aircraft and space vehicles, and several antenna baselines, each separated by large distances, are used to measure direction cosines to an object. From these measurements, space position is computed by triangulation.

TRACK (Navigation). A term used in navigation to describe the direction and distance a ship has moved in a given length of time. If a ship is known to be in a definite position at a given time and to be in another position after the elapse of a definite period of time, then the direction of the rhumb line between the two points is the track, and the length of the lines gives the track distance. Other lines, such as Lambert, great circle, etc., might be computed between the two points, yielding the Lambert track, but the term track is usually restricted to the rhumb line. It should be carefully noted that the positions at the extremes of the track must be accurately known. These may be determined by any reliable fix or, in the case of airplanes, by landmarks beneath the ship. See also **Course**; **Heading**; and **Navigation**.

TRACTION. In a narrow sense, traction refers to the friction developed between a powered surface and one in contact with it. The most common example of traction is the resistance to slipping developed at the point of contact of a driven wheel with the surface on which it rolls. The locomotive driver on its rail, the pneumatic tire on the highway, or the traction engine wheel on earth, illustrate this case. But traction is not confined to wheels, as may be proved by citing another example. The traction sheave of an elevator is a grooved pulley, power-driven, over which passes a rope which is driven by the sheave. The friction between the rope and pulley constitutes the traction.

In a larger sense, traction includes the act of pulling a load over a surface by overcoming the resistance to motion, and the word may also be used descriptively of any vehicle which by its excess of power over its own tractive needs is able to pull other vehicles. In the case of an automobile rolling on a highway, traction is applied through the rear wheels by means of a live axle which delivers a torque to the wheel. The load on the wheel presses it against the road surface enough to develop sufficient friction so that the resistance to motion will be overcome before the wheel slips on the road. If the roadbed be covered with a surface of low frictional power, for example, wet clay, there will be little traction developed because of the low coefficient of friction. The traction of an automobile must overcome wind resistance, wheel bearing friction, rolling resistance, and grade.

TRACTRIX OF HUYGENS.   A higher plane curve of transcendental type. Its equation is the solution of a differential equation resulting from the following problem in mechanics. A rope of length a, attached to a massless particle, moves along the X-axis. The path of the particle in the XY-plane is determined by the differential equation $(y^2 - a^2)p^2 + y^2 = 0$, where $p = dy/dx$. The solution, which is the equation of the tractrix, is

$$\pm x = \sqrt{a^2 - y^2} - a \ln \left\{ \pm \frac{a + \sqrt{a^2 - y^2}}{y} \right\}$$

The evolute of the tractrix is a pair of catenaries, one lying above the X-axis and the other below. There are four branches to the curve, with cusps at $(0, \pm a)$. The X-axis is an asymptote.

See also **Catenary**; and **Curve (Higher Plane)**.

TRADE WINDS.   See **Winds and Air Movement**.

TRAGACANTH.   See **Gums and Mucilages**.

TRAGOPANS (*Aves, Galliformes*).   The tragopans (subfamily *Tragopaninae*) are fairly large, compact, and very colorful *Galliformes*. The tail is short and rounded. There are 18 tail feathers. The tarsi have a short spur in the male. In the male, the throat and side of the face are bare. The male also has a distensible skin flap at the throat which can be inflated in excitement along with hornlike, stalk-shaped erectile bodies on the back of its head. There is 1 genus (*Tragopan*) with 5 closely related species. See Fig. 1.

Fig. 1.   Satyr tragopan (*Tragopan satyra*). (*Sketch by Glenn D. Considine.*)

Like the curassows and guans of tropical America, the tragopans spend a great deal of time on the branches in the crowns of trees where they find much of their food. They mainly eat tender leaves, buds, berries, fruit, and lesser amount of animal food. They are unbelievably shy and have a very acute sense of hearing. At the slightest suspicious sound they slink off like cats, using every bit of cover. Only rarely do they leave the protective darkness of shady forests and, at the most, take a sunbath in lightly wooded areas at noon. Their flight is swift and accompanied by a whirring noise. But in general, tragopans are rather disinclined to fly. See also **Galliformes**.

TRAGULINES (*Mammalia, Artiodactyla*).   A very small group of animals in the order of *Artiodactyla* (even-toed hoofed animals). The Chevrotains, sometimes called Mouse-Deer, appear to be leftovers from an ancient and quite diverse number of ruminating animals. It appears that the ancestors of these small mammals were also the ancestors of other groups in *Artiodactyla*, including the bovines, caprines, cervines, antelopines, and giraffines. Mouse-deer is a misleading term because there is no direct connection between the chevrotains and the deer in particular. Two types of chevrotains are recognized: (1) The Oriental Chevrotains (*Tragulus*); and (2) the Water-Chevrotains (*Hyemoschus*).

There are several species of the Oriental Chevrotains, appearing in southeast Asia, ranging from India to the Philippines. The Indian chevrotain

(*T. meminna*) is a slender delicate animal, about the size of a rabbit. Its fur is red-brown with small white spots, with lighter fur underneath. The animal is nocturnal and prefers the tangled undergrowth. The Water-Chevrotains occur in central Africa. They are semi-aquatic. Some of the Oriental Chevrotains have been domesticated.

TRAJECTORY (of a Path).   A curve in space tracing the points successively occupied by a particle or body in motion. At any given instant the velocity vector of the object is tangent to the trajectory. In the steady-state flow of aggregates of particles, such as parcels of fluid, the trajectories and streamlines of the fluid parcels are identical. Otherwise, the curvature of the trajectory K_T is related to the curvature of the streamline K_S by Blaton's formula

$$K_T = K_S - \frac{1}{V}\frac{\partial \psi}{\partial t}$$

where V is the parcel speed and $\partial \Psi/\partial t$ is the local change of its direction. Note that the use of the term *trajectory* for the path of a body implies that its motion is the result of an externally applied force (i.e., not an engine or propellant in the body itself), and that it usually implies also an open path.

TRALLES SCALE.   See **Specific Gravity**.

TRANSACTINIDE ELEMENTS.   See **Chemical Elements**.

TRANSACTINIUM EARTHS.   A group of chemical elements more frequently termed the Actinides. In order of increasing atomic number, they include actinium, thorium, protactinium, uranium, neptunium, plutonium, americium, curium, berkelium, californium, einsteinium, fermium, mendelevium, nobelium, and lawrencium. See also **Actinide Series**.

TRANSCENDENTAL.   A term applied to numbers, equations, or functions that are not algebraic. The word comes from the Latin, *scandere*, to climb, and it is intended to suggest only that transcendental operations cannot be defined by elementary methods ("*Quod algebrae vires transcendit*"). No special difficulty in understanding their properties is signified nor is there any religious or philosophical connotation to the word.

A transcendental number is not a root of a polynomial in one unknown, with integral coefficients. Typical examples are: e, the base of natural logarithms and π, the area of a circle with unit radius.

Some transcendental functions are: trigonometric and hyperbolic, with their respective inverses; exponential and logarithmic; beta and gamma; elliptic; Bessel; etc. They may be subdivided into a class defined as the solution to a differential equation, which is true for those listed in the previous sentence and one not so defined, like the Riemann zeta function.

Transcendental equations contain one or more transcendental functions. They cannot, in general, be solved by direct analytical means and some approximate method must be used.

See also **Number Theory**.

TRANSDUCER.   1. A device by means of which energy can flow from one or more transmission systems to one or more other transmission systems. The energy transmitted by these systems may be of any form (for example, it may be electric, mechanical, or acoustical), and it may be of the same form or different forms in the various input and output systems. 2. For some purposes, the transducer is defined (more narrowly) as a device capable of being actuated by waves from one or more transmission systems or media, and of supplying related waves to one or more other transmission systems or media. It is sometimes implied that the input and output energies shall be of different forms. For example, an electroacoustic transducer accepts electrical waves and delivers acoustic waves.

Among the types of transducers in addition to those designated by nature of energy change, such as electroacoustic or electromechanical transducers, are:

The *active transducer*, whose output waves are dependent upon sources of power, apart from that supplied by any of the actuating waves, which power is controlled by one or more of these waves.

The *conversion transducer*, an electric transducer in which the input and the output frequencies are different. If the frequency-changing property of a conversion transducer depends upon a generator of frequency different from that of the input or output frequencies, the frequency and voltage or power of this generator are parameters of the conversion transducer.

The *harmonic conversion transducer*, a conversion transducer in which the useful output frequency is a multiple or a submultiple of the input frequency. Either a frequency multiplier or a frequency divider is a special case of harmonic conversion transducer.

The *heterodyne conversion transducer*, a conversion transducer in which the useful output frequency is the sum or difference of the input frequency and an integral multiple of the frequency of another wave.

The *passive transducer*, whose output waves are independent of any sources of power which is controlled by the actuating waves.

The *ideal transducer*, a hypothetical passive transducer that transfers the maximum possible power from a specified source to a load. In linear electric circuits and analogous cases, this is equivalent to a transducer which (a) dissipates no energy and (b) when connected to the specified source and load, presents to each its conjugate.

TRANSFERASES. See **Enzyme**.

TRANSFER CHARACTERISTIC. In electron devices a relation, usually shown by a graph, between the voltage or current of one electrode and the current to another electrode, all other electrode voltages being maintained constant.

TRANSFERENCE NUMBER (or Transport Number). The transport number of a given ion in an electrolyte is the fraction of total current carried by that ion.

TRANSFER FUNCTION. The ratio of output response to input signal or a particular component or system is known as the transfer function. Although this ratio is usually expressed in a mathematical statement, it can be also be expressed in a frequency response plot. In the case of a simple resistive-capacitive type component, the transfer function may be expressed as follows:

$$\frac{E_0(s)}{E_i(s)} = \frac{1}{1 + Ts}$$

This is the Laplace transform of the general equation (see also **Time Constant**) for a first-order system with zero initial conditions. Similar transformations can be developed for more complex systems. In most cases transfer functions are expressed in Laplace notation.

TRANSFER ORBIT. See **Astronautics**.

TRANSFINITE NUMBER. An important concept in the modern mathematical idea of infinity, developed by George Cantor (1845–1918), a German mathematician. Consider the natural numbers 1, 2, 3, ..., which are said to be denumerably infinite and compare them with another set, also containing an infinite number of elements. The latter, too, is denumerably infinite if its members can be matched up one-by-one with the natural numbers. Thus, the even numbers form such a set, for the two sets could be matched up as follows: (1, 2), (2, 4), (3, 6), and so on. All sets of this kind are said to have the cardinal number *aleph-null*, written with the Hebrew character, $\aleph_0$ and this is the first transfinite cardinal number.

However, surprising as it seems at first, given a denumerably infinite set of real numbers between 0 and 1, there are real numbers in this range which are not members of the set. These are the transcendental numbers; the set of all real numbers is not denumerably infinite and Cantor called the cardinal numbers of this set, C, "*the power of the continuum*," which he believed was equal to $\aleph_1$. Mathematicians today, however, are uncertain about this and there may be other transfinite numbers between $\aleph_0$ and C. It is generally agreed that the following are some properties of these strange numbers (n is a finite number): (1) $\aleph_0 + n = \aleph_0$; (2) $\aleph_0 + \aleph_0 = \aleph_0$; (3) $n\aleph_0 = \aleph_0$; (4) $\aleph_0^n = \aleph_0$; (5) $\aleph_0\aleph_0^0 = C$; (6) $\aleph_0 + C = C$; (7) $\aleph_0 C = C$. It is also possible to conceive of cardinal numbers $\aleph_0, \aleph_1, \aleph_2, \ldots, \aleph_n$ where n is a natural number and then $\aleph_w, \aleph_{w+1}, \ldots$ larger in turn than any of the previous ones. See also **Infinity**.

TRANSFORM. 1. If A, B, X are three matrices or three elements of a group then $B = X^{-1}AX$ is the transform of A by X and A, B are conjugate to each other. The complete set of group elements which are conjugate to each other form a class of the group.

2. An integral equation of the first kind,

$$f(y) = \int K(x, y)F(x)\,dx$$

is called the transform of $F(x)$. For certain special cases, see also **Integral Transform**.

TRANSFORMATION (Mathematics). A change of variables in an algebraic expression. Consider a point (x, y) in a plane, referred to a rectangular coordinate system, XOY and let another coordinate system $X'O'Y'$ have its origin at the point (h, k). Then the coordinates of the point in the second system will be (x', y'), where $x = x' + h$, $y = y' + k$. Similarly, if another system $X''OY''$ has the same origin as XOY but the coordinate axes are rotated by the angle θ, the relations are $x = x' \cos\theta - y'' \sin\theta$, $y = x'' \sin\theta + y'' \cos\theta$. A more general case is an affine transformation, $x' = a_1 x + b_1 y + c_1$, $y' = a_2 x + b_2 y + c_2$. If the determinant of the coefficients, $\Delta = a_1 b_2 - a_2 b_1 \neq 0$, special cases of it are translations, rotations, reflections in the axes, strains, elongations, and compressions. Parallel lines or finite points are transformed into similar configurations; the line at infinity remains fixed. If $\Delta = 0$, the transformation is singular.

Matrix notation is frequently useful for discussing transformations. Consider now rectangular coordinate systems in space, $OXYZ$ and $OX'Y'Z'$. The same vector could be described in either system by the relation $\mathbf{x}' = \mathbf{Bx}$, where the elements of the matrix $\mathbf{B}$ are direction cosines for the various pairs of coordinate axes. Alternatively, one may have one coordinate system and rotate the vector $\mathbf{x}$ through the angle θ to obtain a new vector $\mathbf{y}$. A similar matrix equation results, $\mathbf{x} = \mathbf{Ry}$, but the elements of $\mathbf{B}$ and $\mathbf{R}$ are not identical.

Given a transforming matrix like $\mathbf{B}$ or $\mathbf{R}$, it is often the case, that still another coordinate system is desirable so that simpler relations exist between the two vectors. We then seek a transformation of the form $\mathbf{B} = \mathbf{PAQ}$, in which $\mathbf{A}$ and $\mathbf{B}$ are said to be equivalent. Important special transformations are: (a) $\mathbf{P} = \mathbf{Q}^{+1}$, collineatory or similarity transformation (German, *Ähnlichkeits-transformation*); (b) $\mathbf{P} = \tilde{\mathbf{Q}}$, congruent; (c) $\mathbf{P} = \mathbf{Q}\dagger$, conjunctive; (d) $\mathbf{P} = \mathbf{Q} = \mathbf{Q}^{+1}$, real orthogonal; (e) $\mathbf{P} = \mathbf{Q}\dagger = \mathbf{Q}^{+1}$, unitary. Case (c) is identical with (b) if the matrices are real; (d) is also case (b) and case (e) is also (c). For the meaning of the symbols designating the properties of $\mathbf{P}$, see also **Matrix (Mathematics)**.

If w is a function of the complex variable z, the linear fractional transformation is

$$w = (az + b)/(cz - d); (ad - bc) \neq 0$$

where a, b, c, d are constants, real or complex. It is a mapping or transformation in the complex plane. A circle in the z-plane is transformed into a circle in the w-plane. A straight line transforms in the same way, for it may be regarded as a special case of a circle. Since there are only three independent constants in the transformation, any three points z_1, z_2, z_3 are images of w_1, w_2, and w_3. Thus the points in the z-plane become 0, 1, ∞ by the transformation,

$$w = (z_2 - z_3)(z - z_1)/(z_2 - z_1)(z - z_3)$$

Other terms used for this type of transformation are MÄbius, general bilinear, and homographic transformation. See also **Conformal Representation**.

The equation of a polynomial is sometimes transformed in order to change the values of its roots.

TRANSFORMER. A device that raises or lowers the voltage of an alternating current; a static electric device consisting of a winding, or two or more coupled windings, with or without a magnetic core, for introducing mutual coupling between electric circuits (IEEE). Transformers are extensively used in electric power systems to transfer power by electromagnetic induction between circuits at the same frequency, usually with changed values of voltage and current. For example, electricity generated at a central utility station with a voltage of about 22,000 V is transformed to 500,000 V for transmission over high-tension lines. Further downline, the electricity is transformed to about 66,000 V, at which value it may be fed to heavy industries; and still further downline, transformers step the voltage down to about 4000 V for local area distribution to commercial and industrial consumers. Ultimately, transformers (usually pole-mounted) step the voltage down to 230–115 V for use by residential and small business and retail consumers.

The use of transformers by the electric lighting and power industry was commenced more than a century ago (Great Barrington, Massachusetts, 1885; Chicago, Illinois, 1893; Niagara Falls, New York, 1895), during a period when alternating current was winning over direct current as the medium of choice for electric utility systems. As important as transformers are and have been to the electric power distribution system, these devices also find innumerable other applications involving electric circuitry. The wide range of uses include battery chargers, communication systems, electric furnaces, electric tools and toys, and desk lamps, among many others. Instrument transformers for metering electricity consumption (a very difficult problem in the early days of electric utilities) are described in the following article in this encyclopedia.

Stepwise Evolution of the Transformer

The concept of the transformer as ultimately used, unlike many other electrical/electronic inventions (transistor, for example), did not occur as a sudden breakthrough, but rather the transformer developed out of a series of fundamental scientific discoveries. A capsule of the scenario is about as follows: (1) Oersted (1820) demonstrated that an electric current flowing through a conductor creates a magnetic field around that conductor. At that time, Oersted's finding was exciting because electricity and magnetism were considered to be separate, unrelated forces. (The question may have been raised: If, on the one hand, an electric current can create a magnetic field, would a reverse phenomenon occur? i.e., a magnetic field creating an electric current.) (2) Faraday (1831) showed that for a magnetic field to induce a current in a conductor, the field must be a changing one, thus suggesting that the phenomenon would be operative with an interrupted or pulsing direct current, not a steady dc flow, or what later was to be developed, the alternating current, ac. (Tesla did not design the first practical system for generating and transmitting alternating current for electric power until the mid-1880s.) Faraday's experiments are described in some detail in article on **Motor (Electric)**. (3) Although some essential research occurred during the middle 1800s (as by Joseph Henry, whose research on electromagnetic induction independently and unknowingly paralleled that of Faraday), no impelling needs for a transformer-like device existed. It required the economic thrust of ac power generation and distribution, with its almost unlimited commercial potential, for technologists to reduce the principles underlying the transformer to practical, reliable hardware.

Fundamentals of the Transformer

In the transformer, there are no moving parts and no electrical connection between circuits because energy is transferred through a magnetic linkage. Regardless of the voltage, the energy-supply circuit is termed the *primary*; the energy-receiving circuit is the *secondary*. Since so many types of transformers require power to be supplied at changed voltages, and all electric power transmission systems involve changes in voltage, the range of transformer sizes, types, and specifications is very great indeed. In serving a multiplicity of applications, the transformer can range from tiny assemblies the size of a pea or less to massive units weighing 500 tons or more.

In an elementary form, the transformer consists of two coils wound of wire and inductively coupled to each other. When alternating current at a given frequency flows in either coil, an alternating electromotive force (emf) of the same frequency is induced into the other coil. The value of this emf depends upon the degree of coupling and the magnetic flux linking the two coils. The coil connected to a source of alternating emf is usually called the *primary coil*; the emf across this coil is the *primary emf*. The emf induced in the *secondary coil* may be greater than or less than the primary emf, depending upon the ratio of the primary to secondary turns. A transformer is termed a *step-up* or a *step-down* transformer accordingly. Most transformers have stationary iron alloy cores, around which the primary and secondary coils are placed. Because of the high permeability of iron alloys, most of the flux is thereby confined to the core and tight coupling between the coils is thereby obtained. So tight is the coupling between the coils in some transformers that the primary and secondary emfs bear almost exactly the same ratio to each other as the turns in the respective coils or windings. Thus, the turns ratio of a transformer is a common index of its function in raising or lowering potential.

A simple transformer coil and core arrangement is shown in Fig. 1. The primary and secondary coils are wound one over the other on an insulating coil tube or form. The core is laminated to reduce eddy currents. Flux flows in the core along the path indicated, so that all the core flux links both windings. In a circuit diagram, the transformer is represented by the symbol given in Fig. 2.

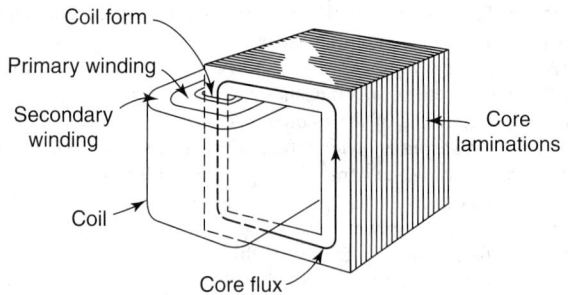

Fig. 1. Transformer coil and core.

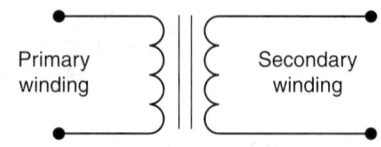

Fig. 2. Simple transformer.

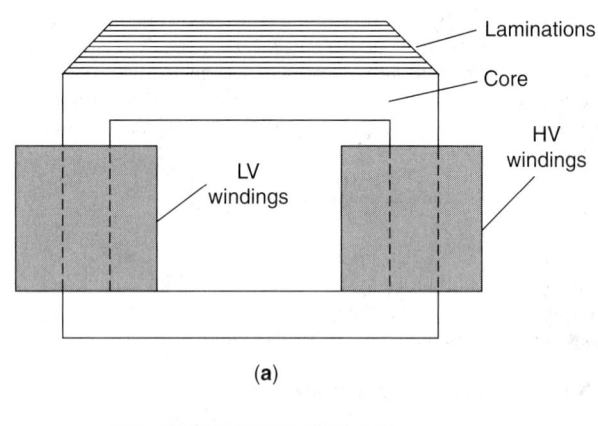

(a)

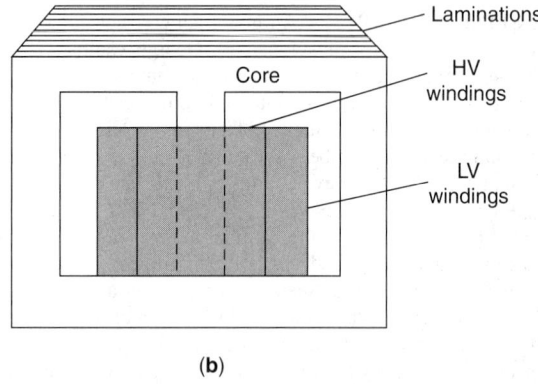

(b)

Fig. 3. Highly schematic diagram of common contemporary transformer configurations; (**a**) *core form*, where the primary encloses one arm of the core and the secondary the other; (**b**) *shell form*, which is made up of E-shaped stampings with the primary and secondary coils nested together on the middle bar. Both cores are constructed from stacked laminations stamped out of iron sheets. In three-phase transformers, the coils are nested on all three bars of (**b**). HV = high voltage; LV = low voltage.

Common contemporary transformer configurations are given in Fig. 3.

In order for a transformer to deliver secondary emf, the primary emf must vary with respect to time. A dc potential produces no voltage in the secondary winding or power in the load. If both varying and dc potentials are impressed across the primary, only the varying part is delivered to the load. This comes about because the emf e in the secondary is induced in that winding by the core flux ϕ according to:

$$e = N \, d\phi \, dt.$$

Reduced to words, the voltage induced in a coil is proportional to the number of turns and to the time rate of change of magnetic flux linking the coil. This change of magnetic flux may be large or small. For a given potential, if the rate of change of flux is small, many turns must be used. Conversely, if a small number of turns is used, a large rate of change of flux is necessary to produce a given potential.

An ideal transformer may be defined as one which neither stores nor dissipates energy. Departures from the ideal transformer are caused by: (1) winding resistance and capacitance; (2) leakage inductance (due to flux which does not link both windings); (3) core hysteresis and eddy current losses; and (4) magnetizing current.

Transformers are required in electronic apparatus to provide the different values of potential for device operation to insulate circuits from each other; to furnish high impedance to alternating current but low impedance to direct current; to change from one impedance level to another; to connect balanced lines to unbalanced loads; and to maintain or modify wave shape and frequency response at different potentials. Electronic transformers differ from power frequency transformers in range of impedance levels, frequencies, size, and weight.

Electronic power transformers are used to supply rectifiers at potentials ranging from 150 V to 750 V. Frequency range transformers are used where the frequency varies, including audio, video, carrier, and control frequencies. Such frequencies vary from a fraction of 1 Hz to uhf (300 to 3000 MHz). Transformers may be wide-band or narrow-band in frequency response, and the core material changes accordingly. In wide-band transformers, the ratio of lowest to highest frequency may be as great as 10^5. For such a wide band, the core material may consist of nickel alloy laminations of high permeability, which maintain uniform secondary voltage at the lowest frequency. This material also makes possible low leakage inductance and winding capacitance, both of which are essential to good response at the highest frequency.

Narrow-band applications use mostly high frequencies and operate over a small percentage of the carrier frequency (e.g., 50 to 55 MHz). At rf frequencies, air-core transformers are used. Pulse transformers are used in radar modulators and computers.

Additional Reading

Gottlieb, I.: "Practical Transformer Handbook: For Electronics, Radio and Communications Engineers," Butterworth-Heinemann, Inc., Woburn, MA, 1998.

Guru, B.S. and H.R. Hiziroglu: "Electric Machinery and Transformers," 3rd Edition, Oxford University Press, New York, NY, 2000.

Heathcote, M. and D.P. Franklin: "The J&P Transformer Book: A Practical Technology of the Power Transformer," 12th Edition, Butterworth-Heinemann, Inc., Woburn, MA, 2000.

Kennedy, B.W.: "Energy Efficient Transformers," The McGraw-Hill Companies, Inc., New York, NY, 1997.

McPherson, G. and R.D. Laramore: "An Introduction to Electrical Machines and Transformers," 2nd Edition, John Wiley & Sons, Inc., New York, NY, 1999.

Pansini, A.J.: "Electrical Transformers and Power Equipment," 3rd Edition, Prentice-Hall, Inc., Upper Saddle River, NJ, 1999.

TRANSFORMER (Instrument). Current and potential transformers extend the ranges of alternating current instruments in the same way as shunts and series resistors work for direct current instruments. See Fig. 1. Taps on the windings permit a single transformer to operate over a wide range. Basic instrument ranges for transformer use are usually 5 amperes or 115 volts. These are nominal secondary ranges for the transformers and are used for computing ratios for the overall range of the meter and transformer combined. In actual practice, the range of the voltmeter may be 150 volts. For relatively high current ranges, the primary often consists of one or more turns passing through the center hole, as shown in Fig. 2. In the case of one turn, the ratio would be 800:5, for two turns, 400:5, and so on. Standard operating procedure and safety for the current transformer requires that the secondary winding always be shorted or connected to the ammeter. Otherwise, dangerous potentials may occur on the secondary of the transformer. Instrument transformers, in addition to permitting measurement of large currents and voltages, perform the important function of insulating the instrument from the line. This affords safety in use. Insulation protection between primary and secondary windings is usually several times the maximum rated voltage of the transformer.

A type of potential transformer arranged with means for connecting the two primary coils in multiple or in series is shown in Fig. 3. This provides for two ranges in a ratio of 2:1. A single secondary winding is used.

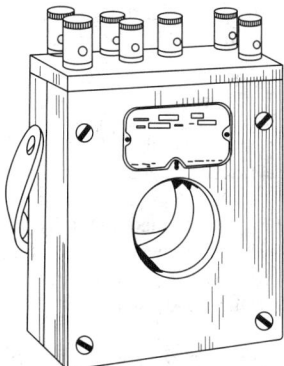

Fig. 1.   External view of instrument transformer.

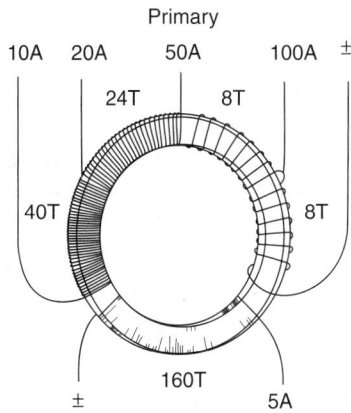

Fig. 2.   Instrument transformer for relatively high current ranges. The primary often consists of one or more turns passing through the center hole.

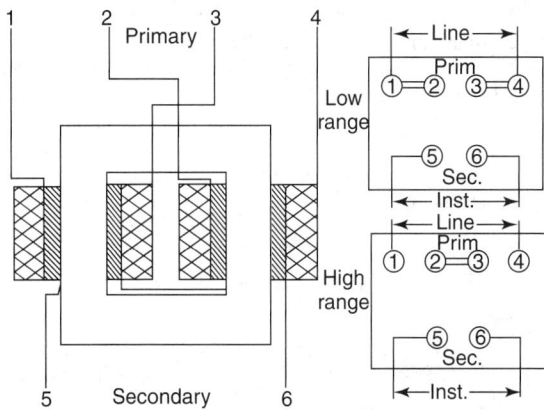

Fig. 3.   Potential transformer arranged with means for connecting the two primary coils in multiple or series.

The chief errors in instrument transformers are due to ratio and phase angle, both of which are controllable in design by using proper flux densities for the magnetic material involved and appropriate construction.

TRANSFUSION (Blood).   See **Blood**.

TRANSIENT (Electrical).   A transient arises when the current in any part of a network changes because of a change of resistance of that part of the circuit, or because the electromotive force operative in its changes. There is an immediate readjustment of the potential differences and currents in the other parts of the circuit. Such readjustments take place very quickly, and the corresponding momentary fluctuations of current are called transients.

Analytically, the general expression for the current in any circuit as a function of time contains both the steady-state terms and the transient terms. The transient terms rapidly become negligible. An example is found in the case of the current in a simple series R-L circuit to which a sinusoidally impressed voltage, $E_{\max}\sin(\omega t + \alpha)$ is applied. The current

$$i = \frac{E_{\max}\sin(\omega t + \alpha + \theta)}{\sqrt{R^2 + \omega^2 L^2}} - \frac{E_{\max}\sin(\alpha + \theta)}{\sqrt{R^2 + \omega^2 L^2}}e^{-Rt/L}$$

The first term on the right of the above relation is the steady-state term. The second term is the transient, which decays with time and eventually becomes very small. The angle θ = phase angle = $-\tan^{-1}\omega L/R$.

The term *transient* is also applied generally to that part of a forced oscillation of a linear system (not necessarily an electric circuit, as defined above) that decays more or less rapidly after the imposition of the force; to be distinguished from the steady state.

TRANSISTOR (Invention and Development). A semiconducting device capable of performing numerous functions in electronic circuits—amplifying electric signals; mixing signals; performing modulation and demodulation, gating, buffering, storing, shifting, and shaping functions, among others. A transistor can operate at high frequencies and has a low power dissipation loss. The transistor is well suited as a switching device for computer and communication systems, as in the areas of transmission, multiplexing, and signal processing. In a very large majority of uses, the transistor has replaced the formerly widely used vacuum tube of the 1940s and 1950s. Transistors are available as discrete units for plug-in boards, but their principal usage in recent years has been in integrated circuits, ranging from small-scale to very-large scale ICs, with a target of incorporating a billion transistors in a single IC by the year 2000 or sooner. The size, speed, and other characteristic advantages of the transistor are described in articles on **Integrated Circuit (IC); Microelectronics;** and **Semiconductors.**

The point-contact transistor was invented in 1948 at the Bell Laboratories. The early device consisted of a piece of n-type germanium about $0.05 \times 0.05 \times 0.005$ inches ($1.3 \times 1.3 \times 0.1$ mm) on which were placed two sharpened points (a beryllium-copper emitter and a phosphor-bronze collector) approximately 0.002 inch (0.05 mm) apart. Each of these points exhibited rectifying diode type current-voltage characteristics. It was then theorized that the operation of the transistor depended on the injection of "holes" into the n-type material through the forward-biased emitter point and the collection at the reverse-biased collector point. The third electrode in the transistor was a low resistance, nonrectifying contact made to the n-type germanium and was designated the base electrode. Power gain was achieved in the structure because the collector current increased at a rate equal to two or three times the emitter current and because the reverse-biased collector terminals could be matched in a circuit to a higher impedance load than the input emitter impedance. The early device is diagrammed in Fig. 1.

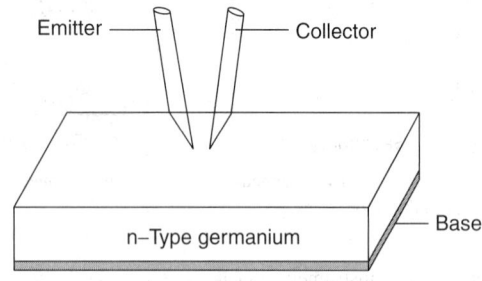

Fig. 1.   Point-contact transistor of the type invented in 1948.

During the intervening years since its invention, the transistor has undergone numerous alterations and refinements, particularly as regards materials of construction (general switchover from germanium to silicon) and the methods used to fabricate the transistor. These refinements represent a continuing process right up to the present day and a process that will continue until sometime in the next century when molecular-scale switching devices emerge. However, it should be stressed that the underlying principles of the solid-state transistor still apply.

Junction Transistor. The junction transistor has numerous advantages over the point-contact transistor of Fig. 1. In a junction transistor, the rectifying characteristics are obtained within the bulk of the semiconductor crystal by placing different types of impurities at different points in the

crystal. The structure of an n-p-n junction transistor is shown in Fig. 2. The emitter and collector regions have an excess of n-type impurities, while the base region has an excess of p-type impurities. The transition from p-type material to n-type material is designated as a p-n junction. In order to achieve transistor action, it is necessary to have the emitter and collector p-n junctions in close physical proximity (fractions of a millimeter and much less). The yardstick of size in the semiconductor industry is the micrometer (micron).

Although in the early days of the transistor germanium was used extensively, the majority of current transistors are made from silicon crystals. The advantages of silicon over germanium include operation at higher temperatures, lower power consumption, and greater surface stability. However, germanium offers advantages in the high-frequency area because electrons and holes travel faster in germanium than in silicon. Gallium arsenide transistors, long under study, are just beginning to take their place commercially. In addition, there are also p-n-p transistors where the type of material used for emitter, base, and collector regions is reversed from the configuration shown for the n-p-n transistor in Fig. 2. The basic difference is that the polarity of all the operating voltages and currents are opposite for these two types of transistors.

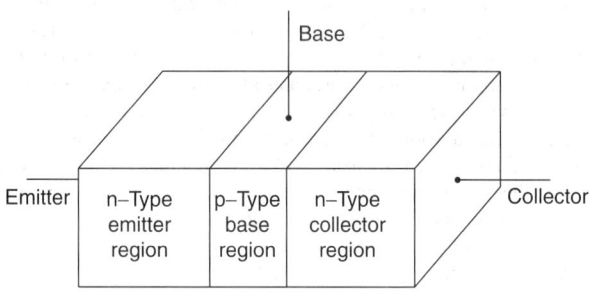

Fig. 2.   Structure of n-p-n junction transistor.

The operation of the n-p-n transistor hinges upon the injection of electrons across the forward biased emitter p-n junction into the base region of the transistor. Once these electrons are injected across the junction they diffuse across the base and arrive at the reverse-biased collector junction. While the emitter junction acts as a source for these electrons, the reverse-biased collector junction acts as a sink for them. We measure the efficiency of such transistors by the current multiplication factor α_F defined as

$$\alpha_F = \frac{\partial I_C}{\partial I_E} \qquad (1)$$

where I_C is the collector current and I_E is the emitter current. In most junction transistors α_F is between 0.95 and 1.00.

The junction transistor can be connected in a circuit so that any one of its three terminals is the common terminal. Figure 3 shows an n-p-n transistor connected in the common emitter configuration with typical operating voltages and currents. Note that in this connection a small incremental increase in base current produces a sizeable increase in collector current. This ratio is defined as

$$h_{FE} = -\frac{\partial I_C}{\partial I_B} = \frac{\alpha_F}{1 - \alpha_F} \qquad (2)$$

Typical values of h_{FE} are between 20 and 200.

There are many types of junction transistors, dependent on the specific method of fabrication chosen. In all cases n-type impurities such as phosphorus, arsenic, or antimony, and p-type impurities such as boron, indium, gallium, or aluminum are introduced into the appropriate regions of the transistor. In a grown-junction transistor, the impurities are introduced at the time that the silicon or germanium crystal is being fabricated. In an alloy-junction structure, the impurities are introduced by melting a pellet of the appropriate doping material on the surface of the semiconductor and regrowing the crystal. Diffused junction transistors are produced by placing the semiconductor material in a furnace at an elevated temperature in an atmosphere containing the desired doping element. The doping material then diffuses into the semiconductor crystal and substitutionally replaces some of the silicon or germanium atoms in the crystal lattice. A region of the opposite type can be created by overcompensating, diffusing into the

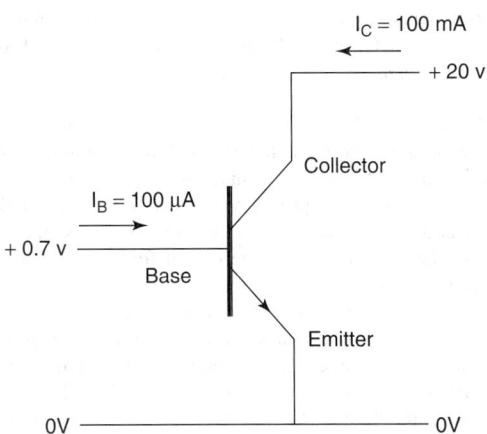

Fig. 3. Circuit symbol and typical potentials for *n*-*p*-*n* silicon-junction transistor.

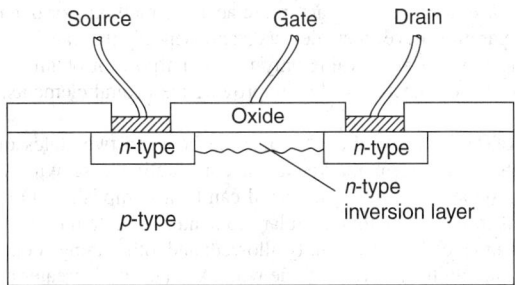

Fig. 5. An *n*-channel enhancement mode MOS transistor.

same region the other type of impurity to a concentration in excess of that previously diffused.

Fig. 4 shows the structure for a silicon planar transistor. This transistor is produced by diffusing boron impurities into *n*-type silicon in order to create the *p*-type base region and later diffusing phosphorus to create the *n*-type emitter region. Note that the surface of the transistor is covered by a layer of silicon dioxide (quartz) which acts as a protection for the exposed *p*-*n* junctions; such layer assures a very high-reliability transistor.

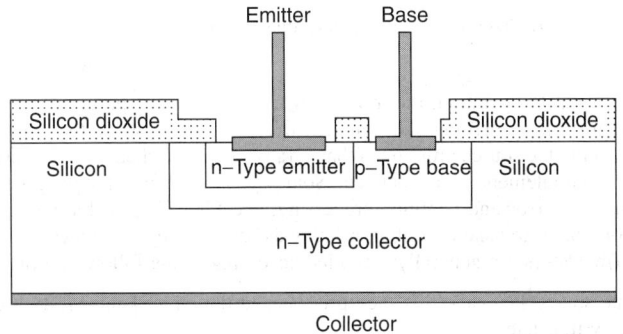

Fig. 4. Cross section of silicon *n*-*p*-*n* planar transistor.

This transistor is also economical to fabricate because mass processing techniques can be utilized. The precise geometrical control achieved by using diffusion techniques results in very reproducible electrical characteristics. See previously mentioned article on microstructure fabrication.

Field effect transistors have been widely used for several years. Of these, the MOS (metal-oxide semiconductor) has been a leading choice of designers. In these transistors, a conducting channel of either electrons or holes is established between two regions, called source and drain, to which ohmic low-resistance contacts are made. Control of the source-to-drain current is achieved by applying a potential to the gate electrode, which affects the width of the conducting channel and/or the number of mobile carriers in the channel. Power gain is achieved between the gate-source input terminals and the drain-source output because of the high input impedance.

In a diffused junction field-effect transistor, the gate is a reverse-biased *p*-*n* junction. In an *n*-channel enhancement mode MOS transistor, the gate electrode is a metallized region separated from the semiconductor by an insulating oxide layer. Application of a positive gate potential to the MOS transistor creates an electron-rich region in the semiconductor (effectively converts the surface of the semiconductor from *p*-type to *n*-type) which serves as a conducting path (*n*-type channel) between the source and the drain. Increasing the gate bias widens the channel and also increases the source-to-drain electron flow.

MOS transistors have been used for integrated circuits because they require a minimum number of fabrication steps and can be produced at high yields. The transistor of Fig. 5, for example, requires only a single *n*-type diffused region of noncritical dimensions. There are also depletion mode transistors, where the *n*-type channel is permanently established and the gate bias (negative) is used to decrease and eventually cut off the drain-to-source current flow.

As amplifiers, transistors are useful up to frequencies of several gigahertz and at power levels of several hundred watts. Transistors have been widely used as switches in computers. The electrical characteristics of a transistor are well suited to this application, in that they closely simulate a relay in its open and closed positions. Computer memory circuits hold tens of thousands of transistors, contained on a small chip of silicon. See also **Memory (Electronic)**.

So-called soft errors become much more important as the transistor becomes smaller. Such errors may be due to the passage of high-energy nuclear particles through a microcircuit, which can cause an error for a short instant, while after the event the circuit works perfectly normally again. Such events arise from the radioactive decay of uranium or thorium, which may be contained in trace quantities in various packaging materials that house the integrated chip. Decay events yield alpha particles, which leave a path of electrical charge as they ionize the silicon atoms during their journey through the microcircuit. Cosmic rays comprise another source of soft error. Among the numerous elementary particles in the cosmic spectrum are neutrons, which interact with silicon atoms causing alpha particles.

Additional Reading

Levinshtein, M. and G. Simin: "Transistors: From Crystals to Integrated Circuits," World Scientific Publishing Company, Inc., River Edge, NJ, 1996.

Neudeck, G.W. and R.F. Pierret: "The Bipolar Junction Transistor," Vol. 3, 2nd Edition, Addison Wesley Longman, Inc., Reading, MA, 1990.

Riordan, M. and L. Hoddeson: "Crystal Fire: The Invention of the Transistor and the Birth of the Information Age," W.W. Norton Company, Inc., New York, NY, 1998.

Tsividis, Y.: "Operation and Modeling of the MOS Transistor," 2nd Edition, McGraw-Hill Higher Education, New York, NY, 1998.

TRANSIT (Astronomy). Both Venus and Mercury revolve about the sun in orbits that lie inside the orbit of the earth about the sun. Accordingly, when the planes of the planetary orbits coincide with the plane of the ecliptic, once during the synodic period of each planet the object will pass between the earth and the sun. Because of the small angular diameter of the planets relative to that of the sun, at the time they pass between the earth and sun they appear as small black spots moving across the brilliant disk of the sun. Such a phenomenon is known as a transit of Venus or a transit of Mercury.

The orbits of both Mercury and Venus are inclined to the plane of the ecliptic with the result that transits do not occur during each synodic period, but only when the planet happens to come to inferior conjunction close to the passage through a node. Transits of Mercury can occur only in May and November, when the earth crosses Mercury's lines of nodes; those of Venus occur only in June and December. Since Mercury is closer to the sun than Venus, transits of Mercury are more common than those of Venus. The last transit of Venus came in 1882 and the next will take place in 2004.

The first recorded observation of a transit of Venus was made in England in 1639 by Horrocks. Four transits have been observed since that time. Since observations of the duration of transit of the planet across the disk of the sun, made from widely separated positions on the earth, provide a method for determination of the important astronomical constant known as solar parallax, expeditions have always been dispatched to observe the phenomena. Within recent years, more accurate methods for determination of solar parallax have been devised, and henceforth, transits of planets across the sun will be of value only for the purpose of obtaining accurate positions of the objects and thus improving the orbital elements.

TRANSITION (Allowed).   A transition between two states of a quantum-mechanical system marked by a comparative ease with which the change in quantum numbers involved can be accomplished. For example, transitions in which the total angular momentum quantum number changes by one in units of h are frequently allowed; and, other things being equal, a transition involving a change of one will take place with greater probability than a competing transition involving a change in this quantum number of two or more.

TRANSITION EFFECT.   A change in the intensity of the secondary radiation associated with a beam of primary radiation as the latter passes from a vacuum into a material medium, or from one medium into another.

TRANSITION LOSS.   In audio systems and components, the transition loss is defined as follows: At any point in a transmission system, the ratio of the available power from that part of the system ahead of the point under consideration to the power delivered to that part of the system beyond the point under consideration. If the input and/or output power consist of more than one component, such as multifrequency signal or noise, then the particular components used and their weighting should be specified. This loss is usually expressed in decibels.

In wave-propagation usage, transition loss is defined in one of two ways, as follows: (1) At a transition or discontinuity between two transmission media, the difference between the power incident upon the discontinuity and the power transmitted beyond the discontinuity which would be observed if the medium beyond the discontinuity were match-terminated. (2) The ratio in decibels of the power incident upon the discontinuity to the power transmitted beyond the discontinuity which would be observed if the medium beyond the discontinuity were match-terminated.

TRANSITION PROBABILITY.   The probability per unit time, symbol w_{if}, that a system in state i will undergo a transition to state f. The quantity

$$w_i = \left[\sum_{\substack{\text{over} \\ \text{all } f}} w_{if} \right]$$

is the total transition probability for state i. The probability that the system is still in state i at the end of time t is then $e^{-w_i t}$. For radioactive transitions, the symbol λ is used instead of w_i, and the total transition probability is then called the disintegration constant.

TRANSITION TEMPERATURE.   1.   An arbitrarily defined temperature within the temperature range in which metal fracture characteristics determined usually by notched tests are changing rapidly such as from primarily fibrous (shear) to primarily crystalline (cleavage) fracture.

2.   The arbitrarily defined temperature in a range in which, the ductility of a material changes rapidly with temperature.

TRANSIT (Surveying).   An instrument used by engineers and surveyors for measuring horizontal angles. Modern transits are usually equipped with a level attached parallel to the telescope, and a graduated vertical circle as well, so that the transit may be employed as a level, and may also be used to read vertical angles. It may also be equipped with stadia wires. An exceptionally accurate transit of the type used for precise surveying is known as a theodolite. See also **Theodolite.**

TRANSLATION.   A motion in which all points of a system have identical displacements. If each point moves on a straight line the translation is rectilinear, if on a curve it is curvilinear. The general motion of a rigid body is a combination of translation and rotation.

TRANSLATION OPERATION (Geometry).   The geometrical process of displacing a body along a straight line, keeping lines fixed in the body always parallel to themselves.

TRANSLATOR.   A network or system having a number of inputs and outputs and so connected that signals representing information expressed in a certain code, when applied to the inputs, cause output signals to appear which are a representation of the input information in a different code. Sometimes called *matrix*.

TRANSLATORS.   A *translator* is a computer program that takes as its input a program written in a particular programming language and transforms it into a functional equivalent program written in another programming language. Two programs are said to be functional equivalent or operation equivalent when, under all circumstances, they exhibit the same behavior and produce the same output for a given set of inputs, if any. The programs themselves are not required to be of the same size or have their operations executed in the same order.

The input to the translator, the *source program* (SP), is written in a *source language* (SL). Likewise, the output of the translator, the *target program* (TP), is written in a *target language* (TL). Since the translator itself is a program, it must be written in some programming language and executed on some computer. The language in which the translator is written is called the *host language* (HL) and the computer on which it is executed is called the *host computer* (HC). The languages involved in the translation process can be illustrated using the T-notation. See Fig. 1.

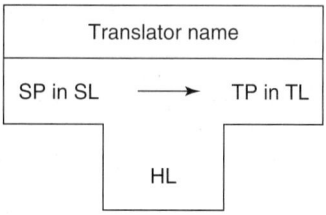

Fig. 1.   Use of the T-notation to illustrate the translation process.

During its operation, the translators may provide the programmer with diagnostics concerning the syntactical or semantic errors that may be present in the source program. What a translator does when it encounters an error is implementation dependent. Some translators notify the programmer about the error and continue processing, checking for possible additional errors; other translators stop translating as soon as they encounter an error.

Translators are generally classified according to the following criteria:

- the source language that they process and the target language to which they translate
- the machine for which they produce the target program.
- the language in which the translator itself is written.
- the completeness of the translation process itself and how they execute the translated programs.
- the number of different activities or phases to carry out the translation process

When classified according to the source language that they process and the target language into which they translate it, translators are called *compilers* if the source language is a high-level language such as C, C++, FORTRAN and the target language is a low-level language such as assembly or machine code. If the source language that they process is assembly language and the target language is machine language the translators are called *assemblers*.

If the target program can be executed on the host computer the translator is called *self-resident*. If the target program cannot be executed on the translator's host computer but in a computer with a different architecture the translator is called a *cross-translator*. Depending on its characteristics a translator may fall into more than one category. For instance, if a translator that runs on an IBM compatible PC translates a source program written in a high-level language into an equivalent target program written in assembly language for an Apple computer, we say that the translator is a cross compiler.

If the translator itself is written in the source language that it processes, the translator is called a *self-translating translator*, for example, a compiler for C++ source programs that is itself written in C++ is a *self-compiling compiler*.

Translators are also classified into two categories based on the completeness of the translation process itself and how they execute the instructions of the target programs. These categories are: compilers and interpreters.

Provided that there are no errors, compilers always translate the entire source programs before executing them. *Interpreters* translate and execute the instructions of the source program one instruction at a time. Interpreters, like any other translator, can also be classified according to the criteria previously explained. In general, interpreters produce programs that are slower to execute and larger in size than their compiled counterparts. However, interpreters can help programmers to develop programs quicker because, in interactive environments, they provide programmers with instant feedback about the syntax of the program as it is being typed. It is not necessary to translate the entire program to realize that the program has a syntax error.

As translators carry out their tasks, the source program undergoes a series of intermediate transformations. The main sequential operations or phases carried out by a translator to transform a source program into a target program are generally called the lexical analysis, syntax analysis, and code generation. During the lexical analysis phase the translator reads the instructions of the source programs and groups the different elements of these instructions into logical units called tokens or lexems, short for lexical elements. A token may be a single character or a sequence of characters that have a particular meaning to the compiler. When processing an instruction the translator ignores comments or white spaces (blanks, carriage return, and tabs). For instance, when a compiler for the programming language Pascal "sees" the statement

if value1 >= value2 then value2 := value2 + 1;

it recognizes the sequence *if* not as two consecutive individual letter but as a single token with special meaning within the language (*keyword*). Likewise, the compiler considers sequence *value1* and *value2* as single tokens and not a sequence of six individual characters. Some symbols that may consist of two or more characters (composite symbols) such as >= and := are also considered single tokens. However, some aspects of the instruction such as the sequence +1; is considered as three individual tokens even if they are written with no space between them. As the individual tokens are recognized by the translator they are classified, in turn, into a larger entity such as a declaration, an assignment statement, etc. To recognize the individual tokens that are present in an instruction, the translator must have the necessary built-in logic. However, what constitutes a token is language dependent and is determined by the grammatical description of the language being translated. The latter is determined by the creator or implementor of the language. The builder of the translator also determines the format of these tokens. Since this phase must read all possible input characters in the source program it sometimes is called a *scanner*. The output of the lexical analysis phase is a sequence of tokens.

The output of the lexical analysis phase is the input for the syntax analysis phase. In this latter phase, the translator checks that the tokens are in the correct order as determined by the grammatical description of the language; that is, that the sequence is grammatically correct. Some authors refer to this activity of the syntactical analyzer as the parsing phase. In addition to checking for the correct order of the token sequence the syntax analyzer may also check that the instructions are semantically correct; that is, that they make sense from an operational point of view. For this reason this phase is sometimes called a semantic analyzer. In some implementations the semantic analyzer is considered a separate phase of the translation process. In general, the output of this syntax analyzer is an abstract data structure called an abstract-syntax tree. The tree is a useful representation of the structure of the operations of the program that, if necessary, can be further operated on to reduce the number of instructions of the target program.

The code generator phase produces as its output the final transformation of the source program into the target program. The input of this phase is the abstract-syntax tree produced by the syntax analysis phase. If necessary, the implementers of the translator may attempt to optimize the code so that it runs faster or so that it occupies a minimum amount of memory when it runs. Most of the time the code generator phase tries to strike a balance between these two frequently incompatible optimal states. The output of this phase can be machine language or assembly language.

In some translators the lexical, syntax, and code generator phases are carried out as separate and individual activities; in some other translators these phases are carried out as a single activity. Depending on the number of these individual phases the translators are called multi-phase or multi-pass or single-phase or single-pass translators.

See also **Computer Operating System**.

Additional Reading

Aho, A.V. and J.D. Ullman: "The Theory of Parsing, Translation and Compiling," Volume I and II. Prentice-Hall, Inc., Upper Saddle River, N.J., 1973.

Barrett, W.A., R.M. Bates, D.A. Gustafson, and J.D. Couch: "Compiler construction: Theory and Practice," Science Research Associates, Inc., Albuquerque, NM, 1986.

Bauer, F.L., et al. "Compiler Construction: An Advanced Course," 2nd Edition, Springer-Verlag Inc., New York, NY, 1976.

Holub, A.I.: "Compiler Design in C," Prentice-Hall Inc., Upper Saddle River, N.J., 1990.

RAMON MATA-TOLEDO, James Madison University, Harrisonburg, VA.

TRANSLUCIDUS. See **Clouds and Cloud Formation**.

TRANSMISSION GRATING. A diffraction grating ruled on a transparent base. Most transmission gratings are plastic replicas of reflection gratings. The radiation is transmitted through, rather than reflected from such a grating.

TRANSMISSION LEVEL. The level of the signal power at any point in a transmission system which is the ratio of the power at that point to the power at some point in the system chosen as a reference point. This ratio is usually expressed in decibels.

TRANSMISSION LIMIT. A limiting wavelength or frequency above or below which a given type of radiation is practically all absorbed by a given medium. If the limits are sharply defined, the medium acts as a sharp cutoff light filter for that radiation.

TRANSMISSION LOSS. The power lost in transmission between one point and another. It is measured as the difference between the net power passing the first point and the net power passing the second. (See also **Standing-Wave Ratio**.) In underwater sound, the transmission loss H is defined to be $H = S - L$, where S is the sound level in decibels, and L is the sound pressure level in decibels above root mean square pressure of 1 dyne per square centimeter.

TRANSMISSION MODE. The various transmission modes in a waveguide are characterized by the electric and magnetic field patterns in a plane normal to the guide axis. When the magnetic vector is perpendicular to the direction of travel, the wave is *transverse magnetic*, or a TM-wave. When the electric vector is perpendicular, it is a *transverse electric*, or TE-wave. If both vectors are perpendicular to the direction of travel, the wave is *transverse electromagnetic*, or a TEM-wave. There are an infinite number of transmission modes in a guide, but each mode possesses a cut-off frequency; waves of lower frequency cannot be propagated by the corresponding mode. The lowest of these frequencies is the absolute cut-off frequency of the guide, and the corresponding mode is the dominant transmission mode. Note that a pair of parallel wires is a waveguide possessing a cut-off frequency of zero, i.e., direct current is propagated.

TRANSMISSION (Pneumatic). Use of compressed air to communicate the values of instrumental-measurement information from one location to the next. Also referred to as pneumatic telemetering, the system is usually confined to relatively short distances within a processing and manufacturing facility, generally not exceeding a thousand feet. The distance over which pneumatic signals may be transmitted is limited by the amount of added lag that the tubing will contribute to the system. The output tubing includes capacitance proportional to the volume of the tube and is distributed along the length of the tubing. The tubing also includes resistance, similarly distributed. Usually transmission tubing is about $\frac{3}{16}$ inch (~5 millimeter) in inside diameter. For tubing of this diameter, the dynamic response for a 350-foot (105-meter) length, terminating in a 1-cubic-inch (~16 cubic centimeters) volume will be approximated by a time constant of 1 second and a dead period of $\frac{1}{3}$ second. For a length of 1,000 feet (300 meters) of tubing, the time constant would be 7 seconds and dead period 1 second.

A pneumatic transmitter provides air output pressure proportional to the value of the variable being measured. Usually, the output-pressure range is from 3 to 15 pounds/square inch (0.2 to 1 atmosphere). Air supply pressure is usually 20 pounds/square inch (1.4 atmospheres). The lowest pressure in the line, corresponding to a minimum instrument reading (often zero on the scale), is 3 psi (0.2 atmosphere). In an instrument calibrated—say from

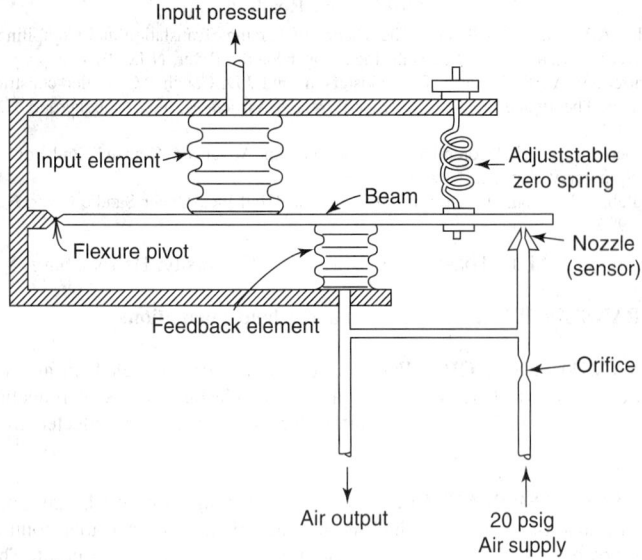

Fig. 1.   Elementary type of pneumatic transmitter.

0° to 100 °C — the transmitted pressure for the high reading would be 15 psi (1 atmosphere). The relationship is reasonably linear. An elementary pneumatic transmitter is shown in the Fig. 1. In this case, it is assumed that a transducer is used to provide an input pressure that is directly related to the variable being measured. As in a pneumatic controller (see also **Pneumatic Controller**), a flapper-nozzle arrangement may be used to vary the output pressure. A feedback element is used to maintain the flapper-nozzle clearance constant. In so doing, more or less air pressure is required, and this is reflected by the transmitter air-output pressure. Where transmission distance is relatively long, a pneumatic relay also will be used in the system.

TRANSMISSOMETER.   An instrument for measuring the extinction coefficient of the atmosphere and for the determination of visual range. Also called *telephotometer*, *transmittance meter*, or *hazemeter*.

TRANSMITTANCE.   Consider a beam of electromagnetic radiation of intensity or radiant power I_0 incident upon a sample and let I be the intensity of the beam transmitted by the sample. Then, the ratio $I/I_0 = T$ is called the transmittance. Experimental measurements of transmittance must usually be corrected for reflection, scattering, and other effects (see also **Bouguer and Lambert Law**). Formerly, such a corrected transmittance result was designated an internal transmittance. Such usage is now discouraged as a reported value of T should be understood as one upon which all appropriate corrections have been made. For solutions, the absorbance $A = \log 1/T$ is more closely related to concentration (see also **Beer's Law**) and many spectrometers record A rather than T.

TRANSMITTANCE (Directional Luminous).   The ratio of the luminance of the surface at which light leaves a diffusing object to the illuminance of the surface at which the light is incident upon the object; provided that (1) the luminance of the emergent surface is expressed in lamberts and the illuminance of the incident surface in lumens/square centimeter, or (2) the luminance of the emergent surface is expressed in foot-lamberts and the illuminance of the incident surface in foot-candles.

TRANSMITTANCY.   The ratio of the internal transmittance of a solution to that of the pure solvent in equivalent thickness. This ratio was supposed to correct for certain unwanted effects in transmittance experiments (see also **Bouguer and Lambert Law**). However, the term is now discouraged since presumably a reported transmittance result has been corrected for all extraneous factors.

TRANSMITTER.   1. A device for converting sound waves into corresponding electrical oscillations, e.g., a microphone or telephone transmitter. 2. In radio communications, the complete group of equipment utilized in converting sound, or an audio-frequency electrical signal into the modulated radio-frequency signal fed to the antenna. The usage varies somewhat and may connote the modulators and radio-frequency stages, or may be applied to more or less of the equipment constituting the complete broadcasting system.

TRANSONIC.   Pertaining to that which occurs or is occurring within the range of speed in which flow patterns change from subsonic to supersonic or vice versa, about Mach 0.8 to 1.2.

TRANSONIC FLOW.   In aerodynamics, flow of a fluid over a body in the range just above and just below the acoustic velocity. Transonic flow presents a special problem in aerodynamics in that neither the equations describing subsonic flow nor the equations describing supersonic flow can be applied in the transonic range.

TRANSPIRATION.   All plants require water for their continued existence and most kinds of plants require it in considerable quantities. In spite of the absolute indispensability of water for their growth and metabolism, plants in general are very inefficient in their use of water. An overwhelmingly large proportion of the water absorbed by most kinds of terrestrial plants escapes from them as water vapor. This process of the loss of water vapor from plants is called transpiration. Although some water vapor loss may occur from any organ of a plant exposed to the atmosphere, in the vast majority of plants most transpirational water loss occurs from the leaves. There are two kinds of foliar transpiration: (1) stomatal transpiration in which water vapor loss occurs through the stomates; and (2) cuticular transpiration, in which evaporation of water takes place directly through the surface of the epidermal cells through the cuticular layer into the atmosphere. In most species, stomatal transpiration represents 90% or more of the total foliar transpiration.

Stomatal transpiration involves the two physical processes of evaporation and diffusion. Evaporation of water takes place from the moist cell wall surfaces into the air-filled intercellular spaces between mesophyll cells. (See also **Leaf**.) If the stomates are closed this soon results in the saturation of the intercellular spaces with water vapor. If, however, the stomates are open, diffusion of water vapor takes place through them into the outside atmosphere except when the vapor pressure of the surrounding atmosphere is equal to that of the intercellular spaces. This latter condition rarely prevails, however, during the daylight hours of clear days when the stomates are most likely to be open. The rate of transpiration of a given plant varies greatly according to the environmental conditions to which it is subjected. The rate of transpiration of most plants is relatively low during the night hours because the stomates are usually closed during that period. During daylight periods when the stomates are open the rate increases, in general, with rise of air temperature or with decrease in atmospheric vapor pressure. Transpiration rates are usually greater in a breeze than in quiet air, but a gentle breeze is almost as effective as a strong wind in its influence on this process. Soil conditions are often even more important in influencing the rate of transpiration than atmospheric conditions. Whenever soil conditions are such that the rate of absorption of water is retarded the rate of transpiration is correspondingly reduced. During droughts, for example, transpiration rates are low or negligible no matter how favorable the atmospheric and light conditions may be to high transpiration rates. See also **Hydrology**.

TRANSPORT EQUATION (Boltzmann).   Consider the evolution of the velocity distribution function f. This evolution proceeds through mechanical changes (flow) and collision.

In gases at low densities, the effect of collisions involving more than two molecules is negligible. If the collisions with the walls can also be neglected, the evolution of the molecular distribution function $f_i(\mathbf{r}, \mathbf{v}, t)$ is given by the Boltzmann integro-differential equation:

$$\frac{\partial f_i(\mathbf{r}, \mathbf{v}, t)}{\partial t} + \mathbf{v}_i \frac{\partial f_i(\mathbf{r}, \mathbf{v}, t)}{\partial \mathbf{r}_i} + \mathbf{X}_i \frac{\partial f_i(\mathbf{r}, \mathbf{v}, t)}{\partial \mathbf{v}_i}$$
$$= 2\pi \iint (f_i' f_j' - f_i f_j) g_{ij} b \, db \, d\mathbf{v}_j \tag{1}$$

b is the impact parameter, g_{ij} is the relative velocity. $\partial f/\partial t$ is the local variation of the distribution function. Equation (1) results from three causes: (1) $\mathbf{v}_i(\partial f_i/\partial \mathbf{r}_i)$, often called the *Streaming Term*, is the variation of the distribution function caused by the movement of the particles in and out of the volume element under consideration; (2) $\mathbf{X}_i(\partial f_i/\partial \mathbf{v}_i)$ is the variation of the distribution function resulting from the external forces X_i acting on the molecules; (3) finally, the term on the right-hand side of Equation (1) expresses the variation of f_i that results from the binary collisions between the molecules, f_i increases when a collision between two molecules having initial velocities $\mathbf{v}'_i$ and $\mathbf{v}'_i$ creates a molecule of the proper velocity $\mathbf{v}_i$. It decreases when a molecule of velocity $\mathbf{v}_i$ collides with a molecule of velocity $\mathbf{v}_j$. The term on the right-hand side of Equation (1) is called the *Collision Integral* of the Boltzmann equation.

TRANSURANIUM ELEMENTS.   The chemical elements with an atomic number higher than 92 (uranium), commencing with 93 (neptunium) and through 103 (lawrencium) frequently are termed Transuranium elements. Any additional elements that may be identified will be a part of this series. See also **Actinide Series**.

TRAP.   1.   A general term for fine-grained, basic and therefore dark-colored igneous rocks, having a relatively high specific gravity. The term is derived from an old Scandinavian word meaning a stair, because these rocks, as particularly well displayed in the 3,000-foot (900-meter) cliffs of the Faroe Islands, develop architectural forms simulating colossal stairways. The term is also used as a synonym for basalt, especially by contractors who use it for road metal.

2.   In electronics, a trap is an absorption filter used to trap or remove an undesired signal from a receiver, as in the sound trap used in the video amplifier of a television receiver to prevent the sound signal from interfering with the picture.

3.   In a diffusion pump system, a trap is a device designed to reduce the effect of vapor pressure of the mercury or oil in the pump, on the high-vacuum side of the system.

4.   Trap is a word also used in connection with the trapping of atoms and ions. See also **Trapped Ions** and also check alphabetical index.

TRAP (Geology).   See **Natural Gas**; and **Petroleum**.

TRAPEZIUM.   See **Quadrilateral**.

TRAPEZOIDAL RULE.   A special case of the Newton-Cotes formula for numerical integration

$$\int_a^b f(x)\,dx = h\left[\frac{y^0}{2} + y_1 + y_2 + \cdots + y_{n-1} + \frac{y_n}{2}\right]$$

where h is the interval between equally spaced values of x, the independent variable. This is the simplest, but least accurate, of the numerical integration rules. It is often used, however, and can give quite satisfactory results.

TRAPPED IONS.   Charged particles, including electrons and atomic ions, can be stored for long periods of time (days are not uncommon) without the usual perturbations associated with confinement (for example, the frequency shifts associated with the collisions of ions with buffer gases in a more traditional optical pumping experiment). Ions that are stored in electromagnetic "traps" can provide the basis for extremely high-resolution spectroscopy. By using lasers, the kinetic energy of the ions can be cooled to millikelvin temperatures, thus suppressing Doppler frequency shifts. Potential accuracies of frequency standards and clocks based on such experiments are anticipated to be better than one part in 10^{15}. The current time standard (cesium clock) has an accuracy of one part in 10^{13} or less. See also **Mass Spectrometry**.

Storage has mainly been achieved in four types of traps: (1) the radio frequency or Paul trap; (2) the Penning trap; (3) the Kingdon electrostatic trap; and (4) the magnetostatic (magnetic bottle) trap. The principles, advantages, and disadvantages of these traps are detailed by D.J. Wineland (*Science*, **226**, 395–400, Oct. 26, 1984).

TRAPPING.   An electron in the conduction band of a semiconductor or insulator may be caught by any irregularity in the crystal lattice and trapped there until it can be released by thermal agitation. The rate of release will depend on $\exp(-E/kT)$ where E is the depth of the trap—hence a material containing large numbers of deep traps may show reduced photoconductivity, for example, and a slow decay as the electrons are released after the light has been removed.

TRAUMA.   A physical injury, the term often signifying a severe wound; or a psychic trauma, representing the psychological consequences of emotional shock.

TRAVELING-WAVE TUBE.   See **Microwave Tubes**.

TRAVERSE.   A series of connected straight lines, forming the outline of a plot of land or the center line of a roadway or railway. A traverse that forms a closed figure is a closed traverse. A traverse that is not closed is an open traverse. The traverse is obtained by means of a survey. Ordinarily, the transit and tape are used, the transit being successively set up over each point constituting the junction of two adjacent lines. The distances between transit stations (the length of the straight lines) are measured by taping with a graduated tape or chain. The bearings of the lines or the angles between the lines, as well as the length of the lines, are noted by the instrument man at the time of the survey in a field book. These field notes are later employed to map the traverse by one of the several methods available, such as tangent offsets, latitudes and departures, coordinates, etc.

A traverse that follows the shoreline of a body of water is a meander line. The general outline of the shore is found by taking offsets from the transit lines at regular intervals and also at points where there is an abrupt change of shape.

TRAVERSE TABLES.   In many problems in surveying and navigation (e.g., the important navigational problem of dead reckoning), the solution of a plane right-triangle becomes necessary. In traverse tables, the plane right-triangle is solved without using tables of trigonometric functions. The tables are constructed by tabulating for successive values of one apex angle the two sides of the triangle as functions of the hypotenuse. The traverse tables, as published for navigational purposes, solve the dead-reckoning problem by tabulating difference of latitude and departure as functions of the distance for each course, either in quarter points of the compass or every degree. Since the tables must contain pages for each value of the apex angle, they are more bulky than ordinary trigonometric tables. However, the rapidity with which the plane triangle may be completely solved more than compensates for the time lost in turning pages, particularly when extreme accuracy is not required in the solution.

See also **Course**; and **Navigation**.

TRAVERTINE.   Carbonated waters dissolve large amounts of calcium carbonate, especially under high temperature. Such waters reaching the earth's surface as hot springs often deposit the calcium carbonate, in great quantities. This material is called travertine from the ancient name for Tivoli, Italy, where a very thick deposit occurs. Travertine may be compact, crystalline, fibrous or, if rapidly deposited, spongy and porous. The less compact varieties are known as tufa. Travertine is being formed at the Mammoth Hot Springs, Yellowstone National Park and at many other localities. A banded travertine used as an ornamental stone is called onyx marble or Mexican onyx.

TREE.   Numerous specific trees are described in this volume—as indicated by the listings of Table 1. One of the most difficult problems in connection with persons conducting literature research on trees involves the naming of trees. Often, material will be listed only under Latin family or genus names. The listings of Table 2 are designed to render assistance in this regard. The names of Table 3 are further expanded to include the major genera within various families, including common name examples for the various genera. Included with many of the specific tree descriptions in this volume are lists of specifications of so-called champion trees in the United States. Over the years, these champion trees have been selected annually by American Forests. Pictures of many of the champion trees also are included. A champion among champions is the record white oak (*Quercus alba*) located at Wye Mills, Maryland. See Fig. 1 on p. 3524.

The magnitude of the millions of trees growing on the earth is dramatized by Table 4 which lists the forested areas of the principal forested nations of the world.

TREE HOPPER   (*Insecta, Homoptera*).   An insect of moderate size with a peculiarly formed thorax, prolonged over the hinder part of the body and in many species with other projections of bizarre form. These insects make up the family *Membracidae*, related to the leaf hoppers and spittle bugs.

TABLE 1. SPECIFIC TREES DESCRIBED IN THIS VOLUME

Title of Entry	Tree Family	Title of Entry	Tree Family
Acacia Tree	*Leguminosae*	Holly Trees and Shrubs	*Aquifoliaceae*
Alder Trees	*Betulaceae*	Hornbeam Trees	*Carpinaceae*
Allspice Tree	*Myrtaceae*	Juniper Trees	*Cupressaceae*
Angiosperms	—	Kentucky Coffee Tree	*Gymnocladus*
Araucarias	*Araucariaceae*	Larch Trees	*Pinaceae*
Arborvitae	*Cupressaceae*	Laurel Family	*Lauraceae*
Ash Trees	*Oleaceae*	Lignum Vitae	*Guaiacum*
Balsa Tree	*Bombacaceae*	Litchi Tree	*Sapindaceae*
Baobab Tree	*Bombaceceae*	Locust Trees	*Leguminosae*
Basswood Trees	*Tiliaceae*	Logwood Tree	*Caesalpiniaceae*
Bayberry Shrubs and Trees	*Myricaceae*	Macadamia Tree	*Proteaceae*
Beech Trees	*Fagaceae*	Magnolia Trees	*Magnoliaceae*
Birch Trees	*Betulaceae*	Mahogany Trees	*Meliaceae*
Bladdernut Tree or Shrub	*Staphleaceae*	Maidenhair Tree	*Ginkgoaceae*
Boswellia Tree	*Burseraceae*	Mangrove Tree	*Rhizophoraceae*
Box Trees and Shrubs	*Aquifoliaceae*	Maple Tree	*Aceraceae*
Brazil-Nut Tree	*Lecythidaceae*	Mesquite Tree	*Prosopis*
Brazilwood	*Caesalpiniaceae*	Mulberry Family	*Moraceae*
Buckeye and Horse Chestnut Trees	*Hippocastanaceae*	Nutmeg Tree	*Myristicaceae*
Buckthorn Shrubs and Trees	*Rhamnaceae*	Nux Vomica Tree	*Loganiaceae*
Cacao Tree	*Sterculiaceae*	Oak Trees	*Fagaceae*
Caesalpinia Tree	*Caesalpiniaceae*	Olive Tree	*Oleaceae*
Cashew and Sumac Trees	*Anacardiaceae*	Palm Trees	*Palmae*
Casuarina Tree	*Casuarina*	Papaya Tree	*Caricaceae*
Catalpa Trees	*Bignoniaceae*	Pepper	*Piperaceae*
Cedar Trees	*Cupressaceae; Pinaceae; Araucariaceae; Simaroubaceae*	Persimmon Trees	*Ebenaceae*
		Pine Trees	*Pinaceae*
Chestnut Trees	*Fagaceae*	Plumeria Tree	*Gentianaceae*
Citrus Trees	*Rutaceae*	Podocarps	*Podocarpaceae*
Clove Tree	*Myrtaceae*	Poinciana	*Delonix*
Coffee Tree and Coffee	*Rubiaceae*	Pomegranate Tree	*Myrtaceae*
Comminphora Tree	*Burseraceae*	Poplar Trees	*Salicaceae*
Conifers		Quassia	*Simaroubaceae*
Coral Tree	*Leguminosae*	Redwood (Coast)	*Taxodiaceae*
Corkwood Tree	*Leitneriaceae*	Redwood (Dawn)	*Taxodiaceae*
Cypress Trees	*Cupressaceae; Taxodiaceae*	Rose Family	*Rosaceae*
Dogwood Shrubs and Trees	*Cornaceae*	Sapodilla	*Sapotaceae*
Elder Trees and Viburnums	*Caprifoliaceae*	Schinus Tree	*Schinus*
Elephant Tree		Silk Cotton Trees	*Bombaceceae*
Elm Trees	*Ulmaceae*	Spruce Trees	*Pinaceae*
Eucalyptus Trees	*Myrtaceae*	Sweetgum Tree	*Liquidambar*
Fir Trees	*Pinaceae*	Sycamore and Plane Trees	*Platanaceae*
Giant Sequoia	*Taxodiaceae*	Teak Tree	*Verbenaceae*
Ginseng	*Araliaceae*	Tung	*Euphorbiaceae*
Guava Trees	*Myrtaceae*	Tupelo Trees	*Davidiaceae*
Hackberry and Zelkova Trees	*Ulmaceae*	Walnut Trees	*Juglandaceae*
Hazelnut Shrubs	*Corylaceae*	Willow Trees	*Salicaceae*
Heather Shrubs and Trees	*Ericaceae*	Witch Hazel	*Hamamelis*
Hemlock Trees	*Pinaceae*	Yew Trees	*Taxaceae*
Hickory and Wingnut Trees	*Juglandaceae*		

TABLE 2. ALPHABETICAL LIST OF FAMILY TREE NAMES — LATIN AND ENGLISH

English	Latin	Latin	English	English	Latin	Latin	English
Basswood	*Tiliaceae*	*Aceraceae*	Maple	Magnolia	*Magnoliaceae*	*Leguminosae*	Pea
Beech	*Fagaceae*	*Anacardiaceae*	Cashew	Mahogany	*Meliaceae*	*Magnoliaceae*	Magnolia
Bignonia	*Bignoniaceae*	*Aquifoliaceae*	Holly	Maple	*Aceraceae*	*Meliaceae*	Mahogany
Birch	*Betulaceae*	*Araucariaceae*	Chile Pine	Mulberry	*Moraceae*	*Moraceae*	Mulberry
Cashew	*Anacardiaceae*	*Betulaceae*	Birch	Myrtle	*Myrtaceae*	*Myrtaceae*	Myrtle
Chile Pine	*Araucariaceae*	*Bignoniaceae*	Bignonia	Oleaster	*Elaegnaceae*	*Oleaceae*	Olive
Citrus	*Rutaceae*	*Caprifoliaceae*	Elder	Olive	*Oleaceae*	*Palmae*	Palm
Cypress	*Cupressaceae*	*Carpinaceae*	Hornbeam	Palm	*Palmae*	*Pinaceae*	Pine
Davidia	*Davidiaceae*	*Cornaceae*	Dogwood	Pea	*Leguminosae*	*Podocarpaceae*	Podocarp
Dogwood	*Cornaceae*	*Corylaceae*	Hazel	Pine	*Pinaceae*	*Proteaceae*	Protea
Ebony	*Ebenaceae*	*Cupressaceae*	Cypress	Podocarp	*Podocarpaceae*	*Rosaceae*	Rose
Elder	*Caprifoliaceae*	*Davidiaceae*	Davidia	Protea	*Proteaceae*	*Rutaceae*	Citrus
Elm	*Ulmaceae*	*Ebenaceae*	Ebony	Quassia	*Simaroubaceae*	*Salicaceae*	Willow
Ginkgo	*Ginkgoaceae*	*Elaegnaceae*	Oleaster	Rose	*Rosaceae*	*Simaroubaceae*	Quassia
Hazel	*Corylaceae*	*Ericaceae*	Heather	Snowbell	*Styracaceae*	*Styracaceae*	Snowbell
Heather	*Ericaceae*	*Fagaceae*	Beech	Swamp Cypress	*Taxodiaceae*	*Taxaceae*	Yew
Holly	*Aquifoliaceae*	*Ginkgoaceae*	Ginkgo	Tea	*Theaceae*	*Taxodiaceae*	Swamp Cypress
Hornbeam	*Carpinaceae*	*Hamamelidaceae*	Witch Hazel	Walnut	*Juglandaceae*	*Theaceae*	Tea
Horse Chestnut	*Hippocastanaceae*	*Hippocastanaceae*	Horse Chestnut	Willow	*Salicaceae*	*Tilliaceae*	Lime or Basswood
Laurel	*Lauraceae*	*Juglandaceae*	Walnut	Witch Hazel	*Hamamelidaceae*	*Ulmaceae*	Elm
Lime	*See Basswood*	*Lauraceae*	Laurel	Yew	*Taxaceae*		

3520

TABLE 3. ABRIDGED ALPHABETICAL LIST OF TREE GENERA VERSUS FAMILIES

| Genus | Name of Family | | Examples[a] |
	English	Latin	
Abies	Pine	*Pinaceae*	Fir
Acacia	Pea	*Leguminosae*	Acacia, cootamundra, wattle, mimosa
Acer	Maple	*Aceraceae*	Maple, sycamore
Aesculus	Horse Chestnut	*Hippocastanaceae*	Buckeye, horse chestnut
Ailanthus	Quassia	*Simaroubaceae*	Tree of heaven
Albizia	Pea	*Leguminosae*	Silk tree
Alnus	Birch	*Betulaceae*	Alder
Amelanchier	Rose	*Rosaceae*	Service-berry
Aralia	Bignonia	*Bignoniaceae*	Japanese angelical tree, Hercules club, Devil's walking stick
Araucaria	Chile Pine	*Araucariaceae*	Parana pine, Chile pine, Norfolk island pine
Arbutus	Heather	*Ericaceae*	Strawberry tree, madrona
Archonotophoenix	Palm	*Palmae*	King palm
Arecastrum	Palm	*Palmae*	Queen palm
Asimina	Laurel	*Lauraceae*	Pawpaw
Athrotaxis	Chile Pine	*Araucariaceae*	Tasmanian cedar, King William pine
Austrocedrus	Cypress	*Cupressaceae*	Chilean incense cedar
Betula	Birch	*Betulaceae*	Birch
Broussonetia	Mulberry	*Moraceae*	Paper mulberry
Buxus	Holly	*Aquifoliaceae*	Common box
Callitris	Cypress	*Cupressaceae*	Cypress pine
Calocedrus	Cypress	*Cupressaceae*	Incense cedar
Camellia	Tea	*Theaceae*	Camellia, tea tree
Carpinus	Hornbeam	*Carpinaceae*	Hornbeam
Carya	Walnut	*Juglandaceae*	Bitternut, pignut, pecan, mockernut, shagbark hickory
Castanea	Beech	*Fagaceae*	Chestnut
Castanopsis	Beech	*Fagaceae*	(Cross between oak and chestnut)
Casuarina	Myrtle	*Myrtaceae*	Dwarf oak or she-oak
Catalpa	Bignonia	*Bignoniaceae*	Catalpa, Indian bean tree
Cedrela	Mahogany	*Meliaceae*	Chinese toon
Cedrus	Pine	*Pinaceae*	True cedars
Celtis	Elm	*Ulmaceae*	Hackberry
Cephalotaxus	Yew	*Taxaceae*	Chinese plum yew; Cow's tail pine; Japanese plum yew
Ceratonia	Pea	*Leguminosae*	Carob
Cercidiphyllum	Witch Hazel	*Hamamelidaceae*	Katsura
Cercis	Pea	*Leguminosae*	Redbud, Judas tree
Chamaecyparis	Cypress	*Cupressaceae*	False cypress
Chamaerops	Palm	*Palmae*	European fan palm
Chionathus	Olive	*Oleaceae*	Fringe tree
Chrysolepis	Beech	*Fagaceae*	Golden chinkapin
Cinnamomum	Laurel	*Lauraceae*	Camphor tree
Citrus	Citrus	*Rutaceae*	Lime, limequat, orange, lemon, grapefruit, mandarin orange, orangequat, sweet orange, Ugli tree
Cladrastis	Pea	*Leguminosae*	Yellow-wood
Clethra	Heather	*Ericaceae*	Clethra
Cocos	Palm	*Palmae*	Coconut palm
Cornus	Dogwood	*Cornaceae*	Dogwood; Cornelian cherry
Corylus	Hazel	*Corylaceae*	Hazel
Cotinus	Cashew	*Anacardiaceae*	Smoke tree
Cotoneaster	Rose	*Rosaceae*	Cotoneaster shrubs
Crataegomespilus	Rose	*Rosaceae*	Bronvaux medlar
Crataegus	Rose	*Rosaceae*	Hawthorn, cockspur thorn
Cryptomeria	Swamp Cypress	*Taxodiaceae*	Japanese cedar
Cunninghamia	Swamp Cypress	*Taxodiaceae*	Chinese fir
Cupressocyparis	Cypress	*Cupressaceae*	Leyland cypress
Cupressus	Cypress	*Cupressaceae*	Cypress
Cydonia	Rose	*Rosaceae*	Quince
Cytisus	Pea	*Leguminosae*	Moroccan broom
Dacrydium	Podocarp	*Podocarpaceae*	Rimu (red pine); Houn pine
Davidia	Davidia	*Davidiaceae*	Dove tree; black gums
Diospyros	Ebony	*Ebenaceae*	Persimmon

(continued)

TABLE 3. (*Continued*)

Genus	English	Latin	Examples[a]
		Name of Family	
Disanthus	Witch Hazel	*Hamamelidaceae*	*Disanthus cercidifolius*
Elaegnus	Oleaster	*Elaegnaceae*	Russian olive
Embothrium	Protea	*Proteaceae*	Chilean firebush
Erica	Heather	*Ericaceae*	Tree heath
Eucalyptus	Myrtle	*Myrtaceae*	Eucalyptus or gum
Eucommia	Elm	*Ulmaceae*	Hardy rubber tree
Eucryphia	Horse Chestnut	*Hippocastanaceae*	*Eucryphia cordifolia*
Euodia	Citrus	*Rutaceae*	Korean euodia
Fagus	Beech	*Fagaceae*	Beech
Ficus	Mulberry	*Moraceae*	Common fig
Fitzroya	Cypress	*Cupressaceae*	Chilean Fitzroya cypress
Fortunella	Citrus	*Rutaceae*	Nagami kumquat
Franklinia	Tea	*Theaceae*	Franklinia
Fraxinus	Olive	*Oleaceae*	Ash (not mountain ash)
Genista	Pea	*Leguminosae*	Mount Etna broom
Ginkgo	Ginkgo	*Ginkgoaceae*	Maidenhair tree
Gleditsia	Pea	*Leguminosae*	Locust
Glyptostrobus	Swamp Cypress	*Taxodiaceae*	Chinese cypress
Gordonia	Tea	*Theaceae*	Similar to camellia
Grevillea	Protea	*Proteaceae*	Silk oak
Gymnocladus	Pea	*Leguminosae*	Kentucky coffee tree
Hakea	Protea	*Proteaceae*	Sea urchin
Halesia	Snowbell	*Styracaceae*	Snowdrop tree; mountain silverbell
Hamemelis	Witch Hazel	*Hamamelidaceae*	Witch hazel
Hippophae	Oleaster	*Elaegnaceae*	Buckthorn
Hoheria	Lime or Basswood	*Tiliaceae*	Ribbonwood
Idesia	Mahogany	*Meliaceae*	Chinese Idesia
Ilex	Holly	*Aquifoliaceae*	Holly, possumhaw, tarajo
Jacaranda	Bignonia	*Bignoniaceae*	Jacaranda
Jubaea	Palm	*Palmae*	Chilean wine palm
Juglans	Walnut	*Juglandaceae*	Walnut
Juniperus	Cypress	*Cupressaceae*	Juniper; pencil cedar
Kalopanax	Bignonia	*Bignoniaceae*	Castor aralia
Koelreuteria	Horse Chestnut	*Hippocastanaceae*	Golden rain tree; Prize of India
Laburnum	Pea	*Leguminosae*	Laburnum
Larix	Pine	*Pinaceae*	Larch
Laurus	Laurel	*Lauraceae*	Laurel; sweet bay; willow-leaf bay
Ligustrum	Olive	*Oleaceae*	Ligustrum lucidum
Liquidambar	Witch Hazel	*Hamamelidaceae*	Sweet gum
Liriodendron	Magnolia	*Magnoliaceae*	Tulip tree
Lithocarpus	Beech	*Fagacae*	Tanbar oak
Livistona	Palm	*Palmae*	Australian palm
Maackia	Pea	*Leguminosae*	*Maackia amurensis; M. sinensis*
Maclura	Mulberry	*Moraceae*	Osage orange
Magnolia	Magnolia	*Magnoliaceae*	Magnolia; cucumber tree; pink tulip tree; swamp bay; sweet bay; umbrella tree
Malus	Rose	*Rosaceae*	Crab apple; orchard apple
Maytenus	Willow	*Salicaceae*	Mayten tree
Mespilus	Swamp Cypress	*Taxodiaceae*	Medlar
Metasequoia	Swamp Cypress	*Taxodiaceae*	Water larch
Morus	Mulberry	*Moraceae*	White, black mulberry
Myrtus	Myrtle	*Myrtaceae*	Myrtle
Nothofagus	Beech	*Fagaceae*	Beeches of southern hemisphere
Nyssa	Davidia	*Davidiaceae*	Sour gum, black gum, tupelo
Olea	Olive	*Oleaceae*	Olive
Ostrya	Hornbeam	*Carpinaceae*	Hop hornbeam
Oxydendrum	Heather	*Ericaceae*	Sorrel tree
Parrotia	Elm	*Ulmaceae*	Parrotia (Persian ironwood)
Paulownia	Bignonia	*Bignoniaceae*	Paulownia; empress tree
Phellodendron	Citrus	*Rutaceae*	Amur cork tree
Phillyrea	Olive	*Oleaceae*	*Phillyrea latifolia*
Phoenix	Palm	*Palmae*	Canary island palm; date palm
Photinia	Rose	*Rosaceae*	Chinese photinia
Picea	Pine	*Pinaceae*	Spruce

TABLE 3. (*Continued*)

Genus	Name of Family		Examples[a]
	English	Latin	
Pinus	Pine	*Pinaceae*	Pine
Pistacia	Cashew	*Anacardiaceae*	Chinese pistachio
Pittosporum	Witch Hazel	*Hamamelidaceae*	Pittosporums
Platanus	Plane	*Platanaceae*	Plane
Podocarpus	Podocarp	*Podocarpaceae*	Podocarp
Poncircus	Citrus	*Rutaceae*	Japanese bitter orange
Populus	Willow	*Salicaceae*	Poplar, cottonwood, aspen
Prunus	Rose	*Rosaceae*	Almond, apricot, gean, mazzard, various cherry, peach, Sloe, blackthorn, Portugal laurel
Pseudolarix	Pine	*Pinaceae*	Golden larch
Pseudotsuga	Pine	*Pinaceae*	Douglas firs
Ptelea	Citrus	*Rutaceae*	Hop tree
Pterocarya	Walnut	*Juglandaceae*	Wing-nut
Pterostyrax	Snowbell	*Styracaceae*	Epaulette tree
Pyrus	Rose	*Rosaceae*	Pear
Quercus	Beech	*Fagaceae*	Oak
Rhododendron	Heather	*Ericaceae*	Rhododendron, rose acacia, great laurel, rose bay
Rhus	Cashew	*Anacardiaceae*	Sumac; varnish tree
Robinia	Pea	*Leguminosae*	Black locust; mop-head acacia clammy-locust
Rosa	Rose	*Rosaceae*	Musk rose
Roystonia	Palm	*Palmae*	Royal palm
Sabal	Palm	*Palmae*	Caribbean cabbage palmetto
Salix	Willow	*Salicaceae*	Willow
Sambucus	Elder	*Caprifoliaceae*	Common elder; elderberry
Sassafras	Laurel	*Lauraceae*	Sassafras
Saxegothaea	Yew	*Taxaceae*	Prince Albert's yew
Schinus	Cashew	*Anacardiaceae*	California pepper tree
Sciadopitys	Pine	*Pinaceae*	Japanese umbrella pine
Sequoia	Swamp Cypress	*Taxodiaceae*	Coast redwood
Sequoiadendron	Swamp Cypress	*Taxodiaceae*	Giant Sequoia ("Big Tree")
Sophora	Pea	*Leguminosae*	Sophora
Sorbus	Rose	*Rosaceae*	Whitebeam; mountain ash; service tree
Stewartia	Tea	*Theaceae*	Stewartias (similar to camellia)
Styrax	Snowbell	*Styracaceae*	Snowbell
Taxoidium	Swamp Cypress	*Taxodiaceae*	Pond cypress; bald cypress; Mexican cypress
Taxus	Yew	*Taxaceae*	Yew
Tetraclinis	Cypress	*Cupressaceae*	North African cypress
Thuja	Cypress	*Cupressaceae*	Arborvitae; white cedar; thuja
Thujopsis	Cypress	*Cupressaceae*	False arborvitae
Tilia	Lime or Basswood	*Tiliaceae*	Basswood; linden; common lime (not to be confused with the citrus lime)
Torreya	Yew	*Taxaceae*	California nutmeg; Japanese kaya
Trachycarpus	Palm	*Palmae*	Chinese windmill palm
Tsuga	Pine	*Pinaceae*	Hemlock
Ulmus	Elm	*Ulmaceae*	Elm
Umbellularia	Laurel	*Lauraceae*	California laurel
Viburnum	Elder	*Caprifoliaceae*	Wayfaring tree; Gueldar rose
Washingtonia	Palm	*Palmae*	Petticoat palm
Yucca	Palm	*Palmae*	Joshua tree
Zelkova	Elm	*Ulmaceae*	Elm zelkova; Japanese zelkova

[a]Examples given in many cases are abridged. Also it is possible for some species of trees given in examples to appear in more than one genera because common names are not always scientifically specific.
Latin names appearing in Example column indicate no common English name.

The buffalo tree hopper, *Ceresa bubalus*, sometimes damages twigs of fruit trees by laying eggs in the bark and is occasionally troublesome in the garden through sucking the juices of young seedlings. Clean cultivation is an effective control.

TREE (Mathematics). A connected subgraph of a connected graph which contains all the vertices of the graph but is free of circuits. Occasionally the phrase "complete" tree is used but the definition given here is more in accord with modern engineering terminology. A graph can possess many trees. In Fig. 1 is an example of a graph with 4 vertices, 6 edges and 16 trees. See also the discussion under **Digraph**.

It can be shown that every finite connected graph G contains at least one tree. A tree contains v vertices and $v - 1$ branches where v is the number of vertices in G.

See also **Graph (Mathematics)**.

Fig. 1.   Record white oak (*Quercus alba*) located at Wye Mills, Maryland. The tree measures 31 feet, 10 inches (9.4 meters, 25 centimeters) in circumference at 41/2 feet (1.4 meters) above the base; is 96 feet (29.4 meters) high; and has a spread of 119 feet (36.3 meters). Selected as champion of species by American Forests in 1996. (*Maryland Department of Economic Development.*)

TABLE 4. PERCENT FORESTATION OF VARIOUS NATIONS

Nation	Percent Land Areain Forest	Percent in Coniferous Trees	Percent in Deciduous Trees	Nation	Percent Land Area in Forest	Percent in Coniferous Trees	Percent in Deciduous Trees
Austria	33	86	14	United Kingdom	15	54	46
Benelux	13	58	42	United States	21	47	53
Canada	23	61	39				
Denmark	10	50	50	Continent/Nation	Approximate Millions of Acres (Hectares)		
Finland	64	81	19	Russia	2,800 (1133)		
France	22	48	52	South America	2,400 (971)		
Germany	28	71	29	Africa	1,900 (769)		
Greece	18	58	42	North America	1,800 (728)		
Italy	20	45	55	Asia	1,300 (536)		
Norway	26	82	18	Europe	350 (142)		
Spain and Portugal	29	66	34	Pacific Area	250 (101)		
Sweden	57	75	25	Central America	180 (73)		
Switzerland	23	75	25				

TREE OF HEAVEN.   See **Mahogany Trees.**

TREE TOAD *(Amphibia, Anura).*   Small animals like the other frogs and toads in form but largely of arboreal habits. The toes are provided with adhesive disks at the tips. The spring peeper and the cricket frog, each about an inch in length, are well known North American species. Our tree frogs deposit their eggs in the water but in some species the larval stage is passed in the egg, which is attached to a leaf. The extensive family *Hylidae* includes tree frogs of every continent.

TRELLIS DRAINAGE.   The type of drainage pattern which is similar to a trellis or fireman's ladder. Trellis drainage is characteristic of the canoe-shaped valleys of the Appalachians.

TREMATODA.   The flukes, a class of flatworms (Phylum *Platyhelminthes*). They are very variable in size, are flattened or rounded in cross section, and are exclusively parasitic. Flukes resemble the free-living flatworms (*Turbellaria*) more closely than tapeworms, but in addition to their parasitic habits they differ in having the mouth usually near the anterior

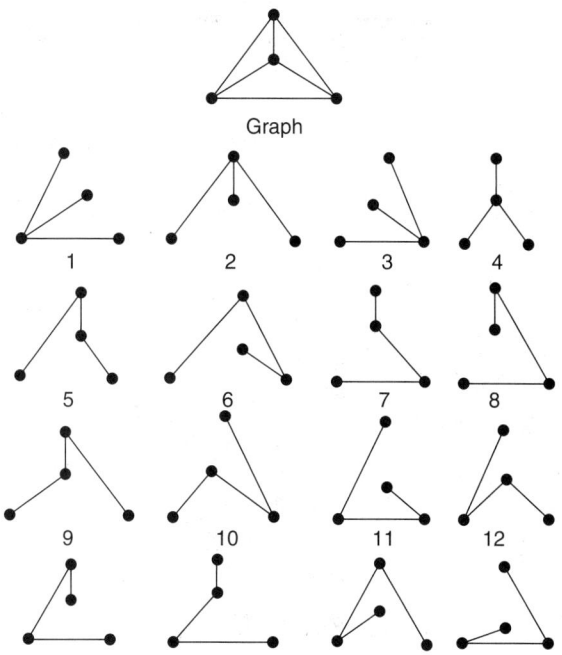

Fig. 1. Examples of graphs.

end of the body and the intestine usually forked and sometimes provided with anal openings. The body is covered with a cuticle and bears suckers and in some species hooks or spines by which the worm attaches itself to the host.

Many flukes hatch from the egg as free-swimming larvae and pass through a life cycle in two or more hosts before attaining maturity. The passage from host to host is associated with particular larval stages.

Some flukes attach themselves to the gills or skin of fishes while others live in the internal organs of various animals. Man is affected by four types, the liver, lung, blood, and intestinal flukes, found chiefly in tropical and Oriental countries. Since some of these parasites pass some stages of their metamorphosis in fishes, the eating of uncooked or imperfectly cooked fish is dangerous in regions where they occur.

TREMBLOR. See **Earth Tectonics and Earthquakes**.

TREMOLITE. The mineral tremolite is a calcium-magnesium silicate corresponding to the formula $Ca_2Mg_5Si_8O_{22}(OH)_2$, belonging to the amphibole group. The replacement of magnesium by ferrous iron causes tremolite to approach actinolite in composition. Tremolite is monoclinic, developing bladed prismatic crystals, but it is frequently found in compact columnar, granular, or fibrous masses. The perfect prismatic cleavage at angles of 56° and 124° typical of this group is to be noted; hardness, 5–6; specific gravity, 2.9–3.1; luster, vitreous to silky; color, varies from white or whitish-gray through shades of green or greenish-yellow; transparent to opaque. Tremolite is formed as a result of contact metamorphism and occurs in marbles, dolomites, and schists. It may alter to talc. Tremolite is found in Switzerland, in the St. Gotthard region, being named for the Tremola Valley, and is common elsewhere in Europe. In the United States it occurs in Maine, Pennsylvania, and New York. In Canada tremolite has been found in Quebec and Ontario.

Hexagonite is a pinkish-purple variety of tremolite which contains a small amount of manganese. So called because it was at first believed to be hexagonal. It has been since shown to be monoclinic, and is found in St. Lawrence County, New York. Some nephrite and asbestos is tremolite.

See also terms listed under **Mineralogy**.

TRENCH FOOT. A serious disease of the foot usually precipitated by lack of attention to the feet as may be encountered by military personnel. Exposure of limbs to low temperatures, particularly under moist conditions, in the range of freezing to about 41°F (5°C) produces a constriction of blood vessels. A net result after prolonged periods of exposure is an edema

and blistering which the victim may not sense because of numbness at the low temperature. Infection or gangrene may follow. Persons who must be exposed in this manner to wet, cold conditions over prolonged periods should wear specially-designed footwear and should warm and thoroughly dry the feet as frequently as possible. In very serious situations, antibiotics may be administered to counter infection, accompanied by rest during which the feet should be elevated. Frequent removal of wet socks and replacing with clean, dry socks is an excellent preventive measure.

R. C. V.

TRENCH (Ocean). See **Earth Tectonics and Earthquakes**.

TREPANNING. Originating a hole by removing an annular ring of material, instead of reducing to chip form the entire volume of the material originally within the hole.

TREPHINE. 1. A hollow, circular, saw-tooth, drill-like instrument that is used for removing a disk of bone from the skull. This opening may then be enlarged or several of these drill holes may be made or connected so as to raise a piece of bone to expose a portion of the brain.

2. To open the skull with a trephine. The skull is usually operated upon for removal of an intracranial tumor, to relieve intracranial pressure, or to drain a brain abscess.

TRIANGLE. A polygon with three sides and three angles. If the three angles all lie in the same plane, the triangle is called plane; if not, so that a spherical polygon is involved, the triangle is also called spherical. The word *triangle*, however, is usually taken to mean a plane triangle. From their properties, they are named as follows: acute, all interior angles acute; congruent, two or more triangles which may be superimposed and which then become equal in all of their properties; equal, with equal areas but not necessarily congruent; equiangular, with three equal interior angles and therefore also equilateral with three equal sides; isosceles, two sides are equal; oblique, containing no right angle; obtuse, containing one obtuse interior angle; right, one angle is $\pi/2$ (the opposite side is the hypotenuse and the other sides are the two legs); scalene, no sides are equal; similar, the angles are equal and the corresponding sides are proportional.

Given a sufficient number of parts of a triangle (sides and angles), the remaining parts may be found, a process called solving the triangle. This is the main concern of plane trigonometry, as taught in elementary courses.

If two arcs of a great circle intersect, a spherical angle is formed. It is measured by the plane angle between the tangents to the circles at their intersection. A polygon with three sides and three spherical angles is a spherical triangle. It may be right, oblique, equilateral, isosceles, or scalene, similar to the corresponding plane triangle. The sum of its angles is greater than π (180°) but less than 3π (540°). The difference between the sum of its angles and π is the spherical excess, E. If the triangle is cut from a sphere of radius R, the area of this spherical triangle is $S = \pi R^2 E/180$, where E is expressed in degrees.

If the vertices of a spherical triangle are taken as poles and arcs of great circles are drawn through them, a second spherical triangle results. It is called the polar triangle of the first.

Spherical triangles, like plane triangles, may be solved. The methods of spherical trigonometry, a generalization of plane trigonometry, are used.

See also **Polygon**; **Trigonometric Function**; and **Trigonometry**.

TRIANGULATION. In surveying, the field work necessary to obtain the angular measurements between the sides of a series of connected triangles, and the length of one or more of the sides, is known as triangulation. The system of connected triangles is called a triangulation system. The topographic surveying of land requires the establishment of control points of known position and elevation. This control usually takes the form of a series of surveyed triangles, with the apices located on prominent points in the area, such as peaks, mounds, cliffs. In laying down a network of triangles, a fairly level region is selected on which to measure a straight base line. The transit is then set over one end of this base line, and the angles between pairs of selected control points taken. The transit is then taken to the other end of the base line, and the angles are again taken to the same points. By occupying, progressively, advancing triangulation points the network may be extended over as large an area as necessary. These angles are read to a high degree of precision by very accurate instruments, using the method of repetition. Even then the angular data

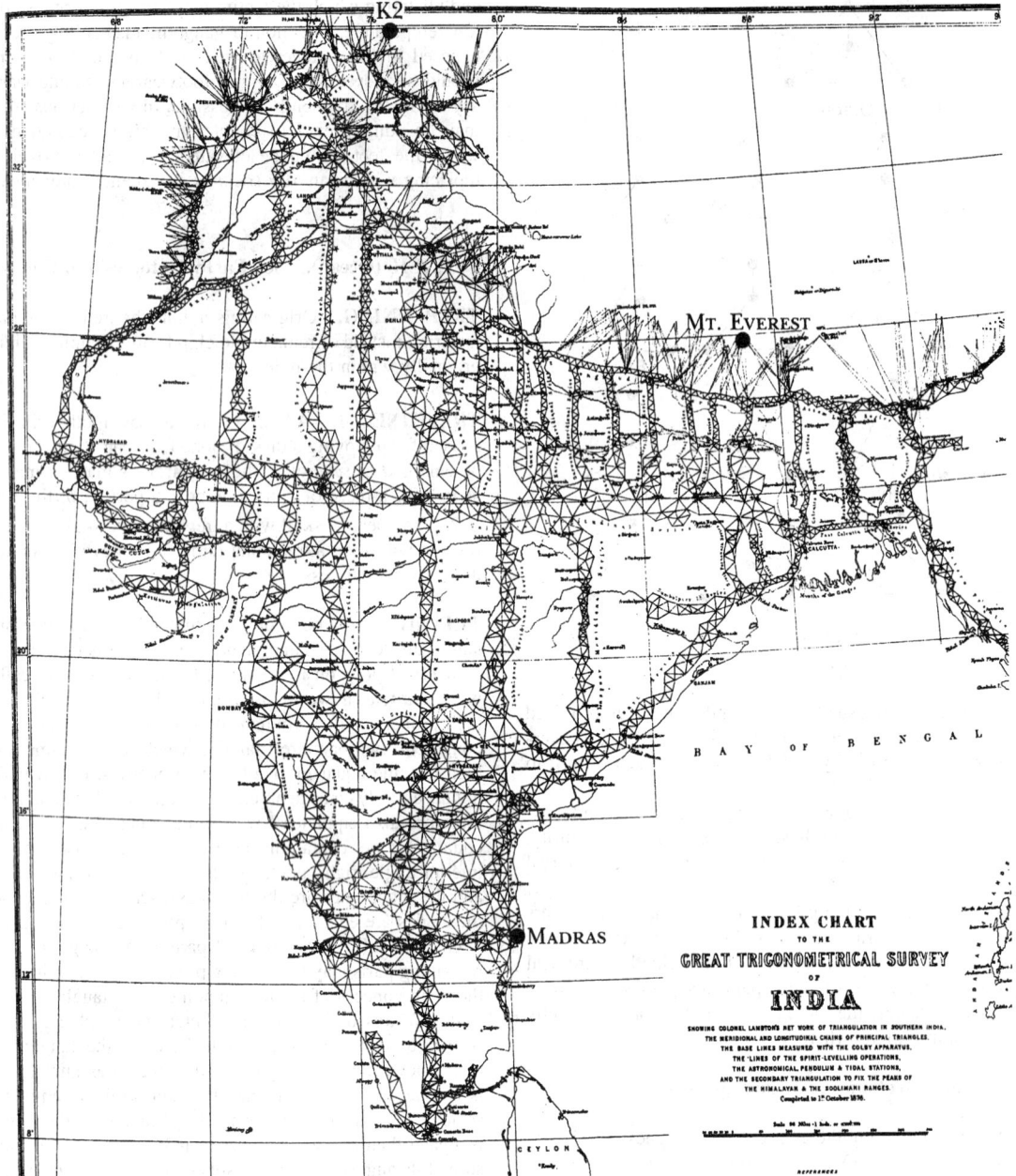

Fig. 1.   Mid-1870s triangulation map of India, referred to as the "Great Trigonometrical Survey." (*Republic of India, New Delhi.*)

as taken in the field will not be consistent, one triangle with another or with the whole. Such small errors or inconsistencies as exist must be adjusted, or balanced, by distributing the error. This is a fairly specialized and intricate process but must be executed to obtain a consistent system of triangles of highest probable accuracy. Upon the completion of this process the position of any triangulation point with respect to any other or with respect to the base line may be very accurately computed. The United States Coast and Geodetic Survey has covered the country with a triangulation system, the sides of the triangles often being many miles in length. The accuracy of triangulation depends upon the accuracy with which the instrument is constructed and read, and the precision with which the base line is measured. The precision of triangulation is classified as first order, second order, third order, etc., first order triangulation being that in which the base line is required to be measured to an accuracy of 1 part in 25,000, and the average triangle closure being only one second. The geodetic surveying calls for first and second order triangulation, but ordinary intermediate work is satisfactory with third or fourth order triangulation.

An early map of India (Fig. 1) exemplifies the "Great Trigonometrical Survey" of India completed in the mid-1870s. The survey commenced in Madras and included the great mountain peaks of the Himalayas. On the

master map, 1 inch (2.54 cm) was equal to 96 miles (154.5 km). (*Republic of India, New Delhi.*)

TRIANGULUM AUSTRALE.   A southern constellation of a triangular configuration and located near Circinus.

TRIANGULUM (the triangle).   A northern constellation that lies between Andromeda and Aries.

TRIASSIC PERIOD.   The earliest, major subdivision of the Mesozoic Era of the geologic time-scale. One of the oldest systemic terms and denoting a three-fold division of the German formations into the lower or Bunter sandstones, the middle or Muschelkalk limestones, and the upper or Keuper copper-bearing shales. The term Triassic was proposed by Alberti in 1834. The period began about 200 million years ago and lasted for about 50 million years. The maximum thickness of formations, 25,000 feet (7,500 meters), occurs in the Alps. In eastern North America, especially in Massachusetts, Connecticut, and New Jersey, occurs a thick series of red sandstones, arkoses, shales, and argillites of freshwater origin, containing the footprints of the earliest known dinosaurs. Interbedded with the sedimentary formations are numerous basalt lava flows and sills. This eastern facies of the Triassic is called the Newark Series after the type locality in New Jersey. In Arizona and New Mexico occur freshwater clastic sediments containing the prostate fossil trunks of ancestral Sequoias in which the original body structures have been perfectly replaced by silica. Some of the tree trunks are over 150 feet (45 meters) long and 3–6 feet (0.9–1.8 meters) in diameter. In the Cordilleran region, the Triassic formations are marine. Triassic rocks also occur in South America, British Isles, Western Europe, Asia, Africa, and Australia. During this period there were great changes in the plants and animals, as disclosed by the fossils. Ferns, cycads, and conifers predominated among the plants. The modern corals, hexacoralla, predominate over the earlier tetracoralla of the Paleozoic, Cystoids and Blastoids have become extinct. The brachiopods are less abundant, their place being taken by the pelecypods. Modern insects begin their development in this period, and the modern (bony) fishes are ascendant. Among the terrestrial animals, the Paleozoic amphibia (*Stegocephalia*) are replaced by the dinosaurs. The highest types of marine invertebrates are ammonites. Marine reptiles, called Ichthyosaurs, and flying reptiles called Pterosaurs, first appeared in Europe. Small reptilian-like mammals also made their first appearance. The economic products of this period are chiefly salt, gypsum, and copper. In eastern North America the period closed with relatively slight uplift and block-faulting of the Newark Series, called the Palisades Disturbance. In the Pacific Coast region there was a withdrawal of the marine waters to mark the close of the period.

TRIBE. See **Taxonomy**.

TRIBOLUMINESCENCE. See **Luminescence**.

TRICHROISM. The property of exhibiting three different colors when viewed in as many different directions.

TRICHROMATIC COEFFICIENTS. Three coefficients, based on the response of the standard eye to the spectral distribution of light from a standard source, which may be used to describe the chromaticity of a source.

TRICKLE IRRIGATION. See **Irrigation**.

TRIDOP. A continuous-wave trajectory measuring system using the Doppler shift caused by a target moving relative to a ground transmitter and three or more receiving stations.

TRIDYMITE. The mineral tridymite is, like quartz, silicon dioxide, SiO_2, but is a high-temperature variety, probably stable above 870 °C. It has a conchoidal fracture; is brittle; hardness, 7; specific gravity, 2.28–2.33; vitreous luster; usually colorless and transparent. It is found chiefly in volcanic rocks of the more acidic types like rhyolite, tachyte, and andesite. It is not a particularly uncommon mineral, occurring in Germany, France, Italy, Japan, the Island of Martinique, Mexico, and in the United States in Wyoming and Washington. Tridymite is hexagonal but when heated to about 1,470 °C passes into an isometric form, cristobalite, which was first noted in the andesitic lavas of the Cerro San Cristobal, Pachuca, Mexico, together with tridymite. Cristobalite has been found also in California and in Germany.

TRIGEMINAL NEURALGIA. See **Neuralgia**.

TRIGLIDS. See **Sea Robins**.

TRIGONOMETRIC CURVE. Plots of the trigonometric functions are shown in Figs. 1–6. See also **Inverse Trigonometric Function**; **Trigonometric Function**; and **Trigonometry**.

TRIGONOMETRIC FUNCTION. If q is one of the acute angles in a right-angle triangle, x is the side of the triangle nearest to θ, y the side opposite the angle, and r the hypotenuse, the trigonometric functions are:

$$\sin \phi = y/r; \cos \phi = x/r$$

$$\tan \phi = y/x; \csc \phi = 1/\sin \phi$$

$$\sec \phi = 1/\cos \phi; \cot \phi = 1/\tan \phi$$

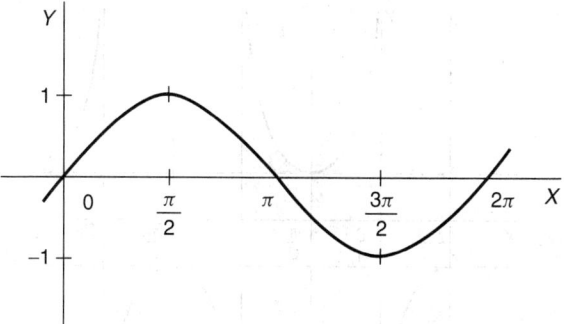

Fig. 1. $y = \sin x$.

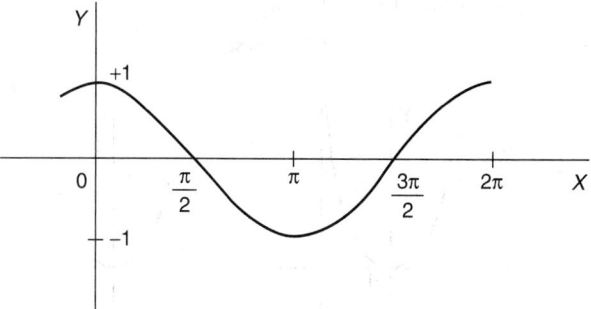

Fig. 2. $y = \cos x$.

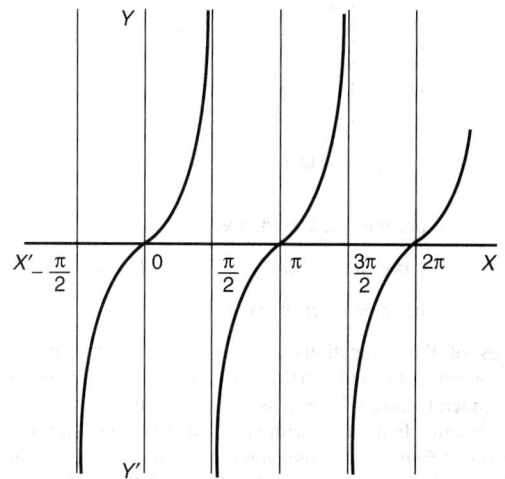

Fig. 3. $y = \tan x$.

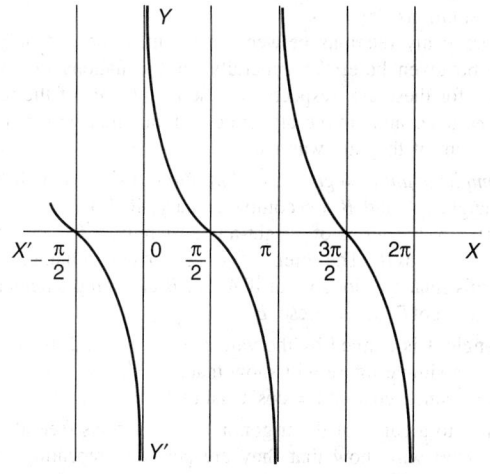

Fig. 4. $y = \cot x$.

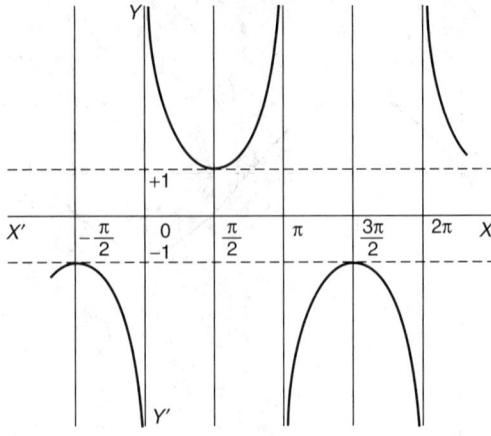

Fig. 5. $y = \csc x$.

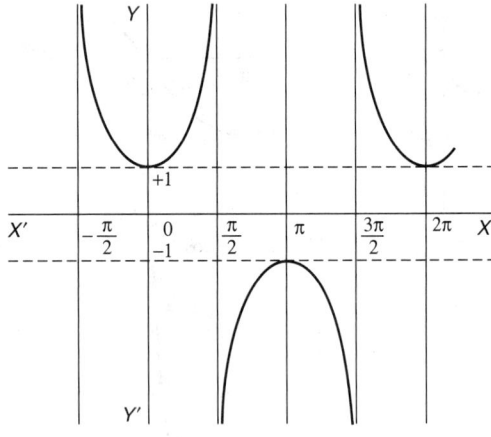

Fig. 6. $y = \sec x$.

Other quantities, but not often used, are:

$$\operatorname{vers}\phi = 1 - \cos\phi; \operatorname{covers}\phi = 1 - \sin\phi$$

$$\operatorname{hav}\phi = \tfrac{1}{2}\operatorname{vers}\phi; \operatorname{exsec}\phi = \sec\phi - 1$$

The names of these functions are, respectively: sine, cosine, tangent, cosecant, secant, cotangent, versine, coversine, haversine, exsecant.

In the general case, where ϕ is not an acute angle in a right-angle triangle, the same definitions, with slight modification, can be used. Place the angle ϕ in a right-handed rectangular coordinate system with its vertex at the origin and its initial side on the positive OX-axis. The terminal side of the angle may lie in any one of the four quadrants but select any point on it (x, y) and draw a radius vector r to the point from the origin. The three letters x, y, r will then determine the trigonometric functions by the previous definitions.

There are many relations between these functions and only a few of them can be given here. We generally omit equations involving $\csc\phi$, $\sec\phi$, $\cot\phi$ for these are, respectively, the reciprocals of the relations for $\sin\phi$, $\cos\phi$, $\tan\phi$ and simple algebraic manipulation will give equations involving them, if they are wanted.

1. *Complementary, Negative, and Multiple Angles*. It will be recalled that two angles, A and B are complementary, if $A + B = 90°$. From the defining relations in terms of a right-angle triangle, it follows that $\sin A = \cos(90° - A) = \cos B$. The prefix "co" in cosine, cotangent, cosecant indicates this relationship, so that if A and B are complementary, $\sin A = \cos B$, $\tan A = \cot B$, $\sec A = \csc B$.

If the angle A is defined by the values of x, y, r and its negative $(-A)$ by x', y', r' a simple figure will show that $x = x'$, $y = -y'$, $r = r'$, hence $\sin(-A) = -\sin A$, $\cos(-A) = \cos A$, $\tan(-A) = -\tan A$.

Reference to a curve of the trigonometric functions (see also **Trigonometric Curve**) will show that they are periodic, repeating their values in multiples of $360°$. In addition, the tangent and cotangent have periods of $180°$. All of the functions also change their signs as the angles move

through these periods. The correct sign can, however, be found by reducing a given angle to the first quadrant where all of the functions are positive. This can be done readily from Table 1, which also contains the information from the first part of this section, as well as numerical values for the functions of some frequently used angles.

TABLE 1. FUNCTIONS OF FREQUENTLY USED ANGLES

	$-A$	$90° \pm A$	$180° \pm A$	$270° \pm A$	$360° \pm A$
sin	$-\sin A$	$\cos A$	$\mp \sin A$	$-\cos A$	$\pm \sin A$
cos	$\cos A$	$\mp \sin A$	$-\cos A$	$\pm \sin A$	$\cos A$
tan	$-\tan A$	$\mp \cot A$	$\pm \tan A$	$\mp \cot A$	$\pm \tan A$

	$0°$	$30°$	$45°$	$60°$	$90°$
sin	0	$1/2$	$\sqrt{2}/2$	$\sqrt{3}/2$	1
cos	1	$\sqrt{3}/2$	$\sqrt{2}/2$	$1/2$	0
tan	0	$\sqrt{3}/3$	1	$\sqrt{3}$	∞

2. *Fundamental Identities*. These relations, eight in number, make it possible to derive all of the equations in the following sections. Easily proved from the definitions of the functions, they are: (a) the three reciprocal relations for $\sin\phi$, $\cos\phi$, $\tan\phi$; (b) the quotients, $\tan\phi = \sin\phi/\cos\phi$ and its reciprocal; (c) the Pythagorean equations, $\sin^2\phi = \cos^2\phi = 1$, $\sec^2\phi = \tan^2\phi + 1$, $\csc^2\phi = 1 + \cot^2\phi$.

3. *The Addition Formulas*. If A and B are any angles $\sin(A \pm B) = \sin A \cos B \pm \cos A \sin B$; $\cos(A \pm B) = \cos A \cos B \mp \sin A \sin B$; $\tan(A \pm B) = \tan A \pm \tan B)/(1 \mp \tan A \tan B)$.

4. *Double-angle Formulas* These are special cases of (3). $\sin 2A = 2 \sin A \cos A$, $\cos 2A = \cos^2 A - \sin^2 A$, $\tan 2A = 2 \tan A/(1 - \tan^2 A)$.

5. *Half-angle Formulas*. These can be obtained from (4). $\sin^2 A/2 = (1 - \cos A)/2$, $\cos^2 A/2 = (1 + \cos A)/2$, $\tan A/2 = (1 - \cos A)/\sin A$.

6. *Product Formulas*. $2 \sin A \cos B = \sin(A + B) + \sin(A - B)$, $2 \sin A \sin B = \cos(A - B) - \cos(A + B)$, $2 \cos A \cos B = \cos(A + B) + \cos(A - B)$.

7. *Sums and Differences*. These may be found from (6). $\sin A \pm \sin B = 2 \sin(A \pm B)/2 \cos(A \mp B)/2$, $\cos A + \cos B = 2 \cos(A + B)/2 \cos(A - B)/2$, $\cos A - \cos B = -2 \sin(A + B)/2 \sin(A - B)/2$.

8. *Series for the Functions*. Taylor series, or other methods, can be used to obtain the infinite series:

$$\sin x = x - \frac{x^3}{3!} + \frac{x^5}{5!} - \frac{x^7}{7!} \pm \cdots; x^2 < \infty$$

$$\cos x = 1 - \frac{x^2}{2!} + \frac{x^4}{4!} - \frac{x^6}{6!} \pm \cdots; x^2 < \infty$$

$$\tan x = x + \frac{x^3}{3} + \frac{2x^5}{15} + \cdots + \frac{2^{2n}(2^{2n} - 1)}{(2n)!}$$

$$\times B_{2n-1} x^{2n-1} + \cdots; x^2 < \pi^2/4?\text{VS}, \%$$

In the last equation, the coefficients are the Bernoulli numbers. If x is small, these series show that approximate values are $\sin x \approx x$, $\cos x \approx 1 - x^2/2$, $\tan x \approx x$, where, as in the three expansions above, x must be expressed in radians.

See also **Inverse Trigonometric Function**; **Spherical Trigonometry**; and **Trigonometry**.

TRIGONOMETRY. The mathematical study of the triangle, normally the plane triangle. The subject seems to have been invented by the Greek Hipparchus, born about 160 B.C., and further developed by Ptolemy of Alexandria, who died in 168 A.D., and by the Hindu and Arabian mathematicians.

The main problem of trigonometry is the finding of unknown parts of a triangle from given data. For a right triangle, two parts must be given in addition to the right angle and one of them must be a side. Let A, B, C be the angles and a, b, c the sides opposite them, respectively. Then if the right angles is C, the remaining parts of the triangle are found from the relation $A + B = 90°$, and the Pythagorean theorem, $a^2 + b^2 = c^2$. Alternatively, suitable combinations of the trigonometric functions may be used.

For an oblique triangle, three of the six parts may be found if three others are given, including one side. One possibility is the division of the oblique triangle into two right triangles and the solution of each of these as described in the preceding paragraph. One could also use:

(1) the law of sines, $a/\sin A = b/\sin B = c/\sin C$; (2) the law of cosines, $a^2 = b^2 + c^2 - 2bc\cos A$; (3) the law of tangents,

$$(a-b)/(a+b) = \frac{\tan(A-B)/2}{\tan(A+B)/2};$$

(4) the half-angle formula, $\tan A/2 = r/(s-a)$, where the perimeter of the triangle, $2s = (a+b+c)$ and r is the radius of a circle inscribed in it, $sr^2 = (s-a)(s-b)(s-c)$. Additional relations can be obtained for (2)–(4) by cyclic permutation of the letters a, b, c and A, B, C.

There are four special cases for an oblique triangle, as follows:

Case	Given	Equation to Use
I	A, B, a	(1)
II	a, b, A	(1)
III	a, b, C	(3) for A, B (1) for c
IV	a, b, c	(2) or (4)

Case II, which is called the ambiguous case, may give two, one, or no solutions. The possibilities are: no solution, $A < 90°$, $a < b\sin A$ or $a \leq b$; one solution, an oblique triangle, $A < 90°$, ab or $A > 90°$, $a > b$, a right triangle, $A < 90°$, $a = c\sin A$; two solutions, oblique triangle, $A < 90°$, $b > a > b\sin A$.

The solution of an oblique triangle is conveniently checked by the Mollweide formulas:

$$(a+b)/c = \frac{\cos(A-B)/2}{\sin C/2}; (a-b)/2 = \frac{\sin(A-B)/2}{\cos C/2}$$

and two other equations, found by cyclic permutation of the letters, a, b, c and A, B, C.

The area of a triangle is given by $2H = \text{base} \times \text{altitude}$. Equivalent expressions are:

(5) $\qquad\qquad H^2 = s(s-a)(s-c);$

(6) $\qquad\qquad 2H = ab\sin C;$

(7) $\qquad\qquad 2H = a^2\dfrac{\sin B\sin C}{\sin(B+C)}.$

Again, other relations may be obtained by cyclic permutations. To find the area in case I, use (7); case III, (6); case IV, (5).

Tables of logarithms are available for the trigonometric functions and triangles are usually solved by logarithmic calculations.

For further relations between angles, see also **Trigonometric Function**. See also **Spherical Trigonometry** for the solution of spherical triangles.

TRIM (Airplane). The trim of an airplane denotes the balance of aerodynamic and thrust forces around a lateral axis through the center of gravity. An airplane would be in trim if it had no tendency to change from its angle of attack with respect to the flight path in steady flight after it has been brought to that angle of attack by a specific angular deflection of the elevator.

TRIMMER CAPACITOR. A small capacitor used to adjust a tuned circuit to the desired resonance frequency. Essential in tracking superheterodyne receivers.

TRIPHYLITE. This mineral is a phosphate of lithium and ferrous iron, $LiFePO_4$. It crystallizes in the orthorhombic system but usually is characterized by large cleavable masses. The hardness is 4.5–5.0; specific gravity, 3.42–3.56, vitreous to resinous luster, translucent, and bluegray color. The mineral occurs as a rare primary mineral in granitic pegmatites and, when available in large quantities, is a source of lithium. Worldwide occurrences include Bavaria, Finland, Sweden, and in the United States, New Hampshire, Maine, and South Dakota.

TRIPLE BOND. See **Chemical Elements**.

TRIPLE POINT. See **Three-Phase Equilibrium**.

TRIPOLITE. A term frequently used synonymously with diatomaceous earth, but more particularly with reference to diatomaceous earth from Tripoli in North Africa. See also **Diatom**; and **Diatomite**.

TRISECTRIX OF MACLAURIN. A higher plane curve, a special case of the cissoid, which is represented in rectangular coordinates by the equation

$$2x(x^2 + y^2) = a(3x^2 - y^2)$$

and in polar coordinates by

$$r\cos\phi/3 = a/2$$

It is also related to the rose curves and thus is a member of the trochoidal type of cyclic curves.

Another trisectrix, that of Longchamps, has the polar equation $r\cos 3\phi = a$. Both can be used in the classical problem of angle trisection.

See also **Cissoid of Diocles**; and **Curve (Higher Plane)**.

TRISMUS. See **Tetanus**.

TRISTIMULUS VALUES. Any sample of light can be matched by the mixture of three different standard stimuli. The magnitudes of the three stimuli needed to match a particular sample are called the tristimulus values of that sample. The standard stimuli must be specified. Those most commonly used are the CIE color mixture functions x, y, z, of which y is the luminosity function.

TRITIUM. The radioactive isotope of hydrogen, with a mass number 3, is termed tritium. It is one form of heavy hydrogen, the other form being deuterium. See also **Nuclear Power Technology**.

TRITIUM (Reactor). See **Nuclear Power Technology**.

TROCHOID. The path described by a point on a line drawn through the center of a circle as it rolls along a straight line. It is a special case of the cyclic curve and its equation can be obtained from that case by the following limiting process:

Let $(R+r)\phi/r = \theta$, $Y = x - R\cos\phi$, $X = y$ and as $R \to \infty$, $R\sin\phi$ becomes $r\theta$ and $\cos\phi$ becomes unity. Finally, replace X and Y by x, y, respectively and r, a by a, b to get the usual form of the equation, $x = a\theta - b\sin\theta$, $y = a - b\cos\theta$.

The given line on which the circle rolls is the X-axis, the generating point is originally on the Y-axis, a is the radius of the circle, b is the distance of the point from the center of the circle, and q is the angle through which the circle has rolled. If $b = a$, the curve is called a cycloid; if $b < a$, the trochoid is curtate and if $b > a$, prolate. Nodes appear in the prolate case, they become cusps for the cycloid, and minima when the trochoid is curtate.

See also **Cycloid**.

TROGON (*Aves, Trogoniformes*). A brilliantly colored and beautifully feathered bird of the Oriental region, tropical Africa, and Central and South America. A single species, the coppery-tailed trogon, *Trogon ambiguus*, enters southern Texas and Arizona. These birds have a moderately large beak, strong and curved. Crests and plumes are highly developed in some of the males. The males also have very long tails. The *quezal* of South America is one of the most widely known of the trogons, probably because it has been pictured numerous times on the postage stamps of Guatemala and partly because its brilliant colors persist in museum specimens. The male is brilliant green with blood-red under parts below the breast. It has a very long tail, drooping plumes over the wings, and the head bears a rounded crest. See also **Trogoniformes**.

TROGONIFORMES (*Aves*). This order which is made up of the Trogons (family *Trogonidae*) are among the most colorful birds we know. They have long, staggered tails and short, powerful beaks which they use to make their brood holes in decaying tree trunks.

Their size ranges from that of a grosbeak to that of a magpie. The breast and the underside of the males are predominantly colored in brilliant shades ranging from red to yellow. They frequently have a magnificent green color above. The plumage of the females is usually not as rich in contrasting colors. The feet are notably weak and the arrangement of the toes is unique:

not only the first but also the second toe is immovable and reversed, while only the third and fourth are directed forward (heterodactyly).

Trogons are usually seen singly or in pairs, but only rarely in small groups. They also occur in higher mountain regions and thus in cooler zones. In the luxuriant tropical forests they are not especially conspicuous despite their brilliant coloring, for they usually sit quietly on a branch or a liana in a strangely upright posture, with the tail pointed directly downward. From these lookout points they take off on short flights to hunt for an insect or to pick fruit in flight. They make their presence known by their loud calls, which are distinctive enough in the individual species to tell them apart.

Although this order, as hole-brooders, make their nesting chambers in decaying tree trunks or stumps, they also use abandoned woodpecker hollows. Some species use the nests of tree termites and even of wasps in which to build their brooding holes. Both sexes share in the building of the nest, but they do not bring nesting materials; the clutch is deposited on the bare floor of the brood chamber. The eggs are nearly globular in shape, white or pastel, but brownish and blue shades also occur. Incubation is shared by both parents.

There are 8 genera with 34 species in the tropics of Central and South America, Africa, and southern Asia. The best known of all trogons is the Quetzal (*Pharomachrus mocino*). This bird was much revered by the ancient civilizations of the Mayas and the Aztecs, and is still the heraldic emblem of Guatemala, whose monetary unit is also named for it. Red and green are beautifully contrasted in the plumage of the quetzal; added to this are the extraordinarily elongated upper tail coverts of the male, forming a magnificent train which can measure up to 1.05 meters ($3\frac{1}{2}$ feet). See also **Trogon**.

TROPHALLAXIS.   An interchange of nourishment between the various members of colonies of insects, reciprocal feeding. The older idea that instinctive parental care is the basis of the concern manifested by ants and termites for the welfare of other individuals of the colony has been substituted by recognition of this interchange. The insects exude secretions at the surface of the body, which are licked off by others. This reward is apparently responsible for much of the solicitude. Even the insects of other species living within such colonies are welcome guests in some cases because of a similar contribution to their hosts. The relationship has complex ramifications which explain other aspects of colonial organization, such as cannibalism, that are difficult to reconcile with the older interpretations of mutual aid within the colony.

TROPHOSOME.   One of several terms used to designate individuals that take in and digest food in colonies of primitive animals, or the stage of development of an individual during which it eats and grows. Trophosome is applied to the polyp stage of hydrozoa, also called gasterozooids. The term trophozooid is applied to an early stage in the development of the coral polyps of the family *Fungidae*, known as mushroom corals, and the term autozooid to individuals that take food in alcyonarian colonies. The root appears again in trophozoite, applied to the growing stage of some parasitic Protozoa (*Sporozoa*).

TROPICAL DISEASES.   These are generally considered to be those ailments occurring mainly, but not necessarily, in the 47 degrees of latitude between the Tropic of Cancer (23.5°N) and the Tropic of Capricorn (23.5°S) of the equator. This area, which presents a large degree of uniformity of warm temperature and only mild seasonal variation in weather patterns, is eminently predisposed to the ecological cultivation of organisms and disease vectors inimical to human health. Of the 4.5×10^9 people populating the earth, some 55% inhabit tropical and subtropical zones and present annually with some 600 million cases of infectious and parasitic diseases, leading to as many as 20 million deaths per year from such diseases as diarrhea, malaria, schistosomiasis, and amebiasis. Malnutrition further exacerbates symptomology in affected individuals and can by itself lead to death.

The climatic conditions of the tropics are so aligned to the development of organisms causing disease in humans that the area has been said to present a boundary to the well-being of humans versus that of nonhumans. Because of these climatically induced dangers to human health, many tropical areas have remained relatively underdeveloped despite their enormous economic potential in lumber, oils, gums, fibers, and fruit, among other important items of commerce.

In tropical countries, diarrheal disease is the major infectious cause of death and this syndrome can be caused by many organisms — for example, *E. coli, V cholerae, Salmonella* sp., *Shigella* sp., and rotavirus. Diarrheas due to amebiasis, giardiasis and intestinal helminths must be considered separately. In developing countries, adults usually suffer from at least one attack of diarrhea annually; children under five years of age have an incidence of two to ten episodes annually. Thus, one can expect to encounter some 4.5 billion cases per year, with a case fatality rate which varies from 1% in the poorest, most malnourished children to less than 0.1% among adults. The effect of this infection rate upon the health of the work force is extensive and the economic effects are high.

Respiratory infections derive from viruses, adenovirus, and bacteria and cause half to two-thirds of the morbidity and mortality in the tropics. Acute tuberculosis is prevalent in approximately 6 to 8 million residents, and of these some 500,000 die annually. Adult men over the age of 25 years, in the economically productive years of life, sustain the highest incidence and death rates for the active disease, thus heavily burdening families dependent upon them for support.

Despite all attempts to control the vector mosquito and all chemotherapeutic measures, malaria still causes an enormous number of deaths. Schistosomiasis infects some 200 million people in the tropics and is spreading as agricultural irrigation projects progress. The extent of water contact largely controls the prevalence and intensity of this parasitic disease, but the majority of individuals remain asymptomatic regardless of the intensity of the infection. As in all tropical infectious diseases, the spread of contagion is facilitated by the usually low socioeconomic status, crowded and insanitary living conditions of many residents.

By the end of this century, 75% of all the people in the world to receive medical attention will be citizens of the tropical developing world. One third of those people now requiring care are in the low-income countries — those whose per capita gross national product amounts to $300 per year or less. Only about 6–10% of any country's GNP (gross national product) is spent on health care. For the poorest countries, this means a total of $6 to $10 per year per person, of which the government provides only about half. When it is realized that the provision of water and sewerage costs between $10 and $32 per capita, the paucity of medical facilities in the tropical field can be placed in perspective.

Over 2.5 million Americans live or travel abroad each year, many returning from regions where dangerous sanitation conditions exist. An estimated 8 to 9 million visitors from foreign countries add to the problems of public health in North America, because 20% of the visitors come from countries where diseases such as typhoid, cholera, yellow fever, or plague are endemic, or even epidemic. With increasing speeds of travel and the number of immigrants from the tropical countries, diseases normally indigenous to the area are appearing with increasing frequency in many of the temperate climate countries from which they were once excluded by virtue of higher standards of hygiene and medical attention. Quarantine is no longer of importance and is rarely used. Because of the speed of travel, epidemics can now span the world almost overnight. An infectious disease acquired, for example, in the tropical zone of Australia, can have its whole incubation period spent in transit over Asia and Europe, to become manifest hours later in North America.

The field of tropical medicine is too wide to be covered in one article in this encyclopedia and reference should be made to specific tropical diseases and their vectors as given in the alphabetical index.

Research on these tropical diseases, their prevention and cure, is being strongly pursued in North America, as for example by the National Institutes of Health in the United States, by the Schools of Tropical Medicine at Liverpool and London in Great Britain, and worldwide, by the World Health Organization (WHO), centered in Switzerland. Most emphasis by these groups is placed on malaria, schistosomiasis, filariasis, and leprosy.

See also **Bacterial Diseases**; and **Infectious Diseases**.

Additional Reading

Arya, O.P.P. and C.A. Hart: "Sexually Transmitted Infections and AIDS in the Tropics," Oxford University Press, Inc., New York, NY, 1999.

Canizares, O.: "A Manual of Dermatology for Developing Countries," 2nd Edition, Oxford University Press, Inc., New York, NY, 1993.

Cherfas, J.: "New Hope for Vaccine Against Schistosomiasis," *Science*, 630 (February 8, 1991).

Cook, G.C.: "Manson's Tropical Diseases," 20th Edition, W.B. Saunders Company, Philadelphia, PA, 1995.

Greenwood, B.M. and K. De Cock: "New and Resurgent Infections: Prediction, Detection, and Management of Tomorrow's Epidemics," John Wiley & Sons, Inc., New York, NY, 1998.

Guerrant, R.L., D.H. Walker, and P.F. Weller: "Essentials of Tropical Infectious Disease," Churchill Livingstone, Inc., Philadelphia, PA, 2001.

Guerrant, R.L., D.H. Walker, and P.F. Weller: "Tropical Infectious Diseases: Principles, Pathogens, and Practice," Volumes 1 & 2, Churchill Livingstone, Inc., Philadelphia, PA, 1999.

Jaret, P., M. Naythions, and S. Franklin: "The Disease Detectives," *Nat'l. Geographic*, 114 (January 1991).

Khusmith, S., et al.: "Protection Against Malaria by Vaccination," *Science*, 715 (May 3, 1991).

Liew, F.Y.: "Vaccination Strategies of Tropical Diseases," CRC Press, LLC., Boca Raton, FL, 1989.

Marshall, E.: "Malaria Vaccine on Trial at Last?" *Science*, 1063 (February 28, 1992).

Neva, F.A.: "Immunotherapy for Parasitic Diseases," *N. Eng. J. Med.*, 55 (January 4, 1990).

Oehen, S., H. Hengartner, and R.M. Zinkernagel: "Vaccinations for Disease," *Science*, 195 (January 11, 1991).

Palmer, P.E.S. and M.M. Reeder: "Imaging of Tropical Diseases: With Epidemiological, Pathological and Clinical Correlation," Volumes 1 & 2, 2nd Edition, Springer-Verlag, Inc., New York, NY, 2001.

Prabhu, S.R., N.W. Johnson, D.F. Wilson, and D.K. Daftary: "Oral Diseases in the Tropics," Oxford University Press, New York, NY, 1992.

Rennie, J.: "Proteins 2, Malaria O," *Sci. Amer.*, 24 (July 1991).

Rennie, J.: "Birds of a Fever," *Sci. Amer.*, 25 (July 1991).

Rose, F.C.: "Recent Advances in Tropical Neurology," Elsevier Science, New York, NY, 1997.

Strickland, G.T., A.J. Magill, and R. Kersey: "Hunter's Tropical Medicine and Emerging Infectious Diseases," 8th Edition, W.B. Saunders Company, Philadelphia, PA, 1997.

Web References

American Society for Tropical Medicine and Hygiene: *http://www.astmh.org/q&a/tropdise.html*

Centers for Disease Control and Prevention: *http://www.cdc.gov/health/diseases.htm*

Royal Society of Tropical Medicine and Hygiene: *http://www.rstmh.org/*

Tropical Medicine and Infectious Diseases: *http://hml.org/WWW/tropical.html*

WHO Division of Control of Tropical Diseases: *http://www.who.int/health-topics/idindex.htm*

ANN C. DeBALDO, Ph.D., College of Public Health, University of South Florida, Tampa, FL.

TROPICAL EASTERLIES. See **Winds and Air Movement**.

TROPICAL FISHES. A majority of the fishes acquired by fanciers for their home aquaria are gathered from numerous lakes, rivers, and lagoons located in the tropics. Several species are described in this encyclopedia, including the following entries: **Characids**; **Cichlids**; **Viviparous Topminnows**; and **Zebra Fish**. An interesting description of several tropical fishes will be found in "The Living Jewels of Lake Malawi," by P. Reinthal and B. Curtsinger, *Nat'l Geographic*, 42 (May 1990).

TROPICAL MONTH. The average period of the revolution of the Moon about the Earth with respect to the vernal equinox, a period of 27 days 7 hours 43 minutes 4.7 seconds, or approximately $27\frac{1}{3}$ days.

TROPICAL STORM. See **Fronts and Storms**.

TROPICAL YEAR. The period of one revolution of the Earth around the Sun, with respect to the vernal equinox. Because of precession of the equinoxes, the tropical year is not 360 degrees with respect to the stars, but 50 minutes 0.3 seconds less. A tropical year is about 20 minutes shorter than a sidereal year, averaging 365 days 5 hours 48 minutes 45.68 seconds in 1955 and is increasing at the rate of 0.005305 second annually. Also called *astronomical, equinoctial, natural*, or *solar year*.

TROPISM (Zoology). An unavoidable response of an animal to some environmental stimulus involving the orientation of the body in relation to the causative factor. In very simple animals this type of reaction is common. Protozoans, for example, may always move toward or away from light, and even animals with organized nervous systems may have some nerve paths associated in such a way that a given condition always evokes the same reaction. This type of response is also called a taxis, a term applied to the responses made without a nervous system and at the other extreme to simple reflexes, which may be nervously intricate although they are automatic.

TROPOPAUSE. The boundary between the troposphere and stratosphere, usually characterized by an abrupt change of lapse rate. The change is in the direction of increased atmospheric stability from regions below to regions above the tropopause. Its height varies from 15 to 20 kilometers in the tropics to about 10 kilometers in polar regions. In polar regions in winter it is often difficult or impossible to determine just where the tropopause lies, since under some conditions there is no abrupt change in lapse rate at any height. See also **Atmosphere (Earth)**.

TROPOSPHERE. That portion of the atmosphere from the Earth's surface to the stratosphere; that is, the lowest 10 to 20 kilometers of the atmosphere. The troposphere is characterized by decreasing temperature with height, appreciable vertical wind motion, appreciable water vapor content, and weather. Dynamically, the troposphere can be divided into the following layers: surface boundary layer, Ekamn layer, and free atmosphere. See also **Atmosphere (Earth)**.

TROUGH. See **Atmosphere (Earth)**.

TROUT (*Osteichthyes*). Of the order *Isospondyli*, family *Salmonidae*, the trout is well known as an excellent food as well as sporting fish. The salmon is also a member of the same family. Trout prefer cold and temperate waters, preferring a range from 50 to 65 °F (10 to 18 °C). They are found in freshwater lakes and streams, but a few species frequent salt water. Trout vary greatly in size and to a slight extent in game qualities. The scales are minute to small and the flesh is superior.

Trout are common in streams of the Northern Hemisphere and have been widely transported for stocking other than their native waters. The brook or speckled trout, *Salvelinus fontinalis*, is the most widely distributed species, ranging from Maine to the Dakotas and northward throughout the continent. Rainbow trout, *Salmo* (*Trutta*) *iridea*, from the West Coast, are now widely stocked in other waters, and, in places where trout fishing is important, imported species, including the European brown trout, *T. trutta*, and the Lochleven trout, are to be expected. The group includes almost two score of species in North America alone, many without common names other than trout while others are called red-fish, green-back, steel-head, blue-back, saibling, and yellow-fin. The lake trouts live in deep water and attain great weight. The mackinaw or common lake trout has been recorded up to 90 pounds (41 kilograms).

In North America the relative merits of the trout and bass as game fishes will probably always remain a matter of opinion. Trout fishing with the fly rod has long been regarded as the aristocrat of piscatorial sports, next to salmon fishing, which is less widely available, but the small-mouth bass has many devotees as a fish for the fly rod and its superiority to the trout is often loyally maintained.

The greenback is a species of trout (*Trutta smaragada*) reported from the headwaters of the Arkansas and South Platte rivers. The Tahoe trout (*Trutta henshawi*), also known as redfish, is found in streams and lakes on the eastern slope of the Sierra Nevada Mountains. The sabling is a fish related to the trout. The name has been applied to a species of charr found in mountain lakes of central Europe (*Salvelinus salvelinus*), and in North America to the introduced European trout (*S. alpinus*), a usage also found to some extent in Europe. A related species (*S. aureolis*), found in lakes of the northeastern states is called the American saibling. The name is also spelled saebling. The charr is a lake fish of the British Isles related to the trout. The trout-perch is a common fish of the Great Lakes and the rivers of the eastern half of the United States, but rare in the South. It attains a maximum length of about 10 inches (25 centimeters), with scales like the perches; otherwise similar to the trout.

The spawning habits of the trout are similar in many respects to those of the salmon. See also **Salmon (Osteichthyes)**. However, whereas the Pacific salmon dies shortly after spawning, many species of trout may spawn for a number of years.

TRUNCATION. The process of approximating the sum of an infinite series by the sum of a finite number of terms in that series. As applied to the dropping of digits in a number, truncation contrasts with rounding.

TRUNCATION ERROR. The error caused by dropping all but a finite number of terms from a possibly infinite series, but often applied to the error due to representing the limit of a sequence by one of its terms, or to representing a function by an interpolation polynomial.

TRUNCUS ARTERIOSUS. The great arterial vessel leading from the heart of vertebrates in the primitive unpaired condition and in the embryo. All blood leaving the heart passes through this vessel, which is differentiated to form the conus arteriosus and the ventral aorta in the fishes and persists as a short trunk in the amphibians. In the reptiles its subdivision to form the aorta and the pulmonary artery is begun and in the birds and mammals this splitting is complete. The truncus is divided by the growth of a pair of ridges in a spiral course along opposite walls. These ridges unite to form a partition.

TRUNK. 1. The body of a vertebrate, bearing the neck and head at one end and the tail at the other, with the two pairs of appendages attached laterally or ventrolaterally near the two ends. The trunk contains the body cavity in which the viscera lie, and is supported by the vertebral column and the ribs. It is divisible into an anterior thorax and a posterior abdomen.

2. The proboscis of the elephant.

TRUSS. A framed structure composed of a series of adjoining triangles which are formed by straight members, all lying in one plane. The members are connected at their points of intersection by pins or gusset plates or by welding. The point of intersection is called a panel point or joint. Since rigidity of the truss is secured by triangles which cannot deform without changing the length of the sides, it is generally assumed that loads applied at the panel points will produce direct stress only. This will be true if the gravity axes of the members meet at the panel points and if the members are considered connected by means of frictionless pins. Actually, since the joints are rigid, bending stresses due to the deflection of the panel points are also developed. These stresses are commonly called secondary stresses. If a load is applied to a member between the panel points it must distribute the load by beam action to the adjacent panel points. The top line of members of a truss is called the top chord and the lower line is known as the lower or bottom chord. The members connecting the top and bottom chords are called web members. The span of the truss is the horizontal distance center to center of end bearings. See Fig. 1.

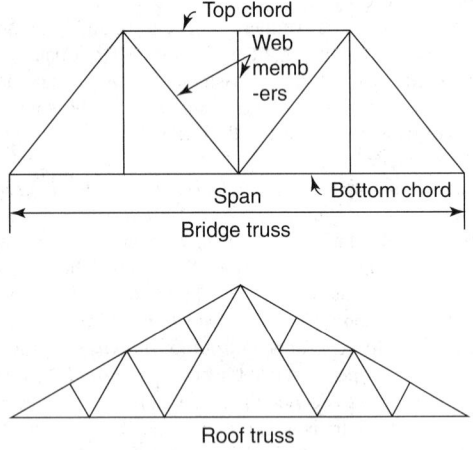

Fig. 1.   Types of trusses.

Truss bridges consist of two or more parallel vertical trusses, which support the floor system. The lateral bracing in the plane of the top and bottom chords of bridges form the web members of horizontal trusses. A roof truss is one that supports the roof of a building. Trusses are also used over auditoriums to carry the loads from the floors above. The members of a truss are usually made of steel or wood, although concrete trusses are sometimes used. Steel has the advantage over wood in that it has far greater strength and is less subject to deterioration, but the latter is frequently used for temporary structures. See also **Bridge (Structural)**.

T SECTION. A two-part network having two series arms and one shunt arm, as shown in Fig. 1. The symmetrical T section is one in which the series arms are similar. The T section two-part may be replaced by the equivalent Pi section two-part, shown in Fig. 2, by means of the following conversion formulas:

$$Y_1 = Z_1/\Delta_z$$
$$Y_2 = Z_2/\Delta_z$$
$$Y_3 = Z_3/\Delta_z$$

where $\Delta_z = Z_1Z_2 + Z_2Z_3 + Z_1Z_3$.

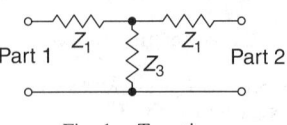

Fig. 1.   T section.

Fig. 2.   Equivalent pi section.

TSETSE FLY (*Insecta, Diptera; Glossina*). The unique viviparous genus Glossina contains some 30 species and subspecies, currently confined to tropical Africa between the latitudes of 15°N and 30°S. An isolated population was once known in southwestern Arabia and this, together with several fossil species, suggests that the distribution of the species was once considerably larger. Even so, the flies occupy roughly half the continent of Africa and are vectors of the trypanosome causing sleeping sickness in humans and various syndromes in other animals.

The tsetse fly is related to the house and stable flies of the North American continent. They are large, brown to grayish, narrow bodied, 6–15 mm long, with a stout proboscis projecting forward, well in front of the head. Except during the act of biting, the proboscis is adjoined laterally by labial palps, which enclose a space, the food canal, through which blood is sucked into the alimentary canal. Although plant juice has been imbibed by the tsetse fly, most of the species feed exclusively on the blood of vertebrates.

Three species groups are recognized—the *fusca*, the *palpalis*, and the *morsitans*. Characteristically, the fusca group is associated with dense, humid tropical forest or forest edges. Palpalis flies are dependent on more or less dense riverine or lacustrine vegetation, but their distribution extends into savanna zones well away from forested areas. The morsitans are the least hygrophilic and occupy bushland and thicket vegetation often far from lakes and rivers.

Tsetse flies have a method of viviparous reproduction, uncommon among higher insects, by which a single larva is produced at a time and is retained and nourished with the body of the female fly to the third instar state, at which time it is extruded by the mother into the ground, where it begins to pupariate. After a quiescent period, the fly, in its teneral state, seeks its first blood meal and then normally feeds at intervals of 3–4 days, dying of starvation if denied blood for longer than 12 days. The average life span of a female *Glossina* is 2 to 3 months and the males have a much shorter life span.

Although flies of the fusca group are vectors of trypanosomes pathogenic to livestock, they never have been associated with the transmission of trypanosomiasis to humans. Species of the palpalis and morsitans groups are active only during daylight hours, with flight and feeding activities decreasing markedly during dull and overcast weather. They hunt their prey mainly by sight, seeking moving, dark colors rather than light, with scent becoming increasingly important at close range.

In natural circumstances where there is no close contact with humans or domestic animals, flies of the palpalis group show preference for feeding on large reptiles, such as monitor lizards and crocodiles. Where humans and animals are available, they are attacked readily and become a major source of food. Sharp outbreaks of sleeping sickness are likely to occur where humans and cattle and small populations of flies tend to depend upon the same limited water supplies.

Infective metacyclic trypanosomes are found in the salivary glands of the tsetse fly. They undergo a complex migration in the fly, which takes about three weeks to complete. It follows that only flies more than three weeks old can be vectors of trypanosomes infective to humans.

The subject of tsetse fly control is a large one and beyond the scope of this article. References given pertain to this. Vaccines have not been effective in treating sleeping sickness because of the ability of the trypanosomes to vary their surface antigens.

See also **African Trypanosomiasis**.

Additional Reading

Leak, S.G.: "Tsetse Biology and Ecology: Their Role in the Epidemiology and Control of Trypanosomosis," CAB International, New York, NY, 1999.

Staff: FAO, "Systematic Approach to Tsetse and Trypanosomiasis Control," Bernan Associates, Lanham, MD, 1995.

Staff: FAO United Nations, "Programme for the Control of African Animal Trypanosomiasis and Related Development: Ecological and Technical Aspects," Bernan Associates, Lanham, MD, 1992.

ANN C. DEBALDO, Ph.D., College of Public Health, University of South Florida, Tampa, FL.

TSUGA.   See **Hemlock Trees**.

TSUNAMI.   (also called Seismic Sea Wave, and popularly, Tidal Wave). An ocean wave produced by a submarine earthquake, landslide, or volcanic eruption. These waves may reach enormous dimensions and have sufficient energy to travel across entire oceans. From the area of the disturbance, the waves will travel outward in all directions, much like the ripples caused by throwing a rock into a pond. The time between wave crests may be from 5 to 90 minutes, and the wave speed in the open ocean will average 450 miles per hour.

Tsunamis reaching heights of more than 100 feet have been recorded. As the waves approach the shallow coastal waters, they appear normal and the speed decreases. Then as the tsunami nears the coastline, it may grow to great height and smash into the shore, causing much destruction.

1. Tsunamis are caused by an underwater disturbance—usually an undersea earthquake. Landslides, volcanic eruptions, and even meteorites can also generate a tsunami.
2. Tsunamis can originate hundreds or even thousands of miles away from coastal areas. Local geography may intensify the effect of a tsunami. Areas at greatest risk are less than 50 feet above sea level and within one mile of the shoreline.

3. People who are near the seashore during a strong earthquake should listen to a radio for a tsunami warning and be ready to evacuate at once to higher ground.
4. Rapid changes in the water level are an indication of an approaching tsunami.
5. Tsunamis arrive as a series of successive "crests" (high water levels) and "troughs" (low water levels). These successive crests and troughs can occur anywhere from 5 to 90 minutes apart. They usually occur 10 to 45 minutes apart.

The Tsunami Warning System, a cooperative international organization and operated by the United States Weather Service, has been in operation since the 1940s. The headquarters of the center is located in Hawaii. An associated Alaska Regional Tsunami Warning System is located in Alaska. Tsunami prediction essentially commences with earthquake monitoring and prediction information. Inputs from these systems (see also **Earth Tectonics and Earthquakes**) are linked with information from a series of tide monitoring installations. Locations of tide stations and of seismograph stations in the Tsunami Warning System are shown in Fig. 1.

When inputs indicate conditions are favorable for a tsunami, a watch is issued for the probable affected area. Warnings are issued when readings from various tidal stations appear to match the seismographic information. Because of the complexity of the factors involved and a large degree of uncertainty nearly always present, there is a tendency to issue watches and warnings as a safety precaution even though a tidal wave of significance may not later develop. Unfortunately, after awhile, persons in likely areas to be affected grow callous to watches and tend to ignore them. A tsunami that hit Hilo, Hawaii in 1960 killed 60 residents even though they had been warned of the coming event an ample 6 hours in advance of the strike.

In the late 1970s, scientists suggested an improved method for making tsunami predictions. For a number of years, specialists have suggested that better analysis and interpretation of seismic waves produced by earthquakes may improve the prediction of tsunamis. Seismic waves range from very short-period waves that result from the sharp snap of rocks under high stress to very long-period waves, due mainly to the slower movements of large sections of the ocean floor. Many researchers believe that tsunamis result mainly from the vertical movement of these large blocks, this leading to a tentative conclusion that the strength of seismic waves of very long period may be the best criterion for an earthquake's ability to generate a tsunami.

Part of the problem is that most seismographs installed in the system are not very sensitive to very long-period waves and thus a given earthquake

Fig. 1.   Network of tide and seismograph stations that are part of the Tsunami Warning System, headquartered in Hawaii. (*National Oceanic and Atmospheric Administration.*)

cannot be analyzed effectively in terms of its potential for producing a tsunami. Equipment has been refined so that today, shorter period waves are used to locate an earthquake and 20-second waves are used to calculate the magnitude. However, some scientists feel that the true magnitude of some earthquakes can only be determined by measuring the characteristics of longer waves, such as 100-second waves. Brune and Kanamori (University of California at San Diego) have observed that the Chilean earthquake of 1960 had a magnitude of 8.3 when calculated on the basis of 20-second waves, but its magnitude was 9.5, or more than 10 times larger in wave amplitude and more than 60 times larger in energy released, when calculated by Kanamori's method, which attempts to include the energy release represented by very long-period seismic waves. Other scientists are coming to the viewpoint that many warnings could be omitted if predictions were based upon longer waves. Two of the first long-period seismographs incorporated in the Tsunami Warning System were installed in Hawaii and on the Russian island of Yuzhno-Sakhalinsk, which is northeast of Vladivostok. Later an installation was made at the Alaska Warning Center in Palmer. Whether or not the latest reasoning proves successful must await a number of years of experience with the earthquakes of the future and the resulting tsunamis.

Continuing Tsunami Research. During the 2000s, research pertaining to the fundamentals of tsunamis and the development of mathematical models of the phenomenon continues. Considerable attention is being directed to specific regions, including the west coasts of Mexico and Chile, the southwestern shelf of Kamchatka (Russia) and, in the United States, the generation of tsunamis in the Alaskan bight and in the Cascadia Subduction Zone off the west coasts of Washington (Puget Sound) and Oregon.

Research also is being directed toward the development of simple and more economic warning systems, particularly in the interest of the developing countries, the coasts of which border on the Pacific Ocean.

The Tsunami Warning System previously described requires millions of dollars for equipment, maintenance, and operation, well beyond the means of some countries. Also, some scientists believe that more localized equipment installations could possibly serve local shore communities better while costing less. These observations, however, do not challenge the need and validity for the larger tsunami network. The National Oceanic and Atmospheric Administration (NOAA) has developed a system costing in the range of $20,000 that can be installed and operated by non experts. The system has been undergoing trials at Valparaiso, Chile, a city that has been struck by nearly 20 tsunamis within the past two centuries. A sensor (accelerometer) is installed in bedrock under the city and can measure tectonic activity in excess of 7.0 on the Richter scale. These measurements are interlocked with level sensors.

Researchers B.F. Atwater and A.L. Moore (University of Washington), in their attempts to model earthquake and tsunami activity in the area over the last thousand years, have reported what they believe to have been a large earthquake on the Seattle fault some time between 1000 and 1100 years ago. The researchers report, "Water surged from Puget Sound, overrunning tidal marshes and mantling them with centimeters of sand. One overrun site is 10 km northwest of downtown Seattle; another is on Whidbey Island, some 30 km farther north. Neither site has been widely mantled with sand at any other time in the past 2000 years. Deposition of the sand coincided — to the year or less — with abrupt, probably tectonic subsidence at the Seattle site and with landsliding into nearby Lake Washington. These findings show that a tsunami was generated in Puget Sound, and they tend to confirm that a large shallow earthquake occurred in the Seattle area about 1000 years ago."

Simulations of tsunamis from great earthquakes on the Cascadia subduction zone have been carried out by M. Ng, P.H. Leblond, and T.S. Murty (University of British Columbia). A numerical model has been used to simulate and assess the hazards of a tsunami generated by a hypothetical earthquake of magnitude 8.5 associated with rupture of the northern sections of the subduction zone. The model indicates that wave amplitudes on the outer coast are closely related to the magnitude of sea-bottom displacement (5 meters). The researchers observe, "Some amplification, up to a factor of 3, may occur in some coastal embayments. Wave amplitudes in the protected waters of Puget Sound and the Straits of Georgia are predicted to be only about one-fifth of those estimated on the outer coast."

Additional Reading

Atwater, B.F. and A.L. Moore: "A Tsunami About 1000 Years Ago in Puget Sound, Washington," *Science,* 1614 (December 4, 1991).

Bernard, E.N., Editor: "Tsunami Hazard: A Practical Guide for Tsunami Hazard Reduction," (Papers from Symposium at Novosibirsk, Russia), International Union of Geodesy and Geophysics, New York, NY, July 1989.

Bryant, E.: "Tsunami: The Underrated Hazard," Cambridge University Press, New York, NY, 2001.

Collins, E.: "Wave Watch," *Sci. Amer.,* 28 (February 1988).

Kubota, I.: "Japan's Weather Service and the Sea," *Oceanus,* 71 (Spring 1987).

Lander, J.F., P.A. Lockridge, and M.J. Kozuch: "Tsunamis Affecting the West Coast of the United States, 1806–1992," DIANE Publishing Company, Collingdale, PA, 1997.

Lander, J.F.: "Tsunamis Affecting Alaska, 1737–1996," DIANE Publishing Company, Collingdale, PA, 1997.

Ng, Max K.-F., P.H. Leblond, and T.S. Murty: "Simulation of Tsunamis from Great Earthquakes on the Cascade Subduction Zone," *Science,* 1248 (November 30, 1990).

Prager, E.J.: "Furious Earth: The Science and Nature of Earthquakes, Volcanoes, and Tsunamis," McGraw-Hill Professional Book Group," New York, NY, 1999.

Soloviev, S.L., K.S. Kim, O.N. Solovieva, et al.: "Tsunamis in the Mediterranean Sea, 2000 B.C.–2000 A.D.," Kluwer Academic Publishers, Norwell, MA, 2000.

Staff: "Long-Wave Runup Models," World Scientific Publishing Company, Inc., River Edge, NJ, 1997.

Tsuchiya, Y. and N. Shuto: "Tsunami: Progress in Prediction, Disaster Prevention and Warning," Kluwer Academic Publishers, Norwell, MA, 1995.

Web References

International Tsunami Information Center (ITIC): *http://www.shoa.cl/oceano/itic/frontpage.html*

National Tsunami Hazard Mitigation Program: *http://www.pmel.noaa.gov/tsunami/*

Tsunami Links: *http://www.pmel.noaa.gov/tsunami-hazard/links.html*

TUATARA (or *Tuatera*).   *Reptilia, Rhynchocephalia.* A primitive reptile, *Sphenodon punctatus,* of New Zealand, superficially like the lizards but different in several anatomical details. It is chiefly noteworthy for the high development of the pineal or median eye.

TUBERCULOSIS. An acute inflammation of the lungs, meninges, pericardium, kidneys, genitals, bones, or lymph system caused by invasion of one or more of these organs by *Mycobacterium tuberculosis,* a long, slender, rod-shaped bacterium. By far the organ most frequently affected is the lung.

Because *M. tuberculosis* is highly adapted to humans and has no reservoir in nature, people are the reservoir for human tuberculosis. Since the virtual elimination of bovine tuberculosis in the United States many years ago, the disease is contracted via person-to-person transmission. Patients with active pulmonary infection shed infected airborne droplets into the environment. Usually close, continued contact is required to spread the disease and hence it tends to occur among members of the same household, schoolmates, persons participating in car pools, and others in similar close conditions. Operators of nursing homes for the elderly must constantly be aware of the presence of *M. tuberculosis.*

With the exception of patients whose immune system is compromised, as in AIDS or infection with the human immunodeficiency virus (HIV), the prognosis of pulmonary tuberculosis is excellent where chemotherapy is commenced as soon as possible after the disease has been positively diagnosed. The principal antitubercular chemotherapeutic agent is *isoniazid,* although rifampin, streptomycin, and ethambutol also are used. Secondary drugs, including para -aminosalicylic acid, pyrazinamide, ethionamide, and cycloserine, are less frequently used. All of these are oral drugs. Also available for parenteral administration are kanamycin, viomycin, and capreomycin.

If the disease is left untreated, very serious complications can occur. Sometimes patients are hospitalized during the initial stages of therapy. The administration of drugs for about two weeks usually markedly reduces the ability of the patient to infect others. Persons with nonpulmonary tuberculosis are considerably less infectious than those with the pulmonary form and thus sometimes can be managed entirely as outpatients.

About 10% of newly recognized cases of tuberculosis are *extrapulmonary.* About half of such patients will have entirely normal chest x-rays; the others will show evidence of inactive pulmonary disease.

The most definitive test for diagnosing tuberculosis is a culture. *M. tuberculosis* grows slowly, and the results of a culture test may require from 3 to 6 weeks. An early morning sputum collection is the best source for performing the test. A number of parallel tests should be made to differentiate *M. tuberculosis* from other mycobacterial species. The initial choice for diagnosis may be a skin test. The Mantoux test involves the careful intradermal administration of a purified protein derivative, which is stabilized by including a polysorbate detergent in the diluent. Palpable induration of over 10 millimeters occurring 48 hours after administration is considered a positive indication of tuberculosis infection, but not necessarily of *active* TB. Also, there are instances where a negative tuberculin test does not exclude a diagnosis of TB.

Unlike the majority of infectious diseases, there is a very long latent period between infection and clinical illness. Initially, the infection usually is found in the lower lung fields. In the alveoli, the bacilli proliferate slowly and, because they do not produce enzymes or toxins, there is little initial inflammatory response. The tuberculin skin test does not become positive for several weeks, at the end of which time cell-mediated immunity has become well established. When this occurs, the majority of bacteria are destroyed, and most persons so infected have a healing of the initial tuberculous lesions. X-rays of such cases may appear normal or indicate focal calcifications. On the other hand, in persons where immunity is incomplete, the bacilli spread the infection. Although the principal site of infection is in the lungs, the bacilli can invade the bloodstream and thus involve other organs, particularly those regions where there is a high blood flow. These include the renal cortex, the vertebral column, and the metaphyseal ends of long bones.

Although many persons with infections may be asymptomatic for a considerable period, the most common symptoms are low-grade fever and nonproductive cough. Notably among patients with reactivated tuberculosis, there is anorexia (loss of appetite), loss of weight, and night sweats. As the disease progresses, there will be low-grade fever, although more severe fever and chills are noted in some cases. Diagnosis can be difficult and requires the assistance of sputum tests and chest x-rays.

Chronology — The Rise, Decline, and Reemergence of TB

Historically, the incidence of tuberculosis has been closely associated with a society's general environment and living conditions. The disease was rampant in crowded city and town conditions of Europe during earlier centuries, where a large percentage of the population was forced to live in squalor and on poor diets. Tuberculosis, particularly the common pulmonary variety, is quite contagious and easily spread by aerosols from the oral and nasal tract. Slowly, during the late 1800s, people were educated by the commonly seen "No Spitting" signs encountered in public places. Because of crowded conditions, TB frequently was encountered in military establishments. The medical profession gradually became aware of how easily TB is spread and consequently public laws required isolation in the form of sanitoriums. Careful regimes for patients were established, and by the 1930s certain surgical procedures were suggested and found limited success. Meticulous care on the part of patients and providers proved successful in reducing the overall mortality of TB.

The first breakthrough occurred with the discovery of antibiotics, several of which were mentioned earlier in this article. These drugs initiated a remarkable decline in mortality, and, at one time, professionals in countries with advanced medical programs were commencing to think in terms of near-eradication of the disease. This optimism, however, did not extend to much of the third world because in many areas sordid living conditions continued, this coupled with inadequate awareness of how contagious TB can be.

The majority of professionals in the field accredited the newly discovered drugs for the success of reducing TB mortality during the 1940s through the 1980s. There is another school of thought that believes that the lower incidence of TB should be attributed more to the buildup of "natural" resistance to *M. tuberculosis* after centuries of exposure to the bacteria. This hypothesis, however, has many more disbelievers than supporters.

The rise of incidence of TB commencing in the mid-1980s generally is attributed to two causes, each of which has had a measurable effect:

1. An increased resistance shown by *M. tuberculosis* to the drugs administered. Current research is illustrating the veracity of this cause.
2. The appearance of the human immunodeficiency virus (HIV) and of AIDS, where the immune system is compromised. This is exacerbated because of the manner in which the presence of TB infection in an otherwise very ill person can be masked for a period of time without positive identification and hence early treatment.

Other possible causes for the increased risk of TB include greater numbers of elderly persons confined in crowded and otherwise unsatisfactory care facilities, the increasing numbers of homeless, or "street people," the immigration of persons from areas where TB is endemic, and the generally unhealthy habits of drug abusers.

TB, HIV, and AIDS. During the 1985–1992 time span, the number of TB cases in the United States rose by nearly 20%. Normally, only 5 to 10% of adults infected with *M. tuberculosis* (as indicated by a positive Mantoux skin test) develop active tuberculosis. Studies have shown, however, that TB, as manifested by a positive skin test, occurs at the rate of 7.9% per year if HIV infection also is present. As pointed out by R. Goodgame, "This is consistent with the observation that active TB is found in 60 percent of Haitians and in over 80 percent of Haitians who have positive skin tests for TB near the time of a diagnosis of AIDS."

Not only does infection with HIV increase the risk of tuberculosis, but there is a growing awareness that the effectiveness of antituberculosis treatment is decreased. Thus, the Centers for Disease Control (CDC) and the American Thoracic Society have recommended that antituberculosis chemotherapy be administered to HIV-infected patients for a minimum of 9 months after organisms have been cleared from the sputum. In a study conducted at the San Francisco General Hospital, it was concluded: "Tuberculosis causes substantial mortality in patients with advanced HIV infection. In patients who comply with the regimen, conventional therapy results in rapid sterilization of sputum, radiographic improvement, and low rates of relapse."

Tuberculosis has been found to be a common result of AIDS in Uganda. This is the result, of course, of the widespread prevalence of *M. tuberculosis* prior to the HIV epidemic. Reactive Mantoux skin tests are noted in from 50 to 60% of unvaccinated adult Ugandans. With over a million adults already HIV-seropositive in the early 1990s, it follows that 50,000 or more cases of HIV-induced active tuberculosis could arise each year, not including the 15,000 or more cases reported before the HIV epidemic. In other areas of the world, other diseases precipitated by HIV and AIDS may include *Pneumocystis carini* rather than TB.

Tuberculosis Meningitis. In 1992, a group of physicians and clinicians in Spain studied the cases of over 2200 patients with tuberculosis. Approximately 25% of the cases had *M. tuberculosis* isolated from the cerebrospinal fluid. The study showed that tuberculous meningitis is a frequent complication of patients with HIV infection in areas of endemic tuberculosis. The study concludes: "HIV-infected patients with tuberculosis are at increased risk for meningitis, but infection with HIV does not appear to change the clinical manifestations or the outcome of tuberculous meningitis."

TB Drug Resistance. In 1992, researchers at London's Hammer-smith Hospital and the Pasteur Institute in Paris described the identification of a mutation that causes *M. tuberculosis* to become resistant to *isoniazid*, the principal drug used to treat the disease. The mechanism described is the deletion of the gene that codes for an enzyme known as *catalase*. The underlying theory is that isoniazid becomes toxic to *M. tuberculosis* only after it has been converted to another compound because of the presence of catalase. One goal then would be that of finding a way to bypass the catalase conversion step. This fundamental discovery has encouraged other bacterial geneticists to investigate the basis for resistance to other drugs by other bacteria. Since 1990, several clusters of multidrug-resistant tuberculosis have been identified among hospitalized patients with AIDS.

See also **Bacterial Diseases**.

Additional Reading

Aldhous, P.: "Genetic Basis Found for Resistance to TB Drug," *Science*, 1038 (August 21, 1992).

Berenguer, J., et al.: "Tuberculous Meningitis in Patients Infected with the Human Immunodeficiency Virus," *N. Eng. J. Med.*, 668 (March 5, 1992).

Bloom, B.R. and C.J.L. Murray: "Tuberculosis: Commentary on a Reemergent Killer," *Science*, 1055 (August 21, 1992).

Cefrey, H., F. Ramen, C. Hayhurst, et al.: "Epidemics: Deadly Diseases throughout History: The Plague, AIDS, Tuberculosis, Cholera, Small Pox, Polio, Influenza, Malaria," The Rosen Publishing Group, Inc., New York, NY, 2001.

Daley, C.L., et al.: "An Outbreak of Tuberculosis with Accelerated Progression Among Persons Infected with the Human Immunodeficiency Virus," *N. Eng. J. Med.*, 231 (January 23, 1992).

Dormandy, T.: "The White Death: A History of Tuberculosis," New York University Press, New York, NY, 2000.

Edlin, B.R., et al.: "An Outbreak of Multidrug-Resistant Tuberculosis Among Hospitalized Patients with the Acquired Immunodeficiency Syndrome," *N. Eng. J. Med.*, 1514 (June 4, 1992).

Evan, S.: "Colorado Scientists Overturn Standard Medical View of TB's Development," *Genetic Engineering News*, 20 (June 1991).

Friedman, L.N.: "Tuberculosis: Current Concepts and Treatment," 2nd Edition, CRC Press, LLC., Boca Raton, FL, 2000.

Goodgame, R.W.: "AIDS in Uganda—Clinical and Social Features (Tuberculosis is Ubiquitous)," *N. Eng. J. Med.*, 383 (August 9, 1990).

Lutwick, L.I.: "Tuberculosis: A Clinical Handbook," Chapman & Hall Medical, New York, NY, 1995.

Ramen, F.: "Tuberculosis," The Rosen Publishing Group, Inc., New York, NY, 2000.

Rangoonwala, R.: "Quinolones in Pulmonary Tuberculosis Management," Marcel Dekker, Inc., New York, NY, 1996.

Rossman, M.D. and R.R. MacGregor: "Tuberculosis: Clinical Management and New Challenges," The McGraw-Hill Companies, Inc., New York, NY, 1995.

Rubin, A.L.: "Tuberculosis Mortality Decline," *Science*, 277 (July 16, 1993).

Small, P.M., et al.: "Treatment of Tuberculosis in Patients with Advanced Human Immunodeficiency Virus Infection," *N. Eng. J. Med.*, 289 (January 31, 1991).

Watson, T.: "A Shot in the Arm for TB Research," *Science*, 886 (February 12, 1993).

Web References
Centers for Disease Control and Prevention: *http://www.cdc.gov/health/diseases.htm*
Johns Hopkins Infectious Diseases: *http://www.hopkins-tb.org/*

TUBERS. See **Solanaceae (Potato Family)**.

TUBFISH. See **Sea Robins**.

TUBULIFLORAE. See **Composite Family**.

TUBULIFORM GLANDS. Silk glands of spiders whose secretion is used in forming the cocoon to contain the eggs. They occur only in the female.

TUCANA. A southern constellation located near Phoenix.

TUFF. Tuff or volcanic tuff is a sedimentary rock, resulting from the partial or complete consolidation of the products of explosive volcanic eruptions. Tuffs may be well sorted and stratified, due to the action of wind or water, or may have an unsorted, heterogeneous character. As particles making up a tuff become coarser the rock grades into an agglomerate.

TULAREMIA. An infectious, communicable disease caused by the gram-negative bacillus, *Francisella tularensis*. The agent is harbored by rabbits, squirrels, skunks, mink, rats, foxes, mice, coyotes, as well as some fowl, snakes, and fish. Dogs and cats, although themselves immune to the disease, may harbor the organism and transmit it to humans by bites and scratches. Vectors include mosquitoes, fleas, and ticks. Blood-sucking insects, usually ticks, are the most commonly identified vectors, but exposure to infected animals is also common. Persons who handle rabbit meat and hides assume a risk of becoming infected.

The disease has been reported in many countries in the Northern hemisphere, with about 300 cases reported in the United States each year. These present a bimodal incidence curve attributed to tick bites in the summer and the hunting of rabbits in winter.

The incubation period is 3 days. The disease takes three forms, depending upon site of infection. (1) In *ulceroglandular tularemia*, an ulcerative papule develops within 3–4 days with regional lymph tissue involvement. The lesion may suppurate (discharge pus). Fever may reach 104–105.8 °F (40–41 °C). Headache may be an accompanying feature. Diseases with somewhat similar symptoms include plague, cat-scratch disease, and infectious mononucleosis. (2) *Pulmonary tularemia*, which can precipitate tracheitis, bronchitis, pneumonia, and plural effusions, is characterized by cough and dyspnea. (3) *Enteric infection* usually causes pharyngitis, sometimes with the formation of a gray membrane. Less commonly, there will be abdominal pain, nausea, vomiting, and diarrhea. When the entry site of infection is not apparent, this may indicate that the infection was waterborne.

The usual drug of treatment is streptomycin administered intramuscularly. Tetracycline and chloramphenicol have also been used successfully. Full recovery depends upon the course followed by the disease and, in some cases, can require several weeks. In treated cases, the mortality is low. In untreated cases, fatality is perhaps 5%, although in the pneumonic and septicemic forms it may reach 30%.

The incidence of tularemia in the United States steadily declined from 1950 (approximately 60 cases/year) to a plateau (about 10 cases/year) in 1970 and since. Recent small outbreaks have been associated with ticks.

Web Reference
University of Maryland Medicine: *http://umm.drkoop.com/conditions/ency/article/000856.htm*

TULIP TREE. See **Magnolia Trees**.

TUMBLE-BUG (*Insecta, Coleoptera*). A beetle of the family *Scarabaeidae*, which forms and buries balls of dung. Numerous species of scavengers in this family have the same habit. They are said to use the dung as a supply of food during periods when they remain underground, and also to bury a ball with an egg attached, to provide food for the developing larva. The balls of dung are often much larger than the beetles themselves, and their clumsy maneuvers in rolling their booty to the place where it is to be buried are responsible for the name tumble-bug.

TUNAS (*Osteichthyes*). *Thunnus* and other genera are part of the mackerel family (see also **Mackerels**), but differ so much in their skeletal structure and circulatory system that they were once classified in their own order. The body of the tuna is similar to that of a typical mackerel. Only the pectoral fins and the part of the body posterior to them are covered with small scales. A corselet extends from the back to beneath the rear end of the dorsal fin, and on the belly to behind the pelvic fins. The tail shaft has a median keel. The back is blue-black, the sides are silver-gray, and the belly is white. The front fins are smoky black, and the rear ones are lighter. The finlets behind the second dorsal fin and anal fin are light yellow, with a dark edge. The eyes are enclosed in bony capsules. There is a pair of deep depressions in front of the bones of the rear aspect of the head. A well-developed subcutaneous circulatory system is connected to the vessels of the lateral musculature. Parts of the lateral muscles located on both sides of the vertebral column have a dark red coloration. Another unusual vessel system is located on the inside of the liver and in the tail. This system is responsible for the "warm bloodedness" of tuna. When excited, tuna can have a body temperature of 10.8–21.6 °F (6–12 °C) higher than that of their surroundings.

The most important species of tuna from a commercial standpoint are the *yellow-fin tuna* (*Thunnus albacares*), which accounts for from 25–36% of the total annual catch, varying from one year to the next; the *skipjack tuna* (*Katsuwanus pelanus*), 20–32% of the total catch; the *albacore* (*T. alalunga*), 20–24% of the catch; the *bigeye tuna* (*T. obesus*), 10–16% of the catch; and the *bluefin tuna* (*T. thynnus*), accounting for 11–15% of the total annual catch worldwide.

The great tuna migrations have long been of interest and the subject of numerous marking experiments. A few tuna caught in the Mediterranean have been found to have hooks in their mouths from the Azores; many caught off the coast of Spain have come from Norway. In Norway, tuna are sometimes caught with harpoons, and these harpoons have been found in Mediterranean specimens. Two tuna marked in 1954 at Martha's Vineyard, Massachusetts were recovered in the Bay of Biscay 5 years later, establishing the fact that at least some American tuna swim across the Atlantic to the shores of Europe.

Bluefin Tuna. This is the largest species of tuna, with lengths ranging up to 16.5 feet (5 meters) and weights up to 1800 pounds (820 kilograms). It is distributed in all warm and temperate seas, including the Mediterranean and the Black Seas, and along Europe's Atlantic coast northward to Tromsö, Norway. It is regularly encountered in the North Sea, and occasionally in the Baltic Sea to Stralsund. On the American East Coast, bluefin tuna occur as far north as Newfoundland. The diet of the bluefin consists of herring, mackerels, and gar, which are usually pursued in small schools for great distances near the surface of the water. Tuna also feed on schooling deep sea fishes. The killer whale is the tuna's chief enemy. During the spawning period in June, tuna migrate by the thousands to the Mediterranean coasts; then they retire again to their feeding grounds. Their small (1–1.2-millimeter) eggs develop quickly. The young (about 4 millimeters in length) hatch after 2 days, and they grow quite rapidly. In the fall of the same year, a great many of these juveniles migrate away from the coastal regions and begin feeding in the open sea. Since the main catch period corresponds with the tuna spawning period (when they are near the coast in greatest numbers), there has been a danger of overfishing them.

Albacore. This fish can easily be recognized by the long, sword-shaped pectoral fin, which extends to underneath the second dorsal fin. The albacore achieves a length up to 43 inches (110 centimeters) and weighs up to 66 pounds (30 kilograms). This species has a distribution somewhat similar to that of the bluefin tuna, but penetrates far to the north with the warm currents and, in winter, moves south again. It is most prevalent in the Pacific Ocean, where it undertakes extensive migrations.

Yellowfin Tuna. This fish is found only in the warmer parts of the ocean. It achieves a length up to 8 feet (2.5 meters) or more and weighs up to 500 pounds (225 kilograms). This species is characterized by a narrow, elongated second dorsal fin, and a similar anal fin. Both fins, plus the gill cover and front of the belly, are brilliant yellow. In one year, a single yellowfin tuna can gain up to nearly 60 pounds (27 kilograms).

Bigeye Tuna. This fish is very similar species to the yellowfin, but it has shorter pectoral fins and larger eyes. Furthermore, it typically inhabits deeper water levels.

Skipjack Tuna. This fish achieves a length of about 3 feet (1 meter) and a weight of 48 pounds (22 kilograms). It is an important commercial species which belongs to a tuna group having stripes instead of spots. It grows very rapidly, but does not live longer than 4 years. The skipjack is especially prevalent in the Pacific, accounting for a high percentage of the total catch in some years. During the summer, skipjack may be found in the North and Baltic Seas. It is recognized by the 4 to 7 dark longitudinal bands on the sides of the body. Skipjack tuna lack a swim bladder. See Fig. 1. Another member of this group, the bonito (*Sarda sarda*) has 8 or 9 longitudinal and dark stripes on its back, extending from the middle of the body upward. Bonitos are widely distributed in the Atlantic. A very similar species, *S. orientalis*, occurs in the Indo-Pacific, while in the northeastern and southeastern Pacific, *S. chillensis* is caught in California and Chile. They are all valued for their firm, tasty meat, and all have local economic importance.

Fig. 1.   Skipjack tuna.

The spanish mackerel, closely related to the other tunas, is described briefly in the entry on **Mackerels**.

Major Changes in Tuna Fisheries. Until about 1961, the majority of the tuna fleet consisted of independent vessels from 300 to 500 gross tons (270 to 450 metric tons). After that time, due to the rapid decline in tuna abundance in other oceans, and also because of the increasing number of overseas bases for operation in the Atlantic (these were acquired by agreement with strategically located countries), a number of smaller vessels were attracted into the area so that the average size of vessel was reduced. The conventional mothership-type of operation, which had proved effective in the earlier years of the central and western Pacific fishery, was not permitted by the Japanese government to operate in the Atlantic, but the modified version of it, the deckboat-carrying mothership, which had first been tested experimentally about 1955 and found good, was permitted to operate there. The first operation of this kind entered the Atlantic in 1961. It was found so effective that by 1963 there were 28 such units fishing in many parts of the Atlantic. Because of its remoteness from the home islands, the Japanese Atlantic tuna fishery developed characteristics that increased efficiency and lowered costs. One such distinction was the development of new and closer markets for their catches in Europe and America. Introduction of Japanese-caught tuna to these countries started in a small way, but ultimately became a large-volume operation. The progressive and intensive development of tuna fishing in the Atlantic continued and dictated changes in the species of tuna caught, and in the seasons and areas the various species could be caught (Kask, 1969).

The seasonal pattern of the Atlantic tuna fishery is quite erratic. Fishing grounds and fishing seasons for each species seem to vary even more in the Atlantic than in other known tuna fishing areas. Yellowfin catches occur in two peaks, one in February–March and the second in August. This catch is mostly made in equatorial waters and is relatively stable, but the rate has fallen in recent years. Albacore catches are made in December in the Southern hemisphere and in July in the Northern hemisphere. In general, albacore are more abundant in the western half of the Atlantic and bigeye in the eastern half. Bigeye catches also show two seasonal peaks, one in summer months in the Northern hemisphere, and the other in the fall months between 5° and 20° south latitude. Bluefin are most abundant in March and April and again in September to November in

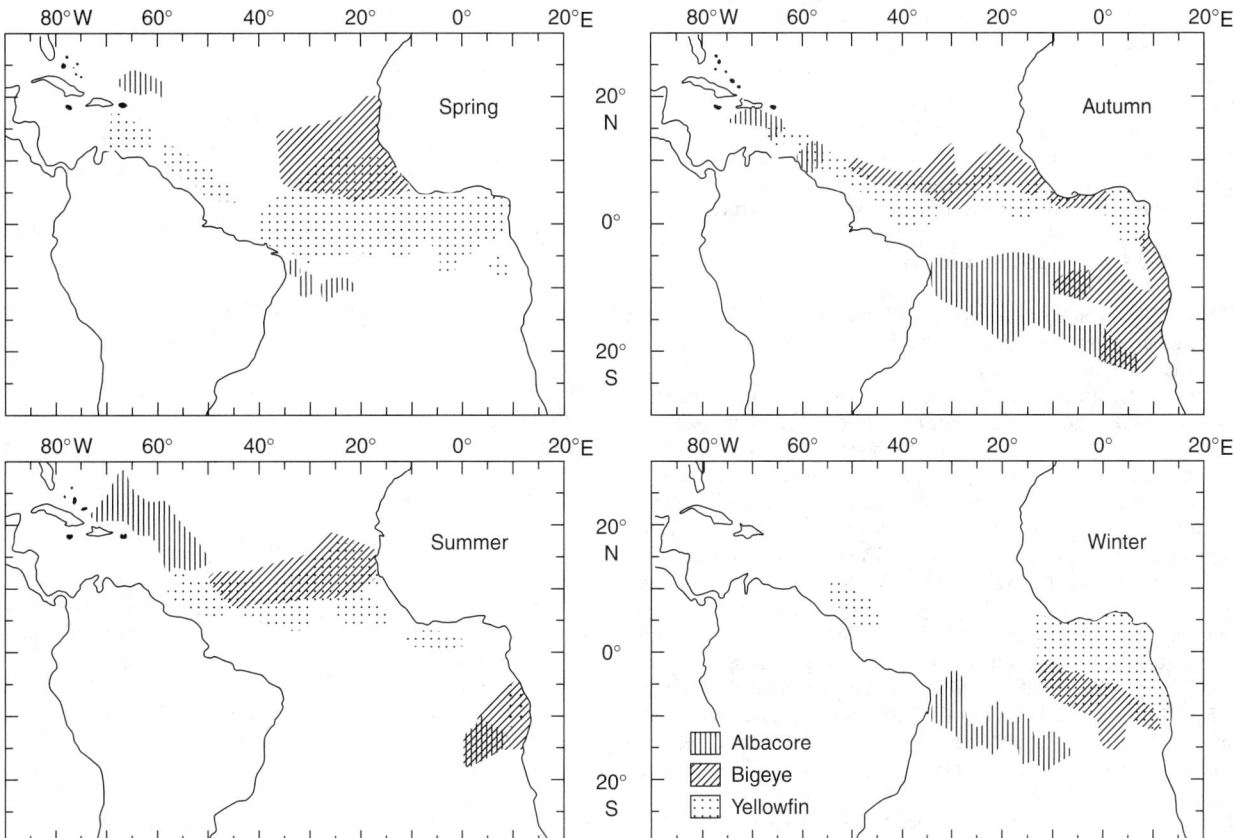

Fig. 2.   Seasonal variations in distribution of longline fishing grounds in the Atlantic Ocean.

the waters off Brazil. The development of the Japanese tuna fishery in the Atlantic has been dramatic, ingenious, and arduous. In recent years, of course, considerable competition for the catch by other countries has taken place. Seasonal variations in the distribution of longline fishing grounds in the Atlantic Ocean are shown in Fig. 2.

See also **Fishes**.

Additional Reading

Bonanno, A. and D. Constance: "Caught in the Net: The Global Tuna Industry, Environmentalism, and the State," University Press of Kansas, Lawrence, KS, 1996.
Fonteneau, A. and J. Marcille: "Resources, Fishing and Biology of Tropical Tunas of the Eastern Central Atlantic," Bernan Associates, Lanham, MD, 1993.
Joseph, J. and J.W. Greenough: "International Management of Tuna, Porpoise, and Billfish," University of Washington Press, Seattle, WA, 1979.
Kask, J.L.: "Tuna Fisheries," in "The Encyclopedia of Marine Resources," (F.E. Firth, editor), Van Nostrand Reinhold, New York, NY, 1969.
Settle, D.M. and C.C. Patterson: "Lead in Albacore: Guide to Lead Pollution in Americans," *Science*, **207**, 1167–1175 (1980).
Sharp, G.D. and A.E. Dizon: "The Physiological Ecology of Tunas," Academic Press, Inc., San Diego, CA, 1979.
Staff: National Research Council, "Dolphins and the Tuna Industry," National Academy Press, Washington, DC, 1992.
Staff: National Research Council, "An Assessment of Atlantic Bluefin Tuna," National Academy Press, Washington, DC, 1994.
NOTE: See also complete list of references at the end of the entry on **Fishes**.

TUNDRA. Treeless plains that lie poleward of the tree line (i.e., the poleward limit of tree growth). The plants thereon are sedges, mosses, and lichens and a few small shrubs. It is mostly underlain by permafrost, so that drainage is bad and the soil may be saturated for long periods. Nearly all of the world's tundra is found in the Northern Hemisphere, where it covers vast expanses of northern North America and Eurasia. In the Southern Hemisphere, only the northern extremities of Antarctica (i.e., Palmer Peninsula) and some surrounding islands contain tundra. See also **Biome**.

TUNG (*Aleurites fordii; Euphorbiaceae*). A tree, native to China and Japan, which is raised for the tung oil of its seeds. It has been introduced to the United States and large acreages are raised along the Gulf Coast. The plant is a tree bearing fruits about 2 inches (5 centimeters) in diameter, each containing 4–5 seeds. The seeds consist of about 65% oil.

Tung oil has been produced for centuries in China, where the oil is extracted by extremely crude methods. In this country, mechanical processing has increased the yield and improved the quality of oil. The fruits are dried, and the seeds are removed. The separated ground seeds are passed through a continuous screw press, which removes the oil. The oil is used in making quick-drying, waterproof varnish, or may be mixed with other oils and resins, to which it imparts desirable properties.

TUNG OIL. See **Cashew and Sumac Trees**.

TUNGSTEN. Chemical element, symbol W, at. no. 74, at. wt. 183.85, periodic table group 6, mp 3390–3420°C, bp 5,660°C, density 19.3 g/cm^3. Two forms of metallic tungsten are known: α-tungsten, which has a body-centered cubic crystal structure, and β-tungsten, which has a face-centered cubic crystal structure. The metal exhibits the phenomenon of passivity so that it is quite resistant to chemical action even though it has a strong reducing action when a fresh surface is exposed, or in potentiometric titrations. Tungsten is a silver-white to steel-gray, brittle, hard metal; not oxidized by air at ordinary temperature, but burns at high temperature, best dissolved by a mixture of hydrofluoric and HNO$_3$ acids. Tungsten has four naturally occurring isotopes ^{180}W, and ^{182}W through ^{184}W. ^{180}W is radioactive with a half-life of approximately 3×10^{14} years. Six other radioactive isotopes have been identified ^{176}W through ^{178}W, ^{181}W, ^{185}W, ^{187}W, and ^{188}W. All have half-lives considerably less than 4 months in length. Tungsten does not occur in the free state and is a relatively scarce element, making up an estimated $7 \times 10^{-3}\%$ of the Earth's crust. The tungsten content of seawater is estimated at about 950 pounds per cubic mile. Although the tungsten mineral *wolframite* (iron manganese tungstate) was described as early as 1574, it was then mistaken as a mineral of tin. The term *tungsten* first appeared about 1758. K.W. Scheele identified tungstic oxide in 1781, after which the calcium tungstate mineral *scheelite* was named. The first metallic tungsten was produced by J.J. d'Elhuyar and F. d'Elhuyar in 1783 by the carbon reduction of the oxide. W.D. Coolidge obtained a patent in 1908 for making ductile tungsten wire for use in incandescent lamps. Divers authorities accredit Scheele or the d'Elhuyar brothers with the discovery of the element. With exception of the United States, the element generally is referred to as *wolfram*. First ionization potential of tungsten is 7.98 eV. Electronic configuration $1s^22s^22p^63s^23p^63d^{10}4s^24p^64d^{10}4f^{14}5s^25p^65d^46s^2$. Ionic radius W^{+4} 0.68 Å, W^{+6} 0.65 Å. Metallic radius 1.3704$_5$ Å. Other important physical properties of tungsten are given under **Chemical Elements**.

Usually tungsten minerals are found in pegmatites, sills, and batholiths. Minerals often accompanying tungsten minerals are cassiterite, quartz, feldspar, sulfides, arsenites, apatite, calcite, molybdenite, and bismuthinite. Several of these minerals are described under separate alphabetical entries. In order of decreasing magnitude, tungsten deposits occur in the People's Republic of China, the United States, Korea, Bolivia, Portugal, Burma, and Australia. Deposits also are found in at least ten other areas. In the United States, the most significant deposits are found in California, Nevada, South Carolina, Idaho, and Colorado. Tungsten concentration in the ores found in the United States run from 0.5% to 3% WO$_3$ (20 pounds 9 kilograms of WO$_3$ contains about 15.9 pounds 7.2 kilograms of W). High-purity tungsten metal is prepared by extracting the tungsten from the ore by use of a strong alkali hydroxide solution at the boiling point. The alkali-metal carbonate or hydroxide thus obtained then is fused to form the water-soluble, alkali-metal tungstate. Where NH$_4$OH is used, the product is ammonium tungstate (NaOH yields sodium tungstate). The compound is reduced to metal powder. Conversion of the powder to massive metal is done by pressing, sintering, and mechanical working at high temperatures. In another process, the ore is fused with sodium carbonate and nitrate to yield sodium tungstate. Reduction of the oxide is accomplished by heating with carbon or hydrogen, whereupon tungsten metal is yielded.

Uses: Approximately one-half of the tungsten produced is in the form of sintered tungsten carbides. These compounds are used for cutting tools and wear-resistant parts. About 15% of production is consumed for making wire used in lamps and also for various shapes used in aerospace and defense products. Another 15% of production is used for high-temperature alloys and powder metallurgy. Approximately 10% goes into high-speed tools. The remaining production goes into a wide variety of applications.

Tungsten carbide, WC, is extremely hard (9.5 on the Mohs scale; diamond = 10) and has a melting point of 2,870°C. This combination of hardness and high-temperature stability makes it an excellent material for cutting tools. Additionally, the wear-resistant properties are excellent, accounting for the use of tungsten carbide for dies for hot and cold working of wire, rod and tubing, mining tools, snow-tire studs, and ball-point pens. For hard carbide tools and dies, tungsten carbide in the form of fine powder (1–10 micrometer particle size) is bonded with cobalt.

Special carbide tools also will often contain various percentages of titanium, tantalum, niobium (columbium), and hafnium carbides, along with the tungsten carbide. Chromium and vanadium carbides are also added to produce special, fine-grain-size grades of cemented tungsten carbide-cobalt materials. See Fig. 1.

For hard-facing applications, fused tungsten carbide is used. Tungsten also forms the ditungsten carbide W$_2$C, which has a melting point of 2,860°C. However, the term *tungsten carbide* usually refers to the mono compound. WC generally is made by combining tungsten metal powder with finely divided lampblack. The mixture then is heated to about 1,500°C. A variety of tungsten powders are made which then are subjected to various powder metallurgy techniques to form numerous shapes with a wide range of characteristics. Tungsten carbide can also be manufactured by a so-called menstrum process, which employs calcium carbide and aluminum metal to reduce scheelite via a thermite reaction, with the tungsten carbide recovered by acid washing.

Tungsten wire, including pure (unalloyed), doped nonsag; (potassium silicate and aluminum chloride or nitrate doped), and thoriated and zirconiated types are used extensively in applications such as filaments for incandescent lamps, thermocouples, arc-lamp electrodes, electrochemical electrodes, and instrument springs. See Fig. 2. Tungsten disks are used for electrical contacts; tungsten is used in glass-to-metal seals, where the coefficient of thermal expansion of tungsten is close to that of hard borosilicate glass; and tungsten pads are used in connection with silicon semiconductors because of the high thermal conductivity of tungsten and good match of the coefficient of thermal expansion of tungsten with that of silicon.

Fig. 1. Cemented tungsten carbide compacts and mud nozzles for oil well drilling. (*Fansteel.*)

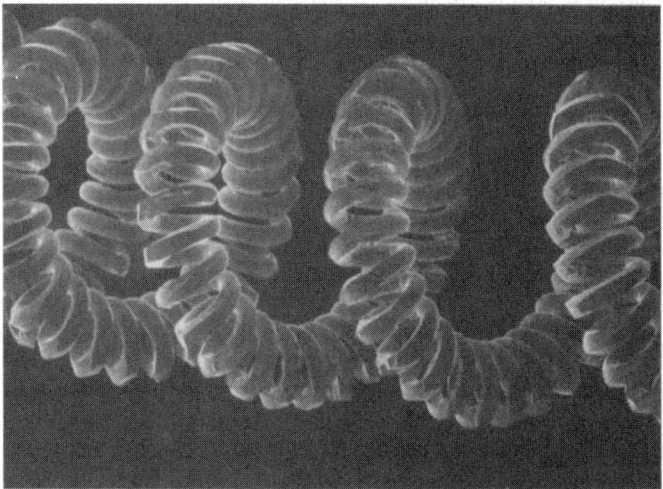

Fig. 2. Scanning electron photomicrograph of a nonsagging tungsten filament taken after several hundred hours of operation at 2500 °C in a 60-watt light bulb. (*After Wittenaur, Nieh, and Wadsworth.*)

Compositions of silver and tungsten and of copper and tungsten find application as electrical contacts where they are subject to severe arcing. As a shield or as containers for radioactive materials, heavy-metal alloys of tungsten alloyed with about 7% nickel and 4% copper are effective. The same alloys also find other uses where high density is required, as in gyroscope rotors, counterweights in aircraft, and self-winding watch parts. Alloys of cobalt, chromium, and tungsten also find use in cutting tools, dies, and wear-resistant parts. The function of tungsten in steel is that of forming stable carbides, strengthening ferrite, and refining the grain size for retaining high hardness at elevated temperatures — a requirement of highspeed steels.

Tungsten chemicals find limited use in inks, paints, enamels, dyes, and glass manufacture. Some tungsten compounds and their derivative phosphors find use in x-ray screens, television picture tubes, and luminescent light sources.

Chemistry and Compounds: In keeping with its $5d^4 6s^2$ electron configuration, tungsten forms many compounds in which its oxidation state is 6+, just as molybdenum does. It forms divalent and tetravalent compounds to about the same extent as molybdenum but its trivalent and pentavalent compounds are somewhat fewer. Its anion chemistry is closely akin to that of molybdenum.

Among the divalent compounds of tungsten, the diiodide, WI_2, dibromide, WBr_2, and the dichloride, WCl_2, are among the most clearly characterized; they all hydrolyze, although the iodide reacts only with warm water. Like molybdenum, tungsten(II) has a complex chloroion, $[W_6Cl_8]^{4+}$ which, however, is much more easily oxidized than its molybdenum analog.

Trivalent tungsten occurs rarely in simple compounds, other than certain high-temperature products such as one of the borides, WB, one of the phosphides, WP, and the complexes. Among the latter is the ion $[W_2Cl_9]^{3-}$ in which the two tungsten atoms participate in a Cl—W—Cl—W—Cl bridging structure, as they do in a W_2Cl_6 structure.

In addition to the tungsten(II) boride mentioned above, the element forms at least two other borides, W_2B and WB_2; it forms a similar series of phosphides, W_2P, WP, and WP_2 as well as WO_2 (brown oxide), W_4O_{11} (blue oxide), and WO_3 (yellow oxide), and two sulfides, WS_2 and WS_3. The tungsten(IV) oxide and sulfide are representative of the simple tetravalent compounds, which also include a tetrabromide, WBr_4, and tetraiodide, WI_4. Like the dihalides, these tetrahalides undergo hydrolysis quite readily.

Among the best known simple pentavalent tungsten compounds are the pentachloride, WCl_5, and the pentabromide, WBr_5. As is true of tungsten(IV), tungsten(V) forms complexes.

By far the greatest number of tungsten compounds are those in which the element is hexavalent. These include all common halides except the iodide, i.e., WF_6, WCl_6, as well as a number of oxyhalides, WOF_4, $WOCl_4$, WO_2Cl_2, $WOBr_4$, and WO_2Br_2, the trioxide, trisulfide, diboride, and diphosphide already mentioned, various complexes and organometallic compounds, and the anions.

Tungsten trioxide dissolves in hot alkali metal hydroxide solutions to yield in more or less hydrated form, the tungstate ion, WO_4^{2-}. However, the ionic species that exist in solution are more complex than mere hydration of the WO_4^{2-} would indicate, and this is especially true of the compounds obtained from such solutions. There are, however, two simple forms of the orthotungstic acid: H_2WO_4, which is precipitated upon addition of HCl to a hot tungstate solution, and $H_2WO_4 \cdot H_2O$, which is similarly obtained from a cold solution. Neutralization of a tungstate solution under most conditions yields, upon crystallization, much more complex salts. The acidic groups condense, with elimination of water, to form complexes, that can be crystallized as salts, which can be regarded as derived from "poly" acids. When such salts have only one kind of metal atom (e.g., W) in their anions, they are called *isopoly acids*; when they have more than one kind, they are called *heteropoly acids*. The latter group comprises an entire field of tungsten chemistry (as well as that of molybdenum and other elements.); the tungstophosphates are important in analysis and other applications. Other examples are the heteropoly acid salts formed by tungsten with oxyanions of boron, silicon, germanium, tin, arsenic, titanium, zirconium, and hafnium. In particular, the 6-series and the 12-series, containing, respectively, 6 and 12 tungsten atoms per molecule, have been extensively investigated.

Other interesting compounds are the "tungsten blues," complex oxides of colloidal nature, obtained by reduction of tungstates in alkaline solution. At higher temperatures, reduction of tungstates (of main group 1 and 2 elements) by alkali metal, hydrogen, zinc, tungsten, or electrolysis yields the semimetallic "tungsten bronzes," formulated as M_nWO_3, where M is the alkali metal and n is less than 1. They have cubical structures with W—O—W groups forming the sides, and the alkali or alkaline earth atom randomly located in the center of some of the cubes. The resulting extra electrons are considered to distribute over the entire structure, giving it metallic properties.

Tungsten forms many other complexes. Of particular interest are the octacyano complexes, containing eight cyanide, CN^- ions coordinated to a single tetravalent or pentavalent tungsten ion, $W(CN)_8^{4-}$ or $W(CN)_8^{3-}$, the latter being exceptionally stable, and forming octacyanotungstic(V) acid, $H_3[W(CN)_8] \cdot 6H_2O$, which is known in salts. Similar complexes are known for molybdenum, rhenium and osmium. The fluorocomplex of tungsten(VI) has the structure $[WF_8]^{2-}$ and forms salts with the higher alkali metals, potassium, rubidium and cesium.

Like molybdenum and chromium, tungsten forms a number of cyclopentadienyl compounds which are also carbonyls, e.g., $C_5H_5W(CO)_3H$, $C_5H_5W(CO)_3CH_3$, $C_5H_5W(CO)_3C_5H_5$, $C_5H_5(CO)\,WW(CO)C_5H_5$, and $C_5H_5(CO)_3WW(CO)_3C_5H_5$. Tungsten(VI) also forms several other organometallic compounds, e.g., $W(OC_6H_5)_6$ and $W(OC_6H_4CH_3)_6$, as well as a simple carbonyl, $W(CO)_6$.

An excellent and comprehensive review of tungsten and its alloys is given in the Wittenauer reference listed.

Additional Reading

Carter, G.F. and D.E. Paul: "Materials Science and Engineering," ASM International, Materials Park, OH, 1991.

Considine, D.M. and G.D. Considine: "Van Nostrand Reinhold Encyclopedia of Chemistry," 4th Edition, Van Nostrand Reinhold, New York, NY, 1984. (A Classic Reference.)

Farrar, L.C. and J.A. Shields, Jr.: "Tungsten and Tungsten-Copper Applications for Coal-Fired Magnetohydrodynamic Power Generation," *J. of Metals* (August 1992).

Greenwood, N.N. and A. Earnshaw: "Chemistry of the Elements," 2nd Edition, Butterworth-Heinemann, Inc., Woburn, MA, 1997.

Krebs, R.E.: "The History and Use of Our Earth's Chemical Elements: A Reference Guide," Greenwood Publishing Group, Inc., Westport, CT, 1998

Lassner, E. and W. Schubert: "Tungsten: Properties, Chemistry, Technology of the Element, Alloys, and Chemical Compounds," Kluwer Academic Publishers, Norwell, MA, 1998.

Lewis, R.J. and N.I. Sax: "Sax's Dangerous Properties of Industrial Materials," 10th Edition, John Wiley & Sons, Inc., New York, NY, 1999.

Lide, D.R.: "CRC Handbook of Chemistry and Physics 2000–2001," 81st Edition, CRC Press, LLC., Boca Raton, FL, 2000.

Shin, K.E., et al.: "High-Temperature Properties of Particle-Strengthened W-Re," *J. of Metals* (August 1990).

Staff: "ASM Handbook—Properties and Selection: Nonferrous Alloys and Pure Metals," ASM International, Materials Park, OH, 1990.

Wittenauer, J.P., T.G. Niehm, and J. Wadsworth: "Tungsten and Its Alloys," *Advanced Materials & Processes*, 28 (September 1992).

Portions of this article were contributed by M. SCHUSSLER, Senior Scientist, Fansteel, North Chicago, IL.

TUNING FORK. An oscillator that has been described as one piece of metal with two moving parts. It is termed a fork because of its obvious forklike shape. The device was invented by John Shore, an English musician, in 1711. Manually operated low-frequency forks still are used to tune musical instruments. Precision tuning forks, electromagnetically driven, are used as frequency standards in wristwatches and clocks; as oscillators in electronic systems for functions similar to those of quartz crystals; and as modulators of laser and other light and energy beams in electro-optical systems. Frequencies range from 10 Hz to more than 25 kHz. Frequency is determined by time (prong) dimensions and varies slightly with amplitude. For high precision, the drive power, transmitted by a magnetic transducer, is held constant by a zener diode, negative feedback, or automatic gain control.

Other fork drives include electrostatic magnetostrictive, pneumatic, piezoelectric and, most efficient of all, electrodynamic. With the latter method, the coils are stationary and the magnet system moves. In the 360 Hz tuning fork watch movement, for example, this drive reduces power requirements to six microwatts.

Normally, a fork is driven to vibrate in a fundamental mode, but highest frequencies are produced by driving it in an overtone mode—which can result in frequencies some six times those for the fundamental mode. In effect, the fork becomes a piece of metal with four moving parts. A 25 kHz fork in this mode is less than an inch (2.5 cm) long.

TUNNEL ENGINEERING. Broadly defined, a tunnel is a subsurface (underground or underwater) passageway. A natural tunnel may exist in some caves, but tunnels of engineering interest are manually-created structures that generally fall within the province of civil engineering.

Tunnels are commonly identified by the purpose for which they serve. Thus, there are railway, highway, and pedestrian tunnels that are located beneath crowded municipalities, that penetrate through mountains, or that rest on the bottom of rivers, lakes, bays, or channels. A tunnel may be as short and as shallow as an underground passage between two buildings or as long as several miles (kilometers), with a large cross section to accommodate several lanes of traffic. It is not uncommon for two tunnels to parallel each other in order to accommodate more traffic. Also, there are tunnels for conveying large quantities of water and sewage. See Table 1.

TABLE 1. OUTSTANDING TUNNELS — WORLDWIDE

Name	Location	Type	Length Mi	Length Km	Completed
Seikan	Tsugara Strait, Japan	RWY	33.1	53.3	1983
Eurotunnel	Folkestone, England–Calais, France	COM	32	50	1994
Simplon (1 and 2)	Alps/Switzerland–Italy	RWY	12.3	19.8	1906/1922
Apennine	Bologna–Florence, Italy	RWY	11.5	18.5	1934
St. Gotthard	Alps/Switzerland	VEH	10.2	16.4	1980
St. Gotthard	Alps/Switzerland	RWY	9.3	14.9	1881
Lötschberg	Alps/Switzerland	RWY	9.1	14.6	1911
Mont Génis	Alps/France	RWY	8.5[a]	13.7	1871
New Cascade	Cascade Mountains, Washington, U.S.	RWY	7.8	12.6	1929
Mount Blanc	Alps/France–Italy	VEH	7.5	12.1	1905
Vosges	Vosges, France	RWY	7.0	11.3	1940
Arlberg	Alps/Austria	RWY	6.3	10.1	1884
Moffat	Rocky Mountains, Colorado, U.S.	RWY	6.2	9.9	1928
Shimuzu	Shimuzu, Japan	RWY	6.1	9.8	1931
Rimutaka	Wairarapa, New Zealand	RWY	5.5	8.9	1955
Mount Ena	Japanese Alps	VEH	5.3	8.5	1976[b]
Great St. Bernard	Alps/Switzerland–Italy	VEH	3.4	5.5	1964
Mount Royal	Montreal, Quebec, Canada	VEH	3.2	5.1	1918
Lincoln	Hudson River, New York City, U.S.	VEH	2.5	4.0	1937
Queensway Road	Mersey River, Liverpool, England	VEH	2.2	3.5	1950
Brooklyn-Battery	East River, New York City, U.S.	VEH	2.1	3.4	1950
Holland	Hudson River, New York–New Jersey, U.S.	VEH	1.7	2.7	1927
Fort McHenry	Baltimore, Maryland, U.S.	VEH	1.7	2.7	1985
Hampton Roads	Norfolk, Virginia, U.S.	VEH	1.4	2.3	1957
Queens-Midtown	East River, New York City, U.S.	VEH	1.3	2.1	1940
Liberty Tubes	Pittsburgh, Pennsylvania, U.S.	VEH	1.2	1.9	1923
Baltimore Harbor	Baltimore, Maryland, U.S.	VEH	1.2	1.9	1957
Allegheny Tunnels	Pennsylvania Turnpike, U.S.	VEH	1.2	1.9	1940[c]

RWY = Railway
COM = Combination
VEH = Vehicular
[a]Lengthened in 1881 to length shown.
[b]Parallel tunnel commenced in 1976.
[c]Parallel tunnel built in 1965; twin tunnel constructed in 1966.
Sources: American Society of Civil Engineers; Bridge, Tunnel and Turnpike Association.

Tunnels date back to the ancient Egyptians, Assyrians, and Indians who constructed tunnels in connection with tombs and temples. It has been determined that a tunnel about 12 feet (3.7 meters) and 15 feet (4.6 meters) long was constructed by the Assyrians to divert water from the Euphrates River. The Romans built a tunnel about 3000 feet (914 meters) long on a roadway between Naples and Pozzuoli. The ends of the tunnel were constructed in the form of conical frustums to permit light to penetrate to the center of the tunnel. Practically all early tunnels were hand-hewn out of rock. Gunpowder for use in tunnel construction was first used circa 1680 in France. The need for tunnels expanded in France and England because of the increasing use of canals. Tunnel construction in sand and wet ground was not undertaken until the early 1800s. A 24-foot (7.3 meters) tunnel was built for the Saint Quentin Canal (France) in 1803. The type of construction required large support timbers and base planking. Such tunnels also required masonry lining to avoid water penetration from above.

Probably the greatest impetus for tunnel building was the introduction of the railway. Two tunnels were constructed on the line of the Liverpool and Manchester Railway in England, and, in the United States, the first railway tunnel was built near Auburn, Pennsylvania, in about 1820. The tunnel measured several hundred feet (meters) in length, was approximately 22 feet (6.7 meters) high and 15 feet (4.6 meters) wide. Numerous tunnels were constructed in the Appalachian mountains and later, as the railways moved west, in the Rocky, Sierra Nevada, and Cascade mountain ranges.

Large highway tunnels were pioneered in Europe (Alps) to shorten the distances between towns as well as to reduce the costs and dangers of winding roads that depended upon natural mountain passes.

Tunnel engineering must consider numerous factors, several of which extend beyond physically creating a passageway:

- A thorough preliminary geological and topological survey of the proposed tunnel location.
- An environmental impact study.
- Design of required drilling and excavating equipment.

- Ample provision for continual earth, rock, and debris removal and selection of an appropriate "dumping" area.
- Selection of acceptable upgrade and downgrade for the types of vehicles that will be using the tunnel.
- Selection of acceptable curves in the tunnel when required.
- Provisions for adequate ventilation.
- Provisions for adequate lighting.
- Creation of a thorough safety program—from the start to the end of construction—and a revised safety program when the tunnel is in operation. All possible contingencies must be considered.
- Provisions for adequate communication—between various stations within the tunnel and to stations on the surface. Communication, of course, plays a large role in the safety program.
- Provisions for expeditious but safe access to and egress from the tunnel. This is particularly important with highway tunnels, which most often require reducing the number of lanes of traffic.

The remainder of this entry is devoted to the *Channel Tunnel*, one of the greatest technological and engineering feats of the 20th century.

The Eurotunnel—Folkestone, England–Calais, France

The first attempt to bore a tunnel under the English Channel was made in 1880, using a machine designed by Colonel Beaumont. The tunnel, which was almost 8 feet (2.4 meters) in diameter, started from an access shaft and was approximately 1.2 miles (2000 meters) long when the project was halted in 1882. A similar tunnel also was started in France from a shaft sunk near Sangatte.

In 1922, a trial bore was made at Folkestone, using a Whitaker machine, which cut a tunnel about 9.8 feet (3 meters) in diameter. Part of this machine has been recovered and is now displayed at the *Eurotunnel* Exhibition Centre in Folkestone.

In 1974, work began on a much larger tunnel system, similar in size and configuration to the *Eurotunnel* scheduled for dedication in 1994. In the United Kingdom, an access tunnel (or adit) was made between the upper

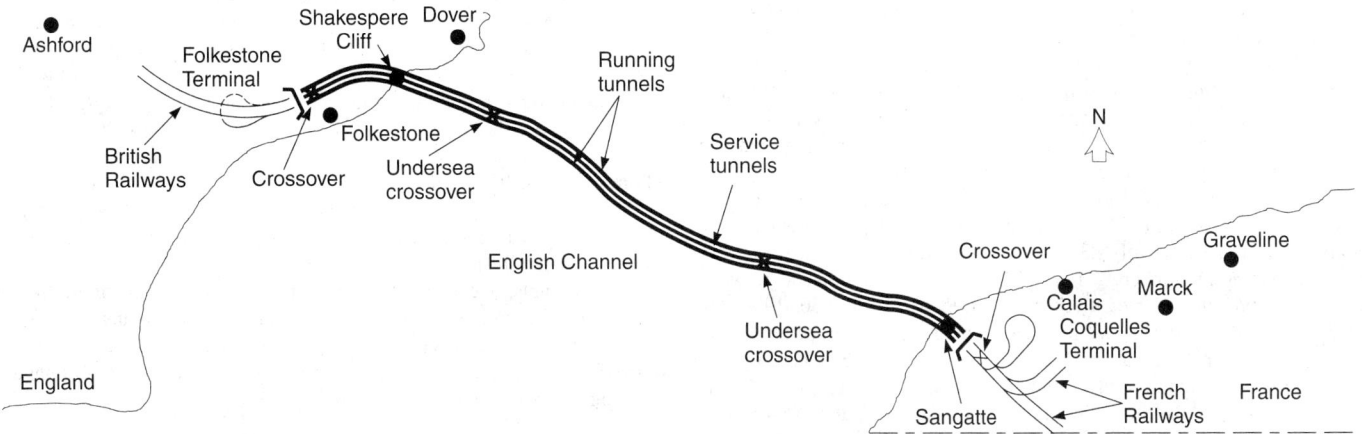

Fig. 1.   Route of *Eurotunnel* as "viewed" from above. (*The Channel Tunnel Group.*)

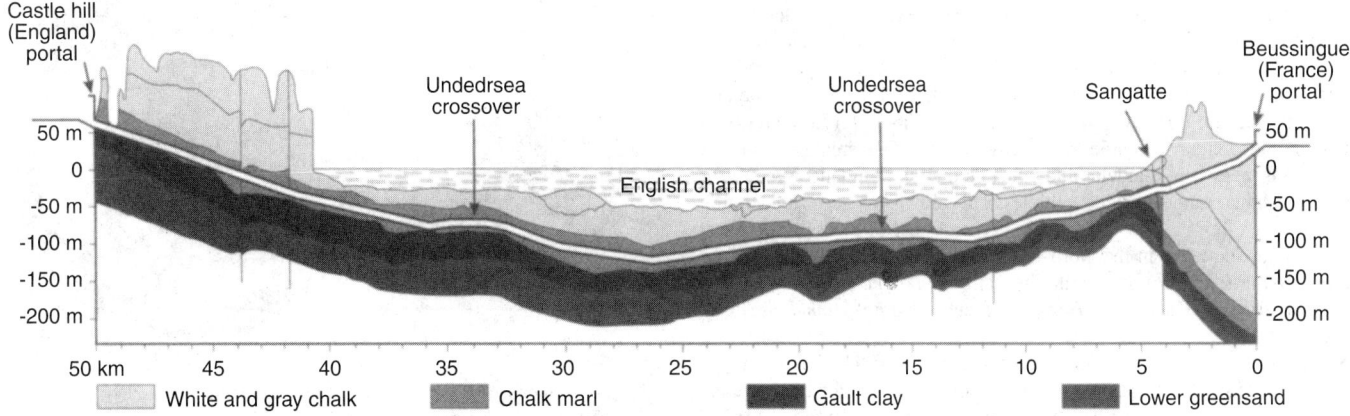

Fig. 2.   Geological cross section of *Eurotunnel* pathway. (*The Channel Tunnel Group.*)

Fig. 3.   The custom designed tunnel boring machines (TBM's) were used to excavate the chalk mari rock which lay beneath the seabed along the tunnel route. Tunneling commenced on December 1, 1987. The rate of advance on the best day was 185 feet (56 meters), in the best week was 966 feet (292.6 meters), and in the best month was 3,649 feet (1,105.7 meters). In all, 287,714,286 cubic feet (8 million cubic meters) of soil was removed during construction at a rate of 2,646 short tons (2,400 tonnes) per hour. (*The Channel Tunnel Group.*)

and lower sites at Shakespeare Cliff near Dover, and a rail access adit was bored from the lower site down to the level of the future rail tunnels. A 1321-foot (400 meters) length of service tunnel was bored before work was abandoned at the start of 1975. During these works, the 1880 Beaumont tunnel and shaft were found to be in good condition. In France, an adit was constructed at Sangatte, but progress was slower because of bad ground conditions. However, a tunnel-boring machine (TBM) had been built and delivered to the site.

The present *Eurotunnel* project was started in the early 1980s, when the British and French governments jointly decided that a fixed link should be built across the Channel. Both governments stipulated, however, that the work would have to be privately financed. In April 1985, bids were invited for the financing, construction, and operation of a fixed link. The *Eurotunnel* proposal was selected in January 1986. The company was awarded the concession for building the Channel Tunnel system and operating it for a period of 55 years beginning in 1987.

On June 15, 1987, the legislative approval for the *Eurotunnel* project was signed by President Mitterrand and received Royal Assent by the British government on July 23, 1987. Actual tunnel construction work commenced on December 1, 1987. The historic first contact between British and French undersea workers occurred on October 30, 1990. On February 10, 1992, *Eurotunnel* directors announced a delay of the tunnel opening date from June 15, 1993, to the summer of 1993. Later, another postponement was announced. Commercial operation began in 1994. The cost of the tunnel at the time of completion was approximately $5.4 billion (£9.0 billion).

The *Eurotunnel* transport company operates every day of the week around the clock to facilitate the movement of people, products, and vehicles across the English Channel through the railway system. The *Channel Tunnel* system consists of three tunnels, each 31.35 miles (50.45 kilometers) long of which 24 miles (38 kilometers) is under the sea. The three tunnels are approximately 100 feet (30 meters) apart and are located under the Channel at an average depth of 148 feet (45 meters) below the seabed. The two larger tunnels are 25 feet (7.6 meters) in diameter and contain a single-track railway line. The smaller tunnel is 16 feet (4.8 meters) in diameter and contains the wire guidance system for custom-designed tunnel service vehicles. The railway system runs shuttle trains between terminals at Folkestone, in the county of Kent in Southeast England and Coquelles in the Nord-Pas-de-Calais region of France. High-speed through-trains connect with the national rail networks of France and Great Britain. The names *Eurotunnel* and *Channel Tunnel* are used interchangeably.

Since the beginning of commercial operation in 1994, the two types of shuttle trains, which transport tourist vehicles or freight trucks, have carried more than 3.5 million cars and coaches and nearly 1 million lorries. The *Eurostar* trains, which carry passengers between London, Paris, and

Brussels, have transported nearly 8 million travelers, while the freight trains have moved nearly 3.9 million short tons (4 million tonnes) of material through the tunnel system.

Geography of Eurotunnel. Contrary to common conceptions, the Eurotunnel is not a tube lying on the sea bed, exposed to the hazards of the sea. Tunnels were bored generally between 82 feet (25 meters) and 148 feet (45 meters) below the sea bed. From Folkestone past Dover and under the Channel for a total distance of about 30 miles (45 kilometers), the tunnels are bored through chalk marl, generally considered to be one of the most consistent and safe tunnelling mediums. As the tunnels approach the Folkestone terminal, they pass through gault clay and other strata and, for the first 3.1 miles (5 kilometers) from the Coquelles (France) terminal, the tunnels pass through more faulted zones and sands and gravels. These materials were the most difficult to bore.

Chalk marl is a soft rock, generally impermeable to water, and is homogeneous. This allows excavation to be carried out at a relatively high rate. The chalk marl is overlain by grey chalk, which generally is a harder and a more fractured material. It is underlain by gault clay, which essentially is impermeable. The vertical and horizontal alignment of the tunnels, therefore, was chosen so that the maximum possible length of the tunnels would be in the chalk marl, taking into account the acceptable gradients (not exceeding 1.1%) and the power consumption criteria applicable to a high-speed railway system, as well as drainage and pumping requirements.

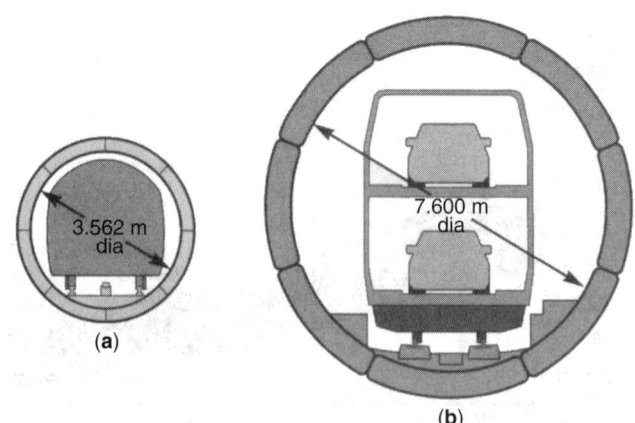

Fig. 4.   Comparison of (**a**) Piccadilly Line (London Underground) subway tunnel with (**b**) one of Eurotunnel's running tunnels. (*The Channel Tunnel Group Limited.*)

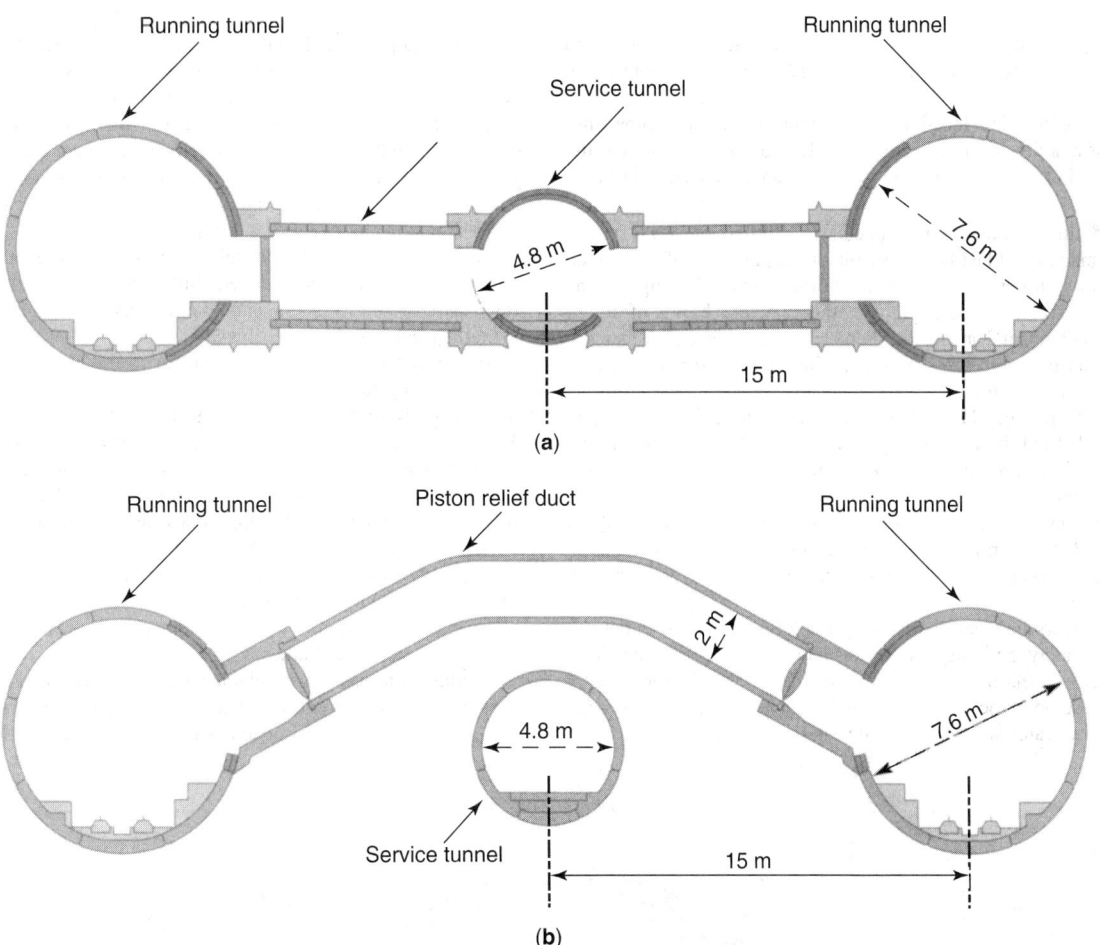

Fig. 5. Typical section through tunnels showing (**a**) cross-passage structure and (**b**) piston relief duct structure. (*The Channel Tunnel Group Limited.*)

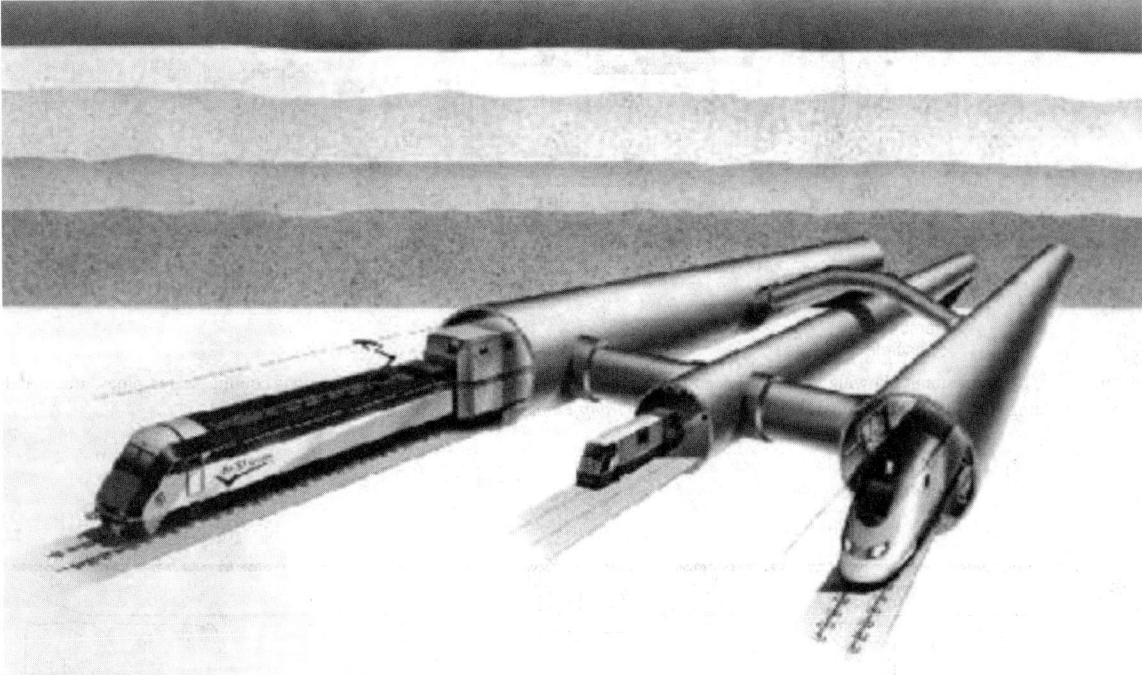

Fig. 6. Three-dimensional sectional view of the three tunnels. The outer tunnels each contain a single railway track, and the central tunnel, which passengers can reach from the rail tunnels, if necessary, is for support services. In normal operation, *Eurotunnel* shuttles and through-trains travel from the UK to France through the southern rail tunnel. Trains from both countries "drive" on the left.

The risk of earthquakes was not overlooked, including very detailed studies of the history of the region. Such studies are on a continuing basis.

In summary, the detailed location of the tunnels was determined not only in terms of a short distance between the European continent and the United Kingdom, but also by favorable geologic conditions. See Figures 1 and 2.

Structure of the Tunnels. The tunnels were bored by very special tunnel-boring machines (TBMs) to permit the installation of reinforced concrete linings or, in certain areas, linings of cast iron. See Fig. 3. The running tunnels have a diameter outside the concrete lining of up to 28.8 feet (8.78 meters) and an inside diameter of 24.9 feet (7.6 meters). The large size can be appreciated by comparing *Eurotunnel* with typical underground tunnels. See Fig. 4.

In addition to boring, the TBMs also erected the tunnel linings. Cement grout is injected behind the linings to fill the voids between them and the chalk marl and to create a solid bond between them. This bond gives the structure extra strength.

Cross passages join the running tunnels and the service tunnel about every 1230 feet (375 meters). Piston relief ducts to permit the equalization of air pressures connect the two running tunnels about every 820 feet (250 meters).

Two rail crossovers between the running tunnels are located under the sea bed approximately one-third of the distance from each portal. Two further sets of crossovers are located close to the portals. In each tunnel crossover, there are doors between the tunnels. Pumping stations along the tunnels handle any water spilled in or entering the tunnels.

Fixed equipment is installed in the service tunnel. See Fig. 5. A three-dimension sectional view of the three tunnels is shown in Fig. 6, and a sectional view of a running tunnel is shown in Fig. 7.

Each running tunnel has a single rail track, overhead line equipment (catenary), and two walkways, one for maintenance and the principal one on the side nearest the service tunnel for any emergency evacuation. The walkways are designed to maintain a shuttle upright and traveling in a straight line in the event of a derailment.

The single rail track through each running tunnel has continuously welded rail on precast concrete rail supports embedded in a concrete track base. Fixed to the sides of the tunnels are cooling pipes, fire mains, signalling equipment, and cables.

The catenary supplies traction power to the shuttles and through-trains in the running tunnels. The line is divided into sections so that only one train is likely to be immobilized at any one time in the event of a fault. *Eurotunnel* has its own standby auxiliary power for emergency situations.

All tunnels and connecting passages have sufficient lighting for operational, maintenance, and emergency purposes. Wherever practical, materials and equipment, including cables, are fire-resistant. All tunnels and connecting passages have sufficient lighting.

A shuttle vehicle for carrying automobiles and passengers riding in the cars is shown in Fig. 8. Cars and vehicles that do not exceed 6 feet (1.85 meters) in height (including top loading racks) are placed on five-vehicle, single-deck shuttles.

In contrast to passenger vehicles, trucks (lorries) and their occupants are separated during the journey. The drivers and passengers, if any, are taken to an air-conditioned rail coach at the front of the shuttle.

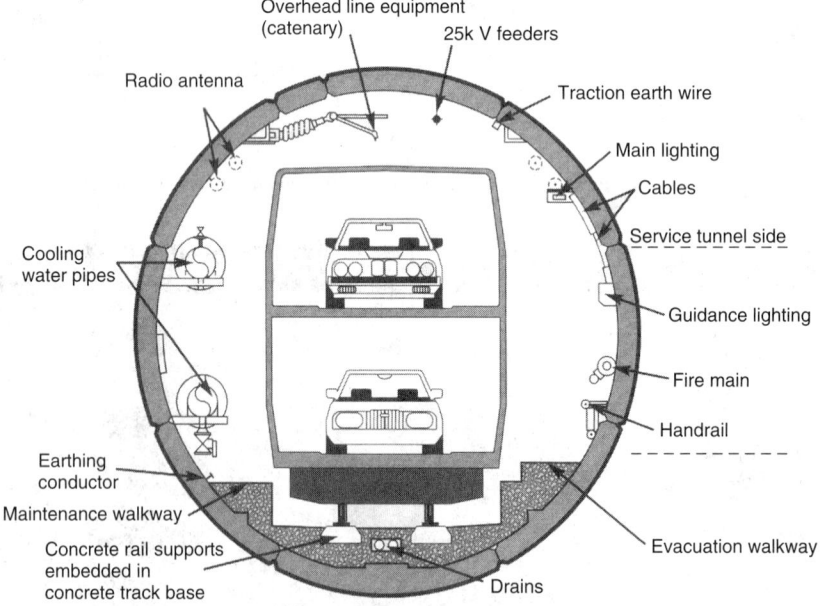

Fig. 7.   Section through a running tunnel showing numerous features, including locomotive overhead power line, cooling water pipes, track and track supports, maintenance walkway, lighting, fire main, and drains. (*The Channel Tunnel Group Limited.*)

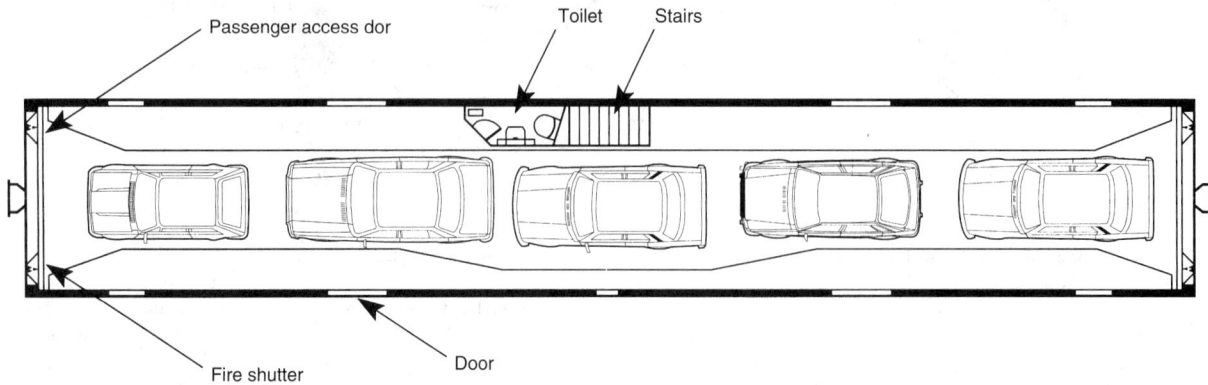

Fig. 8.   Top view of shuttle wagon showing arrangement of five passenger vehicles. Passengers remain with their vehicles. Each wagon features restroom facilities. (*The Channel Tunnel Group Limited.*)

The tunnel journey requires approximately 35 minutes. Procedures at the entrances and exits of the tunnel facility essentially will double that time, making a total trip is about 1 hour and 20 minutes.

Fig. 9. Cross section of a railway tunnel nearing completion in 1991. (*The Channel Tunnel Group Limited.*)

Passenger trains pass through the tunnel much as in passing through any other tunnel. The terminus in England is Waterloo Station, London; in France, it is Gare du Nord or Brussels Midi. An impressive view of one of the running tunnels toward the end of construction in 1991 is shown in Fig. 9.

Additional Reading

Gibb, R.: "The Channel Tunnel: A Geographical Perspective," John Wiley & Sons, Inc., New York, NY, 1994.
Grayson, L.: "Channel Tunnel—the Link to Europe: An Overview and Guide to the Literature," Lawrence Erlbaum Associates, Inc., Mahwah, NJ, 1995.
Guteri, F. and R. Ruthen: "An Undersea Link Between Great Britain and France," *Sci. Amer.*, 22 (January 1991).
Jackson, D.D.: "The Ins and Outs of a Dangerous and Boring Subject," *Smithsonian*, 66 (May 1986).
Kirkland, C.J.: "Engineering the Channel Tunnel, Eurotunnel," Routledge, New York, NY, 1995.
Kuesel, T.R. and E.H. King: "Tunnel Engineering Handbook," Chapman & Hall, New York, NY, 1996.
Mahtab, M.A. and P. Grasso: "Geomechanics Principles in the Design of Tunnels and Caverns in Rocks," Elsevier Science, New York, NY, 1992.
van Dam, L.: "The Big Dig (English Channel Project)," *Technology Review*, 64 (October 1990).
West, G.: "Innovation and the Rise of the Tunnelling Industry," Cambridge University Press, New York, NY, 1988.

Web Reference

Eurotunnel: *http://www.eurotunnel.co.uk/english/explorer_e.htm*

TUPELO TREES. Members of the family *Davidiaceae* (family *Davidia*), tupelos are broadleaf trees found in North America and eastern Asia. Generally, they thrive well in or near swamps. They have alternate leaves and small, regular, greenish flowers in clusters. The drupe is essentially oval and contains a compressed stone. The *Nyssa sylvatica* is known as the sour gum or pepperidge and ranges farther north and farther away from the water than other tupelos. The bark is rough and gray; the wood is light-colored and has a twisted grain, making it difficult to work. It is a desirable wood for some applications, however, because of practically no tendency to split. The wood has been used for the hulls of ships because of its excellent strength when immersed in water and because of its resistance to salt. Record tupelo trees in the United States are listed in Table 1.

TURACOS (*Aves, Cuculiformes, Musophagidae*). This family of birds is confined exclusively to Africa. They range in size from that of a magpie to that of a raven; the length is 40–70 centimeters ($15\frac{1}{2}$–$27\frac{1}{2}$ inches) and the weight is 250–1000 grams ($8\frac{1}{2}$–35 ounces). The lateral foretoe can be turned laterally (reversible toe). Turacos are perching birds of the jungle, the edges of woods, or of the savannas, and their diet is chiefly vegetarian. Usually they live in pairs, occasionally in small flights (4 or 5 birds), but rarely in larger groups. Their calls are characteristic and loud. The nests are shallow, built of twigs in densely foliated trees, and remind one of pigeons' nests. The clutch varies from 1 to 3 white or pastel-shaded unmarked eggs. The young wear a dense down coat at first, and the contour feathers are tardy; the nestling period is about 4 weeks.

The turacos are divided into 4 genera: 1. the Crested Turacos (*Tauraco*); 2. the Violet Plaintain-Eaters (*Musophaga*); 3. the Great Blue Turacos (*Corythaeola*); 4. the Go-Away Birds (*Crinifer*). Altogether they comprise 18 species.

The green color of the plumage of many turaco species does not result, as it does in most other green-plumed birds, from the physical structure of the feathers, but from a green pigment peculiar to turacos called turacoverdin. Turacin, which is a pigment containing copper, is also peculiar to turacos and lends an attractive red to the wing feathers of most of them. See also **Cuculiformes**.

TURBELLARIA. The free-living flatworms, a class of the phylum *Platyhelminthes*. Unlike the parasitic members of this phylum, these worms

TABLE 1. RECORD TUPELO TREES IN THE UNITED STATES[1]

Specimen	Circumference[2] Inches	Centimeters	Height Feet	Meters	Spread Feet	Meters	Location
Black tupelo (typ.) (1998) (*Nyssa sylvatica* var. *sylvatica*)	185	470	78	23.8	64	19.5	New Jersey
Black tupelo (typ.) (1998) (*Nyssa sylvatica* var. *sylvatica*)	182	462	68	20.7	95	29	Connecticut
Ogeechee typelo (1993) (*Nyssa ageche*)	166	422	93	28.3	41	12.5	Florida
Ogeechee typelo (1993) (*Nyssa ageche*)	174	442	81	24.7	48	14.6	Florida
Swamp tupelo (1987) (*Nyssa sylvatica* var. *biflora*)	238	605	102	31.1	57	17.4	Virginia
Water tupelo (1991) (*Nyssa aquatica*)	336	853	105	32	56	17.1	Virginia

[1]From the "National Register of Big Trees," American Forests (by permission).
[2]At 4.5 feet (1.4 meters).

have a cellular ectodermal covering bearing cilia, with the exception of a few parasitic members which resemble the flukes and tapeworms in the absence of cilia. The free-living members of the class also have sense organs located at the anterior end of the body, including a pair of eyes and tentacles. The mouth opens on the ventral surface, either near the head or near the end of the body, and the alimentary tract (see also **Digestive System (Other Life Forms)**) is branched in many forms to extend widely through the body. The terminal portion of the tract forms a protrusible proboscis. The turbellarians are hermaphrodite with very few exceptions.

Flatworms of this group are often common in small streams and ponds, and some are marine. They glide over surfaces by the action of the cilia of the lower surface, aided by the secretion of a trail of mucus, or move by undulations produced by muscular action. Their food consists of small animals, living or dead.

TURBIDIMETRY. The cloudiness in a liquid caused by the presence of finely divided, suspended material is termed *turbidity*. A turbidimeter measures this quality by determining the reduction in transmission of light that is caused by interposing a turbid solution between a light source and a detector, such as the eye or photocell. By using a known volume of solution in comparison with a standard, the instrument makes it possible to determine the mass effect, attributable to the number and size of the particles in the solution, and thus the quantitative amount of material present. Turbidimetric methods are similar to colorimetric procedures in that both involve measurement of the intensity of light transmitted through a medium. They differ in that the light intensity is attenuated by scattering in turbidimetry and by absorption in colorimetry. Both determinations may use similar instrumentation. Several designs of combined nephelometers, turbidimeters, and colorimeters are available.

Turbidimetry is applicable to the determination of suspended material in liquids encountered in nature and in manufacturing processes. Numerous uses are found in water treatment plants, sewage works, power and stream generating plants, beverage bottling plants, and petroleum refineries. The turbidity may be due to a single chemical substance, or it may be due to a combination of several components. For example, silica may be determined in the approximate concentrations of 0.1 to 150 parts per million, expressed as SiO_2. Sometimes, composite turbidities are expressed as equivalent to silica. Higher concentrations may be determined by dilution. Numerous applications are possible in which a turbidity is developed from the test sample under controlled conditions. A widely used determination is that of sulfate, after the addition of barium chloride to an unknown sample to form a suspension of barium sulfate. The procedure is particularly applicable within the concentration range of 0.2 to 100 parts per million sulfate. Routine procedures have been developed for the determination of sulfur in coal, oil, and other organic materials in which the sample first is fused in a sodium peroxide bomb prior to precipitation of the sulfur as barium sulfate.

The *Parr turbidimeter* is an extinction type instrument, which consists of a cylinder to contain the turbid suspension, a lamp filament of fixed intensity at the base, and an adjustable plunger through which visual observation is made. Measurement is made of the depth of turbid medium necessary to extinguish the image of the lamp filament. Standard suspensions are used to prepare a calibration curve, which is a plot of depth *vs.* concentration.

The *Hellige turbidimeter* is also a variable depth type of instrument using visual detection. A combination of vertical and horizontal illumination of the sample and a split ocular permit the eye to function merely to compare the intensities of two images simultaneously appearing in the ocular. This is a special form of double-beam operation, and the intensity emitted by the light source need not be extremely constant from one sample to another. Adjustment is made of a slit in the path of the direct, or vertical, illumination of the sample, and the calibration curve consists of a plot of this slit opening *vs.* concentration.

See also **Analysis (Chemical)**; **Nephelometry**; and **Photometers**.

TURBIDITY CURRENT. See **Ocean**.

TURBINE ENGINE. An engine incorporating a turbine as a principal component; especially, a gas-turbine engine.

TURBINE (Hydraulic). See **Hydroelectric Power**.

TURBINE (Hydroelectric). See **Hydroelectric Power**.

TURBINE (Steam). When steam is expanded adiabatically through a stationary nozzle, it does not retain all the heat it originally had. The heat released during expansion, of course, does not disappear. According to the first law of thermodynamics, it must reappear as work energy in equivalent amounts (778 foot-pounds (108 kilogram-meters) for each Btu). In the case of steam expanding through a stationary nozzle, the moving steam must gain this mechanical energy, with the result that its speed is considerably increased. An ordinary expansion involving a heat drop of about 100 Btus (25 Calories) of steam has the capacity to increase its speed to the large magnitude of 30 miles (48 kilometers)/ minute. This gives some clue as to the reason why so light a fluid medium as steam is capable of producing so much power in a machine of moderate dimensions.

There are two principles of action employed in steam turbines: (1) impulse; and (2) reaction. The impulse principle involves stationary nozzles and moving blades, which absorb the mechanical energy from the steam as it flows over the blades. In the reaction turbine, the nozzles are themselves attached to the shaft. In one case, the motivating force is one of impulse of a stream against a blade; the other, one of a reaction force created by the acceleration of the steam in the moving nozzles. Very high rotative speeds are necessary if the entire adiabatic heat drop for the turbine is released in one set of nozzles. For example, an impulse turbine with a conservative expansion would create a steam speed which, if absorbed on one row of blades revolving in an 18-inch (46-centimeter) circle, would require a rotative speed at the shaft of better than 10,000 revolutions per minute. This gives a clue to the desirability of staging a turbine by sub-dividing the heat drops, the energy being absorbed after each incremental heat liberation. Thus, in a 5-stage impulse turbine, each stage may be called on to absorb only one-fifth of the total heat drop. The subdivision is even greater in the case of reaction turbines, where the Btus liberated per stage rarely exceed 10. In consequence, the number of stages found in the reaction-type turbine is greater than in the impulse-type turbine. Except for areas of nozzles, size of blades, and blade angles, the stages resemble one another.

Large steam turbines fall into three classes: (1) the straight reaction; (2) the straight impulse; and (3) the impulse-reaction turbine. The straight reaction turbine may have numerous stages; the straight impulse turbine rarely more than 20 stages (half that number may be more common). One or two impulse stages preceding a straight reaction section enable reduction of the number of reaction stages to about twenty.

An impulse turbine is pressure staged, i.e., the pressure drop is subdivided among a number of stages, any one stage of which may be, in addition, velocity staged. By this is meant that, following one set of nozzles there may be two rows of moving blades which, together, effect the reduction of velocity and the absorption of energy created in those nozzles. The two rows of moving blades are separated by a row of fixed reversing blades, which receives the steam from the first row and directs it at the proper angle for the second row. A velocity stage within a pressure stage is known as a Curtis stage. A simple pressure stage is a Rateau stage, and the reaction stage often is called Parson's staging. As steam expands through a multistage turbine it increases in volume after each stage, making it necessary either, that the diameter of the circle in which the blades travel be increased, or that the height of the blades be increased to provide sufficient area for the increased volume of flow. Usually both of these expedients are adopted, so that the turbine exhibits, roughly, a somewhat conical shape, being smallest at the high pressure end, and largest at the exhaust.

Essential to the operation of a turbine are a number of auxiliary devices. The impulse turbine rotor receives a moderate end thrust, due to fluid friction on the blades. The reaction turbine has a large end thrust, arising from the drop of steam pressure, across the moving blades, acting on the blading annulus. While the end thrust of an impulse turbine can be accommodated by special thrust bearings, the large forces set up in a reaction turbine necessitate the use of a special device known as a dummy piston. The dummy piston is a circular plate mounted concentric with the axis of the turbine, and having, on one side, high, and on the other, low steam pressure. The pressures are so chosen that the direction of the resultant force on the plate is counter to the end thrust on the blades. A separate piston is used to balance each different diameter of the drum, so that although steam pressure and end thrusts may vary with different loads, the equalizing pressures also similarly vary. A thrust bearing is provided to absorb the small amount of unbalance that may still exist.

The moving parts of the turbine are few. Principally, there are two bearings, one at each end of the turbine, which support the rotor. These are rather heavily loaded so that plain babbitted bearings are usual. Oil is pumped to these bearings and wasted from them to a sump, from which it is withdrawn, filtered, and cooled before being again supplied to the bearings. An auxiliary pump maintains oil pressure during the starting and stopping cycle of the main unit. Where the shaft of the turbine projects through the casing, means are provided for packing against leakage of the steam outward at the high-pressure end, and infiltration of air at the low-pressure end. Packed stuffing boxes are used only on small turbines. This service is performed on large turbines by sealing glands which are built into the turbine, and which seal the shaft with steam or water. The outward leakage from the glands is minimized by labyrinths.

Governing of steam turbines is accomplished by three methods, viz.: (1) throttling at inlet, (2) varying number of inlet nozzles in action, (3) varying duration of full pressure puffs (blasts), of which there are several per second. In addition, some turbines are provided with hand-operated by-pass valves which, by admitting high-pressure steam to low-pressure stages, enable the turbine to carry more overload though, of course, at reduced economy. Of these methods, the first is widely used on small turbines. In the large turbine field, the second is applied to the Curtis type, and the third to the Parsons type.

The losses in a steam turbine, as is the case with the engine, are topped in magnitude by the heat in the exhaust. Unfortunately, this is difficult to reduce because of the troubles attending the use of steam of quality lower than about 85% in the turbine. Wet steam erodes turbine blades, due to the high velocity with which the particles of moisture strike them. The other losses that occur in steam turbines are:

1. Thermodynamic losses:
   a. Leakage. Past shaft gland packings, dummy pistons, diaphragms and blade tips:
   b. Blade and disk friction.
   c. Throttling at the control valves.
   d. Nonstream line flow at other than design conditions.
   e. Leaving loss, and pressure losses in exhaust nozzle.
2. Mechanical losses:
   a. Bearing and stuffing box friction.
   b. Windage action of idle blades.
   c. Gland water-seal power.
   d. Oil pump and governor power.

All the thermodynamic losses of one stage are returned to the steam as it enters the next lower stage, so that one definite advantage of multi-staging is to make available to the next stage the thermal losses of the preceding one. This is one factor in the superior performance of multi-stage turbines. To illustrate the point, consider the case of friction on the nozzles. If the turbine casing is well insulated, the heat that is generated by friction of the steam against the nozzles simply elevates the nozzle temperature until it returns, by conduction to the steam, as much heat as is generated by friction. The greater part of the stage loss is returned to the steam at the low pressure, and in turbine design it is considered that all of the reheat occurs at the low pressure of the stage. Figure 1 shows how an adiabatic heat drop, as in nozzles, followed by a reheat, in which the stage losses are added at the low pressure, results in a steadily increasing entropy in a multistage turbine. This explains the typical expansion line of the steam turbine, which, as pressure is reduced, increases in entropy, whereas an ideal turbine would expand the steam at constant entropy. The increase of entropy corresponds to a decrease of energy availability, and that turbine which has the minimum entropy increase shows the best performance.

It is obvious from the foregoing analysis that all the heat taken out of the steam during its passage from the throttle to the exhaust must have been put onto the rotating shaft as mechanical energy. This is true because of the negligible radiation loss from the casing. Friction and power required to drive the governor and oil pump take their toll of this energy, but they rarely amount to as much as 5% of the heat drop, so that practically all of the heat drop appears at the turbine shaft coupling as useful work.

Trends in Steam Turbine Design. For fossil applications in the 500 to 600 megawatt range, a turbine consisting of a combined high pressure-intermediate pressure turbine element and two double flow low-pressure turbine elements is usually provided. The low-pressure turbine elements may have 25-, $28\frac{1}{2}$, or 31-inch (63.5-, 72.4-, or 78.7-centimeter) long last row blades, depending on the economic optimization with available cooling water, fuel costs and other cycle parameters. The world's first 500 megawatt tandem-compound steam turbine was a unit of this type, placed into service late in 1965. Combined elements, with the high and intermediate pressure blading in a single casing, offer significant advantages. The installed length of such a unit is 18 feet (5.4 meters) less than that required when the high pressure and intermediate pressure elements are contained in separate turbine casings. With one less turbine element, the savings in capital, erection, and maintenance costs are appreciable. A similar arrangement is provided for units in the 150 or 400 megawatt size class, consisting of a combined high pressure-intermediate pressure turbine coupled in tandem with one double flow low-pressure turbine element. The last row blades of the low-pressure turbine element may be 23, 25, $28\frac{1}{2}$, or 31 inches (58.4, 63.5, 72.4, or 78.7 centimeters) long, depending upon the unit rating, cycle, fuel costs, and expected load factor. Many of the improvements developed for the higher-rated units are also available in units rated less than 150 megawatts. An example of this is the single-case, single-exhaust-flow unit for nonreheat applications, such as the 100 megawatt steam turbines designed for some combined-cycle plants.

For larger fossil applications, above 650 megawatts, the complete unit may consist of a separate double-flow high pressure turbine element and a separate double flow intermediate pressure turbine element coupled in tandem with either two or three double-flow low pressure turbine elements. This type unit may utilize either $28\frac{1}{2}$- or 31-inch (72.4- or 78.7-centimeter) long last row blades. Units of both the four casing four flow type and five casing six flow type are in service. Maximum turbine-generator capabilities of units in service are slightly over 800 megawatts and several flow types are projected in the 1,200 megawatt size range.

Prior to the great increases in cost of raw fuel, the story of how the economy of scale in turbine generator unit sizes drove down the cost of electrical power generation is well known. Larger units provide better thermal performance, lower capital costs, and fewer parts to maintain than multiple sets of smaller ratings for the same total capacity. The sharply accelerating trend to even larger unit sizes, which began in the 1950s, continues.

Some recent technological advances that have contributed to improved steam turbines include:

1. Development of an anchored throttle valve-steam chest system, which eliminates the concern for possible turbine casing distortion from main steam inlet pipe reactions. The steam chest, anchored to the foundation, isolates the reactions to the main steam station piping so they are not imposed on the turbine casing.

2. Improvements in the design of internal stationary parts have consisted of innovations to minimize the effects of thermal gradients and thermal cycling. Fossil units in particular are designed with the expectation that they will be called upon to support intermediate peaking or cyclic duty operation. Separate nozzle chambers are used to eliminate locked-in thermal stresses when only a portion of the admission arc is active while operating at reduced load. Inner casings are mounted in the outer casings so they are free to expand radially and axially. The alignment of the inner casing is maintained by supports at the horizontal joint and keys on the vertical center line. All blade rings, dummy rings, and gland rings are separately mounted in the inner or outer casing with their alignment maintained by supports on the horizontal joint and dowel pins at the vertical center line. Each component is free to expand independently, without being restricted or restrained by surrounding parts. Thermal distortion due to temperature gradients and load cycling is practically eliminated. Thermal

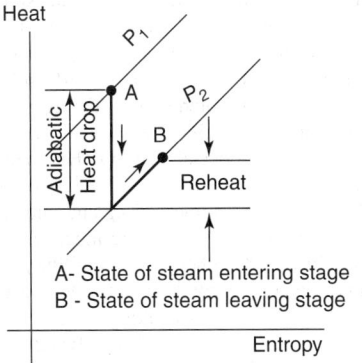

Fig. 1.   Steam conditions in one stage of a turbine.

cracking of casings and nozzle chambers, once a major industry problem, is solved where these features are applied.

3. The first stage in a high-pressure turbine, the control stage, is subjected to the shock load associated with partial arc admission and the flow excitation produced by the nozzle vane wakes. Research has been carried out regarding the operating characteristics of control-stage blading. Troublesome modes of vibration have been identified and their frequencies established. Nozzle vane sections have been developed to provide lower levels of excitation and better control of exciting frequencies.

4. In fossil turbine units, steam temperatures are high enough to affect materially the useful creep life of a turbine rotor at the operating stress level. Considerable design effort has concentrated on reducing the exposed rotor temperature at the two inlet zones of the unit by blanketing these rotor areas with cooling steam. Steam, which has been cooled by expansion through the control stage, blankets the zone between the nozzle chambers and the rotor. For 538 °C inlet temperature, the maximum rotor temperature is limited to about 496 °C. At this reduced temperature, the creep life of the rotor is nearly doubled.

5. Many innovations have been applied to rotating parts to improve thermal performance and reliability. Improved mechanical design in the form of a side-entry blade root has simplified rotor design. This route allows the use of a full row of blades, thereby eliminating the loss in performance and the additional blade loading which exists due to the flow disturbance created when the closing blade of the row must be omitted. This type of blade root was invented by Westinghouse nearly 50 years ago.

6. One of the most significant innovations during the last few years has been the application of sophisticated control techniques made possible by the development of solid-state technology. A particularly valuable feature of digital electrohydraulic control is the ability to switch control between single valve and sequential valve operation. This valve management gives the operator the flexibility to match more favorably temperatures in the first-stage zone of the high-pressure turbine. This minimizes starting and loading periods required for intermediate or cyclic duty application. Automatic startup programs also have been developed.

See also **Cogeneration**.

Additional Reading

Bloch, H.P.: "A Practical Guide to Steam Turbine Technology," McGraw-Hill Professional Book Group, New York, NY, 1994.

King, B.R.: "The Steam Turbine-Generator Today: Materials, Flow Path Design, Repair and Refurbishment: Presented at the 1993 International Joint Power Generation Conference, Kansas City, Missouri, October 17–22, 1993," American Society of Mechanical Engineers, New York, NY, 1993.

Sill, U. and W. Zorner: "Steam Turbine Generators: Process Control and Diagnostics—Modern Instrumentation for the Greatest Economy of Power Plants," John Wiley & Sons, Inc., New York, NY, 1996.

TURBOEXPANDER.   See **Gas and Expansion Turbines**.

TURBOFAN.   A turbojet engine in which additional propulsive thrust is gained by extending a portion of the compressor or turbine blades outside the inner engine case. The extended blades propel bypass air flows along the engine axis but between the inner and outer engine casing. This air is not combusted but does provide additional thrust caused by the propulsive effect imparted to it by the extended compressor blading.

TURBOT.   See **Flatfishes**.

TURBULENCE.   An expression that denotes the presence of irregular eddying motions that are very effective in promoting transport and diffusion of momentum, heat, and matter from one part of the flow to another. For this reason, the existence of turbulence may or may not be welcome. For dispersing pollutants released into the atmosphere, vigorous turbulence is desirable, and its absence above the level of an inversion can lead to smog. In aerodynamics, turbulence increases skin friction but delays separation of flow from aerofoils and increases the stall angle. The definition of turbulent motion is not simple, but the essential characteristics are diffusivity and vorticity.

In general, a turbulent motion consists of a spectrum of eddy sizes, usually identified as Fourier components of the velocity field, which range from scales comparable with the transverse width of the flow to extremely small eddies. The largest eddies set up Reynolds stresses, which extract energy from the mean flow, and most of the bulk properties depend on these eddies. They, in their turn, pass energy to eddies a size smaller, and the process of energy transfer proceeds in a cascade until an eddy size is reached at which energy is lost directly by working against viscous forces. The analytical formulation of these ideas is the theory of local isotropy, initiated by A.N. Kolmogoroff, and since developed to give a good description of the spectrum of turbulence. An important corollary is that the large-scale motion is independent of the fluid viscosity, and so, unless the boundary conditions of the flow involve viscous effects, turbulent flows are similar at all Reynolds numbers. Examples are jets and wakes in which scale effects are unimportant.

The three kinds of turbulent flow may be distinguished:

1. Free turbulence effectively unbounded except by nonturbulent fluid. Characteristically, the effect of the turbulent motion is described by a coefficient of eddy viscosity proportional to the product of the velocity variation and width at the particular section of the flow.

2. Wall turbulence near a solid boundary. The motion is strongly inhomogeneous but is near a condition of local equilibrium. Stress and velocity gradient are related by

$$\tau = k y \frac{du}{dy}$$

3. Convective turbulence in which potential energy of the mean flow is released by the turbulence. Examples are flow between rotating cylinders and heat convection between parallel horizontal planes. The characteristics are strong coupling between different parts of the flow.

Theories of turbulent flow are all, in part, phenomenological, depending on experimental investigations to suggest reasonable approximations and on use of the equations of fluid motion to establish their internal consistency. The most useful instrument for the investigation of the turbulent motion itself is the hot-wire anemometer, a thin wire heated by an electric current whose heat-loss and temperature depends on the fluid velocity past it, but other techniques are also employed. See also **Aerodynamics and Aerostatics**; and **Jet Streams**.

TURBULENCE (Atmospheric).   See **Atmospheric Turbulence**.

TURBULENT FLOW.   A condition of fluid flow in a closed conduit that is above a critical velocity. In this condition, there is a random, irregular motion of the fluid particles in directions transverse to the axis of the main flow. This is converse to the condition of laminar or streamline flow. See also **Laminar Flow**.

At velocities greater than the critical, the fluid velocity profile in the conduit is uniform across the conduit diameter except for a thin layer of fluid at the conduit wall. This *boundary layer* continues to move in laminar flow. In connection with flow measurement, most flowmeters have constant coefficients under turbulent flow conditions. Some flowmeters have the advantage of constant coefficients over Reynolds Number ranges encompassing both turbulent and laminar flows. See also **Fluid Flow (Boundary Layer)**; **Free Turbulent Flow**; and **Reynolds Number**.

TURBULENT FLOW (Air).   See **Aerodynamics and Aerostatics**.

TURING, ALAN (1912–1954).   Turning was a British mathematician and computing pioneer. His theories were important in the development of the digital computer. He studied at Cambridge University and then came to Princeton University in the U.S. for graduate studies. At Princeton, he explored the concept of the modern digital computer. In 1936, Turing produced his first, and best known work, "On Computable Numbers" in which he laid out processes that would be performed by a "universal" computer for solving math problems. The *Turing machine* would judge the "computability" of any problem, solving all mathematical problems provided it was given the proper algorithms. His idea was revolutionary and defined the modern computer.

During WWII, Turing went home to England and worked at Bletchley Park on the ULTRA project. He played a critical role in the development of a machine that helped to decipher German military codes by testing key codes until it found the correct combinations. He constructed relay-driven decoders that shortened the code-breaking time down to hours instead of weeks. Later, in 1945, he went to Great Britain's National Physical Laboratory and worked to help build the Automatic Computing Engine, which was the first British electronic digital computer. Later, at Manchester

University he helped develop the Manchester automatic digital machine. And in 1950, he published a notable paper, which offered what has become know as the *Turing Test* to determine whether machines possessed intelligence and could think.

Turing died June 7, 1954 of an apparent suicide when he ate a cyanide-laced apple. His death was a major loss for the computer science world.

See also **Digital Computer Systems**; and **Turing Machine**.

J. M. I.

TURING MACHINE.   A mathematical abstraction of a device that operates to read from, write on, and move an indefinite tape, thereby providing a model for computerlike procedures. The behavior of a Turing machine is specified by listing an alphabet (i.e., collection of symbols read and written), a set of internal states, and a mapping of an alphabet and internal states which determines what the symbol written and tape motion will be, and also what internal state will follow when the machine is in a given internal state and reads a given symbol. The Turing machine, proposed in 1936 by Alan M. Turing, is the theoretical prototype of real general purpose computers and has had profound impact in many areas of computer science. An excellent description of a Turing machine and its implications is found in "Mathematical Foundations of Programming" by F.S. Beckman, Addison-Wesley, Reading, Massachusetts (1980).

TURKEY *(Aves, Galliformes; Meleagris).*   A large game bird of North and Central America. Wild turkeys are now abundant in some of the protected areas of the eastern states, and in the Southwest they may be found in wild areas. In comparison with their former abundance, however, they are now rare. The eastern species is easily distinguished from the western by the absence of white tips on the feathers of the tail and rump. A third species occurs in Central America. See also **Galliformes**; and **Poultry**.

TURKEY BUZZARD. See **Vulture**.

TURMERIC *(Curcuma longa; Zingiberaceae).*   A name given to both the plant, *Curcuma longa*, and to its derivatives, a dye and a drug, which are obtained from the swollen rhizomes of the plant.

Curcuma, a native of southern Asia, is widely grown in India and other tropical Asian and East Indian lands, where it is used as a drug as well as a condiment and dyestuff. The plant has long smooth pointed leaves, and dull yellow flowers. To prepare it for use, the yellow-brown aromatic rhizome is cleaned and cut into pieces 1–2 inches (2.5–5 centimeters) long. The cut surfaces show the bright yellow color of the substance of the rhizome. These cut pieces are dried in an oven and then ground. In western countries turmeric is sometimes used as a dye; as an ingredient to curry powder; to color and flavor pickles, mustard, and other foods; and in chemistry, where it is a pH indicator. See also **Flavorings**.

TURN-AND-BANK INDICATOR.   An instrument for installation in an aircraft. Also termed a turn-and-slip indicator, the instrument contains a rate gyro mounted with its spin axis athwartship and its spring-restrained gimbal axis aligned fore and aft as shown in Fig. 1. Deflection of this axis from its neutral position is translated to motion of a pointer, calibrated in rate of turn. Below the pointer is a ball bank indicator, which acts as a damped pendulum, showing the dynamic vertical component lying in the plane perpendicular to the fore-and-aft axis. If both pointer and ball are

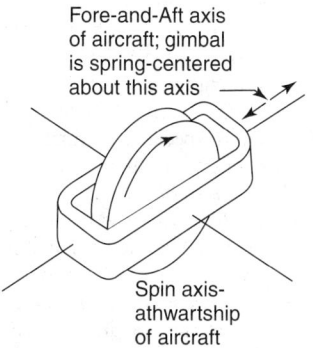

Fig. 1.   Arrangement of a turn gyro as used in turn-and-bank indicator.

centered, the pilot knows the aircraft is flying straight and that the wings are horizontal. The turn-and-bank indicator is also used in making desired rates of turn with proper bank. Both air-drive and electric gyros are used. In a properly designed turn-and-bank indicator, the direction of rotation, moment of inertia, and speed of the gyro wheel are so matched to the centering spring that the gyro spin axis remains nearly horizontal as the plane makes a turn with proper bank at its normal cruising speed. Instead of a rotating gyro wheel, other sensing elements, such as a vibrating reed or a stream of liquid, have been used.

Gyros of the type described are often called *rate gyros* and, when their speed is accurately controlled, they can serve as accurate measures of rate of turn. They find many uses, as in gunsights, bombsights, autopilots, and navigation devices. Instead of driving a pointer, they may mechanically perform some function in a computer or may drive a potentiometer, synchro, switch, or similar device to provide an electrical function of rate of turn.

TURNER'S SYNDROME.   A set of symptoms that develops in a person who has only one *X*-chromosome and no *Y*-chromosome. The person will be female, but will not mature sexually. She will also be stunted physically and may show certain other characteristics, such as a webbing of the skin at the side of the neck and wide spaced nipples on the chest. Some degree of feminine normality can sometimes be induced through the use of female hormone therapy.

TURNIP. See **Brassica**.

TURNSTILE ANTENNA.   An antenna composed of two dipole antennas, normal to each other, with their axes intersecting at their midpoints. Usually, the currents are equal and in phase quadrature.

TURNSTONE. See **Waders, Shorebirds, and Gulls**.

TURPENTINE. See **Resins (Natural)**.

TURQUOISE.   The mineral turquoise is a hydrated phosphate of aluminum and copper. Its exact composition is doubtful, the formula may be expressed $CuAl_6(PO_4)_4(OH)_8 \cdot 5H_2O$; iron is often present. This mineral is found in minute triclinic crystals, but chiefly massive as seams and crusts. The fracture is conchoidal; hardness, 5–6; specific gravity, 2.6–2.8; luster, soft waxy; color, may be various shades of blue, bluish-green and green; essentially opaque. It takes a good polish and the sky-blue varieties have long been used as a gem material. Unfortunately, many beautiful blue stones in time change their color to some greenish hue, usually not attractive, rendering them practically valueless.

For hundreds of years turquoise has been mined in Iran, where it is found with limonite filling crevices in a brecciated trachyteporphyry, and because it found its way into Europe through Turkey, it became known as turquoise, from the French word *turque*, Turkish. Other mines were worked by the Egyptians in ancient times on the Sinai Peninsula. Turquoise is also found in Siberia, Turkestan, Saxony and France; and in the United States in Arizona, California and New Mexico. A blue stone that has passed for turquoise is in reality odontolite, from the Greek meaning tooth, usually fossil teeth or bones colored with iron phosphate. Odontolite is softer than true turquoise, has a somewhat higher specific gravity, 3.0–3.5, and may be distinguished by chemical tests, or by a microscopical examination which will reveal its organic structure.

TURTLES.   Of the class *Reptilia* (reptiles), subclass *Anapsida*, order *Testudines* (turtles). Two suborders of turtles are of principal interest: *Cryptodira* (modern turtles), and *Pleurodira* (side-neck turtles). The organization of the turtles is shown in Table 1.

The characteristic shell sets the turtles so distinctly apart as a separate order that one would find it difficult to confuse them with other animals. Moreover, they are probably the only reptiles that most humans view without prejudice. In fact, the number of people who positively like turtles is surprisingly large. Both terrestrial and aquatic turtles (some of which are called tortoises or terrapins) are often kept as pets in gardens, terrariums, and aquariums.

Turtles are compactly built reptiles with a bony shell, which is a part of the skeleton, protecting both the back and the underside. The animals can more or less completely pull the head and neck, legs, and tail in under

TABLE 1. CLASSIFICATION OF TURTLES
CLASS . . . . REPTILIA (Reptiles)
ORDER . . . . TESTUDINES (Turtles)

Family (Subfamily)	Examples
Suborder: MODERN TURTLES (*Cryptodira*)	
Dermatemydidae	Mexican river turtle (*Dermatemys maivii*)
Chelydridae **Snapping Turtles**	Snapping turtle (*Chelydra serpentina*); Alligator snapping turtle (*Macroclemys temminckii*)
Kinosternidae **Musk Turtles**	Common musk turtle (*Sternotherus odoratus*); Keel-backed musk turtle (*S. carinatus*); Loggerhead musk turtle (*S. minor*); Mud turtle (*Kinosternon subrubrum*); Striped mud turtle (*K. baurii*); Scorpion mud turtle (*K. scorpioides*); Narrow-bridged mud turtle (*Claudius angustatus*); Giant musk turtle (*Staurotypus triporcatus*)
Platysternidae	Big-headed turtle (*Platysternon megacephalum*)
Emydidae	European pond turtle (*Emys orbicularis*); Blanding's turtle (*Emygoidea blandingii*); Chicken turtle (*Deirochelys reticularia*); Painted turtle (*Chrysemys picta*); Pond slider (*C. scripta*); Red-eared turtle (*C. s. elegans*); Cooter or terrapin (*C. floridana*); River cooter (*C. concinna*); Red-bellied turtle (*C. rubriventris*); Ornate slider (*C. ornata*); Brazilian slider (*C. dorbigni*); Jamaican slider (*C. terrapen*); Map turtle (*Graptemys geographica*); False map turtle (*G. pseudogeographica*); Diamondback terrapin (*Malaclemys terrapin*); Box turtle (*Terrapene carolina*); Western box turtle (*T. ornata*); Spotted turtle (*Clemmys guttata*); Wood turtle (*C. insculpta*); Bog turtle (*C. muhlenbergii*); Western or Pacific pond turtle (*C. marmorata*); Caspian terrapin (*C. caspica*); Chinese terrapin (*Clemmys mauremys nigricans*); Japanese terrapin (*C. m. japonica*); Annam turtle (*Annamemys annamensis*); Black pond turtle (*Geoclemys hamiltonii*); Reeve's turtle (*Chinemys reevesii*); Amboina box turtle (*Cuora amboinensis*); Three-lined box turtle (*C. trifasciata*); Annandal's turtle (*Hieremys annandalii*); Diadem turtle (*Hardella thurjii*); Indian roof turtle (*Kachuga tecta*); Painted batugur (*Callagur borneoensis*); Bornean pond turtle (*Orlitia borneensis*)
Testudinidae **Ture Tortoises**	West African hinge-backed tortois (*Kinixys erosa*); Bell's Eastern hinged tortoise (*K. belliana*); Home's hinged tortoise (*K. homeana*); Parrot-beak tortoise (*Homopus areolatus*); Spider tortoise (*Pyxis arachnoides*); Pancake or Tornier's tortoise (*Malacochersus tornieri*); Spur-tailed Mediterranean land tortoise (*Testudo hermanni*) Star tortoise (*Testudo geochelone elegans*); Leopard tortoise (*T. g. pardalis*); African spurred tortoise (*T. g. sulcata*); Geometric tortoise (*Testudo psammobates geometrica*); Tent tortois (*T. p. tentoria*); Bowsprit tortoise (*Testudo chersina angulata*); Radiated tortoise (*Testudo asterochelys radiata*); Aldabra tortoise (*Testudo aldabrachelys gigantea*); Horsfield's tortoise (*Testudo agrionemys horsfieldii*); Red-footed tortoise (*Testudo chelonoidis carbonaria*); Argentine tortoise (*Testudo chilensis*); Yellow-footed tortoise (*T. c. denticulata*); Giant Galápagos tortoise (*T. c. elephantopus*); Gopher tortoise (*Gopherus polyphemus*)
Cheloniidae **Marine Turtles**	Green turtle (*Chelonia mydas*); Hawksbill (*Eretmochelys imbricata*); Loggerhead (*Caretta caretta*); Ridley (*Lepidochelys olivacea*)
Dermochelyidae	Leatherback turtle (*Dermochelys coriacea*)
Carretochelyidae	Pitted-shell turtle (*Carettochelys insculpta*)
Trionychidae (*Cyclanorbinae*) **Soft-shelled Turtles**	Soft terrapin (*Lissemys punctata*); Nubian softshell (*Cyclanorbis elegans*); Senegal softshell (*C. senegalensis*); Aubry's softshell turtle (*Cycloderma aubryi*); Bridled softshell turtle (*C. frenatum*)
(*Trionychinae*) **Soft-Shelled Turtles**	River softshell (*Chitra indica*); Florida softshell (*Trionyx ferox*); Eastern spiny softshell (*T. spiniferus*); Smooth softshell (*T. muticus*)
Suborder: SIDE-NECK TURTLES (*Pleurodira*)	
Pelomedusidae	Helmeted turtle (*Pelomedusa subrufa*); Adanson's mud turtle (*Pelusios adansonii*)
Chelidae **Snake-Necked Turtles**	Matamata (*Chelus fimbriatus*); Otter turtle (*Hydromedusa rectifera*); Murray's river turtle (*Emydura macquarii*)

Note: Some of the better known as well as other species selected at random are included in the various examples given. The examples represent only some of the species of turtles.

the shell. The massive skull is of the anapsid type, having no temporal openings, but rather indentations on the posterior or lower edge. In the species alive today, the jawbones are covered not with teeth but with sharp horny edges. The shell encloses the body so completely that only head, limbs, and tail protrude; it consists of an inner bony capsule and usually an outer covering of large horny plates (sometimes with a thick leathery skin instead).

The basic shell structure is often considerably modified. For example, in the soft-shelled turtles (*Trionychidae*), the pitted-shell turtle, and the leatherneck turtle, in the place of horny scutes there is a thick, leatherlike skin which covers the shell. The marginal bony plates are usually lacking in the soft-shell turtles. The usual bony elements of the shell have almost disappeared in the leatherback turtle, and in their place is a new, tessellated shell, composed of small bony plates, that apparently has nothing to do with the true bony carapace of other turtles. In other species that spend most of their time in the water, there are reductions of parts of the bony shell. Frequently quite large openings (fontanelles) have appeared between the costal plates, so that in certain groups—for example, in the *Cheloniidae*—the ends of the ribs are visible.

Many aquatic and some terrestrial species have evolved single or double traverse hinges in the plastron, places where elastic cartilaginous tissue is interposed between two adjacent plate pairs. In this way, the plastron, ordinarily quite rigid, is divided into movable anterior and posterior sections which the animal can fold upward to close the apertures between the top and bottom of the shell.

The skull of the turtle, in comparison with that of other reptiles, seems short, massive, and strongly arched. There are never true temporal openings, but instead there are characteristic indentations in the temporal region. Not uncommonly the upper jaw terminates anteriorly in a point, curved downward like a hook (as for example in the alligator snapping turtle and the big-headed turtle); the result reminds one of the beak of a bird of prey.

The skin may be either smooth or rough, with scales only on the limbs or the tail; on the upper side of the tail there are sometimes, as in the snapping turtles, scales forming rows of pronounced ridges. On the legs of the *Testudinidae*, too, there are often sturdy lumps, some even ossified at the base. Certain species have special epidermal appendages, the numbers of which can be important in classification. Among these are the barbels on the chin of musk turtles (*Kinosternidae*) and some pleurodires; the warts of the snapping turtles (*Chelydridae*); and the lobes on the head and neck of the matamata. Skin glands are restricted in distribution.

Since turtles can cause the air to flow from one part of the lung to another, the aquatic forms are able to displace their center of gravity at will, so that their lungs act functionally like the swim bladders of fish. But this ability is impaired when the lungs are diseased; for this reason, aquatic turtles with respiratory insufficiencies float at an angle in the water and do not dive. In particularly deep-diving species, the lungs are embedded in bony chambers formed by projections of the inner wall of the shell. It is assumed that this arrangement acts against the effects of high water pressure.

Among the aquatic turtles, there is considerable gas exchange in other parts of the body as well. For example, the soft-shelled turtles are said to take up as much as 70% of their oxygen through the skin that covers the bony shell. Moreover, the mouth cavities of all turtles are richly provided with capillaries. In the marine turtles and soft-shelled turtles, there are also fingerlike villi of skin in the throat through which the blood circulates freely and which serve to remove oxygen from the water. The same function is probably served by the paired vesicles (the anal sacs) attached to the gut near the cloaca; they, too, presumably operate as "physiological gills," taking up a certain amount of oxygen by way of the copious flow of blood through their walls. Thus, the aquatic turtles can actually pass the winter under a completely closed roof of ice. During this period of dormancy, the metabolic rate and thus the oxygen requirement is considerably reduced.

The vertebral column consists of 8 cervical vertebrae, 10 trunk vertebrae, and 18 to 33 tail vertebrae. The form of the cervical vertebrae is extremely variable; it permits the animals to withdraw the head under the shell by bending the neck into an S, either in the vertical (suborder *Cryptodira*) or the horizontal (suborder *Pleurodira*) plane. The limbs usually end in 5 toes, although in one tortoise (*Testudo horsfieldii*), there are only four. The toes are also reduced in the hind limbs of one subspecies of the American box turtles (*Terrapene carolina triunguis*). Freshwater turtles ordinarily have rowing legs somewhat flattened laterally, whose toes are joined by more or less well developed membranes. In contrast, the toes on the columnar (i.e., round in cross-section) legs of the terrestrial forms are not free, but rather form part of a stumpy foot with only the claws sticking out. The legs of cheloniid turtles, soft-shelled turtles, and the pitted-shell turtle have become powerful swimming paddles, very greatly flattened laterally. These animals, too, have no free toes and may have either no claws or as many as three extending from the anterior edge of the paddle. See Fig. 1.

Fig. 1.   Young specimen of a green turtle. When fully grown, it may reach a weight of 272 to 363 kg (600 to 800 pounds). (*New York Zoological Society.*)

With their horny jaws, turtles cut off pieces of food and swallow them whole. As a rule, these blades of horn are exposed, but in the soft-shelled turtles, they are covered with fleshy lips. Although the mouth is well provided with salivary glands, the actual process of digestion probably first takes place in the stomach, which curves to lie crossways in the body cavity. Often it contains stones, picked up voluntarily by both aquatic and terrestrial turtles, presumably as aids to grinding food more thoroughly. The small intestine of the aquatic turtles, predominantly carnivores, is relatively short, but, in the vegetarian land turtles, it can be several times longer than the body. Thirty percent of the cellulose consumed can be digested. The liver is unusually large.

The kidneys excrete a relatively large quantity of urine, which is stored in fluid form in the urinary bladder, present in all turtles. Often turtles squirt a sharp stream of water from the anus when one picks them up. This does not come from the urinary bladder, but rather from the anal sacs.

The turtle brain is small but nevertheless rather highly developed, especially with respect to its centers for the senses of vision, smell, and equilibrium. Thick, movable lids protect the eyes, which have excellent

resolution. Marine turtles and the diamond-backed terrapin have tear glands. Turtles can discriminate distinctly not only forms, but colors as well. Their range of visible wave lengths is shifted considerably toward the red part of the spectrum. Some infrared light waves can be detected. In general, turtles are thought to be deaf, although their hearing organs, however simply constructed, are well developed. Turtles are fundamentally mute, apart from the hoarse cheeping or groaning noises made by testudinids at the peak of excitement during copulation. The sense of smell is well developed. The animals sniff all food extensively, and they can be attracted in masses from some distance by the smell of ripe fruit. In terms of mental capacity, turtles display impressive feats of orientation, learning, and memory. In captivity, they develop a surprising sense for certain regular feeding times.

Turtles live in all parts of the earth and the oceans with a temperate to warm climate, and are particularly numerous in the tropics and subtropics. The aquatic forms are found in swamps, ponds, pools thick with vegetation, large lakes, brooks, rivers, and broad streams. The diamond-backed terrapin is confined to brackish water near the coast. The *Cheloniidae* live exclusively in the open ocean. The *Testudinidae* dwell in warm, dry regions—in steppes and semideserts, on rocky slopes, and in the bush and the savanna. Two South American testudinid species, called *jabutis* by local Indians, inhabit humid tropical forests; the North American gopher tortoises are found in sun-drenched deserts. Some of the *Testudinidae* are semi-aquatic—for example, the hinged tortoises and some of the aquatic turtles, such as the box turtles, at least occasionally spend time on land. Certain aquatic turtles leave the water only to lay eggs or to sun themselves.

In recent years, it has been found that all marine turtles are not carnivorous. The marine green turtle, as an adult, consumes only marine plants, probably accounting for the fact that its flesh does not have an oily taste like other marine turtles. But the alligator snapping turtle, the matamata, the big-headed turtle, and the soft-shelled turtles are pure carnivores. Nor do all terrestrial turtles abstain from animal foods. They show a preference to strong-smelling substances, such as old cheese and sometimes excrement. The Malayan freshwater turtle (*Malayemys subtrijuga*) feeds almost exclusively on mollusks.

All turtles reproduce by way of eggs. There are usually between 2 and 20 eggs in a batch, although the large marine turtles can deposit as many as 100 eggs in one laying. The number of eggs laid usually increases with the age of the female. The enormous masses of eggs laid by the arrau on the sandbanks in the Amazon basin are well known. The eggs are always laid on land, even by marine turtles, and always in about the same manner. Using her hindlegs, the female digs a hole in loose ground; the freshwater turtles can soften the earth as they dig with the fluid they eject from their anal sacs. Observations indicate that female freshwater turtles return to the water several times during the digging process in order to refill these sacs. The nests of marine turtles are built in two stages. First the female prepares a broad trench, throwing the sand out in all directions with all four legs. This large pit functions only to hide the animal during egg-laying. Once the female is thus protected, she begins to dig the real nest chamber at the bottom of the pit.

When the female turtle has completed her preparations, she holds her tail in the entrance to the egg chamber and then lays the eggs, one by one. She catches the eggs first with one hindleg and then with the other, and carefully lets them slide to the floor of the nest. After the eggs are laid, she shoves the excavated earth back into the nest, smooths the surface with her plastron, and even piles leaves or other ground litter over it until the nest site is indistinguishable from its surroundings. The extensive migrations made by marine turtles to their traditional nesting beaches are quite astonishing. After laying the eggs, the female has no further concern over them. Incubation is dependent entirely upon the warmth of the soil. When development within the egg is complete, the young hatch by breaking through the protective coverings with the egg caruncle, a sharp, horny thickening at the tip of the upper jaw, which later falls off.

The life of the turtle follows an annual rhythm, both in the tropics and in temperature latitudes. This yearly cycle, usually determined by temperature, is most evident among species living in cooler regions. When the average daytime and nighttime temperatures fall below a certain value, the animals withdraw to protected places and there spend the inhospitable season in a state of dormancy from which they awaken only when more favorable conditions prevail. European and North American species hibernate in winter. Aquatic turtles in general probably awaken much earlier from hibernation than do land turtles. The period after emergence from hibernation is often the most dangerous for turtles—they cannot withstand sudden cold snaps after emerging. Females lay their eggs shortly after hibernation is over. The dormant period in the tropical regions occurs with most species during summer.

Growth proceeds very slowly in turtles. Turtle size is expressed as the straight-line distance from the anterior to the posterior edge of the shell. One of the smallest species is the bog turtle (*Clemmys muhlenbergi*), which has a length of about 11 centimeters (4.3 inches), whereas the marine leatherback turtle, the giant of the order, reaches a shell length of up to 2 meters (6.6 feet) and a weight of up to 800 kilograms (1764 pounds). Some freshwater turtles, like the arrau and the giant soft-shelled turtle, can attain a length of nearly 1 meter (3.3 feet). The huge land turtles live on two quite isolated and widely separated groups of islands: the Galápagos Islands off western South America; and the Seychelles, north of Madagascar. See Fig. 2. Some of the giant tortoises of the Galápagos have shells up to about 110 centimeters (43.3 inches); those of the subspecies of the Seychelles giant tortoise may measure up to 125 centimeters (49.2 inches).

Fig. 2.   Galapagos tortoises. (*A.M. Winchester.*)

In recent years, there has been considerable concern over the endangerment of certain turtle species, notably sea turtles. An international conference on this topic was held in November 1979 by the U.S. State Department. Participants from 40 nations discussed a conservation strategy for seven species of sea turtle, six of which have been adjudged endangered. In some areas, sea turtle meat is an important food for indigenous populations. The conflict posed here is similar to that which has occurred over the bowhead whale in Alaska. Turtles yield a variety of products of economic value—eggs, skin, meat, calipee (cartilage used for soup), shells, oil, and as trophies. It was pointed out at the conference that the habitats of the sea turtle are being threatened, their eggs are being poached, and adults are being lost in large numbers to shrimp fishing operations, where they are caught in shrimping nets and drowned. The species pointed out at the conference for attention include the hawksbill (supplier of tortoiseshell), the leatherback, the olive (Pacific) ridley, the Kemp's (Atlantic) ridley, and the flatback. It was pointed out that as recently as 1947, some 40,000 Atlantic ridleys were observed nesting on a beach in Mexico. As of the early 1980s, these turtles are at the top of the endangered species list. Some authorities have estimated that there may be no more than 500 to 1000 nesting females left. A committee was appointed to monitor conservation progress well into the 1980s.

Additional Reading

Beardsley, T.M.: "The Sea Turtle's Tale," *Sci. Amer.*, 32 (August 1990).
Epler, R.C.: "Whalers, Whales, and Tortoises," *Oceanus*, 86 (Summer 1987).
Hawley, T.M.: "Galapagos Sea Turtles," *Oceanus*, 34 (Summer 1987).
Holden, C.: "Turtle Navigation," *Science*, 1309 (June 15, 1990).
Klemens, M.W.: "Turtle Conservation," Smithsonian Institution Press, Washington, DC, 2000.
Lewin, R.: "New Look at Turtle Migration Mystery," *Science*, 1009 (February 24, 1989).
Lohmann, K.J.: "How Sea Turtles Navigate," *Sci. Amer.*, 100 (January 1992).
Meylan, A.: "Spongivory in Hawksbill Turtles: A Diet of Glass," *Science*, 393 (January 22, 1988).
Meylan, A.B., B.W. Bowen, and J.C. Avies: "A Genetic Test of the Natal Homing versus Social Facilitation Models for Green Turtle Navigation," *Science*, 724 (May 11, 1990).
News: "Green Sea Turtle Threat," *Nat'l. Geographic*, 138 (April 1991).
Weiss, R.: "Giant Turtles Built for Comfort, Not Speed," *Science News*, 263 (April 26, 1990).
White, C.P.: "Freshwater Turtles," *Nat'l. Geographic*, 40 (January 1986).

TUSSOCK MOTH (*Insecta, Lepidoptera*).   This moth (*Hemerocanoa keycistugna*) is injurious to apple. The moth is yellow and black, with a red head. It attains a length of about 1 inch (2.5 centimeters). The moth prefers feeding on leaves, but on occasion will eat into the fruit. The frothy egg-masses should be collected in fall and winter and destroyed and the trees should be banded to prevent a reinfestation by migrating caterpillars. Other treatment is similar to that for the codling moth, see also **Codling Moth**.

TWILIGHT.   The intervals of incomplete darkness following sunset and preceding sunrise. Twilight is produced primarily by reflection of the light of the sun from the upper atmosphere of the Earth, but effects of scattering of light and refraction also enter into the period of duration of twilight. Morning twilight begins and evening twilight ends when the sun is about 18° below the horizon, although this value varies somewhat with the purity of the atmosphere. The times are determined by arbitrary convection; and several kinds of twilight have been defined and used. Civil twilight, nautical twilight, and astronomical twilight are in common usage, the limiting solar depression angle being, respectively, 6°, 12°, and 18° for those three.

For each of the three kinds of commonly used twilight limits, the duration of the twilight period varies considerably with latitude and calendar date due to the fact that the sun's diurnal path across the observer's celestial sphere meets his horizon at quite different angles and declinations at different latitudes and times of year. When the sun approaches the horizon obliquely, and is at high declination, a longer time (Earth rotation) is required to bring the sun into the limiting depression angle. At very high north or south latitudes, any of the three types of twilight may last as long as 24 hours or may not occur at all.

The so-called *bright segment*, or *twilight arch*, may be observed above the eastern horizon on a clear evening as the sun sets in the west. This is a blue segment bounded by a faintly reddish arc, and is, in reality, the shadow of the Earth cast on the upper atmosphere. It is only a few degrees in vertical angular extent, but discernible over 20 to 30 degrees of horizon. The final disappearance of the bright segment after sunset marks the end of astronomical twilight and represents the beginning of full darkness.

The *anti-twilight arch* is the pink or purplish band of about 3 degrees vertical angular width that lies just above the antisolar point at twilight; it rises with the antisolar point at sunset and sets with the antisolar point at sunrise. See also **Atmospheric Optical Phenomena**; and **Purple Light**.

TWINNING (Crystal).   A process in which a region in a crystal assumes an orientation which is symmetrically related to the basic orientation of the crystal. Usually, layers of atoms within this region are translated with respect to a basic plane (the twinning plane). Each atomic plane is displaced by a distance which is proportional to its distance from the twinning plane. Bands of metal (twin bands) thus assume a lattice structure which is the mirror image of the unchanged portion of the lattice. See also **Mineralogy**.

TWINS.   Two individuals, born at the same birth. The incidence of twins is about 1 to 2% of births. Heredity is a factor in the etiology of multiple pregnancies. Identical twins are twins developing from the cleavage of one fertilized ovum. They are always of the same sex and similar in appearance. Fraternal twins are twins developing from two fertilized ova. These may be of either sex, and may or may not be of similar appearance. See also **Gonads**.

TWO-BODY PROBLEM (Astronomy).   The two-body problem may briefly be stated as follows: given the relative position of two objects at any instant, together with their motions and masses at that instant, predict the positions and motion of those two objects at any other instant. The solution of the motion of bodies in space is the foundation of celestial mechanics. In the simplest case, where only two bodies exist, it is necessary to assume or determine the force law between the two bodies. Kepler's celebrated three laws of planetary motion solved this problem empirically, making the solution necessarily an approximation, as any other solution for the planetary system must be. Newton was the first to attack the two-body problem analytically, and he had to develop his own mathematics to solve it. He was able to derive Kepler's laws and thus explain the motions of Jupiter's satellites, the motion of the moon about the Earth and of the Earth about the sun. However, each of these problems is much more complex than the simple two-body problem. The Earth-moon system is really part of a three-body problem that involves the sun.

The solution to the two-body problem is presented here in a nonhistorical way. Let T be the kinetic energy and V the potential energy. The equations of Lagrange are then written:

$$\frac{d}{dt}\left(\frac{\partial L}{\partial \dot{q}_i} - \frac{\partial L}{\partial q_i}\right) = 0$$

where the qi are the coordinates, L is the Lagrangian function, and the dot over a letter denotes the first derivative with respect to time. The kinetic energy in polar coordinates is

$$T = \frac{m}{2}(\dot{r}^2 + r^2\dot{\theta}^2 r^2 \sin^2\theta \cdot \dot{\phi}^2)$$

If the force field is central, then the potential energy is a function of r only. The Lagrangian is then written:

$$L = T - V = \frac{m}{2}(\dot{r}^2 + r^2\dot{\theta}^2 r^2 \sin^2\theta \cdot \dot{\phi}^2) - V(\dot{r})$$

The Lagrangian is now

$$q_1 = r, \quad q_2 = \theta, \quad q_3 = \phi$$

And from this the equations of motion are obtained:

$$\frac{d}{dt}(m\dot{r}) + \frac{dV}{dr} - mr\dot{\theta}^2 = 0$$

$$\frac{d}{dt}(mr^2\theta) = 0$$

Integrating the last equation gives the law of areas, and substituting this into the first equation leads to the equation of the path.

TYMPANIC MEMBRANE. See **Hearing and the Ear**.

TYNDALL EFFECT. A phenomenon first noticed by Faraday (1857). When a powerful beam of light is sent through a colloidal solution of high dispersity, the sol appears fluorescent and the light is polarized, the amount of polarization depending upon the size of the particles of the colloid. The polarization is complete, if the particles are much smaller than the wavelength of the radiation. See also **Nephelometry**.

TYNDALL, JOHN (1820–1893). John Tyndall was a man of science: draftsman, surveyor, physics professor, mathematician, geologist, atmospheric scientist, public lecturer, and mountaineer. Throughout the course of his Irish and later, English life, he was able to express his thoughts in a manner none had seen or heard before. His ability to paint mental pictures for his audience enabled him to disseminate a popular knowledge of physical science that had not previously existed. Tyndall's original research on the radiative properties of gases as well as his work with other top scientists of his era opened up new fields of science and laid the groundwork for future scientific enterprises.

In January 1859, Tyndall began studying the radiative properties of various gases. Part of his experimentation included the construction of the first ratio spectrophotometer, which he used to measure the absorptive powers of gases such as water vapor, "carbonic acid" (now known as carbon dioxide), ozone, and hydrocarbons. Among his most important discoveries were the vast differences in the abilities of "perfectly colorless and invisible gases and vapors" to absorb and transmit radiant heat. He noted that oxygen, nitrogen, and hydrogen are almost transparent to radiant heat while other gases are quite opaque.

Tyndall's experiments also showed that molecules of water vapor, carbon dioxide, and ozone are the best absorbers of heat radiation, and that even in small quantities, these gases absorb much more strongly than the atmosphere itself. He concluded that among the constituents of the atmosphere, water vapor is the strongest absorber of radiant heat and is therefore the most important gas controlling Earth's surface temperature. He said, without water vapor, the Earth's surface would be "held fast in the iron grip of frost." He later speculated on how fluctuations in water vapor and carbon dioxide could be related to climate change.

Tyndall related his radiation studies to minimum nighttime temperatures and the formation of dew, correctly noting that dew and frost are caused by a loss of heat through radiative processes. He even considered London

as a "heat island," meaning he thought that the city was warmer than its surrounding areas.

Over the course of his life, John Tyndall published numerous papers and essays on his scientific discoveries, as well as literature, religion, mountaineering, and travel. His accomplishments led him to receive five honorary doctorates and become a respected member of thirty-five scientific societies.

See also **Nephelometry**; and **Tyndall Effect**.

TYPHOID FEVER. An acute generalized infection due to the bacillus *Salmonella typhi*. Although this disease can occur worldwide, it is most frequently associated with unsanitary conditions and carriers. In times before there was effective inspection and control over water and milk, as well as other foods, great epidemics of the disease occurred, particularly in the summer. Since the bacillus can survive freezing and drying, it can be carried to food and drink from contaminated sources. In 1984, 380 cases were reported in the United States and 70% of this incidence was imported after being acquired overseas in underdeveloped countries. Typhoid is spread by excreta of persons with the disease. One month after recovery, 50% of the patients will pass some of the infectious organism in their stools; 20% after two months; and 10% after three months. Carriers, approximately 3% of persons who have been infected, will have a chronic infection, usually in the biliary tree or less frequently in the urinary tract. These sites are not usually responsive to chloramphenicol, although penicillin therapy may be effective.

Typhoid fever passes through five stages: (1) The *incubation period* varies from a few days to one week, during which stage the patient will have a mild fever (up to 100 °F; 39.4 °C). In this first stage, there is proliferation of the microorganisms in the gut and penetration into the gut mucosa. Diarrhea is present in only 10–20% of cases. (2) The *active invasion stage* varies in duration from one to two weeks, during which time the patient's temperature will range between 99 and 103 °F (37.2 and 39.4 °C). In this second stage, there is proliferation of the microorganisms within the gut lymphoidal tissue and spreading to regional lymphoidal tissue. Septicemia may be present. Symptoms during this period include headache, malaise, myalgia, loss of appetite, nausea, cough, sore throat, constipation, and sometimes diarrhea. A characteristic rash develops on the abdomen from the seventh to tenth day. The spots are flattened papules, slightly raised, of rose-red color, disappearing on pressure, and 2–4 mm (diameter). Sometimes each is capped by a small vesicle. These spots disappear after 2 to 3 days, leaving a brown stain. (3) During the *established disease stage*, varying from two to four weeks, the temperature will lie between 100 and 103 °F (37.8 and 39.4 °C). During this third stage, reticuloendothelial proliferation occurs. Sometimes the organism will metastasize and form focal points of infection. Necrosis and ulceration may occur at sites of previous lymphoid proliferation. Widespread lesions may occur elsewhere. In the liver, apart from cloudy swelling and small globules of fat, typhoid nodules are a characteristic finding. Symptoms may include toxemia, discomfort in the abdominal region, neuropsychiatric syndromes, intestinal ulceration, bronchitis, relative bradycardia, and possible intestinal perforation with hemorrhage. Symptoms vary in type and degree, depending somewhat upon intensity of initial infection. (4) The *convalescent stage* lasts between four and five weeks. There is a lowering of temperature to a range of 97–100 °F (36.1–37.8 °C). During this period, natural body defenses commence to repair and replace damaged tissue. Relapses of varying duration and intensity may occur. There is also the possibility of late focal complications, including cholecystitis, osteomyelitis, and soft tissue abscesses. Spontaneous abortion and premature labor are common among untreated pregnant patients with typhoid fever.

Recovery from typhoid depends upon a number of factors, including the production of antibodies to the various antigenic compounds of the organism; the development of cell-mediated immunity; and the sensitivity of the strain of *S. typhi* responsible for the disease to the antibiotics chosen by the physician. Cell-mediated immunity appears to be a most important factor in determining the outcome.

Comparatively few antibiotics are effective against typhoid fever. Generally, the drug of choice is chloramphenicol, but strains of *S. typhi* resistant to the drug have started to appear. Amoxicillin has therefore appeared as a true alternative, which may be backed up by ampicillin and trimethoprim-sulfamethazole. Diagnosis is sometimes complicated where persons have recently used oral antibiotics. A small number of patients may relapse several weeks after apparent cure, but this recurrence is usually milder and of shorter duration than the initial infection.

Much work has been done on the preparation of effective vaccines, but none so far produced will give complete protection. See also **Bacterial Diseases**.

Additional Reading

Ackers, M., N. Puhr, R. Tauxe, and E. Mintz: "Laboratory-based Surveillance of Salmonella Typhi Infections in the United States: Antimicrobial Resistance on the rise," *JAMA*; **283**:2668–2673, 2000.

Luby, S., M. Faizan, S. Fisher-Hoch, et al.: "Risk Factors for Infection with Salmonella Typhi in an Endemic Setting, Karachi, Pakistan," *Epidemiology and Infection*, **120**:129–138, 1998.

Mermin, J.H., J.M. Townes, M. Gerber, et al.: "Typhoid Fever in the United States, 1985–1994: Changing Risks of International Travel and Increasing Antimicrobial Resistance," *Archives of Internal Medicine*, **158**:633–38, 1998.

Staff: CDC, "Typhoid Immunization: Recommendations of the Advisory Committee on Immunization Practices," *MMWR* **43** (No. RR-14) 1994.

Web Reference

Centers for Disease Control and Prevention: *http://www.cdc.gov/travel/diseases/ typhoid.htm* or *http://www.cdc.gov/ncidod/dbmd/diseaseinfo/typhoidfever_g.htm*

R.C. VICKERY, M.D., D.Sc., Ph.D., Blanton/Dade City, FL.

TYPHOON.   See **Fronts and Storms**.

TYPHUS.   See **Rickettsial Diseases**.

TYUYAMUNITE.   An ore of uranium with the composition, $Ca(UO_2)_2 (VO_4)_2 \cdot 5-8H_2O$, which occurs in yellow incrustations as a secondary mineral. The mineral is orthorhombic. It occurs as a secondary mineral as incrustations on limestones, and as disseminated impregnations in sandstones. Found abundantly in the Western United States, at Grants, New Mexico, and in Wyoming, Utah, Colorado, Nevada, Arizona and Texas. Also at Tyuya Muyan in Turkestan, the former U.S.S.R.

U

UBAC. A mountain slope so oriented as to receive the minimum available amount of light and warmth from the sun during the day, especially a north-facing slope of the Alps.

UDAD. See **Goats and Sheep**.

UDIC. Said of a soil moisture system that is characterized by not being completely dry for either 90 consecutive days or for 60 consecutive days in the 90-day period following summer solstice, when soil temperature at a depth of 50 cm is above 5 °C.

UGLI TREE. See **Citrus Trees**.

UGRANDITE. A group name for the calcium garnet minerals uvarovite, grossular, and andradite.

UINTAHITE. A black, shiny asphaltite, with a brown streak and conchoidal fracture, which is soluble in turpentine. It occurs primarily in the veins in the Uintas Basin, Utah.

ULBRICHT SPHERE. When a light source is placed at a point inside a sphere with a perfectly diffusing surface, then every part of the surface appears equally illuminated when viewed through an opening in the surface. This principle is the basis of the *Ulbricht sphere*.

ULCER (Acid-Peptic Diseases). The word *ulcer* is sometimes used in lay circles as being synonymous with *sore*. More specifically, an ulcer is a break in the continuity of a surface (skin or mucous membrane) resulting from destruction of underlying tissue and loss of the covering epithelium. Ulcers are found in several parts of the body. Ulcerated tissue is found in ulcerative colitis (not to be confused with the peptic ulcers found in stomach and duodenum). See also **Colitis and Other Inflammatory Bowel Diseases**. Painful ulcers in the mouth may occur singly or in groups and may recur for years. This condition (*aphthous stomatitis*) arises when certain forms of bacteria in the oral cavity multiply, sometimes associated with vitamin deficiency and frequently accompanying ulcerative colitis. Fever or cold sores, which arise in the oral region from Herpes virus, more of these blisters, are described in the entry on **Dermatitis and Dermatosis**. Open ulcerous sores are relatively common in tropical climates and are described toward the end of this entry. *Decubitus ulcers* are typified by the relatively common bedsore, which can be very painful and sometimes quite serious if left untreated over a long period. Decubitus ulcers have been implicated as a causative factor in acute bacterial arthritis. They also can be the source of serious anaerobic or polymicrobial bacteremia, infections which are more commonly found among diabetics. See also **Gangrene**.

Peptic ulcers, caused by action of the digestive juices on the mucosa of the stomach or duodenum, occur with considerable frequency. The size of peptic ulcers varies from $\frac{1}{4}$ inch (6 millimeters) to over 1 inch (2.5 centimeters) in diameter. The ulcer may be deep or shallow, depending largely upon the length of time it has existed. Peptic ulcers may invade gradually deeper and deeper into the stomach or duodenal wall until a large blood vessel is penetrated, causing hemorrhage; or the wall may be perforated. Thus, early treatment is obviously desirable. However, because of the chronic nature of peptic ulcer disease, patients may postpone consulting a physician for months, even years. Pain is the outstanding symptom of peptic ulcer. The pain is believed to result from contact of gastric acids with the base of the ulcer. Pain normally subsides with emptying of the stomach or after neutralizing the acids with antacid preparations.

Duodenal ulcers are rather frequently seen and occur in the duodenum (shortest and broadest part of small intestine, extending from the pyloric end of the stomach to the jejunum). See also **Digestive System (Human)**. Although the incidence of duodenal ulcer disease has declined, it is estimated that during the first half of this century, nearly 10% of the population of advanced, industrialized countries developed this disease at some time during their lifetime. It is believed that the disease has declined since the mid-1950s by at least 50%. The decline has not been satisfactorily explained.

An excellent review article describing the pathogenesis of peptic ulcer and implications for therapy is given in the Soll reference listed. In this article a number of specific questions pertaining to the therapy best suited for gastric and duodenal ulcers are answered.

Men between ages of 45 and 65 years account for about two-thirds of cases. In women, the disease is seen mostly during ages 55 and older. Although genetic connections have not been delineated, there appears to be some familial correlation. Mortality rate of duodenal ulcer disease ranges between 2 to 5 persons per 100,000 population.

In recent years, much rethinking has occurred as regards the therapy of the disease. The role of diet has been reevaluated. The topic remains somewhat controversial even among professionals. At one time, the ingestion of milk and cream at regular intervals throughout the day was considered excellent practice. It has now been established that milk is a poor antacid and actually may promote acid secretion. Bland and soft diets, taken at frequent intervals in small meals, also have been traditionally suggested—yet there is little, if any, hard evidence that this practice is effective. Studies have shown that citrus juices and spicy foods, once believed to be undesirable, probably do not accelerate acid secretion or inhibit the ulcer healing process. Some physicians stress the importance of patient reactions to specific foods, logically eliminating known antagonists. There is general consensus, however, that certain stimulants, such as alcohol and coffee, promote acid secretion and should be avoided. Aspirin antagonizes peptic ulcers in some people. Another point agreed upon is the desirability of avoiding late-evening snacks so that acid secretion will not be stimulated during the night.

The principal therapy is the administration of antacids. Patients vary in the length of time acid is retained in the stomach, in their overall secretory capacity, and in their gastric responses to eating. There is a wide range of antacids available and some experimentation may be required to find the proper preparations and strengths for given individuals. A review of antacids is given in **Antacids**.

Anticholinergics, which competitively inhibit the action of acetylcholine (stimulates gastrin-producing and acid-secreting cells), is a logical therapy, but unfortunately the anticholinergics may produce a number of undesirable side effects (dry mouth; blurred vision; slowness of urination). See also **Choline and Cholinesterase**. This therapy may be used in patients who have hypersecretion or where they do not respond to antacid therapy.

Some *synthetic histamine analogues* have been found effective as inhibitors of histamine-stimulated gastric acid secretion. Coincident with the development of these drugs, researchers have postulated that there may be two types of histamine receptors in tissues, one of these being the H_2 receptor, which can be blocked by specific histamine analogues. Extensive clinical tests have been made of metiamide and cimetidine.

Cimetidine the oral form of which is rapidly absorbed, has been effective in duodenal ulcer disease in many patients. Although the drug has side effects, they appear to be relatively infrequent. Most physicians reserve the use of this drug for difficult situations (recurrent peptic ulcer disease; ulcerative esophagitis; pyloric channel obstruction; bleeding ulcer).

At one time, extensive *gastrectomies* were recommended for persons with continual pain and complications. In recent years, it has been found that relatively limited surgery is effective. The simplest and often effective technique is cutting vagus nerve fibers that go to the upper part of the stomach where acid is secreted.

Diseases sometimes associated with or predispose duodenal ulcer disease include chronic obstructive lung disease, rheumatoid arthritis, cirrhosis of liver, and hyperparathyroidism. In diagnosis of duodenal ulcer disease, the patient who gains relief from abdominal pain within a few minutes after ingesting an antacid preparation is immediately suspected of having the condition. There is an absence of hard evidence to associate stress as a major causative factor in peptic ulcer disease.

Gastric ulcers are peptic ulcers that occur in the stomach. Pain occurring within 30 to 60 minutes after eating is indicative of stomach ulcer, whereas the pain from a duodenal ulcer may not commence for 3 to 4 hours after a meal. Gastric ulcers are somewhat less common than the duodenal ulcer. Since gastric ulcers may occur even in patients with hypochlorhydria (acid secretion is below normal), it is believed that gastric ulcers in many instances are caused, not by excessive acid, but because of a change in the resistance of the stomach mucosa. The great majority of these ulcers occur on the portion of the stomach where orientation of the wall changes from the vertical to the horizontal. A chronic and diffuse gastritis frequently precedes and accompanies gastric ulcer, and it is considered by some authorities to be the contributing factor in ulcer formation.

With a few detailed exceptions, the symptoms of gastric ulcer essentially parallel those for duodenal ulcer. Because gastric carcinoma and benign gastric ulcer are difficult to distinguish, the physician will rely heavily upon laboratory evaluation (x-rays, fiber-optic gastroscopy, possibly biopsy). Gastric ulcers frequently are much larger than duodenal ulcers, accentuating need for early diagnosis and therapy. The recurrence rate for gastric ulcer at the same site is about 50% in the absence of continuing preventive measures. In the United States, duodenal ulcers far outnumber cases of gastric ulcers; the reverse is true in Japan.

D.M. McCarthy (University of New Mexico School of Medicine, Albuquerque) has observed, "The past 10 years have witnessed major changes in our understanding of the pathogenesis and treatment of what are commonly called the acid-peptic diseases. Schwarz's dictum to the effect that 'without acid gastric juice, no peptic ulcer' still applies. Increasingly, however, ulcers have come to be seen as areas of mucosa in which the effects of all noxious influences have exceeded the restorative capacities of all processes favorable to mucosal repair and integrity. A major stimulus to the growth of our knowledge in this area has been the development of drugs that have little or no effect on intragastric acidity, yet that have a clear ability to heal ulcers through mucosal protective effects. Chief among these is *sucralfate*."

Sucralfate is a complex salt of sucrose sulfate and aluminum hydroxide. Several different actions combine to enable sucralfate to prevent acute mucosal injury, reduce inflammation, and heal existing ulcers. Further detail on sucralfate can be found in the McCarthy reference listed.

Misoprostol. The first analogue of prostaglandin E to become commercially available was *misoprostal*. As of 1992, the drug was approved for the treatment of duodenal and gastric ulcers and for prophylaxis against NSAID (defined below) induced ulcers. The drug remains under scrutiny, but a number of promising reports have been published. As observed by R.P. Walt (Queen Elizabeth Hospital, Edgbaston, Birmingham, U.K.), "Much of the published literature about the drug is found in the proceedings of sponsored symposiums rather than in peer-reviewed journals, so caution is warranted in interpreting the data."

Nonsteroidal Anti-inflammatory Drugs (NSAIDS). It has been reported that patients who take NSAIDS may have an increased risk of mucosal damage in the upper gastrointestinal tract. The development of new lesions weeks after the initiation of treatment with NSAIDS has been confirmed in a study. In 1992, M.C. Allison (Royal Infirmary, Glasgow, Scotland) and a group of researchers concluded from their study of 713 patients that: "Patients who take NSAIDS have an increased risk of nonspecific ulceration of the small-intestinal mucosa. These ulcers are less common than ulcers of the stomach or duodenum, but can lead to life-threatening complications."

Helicobacter Pylori: Studies over the last few years have shown that a gram-negative, microaerophilic-curved bacillus, *Helicobacter Pylori*, is responsible for most cases of gastritis not associated with another known primary cause, such as autoimmune gastritis or eosinophilic gastritis and that it also may be a major factor in the pathogenesis of peptic ulcer disease. The probable importance of this microorganism currently

is being scrutinized by a number of researchers. The situation has been summarized by W.L. Peterson (University of Texas Southwestern Medical School, Dallas): "*H. Pylori*" may be the most common cause of human gastrointestinal infection as well as the most frequent cause of gastritis. Furthermore, a persuasive argument can be made that, although most people have no health problems related to *H. pylori* gastritis, some have nonulcer dyspepsia or peptic ulceration. Why some have ulcers but most do not is probably related to other factors thought to be important in the pathogenesis of peptic ulcer (the use of nonsteroidal anti-inflammatory drugs, smoking, and acid hypersecretion, for example). On the other hand, it remains possible that *H. Pylori* gastritis and peptic ulcer are associated, but not casually. Until more data are available, and because antibiotic therapy can lead to side effects (such as pseudomembranous colitis or drug resistance), attempts to treat peptic ulcer or nonulcer dyspepsia by eradicating *H. Pylori* should be limited to randomized, controlled trials."

Tropical Ulcers. Open sores of the skin are quite common in tropical climates. Some are caused by specific bacteria, such as *yaws*; others by parasitic molds (*mandura foot*); a few by single-celled parasites (*leishmaniasis*); still others are the result of secondary invasion of the tissues by various bacteria which gain entrance through wounds and small breaks in the skin. Tropical ulcers develop almost exclusively on the feet and legs, following minor skin breaks and insect bites, the tunnels of itch mites, or the perforations of hookworm larvae. Tiny blisters first appear and then develop shortly into larger, odious sores. The ulcers become tender and painful, and spread rapidly in their early phases, with extensive damage to skin and underlying tissue. Diseased tissue gradually disintegrates and sloughs off, the eroded area becoming increasingly deeper. Unless treated, in some cases the infection may penetrate muscle, tendon, and bone. Fortunately, in a majority of situations, the infection heals spontaneously before it reaches such serious proportions.

Malnutrition is common among persons suffering from tropical ulcers. Severe deficiencies of vitamins, particularly A, B, and C, are considered contributing factors. Antibiotic therapy in recent years has been quite successful in treating a number of these infections.

Additional Reading

Allison, M.C., et al.: "Gastrointestinal Damage Associated with the Use of Nonsteroidal Antiinflammatory Drugs," *N. Eng. J. Med.*, 749 (September 10, 1992).

Dorais, et al.: "Gastrointestinal Damage Associated with Nonsteroidal Antiinflammatory Drugs," *N. Eng. J. Med.*, 1882 (December 24, 1992).

Feldman, M. and M.E. Burton: "Histamine$_2$ Receptor Antagonists: Standard Therapy for Acid-Peptic Diseases," *N. Eng. J. Med.*, 1672 (December 13, 1990).

Gasbarrini, G. and S. Pretolani: "Basic and Clinical Aspects of Helicobacter Pylori Infection," Springer-Verlag, Inc., New York, NY, 1994.

Heatley, R.B. and R.V. Heatley: "The Helicobacter Pylori Handbook," Blackwell Science, Inc., Malden MA, 1996.

Martinez, E. and A. Marcos: "Helicobacter Pylori and Peptic Ulcer Disease," *N. Eng. J. Med.*, 737 (September 5, 1991).

McCarthy, D.M.: "Sucralfate," *N. Eng. J. Med.*, 1017 (October 3, 1991).

Monroe, J.: "Coping with Ulcers, Heartburn, and Stress-Related Stomach Disorders," Rosen Publishing Group, Inc., New York, NY, 2000.

Morison, M.J.: "Prevention and Treatment of Pressure Ulcers," Mosby-Year Book, Inc., St. Louis, MO, 2000.

Morsch, H.H.C.: "Treatment of Peptic Ulcer," *N. Eng. J. Med.*, 998 (October 4, 1990).

Northfield, T.C.C., M. Mendall and P.M. Goggin: "Helicobacter Pylori: Pathophysiology, Epidemiology, and Management," Kluwer Academic Publishers, Norwell, MA, 1993.

Peterson, W.L.: "Helicobacter Pylori and Peptic Ulcer Disease," *N. Eng. J. Med.*, 1043 (April 11, 1991).

Ryan, J., et al.: "Thalidomide to Treat Esophageal Ulcer in AIDS," *N. Eng. J. Med.*, 208 (July 16, 1992).

Soll, A.H.: "Pathogenesis of Peptic Ulcer and Implications for Therapy," *N. Eng. J. Med.*, 909 (March 29, 1990).

Stewart Goodwin, C.S. and B.W. Worsley: "Helicobacter Pylori: Biology and Clinical Practice," CRC Press, LLC., Boca Raton, FL, 1993.

Thagard, P.: "How Scientists Explain Disease," Princeton University Press, Princeton, NJ, 2000.

Thompson, W.G.: "The Ulcer Story: The Authoritative Guide to Ulcers, Dyspepsia, and Heartburn," Perseus Books Group, New York, NY, 1996.

Walt, R.P.: "Misoprostol for the Treatment of Peptic Ulcer and Antiinflammatory-Drug-Induced Gastroduodenal Ulceration," *N. Eng. J. Med.*, 1575 (November 26, 1992).

ULEXITE. This mineral, a hydrated borate of sodium and calcium, $NaCaB_5O_9 \cdot 8H_2O$, is a product of crystallization in arid regions from shallow playas and lakes. Ulexite crystallizes in the triclinic system, but usually occurs in rounded masses of fine-fibered acicular crystals. The hardness is 2.5, specific gravity, 1.96, silky luster and white color. The mineral is found abundantly in Chile and Argentina and in Nevada and California in the United States. Ulexite is a source of boron.

ULMACEAE.   See **Elm Trees**.

ULTIMATE PERIOD.   When the gain of a control system is such that the controlled variable has a continuous oscillation, the period of this oscillation will be the ultimate period. The term is frequently used in process control work and is equal to the reciprocal of the system's natural frequency.

ULTIMATE STRENGTH.   The maximum conventional stress in tension, compression, or shear that a material is capable of undergoing.

ULTRABASIC.   A term proposed by Judd, in 1881, for exceedingly mafic igneous rocks composed largely, if not entirely, of the ferromagnesium minerals such as olivine and pyroxene. The limiting figure of total silica is approximately 45%, or barely sufficient to supply the needs of the basic silicates.

ULTRACENTRIFUGE.   See **Centrifuge**.

ULTRAHIGH VACUUM.   See **Vacuum**.

ULTRAMICROSCOPE.   The ultramicroscope is not an instrument of extraordinary magnifying power, as its name might suggest. The term has reference rather to a special system of illumination for very minute objects. Such objects as colloidal particles, fog drops, or smoke particles are held in liquid or gaseous suspension in an enclosure with an intensely black background (usually of the black-body type). They are illuminated by a convergent pencil of very bright light entering from one side and coming to focus in the field of view—the so-called "Tyndall cone" familiar in experiments on scattering. With this arrangement, objects too small to form visible images in the microscope produce small diffraction ring systems, which appear as minute bright specks on a dark field.

ULTRASONICS.   Sound waves above the frequency normally detectable by the human ear, that is, above 16 to 20 kHz are referred to as *ultrasonic waves*. The particles of matter transmitting a *longitudinal wave* move back and forth about mean positions in a direction parallel to the path of the wave. Alternate compressions and rarefactions in the transmitting material exist along the wave propagation direction. In *shear waves*, the particles move perpendicularly to the direction of wave propagation. In *surface waves*, in seismological studies and in waves through thin stock, the *Rayleigh* and *Lamb waves*, respectively, the particles undergo much more complex vibratory motions than in longitudinal and transverse waves. In most practical applications of ultrasonics, pulses or packets containing a number of oscillation cycles are sent through the solid or liquid under investigation. See also **Cavitation**. A longitudinal wave pulse, when incident on the boundary between two materials having different sound velocities, is transformed into reflected and refracted shear and longitudinal waves. Snell's law governs the angles of reflection and refraction for both types of waves:

$$\frac{\sin \theta}{V} = \text{Constant}$$

where θ is the angle the beam makes with a plane normal to the intervening surface and V is the sound velocity. Therefore, in Fig. 1,

$$\text{Constant} = \frac{\sin \theta_1}{V_{L_1}} = \frac{\sin \theta_2}{V_{L_2}} = \frac{\sin \theta}{V_{S_2}} = \frac{\sin \phi_2}{V_{L_2}}$$

The practical application of ultrasonics requires effective transducers to change electrical energy into mechanical vibrations and vice versa. Transducers are usually piezoelectric, ferroelectric, or magnetostrictive. The application of a voltage across a piezoelectric crystal causes it to deform with an amplitude of deformation proportional to the voltage. Reversal of the voltage causes reversal of the mechanical strain. Quartz and synthetic ceramic materials are used.

Ferroelectric crystals are also electrostrictive. Barium titanate, for example, has an electrical mechanical conversion efficiency about 100

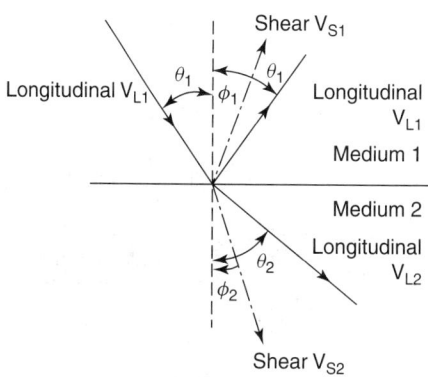

Fig. 1.   Reflection and refraction of ultrasonic waves.

times that of quartz. Unlike the piezoelectric mode of oscillation, in ferroelectric crystals, application of a voltage in either direction across the crystal causes expansion of the crystal. This mode, however, can be converted to the piezoelectric by biasing the expansion in one direction, either by application of a strong dc field, or more commonly by cooling the ferroelectric crystal through its Curie temperature while it is under the influence of a strong electric field (on the order of 10^6 volts/meter).

Transducer crystals are normally cut to a resonant frequency, the thickness being one-half the acoustic wavelength. A bond between the crystal transducer and the specimen matches the acoustic impedance, and carries the acoustic power into the latter. Backing layers may be fixed to the rear surface of the transducer. These layers are selected to reflect power forward into the crystal and specimen in some applications. On the other hand, they may be selected to absorb power so as not to complicate signals received in material testing applications.

Ultrasonic Applications

An appreciation of useful applications of ultrasound dates back to the development of radar and sonar in the late 1930s and early 1940s World War II era. One of the very earliest uses was for detecting flaws in materials. Today ultrasound is one of several effective methods used in the field of nondestructive testing and inspection. Later, the use of ultrasound was extended to include a number of detectors, probes, and transducers for measuring such variables as fluid flow, liquid level, viscosity, density, proximity, and material thickness, among others. Most of these applications are described in other articles in this encyclopedia, notably **Flow Measurement**; **Nondestructive Testing (NDT)**; and **Thickness Measurement and Gaging Systems**. Ultrasonic devices also are used in industrial processing applications, such as cleaning, drilling, emulsifying, soldering, and welding. Possible the most noteworthy of ultrasonic devices are found in the medical field for use in diagnostics where ultrasound vies with x-ray and other diagnostic procedures. Ultrasound is also applied in some security systems to detect intruders.

Some industrial instrumentation applications for ultrasound not described elsewhere are included in the following paragraphs to illustrate the basic principles in applying ultrasonics for a variety of measurements.

Ultrasonic Density Sensors.   A useful application is found in the density measurement of lime slurries for the purpose of adjusting the pH in acid water neutralization processes. Where traditional measurement methods are not suited, control engineers will often turn to more sophisticated techniques, such as ultrasound.

Slurries are difficult to handle, have a strong tendency to settle out and to coat equipment with which they come in contact. An ultrasonic density control sensor can be fully immersed in an agitated slurry, thus avoiding coating and clogging. Use of the ultrasound device for controlling the specific gravity of slurries within ±0.01% of the desired value has been demonstrated. In the usual application, the specific gravity ranges between 1.05 and 1.10. Actually, the ultrasonic sensor measures the percentage of suspended solids in the slurry, thus providing a very close approximation of the true specific gravity. The suspended solids present in the slurry attenuate the ultrasonic beam, with the resulting electronic signal being proportional to the solids in suspension.

Ultrasonic Level Detectors.   In the ultrasound system shown in Fig. 2, one transmitting sensor creates a sonic beam, the waves of which are picked

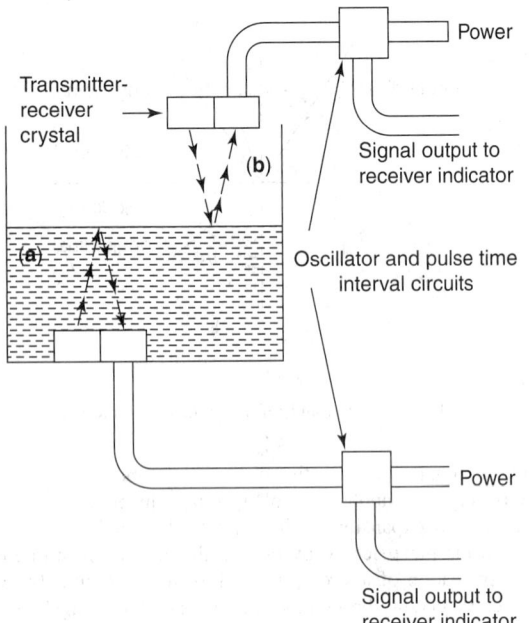

Fig. 2. Continuous sonic-type level measuring device: (**a**) Liquid phase; (**b**) vapor phase.

up by a receiving sensor. This can be accomplished by a direct path or by the reflective waves from the transmitter, which strike a fluid surface and are reflected back to the receiving sensor. Sensors can be spaced as close as $\frac{1}{4}$ in. (6 mm) or as far apart as 10 ft (3 m) in a direct beam path. They can pipe their beams through tubing where sensor beams cannot be direct, but generally the tubing length is limited to several feet (few meters). The sensor's sound beam is unaffected by mist, smoke, dust, or fumes, since it is interrupted only by a solid or fluid entering the beam path. This type of system is often used in connection with measuring the level of dry bulk solids.

Ultrasonic Thermometers. These are usually designed to respond to the temperature dependence of sound speed. In special cases where only one particular temperature is of interest, such as the temperature of a phase change, or the recrystallization temperature of a substance, the temperature dependence of attenuation may be utilized. Ultrasonic thermometers have found applications in the range -80 to $+250\,°C$, where the so-called quartz thermometer offers resolution of 0.1 millidegree and linear superiority to platinum resistance thermometers.

Ultrasonic Pressure Transducers. Advantage is taken of the fact that pressure influences sound propagation in solids, liquids, and gases, but in different ways. In solids, applied pressure leads to so-called stress-induced anisotropy. In liquids, the effects of pressure are usually small (relative to effects in gases), but the frequency of relaxation peaks can be shifted significantly.

Sonochemistry. Where liquids and solids must react, researchers are finding that energy in the ultrasound range often promotes such reactions. K.S. Suslick (University of Illinois), who has pioneered in the field, observes, "Ultrasound causes high-energy chemistry. It does so through the process of acoustic cavitation; the formation, growth and implosive collapse of bubbles in a liquid. During cavitational collapse, intense heating of the bubbles occurs. These localized hot spots have temperatures of roughly $5000\,°C$, pressures of about 500 atmospheres, and lifetimes of a few microseconds. Shock waves from cavitation in liquid-solid slurries produce high-velocity interparticle collisions, the impact of which is sufficient to melt most metals." In recent research, Suslick also reports that ultrasound creates clean, highly reactive surfaces on metals. In both homogeneous and heterogeneous situations, ultrasound assists in initiating and enhancing a number of catalytic reactions.

In investigating the phenomenon of acoustic cavitation in homogeneous liquids, researchers find that the velocity of sound in liquids typically is about 1500 m/s. Ultrasound spans the frequencies of approximately 15 kHz to 10 MHz, with acoustic wavelengths of 10 to 0.01 cm. Researchers in sonochemistry stress that these are not molecular dimensions and thus no direct coupling of the acoustic field with chemical species takes place on

a molecular level. Rather, the effects of ultrasound arise from a number of physical mechanisms, including the production of cavitation. Cavitation, originally investigated by Lord Rayleigh (1895), can produce inordinate local temperatures of 10,000 K and pressures up to 10,000 atmospheres when cavities collapse. These physical parameters indeed are conducive to promoting chemical reactivity. Considerable research over the years has been directed toward investigating the chemical effects of ultrasound on inorganic liquids, notably water. Little effort to date has been directed toward organic liquids.

In his summary of an excellent review article on sonochemistry (reference listed), Suslick points out, "Chemical applications of ultrasound are just beginning to emerge. The very high temperatures and very short times of cavitational collapse makes sonochemistry a unique interaction of energy and matter. In addition, ultrasound is well suited to industrial applications. Since the reaction liquid itself carries the sound, there is no barrier to its use with large volumes. In fact, ultrasound is already heavily used industrially for the physical processing of liquids, such as emulsification, solvent degreasing, solid dispersion, and sol formation. It is also extremely important in solids processing, including cutting, welding, cleaning and precipitation." See also **Cavitation**.

Ultrasonic Nondestructive Characterization (NDC) of Materials. The use of ultrasonics for the nondestructive characterization (testing) of metals and other materials was one of the first practical applications for ultrasound and dates back to the 1950s. The detailed properties of materials that can be measured with ultrasound include microstructure, surface characteristics, elastic properties, density, porosity, mechanical properties, process characteristics, and overt flaws. Ultrasound waves transmitted may take the form of longitudinal, shear, and surface waves. Wave characteristics include ultrasonic velocities and frequency dependence of ultrasonic attenuation/absorption. Specific reactions occurring within a tested material may include ultrasound reflection, transmission, refraction, diffraction, interference, scattering, and absorption. In applying ultrasound to testing, the characteristics of the radiation source that can be manipulated include wave type, frequency, bandwidth, pulse shape, and pulse size.

By comparison with other NDC methods, such as liquid penetrant examination, magnetic particle, eddy current testing, and radiography, the ultrasonic method is the only technique that is applicable to a wide range of materials.

The success of an ultrasonic NDC application depends upon the selection of the best-qualified transducer (i.e., one with optimum frequency response, pulse width and shape). Transducer characteristics can be customized through the use of the best-suited piezoelectric material, such as lead zirconate-lead titanate, lead metaniobates, polymer piezoelectrics, and other advanced ferro-electric materials.

Other Transducers. Ultrasound also has been used for the measurement of force, vibration, acceleration, interface location, position changes, differentiation between the composition of differing materials, grain size in metals, and evaluation of stress and strain and elasticity in materials. Sonic devices can used to detect gas leaks, and to count discrete parts by means of an interrupted sound beam. Frequently, an ultrasonic device can be applied where photoelectric devices are used. Particularly in situations where light-sensitive materials are being processed (hence presence of light must be avoided), ultrasonic devices may be the detectors of choice.

Ultrasonic Flowmeters. These instruments are described in **Flow Measurement (Fluids)**.

Ultrasonic Applications in Medicine

Ultrasonic imagery is one of the earlier tools used in the medical field for noninvasively probing body organs and tissues. Experience with this technique dates back at least to the 1970s.

The nonionizing character of ultrasonic radiation is particularly attractive, permitting the use of the technique repeatedly with a given patient at low risk. The sound frequency used for most diagnostic instruments ranges from 1 to 10 MHz. These frequencies are generated by piezoelectric transducers that reversibly convert electrical to vibratory mechanical (sound) energy. Short sound pulses propagate through the body, but a small portion of the energy is reflected back to the instrument where there are interfaces between tissues having different acoustic impedances.

Ultrasonic examination has become a well-established diagnostic tool in many abdominal diseases. In recent years, the degree of resolution attainable with ultrasonic equipment has been improved considerably, making more accurate diagnoses possible. As a consequence, transabdominal sonography has become increasingly important for diagnosing diseases of the gastrointestinal tract. It is noteworthy to observe that even acute appendicitis usually can be diagnosed by ultrasonic examination.

As noted by B. Limberg (University of Frankfurt), "With the use of sonography alone, it is impossible to detect diseases of the colon reliably, since the large bowel cannot be visualized in its entirety and detailed evaluation of wall structures and intra-aluminal lesions is difficult. The method can be improved considerably, however, by the retrograde instillation of water into the colon, a method known as *hydrocolonic sonography*. Studies have shown that this method permits not only the diagnosis of colonic tumors, but also that of inflammatory colonic diseases, such as Crohn's disease and ulcerative colitis." The detailed test of 300 patients in which conventional sonography and hydrocolonic sonography were used are given in the Limberg reference listed.

In order to facilitate surgery, preoperative imaging procedures, such as ultrasonography, computed tomography (CT), and angiography, have been used to localize the primary tumor in the case of pancreatic endocrine tumors. Because of their size, insulinomas and gastrinomas cannot be identified in up to 30% of patients. As pointed out by T. Rösch (Technical University of Munich), a large team made a comparative study involving some 37 patients. Some of the study results included: "The introduction of endoscopic ultrasonography has allowed high-resolution imaging of the pancreas that can distinguish structures as small as 2 to 3 mm in diameter. The accuracy of this procedure in diagnosing small pancreatic carcinomas has been reported to be close to 100 percent. The method also seems to be useful for the preoperative localization of small endocrine tumors of the pancreas, given the experience in this research. We report here the collective experience at six centers where endoscopic ultrasonography has been used for the preoperative localization of small tumors of the pancreas. Only patients with normal results on transabdominal ultrasonography and CT were included."

Ultrasound in Surgery. The use of extracorporeal shock waves in a procedure known as *lithotripsy* for breaking up gall stones into fragments is described in an article on **Gallbladder and Biliary Tract Diseases**. Percutaneous lithotripsy to remove stones and fragments from the kidney and urinary tract are described in an article on **Kidney and Urinary Tract**.

Ultrasound May Reduce Need for Amniocentesis. In many countries ultrasonography is routinely performed between the 16th and 24th weeks of gestation and can identify over 94% of fetuses with spina bifida, encephalocele, gastroschisis, or omphalocele. Therefore, the risk of one of these anomalies is substantially reduced for a given maternal serum alphafetoprotein level if the results of ultrasonography are normal. Where abnormalities may appear, it is not uncommon to perform a Level 2 ultrasonography, prior to proceeding with the more complex, higher-risk, and greater-cost procedure, amniocentesis.

Ultrasound Assists Drug Implants. Plastic materials infused with drugs are used for a number of purposes where drugs are gradually diffused into the bloodstream. Such implants are used for contraceptions and various chemotherapies. The effects of exposure to ultrasound of biodegradable implant materials have been studied at several teaching hospitals with interesting results.

Human Perception of Ultrasonic Speech. The upper range of human air-conduction hearing has been estimated to be no higher than approximately 24,000 Hz, although there have been a conservative number of reports that indicate hearing well into the ultrasonic range, but only when signals are delivered by bone conduction. M.L. Lenhardt (Medical College of Virginia) and a team of researchers has reported that "Bone-conducted ultrasonic hearing has been found capable of supporting frequency discrimination and speech detection in normal, older hearing-impaired and profoundly deaf human subjects. When speech signals were modulated into the ultrasonic range, listening to words resulted in the clear perception of the speech stimuli and not a sense of high-frequency vibration. These data suggest that ultrasonic bone-conduction hearing has potential as an alternative communication channel in the rehabilitation of hearing disorders." Further details are given in the Lenhardt reference listed.

Additional Reading

Abramov, O.V.: "High-Intensity Ultrasonics: Theory and Industrial Applications," Gordon & Breach Publishing Group, Newark, NJ, 1998.

Doktycz, S.J. and K.S. Suslick: "Interparticle Collisions Driven by Ultrasound," *Science*, 1067 (March 2, 1990).

Edward, I., E.I. Bluth, P.W. Ralls, P. Arger, and C. Benson: "Ultrasound: A Practical Approach to Clinical Problems," Thieme Medical Publishers, Inc., New York, NY, 1999.

Evans, D.H. and N. McDicken: "Doppler Ultrasound: Physics, Instrumentation, and Signal Processing," 2nd Edition, John Wiley & Sons, Inc., New York, NY, 2000.

Harness, J.K.: "Ultrasound in Surgical Practice: Basic Principles and Clinical Applications," John Wiley & Sons, Inc., New York, NY, 1999.

Lenhardt, M.L., et al.: "Human Ultrasonic Speech Perception," *Science*, 82 (July 5, 1991).

Limberg, B.: "Diagnosis and Staging of Colonic Tumors by Conventional Abdominal Sonography as Compared with Hydrocolonic Sonography," *N. Eng. J. Med.*, 65 (July 9, 1992).

Nadel, A.S., et al.: "Absence of Needs for Amniocentesis in Patients with Elevated Levels of Maternal Serum Alpha-Fetoprotein and Normal Ultrasonographic Examinations," *N. Eng. J. Med.*, 557 (August 30, 1990).

Papadakis, E.P.: "Ultrasonic Instruments and Devices," Academic Press, Inc., San Diego, CA, 2000.

Rennie, J.: "Ultrasound Speeds the Release of Drugs from Medical Implants," *Sci. Amer.*, 30 (April 1990).

Rifkin, M.D., et al.: "Comparison of Magnetic Resonance Imaging Ultrasonography in Staging Early Prostate Cancer," *N. Eng. J. Med.*, 621 (September 6, 1990).

Rösch, T., et al.: "Localization of Pancreatic Endocrine Tumors by Endoscopic Ultrasonography," *N. Eng. J. Med.*, 1721 (June 25, 1992).

Rose, J.L.: "Ultrasonic Waves in Solid Media," Cambridge University Press, New York, NY, 1999.

Schmerr, L.W.: "Fundamentals of Ultrasonic Nondestructive Evaluation: A Modeling Approach," Perseus Publishing, Boulder, CO, 1998.

Suslick, K.S., Editor: "Ultrasound. Its Chemical, Physical, and Biological Effects," VCH Publishers, New York, NY, 1988.

Suslick, K.S.: "The Chemical Effects of Ultrasound," *Sci. Amer.*, 80 (February 1989).

Suslick, K.S.: "Sonochemistry," *Science*, 1439 (March 23, 1990).

Thurston, R.N., E. Papadakis: "Ultrasonic Instruments and Devices I: Reference for Modern Instrumentation, Techniques, and Technology," Vol. 23, Academic Press, Inc., San Diego, CA, 1999.

Thurston, R.N., E.P. Papadakis, and A.D. Pierce: "Ultrasonic Instruments and Devices II: Reference for Modern Instrumentation, Techniques, and Technology," Vol. 24, Academic Press, Inc., San Diego, CA, 1999.

Web Reference

Ultrasonic Industry Association: *http://www.ultrasonics.org/*

ULTRASONIC TESTING.   See **Nondestructive Testing (NDT)**.

ULTRASONIC THICKNESS GAGE.   See **Thickness Measurement and Gaging Systems**.

ULTRASONIC WELDING.   See **Welding**.

ULTRAVIOLET ASTRONOMY.   For many decades, astronomers postulated that ultraviolet and extreme ultraviolet radiation emanating from the stars and other objects in the cosmos could yield information that could answer some fundamental questions: "How are galaxies formed? What material makes up the ubiquitous clouds of interstellar dust grains? How does the grouping of galaxies into clusters influence the birth rate of new stars?" And, with the answers to these questions on hand, one could be led to scores of answers pertaining to other aspects of the cosmos.

Earnest research dates back to the 1960s and is reviewed briefly later in this article. The astronomical problem of viewing the cosmos in the UV region of the spectrum is not as simple as the problem that, at one time, plagued optical telescopes, which operate in the visual light range—namely, overcoming the effects of the Earth's atmosphere. Orbiting telescopes, a gross improvement in telescope optics, and the use of adaptive optics have markedly overcome that problem. See also **Telescope (Astronomical-Optical)**.

When searching the cosmos with a UV and extreme UV instrument, there is the so-called "haze of hydrogen atoms," which, at one time, was believed to be present throughout space. Further, it was reasoned that photons at wavelengths of between 100 and 912 angstroms have just sufficient energy to ionize and thus are swept up by neutral hydrogen. But a remarkable discovery was made in 1975 by the Apollo-Soyuz mission, which carried an extreme ultraviolet telescope designed by Stuart Bowyer (University of California, Berkeley). Although it is reported that some scientists had expected a patchy medium when observed, the telescope detected four distinct stars other than the sun. This marked the first time that extreme UV light had been detected beyond the solar system. Without immediate analysis, this event showed that certainly, under certain circumstances, extreme ultraviolet astronomy could be done.

It was not until Priscilla Frish and Donald York (University of Chicago) reported that the sun resides in a void that has been swept clear of neutral hydrogen. One explanation offered was that of nearby supernova explosions. The void (clear of hydrogen) stretches for some 100 light years in most directions. In more recent findings, Barry Welsh (University of California, Berkeley) determined that the void is elongated into an interstellar "tunnel" at least 1000 light-years long. Thus, there is a window for future valuable UV exploration. This is touched on later in this article.

Chronology — UV Astronomy

Ultraviolet observations have been carried out since the beginning of the 1960s, mainly of the sun, using sounding rockets. The first UV space observatory was the *Orbiting Astronomical Satellite* (*OAO-2*), a photometric instrument launched in the late 1960s. Its primary function was ultraviolet filter photometry and low dispersion spectroscopy, largely concentrating on the ultraviolet continuum of stars and the details of dust absorption in the interstellar medium. Its major contribution was the determination of the properties of the enhanced absorption from interstellar dust grains at l = 2200 Å.

The next major instrument in this series was *COPERNICUS*, launched in 1973, which was designed to perform high resolution spectroscopic observations of interstellar absorption lines at wavelengths less than l = 1200 Å. The major achievements of this instrument included the discovery of the hot phase of the diffuse interstellar gas, and the first determination of the abundance of deuterium in the interstellar medium. It was also capable of studying for the first time absorption from hydrogen (H_2), the molecular component of the diffuse gas. Its high wavelength resolution permitted study, to an unprecedented accuracy, of the dynamics of the gas in the interstellar medium. It also provided considerable information about the dynamics of the atmospheres of luminous hot stars.

The *Netherlands Astronomical Satellite* (*ANS*) and the *TD-1* satellite have performed both photometric and spectrophotometric surveys of large numbers of stars.

Astronomers have also used manned flights as an opportunity to carry out ultraviolet imaging experiments. There were several UV cameras used during the *Apollo* lunar missions. During the *Skylab* mission, in addition to high resolution far-UV observations of the sun, objective prism spectra were obtained during the S-019 experiment and UV imaging was performed during the S-301 project. Imaging was also obtained during the *Apollo-Soyuz* mission.

Copernicus. The third Orbiting Astronomical Observatory the *Copernicus* satellite, otherwise known as the Orbiting Astronomical Observatory 3 (OAO-3), was launched by NASA from Cape Canaveral on August 21, 1972 and operated until February 1981. At that time, the spacecraft was the heaviest (2.2 tonnes) unmanned space observatory launched. The satellite, named *Copernicus*, was designed to seek answers to some of the fundamental questions concerning stars and interstellar matter by obtaining high resolution ultraviolet spectra, especially in the 900–1200 Å wavelength region. The principal experiment was a 32-inch (0.8 m) diameter reflecting telescope and ultraviolet spectrometer system designed, built, and operated by the Princeton University Observatory. Between August 1972 and February 1981 a total of 549 different objects were observed (plus measurements of air-glow and geocoronal Lyman-alpha emission), and 687,960 separate scans were obtained.

Copernicus also contained a set of X-ray detectors co-aligned with the ultraviolet telescope. Further information about this set of experiments is available from Goddard's High Energy Astrophysics Science Archive Research Center (HEASARC).

International Ultraviolet Explorer Observatory (IUE). Space UV astronomy was revitalized by the launch of the IUE observatory, a joint project of the National Aeronautics and Space Administration (NASA) and the European Space Agency (ESA), as well as of the Science and Engineering Research Council of the United Kingdom (SERC). The IUE satellite was launched on January 26, 1978. It had an expected lifetime of 3 years, with a goal of 5 years, but exceeded that beyond anyone's wildest dreams. When it was shut down on September 30, 1996, it had been in continuous operation for 18 years and 9 months. It was the most successful scientific satellite to date.

The small (45 cm) telescope operated as a photometrically accurate spectrograph from l = 1150–3300 Å. It provided a low-resolution spectrophotometer (7 Å resolution) and a high-dispersion spectrograph (0.15 Å

resolution) over this entire wavelength range. The satellite was in a geostationary orbit which was under direct NASA control for 16 hours per day and under ESA-SERC control for the remaining eight hours.

IUE contributed to the understanding of virtually all types of astronomically interesting objects, from planets to quasars. Here it is possible only to highlight some of its major contributions. It was used during the International Halley Watch to monitor the UV spectrum of Comet Halley during the 1986 close satellite encounters. It revealed the long-term structure of the plasma in the vicinity of Jupiter (the Iotorus), and provided information about the UV reflectivity of the major planets.

It is known from optical observations of solar-type stars that the outer atmospheres of main sequence stars with temperatures less than about 8000 K are extremely hot, tenuous plasmas. These regions, the chromospheres, reach temperatures of over 105 K. The longevity of *IUE* permitted the determination of temperature and structural changes in these atmospheres during the starspot cycles of many cool stars. It also was used to study the emission from regions connected with large-scale magnetic loop structures in close, active binary stars like RS Canes Venaticorum stars. It was also the first instrument to demonstrate the existence of active chromospheres in pre-main sequence T Tauri stars.

Following up on the discovery by *COPERNICUS* of the hot (105–106 K) component of the galactic interstellar medium, IUE played a key role in mapping out the extent of the galactic corona, and in tying the observations of the diffuse interstellar medium of our galaxy into a cosmic context. It is now possible to compare the structure of the interstellar gas along many lines of sight with that detected by observations of extragalactic systems. It is also possible to add to the *COPERNICUS* data a large number of previously inaccessible elements whose abundances in the diffuse gas between the stars can now be determined.

COPERNICUS showed that most luminous stars have strong mass outflows, reaching terminal velocities of thousands of kilometers per second. With *IUE*, a greater sample of these stars, both in our Galaxy and the neighboring Magellanic Clouds, have been studied. These observations support the idea that radiation, being absorbed and scattered in the stellar envelope, is responsible for transferring sufficient momentum to the gas that it is expelled from the star. The more luminous the star, the stronger its wind, among stars hotter than about 10,000 K. *IUE* observations have also showed that these winds are not stable, and that they display waves or blobs which can be traced for several hours to days through the wind by the absorption they produce against the light emitting surface of the star. Because the ground state transitions of a number of important atomic species occurred in the *IUE* wavelength region, like C IV and Si IV, it was possible to obtain information over a greater radial extent of stellar winds than can be understood by using optical emission profiles. *IUE* also was instrumental in studying the structure and dynamics of extended envelopes around emission line stars. It was the first satellite to be used in the same manner as a ground-based observatory would be — that is, implemented to carry out both survey and monitoring programs.

FAUST Telescope. As early as 1983, the Far Ultraviolet Space Telescope, designed to observe the far ultraviolet emanations of stars and galaxies over great swathes of sky, was ready for launching. Because of shuttle problems and the priorities given to other science projects, FAUST was not launched until March 30, 1992. By March 30, investigators had completed 20 of the 34 planned observations, at which time a faulty fuse (not reparable in orbit) malfunctioned and put the telescope completely out of service. Although very disappointed, researchers claim that they have sufficient data to analyze over the next few years.

Astro-1 and 2 Space Shuttle Mission. In February 1978, NASA issued an Announcement of Opportunity for instruments that could travel aboard the Space Shuttle and utilize the unique capabilities of Spacelab. Many teams responded, and over 40 proposals were selected for further study. Three telescopes, Hopkins Ultraviolet Telescope (HUT), Ultraviolet Imaging Telescope (UIT), and Wisconsin Ultraviolet Photo-Polarimeter Experiment (WUPPE), evolved as a single payload and the mission was assigned to Goddard Space Flight Center (GSFC). Because the Instrument Pointing System (IPS) and other Spacelab facilities were needed for the mission, management of the payload was moved to Marshall Space Flight Center (MSFC) in 1982, where it was given the name Astro.

The ASTRO-1 mission was flown on the space shuttle Columbia during 2–10 December, 1990. The same three instruments were later flown on the space shuttle Endeavor from 3–17 March, 1995 as part of the ASTRO-2 mission.

The *Hopkins Ultraviolet Telescope Project* (HUT) was conceived, designed, and built by astronomers and engineers at JHU to perform astronomical observations in the far-ultraviolet portion of the electromagnetic spectrum, wavelengths of light that are inaccessible to ground-based telescopes. HUT consists of a 90-centimeter (36-inch) f/2 mirror that focuses light from celestial sources onto a prime focus spectrograph.

Covering the 825–1850 Å region with about 3 Å resolution, HUT opened the astrophysically important 912 to 1200 Å window to detailed scrutiny for the first time. In typical 1800 s integrations, HUT observed faint astronomical objects with visual magnitudes of about 16.

Originally designed to explore the far- and extreme-ultraviolet ranges on Astro-1, HUT was modified for Astro-2 to concentrate on the far-ultraviolet. The changes made to HUT for Astro-2 included a new detector system and new silicon carbide coatings on the mirror and grating which replaced the original iridium and osmium. These improvements provided a factor of 2.3 increase in sensitivity in the primary operating range of 825 to 1850 angstroms, especially in the 912- to 1200-angstrom region unique to HUT.

The telescope has been used to observe objects as diverse as planets in our own solar system, individual and binary stars, star clusters, gaseous nebulae, normal and active galaxies, and even the enigmatic quasars (the most distant objects in the Universe!).

The exciting findings of the Astro-1 mission are reported in some detail by Davidsen. (See reference.) Particularly with reference to the *HUT*, Davidsen reports, "Among the new insights concerning the origin of the ultraviolet light from the old stellar population in elliptical galaxies, new evidence for a hot, gaseous corona surrounding the Milky Way was found as well as improved views of the physical conditions in active galactic nuclei, and a measurement of the ionization state of the local interstellar medium."

The *Ultraviolet Imaging Telescope* (UIT) consisted of an f/9 Ritchey-Chretien telescope with an aperture of 38 cm. Two detectors covered the near-ultraviolet and far-ultraviolet wavelength ranges from 1200 to 3300 Å. Six filters were available for each camera. Magnetically two-stage image intensifiers produced images that were recorded on 70 mm film (Kodak IIa-O). The resulting images cover a 40 arcmin field of view, with a resolution of 3 arcsec. For comparison, the Wide Field Camera #2 on HST has much smaller field of view, 2.5 arcmin × 2.5 arcmin, with a higher resolution of about 0.1 arcsec. UIT could therefore obtain ultraviolet imaging of much larger, more extended objects, than HST. In addition, a diffraction grating was used for full field low resolution spectroscopy. The film images were digitized, linearized and flat fielded through a standard reduction system.

Further details about UIT may be obtained from The Ultraviolet Imaging Telescope: Instrument and Data Characteristics, an article written by UIT mission personnel describing the design, operation, data reduction, and calibration of the instrument. *http://adsbit.harvard.edu/cgi-bin/nph-iarticle_query?bibcode=1997PASP..109..584S*

The *Wisconsin Ultraviolet Photo-Polarimeter Experiment* (WUPPE) was a pioneering effort to explore polarization and photometry in the ultraviolet (UV) spectrum. It was the first and most comprehensive effort to exploit the unique powers of polarimetry at wavelengths not visible on Earth. The instrument was designed and built at the University of Wisconsin Space Astronomy Laboratory in the 1980's. WUPPE flew on two NASA Space Shuttle missions: ASTRO-1 and ASTRO-2.

WUPPE obtained medium resolution spectropolarimetry to study the interstellar medium, hot stars, stars with circumstellar material, interacting binaries, novae, solar system objects, and active galaxies. It was a unique instrument, obtaining a large amount of the ultraviolet spectropolarimetry data currently available.

WUPPE obtained simultaneous spectral and polarization measurements from 1400 to 3300 Å. The instrument included a 0.5 m f/10 Cassegrain telescope and a modified Monk-Gilleson spectropolarimeter. Two rotating wheels were used to select the focal plane aperture and the polarimetric analyzer. A magnesium fluoride Wollaston polarizing beam-splitter placed between the aperture and the relay mirror split the beam into two orthogonally-polarized spectra. The detector consisted of dual Reticon arrays of 1024 pixel photodiodes coupled by fiber optics to a microchannel plate intensifier. A set of halfwave plates at 6 different angles provided spectropolarimetric modulation with 16 Å spectral resolution. In addition, a Lyot analyzer was used to provide low resolution (50–100 Å) spectropolarimetry on fainter objects. However the Lyot had calibration problems and no useful polarimetry data were obtained from that mode.

During Astro-1, WUPPE obtained 98 observations of 75 targets. Thirty-three of the observations resulted in usable polarimetric data (as well as spectroscopy). Forty-two of the pointings resulted in only UV spectra. During Astro-2, WUPPE obtained 369 observations of 254 targets. One hundred-fifty of the pointings resulted in polarimetric data. Forty-seven pointings produced only UV spectra.

For more information about WUPPE, please see Web References.

Hubble Space Telescope (HST). The HST is an orbiting astronomical observatory operating from the near-infrared into the ultraviolet. Launched in 1990 and scheduled to operate through 2010, HST carries and has carried a wide variety of instruments producing imaging, spectrographic, astrometric, and photometric data through both pointed and parallel observing programs.

The Hubble Space Telescope is a cooperative program of the European Space Agency (ESA) and the National Aeronautics and Space Administration (NASA) to operate a long-lived space-based observatory for the benefit of the international astronomical community. HST is an observatory first dreamt of in the 1940s, designed and built in the 1970s and 80s. Since its preliminary inception, HST was designed to be a different type of mission for NASA—a long term space- based observatory. To accomplish this goal and protect the spacecraft against instrument and equipment failures, NASA had always planned on regular servicing missions. Hubble has special grapple fixtures, 76 handholds, and stabilized in all three axes. HST is a 2.4-meter reflecting telescope which was deployed in low-Earth orbit (600 kilometers) by the crew of the space shuttle Discovery (STS-31) on 25 April 1990.

Responsibility for conducting and coordinating the science operations of the Hubble Space Telescope rests with the Space Telescope Science Institute (STScI) on the Johns Hopkins University Homewood Campus in Baltimore, Maryland. STScI is operated for NASA by the Association of Universities for Research in Astronomy, Inc. (AURA). See also **Hubble Space Telescope (HST)**.

ROSAT, the ROentgen SATellite. This satellite was an X-ray observatory developed through a cooperative program between Germany, the United States, and the United Kingdom. The satellite was designed and operated by Germany, and was launched by the United States on June 1, 1990. It was turned off on February 12, 1999. The ROSAT mission began with a 6-month, all sky Position Sensitive Proportional Counter (PSPC) survey, after which the satellite began a series of pointed observations that continued for the duration of the project. See also **ROSAT (Roentgen Satellite)**.

Extreme Ultraviolet Explorer (EUVE). This satellite, mentioned earlier, was launched in June 1992. The EUVE conducted the first extreme ultraviolet (70–760 Angstroms) survey of the sky and subsequently began a Guest Observer Program of pointed spectroscopy, that ended on January 31, 2001. The satellite has four photometric imaging systems and a three-channel EUV spectrometer. The imaging instruments were used to complete the sky survey. The spectrometers were used for the pointed spectroscopic programs, which collected data from over 350 unique astronomical targets.

At any time, *EUVE* can shift to a nonsurvey mode, and target specific sources and record detailed spectroscopic data. *EUVE* has taken the first extreme UV snapshots of the moon and the Cygnus loop, a circular glow produced by the outrushing gaseous shell from an exploding supernova careening into the interstellar medium. *EUVE* has proved that the interstellar medium can be probed by peering through gaps in the medium.

The Far Ultraviolet Spectroscopic Explorer (FUSE). FUSE is a NASA-supported astronomy mission that was launched on June 24, 1999, to explore the Universe using the technique of high-resolution spectroscopy in the far-ultraviolet spectral region. The Johns Hopkins University has the lead role in developing and now operating the mission, in collaboration with The University of Colorado at Boulder, The University of California at Berkeley, international partners the Canadian Space Agency (CSA) and the French Space Agency (CNES), and corporate partners. FUSE is part of NASA's Origins Program under the auspices of NASA's Office of Space Science. FUSE is planned for a 3 year lifetime with funding for an additional 2 years expected.

FUSE will cover the 905–1187 Å spectral region and will obtain high resolution spectra of hot and cool stars, AGNs, supernova remnants, planetary nebulae, solar system objects as well as perform detailed studies of the interstellar medium. FUSE will be able to observe sources 10,000 times fainter than *Copernicus*, an early FUV mission, and has superior resolving power than the Hopkins Ultraviolet Telescope (HUT) and the Orbiting Retrievable Far and Extreme Ultraviolet.

The FUSE satellite consists of two primary sections, the spacecraft and the science instrument. The spacecraft contains all of the elements necessary for powering and pointing the satellite: the attitude control system, the solar panels, communications electronics, and antennas. The science instrument collects the light of distant objects and contains the equipment necessary to disperse and record the light: the telescope mirrors, the spectrograph (and its electronic detectors), and an electronic guide camera called the *Fine Error Sensor* (or FES). The spacecraft and the science instrument each have their own computers, which together coordinate the activities of the satellite.

Astronomers will view the Universe in a whole new light using the unique data obtained with FUSE. In particular, they seek answers to long-standing questions such as: *"What were the conditions like in the first few minutes after the Big Bang?"*, *"How are the chemical elements dispersed throughout galaxies, and how does this affect the way galaxies evolve?"*, and *"What are the properties of the interstellar gas clouds out of which stars and solar systems form?"* All of these questions, and many others, can be addressed by observing the far ultraviolet light from stars, interstellar gas, and distant galaxies with FUSE.

The scientific approach of the FUSE mission is special because a science team has been charged by NASA with providing answers, or at least partial answers, to intriguing questions like those posed above. Toward this end, the FUSE science team will undertake a comprehensive study of the cosmic abundance of deuterium, a rare form of "heavy hydrogen" formed only in the Big Bang. The team will also study the hot gas content of our galaxy, the Milky Way, and its nearest neighbor galaxies, the Magellanic Clouds. To conduct these large studies, the FUSE science team will observe hundreds of astronomical objects, using about half of the observing time during the three-year mission. The remaining observing time is devoted to a Guest Investigator program where NASA selects scientific investigations proposed by astronomers world-wide. For additional information on FUSE see *http://fusewww.gsfc.nasa.gov/*.

Additional Reading

Bowyer, S. and R.F. Malina: "Astrophysics in the Extreme Ultraviolet," Kluwer Academic Publishers, Norwell, MA, 1996.

Brumby, S.P.: "Small Missions for Energetic Astrophysics: Ultraviolet to Gamma-Ray," American Institute of Physics, College Park, MD, 1999.

Croswell, K.: "A Long Look in the Extreme Ultraviolet," *Science*, 32 (October 4, 1993).

Davidsen, A.F.: "Far-Ultraviolet Astronomy on the Astro-1 Space Shuttle Mission," *Science*, 327 (January 15, 1991).

Drew, J.E.: "New Developments in X-Ray and Ultraviolet Astronomy," Elsevier Science, New York, NY, 1995.

Hamilton, D.: "Leaky Pipes Delay Astronomy Mission: Astro-1," *Science*, 1486 (June 22, 1990).

Henry, R.C., R. Wilson, J. Murthy, et al.: "Atlas of the Ultraviolet Sky," Johns Hopkins University Press, Baltimore, MD, 1988.

Holden, C.: "FAUST Blows a Fuse," *Science*, 175 (April 10, 1992).

Morse, J., J.M. Shull, and A. Kinney: "Ultraviolet-Optical Space Astronomy beyond HST," Vol. 164, Astronomical Society of the Pacific, San Francisco, CA, 1999.

Tatarewicz, J.N.: "Space Technology and Planetary Astronomy," Indiana University Press, Bloomington, Indiana, 1990.

Taubes, G.: "Physicists Explore the Driplines: Clues to the Structure of the Nucleus and the Metabolism of Stars," *Science*, 1874 (June 25, 1993).

Travis, J.: "Probing the Cosmos in Extreme Ultraviolet," *Science*, 1250 (February 26, 1993).

Web References

Astronomy and Astrophysics Flight Project Information at the NSSDC: *http://nssdc.gsfc.nasa.gov/astro/flight_projects.html#E*

The Hubble Space Telescope: *http://www.stsci.edu/hst/*

The Space Telescope Science Institute (STScI): *http://archive.stsci.edu/euve/*

The Hopkins Ultraviolet Telescope Project: *http://praxis.pha.jhu.edu/hut.html*

WUPPE Mission: *http://www.sal.wisc.edu/WUPPE/*

ULTRAVIOLET PHOTOGRAPHY AND IMAGERY. See **Photography and Imagery**.

ULTRAVIOLET RADIATION. This region of the electromagnetic spectrum is subdivided into: (1) the near-ultraviolet, 4,000 to 3,000 Å, present in sunlight, producing important biological effects, but not detectable by the human eye; (2) the middle-ultraviolet, 3,000 to 2,000 Å, not present in sunlight as it reaches the Earth's surface, but well transmitted through air; and (3) the long- or extreme-ultraviolet (XUV), 2,000 to 100 Å. The latter borders on x-radiation. The latter is also called the far-ultraviolet and it is not transmitted through air. The region between 2,000 and 1,350 Å is sometimes referred to as the Schumann region after its discoverer. The boundary between far-ultraviolet and x-rays is arbitrary.

Ultraviolet radiation is emitted by nearly all light sources to some degree. Generally, the higher the temperature of the source, or the more energetic the excitation, the shorter are the wavelengths produced. Tungsten lamps in quartz envelopes radiate in the ultraviolet in accordance with Planck's law, slightly modified by the emissivity function of tungsten. Because of its high temperature (3,800 K), the crater of an open carbon arc is an excellent source of ultraviolet radiation, extending to the air cutoff. Electrical discharges through gases produce intense ultraviolet emission, mainly in lines and bands. A widely used source is the quartz mercury arc. Magnetically compressed plasma, as produced by devices such as zeta and theta pinch and which reach an extremely high temperature, are also sources of highly ionized atoms and emission lines in the far-ultraviolet. Such radiation is also produced by a synchrotron.

Solids, liquids, and gases normally transmit effectively in the near-ultraviolet range, but become opaque in the middle or extreme ultraviolet range. For constructing lenses and prisms used in ultraviolet instruments, the unusual transmittance of crystal and fused quartz, fluorite, and lithium fluoride are an immense advantage. Gases vary considerably in their absorption characteristics. Oxygen molecules cause air to become opaque below about 1,850 Å. Molecular nitrogen is relatively transparent down to 1,000 Å. Hydrogen absorbs in the Lyman series lines, and in an ionization continuum beyond the series limit, 911.7 Å. Helium is the most ultraviolet-transparent of all gases. Absorption first takes place in the resonance lines, the longest lying at 584 Å, and in a continuum beyond the series limit, 504 Å. The more complex gaseous molecules, such as CO_2, NO, and N_2O are rather opaque throughout most of the far-ultraviolet. Water vapor commences to absorb at wavelengths below 1,850 Å.

As with all visible radiation, reflection occurs for ultraviolet. Reflectance becomes less as the wavelength decreases. Aluminum is the best reflector over much of the long-wavelength region, reaching 90% down to 2,000 Å (when the metal is properly prepared), and 80% down to 1,200 Å when the aluminum is coated with a thin layer of magnesium fluoride to prevent growth of aluminum oxide. Platinum is the best reflector below 1,000 Å, achieving 20% at 600 Å, but only about 4% at 300 Å.

The simplest way to detect and measure ultraviolet radiation is by using the fluorescence process, converting the ultraviolet into radiation that can be seen; or into the near-ultraviolet range, which can be easily photographed or measured with conventional photomultipliers. Materials used for the extreme ultraviolet include oil and sodium salicylate, the latter particularly valuable because its quantum efficiency of fluorescence is high and nearly independent of wavelength. Thus, an ordinary photomultiplier with a sodium salicylate coated glass window becomes a sensitive radiometer for use throughout the entire ultraviolet region.

Ionization chambers and Geiger counters can be used for detecting extreme ultraviolet radiation. Knowledge of the ionization efficiency of the gas makes it possible to use them for measurement of absolute energy. Ultraviolet radiation also can be detected with a thermocouple, thermopile, or bolometer.

The Sun emits strongly throughout the ultraviolet, but only the near-ultraviolet reaches the Earth's surface, wavelengths shorter than 2,900 Å being absorbed by a layer of ozone in the atmosphere. See also **Oxygen**. Possible disturbance of the ozone layer by various air-polluting chemicals is considered of major importance because of the possibility of eliminating this effect and thus exposing the Earth's surface to the shorter, more dangerous ultraviolet radiation. In addition to producing sunburn, exposure of the eye to ultraviolet can cause a painful burn of the cornea and conjunctivitis. Snow blindness is caused by reflection of intense sources of middle-ultraviolet radiation from snowfields and glaciers. Ultraviolet radiation also enters into the photochemical processes which contribute to the production of smog.

The use of ultraviolet lamps has been practiced for a number of years in some hospitals, schools, and factories to check the spread of respiratory infections, but their effectiveness is inconclusive. Possibly the most effective use of the characteristics of ultraviolet radiation is in optical and instrumentation applications. See also **Ultraviolet Spectrometers**.

For the use of ultraviolet radiation in astronomy, see also **Ultraviolet Astronomy**. See also specific planets. For the use of ultraviolet lasers in chemistry, see also **Photochemistry and Photolysis**.

Additional Reading

Attwood, D.T.: "Soft X-Rays and Extreme Ultraviolet Radiation: Principles and Applications," Cambridge University Press, New York, NY, 1999.

Cockell, C. and A.R. Blaustein: "Ecosystems, Evolution, and Ultraviolet Radiation," Springer-Verlag, Inc., New York, NY, 2001.

Huffman, R.E.: "Atmosphere Ultraviolet Remote Sensing," Academic Press, Inc., San Diego, CA, 1992.

Nilsson, A.: "Ultraviolet Reflections: Life under a Thinning Ozone Layer," John Wiley & Sons, Inc., New York, NY, 1996.

ULTRAVIOLET SPECTROMETERS. Ultraviolet instruments are based upon the selective absorbance of ultraviolet radiation by various substances. The absorbance of a substance is directly proportional to the concentration of the substance which causes the absorption in accordance with the Lamber-Beer law (or simply Beer's law):

$$A = abc = \log \frac{I_0}{I} \log \frac{1}{T}$$

where A = absorbance; a = molar absorptivity, 1/(mole)(centimeter); b = path length, centimeters; c = concentration, moles/1; I_0 = intensity of radiation striking detector with nonabsorbing sample in light path; I = intensity of radiation striking detector with concentration c of absorbing sample in light path b; and T = transmittance = I/I_0.

For the vapor phase,

$$A = \frac{abc'}{2{,}450}$$

at 25 °C and 760 torr pressure, where c' = volume percent or mole percent, or

$$A = \frac{abc'}{2{,}450} \times \frac{P + 14.7}{14.7} \times \frac{298}{t + 273}$$

at any temperature or pressure, where P = pressure, psig; and t = temperature °C.

For the liquid and solid phases.

$$c = \frac{c'' \times d}{\text{M.W.}} \times 10 = \text{moles/1}$$

where c'' = weight percent in liquid; d = density of liquid; and M.W. = molecular weight of material to be measured.

$$A = \frac{10abc''d}{\text{M.W.}}$$

The fundamental elements of an ultraviolet-absorption analyzer include: (a) a radiation source; (b) suitable optical filters; (c) a sample cell; and (d) an output meter. A transmittance measurement is made by calculating the ratio of the reading of the output with the sample in the cell to the reading with the cell empty (of ultraviolet-absorbing materials). The concentration can be calculated from the known absorptivity of the substances as previously demonstrated by the equations; or it may be determined by comparison with known samples.

Sources of ultraviolet radiation include: (a) tungsten-filament incandescent lamps; (b) tungsten-iodine cycle lamps with quartz envelopes; (c) mercury-vapor lamps; and (d) the zinc discharge lamp. Other types are available, but enjoy only limited application. The hydrogen or deuterium lamps are used in the laboratory, but are delicate and costly for process uses.

The analytical radiation in an ultraviolet analyzer must be as nearly monochromatic as possible in the interest of high linearity, long-time stability, and sustained accuracy. Monochromatic radiation is obtained by proper selection of sources, filters, and phototubes, each of which is selective in regard to the wavelengths that it respectively emits, transmits, or responds to. For lenses and windows, fused quartz is the most commonly used for windows and for some lenses. Corning 7910 glass and synthetic sapphire transmit throughout the near-ultraviolet range. Mirrors of rhodium or special alloys may be less reflecting in certain regions, but are preferred over silver and aluminum because of their high resistance to scratching and corrosion. Semitransparent mirrors for beamsplitting are made of special alloys or chromium coatings evaporated upon quartz or glass.

Vacuum phototubes are preferred as detectors over barrier-layer photocells because of their higher signal-to-noise ratio, greater stability, longer life, and freedom from fatigue. Simple tubes are preferred over multiplier types because they are less costly, are more stable, and can be used in simpler circuits.

The simplest ultraviolet-absorption analyzer is the *single beam* type. The output of this type of instrument will be affected by fluctuations and drift of the light source, dirt or bubbles in the sample cell, and any drift in the detector or detector circuit. Thus, single-beam instruments operate on relatively low sensitivity (high absorbance) levels to provide reasonably stable analyses. Improved single-beam instruments have found extensive applications where only a "go-no-go" or broad range measurement will suffice. The *split-beam* analyzer overcomes most of the foregoing shortcomings and is based upon a differential absorption measurement at two wavelengths. The optical diagram of a double-beam grating ultraviolet spectrophotometer is shown in Fig. 1. Each band of wavelengths, isolated by the grating monochromator, is split into two beams, which pass

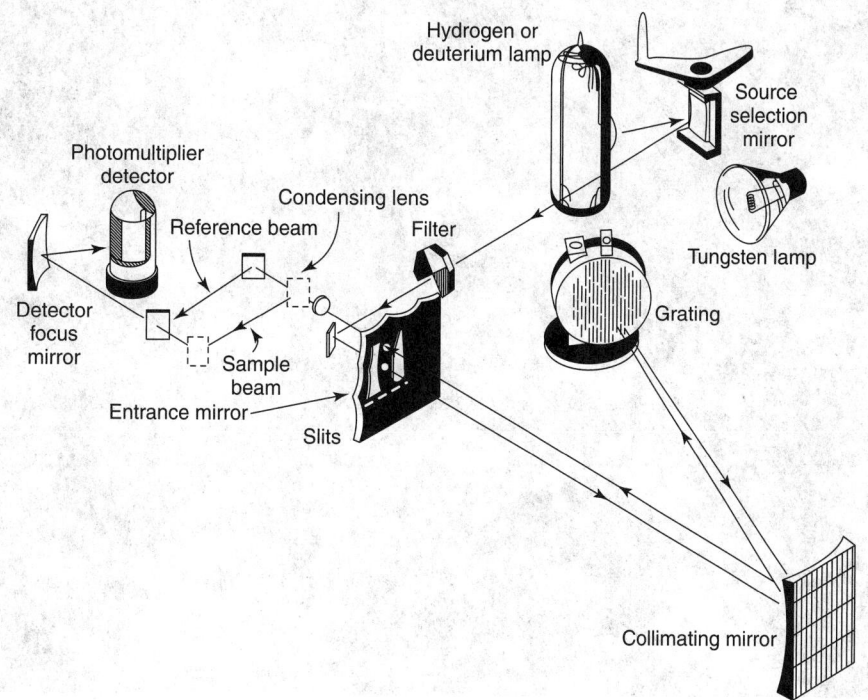

Fig. 1.   Optical path of double-beam grating ultraviolet spectrophotometer. (*Beckman.*)

alternately through the sample and reference paths. The two beams are then recombined along a single path, but separated in time. Thus, the detector receives an alternating optical signal consisting of radiant power P through the sample and P_0 through the reference. This output is converted into an electrical signal that is related to the transmittance of the sample P/P_0.

One of the more important areas of use of ultraviolet instruments is the identification and determination of biologically active substances. Many components in body fluids can be determined either directly or through colorimetric methods. Drugs and narcotics can be measured both in the body as well as in formulations. Vitamin assay is another related activity. Nearly all metals and nonmetals can be determined through their ultraviolet absorption or by colorimetric methods. In recent years, ultraviolet instruments have been used extensively for the determination of air and water pollutants, such as aldehydes, phenolics, and ozone; for the analysis of dyestuffs; for studies on polyaromatics and other carcinogenic substances; for the determination of food additives; for the analysis of petroleum fractions; and for the detection and determination of pesticide residues.

ULYSSES MISSION. *Ulysses* is a joint United States National Aeronautics and Space Administration (NASA) and European Space Agency (ESA) mission to explore and accurately assess the total solar environment. The primary mission of the *Ulysses* spacecraft is to characterize the heliosphere as a function of solar latitude. The heliosphere is an immense magnetic area of interplanetary space occupied by the Sun's atmosphere and dominated by the outflow of the solar wind. Solar wind may be defined as the stream of charged particles that flows continuously away from the Sun in all directions at speeds of up to one million miles per hour (up to 600–800 kilometers per second).

The *Ulysses* spacecraft was launched on October 6, 1990 by the space shuttle *Discovery* with two upper stages. See Fig. 1. Dornier Systems of Germany built the spacecraft for ESA, which is responsible for the launch and on-orbit operations of the spacecraft. In addition to the spacecraft, ESA provided five of the nine instruments and was responsible for instrument integration. NASA provided the space shuttle *Discovery* as the launch vehicle, and the IUS and (PAM) upper stages along with four of the science instruments and the mission operations facilities. In addition, NASA provided the Radioisotope Thermoelectric Generator (RTG) power source, which was built for the U.S. Department of Energy by the General Electric Company. Jet Propulsion Laboratory (JPL) manages the U.S. portion of the mission for NASA's Office of Space Science. *Ulysses* is

being tracked and data is being gathered by NASA's Deep Space Network, which is operated by JPL. See also **Antenna (Communications)**; and **Deep Space Network**. Investigation teams both in Europe and the USA provided scientific experiments. Spacecraft operations and data analysis are being performed at JPL by a joint ESA/JPL team.

Mission Objectives. The primary scientific objectives of the *Ulysses* mission are to investigate as a function of solar latitude:

1. the properties of the solar corona;
2. the solar wind;
3. the structure of the Sun–wind interaction;
4. the heliospheric magnetic field;
5. solar and non-solar cosmic rays;
6. solar radio bursts and plasma waves;
7. solar x-rays; and
8. interstellar/interplanetary neutral gas and dust.

The *Ulysses* Mission is the first spacecraft to explore interplanetary space at high solar latitudes. It will enable measurements to be made far out of the ecliptic plane over the solar poles. Specifically, the space physics goals are to:

1. Gain knowledge of the interplanetary medium at high solar latitudes due to the solar chromospheric and photospheric conditions, sunspots, solar flares, prominences, and coronal holes.
2. Characterize various particles from the ecliptic to the Sun's poles, including solar wind, galactic cosmic rays, interplanetary dust, and solar energetic particles.
3. Analyze the heliospheric medium from the ecliptic to the poles, including plasma waves, solar radio waves, and heliospheric magnetic fields.

To reach high solar latitudes, the spacecraft was aimed close to Jupiter so that Jupiter's huge gravitational field would accelerate *Ulysses* out of the ecliptic plane to high latitudes; a feat no man-made launch vehicle alone could accomplish. Such a tremendous amount of energy is required that even with the powerful launch vehicles available today, direct launch to high latitudes from the Earth cannot be achieved. This is because the Earth itself orbits the Sun at a speed of 65,000 miles per hour (30 kilometers per second) in a plane perpendicular to the desired solar polar orbit. The energy imparted to a space probe must cancel out this motion in addition to providing the correct polar trajectory.

Fig. 1.   The Sun as seen by a X-ray telescope, the *Ulysses* spacecraft, and the Earth–Moon system. (*NASA, JPL.*)

Encounter with Jupiter occurred on February 8, 1992 (See also **Jupiter**), where a gravity-assist maneuver placed the spacecraft in a unique solar polar orbit, allowing it to fly over the south pole of the Sun in 1994 and over the north pole in 1995. Since then *Ulysses* traveled to higher latitudes. See Fig. 2. The periods of primary scientific interest are when *Ulysses* is at or higher than 70 degrees latitude at both the Sun's south and north poles. On June 26, 1994, *Ulysses* reached 70 degrees south. There it began a four-month observation from high latitudes of the complex forces at work in the Sun's outer atmosphere, the corona. Maximum southern latitude of 80.2 degrees was achieved on September 13, 1994. *Ulysses* traveled through high northern latitudes during June through September of 1995. The high latitude observations were obtained during the quiet (minimum) portion of the 11-year solar cycle. The spacecraft's orbital period is six years and, if kept in operation, high latitude observations will be obtained during the active (maximum) portion of the solar cycle. Prior to the *Ulysses* mission, studies of the Sun were made from Earth using Earth-based sensors. More recently, solar studies have been conducted from space stations; however, these investigations have been mostly from the ecliptic plane (the plane in which most of the planets travel around the Sun). No previous spacecraft have been able to reach solar latitudes higher than 32 degrees. Now that *Ulysses* high latitude data is available, scientists are obtaining a new understanding of the processes occurring at high solar latitudes.

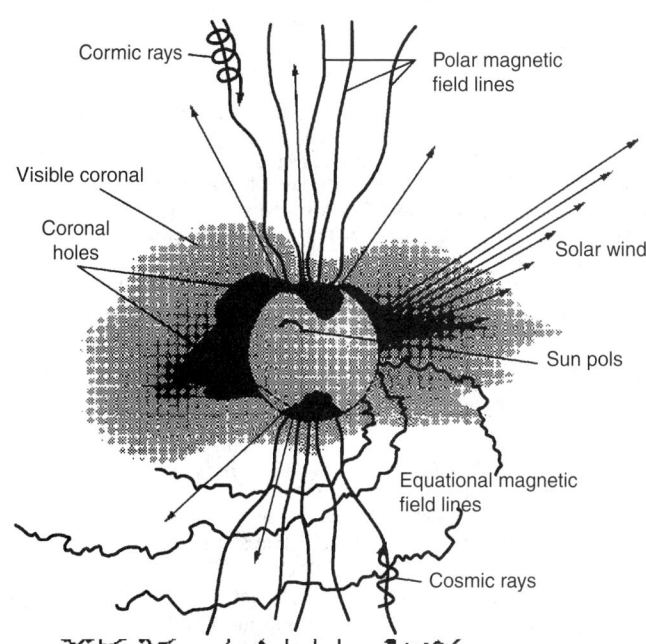

Fig. 3. A composite of the Solar phenomena being investigated by the *Ulysses* mission. Previous Earth-based observations and theory indicated that there would be quite significant differences between the low latitude regions and the poles. (*NASA, JPL.*)

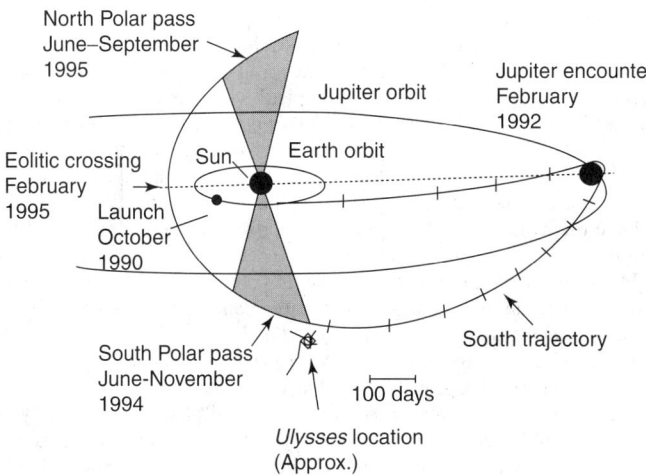

Fig. 2. Trajectory of the *Ulysses* spacecraft. (*NASA, JPL.*)

Scientists have long been aware of differences between the Polar Regions of the Sun and lower latitudes. For example, sunspots are only seen at lower latitudes, and photographs of the solar corona taken during solar eclipses often show dark regions over the poles.

Solar Corona. The solar corona consists of hot gasses at such high temperatures that the gravitational field of the Sun can not prevent escape of coronal gas as the solar wind. However, the Sun has a global magnetic field. Many of the solar magnetic field lines that leave the solar surface return to the surface, but some of the field lines, particularly those over the poles, extend deep into interplanetary space. The solar wind expands into interplanetary space along these field lines, and the regions of the *corona* from which the hot gas escapes are dark because of the low gas density. These regions are known as coronal holes. See Fig. 3.

The properties of the Sun's polar magnetic field are poorly understood, although the magnetic field has an important influence on the escape of the solar wind. In addition, the complex processes that serve to both heat and accelerate the solar wind are not well understood. *Ulysses* observations over the poles should provide important new information on how the solar wind expands from the Sun, which will aid scientists in understanding these processes. The magnetic field also has a tremendous influence on matter approaching the Sun from the Milky Way galaxy and from the nearby interstellar medium. Incoming cosmic rays are subjected to forces exerted by the magnetic field. Scientists believe that the structure of the Sun's magnetic field is such that entry of cosmic

rays occurs by way of the Sun's Polar Regions. Scientists hope that *Ulysses* findings can provide new information as to the extent to which the galactic cosmic rays use this route, and on the ways in which their properties are modified as a result. Scientists also hope to gain more knowledge of the intensity and properties of the cosmic rays far from the Sun.

With the first phase of its mission successfully completed, *Ulysses* has now embarked on a second orbit of the Sun; it will for the first time investigate the high latitude properties of the solar wind during the maximum of the solar activity cycle. The spacecraft and its scientific payload are in excellent condition. The ultimate goal of this next phase of the mission is the study of the Sun's polar regions under conditions of high solar activity, culminating in polar passes in 2000 and 2001 (South Polar pass: September 2000 to January 2001; North Polar pass: September to December 2001). Much before this, however, with *Ulysses* descending slowly in latitude after the northern polar pass, there is a unique opportunity to make coordinated observations with ESA's SOHO spacecraft, which carries an extensive complement of experiments dedicated to studying the Sun's corona and the solar wind. The period around aphelion (1997–98) was also of great interest. During this interval *Ulysses* spent many months close to the ecliptic at almost constant radial distance (~5 AU) from the Sun, enabling the scientists to study the evolution and development of many interplanetary phenomena, free of concern about spatial variations. The out-of-ecliptic orbit of *Ulysses* is expected to cover a period of 6.2 years, corresponding to approximately half a solar cycle.

As noted above, the high-latitude passes of the second solar orbit will occur close to the maximum of solar cycle 23. The conditions in the Polar Regions are expected to be dramatically different from those encountered during the prime mission. In particular, the rather simple configuration of the corona found at solar minimum, with large coronal holes over the polar caps, will have been replaced by a much more complex arrangement, probably including high-latitude streamers. Transient events (solar flares, coronal mass ejections, etc.) related to the increase in solar activity will dominate, greatly disturbing the underlying structure of the solar wind and influencing the transport of cosmic rays and energetic solar particles. *Ulysses* is clearly in a unique position to study the evolution of the three-dimensional heliosphere from the current solar minimum to maximum activity conditions. No other space mission in the foreseeable future will address these goals.

Ulysses Mission fast facts are given in Table 1.

TABLE 1. ULYSSES MISSION FAST FACTS

Launch date	October 6, 1990
Payload	11 instruments
Orbit	Heliocentric, 5 by 1.3 astronomical units
Design life	5 years (prime mission)
Length	7.3 feet (2.2 meters)
Weight	805 pounds (366 kilograms)
Diameter	13.5 feet (4.1 meters), including booms
Launch vehicle	Space Shuttle/inertial upper stage/payload assist module
International participation	United Kingdom, Germany, Italy, Switzerland, United States

A complete descriptions of the following *Ulysses* Scientific Instruments can be found at *http://ulysses.jpl.nasa.gov/ULSHOME/ESAULS/instruments. html*. See Fig. 4.

- Magnetometer (VHM/FGM)
- Solar Wind Plasma Experiment (SWOOPS)
- Solar Wind Ion Composition Instrument (SWICS)
- Unified Radio and Plasma Wave Instrument (URAP)
- Energetic Particle Instrument (EPAC)
- Low-Energy Ion and Electron Experiment (HISCALE)
- Cosmic Ray and Solar Particle Instrument (COSPIN)
- Solar X-ray and Cosmic Gamma-Ray Burst Instrument (GRB)
- Dust Experiment (DUST)
- Coronal-Sounding Experiment (SCE)
- Gravitational Wave Experiment (GWE)

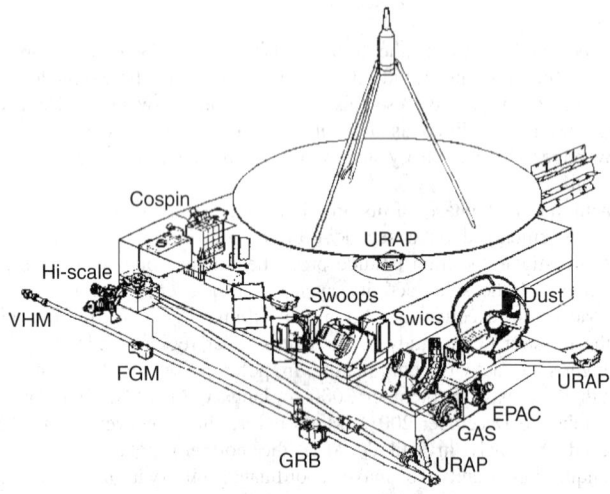

Fig. 4.   *Ulysses* Scientific Instruments (*NASA, JPL*).

Additional Reading

Balogh A. et al.: "Magnetometer (VHM/FGM): The Magnetic Field Investigation on the Ulysses Mission: Instrumentation and Preliminary Scientific Results," *Astron. Astrophys. Suppl. Ser.* **92**, 221–236 (1992).

Bame S.J. et al.: "Solar Wind Plasma Experiment (SWOOPS): The Ulysses Solar Wind Plasma Experiment," *Astron. Astrophys. Suppl. Ser.* **92**, 237–265 (1992).

Bertotti B. et al.: "Gravitational Wave Experiment (GWE): The Gravitational Wave Experiment," *Astron. Astrophys. Suppl. Ser.* **92**, 431–440 (1992).

Bird M.K. et al.: "Coronal-Sounding Experiment (SCE): The Coronal-Sounding Experiment," *Astron. Astrophys. Suppl. Ser.* **92**, 425–430 (1992).

ESA Bulletin No. 67, August 1991, for a report on the first scientific results from the in-ecliptic phase of the mission.

For information on *Ulysses* Scientific Instruments, (See Fig. 4) and the following papers:

Gloeckler G. et al.: "Solar Wind Ion Composition Instrument (SWICS): The Solar Wind Ion Composition Spectrometer," *Astron. Astrophys. Suppl. Ser.* **92**, 267–289 (1992).

Gruen E. et al.: "Dust Experiment (DUST): The Ulysses Dust Experiment," *Astron. Astrophys. Suppl. Ser.* **92**, 411–423 (1992).

Hurley K. et al.: "Solar X-ray and Cosmic Gamma-Ray Burst Instrument (GRB): The Solar X-Ray/Cosmic Gamma-Ray Burst Experiment Aboard Ulysses," *Astron. Astrophys. Suppl. Ser.* **92**, 401–410 (1992).

Keppler E. et al.: "Energetic Particle Instrument (EPAC): The Ulysses Energetic Particle Composition Experiment (EPAC)," *Astron. Astrophys. Suppl. Ser.* **92**, 317–331 (1992).

Lanzerotti L. et al.: "Low-Energy Ion and Electron Experiment (HISCALE): Heliosphere Instrument for Spectra, Composition and Anisotropy at Low Energies," *Astron. Astrophys. Suppl. Ser.* **92**, 349–363 (1992).

Marsden, R.G. and K.P. Wenzel, *ESA Bulletin No. 72*, November 1992, ESA Space Science Department, ESTEC, Noordwijk, The Netherlands.

Simpson J.A. et al.: "HISCALE (Heliosphere Instrument for Spectra, Composition, and Anisotropy and Low Energies) energetic particle experiment at JHU/APL," *Astron. Astrophys. Suppl. Ser.* **92**, 365–399 (1992).

Simpson J.A. et al., "Solar X-ray and Cosmic Gamma-Ray Burst Instrument (GRB): The Solar X-Ray/Cosmic Gamma-Ray Burst Experiment Aboard Ulysses," *Astron. Astrophys. Suppl. Ser.* **92**, 365–399 (1992).

Stone R.G. et al., "Unified Radio and Plasma Wave Instrument (URAP): The Unified Radio and Plasma Wave Investigation," *Astron. Astrophys. Suppl. Ser.* **92**, 291–316 (1992).

Web References

European Space Agency: *http://sci.esa.int/ulysses/*
Ulysses Mission: *http://ulysses.jpl.nasa.gov/mission/mission.html*
Ulysses Mission Operations Home Page: *http://ulysses-ops.jpl.esa.int/*
Ulysses Spacecraft: *http://nssdc.gsfc.nasa.gov/nmc/tmp/1990-090B.html*

UMBELLIFERAE (Carrot Family).   The plants of this family, mostly found in the north Temperate Zone, are nearly all annual or biennial herbs, with a few trees or shrubs. They are characterized by having stems hollow except at the nodes, alternate leaves that are either pinnately or ternately compound, and large numbers of small flowers produced in simple or compound umbels. The distinctive umbel is an influorescence in which the pedicels of the individual flowers arise from a very short axis and are of nearly equal length, producing a flat-topped or rounded cluster.

The single flowers are small and regular, the calyx being either absent or fused to the ovary, and five-toothed, the corolla composed of five separate petals, the stamens five and the single pistil with inferior ovary, with two one-seeded carpels. The flowers are insect-pollinated. The fruit is a form known as a schizocarp, a dry fruit having two carpels which separate at maturity into two mericarps, each of which usually has five longitudinal ridges. Between the ridges are found longitudinal oil canals containing the volatile oils which give the odors characteristic of many members of this family.

Many members of this family are important food plants to people and domestic animals. Among these are plants grown for their roots, the carrot, *Daucus carota*, and parsnip, *Pastinaca sativa*, both natives of Europe; also celery, *Apium graveolens*, grown for its leaf stalk, and parsley, *Petroselinum crispum*, grown for the much dissected leaves which are used as a garnish and for flavoring. A large number of plants in this family are frequently cultivated, especially in Europe, for flavorings and medicinal use, though possibly their therapeutic value is rather overrated. Among these are anise (*Pimpinella anisum*), caraway (*Carum carvi*), coriander (*Coriandrum sativum*), fennel (*Foeniculum officinale*), and dill (*Anethum graveolens*). Many of these are also used for flavoring foods. *Ferula asafoetida* produces a drug, asafoetida.

In contrast to these, which are useful to humans, are many members of the family which contain violent poisons. *Conium maculatum*, the Poison Hemlock, is reputed to have been the source of the poison which Socrates drank. *Cicuta maculata* and related species are sometimes eaten by livestock with fatal results.

UMBILICAL CORD.   Any of the servicing electrical or fluid lines between the ground or a tower and an uprighted rocket vehicle before the launch. Often shortened to umbilical.

UMBILICAL TOWER.   A vertical structure supporting the umbilical cords running into a rocket in launching position.

UMBILICUS.   The scar on the ventral surface of the abdomen where the umbilical cord is severed at birth in the placental mammals. The navel.

UMBRA.   1. The darkest part of a shadow in which light is completely cut off by an intervening object. A lighter part surrounding the umbra, in which the light is only partly cut off, is called the *penumbra*.

2. The darker central portion of a sun spot, surrounded by the lighter penumbra. See also **Eclipse**.

UMKEHR EFFECT.   Due to the presence of the ozone layer, an anomaly of the relative zenith intensity of scattered sunlight at certain wavelengths in the ultraviolet as the sun approaches the horizon.

UNAVAILABLE ENERGY. When an irreversible process takes place, the effect on the universe is the same as that which would be produced if a certain quantity of energy in a form completely available for work were converted to a form in which it was completely unavailable for work. This amount of energy is called the unavailable energy.

UNCERTAINTY PRINCIPLE. Also sometimes referred to as the *indeterminancy principle*, this was first stated by Heisenberg in connection with the position and momentum of an electron. In essence, the postulate states that it is impossible to determine simultaneously both the exact position and the exact momentum of an electron and thus these values must be expressed as a probability. If, for example, one determines the precise location of an electron, all information about the electron's velocity is lost. On the other hand, knowledge of the electron's velocity can be obtained only at the expense of knowledge of its location.

Classical Newtonian mechanics assumes that a physical system can be kept under continuous observation without thereby disturbing it. This is reasonable when the system is a planet or even a spinning top, but is unacceptable for microscopic systems, such as an atom. To observe the motion of an electron, it is necessary to illuminate it with light of ultrashort wavelength (gamma rays); momentum is transferred from the radiation to the electron and the particle's velocity is, therefore, continuously disturbed. The effect upon a system of observing it can not be determined exactly, and this means that the state of a system at any time cannot be known with complete precision. As a consequence, predictions regarding the behavior of microscopic systems have to be made on a probability basis and complete certainty can rarely be achieved. This limitation is accepted and is made one of the foundation stones upon which the theory of quantum mechanics is constructed.

This principle can be extended to other phenomena of a like nature, that is, in the simultaneous determination of the values of two canonically conjugated variables, the product of the smallest possible uncertainties in their values is of the order of magnitude of the Planck constant h. If Δq is the range of values that might be found for the coordinate q of a particle, and Δp is the range in the simultaneous determination of the corresponding component of its momentum p, then $\Delta p \cdot \Delta q \geq h$. Similarly, if ΔE and Δt are the uncertainties in the simultaneous determination of the energy and the time, $\Delta E \cdot \Delta t \geq h$. In the same way, the principle applies to any other pair of canonically conjugated variables. See also **Quantum Mechanics**.

UNCONFORMITY (Geological). If deposition in a given area is interrupted for a time by erosional processes, then renewed, a dissected surface will separate the two groups of beds, which are then said to be unconformable, and the erosion surface marking their contact is called an unconformity. If this unconformable relationship of two rock groups is of limited extent, the unconformity is then said to be local; if of wide extent it is called a regional unconformity. If two groups of sedimentary rocks are separated by an unconformity on tilted beds, it is then called a nonconformity or angular unconformity. When the plane of erosion occurs between relatively horizontal formations it is called a disconformity. Principal types of unconformities are illustrated in Fig. 1. *A* represents a nonconformity having an irregular erosion surface, developed on the upturned edges of different formations, and covered with a basal conglomerate which grades through sandstone and shale into limestone. In *B* the nonconformable series has also been deformed together with the second wave of deformation which affected the already deformed beds beneath the unconformity. *C* represents a major nonconformity in which the erosion surface truncates highly metamorphosed formations and batholithic intrusives. The basal conglomerate rests upon a plane of erosion which, in itself, must represent a great physical hiatus or lack of stratigraphic record for this region. *D* and *E* represent disconformities that have been developed upon a prelithified surface. In *E* the disconformable formations have been subsequently deformed (tilted). *F* represents a graduational contact, or one in which erosion and depositions have been relatively continuous. *G* represents a graduational contact developed in plastic sediments, with the development of mud cracks and intraformational conglomerates. *H* represents a disconformity in which the basal breccia is composed of the lithified fragments of the older formations, plus

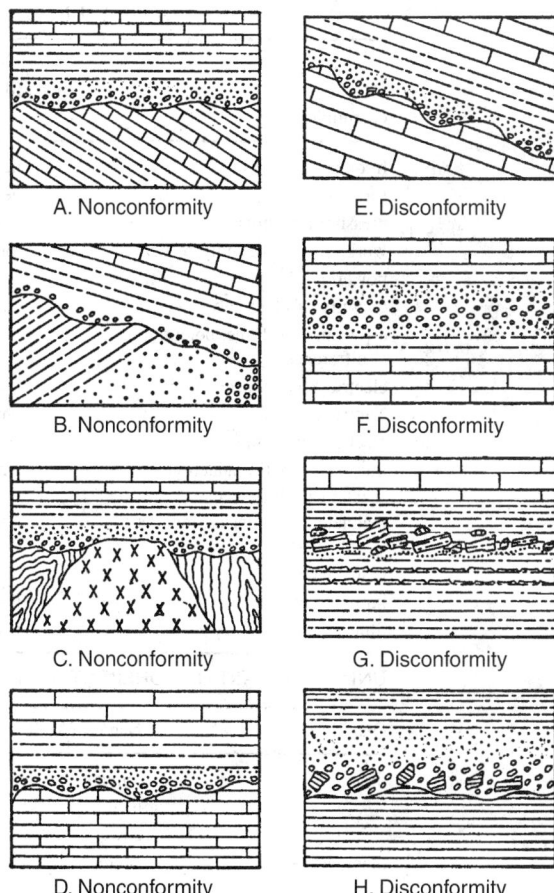

Fig. 1.   Types of geological unconformities. (*After Field.*)

foreign clastic material. As the fossils above and below the disconformity are the same, the disconformity, though pronounced, signifies slight, if any, hiatus (diastem). Disconformities of types *F*, *G*, and *H* may represent either slight or great hiatus. Note: The true amount of hiatus can only be measured by paleontological means. With the exception of a major nonconformity, the physical evidence of hiatus (unconformity) is seldom a safe criterion alone. Thus the terms nonconformity and disconformity have structural but not necessarily stratigraphic, or time, significance.

UNDERDAMPED.   See Damping.

UNDERGROUND MINING.   See Coal.

UNDERWATER SOUND.   See Sonar.

UNDERWING (*Insecta, Lepidoptera*). A moth whose fore wings are colored in dull shades of gray or brown, but whose hind wings are in most species brightly banded with black and some shade of yellow, orange or red. The name is used for the large moths of the family *Noctuidae* and the genus *Catocala*, although members of other genera are colored in a similar way. The genus *Catocala* is so extensive, with more than 100 species in North America alone, that it has attracted the attention of specialists and has been studied in detail. For the same reason the moths are very likely to be seen without special search. They hide about buildings in the daytime, as well as on the trunks of trees.

UNIMODULAR GROUP.   See Lie Group.

UNIT FUNCTION.   See Laplace Transform.

UNIT RAMP FUNCTION.   See Laplace Transform.

TABLE 1. UNITS DERIVED FROM BASE SI QUANTITIES

Quantity	Name of SI Derived Unit	Symbol	Expressed in Terms of SI Base or Derived Units
frequency	hertz	Hz	$1 \text{ Hz} = 1 \text{ s}^{-1}$
force	newton	N	$1 \text{ N} = 1 \text{ kg} \cdot \text{m/s}^2$
pressure and stress	pascal	Pa	$1 \text{ Pa} = 1 \text{ N/m}^2$
work, energy, quantity of heat	joule	J	$1 \text{ J} = 1 \text{ N} \cdot \text{m}$
power	watt	W	$1 \text{ W} = 1 \text{ J/s}$
quantity of electricity	coulomb	C	$1 \text{ C} = 1 \text{ A} \cdot \text{s}$
electromotive force, potential difference	volt	V	$1 \text{ V} = 1 \text{ W/A}$
electric capacitance	farad	F	$1 \text{ F} = 1 \text{ A} \cdot \text{s/V}$
electric resistance	ohm	Ω	$1 \text{ }\Omega = 1 \text{ V/A}$
electric conductance	siemens	S	$1 \text{ S} = 1 \text{ }\Omega^{-1}$
flux of magnetic induction, magnetic flux	weber	Wb	$1 \text{ Wb} = 1 \text{ V} \cdot \text{s}$
magnetic flux density, magnetic induction	tesla	T	$1 \text{ T} = 1 \text{ Wb/m}^2$
inductance	henry	H	$1 \text{ H} = 1 \text{ V} \cdot \text{s/A}$
luminous flux	lumen	lm	$1 \text{ lm} = 1 \text{ cd} \cdot \text{sr}$
illuminance	lux	lx	$1 \text{ lx} = 1 \text{ lm/m}^2$

UNITS, NOT PART OF COHERENT SYSTEM, BUT GENERALLY ACCEPTED FOR USE WITH SI UNITS

Quantity	Name of Unit	Unit Symbol	Magnitude in SI Units
time	minute	min	60 s
	hour	h	3600 s
	day	d	86400 s
plane angle	degree	°	$\pi/180$ rad
	minute	′	$\pi/10\,800$ rad
	second	″	$\pi/648\,000$ rad
volume	litre	l	$1 \text{ l} = 1 \text{ dm}^3$
mass	tonne	t	$1 \text{ t} = 10^3 \text{ kg}$
energy	electronvolt	eV	approx. 1.60219×10^{-19} J
mass of an atom	atomic mass unit	u	approx. 1.66053×10^{-27} kg
length	astronomical unit	AU	149600×10^6 m
	parsec	pc	approx. 30857×10^{12} m

UNITS AND STANDARDS.

The General Conference on Weights and Measures, to which the United States adheres by treaty, has established the International System of Units, called SI units. The base quantities of this system and the corresponding units and symbols are:

Length	meter	m
Mass	kilogram	kg
Time	second	s
Electric current	ampere	A
Thermodynamic temperature	kelvin	K
Luminous intensity	candela	cd
Amount of substance	mole	mol

The units radian (rad) for plane angle and steradian (sr) for solid angle are described as supplementary units and are normally treated as though they were base units, although the corresponding quantities may be treated as dimensionless. The coherent SI unit system consists of the foregoing, plus all of the units derived from them by multiplication and division without introducing numerical factors.

A number of derived units have been given special names and symbols. See Table 1.

Decimal multiples of the coherent base and derived SI units are formed by attaching to these units the prefixes shown in Table 2.

The following units were accepted by the International Committee of Weights and Measures for use with the SI units for a transitional period: angstrom ($1 \text{ Å} = 10^{-10}$ m), barn ($1 \text{ b} = 10^{-28} \text{ m}^2$), bar ($1 \text{ bar} = 10^5$ Pa), standard atmosphere ($1 \text{ atm} = 101\,325$ Pa), curie ($1 \text{ Ci} = 3.7 \times 10^{10} \text{ s}^{-1}$), roentgen ($1 \text{ R} = 2.58 \times 10^{-4}$ C/kg), rad ($1 \text{ rad} = 10^2$ J/kg).

TABLE 2. STANDARD PREFIXES USED WITH SI UNITS

Factor by Which the Unit Is Multiplied	Prefix Name	Prefix Symbol
10^{12}	tera	T
10^9	giga	G
10^6	mega	M
10^3	kilo	k
10^2	hecto	h
10	deca	da
10^{-1}	deci	d
10^{-2}	centi	c
10^{-3}	milli	m
10^{-6}	micro	m
10^{-9}	nano	n
10^{-12}	pico	p
10^{-15}	femto	f
10^{-18}	atto	a

The *cgs system of units*, based on the centimeter, gram, and second as units in mechanics, is a metric system which continues to be used in some branches of physics. In daily life, the customary units in the United States are those based on the foot, pound-force, and second, but these units are rarely used in physics except for the description of equipment (e.g., "a 2-inch pipe"). Considerable effort has been going forth in the United

TABLE 3. PRINCIPAL UNITS — SYMBOLS, DEFINITIONS, DIMENSIONS

AMPERE (A). *The constant current that, if maintained in two straight parallel conductors that are of infinite length and negligible cross section and are separated from each other by a distance of 1 meter in a vacuum, will produce between these conductors a force equal to 2×10^{-7} newton per meter of length.*

(The SI unit of electric current.)

AMPERE PER METER (A/m). *The magnetic field strength in the interior of an elongated uniformly wound solenoid which is excited with a linear current density in its winding of 1 ampere per meter of axial distance.*

(The SI unit of magnetic field strength.)

AMPERE-HOUR (Ah). *The quantity of electricity represented by a current of 1 ampere flowing for 1 hour.*

ANGSTROM Å. *A unit of length equal to 10^{-10} meter.* *

APOSTILB (asb). *A unit of luminance. One lumen per square meter leaves a surface whose luminance is 1 apostilb in all directions within a hemisphere.*

(The candela per square meter is the preferred unit of luminance.)

ATMOSPHERE, STANDARD (atm). *A unit of pressure. One standard atmosphere equals 101,325 newtons per square meter.*

ATOMIC MASS UNIT, UNIFIED (u). *The atomic mass unit (unified) is 1/12th of the mass of an atom of the ^{12}C nuclide.*

(Use of the prior atomic mass unit (amu), defined by reference to oxygen, is no longer preferred.)

BAR (bar). *A unit of pressure. One bar equals 100,000 newtons per square meter.*

BARN (b). *A unit of nuclear cross section. One barn equals 10^{-28} square meter.*

BARREL (bbl). *A unit of volume. One barrel equals 9,702 cubic inches; or 0.15899 cubic meters.*

(This is the standard barrel used for petroleum, etc. A different standard barrel is used for fruits, vegetables, and dry commodities.)

BAUD (Bd). *A unit of signaling speed. One baud equals one element per second.*

(The signaling speed in bauds is equal to the reciprocal of the signal element length in seconds.)

BEL (B). *A dimensionless unit for expressing the ratio of two values of power, being the logarithm to the base 10 of the power ratio.*

(The more commonly used unit, decibel (dB), is 10 times the logarithm to the base 10 of the power ratio. A bel is 10 decibel.)

BIT (b). *A unit of information, generally represented by a pulse. A bit is a binary digit, i.e., a 1 or 0 in computer technology.*

(In information theory, the bit is the smallest possible unit of information.)

BIT PER SECOND (b/s). *A unit of signaling speed. A transference rate of 1 bit per second.*

BRITISH THERMAL UNIT (Btu). *A unit of heat. The heat required to warm 1 pound of pure water through an interval of 1 degree Fahrenheit.*

CALORIE (International Table) (cal$_{It}$). *A unit of heat. One International Table calorie equals 4.1868 joules.*

(The 9th Conférence Générale des Poids et Mesures adopted the joule as the unit of heat.)

CALORIE (Thermochemical Calorie) (cal). *A unit of heat. One calorie equals 4.1840 joules.*

(See foregoing note.)

CANDELA (cd). *The luminous intensity of 1/6,000,000 of a square meter of a radiating cavity at the temperature of freezing platinum (2042 K).*

(The SI unit of luminous intensity. The unit formerly was called the *candle*.)

CIRCULAR MIL (cmil). *The area of a circle whose diameter is 0.0001 inch. One circular mil equals $\pi/4 \cdot 10^{-6}$ square inches.*

COULOMB (C). *The quantity of electric charge which passes any cross section of a conductor in 1 second when the current is maintained constant at 1 ampere.*

(The SI unit of electric charge.)

CURIE (Ci). *The unit of activity in the field of radiation dosimetry. One curie equals 3.7×10^{10} disintegrations per second.*

(The activity of 1 gram of ^{226}Ra is slightly less than 1 curie.)

CYCLE (c). *An interval of space or time in which is completed 1 round of events or phenomena.*

CYCLE PER SECOND (Hz, c/s). *The number of cycles per second.*

(The name hertz (Hz) is the accepted international term. The abbreviation Hz is preferred to c/s.)

DARCY (D). *A unit of permeability of a porous medium. One darcy equals 1 cP (cm/s)(cm/atm) equals 0.986923 square micrometers.*

(A permeability of 1 darcy will allow the flow of 1 cubic centimeter per second of fluid of 1 centipoise viscosity through an area of 1 square centimeter under a pressure gradient of 1 atmosphere per centimeter.)

DAY (d). *A unit of time, the exact definition of which is dependent upon which system of time measurement is referred to, i.e., apparent solar time, mean solar time, universal time, apparent sidereal time, ephemeris time, or atomic time.* See **Time**. With exception of atomic time, the time base is referenced to rotation of the Earth. For general purposes, a day is considered the period taken for 1 revolution of the Earth about its axis.

DEGREE CELSIUS (°C). *One unit of temperature on the Celsius temperature scale, which is derived from the thermodynamic of Kelvin scale of temperature and related by: Temperature (degrees Celsius) equals Temperature (Kelvin units) minus 273.15. See Temperature.*

DEGREE FAHRENHEIT (°F). *One unit of temperature on the Fahrenheit temperature scale, which is related to the Celsius temperature scale by: Temperature (degrees Fahrenheit) equals 1.8 × (degrees Celsius) plus 32. See Temperature.*

DEGREE RANKINE (°R). *One unit of temperature on the Rankine temperature scale, which is related to the Fahrenheit temperature scale by: Temperature (degrees Rankine) equals Temperature (degrees Fahrenheit) plus 459.69. See* **Temperature.**

DYNE (dyn). *A unit of force. One dyne equals the force necessary to give 1 gram mass an acceleration of 1 centimeter/(second)(second).*

(The dyne is the unit of force in the CGS system.)

ELECTRONVOLT (eV). *A unit of energy. One electronvolt equals the energy acquired by an electron when it passes through a potential difference of 1 volt in a vacuum.*

(One electronvolt equals 1.602×10^{-12} erg.)

ERG (erg). *A unit of energy. One erg equals 10^{-7} joule.*

(Also, 1 erg equals the work done when a force of 1 dyne is applied through a distance of 1 centimeter. One foot-pound equals 13,560,000 ergs.)

FARAD (F). *The capacitance of a capacitor in which a charge of 1 coulomb produces a potential difference of 1 volt between the terminals.*

(The SI unit of capacitance.)

FOOTCANDLE (fc). *A unit of luminance. One footcandle equals 1 lumen per square foot.*

(The name *lumen per square foot* is recommended for this unit. The SI unit, lux (lumen per square meter), is preferred.)

FOOTLAMBERT (fL). *A unit of luminance. One lumen per square foot leaves a surface whose luminance is 1 footlambert in all directions within a hemisphere.*

(If luminance is measured in English units, the candela per square inch is preferred. However, use of the SI unit, the candela per square meter, is generally accepted.)

GAL (Gal). *A unit of acceleration. One Gal equals 1 centimeter per second per second.*

GALLON (gal). Because the gallon, quart, and pint differ in the United States and the United Kingdom, the use of this unit and term is generally discouraged for scientific purposes. An imperial gallon equals 1.20095 U.S. gallons. One U.S. gallon equals 3.785×10^{-3} cubic meter.

GAUSS (G). *A unit of magnetic flux density, or magnetic induction. The ratio of the flux in any cross section to the area of that cross section, the cross section being taken normal to the direction of flow. One gauss equals 1 maxwell per square centimeter.*

(The gauss is a unit of the CGS system. Use of the SI unit, the *tesla*, is preferred.)

GILBERT (Gb). *A unit of magnetomotive force. One gilbert equals 0.4 π (ni), where (ni) is an ampere-turn.*

(The gilbert is a unit of the CGS system. Use of the SI unit, the ampere (or ampere-turn), is preferred.)

GRAIN (gr). *A unit of mass. One grain equals 0.06480 gram.*

(One ounce, avoirdupois, equals 437.5 grains; 1 ounce, troy, equals 480 grains; 1 ounce, apothecaries', equals 480 grains. One pound, avoirdupois, equals 7,000 grains.)

GRAM (g). *A unit of mass. One gram equals 1/1,000th kilogram.*

(See also KILOGRAM in this list.)

HENRY (H). *A unit of inductance. The inductance of a circuit in which a current of 1 ampere induces a flux linkage of 1 weber.*

(The SI unit of inductance.)

HERTZ (Hz). *A unit of frequency. One hertz equals a frequency of one cycle per second.*

(The SI unit of frequency.)

HORSEPOWER (hp). The horsepower is considered an anachronism in science and technology. Use of the SI unit of power, the watt, is preferred. When used, 1 horsepower equals (1) 42.44 Btu/minute; (2) 33,000 foot-pounds/minute; or (3) 550 foot-pounds/second.

HOUR (h). *A unit of time. One hour equals 60 minutes, or 3,600 seconds.*

INCH (in). *A unit of length. One inch equals 2.540×10^{-2} meter.*

INCH OF MERCURY (inHg). *A unit of pressure. One inch of mercury equals 3,386.4 newtons per square miter.*

(An inch of mercury also equals (1) 0.03342 atmosphere; (2) 1.133 feet of water; (3) 345.3 kilograms/square meter; (4) 70.73 pounds/square foot; or (5) 0.4912 pounds/square inch.)

INCH OF WATER (in H$_2$O). *A unit of pressure. One inch of water equals 249.09 newtons per square meter.*

(An inch of water also equals (1) 2.458×10^{-3} atmosphere; (2) 0.07355 inch of mercury; (3) 2.540×10^{-3} kilogram/square centimeter; (4) 0.5781 ounce/

(continued)

TABLE 3. (*Continued*)

square inch; (5) 5.204 pounds/square foot; or (6) 0.03613 pound/square inch. The latter Figures hold for a temperature of 4 °C.)

JOULE (J). *A unit of energy. The work done by 1 newton acting through a distance of 1 meter.*
(The SI unit of energy. One joule equals 1 watt-second; equals 10^7 ergs; equals 10^7 dyne-centimeters.)

JOULE PER KELVIN (J/K). *A unit of heat capacity and entropy.*

KELVIN (K). *The basic unit of thermodynamic temperature. One kelvin is the fraction 1/273.16 of the thermodynamic temperature of the triple point of water.*
(The term *degree Kelvin* was officially dropped in 1967. Thus, the symbol is K and not °K. Relationship of the Kelvin scale to the Celsius scale is given earlier in this list under DEGREE CELSIUS.)

KILOGRAM (kg). *A unit of mass and is based upon a cylinder of platinum-iridium alloy kept by the International Bureau of Weights and Measures at Paris.* A duplicate in the custody of the National Bureau of Standards at Washington is the mass standard for the United States. The kilogram is the only base unit still defined by an artifact.
(A kilogram equals (1) 1,000 grams; (2) 2.205 pounds; (3) 9.842×10^{-4} long tons; or (4) 1.102×10^{-3} short tons.

KNOT (kn). *A unit of speed. One knot equals 1 nautical mile per hour.*
(A knot also equals 6,080.2 feet/hour; or 1.151 statute miles/hour.)

LAMBERT (L). *A unit of luminance. One lumen per square centimeter leaves a surface whose luminance is 1 lambert in all directions within a hemisphere.*
(The candela per square meter is the preferred unit of luminance.)

LITER (l). *A unit of volume. One liter equals 10^{-3} cubic meter.*
(A liter also equals (1) 1,000 cubic centimeters; (2) 0.03531 cubic foot; (3) 61.02 cubic inches; (4) 1.308×10^{-3} cubic yard; (5) 0.2642 U.S. liquid gallon; (6) 1.057 U.S. liquid quarts; or (7) 0.22 Imperial gallon.

LUMEN (lm). *A unit of luminous flux. The flux through a unit solid angle (steradian) from a uniform point source of 1 candela.*
(The SI unit of luminous flux.)

LUMEN PER SQUARE FOOT (lm/ft^2). *A unit of illuminance and also a unit of luminous excitation.*
(Use of the SI unit, lumen per square meter, is preferred.)

LUMEN PER SQUARE METER (lm/m^2). *A unit of luminous excitation.*
(The SI unit of luminous excitation.)

LUMEN PER WATT (lm/W). *A unit of luminous efficacy.*
(The SI unit of luminous efficacy.)

LUMEN SECOND (lm · s). *A unit of quantity of light.*
(The SI unit of quantity of light.)

LUX (lx). *A unit of illuminance. One lux equals 1 lumen per square meter.*
(The SI unit of illuminance.)

MAXWELL (Mx). *A unit of magnetic flux. The flux through a square centimeter normal to a field at 1 centimeter from a unit magnetic pole.*

METER (m). *A unit of length. Defined as 1,650,763.73 wavelengths in vacuum of the orange-red line of the spectrum of 86 Kr (krypton).*
(The SI unit of length. A meter also equals (1) 100 centimeters; (2) 3.281 feet; (3) 39.37 inches; (4) 0.001 kilometer; (5) 5.396×10^{-4} nautical mile; (6) 6.214×10^{-4} statute mile; or (7) 1.094 yards.)

MHO (mho). *A unit of conductance (and of admittance). The conductance of a conductor whose resistance is 1 ohm.*
(The name *siemens* (S) also is used for this quantity.)

MICROMETER (μm). *A unit of length. One micrometer equals one-millionth of a meter.*
(The term *micron* formerly used for this unit no longer is preferred.)

MICRON. See MICROMETER above.

MIL (mil). *A unit of length. One mil equals one-thousandth of an inch.*

MILE, STATUTE (mi). *A unit of length. One mile equals 5,280 feet.*
(One statute mile also equals (1) 1.609 kilometers; (2) 1,760 yards; (3) 6.336×10^4 inches; or (4) 0.8684 nautical mile.)

MILE, NAUTICAL (nmi). *A unit of length. One nautical mile equals 1.1516 statute miles.*
(One nautical mile also equals (1) 6,080.27 feet; (2) 1.853 kilometers; or (3) 2,027 yards.)

MINUTE, TIME (min). *A unit of time. One minute equals 60 seconds.*
(Time also may be designated by means of superscripts, as in $9^h46^m30^s$, where there otherwise will be no confusion with abbreviations.)

MOLE (mol). *A unit of amount of substance. One mole is an amount of a substance, in specified mass units, equal to the molecular weight of that substance.*
(The SI unit for amount of substance. Examples are the gram mole or the pound mole.)

NEPER (Np). *A dimensionless unit for expressing the ratio of two voltages, two currents, or two power values in a logarithmic manner. The number of nepers is the natural (Napierian) logarithm of the square root of the ratio of the two values being compared.* Thus, the neper uses the base of 2.71828 in contrast

with the bel (or decibel) which uses the common-logarithm base of 10. One neper equals 8.686 decibels.

NEWTON (N). *A unit of force. One newton is the force that will impart an acceleration of 1 meter per second per second to a mass of 1 kilogram.*
(The SI unit of force. One newton equals 10^5 dynes.)

NIT (nt). *A unit of luminance and is synonymous with candela per square meter.*

OERSTED (Oe). *A unit of magnetic field strength. The magnetic field produced at the center of a plane circular coil of 1 turn and of radius 1 centimeter, which carries a current of $(\frac{1}{2}\pi)$ abamperes.*
(An abampere equals 10 amperes. The oersted is the CGS unit of magnetic field strength. Use of the SI unit, the ampere per meter, is preferred.)

OHM (Ω). *A unit of resistance (and of impedance). The resistance of a conductor such that a constant current of 1 ampere in it produces a voltage differences of 1 volt between its ends.*
(The SI unit of resistance.)

PASCAL (Pa). *A unit of pressure or stress. One pascal equals 1 newton per square meter.*

PHON (phon). *A unit of loudness level. The pressure level in decibels of a pure 1,000 Hz tone.*

PHOT (ph). *A unit of illuminance. One phot equals 1 lumen per square centimeter.*
(The phot is the CGS unit of illuminance. Use of the SI unit, the lux, is preferred.)

PINT (pt). Because the gallon, quart, and pint differ in the United States and the United Kingdom, the use of this unit and term is generally discouraged for scientific purposes. One U.S. pint equals: (1) 473.2 cubic centimeters; (2) 0.01671 cubic foot; (3) 28.87 cubic inches; (4) 4.732×10^{-4} cubic meter; (5) 6.189×10^{-4} cubic yard; (6) 0.125 U.S. gallon; (7) 0.4732 liter; or (8) 0.5 liquid U.S. quart.

POISE (P). A unit of dynamic viscosity. The unit is expressed in dyne second per square centimeter. The centipoise (cP) is more commonly used. The formal definition of viscosity arises from the concept put forward by Newton that under conditions of parallel flow, the shearing stress is proportional to the velocity gradient. If the force acting on each of two planes of area A parallel to each other, moving parallel to each other with a relative velocity V, and separated by a perpendicular distance X, be denoted by F, the shearing stress is F/A and the velocity gradient, which will be linear for a true liquid, is V/X. Thus, $F/A = \eta V/X$, where the constant η is the viscosity coefficient or dynamic viscosity of the liquid. The poise is the CGS unit of dynamic viscosity.

POUNDAL (pdl). *A unit of force. One poundal equals the force required to give a standard 1-pound body an acceleration of 1 foot per second per second.*

QUART (qt). Because the gallon, quart, and pint differ in the United States and the United Kingdom, the use of this unit and term is generally discouraged for scientific purposes. One U.S. quart equals: (1) 946.4 cubic centimeters; (2) 0.03342 cubic foot; (3) 57.75 cubic inches; (4) 9.464×10^{-4} cubic meter; (5) 1.238×10^{-3} cubic yard; (6) 0.25 U.S. gallon; or (7) 0.9463 liter.

RAD (rd). *A unit of absorbed dose in the field of radiation dosimetry. One rad equals the absorption of energy in any medium of 100 ergs per gram.*

RADIAN (rad). *A unit of plane angle. One radian equals the angle subtended at the center by a circular arc which is equal in length to the radius of the circle.*
(The SI unit of plane angle.)

REM (rem). *A unit of dose equivalent in the field of radiation dosimetry. One rem equals the amount of ionizing radiation of any type which produces the same damage to humans as 1 roentgen of approximately 200 kilovolts x-radiation.*
(The unit is abbreviation of Roentgen Equivalent Man.)

REVOLUTION PER MINUTE (r/min). Although use of rpm as an abbreviation is common, it should not be used as a symbol.

ROENTGEN (R). A unit of exposure in the field of radiation dosimetry. That quantity of x- or gamma-radiation such that the associated corpuscular emission per 0.001293 gram of dry air (equals 1 cubic centimeter at 0 °C and 769 millimeters of mercury pressure) produces in air ions carrying 1 esu of quantity of electricity of either sign.
(The emu (electrostatic unit) is a unit in the CGS system in which the statcoulomb is the charge that repels an exactly similar charge in a vacuum with a force of 1 dyne. One statcoulomb equals 3.3356×10^{-10} coulomb.)

SECOND (s). *A unit of time. The duration of 9,192,631,770 periods of the radiation corresponding to the transition between the two hyperfine levels of the ground state of the ^{133}Cs (cesium) atom.*
(The SI unit of time.)

SLUG (slug). *A unit of mass. One slug equals 14.5959 kilograms.*

STERADIAN (sr). *A unit of solid angle. One steradian equals the solid angle subtended at the center by $\frac{1}{4}\pi$ of the surface area of a sphere of unit radius.*

STILB (sb). *A unit of luminance. One stilb equals 1 candela per square centimeter.*

STOKES (St). A unit of kinematic viscosity. The centistokes (cSt) is more commonly used. Kinematic viscosity is the dynamic viscosity divided by the density. See POISE given previously in this list.

TABLE 3. (*Continued*)

TESLA (T). *A unit of magnetic flux density (magnetic induction). The magnetic flux density of a uniform field that produces a torque of 1 newton-meter on a plane current loop carrying 1 ampere and having a projected area of 1 square meter on the plane perpendicular to the field.* T = N/A · m.
(The SI unit of magnetic flux density.)

THERM (thm). *A unit of heat. One therm equals 100,000 British thermal units.*

TON (ton). *A unit of weight. If not otherwise specified, a short ton equal to 2,000 pounds is assumed. A long ton equals 2,240 pounds. A metric ton equals 1,000 kilograms (2,205 pounds), also called tonne (t).*

VAR (var). *A unit of reactive power. The reactive power at the port of entry of a single-phase two-wire circuit when the product of (a) the rms (root mean square) value in amperes of the sinusoidal current, (b) the rms value in volts of the voltage, and (c) the sine of the angular phase difference by which the voltage leads the current is equal to 1.*
(The SI unit of reactive power.)

VOLT (V). *A unit of voltage. The voltage between 2 points of a conducting wire carrying a constant current of 1 ampere, when the power dissipated between these points is 1 watt.*
(The SI unit of voltage.)

VOLTAMPERE (VA). *A unit of apparent power. The apparent power at the port of entry of a single-phase two-wire circuit when the product of (a) the rms (root mean square) value in amperes of the current and (b) the rms value in volts of the voltage is equal to 1.*

(The SI unit of apparent power.)

WATT (W). *A unit of power. The watt equals 1 joule per second.*
(The SI unit of power.) One watt equals: (1) 3.4192 Btu/hour; (2) 0.05688 Btu/minute; (3) 10^7 ergs/second; (4) 44.27 foot-pounds/minute; (5) 0.7378 foot-pounds/second; (6) 1.341×10^{-3} horsepower; (7) 1.360×10^{-3} metric horsepower; (8) 0.01433 kilogram-calories/minute; or (9) 0.001 kilowatt.

WATT PER METER KELVIN (W/m · K). The SI unit of thermal conductivity.

WATT PER STERADIAN (W/sr). The SI unit of radiant intensity.

WATT PER STERADIAN SQUARE METER (W/Sr · m^2). The SI unit of radiance.

WATTHOUR (Wh). *A unit of energy. One watthour equals 3,600 joules.*
(One watthour equals: (1) 3.413 Btu; (2) 3.60×10^{10} ergs; (3) 2,656 footpounds; (4) 859.85 gram-calories; (5) 1.341×10^{-3} horsepower-hour; (6) 0.8598 kilogram-calorie; (7) 367.2 kilogram-meters; or (8) 0.001 kilowatt-hour.

WEBER (Wb). *A unit of magnetic flux. The magnetic flux passing through an area of 1 square meter placed normal to a uniform magnetic field of magnetic flux density equal to 1 tesla.* Wb = T · m^2.
(The SI unit of magnetic flux.) If the flux linked by a circuit changes at a uniform rate of 1 weber per second, a voltage of 1 volt is induced in the circuit. Wb = V · s.

*Although, officially, A (without small circle over it) may be used as an abbreviation or symbol for angstrom, to avoid possible confusion with the use of A for ampere, the Å symbol is used throughout this text.

TABLE 4. COMMON EQUIVALENTS AND CONVERSIONS

Approximate Common Equivalents		Conversions Accurate to Parts Per Million	
1 inch	25 millimeters	inches × 25.4*	millimeters
1 foot	0.3 meter	feet × 0.3048*	meters
1 yard	0.9 meter	yards × 0.9144*	meters
1 mile	1.6 kilometers	miles × 1.60934	kilometers
1 square inch	6.5 square centimeters	square inches × 6.4516*	square centimeters
1 square foot	0.09 square meter	square feet × 0.0929030	square meters
1 square yard	0.8 square meter	square yards × 0.836127	square meters
1 acre	0.4 hectare	acres × 0.404686	hectares
1 cubic inch	16 cubic centimeters	cubic inches × 16.3871	cubic centimeters
1 cubic foot	0.03 cubic meter	cubic feet × 0.0283168	cubic meters
1 cubic yard	0.8 cubic meter	cubic yards × 0.764555	cubic meters
1 quart (liquid)	0.9463 liter	quarts (liquid) × 0.946353	liters
1 gallon	0.004 cubic meter	gallons × 0.00378541	cubic meters
1 ounce (avoirdupois)	28 grams	ounces (avoirdupois) × 28.3495	grams
1 pound (avoirdupois)	0.45 kilogram	pounds (avoirdupois) × 0.453592	kilograms
1 horsepower	0.75 kilowatt	horsepower × 0.745700	kilowatts
1 millimeter	0.04 inch	millimeters × 0.0393701	inches
1 meter	3.3 feet	meters × 3.28084	feet
1 meter	1.1 yards	meters × 1.09361	yards
1 kilometer	0.6 mile	kilometers × 0.621371	miles
1 square centimeter	0.16 square inch	square centimeters × 0.155000	square inches
1 square meter	11 square feet	square meters × 10.7639	square feet
1 square meter	1.2 square yards	square meters × 1.19599	square yards
1 hectare	2.5 acres	hectares × 2.47105	acres
1 cubic centimeter	0.06 cubic inch	cubic centimeters × 0.0610237	cubic inches
1 cubic meter	35 cubic feet	cubic meters × 35.3147	cubic feet
1 gram	0.035 ounce (avoirdupois)	grams × 0.0352740	ounces (avoirdupois)
1 kilogram	2.2 pounds (avoirdupois)	kilograms × 2.20462	pounds (avoirdupois)
1 kilowatt	1.3 horsepower	kilowatts × 1.34102	horsepower

States and a number of other countries that are not accustomed to using the metric system to ultimately adopt it.

About one hundred of the most frequently used units are defined in Table 3. Common equivalents and conversions are given in Table 4.

UNIVALVES. See Mollusks.

UNIVERSAL TIME (CT). See Time.

UNIVERSE (The). Approximately 15 billion years ago in cosmic history, the first galaxies took shape from vast clouds of early chemical elements.

In the furnace of stars, life-sustaining chemicals such as carbon and oxygen came into being. Then, in awe-inspiring blasts from dying stars,

life's chemicals blew out into space, only to condense anew into stars like our sun and planets like Earth.

Through the mixing of these vital chemicals and energy, the living Universe blossomed with the earliest self-replicating organisms and the profusion of life on our planet.

Timeline of the Universe

The 15-billion-year-long history of the Universe shown in Fig. 1 illustrates the major chain of events that eventually led to life on Earth. The sequence starts at upper right with the Big Bang and proceeds counter-clockwise following the arrows to the Chemistry of Life.

The Big Bang

Scientists believe that the universe was created about 15 billion years ago in a single violent event known as the Big Bang. All the space,

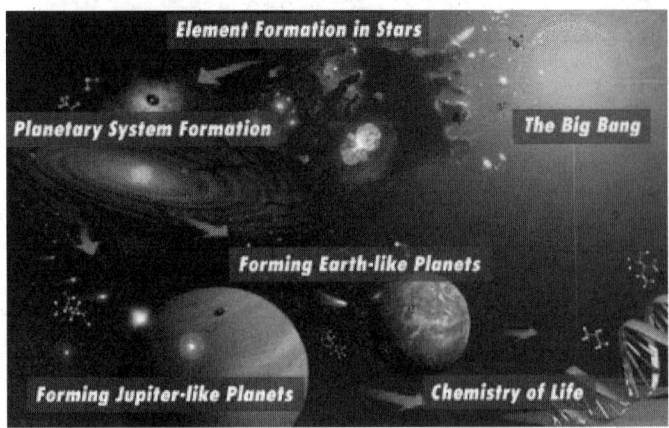

Fig. 1.   Timeline of the Universe. (*NASA/Jet Propulsion Laboratory, Pasadena, CA.*)

time, energy, and matter that constitute today's universe originated in the Big Bang. The early universe was extremely small, dense, and hot. For the first fraction of a second, only energy existed. See also **Cosmology**.

As the universe expanded and cooled, the four fundamental forces (gravity, electromagnetism, and the strong and weak nuclear forces) became distinct. Quarks, then atomic particles and their antimatter partners, appeared. As matter and antimatter met, they annihilated each other, leaving behind energy and a slight excess of ordinary matter—almost exclusively the lightest elements, hydrogen and helium. The faint residual heat from the Big Bang can be observed coming from everywhere in the sky.

Galaxies.   The young universe did not have a perfectly even distribution of energy and particles. These irregularities allowed forces to start to collect and concentrate matter. Accumulations started to develop ever more complicated structures. Concentrations of matter formed into clouds, then condensed into stars and the collections of stars we call galaxies. The way in which galaxies spin indicates that their visible portions of stars and

Fig. 2.   Distant Galaxies and the Hubble Deep Field. The Hubble Deep Field is our deepest and most detailed look at the horizon of the visible universe. This dark patch of sky was selected by astronomers to be as empty as possible of foreground stars and known clusters of galaxies. This image was constructed from 342 separate Wide-Field Planetary Camera-2 (WFPC2) exposures taken in ultraviolet, blue, red, and infrared light during ten consecutive days of observing in December 1995 by the Hubble Space Telescope (HST). The Hubble Deep Field is located near the handle of the Big Dipper, and its size in the sky would appear to the naked eye about equal to the size of a grain of sand held at arm's length.

The Hubble Deep Field shows over 3000 galaxies at various distances and stages of evolution. There are only four obvious stars visible in the image (they appear as point-like objects with diffraction spikes, can you find them?). The small number of stars is a consequence both of the tiny field of view and of the fact that we are looking up out of the plane of the Milky Way Galaxy. Many different types of galaxies are visible, including spirals like our own, almost featureless ellipticals, and many disturbed-looking "oddballs." Some of these oddball galaxies may be in the midst of titanic collisions with other galaxies, while others are still in the star-forming exuberance of youth.

Astronomers are using the largest telescopes in the world to determine which of the galaxies in the image are relatively nearby and faint, and which are truly at the edge of the visible universe. The light from these farthest galaxies took many billions of years to cross the vast expanse of the universe, and so we are seeing them as they appeared very shortly after they and the universe were born. The Hubble Deep Field therefore promises to become the Rosetta Stone of cosmology, allowing astronomers to answer fundamental questions about the age, size, and composition of the universe. Whatever the answers, the Hubble Deep Field will rank among the greatest scientific treasures of the twentieth century. (*Robert Williams (Space Telescope Science Institute), the Hubble Deep Field Team, and NASA.*)

diffuse gas and dust clouds known as nebulae constitute only one tenth of the total mass. The so-called "missing mass" could hold the key to the ultimate fate of the universe—that is whether it expands forever or is pulled back together by the combined gravitational attraction of all of its mass.

From the standpoint of the development of life, what matters is that each galaxy is a stellar factory, producing stars out of giant gas clouds; each star is a chemical factory, transmuting simple elements into heavier, more complex ones; and life is a collection of some of these complex molecules. See Fig. 2. Visible matter comes in a wonderful variety of galaxy forms, characterized by their distributions of stars and glowing or dark nebulae. See also **Galaxy**; and **Nebula**.

Giant Molecular Clouds. The largest inhabitants of galaxies are giant clouds of molecules that contain the raw material for stars and planets. A cloud with a diameter of 300 light years (1 light year is equal to about 10 trillion kilometers) contains enough mass to manufacture 10,000 to a million stars, each with the mass of our Sun. However, only about 10 percent of the cloud will be in clumps dense enough for stars to form—enough to produce a few hundred to a few thousand new stars. Giant molecular clouds last for 10 to 100 million years before they dissipate.

Element Formation in Stars

Clouds and Star Birth. Gravity acts on individual particles to form collections that attract still more particles. Under the right conditions, gravity can overcome the disruptive forces of heat and turbulence to create spheres of gas that are hot enough and dense enough at their centers so that hydrogen can fuse into helium—creating a star. See Fig. 3. But this new star will probably not yet be apparent in visible light. The young star is surrounded by a dense, opaque shroud of dust. As the star heats the dust, the star becomes detectable by infrared telescopes as a "hot spot" within a large, dense molecular cloud. Winds from the star will eventually blow away residual gas and dust and the star will become visible in optical telescopes.

Mature Stars and Nucleosynthesis. Young stars grow and shrink as they try to strike an evolving balance between gravity, which tries to compress the star, and the pressure from the fusion reactions that try to make the star expand. **Mature stars** have achieved that delicate balance and spend almost their entire lives that way.

A star's size, color, brightness, and life span are the consequence of the total amount of its mass. Stars with only small amounts of material (a few tenths the mass of our Sun) become cool "red dwarfs" that live for many billions of years. Stars with the mass of our Sun last for about 10 billion years. Giant stars, with a few tens of the mass of our Sun, consume their fuel furiously and burn, white-hot, for only a few million years.

Over its entire lifetime, a star's hydrogen is being fused into helium. Late in the star's life, its helium mass becomes great enough to reach the necessary pressure and temperature, and the helium begins to fuse into still heavier elements. Shells of fusion, each requiring higher and higher pressures and temperatures, form from the ashes of the previous reaction and create new elements in the process known as **nucleosynthesis**. The additional heat produced in the core causes the star to swell. See also **Star**.

Star Death and the Distribution of Elements for New Star Systems. Eventually, all stars run out of fuel in their cores. They lose their equilibrium as the force of gravity comes to dominate. Different-mass stars end their lives differently. Low-mass stars die quietly as their nuclear fires dwindle. The core in a Sun-like star collapses rapidly into an Earth-size white dwarf. The star's outer layers, containing atoms formed in the fusion process, are left as expanding bubbles or jets of material that expand out into the universe. A massive star's core collapses almost instantaneously. It rebounds outward and strikes other material falling inward. This collision occurs with so much energy that it creates all of the naturally occurring elements and blows the star apart. This explosion, a **supernova**, is the source of all the heavy elements that are found in nebulae,

Fig. 3. Star Birth in the Eagle Nebula. These magnificent, pillar-like structures are vast columns of gas and dust in the Eagle Nebula, within which new stars have recently formed. The pillars protrude from the interior wall of a dark molecular cloud like stalagmites from the floor of a cavern. The tallest of the pillars (at left) is about one light year in length from base to tip. The Eagle Nebula is a star-forming region 7000 light years away in the constellation Serpens.

The pillars are in some ways akin to buttes in the desert, where basalt and other dense rock have protected a region from erosion while the surrounding landscape has been worn away. In this celestial setting, especially dense clouds of dust and gas have survived longer than their surroundings as they are flooded by intense ultraviolet light from nearby hot, massive, newborn stars in an erosive process called photoevaporation. The ultraviolet light is also responsible for illuminating the convoluted surfaces of the pillars and the ghostly streamers of gas boiling away from their surfaces.

As the pillars are slowly eroded by the ultraviolet light, small globules of even denser gas and dust buried within the pillars are uncovered. These globules have been named evaporating gaseous globules (EGGs). The term describes their nature, for forming inside some of the EGGs are embryonic stars, which abruptly stop growing when the EGGs are uncovered and separated from the larger reservoir of gas and dust from which they were drawing mass. Eventually, the new-born stars emerge as the EGGs themselves succumb to photoevaporation.

This image has been constructed from three separate WFPC2 exposures taken in the light of emissions from different atomic constituents: red shows emission from singly ionized sulfur atoms, green is emission from atomic hydrogen, and blue shows emission from doubly ionized oxygen atoms. (*J. Hester and P. Scowen (Arizona State University), and NASA.*)

Fig. 4.   Artist's conception of the formation of planetesimals within a photoplanetary disk. (*Pat Rawlilngs, for the Jet Propulsion Laboratory.*)

stars, planets, and interstellar space. See also **Supernova 1987A and 1993J.**

Deep in cold, interstellar space, elements such as carbon, oxygen, and nitrogen can combine with primordial hydrogen to form complex molecules, particularly in dense condensations of gas called molecular clouds, where collisions between gas atoms and dust grains are possible. A large number of complex molecules, particularly those involving carbon atoms, have been detected in interstellar space.

Planetary System Formation

Photoplanetary Disks. Forming planetary systems may appear as dark or luminous disks silhouetted against a glowing nebula. Others, deeply embedded in their natal clouds, can be seen only in infrared light. Still others show knots of material with long, comet-like streamers formed as an interstellar wind blows the area clear.

These protoplanetary regions are up to 20 times the diameters of our solar system. All the material in a protoplanetary disk spins in the same direction around the star. The disk includes the complex molecules found in the original nebula, plus others that may form as the density and temperature change in the dense regions of the nebula surrounding the star.

Planetesimals Form and Photoplanets Condense. Within the spinning protoplanetary disk, gravity allows clumps to form and grow—creating objects called planetesimals. Heavy metals and silicates can survive the winds and high temperatures found close to the star, but lighter, volatile materials such as water and hydrogen gas survive only in the outer parts of the disks.

Clumps of solid material begin to solidify as they accumulate enough mass; growing larger as the result of collisions. Eventually, a few large objects—protoplanets—will begin to dominate the nebula, accreting more and more of the nebular material as the amount of free dust and gas is steadily reduced. See Fig. 4.

Forming Jupiter- and Earth-like Planets

Very subtle differences in the way mass is distributed in the protoplanetary disk will determine where planets will form and how big they will be.

The rocky and metallic planetesimals in the inner solar system form into Earth-like planets with molten interiors. As these Earth-like planets radiate the heat of compression into space, they form hard crusts. Over long periods of time, they may become solid all the way through. Bombardment by rocky and icy planetesimals disrupts the surface but also delivers elements and molecules, including, most critically from the standpoint of the evolution of life, water.

Predominantly icy objects in the outer solar system form Jupiter-like planets. With or without a rock and metal core, these planets are mostly

liquid surrounded by thick gaseous layers; the composition of a Jupiter-like planet is similar to that of its star. These planets, too, are subject to frequent impact by icy and rocky objects.

Chemistry of Life

Found in interstellar space, and therefore in planet-forming nebulae, are complex carbon molecules and amino acids—the building blocks of life. The universe appears to be very well populated with the raw materials to manufacture the deoxyribonucleic (DNA) molecule, the blueprint of all life on Earth. However, the method for ensuring that all of the right components find each other in the right quantities and under the right circumstances has yet to be identified. The fact that it happened once, and with such prolific consequences, suggests that in the chain of events described here, the opportunity exists for repeated occurrences of life elsewhere in the universe.

See also **Astrobiology**; **Hubble Space Telescope (HST)**; and **Origins Program**.

Web References

History Of The Universe: *http://www.historyoftheuniverse.com/*
NASA/Jet Propulsion Laboratory: *http://origins.jpl.nasa.gov/science/science.html*

NASA/Jet Propulsion Laboratory, Pasadena, CA.

UNSYMMETRICAL BENDING.   The condition that exists at any cross section of a flexural member, (see also **Flexure**) when the plane of the loads contains the shear center, but does not coincide with either of the two principal planes of bending, is known as unsymmetrical bending. Under these conditions the flexure formula is not applicable because the neutral axis is not perpendicular to the plane of the loads although it does pass through the center of gravity of the cross section.

The flexural stress due to unsymmetrical bending is found from the formula

$$s = \frac{M \cos\theta y}{I_x} + \frac{M \sin\theta x}{I_y}$$

where
$s =$ Flexural unit stress at a point.
$M =$ Bending moment.
$x, y =$ Coordinates of the point taken with respect to the principal axes.
$I_x, I_y =$ Moments of inertia referred to the principal axes.
$\theta =$ Angle which the plane of the loads, and hence the plane of the bending moment, makes with the y-axis.

It should be noted that this formula, which is applicable if the stress does not exceed the proportional limit, may be obtained by resolving the bending moment into components parallel to the two axes. The flexure formula is then applied to each of the principal axes and the results combined by the theory of superposition. The simplest way to determine the sign of the resultant stress (+ for tension and − for compression) is to note the sign for bending about each axis and use this sign when combining the results in the formula.

UPLIFT (Hydraulic). If water from a reservoir should work its way between the base of a dam and its foundation, it would exert a pressure upwards against the base of the dam. It would, in the extreme case, equal the full hydrostatic head corresponding to the height of the water surface above the base of the dam. This water pressure is known as uplift, and could possibly overturn a gravity type dam if it were not prevented, or allowed for in the design. Sometimes $\frac{2}{3}$ of the static head is assumed to be acting as uplift. Uplift might be considered to be maximum at the upstream edge of the base, decreasing from that to zero or to a pressure corresponding to the pool elevation at the downstream edge. Where measures are taken to prevent uplift, the foundation must be very thoroughly grouted, cutoff walls must be let into the foundation near the upstream edge of the base, and drains provided to relieve any pressure which might be built up by a slow seepage.

UPPER AIR OBSERVATIONS. To some extent, the practical application of weather satellites has altered the former full dependence upon rockets, balloons, and high-flying aircraft for observations of the meteorological conditions existing in the upper air. However, rockets and balloons still play an important role. Balloons are released at regular intervals by a number of weather stations, many of which are equipped with radiosonde instrumentation. See also **Radiosonde**; and **Weather Technology**.

Meteorological rockets are designed primarily for routine upper-air observation (as opposed to research) in the portion of the atmosphere that is inaccessible to balloons, from about 20 to 60 miles (32 to 96 kilometers) altitude. Upper-air measurements by sounding rocket are made, in the main, by the use of three techniques: (1) Rocket grenades are ejected and exploded at ascending altitudes. Atmospheric temperature and wind characteristics at the successive altitudes are determined from the time and distance data gathered by means of microphones, optical instruments, and radar equipment on the ground. (2) Tracer elements, such as sodium, are ejected into the air at high altitudes to form luminous clouds or trails, from which observations are made of wind direction and velocity, and wind shear effects. (3) A pitot-static tube, a modification of the pitot tube, is carried at the tip of a sounding rocket. Reactions of various sensitive parts of the tube to atmospheric pressures, ionization, and radiation are telemetered to the ground, and these data are used to calculate atmospheric pressure and density at various altitudes.

UPSIDE-DOWN CATFISH. See **Catfishes**.

UPSILON PARTICLE. As of 1977, when the upsilon particle was discovered at the Fermi National Accelerator Laboratory, the particle was the heaviest to be identified. Discovery of upsilon prompted physicists to introduce a massive new quark, raising the number of quarks from four to five (but probably six). The upsilon has a mass three times greater than any subatomic entity previously identified. It was discovered in energetic collisions between protons and copper nuclei. With a mass at its lower energy state equivalent to 9.0 GeV and masses in excited states equivalent to 10 and 10.4 GeV, the upsilon particle has been interpreted by scientists as consisting of a massive new quark (fifth) bound to its antiquark. The experiment was later reinforced by research at the Deutsches Elektronen-Synchrotron (DESY), located near Hamburg. At Fermilab, the excited upsilon particle appeared as a resonance in the yield of muons generated in collisions between protons and nuclei. A discussion of the upsilon experiment is described by a principal scientist of the project, L.M. Lederman (*Sci. Amer.*, **239**(4), 72–80, 1978). See also **Particles (Subatomic)**.

URACIL. See **Cell (Biology)**; and **Nucleic Acids and Nucleoproteins**.

URALITE. A metamorphic mineral. It is well established that pyroxene rocks may be metamorphosed into hornblende rocks. If the hornblende thus produced is fibrous and retains the original form of the pyroxene, it is called uralite, and the process by which the change is brought about chemical process which in many cases is accompanied by the generation chemical process which in many cases is accompanied by the generation of new minerals such as calcite, epidote, and magnetite. Uralite was first observed in rocks from the Ural Mountains, hence its name. See also **Hornblende**.

URANINITE. A mineral approximating the composition UO_2, but containing besides the higher oxide of uranium, UO_3, and oxides of lead, thorium, and rare earths. The uraninite usually occurs as cubic or cubo-octahedral crystals of specific gravity 7.5–10; when in masses of pitchy luster it is called pitchblende, specific gravity 6.5–9. All uraninites and pitchblende contain a minute amount of radium. It was in pitchblende obtained from the Joachimsthal in Czechoslovakia that Mme Curie discovered radium. Other localities for uraninite are in Saxony, Rumania, Norway, Cornwall, East Africa, and in the United States in the pegmatites of Connecticut Grafton Center, New Hampshire, North Carolina, and South Dakota, and in Gilpin County, Colorado. An important occurrence of pitchblende is at Great Bear Lake, Northwest Territories, Canada, where it has been found in large quantities associated with silver.

URANIUM. Chemical element symbol, U, at. no. 92, at. wt. 238.03, periodic table group (Actinides), mp 1,131 to 1.133 °C, bp 3,818 °C, density 18.9 g/cm³ (20 °C). Uranium metal is found in three allotropic forms: (1) *alpha phase*, stable below 668 °C, orthorhombic; (2) *beta phase*, existing between 668 and 774 °C, tetragonal; and (3) *gamma phase*, above 774 °C, body-centered cubic crystal structure. The gamma phase behaves most nearly that of a true metal. The alpha phase has several nonmetallic features in its crystallography. The beta phase is brittle. See also **Chemical Elements**.

Prior to the production of artificially-created elements, uranium was the highest in terms of atomic number and atomic weight. It was difficult to locate uranium in the periodic table of the elements, although chemically uranium resembles the elements of group 6b, namely, chromium, molybdenum, and tungsten. Subsequent to the production of the transuranium elements (atomic numbers 93 through 103), these elements, along with actinium (89), thorium (90), protactinium (91), and uranium (92) have been placed into the Actinide group of transition elements. They are similar in their mutual relations to the rare-earth group lanthanum (57) to lutetium (71). See also **Periodic Table of the Elements**.

Earthly abundance of uranium will be described shortly. In terms of presence in seawater, no significant concentrations have been reported. In terms of cosmic abundance, uranium also is very scarce. The study by Harold C. Urey (1952), in which silicon was given a base figure of 10,000, the concentration of uranium was represented by a figure of 0.0002.

Uranium is a white metal, ductile, malleable, and capable of taking a high polish, but tarnishes readily on exposure to the atmosphere. Finely divided uranium burns upon exposure to air, and the compact metal burns when heated in air at 170 °C. Uranium metal slowly decomposes water at ordinary temperatures and rapidly at 100 °C; is soluble in HCl and in HNO_3; and is unattacked by alkalis. Chemically related to chromium, molybdenum, and tungsten; and, like thorium, is radioactive. In the radioactive decomposition radium is formed. Discovered by Klaproth in 1789.

The element uranium found in nature consists of the three isotopes of mass numbers 238, 235, 234 with relative abundances 99.28, 0.71, and 0.006%, respectively.

The isotope ^{238}U is the parent of the natural uranium $4n + 2$ radioactive series, and the isotope ^{235}U is the parent of the natural actinium $4n + 3$ radioactive series.

The isotope ^{235}U has great importance because it undergoes the nuclear fission reaction with slow neutrons, and it has been separated in substantial amounts in nearly 100% isotopic composition.

Electronic configuration is $1s^2 2s^2 2p^6 3s^2 3p^6 3d^{10} 4s^2 4p^6 4d^{10} 4f^{14} 5s^2 5p^6 5d^{10} 5f^3 6s^2 6p^6 6d^1 7s^2$. Ionic radii U^{4+} 0.89 Å; U^{3+} 1.04 Å (Zachariasen). Metallic radius 1.4318 Å (805 °C). Oxidation potential $U + 2H_2O \rightarrow UO_2^{2+} + 4H^+ + 6e^-$, 0.82 V.

Uranium Reserves

Uranium has been known to be a distinct element since 1789. Apart from the small amount of its salts used in yellow pottery glazes, however, it remained more or less a laboratory curiosity until the 1920s. Then, the treatment of uranium ore, for the recovery of its radium (for the treatment of cancer), began in Eastern Europe and Zaire, followed by Canada in 1933. The separated uranium was mostly stockpiled or discarded.

After the development and successful explosion of the atomic bomb toward the end of World War II, an urgent search for workable uranium deposits was set in motion all over the world. The only high-grade deposits known to the western world were those in the countries just named as radium sources, but in view of the limited demand previously, serious exploration for uranium had never been undertaken. However, the offer of contracts by the U.S. Atomic Energy Commission, for fixed quantities at stated prices stimulated exploration for this hitherto largely ignored material.

Uranium is rather widely distributed throughout the world. See Table 1. Deposits vary markedly in richness.

TABLE 1. TYPES OF NATURAL URANIUM RESOURCES

Type of Resource	Ore Grade (Ppm Uranium)	Principal Known Locations
Vein deposits	10,000–30,000	Canada (near Great Bear Lake in the Northwest Territory) Western United States France Germany Russia Africa Australia China
Vein deposits (pegmatites, unconformity deposits)	2,000–10,000	Canada (Saskatchewan) Russia. Australia
Fossil placers, sandstones	200–2,000	Canada (Ontario) Western United States Brazil Chile Russia Japan Australia Africa
Shales, phosphates	10–100	United States (Florida) Morocco Sweden Russia
Pegmatites, other igneous and metamorphic	1–10	Canada (Ontario) deposits Greenland Brazil Spain Russia India Africa Australia

Uranium Reserves. During the past few decades, the construction of new nuclear power reactors in the United States has been limited, although this does not hold for France and some other countries. As with any nonrenewable fuel, there is concern over some ultimate date in the future when the supply may approach exhaustion. As of the early 1990s, with a new period of calm prevailing pertaining to the need for construction of nuclear weaponry, forecasts of useable uranium reserves made in earlier years no longer hold, and revised forecasts are lacking. In terms of current usage of uranium, there are several avenues available for conserving the fuel.

(1) *Improvement of uranium efficiency in thermal reactors* which consume more fissionable material than they breed. If it is assumed that on average a typical light-water reactor (LWR) is operated at 75% total capacity (load factor = 0.75), it will consume about 6000 tons of U_3O_8 per gigawatt of electrical output GWe over an expected 30-year reactor life. During this period, the reactor will produce about 5 tons of fissile plutonium, which is discharged in the spent fuel. In the past, minimal consumption of uranium has not been a major design goal. Improvements in reactors (not requiring major alterations) could effect savings of from 10 to 15% in uranium consumption even without considering fuel recycling. For example, simply enriching the level of U-235 from 3 to 4.2% would allow the fuel to remain in the reactor for 5 years instead of the usual 3 years. As pointed out by Hafenmeister (California Polytechnic University), the longer residence time and the higher enrichment would allow a greater fraction of the U-235 to be used, increasing the in situ generation and burning of plutonium, while at the same time reducing the discharge of plutonium by 30% (from 5 to 3.5 tons over lifetime of the reactor). If the burn-up of fuel is increased from the traditional 30,000 to 50,000 megawatt-days per ton, the level of U-235 in the discharged fuel is reduced from 0.85% to 0.71%, and refueling can be considered either on a 12- or an 18-month cycle.

(2) *Design changes in new reactors can conserve uranium.* Traditional LWRs use control poisons such as boric acid in the reactor coolant as a means to reduce reactivity. This practice results in a waste of from 5 to 10% of available neutrons. Newer designs which would allow faster and more frequent refueling could reduce the need for such poisons and consequent loss of neutrons. Such changes could result in a saving of some 25% of the fuel required.

(3) *Reprocessing of spent fuel and blanket materials*, with the recovery of purified uranium and plutonium, could effect another fuel savings of nearly 20%. Such reprocessing not only would conserve the U_3O_8 supply, but would also alleviate a severe nuclear waste problem. As described in the entry on **Nuclear Power Technology**, the spent-fuel storage facilities at existing nuclear power facilities already have reached or are rapidly approaching their full capacity, and new storage systems must be developed. Processes for spent-fuel recovery are described in the next section of this article. The principal deterrent to reprocessing is concern with "nuclear proliferation." In the most straightforward reprocessing scheme, plutonium is recovered along with uranium. Plutonium is the primary ingredient of nuclear weapons. Contemporary nuclear power plants operating in a number of countries do not yield plutonium in a form useful for weapons production, whereas a fuel reprocessing scheme would.

(4) *Extending uranium supplies by using thorium in LWRs.* Instead of using boron as a poison in the coolant, 40% heavy water (D_2O) could be used along with thorium. Hafemeister observes that the heavier D_2O in this "spectral shift control reactor" would result in breeding and then burning more plutonium from the fertile U-238. It is estimated that the savings would be about 12% of the U_3O_8 when compared to the LWR on a conventional uranium cycle; and about 20% of the U-233 required for a LWR on a thorium cycle. The Canada Deuterium Uranium (CANDU) power reactors are described under "Heavy Water Reactor" in the entry on **Nuclear Power Technology**. This means of uranium conservation, of course, would require large capital expenditures.

(5) *Fast breeder reactors.* These reactors (FBRs) produce more fissionable material during operation than is originally furnished. A number of nations are interested in the FBR as a means of extending their available uranium. Plutonium is more valuable in an FBR because it raises breeding ratios by 10 to 20% when compared with U-235. But, as pointed out in the entry on **Nuclear Power Technology**, the FBR essentially remains in the design and development stage. In terms of nuclear proliferation, uranium has an advantage over plutonium because isotopic enrichment is necessary to obtain weapons-usable material from a fuel containing both U-233 and U-238, whereas only chemical separations are required to purify plutonium from a fuel containing plutonium and uranium.

Uranium Spent-Fuel Reprocessing

In the PUREX process, the spent fuel and blanket materials are dissolved in nitric acid to form nitrates of plutonium and uranium. These are separated chemically from the other fission products, including the highly radioactive actinides, and then the two nitrates are separated into two streams of partially purified plutonium and uranium. Additional processing will yield whatever purity of the two elements is desired. The process yields purified plutonium, purified uranium, and high-level wastes. See also "Radioactive Wastes" in the entry on **Nuclear Power Technology**. Because of the yield of purified plutonium, the PUREX process is most undesirable from a nuclear weapons proliferation standpoint.

In a modified PUREX process (sometimes called coprocessing), the plutonium and uranium are not handled separately as their nitrates, but rather they are processed together, yielding plutonium diluted with uranium. The diluted product is useful for nuclear reactors, but not directly usable for nuclear weapons.

In the CIVEX fuel reprocessing system, most of the uranium is separated and purified from the spent nuclear reactor fuel, while some of the fission fragments, some of the uranium, and all of the plutonium are handled together. The system thus yields a fuel containing some of all three materials. The fission products render the fuel unsafe for any but sophisticated handling for a period of at least one year. The plutonium does not exist in high concentrations. Hafemeister (1979) observes that the costs of remote fabrication for CIVEX make it potentially attractive only for nations with large breeder reactor programs. In addition, the fission products in CIVEX fuels would, to some extent, act as poisons in thermal reactors, reducing the flow of neutrons.

Processing of Uranium Ores

Preconcentration of uranium ores by methods based on gravity, magnetic, or electrical properties, or by flotation have not been generally successful because of the softness of the raw materials and the absence of differing physical properties among the constituents. Thus, treatment has involved the whole ore wherein an extractant has been used. The uranium is recovered from solution after appropriate solid-liquid separations. Either an acid or alkaline extractant can be used. The extraction is enhanced by use of fine grinding of the ore, by increasing the concentration of extractant, and by using higher temperatures and oxidizing agents.

The uranium can be separated from the inert material by solid-liquid separation, filtration, or countercurrent washing. The uranium is recovered from solution by treatment with ion-exchange resin or solvent extraction. Alkaline leaching predominates over the acid process. The process is illustrated and briefly described in the Fig. 1. The uranium concentrate produced requires additional refining before fabrication into a fuel for nuclear reactors. Generally, this is accomplished by dissolving the product in HNO_3, filtering, and treating the uranyl nitrate solution by solvent extraction methods.

Chemistry of Uranium

Uranium has the four oxidation states, (III), (IV), (V), and (VI); the ions in aqueous solution are usually represented as U^{3+}, U^{4+}, UO_2^+, and UO_2^{2+}. The oxidation-reduction scheme, on the hydrogen scale (in which the potential for $\frac{1}{2}H_2 \rightarrow H^+$ is taken as zero) is indicated in Fig. 2.

The UO_2^+ ion is unstable in solution and undergoes disproportionation to U^{4+} and UO_2^{2+}. A few solid compounds of this oxidation state are known, as for example, UF_5 and UCl_5.

The ion U^{3+} forms intense red solutions in H_2O and is oxidized by water at an appreciable rate. The rate of oxidation appears to increase with

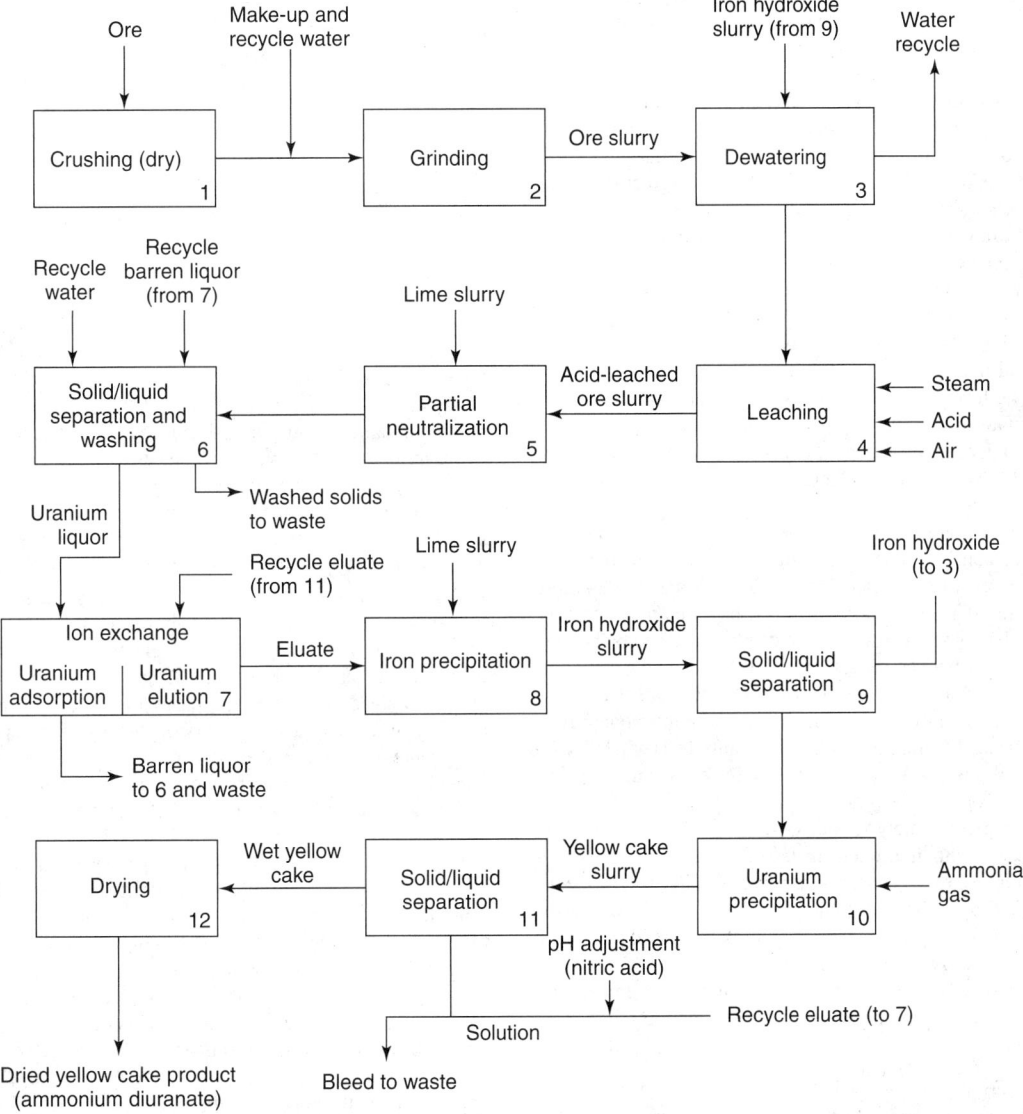

Fig. 1. Schematic flowsheet of uranium processing (acid leach and ion exchange) operation. Numbers refer to the numbers that appear in the boxes on the flowsheet. Operations (3), (6), (9), and (11) may be done by thickening or filtration. Most often, thickeners are used, followed by filters. The pH of the leach slurry (4) is elevated to reduce its corrosive effect and to improve the ion-exchange operation on the uranium liquor subsequently separated. In the ion exchange operation (7), resin contained in closed columns is alternately loaded with uranium and then eluted. The resin adsorbs the complex anions, such as $UO_2(SO_4)_3^{4-}$, in which the uranium is present in the leach solution. Ammonium nitrate is used for elution, obtained by recycling the uranium filtrate liquor after pH adjustment. Iron adsorbed with the uranium is eluted with it. Iron separation operation (8) is needed inasmuch as the iron hydroxide slurry is heavily contaminated with calcium sulfate and coprecipitated uranium salts. Therefore, the slurry is recycled to the watering stage (3). Washed solids from (6), the waste barren liquor from (7), and the uranium filtrate from (11) are combined. The pH is elevated to 7.5 by adding lime slurry before the mixture is pumped to the tailings disposal area. (*Rio Algom Mines Limited, Toronto*.)

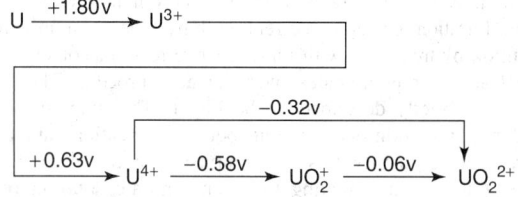

Fig. 2.   Oxidation-reduction potentials of uranium ions (in 1-molar hydrochloric acid).

increasing ionic strength, although concentrated solutions are said to be stabilized by strong acids such as hydrochloric. Solutions of uranium(IV) are green, and uranium(VI) solutions, yellow.

The (IV) and (VI) are the important oxidation states and therefore the more important phases of the chemistry of uranium may be related to the two oxides UO_2 and UO_3, uranium dioxide and uranium trioxide. A series of salts such as the chloride and sulfate, UCl_4 and $U(SO_4)_2 \cdot 9H_2O$ is obtained from UO_2. The more common uranyl salts as $UO_2(NO_3)_2 \cdot 6H_2O$, UO_2Cl_2, and $UO_2SO \cdot nH_2O$ in which the UO_2^{2+} (uranyl ion) acts as a radical, are derived from UO_3. UO_3 is amphiprotic and forms a series of alkali and double alkali uranates and polyuranates of limited solubility, such as $Na_2U_2O_7$, $NaZn(UO_2)_3(C_2H_3O_2)_9 \cdot 6H_2O$, $NaMg(UO)_2)_3(C_2H_3O_2)_9 \cdot 6H_2O$, etc.

The element uranium also exhibits a formal oxidation number of (II) in a few solid compounds, semimetallic in nature, such as UO and US. No simple uranium ions of oxidation state (II) are known in solution.

In addition to three oxides, UO_2 (brown, cubic), U_3O_8 (greenish black, orthorhombic) and UO_3 (orange, hexagonal), which have been known for a long time, there are known to exist the monoxide, UO, and the pentoxide U_2O_5. There is also some evidence for the existence of U_4O_7 and U_6O_{17}. The phase relationships in the uranium-oxygen system are very complex because solid solutions are readily formed, so that it is possible to obtain uranium "oxides" with practically any composition intermediate between UO and UO_3, and with many crystal structures.

Uranyl peroxide, the formula of which is usually given as $UO_4 \cdot 2H_2O$, is formed by precipitation from solutions of uranyl nitrate by hydrogen peroxide. Alkali hydroxides, hydrogen peroxide, and sodium peroxide form soluble peroxyuranates, $Na_2UO_6 \cdot 4H_2O$ and $Na_4UO_8 \cdot 8H_2O$, when added to solutions of uranyl salts.

Two uranium carbides are known, the monocarbide, UC, and the dicarbide, UC_2. These can be prepared by direct reaction of carbon with molten uranium, or by reaction of carbon monoxide with metallic uranium at elevated temperatures. The sesquicarbide, U_2C_3, has been found to exist as a stable compound below about $1800\,°C$ and can be produced by heating a mixture of UC and UC_2 between 1,250 and $1,800\,°C$.

Uranium and nitrogen form an extensive series of compounds that can be prepared by direct action of nitrogen on the metal. Uranium mononitride, UN, is the lowest nitride of uranium. If the mononitride is treated with more nitrogen at atmospheric pressure, U_2N_3 is formed. With nitrogen under high pressure UN_2 can be prepared, but it is difficult to obtain samples of UN_2 that are completely free of UN.

Uranium metal reacts with hydrogen at $250-300\,°C$ to form a well-defined hydride, which resembles the rare-earth hydrides in many respects. The formula of this substance has been shown to be $UH_{3.00}$. The hydride undergoes decomposition with increasing temperature; the dissociation pressure of UH_3 is one atmosphere at $436\,°C$.

Uranium tetrafluoride serves as a starting material for the preparation of the other fluorides. It is best prepared by hydrofluorination of uranium dioxide:

$$UO_2 + 4HF \xrightarrow{500\,°C} UF_4 + 2\,H_2O$$

Uranium trifluoride can be prepared by reduction of UF_4 with hydrogen at $1,000\,°C$. Uranium hexafluoride, UF_6, white and orthorhombic, is best obtained by direct fluorination of UF_4, green and monoclinic, although any uranium compound will yield UF_6 by reaction with fluorine at elevated temperatures:

$$UF_4 + F_2 \xrightarrow{350\,°C} UF_6$$

The hexafluoride can also be prepared by the interesting reaction:

$$2\,UF_4 + O_2 \xrightarrow{900\,°C} UF_6 + UO_2F_2$$

The intermediate fluorides $U_2F_9(UF_{4.5})$, $U_4F_{17}(UF_{4.25})$ and UF_5 are prepared by reaction of solid UF_4 and gaseous UF_6 under appropriate conditions of temperature and pressure.

Uranium hexafluoride is probably the most interesting of the uranium fluorides. Under ordinary conditions, it is a dense, white solid with a vapor pressure of about 120 mm at room temperature. It can readily be sublimed or distilled, and it is by far the most volatile uranium compound known. Despite its high molecular weight, gaseous UF_6 is almost a perfect gas, and many of the properties of the vapor can be predicted from kinetic theory.

Uranium tetrachloride can be prepared by direct combination of chlorine with uranium metal or hydride; it can also be obtained by chlorination of uranium oxides with carbon tetrachloride, phosgene, sulfur chloride, or other powerful chlorinating agents. The trichloride is obtained by reaction of UCl_4 with hydrogen and the higher chlorides by reaction of UCl_4 and Cl_2. Uranium hexachloride, UCl_6 is a rather volatile, somewhat unstable substance. All of the uranium chlorides dissolve in or react readily with water to give solutions in which the oxidation state of the ion corresponds to that in the solid. All of the solid chlorides are sensitive to moisture and air.

The trichloride, tribromide and triiodide of uranium are obtained either by reaction of the elements or by treatment of UH_3 with the appropriate halogen acid. The thermal stability of the halides decreases as the atomic number of the halogen increases. No higher uranium bromides or iodides are known.

A series of oxyhalides of the type UO_2F_2, $UOCl_2$, UO_2Br_2, etc., are known. They are all water-soluble substances which become increasingly less stable in going from the oxyfluoride to the oxyiodide.

Uranyl ion forms complexes with many oxy anions. Both U(VI) and U(IV) compounds dissolve in alkali carbonate solutions with formation of carbonato complexes. Those of the larger alkali cations are only slightly soluble: $K_{sp} = 6 \times 10^{-5}$ for both $K_4[UO_2(CO_3)_3]$ and $(NH_4)_4[UO_2(CO_3)_3] \cdot 2H_2O$.

Aqueous solutions of uranium(III), uranium(IV), and uranium(VI) are readily obtained. Solutions of uranium(III) are blood-red in appearance; hydrogen is slowly evolved with the formation of uranium(IV) is a strong reducing agent and is easily oxidized to uranyl ion by oxygen, peroxide, and numerous other oxidizing agents. Uranyl solutions in turn may be reduced to uranium(IV) with sodium dithionite, zinc or cadmium amalgams, or by electrochemical or photochemical means.

Separation of Isotopes. Several methods are available for the separation of isotopes, including gaseous diffusion, centrifugation, electromagnetic methods, thermal diffusion, electrolytic methods, distillation, and chemical-exchange methods. In the late 1960s, another, radically different process was added to the technology of isotope separation. This is known as laser enrichment and is described shortly. The separation of ^{235}U from ^{238}U represented the first large-scale isotope-separation operation and, after considerable study, the principal plant utilized gaseous diffusion.

The gaseous diffusion method of isotope separation is based upon the difference in the rate of diffusion of gases that differ in density. Since the rate of diffusion of a gas is inversely proportionate to the square root of its density, the lighter of two gases will diffuse more rapidly than the heavier. Therefore, the result of a partial diffusion process will be an enrichment of the partial product in the lighter component.

To separate isotopes by this process, they must be in the gaseous form. Therefore, the separation of isotopes of uranium required the conversion of the metallic uranium into a gaseous compound, for which purpose the hexafluoride, UF_6, was chosen. Since the atomic weight of fluorine is 19, the molecular weight of the hexafluoride of ^{235}U is $235 + (6 \times 19) = 349$, and the molecular weight of the hexafluoride of ^{238}U is $238 + (6 \times 19) = 352$. Since the rate of diffusion of a gas is inversely proportional to the square root of its density (mass per unit volume), the maximum separation factor for one diffusion process of the uranium isotopes is $\sqrt{352/349} = 1.0043$. Since only part of the gas can be allowed to diffuse, the actual separation factor is even less than this theoretical maximum.

From this small figure, it is apparent that many diffusion stages are necessary in the separation of ^{235}U from ^{238}U. The number originally calculated for the Oak Ridge plant was about 4,000. Other reasons are the small apertures demanded by diffusion processes (in this case less than. 00001 centimeter in diameter), which reduce the rate of gas flow and demand a great barrier area for appreciable production.

The **centrifugal method of isotope separation** consists essentially of the passage of the mixture through a rapidly rotating force field, such as that of a rotating cylinder. If a current of mixed gases is passed into such a cylinder, moving parallel to the axis of rotation, the lighter gas will tend to concentrate near the axis, and the heavier gas, near the periphery. This is the principle of the cream separator; its successful application to separation of isotopes in the gaseous phase requires apparatus operating at very high speeds of rotation.

The **electromagnetic method of isotope separation** is based upon the principle of the mass spectrograph. As in that apparatus, a stream of charged particles is passed through a system of electric and magnetic fields. If the particles are ions of two or more isotopes of the same element, all bearing the same charge, the deflections produced by the fields will vary with the masses of the particles, and will thus provide a means for their separation. This method is especially effective for the separation of particles of a number of masses, and has been widely used for that purpose in research studies and in production-separation operations. The method is also used extensively in a number of research laboratories, particularly those of northern Europe, for the isotopic separation of individual radioactive nuclides that are to be used as sources in instruments, such as beta- and gamma-ray spectrometers, in which measurements are made of the characteristics of ionizing radiations.

The **thermal diffusion method of isotope separation** has broad application to liquid-phase as well as gaseous-phase separations. The apparatus widely used for this purpose consists of a vertical tube provided with an electrically heated central wire. The gaseous or liquid mixture containing the isotopes to be separated is placed in the tube, and heated by means of the wire. In such an apparatus two effects act to separate the isotopes. Thermal diffusion tends to concentrate the heavier isotopes in the cooler outer portions of the system, while the portions near the hot wire are enriched in the lighter isotopes. At the same time, thermal convection causes the hotter fluid near the hot wire to rise, while the cooler fluid in the outer portions of the system tends to fall. The overall result of these two effects causes the heavier isotopes to collect at the bottom of the tube and the lighter at the top, whereby both fractions may be withdrawn.

The **electrolytic method of isotope separation** is of importance not only because of its present day uses, but also because of its historical interest. It was by this method that G.N. Lewis and his co-workers at the University of California obtained practically pure deuterium. Since deuterium oxide had been shown to be present in ordinary water, the conclusion was drawn that water (or rather the dilute aqueous solution) from electrolytic cells used for the production of hydrogen and oxygen by continuous electrolysis of water, should be richer in the heavier isotope (deuterium having a mass number of 2, as against 1 for protium). Starting with such residual water from an electrolytic cell, it was found that by repeated electrolysis a small residue consisting almost entirely of deuterium oxide (D_2O) was obtained. This process is still used for the separation of pure hydrogen isotopes, as well as for other purposes.

The **distillation method of isotope separation** has also been the basis of important research contributions. In the work on the hydrogen isotopes, it preceded the electrolytic separation methods discussed above. Following the suggestion by Birge and Menzel, of the possible presence of deuterium in ordinary hydrogen to the extent of 1 part in about 4,500, Urey, Brickwedde and Murphy, in 1931, began their search for this isotope. By evaporating about 4,000 milliliters of liquid hydrogen to a volume of about 1 milliliter, Brickwedde obtained a residue that gave conclusive spectroscopic evidence of the presence of deuterium. The distillation method of separation has been responsible for many other research contributions. Among them may be mentioned separation of the isotopes of oxygen, of mercury, zinc, potassium and chlorine.

The **chemical exchange methods of isotope separation** are of value, not only for that purpose, but also because they provide a direct means for the study of chemical reactions. A well-known example of an isotopic chemical exchange is the heavy water equilibrium:

$$H_2 + D_2O \rightleftharpoons D_2 + H_2O$$

In this equation, the formula H_2 is used for the hydrogen isotope of mass number 1, which constitutes all but a small fraction of ordinary hydrogen; H_2O is the corresponding "light water"; D_2 is hydrogen of mass number 2 (deuterium) and D_2O is the corresponding heavy water. The double arrows indicate an equilibrium reaction, whereby under suitable conditions ordinary hydrogen reacts with heavy water to produce hydrogen of mass number 2 and light water. If one were to start either with the two reactants on the left of the arrows, or with the two on the right, the system at equilibrium would have all four present. However, in this system at equilibrium the reverse reaction predominates, so that the ratio of ^2H to ^1H (that is, the ratio of D_2O to H_2O) in the liquid phase is about three times as great as in the gas. Because of this differential reactivity, this method is useful in the separation of the two hydrogen isotopes.

Another equilibrium system is useful in the separation of ^{14}N and ^{15}N. It is represented by the equation:

$$^{15}NH_3 + {}^{14}NH_4NO_3 \rightleftharpoons {}^{14}NH_3 + {}^{15}NH_4NO_3$$

This exchange reaction is conducted by the countercurrent flow of ammonia gas and ammonium nitrate solution (in water). The forward reaction is favored, resulting in the concentration of the ^{15}N in the ammonium nitrate in solution. The multistate conduct of this reaction that is necessary for effective operation is accomplished by arranging later stages in which the enriched ammonium nitrate solution is divided into two parts. One part is treated with caustic soda to displace the enriched NH_3, which is then used in a second stage of the process with the other part of the NH_4NO_3 solution. Three or more stages may thus be used, until the desired concentration of ^{15}N has been effected.

Another method of isotope separation is by ion mobility, a process based on the difference in mobility of the ions in an electrolytic solution, under the influence of an electric field.

The **laser process for enrichment of uranium** dates back to the early 1970s. In addition to early and continuing development of this process at the Lawrence Livermore National Laboratory, pioneering efforts were made by Exxon Nuclear Corporation and Avco Corporation in early 1971. These efforts concentrated on an atomic vapor laser isotope separation approach. The process now in a reasonably late stage of development at the Lawrence laboratory is known by the acronum AVLIS (for atomic vapor laser isotope separation). The process differs radically from all other uranium enrichment approaches. The AVLIS processes uses a bank of very finely tuned lasers to create an electrical charge on uranium-235 atoms while leaving nonfissle uranium-238 atoms unchanged. Fundamentally, the process involves the firing of light from high-powered copper-vapor lasers into a stream of uranium atoms. Through tuning of the lasers, electrons will be stripped from some of the atoms, thus leaving positively charged uranium-235 ions. These are drawn to negatively charged plates. The uncharged uranium-238 atoms pass through the process unaffected. Advantages claimed for the process include a smaller capital investment to build and less costly to operate and maintain. Also, the construction of laser plants can be smaller and modular as compared with the former huge diffusion, centrifuge, and magnetic plants.

Additional Reading

Bothwell, R.: "Nucleus: The History of Atomic Energy Limited," University of Toronto, Toronto, Canada, 1988.

Golay, M.W. and N.E. Todreas: "Advanced Light-Water Reactors," *Sci. Amer.*, **82** (April 1990).

Golay, M.W.: "Longer Life for Nuclear Plants," *Technology Review (MIT)*, 25 (May/June 1990).

Goldschmidt, B.: "Atomic Rivals," Rutgers University Press, New Brunswick, NJ, 1990.

Grenwood, N.N. and A. Earnshaw: "Chemistry of the Elements," 2nd Edition, Butterworth-Heinemann, Inc., Woburn, MA, 1997.

Hafemeister, D.W.: "Nonproliferation and Alternative Nuclear Technologies," *Technology Review (MIT)*, **81**(3), 58–62 (1979).

Hotta, H.: "Recovery of Uranium from Seawater," Oceanus, 30 (Spring 1987).

Lewis, R.J. and N.I. Sax: "Sax'x Dangerous Properties of Industrial Materials," 10th Edition, John Wiley & Sons, Inc., New York, NY, 1999.

Lide, D.R.: "CRC Handbook of Chemistry and Physics 2000–2001," 81st Edition, CRC Press, LLC., Boca Raton, FL, 2000.

Marshall, E.: "Counting on New Nukes," *Science*, 1024 (March **2**, 1990).

Slovac, P., J.B. Flynn, and M. Layman: "Perceived Risk, Trust, and the Politics of Nuclear Waste," *Science*, 1603 (December 13, 1991).

Spinard, B.I.: "U.S. Nuclear Power in the Next Twenty Years," *Science*, 707 (December 12, 1988).

Suzuki, T.: "Japan's Nuclear Dilemma," *Technology Review (MIT)*, 41 (October 1991).

URANIUM FUEL.   See **Nuclear Power Technology**.

URANIUM SERIES.   See Radioactivity.

URANUS.   Seventh planet outward from the Sun, Uranus is estimated to have a diameter of about 4 times that of the Earth. To the naked eye, Uranus is barely visible as a sixth-magnitude star, but in a telescope of moderate aperture, the body appears as a disk with a bluish-green color. The latter is now known to be caused by methane in the outer layer of the planet. Methane absorbs in the red. Uranus, like Neptune, is intermediate between the terrestrial planets (Mercury, Venus, Earth, and Mars), which are made up mostly of metals and their oxides, and the Jovian planets that consist mostly of water, ammonia, and methane ices. See Fig. 1.

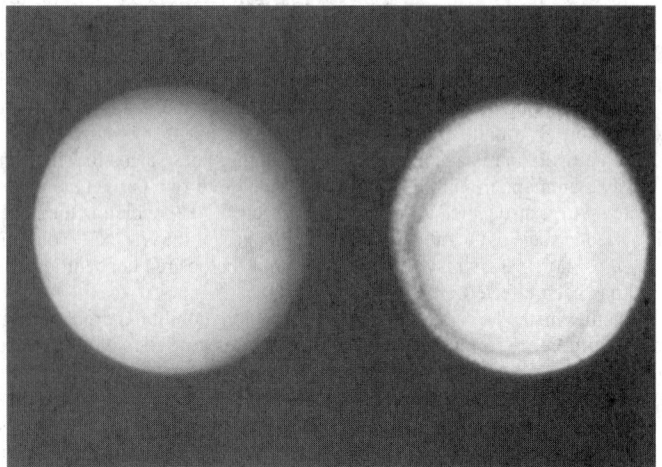

Fig. 1.   Views of Uranus made from *Voyager* 2 on January 17, 1986, when the spacecraft was about 9.2 million kilometers (5.7 million miles) from the planet. (*Left*) Black-and-white rendition of the original grennish-blue image of Uranus. (*Right*) Black-and-white replica of the false-color image, indicating features of the planet not obtainable without use of color filters. (*Jet Propulsion Laboratory, Pasadena, California.*)

The most distinctive feature of Uranus is its unusual rotational position, tipped over on its axis. Some scientists suggest that early in the history of the planet, a collision with another, smaller body may have tilted the planet from a vertical or near-vertical axis to its present orientation. In the 84-year long orbit of Uranus, one pole is in sunlight for 42 years while the other is in darkness. For many years, the length of a Uranian day had been estimated between 16 and 24 hours. Data from *Voyager* 2 (1986) confirmed a rotational period of 17.24 hours. However, features at different latitudes were found to circle at different rates, with periods ranging from 14 to 18 hours. The 17.24-hour period was determined by observing individual convective plumes of methane ice, using periodic radio emissions.

Uranus has 15 confirmed satellites, five of which are considered major. Ten additional satellites were discovered by *Voyager* 2, one during its approach to the planet in 1985, and the other nine moons found during the encounter. The orbital planes of these satellites, excepting Miranda, lie close to the plane of the planet's equator and hence are nearly perpendicular to the plane of the ecliptic. Further data on the satellites are given in this article.

Uranus in Perspective

Uranus was found accidentally by Herschel, in 1781, while sweeping the sky with a 7-inch reflecting telescope of his own manufacture. His discovery stirred up a tremendous amount of popular interest in astronomy, and history relates that during the weeks following the discovery, the streets in front of Herschel's house were crowded with people eager to get a view of the telescope and a glimpse of the discoverer. Herschel named the planet Georgium Sidus (star of the Georges) in honor of the then reigning king of England, George III, but the name was never adopted on the continent. Many Europeans called the planet Herschel, in honor of the discoverer, but the name Uranus, proposed by Bode, is the one that has survived.

Early ground observations showed that the planet is very much flattened at the poles, which provides evidence of a relatively high rotational speed. In 1912, Lowell and Slipher, using the Doppler principle, indicated a period of rotation of about 19.75 hours (now considered an inaccurate figure).

Several years ago, Dunham concluded the presence of large amounts of methane in the planet's atmosphere. On March 10, 1977, Uranus occulted a 9th-magnitude star and scientists at four observatories (Kuiper Airborne Observatory, Perth Observatory, South African Astronomical Observatory, and Kalvalur, India observes) photoelectrically recorded the star as it disappeared behind the planet and reappeared again. They also detected a series of fluctuations in the star's light. At first believed to be caused by moonlets, the final conclusion drawn was that the fluctuations were caused by the presence of rings. At a somewhat later date, additional rings were found, bringing the total to nine. The latter number was confirmed by *Voyager* 2 data. The finding of these rings erased Saturn's former distinction as being the only ringed planet in the solar system. Later, Neptune was also found to have rings.

During the late 1700s, when Herschel first noted a definite polar flattening of Uranus, he also suggested that there may be two rings at right angles to each other, and, in fact, after numerous observations, appeared somewhat convinced that a ring about the planet was the cause of a bulge at the equator. Herschel's ring observations were not taken seriously in the scientific community, which attributed the suspicion to inadequate instrumentation—that is, until the 1977 observations.

Voyager 2 Encounter with Uranus.   On January 24, 1986, *Voyager* 2, which previously had visited Jupiter on March 5, 1979 and was in the vicinity of Saturn on August 25, 1981, encountered Uranus. It should be pointed out that *Voyager* 2 later continued on to Neptune. The path of *Voyager* 2 is sketched in the article on Neptune (Fig. 1). Details of *Voyager* 2's flight path past Uranus are given in Fig. 2. The Uranus encounter is quite remarkable, considering that the spacecraft was launched in 1977. Interesting statistics pertaining to the Uranian encounter include:

- One-way light time, Earth to Uranus, is 2 hours, 44 minutes, 50 seconds.
- Distance of *Voyager* 2 from Earth at time of encounter, 2,965,400,000 km (1,842,610,000 mi).
- Closest distance of *Voyager* 2 to Uranus cloudtops, 81,500 km (50,600 mi)
  - to center of planet, 107,000 km (64,500 mi).
  - to satellite Miranda, 29,000 km (18,000 mi).
  - to satellite Ariel, 127,000 km (79,000 mi).
  - to satellite Umbriel, 325,000 km (202,000 mi).
  - to satellite Titania, 365,200 km (227,000 mi).
  - to satellite Oberon, 470,600 km (300,000 mi).
- Velocity of *Voyager* 2, Geocentric: 36 km/second (80,200 mph)
  Heliocentric: 22 km/second (49,000 mph)

The increasing distance between the spacecraft and tracking stations on Earth, as compared with earlier planetary encounters, required innovative telecommunications techniques. For example, at Jupiter, data rates of 115,200 bits/second were possible; at Saturn, the rate dropped to 44,800 bits/second. At Uranus, the spacecraft signal was considerably weaker, but the same signal, received at two or more antennas (different groupings of 34- and 64-meter antennas) at each of three deep space network complexes were combined in a technique known as *arraying*. This technique reinforced the strength of the signal received by electronically combining the same telemetry recorded at multiple stations. Thus, arraying made possible a data rate of 21,600 bits per second at Uranus. Without arraying, fewer than half the observations planned could have been performed and satisfactory data returned to Earth.

Image-motion compensation was required in connection with *Voyager* 2's camera equipment. The high velocity of the spacecraft as well as the light levels in the Uranian system required this special instrumentation to prevent image smear and assure good image resolution. Much engineering of this type was performed on the spacecraft by ground-based personnel. It should be pointed out that with Uranus being twice as far from the sun as Saturn (the previous *Voyager* 2 encounter), the light levels at Uranus were four times lower and the targets were inherently darker. It has been estimated that the sunlight received on the illuminated portions of Uranus do not exceed the light received by Earth during a total eclipse of the sun.

Voyager 2 investigations and instrumentation systems used to observe the Uranian system included: Imaging system; photopolarimetry; infrared interferometer spectroscopy; ultraviolet spectroscopy; radio science, magnetometry; plasma; low-energy charged particles; cosmic ray, planetary radio astronomy; and plasma waves.

Considerable information on the *Voyager* 2 spacecraft is given in separate article on *Voyager* **Missions to Jupiter and Saturn.**

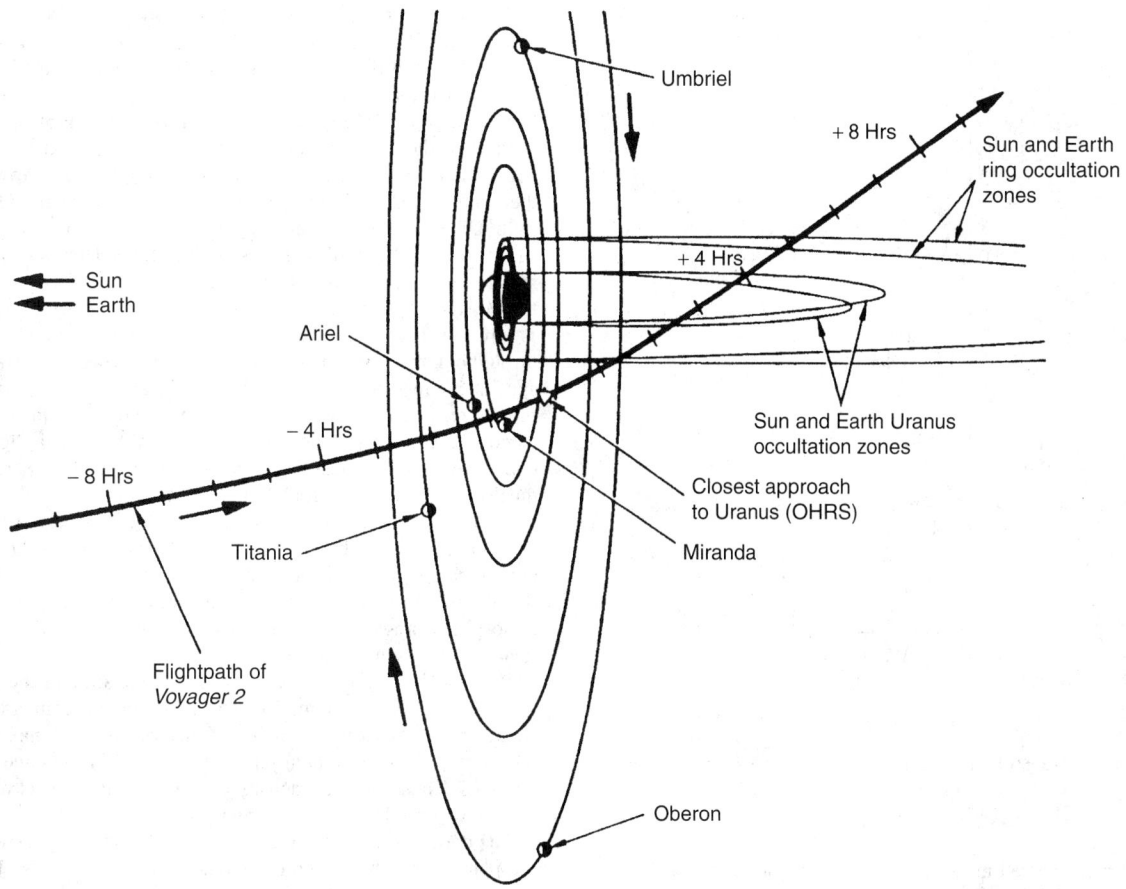

Fig. 2.   The flight path of *Voyager 2* past Uranus (January 24, 1986). This view is vertical to plane of flight path. (*Jet Propulsion Laboratory, Pasadena, California.*)

The Uranian Atmosphere

Several months after *Voyager 2*'s encounter with the Uranian system, scientists prepared preliminary summaries of the data received. The process of refining these data will continue for several years. Reporting on the planet's atmosphere, E.C. Stone (California Institute of Technology) and E.D. Miner (Jet Propulsion Laboratory) provided the following highlights:

Uranus apparently does not have a large internal heat source. Although the sun directly heats the polar regions, the atmosphere of Uranus resembles those of Jupiter and Saturn. See also **Jupiter**; and **Saturn**.

At a pressure of 600 mbar, temperatures at the planet's poles and the equator are close to the same. This indicates some means of dynamical redistribution of the solar energy received in the polar regions. (Slightly colder bands were found at about 25 °S and 40 °N, a condition contrasting with the findings for Jupiter and Saturn.)

At higher altitudes (pressure of 100 mbar), the temperature drops to a minimum of 52 ± 2 K. The temperature reaches a maximum in the extreme upper atmosphere, 750 K.

At an altitude of 2000 to 3500 km (1250 to 2185 mi) beyond the 100 mbar region, a multilayer ionosphere exists. It is estimated that this layer may extend to an altitude of 10,000 km (6,240 mi). A hydrogen corona extends from the planet outward beyond the rings. The corona has an estimated density of 100 cm^{-3} at the epsilon ring.

Helium presence in the atmosphere (range of 0.15 ± 0.05 mole fraction) is somewhat greater than that at Jupiter and Saturn, but consistent with solar helium abundance. The upper atmosphere also contains a trace of methane. (Methane absorbs in the red, thus giving the planet its blue-green color.)

Radio occultation information indicated a cloud deck of methane ice at pressures of 900 to 2300 mbar. These measurements indicated a methane mole fraction of about 2% deeper in the atmosphere. Although this value is some twenty times that of solar carbon abundance, it was not surprising because of the ice-rich material from which Uranus is presumed to have formed.

At about 50 °S, methane clouds are noted as band with a 700-km (440-mi) latitudinal scale. These clouds suggest discrete sources and little latitudinal diffusion. Convective plumes of methane ice having prograde

velocity of from 40 to 160 m sec^{-1} relative to the interior of the planet were noted. These winds flow in a direction opposite to that of the thermal winds that may be expected from the observed latitudinal temperature gradient. Thus, the initial conclusion is that other dynamical processes must be present.

At higher altitudes, intense ultraviolet light was found emitting almost uniformly from the sunlit hemisphere, this emission resulting from *electroglow*.[1]

Auroral emissions were noted on the dark side of Uranus. These emissions are produced by the excitation of H_2 by approximately 10-keV electrons and form an auroral region 15° to 20° in diameter about the magnetic pole.

As noted by Ingersoll, in a three-layer model of Uranus, melted ices form a liquid "ocean" between a rocky core and a gaseous hydrogen-helium atmosphere. *Voyager 2* data, however, suggest a two-layer model in which the gases and ices are mixed in a dense atmosphere. Near the visible surface layers of the atmosphere of Uranus, ammonia, ammonium hydrosulfide (NH$_4$HS) and water are believed by some researchers to condense to form icy clouds. The condensation sequence would be determined by the condensation temperatures of the constituent materials. Uranus is sufficiently cold to permit methane to condense above the other clouds. Thus the diagram of Fig. 3 is suggested.

The Uranian Ring System

The existence of rings around Uranus was not confirmed until 1977. The rings are dark, narrow, and dissimilar to the broad, bright rings of Saturn. Uranian rings, in some cases, are noncircular and vary in width. There are vast reaches of space between the rings. The rings of Uranus were discovered during a stellar occultation, in which the brightness of a star was being monitored as Uranus passed between the star and the Earth. The star involved in the 1977 observation was SAO 15867. The observers

[1] Electroglow is a phenomenon first noted on Jupiter and Saturn. The emission consists of both molecular and atomic hydrogen emissions.

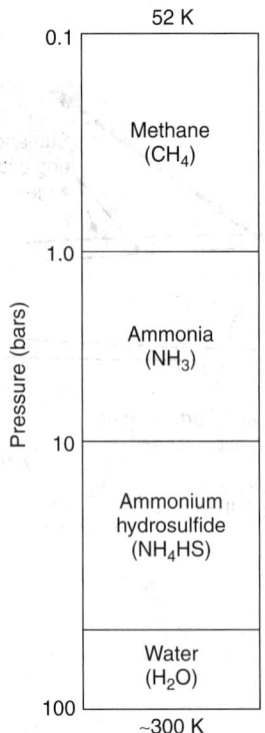

Fig. 3.   Estimated variation of the atmosphere of Uranus versus pressure (bars). (*After Ingersoll.*)

Two new rings were discovered by *Voyager 2*, identified as 1986UIR and 1986U2R. From ground-based occultation data, there may be other minor or partial rings (arcs), but these were not confirmed by *Voyager 2*.

From *Voyager* data analyses thus far (early 1988), the physical data pertaining to the Uranian rings are summarized in Table 1. See also Fig. 4.

Prior to the **Voyager 2** encounter, it was expected that as many as 18 small satellites (moons or shepherds) confining the narrow rings between them might be confirmed. However, only two such shepherd satellites (1986U7 and 1986U8) were found on either side of the epsilon ring. It is believed that shepherd satellites for the other Uranian rings may be too small (less than 14 km; 10 mi in diameter) and dark for observation by the *Voyager 2* instrumentation.

Voyager 2 data revealed quantities of microscopic dust widely spread throughout the rings. Cuzzi and Esposito observe that such dust is of special interest because tiny particles in and around planetary rings are rapidly removed as a result of erosion by micrometeoroids and radiation-belt electrons. The existence of the dust implies that a local, long-lived source consisting of more massive particles must exist. However, relatively little dust (less than 1%) was found in any of the main rings. The dustiest feature was found in one of the most prominent of the new rings (1986U1R). Researchers found the lack of dust in the main rings rather surprising, as compared to what would have been expected from extrapolations of Saturn ring data. As currently explained, the planet's considerably denser atmosphere drags microscopic dust particles out of their orbits within a matter of a few hundred years.

Voyager 2 instrumentation showed that the sizes of the largest particles in the rings average about 2 meters and thus are comparable to particles found in Saturn's rings. Unlike the material in Saturnian rings, however, the rings of Uranus contain relatively few millimeter- and centimeter-size particles. Thus, the mass density of the Uranian rings is markedly greater than, for example, Saturn's B ring.

The darkness and lack of color of the Uranian rings essentially matches the planet's moons. Several hypotheses concerning this observation have been proposed and described in the literature. In particular, see the Cuzzi/Esposito 1987 reference.

Satellites of Uranus

Prior to the *Voyager 2* encounter, studies of the Uranian satellites were confined to dynamical analyses and disk-integrated photometry, radiometry, and spectroscopy. The orbits of the five largest satellites are all of low inclination, with the exception of Miranda with an inclination 4 degrees. As described in the following paragraphs, while the Uranian satellites are similar in some respects, in other qualities they differ radically from one another. *Voyager 2* data requires a different model for each of the satellites. As observed by the *Voyager 2* imaging team (May 1986), caution must be exercised in interpreting the *Voyager* observations of the satellites as being representative of their entire surface. Because the subsolar points on Uranus and its satellites are currently close to their south poles, Voyager images covered only the southern hemisphere. Also, the terrain types of many objects in the solar system display global dichotomies (for example, the southern highlands and the northern lowlands of Mars and the dark

noted that as Uranus passed in front of the star, the starlight was blocked sharply during several short intervals as the planet passed through the path. The occultations (as observed by several researchers) on the one side of Uranus were matched by those on the other side of the planet. Thus, it was logical to conclude that a number of rings around the planet were the cause. Such observations continued. They involved well over 200 stellar occultations by Uranus and the presence of nine narrow rings was well established. In order of increasing distance from Uranus, the rings are identified as 6, 5, 4, alpha, beta, eta, gamma, delta, and epsilon. In addition to some noncircular paths taken by the rings, not all of the rings lie in the plane of the planet's equator. Widths vary from only 2 km (1 1/4 mi) to almost 100 km (62 1/2 mi). The largest and outermost ring is epsilon. Its width ranges from 20 to 96 km (12 1/2 to 60 mi). As noted by Cuzzi and Esposito, the variations in the width of epsilon are not random. The width increases in proportion to the distance of the ring material from Uranus; i.e., at its closest point to the planet, the epsilon ring is narrowest and nearly opaque; at its farthest point from the planet, it is 5 times wider and 5 times more transparent. Similar behavior is exhibited by the alpha and beta rings.

TABLE 1. RING SYSTEM OF URANUS

Feature	Distance from Uranus Center (10^2 KM)	Eccentricity (10^{-3})	Inclination (10^{-3} Degrees)	Width (km)[†]	Mean Visual Optical Depth[†]	Maximum Radio Optical Depth[†]
1986U2R	37–39.5	0?	0?	~2500	0.001–0.0001	
6 ring	41.85*	1.0*	63*	1–3	0.2–0.3	0.8–1.1
5 ring	42.24*	1.9*	52	2–3	0.5–0.6	1.2–2.3
4 ring	42.58*	1.1*	32*	2–3	0.3	1.2–1.5
Alpha	44.73*	0.8*	14*	7–12	0.3–0.4	1.0–2.5
Beta	45.67*	0.4*	5*	7–12	0.2	0.5–1.0
Eta	47.18*	(0)*	(2)*	0–2	0.1–0.4	0.7–1.1
Gamma	47.63*	(0)*	(11)*	1–4	1.3–2.3	3.8–6.9
Delta	48.31*	(0)*	4*	3–9	0.3–0.4	1.2–4
1986U1R	50.04	0?	0?	1–2	0.1	
Epsilon	51.16*	7.9*	(1)*	22–93	0.5–2.1	2.5–8

[†]Ring widths and optical depths vary greatly with ring longitude.
*Estimate from early evaluation of data.
Source: Jet Propulsion Laboratory, Pasadena, California.

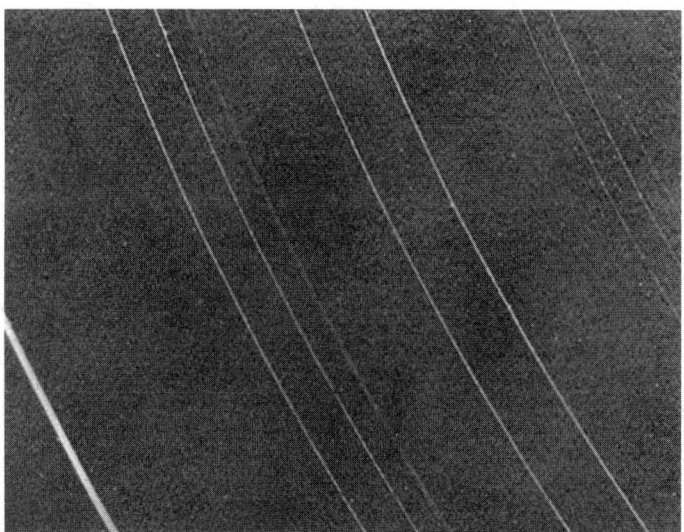

Fig. 4. Rings of Uranus shown in image taken by *Voyager 2* on January 23, 1986 at a distance of approximately 1,110,000 km (690,000 mi) as the spacecraft approached the Uranian system. (*Jet Propulsion Laboratory, Pasadena, California.*)

leading and bright trailing hemispheres of Iapetus (a Saturnian moon). A summary of the physical properties of the Uranian satellite system is given in Table 2. See also Fig. 5.

Oberon. The outermost and second largest of the moons of Uranus is of a gray, rather uniform appearance. The surface is generously covered with large, old craters. They range in size from quite small, less than 12 km (7.5 mi) up to large, exceeding 100 km (60+ mi). The large craters do not appear to have been covered over by fresh material and thus most likely represent the formation by heliocentric bombardment some 4 billion years ago. There is very dark material on a few of the largest craters, of a type familiar to that found on 1985U1. Thus, it is likely that Oberon was somewhat active in the past. However, with present knowledge, it appears that Oberon mainly has been a passive target for occasional incoming projectiles. A high-resolution image of Oberon is shown in Fig. 6.

Titania. The second-outermost and the largest of the moons of Uranus (only slightly larger than Oberon) displays evidence of global tectonics and thus is in marked contrast with Oberon. The surface is heavily cratered, indicating bombardment by impactors from a second group of planetocentric objects, as well as by earlier heliocentric impactors. Unlike Oberon, heliocentric bombardment of Titania is not so evident, probably

because of the obliteration of such craters by some poorly understood resurfacing process. Models constructed by researchers suggest that the resurfacing resulted from extensive volcanic extrusion of material onto the surface. The model assumes that such volcanic activity commenced even before the ending of the heliocentric bombardment. In this scenario, large craters would have been flooded, thence frozen in a relatively smooth ice surface, or because the planet was relatively warm, the crust would have been softened and simply collapsed. Later tectonic activity produced a network of extensional faults, with blocks of crust dropping down along the faults, thus creating grabens. It has been estimated that Titania has been a quiet body for the past 3 billion years, with only possible interruptions caused by the impacts of a relatively few comets. In another possible scenario, E.M. Shoemaker (U.S. Geological Survey), an authority on planetary cratering, suggests that the early cratering of Titania by heliocentric objects was so intense that a large, late-arriving impactor may have blasted the moon apart. Pieces of material resulting from this shattering would have remained in an orbit about Titania, from which they were later reassembled into a "new" moon, which we now call Titania. Obviously this recreated moon would have a new surface without any trace of the early heliocentric bombardment. Because Oberon and Titania have similar bulk properties and yet are so different in appearance, this scenario offers an explanation of their striking differences. Further, it has been suggested that inasmuch as Titania is closer to Uranus, the probable gravitational focus of an impacting object would have been closer to Titania than to Oberon. A high-resolution image of Titania is given in Fig. 7.

Umbriel. The third outermost and third-largest of the moons of Uranus is described as the least distinctive of the planet's moon and has been described as some observers as "very bland." The satellite lacks global-scale variations in brightness and is not marked by bright ray craters. High-resolution images of the moon (made during *Voyager 2's* closest approach) reveal that the surface of Umbriel is dominated by large but rayless craters, attributed to very early bombardment. Researchers to date have found it difficult to explain satisfactorially the bland character of the moon. One hypothesis suggests that the bright rays around Umbriel's craters may have been erased by micrometeoroids, which not only ravaged the surface in great numbers, but also mixed dark underlying material with debris from the bombardment. This process is sometimes referred to as "impact gardening." Another scenario postulates that the rays of the craters contained methane that has become darkened by energetic radiation. Neither of these postulations, however, explain why rays are absent on Umbriel, but present on the other moons of Uranus. Possibly, a more logical explanation may be that rays never did exist on Umbriel. Perhaps dark material on the moon blankets the moon for several kilometers in depth and hence the ejecta from impacts would be dark and thus light rays would not be observable. A moon of this type, Ganymede, was discovered in the Jovian system. The reflectance of Umbriel is only about 20%, indicating that the material near its surface is composed of dark, carbon-rich substances.

TABLE 2. PHYSICAL DATA OF URANIAN SATELLITE SYSTEM

Satellite Name	Diameter (km)	Distance from Uranus Center (10^3 km)	Orbital Period (Hours)	$G \times Mass$ (km^2 sec^{-2})	Density (G cm^{-3})	Normal Albedo
Titania	1610 ± 10	436.3	208.9	232 ± 12	1.59 ± 0.09	0.28 ± 0.02
Oberon	1550 ± 20	583.4	323.1	195 ± 11	1.50 ± 0.10	0.24 ± 0.01
Umbriel	1190 ± 20	266.0	99.5	85 ± 16	1.44 ± 0.28	0.19 ± 0.01
Ariel	1160 ± 10	190.9	60.5	90 ± 16	1.65 ± 0.30	0.40 ± 0.02
Miranda	484 ± 10	129.9	33.9	5.0 ± 1.5	1.26 ± 0.39	0.34 ± 0.02
1985U1	170 ± 10	86.0	18.3			0.07 ± 0.02
1986U1	~80*	66.1	12.3			<0.1*
1986U2	~80*	64.6	11.8			<0.1*
1986U3	~60*	61.8	11.1			<0.1*
1986U4	~60*	69.9	13.4			<0.1*
1986U5	~60*	75.3	14.9			<0.1*
1986U6	~60*	62.7	11.4			<0.1*
1986U7	~40*	49.7	8.0			<0.1*
1986U8	~50*	53.8	9.0			<0.1*
1986U9	~50*	59.2	10.4			<0.1*

*Minor satellite diamers are upper limits. Corresponding normal albedos are calculated from these diameters (and thus lower limits) and are probably lower than the listed upper limits of 0.1).
Source: Jet Propulsion Laboratory, Pasadena, California.

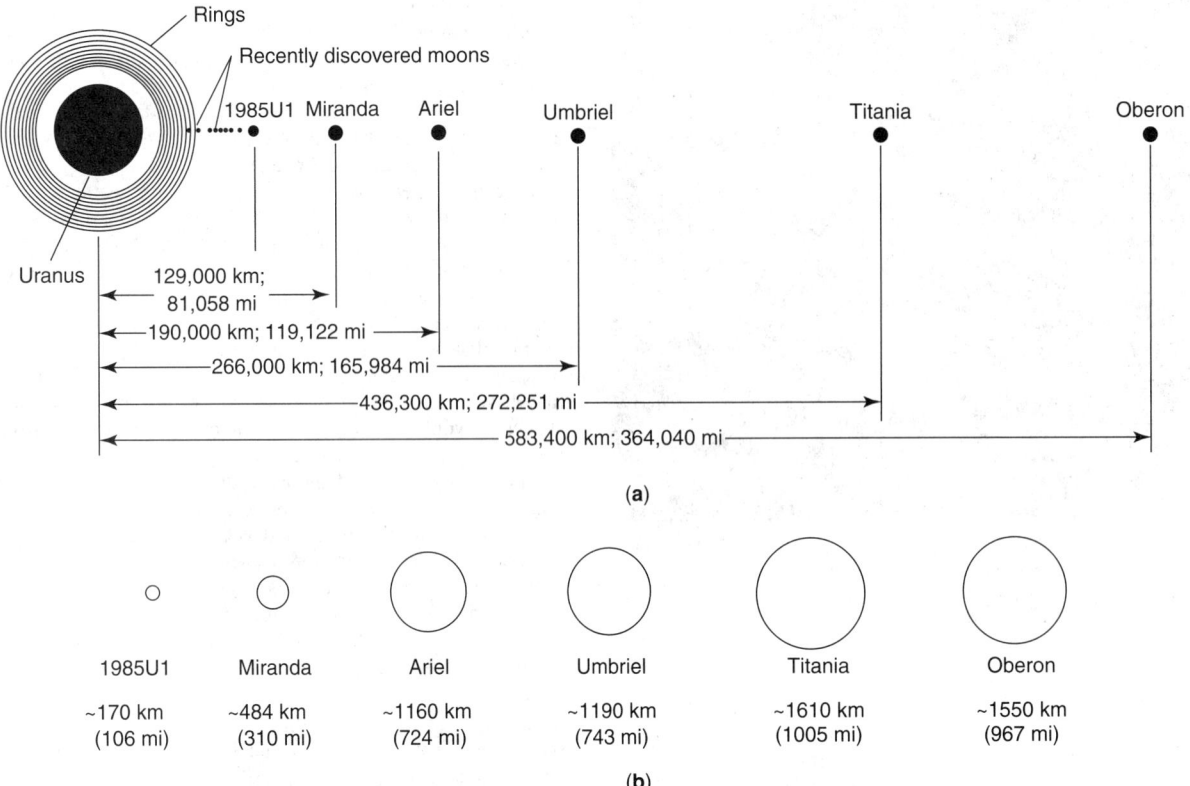

Fig. 5.   The five principal satellites of Uranus and the recently discovered small moon 1985U1: **(a)** distance from center of Uranus to lunar orbits. Several very small satellites were discovered by *Voyager 2* as it approached the planet, one or more of which has been designated as a shepherding moon within the ring structure. **(b)** relative size of satellites (diameters indicated).

Fig. 6.   Best image of Oberon obtained by *Voyager 2* on January 24, 1986. Note cratering and large peak on moon's lower limb. Range: 660,000 km (410,000 mi). (*Jet Propulsion Laboratory, Pasadena, California.*)

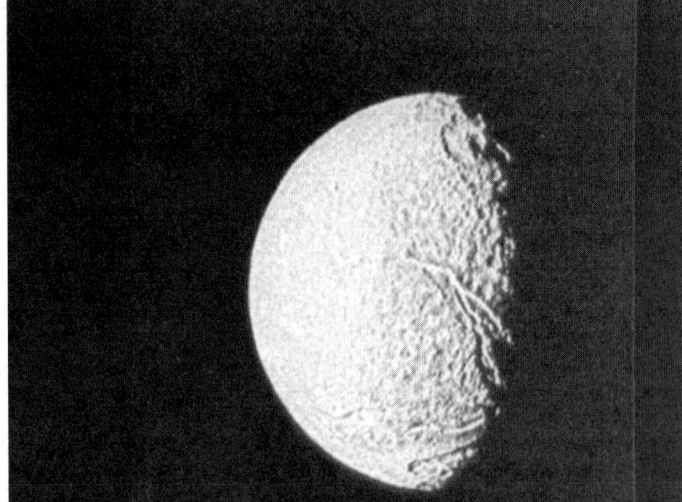

Fig. 7.   High-resolution image of Titania obtained by *Voyager 2* on January 24, 1986. Range: 368,000 km (229,000 mi). Note prominent fault valleys nearly 1600 km (1000 mi) in length. (*Jet Propulsion Laboratory, Pasadena, California.*)

To compound the current mysteries of Umbriel, there are two outstanding bright features near the equator of the satellite. One of these is a ring some 80 km in diameter. It appears to cover the floor of an impact crater. Some researchers have proposed that the bright material in this feature probably originated below the surface of the body. Umbriel also features a spot on the peak of another large crater. With reference to the aforementioned underlying dark material, these latter two features may be exceptions to the rule. A high-resolution image of Umbriel is given in Fig. 8.

Ariel. Not counting the recently discovered very small moon, 1985U1, Ariel is the second moon outward from the planet. It is very slightly (some 30 km) smaller (diameter) than Umbriel. This satellite appears to have been quite active, with features indicating crustal rifting and icy volcanic

flows. Ariel is the brightest of the Uranian moons, quite in contrast with Umbriel, the darkest. Reflectance is approximately 40%. Researchers have concluded thus far that Ariel may be the youngest and least cratered of the Uranian satellites. Ariel has been compared with the Jovian satellites, Europa and Ganymede, and Saturn's Enceladus. It has been proposed that Ariel's history may parallel that of Titania, previously described, with the exception that the geologic activity on Ariel has been more pronounced and occurred over a longer time span. There is little evidence of heliocentric bombardment of Ariel, featuring only about a third of the number of such craters found on Titania. Thus, Ariel's surface has been restructured over a broader area and for a longer period. Ariel features a global network of fault valleys, some of which are estimated to be tens of kilometers in

Fig. 8.   The most detailed image of Umbriel made by *Voyager 2* on January 24, 1986. Range: 557,000 km (346,000 mi). The surface is a dark, nearly uniform gray coloration. There is an abundance of large, ancient craters. Note the two brightest features: (1) A bright ring at a longitude of 270° (top of view) which appears to lie in an 80 km (50 mi) crater. (2) Bright spot on the central peak of another crater at a longitude of about 310° (central left of view). (*Jet Propulsion Laboratory, Pasadena, California.*)

depth. Images have provided strong evidence that Ariel's resurfacing has been accomplished by way of volcanic activity. Considerable upwelling of material on Ariel has occurred. However, most researchers do not believe that the material was molten rock, but rather a relatively warm, plastic mixture of ice and rock that behaved much like a glacier on Earth. An image of Ariel is given in Fig. 9.

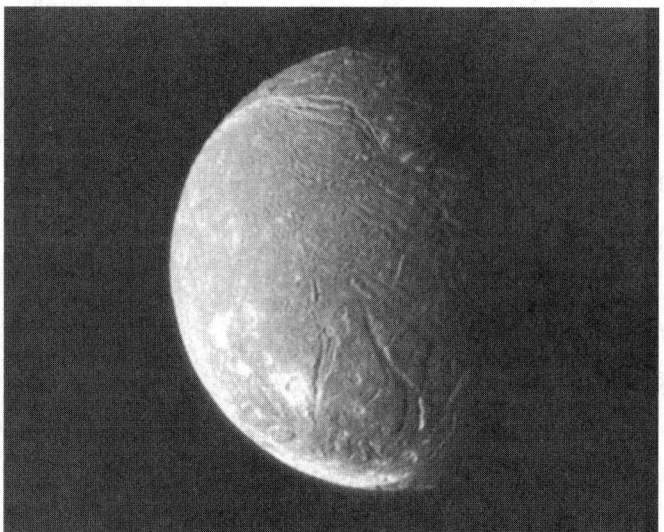

Fig. 9.   The most detailed view (mosaic) of Ariel obtained by *Voyager 2* on January 24, 1986. Range: 128,700 km (80,000 mi). Image shows numerous faults and valleys. Most prominent features are the broad, crisscrossing grabens, or fault valleys, along the equator (near right of photograph). The supposition that much of Ariel was re-covered (resurfaced) after the heliocentric bombardment is based upon the almost total absence of large craters. (*Jet Propulsion Laboratory, Pasadena, California.*)

Miranda. The smallest of the five principal Uranian moons (about 3% of the diameter of Titania) has proved to be one of the most interesting of the planet's moons, as determined by *Voyager 2* data. The most ancient geologic feature on the satellite is a series of densely cratered rolling plains packed by the planetocentric bombardment. Most of the surface visible to *Voyager 2* is covered in this manner, but there are a few very fascinating features superposed on the rolling plains. Called *ovoids* by some researchers, there are three huge, oval-to-trapezoidal regions,

which range between 200 and 300 km (125 and 187 mi) in diameter. Considering the diameter of Miranda (484 km; 310 mi), these indeed are immense features. Within the perimeters of the ovoids, numerous craters (considered much younger than the plains) are found. As described by Johnson, et al. (see reference), the ovoids consist of locally parallel belts of ridges, grooves, and scarps that abut one another at odd angles. Bright material as well as dark material is exposed along the scarps and in fresh craters inside the ovoids. In contrast, the rolling plains have a fairly uniform reflectance. There is an additional fascinating feature. The ovoids and the plains are cut by huge fracture zones that circle the satellite, creating fault valleys whose steep, terraced cliffs are as much as 10 to 20 km (~6 to 12 mi) high.

As with the strange features on Titania, early postulations concerning Miranda have developed. One researcher suggests that Miranda was blasted apart, as Titania, several times instead of just once for Titania. After each occasion, the moon reassembled itself, during which time rock and ice became partially separated. Then, when the reassembled moon was fractured again, the pieces that were generated consisted, not of a mixture of rock and ice, but of one or the other materials. Because the satellite, at that time, still had some internal heat, viscous flow in its interior was possible. Thus rock masses commenced to sink again toward the center of the body, with the lighter ice rising and flowing into the space behind them. This caused the surface to crack along concentric stress lines. Thus, the surface disturbances evidenced by the ovoids were left by sinking rock masses. To some researchers, this postulation does not take into account the probability that large pieces of rock would not retain their identity in the type of cataclysm which normally would be presumed to create very small pieces of resulting material.

The other scenario suggests that when Miranda accreted, it consisted of a relatively uniform mixture of rock and ice and that from that point, the satellite commenced to differentiate. The rock sand moved toward the center and ice masses rose toward the surface. Since the patterns of the ovoids are roughly similar, it is suggested that some "internal organization" to the flow (a form of convection cell) created the "typical" ovoidal pattern.

Some authorities suggest that the ovoids are only about a billion years old; others suggest that the evolution of the moon ceased some 3 to 4 billion years ago. Three striking images of Miranda are given in Fig. 10.

Editorial limitations do not permit presenting further details concerning the analysis of *Voyager 2* data in connection with the Uranian satellites. Considerably more detail can be found in the excellent Johnson, et al. reference (1987). As more time permits further modeling of the moons and scientific exchanges of viewpoints, some of the current hypotheses will gain increased acceptance, while others will be disproved or at least become more speculative.

Newly Discovered Uranian Satellites. The five major satellites of Uranus were discovered and briefly described from Earth observations prior to receiving information from *Voyager 2*. On its early approach to the planet in 1985, *Voyager 2* discovered the satellite 1985U1. The satellite appears to be made of the same primordial ice-rock material of the other moons, but surprisingly it is not so bright as the other moons. Some researchers suggest that 1985U1 is a homogeneous mixture of ice and rock, in contrast with the major moons which have undergone differentiation. It is assumed that much of the dense rock on the larger moons has settled toward their center, with brighter ice rising to the surface and thus the lighter appearance. At the time of the encounter with the planet, *Voyager 2* discovered nine additional small satellites (named, respectively, 1986U1 through 1986U9). The diameters of these smaller satellites range from about 1/3 to 1/4 the size (diameter) of 1985U1, which has a diameter of 170 ± 10 km (~106 mi). The ten new satellites orbit between Uranus and Miranda. As with the major satellites, the recently discovered satellites are in synchronous rotation, with one side always facing Uranus. It is presumed that the newly found satellites are also in locked rotation. Two of the ten smaller satellites (1986U7 and 1986U8) are shephering moons on each side of the planet's epsilon ring and it is suggested that the two moons keep the ring particles in place by exerting gravitational forces on them. Data from *Voyager 2* did not indicate any shepherding moons near any of the other eight rings, although they may exist and are too small to detect.

Magnetosphere of Uranus

The Uranian magnetic field was first detected by radio emissions five days prior to *Voyager's* closest approach to the planet — at a distance of about 275 Uranian radii (R_U). As reported by Stone and Miner, *Voyager*

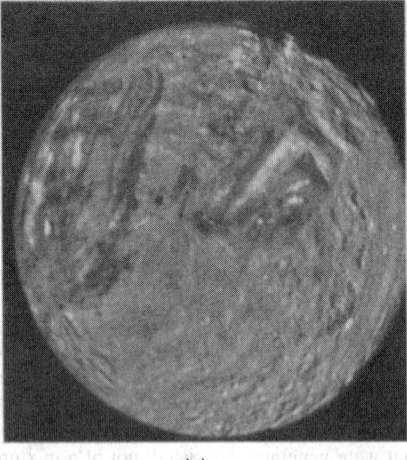
(a)

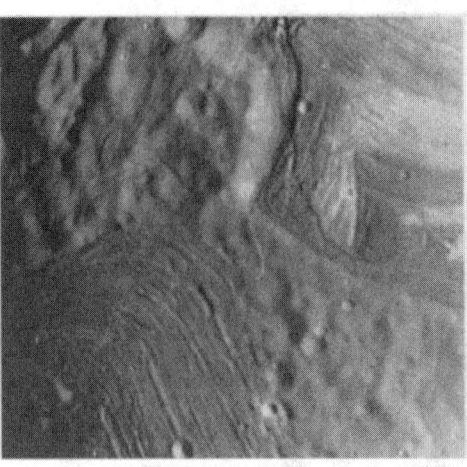
(b)

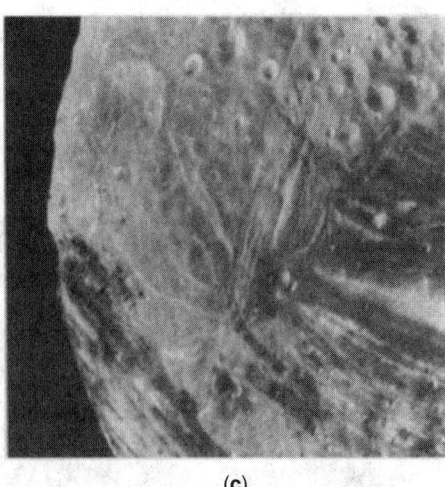
(c)

Fig. 10.   Images of Miranda taken by *Voyager 2* on January 24, 1986. (**a**) A computer mosaic of the satellite shows varied geologic regions at high resolution. Range: 18,700 to 30,090 km (25,000 to 40,225 mi). (**b**) Unusual "chevron" feature seen on *Voyager 2's* approach to Miranda. Range: 41,830 km (26,000 mi). (**c**) Image taken shortly before closest approach by *Voyager 2*. Range: 30,570 km (19,000 mi). (*Jet Propulsion Laboratory, Pasadena, California.*)

subsequently crossed a well-defined, detached bow shock at $23.5R_U$ and entered a fully developed magnetosphere at $18R_U$, revealing a magnetic dipole field with an axis at an unexpected large angle of $60°$ with respect to the rotation axis of the planet and offset from the center of Uranus by $0.3R_U$. The dipole moment was measured at $0.23GR_U$, giving a surface magnetic field ranging from 0.1 to 1.1 G. The intensity of the field and its offset suggested that it is generated at an intermediate depth where water may be under sufficient pressure to be electrically conductive. The rotation period of the magnetic field, presumed to be that of this interior region, was measured at 17.24 ± 0.01 hours. Researchers report that the dipole field appears to be deformed by the incident solar wind, resulting in a magnetotail geometry similar to that at Earth. The tail has a radius of $42R_U$ at a distance of $67R_U$ behind Uranus and has a plasma sheet about $10R_U$ thick. In as much as the planetary rotation axis is directed virtually sunward at the time of the Uranian year when measured, the magnetotail is estimated to rotate about the antisun line with the same period as the planet.

It has been reported that within the magnetosphere there is an extensive distribution of charged particles, mainly hydrogen ions and electrons. Two main plasma ion populations were determined: (1) a warm ($\sim$10 eV) component within about $7R_U$, with a maximum density of about 2 cm^{-3} planetward of Miranda's orbit; and (2) a hot (1000 eV) component that is confined outside approximately $5R_U$. It is suggested that ionization of the extended hydrogen corona may be the principal source of the plasma. However, the ionosphere and the solar wind also may be sources. The energy density of the plasma is small as compared with that of the magnetic field.

A Jet Propulsion Laboratory (JPL) report of May 1986 observes that the orientation of the rotation axis results in sunward convection of the plasma. This is quite different from that in the magnetosphere of Jupiter and Saturn, in which radial transport is dominated by diffusion. It is suggested that the rapid convective time scale of approximately 40 hours probably precludes the accumulation of a significant density of heavy ions sputtered from the surfaces of the moon, Miranda.

The trapped ion population at higher energies outside Miranda's orbit was found to be dominated by protons, with spectra characterized by temperatures of 4 to 30 keV. The proton fluxes are too small to cause significant distortion of the magnetic field, but are sufficient to modify and darken any methane on the surfaces of the satellite in a span of less than 100,000 years. The JPL report also notes that energetic electrons with temperatures greater than 20 keV were observed throughout the magnetosphere, terminating abruptly at about $18R_U$ on the dark side of the planet where the magnetotail commences. The approximately 1-MeV electron fluxes peak inside Miranda's orbit because of the adiabatic acceleration that occurs as the electrons diffuse radially inward from the outer magnetosphere. Intense whistler-mode hiss and chorus emissions were found inside $8R_U$. These radio emissions cause electron precipitation, which, in turn, may contribute to nightside ultraviolet emissions.

Radio emissions from Uranus also occur, but they are of lower average power than found at Saturn and are detectable only near the planet.

Additional Reading (Reports by *Voyager 2* Encounter Team)

Bridge, H.S., et al.: "Plasma Observations Near Uranus: Initial Results from Voyager 2," *Science*, **233**, 89–92 (1986).

Broadfoot, A.L., et al.: "Ultraviolet Spectrometer Observations of Uranus," *Science*, **233**, 74–78 (1986).

Gurnett, D.A., et al.: "First Plasma Wave Observations at Uranus," *Science*, **233**, 106–109 (1986).

Hanel, R., et al.: Infrared Observations of the Uranian System," *Science*, **233**, 70–73 (1986).

Krimigis, S.M., et al.: "The Magnetosphere of Uranus: Hot Plasma and Radiation Environment," *Science*, **233**, 97–102 (1986).

Lane, A.L., et al.: "Photometry from Voyager 2: Initial Results from the Uranian Atmosphere, Satellites, and Rings," *Science*, **233**, 65–69 (1986).

Ness, N.F., et al.: "Magnetic Fields at Uranus," *Science*, **233**, 85–89 (1986).

Smith, B.A., et al.: "Voyager 2 in the Uranian System: Imaging Science Results," *Science*, **233**, 43–64 (1986).

Stone, E.C., et al.: "Energetic Charged Particles in the Uranian Magnetosphere," *Science*, **233**, 93–97 (1986).

Stone, F.C. and E.D. Miner: "The Voyager 2 Encounter with the Uranian System," *Science*, **233**, 39–43 (1986).

Tyler, G.L., et al.: "Voyager 2 Radio Science Observations of the Uranian System: Atmosphere, Rings, and Satellites," *Science*, **233**, 79–84 (1986).

Warwick, J.W., et al.: "Voyager 2 Radio Observations of Uranus," *Science*, **233**, 102–106 (1986).

(Other References)

Bergstralh, J.T., E.D. Miner, and M.S. Matthews: "Uranus," University of Arizona Press, Tuscon, AZ, 1997.

Brown, R.H. and D.P. Cruikshank, "The Moons of Uranus, Neptune, and Pluto," *Sci. Amer.*, 38–47 (July 1985).

Burns, J.A. and M.S. Matthews: "Satellites," University of Arizona Press, Tucson, AZ, 1986.

Cuzzi, J.N.: "Ringed Planets: Still Mysterious," *Sky and Telescope*, 511–515 (December 1984); 19–23 (January 1985).

Cuzzi, J.N. and L.W. Esposito: "The Rings of Uranus," *Sci. Amer.*, 52–66 (July 1987).

Eberhart, J.: "Locating Uranus' Auroras," *Science News*, 248 (October 20, 1990).

Elliot, J. and R. Kerr: "Rings: Discoveries from Galileo to Voyager," The MIT Press, Cambridge, MA, 1987.

Fridman, A.M. and N.N. Gorkavyi: "Physics of Planetary Rings," Springer-Verlag, Inc., New York, NY, 1999.

Gore, R.: "Uranus — Voyager Visits a Dark Planet," *National Geographic*, 178–194 (August 1986).

Greenberg, R. and A. Brahic: "Planetary Rings," University of Arizona Press, Tucson, Arizona, 1984.

Hunt, G., Editor: "Uranus and the Outer Planets," Cambridge University Press, New York, NY, 1982.

Ingersoll, A.P.: "Uranus," *Sci. Amer.*, 38–45 (January 1987).

Ingersoll, A.P.: "Atmospheric Dynamics of the Outer Planets," *Science*, 308 (April 20, 1990).

Jankowski, D.G. and S.W. Squryes: "Solid-State Ice Volcanism on the Satellite of Uranus," *Science*, 1322 (September 9, 1988).

Johnson, T.V., R.H. Brown, and L.A. Soderblom: "The Moons of Uranus," *Sci. Amer.*, 48–60 (April 1987).

Laeser, R.P., W.I. McLaughlin, and D.M. Wolff: "Engineering Voyager 2's Encounter with Uranus," *Sci. Amer.*, 36 (November 1986).

Lunine, J.I.: "Origin and Evolution of Outer Solar Systems Atmospheres," *Science*, 141 (July 14, 1989).

Melosh, H.J. and D.M. Janes: "Ice Volcanism on Ariel," *Science*, 195 (July 14, 1989).

Miner, E.D.: "Uranus: The Planet, Rings and Satellites," 2nd Edition, John Wiley & Sons, Inc., New York, NY, 1998.

Nellis, W.J. et al.: "The Nature of the Interior of Uranus Based on Studies of Planetary Ices at High Dynamic Pressure," *Science*, 779 (May 6, 1988).

Web Reference

Uranus Information & Facts: *http://search.jpl.nasa.gov/cgi-bin/query?*mss = search & q = Uranus

EDITOR'S NOTE: The staff of this encyclopedia is most appreciative of the cooperation extended by A.S. Woods and his associates at the Jet Propulsion Laboratory, Pasadena, CA in furnishing data for this article.

UREA. $H_2N \cdot CO \cdot NH_2$, formula weight 60.06, colorless crystalline solid, mp 132.7 °C, sublimes unchanged under vacuum at its melting point, sp gr 1.335. Heating above the mp at atmospheric pressure causes decomposition, with the production of NH_3, isocyanic acid HNCO, cyanuric acid $(HNCO)_3$, biuret $NH_2CONHCONH_2$, and other products. Also known as carbamide, urea is very soluble in H_2O, soluble in alcohol, and slightly soluble in ether. The compound was discovered by Rouelle in 1773 as a constituent of urine. Historically, urea was the first organic compound to be synthesized from inorganic ingredients, accomplished by Wöhler in 1828. However, a century passed before the compound was manufactured on a large scale.

Because of the reactivity and versatility of its derivatives, urea is a very high-tonnage chemical. The compound and its derivatives are widely used in fertilizers, pharmaceuticals (e.g., barbiturates), and synthetic resins and plastics (urethanes). Although there are several chemical engineering approaches to the synthesis of urea, the principal reaction is that of combining NH_3 with CO_2 in a first step to form ammonium carbamate. In a second step, dehydrating the ammonium carbamate to yield urea: (1) $2NH_3 + CO_2 \rightarrow NH_2COONH_4$, (2) $NH_2COONH_4 \rightarrow NH_2CONH_2 + H_2O$. The processing is complicated because of the severe corrosiveness of the reactants, usually requiring reaction vessels that are lined with lead, titanium, zirconium, silver, or stainless steel. The second step of the process requires a temperature of about 200 °C to effect the dehydration of the ammonium carbamate. The processing pressure ranges from 160 to 250 atmospheres. Only about one-half of the ammonium carbamate is dehydrated in the first pass. Thus, the excess carbamate, after separation from the urea, must be recycled to the urea reactor or used for other products, such as the production of ammonium sulfate.

Some of the reactions of urea and derivatives include: (1) as a weak mono-acid base, urea forms stable salts, such as urea nitrate $CO(NH_2)_2 \cdot HNO_3$ and urea oxalate $2CO(NH_2)_2 \cdot H_2C_2O_4$; (2) urea reacts with malonic acid to form barbituric acid $CO(NHCO)_2CH_2$, the derivatives of which are barbiturates (sedative drugs); (3) with alcohols, urea reacts to form urethanes; (4) with formaldehyde, urea forms ureaforms which can be used as slow-release fertilizers and also as ingredients for adhesives and plastics; (5) with hydrogen peroxide, urea forms a useful crystalline oxidizing agent; (6) with straight-chain alkanes, urea forms crystalline complexes (clathrates) which are used in the petroleum industry for separating straight- and branched-chain hydrocarbons; (7) when heated rapidly to about 350 °C in a fluidized bed at atmospheric pressure, urea decomposes to isocyanic acid and NH_3. The latter products, when passed

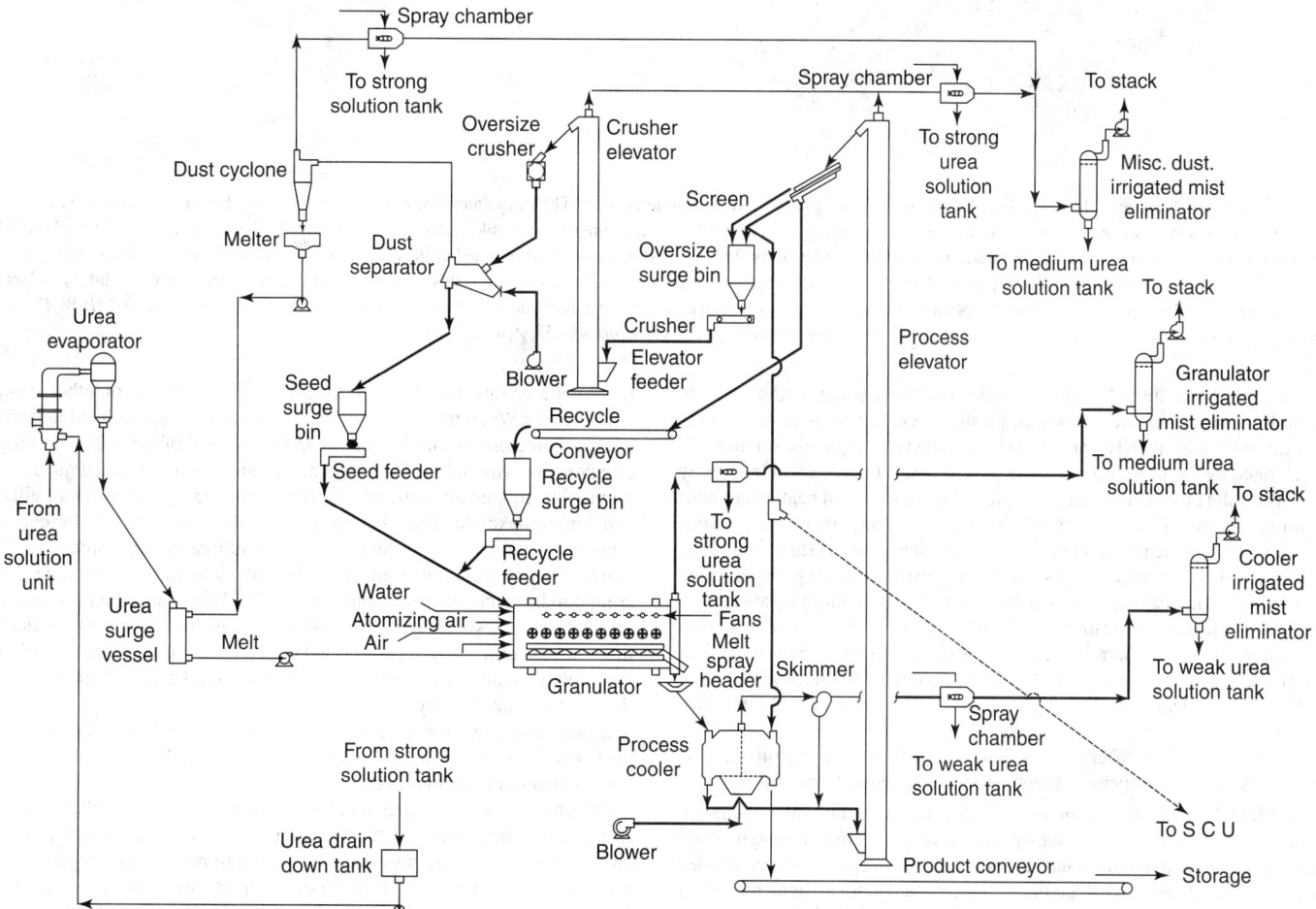

Fig. 1. A new process (*Urea Technologies*) developed for the Tennessee Valley Authority operates at considerable energy savings. Urea is produced in an overall exothermic reaction of ammonia and carbon dioxide at elevated pressure and temperature. In a highly exothermic reaction, ammonium carbamate is first formed as an intermediate compound, followed by its dehydration to urea and water, which is a slightly endothermic reaction. The conversion of CO_2 and NH_3 to urea depends on the ammonia-to-carbon dioxide ratio, temperature, and water-to-carbon dioxide ratio, among other factors. The new process makes maximum use of the heat created in the initial reaction, including heat recycling. (*Urea Technologies and Tennessee Valley Authority.*)

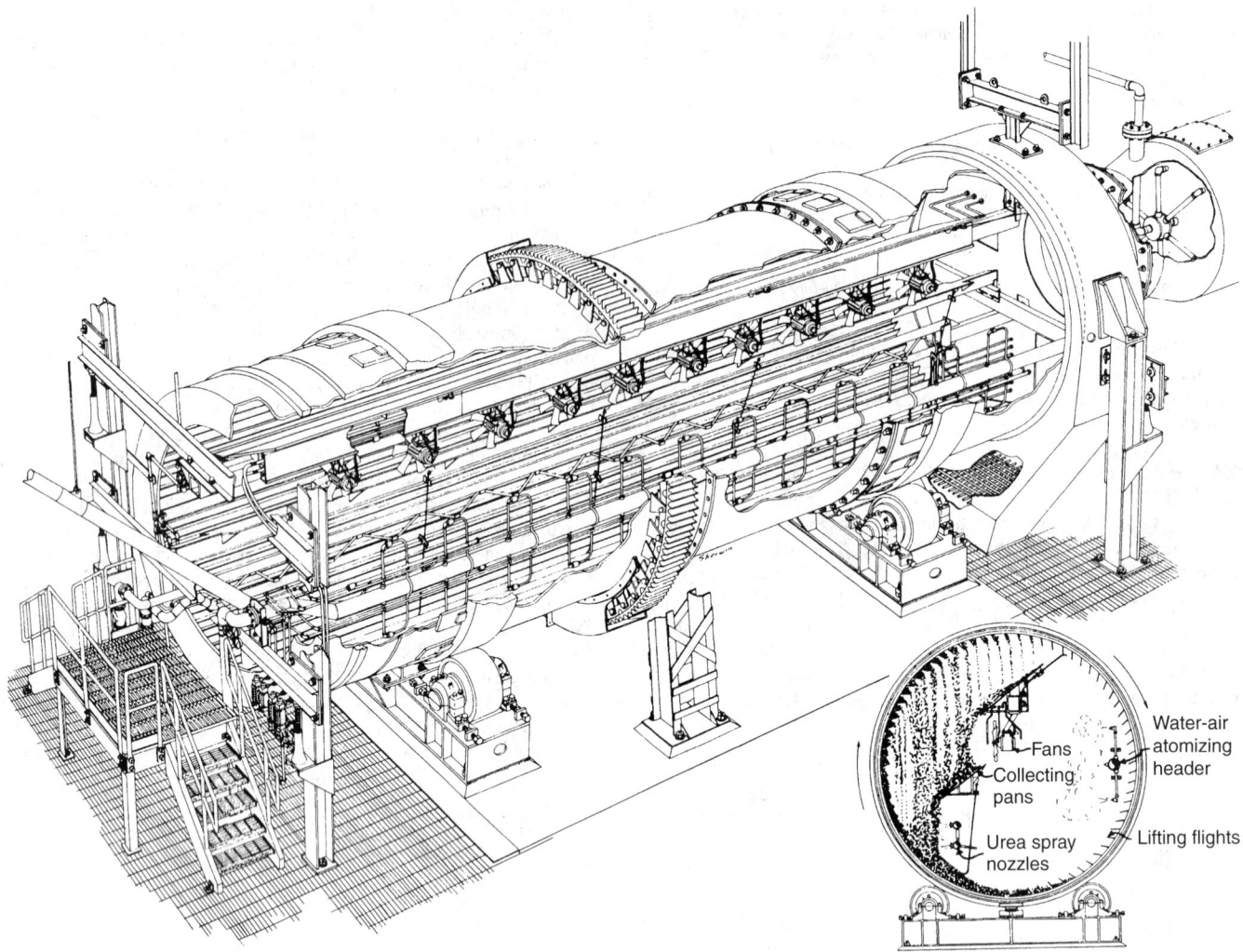

Fig. 2. In the urea process shown in Fig. 1, a falling-curtain granulation technique is used. The granulator drum is shown here. The key to the process is the rotary drum, with specially designed components, including collecting pans, air-circulation fans, urea spray header, and water spray header. As the drum rotates, seed particles, recycled undersize granules, and intermediate-size granules are lifted by flights and discharged onto inclined collecting pans. The material forms a dense falling curtain of granules as it slides from the collecting pans. The urea melt is sprayed through hydraulic atomizing nozzles onto this falling curtain and quickly solidifies. Controls over several variables in the process make it possible to form granules within narrow size specifications. Further details of process are given by Kirkland (*R.W. Kirkland,* "*Energy-efficient route to granular urea,*" *Chemical Engineering Progress, April, 1984, pp. 49–53, April 1984*).

over a catalyst at 400 °C, yield melamine $(NCH_2)_3$ which is the triamide of cyanuric acid and widely used in plastics; (8) with acids or bases, urea hydrolyzes, yielding NH_3 and CO_2. Hydrolysis in aqueous solutions is accelerated by the presence of urease (an enzyme). This reaction frequently is used for the quantitative determination of urea; (9) upon heating aqueous solutions of urea, biuret is formed. When crystallizing urea from aqueous solutions, the presence of about 5% biuret alters the crystals from long needles to short rhombic prisms, the latter greatly enhancing the handling properties of the final product. A content of up to 1.5% biuret is satisfactory for most fertilizer applications, although for citrus fruits, coffee plants, and cherry trees, the biuret content must remain below 0.3%. As a feed supplement for ruminants, pure biuret has proved advantageous because of the slower release rate of NH_3 from biuret as compared with urea. See also **Fertilizer**.

There are several different processes for making urea fertilizer. One process designed for energy savings is shown in Figs. 1 and 2.

Arginine-Urea Cycle (Ornithine Cycle). In adult animals, including humans, the characteristic tissue-specific levels of different enzymes are maintained by a dynamic balance between the independently controlled rates of biosynthesis and degradation of each enzyme. A dynamic rather than a static system most likely emerged because it enables organisms to adapt to widely different nutritional conditions and other environmental changes. Depending upon the physiological state of the animal at a given moment, amino acids derived from the hydrolysis of exogenous or endogenous protein may be predominantly utilized for *synthesis* of tissue-specific proteins. Or, their carbon chains may be *metabolized* further

to provide energy (ATP) or intermediates for synthesis of other cellular constituents. When the carbon chains of amino acids are utilized to provide energy, some provision must be made for disposal of the reduced nitrogen components. Animal tissues in general cannot tolerate accumulation of ammonia. Aquatic animals, which are surrounded by a convenient diluent, can simply excrete ammonia as rapidly as it is formed. In contrast, land-based animals have devised other solutions to this problem. They convert amino acid nitrogen and ammonia into nitrogen-rich, nontoxic compounds, such as *urea* and *uric acid*. These are then excreted at intervals. Synthesis of urea, the primary nitrogenous excretory product of mammals, is efficiently accomplished in the liver by combining a portion of the already established pathway of arginine biosynthesis with the hydrolytic degradative enzyme, arginase.

Some of the enzymes required are widely distributed, but *ornithine carbamoyltransferase* occurs only in the liver and thus the complete urea cycle occurs only in that organ.

In human liver, a given molecule of arginine has four possible metabolic fates. It can be converted to (1) argininosuccinic acid, or (2) argininyl-sRNA, or (3) ornithine plus urea, or (4) ornithine plus glycocyamine, the precursor of creatine. The flow along the pathway to (4) is regulated by feedback repression in which the steady-state level of the enzyme involved (arginine:glycine amidinotransferase) is regulated by the concentration of liver creatine. Thus, runaway synthesis of creatine is prevented. The flow along the pathway is regulated by supply and demand.

Experimentally, it has been observed that above a certain basal level, the quantity of urea excreted is proportional to the amount of ingested protein.

The enzyme urease was not discovered until 1926 (by Sumner). It was the first enzyme to be isolated as a crystalline protein. Sumner's accomplishment confirmed the then growing belief that enzymes, the biological catalysts, were indeed from the chemical standpoint protein molecules. Urease catalyzes the cleavage of urea to ammonia and carbon dioxide.

For related information and references, see articles on **Fertilizer**; and **Gene Science**.

UREA (Fertilizer).   See **Fertilizer**.

UREA-FORMALDEHYDE RESINS.   See **Amino Resins**.

URIC ACID.   See **Gout**; **Purines**; and **Water**.

URIC ACID STONES.   See **Gout**; and **Kidney and Urinary Tract**.

URINE.   The fluid secreted from the blood by the kidneys, stored in the bladder, and discharged by the urethra. In health, it is amber colored. About 1,250 milliliters of urine are excreted in 24 hours by normal humans, with specific gravities usually between 1.018 and 1.024 extremes: 1.003–1.040). Flow ranges from 0.5–20 milliliters/minute with extremes of dehydration and hydration. Maximum osmolar concentration is 1,400, compared to plasma osmolarity of 300. In diabetes insipidus, characterized by inadequate antidiuretic hormone (ADH) production, volumes of 15–25 liters/day of dilute urine may be formed. In addition to the substances listed in Table 1, there are trace amounts of purine bases and methylated purines, glucuronates, the pigments urochrome and urobilin, hippuric acid, and amino acids. In pathological states, other substances may appear: proteins (nephrosis); bile pigments and salts (biliary obstruction); glucose, acetone, acetoacetic acid and betahydroxybutyric acid (diabetes mellitus). The U/P ratios of the substances in the table vary widely because of differential handling by the kidney. Quantitative knowledge of glomerular filtration, tubular reabsorption, and secretion of these requires an understanding of the concept of renal plasma clearance.

TABLE 1. COMPOSITION OF 24-HOUR URINE IN THE NORMAL ADULT[a]

Substance	Amount (Grams)	U/Pb[b]
Urea	6.0–180.0 (nitrogen)	60.0
Creatinine	0.3–0.8 (nitrogen)	70.0
Ammonia	0.4–1.0 (nitrogen)	—
Uric acid	0.08–0.2 (nitrogen)	20.0
Sodium	2.0–4.0	0.8–1.5
Potassium	1.5–2.0	10.0–15.0
Calcium	0.1–0.3	—
Magnesium	0.1–0.2	—
Chloride	4.0–8.0	0.8–2.0
Bicarbonate	—	0.0–2.0
Phosphate	0.7–1.6 (phosphorus)	25.0
Inorganic sulfate	0.6–1.8 (sulfur)	50.0
Organic sulfate	0.06–0.2 (sulfur)	—

[a]Based upon data by White, Handler, Smith, and Stetten.
[b]U/P ratio = ratio of urinary to plasma concentration.

The rate at which a substance (X) is excreted in the urine is the product of its urinary concentration, U_x (milligram/milliliter), and the volume of urine per minute, V. The rate of excretion $(U_x V)$ depends, among other factors, upon the concentration of X in the plasma, P_x (milligram/milliliter). It is therefore reasonable to relate $U_x V$ to P_x and this is called the clearance ratio: $(U_x \cdot V)/P_x$, or more generally, UV/P. This has the dimensions of volume and is in reality the smallest volume from which the kidneys can obtain the amount of X excreted per minute. The kidneys do not usually clear the plasma completely of X, but clear a larger volume incompletely. The clearance is therefore not a real, but a virtual volume. When substances are being cleared simultaneously, each has its own clearance rate, depending upon the amount absorbed from the glomerular filtrate or added by tubular secretion. The former will have the lower clearance, the latter the higher. Those cleared only by glomerular filtration will be intermediate, and their clearance will in effect measure the rate of glomerular filtration in milliliters/minute.

The best-known substance that can be infused into blood to provide a clearance equal to glomerular filtration rate is *inulin*, a polymer of fructose containing 32 hexose molecules (molecular weight 5,200). Strong evidence indicates that it is neither reabsorbed nor secreted, is freely filterable, is not metabolized, and has no physiological influences. Its clearance in humans is 120–130 milliliters/minute. This is taken to be the glomerular filtration rate (*GFR*) or C_F (amount of plasma water filtered through glomeruli/minute). Besides inulin in the dog and other vertebrates, creatinine, thiosulfate, ferrocyanide, and mannitol also fulfill these requirements.

Knowing the glomerular filtration rate permits quantification of the amount of any substance freely filtered (C_F (milliliters/minute) × P_x (milligrams/milliliter)). Subtracting from this one minute's excretion, $U_x V$, would give the amount reabsorbed in milligrams/minute. A classical example is the glucose mechanism. At normal plasma concentrations, none or a trace appears in the urine. When plasma glucose is elevated to about 180–200 milligram percent (the "threshold"), the amount appearing in the urine begins to increase. As concentration is raised more, the nephrons become progressively saturated until the rate of reabsorption becomes constant and maximal. This indication of saturation of the transport system is referred to as the T_m ("tubular maximum — T_{mG}"). In humans, T_{mG} has the value of 340 milligrams/minute. Absorption occurs in the proximal convoluted tubules. See also **Excretory System**; and **Kidney and Urinary Tract**.

Additional Reading

McBride, L.J.: "Textbook of Urinalysis and Body Fluids: A Clinical Approach," Lippincott-Raven Publishers, Philadelphia, PA, 1997.

Ringsrud, K.M. and J.J. Linne: "Urinalysis and Body Fluids: A Colortext and Atlas," Mosby-Year Book, Inc., St. Louis, MO, 1995.

Stamey, T.A. and R.W. Kindrachuk: "Urinary Sediment and Urinalysis: A Practical Guide for the Health Science Professional," W.B. Saunders Company, Philadelphia, PA, 1996.

UROLOGY.   That branch of medicine dealing with the diseases of the urogenital tract in the male and the urinary tract in the female. See also **Kidney and Urinary Tract**.

URSA MAJOR (the greater bear).   A constellation best known in the United States as the Big Dipper and in England as the Plow or the Wagon. Ursa Major is circumpolar for both Europe and North America, and two of its stars, known as the pointers, are very useful in locating the bright star Polaris, since the line joining them, if extended, will pass close to the celestial pole. The star Mizar (ζ Ursae Majoris) at the bend of the handle of the dipper is an easy visual double star. (See map accompanying entry on **Constellations**.)

URSA MINOR (the smaller bear).   A constellation frequently referred to as the Little Dipper. Ursa Minor is best known for the fact that the bright star Polaris at the end of the handle of the asterism is, at present, the closest bright star to the north celestial pole of rotation. (See map accompanying entry on **Constellations**.)

URSINES.   See **Bears**.

URTICARIA.   Commonly called *hives*, this is an allergic response characterized by the appearance on the skin of firm, elevated, circumscribed lesions, varying from the size of a small pea to several inches in diameter. Usually, the eruption is characterized by intense itching. Each lesion undergoes its own cycle, ranging from redness to swelling, followed by formation of a pink halo, and then gradually fading. Several lesions in different stages of the cycle sometimes give a blotchy appearance to the affected area. Hives may last only several hours, or they may persist for many months. The condition stems from a variety of causes, some of which are not well understood. In general, urticaria occurs when a hypersensitive individual comes in contact with a specific protein substance. Certain protein foods, serum or blood injections, or certain drugs may cause this type of eruption. Hives of large size, sometimes termed *angioneurotic edema*, may cause secondary problems when certain areas are affected — as impairment of vision when an eyelid may be involved, or impairment of swallowing and breathing where the throat and larynx may be involved. While the most certain route to successful treatment is that of patiently determining the causative substance and then avoiding future contact with it, more immediate measures are required for angioneurotic edema and for alleviating the discomfort where large areas are covered by smaller hives, particularly when present in the groin and on the hands and feet.

Drugs found useful in controlling severe urticaria attacks include epinephrine, aminophylline, intravenous and intramuscular antihistamines, such as diphenylhydramine, and intravenous and intramuscular hydrocortisone. For milder forms of the disease, antihistamines usually suffice. Topical agents include calamine lotion, corticosteroids, or menthol, phenol, and camphor in alcohol. These remedies reduce itching, but do not control the condition.

It is interesting to note that an individual may suddenly develop an allergic reaction in the form of hives to a substance (food, drug, serum, etc.) to which he has been exposed over a number of years without any reaction. In the first occurrence of hives in reaction to a given substance, the hives may not appear for several days after exposure. Subsequent exposures may produce hives within a few minutes or hours. In addition to foods and drugs as offending materials, hives may result from prolonged exposure to light, extremes of heat or cold, animal dander, plants, fungi, bites, and serums. *Papular urticaria* is a form of hives that occurs in young children, normally in the spring, and generally attributed to insects. The eruptions appear mainly on the arms and legs and are small, hard papules.

See also **Alkaloids**.

Additional Reading

Basketter, D.: "Toxicology of Contact Dermatitis: Allergy, Irritancy and Urticaria," John Wiley & Sons, Inc., New York, NY, 1999.

Henz, B.M., E. Monroe, T. Zuberbier, and J. Grabbe: "Urticaria: Clinical, Diagnostic and Therapeutic Aspects," Springer-Verlag, Inc., New York, NY, 1997.

Lahti, A. and H.I. Maibach: "Contact Urticaria Syndrome," CRC Press, LLC., Boca Raton, FL, 1997.

UTERUS. See **Embryo**; **Gonads**; and **Hormones**.

UTISOLS. See **Soil**.

UTTER CHAOS. Since the formative years of science, the precepts of classical mechanics were entrenched firmly in the pursuit of dynamic systems and guided by the unwavering notion that the behavior of complex systems could be predicted accurately provided that one had enough information and intelligence. The concept (or theory) of chaos has challenged this historic approach. The ground rules are changing!

The "sufficient information" doctrine first was challenged at the atomic level by quantum mechanics in the 1920s. In the 1980s, prior tenets received another setback with the emergence of chaos theory. This theory holds that for microscopic or macroscopic systems, tiny variations in initial conditions sometimes may create unexpected, radically different outcomes, seemingly making it impossible to predict fully the behavior of some systems. Perhaps most startling of all, such behavior can arise in relatively simple systems governed by a few uncomplicated equations. Thus, relatively simple or highly complex systems can exhibit chaos. During the course of the first score of years of its existence, chaos theory generated wide interest in academia, but relatively few practical examples. However, quite recently, the science of system dynamics has entered a new era, one that is comparable to the time frame when quantum mechanics was "fleshing out."

A physicist at the Electric Power Research Institute recently observed, "With chaos, we're on the brink of a new classical dynamics and people thought that classical physics was dead." Another scientist has observed, "It's called the curse of dimensionality"—the amount of data you need to understand a system rises exponentially with the system's dimensionality, that is, the number of independent variables or degrees of freedom needed to describe it. Some of the projects involving what we thought would be simple questions have turned out to be very difficult. And, of course, there's the problem of noise. In many cases, it may be very hard to get data sets that are sufficiently tidy for understanding chaos. On the other hand, chaos theory can help us learn the limits of predictability for very complex systems, such as the weather, and may even give us new tools for controlling these systems.

The implications of chaos theory for electric power equipment and networks are both disturbing and exciting. On the one hand, an unsuspected potential for instability may lurk among the operating conditions of systems thought to be well understood. Sudden voltage collapses on power grids, for example, may indicate the presence of underlying chaotic dynamics. On the other hand, understanding chaos may provide unprecedented control over some of the most complex and elusive natural processes, such as combustion, corrosion, and superconductivity.

Researchers observe, "The problem is how to distinguish 'deterministic chaos' from stochastic, or totally random behavior. Chaos has an underlying order, a pattern that's not periodic, but is not completely random either. In any real system, however, some stochastic processes are also likely to be present as noise. It's like looking for a fuzzy pattern through a fog."

As early as 1899, Jules-Henri Poincaré (France) recognized the possibility for chaotic behavior in dynamics systems. However, it was not until 1961 that the meteorologist, Edward Lorenz, observed the phenomenon when he was attempting to construct a simple computer model of weather on the basis of convection currents in the earth's atmosphere.

Lorenz mapped a three-dimensional pattern (called a butterfly) that commonly appears when plotting chaotic data. The development of chaos science to where it is today is exquisitely summarized: "It took more than a decade and a half for this phenomenological pattern to gain enough recognition to be named and it took even longer for investigation of chaos to earn scientific respectability."

Processes currently under investigation with reference to chaos include fluidized-bed combustion, electric power grids, and chaos as related to fractal geometry.

See **Mathematics (State of the Art Reviews)**.

UVEITIS. Uveitis refers to an inflammation of the middle layer of the eye, the uvea the part of the eye that supplies blood and nutrients to the eye. Any damage to this part of the eye can affect vision. Blood vessels must remain out of the way of the cornea and lens in order for people to have clear vision. When the vessels become irritated or inflamed, blood cells and other matter can leak out and cause cloudiness, blurring, redness and other symptoms.

There are two forms of uveitis:

Anterior uveitis, the most common form, occurs in the iris. It is often associated with autoimmune diseases, such as rheumatoid arthritis and Lupus, and Lyme disease. Anterior uveitis usually occurs in people 20 to 50 years old, and it can affect one or both eyes. It usually lasts a few days to a few weeks even with treatment, and recurrences are common. See also **Lyme Disease**.

Posterior uveitis refers to the inflammation of the uvea that is in the back portion of the eye. Posterior uveitis is often associated with body infections, but many systemic diseases can be the cause. Uveitis is one of the leading causes of blindness in the United States.

It may be caused with an underlying systemic disease, or it may occur as a result of trauma to the eye.

Uveitis can either be acute or chronic. Acute uveitis is a result most often associated with blunt trauma. The chronic form is more often associated with other systemic disorders.

It is important to note, however, that many patients do not have obvious underlying causes for their uveitis and must undergo a comprehensive evaluation to determine the cause, if possible.

Symptoms of uveitis can include pain, redness, sensitivity to light and blurring. It can occur suddenly. Some patients complain of a deep, dull aching of the eye.

The first step in treating uveitis is detection. When treating anterior uveitis, eye care professionals aim at first reducing the inflammation and any discomfort, and second, preventing worsening of the condition. Treatment varies from patient to patient based, of course, on the patient's history and severity of the condition. Frequently, treatment includes the use of a steroid eye medicine to reduce the inflammation, a pain medicine for comfort, and possibly other medicines depending upon other present conditions. Patients are generally re-evaluated by their eye care professional within a range of one to seven days after beginning treatment.

To treat posterior uveitis, the physician must first discover the systemic disease that may be causing the inflammation. Then the disease can be treated. Antiinflammatory medications, steroids, and even in some cases, chemotherapy can help decrease inflammation in the eye.

Posterior uveitis usually lasts much longer than anterior uveitis, continuing for months to years, and it can cause permanent vision loss despite treatment.

As with many vision problems, uveitis can result in other complications such as glaucoma, retinal damage and vision loss if left untreated. Patients with symptoms should consult their eye care professional immediately. See also **Glaucoma**; **Retina**; and **Vision and the Eye**.

Vision Rx, Inc., Elmsford, NY.

V

V ANTENNA.   A V-shaped arrangement of conductors, balanced-fed at the apex, and with included angle, length, and elevation proportioned to give the desired directivity.

VACCINATION.   The process of introducing a **vaccine** in the body for the purpose of immunization against disease. The immunity is the result of the stimulation of **antibody** production against the organism contained in the vaccine. Vaccination was first used against smallpox, but is now carried out as a method of immunizing against many diseases including Anthrax, Diphtheria/Tetanus/Pertussis (DTaP), Hepatitis A, Hepatitis B, *Haemophilus Influenzae* type b (Hib), Influenza, Lyme Disease, Measles/Mumps/Rubella (MMR), Meningococcal, Pneumococcal Polysaccharide, Polio, and Varicella (Chickenpox). See also **Anthrax; Chickenpox; Diphtheria; Liver; Influenza; Lyme Disease; Measles; Meningitis; Mumps; Pertussis (Whooping Cough); Poliomyelitis; Rubella;** and **Tetanus**.

VACCINES.   A preparation which on injection will induce and active immunity in the body. Vaccines are made up of dead or attenuated infectious agents, bacteria or viruses, and each one is specific: the intracutaneous inoculation of tetanus vaccine protects the individual against tetanus, and the rabies vaccine protects against rabies, etc.

Perhaps the greatest success story in public health is the reduction of infectious diseases resulting from the use of vaccines. Routine immunization has eradicated smallpox from the globe and led to the near elimination of wild polio virus. Vaccines have reduced preventable infectious diseases to an all-time low and now few people experience the devastating effects of measles, pertussis and other illnesses. Prior to approval by the Food and Drug Administration (FDA), vaccines are extensively tested by scientists to ensure that they are effective and safe. Vaccines are the best defense we have against infectious diseases. However, no vaccine is 100% safe or effective. Differences in the way individual immune systems react to a vaccine account for rare occasions when people are not protected following immunization or when they experience side effects.

As infectious diseases continue to decline, some people have become less interested in the consequences of preventable illnesses like diphtheria and tetanus. Instead, they have become increasingly concerned about the risks associated with vaccines. After all, vaccines are given to healthy individuals, many of whom are children, and therefore a high standard of safety is required. Since vaccination is such a common and memorable event, any illness following immunization may be attributed to the vaccine. While some of these reactions may be caused by the vaccine, many of them are unrelated events that occur after vaccination by coincidence. Therefore, the scientific research that attempts to distinguish true vaccine side effects from unrelated, chance occurrences is crucial. This knowledge is necessary in order to maintain public confidence in immunization programs. As science continues to advance, we are constantly striving to develop safer vaccines and improve delivery in order to better protect ourselves against disease.

More than two hundred years ago, Edward Jenner, a country physician practicing in England, noted that milkmaids rarely suffered form smallpox, a disease that was known to kill up to 40 percent of those who contracted it. The milkmaids often did get cowpox, a related but far less serious disease, and those who did, never became ill with smallpox. In an experiment that was to prove a revelation, Jenner took a few drops of fluid from a skin sore of a woman who had cowpox and injected the fluid into the arm of healthy young boy who had never had cowpox or smallpox. Six weeks later, Jenner injected the boy with fluid from a smallpox sore, but the boy remained free of the dreaded smallpox. Dr. Jenner had discovered one of the fundamental principles of immunization. He had used a relatively harmless foreign substance to evoke an immune response that would protect someone from a disease-causing *microbe*.

In those days, a million people died from smallpox each year in Europe alone, most of them children. Those who survived were often left with grim reminders of their ordeals: blindness, deep scars, deformities. When Jenner laid the foundation for modern vaccines in 1796, he started on a course that would ease the suffering of people around the world. By the beginning of this century, vaccines for rabies, diphtheria, typhoid fever and plague were in use, in addition to the vaccine for smallpox.

Yet vaccination was not immediately accepted. The idea of deliberately introducing a potentially harmful microbe into people was met with suspicion and even outrage by many in the medical and scientific communities and public opinion was bitterly divided over the merits of vaccination. It took some time to convince people that the benefits of vaccination outweigh the few risks. Today vaccines are far safer and more protective than those early vaccines. And as science advances, we are developing even better vaccines to protect ourselves from disease.

Benefits

Disease prevention is the key to public health. Vaccines benefit in particular the people who receive them, and in turn those people cannot spread disease to others who have not been vaccinated. Infection cannot spread if it never gains a foothold. Infectious diseases cause enormous suffering, strain the capabilities of our health care system, and deplete financial resources. For the individual, the health care provider, and in the interest of conserving human and financial resources, it is always better to prevent a disease than to treat it.

Veterinary vaccines benefit people, too. Some diseases, such as rabies, anthrax, certain types of encephalitis, and Rift Valley fever, are readily transmissible from animal species to humans. In many instances, livestock and pets are vaccinated not only for their own health, but also for that of their owners.

In the United States, federal and state public health programs help assure that children receive vaccines. Many childhood diseases that were a normal part of growing up just 50 years ago are now preventable. Measles, rubella (German measles), mumps, pertussis, (whooping cough), and chickenpox were almost unavoidable. Most people did not reach adulthood without their families or circle of friends being touched by a serious illness or death caused by an infectious disease. For the most part, children suffered through the course of the disease and were left with *naturally acquired immunity*, some schoolwork to catch up on, and perhaps a little pockmark somewhere on their skin. However, in some cases, children died, or they were left with permanent loss of hearing or sight or other tragic effects of serious infections.

Adult Immunization

Although most of us receive the great majority of our immunizations during childhood, it is important to remember that vaccines are not just for young children. Adolescents and adults should keep up-to-date on tetanus and diphtheria immunizations. Adults, who have not had diseases such as measles or chickenpox during childhood, or the vaccines to prevent them, should consider being immunized. Ironically, childhood diseases such as measles, mumps, and chickenpox can be far more serious in adults.

People who travel overseas should determine, together with their physicians or at international travel clinics, which vaccines would be appropriate based on their destinations. Effective vaccines are available to prevent yellow fever, polio, typhoid fever, hepatitis A, cholera, and other bacterial and viral diseases that are more prevalent abroad than in the United States.

Each year, as we prepare for winter and the flu season, many adults should consider the benefits of the flu vaccine. In addition to flu vaccine, immunizations for pneumococcal pneumonia, hepatitis A, and hepatitis B are recommended for people who may be at risk.

Evaluating a Vaccine

Variations in individuals and their immune systems are many and subtle; thus no vaccine is totally effective. In the United States, a vaccine is approved for general use if it fulfills several stringent requirements.

- The vaccine must be safe. Although it is quite unlikely that a vaccine will ever be 100 percent safe, it must produce *protective immunity* with only minimal side effects (such as redness and soreness at the vaccination site) for the overwhelming majority of those who receive it. More discomfort in side effects can be acceptable, however, depending upon the severity of the disease the vaccine is designed to prevent. For example, most people would consider vaccine side effects that mimicked the symptoms of a bad cold acceptable if the vaccine protected them from HIV disease.
- The vaccine must be *immunogenic*, that is, it must cause a strong and measurable immune response. Vaccines usually contain *antigens*, bits of material, sometime form the disease-causing microbe itself, which can stimulate the immune system to respond and fight off a potential infection. When a vaccine is immunogenic, it primes the recipient's immune system to recognize the disease-causing microbe and launch a counterattack before illness can occur. In addition, the vaccine must induce the right type of immunity. When microbes invade, they cause disease in different ways, and different parts of the immune system respond to fight them. Vaccines must stimulate the specific parts of the immune system that protect against a particular kind of organism.
- The vaccine must be stable during its shelf life, that is to say, its *potency* must remain at the proper level for the vaccine to evoke an immune response. Many *inactivated vaccines* are simple to store, since they are in powdered form and are reconstituted with the appropriate fluid before they are given. *Live, attenuated vaccines*, however, require refrigeration from manufacturer to clinic to maintain stability and potency.

All approaches to vaccine development focus on the immune system and the body's natural defenses against foreign invaders. To understand something of how vaccines work, it is best to start with the immune system. See **The Immune System** discussed later in the entry. Together, your immune system and vaccines are powerful allies in the fight against disease.

Protection Against Microbes

The body has an arsenal of defenses that it uses to ward off foreign invaders. Healthy immune systems recognize as foreign many things that are not self.

Most of the foreign invaders that confront the human immune system are microscopic. Fungi, parasites, bacteria, and viruses populate our planet in far greater numbers than any other living organisms. Some are beneficial. We coexist quite happily with certain types of bacteria that live in our gastrointestinal and genital tracts. These bacteria help prevent infection by harmful organisms. We are even dependent on some microorganisms that inhabit our gastrointestinal tracts for their help in digesting food. Whether beneficial or harmful, however, all foreign microbes in the human body display special markers on antigens called *antigenic determinants* or *epitopes*. It is these antigens and their determinants that the immune system may recognize as harmful and identify for destruction.

The Immune System

Organs of the Immune System. The immune system is complex of organs—highly specialized cells and even a circulatory system separate from blood vessels—all of which work together to clear infection.

The organs of the immune system, positioned throughout the body, are called *lymphoid organs*. The word "lymph" in Greek means a pure, clear stream—and appropriate description considering its appearance and the purpose of our immune system.

Lymphatic vessels and *lymph nodes* are the parts of the special circulatory system that carries lymph, a transparent fluid containing white blood cells, chiefly *lymphocytes*. Lymph bathes the tissues of the body, and the lymphatic vessels collect and move it eventually back into the blood circulation. Lymph nodes dot the network of lymphatic vessels and provide meeting grounds for the immune system cells that defend against invaders. The *spleen*, at the upper left of the abdomen, is also a staging ground and a place where immune system cells confront foreign microbes.

Both immune cells and foreign molecules enter the lymph nodes via blood vessels or lymphatic system and eventually return to the bloodstream. Once in the bloodstream, lymphocytes are transported to tissues throughout the body, where they act as sentries on the lookout for foreign antigens.

How the Immune System Works. Cells will grow into the many types of more specialized cells that circulate throughout the immune system are produced in the bone marrow. This nutrient-rich, spongy tissue is found in the center shafts of certain long, flat bones of the body, such as the bones of the pelvis. The cells most relevant for understanding vaccines are the lymphocytes, numbering close to one trillion.

The two major classes of lymphocytes are *B cells*, which grow to maturity in the bone marrow, and *T cells*, which mature in the thymus, high in the chest behind the breastbone.

B cells produce *antibodies* that circulate in the blood lymph streams and attach to foreign antigens to mark them for destruction by other immune cells. B cells are part of what is known as *antibody-mediated* or *humoral immunity*, so called because the antibodies circulate in the blood and lymph, which the ancient Greeks called, the body's "humors." Certain T cells, which also patrol the blood and lymph for foreign invaders, can do more than mark the antigens; they attack and destroy diseased cells they recognize as foreign. T cell lymphocytes are responsible for *cell-mediated immunity* (or *cellular immunity*). T cells also orchestrate, regulate and coordinate the overall immune response. T cells depend on unique cell surface molecules called the *major histocompatibility complex* (MHC) to help them recognize antigen fragments.

Antibodies. The antibodies that B cells produce are basic templates with a special region that is highly specific to target a given antigen. Much like a car coming off a production line, the antibody's frame remains constant, but through chemical and cellular messages, the immune system selects a green sedan, a red convertible or a white truck to combat this particular invader. However, in contrast to cars, the variety of antibodies is very large. Different antibodies are destined for different purposes. Some coat the foreign invaders to make them attractive to the circulating scavenger cells, *phagocytes*, which will engulf an unwelcome microbe. When some antibodies combine with antigens, they activate a cascade of nine proteins, known as *complement*, that have been circulating in inactive form in the blood. Complement forms a partnership with antibodies, once they have reacted with antigen, to help destroy invaders and remove them form the body. Still other types of antibodies block viruses for entering cells.

T Cells. T cells have two major roles in immune defense. Regulatory T cells are essential for orchestrating the response of an elaborate system of different types of immune cells. *Helper T cells*, for example, also known as CD4 positive T cells (CD4+ T cells), alert B cells to start making antibodies; they also can activate other T cells and immune system scavenger cells called *macrophages* and influence which type of antibody is produced. Certain T cells, called CD8 positive T cells (CD8+ T cells), can become killer cells that attack and destroy infected cells. The *killer cells* are also called *cytotoxic T cells* or CTLs (cytotoxic lymphocytes).

Naturally Acquired Immunity. As early as 2500 years ago in Greece, some people understood enough about contagion to know that a person who had recovered from plague would not get it again. Later, physicians recognized that a person acquires immunity to many diseases in this way.

This protection comes for another special cell of the immune system. Whenever B cells and T cells are summoned, they transform some of their numbers into *memory cells*. Although the army of antibodies necessary to destroy an infectious agent does not remain, the memory cells do. If the memory cells recognize the invader again, the immune system quickly mounts a defense and defeats the interlopers before illness can occur. This type of immunity is naturally acquired. Most of us benefit from naturally acquired immunity from our earliest days. Certain types of immune cells are passed form mother to fetus during pregnancy. Thus, most newborns, though vulnerable in many ways, have a head start in fighting disease.

Artificially Acquired Immunity. Artificially acquired immunity can be either passive or active. Artificially acquired passive immunity results when antibodies produced by another animal or human are given to someone to prevent or treat disease. For example, administering tetanus antitoxin or rabies immune globulin to someone is a way of conferring passive immunity. This type of immunization is effective very quickly, but since it lasts only a short time, it is used to protect people when they are particularly vulnerable, such as immediately after exposure to a serious disease.

Artificially acquired active immunity is achieved through safe and effective vaccines. Traditional vaccines are preparations of killed or weakened bacteria or viruses, or parts of these microbes, or *inactivated toxins* from the disease-causing agent. Recently, innovative vaccine technologies have revealed many more ways to give people active immunity to a disease. These include *subunit vaccines, conjugate vaccines* and *naked DNA vaccines.*

Different Types of Vaccines

When a new disease emerges or a familiar one becomes a more significant health threat than it has been in the past, scientists, physicians and public health workers recognize the need for a new way to prevent the disease. Once scientists have identified the organism or toxin that causes the illness, they pursue a number of approaches to develop a vaccine.

Vaccine development has its early roots in the work of Edward Jenner, who discovered how to protect people from smallpox, and Louis Pasteur, who developed a vaccine to protect from rabies. Those pioneering efforts subsequently led to vaccines for diseases that once claimed millions of lives worldwide.

The purpose of a vaccine is to bring about active immunity by provoking a response from a person's immune system — marshaling B and T cells to swing into action — and creating a memory within the immune system so that exposure to the active disease agent will stimulate an already primed immune system to fight the disease. Some vaccines are combinations that protect against several diseases. Most of us are familiar with the DTP (diphtheria, tetanus, pertussis) and MMR (measles, mumps, rubella) vaccines that children in the United States receive. Scientists extensively test these combination vaccines to make sure that none of the antigens detracts from the immune priming effect of the others. Thus the vaccines can provide triple protection, the recipients are spared extra needle sticks, and the public health costs are reduced.

Vaccines Licensed in the United States include:

- Adenovirus vaccine.
- Anthrax vaccine.
- Bacille Calmette-Guérin vaccine.
- Cholera vaccine.
- Diphtheria toxoid.
- Diphtheria and tetanus toxoids (DT).
- Diphtheria and tetanus toxoids and acellular pertussis vaccine (DTP).
- Diphtheria and tetanus toxoids and whole cell pertussis vaccine.
- Diphtheria and tetanus toxoids and pertussis and *Haemophilus influenzae* b (Hib) conjugate vaccine.
- Hib conjugate vaccine (with diphtheria, meningococcal, and tetanus conjugates).
- Hepatitis A vaccine.
- Hepatitis B vaccine.
- Influenza virus vaccine.
- Japanese encephalitis virus vaccine.
- Measles virus vaccine.
- Measles and mumps virus vaccine.
- Measles, mumps, and rubella virus vaccine (MMR).
- Meningococcal vaccine, Group A.
- Meningococcal vaccine, Group C.
- Meningococcal vaccine, Groups A and C.
- Meningococcal vaccine, Groups A, C, Y, and W-135.
- Mumps virus vaccine.
- Pertussis vaccine (acellular).
- Pertussis vaccine (whole cell).
- Plague vaccine.
- Pneumococcal vaccine.
- Poliovirus vaccine — inactivated
- Poliovirus vaccine — live, attenuated.
- Rabies vaccine.
- Smallpox vaccine.
- Tetanus toxoid.
- Typhoid vaccine.
- Varicella (chickenpox) virus vaccine.
- Yellow fever vaccine.

Based on the biological and chemical characteristics of the disease-causing agent and on what type of immunity is desired, researchers begin to develop one of the following types of vaccines. Vaccines can be produced from 10 inactivated (killed), 2) live, attenuated (weakened), or 3) synthetic (laboratory-made) microbial materials.

Traditional Vaccines

Inactivated Vaccines. Inactivated vaccines are produced by killing the disease-causing microorganism with chemicals or heat. Such vaccines are stable and safe; they cannot revert to the *virulent* (disease-causing) form. They often do not require refrigeration, a quality that makes them accessible to the people of many developing countries, as well as practical for vaccinating people who are highly mobile, such as members of the armed forces. However, most inactivated vaccines stimulate a relatively weak immune response and must be given more than once. A vaccine that requires several doses (boosters) has a limited usefulness, especially in areas where people have less access to regular health care.

The flu shot is an inactivated vaccine, as are the vaccines for cholera, plague, and hepatitis A.

Live, Attenuated Vaccines. To make a live, attenuated vaccine, the disease-causing organism is grown under special laboratory conditions that cause it to lose its virulence, or disease-causing properties. Although live vaccines require special handling and storage in order to maintain their potency, they produce both antibody-mediated and cell-mediated immunity and generally require only one boost, or additional dose. Most live vaccines are injected; some, however, such as the polio vaccine, administered in the nose, show promise in preventing flu.

While there are advantages to live vaccines, there is one caution. It is the nature of living things to change, to *mutate*, and the organisms used in live vaccines are no different. There is a remote possibility that the organism may revert to a virulent form and cause disease. It is for this reason that live vaccines continue to be carefully tested and monitored.

For their own protection, people with compromised immune systems — such as people who are taking *immunosuppressive* drugs, people who have cancer or people living with HIV — are usually not given live vaccines.

The vaccines for yellow fever, measles, rubella, and mumps are all produced from live, attenuated organisms.

Toxoids. A *toxoid* is an inactivated toxin, the harmful substance produced by a microbe. Many of the microbes that infect people are not themselves harmful. It is the powerful toxins they produce that can cause illness. For example, the bacterium that causes tetanus is found everywhere in nature, and in an environment with plenty of oxygen, it is harmless. If that same organism is put into an environment without oxygen, however, the organism starts to change and produce tetanus toxin, a substance far more potent than the well-known poison sodium cyanide. To inactivate such powerful toxins, vaccine manufacturers treat them with materials known to completely cripple any disease-causing ability. *Formalin*, a solution of formaldehyde and sterile water, is most often used to inactivate toxins and produce toxoids.

Toxoids are used to immunize people against tetanus and diphtheria.

New and Second-Generation Vaccines

Scientists are using new technologies to improve traditional vaccines. These new second-generation vaccines, as well as vaccines for diseases that had not been preventable very long ago, are made using powerful techniques such as *recombinant genetic engineering* (also called recombinant DNA technology).

Conjugate Vaccines. The bacteria that cause some diseases, such as pneumococcal pneumonia and certain types of meningitis, have special outer coats. These coats disguise antigens so that the immature immune systems of infants and younger children are unable to recognize these harmful bacteria. In a conjugate vaccine, proteins or toxins from a second type of organism, one that an immature immune system can recognize, are linked to the outer coats of the disease-causing bacteria. This enables a young immune system to respond and defend against the disease agent.

Currently, conjugate vaccines are available to protect against a type of bacterial meningitis caused by *Haemophilus influenzae* type b (Hib). Meningitis, an inflammation of the fluid-filled membranes that protect the brain and spinal cord, can be fatal or it can cause severe, life-long disabilities such as deafness and mental retardation. Since Hib vaccines have been widespread use in the United States, Hib meningitis has nearly disappeared among babies and young children.

Subunit Vaccines. Sometimes vaccines developed form antigenic fragments are able to evoke an immune response, often with fewer side effects than might be caused by a vaccine made from the whole organism. Subunit vaccines can be made by taking apart the actual microbe, or they can be made in the laboratory using genetic engineering techniques.

Today, subunit vaccines are used to protect against pneumonia caused by *Streptococcus pneumoniae* and against a type of meningitis.

A *recombinant subunit vaccine* for hepatitis B virus infection is now licensed for use in the United States. The recombinant vaccine is made by inserting a tiny portion of the hepatitis B virus' genetic material into common baker's yeast. The process induces the yeast to produce an antigen, which is then purified. The purified antigen, when combined with an *adjuvant*, a substance that stimulates the immune system, results in a safe and very effective vaccine.

Recombinant Vector Vaccines. A vaccine *vector*, or carrier, is a weakened virus or bacterium into which harmless *genetic material* from another disease-causing organism can be inserted.

The vaccinia virus, the virus that caused cowpox, is now used to make *recombinant vector vaccines*. In the submicroscopic world of viruses, vaccinia is relatively large and has ample room to accept additional genetic fragments. A vaccinia virus with several genes from the human immunodeficiency virus (HIV) is currently being tested as a vaccine for acquired immune deficiency syndrome (AIDS). In addition, a close relative of vaccinia, canarypox virus, engineered with harmless fragments of HIV, is being tested in human volunteers as a vaccine for AIDS.

Similarly, scientists are testing a weakened bacterium — salmonella — to carry portions of such microbes as the hepatitis B virus. Currently no recombinant vector vaccines are licensed for general use in the United States.

Vaccine Development and Testing

To be approved for general use, a candidate vaccine must go through a long period of testing and validation. The time between discovery of a disease agent and production of a widely available vaccine has been as long as 50 years. Today, with biological synthesis and recombinant vaccine development techniques, the length of time from basic research to availability of licensed product can sometimes be greatly reduced.

Basic Research and Development. Basic research focuses on biochemistry and physiology, and on mechanisms that disease-causing microbes use to cause damage. Such research also takes into account the biophysical characteristics of the organisms that might be used in vaccines or drugs to prevent or interrupt the disease process.

To develop a candidate vaccine, scientists test vaccine preparations in cell-culture, and often eventually in animals such as mice, guinea pigs, or even monkeys. In some cases, computers can help researchers visualize the vaccine candidates in three dimensions to predict how vaccine antigens will interact with the immune system. If the vaccine candidate is shown to be promising throughout the *preclinical* evaluations, it can become an *investigational vaccine*.

An investigational vaccine is one that successfully has gone through basic research and developmental processes, often including preclinical trials in animals, and has been approved by the U.S. Food and Drug Administration (FDA) for use in human volunteers in clinical trials.

Clinical Studies. Clinical studies rely entirely upon the participation of volunteers, people who contribute their time and energy for the advancement of science and improved health care for all. Thousands of volunteers of all ages and from all walks of life have participated in these studies. A typical volunteer in a vaccine study agrees to be given the vaccine, makes frequent visits to a clinic for evaluation, participates in medical testing, and provides blood or tissue samples that will be used in assessing the vaccine's safety and potential effectiveness. Unlike the boy vaccinated by Dr. Jenner 200 years ago, volunteers today must sign an informed consent document indicating their understanding of the study, its risks, and their willingness to participate.

A candidate vaccine undergoes three phases of clinical trials before it can be licensed for public use. Phase I trials, to determine the safety of various doses of the vaccine, usually begin with small numbers of volunteers, and then expand to include more volunteers if the vaccine appears to be safe.

Phase II trials, to determine whether the vaccine is safe and immunogenic, are open to hundreds of volunteers. The vaccine is tested for safety, for its ability to evoke an immune response, and for its potential to prevent disease.

Phase III trials are large-scale efficacy studies, often in thousands of individuals, to confirm that the vaccine safely prevents disease. A vaccine is considered successful if its overall effect is beneficial; it should prevent disease, and any side effects should be minimal. If the disease the vaccine is designed to prevent is rare in the United States, Phase III trials may be conducted in a country where the disease is prevalent. In the cases of such international cooperation, each government signs an agreement and expects its citizens to benefit from the study.

Side Effects and Adverse Reactions. The most common side effects of vaccines include low-grade fever, soreness and redness at the site of the injection, or sometimes body aches for up to 24 hours after vaccination. These minor side effects of a vaccine are far preferable to having the disease.

The extensive testing that a vaccine undergoes before it is licensed for public use is conducted, in part to assure safety as much as possible by closely observing large numbers of volunteers for harmful side effects. But no matter how thorough the testing, it is impossible to allow completely for the extensive variation among individuals, their immune systems, and their reactions to the introduction of new substances into their bodies. Serious systemic reactions to vaccines can occur, although they are very rare. The FDA and the Centers for Disease Control and Prevention monitor vaccine distribution and use. Information about adverse reactions to a vaccine is collected even after the vaccine is licensed for general use. Both organizations record any incident of a serious reaction and follow up on any re-evaluation of the vaccine that is necessary.

Basic and Clinical Research

Basic research conducted and supported by the National Institute of Allergy and Infectious Diseases (NIAID's) is helping scientists understand more about infectious microbes and human immune responses. From this understanding has come the development of vaccines and other tools needed to prevent many infectious diseases and immune system disorders. The goal of these efforts is not only to protect individuals from serious infections; but eventually to eradicate diseases, as we have seen with smallpox.

In clinical research, NIAID revolutionized the classical but cumbersome, piecemeal approach to clinical vaccine studies by designing a network of vaccine and treatment evaluation units, and more recently, an international network for testing vaccines to prevent infection with HIV. Testing sites are based at leading university medical research centers, public health departments, and community clinics. Investigators working within these networks and other NIAID-supported researchers played major roles in the clinical studies required for licensing of vaccines for Hib meningitis and or new *acellular* pertussis vaccines.

The Children's Vaccine Initiative

NIAID has been recognized by Congress as the lead agency to provide the scientific and programmatic direction for the Children's Vaccine Initiative (CVI). The CVI, a worldwide public health effort, combines the efforts of many international scientists and representatives of the United Nations International Children's Emergency Fund, the World Health Organization, the World Bank, the United Nations Development Program, the Rockefeller Foundation and vaccine manufactures, biotechnology firms, and national research agencies. The CVI has made impressive strides towards its goal of universal immunization. Already 80 percent of all children in the United States receive their initial immunizations by their first birthdays. NIAID is making progress toward achievable goals, such as making existing vaccines safer and more effective and improving vaccines so that fewer doses will be needed.

Vaccines of the Future

Currently scientists are pursuing many promising new strategies in vaccine development and exploring novel ways to administer vaccines. The following descriptions of just a few of these innovative ideas provide a preview of safer, more effective ways to fight disease.

Exposing *mucosal membranes* to vaccines is a strategy that can produce an immune response in a less-stressful and better-targeted manner. Mucosal membranes are located throughout our bodies, but are most accessible in the lungs, nose, mouth, throat, gastrointestinal tract, rectum and vagina. The oral polio vaccine, in use since the 1950's, is an early example of the effectiveness of this strategy. Another possible mucosal route of administration is through the nose, and flu vaccines may soon be widely available in a nasal spray. Researchers have shown that the route of entry a disease-causing organism takes is often an effective vaccine route as well.

Many vaccine improvements may result from progress in designing better adjuvants. At present, only an aluminum salt called *alum* is approved as an adjuvant by the FDA, but scientists are studying many new natural and synthetic compounds.

Scientists are also looking at new ways of presenting the vaccine to the immune system. *Microspheres*, tiny spheres containing bits of antigenic material, show promise in that they can release small doses of vaccine over extended periods of time as the microspheres gradually dissolve in the body. This means that someone may be able to receive two or three doses in just one administration of vaccine.

Perhaps the most exciting new vaccine technique is introducing pure genetic material directly into the body. This genetic material, called "naked DNA," encodes a few proteins from a disease-causing organism. The DNA is then incorporated into the body's own cells, which make the proteins encoded by the new DNA. It is these proteins that are recognized as foreign and stimulate the immune system. In this way, the DNA will have an effect similar to that of a live, attenuated vaccine. In effect, the DNA will produce antigens for years and induce strong, long-lasting immunity. At the same time, the exclusion of genes that are critical to the disease-causing organism's survival will assure that the vaccines are safe and do not actually cause disease.

Researchers are also exploring ways to create edible vaccines. By genetically engineering plants to incorporate synthetic antigens, scientists may be able to develop a banana or potato, for example, that will produce protective immunity when eaten. Obviously, such a vaccine technique would greatly simplify immunization for many people of the world.

Vaccines remain among the most powerful tools we have for disease prevention, and advances in biotechnology have ushered in a new era in vaccine development that holds even more promise for improving public health. NIAID remains a leader in the discovery and testing of new and improved vaccines and will continue to nourish this exciting renaissance in vaccine development.

See also **Immune System and Immunology**.

An abridged glossary of terms used to describe basic vaccine science and the immune system would include:

Acellular vaccine: Devoid of whole cells. Acellular vaccines contain only portions of cellular material.

Active immunity: Immunity produced by the body in response to stimulation by a disease-causing organism (naturally acquired active immunity) or by a vaccine (artificially acquired active immunity).

Adjuvant: A substance sometimes included in a vaccine formulation to enhance the immune-stimulating properties of a vaccine.

Antibody: Soluble protein molecule produced and secreted by B cells in response to an antigen and capable of binding to that specific antigen.

Antibody-mediated immunity (humoral immunity): Immune protection provided by B cells, which secrete antibodies in response to antigen (as distinct from that provided by the direct action of immune cells or cellular immunity).

Antigen: A substance that provokes an immune response.

B cells: Small white blood cells crucial to the immune defenses. Also known as B lymphocytes, they are derived from bone marrow and develop into plasma cells, which produce antibodies.

Bone marrow: Soft tissue located in the cavities of the bones. The bone marrow is the source of all blood cells.

Cell-mediated immunity (cellular immunity): Immune protection provided by the direct action of immune cells (as distinct from that provided by soluble molecules such as antibodies).

Complement: A complex series of blood proteins whose action "complements" the work of antibodies. Complement destroys antibody-coated cells, produces inflammation, and regulates immune reactions.

Conjugate vaccine: A vaccine in which proteins that are easily recognizable to the immune system are linked to the outer coat of the disease-causing organism to promote an immune response.

Cytotoxic T cells: A subset of T lymphocytes that can kill other cells infected by viruses, fungi or certain bacteria, or cells transformed by cancer.

Efficacy: In vaccine research, the ability of a vaccine to produce a desired clinical effect, such as protection against a specific disease, at the optimal dosage and schedule in a given population. A vaccine may be tested for efficacy in Phase III studies if it appears to be safe and shows promise in smaller Phase I and II studies.

Epitope: A unique shape or marker carried on an antigen's surface, which triggers a corresponding antibody response.

Formalin: A solution of water and formaldehyde, used as an antiseptic, disinfectant or fixative.

Genetic material: Deoxyribonucleic acid (DNA) that carries the directions a cell uses to perform a specific function, such as making a given protein.

Haemophilus influenzae type b (HIB): A bacterium found in the respiratory tract that causes acute respiratory infections, including pneumonia, and other diseases such as meningitis.

Helper T cells: A subset of T cells that typically carry the CD4 marker and are essential for turning antibody production on, activating cytotoxic T cells, and initiating many other immune responses.

Humoral immunity: Immune protection provided by soluble factors such as antibodies, which circulate in the body's fluids, primarily serum and lymph. See **Antibody-mediated immunity**.

Immunogenic: Capable of stimulating an immune response (immunogenicity).

Immunosuppressive: Capable of reducing immune responses. For instance, drugs given to prevent transplant rejection are immunosuppressive.

Inactivated toxins: Organic toxins, such as those produced by bacteria and viruses, that have been killed by chemical means, heat, or irradiation, and are no longer capable of causing disease.

Inactivated vaccine (Killed vaccine): A vaccine made from a whole virus or bacterium whose biological ability to grow or reproduce is ended.

Investigational vaccine: A vaccine that has been approved by the Food and Drug Administration for testing in humans, but that has not yet completed evaluation and been accepted for licensure and public use.

Killer T cells (cytotoxic T cells, cytotoxic lymphocytes, CTLs): A subset of T cells that can kill cancer cells and cells infected with viruses, fungi or certain bacteria.

Live, attenuated vaccine: A vaccine whose biological activity has not been inactivated, but whose ability to cause disease has been weakened.

Lymph: A transparent, slightly yellow fluid that carries lymphocytes, bathes the body tissues, and drains into the lymphatic vessels.

Lymph nodes: Small bean-shaped organs of the immune system, distributed widely throughout the body and linked by lymphatic vessels. Lymph nodes are gathering sites of B, T, and other immune cells.

Lymphatic vessels: A body-wide network of channels, similar to the blood vessels, that transport lymph to the immune organs and into the bloodstream.

Lymphocytes: Small white blood cells produced in the bone marrow and thymus that are essential in immune defense.

Lymphoid organs: Organs of the immune system, where lymphocytes develop and congregate. They include the bone marrow, thymus, lymph nodes, spleen, and various other clusters of lymphoid tissue, such as the appendix. The blood vessels and lymphatic vessels also can be considered lymphoid organs.

Major histocompatibility complex (MHC): A large set of cell surface molecules in each individual, encoded by genes. MHCs serve as unique biochemical markers of individual identity.

Macrophage: A large and versatile immune cell that acts as a microbe-devouring phagocyte, and antigen-presenting cell, or an important source of immune secretions.

Memory cell: A subset of T cells and B cells that have been exposed more readily when the immune system encounters the same antigens again.

Microbe: A minute living organism, such as a bacterium or virus.

Microspheres: Tiny, microscopic spheres that can carry vaccines or drugs and can pass easily through the body's tissues.

Mucosal membranes: The moist tissues lining body cavities or passages that have an opening to the external world, such as the mouth, nose, rectum or vagina. Mucosal immunity depends on immune cells and antibodies being present in the linings of the reproductive, respiratory, gastrointestinal tracts, and other mucosal membranes.

Mutate: To change a gene or unit of hereditary material that results in a new inheritable characteristic.

Naked DNA vaccine: Vaccine made up of deoxyribonucleic acid that is not encased or encapsulated. In naked DNA vaccines, genetic material is injected directly into the vaccine recipient.

Naturally acquired immunity: Immunity produced by immune cells passed from mother to fetus or by the body in response to exposure to a disease-causing organism.

Passive immunity: Immunity resulting from the transfer of cells or antibodies or antiserum produced by another individual.

Peyer's patches: A collection of lymphoid tissues in the intestinal tract.

Phagocyte: An immune cell that is able to ingest and destroy microbes and other foreign matter.

Potency: The strength of substance. In vaccines, one of the qualities, which makes a vaccine protective.

Preclinical: A phase of study of a vaccine or drug that is completed before clinical studies are carried out in people. Preclinical studies may be conducted in cells or in animals.

Protective immunity: Complete resistance to disease, whether long-lasting or temporary.

Recombinant DNA technology: The technique by which genetic material from one organism is inserted into a foreign cell or another organism in order to mass-produce the protein encoded by the inserted genes. (Also called recombinant genetic engineering.)

Recombinant subunit vaccine: A subunit vaccine made using recombinant DNA technology. See **subunit vaccine**.

Recombinant vector vaccine: A vaccine that combines a vector—a harmless bacterium or virus used to transport an antigen into the body to stimulate protective immunity—and an antigen or immunogen from an organism other than the vector.

Subunit vaccine: A vaccine that uses one or more components of a disease-causing organism, rather than the whole, to stimulate an immune response.

T cells: Small white blood cells that orchestrate or directly participate in immune defenses. Also known as T lymphocytes, they mature in the thymus.

Thymus: A primary lymphoid organ, high in the chest, where T lymphocytes proliferate and mature.

Toxoid: An inactivated or killed organic toxin used to immunize against specific bacteria.

Vector: In vaccine technology, a bacterium or virus that does not cause disease in humans and is used in genetically engineered vaccines to transport genes coding for antigens into the body to induce an immune response.

Virulent: Toxic, causing disease.

National Institute of Allergy and Infectious Diseases, National Institutes of Health: Bethesda, MD.

Additional Reading

Ada, G.L. and A.J. Ramsay: "Vaccines, Vaccination and the Immune Response," Lippincott-Raven Publishers, Philadelphia, PA, 1996.

Bazin, H. and E. Jenner: "The Eradication of Smallpox," Academic Press, Inc., San Diego, CA, 2000.

Kiyono, H., P.L. Ogra, and J.R. McGhee: "Mucosal Vaccines," Academic Press, Inc., San Diego, CA, 1996.

Levine, M.M., G.C. Woodrow, J.B. Kaper, and G.S. Cobon: "New Generation Vaccines," 2nd Edition, Marcel Dekker, Inc., New York, NY, 1997.

O'Hagan, D.T.: "Vaccine Adjuvants: Preparation Methods and Research Protocols," Vol. 42, Humana Press, Totowa, NJ, 2000.

Orenstein, W.A. and R. Zorab: "Vaccines," Harcourt Brace & Company, San Diego, CA, 1999.

Stanberry, L.R. and D. Bernstein: "Sexually Transmitted Diseases: Vaccines, Prevention and Control," Academic Press, Inc., San Diego, CA, 2000.

Web References

American Public Health Association (APHA): *http://www.apha.org/*

Food and Drug Administration (FDA): Center for Biologics Evaluation and Research: *http://www.fda.gov/cber/*

Institute for Vaccine Safety: Johns Hopkins School of Public Health: *http://www.vaccinesafety.edu/*

National Institute of Allergy and Infectious Diseases: *http://www.niaid.nih.gov*

National Institute of Health (NIH): *http://www.nih.gov*

The Centers for Disease Control and Prevention: National Immunization Program: *http://www.cdc.gov/nip/*

The Immunization Gateway: Your Vaccine Fact-Finder: *http://www.immunofacts.com/*

VACCINIA (Cowpox). The skin reaction which follows the introduction of the virus of vaccinia into the body, and which is followed by the development of immunity against vaccinia itself and smallpox. This process may be effected, accidentally but is commonly deliberate, in the form of vaccination, in which the virus derived from lesions on the skin of calves or sheep is inoculated intradermally. The strains of virus so used have been artificially attenuated in virulence by repeated passage through animals; in consequence the lesion produced is of minimal size consistent with effective immunization. Vaccinia is occasionally complicated, especially by encephalitis, which may be fatal, and it is in part due to this that compulsory vaccination is on the decline.

VACUOLE. A small globule of clear fluid in the cytoplasm of a cell. In preparations for microscopic study the contents of the vacuole are usually dissolved away, so that an open space alone remains, but even vacuoles whose contents are undisturbed usually appear vacant because of their transparency. In the multicellular body fat cells afford a good illustration of vacuoles as globules of fat accumulate in them. One-celled animals also offer a good example in the contractile or pulsating vacuole. This is a globule of clear liquid that forms and discharges periodically, sometimes at a fixed point in the cell. It is interpreted as an organ for the removal of surplus water with dissolved wastes from the protoplasm. See also **Cell (Biology)**.

VACUUM. According to definition, a space entirely devoid of matter. The term is used in a relative sense in vacuum technology to denote gas pressures below the normal atmosphere pressure of 760 torr (1 torr = 1 millimeter of mercury). The degree or quality of the vacuum attained is indicated by the total pressure of the residual gases in the vessel that is pumped. Table 1 shows generally accepted terminology for denoting various degrees of vacuum, together with pertinent pressure ranges; the calculated molecular density (from the equation $p = nkT$, where p is the pressure n is the molecular density, i.e., number of molecules per cubic centimeter; k is the Boltzmann's constant; and T is the absolute temperature taken to be 293 K (20 °C); and the mean free path λ from the approximate equation for air: $\lambda = 5/p$ centimeters, where p is the pressure in millitorr).

In the quantum field theories that describe the physics of elementary particles, the vacuum becomes somewhat more complex than previously defined. Even in empty space, matter can appear spontaneously as a result of fluctuations of the vacuum. It may be pointed out, for example, that an electron and a positron, or antielectron, can be created out of the void. Particles created in this way have only a fleeting existence; they are annihilated almost as soon as they appear, and their pressure can never be detected directly. They are called *virtual particles* in order to distinguish them from real particles. Thus, the traditional definition of vacuum (space with no real particles in it) holds. In their excellent paper, the aforementioned authors discuss how, near a superheavy atomic nucleus, empty space may become unstable, with the result that matter and antimatter can be created without any input of energy. The process may soon be observed experimentally.

Even when all matter and heat radiation have been removed from a region of space, the vacuum of classical physics remains filled with a distinctive pattern of electromagnetic fields. The discovery of a connection between thermal radiation and the structure of the classical vacuum reveals an unexpected unity in the laws of physics, but it also complicates the view of what was once considered simply *empty space*. But, even with its pattern of electric and magnetic fields, the vacuum remains the simplest state of

TABLE 1. VARIOUS QUALITIES OF VACUUM AND PRESSURE RANGES

Quality of Vacuum	Pressure Range (torr)	Molecular Density, n (molecules/cubic centimeter)	Mean Free Path, λ (centimeters)
Coarse or rough vacuum	760–1	$2.69 \times 10^{19} - 3.5 \times 10^{16}$	$6.6 \times 10^{-6} - 5 \times 10^{-3}$
Medium vacuum	$1-10^{-3}$	$3.5 \times 10^{16} - 3.5 \times 10^{13}$	$5 \times 10^{-3} - 5$
High vacuum	$10^{-3} - 10^{-7}$	$3.5 \times 10^{13} - 3.5 \times 10^{9}$	$5 - 5 \times 10^{4}$
Very high vacuum	$10^{-7} - 10^{-9}$	$3.5 \times 10^{9} - 3.5 \times 10^{7}$	$5 \times 10^{4} - 5 \times 10^{6}$
Ultrahigh vacuum	$<10^{-9}$	$<3.5 \times 10^{7}$	$>5 \times 10^{6}$

nature. Perhaps this statement reflects more on the subtlety of nature than it does on the simplicity of the vacuum.

Vacuum Pumps

Two widely used vacuum pumps are the mechanical rotary oil-sealed pump and the vapor pump. The former provides a medium vacuum and works relative to the atmosphere. The vapor pump, on the other hand, provides a high or very high vacuum and operates relative to a medium vacuum provided by a rotary pump, referred to as a backing pump in this connection. Thus, the most widely used high-vacuum system able to establish an ultimate pressure of about 10^{-6} torr or below consists of a vapor pump backed by a rotary pump.

Several patterns of rotary oil-sealed pumps exist, but they have in common the fact that the volume between a rotor (or rotating plunger) and a stator is divided into two crescent-shaped sections, which are isolated from one another as regards the passage of gas. Further, they are furnished with an intake port and a discharge outlet valve to the atmosphere. One revolution of the rotor (speeds of 450 to 700 revolutions per minute are used), gas is swept from the intake port, compressed, and discharged to the atmosphere via the one-way outlet valve. The mechanism is immersed in a low-vapor-pressure oil for sealing and lubrication in a small pump; larger units have a separate oil reservoir and feed device. A spring-loaded vane type of rotary oil-sealed pump is shown in Fig. 1. A single-stage pump of this kind provides an ultimate pressure of about 10^{-2} torr; a two-stage one with two units in cascade will give an ultimate of about 10^{-4} torr. Rotary pumps with speeds from 20 to 20,000 liters/minute are commercially available, the smallest being driven by an 1/8-hp motor, the largest requiring a 40-hp motor.

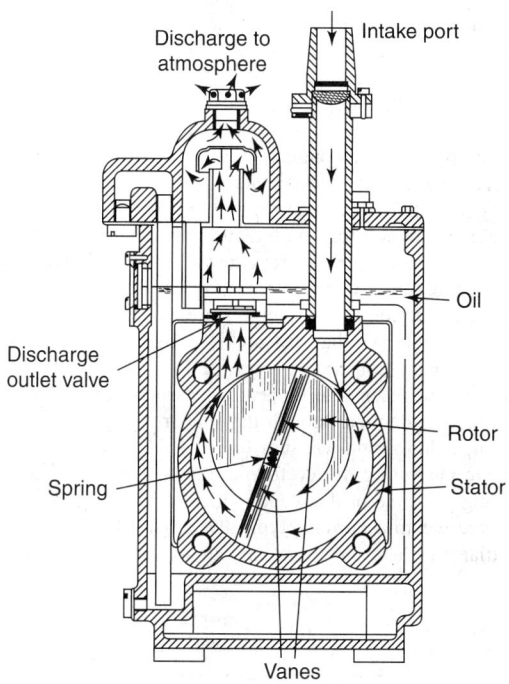

Fig. 1.   Spring-loaded vane-type rotary oil-sealed vacuum pump.

These pumps handle permanent gases efficiently. Condensable vapors, e.g., water vapor, are not satisfactorily pumped because they may liquefy during the compression part of the rotation. To prevent this, gas ballast is a common provision whereby air from the atmosphere is admitted to the pump through a simple, adjustable screw valve to the region between the rotor and stator just before the discharge outlet valve. The amount of extra air admitted is readily adjusted to provide a compressed gas-vapor mixture, which opens the discharge valve before vapor condensation occurs. Gas ballasting will clearly increase significantly the ultimate pressure provided by the pump, but this is not important since the gas-ballast valve can be closed after initial pumping has removed most of the water vapor.

Vapor pumps are of two main types: vapor diffusion pumps and vapor ejector pumps. Both employ vapor (of either mercury or a low-vapor-pressure oil) issuing from a jet as a means of driving gas in the direction

from the intake port to the discharge outlet, which is maintained at a medium vacuum by a backing rotary pump. In the diffusion pump (a two-stage design utilizing oil as the pump fluid), the vapor issuing from the top, first-stage jet is directed downward towards the backing region. Gas molecules from the intake port diffuse into the streaming vapor. The directed oil molecules collide with the gas molecules to give them velocity components toward the backing region. A large pressure gradient is thereby established in the pump so that the intake pressure may be over 100,000 times less than the backing pressure. The intake pressure may therefore be 10^{-6} torr or lower with a backing pressure of 10^{-1} torr.

In the diffusion pump, the vapor stream is not essentially influenced by the gas pumped. In the vapor ejector pump, however, the vapor stream is enabled by a higher boiler pressure to be denser and of greater speed with a higher intake pressure, so that the gas is entrained by the high-speed vapor. Viscous drag and turbulent mixing now carry the gas at initially supersonic speeds down a pump housing of diminishing cross section. The ejector pump is designed to operate with a maximum pumping speed at an intake pressure of 10^{-1} to 10^{-3} torr and with a backing pressure of 0.5 to 1 torr or more. The diffusion pump, on the other hand, is designed to have a fairly constant speed from 10^{-3} torr down to an ultimate 10^{-6} torr or much lower in a modern, bakeable stainless steel system.

An important mechanical pump, which operates in the same pressure region as the oil ejector, is the Roots pump, capable of very great speeds and requiring backing by a rotary oil-sealed pump.

A vacuum system may consist of a diffusion pump and a backing pump, together with baffles, cold traps and isolation valves. The cold trap is essential if a mercury vapor diffusion pump is used and is best filled with liquid nitrogen ($-196\,°C$); otherwise, the system will be exposed to the mercury vapor pressure, which is 10^{-3} torr at $18\,°C$.

Ultrahigh-vacuum systems with stainless steel traps and metal sealing gaskets, and bakeable (except for the pumps) for several hours to $450\,°C$, may be constructed to provide an ultimate pressure of 10^{-9} to 10^{-10} torr.

Other vacuum pumps include the sorption type based on the high gas take-up of charcoal or molecular sieve material at liquid nitrogen temperatures. Sorption pumps may be used in place of rotary pumps, with a desirable freedom from rotary pump oil vapor, especially in systems where the amount of gas to be handled is limited.

The chief rival to the vapor diffusion pump at present is the getterion pump of the Penning discharge type, sometimes called the sputter-ion pump, with electrodes of titanium metal. The principle of operation is illustrated by Fig. 2 where an egg-box type anode is situated between plane cathodes. The anode-cathode operating potential difference is of the order of 2 to 10 kV, and the magnetic flux density is about 3,000 gauss. The chief pumping action with active gases such as hydrogen, nitrogen and oxygen is to the anode which receives deposited titanium (which has very high gas affinity) sputtered from the cathodes under the action of the positive ion bombardment. Some gas, especially the inert gases like argon, is pumped to the cathodes.

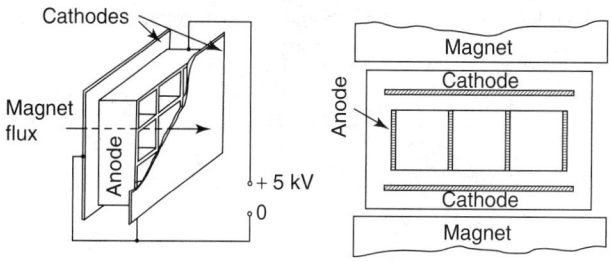

Fig. 2.   Sputter-ion vacuum pump.

A typical multicell pump of moderate size of this type has a pumping speed of about 250 liters/second. Much larger pumps with speeds of up to 5,000 liters/second are commercially available, as are small single-cell units with speeds of some 2 liters/second.

The sputter-ion pumps provide a vapor-free system giving a so-called dry vacuum, and they are often incorporated in plant with molecular sieve sorption as the backing pump. For medium-size laboratory plant able to provide ultrahigh vacuum they are most attractive. Probably their chief disadvantage is that the life of the pump is only about 40 hours at 10^{-3} torr, but this increases inversely with the pressure, so that it is 40,000 hours at 10^{-6} torr. At present, they are therefore strong rivals to the diffusion pump for plant where moderate amounts of gas are handled in the lower pressure ranges.

In cryogenic pumping, the provision is, within an initially evacuated system, a surface which is at such a low temperature that gas impinging on the surface will be condensed. For example, if the surface is maintained at the temperature of liquid helium ($-269\,°C$), all other gases have insignificantly low vapor pressures at this temperature and molecules of these gases impinging on the surface would remain there. A pumping speed for nitrogen of nearly 12 liters sec^{-1} cm^{-2} of cooled surface is hence theoretically possible. Liquid nitrogen, together with molecular sieve and other sorbent surfaces, and also liquid-hydrogen ($-253\,°C$, at which the vapor pressure of solid nitrogen is 10^{-10} torr) and liquid-helium cooled metallic surfaces, are being actively investigated for the possibility of providing very high pumping speeds (10^6 liters/sec is not out of question) in space simulators and other plants.

Vacuum Measurement

Subatmospheric pressure usually is expressed in reference to perfect vacuum or absolute zero pressure. Like absolute zero temperature (the concept is analogous), absolute zero pressure cannot be achieved, but it does provide a convenient reference datum. Standard atmospheric pressure is 14.695 psi absolute, 30 inches of mercury absolute, or 760 mmHg of density 13.595 g/cm^3 where acceleration due to gravity is $g = 980.665$ cm/s^2. 1 mmHg, which equals 1 torr, is the most commonly used unit of absolute pressure. Derived units, the million or micrometer, representing 1/1000 of 1 mmHg or 1 torr, are also used for subtorr pressures.

In the MKS system of units, standard atmospheric pressure is 750 torr and is expressed as 100,000 Pa (N/m^2) or 100 kPa. This means that 1 Pa is equivalent to 7.5 millitorr (1 torr = 133.3 pascal). Vacuum, usually expressed in inches of mercury, is the depression of pressure below the atmospheric level, with absolute zero pressure corresponding to a vacuum of 30 inches of mercury.

When specifying and using vacuum gages, one must constantly keep in mind that atmospheric pressure is *not* constant and that it also varies with elevation above sea level.

Vacuum gages can be either direct or indirect reading. Those that measure pressure by calculating the force exerted by incident particles of gas are direct reading, while instruments that record pressure by measuring a gas property that changes in a predictable manner with gas density are indirect reading.

The range of operation for these two classes of vacuum instruments is given in Table 2. Since the pressure range of interest in present vacuum technology extends from 760 to 10^{-13} torr (over 16 orders of magnitude), there is no single gage capable of covering such a wide range. The ranges of vacuum where specific types of gages are most applicable are shown in Fig. 3.

The operating principles of some vacuum gages, such as liquid manometers, bourdon, bellows, and diaphragm gages involving elastic members,

TABLE 2. RANGE OF OPERATION OF MAJOR VACUUM GAGES

Principle	Gage type	Range, torr
Direct reading	Force measuring:	
	Bourdon, bellows, manometer (oil and mercury),	$760–10^{-6}$
	McLeod capacitance (diaphragm)	760×10^{-6}
Indirect reading	Thermal conductivity:	
	Thermocouple (thermopile)	$10–10^{-3}$
	Pirani (thermistor)	$10–10^{-4}$
	Molecular friction	$10^{-2}–10^{-7}$
	Ionization:	
	Hot filament	$10–10^{-10}$
	Cold cathode	$10^{-2}–10^{-15}$

were described earlier in this article. The remaining vacuum measurement devices include the thermal conductivity (or Pirani and thermocouple-type gages), the hot-filament ionization gage, the cold-cathode ionization gage (Philips), the spinning-rotor friction gage, and the partial-pressure analyzer.

Pirani or Thermocouple Vacuum Gage. Commercial thermal conductivity gages ordinarily should not be considered as precision devices. Within their rather limited but industrially important pressure range, they are outstandingly useful. The virtues of these gages include low cost, electrical indication readily adapted to remote readings, sturdiness, simplicity, and interchangeability of sensing elements. They are well adapted for uses where a single power supply and measuring circuit is used with several sensing elements located in different parts of the same vacuum system or in several different systems.

The working element of the gages consists of a metal wire or ribbon exposed to the unknown pressure and heated by an electric current. See Fig. 4. The temperature attained by the heater is such that the total rate of heat loss by radiation, gas convection, as thermal conduction, and thermal conduction through the supporting leads equals the electric power input to the element. Convection is unimportant and can be disregarded, but the heat loss by thermal conduction through the gas is a function of pressure. At pressures of approximately 10 torr and higher, the thermal conductivity of a gas is high and roughly independent of further pressure increases. Below about 1 torr, on the other hand, the thermal conductivity decreases with decreasing pressure, eventually in linear fashion, reaching zero at zero pressure. At pressures above a few torr, the cooling by thermal conduction limits the temperature attained by the heater to a relatively low value. As the pressure is reduced below a few hundred millitorr, the heater temperature rises, and at the lowest pressures, the heater temperature reaches an upper value established by heat radiation and by thermal conduction through the supporting leads.

Hot-Filament Ionization Vacuum Gage. This gage is the most widely used pressure-measuring device for the region from 10^{-2} to 10^{-11} torr. The operating principle of this gage is illustrated in Fig. 5.

A regulated electron current (typically about 10 mA) is emitted from a heated filament. The electrons are attracted to the helical grid by a

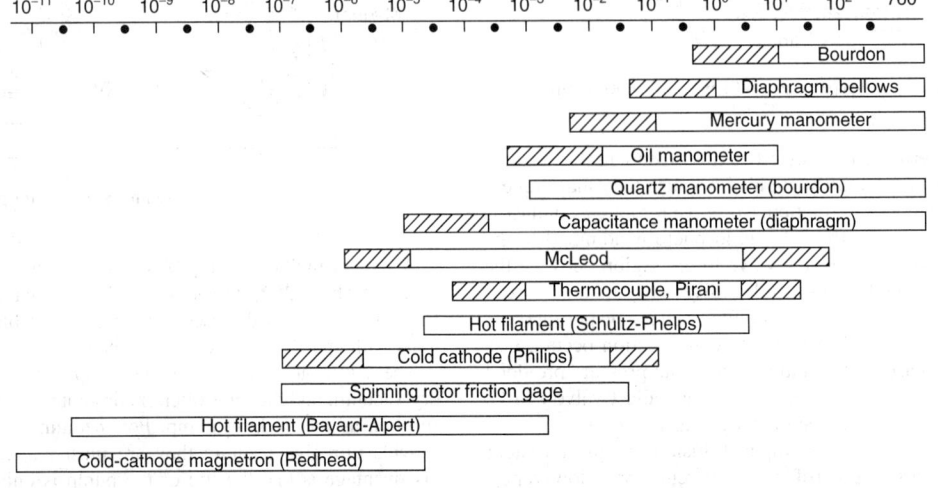

Fig. 3.   Vacuum gages versus measurement range.

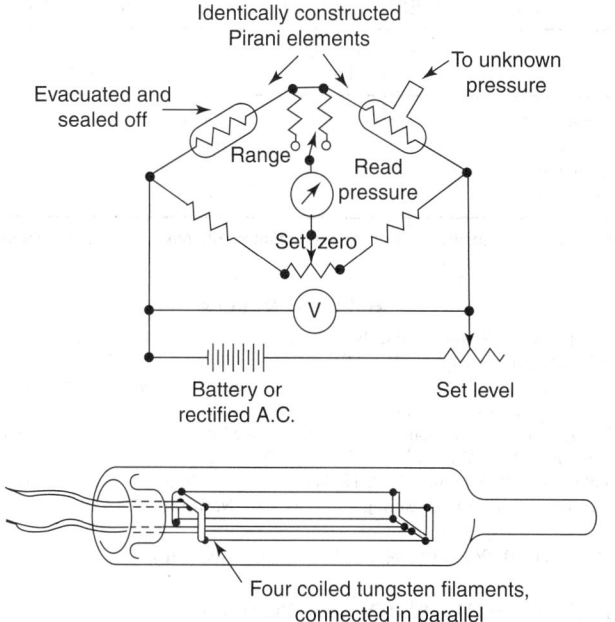

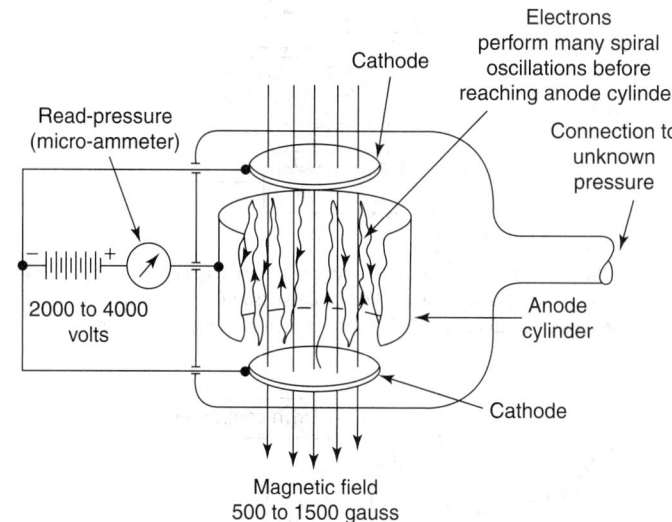

increased opportunity to encounter and ionize molecules of gas in the interelectrode region, even though this gas may be extremely rarefied. It has been found possible by this use of a magnetic field and appropriately designed electrodes, as indicated in Fig. 6, to maintain an electric discharge at pressures below 10^{-9} torr.

Fig. 6.   Philips cold-cathode ionization vacuum gage.

Comparison with the hot-filament ionization gage reveals that, in the hot-filament gage, the source of the inherently linear relationship between gas pressure (more exactly molecular density) and gage reading is the fact that the ionizing current is established and regulated independently of the resulting ion current. In the Philips gage this situation does not hold. Maintenance of the gas discharge current involves a complicated set of interactions in which electrons, positive ions, and photoelectrically effective x-rays all play a significant part. It is thus not surprising that the output current of the Philips gage is not perfectly linear with respect to pressure. Slight discontinuities in the calibration are also sometimes found, since the magnetic fields customarily used are too low to stabilize the gas discharge completely. Despite these objections, a Philips gage is a highly useful device, particularly where accuracy better than 10 or 20% is not required.

The Philips gage is composition-sensitive, but, unlike the situation with the hot-filament ionization gage, the sensitivity relative to some reference gas, such as air or argon, is not independent of pressure. Leak hunting with a Philips gage and a probe gas or liquid is a useful technique. Unlike the hot-filament ionization gage, the Philips gage does not involve the use of a high-temperature filament and consequently does not subject the gas to thermal stress. The voltages applied in the Philips gage are on the order of a few thousand volts, which is sufficient to cause some sputtering at the high-pressure end of the range. This results in a certain amount of gettering or enforced takeup of the gas by the electrodes and other part of the gage. Various design refinements have been used to facilitate periodic cleaning of the vacuum chamber and electrodes, since polymerized organic molecules are an ever-present contaminant.

The conventional cold-cathode (Philips) gage is used in the range from 10^{-2} to 10^{-7} torr. Redhead has developed a modified cold-cathode gage useful in the 10^{-6} to 10^{-12} torr range. See Fig. 7. The operating voltage is about 5000 volts in a 1-kG magnetic field.

Spinning-Rotor Friction Vacuum Gage. Although liquid manometers (U-tube, McLeod) serve as pressure standards for subatmospheric measurements (760 to 10^{-5} torr), and capacitance (also quartz) manometers duplicate this range as useful transfer standards, calibration at lower pressures depends on both volume expansion techniques and presumed linearity of the measuring system. The friction gage allows extension of the calibration range directly down to 10^{-7} torr. It measures pressure in a vacuum system by sensing the deceleration of a rotating steel ball levitating in a magnetic field. See Fig. 8.

Partial-Pressure (Residual Gas) Analyzers. Many applications of high-vacuum technology are more concerned with the partial pressure of

Fig. 4.   Pirani gage: (*Top*) Gage in fixed-voltage Wheatstone bridge. (*Bottom*) Sensing element.

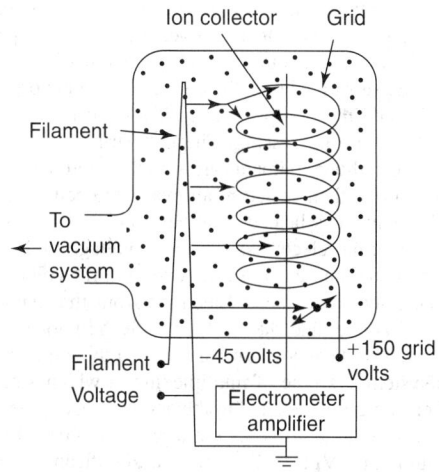

Fig. 5.   Hot-filament ionization gage. (*Bayard-Alpert type.*)

dc potential of about +150 volts. In their passage from filament to grid, the electrons collide with gas molecules in the gage envelope, causing a fraction of them to be ionized. The gas ions formed by electron collisions are attracted to the central ion collector wire by the negative voltage on the collector (typically −30 volts). Ion currents collected are on the order of 100 mA/torr. This current is amplified and displayed using an electronic amplifier.

This ion current will differ for different gases at the same pressure — that is, a hot-filament ionization gage is composition-dependent. Over a wide range of molecular density, however, the ion current from a gas of constant composition will be directly proportional to the molecular density of the gas in the gage.

Cold-Cathode Ionization Vacuum Gage. This ingenious gage, invented by Penning, possesses many of the advantages of the hot-filament ionization gage without being susceptible to burnout. Ordinarily, an electrical discharge between two electrodes in a gas cannot be sustained below a few millitorr pressure. To simplify a complicated set of relationships, this is because the "birthrate" of new electrons capable of sustaining ionization is smaller than the "death rate" of electrons and ions. In the Philips gage this difficulty is overcome by the use of a collimating magnetic field, which forces the electrons to traverse a tremendously increased path length before they can reach the collecting electrode. In traversing this very long path, they have a correspondingly

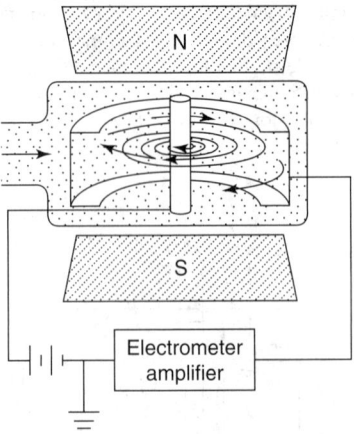

Fig. 7.   Inverted magnetron, a cold-cathode gage, produces electrons by applying a high voltage to unheated electrodes. Electrons spiraling in toward the central electrode ionize gas molecules, which are collected on the curved cathode.

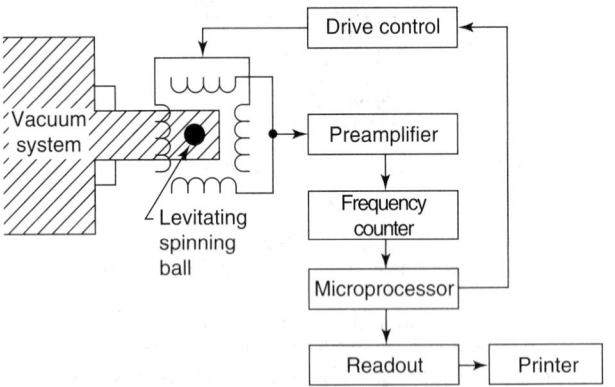

Fig. 8.   Schematic diagram of spinning-rotor friction gage.

particular gas species than with total pressure. Also, "total" pressure gages generally give accurate readings only in pure gases. For these reasons partial-pressure analyzers are finding increasing application. These are basically low-resolution, high-sensitivity mass spectrometers that ionize a gas sample in a manner similar to that of a hot-filament ionization gage. The resulting ions are then separated in an analyzer section, depending on the mass-to-charge ratio of each ion. The ion current corresponding to one ion type is then collected, amplified, and displayed. Partial-pressure gages are very valuable diagnostic tools in both research and production work. The major types are listed in Table 3.

TABLE 3. CHARACTERISTICS OF PARTIAL-PRESSURE ANALYZERS

Type	Minimum partial pressure, torr	Resolution,* au	Magnetic field
Magnetic sector	10^{-11}	20–150	Yes
Cycloidal	10^{-11}	100–150	Yes
Quadrupole	10^{-12}	100–300	No
Time of flight	10^{-12}	200	No

*Maximum mass number at which a mass number difference of 1 can be observed.

Additional Reading

Chanbers, A., R.K. Fitch, and B.S. Halliday: "Basic Vacuum Technology," Iop Publishing, Philadelphia, PA, 1998.

Hoffman, D.M., J.H. Thomas, and B. Singh: "Handbook of Vacuum Science and Technology," Morgan Kaufmann Publishers, Orlando, FL, 1997.

Lafferty, J.M.: "Foundations of Vacuum Science and Technology," John Wiley & Sons, Inc., New York, NY, 1997.

Santeler, D.J., D.W. Jones, D.H. Holkeboer, and F. Pagano: "Vacuum Technology and Space Simulation," Springer-Verlag, Inc., New York, NY, 1997.

VACUUM DEPOSITION (Thin-Film).   See **Thin Films.**

VACUUM DISTILLATION.   See **Petroleum.**

VACUUM FLUORESCENT DISPLAYS.   A vacuum fluorescent display (VFD) is essentially a flat CRT that uses multiple cathodes and a matrix deflection system. The vast majority of VFDs are segmented types; high-resolution dot matrix displays have been developed over a number of years, but they are expensive to manufacture, so market penetration has not occurred. The basic structure of a VFD is shown in Figure 1.

The glass anode substrate holds the connector conductors, the insulation layer, the anode, and the phosphor layer. These structures are formed by thin-film and thick-film processing. The grid structure is positioned over the anode, and the cathode filaments are stretched and held above the grid. The entire structure is then sealed in an evacuated cell.

The cathode filament is heated to about $600\,°C$ ($1,100\,°F$) to facilitate the emission of thermal electrons. An anode voltage of 10 to 50 volts is supplied to the anode, and at the same time the grid voltage is applied to the grid of that selected segment. Electrons from the filament accelerate past the grid and anode, and they collide with the phosphor, resulting in photon emission. The display can be driven in a dynamic or static system. The dynamic system is a type of multiplexing in which a pulse is applied to the grid of each digit or row and to the anode of each selected digit. The multiplexing of very large arrays is not a problem with VFD technology.

The front luminous VFD (FLVFD) is a significant advance that has improved the appearance of the basic structure. This display uses a phosphor deposited on a transparent conductive film of ITO. The emission is observed as one looks through the phosphor, formed as a thin film to increase the amount of light transmitted. Narrow lines of aluminum with

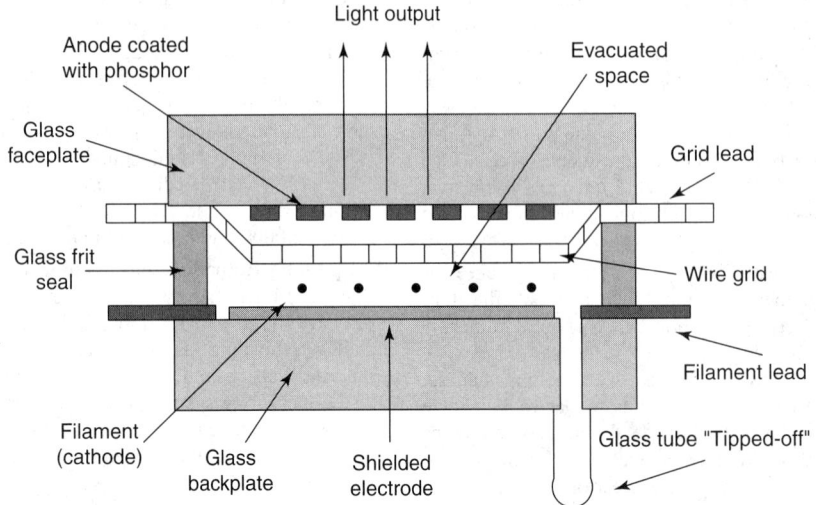

Fig. 1.   Cross section of vacuum fluorescent display.

wide spacing are deposited on the glass as conductors to help boost the brightness of the phosphor.

The single-matrix VFD is the simplest in construction and lowest in manufacturing costs. The simplicity of the design allows the displays to be made with the highest resolution. The design has a major shortcoming: The increased number of lines leads to a lower duty cycle, requiring a higher anode voltage in order to maintain brightness. As mentioned in previous sections, high-voltage drivers are expensive.

Futaba Corporation (*http://www.futaba.com/*) has built displays of this type with formats of 320×300 and 640×400 pixels. The firm also demonstrated a 640×400 pixel, graphic VFD with 16 gray levels. The development of a driver IC (32-bit, 64 levels) for the plasma display has allowed the grayscale to be achieved by the pulse-width control method. Using this IC, images with grayscale were displayed on the VFD when NTSC signals were applied. A front luminous VFD was used; it had 640×200 pixels with a pixel pitch of 0.28×0.56 mm. This led to a panel with 640×400 pixels using electrostatic deflection.

Despite the relative simplicity of the character type displays, dot matrix displays with a high pixel count are still very difficult to manufacture, especially at a low cost. Large-area graphic VFDs are not considered competitive with LCDs, PDPs, or TFEL displays and have not been widely used.

Trends in VFD manufacturing include higher brightness, reduced operating voltages, larger sizes, longer life, improved control circuitry, single voltage power source, and a larger selection of colors. Although the majority of the VFDs on the market are blue-green monochrome versions, some dual-color standard models are available, and three-color models can be made on a custom basis. Suppliers are working on full-color (16.77 million colors) versions, but the large dot pitch of VFDs limits the practicality of full-color displays. Most VFDs are custom designed, and good engineering support and customer service are essential in this market segment.

Manufacturers are also trying to improve the brightness and longevity of the three primary colors and reduce the dot pitch to <0.6 mm. Improvements in phosphors, device design, and cathodes are intended to increase the performance of graphic VFDs. Technological development focuses on the following: more colors, higher luminance, larger sizes, and longer life expectancy. Current trends related to display tube manufacturing include rib-grid fluorescent displays, color fluorescent displays, LCD-compatible character modules for easy equipment installation, and integration of memory function on displays.

Rib-grid VFDs use thick-film printed grids instead of metal mesh grids, allowing for greater freedom in pattern design. A VFD in this format has partition walls (ribs) with a width of 0.1 to 0.14 mm. This frees the VFDs from various restrictions required by the metal mesh. Usually, no limitations are placed on the shape and size of rib-grid VFDs. Rib-grid fluorescent display tubes are primarily custom designed because of the ease of laying out display patterns and producing higher resolutions (pixel densities). The higher pixel density can be obtained by printing the grid electrodes. This allows the display to show a large amount of information in a limited space. Because the spacing between VFD segments developed using thick-film printed grids can decrease in size when compared with spacing required for metal mesh grids, the spacing between the adjacent characters narrows. At the same time, the width of each character widens, allowing for designs with easy-to-read displays.

VFDs are increasing in luminance and viewing area, while at the same time using reduced voltages. In cases in which an LCD module does not provide the appropriate brightness or viewing angle, a U-version small character module uses the interface specifications that are virtually the same as those of LCDs. Such VFDs use control codes that are almost the same as those used by LCD controllers. This module is identical in overall thickness to a 2-line × 20-character LCD with backlighting.

Small character modules capable of displaying 20 × 2 lines or 24 × 2 lines of the 5 × 7 dot fonts are coming to the market. These modules are designed as replacements for LCD modules. Their power consumption is reduced, and a power-saving design is adopted to suspend the converters' operation in the display off state. A 2-line × 16-character U-version VFD has also been developed. It is compatible with the 2 × 16 miniature LCD character module. See also **Liquid Crystal Display Technology**.

Color Vacuum Fluorescent Displays

Color VFD tube modules adopt two different luminescent materials, each placed next to one another, with minimal space between them in order to emit three colors. Futaba has shown a system with a green tube that offers 500 cd/m^2 of luminance and a red tube with 250 cd/m^2 of luminance; the company claims lifetimes of 10,000 hours when used as a text display. Blue may be added to the color capability. It may take two years before a blue phosphor is ready for commercial use. These products are a long way from full-color capability.

Chip-Lighting Vacuum Fluorescent Display (CLVFD)

The chip-lighting VFD is a blend of semiconductor technology and conventional VFD technology. The display is a phosphor dot matrix on top of a 5.4×6 mm semiconductor chip that integrates memory function and display driver circuits. The basic structure of this display device is similar to the conventional VFD except for the structure of the anode. In a typical VFD, wiring, insulation, and phosphor layers are fabricated directly on the glass plate and no semiconductor devices are arranged internally. With CLVFD, phosphor matrices are fabricated on a semiconductor chip and arranged on a glass plate. The chip and outer leads are connected by wire bonding. These types of graphic displays contain semiconductor driver ICs with embedded memory. Claimed benefits of CLVFDs over conventional VFDs include low noise, longer life, higher brightness, and stable operation under varied conditions.

Chip-In-Glass (CIG) Vacuum Fluorescent Display

In chip-in-glass (CIG) technology, slim chips are hidden in the glass envelopes to provide a smaller and more functional display. The main feature of CIG design is the elimination of special external VFD drivers, along with many of the lead pins. Since CIGs do not require chip-mounting space, as do conventional VFDs, they can potentially be made in smaller packages.

Ise Electronics and NEC have used CIG technology to produce dot matrix VFDs. Production costs are lower than for conventional VFDs, and the devices are more compact. Graphic displays for presenting Chinese or Japanese Kanji script have also been developed. The use of static drives allows operation at low voltages (about 10 volts) and lower power consumption, a longer product life, and less noise (compared to dynamic drives) when displaying the same kind of patterns. However, static drive operation has the disadvantage of increasing the number of driver chips with higher segment counts. See also **Cathode-Ray Tube**; and **Television (TV)**, and the family of articles catalogued under **Flat panel Display Technology**.

For additional reading, refer to Flat Panel Display Technology entry.

Stanford Resources, Inc., San Jose, CA.

VAGINA. See **Gonads**; and **Hormones**.

VALENCE. See **Chemical Elements**; **Molecule**.

VALENCE ELECTRON. See **Electron**.

VALINE. See **Amino Acids**.

VALLEY FEVER. See **Coccidioidomycosis**.

VALLEY WIND. See **Winds and Air Movement**.

VALVE (Control). The final control element that directly changes the value of the controlled variable in an automatic control system by varying the rate of flow of some medium, such as steam, water, reactant, etc. The control valve has been defined as a continuously variable orifice in a fluid-flow line. The complete control valve consists of an actuator and a valve body. The actuator provides the power to vary the port opening in the valve body. The valve body assembly consists of a pressure-tight chamber that is screwed, flanged, or welded into a pipeline. Control valves frequently are classified on the basis (1) of the *means of actuation*, notably pneumatically or electrically, although there are also hydraulically operated valves as well as electromechanical, electropneumatic, and electrohydraulic modes of operation; (2) of the style of *valve-body design*, which follows to a large degree the valve body styles of manually operated valves, notably globe valves and rotary-stem valves, as well as the lesser used butterfly valves, slide valves, 3-way valves, angle valves, Saunders patent valves, etc.; and (3) of the *characteristic* of the valve, i.e., the flow behavior as the valve operates through its rated stroke, such as a quick-opening characteristic, an equal-percentage characteristic, a parabolic characteristic, etc. Control valves also may be classified in terms of the service for which they are intended—high temperature, high pressure, cryogenic and vacuum ranges, etc. Valve bodies are obtainable in numerous materials and sizes. Depending upon quality of service, the materials used include carbon steel, alloy steel, stainless steel, bronze, and corrosion-resistant alloys, such as Monel, Hastelloy, Inconel, Duriron, etc.

A diaphragm-actuated (pneumatic) control valve with cage-guided valve plug is shown in Fig. 1. An increasing pneumatic error signal operates in the upper chamber of the valve to push the stem downward against spring pressure. The standard range of pneumatic signals is from 3 to 15 pounds per square inch (0.2 to ~1 atmosphere). In the high-pressure double-port

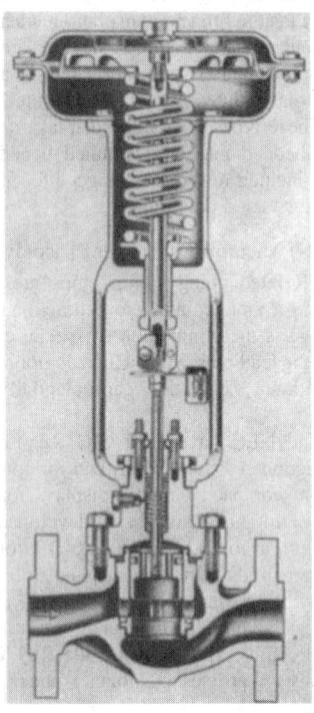

Fig. 1.   Section of diaphragm-actuated control valve with cage-guided valve plug.

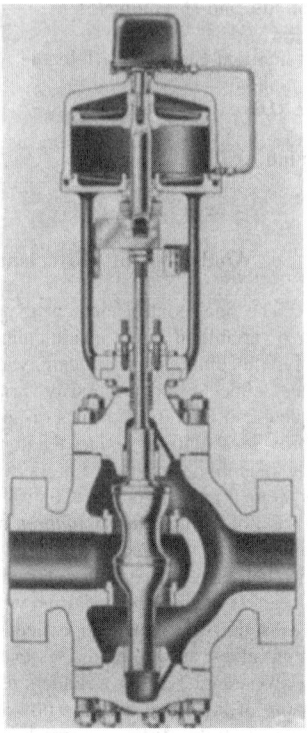

Fig. 2.   Section of piston-actuated, high-pressure, double-port control valve.

control valve of Fig. 2, a pneumatic piston operator is used instead of a diaphragm-type motor.

Electric actuators for control valves fall into two basic categories: (1) inherently two-position actuators, including solenoid and relay actuation; and (2) inherently infinite-position actuators, including reversible electric motor drives, saturable-core reactors, silicon-controlled rectifiers, variable-speed electric motor drives and electropneumatic converters.

For a comprehensive discussion of automatically controlled valves, reference to the "Process/Industrial Instruments and Controls Handbook," Considine, D.M. and G.K. McMillan: 5th Edition, McGraw-Hill Companies, Inc., 1999, is suggested.

Additional Reading

Borden, G.: "Control Valves: Practical Guides for Measurement and Control," ISA, Research Triangle Park, NC, 1998.
Goodman, R.B.: "A Primer on Pneumatic Valves and Controls," Krieger Publishing Company, Melbourne, FL, 1997.

VAMPIRE BAT.   See **Bats**.

VANADINITE.   The mineral vanadinite corresponds to the formula $Pb_5(VO_4)_3Cl$, being composed of lead chloride and lead vanadate in the proportion of 90.2% of the former and 9.8% of the latter. It crystallizes in the hexagonal system, is usually prismatic, but the crystals are often skeletal or cavernous; it may be found in crusts. Its fracture is uneven; brittle; hardness, 2.75–3; specific gravity, 6.86; fresh fractures show a resinous luster; color, yellow, yellowish-brown, reddish-brown, and red; streak, white to yellowish; translucent to opaque. Vanadinite, not a common mineral, occurs as an alteration product in lead deposits. It is found in the Ural Mountains, Austria, Spain, Scotland, Morocco, the Transvaal, Argentina, and Mexico. In the United States it occurs in Arizona, New Mexico, and South Dakota. It is used as an ore of vanadium and to some extent of lead as well. It is interesting to note that this mineral was first described as a chromate upon its discovery in Mexico in 1801. It was not until the discovery of the element vanadium in 1830 that the true nature of this compound was known.

VANADIUM.   Chemical element symbol V, at. no. 23, at. wt. 50.942, periodic table group 5, mp 1880–2000 °C, bp 3380 °C, density 6.10 g/cm³. Elemental-vanadium has a body-centered cubic crystal structure. Vanadium is a silver-white, very hard (7 on the Mohs scale), oxidizes upon exposure to air, burns upon ignition to form the pentoxide V_2O_5, insoluble in HCl, slowly dissolves in hydrofluoric, HNO_3, or H_2SO_4 (hot, concentrated), or aqua regia. Insoluble in HaOH solution. The element was first reported by Andrés Manuel del Rio in 1801; later and separately reported by Nils Gabriel Sefstr "om in 1830. There are two naturally occurring isotopes ^{50}V (radioactive and with a half-life something greater than 10^{14} years and only present to the extent of 0.24% in natural substances) and ^{51}V (99.76 abundance percentage). Five other radioactive isotopes have been identified ^{46}V through ^{49}V and ^{52}V. ^{49}V has a half-life of 330 days. ^{48}V has a half-life of 16.1 days; the others have half-lives measured in seconds or minutes. Vanadium ranks 22nd among chemical elements occurring in the Earth's crust. An average composition of igneous rocks contains 0.017% V. It is estimated that vanadium occurs in seawater to the extent of about 9.5 tons per cubic mile (2.1 metric tons per cubic kilometer). First ionization potential 6.74 eV; second, 14.7 eV; third 29.6 eV; fourth 48.3 eV; fifth, 68.64 eV. Oxidation potentials: $V \rightarrow V^{2+} + 2e^-$, 1.5 V, $V^{2+} \rightarrow V^{3+} + e^-$, 0.255 V, $VO^{2+} + H_2O \rightarrow VO_2^+ + 2H^+ + e^-$, −1.00 V. Electronic configuration $1s^2 2s^2 2p^6 3s^2 3p^6 3d^3 4s^2$.

Other important physical characteristics of vanadium are given under **Chemical Elements**.

Vanadium occurs as patronite, containing vanadium pentasulfide, in Peru, as carnotite, potassium uranyl vanadate, in Colorado and Utah, as vanadinite, lead vanadate, in Arizona, New Mexico, the Republic of South Africa and Zambia. See also **Carnotite**; and **Vanadinite**. Ships burning Venezuelan or Mexican petroleum fuel oil recover vanadium oxide from the boiler and stack dust. In Italy the refining of bauxite ore (for aluminum) yields vanadium, and in Germany some iron ores contain vanadium. Vanadium and radium ore is found in southwestern Colorado and southeastern Utah; in Arizona a complex ore of gold, silver, and lead contains vanadium and molybdenum; and extensive deposits of phosphate rock in Idaho yield tonnage quantities of vanadium. The sulfide ore is roasted to remove sulfur, and the residue fused with sodium carbonate, forming sodium vanadate. This last is extracted with H_2O and excess of H_2SO_4 is added, causing precipitation of vanadium pentoxide, which is later reduced by carbon or aluminum at high temperatures.

Uses: Most vanadium produced is consumed as ferrovanadium. Ferrovanadium is made by the aluminum or silicon reduction of the oxide in the presence of iron in an electric-arc furnace. The product contains about 85% vanadium, 12% carbon, and 2% iron. Ductile vanadium metal also is produced in significant quantities, mainly by the calcium reduction of the oxide in a process developed by McKechnie and Seybolt. Pure vanadium oxide, calcium metal, and iodine are charged into a heavy-walled steel cylinder, excluding all moisture. After evacuation, heat is applied to initiate the reaction. Molten droplets of vanadium collect beneath the calcium oxide-calcium iodide slag and there form a single button or regulus. Ductile vanadium metal produced in this manner has an analysis of 99.7% vanadium, 0.10% oxygen, 0.04% nitrogen, 0.008% hydrogen, 0.04% iron, and 0.03% carbon. The metal is soft and ductile, can be hot- and cold-worked easily. Any heating must be done in vacuum or an inert atmosphere because the metal oxidizes readily. With exception of HNO_3, the metal withstands other acids and aerated saltwater better than most stainless steels. It has a comparatively low cross section for neutron capture and is of interest in the nuclear field. The density is 22% less than iron and 28% greater than titanium. The coefficients of thermal and electrical conductivity are higher than those of titanium.

Most vanadium is used by the steel industry. The addition of vanadium to steel causes the formation of vanadium carbide. The carbides are very hard and wear-resistant; they maintain a fine dispersion. The addition of very small quantities (0.02–0.08%) of vanadium enhances strength and toughness of the resulting steel. Many structural, plate, bar, and pipe steels contain vanadium in these amounts. For little additional cost, in comparison with plain carbon steels, there is a marked increase in performance, including higher strength in the as-rolled condition without heat treatment. Often, manganese and copper are added in small quantities along with the vanadium. Some sheet steels that are used for deep-drawing as in auto and home-appliance parts contain vanadium to suppress aging. For such steels, ferrovanadium is added to rimming steels, resulting in a good, nonaging, deep-drawing steel at a smaller cost than for aluminum-killed deep-drawing steel. Some large steel forgings contain vanadium to the extent of 0.5 to 0.15% with the object of improving the mechanical properties of the forgings. Vanadium is particularly effective in raising the strength and ductility of large steel castings and forgings when added in a small percentage. A large number of tool steels contain vanadium to the extent of 0.10 to 5.00%. In these steels, vanadium insures the retention of hardness and cutting ability at the high temperatures resulting from the rapid cutting of metals. The use of vanadium in cast iron controls the size and distribution of graphite flakes and thus improves strength and wear resistance. One of the most popular of the titanium-base alloys contains 4% vanadium and 6% aluminum. The addition ingredient for this titanium alloy, produced in large quantities, is a base 40:60 vanadium-alloy.

The use of certain vanadium compounds as catalysts has been increasing. Vanadium oxytrichloride is a catalyst in making ethylene-propylene rubber. Ammonium metavanadate and vanadium pentoxide are used as oxidation catalysts, particularly in the production of polyamides, such as nylon, in the manufacture of H_2SO_4 by the contact process, in the production of phthalic and maleic anhydrides, and in numerous other oxidation reactions, such as alcohol to acetaldehyde, anthracene to anthraquinone, sugar to oxalic acid, and diphenylamine to carbazole. Vanadium compounds have been used for many years in the ceramics field for enamels and glazes. Colors are produced by various combinations of vanadium oxide and silica, zirconia, zinc, lead, tin, selenium, and cadmium. Vanadium intermediate compounds also are used in the making of aniline black used by the dye industry.

Chemistry and Compounds. The common oxidation states are: vanadous, V^{2+}; vanadic, V^{3+}; vanadyl, VO^{2+} or VO^{3+}; pervanadyl, VO_2^+; metavanadate, VO_3^+. There are also orthovanadates, VO_4^{3-}, pyrovanadates, $V_2O_7^{4-}$; and complex polyvanadates. The latter group includes di, tri-, tetra-, and octavanadates. The hexavanadates are regarded as oxyvanadium(V) pentavanadates, containing the group $V_5O_{16}^{7-}$. Vanadium ions also form heteropolyacids with acids of molybdenum, tungsten, arsenic, phosphorus, silicon and tin. The peroxy vanadium ions include the diperoxyortho-vanadate ions $[VO_2(O_2)_2]^{3-}$ and the peroxovanadium(V) ions, $[V(O_2)]^{3+}$.

Among the simpler compounds of vanadium are the following:

Fluorides: Vanadium trifluoride VF_3, green crystalline solid; vanadium tetrafluoride VF_4, brownish-yellow crystalline solid; vanadium pentafluoride VF_5, brownish-yellow crystalline solid, sublimes on heating.

Chlorides: Vanadium dichloride VCl_2, green crystalline solid, a strong reducing agent; vanadium trichloride VCl_3, pink crystalline solid; vanadium tetrachloride VCl_4, reddish-brown liquid, bp 148 °C.

Bromide: Vanadium tribromide VBr_3, green crystalline solid.

Iodides: Vanadium diiodide VI_2, usually hydrated, green crystalline solid; vanadium triiodide VI_3, brown crystalline solid.

Oxyhalides of vanadium are common, including VOF_2, VOF_3, $VOCl$, $VOCl_2$, $VOCl_3$, $VOBr$ and $VOBr_3$.

Hydroxides: Vanadium dihydroxide $V(OH)_2$, brown precipitate by reaction of NaOH solution with hypovanadous acid (one of the most powerful of reducing agents) lavender solution; vanadous hydroxide $V(OH)_3$, green precipitate by reaction of NaOH solution with vanadous salt, green solution.

Oxides: Vanadium monoxide VO, gray solid; vanadium trioxide V_2O_3, black solid; vanadium dioxide VO_2, dark blue solid; vanadium pentoxide V_2O_5, orange to red solid. The last is the most important oxide; formed by the ignition in air of vanadium sulfide, or other oxide, or vanadium; used as a catalyzer, e.g., the reaction SO_2 gas plus oxygen of air to form sulfur trioxide, and the oxidation of naphthalene by air to form phthalic anhydride.

Bismuth vanadate, $BiVO_4$, exhibits a ferroelastic-paraelastic phase transition and has been the object of considerable investigation. Lattice dimensions of the compound have been determined under numerous different high-pressure or high-temperature conditions and all confirm this transition. R.M. Hazen and J.W.E. Mariathasan, key investigators in this area, have suggested that *in situ* determination of lattice parameters and crystal structures at combined temperature and pressure should be especially valuable in the documentation of nonquenchable, reversible phase transitions, such as that displayed by bismuth vanadate.

Sulfides: Vanadium monosulfide VS; vanadium trisulfide V_2S_3, most stable; vanadium pentasulfide V_2S_5.

Biological Systems: Very little is known of what roles V may play in biological systems. In 1984, scientists at the University of British Columbia found that vanadates stimulate glucose oxidation and transport in adipocytes, enhance glycogen synthesis in liver and diaphragm, and inhibit hepatic glucogenesis and intestinal glucose transport. Working with diabetic rats, the investigators found that vanadates appear to control the high blood glucose and prevent the decline in cardiac performance due to diabetes.

Additional Reading

Carter, G.F. and D.E. Paul: "Materials Science and Engineering," ASM International, Materials Park, OH, 1991.

Greenwood, N.N. and A. Earnshaw: "Chemistry of the Elements," 2nd Edition, Butterworth-Heinemann, Inc., Woburn, MA, 1997.

Hazen, R.M. and J.W.E. Mariathasan: "Bismth Vanadate: A High-Pressure, High-Temperature Crystallographic Study of the Ferroelastic-Paraelastic Transition," *Science*, **216**, 991–992 (1982).

Lide, D.R.: "CRC Handbook of Chemistry and Physics 2000–2001," 81st Edition, CRC Press, LLC., Boca Raton, FL, 2000.

Niriagu, J.O.: "Vanadium in the Environment: Health Effects," Vol. 2, John Wiley & Sons, Inc., New York, NY, 1998.

Staff: "ASM Handbook—Properties and Selection: Nonferrous Alloys and Pure Metals," ASM International, Materials Park, OH, 1990.

Staff: "Properties and Selection: Iron, Steels, and High-Performance Alloys," ASM International, Materials Park, OH, 1990.

Tracy, A.S. and D.C. Crans: "Vanadium Compounds: Chemistry, Biochemistry, and Therapeutic Applications," Oxford University Press, Inc., New York, NY, 1998.

VAN ALLEN, JAMES (1914–1994). Van Allen was an American physicist who earned his Ph.D. in physics from the State University of Iowa. Between the years of 1939 and 1942, he researched at the Carnegie Institution's Department of Terrestrial Magnetism. In 1942, he began weapons research for the U.S. Navy at Johns Hopkins University. He is known for developing the radio proximity fuse that increased missile efficiency.

In 1946, Van Allen became the administrator and coordinator of the U.S. Army's V-2 research program and designed payloads to be sent into the atmosphere. In 1951, he returned to the State University of Iowa and became head of the department of physics and astronomy. During this research time, he discovered high-altitude radiation belts. This discovery is now named the Van Allen radiation belts.

See also **Van Allen Radiation Belts**.

J. M. I.

VAN ALLEN RADIATION BELTS.   These are belts of charged particles (electrons and protons) that are trapped by the Earth's external magnetic field and which circle the Earth at altitudes of approximately 1000 to 6000 kilometers. The paths of the particles are determined by the direction of the external lines of force of the Earth's magnetic field. The particles migrate from the region above the Earth's equator toward the North Pole, then toward the South Pole, and then return to the region above the equator. This is a definition essentially developed by the IEEE Communications Committee.

Although the belts were predicted a quarter-century earlier by James Van Allen, they were not confirmed until 1958 by instruments aboard the U.S. Explorer 1 satellite. The Van Allen belts are well known for single-event upsets that occur when a single, heavy ion shoots through a satellite-based semiconductor, causing a memory bit to change. Radhardening of electronic devices for Van Allen radiation sources as well as other radiation found in outer space (or in the terrestrial atmosphere as the result of a nuclear event) is extremely important to military and aerospace equipment. See also **Radiation Hardening (Electronics)**.

VAN DE GRAAF GENERATOR.   See Particles (Subatomic).

VAN DER WAALS EQUATION.   A form of the equation of state, relating the pressure, volume, and temperature of a gas, and the gas constant. Van der Waals applied corrections for the reduction of total pressure by the attraction of molecules (effective at boundary surfaces) and for the effect of reduction of total volume by the volume of the molecules (but not the actual molecular volume). The equation takes the form

$$\left(P + \frac{a}{V^2}\right)(V - b) = RT$$

in which P is the pressure of the gas, V is the volume, T is the absolute temperature, R is the gas constant, and a and b are correction terms which have been evaluated and reported for many gases. See also **Characteristic Equation**.

VAN DER WAALS FORCES.   Interatomic or intermolecular forces of attraction due to the interaction between fluctuating dipole moments associated with molecules not possessing permanent dipole moments. These dipoles result from momentary dissymmetry in the positive and negative charges of the atom or molecule, and on neighboring atoms or molecules. These dipoles tend to align in antiparallel direction and thus result in a net attractive force. This force varies inversely as the seventh power of the distance between ions.

VAN DER WAALS, JOHANNES DIDERIK (1837–1923).   Van der Waals was a Dutch physicist, he was a professor of physics at the University of Amsterdam from 1877 to 1907. He was especially interested in thermodynamics and he won the 1910 Nobel Laureate in Physics for his work on gases and liquids, his most well known expressed as Van der Waals Equation. He also studied the attractive forces, which hold the atoms of molecules together and is named now Van der Waals forces. Van der Waals bonding, a crystalline structure with the weakest bonding, is also named in his honor.

See also **Characteristic Equation**; **Chemical Elements**; **Van Der Waals Equation**; and **Van Der Waals Forces**.

J. M. I.

VANE.   1. A thin and more-or-less flat object intended to align itself with a stream or flow in a manner similar to that of the common weathercock, as: (a) a device that project ahead of an aircraft to sense gusts or other actions of the air so as to create impulses or signals that are transmitted to the control system to stabilize the aircraft; (b) a fixed or movable surface used to control or give stability to a rocket.

2. A blade or paddlelike object, often fashioned like an airfoil and usually one of several, that rotates about an axis, either being moved by a flow or creating a flow itself, such as the blade of a turbine, of a fan, of a rotary pump or air compressor, etc.

3. Any of certain stationary blades, plates, or the like that serve to guide or direct a flow, or to create a special kind of flow, as: (a) any of the blades in the nozzle ring of a gas-turbine engine; (b) any of the plates or slatlike objects that guide the flow in a wind tunnel; (c) a plate or fence projecting from a wing to prevent spanwise flow.

VANILLA BEAN.   See **Bean**.

VAN SLYKE REACTION.   See **Amino Acids**.

VAN'T HOFF EQUATION.   A relationship representing the variation with temperature (at constant pressure) of the equilibrium constant of a gaseous reaction in terms of the change in heat content, i.e., of the heat of reaction (at constant pressure). It has the form:

$$\frac{d \ln K_p}{dT} = \frac{\Delta H}{RT^2}$$

in which K_p is the equilibrium constant at constant pressure, T is absolute temperature, R is the gas constant, and ΔH is the standard change in heat content, or, for ideal gases, the change in heat content.

VANT'T HOFF, JACOBUS HENDRICUS (1852–1911).   Van't Hoff was born in Rotterdam, the Netherlands. In 1869 he entered the Polytechnic School at Delft and obtained his technology diploma in 1871. His decision to follow a purely scientific career, however, came soon afterwards during vacation–work at a sugar factory when he anticipated for himself a dreary profession as a technologist. After having spent a year at Leyden, mainly for mathematics, he went to Bonn to work with A.F. Kekulé from autumn 1872 to spring 1873. This period was followed by another in Paris with A. Wurtz, when he attended a large part of the curriculum for 1873–1874. He returned to Holland in 1874 and received his Ph.D. from the University of Utrecht under E. Mulder. He did research on chemical dynamics and equilibrium. He was awarded the first Nobel Prize in Chemistry in 1901.

In his later years, Van't Hoff spent his research time trying to understand the action of enzymes as biological catalysts.

See also **Van't Hoff Equation**; and **Van't Hoff Law**.

J. M. I.

VAN'T HOFF LAW.   A dissolved substance has the same osmotic pressure as the gas pressure it would exert in the form of an ideal gas occupying the same volume as that of the solution.

VAPOR.   A substance in the gaseous state, but below its critical temperature, is called a vapor. If a pure liquid partly filling a closed container is allowed to stand, the space above it becomes filled with the vapor of the liquid, which then develops a pressure. This vapor pressure increases up to a certain limit, depending upon the temperature, where it becomes constant, and the space is then said to be saturated.

Such a body of vapor is not subject to all of the laws of gases. If the space occupied by it is diminished without change of temperature, there is no increase in pressure, but instead part of the vapor condenses. And if the temperature is raised, the pressure goes up not at a uniform but at an increasing rate, because of both the expansion of the liquid and the further evaporation from it. The relation of vapor to liquid takes on a curious aspect as the critical state is approached, in which the vapor and the liquid have equal density.

See also **Boiler (Steam Generator)**; **Boiling**; and **Supersaturated Vapor**.

VAPOR-ABSORPTION REFRIGERATION.   See **Refrigeration**.

VAPOR DENSITY.   The density of a gas referred to the density of hydrogen or air as unity. If the density of hydrogen is taken as 2, the vapor density is approximately the molecular weight; if it is taken as one, the vapor density equals about half the molecular weight.

VAPOR-FILLED THERMOMETER.   See **Thermometer (Filled-System)**.

VAPORIZATION.   The change of a substance from the liquid or solid state to the gaseous state.

VAPORIZATION (Heat of). The evaporation of a given mass of any liquid requires a definite quantity of heat, dependent upon the liquid and upon the temperature at which it evaporates. The quantity required per unit mass at a fixed temperature is called the heat of vaporization of the substance at that temperature. It may be measured by allowing the vapor to condense in a suitable calorimeter, the heat thus evolved, corrected for fall of temperature before and after condensation, being observed. (The heat evolved in condensing is equal to that absorbed when the liquid evaporates.) The result is often surprising. For example, the evaporation of water at the boiling point requires about 540 calories per gram, or more than five times the heat required to raise its temperature from freezing to boiling. The explanation is the large amount of energy necessary to separate the molecules against their cohesion, and the much smaller amount (about 7.4% of the whole) which is used in expanding the vapor against atmospheric pressure. At lower temperatures the value is still greater, because the cohesion is then more effective; with water, for each degree below the normal boiling point, about 0.6 calorie per gram must be added to the heat of vaporization. Trouton found that the heat of vaporization per mole for different liquids bears a nearly constant ratio to the absolute temperature of the boiling point.

A number of methods have been developed for measuring the heat of vaporization. In the Awberg and Griffith's method, the flow is continuous; the heat of evaporation of the liquid under investigation is transmitted to a stream of water flowing at constant rate and its increase in temperature is measured.

VAPOR LOCK. Volatility of a fuel such as gasoline makes for easier starting of an engine using it, but creates one undesirable feature, namely, vapor lock. This phenomenon is associated with the fuel supply to internal combustion engines. It occurs mainly in the fuel lines conveying a volatile fuel. If the fuel line from supply tank to engine has many bends and fittings, if it is exposed to engine heat, or if parts of it are under vacuum, bubbles of fuel vapor may form in the lines, causing irregular engine operation, or even stoppage. See also **Petroleum**.

VAPOR PRESSURE. The vapor pressure of a substance (solid or liquid) is the pressure exerted by its vapor when in equilibrium with the substance. For pure substances it depends only on the temperature. The simplest way to measure the vapor pressure of a substance is to introduce a small amount of it into the closed end of a barometer tube and note the decrease in the height of the barometer.

The vapor pressure of a solvent is lowered on dissolving the solute in it. This lowering for dilute solutions is proportional to the mole fraction of the solute (Raoult's Law). The lowering of the vapor pressure of the solution can be related to the lowering of the freezing point and the elevation of the boiling point. These phenomena serve as a basis for molecular weight determinations. If both components of the solution are volatile, each lowers the vapor pressure of the other and the ratios of the two substances in the liquid and vapor phase are not necessarily the same. Use is made of this fact to separate the two substances by distillation.

Equilibrium vapor pressure is the vapor pressure of a system in which two or more phases or a substance coexist in equilibrium. In meteorology, the reference is to water substance, unless otherwise specified. If the system consists of moist air in equilibrium with a plane surface of pure water or ice, the more specialized term *saturation vapor pressure* is usually employed, in which case, the vapor pressure is a function of temperature only. In the atmosphere, the system is complicated by the presence of impurities in liquid or solid water substance (see also **Raoult's Law**), drops or ice crystals or both, existing as aerosols; and, in general, the problem becomes one of nucleation. For example, the difference in vapor pressure over supercooled water drops and ice crystals is the basis for the Bergeron-Findeisen theory of precipitation formation.

VAPOR-TYPE VACUUM PUMP. See **Vacuum**.

VARACTOR. See **Semiconductor**.

VARIABLE-AREA FLOWMETER. See **Flow Measurement**.

VARIABLE-FOCUS LENS. A lens system, part of which is movable, and so designed as to have correction for lens aberrations, continual sharp focusing of the image on the receiving film and constant value in the F/System as the focal length is changed. Such a lens gives the effect of moving the camera towards or away from the object. Variable focus lenses are used in motion picture and television cameras; commonly called Zoomar lenses.

VARIABLE (Process). The quantity or characteristic that is the object of measurement in an instrumentation or automatic control system. Other terms used include measurement variable, instrumentation variable, and process variable. The latter term is commonly used in the manufacturing industries. Numerous ways to classify variables have been proposed — by methods of measurement, by end-measurement objectives, and so on. One of the most convenient and meaningful classifications is the physical and/or chemical nature of the variable, as follows:

Thermal Variables. These variables relate to the condition or character of a material dependent upon its thermal energy. Variables included are: *temperature, specific heat, thermal-energy variables* (enthalpy, entropy, etc.), and calorific value.

Radiation Variables. These variables relate to the emission, propagation, and absorption of energy through space or through a material in the form of waves; and by extension, corpuscular emission, propagation, and absorption. Variables included are: *nuclear radiation*; *electromagnetic radiation* (radiant heat, infrared, visible, and ultraviolet light; x- and cosmic rays; gamma radiation).

Force Variables, including: *total force, moment or torque,* and *force per unit area,* such as pressure, vacuum, and unit stress.

Rate Variables. These variables are concerned with the rate at which a body is moving toward or away from a fixed point. Time always is a component of a rate variable. Variables included are: *flow, speed, velocity,* and *acceleration*.

Quantity Variables. These variables relate to the total quantity of material that exists within specific boundaries. Variables include: *mass* and *weight*.

Physical Property Variables. These variables are concerned with the physical properties of materials with the exception of those properties which are related to chemical composition and direct mass and weight. Variables included are: *density* and *specific gravity, humidity, moisture content, viscosity, consistency,* and *structural characteristics,* such as hardness, ductility, and lattice structure.

Chemical-Composition Variables. These variables relate to the chemical properties and analysis of substances. A very abridged list of analysis variables would include: Identification and concentration of carbon dioxide, carbon monoxide, hydrogen, nitrogen, oxygen, water, hydrogen sulfide, nitrogen oxides, sulfur oxides, methane, ethylene, alcohol, and so on. Also included in this category is measurement of pH (hydrogen ion concentration) and redox measurements. See also **Analysis (Chemical)**.

Electrical Variables. Included here are those variables which are measured as the "product" of a process, as in the case of measuring the current and voltage of a generator, and also as part of an instrumentation system. Numerous transducers, of course, yield electrical signals that represent by inference some other variable quantity, such as a temperature or pressure. Variables in this class include: *electromotive force, electric current, resistance, conductance, inductance, capacitance,* and *impedance*.

Geometric Variables. These variables are related to position or dimension and relate to the fundamental standard of length. Variables include: *position, dimension, contour,* and *level* (as of a material in a tank or bin).

See also **Control System**.

VARIABLE SAMPLING FRACTION. See **Sampling (Statistics)**.

VARIABLE STAR. Any star whose brightness changes, for whatever reason, is designated a *variable star*. Although oriental records indicate that novae and supernovae were recorded as early as the fourth century A.D., the first systematic observations of variability date from the late sixteenth century, with the discovery of Mira (*o* Ceti) by Fabricius. The first periodic variable to be well studied, Algol, was identified by Goodricke in the late eighteenth century, and modeled as an eclipsing variable by Goodricke and soon afterwards by Herschel. At the time, all periodic variability was assumed to be due to a dark companion producing eclipses of the visible star. By the beginning of the 20th century, it was recognized that this model would not explain the range of behavior attributed to these stars. The pulsation mechanism, developed primarily by Eddington and Rosseland, was eventually adopted as an alternative source of periodic fluctuation. This has lead to the separation of variable stars into two broad categories: (1) those which are eclipsing binaries and (2) those which are intrinsic variables. In addition there are cataclysmic variables, which are discussed in the entry on **Nova and Supernova**; and the dMe or *flare stars*. See also **Flare Stars**.

VARIANCE. In statistics, the variance of a population is the second moment about the mean, that is to say, the average of the square of deviations from the mean. It is the most commonly used measure of dispersion. In mechanics, the term refers to the number of degrees of freedom of a system. In physics and chemistry, the variance is the number of degrees of freedom of a system, or the degrees of freedom themselves.

VARIANCE (Analysis of). See **Analysis of Variance**.

VARIATE. See **Random Variable**.

VARIATE DIFFERENCE METHOD. If the nonrandom part of a time-series can be represented locally by a polynomial, this part can be removed by taking successive differences of the series. If the variances of the successive difference series are calculated, and the rth variance divided by $(2r)!/(r!)^2$, the resulting numbers should decrease until the trend has been eliminated and then remain at a roughly constant value which estimates the variance of the random part of the original series. A similar technique can be used to estimate the correlation between the random parts of two series.

VARIATION (Coefficient of). The coefficient of variation of a distribution, V, is defined as the standard deviation divided by the mean,

$$V = \frac{\sigma_x}{\bar{x}}$$

Some authors define it as

$$V = 100\frac{\sigma_x}{\bar{x}}$$

in order to avoid decimals.

VARIATION (Mathematics). If two variables x and y are related by the power function relationship $y = kx^n$, where k and n are constants, we say that y varies as x^n, or y varies as the nth power of the x, or that y is proportional to x^n. The factor k is called the constant of variation, or the constant of proportionality, or the proportionality factor. For another use in mathematics, see also **Variations (Calculus of)**.

VARIATIONS (Calculus of). The problem of the calculus of variations is to determine functions in such a way that a definite integral depending upon them and their derivatives may assume an extreme value. In the simplest case the problem is to minimize the value of the integral

$$I[y] = \int_a^b F(x, y, y')\,dx$$

with given a, b, $y(a)$, $y(b)$. Thus, if $y = f(x)$ is the desired minimizing function and if we consider neighboring functions $y + \varepsilon\eta(x)$, where $\eta(x)$ is any fixed function of x with $\eta(a) = \eta(b) = 0$ and ε is a real number, we may regard $I[y + \varepsilon\eta]$ as a function of ε alone with a minimum at $\varepsilon = 0$, so that

$$\left.\frac{dI}{d\varepsilon}\right|_{\varepsilon=0} = 0$$

But

$$\frac{dI}{d\varepsilon} = \int_a^b \left(\frac{\partial F}{\partial y}\eta + \frac{\partial F}{\partial y'}\eta'\right)$$

so that, by integration by parts,

$$\int_a^b \eta\left\{\frac{\partial F}{\partial y} - \frac{d}{dx}\left(\frac{\partial F}{\partial y'}\right)\right\} = 0$$

and from the fundamental lemma (see definition below).

$$\frac{\partial F}{\partial y} - \frac{d}{dx}\left(\frac{\partial F}{\partial y'}\right) = 0$$

which is called the Euler-Lagrange equation of the original problem. It represents a necessary condition for an extreme value of the definite integral.

The fundamental lemma referred to above may be stated thus. If $\phi(x)$ is continuous in the closed interval $[a, b]$ and $\int_a^b \eta(x)\phi(x)dx = 0$ for all $\eta(x)$ with continuous first derivatives and such that $\eta(a) = \eta(b) = 0$, then $\phi(x)$ is identically zero.

The above remarks can be generalized directly to deal with integrands involving derivatives of higher than first order and with the case where the integrand depends upon more than one function. The function $\varepsilon\eta(x)$ is called the variation of y, and $\varepsilon(dI/d\varepsilon)$ is the variation of I. (Compare the definition of the differential of a dependent variable.)

VARICELLA. See **Chickenpox**.

VARICOCELE. See **Arteries and Veins**.

VARIOLA. See **Smallpox**.

VARIOLITE. A fine-grained basic rock that contains spherulites made up of fibers of feldspar and augite in radial development.

The spherulites themselves are known as varioles, and the texture of such rocks is said to be variolitic.

VARIOMETER. An instrument for comparing magnetic forces, especially of the earth's magnetic field.

VARLEY BRIDGE. See **Bridge Circuits (Electrical/Electronic)**.

VARVES. The annual layers of sediment deposited in lakes and fiords by melt-water from glaciers. Each layer consists of two parts deposited at different seasons and differing in color and texture so that the layers can be measured and counted. If the series is complete, the number of layers gives the date on which the ground was vacated by the retreating ice.

VASCULAR HEADACHE. See **Headache**.

VASCULAR SYSTEM (Plants). A complex system of cells and tissues called xylem and phloem, serving to conduct water, mineral salts, and food through the plant. The vascular system also gives strength and support to the plant. It composes the bulk of the tissues of roots and stems of woody plants.

VASCULAR TISSUE. Blood, lymph, and related fluids of the body. These liquids are regarded as tissues because they consist of characteristic cells lying in an intercellular substance. The liquid condition of the intercellular substance is responsible for their fluidity.

VAS DEFERENS. See **Gonads**.

VASOMOTOR NERVES. See **Arteries and Veins**.

VASOPRESSIN. See **Central and Peripheral Nervous Systems**.

VAST GALAXY DRIFT. Two astronomers have discovered that our own Milky Way Galaxy and most of its neighboring galaxies contained within a huge volume of the universe, one billion light years in diameter, are drifting with respect to the more distant universe. This startling result may imply that the universe is "lumpier" on much larger scales than can be readily explained by any current theory. The new observations thus challenge our understanding of how the universe evolved.

This surprising conclusion comes from the deepest systematic survey of galaxy distances to date, conducted by Dr. Tod R. Lauer of National Optical Astronomy Observations (NOAO) in Tucson, Arizona, and Dr. Marc Postman of the Space Telescope Science Institute (STScI) in Baltimore, Maryland. The two astronomers used NOAO telescopes at Kit Peak National Observatory, new Tucson, Arizona, and at Cerro Tololo Inter-American Observatory, near La Serena, Chile, to study galaxy motions over the entire sky out to a distance of over 500 million light years, thus exploring a volume of space about thirty times larger than has been surveyed by the National Astronomy Observatories, Tucson, AZ.

The expansion of the universe causes all galaxies to be moving away from us. Galaxies at the far edge of the volume surveyed by Lauer and Postman are receding from us at 5 percent of the speed of light. The large flow that the astronomers discovered comes from looking at the galaxy motions "left over" once the expansion of the universe has been taken into account. This flow means that the nearby universe appears to be drifting in a particular direction with respect to the move distant universe, as well as expanding.

Lauer and Postman have measured the drift of the Milky Way with respect to 119 clusters of galaxies located all over the sky at distances as far as 500 million light years. The galaxy clusters are at a variety of distances from us, and galaxies in the distant clusters appear dimmer than the ones in nearby clusters. However, once the various distances are accounted for, the brightest galaxy in each cluster is always found to give off roughly the same amount of light. Astronomers refer to such objects as "standard candles." In a uniformly expanding universe, the distances to the clusters are estimated by how fast they are moving away from us. If the Milky Way Galaxy is drifting, however, its motion makes measurement of the expansion speed depend on the direction we are looking, and the "standard candles" galaxies will appear to vary slightly in brightness in a smooth pattern across the sky. Lauer and Postman used images of the cluster galaxies to detect this pattern and determine the motion of our own galaxy.

If the motion of the Milky Way is caused by galaxies closer in than the set of clusters, its motion with respect to the distant clusters should be essentially identical to that with respect to the microwave back ground radiation. But the motion of the Milky Way that Postman and Lauer measured from the distant clusters is in a completely different direction from the inferred from the microwave background. The most likely solution to this dilemma is that the clusters themselves are moving with respect to the microwave background with an average velocity of 425 miles per second toward that direction of the constellation of Virgo. Because of the enormous size of the volume containing the clusters, however, this result would imply the existence of even more distant and massive concentrations of matter if the motions are caused by gravitational force.

See **Cosmology**.

V-BAND. A frequency band used in radar extending approximately from 46 to 56 gigacycles per second.

V CENTER. One of the simple types of color centers, originally considered to be a positive ion vacancy with a bound positive hole. In the alkali halides the absorption band occurs in the ultraviolet part of the spectrum.

VECTOR. 1. An element of a *linear* or *vector space*. See also **Linear Space**. 2. A directed line segment in Euclidean n-space. If $(\lambda_1, \lambda_2, \ldots, \lambda_n)$ is a point in Euclidean n-space, then the line segment from the origin $(0, 0, \ldots, 0)$ to $(\lambda_1, \lambda_2, \ldots, \lambda_n)$ is the directed line segment associated with the vector $(\lambda_1, \lambda_2, \ldots, \lambda_n)$.

A vector is often indicated graphically by means of an arrow (technically called a *stroke*). The length of the arrow is proportional to the scalar magnitude of the vector, and the direction in which the arrow points is the direction of the vector. The tail or initial point of the arrow is its *origin*; the head or final point is its *terminus*.

A vector of unit length, drawn in the positive direction and tangential to a coordinate system, is a unit vector. It is not necessary that the system be orthogonal. In the common case, a rectangular Cartesian coordinate system is used and the unit vectors along OX, OY, OZ axes are called **i**, **j**, **k**, respectively.

For an arbitrary vector, scalar quantities called the components are required to determine it numerically. In three dimensions, they are directed lines, parallel to the axes of a coordinate system. Thus, if a rectangular Cartesian system is used, with unit vectors **i**, **j**, **k**, any vector may be written as $\mathbf{A} = \mathbf{i}Ax + \mathbf{j}Ay + \mathbf{k}Az$, where (Ax, Ay, Az) are its three components. In the more general case of an n-dimensional vector the components of the vector are the n matrix elements of a column or row matrix. If, in a rectangular coordinate system, a point has coordinates (x, y, z) then its *position vector* is one drawn from the coordinate origin to the point. It may be written as $\mathbf{R} = \mathbf{i}x + \mathbf{j}y + \mathbf{k}z$. In polar coordinates or in spherical polar coordinates, a vector drawn from the origin of the coordinate system to a point is a *radius vector*.

As previously seen, a vector is commonly indicated by a boldface letter such as **A**, which stand for its three scalar components (A_1, A_2, A_3) referred to some coordinate system. In the Gibbs notation, scalar and vector products (see also **Vector Multiplication**) are shown with dots and crosses, respectively. Thus, if C is a scalar and **V**, **A**, **B** are vectors, then $C = \mathbf{A} \cdot \mathbf{B}$ and $\mathbf{V} = \mathbf{A} \times \mathbf{B}$. Less commonly used symbols have been proposed by Hamilton, Grassmann, Heaviside, and others. They include: $T\mathbf{A}$ (T for tensor), $|\mathbf{A}|$ for the magnitude of a vector; $S\mathbf{AB}$, $(\mathbf{AB})$ for the scalar product; $V\mathbf{AB}$, $\mathbf{A}V\mathbf{B}$, and $[\mathbf{AB}]$ for the vector product.

A more precise definition of a vector is often required. Suppose that a point located in a rectangular coordinate system has components (x_1, x_2, x_3). The same point, however, could also be described in other coordinate systems, obtained from the first one by translation of the origin and rotations about the coordinate axes. If the components of the point in the second system are (x'_1, x'_2, x'_3), assumed for convenience to have the same origin as that of the first system, then the relation between the components, called a linear transformation, is

$$x'_i = \sum_{j=1}^{3} c_{ij} x'_j. \quad i = 1, 2, 3$$

where the c_{ij} are the nine direction cosines between the various coordinate-axis pairs. Matrix notation may also be used to write $\mathbf{x}' = \mathbf{Rx}$, where x' and x are column vectors; $\mathbf{R}$ is the orthogonal matrix of the direction cosines. If this transformation law holds, one speaks of a *polar, proper*, or *localized vector*; if the law does not hold, a pseudovector or *axial vector*.

The concept of vector may be generalized extensively. A *four-vector* has four components. One type is called a quaternion, another, used principally in relativity theory, has for its components (x, y, z, ict), where x, y, z are positional coordinates, $i = sqrt-1$, c is the velocity of light, and t is the time. The components of such a vector in one coordinate system are related to the components in another system by a *Lorentz transformation*. See also **Tensor**.

If (x', y', z') are functions of (x, y, z), then the vector $\mathbf{V}' = \mathbf{i}x' + \mathbf{j}y' + \mathbf{k}z'$ is a vector function of the vector $\mathbf{V} = \mathbf{i}x + \mathbf{j}y + \mathbf{k}z$, where $(\mathbf{i}, \mathbf{j}, \mathbf{k})$ are unit vectors. The function is a linear vector function if $f(\mathbf{A} + \mathbf{B}) = f(\mathbf{A} + f(\mathbf{B}))$, for all vectors $\mathbf{A}, \mathbf{B}$ and $f(k\mathbf{V}) = kf(\mathbf{V})$, where k is a scalar. More generally, suppose the components of $\mathbf{V}$ and $\mathbf{V}'$ are (V_1, V_2, V_3) and (V'_1, V'_2, V'_3) respectively and that the relation between the two vectors in matrix form is $\mathbf{V} = \mathbf{MV}'$, where $\mathbf{M}$ is a (3×3)-matrix with elements M_{ij}. Then the function is a linear vector function, for $f(k\mathbf{V}) = kf(\mathbf{V})$ and $f(\mathbf{V} + \mathbf{V}') = f(\mathbf{V}) + f(\mathbf{V}')$, as before.

The vector $\mathbf{V}$ can be written as the sum of a symmetric and an antisymmetric linear vector function, $\mathbf{V} = \mathbf{S} + \mathbf{A}$. If $\mathbf{S} + \mathbf{QV}'$, then $Q_{ii} + M_{ii}Q_{ij} + Q_{ji} + (M_{ij} + M_{ji})/2$, $i \neq j$. Similarly, $\mathbf{A} = \mathbf{TV}'$, $T_{ii} = 0$, $T_{ij} + (M_{ij} - M_{ji})/2$. The antisymmetric function can also be written as a vector product, $\mathbf{A} + \mathbf{T} \times \mathbf{V}'$.

See also **Dyadic**.

VECTOR ADDITION. If **A**, **B** are vectors with components A_x, A_y, A_z and B_x, B_y, B_z, respectively, their sum is a new vector $\mathbf{C} = \mathbf{A} + \mathbf{B}$, with components $A_x + B_x, A_y + B_y, A_z + B_z$. Vector addition obeys the commutative and associative laws of algebra: $\mathbf{A} + \mathbf{B} = \mathbf{B} + \mathbf{A}$; $(\mathbf{A} + \mathbf{B}) + \mathbf{C} = \mathbf{A} + (\mathbf{B} + \mathbf{C})$. To subtract a vector $\mathbf{B}$ from a vector $\mathbf{A}$, take the negative of $\mathbf{B}$ and add $-\mathbf{B}$ to $\mathbf{A}$.

VECTOR DERIVATIVE. If a vector $\mathbf{R}$ is a function of a single scalar variable t, there are three possible ways in which $\mathbf{R}$ may vary with t, for if $\mathbf{R}_1$ and $\mathbf{R}_2$ refer to t_1 and t_2, respectively, then $\mathbf{R}_2$ may differ from $\mathbf{R}_1$: in magnitude only; in direction only; in both magnitude and direction. Since even the general case is relatively simple, assume that a curve is traced by the terminus of the continuously varying vector $\mathbf{R}$, the origin of the vector being kept fixed at the origin of a coordinate system. Let A and B be two neighboring points on this curve and let $\mathbf{R}_1$ and $\mathbf{R}_2$ be their position vectors, then the vector $\Delta\mathbf{R} = \mathbf{R}_2 - \mathbf{R}_1$ has the direction of the secant AB, which approaches the tangent to the curve at A as $\Delta t = t_2 - t_1$ approaches zero. The quotient $\Delta\mathbf{R}/\Delta t$ is the average rate of change of $\mathbf{R}$ in the interval between t_1 and t_2. The derivative is defined as

$$\lim_{t \to 0} \Delta\mathbf{R}/\Delta t = d\mathbf{R}/dt$$

In terms of unit vectors, and with the use of primes for differentiation, $\mathbf{R} = \mathbf{i}R_x + \mathbf{j}R_y + \mathbf{k}R_z$, $\mathbf{R}' = \mathbf{i}R'_x + \mathbf{j}R'_y + \mathbf{k}R'_z$, $\mathbf{R}'' = \mathbf{i}R''_x + \mathbf{j}R''_v + \mathbf{k}R''_z$. For a composite function of two or more vectors, each depending on a single scalar t, the usual rules of differentiation hold, except that the order of the vectors must be retained if vector products are involved.

There are also several differential vector operators. See also **Curl**; **Del**; **Divergence (Mathematics)**; **Gradient (Mathematics)**; and **Laplacian**.

VECTOR FIELD. An assignment of a vector to each point of region of space. Thus, in three dimensions, the vector assigned to each point is described by three quantities, the components of the vector along the coordinate axes. Examples are wind velocities in the atmosphere, electrostatic or electromagnetic field. See also **Scalar Field**.

VECTOR INTEGRAL. A kind of inverse operation to differentiation. Corresponding to ordinary definite integrals (see also **Multiple Integral**) there are *line integrals, surface integrals*, and *volume integrals* of vector functions.

(a) *Line integral.* Suppose $\mathbf{r} = \mathbf{r}(t)$ determines a curve C in space and that $d\mathbf{r}$ is an infinitesimal line element of this curve. Three different line integrals may then be formed, using a scalar ϕ or a vector $\mathbf{V}$:

$$(1) \quad \int_C \phi \, d\mathbf{r}; \quad (2) \quad \int_C \mathbf{V} \cdot d\mathbf{r}; \quad (3) \quad \int_C \mathbf{V} \times d\mathbf{r}$$

The results of integration are a vector, a scalar, a vector, respectively.

In each case, the line integral may be reduced to a sum of definite integrals and evaluated by the usual methods of integral calculus. The line integral may be generalized for the complex variable and the result is a contour integral.

(b) *Surface integral.* There are three possible cases:

$$(1) \quad \int_S \phi \, d\mathbf{S}; \quad (2) \quad \int_S \mathbf{V} \cdot d\mathbf{S}; \quad (3) \quad \int_S \mathbf{V} \times d\mathbf{S}$$

which give a vector, a scalar, a vector. It is convenient to write only one integral sign, in general, and to understand by the symbol S, attached to the integral sign, that the limits of integration are suitably chosen. The surface element $d\mathbf{S} = d\mathbf{x} \times d\mathbf{y}$.

In case (2), if $\mathbf{V}$ is the product of density and velocity of a fluid (or electric, magnetic, gravitational force; heat, etc.), the integral is the *flux* of $\mathbf{V}$ through the surface. See also **Area**.

(c) *Volume integral.* In vector notation, the element of volume $d\tau = dx\,dy\,dz$ is a scalar. There are thus two possible volume integrals:

$$(1) \quad \int \phi \, d\tau, \quad (2) \quad \int_\tau \mathbf{V} d\tau$$

The integrals are, respectively, a scalar and a vector. As is frequently the custom, only one integral sign is used, and the symbol τ is a reminder that the integration is triple and that appropriate limits of integration are to be supplied. See also **Volume (Geometry)**.

A circulatory integral is a vector function

$$\int \mathbf{V} \cdot d\mathbf{r}$$

over a closed contour. It is a measure of the tendency of lines of force to close up. If $\mathbf{V}$ refers to a fluid, then this integral is a measure of the flow around the path chosen. When the vector field has a potential, this integral is zero and the field is said to be irrotational, thus its curl must vanish.

If $\mathbf{V}$ describes a vector field, for example, the velocity of an incompressible fluid, then the total flux through a surface S in the field is given by

$$\int_S \mathbf{V} \cdot d\mathbf{S}$$

The vector $\mathbf{V}$ may refer to electric, magnetic, or gravitational force; heat or a fluid, etc. The surface integral may be converted to a volume integral by Gauss's theorem.

For other relations between vector integrals see also **Green Function**; and **Stokes Theorem**.

VECTOR MULTIPLICATION. There are two distinct kinds of products of two vectors: the scalar product and the vector product (see also **Pseudovector**). They are also sometimes called inner and outer products but these terms more commonly refer to tensor products. There are also several possibilities for the product of three or four vectors, as the subsequent discussion will show.

1. *Scalar Product.* If $\mathbf{A}$ and $\mathbf{B}$ are two vectors, of magnitude A, B, respectively, their scalar product is $\mathbf{A} \cdot \mathbf{B} = AB\cos\theta$, where θ is the angle between the two vectors. This product, which is a scalar quantity, is also known as the dot product. If the vectors are complex, the result of multiplication is the Hermitian scalar product (see also **Linear Space**).

The scalar product of two vectors obeys the commutative and distributive laws: $\mathbf{A} \cdot \mathbf{B} = \mathbf{B} \cdot \mathbf{A}$; $\mathbf{A} \cdot (\mathbf{B} + \mathbf{C}) = \mathbf{A} \cdot \mathbf{B} + \mathbf{A} \cdot \mathbf{C}$. If $\mathbf{A}$ is perpendicular to $\mathbf{B}$, then $\mathbf{A} \cdot \mathbf{B} = 0$, and consequently if $\mathbf{A} \cdot \mathbf{B} = 0$, then $\mathbf{A}$ is perpendicular to $\mathbf{B}$ and the two vectors are said to be orthogonal. If $\mathbf{A}$ is parallel to $\mathbf{B}$, then $\mathbf{A} \cdot \mathbf{B} = AB$. Consequently, $\mathbf{A} \cdot \mathbf{A} = A^2$, the square of the length of $\mathbf{A}$.

2. *Vector Product.* The vector product of $\mathbf{A}$ and $\mathbf{B}$ (also called skew or cross product) is of length $C = AB\sin\theta$ and its direction is perpendicular to the plane determined by $\mathbf{A}$ and $\mathbf{B}$. In Cartesian coordinates, with unit vectors i, j, k, one has

$$\mathbf{C} = \mathbf{A} \times \mathbf{B} = (A_yA_z - A_zB_z)\mathbf{i} + (A_zB_z - A_zB_z)\mathbf{j}$$

$$+ (A_xB_y - A_yB_x)\mathbf{k} = \begin{vmatrix} \mathbf{i} & \mathbf{j} & \mathbf{k} \\ A_x & A_y & A_z \\ B_x & B_y & B_z \end{vmatrix}$$

Vector multiplication is not commutative, in this case, for $\mathbf{A} \times \mathbf{B} = -\mathbf{B} \times \mathbf{A}$ but the distributive law of multiplication still holds.

3. *Products of Unit Vectors.* For the two kinds of vector multiplication described in (1) and (2), the results for unit vectors are: $\mathbf{i} \cdot \mathbf{j} = \mathbf{j} \cdot \mathbf{i} = \mathbf{i} \cdot \mathbf{k} = \mathbf{k} \cdot \mathbf{i} = \mathbf{j} \cdot \mathbf{k} = \mathbf{k} \cdot \mathbf{j} = 0$; $\mathbf{i} \cdot \mathbf{i} = \mathbf{j} \cdot \mathbf{j} = \mathbf{k} \cdot \mathbf{k} = i^2 = j^2 = k^2 = 1$; $\mathbf{i} \times \mathbf{j} = -\mathbf{j} \times \mathbf{i} = \mathbf{k}$; $\mathbf{j} \times \mathbf{k} = -\mathbf{k} \times \mathbf{j} = \mathbf{i}$; $\mathbf{k} \times \mathbf{i} = -\mathbf{i} \times \mathbf{k} = \mathbf{j}$; $\mathbf{i} \times \mathbf{i} = \mathbf{j} \times \mathbf{j} = \mathbf{k} \times \mathbf{k} = 0$.

4. *Triple Products of Vectors.* Three vectors $\mathbf{A}$, $\mathbf{B}$, $\mathbf{C}$ may be combined to form products with meaning in several ways: (a) $\mathbf{A}(\mathbf{B} \cdot \mathbf{C})$, a vector with the same direction as $\mathbf{A}$ and magnitude ABC $\cos\theta$, where θ is the angle between $\mathbf{B}$ and $\mathbf{C}$. (b) $\mathbf{A} \cdot (\mathbf{B} \times \mathbf{C})$, the scalar triple product, giving the volume of a parallelepiped with edges $\mathbf{A}$, $\mathbf{B}$, $\mathbf{C}$. It is frequently indicated by the symbol [**ABC**] and if the three vectors all lie in the same plane [**ABC**] $= 0$. It may be written in terms of its components as a determinant

$$[\mathbf{ABC}] = \begin{vmatrix} A_x & A_y & A_z \\ B_x & B_y & B_z \\ C_x & C_y & C_z \end{vmatrix}$$

Its properties include [**ABC**] = [**BCA**] = [**CAB**] = $-$[**ACB**] = $-$[**BAC**] = $-$[**CBA**]. (c) The vector triple product, $\mathbf{V} = \mathbf{A} \times (\mathbf{B} \times \mathbf{C})$ is perpendicular to both $\mathbf{A}$ and the vector ($\mathbf{B} \times \mathbf{C}$). It therefore lies in the plane determined by $\mathbf{B}$ and $\mathbf{C}$. Its properties include:

$$\mathbf{A} \times (\mathbf{B} \times \mathbf{C}) = \mathbf{B}(\mathbf{A} \cdot \mathbf{C}) - \mathbf{C}(\mathbf{A} \cdot \mathbf{B}) = -\mathbf{A} \times (\mathbf{C} \times \mathbf{B})$$

$$= (\mathbf{C} \times \mathbf{B}) \times \mathbf{A} = -(\mathbf{B} \times \mathbf{C}) \times \mathbf{A}.$$

5. *Quadruple Products of Vectors.* If $\mathbf{A}$, $\mathbf{B}$, $\mathbf{C}$, $\mathbf{D}$ are any four vectors, two types of quadruple products can occur:

(a) $(\mathbf{A} \times \mathbf{B}) \cdot (\mathbf{C} \times \mathbf{D}) = (\mathbf{A} \cdot \mathbf{C})(\mathbf{B} \cdot \mathbf{D}) - (\mathbf{A} \cdot \mathbf{D})(\mathbf{B} \cdot \mathbf{C})$;
(b) $(\mathbf{A} \times \mathbf{B}) \times (\mathbf{C} \times \mathbf{D}) = (\mathbf{A} \cdot \mathbf{C} \times \mathbf{D})\mathbf{B} - (\mathbf{B} \cdot \mathbf{C} \times \mathbf{D})\mathbf{A}$.

The latter equation is also conveniently written in the equivalent form [**ABC**]**B** $-$ [**BCD**]**A** = [**ABD**]**C** $-$ [**ABC**]**D**, where the bracket signifies the scalar triple product. See also **Reciprocal Vector System**.

Products of more than four vectors can always be reduced to combinations of one or more of the preceding types.

VECTOR SPACE. See **Linear Space (Vector Space)**.

VEERING WIND. See **Winds and Air Movement**.

VEERY. See **Thrush**.

VEGA (α Lyrae). Ranking fifth in apparent brightness among the stars, Vega has a true brightness value of 55 as compared with unity for the sun. Vega is a white, spectral type A star and is located in the constellation Lyra north of the ecliptic. Estimated distance from the Earth is 27 light years. In the northern latitudes, Vega is visible during some portion of every night throughout the year, and dominates the summer skies. Because of its distinctly bluish tinge, it is one of the most beautiful stars of the northern skies, and references to it are found in ancient literature. At one time, Vega was the pole star, and because of precession, the pole will be close to Vega about 11,500 years hence.

VEGETABLE OILS (Edible).

Vegetable oils are prepared from at least ten major oilseeds, plus a few other sources that may develop into high volume at some future date. There are both edible and nonedible vegetable oils. Technical and industrial vegetable oils, such as castorseed and tung oils, are used in nonfood applications. Some oils, such as linseed (flaxseed) and olive oil, depending upon the manner in which they have been treated and refined, find both food and nonfood applications. Other oils, such as soybean, sunflower, rapeseed, sesame, and safflower oils, are used predominantly in food processing and in the production of feedstuffs. Usually, in considering edible vegetable oils, the availability of animal fats, such as butterfat, lard, tallow, etc., and of fish and other marine oils, is noted because there is surprising versatility among all oils and fats, allowing significant substitutions in the same end-product. Perhaps this is best illustrated by the fact that coconut oil, corn (maize) oil, cottonseed oil, palm oil, groundnut (peanut) oil, safflower oil, and soybean oil all are or have been used in commercial margarines for the retail market. The processor of a given brand of margarine will select the oil or a combination of oils, considering price and availability, as well as desirable end-product characteristics (stick or brick; soft tub; diet or imitation; etc.), among other factors. The pressed cakes and meals remaining after extraction of oil from beans and seeds finds wide use in animal feedstuffs.

The consumption of fats and oils by persons in the United States has increased steadily since the early 1960s. In 1963, 46.3 pounds (21 kilograms) were consumed per capita per year, rising to 54.4 pounds (24.7 kilograms) in the late 1970s. During this period, butter dropped from 13.9% of total fats and oils to 7.7%; lard dropped from 12.9% to 4%; margarine increased from 19.4% to 20.2%; shortenings of various kinds increased from 27.2% to 30.6%; and salad and cooking oils increased markedly from 26.6% to 37.5%.

On a worldwide basis, as of the early 1980s, the various sources of edible oils, in terms of volume of production, rank as follows: Soybeans, 50.3%; cottonseed, 16.2%; groundnuts (peanuts), 11.4%; sunflower seeds, 7.9%; rapeseed, 6.9%; copra (coconut), 2.9%; flaxseed (linseed), 1.8%; sesame seed, 1.1%; palm kernel, 0.9%; and safflower seed, 0.6%. See also **Diet**.

Principal edible uses of these oils are found in cooking and salad oils; frying oils; margarine, mayonnaise, and salad dressings; bakery, cake mix, and pie shortenings; and whipped topping and other nondairy products, such as coffee creamers.

Definitions

Some of the common terms used in describing edible oils and fats are:

Saturated — the state in which all available valence bonds of an atom, especially carbon, are attached to other atoms. The straight-chain alkyls (paraffins) are typical saturated compounds. Several fatty acids found in oils and fats are listed in Table 1. Where no double- or triple-bonds are available, such compounds cannot be hydrogenated because there are no places for attaching additional hydrogen atoms.

Monosaturated — the state where one double bond is present in the compound. Under proper conditions of hydrogenation, an additional hydrogen atom can be added, thus resulting in a saturated compound.

Polyunsaturated — the state where two or more double bonds are present in the compound. Again, under proper conditions of hydrogenation, one or more additional hydrogen atoms can be added, thus resulting in a compound that is less unsaturated or that is saturated, depending upon the number of double bonds originally available and the degree of hydrogenation effected.

Hydrogenation — any reaction of hydrogen with an organic compound. In the case of fats and oils, this is the direct addition of hydrogen to double bonds of unsaturated molecules, resulting in a partially or a fully saturated product.

Hydrogenation can be accomplished with gaseous hydrogen under pressure in the presence of a catalyst (nickel, platinum, or palladium). The degree of saturation of an oil or its fatty acid substituents affects its fluidity (melting or softening point), density, refractive index, and general reactivity. Frequently, the degree of hydrogenation of an oil will be determined by measuring its refractive index. Vegetable and fish oils can be hardened or solidified by catalytic hydrogenation. Partial hydrogenation clarifies some oils and makes them odorless. Fatty oils, such as oleic acid, are converted into stearic acid by hydrogenation. Coconut oil, groundnut (peanut) oil, and cottonseed oils can be made to appear, taste, and smell like lard; or they can be made to resemble tallow. Sometimes, hydrogenated oils are referred to as synthetic shortenings. Generally, hydrogenated oils have higher melting points and lower iodine values than the natural, untreated oils.

For many years, hydrogenation was a batch operation, but in recent years, continuous hydrogenation processes have been displacing the batch methods, especially in large-scale facilities. Hydrogenation pressures vary as well as temperatures, but the latter are usually in the 93–135 °C range. It is not uncommon to mix two or more oils prior to hydrogenation. Considerably more detail on the hydrogenation process is given in the entry on **Hydrogenation**.

Deodorizing — crude oils tend to have undesirable odors and tastes. These are due to the presence of various volatile components. These are removed by passing steam through the heated oil under diminished pressure.

Winterizing — because of widespread use of refrigeration, salad oils and other end-products prepared from vegetable oils must not become cloudy or solidify at relatively low storage temperatures. Thus, most salad oils differ from cooking oils, in that the latter, if stored under refrigeration, may slowly solidify and be difficult to pour from a container. To prevent cloudiness and solidification at low temperatures, the oils are "winterized." This involves subjecting the oils to low temperatures in stages, during which time crystals of high-melting-point fat are formed. A final low temperature of about 5.5 °C is reached in the process, at which temperature the oil is allowed to stand for a considerable time, during which considerable crystallization occurs. Slow cooling and gentle agitation ensure maximum removal of the high-melting-point fats. The crystalline material remaining after filtration is essentially stearine, with traces of wax, gums, and soaps. Improvements in the winterizing process during the past several years have reduced the time required from 3 to 6 days down to about 5 hours. The more recent processes involve solvent extraction and centrifugation.

Iodine Value (or Number) — the percentage of iodine that will be absorbed by a chemically unsaturated substance, such as vegetable oil, in

TABLE 1. CHARACTERISTICS OF VEGETABLE OIL FATTY ACIDS

Principal Fatty Acid	Formula Weight	Number of Double Bonds	Position Of Bonds	Number of Carbon Atoms	Formula
SATURATED COMPOUNDS					
Caproic	116.09	0	—	6	$C_5H_{11}COOH$
Caprylic	144.21	0	—	8	$C_7H_{15}COOH$
Capric	172.26	0	—	10	$C_9H_{19}COOH$
Lauric	200.31	0	—	12	$C_{11}H_{23}COOH$
Myristic	228.36	0	—	14	$C_{13}H_{27}COOH$
Palmitic	256.42	0	—	16	$C_{15}H_{31}COOH$
Stearic	258.47	0	—	20	$C_{19}H_{39}COOH$
UNSATURATED COMPOUNDS					
Palmitolenic	254.42	1		16	$C_{15}H_{29}COOH$
Linolenic	278.42	3	cis-9, cis-12, cis-15	18	$C_{17}H_{29}COOH$
Linoleic	180.44	2	cis-9, cis-12	18	$C_{17}H_{31}COOH$
Oleic	282.45	1	cis-9	18	$C_{17}H_{33}COOH$

TABLE 2. CHARACTERISTICS AND PROPERTIES OF MAJOR VEGETABLE OILS

Property	Coconut Oil	Cottonseed Oil	Groundnut (Peanut) Oil	Pal Oil	Rapeseed Oil
Color	White	Pale-yellow or yellowish-brown to dark ruby-red, or black-red.	Yellow to greenish-yellow	Yellow-brown	Brown (raw); yellow (refined)
State (at room temperature)	Semisolid (nondrying oil)	Liquid (semidrying oil)	Liquid (nondrying)	Solid (buttery)	Liquid (viscous)
Odor	Slight	Slight (when refined)	Nutlike	Agreeable	Characteristic
Melting point	77°–90°F; 25°–32°C	Slightly below 32°F; 0°C# (before winterizing process)	23°–37°F; −5°–+3°C@	86°F; 30°C	328°F; 0°C$
Specific gravity	0.92	0.915–0.921	0.912–0.920	0.952	0.913–0.916
Saponification value	250–264	190–198	186–194	247.6	174
Iodine value	7–10	109–116	88–98	13.5	100.3
Principal constituents	Glycerides of fatty acids of approximate composition: Lauric acid 45–48% Myristic acid 17–20% Capric acid 6–8% Palmitic acid 5–9% Caprylic acid 5–7% Oleic acid 4–8% Stearic acid 2–5% Arachidic acid 1%± Palmitoleic acid 0.4%± Unsaturated acids 8% (Approximate) Saturated acids 92% (Approximate)	Glycerides of fatty acids of approximate composition: Linoleic acid 47–50% Palmitic acid 26–27% Oleic acid 18–19% Stearic acid 2%+ Palmitoleic acid 1%± Myristic acid 1%− Capric acid 0.5% Linolenic acid 0.4%± Lauric acid 0.4% Arachidic acid 0.2%± Unsaturated acids 73%	Glycerides of fatty acids of approximate composition: Oleic acid 40–52% Linoleic acid 25–37% Palmitic acid 8.5–10.5% Behenic acid 2.5%± Lignoceric acid 1.5%± Stearic acid 1.5–3.0% Arachidic acid 1.5–2.5% Myristic 0.2%± Unsaturated acids 82% (Approximate) Saturated acids 18% (Approximate) 27%	Triglycerides of fatty acids of approximate composition: Palmitic acid 37–47% Oleic acid 31–44% Lauric acid 4–8% Stearic acid 2.5–5.5% Myristic acid 0.8–1.5% Linoleic acid 1–3% Palmitoleic acid 0.2–0.4% Arachidic acid 0.2–0.4% Unsaturated acids 50% Saturated acids 50% (Approximate) There are some significant differences in composition of palm oil (from fleshy fruit pulp) and palm kernel oil.[symbol]	High in unsaturated acids, especially oleic, linoleic, and erucic acids, the latter being 40% in older varieties, but more recent varieties have a lower erucic acid content.

3610

Soluble in	Alcohol, carbon disulfide, chloroform, ether. Insoluble in water.		Carbon disulfide, chloroform, ether, petroleum ether. Insoluble in alkalies, but saponified by alkali hydroxides with formation of soaps. Slightly soluble in alcohol. Insoluble in water.	Benzene, carbon disulfide, chloroform, ether. Slightly form, soluble in alcohol.	Alcohol, carbon disulfide, chloroform, ether. mmiscible in water.
Grades	Crude, refined		Crude, refined, edible, USP.	Crude, refined, prime summer yellow, bleachable, USP.	Crude, refined, Ceyon, Manila
Derivation	Separation of fat from palm fruits by expression or centrifugation.		By pressing ground groundnut (peanut) meats or by extraction with hot or cold solvents. Purified by bleaching with fuller's earth or carbon. Hot-pressed oil may be allowed to stand to deposit stearin, prior to filtering.	From cotton seeds by hot-pressing or solvent extraction.	Hydraulic press or expeller extraction from coconut meat, followed by alkali-refining, and bleaching.
Food uses	Salad dressings, margarine, substitute for soybean oil.	Food shortenings, margarines, competes with soybean oil.	Salad oils, mayonnaise, margarine Sometimes a substitutefor olive oil. Cooking oil.	Margarine, shortening, salad oils and dressings, stabilizers.	Margarine, hydrogenated shortenings, synthetic cocoa dietary supplements.
Special notations	$Flash point is 325 °F (617 °F) Autoignition temperature is 836 °F (447 °C).	⊘Approximate composition of palm kernel oil is: 50% lauric acid, 15% oleic acid, 16% palmitic acid, 7% myristic acid with lesser amounts of capric and caprylic acids. See also **Palm Oil**.	@Flash point is 540 °F (282 °C) See also **Groundnut (Peanut)**.	#Solidification range is 88°–95 °F (31°–35 °C) Flashpoint is 486 °F (252 °C) See also entry on **Cottonseed**.	See also entry on **Coconut**.

(continued)

TABLE 2. (*Continued*)

Property	Safflower Seed Oil	Sesame Seed Oil	Soybean Oil	Sunflower Oil
Color	Straw (Nonyellowing)	Yellow	Pale Yellow	Pale Yellow
State (at room temperature)	Liquid	Liquid	Liquid (fixed drying oil)	Liquid (semidrying)
Odor	Almost odorless	Almost odorless	Characteristic	Pleasant
Melting point	0°–+13°F; −18°– −25°C	68°–77°F; 20°–25°C	77°–88°F; 22° 31°	3° + °F; −16°– −18°C
Specific gravity	0.923–0.927	0.9187	0.924–0.929	0.924–0.926
Saponification value	186–193	188–193	190–193	186–194
Iodine value	140–152	103–114	137–143	130–135
Principal constituents	Basic fatty acid composition: Linoleic acid 78%± Oleic acid 13%± Stearic acid 3%± Palmitic acid 6%± Unsaturated acids 90% Saturated acids 10% (Approximate)	Basic fatty acid composition: Oleic acid 40%± Linoleic acid 44%± Other saturates 1%± Palmitic acid 9%± Stearic acid 4%± Other saturates 2%± Unsaturated acids 85% Saturated acids 15% (Approximate)	Basic fatty acid composition: Oleic acid 20–25% Linoleic acid 48–53% Linoleic acid 5–9% Stearic acid 3–5% Palmitic acid 9–12% Myristic acid 0.2%± Arachidic acid 0.2%± Palmitoleic acid 0.3% Lauric acid 0.1% Unsaturated acids 84% Saturated acids 16% (Approximate)	Basic fatty acid composition: Linoleic acid 60–75% Oleic acid 17–32% Palmitic acid 7–10% Stearic acid 4–8% Myristic acid 0.1% Arachidic acid 0.3% Linolenic acid 0.3%± Unsaturated acid 90% Saturated acids 10% (Approximate)
Soluble in		Benzene, carbon disulfide, chloroform, ether. Slightly soluble in alcohol.	Alcohol, carbon disulfide, chloroform, ether	Alcohol, carbon disulfide, chloroform, ether
Grades	Crude, refined	Edible should contain less than free fatty acids. Semi-refined, coast, USP.	Coast, refined (salad), crude, foots (for soapstock), clarified.	Crude, refined
Derivation	Hydraulic press or solvent extraction of seeds.	Pressing of seeds.	Oil for edible purposes is bleached with fuller's earth.	Expression from seeds.
Food uses	Dietetic foods, margarine, hydrogenated shortenings	Shortenings, salad oil, margarine	High-protein foods, margarine, salad dressings	Margarine, shortening
Special notations	Considered by some authorities as the most natural, nutritionally sound vegetable oil. See also **Safflower Seed Oil.**	≠Solidifying point is 23°F (−5°C) See also **Sesame Seed Oil.**	See also **Soybean Processing.**	See also **Safflower Seed Oil.**

a given time under specified conditions. The iodine value is a measure of degree of unsaturation. Two methods are described in the "Food Chemicals Codex:" the Hanus method and the Wijs method.

Saponification Number — the number of milligrams of potassium hydroxide required to hydrolyze one gram of sample of an ester (glyceride, fat) or mixture. This test is also described in the "Food Chemicals Codex."

Margarines

Regular margarines (by government regulations) contain at least 80% fat. The remaining content is ~16% water and small amounts (about 4%) of skim or nonfat dry milk and salt. There are several unsalted margarines available. Small amounts of emulsifying agents consisting of lecithin and/or monoglycerides and diglycerides are contained in a number of commercially available margarines. The preservatives most commonly used include sodium benzoate, potassium sorbate, calcium disodium EDTA, isopropyl citrate, and citric acid. Most margarines in the United States are fortified with about 15,000 USP units of vitamin A and a few margarines also contain up to 2000 USP units of vitamin D.

Margarines produced in the United States resemble butter in their proximate composition. The great majority of margarines provide more polyunsaturated fatty acids than an equivalent weight of butter. Diet and imitation margarines were introduced about a decade ago and contain less than 80% fat. The fat content of light blends is about 60% and that of diet imitation margarines is about 40%. Lower fat content in these products is compensated for by higher water content.

A survey of some 40 margarines made in the United States in the late 1970s showed that stick margarines made from partially hydrogenated soybean and cottonseed oil were by far the largest category. Stick margarines made from corn (maize) oil and partially hydrogenated corn oil, and from partially hydrogenated soybean oil and liquid cottonseed oil ranked second and third, respectively. The survey showed that margarines with formulations that included liquid cottonseed oil or liquid and/or partially-hardened palm oil usually contained higher amounts of palmitic acid than did other margarines of the same type. The stearic acid content was reasonably constant for most hard and soft margarines. Margarines made with coconut oil were extremely saturated and contained large amounts of lauric and myristic acids.

Properties of Major Edible Vegetable Oils

The major properties of nine of the principal edible vegetable oils are summarized in Table 2. For descriptions of the constituent acids, see also **Arachidic Acid**; **Caproic Acid**; **Capric Acid**; **Lauric Acid**; **Linoleic Acid**; **Linolenic Acid**; **Myristic Acid**; **Oleic Acid**; **Palmitic Acid**; and **Stearic Acid and Stearates**.

Unconventional Sources of Oils and Proteins

Research efforts continue to find new sources of both food and nonfood oils. Some investigators have reported on the prospects of various gourds indigenous to western North America. Such cucurbits are vigorous and highly drought resistant and produce large quantities of foliage and fruit containing seeds rich in protein and oil, with an extensive system of large, fleshy storage roots which contain starch. Varieties studied have included *Cucurbita foetidissima* and *C. digitata*. Also studied have been varieties of hibiscus (okra). As pointed out by the researchers, the non-conventional protein sources as well as oils would be welcome in a time of protein deficits.

Investigators also have described the potential of the *jojoba plant*, a native of the Sonoran desert of Mexico, Arizona, and California. The desert shrub produces seeds about the size of groundnuts (peanuts) which contain a liquid wax, frequently called *jojoba oil*. The oil is similar to sperm whale oil in its suitability for a number of industrial applications. Sperm whale was placed on the endangered species list by a number of countries, including the United States, in 1971. Prospects of an edible oil from jojoba are undetermined, but the meal remaining after pressing jojoba seeds contains 30–35% protein, and may have potential as a livestock feed. A new jojoba industry would mean a new, renewable resource that has not been exploited before and would offer economic relief to the people of the Sonoran desert region. Other researchers have recently shown a renewed interest in the lupine as a source of edible oils and proteins.

Additional Reading

Bockisch, M.: "Fats and Oils Handbook," AOCS Press, Champaign, IL, 1998.
Gunstone, F.D. and D. Firestone: "Scientia Gras: A Select History of Fat Science and Technology," AOCS Press, Champaign, IL, 2000.
O'Brien, R.D., W.E. Farr, and P.J. Wan: "Introduction to Fats and Oils Technology," 2nd Edition, AOCS Press, Champaign, IL, 2000.
Przybylski, R. and B.E. McDonald: "Development and Processing of Vegetable Oils for Human Nutrition," AOCS Press, Champaign, IL, 1995.
Widak, N.: "Physical Properties of Fats, Oils, and Emulsifiers," AOCS Press, Champaign, IL, 2000.

VEIN (Geology). A small or large fissure that has been filled with mineral matter by deposition from aqueous solutions, including "liquors" and gaseous emanations from magmas. Lode means essentially the same thing, the latter being an old Cornish mining term referring to the formations that would "lead" or direct the miner to the desired minerals.

VELA. A southern constellation which once, with Carina and Puppis, was part of a superconstellation known as Argo Navis.

VELOCIMETER. A continuous-wave reflection Doppler system used to measure the radial velocity of an object.

VELOCITY AND SPEED MEASUREMENT. Velocity and speed are considered to be rate variables. See also **Variable (Process)**.

Velocity

Velocity is the time rate of change of position. (Unless angular velocity is specified, this term is understood to refer to linear motion, which may be emphasized by the expression "linear velocity.") Strictly, the velocity of a moving point must specify both the speed and the direction of the motion, and is therefore a vector; though the term is sometimes more loosely used as merely synonymous with speed. The velocity of a point is the time rate of the distance s from a fixed origin O, expressed as the vector derivative of s with respect to the time, ds/dt (Fig. 1); while the speed is the magnitude of the velocity and is not a vector. If the direction of motion is constant, so that the motion is in a straight line (but not necessarily with constant speed), and if the line of motion is clearly understood, it is convenient to treat the distance s and the velocity ds/dt as scalars with respect to some zero point on that line and with appropriate algebraic signs (Fig. 2); otherwise they must be regarded as vectors. If the velocity is variable, account must be taken of the acceleration. Examples of both curved and rectilinear motion are treated under kinematics.

Angular Velocity. A quantity relating to rotational motion. While the use of the term "angular velocity" may be extended to any motion of a point with respect to any axis, it is commonly applied to cases of rotation. Its instantaneous value is defined as the vector, whose magnitude is the time rate of change of the angle θ rotated through, for example, $d\theta/dt$, and whose direction is arbitrarily defined as that direction of the rotation axis for which the rotation is clockwise. The usual symbol is ω or Ω.

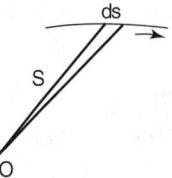

Fig. 1.   Velocity expressed as a vector derivative.

The concept of angular velocity is most useful in the case of rigid body motion. If a rigid body rotates about a fixed axis and the position vector of any point P with respect to any point on the axis as origin is **r**, the velocity $d\mathbf{r}/dt$ of P relative to this origin is $d\mathbf{r}/dt = \boldsymbol{\omega} \times \mathbf{r}$, where $\boldsymbol{\omega}$ is the instantaneous vector angular velocity. This indeed may serve as a definition of $\boldsymbol{\omega}$.

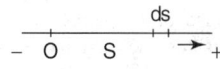

Fig. 2.   Treatment of distance and velocity as scalars.

The average angular velocity may be defined as the ratio of the angular displacement divided by the time. In general, however, this is not a vector, since a finite angular displacement is not a vector. The instantaneous angular velocity is more widely used.

Angular velocities, like linear velocities, are vectorially added, for example, if a top is spinning about an axis which is simultaneously being tipped over toward the table, the resultant angular velocity is the vector sum of the angular velocities of spin and of tipping. This enters into the theory of precession. The derivatives of the Eulerian angles are sometimes very useful in describing the angular motion of a rigid body which has components of angular velocity about all its principal axes.

Speed

Speed is a scalar quantity equal to the magnitude of velocity. Velocity is a vector quantity denoting both the direction and the speed of a linear motion, or denoting the direction of rotation and the angular speed in the case of rotation. Industrially, linear speeds are frequently inferred from rotational measurements simply because of the manner in which most machines are designed — with rotating shafts, wheels, and gears to which speed transducers can be conveniently attached.

As with most other areas of industrial instrumentation, analog-type sensors of speed no longer exclusively predominate the technology. Commencing on a small scale in the mid-1950s, digital speed sensors have been developed and, for many applications, digital sensors are preferred because of the ease with which they can be integrated into overall digital control systems. However, it should be stressed that analog-type tachometers are still widely used. The word *tachometer*, which once denoted the conventional direct-current and alternating-current tachometers, has been extended in interpretation to include numerous methodologies, including analog and digital types.

Direct-Current Tachometers. These instruments operate on the principle of the voltage generated when a magnet is rotated with reference to a conductor (as shown in Fig. 3) or vice versa. Polarity of voltage and thus the direction of current flow will depend upon the polarity of the field and the direction of the conductor motion. A dc tachometer system is comprised of a simple dc generator and a dc indicator or recorder. Generator speeds range from 100 to 5000 revolutions per minute (rpm).

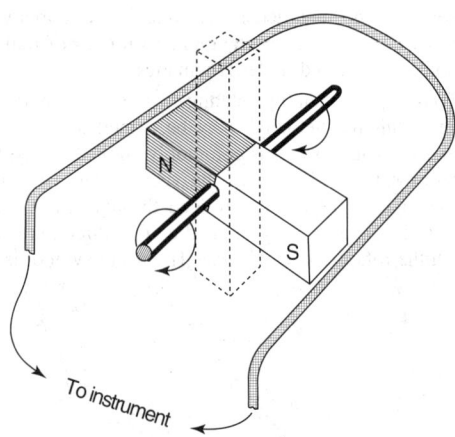

Fig. 3. Rotating magnet within fixed conductor generates voltage in direct-current-type tachometer.

DC tachometers are of two basic types: (1) brush type; and (2) brushless type. Advantages of either type include linear output through zero speed, acceptable accuracy, and refinement of design as the result of many years of experience. Generally, the characteristics of the tachometer can be matched to the equipment whose speed is being determined by employing suitable gearing for effecting speed reduction or multiplication. Brush-type tachometers use either (1) an iron core, or (2) a moving coil. In essence, the iron core type is a simple dc motor that is used as an electrical generator. The moving-coil brush-type dc tachometer uses a winding in the form of a shell or cup. Because of problems with brushes in certain kinds of environments, the brushless dc tachometer is frequently specified. In this design, the positions of the winding and magnet are reversed — with the magnet rotating, while the winding is stationary. Thus, there are

no electrical connections required to the rotating member. Additional electrical circuitry is required to sense the position of the rotor and provide appropriate switching. These features add significantly to the cost of the tachometer.

Alternating-Current Tachometers. Fundamentally, an ac tachometer is a three phase electric generator with a three-phase rectifier on the output. There are two principal formats. In the *voltage-responsive tachometer system*, an ac generator and a rectifier-type indicator are used. These instruments can be used where the generator speed for full scale is not less than 500 nor greater than 5000 rpm. With adequate attention to bearings, conventional ac generators may be used at speeds up to 10,000 rpm. The ac tachometer generator embodies a stator surrounding a rotating Alnico permanent magnet. The output of the generator for voltage-responsive systems is temperature compensated and is proportional to speed. In the *frequency-responsive tachometer system*, the system consists of a dc indicator or recorder, a frequency-responsive network which may be contained in a recorder or separate transformer box, and an ac tachometer generator of either the conventional or bearing-less form.

Bearingless tachometer generators are ac generators of the most basic form consisting only of a permanent magnet rotor and a stator. The devices have no bearing or brushes. They are designed to be impervious to oil, grease, and relatively high temperatures and, consequently, may be installed in inaccessible areas, such as gearboxes, which permit saving of space. They have very low torque burdens of less than one ounce-inch and are capable of speeds up to 100,000 rpm.

In general, when a bearingless generator is used, the frequency-responsive approach is used. Since the system is solely dependent upon the frequency output of the generator, voltage variations caused by reductions in the magnetic strength of the rotor due to handling, poor alignment of stator and rotor, or axial travel of the rotor with respect to the stator will not affect the overall accuracy.

Capacitor-Type Tachometer. In the instrument shown in (Fig. 4), the charging current of a capacitor is utilized. The pickup head usually contains a reversing switch, operated from a spindle, which reverses twice with each revolution. Thus, battery potential is applied to the capacitor in each direction, and, with each impulse, a current is passed through the milliammeter. The indicator responds to the average value of these impulses. Therefore, the indications are proportional to the rates of the pulses, which, in turn, are proportional to the rates of the spindle revolutions. No current is drawn from the battery when the spindle is not revolving. A high-accuracy instrument is available wherein the capacitor and reversing switch are connected to one leg of a bridge circuit. The pulse from the periodically charging capacitor upset the balance of the bridge and thus cause an indication by the milliammeter.

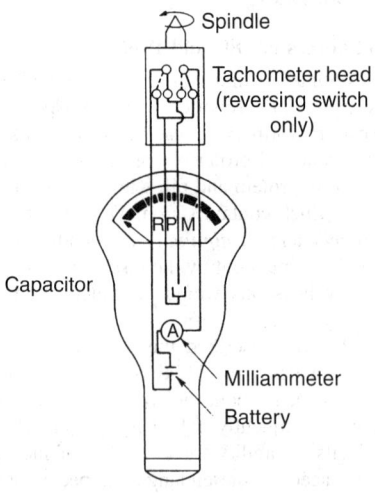

Fig. 4. Capacity-type impulse tachometer.

Eddy-Current Tachometer. For many years, this type of instrument has been used for automobile speedometers, in which case a flexible shaft arrangement is utilized, but it also finds extensive industrial use. In its basic form, shown in (Fig. 5), the drag-type instrument uses a permanent magnet which is revolved by the source being measured. Close to the revolving

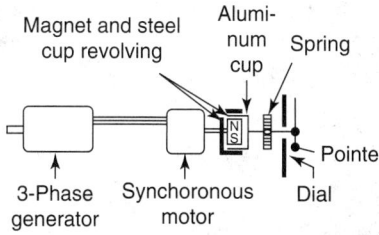

Fig. 5.   Drag-type eddy-current tachometer.

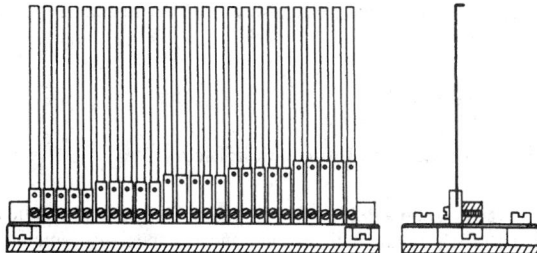

Fig. 6.   Vibrating-reed tachometer.

magnet is an aluminum disk, pivoted so as to turn against a spring. A pointer attached to the pivoted disk is associated with a calibrated scale. As the permanent magnet is revolved, eddy currents are set up in the disk. The magnetic fields caused by these eddy currents produce a torque which acts in a direction to resist this action and turns the disk against the spring. The disk turns in the direction of the rotating magnetic field and will turn (or be dragged) until the torque developed equals that of the spring. This torque is proportional to the speed of the rotating magnet. The instrument has a uniform scale. This instrument has been highly refined over the years and is available in a number of formats. Remote indication, for example, is obtainable with one form of this tachometer. A three-phase generator is driven from the shaft whose speed is to be measured. The generator output is connected to a three-phase synchronous motor, attached to the indicator, which rotates the magnetic field. Several indicators, each with its own synchronous motor, may be connected to the three-phase generator and will indicate in proportion to the speed of the generator.

Since the indications of this type of instrument are dependent on the frequency of the three-phase generator and not upon its voltage, a high degree of accuracy is attained. These instruments are considered accurate to 10 rpm in a full-scale range of 3000 rpm. The indicator pointer may be revolved through as much as 1080°, in which case a secondary pointer indicates the number of revolutions of the main pointer.

Photoelectric Tachometers. In one instrument of this type, designed to measure speeds up to 3 million rpm, the movable part subject to measurement is arranged to provide reflecting and absorbing areas. The interrupted reflected light produces, by means of a photocell, electric impulses which are applied to a frequency meter which generates a square wave from the pulse voltage and applies it to a discriminating circuit. A fixed current pulse at each half cycle is produced. These pulses are rectified and applied to a dc milliammeter which indicates the average value. Thus, the meter readings are proportional to the number of pulses per second, or the frequency.

In a recent *laser velocimeter* developed for measuring linear velocity, as of a jet engine exhaust, where no contact can be permitted, a laser, mirror, beam splitter, and lens system, and phototube sensor take advantage of the Doppler effect in producing interference fringes which are a measure of the velocity.

Variable-Reluctance Tachometer. In this type of electronic tachometer, pulses produced by the pickup are proportional to speed and are amplified, rectified, and control the direct current to a milliammeter. This type of instrument is rated at 10,000 to 50,000 rpm, with an accuracy of ±1/2% of full-scale reading. The pickup is rated to withstand ambient temperatures from −51 to +160 °C.

Vibrating-Reed Tachometer. Vibrating reeds provide a natural means of measuring the frequency of vibrating or revolving equipment. The reeds are of various lengths, in accordance with their natural period of vibration. They are mounted on a base with a reference scale so that observation of the reed that is vibrating forms a means of measuring the frequency of vibration. Purely electrical frequency is measured by using a solenoid valve associated with the reeds. The frequency of the output of a generator may be measured by simply touching the generator housing with the case of the indicator. Mechanical frequencies, oscillations, and rotations may be measured by bringing the member to be measured in mechanical contact with the instrument by touching it to the bearing support, case, or frame of the rotating device. Once the reeds are correctly adjusted and aged, they retain adjustment for long periods. Individual reeds are adjusted to ±0.3%. A change in temperature of ±25 °C may result in an error of about 0.5%. See Fig. 6.

Pneumatic Tachometer. Applicable to speed determinations from 0 to 6500 rpm with accuracies rated at ±0.5% of span, this instrument operates on the force-balance principle with magnetic actuation of a standard pneumatic circuit. See also **Pneumatic Controller**. The input shaft carries an eight-pole permanent magnet. A nonmagnetic alloy disk is held in position by flexure mounts between the keeper plate and the poles of the magnet. As the input shaft rotates the magnet, the magnetomotive pull, due to eddy currents generated in the disk, tends to turn it in the same direction. As reported by the manufacturer (Foxboro), this pull positions the force bar attached to the disk in relation to the nozzle, thereby causing an increase in the back pressure in the air flow through the nozzle. This back pressure is amplified by a relay which produces an output pressure proportional to speed.

Optical Shaft Encoders. Used mainly to sense the exact position of a machine member, optical shaft encoders (digital) can also be used for the measurement of speed.

Other Methods. The measurement of wind and air velocities is described in the entry on **Wind and Air Velocity Measurement**. Governors used for controlling the speed of large rotating equipment are described under **Governor**. Stroboscopes can be used as tachometers provided they have their own flashing-rate control circuits and calibration display. See also **Stroboscope**. Doppler radars are commonly used for measuring linear speeds in auto traffic control. See also **Radar**.

VELOCITY CURVE (Stellar).   A plot of radial velocity of a star as ordinates against time as abscissae is known as the velocity curve for the star. The method of formation of a mean velocity curve is similar to that described for the determination of a mean light curve. For the use of the velocity curve in determining the orbit of a spectroscopic binary, see also **Spectroscopic Binaries**.

VELOCITY LIMITING CONTROL.   In an automatic control system, control action in which the rate of change of a specified variable will not exceed a predetermined limit.

VELOCITY METER.   See **Accelerometer**.

VELOCITY MODULATION.   This is a form of electron modulation in which the electrons passing through a resonant cavity in a tube such as the klystron are acted upon by a modulating field in such a manner that their velocities cause them to pass through the collector cavity in groups.

VELOCITY PROFILE.   The graphical representation of the variation with displacement normal to the general direction of flow of the mean flow velocity in a shear flow. For example, the velocity profile of laminar flow through a circular tube is parabolic.

VELVET ANT (*Insecta, Hymenoptera*).   A wasp of the family *Multillidae*. The females of these insects are wingless and consequently are antlike in form, but they differ in the absence of the dorsal prominence on the slender waist that characterizes the ants. They are densely hairy insects, usually brightly banded with some shade of red or yellow, black, and sometimes white. The males differ in other details than the presence of wings, hence it is difficult to associate the sexes unless they are taken together. Velvet ants are parasitic in the nests of other insects and some have been reported as parasites on the tsetse fly.

VENA CONTRACTA.   A term frequently used in connection with orifice-type flowmeters. As the flow passes through an orifice plate, the flow achieves its narrowest cross section somewhat downstream of the plane of the plate. This location is called the vena contracta, meaning the narrowest jet cross section. See also **Flow Measurement (Liquids and Gases)**; and **Orifice**.

VENEER.   See **Wood**.

VENN DIAGRAM.   See **Information Theory**.

VENN, JOHN (1834–1923).   Venn was a British mathematician who developed Boole's mathematical logic and is especially remembered for Venn diagrams. He taught logic and probability theory at Cambridge University. He wrote *Logic of Chance, Symbolic Logic,* and *The Principles of Empirical Logic,* all of which influenced the development of the theory of statistics and mathematical logic.

Venn also had an interest in history and compiled a history of Cambridge University, which was published in 1922.

J. M. I.

VENOM (Snake).   See **Snakes**.

VENTIFACT.   A stone whose form has been modified by the sand-blasting effect of wind-carried sediments. See also **Dreikanter**.

VENTILATING FAN.   See **Fan**.

VENTRICLE (Brain).   A cavity of the vertebrate brain. The first and second ventricles are in the cerebral hemispheres. The third is formed of the persisting median cavity of the first primitive brain vesicle. The fourth ventricle is the cavity of the third primary vesicle, and all other remnants of the original cavities become narrowed passages. The ventricular system contains cerebrospinal fluid. See also **Central and Peripheral Nervous Systems**.

VENTURI FLOWMETER.   See **Flow Measurement**.

VENUS.   The second planet from the sun, traveling in an orbit between that of Mercury and the Earth. The mean distance of Venus from the sun is 108.21 million kilometers (67.24 million miles). Venus comes closer to the Earth than other neighboring planets. At inferior conjunction, when Venus lies between the Earth and the sun, it is about 40 million kilometers (25 million miles) distant. Because only a crescent of the planet is viewed from the Earth at inferior conjunction, the planet does not appear brighter than when the full disk is viewed at the time of superior conjunction, a distance of some 259 million kilometers (161 million miles), when the planet lies behind the sun. A full cycle of phases, as seen from the Earth, requires 584 days (the synodic period). Venus makes one complete revolution around the sun in 225 days. The axis of rotation of the planet is nearly perpendicular to the plane of its orbit. In past years, the period of rotation of the planet about its own axis has been variously estimated. One accepted value is 243.09 ± 0.18 earth days. Observations have indicated that Venus rotates in a retrograde sense with respect to the behavior of all other planets excepting Uranus. The diameter of the planet is estimated at 12,102.8 kilometers (7521 miles) at the equator. The orbit of Venus around the sun is nearly a perfect circle (eccentricity = 0.0068). Its orbital velocity is estimated at about 35 kilometers (21.8 miles) per second. The density is estimated at 5.431 grams per cubic centimeter. There are no known satellites.

Both Venus and Mercury revolve about the sun in orbits that lie inside the orbit of the Earth about the sun. Accordingly, when the planes of the planetary orbits coincide with the plane of the ecliptic, once during the synodic period of each planet, the object will pass between the Earth and the sun. Because of the small angular diameter of the planets relative to that of the sun, at the time they pass between the Earth and the sun, they appear as small black spots moving across the brilliant disk of the sun. Such a phenomenon is known as a transit of Venus or a transit of Mercury. The first recorded observation of a transit of Venus was made by Horrocks in England in 1639. See also **Transit (Astronomy)**.

An early view of Venus from an earth-launched spacecraft is shown in Fig. 1. A chronology of missions to Venus, as of mid-1993, is given toward the end of this article and includes American and Russian ventures to the planet.

Fig. 1.   Venus as viewed from a distance of 720,000 km (450,000 mi) when *Pioneer Venus* was approaching the planet in 1978. A whirlpool-like vortex in the polar clouds that provides a downward motion of the atmosphere was noted. The Venus spacecraft, scheduled to operate for 243 Earth-days (the time required by Venus to turn on its axis) performed so well that the life of the mission was extended. (*NASA Ames Research Center, Moffett Field, California.*)

Venus Unlike Earth

For many decades, scientists, including the very early astronomers, regarded Venus as a sister planet of Earth. There were many reasons for this comparison. Left unscrutinized, both planets appeared to have heavy vaporous atmospheres; the two planets have comparable masses and diameters, especially considering the large differences that exist between any two other planets in the solar system; and, in fact, its comparative nearness to Earth was a qualitative factor among some early scientists. By way of earthbound observatories, including spectroscopic observations, these similarities faded in recent years, but only since the late 1970s have they been discounted.

Prior to obtaining excellent images of Venus, there was an opinion among some experts that, although the two planets are quite different today, perhaps they did evolve via similar mechanisms. *Magellan* and *Galileo* findings refute this possibility. The active geologic period that Venus currently is experiencing has no parallel in the evolution of Earth.

This departure of prior entrenched opinions is giving Venusian exploration renewed vigor as the experts now attempt to explain why Earth and Venus did not develop as "twin" planets.

One supposition emanating out of the data thus far embraces the concept that a tremendous impact on Earth created the moon and removed much of its atmosphere, whereas Venus did not experience such an event, and thus the atmospheres of the two planets as we witness them today evolved differently. Another concept involves the lack of a hydrosphere (oceans) on Venus that prevents volatiles from recycling, inhibiting subduction and thus creating a more voluminous and thicker and heavier crust on Venus than on Earth. Observations suggest that the Venusian crust differs from Earth's because of the absence of volatiles and a significant energy source within the interior of the planet. This also matches magnetic measurements made thus far.

It has been proposed that the Venusian atmosphere, unlike Earth's, increases the planet's surface temperature (up to 730 K). Thus, Venus contrasts with Earth's oceans and its disequilibrium atmosphere.

These and other concepts are being reviewed, based upon the analysis of data from *Magellan* and *Galileo*. At a minimum, most researchers believe that it will require 5 additional years to develop cause-and-effect scenarios to match the data and to be acceptable to a majority of researchers who are pursuing the numerous aspects of Venusian evolution.

Surface Features of Venus. Sufficient information has not been analyzed thus far to explain the details of the various surface features, as shown in Figs. 2–10. One of the most publicized photos of Venus is shown in Fig. 11. This is a false-color perspective of the volcano *Sif Mons*. A radar altimetry map of a portion of the Venusian surface is shown in Fig. 12. Based upon data received from **Magellan**, scores of maps like this are in preparation.

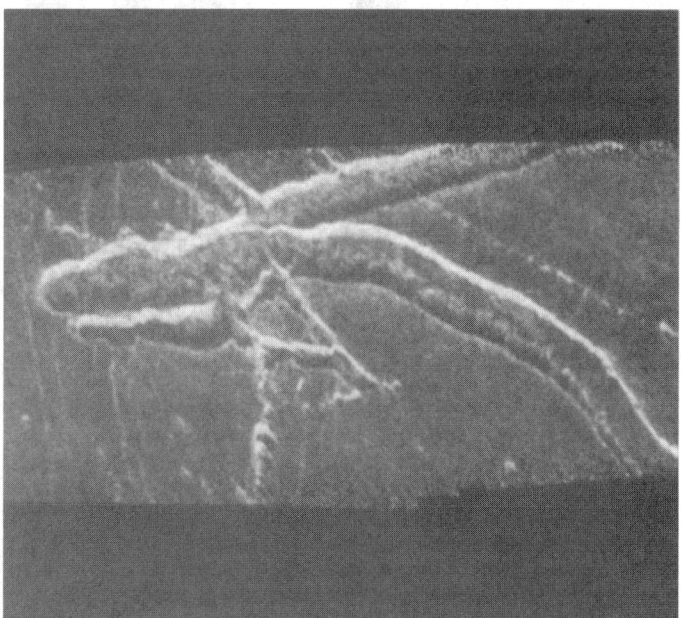

Fig. 2. Trough features on Venus, 28 km (17 mi) wide. Location: 60 S; 347 E. (*Jet Propulsion Laboratory.*)

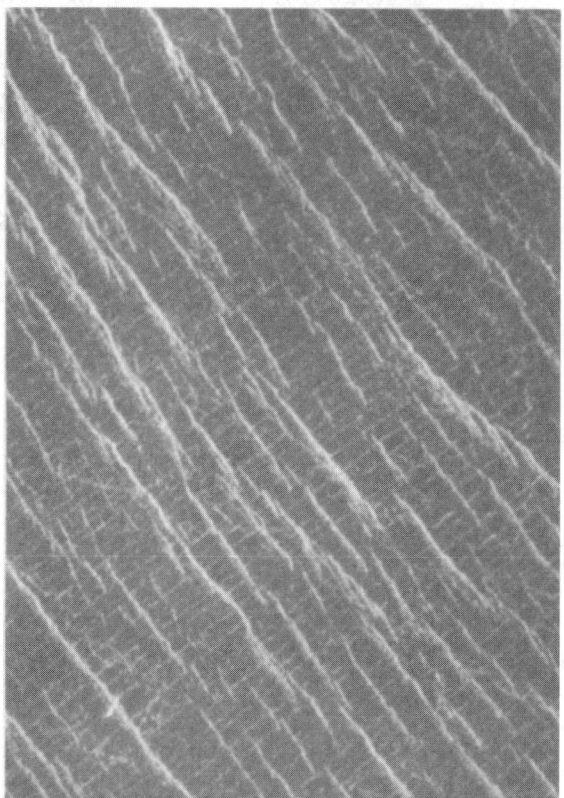

Fig. 3. Gridded plains, 37 km (23 mi) wide. Location: 30 N; 333 E. (*Jet Propulsion Laboratory.*)

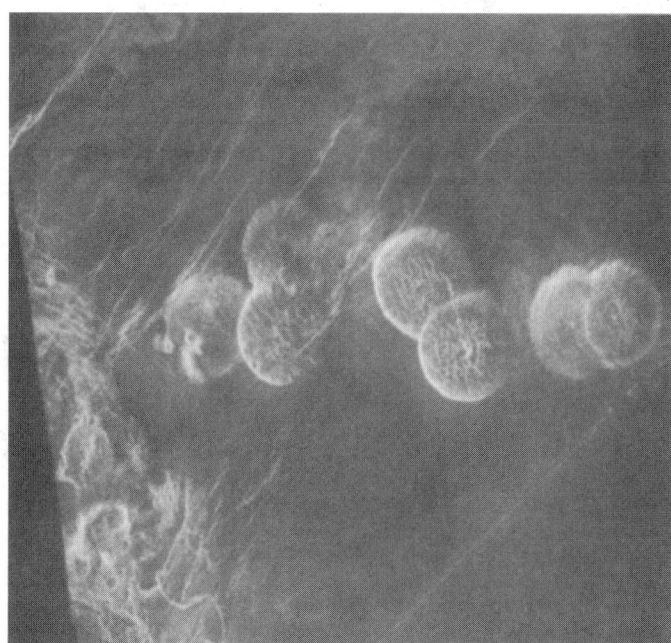

Fig. 4. "Pancake" volcanic dome on Venus, average 25 K (15.5 mi) in diameter. Location: 30 S; 11.8 E. (*Jet Propulsion Laboratory.*)

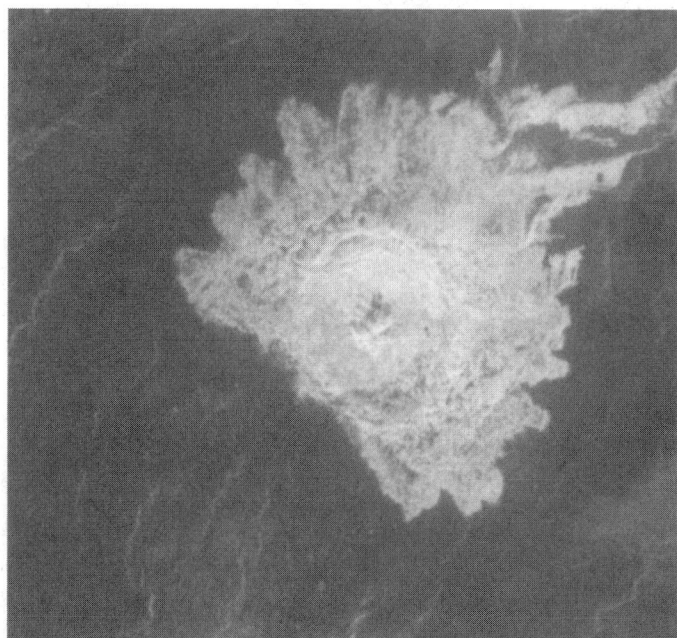

Fig. 5. Impact crater Aurelia, 32 km (20 mi) in diameter. Location: 20.3 N; 331.8 E. (*Jet Propulsion Laboratory.*)

Early Surface Theories. Because Venus is shrouded in clouds, observations of the planet's surface had to await the perfection of Earth-based radar techniques and, later, the probes which landed on the surface and orbiters equipped with radar and other instrumentation that would provide leads pertaining to the surface. The thick atmosphere also prevented traditional telescopes from determining the rate at which Venus rotates. Among the first radar observations of Venus were those made by the Lincoln Laboratory (Massachusetts Institute of Technology) in 1961. At that time, W.D. Smith and others sought evidence of frequency broadening in the returned radar echoes, assuming that if the planet were rotating, a differential Doppler shift would be noted in the signals reflected from different areas of the surface. Signals striking an approaching surface feature would be increased in frequency, whereas those signals striking a receding surface would be decreased in frequency. The net result is that the frequency spectrum of the echoes would be broadened. While information from the early experiments were fairly limited, Smith et al. did suggest (and was

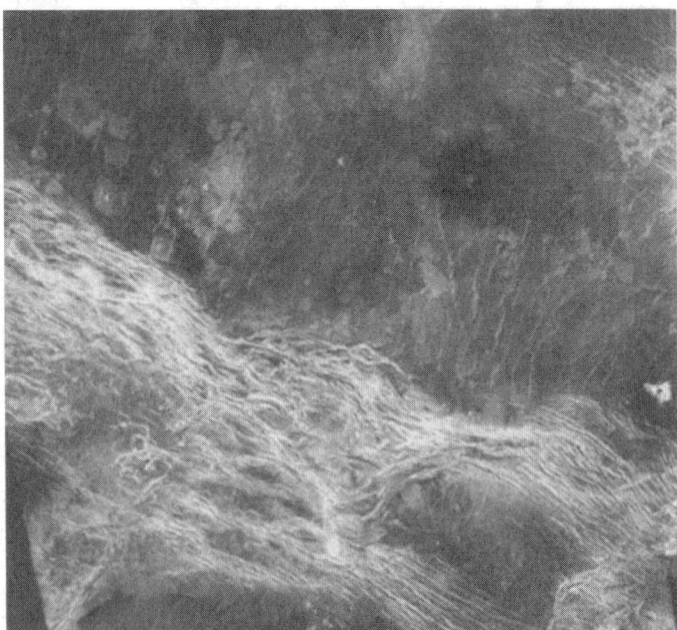

Fig. 6. Ridge belts in Lavinia Region, 615 km (382 mi) wide. (*Jet Propulsion Laboratory.*)

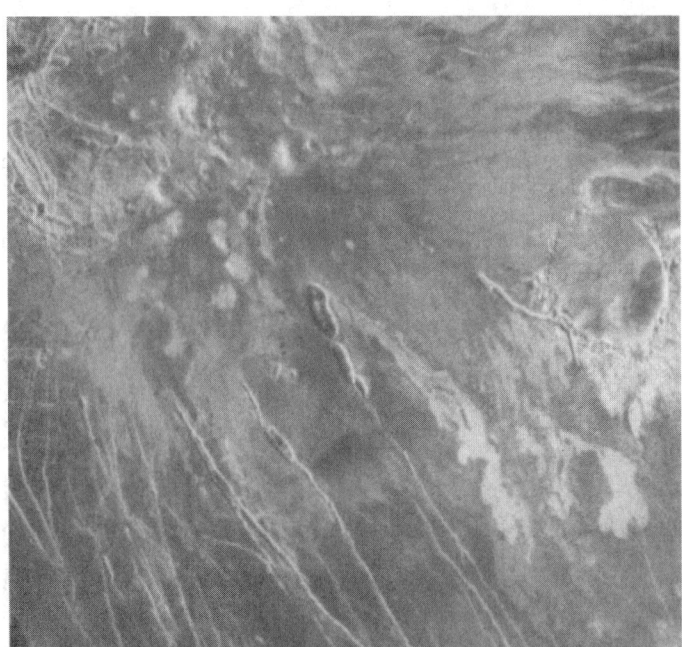

Fig. 8. Fractures and lava-flooded crater, 300 km (185 mi) wide. Location: 60 S; 352 E. (*Jet Propulsion Laboratory.*)

Fig. 7. "Turtle-Back" fractured dome in Freyja Montes, 70 km (43 mi) wide. Location: 72 N; 342 E. (*Jet Propulsion Laboratory.*)

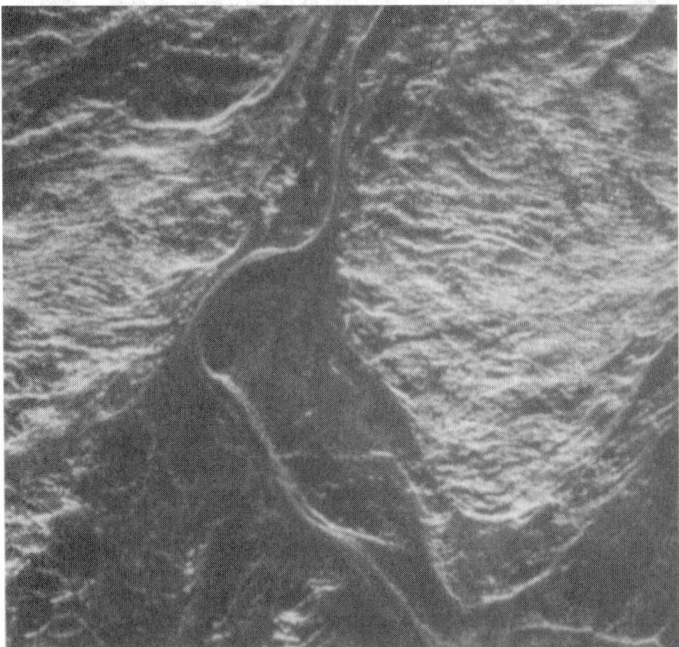

Fig. 9. Sinuous volcanic channel north of Freyja Montes, 300 km (185 mi) wide. Location: 60 S; 352 E. (*Jet Propulsion Laboratory.*)

later proved to be correct) that Venus' orbital motion may be relatively slow and retrograde. These findings were soon confirmed by researchers at the Jet Propulsion Laboratory (Pasadena, California) in 1962, at which time a period of rotation (retrograde) of 240 earth days was established for Venus.

In the middle and later 1960s, additional radar astronomy facilities studied Venus. These included the Arecibo installation in Puerto Rico, the deep-space tracking antenna at Goldstone, California, and the Haystack radar system in Massachusetts. With improvements in instrumental techniques, the accuracy of radar ranging was better than 1 kilometer (0.62 mile). As pointed out by Pettingill, Campbell, and Masursky (1980), it was eventually possible to observe variations in the radius of Venus as the reflecting region migrated around the equator of the planet. A correct value for radius is important in assessing other properties of the planet. For example, using measurement of refraction of the *Mariner 5* spacecraft's radio signal by the

Venusian atmosphere, scientists calculated atmospheric pressure and temperature at the surface as a function of distance from the planet's center of mass.

Radar findings of the Venusian surface obtained by *Pioneer* Venus indicate that the planet is largely a lowland plain, which extends uninterrupted for thousands of kilometers, covering 95% of the surface mapped to date. Venus may be the flattest body in the inner solar system. The crusts of all other inner planets, including the earth's moon, exhibit considerable relief and owe this principally to the presence of two very different kinds of rock. The relatively dense basalt (see also **Basalt**) forms vast, low plains, as found in the oceanic basins of the earth and the maria of the moon; and the less dense granite and anorthosite (see also **Anorthosite**; and **Granite**) form elevated continental regions. Scientists have found that less than 5% of the surface of Venus is elevated to the extent that it might be considered continental, whereas on earth over 30% of the surface is composed of continents (including the continental shelves). There are two

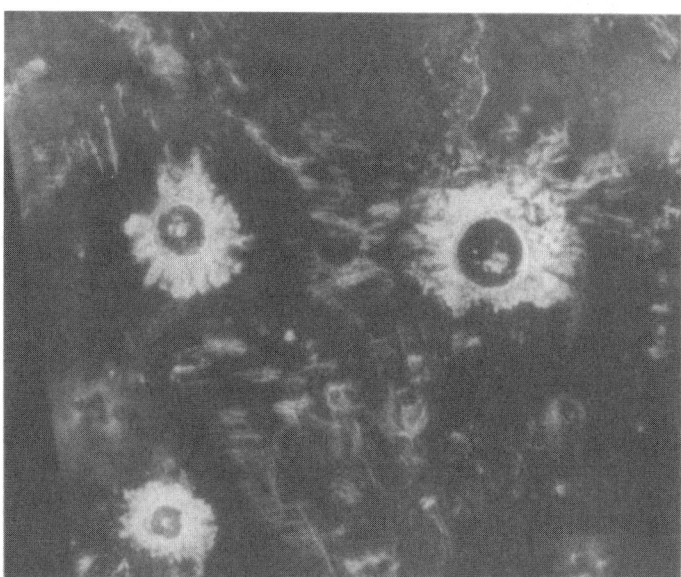

Fig. 10.   Mosaic of three impact craters and fractured plains in Lavinia Planitia, 500 km (800 mi) wide. Location: S7 D; 339 E. (*Jet Propulsion Laboratory.*)

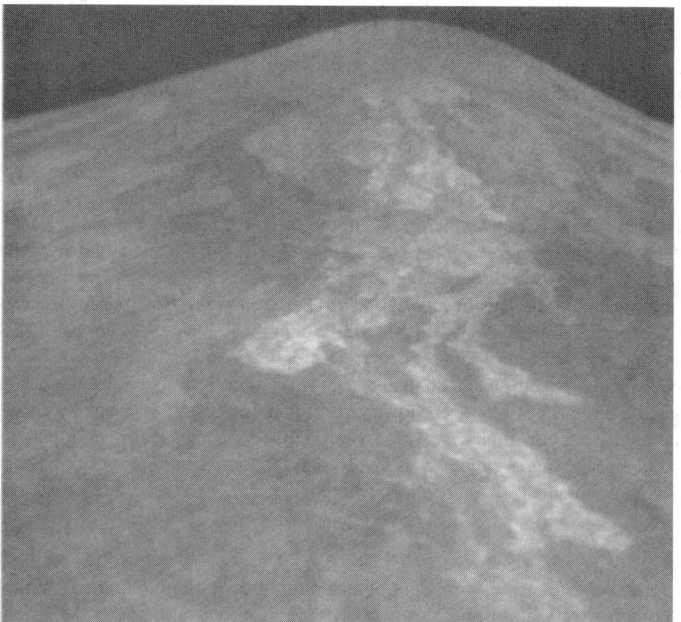

Fig. 11.   False-color perspective of the volcano of *Sif Mons* on Venus. (*Jet Propulsion Laboratory.*)

small regions (about the size of Ohio) on Venus that are depressed over 1.5 kilometers (1 mile) below the mean radius altitude. It cannot readily be assumed that the vast lowlands of Venus are basalt, another surprise to researchers, because the Soviet *Venera 8* lander showed that radioactivity in the lowlands compared well with that of terrestrial granite. Another puzzling find is that the lowlands appear to contain craters — not the highlands. The craters appear to be shallow and thus would indicate a strong planetary crust. One hypothesis is that the very high temperatures at the planet's surface have weakened and possibly melted the crustal rock, a process that would decrease relief.

As reported by the *Pioneer* Venus orbiter mission team in 1979, the topographic results obtained by the orbiter radar operating in the altimetry mode indicated that the radar bright feature called Maxwell is a broad plateau with a rough surface 1000 kilometers (621 miles) by 700 kilometers (435 miles) in extent and rising some 6 kilometers (3.7 miles) above the surrounding plain. A ridge some 6 kilometers (3.7 miles) higher runs parallel to the northwest-southeast plateau margins. Actually, a series of ridges in this direction gives the feature a grainy pattern. A dark, circular spot has been observed and may be a crater partly

filled with lava. The mountain ridges have proved thus far to be the highest features observed on the planet. See Fig. 12.

A smooth pear-shaped region west of Maxwell is a plateau 1000 kilometers (621 miles) across and elevated to 5 kilometers (3.1 miles) above the plains to its south and west. This feature appeared to be tectonic, whereas Maxwell may be either tectonic or of complex volcanic origin. The radar bright features known as Gauss and Hertz may be volcanic. The first data, obtained in the imaging mode at a longitude of about 340° in the Northern hemisphere, appeared to show impact craters about 250 kilometers (155 miles) in diameter with only about 600 meters (1969 feet). Scientists are somewhat puzzled by the ring-shaped features observed thus far on Venus. At this time, caution is followed in positively identifying some of these features as true impact craters. Some of the rims are not quite circular and the radial extent of small-scale roughness near them is greater than that of impact craters observed on other planetary bodies. Surface resolution is only about 10 kilometers (6.2 miles), so that detailed structure cannot be ascertained, and this, of course, is critical to the interpretation of the nature of a feature and an understanding of its origin and evolution.

Altimeter data gained thus far have revealed a significant difference between the topography of Venus and that of the Earth. Because Venus rotates so slowly, there is little if any bulge at the equator. Most radar mapping to date indicates that a high percentage of the surface of Venus lies within a range of one kilometer (3281 feet) of the mean surface altitude (equivalent to a radius of 6051.4 kilometers (3760.4 miles). Observers have indicated that Venus does have "continents" of a sort, that is, areas that rise over 2 kilometers (1.2 miles) above the mean surface altitude. At least two so-called continental areas have been found to date — the Lakshmi plateau to the west of aforementioned Maxwell and another (Aphrodite) just south of the equator between longitudes 70 and 140 degrees east.

Burns (Arecibo Observatory) suggests that the surface of Venus exhibits close to the number of craters with a diameter larger than about 80 kilometers (496 miles) that cratering models predict should have been made within the past 600 million to one billion years. Because of the planet's dense atmosphere, the number of smaller craters (up to about 20 kilometers; 12.5 miles across) would be considerably less than in most other planetary bodies because such meteorites would have been consumed in the Venusian atmosphere. Unfortunately, limitations on radar techniques (either instrumental or distance from the planet) have made it impossible to confirm this prediction to date. As of the late 1980s, investigators are aggressively seeking answers to the question — What are the relative roles of three major processes that may have determined the Venusian land mass: meteorite impact, volcanism, and tectonism? Scientists are eagerly looking forward to the **Magellan** mission previously described.

Comments pertaining to past or active volcanism, the possible presence of lightning in the atmosphere, among other important and intriguing phenomena on Venus essentially remained ungelled at this juncture. Several of the references listed here, however, contain numerous hypotheses of interest.

Atmosphere of Venus — Early Theories

(1) *Missing Oxygen.* Some theorists have suggested that ancient Venus had an ocean, something like that of Earth. This has been part of the "sister" or "twin" planet concept which has persisted for decades. For this early chapter in a scenario, two fundamental questions require answers: (a) Where is the missing hydrogen? (b) Where is the missing oxygen from the H_2O that would have made up that ocean? The present water in the desiccated Venusian atmosphere has been detected to be only 0.0014% of a terrestrial ocean. Conditions at the Venusian surface appear to rule out any underground water. The mass spectrometer of Pioneer Venus detected in a cloud droplet of sulfuric acid a hydrogen sample whose isotopic ratio showed that at one time Venus had at least as much water as a few tenths of a percent of the Earth's present oceanic volume. Hydrogen from the missing water could easily have escaped to space. Thus, a disproportionate amount of deuterium would have remained. It has been suggested that perhaps the oxygen combined with dark, chemically reduced basalt on the surface, resulting in reddish, oxidized rock having the color of volcanic cinder cones found in Hawaiian volcanoes. Researchers at Brown University and the Russian Academy of Sciences made an intense study of color pictures of the Venus surface as returned by the *Venera 13* and *14* landers. The orange sunlight component had to be filtered out and corrections had to be made for discoloration of Veneras' camera

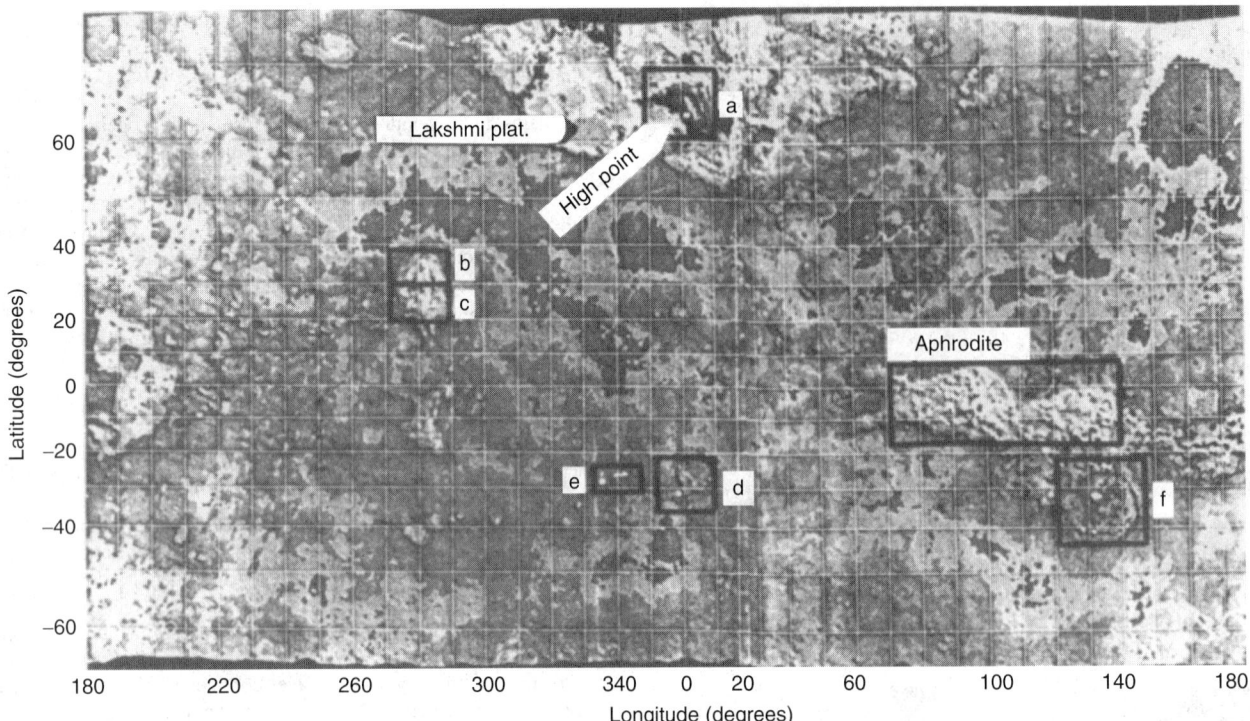

Fig. 12.   Radar altimetry map of Venus from −65 °S to +70 °N latitude and covering 360 degrees of longitude. Purely black areas along upper and lower edges and the small black spot at about 60 °N and 260 ° are unmapped areas. Principal features of interest to scientists are (a) the Maxwell region (named after James Clerk Maxwell) and first noted by earth-based, low-resolution radar observations. This region contains the highest point on the Venusian surface (above surface calculated by using mean radius of planet). Arrow points to location of Montes Maxwell which rises some 10,670 meters (35,000 feet) above the lowland plain. Comparable in height to Mount Everest, the Venusian peak differs in that it appears to rise alone from the surrounding plains and plateaus. Large mountains on the earth usually are associated with somewhat greater gravitational pull than lowlands. Data taken thus far in this region indicate no such gravity anomaly for Montes Maxwell. (b) and (c) comprise the Beta region. Rhea Mons is located in area (b); Theia Mons in area (c). The Beta region (Regio Beta) is a highland with the two aforementioned shield volcanoes (so-named because they resemble low, expansive shields of rock). They rise along a fault zone from a plateau about 1600 kilometers (1000 miles) long. Extensive flat sheets typically formed by basalt flows (an example on earth is the Snake River Plateau in Idaho) surprisingly have not been detected in the Beta region. This has led to speculation that a water-poor Venusian flow may have been more viscous than other basalts and thus maintained their relief even at greater distances from the vent. (d) The Alpha region reveals a striking northeast-to-southwest grain similar to faulted basin ranges found in Nevada. Within this region, a feature called Eve has been observed. This bright central spot lies within a prominent ring-shaped feature that resembles many impact craters seen on the moon and Mercury. (e) Three craterlike features have been observed in the region to the left of (d). (f) In this region, a circular feature some 1800 kilometers (1116 miles) across has been observed and believed to be the remnant of a gigantic impact crater. The Lakshmi plateau (Planum Lakshmi) is a pear-shaped plateau that is larger and higher than any plateau on the earth. Some scientists suggest that it was uplifted with relatively little horizontal motion. A similar process was involved in the formation of the major plateaus of the earth. Some scientists suggest that it was uplifted with relatively little horizontal motion. A similar process was involved in the formation of the major plateaus of the earth. Another major feature and suggested "continental" area of Venus is known as Aphrodite. (*Based upon a computer colored map prepared by Eric Eliason, U.S. Geological Survey.*)

coloration chips which had been seared by the 500 °C (932 °F) Venusian surface temperature. With all corrections made, the rocks were shown to be dark and nearly colorless. The researchers heated red, oxidized basalt to the same temperature and the sample appeared essentially the same as the reprocessed pictures from Venus. It was found, however, that oxidized basalt reflects more light than hot unoxidized basalt outside the visible range (wavelength of 1 micrometer). This led to the conclusion that the imaged Venusian rocks are indeed dark and *oxidized*. This yielded a clue to the effect that the missing oxygen is now part of the rock structure. Hunter of the University of Arizona estimated that the minimum amount of water required by the isotopic analysis could have formed an "ocean" 10 meters deep. Losing it would have required that oxygen combine with an amount of rock equivalent to a layer 40 meters thick that was then buried by volcanism. By comparison, a full terrestrial ocean would have required rock some 10 km thick in order to combine with the oxygen of a former ocean.

(2) *Missing Hydrogen.* McElroy, Prather, and Rodriguez (Harvard University) reported on the escape of hydrogen from Venus. The planet contains quantities of carbon and nitrogen similar to Earth, but hydrogen is deficient. The scenario: Recombination of O_2^+ represents a source of fast oxygen atoms in Venus' exosphere, and subsequent collisions of oxygen atoms with hydrogen atoms led to the escape of about 10^7 hydrogen atoms per square centimeter per second. The escape of deuterium atoms is considered negligible, with an expected increase of the ratio of deuterium to hydrogen over time. It is suggested that the mass-2 ion observed by

Pioneer Venus is D^+, which implies a ratio of deuterium to hydrogen in the contemporary atmosphere of about 10^{-2}, an initial ratio of 5×10^{-5}, and an original H_2O abundance not less than 800 grams per square centimeter. The researchers derived a value for the primordial D/H ratio on Venus similar to ratios observed for Jupiter and Earth. The analysis implies an initial abundance for H_2O on Venus of 8×10^2 g cm^{-2}, or approximately 300 times less than the terrestrial value.

(3) *Atmospheric Composition.* Based upon data returned from *Venera 13* and *14*, it is estimated that carbon dioxide (CO_2) constitutes about 96% of the atmosphere and is 10 times more abundant than nitrogen (N_2) and about 10,000 times more abundant than sulfur dioxide (SO_2), which is present at a concentration of about 150 ppm. Venera data found higher levels of hydrogen sulfide (H_2S) and of carbonyl sulfide (OCS). The atmosphere also contains sulfuric acid cloud particles and traces of hydrochloric acid (10^{-1} ppm) and hydrofluoric acid (10^{-3} ppm). The presence of sulfur and acidic compounds poses the basic question — whence do they come?

(4) *Sulfur Compounds.* Prinn (Massachusetts Institute of Technology) reports (see 1985 reference) that sulfur is a minor atmospheric constituent on Venus. The gaseous progenitors of the clouds make up 0.02% and the cloud particles themselves constitute only 0.00002% of the atmosphere. Of importance, however, is that these sulfur-bearing clouds markedly affect the planet's climate. For example, they reflect nearly 80% of the total incident sunlight (notably at red and yellow wavelengths). Thus, even though Venus is closer to the sun than Earth, it receives much less solar energy. Two-thirds of that energy is retained in the Venusian clouds because

the clouds are absorptive in the ultraviolet and near-infrared regions. Thus, about one-third of the incident solar energy reaches the Venusian surface. This is essentially opposite of the situation on Earth.

Prinn reports that the dominant constituent of the clouds is concentrated sulfuric acid, estimated at 75% by mass. The acid is in the form of liquid droplets. Polarization studies exclude other candidate substances and, in particular, they exclude water. Sulfur dioxide and water vapor present in the lower cloud region are believed to chemically react, forming sulfuric acid. The sulfuric acid component of Earth's atmosphere is easily explained, resulting from two main sources — fossil fuel burning and to a much lesser extent, volcanoes. The surface conditions of Venus (sometimes described as a sort of Hell) seem to favor volcanism as the source. There is, however, another view. Lewis (University of Arizona) has suggested that these gases are "cooked" out of surface rocks by intense heat. It is further suggested that the gases are added to the atmosphere at the same rate at which they are removed by reactions with the surface — thus no net flow of a compound into or out of the atmosphere. In essence, the sulfur gases, hydrogen chloride, and hydrogen fluoride in the atmosphere are in chemical equilibrium with minerals in the crust.

(5) *Outgassing Comparison — Venus and Earth*. Prinn observes that the Venusian atmosphere is 90 times as massive as that of Earth. The total amount of carbon dioxide and nitrogen on Venus is estimated to be about 30% less than that on Earth. The concentration of argon-40 is roughly one-third of the terrestrial value, while the abundance in the crust of the argon source (radioactive potassium 40) is about the same on both planets. This may indicate that comparable amounts of outgassing and thus comparable levels of volcanism may have occurred on the planets in the past. It should be emphasized, however, that the aforementioned gases are relatively stable and can survive in an atmosphere over geologic periods of time. Thus, they do not necessarily indicate outgassing on Venus in recent times. Evidence for recent volcanism on Venus lies in the clouds and in the sulfur gases that produce them.

Chronology of Missions to Venus

Interest in exploring Venus was manifested very early in the space programs of the United States, Russia, and other countries that have funded space ventures. Major missions to Venus have included:

Sputnik 7 (4, February 1961). This was the first Soviet attempt at a Venus probe. The probe was successfully launched into Earth orbit with a SL-6/A-2-e launcher. The launch payload consisted of an Earth orbiting launch platform (Tyazheliy Sputnik 4) and the Venera probe. The fourth stage (a Zond rocket) was supposed to launch the Venera probe towards a landing on Venus after one Earth orbit but ignition failed, probably due to a faulty timer, and the spacecraft remained in Earth orbit. Because of its large size 14,300 pounds (6,483 kilograms), the mission was originally thought by non-Soviet observers to be a failed manned mission, and later was described as a test of an Earth orbiting platform from which an interplanetary probe could be launched.

Venera 1 (12, February 1961). Venera 1 was the first spacecraft to fly by Venus. The probe consisted of a cylindrical body topped by a dome, totaling 2 meters in height. Two solar panels extended radially from the cylinder. A large (over 6.5 feet (2 meter diameter) high-gain net antenna was used to receive signals from the ground. This antenna was attached to the cylinder. A long antenna arm was used to transmit signals to Earth. The probe was equipped with scientific instruments including a magnetometer attached to the end of a 2 meter (6.5 feet) boom, ion traps, micrometeorite detectors, and cosmic radiation counters. The dome contained a pressurized sphere which carried a Soviet pennant and was designed to float on the putative Venus oceans after the intended Venus impact. Venera 1 had no on-board propulsion systems. Temperature control was achieved with thermal shutters.

Venera 1 was launched along with an Earth orbiting launch platform (Tyazheliy Sputnik 5 (61-003C)) with a SL-6/A-2-e launcher. From a 229×282 km orbit, the Venera 1 automatic interplanetary station was launched from the platform towards Venus with the fourth stage Zond rocket. On 19 February, 7 days after launch at a distance of about two million km from Earth, contact with the spacecraft was lost. On May 19 and 20, 1961, *Venera 1* passed within 100,000 km of Venus and entered a heliocentric orbit.

Mariner 1 (22, July 1962). This was to be the first Mariner mission. It was intended to perform a Venus flyby. The vehicle was destroyed by the Range Safety Officer 293 seconds after launch at 09:26:16 UT

when it veered off course. The booster had performed satisfactorily until an unscheduled yaw-lift (northeast) maneuver was detected by the range safety officer. Faulty application of the guidance commands made steering impossible and were directing the spacecraft towards a crash, possibly in the North Atlantic shipping lanes or in an inhabited area. The destruct command was sent 6 seconds before separation, after which the launch vehicle could not have been destroyed. The radio transponder continued to transmit signals for 64 seconds after the destruct command had been sent.

Sputnik 19 (25, August 1962). Sputnik 19 was a Venera-type spacecraft intended to make a landing on Venus. The SL-6/A-2-e launcher put the spacecraft into Earth orbit, but the escape stage failed and the probe remained in geocentric orbit for three days until the orbit decayed on 28 August and it re-entered Earth's atmosphere.

This spacecraft was originally designated Sputnik 23 in the U.S. Naval Space Command Satellite Situation Summary.

Mariner 2. (27, August 1962). The Mariner 2 spacecraft was the second of a series of spacecraft used for planetary exploration in the flyby, or nonlanding, mode and the first spacecraft to successfully encounter another planet. *Mariner 2* was a backup for the Mariner 1 mission, which failed shortly after launch to Venus. The objective of the Mariner 2 mission was to fly by Venus and return data on the planet's atmosphere, magnetic field, charged particle environment, and mass. It also made measurements of the interplanetary medium during its cruise to Venus and after the flyby.

After launch and termination of the Agena first burn, the Agena-Mariner was in a 118 km altitude Earth parking orbit. The Agena second burn some 980 seconds later followed by Agena-Mariner separation injected the *Mariner 2* spacecraft into a geocentric escape hyperbola at 26 minutes 3 seconds after lift-off. Solar panel extension was completed about 44 minutes after launch. On 29 August 1962 cruise science experiments were turned on. The midcourse maneuver was initiated at 22:49:00 UT on 4 September and completed at 2:45:25 UT 5 September. On 8 September at 17:50 UT the spacecraft suddenly lost its attitude control, which was restored by the gyroscopes 3 minutes later. The cause was unknown but may have been a collision with a small object. On October 31 the output from one solar panel deteriorated abruptly, and the science cruise instruments were turned off. A week later the panel resumed normal function and instruments were turned back on. The panel permanently failed on 15 November, but *Mariner 2* was close enough to the Sun that one panel could supply adequate power. On December 14 the radiometers were turned on. *Mariner 2* approached Venus from 30 degrees above the dark side of the planet, and passed below the planet at its closest distance of 34,773 km at 19:59:28 UT 14 December 1962. After encounter, cruise mode resumed. Spacecraft perihelion occurred on 27 December at a distance of 105,464,560 km. The last transmission from *Mariner 2* was received on 3 January 1963 at 07:00 UT. Mariner 2 remains in heliocentric orbit.

Scientific discoveries made by *Mariner 2*; included a slow retrograde rotation rate for Venus; hot surface temperatures and high surface pressures; a predominantly carbon dioxide atmosphere; continuous cloud cover with a top altitude of about 60 km; and no detectable magnetic field. It was also shown that in interplanetary space the solar wind streams continuously and the cosmic dust density is much lower than the near-Earth region. Improved estimates of Venus' mass and the value of the astronomical unit were made.

Total research, development, launch, and support costs for the Mariner series of spacecraft (Mariners 1 through 10) was approximately $554 million.

Sputnik 20 (1, September 1962). Sputnik 20 (1962 Alpha Tau 1) was intended to be a Venus landing mission. The Venera-type spacecraft was successfully inserted into geocentric orbit by the SL-6/A-2-e launcher. The escape stage failed and the spacecraft was stranded in Earth orbit until it re-entered the Earth's atmosphere 5 days later.

This spacecraft was originally designated Sputnik 24 in the U.S. Naval SpaceCommand Satellite Situation Summary.

Sputnik 21 (12, September 1962). Sputnik 21 was an attempted Venus flyby mission. The SL-6/A-2-e launcher put the craft into Earth orbit, but the third stage exploded, destroying the spacecraft.

This spacecraft was originally designated Sputnik 25 in the U.S. Naval Space Command Satellite Situation Summary.

Cosmos 21 (11, November 1963). This mission has been tentatively identified as a technology test of the Venera series space probes. It may have been an attempted Venus flyby, presumably similar to the later

Cosmos 27 mission, or it may have been intended from the beginning to remain in geocentric orbit. In any case, the spacecraft never left Earth orbit after insertion by the SL-6/A-2-e launcher. The orbit decayed on 14 November, three days after launch.

Beginning in 1963, the name Cosmos was given to Soviet spacecraft which remained in Earth orbit, regardless of whether that was their intended final destination. The designation of this mission as an intended planetary probe is based on evidence from Soviet and non-Soviet sources and historical documents. Typically Soviet planetary missions were initially put into an Earth parking orbit as a launch platform with a rocket engine and attached probe. The probes were then launched toward their targets with an engine burn with a duration of roughly 4 minutes. If the engine misfired or the burn was not completed, the probes would be left in Earth orbit and given a Cosmos designation.

Venera 1964A (19, February 1964). Attempted Venus Flyby (Launch Failure).

Venera 1964B (1, March 1964). Attempted Venus Flyby (Launch Failure).

Cosmos 27 (27, March 1964). This mission was intended as a Venus flyby. The SL-6/A-2-e launcher successfully achieved Earth orbit, but the spacecraft failed to escape orbit for its flight to Venus, and was designated Cosmos 27.

Zond 1 (2, April 1964). Zond 1 was launched from an earth orbiting platform Tyazheliy Sputnik (64-016A) towards Venus. It flew by Venus on July 14, 1964, at a distance of 100,000 km and entered a heliocentric orbit. The announced mission objectives were space research and testing of onboard systems and units. Communications from the spacecraft failed soon after May 14, 1964.

Venera 2 (12, November 1965). Venera 2 was launched from a Tyazheliy Sputnik (65-091B) towards the planet Venus and carried a TV system and scientific instruments. On February 27, 1966, the spacecraft passed Venus at a distance of 24,000 km and entered a heliocentric orbit. The spacecraft system had ceased to operate before the planet was reached and returned no data.

Venera 3 (16, November 1965). Venera 3 was launched from a Tyazheliy Sputnik (65-092B) towards the planet Venus. The mission of this spacecraft was to land on the Venusian surface. The entry body contained a radio communication system, scientific instruments, electrical power sources, and medallions bearing the coat of arms of the U.S.S.R. The station impacted Venus on March 1, 1966. However, the communications systems had failed before planetary data could be returned.

Cosmos 96 (23, November 1965). This mission was intended as a Venus lander, presumably similar in design to the *Venera 3*, which had launched a week earlier. The spacecraft attained Earth orbit and the main rocket body (65-094B) separated from the orbiting launch platform. It is believed an explosion (perhaps during ignition for insertion of the spacecraft into a Venus transfer orbit) damaged the platform, resulting in at least six additional fragments (designated 65-094 C–H). The damaged spacecraft remained in orbit for 16 days and reentered the Earth's atmosphere on 9 December 1965.

Venera 1965A (23, November 1965). *Venera 1965A* was an attempted Venus flyby mission, possibly similar to the Venera 2 flyby mission launched two weeks earlier. It is believed the SL-6/A-2-e launcher failed.

Venera 4 (12, June 1967). *Venera 4* was launched from a Tyazheliy Sputnik (67-058B) towards the planet Venus with the announced mission of direct atmospheric studies. Part of a series of exploratory programs of Venus mounted by Russia, this spacecraft arrived at Venus within just a few hours of *Mariner 5's* rendezvous with the planet. *Venera 4* carried an instrument package designed to land on the planet's surface after a descent by parachute. Within about 75 minutes, after registering a temperature well above 260 °C (500 °F), data transmission ceased. Scientists later estimated that when transmission ceased the package was at an altitude some 30 km (20 mi) above the Venusian surface.

Mariner 5. On 14, June 1967 the Mariner 5 spacecraft was launched. *Mariner 5* was the fifth in a series of spacecraft used for planetary exploration in the flyby mode. Mariner 5 was a refurbished backup spacecraft for the Mariner 4 mission and was converted from a Mars mission to a Venus mission. The spacecraft was fully attitude stabilized, using the sun and Canopus as references. A central computer and sequencer subsystem supplied timing sequences and computing services for other spacecraft subsystems. The spacecraft passed 4,000 km from Venus on October 19, 1967. The spacecraft instruments measured both interplanetary

and Venusian magnetic fields, charged particles, and plasmas, as well as the radio refractivity and UV emissions of the Venusian atmosphere.

Marner 5 traveled around the dark side of Venus at a distance of about 9655 km (6000 mi). Again, data confirmed that no significant magnetic field existed. Radio signals, passed through the Venusian atmosphere (once from the day side; once from the night side of the planet) evidenced a very heavy concentration of carbon dioxide (CO_2) in the atmosphere. The atmosphere was estimated to exert a pressure at the planet's surface of about 100 times the normal sea level pressure (1 atm) on Earth. The mission was termed a success.

Cosmos 167 (17, June 1967). This mission was intended to be a Venus lander, similar in design to the *Venera 4* spacecraft launched 5 days earlier on 12 June. The spacecraft became stranded in Earth orbit and was designated *Cosmos 167*. Its orbit decayed and it re-entered Earth's atmosphere 8 days after launch.

Venera 5 (5, January 1969). Venera 5 was launched from a Tyazheliy Sputnik (69-001C) towards Venus to obtain atmospheric data. The spacecraft was very similar to Venera 4 although it was of a stronger design. When the atmosphere of Venus was approached, a capsule weighing 405 kg and containing scientific instruments was jettisoned from the main spacecraft. During satellite descent towards the surface of Venus, a parachute opened to slow the rate of descent. For 53 min on May 16, 1969, while the capsule was suspended from the parachute, data from the Venusian atmosphere were returned. The spacecraft also carried a medallion bearing the coat of arms of the U.S.S.R. and a bas-relief of V.I. Lenin to the night side of Venus.

Venera 6 (10, January 1969). Venera 6 was launched from a Tyazheliy Sputnik (69-002C) towards Venus to obtain atmospheric data. The spacecraft was very similar to Venera 4 although it was of a stronger design. When the atmosphere of Venus was approached, a capsule weighing 405 kg was jettisoned from the main spacecraft. This capsule contained scientific instruments. During descent towards the surface of Venus, a parachute opened to slow the rate of descent. For 51 min on May 17, 1969, while the capsule was suspended from the parachute, data from the Venusian atmosphere were returned.

Venera 7 (17, August 1970). Venera 7 was launched from a Tyazheliy Sputnik in an earth parking orbit towards Venus to study the Venusian atmosphere and other phenomena of the planet. Venera 7 entered the atmosphere of Venus on December 15, 1970, and a landing capsule was jettisoned. After aerodynamic braking, a parachute system was deployed. The capsule antenna was extended, and signals were returned for 35 min. Another 23 min of very weak signals were received after the spacecraft landed on Venus. The capsule was the first man-made object to return data after landing on another planet.

Cosmos 359 (22, August 1970). This mission was an attempted Venus flight, perhaps a lander similar to the Venera 7 mission launched 5 days earlier on 17 August. The SL-6/A-2-e launcher successfully brought the spacecraft to Earth orbit and the spacecraft payload was separated from the Tyazheliy Sputnik, but the escape stage failed during firing, putting the payload into a slightly more elliptical geocentric orbit. The mission was designated Cosmos 359.

Venera 8 (27, March 1972). Venera 8 was a Venus atmospheric probe. Its instrumentation included temperature, pressure, and light sensors as well as radio transmitters. The spacecraft took 117 days to reach Venus, entering the atmosphere on 22 July 1972. Descent speed was reduced from 41,696 km/hr to about 900 km/hr by aerobraking. The 2.5 meter diameter parachute opened at an altitude of 60 km, and a refrigeration system was used to cool the interior components. Venera 8 transmitted data during the descent and continued to send back data for 50 minutes after landing. The probe confirmed the earlier data on the high Venus surface temperature and pressure returned by Venera 7, and also measured the light level as being suitable for surface photography, finding it to be similar to the amount of light on Earth on an overcast day.

Cosmos 482 (31, March 1972). This mission has been identified as an attempted Venus probe which failed to escape low Earth orbit. It was launched by an SL-6/A-2-e launcher 4 days after the *Venera 8* atmospheric probe and may have been similar in design and mission plan. After achieving an Earth parking orbit, the spacecraft made an apparent attempt to launch into a Venus transfer trajectory. It separated into four pieces, two of which remained in low Earth orbit and decayed within 48 hours, and two pieces (presumably the payload and detached engine unit) went into a higher 210 × 9800 km orbit. It is thought that a malfunction resulted in an engine burn which did not achieve sufficient velocity for the Venus transfer and left the payload in this elliptical Earth orbit.

Mariner 10 (4, November 1973). Mariner 10 was the seventh successful launch in the Mariner series and the first spacecraft to use the gravitational pull of one planet (Venus) to reach another (Mercury). The spacecraft flew by Mercury three times in a heliocentric orbit and returned images and data on the planet. Mariner 10 is the only spacecraft to have visited Mercury.

The spacecraft structure was an eight-sided framework with eight electronics compartments. It measured 1.39 m diagonally and 0.457 m in depth. Two solar panels, each 2.7 m long and 0.97 m wide, were attached at the top, supporting 5.1 sq m of solar cell area. The rocket engine was liquid-fueled, with two sets of reaction jets used to stabilize the spacecraft on three axes. It carried a low-gain omnidirectional antenna, composed of a honeycomb-disk parabolic reflector, 1.37 m in diameter, with focal length 55 cm. Feeds enabled the spacecraft to transmit at S-and X-band frequencies. The spacecraft carried a Canopus star tracker, located on the upper ring structure of the octagonal satellite, and acquisition sun sensors on the tips of the solar panels. The interior of the spacecraft was insulated with multilayer thermal blankets at top and bottom. A sunshade was deployed after launch to protect the spacecraft on the solar-oriented side.

Instruments on-board the spacecraft measured the atmospheric, surface, and physical characteristics of Mercury and Venus. Experiments included television photography, magnetic field, plasma, infrared radiometry, ultraviolet spectroscopy, and radio science detectors. An experimental X-band, high-frequency transmitter was flown for the first time on this spacecraft.

Mariner 10 was placed in a parking orbit after launch for approximately 25 minutes, then placed in orbit around the Sun en route to Venus. The orbit direction was opposite to the motion of the Earth around the Sun. Midcourse corrections were made. The spacecraft passed Venus on February 5, 1974, at a distance of 4200 km. It crossed the orbit of Mercury on March 29, 1974, at 2046 UT, at a distance of about 704 km from the surface. The TV and UV experiments were turned on the comet Kohoutek while the spacecraft was on the way to Venus. A second encounter with Mercury, when more photographs were taken, occurred on September 21, 1974, at an altitude of about 47,000 km. A third and last Mercury encounter at an altitude of 327 km, with additional photography of about 300 photographs and magnetic field measurements occurred on March 16, 1975. Engineering tests were continued until March 24, 1975, when the supply of attitude-control gas was depleted and the mission was terminated.

Venera 9 (8, June 1975). On October 20, 1975, this spacecraft was separated from the Orbiter, and landing was made with the sun near zenith at 0513 UT on October 22. A system of circulating fluid was used to distribute the heat load. This system, plus precooling prior to entry, permitted operation of the spacecraft for 53 min after landing. During descent, heat dissipation and deceleration were accomplished sequentially by protective hemispheric shells, three parachutes, a disk-shaped drag brake, and a compressible, metal, doughnut-shaped, landing cushion. The landing was about 2,200 km from the *Venera 10* landing site. Preliminary results indicated: (A) clouds 30–40 km thick with bases at 30–35 km altitude, (B) atmospheric constituents including HCl, HF, Br, and I, (C) surface pressure about 90 (earth) atmospheres, (D) surface temperature 485 deg C, (E) light levels comparable to those at earth midlatitudes on a cloudy summer day, and (F) successful TV photography showing shadows, no apparent dust in the air, and a variety of 30–40 cm rocks which were not eroded.

Venera 10 (14 June 1975). On October 23, 1975, this spacecraft was separated from the Orbiter, and landing was made with the sun near zenith, at 0517 UT, on October 25. A system of circulating fluid was used to distribute the heat load. This system, plus precooling prior to entry, permitted operation of the spacecraft for 65 min after landing. During descent, heat dissipation and deceleration were accomplished sequentially by protective hemispheric shells, three parachutes, a disk-shaped drag brake, and a compressible, metal, doughnut-shaped, landing cushion. The landing was about 2,200 km distant from *Venera 9*. Preliminary results provided: (A) profile of altitude (km)/pressure (earth atmospheres)/temperature (deg C) of 42/3.3/158, 15/37/363, and 0/92/465, (B) successful TV photography showing large pancake rocks with lava or other weathered rocks in between, and (C) surface wind speed of 3.5 m/s.

Pioneer Venus. This was a major mission and arrived at Venus in December 1978. This was the first United States mission directed specifically at investigating the atmosphere of Venus on a planetary scale and radar mapping the planet's surface features. This was a two-spacecraft

mission, consisting of an orbiter and a multiprobe spacecraft. The orbiter was launched (20, May 1978) and scheduled to arrive at Venus 5 days prior to the probe-carrying spacecraft. To effect this timing, the orbiter was launched on a trajectory that took it more than 180 degrees around the solar system in 8 months. The probes, launched at a later date (8, August 1978) and followed a direct-path trajectory from Earth to Venus, requiring just the right time span for the planned rendezvous. Data from *Pioneer Venus* was studied intensely and, in fact, continues today as a major part of the database used in studying, hypothecating, and building models of the physical and chemical features of the planet.

The probes confirmed the expected high surface temperature and the presence of high winds (320 km; 200 mph), these possibly accounting for the transfer of heat into the night side in spite of the planet's low rotational speed. Atmosphere and cloud chemistry were examined in detail. The probes detected four layers of clouds and more light on the surface than might be expected solely from sunlight. (This light enabled a later *Venera* mission to obtain images of rocks on the surface.) Investigators, using *Pioneer Venus* data, tentatively concluded that sulfur plays a major role and that sulfur could be responsible for the light glow at the surface. The *Pioneer* orbiter confirmed the Venusian cloud patterns and circulation previously noted by *Mariner 10*. A large variability in the ionosphere was noted.

Details of the mechanics of the *Pioneer Venus* mission are given in Figs. 13, 14, 15, 16, and 17.

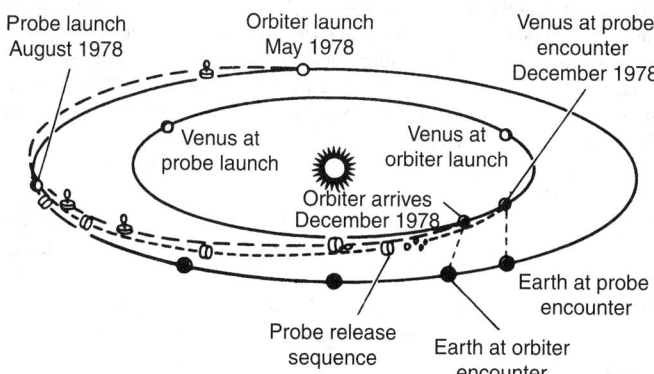

Fig. 13. Orbits of *Pioneer Venus* orbiter and probe bus. The program consisted of two scientifically related experiments and two spacecraft, an *orbiter* and a *multiprobe*. The orbiter was launched on May 20, 1978 and was inserted into a highly eccentric, near-polar orbit around Venus on December 4, 1978 after an interplanetary journey of 198 days. The multiprobe was launched on August 8, 1978 and reached Venus on December 9, 1978, just 5 days after orbital insertion. The orbiter trajectory and multiprobe impact locations matched nearly perfectly the conditions specified prior to launch. (*NASA Ames Research Center, Moffett Field, California.*)

The *Pioneer Venus Orbiter* carried 17 experiments (with a total mass of 45 kg (99.2 pounds):

- A cloud photopolarimeter to measure the vertical distribution of the clouds.
- A surface radar mapper to determine topography and surface characteristics.
- An infrared radiometer to measure IR emissions from the Venus atmosphere.
- An airglow ultraviolet spectrometer to measure scattered and emitted UV light.
- A neutral mass spectrometer to determine the composition of the upper atmosphere.
- A solar wind plasma analyzer to measure properties of the solar wind.
- A magnetometer to characterize the magnetic field at Venus.
- An electric field detector to study the solar wind and its interactions.
- An electron temperature probe to study the thermal properties of the ionosphere.
- An ion mass spectrometer to characterize the ionospheric ion population.
- A charged particle retarding potential analyzer to study ionospheric particles.
- Two radio science experiments to determine the gravity field of Venus.
- A radio occultation experiment to characterize the atmosphere.

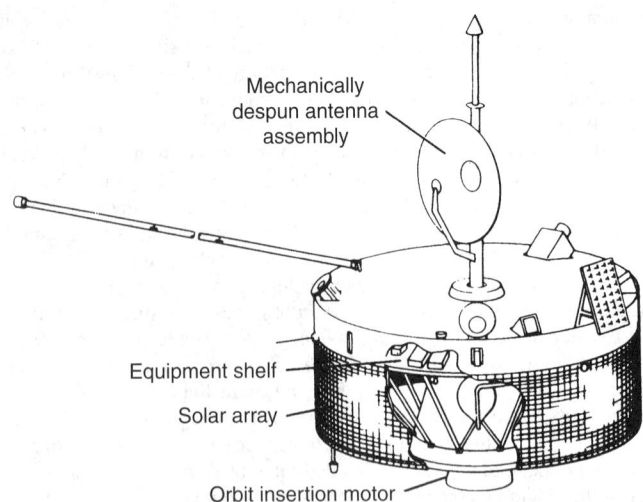

Fig. 14. *Pioneer Venus orbiter*. The orbiter was spin-stabilized in the range of 4.90 to 4.99 revolutions/minute in orbit, and the positive-spin-axis pointed in the direction of the planet's south ecliptic pole. Parameters of the orbiter included: Periapsis altitude, 150 to 260 km; apoapsis altitude, 66,900 km; eccentricity, 0.843; average period, 24.03 hours; inclination to equator, 105.6°; latitude of periapsis, 17°; longitude of periapsis, 170.2°; and original primary mission duration of 243 days. Despite its initial relatively short design life, as of 1987, the 8-year-old orbiter was still providing scientists with information about Venus and, in fact, in 1986, the orbiter was diverted from Venus to look at Halley's comet, the only United States spacecraft to observe the comet. (*NASA Ames Research Center, Moffett Field, California.*)

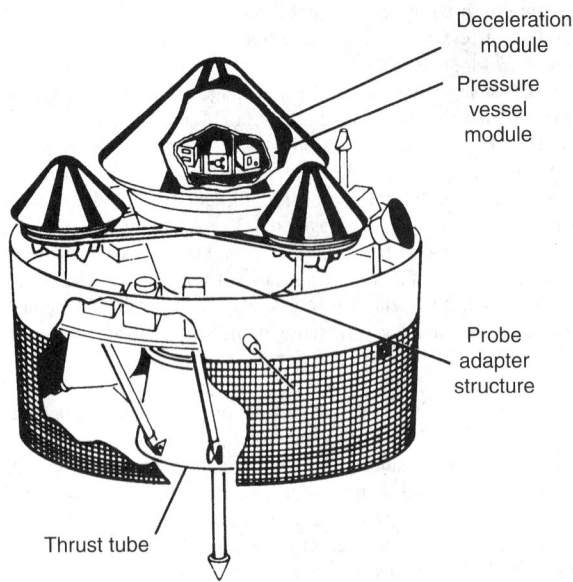

Fig. 16. *Pioneer Venus* multiprobe spacecraft. (*NASA Ames Research Center, Moffett Field, California.*)

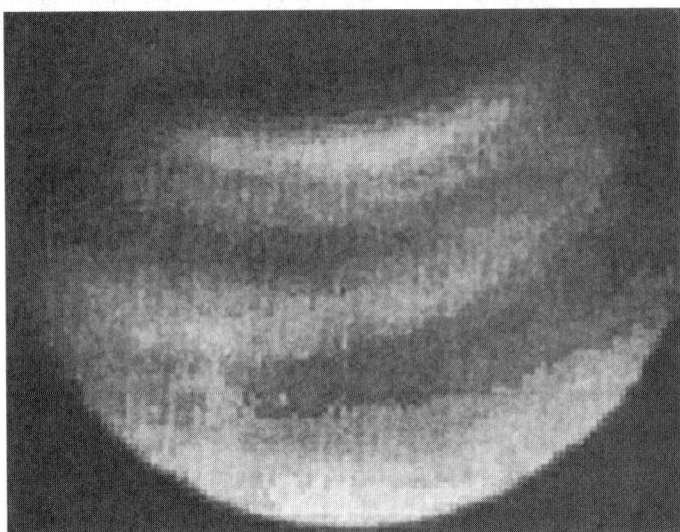

Fig. 17. Computer enhanced ultraviolet image of Venus made by *Pioneer Venus* orbiter on January 31, 1979. Probe instruments detected several hundred times more primordial argon and neon gases than on the Earth. The probes also detected an unexpected glow, perhaps caused by reactions of sulfur compounds in extreme heat near the surface. One of the mission probes survived for more than 67 minutes before succumbing to the searing environment. The Venusian atmosphere and surface databases were subsequently augmented and modified as the result of additional information obtained by the *Venera 13* and *14* missions. (*NASA Ames Research Center, Moffett Field, California.*)

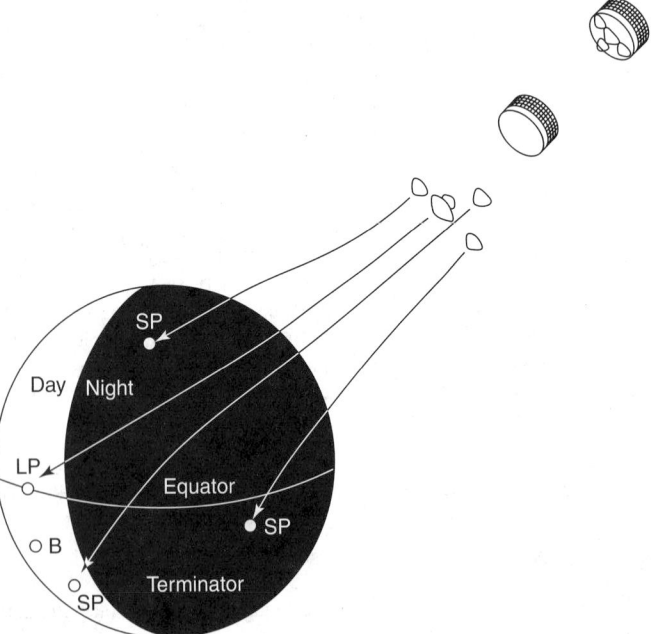

Fig. 15. Approximate entry points for *Pioneer Venus* atmosphere probes. The multiprobe during interplanetary cruise consisted of a bus transporting a large probe and three small probes, named north, day, and night. The spin-stabilized bus was oriented, and the large probe was released toward its planned Venus entry location on November 16, 1978. About 4 1/2 days later (November 20, 1978), the three small probes were released toward their planned entry locations. All probes were then silent before entry on December 9, when they were activated by pretimed internal sequencers. (*NASA Ames Research Center, Moffett Field, California.*)

- An atmospheric drag experiment to study the upper atmosphere.
- A radio science atmospheric and solar wind turbulence experiment and
- A gamma ray burst detector to record gamma ray burst events.

The *Pioneer Venus Multiprobe* consisted of a bus which carried one large and three small atmospheric probes. The large probe was released

on November 16, 1978 and the three small probes on November 20. All four probes entered the Venus atmosphere on December 9, followed by the bus.

The Pioneer Venus large probe was equipped with 7 science experiments, contained within a sealed spherical pressure vessel. This pressure vessel was encased in a nose cone and aft protective cover. After deceleration from initial atmospheric entry at about 11.5 km/s (37,700 feet/s) near the equator on the Venus night side, a parachute was deployed at 47 km (29.2 miles) altitude. The large probe was about 1.5 m (4.92 feet) in diameter and the pressure vessel itself was 73.2 cm (2.4 feet) in diameter. The science experiments were:

- A neutral mass spectrometer to measure the atmospheric composition.
- A gas chromatograph to measure the atmospheric composition.
- A solar flux radiometer to measure solar flux penetration in the atmosphere.

- An infrared radiometer to measure distribution of infrared radiation.
- A cloud particle size spectrometer to measure particle size and shape.
- A nephelometer to search for cloud particles.
- Temperature, pressure, and acceleration sensors.

The three small probes were identical to each other, 0.8 m (2.62 feet) in diameter. These probes also consisted of spherical pressure vessels surrounded by an aeroshell, but unlike the large probe, they had no parachutes and the aeroshells did not separate from the probe. Each small probe carried a nephelometer and temperature, pressure, and acceleration sensors, as well as a net flux radiometer experiment to map the distribution of sources and sinks of radiative energy in the atmosphere. The radio signals from all four probes were also used to characterize the winds, turbulence, and propagation in the atmosphere. The small probes were each targeted at different parts of the planet and were named accordingly. The North probe entered the atmosphere at about 60 degrees north latitude on the day side. The night probe entered on the night side. The day probe entered well into the day side, and was the only one of the four probes which continued to send radio signals back after impact, for over an hour.

The *Pioneer Venus* bus also carried two experiments, a neutral mass spectrometer and an ion mass spectrometer to study the composition of the atmosphere. With no heat shield or parachute, the bus survived and made measurements only to about 110 km altitude before burning up. The bus was a 2.5 m (8.2 feet) diameter cylinder weighing 290 kg (639 pounds), and afforded us our only direct view of the upper Venus atmosphere, as the probes did not begin making direct measurements until they had decelerated lower in the atmosphere.

Venera 11 (9, September 1978). Venera 11 was part of a two-spacecraft mission to study Venus and the interplanetary medium. Each of the two spacecraft,

Venera 11 and *Venera 12*, consisted of a flight platform and a lander probe. Identical instruments were carried on both spacecraft. The flight platform had instruments to study solar-wind composition, gamma-ray bursts, ultraviolet radiation, and the electron density of the ionosphere of Venus. The lander probe carried instruments to study the characteristics and composition of the atmosphere of Venus.

The *Venera 11* descent craft carried instruments designed to study the detailed chemical composition of the atmosphere, the nature of the clouds, and the thermal balance of the atmosphere. Separating from its flight platform on December 23, 1978 it entered the Venus atmosphere two days later at 11.2 km/sec. During the descent, it employed aerodynamic braking followed by parachute braking and ending with atmospheric braking. It made a soft landing on the surface at 06:24 Moscow time on 25 December after a descent time of approximately 1 hour. The touchdown speed was 7–8 m/s. Information was transmitted to the flight platform for retransmittal to earth until it moved out of range 95 minutes after touchdown. It is unknown whether the Lander Probe carried an imaging system. No mention of it occurs in the Soviet literature examined by the author. Two other experiments on the Lander did fail, and their failure was acknowledged by the Soviets. Some U.S. literature on the subject notes that the imaging system "failed" but did return some data. Among the instruments on board was a gas chromatograph to measure the composition of the Venus atmosphere, instruments to study scattered solar radiation and soil composition, and a device named Groza which was designed to measure atmospheric electrical discharges. Results reported included evidence of lightning and thunder, a high Ar_{36}/Ar_{40} ratio, and the discovery of carbon monoxide at low altitudes.

Venera 12 (14, September 1978). The Venera 12 descent craft carried instruments designed to study the detailed chemical composition of the atmosphere, the nature of the clouds, and the thermal balance of the atmosphere. Separating from its flight platform on December 19, 1978, it entered the Venus atmosphere two days later at 11.2 km/sec. During the descent, it employed aerodynamic braking followed by parachute braking and ending with atmospheric braking. It made a soft landing on the surface at 06:30 Moscow time on 21 December after a descent time of approximately 1 hour. The touchdown speed was 7–8 m/s. Information was transmitted to the flight platform for retransmittal to earth, until it moved out of range 110 minutes after touchdown. It is unknown whether the Lander Probe carried an imaging system. No mention of it occurs in the Soviet literature examined by the author. Two other experiments on the Lander did fail, and their failure was acknowledged by the Soviets. Some U.S. literature on the subject notes that the imaging system "failed" but did return some data.

Venera 13 and 14. *Venera 13* was launched on 30, October 1981, and *Venera 14* was launched on 4, November, 1981. In March 1982, these Russian spacecraft explored the surface of Venus. The spacecraft landed on the surface, sampled the rock, and determined its chemical composition. Each spacecraft carried a miniature drilling rig for taking samples. These were conveyed through an air lock to a measuring cell. There each sample was bathed in radioactivity emitted by plutonium-238 and iron-55. The plutonium isotope elicited the emission of x-rays at characteristic frequencies from magnesium, aluminum, and silicon atoms; the iron isotope—for potassium, calcium, and titanium. The X-radiation was analyzed by on-board spectrometers, which were recalibrated when empty. Approximately 4 minutes after landing, the first soil sample was retrieved and analyzed.

Venera 13 landed on an upland rolling plain; *Venera 14* on a lowland. From prior radar mapping made by *Pioneer Venus*, it was estimated that the *Venera* samplings were representative of about 92% of the planet's surface. In total, *Venera 13* acquired 38 samples; *Venera 14*, 20 samples. See Table 1.

TABLE 1. SURFACE COMPOSITION OF VENUS AS DETERMINED BY *VENERA* SPACECRAFT

Element	Venera 13	Venera 14
Magnesium (MgO)	11.4 ± 6.2	8.1 ± 3.3
Aluminum (Al_2O_3)	15.8 ± 3.0	17.9 ± 2.6
Silicon (SiO_2)	45.1 ± 3.0	48.7 ± 3.6
Potassium (K_2O)	4.0 ± 0.6	0.2 ± 0.1
Calcium (CaO)	7.1 ± 1.0	10.3 ± 1.2
Titanium (TiO_2)	1.6 ± 0.5	1.3 ± 0.4
Manganese (MnO)	0.2 ± 0.1	0.2 ± 0.1
Iron (FeO)	0.3 ± 2.2	8.8 ± 1.8
Sulfur (SO_3)	1.6 ± 1.0	0.9 ± 0.8
Others	3.9	3.6

The Russian team reported that the *Venera* landers transmitted almost complete panoramic views of their landing sites, made possible by the aforementioned sulfur (?) light glow. Image analysis showed the presence of rock formations undergoing geomorphic degradation. The formations displayed ripple marks, thin layering, differential erosion, and curvilinear fracturings. Shortly after the mission, some researchers suggested that some of the rocky material observed could be interpreted as lithified elastic sediments. The lithification could have occurred at depth, or at the surface, resulting in a type of duricrust. The origin of the sediments remains unknown, but it could be aeolian, volcanic, or related to impacts, or possibly to turbidity currents. A number of hypotheses and scenarios have been developed, but none proved convincingly.

Venera 15 and 16. *Venera 15* was part of a two spacecraft mission (along with *Venera 16*) designed to use 8 cm band side-looking radar mappers to study the surface properties of Venus. *Venera 15* was launched on 2, June 1983 and *Venera 16* on 7, June 1983. The two spacecraft were inserted into Venus orbit a day apart with their orbital planes shifted by an angle of approximately 4 degrees relative to one another. This made it possible to reimage an area if necessary. Each spacecraft was in a nearly polar orbit with a periapsis at 62 N latitude. Together, the two spacecraft imaged the area from the north pole down to about 30 degrees N latitude over the 8 months of mapping operations.

The *Venera 15* and *16* spacecraft were identical and were based on modifications to the orbiter portions of the Venera 9 and 14 probes. Each spacecraft consisted of a 5 m long cylinder with a 6 m diameter, 1.4 m tall parabolic dish antenna for the synthetic aperture radar (SAR) at one end. A 1 meter diameter parabolic dish antenna for the radio altimeter was also located at this end. The electrical axis of the radio altimeter antenna was lined up with the axis of the cylinder. The electrical axis of the SAR deviated from the spacecraft axis by 10 degrees. During imaging, the radio altimeter would be lined up with the center of the planet (local vertical) and the SAR would be looking off to the side at 10 degrees. A bulge at the opposite end of the cylinder held fuel tanks and propulsion units. Two square solar arrays extended like wings from the sides of the cylinder. A 2.6 m radio dish antenna for communications was also attached to the side of the cylinder.

VEGA 1 and *2* Venus Balloon Experiment. In June 1985, this mission was undertaken jointly by France and Russian, with major cooperation by the United States. Several other countries participated in the establishment of a worldwide balloon-tracking network. A total of 20 tracking stations participated, ranging from one in British Columbia, Canada, 5 in the United States, 6 in Russia, and one each in Puerto Rico, Brazil, Spain, England, Germany, Sweden, South Africa, and Australia. Very long baseline interferometry (VLBI) technology was used.

The experiment was first proposed in 1967 by Blamont (Centre National d'Etudes Spatiales (CNES)). The initial proposal was studied and refined by the Space Research Institute (IXI) of Russia, with cooperation of NASA (United States), particularly in organizing the international network of radio telescopes for tracking.

This spacecraft mission combined a Venus swingby and a Comet Halley flyby. Two identical spacecraft, Vega 1 and Vega 2, were launched December 15 and 21, 1984, respectively. After carrying Venus entry probes to the vicinity of Venus (arrival and deployment of probes were scheduled for June 11–15, 1985), the two spacecraft were retargetted using Venus gravity field assistance to intercept Comet Halley in March 1986. The first spacecraft encountered Comet Halley on March 6, 1986, and the second three days later. The flyby velocity was 77.7 km/s. Although the spacecraft could be targetted with a precision of 100 km, the position of the spacecraft relative to the comet nucleus was estimated to be known only to within a few thousand kilometers. This, together with the problem of dust protection, led to estimated flyby distances of 10,000 km for the first spacecraft and 3000 km for the second.

The spacecraft was three-axis stabilized. Its main features were large solar panels, a high-gain antenna dish, and an automatic pointing platform carrying those experiments that required pointing at the comet nucleus. The automatic platform could rotate through + or −110 deg and + or −40 deg in two perpendicular directions with a pointing accuracy of 5 arc-min and a stability of 1 arc-min/s. It carried the narrow- and the wide-angle camera, the three-channel spectrometer, and the infrared sounder. All other experiments were body-mounted, with the exception of two magnetometer sensors and various plasma probes and plasma wave analyzers which were mounted on a 5-m boom. The total scientific payload weighed 125 kg and had a data rate of 65 kbs in fast telemetry mode for encounter. There was also a slow telemetry mode for the cruise mode. The comet-encounter science data-take was from 2.5 h before until 0.5 h after the closest approach, with several periods of data-take before and after, each lasting about 2 h. Continuous coverage for plasma and dust instruments was provided by an onboard memory (5-megabit tape recorder). The spacecraft was shielded from hypervelocity dust impacts by a shield consisting of a 100-micrometer multilayer sheet 20 to 30 cm from the spacecraft, and a 1-mm Al sheet 5 to 10 cm from the spacecraft. Approximately half of the VEGA spacecraft was devoted to the Halley module, and half to the Venus lander package. The total scientific payload weight was 144.3 kg.

The Venus package consisted of a sphere 240 cm in diameter, which separated two days before arrival at Venus and entered the planet's atmosphere on an inclined path, without active maneuvers, as was done on previous Venera missions. The lander probe was identical to those of *Venera 9* through *14* and similarly had two objectives, the study of the atmosphere and the study of the superficial crust. In addition to temperature and pressure measuring instruments, the descent probe carried a UV spectrometer for measurement of minor atmospheric constituents, an instrument dedicated to measurement of the concentration of H_2O, and other instruments for determination of the chemical composition of the condensed phase: a gas-phase chromatograph; an X-ray spectrometer observing the fluorescence of grains or drops; and a mass spectrograph measuring the chemical composition of the grains or drops. The X-ray spectrometer separated the grains according to their sizes using a laser imaging device, while the mass spectrograph separated them according to their sizes using an aerodynamical inertial separator. After landing, a small surface sample near the probe was to be analyzed by gamma spectroscopy and X-ray fluorescence. The UV spectrometer, the mass spectrograph, and the pressure- and temperature-measuring instruments were developed in cooperation between French and Soviet investigators.

In addition to the lander probe, a constant-pressure instrumented balloon aerostat was deployed immediately after entry into the atmosphere at an altitude of 54 km (35 mi). The 3.4 meter diameter balloon supported a total mass of 25-kg. A 5-kg payload hung suspended 12 meters below the balloon. It floated at approximately 50 km altitude in the middle, most active layer of the Venus three-tiered cloud system. Data from the balloon instruments were transmitted directly to Earth for the 47-hr lifetime of

the mission. (The batteries had a lifetime of 60 hrs.) Onboard instruments were to measure temperature, pressure, vertical wind velocity, and visibility (density of local aerosols). Very long baseline interferometry was used to track the motion of the balloon to provide the wind velocity in the clouds. The tracking was to be done by a 6-station network on Soviet territory and by a network of 12 stations distributed world-wide (organized by France and the NASA Deep Space Network). After two days and 9000 km, the balloon entered the dayside of Venus and expanded and burst due to solar heating.

Magellan Mission. The **Magellan** spacecraft was launched on May 4, 1989, from aboard the space shuttle *Atlantis* and went into orbit around Venus on August 10, 1990. The spacecraft was designed to complete one orbit around Venus every 3 hours and 15 minutes, passing as close to the planet as 294 km (183 miles) and as far away from Venus as 8472 km (5265 miles). The basic scientific instrument aboard the spacecraft is a synthetic aperture radar (SAR), which can look through the thick clouds that perpetually shield the surface of the planet. See Fig. 18. The radar provided images of 98 percent of the planet's surface with a resolution of 100 meters (330 feet).

Fig. 18.   During its primary mission at the planet of 243 days, approximately one Venusian year, the *Magellan* Spacecraft used synthetic aperture radar to penetrate Venus' thick cloud cover and return maps of about 98 percent of the planet's surface. Other experiments include measuring heights of surface features on Venus with a radar altimeter as well as studies of the planet's gravitational field. Designed as a low-cost scientific mission with a design life of four years, *Magellan* was manufactured by Martin Marietta Aerospace Company with spare parts from other JPL flight projects including Voyager and Galileo. (*Jet Propulsion Laboratory.*)

At the same excellent images of Venus were returned to Earth as the spacecraft continued on its approach in late August 1990, a series of electronic system disruptions occurred. On August 23, the spacecraft's signal faded for a period of 21 hours before engineers at the space center could reestablish it. Thereafter, *Magellan* operated in accordance with specifications and completed its initial mission.

Venus completes one turn around its axis in 243 Earth days. That period of time (one Venus rotation) was specified as the time span for *Magellan's*

Fig. 19.   This map of the topography of Venus was obtained by the *Magellan* radar altimeter during its 24 months of systematic mapping. Color is used to code elevation (see color bar), and simulated shading to emphasize relief. Red corresponds to the highest, blue to the lowest elevations. The upper image shows the portion of the planet between 69 degrees north and 69 degrees south latitude in Mercator projection; beneath it are the two polar regions, covering latitudes above 44 degree in stereographic projection. Height accuracy is better than 50 meters (165 feet); horizontal (footprint) resolution of the surface depends on spacecraft altitude, with a resolution of about 10 kilometers (6 miles) near the equator and as much as 25 kilometers (15.5 miles) at higher latitudes. The *Magellan* altimeter acquired topography data over 98 percent of the planet's surface. Gray areas show the coarser results from the *Pioneer Venus* (1978) and *Venera* 15/16 (1983) radar altimeters, and indicate where data was not obtained by *Magellan*. The elevated region in the north is Ishtar Terra, dominated by Maxwell Montes (the planet's highest mountains) which rise 11 kilometers (36,000 feet) above the planetary mean elevation. Southwest of Ishtar are the highlands of Beta Regio and Phoebe Regio, which are bisected by a major north-south trending rift zone. The scorpion-shaped feature extending along the equator between 70 and 210 degrees longitude is Aphrodite Terra, a continent-like highland that contains several spectacular volcanoes at its eastern limit: Maat, Ozza and Sapas Montes. The altimetric data shown here were compiled and analyzed at the Center for Space Research, Massachusetts Institute of Technology. (*Jet Propulsion Laboratory*.)

primary mission. During that period, *Magellan* mapped approximately 80% of the surface of the planet. Thus, the primary mission of *Magellan* was completed by the spring of 1991. During the early days of *Magellan's* orbiting experience around the planet, sensational images of Venus were returned to Earth. Even though some criticism has been expressed over enhancement of some of the views, the images provided for the first time a comprehensive picture of the planet that exists under its cloud cover. The smallest visible objects in some views measure approximately 120 meters (400 feet).

The mission of the Magellan program was to develop an understanding and knowledge of the geological structure of Venus, including its density and distribution. During the four-year orbital tour of *Magellan* (1990–1994), detailed topographical images of the surface of Venus were obtained. In addition to the topographic maps, measurements of surface altitude were obtained using radar altimetry, and measurements of the planets gravitational field were obtained using precision radio tracking. See Figs. 19–24. At the end of its useful life, Magellan was directed into a gradual dive, which was designed to maximize information gathering on the atmospheric properties of Venus during the final days of the program. Magellan burned up in the Venusian atmosphere on October 12, 1994. See also **Magellan Mission to Venus**.

Galileo's Flyby of Venus. The primary mission of *Galileo*, launched on October 18, 1989, is that of exploring Jupiter, where it is scheduled to arrive on December 7, 1995. The spacecraft has encountered several difficulties, which are described in article on **Jupiter**. A part of *Galileo's* mission plan was that of flying by Venus in February 1990. This flyby was accomplished quite successfully. Important new information pertaining to Venus was returned, including the possible discovery of lightning in the atmosphere of Venus, magnetic field studies of the solar wind, plasma wave observations, determination of energetic particles, images of the Venusian cloud deck, middle infrared thermal maps of the planet, and ultraviolet spectrometric observations.

The *Galileo* spacecraft is described in more detail in article on **Jupiter**. See also **Galileo Mission to Jupiter**.

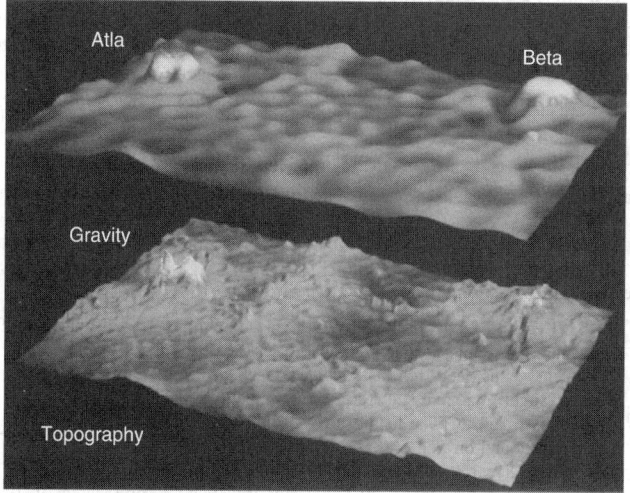

Fig. 20.   The computer-generated perspective shown here compares gravity and topography over a region 12,700 kilometers by 8,450 kilometers (7,860 miles by 5,240 miles) and extends from longitudes 180 degrees east to 300 degrees east and latitudes 40 degrees north to 40 degrees south. The highlands of Beta Regio and Atla Regio, sites of rifting and large volcanoes, have corresponding high topography and high gravity. These areas are interpreted to be sites where hot mantle material is upwelling, forming "hot spots," similar to areas on Earth such as Hawaii. The two gravity highs at Atla correspond to the volcanoes Maat Mons, the higher of the two peaks, and Ozza Mons respectively. Gravity anomalies at Atla and Beta are the largest on Venus and these may be the sites of relatively young geologic features. In contrast to Earth, it is also seen that there is a near-perfect correlation between gravity and topography with anomalies, both positive (light blue) and negative (dark blue), being correlated with topographic highs and lows. This correspondence is interpreted to indicate that, relative to Earth, the formation of features on Venus is more strongly linked to fluid motions in the mantle. (*Jet Propulsion Laboratory*.)

Fig. 21. This global view of the surface of Venus is centered at 90 degrees east longitude. *Magellan* synthetic aperture radar mosaics from the three eight-month cycles of *Magellan* radar mapping are mapped onto a computer-simulated globe to create this image. *Magellan* obtained coverage of 98 percent of the surface of Venus. Remaining gaps are filled with data from previous Venus missions — the *Venera 15* and *16* radar and *Pioneer-Venus* Orbiter altimetry — and data from Earth-based radar observations from the Arecibo radio telescope. Simulated color is used to enhance small-scale structures. The simulated hues are based on color images obtained by the *Venera 13* and *14* landing craft. The bright feature near the center of the image is Ovda Regio, a mountainous region in the western portion of the great Aphrodite equatorial highland. The dark areas scattered across the Venusian plains consist of extremely smooth deposits associated with large meteorite impacts. The image was produced by the Solar System Visualization Project and the *Magellan* Science team at the Jet Propulsion Laboratory Multimission Image Processing Laboratory. (*Jet Propulsion Laboratory*.)

Additional Reading

Hamilton, D.P.: "Magellan Mission Extension Gets a Boost," *Science*, 1383 (June 5, 1992).
Kaula, W.M.: "Venus: A Contrast in Evolution to Earth,"*Science*, 1191 (March 9, 1990).
News and Progress Reports — Science Magazine:

"NASA Keeps Its Fingers Crossed While Magellan Shines," 27 (October 6, 1990).
"Magellan: No Venusian Plate Tectonics Seen," (April 12, 1991).
"A Violent Venus Seen from a Troubled Magellan," 977 (August 31, 1990).
"Venus Caught in a Geologic Act," 1208 (September 13, 1991).
"Do NASA Images Create Fantastic Voyages?" 1632 (March 27, 1992).
"Lightning Found on Venus at Last?" 1492 (September 27, 1992).
"Did Venus Hiccup or Just Run Down?" 1400 (March 5, 1993).
"More Venus Science, or the Off Switch for Magellan?" 1966 (March 19, 1993).

Staff: "The Volcanoes of Venus," Technology Review (MIT), 15 (January 1993).

Reports by the *Magellan* Team

Arvidson, R.E., et al.: "Magellan: Initial Analysis of Venus Surface Modification," *Science*, 270 (April 12, 1991).
Head, J.W., et al.: "Venus Volcanism: Initial Analysis from Magellan Data," *Science*, 276 (April 12, 1991).
Pettengill, G.H., et al.: "Magellan: Radar Performance and Data Products," *Science*, 260 (April 12, 1991).
Phillips, R.J., et al.: "Impact Craters on Venus: Initial Analysis from Magellan," *Science*, 288 (April 12, 1991).
Saunders, R.S. and G.H. Pettengill: "Magellan: Mission Summary," *Science*, 247 (April 12, 1991).
Saunders, R.S., et al.: "An Overview of Venus Geology," *Science*, 249 (April 12, 1991).
Solomon, S.C. and J.W. Head: "Fundamental Issues in the Geology and Geophysics of Venus," *Science*, 252 (April 12, 1991).

Solomon, S.C., et al.: "Venus Tectonics: Initial Analysis from Magellan," *Science*, 297 (April 12, 1991).
Tyler, G.L., et al.: "Magellan Electrical and Physical Properties of Venus' Surface," *Science*, 265 (April 12, 1991).

Reports by the *Galileo* Flyby Team

Belton, M.J.S., et al.: "Images from Galileo of the Venus Cloud Deck," *Science*, 1531 (September 27, 1991).
Carlson, R.W., et al.: "Galileo Infrared Imaging Spectroscopy Measurements at Venus," *Science*, 1541 (September 27, 1991).
Crisp, D., et al.: "Ground-Based Near-Infrared Imaging Observations of Venus During the Galileo Encounter," *Science*, 1538 (September 27, 1991).
Frank, L.A., et al.: "Plasma Observations at Venus with Galileo," *Science*, 1528 (September 27, 1991).
Gurnett, D.A., et al.: "Lightning and Plasma Wave Observations from the Galileo Flyby of Venus," *Science*, 1522 (September 27, 1991).
Hord, C.W., et al.: "Galileo Ultraviolet Spectrometer Experiment: Initial Venus and Interplanetary Cruise Results," *Science*, 1548 (September 27, 1991).
Johnson, T.V., et al.: "The Galileo Venus Encounter," *Science*, 1516 (September 27, 1991).
Frank, L.A., et al.: "Plasma Observations at Venus with Galileo," *Science*, 1528 (September 27, 1991).
Orton, G.S., et al.: "Middle Infrared Thermal Maps of Venus at the Time of the Galileo Encounter," *Science*, 1536 (September 27, 1991).
Williams, D.J., et al.: "Energetic Particles at Venus: Galileo Results," *Science*, 1525 (September 27, 1991).

Pre-1988 References

Alexandrov, Y.N.: "Venus: Detailed Mapping of Maxwell Montes Region," *Science*, **231**, 1271–1273 (1986).
Borucki, W.J.: "Lightning on Venus — An Alternative View," *The Planetary Report*, 6 (July/August 1987).
Campbell, D.B., et al.: "Venus: Identification of Banded Terrain in the Mountains of Ishtar Terra," *Science*, **221**, 644–647 (1983).
Campbell, D.B., et al.: "Venus: Volcanism and Rift Formation in Beta Regio," *Science*, **226**, 167–169 (1984).
DeMore, W.B. and Y.L. Yung: "Catalytic Processes in the Atmospheres of Earth and Venus," *Science*, **217**, 1209–1213 (1982).
Donahue, T.M., et al.: "Venus Was Wet: A Measurement of the Ratio of Deuterium to Hydrogen," *Science*, **216**, 630–633 (1982).
Esposito, L.W.: "Sulfur Dioxide: Episodic Injection Shows Evidence for Active Venus Volcanism," *Science*, **223**, 1072–1074 (1984).
Florensky, C.P., et al.: "Venera 13 and Venera 14 : Sedimentary Rocks on Venus?" *Science*, **221**, 57–59 (1983).
Ford, P.B. and G.H. Pettengill: "Venus: Global Surface Radio Emissivity," *Science*, **220**, 1379–1380 (1983).
Hazard, C.: "Hughes Radar to Unravel Venusian Cloud Mystery," *Hughesnews*, 1, 8 (May 22, 1987).
Hunter, D.M., et al.: "Venus," University of Arizona Press, Tucson, Arizona, 1983.
Kerr, R.A.: "Venus Highlights," *Science*, **215**, 278–279 (1982).
Lewis, J.S. and B. Fegley, Jr.: "Venus: Halide Cloud Condensation and Volatile Element Inventories," *Science*, **216**, 1223–1225 (1982).
Lewis, J. and R. Prinn: "Planets and Their Atmospheres: Origin and Evolution," Academic Press, Orlando, Florida, 1984.
McElroy, M.B., M.J. Prather, and J.M. Rodriguez: "Escape of Hydrogen from Venus," *Science*, **215**, 1614–1615 (1982).
Munroe, K.: "Hughes Radar "Eyes" of New Venus Mapper," *Hughesnews*, 1, 9 (January **20**, 1984).
Nozette, S. and J.S. Lewis: "Venus: Chemical Weathering of Igneous Rocks and Buffering of Atmospheric Composition," *Science*, **216**, 181–183 (1982).
Pettengill, G.H., P.G. Ford, and S. Nozette: ";Venus: Global Surface Radar Reflectivity," *Science*, **217**, 640–642 (1982).
Phillips, R.J., et al.: "Tectonics and Evolution of Venus," *Science*, **212**, 879–887 (1981).
Prather, M.J. and M.B. McElroy: "Helium on Venus: Implications for Uranium and Thorium," *Science*, **220**, 410–411 (1983).
Prinn, R.G.: "The Volcanoes and Clouds of Venus," *Sci. Amer.*, 46–53 (March 1985).
Schubert, G. and C. Covey: "The Atmosphere of Venus,"*Sci. Amer.*, 66–74 (July 1981).
Taylor, H.A., Jr.: "Auroras at Venus?" *The Planetary Report*, 4–6 (July/August 1987).

Web References

Galileo Project Information: *http://nssdc.gsfc.nasa.gov/planetary/galileo.html*
Images of Venus: *http://nssdc.gsfc.nasa.gov/imgcat/html/group_page/VN.html*
Magellan Mission: *http://nssdc.gsfc.nasa.gov/planetary/magellan.html*
Venus: *http://nssdc.gsfc.nasa.gov/planetary/planets/venuspage.html*

VENUS' GIRDLE (*Ctenophora, Cestida*).   A transparent marine animal of ribbon-like form. The longitudinal axis of the body lies across the width

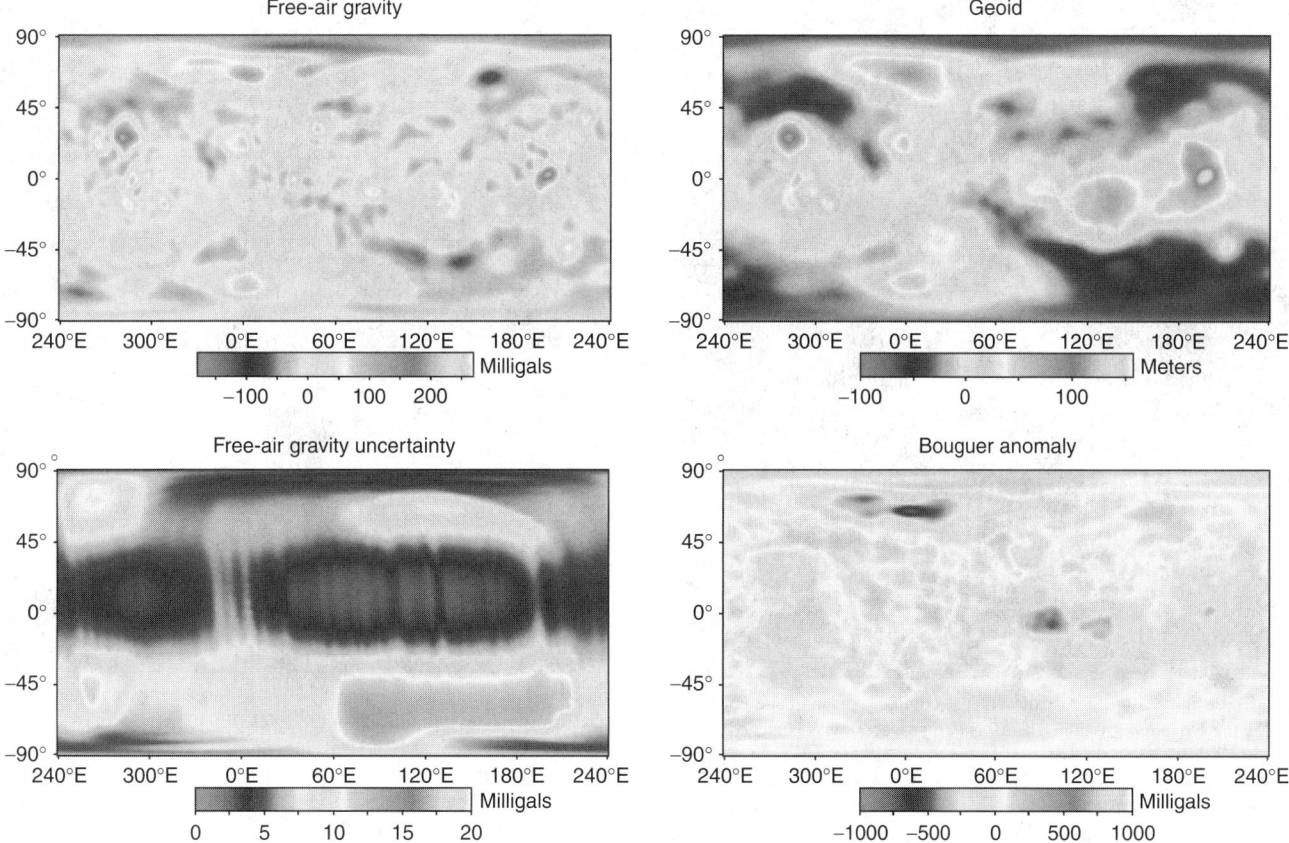

Fig. 22.   These four images summarize the results of the *Magellan* gravity experiment on Venus.

FREE AIR GRAVITY MAP: "Free Air" means raw gravity measured in milligals (thousandths of a gal. One-gal = one centimeter per second squared). These variations are measured relative to the Venus average of 800 gals. The largest positive gravity effects are found at Beta Regio, latitude 25-north, longitude 281 east, and Atla Regio, 2 north, 199 east. The measurements were 200 milligals and 203 milligals, respectively. The largest negative effects are all in the low lands, ranging from 40 to 80 milligals. The differences are similar to those of the Earth, but unlike those of the moon and Mars, which are much larger. On Venus, however, there is almost perfect correlation of topography with gravity—highs with highs and lows with lows—which is not the case on Earth where some gravity highs occur in the oceans and some gravity lows occur over continents. Earth's major gravity features are produced by density variations deep in the mantle. That also is true for Venus, except on Venus the surface highs and lows are also tied to variations in the mantle.

FREE AIR GRAVITY UNCERTAINTY MAP: The smallest uncertainty in the gravity field solution is shown by the small values stretching across the equatorial zone. This is the result of excellent data from Magellan cycle 4 coverage from September 1992 to May 1993. Another block of excellent data was acquired during cycle 5 after a successful aerobraking, which placed the spacecraft in a near-circular orbit and provided data for the polar regions. The largest uncertainties between 90 east and 215 east are due to the fact that there was no coverage during cycle 5 in that area.

GEOID MAP: The surface geoid (theoretically, continuously level surface) shows height variations in a gravity surface that is equipotential, or has the same measured value. The peak measurements are again over Beta Regio (113.7 meters or 375.2 feet) and Atla Regio (141.4 meters or 466.6 feet).

BOUGUER ANOMALY MAP: (Named for 18th Century French scientist Pierre Bouguer) A Bouguer map is produced by subtracting from the observed free-air gravity a theoretical gravity based on the topography. If the observed gravity was produced from rigid surface topography only, the difference would be zero. There are very large differences, especially in the Aphrodite region, 70 east to 150 east, near the equator and the Ishtar Terra region, 65 north latitude. Those differences indicate that the interior of Venus contains variations in density, which may be related to large-scale motions or temperature differences in the planet's mantle. Bouguer anomalies are usually interpreted as variations in crustal thickness. Negative anomalies would indicate thickened crust and are seen in the Venus highlands. This also is the case for Earth where the most negative Bouguer anomalies occur in mountainous regions. (*Jet Propulsion Laboratory.*)

of the ribbon, hence the body is short and is greatly elongated on one transverse axis and very short on another.

VERMICULITE INSULATION.   See **Insulation (Thermal)**.

VERNAL EQUINOX.   1. That point of intersection of the ecliptic and the celestial equator, occupied by the Sun as it changes from south to north declination, on or about March 21. Also called *March equinox, first point of Aries.*

2. That instant the Sun reaches the point of zero declination when crossing the celestial equator from south to north.

VERNALIZATION.   Low-temperature treatment of seeds before sowing, which shortens the time to flowering of the plants which develop from them.

VERNIER.   A scale or control used for fine adjustment to obtain a more precise reading of an instrument or closer adjustment of any equipment.

VERNIER ENGINE.   A rocket engine of small thrust used primarily to obtain a fine adjustment in the velocity and trajectory of a rocket vehicle just after the thrust cutoff of the last sustainer engine, and used secondarily to add thrust to a booster or sustainer engine. Also called *vernier rocket.*

VERSINE.   See **Trigonometric Function**.

VERTEBRA.   One of the 33 bones making up the spinal column. In humans, there are 7 cervical, 12 dorsal, 5 lumbar, 5 sacral, and 4 coccygeal vertebrae. See Fig. 1. See also **Skeletal System**.

VERTEX.   An endpoint of an edge. Point, O-cell, and node are other names used in the literature. As a rule, an isolated point is not considered a vertex. The term vertex also signifies the point common to the two straight lines forming an angle.

Vertices, adjacent. Two vertices of a graph G are adjacent if they are endpoints of an edge to G.

Vertex (cut). Let G be a connected separable graph. Then, by definition, there exists at least one subgraph G_s which has only one vertex β_c in

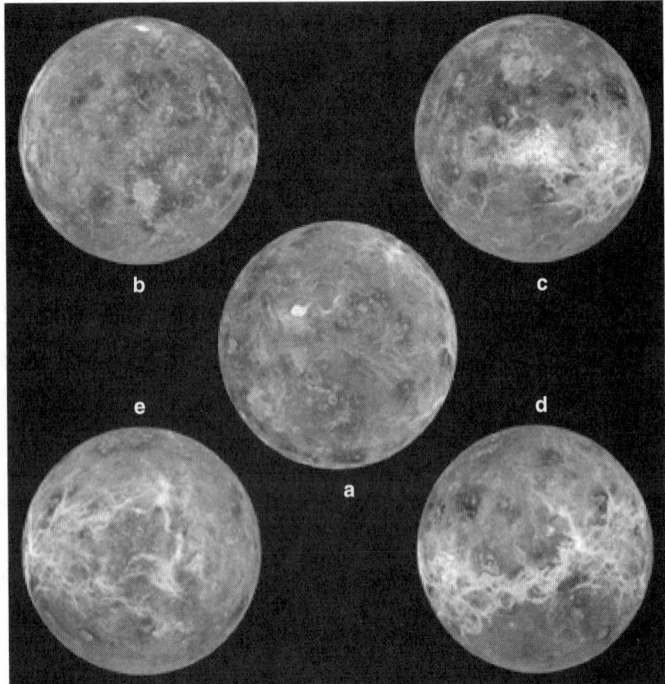

Fig. 23. These images are composites of the complete radar image collection obtained by the *Magellan* mission. The *Magellan* spacecraft was launched aboard space shuttle Atlantis in May 1989 and began mapping the surface of Venus in September 1990. The spacecraft continued to orbit Venus for four years, returning high-resolution images, altimetry, thermal emissions and gravity maps of 98 percent of the surface. *Magellan* spacecraft operations ended on October 12, 1994, when the radio contact was lost with the spacecraft during its controlled descent into the deeper portions of the Venusian atmosphere. The surface of Venus is displayed in these five global views. The center image (A) is centered at Venus's north pole. The other four images are centered around the equator of Venus at (B) 0 degrees longitude, (C) 90 degrees east longitude, (D) 180 degrees and (E) 270 degrees east longitude. *Magellan* synthetic aperture radar mosaics are mapped onto a rectangular latitude-longitude grid to create this image. Data gaps are filled with *Pioneer-Venus* Orbiter altimetric data, or a constant mid-range value. Simulated color is used to enhance small-scale structure. The simulated hues are based on color images recorded by the Soviet *Venera* 13 and 14 spacecraft. The bright region near the center in the polar view is Maxwell Montes, the highest mountain range on Venus. Ovda Regio is centered in the (C) 90 degrees east longitude view. Atla Regio is seen prominently in the (D) 180 east longitude view. The scattered dark patches in this image are halos surrounding some of the younger impact craters. This global data set reveals a number of craters consistent with an average Venus surface age of 300 million to 500 million years. The image was produced by the Solar System Visualization Project and the *Magellan* science team at the Jet Propulsion Laboratory's Multimission Image Processing Laboratory. (*Jet Propulsion Laboratory.*)

Fig. 24. This hemispheric view of Venus, as revealed by more than a decade of radar investigations culminating in the 1990–1994 *Magellan* mission, is centered at 0 degrees east longitude. The *Magellan* Spacecraft imaged more than 98 percent of Venus at a resolution of about 100 meters (330 feet); the effective resolution of this image is about 3 kilometers (1.8 miles). A mosaic of the Magellan images (most with illumination from the west) forms the image base. Gaps in the *Magellan* coverage were filled with images from the Earth-based Arecibo radar in a region centered roughly on 0 degrees latitude and longitude, and with a neutral tone elsewhere (primarily near the south pole). The composite image was processed to improve contrast and to emphasize small features, and was color-coded to represent elevation. Gaps in the elevation data from the *Magellan* radar altimeter were filled with altimetry from the Venera spacecraft and the U.S. *Pioneer Venus* missions. An Orthographic projection was used, simulating a distant view of one hemisphere of the planet. The *Magellan* mission was managed for NASA by the Jet Propulsion Laboratory (JPL), Pasadena, CA. Data processed by JPL, the Massachusetts Institute of Technology, Cambridge, MA, and the U.S. Geological Survey, Flagstaff, AZ. (*Jet Propulsion Laboratory.*)

common with its complement; β_c is a cut vertex. A necessary and sufficient condition for a graph G to be nonseparable is that it have no cut vertex.

Vertex degree. The number of edges incident at the vertex.

Vertex (final). The vertex of the last edge of an edge sequence not shared with the previous edge.

Vertex (initial). The vertex of the first edge of an edge sequence which is not common to the second edge.

Vertex (internal). A vertex of an edge sequence which is not terminal.

Vertex (terminal). The initial and final edges of an edge sequence.

VERTEX MATRIX. The vertex matrix $\mathbf{A}_a = (a_{ij})$ of a linear oriented graph G possessing v vertices β_i, $(i = 1, 2, \ldots, v)$ and e elements ε_j, $(j = 1, 2, \ldots, e)$, is a matrix with v rows and e columns such that

1. $a_{ij} = 1$ if element ε_j is incident at vertex β_i and oriented away from β_i;
2. $a_{ij} = -1$ if element ε_j is incident at β_i and oriented toward β_i;
3. $a_{ij} = 0$ if element ε_j is not incident at vertex β_i.

Fig. 1. Vertebral column of human.

The vertex matrix has exactly two non-zero elements, one $+1$ and one -1 in each column and at least one non-zero element in each row. The rank of the vertex matrix of a connected graph is $v - 1$.

For a nonoriented graph the vertex matrix is defined similarly:

1. $a_{ij} = 1$ if element ε_j is incident at vertex β_i and
2. $a_{ij} = 0$ otherwise.

A graph is completely defined by its vertex matrix.

VERTICAL CIRCLE.   Any great circle on the celestial sphere that passes through the zenith and nadir. Since the zenith and nadir are poles of the horizon, a vertical circle must be perpendicular to the horizon. The vertical circle that cuts the horizon at the north and south points is known as the local meridian, and the vertical circle that cuts the horizon at the east and west points is known as the prime vertical. See also **Celestial Sphere and Astronomical Triangle**.

VERTIGO.   True vertigo arises as the result of a disturbance of the semicircular canals of the inner ear (auditory organ). *Objective vertigo* is the sensation that the external world is revolving about the patient; *subjective vertigo* is the sensation that the patient is himself moving in space. The disturbance may arise from a variety of causes: various diseases of the central nervous system; certain infectious diseases; the influence of certain toxic substances (such as alcohol, various sedatives, and some antibiotics); diseases of the visual system and of the blood; environmental factors (sunstroke, motion sickness); and, notably, Menière's disease. See also **Hearing and the Ear**.

VERTISOLS.   See **Soil**.

VESICANT.   A chemical substance that causes blisters to form on the skin. See also **Chlorinated Organics**.

VESICULAR (Petrology).   Said of the texture of a rock, especially a lava, characterized by abundant vesicles (holes, pores, etc.) formed as a result of the spansion of gases during the fluid stage of the lava. A less-preferred synonym is *cellular*.

VESTIBULAR.   Of or pertaining to the nature of a vestibule. In anatomy and zoology, a vestibule is a cavity or space that serves as an entrance to another cavity or space. The vestibule of the inner ear, for example, leads into the cochlea.

VESTIBULAR APPARATUS (Ear).   See **Hearing and the Ear**.

VESTIGIAL STRUCTURES.   Elements appearing in various life forms which, although often quite underdeveloped, are no longer needed or functional and represent a carry-over from more primitive forms. The human appendix is an example.

VESUVIANITE.   The mineral vesuvianite is a very complex silicate of calcium and aluminum with fluorine which may also contain varying amounts of boron, iron, lithium, magnesium, manganese, potassium, sodium and titanium. A suggested formula is $Ca_{10}Mg_2Al_4(SiO_4)_5$ $(Si_2O_7)_2(OH)_4$. Its tetragonal crystals are usually short, somewhat stoutish prisms, sometimes pyramids, but columnar to massive varieties are common. It is essentially without cleavage; fracture, uneven; brittle hardness, 6–7; specific gravity, 3.3–3.5; luster, vitreous to greasy resinous; color, commonly some shade of brown, green, or white, may be reddish, bluish or yellowish; may be transparent, but is usually translucent.

This mineral was formerly called idocrase, having been named Haüy from the Greek words meaning *form* and *mixture* because it resembled crystals of other species, scarcely a valid distinction. Werner gave it the name vesuvianite from Mt. Vesuvius where it was first found blocks of limestone appearing as inclusions in the lava. Vesuvianite not a constituent of the igneous rocks, but rather a contact metamorphic mineral resulting from the alteration of impure limestones and dolomites It is usually associated with diopside, wollastonite, epidote, grossularite and garnet. There are many localities worthy of mention among which are: The Urals, Central Europe, Rumania, Trentino and Monzoni in Italy, as well as at Mt. Vesuvius and Mt. Somma, Switzerland, Mexico and Japan. In the United States vesuvianite is found in Androscoggin and York Counties in Maine; Orange and Warren Counties, New York; Sussex County, New Jersey; Garland County, Arkansas; and Riverside and Tulare Counties in California.

VETCH.   Of the family *Leguminoseae* (pea family), genus *Vicia*, vetch is the common designation for numerous herbs that are cultivated in several regions of the world, mainly for fodder, green manure, and as a cover, particularly for orchards. Vetch has the important advantage of all leguminous plants, namely, the ability to enrich poor soils with nitrogen, a factor that makes possible the planting of vetch in regions where other nutritious plants will not thrive.

VIBRATING SYSTEM (Oscillation).   See **Oscillation**.

VIBRATION.   Generally, any form of sustained motion, characteristic of a finite system, in which each particle or element of the system moves to-and-fro about an equilibrium position. In the simplest situation, such a motion possesses a unique "periodic time." The inverse of this quantity, the "frequency," is the number of complete to-and-fro excursions per unit time. A simple pendulum is an example of a form of vibration in which the motions are slow. See also **Pendulum**. Often one thinks of vibration in terms of much higher frequencies. Historically, pendulum motion was the first simple harmonic motion to be understood. Simple harmonic motion is referred to as "isochronous" because, in a given case, the periodic time is the same whatever the amplitude.

In vibratory motion, the restoring forces responsible for the individual motions of the constituent elements of a vibrating system may arise from *tensions* externally applied to the system—as with stretched strings and membranes; from *elastic stresses* developed internally as the system is deformed—as with rods vibrating longitudinally, transversely, or torsionally, or with air columns in organ pipes; or in *other ways*, such as *surface tension* forces—as involved in the pulsation vibrations of thin films of liquid.

The complete period for a small amplitude for a simple pendulum is

$$T = 2\pi\sqrt{\frac{l}{g}}; \quad g = 4\pi^2\frac{l}{T^2}$$

where l is length in centimeters; g is acceleration due to gravity in centimeters per sec_2 and T is time in seconds.

If the period is P for an arc θ, the time of vibration in an indefinitely small arc is approximately

$$T = \frac{P}{1 + \frac{1}{4}\sin^2\frac{\theta}{4}}$$

For a compound pendulum, where the body of mass m is suspended from a point about which its moment of inertia is I, with its center of gravity a distance h below the point of suspension, the period is

$$T = 2\pi\sqrt{\frac{I}{mgh}}$$

The frequency of vibration of a closed pipe or air column, for the fundamental and first three overtones respectively, is

$$n_0 = \frac{V}{4l}, \quad n_1 = \frac{3V}{4l}, \quad n_2 = \frac{5V}{4l}, \quad n_3 = \frac{7V}{4l}$$

where l is the length of the column and V is the velocity of sound in air.

For an open pipe, the frequency of vibration is

$$n_0 = \frac{V}{2l}, \quad n_1 = \frac{2V}{2l}, \quad n_2 = \frac{3V}{2l}, \quad n_3 = \frac{4V}{2l}$$

The fundamental frequency of vibrating strings (stretched) is given by

$$n = \frac{1}{2l}\sqrt{\frac{T}{m}}$$

where l is the length; T is the tension; and m is the mass per unit length.

With essentially linear systems, such as strings and rods and organ pipes, when the conditions at the two ends of the system are the same as they are with a stretched string, and there is no internal constraint, the normal modes in general constitute a harmonic series. When the end conditions are different, as in a pipe which is closed at one end and open at the other, in general the even-numbered harmonics are not represented. In all cases, however, the theorem concerning the decomposition of the most complex sustained motion in terms of the normal modes applies.

As an example of elastic vibrations, consider the transverse vibrations of a thin, uniform rod, which is clamped at one end. For a displacement profile of arbitrary shape, the elastic forces are brought into play across any section of the rod at that section, and the unbalanced restoring force acting on a small finite element of the rod is similarly proportional to its displacement. But the proportionality constant relating restoring force per unit length to displacement will, in general, vary along the rod. Only for displacement profiles of certain shapes will this constant be the same overall. These are the shapes of displacement profile corresponding to the normal mode vibrations, and it is the object of theory to identify them and to deduce the corresponding frequencies. Even an approximate theory is complex. All that can be done in the interest of simplicity is to quote the ratios of the frequencies of the first three normal mode vibrations, i.e., 1:6.27:17.55. Very clearly, the normal modes in this case do not constitute a harmonic series; in musical parlance, the first overtone is some 2.5 octaves higher than the fundamental.

Vibration testing is important to many technologies, as for example, testing building structures, aircraft, missiles, and vehicles of all kinds. Testing techniques have become increasingly precise and are extended to deal with problems in such fields as the design of machines, construction of bridges, the development of new materials, and the study of earthquakes. Vibration can be measured in several ways. The linear piezoelectric accelerometer is widely used for measuring vibration. The piezoelectric element, which is acted on by the inertial force of the mass, may be made of polycrystalline ceramics, such as barium titanate or lead zirconate. See Fig. 1. The resonance is sharply underdamped. Over the useful frequency range, the equivalent circuit of the accelerometer is shown in Fig. 2. The frequency response of the accelerometer is shown in Fig. 3. Because of its high impedance, the accelerometer must be isolated from the data-conditioning equipment, and the signal must be amplified.

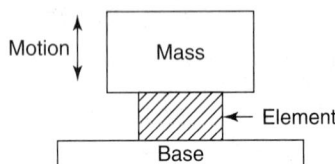

Fig. 1.   Simplified representation of accelerometer.

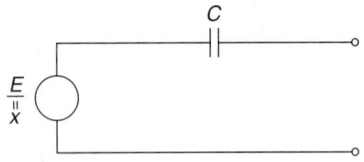

Fig. 2.   Equivalent circuit of accelerometer over useful frequency range.

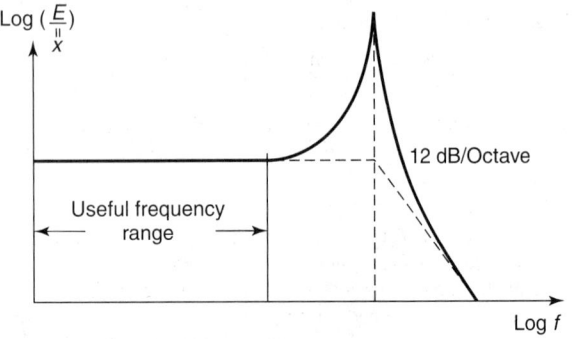

Fig. 3.   Frequency response of an accelerometer.

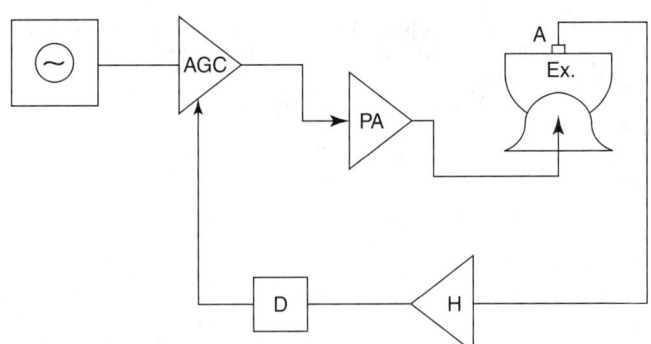

Fig. 4.   Application of an accelerometer to closely controlled sine-wave vibration.

The manner in which an accelerometer is used to closely control sine-wave vibration testing is shown in Fig. 4. An electrodynamic vibration exciter is used. An oscillator puts out a sinusoidal signal which is fed via a variable gain (AGC) to a power amplifier (PA). The latter, in turn, drives the vibration exciter (EX). The accelerometer A has its output amplified by the signal-conditioning circuit H. The signal then is detected by D, and a dc voltage proportional to signal level is used to control the gain of the AGC stage. Thus, independent of the usual resonant effects at the exciter or specimen, the vibration level at A can be carefully controlled.

Use of an accelerometer to control a random vibration is shown in Fig. 5. An automatic random equalizer is used. A random-noise generator supplies the input to a set of contiguous bandpass filters. In some apparatus, these filters are crystal in nature. Each filter output is controlled by an AGC stage so as to maintain its spectral power at a desired level. The reference voltage at each servoloop summing point is used to determine the vibration level of the corresponding filter channel. A large loop gain makes it possible to calibrate the potentiometer supplying the reference voltage so as to follow any desired spectral-density plot. There is one such control for each filter channel.

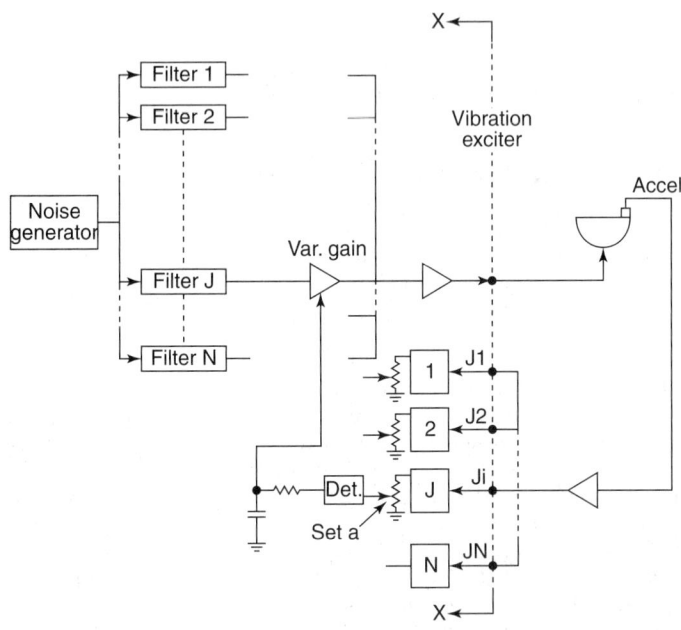

Fig. 5.   Use of an accelerometer to control a random vibration.

Vibration testing using an accelerometer can be extended to force impedance testing or mechanical-impedance testing, using an impedance head. The latter consists of an accelerometer plus a force gage, measuring at the same point in the same direction. The force gage consists of a piezoelectric element between two plates and provides an output proportional to the pressure between the plates. The stiffness of the impedance head should be kept large, and the mass of the impedance head should be kept small.

Commercial accelerometers usually are small and cylindrical in form, weighing between 0.001 and 0.2 pound (0.5 and 91 grams). Accelerometer sensitivity generally increases with size. The resonant frequency commonly is from 25,000 to 100,000 Hz, a point two to three times the maximum operating frequency. See also **Accelerometer**.

A velocity vibration pickup is a self-generating transducer having an output voltage proportional to velocity. The device senses the relative velocity between a coil and the magnetic field imposed on the coil by a self-contained magnetic source. In some types, a seismically suspended coil moves relative to a uniform magnetic field. A variation of this is a seismic magnet that moves relative to an anchored coil. Other types function by means of a coil that senses variations in the magnetic field. A typical pickup is shown in Fig. 6. The output signal of a velocity pickup normally is large enough so as not to require amplification. The velocity signal can be integrated and differentiated to obtain displacement and acceleration, respectively. The natural frequency of the suspended moving element usually is low, ranging from 2.5 to 10 Hz in various models. This natural frequency establishes the lower operating-frequency limit. Low velocities associated with high-frequency vibration and pickup structural resonances limit the upper frequency to approximately 2,000 Hz. The velocity pickup is useful over a range of amplitudes determined by the sensitivity and friction at low displacements and by space limitation on moving parts at high displacements.

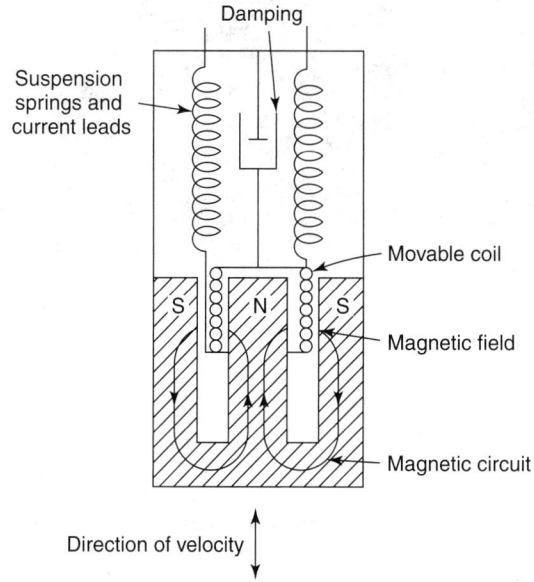

Fig. 6.   Schematic representation of a velocity vibration pickup.

Additional Reading

Inman, D.J.: "Engineering Vibration," Prentice-Hall, Inc., Upper Saddle River, NJ, 2000.
Meirovitch, L.: "Fundamentals of Vibrations," McGraw-Hill Higher Education, New York, NY, 2000.
Ramamurti, V.: "Mechanical Vibration Practice with Basic Theory," CRC Press, LLC., Boca Raton, FL, 2000.
Steinberg, D.S.: "Preventing Thermal Cycling and Vibration Failures in Electronic Equipment," John Wiley & Sons, Inc., New York, NY, 2001.

VIBRATION (Music).   See Musical Sound.

VIBRATION ANALYSIS.   See Spectrum Analysis.

VIBRATION EXCITER. A transducer used in simulation of predetermined mechanical vibration environments. A mounting or table surface is usually provided for threaded attachment of the item to be vibrated for testing. Vibration exciters can be classified into four main groups: (1) mechanical, (2) electromechanical, (3) electrohydraulic, and (4) pneumatic, depending upon the source of vibration-producing energy.

The least sophisticated types are pneumatic and electromagnetic. These types are limited in their capability to reproduce vibration over a wide and controllable range. Piezoelectric exciters generally are useful at high

frequencies only. Hence, these devices are limited in application to calibration of instruments and ultrasonic apparatus. Mechanical exciters are driven by electric motors which generate vibration by use of cranks or cams (direct drive), or by use of eccentric weights (reaction).

These exciters are useful over a relatively narrow range of controllable frequencies and amplitudes. They are relatively low cost and hence are widely used wherever they can be applied.

Electrodynamic exciter construction is allied to that of a loudspeaker. Through the use of electronic amplifiers and accessories, a wide range of input signals can be converted to mechanical vibration output with excellent fidelity. The input signals commonly used are sinusoidal, complex, random, and tape recording, all with a high degree of controllability within the exciter system specifications.

VIBURNUMS.   See Elder Trees and Viburnums.

VICUNA.   See Camels and Llamas.

VIDICON. A television pickup tube utilizing a photoconductor as the sensing element. In conjunction with a telescope this is known as a *vidicon telescope*.

VIGNETTING EFFECT. The falling-off in brightness toward the margin of an illuminated field, due to the mutilation of the more oblique bundles of light by the combined effects of diaphragm and lens aperture. This is a source of trouble in photographic objectives when they are used at large apertures, or close to the limit of their fully illuminated fields.

VIKING MISSION TO MARS. NASA's *Viking* Project was the culmination of a series of missions to explore Mars that had begun in 1964 with *Mariner 4*, and continued with the *Mariner 6* and *7* flybys in 1969, and the *Mariner 9* orbital mission in 1971 and 1972.

Viking found a place in history when it became the first mission to land a spacecraft successfully on the surface of another planet. Two identical spacecraft, each consisting of a lander and an orbiter, were built. Each orbiter-lander pair flew together and entered Mars orbit; the landers then separated and descended to the planet's surface.

Mission Design

Both spacecraft were launched from Cape Canaveral, Florida — *Viking 1* on August 20, 1975, and *Viking 2* on September 9, 1975. The landers were sterilized before launch to prevent contamination of Mars with organisms from Earth. The spacecraft spent nearly a year cruising to Mars. *Viking 1* reached Mars orbit June 19, 1976; *Viking 2* began orbiting Mars August 7, 1976. See Fig. 1.

After studying orbiter photos, the Viking site certification team considered the original landing site proposed for *Viking 1* unsafe. The team examined nearby sites, and *Viking 1* landed on Mars July 20, 1976, on the western slope of Chryse Planitia (the Plains of Gold) at 22.3 degrees north latitude, 48.0 degrees longitude.

The site certification team also decided the planned landing site for *Viking 2* was unsafe after it examined high-resolution photos. Certification of a new landing site took place in time for a Mars landing September 3, 1976, at Utopia Planitia, at 47.7 degrees north latitude and 48.0 degrees longitude.

The *Viking* mission was planned to continue for 90 days after landing. Each orbiter and lander operated far beyond its design lifetime. *Viking Orbiter 1* exceeded four years of active flight operations in Mars orbit.

The *Viking* project's primary mission ended November 15, 1976, 11 days before Mars' superior conjunction (its passage behind the Sun). After conjunction, in mid-December 1976, controllers reestablished telemetry and command operations, and began extended mission operations.

The first spacecraft to cease functioning was *Viking Orbiter 2*, on July 25, 1978; the spacecraft had used all the gas in its attitude-control system, which kept the craft's solar panels pointed at the Sun to power the orbiter. When the spacecraft drifted off the Sun line, the controllers at Jet Prapulane Laboratory (JPL) sent commands to shut off power to *Viking Orbiter 2's* transmitter.

Viking Orbiter 1 began to run short of attitude-control gas in 1978, but through careful planning to conserve the remaining supply, engineers found it possible to continue acquiring science data at a reduced level for another two years. The gas supply was finally exhausted and *Viking*

Fig. 1. *Viking* Orbiter.

Orbiter 1's electrical power was commanded off on August 7, 1980, after 1,489 orbits of Mars.

The last data from *Viking Lander 2* arrived at Earth on April 11, 1980. *Viking Lander 1* made its final transmission to Earth Nov. 11, 1982. Controllers at JPL tried unsuccessfully for another six and one-half months to regain contact with *Viking Lander 1*. The overall mission came to an end May 21, 1983.

Viking Orbiters

The *Viking* spacecraft consisted of two large orbiters, each weighing 2,325 kilograms (5,125 pounds) with fuel. Each orbiter carried a lander, weighing 576 kilograms (1,270 pounds), and their design was greatly influenced by the size of these landers.

The orbiters were a follow-on design to the *Mariner* class of planetary spacecraft with specific design changes for the 1976 surface mission. Operational lifetime requirements for the orbiters were 120 days in orbit and 90 days after landing.

The combined weight of the orbiter and lander was one factor that contributed to an 11-month transit time to Mars, instead of the five months for *Mariner* missions. The longer flight time then dictated an increased design life for the spacecraft, larger solar panels to allow for longer degradation from solar radiation and additional attitude control gas.

The basic structure of the orbiter was an octagon approximately 2.4 meters (8 feet) across. The eight sides of the ring-like structure were 45.7 centimeters (18 inches) high and were alternately 1.4 by 0.6 meters (55 by 22 inches).

Electronic bays were mounted to the faces of the structure, and the propulsion module was attached at four points. There were 16 bays, or compartments, three on each of the long sides and one on each short side.

The orbiter was 3.3 meters (10.8 feet) high and 9.7 meters (32 feet) across the extended solar panels. With fuel, the orbiters weighed in excess of 2,300 kilograms (5,000 pounds).

Combined area of the four panels was 15 square meters (161 square feet), and they provided both regulated and unregulated direct current

power; unregulated power was provided to the radio transmitter and the lander.

Two 30-amp-hour, nickel–cadmium, rechargeable batteries provided power when the spacecraft was not facing the Sun during launch, correction maneuvers, and Mars occultation.

The orbiter was stabilized in flight by locking onto the Sun for pitch and yaw references and onto the star Canopus for roll reference. The attitude control subsystem kept this attitude with nitrogen gas jets located at the solar panel tips. The jets would fire to correct any drift. A cruise Sun sensor and the Canopus sensor provided error signals. Before Sun acquisition four acquisition Sun sensors were used and then turned off.

The attitude control subsystem also operated in an all-inertial mode or in roll-inertial with pitch and yaw control, still using the Sun sensors. During correction maneuvers, the attitude control subsystem aligned the vehicle to a specified attitude in response to commands from the on-board computer. Attitude control during engine burns was provided in roll by the attitude control subsystem and in pitch and yaw by an autopilot that commanded engine gimballing.

If Sun lock was lost, the attitude control subsystem automatically realigned the spacecraft. In loss of Canopus lock, the subsystem switched to roll-inertial and waited for commands from the spacecraft computer. The nitrogen gas supply for the subsystem could be augmented by diverting excess helium gas from the propulsion module, if necessary.

Two on-board general purpose computers in the computer command subsystem decoded commands and either ordered the desired function at once or stored the commands in a 4,096-word plated-wire memory. All orbiter events were controlled by the computer command subsystem, including correction maneuvers, engine burns, science sequences and high-gain antenna pointing.

The main orbiter communications system was a two-way, S-band, high-rate radio link providing Earth command, radio tracking and science and engineering data return. This link used either a steerable 1.5 meters (59 inches) dish high-gain antenna or an omni-directional low-gain antenna, both of them on the orbiter. The low-gain antenna was used to send and receive near Earth, and the high-gain antenna was used as the orbiter journeyed farther from Earth.

S-band transmission rates varied from 8.3 or 33.3 bits per second for engineering data to 2,000 to 16,000 bits per second for lander and orbiter science data.

Relay from the lander was achieved through an antenna mounted on the outer edge of a solar panel. It was activated before separation and received from the lander through separation, entry, landing and surface operations. The bit rate during entry and landing was 4,000 bits per second; landed rate was 16,000 bits per second.

Data were stored aboard the orbiter on two eight-track digital tape recorders. Seven tracks were used for picture data and the eighth track for infrared data or relayed lander data. Each recorder could store 640 million bits.

Data collected by the orbiter, including lander data, were converted into digital form by the flight data subsystem and routed to the communications subsystem for transmission or to the tape recorders for storage. This subsystem also provided timing signals for the three orbiter science experiments.

Viking Lander

The lander spacecraft was composed of five basic systems: the lander body, the bioshield cap and base, the aeroshell, the base cover and parachute system and lander subsystems. See Figs. 2 and 3.

The completely outfitted lander measured approximately 3 meters (10 feet) across and was about 2 meters (7 feet) tall. It weighed about 576 kilograms (1,270 pounds) without fuel.

The lander and all exterior assemblies were painted light gray to reflect solar heat and to protect equipment from abrasion. The paint was made of rubber-based silicone.

The body was a basic platform for science instruments and operational subsystems. It was a hexagon-shaped box with three 109 centimeters (43 inches) side-beams and three 56-centimeter (22-inch) short sides. It looks like a triangle with blunted corners.

The box was built of aluminum and titanium alloys, and was insulated with spun fiberglass and dacron cloth to protect equipment and to lessen heat loss. The hollow container was 1.5 meters (59 inches) wide and 46 centimeters (18 inches) deep, with cover plates on the top and bottom.

The lander body was supported by three landing legs, 1.3 meters (51 inches) long, attached to the short-side bottom corners of the body. The legs gave the lander a ground clearance of 22 centimeters (8.7 inches).

Each leg had a main strut assembly and an A-frame assembly, to which was attached a circular footpad 30.5 centimeters (12 inches) in diameter. The main struts contained bonded, crushed aluminum honeycomb to reduce the shock of landing.

The two-piece bioshield was a pressurized cocoon that completely sealed the lander from any possibility of biological contamination until *Viking* left Earth's atmosphere.

The two bioshield halves generally resembled an egg, and the shield's white thermal paint heightened the resemblance. It measured 3.7 meters (12 feet) in diameter and was 1.9 meters (6.4 feet) deep. It was made of coated, woven fiberglass, 0.13 millimeters (0.005 inches) thin, bonded to an aluminum support structure.

The bioshield was vented to prevent over-pressurization and possible rupture of its sterile seal. The aeroshell was an aerodynamic heat shield made of aluminum alloy in a 140-degree, flat cone shape and stiffened with concentric rings. It fit between the lander and the bioshield base. It was 3.5 meters (11.5 feet) in diameter and its aluminum skin was 0.86 millimeters (0.034 inches) thin.

Bonded to its exterior was a lightweight, cork-like ablative material that burned away to protect the lander from aerodynamic heating at entry temperatures which may have reached 1,500 °C (2,730 °F).

The interior of the aeroshell contained twelve small reaction control engines, in four clusters of three around the aeroshell's edge, and two

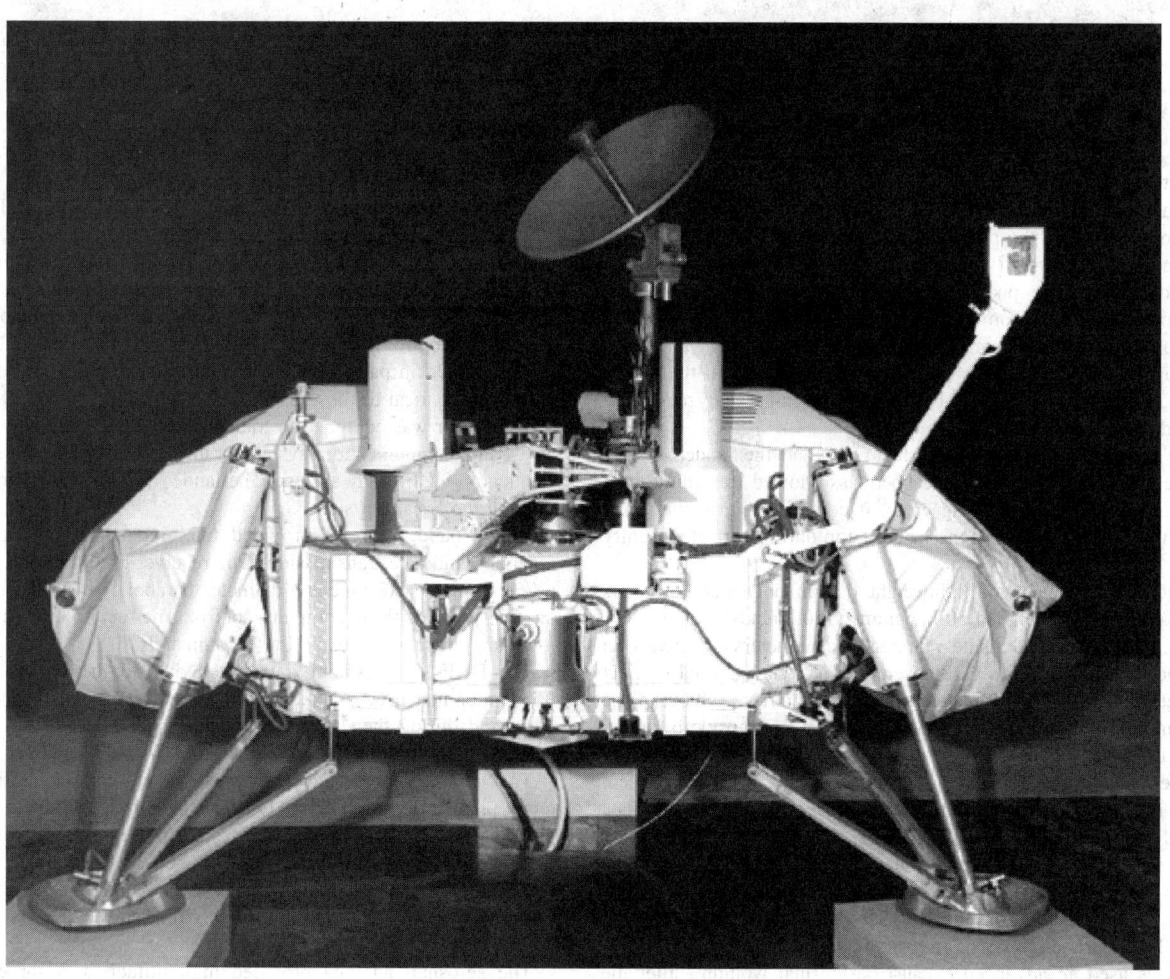

Fig. 2. *Viking* Lander 1.

Fig. 3. *Viking* Lander 2.

spherical titanium tanks that contained 85 kilograms (188 pounds) of hydrazine mono-propellant.

The engines controlled the pitch and yaw to align the lander for entry, help slow the craft during early entry, and maintain roll control.

During the long cruise phase, an umbilical connection through the aeroshell provided power from the orbiter to the lander; housekeeping data also flowed through this connection.

The aeroshell also contained two science instruments—the upper atmosphere mass spectrometer and the retarding potential analyzer—plus pressure and temperature sensors.

The base cover fit between the bioshield cap and the lander. It was made of aluminum and fiberglass; the fiberglass allowed transmission of telemetry data to the orbiter during entry. It covered the parachute and its ejection mortar, and protected the lander's top during part of the entry phase.

The parachute was made of lightweight dacron polyester 16 meters (53 feet) in diameter. It weighed 50 kilograms (110 pounds).

The parachute was packed inside a mortar 38 centimeters (15 inches) in diameter and mounted into the base cover. The mortar was fired to eject the parachute at about 139 kilometers per hour (75 miles per hour). The chute had an extra-long suspension line that trailed the capsule by about 30 meters (100 feet).

The lander subsystems were divided into six major categories: descent engines, communications equipment, power sources, landing radars, data storage, and guidance and control.

Three terminal descent engines provided attitude control and reduced the lander's velocity after parachute separation. The 2,600-newton (600-pound) throttleable engines were located 120 degrees apart on the lander's sidebeams. They burned hydrazine monopropellant.

The engines used an advanced exhaust design that wouldn't alter the landing site environment. An unusual grouping of 18 small nozzles on each engine would spread engine exhaust over a wide angle that wouldn't alter the surface or unduly disturb the chemical and biological experiments.

Two spherical titanium tanks, attached to opposite sides of the lander body beneath the RTG wind covers, fed the descent engines from an 85 kilograms (188 pounds) hydrazine propellant supply.

Four small reaction control engines used hydrazine mono-propellant thrusters to control lander roll attitude during terminal descent. The engines were mounted in pairs on the terminal descent engines' propellant tanks and were identical to those used on the aeroshell.

The lander was equipped to transmit information directly to Earth with an S-band communications system, or through the orbiter with an ultra-high frequency (UHF) relay system. The lander also received Earth commands through the S-band system.

Two S-band receivers provided total redundancy in both command receiving and data transmission. One receiver used the high-gain antenna, a 76 centimeters (30-inches) diameter parabolic reflector dish that could be pointed to Earth by computer control. The second receiver used a fixed low-gain antenna to receive Earth commands.

The UHF relay system transmitted data to the orbiter with a radio transmitter that used a fixed antenna. The UHF system operated during entry and during the first three days of landed operations. After that, it was operated only during specific periods.

The radar altimeter measured the lander's altitude during the early entry phase, alerting the lander computer to execute the proper entry commands. The radar was a solid-state pulse radar with two specially designed antennas: one was mounted beneath the lander and one was mounted through the aeroshell.

Altitude data were received from 1,370 kilometers down to 30.5 meters (740 miles to 100 feet).

The aeroshell antenna provided high-altitude data for entry science, vehicle control and parachute deployment. The lander antenna was switched into operation at aeroshell separation and provided altitude data for guidance and control, and for terminal descent engine ignition.

The terminal descent landing radar measured the horizontal velocity of the lander during the final landing phase. It was located directly beneath the lander and was turned on at about 12 kilometers (4,000 feet) above the surface. It consisted of four continuous-wave Doppler radar beams that could measure velocity to an accuracy of plus or minus one meter per second.

Both radars were essential for mission success, so the terminal descent landing radar could work with any three of its four beams, and identical sets of radar altimeter electronics could be switched to either of the radar antennas.

The "brain" of the lander was its guidance control and sequencing computer. That computer commanded everything the lander did through software (computer programs) stored in advance or relayed by Earth controllers.

The computer was one of the greatest technical challenges of *Viking*. It consisted of two general-purpose computer channels with plated-wire memories, each with an 18,000-word storage capacity. One channel would be operational while the other was in reserve.

Among other programs, the computer had instructions stored in its memory that could control the lander's first 22 days on Mars without any contact from Earth. These instructions would be updated and modified by Earth commands once communications had been established.

Basic power for the lander was provided by two SNAP 19-style, 35-watt radioisotope thermoelectric generators (RTGs) developed by the then U.S. Energy Research and Development Administration. They were located atop the lander and were connected in series to double their voltage and reduce power loss.

The SNAP 19 *Viking generator* was 147 centimeters (23 inches) across the housing fin tips, 96 centimeters (15 inches) in length and weighed 15.3 kilograms (34 pounds).

The first isotopic space generator was put into service in June 1961, on a Navy navigational satellite. Advances in SNAP systems were made with the development and flight of SNAP 19 aboard Nimbus III, launched in April 1969. This use of SNAP 19 represented a major milestone in the development of long-lived, highly reliable isotope power systems for space use by NASA. The SNAP 27 generator was developed to power five science stations left on the Moon by the *Apollo 12, 14, 15, 16* and *17* astronauts. The continuing operation of these generators provided new dimensions of data about the Moon and the universe. Four SNAP 19 nuclear generators later provided the electrical power for each of two NASA pioneering Jupiter flyby spacecraft known as *Pioneers 10* and *11*.

The generators provided a long-lived source of electricity and heat on Mars, where sunlight is half as strong as on Earth, and is non-existent during the Martian night, when temperatures can drop as low as $-120\,^{\circ}C$ ($-184\,^{\circ}F$).

The generators used thermoelectric elements to convert heat from decaying plutonium-236 into 70 watts of electrical power.

Waste heat or unconverted heat was conveyed by thermal switches to the lander's interior instrument compartment, when required. Covers over the RTGs pre-vented excess heat dissipation into the environment.

Four nickel-cadmium, rechargeable batteries helped supply lander power requirements in peak activity periods. The batteries, mounted in pairs inside the lander, were charged by the RTGs with power available when other lander power requirements were less than RTG output.

This equipment collected and controlled the flow of lander scientific and engineering data. It consisted of a data acquisition and processing unit, a data storage memory and a tape recorder.

The data acquisition and processing unit collected the science and engineering information and routed it to one of three places: to Earth through the S-band high-gain antenna, to the data storage memory or to the tape recorder.

Information was stored in the data storage memory for short periods. Several times a day the memory would transfer data to the tape recorder or back to the data acquisition and processing unit for further transmission. The memory had a storage capacity of 8,200 words.

Data were stored on the tape recorder for long periods. The recorder could transmit at high speed back through the data acquisition and processing unit, and the UHF link to an orbiter passing overhead. The recorder could store as many as 40 million bits of information and it could record at two speeds and play back at five.

Science Experiments

With a single exception, the seismic instruments, the science instruments acquired more data than expected. The seismometer on *Viking Lander 1* would not work after landing, and the seismometer on *Viking Lander 2* detected only one event that may have been seismic. Nevertheless, it provided data on wind velocity at the landing site to supplement information from the meteorology experiment, and showed that Mars has very low seismic background.

The three biology experiments discovered unexpected and enigmatic chemical activity in the Martian soil, but provided no clear evidence for the presence of living microorganisms in soil near the landing sites. According to mission biologists, Mars is self-sterilizing. They believe the combination of solar ultraviolet radiation that saturates the surface, the extreme dryness of the soil and the oxidizing nature of the soil chemistry prevent the formation of living organisms in the Martian soil. The question of life on Mars at some time in the distant past remains open.

The landers' gas chromatograph/mass spectrometer instruments found no sign of organic chemistry at either landing site, but they did provide a precise and definitive analysis of the composition of the Martian atmosphere and found previously undetected trace elements. The X-ray fluorescence spectrometers measured elemental composition of the Martian soil.

Viking measured physical and magnetic properties of the soil. As the landers descended toward the surface they also measured composition and physical properties of the Martian upper atmosphere.

The two landers continuously monitored weather at the landing sites. Weather in the Martian midsummer was repetitious, but in other seasons it became variable and more interesting. Cyclic variations appeared in weather patterns (probably the passage of alternating cyclones and anticyclones). Atmospheric temperatures at the southern landing site (*Viking Lander 1*) were as high as $-14\,^{\circ}C$ ($7\,^{\circ}F$) at midday, and the predawn summer temperature was $-77\,^{\circ}C$ ($-107\,^{\circ}F$). In contrast, the diurnal temperatures at the northern landing site (*Viking Lander 2*) during midwinter dust storms varied as little as $4\,^{\circ}C$ ($7\,^{\circ}F$) on some days. The lowest predawn temperature was $-120\,^{\circ}C$ ($-184\,^{\circ}F$), about the frost point of carbon dioxide. A thin layer of water frost covered the ground around *Viking Lander 2* each winter.

Barometric pressure varies at each landing site on a semiannual basis, because carbon dioxide, the major constituent of the atmosphere, freezes out to form an immense polar cap, alternately at each pole. The carbon dioxide forms a great cover of snow and then evaporates again with the coming of spring in each hemisphere. When the southern cap was largest, the mean daily pressure observed by *Viking Lander 1* was as low as 6.8 millibars; at other times of the year it was as high as 9.0 millibars. The pressures at the *Viking Lander 2* site were 7.3 and 10.8 millibars. (For comparison, the surface pressure on Earth at sea level is about 1,000 millibars.)

Martian winds generally blow more slowly than expected. Scientists had expected them to reach speeds of several hundred miles an hour from observing global dust storms, but neither lander recorded gusts over 120 kilometers (74 miles) an hour, and average velocities were considerably lower. Nevertheless, the orbiters observed more than a dozen small dust storms. During the first southern summer, two global dust storms occurred, about four Earth months apart. Both storms obscured the Sun at the landing sites for a time and hid most of the planet's surface from the orbiters' cameras. The strong winds that caused the storms blew in the southern hemisphere.

Photographs from the landers and orbiters surpassed expectations in quality and quantity. The total exceeded 4,500 from the landers and 52,000 from the orbiters. The landers provided the first close-up look at the surface, monitored variations in atmospheric opacity over several Martian years, and determined the mean size of the atmospheric aerosols. The orbiter cameras observed new and often puzzling terrain and provided clearer detail on known features, including some color and stereo observations. *Viking's orbiters* mapped 97 percent of the Martian surface.

The infrared thermal mappers and the atmospheric water detectors on the orbiters acquired data almost daily, observing the planet at low and high resolution. The massive quantity of data from the two instruments will require considerable time for analysis. Understanding of the global meteorology of Mars *Viking* also definitively determined that the residual north polar ice cap (that survives the northern summer) is water ice, rather than frozen carbon dioxide (dry ice) as once believed.

Analysis of radio signals from the landers and the orbiters—including Doppler, ranging and occultation data, and the signal strength of the lander-to-orbiter relay link—provided much valuable information. See also **Mars**.

Other Significant Discoveries

Other significant discoveries of the *Viking* mission included:

- The Martian surface is a type of iron-rich clay that contains a highly oxidizing substance that releases oxygen when it is wetted.
- The surface contains no organic molecules that were detectable at the parts per billion level, less in fact, than soil samples returned from the Moon by Apollo astronauts.
- Nitrogen, never before detected, is a significant component of the Martian atmosphere, and enrichment of the heavier isotopes of nitrogen and argon relative to the lighter isotopes implies that atmospheric density was much greater than in the distant past.
- Changes in the Martian surface occur extremely slowly, at least at the Viking landing sites. Only a few small changes took place during the mission life-time.
- The greatest concentration of water vapor in the atmosphere is near the edge of the north polar cap in midsummer. From summer to fall, peak concentration moves toward the equator, with a 30% decrease in peak abundance. In southern summer, the planet is dry, probably also an effect of the dust storms.
- The density of both of Mars' satellites is low, about two grams per cubic centimeter, implying that they originated as asteroids captured by Mars' gravity. The surface of Phobos is marked with two families of parallel striations, probably fractures caused by a large impact that may nearly have broken Phobos apart.
- Measurements of the round-trip time for radio signals between Earth and the *Viking* spacecraft, made while Mars was beyond the Sun (near the solar conjunctions), have determined delay of the signals caused by the Sun's gravitational field. The result confirms Albert Einstein's prediction to an estimated accuracy of 0.1%. This is 20 times greater than any other test.
- Atmospheric pressure varies by 30% during the Martian year because carbon dioxide condenses and sublimes at the polar caps.
- The permanent north cap is water ice; the southern cap probably retains some carbon dioxide ice through the summer.
- Water vapor is relatively abundant only in the far north during the summer, but subsurface water (permafrost) covers much if not all of the planet.
- Northern and Southern Hemispheres are drastically different climatically, because of the global dust storms that originate in the south in summer.

VINEGAR EEL (*Nematoda*). A minute roundworm, *Anguillula (Turbatrix) aceti*, found in the "mother" of vinegar. It reaches a length of 2 millimeters. The worms have been found in other situations, including the human bladder.

VINYL CHLORIDE. See **Chlorinated Organics**.

VINYL ESTER RESINS. The vinyl ester resins are a relatively recent addition[1] to thermosetting-polymer-chemistry. Superficially, they are similar to unsaturated polyester resins insofar as they contain ethylmic unsaturation and are cured through a free-radical mechanism, usually in the presence of a vinyl monomer, such as styrene. However, close examination of the chemistry and structure of the vinyl ester resins demonstrates several basic differences which lead to their unique characteristics.

Vinyl ester resins are manufactured through an addition reaction of an epoxy resin with an acrylic monomer, such as acrylic acid, methacrylic acid, or the half-ester product of an hydroxyalkyl acrylate and anhydride. In contrast, the polyester resins are condensation products of dibasic acids and polyhydric alcohols. The relatively low-molecular-weight precise polymer structure of the vinyl ester resins is in contrast to the high-molecular-weight random structure of the polyesters.

Of particular importance in describing the difference between these two families of resins are the locations of the reactive unsaturation. In the polyester resin, these groups are located along the backbone of the polymer with terminal hydroxyl or carboxylic acid groups. In contrast, the vinyl ester resins contain no significant acidity but terminate in reactive vinyl ester groups. Because of the location of these reactive sites, the vinyl ester resins will homopolymerize as well as coreact with various vinyl monomers.

Resin Properties. Vinyl esters, because of their relatively low molecular weight and precise structure, can be characterized as low-viscosity, fast-wetting, consistent-reactivity products. Typical property profiles for some uncured vinyl ester resins are given in Table 1. Properties of some cured resins are given in Table 2. Typically, 6-month stability can be expected at 25°C (77°F) with decreased storage life under elevated temperatures. In general, anyone familiar with the proper storage and handling of unsaturated polyester resins and styrene monomer experiences no difficulty with these materials.

TABLE 1. PROPERTIES OF TYPICAL UNCURED VINYL ESTER RESINS

Property	Standard Resin	Low Viscosity Resin
Monomer type	styrene	styrene
Level, %	45	45
Viscosity at 77°F (25°C), centipoises	550	200
Acid number	5	5
Specific gravity	1.04	1.04
SPI gel time (1% benzoyl peroxide), minutes		
at 82°C (180°F)	10	12
at 121°C (250°F)	1.4	1.5
Flash point (Tag open cup)		
°C	34	34
°F	93	93

Source: The Dow Chemical Company.

TABLE 2. PROPERTIES OF TYPICAL CURED VINYL ESTER RESINS

CLEAR-CASTING PROPERTIES	
Tensile strength, psi	12,000
megapascals	83
Tensile modulus, psi	500,000
megapascals	3,447
Ultimate elongation, %	5
Flexural strength, psi	18,000
megapascals	124
Flexural modulus, psi	450,000
megapascals	3,103
Yield compressive strength, psi	17,000
megapascals	117
Compressive modulus, psi	350,000
megapascals	2,413
Deflection at yield, %	7
Heat-distortion temperature, °C	101.7
°F	215.0
Barcol hardness	35
GLASS-REINFORCED LAMINATE PROPERTIES	
Laminate thickness, inch	0.25
millimeters	6.3
Fiber-glass content, %	30
Tensile strength, psi	19,000
megapascals	131
Tensile modulus, psi	1,400,000
megapascals	9,653
Flexural strength, psi	22,000
megapascals	152
Flexural modulus, psi	1,000,000
megapascals	6,895

Source: The Dow Chemical Company.

Cure Mechanism. The free-radical cure mechanism of the vinyl ester resins is well understood. In most respects, it is similar to that of the unsaturated polyester resins. To initiate the curing process, it is necessary to generate free radicals within the resin mass. Organic peroxides are the most common source of free radicals. These peroxides will decompose under the influence of elevated temperatures or chemical promoters, e.g., organometallics or tertiary amines, to form free radicals. Generation of

[1] The first literature reference was a patent issued in 1962 for a tooth-filling compound. Commercialization did not start until the late 1960s.

free radicals also can be effected by ultraviolet or high-energy radiation applied directly to the resin system. The free radicals thus formed react to open the double bond of the vinyl group. Once opened, the resin vinyl group is highly reactive and rapidly combines with several more vinyl groups available from both the unreacted resin and the monomer. This exothermic reaction is rapidly carried to completion, forming a 3-dimensional thermosetting network.

Applications. As might be expected from the wide variation in resin properties that can be built into the molecule, the vinyl ester resins find many applications. Chief among these are fiberglass-reinforced plastics, where the inherent characteristics of the vinyl ester resin provide a cost and performance advantage over other materials. The largest application is in the manufacture of corrosion-resistant reinforced plastic structures. Because of the reduced number of ester groups within the resin structure, the corrosion-resistant vinyl ester resins are less prone to attack by hydrolysis than the bisphenol A-fumaric acid polyesters. In addition, the resilience of the vinyl ester resins (4–6% ultimate tensile elongation) results in a fabricated part which is less prone to damage during shipping, field erection, and service.

All fabrication methods commonly used in the manufacture of reinforced plastics can be used with the vinyl ester resins. In those applications, such as filament winding and bag molding, where fast wetting is important, significant increases in output can be realized. One technique to which vinyl ester resins are particularly suited is the use of sheet molding compound (SMC). Here, the resin system (usually with a high filler loading) is combined in sheet form with the glass reinforcement and chemically thickened through the use of metal oxides, such as MgO. The SMC then is molded, usually in a matched-die molding operation, to give the desired final product, e.g., automotive parts, appliance housings, electrical structures, and panel configurations. Vinyl ester resins diluted with vinyl toluene monomer are used in the production of high-temperature electrical laminating systems. See Structures 1 and 2.

P.H. COOK, Dow Chemical U.S.A., an operating unit of The Dow Chemical Company, Freeport, TX.

VINYLPYRIDINES.   See **Pyridine and Derivatives**.

VIPER FISHES *(Osteichthyes)*.   Of the order *Isospondyli*, family *Chauliodontidae*, viper fishes have an unattractive, snake-like appearance, which is dramatized by their large fanglike teeth. The three species all inhabit deep ocean waters, ranging from very cold to tropical. The longest of these species attains a length of about 10 inches (25 centimeters). In catching its prey, the fish swims with its mouth wide open and upper fangs extended so as to spear the victim. Studies have shown that these fishes engage in vertical migrations, moving upward some 1500 feet (450 meters) at night, but not known to surface.

VIPERINE SNAKES.   See Snakes.

VIRAL HEMORRHAGIC FEVERS.   The term viral hemorrhagic fever (VHF) refers to a group of illnesses that are caused by several distinct families of viruses. While some types of hemorrhagic fever viruses can cause relatively mild illnesses, many of these viruses cause severe, life-threatening disease.

The Special Pathogens Branch (SPB) primarily works with hemorrhagic fever viruses that are classified as biosafety level four (BSL-4) pathogens. A listing of these viruses appears in Table 1. The Division of Vector-Borne Infectious Diseases, also in the National Center for Infectious Diseases, works with the non-BSL-4 viruses that cause two other hemorrhagic fevers, dengue hemorrhagic fever and yellow fever.

VHFs are caused by viruses of four distinct families: arenaviruses, filoviruses, bunyaviruses, and flaviviruses. Each of these families share a number of features:

- They are all RNA viruses, and all are covered, or enveloped, in a fatty (lipid) coating.
- Their survival is dependent on an animal or insect host, called the natural reservoir.
- The viruses are geographically restricted to the areas where their host species live.
- Humans are not the natural reservoir for any of these viruses. Humans are infected when they come into contact with infected hosts. However, with some viruses, after the accidental transmission from the host, humans can transmit the virus to one another.

Structure 1.   Typical Vinyl Ester Resin.

Structure 2.   Typical Unsaturated Polyester Resin.

TABLE 1. HEMORRHAGIC FEVER VIRUSES

Arenaviruses	Bunyaviruses	Filoviruses	Flaviviruses
Argentine	Crimean-Congo hemorrhagic fever (CCHF)	Ebola hemorrhagic fever	Tick-borne
Hemorrhagic fever			Encephalitis
Bolivian	Rift Valley fever	Marburg hemorrhagic fever	Kyasanur Forest disease
Hemorrhagic fever			
Sabia-associated	Hantavirus Pulmonary syndrome (HPS)		Omsk hemorrhagic fever
Hemorrhagic fever			
Lassa fever	Hemorrhagic fever with renal syndrome (HFRS)		
Lymphocytic Choriomeningitis (LCM)			
Venezuelan hemorrhagic fever			

- Human cases or outbreaks of hemorrhagic fevers caused by these viruses occur sporadically and irregularly. The occurrence of outbreaks cannot be easily predicted.
- With a few noteworthy exceptions, there is no cure or established drug treatment for VHFs.

In rare cases, other viral and bacterial infections can cause a hemorrhagic fever; scrub typhus is a good example.

Viruses associated with a majority of the VHFs are zoonotic. This means that these viruses naturally reside in an animal reservoir host or arthropod vector. They are totally dependent on their hosts for replication and overall survival. For the most part, rodents and arthropods are the main reservoirs for viruses causing VHFs. The multimammate rat, cotton rat, deer mouse, house mouse, and other field rodents are examples of reservoir hosts. Arthropod ticks and mosquitoes serve as vectors for some of the illnesses. However, the hosts of some viruses remain unknown, Ebola and Marbur viruses are well-known examples.

The viruses that cause VHFs are distributed over much of the globe. However, because each virus is associated with one or more particular host species, the virus and the disease it causes are usually seen only where the host species live(s). Some hosts, such as the rodent species carrying several of the New World arenaviruses, live in geographically restricted areas. Therefore, the risk of getting VHFs caused by these viruses is restricted to those areas. Other hosts range over continents, such as the rodents that carry viruses which cause various forms of Hantavirus pulmonary syndrome (HPS) in North and South America, or the different set of rodents that carry viruses which cause hemorrhagic fever with renal syndrome (HFRS) in Europe and Asia. A few hosts are distributed nearly worldwide, such as the common rat. It can carry Seoul virus, a cause of HFRS; therefore, humans can get HFRS anywhere where the common rat is found.

Although people usually become infected only in areas where the host lives, occasionally people become infected by a host that has been exported from its native habitat. For example, the first outbreaks of Marburci hemorrhagic fever, in Marburg and Frankfurt, Germany, and in Yugoslavia, occurred when laboratory workers handled imported monkeys infected with Marburg virus. Occasionally, a person becomes infected in an area where the virus occurs naturally and then travels elsewhere. If the virus is a type that can be transmitted further by person-to-person contact, the traveler could infect other people. For instance, in 1996, a medical professional treating patients with Ebola hemorrhagic fever (Ebola HF) in Gabon unknowingly became infected. When he later traveled to South Africa and was treated for Ebola HF in a hospital, the virus was transmitted to a nurse. She became ill and died. Because more and more people travel each year, outbreaks of these diseases are becoming an increasing threat in places where they rarely, if ever, have been seen before.

Viruses causing hemorrhagic fever are initially transmitted to humans when the activities of infected reservoir hosts or vectors and humans overlap. The viruses carried in rodent reservoirs are transmitted when humans have contact with urine, fecal matter, saliva, or other body excretions from infected rodents. The viruses associated with arthropod vectors are spread most often when the vector mosquito or tick bites a human, or when a human crushes a tick. However, some of these vectors may spread virus to animals, livestock, for example. Humans then become infected when they care for or slaughter the animals.

Some viruses that cause hemorrhagic fever can spread from one person to another, once an initial person has become infected. Ebola, Marburg, Lassa, and Crimean-Congo hemorrhagic fever viruses are examples. This type of secondary transmission of the virus can occur directly, through close contact with infected people or their body fluids. It can also occur indirectly, through contact with objects contaminated with infected body fluids. For example, contaminated syringes and needles have played an important role in spreading infection in outbreaks of Ebola hemorrhagic fever and Lassa fever.

Specific signs and symptoms vary by the type of VHF, but initial signs and symptoms often include marked fever, fatigue, dizziness, muscle aches, loss of strength, and exhaustion. Patients with severe cases of VHF often show signs of bleeding under the skin, in internal organs, or from body orifices like the mouth, eyes, or ears. However, although they may bleed from many sites around the body, patients rarely die because of blood loss. Severely ill patients cases may also show shock, nervous system malfunction, coma, delirium, and seizures. Some types of VHF are associated with renal (kidney) failure.

Treatment

Patients receive supportive therapy, but generally speaking, there is no other treatment or established cure for VHFS. Ribavirin, an antiviral drug, has been effective in treating some individuals with Lassa fever or HFRS. Treatment with convalescent-phase plasma has been used with success in some patients with Argentine hemorrhagic fever.

With the exception of yellow fever and Argentine hemorrhagic fever, for which vaccines have been developed, no vaccines exist that can protect against these diseases. Therefore, prevention efforts must concentrate on avoiding contact with host species. If prevention methods fail and a case of VHF does occur, efforts should focus on preventing further transmission from person to person, if the virus can be transmitted in this way.

Because many of the hosts that carry hemorrhagic fever viruses are rodents, disease prevention efforts include: controlling rodent populations; discouraging rodents from entering or living in homes or workplaces; and encouraging safe cleanup of rodent nests and droppings.

For hemorrhagic fever viruses spread by arthropod vectors, prevention efforts often focus on community-wide insect and arthropod control. In addition, people are encouraged to use insect repellant, proper clothing, bednets, window screens, and other insect barriers to avoid being bitten.

For those hemorrhagic fever viruses that can be transmitted from one person to another, avoiding close physical contact with infected people and their body fluids is the most important way of controlling the spread of disease. Barrier nursing or infection control techniques include isolating infected individuals and wearing protective clothing. Other infection control recommendations include proper use, disinfection, and disposal of instruments and equipment used in treating or caring for patients with VHF, such as needles and thermometers.

In conjunction with the World Health Organization, the CDC has developed practical, hospital-based guidelines, titled *Infection Control for Viral Haemorrhagic Fevers In the African Health Care Setting*. The manual can help health-care facilities recognize cases and prevent further hospital-based disease transmission using locally available materials and few financial resources.

Scientists and researchers are challenged with developing containment, treatment, and vaccine strategies for these diseases. Another goal is to develop immunologic and molecular tools for more rapid disease diagnosis, and to study how the viruses are transmitted and exactly how the disease affects the body (pathogenesis). A third goal is to understand the ecology of these viruses and their hosts in order to offer preventive public health advice for avoiding infection. See also **Arenaviruses**; **Dengue Fever and Dengue Hemorrhagic Fever**; **Ebola Hemorrhagic Fever**; **Filoviruses**; **Lassa Fever (African Hemorrhagic Fever)**; **Lymphocytic Choriomeningitis (LCM)**; **Marburg Hemorrhagic Fever**; **Rift Valley Fever**; and **Virus**.

An alphabetical listing of common terms that frequently occur in epidemiological and health prevention articles is listed below.

Aerosol: A fine mist or spray that contains minute particles.

Antibody: Proteins produced by an organism's immune system to recognize foreign substances.

Antigen: Any substance that stimulates an immune response by the body. The immune system recognizes such substances as being foreign, and produces cellular antibodies to fight them. Antigen/antibody response is an important part of a person's immunity to disease.

Assay: A quantitative or qualitative evaluation, or test, of a substance. Frequently used to describe tests of the presence or concentration of infectious agents, antibodies, etc.

Biosafety level: Specific combinations of work practices, safety equipment, and facilities, which are designed to minimize the exposure of workers and the environment to infectious agents. Biosafety level 1 applies to agents that do not ordinarily cause human disease. Biosafety level 2 is appropriate for agents that can cause human disease, but whose potential for transmission is limited. Biosafety level 3 applies to agents that may be transmitted by the respiratory route that can cause serious infection. Biosafety level 4 is used for the diagnosis of exotic agents that pose a high risk of life-threatening disease, which may be transmitted by the aerosol route and for which there is no vaccine or therapy.

Carrier: A person or animal that harbors a specific infectious agent without visible symptoms of the disease. A carrier acts as a potential source of infection.

Case-fatality proportion: The number of cases of a disease ending in death compared to the number of cases of the disease. Usually expressed as a percentage. Deaths from other diseases are often expressed as mortality rates; SPB normally uses case-fatality proportions. This is due to the fact that rates include a time determinant, for example, 100 deaths per 1000 cases per year. However, the diseases with which SPB works break out sporadically and occur as brief epidemics.

Case-to-infection ratio or proportion: The number of cases of a disease (in humans) compared to the number of infections with the agent that causes the disease (in humans).

Cotton rat (Siqmodon hispidus): Typically found in the southeastern United States and way down into Central and South America. The cotton rat has a bigger body than the deer mouse: head and body about 5–7 inches, and another 3–4 inches for the tail. The hair is longer and coarser, of a grayish brown color, even grayish black. The cotton rat prefers overgrown areas with shrubs and tall grasses.

Deer mouse (Peromyscus maniculatus): A deceptively cute animal, with big eyes and big ears. Its head and body are normally about 2–3 inches long, and the tail adds another 2–3 inches in length. A variety of colors, from gray to reddish brown, depending on its age may be seen. The underbelly is always white and the tail has sharply defined white sides. The deer mouse is found almost everywhere in North America. Usually, the deer mouse likes woodlands, but also turns up in desert areas.

Disease: Formally speaking, a disease is the condition in which the functioning of the body or a part of the body is interfered with or damaged. In a person with an infectious disease, the infectious agent that has entered the body causes it to function abnormally in some way or ways. The type of abnormal functioning that occurs is the disease. Usually the body will show some signs and symptoms of the problems it is having with functioning. Disease should not be confused with infection.

ELISA (enzyme-linked-immunosorbent serologic assay): A technique that relies on an enzymatic conversion reaction. It is used to detect the presence of specific substances, such as enzymes, viruses, antibodies or bacteria.

Endemic: Disease that is widespread in a given population.

Enzootic: A disease that is constantly present in the animal community, but only occurs in a small number of cases.

Epidemic: The occurrence of cases of an illness in a community or region which is in excess of the number of cases normally expected for that disease in that area at that time.

Epizootic: An outbreak or epidemic of disease in animal populations.

Host: An organism in which a parasite lives and by which it is nourished.

IgG: One of many antibodies present in blood serum that is usually indicative of a recent or remote infection. IgG is most prevalent about three weeks after an infection begins.

Igm: One of many antibodies present in blood serum that is usually indicative of an acute infection.

Immunohistochemistry: A type of assay in which specific antigens are made visible by the use of fluorescent dye or enzyme markers.

Infection: The entry and development of an infectious agent in the body of a person or animal. In an apparent "manifest" infection, the infected person outwardly appears to be sick. In an inapparent infection, there is no outward sign that an infectious agent has entered that person at all. For example, although humans have become infected with Ebola-Reston, a species of Ebola virus, they have not shown any sign of illness. By contrast, in recorded outbreaks of Ebola hemorrhagic fever caused by Ebola-Zaire, another species of Ebola virus, severe illness followed infection with the virus, and a great proportion of the case-patients died. Infection should not be confused with disease.

Multimammate rat (M. elythroleucus or M. huberti): Any of several species of the genus *Mastomys*. *Mastomys* rodents breed very frequently, producing large numbers of offspring. They are numerous in the savannas and forests of West, Central, and East Africa. Some species, like the *M. huberti* prefer to live in human homes.

Multimammate rats are known carriers of Lassa virus. See also **Lassa Fever (African Hemorrhagic Fever)**.

Nosocomial infection: An infection occurring in a patient which is acquired at a hospital or other healthcare facility. Commonly called a cross infection.

Report of a disease: An official report that notifies an appropriate health authority of the occurrence of a disease in a human or in an animal. Human diseases usually are reported first to the local health authority, such as a county health department.

Reservoir: Any person, animal, arthropod, plant, soil or substance in which an infective agent normally lives and multiplies. The infectious agent primarily depends on the reservoir for its survival.

Rice rat (Orvzomys Palustris): Slightly smaller than the cotton rat, having a head and body 5–6 inches long, plus a very long, 4- to 7-inch tail. The Rice rat has short soft grayish brown fur on top, and gray or tawny underbellies. Their feet are whitish. As you might expect from the name, this rat likes marshy areas and is semiaquatic. Typically the Rice rat can be found in the southeastern United States and in Central America.

Risk: (1) The chance of being exposed to an infectious agent by its specific transmission mechanism. (2) The chance of becoming infected if exposed to an infectious agent by its specific transmission mechanism.

RT-PCR (reverse transcriptase polymerase chain reaction): Powerful technique for producing millions of copies of specific parts of the genetic code of an organism so that it may be readily analyzed. More specifically, RT-PCR produces copies of a specific region of complementary DNA that has been converted from RNA. The technique is often used to help in the identification of an infectious agent.

Surveillance of disease: The ongoing systematic collection and analysis of data and the provision of information which leads to action being taken to prevent and control an infectious disease.

Transmission of infectious agents (such as a virus): Any mechanism through which an infectious agent, such as a virus, is spread from a reservoir (or source) to a human being. Usually, each type of infectious agent is spread by only one or a few of the different mechanisms. There are several types of transmission mechanism:

Direct transmission: This type of transmission is, at base, immediate. The transfer of the infectious agent is, as the name implies, directly into the body. Different infectious agents may enter the body using different routes. Some routes by which infectious diseases are spread directly include personal contact, such as touching, biting, kissing, or sexual intercourse. In these cases the agent enters the body through the skin, mouth, an open cut or sore, or sexual organs. Infectious agents may spread by tiny droplets of spray directly into the conjunctiva (the mucus membranes of the eye), or the nose or mouth during sneezing, coughing, spitting, singing or talking (although usually this type of spread is limited to about within one meter's distance.) This is called droplet spread.

Indirect transmission: Indirect transmission may happen in any of several ways:

Vehicle-borne transmission: In this situation, a vehicle i.e., an inanimate object or material called in scientific terms a "fomite,"-becomes contaminated with the infectious agent. The agent, such as a virus, may or may not have multiplied or developed in or on the vehicle. The vehicle contacts the person's body. It may be ingested (eaten or drunk), touch the skin, or be introduced internally during surgery or medical treatment. Examples of vehicles that can transmit diseases include cooking or eating utensils, bedding or clothing, toys, surgical or medical instruments (like catheters) or dressings. Water, food, drinks (like milk) and biological products like blood, serum, plasma, tissues or organs can also be vehicles.

Vector-borne transmission: When researchers talk about vectors, often they are talking about insects, which as a group of invertebrate animals carry a host of different infectious agents. (However, a vector can be any living creature that transmits an infectious agent to humans.)

Vectors may mechanically spread the infectious agent, such as a virus or parasite. In this scenario the vector, for instance a mosquito, contaminates its feet or proboscis ("nose") with the infectious agent, or the agent passes through its gastrointestinal tract. The agent is transmitted from the vector when it bites or touches a person. In the case of an insect, the infectious agent may be injected with the insect's salivary fluid when it bites. Or the insect may regurgitate material or deposit feces on the skin, which then enter a person's body, typically through a bite wound or skin that has been broken by scratching or rubbing.

In the case of some infectious agents, vectors are only capable of transmitting the disease during a certain time period. In these situations, vectors play host to the agent. The agent needs the host to develop and mature or to reproduce (multiply) or both, (called cyclopropagative). Once the agent is within the vector animal, an incubation period follows during which the agent grows or reproduces, or both, depending on the type of agent. Only after this phase is over does the vector become infective. That is, only then can it transmit an agent that is capable of causing disease in the person.

Airborne transmission: In this type of transmission, infective agents are spread as aerosols, and usually enter a person through the respiratory tract. Aerosols are tiny particles, consisting in part or completely of the infectious agent itself, which become suspended in the air. These particles may remain suspended in the air for long periods of time, and some retain their ability to cause disease, while others degenerate due to the effects of sunlight, dryness or other conditions. When a person breathes in these particles, they become infected with the agent-especially in the alveoli of the lungs. How do infectious aerosols get into the air? Small particles of many different sizes contaminated with the infective agent may rise up from soil, clothes, bedding or floors when these are moved, cleaned, or blown by wind. These dust particles may be fungal spores (infective agents themselves) tiny bits of infected feces, or tiny particles of dirt or soil that have been contaminated with the agent.

Droplet nuclei can remain in the air for a long time. Droplet nuclei are usually the small residues that appear when fluid emitted from an infected host evaporates. In the case of the virus causing hantavirus pulmonary syndrome, the rodent carriers produce urine. The act of spraying the urine may create the aerosols directly, or the virus particles may rise into the air as the urine evaporates. In other situations, the droplets may occur as an unintended result of mechanical or work processes or atomization by heating, cooling, or venting systems in microbiology laboratories, autopsy rooms, slaughterhouses, or elsewhere.

Both kinds of particles are very tiny. Larger droplets or objects that may be sprayed or blown, but that immediately settle down on something rather than remaining suspended, are not considered to belong to the airborne transmission mechanism. Such sprays are considered direct transmission.

Vector: A carrier which transmits infective agent from one host to another.

Virus: A virus is an extremely tiny infectious agent that is only able to live inside a cell. Basically, viruses are composed of just two parts. The outer part is a protective shell made of protein. This shell is often surrounded by another protective layer or envelope, made of protein or lipids (fats). The inner part is made of genetic material, either RNA or DNA.

A virus does not have any other structures (called organelles) that living cells have, like a nucleus or mitochondria. These organelles are the tiny organs that maintain a cell's metabolism (life processes). A virus has no metabolism at all.

Because a virus lacks organelles, it cannot reproduce by itself. To reproduce, a virus invades a cell within the body of a human or other creature, called the host. Each type of virus has particular types of host creatures and host cells that it will invade successfully.

Once within the host cell, the virus uses the cell's own organelles to produce more viruses. In essence, the virus forces the cell to replicate the virus' own genetic material and protective shell. Once replicated, the new viruses leave the host cell and are ready to invade others. See also **Cell (Biology)**; and **Virus**.

White-footed mouse (Peromyscus leucopus): This mouse is hard to distinguish from the deer mouse. The head and body together are about four inches long. The tail is normally shorter than its body (about 2–4 inches long). The fur White-footed mouse ranges from pale brown to reddish brown, while its underside and feet are white. The white-footed mouse is found through southern New England, the Mid-Atlantic and Southern states, the Midwestern and Western states, and Mexico. It prefers wooded and brushy areas, although sometimes it will live in more open ground.

Zoonotic disease or infection: An infection or infectious disease that may be transmitted from vertebrate animals (such as a rodent) to humans.

Additional Reading

Fields, M.B.N., D.M. Knipe, P.M. Howley, and R.M. Chanock: "Fields Virology," 3rd Edition, Lippincott Williams & Wilkins, Philadelphia, PA, 1996.

Galasso, G.J., T.C. Merigan, and R.J. Whitley: "Antiviral Agents and Human Viral Diseases," 4th Edition, Lippincott Williams & Wilkins, Philadelphia, PA, 1997.

Gubler, D.J. and G. Kuno: "Dengue and Dengue Hemorrhagic Fever," CAB International, New York, NY, 1998.

Leland, S.S. and S. Ozmat: "Clinical Virology," W.B. Saunders Company, Philadelphia, PA, 1996.

Love, C.B. and P.B. Jahrling: "Viral Hemorrhagic Fever," DIANE Publishing Company, Collingdale, PA, 1996.

Pattison, J.R., J.E. Banatvala, and A.J. Zuckerman: "Principles and Practice of Clinical Virology," 4th Edition, John Wiley & Sons, Inc., New York, NY, 2000.

Richman, D.D., R.J. Whitley, and F.G. Hayden: "Clinical Virology," Harcourt Brace & Company, San Diego, CA, 1998.

Salvato, M.S.: "The Arenavirdiae," Kluwer Academic Publishers, Norwell, MA, 1993.

Voyles, B.A.: "The Biology of Viruses," Mosby-Year Book, Inc., St. Louis, MO, 1993.

Center for Disease Control and Prevention (CDC), Atlanta, GA.

VIRGA.   See **Precipitation and Hydrometeors**.

VIRGO.   One of the earliest named constellations; the sixth sign of the zodiac. References to Virgo are found in every known literature and are always connected in some manner with a maiden and the harvest. Among the Egyptians, Virgo was associated with Isis, who was said to have formed the Milky Way by dropping innumerable wheat heads in the sky.

Astronomically, the constellation is famous for the large cluster of galaxies found in it. Sir William Herschel found no less than 323 galaxies in this part of the sky, and more recent observations have raised the number to more than 500. It also contains a large number of variable stars and the well-known bright star, Spica. (See map accompanying entry on **Constellations**.) See also **Galaxy**.

VIRTUAL WORK.   See **Work (Virtual)**.

VIRUS.   Viruses are considered to be the smallest infectious agents capable of replicating themselves inside eukaryotic or prokaryotic cells. The majority of these extremely small infectious particles fall within a size range of about 0.02–0.25 micrometer and can only be visualized directly with the aid of an electron microscope.

In 1898, Loeffler and Frosch demonstrated that foot-and-mouth disease of cattle could be transferred by material passed through a filter capable of excluding bacteria. This new group of "organisms" subsequently became known as *filterable viruses*. Years of debate have centered around the question of whether viruses are living or nonliving and, although resolution of this is now considered to be simply a problem of semantics, several fundamental differences distinguish viruses from other organisms. Viruses, unlike true microorganisms, are not cells, do not replicate by binary fission, and contain a genome consisting of only one type of nucleic acid (DNA or RNA, double or single stranded). They contain no organelles, such as mitochondria or ribosomes (except for the Arena viruses, which contain cellular ribosomes). Some virions contain special enzymes, such as transcriptase required for initiation of the vital growth cycle, not present in host cells.

Smaller than viruses, however, are the particles known as *viroids*, which are nothing more than very short strands of RNA uncoated by protein as is a normal virus. Viroids are known to be responsible for several plant diseases and may also be involved in animal and rare human nerve diseases. Very little is known about viroids except that when they are introduced into a host cell they replicate without the assistance of a helper virus. They are not translated into proteins and, presumably, their replication must rely entirely upon the enzyme systems of the host.

Viruses are obligatory intracellular parasites and, as such, cannot replicate in cell-free media. Therefore, the study of viruses is normally carried out with cultured cells. These are classified as: (1) *Primary cell cultures*; (2) *diploid cell strains*; and (3) *continuous cell lines*. Primary cell cultures are developed on tissue freshly removed from a plant or animal, contain several cell types capable of supporting replication of a wide range of viruses, but are limited to only a few cycles of cell division in vitro. Diploid cell strains contain cells of a single type which retain their original diploid chromosome number and are capable of up to about 100 divisions in vitro. Continuous cell lines consist of transformed or dedifferentiated cells of a single type which bear little resemblance to normal cells of that type. Continuous cell lines are capable of indefinite propagation in vitro.

Viral Structure. The mature virus particle, referred to as the *virion*, consists of a nucleic acid molecule(s) surrounded by a protein coat, the *capsid*. The capsid is composed of a number of *capsomeres* comprising one or more polypeptide chains and in some viruses surrounds a protein *core*. The capsid and enclosed nucleic acid together constitute the *nucleocapsid*. The viral capsid symmetry is characteristic of groups of virus and may be icosahedral (cubic) or helical. The icosahedron has 12 vertices and 20 faces, each an equilateral triangle. Nucleocapsids exhibiting helical symmetry consist of capsomeres and nucleic acid wound together into a spiral or helix. However, regardless of capsid symmetry, the actual virion may appear to be round, brick, or bullet-shaped. Some icosahedral and all helical viruses are enclosed in an outer *envelope* composed of lipoprotein, which is derived directly from the virus-modified cellular membrane during release of the nucleocapsid from the infected cell by a process called *budding*. Enveloped viruses can usually be inactivated by ether, chloroform, or bile salts.

Nucleic acid extracted from purified virus using phenol or dodecyl sulfate is easily destroyed by the homologous nucleases present in normal sera or tissues. DNA is destroyed by the enzyme deoxyribonuclease; RNA by ribonucleases. This provides one means of identifying the type of nucleic acid. The intact virus is not affected by these enzymes.

Bacteriophage, a virus infecting bacterial cells, has a structure somewhat different from those previously described. A *head* contains the nucleic acid and the viral DNA passes through a *tail* during the infection process. In the T-even phages (Fig. 1), the tail consists of a tube surrounded by a *sheath* and is connected to a thin *collar* at the head end and a plate at the tip end. The sheath is capable of contraction and the plate possesses *pins* and *tail fibers*, which are the organs of attachment of the bacteriophage to the wall of the host cell.

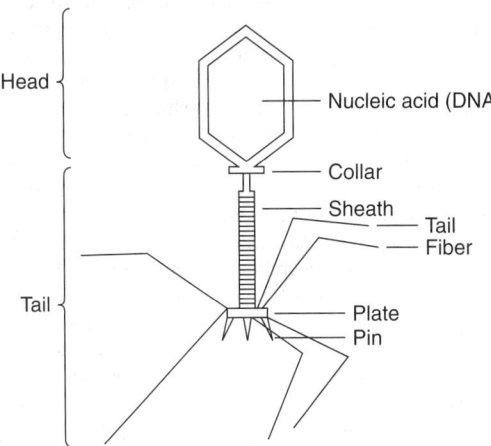

Fig. 1. T-even bacteriophage.

Some strains of the bacterium *Escherichia coli* harbor a dormant virus called lambda, which consists of a long molecule of DNA enclosed in protein. Exposure of such infected bacteria to ultraviolet light suddenly "switches on" these inactive lambda. The viruses proliferate and some 45 minutes after irradiation the bacteria burst, yielding a crop of new virus particles. If the bacteria are not irradiated, they grow normally, and rarely give rise spontaneously to viruses.

Viral Replication. In contrast to eukaryotic and prokaryotic cells, which multiply by binary fission, viruses multiply by synthesis of their separate components, followed by assembly. Several stages are involved in viral replication:

(1) *Attachment or adsorption*. The virus becomes attached to the cell via specific receptors. Thus, cells lacking the receptors are resistant to attack.

(2) *Penetration*. With enveloped viruses, this step occurs when the virion's envelope fuses with the cellular membrane. Naked virions penetrate intact through the cellular plasma membrane and into the cell cytoplasm. Viruses may also enter the cell by cellular phagocytosis. Ordinarily, without a protein coat, viral nucleic acid is incapable of entering a cell, showing the importance of the coat in infectivity. The efficiency of infection with naked nucleic acid can be increased by the presence of basic polymers, such as DEAE-dextran, or by pretreating cells with hypertonic salt solution. Even under the most favorable conditions, however, the efficiency of infection is not more than 1% that of the corresponding intact virions.

(3) *Uncoating and eclipse*. Uncoating is detected by the lability of viral nucleic acids to nuclease after the artificial disruption of the cell. Eclipse is recognized by loss of infectivity of intracellular virions recovered from disrupted cells.

Once inside the cell, virulent viruses turn off cellular macromolecular synthesis and disaggregate cellular polyribosomes, thus favoring a shift to viral synthesis. These viruses cause the ultimate destruction of the infected cell. In contrast, moderate viruses may stimulate host DNA, mRNA, and protein synthesis—a phenomenon which may be of considerable importance in viral carcinogenesis.

In general, the DNA viruses multiply in the nucleus of the host cell. The viral DNA is transcribed in the nucleus and the resultant mRNA translated into proteins on cytoplasmic ribosomes. Depending upon the virus type,

"early" or "late" proteins may be synthesized. These proteins may function as enzymes in replication of the viral DNA, as structural components of progeny virions, or as regulatory proteins. Replication of the viral DNA is semiconservative and, in general, depends upon viral proteins.

RNA viruses usually replicate in the cytoplasm and can be divided into five classes according to the nature of the RNA in the virion. *Class I* viruses contain a molecule of single-stranded RNA which acts as mRNA to be translated into viral proteins. The RNA is said to have plus *strand polarity*. The picornaviruses are an example. *Class II* viruses (e.g., paramyxoviruses) have a molecule of single-stranded RNA which cannot act as mRNA (minus strand polarity). A virion transcriptase synthesizes several complementary messenger molecules from which viral proteins are translated. *Class III* viruses (e.g., myxoviruses) contain single-stranded RNA of minus strand polarity, present in seven or more segments. A virion transcriptase transcribes each segment into a complementary messenger. *Class IV* viruses contain ten segments of double-stranded RNA which is transcribed into mRNA by a viral transcriptase. Representative of this group are the reoviruses. *Class V* viruses (e.g., leukoviruses) contain segmented single-stranded RNA of messenger polarity. Each RNA segment is transcribed into DNA by a *reverse transcriptase* present in the virion; mRNA is then transcribed from the DNA.

(4) *Assembly and release*. The assembly of the capsid and its association with nucleic acid is then followed by release of the virus from the cell. This may occur in different ways, depending upon the nature of the virus. Naked viruses may be released slowly and extruded without cell lysis, or released rapidly by disruption of the cell membrane. DNA viruses, which mature in the nucleus, tend to accumulate within infected cells over a long period. Enveloped viruses generally acquire their envelope and leave the cell by budding through the nuclear or cytoplasmic membrane at a point where virus-specified proteins have been inserted. The budding process is compatible with cell survival.

Viral Classification

Several methods of viral classification are in use. Classification based upon epidemiological criteria, such as enteric or respiratory viruses, is useful, but of more significance are schemes based upon the morphology of the virion (symmetry, envelope, etc.) and type of nucleic acid (DNA, RNA, number of strands, polarity, etc.).

The two groups of viruses, RNA and DNA, are further divided according to size, morphology, and biological and chemical properties. Thus, the icosahedral RNA viruses that are ether stable are divided into the picornaviruses and the reoviruses. The name picornavirus comes from *pico* (meaning very small) and *rna* (indicating the type of nucleic acid). Included in the group are enteroviruses, such as polio, Coxsackie, foot-and-mouth, and echoviruses, among others, and also the rhinoviruses. The picornavirus capsid consists of 60 subunits each made up of four proteins, which change by mutation to yield antibody-resistant strains of cold and polio viruses. The reoviruses (*r*espiratory *e*nteric *o*rphan virus) cause inapparent infection in humans and other animals, and their relationship to spontaneous disease is uncertain. They are morphologically similar to the wound-tumor virus of clover, and a small cross-activity with this virus by means of complement fixation has been reported.

The arboviruses are those which multiply in both vertebrates and arthropods. The former serve as reservoirs and the latter primarily as vectors. Arbovirus is a somewhat arbitrary epidemiological classification, which contains several heterogeneous groups. The togaviruses contain such entities as Eastern and Western equine encephalitis viruses (EEE and WEE) and dengue and yellow fever viruses, which have mosquitoes as vectors. The arenaviruses comprise such agents as Lassa, Tacaribe, and lymphocytic choriomeningitis viruses. Arboviruses are dangerous and difficult to study. They appear to contain single-stranded RNA (positive polarity), are ether sensitive, and are relatively unstable. The capsids are suggestive of icosahedral symmetry.

Myxoviruses, orthomyxovirus, and paramyxoviruses are spherical or filamentous, enveloped single-stranded RNA viruses. The myxovirus group contains the influenza viruses which, in turn, have been separated into three distinct antigenic types, designated A, B, and C. The genome of myxoviruses, unlike that of the paramyxoviruses, is segmented. It is this characteristic that is responsible for the devastating influenza pandemics which have occurred periodically. Influenza A viruses have undergone three major antigenic shifts since 1933, and each new variant is able to successfully infect populations of individuals immune only to preexisting types. In the influenza pandemic of the winter of 1917–1918, over 20 million persons died worldwide, with better than one-half million fatal cases in the United States. Over 50 million cases of influenza were reported in the United States in the 1968–1969 winter. These cases were attributed to a hitherto unknown variant, first isolated in Hong Kong (hence named "Hong Kong flu"). Some 20,000 and possibly as many as 80,000 deaths resulted from this influenza invasion and the side effects it produced. In the 1972–1973 winter season, a much milder and minor variant, called the "London flu," caused well over 2000 deaths in the United States, particularly from complications such as pneumonia. When combined with influenza, pneumonia is the fifth most serious public health problem in the United States. In terms of absenteeism, it is the number one problem.

Only Type A influenza virus has been found to be capable of producing *pandemics*. The influenza A virus is identified as a medium-size RNA virus, some 110 nanometers in diameter and delimited by a membrane of lipids and polysaccharides derived from the host cell and virus-specific protein. Five distinct proteins have been identified, three of which are inside the virion. A schematic representation of the influenza virus emerging from a cell is given in Fig. 2.

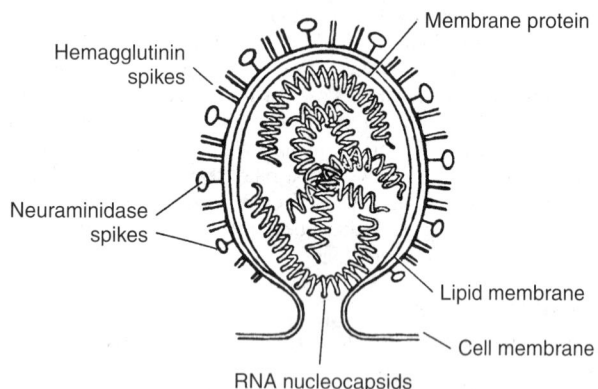

Fig. 2.   Schematic representation of an influenza virus emerging from a cell. (*After Kilbourne.*)

It has been reported that the antigenic shifts are manifested in hemagglutinin and neuraminidase, two glycoproteins found on the surface of the influenza virion. It is suggested that the hemagglutinin binds the virus to t he target cell and when the hemagglutinin function is inhibited (as by an antibody), the virus is no longer infective. It is believed that the neuraminidase cleaves a glycoside bond in the host membrane. This action frees the newly formed virus from the cell and, if inhibited, will not reduce the infectivity of the virus, but will deter the spread of virus particles to other cells.

The emergence of new influenza subtypes appears to be too abrupt to be explained fully by conventional concepts of mutation and the full story may rest in the very nature of the segmented viral genome. If a host cell is infected by two different subtypes of influenza virus at the same time, the genes from the subtype may undergo random reassortment in the cell, resulting not only in production of the two original subtypes, but in production of one or several other subtypes as well. Each hybrid, of course, will have a different but full set of genes and recombination within the infected host can explain the large mutations that occur about once every decade. However, it is well established that only one influenza subtype can exist in humans at any given time. Also, that the emergence of a new subtype, such as the Hong Kong strain, is usually accompanied by the abrupt disappearance of the antecedent subtypes—thus allowing little, if any, opportunity for recombinations to occur within human cells. Of considerable interest, however, is the fact that several virus strains can exist simultaneously within animal hosts. In animals, the appearance of a new influenza virus strain is not necessarily accompanied by the disappearance of previously recognized strains. It has been established that there are at least two discrete subtypes of equine influenza, eight or more avian strains and two subtypes in swine. Thus, the postulation that recombination occurs in animals which share the general environs with humans. Some evidence of this may derive from the fact that most new subtypes appear to originate in Asia, where animals and humans commonly inhabit the same building.

The paramyxovirus group includes the causative viruses of mumps, measles, parainfluenza, Newcastle disease, canine distemper, and several other diseases. These viruses are generally larger than the myxoviruses, are enveloped and pleomorphic, and contain one molecule of single-stranded RNA.

The rhabdoviruses, causative agents of rabies in humans and other animals, are also enveloped and contain a single strand of RNA. A peculiar bullet-shaped morphology disguises the helical nucleocapsid.

The human immunodeficiency virus (HIV) which induces AIDS is a retrovirus. Its genetic material is RNA and it carries with it a reverse transcriptase which catalyzes transcription of viral RNA into double helical DNA that then integrates into the genome of the infected cell where it is known as a *provirus*. Transcription of this provirus produces new viral RNA and proteins. One characteristic of HIV is that its genome is significantly more complex than that of other known retroviruses; it possesses at least seven types of genes instead of the normal three. The virus replicates by budding off from a T lymphocyte to become a free infectious virus. See also **Immune System and Immunology.**

The chemical nature of many viruses, which either do not grow well or do not lend themselves to purification, is unknown. The Riley lactic dehydrogenase virus is a nonpathogenic virus, which is recognized only by an increase in lactic dehydrogenase in the blood of infected mice. A lipovirus described by Chang causes marked degradation of infected cells and releases a lipogenic toxin dissociable from infectivity, which is capable of inducing fatty degeneration in other uninfected cells. A marked increase in the gamma globulin fraction of blood serum of mink infected with Aleutian mink disease is an indication of infection with a virus which causes a color change in the fur and often sickness and death.

Several groups of viruses, of importance in human disease, contain DNA. The adenoviruses, named for their original isolation from adenoid tissue, contain double-stranded DNA, have icosahedral symmetry and lack and envelope. This group of viruses that multiply in the nucleus of infected cells is usually associated with respiratory tract and eye infections, although it is now apparent that adenoviruses are not the etiological agents for the majority of acute viral respiratory infections. Although adenoviruses exhibit marked oncogenic (tumor causing) potential in animals, they are probably not oncogenic for humans.

The adenoviruses contain at least three protein moieties, and certain types are capable of inducing one or more new host antigens, such as tumor (T) antigens, the chemistry of which is presently unknown. The viral proteins can be separated by gel diffusion and correlated with results obtained by complement fixation. One moiety is the toxic protein that causes the host cell to degenerate. Another corresponds to the group antigen common to all 31 types of adenoviruses and the third is the type-specific protein.

The papovaviruses (*pap*-illoma, *po*-lyoma, *va*-cuolating agent, SV40) are small, nonenveloped, icosahedral viruses which also replicate in the cell nucleus. The virion contains double-stranded DNA. Apart from causing several forms of warts, this group of viruses is of interest as models for understanding mechanisms of viral carcinogenesis.

The major herpesviruses (Gr: *herpein* = to creep) that infect humans are herpes simplex (Type I: fever blisters; Type II: genital lesions), varicella (chickenpox), zoster (shingles), cytomegalovirus, and Epstein-Bar viruses. A number of viruses that infect lower animals also belong to this group of enveloped, icosahedral, double-stranded DNA viruses. Most members of this group tend to produce latent infections with periodic recurrent disease. Two examples are fever blisters caused by herpes simplex I, and shingles, the recurrent form of chickenpox. Cytomegalovirus causes a severe, often fatal, illness of newborns, usually affecting the salivary glands, brain, lungs, kidneys, and liver. Surpassing rubella virus, this is the most common viral cause of mental retardation. It has been estimated that cytomegalovirus (CMV) causes serious mental retardation of more than 3000 infants annually in the United States alone. In addition to mental retardation, the disease in infants may cause blindness and deafness. In about 90% of infants infected with CMV, the disease can be detected only through urine examination. In about 10% of the cases, the disease is typified by enlargement of the spleen and liver, blood abnormalities, and hepatitis. Microcephaly (abnormally small head) is also sometimes an indication. CMV causes enlargement of the affected cells (cytomegaly). The disease is found throughout the world and it is believed that congenital infections result from a primary infection of the mother during pregnancy. CMV, like herpesviruses, probably persist in a latent stage for long spans of time. Immunosuppressed patients, such as those suffering from cancer or recipients of organ transplants, are also prone to infections with CMV.

The Epstein-Barr viruses play an etiological role in infectious mononucleosis, an acute infectious disease that affects lymphoid tissue throughout the body. A strong association of this virus with Burkitt's lymphoma and perhaps nasopharyngeal carcinoma also has been observed.

The poxviruses are the largest and most complex viruses of vertebrates and contain a large, double-stranded DNA molecule. The virions are complex, brick-shaped particles, covered by several membrane layers of viral origin. Unlike other DNA viruses of mammals, poxviruses multiply in the cell cytoplasm. They can be divided into several groups on the basis of specific antigens, morphology, and natural hosts. *Group I* consists of mammalian viruses, such as variola (smallpox), vaccinia, cowpox, ectromelia, and monkeypox. Of this group, variola or smallpox has caused the greatest human morbidity and mortality. However, because the virus has no animal reservoir, and is spread chiefly by human contact, the World Health Organization was able to announce in 1980 that, because of massive immunization campaigns, smallpox has been completely eradicated. Since that announcement, remaining stores of the virus have been destroyed to prevent laboratory accidents, such as the one in 1979, which took the life of a scientist. *Group II* comprises the tumor-producing viruses, the fibroma and myxoma viruses.

The hepatitic viruses appear to fall into two different groups of small, icosahedral DNA viruses. *Type A* causes infectious hepatitis and is transmitted through the oral-intestinal route. *Type B* is transmitted by injection, usually of infected blood or its products.

Slow Viruses

During the last decade or two, there has been increasing speculation and some tentative evidence that so-called *slow viruses* may be operative and may be the underlying causes of a number of degenerative diseases, long poorly understood, such as multiple sclerosis and rheumatoid arthritis, among others. More recently, there have been increasing postulations of an association between viruses and diabetes. In fact, rather positive identification of slow viruses with some rare diseases has been established. Most investigators caution that the term "slow" should not necessarily be fully interpreted in terms of a virus per se, but equally if not completely with the manifestations of the virus. So-called slow virus infections are characterized by a long incubation period, followed by a protracted course of disease. The slowness may arise in some cases from the virus itself, but the slow pace also may be the result of weak but prolonged interactions between the virus and the host's immune system. It is also possible that these characterizations of slowness may not be attributable to viruses at all, but to some other unknown causative factors. Obviously as of this juncture, investigators are following a source of suspicion rather than a chain of hard evidence. Nevertheless, the case for the slow viruses is becoming increasingly convincing. The causative agents for at least four rare diseases, two in humans and two in animals, are sometimes referred to as "unconventional viruses."

One of these diseases in humans is *kuru*, encountered only in the Fore people and their neighbors in New Guinea. The disease for many years was considered a genetic disease. However, it has been established that the disease can be transmitted to chimpanzees by injection of extracts from the brains of human kuru victims into the brains of chimpanzees. Kuru is a neurological disease with brain lesions located mainly in the gray matter. The cerebral cortex takes on a spongy appearance. The other human disease is Creutzfeld-Jakob disease, rare but of worldwide distribution. It involves the premature development of the mental deterioration sometimes seen in old age. It also has been established that it is caused by a transmissible agent that can infect chimpanzees and lower primates. One of the animal diseases referred to is *scapie*, known for over two centuries as a fatal disease among sheep. The other animal disease is *transmissible mink encephalopathy*, first discovered in Wisconsin in the late 1940s. A puzzling aspect of the unconventional slow viruses is the fact that they cannot be observed with an electron microscope. Another puzzling aspect is their apparent lack of antigenicity. Although it has not been possible to demonstrate that any of these four "agents" will evoke production of antibodies, recent work has found fibrils in the brains of infected animals that are believed to be specific markers for the "unconventional" slow viruses and may indeed be the etiological agent. These unconventional slow viruses are not destroyed by ultraviolet radiation, and they are highly resistant to treatment with formalin or heat, but infectivity is destroyed by phenol or

ether. Some investigators believe that these agents may incorporate a very small nucleus of the size range of the viroids (self-replicating infectious RNA molecules known to produce certain plant diseases).

Two slow infections of the human central nervous system—*progressive multifocal leukoencephalopathy* (PML) and *subacute sclerosing panencephalitis* (SSPE) are thought to be associated with conventional viruses. Although PML does not cause inflammation of the brain, it does produce demyelination, i.e., destruction of the layers of membranes surrounding nerve axons. Some investigators believe that the virus is a papovavirus (group of small viruses including human wart virus, simian virus 40, and the polyoma virus of mice). It is reasoned that in PML the virus destroys the cells needed for formation and maintenance of the myelin sheath. A conventional virus has been isolated from the brains of persons suffering from SSPE. An association between measles (in patients under two years of age) or immunization with a live measles virus vaccine and later development of SSPE has been shown. SSPE patients have unusually high titres of measles antibodies and affected brain cells have inclusions similar to those seen in measles infections.

Slow viruses are becoming increasingly suspect in the instances of much more common diseases, particularly the autoimmune diseases. An autoimmune disease may be defined as a disease wherein the immune system of the body does not direct its attack on an invading foreign substance, but instead at the body's own tissue. Many authorities consider rheumatoid arthritis and multiple sclerosis as autoimmune diseases. The precise causes of these diseases have remained obscure. Multiple sclerosis is a demyelinating disease and has variously been described as an autoimmune disease, a viral disease, or an autoimmune disease provoked by a virus. Epidemiological studies indicate that from 3 to 23 years may elapse between the time of exposure to the virus and the onset of symptoms. Further evidence points to involvement of a myxovirus. Measles virus is of this kind.

Possible Viral Connection to Diabetes. A Norwegian physician (J. Stang) in 1864 noted that diabetes developed in one of his patients within a short period after a mumps infection and was probably the first person to indicate a possible connection between viruses and diabetes. Over the years, numerous other connections have been attempted to relate diabetes with mumps, hepatitis, rubella, coxsackie, and influenza viruses, adenoviruses, enteroviruses, and cytomegalovirus. One of the presumptions made is that viruses are understood to replicate in the pancreas. Commencing in the late 1950s, more substantive evidence has been given. Reports from Sweden in 1958 link juvenile diabetes with mumps infection. Reports from New York State in 1974 relate closely the cycles of incidence of mumps and those of juvenile diabetes. The study was based upon investigation of records for the period 1946–1971. Tentative conclusions indicate an average lag period of about 3.8 years between onset of diabetes and exposure to mumps and it is reasoned that this represents the time required for the virus to produce permanent damage to the pancreas. Other investigators have statistically linked diabetes to rubella (German measles). Some authorities suggest that the pancreas, along with other embryonic organs, may be damaged by the virus that causes congenital rubella. The records of nearly 3,000 juvenile diabetics treated at King's College Hospital in London (1955–1968) have been studied, and they reveal a seasonal pattern on the onset of juvenile diabetes, striking a low incidence in June and a high incidence in October. Without presenting the details, conclusions are suggested that an association of viral infections with the juvenile form of diabetes is evident. However, the relationship, if any, has not been determined in the case of the maturity-onset form of diabetes.

Viral Diagnosis and Vaccination

Viral Diagnosis. Three major approaches to identification of viruses are commonly used: (1) *Microscopy*. Viruses may be observed directly by electron microscopy; viral antigens may be recognized in infected tissue by immunofluorescence, using virus-specific antisera; virus-induced pathology may be identified by light microscopy. (2) *Virus isolation*. Provisional viral identification may be based upon cytopathic effects produced in cell cultures infected with virus present in tissues or secretions of the patient. (3) *Serology*. Antibodies specific for a particular virus may be identified in a patient's serum. A very sensitive, accurate, and recently developed diagnostic approach is radioimmunoassay, which involves the use of an isotope-labelled antibody or antigen.

Viral Vaccination. Vaccines, agents that elicit a specific antiviral immune response, have been very successful against smallpox, measles, rubella, poliomyelitis, and yellow fever, all of which are generalized diseases. Vaccines against diseases caused by respiratory tract viruses, where great antigenic diversity is found, have been less effective.

Vaccines may be prepared by rendering viruses harmless without affecting their immunogenicity. This can be done by either inactivating the virus, or by selecting avirulent mutants. The most successful vaccines are "living" avirulent viruses, which possess the advantage of multiplying in the host and which usually require only a single dose to be effective. This leads to prolonged immunological stimulation similar to that which occurs in natural infection. Live vaccines, however, are subject to a number of problems, such as genetic instability and contamination by extraneous viruses. Inactivated viruses are usually produced by treatment with formaldehyde, which destroys their infectivity. The major difficulty with inactivated viruses is the administration of sufficient viral antigen to induce a lasting immunity. In many cases, several injections must be given over a substantial period of time. The only inactivated viral vaccine in widespread use in humans is the influenza vaccine. The inactivated Salk polio vaccine has been largely replaced by the attenuated live-virus Sabin vaccine.

Interferon. Interferons (IFN) are proteins that evert virus nonspecific antiviral activities in cells through metabolic processes involving synthesis of both DNA and protein. The number of interferon-inducing substances has increased to include not only all of the major virus groups, but also bacterial and fungal products, nucleic acids, polymers, mitogens, and various low-molecular-weight substances. However, as interferons are induced by viruses and inhibit viral replication, viruses are usually considered to be natural inducers. The ability of viruses to induce interferon production depends upon the virus type. Some viruses, such as that responsible for Newcastle disease, are good inducers, while others, such as the adenoviruses, are regarded as poor inducers. Further, the type of cell used presents another factor in interferon production. In the whole animal, cells of the reticuloendothelial system are generally considered to be the major interferon producers. Recently, interferons have been classified into types on the basis of their antigenic specificities. Alpha and beta interferons (formerly called leukocyte and fibroblast, respectively) are acid stable and correspond to what have been called *Type I* IFNs (interferons). Gamma interferons (formerly called Immune) are acid labile and correspond to *Type II* IFNs.

Although interferon has been studied extensively for over a decade, the mechanism of its antiviral activity remains unclear. Considerable evidence exists to support the concept that interferon inhibits virus-specific protein synthesis, thus blocking viral replication in cells adjacent to the infected cell producing the interferon. There is no established reason to conclude, however, that interferon exerts antiviral action through a single mechanism. Interferon is probably one of the most important early determinants of recovery from a number of viral diseases.

Recent work has centered upon use of interferon as a therapeutic agent in humans and animals. In humans, local application of monkey interferon is effective in reducing the severity of vaccinia virus skin infections. Recent results with herpes keratitis and chronic hepatitis are promising. Interferon appears to be active against oncogenic viruses in the treatment of such cancers as osteogenic sarcoma, and at present it is only the limited availability of interferon that prevents more extensive testing.

A number of virus diseases and virus related topics are described in this encyclopedia. Check alphabetical index for antiviral drugs, cancer research, chickenpox, common cold; coxsackie virus, dengue (breakbone fever), hepatitis, infectious mononucleosis, influenza, measles, mumps, Norwalk virus, poliomyelitis, rabies, Rift Valley fever, vaccinia, virus diseases (plants), and yellow fever. See also **Acquired Immune Deficiency Syndrome (AIDS)**; **Arenaviruses**; and **Filoviruses**.

Additional Reading

Ahmed, R. and I.S.Y. Chen: "Persistent Viral Infections," John Wiley & Sons, Inc., New York, NY, 1999.

Campbell, I. and M. Buchmeier: "Neurovirology: Virus and the Brain (Advances in Virus Research," Vol. 56, Academic Press, Inc., San Diego, CA, 2001.

Cann, A.J.: "Virus Culture: A Practical Approach," Oxford University Press, Inc., New York, NY, 2000.

Goode, J.: "Gastroenteritis Viruses," Vol. 238, Novartis Foundation Symposium, John Wiley & Sons, Inc., New York, NY, 2001.

Gosztonyi, I.G., M. Cooper, and R.W. Compans: "Mechanisms of Neuronal Damage in Virus Infections of the Nervous System," Springer-Verlag, Inc., New York, NY, 2001.

Maramorosch, K., F.A. Murphy, and A.J. Shatkin: "Advances in Virus Research," Vol. 54, Academic Press, Inc., San Diego, CA, 1999.

Montagnier, L.: "Virus: The Co-Discover of HIV Tracks Its Rampage and Charts the Future," W.W. Norton Company, Inc., New York, NY, 1999.

Nowak, M.A. and R. May: "Virus Dynamics: Mathematical Principles of Immunology and Virology," Oxford University Press, Inc., New York, NY, 2000.

Wagner, E.K. and M. Hewlett: "Basic Virology," Blackwell Science, Inc., Malden, MA, 1999.

Zuckerman, A.J., J.R. Pattison, and J.E. Banatvala: "Principles and Practice of Clinical Virology," 4th Edition, John Wiley & Sons, Inc., New York, NY, 2000.

ANN C. DEBALDO, Ph.D., Assoc. Prof., College of Public Health
University of South Florida, Tampa, FL.

VISCERA. Organs lying more or less freely in the cavities of the body. Usually applied to the heart and lungs as thoracic viscera and to the stomach, intestines, spleen, liver and pancreas, and some of the reproductive organs as abdominal viscera.

VISCOELASTICITY. Mechanical behavior of material which exhibits viscous and delayed elastic response to stress in addition to instantaneous elasticity. Such properties can be considered to be associated with rate effects—time derivatives of arbitrary order of both stress and strain appearing in the constitutive equation—or hereditary or memory influences which include the history of the stress and strain variation from the undisturbed state. See also **Rheology**.

VISCOSITY. A property of fluids, which appears as a dissipative resistance to flow. A solid subjected to external forces can attain a condition of elastic equilibrium with elastic stresses balancing the applied forces. A fluid subject to external forces can be in static equilibrium only if the forces are derivable from a potential function and the usual result is steady flow resisted by viscous stresses set up by distortion of fluid elements. The mechanism and nature of the viscous effect may be very different in gases and in liquids.

In a gas, viscous stresses arise from migration of molecules, which carry with them momentum relative to their starting-points. To the extent that the gas may be treated as a continuous fluid, i.e., on scales large compared with the mean free path and for time intervals large compared with the collision frequency, Newton's law of fluid friction is obeyed. The elementary form of the law is that, in simple shearing motion, the shear stress is proportional to the rate of shear. More generally, the stress tensor p_{ij} is linearly related to the rate of deformation tensor

$$S_{ij} = \frac{\partial u_i}{\partial x_j} + \frac{\partial u_j}{\partial x_i}.$$

The kinetic theory of gases leads to

$$p_{ij} = \eta \left(\frac{\partial u_i}{\partial x_j} + \frac{\partial u_j}{\partial x_i} \right) - \delta_{ij} \left(\frac{1}{3} \eta \frac{\partial u_j}{\partial x_j} + p \right)$$

where

u_i is the velocity at position x_i
p is the hydrostatic pressure
$\eta = A\rho\tilde{c}\lambda$ is the coefficient of (dynamic) viscosity
ρ is the density
$\tilde{c}$ is the mean speed of the molecules
λ is the mean free path

The constant A depends on the forces between the molecules and is nearly 0.5 for hard spherical molecules. For low densities, the viscosity is independent of pressure and proportional to the square root of the absolute temperature. For large densities, the Enskog relation describes the variation with density,

$$\frac{\eta}{\rho} = \frac{1}{2545} \left| \frac{v}{b} + 0.8000 + 0.7614 \frac{b}{v} \right| \left(\frac{\eta}{\rho} \right)_{\min}$$

where v and b mean the same as in van der Waals equation. The variation with temperature is given by Sutherland's formula,

$$\eta = \eta_0 \left(\frac{T}{273.2} \right)^{3/2} \frac{c + 273.2}{c + T}$$

where c is a constant and η_0 is the viscosity at $0\,^\circ\text{C}$.

The viscous stresses in liquids arise more from intermolecular attraction than from molecular migration and more complicated kinds of behavior are found. The concept of a liquid as an imperfect solid suggests that a liquid behaves as an elastic solid with considerable thermal creep. A simple model has a relaxation time λ, and relates stress to strain by

$$p_{ij} + \lambda \frac{\partial p_{ij}}{\partial r} = \eta_1 \left(\frac{\partial u_i}{\partial x_j} + \frac{\partial u_j}{\partial x_i} \right) - \delta_{ij} \left(\frac{1}{3} \eta_1 \frac{\partial u_j}{\partial x_j} - p \right)$$

describing a simple visco-elastic fluid. For some liquids, the relaxation time is small compared with time scales of ordinary flows, and the liquid exhibits Newtonian behavior. Many liquids, such as solutions of proteins and polymers, are non-Newtonian in behavior and show relaxation effects and even nonlinear response to deformation. The variation of viscosity of Newtonian liquids with temperature and pressure is described by the Arrhenius formulas. Characteristically, the viscosity of a liquid decreases rapidly with increase of temperature and increases with pressure. An exception is liquid helium I whose viscosity at its own vapor pressure increases with temperature.

The Newtonian coefficient of viscosity has dimensions $ML^{-1}T^{-1}$ and the cgs unit is the poise (after Poiseuille). The centipoise is frequently used for mobile liquids. English units are lbm ft^{-1} sec^{-1} and lbf ft^{-2} sec. Special units are used in engineering practice, related to specific instruments for the measurement of viscosity. For problems involving the action of viscous stresses on the fluid itself, it is convenient to use the kinematic viscosity, $v = \eta/\rho$. It may be regarded as the diffusivity of momentum.

See also **Units and Standards**.

In view of the importance of viscosity as an index of the suitability of liquids for practical purposes such as lubrication, many methods of measurement have been devised. There are four chief kinds of absolute methods, as follows:

1. *Stokes' law methods.* These involve measurement of the rate of fall of a small sphere in the liquid. The viscosity is given by

$$\eta = \frac{2gr^2(\rho - \rho_l)}{qV}$$

where r is the radius of the sphere, ρ is the density of the sphere, ρ_l is the density of the fluid and V the rate of fall. It is essential that the Reynolds number of flow $V + \rho/\eta$ should not exceed 0.1, i.e.,

$$V^2 < 0.02 \frac{\rho - \rho_l}{\rho} gr$$

2. *Capillary tube methods.* These apply the Poiseuille equation to the measurement of viscosity. Allowance must be made for end effects, and the Reynolds number of flow must be small enough to ensure laminar flow.

3. *Rotating cylinder methods.* The torque transmitted to a stationary inner cylinder when the outer one is rotated at constant speed is given by a simple equation involving the viscosity, provided the speed of rotation is not too high. The method is wellsuited for precision measurement as end-effects may be eliminated by guard-rings.

4. *Oscillating disk method.* A disk oscillating in a plane parallel and close to a plane surface undergoes a damping due to the viscosity of the intervening fluid. From the dimensions of the apparatus and the rate of damping, the viscosity may be calculated.

Most of the absolute methods have been applied in simplified form, i.e., operationally convenient but subject to corrections that must be determined by use with a fluid of known viscosity. Thus, the Engler and Saybolt viscosimeters use the time taken for a sample of liquid to flow past a constriction in a tube as a measure of its viscosity. See also **Rheology**.

VISIBILITY. In United States weather-observing practice, the greatest distance in a given direction at which it is just possible to see and identify with the unaided eye (a) in the daytime, a prominent dark object against the sky at the horizon, and (b) at night, a known, preferably unfocused, moderately intense light source. Daytime estimates of visibility are subjective evaluations of atmospheric attenuation of contrast, while nighttime estimates represent attempts to evaluate something quite different, namely attenuation of flux density.

According to United States weather-observing practice, after visibilities have been determined around the entire horizon, they are resolved into a single value of *prevailing visibility*, which is the greatest horizontal visibility equaled or surpassed throughout half, though not necessarily a continuous half, of the horizon circle. Visibility determined from a point on the ground is *surface visibility*, as opposed to *control tower visibility*

observed from an airport control tower. *Vertical visibility* is the distance that can be seen vertically into a surface-based obscuring phenomenon such as fog, rain, or snow.

Transmissometers are used to measure visibility at airports along the active runway. A transmissometer consists of a narrow light beam focused into a photoelectric cell 250 feet (75 meters) away. The output of the photocell is a measure of the light received from the spotlight, i.e. the amount of light transmitted. The attenuation of the light beam is a measure of the transmissivity of the atmosphere and is related to what the human eye (aircraft pilot's) can see. If there is no attenuation the visibility is "unlimited." If there is 100% attenuation the visibility is less than 250 feet (75 meters), the baseline distance.

Visibility obtained by the use of a transmissometer is called *Runway Visual Range* and it is intended to be the distance a pilot can see at touchdown for landing.

See also **Fog and Fog Clearing**; and **Weather Technology**.

VISION AND THE EYE. The complete visual system requires light, the eye, and a conscious observer. In its gross aspects, the visual system includes the eyes, the *extraocular muscles* which control eye position in the *bony orbit* (eye socket), the optic and other nerves that connect the eyes to the brain, and those several areas of the brain that are in neural communication with the eyes. This summary will stress the informational aspects of human vision. It should be realized, however, that no visual system could function without its *protective mechanisms* (tears and eyelids, especially) or if its normal metabolism (mediated through the vascular supply of eye and brain) should malfunction.

The visual system is particularly well adapted for the rapid and precise extraction of spatial information from a more-or-less remote external world. It does this by analyzing, in ways that are as yet imperfectly understood, the continuously changing patterns of radiant flux impinging upon the surfaces of the eyes. Much of this light is reflected from objects that must be discriminated, recognized, attended to, and/or avoided in the environment. This ability transcends enormous variations in intensity, quality, and geometry of illumination as well as vantage point of the observer. A block diagram of the visual system is given in Fig. 1.

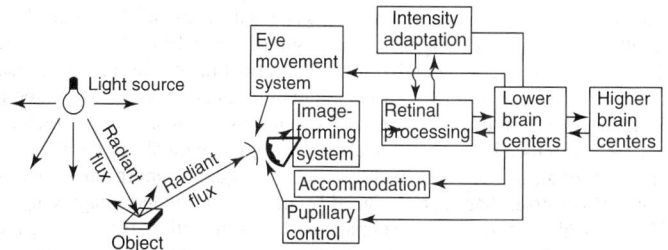

Fig. 1.   Functional components of the visual system.

Although image formation in the eye is importantly involved, the analogy between eye and photographic camera has been badly overworked and tends to create the erroneous impression that little else is needed to explain how we see. Image formation is greatly complicated by the movement of the eyes within the head, and of both eyes and head relative to the external sea of radiant energy. Such visual input is ordinarily sampled by discrete momentary pauses of the eyes called *fixations*, interrupted by very rapid ballistic motions known as *saccades* which bring the eyes from one fixation position to the next. Smooth movements of the eyes can occur when an object having a predictable motion is available to be followed. A large body of evidence suggests that the visual input is processed by the brain in "time frames" of about 100 milliseconds, although the peripheral parts of the visual system operate with much shorter time constants than this.

Each eye controls many important functions within one mobile housing; it is a device to form an image upon a vast array of light sensitive *photoreceptors*, but it also contains systems to dissect, encode, and transmit information derived therefrom. A cross section of the human eye is shown in Fig. 2. The primary refracting surface is the *cornea*, a complex yet transparent structure, which admits light through the anterior part of the outer surface of the eye. The *iris* contains muscles that alter the size of the entrance port of the eye, the *pupil*. The *crystalline lens* has a variable shape,

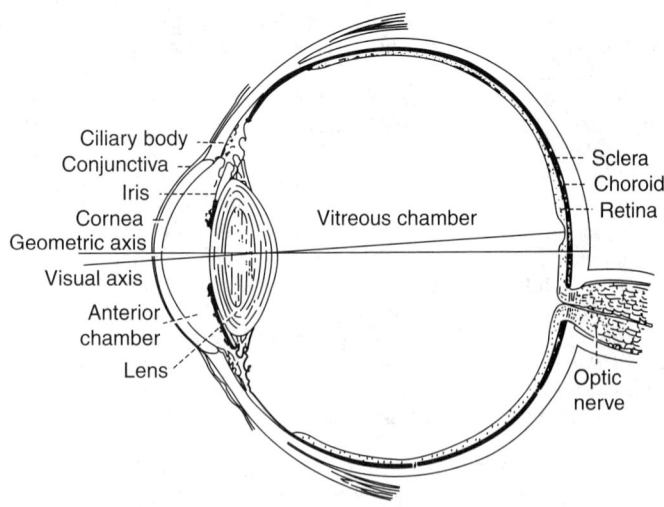

Fig. 2.   Horizontal cross section of the right eye of the human.

under the indirect control of the ciliary muscle. Since it has a refractive index higher than the surrounding media, it gives the eye a variable focal length, allowing *accommodation* to objects at varying distances from the eye. The iris muscles and the ciliary muscle, known collectively as the *intraocular* muscles of the eye, are controlled by impulses having their origins in separate but interacting centers in the brain stem. These brain centers also receive nerve impulses from the eye. These loops, and those involving the extraocular musculature and thus eye position, have some of the properties of nonlinear servosystems, and have been actively investigated as such.

Much of the remainder of the eye is filled with fluids and materials under pressure, which help the eye maintain its shape. The *aqueous humor* — thin, watery, and continuously being replaced — fills the *anterior chamber* between cornea and lens. The *vitreous humor* — thinly jellylike and of very low metabolism — fills the majority of the eye's volume. The image produced through these structures is formed upon the *retina* at the back of the eye. The retinal image is very small, because the eye itself is small and has a short posterior focal length of about 19 to 23 millimeters, depending upon accommodative state. The retinal image has a point-spread function on the order of two to three minutes of arc, corresponding to about 10 micrometers on the retina for ideal conditions. These conditions include a 2 to 3 millimeters pupil, monochromatic light, optimal accommodation and a normal, young, and healthy eye. This quality approaches, but is somewhat worse, than that produced by diffraction-limited imagery in an ideal optical system. The retinal image is always in motion. Even during the best efforts at steady fixation, there exists an irreducible tremor of the eye whose high-frequency components are in the 20 to 30 second-of-arc range, with larger drifting and saccadic movements up to 5 minutes of arc. It is possible to eliminate this residual motion by various optical techniques. Such stabilization usually results in a total loss of vision, providing an elegant demonstration that the visual system responds primarily to *changes* in light patterns, rather than to steady states. Electrophysiological evidence from animals amply confirms this.

The retina is a thin structure of extreme complexity. It is considered embryologically to be a displaced part of the brain, and it is of clinical importance as the only part of the central nervous system that can be directly observed in the intact living subject. The receptors, the *rods* and *cones*, line the back surface of the retina, in immediate contact with a dark layer (the choroid) which helps to nourish the receptors and to prevent multiple reflection of light. There are about 125,000,000 receptors in each human eye, of which only about 5% are cones. The cones are, however, of an importance disproportionate to their relative number. In particular, there is a small central bouquet of about 2,000 cones located in a rod-free depression of the retina known as the *fovea centralis*. There, they are packed together into a hexagonal array having a density of about 150,000/square millimeter. These are capable of dissecting the finest details of the optimal retinal image. This process is aided by the lateral displacement of other retinal structures through which light must pass to reach the cone receptors. Moreover, this is the area of the retina where images have the highest attention value and which "projects" to a disproportionately huge area of the visual brain. The extraocular

muscles move the eye more or less automatically, in the act of fixation, to put object of interest into this region, where their details can be most critically appreciated, while the accommodative mechanism alters the shape of the lens to produce the sharpest possible image in this region. The cones, including those in the fovea, function only at high luminance levels (approximately, above .01 candela/square meter), below which are functionally blind and the rods take over. Thus the retina contains two systems intermixed: (a) the cone system (photopic), good for high-acuity vision, which also mediates all color vision; (b) the achromatic rod system, which has relatively poor spatial resolving power, but very high sensitivity.

The rods and cones are synaptically connected to the *bipolar cells*, which in turn relate to the *ganglion cells*, whose axons constitute the optic nerve fibers. There are also rich horizontal connections among the receptors, among the bipolar cells, and among the ganglion cells. In addition, there is a high degree of convergence: the 125,000,000 receptors ultimately feed into only 1,000,000 nerve fibers of the flexible optic nerve, which therefore constitutes the principal "bottleneck" of information flow in the visual system. The convergence ratio for the fovea is about 1:1, helping to preserve the high-detail vision of this region, while in the peripheral retina this ratio is many thousands to one, leading to high sensitivity at the sacrifice of resolving power.

The pathways from retina to brain are by no means independent, including those emerging from the central fovea. The horizontal interconnections are utilized to allow inhibitory processes to sharpen the "neural image" by a process of border enhancement, but much more complicated preprocessing of information occurs also. It is abundantly clear that the brain does not receive a replica of what is on the retina, although a spatial isomorphism between retina and brain does exist. Rather, the messages sent to the brain tend to carry information that is already processed in complex ways to make efficient use of the limited communications pathways between eye and brain in an adaptively significant manner.

Because the two eyes are located in slightly different places in space, a disparity of the two retinal images results. Rather than to produce a blurred or confused picture, this *retinal disparity* results in the appearance of *stereoscopic depth*. Such depth judgments are remarkably precise, consistent with the findings that all but the smallest eye movements and accommodative adjustments are highly correlated between the two eyes, and that neural units in the visual brain are precisely connected, by way of intermediate synapses, to optically corresponding areas.

The normal eye exhibits a large amount of chromatic aberration, which is not normally noticeable. There are at least two reasons for this: (a) the cone receptors exhibit a directional sensitivity, which reduces the visual effectiveness of light entering the marginal zones of the pupil; (b) the visual system has a remarkable capacity to adapt to systematic distortions, of almost any kind, which do not carry useful information from the external world. For example, observers learn with practice to compensate for the effects of gross visual displacement caused by prisms placed before the eyes, and are not longer able to see the chromatic fringes produced by such prisms. Removal of the prisms produce reappearance of chromatic fringes and an apparent displacement in the opposite direction. The explanation of such effects is not simple: in this example, the adaptation to displacement is probably kinesthetic rather than visual, but the adaptation to fringes is almost certainly confined to the visual system. Related to this are many entoptic phenomena that are seldom perceived: (a) the shadows of the retinal blood vessels, which are in front of the receptors; (b) the blind spot in the visual field, caused by the receptor-free optic disc, large enough to contain 200 images of the moon; (c) "floaters," usually shadows of debris in the vitreous humor, clearly visible if attended to against bright, uniform surfaces such as the sky; (d) fleeting specks of light probably caused by the movements of corpuscles within the retinal blood vessels; (e) Maxwell's spot, probably corresponding to the region of the macular pigment, and many others.

The initial nonoptical event in the visual process is the absorption of single light quanta by single molecules of visual photopigment, of which millions are located in each rod or cone. Under ideal conditions, as few as a half-dozen of these elemental events within fairly broad bounds of time and area, are sufficient to lead to a visual sensation. The visual photopigment contained in the rods is *rhodopsin*, having a peak sensitivity at about 505 nanometers. It has been much studied and is found in most animals including man. Absorption of light by rhodopsin probably produces graded potentials at the receptors that trigger all-or-none nerve impulses by the time the ganglion cells are activated, if not before. The exact mechanisms whereby light absorption gives rise to receptor potentials (and these to nerve impulses), although under active investigation, cannot be said yet to be satisfactorily understood.

Modern neurophysiology has learned much about the operation of the individual nerve cell, but rather little about the meaning of the circuits they compose in the brain. An initial approach to visual information processing is described by these authors. See also the entry on **Central and Peripheral Nervous Systems**.

Color vision depends upon the existence of three classes of visual photopigment, all different from rhodopsin, housed in different proportions in different classes of cone receptors. When two fields of light that are physically different look exactly alike in color (*metameric* matches), it is probable that the rate at which light is being absorbed in the three classes of cone photopigment is the same from both fields, although this is not yet definitely established. The perception of color clearly involves the higher levels of the visual system as well.

Another important property of the eye is its adaptation to intensity by means of which the eye changes its gain and other characteristics, enabling it to respond discriminatively over a stimulus intensity range of about ten billion to one. At one time, it was felt that the bleaching of photopigments was primarily involved in this process. Recent evidence indicates, however, that this plays only a minor role and the true mechanisms are numerous and include changes in the organizational properties of retinal networks.

Physical Disorders of the Eye

Near-and Farsightedness. Ideally, the lens of the eye receives light from the outside and bends it in such a way that an image is resolved upon a small point of the retina. In order to maintain focus on the retina, the lens must change its shape when objects are viewed from different distances. However, the cornea has more than twice the focusing power of the lens. If the cornea-lens combination focuses the light rays at a point in *front of the retina*, the person is *near-sighted* (myopia) and *cannot see distant objects* clearly without glasses. If, on the other hand, the light rays are focused at a point *behind the retina*, the person is *farsighted* (*hyperopia*). *Presbyopea*, a decrease in the ability of the eye to accommodate to various distances, is a normal concomitant of the aging process. It is common in persons over 40 years. Unless caused by disease, accommodation is afforded by the use of glasses.

The manner in which light rays from distant objects are focused on the retina in a normal eye is shown in Fig. 3. *Myopia* may occur when the eyeball has grown too long for the normal focus of the cornea and the lens, so that the image projected onto the retina is blurred and out of focus. See Fig. 4. An eye of normal size can also develop the problem if the curvature of the cornea and lens increase, causing greater refractive (bending of light) power. The focus falls short and a blurred image is transmitted to the retina. See Fig. 5.

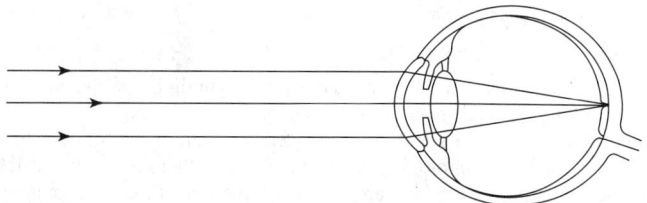

Fig. 3.   Pattern of light rays from distant objects as focused on the retina of a normal eye. (*American Association of Ophthalmology.*)

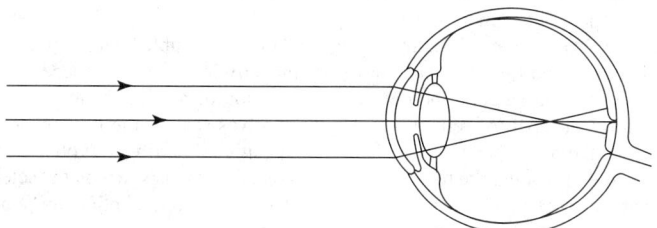

Fig. 4.   Pattern of light rays from distant objects as focused in a myopic eye due to abnormal length of the eyeball. The rays are focused short of the retina by the normal curvatures of the cornea and lens. (*American Association of Ophthalmology.*)

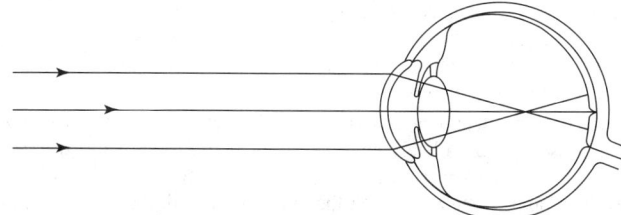

Fig. 5.   Pattern of light rays from distant objects as focused in a myopic eye due to increased curvatures of the cornea and lens. The rays are focused short of the retina. (*American Association of Ophthalmology.*)

A distinction is usually made between *simple* and *pathologic myopia*. Simple myopia is the most common form and is not attributable to eye disease. The cause of pathologic myopia is not well understood. Various hypotheses, including deficiency of vitamins and eyestrain due to reading, have been put forth. There is no evidence that reading (even in poor light) will influence myopia, or that it can be changed by eye exercises. The myopic person can have vision as sharp and clear as anyone by wearing the proper eyeglasses or contact lenses. See Fig. 6. A surgical procedure for correcting nearsightedness in some patients was introduced a few years ago. Known as *radial keratotomy*, the procedure involves making radial incisions in the cornea, permitting peripheral bowing which causes central flattening of the cornea.[1]

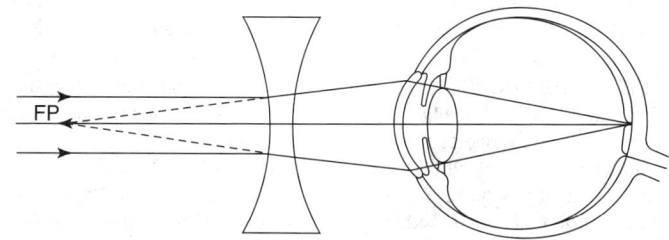

Fig. 6.   The myopic eye has a fixed far point (F.P.) beyond which objects cannot be seen clearly. To correct this problem, the proper diverging lens placed before the eye will permit light rays from distant objects to be focused sharply on the retina. (*American Association of Ophthalmology.*)

In *hyperopia* (*hypermetropia* or *farsightedness*), there is a defect in refraction of light into the eye which diminishes the ability to see at close range. Rays of light coming from nearby objects cannot be focused on the retina, but those coming from distant objects can be properly refracted. As a result, close images are blurred, while distant ones are clear. The cause may be too shallow an eyeball, or insufficient convexity of the lens, or changes in the refractive media of the eye with increasing age of the individual. The condition usually is easily correctible by use of glasses having convex lenses, or the equivalent contact lenses. See Fig. 7. A surgical procedure for correcting farsightedness in some patients is also

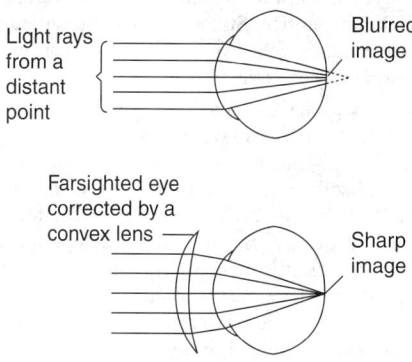

Fig. 7.   Path of light rays (above) in farsighted eye, in normal eye (below).

[1] Some of the more recent surgical procedures described have stirred sone controversy among experts. (See *Science*, **35**, February 1, 1985.)

relatively new. Known as *epikeratophakia*, the procedure involves suturing a corneal button to the front of the eye. If unsuccessful, the lens (intended to serve as a permanent contact lens, in essence) can be removed and vision returned to preoperative levels.

Astigmatism. This is a common optical defect, but in most cases it is so slight as to go completely unnoticed. It is manifested by a distortion of vision. Thus, in looking at an object, a straight line in the vicinity of the object may appear curved. When the eyes are moved, a motionless object may seem to move as it passes through the distorted area of the field of vision. See also **Astimagtism.**

Nystagmus. This is an involuntary rapid movement of the eyeball, which may be from side to side, up and down, circular, or a combination of these movements. It may be due to incoordination of the eye muscles, disturbances of the nerves of the eye, or to disturbances of the vestibular canals in the ear.

Night Blindness. The greatest concentration of nerve endings in the retina is found in a particular area called the *fovea*, which is located near the center of the retina. The nerve endings in this area are called *cones* and are responsible for direct vision and detection of both intensity and color of light. Scattered near the fovea and distributed in greater numbers elsewhere in the retina are other types of nerve endings called rods. These nerves have little ability to detect color, but are extremely sensitive to light. In daylight, they are almost unable to operate, but when the light is dimmed, their sensory ability returns (*dark adaptation*). The inability to see in dim light is called night blindness. About 30 minutes are required for dark adaptation to reach its maximum. Since only a few of the rods are located in the foveal region, one does not look directly at objects to view them in the dark; instead, the individual looks slightly to the side or just above or below the object.

Color-Blindness. There are several hereditary conditions that may cause an individual to be color-blind. In most instances, the cones of the fovea are operative, but do not distinguish colors. The most common type of color-blindness is termed "red-green" color-blindness. The individual has defective recognition of reds and greens. Rarer types prevent differentiation of other pairs of colors. Total color blindness is the rarest type; the individual can distinguish only shadings of gray and black. The inheritance of red-green color blindness usually follows a sex-linked pattern. Briefly, the disability appears more frequently in men than in women; although only one girl in 100 will be color-blind, the disability will appear in one out of every 10 or 12 boys. The disorder will not appear in a color-blind man's son unless the boy's mother is either color-blind, or is a carrier of the gene for color-blindness. Consequently, the disorder only rarely is transmitted from father to son. More commonly, color-blindness is transmitted from an affected man through his daughters (in whom the condition usually is not expressed) to about one-half of his grandsons. Color-blindness also can be caused by disease or injury of the retina, the optic nerve, or the conduction paths of the eye to the brain. See also **Color Blindness.**

Strabismus. Normal binocular vision is the ability of each eye to look at the same point in distance. It is the result of balanced muscular coordination, allowing proper convergence of the eyes to take place. When one eye cannot achieve binocular vision with the other because it deviates inward ("cross-eyes"), outward ("wall-eyes"), upward, or downward, strabismus is said to be present. Strabismus may be caused by paralysis of one or more ocular muscles, or by congenital imbalance of the muscles. See also **Strabismus (Cross-Eyes).**

Retinal Disorders. The retina is a delicate sheet of tissue made up of ten layers. The pigmented outermost layer normally lies snug against the back wall of the eye. The nine inner layers, which pick up light rays, have in front of them a semi-solid transparent gel called the *vitreous*. This vitreous gel occupies most of the eye cavity. See Fig. 8. The retina's function is like that of film in a camera. By means of a complex chemical process it transforms light into electrical energy and transmits it to the brain. The brain, in turn, interprets these electrical impulses as visual images. Only when the retina is intact and in proper position can vision be normal.

During normal aging, the gel becomes liquified. It collapses in a significant percentage of the total population by the age of 65 years. The breakdown of the gel or liquification is a prolonged process rather than a single episode in the patient's life. During the collapse of the gel, if a retinal adhesion is present, it creates a traction force tearing a hole in the retina. The patient may be aware of flashing lights and a shower of

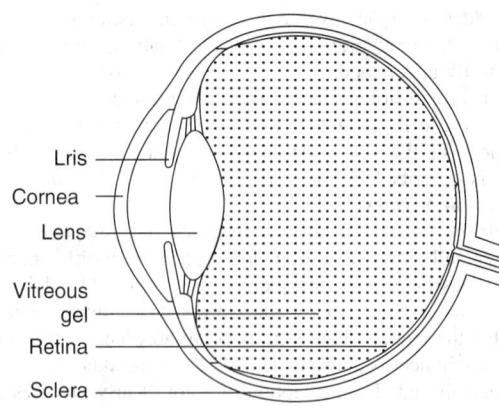

Fig. 8.   Sectional view of human eye showing large amount of space taken up by vitreous gel between the lens and the retina. The retina, made up of ten layers of delicate tissue, may be torn, develop holes, and sometimes become detached. (*Wills Eye Hospital, Philadelphia, Pennsylvania.*)

spots during this period. The vitreous collapse continues and fluid may seep through the hole in the retina, lifting or detaching the retina, much as wallpaper falling from a ceiling when there is a leak in the roof. A black visual curtain results, with the loss of sight. See Fig. 9.

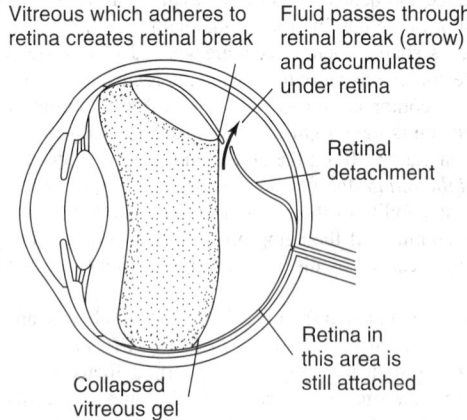

Fig. 9.   When the retina detaches it becomes separated from the eye wall by a layer of fluid. Most retinal detachments occur spontaneously when the vitreous gel shrinks and tugs on the retina causing one or more holes or tears to occur. Liquefied vitreous fluid then seeps through the hole and accumulates behind the retina, detaching it from the back wall of the eye. (*Wills Eye Hospital, Philadelphia, Pennsylvania.*)

Careful examination with the use of a bright light of the binocular ophthalmoscope will reveal a tear before the detachment occurs. In some cases, when tears and holes are detected before the retina actually detaches, the ophthalmologist may be able to seal them by means of laser photocoagulation or cryotherapy. Both of these methods are relatively simple and safe, requiring no general anesthetic, and are usually performed on an outpatient basis. See Figs. 10 and 11.

Early diagnosis and treatment of retinal problems is very important because the progression from hole to *actual detachment* may be rapid and, once detachment occurs, major eye surgery is required. Recent advances in diagnostic and surgical techniques, coupled with improved silicone supporting materials, have significantly increased the success rate in detachment surgery and decreased the period of hospitalization.

Surgical repair of retinal detachment involves sealing the hole or holes so that the retina can return to its normal position. Scarring is produced by freezing or heating the tissue around the hole. A patch is sewn on the sclera over the hole to support the tear by reducing vitreous traction. In some cases, a silicone rubber sponge is used. In other cases, solid silicone rubber patches are used. An encircling band of silicone rubber also may be used to provide support during the period of scar formation. See Fig. 12. As a rule, the band remains permanently in place, in the hopeful expectation that it will prevent a recurrence of the detachment.

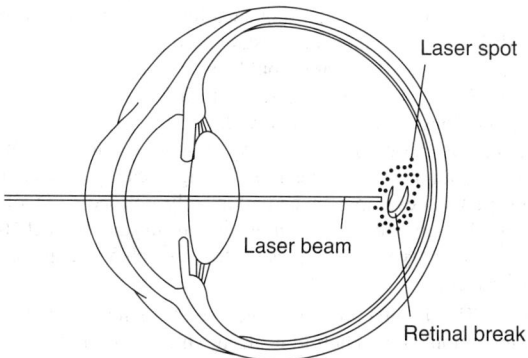

Fig. 10. Laser photocoagulation is a process in which split-second bursts of intense, precisely focused light are directed through the pupil onto the retina. This procedure is successfully used to repair the retina prior to it becoming fully detached. (*Wills Eye Hospital, Philadelphia, Pennsylvania.*)

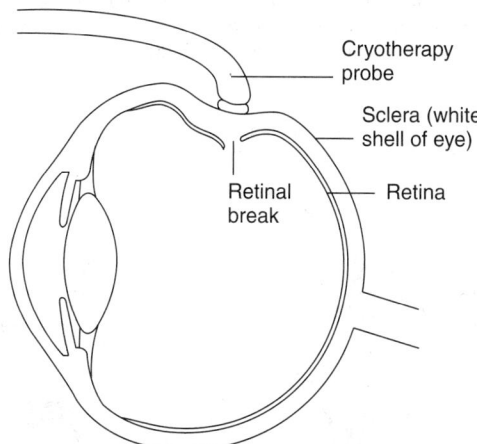

Fig. 11. Cryotherapy is a process in which a freezing probe (which looks similar to a ballpoint pen) is applied briefly to the outside of the eye wall. The probe freezes through to the retinal hole and promotes growth of sealing scar tissue.

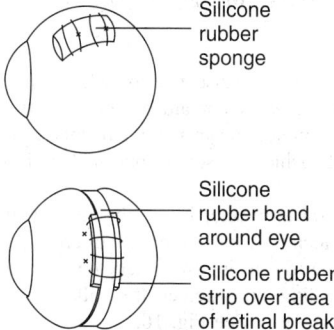

Fig. 12. Surgical repair of retinal detachment. Holes in the retina are sealed by scarring produced by freezing or heating the tissue around the hole to support the tear by reducing vitreous traction. In some cases, a silicone rubber sponge is used, as shown in top of diagram. In other cases, an encircling band of silicone rubber may be used to provide support during the period of scar formation. as shown in bottom of diagram. (*Wills Eye Hospital, Philadelphia, Pennsylvania.*)

Vitrectomy. The vitreous gel may undergo various other changes which will cause a decrease or loss of vision. These changes may or may not be related to retinal detachment. In some patients, especially diabetics, the vitreous may become permanently filled with blood, which acts like a cloud, preventing light from reaching the thin, sensitive retinal layer. In one type of surgery, instruments that look like small metal straws are inserted into the vitreous cavity. See Fig. 13. Bloody vitreous is removed by suction, in a procedure called vitrectomy. As the flood-filled vitreous is withdrawn, a clear fluid is automatically infused through a third opening in the eye. This exchange of clear fluid for opaque fluid permits light to reach the retina, returning improved vision.

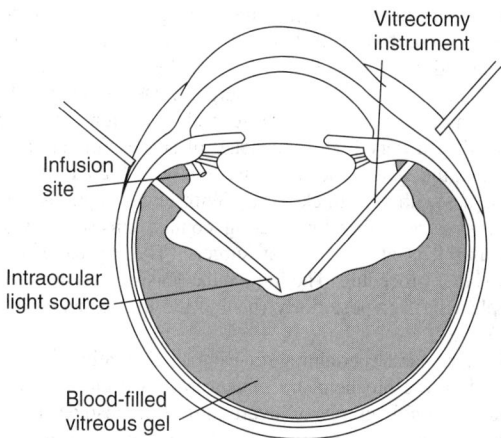

Fig. 13. Vitrectomy, a surgical procedure for exchanging blood-filled vitreous with a clear fluid. Instruments that look like small metal straws are inserted into the vitreous cavity and the blood vitreous is removed by suction, after which a clear fluid is automatically infused through a third opening in the eye. (*Wills Eye Hospital, Philadelphia, Pennsylvania.*)

In some patients, scar tissue may develop in the vitreous gel. As the vitreous shrinks, this scar tissue pulls on the retina and causes it to separate from the eye wall. This is known as *traction retinal detachment*. See Fig. 14. In this complex situation, vitrectomy is combined with retinal detachment surgery. The scar tissue is first removed with the vitrectomy instrument and automated scissors to release traction on the retina.

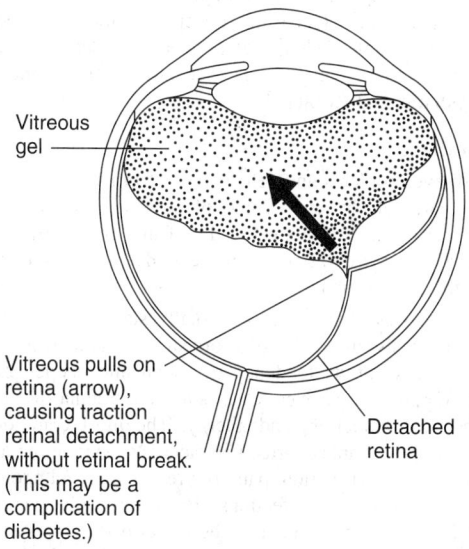

Fig. 14. Complex eye surgery in which vitrectomy is combined with retinal detachment surgery. Scar tissue is first removed with the vitrectomy instrument and automated scissors to release traction on the retina. (*Wills Eye Hospital, Philadelphia, Pennsylvania.*)

Cataract. The loss of sight due to cataract has been known since antiquity. Cataract is defined as any opacity (clouding) of the lens of the eye that alters or blocks the passage of rays of light through it, and thus interferes with vision. The lens of the eye is normally a transparent body that lies directly behind the iris and pupil. Cataract is not a film that grows over the lens, but a cloudiness of the lens itself. Cataracts are not growths and are not contagious. Among known causes of cataract are some general diseases, such as diabetes. A child may be born with cataract from hereditary influences or because the mother had rubella (German measles) during the first trimester of pregnancy. Radiation and dietary deficiencies also may produce cataract, as can chemical camage or mechanical injury.

In some cases, a cataract may not necessarily become progressively worse and visual acuity can be improved by specially prescribed spectacles. Occasionally, when the cataract is small and centrally located, drops can be given to keep the pupil dilated and allow the person to see around the cataract, thus making surgery unnecessary, or at least postpone it. An individual should not self-diagnose a cataract and allow it to "ripen" prior to consulting a physician. However, the majority of cataracts are progressive and eventually require surgery. Development of advanced surgical techniques during the last several years for removing the lens have resulted in a very high percentage of successful cataract surgeries. At one time, the natural lens of the eye was "replaced" by wearing aphakic glasses with very thick lenses. With few exceptions, these glasses have been replaced either by wearing contact lenses or by implanting an intraocular lens at the time of surgery. The current major trend is toward the latter procedure. The intraocular lens is a clear plastic implant, which is placed either posteriorly (behind the iris) or anteriorly (in front of the iris).

Most lens research pertaining to cataract formation in the past has concentrated on the biochemistry of the processes that may be causative factors. More recently, groups of investigators have also been concentrating on the biophysics of the problem. For example, scientists at Rush University (Chicago) have reported on a new and functional theory of how the lens can maintain its volume and transparency. These investigators found that the lens has a highly unusual structure—the fiber cells in its interior have extremely impermeable membranes. The researchers have proposed that impermeability is a rigid requirement if small ions are not to penetrate the lens and cause cataract. Other scientists as well have been constructing mathematical models of the electrical properties of the lens. Out of this work came the concept that the lens has two kinds of cells—epithelial cells on the outer surface and fiber cells on the interior. It has been shown that the membranes of most cells of the body are far more permeable to sodium than are lens fiber cell membranes. The concept of a "sodium pump" in the lens, which would extrude any sodium ions that might wander into the lens, has been ruled out, because such pumps require energy, and because the lens does not provide facilities for metabolic energy input. Hence the concept that the membranes must be absolutely impermeable. When that permeability deteriorates, a cataract forms according to this hypothesis. See also **Central and Peripheral Nervous Systems**; and **Cataract**.

Diseases of the Eye

A number of eye problems and defects result from various diseases, which affect vital parts of the eye. Some of these conditions are described in separate entries in this encyclopedia. Thus, see **Conjunctivitis**; and **Glaucoma**. For diseases affecting the eyelids, see also **Eyelid** and the glossary at the end of the entry.

Corneal Diseases. *Corneal ulcer* usually starts as a small, gray area of localized necrosis (tissue death). Corneal ulcer is considered a medical emergency because of its tendency to widen and deepen rapidly, until much of the cornea is destroyed. Causes include trauma (usually a foreign body in the eye), infection, and allergy. The most common infectious agent is the herpes simplex virus. Characteristic symptoms are reddened eyes, discharge and lacrimation, pain, blurred vision, and photophobia, but individual symptoms vary. Infections are treated with specific antibiotic therapy, and steroid compounds may be prescribed.

Keratitis is inflammation of the cornea caused by infection, trauma, or chemical irritation. There are many types of keratitis. *Interstitial keratitis* is an inflammation of the deep layers of the cornea, and it occurs most commonly among children afflicted with congenital syphilis, and appears between the ages of 5 and 15. The cornea becomes progressively grayish and opaque. Eventually, both eyes are usually involved. If drug therapy fails, corneal transplantation may be considered in suitable cases. *Industrial keratitis* takes many forms. Almost always, the source of inflammation is physical trauma or chemical irritation. See also **Keratitis**.

Keratopathy is deterioration of the cornea with aging, or in the presence of other eye diseases, the production of outgrowths of hyaline tissue on the back surface of the cornea. These transparent growths interfere with vision by scattering incident light. The condition is called *guttate keratopathy*. The cornea also may develop blisters on its front surface (*bullous keratopathy*) which interfere with vision and may be very painful.

Corneal transplantation. Surgical techniques make it possible to replace all or part of a diseased human cornea by a corresponding segment of a clear human cornea obtained from another individual. In addition to damage from corneal diseases, the transparency of the cornea may have been scarred by injury or burn. For the corneal transplant to be successful in restoring sight, the other parts of the eye must be in good condition.

Because of difficulties of acquiring healthy corneas for transplantation when needed, eye banks have been established. Persons wishing to donate the eyes immediately after death should best contact an eye bank or inform a physician. Simply mentioning this desire in one's will is not sufficient because of the long time factor involved.

Donor corneas can be frozen for long periods of time without apparent damage of any kind of the corneal tissue. Significant advances have been made toward the development and ultimately common use of artificial corneas. Problems involve rejection of foreign substances, such as silicone rubber, synthetic polymers, and plastics from which artificial corneas have been constructed.

Retinitis. This is inflammation or edema of the retina, often associated with inflammation of the choroid coat, in which case the term *choroiditis* may be used.

Uveitis. The eye has three coats—an outer, tough protecting layer called the *cornea/sclera*; an inner, light-sensitive layer called the retina; and a middle, pigmented, nourishing layer called the *choroid*. See Fig. 15. If the outer and inner layers of the eye are removed, the remaining tissue—darker and full of blood vessels—resembles a grape. If this is inflamed, the condition is called *uveitis*. This is a fairly common condition and results in approximately 10% of all blindness.

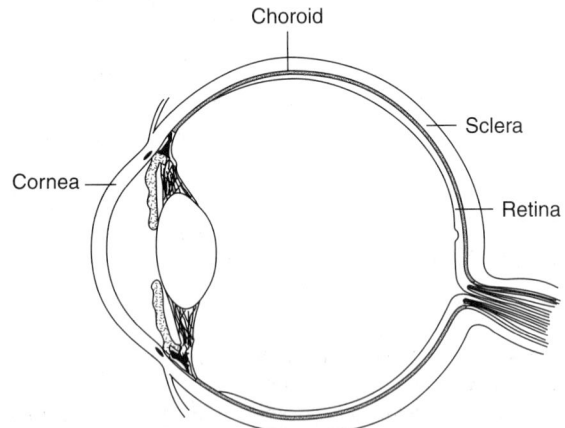

Fig. 15.   Coats of the eye. (*Wills Eye Hospital, Philadelphia, Pennsylvania.*)

Many systemic diseases cause uveitis. More frequently, an attack of inflammation in the uvea is not attributable to any known disease and is believed to be immunologic in nature. In this case, the uvea reacts in a highly sensitized fashion to some abnormal or foreign protein within the body.

Uveitis is a general term referring to the entire uveal tract, but usually this condition is localized or confined to areas that are more often involved, such as the iris (iritis), the ciliary body (cyclitis), or the choroid (choroiditis), depending on whether the inflammation is in the front, middle, or back of the eye. See Fig. 16.

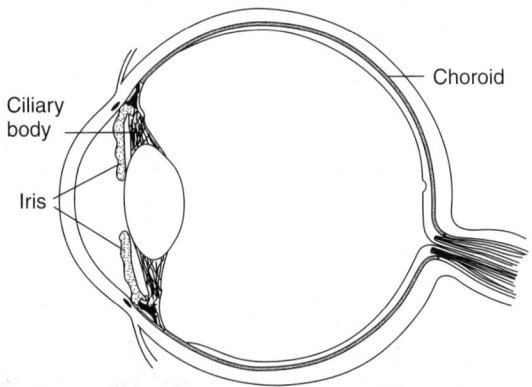

Fig. 16.   Areas of the eye that may be involved in uveitis. (*Wills Eye Hospital, Philadelphia, Pennsylvania.*)

Diseases in which uveitis may be present include ankylosing spondylitis, Behcet's syndrome (rare), epidemic typhus, erythema multiforme, gonorrhea, leptospirosis, meningococcal disease, Reiter's syndrome (rare), and Whipple's disease (rare), among others. See also **Uveitis.**

Disorders of the Iris. Other than certain congenital malformations, *iritis* is the most common condition affecting the iris. This is an acute or chronic inflammation of the iris, due to any of a variety of causes. When the ciliary body, which lies behind the iris, is also involved, the condition is termed *iridocyclitis*. In iritis, the iris looks muddy, dull, and swollen. Symptoms usually include throbbing eye pain, blurred vision, photophobia, and sometimes swelling of the upper lid. Prompt medical care is essential because of the danger of secondary glaucoma ending in blindness. Often atropine drops are prescribed to keep the pupil dilated and adrenocorticoid steroids to shorten the course of the disease.

Tumors of the Eye. These may be benign or malignant. They can involve outer or inner ocular structures. Although they seldom occur, benign tumors can cause serious eye damage and malignant tumors are a threat to life. Tumors of the eyelids closely resemble tumors of the skin elsewhere. However, they may interfere with vision and irritate the eyeball by friction or pressure. Tumors that arise within the eyeball may cause increased ocular pressure (glaucoma), bulging of the affected eye from its socket (exophthalmos), pain, defects in vision, and other symptoms. In the early stages of tumor growth, the patient may experience no symptoms. The two principal malignant tumors are *malignant melanoma* and *retinoblastoma*. Melanoma occurs almost exclusively in adults, usually between the ages of 40 and 60 and involves only one eye and is first noticed as a defect in the visual field. Eventually, retinal detachment takes place around the tumor.

Retinoblastoma is a congenitally acquired cancer. It occurs in children under five years of age, usually in one eye, but sometimes in both. Occasionally several children in the same family have the condition. It has been known to occur in the offspring of adults who were cured of retinoblastoma in childhood.

Researchers at the Massachusetts Eye and Ear Infirmary and the Whitehead Institute announced in 1986 the isolation of the entire retinoblastoma gene. The retinoblastoma gene is thought to be a rare mutation of a normal gene. Children develop retinoblastoma when they inherit one copy of the gene and then, by chance, the homologous segment of their other chromosome in one or more of their retinal cells is deleted or mutated. The result of this sequence is that the individual has no normal gene to counter the effects of the retinoblastoma gene and cancer develops. Search for the retinoblastoma gene stems back to the 1960s to a researcher at the University of Minnesota who reported a patient with an eye tumor who also had a deletion on chromosome 13. This case provided the first hint of where the retinoblastoma gene might be located. Researchers are now searching for the protein that the gene normally codes for. This finding also has a broader implications. It is the first recessive cancer-causing gene ever isolated. All other oncogenes are dominant, that is, one copy of the gene leads to cancer. Researchers at the Muscular Dystrophy Association observe that the dominant genes code for products that apparently direct cells to grow uncontrollably. Because a recessive gene only causes cancer where there is no good copy of the homologous gene present, the interpretation is that these genes may be coding for substances that tell a cell to stop proliferating.

Blindness

Blindness may be caused by any of a number of conditions, including severe instances of the various diseases and disorders previously described. Emotional upsets and hysteria may produce temporary blindness. In addition to the conditions described, permanent blindness can be caused by pressure on the optic nerve by tumors, brain damage by skull fracture or loss of adequate blood supply, and, of course, eye injuries. Comparatively few years ago, most blind persons were considered incurable, but rapid advancements in medical science have greatly altered the situation.

Blindness affects all age groups, but most of those so afflicted are elderly, although about 60% of the blind are not sightless in the real sense of the word. In some states, a person is considered "industrially blind" when the visual acuity is less than 20/200 in the better eye with properly fitted glasses. These Figures mean that the person taking the test for visual acuity can only see at 20 feet (6 meters) what one with normal vision can see at 200 feet (60 meters). Normal vision is defined as 20/20 vision or better. Some persons classified as blind can see clearly, but over only a

small area. That is, their vision may be limited to an area approximately the size of that to be seen by looking through the barrel of a gun.

Reading Aids. The usual reading aids for the blind will not be discussed here, but a few of the electronic reading aids will be described briefly.

1. A letter-recognition device for reading short sections of print to make it possible to check bank statements, proof typed letters, or recognize a particular book or magazine by reading its title. The device consists of a probe with 8 to 11 photocells arranged in a linear array, each individually connected to an oscillator the frequency of which increases in pitch from the bottom to the top of the array. When a photocell "sees" dark, it keys on its oscillator. Thus, as the probe is moved along a line of letters, each letter generates a characteristic sequence of chords, which can be learned, but with considerable difficulty.

2. A character recognition device which takes the outputs from an array of specially placed photocells, puts them through a recognition matrix, and triggers a section of magnetic tape that speaks the appropriate letter. Because of slow speed, such a device is not geared to reading for pleasure, but can facilitate reading for information.

3. A device that is a complete automatic computer-controlled print-to-speech converter, which does the equivalent of a human being reading printed material onto magnetic tape. The device is costly and before available on a mass basis, the logistics of a communication and distribution system will have to be worked out, probably utilizing local libraries and telephone lines.

Although in development for a number of years, convenient, low cost reading aids remain to be marketed. Since programs of this type are of a dynamic nature, details cannot be presented here. The Veterans Administration, which has sponsored much development along these lines, is an appropriate contact for additional information.

See also **Braille System**; **Lasers (Eye Surgery)**; and **Retina.**

An abridged glossary of terms used to describe all aspects of vision and the eye would include:

Aberration. Blurred or distorted image quality that results from inherent physical properties (shape, curvature, density) of an optical device such as a lens or prism.

Accommodation. The ability of the eye to change its focus from distant to near objects as well as from near to distance objects. This process is achieved by the crystalline lens changing its shape.

Achromatopsia. Congenital absence of the ability to see colors. Caused by absence or defects in light-sensitive retinal receptor cells (cones) that provide sharp visual acuity and color discrimination. See also **Achromatopsia**; and **Color Blindness.**

Afterimage. A sensation of seeing an object that is no longer in your sight. Also called *after sensation, after vision.*

Age-related macular degeneration (AMD). Degeneration of the photoreceptors in the macula or central region of the retina. This area of the retina is responsible for central vision, used for reading, seeing faces, and so on. Often associated with aging. See also **Age-Related Macular Degeneration**; and **Macula.**

AIDS-related eye disorders. Anyone who is HIV positive should have regular eye examinations by an ophthalmologist, especially if their T-cell count falls under 250 or if vision changes are obvious. The most common eye problem from AIDS is called Cotton Wool Spots. A more serious problem is an eye infection caused by cytomegalovirus (CMV) which can permanently impair vision. Other AIDS-related eye disorders include red eye, shingles, detached retina and Kaposi's sarcoma, a slow growing tumor on the eyelid or spot on the white of the eye.

See also **AIDS-Related Eye Disorders.**

AK. Acronym for Astigmatic Keratotomy. A surgical procedure used to correct moderate cases of astigmatism. Often performed at the same time as the Radial Keratotomy (RK) procedure for correcting nearsightedness. Both AK and RK have been largely replaced by newer refractive surgery procedures. See also **Astigmatic Keratotomy (AK)**; and **Refractive Eye Surgery.**

ALK. Automated Lamellar Keratoplasty. See also **Automated Lamellar Keratoplasty (ALK)**; and **Refractive Eye Surgery.**

Allergies of the eye. Allergic reactions triggered by the body's immune system to protect eyes from injury. Ocular allergies can be seasonal, such as those caused by pollens, or caused by bacteria's, food sensitivities, cosmetics, fabrics, soaps and other substances. Visible symptoms of eye allergies may include swelling, hives, itching, watering, eye pain and sensitivity to light.

Allyl resin plastic lens. The lightest of all vision-correction lenses. But they are thicker than other plastic lenses, are less scratch resistant, and do not block as much ultraviolet light as polycarbonate and glass lenses.

Amacrine cell. A type of neuron seen in the retina.

Amblyopia. Also called lazy eye, this is a condition of decreased vision in one or both eyes that occurs without detectable structural abnormalities or disease in the eye or visual pathways. See also **Amblyopia**.

Amsler grid. Home eye test featuring equally spaced horizontal and vertical lines in a grid pattern. Used for determining the presence of macular degeneration and other defects that affect central vision. See also **Age-related Macular Degeneration**; and **Amsler Grid**.

Angiography. Test used to examine blood vessels in the retina, choroid and iris of the eye. See also **Fluorescein Angiography**.

Angioma. Benign tumor consisting of small blood vessels. Can affect eye with tumors beneath or on top of the retina, which may lead to blood leakage and possible retinal detachment. See also **Von Hippel-Lindau Disease (VHL)**.

Aniridia. A birth defect, in which a child is born without an iris, so there is no way to control the amount of light that enters the eye. The only treatment is to use colored eye lenses to reduce the amount of light entering the eye.

Anisometropia. A condition that occurs when each of a person's eyes have a different refractive power.

Anterior segment. Front third of eyeball, including the cornea, anterior chamber, iris and ciliary body.

Anterior capsulotomy. Cataract surgery technique used to make a round opening in the front (anterior) of the capsule, which encases the eye's natural lens.

Aphakia. Absence of the eye's natural crystalline lens, usually after cataract extraction.

Astigmatism. Uneven curvature of the cornea in which refractive light rays are bent out of focus resulting in distorted vision. Those people with astigmatism are usually born with the disorder and it does not worsen with age. Often occurs in conjunction with nearsightedness or farsightedness. See also **Astigmatism**.

Asthenopia. The term for eye strain.

Automatic keratometer. Instrument used to measure curvature of the front surface of the cornea.

Automatic refractor. Electro mechanical device, that provides an "objective" measurement of an eye's refractive error.

Bardet-Biedl Syndrome. A genetic disorder characterized by degeneration of the light sensitive cells in the periphery of the retina causing night blindness, tunnel vision, decreased visual acuity, and photophobia. Other conditions include obesity, developmental delays, spastic paraplegia and renal disorders. See also **Bardet-Biedl Syndrome (BBS)**.

Basil cell nevus syndrome. An inherited group of multiple defects that result in an unusual facial appearance and a predisposition for skin cancer. The disorder affects the eyes as well as the skin, nervous system, endocrine glands and bones. Children with the disease have wide-set eyes, a broad nose and heavy protruding brow. The disorder results in the appearance of skin cancers, particularly around the cheekbones, upper lip and eyes. See also **Basal Cell Nevus Syndrome**.

Batten Disease. A group of inherited neurological disorders that affect children and are marked by progressive vision failure and other disorders that affect the brain. Most prominent in families of Northern European-Scandinavian ancestry. See also **Batten Disease**.

Bell's Palsy and the eye. Paralysis of muscles controlled by the facial cranial nerve. Eyelid on affected side does not close properly, so corneal drying may become a problem.

Binocular vision. The ability to maintain visual focus on an object with both eyes, creating a single visual image. Lack of binocular vision is normal in infants. Adults without binocular vision experience distortions in depth perception and visual measurement of distance.

Blepharitis. Common, persistent and sometimes chronic inflammation of the eyelids, resulting from bacteria, that reside on the skin. In certain individuals, these bacteria thrive in the skin at the base of the eyelashes or even in the oil glands near the eyelid, resulting in dandruff-like scales and particles. See also **Blepharitis**.

Blepheroplasty. Surgery to improve the appearance of the eyelids. In addition to detracting from overall appearance, drooping eyelids may cause functional problems such as impaired field of vision or difficulty wearing glasses. See also **Blepharoplasty**.

Blepharospasm. Condition characterized by uncontrollable, forcible closure of the eyelids caused by a progressive dysfunction of the nerve that controls muscles around the eye. Less serious form may cause eyelid twitches or tics. See also **Blepharospasm**.

Blind spot. 1. A small area of the retina where the optic nerve enters the eye, this type of blind spot occurs normally in all eyes. Also called optic disk. 2. Any gap in the visual field corresponding to an area of the retina where no visual cells are present; this type of blind spot is associated with eye disease.

Blindsight. A phenomenon reported in individuals who have damage to the primary visual cortex resulting in blindness. Individuals with blindsight report that they are unable to see, yet when examined are able to indicate the presence and location of objects.

Braille. A system of raised-dot writing devised by Louis Braille (1809–1852). Each braille character or "cell" is made up of 6 dot positions that are arranged in a rectangle comprising 2 columns of 3 dots each. A dot may be raised at any of the 6 positions, and each combination of raised dots corresponds to a letter of the alphabet, a punctuation mark, and another symbol. See also Braille **System**.

B-Scan. Ultrasound procedure in which high frequency waves are reflected by eye tissues and orbital structures and then converted into electrical pulses, which are displaced as bright spots on a black background. Provides a cross-sectional view of tissues used for evaluating structures that cannot be seen directly.

Calcarine sulcus. The portion of the occipital lobe of the brain where sight takes place. The central visual field is represented in the posterior calcarine sulcus. The peripheral visual field is represented in the anterior portion of the calcarine sulcus.

Cauterization. The use of heat to destroy abnormal cells. Also called *diathermy* or *electrodiathermy*.

Chalazion of the eyelid. Inflammatory lump in a meibomian gland of the eyelid. This gland, located just inside the eyelid, supplies the outer portion of the tear film, preventing rapid tear evaporation and overflow and providing airtight eyelid closure.

Choroid. Layer of major blood vessels, which lies between the retina and the sclera (white of the eye), that provides oxygen and nutrients to the retina.

Choroidal hemangioma. A noncancerous growth within the "choroid" blood vessel layer, which lies beneath the retina. Many choroidal hemangiomas never grow or leak fluid, and the ophthalmologist observes them without treatment. However, if the growth is located in the macula (center of vision) or if it leaks fluid, it requires treatment because it can cause a retinal detachment or other vision problem.

Choroidal melanoma. Relatively rare malignant melanomas that form in the choroid, the blood vessel layer beneath the retina. Small melanomas can be treated and are usually watched for growth prior to treatment Medium to large sized melanomas are usually treated with either radiation therapy or removal of the eye.

Choroidal metastasis. Sometimes, malignant tumors from other parts of the body can spread to areas in and around the eye. In women, these are usually from the breast, while in men they are more often from the lung. Chemotherapy can sometimes be used to treat these tumors, although radiation is usually a more definitive treatment.

Choroidal nevus. Rare tumors that can grow into a malignant melanoma. Like a raised freckle on the skin, nevi can occur inside the eye and should be seen by an eye doctor.

Choroidal osteoma. Benign bony tumors that can arise within the choroid blood vessel layer beneath the retina. They are usually located near the optic nerve and can cause vision loss.

Closed-angle glaucoma. Serious form of glaucoma, that can result in a sudden rise in intraocular pressure. Condition develops when the drainage angle, which allows the aqueous (fluid behind the iris) to flow out of the eye, is suddenly blocked. Symptoms of a partial blockage can include blurred vision, halos around lights, pain and redness of the eye. A sudden blockage can result in severe pain. To avoid permanent damage to the eyes, an ophthalmologist should treat the symptoms immediately. See also **Glaucoma**.

Collagen. In referring to the eye, collagen is the protein fibrils within the corneal tissue that help sustain its shape. Some new refractive vision correction procedures heat these fibrils with a special laser causing them to shrink and change the shape of the cornea.

CO_2 laser. High-energy laser, that can be used for making surgical incisions and for skin resurfacing. Emits short, intense bursts of energy that vaporize the outer layer of skin without damage to the underlying skin. See also **Lasers (Eye Surgery)**.

Color deficiency. Partial or total inability to distinguish specific colors. See also **Color Blindness**.

Computer vision syndrome (CVS). Collection of symptoms brought about by eyestrain associated with prolonged use of the computer. Symptoms include eyestrain, blurred vision, headaches, flickering or flashing sensations, loss of appetite, nausea and dry or irritated eyes. See also **Computer Vision Syndrome (CVS)**.

Cone. A type of specialized light-sensitive cell (photoreceptor) in the retina that provides sharp central vision and color vision. Highly concentrated in fovea. Three classes of cones exist: short, medium, and long wavelength cones.

Conjunctiva. Thin, transparent membrane that lines the inner surface of the eyelid and covers the sclera where it becomes the white of the eye. See also **Conjunctiva**.

Conjunctival tumor. Malignant cancers, that grow on the surface of the eye. The most common conjunctival cancers are squamous carcinoma, malignant melanoma and lymphoma. See also **Conjunctival Tumors**.

Conjunctivitis. Inflammation of conjunctiva or membrane, that covers the white of the eye and inner surfaces of the eyelid. Characterized by discharge, grittiness, redness and swelling. May result from virus, bacteria, allergens, chemical exposure or ultraviolet light exposure and, depending on cause, can be contagious. Sometimes called *pink eye*. See also **Conjunctiva**; and **Conjunctivitis**.

Contact lens. Thin plastic or glass lens designed to fit over the surface of the cornea, usually for correction of a refractive error, but can also be cosmetic. See also **Contact Lenses**.

Contrast sensitivity. Measure of visual ability to distinguish details between an object and its background under varying degrees of contrast.

Convergence. Coordinated movement of the two eyes so that the images of a single point fall on corresponding points of the two retinas.

Cornea. Clear structure that covers the front part of the eye including the iris and pupil. The cornea provides most of the eye's optical power, while the crystalline inner lens, located behind the iris serves to "fine tune" the focus of the images. When both are working properly, a sharp image is focused on the retina and transmitted through the optic nerve to the brain. See also **Cornea**.

Corneal abrasion. A tearing, scrape or puncture of the cornea, sometimes accompanied by a loss of epithelium, the outer tissue layer of the cornea. Although a corneal abrasion can be very painful, the corneal layer of the eye heals more quickly than most other tissues in the human body. See also **Corneal Abrasion**.

Corneal dystrophy. Any of a number of rare hereditary abnormalities, that are characterized by an accumulation of abnormal material in the cornea. This accumulation may occur later in life and result in cloudiness of the cornea and reduction in vision. See also **Corneal Dystrophy**.

Corneal mapping. Procedure whereby a detailed map of the corneal surface is drawn by a corneal topography instrument. Maps are used to evaluate the cornea prior to treatment and are especially valuable as a tool in preparing for refractive vision correction.

Corneal ring. Plastic half-ring segments used in a particular type of refractive surgery. This procedure involves implantation of the ring segments in the peripheral area of the cornea in order to change its contour to the shape desired to correct cases of nearsightedness. These rings may later be removed or replaced.

Corneal topography. Procedure that creates detailed maps of the surface of the eye using an instrument that combines a computer and video camera. The maps are used to evaluate the cornea prior to treatment and are especially valuable as a tool in preparing for refractive vision correction. See also **Corneal Topography (Videokeratography)**.

Corneal ulcers. Wound on the surface of the eye similar to a scrape or cut on the skin. The cornea is covered by a layer of tissue called the epithelium which, when damaged can result in infection if left untreated. Corneal ulcers have many causes including injury, abnormal eyelashes, absence of tear production and infections. Persistent irritation or redness is reason to see an eye doctor.

Cylindrical surface. With respect to eyeglass lenses, is not evenly curved, but more like an egg or a football.

Diabetic retinopathy. Potentially serious complication of diabetes that results in the weakening of tiny blood vessels that nourish the retina. See also **Diabetic Retinopathy**.

Diopter. Unit that measures the degree of refractive error in an eye or the light-bending power of a lens. In an eyeglass or contact lens prescription, a negative number refers to nearsightedness, while a positive number refers to farsightedness. For instance, a -8.00 diopter lens is very nearsighted, while a $+0.75$ is slightly farsighted. Astigmatism can be measured as either a positive $(+)$ or a negative $(-)$ but will be accompanied by another number such as "@ 90" indicating the direction (axis) of the astigmatism.

Diplopia. The condition whereby a single object appears as two objects. Occurs because the eye muscles do not act equally. Also called double vision.

Drooping eyelids. Condition (usually hereditary), in which the upper eyelid(s) sag. May be congenital or caused by a later problem associated with a nonfunctioning levator muscle.

Drusen. Tiny yellow or white deposits in the retina of the eye or on the optic nerve head, visible to an eye care specialist during an eye examination. One of the most common early signs of age-related macular degeneration (ARMD). The presence of drusen alone does not indicate disease, but it may mean that the eye is at risk for developing more severe ARMD.

Dry eye. Condition due to a deficiency in the production and/or composition of tears by the eye's lacrimal glands. Symptoms are redness, swelling and irritation, often accompanied by excessive watering of the affected eye. See also **Dry Eye**.

Ectropion. An abnormal turning out of an eyelid.

Electro-oculogram. A record of the standing voltage between the front and back of the eye that is correlated with eyeball movement, as in rapid eye movement (REM) sleep, and obtained by electrodes suitably placed on the skin near the eye.

Electroretinography. A test that measures the electrical impulses of the retina when the eye is exposed to light. For an ERG, an electrode is placed on the cornea at the front of the eye. The electrode measures the electrical response of the rods and cones, the visual cells in the retina at the back of the eye. An abnormal ERG is found in conditions such as arteriosclerosis of the retina, detachment of the retina, and temporal arteritis with eye involvement. The instrument used to do electroretinography is an electroretinograph, and the resultant recording is called an *electroretinogram*.

Emmetropia. Refractive condition, in which no refractive error is present. In other words, perfect 20/20 vision. Distant images are focused sharply on the retina without the need for corrective lenses.

Enucleation. Surgical removal of a diseased or damaged eyeball, leaving eye muscles and remaining orbital contents intact. See also **Enucleation**.

Epiphora. Overflow of tears down the cheek caused by defective tear drainage system or by excessive flow of tears. See also **Epiphora**.

Erbium laser. Laser developed as a refinement to the CO_2 laser used in skin resurfacing for acne removal and removal of shallow wrinkles. Erbium laser shows promise in a new procedure to emulsify the lens nucleus during cataract surgery.

Esophoria. A tendency of one eye to turn inward.

Esotropia. Misalignment of the eyes in which one eye deviates inward toward the nose while the other fixates normally. See also **Esotropia**.

Exophoria. A tendency of one eye to turn outward from the nose.

Exotropia. Misalignment of the eyes in which one eye deviates outward away from the nose while the other fixates normally. See also **Exotropia**.

Extracapsular cataract extraction. A surgical procedure that involves removing a cloudy or opaque lens (cataract) while leaving the back lens capsule intact. See also **Extracapsular Cataract Extraction (ECCE)**.

Field analyzer. Automated projection perimeter instrument, which uses projected points of light to determine central or peripheral field of vision. Used for early diagnosis, treatment and management of diseases resulting in visual field loss.

Fixation. The act, or an instance of focusing the eyes on an object.

Fixation point. The point in the visual field on which the two eyes focus in normal vision, and for each eye is the point that directly stimulates the fovea of the retina.

Flashers. Bright bursts of light, that are sometimes seen when the eyes are closed. These occur normally as the eye ages and the vitreous humor, which is jelly-like substance that fills center cavity of eye, begins pulling away from the retina. The flashes can appear off and on for several weeks or months. If there is a sudden appearance of many light flashes, an ophthalmologist should be consulted immediately to see if the retina has been torn.

Floaters. Particles that float in the vitreous, which is jelly-like substance that fills center cavity of eye, and cast shadows on the retina. The particles appear to be strung together with a web-like thread. See also **Floaters**.

Fluorescein angiography. Test used to examine blood vessels in the retina, choroid and iris of the eye. Fluorescein dye is injected into an arm vein and rapid, sequential photographs are taken of the eye as the dye circulates. See also **Fluorescein Angiography**.

Fovea. A small rodless area of the retina that affords the sharpest vision because the layers of the retina spread aside to let light fall directly on the cones, which are the cells that give the clearest vision.

Glaucoma. Condition caused by excessive buildup of fluid inside the eye putting pressure on the retina. Glaucoma is the leading cause of blindness among adults in the United States. There are two types of glaucoma; the most common being open angle and the less common, but more serious, called *narrow angle*. Glaucoma has few if any symptoms, but a simple, painless eye test detects the problem. If untreated, glaucoma can result in gradual, painless, irreversible loss of vision. See also **Glaucoma**.

Graves disease. Condition caused by over activity of the thyroid gland. Patients with Graves disease may experience these eye symptoms: inflammation of the eyes, swelling of the tissues around the eyes, and protrusion of the eyes. See also **Thyroid Gland**.

Heterophobia. A constant tendency of one eye to deviate in one or another direction due to imperfect balance of ocular muscles.

Histoplasmosis. Disease caused when airborne spores of the fungus Histoplasma Capsulatum are inhaled into the lungs. This fungus is found throughout the world where bird or bat droppings accumulate. Initial symptoms are usually mild, similar to a common cold. Even mild cases can later cause a serious eye disease called ocular histoplasmosis syndrome, a leading cause of vision loss in Americans ages 20 to 40. Although most people infected with the fungus never develop the disease, people who have lived in an area with a high rate of histoplasmosis should have their eyes examined regularly. See also **Histoplasmosis**.

Horopter. For a given fixation point, the set of points on the retina that are perceived to have zero retinal disparity. The points sweep out an arc in space that intersects a fixated point in space.

Hyperopia (farsightedness). Vision that results when there is too short a distance from the cornea to the retina. See also **Hyperopia (Farsightedness)**.

Infant vision development. At birth, infants can see patterns of light and dark but have blurred vision of specific objects. During the first 4 months, the two eyes will begin working together, with visual horizon expanding, vision becoming clearer and color vision developing. It is believed that by 4 months, an infant's color vision is similar to an adult's. By 6 months of age, babies acquire eye movement control and eye-hand coordination skills. It is normal at this stage for an infant's eyes to appear at times as crossed or out of alignment. Persistent misalignment should be reported to an eye care professional.

Inferior. In eyecare terminology, referring to the lower half of the eye.

Intracapsular Cataract Extraction (ICCE). Older method of cataract surgery, that involves removal of the entire lens. Some surgeons may still use this method in selected cases. See also **Intracapsular Cataract Extraction (ICCE)**.

Intraocular lens (IOL). Plastic lens surgically implanted to replace the focusing power of the natural crystalline lens following cataract extraction or during a refractive surgery procedure called Clear Lens Extraction. See also **Intraocular Lenses (IOLs)**.

Intraocular pressure (IOP). Measurement of fluid pressure inside the eye. All eye exams include a measurement of this eye pressure with an instrument called a tonometer. Glaucoma, the leading cause of blindness in adults, is caused by a buildup of pressure inside the eye and often occurs without noticeable symptoms. This buildup puts pressure on the optic nerve, which can permanently damage eyesight or even cause blindness. See also **Tonometer**.

Iritis. Inflammation of the iris causing pain, tearing, blurred vision, small pupil, and a red congested eye. See also **Iritis**.

Isopter. A contour line in a representation of the visual field around the points representing the macula lutea that passes through the points of equal visual acuity.

Itchy eyes. Condition that can be caused by many factors such as pollutants in the air, allergies, chemical exposure (swimming pool for instance), sun glare, viral and bacterial infections and conjunctivitis (pink eye). See also **Itchy Eyes**.

Keratitis. Infection or inflammation of the cornea which can be caused by a variety of conditions, including infections, dry eyes, foreign objects, contact lenses, intense light, vitamin A deficiency or allergies. Usual treatment is with antibiotic or anti-viral eye drops and ointments. See also **Keratitis**.

Keratoconus. Hereditary, degenerative corneal disease characterized by generalized thinning and cone-shaped protrusion of the corneal area usually affecting the vision of both eyes. See also **Keratoconus**.

Keratometer. Device for measuring the curvature of the cornea and for detecting and measuring astigmatism. The keratometer measures the frontal curvature, or steepness, of the cornea and compares high and low points to determine if a refractive problem exists.

Keratometry. Measurement of the form and curvature of the cornea.

Lacrimal system. Orbital structures of the eye responsible for tear production and drainage. Tears are produced in the lacrimal gland above the outer corner of the eye. See also **Lacrimal System**; and **Tears**.

Laser. Acronym for Light Amplification by Simulated Emission of Radiation. A very narrow, hi-intensity light, which can vaporize tissue and/or join structures.

Laser: Excimer. A high-energy, cold laser that is used in the Photorefractive Keratectomy (PRK) and Laser In-Situ Keratomileusis (LASIK) procedures to sculpt the central zone of the cornea to correct nearsightedness, farsightedness or astigmatism. See also **Laser In-situ Keratomileusis (LASIK)**; **Photoreactive Keratectomy (PRK)**; and **Refractive Eye Surgery**.

Laser: CO_2. High-energy laser that can be used for making surgical incisions and for skin resurfacing. Emits short, intense bursts of energy that vaporize the outer layer of skin without damage to the underlying skin.

Laser: Erbium. Laser developed as a refinement to the CO_2 laser used in skin resurfacing for acne scarring and wrinkle removal for shallow wrinkles. Erbium laser shows promise in a new procedure to emulsify the lens nucleus during cataract surgery.

Laser: Holmium. Infrared (thermal) laser developed in the late 1980s. The laser's beam is cool and can remove small areas of tissue without affecting surrounding tissue. In a refractive surgery procedure called Laser Thermal Keratoplasty (LTK), the Holmium laser is used to shrink the peripheral area of the cornea. See also **Laser Thermal Keratoplasty (LTK)**; and **Refractive Eye Surgery**.

Laser In-Situ Keratomileusis (LASIK). Currently the most widely used refractive eye surgery procedure. Combines the minimal post-operative discomfort of Automated Lamellar Keratoplasty (ALK) with the computer-controlled precision of Photorefractive Keratectomy (PRK). See **Automated Lamellar Keratoplasty (ALK)**; **Laser In-situ Keratomileusis (LASIK)**; **Photoreactive Keratectomy (PRK)**; and **Refractive Eye Surgery**.

Laser pointers. Pointing devices that use a laser beam. Light energy that some laser pointers can deliver into the eye may be more damaging than staring directly into the sun. Even momentary exposure can cause discomfort and temporary vision impairment.

Laser Vision Correction. Any of several surgical vision correction techniques that use a computer-controlled laser to gently sculpt the corneal tissue to correct vision error caused by corneas that are too steep (nearsighted), too flat (farsighted) or uneven (astigmatic). See also **Refractive Eye Surgery**.

Laser: Yag. Short pulsed, high-energy light beam that can be precisely focused by computer to optically cut, perforate, or fragment tissue.

Lateral. In eye care terminology, denoting a position farther from the median plane or midline of the eye.

Lateral geniculate nucleus. Structure in the thalamus, which is a major recipient of axons from the retina.

Lazy eye. Condition of decreased vision in one or both eyes without detectable structural abnormalities or disease in the eye or visual pathways. Also known as amblyopia. See also **Amblyopia**.

Lens: aspheric. Premium contact lens for borderline astigmatic patients and emerging presbyopes.

Lens: bifocal. Eyeglass lens made up of a main lens on top for distance vision and an additional lens on the bottom for near vision.

Lens: bitoric. Toric contact lens used to correct residual astigmatism.

Lens: crystalline. Natural lens of the eye that lies behind the iris and helps bring light rays to focus on the retina. Muscular elastic fibers contract or expand the lens so that it can focus on objects at varying distances. In this manner, the lens fine-tunes the focus of the light rays refracted by the cornea so that a sharp, clear and colorful image is focused on the retina. The lens of the eye is often compared to the lens of a camera.

Lens: meter. Device used for determining the refractive power of an eyeglass or contact lens.

Lens: multifocal. Eyeglasses or contact lenses, that enable the wearer to focus through two or more prescriptions for different distances on one lens. Bifocals have two points of focus, one for distance and the other for near, while trifocals have three points of focus—distance, intermediate and near. Progressive "no-line" eyeglass lenses offer a continuous range of focus from top to bottom.

Lens: progressive. "No-line" multi-focal eyeglass lenses with progressive powers that graduate from distance to reading power. Although progressive lenses often require a greater period of adjustment, these are the most versatile of all multi-focal designs because of the continuous range of focus.

Lens: spheric. Eyeglass or contact lens with a single continuous curve across the entire front surface.

Lens: toric. Contact lens that contains both a spherical and cylinder component to correct astigmatism. Thickness of lens may be modified from one meridian to another, thereby enabling the lens to maintain correct orientation on the eye.

Lens: trifocal. Eyeglass or contact lens design that includes three focal areas: usually a reading lens, a lens for faraway viewing, and a lens for mid-distance viewing.

Lensectomy. Surgery to remove the natural crystalline lens of the eye. During most cataract surgery, the cloudy cataract lens is removed and replaced with a plastic prescription lens. In a procedure called "Clear Lens Extraction" or "Clear Lensectomy," the eye's natural lens is also removed and replaced with a plastic prescription lens in order to correct refractive disorders. The word "clear" denotes that the natural lens that is removed is clear, rather than a cloudy cataract lens. See also **Cataract**.

Lensometer. Device used for determining the refractive power of an eyeglass or contact lens.

Lenticular. Special noncataract lenses for patients who have cataracts.

Light adaptation. The automatic adjustment of the pupil or the retina to the amount of light that is present at any given time.

Low vision. The condition that exists when ordinary eyeglasses, contact lenses, intraocular lens implants or refractive surgery do not provide clear vision. Should not be confused with blindness. People with low vision have useful vision that can be improved with vision devices. The terms "legally blind" or "partially sighted" are often used in association with low vision. A person is considered legally blind when the best-corrected vision in the better eye is no more than 20/200 and/or the field of view is less than 20 degrees. A person with best-corrected vision of no more than 20/70 in the better eye is considered partially sighted or visually impaired.

Lutein. Yellow carotenoid pigment found in body fats, egg yolks and green plants that promotes healthy eyes. One of two primary pigments found in the central part of the retina which helps filter out damaging light. See also **Lutein**.

Lysozyme. A basic protein that is present in egg white, saliva, and tears.

Macula. Small centralized area of the retina responsible for acute central vision. Damage to this portion of the retina severely limits a patient's ability to read, recognize faces and perform any other task that requires straight-ahead vision. See also **Macula**.

Macula lutea. The cone-rich area of the human eye, that contains the fovea. Also called yellow spot.

Macular degeneration. Degeneration of the photoreceptors in the macula or central region of the retina. This area of the retina is responsible for central vision, used for reading, seeing faces, and so on. Often associated with aging. See also **Age-related Macular Degeneration**.

Meibomian gland. Little glands in the eyelids, that make a fatty lubricant which they discharge through tiny openings in the edges of the lids. The meibomian glands can become inflamed, a condition termed meibomianitis or meibomitis. Chronic inflammation of the meibomian glands leads to a cyst, also called *chalazion*, which is a pimple in the margin of the eyelid. The meibomian glands are also known as the *palpebral glands, tarsal glands*, or *tarsoconjunctival glands*.

Melanin. Black pigment in the pigment epithelium cells that absorbs light not captured by the retina, thereby preventing the light from being reflected off the back of the eye.

Microkeratome. Sophisticated surgical device used to shave a very thin amount of the cornea at a predetermined depth. Used in LASIK refractive vision correction procedure to create the corneal "flap" which is lifted. Then the Excimer laser sculpts the underlying corneal tissue See also **Laser In-situ Keratomileusis (LAISK)**.

Microspectrophotometry. A procedure that involves passing a narrow measuring beam through the outer segments of individual photoreceptors to measure absorbancy in excised retinas.

Miosis. Condition in which the pupil is constricted. Occurs as a normal response to a bright light stimulus, to focusing on a near object (known as accommodation), or to administer certain drugs.

Monochromat. An individual who is completely color-blind.

Multi-focal. Eyeglasses or contact lenses, that allow the wearer to focus through different prescriptions for different distances on the same lens. Bifocals have two points of focus, one for distance and the other for near, while trifocals have three points of focus—distance, intermediate and near. Progressive "no-line" eyeglass lenses offer a continuous range of focus from top to bottom.

Multiple sclerosis. Chronic central nervous system disorder in which there is loss of the protective myelin sheath surrounding nerve tissue. Effects on the eye include optic nerve inflammation with reduced vision, double vision and involuntary eye oscillations.

Mydriasis. Increase in pupil size (dilation) occurring normally in the dark. May occur artificially through the use of drugs.

Mydriatic. A drop that dilates the pupil.

Myope. Medical term for nearsighted person. A person with good reading vision, but who has difficulty seeing distant objects.

Myopia. Also called nearsightedness. The front curvature of the cornea is too steep in a nearsighted person, causing good reading vision but poor distance vision. See also **Myopia (Nearsightedness)**.

Nasal. When referring to the eye, this term indicates inward direction toward the nose. Also refers to that half of the eye or visual field from the middle of the eye inward.

Nasal lacrimal system. That portion of the lacrimal (tear producing) system that includes ducts which drain tears from the eye into the nose.

Nasal lacrimal duct. Tear drainage channel that extends from lacrimal sac to opening in mucous membrane of nose.

Occlusion. The transient approximation of the edges of a natural opening, i.e, occlusion of the eyelids.

Ocular. Of, pertaining to, or affecting the eye.

Ocular hypertension. Condition in which the intraocular pressure of the eye is elevated above normal without any obvious optic nerve damage or visual field defects. Over time, ocular hypertension may develop into glaucoma.

See also **Ocular Hypertension**.

Ocular motility. The movement of the eye.

Onchocerciasis. River blindness, a disease caused by a parasitic worm (Onchocerca volvulus) transmitted by biting blackflies (buffalo gnats) that breed in fast-flowing rivers. The adult worms can live for up to 15 years in nodules beneath the skin and in the muscles of infected persons, where they produce millions of worm embryos (microfilariae) that invade the skin and other tissues including the eyes. About 18 million persons are affected, mostly in Africa and also in Yemen and Latin America. Both living and dead microfilariae cause severe itching in the skin and sometimes blindness after many years. Since1987, the drug ivermectin (brand name: Stromectol) has been provided by the manufacturer (Merck) free of charge. A single oral dose administered once a year prevents the accumulation of microfilariae in persons at risk. No drug suitable for mass treatment can kill the adult worms in the body, and therefore, onchocerciasis cannot be wiped out. The blindness, however, can be eliminated.

Open-angle glaucoma. The less serious of two types of glaucoma, a condition caused by excessive buildup of fluid inside the eye putting pressure on the retina. Sometimes called chronic glaucoma, it is caused by a gradual blocking of aqueous outflow from the eye. It can develop slowly with no noticeable symptoms. If untreated, open-angle glaucoma results in a gradual, painless, irreversible loss of vision. See also **Glaucoma**.

Ophthalmologist. A medical doctor (M.D.) with education, training and experience in medical and surgical treatment of eye diseases and disorders.

Ophthalmoscope. An instrument for viewing the interior of the eye. It comprises a concave mirror with a hole in the center through which the observer examines the eye, a source of light that is reflected into the eye by the mirror, and lenses in the mirror that can be rotated into the opening in the mirror to neutralize the refracting power of the eye being examined. It makes the image of the retina clear.

Ophthalmology. Branch of medical science that deals with the structure, functions and diseases of the human eye.

Optic disc. The circular area in the back of the inside of the eye where the optic nerve connects to the retina. Also called the optic nerve head. Contains no photoreceptors and therefore creates a blind spot in the visual field.

Optic nerve. Connects the eye to the brain. It carries the impulses formed by the retina, the nerve layer that lines the back of the eye and senses light and creates impulses. The brain interprets the images. Using an ophthalmoscope, the head of the optic nerve can be seen. It can be viewed as the only visible part of the brain (or extension of it).

Optic nerve head. The circular area in the back of the inside of the eye where the optic nerve connects to the retina.

Optic nerve pathways. The left and right branches of the optic nerves join behind the eyes, just in front of the pituitary gland, to form a cross-shaped structure called the *optic chiasma*. Within the optic chiasma, some of the nerve fibers cross. The fibers from the nasal (inside) half of each retina cross over, but those from the temporal (outside) half do not. Specifically, the fibers from the nasal half of the left eye and the temporal half of the right eye form the right optic tract; and the fibers from the nasal half of the right eye and the temporal half of the left form the left optic tract. The nerve fibers then continue along in the optic tracts. Just before they reach the thalamus of the brain, a few of the nerve fibers leave to enter nerve nuclei that function in visual reflexes. Most of the nerve fibers enter the thalamus, forming a junction (synapse) in the back of the thalamus. From there the visual impulses enter nerve pathways called the optic radiations, which lead to the visual (sight) cortex of the occipital (back) lobes of the brain.

Optic neuritis. Inflammation of the optic nerve. Characterized by rapid onset of decreased vision and usually accompanied by discomfort upon eye movement and central visual field defect. See also **Optic Neuritis**.

Optic neuroma. A benign tumor of the optic nerve.

Optic tract. The portion of each optic nerve between the optic chiasma and the diencephalon proper.

Optometrist. Doctors of Optometry (O.D.) are primary health care providers who examine, diagnose, treat and manage diseases and disorders of the eye and associated structures. In accordance with state law, optometrists prescribe, fit and dispense ocular medications, glasses and contact lenses. An optometrist cannot perform surgery but often works with an ophthalmologist on pre-and post-surgical care.

Optometry. The profession of examining the eyes, measuring vision and prescribing corrective lenses, diagnosing diseases of the eye and treating certain conditions that do not require license as a medical doctor.

Orbit. Pyramid-shaped cavity in the skull containing the eyeball, its muscles, blood supply, nerve supply and fat.

Orbital tumor. Process by which a living organism assimilates food and uses it for growth and tissue replacement. In the prevention of ocular disease, the role of vitamins and minerals has taken on a major role. To maintain the health of the eye, a high-fiber, high-carbohydrate, high-antioxidant, low-fat, low-protein diet is best.

Orthokeratology. Controversial nonsurgical contact lens procedure designed to eliminate nearsightedness and astigmatism. See also **Orthokeratology**.

Overconvergence. Condition in which the eyes come too far inward when focusing on a near object, resulting in blurring.

Pachymeter. Ultrasound machine used in measuring thickness of the cornea. Especially important in determining treatment depth in refractive surgery.

Pachometer. Instrument that uses optical principle of split images to measure corneal thickness or anterior chamber depth.

Palpebral glands Little glands in the eyelids, that make a fatty lubricant which they discharge through tiny openings in the edges of the lids. The meibomian glands can become inflamed, a condition termed meibomianitis or meibomitis. Chronic inflammation of the meibomian glands lead to a cyst, also called *chalazion*, which is a pimple in the margin of the eyelid. The meibomian glands are also known as the *meibomian glands, tarsal glands,* or *tarsoconjunctival glands.*

Perimeter. Instrument used for determining central or peripheral field of vision.

Perimetry. Method of charting extent of visual field as seen by the stationary eye. Aids in the detection of damage to the sensory visual pathways. See also **Perimetry**.

Peripheral vision. Side vision; vision not in the straight-ahead direction.

Phacoemulsification. Technique for removing eye's natural crystalline lens in cataract surgery or clear lens extraction. See also **Pacoemulsification**.

Phaco machine. Ultrasonic instrument used in phacoemulsification technique for removing the natural lens of the eye during cataract or clear lens extraction surgery. Using a hand piece connected to the phaco machine, the surgeon dissolves the lens into small fragments, aspirates the contents and irrigates the eye.

Phoropter. An instrument used by the eye doctor to determine the degree of myopia (nearsightedness), hyperopia (farsightedness) or astigmatism that is present in the patient's eye.

Photocoagulation. A surgical process of sealing off, or clotting, tissue by means of a laser beam. It is used in cancer treatment to destroy blood vessels entering a tumor and deprive it of nutrients, in the treatment of a detached retina, to destroy abnormal blood vessels in the retina, to treat tumors in the eye; etc.

Photophobia. Abnormal sensitivity to, and discomfort from, light. Frequently associated with excessive tearing and often due to inflammation of iris or cornea.

Photorefractive Keratectomy (PRK). Refractive eye surgery procedure that employs a computer-controlled Excimer laser system to sculpt by ablation (vaporize) the central corneal zone, or visual axis. By changing the shape of the patient's cornea, the eye care surgeon can correct some cases of nearsightedness, farsightedness and astigmatism. See **Photorefractive Keratectomy (PRK)**; and **Refractive Eye Surgery**.

Pigmentary retinopathy. Any of several hereditary progressive degenerative diseases of the eye marked by night blindness in the early stages, atrophy and pigment changes in the retina, constriction of the visual field, and eventual blindness. Also called *retinis pigmentosa.*

Pink eye. Also known as conjunctivitis. Inflammation of conjunctiva or membrane that covers the white of the eye and inner surfaces of the eyelid. Characterized by discharge, grittiness, redness and swelling. May result from virus, bacteria, allergens, chemical exposure, or ultraviolet light exposure. Can be contagious. See also **Conjunctiva**; and **Conjunctivitis**.

Plano. Term used by eye care professionals to describe lenses with no focusing power. The term is most often applied to nonprescription sunglasses or contact lenses that are worn for cosmetic purposes only.

Polarized lenses. Eyeglass lenses designed to protect the eye against UV rays and reduce the glare of reflected light. May be constructed of glass, plastic or lightweight, hi-index plastic.

Pole. Posterior pole refers to back curvature of eyeballs, usually to the retina between optic nerve and macular area. Anterior pole refers to center of front surface of cornea.

Posterior chamber. In referring to the eye, that space between the back which is filled with aqueous fluid between the iris and front face of the vitreous.

Presbyope. Person who has difficulty reading print and seeing nearby objects because of age-related loss of elasticity of the eye's natural crystalline lens.

Presbyopia. Deterioration in the ability of the eye's natural crystalline lens to expand or contract in order to focus on close objects. See also **Presbyopia**.

Prosopagnosia. The inability to recognize faces.

Ptosis. Condition in which, the upper eyelid(s) sag. May be congenital or caused by a later problem associated with a nonfunctioning levator muscle. Usually hereditary.

Punctum plug. Small, nondissolvable silicone plugs inserted in the tear draining ducts to close the openings of the tear draining system in order to slow drainage and loss of tears. Used to provide dry eye relief and/or reduce or eliminate the major cause of contact lens discomfort. See also **Tears**.

Pupil dilation. Enlarged pupil resulting from contraction of dilator muscle or relaxation of iris sphincter. Normally occurs in dim illumination conditions. Comprehensive examination of the interior of the eye requires pupil dilation through the administration of appropriate eye drops by the examining eye professional. Dilation can also be caused by injuries, including blunt trauma.

Radial Keratotomy (RK). Surgical procedure where patterned surgical incisions are made in the peripheral area of the cornea. As these incisions heal, the cornea is flattened to the degree required to make the desired refractive error correction. This early refractive vision correction procedure has been largely replaced by newer, more accurate procedures such as LASIK. See also **Refractive Eye Surgery**.

Red eye. Any condition, that causes the white part of the eye to look red. Redness usually is because of engorged blood vessels on the surface of the eye or hemorrhaging on the surface. In making a diagnosis, an eye doctor will pay close attention to the location and pattern of redness.

Refraction: objective. Eye test utilizing an automated device called an Auto Refractor that instantly measures the power of the eye and the outer shape of the eye. The device needs no input from the patient such as that required by the "Which is better? 1 or 2" question used in a phoropter exam.

Refraction: subjective. Eye test using a phoropter where patient is given choices between lens 1 and 2, providing the examining eye doctor with "subjective" information regarding which prescription will make the patient more comfortable.

Refractive error. Optical defect of the eye that causes light rays to focus in front of the retina (nearsighted), behind the retina (farsighted), or in several different places on the retina (astigmatic), resulting in less than perfect vision. These defects can normally be corrected with eyeglasses, contact lenses or refractive eye surgery.

Refractive surgery. Elective eye surgery which corrects optical defects of the eye by either changing the shape of the cornea or by inserting a plastic lens to supplement the eye's natural focusing ability. See also **Refractive Eye Surgery**.

Refractor. Instrument that aids an eyecare professional in determining the proper corrective prescription for a patient with less-than-perfect vision.

Retinal detachment. Painless disorder, where patient may notice gradual raising or lowering of a curtain over the visual field of the affected eye. See also **Retinal Detachment**.

Retinal hole. With age, the retina starts to thin and weaken especially near its attachments with the front of the eye (periphery). In addition, the jelly-like ball that fills most of the eye behind the lens, called the *vitreous*, changes from a firm substance to a loose fluid. When the vitreous becomes fluid, it can easily move and tug on its attachments and become detached, pulling a small bit of retina with it. When this happens, a hole is left in the retina.

Retinal rivalry. The perception of first one then the other of two visual stimuli, which differ in color or form when they are presented at the same time to congruent areas of both eyes.

Retinitis pigmentosa. Any of several hereditary progressive degenerative diseases of the eye marked by night blindness in the early stages, atrophy and pigment changes in the retina, constriction of the visual field, and eventual blindness. Also called *pigmentary retinopathy*. See also **Retinitis Pigmentosa**.

Retinoblastoma. Hereditary, malignant intraocular tumor, that develops from retinal cells. See also **Retinoblastoma**.

Rhabdomyosarcoma. A malignant tumor composed of striated muscle fibers.

Rhodopsin. The visual pigment in rod cells. Also called visual purple.

Sclera. Tough outermost layer of the eye, that is visible as the white of the eye.

Scleral buckle. Surgical procedure that repairs a retinal detachment by indenting, or buckling the sclera inward, usually by sewing a piece of preserved sclera or silicone rubber to the scleral surface. This pushes choroid and pigment epithelium closer to the retina and helps relax the vitreous tug on the retinal surface. See also **Scleral Buckle**.

Scotoma. Nonseeing area, or blind spot, within visual field, resulting from damage to visual pathways or to the retina. Blind spots exist normally in all eyes and mark the site of the optic nerve.

Scotopic vision. Dim light conditions where only rods are functional. Also called *twilight vision.*

Slit lamp. Microscope used in eye examination that projects a thin, intense beam of light into the eye through a controlled diaphragm. Aids in the diagnosis of diseases or trauma, which affect the structural properties of the anterior eye segment. See also **Slit Lamp**.

Snellen Eye Chart. Test chart for assessing visual acuity. Rows of letters, numbers and symbols in standardized graded sizes, with a designated distance at which each row should be legible to a normal eye. See also **Snellen Eye Chart**.

Spherical. Single prescription contact lenses with smooth spherical surfaces that bend light rays equally in all directions (360 degrees).

Sports vision. Growing specialty in many eye care offices, usually focusing on the safety aspects of sports, but may also include vision therapy designed to strengthen eye coordination and/or overcome congenital weaknesses. Sports safety usually includes ultra-violet and sun protection, glare protection and safety from direct impact.

Spots. Particles that float in the vitreous, which is jelly-like substance, that fills the center cavity of the eye, and cast shadows on the retina. Some spots are formed before birth while others occur normally with aging. The sudden appearance of many spots can be an indication of a serious eye disorder and should be checked by your eye doctor.

Stereopsis. Perception of depth, depending on the differences in the images projected on the retinas of the two eyes.

Stye. Inflammation of one or more sebaceous, or fluid-producing, glands of an eyelid. See also **Stye (Hordeolum)**.

Stiles-Crawford effect. An optical phenomenon in which light passing through the center of the pupil is perceived as more intense than light passing through the periphery of the pupil.

Superior. In eye care terminology, referring to the upper half of the eye.

Superior colliculus. Part of the brain that constitutes a primitive center for vision. Also called *optic lobe, optic tectu m.*

Sympathetic ophthalmia. Inflammation of one eye following inflammation in the other eye.

Tapetum. Silvery lining behind the retina in some animals active in dim light. Reflects light back through the eye and allows the photoreceptors a second chance to absorb photons.

Tarsoconjunctival glands. Little glands in the eyelids, that make a fatty lubricant which they discharge through tiny openings in the edges of the lids. The meibomian glands can become inflamed, a condition termed *meibomianitis* or *meibomitis*. Chronic inflammation of the meibomian glands lead to a cyst, also called *chalazion*, which is a pimple in the margin of the eyelid. The meibomian glands are also known as the *meibomian glands*, or *palpebral glands*.

Tay-Sachs disease (TSD). A progressive neurodegenerative disorder affecting babies. The child with TSD usually develops normally for the first few months of life. Then the baby gets an exaggerated startle reaction, and loses head control by 6 to 8 months of age. The infant cannot roll over or sit up, spasticity and rigidity develop, and excessive drooling and convulsions become evident. Blindness and head enlargement set in by the second year. It is fatal by age 5. TSD results from an enzyme deficiency, which in turn allows a fat, named lipid, to be deposited in the brain. It occurs primarily in Ashkenazi Jews of European origin, a group comprising 95 percent of the Jews in the United States.

Tears. Watery, slightly salty secretion of the lacrimal glands that serve to lubricate the front of the eye and wash away particles and foreign bodies. Natural tears are composed of three layers: the outer oily layer, the middle watery layer, and the inner mucus layer. See also **Tears**.

Thermokeratoplasty. Refractive eye surgery procedure that involves use of the Holmium laser to heat and shrink tissues in the peripheral area of the cornea in order to change the shape of the cornea to correct farsightedness and/or some cases of astigmatism. See also **Refractive Eye Surgery**.

Tonometer. Instrument used by the eye care professional to determine the intraocular or internal pressure of the eye. See also **Tonometer**.

Toric. Contact lenses that contain a cylinder component to correct astigmatism by bearing two different optical powers at right angles to each other. These lenses may be thicker in one meridian to enable the lens to maintain proper orientation on the eye. See also **Contact Lenses**.

Trachoma. Severe, chronic, contagious conjunctival eyelid and corneal infection caused by a virus. Leads to corneal blood vessel formation, corneal clouding, conjunctival and eyelid scarring and dry eyes. Leading cause of blindness in the world. See also **Trachoma**.

Trifocals. Eyeglass or contact lens design that includes three focal areas: usually a reading lens, a lens for faraway viewing, and a lens for mid-distance viewing.

Ultrasound: A-Scan. Type of ultrasound device that emits very high frequency waves that are reflected by the ocular structures and converted into electrical impulses. Used for differentiating normal and abnormal eye tissue or for measuring length of eyeball.

Ultrasound: B-Scan. Ultrasound procedure in which high frequency waves are reflected by eye tissues and orbital structures and then converted into electrical pulses that are displaced as bright spots on a black background. Provides a cross-sectional view of tissues used for evaluating structures that cannot be seen directly.

Ultrasound: eye. Transmission into the eye of high frequency sound waves that are reflected by ocular tissues and displayed on a screen so the internal structures can be visualized. Aids in diagnosis of eye and orbital problems.

Visual acuity. Measure of eye's ability to distinguish object details and shape. Assessed by smallest identifiable object that can be seen at a specified distance, usually 20 feet for distance vision and 16 inches for near vision. See also **Visual Acuity**.

Visual field. Extent of space visible to an eye as it looks (fixates) straight ahead. Measured in degrees away from fixation. See also **Visual Field**.

Vitreous. A clear, jelly-like liquid, that fills the middle of the eye. Also called the *vitreous humor*.

Vitreous detachment. Separation of vitreous, which is the jelly-like substance that fills the eye behind the lens, from the retinal surface. Frequently occurs with aging, but may occur in diseases such as diabetes and severe myopia. Usually harmless, but can create retinal tears, which may in turn lead to retinal detachment. See also **Vitreous Detachment**.

Von Hippel's angioma. Also known as Von Hippel-Lindau disease, this hereditary disorder is characterized by tumors of the retina, central nervous system, and visceral organs. See also **Von Hippel-Lindau Disease (VHL)**.

For a free eye test visit Vision Rx. Com: *http://www.visionrx.com/et/begin.asp?frombc=1*

Additional Reading

Abelson, M.B.: "Allergic Diseases of the Eye," W.B. Saunders Company, Philadelphia, PA, 2000.

Bahn, R.S.: "Thyroid Eye Disease," Kluwer Academic Publishers, Norwell, MA, 2001.

Barlow, R.B., Jr.: "What the Brain Tells the Eye," *Sci. Amer.*, 90 (April 1990).

Barns, M.W.: "Laser Surgery," *Sci. Amer.*, 84 (June 1991).

Batterbury, M. and B. Bowling: "Ophthalmology: An Illustrated Colour Test," *Harcourt Health Sciences*, San Diego, CA, 1999.

Cassel, G.H., M.D. Billig, and H.G. Randall: "The Eye Book: A Complete Guide to Eye Disorders and Health," Johns Hopkins University Press, Baltimore, MD, 2000.

Chawla, H.B.: "Ophthalmology," Butterworth-Heinemann, Inc., Woburn, MA, 1998.

Dryja, T.P., et al.: "Mutations within the Rhodopsin Gene in Patients with Autosomal Dominant Retinitis Pigmentosa," *N. Eng. J. Med.*, 1302 (November 8, 1990).

Dutton, J.J., S.F. Byrne, and A.D. Proa: "Diagnostic Atlas of Orbital Diseases," W.B. Saunders Company, Philadelphia, PA, 2000.

Eagle, R.C. and R. Lampert: "Eye Pathology: An Atlas and Basic Text," W.B. Saunders Company, Philadelphia, PA, 1999.

Easty, D.L. and J.M. Sparrow: "Oxford Textbook of Ophthalmology," Vol. 2, Oxford University Press, Inc., New York, NY, 1999.

Efron, N.: "Contact Lens Complications," Butterworth-Heinemann, Inc., Woburn, MA, 1999.

Evans, B.: "Pickwell's Binocular Vision Anomalies: Investigation and Treatment," Butterworth-Heinemann, Inc., Woburn, MA, 2001.

Evans, B. and S. Doshi: "Binocular Vision Disorders Assessment Investigation and Management," Butterworth-Heinemann, Inc., Woburn, MA, 2001.

Flam, F.: "Physicists Take a Hard Look at Vision," *Science*, 982 (August 20, 1993).

Galloway, N.R. and W.M. Amoaku: "Common Eye Diseases and Their Management," Springer-Verlag, Inc., New York, NY, 1999.

Glaser, J.S.: "Neuroophthalmology," 3rd Edition, Lippincott Williams & Wilkins, Philadelphia, PA, 1999.

Glickstein, M.: "The Discovery of the Visual Cortex," *Sci. Amer.*, 118 (September 1988).

Gold, D.H. and T.A. Weingeist, Editors: "The Eye in Systemic Disease," J.B. Lippincott, Philadelphia, PA, 1990.

Gold, D.H.: "Color Atlas of the Eye in Systemic Disease," Lipponcott-Raven, Philadelphia, PA, 2000.

Gross, D.A.: "Introduction to the Optics of the Eye," Butterworth-Heinemann, Inc., Woburn, MA, 2001.

Grossman, M. and G. Swartwout: "Natural Eye Care: An Encyclopedia," Keats Publishing, Inc., Chicago, IL, 1999.

Guyer, D.R., S. Chang, and W.R. Green, et al.: "Retina-Vitreous-Macula," Vol. 1, Harcourt Brace & Company, San Diego, CA, 1998.

Henson, D.B.: "Visual Fields," 2nd Edition, Butterworth-Heinemann, Inc., Woburn, MA, 2000.

Hubel, D.H.: "Eye, Brain, and Vision," W.H. Freeman Company, New York, NY, 1995.

Isbert, R.: "Fluorescein and ICG Angiography," Thieme Medical Publishers, Inc., New York, NY, 1998.

Jaffe, N.S., M.S. Jaffe, and G.F. Jaffe: "Cataract Surgery and Its Complications," 6th Edition, Mosby-Year Book, Inc., St. Louis, MO, 1997.

Jones, D.W. and L. Jones: "Common Contact Lens Complications: Their Recognition and Management," Butterworth-Heinemann, Inc., Woburn, MA, 2000.

Koertz, J.F. and G.H. Handelman: "How the Human Eye Focuses," *Sci. Amer.*, 92 (July 1988).

Lam, D.M. and C.J. Schatz: "Development of the Visual System," MIT Press, Cambridge, MA, 1991.

Leitman, M.W.: "Manual for Eye Examination and Diagnosis," 5th Edition, Blackwell Science, Inc., Malden, MA, 2000.

Lightman, S.: "HIV and the Eye," World Scientific Publishing Company, Inc., Riveredge, NJ, 2000.

Loewenfeld, I.E. and O. Lowenstein: "The Pupil," Butterworth-Heinemann, Inc., Woburn, MA, 1999.

Mahowald, M.A. and C. Mead: "The Silicon Retina," *Sci. Amer.*, 76 (May 1991).

Marx, J.: "How the Retinoblastoma Gene May Inhibit Cell Growth," *Science*, 1492 (June 14, 1991).

Merimee, T.J.: "Diabetic Retinopathy," *N. Eng. J. Med.*, 978 (April 5, 1990).

Miller, K.D., J.B. Keller, and M.P. Stryker: "Ocular Dominance Column Development: Analysis and Simulation," *Science*, 605 (August 11, 1989).

Nathans, J.: "The Genes for Color Vision," *Sci. Amer.*, 42 (February 1989).

Nathans, J., et al.: "Molecular Genetics of Human Blue Cone Monochromacy," *Science*, 831 (August 25, 1989).

Packer, A.J.: "Manual of Retinal Surgery," 2nd Edition, Butterworth-Heinemann, Inc., Woburn, MA, 2000.

Parrish, R.K.: "The Atlas of Ophthalmology," Butterworth-Heinemann, Inc., Woburn, MA, 1999.

Ramachandran, V.S.: "Form, Motion, and Binocular Rivalry," *Science*, 950 (February 22, 1991).

Rootman, J.: "Disease of the Orbit," Lippincott Williams & Wilkins, Philadelphia, PA, 2001.

Rudnicka, A. and J. Birch: "Diabetic Eye Disease: Identification and CO-Management," Butterworth-Heinemann, Inc., Woburn, MA, 2000.

Schoenlein, R.W., et al.: "The First Step in Vision: Femtosecond Isomerization of Rhodopsin," *Science*, 412 (October 18, 1991).

Schwarz, U., C. Busettini, and F.A. Miles: "Ocular Responses to Linear Motion are Inversely Proportional to Viewing Distance," *Science*, 1394 (September 22, 1989).

Shingleton, B.J.: "Eye Injuries," *N. Eng. J. Med.*, 408 (August 8, 1991).

Steinert, R.F. and C.A. Puliafito: "Nd-YAG Laser in Ophthalmology: Principles and Clinical Application of Photodisruption," W.B. Saunders Company, Philadelphia, PA, 1998.

Traboulsi, E.I.: "Genetic Diseases of the Eye," Oxford University Press, Ilnc., New York, NY, 1999.

Web References

International Academy of Sports Vision: *http://www.iasv.net/*

Karolinska Institutet: *http://www.mic.ki.se/Diseases/c11.html*

St LukesEye.com: *http://www.stlukeseye.com/eye_diseases.htm*

Vision Rx: *http://www.visionrx.com/home.asp*

VISUAL ACUITY.   Visual acuity is the medical term for sharpness of vision. It deals with the sharpness, or discrimination, of central vision, rather than the extent or clarity of peripheral vision. Refractive errors, which can be corrected with eyeglasses, are the most common cause of poor visual acuity. These include myopia, or nearsightedness; hyperopia, or farsightedness; and astigmatism. Myopia is a reduced ability to see distant objects clearly, hyperopia is a condition that initially causes difficulty in seeing nearby objects and progresses to affect distance vision, and astigmatism is blurred vision caused by abnormal curvature of the front surface of the cornea.

Overall visual acuity is measured by using the Snellen Eye Chart, with the large E at the top followed by rows of letters where each row is smaller than the previous one. A chart using the letter E facing up, down, left, and right is used for children and those who do not read. A person whose vision can be corrected to 20/200 vision in the better eye is considered legally blind.

Near visual acuity is measured with the Jaeger card, which has print samples of different sizes. The card is held 14 inches from the person's eye for the test. A result of 14/20 means that the person can read at 14 inches what someone with normal vision can read at 20 inches. The results of visual acuity tests are used to prescribe eyeglasses or other corrective measures. See also **Snellen Eye Chart**.

Vision Rx, Inc., Elmsford, NY.

VISUAL BINARIES.   The visual binary is a pair of stars (i.e., a double star that can be separated in a telescope). The larger the aperture of a telescope, the greater the resolving power and hence the greater the ability to see two point sources as separated. The limiting factor in observing visual binaries is not the aperture of the telescope, but the unsteadiness of the Earth's atmosphere, a phenomenon called "seeing."

Algol (Beta Persei), also known as the winking demon star, is one of the best observed of the visual binaries. See also **Algol (β Persei)**.

VISUAL FIELD.   The visual field is the total area in which perception is possible while an individual is looking straight ahead. For a person with normal vision, the visual field usually extends outward over an approximately 90-degree angle on each side of the vertical midline of the face. But the angle is smaller above and below the midline, especially for a person whose eyes are deep-set or who has prominent eyebrows. Because the visual fields of the two eyes overlap to a large extent, a defect in the field of one eye may not be evident when both eyes are open. Thus, the visual field can be divided into the area that is visible only to the right eye, the area that is visible to the left eye, and the middle region of binocular vision.

All the light from the visual fields that are left of center of both eyes falls on the right sides of the retinas of both eyes. This information is transmitted by the optic nerves to the right visual cortex of the brain. Information about the right fields of vision is transmitted to the left cortex, and information about the region of binocular vision is transmitted to both the right and left cortex.

An examination of the visual field measures the responsiveness of the peripheral area of the retina. That area does not have as many information-carrying nerve fibers as the center of the retina. The photoreceptors in the peripheral part of the retina are rods, which are less sensitive to light than the cones in the center of the retina, and so only larger and brighter objects are seen on the periphery of the field of vision. It is impossible to read fine print that is as little as 5 degrees to one side of a point on which the vision is fixed. The visual field examination generally is done by using electronic equipment that prints a computerized reading of an individual's field of vision and that can measure an individual's attention span and the consistency, accuracy and pattern of his or her responses to visual stimuli.

Peripheral vision can be lost because of glaucoma or a stroke. Although the loss caused by a stroke is sudden and obvious, the damage done by chronic glaucoma usually is gradual. Therefore, it is not apparent to the affected individual until considerable vision has been lost. Periodic tests of the visual field thus are important as people mature, because the incidence of glaucoma increases with age.

Loss of peripheral vision is obvious to anyone with macular degeneration, a progressive condition in which deterioration of the macula, the central part of the retina, produces a growing blind spot in the center of the field of vision.

Vision Rx, Inc., Elmsford, NY.

VISUAL FUNCTION (Eye).   The eye works like a camera with two lenses. The first lens is the cornea, a clear membrane that covers the front of the eye. The second lens is the eye's natural crystalline lens, which is located behind the pupil. The cornea is responsible for about 70% of the eye's focusing power, while the natural lens fine-tunes the image before it is focused on the retina, at the back of the eye. The retina works like the film in a camera, receiving light images and sending them through the optic nerve to the brain. If both lenses are working properly, the image is focused precisely on the surface of the retina and the result is perfect 20/20 vision.

Just as people are born with different sizes and shapes of hands, their eyes also vary in form and proportion. A perfect eye has an evenly rounded cornea that allows light to fall exactly on the retina resulting in perfect vision.

See also **Vision and the Eye**.

Vision Rx, Inc., Elmsford, NY.

VITAMIN.   An organic compound that performs specific and necessary functions in humans, livestock, and other living organisms—even when it may be present in very small concentrations, at the milligram or microgram per 100 gram levels. The term *vitamine* was proposed by a Polish biochemist (Casimir Funk) in 1912 to designate substances required in trace amounts in the diet to prevent various nutritional-deficiency diseases.

Nearly all vitamins are associated in some way with the normal growth function as well as with the maintenance and efficiency of living things. Various species are capable of synthesizing some of the vitamins from precursors that are present in the body. Synthesis is frequently by way of intestinal bacteria. In the case of vitamin D, substances in the skin combine with ultraviolet radiation from sunlight to yield the essential substance. Some vitamins, such as vitamin C, are specific, singular substances—in this case ascorbic acid. With other vitamins, there is a range of related compounds, as exemplified by the D, E, and K vitamins.

Because of inconsistencies in nomenclature, the B vitamins are not closely related as one might suspect. The B vitamins are different specific substances, and the use of the letter B to designate them indicates a degree of commonality that actually is not the case. Vitamin B_1 is thiamine, vitamin B_2 is riboflavin, vitamin B_6 is pyridoxine, vitamin B_{12} is cobalamin. Vitamins B_6 and B_{12}, for example differ markedly in function and structure. The alphabetical method of designation became complex and somewhat confusing as the various vitamins were recognized and studied over many years. During this period some substances were found to be identical with previously announced and described vitamins; or some substances were found not to be vitamins at all. Thus, over the years, the International Union of Pure and Applied Chemistry (I.U.P.A.C.) assigned new names to several of the vitamins.

The major vitamins are described in separate alphabetical entries in this book. Titles used for these entries have been selected on the basis of the most frequently used designations as of the early 1980s. In alphabetical order, the vitamins described in this book are: **Ascorbic Acid (Vitamin C); Biotin; Choline; Folic Acid; Inositol; Niacin; Pantothenic Acid; Riboflavin (Vitamin B_2); Thiamine (Vitamin B_1); Vitamin A; Vitamin B_6 (Pyridoxine); Vitamin B_{12} (Cobalamin); Vitamin D; Vitamin E (Tocopherols); and Vitamin K.**

The daily requirements of vitamins by humans are summarized in the entry on **Diet**. The relationship between hormones and vitamins is described in the entry on **Hormones**. Vitamins also are mentioned

TABLE 1. COMPARATIVE LOSSES OF VITAMINS FROM VEGETABLES (CANNING AND FREEZE PROCESSING)

Method of Preservation	Value	Loss of Vitamins as Compared with Values of Fresh Cooke				
		Vitamin A	Thiamine (B$_1$)	Riboflavin (B$_2$)	Niacin	Ascorbic Acid (C)
Frozen, cooked (boiled), drained	mean	12%	20%	24%	24%	26%
	range	0–50%	0–61%	0–45%	0–56%	0–78%
Canned, drained solids	mean	10%	67%	42%	49%	51%
	range	0–32%	56–83%	14–50%	30–65%	28–67%

TABLE 2. COMPARATIVE LOSSES OF VITAMINS FROM FRUITS

Method of Preservation	Value	Loss of Vitamins as Compared with Values of Fresh Products				
		Vitamin A	Thiamine (B$_1$)	Riboflavin (B$_2$)	Niacin	Ascorbic Acid (C)
Frozen (not thawed)	mean	37%	29%	17%	16%	18%
	range	0–78%	0–66%	0–67%	0–33%	0–50%
Canned, solids and liquids	mean	39%	47%	57%	42%	56%
	range	0–68%	22–67%	33–83%	25–60%	11–86%

frequently in descriptions of various fruits, vegetables, and other food-stuffs throughout the book. Vitamins also figure prominently in discussions of some of the diseases, scores of which are described in this book.

Loss of Vitamins in Processing. During the last several years, much research has gone into determining the loss in effectiveness of vitamins as various foods are processed. It is a common tendency on the part of consumers to regard any fresh food as representing perfection in terms of nutritive, including vitamin, value and, conversely, to regard processed foods as nutritionally inferior. Under normal circumstances, these observations are true. Because fresh foods frequently are stored for several days at temperatures well above their freezing points, there are vitamin losses in unprocessed produce. Ascorbic acid content in vegeTables, for example, can severely degrade during improper storage. The degradation of vitamin values depends a great deal upon the type of food substance, the particular vitamin, and the manner in which the raw food is processed. The consumer today also is protected by vitamin-fortified foods, where vitamins have been added to compensate for losses during processing, or, in some cases, to generally enrich the foods nutritionally. Losses of vitamins from fruits and vegeTables during processing are tabulated in Tables 1 and 2.

Additional Reading

Ball, G.F.M.: "Fat-Soluble Vitamin Assays in Food Analysis: A Comprehensive Review," Elsevier, New York, NY, 1989.

Ball, G.F.M.: "Bioavailability and Analysis of Vitamins in Foods," Chapman & Hall, New York, NY, 1997.

Barinaga, M.: "Vitamin C Gets a Little Respect," *Science*, 374 (October 18, 1991).

Beardsley, T.: "Vitamin A and Its Cousins are Potent Regulators of Cells," *Sci. Amer.*, 16 (February 1991).

Blomhoff, R., et al.: "Transport and Storage of Vitamin A," *Science*, 399 (October 19, 1990).

Gaby, S.K., et al.: "Vitamin Intake and Health: A Scientific Review," Marcel Dekker, New York, NY, 1991.

Lipkowitz, M.A. and T. Navarra: "Encyclopedia of Vitamins, Minerals and Supplements," Facts on File, Inc., New York, NY, 1996.

Staff: Academic Press: "Vitamins and Hormones," Vol. 63, Academic Press, IInc., San Diego, CA, 2001.

Suttie, J.W., D.B. McCormic, and R. Rucker: "Handbook of Vitamins," 3rd Edition, Marcel Dekker, Inc., New York, NY, 2001.

VITAMIN A. This substance also has been referred to as retinol, axerophthol, biosterol, vitamin A$_1$, anti-xerophthalmic vitamin, and anti-infective vitamin. The physiological forms of the vitamin include: Retinol (vitamin A$_1$) and esters; 3-dehydroretinol (vitamin A$_2$) and esters; 3-dehydroretinal (retinine-2); retinoic acid; neovitamin A; neo-b-vitamin A$_1$. The vitamin is required by numerous animal species. All vertebrates and some invertebrates convert plant dietary carotenoids in gut to vitamin A$_1$, which is absorbed. Most animal species store appreciable amounts of the vitamin in their livers, have low concentrations in the blood, and undetectable quantities in most other tissues. A deficiency of the vitamin produces a variety of symptoms, the most uniform being eye lesions, nerve degeneration, bone abnormalities, membrane keratinization, reproductive failure, and congenital abnormalities. Toxic symptoms from large doses of vitamin A are readily produced in animals and humans. Overdosage may cause irritability, nerve lesions, fatigue, insomnia, pain in bones and joints, exophthalmia, and mucous cell formation in keratinized membranes.

The principal physiological functions of this vitamin include growth, production of visual purple, maintenance of skin and epithelial cells, resistance to infection, gluconeogenesis, mucopolysaccharide synthesis, bone development, maintenance of myelin and membranes, maintenance of color and peripheral vision, maintenance of adrenal cortex and steroid hormone synthesis. Specific vitamin A deficiency diseases include xerophthalmia, nyctalopia, hemeralopia, keratomalacia, and hyperkeratosis.

In the rods of the retina, retinal is found combined with the protein *opsin*, the complex being called *rhodopsin* (visual purple). Although the entire series of reactions involved in dark vision has not been entirely worked out, the major steps in the cycle are quite clear. All-*trans*-retinol from the blood is oxidized by alcohol dehydrogenase (with NADP, nicotinamide adenine dinucleotide phosphate) to retinol which, in turn, is isomerized in the retina to 11-*cis*-retinal. This combines with opsin to form rhodopsin. On exposure to light, rhodopsin undergoes a sequence of changes with the eventual splitting off of retinal, which now has the all-*trans* configuration. This presumably can be reutilized in the retina by isomerization, or it can be reduced to retinol by alcohol dehydrogenase and returned to the circulation either as the free alcohol or as an ester.

The relatively recent observation, that retinoic acid can replace retinol or retinal for normal growth of animals, gave rise to further concepts in the biochemistry of vitamin A. Although retinoic acid cannot be demonstrated to be present normally in animal tissues, its formation by liver aldehyde dehydrogenase (NAD) and aldehyde oxidase has been accomplished, so that the molecule must be considered in the general scheme of vitamin A metabolism. When retinoic acid is given to animals as the only form of vitamin A, growth is normal, but the animals eventually become sterile and blind. This had led to the consideration that vitamin A may have at least three independent functions: (1) growth; (2) vision; and (3) reproduction.

The reversal of the oxidative pathway of vitamin A (retinol → retinal → retinoic acid) does not occur in the body. When retinoic acid is fed to animals, even in relatively large doses, there is no storage and, in fact, the molecule is rapidly metabolized and cannot be found several hours after administration. The metabolic products have not been fully identified. Several fractions from liver or intestine, isolated after administering retinoic acid marked with carbon-14, have been shown to have biological activity.

In 1912, Hopkins reported a factor in milk needed for the growth of rats. In 1913, Osborne and Mendel demonstrated that milk factor is fat soluble, and present in other fats also. McCollum and Davis, in 1913–1915, identified milk factor (fat-soluble A) in butter and egg yolk. In 1917, McCollum and Simmonds found xerophthalmia in rats due to lack of fat-soluble A. In 1920, Drummond renamed fat-soluble A, vitamin A. In 1930, Moore determined that carotene is a precursor of vitamin A. See also **Carotenoids**. During 1930–1937, Karrer et al. isolated and synthesized vitamin A. In 1935, Wald reported visual purple in retina to be a complex of protein and vitamin A.

Distribution and Sources. Provitamin carotenoids are contained in numerous foods, but of varying concentrations.

High vitamin A and procarotenoids content (10,000–76,000 I.U./100 grams).[1] Carrot, dandelion green, kohlrabi, liver (beef, calf, chicken, pig, sheep), liver oil (cod, halibut, salmon, shark, sperm whale), mint, palm oil, parsley, spinach, turnip greens.

Medium vitamin A and procarotenoids content (1,000–10,000 I.U./100 grams). Apricot, beet greens, broccoli, butter, chard, cheese (except cottage), cherry (sour), chicory, chives, collards, cream, eel, egg yolk, endive, fennel, kale, kidney (beef, pig, sheep), leek greens, lettuce (butterhead and romaine), liver (pork), mango, margarine, melons (yellow), milk (dried), mustard, nectarine, peach, pumpkin, squash (acorn, butternut, hubbard), sweet potato, tomato, watercress, whitefish.

Low vitamin A and procarotenoids content (100–1,000 I.U./100 grams). Artichoke, asparagus, avocado, banana, bean (except kidney), berry (black-, blue-, boysen-, goose-, logan-, rasp-), Brussels sprouts, cabbage, carp, cashew, celery, cherry (sweet), clam, corn (maize), cowpea, cucumber, currant (red), grape, groundnut (peanut), hazelnut, herring, kumquat, leek, lentil (dry), lettuce, milk, okra, olive, orange, oyster, pea, pecan, pepper (sweet), pineapple, pistachio, plum, prune, rhubarb, rutabaga, salmon, sardine, squash (summer and zucchini), tangerine, walnut (black).

In higher plants, carotenoids are produced in green leaves. In animals, conversion of carotenoids to vitamin A occurs in the intestinal wall. Storage is in the liver; also kidney in rat and cat. Target tissues are retina, skin, bone, liver, adrenals, germinal epithelium. Commercial Vitamin A supplements are obtained chemically by extraction of fish liver; or synthetically from citral or β-ionone.

Bioavailability of Vitamin A. Factors which may cause a decrease in the availability of vitamin A include: (1) liver damage; (2) impaired intestinal conversion of carotenes; (3) impaired absorption (low bile); (4) loss in food preparation (cooking and frying — heat oxidation); (5) presence of antagonists; (6) illness, causing increased destruction and excretion of the vitamin. Increases in availability may result from: (1) storage in body (liver); (2) factors which stimulate intestinal conversion of carotenes — tetraiodothyronine (thyroxine), insulin; (3) absorption aids — bile, fat; and (4) dietary protein which mobilizes vitamin A from storage in liver.

Antagonists of vitamin A include sodium benzoate, bromobenzene, citral, oxidized derivatives of vitamin A, excessive concentrations of thyroxine, estrogens, vitamin E (as regards membrane permeability). Synergists include vitamins B₂, B₁₂, and E, ascorbic acid, thyroxine, testosterone, melanocyte-stimulating hormone (MSH), and somatotrophin growth hormone.

Unusual features of vitamin A as observed by some investigators include: (1) decreases serum cholesterol in large-quantity administration (chicks); (2) dietary protein required to mobilize liver reserves of vitamin A; (3) decreased quantities in tumors; (4) coenzyme Q_{10} accumulates in A-deficient rat liver; (5) Ubichromenol-50 accumulation in A-deficient rat liver; (6) retinoic acid functions as vitamin A except for visual and reproductive functions; (7) anti-infection properties and anti-allergic properties; (8) decreases basal metabolism; (9) detoxification of poisons in the liver aided by vitamin A; and (10) vitamin A is involved in triose → glucose conversions.

Additional Reading

Ball, G.F.M.: "Bioavailability and Analysis of Vitamins In Foods," Chapman & Hall, New York, NY, 1997.

Combs, G.F. Jr.: "The Vitamins: Fundamental Aspects in Nutrition and Health," 2nd Edition, Academic Press, Inc., San Diego, CA, 1998.

Eitenmiller, R.R. and W.O. Landen: "Vitamin Analysis for the Health and Food Sciences," CRC Press, LLC., Boca Raton, FL, 1998.

Lipkowitz, M.A. and T. Navarra: "Encyclopedia of Vitamins, Minerals and Supplements," Facts on File, Inc., New York, NY, 1996.

McDowell, L.R.: "Vitamins in Animal and Human Nutrition," 2nd Edition, Iowa State University Press, Ames, IA, 2000.

Nau, H. and W.S. Blaner: "Retinoids: The Biochemical and Molecular Basis of Vitamin A and Retinoid Action," Vol. 139, Springer-Verlag, Inc., New York, NY, 1999.

[1] One I.U. = 0.344 microgram vitamin A acetate = 0.3 microgram retinol.

Newstrom, H.: "Nutrients Catalog: Vitamins, Minerals, Amino Acids, Macronutrients — Beneficial Use, Helpers, Inhibitors, Food Sources, Intake Recommendations, and Symptoms of over or under Use," McFarland & Company, Inc., Publishers, Jefferson, NC, 1993.

VITAMIN B₆ (Pyridoxine). Infrequently called adermine or pyridoxol, this vitamin participates in protein, carbohydrate, and lipid metabolism. The metabolically active form of B₆ is pyridoxal phosphate, the structures of which are:

Pyridoxal phosphate enzymes mediate the nonoxidative decarboxylation of amino acids. This mechanism is of primary importance in bacteria, but it may be essential to proper function of the nervous system in humans by providing a pathway for the synthesis of a nerve impulse inhibitor, γ-amino-butyric acid from glutamic acid:

Pyridoxal phosphate is also a cofactor for transamination reactions. In these reactions, an amino group is transferred from an amino acid to an α-keto acid, thus forming a new amino acid and a new α-keto acid. Transamination reactions are important for the synthesis of amino acids from non-protein metabolites and for the degradation of amino acids for energy production. Since pyridoxal phosphate is intimately involved in amino acid metabolism, the dietary requirement for vitamin B₆ increases as the protein content of the diet increases.

The coenzyme especially participates in gluconeogenesis, production of neural hormones, bile acids, unsaturated fatty acids, and porphyrins.

A deficiency of the vitamin can result in lymphopenia, convulsions, dermatitis, irritability, and nervous disorders in humans. A deficiency in monkeys may cause arteriosclerosis, while in rats, acrodynia. Research indicates that all animals require vitamin B₆. Bacteria in intestines generate some of this vitamin, but relatively little is available to humans in this form. Endogenous sources are available to plants, fungi, and some bacteria.

In 1934, György cured a dermatitis in rats (not due to vitamins B₁ or B₂) with a yeast extract factor. In 1938, Lepkovsky isolated a similar factor from rice bran extract. In that same year, Keresztesy and Stevens isolated and crystallized pure B₆ from rice polishings. Also, in the same year, Kohn, Wendt, and Westphal synthesized pyridoxine and gave pyridoxine its present name. In the following year (1939), Stiller, Keresztesy, and Stevens established the structure of the vitamin. In 1945, Snell observed pyridoxal and pyridoxamine. The recognition of and establishment of B₆ requirements in humans was not achieved until 1953, by Snyderman et al.

In plants, the vitamin is present as pyridoxol-5-phosphate, pyridoxal-5-phosphate, or pyridoxamine-phosphate. In plants, production sites are found in fungi, cereal germ, and seeds.

Commercially, the vitamin is available as a dietary supplement in the compound pyridoxine hydrochloride. The compound can be synthesized by condensing ethoxy acetylacetone with cyanoacetamide (method of Harris and Folkers); or from oxazoles.

Distribution and Sources. Most fruits and vegetables are low in pyridoxine content, although most nuts are quite high. Cereals and a number of other substances have low-to-medium content.

High pyridoxine content (1,000–10,000 micrograms/100 grams). Groundnut (peanut), herring, liver (beef, calf, pork), molasses (black strap), rice (brown), salmon, walnut, wheat germ, yeast.

Medium pyridoxine content (100–1,000 micrograms/100 grams). Avocado, banana, barley, beef, Brussels sprouts, butter, cabbage, carrot, cauliflower, cod, corn (maize), eggs, flounder, grape, halibut, kale, lamb, mackerel, oats, pea, pear, pork, potato, rye, sardine, soybean, spinach, tomato, tuna, turnip, veal (brain, heart, kidney), whale, wheat, yam.

Low pyridoxine content (10–100 micrograms/100 grams). Apple, asparagus, bean, beet greens, cantaloupe, cheese, cherry, currant (red), grapefruit, lemon, lettuce, milk, onion, orange, peach, raisin, strawberry, watermelon.

Bioavailability of Pyridoxine. Factors which tend to decrease bioavailability of pyridoxine include: (1) Administration of isoniazid; (2) loss in cooking (estimated at 30–45%)—vitamin is water-soluble; (3) diuresis and gastrointestinal diseases; (4) irradiation. Availability can be increased by stimulating intestinal bacterial production (very small amount), and storage in liver. The target tissues of B_6 are nervous tissue, liver, lymph nodes, and muscle tissue. Storage is by muscle phosphorylase (skeletal muscle—small amount). It is estimated that 57% of the vitamin ingested per day is excreted. The vitamin exerts only limited toxicity for humans.

Precursors for biosynthesis of the vitamin include glycine, serine, or glycolaldehyde, although further research is required for further confirmation of these substances. Intermediates have not been identified. Antagonists of B_6 include 4-deoxypyridoxine, 4-methoxypyridoxine, toxopyrimidine, penicillamine, semicarbazide, and isoniazid. Synergists include ascorbic acid, biotin, epinephrine, folic acid, glucagon, niacin, norepinephrine, somatotrophin (growth hormone), and vitamins B_1, B_2, and E.

Determination of Vitamin B_6. As pointed out by investigators Gregory and Kirk (Department of Food Science and Human Nutrition, Michigan State University, East Lansing, Michigan), development of an adequate chemical procedure for the determination of biologically active forms of vitamin B_6 in foods has been a complex problem. Basic studies by Bonavita (1960), Toepfer et al. (1961), and Polansky et al. (1964) have demonstrated the feasibility of fluorometric measurement of pyridoxal (PAL), pyridoxamine (PAM), and pyrodixine (PIN) by conversion to PAL and reaction with potassium cyanide, forming the fluorophore 4-pyridoxic acid lactone. Various fluorometric methods have been applied to vitamin B_6 compounds in biological materials (Fujiita et al., 1955; Contractor and Shane, 1968; Loo and Badger, 1969; Takanashi et al., 1970; Fieldlerova and Davidek, 1974; Chin, 1975). The results of Chin suggested that interfering compounds may be present in the PAL fraction after column chromatographic separation of the B_6 analogs by the procedure of Toepfer and Lehmann (1961). In the Gregory-Kirk (1977) study, methods for improving chromatographic separation and fluorometric determination of vitamin B_6 compounds in foods were investigated. Their findings are presented in the reference indicated.

Traditionally, B_6 compounds also have been determined by bioassay, including rat and chicken growth assays.

Additional Reading

Ball, G.F.M.: "Bioavailability and Analysis of Vitamins In Foods," Chapman & Hall, New York, NY, 1997.

Combs, G.F. Jr.: "The Vitamins: Fundamental Aspects in Nutrition and Health," 2nd Edition, Academic Press, Inc., San Diego, CA, 1998.

Eitenmiller, R.R. and W.O. Landen: "Vitamin Analysis for the Health and Food Sciences," CRC Press, LLC., Boca Raton, FL, 1998.

Lipkowitz, M.A. and T. Navarra: "Encyclopedia of Vitamins, Minerals and Supplements," Facts on File, Inc., New York, NY, 1996.

McDowell, L.R.: "Vitamins in Animal and Human Nutrition," 2nd Edition, Iowa State University Press, Ames, IA, 2000.

Newstrom, H.: "Nutrients Catalog: Vitamins, Minerals, Amino Acids, Macronutrients—Beneficial Use, Helpers, Inhibitors, Food Sources, Intake Recommendations, and Symptoms of over or under Use," McFarland & Company, Inc., Publishers, Jefferson, NC, 1993.

VITAMIN B₁₂ (Cobalamin). Sometimes also called cyanocobalamin, this vitamin is one of the more recent of the major B complex vitamins to be fully identified, with its structure not definitized (by Hodkin

et al.) until 1955. The vitamin is required by most vertebrates, some protozoa, bacteria, and algae. Principal physiological functions include: (1) Coenzyme in nucleic acid, protein, and lipid synthesis; (2) maintains growth; (3) participates in methylations; (4) maintains epithelial cells and nervous system (myelin sheath); (5) erythropoiesis (with folic acid); (6) leukopoiesis. Deficiency diseases or disorders include retarded growth; pernicious anemia; megaloblastic anemia; macrocytic, hyperchromic anemia; glossitis; spinal cord degeneration; and sprue. The major physiological forms of B₁₂ available include hydroxocobalamin (vitamin B₁₂ₐ) and aquocobalamin (vitamin B₁₂c).

In 1926, Minot and Murphy controlled pernicious anemia using liver. In 1944, Castle demonstrated intrinsic factor needed to control pernicious anemia with liver. Rickes et al., in 1948, isolated and crystallized factor in liver controlling pernicious anemia. In that same year, Smith and Parker crystallized and designated liver factor as vitamin B₁₂. West demonstrated, in 1948, clinical activity of vitamin B₁₂, and, in 1955, Hodgkin et al. determined the structure of the vitamin. This is shown in Structure 1. Vitamin B₁₂ is the only vitamin with a metal ion—in this case, cobalt. Surrounding the cobalt is a macrocyclic corrin ring that is comprised of four nitrogen-containing, five-membered rings joined through three methylene bridges. There is a similarity between this corrin ring and the dihydroporphyrin (chlorin) ring of chlorophyll.

Structure 1.   Vitamin B₁₂.

Absorption of Vitamin B_{12}. This vitamin is not synthesized in animals, but rather it results from the bacterial or fungal fermentation in the rumen, after which it is absorbed and concentrated during metabolism. Among the known vitamins, this exclusive microbial synthesis is of great interest. One of the major results of vitamin B₁₂ deficiency is pernicious anemia. This disease, however, usually does not result from a dietary deficiency of the vitamin, but rather by an absence of a glycoprotein ("gastric intrinsic factor") in the gastric juices that facilitates absorption of the vitamin in the intestine. Control of the diseases hence is either by injection of B₁₂ or by oral administration of the intrinsic factor, with or without the vitamin injection. See also **Anemias**.

There are two separate and distinct mechanisms for absorption of vitamin B₁₂. One mechanism is active, the other passive; both operate simultaneously. The active process is physiologically more important, since it is operative primarily in the presence of the small (1–2 micrograms) quantities of vitamin B₁₂ made available for absorption from the average meal. This special mechanism, perhaps uniquely necessary for vitamin B₁₂ because of its large size and polar properties, operates as follows. The normal gastric mucosa secretes a substance, called the *intrinsic factor of Castle*, which combines with free vitamin B₁₂. The complex travels down the intestine to the ileum, where, in the presence of calcium and pH above 6, it attaches to "receptors" lining the wall of the ileal mucosa.

Vitamin B$_{12}$ is then freed from intrinsic factor via a "releasing factor" mechanism of unknown nature, operating either at the surface of or within the ileal mucosal cell, and passes into the bloodstream. Thus, important requirements for normal absorption of vitamin B$_{12}$ from food are: (1) the vitamin must be freed from its peptide bonds in food; (2) the gastric mucosa must secrete an adequate quantity of intrinsic factor; (3) the ileal mucosa must be sufficiently normal both structurally and functionally so that vitamin B$_{12}$ may be absorbed across it.

Intrinsic factor is believed to be a glycoprotein or mucopolysaccharide with a molecular weight in the range of 50,000 and an end-group conformation like that of partly degraded blood group substance. The sole known role of intrinsic factor is to facilitate the transport of the large (molecular weight = 1,355) vitamin B$_{12}$ molecule across the wall of the ileal mucosa and into the bloodstream. Antibodies to intrinsic factor exist in the serum of approximately half of all patients with pernicious anemia.

The second mechanism for vitamin B$_{12}$ absorption is operative primarily in the presence of quantities of vitamin B$_{12}$ greater than those made available for absorption from the average diet (*i.e.*, quantities greater than about 30 micrograms). This mechanism is a passive one, probably diffusion, and most likely occurs along the entire length of the small intestine. It operates when patients with pernicious anemia (vitamin B$_{12}$ deficiency due to inadequate or absent intrinsic factor secretion of unknown cause) are treated with large quantities (500 micrograms or more daily) of oral vitamin B$_{12}$. Such treatment is probably better than treatment with oral hog intrinsic factor, to which refractoriness often develops, but it is not as certain as treatment with monthly injections of vitamin B$_{12}$.

Deficiency Effects. Further elucidating on the physiologic functions and deficiency disorders of vitamin B$_{12}$, this vitamin is required for DNA (deoxyribonucleic acid) synthesis and, therefore, is necessary in every reproducing cell in humans for maintenance of the ability to divide. The vitamin functions coenzymatically in the methylation of homocysteine to methionine. It is important in several isomerization reactions, and as a reducing agent, and is probably of special importance in enzymatic reduction of ribosides to deoxyribosides. It is involved in protein synthesis, partly via its role in the conversion of homocysteine to methionine; in fat and carbohydrate metabolism, partly via its role in the isomerization of succinate to methylmalonate (which then may be decarboxylated to propionate), and in folate metabolism. Where these two vitamins interrelate, vitamin B$_{12}$ appears to serve as a coenzyme and folate as a substrate; such is true in the vitamin B$_{12}$-mediated transfer of a methyl group from N^5-methyltetrahydrofolic acid to homocysteine, which is thereby converted to methionine.

Vitamin B$_{12}$ is one of the most potent nutrients known; the minimal daily requirement for absorption by the normal adult is probably in the range of 0.1 microgram. This equals, for example, 1/500th of the minimal daily adult folate requirement, which is in the range of 50 micrograms.

As with all nutritional deficiencies, lack of vitamin B$_{12}$ may arise from inadequate ingestion, absorption, or utilization, and from increased requirement or increased excretion. Deficiency of vitamin B$_{12}$ produces megaloblastic (large germ cell) anemia, damage to the alimentary tract (glossitis being the most striking feature), and neurologic damage. The most classic neurologic sign of vitamin B$_{12}$ deficiency is decreased ability to perceive the vibration of a tuning fork pressed against the ankles. This finding is associated with damage to the posterior and lateral columns of the spinal cord, and also with damage to the peripheral nerves. This damage occurs because vitamin B$_{12}$ deficiency results in gradual deterioration of the myelin sheath, which is followed by deterioration of the axon. These processes occur slowly over months to years, and during this stage are reversible by treatment with vitamin B$_{12}$. However, when the nerve nucleus finally deteriorates, the neurologic damage becomes irreversible.

Distribution and Sources of Vitamin B$_{12}$. Vegetables, fruits, seeds, and nuts have a very low content of this vitamin.

High vitamin B$_{12}$ content (50–500 micrograms/100 grams). Brain (beef), kidney (beef, lamb), liver (beef, calf, lamb, pork).

Medium Vitamin B$_{12}$ content (5–50 micrograms/100 grams). Clam, crab, egg yolk, heart (beef, chicken, rabbit), kidney (rabbit), liver (chicken, rabbit), oysters, sardine, salmon.

Low Vitamin B$_{12}$ content (0.5–5 micrograms/100 grams). Beef, cod, cheeses, chicken, eggs, flounder, haddock, halibut, lamb, lobster, milk, pork, scallops, shrimp, swordfish, tuna, whale.

Vitamin B$_{12}$ dietary supplements are often prepared commercially by the fermentation of *S. griseus, S. aureofaciens, Propionibacterium*; or as a by-product of antibiotic production.

Certain species of bacteria and actinomycetes biosynthesize vitamin B$_{12}$. Precursors for this synthesis include glycine-corrin nucleus; δ-aminolevulinic acid-corrin nucleus; and methionine-corrin nucleus. Intermediates during the synthesis include porphobilinogen, α-D-ribosides of benzimidazole; 5,6-dimethylbenzimidazole; and α-ribazole. Antagonists of vitamin B$_{12}$ include methylamide, ethylamide, anilide, lactone derivatives, pteridine, nicotinamide. Synergists include ascorbic acid, biotin, folic acid, pantothenic acid, thiamine, and vitamins A and E.

Bioavailability of Vitamin B$_{12}$. Factors which tend to decrease the availability of this vitamin include: (1) cooking losses, since the vitamin is heat labile; (2) cobalt deficiency in ruminants; (3) intestinal malabsorption or parasites; (4) lack of intrinsic factor; (5) intestinal disease; (6) aging; (7) vegetarian diet; (8) excretion in feces; (9) gastrectomy. Factors which help to increase availability include: (1) administration of sorbitol; (2) synthesis by intestinal bacteria (not normally); (3) reduced temperature; and (4) presence of food in the stomach.

Although vitamin B$_{12}$ is essentially considered nontoxic, polycythemia has been reported from excessive dosages. From 30 to 60% of the vitamin is stored in the liver; the remainder is found in the kidneys, lungs, and spleen. Target tissues are the central nervous system, kidneys, myocardium, muscle, skin, and bone.

Unusual features of vitamin B$_{12}$ observed by some investigators include: (1) the cyanide group is an artifact of preparation; (2) the only vitamin synthesized in appreciable amounts only by microorganisms (possible in tumors); (3) only vitamin with a metal ion; (4) works with glutathione; (5) glutathione content decreased on B$_{12}$ deficiency; (6) mitosis retarded in B$_{12}$ deficiency; (7) requires intrinsic factor (enzyme) for oral activity; (8) increases tumor size (Rous sarcoma); (9) diamagnetic properties; (10) no acidic or basic groups revealed on titration (no pKa).

Additional Sources of B$_{12}$. Fermented soybean and fish products have been found to contain B$_{12}$ (Lee et al., 1958). Nutritionally significant amounts of B$_{12}$ also were found in the Indonesian fermented products, *ontjom* and *tempeh* (Liem et al., 1977). The microbial production of vitamin B$_{12}$ in *kimchi*, Korean fermented vegetables, including cabbage, has been reported (Lee et al., 1958; Kim et al., 1960). The strain producing the vitamin during the fermentation was identified as *Bacillus megaterium*. As reported by Ro, Woodburn, and Sandine (1979), Foods and Nutrition Department and Department of Microbiology, Oregon State University, Corvallis, Oregon, inoculation of fermented foods with strains known to produce vitamin B$_{12}$ has been evaluated as a vitamin enrichment method. Soybean paste inoculated with *Bacillus megaterium* and fermented was found to contain increased vitamin levels (Choe et al., 1963; Ke et al., 1963). Propionibacterium species widely used in the industrial production of vitamin B$_{12}$ (Wuest and Perlman, 1968) have been recommended for vitamin fortification of some dairy products. Karlin (1961) fortified *kefir* with vitamin B$_{12}$ by the addition of *Propionibacterium* to the kefir grains. Kruglova (1963) prepared vitamin-enriched curds from pasteurized cow's milk by fermentation with equal parts of cultures of lactic acid and propionic acid bacteria (2.5% each). The curds had approximately 10 times more vitamin B$_{12}$ than when produced in the usual way with only lactobacilli. In 1979, Ro, Woodburn, and Sandine undertook to increase the vitamin B$_{12}$ content in the production of kimchi. Changes in the ascorbic acid content during the kimchi fermentation were also observed.

Determination of Vitamin B$_{12}$. Microbial (using *L. leichmanii, O. malhamensis, E. gracilis*, etc.) bioassay methods are used, as are checking the effects of curative doses on experimental animals (chick, rat, etc.). Physicochemical methods used include spectrophotometry, polarography, and isotope dilution.

Additional Reading

Ball, G.F.M.: "Bioavailability and Analysis of Vitamins In Foods," Chapman & Hall, New York, NY, 1997.

Combs, G.F., Jr.: "The Vitamins: Fundamental Aspects in Nutrition and Health," 2nd Edition, Academic Press, Inc., San Diego, CA, 1998.

Eitenmiller, R.R. and W.O. Landen: "Vitamin Analysis for the Health and Food Sciences," CRC Press, LLC., Boca Raton, FL, 1998.

Lipkowitz, M.A. and T. Navarra: "Encyclopedia of Vitamins, Minerals and Supplements," Facts on File, Inc., New York, NY, 1996.

McDowell, L.R.: "Vitamins in Animal and Human Nutrition," 2nd Edition, Iowa State University Press, Ames, IA, 2000.

Newstrom, H.: "Nutrients Catalog: Vitamins, Minerals, Amino Acids, Macronutrients—Beneficial Use, Helpers, Inhibitors, Food Sources, Intake Recommendations, and Symptoms of over or under Use," McFarland & Company, Inc., Publishers, Jefferson, NC, 1993.

VITAMIN D. Although the term "Vitamin D" is convenient to use in discussions of nutrition, this singular term is unsatisfactory when used in a strict biochemical context—because there are different substances, each of which is capable of performing vitamin D nutritional functions, namely, that of promoting growth, including bone growth, and preventing rickets in young animals. With reference to generalized terms used over the years in the development and refining of knowledge of related substances, such terms as *antirachitic vitamin, rachitamin, rachiasterol, cholecalciferol, activated 7-dehydrocholesterol,* etc., have been used. As pointed out later, some of these terms remain quite appropriate.

As a brief introductory summary, vitamin D substances perform the following fundamental physiological functions: (1) promote normal growth (via bone growth); (2) enhance calcium and phosphorus absorption from the intestine; (3) serve to prevent rickets; (4) increase tubular phosphorus reabsorption; (5) increase citrate blood levels; (6) maintain and activate alkaline phosphatase in bone; (7) maintain serum calcium and phosphorus levels. A deficiency of D substances may be manifested in the form of rickets, osteomalacia, and hypoparathyroidism. Vitamin D substances are required by vertebrates, who synthesize these substances in the skin when under ultraviolet radiation. Animals requiring exogenous sources include infant vertebrates and deficient adult vertebrates. Included there are vitamin D_2 (calciferol; ergocalciferol) and vitamin D_3 (activated 7-dehydrocholesterol; cholecalciferol).

The most important or at least the best-known members of the family of D vitamins are vitamin D_2 (calciferol), which is indicated in abbreviated form in Structure 1 and can be produced by ultraviolet irradiation of ergosterol, and vitamin D_3 [Structure 2], which may be produced by the irradiation of 7-dehydrocholesterol.

Structure 1.   Vitamin D_2.

Structure 2.   Vitamin D_3.

Nomenclature. Subscript numerals have a different connotation in connection with vitamin D substances than is true, for example, with B vitamins. Vitamins B_1, B_2, B_6, B_{12}, etc., represent individual substances which have little or no chemical resemblance to each other and perform different metabolic functions. The various vitamin D's, however, have very similar structures, differing only in the side chains, and perform the same functions.

Biochemical Requirements. There are several unique features exhibited by the D vitamins. First, they are not required nutritionally at all if the organism has access to ultraviolet light (which is present in sunlight). Some animals, kept away from ultraviolet light, require so little D vitamins that the need cannot be demonstrated using ordinary diets. Rats, for example, exhibit a need for D vitamins when the calcium/phosphorus ratio in the diet is about 5:1 but not when it is the more usual 1:1. Chickens, on the other hand, exhibit a need even when the calcium/phosphorus ratio is "normal" (1.5:1).

Different species of animals respond distinctively to the different members of the vitamin D family. The most striking example of this is the fact that vitamin D_2 (calciferol) has practically no vitamin D activity for chickens. Rats respond about equally to D_2 and D_3. Human beings respond both to D_2 and D_3. Information as to how various animals react to the other less known forms of vitamin D is largely lacking and for practical reasons is not sought after.

Members of the vitamin D family are extremely difficult to isolate and identify in pure form from any source. Fish liver oils are rich sources, and vitamins D_2 and D_3 have been isolated from them. Most ordinary foods are such poor sources in terms of amounts present, that the presence of D vitamins in them has not been demonstrated. Sterols that can be converted into some form of vitamin D by ultraviolet light are, however, widespread, and it may be inferred that D vitamins are often present even when their presence has never been demonstrated.

The requirements of animals for D vitamins in terms of actual weight are extremely small. It is estimated that human beings need about 400 international units of vitamin D per day. Since an international unit of vitamin D corresponds to 0.025 microgram of crystalline vitamin D, this means that the daily human requirement is about 0.01 milligram. Foods can contain as little as 0.02 parts per million of vitamin D and yet furnish an ample supply on the basis of the foregoing estimate.

Excessive dosages of D vitamins have caused excessive calcification and damage (hypervitaminosis). The full story of vitamin D dosage remains obscure. It has been observed, for example, that some "susceptible" children do not respond to the usual doses, but require 5,000–10,000 units per day to keep them free from rickets. There are other children that are afflicted with "vitamin D-resistant rickets" who do not respond even to these high doses, but may do so when doses of the order of 500,000–1,000,000 units are administered. Although unclear, it would seem that in some individuals the vitamin D has difficulty in getting through to where it is needed.

For many years it has been recognized that all cells need calcium to function because their growth and development is related to changes in their intracellular calcium content. Reasoning further, it has been postulated that calcium may serve as a cellular regulatory agent. Growing interest has been shown by investigators, in a steroid that is derived from vitamin D and that regulates the amount of calcium in the animal's blood. This substance has been referred to as a hormone. It is 1,25-dihydroxyvitamin D_3 and is metabolized from vitamin D. In response to a skeletal need for calcium, the hormone is secreted by the kidney and transported to the intestine and bones. Many authorities believe that parathyroid hormone is involved in signaling the kidney to release 1,25-$(OH)_2D_3$. Hypoparathyroid patients lack parathyroid hormone and fail to make 1,25-$(OH)_2D_3$. The result is an abnormally low concentration of calcium in the blood, producing severe bone disease. DeLuca and associates have used 1,25$(OH)_2D_3$ along with calcium to correct deficits in serum calcium concentrations of a limited number of patients. Corticosteroid therapy of long duration is known to produce bone disease. Corticosteroids are frequently administered to persons with rheumatoid arthritis, systemic lupus erythematosis, and asthma, in addition to persons who have received transplants. Some investigators have found that large doses of vitamin D tend to overcome the adverse effects of the corticosteroids. Findings to date essentially are the results of clinical applications rather than based upon a more detailed knowledge of the molecular mechanisms that operate in the metabolism of D vitamins.

Chronology of Vitamin D Substances. In 1918, Mellanby produced experimental rickets in dogs. In 1919, Huldschinsky ameliorated rachitic symptoms in children with ultraviolet radiation. Hess, in 1922, showed that liver oils contain the same antirachitic factor as sunlight. In that same year, McCollum increased calcium deposition in rachitic rats with cod liver oil factor. In 1924, Steenbook and Hess demonstrated irradiated foods have antirachitic properties. It was in 1925 that McCollum named antirachitic factor as vitamin D. In 1931, Angus isolated crystalline vitamin D (calciferol). In 1936, Windaus isolated vitamin D3 (activated 7-dehydrocholesterol).

Rickets. Vitamin D deficiency (also calcium deficiency) produces a condition known in children as *rickets* and in adults as *osteomalacia*. The bones and teeth of children with rickets are poorly formed and soft. A child with rickets frequently has malformed limbs, especially bowlegs. Blood clotting may be impaired, and, in extreme cases, there may be disturbances of the nervous system. An improvement in the level of calcium in the diet, along with vitamin D or parathyroid extract when required, brings about a hardening of the bones, but leaves them misshapen if deformity has already occurred. Adults, particularly pregnant or nursing women, also require vitamin D because calcium and phosphorus are continually dissolving from bones; and vitamin D is necessary for their utilization. Rickets is not to be confused with the entirely unrelated Rickeetsial group of diseases (Rocky Mountain fever, etc.) that are of virus origin.

Distribution and Sources. Fruits, nuts, and grains are not sources of vitamin D. Animal sources predominate.

High Vitamin D content (1,000–25 × 10⁶ I.U./100 grams).[1] Liver oils from: Bonito, cod, halibut, herring, lingcod, sablefish, sea bass, soupfin shark, swordfish, tuna.

Medium Vitamin D content (100–1,000 I.U./100 grams) Egg yolk, herring, kippers, lard, mackerel, margarine, pilchards, salmon, sardine, shrimp, tuna.

Low Vitamin D content (10–100 I.U./100 grams). Beef, butter, cheeses, cod roe, cream, eggs, grain oils, halibut, horse meat, liver (beef, calf, lamb, pork), milk (vitamin D fortified),[2] veal, vegetable oils.

Bioavailability. Factors which tend to cause a decrease in available vitamin D substances include: (1) liver damage; (2) presence of antagonists; (3) presence of phytin in gut; (4) low bile salts in gut; (5) high pH in gut; (6) destruction of intestinal flora; and (7) excretion in feces. Factors that enhance availability include: (1) storage in liver and skin; (2) absorption aids, such as bile salts; (3) decrease in pH of lower intestine; and (4) irradiation by ultraviolet. Antagonists of vitamin D include toxisterol, phytin, phlorizin, cortisone, cortisol, thyrocalcitonin, and parathormone. Synergists include niacin, parathormone (concentration dependent), and somatotrophin (growth hormone).

Dosages exceeding 4000 I.U./day may cause varying degrees of toxicity in humans. Symptoms include anorexia, nausea, thirst, and diarrhea. There also may be polyuria, muscular weakness, and joint pains. Serum calcium increases and calcification of soft tissues (arteries, muscle) may commence. Arterial lesions and kidney injury have been noted in rats.

In the biosynthesis of vitamin D substances, precursors include cholesterol (skin + ultraviolet radiation) in animals; ergosterol (algae, yeast + ultraviolet radiation). Intermediates in the biosynthesis include preergocalciferol, tachysterol, and 7-dehydrocholesterol. Provitamins in very small quantities are generated in the leaves, seeds, and shoots of plants. In animals, the production site is the skin. Target tissues in animals are bone, intestine, kidney, and liver. Storage sites in animals are liver and skin.

Commercial vitamin D dietary supplements are prepared by the irradiation of ergosterol, 7-dehydrocholesterol; or by extraction of fish liver oils.

Unusual features of vitamin D substances noted by some investigators include: (1) vitamin has hormonal qualities due to internal synthesis; (2) vitamin D2 has little activity for chickens — various species differ in response to the vitamin; (3) vitamin D substances may play a role in aging calcification phenomena, especially in skin; (4) the vitamin can mimic rickets with a high-calcium–low-phosphorus diet; (5) the vitamin can mimic osteomalacia under same conditions; (6) the vitamin is absorbed through skin; (7) the vitamin activates transport of heavy metals by intestinal cells; (8) the vitamin has an exceptionally long half-life (days to weeks); (9) furred and feathered animals obtain some vitamin D as the result of grooming and licking; (10) fishes are believed to obtain vitamin D from marine invertebrates; (11) the vitamin has been found useful in the treatment of lead poisoning.

Determination of Vitamin D. Bioassay techniques involve testing rats on antirachitic qualities. An important physicochemical method involves reaction with antimony trichloride.

See also entries on **Calcium**; **Hormone**; **Lipids**; and **Phosphorus**.

Additional Reading

Ball, G.F.M.: "Bioavailability and Analysis of Vitamins In Foods," Chapman & Hall, New York, NY, 1997.

[1] One I.U. = 0.025 microgram vitamin D_3.

[2] Milk is normally a poor source of vitamin D. Since mild forms a major part of the diet in many countries, particularly for children, the product is commonly fortified with vitamin D substances.

Combs, G.F., Jr.: "The Vitamins: Fundamental Aspects in Nutrition and Health," 2nd Edition, Academic Press, Inc., San Diego, CA, 1998.

Eitenmiller, R.R. and W.O. Landen: "Vitamin Analysis for the Health and Food Sciences," CRC Press, LLC., Boca Raton, FL, 1998.

Lipkowitz, M.A. and T. Navarra: "Encyclopedia of Vitamins, Minerals and Supplements," Facts on File, Inc., New York, NY, 1996.

McDowell, L.R.: "Vitamins in Animal and Human Nutrition," 2nd Edition, Iowa State University Press, Ames, IA, 2000.

Newstrom, H.: "Nutrients Catalog: Vitamins, Minerals, Amino Acids, Macronutrients — Beneficial Use, Helpers, Inhibitors, Food Sources, Intake Recommendations, and Symptoms of over or under Use," McFarland & Company, Inc., Publishers, Jefferson, NC, 1993.

VITAMIN E. Sometimes referred to as the *antisterility vitamin*, factor X (an earlier designation), chemically vitamin E is alpha-tocopherol, the structure of which is:

Alpha-tocopherol

Active analogues and related compounds include: dl-α-Tocopherol; 1-α-tocopherol; esters (succinate, acetate, phosphate), and β, ζ_1, ζ_2-tocopherols. The principal physiological forms are D-a-tocopherol, tocopheronolactone, and their phosphate esters.

The physiological functions of vitamin E substances include: (1) biological antioxidant; (2) normal growth maintenance; (3) protects unsaturated fatty acids and membrane structures; (4) aids intestinal absorption of unsaturated fatty acids; (5) maintains normal muscle metabolism; (6) maintains integrity of vascular system and central nervous system; (7) detoxifying agent; and (8) maintains kidney tubules, lungs, genital structures, liver, and red blood cell membranes.

In livestock and laboratory animals, a deficiency of vitamin E substances may cause degeneration of reproductive tissues, muscular dystrophy, encephalomalacia, and liver necrosis. Considerable research is required to fully determine supplementation of livestock diets unless typical symptoms of a deficiency appear. Symptoms have appeared where there are selenium deficiencies in the soil and where there are excessive levels of nitrates in the soil. "White muscle" is the term used to describe a condition of muscular dystrophy in cattle.

In 1922, Evans and Bishop reported dietary factor "X" needed for normal rat reproduction. In that same year, Matill found dietary factor "X" in yeast and lettuce. Evans et al., in 1923, found factor "X" in alfalfa, butterfat, meat, oats, and wheat. The designation *factor "X"* was changed to *vitamin E* by Sure in 1924. In 1936, Evans et al. demonstrated that vitamin E belongs to the tocopherol family of compounds. During that year, these researchers isolated several active tocopherols and found a-tocopherol to be the most active of the number. Fernholz, in 1938, determined the structure of vitamin E. It was first synthesized by Karrer during that same year. During the interim between 1938 and 1956, several tocopherols were identified and studied. It was in 1956 that Green observed the eighth in the family of tocopherols.

The tocopherols were identified as naturally occurring oily substances and the first three were characterized as alpha, beta, and gamma forms, the biological activity of which decreased in that order.

Vitamin E substances are necessary for the normal growth of animals. Without vitamin E, the animals develop infertility, abnormalities of the central nervous system, and myopathies involving both skeletal and cardiac muscle. The antioxidant activity of the tocopherols is in reverse order to that of their vitamin activity. Muscular tissue taken from a deficient animal has an increased rate of oxygen utilization. The tocopherols are so widely distributed in natural foods that a spontaneous deficiency is infrequent unless diseases of the gastrointestinal or biliary system hinder absorption. Symptoms indicating a vitamin E deficiency include: (1) Red blood cell hemolysis, creatinuria, xanthomatosis and cirrhosis of gall bladder, steatorrhea (in young), cystic fibrosis of pancreases (in young), poorly developed muscles. Rats, dogs, monkeys, and chickens display muscular dystrophy; myocardial degeneration is observed in dogs and rabbits; resorption of fetus, degeneration of germ epithelium, disturbance of estrus cycle are observed in rats; hepatic necrosis is shown in rats; encephalomalacia and vascular degeneration is manifested in chickens.

Role of Vitamin E in Humans. The fundamental needs for vitamin E in humans have long been established. There are factors associated with this vitamin, however, that have created controversy and disagreement among highly qualified professional people. Although nearly every vitamin, at one time or other, has been used unwisely (in retrospect) in the treatment of human diseases, perhaps no other vitamin substance has aroused more discussion among clinicians than vitamin E. Because deficient animals develop a form of myopathy, it was natural to test the therapeutic efficacy of vitamin E in various forms of progressive muscular dystrophy and in diseases of the reproductive system. Enthusiastic claims have been made, and refuted, by investigators. From the standpoint of solid evidence, as of the early 1980s, the principal advantage of administering vitamin E lies exclusively in those instances where a vitamin E malabsorption syndrome exists. Associated with this fundamental situation are hemolytic anemia of premature infants; diseases caused by poor fat and oil absorption, and intermittent claudication (limping). A 1979 Institute of Food Technologists "Food Safety and Nutrition Panel" reported no incidence of vitamin E deficiency. Three underlying reasons were cited for this: (1) ample storage in adipose tissue; (2) slow elimination from the body; and (3) prevalence in foods. Significant amounts are present in vegetable oils and margarine (70% of the average daily intake), cereal products, fish, meat, eggs, dairy products, and leafy green vegeTables.

Cure-all claims for the vitamin appear to stem from the vitamin's antioxidant properties and subsequent ability to neutralize harmful free radical products of oxidation. This had led to vitamin E administration for diseases of the circulatory, reproductive, and nervous systems, increased athletic and sexual endurance, and protection against aging and air pollution effects. Although some claims have been verified by animal studies, evidence is not conclusive for humans. Elderly individuals have resorted to vitamin E in hopes of slowing the aging process. The idea is not unfounded, for in the laboratory, the nutrient neutralizes radicals normally contributing to aging pigment formation. Neutralization within humans, however, remains unproven.

Distribution and Sources. Oily substances are, by far, the best natural sources of vitamin E.

High vitamin E content (50–300 milligrams/100 grams). Corn (maize) oil, cottonseed oil, margarine, safflower oil, soybean oil, wheat germ oil.

Medium vitamin E content (5–50 milligrams/100 grams). Alfalfa, apple seeds, asparagus, barley, cabbage, chocolate, coconut oil, groundnut (peanut), groundnut (peanut) oil, olive oil, rose hips, soybean (dry), spinach, wheat germ, yeast.

Low vitamin E content (0.5–5 milligrams/100 grams). Apple, bacon, bean (dry navy), beef, beef liver, blackberry, Brussels sprouts, butter, carrot, cauliflower, cheeses, coconut, corn (maize), corn (maize) meal, eggs, flour (whole wheat), kale, kohlrabi, lamb, lettuce, mustard, oats, oatmeal, olive, parsnip, pea, pear, pepper (sweet), pork, rice (brown), rye, sweet potato, turnip greens, veal, wheat.

Production sites for vitamin E biosynthesis occur in nuts, seeds, cereal germ, green leaves, legumes. Biosynthesis also occurs in some microorganisms. Precursors for biosynthesis include mevalonic acid and phenylalanine (probably these compounds with side chains). Considerably more research is required to pinpoint the exact precursors. Tocotrienol occurs as an intermediate in the biosynthesis.

Commercial production of vitamin E tocopherols is by way of molecular distillation from vegetable oils.

Antagonists of the tocopherols include α-tocopherol quinone, oxidants, cod liver oil, and thyroxine. Synergists include ascorbic acid, estradiol, somatotrophin (growth hormone), testosterone, and vitamins A, B_6, B_{12}, and K.

Bioavailability of Vitamin E. Factors which tend to reduce availability of the vitamin include: (1) presence of antagonists; (2) mineral oil ingestion; (3) presence of vitamin E oxidation products; (4) occurrence with other less active analogues; (5) excessive excretion in feces; (6) impaired fat absorption; (7) chemical binding in foods; (8) cooking losses (vitamin is heat and oxygen labile); (9) losses in frozen storage, steatorrhea, and variability of natural sources. Factors which may increase absorption include: (1) Storage of vitamin in adipose and muscle tissues; (2) esterification, which increases stability; (3) use of unprocessed fresh food sources; and (4) absorption aids, such as bile salts.

Storage sites for the tocopherols in the body include muscle and adipose tissues and the liver. Target tissues include the adrenals, pituitary, kidney, genital organs, muscles, liver, lungs, and bone marrow.

Unusual features of vitamin E substances as observed by various investigators include: (1) the vitamin may be involved in aging mechanisms by protecting unsaturated fatty acids and membranes against free radicals; (2) only D-isomers occur naturally; (3) vitamin E is replaceable by selenium salts in therapy of rat and pig liver necrosis, and chick exudative diathesis; (4) vitamin E is replaceable by coenzyme Q (see also **Coenzymes**) and antioxidants for certain symptoms of vitamin E deficiency, but not for all, e.g., red blood cell hemolysis, resorption gestation not affected; (5) species differences in response to vitamin E treatment of similar symptoms, e.g., muscular dystrophy—positive in rabbits, negative in humans; (6) other tocopherols are only slightly active as compared with vitamin E; (7) vitamin content is decreased in tumors.

Alpha-Tocopherol and Nitrosamine Formation. Because of the growing concern, commencing in the late 1970s, as regards the formation of N-nitrosamines, such as dimethylnitrosamine and N-nitrosopyrrolidine, upon cooking of certain meat products cured with sodium nitrite, a number of investigators began studies to find materials that may inhibit nitrosamine formation. Reporting in late 1978, W.J. Mergens and a team of investigators (Hoffmann–LaRoche Inc., Nutley, New Jersey) observed that N-nitrosopyrrolidine has been found in fried bacon, but not in raw bacon (Fazio et al., 1973; Fiddler et al., 1974), apparently because of the influence of heat in accelerating the reaction of nitrite with the amine group of proline or its decarboxylated product, pyrrolidine, formed in frying (Archer et al., 1976; Hwang and Rosen, 1976). The effect of ascorbic acid in inhibiting nitrosamine formation has been demonstrated by various workers both in vitro and in vivo (Mirvish et al., 1972, 1973; Kamm et al., 1973, 1975; Greenblatt, 1973; Ivankovic et al., 1973).

The promising contribution of adding tocopherol to bacon, along with sodium ascorbate, to inhibit nitrosamine formation undertaken by Mergens and associates is reported in detail in the Mergens et al. reference (1978).

Determination of Vitamin E. Bioassay methods include measurements of quantity required to prevent fetal resorption; and for red blood cell hemolysis (in rat). Measurements also are made of liver storage in the chick. Physicochemical methods used include colorimetric two-dimensional paper chromatography.

Additional Reading

Archer, M.C., et al.: "Nitrosamine Rofmation in the Presence of Carbonyl Compounds," IARS Scientific Publication 14, International Agency for Research on Cancer, Lyon, France, 1976.

Ball, G.F.M.: "Bioavailability and Analysis of Vitamins in Foods," Chapman & Hall, New York, NY, 1997.

Combs, G.F., Jr.: "The Vitamins: Fundamental Aspects in Nutrition and Health," 2nd Edition, Academic Press, Inc., San Diego, CA, 1998.

Cort, W.M., W. Mergens, and A. Greene: "Stability of Alpha-and Gamma-Tocopherol: Fe^{3+} and Cu^{2+} Interactions," *J. Food Sci*, **43**, 3,797–802 (1978).

Eitenmiller, R.R. and W.O. Landen: "Vitamin Analysis for the Health and Food Sciences," CRC Press, LLC., Boca Raton, FL, 1998.

Fazio, T., et al.: "Nitrosopyrrolidine in Cooked Bacon," *J. Assoc. Offic. Anal. Chem.*, **56**, 919 (1973).

Fiddler, W., et al.: "Some Current Observations on the Occurrence and Formation of N-nitrosamines," Proc., 18th Meeting Meat Res. Workers, Guelph, Ontario, Canada, 1972.

Greenblatt, M.: "Ascorbic Acid Blocking of Aminopyrine Nitrosation in NZO/BI Mice," *J. Nat. Cancer Inst.*, **50**, 1055 (1973).

Hwang, L.S. and J.D. Rosen: "Nitrosopyrrolidine Formation in Fried Bacon," *J. Agric. Food Chem.*, **24**, 1152 (1976).

Ivankovic, S., et al.: "Verhutung van Nitrosamidbedingtem Hydrocephalus durch Ascorbinsaure noch praenataler Gabe von Aethylharnstoff und Nitrite an Ratten," *Z. Krebsforsch.*, **79**, 145 (1973).

Kamm, J.J., et al.: "Protective Effect of Ascorbic Acid on Hepatotoxicity Caused by Sodium Nitrite plus Aminopyrine," *Proc., Nat. Acad. Sci.*, **70**, 747 (1973).

Kamm, J.J., et al.: "Inhibition of Amine-Nitrate Hepatotoxicity by Alpha-Tocopherol," *Toxical. Appl. Pharmacol.*, **41**, 575 (1977).

Lipkowitz, M.A. and T. Navarra: "Encyclopedia of Vitamins, Minerals and Supplements," Facts on File, Inc., New York, NY, 1996.

McDowell, L.R.: "Vitamins in Animal and Human Nutrition," 2nd Edition, Iowa State University Press, Ames, IA, 2000.

Mergens, W.J., et al.: "Stability of Tocopherol in Bacon," *Food Technol.*, **32**, 11, 40–44, 52 (1978).

Newstrom, H.: "Nutrients Catalog: Vitamins, Minerals, Amino Acids, Macronutrients — Beneficial Use, Helpers, Inhibitors, Food Sources, IIntake Recommendations, and Symptoms of over or under Use," McFarland & Company, Inc., Publishers, Jefferson, NC, 1993.

VITAMIN K. Sometimes referred to as the *antihemmorhagic vitamin*, and, earlier in its development, the prothrombin factor or Koagulationsvitamin, vitamin K is a substituted derivative of naphthoquinone and occurs in several forms. The designation *phylloquinone*, or K_1, refers to 2-methyl-3-phytyl-1,4 naphthoquinone; the designations *farnoquinone* and *prenylmenaquinone*, or K_2, refer to 2-difarnesyl-3-methyl-1,4-naphthoquinone. *Menadione*, sometimes called oil-soluble vitamin K3, is 2-methyl-1,4-naphthoquinone. The structure of phylloquinone is:

Generally, when vitamin K substances are absent or deficient in the diet of animals, including humans, a hemorrhagic disorder will appear. Young fowls that are allowed to continue on a deficient diet for extended periods will ultimately die of internal hemorrhage, or from extensive bleeding from small external wounds. Fowls experience difficulty in absorbing vitamin K from the intestine, whereas humans, rats, and dogs absorb it readily and normally obtain their requirement form intestinal bacteria without need of dietary supplementation. If, however, bacterial synthesis is inhibited by the use of sulfa drugs or certain antibiotics, the disease will develop, unless the diet is supplemented with some form of vitamin K. When there is a decrease in the amount of bile salts in the intestine, as in obstructive jaundice, vitamin K is absorbed in such small amounts that the disease will also ensue. The use of vitamin K also is suggested to control and prevent the disease in premature babies. Vitamin K_1 is also able to reverse the hemorrhagic condition resulting from the administration of dicumarol to animals.

It has been reported that vitamin K_1 and several of the vitamin K_2 homologues are capable of restoring electron transport in solvent-extracted or irradiated bacterial and mitochondrial preparations. Other reports suggest that vitamin K is concerned with the phosphorylation reactions accompanying oxidative phosphorylation. The capacity of these compounds to exist in several forms, e.g., quinone, quinol, chromanol, etc., appears to strengthen the proposal that links them to oxidative phosphorylation. Information has suggested that vitamin K acts to induce prothrombin synthesis. Since prothrombin has been shown to be synthesized only by liver parenchymal cells in the dog, it would appear that the proposed role for vitamin K is not specific for only prothrombin synthesis, but applicable to other proteins.

In 1929, Dam reported chicks on a synthetic diet develop hemorrhagic conditions. In 1935, Dam named vitamin K as the missing factor in synthetic diets. In that same year, Almquist and Stokstad demonstrated the presence of vitamin K in fish meal and alfalfa. In 1939, Dam and Karrer isolated vitamin K from alfalfa; and, in that same year, Doisy isolated K_1 from alfalfa, K_2 from fish meal, and demonstrated differences of the two substances. Also, in 1939, MacCorquodale, Cheney, and Fieser determined the structure of vitamin K_1. In that same year, Almquist and Klose synthesized vitamin K_1 for the first time. In 1941, Link et al. discovered dicoumarol, an anticoagulant and antagonist of vitamin K.

In addition to compounds previously mentioned, active analogues and related compounds include menadiol diphosphate, menadione bisulfite, phthicol, synkayvite, menadiol (vitamin K_4), and compounds designated as vitamins K_5, K_6, and K_7. The vitamin is frequently administered to poultry via feedstuffs. Intestinal bacteria, normally functioning, supply the vitamin to the human body.

In the therapy of deep venous thrombosis, heparin is commonly administered. This drug takes effect immediately to prevent further thrombus formation. However, heparin is regarded as a hazardous drug and possibly may be the leading cause of drug-related deaths in hospitalized patients who are relatively well. Usually administered intravenously, preferably by pump-driven infusion at a constant rate rather than by intermittent injections, it sometimes may cause major bleeding, which

is particularly hazardous if it is intracranial. The action of heparin can be terminated almost immediately by intravenous injection of protamine sulfate, but where there may be less urgency, vitamin K_1 may be used. The vitamin preparation may be administered intravenously, intramuscularly, or subcutaneously.

Vitamin K is also an antagonist of warfarin, which is sometimes used in rodenticides. Pets that have been exposed to warfarin-containing poisons may be saved from death by internal hemorrhaging through the immediate administration of vitamin K.

Vitamin K is sometimes used in the treatment of viral hepatitis.

It has been found that vitamin K analogues possess an ability to insert themselves into the oxygen-binding cleft of hemoglobin. This may result in hemolysis (dissolution of red blood corpuscles with liberation of their hemoglobin).

See also **Anticoagulants**.

Distribution and Sources. Some fruits, vegetables, and nuts, as well as meat products, contain good sources of K vitamins. Intestinal bacteria, M. phlei, synthesize it.

> *High vitamin K content (100–300 micrograms/100 grams).* Beef kidney, beef liver, cabbage, cauliflower, pork, soybean, spinach.
> *Medium vitamin K content (10–100 micrograms/100 grams).* Alfalfa, egg yolk, pine needles, potato, strawberry, tomato, wheat (bran, germ, whole).
> *Low vitamin K content (0–10 micrograms/100 grams).* Carrot, corn (maize), milk, mushroom, parsley, pea.

Commercial production of vitamin K is by column chromatography of fish meal extracts. In biosynthesis, precursors include polyacetic acid (ring); acetate (side chain). Intermediates include dehydroquinic acid (ring); farnesol (side chain).

Bioavailability of Vitamin K. Factors which decrease availability of the vitamin include: (1) biliary obstruction; (2) liver damage — cirrhosis, toxins; (3) poor food preparation (vitamin is strong-acid, alkali, light, and reduction labile); (4) impaired lipid absorption in gut; (5) presence of antagonists; (6) ingestion of mineral oil; (7) sterilization of gut with antibiotics and sulfa drugs; and (8) excessive excretion in feces. Availability may be increased by way of storage in the liver and absorption aids, such as bile salts.

Antagonists of vitamin K substances include dicoumarol, sulfonamides, antibiotics, α-tocopherol quinone, dihydroxystearic acid glycide, salicylates, iodinin, warfarin. Synergists include ascorbic acid, somatotrophin (growth hormone), and vitamins A and E.

General symptoms of a vitamin K deficiency include hypoprothrombinemia, increased bleeding and hemorrhage, increased clotting time, and neonatal hemorrhage. Internal hemorrhage is a symptom in chicks. Usually the vitamin is nontoxic, but, in humans, very excessive dosages can cause thrombosis, vomiting, and porphyrinuria. Target tissues are liver and vascular system. Small quantities are stored in liver.

Determination of Vitamin K. A vitamin K deficient chick assay may be made; or physicochemical techniques, including polarographic methods, spectrophotometry of pure solutions, and prothrombin time determinations, may be used.

Additional Reading

Ball, G.F.M.: "Bioavailability and Analysis of Vitamins In Foods," Chapman & Hall, New York, NY, 1997.
Combs, G.F., Jr.: "The Vitamins: Fundamental Aspects in Nutrition and Health," 2nd Edition, Academic Press, Inc., San Diego, CA, 1998.
Eitenmiller, R.R. and W.O. Landen: "Vitamin Analysis for the Health and Food Sciences," CRC Press, LLC, Boca Raton, FL, 1998.
Lipkowitz, M.A. and T. Navarra: "Encyclopedia of Vitamins, Minerals and Supplements," Facts on File, Inc., New York, NY, 1996.
McDowell, L.R.: "Vitamins in Animal and Human Nutrition," 2nd Edition, Iowa State University Press, Ames, IA, 2000.
Newstrom, H.: "Nutrients Catalog: Vitamins, Minerals, Amino Acids, Macronutrients — Beneficial Use, Helpers, Inhibitors, Food Sources, Intake Recommendations, and Symptoms of over or under Use," McFarland & Company, Inc., Publishers, Jefferson, NC, 1993.

VITREOUS DETACHMENT. Vitreous detachment, often called posterior vitreous detachment, occurs when the vitreous humor, the gel that fills the inside of the eye behind the lens, shifts and separates from the retina. This shift is often harmless, although it does produce visible symptoms.

One is the appearance of floaters, particles floating in the vitreous humor that appear to be circular spots or small spots strung together with a weblike thread. Floaters are most noticeable in bright light or when looking at distant objects. These floaters occur because of the formation of opaque spots in the vitreous. They sometimes disappear in weeks, but can persist for much longer periods, months or years.

Vitreous detachment can also result in flashes; quick bursts of light caused when the vitreous rubs against the retina. These generally are short-lived.

If the floaters increase in number or the flashing becomes continuous, an immediate visit to an ophthalmologist is recommended because these can be symptoms of a retinal detachment or other problem resulting from the shift of the vitreous. Left untreated, a retinal detachment can lead to permanent loss of vision.

A major warning sign of a possible retinal detachment is that the flashers and floaters occur in just one eye. (Flashers in both eyes are unusual, and generally are caused by migraine headaches.) Floaters in one eye can also be symptoms of a hemorrhage in the eye or an infection, which also require quick treatment. Other warning signs include blurred or distorted vision, the appearance of cobwebs or veils, and the feeling that a curtain is descending over the visual field.

Persons who are at higher risk of retinal detachment caused by a vitreous detachment include those who are nearsighted because myopia is associated with a thinner-than-normal retina, those who have undergone eye surgery such as a cataract operation, and those with eye inflammation (uveitis) or a previous eye injury. The occurrence of vitreous detachment in one eye indicates an increased risk for the same detachment to occur in the other eye.

Fortunately, most people who experience vitreous detachment will not require medical care. If the floaters are persistent or very bothersome, the only treatment available is vitrectomy surgery during which the vitreous gel is surgically removed and replaced with a saline solution. Although the risks associated with vitrectomy surgery are relatively small, they usually outweigh the benefits as a treatment for floaters. See also **Floaters**; **Retinal Detachment**; and **Uveitis**.

Vision Rx, Inc., Elmsford, NY.

VITREOUS STATE.   When certain liquids are cooled fairly rapidly, crystals do not form at a definite temperature, but the viscosity of the liquid increases steadily until a glassy substance is obtained. A glass may be thought of as a disordered amorphous solid, or as a supercooled liquid, which only devitrifies into the crystalline state after extremely long standing. Glasses are optically isotropic, which explains their value in optical instruments. The property of forming a glass is possessed particularly by the oxides of silicon, boron, germanium, arsenic, phosphorus, etc., and by many organic compounds, especially those containing several hydroxyl groups per molecule. See Fig. 1.

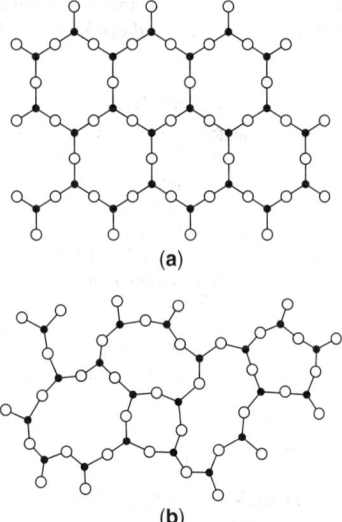

(a)

(b)

Fig. 1.   Two-dimensional diagram showing **(a)** an oxide of composition X_2O_3 in the crystalline form; and **(b)** the same oxide in the vitreous state.

VITROPHYRE.   A volcanic glass carrying sporadic distinct crystals of feldspar and other minerals; in short, a porphyritic glass.

VIVERRINES *(Mammalia, Carnivora).*   A family of carnivores, sometimes referred to as the civet family. The general organization of the viverrines is indicated in the Table 1. Civets are closely related to the hyenas and the cats. They have relatively coarse hair and, in some species, an erectile mane. The teeth are hyena-like, but the claws and tail are cat-like. Civets range from 2 to 3 feet (0.6 to 0.9 meter) in length and possess black and white stripes and spots. They are found in the more tropical parts of the Old World, particularly in Africa and Malaya. They are not found in the Australian region. Dietary items include birds' eggs, smaller mammals, snakes, and lizards. They are known to devour large quantities of crocodile eggs along the Nile and hence are considered of economic value. They possess an odorous and fatty substance in a pouch connected with the sexual organs. The material can be used for compounding perfumes when properly treated. This material is known as *civet* when purified. The so-called civet-cat in the Americas is not a true civet, but is related to the raccoons. The *rasse* is a civet of the Oriental region. *Genets* have more slender bodies and shorter legs than the civets, but are closely related. Several species are found in Africa, Asia, and Europe. *Linsangs*, also closely related to civets, are slender predacious animals with short legs and very long tails. They occur in the Oriental region and to a lesser extent in Africa.

TABLE 1. GENERAL ORGANIZATION OF THE VIVERRINES

Civets *(Viverrinae)*	Galidines *(Galidiinae)*
True Civets *(Viverra* and *Civettictis)*	FOSSAS *(Cryptoproctinae)*
Rasse *(Viverricula)*	
Genets *(Genetta)*	MONGOOSES *(Herpestinae)*
African Linsang *(Poiana)*	
Linsangs *(Prionodon. ...)*	True Mongooses *(Herpestes)*
Water-Civet *(Osbornictis)*	Banded Mongoose *(Mungos)*
Palm-Civets *(Paradoxurinae)*	Dwarf Mongooses
	(*Helogale*)
Musangs *(Paradoxurus)*	Marsh Mongooses *(Atilax)*
Masked Palm-Civets *(Paguma)*	Cusimanses *(Crossarchus)*
Small-toothed Palm-Civets *(Arctogalidia)*	White-tailed Mongooses
	(*Ichneumia*)
Celebesean Palm-Civet *(Macrogalidia)*	Bushy-tailed Mongooses
	(*Cynictis*)
Binturong *(Arctictis)*	Dog-Mongooses *(Bdeogale)*
West African False Palm-Civet *(Nandinia)*	Xenogales *(Xenogale)*
	Meerkat *(Suricata)*
HEMIGALES *(Hemigalinae)*	
Hemigales *(Hemigale, ...)*	
Otter-Civet *(Cynogale)*	
Fanaloka *(Fossa)*	
Anteater-Civet *(Eupleres)*	

The *mongoose* is a slender animal with short legs and long tail, related to the civets, but it possesses no scent glands. These animals live in Africa and the Oriental region and have been introduced successfully into the West Indies and other regions. Mongooses kill many small animals and are valuable for destroying rats and other vermin, but they will also attack useful animals when the supply of vermin is low. Mongooses are particularly noted for their ability to kill snakes. The Indian mongoose (*Herpestes mungo*) is readily tamed and is frequently kept to free a household of undesirable pests. The Egyptian mongoose (*H. ichneumon*) is also called the *ichneumon*.

The *meerkat* is a South African animal of the civet family, related to the mongooses. The name is sometimes applied to the penciled mongoose. The *suricate* is an animal of southern Africa (*Suricata tetradactyla*) related to the mongooses and a part of the civet family. The suricate is of moderate size, gray with transverse dark bands on the back and a whitish crown, and has rather short legs. The animal can make an interesting pet. This species shares the name *meerkat* with one of the mongooses.

See Fig. 1.

VIVIANITE.   The mineral vivianite is a hydrous iron phosphate, $Fe_3(PO_4)_2 \cdot 8H_2O$, its monoclinic crystals are usually prismatic or blade-like but may be in massive forms. Vivianite has one perfect cleavage;

(a)

(b)

Fig. 1.   Examples of viverrines: (**a**) Young genet about to invite its mother to play suddenly reacts to a suspicious noise; (**b**) yellow mongoose (left) in sitting position; (right) in standing guard position.

hardness 1.5–2; specific gravity 2.58–2.68; luster, pearly on cleavage; faces, otherwise vitreous; colorless, when freshly exposed, but becoming blue or brownish with the alteration of the ferrous to ferric iron; transparent to translucent. Vivianite is an associate of pyrrhotite, pyrite and copper and tin ores. It is found also in clay beds forming the so-called blue iron earth which is common and of wide distribution in peat bogs. Vivianite is found in Rumania; Bavaria; Cornwall in England; and elsewhere in Europe; Australia; Bolivia; and Greenland. In the United States it occurs in New Jersey, Delaware, and Colorado. This mineral was named by Werner after the English mineralogist J.G. Vivian, its discoverer.

VIVIPAROUS REPRODUCTION.   The production of live offspring following a period of internal nourishment from the mother by means of a placenta. Only the mammals have this type of reproduction, although a few simple mammals, such as the marsupials and the prototheria still produce eggs with coverings, so that the embryo forms no connection with the mother before birth.

VIVIPAROUS TOPMINNOWS (*Osteichthyes*).   Of the order *Microcyprini*, family *Poeciliidae*, probably the most famous of this family of beautiful, streamlined little fishes is the "guppy," named for its discoverer, R.J.L. Guppy, who first noted the fish in 1866 in streams on the island of Trinidad. The *Lebistes reticulatus* remains a favorite of tropical-fish hobbyists, and literally millions have found their way to home fish tanks. The guppy, of course, is well known for the live bearing of its young. Fully grown females reach 2–21/2 inches (5–6.4 centimeters) in length;

the males are considerably shorter. The natural habitat of the guppy is the southern Caribbean extending as far south as southern Brazil.

There are some 45 genera in the topminnow family, and their natural range is quite wide, but all in the western hemisphere. In addition to the guppy, of course, numerous other species are found in small aquariums. For example, the *Gambusia affins* (mosquito fish) was well known by aquarists prior to the discovery of the guppy. It is somewhat larger, ranging up to 3 inches (7.5 centimeters) in length. This fish is welcome in the tropics because of its ability to destroy mosquito larvae and pupae. The dwarf mosquito fish (*Heterandria formosa*) reaches only about 1 inch (2.5 centimeters) in length and inhabits the waters from North Carolina to Florida. Another very popular fish with hobbyists is *Mollienesia latipinna* (mollies). This species attains a length of about 4 inches and ranges from South Carolina to Mexico. Through selective breeding, all black mollies have been produced. Other popular varieties include the "platys" and "swordtails." These varieties are of the genus *Xiphophorus*. Topminnows have been used extensively in medical research, notably in genetic investigations.

VODKA.   A colorless, essentially tasteless and odorless, highly rectified alcoholic beverage that until after World War II was essentially limited in its production and distribution to the Slavic nations (Russia, Poland, etc.) of eastern Europe and to a few mideastern countries (Iran, Turkey). Vodka has been made and consumed in these countries for hundreds of years. In the western countries, the drink probably most reminiscent of vodka is a high-quality gin. In the late 1940s and early 1950s, vodka was introduced (on a commercially sponsored scale) to several of the western countries, including western Europe, the United Kingdom, and the United States, and received wide and early acclaim. It was no longer regarded as a curiosity in these countries. As of the early 1980s, vodka is rapidly crowding into second place in American alcoholic beverage consumption, with Bourbon whiskeys and blends still maintaining an impressive lead. Generally, traditionalists in the field of alcoholic beverage production and distribution were taken by surprise at the rapid spread of vodka's acceptance and, during the interim, have offered a number of explanations. Surprise was registered particularly because vodka is such a neutral and "bland" beverage, with no real character other than its "lack of character" so to speak. However, some authorities believe that it was this very lack of character that made it a superior substitute for gin in numerous cocktail recipes, particularly in the United States where cocktails are very popular. Some authorities also believe that vodka, because of its lack of a characteristic whiskey, gin, or wine taste, may have increased overall liquor consumption because of the objections of some consumers to characteristic tastes and aromas. For a while, it appeared that vodka was probably a fad, but this has long since been disproved. For example, vodka has been accepted in Scotland, where native whiskies had been essentially regarded as the only strong alcoholic beverage to consider.

Traditionally, vodka has been prepared from a variety of starch-or sugar-containing raw materials (grains, molasses, grapes, potatoes, etc.). In the eastern European countries, a source that was in surplus was usually selected at any given time for making vodka. Up to and including the fermenting process, vodka is essentially prepared like whiskies and gins. It is the rectification that is different. The vodka distiller uses a series of columnar fractionators to produce an effluent of approximately 190° proof that is essentially neutral—a very strong ethanol-water solution. To remove any substances that would add to flavor or aroma, the distillate is then usually filtered through activated charcoal. In contrast with gin, no artificial or natural flavorings or colorings are added to vodka. In the United States, vodka is sold as 80° and 100° proof.

Although most usually mixed with water (or in cocktails), the eastern Europeans frequently prefer drinking vodka *straight* (without water). Vodka also is available, mainly in Europe, with some flavoring and sweetening.

VOICE AND SOUND PRODUCTION.   The incidental production of sound waves in air or similar vibrations in the water by the movements of living things is so common that a sense of hearing serves animals as a valuable means of detecting the presence of enemies, prey, or members of their own species. Together with this sense, the ability to produce sound voluntarily serves many uses, as we know from our own experience. In some cases the act involves only the special use of common organs, as when the beaver slaps the water with his broad tail in a warning signal.

Sound production immediately brings to mind, however, organs such as the human larynx with its great range of activity and other organs specially developed for sound production.

Among the vertebrates, the production of sound is very commonly associated with the respiratory system, since the air-breathing forms have an opportunity to vibrate special structures by controlled currents of air expelled from the lungs. In the birds the apparatus is the syrinx, located at the inner end of the trachea, and, in mammals, it is the larynx, near the outer end of the same passage. The frog also has a larynx with vocal cords in its very short trachea. Vocal cords are lacking in the reptiles, although some of these animals produce sound by the expulsion of air from the respiratory system, as in the hissing of snakes and the so-called bellow of the alligator. Whispering of human beings is in the same category. Some fishes produce sound by means of the swim bladder. A noteworthy example is the croaker (*Micropogon undulatus* Linn.) of the eastern and southern coastal waters of the United States.

Of all vertebrates, the mammals and birds are most generally equipped with elaborate vocal organs; hence, the syrinx and larynx of these classes claim special attention. The two organs are quite different in origin and structure.

The syrinx is a chamber located at the junction of the trachea and the two major bronchi. It is supported by modified cartilages and contains a projecting flexible membrane, which is thrown into vibration by the expulsion of air from the lungs. Muscular control of the tension of this membrane permits the bird to control the pitch of the sounds produced, which vary from comparatively simple calls to elaborate songs. Some birds are able to mimic sounds very faithfully, hence modulation of the sounds produced by the syrinx is involved, as in the case of human speech. The mocking bird (*Mimus polyglottos*) affords an excellent example of mimicry of the songs of other species, while parrots and mynas extend their virtuosity to mimicry of human speech. The mynas, especially, enunciate words with amazing delicacy of inflection.

The larynx is essentially a chamber at the pharyngeal end of the trachea, supported by modified cartilages and in most species of mammals containing vocal folds, or cords, whose tension is regulated by muscles acting on the laryngeal wall. The human larynx is among the most highly developed organs of the kind. See Fig. 1. A slit-like opening, the glottis, leads into it from the pharynx. This opening is flanked by lobes of the cricoid cartilage and by ligamentous folds, the vocal cords, imbedded in the mucous lining. The larynx is further supported by large right and left quadrangular plates of the thyroid cartilage whose ventral junction forms the "Adam's apple." A flap-like epiglottis closes down over the glottis during the act of swallowing. During ordinary breathing the glottis is V-shaped, becoming more rounded in deep inspiration. When high tones are produced, the tension of the vocal cords closely approximates them and narrows the opening to a mere slit. Although the production of sound by this organ is due simply to the vibration of the cords by the properly controlled ejection of air, sound may be produced, in whispering, without the vibration of the cords. In either case, an important modulating effect

is produced by the mouth and tongue. The oral, nasal and pharyngeal cavities also act as resonators. Individual differences in the pitch of the human voice depend on the length, thickness and elasticity of the cords. In any one individual, pitch is varied by changing the tension of the cords.

Much research in recent years has been directed toward thoroughly understanding how the human vocal system functions, not simply in terms of generalities which are relatively well understood, but in terms of specific parameters, so that computer models can be constructed and manipulated. See Fig. 2. Such research will one day become important in terms of human communication between computers and machines, but an even more practical and immediate priority is finding efficient and realistically acceptable ways of digitizing human speech for distant transmission and, in turn, converting the received signals into voice sounds. Very important is the rate at which such signals can be transmitted, a factor in determining cost and offering numerous other advantages.

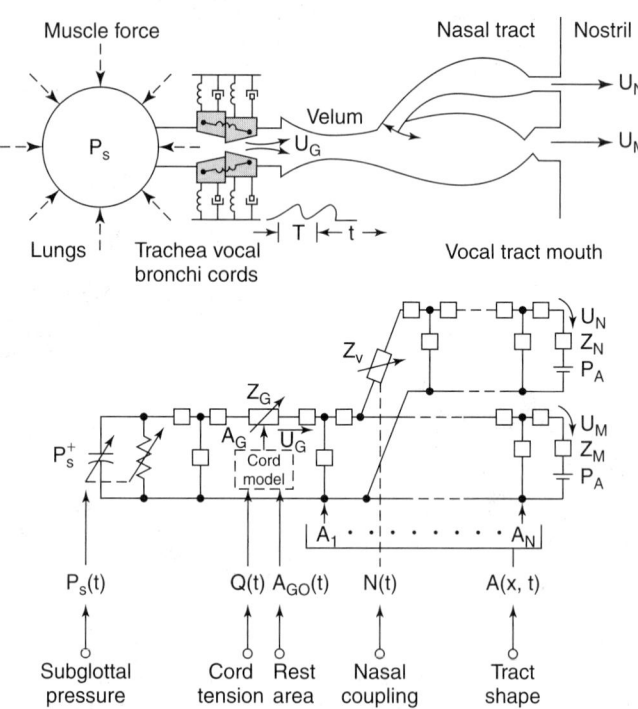

Fig. 2.   Speech synthesizer based upon computer models for vocal cord vibration, sound propagation in a yielding-wall tube, and turbulent flow generation at places of constriction. The control inputs are analogous to the human physiology and represent subglottal air pressure in the lungs (P_s); vocal-cord tension (Q) and area of opening at rest (A_{g0}); the cross-sectional shape of the vocal tract, [$A(x)$]; and the area of coupling to the nasal tract (N). (*AT&T Bell Laboratories.*)

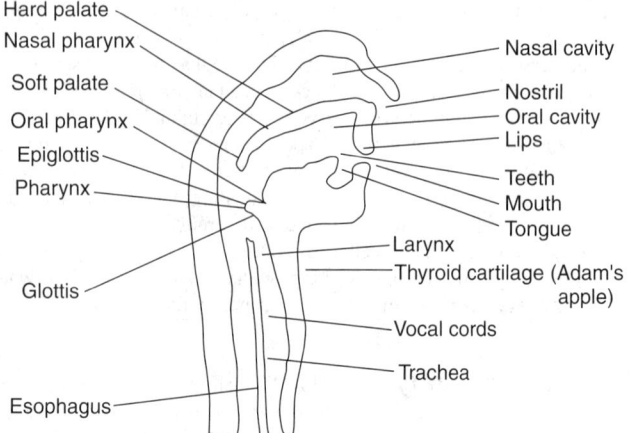

Fig. 1.   The human vocal tract is basically an acoustic tube with the lips at one end and the vocal cords at the other. During speech production, the shape of the tract changes continuously, due to movements of the tongue, jaw, and lips. Vocal tract models suitable for speech coding have been under study for several years. (*AT&T Bell Laboratories.*)

Supplementary resonating structures occur in some mammals, such as the enormously dilated hyoid bone and cavernous diverticula of the larynx in the howler monkeys (*Alouata* sp.). In the bats the voice and the highly developed ears serve to detect obstacles in flight by the reflection of sounds.

Among the insects sound production is highly specialized. Here it is of the type called stridulation. The apparatus varies from a comparatively simple roughening of apposed surfaces, which produce vibrations when rubbed together, to the elaborate organs of the male cicada. A common example of the simple form is found in the milkweek beetles (*Tetraopes*). By moving the prohorax up and down, its ventral sclerite is rubbed against the mesosternum, producing a squeaking sound audible to the human ear at very short distances. A more elaborate stridulation occurs in the order Orthoptera, exemplified by the crickets, katydids and grasshoppers. The crickets have parts called scrapers and files developed on the basal portion of the wings. When rubbed together they throw the entire wing membrane into vibration, producing the shrill singing so familiar to everyone. The short-horned grasshoppers set the wings into vibration by means of a row of fine projections on the inner surface of the hind legs.

The sound-producing organs of the male cicada consist of a pair of deep depressions at the base of the abdomen, covered with thin plates. Within the cavities are membranes that are thrown into vibration by direct muscular action, thus producing sound, and others which serve as resonators. Although the sound produced by cicadas of different species varies greatly in volume and quality, probably no insects exceed in volume or incisiveness the shrill vibrant singing of some of the common cicadas of the United States.

Auditory organs occur in many insects with special stridulating organs, but in others, such as some members of the order *Diptera*, their presence suggests that the rapid movement of the wings in flight produces the sound that the insect hears, thus providing for recognition within the species. Certainly the wing beat may have such an effect, since it is quite evident to us in the humming of mosquitoes and midges.

Voice and Speech Coding

Researchers have and are continuing to explore efficient digital speech coding algorithms to create natural-sounding voice signals at low bit rates. The traditional high quality of voice transmission will continue to be maintained in a majority of telephone networks, but the methods of encoding used most likely will change within the relatively near future. Instead of transmitting the acoustic voice signal in the conventional manner at 64,000 bits per second, the target is to deliver a more efficiently digitized telephone voice signal using as low a bit rate as possible, while the speech the listener hears is comparable in quality to speech produced by high-bit-rate coding. Speech coding at low bit rates most likely will play an important part in providing future capabilities for telephone communication systems, including voice encryption, transmitting voice mail over telephone networks, and flexible management of voice and data over packet networks.

Various Speech Coding Methods[1]. The conventional method of digital encoding, known as pulse code modulation (PCM) and first standardized in the 1960s, uses *waveform coding*. Limited to a bandwidth of 4 kHz, the analog voice signal is sampled 8000 times a second, and each sample is converted into an 8-bit digital code. The resulting bit rate is 64 kilobits per second.

Introduced at a later date was a waveform coding technique, known as Adaptive Differential Pulse Code Modulation (ADPCM), used for encoding toll-quality speech at 32 kilobits per second. A new concept, now under intense investigation, is based on *source coding* concepts. This approach generates the digitized voice signal by analyzing how human speech is produced and how the ears perceive the synthesized version. Coding and decoding are done at 9.6 kilobits per second, which is only one-sixth the rate of the most widely used standard rate of 64 kilobits per second.

Low-Bit-Rate Speech Coding. Low-bit-rate coding can offer several advantages. First, carriers may be able to increase the number of channels handled by a given system. For example, a T1 carrier rated for 1.544 megabits per second can carry up to 24 voice-grade channels at 64 kilobits per second each, or up to 47 channels at 32 kilobits per second each, but the same digital carrier could be upgraded to as many as 160 voice channels at 9.6 kilobits per second each. Some of these channels could be assigned for use as specific signaling channels used by the telecommunications carrier.

Second, speech coding at low bit rates eventually may help to implement the use of digitized voice signals in public switched loop circuits that presently are unable to carry high-quality, digitally encoded speech. Digitized speech must be transmitted over unconditioned, ordinary loop circuits before end-to-end digital voice and data transmission through the public switched network can become a reality.

The highest bit rate considered acceptable for an unconditioned circuit in the public loop plant today is about 4.8 kilobits per second, which is used by voiceband data modems. Digital transmission rates higher than 4.8 kilobits per second on a dial-up public line could experience significant pulse distortions, which would increase error rates. Signal errors on a digitized voice line could introduce unacceptable speech distortions, depending upon the type of speech decoder used.

A digital voice coder capable of transmitting intelligible digitized speech over dial-up lines, known as a vocoder (voice coder), was introduced a few years ago. It typically operates at 2.4 kilobits per second, but the speech signal sounds mechanical and artificial. Continuing research has shown that *synthesized speech* encoded at 9.6 kilobits per second produces voice

[1] Information furnished by B.S. Atal and R.Q. Hofacker, Jr., AT&T Bell Laboratories.

quality far more natural than that created by the vocoders, but toll quality continues to be sought.

Past efforts toward source coding were impeded by the lack of high-speed microchips. Research on the complex behavior of human vocal and aural organs as they produce and listen to speech signals has revealed how complex the speech communication process is. It is very difficult to synthesize the human voice with artificial devices, especially in real time.

When algorithms were designed around relatively unsophisticated hardware, the low-bit-rate synthesized voice could not equal a normal telephone voice transmitted by analog or PCM digital systems. However, with the improvement of available hardware, particularly in speed, high-speed digital signal processing could markedly improve the source coding system and thus achieve high-quality voice synthesis in real time. Alternatively, the coded bit rate could be reduced even further—perhaps as low as 2.4 kilobits per second, while generating a high-quality voice signal. Further advances can be made from learning more about the physics of speaking and hearing. For example, speech scientists know that average speech is burdened with a tremendous load of redundant information.

Two speech signals need not be *physically* identical to conclude that they are *perceptually identical*. This means that there is some latitude when the signal is reproduced. The key to success in speech coding lies in the proper appreciation of *what we hear* and *what we do not*. Nature requires redundant components in speech signals to reinforce speech sounds so they will survive passage through the air in noisy environments. These redundant components are not required when speech is being transmitted across a telecommunications network. They take up valuable transmission capacity. Source coding research seeks ways to eliminate such redundant speech components before the voice is transmitted over the digital channel, restoring only the necessary components at the receiving end. See Fig. 3.

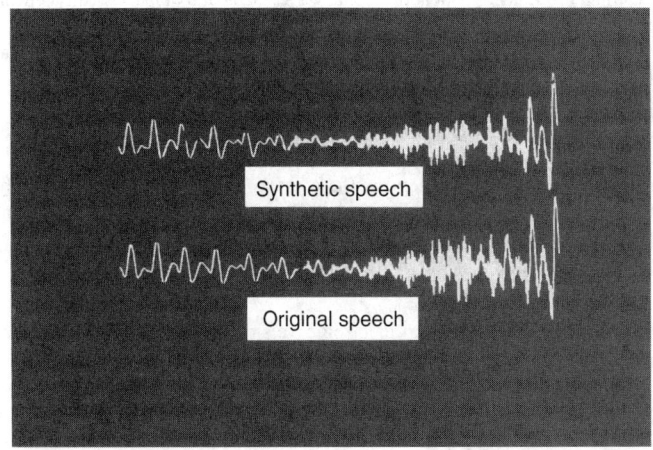

Fig. 3. Most of the average speech signal is full of redundant components. By identifying and removing those redundancies during coding for telephone transmission, the efficiency of low-bit-rate digitized voice systems can be greatly improved with minimum loss of quality.

Minimizing the Information Rate. Human speech *perception* is complex. It consists of a sequence of transformations occurring at several levels in the ear and the brain. An extremely important transformation occurs in the ear's cochlea, where the acoustic signal is converted into neural signals. These are transmitted by the auditory nerve to the higher auditory centers of the brain. See Fig. 4. For efficient coding, we need to know just how much of the speech signal's information is processed in the acoustic-to-neural transformation. If certain irrelevant details are going to be lost during the conversion anyway, why transmit them through the digital communication channel?

Coding speech efficiently therefore requires that we determine just how closely the synthesized signal matches the natural speech signal in a perceptual sense. This continuous comparison of the two signals is done by including both a speech synthesis model and a model of auditory perception in the encoding process.

Teaching Machines to Talk. Speech coders are guided by a speech synthesis model. A properly developed speech synthesis model plays a key role in generating high-quality voice signals at low bit rates. The most commonly used speech synthesis model in source coding is the

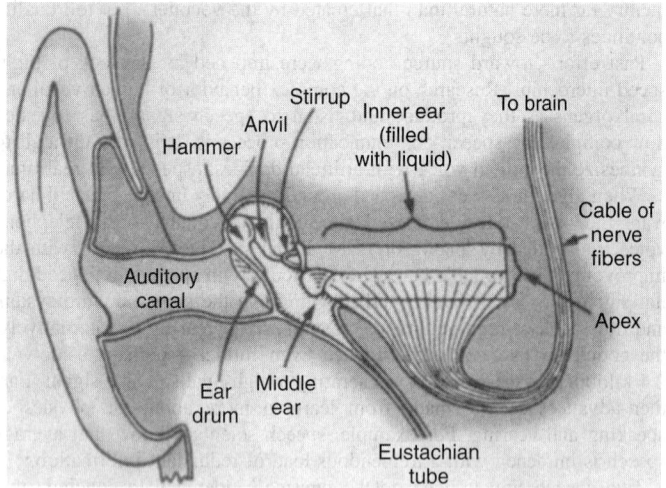

Fig. 4. Schematic of human ear shows snailshell-shaped cochlea as a flattened "inner ear" structure that houses the basilar membrane, which connects to thousands of nerve endings. The membrane acts as a peripheral frequency analyzer, selecting frequency bands and transforming acoustical energy into nerve impulses when stimulated by signal waves traveling through the inner ear's fluid. Not all of the signal detail that enters the cochlea is changed into nerve impulses that the brain understands as sound. If the signal components are ignored by the acoustic-to-neural transformation, then these signals need not be transmitted through a digital telecommunications system. This would potentially increase digital transmission capacity, offer more flexibility in system utilization, and lower costs for the analog/digital conversion hardware.

During the transformation of sound waves to neural impulses, the delicate components of the ear perform much like a bank of bandpass filters. Sounds are grouped together in frequency bands, some of which are broader than others. An important factor is *masking*, the inability of the human hearing mechanism to detect weak sounds in the presence of strong ones. Traditionally, a communications problem involves a signal being masked by noise. In speech coding, the opposite can be true. Every coder introduces errors during the encoding and decoding processes. These errors (or noise) can be masked by clever manipulation of the speech signal. By taking these factors into account, the digitized speech signal sent over telephone transmission circuits could be trimmed substantially. (*AT&T Bell Laboratories.*)

time-varying linear filter, an electronic version of the human vocal tract. This is excited by either a periodic pulse train (to produce voiced sounds), or a white-noise source (to generate unvoiced sounds). Voiced sounds, such as vowels, semivowels, voiced stops, and nasals, come from vocalcord vibrations. Unvoiced sounds, including fricatives (like "f", "s", "sh", etc.), are made without using the vocal cords, but by generating turbulence in

the vocal tract. The voice synthesis model requires information about the characteristics of the speech signal we want to code — whether it is voiced or unvoiced, its pitch period, and its linear filter parameters. Present low-bit-rate speech coders, such as the *linear predictive code* (LPC) vocoders, use this synthesis model. Their machine-like sound is a major drawback. See Fig. 5.

A more natural-sounding speech signal is produced when the newer *multi-pulse* linear predictive coder (MPLPC) concept[2] is used. Instead of pacing the excitation pulses at only one per pitch period as speech proceeds, MPLPC uses many pulses per period and positions them dynamically according to variations in the speech signal. This approach does not require the knowledge of speech characteristics, such as pitch period, or whether the speech signal is voiced or unvoiced (often used in the older vocoders and which are often difficult to determine). Multi-pulse LPC is sufficiently flexible to achieve any desired performance level within hardware limitations. See Fig. 6.

Average Versus Peak Bit Rates. Information in any speech signal is not spread evenly in time. Some speech segments are more important than others. Although this is well known, waveform speech coding systems in the past often allocated the same number of bits for each and every sample of the speech signal, with the result that the average bit rate equalled the peak bit rate.

An early approach to utilization of the variations in speech patterns was Time Assignment Speech Interpolation (TASI). Noting that voice conversations were formed of "talk spurts" with pauses between them, researchers developed a high-speed switch that interleaved various conversations on the same channel.

From a listener's viewpoint, what is most important in voice synthesis is constant voice quality during a conversation. By functioning as a variable rate coder, the new multipulse coder provides much more flexibility in assigning the bits needed as the different portions of the speech signal change.

Many of the newer applications of speech coding, such as voice mail and packetized voice transmission, can achieve superior efficiency by taking advantage of the difference between average and peak bit rates. In voice mail applications, for example, the total message-storage capacity is determined by the *average* bit rate, not the *peak* bit rate. By exploiting the variable nature of the speech signal, a hypothetical coder operating at a peak bit rate of only 9.6 kilobits per second might achieve acceptable performance with an average bit rate of only 9.6 kilobits per second. That means four times more digital storage capacity than with a fixed bit rate of 32 kilobits per second. Moreover, for packetized voice transmission, a programmed speech coder could help to reduce the transmission delays inherent in variable rate coding and control loss of important packets during transmission. Low-bit-rate voice coders, such as multi-pulse LPC, have the built-in capabilities to provide this control.

[2] An invention of AT&T Bell Laboratories.

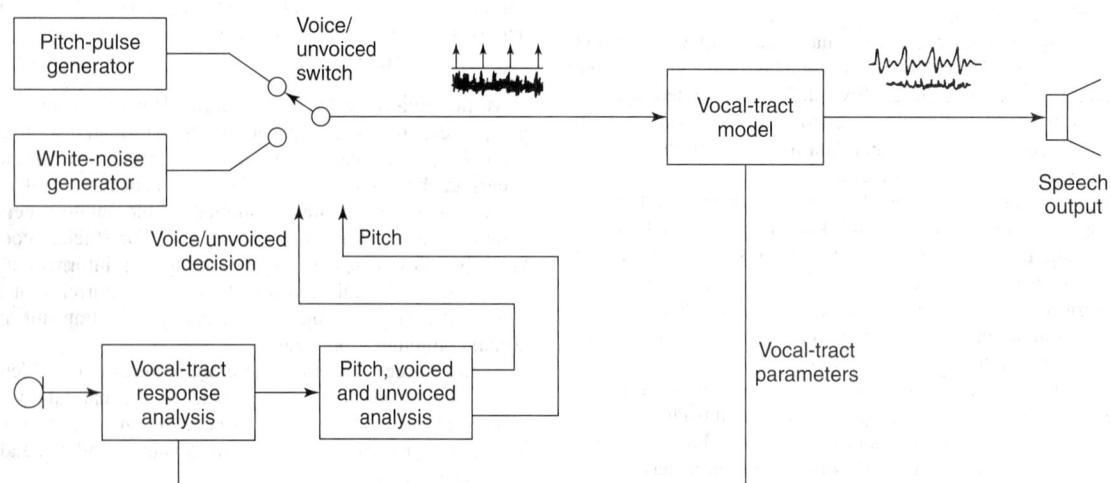

Fig. 5. Vocoder model for speech synthesis, useful for synthesizing digitally encoded speech at a bit rate sufficiently low for use on loop transmission facilities. The device operates with primtive source coding to digitally generate speech at 2.4 kilobits per second. Because its excitation pulses are paced at only one per pitch period, the synthesized speech sounds mechanical. (*AT&T Bell Laboratories.*)

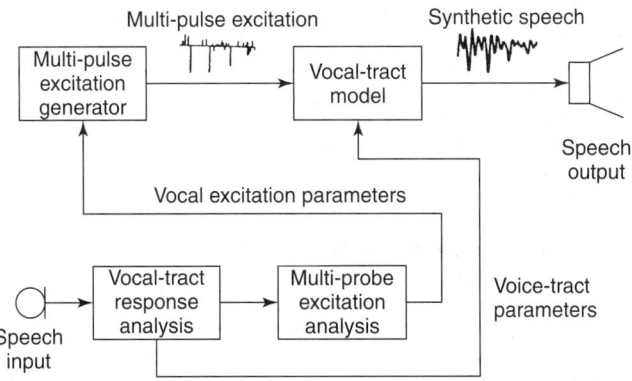

Fig. 6. Multi-pulse linear predictive coding (MPLPC) model for speech synthesis. By using a number of excitation pulses in every pitch period (usually about 10 milliseconds in an adult male voice), the MPLPC produces natural-sounding speech at bit rates as low as 9.6 kilobits per second. Normally, only a few pulses — typically, 8 to 16 pulses every 10 milliseconds — are needed to generate speech dramatically superior to that of the vocoder. (*AT&T Bell Laboratories.*)

Another application may be interactive business transaction that use synthesized voice prompts. A customer could key in data on the telephone keypad, while following instructions, confirmations, and summaries provided by the synthesized voice prompts.

Additional Reading

Atal, B.S. and R.Q. Hofacker, Jr.: "The Telephone Voice of the Future," Record, 4–10, AT&T Bell Laboratories, Short Hills, NJ, (July 1985).

Balentilne, B. and D.P. Morgan: "How to Build Speech Recognition Application: A Style Guide for Telephony Dialogues," Enterprise Integration Group, New York, NY, 1999.

Bates, R.J. and D.W. Gregory: "Voice and Data Communications Handbook," 3rd Edition, Mc-Graw-Hill Professional Book Group, New York, NY, 2000.

Becchetti, C.: "Speech Recognition: Theory and C++Implementation," John Wiley & Sons, Inc., New York, NY, 1998.

Furui, S.: "Digital Speech Processing, Synthesis, and Recognition," 2nd Edition, Marcell Dekker, Inc., New York, NY, 2000.

House, A.S.: "Recognition of Speech by Machine," Academic Press, Inc., San Diego, CA, 1990.

Jurafsky, D. and J.H. Martin: "Speech and Language Processing: An Introduction to Natural Language Processing, Computational Linguistics, and Speech Recognition," Prentice-Hall, Inc., Upper Saddle River, NJ, 2000.

Lee, Chin-Hui, F.K. Soong, and K.K. Paliwal: "Automatic Speech and Speaker Recognition: Advanced Topics," Kluwer Academic Publishers, Norwell, MA, 1996.

Rabiner, L.R. and B.H. Juang: "Fundamentals of Speech Recognition," Prentice-Hall, Inc., Upper Saddle River, NJ, 1993.

VOLATILE. Having a low boiling or subliming temperature at ordinary pressure; in other words, having a high vapor pressure, as ether, camphor, naphthalene, iodine, chloroform, benzene; or methyl chloride.

VOLATILE OILS. The volatile oils are distinguished from the fixed oils by the fact that a drop of one of the former does not leave a spot on paper. Members of certain plant families, such as the *Labiatae*, contain a larger percentage of such oils than do other families. But volatile oils are in no sense restricted to any small group, nor are they found only in certain tissues. Sometimes, certain parts may be principally used for the oils, as the seeds of the *Umbelliferae*.

Various methods are used in extracting the oils from the plant tissue. Many are distilled with water or steam, the oil being carried over with the distillate. In others, as for example oil of bitter almonds, the oil develops in the tissues only after fermentation. It is then obtained by distillation. Another method, and one especially used for more delicate and valuable oils, is called "enfleurage." In this method the flowers containing the oil are spread as a thin layer over a layer of lard or olive oil. The latter absorbs the delicate oil in the flowers, after which distillation may separate the volatile oil from the other.

VOLATILITY PRODUCT. The product of the concentrations of two or more ions or molecules that react to produce a volatile substance. The volatility product is analogous to the solubility product, except that, when it is exceeded, the substance escapes from the system by volatilization rather than precipitation. As with the solubility product, if any of the reacting ions or molecules have a numerical coefficient greater than one, then the concentration term of that ion or molecule is raised to the corresponding power.

VOLCANIC MAGMA.   See **Lava**.

VOLCANO.   A generalized definition of a volcano would be, "A vent in the surface of the earth through which magma and associated gases and ash erupt." The word is also used to designate the form or structure, usually conical, that is produced by the ejected material (ejecta). There are active and dormant volcanoes on Earth, and volcanoes also have been recognized in imagery received from exploring some of the other planets in the solar system, notably Venus, and Jupiter's moon Io, on which at least eight active volcanoes have been identified. As is explained later, volcanoes on earth are the result of tectonic activity, which provides openings in the earth's crust, allowing the release of energy and material from below the planet's lithosphere. The concept of plate tectonics is fundamental to understanding the formation of volcanoes, as it also is in connection with earthquakes. Both phenomena reflect the subterranean forces that do not attain permanent equilibria, but that are constantly shifting, if ever so slowly. A rather comprehensive table listing major volcanoes on the Earth known since several centuries B.C. is given in article on **Earth**.

Most geothermal regions where underground hot water and steam are extracted for commercial heating and electric power generation are found near active or dormant volcanoes. Examples include the Larderello field in Italy, where electric power has been generated since 1913; the terrain under Reykjavik, Iceland; the Wairakei geothermal fields on the North Island of New Zealand; the Geysers area of California, discovered in 1847, but not exploited commercially until 1970. See also **Geothermal Energy**.

Ill-Famed Volcanoes

A thorough inventory of volcanoes, ranging from active to dormat and "extinct" would number in the thousands. Volcanoes range widely in their physical dimensions and in their behavior. Only a few can be described here.

Mt. Etna (Sicily). The activity of this volcano dates back to the beginning of recorded history, with more than 11 eruptions known to have occurred prior to 1 A.D. About 70 eruptions have been recorded since then, several entailing high losses of life and property. A relatively major eruption occurred in November 1979, followed by activity in 1989 and a major eruption on December 14, 1991. This followed nearly 2 years of inactivity. Extensive measures were taken to combat the volcano in 1991 in an effort to protect the village of Zafferana Etnea, with a population of 7000 people, at the base of the volcano. As reported by Gasparini (University of Naples), "A 234-meter (768-foot) long and 21-meter (69-foot) high dam was constructed of earth, scoriae, and stones. The dam contained the lava and was breached after about one month. Three additional smaller earthen barriers were built to slow the flow toward Zafferana. Another effort consisted of several attempts to stop the advance of the lava front by the diversion of the flow out of its natural and extensively tunneled channel. The main intervention was made at an elevation of 2,000 meters (6,532 feet) in an almost inaccessible zone with the extensive use of helicopters." In the end, these and additional efforts saved the village. The eruption ended on March 30, 1993, culminating 473 days of continuous lava flows. It has been estimated that this probably was the largest eruption of Etna in the last 300 years, covering about 7 square kilometers (2.6 square miles) with in excess of 250 million cubic meters (8.8 billion cubic feet) of lava.

Cone-shaped Mt. Etna rises some 3300 meters (10,380 feet), with a circumference at the base of about 145 kilometers (90 miles).

In 1991, a group of French scientists found that Mt. Etna is one of the Earth's most potent sources of carbon dioxide (CO_2), estimating that even during quiescent periods, the volcano releases some 25 million tons of CO_2 into the atmosphere each year. This is many more times the amount released by the more spectacular Hawaiian volcanoes. Because of difficulties in measuring CO_2 directly, the researchers measured sulfur dioxide (SO_2) and used a ratio function for estimating the relative abundance of the two gases. The investigators found that not only do large amounts of CO_2 emanate from the crater, but that the atmosphere surrounding the base of the volcano also is inordinately heavy with the gas. The basic cause of the CO_2 has not been determined precisely, but one investigator suspects that the CO_2 is released from carbon-rich rocks around the volcano. Another researcher attributes the relatively alkaline lava as the principal source. See description of the Lake Nyos disaster (1976) described later in this article.

Mt. Vesuvius (Italy). Ash and lava from this dreaded volcano literally buried thousands of people and their homes in the cities of Pompeii and Herculanum on August 24, 79 A.D. Although Mt. Vesuvius has not erupted since 1944 and currently does not indicate renewed activity, the Phlegraaean Fields caldera in nearby geothermal areas has shown signs of unrest since 1970. This area has produced sporadic earthquake swarms, often regarded as precursors of volcanic eruptions.

Charles Dickens in his "Pictures from Italy" described a visit to Pompeii and Herculanum and his ascent and descent of Mt. Vesuvius in 1844.

Mt. Vesuvius and the Phlegraean Fields are of major concern to the Italian government because of the proximity of Naples, a city of 2 million people. In fact, in recent years European scientists have shown an increasing interest in the possible future threats of volcanoes in western and Eastern Europe. During the early 1990s, the European Science Foundation created the European Volcanological Project (EVOP), which is chartered to gather the knowledge of volcanologists all across Europe. Although not within the "Ring of Fire" mentioned later, there are volcanoes designated as active in Greece, Turkey, the Canary Islands (under Spanish control), the Azores (under Portuguese control), and Iceland. It is interesting to note that the volcanoes of Italy and Greece have erupted about 140 times since the great eruption of Santorini in the Aegean Sea in 1500 B.C. The disappearance of the Minoan civilization on the island of Crete is attributed to that eruption.

Krakatau (Indonesia). This volcano is remembered most in the literature as the "volcanic island that literally disappeared into the sea." This eruption occurred in late August 1883, following unrest that commenced in May of that year. Actually, one-third of the island remained above sea level, with the remainder of the island sinking well below sea level. The remaining island was about 15.5 square kilometers (6 square miles). The power released by the volcano has been estimated as being equal to or greater than one of the early military nuclear bombs. Essentially all life of any form was consumed by the blast. A submarine earthquake accompanying the blast created ocean waves as high as 15 meters (50 feet) that traveled several thousand miles. The waves killed more than 36,000 people along the coasts of Java and Sumatra and destroyed incalculable amounts of property. Historians recorded reports of the blast being heard by persons some 4800 kilometers (about 3000 miles) distant. The dust cloud produced by Krakatau was noted in the upper atmosphere for about 3 years.

As reported by D. and M. Plage, who visited the wildlife reserve of Ujung Kulon Peninsula on the large island of Java, "Thirty-five miles across the sea from Ujung Kulon, the island of Krakatau ripped itself apart in 1883. Tsunamis reaching the Javanese shore washed away the few villages that dotted this isolated outpost. The people never returned; the wildlife did, and it is protected today by the Ujung Kulon National Park. We spent a year there, examining the regeneration of life on both the peninsula and Krakatau a century after eruption."

El Chichón Volcano, Chiapas, Mexico. Between March 28 and April 4, 1982, this volcano in southern Mexico, even though regarded as a relatively small volcanic event, injected a fine mist of sulfuric acid droplets into the stratosphere considered denser than any volcanic cloud since the great eruption of Krakatau in 1883. The blast tore away the upper 200-meter (656-ft) portion of the old volcanic cone, after approximately 600 years of dormancy. The crater left by the 1982 eruption lies within the old crater, which is masked by new volcanic deposit. Satellite observations made shortly after the volcano erupted for the first time indicated that an aerosol cloud moved rapidly westward and, within a few weeks, a veil of fine material reached around the globe. As reported by Rampino (NASA) and Self (University of Texas), in less than a year the stratospheric cloud (some 25 km; 15.6 mi aloft) had blanketed the entire Northern Hemisphere and much of the Southern Hemisphere. El Chichón ejected a volume of ash comparable to that of Mount Saint Helens in 1980, but the cloud from El Chichón was estimated to be at least 100 times denser. It has been suggested that the sulfur, from which the sulfuric acid aerosol was formed through photochemical reaction, originated either from sedimentary deposits underlying the volcano, or from sulfide deposits on one tectonic plate of the Earth's crust that is plunging under another and melting.

Tectonically, El Chichón is related to the Guatemalan volcanic belts. Three major plates meet near this region — the Caribbean plate is sliding past the North American plate along a series of faults in Guatemala; the Cocos plate is being subducted under the North American and Caribbean plates at the Middle American trench off the coast of Mexico. See the map of plates in the article on **Earth Tectonics and Earthquakes.** Acidity of ice in the Greenland ice sheet were found to correlate with the eruption of El Chichón, as did those of an earlier similar eruption of the Agung volcano in Indonesia in 1963. Effects on the climate as caused by the El Chichón eruption are still being studied. The first description of El Chichón in the literature did not appear until 1928. Carbon-14 dating indicates that the last prior eruption of the volcano occurred between A.D. 1350 and 1400, and that the magnitude of that eruption was probably ten times greater in magnitude than that of 1982. El Chichón is located in a very remote and relatively inaccessible area of Mexico. For comparison purposes, scientists had only a few photographs that were made of the volcano prior to 1982.

Mt. Pinatubo (The Philippines). There are approximately 100 eruptive centers in the Philippine Islands. There was a major eruption of Mt. Taal on Luzon in 1965, costing 190 lives as well as property damage. This volcano also erupted in 1968, but on a smaller scale. Mt. Pinatubo on Luzon erupted on June 10, 1991, after not having erupted for over 500 years (1380). As with numerous volcanoes, an eruption may be "expected" at most any time, but the enormity of the 1991 eruption was surprising even to most experts in the field. As observed by one scientist at the National Aeronautics and Space Administration, "It's no Krakatau, but this is possibly the largest eruption of this century." Of major significance is the possible effects the eruption may have on global weather. In this regard, Mt. Pinatubo is sometimes compared with the 1982 eruption of El Chichón (Mexico).

Detailed studies of the 1992 and 1993 weather data remain to be completely analyzed, and it may require a few more years to study and build models of the data accumulated before definite conclusions on the volcano's effects can be announced with confidence. The NASA spaceborne Earth Radiation Budget Experiment has been used in connection with these studies. The change in the Earth's radiation budget that is initiated by the eruptions is termed *volcanic aerosol forcing*, or, simply, *volcanic forcing*. Needless to say, fundamental studies of the "greenhouse effect," or global warming, become more complex each time there is a major volcanic eruption of the enormit Pinatubo buto or El Chichón.

Mt. Unzen (Japan). After a quiescent period of about 200 years, Mt. Unzen, located near Nagasaki, Japan, collapsed, sending a half-million cubic meters of lava and hot ash traveling at high speed down the Mizunashi River and into the nearby town of Kamikoba. A total of 43 peopled were killed, including a large number of volcanologists from several countries. The group of scientists was engaged in research. There also were civilian deaths and property damage near Kamikoba. Since the disaster, which occurred on the afternoon of June 3, 1991, the Japanese government has been investigating ways and means to counter the volcano's destructive capabilities. The volcano remains reasonably active. For example, an eruption on June 23, 1993, produced damage in a previously untouched river valley on the volcano's northeast slope. This caused one death and destroyed 80 structures.

Despite a willingness on the part of the Japanese government to invest heavily in protecting lives and property at the base of the volcano (reminiscent of the previously described success at Mt. Etna), some specialists believe that the volcano is entirely too fickle to justify strategic protective plans. One volcanologist at the Kyusu University's Simabara Earthquake and Volcano Observatory has indicated that the best solution most probably would be that of moving the people who live in dangerous areas to safer places.

Mt. Pelée (Lesser Antilles). Located on the island of Martinique in the Caribbean northeast of Venezuela, Mt. Pelée erupted in 1902, destroying Saint-Pierre, formerly the largest city on the island. An estimated 36,000 persons were killed as the result of the blast.

Nevado del Ruiz Volcano, Colombia. After an inactive period of some 140 years, Nevado del Ruiz, a volcano located in northwestern Colombia, became seismically active in November 1984 and erupted explosively on November 13, 1985. Earlier eruptions between 1828 and 1833 and again in 1845 were relatively minor, but the latter did create a volcanic mudflow (lahar) that traveled down the 5300-meter (17,390 ft) high mountain on its eastern flank, reaching the site of Amero (a distance of over 50 km; 31 mi) and causing over a thousand casualties. A major eruption occurred in 1595, producing an ash and lapilli (pyroclastic particles) and lahars.

In terms of lives lost, the 1985 eruption of Ruiz was a major disaster. Pyroclastic flows and surges, generated during the initial stage of the eruption, triggered surface melting of approximately 10% of the volcano's ice cap, thus resulting in meltwater flooding. The erosive floods swept soils and loose sediments from the volcano's flanks and developed mud flows, which claimed at least 25,000 lives. Volcanologists regard the 1985 eruption of Ruiz as the largest volcanic disaster since the 1902 eruption of Mount Pelée in Martinique that killed some 28,000 people. Investigators reported that Ruiz released a small volume of pyroclastic material and a disproportionately large volume of volcanic gases.

Before the eruption, summit fumarole gases became less water-rich and the sulfur/chlorine ratio increased. Remote measurements of sulfur dioxide (SO_2) flux after the eruption indicated active degassing at levels associated with eruptive or inter-eruptive stages of other volcanoes. Thermal water analyses revealed increases in magnesium, calcium, and potassium and an increase in the magnesium/chlorine ratio, suggesting that these elements may have been leached from new magma. Water from the lahar contained high concentrations of sulfate and had a sulfur/chlorine ratio of 4.67, suggesting to experts that water ejected from the crater lake and turbulent mixing of pyroclasts and glacial ice triggered the lahar. The uniform composition of the pumices and the unusually high ratio of gas to magma suggest that, although a new batch of magma triggered the eruption, the pumice that erupted may have been old.

In June 1993, D.L. López (University of British Columbia) and S.N. Williams (Arizona State University) observed, "Catastrophic volcanic collapse, without precursory magmatic activity, is characteristic of many volcanic disasters. The extent and locations of hydrothermal discharges at Nevado del Ruiz suggest that many volcanoes' collapse may result from the interactions between hydrothermal fluids and the volcanic edifice. Rock dissolution and hydrothermal mineral alteration, combined with physical triggers such as earthquakes, can produce volcanic collapse. Hot spring water compositions, residence times, and flow paths through faults were used to model potential collapse at Ruiz. Caldera dimensions, deposits, and alteration mineral volumes are consistent with parameters observed at other volcanoes." Details are given in this scholarly paper. See reference listed.

Kilauea Volcano, Hawaii. Volcanic activity in this region of the world might be described as semi-continuous, much of which does not reach the newspaper headlines. In 1983, an eruption of Kilauea occurred along the volcano's east rift zone. The eruption commenced on January 2, 1983 and continued for about two weeks, during which a number of eruptive episodes produced several lava flows. Additional lava flows occurred during eruptions in February of that year. Scientists found the active laval flows useful for running field verification measurements of induced convective heat flow in basaltic lava. Convective heat flow in lava and magma has been studied because of its importance to energy extraction from molten magma and because of its relation to tectonic processes. As reported by Hardee (Geophysics Research Division, Sandia National Laboratories), eight field measurements of induced natural convection were made, giving heat flux values that ranged from 1.78 to 8.09 kW/cm^2 at lava temperatures of 1088 and 1128 °C (1993 and 2064 °F), respectively. The field measurements agreed with prior laboratory measurements in furnace-melted samples of molten lava.

The Hawaiian Islands, of volcanic origin, were not created simultaneously. The island of Kauai, for example, is estimated to be 5.1 million years old; Maui, 1.3 million years old; and Hawaii, 800,000 years old. Considerably further distant, the Midway islands are estimated to be 28 million years old. As the seamounts rose above sea level, numerous life forms commenced to occupy the islands. The process of occupation by living forms remains poorly understood. The Scottish biologist, Sir John Arthur Thomson, commented in 1920, "Living creatures press up against all barriers; they fill every possible niche all the world over.... We see life persistent and intrusive—spreading everywhere, insinuating itself, adapting itself, resisting everything, defying everything, surviving everything!"

The Hawaiian Islands resulted from plate tectonic activity, but, according to many scientists, were not directly influenced by continental landmasses.

G. Sen (Florida International University) and R.E. Jones (University of California, Los Angeles) in their studies of the origination of Hawaiian volcanism have reported, "The maximum depth at which large (greater than 1000 km^3) terrestrial mafic magma chambers can form has generally been thought to be the Moho, which occurs at a mean depth of about 35 km beneath the continents and 8 km beneath the ocean basins. However, the presence of layers of cumulus magnesium-rich spinel and olivine and intercumulus garnet in an unusual mantle xenolith from Oahu, Hawaii, suggests that this rock is a fragment of a large magma chamber that formed at a depth of about 90 km; Hawaiian shield-building magmas may pond and fractionate in such magma chambers before continuing their ascent. This depth is at or near the base of the 90-million-year-old lithosphere beneath Oahu; thus, rejuvenated stage alkalic magmas containing mantle xenoliths evidently also originate below the lithosphere."

The researchers report in exquisite detail their findings pertaining to the evolution of a typical Hawaiian volcano, indicating that this is a four-stage process.

System studies of Kilauea began as early as 1912, with the establishment of the Hawaiian Volcano Observatory, and continue to the present. These studies have revealed much about the physical and chemical processes that trigger and sustain volcanic eruption and have served as models for investigations of other, often more violent volcanoes in many parts of the world. The Observatory is managed by the U.S. Geological Survey. A detailed description of the Observatory's findings is given in the Delaney reference listed.

It is interesting to note that some of the world's largest optical telescopes are located atop Mauna Kea, long considered to be an "extinct" volcano.

Further reference to article on **Earth Tectonics and Earthquakes** is suggested.

Oldoinyo Lengai (Tanzania). This volcano is famous for its unusual alkali-rich magma, termed *natrocarbonatite*, and is believed to be the only active *carbonatite* volcano on Earth. The volcano has a steep cone about 2900 meters (9515 feet) high. It is located just a few km south of Lake Narron in the Tanzanian part of the Gregory Rift Valley. Early explorers reached the area in 1880. Since then, about ten eruptions have been noted, most of them of an explosive nature. However, until the 1960 lava eruption, the carbonatitic composition of the lava had not been recognized. Another explosive eruption occurred in 1966. No additional activity was reported until an explosive eruption in January 1983. Activity during 1988 was purely effusive.

The lava chemistry of this volcano has attracted the interests of volcanologists, even though the volcano is very difficult to reach. Reporting on recent findings, M. Krafft (Centre Vulcain, Cernay, France) and J. Keller (Mineralogisch-Petrographisches Institut der Universität, Freiburg, Germany) observe, "The petrogenesis of carbonatites has important implications for mantle processes and for the magmatic evolution of mantle melts rich in carbon dioxide. Oldoinyo Lengai has highly alkalic, sodium-rich lava. Although different in composition from the more common calcium-rich carbonatites, this provides the opportunity for observation of the physical characteristics of carbonatite melts. Temperature measurements on active carbonatitic lava flows and from carbonatitic lava lakes were carried out during a period of effusive activity in June 1988. Temperatures ranged from 491° to 519 °C. The highest temperature, measured from a carbonatitic lava lake, was 544 °C. These temperatures are several hundred degrees lower than measurements from any silicate lava. At the observed temperatures, the carbonatite melt had lower viscosities than most fluid basaltic lavas. The unusually low magmatic temperatures were confirmed with 1-atmosphere melting experiments on natural samples."

Soufrière Volcano, St. Vincent, Lesser Antilles. A swarm of local earthquakes on the island of St. Vincent occurred just a few hours before the eruption of this volcano got underway in the spring of 1979. The eruption consisted of two phases. As reported by Fiske (Smithsonian Institution) and Sigurdsson (University of Rhode Island), during the initial vulcanian explosive phase (April 13 to 26), a series of vertical explosions blasted a new vent through the lava island (created in 1971–1972) in the middle of the crater lake (1 km; 0.6 mi wide), causing steam and tephra to rise 8 to 20 km (5 to 12$\frac{1}{2}$ mi) into the atmosphere. This phase also added layers of air-fall debris on the island and surrounding sea. Also, during this period, small pyroclastic and mud flows traveled down the slopes. In a few days, the crater lake flooded the vent of the volcano. During the next few days, the lake completely disappeared. The second phase consisted of the quiet extrusion of viscous basaltic andesite lava, resulting in the growth of a dome over the vent. It was estimated that the total mass of rock produced during the eruption was 2.4×10^{11} kg as compared with 3.8×10^{11} kg produced by the same volcano in a destructive eruption that occurred in 1902.

Scientific coverage of the event was exceptional, including observations from the ground, aircraft, and satellites. Because of intense interest and the large amounts of data collected, the Smithsonian Institution hosted a "Soufrière Conference" in September 1979. Meteorologists concluded that the atmospheric plume created was not an important climate modifier. Fascinating "skirt" cloud formations were noted during the eruption. These clouds had a nearly vertical orientation. They are described in the Barr reference listed.

Alaska's Katmai Region. What was described as a colossal volcanic eruption occurred at Mt. Katmai on June 6, 1912. At the time, the region was very sparsely populated and observations were few. No casualties or property damage were reported. Mt. Katmai is located in the Alaskan region denoted as the "Valley of Ten Thousand Smokes." Mt. Katmai is about 2042 meters (6700 feet) high. There are numerous volcanoes in Alaska and the nearby Aleutian Islands. Among these are Mt. Wrangell, 4317 meters (14,163 feet) high, in Alaska and over 30 active vents and numerous inactive cones in the Aleutians. A volcano on Akutan Island, 1220 meters (4000 feet) high, exploded in 1974, with ash and debris rising several hundred meters. The volcano Great Sitkin, 1750 meters (5741 feet) high displayed explosive activity in the spring of 1974. A volcano on Augustine Island, 1220 meters (4000 feet) high, erupted in 1976 and again on March 17, 1986.

The area around Mt. Katmai was designated a national monument in 1918 and presently is part of Katmai National Park. Commencing in 1989, scientists have been making a geophysical study of the region's immense volcanic system. As of 1991, plans called for drilling approximately 1220 meters (4000 feet) beneath the surface to determine the underground mechanics of the volcanic system. Site of the drilling is the dome-plugged volcanic throat of Novarupta in Katmai. Also, it is hoped that details of the Katmai eruption of 1912 can be understood better by analyzing core samples. It has been estimated that when Katmai erupted it produced 30 times the ash of Mt. St. Helens. Careful regard for the environment will be a predominant consideration.

Cascade Range (Northwestern United States). From the standpoint of plate tectonics, this region is ideally located in terms of volcanic activity. Prior to the eruption of Mt. St. Helens in southwest Washington on May 18, 1983, the last prior eruption in the region was a period of activity at Lassen Peak (California) during the 1914–1917 time span. The Mt. St. Helens eruption is described later in this article. Some 7000 years ago, Mazama, a 3000-meter (9843-feet) high volcano in southern Oregon erupted violently, ejecting about 40 cubic kilometers (9.6 cubic miles) of ash and lava. The ash spread over the entire northwestern United States and as far as Saskatchewan (Canada). During the eruption, the top of the mountain collapsed, leaving a caldera about 10 kilometers (6.2 miles) across and about one kilometer (0.62 mile) deep, which filled with rainwater to form what is now called Crater Lake.

Principal Types of Volcanoes[1]

The word "volcano" is derived from the name of a small island, *Vulcano*, situated in the Mediterranean Sea off Sicily. Centuries ago, the people living in the area believed that Vulcano was the chimney of the forge of Vulcan—that is, the blacksmith of the Roman gods. They thought that the hot lava fragments and clouds of dust erupting from Vulcano came from Vulcan's forge. Modern geologists generally put volcanoes into four generalized groups, but obviously there could be other classifications and numerous subclassifications.

Cinder Cones. These are the simplest type of volcano. They are built from particles and "blobs" of congealed lava ejected from a *single* vent. As the gas-charged lava is blown violently into the air, it breaks into small fragments that solidify and fall as cinders around the vent, forming a circular or oval dome. Most cinder cones have a bowl-shaped crater at the summit and rarely rise more than about 300 meters (~1000 feet) above their surroundings. Cinder cones are numerous in western North America as well as throughout other volcanic terrains of the world.

Composite Volcanoes. Some of the Earth's grandest mountains are *composite* volcanoes (sometimes called *stratovolcanoes*). Typically, they are steep-sided symmetrical cones of large dimension built of alternating layers of lava flows, volcanic ash, cinders, and blocks of material and may rise as much as 2400 meters (~8000 feet) above their bases. Some of the most conspicuous and beautiful mountains in the world are composite volcanoes, including Mt. Fuji (Japan), Mt. Cotopaxi (Ecuador), Mt. Shasta (California), Mt. Hood (Oregon), Mt. St. Helens (Washington), and Mt. Rainier (Washington).

Most composite volcanoes have a crater at the summit that contains a central vent or a clustered group of vents. Lava either lows through breaks in the crater wall or issues from fissures on the flanks of the cone. Lava, solidified within the fissures, forms dikes that act as ribs that greatly strengthen the cone.

The essential features of a composite volcano is a conduit system through which magma from a reservoir deep in the Earth's crust rises to the surface. The volcano is built up by the accumulation of material erupted through the conduit and increases in size as lava, cinders, ash, and so forth are added to the slopes.

Shield Volcanoes. These volcanoes are built almost entirely of fluid lava flows. Flow after flow pours out in all directions from a central summit vent or group of vents, building a broad, gently sloping cone of flat, domical shape, with a profile much like that of a warrior's shield. They are built up slowly by the accretion of thousands of flows of highly fluid basaltic[2] lava that spread widely over great distances and then may cool as thin, gently dipping sheets. Lavas also commonly erupt from vents along fractures (rift zones) that develop on the flanks of the cone. Some of the larger volcanoes in the world are shield volcanoes. In northern California and Oregon, many shield volcanoes have diameters of 4.8 to 6.4 kilometers (~3 to 4 miles) and heights of 460 to 610 meters (1500 to 2000 feet). The Hawaiian Islands are composed of linear chains of these volcanoes, including Mt. Kilauea and Mt. Loa on the island of Hawaii.

In some shield-volcanic eruptions, basaltic lava pours out quietly from long fissures instead of central vents, flooding the surrounding countryside with lava flow upon lava flow and forming broad plateaus. Lava plateaus of this type can be seen in Iceland, southeastern Washington, eastern Oregon, and southern Idaho.

Lava Dome Volcanoes. Volcanic or lava domes are formed by relatively small, bulbous masses of lava too viscous to flow any great distance. Consequently, upon extrusion, the lava piles over and around its vent. A dome grows largely by expansion from within. As it grows its outer surface cools and hardens and then shatters, spilling loose fragments down its sides. Some domes form craggy knobs or spines over the volcanic vent, whereas others form short, steep-sided lava flows known as "coulees." Volcanic domes commonly occur within the craters or on the flanks of large composite volcanoes. The nearly circular Novarupta Dome that formed during the 1912 eruption of Mt. Katmai (Alaska) measures 274 meters (~900 feet) across and is 61 meters (~200 feet) high. The internal structure of this dome, defined by layering of lava fanning upward and outward from the center, indicates that it grew largely by expansion from within. Mount Pelée in Martinique and Lassen Peak and Mono dome in California are examples of lava domes.

Submarine Volcanoes. These volcanoes and volcanic vents are common features on certain zones of the ocean floor. Some are active at the present time and, in shallow water, disclose their presence by blasting steam and rock debris high above the surface of the sea. Many others lie at such great depths that the tremendous weight of the water above them results in high, confining pressure and prevents the formation and release of steam and gases. Even very large, deepwater eruptions may not disturb the ocean floor.

The famous black sand beaches of Hawaii were created virtually instantaneously by the violent interaction between hot lava and seawater. Underwater mountains sometimes are referred to as *seamounts*.

Macdonald Seamount. When surveyed in 1975, the peak of this seamount was found to be only 49 meters (161 ft) below the surface of the Pacific. Since then, this undersea volcano (about 1600 km; 1000 mi southeast of Tahiti) has been erupting and growing fast. Since 1977, seismic stations on several Polynesian islands have detected a total of 12 seismic swarms, directly attributed to eruptions at Macdonald. French divers, in 1983, explored the plateau around the summit and found evidence of recent volcanism—lava spatter cones, free of algae, flanking a fissure with fresh walls. Soundings then indicated that the summit of the volcano had risen to just 27 meters (88 ft) below sea level. Macdonald may be an island "in the making."

---

[1] Source of Information: *U.S. Geological Survey.*

[2] From *basalt*, a hard, dense, dark volcanic rock.

The Mechanisms of Volcano Origination

As viewed from the perspective of *plate tectonics*, volcanoes appear to originate in three ways:

(1) *Plate Collisions.* Volcanoes which are associated with the coming together of two tectonic plates and, in particular, where one plate commences advancement under an adjacent plate. This process is called *subduction* which may be defined as the process of one crustal block descending beneath another, by folding or faulting or both. The concept was originally used by Alpine geologists. Movement along the subduction zone (the "line" of contact between adjacent crustal blocks) will range from a very few to several centimeters per year and there is some correlation between these rates of movement and the activity of volcanoes on one of these blocks. According to current plate tectonics theory, there are several large plates that "float" on viscous underlayers in the Earth's mantle, such as the Pacific Plate, North American Plate, Eurasian Plate, Antarctic Plate, etc. There are usually several smaller plates or "plate regions" associated with these large main plates. As shown by Fig. 1, the Juan de Fuca Plate which lies between the Pacific Plate and the North American Plate has been responsible for the actions of Mount St. Helens in Washington (U.S.A.). It has been estimated that movement along the subduction zone between these two plates is in the range of 2–3 centimeters (0.8–1.2 inches) per year. The underthrusting of one plate relative to another generates high pressures and temperatures and disturbs the magma chamber, a reservoir of magma in the shallow part of the lithosphere (to a few thousand meters), as well as exposing pathways for the flow of magma upward to the surface of the Earth. Hays (1979) suggested that increased volcanism will occur when the underthrust oceanic plate reaches a depth of between 100 and 200 meters (328 and 656 feet), where conditions are suitable for the production of andesitic magma. The vent of a volcano is first formed because it provides a path of least resistance to flow to the surface. This vent, once formed, serves during periods of dormancy as a solidified rock plug. However, the pathways can change, as witness the number of supplementary vents found on some volcanoes. The vent of a volcano is, in essence, one end of a natural pipeline. See Fig. 2.

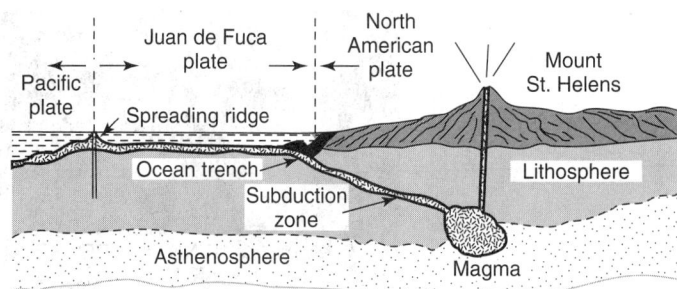

Fig. 2. Generalized arrangement of principal elements of a subduction-process-associated volcano as visualized by some scientists.

As observed by Hékinian (Centre Océanologique de Bretagne, Brest) in an excellent paper on undersea volcanoes (see reference), the volcanic activity that continually remakes the surface of the Earth takes place almost entirely at the bottom of the ocean, far beyond the reach of the traditional investigative tools of the volcanologist. Magma wells up from the mantle in two characteristic locations: along the active midocean ridges, where the spreading tectonic plates grow by the steady accretion of solidifying mantle material; and at the isolated volcanic structures called seamounts, which typically are strung out in chains across the interior of the plates. Sonar, special lighting for photography, seismology, and submersible vessels, manned and unmanned, have contributed much to the knowledge of undersea volcanic activity during the past few years. Some of the volcanoes observed appear in sonar images much the same as an aerial view of volcanoes located above sea level. An early sonar image, using side-looking sonar (system name, GLORIA) was made by the Institute of Oceanographic Sciences (Wormley, England) in 1980. It showed a volcano some 1.5 km (0.9 mi) high and about 9.5 km (6 mi) in diameter at the base. Volcanic activity is associated with hydrothermal vents ("black smokers") that vent noxious gases and the spiral-like tower structure of which contain metal-rich sulfide deposits. These are other volcanic features of the ocean floor, such as ocean floor fissures, pillow lava, and sheet lava, are described and illustrated in the article on **Ocean**.

The eruptive behavior of a volcano is directly related to the character of the magma that feeds it and, in particular, to the viscosity and innate strength of the magma.

Volcanoes associated with the subduction process occur all around the so-called Ring of Fire, which is associated with the Pacific Plate and its subplates. See Fig. 3. The numerous volcanoes found in the Aleutians, Alaska, and along the Pacific coast of the United States are a part of this ring. With exception of some volcanoes in the Aleutians and Alaska and, more recently, of Mt. Saint Helens in Washington, volcanoes in the North American portion of the ring have been dormant for several decades. The volcanoes of Indonesia, Java, the Philippines and other parts of Southeast Asia and the South Pacific, which are much more active, as well as those volcanoes located along the western edges of South and Central America, are also a part of this ring. Because of this, the subduction-associated volcanoes are sometimes called Ring of Fire volcanoes. The magma of these volcanoes is siliceous, with a relatively high viscosity and a high content of dissolved gases. As the magma rises toward the Earth's surface, a combination of high temperature and lessening pressure cause some of the gases to come out of solution and form bubbles. However, because of a combination of factors, including the mechanical strength of the bubble envelope materials, the bubbles do not reach an equilibrium size. Rather, they remain somewhat analogous to little grenades that confine pressures up to several hundred atmospheres. Thus, instead of a smooth lava flow rising and flowing from the vent, the gas bubbles, which also contain solid particles, retain their pent-up energy until a final massive eruption takes place, freeing them to a low pressure of one atmosphere. As they emerge and as they rise to heights in the atmosphere, they explode, thus releasing not only gaseous vapors, but much entrained solid material as well. Because the siliceous magma is very viscous, it forms pyroclastic avalanche flows rather than smooth-flowing rivers of lava. The 1980 eruption of Mt. Saint Helens, described a bit later, was characteristic of the behavior of this kind of volcano.

(2) *Divergence of Plates.* A second category of volcanoes, as viewed from plate tectonics, is the *rift volcano*, which occurs where plates are

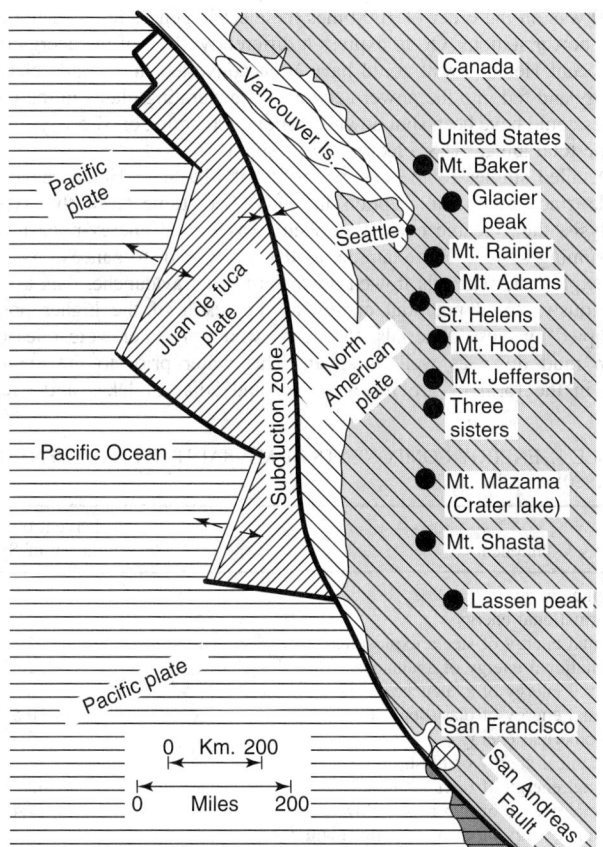

Fig. 1. Chain of volcanoes in the western United States, extending from Mt. Baker in northern Washington to Lassen Peak in northern California. Diagram shows position of the Juan de Fuca Plate with reference to the North American and Pacific Plates.

Fig. 3.   Location of "Ring of Fire" as related to major tectonic plates of the world.

diverging rather than converging. Their eruptions are more effusive,[3] yielding a basaltic magma which has relatively low viscosity and low content of dissolved gases. See also **Basalt**. However, when these volcanoes erupt in shallow water or through continental crust, they can be explosive.

(3) *Island Volcanoes.* A third category, the *hot-spot volcano*, penetrates the Earth's crust, but how directly it is associated with plate mechanics is subject to some debate, particularly if the view is taken that *all* plates are coupled globally. The oceanic island volcanoes are usually of this type and are exemplified by the volcanoes of the Hawaiian Islands. Particularly when these volcanoes penetrate oceanic crust, they are effusive, but under certain conditions also can produce violent explosions. These volcanoes also yield basaltic magmas and may exhibit a most unusual and beautiful behavior when they are active. They may be characterized by numerous vents and large calderas. Molten lava in the caldera may light up the night sky for long periods and steam and heat combine to generate cumulonimbus clouds overhead. Although the gas content of basaltic magma is less than that of siliceous magma, nevertheless carbon dioxide, water, and other gases come out of solution as the magma rises upward. The bubbles expand easily against low viscous forces and frequently, during just the last few meters of their escape, form beautiful and dramatic incandescent fountains of lava. These volcanoes, however, do not always behave in an effusive manner. For example, Kilauea Volcano on the island of Hawaii suddenly erupted in 1790 with a severe explosion. A similar, but lesser explosion occurred in 1924. Otherwise, the 44 eruptions of the volcano recorded since 1900 have been of the gentle kind, producing prodigious quantities of lava that change some part of the island's surface every few years. This volcano has been the object of continuing research for many years.

Energy and Ejecta of Volcanoes

Estimates of prehistoric volcanic eruptions, which produced quantities of ejecta (gas, ash, debris) in excess of 1000 cubic kilometers (240 cubic miles), tend to dwarf the eruptions which have occurred during the period of recorded history. As the result of paleogeological findings of recent years, it is believed that very early volcanic activity was concentrated largely in Japan, New Zealand, Central America, and the western United States. Geologists have estimated that volcanic activity experienced during the lifetime of the Earth to date is but a fraction of what could occur under the most favorable conditions for such eruptions, i.e., there is no indication of any diminution of thermal energy below the Earth's crust.

In describing volcanic eruptions, the news media are prone to compare the power of a volcano with that of a bomb. For example, the 1980 explosion of Mt. Saint Helens was likened to a 400-megaton nuclear bomb

(over 8 times more powerful than any nuclear device detonated to date). This kind of comparison is misleading because it does not give proper recognition to the *rate of energy release*. A bomb made of TNT or of nuclear components releases energy essentially in an instant, whereas the energy of a volcanic eruption is spread over a much longer time span. It is estimated that Mt. Saint Helens released 1.7×10^{18} joules over a 9-hour period, whereas a 400-megaton bomb would release only 4.2×10^{15} joules. Thus, the sustained power output exhibited by Mt. Saint Helens was approximately equivalent to detonating 27,000 Hiroshima-size nuclear bombs at a rate of one detonation per second over a 7.5-hour period. Decker (1981) estimated the power generation by Mt. Saint Helens on May 18, 1980 was on the order of 100 times the electric generating capacity of all U.S. power stations. Or, to express it differently, if all of the energy of the Mt. Saint Helens event could have been converted into usable electrical energy and fed into the power network, all generating plants in the United States could have been shut down for about 900 hours, or 37.5 days (10.3% of the total annual U.S. electric power supply).

Detailed studies of the energy profile of Mt. Saint Helens have been made since the main eruption in 1980. See Table 1. In this type of analysis, the volcanic event is looked at as a total energy-releasing and consuming system. For example, as the result of disturbances created by the explosion, materials flowed downward in an avalanche, thus utilizing previously stored potential energy as the result of the higher position occupied by these materials. Similarly, stores of water were released to form mud flows as well as steam. This water prior to the explosion was in the form of groundwater, ice, and adjacent lakes and streams.

TABLE 1. ENERGY COMPONENTS OF MT. SAINT HELENS EVENT (MAY 18, 1980)

Energy Component	Percent of Total Energy of System
Seismic energy	*
Deformational energy	1.0
Thermal energy of steam explosions	*
Gravitational energy of avalanche	2.6
Thermal energy of avalanche	10.5
Thermal energy of blast deposits	10.5
Hydrothermal energy	5.3
Mechanical energy of blasts	*
Thermal energy of high ash cloud	43.7
Thermal energy of pyroclastic flows	17.4
Thermal energy of eruptions after main event	8.9
All other	0.1
Total	100.0

*Insignificant on this scale. However, as explained in text these were nevertheless large amounts of energy.

[3] As contrasted with being violently explosive, the flow of magma from an effusive volcano is much smoother and gentle and sometimes has been likened to t "toothpaste flowing out of a tube."

Although some of the energy components, as a percent of the total energy system, appear insignificant, they were nevertheless very large, as evidenced by the destruction which they wrought. The thermal energy of steam explosions, not shown on table, was estimated at 10^{14} joules; the mechanical energy of blasts, 3×10^6 joules; and the pre-eruption seismic energy, 1.8×10^{13} joules.

The energy profile and the amount and characteristics of the ejecta are affected by the kind of magma, dissolved gas, groundwater, and originating pressure. Volcanic eruptions are commonly evaluated in terms of the tonnage of ejecta produced. However, there is no direct relationship between ejecta and damage. As shown by Fig. 4, the eruption of Mt. Saint Helens in 1980 was very low in terms of ejecta produced by a number of other volcanoes and including a prior eruption of Mt. Saint Helens about 1900 B.C., which produced about 4 times the quantity of ejecta produced in the 1980 event. Total damage is related to numerous factors in addition to production of ash, which destroys by rapidly covering over (literally burying) structures and people or by chemical poisoning from gases evolved. As in the case of Mt. Saint Helens, the ejecta principally posed a severe inconvenience over a relatively limited region, with the majority of damage stemming from the thermal and blast energy that destroyed large areas of forest and recreational grounds; and from the disturbance of water flow and the clogging of streams and reservoirs with ash-created mud; and the destruction of adjacent land by the mud and avalanche flow. This type of event is contrasted with the notable event of Vesuvius, which in A.D. 79 literally covered Herculaneum and Pompeii with ash.

As with earthquakes, the relative importance of volcanic events is usually equated much more with loss of life and property statistics rather than with measurements of a strictly scientific nature.

The rather different nature of the eruption of Krakatoa (1883), previously mentioned, has intrigued geologists and volcanologists for many years. In recent years, it has been suggested that possibly underwater volcanic eruptions, such as Krakatoa, may be accompanied by enormous vapor explosions. As pointed out in the entry on **Natural Gas**, vapor phase explosions are known to occur, but to date there is no established explanation for their initiation and propagation (Fowles, 1979).

The effect of gases and aerosols (particles) in the stratosphere and troposphere upon the amount of solar radiation reaching the surface of the earth is not thoroughly understood. In the case of the Mt. Saint Helens 1980 eruption, due to gravitational sedimentation, larger particles were found in the tropospheric samples than in the stratospheric ones. Most of the particles in the stratosphere were several tenths of a micrometer in size, whereas a strong enhancement of micrometer-sized particles was observed in the tropospheric samples. Sulfuric acid was present chiefly as a coating on the ash particles in the troposphere and as both separate small particles and large composite particles in the stratosphere (Chuan et al., 1981; Farlow et al., 1981). The concentrations of particles less than 10 micrometers in diameter in the ash emissions from Mt. Saint Helens

were more than 1000 times greater than those in the ambient air. Mass loadings of particles less than 2 micrometers in diameter were generally several hundred micrograms per cubic meter. Mineralogy of the particles varied considerably with time after eruption. The volume percent of glass ranged from 37 to 100%; of plagioclase, 0 to 32%; hornblende, 0 to 21%; and pyroxene, 0 to 22%.

Increased burdens of atmospheric aerosols can lead to either a cooling or warming of the Earth, depending upon the optical properties of the particles. Samples of particles from Mt. Saint Helens were collected in both the stratosphere and troposphere for measurement of the light absorption coefficient. Results indicate that the stratospheric dust had a small but finite absorption coefficient, ranging up to 2×10^{-7} per meter at a wavelength of 0.55 micrometer, which is estimated to yield an albedo for single scatter of 0.98 or greater. Tropospheric results showed similar high values of an albedo for single scatter.

Gaseous emissions vary greatly from one volcano to the next and, in fact, from one eruption to the next. During the steady-state period of activity of Soufrière Volcano (St. Vincent) in 1979, the mass emissions of sulfur dioxide into the troposphere amounted to a mean value of 339 ± 126 metric tons per day. This value is similar to the sulfur dioxide emissions of other Central American volcanoes, but less than those measured at Mount Etna, which is an exceptionally strong volcanic source of sulfur dioxide.

Volcanic Introduction of Chlorine into the Stratosphere. Some researchers (Johnston, 1980) now suspect that considerably more hydrogen chloride (HCl) is injected into the stratosphere by certain volcanic eruptions than previously estimated. In fact, this could amount to as much as 17–36% of the worldwide production of industrial production of chlorine in fluorocarbons—as of 1975 prior to severe reductions in such production brought about by concern with degradation of the protective ozone layer in the stratosphere. Under normal conditions, hydrogen chloride emanating from anthropogenic sources is not a threat to the ozone layer simply because the HCl is soluble in water and thus is removed in rain prior to reaching the stratosphere. Large volcanic eruptions inject HCl directly into the stratosphere forcefully and in significant quantities and thus bypass the absorptive effects of the lower atmosphere. It has been estimated that eruptions of this nature occur at least once per year. As pointed out by Johnston, the moderately to highly silicic magmas of volcanoes along the continental and island arcs, because of their high chlorine content, extreme explosivity, and frequent eruptions are likely to have the greatest atmospheric impact. These observations suggest that the impact of anthropogenic production of chlorine in fluorcarbons should once again be reviewed against the backdrop of disturbance of the ozone layer that may arise from natural, volcanic causes. Researchers have observed, for example, that the Augustine Volcano (Alaska), which erupted in 1976, may have injected 289×10^9 kilograms of HCl into the stratosphere. This quantity is about 570 times the 1975 world industrial production of chlorine and fluorocarbons.

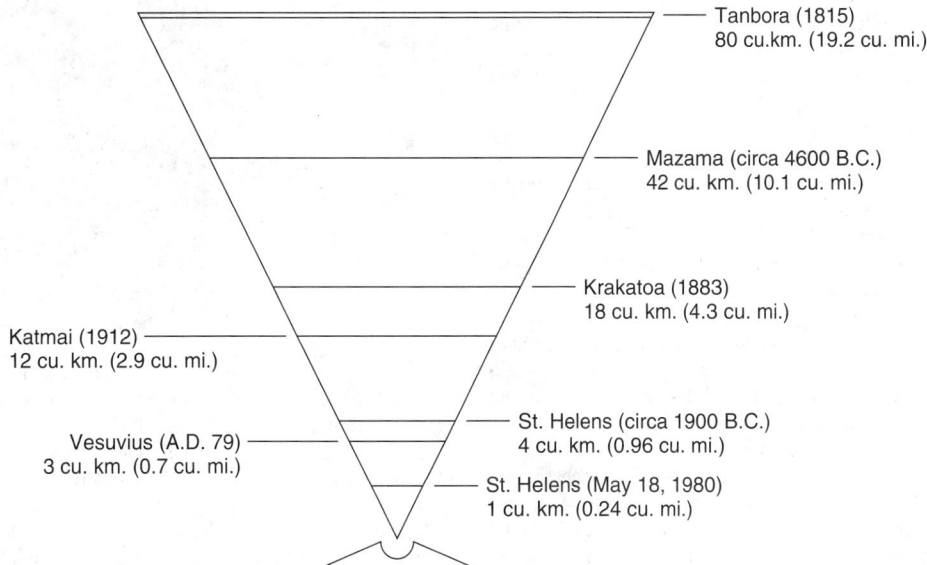

Fig. 4.   Relative quantities of ejecta from major and well-known volcanic eruptions.

Volcano-Climate and Climate-Volcano Relationships. Although it is readily apparent that volcanic eruptions can affect weather, at least over the short term, the effect of climatic change as a precursor of volcanic eruption is less apparent and still rather poorly understood. Research along these lines has been active in recent years (Ninkovich and Donn, 1976, among others). The general theme is that variations in climate lead to stress changes in the Earth's crust as, for example, by loading and unloading of ice and water masses and by axial and global spin-axis rate changes (due to change in symmetry of mass), factors that may augment volcanic and seismic potential. Relatively recent research has indicated that active volcanic periods (sometimes called "volcanic pulses") which, in past theory, were supposed to have triggered glacial advances, actually occurred hundreds of years after glacial activity. Further, major ice-sheet expansions that have occurred during the past two million years appear to have lagged volcanic triggering events by as much as 70,000 years. However, other research points to some correlation of past major volcanic eruptions and cooling periods. A major part of the problem in sorting out the sequence of events arises from inaccuracies in the dating of ancient volcanic and glacial events.

Rampino, et al. (1979) observe that, although some of the geological dates may be questioned, the instrumental temperature records for the past 200 years show that the major historical eruptions associated with coolings actually occurred after decade-long temperature decreases had been initiated. The examples cited include Tambora (1815), Krakatoa (1883), the 1902 series of eruptions (Santa Maria, Mount Pelée, etc.), and Katmai (1912). Rampino et al. also observe that probably one of the greatest single explosive events of the last 2 million years was the eruption of Toba in Sumatra; its vast ash shower extended from Malaysia to India. Dated at 75,000 years ago, it apparently followed the initiation of the cooling that began with the end of the Brørup (Saint Pierre) interstadial by several thousand years.

Ninkovich and Donn (1976) point out that during historic times, the atmospheric temperature quickly recovered after relatively large eruptions. Only an unusually high frequency of eruptions could have a long-term climatic effect. During historic times, this has not occurred.

As pointed out in the entry on **Climate,** the weather (over a period of a year or so) and the climate (over several centuries or more) are two different considerations. The effects of volcanic ash in the atmosphere on weather have been well documented in a number of instances. One of the most thoroughly researched is the so-called Year without a Summer, which adversely affected New England in 1816, following the eruption of Tambora in 1815. In New England in 1816, it snowed in June, with killing frosts continuing through August. The economic and social effects of this weather change are well portrayed by Stommel and Stommel (1979).

Because of its recent interest, it should be pointed out that, in the observations of one scientist, Mt. Saint Helens just did not have what it takes to thicken the light haze in the stratosphere sufficiently to block off much sunlight and thus cool the air near the ground. Little or no detectable effects were observed after the May 18, 1980 eruption.

Tsunamis Generated by Volcanic Eruptions. As reported by Kienle and Kowalik (University of Alaska-Fairbanks) and Murty (Institute of Ocean Sciences, British Columbia), during an eruption of the Alaskan volcano Mount St. Augustine in the spring of 1986, concern developed over the possibility that a tsunami (large water wave induced by sudden upheaval or subsidence of the sea floor) might be generated by the collapse of a portion of the volcano into the shallow water of Cook Inlet. A similar edifice collapse of the volcano and ensuing sea wave occurred during an eruption in 1883. Other sea waves resulting in large loss of life and property have been generated by the eruption of coastal volcanoes in various parts of the world. Fortunately, this did not occur in 1986.

Tsunamis can travel across the sea over long distances and can be profoundly hazardous, sometimes arriving unexpectedly, long distances from the source. The researchers estimate that tsunamis of volcanic origin (not including earthquakes and submarine landslides) have killed about 25% of the people who have experienced the direct effects of catastrophic volcanic eruptions since A.D. 1000. See also **Tsunami.**

Because progress toward fully understanding how volcanic processes can trigger tsunamis has been slow, and because few modern field observations are available to study the problem of wave generation, propagation, and run-up, the investigators targeted on the case of Mount St. Augustine in creating numerical simulations and models. Details are beyond editorial limitations here. See reference listed.

Mount Saint Helens. The eruption of Mount St Helens in 1980 has been one of the most analyzed and discussed of volcanic events in recent decades. The eruption was the first such eruption in the lower 48 United States since the milder eruptions of Lassen Peak in California in 1914–1917. In the 1980 event, 62 persons were killed or reported missing. Property damage exceeded $1 billion.

About 2.6 cubic kilometers (0.65 cubic mile) of volcanic rock were displaced. About 0.5 cubic kilometer (0.12 cubic mile) of liquid rock magma were produced. An area of over 500 square kilometers (193 square miles), essentially mountainous and forested terrain, was devastated. One of the largest avalanches in recorded history resulted.

Although comparatively small on the long-term scale of volcanic power, the eruption was the fourth largest to occur in the past 100 years, surpassed (in terms of ejecta) only be Krakatoa (1883), Santa Maria (Guatemala) (1902), and Katmai (Alaska) (1912).

Mount Saint Helens was last active in 1831–1857. For years, it has been considered by geologists to have a "bad record," dating back some 4500 years, with at least 20 separate events.

The volcano gave its first warning of recent activity by seismographic measurements on April 20, just short of one month prior to the big eruption. Heightened seismographic activity was followed on March 17 by small steam and ash eruptions. The column of gases and aerosols from the main eruption reached a height of some 4.9 kilometers (3 miles). The vapor released contained small amounts of carbon dioxide, sulfur dioxide, hydrogen sulfide, and hydrogen chloride, along with much steam. A prior small crater progressively changed to an oval basin (500 × 300 meters; 1640 × 984 feet) and a depth of 200 meters (656 feet). See Fig. 5.

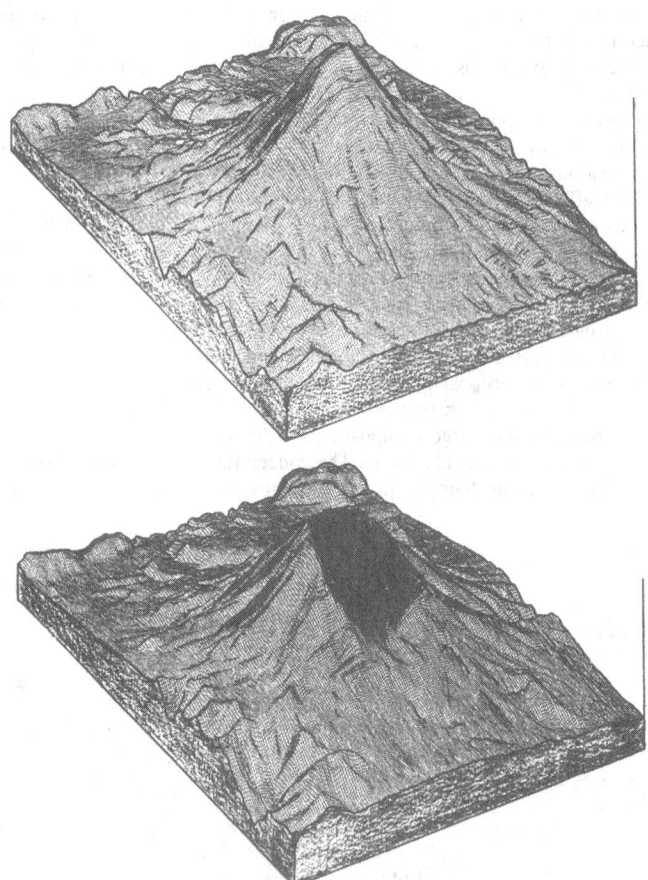

Fig. 5.   Alteration of the peak of Mount Saint Helens by explosive eruption of May 18, 1980. Upper view, before explosion; lower view, after. Some 400 meters (1312 feet) of the top were removed, forming a crater of about 750 meters (2461 feet) deep. (*Drawings based on digitally prepared map created at the Western Mapping Center of the U.S. Geological Survey.*)

Geologists estimate that, during the preliminary period of activity, energy dissipated by the volcano amounted to only 1/17,000th of the energy release during the great eruption. Prior to the large eruption, seismic activity increased to a rate of about 50 earthquakes of magnitude 3+ (known as an "earthquake swarm") per day. A bulge in the mountain appeared. In early May, the bulge expanded at about a rate of 1.5 meters (about 5 feet) per day. The diameter of the bulge reached about 2 kilometers (1.2 miles) and had moved outward from the original profile of the mountain by about 100 meters (382 feet). See Fig. 6.

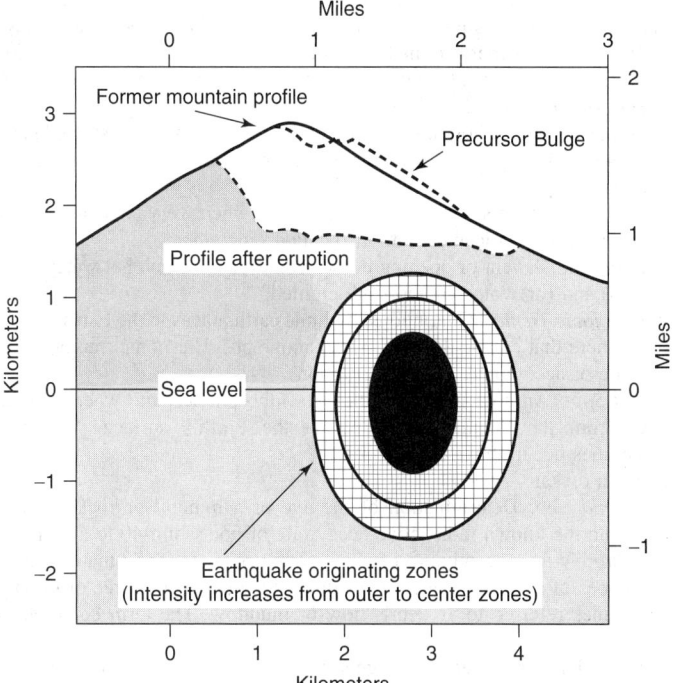

Fig. 6.   Sectional profile of Mount Saint Helens showing outline of mountaintop before May 18, 1980 eruption. Note the precursor bulge which commenced to form in April and the profile after eruption. The large oval depicts assumed originating zones of "earthquake swarms." (*Drawing based on data gathered by the U.S. Geological Survey and the Department of Geophysics, University of Washington. Portrayal suggested by Decker and Decker, 1981.*)

The major eruption occurred at 8:32 AM (May 18, 1980). In addition to the explosion the avalanche produced was catastrophic. Some 2 cubic kilometers (0.5 cubic mile) of crushed rock and glacier ice plunged into nearby Spirit Lake. Mud and avalanche flows proceeded in all directions from the center of the eruption. Estimates suggest that the avalanche at one point accelerated to a velocity of about 250 kilometers (155 miles) per hour. Another path of the avalanche took it over a nearby mountainous ridge that was some 360 meters (1181 feet) high. Ultimate measurements showed that in just one direction the avalanche (pyroclastic flow) deposited debris over a length of 21 kilometers (13 miles) and from 1 to 2 kilometers (0.6 to 1.2 miles) wide and about 150 meters (492 feet) deep.

The lateral blast of the explosion is estimated to have reached a temperature of 300 °C (572 °F) with steam-and-debris-filled clouds traveling at a rate of 100–400 kilometers (62–248 miles) per hour (well below the speed of sound). This blast destroyed some 550 square kilometers (212 square miles) of mountain topography in the northeast-northwest arc out of the volcano. Large trees were burned and uprooted in an area some 10–15 kilometers (6.2–9.3 miles) across and beyond this, in a blowdown zone.

The vertical plume from Saint Helens reached a high altitude and due to high-altitude winds commenced to fall on central Washington within a few hours. At one location (Ritzville, Washington), located about 330 kilometers (205 miles) from the volcano, the ash fall accumulation measured 70 millimeters (2.75 inches) in thickness. The ash fall was comprised of sand-sized fragments of dark volcanic rock with feldspar crystals of a lighter color. A bit later, a deposit of saltlike volcanic glass followed.

Mud flows were a serious consequence of the eruption. The Toutle and Cowlitz rivers were clogged. Historic highwater marks were exceeded by some 9 meters (29.5 feet).

The explosion of Mt. Saint Helens has been compared technically with that of the eruption of Bezymianny in Kamchatka (1955–1956). The

latter event also went through five similar phases. The earthquake swarm persisted 23 days instead of only a few days in the case of Mt. Saint Helens. Seismic monitoring gave geologists some warning of the probable nature of the eruption, although its extent exceeded expectations.

Ash from the eruption fell over a diverse agricultural area with some deposits up to 30 kilograms per square meter (6.1 pounds per square foot). Crop losses were estimated in eastern Washington at about $100 million, or about 7% of the annual crop. Ash on plant leaves reduced photosynthesis by up to 90%, but most plants shed the ash through actions of wind and rain. With the possible exception of sulfur, there was no nutrient value of the ash. Insect loss (destructive and beneficial) was significant, but no populations were fully destroyed.

Since 1980, magma has worked its way upward to form a new dome, measuring approximately 320 meters high and 1 km wide (1050 ft; 0.6 mi), which rests on the crater floor. Mount Saint Helens has become a real-time laboratory for the U.S. Geological Survey and other scientists interested in volcanology. The USGS operates the Cascades Volcano Observatory (CVO) near Mount Saint Helens. The scientists at CVO are reasonably confident that they will be able to warn days or even weeks in advance of the next dome-building eruption. Numerous seismometers and tiltmeters are placed on the dome. Other instrumentation includes laser distance-measuring equipment for monitoring "imperceptible" dome movements. Studies to date indicate that slow but accelerating outward movement of the dome will occur before an actual eruption. CVO scientists have reported that the precursory dome movement is a sluggish version of one way that the dome grows during a full-blown explosion. A model of dome growth has been called the "ruptured onion": as new magma enters a molten core within the dome, it adds to the dome and causes it to expand laterally to form a layered, onionlike structure. Some highly viscous magma may also escape as lava as it nears the top of the dome. The driving force appears to be magma that exists in temporary storage that is about 2 km (1.2 mi) below the crater floor. This underground magma pushes the less buoyant and more viscous magma ahead by way of a conduit that is only 5 to 10 meters (15 to 30 ft) wide. As increasing amounts of the volcano's "throat" is cleared, the flow increases, thus causing the characteristic accelerating dome growth.

It is interesting to note the correlation between eruption precursors and the moon, a relationship that may prove useful for forecasting the volcano's behavior. It is assumed that additional stress resulting from the lunar gravitational distortion of the Earth's crust affect the volcano's behavior and could, when coupled with other forces, trigger a dome-building eruption (to be distinguished from a major explosive eruption). The correlation still is regarded speculatively by some scientists.

In considering what factors may cause the next explosive eruption of Mount Saint Helens, three are regarded most likely: (1) a catastrophic landslide that may fall off the dome; (2) a new injection of magma into the system; or (3) blockage of the volcano's throat by congealing magma that would result in a great build-up of pressure in the system. Because CVO scientists are confident in predicting short-term behavior of the volcano, the general area of Mount Saint Helens has been reopened to the public.

Most everyone agrees with Herbert E. McLean, who in writing for American Forests magazine in 1991, stated, "In the 10 years since it blew its top, the world's most monitored mountain has become a monumental showplace and schoolroom."

Lake Nyos Disaster—Not of Volcanic Origin, But Related. On August 21, 1986, a sudden catastrophic release of gas from Lake Nyos (Cameroon, West Africa) caused the deaths of at least 1700 people. Inasmuch as this is a crater lake, there was strong initial speculation that the disaster was related to volcanic activity. This impression among the lay public has persisted, although official reports of investigations of the incident have since disproved a direct volcanic connection. Testimonies of survivors and autopsies of the victims conclusively support the cause of death (including over 3000 cattle) was a gas cloud composed mainly (estimated 98 to 99%) of carbon dioxide (CO_2), which resulted in CO_2 asphyxiation. Because CO_2 is denser than ambient air and because its density would have increased as it cooled upon expansion, the cloud tended to maintain its integrity as it spilled over the crater rim into low-lying areas. Typical volcanic gases—hydrogen, hydrogen sulfide, carbon monoxide, hydrogen fluoride, and sulfur dioxide—were below detection limits of a few parts per million 12 days after the disaster. Carbon dioxide in the lake water was estimated at least 100 times more abundant relative to sulfur than in volcanic gases. A further clue to the existence of an overwhelming

CO_2 cloud, as contrasted with that which might have resulted from a volcanic eruption, is the observation that foliage was not harmed during the incident.

It is interesting to note that a similar catastrophe occurred in Cameroon in 1984, when 37 persons died along the shore of Lake Monoun, another typical deep, tropical lake, perennially warmed at the top to form a lid shutting in deeper, denser waters. Decomposition of organic matter in such lakes consumes all available oxygen in its deep waters, resulting in production of CO_2 and some methane. Investigators of the 1984 disaster also concluded that the cause of fatalities was an overwhelming presence of carbon dioxide.

An official report summarizes the scenario of the disaster as follows. Chemical, isotopic, geologic, and medical evidence supports the hypotheses that (1) the bulk of gas released was CO_2 that had been stored in the lake's hypolimnion (the lowermost layer of water in a lake, characterized by an essentially uniform temperature that is generally colder than elsewhere in the lake and often by relatively stagnant or oxygen-poor water); (2) the victims exposed to the gas cloud died of CO_2 asphyxiation; (3) the CO_2 was derived from magmatic sources; and (4) there was no significant, direct volcanic activity involved. The limnological nature of the gas release suggests that hazardous lakes may be identical and thus should be monitored to warn of future incidents of this type. See Kling, et al. reference listed.

An abridged glossary of terms used to describe Volcano activity and features would include:

Aa: Aa (pronounced "ah-ah" -a Hawaiian term), is lava that has a rough, jagged, spiny, and generally clinkery surface.

Active volcano: A volcano that is currently erupting, or has erupted during recorded history.

Aerosol: Fine liquid or solid particles suspended in the atmosphere. Aerosols resulting from volcanic eruptions are tiny droplets of sulfuric acid — sulfur dioxide that has picked up oxygen and water.

Airfall: Volcanic ash that has fallen through the air from an eruption cloud. A deposit so formed is usually well sorted and layered. Also called *ashfall*.

Andesite: A medium-colored dark gray volcanic rock containing 53–63 percent silica with a moderate viscosity when in a molten state. Intermediate in color, composition, and eruptive character between basalt and dacite.

Ash (volcanic): Fragments less than 2 millimeters (about 1/8 inch) in diameter of lava or rock blasted into the air by volcanic explosions.

Ash cloud: The fine material that is generated by a pyroclastic flow and rises above it.

Ash flow: A pyroclastic flow consisting predominantly of ash-sized (less than 4 millimeters in diameter) particles. Also called a *glowing avalanche* if it is of very high temperature.

Atmospheric shock wave: Strong compressive atmospheric wave driven by volcanic ejecta.

Ballistic fragment: An explosively ejected rock fragment that follows a ballistic trajectory.

Basalt: Dark-colored, low-silica (less than 53 percent SiO_2), low viscosity volcanic rock that is relatively fluid when molten; eruptions of basalt are generally nonexplosive and tend to produce relatively long thin lava flows like those common in Hawaii.

Base surge: Turbulent, low-density cloud of rock debris and water and (or) steam that moves over the ground surface at high speed.

Black Sand Beach: The famous "black sand" beaches of Hawaii were created virtually instantaneously by the violent interaction between hot lava and sea water.

Blowdown: Trees felled by a volcanic blast.

Caldera: A large volcanic collapse depression, commonly circular or elliptical when seen from above.

Cinders: Cinders are lava fragments about 1 centimeter (about 1/2 inch) in diameter.

Cinder cone: A steep-sided volcano formed by the explosive eruption of cinders that form around a vent.

Conduit (volcanic): A subterranean passage through which magma reaches the surface during volcanic activity.

"Continental" Volcanoes: In the typical "continental" environment, volcanoes are located in unstable, mountainous belts that have thick roots of granite or granitelike rock. Magmas, generated near the base of the mountain root, rise slowly or intermittently along fractures in the crust.

During passage through the granite layer, magmas are commonly modified or changed in composition and erupt on the surface to form volcanoes constructed of nonbasaltic rocks.

Crater: A steep-sided, usually circular depression formed by either explosion or collapse at a volcanic vent.

Dacite: Typically light-colored, fairly silica-rich (63 to 68 percent SiO_2) volcanic rock with a high viscosity when in a molten state; eruptions are commonly explosive (e.g., Mount St. Helens' eruption of May 18, 1980) and may produce voluminous tephra, pyroclastic flows, and lava domes.

Diatreme: A general term for a volcanic vent or pipe drilled through enclosing rocks (usually flat-lying sedimentary rocks) by the explosive energy of gas-charged magmas.

Directed blast: A hot, low-density mixture of rock debris, ash, and gases that moves at high speed along the ground surface, and generated by explosions.

Dome: A steep-sided mount that forms when very viscous lava is extruded from a volcanic vent. Also called *Lava dome*.

Fumarole: A vent or opening in the ground from which hot water vapor (steam) and (or) volcanic gases are emitted.

Harmonic Tremor: Continuous rhythmic earthquakes in the Earth's upper lithosphere that can be detected by seismographs. Harmonic tremors often precede or accompany volcanic eruptions.

Hot Spot: An area in the middle of a lithospheric plate where magma rises from the mantle and erupts at the Earth's surface. Volcanoes sometimes occur above a hot spot.

Igneous rocks:

K-Ar dating: Determination of the age of a mineral or rock in years based on the known radioactive decay rate of potassium-40 to argon-40.

Lahar: A flowing mixture of water-saturated rock debris that forms on the slopes of a volcano, and moves downslope under the force of gravity, sometimes referred to as debris flow or mudflow. The term comes from Indonesia.

Lava: The term used for magma once it has erupted onto the Earth's surface.

Lava tube: During long-lived eruptions, lava flows tend to become "channeled" into a few main streams. Overflows of lava from these streams solidify quickly and plaster on to the channel walls, building natural levees or ramparts that allow the level of the lava to be raised. Lava streams that flow steadily in a confined channel for many hours to days may develop a solid crust or roof and thus change gradually into streams within lava tubes. Tube-fed lava can be transported for great distances from the eruption sites.

Lithic (volcanic): Pertains to pyroclastic deposits that contain abundant fragments of previously-formed rocks and/or dense fragments.

Loess: A well-sorted deposit of windblown silt-sized particles that forms a blanket over the landscape.

Maars: Also called "tuff cones", maars are shallow, flat-floored craters formed above diatremes as a result of a violent expansion of magmatic gas or steam. Maars range in size from 200 to 6,500 feet across and from 30 to 650 feet deep, and most are commonly filled with water to form natural lakes.

Magma: Molten rock containing liquids, crystals, and dissolved gases that forms within the upper part of the Earth's mantle and crust. When erupted onto the Earth's surface, it is called lava.

Mantle: A zone in the Earth's interior between the crust and the core that is 2,900 kilometers (1,740 miles) thick. (The lithosphere is composed of the topmost 65–70 kilometers (39–42 miles) of mantle and the crust).

Mudflow: The flowing mixture of water and debris (intermediate between a volcanic avalanche and a water flood) that forms on the slopes of a volcano. Sometimes called a *debris flow* or *lahar*, a term from Indonesia where volcanic mudflows are a major hazard.

Pahoehoe: Pahoehoe (pronounced "pah-hoy-hoy" -a Hawaiian term), is lava that in solidified form is characterized by a smooth, billowy, or ropy surface.

Pillow Lava: Fluid lava erupted or flowing under water may form a special structure called *pillow lava*. Such structures form when molten lava breaks through the thin walls of underwater tubes, squeezes out like toothpaste, and quickly solidifies as irregular, tongue-like protrusions. This process is repeated countless times, and the resulting protrusions stack one upon another as the lava flow advances underwater. The term pillow comes from the observation that these stacked protrusions are sack-or pillow-shaped in cross section. The bulk of the submarine part of a Hawaiian volcano is composed of pillow lavas.

Pluton: Pertaining to igneous rock bodies that form at great depth.

Pumice: A light-colored, frothy volcanic rock, usually of dacite or rhyolite composition, formed by the expansion of gas in erupting lava. Commonly perceived as lumps or fragments of pea size and larger but can also occur abundantly as ash-size particles. Because of its numerous gas bubbles, pumice commonly floats on water.

Pyroclastic: Pertaining to fragmented (clastic) rock material formed by a volcanic explosion or ejection from a volcanic vent.

Reticulite: During the exceptionally high fountaining episodes of some eruptions, an extremely vesicular, feathery light pumice, called *reticulite* or thread-lace scoria, can form and be carried many miles downwind from the high lava fountains.

Rhyolite: Typically a light-colored crystalline or black glassy volcanic rock or magma, containing more than 68 percent silica with a very high viscosity when in a molten state.

Satellite vent: A secondary vent of flank vent at a volcanic center.

Shield volcano: A volcano that resembles an inverted warrior's shield. It has long gentle slopes produced by multiple eruptions of fluid lava flows.

Spreading Ridges: Places on the ocean floor where lithospheric plates separate and magma erupts. About 80 percent of the Earth's volcanic activity occurs on the ocean floor.

Subduction Zone: The place where two lithosphere plates come together, one riding over the other. Most volcanoes on land occur parallel to and inland from the boundary between the two plates. (Teacher's Packet)

Tephra: Solid material of all sizes explosively ejected from a volcano into the atmosphere.

Tuff: Used loosely as a collective term for all consolidated pyroclastic rocks.

Vent: The opening at the Earth's surface through which volcanic materials (lava, tephra, and gases) erupt. Vents can be at a volcano's summit or on its slopes; they can be circular (craters) or linear (fissures).

Additional Reading

1994–2000 References

Lentz, H.M. III: "The Volcano Registry: Names, Locations, Descriptions and History for over 1500 Sites," McFarland & Company, Inc., Publishers, Jefferson, NC, 1999.

McBirney, A.R. and Jacques-Marie Bardintzeff: "Volcanology," Jones & Bartlett Publishers, Inc., Sudbury, MA, 2000.

Scarpa, R. and R.I. Tilling: "Monitoring and Mitigation of Volcano Hazards," Springer-Verlag, Inc., New York, NY, 1996.

Shoji, S., M. Nanzyo, and R. Dahlgren: "Volcanic Ash Soils: Genesis, Properties and Utilization," Elsevier Science, New York, NY, 1994.

Web References

Earth Sciences Organizations: *http://www-vl-es.geo.ucalgary.ca/VL/html/es-orgs-by-location.html*
U.S. Geological Survey: *http://vulcan.wr.usgs.gov/Glossary/framework.html*
Volcano Information Center: *http://www.geol.ucsb.edu/~fisher/*
Volcano World: *http://volcano.und.nodak.edu/vw.html*

1989–1993 References

Amato, I.: "Mt. Unzen," *Science*, 827 (August 13, 1993).
Amos, W.H.: "Hawaii's Volcanic Cradle of Life," *Nat'l Geographic*, 70 (July 1990).
Cashman, K.V. and R.S. Fiske: "Fallout of Pyroclastic Debris from Submarine Volcanic Eruptions," *Science*, 275 (July 19, 1991).
Courtillot, V.E.: "A Volcanic Eruption," *Sci. Amer.*, 85 (October 1990).
Crumpler, L.S., J.W. Head, and J.C. Aubele: "Relation of Major Volcanic Center Concentration on Venus to Global Tectonic Patterns," *Science*, 591 (July 30, 1993).
Delaney, P.T., et al.: "Deep Magma Body Beneath the Summit and Rift Zones of Kilauea Volcano, Hawaii," *Science*, 1311 (March 16, 1990).
Dvorak, J.J., C. Johnson, and R.I. Tilling: "Dynamics of Kilauea Volcano," *Sci. Amer.*, 46 (August 1992).
Ellis, M. and G. King: "Structural Control of Flank Volcanism in Continental Rifts," *Science*, 839 (November 8, 1991).
Fisher, R.V. and H.-U. Schmincke: "Pyroclastic Rocks," Springer-Verlag, Inc., New York, NY, 1989.
Flam, F.: "Volcano Claims Scientists' Lives," *Science*, 1488 (June 14, 1991).
Fryer, P.: "Mud Volcanoes of the Marianas," *Sci. Amer.*, 46 (February 1992).
Gasparini, P.: "Research on Volcanic Hazards in Europe," *Science*, 1759 (June 18, 1993).
Holden, C.: "Doing Science at the Gates of Hell," *Science*, 475 (August 3, 1990).
Ingebritsen, S.E., D.R. Sherrod, and R.H. Mariner: "Heat Flow and Hydrothermal Circulation in the Cascade Range, North-Central Oregon," *Science*, 1458 (March 17, 1989).

Kerr, R.A.: "Volcanoes Can Muddle the Greenhouse," *Science*, 127 (July 14, 1989).
Kerr, R.A.: "Did a Burst of Volcanism Overheat Ancient Earth?" *Science*, 746 (February 15, 1991).
Kerr, R.A.: "Did a Volcano Help Kill Off the Dinosaurs?" *Science*, 1496 (June 14, 1991).
Kerr, R.A.: "Volcanologists Ponder a Spate of Deaths in the Line of Duty," *Science*, 289 (April 16, 1993).
Krafft, M. and J. Keller: "Temperature Measurements in Carbonatite Lava Lakes and Flows from Oldoinyo Lengai, Tanzania," *Science*, 168 (July 14, 1989).
Latter, J.H., Editor: "Volcanic Hazards: Assessment and Monitoring," Springer-Verlag, New York, NY, 1989.
Liu, M., et al.: "Development of Diapiric Structures in the Upper Mantle Due to Phase Transitions," *Science*, 1836 (June 28, 1991).
Lopez, D.L. and S.N. Williams: "Catastrophic Volcanic Collapse: Relation to Hydrothermal Processes," *Science*, 1794 (June 18, 1993).
McClelland, L., et al.: "Global Volcanism," Prentice-Hall, Inc., Upper Saddle River, NJ, 1989.
McDowell, R.: "Eruption in Colombia," *Nat'l. Geographic*, 640 (May 1986).
McLean, H.E.: "The Amazing Recovery of Mt. St. Helens," *Amer. Forests*, 26 (March/April 1991).
Minnis, P., et al.: "Radiative Climate Forcing by the Mount Pinatubo Eruption," *Science*, 1411 (March 5, 1993).
Mohlenbrock, R.H.: "Mount St. Helens, Washington," *Natural History*, 26 (June 1990).
Powell, C.S.: "Greenhouse Gusher," *Sci. Amer.*, 20 (October 1991).
Propp, M.V. and V.G. Tarasov: "Caldron in the Sea," *Oceanus*, 54 (Winter 1989/1990).
Rampino, M.R. and R.B. Stothers: "Flood Basalt Volcanism During the Past 250 Million Years," *Science*, 663 (August 5, 1988).
Sakai, H., et al.: "Venting of Carbon Dioxide–Rich Fluid and Hydrate Formation in Mid-Okinawa Trough Backarc Basin," *Science*, 1093 (June 1, 1990).
Sen, G. and R.E. Jones: "Cumulate Xenolith in Oahu, Hawaii: Implications for Deep Magma Chambers and Hawaiian Volcanism," *Science*, 1154 (September 7, 1990).
Staff: "Drilling to the Heart of an Alaska Volcano," *Nat'l. Geographic*, News Section (May 1991).
Tarduno, J.A., et al.: "Rapid Formation of Ontong Java Plateau by Aptian Mantle Plume Volcanism," *Science*, 399 (October 18, 1991).
Valentine, G.A. and R.V. Fisher: "Glowing Avalanches: New Research on Volcanic Density Currents," *Science*, 1130 (February 19, 1993).
White, R.S. and D.P. McKenzie: "Volcanism at Rifts," *Sci. Amer.*, 62 (July 1989).
Wohletz, K. and G. Heiken: "Volcanology and Geothermal Energy," University of California Press, Berkeley, CA, 1992.

Pre-1989 References

Aydin, A. and J.M. DeGraff: "Evolution of Polygonal Fracture Patterns in Lava Flows," *Science*, 239, 471–476 (1988).
Barr, S. and J.L. Heffter: "Meteorological Analysis of the Eruption of Soufrière in April 1979," *Science*, 216, 1109–1111 (1982).
Barr, S.: "Skirt Clouds Associated with the Soufrière Eruption of 17 April 1979," *Science*, 216, 1111–1112 (1982).
Beget, J.E.: "Recent Volcanic Activity at Glacier Peak," *Science*, 215, 1389–1390 (1982).
Blong, R.J.: "The Time of Darkness: Local Legends and Volcanic Reality in Papua New Guinea," Univ. of Washington Press, Seattle, Washington, 1982.
Casadevall, T., et al.: "Gas Emissions and the Eruption of Mount St. Helens through 1982," *Science*, 221, 1383–1385 (1983).
Cashman, K.V. and J.E. Taggart: "Petrologic Monitoring of 1981 and 1982 Eruptive Products from Mount St. Helens," *Science*, 221, 1385–1387 (1983).
Chadwick, W.W., Jr., et al.: "Deformation Monitoring at Mount St. Helens in 1981 and 1982" *Science*, 221, 1379–1380 (1983).
Chester, A.M., et al., Editors.: "Mount Etna: The Anatomy of a Volcano," Stanford Univ. Press, Stanford, CA, 1986.
Chuan, R.L., et al.: *Science*, 211, 830 (1981).
Cronn, D.R. and W. Nutmagul: "Volcanic Gases in the April 1979 Soufrière Eruption," *Science*, 216, 1121–1123 (1982).
Decker, R. and B. Decker: "Volcanoes," W.H. Freeman, San Francisco, CA, 1981.
Dzurisin, D., J.A. Westphal, and D.J. Johnson: "Eruption Prediction Aided by Electronic Tiltmeter Data on Mount St. Helens," *Science*, 221, 1381–1383 (1983).
Edmond, J.M. and K. Von Damm: "Hot Springs on the Ocean Floor," *Sci. Amer.*, 78–93 (April 1983).
Farlow, N.H., et al.: *Science*, 211, 839 (1982).
Fedotov, A. and Ye K. Markhinin, Editors: "The Great Tolbachik Fissure Eruption," Cambridge University Press, New York, NY, 1983.
Fiske, R.S. and J.B. Shepherd: "Deformation Studies on Soufrière, St. Vincent, Between 1977 and 1981," *Science*, 216, 1125–1126 (1982).
Fiske, R.S. and H. Sigurdsson: "Soufrière Volcano, St. Vincent: Observations of Its 1979 Eruption from the Ground, Aircraft, and Satellites," *Science*, 216, 1105–1106 (1982).
Fowles, G.R.: "Vapor Phase Explosions: Elementary Detonations?," *Science*, 204, 168–169 (1979).

Francheteau, J., et al.: "Birth of an Ocean: The Crest of the East Pacific Rise," *Centre National pour l'Exploitation des Océans*, 1980.

Francis, P.: "Giant Volcanic Calderas," *Sci. Amer.*, 60–70 (June 1983).

Francis, P. and S. Self: "The Eruption of Krakatau," *Sci. Amer.*, 172–187 (November 1983).

Hardee, H.C.: "Heat Transfer Measurements of the 1983 Kilauea Lava Flow," *Science*, **222**, 47–48 (1983).

Heiken, G. and K. Wohletz: "Volcanic Ash," University of California Press, Berkeley, CA, 1986.

Hékinian, R.: "Petrology of the Ocean Floor," Elsevier Science, New York, NY, 1982.

Hékinian, R.: "Undersea Volcanoes," *Sci. Amer.*, 45–55 (July 1984).

Hoffer, J.M., G. Gomez, and P. Muela: "Eruption of El Chichón Volcano, Chiapas, Mexico, 28 March to 7 April 1982," *Science*, **218**, 1307–1308 (1982).

Hofmann, D.J. and J.M. Rosen: "Sulfuric Acid Droplet Formation and Growth in the Stratosphere After the 1982 Eruption of El Chichón," *Science*, **222**, 325–327 (1983).

Johnston, D.A.: "Volcanic Contribution of Chlorine to the Stratosphere: More Significant to Ozone than Previously Estimated?" *Science*, **209**, 491–493 (1980).

Keen, R.A.: "Volcanic Aerosols and Lunar Eclipses," *Science*, **222**, 1011–1013 (1983).

Kerr, R.A.: "Lake Nyos Was Rigged for Disaster," *Science*, **235**, 528–529 (1987).

Kienle, J., Z. Kowalik, and T.S. Murty: "Tsunamis Generated by Eruptions from Mount St. Augustine Volcano, Alaska," *Science*, **236**, 1442–1447 (1987).

Kling, G.W. et al.: "The 1986 Lake Nyos Gas Disaster in Cameroon, West Africa," *Science*, **236**, 169–175 (1987).

Kotra, J.P., et al.: "El ChichÄn: Composition of Plume Gases and Particles," *Science*, **222**, 1018–1020 (1983).

Lipman, P.W. and D.R. Mullineaux, Editors.: "The 1980 Eruptions of Mount St. Helens, Washington," U.S. Geological Survey, Reston, VA, 1982.

Malin, M.C. and M.F. Sheridan: "Computer-Assisted Mapping of Pyroclastic Surges," *Science*, **217**, 637–639 (1982).

Malone, S.D., C. Boyko, and C.S. Weaver: "Seismic Precursors to the Mount St. Helens Eruptions in 1981 and 1982," *Science*, **221**, 1376–1378 (1983).

McCormick, M.P., et al.: "Stratospheric Aerosol Effects from Soufrière Volcano as Measured by the SAGE Satellite System," *Science*, **216**, 1115–1118 (1982).

Melson, W.G.: "Monitoring the 1980–1982 Eruptions of Mount St. Helens: Compositions and Abundances of Glass," *Science*, **221**, 1387–1391 (1983).

Naranjo, J.L., et al.: "Eruption of the Nevado del Ruiz Volcano, Colombia, On 13 November 1985: Tephra Fall and Lahars," *Science*, **233**, 961–963 (1986).

Ninkovich, D. and W.L. Donn: "Explosive Cenozoic Volcanism and Climatic Implications," *Science*, **194**, 899–906 (1976).

Plage, D. and M. Plage: "In the Shadow of Krakatau: Return of Java's Wildlife," *Nat'l. Geographic*, 750 (June 1985).

Prinn, R.G.: "The Volcanoes and Clouds of Venus," *Sci. Am.*, 46–53 (March 1985).

Patrusky, B.: "Mass Extinctions: Volcanic, or Extraterrestrial Causes, or Both?" *Oceanus*, 40–48 (Fall 1987).

Rampino, M.R., S. Self, and R.W. Fairbridge: "Can Rapid Climatic Change Cause Volcanic Eruptions?" *Science*, **206**, 826–828 (1979).

Rampino, M.R. and S. Self: "The Atmospheric Effects of El Chichón," *Sci. Amer.*, 48–57 (January 1984).

Robock, A.: "The Mount St. Helens Volcanic Eruption of 18 May 1980: Minimal Climatic Effect," *Science*, **212**, 1383–1384 (1981).

Sano, Y., et al.: "Helium-3 Emission Related to Volcanic Activity," *Science*, **224**, 150–151 (1984).

Stommel, H. and E. Stommel: "The Year without a Summer," *Sci. Amer.*, **240**, 5, 176–186 (1979).

Walker, D.A., S. McCreery, and F.J. Oliveira: "Kaitoku Seamount and the Mystery Cloud of 9 April 1984," *Science*, **227**, 607–611 (1985).

Woods, D.C. R.L. Chuan, and W.I. Rose: "Halite Particles Injected into the Stratosphere by the 1982 El Chichón Eruption," *Science*, **230**, 170–172 (1985).

VOLTAGE DIVIDER. The ordinary three-terminal resistance may be used as a voltage divider. If no current is taken from the intermediate tap the voltages will be in proportion to the resistance included between the taps. In many applications a potentiometer is used as a voltage divider so the voltage may be adjusted by varying the position of the tap. If current is drawn from the tap the exact distribution of the voltage is altered but the total voltage applied across the divider is still divided between the various sections. Figure 1 shows some typical arrangements.

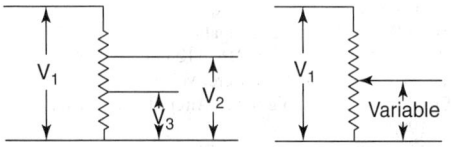

Fig. 1.   Voltage divider.

VOLTAGE DOUBLER. This is a connection of capacitance and rectifiers across an ac source which gives a dc output voltage approximately twice that of the normal connection. The exact value of the output voltage depends upon the load and for very high loads may not even approach the theoretical double value. A circuit is shown in the accompanying figure. On one half-cycle C_1 charges through rectifier 1 and on the other half-cycle C_2 charges through rectifier 2, the polarities being as shown in Fig. 1. It is seen then that the output voltage is the sum of the two capacitor voltages.

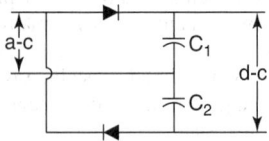

Fig. 1.   Voltage doubler.

VOLTAGE-TO-FREQUENCY A/D CONVERTER. This type of analog-to-digital converter converts the magnitude of the analog-input signal (usually a current or a voltage) into a frequency which can be measured digitally. As an example, the analog-input signal could control the bias of a varacter diode. If the diode is used as the frequency-determining component of a resonant circuit in an oscillator, then the frequency of oscillation would depend on the value of the input signal.

In the Fig. 1, the voltage-to-frequency A/D converter uses an integrator. At commencement of conversion, the counter is returned to zero and the integrator is reset. An input signal V_S is integrated until an output value V_r is reached by the integrator. A change in state of the comparator output detects this point. Upon this change of state of the comparator, the integrator is reset and the process repeated.

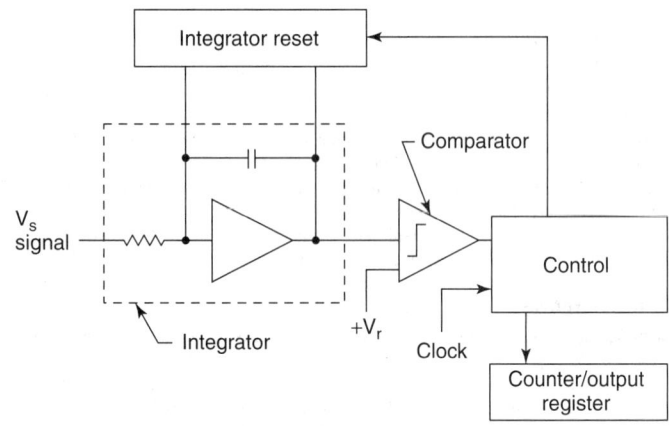

Fig. 1.   Voltage-to-frequency analog-to-digital converter.

The rate of integration depends upon the value of the input signal. Thus, the frequency of the sawtooth waveform that is generated by the integrator is a function of the input signal. If the number of integration cycles that take place in a fixed time period are counted, a digital representation of the input signal is yielded. Clock pulses in a separate counter determine the counting interval.

An advantage of the voltage-to-frequency A/D converter is that it provides a digital representation of the average value of the input signal during the conversion interval. Thus, input-noise transients have minimal effect on performance. As the result of serial counting, this A/D converter is comparatively slow. Speed usually is less than 1,000 samples/second at a resolution of 10 bits or less.

THOMAS J. HARRISON, International Business Machines Corporation, Boca Raton, Florida.

VOLTERRA, VITO (1860–1940). Volterra was an Italian mathematician and physicist. It is believed that at about age thirteen he was so inspired by the Jules Verne's novel, *From the Earth to the Moon*, that he tried to determine the trajectory of a projectile in the combined gravitational fields of the Earth and moon. Volterra made important contributions to areas of higher analysis and mathematical physics. His name is associated with the creation of the theory of functionals and with the theory of integral and integro-differential equations. He also made important contributions in many areas of applied science. He did significant work on the mathematics of population growth, arguing that cooperation among people would encourage increases in the food supply and thus the Earth could accommodate a greater increase in population.

In 1917, Volterra established the Office for War Inventions in Italy and promoted scientific and technical collaboration among the Allies.

Volterra was a brilliant scientist who became a member of almost every major scientific and mathematical society. He was elected to the Pontifical Academy of Sciences in 1936. He was also honored at many universities as a gifted teacher.

See also **Functional Analysis**; and **Integral Equation**.

J. M. I.

VOLUME (Critical).   The volume occupied by one gram of a liquid or gaseous substance at its critical temperature and critical pressure.

VOLUME (Geometry).   The volume of a parallelepiped is often represented as a scalar triple product of three vectors, the edges of the solid. In calculus, the volume of a solid may be found by means of definite integrals, as follows: (a) If a plane perpendicular to the X-axis at a distance x from the origin cuts, from a given solid, a section whose area is $A(x)$, then the volume of that part of the solid between $x = a$ and $x = b$ is given by the definite integral

$$V = \int_a^b A(x)\,dx$$

(b) Suppose the solid is that part of a right circular cylinder included between the XOY-plane and the surface $z = f(x, y)$, then its volume is given by the definite double integral

$$V = \int_{a1}^{a2} \int_{b1}^{b2} f(x, y)\,dx\,dy$$

where the base of the cylinder is bounded by the curves $x = a_1$, $x = a_2$, $y = b_1$, $y = b_2$. (c) If the solid is divided into volume elements, $d\tau = dx\,dy\,dz$ which are right prisms with base $dx\,dy$ and altitude dz, the volume of the solid is given by the triple definite integral

$$V = \int_{x1\,y1\,z1}^{x2\,y2\,z2} dx\,dy\,dz$$

where the limits are determined by the boundary of the given solid. In spherical polar coordinates $d\tau = r\sin\theta\,dr\,d\theta\,d\phi$ and in cylindrical coordinates $d\tau = p\,dp\,d\phi\,dz$.

Vector methods are often useful for expressing the volume of a solid (see also **Vector Integral**).

See also **Calculus**; and **Parallelepiped**.

VOLUME (Standard).   The volume occupied by one gram molecular weight of a gas at $0\,°C$ and a pressure of 1 standard atmosphere.

VON BRAUN, WERNHER (1912–1977).   Wernher von Braun is, without doubt, the greatest rocket scientist in history. His crowning achievement, as head of NASA's Marshall Space Flight Center, was to direct the mission to land the first men on the Moon in July 1969.

Wernher von Braun was born in 1912 in Wirsitz, Germany (now part of Poland), into an aristocratic Prussian family. His father was Baron Magnus von Braun and his mother was a direct descendent of Valdemar I of Denmark (1131–1182). At an early age Wernher developed a fascination for rockets, inspired by the ancient Chinese who invented fireworks. At the age of 12, he tried his first practical rocket experiment. He strapped six rockets to a small wagon and lit them up. The wagon performed beyond his wildest dreams and careened about crazily, trailing a tail of fire like a comet. When the rockets finally burned out, ending their sparkling performance with a magnificent thunderclap, the wagon rolled majestically to a halt. The police, who arrived late for the beginning of his experiment, but in time for the grand finale, were unappreciative. They took young Wernher into custody. Fortunately, no one was injured and he was released to the Minister of Agriculture, his father. So began a career in rocketry that changed human history.

Von Braun attended the French Gymnasium School in Berlin, but was not a star pupil. He spent much of his time building an automobile in his father's garage instead of studying books. Von Braun's grades improved after his father transferred him to a boarding school. Part of his education there involved working in small groups to develop technical skills. He would draw upon these lessons many times later in his career when working in teams. Before bedtime he was permitted to examine the stars with a small telescope that his mother bought him. Thus began his interest in astronomy.

One day in 1925, von Braun saw an ad in an astronomy magazine about a book called *The Rocket to the Interplanetary Spaces*, by Professor Hermann Oberth. He ordered the book at once and, when it arrived, opened it breathlessly. To his consternation, he could not understand a word. Its pages were a baffling conglomeration of mathematical symbols and formulae. Rushing to his teacher, he cried, "How can I understand what this man is saying?" To von Braun's dismay, his teacher told him to study mathematics and physics, but with the glamorous prospect of a life devoted to space travel, these subjects took on a new meaning. Von Braun was determined to master them and he began to bury himself in their mysteries, and after a few years, he succeeded in graduating a year ahead of his class.

After graduating from school, von Braun became a student at the Berlin Institute of Technology and worked in his spare time as an assistant for Professor Oberth at the German Society for Space Travel. Oberth was trying to prove that liquid fuels, instead of solids, offered the best approach to powering rockets for space vehicles. Oberth's other two assistants were Klaus Riedel and Rudolf Nebel. Their equipment was crude and the ignition system was perilous. Riedel would toss a flaming gasoline-soaked rag over the gas-spitting motor and duck for cover before Oberth opened the fuel valves, and then the motor would start with a roar!

Oberth and his assistants were allowed to conduct experiments as guests on the proving grounds of the Chemical and Technical Institute, the German equivalent of the U.S. Bureau of Standards. In August 1930, Oberth's little rocket engine succeeded in producing a thrust of seven kilograms for 90 seconds, burning gasoline and liquid oxygen. An official of the Institute certified the demonstration and the liquid-fueled rocket motor was, thus, recognized for the first time in Germany as a respectable member of the family of internal-combustion engines. This was a tremendous step forward but, because he had to support a large family, Oberth was forced to return to his teaching job in Romania.

After Oberth's departure, the team's guest status at the proving grounds expired and a new place to conduct the experiments had to be found. Nebel secured a lease on an abandoned 300-acre ammunition storage depot on the outskirts of Berlin. He persuaded the city fathers to let them use the site free-of-charge and for an indefinite period. One of the blockhouses was used as the laboratory and on this building was hung the sign "Raketenflugplatz Berlin" (Berlin Rocket Field).

Nebel did an amazing job of scrounging free materials, which were swapped for skilled labor, such as tin bending or welding. Riedel sketched out a design for a "Minimum Rocket," which they started to build. The motor was located in the nose, not for any scientific reason, but simply because Nebel had scrounged a truckload of aluminum tubing which could only be used if the motor dragged the tanks by the fuel lines.

In 1931, von Braun interrupted his studies at the Institute of Technology in Berlin to study for a semester at the Federal Institute of Technology in Zurich. He returned to Berlin in October for the first public firing of Riedel's rocket. Several local industrialists had been persuaded by Nebel and Riedel to pay one mark each to witness the demonstration. When the moment of truth came, the rocket moved halfway up the launcher tracks, then settled peacefully back on the pad. What an embarrassment, but the admission fees were not returned! The pressurization of the rocket's fuel tanks was unreliable and this problem was soon corrected. Within a few weeks, successful launchings became commonplace and the rocket reached an altitude of 1,000 feet. A small parachute carried the tail section gently back to Earth. Riedel would dash across the field in an old car, jump out, and sometimes catch the rocket before it struck the ground. After such a lucky "hand recovery," they could fire the rocket again immediately.

While taking part in these exciting activities in his spare time, von Braun continued with his formal studies, and graduated from the Berlin Institute of Technology with a bachelor's degree in aeronautical engineering in 1932. Von Braun's exposure to rocketry convinced him that the exploration of space would require far more than just applications of the current engineering technology. To this end he enrolled as a graduate student at the University of Berlin and gained his Ph.D. in physics in 1934. His thesis was about liquid rocket propulsion. Solid propellant rockets had been used for centuries, but liquid propulsion was new. Only miniature motors had been built and tested that used a liquid oxygen/alcohol propellant combination. Von Braun wanted to analyze some of the puzzling phenomena that take place in a rocket engine, such as atomization, combustion, and expansion of gases. Experimentation would be costly and von Braun considered himself fortunate when the research department of the German Army Ordnance Corps sponsored his research and permitted him to conduct dangerous experiments at the Kummersdorf Army Proving Ground.

After gaining his Ph.D., von Braun became a civilian employee of the Army and continued with this work. He designed the V-2 rocket that was used so effectively against Britain during World War II. At the end of the war, the von Braun team at Peenemunde on the Baltic Sea headed south and surrendered to U.S. forces rather than risk capture by the Soviet army. With 120 of his associates, von Braun was brought to the U.S. as part of "Operation Paperclip" in order to demonstrate their achievements with V-2 rockets. He arrived in the U.S. in September 1945 under contract to the U.S. Army. During the following five years, he directed high-altitude firings of V-2 rockets at the White Sands Missile Range in New Mexico and was project director of the guided missile development unit at Fort Bliss, Texas.

In 1950, the Fort Bliss rocket development group was transferred to Huntsville, Alabama, where the Army centered its rocket development activities. In Huntsville, the von Braun team worked on the ballistic rockets called Redstone, Jupiter C, Juno, and the Saturn 1B.

After the Soviets' Sputnik went into orbit in 1957 and the Navy's Vanguard rocket blew up on its pad, a version of von Braun's Redstone rocket, called Jupiter-C, put Explorer 1 into orbit in 1958, and another version carried Alan Shepard on the first U.S. sub-orbital flight in 1961.

Von Braun and his team were transferred from the Army's control to NASA's when the National Aeronautics and Space Administration was established in 1958. Von Braun became director of NASA's Marshall Space Flight Center (MSFC) in Huntsville, which was dedicated on July 1, 1960. The von Braun team members were given a choice whether to go to NASA or stay with the Army Ballistic Missile Agency. All elected to be transferred to MSFC on this same date.

Von Braun's management style was to let everybody on the team know that they were important to its success and that any work they did was a reflection on the team. The result was that every team member did his best.

When President Kennedy called for a Moon landing within the decade, von Braun was asked to lead the effort to design and build the rocket. The Saturn V, developed by MSFC, won the race with the Soviet Union to put the first men on the Moon in 1969. After the Moon landings, at the insistence of the NASA Administrator, von Braun, in 1970, went to NASA Headquarters in Washington, DC to serve as Deputy Associate Administrator and to promote space activities. But public interest and support had declined, and von Braun resigned in May 1972 to become vice president for engineering and development at Fairchild Industries, Inc.

In 1975, he founded and became the first president of the National Space Institute, a private group designed to increase public understanding and support of space activities.

See also **Goddard, Robert H.**; **Moon (Earth's)**; **Rocket Engine**; **Rocket Propellants**; and **Rocketry**.

Additional Reading

Stuhlinger, E. and F. Ordway: "Wernher von Braun, Crusader for Space, A Biographical Memoir," Krieger Publishing Company, Melbourne, FL, 1996.

Web Reference

NASA Marshall Space Flight Center's Web site on Wernher von Braun: *http://history.msfc.nasa.gov/vonbraun/*

VON FRISCH, KARL (1886–1982). Von Frisch was an Austrian ethologist and zoologist born in Vienna. His affiliation was Germany's Munich University Institute of Zoology. He did pioneering work in comparative behavioral psychology. His studies dealt with complex communication between insects particularly under natural conditions. In the 1940's, von Frisch first discovered the significance of the honeybees' dances. Von Frisch and his colleagues made careful accounts of the dance language. He noted with a stopwatch the time that movements lasted, the direction the bee faced while waggling, and how fast the bee dancer

finished her circuits. He concluded that the honeybees' dance functioned as a language. Karl von Frisch won the 1973 Nobel Laureate in Medicine for discoveries concerning organization and elicitation of individual and social behavior patterns.

See also **Honeybees**.

J. M. I.

VON HIPPEL-LINDAU DISEASE (VHL). A genetic disorder, of abnormal growth of blood vessels that develop into angiomas or tumors. Angiomas can occur in various parts of the body but generally occur in areas rich in blood vessels, such as in the eye. Problems occur when the angioma grows causing the walls of the blood vessels to weaken and potentially leak. When leakage occurs, damage may result in surrounding tissues and affect organs. Overall, approximately one in 32,000 people in the world has VHL, but the disorder does not favor specific ethnic groups.

VHL is named after two doctors. Dr. Eugene Von Hippel described the genetic condition of the abnormal growth of blood vessels that develop into angiomas or tumors. His particular area of interest and research involved the eye, specifically angiomas in the retina. Dr. Arvid Lindau identified tumors of the spine and brain and is usually associated with VHL in the central nervous system.

As every person is different, so is the severity and onset of VHL, although it usually occurs in adulthood. Anyone with a history of VHL in his or her family should consult with a medical professional about what to do should the symptoms arise. Sometimes, VHL shows no signs or symptoms and the only definitive diagnosis is through DNA testing for the genetic marker, which is important because children can be affected.

An angioma in the eye may be one of the earliest signs of VHL disease. Such angiomas can occur beneath and on top of the retina and are identified by eye care professionals as lesions. If an angioma or blood vessels leak, the retina can detach and result in blindness. If no leakage is identified, careful monitoring of the tumor and its growth is essential in order to prevent retina detachment. Angiomas in the retina or optic nerve are usually not cancerous. These tumors tend to progress slowly, so early detection is key to preventing retina detachment.

The treatment of the retinal angioma depends on its severity and the diagnosis pertaining to any present lesions. Small lesions are easy to treat with laser surgery. For large tumors, eye care specialists may recommend cryotherapy. People with a history of VHL may undergo CT or MRI imaging scans to monitor tumor growth both in the eye as well as other parts of the body. So far, there is no cure for VHL. Research and studies continue in order to help people and professionals better manage VHL.

Vision Rx, Inc., Elmsford, NY.

VON KLITZING, KLAUS (1943). Von Klitzing is a German physicist who began his career in physics with a dissertation of the electrical properties of indium antimonide, a compound of two semiconducting elements, indium and antimony. He was particularly interested in the effects of magnetic fields on the conducting properties of semiconductors and was awarded his Ph.D. for research in this field. He went to Oxford University because of the availability of superconducting magnets and then went to the High-Field Magnet Laboratory in Grenoble, France where he made his famous discovery of the quantized Hall effect. He was awarded the 1985 Nobel Prize in physics for his discovery. His discovery has made it possible to establish a new international standard for the ohm.

See also **Hall Effect and Quantized Hall Effect**.

J. M. I.

VON NEUMANN, JOHN (1903–1957). Von Neuman is recognized as an American mathematician. He was born in Budapest, Hungary. He received a Ph.D. in mathematics at the University of Budapest in 1926. He came to America in 1930 as a visiting professor at Princeton University and in 1933 he became professor of mathematics at the Institute for Advanced Study at Princeton, New Jersey, and remained in this position until his death. From 1943 until 1955 he was also associated with Los Alamos Scientific Laboratory. von Neuman was U.S. Atomic Energy Commissioner in 1954 and he received the Enrico Fermi Award of the Atomic Energy Commission in 1957.

Much of von Neuman's early work was in quantum theory. He made contributions to the study of periodic functions, the ergodic theorem, and the algebra of bounded operators. In 1928, he gave a presentation, which was the beginning of a new branch of science referred to as the theory of games. Later in 1944, he developed the analogy between games of strategy and situations in economics. His theory of games has found a wide range of applications in social sciences.

Neu also remembered for his work in the pioneering of computer theory and design. In 1952, von Neumann and his colleagues designed and built the first computer, which was able to use a flexible stored program.

See also **Game Theory**; **Digital Computer Systems**; and **Mathematics (State of the Art Reviews)**.

J. M. I.

VORTEX LAWS.　A vortex is defined in meteorology as a mass of fluid in which the flow is circulatory.

The filament or thread of the vortex is the locus of the centers of circulation. Hydrodynamic analysis of fluid flow, attributed to Helmholtz, has produced the following laws governing vortex flow:

1. The strength C of a vortex is constant along the filament.
2. The identity of the fluid in a vortex does not alter during the life of the vortex.
3. Filaments have no ending; they are either closed paths, or the ends extend to infinity.

VOYAGER MISSIONS TO JUPITER AND SATURN.　The *Voyager* mission comprised a major thrust in the strategy of the U.S. National Aeronautics and Space Administration (NASA) for obtaining information on the outer solar system. The mission was officially approved in May 1972 and has returned more new knowledge about the outer planets than had existed in all of the preceding history of astronomy and planetary science. Objectives of the *Voyager* missions included the study of Jupiter and Saturn, including their satellites, as well as observations of interplanetary media between the Earth and these planets. The *Voyager* spacecraft were designed to take advantage of a rare geometric arrangement of the outer planets that occurs only once every 176 years. This configuration allows a single spacecraft to swing from one planet to the next without the need for large onboard propulsion systems. The two spacecraft were launched in 1977. *Voyager 2* was launched first, on August 20, 1977, followed by *Voyager 1*, which was put on a faster shorter trajectory to Jupiter on September 5, 1977. Both *Voyager* launches took place at the NASA Kennedy Space Center at Cape Canaveral, Florida. See Fig. 1. Eighteen months after launch, *Voyager 1* reached Jupiter, 650 million kilometers (400 million miles) away. The spacecraft made its closest approach on March 5, 1979, and *Voyager 2* followed with its closest approach occurring on July 9, 1979. The first spacecraft flew within 206,700 kilometers (128,400 miles) of the planet's cloud tops, and *Voyager 2* came within

Fig. 1.　OUTWARD BOUND *VOYAGER*—A *Titan-Centaur* launch vehicle hurls *Voyager 1* from Cape Canaveral toward its rendezvous with Jupiter and Saturn. The launch took place at 5:56 A.M. (PDT) September 5, 1977. (*Jet Propulsion Laboratory*.)

570,000 kilometers (350,000 miles). On November 12, 1980 *Voyager 1* made its closest approach to Saturn followed by *Voyager 2* on August 25, 1981. *Voyager 1* flew within 64,200 kilometers (40,000 miles) of the cloud tops, while *Voyager 2* came within 41,000 kilometers (26,000 miles).

These spacecraft represented the technology of the 1970's. The very best hardware was incorporated into the spacecraft, not only to successfully complete the missions to Jupiter and Saturn, but also with the hopeful expectations that the systems would continue functioning satisfactorily and thus enable visits to Uranus and Neptune. After *Voyager 2's* successful Saturn encounter, it was shown that *Voyager 2* would likely be able to fly on to Uranus with all instruments operating. NASA provided additional funding to continue operating the two spacecraft and authorized the Jet Propulsion Laboratory (JPL) to conduct a Uranus flyby. Subsequently, NASA also authorized the Neptune leg of the mission, which was renamed to the *Voyager* Neptune Interstellar Mission. On January 24, 1986 *Voyager 2* made its closest approach to Uranus, coming within 81,500 kilometers (50,600 miles) of the planet's cloud tops. *Voyager 2* made its closest approach to Neptune, 5,000 kilometers (3,000 miles) on August 25,1989. Following *Voyager 2's* closest approach to Neptune, the spacecraft flew southward, below the ecliptic plane and onto a course that will take it to interstellar space. Reflecting the *Voyagers'* new transplanetary destinations, the project is now known as the *Voyager* Interstellar Mission. See also **Neptune (Planet)**; and **Uranus**.

Voyager 1 is now leaving the solar system, rising above the ecliptic plane at an angle of about 35 degrees at a rate of about 520 million kilometers (about 320 million miles) a year. *Voyager 2* is also headed out of the solar system, diving below the ecliptic plane at an angle of about 48 degrees and a rate of about 470 million kilometers (about 290 million miles) a year.

The Spacecraft

The *Voyager* mission module after injection weighed 826 kilograms (1,820 pounds), including a 105-kilogram (231-pound) scientific instrument payload. The propulsion module, with its large solid propellant rocket motor, weighed 1220 kilograms (2,960 pounds). The spacecraft adaptor joined the spacecraft with the *Centaur* stage of the launch vehicle. It weighed 47.2 kilograms (104 pounds) and thus the total launch weight was 2100 kilograms (4,360 pounds). To assure proper operation for the four–year flight to Saturn and beyond, the mission module sub-systems were designed with high reliability and extensive redundancy. Like the *Mariners* that explored the solar system's inner planets and the *Viking*

Mars Orbiters, the *Voyagers* are stabilized on three axes, using the sun and Canopus (a star) as celestial reference points. See Fig. 2.

Three engineering subsystems are programmable for onboard control of spacecraft functions. Only trajectory correction maneuvers must be enabled by ground command. These subsystems are the computer command subsystems (CCS), the flight data subsystems (FDS), and the attitude and articulation control subsystems (AACS). The memories of these units can be updated or modified by ground command at any time. Hot gas jets provide thrust for attitude stabilization as well as for trajectory correction maneuvers.

The science instruments required to view the planets and their moons are mounted on a two-axis scan platform at the end of the science boom for precise pointing. Other body-fixed and boom-mounted instruments are aligned for proper interpretation of their measurements. Data storage capacity on the spacecraft is about 536 million bits of information-the equivalent of about 100 full-resolution photos. Dual frequency communication links, S-band and X-band, provide accurate navigation data and large amounts of scientific information during planetary encounter periods (up to 115,200 bits per second at Jupiter; and 44,800 bits per second at Saturn). The dominant feature of the spacecraft is the 3.66-meter (12-foot) diameter high-gain antenna which points toward the Earth continually. While the high-gain antenna dish is white, most visible parts of the spacecraft are black-blanketed or wrapped for thermal control and micrometeoroid protection. A few small areas are finished in gold foil or have polished aluminum surfaces.

The basic mission module structure (see Fig. 3) is a 29.5-kilogram (65-pound), ten-sided aluminum framework with ten electronics packaging compartments. The structure is 47 centimeters (18.5 inches) high and 1.78 meters (5.8 feet) across. The electronics assemblies are structural elements of the ten-sided box.

Three Radioisotope Thermoelectric Generators (RTGs) are assembled in tandem on a deployable boom hinged on an outrigger arrangement of struts attached to the basic structure. See Fig. 4. The generators convert the heat released by the isotopic decay of plutonium-238 into electricity. Each isotope heat source has a capacity of 2400 thermal watts with a resultant maximum electrical power output of 160 watts at the start of the mission. Due to the natural radioactive decay of the Plutonium fuel source, the electrical energy provided by the RTGs is continually declining. By the time the spacecraft passed Saturn, the total power supply was reduced to about 384 watts. At the beginning of 1997, the power generated by *Voyager 1* has dropped to 334 watts and 336 watts for *Voyager 2*. Both

Fig. 2.   Seen here is a full-scale model of one of the twin *Voyager* spacecraft, which was sent to explore the giant outer planets in our solar system. *Voyager 2* was launched August 20, 1977 followed by the launch of *Voyager 1* sixteen days later. (*Jet Propulsion Laboratory, Pasadena, California/NASA.*)

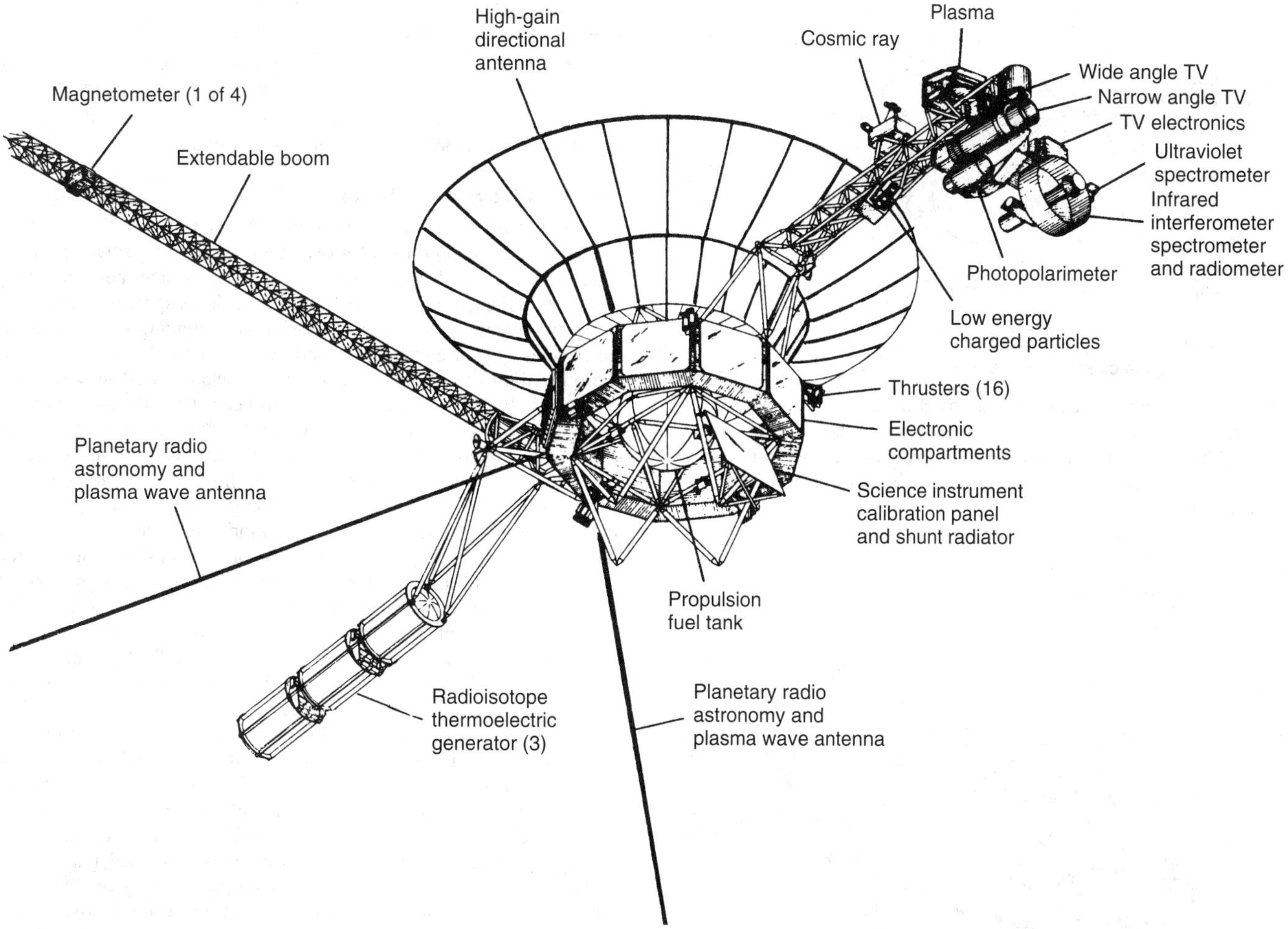

Magnetometer (1 of 4)

Extendable boom

High-gain directional antenna

Planetary radio astronomy and plasma wave antenna

Cosmic ray

Plasma

Wide angle TV

Narrow angle TV

TV electronics

Ultraviolet spectrometer

Infrared interferometer spectrometer and radiometer

Photopolarimeter

Low energy charged particles

Thrusters (16)

Electronic compartments

Science instrument calibration panel and shunt radiator

Propulsion fuel tank

Radioisotope thermoelectric generator (3)

Planetary radio astronomy and plasma wave antenna

Fig. 3.   Principal features of *Voyager* spacecraft. (*Jet Propupulsion Laboratory.*)

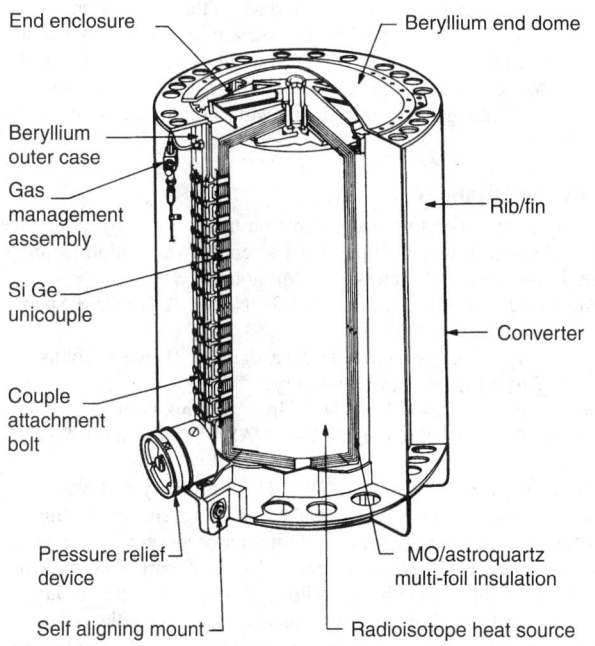

End enclosure

Beryllium end dome

Beryllium outer case

Gas management assembly

Si Ge unicouple

Rib/fin

Converter

Couple attachment bolt

Pressure relief device

Self aligning mount

MO/astroquartz multi-foil insulation

Radioisotope heat source

Fig. 4.   Radioisotope thermoelectric generator aboard *Voyager* spacecraft. (*Jet Propulsion Laboratory.*)

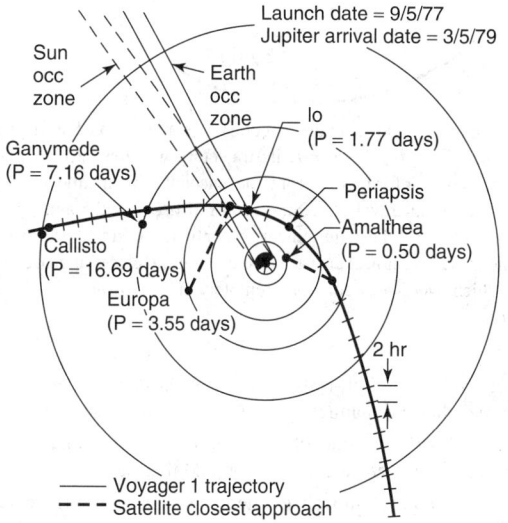

Launch date = 9/5/77
Jupiter arrival date = 3/5/79

Sun occ zone

Earth occ zone

Io (P = 1.77 days)

Ganymede (P = 7.16 days)

Periapsis

Amalthea (P = 0.50 days)

Callisto (P = 16.69 days)

Europa (P = 3.55 days)

2 hr

——— Voyager 1 trajectory
- - - - Satellite closest approach

Fig. 5.   *Voyager 1* Jupiter encounter geometry. This view from the planet's north pole shows the trajectory closest approach geometry to the planet and selected satellites, satellite periods, 2-hour tick marks along the trajectory path, and the Earth and Sun occultation zones through which the spacecraft proceeded. (*Jet Propulsion Laboratory.*)

The encounter geometry of *Voyager 1* and *2* with Jupiter are indicated in Figs. 5 and 6; that for Saturn is shown in Fig. 7.

Scientific Experiments Aboard *Voyagers*

The principal scientific investigations programmed for *Voyager* encounters included the following:

of these power levels represent performances higher than the pre-launch predictions. At the present time both spacecraft have adequate electrical power and attitude control propellant to continue operating until around 2020 when the available electrical power will no longer support science instrument operation.

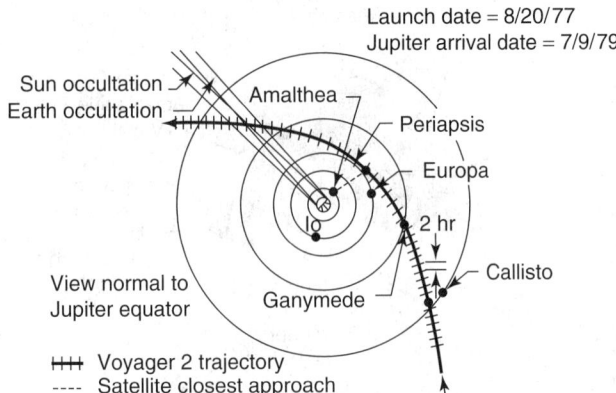

Fig. 6. *Voyager 2* Jupiter encounter geometry. This view from the planet's north pole shows the trajectory closest approach geometry to the planet and selected satellites, 2-hour time interval tick marks along the trajectory path, and the Earth and Sun occultation zones through which the spacecraft proceeded. (*Jet Propulsion Laboratory.*)

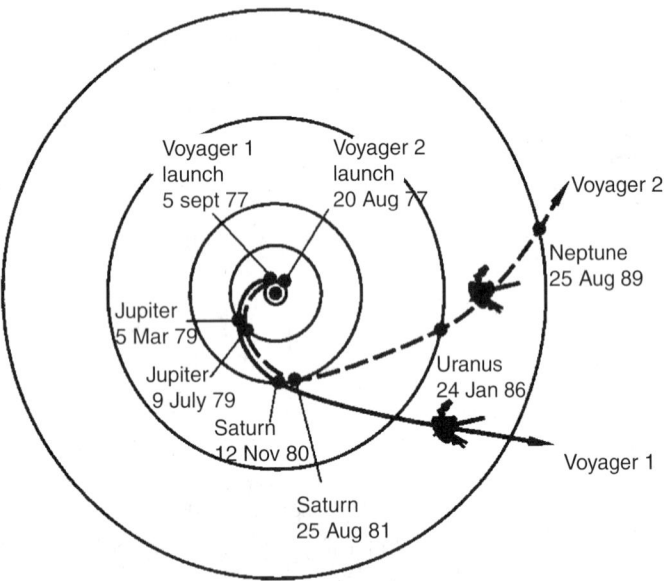

Fig. 7. *Voyagers 1* and 2 Saturn encounter geometry. As Saturn was the final planetary encounter for *Voyager 1*, its trajectory was designed to maximize the return of scientific information from the encounter with little regard to where the spacecraft would travel after Saturn. For *Voyager 2*, the aim point at Saturn was defined solely by the requirement to continue the trajectory to Uranus. The arrival time at Saturn, however was selected to allow closer approaches to several satellites which were viewed more remotely from *Voyager 1*. (*Jet Propulsion Laboratory.*)

Imaging Science. High-resolution reconnaissance over large phase angles; atmospheric dynamics; geologic structure of satellites.

Infrared Radiation. Atmospheric composition, thermal structure, and dynamics; satellite surface composition and thermal properties.

Photopolarimetry. Atmospheric aerosols; satellite surface texture and sodium cloud.

Radio Science. Atmospheric and ionospheric structure, constituents, and dynamics.

Ultraviolet Spectroscopy. Upper atmospheric composition and structure; auroral processes; distribution of ions and neutral atoms in planetary system.

Magnetic Fields. Planetary magnetic field; magnetospheric structure; Io flux tube currents.

Plasma Particles. Magnetospheric ion and electron distribution; solar wind interaction with planer; ions from satellites.

Plasma Waves. Plasma electron densities; wave-particle interactions; low-frequency wave emissions.

Planetary Radio Astronomy. Polarization and spectra of radio frequency emissions; Io radio modulation process; plasma densities.

Low-Energy Charged Particles. Distribution, composition, and flow of energetic ions and electrons; satellite-energetic particle interactions.

Cosmic-ray Particles. Distribution, composition, and flow of high-energy trapped nuclei; energetic electron spectra.

Voyager's Interstellar Mission

At the start of the *Voyager* Interstellar Mission (VIM), the two *Voyager* spacecraft had been in flight for over 12 years. The objective of the (VIM) is to extend the NASA exploration of the solar system beyond the region of the outer planets to the outer limits of the Sun's sphere of influence, and possibly beyond. This extended mission is continuing to characterize the outer solar system environment and search for the heliopause boundary, the outer limit of the Sun's magnetic field and outward flow of the solar wind. Exactly where the heliopause is has been one of the great unanswered questions in space physics. By studying the radio emissions, scientists now theorize the heliopause exists some 90 to 120 astronomical units (AU) from the Sun. One AU is equal to 150 million kilometers (93 million miles).

We should consider the VIM as having three notable phases: the termination shock, heliosheath exploration, and interstellar exploration phases. The two *Voyager* spacecraft began the VIM operating, and are still operating, in an environment controlled by the Suns magnetic field with the plasma particles being dominated by those contained in the expanding supersonic solar wind. This is the characteristic environment of the termination shock phase. At some distance from the Sun, the supersonic solar wind will be held back from further expansion by the interstellar wind. One of the features to be encountered by a spacecraft as a result of this interstellar wind/solar wind interaction will be the termination shock where the solar wind slows from supersonic to subsonic speed and large changes in plasma flow direction and magnetic field orientation occur. Passage through the termination shock ends the termination shock phase and the heliosheath exploration phase begins. Since the exact location of the termination shock is not known, it is estimated that *Voyager 1* will complete the termination shock phase about the year 2000 or 2001 when it will be about 80 AU from the Sun. Once final passage through the termination shock, the spacecraft will be operating in the heliosheath environment which is dominated by the Suns magnetic field and particles contained in the solar wind. When the heliosheath exploration phase ends with passage through the heliopause which is the outer extent of the Suns magnetic field and solar wind. The thickness of the heliosheath is uncertain and could be tens of AU thick taking a couple of years to traverse. Passage through the heliopause begins the interstellar exploration phase with the spacecraft operating in an interstellar wind dominated environment. See Fig. 8.

Science Investigations

There are seven operating instruments on-board each *Voyager* spacecraft. Five of these instruments support the science investigation teams participating in the *Voyager's* Interstellar Mission (VIM).

Magnetic Field Investigation (MAG). NASA & Goddard Space Flight Center.

Low Energy Charged Particle Investigation (LECP). Johns Hopkins University/Applied Physics Laboratory.

Plasma Investigation (PLS). MIT Space Plasma Group.

Cosmic Ray Investigation (CRS). NASA & Goddard Space Flight Center.

Plasma Wave Investigation (PWS). The University of Iowa

These science teams are currently collecting and evaluating data on the strength and orientation of the Sun's magnetic field; the composition, direction and energy spectra of the solar wind particles and interstellar cosmic rays; the strength of radio emissions that are thought to be originating at the heliopause, beyond which is interstellar space; and the distribution of hydrogen within the outer heliosphere. In addition there are two science instruments that do not have official science investigation teams associated with them. These instruments are: *Planetary Radio Astronomy Subsystem* (PRA) and the *Ultraviolet Spectrometer Subsystem* (UVS). However the data captured by these instruments is made available to interested scientists.

VULTURE *(Aves, Falconiformes).* A large, flesh-eating bird with a hooked beak but with claws less strongly developed than those of the

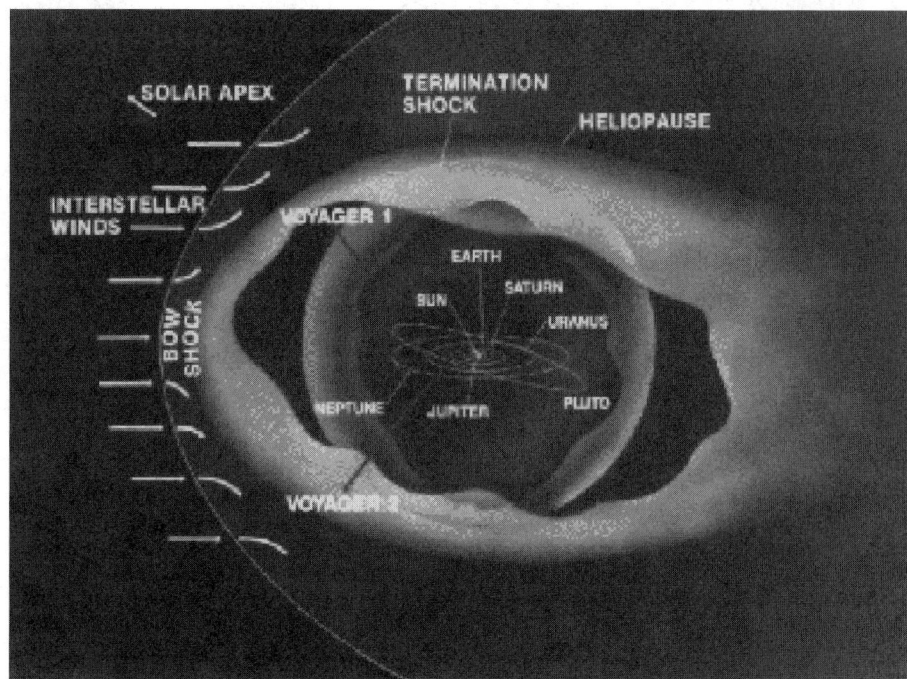

Fig. 8.   Illustration showing the interstellar exploration by *Voyager*. (*Jet Propulsion Laboratory Pasadena, California/NASA.*)

eagles, hawks, and owls. Vultures feed largely on carrion, but many species are also known to attack living animals.

The Old World vultures belong to a family distinct from the New World species. The latter differ in having the nostrils confluent, so that the beak is perforated transversely. With the exception of the lammergeier, an Old World species, all vultures have the head and neck almost bare of feathers and sometimes brightly colored.

The turkey vultures or turkey buzzards (Fig. 1), *Cathartes aura*, of North America are the most widely distributed representatives of the group, and the condor of South America is probably most widely known for its enormous size. It has been recorded with a length of 4 feet (1.2 meters) and a wingspread of 9 feet (2.7 meters). The California vulture or condor, *Gymnogyps californianus*, of the southwestern states and Lower California, has also been recorded with a maximum length of 4 feet (1.2 meters) or more, and its wingspread is said to reach almost 11 feet (3.3 meters). All these birds are magnificent fliers, soaring for long periods without flapping a wing. See also **Falconiformes**.

Fig. 1.   Turkey buzzard (*Cathartes aura septembrionales*). Black with brown edging to feathers. Skin of head and neck bare and red.

W

WAD. The mineral wad, sometimes called *bog manganese*, occurs in amorphous masses, and consists of mixtures of manganese oxides, MnO_2 and MnO, and oxides of other metals such as copper, lead, cobalt, and iron. It is bluish- to brownish-black, usually soft enough to soil the fingers and often porous and light. It is not a distinct mineral species.

WADDEN ZEE. An inlet of the North Sea located in the Netherlands.

WADERS, SHOREBIRDS, AND GULLS *(Aves, Charadriiformes).* Among the many shore and wading birds making up this large order are the various species of auks, guillemots, and puffins; the sandpipers, ruffs, snipes, and woodcock; the plovers and killdeers; the gulls, skimmers and terns; the jacana, phalarope, water pheasant, and wrybill.

The *auks* have large, compressed beaks with oblique grooves toward the tip. One species, the razorbill (*Alca torda*) occurs on both sides of the Atlantic, nesting on rocky ledges. This species is related to the extinct great auk and to the murres, puffins, and guillemots, and is also known as the razor-bill auk. There are over 20 species of auks, which range from 7 to 30 inches (18–76 centimeters) in length. As a group, they are principally found in the North Atlantic and the Arctic regions. The color varies with particular species. Most are grayish, but some are black. Some species have ornamental head feathers. The feet are webbed and the legs are relatively short. The auk is considered to be quite gregarious. The garefowl (*Pinguinus (Alca) impennis*) is now extinct. *Guillemots* are birds with short legs, webbed feet, and upright posture. The several species are found chiefly about the northern oceans. The dovekie is a black guillemot (*Uria grylla*), a marine bird of the north Atlantic (also spelled dovekee and dovekey). The murres are moderately large marine birds closely related to the auks. To some extent, the name is synonymous with guillemot, but different species have been named as one or the other by ornithologists. The murres are powerful swimmers and divers. The murrelets are of a small species of marine diving birds and related to the murres. *Puffins* are marine birds of several species whose thick compressed beaks are almost as large as their heads. They are found on both sides of the Atlantic, nesting in northern latitudes and wintering south to New York and the Mediterranean. The related species of the Pacific are commonly called *shearwaters*.

Sandpipers are long-legged shore birds related to the snipes and plovers. They are named from their association with the bare margins of streams and ponds, where their clear piping calls are a familiar sound. There is a score or more of species in North America, some bearing other names, such as sanderling and knot. Other continents have representatives of the group during at least part of a year. *Snipes* are long-legged and long-beaked wading birds related to the sandpipers and woodcock. Snipes are found in marshy ground and occur on all continents. The principal species in North America is the Wilson snipe (*Capella delicata*) or jack snipe, which is regarded to a limited extent as a game bird. Its erractic flight makes it a severe test of marksmanship. The curlews are shore birds of several species with long legs, a moderately long neck, and a long, curved beak. They are related to the snipes. The *dunlin* (*Tringa*) is a common European shore bird, a type of sandpiper. It is also called the ox bird. The *godwits* are wading birds allied to the sandpipers, with long legs and a long slender beak. The several species nest in the far north, but migrate to the Southern Hemisphere in winter. The *greenshank* is a European sandpiper (*Tringa nebularia*), related to the willets of North America. The greenshank migrates into South Africa and Australia. The *redshank* is a Eurasian bird (*Tringa totanus*) related to the snipes and sandpipers, named from the red color of the bare part of the legs. The name is also applied to a larger related species of the European *T. erythropus*. The *ruff* is a European bird (*Philomachus (Machetes) pugnax*) related to the sandpipers. The name comes from the seasonal development of a large ruff about the neck of the male. The females are called *reeves*. Scales develop on the toes during winter months. The scales are long and help to support the bird as it walks on the snow. *Seed snipes* are South American birds resembling quails, but classes as intermediate between true snipes and the gulls. They are closely related to the *sheathbills*. The *tatler* is a North American sandpiper, *Heterocelus* (*Heteractitis icanus*), or wandering tatler, found along the Pacific coast of North America where it breeds in northern latitudes and on some of the Pacific islands. The *woodcock* is a woodland bird, *Scolopax minor*, of the northern and eastern part of the United States and related to the snipes. The woodcock has a long straight beak and short neck, tail, and legs, and its mottled brown plumage is a fine example of protective coloration. The woodcock is of limited interest as a game bird. Yellow-legs is an American bird related to the snipes and sandpipers. Two species, the greater, *Totanus* (*Tringa*) *melanoleucus*, and the lesser, *T.* (*Tringa*) *flavipes*, are widely distributed over North America and migrate into South America. The latter is sometimes seen in Europe.

The *plovers* are wading birds with moderately long legs and a moderate beak. Although usually found near water, many species frequent dry ground, such as meadows and pastures. Plovers average about 10 inches in length. In some species, the white plumage of the neck features two beautiful rings of black down. The common coloration for most species is black, with white and brown. Plovers are known for their rather raucous voices and also for their tricks when in trouble. They sometimes appear as though falling in flight as the result of a broken wing, or possibly feigning a broken leg when faced by intruders. The *killdeer, Oxyechus vociferus,* well named for its loud voice is the most common of several North American species of plovers. The *golden plover, Pluvialis dominica,* is remarkable for its long migrations between arctic breeding grounds and its winter home in Patagonia. The courser is a long-legged bird related to the plovers and is found in Africa and southern Europe. One species is a desert bird. *Dotterels* (*Eudromias*) represent several species of Old World birds related to the plovers. *Lapwings* (*Vanellus*) are birds of several species resembling plovers both in appearance and habit. The common European lapwing (*Vanellus vanellus*) also is known as the *peewit*. The *oyster catchers* are birds of few species, but of worldwide distribution. They are related to the *avocets*, from which they differ in being more stoutly built. Oyster catchers frequent the coasts and eat bivalve mollusks, among a much larger variety of food. The American oyster catcher is *Haematopus palliatus*. *Sheathbills* are related to the oyster catchers and derive their name from the horny sheath enclosing the base of the beak. Two species are known, both found in extreme southern latitudes and the Antarctic. *Turnstones* are shore birds related to the plovers and oyster catchers. They are found chiefly in the far north, migrating southward, chiefly along the coasts, in both the Old and New Worlds. They winter as far south as Patagonia. The *whimbrels* are northern shore birds with long legs and a long curved beak. They are related to the curlews and plovers. The term whimbrel applies to the European species. Species occurring in North America are called the *Eskimo, Phaeopus borealis,* and the *Hudsonian, P. hudsonicus,* curlews.

Gulls are water birds with webbed feet and long narrow wings. Their flight is powerful and easy, and they are more often seen in the air than on the water. They are common along the seashore and on larger bodies of fresh water, but Franklin gulls are often seen far from water, even following the plow to pick up insects. With exception of the sabine gully, they may be distinguished from the closely related terns by the square end of the tail. The *kittiwake* (*Rissa*) is a related species of the north Pacific and Bering Sea. The *jaegers,* also called *skuas,* are birds closely related to the gulls, from which they differ in the elongate middle tail feathers. Four species nest in the Arctic regions and migrate into Europe, the southern United States, and some to the Southern Hemisphere. One species is found in the Antarctic and another lives along the west coast of South America. These birds hunt and fish like gulls. They range up to 2 feet (0.6 meter) in length. They fly with great speed and are known for rather nasty dispositions, even attacking people when too close to their nests. *Skimmers* are birds related to the terns and gulls, but distinguished by a peculiar beak. The entire beak is long and compressed and the lower mandible is much longer than the upper. The bird dips this lower mandible into the water as it flies. Of the few species, only one, the black skimmer, *Rhynchops nigra*, is known in North America. *Terns* resemble the gulls, but have a more slender and tapering beak. The tail is forked. Terns are very much like gulls in habit and are usually seen in flight over water, both on the coasts and near ponds and lakes. About 10 species are normally found in the United States. The common tern, *Sterna hirunda*, and the Arctic tern, *S. paradisaea*, breed within areas of the Arctic Circle. Four of the relatively few species breed in the Antarctic. Terns are well known for their very long-distant migrations. A tern banded in Labrador was recovered 11,000 miles (17,700 kilometers) away in Africa, only 3 months after the bird had learned to fly. Many of the terns have black heads with white bodies. The tail is deeply forked, wings long and pointed. They deposit 3 white eggs with dark flecks in a small cup rolled out in the sand. The eggs often are covered with sand. During courtship, the terns go through a ritual something like that of the *bower bird*. The tern is a beautiful and most graceful bird.

The *jacanas* are shore birds of South America, Africa, and the Oriental region. They have long legs and tail and very long toes. Depending upon species, they range in length from 6 to 12 inches (15–30 centimeters). Spurs on the wings are sometimes very sharp. In some species, the toes have nails up to 4 inches (10 centimeters) long. They help to hold them on to rafts of lilies or plants floating on the surface of ponds and lakes. Unlike most birds, the female jacana possesses the elaborate plumage instead of the male. Generally, the roles of sexes are reversed. The male makes the nest and cares for and incubates the eggs and tends to most of the rearing of the young. The female is also the more aggressive member of the family. The female entices the male during courtship through dances akin to the bower bird. Jacanas have a catlike cry. They have a red maroon coloration with a black head. They fly in beautiful pack formations. The eggs are 4 in number and often are submerged in warm sunlit water. Incubation period ranges from 22 to 24 days. The nails are long at birth. The young soon follow the male onto rafts for training in flying and hunting for food.

The *phalaropes* are wading birds of several species, with long legs, lobed toes, and a moderately long beak. The red (*Phalaropus fulicaris*) and northern (*Lobipes lobatus*) phalaropes breed throughout the northern part of the northern hemisphere. *Wilson's phalarope, Steganopus tricolor*, of North America is one of the most beautiful of wading birds. These birds are also unusual in the reversal of sex roles, the female being larger, more brightly colored, and taking the aggressive role in courtship.

The *water pheasant* is a large and beautiful water bird of India and Ceylon. Its nest floats on the water or is anchored to water plants. The bird is related to the jacana. The *wrybill* is a New Zealand bird whose beak is asymmetrical. The terminal half of the organ bends to the right. The bird is said to seek its food, consisting of insects and other small creatures, by reaching under the edges of stone. This habit is apparently well served by the peculiar adaption of the beak.

Pratincoles are birds of several species found on all continents of the Old World. Their long wings and forked tail give them the appearance of swallows when in flight, but their moderately long legs are a point of resemblance with the plovers. They live chiefly near water, but are insectivorous in habits.

Avocets are wading birds of several species found in the Old and New World. The beak is curved upward at the tip and the feet are fully webbed. Avocets are about 18 inches (46 centimeters) in length. Nests are found on the ground near water. Shallow pounds or muddy seacoasts are preferred habitats. The avocet catches its food by flinging its head from one side to the other with mouth open while in the water. It also gains food while wading. See also **Charadriiformes**.

WADI.   A term used in desert regions for a stream bed or channel, a steep-sided and bouldery ravine, gully, or valley, or a dry wash that is usually dry except during the rainy season and that often forms an oasis. The word also may be used to designate an intermittent and torrential stream that flows through a wadi and ends in a closed basin, or a shallow, usually sharply defined closed basin in which a wadi terminates. Also spelled *wady* or *waddy*. See also **Arroyo**.

WADING BIRDS.   See **Waders, Shorebirds, and Gulls**.

WAHOO.   See **Mackerels**.

WAKE.   The disturbed region behind a body placed in a stream of fluid. The flow derives its energy and momentum from the drag force on the body and measurements in the wake can be used to infer the force.

WALKING FERN.   See **Ferns**.

WALKINGSTICK (*Insecta, Orthoptera*).   A slender wingless insect related to the grasshoppers. The walkingsticks are elongated in every part and closely resemble the twigs or stalks of the vegetation on which they live. Together with winged species found in the warmer regions of the world they make up the family Phasmidae. Some of the winged species resemble leaves. In the tropics, some of these insects measure 15 inches (38 centimeters) from end to end.

WALLABY.   See **Marsupialia**.

WALLEYE.   See **Perches and Darters**.

WALNUT TREES.   Members of the family *Juglandaceae* (walnut family), these trees are of several species. They are all termed walnut trees with exception of the butternut tree. Collectively, the Juglans may be described as deciduous trees, although some occur as large shrubs. They are best known for their hard-shelled nut which can be harvested in the autumn. See Fig. 1. Important species of the walnut family include:

American walnut	*Juglans nigra*
Black walnut	See American walnut
Butternut	*J. cinerea*
Chinese walnut	*J. cathayensis*
Circassian walnut	*J. regina*
Cut-leaved walnut	*J. regia "Laciniata"*
English walnut	*J. regia*
Manchurian walnut	*J. mandshurica*
Persian walnut	*J. regia*
Texas walnut	*J. rupestris*

The black walnut or American walnut is native to the United States and is found well distributed along the eastern seacoast and west to Michigan and throughout most of the Midwestern states. It is a large tree, with a height of 30 to as much as 65 feet (9 to 19.5 meters). The trunk is about 21/2 to 3 feet (0.8 to 0.9 meter) in circumference. As shown by Table 1, some specimens attain much larger dimensions. The branches are stout. The bark is sepia in coloration, rough with deep furrows that are perpendicular and short. The leaf is compound and lance-shaped, with from 9 to 19 leaflets per leaf. They are pointed and downy. The nut is round, approximately 4 to 6 inches (10 to 15 centimeters) in circumference, with an outer thick hull. The hull is divided into 6 lobes and is dark green-brown in color. The sweet and oily kernel is inside a tough shell protected by the hull.

It requires about 80 years for walnut wood to mature and be used for timber. The specific gravity when kiln dried is 0.56; the shearing strength parallel to the grain is 1,000 psi (6.9 MPa). Compressive strength perpendicular to the grain is 1,730 psi (11.9 MPa). The strong, durable wood is rich brown in color, darker than European walnut and more uniform in color. The wood is closed-grained and aromatic. The wood is sometimes referred to as "Queen Ann's cabinet wood," and is particularly suited for carving, and especially valuable for furniture making and a leading wood for gun stocks.

Old walnut trees (a half century or more of age) are quite valuable even though facing death through diseases, such as *walnut dieback*, or suffering the effects of a lightning bolt. Owners of dying walnut trees will frequently send notices to walnut buyers asking for bids. For large trees (2 feet; 0.6 meter or more in diameter), the value will be in the thousands of dollars. The larger logs are frequently used for veneer. For example, the butt log from a large tree may have a volume of 1050 board feet,[1] which will produce approximately 80,000 square feet (7432 sq meters) or enough veneer to cover a wall 8 feet (2.4 meters) high and 13/4 miles (2.8 km) long. In processing, the log will be sawn into two pieces, lengthwise. Then, each half will be sawn in half (also lengthwise), making four quarters, known as *flitches*, from the single log. Each flitch is placed on a machine

---

[1] One board foot equals 2.36 cubic meters.

(a)   (b)

Fig. 1.   (**a**) English or Carpathian walnut (*Juglans regia*); (**b**) Black walnut (*Juglans nigra*). (*USDA photos.*)

TABLE 1. RECORD WALNUT TREES IN THE UNITED STATES[1]

Specimen	Circumference[2]		Height		Spread		Location
	Inches	Centimeters	Feet	Meters	Feet	Meters	
Arizona walnut (1999) (*Juglans major*)	225	572	85	25.9	98	29.9	New Mexico
Black walnut (1991) (*Juglans nigra*)	278	706	130	39.6	140	42.7	Oregon
Butternut walnut (1998) (*Juglans cinerea*)	259	658	78	23.8	76	23.2	Connecticut
Little walnut (1980) (*Juglans microcarpa*)	160	406	50	15.2	80	24.4	Texas
Little walnut (1986) (*Juglans microcarpa*)	160	406	53	16.2	65	19.8	New Mexico
Northern California walnut (1986) (*Juglans hindsii*)	290	737	115	35.1	106	32.3	California
Southern California walnut (1973) (*Juglans californica*)	241	612	116	35.4	95	29	California

[1]From the "National Register of Big Trees," American Forests (by permission).
[2]At 4.5 feet (1.4 meters).

with a long knife blade that slices off the thin sheet of veneer. Much more detail on this topic is given by J. Birkemeir (*American Forests*, 13–15, September 1980). The processing of walnut wood and nuts is well covered by J.P. Jackson (*American Forests*, 18–58, October 1981).

Oil is obtained from pressing the nut kernel. The oil is fast-drying and yellow in color, and used for artists' paints. A walnut-shell flour made from walnut shells in California is used as a filler in plastics, providing strength.

The butternut tree is found from Maine to Minnesota and throughout the New England States. However, it does not occur in larger numbers. The tree is short-lived. The usual height is from 50 to 60 feet (15 to 18 meters), with a trunk of about 1 to 2 feet (0.3 to 0.6 meter) in diameter. The bark is furrowed and dark gray. The fruit is about 1 1/2 inches (3.8 centimeters) long, oval-shaped, with a brown, tough husk. The husk contains a yellowish liquid. The kernel is oily. Butternut wood takes a high polish and is considered quite valuable by cabinet makers.

WALRUS.   See **Sea Lions and Seals**.

WANDERROO.   See **Monkeys and Baboons**.

WARBLE FLY (*Insecta, Diptera; Hypoderma*).   A bot fly whose larva migrates through the connective tissues of cattle to complete their development in small abscesses called warbles opening through the skin of the animal's back. The adult flies attach their eggs to the hairs of cattle and the newly hatched larva enters the skin by way of the hair follicle. During its development it migrates extensively before reaching its final position under the skin of the back. The perforations leading into the warbles damage the best part of the hide, and the insect is sometimes a source of economic loss as a cause of illness in cattle. The maggots can be pressed out of the warbles when they once become evident or can be destroyed by smearing an ointment over the openings in the skin.

WARBLER *(Aves, Passeriformes)*.   A small bird related to the thrushes. The warblers of the Old World are an extensive family (*Sylviidae*) represented in North America only by the *kinglets* and *gnatcatchers*. The birds commonly called warblers in America are more accurately distinguished as wood warblers and make up the family *Mniotiltidae*, more closely related to the vireos.

Both groups include species whose common names do not indicate their association. Among European examples are the *whitethroat*, the *hedge sparrow*, the *firecrest*, and among the American warblers are the *oven bird, water thrushes, chats,* and *redstarts*.

Many of these birds are beautiful. Their numerous species are a delight to bird lovers during the spring migration in the United States, and in the fall, due to the great variation of patterns and colors between the sexes and the immature individuals, they are as much a puzzle as a pleasure. See Fig. 1.

Fig. 1.   Hooded warbler (*Wilsonia ciatrina*). Olive green above; yellow below. The male has a black hood covering the top of the head and running around the neck to the throat, giving the effect of a yellow mask across the face.

The *vireo* is a small, quietly colored bird of a family related to the warblers. The vireos are mostly gray to olive gray above and white below. Some species show traces of contrasting black or yellow. They construct beautifully cupped nests, suspended in the crotch of a twig.

The *whitethroat* is a European warbler. The common species, *Sylvia cinerea*, is a gray-brown color above and whitish below, and the lesser whitethroat, *S. curruca*, is gray above with dark brown ear coverts and tinged with pink on the breast. The *yellowthroat* is a North American warbler distinguished by its bright yellow front with black patch through the eyes and along the side of the head. Known as the Maryland yellowthroat in the eastern and central states and in its western varieties as the western, Pacific, tule, or salt-marsh yellowthroat. The bird frequents low thickets and marshy places.

WARM FRONT.   See **Fronts and Storms**.

WART (*Verrucae*).   Circumscribed, raised skin lesions, gray, tan, brown, or black in color, having a piled up or horny surface. *Verruca vulgaris* commonly occurs on the hands of children, as multiple, small warts over the fingers and back of the hand. Other forms are plantar warts on the soles of the feet, and venereal warts on the genitalia. It has been established that the types of warts appearing on children are caused by an identifiable filtrable virus. However, there is much to learn concerning the basic chemistry and mechanics of the behavior of these masses. Sometimes, on the one hand, they may disappear within a day or two after no attempt whatever to treat them. Other times, they stubbornly resist all attempts to remove them and, once removed, they may return to the exact spot on the body. Many warts disappear after varying lengths of time and thus do not require treatment. For stubborn cases where for cosmetic purposes it is highly desirable to remove the wart, various methods have been used with varying degrees of success. These include application of liquid nitrogen with a cotton-tipped applicator for freezing the lesion. This causes blisters and subsequent dissolution of the wart. Electrodesiccation and curettage are effective for persistent or recurrent lesions, but care must be exercised to avoid scarring.

The *Verruca plana* (skin-colored, flat wart), commonly seen on the face and back of the hand, is more difficult to cure. Freezing with liquid nitrogen, applying trichloracetic acid, or painting the lesions daily with 10% salicylic acid and 10% lactic acid in flexible collodion may be effective.

The *Verruca plantaris* (a more complex, mosaic-like lesion, often consisting of multiple discrete, or confluent lesions) is the most difficult to treat. These warts can be quite painful and, depending upon location, can be debilitating. After the wart is pared, 50% bichloracetic or trichloroacetic acid may be applied, or nightly applications of 40% salicylic acid plasters may be used. In unresponsive cases, surgical dissection and removal of the wart may be required.

Because of the similar appearance of some warts to much more serious skin lesions, the untrained person should not participate in diagnosis or treatment. Procedures such as those described should be limited to the physician.

WARTHOG.   See **Swine**.

WART SNAKES.   See **Snakes**.

WASP (*Insecta, Hymenoptera*).   An insect related to the ants and bees. Some species are solitary and some social. Most species have four membranous wings, the front wings much larger than the hinder pair and both with few veins, joined to form closed cells. Some wasps burrow, some build nests of mud, and some use a coarse paper made by chewing wood from weathered surfaces. The last species are more commonly called hornets. Wasps are well known for their ability to sting; the more commonly known species inflict painful wounds because of their large size. See Fig. 1.

Fig. 1.   Wasps.

From the scientific point of view, the term wasp is almost without value, since it applies to many different groups of insects. The members of one superfamily, the *Vespoidea*, include the spider wasps, the cuckoo wasps, the velvet ants, the true ants, the true wasps, and the potter wasps, each making up a different family, while the hornets and yellow-jackets, most familiar of wasps, are only one family of the group. Scientifically another superfamily, *Sphecoidea*, is made up entirely of insects called wasps according to some writers, but includes the bees according to others. The wasps of this division are represented by the giant cicada-killer, the largest of North American wasps, and the thread-waisted wasps.

Wasps have been extensively studied because of their complex behavior in making and provisioning nests for their young. In the works of the Peckhams and the Raus some remarkable and interesting records are preserved. See also **Dried-Fruit Insects**; **Hornet**; and "Fig Trees" in the entry on **Mulberry Family**.

WASTES AND POLLUTION.   The approach to reducing waste and abating pollution is complex and sometimes even controversial among the experts. Differences arise because not all of the facts are on hand. The basic chemistry and physics of the Earth's three waste "sinks" — the atmosphere, the hydrosphere, and the lithosphere — remain poorly understood. Although there is consensus among both technologists and the lay public that serious environmental pollution problems exist and indeed are worsening, there are differences pertaining to details and priorities. Environmental scientists and engineers are influenced by factors that are not exclusively scientific, but that are of a societal and economic nature as well. The populace in general is slow to accept changes in life-style preferences and habits, and commerce and business interests frequently resist the concept that environmental costs must be added to the other costs of doing business. Such conflicting factors minimize the key ingredient of attaining success — *dedication*.

Even in view of the aforementioned difficulties, impressive pollution abatement successes have been made, but unfortunately the pace of these programs has not kept up with the rate of the worsening environment. To illustrate the partial successes to date, one need only to compare the environmental status in the advanced industrial nations with the environmental damage found in the former Soviet Bloc, where environmental problems essentially were ignored.

Pathways to Environmental Correction

From an idealistic viewpoint, one may place the efforts for restoring the purity of the natural environment as falling along three pathways. This is an approximation, and the pathways are not mutually exclusive. Table 1 essentially is included as a checklist of the numerous actions that are being taken.

Pathway 1 includes those actions that are directed more toward eliminating or drastically reducing waste production — that is, efforts made to correct the cause (waste) rather than the effect (pollution). Had actions along these lines been taken years ago, the massive pollution experienced today most likely would not have occurred.

Pathway 2 actions recognize that, with current technology and societal attitudes coupled with economic factors, a considerable amount of pollution must be accepted. Actions along this pathway, however, can reduce the amount of wastes produced and hence resulting pollution.

Pathway 3 actions recognize the current inevitability of mass pollution, but which are directed toward reducing the long-term effects of waste disposal. In the past, pollution has occurred in somewhat of a step-like fashion (i.e., creation of the wastes in the first place, sometimes followed by unscrupulous means taken to "hide" or simply "forget" abandoned wastes, followed by waste site clean up). These problems occur simply because wastes have been or are being disposed of improperly.

Establishment of Regulations

When it became apparent that environmental protection could not be accomplished strictly on a voluntary basis, several of the advanced industrial countries were forced to take regulatory actions (circa mid-1960s). The problem was too complex and not sufficiently understood at the outset to institute complete legislation at one time. Consequently in the United States, for example, numerous special acts were passed but stretched out for several years. See Table 2 on p. 3701. This has resulted in difficult compliance and enforcement procedures.

Numerous scholars of the environment have noted several major differences between the way some of the advanced European industrial nations and Japan approach the problems of waste reduction and pollution abatement, and the general approach adopted by the United States over the years. The European countries referred to here notably are the Netherlands, Sweden, and Germany and, of course, do not include the former Soviet Bloc countries.

In the United States, *regulation* is by far the predominant controlling tool. The U.S. system is highly legalized. In the European countries and Japan, regulation is but one tool used.

Government regulators in the United States infrequently provide specific professional assistance to a firm with pollution problems. In the European countries mentioned and in Japan, regulatory personnel frequently work closely and cooperatively with industrial personnel in seeking solutions. In fact, in some cases, government grants are made available for remedial implementation.

The principal incentive for regulators in the United States is that of *developing* and *enforcing* regulations. Less emphasis is directed toward finding better methods for achieving improved results. In connection with the Superfund program, regulators are credited when pollution sites are cleaned up, but emphasis is given to *initiating* the action, with less accountability required where the cleanup has been found inadequate.

In Japan and the European countries mentioned, regulators are rewarded for eliminating waste streams and cleaning up pollution. Regulators contribute technical expertise. In some cases, research is conducted in government laboratories in an effort to solve a particular pollution problem. The emphasis is on cooperation, rather than adjudication.

In the United States, regulations are highly detailed, sometimes overburdened with detail. They are drawn up so that they can withstand litigation and with little practical latitude in enforcement. Industry has no greater access to the regulators than does any other interest. This, unfortunately, tends to create a climate of confrontation rather than one of cooperation.

In the European countries mentioned and in Japan, polluting firms are not required to pay for past practices that did not break former law, with the exception of sites that the polluter continues to own. In the United States, industrial polluters are liable for cleanup costs for sites to which they have contributed waste, even though no laws were broken when the pollution was created.

A more thorough examination of these policy differences is given in the Beecher/Rappaport reference listed.

Particular Attention Given to Hazardous Wastes. In addition to toxicity, hazardous wastes include materials that may become chemically reactive, including ignitability and explosibility, or that may be corrosive. Some toxic materials require extensive pretreatment prior to dumping. See Table 3 on p. 3701.

With the growing use of throwaway products in the hospital and medical field, there is an increasing danger stemming from *infectious wastes*. The

---

TABLE 1. WASTE-POLLUTION OPTIONS AND STRATEGIES (Abridged)

PATHWAY 1
ELIMINATE OR GROSSLY REDUCE WASTES

Increase Life Span of Products
Reduce "junking" frequency. Design for corrosion and wear resistance; easy maintainability. Discourage frequent styling changes simply in interest of increasing marketing appeal.

Design Energy Efficiency into Products and Processes
Traditional processes for converting energy resources (fuels) into electricity, for example, are major polluters. Thus, end products and end processes should consume minimum energy to overcome "hidden" pollution costs. Considerable progress has been made to design more efficient heating and cooling systems (at manufacturing and consuming levels). New electric motor designs consume less energy. Electric lamp efficiency is increasing.

Use Less Pollutive Energy Sources
Check fuel BTU content vs. pollution generated. Also, pretreat fuels and design equipment to increase combustion efficiency.

Evaluate Nontraditional Energy Sources and Conversion Processes
Select least pollutive of common fossil fuels. Consider the feasibility of geothermal, hydro, solar, biomass, and other substitute energy resources. Nuclear power scores high as a non-polluter, with the exception of the radioactivity wastes. A new generation of nuclear reactors is underway that will increase safe operation manyfold. Much research on nontraditional energy resources continues, but generally technical problems have slowed the pace of progress. These efforts are addressed elsewhere in this encyclopedia. Check index.

Search for New Products and Processes (Substitutes)
Many existing products either generate excessive pollutants during their manufacture, require large amounts of energy (hence hidden polluters), or are adversely pollutive in their own right. Some agricultural chemicals and refrigerants exemplify the latter property. The techniques of organic synthesis provide an avenue to substitute product and process development. The recent development of new refrigerants to replace chlorofluorocarbons (ozone problem) and biological insecticides are examples. Some progress has been made in developing less-pollutive fuels for internal combustion engines and the substitution of new technology for traditional motive power, such as electric- and solar-powered vehicles.

De-emphasize Throw-Away Products
Although somewhat justifiable for certain hospital and medical products, these are very hazardous in wastes. This practice should be reevaluated. Other throw-away items, such as pens and cameras are designed because of marketing motivations and contribute to the waste-disposal load.

Eliminate Product Frills and Frivolous Products
Packaging engineers in recent years have contributed tremendously to waste creation. A substantial portion of household and restaurant wastes, for example, consist of packaging and shipping materials. The marketplace is full of junk merchandise.

Safely Transport Products
Moving polluting products from the manufacturing source to a consuming destination poses an environmentally damaging threat. Product containers must be designed to withstand forceful damage. In connection with petrochemical products, containers range in size from oil drums and chemical-containing carboys to ocean-going oil tankers.

J.S. Hirschhorn (Congressional Office of Technology Assessment) as early as 1988 presented a scholarly summary of waste reduction as the ultimate key to pollution abatement, "Waste reduction is the only way to save industry some of the escalating costs of the current waste-management system." The direct costs of waste disposal have increased some 50 times just over the past few years. Hirschhorn listed six steps to waste reduction:

1. Transfer the economic motivation for waste reduction to those engaged in the manufacturing process.
2. Motivate employees by crediting their performance records by meeting waste-reduction timetables established by management and for proposing waste-reduction concepts.
3. Seek technical assistance from outside sources to gain new viewpoints and incentives.
4. Conduct and maintain a waste-reduction audit.
5. Make waste reduction a lasting part of corporate culture. Approach waste-reduction goals today as energy conservation was stressed a decade or so ago.
6. Initiate a corporate-wide waste-reduction educational program.

TABLE 1. (*Continued*)

PATHWAY 2

RECYCLE WASTE MATERIALS

Design Containers for Recycling

Although not always desirable from a marketer's or consumer's viewpoint, throw-away containers of all kinds contribute massively to the waste-handling problem and to pollution.

Design Products and Components for Recycling

Whereas the aluminum beverage container cannot be reused as such, the aluminum in the can may be reprocessed. The recycling of aluminum is one of the current successes along these lines. The production of raw aluminum from ores consumes enormous amounts of electricity and thus contributes to pollution. Similarly, for years scrap and junk yards have specialized in recycling other metals and all manner of machine parts. From the viewpoint of pollution, this is an excellent practice. Nonmetallics have proved to be more difficult to recycle (most plastics, for example), but much technical progress in this area is underway. Recycled wood fibers in paper products has enjoyed much success. Product designers are in an excellent position to consider the recycling potential of materials after the useful life of the product itself has expired.

Design Processes for Recycling

Cooling water is a prime utility in manufacturing and most notably in the chemical and petrochemical industries. Excellent progress has been made in recycling water instead of continuously dipping into natural water reservoirs. Use is made of cooling ponds and relatively simple water treating at the plant site, thus bypassing pollutive procedures. This also avoids thermal pollution of water source.

Much more complex substances than water should be considered for recycling. These would include the reuse of solvents, cleaning compounds, and, in some instances, using traditional waste components as sources of raw materials. Recovery of valuable materials from wastes may prove less expensive than procuring the same substance from a supplier.

Consider Wastes as Energy Resources

Some industries that produce large amounts of combustible solid wastes have used such materials to augment solid fuels, such as coal, to generate utility steam and hot water. See article on **Wastes As Energy Sources**.

PATHWAY 3

NOTE: The actual disposition of waste into one of the Earth's "waste sinks" is the least attractive of pollution handling procedures. To date, however, dumping of waste remains the most widely used practice. Progress along Pathways 1 and 2 contribute to a progressive reduction in the tonnage of waste to be disposed and in the long term will alleviate the pollution problem in a major way.

In the advanced countries of the world, waste disposal is no longer a simple matter of venting gases and vapors into the atmosphere, or of finding the nearest creek or river, or of creating a landfill.

The carefree disposal of waste that took place several years ago resulted in a public outcry and the creation of numerous, often quite complex regulations. Thus, the polluting source today must take a number of costly actions prior to the ultimate disposal of the waste.

WASTE DISPOSAL — RELOCATING THE WASTE

Classify and Characterize the Waste

Regulations vary considerably, depending upon the nature of the waste. Hazardous (toxic) wastes are treated as a separate category by federal, state, provincial, and municipal regulatory agencies.

Pretreat Wastes Prior to Disposal

In addition to rigid requirements for hazardous wastes, other regulations may require various forms of pretreatment, such as sorting wastes into various categories in the interest of handling efficiency at waste sites, incinerators, and so on. Although they require handling as wastes, biodegradable wastes pose a lesser threat to long-term pollution and often require less stringent regulation.

Select an Appropriate Depository

Gases, vapors, and airborne particulates, unless present in minor amounts, generally will require postproduction treatment before venting to the atmosphere. Post-treatment is also frequently required for the disposal of liquid and solids. The topic of disposal site selection for liquids and solids is complex because there are so many classes of materials, including such diverse wastes as sewage, public building and household wastes, medical and hospital wastes, packaging wastes, transportation vehicle wastes, office wastes, and agricultural wastes. The lists numbers into the hundreds of categories.

In the case of fluids and solids, the Earth's hydrosphere or lithosphere are the only sinks available. For solids, some geological formations are much more appropriate than others. Some areas may appear suitable, but are found to exist over an aquifer (essentially an underground stream), and hence pollution can occur in the lithosphere and then pass along to the hydrosphere.

In the past, a number of polluters have used temporary waste storage means, such as aboveground tanks. Storage of radioactive wastes at nuclear power facilities is another example. In-plant storage or nearby polluter-owned sites must meet all current pollution regulations. These practices have been costly in retrospect. They have comprised many of the targets of the so-called Superfund.

Consider a Professional Waste-Handling Firm

Expert assistance (applying mainly to liquid and solid wastes) is available, but extreme caution must be taken in selecting such assistance. There have been several instances of fraudulent practices that have led to disastrous pollution and resulted in strict legal judgments against the initiating polluter.

Run Continuous Checks on Waste Disposal Costs

Where costs continue to spiral, this may provide the incentive to initiate actions along Pathways 1 and 2, which can lower pollution costs in the long term.

Consider the Inevitable Conflicts

Severe regulations are reasonably clear in terms of what a polluter can and cannot do. But tradeoffs do remain. A basic triangle of conflicting forces — that is, Energy vs. Economy vs. Environment — is described in article on Electric Power Production and Distribution.

Louis J. Thibodeaux (Louisiana State University and Director, EPA-Sponsored Hazardous Waste Research Center), in 1990, outlined the four *natural laws* of hazardous waste:

1. In converting thermal energy to useful work, a certain amount of waste (thermal) energy must be discharged into the environment.
2. It is impossible to recycle waste completely. Recycling is one aspect of waste minimization, not a solution.
3. Some fraction of the energy and material needed to drive processes and make products will always be degraded to waste that will have to be disposed of in an environmentally acceptable manner, such as incineration or some version of the solidification/fixation process.
4. Small waste leaks are unavoidable and acceptable. Ecosystems can handle small infusions of hazardous substance. Such discharges must, however, be made small so that there will be no harmful effects either locally or globally.

---

ultimate disposal of millions of needles and condoms becomes a part of municipal wastes. The virulence of microorganisms under such conditions is poorly understood. Throwaway diapers not only contribute immensely to the volume of wastes to be handled, but also contain hosts of living microorganisms, the survival rate of which, under disposal conditions, have not been documented. The common childhood intestinal pathogens, such as retoviruses, hepatitis A virus, and the protozoans *Giardia* and *Cryptospordium*, have not been ruled out.

Incineration of Hazardous Wastes

One authority has commented that the greatest public health danger with medical waste in the United States is *substandard* incineration practices at local hospitals. It has been estimated that in the early 1990s there were about 6000 substandard medical-waste incinerators throughout the United States. The congressional Office of Technology Assessment (OTA)

has estimated that air emissions of dioxin and heavy metals from these incinerators average from 10 to 100 times more per gram of waste burned than emissions from well-controlled municipal waste incinerators. Also, hospital incinerators produce toxic remains in the ashes that can contaminate surface and groundwater when dumped in landfills.

The problem is exacerbated in large and crowded communities. For example, there are well over 50 hospital incinerators in New York City. Most local hospital incinerators are not equipped with acid-gas scrubbers, which convert harmful airborne substances into harmless calcium salts. Nor are most incinerators equipped with electrostatic precipitators to capture particles that have adsorbed toxic flue gases.

Authorities report that quite the contrary conditions exist in several parts of Europe, notably Switzerland and Germany. Legislation in the early 1980s in Germany mandated the closing of hospital incinerators, requiring that medical wastes be sent to regional facilities, at which the latest in

TABLE 2. WASTES AND POLLUTANTS REGULATORY STATUTES

(United States — Partial List)

Clean Air Act
Clean Water Act
Comprehensive Environmental Response, Compensation, and
  Liability Act (popularly known as the Superfund)
Federal Insecticide, Fungicide, and Rodenticide Act
Food, Drug, and Cosmetic Act
Hazardous Materials Transportation Act
National Environmental Policy Act
Occupational Safety and Health Act
Resource Conservation and Recovery Act
Safe Drinking Water Act
Superfund Amendments and Reauthorization Act
Toxic Substances Control Act

Note: Most of these statutes have been enacted since the early 1970s.

TABLE 3. REGULATOR'S CHARACTERIZATION OF HAZARDOUS WASTES

Toxicity

Definition of Extract: The liquid component of a solid waste and deionized water at a pH 5.0 that has been in continuous contact with the solid phase of the waste for a minimum of 24 hours.

Permissible Upper Limit of Contaminant:

	Milligrams/Liter
Arsenic	5.0
Barium	100.0
Cadmium	1.0
Chromium	5.0
Lead	5.0
Mercury	0.2
Selenium	1.0
Silver	5.0
Endrin insecticide	0.02
Lindane insecticide	0.4
Methoxychlor insecticide	10.0
Toxaphene insecticide	0.5
2,4-D (2,4-Dichlorophenoxyacetic acid)	10.0
Silvex 2-(2,4–5 Trichlorophenoxy propionic acid)	1.0

Reactivity

Any substance that:

- Is normally unstable and readily undergoes violent changes with detonations.
- Reacts violently with water.
- Reacts with water to generate toxic gases, vapors, or fumes in a quantity that is dangerous to human health or the environment.
- Is capable of detonating or undergoing an explosive reaction when subjected to a strong initiating source or if heated when confined.
- Is capable of detonation or explosive decomposition or reaction at standard temperatures and pressures.
- Is normally considered explosive and that meets transportation regulations.
- Forms potentially explosive mixtures with water.
- Is a cyanide or a sulfide-bearing material that, when exposed to a pH between 2 and 12.5, can generate toxic gases, vapors, or fumes that are dangerous to human health or the environment.

Ignitability

Any substance that:

- Is a liquid with a flash point of less than 60 °C (140 °F).
- Is a solid and is capable of causing fire through friction, absorption of moisture, or spontaneous chemical changes that, when ignited, burns so vigorously as to create a hazard.
- Is a compressed gas or oxidizer that does not meet transportation regulations.

Corrosiveness

Any aqueous liquid with a pH less than or equal to 2.0 or greater than or equal to 12.5.
Any liquid that corrodes steel at a rate greater than 0.006 meter (1/4-inch) per year.

technology is deployed. Incinerator operators are given special training and require certification of their skills. Particular precautions are used in feeding such incinerators to protect plant workers. The incinerated remains are disposed of in specially lined landfills.

Recently, the technology of autoclaving instead of incineration has been proposed. Disinfection is achieved through the use of high-pressure steam, which assists in breaking the refuse down and ready for compacting.

German authorities also mandate strict regulations over the transport of medical waste, disallowing the transport of foodstuffs in trucks that also are used to handle medical wastes. Hospitals also are required to "tag" all refuse, designating such categories as "office, cafeteria, and general," "infectious" (including pathological body parts, syringes, needles), "radioactive," and "dangerous to handle" (such as scalpels, which must be placed in unopenable containers).

Considerable design effort has been invested in the improvement of incineration systems. D.A. Tillman (Ebasco Environmental) and associates report that "Rotary kilns have become the incinerators of choice for eliminating hazardous wastes in accordance with the U.S. Resource Conservation and Recovery Act, Superfund, and related legislation."

The Ebasco team points out, "A rotary kiln is basically a rotating cylinder that, typically, is refractory lined. The cylinder is tilted slightly (i.e. 3°) and the feed material goes into the upper end. A heat source is applied to the material, usually by combusting liquid or gaseous fuel within the kiln. Gravity moves the material through the cylinder, and it is discharged from the lower end. When rotary kilns are used for hazardous waste incineration, the gaseous products from the kiln are ducted to a secondary combustion chamber for subsequent destruction." Several overall system design configurations are possible. Generalized configurations are illustrated in Figs. 1 and 2. See Tillman reference listed.

J.F. Mullen (Dorr-Oliver Inc.) reports that fluidized bed incinerators have been used for municipal sludge and industrial waste incineration since the early 1960s for a variety of wastes (petroleum tank bottoms, sludge from pharmaceutical, pulp and paper, and nylon manufacturing operations), waste plastics, waste oils, and solvents. Fluid beds were first considered for incinerating hazardous wastes in the 1980s.

Advantages claimed for fluid bed technology include efficient combustion, ease of control for handling a variety of feeds, and reasonably low capital and operating costs. Specific advantages claimed for incinerating hazardous wastes include fuel savings and lower emissions of nitrogen oxides (NO_x) and metals. The principles of a fluid bed incinerator are illustrated in Fig. 3. Among the primary organic hazardous constituents (POHCs) of waste, the fluid-bed has successfully demonstrated its effectiveness in achieving 99.99% efficiency in removal of aniline, carbon tetrachloride, chloroform, chlorobenzene, cresol, para -dichlorobenzene, methyl methacrylate, naphthalene, perchloroethylene, phenol, tetrachloroethane, 1,1,1-trichloroethane, trichloroethylene, and toluene.

Soil-Washing Technology

Tons of earth per hour can be sifted and scrubbed clean of hazardous materials by using what may be called a heavy-duty industrial *washing machine*. Soils contaminated with hazardous or radioactive materials may be removed by washing. Contaminating material that "comes out in the wash" so to speak, such as copper, then can be recycled and reused by industry. Although not widely publicized, soil washing is not a new technology. It has been estimated that nearly 100,000 tons of contaminated soils are remedied in this manner each year in Europe. The bulk of these soils are sands contaminated with hydrocarbons and/or heavy metals. Westinghouse has developed a commercial unit.

The soil-washing process itself is environmentally benign. It is a closed-loop system, and thus no contaminants are discharged into the air or land. It is a permanent solution to many hazardous problems because it physically removes contaminants from the soil. Examples of tests made on the process are graphed in Fig. 4. Soil washers are mounted on truck trailers for mobility. A flow sheet of the process is given in Fig. 5.

In a particular application, Westinghouse unveiled in late 1992 equipment for removing lead particles from the residue created when bridges and other metal structures are sand-blasted to remove old lead-containing paint. This problem is commonly encountered by highway construction and maintenance personnel, but the system is not limited to such applications. It has been estimated that the method has achieved a two-thirds reduction in disposal costs as compared with burial or smelting

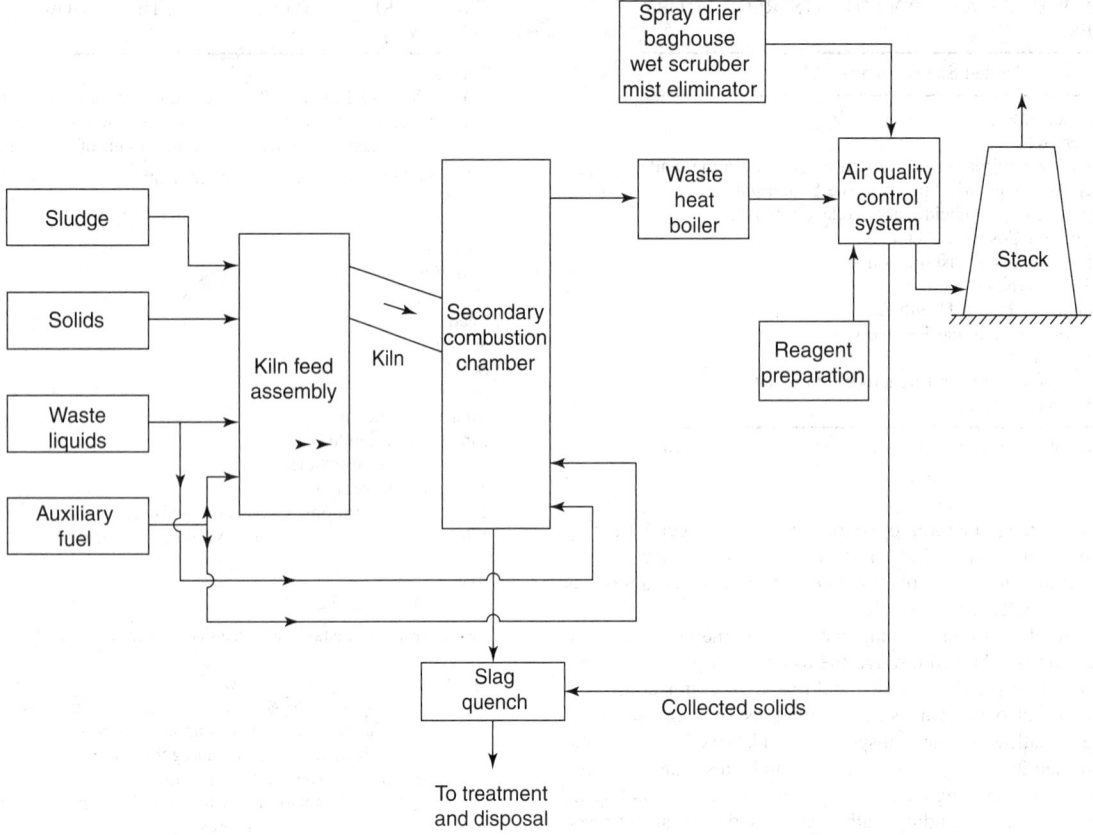

Fig. 1.   Simplified flowsheet of incineration train in a hazardous waste treatment complex. (*After Tillman.*)

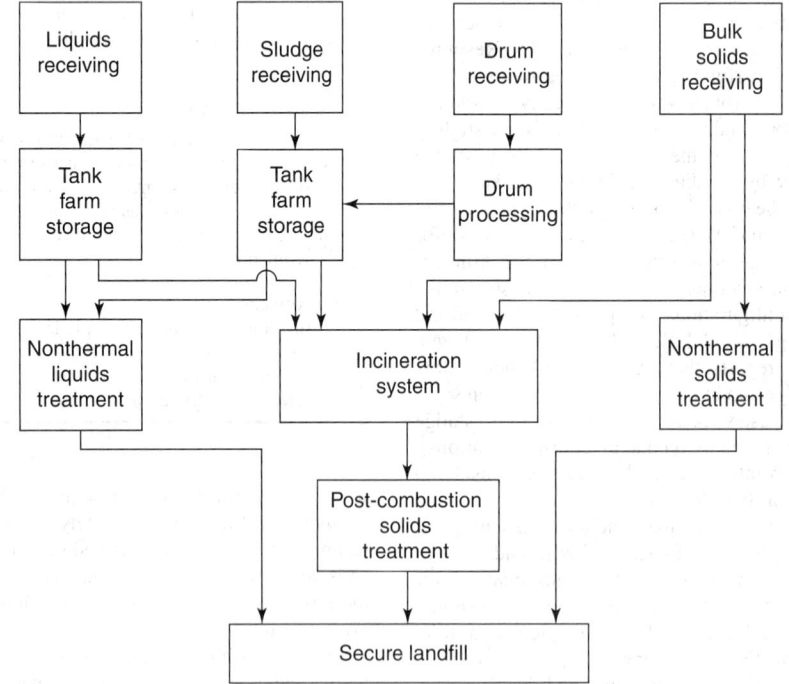

Fig. 2.   Generalized flowsheet of a hazardous waste treatment system complex that utilizes a rotary kiln incineration system. (*After Tillman.*)

without pretreatment. One of the first large-scale tests of the equipment was conducted by the Minnesota Department of Transportation. A portable soil-washing machine for highway work is shown in Figs. 6 and 7.

Soil washing is based on the use of water-based leachates, which are recycled continuously in the machine. The end products are a relatively clean, coarser soil fraction and a wash water containing finer soil particles and most contaminants. The small levels of organic contamination remaining in the coarser materials are removed readily with heat. The wash water, depending upon the specific contaminants, responds to various traditional treatment methods, such as bioremediation, air stripping, chemical precipitation, or membrane separation.

Recycling Plastic Wastes

Although plastic wastes in refuse are highly visible to the public, and thus have caused considerable consumer pressures on plastics manufacturers to take anti-pollution measures, disposal of plastics creates special problems

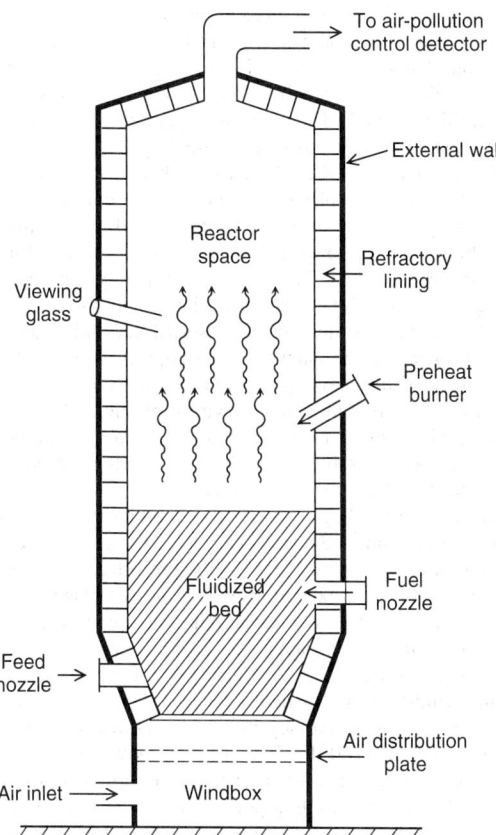

Fig. 3. Schematic diagram of a fluid-bed incinerator. Waste is injected into a bed of inert material that is fluidized by large quantities of air flowing upward through the unit. (*After Mullen.*)

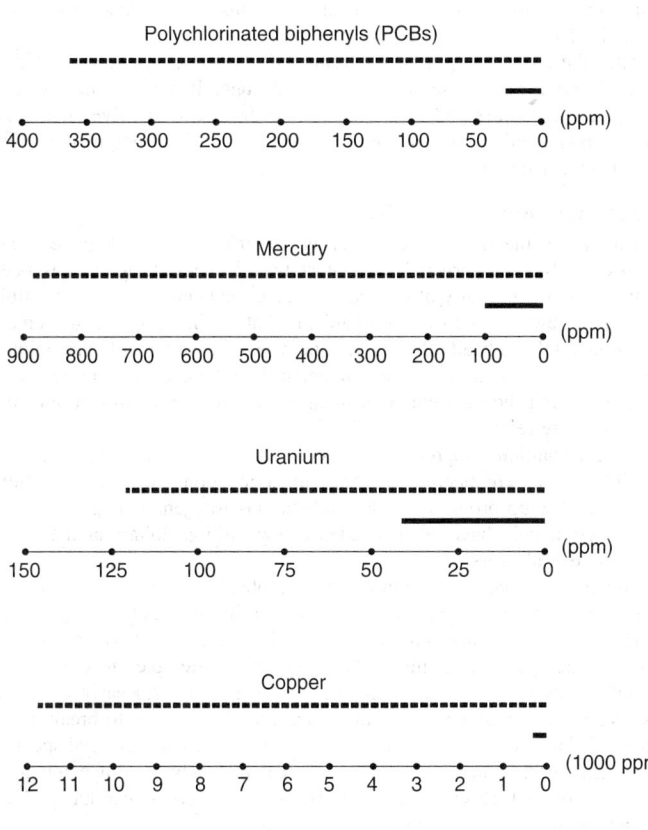

Fig. 4. Examples of performance of soil-washing process. (*Westinghouse Electric Corporation.*)

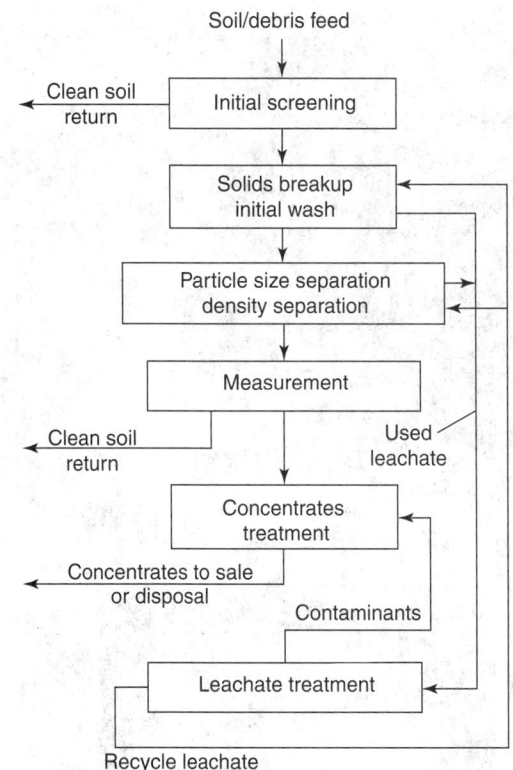

Fig. 5. Soil-washing process flowsheet. (*Westinghouse Electric Corporation.*)

Fig. 6. Portable soil-washing machine shown here at a highway maintenance site used for removal of lead particles from the residue created when bridges and other metal structures are sand-blasted to remove old paint containing lead. There are numerous applications for such portable plants. (*Westinghouse Electric Corporation.*)

in municipal incinerators because of the formation of some toxic gases. In terms of dumpsite disposal, the absence of biodegradability results in long-term solids buildup. Actually, on a weight basis, however, plastics only comprise 7% of municipal wastes, as shown in Table 4.

The principal plastics that show up in municipal wastes are the polyethylenes, polystyrenes, and polypropylenes. These include polyethylene terephthalate (PET) used in soft drink containers, high-density polyethylene (HDPE), used in milk jugs, and polystyrene, used in fast-food containers, which, incidentally, were first banned in Oregon (1989).

As early as 1989, 7 billion plastic soft drink containers were produced in the United States and nearly an equivalent tonnage in Europe. As of early 1991, it was reported that about 28% of the PET bottles produced in the United States were recycled, yielding over 20 million pounds of PET for subsequent use in making carpet yarn, fiberfill for clothing, nonfood

Fig. 7.   Operator at controls of portable soil-washing machine used at highway reconstruction site. (*Westinghouse Electric Corporation.*)

TABLE 4. MAJOR MATERIALS IN MUNICI-PAL WASTES

Material	Percent by Weight
Paper products	40
Yard waste	18
Food waste (garbage)	12
Glass	8
Plastics	7
Steel/metals	7
Other	8
	100
Principal Classes of Plastics	
Low-density polyethylene (LDPE)	24
High-density polyethylene (HDPE)	19
Polystyrene	14.5
Polypropylene	14.5
Polypropylene terephthalate (PET)	4.5
Other plastics	23.5
	100.0

containers, automobile parts, fencing, and industrial strapping. A doubling of these amounts was expected by 1995.

A principal plastic reclaiming process used was developed in the Netherlands by a resin manufacturing firm. In this process, after removal of labels and base cups by machine, the bottles (clear or green) are color sorted and granulated. The PET flake is then washed to remove glue. Closure material is separated by flotation from the PET. The remaining PET is dried and sifted for fines and then is ready for reuse.

In 1989, seven of the leading plastics producers in the United States formed the National Polystyrene Recycling Company, with a 1995 target for achieving a minimum of a 25% recycling rate for polystyrene. A pilot recycling center was set up in Leominster, Massachusetts.

A joint venture of a leading plastics manufacturer and a major waste management firm was established in 1990 to recycle PET and HDPE materials. The plan was that two existing plants would be joined by three additional recycling operations by 1994. Jointly, the plants would recycle about 200 million pounds of these plastics per year. Taken in perspective, however, this is a small quantity of the total of 1.5 billion pounds (PET) and 6.5 billion pounds (HDPE) disposed each year in the United States.

Still another leading plastics manufacturer commenced operation recently for recycling 400 million pounds/year of plastic film and rigid containers. Following a pilot-plant test run, a wide variety of polyethylene materials can be processed. These include polyethylene wrap, lawn and grocery bags, and containers (detergent, bleach, and motor oil), as well as plastic milk and juice bottles and PET soft drink and liquor containers. When in full operation, the plant will serve a 500-mile (~800-km) radius area. It is projected that new applications for the recycled materials will displace some of the requirements for virgin plastic and nonplastic materials. Food applications of recycled material are not in the current plans because of the special problems involved in altering the color and other physical properties of the recycled resins.

Plastic materials recycling poses a serious problem when plastic refuse is received in a commingled state. Because of the great variety of plastic products, only a comparatively few can be presorted before delivery to a recycling plant. These few exceptions would include plastic milk and other beverage containers. Impressive research has been underway at Rensselaer Polytechnic Institute, and a patent has been obtained for a process that dissolves shredded plastics (as described by a researcher), "one polymer at a time in a chip-filled vat" where a solvent (xylene) is used to dissolve five groups of plastics at five separate temperatures, ranging from room temperature to 138 °C (280 °F). For example, polystyrene dissolves first, while the other plastics remain unaffected. The top solubility temperature required is well below the boiling point of the xylene solvent. As pointed out by the researchers, the xylene polymer solution (for a specific polymer) drains to a separate part of the system, where it is heated under pressure to near the boiling point of xylene. Pressure is required to keep the xylene-polymer solution in liquid form. The solution then is sent through a valve into a vacuum chamber to undergo flash devolitization. The sudden change from high to low pressure, researchers say, causes the xylene to vaporize instantly, leaving behind the pure polymer.

After the recovered polymer is removed, the same xylenes are recompressed and cooled to return to its liquid state. It again is heated to a different temperature and reintroduced into the vat to dissolve another of the polymers, and the interim steps are repeated until all polymers present have been separated.

Genetic Engineering and Pollution

Research on the genetic engineering of microbes to degrade toxic wastes has been underway for at least two decades. Progress has been relatively slow, partially attributed to societal concerns over the possible release of new, untested microorganisms that in themselves could create environmental and health threats. As one researcher in the field has pointed out, "No one wants to release organisms before the possible consequences are known, but the possible consequences will remain unknown until the organisms are released."

A very cautious approach has been taken thus far concerning the possible use of engineered microbes in connection with the Superfund dumpsite cleanup program. Budget allocations for genetic engineering in the pollution field have not exceeded a few million dollars annually for the past several years.

Pseudomonas putida, a common soil microbe, has been the target of several researchers. A few years ago, scientists at the University Medical Center (Geneva, Switzerland) created a a microbe that eats 4-ethylbenzoate (4-EB), a toxic synthetic chemical. The researchers attempted to find all the "right" genes and combine them into one organism. As early as 1980, researchers in the United States designed a microorganism to break down crude oil. This also involved research with *Pseudomonas* bacterial species. Other engineered microbes have been created that break down much of the sulfur in coal. Research at Johns Hopkins University also has yielded engineered microbes that can metabolize sulfur.

Microorganisms have been used successfully to clean up contaminated wastewater of such substances as pentachlorophenols (PCPs), polychlorinated biphenyls (PCBs), iron cyanide, and leachates containing chlorinated compounds, phenols, and formaldehydes. A dry cleaning solvent, tetrachlorodiphenylethane (TCE), which is a suspected carcinogen, has been degraded through the use of enzymes released by a strain of the bacterium, *Pseudomonas cetacia*. A wastewater treatment plant has used a mutant strain of *Pseudomonas*, which consumes mine-generated cyanide wastes. A process has been developed that uses naturally occurring microbes to concentrate phosphorus for easy removal. It has been estimated that 99.3% of phosphorus in wastewaters can be removed by the process. One firm has developed a bioreactor that uses an aerobic microorganism (naturally occurring in white-roto fungus) to break down toxic substances, such as 2-chlorophenol.

Other researchers have designed a microorganism for breaking down crude oil, and some experience has been gained from testing the product for cleaning up oil tanker spills. Several proprietary engineered microbes have been announced, but with little detail given. Research firms in the field are delaying commercialization, pending clarification of federal, state, and provincial regulations. See also **Water Pollution**.

Fermentation. In 1990, a bioprocessing research center was opened at Penn State University. A pilot plant demonstrates various processes for converting (recycling) agricultural and food processing wastes into dietary supplements for animals. The pilot plant is designed with versatility and flexibility for testing and processing a wide variety of substances. One example of agricultural and livestock wastes that can occur during abnormal situations is an estimated 1 million tons/day of chickens that die of natural causes and are buried at a site on the Delmarva Peninsula, which borders on the Chesapeake Bay. In extremely hot weather, this figure can increase to 4 million tons/day. Also, it has been estimated that 600 tons/day of fish wastes are dumped off Kodiak Island. The Netherlands is estimated to produce 100 million tons/year of poultry manure, twice the amount that the land can absorb naturally. Researchers at Penn State have applied for patents on one of the fermentation processes developed. A type of marine yeast converts protein-bearing waste into a slurry of water-soluble proteins. The process destroys all pathogens. Thus, when the slurry is dried, it is suitable as a dietary supplement. Researchers note that the same principles could be applied to cesspool wastes. Although centralized sewer systems collect *un*sanitary wastes in most urban areas today, there remain multi-thousands or millions of cesspools used in less-populated areas of the country.

Recycling/Regenerating Paper Wastes

Considerable progress has been made during the past decade for recycling paper wastes, which according to Table 4 constitute well over one-third of the typical municipal waste produced.

Investigators at Texas A&M University (Austin, Texas) are targeting on improved processes for regenerating paper wastes. One of the key problems is the cellulose content in newspapers and paper products. Cellulose is very difficult to "digest." The researchers have developed an ammonia fiber explosion (AFEX) technique that more efficiently utilizes enzymatic digestion of cellulose. Researchers explain that ground-up municipal waste is placed in a tank and soaked with ammonia for about one-half hour, after which high pressure is applied. When this pressure is released abruptly, cellulose fibers are literally blown apart, making it much easier for enzymes to digest them. The enzymes break down the cellulose into individual glucose molecules, after which yeast can convert them to ethanol. The researchers have noted an improvement of up to 150% in digestion as the result of the AFEX process. Researchers also forecast that 180 billion tons of municipal wastes could be converted into over 8 billion gallons of ethanol fuel for automotive consumption.

In many communities, prescribed routines for consumers to separate paper from other items of trash have been highly successful and, in fact, in recent years there has been more paper ready to recycle than there are facilities to process it. Consequently, a lot of newsprint and other paper products still go to the dumpsite. Paper in the household trash is a highly visible waste to the average consumer—hence, much cooperation from the populace has been evidenced. In recognition of an imminent "glut" of newsprint to recycle, some municipalities and states have lowered their targets. One example is Wisconsin, which had set a goal of 50% recycled fiber by 1995 to 17% by 2001.

As mentioned earlier in this article, there is a continuing conflict (triangle) of forces that come into play in numerous decisions that interlock the factors of energy, environment, and economy. Newsprint as of the early 1990s is an example. Publishers, particularly in the northeastern United States, can obtain paper of high quality and made of virgin fibers from nearby Canada at an attractive cost, not much in excess of the cost of recycled newsprint. Canadian paper is imported duty-free. Approximately 58% of the paper consumed in the United States (overall) is imported from Canadian mills from trees grown in Canada. Expected lower timber prices in the southeastern United States may make that region more competitive with Canada over the next decade or two.

Average use by newspapers of recycled fibers seldom exceeds 50% of the total. However, the publishers of the *Los Angeles Times* use approximately 80% recycled fiber.

Recycled newsprint does pose problems to publishers, notably those of books and magazines. Some virgin pulp is required to hold the paper together. Magazine publishers usually purchase what recycled paper they do use from pulp-substitute suppliers—that is, firms that process only selected used paper, such as envelope trim and cuttings, ledgers, business forms, and computer printout paper. A long-range recycling program also poses the problem of dealing with shorter fibers. Each time paper is reprocessed, the fibers are shortened. The present use of recycled fiber in newspapers is, on the average, about 20%.

The processing of recycled paper is essentially the same as the production of paper from virgin fibers once the feed slurry is made. See article on **Papermaking and Finishing**; and **Pulp (Wood) Production and Processing**.

In preparing the slurry for recycled newsprint, first all trash must be removed from the waste paper (some manual labor assisted by machine metal detectors, etc.). The paper then passes to a pulper, where water and some reagents are used to accomplish de-inking. In the pulper, the waste paper is shredded by rotating cutting blades to produce a slurry. This slurry is passed through a continuous pulper, similar to that used in making virgin paper slurry after the natural fibers have been processed in a digester. The slurry then passes through screens and onto a three-stage washer, where ink particles are fully removed and the paper adjusted for the proper consistency prior to being introduced onto the paper machine.

Considering the numerous errors that have appeared in the environmental literature, Stephen Strauss, a science writer for the *Toronto Globe and Mail*, points out how easy it is to draw conclusions regarding environmental measures when raw data have not been gathered, calculated, or presented with exacting care. Strauss makes the serious but amusing observation, "When historians of technology reflect on the final quarter of the twentieth century, they may well surmise that the archetypal public debate (over pollution) centered around the throw-away (paper or foam) cup."

See Strauss reference listed.

Several other articles in this encyclopedia address the topic of waste and pollution. See also **Pollution (Air)**; **Water Pollution**; and alphabetical index. Global warming concepts are described in article on **Climate**; radon is covered under **Radon**; and ozone is described in article on **Polar Research**.

Additional Reading

Abelson, P.H.: "Remediation of Hazardous Waste Sites," *Science*, 901 (February 21, 1992).

Beecher, N. and A. Rappaport: "Hazardous Waste Management Policies Overseas," *Chem. Eng. Progress*, 30 (May 1990).

Bishop, P.L.: "Pollution Prevention: Fundamentals and Practice," McGraw-Hill Higher Education, New York, NY, 1999.

Boerner, D.A.: "Recycling the Paper Forest," *Amer. Forests*, 37 (July/August 1990).

Bumble, S.: "Computer Simulated Plant Design for Waste Minimization/Pollution Prevention," Lewis Publishers, Boca Raton, FL, 2000.

Cezeaus, A.: "East Meets West (Germany) to Look for Toxic Waste Sites," *Science*, 620 (February 8, 1991).

Crouch, M.S.: "Check Soil Contamination Easily," *Chem. Eng. Progress*, 41 (September 1990).

Davenport, G.B.: "The ABCs of Hazardous Waste Legislation," *Chem. Eng. Progress*, 45 (May 1992).

Davis, M.L. and D.A. Cornwell: "Introduction to Environmental Engineering," 3rd Edition, The McGraw-Hill Companies, Inc., New York, NY, 1997.

Davis, W.T. and A.J. Buonicore: "Air Pollution Engineering Manual," John Wiley & Sons, Inc., New York, NY, 1997.

Davis, W.T.: "Air Pollution Engineering Manual," 2nd Edition, John Wiley & Sons, Inc., New York, NY, 2000.

Dupont, R.R., K. Ganesan, and L. Theodore: "he Pollution Prevention the Waste Management Approach to the 21st Century," Lewis Publishers, Boca Raton, FL, 1999.

Erb, J., E. Ortiz, and G. Woodside: "On -Line Characterization of Stack Emissions," *Chem. Eng. Progress*, 40 (May 1990).

Evanoff, S.P.: "Hazardous Waste Reduction in the Aerospace Industry," *Chem. Eng. Progress*, 51 (April 1990).

Garg, S.: "Introduction of Recombinant DNA-Engineered Organisms into the Environment: Key Issues," National Academy Press, Washington, DC, 1987.

Garg, S. and D.P. Garg: "Genetic Engineering and Pollution Control," *Chem. Eng. Progress*, 46 (May 1990).

Gibbons, A.: "Making Plastics that Biodegrade," *Technology Review (MIT)*, 69 (February 1989).

Greenberg, R.A.: "Workshop Participants Focus on (Food) Packaging Waste Management," *Food Technology*, 42 (January 1991).

Hershkowitz, A.: "Without a Trace: Handling Medical Waste Safely," *Technology Review (MIT)*, 35 (August/September 1990).

Higgins, T.E.: "Pollution Prevention Handbook," Lewils Publishers, Boca Raton, FL, 1995.

Hodge, C.A., N.N. Popovici: "Pollution Control in Fertilizer Production," Marcel Dekker, Inc., New York, NY, 1994.

Hooker, L.: "Danger Below (Underground Aquifers)," *Chem. Eng. Progress*, 52 (May 1990).

Kamrin, M.A.: "Toxicology: A Primer," Lewis Publishers, Boca Raton, FL, 1988.

Leaf, D.A.: "Acid Rain and the Clean Air Act," *Chem. Eng. Progress*, 25 (May 1990).

Lecomte, P., C. Mariotti: "Handbook of Diagnostic Procedures for Petroleum-Contaminated Sites," John Wiley & Sons, Inc., New York, NY, 1999.

Lohr, L.: "Managing Solid Byproducts of Industrial Food Processing," *Food Review*, 21 (April–June 1991).

Loupe, D.E.: "To Rot or Not; Landfill Designers Argue the Benefits of Burying Garbage Wet vs Dry," *Science News*, 218 (October 6, 1990).

Majumdar, S.B.: "Regulatory Requirements and Hazardous Materials," *Chem. Eng. Progress*, 17 (May 1990).

Martin, A.M. et al.: "Control Odors from Chemical Process Industries," *Chem. Eng. Progress*, 51 (December 1992).

Morrow, D.R.: "Recycling of Plastic Packaging Materials," *Food Technology*, 89 (December 1989).

Mullen, J.F.: "Consider Fluid-Bed Incineration for Hazardous Waste Destruction," *Chem. Eng. Progress*, 50 (June 1992).

Nathanson, A.A.: "Basic Environmental Technology: Water Supply, Waste Management, and Pollution," 2nd Edition, Prentice-Hall, Inc., Upper Saddle River, NJ, 1996.

Nathanson, J.A.: "Basic Environmental Technology," 3rd Edition, Prentice-Hall, Inc., Upper Saddle River, NJ, 1999.

Nemerow, N.L.: "Zero Pollution for Industry: Waste Minimization through Industrial Complexes," John Wiley & Sons, Inc., New York, NY, 1995.

Ostler, N.K.: "Introduction to Environmental Technology," Prentice-Hall, Inc., Upper Saddle River, NJ, 1995.

Ostler, N.K., M. Malachowski, and T.A. Byrne: "Health Effects of Hazardous Materials," Prentice-Hall, Inc., Upper Saddle River, NJ, 1996.

Ostler, N.K., J.T. Nielsen: "Waste Management Concepts," Prentice-Hall, Inc., Upper Saddle River, NJ, 1997.

Powell, C.S.: "Plastic Goes Green (Recycled Plastics)," *Sci. Amer.*, 101 (August 1990).

Pszozola, D.E.: "Bottle Manufacturer Operates Plastic Recycling Plant," *Food Technology*, 54 (January 1991).

Rathje, W.L., L. Psihoyos: "Once and Future Landfills," *Nat'l. Geographic*, 116 (May 1991).

Renko, R.J.: "Minimize Operating Costs in Meeting Fume Emission Control Standards," *Chem. Eng. Progress*, 47 (October 1990).

Staff: "Environmental Protection, Safety, and Hazardous Waste Management," *Chem. Eng. Progress*, 15 (December 1988).

Staff: "Elements of Toxicology," *Chem. Eng. Progress*, 37 (August 1989).

Staff: "Plastic Recycling Plant in Philadelphia," *Chem. Eng. Progress*, 10 (February 1990).

Staff: "Nylon Meshes Well with the Environment," *Advanced Materials & Processes*, 6 (July 1990).

Staff: "Penn State Opens Pilot Plant for Biotechnology Companies," *Chem. Eng. Progress*, 9 (September 1990).

Staff: "Effective Management of Food Packing: From Production to Disposal," *Food Technology*, 225 (May 1991).

Staff: "Process Pushes the Upside of Garbage," *Chem. Eng. Progress*, 12 (October 1991).

Staff: "Solvent Sorts Out Plastics," *Chem. Eng. Progress*, 22 (November 1991).

Strauss, W.: "The Haze Around Environmental Audits," *Technology Review (MIT)*, 19 (April 1992).

Testin, R.F., P.J. Vergano: "Food Packaging," *Food Review*, 31 (April–June 1991).

Theodore, L., Y.C. McGuinn: "Pollution Prevention," John Wiley & Sons, Inc., New York, NY, 1997.

Thibodeaux, L.G.: "The Four Natural Laws of Hazardous Waste," *Chem. Eng. Progress*, 7 (May 1990).

Tillman, D.A., A.J. Rossi, and K.M. Vick: "Rotary Incineration Systems for Solid Hazardous Wastes," *Chem. Eng. Progress*, 19 (July 1990).

Woodard, F.: "Industrial Waste Treatment Handbook," Butterworth-Heinemann, Inc., Woburn, MA, 2001.

WASTES AS ENERGY SOURCES. Initially, *biomass* was defined as the amount of living organisms in a particular area, stated in terms of the weight or volume of organisms per unit area or of the volume of the environment. This definition still applies very well to ecological and geophysical assessments of land areas or regions and depths of the seas and lakes.

In modern technology, the term also may be used to describe the exploitation of living terrestrial materials, such as plants or marine plants, all or parts of which may be combusted directly for the thermal energy that they yield or, more indirectly, as raw materials for processes that can convert the biomass into fuels.[1] Very generally, biomass may be considered the total amount of living matter within a given unit of area, volume, or mass. When biomaterials serve as foods or provide fibers and items of construction, for example, waste is created. These materials are typified by straw, sawdust, sewage sludge, and so on, which possess value as energy sources. That aspect of biomass is the topic of this article.

Generally, the pace of research and construction of facilities for transforming solid wastes into energy forms, such as methane, or to recover heat from combusting the wastes, slowed during the late 1980s and early 1990s. There are several causative factors, but of course a breakthrough could reverse these trends. In the 1970s, during the time of the oil embargo and energy crisis, finding new sources of energy was a major incentive. The fervor of the former energy programs has largely deteriorated in the presence of what is now considered by many a severe environmental crisis. This latter incentive has not been sufficient to foster extensive research programs in transforming garbage and other trash into energy because the *primary* advantages remain as an energy source and not as a means of pollution abatement. Nevertheless, progress has been made. Simply combusting municipal rubbish as a means of waste disposal is described in the preceding article.

Considerable research is being conducted on a variety of biomass feedstocks that contain cellulose, from which ethanol can be produced. Again, ethanol as a transportation fuel component has not been widely accepted, even though aggressively promoted in some areas. Brazil usually is cited as a prime example of progress in this area. L.R. Lynd (Dartmouth College) and a team of researchers have been studying the impacts of alternative fuel use on carbon dioxide (CO_2) accumulation, energy security, and economic effects on the United States as a whole. The team observes, "Production of ethanol from cellulosic biomass is believed to be an emerging energy technology with particularly great potential for the U.S. transportation sector. Research to improve conversion processes and to develop cellulosic energy crops is necessary to reduce costs and to increase production potential. Success can reasonably be expected in both these areas in light of the immature state of current technology and the powerful approaches available." In terms of their potential for yielding ethanol, in order of diminishing production potential, are agricultural wastes, forest sources, and municipal solid waste.

The majority of agricultural, commercial, industrial, and urban or municipal wastes are of a biological rather than mineral nature and thus fall under the umbrella of biomass. The simple burning of wood for heat illustrates one of the simplest ways to convert biomass to energy. All biomass represents an indirect form of solar energy. Biomass, as a source of energy, differs from coal, natural gas, and petroleum in one major way—biomass is renewable. Some potential biomass energy crops can be renewed as frequently as two or three times per year, depending upon location, while other materials such as trees have a renewable cycle of several years. Anthropogenic wastes are renewed on a daily basis. Interest in biomass over the last several years has stemmed from the overall concern with ultimate exhaustion of nonrenewable energy sources, as well as gaining a degree of political independence by many nations that either do not have any fossil fuel resources, or that have insufficient supplies to maintain a strong economic and industrial position.

Urban Wastes as Energy Sources

Urban waste includes household, sewage, commercial, institutional, manufacturing, and demolition waste. The availability of this waste is directly

---

[1] The generation of biomass from carbon dioxide is called "primary production" because it is the first fundamental step in turning inorganic material into organic compounds and cell constituents. This reduction of carbon dioxide uses sunlight as the source of energy. See also **Photosynthesis**.

related to the population living in urban areas of adequate size to support a given size system.

Manufacturing and processing wastes include all residuals generated from material inputs that leave the plant as product output. Office and packaging wastes associated with this sector are included in the urban waste sector. The majority of these wastes are from pulp and paper manufacturing, primary and secondary wood manufacturing, and the construction industry.

The energy recovery system selected dictates the extent that solid waste must be prepared. Some systems require nothing more than the removal of massive noncombustibles, such as kitchen appliances from the refuse, while other processes require extensive shredding, air classification, reshredding, and drying. In conjunction with fuel preparation, it is usually worthwhile to reclaim metals and glass for recycling.

One-stage shredding is often used to reduce waste to a nominal size as small as 1 inch (2.5 centimeters). When finer-sized fuel is required, a second shredding step is usually used after air classification has removed many of the noncombustibles. Both vertical and horizontal air classifiers depend on the heavy noncombustibles settling out by gravity in a moving air stream, while the lighter combustibles are pneumatically transferred through the air classifier. Denser combustibles, such as rubber and leather, may be removed with the heavy fraction, while some of the fine glass and metal foils are carried with the combustibles. Thus, desired separation may not always be achieved on one pass through an air classifier.

Some energy recovery systems require drying to remove excess moisture in the waste. This is required when sewage sludge is used as a fuel. Usually, waste heat from the total process can be used for the drying system.

Pyrolysis. In one system, municipal refuse is charged at the top of a shaft furnace and is pyrolyzed as it passes downward through the furnace. Oxygen enters the furnace through tuyeres near the furnace bottom and passes upward through a 1425 to 1650 °C combustion zone. The products of combustion then pass through a pyrolysis zone and exit at about 93 °C. The off-gas then passes through an electrostatic precipitator to remove flyash and oil formed during pyrolysis. The latter are recycled to the furnace combustion zone. The gas then passes through an acid absorber and a condenser. The clean fuel gas has a heating value of about 300 Btu/cubic foot (2670 Calories/cubic meter) and a flame temperature equivalent to that of natural gas. The solid waste that remains is a slag at the furnace bottom.

Biological Methane Production. This process involves the anaerobic digestion of a solid waste and water or sewage sludge slurry at 60 °C for five days to produce a methane-rich gas. Solid waste is prepared by shredding and air classification, followed by blending with water to produce a mixture of 10 to 20% solids concentration. The slurry is heated and placed in a mixed digester at 60 °C for 5 days detention. The digester gas is drawn off and separated into carbon dioxide and methane. The spent slurry from the digester is pumped through a heat exchanger to partially heat the incoming slurry prior to filtration. The filtrate is returned to the blender and the sludge is used for landfill. Heat addition to the refuse slurry is required to maintain the required digester temperature. The process is well suited for use on sewage sludge, animal manures, and other high-moisture-content solid wastes. It is estimated that the process can reduce the volume of volatile solids by 75%, while producing about 3000 cubic feet (85 cubic meters) of methane per ton of incoming solid waste. The major residue is used for landfill or incinerated. About 10% of the methane is required to heat the digester feed.

Direct Steam Process. One process uses a rotary kiln pyrolizer followed by an afterburner and boiler to produce steam from shredded waste. The pyrolysis process in a kiln is operated countercurrently. Solid waste enters at one end and pyrolyzed residue is discharged at the other. External fuel and air are introduced at the residue discharge area and combustion products and pyrolysis gases leave the kiln at the feed opening. This arrangement causes the solid waste to be exposed to progressively higher temperatures as it passes through the kiln. The kiln off-gases pass through a refractory-lined afterburner into which air is introduced to allow complete combustion prior to passing through the waste heat boiler. A wet scrubber is used for air pollution control, while an induced draft fan is used to draw the gases through the system. One ton of solid waste, augmented by 1.25 million Btu (0.3 million Calories) from auxiliary fuel and 55 kilowatt-hours of electricity will produce about 4800 pounds (2177 kilograms) of steam at 330 psig (22.4 atmospheres) along with 200 pounds (91 kilograms) of char.

Waterwall Incinerators. These devices generate steam by burning unprepared solid waste on a grate and passing the hot products of combustion through a boiler. Numerous waterwall incinerators have been built in Europe and the United States. Unprepared refuse is taken from storage pits and charged directly into the incinerator feed hopper. From there, the refuse drops onto a feed chute and then is fed automatically onto the stoker by means of a hydraulic feed ram. Temperatures in the 870 °C range effectively burn the solid waste. Before the flue gas enters the boiler, secondary air is added to produce a temperature near 1090 °C. The boiler is constructed of membrane waterwalled tubes with extruded fins. After passing through the boiler, the gases travel through an economizer section and then into an electrostatic precipitator for particle removal. A typical 1000 tons/day (900 metric tons/day) waterwall incinerator produces about 300,000 pounds (136,080 kilograms) of steam per hour.

Additional engineering is required where toxic wastes may be present.

Principal Biomass Materials for Energy

Biomass-to-energy systems fall into two principal categories: (1) materials for direct combustion that will generate heat for processing, for warming living and working spaces, for steam and hence also for generating electricity; and (2) materials from which both fuels and chemicals can be obtained through biochemical or thermochemical conversion processes. Resulting fuels must have ample caloric content per unit of weight and thus rich in carbon and hydrogen and poor in the content of atoms, such as oxygen and nitrogen, which do not contribute to the caloric value of the fuel. In searching for new biomass raw materials, scientists have found it helpful to study the various biosynthetic pathways followed by plants from seed to maturation.

Direct Wood Burning. Wood was the major fuel of the United States until about 1886 when the consumption of coal equaled that of wood. Oil did not appear on the chart until about 1900 and gas in 1910–1920. The use of wood tapered off while other fuels climbed at amazing rates, but wood never ceased as a factor, even if small. It is interesting to note that about a million modern woodburning stoves are in use and that about 40% of the wood products industry is furnished by combusting bark and mill wastes. This amounts to about 1 quad (10^{15} Btu). Wood burning has been sufficiently extensive during the past few years to cause environmental concerns in some regions. Although wood has a low sulfur content and produces minimal amounts of nitrogen oxides even by the hottest fires (1370 °C; 2500 °F), it contains air pollutants in the form of particulates, gases, and tars. Environmentalists in the New England region have estimated that the 300,000–400,000 tons of wood burned per year (New Hampshire only), if the fuel is very dry red oak, will add 1000 tons of particulates to the air; if dry white pine is burned, the total may be over 5000 tons. Since a mixture of woods usually is used, the figure lies somewhere between the two aforementioned quantities.

Agricultural Wastes. In the absence of an energy crisis, with a few exceptions, attention to the use of agricultural wastes has waned. These materials continue to represent an essentially untouched source of energy in most countries. In the course of extensive scientific research in the early 1970s, at a time when renewable energy sources were being aggressively sought, evaluations were made of various materials for their caloric content and availability. Included were corn (maize), sorghum, wheat, sugar beats, sugarcane, pineapple, and cassava, among others. Several of these crops were considered as sources for the production of alcohol for admixture with gasoline (gasahol) as an automotive fuel. Brazil has had considerable success in this regard with sugarcane as a raw material.

Although burning and incineration offers means for disposing some waste products, these processes do contribute to air pollution and thus must meet the requirements now followed for the common fossil fuels.

Additional Reading

Abelson, P.H.: "Improved Yields of Biomass," *Science*, 1469 (June 14, 1991).

Corcoran, E.: "Dirty Business: How Companies are Seeking Their Fortunes in Garbage," *Sci. Amer.*, 98 (September 1989).

Klass, D.L.: "Biomass for Renewable Energy, Fuels, and Chemicals," Harcourt Brace & Company, San Diego, CA, 1998.

Kumer, R., J.K. Van Sloun: "Purification (of Methane) by Adsorptive Separation," *Chem. Eng. Progress*, 34 (January 1989).

Lynd, L.R. et al.: "Fuel Ethanol from Cellulosic Biomass," *Science*, 1318 (March 15, 1991).

Monoastersky, R.: "Biomass Burning Ignites Concern," *Science News*, 196 (March 31, 1990).

Rowell, R.M., T.P. Schultzand, and R. Narayan: "Emerging Technologies for Materials and Chemicals from Biomass," American Chemical Society, Washington, DC, 1992.

Saha, W.C., J. Woodward, and B.C. Saha: "Fuels and Chemicals from Biomass, Vol. 666, American Chemical Society, Washington, DC, 1997.

Staff: "Waste Management: Technology Cuts through Emotional Myths," *Westinghouse Technology*, 15 (October 1988).

Turbak, G.: "Woodburning's New Age," *Amer. Forests*, 52 (November–December 1989).

Wereko-Brobby, C.Y., E.B. Hagen: "Biomass Conversion and Technology," John Wiley & Sons, Inc., New York, NY, 1996.

Wright, J.D.: "Ethanol from Biomass by Enzymatic Hydrolysis," *Chem. Eng. Progress*, 62 (August 1988).

Wyman, C.E.: "Handbook on Bioethanol: Production and Utilization," Taylor & Francis, Inc., Philadelphia, PA, 1996.

WASTEWATER TREATMENT.　See **Water Pollution**.

WATER.　A colorless (blue in thick layers) liquid, H_2O, odorless, tasteless, melting point $0\,°C$ (one of the standard temperature points), boiling point $100\,°C$ at 760 millimeters of mercury pressure (another standard temperature point).

The boiling point of water increases with increasing pressure: $100.366\,°C$ at 770 mm; $120.6\,°C$ at 1,520 mm; $180.5\,°C$ at 7,600 mm. The boiling point decreases with decreasing pressure: $99.360\,°C$ at 750.0 mm; $99.255\,°C$ at 740.0 mm; $98.877\,°C$ at 730.0 mm; $81.7\,°C$ at 380 mm; $46.1\,°C$ at 76.0 mm.

At $0\,°C$, the density of water is 0.99987 gram per milliliter. At $8\,°C$, 0.99988; at $15\,°C$, 0.99913; at $16\,°C$, 0.99897; at $17.5\,°C$, 0.99871; at $20\,°C$, 0.99823; at $25\,°C$, 0.99707; at $40\,°C$, 0.99224; at $50\,°C$, 0.99807; at $75\,°C$, 0.97489; at $100\,°C$, 0.95838; at $120\,°C$, 0.9434.

The critical temperature of water is $374.15\,°C$; critical pressure, 218.4 atmospheres; critical density, 0.323 gram per cubic centimeter.

The viscosity at $0\,°C$ is 0.01792 poise (dyne second per square centimeter), specific viscosity, 1.000. At $20\,°C$ the viscosity is 0.01005 poise, specific viscosity, 0.561. At $50\,°C$, the viscosity is 0.00549 poise, specific viscosity, 0.307. At $75\,°C$, the viscosity is 0.00380 poise, specific viscosity, 0.212. At $100\,°C$, the viscosity is 0.00284 poise, specific viscosity, 0.158.

The surface tension of water against air at $0\,°C$ is 75.6 dynes per centimeter. At $10\,°C$, the surface tension is 74.22; at $20\,°C$, 72.75; at $30\,°C$, 71.18; at $60\,°C$, 66.18; and at $100\,°C$, 58.9.

The specific heat of water is 1.000000 at $15\,°C$ (standard of specific heat). At $0\,°C$, the specific heat is 1.00874; at $25\,°C$, 0.99765; at $35\,°C$, 0.99743 (minimum); at $50\,°C$, 0.99829; at $65\,°C$, 1.00001; at $80\,°C$, 1.00239; at $100\,°C$, 1.00645; at $120\,°C$, 1.016; at $180\,°C$, 1.04.

The electrical conductivity of water at $18\,°C$ is 0.04×10^{-6} reciprocal ohms (measurements of Kohlraush and Heydweiller, 1902); of pure water in equilibrium with air, 0.8×10^{-6}; of ordinary distilled water, about 5×10^{-6}.

The dielectric constant of water (specific inductive capacity) is 81.07 at $18\,°C$.

Pure water, when free of dissolved gases, may be heated above $100\,°C$ (even up to $180\,°C$) without boiling, but upon further heating, boiling with explosive violence may occur. Steam at $100\,°C$ occupies a volume 1,700 times greater than water at $100\,°C$. Pure water, when not agitated, may be cooled somewhat below $0\,°C$ without freezing, but upon further cooling it congeals with an increase of volume (density of ice, 0.917) exerting great force, when confined, but if in intimate contact with water at atmospheric pressure, the freezing temperature is $0\,°C$. The vapor pressure of ice and water is 4.579 millimeters at 1 atmosphere pressure and $0\,°C$. The triple point (ice, water, and water vapor) occurs at a saturation vapor pressure of 6.11 millimeters and at a temperature of 273.16 K. When water is compressed to about 20,000 atmospheres and then cooled, other varieties of ice, all denser than water, are formed. Ice II is 12% denser; Ice III is 3% denser. At least six varieties of ice are known.

In ice I, ice II, and ice III, each oxygen atom is surrounded tetrahedrally by four other oxygen atoms, the difference between the three forms being largely in some distortion of the linkages, since the O—O distance varies little from the 2.76 value of ice I. Each oxygen atom has 2 hydrogen atoms quite close (about 1 Å) to it. At lower temperatures, and presumably higher pressures (forms V, VI, and VII, the water molecules with four hydrogen bonds are more in evidence.

Liquid water exhibits the same tendency toward increased bonding at lower temperatures. While individual water molecules have a nonlinear structure, there is association between H_2O molecules by hydrogen bonding, the degree of association being greater at lower temperatures. Based upon the statistical mechanical treatment of Frank and Wen, liquid water may be regarded as a mixture of hydrogen bonded clusters and unbonded molecules. Other research characterizes this model in terms of 5 species: unbonded molecules, tetrahydrogen bonded molecules in the interior of cluster; and surface molecules connected to the cluster by 1, 2, or 3 hydrogen bonds.

The chemical properties of water change with temperature at high temperatures. The reaction, $2H_2O \leftrightarrow 2H_2 + O_2$, shows an appreciable shift to the right, reaching 0.8% at $2,000\,°C$, and increasing rapidly above that temperature. At ordinary temperatures, the equilibrium, $2H_2O \leftrightarrow H_3O^+ + OH^-$, is important because it enables water to act either as a proton donor or acceptor. With stronger acids, water can act as a proton acceptor:

$$HCl + H_2O \longleftrightarrow H_3O^+ + Cl^-$$

$$HBr + H_2O \longleftrightarrow H_3O^+ + Br^-$$

$$HSO_4^- + H_2O \longleftrightarrow H_3O^+ + SO_4^{2-}$$

With stronger bases, water can act as a proton donor:

$$H_2O + CO_3^{2-} \longrightarrow HCO_3^- + OH^-$$

$$H_2O + NH_3 \longrightarrow NH_4^- + OH^-$$

Although the ions shown are written as CO_3^{2-}, HCO_3^-, Cl^-, etc., they of course are more or less solvated by the water, i.e., they have water molecules attached to them by ion-dipole bonds, since water is a polar compound. One of the most strongly marked properties of water is its behavior as an electrolytic solvent, which is due to its high dielectric constant. The energy of separation of two ions is an inverse function of the dielectric constant of the solvent. Some of the parameters of water are shown in diagram of Fig. 1.

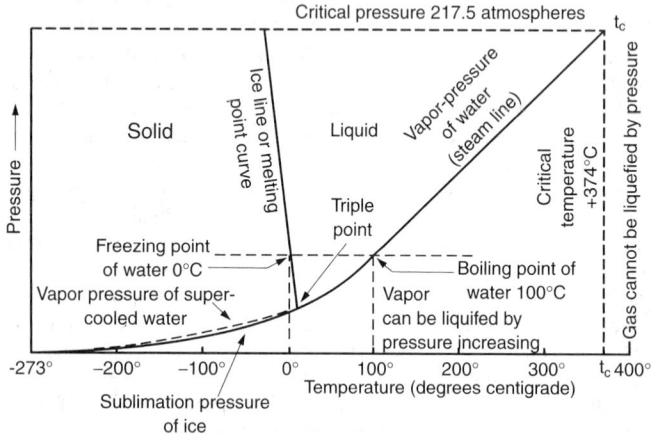

Fig. 1.　Pressure-temperature diagram for water.

Molecular Architecture. Greatly oversimplified, the water molecule may appear as shown in Fig. 2, which indicates the equilibrium position of the oxygen atom and the hydrogen atoms, i.e., the equilibrium position of the positive and negative charges of the molecule. Because of this orientation, the water molecule has a strong tendency to be oriented in an electrical field. The dipole moment depends upon the magnitude of the charge separation within the molecule, and in the water molecule, the separation is large. Thus, water may be described as having an exceptionally large dipole moment and consequently a large dielectric constant. On the basis of ascribing a dielectric constant of 1 for a vacuum, the dielectric constant of water is 80; i.e., in water, 2 electrical charges will attract or repel each other with only 1/80th as much strength as would be the case in a vacuum. This accounts, at least in part, for the remarkable ability of water to dissolve substances, particularly materials whose molecules are held together primarily by ionic bonding. The bonding arrangements within the water molecule also account for the exceptional cohesive power exhibited in water's high surface tension and the outstanding ability of water to adhere strongly to a variety of materials (the property of wetting). Bonding also accounts for the manner in which water crystals, e.g., snowflakes, are formed and for the maximum density of water (4 °C), below which water assumes less dense forms, causing ice to float. See also **Clouds and Cloud Formation**; and **Precipitation and Hydrometeors**. Bonding is responsible for the exceptional heat capacity and exceptionally high latent heats of fusion and evaporation of water.

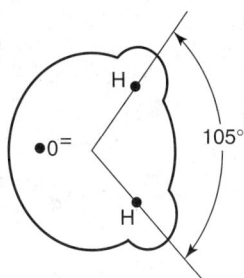

Fig. 2.   Schematic representation of arrangement of electrical charges in water molecule.

The molecular behavior of various molecular types in electrolytes is shown in Fig. 3. This behavior is of particular significance to the role of water in biological systems. See also **Molecule**.

Although of continuing interest, research on the structure of water is exceptionally difficult. X-ray crystallography has provided a good picture of the stable hydrogen-bonded structure of ice, but thus far the structural imaging of liquid water has escaped investigators, and hence blackboard theorizing and computer modeling are the principal research techniques followed. In the *Journal of the American Chemical Society* (*JACS*) of May 20, 1992, water scientists present a complex new theory that may account for so many of the physical properties of liquid water.

Heavy water, also known as deuterium oxide, D_2O, is water in which the hydrogen of the water molecule consists entirely of the heavy-hydrogen isotope having a mass number of 2. The density of heavy water is 1.1076 at 20 °C. Heavy water has been used as a moderator in nuclear reactors as well as a coolant.

Uses. Water is such a common substance that its importance and versatility are usually taken for granted. Included among the major ways in which water is important would be: (1) as a raw material for incorporation into final products without chemical change; (2) as a raw material for undergoing chemical change; (3) as a transport and conveyance medium with water acting as a solvent or carrier of solutions and suspensions in and out of reactions and physical-change operations—at an industrial as well as biochemical level; (4) as a heating and cooling medium over the wide temperature range from below normal freezing temperature (brine solutions, for example) to those of superheated steam; (5) as an energy-storage medium; (6) as a gathering medium for waste products; (7) as a cleaning medium; (8) as a shield against heat and nuclear radiation (heavy water); (9) as a convenient standard in terms of temperature, density, viscosity, and other units; and (10) with exception of a few situations where the presence of water is hazardous, as a fire-fighting medium.

Water Metabolism in Vertebrates. Those vertebrates that now inhabit land, seas, brackish and fresh waters have survived because they have developed homeostatic mechanisms that enable them to cope with considerable variation in the content and availability of water, sodium, potassium, and chloride in their external environment. These mechanisms prevent life-threatening changes in their internal environment by (1) assuring that the cells are bathed by fluid with the same osmotic concentrations as themselves; and (2) by preventing major qualitative changes in the intra- and extracellular content of these ions or water. Regardless of species, one is impressed not by the differences, but by the similarities in the ionic composition of their intra- and extracellular fluids. The water content of the fat-free tissues of all vertebrates ranges between 70 and 80%. Water diffuses freely along its concentration gradient (osmosis) throughout all body tissues. Therefore, any deviation of the osmotic pressure of intra- or extracellular fluids, by either withdrawal or addition of water, causes an immediate movement of water from the more dilute to the more concentrated solution until osmotic equilibrium is reestablished.

Water is lost from the body of mammals by evaporation across the skin and in the expired air, urine, and feces. The more arid the environment, the more a mammal must be able to reduce water loss and tolerate longer periods of water dehydration and hypertonicity of its body fluids.

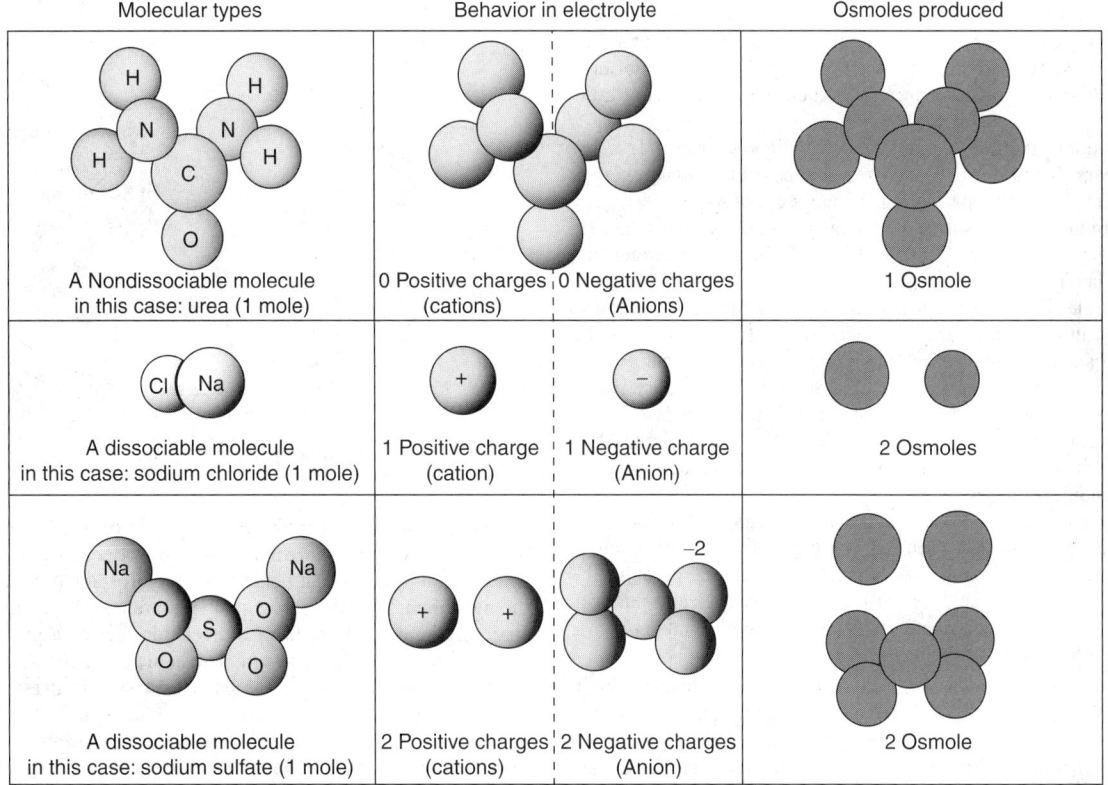

Fig. 3.   Behavior of various molecular types in electrolytes. (*After Maffy.*)

According to Chew, vertebrates fall into several groups in terms of how they maintain their water balance. *Fishes and amphibians* in fresh water are very hypertonic to their medium and must counteract a continual dilution of their body fluids. Water influx is reduced by the relative impermeability of the skin, and is balanced by diuresis. Electrolytes lost in this urine are replaced in food eaten and by absorption through gill surfaces (fishes) and skin (amphibians). *Marine elasmobranches* are unique in maintaining themselves slightly hypertonic to sea water by retention of urea (2±%) in their body fluids, making their osmotic pressure more than twice that of marine teleosts. *Marine teleosts* and terrestrial tetrapods face the continual problem of counteracting desiccation due to osmotic loss to a hypertonic medium or to evaporation. Mammals are the most effective of vertebrates in conserving urine water by concentrating the urine, which is achieved by reabsorption of water in the kidney tubules.

Terrestrial tetrapods adjust by avoidance of evaporative stress, reduction of evaporative and urinary water losses, and temporary toleration of hyperthermia or hypernatremia. Antidiuretic hormone (ADH) from the neurapophysis is very important in enhancing uptake of water through the skin (amphibians), reduction in glomerular filtration (amphibians, reptiles, birds), and increase in tubular reabsorption of water (mammals).

Water balance processes are best developed in species inhabiting deserts, where little drinking water is available and climatic conditions accentuate evaporation.

Certain toads and frogs survive in deserts, needing open water only for breeding, largely by remaining dormant during dry periods. Evaporation is greatly retarded in a cool damp burrow, and urine volume is reduced by 98–99% (filtration antidiuresis), but urine remains hypotonic. Urinary water may be recycled through the body by reabsorption from the bladder. Dormant animals tolerate a loss of 50–60% of their body water. They emerge during rains, and in their dehydrated state quickly reabsorb water through the skin.

Terrestrial reptiles also avoid considerable evaporation by being quiescent in burrows much of the time. Also, their skin is more impermeable than that of amphibians, although water is still lost in expired air. Hydrated lizards have low urine filtration rate (urine always hypotonic), and may become almost anuric when dehydrated. During dehydration, electrolyte wastes are retained in the body and tolerated in concentrations fatal to birds and mammals, until water is available for their excretion. A carnivorous diet (70±% water) provides adequate water intake while food is available. Water can be reabsorbed osmotically from the cloaca, reabsorption being particularly effective because of the nature of the principal nitrogenous waste, uric acid, which has a very low solubility. As uric acid precipitates in the cloaca, its osmotic effect is removed, and further water can then be absorbed by osmosis. This is probably the major value of uric acid excretion. Precipitated wastes are excreted en masse, with very little fluid loss.

Birds, being homeothermic, cannot reduce their evaporative loss by becoming dormant. Being diurnally active and exposed to radiant energy, they must often expend water for cooling, by panting. Consequently, in arid regions the distribution of birds is limited to areas within flying distance of water. Some water expenditure is avoided by allowing hyperthermia (up to 3 °C) in the daytime.

Desert rodents lead the most water-independent life of all vertebrates. Kangaroo rats can so reduce their evaporation that they are able to maintain water balance on only metabolic water. Other species survive on only metabolic water plus free water in air-dry seeds. Respiratory water loss is reduced by cool nasal mucosal surfaces, which condense water from warm air coming from the lungs, before it can be expired. Skin impermeability involves a physical vapor barrier in the epidermis, plus unknown physiological factors.

Many larger mammals are exposed to daytime radiant energy and need to dissipate heat by sweating, panting or wetting themselves with saliva (marsupials). These water expenditures must be balanced periodically by drinking. A dehydrated camel is particularly physiologically adapted to store heat (rather than dissipate it by evaporation), undergoing a temperature rise of up to 6° in the daytime.

Water in the Human Body. The adult male human body contains about 60% (weight) of water and the adult female body about 50%. The large amount of body water is compartmentalized, each compartment being bounded by membranes. It has been estimated (Edelman and Leibman, 1959) that 55% of this water is contained within cells and that it is bounded by cell membranes. The remaining, extracellular water or fluids

(ECF) is made up of a relatively small volume of plasma (7.5% of total body water in the vascular tree), with the remaining 37.5% in nonplasma and located outside the vascular tree. The latter includes interstitial water (20%), another 15% in bone and dense connective tissue, and 2.5% in secretions. These numbers are shown graphically in Figs. 4 and 5.

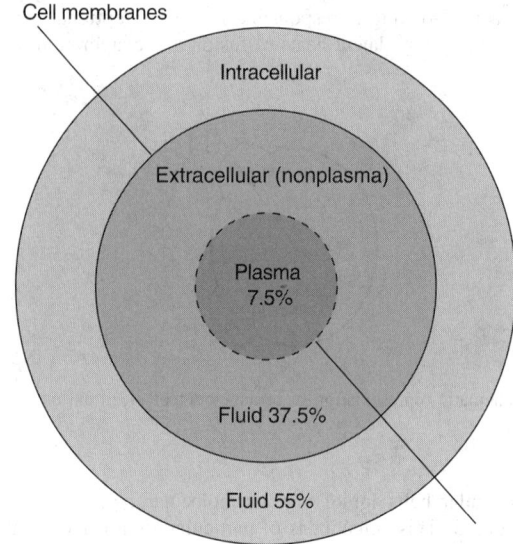

Fig. 4.   The three principal categories of body water.

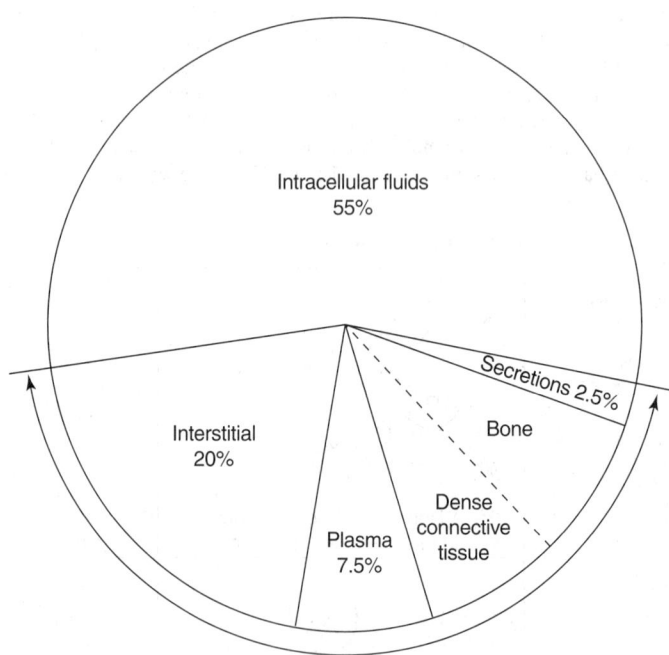

Fig. 5.   Pie chart showing approximate volumetric proportions of body fluids.

Two driving forces control the movement of water in the body, namely, hydrostatic pressure and osmotic pressure. Because transmembrane pressures are so low, it is not believed that hydrostatic pressure plays a role in the movement of water across cell membranes. On the other hand, hydrostatic pressure resulting from heart action creates a gradient of about 20 millimeters of mercury pressure across the capillary walls. The principles of osmosis are described in the entry on **Osmotic Pressure**. Normally, in describing osmotic pressure, reference is made to salt solutions of differing concentrations separated by a membrane. Concentrations are expressed in terms of solute in solvent (water). When thinking in terms of body water, one usually considers the addition of solute to the water as a dilution of the pure water rather than as an increase in solute content of the water. For example, pure water contains 55.5 moles per kilogram (about 55,500 mmoles/liter). Body fluids, that is, intracellular fluid (ICF) and extracellular fluid (ECF) contain approximately 99.5% water molecules and 0.5% solute molecules. This equals about 55,200 mmoles of water per liter; and 300 mmoles of solute per liter. Osmolality is a measure of the concentration of osmoles present in the water. If the osmolality is low, the concentration of water is high; if the osmolality is high, the concentration of water is low. Movement of water across the cell membranes occurs because of differences in concentration, the movement always being from a phase of high water concentration to one of lower concentration. Movement is "downhill" so to speak and the motivating force is called osmotic pressure. This movement can be counteracted by the application of an opposing force, notably hydrostatic pressure.

It is interesting to note that the main cation present in ICF is the potassium ion, whereas the principal cation in ECF is the sodium ion. The role of potassium and sodium ions in the biological system is described in the entry on **Potassium and Sodium (In Biological Systems)**.

Water Intoxication. Either an increased intake of water or a decreased output of water can cause an excess of body water. Because healthy kidneys have outstanding ability to increase water excretion, a condition of water intoxication usually occurs because of a disability to excrete water (*hyponatremia*). Frequently impairment may not be the result of disease or damage to the kidneys per se, but rather due to faulty processing of stimuli by the kidneys. Excessive renal reabsorption of water can result from the action of antidiuretic hormone (ADH) or by the excessive reabsorption of sodium in the proximal tubules. Under such conditions, the excretion of water and sodium will be low. Retention of salt and water causes an expansion of the ECF, usually resulting in edema and effusions. See also **Kidney and Urinary Tract**.

Water Deficiency. This condition occurs when water output exceeds intake. Water is continually lost by way of the lungs, skin, and kidneys and thus a deficiency of body water will occur if a critical minimal supply is not maintained. Decreased intake when water is available is uncommon. Very rarely, a brain malfunction may interfere with one's sense of thirst. Increased output of water can result from many causes. For example, a person with diabetes insipidus who lacks ADH (antidiuretic hormone) or a person whose kidneys do not respond normally to ADH, as in instances of nephrogenic diabetes insipidus, will increase water output. Other diseases which may cause excess excretion of water include osmotic diuresis, hypercalcemia, hypokalemia, chronic pyelonephritis, and sickle cell anemia, among others. Excessive water losses are also experienced in some cases with advanced age and in some burn cases. Two clinical features are good measures of dehydration—weight loss of the patient and an elevation of the serum sodium concentration. In situations of dehydration, the body initiates mechanisms which manipulate the transfer of water from one compartment to the next, retaining water in those cells and organs where it is most needed.

In cases of severe dehydration, rehydration must be brought about carefully and in steps. The usual practice is to make an initial estimate of water or electrolytes that require replacement and then to administer one-half the amount of the deficit, after which another series of measurements is made, followed by replacement of one-half of the estimate—until a satisfactory ultimate balance has been attained. If rehydration is not gradual, some organs, such as the brain, may take up water beyond normal requirements and this can result in cerebral edema. Also, in the case of acute dehydration, permanent brain damage can occur as the result of a shrinking brain tearing away vessels, causing cerebral hemorrhages.

Water and Macromolecules. Over eons of time, processes have appeared for making the great variety of macromolecules required by living organisms. Most of these developments have occurred in a water medium, or in one having a high water vapor content. So it is not surprising that the majority of the macromolecules involved in the life processes are hydrophilic, in different degrees. One may find minor exceptions in the cases of certain fats and lipids. It should be noted that in living organisms, the hydrophilic macromolecules of one kind often join with those of other kinds to produce the useful structures required in the life process, i.e., membranes.

Decades ago, biochemists recognized that it was difficult to remove all water from a large number of macromolecular materials. The term "bound water" was coined to explain the great affinity many of these materials showed for water, particularly the proteins. Biochemists were convinced that biological behavior, at least in part, resulted from the amount of "bound water" contained in the macromolecular structure. As an example, water held in plant structures so that it did not freeze in below-freezing temperatures was considered to be bound. At the time these ideas found favor, the method of lyophilization or quick freeze-drying had not been perfected. Lyophilization removes the bulk of the water held in biological substances without destroying their structures or their activity.

In more recent years, lyophilized proteins have been further dried to a constant weight in a high vacuum and then studied as they adsorbed water vapor. The heats of adsorption of the first water vapor molecules were considerably higher than the values obtained as the adsorption approached that of the saturated vapor. These results indicated that the first molecules to be adsorbed were on the most water-loving or active sites. Such adsorbed molecules on the higher-energy sites would be desorbed last in a high vacuum. Without going into considerable detail, the results of this line of research led to the conclusion that the earlier concept of "bound water" was unfounded.

Many biological systems depend in part on their degree of hydration. Most biological membranes are hydrophilic, but should the membrane be a multi-layered one, the hydration of the different layers may well differ markedly. A great many of the membranes used by living organisms are known to be selective in what passes through them. For many years the theory of the role of some membranes in pumping water into the cells they surround even against high osmotic gradients due to salt concentration has maintained among most biologists. The fascinating field of membrane biophysics has shed much light concerning the hydration of biological macromolecules.

As pointed out by Kolata, a primary difference between living and dead cells is that living cells selectively retain certain ions, such as potassium, and exclude others, such as sodium. The water in dead cells reflects the conditions of the solution around them. The conventional explanation is that this difference is due to ion "pumps" in membranes, pumps purported to use cell energy to transport some ions into and other ions out of the cell. There is another school of thought, however, which denies that such pumps exist. This school claims that ions are excluded from cells on the basis of their low solubilities in cellular water, except when specific charged sites with which the ions can associate are available. It is maintained that cell water has a different structure than either liquid water or ice and that it is this special structure that affects the solubility of various ions in it. This school also suggests that there is evidence that ion pumps are thermodynamically impossible, requiring more energy than is available to the cell. The school also claims that nuclear magnetic resonance (NMR) studies show that cell water is more structured than liquid water and less structured than ice. This, it is believed, would affect the solubilities of ions in the cell and could account for selective ion exclusions. A number of investigators, including advocates of pumps, have agreed that cell water may have some ordered structure that makes it different from liquid water. Most investigators do not question that cell water is likely to be structured, but do ask to what extent it is structured and what the physiological importance of this structure may be.

Degassed Water. It is interesting to note that during the past several years, Russian scientists have raised some intriguing points concerning water, not all of which have held up under intensive scrutiny. Some readers may recall the discussions of the early 1970s concerning the "discovery" of polywater or so-called anomalous water by some Russian scientists. Observations of this water did not meet with accepted criteria for physical properties. For a while, it was considered by some scientists to be of a polymeric nature, but as the result of subsequent numerous exchanges of views and a careful scrutiny of the water it was found to be ordinary water which had a large concentration of dissolved minerals.

In 1978, other Russian scientists proposed that meltwater (from freshly melted snows) carries certain biological properties not known in ordinary water. The Russian scientists theorized that meltwater retains some of the order that is characteristic of frozen water and that this increased order alters vital reaction rates within cells. This tends to tie in with the concept of structured water discussed briefly later.

In the course of investigation at the Institute of Fruit-Growing and Vine-Growing (Kazakhstan, Russia), investigators tested the relative growing rates, qualities, and yields of various plants subjected to meltwater, tap water, and boiled water. Although the plants responded in a superior way to the meltwater, as compared with tap water, it was found (more or less accidentally) that the plants responded even better to quickly cooled boiled water. Thus, the experimenters concluded that the superior action on plants derived from the fact that some of the waters tested (meltwater and quickly-cooled boiled water) contained less dissolved gases. In other words, the claim was made that any degassed water is superior when administered to plants. Igor Zelepukhin suggests that the conductivity of degassed water is decreased considerably and there are comparative increases in density, viscosity, surface tension, energy of intermolecular interaction, and internal pressure, factors that may enhance the value of water. Zelephuhkin thus observed further that degassed water bears a closer resemblance to the fluid in cells than does ordinary tap water. Experiments in utilizing degassed water for cattle and livestock are proceeding. Other Russian investigators have observed that concrete prepared with degassed water is from 8 to 10% stronger than when ordinary water is used. It should be stressed that, as of the early 1990s, these observations have not as yet been accepted by the general scientific community. Considerably more research is required toward the development of hard, convincing data.

Raw Water and Treatment

Aside from desalinators used in some regions of the world that have severe insufficiencies of rain and freshwater and thus must depend upon purified saline waters, drinking water and the water required by industry for a plurality of reasons come from two classes of natural sources. The first is *surface waters* from ponds, streams, rivers, lakes, waterfalls, and glaciers. Natural water precipitation (rain and snow) is the result of Earth's natural hydrologic cycle. Precipitation also reaches below the surface to collect or flow in aquifers and thus is referred to as groundwater. See also **Climate**; **Desalination**; and **Hydrology**.

Prior to the great increases in world population and the development of extensive industrialization and modern agriculture, nature provided a number of built-in processes by which flowing streams essentially can be self-purifying, provided, of course, that pollution of any significant magnitude or intensity did not reach the natural water source. Today, with the exception of very few locales, such pristine conditions do not occur, and, in fact, some form of raw water treatment has existed for nearly five centuries. Traditionally called *waterworks*, or *pumping plants*, the earliest plant is believed to have been one designed by Peter Maurice and located near London Bridge in 1562. Its capacity exceeded 300,000 gallons (1136 cubic meters) for furnishing water to London. The first municipal plant installed in the United States was built in Boston in 1652. Only 15 more plants had been built in the United States by 1800, all located in the northeastern part of the country. The presence of bacteria in public water supplies was the major incentive for treating as well as filtering and pumping water to large municipalities. In terms of the treating operations performed at water treating plants, treating chemicals (including chlorine introduction methods) have grossly improved, and the various processes have increased in size and numbers to handle much greater quantities of water. One of the major factors that distinguishes the water treating plants of the last few decades from their earlier counterparts is the availability of much more sensitive instruments to measure water quality and the automatic control means introduced to accelerate processing reaction time. Public concern and the consequent great expansion of regulatory requirements are the result of raw water supply pollution. Water consumers today also are much more demanding for controlling water taste and odor. Controls over the introduction of toxic wastes (see also **Wastes and Pollution**) have been mandated.

In an average situation, water will be pumped from a natural source, passed through bar screens to remove any large pieces of debris, and moved on to a raw water storage tank for temporary holding. Treating boiler feedwater requires customized methods. See also **Feedwater (Boiler)**.

Raw water treatment plants are found in nearly all municipalities throughout the U.S., Canada, and other advanced countries. Traditionally, these plants have operated under strict regulations and guidelines. Additionally, some industrial firms will re-treat municipal water for a variety of special reasons. Ultrapure water, for example, is required by the semiconductor industry, and means used to obtain ultrapurity include reverse osmosis processes, decarbonators, distillation, deionization, ultraviolet sterilization, and ultrafiltration.

Principal Impurities. Substances that must be removed in order to meet required standards for municipal water fall into three general categories:

1. Suspended matter, color, and organic matter — sediment, turbidity, microorganisms, taste, odor, and other organic matter.
2. Dissolved mineral matter — bicarbonates, sulfate, chlorides of calcium, magnesium, and sodium. Small amounts of silica and alumina

commonly are present. Other constituents frequently present are iron, manganese, fluorides, nitrates, potassium, and sulfuric acid. (The presence of mine drainage components [common in some regions] and components of trade wastes are less frequently encountered today because of severe pollution restrictions. Severely toxic wastes also are controlled, with current regulations in most advanced industrialized nations mandating pretreatment to eliminate or neutralize all such substances prior to disposing of them in public reservoirs. However, municipal water treatment plant managers require the analysis of incoming raw water for the presence of toxic substances.)
3. Dissolved gases — Usually, these are present as oxygen, nitrogen, carbon dioxide, and, less frequently, hydrogen sulfide and methane.

Major Water Treating Operations

The principal operations required to remove and alleviate the undesirable components of raw water include the following.

Sedimentation. With waters containing large amounts of coarse, easily settled, suspended matter, sedimentation (plain sedimentation) is often of value in reducing the load on the filters and effecting economies in amounts of chemicals used for coagulation. Sedimentation may be carried out in sedimentation tanks or basins or in reservoirs. Detention periods vary over a wide range — from a few hours up to one or more months.

Coagulation. Coagulation is employed to form, by cataphoresis and entanglement, larger aggregates with the turbidity, color, microorganisms, and other organic matter present in the water. These larger particles, known as the "floc," may then be removed by filtration through a sand or Anthrafilt filter or by settling and filtration. The coagulant employed is either an aluminum or iron salt, usually the surface. Aluminum sulfate is the most widely used coagulant. Others are ferric sulfate, ferrous sulfate (must be oxidized by air or chlorine) and sodium aluminate. Most favorable pH values for aluminum coagulants usually range from 5.5 to 6.8 and for iron from 3.5 to 5.5 and above 9.0 but there are exceptions. Coagulation aids are ground clay (not too finely pulverized) and activated silica.

Filtration. Filtration is effected by flowing the coagulated or coagulated and settled water downward through a bed of fine filter sand or Anthrafilt is either a pressure type or gravity type filter. Flow rates in industrial practice range up to 3 gpm per sq. ft. (122 liters/min/sq. meter) of filter bed area while in municipal practice maximum flow rate is usually 2 gpm per sq. ft. (81.5 liters/min/sq. meter).

Chlorination. Chlorination is the most widely used disinfecting or sterilizing process. Where daily water requirements are not large, it is common practice to use a hypochlorite, but for large plants liquefied chlorine gas is used. Chlorination may be practiced before filtration (prechlorination), after filtration (post-chlorination), or both before and after.

Taste and Odor Removal. Except for sulfur waters, most tastes and odors are organic in nature. Activated carbon is widely used for their removal. In powdered form, it may be added to the water being treated in coagulation and settling equipment. In such installations, aeration is frequently used as preliminary treatment. In granular form, it is used in filters (activated carbon filters or purifiers). As substances producing tastes and odors are usually extremely small in amount, activated carbon filters are frequently operated for 6 months to one or more years before replacement of bed is necessary.

Optional Treating Operations

These may include improving water quality for domestic and industrial purposes.

Hardness Removal (Water Softening). *Sodium Cation Exchanger (Zeolite) Process.* This is the most widely used water-softening process in industrial, commercial, institutional and household applications. Hard water is softened by flowing it, usually downward, through a bed (2 feet to over 8 feet in thickness) of a granular or bead type sodium cation exchanger in either a pressure-type (most widely used) or gravity-type water softener. As water comes in contact with sodium cation exchanger, hardness (calcium and magnesium ions) is taken up and held by the exchanger which gives up to the water an equivalent amount of sodium ions. At end of softening run (4 to over 24 hours in industrial practice and 1 to over 2 weeks in household use), softener is cut out of service, regenerated and returned to service (1/2 to 11/2 hours). Regeneration is effected in 3 steps: (1) backwashing to cleanse and hydraulically regrade the bed (2) salting with specified amount of common salt (sodium chloride) solution, usually 10 to 15% in strength, which removes calcium and magnesium from the exchanger and restores sodium to it and (3) rinsing to remove calcium and magnesium chlorides and excess salt.

Hydrogen Cation Exchanger Process. Calcium, magnesium, sodium and other cations are removed by flowing water (usually downward) through a bed (2 feet to over 8 feet in thickness) in an acid-proof pressure-type (most widely used) or gravity-type shell. As water comes in contact with hydrogen cation exchanger, calcium, magnesium, sodium and other cations are taken up by the exchanger which gives up to the water an equivalent amount of hydrogen ions. At the end of operating run (4 to over 24 hours), unit is cut out of service, regenerated and returned to service (11/4 to 2 hours). Regeneration is effected in three steps: (1) backwashing to cleanse and hydraulically regrade the bed; (2) acid treatment with sulfuric or hydrochloric acid which removes metallic cations from the bed and restores hydrogen to it; and (3) rinsing to remove salts (sulfates or chlorides) and excess acid. The carbon dioxide formed from the bicarbonates may be reduced to below 5 to 10 ppm by aeration. The sulfuric and hydrochloric acids formed from chlorides and sulfates may be (1) neutralized with an alkali (usually caustic soda), (2) neutralized by sodium bicarbonate content of a sodium cation exchanger softened water (in which case aeration follows neutralization), or (3) removed by an anion exchanger.

Cold Lime (or Lime Soda) Process. Chemicals used may be (1) lime plus a coagulant or (2) lime plus soda ash plus a coagulant. Dosages vary according to composition of raw water and result desired such as (a) calcium alkalinity reduction, (b) calcium and magnesium alkalinity reduction, (c) reduction of total hardness without excess chemicals and (d) excess chemical treatment. Precipitates produced are calcium carbonate and magnesium hydroxide. Rated residuals without excess chemicals are 35 ppm for calcium and 33 ppm for magnesium, both expressed as calcium carbonate. Operating results will range between these and theoretical solubilities. With excess chemicals, total hardness may be lowered to 16 ppm. The process is best carried out in the sludge blanket-type of equipment in which the treated water is filtered upward through a suspended blanket of previously formed sludge. Detention periods range from one to two hours. Usually, treated water is filtered before going to service but where small amounts of turbidity are unobjectionable, filters may be omitted.

Hot Lime Soda Process. In this process, treatment with lime and soda ash is carried out at temperatures around the boiling point in closed, steel pressure tanks. Heating is usually accomplished with exhaust steam and pressures most widely used range from 5 to 10 psig, but higher pressures up to but seldom above 20 psig are also used. At these temperatures, the reactions proceed swiftly and precipitates formed are larger than those in cold lime soda process so no coagulant is needed. Detention period is one hour and de-aeration effected in primary heater is sufficient to lower dissolved oxygen content to 0.3 milliliter per liter which is sufficient for low-pressure boilers. For high-pressure boilers, either an integral or separate de-aerator is used and this will bring the dissolved oxygen down to less than 0.005 ml/l. With 20 to 30 ppm excess soda ash, the hardness will be reduced to 25 ppm. Softening to practically zero hardness may be effected by either (1) two-stage hot lime soda phosphate treatment in which effluent from hot lime soda softener is treated with sodium phosphate, or (2) the filtered effluent is passed through a sodium cation exchanger. Anthrafilt filters are usually employed with hot process softeners.

Fluoridation. See entry on **Fluorine**.

Demineralization (Deionization). Metallic cations are removed by a hydrogen cation exchanger. Anions are removed by an anion exchanger. Depending on the hookup used, the carbon dioxide formed from the bicarbonates may be removed mechanically by an aerator, degasifier or vacuum de-aerator, or chemically by a strongly basic anion exchanger. Strongly basic anion exchangers will remove both strongly ionized acids, such as sulfuric and hydrochloric, and weakly ionized acids, such as silicic and carbonic. Weakly basic anion exchangers will remove only strongly ionized acids.

Iron and Manganese Removal. In clear, deep ground waters, iron and/or manganese may occur as soluble, colorless, divalent bicarbonates. These may be removed (1) by oxidation plus settling (if necessary) plus filtration (2) by cation exchange with sodium or hydrogen cation exchangers or (3) filtration through an oxidizing (manganese zeolite) filter. In (1) addition of an alkali or lime may be needed to build up the pH value

so as to speed up the oxidation. Iron and/or manganese in organic (chelated) form may usually be removed by coagulation, settling and filtration. In acid waters, these metals may be removed by neutralization (plus increase of pH), aeration, settling and filtration.

Fluoride Removal. Fluorides may be reduced to below 1 ppm by filtration through a bed of a specially prepared, granular bone char (bone black). Regeneration is effected with caustic soda solution followed by treatment with dilute phosphoric acid.

Dissolved Gases. *Oxygen and Nitrogen* may be removed (1) hot in a deaerating heater (de-aerator) or (2) cold in a vacuum de-aerator. *Carbon dioxide* may be removed in (1) an aerator, (2) a de-aerating heater or (3) a vacuum de-aerator or it may be neutralized with lime or an alkali or by filtration through a bed of granular calcite. *Hydrogen sulfide* may be removed by (1) aeration followed by chlorination, (2) treatment with flue gas plus aeration followed by chlorination or (3) filtration through an oxidizing manganese zeolite filter (household use). If sulfur content and pH values are high, (1) may effect but little removal but, in some cases, with fairly long detention periods, sulfur bacteria may effect notable reductions.

Treatment of Wastewater (Sewage)

Wastewater may be defined as the spent or used water from a community or industry that contains dissolved or suspended matter. Toxic wastes must receive pretreatment by the polluter prior to introduction into a water reservoir or municipal used-water return lines. Most wastewater is 99.94% water by weight. The remaining 0.06% is material dissolved or suspended in the water. Water chemists differentiate *suspended solids* and *dissolved contaminants*. The concentration of dissolved or suspended matter usually is expressed as milligrams of pollutants per liter of water (mg/l), or as parts per million (ppm) (weight). On another scale, 1 ppm can be visualized as being equivalent to 1 minute of time in 1.9 years. Although pollutants may be so minute, innumerable studies have shown their adverse effects on human and other animal life.

A generally accepted estimate is that each individual, on a national (U.S.) average, contributes approximately 265 to 568 litres of water per day to a community's wastewater flow. While most people think of wastewater as only "sewage," wastewater also comes from other sources—commercial, industrial, and storm and ground water. Generally, each house or business has a pipe or sewer that carries the wastewater to the wastewater treatment plant. **Sanitary sewers** carry only domestic and industrial wastewater, while **combined sewers** carry wastewater and storm water runoff. Every reasonable effort is made to exclude storm (inflow) and ground (infiltration) water from the sanitary sewer system. These efforts are usually less successful on older sewer systems that leak.

The wastewater from the sewer system either flows by gravity or is pumped into the treatment plant. Usually, treatment consists of two major steps, primary and secondary, along with a process to dispose of solids removed during the two steps.

In **primary treatment**, the objective is to physically remove suspended solids from the wastewater, either by screening, settling, or floating. The major goal of **secondary treatment** is to biologically remove contaminants that are dissolved in wastewater. In secondary treatment, air is supplied to encourage the natural processes of growth of bacteria and other biological organisms to consume most of the waste. These organisms and other solids are then separated from the wastewater. Before discharge to the **receiving stream**, the water usually passes through a tank where a small amount of chemical (usually chlorine) is added to disinfect the treated water.

In primary and secondary treatment, solids are settled and removed for further processing. Solids, usually referred to as sludge, are normally processed in three steps—digestion, dewatering, and disposal. The digestion step reduces the volatile solids and prepares the sludge for further processing. Dewatering involves the application of a variety of processes that reduces the water content of the sludge and, in turn, its volume. The final step is the ultimate management of this treated material, or biosolids, which can be used beneficially through methods such as land application. See Figs. 6 and 7.

Sludge Handling. Treatment professionals refer to solids in general as **sludge**. Beneficial sludges are called "biosolids." But these solutions are not the thick, molasses-like substances that most people think of when they hear the word "sludge." Wastewater sludges are slurries of water and solids that are roughly 100 times more concentrated than untreated wastewater. That is, they contain about 3% solids compared to the .03% (or less) concentration of the initial flow into a treatment plant. The various

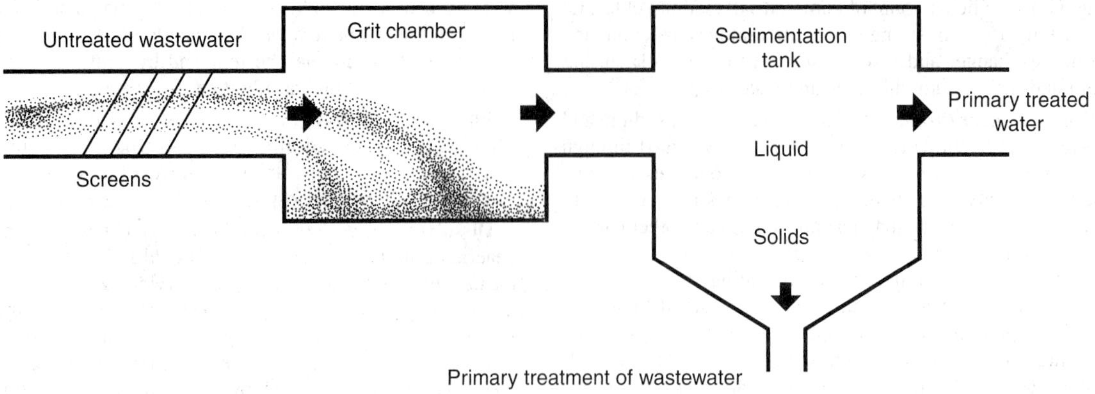

Untreated wastewater   Grit chamber   Sedimentation tank

Screens

Primary treated water

Liquid

Solids

Primary treatment of wastewater

Fig. 6.   Wastewater entering a treatment plant receives primary treatment first. In this state, a series of operations removes most of the solids that can be screened out, will float, or will settle.

Screening removes large floating objects from the incoming wastewater stream. Treatment plant screens are sturdily built to withstand the flow of untreated wastewater for years at a time. Rags, wood, plastics, and other floating objects could clog pipes and disable treatment plant pumps if not removed at this point. Typically, screens are made of steel or iron bars set in parallel about one-half inch apart. Some treatment plants use a device known as a comminuter, which combines the functions of a screen with that of a grinder.

Sand, grit, and gravel flow through the screens to be picked up in the next stage of primary treatment—the grit chamber. Grit chambers are large tanks designed to slow the wastewater down just long enough for the grit to drop to the bottom. Grit is usually washed after its removal from the chamber and buried in a landfill.

After the flow passes out of the grit chamber, it enters a more sophisticated settling basin called a sedimentation tank. Sedimentation removes the solids that are too light to fall out in the grit chamber. Sedimentation tanks are designed to hold wastewater for several hours. During that time the suspended solids drift to the bottom of the tank, where they can be pushed into a large mass by mechanical scrapers and pumped out of the bottom of the tank. The solids removed at this point are called primary sludge. The primary sludge is usually pumped to a sludge digester for further treatment. During the sedimentation process, floatable substances, such as grease and oil, rise to the surface and are removed by a surface skimming system. The skimmed materials are either sent to the sludge digester for treatment along with the primary sludge or are incinerated. Sedimentation marks the end of primary treatment. At this point, most of the solids in the stream that can be removed by the purely physical processes of screening, skimming, and settling have been collected. An additional set of techniques using biological processes must be employed next. (*After Water Environment Federation.*)

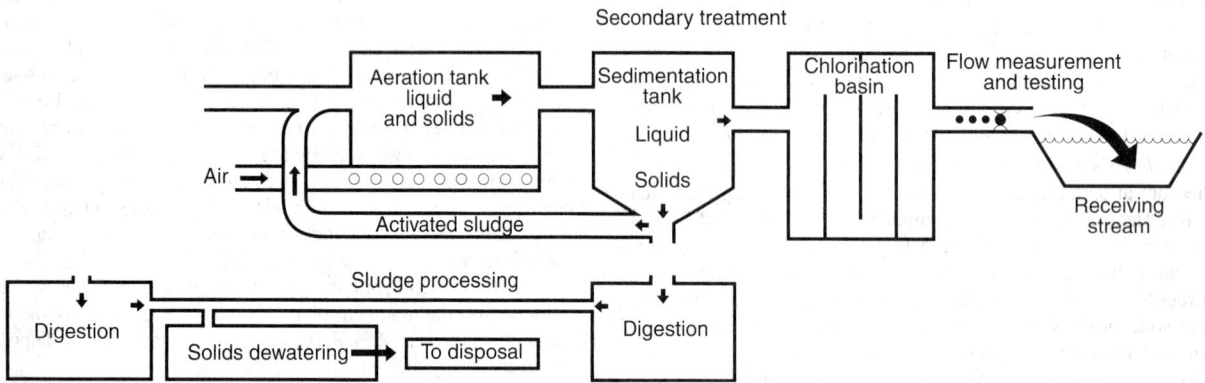

Fig. 7.   Wastewater flowing out of primary treatment still contains some suspended solids and other solids that are dissolved in the water. In a natural stream, such substances are a source of food for protozoa, fungi, algae, and hundreds of varieties of bacteria. The secondary treatment stage is a highly controlled artificial environment in which these same microscopic organisms are allowed to work as fast and efficiently as they can. The microorganisms biologically convert the dissolved solids in the wastewater to suspended solids that will physically settle out at the end of secondary treatment.

There are several different ways to optimize biological conversion. Secondary treatment promotes the growth of millions of microorganisms, bringing them into close contact with the wastewater on which they feed. Care is taken to make sure that the temperature, oxygen level, and contact time support rapid and complete consumption of the dissolved wastes. The final products are carbon dioxide, water and more microorganisms. Three widely employed types of secondary treatment are common: activated sludge, trickling filters, and lagoons.

The most common is the activated sludge process. Activated sludge processes are much more tightly controlled than either trickling filters or lagoons. In this form of treatment, wastewater and microorganisms are mixed for a few hours in a large tank by constant aeration and agitation. Once the aeration is complete, the mixture of water and microorganisms flows to a sedimentation tank similar to the one used in primary treatment. The microorganisms and other solids settle to the bottom of the sedimentation tank. Since activated sludge is a continuous process, a portion of the settled solids (return activated sludge) are circulated back to the beginning of the process to serve as "seed" organisms. The part not needed for "seed" is commonly called waste activated sludge and is sent to a sludge digester for further treatment. (*After Water Environment Federation.*)

techniques for handling these flows are designed to increase the solids concentration even further, to as much as 50%.

As a rule of thumb, higher degrees of wastewater treatment produce larger volumes of sludge. For example, primary treatment usually produces 2500 to 3500 gallons of sludge for every 1 million gallons of wastewater. Secondary treatment usually produces 15,000 to 25,000 gallons for every 1 million gallons treated. To try to dispose of or recycle such volumes of waste is practical. Thus, sludge handling methods are designed to remove as much water from the mixture as possible.

The spectrum of sludge handling techniques is divided into processes that condition, thicken, stabilize, and dewater the sludge flow. Conditioning operations usually employ chemicals or heat to make the sludge release water more easily. Thickening techniques use gravity, flotation, and chemicals to separate water from the solids. Conditioning and thickening are usually the first steps in handling primary and secondary sludges.

Stabilization converts the organic matter in the sludge so that the biosolids can be disposed of or used as a soil conditioner without posing a health hazard in the general environment. Sludge stabilization can occur

although they are closely related. The water measurers make up the family *Hydrometridae*, and the other insects are water striders of the family *Gerridae*.

WATER MOCCASIN. See **Snakes**.

WATER PENNY *(Insecta, Coleoptera)*. A small flattened oval insect found chiefly on the underside of rocks in running water. Water pennies are the larvae of beetles of the family *Psephenidae*. They resemble crustaceans and were originally described as such.

WATER PHEASANT. See **Waders, Shorebirds, and Gulls**.

WATER PLANTS. See **Hydrophytes**.

WATER POLLINATION. See **Pollination**.

WATER POLLUTION. Means for treating polluted waters are described in a prior entry on **Water**. This article is devoted to the current state of water pollution in the United States and some European countries, sources of pollution, and ways and means for preventing water pollution. This article also relates directly to articles on **Wastes and Pollution**; and a later article on **Water Resources**.

Water Pollution in the United States[1]

For an abridged analysis of the water pollution problem in the United States, a concerted look is taken at (1) the rivers and (2) groundwater.

[1] It is interesting to note how interrelated the topics of water, air, and solids (soil) pollution are. Acid rain, for example, commences as an air pollutant and ends up as a soil and water pollutant. See also **Pollution (Air)**. Thus, water pollution may be direct or indirect. Because they have a mass, air pollutants ultimately fall to Earth's surface and thus pollute the oceans, bodies of freshwater, and the land.

Water Quality in the Rivers. River water serves as an excellent overall index of the pollution problem, because ultimately, most water, even consumed groundwater and a vast majority of the lakes, ponds, etc., in the long run ends up in a river (or series of rivers), and thence flows to the oceans. The majority of rivers are accessible with relative ease and thus add to the convenience of obtaining water samples for analysis.

Probably the most complete study of river water quality was completed by the U.S. Geological Survey, released in early 1987 and periodically updated. The initial survey was coordinated by Smith and Alexander (USGS) and Wolman (The Johns Hopkins University), including water quality records from two nationwide sampling networks. The network included over 300 locations on the major rivers of the United States. Twenty-four water quality parameters are measured. Originally, the two networks were comprised of: (1) the National Stream Quality Accounting Network (NASQUAN) and (2) the National Water Quality Surveillance System (NWQSS). Locations of stations are shown on the map in Fig. 1. The measured water-quality indicators include:

pH (hydrogen ion concentration)	Trace Elements
Alkalinity ($CaCO_3$)	Arsenic
Sulfate (SO_4)	Cadmium
Nitrate (total as N) (TN)	Chromium
Phosphorus (total as P) (TP)	Lead
Calcium	Iron
Magnesium	Manganese
Sodium	Mercury
Potassium	Selenium
Chloride	Zinc
Suspended sediment (SS)	
Fecal coliform bacteria	
Fecal Streptococcal bacteria	
Dissolved oxygen	
Dissolved-oxygen deficit (DOD)	

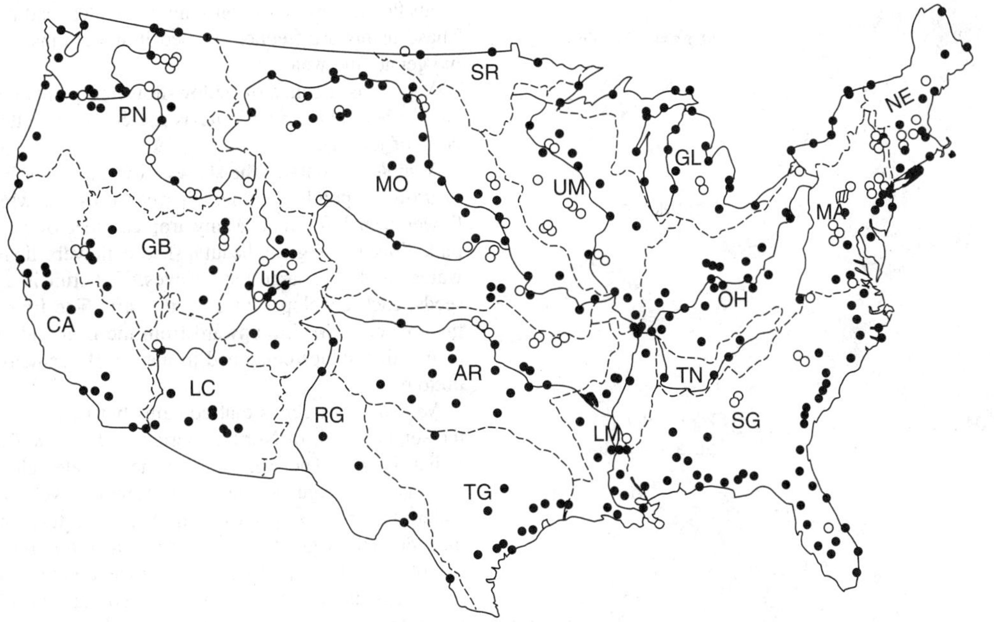

Fig. 1. Analysis of major river waters based upon a sampling network comprised of over 300 stations and involving two systems:
(1) The National Stream Quality Accounting Network (NASQAN), indicated by solid black dots in map; and (2) the National Water Quality Surveillance System (NWQSS), indicated by open black circles on map. Shown in outline are regional drainage basins. Abbreviations used for these basins are:

NE New England	UM Upper Mississippi	RG Rio Grande
MA Mid-Atlantic	LM Lower Mississippi	LC Lower Colorado
SG Southeast-Gulf	TG Texas-Gulf	UC Upper Colorado
TN Tennessee	AR Arkansas-Red	GB Great Basin
OH Ohio	MO Missouri	CA California
GL Great Lakes	SR Souris-Red-Rainy	PN Pacific Northwest

The largest rivers are shown as solid black lines. The NASQAN stations are located and associated with these rivers and their tributaries; the NWQSS stations are located along smaller rivers and usually near agricultural areas and some urban communities. (*U.S. Geological Survey.*)

with (**aerobic**) or without (*anaerobic*) oxygen in special tanks called digesters. Sometimes chemicals such as lime are used for stabilization.

Dewatering is done by mechanical means. Filters, centrifuges, and presses remove even more water from the biosolids. Biosolids dewatered by such equipment have the consistency of wet mud and can have a solids concentration of up to 20%. Other techniques, such as drying beds or special presses, can be used to dewater the sludge, producing up to 50% solids—about the consistency of dry soil. At the end of a sludge handling process, the concentrated solids can be placed in landfills, incinerated, applied to land, or composted for use as a soil conditioner.

Alternatives to Sludge Process. For some wastewater treating situations, the complexities of the sludge process may not be required. In such instances, trickling filters or lagoons may be used.

Trickling filters. are large beds of coarse, loosely packed material—rocks, wooden slats, or shaped plastic pieces—over which the wastewater is sprayed or spread. The surfaces of the filter material (also known as the "medium") become breeding grounds for the microorganisms that consume the wastes. A common trickling filter is a bed of stones 3 to 10 feet deep. Under the bed, a system of drains collects the treated wastewater and diverts it to a sedimentation tank or back over the filter medium for additional treatment. In the sedimentation tank, suspended solids settle and are pumped to a sludge digester. Trickling filters are relatively simple to construct and operate. Many communities in the United States rely on them for secondary treatment.

Lagoons are used by some communities to achieve secondary treatment. Lagoons generally treat the total wastewater from a community until the biological oxidation processes have consumed and converted most of the wastes present. This form of treatment depends heavily on the interaction of sunlight, algae, and oxygen. Sometimes the wastewater is aerated to speed the process, since these interactions are relatively slow. There is usually no sedimentation tank associated with a lagoon. Suspended solids settle to the bottom of the lagoon, where they remain or are removed every few years. Generally speaking, lagoons are simpler to operate than other forms of secondary treatment, but are less efficient.

At the end of a secondary treatment process, the wastewater is disinfected to remove disease-causing organisms. Usually, an agent such as chlorine is added to the stream of wastewater before it is discharged to receiving waters. Sometimes other techniques are used if the receiving waters are sensitive to the addition of chlorine.

An alternate to secondary and higher levels of treatment is land application of wastewater. The wastewater is usually sprayed over natural or specially sloped and seeded land. The wastewater seeps into the soil where natural solid microorganisms consume the wastes. The treated water is either used by plants, stays in the ground, or is collected and routed to a receiving stream.

Role of Sunlight as a Detoxifying Agent

In the late 1980s, Sandia National Laboratories announced the development of a solar-powered reactor to generate low-cost electrical power and recently have found an alternative use for it, namely, for the detoxification of polluted water. As stated by the project leader, C. Tyner, "We believe this process will destroy most organic materials, including industrial solvents, pesticides, dioxins, PCBs, and munitions chemicals." The process breaks down toxic chemicals into smaller, safer molecules. Current methods remove organic wastes from water by bubbling air through the water and thus volatilizing them for release into the atmosphere (not attractive) or by running the polluted water through carbon filters. The researchers have set up a troughlike arrangement (similar to sunlight collectors used for solar furnaces) along which runs a radiation-transparent tube holding the flowing water, which thus receives a maximum concentration of solar radiation. The future of the project will be determined largely by cost considerations.

See also **Wastes and Pollution**; **Water Pollution**; and **Water Resources**.

Additional Reading

Amato, I.: "A New Blueprint for Water's Architecture," *Science*, 1764 (June 26, 1992).
Beardsley, T.: "Mr. Clean: Sunlight Can Destroy Dangerous Chemicals," *Sci. Amer.*, 83 (June 1989).
Chew, R.M.: "Water Metabolism in Mammals," in "Physiological Mammalogy" (W.W. Mayer and R.G. Van Gelder, Eds.) Academic Press, Orlando, Florida, 1963.
Corbitt, R.A.: "Standard Handbook of Environmental Engineering," 2nd Edition, The McGraw-Hill Companies, Inc., New York, NY, 1998.
Crompton, T.R.: "Determination of Metals in Natural and Treated Water," Taylor & Francis, Inc., Philadelphia, PA, 2001.
Dale, J.E.: "Plants and Water," Cambridge University Press, New York, NY, 2001.
Hauser, B.: "Fundamentals of Drinking Water," Lewis Publishers, Boca Raton, FL, 2001.
Herschy, R.W. and R.W. Fairbridge: "Encyclopedia of Hydrology and Water Resources," Chapman & Hall, New York, NY, 1998.
Horan, N.: "Biological Waste Water Treatment Systems," 2nd Edition, John Wiley & Sons, Inc., New York, NY, 2001.
Kolata, G.B.: "Water Structure and Ion Binding: A Role in Cell Physiology?" *Science*, **192**, 1220–1222 (1976).
Martin, A.M.: "Use Biomonitoring Data to Reduce Effluent Toxicity," *Chem. Eng. Progress*, 43 (September 1992).
McLaughlin, L.A., H.S. McLaughlin, and K.A. Groff: "Develop an Effective Wastewater Treatment Strategy," *Chem. Eng. Progress*, 34 (September 1992).
Morra, M.: "Water Biotechnological Surface Science," John Wiley & Sons, Inc., New York, NY, 2001.
Nollet, L.M.L.: "Handbook of Water Analysis," Marcel Dekker, Inc., New York, NY, 2000.
Okoniewski, B.A.: "Remove VOCs from Wastewater by Air Stripping," *Chem. Eng. Progress*, 89 (December 1992).
Pankow, J.F.: "Aquatic Chemistry Concepts," Lewis Publishers, Inc., Boca Raton, FL, 1991.
Patra, K.C.: "Hydrology and Water Resources Engineering," CRC Press, LLC., Boca Raton, FL, 2000.
Schultz, G.A., E.T. Engman: "Remote Sensing in Hydrology and Water Management," Springer-Verlag, Inc., New York, NY, 2000.
Staff: "Sun-Powered Chemical Reactor," *Chem. Eng. Progress*, 12 (September 1990).
Staff: "Structural Ice," *Advanced Materials & Processes*, 4 (December 1991).
Staff: Water Amer, American Waterworks Association, "The Drinking Water Dictionary," McGraw-Hill Professional Book Group, New York, NY, 2001.
van der Leeden, F., F.L. Troise, and D.K. Todd: "The Water Encyclopedia," 2nd Edition, Lewis Publishers, Inc., Boca Raton, FL, 1990.
WEF: "About Wastewater Treatment," Water Environment Federation, Alexandria, VA, 1993.
WEF: "Clean Water for Today: What is Wastewater Treatment?" Water Environment Federation, Alexandria, VA, 1994.
Weinberg, C.J., R.H. Williams: "Energy from the Sun (Wind and Biomass)," *Sci. Amer.*, 147 (September 1990).

WATER BALANCE (Fishes). See **Fishes**.

WATER BOATMAN *(Insecta, Hemiptera)*. An aquatic bug of the family *Corixidae*. These insects are flattened, broad at the head and tapering bluntly at the opposite end of the body. The fringed posterior legs project like a pair of oars and are used in swimming. Water boatmen breathe air but they are able to descend to considerable depths, carrying a film of air on the ventral surface of the body. They feed on ooze containing plant matter and minute animals.

WATERBUCK. See **Antelope**.

WATER CLOUDS. See **Clouds and Cloud Formation**.

WATER CYCLE. See **Hydrology**.

WATER (Desalination). See **Desalination**.

WATER (Electrolysis). See **Hydrogen (Fuel)**.

WATER FLEA *(Crustacea, Cladocera)*. Minute aquatic crustaceans of compact form, usually transversely compressed and provided with a bivalve carapace. They are superficially like fleas in form.

WATERFOWL *(Aves, Anseriformes)*. This large order of birds includes the *ducks, geese, mergansers, swans*, and related species. Ducks are swimming birds of moderately large size with heavy bodies, short legs, webbed feet, and broad flattened beaks with sieve plates at the sides. In common usage, the term includes the closely related teals, sheldrakes, and mergansers. With exception of the mergansers, which have narrow beaks with serrate edges and live chiefly on fish, the ducks eat rice and other plant products, with some insects and other small animals. Ducks are among the leading game birds and many wild species have very palatable flesh. Many species have been domesticated and interbred with the common domestic duck.

The *baldpate* is a North American duck (*Mareca americana*) and is not to be confused with a dove of the same common name found in the West Indies. The baldpate is about 18 inches (45 centimeters) long, with a bill rounded at the tip. The wings are long and the tail is wedge-shaped. The head is mainly black and gray with white on the very top. Green plumage surrounds the eye. Baldpates are swift, but irregular in flying pattern. These birds breed in Canada and winter in Central America. They feed on insects, grain, tender green plants, leeches, and worms. When feeding, the head is usually completely immersed below the surface of the water. The birds nest on the ground and have from 8 to 10 eggs at incubation time. The eggs are white and close to the size of a chicken egg. The European species, known as the *widgeon* (*M. penelope*) is smaller and is sometimes seen along the American coast line.

The *bufflehead, Charitonetta albeola*, is a small North American duck of wide distribution, particularly of inland waters. The bird ranges up to 24 inches (61 centimeters), wing tip to wing tip and is about 10 inches (25 centimeters) long. The color is white and black, crested with rich silky green plumage and white streaks across the top of the head. The bufflehead nests in old stumps, burrows, or holes in trees. The eggs are dark and number from 10 to 12. These birds fly very fast. They range from the Great Lakes north to the Arctic Circle.

The *canvas-back* is a large North American duck, *Nyroca vallisneria*, similar to the redhead, but with a dusky crown and face. Its flesh at the proper season is considered superior to that of other ducks.

Eider ducks are of several species that breed in the far north along rocky coasts. They are noted for their fine down with which they line their nests. This material has a high commercial value and is collected for market and used in pillows and comforters.

The *gadwall* is a North American duck, *Anas strepera*. The *goosander* is a merganser duck of the Northern Hemisphere, *Mergus merganser*. Old squaw or old wife is the name of a small duck, *Clangula hymenalis*, of the latitudes of most of the Northern Hemisphere, migrating into the United States in winter. It is chiefly black and white and the male has a long slender tail. It is also called the long-tailed duck. The *scaup* or *bluebill*, *Aythya* (*Nyroca*) *marila*, is a duck found throughout the Northern Hemisphere and the lesser scaup, *N. affinis*, is common in North America. Both are principally black and light gray or white. The head of the scaup duck is glossed with green and that of the lesser scaup with purple.

The *scoter* is a duck found chiefly along the seacoasts and less often on fresh waters. The American scoter, *Oidemia americana*, is a bird of the northern Atlantic, breeding from Alaska to Labrador. The *white-winged scoter, Melanitta deglandi*, breeds from the northern United States northward and is fairly common on the Pacific Coast. The *surf scoter, M. perspicillata*, also breeds northward into the Arctic regions and is more generally distributed on both coasts and in inland waters. All scoters are dark-colored birds with a few light areas.

The *shoveller* is a widely distributed duck, *Spatula clypeata*, of the Northern Hemisphere. The male has a dark green head and is beautifully marked with blue, white, chestnut, and orange. The female is brownish but for the blue wing patch. The species is named from the greatly broadened beak. Also called the *spoonbill*.

The *teal* is a small duck whose beak has almost parallel sides. The several species are beautifully marked. Teals are found throughout Europe, Asia, and North America. The common European species migrates into Africa and is sometimes found in northern and eastern North America, while other species have a much more limited range. In North America, the *blue-winged teal, Querquedula discors*, is common east of the Rocky Mountains; the *green-winged teal, Nettion carolinense*, throughout the continent; and the *cinnamon teal, Q. cyanoptera*, west of the Mississippi. The *garganey, Anas querquedula*, is the summer teal of Europe.

Pochard is a name applied in the Old World to ducks of several species related to the *scaup ducks*, the *redhead*, and the *canvas-back* of North America. Smew is a European name for a small merganser, *Mergellus albellus*. The name also is applied to widgeons and pochards, and said to be used for the pintail duck, although pintail is by far the more common name.

Geese are large swimming birds with webbed feet and thick, strong beaks. They live chiefly on vegetation. Geese are found on all continents. They are strong fliers, and some species migrate from their nesting grounds in the north to the Southern Hemisphere. Among the several North American species, the *Canada goose, Branta canadensis*, is sometimes called *brant*, and a related species is the *black brant, B. nigricans*. Geese are excellent game birds and good food. The domestic goose is also valuable for its smaller feathers and down for filling pillows and cushions.

The *swan* is a large bird related to the geese and of similar form and habits. The swan differs in the length of the neck, which is at least as long as the body. Although the more familiar species are white, some swans are marked with black and in Australia a species with almost entirely black plumage occurs. The *mute swan, Cygnus olor*, is habitually silent. The two North American species, the *whistling, Cygnus columbianus*, and *trumpeter, C. buccinator*, swans, breed far to the north and are not often seen. The whistling swan also bears the name *whooper, C. cygnus*, in Europe. See also **Anseriformes**; and **Poultry**.

WATER HAMMER. Sudden stoppage of water flow in a long pressure conduit caused by the closing of valves can, if the rate of closure be rapid enough, cause the conduit to be subjected to a sharp, hammerlike blow from a steep front pressure wave. Water moving in a long pipe-line has considerable mass. To decelerate a mass requires a force equal to the mass times the deceleration (negative acceleration). If the rate of deceleration is large, the force will be large. Hence if a valve or gate is suddenly closed, the water has high deceleration, and a large force is set up. Due to the elastic nature of conduit material, this force acts expansively, slightly stretching the pipe. When the inertia force has disappeared, the pipe regains its original girth and produces secondary pressure waves. A calculation shows that the power required to decelerate water in a 5-foot (1.5 meter) pipe 2,000 feet (600 meters) long is 1,400 horsepower (1044 kilowatts). This calculation was based on an assumption of 10 feet (3 meters) per second water velocity, and 5 seconds was the time taken to close the valve, and it should acquaint one with the magnitude of power behind the water hammer. To cushion all or parts of the water conduit against the destructive effect of water hammer forces, relief valves, bursting plates, and surge tanks have been used. Water hammer may also be caused by the sudden collapse of steam bubbles upon entering cold water, as when the steam is turned into a cold radiator partly filled with water.

WATER HYACINTH (*Eichhornia crassipes; Pontederiaceae*). This plant occurs widespread in tropical and subtropical regions, where it often becomes a troublesome weed. In Florida it sometimes forms floating masses so dense as to become a serious hindrance to river navigation. Very noticeable are the leaves, the petioles of which are swollen in bladder-like enlargements containing many air spaces. These cause the plant to remain floating at the surface of the water. Because of the broad shining green blades, the plants are easily blown about on the surface by the wind. The dark-colored roots form a dense mass beneath the water surface. The root cap at the tip of each rootlet is a very conspicuous structure. The flowers are showy and pale lavender in color. They are trimorphic, there being three different lengths of styles. The plant is frequently found in cultivation in northern regions, where, however, it is not hardy.

WATERJET CUTTING. The power of a pressurized waterjet has been known for decades. For example, pressurized jet has been used in the timber, lumber, and pulpwood industries for many years as a means of debarking, in less than one minute, logs from huge trees. The concept of using an abrasive waterjet for machining metals was first developed in 1974, but only since the mid-1980s has practical equipment become available for use in precision machining. Testing of automated abrasive waterjet cutting systems was undertaken by Gerin Sylvia (University of Rhode Island) in early 1985. Manufacturers of equipment now claim that the process can be used to cut "everything" from simple gray cast iron to 2-inch (50 mm) thick armor plate and boron-reinforced aluminum. See Table 1. The relationship between jet pressure and jet velocity is shown by Table 2.

TABLE 1. REPRESENTATIVE MATERIALS CUT BY ABRASIVE WATERJET (Thickness = 1 inch; 25 mm)

Material	Cutting Speed In. min	Cutting Speed Mm/min
17-4PH Stainless steel	2.0	51
HY-80 High-strength steel	2.0	51
6A1-4V Titanium	2.0	51
Ni-Cr Superalloy (UNS NO 7718)	2.0	51
Aluminum	4.0	102
Lead	18.0	457
Glass	18.0	457

Source: Flow Systems Inc.

TABLE 2. PRESSURE VERSUS VELOCITY OF ABRASIVE WATERJET

Pressure Psi	Pressure MPa	Jet Velocity Ft/sec	Jet Velocity M/sec
60 000	415	2985	910
50 000	345	2726	830
40 000	275	2438	745
30 000	205	2111	645
20 000	140	1724	525

Note: A doubling of system pressure produces only about a 50% increase in jet velocity, thus illustrating how minor pressure fluctuations do not affect cutting ability (velocity) of the jet.
Source: Flow Systems Inc.

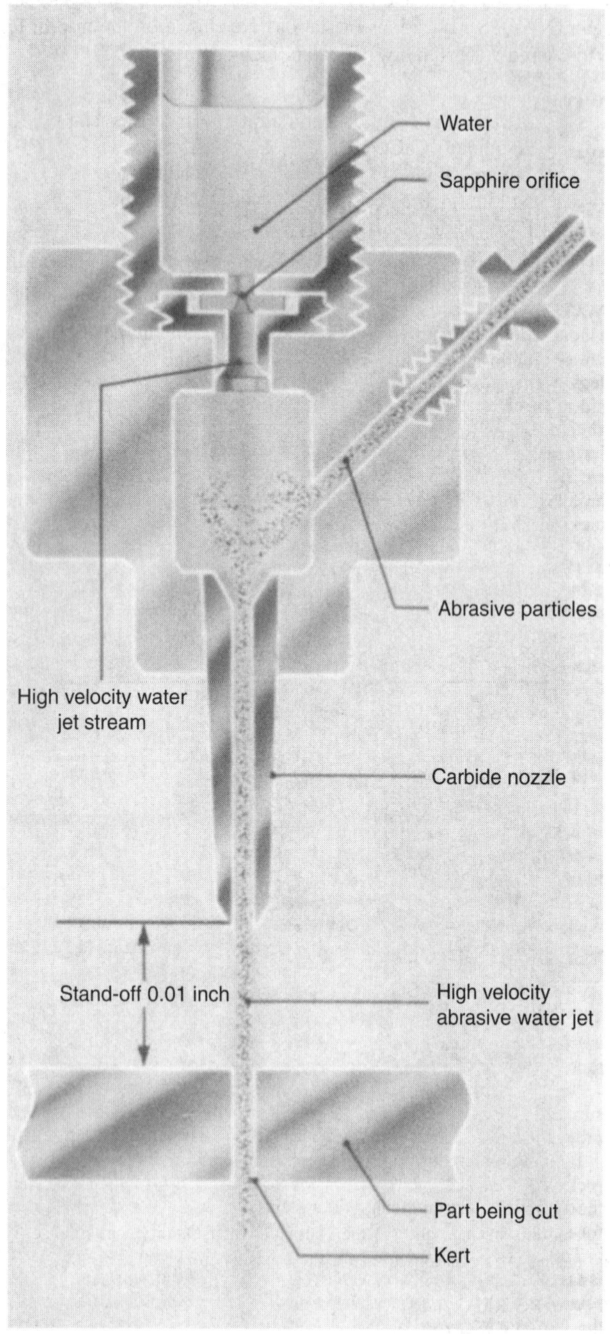

Fig. 1. Cross section of abrasive waterjet cutter, showing how water and abrasive material mix and travel through the nozzle. (*Aerospace Engineering*.)

Abrasive waterjet cutting produces no heat that can degrade metallurgical properties. Because of the smooth edges produced, often no postmachining is required. Generally, the fresh-cut edge has a surface roughness of 150 to 250 microinches (2.5 to 6.3 micrometers). Although there is no surface deformation because of heat, as is common with plasma cutting or grinding, the jet usually cuts at slower travel speeds. However, with reduced or no postcutting operations, net speeds may be increased. Maneuverability is also an excellent feature of jet cutting. Saws, for example, are mainly limited to cutting straight lines or essentially small arces of circular cuts, whereas the jet can be guided to cut curves, holes, and complex shapes. In addition to metals and glass, jet cutting can be used to cut through large sections of concrete.

Abrasives used are fine rather than coarse. It has been found that the harder abrasives, such as garnet, cut more effectively and up to 30% faster than foundry sand. However, where practical, sand is often favored because of its much lower cost.

As pointed out by Brahney (*Aerospace Engineering*), particularly in connection with aerospace structures, it is becoming increasingly apparent that traditional machining methods, such as routing and abrasive wheels, will not be acceptable. Although some individual methods can provide satisfactory results for specific tasks, they are not always an effective solution to the accuracy and flexibility demanded by complex cutting and trimming operations. For this reason, engineers have studied the feasibility of combining the special capabilities, including enhanced repeatability, afforded by computer automation with the power and precision offered by abrasive waterjet cutting technology.

Aerospace engineers have found that the waterjet process is capable of cutting aluminum up to 6 inches (152 mm) thick and composites up to 4 inches (102 mm) thick in a single pass.

A cross section of an abrasive waterjet cutter is given in Fig. 1.

WATER LILIES (*Nymphaeaceae*). The water lilies form a small family of water or marsh plants. The leaves may be submerged or floating or carried well above the surface of the water on stiff petioles, as in the Lotus. The flowers are usually large and solitary. The principal genera are *Cabomba, Nuphar, Nymphaea*, and *Nelumbo*.

Cabomba is a genus of tropical American water lilies having two types of leaves; some are submerged and much-divided, into linear segments, while others are entire and floating with the petiole centrally attached. The small flowers are borne on long peduncles, and have their parts in threes. These plants are frequently used in aquariums, both for ornament and to oxygenate the water.

Nuphar is a genus of yellow-flowered plants occurring in the Northern Hemisphere. *Nuphar advena* is the common yellow water lily or spatter-dock of the marshes.

Nymphaea contains the showy-flowered water lilies so frequently grown in artificial ponds. Northern hardy forms are white- or sometimes pink-flowered and fragrant. Many tropical species have red, yellow, blue, or pink flowers of great beauty. These flowers float on the surface of the water, as do the large cleft leaves. The fruit is a berry containing many seeds, each enveloped in a spongy aril. The fruit ripens under water, the mature seeds floating upward from the fruit and drifting about, by means of the air bubbles contained in the aril. Eventually each seed sinks to the bottom.

Nelumbo is a genus that contains but two species, *N. lutea*, a native of the southern half of North America, and *N. Nucifera*, of Asia and the East Indian Islands. The American species is pale yellow-flowered, the flowers and also the large peltate leaves standing well above the surface of the water. The Asiatic species is the Sacred Lotus, which has showy fragrant pink flowers of great beauty. The fruit of the lotus is a curious obconical receptacle in the top of which are embedded the many carpels. At maturity the receptacle is very light and dry, so that when broken from its stalk it floats on the water, carrying the seeds about until it breaks apart. The seeds of the lotus are used as food by many peoples, especially in Asia.

Victoria regia, the giant water lily of the Amazon, is related to *Nymphaea*. It is a plant of tremendous size, the floating leaf with its upturned rim often having a diameter of six feet or more. The flowers are likewise very large. The seeds are used as food in the Amazon valley, where the plant is native. See also **Hydrophytes**.

WATER LOSS (Plants). See **Guttation**.

WATER MEASURER (*Insecta, Hemiptera*). A long slender bug that creeps slowly on the surface of water. Also called the marsh treaders. These are not the common insects that skate rapidly on the water,

Particularly noteworthy are widespread decreases in fecal bacteria and lead concentrations and widespread increases in nitrate, chloride, arsenic, and cadmium concentrations. Recorded increases in municipal waste treatment, use of salt on highways, and nitrogen fertilizer application, along with decreases in leaded gasoline consumption and regionally variable trends in coal production and combustion during the period, appear to be reflected in water-quality changes. In addition to data from the network of sampling stations, the researchers depended upon considerable ancillary data in their interpretation of the sampling station trend results.

Because of the passage of restrictive legislation, mainly over the past decade, improvements in river water quality can and should be expected. As discussed later, indirect pollution of landfills as they reach underground aquifers is an even more important concern and is less visible and hence less easy to police and control.

Since the passage of the Clean Water Act by the U.S. Congress, numerous improvements have been evidenced and thus tend to reinforce confidence that in many aspects the environment can be improved.

Trends in River Water Pollution

Biological Oxygen Demand (BOD). The Clean Water Act was passed by the U.S. Congress in 1972. In the decade that followed, municipal loads of BOD decreased an estimated 46% and industrial BOD loads decreased at least 71% nationwide. These reductions are impressive, especially in light of an increase in population (up 11% during the period) and an increase in the gross national product (GNP, adjusted for inflation) of 25%. Federal funding for upgrading municipal facilities peaked in 1980 and amounted to $35 billion for the decade.

Dissolved Oxygen Deficit (DOD). The sampling station report indicated a net decrease in DOD (thus an improvement in dissolved oxygen conditions). DOD declines were reported in the New England, Mid-Atlantic, Ohio, and Mississippi regional basins; increases were most frequently found in the Southeast. These data appear to confirm the success of point-source control efforts. Decreases in DOD were found most often (beyond expectations) where point sources dominated; increases where nonpoint sources prevailed. The chronology of the station data also indicates that gains from industrial water treatment preceded those from modernization of municipal facilities.

Fecal Bacteria (coliform, FC; streptococcal, FS). During the study period, decreases in FC and FS were widespread. Major decreases were particularly evident in parts of the Gulf Coast, central Mississippi, and Columbia basins; significant decreases were also noted in the Arkansas-Red basin and along the Atlantic Ocean. During the study period, major efforts were made to control both municipal and agricultural sources of fecal bacteria.

Suspended Sediment (SS). These impurities, originating from nonpoint sources (notably agricultural) are probably the most damaging nonpoint sources of water pollution. E.H. Clark estimates the cost of the hydrologic impacts of soil erosion and related nutrients on aquatic ecosystems may increase. Fertilizer application rates increased by 68% during most of the testing period described here. The long-term history of fertilizer in the United States has been one of almost continuous increases in nitrogen and phosphorus application rates. As reported by Smith, et al., the extent to which changes in agricultural practice are reflected in trends of SS, phosphorus, and nitrogen concentrations in the nation's rivers has largely been a matter of conjecture because of the lack of systematic long-term studies.

Certain forms of land use, such as logging, traditionally have been associated with high rates of soil erosion. Thus, the study indicated large increases in SS in the Columbia (logging) and in the Arkansas-Red and Mississippi basins (agriculture) during the test period. Significant amounts of SS were also found along the Texas Gulf Coast, northern Florida, and scattered locations in northern Ohio, New England, and northern Minnesota. Declining rates of SS were indicated in the Missouri River basin and have been clearly traced to the effects of reservoir construction throughout that basin during the 1950s and 1960s.

Total Phosphorus (TP). Mainly as the result of controls over point sources of pollution, decreasing trends in TP were found in the Great Lakes and Upper Mississippi regions. Significant correlations in TP increases were found in connection with nonpoint sources (mainly agriculture) in several regions.

Total Nitrate (TN). Increases in TN were found to be widespread. Although also found to be significant in the Pacific Northwest and

along the California coast, the great majority of increases in TN were found east of the 100th meridian (southward from North Dakota to southern Texas). Increases of TN were strongly associated with agricultural activity including fertilized acreage (as a percentage of basin area), livestock population density, and feedlot activity. Atmospheric deposition, in addition to agricultural runoff, was found to be a major source of nitrate in surface waters, particularly in the forested basins of the East and northern Midwest. Separate studies of NOx emissions showed a general pattern of increasing rates, notably in the Ohio, Mid-Atlantic, Great Lakes, and Upper Mississippi basins. The river water quality study correlated well with the emission studies.

It is interesting to note that the differences in nitrogen and phosphorus trend patterns appear to be the result of three factors: (1) atmospheric deposition seems to play a large role in the high frequency of nitrate trends; (2) the low frequency of, and strong association patterns between phosphorus and SS trends suggests that increases in TP resulting from rises in agricultural activity have been moderated or delayed by temporary storage of sediment-ground TP in stream channels; and (3) during the study, point-source control efforts were focused much more intensely on TP than on TN because phosphorus was considered more limiting to eutrophication in freshwater ecosystems. The greatest consequence of the differences in TN and TP were seen in changes in the delivery of nutrients to coastal areas. Nitrate loads to East Coast estuaries, the Great Lakes, and the Gulf of Mexico have increased markedly, while phosphorus loads to coastal areas have changed little or have declined. Exceptions are the Gulf Coast and the Pacific Northwest basins, in which instances phosphorus loads to estuaries have increased in association with substantial increases in sediment loads. Researchers have stressed that increased delivery of nitrate to estuaries is of major concern because of the tendency for nitrogen to be limiting to eutrophication in many estuarine environments. See Table 1.

TABLE 1. CHANGES AND TREND PATTERNS OF DELIVERY OF NUTRIENTS TO COASTAL AREAS OF THE UNITED STATES

| | Change in Load | |
Region	Total Nitrate (%)	Total Phosphorus (%)
Northeast Atlantic Coast	32	−20
Long Island Sound/New York Bight	26	−1
Chesapeake Bay	29	−0.5
Southeast Atlantic Coast	20	12
Albemarle/Pamlico Sound	28	0
Gulf Coast	46	55
Great Lakes	36	−7
Pacific Northwest	6	34
California	−5	−5

Source: Smith/Alexander/Wolman 1987 reference listed.

Salinity. Increasing trends in chloride, sulfate, and sodium were found in a large majority of the rivers tested. An average increase of 30% in salinity was found. Researchers attribute this increase to the following factors. (1) Increases in population in the basins surveyed—human wastes are a major source of chloride in populated basins. (2) Use of salt on highways increased by a factor of 12 between 1950 and 1980 and is considered a major contributor to total stream salinity. (The rates of highway salt use were particularly significant in the Ohio, Tennessee, lower Missouri, and Arkansas-Red basins.) (3) In the Missouri, Arkansas, and Tennessee basins, salinity increases were accurately correlated with changes in surface coal production during the testing period. By contrast, salinity decreased in the Upper Colorado Basin, an area historically plagued with salt problems. Decreases are attributed to control efforts and the effects of reservoir filling during the early 1970s.

Trace Elements. Moore and Ramamoorthy reported that trends in toxic element concentrations in surface waters have remained largely unknown despite rapidly increasing knowledge of the potential sources of toxic substances in aquatic systems. The study reported on analyses of several trace elements, including arsenic, cadmium, chromium, lead, iron, manganese, mercury, selenium, and zinc. Researchers found increases in dissolved arsenic and cadmium, particularly in those basins in the industrial Midwest. Atmospheric deposition is considered more significant than terrestrial sources of trace element contamination. Heit et al. confirmed this conclusion as based upon the results of lake sediment analyses in regions with high deposition of fossil-fuel combustion products. At many network sampling stations, dissolved lead concentrations showed a decrease, but despite significant declines in gasoline lead consumption, some sampling stations showed no decline. Lead concentrations were found particularly heavy in the Texas Gulf Coast region.

In deference to the excellence of the sampling study, researchers point out that additional sampling in selected smaller basins is needed to improve the ability to determine the effects of changes in point-source pollution. Although the effects of improved sewage treatment on dissolved oxygen levels appear to be more localized than previously thought, it is possible that the ecological and social benefits of water-quality improvements have been large in proportion to their spatial extent. As pointed out by W.M. Leo, et al., individual case studies have demonstrated local effects of point-source pollution controls, but they do not provide an adequate national sample on which to base an assessment of the benefit of pollution abatement programs. Smith/Alexander/Wolman, in their report, suggest that in designing water-quality monitoring programs for the future, we should recognize the growing number of both point- and nonpoint-source issues that our economic and political systems must address.

Groundwater Quality and Contamination

Groundwater as used by humans consists of subsurface water, which occurs in fully saturated soils and geological formations. Nearly half the population of the United States uses groundwater from wells or springs as a primary source of drinking water. An estimated 75% of major cities depend upon groundwater for most of their supply. Total fresh groundwater withdrawals (1980) were estimated at 88.5 billion gallons per day, of which 65% was used for agricultural irrigation. See also the article on **Groundwater**. Numerous regulatory acts have been passed by the U.S. Congress in recent years to protect groundwater from pollution. These various acts are listed in article on **Wastes and Pollution**.

As early as the 1980s, the U.S. Environmental Protection Agency (EPA) proposed a national groundwater strategy—with the emphasis of targeting on prevention as contrasted with taking remedial action in treating known contaminated sites. Transport of groundwater is very slow. Flow rates are governed by hydraulic gradients and aquifer permeability, thus flows may range from a few centimeters to a meter or so per day. Contaminants usually mix with the water at a low rate. Sometimes natural ingredients tend to retard the actions of contaminants; in other cases the concentrations of contaminants in groundwater may exceed those found in surface water. Groundwater contamination occurs out of view and the effects of pollution may not be noted for weeks or months (even years in sparsely settled communities) after the initial cause(s). A wide variety of substances are involved in groundwater contamination. These include inorganic ions (chloride, nitrate, heavy metals), complex synthetic organic chemicals, and pathogens (viruses and bacteria). Thus, rarely does one find the same combination of variables when attacking a groundwater contamination problem.

The extent of groundwater is statistically very impressive: (1) of the water in the hydrologic cycle, 4% is groundwater, exceeded only by the seas and oceans; (2) groundwater volume at any given time exceeds that of the combined fresh water in lakes, streams, and rivers, of which an estimated 30% of the volume is furnished by groundwater; (3) during dry spells, groundwater furnishes a large majority of the low water flow of streams.

The manner in which waste disposal practices may interact with and contaminate the groundwater system is illustrated schematically in Fig. 2.

A study of groundwater contamination in ten selected states, sponsored by the Environmental Assessment Council of the Academy of Natural Sciences (Philadelphia) provides an excellent view of the sources and kinds of contamination that have been found in known, reported incidents of contamination. See Table 2.

One of the most serious cases of groundwater pollution occurred above the Great Miami Aquifer, which is a 2-mile (3.2-km) wide and 80-mile (129-km) long water basin that essentially follows the Great Miami River in southern Ohio and a section of which is located directly under the city of Hamilton. This aquifer is considered to be one of the Midwest's most productive sources of groundwater, making up about one-third of Ohio's groundwater. At points, the aquifer rises to within 20 feet (6 m) of ground level. A small, private firm established a waste-disposal operation in the 1970s at a time when regulations were minimal.

Cleaning up the severely polluting operation commenced in May 1983 and was one of the first projects selected by the EPA to clean up after passage of the Superfund Act in late 1980. Cleanup contractors found 8600 drums, 30 storage tanks, and two open-top tanks on an approximately 10-acre (40,470–square-meter) area. Containers were found holding and leaking some 300,000 gallons (1,135,000 liters) of toxic wastes, including pesticides, rodenticides, waste oils, plastics and resins, acids, arsenic, and cyanide sludges. Among the chemicals found were DDT and PCBs. It was estimated in 1990 that environmental correction actions will require an additional 10 years at a multimillion dollar cost. Professionals in the cleanup field have estimated the existence of about 400 additional sites of serious groundwater pollution of the extent and nature of the site just described.

Remediation of Poisoned Aquifers. Means for correcting polluted underground water systems are limited and costly. To the knowledgeable would-be polluter, the costs and time involved serve as incentives for preventing such damage in the first place.

Containment. Once subterranean pollution has been detected for the first time, not only must the source(s) be located, but the extent of damage must be determined. This often involves the meticulous analysis of core samples and thorough geologic and hydrologic mapping of the poisoned volume of earth and groundwater affected. Effective permanent remedial actions require an extensive and reliable data base. In some cases it may be found that, by way of constructing underground barriers, the polluted underground sections can be isolated and an aquifer rerouted. In such rerouting, it may be found that gravity flow underground may be insufficient, thus requiring pumps to assure flow through what, in essence, is a new "artery" for groundwater flow.

Where containment (isolation) procedures are not practical or fully adequate, extraction of pollutants may be required.

Extraction. In this process, sometimes referred to as the "pump and treat" method, polluted water is pumped to the surface, after which it is treated to remove toxic materials. Then the treated water is returned to the aquifer. Depending upon the extent of pollution, a large water-treating facility may be required adjacent the cleanup site, even though the facility may be required only for a comparatively short time of a few to several months.

Bioremediation. This process is gaining acceptance among site cleanup professionals. Wastes that have been pumped out of the aquifer are transferred to specially constructed tanks, to which fungi, bacteria, and other microbes are added, along with hydrogen peroxide, which furnishes oxygen. Also added are small amounts of nitrogen fertilizer, which acts as a nutrient for the microbes. Under these ideal conditions, the microorganisms secrete enzymes that transform immense quantities of substances such as benzene and trichloroethylene into harmless salts. A delicate balance must be maintained in providing just enough oxygen and nutrients, thus avoiding repollution of the wastes.

For cleaning up sites less extensive than the one just described, and thus requiring aboveground treating facilities for less time, portable plants are receiving consideration. New techniques also are being evaluated. In one system, a *wet oxidation* process converts waste sludge into clean water and sterile ash. Organic pollutants also are being subjected to a laser oxidation process that breaks molecular bonds in a photochemical reaction chamber. Low-frequency soundwaves and electric charges also are being tested to free pollutants from soils. In this process, after a positively charged anode is introduced to the ground, positively charged water and contaminants migrate toward a negatively charged withdrawal well, which acts as a cathode.

Above- and Underground Storage of Products and Wastes

For practical economic reasons, some potentially polluting products must be stored for indeterminate periods above ground. Chemical, petroleum, and petrochemicals are examples of the need for temporary holding. In the blending of gasolines, for example, refineries pump different grades of fuel to a "gasoline pool," out of which quantities can be drawn and blended to achieve the desired specifications for a given gasoline. In producing commercial chemicals where batch or semibatch processing is used, a supply for several weeks or months may be produced within a short time frame and then stored in inventory to fill orders over a period much longer than that required to make the product. Thus, numerous products that are considered toxic must be inventoried. Frequently, at the manufacturing level, aboveground tanks are preferred. In terms of gasoline,

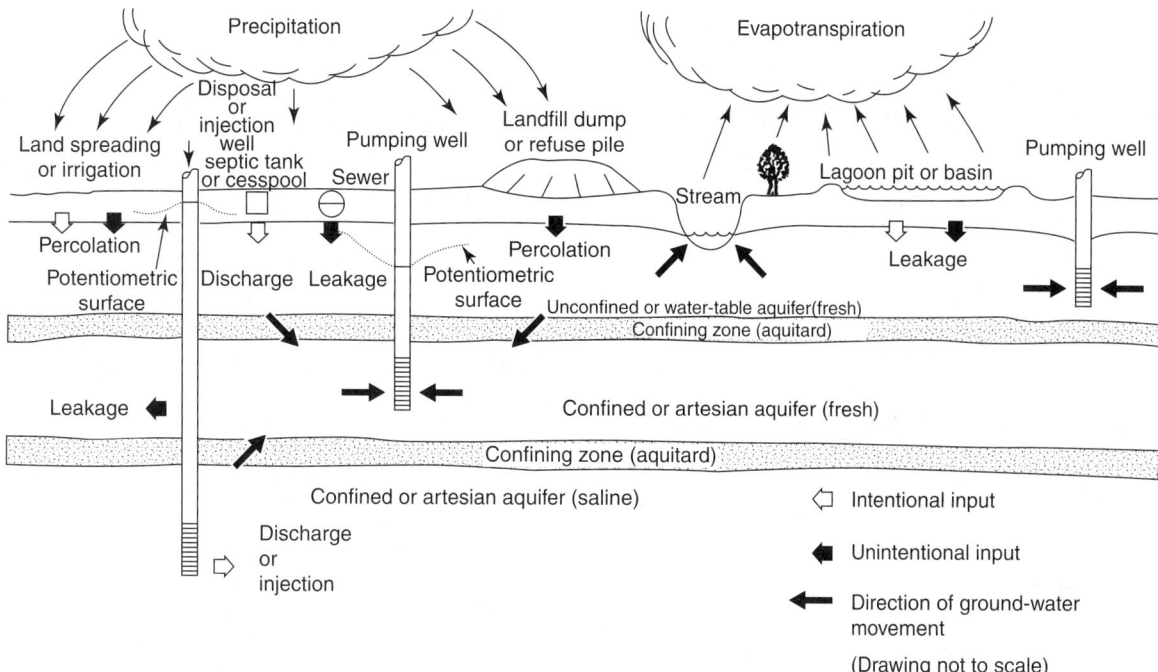

Fig. 2. Processes by which wastes can reach and contaminate groundwater systems. (*U.S. Environmental Protection Agency.*)

TABLE 2. REPORTED INCIDENTS OF GROUNDWATER CONTAMINATION IN TEN SELECTED STATES (Academy of Natural Sciences, Philadelphia)

State	Groundwater Consumed (Million Gallons/Day)	Source of Contamination
Arizona	4800	Industrial wastes; landfill leachate; human and animal wastes
California	13,400–19,000	Saltwater intrusion; nitrates from agriculture; brines and other industrial and military wastes
Connecticut	116	Industrial wastes, petroleum products, human and animal wastes
Florida	3000	Chlorides from saltwater intrusion and agricultural return flow; industrial wastes; human and animal wastes
Idaho	5600	Human and animal wastes; industrial wastes; radioactive wastes
Illinois	1000	Human and animal wastes; landfill leachate; industrial wastes
Nebraska	5900	Irrigation and agriculture; human and animal wastes; industrial wastes
New Jersey	790	Industrial wastes; petroleum products; human and animal wastes
New Mexico	1500	Oil field brines; human and animal wastes; mine wastes
South Carolina	200	Petroleum products; industrial wastes; human and animal wastes

Note: Probably the most newsworthy, publicized incidents have stemmed from public and industrial dumping sites, covered in this summary under "landfill leachate." A class of pesticides most commonly found in groundwater is nematocides. They are particularly difficult because manufacturers design them to be both persistent and toxic. DBCP (1,2-dibromo-3-chloropropane) is a representative nematocide.

at the consuming level, safety regulations and sheer convenience require that fuels be dispensed from underground tanks. Unfortunately, over the years, tank leaks have become an important cause of land pollution and ultimately the poisoning of aquifers.

Other toxic products that require some form of storage include solvents, chemical raw materials (fluid) and wastes that are "waiting" for final disposal/destruction.

Pollutant Pathways When Leaked Follow Different Pathways Underground. Relatively recently, cleanup specialists have made the surprising discovery that organic substances generally differ from inorganic substances when they are leaked underground. See Fig. 3. This information is helpful for preventing future leaks and for remedying situations where pollution already has occurred.

Means for Preventing Tank Leakage. In addition to the "common sense" approaches, such as selecting corrosion, weather, and moisture-resistant materials of construction, providing excellent foundations, and leak-detection instrumentation, there are additional preventive measures that can be taken. Cathodic protection, for example, is described in an article on **Corrosion**.

Secondary containment of underground storage tanks is another means. EPA regulations now require secondary containment systems for hazardous substance underground storage. Many industrial firms now are turning to flexible membrane systems, which offer a number of advantages over their rigid counterparts. EPA regulations for new tank construction permit four options:

1. Fiberglass-reinforced plastic.
2. Steel with cathodic protection, using either dielectric coating, field installation by a corrosion expert, or impressed current system.
3. Metal without cathodic protection if site is determined to be noncorrosive.
4. Steel-fiberglass–reinforced plastic composite.

Options for existing tanks include:

1. Interior lining.
2. Cathodic protection if internal inspection is conducted and tank is less than 10 years old and is monitored monthly.

Landvaults. Approval of aboveground permanent landvaults sometimes can be obtained. They have a number of advantages:

1. A double or triple composite liner system provides maximum protection.
2. They permit future retrieval of wastes for improved treatment when such a process may be developed, and, inasmuch as the vault is fully sealed, the underground geology need only be strong, stable, and free of any tectonic history.

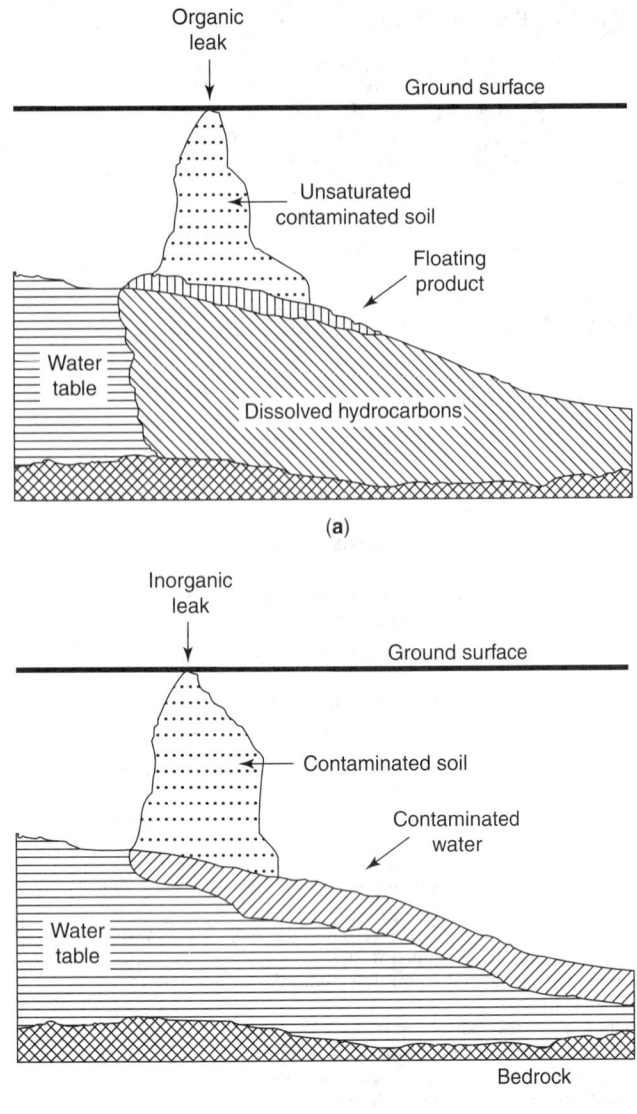

Organic
leak
↓
Ground surface

Unsaturated
contaminated soil

Floating
product

Water
table

Dissolved hydrocarbons

(a)

Inorganic
leak
↓
Ground surface

Contaminated soil

Contaminated
water

Water
table

Bedrock

(b)

Fig. 3. Schematic diagrams illustrate how organic and inorganic fluids differ in how they permeate ground and water table: (**a**) Organic substances contaminate the soil to the surface of the groundwater table. If contaminant is lighter than water and only slightly soluble, such as a petroleum product, a substantial layer of product may be found floating on the groundwater surface. The groundwater will contain varying concentrations of dissolved contaminants, depending upon solubility. (**b**) Inorganic compounds, such as acidic wastes containing metals, may be attenuated in their flow. Soils tend to attenuate movement of positively charged ions (cationic) metals at various rates, depending on the cation exchange capacity of the soil. Clays have a high cation exchange capacity. Thus, extensive contamination of soils by metals is infrequent. In contrast, acids can increase movement through the soil.

Pollution of the Seas and Oceans

Over the centuries, populations living along the tens of thousands of miles of coastline of the oceans and seas worldwide have dumped their refuse into these sinks of saline water. This was done with little if any reluctance or concern prior to the early 1900s. Ocean waters, without desalination, could not serve as drinking or irrigation water, and, inasmuch as these bodies of water are so tremendous as compared with the amounts of waste that may be added to them, there was no real sense of guilt when polluting them. Only within the last few decades has a public awareness of pollution of all types, including the oceans, developed. Modern means for exploring Earth's hydrosphere, from the surface to great depths, are relatively recent achievements. The American explorer, Charles Beebe, did not accomplish his remarkable descent to ocean depths until 1934, when he reached a depth of 3028 feet (923 m). Progress in the ocean sciences is described in several articles devoted to the ocean. See also **Ocean.**

The effects of polluted waters on the proliferation and edibility of fishes and other forms of ocean life have been researched extensively in recent years, and this knowledge has created a vital incentive for reducing pollution. Just a few years ago lightweight refuse that rises to the ocean's surface and littered beaches created an environmental crisis and another incentive for ceasing ocean dumping of sewage and solid wastes into the oceans.

The U.S. Congress passed the Ocean Dumping Act in 1972. To continue with such dumping, permits were required and the polluter had to demonstrate to the Environmental Protection Agency that the materials to be dumped would not "unreasonably degrade or endanger human health or the marine environment." Radiological, chemical, and biological warfare agents and high-level radioactive wastes were fully banned. The dumping of dredged materials from navigable waters was put under the regulation of the U.S. Army Corps of Engineers. Dredged material is comprised of a mixture of sand, silt, and clay, but can include rock, gravel, organic matter, and contaminants derived from a wide range of agricultural, urban, and industrial sources. As reported by R.M. Engler (U.S. Army Corps of Engineers), "Contaminated or otherwise unacceptable dredged material accounts for only a small fraction of the total — less than 10 percent in the U.S. and globally." Clean dredged waste has a number of positive uses, including the enhancement of wetlands and aquatic and wildlife habitats, beach nourishment, offshore mound and island construction, agriculture mariculture uses, and as a construction aggregate. All requests for the ocean dumping of dredged materials had to be accompanied by a list of suggested alternative actions.

As will be noted in the article on **Wastes and Pollution,** much emphasis is given to reducing pollution by creating less waste. "No Waste," of course, is an impractical target, but "Much Less Waste" can be achieved globally within the next several years, even considering major population increases. But, even then, from an economical and practical viewpoint, some scientists do not believe that the oceans can escape, in the long term, playing a major role as a "waste sink." As of the mid-1990s, both scientific and lay opinions were strongly polarized. The oceans can accommodate huge annual volumes of nonhazardous, nonfloatable wastes. Much scientific research and engineering remains to be done in removing toxic materials prior to dumping and thus protect ocean life forms and the people who eat seafood. Detoxification is mandated now prior to land dumping. Also, much more knowledge of ocean characteristics is required in the way of selecting dumping sites.

Oceanologists and waste-handling professionals presently are attempting to answer the question, "How can society use the oceans for waste disposal without harming the marine environment or fisheries resources?" Much research remains, and numerous other questions remain unanswered.

The Boston Harbor Project

For nearly a century, scores of villages (later to become cities) developed around the city of Boston, Massachusetts. The Boston Harbor and associated rivers (Charles, Mystic, and Neponset) traditionally have been used for the disposal of sewage wastes. In 1904, Boston established a centralized area sewage disposal system, which because of population expansion, has multiplied in capacity manyfold. The cities that cluster around Boston found it easier and less costly to hook into the Boston central sewage system than to build their own treating and disposal facilities. The result over the past 20 years or so has been no surprise, and Boston Harbor became known as one of the most polluted in the world. Reconstruction of the Boston sewage system commenced during the same time frame as increased state and federal regulations governing sewage disposal. Although construction of a 9.5-mile (25-km) tunnel to carry sewage wastes out of the immediate Boston harbor area to Massachusetts Bay was nearing completion as of 1994, this project serves as an interesting example of how a major environmental project can become entangled in legislation and regulatory red tape and the involvement of citizen groups in environmental issues. An excellent background description is contained in the Spring 1993 issue of *Oceanus* magazine.

The waste-transporting tunnel was bored through bedrock that supports the seafloor of Boston Harbor and Massachusetts Bay. With exception of the tunnel bored under the English Channel (see also **Tunnel Engineering**), it was the major tunneling project to take place during the 1980s and 1990s. Toward the end of the tunnel, a series of 55 vertical effluent discharge pipes release the sewage into the bay.

Further Quantification of Ocean Data Needed

Factors[2] that remain to be studied concerning the oceans in terms of waste disposal include:

Physical — Diffusion, advection, sedimentation.
Chemical — Volatilization, neutralization, precipitation, flocculation, adsorption, desorption, dissolution, oxidation.
Benthic — Geochemical, biological.

Over the past several decades, oceanographers have found that the oceans and seas are not as uniform as once contemplated, ranging with regard to their latitudinal and longitudinal locations and also with depth. This suggests, then, that various areas of the ocean may be more suited to accept anthropogenic debris than others. For example, temperature profiles, depths, thermal gradients, and numerous other physical, biochemical, and life-sustaining qualities are known to exist.

An early appreciation of natural oceanic detritus is given by biologist Rachel Carson, who in 1950 observed in her book, *The Sea Around Us*, "When I think of the floor of the deep sea, the single overwhelming fact that possesses my imagination is the accumulation of sediments. I see always the steady, unremitting, downward drift of materials from above, flake upon flake.... For the sediments are the materials of the most stupendous 'snowfall' the earth has ever seen." Fine particles are sinking into the global oceans every microsecond of the day and night. Such particles may include "shreds and motes and globs of stuff" — dead and decayed remains of plants and animals, meteorites and other cosmic dust, old lobster molts, volcanic fallout, radioactive fallout, the pollen from flowers, grains of sand from the deserts — in summary, just about everything that is airborne contributes to oceanic detritus. The rate of passage of suspended material from the ocean's surface to the bottom of the sea varies widely, but some scientists estimate that for some particles it may take thousands of years to reach the ultimate deep-sea graveyard. It is only in recent years that some oceanologists have studied this phenomenon seriously. Some detritus never may reach the sea floor because it is consumed by various forms of ocean life as food. A. Aldredge and M. Silver (University of California) observe, "Marine snow particles are typically smaller than a pinhead. However, in Monterey Bay off California (and probably elsewhere), the particles may be 4 inches (10 cm) across. These particles are usually aggregates of many smaller particles that stick together, often with mucus.... The largest type of marine snow is the *giant house*, which can be 6 feet (1.8 meter) across. Each house is a blob of mucus that has been secreted by a zooplankter. Some investigators refer to them as "floating islands.... While a single zooplankter inhabits its balloon of mucus, there may be hundreds of copepods on its exterior, apparently grazing on the nourishing tidbits of *marine snow*.... We find typically that *marine snow* hosts organisms in very high concentrations.... There are certain groups of organisms that seem to be found only on these.... Among the unusual characters are rare species of copepod, certain protozoans and a unique assemblage of bacteria.... *Marine snow* is a major means by which material reaches the ocean floor, a major vehicle for the transport of organic matter down to the ocean's interior.... As these particles sink through the water, they change continuously, as do the communities living on them."

Scientists have observed that *marine snow* can be a nuisance. Because in some cases gas may be produced, particles rise and create a scum, and this can be dried by the sun to produce a surface of sufficient strength to permit seagulls to walk upon it. Such scum extends for many thousands of acres (hectares) in the Adriatic Sea, where it has become a menace to fishermen and the tourist trade. Such scums were reported as early as the 1700s.

Natural Pollution of the Oceans. Frequently overlooked is what may be termed "natural" pollution, which, when coupled with artificial (anthropogenic) pollution, contributes to the sum total of all pollutants found in fresh and ocean waters worldwide. Deep fissures in the ocean floor, fumaroles, and seamounts (underwater volcanoes) release megatons of sulfur-laden and other noxious gases into ocean water; other discontinuities in the ocean basins release vast quantities of crude oil and other hydrocarbons. Surface volcanoes are major contributors to atmospheric pollution, much of which ultimately affects Earth's hydrosphere. The present dissolved solids content of the oceans represents natural water pollution that has taken place ever since the land masses rose above sea level — through a constant erosion of soil.

[2] As suggested by I.W. Duedall (Florida Institute of Technology).

Oil and Hydrocarbon Pollution of the Oceans

Petroleum is not a substance foreign to the marine environment. Natural seeps have been discharging petroleum hydrocarbons into the marine environment for millions of years, in amounts substantially greater than those resulting, for example, from present offshore production activities. About 200 submarine oil seeps have been identified worldwide. There is little doubt that many more exist. Petroleum has also continuously entered the seas as a result of erosion of uplifted sedimentary rocks containing trace amounts of petroleum hydrocarbons. There is also evidence of organisms living in the ocean that biologically produce hydrocarbons, ranging from gases (methane, ethane) through liquids, to solid paraffin waxes of high molecular weight.

These substances enter the marine environment from a variety of sources, both through natural phenomena and anthropogenic activities. In 1985, the National Academy of Sciences (NAS) published an assessment of petroleum pollution of the world's oceans and estimated that between 1.7 and 8.8 million metric tons per annum (mta) of oil enter the oceans. Within this range, 3.2 mta is regarded as the best single estimate — equivalent to about 0.1% of the total oil produced annually worldwide (about 3 billion metric tons).

Table 3 compares these estimates with those published by NAS in 1972. All categories except spills show about a 50% decrease. According to the NAS, part of the reduction in most categories can be attributed to refinements in estimating techniques. Other reductions, however, are attributed to efforts to reduce oil pollution, such as the international program to reduce tanker operational discharges. Because annual Figures for marine accidental spills vary significantly, an average over a number of years was used for both the 1975 and 1985 reports. The increase shown for such spills in the 1985 study reflects the influence of major incidents — such as the loss of the 220,000 dead-weight-ton tankers *Amoco Vadiz*, which occurred in the period (1975 through 1980) — used to calculate its annual average. The Alaskan oil spill (1989) is discussed later.

TABLE 3. PETROLEUM INPUTS TO THE OCEANS FROM DIFFERENT SOURCES

Source	1975 Study[1]	1985 Study[2]
	(Million Metric Tons Annually)	
Municipal and Industrial Wastewater		
Discharges and Runoffs	2.5	1.0
Refinery Wastewater Discharges	0.2	0.1
Offshore Oil Production	0.08	0.05
Marine Transportation		
Tanker Operations	1.3	0.7
Accidental Spills	0.2	0.4
Other Maritime Activities	0.6	0.4
Natural Seeps and Erosion	0.6	0.25
Atmospheric Fallout	0.6	0.3
	6.1	3.2

[1] Petroleum in the Marine Environment, National Academy of Sciences, 1975.
[2] Oil in the Sea: Inputs, Fates, and Effects, National Academy of Sciences, 1985.

Most surface and near-surface open ocean waters contain petroleum hydrocarbons in the range of about 1 to 10 parts per billion (ppb), according to the NAS 1985 study. In deeper open ocean waters, the concentrations are 1 ppb or less. (One ppb is equal to about one drop of oil in 100,000 quarts of water.) Coastal waters, particularly those near populated and industrialized areas where the presence of oil is most likely, show higher levels (up to 100 ppb). Laboratory experiments conducted to determine the toxicity of crude oils or petroleum products indicate that concentrations from 10 to 100 times greater than those of coastal waters are required before measurable effects on marine organisms can be detected.

The process whereby an oil slick from a sudden spill is ultimately disposed has been the subject of intense research. Figure 4 depicts the processes occurring in the water, the overlying atmosphere, and the underlying bottom sediments. Figure 5 relates the time following discharge of oil into the sea to the various processes of movement and degradation. These processes include:

Drift. The movement of the center of a mass of oil on the surface of the ocean, drift, is caused by the combined action of wind, surface currents,

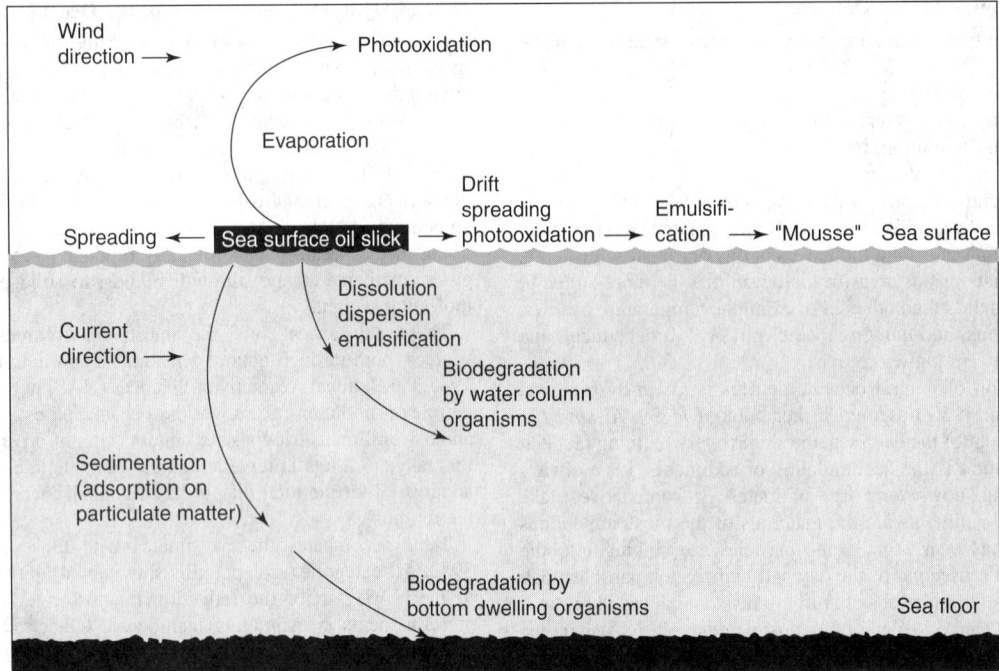

Fig. 4.   Processes that act upon an oil slick.

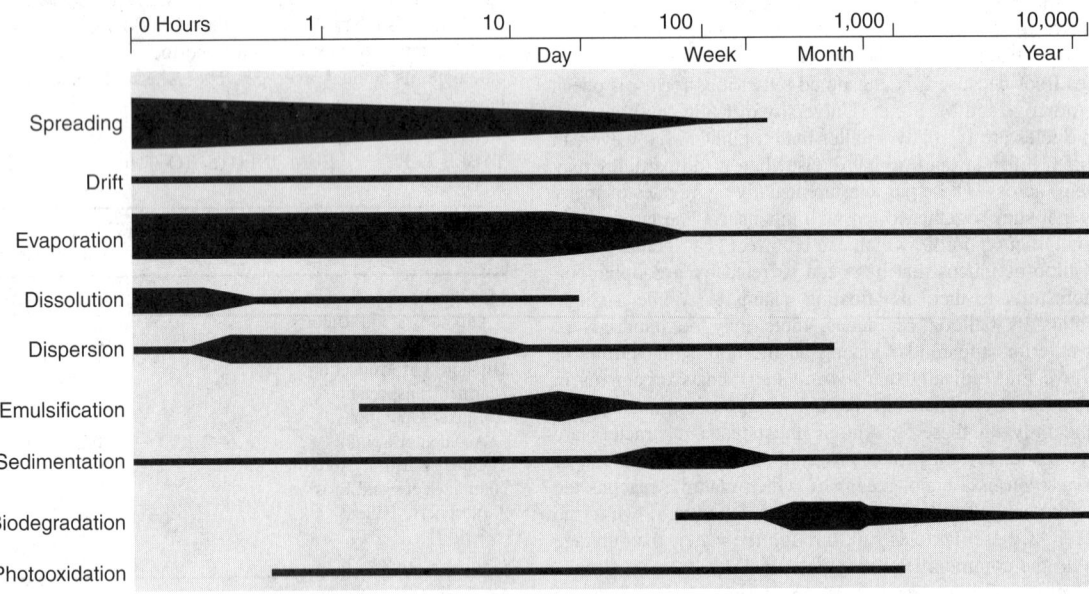

Line length-probable time span of any process.
Line width-relative magnitude of the process both through time and in relation
to other contemporary processes.

Fig. 5.   Time span and relative magnitude of processes acting on spilled oil.

waves, and tides. With larger amounts of oil, such as occur with accidental spills, the oil drift is largely independent of spill volume, spreading, or weathering. Since the "thick" portion of the slick drifts faster than the "thin" portion, a heavy oil accumulation forms the leading edge of an advancing slick. The drift process is always active—from the moment oil is released into the sea until the oil disappears from the surface. It is difficult to predict precisely all the complex interactions of oceanographic and meteorological factors that influence drift of an individual oil slick. However, field and laboratory observations of oil slick drift are surprisingly consistent.

Evaporation. The primary weathering process involved in the natural removal of oil from the sea is evaporation. It is particularly dominant soon after oil is released. Evaporation involves the transfer of hydrocarbon components from the liquid oil phase to the vapor phase. Estimates from

major spills as well as experimental data indicate that evaporation may be responsible for the loss of up to 50% of a surface oil slick's volume during its life. Evaporation rates of oil at sea are determined by wind velocity, water and air temperatures, sea roughness, and oil composition. Some of the light, low-boiling hydrocarbons, such as benzene, toluene, and xylenes, which are rapidly lost through evaporation, are the most toxic. Thus, their removal decreases toxicity to marine life of the oil remaining on the surface.

Photooxidation. Much of the oil that evaporates is photo-oxidized in the atmosphere, but in the absence of good analytical data the contribution of this process cannot be estimated. It is reasonable to assume that some of the oil returns to the seas as atmospheric fallout.

Dissolution. This is another early process acting on spilled oil. Dissolution involves the transfer of oil compounds from a floating slick, or from dispersed oil droplets, into solution in the water phase. Lower-molecular-weight compounds tend to be the most soluble. It is unlikely, however, that dissolution significantly affects slick weathering as only a very small fraction of the oil dissolves. Evaporation proceeds much more rapidly than dissolution.

Dispersion. Oil also enters the water in forms larger than dissolved molecules. In natural dispersion, small droplets of oil (ranging in diameter from very small fractions of a millimeter to a few millimeters) are incorporated into the water in the form of a dilute oil-in-water suspension. Natural dispersion reaches a maximum rate in only a few hours (4 to 10) following a spill, but continues for some time. Oil dispersion is influenced by oil composition (e.g., wax and asphaltene content), density, viscosity, oil-water interfacial tension, and water turbulence. Crude oils and many petroleum products contain trace amounts of nitrogen, sulfur, and oxygen-bearing organic compounds, which can act as natural surfactants. These surfactants reduce the oil-water interfacial tension, allowing the oil to break up and to disperse into droplets more readily. Chemical dispersants may be applied to an oil slick to supplement the natural surfactants, thereby enhancing the dispersion process. The primary purpose for removing oil from the surface through dispersion is to enhance the degradation process. The increase in the surface area of dispersed oil droplets resulting from surfactant action accelerates the degradation of oil. Accelerating the dispersion process through the use of chemical dispersants also reduces the threat of floating oil stranding on a shoreline, where it can damage biota and property.

Emulsification. This is a water-in-oil process in which water is incorporated into the floating oil. Such emulsions, which may contain from 20 to 80% water, are often very viscous and referred to as "mousse." Mousse formation is highly dependent on oil composition. High levels of asphalt-type compounds, as well as waxes, appear to promote the formation of these emulsions. Ocean turbulence also accelerates mousse formation, although a fully developed, stable emulsion may be formed from some oils under relatively quiescent open-water conditions. Early treatment of spilled oil with chemical dispersants is an excellent way to prevent emulsification.

Sedimentation. Some organisms may ingest dispersed oil droplets in the water column and subsequently deposit them as fecal pellets. In some instances, this has been estimated to be a significant form of sedimentation.

Biodegradation. This is an important process for removing petroleum hydrocarbons from the marine environment. All surface waters, fresh or marine, contain natural populations of bacteria, yeast, and fungi capable of metabolizing and chemically degrading hydrocarbons through their normal life processes. These organisms are also primarily responsible for degrading most of the biologically produced hydrocarbons in the ocean. The rate and extent of biodegradation depend on the abundance and variety of existing microorganisms, their predators, available oxygen and nutrients, temperature, and oil composition. Hydrocarbons, dissolved or dispersed in water, are the most easily degraded. Degradation of hydrocarbons contained in bottom sediments also occurs if oxygen is present. Emulsified oil (mousse) is slow to degrade because water is trapped within the emulsion and the nutrients and oxygen essential to biodegradation are kept out.

Invention of Oil-Eating Bacterium. It is interesting to note that the concept of genetically altering bacteria to transform crude oil into cattle feed was proposed by A.M. Chakrabarty (General Electric Co.) in the late 1960s. A patent was not granted until 1980 after considerable litigation. The concept is considered by many as the cornerstone of the biotech industry. Chakrabarty as of the 1990s continues in this field. Although applications in other areas of the petroleum industry may find applications for such bacteria, the primary interest in recent years has been in connection with oil tanker spill cleanups.

After the Alaskan oil spill of March 1989, approximately 70 miles (113 km) of beaches around Prince William Sound were sprayed with a fertilizer (*Inipol*) that had been invented by a French petroleum company (*Elf Aquataine*). The effort was made to stimulate the growth of naturally occurring bacteria (i.e., microorganisms that eat petroleum). This was the first large-scale test of bacteria for cleaning up an oil spill. Improvement was noted within a couple of weeks after the spraying. Research continues with the material to determine its effectiveness on shoreline rocks and pebbles. Early findings indicated that oil caught beneath the surface of rocks was consumed in 6 to 7 weeks. Researcher C. Oppenheimer (University of Texas) also developed microbial strains for producing fatty acids from oil, the product of which is more soluble in water. These compounds serve as food for plankton and other organisms.

Oil Tanker Spills

Usually the most publicized and one of the most dramatic examples of ocean water (and adjacent shoreline) pollution involves oil tanker or barge accidents. Most often, the saline waters of the oceans and seas are polluted, although there are instances where such accidents have occurred in fresh and brackish waters.

Oil spills present many different variations in the manner in which they develop and react to cleanup efforts. The Alaskan spill that occurred on March 24, 1989, in Prince William Sound was the most extensively researched to date. The cost of cleaning up the spill also exceeded all other tanker spill expenses to date. A study reported by the U.S. Forest Service estimates the final fate of the 10 million gallons as follows: evaporated, 35%; recovered, 17%; burned, 8%; biodegraded, 5%; and dispersed, 5%. The total in the form of oil slicks on the Sound amounted to about 10% of the original spill and that on the shoreline about 18%. For many weeks after the spill, a fleet of specially equipped vessels mopped up about 120,000 gallons of crude oil per day. Other tanker spills have carried crude oil of somewhat different composition. The state of the sea, temperature, winds, and presence or absence of sunshine are among other variables that make each spill unique.

Tanker Design. One ship designer has observed that modern tankers are uniquely fragile and unwieldy vessels. The goal of tanker design is "to get as much cargo as you can into as little steel as possible and still have economical propulsion." Thus, the larger the ship, the easier to meet these specifications. Ultra-large crude carriers (ULCCs) required a distance of 3 miles (4.8 km) and about 20 minutes to stop from a top speed of 15 to 16 knots. Oil tankers have steadily increased in capacity and length over the last few decades. See Fig. 6.

Because of several major tanker incidents over the last few years and highlighted by the Alaskan spill, much design thought has gone into the "double-hull" concept. Indeed, this is not a new concept because two "skins" are used on the nearly 60,000 merchant vessels afloat — with the important *exception* of oil tankers. Double hulls have been standard on liquefied natural gas (LNG) ships for many years, and the design has been credited with preventing disasters at sea. The double-hull design enabled one LNG ship to sail many miles at its highest speed to the nearest port even though the outer hull had been torn open under several of the cargo tanks. None of the highly volatile cargo escaped.

The oil industry has objected to two hulls for the following reasons:

1. If oil leaked from the inner to the outer shell, the space in between could generate a vapor and thus be an explosion hazard.
2. When the outer hull was breached by an accident, water would fill the void, causing the ship to lose buoyancy and possibly go aground.

Even with these objections, there are some 530 oil tankers with double hulls, and these have been accident-free thus far.

Other improvements in tanker design have included more precise navigation systems that are customized to the vessel and the coarse of travel that it usually follows. In a single video readout are shown the ship's location and course with respect to shoreline, bottom contours, buoys, markers, and other ships. With another system, which includes radar reflectors located along the shoreline, the ship location can be determined within less than about 6 feet (1.8 meters). Another concept embraces use of a funnel that can be lowered from the ship immediately when a spill occurs and sucks the spilled oil back into the ship. A design of this kind has been under test in the Gulf of Mexico.

Radioactive Waste Dumping. Numerous proposed solutions for dumping radioactive wastes, including ocean burial, are described in the article on **Nuclear Power**.

Additional Reading

Abel, P.D.: "Water Pollution: Biology," 2nd Edition, Taylor & Francis, Inc., Philadelphia, PA, 1996.
Abelson, P.H.: "Oil Spills," *Science*, 629 (May 12, 1989).
Aubrey, D.G. and M.S. Connor: "Boston Harbor: Fallout Over the Outfall," *Oceanus*, 61 (Spring 1993).
Barinaga, M.: "Alaska Oil Spill: Health Risks Uncovered," *Science*, 463 (August 4, 1989).
Battle, J.B. and M. Lipeles: "Water Pollution," 3rd Edition, Anderson Publishing Company, Cincinnati, OH, 1998.

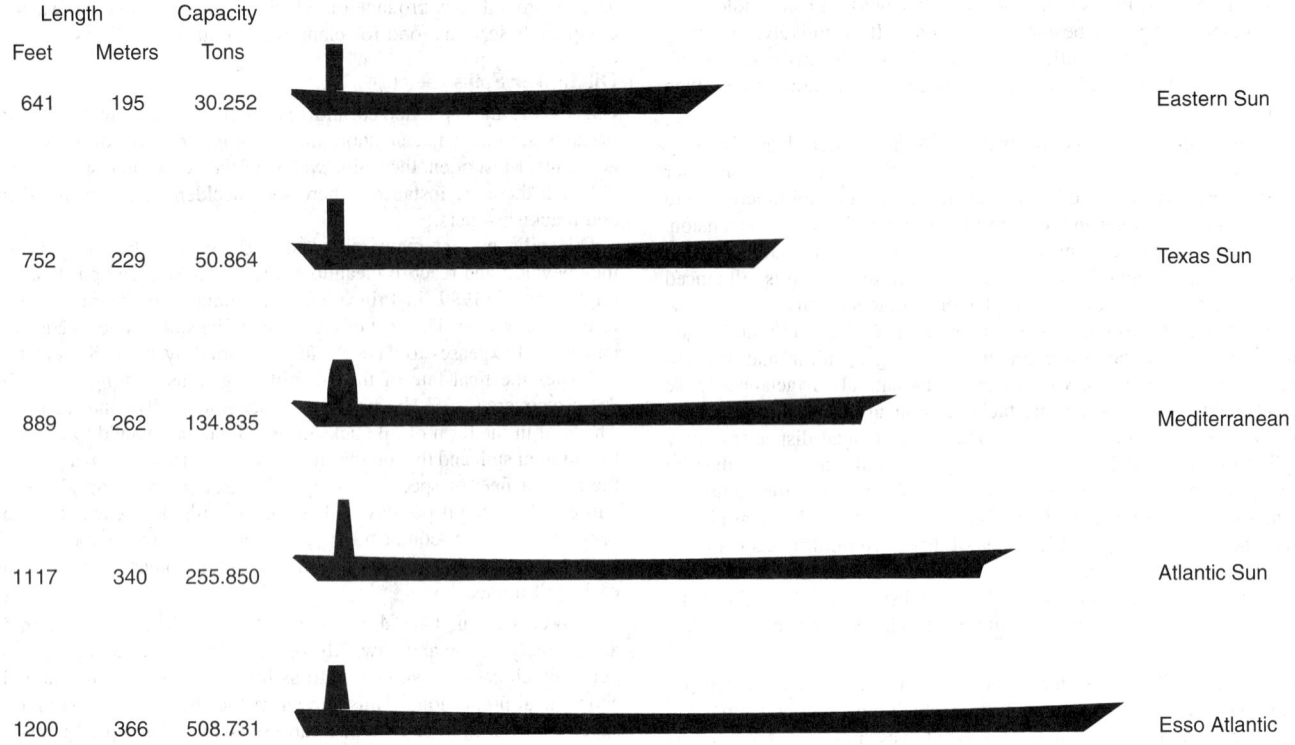

Length		Capacity		
Feet	Meters	Tons		
641	195	30.252		Eastern Sun
752	229	50.864		Texas Sun
889	262	134.835		Mediterranean
1117	340	255.850		Atlantic Sun
1200	366	508.731		Esso Atlantic

Fig. 6.   Trends in oil tanker design characteristics. (*After A. Dane.*)

Broadus, J.M.: "Tailoring Waste Disposal to Economic Realities," *Oceanus*, 707 (Summer 1990).

Cadwallader, M.: "Above-Ground Landvaults for Waste Containment," *Chem. Eng. Progress*, 9 (August 1989).

Capuzzo, J.E.M.: "Effects of Wastes on the Ocean: The Coastal Example," *Oceanus*, 39 (Summer 1990).

Clarke, E.H., II, A. Haverkamp, and W. Chapman: "Eroding Spo's" The Off-Farm Impacts, Conservation Foundation, Washington, DC, 1985.

Crawford, M.: "Bacteria Effective in Alaska Cleanup," *Science*, 1537 (March 30, 1990).

Curtis, C.E.: "Protecting the Oceans," *Oceanus*, 19 (Summer 1990).

Dane, A.: "America's Oil Tanker Mess," *Popular Mechanics*, 51 (November 1989).

Dane, A.: "Oil Slick Buster," *Popular Mechanics*, 58 (May 1990).

Dane, A.: "Learning from Disaster," *Popular Mechanics*, 94 (September 1991).

Duedall, I.W.: "A Brief History of Ocean Disposal," *Oceanus*, 29 (Summer 1990).

Eckenfelder, W.W., Jr.: "Industrial Water Pollution Control," 3rd Edition, The McGraw-Hill Companies, Inc., New York, NY, 1999.

Erickson, D.: "Oil-Eating Bacterium that Spawned an Industry," *Sci. Amer.*, 88 (June 1990).

Grassle, F.: "Sludge Reaching Bottom at the 106 Site, Not Dispersing as Plan Predicted," *Oceanus*, 61 (Summer 1990).

Haberl, R., P. Cooper, R. Perfler, and J. Laber: "Wetland Systems for Water Pollution Control 1996," Elsevier Science, New York, NY, 1997.

Hawley, T.M.: "Herculean Labors to Clean Wastewater," *Oceanus*, 772 (Summer 1990).

Helmer, R., I. Hespanhol: "Water Pollution Control: A Guide to the Use of Water Quality Management Principles," Routledge, New York, NY, 1997.

Higgins, T.E. and W.D. Byers: "Leaking Underground Tanks: Conventional and Innovative Clean-Up Techniques," *Chem. Eng. Progress*, 12 (May 1989).

Hodgson, B. and N. Forbes: "Alaska's Big Spill: Can the Wilderness Heal?" *Nat'l. Geographic*, 5 (January 1990).

Hollister, C.D.: "Options for Waste: Space, Land, or Sea?" *Oceanus*, 13 (Summer 1990).

Holloway, M.: "Soiled Shores," *Sci. Amer.*, 102 (October 1991).

Holloway, M.: "Abyssal Proposal (Ocean Depths and Sewage Sludge)," *Sci. Amer.*, 30 (February 1992).

Hooker, L.: "Danger Below (Underground Aquifers)," *Chem. Eng. Progress*, 52 (May 1990).

Kistos, T.R., J.K.M. Bondareff: "Congress and Waste Disposal at Sea," *Oceanus*, 23 (Summer 1990).

Leo, W.M. et al.: Before and After Case Studies, EPA-430/9-007, Environmental Protection Agency, Washington, DC, 1984.

Levy, P.F.: "Sewer Infrastructure," *Oceanus*, 53 (Spring 1993).

Liptbak, B.G. and D.H. Liu: "Groundwater and Surface Water Pollution," Lewis Publishers, Boca Raton, FL, 1999.

Marshall, E.: "Valdez: The Predicted Oil Spill," *Science*, 20 (April 7, 1989).

Mayer, J., S. McClurg: "Water Pollution," Water Education Foundation, Sacramento, CA, 1996.

Moore, J.W., S. Ramamoorthy: "Heavy Metals in Natural Waters," Springer-Verlag, Inc., New York, NY, 1984.

Noll, K.E., V. Gounaris, and Wain-sun Hou: "Adsorption Technology for Air and Water Pollution Control," Lewis Publishers, Boca Raton, FL, 1991.

Peterson, S.: "Alternatives to the Big Pipe (Boston)," *Oceanus*, 71 (Spring 1993).

Schmitz, R.J.: "Introduction to Water Pollution Biology," Buterworth-Heinemann, Inc., Woburn, MA, 1995.

Semonelli, C.T.: "Secondary Containment of Underground Storage Tanks," *Chem. Eng. Progress*, 78 (June 1990).

Smith, R.A., R.B. Alexander, and M.G. Wolman: "Water-Quality Trends in the Nation's Rivers," *Science*, **235**, 1607–1615 (1987).

Spencer, D.W.: "The Ocean and Waste Management," *Oceanus*, 5 (Summer 1990).

Staff: "Prevention of Water Pollution by Agriculture and Related Activities," Bernan Associates, Lanham, MD, 1993.

Staff: National Research Council, Ocean Studies Board, and Water Science Technology Staff, "Clean Coastal Waters: Understanding and Reducing the Effects of Nutrient Pollution," National Academy Press, Washington, DC, 2000.

Stegeman, J.J.: "Detecting the Biological Effects of Deep-Sea Waste Disposal," *Oceanus*, 54 (Summer 1990).

Stone, R.: "Icy Inferno: Researchers Plan Oil Blaze in Arctic," *Science*, 1203 (September 13, 1991).

Viessman, W. and M.J. Hammer: "Water Supply and Pollution Control," 6th Edition, Addison Wesley Longman, Inc., Reading, MA, 1998.

Wolman, M.G.: *Science*, **174**, 905 (1971).

WATER RESOURCES. As pointed out by various authorities, major problems pertaining to water, in addition to pollution, include: (1) the heavy consumption, which began some years ago to limit the growth of various cities and of agriculture, particularly in the southwestern United States; (2) evaporation losses from reservoirs and storage ponds, particularly important in the arid western and west central sections of the United States, the control of which may be found to be more economic than that of developing new water sources; (3) lowering water Tables, again of major concern to the southwest and Pacific coastal regions of the United States, but also becoming a major factor near larger cities, and to some middle-Atlantic areas of the United States, a situation which is exerting considerable influence on the choice of new irrigation areas; (4) long-distance water transmission systems, which in the future may not be confined to the western United States; (5) waste water return to the oceans, which can become the largest and least expensive potential secondary water source and a very attractive source for many industrial uses; (6) salt water encroachment, which already is destroying some water sources and land; (7) watershed trash vegetation, the eradication of which can increase water yield, particularly in the southwestern United States; and (8) storage of seasonal and flood flows, which has been practiced for many years in most areas that are away from good lake or groundwater supplies, but which will require extension as the water problem becomes more severe in the less-arid areas. See also **Dam**.

The water supply problems of the United States image those of many other areas of the world. As is evident from Fig. 1, the eastern one-third of the United States, excepting a few areas and several of the major cities, does not generally run a seasonal water deficiency. Somewhat over one-third of the western United States, however, is characterized by no available surplus, and other areas by summer deficiency and winter surpluses.

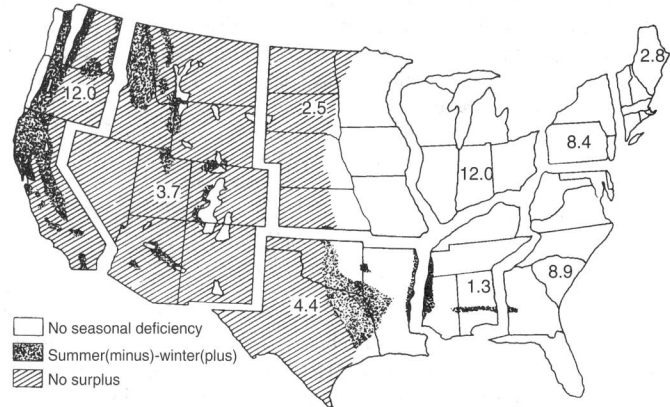

Fig. 1. Water supply versus population increases. Figures indicate expected increases (millions of people) by the year 2000. (*U.S. Department of the Interior.*)

Water resources must be evaluated in the long term because unusually wet or dry winters (such as 1980–1981), while of near-crisis proportions, do not represent the average of a decade or more. What is important is to watch trends that appear to be consistent.

Widespread flooding in the midwestern United States during the summer of 1993 is regarded by some experts as a 1 in 100 year experience. Every century experiences a few weather phenomena that statistically contribute to the minimums and maximums of data, but do not necessarily predict any permanent changes in the weather picture for a given region. Final analyses of data from the 1993 flooding will not be completed for another year or two.

The environmental problem of finding additional sources of clean water have some parallels with the problem of finding new ways for disposing of wastes. These are indeed limited. Again, the subject of *waste reduction* enters into the picture. Whereas, fundamentally, pollution equates with cutting back on the production of wastes, water shortages equate markedly with reducing the waste of water. This, too, presents a variety of sociological preferences and concerns. Persons who live in communities where water shortages approach near-crisis proportions are fully aware of how inconvenient it is to cut back on water consumption. Over the years, a few imaginative suggestions have been made. Weather alteration in an effort to produce rain essentially has been abandoned by technologists. Dam building, in addition to creating hydroelectric power, also contributes to smoothing out the water supply for many regions. In recent years, however, a substantial citizen's movement against creating dams has arisen.

Although nearing the fantasy level, icebergs have not been ruled out technically as "portable and potable" sources of excellent water.

While desalination now serves a number of the arid regions of the world (see also **Desalination**) and the costs of desalination have been reduced because of improved processes, the economics of processing discourage the use of desalination in the less arid regions of the world. However, if the world's population continues to expand geometrically, the price per gallon or liter of drinking water, for example, may rise beyond belief.

Water from Icebergs. The concept of obtaining fresh water from icebergs dates back many years. One of the earlier proponents was Isaacs (Scripps Institute of Oceanography) who described the concept in the 1940s. The concept has been characterized by a cycling of interest over the years, but with little follow-through among world scientists and planners. In

1977, the first conference on iceberg utilization of major proportions was held at Iowa State University, with over 200 scientists, consultants, and representatives of private firms from 18 nations present to consider the technical, economic, environmental, and legal problems that may be involved in transporting and exploiting icebergs. In discussions over the last several years, most attention has been given to icebergs from the Antarctic rather than the Arctic region, principally those icebergs that break away from the Ross Ice Shelf.

Melted water from an iceberg is extremely pure, with only traces of rock fragments and with almost complete absence of trapped organic matter. Contamination has been estimated at one part per billion and thus far superior to other fresh water unless it is distilled. The Iowa State conference was financed in part by the National Science Foundation and Saudi Arabia. In contemplating the possible use of icebergs as a future source of fresh water, it is interesting to note that although there are about 1.4 billion cubic kilometers of water on Earth, only about 9 million cubic kilometers (six-tenths of one per cent) of the total is both liquid and fresh. The useful supply tends to be about 10–20% of the total precipitation per year, or about 10,000–20,000 cubic kilometers.

The ice sheet in the Antarctic accumulates an input of precipitation equivalent to about 2000 cubic kilometers of water per year. About half of this total forms into tabular (flat) icebergs, each equivalent to about 200 cubic kilometers of water. The slabs are from 200 to 250 meters thick and up to about 0.5 kilometer wide. Greenland icebergs, in contrast, are more irregular in shape and size. Antarctic icebergs would have to be transported about 4,000 miles (6,436 kilometers) to the southwestern coast of Australia; about 9,500 miles (15,286 kilometers) to southern California; and about 12,000 miles (19,308 kilometers) to the Middle East. Several studies have been made as regards towing icebergs and the consensus is that the technology for this already exists. Towing of some icebergs already has been done in connection with oil drilling activities, where tugs are used to deflect Greenland-originated icebergs from the path of drillships and platforms. Also, many attempts have been made to destroy (break up) icebergs through the use of explosives to prevent the icebergs from drifting into major sea lanes. These experiments have largely been unsuccessful because one kilogram of explosive will break up only about 4 cubic meters of ice. Even though ice possesses a low strength and density, it reacts to blasting much like ordinary rock.

During the 1990s, the use of icebergs as a source of fresh water remained a topic of considerable conjecture. Topics discussed at the Iowa State Conference included: (1) use of earth resources satellites to find the most suitable icebergs for this purpose; (2) study of various transportation modes, such as tugs for pulling, semi-submersibles and submarines for pushing, and propellers mounted on icebergs to make them self-transportable; (3) ways to minimize melting during transport, e.g., stretching plastic sheets over the iceberg or spraying its surface with urethane foam; (4) investigation of legal problems relating to ownership of Antarctic ice; (5) the numerous environmental effects which such actions might provoke; (6) possible use of icebergs as energy sources, through harnessing of thermal and salinity gradients. See also **Ocean Resources (Energy)**. If the concept takes hold, it would seem that the first tests would be conducted in the Southern Hemisphere.

For further details, see "The Iceberg Cometh," by W.F. Weeks and Malcolm Mellor, *Technol. Rev. (MIT)*, pp. 66–75 (August/September, 1979), and *Science*, **198**, 274–276 (1977).

Many of the scientific aspects of water resources are described in several entries in this encyclopedia. See also **Climate**; **Connate Water**; **Desalination**; **Drainage Systems**; **Estuary**; **Groundwater**; **Hydrology**; **Precipitation and Hydrometeors**; **Soil**; **Water Pollution**; and **Watershed**.

WATER SCORPION (*Insecta, Hemiptera*). Moderately large water bugs of oval or slender and elongate form, with a long breathing tube at the end of the body. The front legs are adapted for catching small prey.

WATERSHED. The area that supplies water to a stream and its tributaries by direct runoff and by ground water runoff is the drainage area or watershed for the stream. The yield of the watershed is the direct runoff plus the ground water runoff for a given period of time. That part of yield used for industrial or domestic purposes is called the draft. When the draft approaches the yield, the latter may be adjusted by the construction of an impounding reservoir. See also **Drainage Systems**; and **Hydrology**.

WATER SNAKES. See **Snakes**.

WATERSPOUT.   See **Fronts and Storms**.

WATER STRIDER (*Insecta, Hemiptera*).   A moderately large bug whose two posterior pairs of legs are modified for locomotion on the surface film of water. The claws are set back from the tip of the leg and the hairs of the dense covering are turned under so that the tip of the leg rests on their curved surfaces. These bugs are common on streams and ponds. The members of one genus, *Halobates*, are the only truly marine insects. They are found in tropical waters, often far from land, and their eggs have been found attached to floating feathers.

WATER TEMPERATURE.   See **Atmosphere-Ocean Interface**.

WATSON, JAMES DEWEY (1928).   James Watson, an American chemist, was a brilliant student who began college at the University of Chicago at the young age of 15. Since he was young, Watson had a passion for bird watching and this interest was probably a factor in his fascination with genetics. He is known for his research contributions in the field of genetics.

He began working with Francis Crick in 1950 at the Cavendish laboratories. In 1953, Watson and Crick, using the photographs of Rosalind Franklin, which exposed crystallized molecules from the nucleus, identified the material that biologists were viewing in the nucleus as DNA. Watson and Crick created a three-dimensional structure DNA model, which provided scientists with a valuable tool in the study of heredity. In 1962 Watson-Crick were awarded the Nobel Prize for their work.

Watson taught at Harvard and CalTech and in 1968 he became the directorship for Cold Spring Harbor Laboratory. Under his directorship the lab became a leading research center in the world for molecular biology. In 1988, Watson became head of the Human Genome Project.

See also **Crick Francis**; and **Genetics and Gene Science**.

J. M. I.

WATTLE TREE.   See **Acacia Trees**

WAVE ANALYZER.   See **Spectrum Analysis**.

WAVE CYCLONE.   See **Fronts and Storms**.

WAVE ENERGY.   See **Ocean Resources (Energy)**; and **Tidal Energy**.

WAVE EQUATION.   A partial differential equation of classical mathematical physics. In vector form it is

$$c^2 \nabla^2 \phi = \partial^2 \phi / \partial t^2$$

where ϕ is a quantity characterizing the propagated disturbance, c is the velocity of the wave, and t is the time. It applies also in the electromagnetic theory of light and in the theories of electric waves, elastic vibrations, and sound. In solving it by the method of separation of variables, assume $\phi(x, y, z, t) = F(x, y, z)T(t)$. The result is $T(t) = e^{\pm ikct}$ and $\nabla^2 F + k^2 F = 0$, where k is an arbitrary constant. In most cases, some other curvilinear coordinate system (u, v, w) is more suitable than (x, y, z). If the partial differential equation again separates there will be three ordinary differential equations in $f_1(u)$, $f_2(v)$, $f_3(w)$ and the solution of the wave equation becomes $\phi = f_1(u)f_2(v)f_3(w)T(t)$. In cylindrical coordinates, Bessel functions occur in this solution; in spherical polar coordinates, Bessel and associated Legendre functions.

In quantum mechanics, the Schrödinger wave equation for a single particle of mass m and potential energy $V(\mathbf{r}, t)$ is

$$\frac{h^2}{2m} \nabla^2 u = Vu - ih \frac{\partial u}{\partial t}$$

where $h = h/2\pi$ and h is Planck's constant. See also **Schrödinger Equation**.

WAVE (Estuary).   See **Estuary**; and **Tidal Energy**.

WAVE (Gravitational).   See **Gravitation**.

WAVEGUIDE.   A structure with the capability of confining and supporting the energy of an electromagnetic wave to a specific relatively narrow and controllable path. Waveguides are altered to suit the needs of the

medium communicated, such as microwave guides, coaxial cables, and optical fiber guides.

WAVE INTERFERENCE.   The phenomenon that results when waves of the same or nearly the same frequency are superposed; characterized by a spatial or temporal distribution of amplitude of some specified characteristic differing from that of the individual superposed waves. Also called *interference*.

WAVELENGTH.   Of a periodic wave in an isotropic medium, the perpendicular distance between two surfaces of equal phase in which the displacements have a difference in phase of one complete period. Otherwise phrased, the wavelength is the "space period" of a wave, i.e., the least translation distance that leaves the wave invariant.

In a harmonic wave, the wavelength is the distance between any two points at which the phase at the same instant differs by 2π, specifically if the wave is a plane harmonic wave in the x-direction with angular frequency ω, so that the disturbance has the form

$$\sin(\omega t - kx)$$

the wavelength is

$$\lambda = 2\pi/k = 2\pi V/\omega$$

where V is the wave velocity.

WAVELLITE.   The mineral wavellite is a hydrous phosphate of aluminum, formula $Al_3(PO_4)_2(OH)_3 \cdot 5H_2O$. It is orthorhombic but crystals are of rare occurrence as it is ordinarily found in crusts or radial aggregates, sometimes fibrous. Its hardness is 3.25–4; specific gravity, 2.36; may be of various colors, gray, blue, green, yellow, black, or colorless. It has a vitreous luster, and is translucent. This mineral is of secondary origin, probably formed by waters bearing phosphoric acid which have acted on aluminum minerals. Wavellite is found in Saxony, Bavaria, Devonshire, from where it was originally described; and in the United States in Chester and Cumberland Counties, Pennsylvania; and Montgomery and Garland Counties, Arkansas. It was named after its discoverer, Dr. Wavel.

WAVE (Meteorology).   See **Atmosphere (Earth)**.

WAVE MOTION (Sound).   See **Acoustics**.

WAVE NORMAL.   The curves of a family normal to the equiphase surface of a wave are called wave normals. The equiphase surfaces are those satisfying $\Phi(x, y, z) = $ constant for the wave $A(x, y, z)e^{j[\omega t - \Phi(x, y, z)]}$. Wave normal may also be defined as a unit vector normal to an equiphase surface, with its positive direction taken on the same side of the surface as the direction of propagation. In isotropic media, the wave normal is in the direction of propagation.

WAVE NUMBER.   The reciprocal of the wavelength in a harmonic wave. Some authors use $2\pi/\lambda$ instead of $1/\lambda$ in this sense. In this case, it is often symbolized by k and called the wave parameter.

WAVE (Ocean).   See **Atmosphere-Ocean Interface**; and **Ocean**.

WAVE PERIOD.   In a harmonic wave, the time between attainment of successive maximum disturbances at the same place. The reciprocal of the frequency.

WAVE PHASE.   The argument in the wave function. Thus in the general arbitrary progressive wave function $f(x - Vt)$ the phase is $x - Vt$. For a harmonic plane wave in which the disturbance has the form $\sin(\omega - kx)$, the phase is $\omega t - kx$. Here ω is the angular frequency $= 2\pi f$, where f is the actual frequency, $k = 2\pi/\lambda$, where λ is the wavelength.

WAVE POLE (or Wave Staff).   A device for measuring sea-surface waves. It consists of a weighted pole below which a disk is suspended at a depth sufficiently deep for the wave motion associated with deep-water waves to be negligible. The pole will then remain nearly as if anchored to the bottom, and wave height and period can be ascertained by observing or recording the length of the pole that extends above the surface.

WAVE PROPAGATION (Huygens' Principle). A well-known method of analysis applied to problems of wave propagation. It recognizes that each point of an advancing wave front is in fact the center of a fresh disturbance, and the source of a new train of waves; and that the advancing wave as a whole may be regarded as the resultant of the secondary waves arising from points in the medium already traversed. This view of wave propagation facilitates the study of various phenomena, such as diffraction. For example, if two rooms are connected by an open doorway, and sounds are produced in a remote corner of one room, a person in any part of the other will hear the sounds as proceeding from the doorway, which is indeed the case. So far as the second room is concerned, the vibrating air in the doorway is the source, and from it sound waves enlarge in all directions through the second room. The same is true of light reaching a slit or passing the edge of an obstacle, though this is not quite so easily observed because of the short wavelength. The interference of light from variously distant areas of the moving wave front accounts for the maxima and minima observable as diffraction fringes.

WAVE PROPAGATION (Transition Loss). See **Transition Loss**.

WAVES AND WAVE MECHANICS. A wave is a disturbance propagated in a medium in such a manner that at any point in the medium, the displacement is a function of the time, while at any instant the displacement at a point is a function of the position of the point. Any physical quantity having the same relationship to some independent variable (usually time) that a propagated disturbance has, at a particular instant, with respect to space, may be called a wave. In this definition, displacement is used as a general term, indicating not only mechanical displacement, but also electric displacement, etc. In short, a wave is a time-varying quantity that is also a function of position; for example, any time-varying voltage or current in a network is often called a wave. Other examples of waves are: (a) a wave on the surface of a liquid, in which the disturbance is the displacement of any particle in the surface from its equilibrium position; (b) an acoustic wave, in which the disturbance is the change in pressure from its equilibrium value at any point in a material medium (fluid or solid); (c) an electromagnetic wave, in which the disturbance is the change in the electric and magnitude field intensities from their equilibrium values in space.

Wave Mechanics. This is a term and field that essentially grew out of the quantum theory. Wave mechanics in the modern sense is an integral part of quantum mechanics. The fact that radiant energy (light, X-rays, etc.) is certainly emitted by atoms or molecules and is as certainly done up in parcels, called quanta, the magnitude of each of which is definitely associated with a vibration or wave frequency of some kind (see also **Planck Law**), leads one to inquire what there is about an atom or the electrons in it that has to do with vibrations or waves. The now classic Davisson-Germer experiment gave most conclusive evidence that electrons actually do have wave characteristics even when flying freely through space (or at least when they strike and rebound from something like a crystal), and that, again, the energy of their motion is expressible in terms of a wave or vibration frequency. Even whole atoms are reflected by crystals as if they were waves, as shown by the experiments of Ellett, Olson, and Zahl.

Such facts have given rise to the idea that perhaps all physical processes are, in the last analysis, wave processes, with frequencies or wavelengths appropriate to the quanta into which the energy divides itself. Instead of being particles which revolve in orbits like planets, the electrons in the atom, according to this conception, become wave trains reverberating like sound in a closed room, and setting up stationary interference patterns corresponding to the stationary quantum states. It is of such boldly revolutionary concepts that the new wave mechanics is built. The mathematical formulation of the theory has been developed largely by de Broglie and by Schrödinger.

As pointed out in the entry on **Relativity and Relativity Theory**, Einstein played an invaluable role in establishing concepts that later led to the development of quantum mechanics. Rather than duplicate information here, the reader is referred, in particular, to three entries in this encyclopedia—**Gravitation; Quantum Mechanics**; and the aforementioned entry on relativity.

Wave Motion. The motion of waves can be said to be the most common and the most important type of motion that we know. It is through wave motion that sounds come to our ears, light to our eyes, and electromagnetic waves to our radios and television sets; tidal waves and earthquakes, among

other phenomena, are also characterized by wave motion. Wave motion can be defined as that mechanism by which energy is transported from a source to a distant receiver without the transfer of matter between the two points.

Waves can be classified according to the manner of their production, namely, a vibrating material object, or in the case of electromagnetic waves, sources such as electrical oscillations in an aerial. The wind blowing across water causes surface waves; a piezoelectric quartz crystal vibrating under an applied electric field can generate underwater motion. Waves can also be classified according to the medium in which they travel. The most useful classification, however, involves the direction of motion of the particles of a medium (or of an electric or magnetic field in the case of electromagnetic waves) relative to the direction in which the energy of the wave is itself propagated. Such a classification is useful because wave motions falling into the same class according to the selected criterion will have other similar properties.

Wave motion can be most easily understood if one considers first, as an example, wave motion of a horizontal, stretched string and then, by analogy, other types of wave motion. If one end of such a string is moved up and down, a rhythmic disturbance travels along the string. Each particle of the string moves up and down, while at the same time, the wave motion moves along the length of the string. It is the state of the particles that advances, the medium as a whole returning to its initial condition after the disturbance has passed. Such a wave motion, one in which the vibratory motion of the medium is at right angles, or essentially at right angles, to the direction of propagation, is called *transverse*. Surface waves on liquids are transverse; so also are electromagnetic waves (X-rays, visible light, radio waves, etc.), but here, since electromagnetic waves can travel in a vacuum, we must think of the electric and magnetic fields associated with such waves as changing in intensity in a direction at right angles to the direction of propagation.

Another type of wave motion, termed *longitudinal* or *compressional*, can occur only in material media. In this type of motion, the particles of the medium move forward and backward along the direction of propagation of the wave. Compressional waves are exemplified by sound waves in air, in which a volume in the path of the wave is alternately compressed and rarefied. These variations in pressure are very small. Even for the loudest sounds that an ear can tolerate, the pressure variations are of the order of 280 dynes per square centimeter (above and below atmospheric pressure of about 1 million dyes per square centimeter).

Yet another type of wave motion is the *torsional wave*, which can take place only in solids. Less frequently seen, this type can be demonstrated by a long helical spring supported on a flat surface. As one end of the spring is given a quick, momentary twist about the axis of the spring, a pulse travels down the spring.

Whatever the type of wave, certain useful definitions can be set forth and general statements made. *Phase* describes the relative position and direction of movement of a particle in its periodic motion as it participates in a wave motion, or the relative intensity at a point of the electric field accompanying electromagnetic waves. *Frequency* is the number of complete vibrations performed per unit of time by a particle (or field) through which a wave passes. *Period* is the time required for one complete vibration of a particle participating in wave motion. *Wavelength* is the distance between any two points that are in phase on successive waves or pulses. The *velocity* of a wave is the product of the frequency and the wavelength. All waves except electromagnetic waves require a medium for propagation.

Wave motions may vary in the energy they transport per unit of time. This property depends upon the amplitude. The *amplitude* of a wave in a string, to take again an example, is the maximum displacement experienced by the particles of the string as they move from their equilibrium positions. The *intensity* of the wave is the power (energy per second) passing through a square centimeter perpendicular to the wave front, and it is related to the square of the amplitude. In the case of sound, intensity is related to loudness.

Two or more waves crossing one another's paths will not cause any change in the direction, frequency, or intensity of any of them. The displacement effects of two or more waves of the same kind passing through a medium are additive at any point. And, at any moment, the displacement at a point is the vector sum of the separate displacements caused by the separate waves. Two transverse wave motions passing at right angles through a point in a medium cause a particle at that point to perform a path called a *Lissajous figure* (see also **Frequency**

Measurement), whose form depends upon the amplitudes and frequencies of the two waves and upon their phase relationship.

Standing (stationary) waves are produced by combining two similar wave trains moving in opposite directions. Not themselves waves, they are patterns of vibration that simulate waves standing still. An example is exhibited by a string, one end of which is fastened rigidly and the other vibrated transversely at a constant frequency. Waves traveling down the string from the source meet waves reflected from the fixed end. If the tension in the string is adjusted properly, the string can be made to display *nodes*, points where the string does not move transversely because at those points the two waves cancel the effects of one another. Between the regularly spaced nodes are found the *antinodes* or loops, where the two waves reinforce one another.

Wave motion is also characterized by the phenomena of absorption, reflection, refraction, interference, diffraction, beats, resonance, and polarization. (Longitudinal waves do not exhibit polarization.)

Complex waves can be analyzed into sets of simple waves according to the principles of Fourier analysis, where by "simple waves" are meant waves whose variations of displacement with time can be represented by sine curves.

Compressional waves require about 5 seconds to travel a mile in air, 1 second to travel a mile in water, and $\frac{1}{3}$ second to travel a mile in iron. To travel a kilometer a compressional wave takes 3.1 seconds in air, 0.6 second in water, and 0.2 second in iron. Although varying in speed from material to material, low-frequency compressional waves travel with the same speed in a particular medium, i.e., they do not exhibit dispersion. Small variations in speed sometimes found at high frequencies are due to relaxation phenomena.

Compressional waves in a fluid have a speed v that depends only on the density ρ and the adiabatic bulk modulus β of the medium according to the relation

$$v = \sqrt{\beta/\rho}$$

In solids, the speed of compressional waves is given by the relation

$$v = \sqrt{Y/\rho}$$

where Y is Young's Modulus. Thus, it may be seen that a study of the propagation of waves in a medium gives important information about the medium.

Transverse waves on the surfaces of liquid do *not* travel with a fixed speed that is dependent only upon the properties of the liquid. Their speed depends upon their wavelength and amplitude, the depth of the liquid, and whether the surface is confined, as in a canal. Surface tension waves on the surface of water have wavelengths less than 1.7 centimeters, while gravity waves on the surface of water have wavelengths greater than 1.7 centimeters. Ripples on water often move only 30 centimeters per second. They have higher velocity as their wavelength becomes smaller. In contrast, for example, ocean waves measuring 244 meters (800 feet) from crest to crest have been found to travel 20 meters per second (45 miles per hour).

Transverse waves in strings (or wires) travel with a speed that depends only on the tension in the string T and the mass per unit length of the string m, according to the formula

$$v = \sqrt{T/m}$$

Electromagnetic waves of all frequencies travel with the same speed in a vacuum (2.9979×10^{10} centimeters/second). In a particular medium, however, different frequencies (colors in the case of visible light) travel with different speeds. The speeds at particular frequencies also vary with the media.

The observed wavelength of a wave motion, whether it be longitudinal or transverse, depends on whether the source and the receiver are moving relative to one another, a phenomenon known as the Doppler effect. In the case of electromagnetic waves, the Doppler effect depends only on the relative velocity. In the case of a compressional wave, as for instance, sound, the magnitude of the effect depends not only on the relative velocity of the source and receiver, but also the effect depends upon whether the receiver or the source is in motion with respect to the transmitting medium. In both cases, the wavelength of the wave is increased if the source and receiver move away from one another and the wavelength is decreased if they move toward one another.

In some uses of wave motion, care must be taken to differentiate between "phase" and "group" velocity. The group velocity of a wave is the velocity usually observed, and the energy in a wave is transmitted with the group velocity. For example, measurements of the speed of light wherein the time for "chopped" pulses to travel a known distance is determined, result in values of the group velocity. On the other hand, if one carefully observes the expanding group of ripples when a stone is dropped into water, the group travels with one velocity, the group velocity, while a particular wave crest will advance through the group to the outer leading edge and exhibit the phase velocity. The difference in the two velocities depends on the wavelength in the material medium and on the dispersion, i.e., on the change in phase velocity with wavelength. In a vacuum, such an interstellar space, the two velocities are the same for light, whatever the color.

See also **Quantum Mechanics**.

Additional Reading

Boccotti, P.: "Wave Mechanics for Ocean Engineering," Elsevier Science, New York, NY, 2000.
Greiner, W.: "Relativistic Quantum Mechanics: Wave Equations," Springer-Verlag, Inc., New York, NY, 2000.
Mehra, J. and H. Rechenberg: "Erwin Schrodinger and the Rise of Wave Mechanics: The Creation of Wave Mechanics; Early Response and Applications 1925–1926," Springer-Verlag, Inc., New York, NY, 2000.
Pauli, W. and C.P. Enz: "Wave Mechanics," Dover Publications, Inc., Mineola, NY, 2000.
Schroedinger, E.: "Collected Papers on Wave Mechanics," 3rd Edition, American Mathematical Society, Providence, RI, 1997.

WAVE (Supersonic).   See **Supersonic Aerodynamics**.

WAVE TRAIN.   The series of waves produced by a "vibrating" body is called a train of waves.

WAVE VECTOR.   A vector, usually represented as **k**, which, at each position in space, points in the direction of propagation of the wave under consideration. The magnitude assigned to the wave vector associated with a wave of wavelength λ is not unique, but with present usage it is almost certain to be either unity, $1/\lambda$, or $2\pi/\lambda$. The magnitude of the wave vector is called the wave number when $|k| = 1/\lambda$ and sometimes when $|k| = 2\pi/\lambda$.

WAVE VELOCITY (or Phase Velocity).   A plane wave has a velocity uniquely defined as the reciprocal of the phase slowness. The wave velocity is thus the velocity with which the displacement profile of a sinusoidal progressive wave travels. In three dimensions, the reciprocal of the magnitude of the (vector) phase slowness is the speed of the wave along a wave normal, and is often called phase velocity.

WAVE VELOCITY SURFACE.   An elliptical surface used to indicate the difference in the velocity of the ordinary and the extraordinary ray of radiation in different directions in a double-refracting crystal.

WAX MOTH (*Insecta, Lepidoptera*).   A small moth whose larva lives in the combs of bee hives, spinning silken tunnels as it burrows through them. Although it eats the wax of which the combs are built and will attack the pure wax in stored comb foundation, careful studies have shown that it does not thrive on a pure wax diet. The other materials in old combs are a necessary source of nitrogen, without which the caterpillar may live but cannot grow and develop normally.

The moth does not become a serious pest in strong colonies of bees but it is sometimes a cause of serious damage in weak colonies and stored combs.

WAX MYRTLE.   See **Bayberry Shrubs and Trees**.

WAXWING (*Aves, Paseriformes*).   A bird of the Northern Hemisphere, smoothly gray and brown, with limited yellow shading and black marks, and red tips of horny material on some of the wing feathers. The name waxwing refers to these tips. The head bears a sharp crest. One species, the cedar waxwing, or cedar bird, *Bombycilla cedrorum*, is peculiar to the United States. The Bohemian waxwing, *B. garrula*, nests in northern latitudes in Europe and North America and is occasional through the northern half of the states. A third species, the Japanese waxwing, *B. japonica*, breeds in eastern Asia.

WAYFARING TREE. See **Elder Trees and Viburnums**.

WEAKFISH. See **Croakers**.

WEAK INTERACTION. See **Particles (Subatomic)**.

WEASEL. See **Mustelines**.

WEATHERING. The processes by which the atmospheric agencies, commonly associated with the weather, mechanically disintegrate or chemically decompose the rocks at or near the Earth's surface. Mechanical weathering includes the effects produced by changes of temperature, the action of frost, etc. Chemical weathering includes the solvent action of water, the union of atmospheric oxygen with rock materials — oxidation, union with atmospheric carbon dioxide — carbonation, and the chemical combination of substances with water — hydration.

WEATHER TECHNOLOGY. The practical application of the disciplines of meteorology, hydrology, and climatology, greatly assisted by the computer and communication sciences, to short- and medium-term forecasting of the weather. Forecasting responsibilities in the United States fall under the umbrella of the National Oceanic and Atmospheric Administration (NOAA, pronounced *Noah*), of which the National Weather Service (NWS) is a part. Day-to-day operations for NWS are handled by the National Meteorological Center (NMC), located in Camp Springs, Maryland.

NMC confines its "long-range" weather forecasts to the comparatively brief time span of 10 days. Even with the aid of supercomputers, which process multi-tiered data from about 10,000 sources every 6 hours, extrapolating these data for each of the next 10 days requires a very large quantity of computer time. As pointed out by one NMC scientist, "We've seen situations where the model has had substantial skill out to 25 or 30 days. We've also seen situations where it fails to anticipate a change just three days down the road." Efforts are being made to see if a reliable prediction can be distinguished from a bad prediction in advance. Perhaps there are combinations of weather factors that favor accurate forecasts, as contrasted with other weather-factor combinations that destroy any degree of predictive accuracy.

Weather forecasting of the kind carried out by NMC also is pursued at other world locations, of which the European Center for Medium Range Weather Forecasts (Reading, England) is one.

Centers that engage more in long-term climatological trends are exemplified by the University Corporation for Atmospheric Research (sponsored by the National Science Foundation) and whose main facility, the National Center for Atmospheric Research (UCAR) is located in Boulder, Colorado.

Many professional weather forecasters today acknowledge that long-term weather forecasts are indeed tenuous at best — and for at least two fundamental reasons:

1. An inadequate database and the inherently small errors present in measurements. Recalling that the surface of the Earth is about 200 million square miles (581 km^2), it is immediately obvious that the Earth's largest, interactive dynamic system is not adequately measured — because weather formation is not regional, but is worldwide. With expansion and refinement of weather measurements and aided by supercomputers, meteorologists, until the last decade or two, envisioned that reasonably accurate, long-term weather forecasting could become a reality. But more and better instrumentation and computer power no long appear to offer the ultimate solution. This goal may be theoretically impossible.

2. The concept of chaos, presently accepted by many weather scientists, has demonstrated that very small differences in model input data, as may be caused by instrumental errors or intuitive biases introduced into a computer simulator, can grow into greatly exaggerated errors with the passage of time. Findings thus far have shown that, with weather systems, these tiny errors do not create marked inaccuracies for about two weeks, but after that any forecast may be skewed to one of sheer inaccuracy and thus is worthless.

Weather satellites have been invaluable in terms of forecasting in the immediate and short term. They have been very helpful in tracking hurricanes, but unfortunately the prediction of landfall remains largely a judgmental call on the part of knowledgeable hurricane specialists who have years of experience. Hurricanes and tornadoes are described in article on **Fronts and Storms**.

Computer models continue to be developed, if not for long-term weather forecasting, then in the interest of developing a better understanding of the processes that go on in very-short-term weather phenomena, such as tornadoes. In 1989, Kelvin Droegermeier (University of Oklahoma), a meteorologist, introduced the Piecewise Parabolic Method (PPM) to "build into" a weather problem a knowledge of physics and an understanding of fluid flows. PPM initially was developed by P. Goodward (University of Minnesota) and P. Colella (University of California, Berkeley). The concept is based upon the work of S.K. Godunov (Russia) and Bram van Leer (Netherlands). They desired a numerical technique that could process the shock discontinuities found in supersonic fluid flow problems.

Woodward observes that the equations originally developed were for astrophysical applications (jets from the nuclei of galaxies and motions of fluids in stars) also describe some weather phenomena on Earth. Instead of the standard techniques where an attempt is made to keep track of numerous variables and using a finite set of grid points, PPM tracks the average of a variable at each grid cell. A given variable at each grid cell is

---

Chaos as Related to Weather Forecasting

Much has been written pertaining to the concept of *chaos* in recent years. As early as 1963, Edward Lorenz (Massachusetts Institute of Technology), when speaking before an assembly of meteorologists, described what he termed, "The Butterfly Effect." The speaker related that when he thought about weather forecasting (which he had done for some years), he thought about butterflies. Lorenz observed, "Imagine a butterfly stirring deep in the Amazon forest. Its delicate motion alters tiny air currents that influence large eddies ever so slightly. Can those eddies transform a Texas tornado days later?" His comment was taken seriously by professionals in the field of weather predicting. Aksel Winn- Nelson (European Centre for Medium Range Weather Forests) responded, "If the weather is sensitive to such minute influences, no one could predict its course."

Lorenz initiated a new way for meteorologists to look at the weather system. Prior theories had been based on the traditional "clockwork" universe, which was championed by Laplace and Descartes in the 18th century, to the effect that a full comprehension of each part of a system can yield a prediction of the future of that system. This led to a conclusion that, within reason and tempered with judgment, supercomputers could assist in predicting the weather. Winn-Nelson observed, "We thought if you just knew the state of the atmosphere sufficiently well and if you built the right models with powerful enough computers, there should be no limit in predicting the weather. Lorenz's work came as quite a shock."

Chaos is an important factor in weather systems. As explained in the article on *Chaos*, it is a topic that also has been applied to the study of numerous other complex phenomena, one being turbulence, which, prior to his announcement in 1963, Lorenz had attempted to create computer models for, long-term weather prediction. In one experiment, he rounded off the input numbers and found that this produced very serious changes in the end results. Lorenz learned a fundamental of chaos, namely, that tiny changes in the input information grossly exaggerate the final result.

Weather scientists, as well as researchers in other disciplines, have found some "order" in chaos. Characteristic patterns of different phenomena appear repetitive and represent what may be termed a "signature" of various unique kinds of behavior. They are called *strange attractors*. A strange attractor may be described as a system's "preference" for behavior. As explained by Robert Pool (see reference), "The most important quantity associated with a strange attractor is its dimensions, which indicates how complicated the pattern is and gives a rough indication of how many variables it takes to describe the behavior of the system. If a strange attractor is low dimensional, it indicates that the weather under consideration is simple enough to model with only a few variables. Also, since the strange attractor contains the system's "preference" for behavior, it is possible to use the strange attractor to predict the system's future actions."

This, then, may become an index to those weather systems that may be predicted correctly over longer periods than others whose index is greater."

represented by a parabola. Some details of the PPM method are described by Cipra (see reference listed).

Forecasting Success in Britain in 1988. A storm in October 1987 that struck southern England with wind gusts of up to 99 mph (160 km/hour) was not predicted much in advance by British forecasters. It was labeled a "once-in-a-lifetime" storm, and warning was just a matter of a few hours. Not so in the case of another major storm that struck Britain less than 2 years later in March 1988. The Meteorological Office alerted the populace more than 1 day ahead of the storm. During the interval the Office had acquired a computer eight times more powerful than was in place at the time of the 1987 storm. Probably the increased computer power helped, but most likely the second storm fell into the more easily predictable category and thus exemplified a storm behavioral pattern related to the strange attractor of chaos theory as previously described. A spokesman for the Meteorological Office simply observed, "This was a better behaved storm."

Mesoscale and Microscale Meteorology

Considerable research is being undertaken as of 1994 toward understanding atmospheric phenomena on space scales of millimeters to a megameter and on time scales of seconds to a day. Key research activities include the study of:

1. The physical mechanisms that govern the behavior of mesoscale weather systems and the factors that determine their predictability.
2. The basic nature of moist atmospheric convection, its interaction with topography and other lower boundary effects, and how cloud and precipitation microphysics and large-scale dynamical factors influence its behavior.
3. The fundamental atmospheric processes that operate on scales of up to a few kilometers, including boundary-layer processes, turbulence, and cloud and precipitation microphysics.

Such programs, as conducted at the National Center for Atmospheric Research laboratory and at university laboratories in the United States and Europe, include the conduction of major field experiments and numerical model development.

Accomplishments have included the following:

- Large-eddy simulations have shown how radiative cooling generates turbulence in cloud-topped mixed layers, which has important implications for the dissipation of stratiform clouds through the entrainment of warm, dry air from above.
- Observational studies, using satellites and airborne data gathering, have revealed that horizontal inhomogeneity in marine stratiform cloud regimes is commonly observed and has a significant effect on cloud formation and dissipation.
- Measured changes in the size spectra of sulfuric acid droplets in the arctic stratosphere indicate, contrary to previous assumptions, that the majority of the droplets are supercooled at the temperature at which polar stratispheric clouds could begin to form, but that they freeze homogeneously at slightly colder temperatures. This observation alters the view of the mechanisms important in the formation of polar stratospheric clouds that are directly involved in polar ozone loss.
- High-resolution mesoscale model simulations have reproduced the remarkable structure of recently documented warm-core marine cyclones that are in contradiction to the classical Norwegian cyclone model.
- Studies of how echoes and downbursts identify mechanisms producing these significant storm features.
- Initiation of convection along the Colorado Front Range in the presence of southerly flow has been shown to occur through complex nonlinear interactions involving boundary-layer instabilities, tropospheric gravity waves, and nocturnal low-Froude-number flow effects.
- Momentum transport by organized convection in both idealized models and the real atmosphere has been represented by a comprehensive nonlinear theory and applied to represent momentum transport in general circulation models.
- Entrainment and mixing along the interface between the cloud and its environment have been shown to occur through convective and shearing instabilities and vorticity generation.
- Techniques have been developed to assimilate precipitable water measurements into a mesoscale model. The assimilation of precipitable water produces a much-improved estimate of the vertical structure of water

vapor than statistical methods based on climatology and improves the precipitation forecast of the model.

Computer Modeling. The principle behind modeling is that many natural processes can be described by mathematical equations. These equations, which describe everything from basic tenets (such as Newton's laws) to detailed processes (such as how solar radiation is absorbed or reflected by the Earth and clouds), can be solved using a computer. Obviously, to perform accurately a model requires both good equations that describe the world realistically and good data. Models come in all scales, from a single cloud to the entire planet and the atmosphere above it, and in many levels of complexity. Each is designed for a certain research target. See Fig. 1.

Fig. 1.   Computer modeling enables scientists to visualize the processes they are investigating. Air motions in and around a thunderstorm are shown schematically in this diagram. (*National Center for Atmospheric Research.*)

Computer modeling does have limitations. For example, climate modelers would not use a weather-forecasting model because such a model does not have to account for long-term chances, such as variations in the extent of the polar ice caps. In contrast, forecasters would not use a climate model because it might average or ignore other short-term variations that affect weather. A cloud physicist or solar scientist would not use a climate nor a weather model because these models would not include the specialized equations needed. The impracticality of using weather models for time spans exceeding several days is mentioned earlier in this article. More detail on climate modeling is given in article on **Climate**.

Weather Forecasting

The data base for a weather projection consists of as complete a description as possible of all the interrelated parameters that define the atmosphere and, in particular, its relation to human activities. These parameters are obtained from meteorological observations at fixed times, usually known as "time zero." The basic structure of the atmosphere is projected stepwise by integrating the time rate of change of the descriptive parameters. For many years, the stepwise project was done manually by use of linear or accelerated extrapolation. In recent years, these tasks have been assigned to computers. Most forecasting in the early 1990s is a mix of computer outputs with human refinements. The trend is to turn over almost all of the task to computers.

Forecasts deviate from actual weather increasingly as the increments of time increase. Forecasts for 12 hours are reasonably reliable. Forecasts for 24 hours verify within the range of 75 to 85%. Forecasts for 96 hours are nearer the 60% level of validity. There are two causes for deterioration: (1) The original data base is faulty and incomplete; and (2) the projection formula does not contain all the terms that actually exist in the atmosphere. When the data base is complete and sufficient and the time rate of change equations are in a one-to-one correspondence with nature, the forecasts are accurate. Tide and moonrise/moonset forecasts are an example, of course, of accurate forecasting. Future trends are toward increasing accuracy in weather forecasting because the data base is constantly improving and projection formulas are undergoing continual modification and refinement. A factor frequently reported today is the wind-chill factor. See Table 1.

Types of Forecasts. There are several categories of forecasts, divided according to the period of time, the geographical areas, and the special

TABLE 1. WIND-CHILL FACTOR

Wind Speed (mph)	Actual Temperature (°F)									
	50	40	30	20	10	0	−10	−20	−30	−40
5	48	37	27	16	6	−5	−15	−26	−36	−47
10	40	28	16	4	−9	−24	−33	−46	−58	−70
15	36	22	9	−5	−18	−32	−45	−58	−72	−85
20	32	18	4	−10	−25	−39	−53	−67	−82	−96
25	30	16	0	−15	−29	−44	−59	−74	−88	−104
30	28	13	−2	−18	−33	−48	−63	−79	−94	−109
35	27	11	−4	−21	−35	−51	−67	−82	−98	−113
40	26	10	−6	−21	−37	−53	−69	−85	−100	−116
	Equivalent chill temperature (°F)									

Note: °C = (°F − 32) × 5/9.
kph = 0.6214 mph.

purposes for which they are made, as well as by the considerations or techniques upon which they are based.

In categorizing weather forecasts according to geographic areas, the broadest categories may be thought of in terms of local or *microforecasts* (highly detailed forecasts of conditions over relatively limited areas, such as cities or airports) and district or *macroforecasts* (general forecasts for conditions over relatively large areas of tens- or hundreds-of-thousands of square miles). In terms of periods of time, forecasts, in general, may be thought of as daily or *short-range forecasts*, extending from 12 to 48 hours in advance, and being usually detailed for specific geographic areas; *medium-range forecasts*, extending from about 2 days to a week in advance; and *long-range* or *extended forecasts* for periods greater than 2 days or a week in advance. The longest-range and the most general of all types is the *climatic forecast*, which predicts future climate and general weather conditions for a specified region over a period of years.

Aviation Forecasts. These reports are required for periods of several hours to 1 day. Most aircraft flights are of less than 6 hours duration; commercial airliners seldom remain aloft more than 12 hours. Forecasts of fog and low clouds, turbulence, thunderstorms and accompanying severe weather, freezing rain and snow, aircraft icing, and of winds aloft at cruise levels are paramount to aviation. In addition to maps, updated aviation weather information is available from specific stations via sequence reports. See also **Precipitation and Hydrometeors**. See also Figs. 2–5.

Terminal Forecasts. These contain information for specific airports on expected ceiling, cloud heights, cloud amounts, visibility, weather and

Fig. 2. Airports utilize a number of local weather sensors. Shown here is a fully automated runway visual range measurement and reporting system. The system consists of a light projector, a background luminance sensor, a light receiver, and a data-processing unit. Measuring range is 0–100% transmittance. Resolution is 0.02% of transmittance. Accuracy is ±1% of transmittance value through whole measuring range verified by calibrated optical filters. (*Vaisala.*)

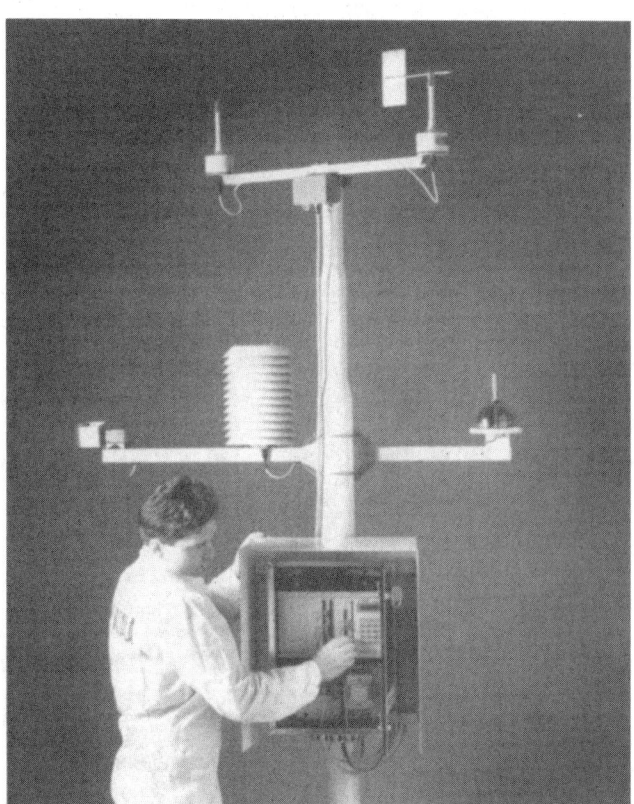

Fig. 3. Automatic weather station suitable for airport use or any local site. The instrument measures up to 60 weather parameters. Available with a wide range of weather data inputs. Data may be analog in several ranges of sensitivity. Or digital and pulse frequency inputs or serial inputs from intelligent sensors can be accommodated. Customized software is available. (*Vaisala.*)

obstructions to vision and surface wind. They are issued 3 times/day and are valid for 24 hours. The last 6 hours of each forecast are covered by a categorical statement indicating whether VFR (Visual Flight Rules), MVFR (Marginal Visual Flight Rules), IFR (Instrument Flight Rules), or LIFR (Low Instrument Flight Rules) conditions are expected. Terminal forecasts are written in the following form:

Ceiling: Identified by the letter C.
Cloud Heights: In hundreds of feet above the station (ground).
Cloud Layers: Stated in ascending order of height.
Visibility: In statute miles, but omitted if over 6 miles.
Weather and Obstruction to Vision: Standard weather and obstruction to vision symbols are used.
Surface Wind: In tens of degrees and knots; omitted when less than 10.

Area Forecasts. These are 18-hour aviation forecasts plus a 12-hour categorical outlook prepared 2 times/day giving general descriptions of cloud cover, weather, and frontal conditions for an area the size of several states. Heights of cloud tops and icing are referenced ABOVE SEA LEVEL (ASL); ceiling heights, ABOVE GROUND LEVEL (AGL); bases of cloud layers are ASL unless indicated. Each SIGMET or AIRMET affecting an FA (Flight Advisory) area will also serve to amend the Area Forecast.

SIGMET or AIRMET[1] messages warn aircraft personnel in flight of potentially hazardous weather, such as squall lines, thunderstorms, fog, icing, and turbulence. SIGMET concerns severe and extreme conditions of importance to all aircraft. AIRMET concerns less severe conditions which may be hazardous to some aircraft or to relatively inexperienced pilots. Both are broadcast by FAA (Federal Aviation Agency) on NAVAID (Navigational Aid) voice channels.

Winds and Temperatures Aloft (FD) Forecasts.[2] These are 12-hour forecasts of wind direction (nearest 10 true North) and speed (knots) for selected flight levels. Temperatures aloft (°C) are included for all but the 3000-foot level.

---

[1] SIGMET is an acronym for Significant Meteorological Information; AIRMET, Airman Meteorological Information.

[2] FD is an arbitrary designation for these forecasts.

Fig. 4. UHF telemetry antenna designed for reception of radio signals in the 403 MHz band. Antenna ensures reliable operation for both low- and high-altitude radiosonde observations even under the highest of wind conditions. Operating range is from 0 to 200 km. (*Vaisala.*)

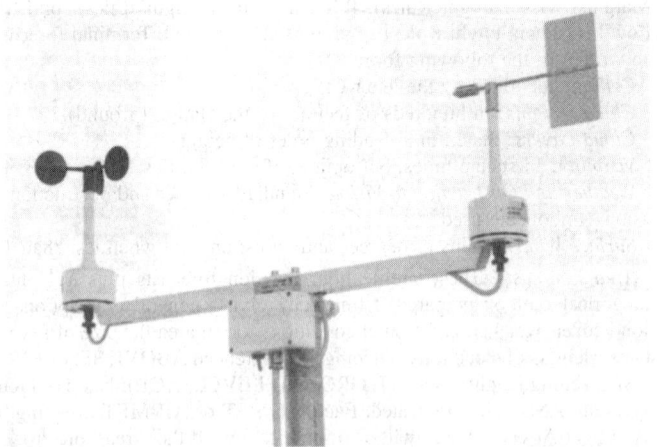

Fig. 5. Microprocessor-based wind-measurement system useful for airport or other local weather stations. Minimum and maximum displays provide instant, 5-second, and 2- or 10-minute average maximum and minimum wind speeds and direction. Meters per second, km/h, knots, and mph can be selected for display. The system also incorporates a wind speed alarm in both audible and visual means. Accuracy: ±1.5% full scale (*Vaisala*).

The Impact of Computed Weather Forecasts on Accuracy. The weather forecasts issued by the National Weather Service have been much more accurate in the last ten years than those that were issued previously. This is due to improvements, in the several models programmed to make numerical weather predication forecasts and the application of the methods of ensemble of forecasting to several models to get the official forecast.

Ensemble forecasts are made with several models, each of which are very good at forecasting but suffers from the universal curse of being a chaotic system. This is due to the fact that some time inside or outside the life of the official forecast, and in some areas, it has a disturbance that does not appear in the verifying weather or a disturbance that does not occur that does appear in the verifying weather map.

The ensemble forecast in its simplest form uses five to more than ten forecasts computed with different models and most of the time the pattern that the majority of the forecasts have displayed can be chosen as the forecasts. This method has improved the forecasts and stretched out the time that they are useful. Sometimes it uses forecasts that start on day 1 and compare these with forecasts from the same model that star on day 2.

Contrast this with the condition 30 to 40 years ago when a forecast for this afternoon was all right but any forecast for tomorrow had about a 50/50 chance of being right. A five-day forecast was experimental and gave very little useful information.

The use of ensemble forecasts is a direct result of improved computing power. As computing power has grown, the number of different models that can be run has grown. Computing power has been growing due to the number of weather services that perform daily weather forecasts and develop models; and the computing capacity at one computing establishment being capable of running two or three simultaneous computations with three or more models.

Commercial Forecasts. These forecasts are directed to commercial and public interests.

Agricultural weather forecasting requires a varied time span — from a day or less to a whole season. Day-to-day agricultural operations, such as planting or harvesting specific crops, require a 24-hour forecast. Other agricultural operations require longer advance planning and forecasting of relevant weather conditions must extend to 72 hours or longer. Longer term planning has to be based upon a climatological forecast. The principal elements of concern to agriculture, of course, are precipitation, temperature, sunshine, and often, wind.

Outdoor construction weather forecasts fall mainly within the time frame of 12 to 72 hours. About a half-day advance warning is needed to guard against inclement conditions of wind, precipitation, and storms. Depending upon the nature of the construction project and accompanying economic and labor factors, reliable forecasting up to 48 or 72 hours is desired.

Forest and grassfire weather forecasts require lead times of from 24 to 72 hours. Elements of prime interest are wind, relative humidity, and temperature, i.e., the principal factors that contribute to producing a favorable combustible state for grass, brush, and trees. Precipitation is important because it virtually eliminates combustibility for the immediate period. Thunderstorms are significant because of the factor which lightning plays in starting such fires.

Public weather forecasts range from a few hours to a day. In addition to providing information for personal and family activity planning, warnings in short time frames are needed for prediction of actual or potential thunderstorms, hail, tornadoes, flash floods, excessive rainfall, hurricanes, and other tropical storms. A variety of public interests are involved wherein low or high temperatures, strong winds, snow, and glaze icing are relevant. Outdoor recreational weather forecasts are needed primarily in the time frame of 24 to 96 hours to assist planning by recreational facility operators and the using public. Weather forecasts are also important to merchandisers who sell weather-oriented gear, such as rainwear, air conditioners, etc.

Satellite imagery, computer graphics, computer simulations, and the ability to stop, reverse, speed up, or slow down images literally has combined serious meteorological information with showmanship and entertainment. Laypeople are much better acquainted with the rudimentary principles of meteorology than they were just a decade ago, mainly because of the interesting way weather is reported by television and the amount of television time devoted to covering it. Special TV weather channels made their appearance in the late 1970s, and there were many doubts that a weather channel could find sufficient commercial advertising sponsorship. These doubts were disproved.

Weather data reporting and forecasting requires a vast array of sensors and an extensive telecommunications network. The system undergoes continuous modernization in an effort to keep up with technological improvements. The consequence at any given time is a mixture of traditional and advanced support facilities.

Special Weather Alerts. The National Weather Service in the United States issues special alerts and reports pertaining to particularly hazardous conditions, such as flood watches and hurricane, thunderstorm, and tornado alerts, usually on a local or regional basis, as required. A special radio communications band has been established for both routine and alert reporting that operates continuously on a 24-hour basis. Local radio and television channels also will interrupt regular broadcasts to carry such information.

These topics are described in much more detail in article on **Fronts and Storms**.

Types of Weather Observations

The principal categories of weather observations are: (1) *surface observations*, and (2) *upper air observations*.

Surface Observations. These data are used to evaluate the state of the atmosphere as observed from one or several points on the surface of the Earth. Usually the term applies to observations that are made primarily for the purpose of preparing surface synoptic charts. There are several subcategories of surface weather observations:

Synoptic Weather Observations. These are made, usually, at 3-hour and 6-hour intervals, of sky cover, state of the sky, cloud height, atmospheric pressure reduced to sea level, temperature, dew point, wind speed and direction, amount of precipitation, hydrometeors and lithometeors, and special atmospheric phenomena that prevail at the time of the observation or have been observed since the previous specified observation.

Aviation Weather Observations. These pertain to the operation of aircraft. Included are the cloud height or vertical visibility, sky cover, visibility, obstructions to vision, precipitation and thunderstorms, and wind speed and direction that prevail at the time of the observation. Complete observations include the sea-level pressure, temperature, dewpoint temperature, and altimeter setting, and are made at regular, equal intervals, usually on the hour. Local extra observations may be taken every 15 minutes when there are impending aircraft operations and when weather conditions are near certain operational weather limits established as safety minimums for aircraft landings and take-offs. See also articles on **Gust Front**; and **Wind Shear**.

Marine Weather Observations. The weather on land is grossly affected by the weather over the oceans. See also **Atmosphere-Ocean Interface**. Marine observations include those made from a ship at sea or from a buoy that is equipped to telemeter instrumental reading. They may include total cloud amount; wind direction and speed; visibility; weather, pressure; temperature; selected cloud-layer data (amount, type, and height); pressure tendency; seawater temperature; dew-point temperature; state of the sea (waves); and sea ice.

Climatological Weather Observations. These data essentially serve as part of the database used for the science of climatology, including very long period weather forecasting. Measurements are made at least once daily from climatological substations and include the evaluation of minimum and maximum temperature, of total precipitation since the previous observation, of weather (cloudy, clear, etc.), and of atmospheric phenomena. Such observations comprise the bulk of climatological data in the United States and in much of the world. There are about 10,000 climatological substations in the United States. See also **Climate**.

Upper Air Observations. These are observations of atmospheric conditions that are beyond the effective range of a surface weather observatory. The term is usually applied to those observations that are used in the analysis of upper air charts and the display of upper air conditions. The first data level of upper air observations is usually 850 millibars pressure (approximately 5000 feet: 1524 meters). Upper air observations by balloon-carried instruments regularly reach altitudes of 100,000 feet (30,480 meters). Observations by rockets reach as high as 250 miles (402 kilometers). The usual elements observed include temperature, water content of the air, wind direction and speed, and altitude and pressure. See also **Balloon**; and **Upper Air Observations**.

Weather Data Collection

Most of the weather observatories collect surface data, but some sophisticated installations collect upper air and radar data. Inasmuch as an average of at least six elements is observed and recorded at each observation, the total amount of data observed and recorded in any given 24-hour period is monumental. Yet, compared with the size of the atmosphere and the ocean areas, the data collected are entirely too few to adequately describe the whole atmosphere.

Weather data are collected by numerous means, some of which are described in the following paragraphs.

Radar Meteorology. *Weather radar* is applied to any radar suitable for the detection of precipitation or clouds. The general qualifications for weather radars are: (a) wavelength between one and 30 centimeters; (b) pulse transmission with high peak power (several megawatts); (c) relatively narrow beam widths; (d) pulse lengths of a few microseconds or less; (e) pulse repetition frequencies that are hertz of several hundred; and (f) automatic azimuth and/or elevation angle scanning. Electronic circuits that permit the quantitative measurement of the signal strength of the returned signal are necessary for more comprehensive studies. Since hydrometeors can scatter radio energy, weather radars operating on certain frequency bands can detect the presence of precipitation at distances up to several hundred miles from the radar, depending upon meteorological conditions and the type of radar. Evaluation of the echoes that appear on the indicator of a weather radar are made in terms of orientation, coverage, intensity, tendency of intensity, height, movement, and unique characteristics of echoes, which may be indicative of certain types of severe storms (such as hurricanes, tornadoes, or thunderstorms).

Cloud-detection radar is a type of weather radar designed specifically for the detection of clouds. It operates on radio frequencies near 30,000 megahertz/sec. corresponding to a wavelength of about one cm. (K band). Short wavelengths are essential to these radars in order to obtain appreciable scatter from the small drops of which clouds are formed. The beam is usually directed toward the zenith: thus, only clouds and precipitation passing directly above the radar are detected. It is capable, in its present form, of detecting clouds in multiple layers to an altitude of about 50,000 feet (15,240 meters).

Height-finding radar, designed specifically for the accurate measurement of target altitudes, is extremely useful as a weather radar in studying vertical cross-section of clouds and precipitation areas. The beam width in the vertical is generally significantly smaller than in the horizontal. Target information is presented on a *range-height indicator scope*, a type of radarscope on which echoes are displayed in coordinates of slant range and elevation angle, simulating, thereby, a vertical cross-section of the atmosphere along some azimuth from the radar.

Certain radars are used in connection with the balloon-borne radiosonde observations, by means of which meteorological data is measured and transmitted. Also useful as weather radars are the radar signal spectrograph, and, as mentioned above, the radarscope.

A considerable number of weather surveillance radar observatories have been established in the United States and nearly all of the airspace comes under the umbrella of radar surveillance. The equipment used operates near the 10 centimeter wavelength. The observational range is 250 miles (402 kilometers), although at this distance the radar beam passes above some target material because of the curvature of the earth. Within an operational range of 125 miles (201 kilometers), the weather surveillance radar is very effective in detecting air space that contains liquid and solid water particles larger than small cloud droplets. The radar hardware is also able to stand at one azimuth to scan vertically to depict the height of echoes. Most radar installations are equipped with electronic capability to transfer the scope picture to the carrier wave of good quality communications land lines.

Airborne Radar. Weather detecting radars operating in the wavelength range of 3 to 5.5 centimeters have been installed in most commercial airlines. The range of these sets varies from 150 to 300 miles (241 to 483 kilometers), which allows ample time for weather detection and avoidance planning. The scope viewing area is an arc of 45 to 60° directly ahead of the plane. These installations have uptilt and downtilt capability to provide for vertical scanning. Most aircraft radars have the capability to display two levels of echo intensity. The lowest level covers the range of precipitation from one through three on a scale of seven. The upper level cuts in at the level four and is related to moderate and heavy precipitation. The purpose of the upper level is to "contour" areas of more intense precipitation because there is a good correlation between turbulence and precipitation intensity in storms. The scope depiction, in black-and-white, is one of a whitish area where the lower-level intensity is detected. The upper level of heavier precipitation is dark, surrounded by the whitish area. The heavy precipitation thus is "contoured." Scope depiction in color utilizes two or more colors to outline the relative intensity of precipitation.

The primary purpose of airborne radar on airliners as or weather avoidance. The automated lateral scan and the uptilt/downtilt permits the crewmen to analyze a storm and plan an avoidance flight path that will keep the plane away from the storm.

Airborne radar can also be used for mapping and navigation. Bodies of water, shorelines, and islands are clearly depicted even though the plane may be well above a thick cloud layer.

Weather Surveillance Radars. These systems depict several levels of echo intensity. In most installations, the electronic circuitry is arranged to distinguish up to seven levels of intensity, ranging from very light precipitation to very intense torrential downpours. The levels of intensity are separated by the degree of shading of the scope, the least intense being very light. Successive levels repeat the shading from white to black. Surveillance radar thus can depict the core of thunderstorms, can evaluate the rate of rainfall, and determine the total rainfall over specific areas. The radar is also very effective in depicting the leading edge of rain or snow and the ending edge as a storm moves along. Weather surveillance radar is able to observe all the air space between the standard widely separated observatories and thereby provide complete surveillance of precipitation.

Inadequate funding in recent years has slowed improvements in weather radar. One meteorologist has observed that "Radars that now track violent storms, for example, are so old (most were installed in 1957) that they rely on vacuum tubes. The systems fail frequently and can be put of commission for days, even weeks. Radars have been down when storms ripped through areas with disastrous consequences." After a disastrous experience in parts of North Carolina, where radar had been out of service for ten days, federal investigators were prompted to upgrade and replace the aged U.S. system of weather radars. Over 160 modern technology weather radars are scheduled for the NWS by 1997.

Weather Satellite Technology

The first weather-observing satellite launched in the United States was placed in orbit on April 1, 1960. This satellite was a television infrared observing satellite (TIROS I). Pictures of clouds taken in the infrared were telemetered to earth stations where they were analyzed and reconstructed in the form of cloud-cover maps. A total of ten TIROS satellites were placed in orbit. A NIMBUS series of satellites was commenced about two years after TIROS I. The NIMBUS satellite observatories were placed in a sun-synchronous polar orbit. This series introduced the automatic picture transmission (APT) from its television cameras. They also introduced an advanced videcon camera system. The ESSA series of weather-observing satellite was begun in 1966. This series embraced all of the advanced features of earlier satellites, plus newer technological innovations. Infrared scanning, as represented by the ESSA series, had become a fine technological art during a comparatively short period of eight years. See Fig. 6.

A series of "applications technology satellites" was commenced in Pictures of clouds taken in the infrared were telemetered to earth stations where they were analyzed and reconstructed in the form of cloud-cover maps. A total of ten TIROS satellites were placed in orbit. A NIMBUS series of satellites was commenced about two years after TIROS I. The NIMBUS satellite observatories were placed in a sun-synchronous polar orbit. This series introduced the automatic picture transmission (APT) from its television cameras. They also introduced an advanced videcon camera system. The ESSA series of weather-observing satellite was begun in 1966. This series embraced all of the advanced features of earlier satellites, plus newer technological innovations. Infrared scanning, as represented by the ESSA series, had become a fine technological art during a comparatively short period of eight years.

A series of "applications technology satellites" commenced in 1966. The satellites, known as ATS, were placed in orbit near the equator at approximately 23,000 miles (37,000 kilometers). At this altitude, the eastward speed of the satellite matched the rotational speed of the earth, thus making the satellite to appear to "stand" over a fixed point on the earth. The cameras on the ATS were timed to scan the earth below and return the picture within a little less than one-half hour. Successive pictures of cloud structures displayed cloud movements and for the first time these could be pieced together to make, in essence, a "movie" of cloud movements.

Satellite weather observation continued to advance. By using a number of select, narrow-band frequencies in the infrared, satellite hardware is now able to measure temperatures down to the Earth's surface. Data are telemetered to earth receiving stations for analysis and display in graphic form as maps or in alphanumeric form as desired. Within the realm of present technological capability limitations is a weather observing satellite

Fig. 6. View of Western Hemisphere taken by an early weather satellite. (*National Aeronautics and Space Administration.*)

that can scan the earth below for cloud cover, measure temperature and water vapor content of the air, and provide a readout of the cloud tops. A series of polar orbiting satellites, fully equipped, can put the whole atmosphere under surveillance continuously.

Weather-observing satellites are powered by solar energy and, therefore, can continue to function so long as their energy systems function properly. The longevity of the systems continues to improve with each newly added orbiting satellite. Most of the earlier weather satellites have failed or fallen back into the Earth's atmosphere where they were destroyed.

Geostationary Operational Environment Satellites (GOES). This series of weather satellites commenced in the early 1980s. After a failure of one of two satellites, in July 1984, another GOES was launched by a Delta rocket on February 26, 1987 after several postponements. With coverage by only one performing satellite during the period, it was necessary to shift the satellite position so that the eastern and western portions of North America could be under surveillance by giving up portions of coverage to the east and west of the continent. Loss of another satellite again reduced the system to one GOES-7 satellite in the early 1990s.

With the expiration of GOES-7 in the early 1990s, arrangements were made to utilize other available satellites for procuring weather information. The first of the GOES-NEXT series of weather satellites is scheduled for launching in the spring of 1994, with plans to complete the full series in 1994–1995.

Liquid water and ice clouds are routinely tracked on imagery from all geostationary satellites around the world at least twice per day, and the resultant wind data are fed into global weather forecast models. Geostationary satellites operated by the United States, the European Space Agency (ESA), Japan, and India are used to obtain the cloud wind data.

The GOES-NEXT Program. Plans for this next series of weather satellites commenced in 1985, but as of early 1994 the program had not come to fruition. The GOES-NEXT series was designed to simultaneously record visual images and take soundings of both temperature and moisture. See Fig. 7. The GOES satellite can accomplish both of these operations, but not simultaneously. The program has experienced numerous difficulties. NOAA gave NASA the responsibility to run the program, and NASA selected contractors as early as 1985. Total cost of the program has risen from about $640 mil to $1.7 bil, largely because of cost overruns and technical problems. See also **Satellites (Scientific and Reconnaissance)**.

Sunspots and Weather

For several years, some meteorologists have claimed a strong connection between sunspots and weather events on Earth. There may be a connection,

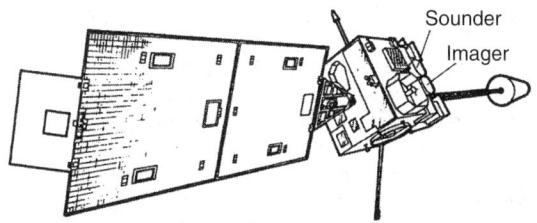

Fig. 7.   Artist's concept of a GOES-NEXT satellite. (*National Aeronautics and Space Administration.*)

but not of the magnitude once considered. There was strong statistical evidence of the connection over a 40-year period—that is, until the winter of 1989, at which time a close correlation was tested. There was unusually heavy sunspot activity during that time. There should have been an exceptionally cold winter in the east-central United States. Actually, the exact opposite occurred. But, at a conference on the Climate Impact of Solar Variability sponsored by NASA in April 1990, one professional observed, "The jury is still out on the 110-year solar effect. We're going to have to wait longer." Part of the reasoning is that any impact of solar activity on the winter of 1989 was overwhelmed by the effects of the El Niño cycle. Searching through past records relative to El Niño and sunspot activity has shown that a week case for sunspot effects on the weather may remain valid. See also **Atmosphere-Ocean Interface**; and **El Niño**.

Testing the Effects of Weather (Weatherability)

Numerous artificially created weather-condition test chambers are used by manufacturers of various machines and systems, notably in the transportation vehicle field. Considered the largest of such facilities, the Centre Scientifique et Technique du Batiment (CSTM), located in Nantes, France, was dedicated in October 1990. Sometimes referred to as the Jules Verne wind tunnel, the 54,000-square-foot (5000 m²) facility is available to meet long-unfulfilled needs of researchers in government laboratories, universities, and industry for conducting tests under simulated extreme climatic conditions. The building is equipped with six fans, having a total power capability of 3.2 MW. These fans are capable of generating winds up to a top speed of 195 mph (310 km/hr). The fans are designed to operate independently and simulate sudden changes in wind speed. The facility also has a sand-input system for simulating sandstorms at speeds up to 60 mph (100 km/hr). Rain also can be simulated up to a precipitation level of 0.1 in/min (3 mm/min) in the presence of winds up to 60 mph (100 km/hr). Sunlight can be simulated through use of metallic-halogen lamps for sunlight effects up to 1.1 kW/square meter.

Additional Reading

Albright, B.A.: "Aerosols, Cloud Microphysics, and Fractional Cloudiness," *Science*, 1227 (September 15, 1989).
Aquado, E., J.E. Burt: "Understanding Weather and Climate," 2nd Edition, Prentice-Hall, Inc., Upper Saddle River NJ, 2000.
Beer, T.: "The Applied Environmetrics Meteorological Tables," CRC Press, LLC., Boca Raton, FL, 1990.
Bjerklie, D.: "A Modern Weather Service," *Technology Review (MIT)*, 18 (January 1992).
Cess, R.D. et al.: "Interpretation of Cloud-Climate Feedback as Produced by 14 Atmospheric General Circulation Models," *Science*, 513 (August 4, 1989).
Chiles, J.R.: "NASA's Giant Research Balloons," *Smithsonian*, 82 (January 1987).
Cipra, B.A.: "An Astrophysical Guide to the Weather on Earth," *Science*, 212 (October 13, 1989).
Cotton, W.R., R.A. Anthes: "Storm and Cloud Dynamics," Academic Press, Inc., San Diego, CA, 1997.
Cowen, R.: "Launch Delays Jeopardize Weather Forecasts," *Science News*, 5 (July 6, 1991).
Cowen, R.: "Weather Report: NASA GOES Astray," *Science News*, 68 (August 3, 1991).
Dunlop, S.: "A Dictionary of Weather," Oxford University Press, Inc., New York, NY, 2001.
Friedman, R.M.: "Appropriating the Weather. Vilhelm Bjerknes and the Construction of a Modern Meteorology," Cornell University Press, Ithaca, New York, NY, 1993.
Gibbons, A.: "Chaos and the Real World," *Technology Review (MIT)*, 12 (July 1988).
Hamilton, D.: "Will GOES-NEXT Go Next?" *Science*, 133 (July 12, 1991).
Hamilton, D.P.: "Giving Up On GOES-NEXT," *Science*, 499 (August 2, 1991).
Holden, C.: "Earthwinds Around the World," *Science*, 964 (May 25, 1990).
Horgan, J.: "Pinning Down Clouds," *Sci. Amer.*, 22 (May 1989).
Kerr, R.A.: "Sunspot-Weather Link is Down but Not Out," *Science*, 684 (May 11, 1990).
Kerr, R.A.: "NOAA Revived for the Green Decade," *Science*, 1177 (June 8, 1990).
Kerr, R.A.: "A Military Navigation System Might Probe Lofty Weather," *Science*, 318 (April 17, 1992).
Kjelgaard, M.J.: "Engineering Weather Data," The McGraw-Hill Companies, Inc., New York, NY, 2001.
List, R.J.: "Smithsonian Meteorological Tables," Smithsonian Institution Press, Washington, DC, 2000.
Monastersky, R.: "Forecasting Into Chaos: Meteorologists Seek to Foresee Unpredictability," *Science News*, 280 (May 5, 1990).
Monmonier, M.: "Air Apparent: How Meteorologists Learned to Map, Predict, and Dramatize Weather," University of Chicago Press, Chicago, IL, 2000.
Pool, R.: "Is Something Strange About the Weather?" *Science*, 1290 (March 10, 1989).
Potter, T.D.: "Handbook of Weather, Climate, and Oceans," The McGraw-Hill Companies, Inc., New York, NY, 2001.
Reynolds, R.: "The Cambridge Guide to Weather," Cambridge University Press, New York, NY, 2000.
Ross, P.E.: "Lorenz's Butterfly: Weather Forecasters Grapple with the Limits of Accuracy," *Sci. Amer.*, 42 (September 1990).
Spencer, R.W., J.R. Christy: "Precise Monitoring of Global Temperature Trends from Satellites," *Science*, 1558 (March 30, 1990).
Staff: "Looking for a Change in the Weather?" *Advanced Materials & Processes*, 6 (August 1991).
Staff: "Weather Watchers to get Hughes' Help," *Hughesnews*, 3 (July 10, 1992).
Staff: Diagram Group, "Weather and Climate on File," Facts on File, Inc., New York, NY, 2000.
Stein, P.: "MacMillan Encyclopedia of Weather," Simon & Schuster Trade, New York, NY, 2001.
Tippie, V.K., J.H. Cawley: "Modernizing NOAA's Ocean Services," *Oceanus*, 84 (Spring 1991).
Note: Numerous helpful publications are available from:
NASA—National Aeronautics and Space Administration, Washington, DC
NCAR—National Center for Atmospheric Research, Boulder, Colorado.
NOAA—National Oceanic and Atmospheric Administration, Washington, DC

Web References

American Meteorological Society: *http://www.ametsoc.org/AMS/index.html*
NOAA: *http://www.noaa.gov/*

Dr. JOHN C. FREEMAN, Certified Consulting Meteorologist, Weather Research Center, TX.

WEAVERBIRD *(Aves, Passeriformes).*   A bird of a large group found in Africa, Australia, and tropical Asia. They build remarkable nests, weaving their materials intricately and sometimes in complex forms. Some species are gregarious, building large nests for the entire colony. Many species of weavers are brilliantly colored. The group includes the ox birds, whidah birds, bishop birds, munias, and weaver finches.

The blood weaver-finches represent a group of weaverbirds of Arabia and Africa, named from the prevalence of scarlet in their plumage.

WEBER EQUATION.   The second-order equation

$$y'' + (n + \tfrac{1}{2} - \tfrac{1}{4}x^2)y = 0$$

It has an irregular singular point at ∞ and occurs in solving the wave equation in parabolic cylindrical coordinates. It is a special case of the confluent hypergeometric equation and its solutions are confluent hypergeometric series.

WEBER-FECHNER LAW.   An approximate psychophysical law relating the degree of response or sensation of a sense organ and the intensity of the stimulus. The law asserts that equal increments of sensation are associated with equal increments of the logarithm of the stimulus, or that the just noticeable difference in any sensation results from a change in the stimulus which bears a constant ratio to the value of the stimulus. Also called *Weber law*.

WEBWORM *(Insecta, Lepidoptera).*   A caterpillar that surrounds the site of its work on plants with a mixture of silk and debris. Several insects belonging to different families are known as webworms. The burrowing webworms (*Acrolophidae*) eat the roots of grass and may damage corn when planted on sod ground. The sod webworms work at the base of the stem and damage grasses, cereals, and other plants. They belong to the subfamily Crambinae of the family *Pyralidae*, which also contains the cabbage and garden webworms, members of a different subfamily, Pyraustinae. The former affects cabbage and related plants and the latter attacks corn, cotton, and various species of garden plants. The European corn borer is closely related to the last two species. See also **European Corn Borer**.

Cultural methods are the most important in controlling these pests. They vary according to the species, the crop, and the conditions.

Some of the specialists among the webworms are described in the rest of this entry.

Alfalfa webworm (Loxostege commixtalis, Walker). A small webbing-type caterpillar that can destroy an entire field of alfalfa within a very short period if not effectively controlled. Sometimes this insect marches as an army, much as the armyworm (see also **Armyworm**). Although a crop specialist, the alfalfa webworm will consume grains and garden crops.

Beet webworm (Loxostege stricticalis, Linne). Very similar to the alfalfa webworm. This insect is known to destroy many thousands of acres of sugarbeet in some years. The pest occurs from the Mississippi Valley westward to the Rocky Mountains. The insect winters as a larva in cells or tubes in the soil that are lined with silk. See accompanying illustration. The adults emerge in March and through June and are active at night. The wingspread ranges from 1 to 1.5 inches (2.5 to 4 centimeters). The insect devours leaves of sugarbeet and while doing this also spins a web that draws leaves together to form a tube that can be several inches long. This heads to their hiding place, often a lump of soil near the base of the plant. There may be 3 generations of beet webworm per year.

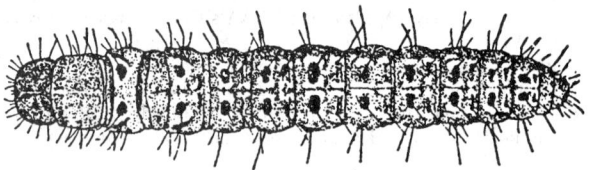

Fig. 1.   Sod webworm larva.

Corn root webworm or sod webworm (Grambus sp.). This is a gray or brown caterpillar approximately $\frac{1}{2}$-inch (12 millimeters) long which habitates in silk-lined burrows in the soil near the base of the corn (maize) or other plant attacked. See Fig. 1. This insect thrives in grassland. It cuts off plants near the surface of the soil in a manner quite similar to various cutworms. Often parts of the injured plant will be dragged into the tunnel or burrow. In addition to injuring corn and other crops, the insects are very destructive to lawns. Frequently, the first sign of an invasion by these webworms will be the presence of various birds vigorously feeding on the worms.

Fruit webworm or tent caterpillar (Malacosoma americanum, Fabricius). This insect occurs throughout the United States and attacks a number of fruit trees, such as apple, peach, plum, and wild cherry, as well as certain forest and shade trees, such as beech, birch, oak, poplar, and willow. This insect was first noted in North America in 1646 and tends to occur in cycles of about every 10 years. The adult is a white-to-brown moth and very active near lights at night in June and July. The larva is a hairy caterpillar about 1 to 2 inches (2.5 to 5 centimeters) long. The caterpillar feeds on leaves, sometimes nearly fully defoliating a tree. It is characterized by the construction of tents of webbing on tree branches.

Nonchemical methods of control include removing the tents of webbing from branches by using a pole that has a cone-shaped brush or several nails on its end. The webs can be wound on the end of the pole and then should be immediately burned. The webs should not be burned on the tree because this almost always causes injury to the tree.

The eastern caterpillar lays its eggs on twigs of the tree during the summer. Eggs are laid in bands around the twigs and cover the twigs with a foamy secretion. This secretion dries to a firm, brown covering that looks like an enlargement of the twig. Larvae develop inside the egg, but do not hatch until spring. These egg masses can easily be removed from the branches and destroyed before they hatch and cause damage.

Garden webworm (Loxostege similalis, Guenee). This species of webworm may be described as a universal or general feeder, attacking a wide variety of food crops as well as undesired weeds. Although it does not specialize on alfalfa or sugarbeet, as do the specialists previously described, the garden webworm will also attack these crops. However, it is not destructive of small grains and grasses. Distribution of the insect

is throughout North America and much of South America, but notably Mexico. The insect winters in the pupal stage in the soil in the vicinity of the last plants to be attacked during the fall. In very southern climes, some larvae may pass the winter in that form. Quite early in the spring, the adult moth emerges from the pupa. Masses of eggs, ranging from just a few up to about 50, are laid by the females. The moths are active at night. Only 3 days to a week are required for the eggs to hatch, and the worms are fully grown in from 3 to 5 weeks. If there is a shortage of food supply, the garden webworms have been known to march as an army much like the armyworm.

Parsnip webworm (Depressaria heracliana, Linne). This insect is found in the northern United States and east of the Mississippi River; as well as in southern Canada. Preferred plants for attack include parsnip, celery, wild carrot, and certain weeds. The insect particularly impedes the production of celery and parsnip seed because it invades the flower heads of the plants, engulfing them with silken threads. The caterpillar is small, ranging from yellow to green to gray, with a covering of short hairs and tiny black spots. The adult grayish moth winters in loose bark and debris. The female lays eggs in late spring. When its destructive feeding on buds and seeds is complete, the worm seeks out a suitable place to pupate, emerging as an adult in late summer.

WEDDLE RULE.   A procedure for numerical integration. If $y_0, y_1, y_2, \ldots, y_6$ are the values of $y = f(x)$ at seven equally-spaced values of $x_0, x_1, \ldots, x_6$ and $h = x_i - x_j$, then

$$\int_{x_1}^{x_6} f(x)\,dx = \frac{3h}{10}(y_0 + 5y_1 + y_2 + 6y_3 + y_4 + 5y_5 + y_6) + R$$

where R is a remainder term, giving the error in the value of the integral.

WEDGE.   See **Machine (Simple)**.

WEED CONTROL.   See **Herbicide**.

WEEPING SPRUCE.   See **Spruce Trees**.

WEEVERFISHES *(Osteichthyes).*   Of the family *Trachinidae*, weevers are found in the coastal waters of Europe, along the coast of Chile, and the waters of northern and western Africa. These fishes are probably best known for their poisonous spines, which contain a venomous nerve toxin. *Trachinus draco* prefers relatively deep waters and attains a length of about 17 inches (43 centimeters). Its habitat is from the Baltic to the Black Sea. The smaller *Trachinus viper* seldom exceeds 6 inches in length and is reasonably abundant in the North Sea southward to the Mediterranean. These fishes have the bad habit of burrowing in sandy bottoms and, because of their poisonous nature are a hazard to skin divers.

WEEVIL *(Insecta, Coleoptera).*   Most weevils are members of a large suborder of *Coleoptera* known as Polyphaga. These insects, with some exceptions, have the head prolonged into a snout which bears the small mandibles at its tip. The snout is most conspicuous in the nut weevils, where it exceeds the length of the body and is very slender. Weevils are frequently termed *snout beetles*. See Fig. 1.

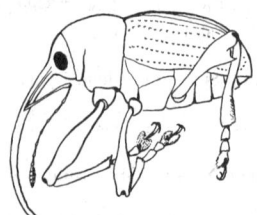

Fig. 1.   Typical snout beetle.

A number of the weevils are important economic pests, notably the cotton boll weevil, the curculio (pear and plum), and the granary weevil. The first of these lives as a larva in the squares and bolls of the cotton plant, preventing the normal development of seeds and lint, and the last is found chiefly in stored grains. The grain weevil can live for hours in an atmosphere of carbon dioxide. Although the carbon dioxide acts as an anesthetic, the insect can survive because of the reserves of oxygen in its body. Other important economic weevils include the alfalfa weevil, the potato weevil, and the rice water weevil.

See also **Grain-Storage Insects**.

WEGENER, ALFRED LOTHAR (1880–1930). Alfred Wegener was born in Berlin on November 1, 1880. He studied the natural sciences at the University of Berlin, receiving a doctorate in astronomy in 1904. He did not pursue a career in astronomy, however, but turned instead to meteorology, where the telegraph, Atlantic cable, and wireless were fostering rapid advances in storm tracking and forecasting.

In 1905 Wegener went to work at the Royal Prussian Aeronautical Observatory near Berlin, where he used kites and balloons to study the upper atmosphere. He also flew in hot air balloons; indeed, in 1906 he and his brother Kurt broke the world endurance record by staying aloft for more than 52 hours.

Thanks to his upper-air work, Wegener was invited to join a 1906 Danish expedition to Greenland's unmapped northeast coast. He was thrilled: As a youth he had dreamed of exploring the Arctic, attracted by both the scientific and physical challenges. During this expedition Wegener became the first to use kites and tethered balloons to study the polar atmosphere.

When he returned to Germany, Wegener's Arctic research earned him a position at the small University of Marberg where, beginning in 1909, he lectured on meteorology, astronomy, and "astronomic-geographic position-fitting for explorers."

Both students and professors were impressed by the clarity of the young meteorologist's thinking, by his ability to explain difficult concepts in simple terms, and by the intuitive leaps of his nimble mind.

"With what ease he found his way through the most complicated work of the theoreticians, with what feeling for the important point!" a somewhat-awed colleague, physics professor Hans Benndorf, would later write.

In 1911, still only 30, Wegener collected his meteorology lectures into a book, *The Thermodynamics of the Atmosphere*, which soon became a standard text throughout Germany. After reading it, the distinguished Russian climatologist Alexander Woeikoff wrote that a new star had risen in meteorology.

In 1912, the year of his continental-drift presentations, Wegener again answered the siren call of Greenland. His four-man expedition "escaped death only by a miracle" while climbing a suddenly calving glacier on the northeast coast, then became the first to overwinter on the ice cap. The following spring, they barely survived the longest crossing of the great ice sheet ever made, traversing 750 miles of barren snow and ice rising to heights of 10,000 feet.

During these perilous adventures, Wegener collected volumes of unique scientific data. The resulting publications established him as one of the world's leading experts on polar meteorology and glaciology. According to fellow meteorologist and Greenland explorer Dr. Johannes Georgi, Wegener was the first to trace storm tracks over the ice cap.

When he returned to Marberg, Wegener resumed work on continental drift, marshaling all the scientific evidence he could find to support his theory. Using this pioneering interdisciplinary approach, Wegener wrote one of the most influential and controversial books in the history of science: *The Origin of Continents and Oceans*, published in 1915. Because of the First World War, Wegener's book went unnoticed outside Germany. In 1922, however, a third (revised) edition was translated into English, French, Russian, Spanish, and Swedish, pushing Wegener's theory of continental drift to the forefront of debate in the earth sciences.

Wegener began by demolishing the theory that large land bridges had once connected the continents and had since sunk into the sea as part of a general cooling and contraction of the Earth. He pointed out that the continents are made of a different, less dense rock (granite) than the volcanic basalt that makes up the deep-sea floor in which Wegener proposed that the continents floated somewhat like icebergs in water. Wegener also noted that the continents move up and down to maintain equilibrium in a process called *isostasy*. As an example he cited the sinking of Northern Hemisphere lands under the weight of continental ice sheets in the last ice age, and their rise since the ice melted some 10,000 years ago.

Given the difference in density between continents and sea floor, plus the process of isostasy, Wegener reasoned that if continent-size land bridges had existed and somehow been forced to the ocean bottom, they would have "bobbed-up" again when the force was released. Therefore, since fossil and geological evidence clearly showed the continents were once

connected, the only logical alternative was that the continents themselves had been joined and had since drifted apart. See Fig. 1.

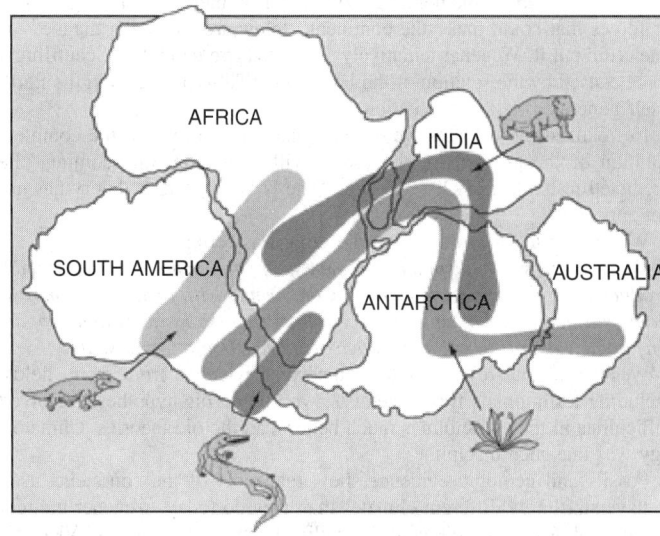

Fig. 1. Certain fossils appear in continuous bands across continents that are now separated by thousands of miles of ocean. Wegener believed this fact was one of the strongest pieces of evidence for his theory. (Map courtesy of the *United States Geological Survey*.)

Wegener also offered a more plausible explanation for mountain ranges. According to the cooling, contracting-Earth theory, they formed on the Earth's crust as wrinkles form on the skin of a drying apple. If this were so, however, they should be spread evenly over the Earth; instead mountain ranges occur in narrow bands, usually at the edge of a continent. Wegener said they formed when the edge of a drifting continent crumpled and folded—as when India hit Asia and formed the Himalayas.

He also noted that when Africa and South America are fitted together, mountain ranges (and coal deposits) run uninterrupted across both continents, writing:

It is just as if we were to refit the torn pieces of a newspaper by matching their edges and then check whether the lines of print ran smoothly across. If they do, there is nothing left but to conclude that the pieces were in fact joined in this way.

By his third edition (1922), Wegener was citing geological evidence that some 300 million years ago all the continents had been joined in a supercontinent stretching from pole to pole. He called it Pangaea (all lands), and said it began to break up about 200 million years ago, when the continents started moving to their current positions.

Perhaps the best summary of Wegener's revolutionary theory was provided by countryman Hans Cloos: "It placed an easily comprehensible, tremendously exciting structure of ideas upon a solid foundation. It released the continents from the Earth's core and transformed them into icebergs of gneiss [granite] on a sea of basalt. It let them float and drift, break apart and converge. Where they broke away, cracks, rifts, trenches remain; where they collided, ranges of folded mountains appear."

Except for a few converts, and those like Cloos who could not accept the concept but was clearly fascinated by it, the international geological community's reaction to Wegener's theory was militantly hostile. American geologist Frank Taylor had published a similar theory in 1910, but most of his colleagues had simply ignored it. Wegener's more cogent and comprehensive work, however, was impossible to ignore and ignited a firestorm of rage and rancor. Moreover, most of the blistering attacks were aimed at Wegener himself, an outsider who seemed to be attacking the very foundations of geology.

Because of this abuse, Wegener could not get a professorship at any German university. Fortunately, the University of Graz in Austria was more tolerant of controversy, and in 1924 it appointed him professor of meteorology and geophysics.

In 1926 Wegener was invited to an international symposium in New York called to discuss his theory. Though he found some supporters, many speakers were sarcastic to the point of insult. Wegener said little. He just sat smoking his pipe and listening. His attitude seems to have mirrored that of Galileo who, forced to recant Copernicus' theory that the Earth moves around the sun, is said to have murmured, "Nevertheless, it moves!"

Scientifically, of course, Wegener's case was not as good as Galileo's which was based on mathematics. His major problem was finding a force or forces that could make the continents "plow around in the mantle," as one critic put it. Wegener tentatively suggested two candidates: centrifugal force caused by the rotation of the Earth, and tidal-type waves in the Earth itself generated by the gravitational pull of the sun and moon.

He realized these forces were inadequate. "It is probable the complete solution of the problem of the forces will be a long time coming," he predicted in his last (1929) revision. "The Newton of drift theory has not yet appeared."

Wegener noted, however, that one thing was certain:

The forces, which displace continents, are the same as those, which produce great fold-mountain ranges. Continental drift, faults and compressions, earthquakes, volcanicity, [ocean] transgression cycles and [apparent] polar wandering are undoubtedly connected on a grand scale.

Wegener's final revision cited supporting evidence from many fields, including testimonials from scientists, who found his hypothesis resolved difficulties in their disciplines much better than the old theories. Climatology was one such discipline.

Fossils and geologic evidence show that most of the continents used to have startlingly different climates than they do today. Wegener thought continental drift was the key to these climatic puzzles, so he and Vladimir Koppen plotted ancient deserts, jungles, and ice sheets on paleogeographic maps based on Wegener's theory. Suddenly the pieces of the puzzles fell into place, producing simple, plausible pictures of past climates. Evidence of the Permo-Carboniferous ice-age era that peaked some 280 million years ago, for example, was scattered over almost half the Earth, including the hottest deserts. On Wegener's map, however, it clustered neatly around the South Pole, because Africa, Antarctica, Australia, and India had once comprised a Southern Hemisphere supercontinent (Gondwanaland).

Wegener considered such paleoclimatic validation one of the strongest proofs of his theory. Conversely, continental drift has since become the organizing principle of paleoclimatology and other paleosciences.

Unfortunately, though Wegener's explanation of the Permo-Carboniferous ice age impressed even his critics, the merit of much of the rest of his supporting evidence was not widely recognized at the time. As a result, most geologists eventually dismissed his theory as a fairy tale or "mere geopoetry."

Despite general rejection, Wegener's compelling concept continued to attract a few advocates over the next several decades. Then, beginning in the mid-1950s, a series of confirming discoveries in paleomagnetism and oceanography finally convinced most scientists that continents do indeed move. Moreover, as Wegener had predicted, the movement is part of a grandscale process that causes mountain-building, earthquakes, volcanic eruptions, sea-level fluctuations, and apparent polar wandering as it rearranges Earth's geography.

Geologists call the process "plate tectonics," after the large moving plates that form the planet's outer shell. These plates carry both continents and sea floor, but unlike the sea floor, the less-dense, buoyant continents resist subduction into the mantle. Thus, despite significant differences in detail, Alfred Wegener was right in most of his major concepts. Plate tectonics also confirms the accuracy of many of his paleogeographic reconstructions.

Ironically, though the lack of a credible driving force was the main objection to Wegener's theory, plate tectonics has been almost universally accepted despite the absence of scientific consensus as to its cause. Convection currents in the molten magma of the upper mantle are the favorite candidate; Wegener discussed this possibility in his 1929 revision.

During the last few decades, Alfred Wegener has finally gotten the recognition he deserves. Unfortunately, as with most visionaries, it must be posthumous praise. See also **Ocean**; and **Earth Tectonics and Earthquakes**.

Additional Reading

Koppen, V. and A. Wegener: "The Climates of the Geological Past," D. Van Nostrand, London, England, 1863.
Schwarzbach, M. and A. Wegener: "The Father of Continental Drift," Science Tech, Madison, WI, 1986.
Wegener, A.: "The Origin of Continents and Oceans," Dover Publications, Mineola, NY, 1966.

PATRICK HUGHES, NASA.

WEIBULL DISTRIBUTION. The cumulative distribution function $F(t) = 1 - e^{-\alpha(t-\gamma)^\beta}$ where $t > \gamma$, $\alpha > 0$, and $b > 0$ is referred to as the three-parameter Weibull distribution. The two-parameter Weibull distribution is obtained by setting $\gamma = 0$. The density function is $f(t) = F'(t) = \alpha\beta(t - \gamma)^{\beta-1}e^{-(t-\gamma)^\beta}$. This distribution has become popular in statistical reliability problems. The moments of the Weibull distribution are given by

$$E(t) = \bar{t} = \frac{a}{\alpha} + \gamma$$

$$\sigma^2 = (b - a^2)/\alpha^2$$

$$\in_3 = (c - 3ab + 2a^3)/\alpha^3$$

where

$$a = \Gamma\left(1 + \frac{1}{\beta}\right)$$

$$b = \Gamma\left(1 + \frac{2}{\beta}\right)$$

$$c = \Gamma\left(1 + \frac{3}{\beta}\right)$$

and $\in_3$ is the third moment about the mean. In practice, the two parameter ($\gamma = 0$) distribution is simpler to use. In reliability problems the instantaneous failure rate $Z(t)$ is given by

$$Z(t) = F'(t)/(1 - F(t)) \quad t > 0$$

For the two parameter Weibull distribution $Z(t) = \alpha\beta t^{\beta-1} t > 0$. It is easily seen that if $\beta < 1$ the failure rate decreases with time, if $\beta > 1$ it increases with time and if $\beta = 1$ the failure rate equals α and the Weibull distribution is equal to the exponential distribution. It can be seen that $\ln\ln 1/(1 - F(t)) = \ln\alpha + \beta\ln t$. The right hand side is linear in $\ln t$. Special graph paper having its scales transformed so that the divisions on the horizontal axis are proportional to $\ln t$ and those on the vertical scale are proportional to $\ln\ln 1/(1 - F(t))$ can be obtained. To estimate α and β by graphical methods, one may use the following method:

If the ith unit fails at time t_i, estimate $F(t_i)$ by using the estimator

$$\hat{F}(t_i) = \frac{i - \frac{1}{2}}{n}$$

Plot the point corresponding to t_i, $\hat{F}(t_i)$ on Weibull paper. Draw a straight line fit to the points. If the points fall reasonably close to a straight line, it can be assumed that the underlying failure-time distribution is of the Weibull type. The graphically obtained answers will suffice for many applications. The parameters α and β may also be estimated by applying linear regression methods to fit a straight line to the transformed data.

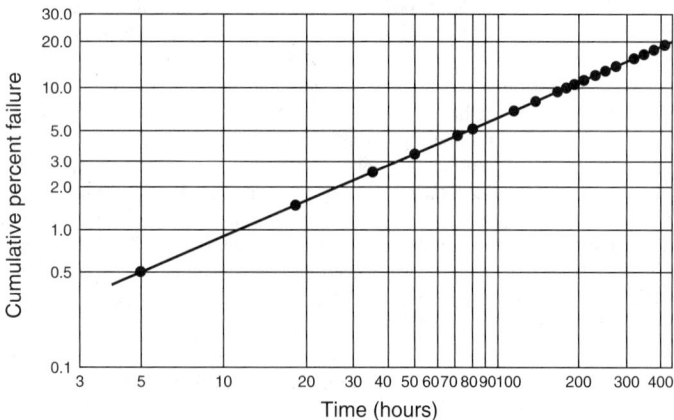

Fig. 1.   Weibull failure-time distribution.

In some applications, failure times are observed independently. One orders the observations by ascending time to failure and then uses the Weibull distribution in the same manner as if zero time were identical for all observations. Figure 1 shows the use of the Weibull distribution. It can be seen that 5 hours is a good estimate for the time within which $\frac{1}{2}\%$ of the units of this type will fail.

DONALD R. HODGE, The BDM Corporation, Vienna, VA.

WEIERSTRASS DEFINITION.   See **Gamma Function**.

WEIERSTRASS FUNCTION.   This function $\rho(z) = \rho(z/\omega, \omega')$ is an elliptic function of periods $2\omega, 2\omega'$ which: (1) is of order two; (2) has a double pole at $z = 0$, the principal part of the function at the pole being z^{-2}; (3) makes $\rho(z) - z^{-2}$ analytic in a neighborhood of, and vanish at $z = 0$. In this definition, the order of an elliptic function is its number of poles, each counted according to its multiplicity, in the parallelogram formed for fixed z_0 by the points

$$z = z_0 + 2\zeta\omega + 2\eta\omega', \quad 0 \leq \zeta < 1; \quad 0 \leq \eta < 1$$

Weierstrass' zeta function is a certain meromorphic function with simple poles, derivable from Weierstrass' function, but not itself doubly-periodic and hence not an elliptic function.

See also **Elliptic Integral**; and **Gamma Function**.

WEIGHING.   A balance or a scale is an instrument with which the mass of a body is measured, where the force of gravity on the body — weight — is used as a basic for comparison. However, the strict physical definition of mass, i.e., the properties of gravity and inertia, is not stressed in weighing. The mass determined by weighing, in the broadest sense, serves more as measure for the quantity of substance in a body, from which are derived such practical measurements as market value, relative proportions in compounds, reaction equivalents, etc. Therefore, the result read from the balance is generally taken as a measure of the quantity weighed; corrective calculations (e.g., for air buoyancy) are done only in exceptional cases.

Mass and Weight. In everyday speech, weight is the preferred term for a mass determined by weighing. In scientific usage, however, there is adherence to the stricter physical understanding of the terms mass and weight.

By *mass* we understand that basic characteristics of matter which manifests itself on the one hand as inertia and on the other hand as mutual attraction or gravitation. The basic unit of mass is the kilogram, which is embodied in a standard (kilogram prototype) at the Bureau International des Poids et Mesures in Paris.

Weight, in a physical sense, is the force exerted on a body by the gravitational field of the earth. Defined this way, weight is the product of mass with local gravitational acceleration g, expressed by the force unit Newton (N), which is short for kg m/s^2.

Since it depends on geographic latitude, altitude above sea level and the density of the earth at the location, gravitational acceleration is not a constant. Expressed in m/s^2, g is equivalent to, for example: 9.80943 in Paris, France; 9.80267 in New York; 9.79524 in Atlanta, Georgia; and 9.79965 in San Francisco, California.

For this reason, the sensitivity of a balance should be fine adjusted on location by using a standard mass, as long as the weighing principle does not involve direct comparisons of mass.

In another context, weight — also called weight piece or weight stone — is the embodiment of a given amount of mass.

Classification of Balances and Scales

Balances and scales are differentiated in accordance with their (1) design, (2) weighing principle, (3) application, and (4) metrological criteria, including sensitivity, accuracy, resolution, and maximum load.

A laboratory instrument is commonly referred to as a "balance," but the term "scale" also is appropriate. They are designed for light loads, usually less than 1 pound or 0.5 kilogram. Industrial scales for accepting light to very heavy loads are usually referred to as *scales*.

Laboratory Balances

Figure 1 illustrates how balances are categorized into four accuracy classes, as recommended by the Organisation Internationale de Métrologie Légale (OIML). With laboratory balances, there is also the following conventional nomenclature:

Ultramicroanalytical balances	$d_d = 0.1$ μg
Microanalytical balances	$d_d = 1$ μg
Semimicroanalytical balances	$d_d = 0.01$ mg
Macroanalytical balances	$d_d = 0.1$ mg
Top-loading precision balances	$d_d \geq 1$ mg

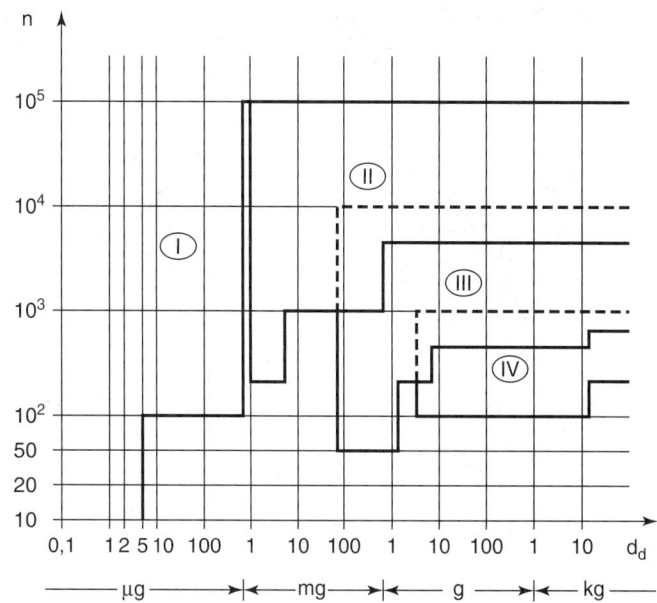

Fig. 1.   Classification of balances into four accuracy groups: (I) special accuracy, (II) high accuracy, (III) medium accuracy, and (IV) ordinary accuracy.

With top-loading precision balances, the display range is 10^4 to 10^5 scale divisions, and with some analytical balances it can be over 10^7 scale divisions.

During the past few decades, hand-manipulated laboratory balances largely have been replaced by electronic designs. However, numbers of the older style balances still can be found in locations where weighing is not frequently required.

Electronic Balance Operating Principles.   In a weighing system, there are three basic functions, as shown by Fig. 2. (1) The weight of the object to be weighed, in the form of a randomly distributed pressure p on the weighing pan, is the unknown input quantity. The *load transfer mechanism*, composed of mechanical levers and guides, translates the weighing load into a measurable single force F. (2) The measuring transducer (called a *load cell* in balances) produces an output signal proportional to the initial force, e.g., a voltage or frequency change. (3) The load cell signal is processed through analog and digital electronic circuitry where it undergoes certain computations and is finally displayed as a weighing result in digital form.

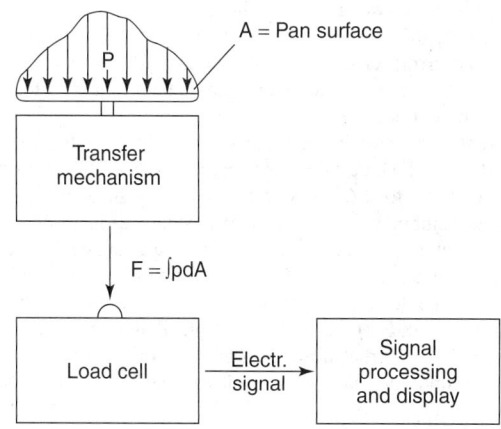

Fig. 2.   Fundamental elements of a balance. (*Mettler.*)

With *hanging pan* balances (Fig. 3), the resulting single force F is generated by an inner reaction in the suspension bearing and is led to the load cell via the balance beam. Instead of the knife bearings shown, newer balances also make use of torsion band or crossflex bearings. All things being equal, the hanging pan design provides the highest accuracy and highest resolution of small weight differences. It is used in those cases where technical weighing capabilities are stretched to their limit, and the disadvantage of a pendulous pan must therefore be accepted.

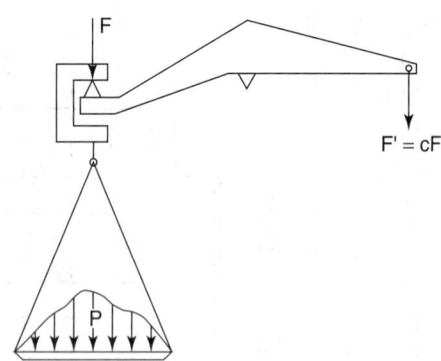

Fig. 3.   Hanging pan showing weight transfer. (*Mettler.*)

The principle of *top pan* weighing (Fig. 4) is generally used in top-loading precision balances, but has recently been introduced in analytical balances as well. The support column of the weighing pan is held vertical by two pairs of guides with elastic flex bearings. The weight force can be taken from any desired place by the support column and is led to the load cell either directly or by lever action.

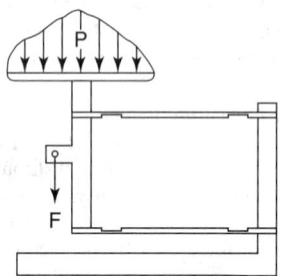

Fig. 4.   Top pan, showing weight transfer. (*Mettler.*)

A basic requirement of this system is the parallelism of the guides with an accuracy of up to ±1 μm. Parallelism deviations cause so-called side-load errors. By that, we mean deviations in weighing results occurring when a test weight is moved from the center to the edge of the weighing pan. Side load errors can be caused by sharp blows to the pan and by overstressing the guides. As a result, many balances are protected against such blows and overstresses to the weighing pan.

Load-measuring cells in laboratory balances operate, for the most part, according to the electromagnetic force compensation principle. With a resolution of up to 1,600,000 scale divisions in analytical balances, this principle combines the accuracy of mechanical instruments with the advantage of an automatic reading of the weighing result. See Fig. 5. The movable system part (weighing pan, support column, coil, position indicator) and the load are held in constant floating equilibrium by a compensating force F equal to the weight. This compensating force F is generated by the current flowing through a coil which is located in the air gap of a cylindrical magnet, according to the following equation:

$$F = I \cdot l \cdot B$$

where I equals current, l equals total length of coil wire, and B equals the magnetic flux density in the air gap.

The following processes take place in the control loop of the cell:

1. With a change of the load, the movable system part at first responds to the weight change and moves vertically, usually by only fractions of a millimeter.

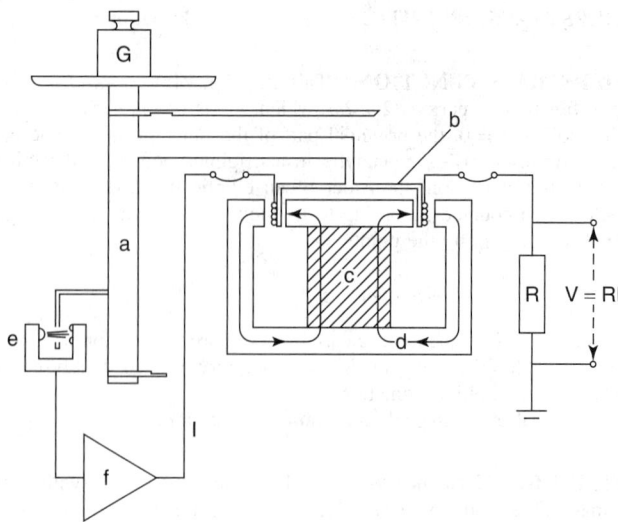

Fig. 5.   Electromagnetic force compensation principle: (**a**) Movable system part, (**b**) coil (cross section), (**c**) cross section of cylindrical magnet with permanent magnet core and soft iron jacket, (**d**) field lines of force, (**e**) photoelectric position indicator, (**f**) servo amplifier. (*Mettler.*)

2. The light gate, which is part of the movable system component, changes the amount of light received by the photoelectric position indicator. An electrical signal is then passed on to the servo amplifier.
3. The amplifier changes the current flowing through the coil only as much as is required to bring the movable system component back to its state of equilibrium.

Weight now results from a measure of current I, e.g., by measuring voltage $V = R \cdot I$ across a precision resistance switched in series to the coil.

With another form and arrangement of coil and magnet, we also find this principle as a torque-compensation system, especially in micro balances. In a practical arrangement, such a system resembles conventional moving-coil instruments for measuring current and voltage. Here the balance beam is connected to the moving coil, which is suspended on a torsion band. The torque caused by the load on the balance beam is compensated by an opposite torque, generated by electromagnetic interaction between the current through the coil and the field of the permanent magnet that surrounds the coil. The coil current is regulated through the control loop in such a way that the beam stays in its prescribed horizontal position. This is analogous to the force compensation system, but here the current is proportional to the torque and therefore to the load. The previously described steps of the signal evaluation apply.

In the *pulse width modulation* principle, square-wave direct-current impulses of a constant amplitude I_0 and frequency v, flow through the coil, with a pulse duration t' as a control variable (Fig. 6). The compensation force is determined here by the time average of the current $I = I_0 \cdot v \cdot t'$ and is thus proportional to pulse duration t'. Pulse duration t' is added up over a large number of pulses, which increases the measuring resolution.

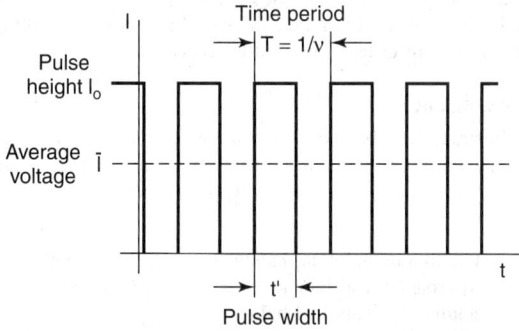

Fig. 6.   Principle of pulse width modulation of compensating current. (*Mettler.*)

Finally, it should be noted that in addition to electromagnetic force compensation, the *oscillating string principle* is also sometimes used. The relationship between tension and oscillation frequency in a string

$$f_n = n/2l \cdot \sqrt{F/\mu} \quad (n = 1, 2, 3 \ldots)$$

provides the basis for the force/frequency converter. Here, n stands for the harmonic order of oscillation, l for the length of the string, and m for the string mass per unit length.

Electronic Signal Processing Part. By this term we encompass that function group which transforms the output of the measuring transducer into a numeric display or a data output. For the user, the most important items here are the control and calculating operations—depending on the degree of balance complexity.

Push-Button Tare. Resetting to zero of the display with any load within the weighing range.

Adjustable Integration Time. The weight values are averaged over a preselected time period. This is useful in weighing laboratory animals as well as for weighing under unstable conditions in general, e.g., with building vibrations or air drafts.

Rounding Off the Result. The digital electronics of Mettler balances produce internal weighing values of a higher resolution than are indicated by the display. In this way, an exact centering of the balance zero point is achieved when the balance is tared. The net internal value is numerically rounded off for display.

Stability Detector. Sequential weighing results are compared with one another. When a result stays the same, the data output is released and the validity of the display value is confirmed, e.g., when the weight unit symbol lights up.

In addition, with some of the newest balances, other useful values can be directly determined from the weighing result, e.g., parts count, percentage value, net total and target value for fill weighings. Necessary calculations are carried out by a built-in microprocessor; here the user enters control instructions and additional data into the balance via a push-button input instrument.

Special Applications. In a number of experimental processes, mass or a change in mass must be determined under special environmental conditions, or the effect of a force field superimposed on a weight must be measured. For these reasons, weighings are sometimes carried out under vacuum, under high- or low-temperature conditions, in special atmospheres, in magnetic or electrostatic fields, and in liquids.[1]

Industrial Scales

There are numerous industrial and commercial scales, ranging from those of small capacity, as may be found in retail operations, through multi-tonnage scales found in the process industries for weighing bulk materials or large discrete objects, such as a reel of wire. In terms of method, there are gravimetric scales in which other masses (weights) are used, usually via a mechanical lever system, to compare the mass of the unknown. There are spring scales, in which the gravitational force on the unknown is matched by a spring force of known constant; and, as with the much smaller balances previously described, there are electronic load-cell configurations. Pneumatic and hydraulic load cells are also used as the weight (force) sensors. The majority of scales operate on a batch, i.e., one-load-at-a-time basis. Continuous gravimetric feeders are described in **Feeder (Gravimetric)**. In batch weighing, the container for the load may range from a simple platform, as found in a retail operation, to numerous industrial configurations, including tank scales, floor scales, hopper scales, railway track, and highway vehicular scales.

Early scales required the manual movement of a sliding poise on a weigh beam that was connected directly to the load-reducing mechanical lever system. Industrial scales, over the years, have been built for rugged service and maintain their useful life for many years. Consequently, scales of old design still are found in parts of the world, frequently for the weighing of commodities in the third world countries. The simple steelyard still is

[1] Portions of the preceding part of this article were prepared by Dr. Walter Kupper, Research and Development Manager, Mettler Instrument Corporation, Hightstown, New Jersey.

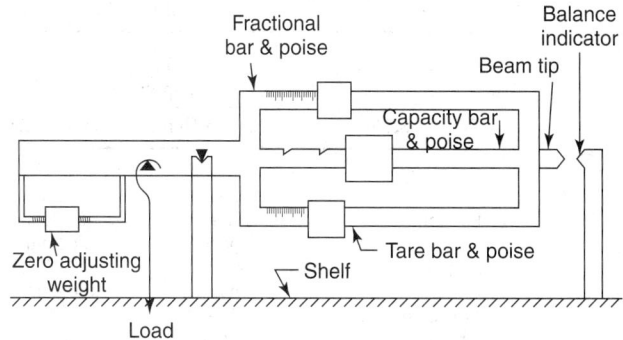

Fig. 7. Weigh beam with sliding poises.

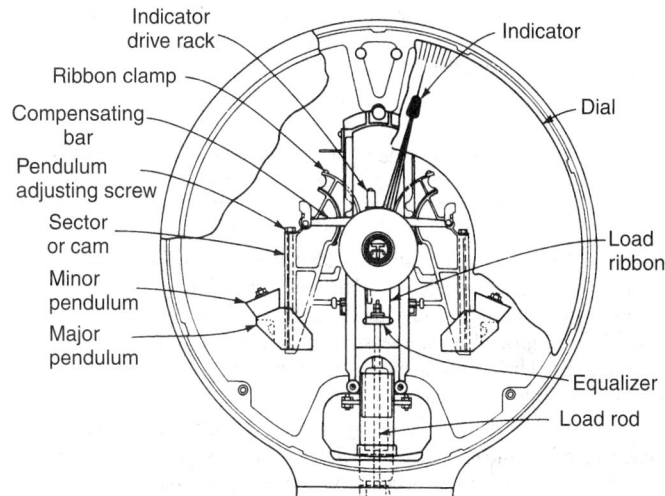

Fig. 8. Double-pendulum dial indicator, a classic design for industrial scales for many years.

found in isolated areas. A weigh beam with sliding poises is shown in Fig. 7.

Direct indication of weight, without manual manipulation of poises, was always available with spring scales for relatively small loads—because the spring "automatically" extended without any manual intervention. The first automatic pendulum-type scale appeared many decades ago and this largely replaced the weigh beam with its sliding poises. A comparison of the two methods is given in Figs. 7 and 8. Once a means was devised to achieve "automatic" positioning of the pointer to indicate weight directly, it was relatively easy to add accessory devices in the form of photoelectric pickups, mercury magnetic switches, etc. to sense the pointer position and then actuate automatically a printing device, or to control automatic feeding (unloading, loading) equipment. In other words, semicontinuous weighing could be achieved by automatically interlocking a series of batch weighings. Fully continuous weighing is achieved by belt scales in which a section of belt rides over weight-sensing devices and these are interlocked in numerous ways to effect a control over weight-flow. Frequently, a series of such weigh-belts will be interlocked to achieve a gravimetric blending of several ingredients, as may be encountered in the glass, rubber, feed, and other industries, where processing or final products are based upon closely maintaining proportions of materials on a weight basis.

The principal types of industrial scales are listed in Table 1.

Load Cells Revolutionize Industrial Scales. Most scales installed within the last few decades, instead of depending upon mechanical lever systems and counterbalancing systems, measure the deflection of supporting beam structures with hydraulic, pneumatic, or electronic (strain-gage) systems. Surprisingly, these substitutions were not made to achieve greater accuracy, but rather to reduce the bulk, weight, and iron content of scales and to provide means for remote weight data displays and ways to automatically control scale operations (as in batch processes) as one can manage any other sensor. At one time, the lever-type scale was one of the most accurate of sensors, though cumbersome.

TABLE 1. PRINCIPAL TYPES OF INDUSTRIAL SCALES

Type of Scale	Capacity Range		Load Receiving Element		
	English Units	Metric Units	Type	Size	
				Feet	~meters
Bench scale	5–300 pounds	2.3–136 kilograms	platform	5 × 5 to 24 × 24	1.5 × 1.5 to 7.3 × 7.3
Portable scale	100–5000 pounds	45–2268 kilograms	platform	2 × 2 to 6 × 6	0.6 × 0.6 to 1.8 × 1.8
Floor/warehouse scale	500–20,000 pounds	227–9072 kilograms	platform	3 × 3 to 10 × 10	0.9 × 0.9 to 3 × 3
Motor truck scale (axle load)	10–75 tons	9–68 metric tons	platform	8 × 10 to 10 × 100	2.4 × 3 to 3 × 30
Cement scale	20–100 tons	18–91 metric tons	platform	10 × 20 to 16 × 100	3 × 6.1 5 × 30
Railway track scale	100–200 tons	91–182 metric tons	track span	4 × 110	1.2 × 33.5
Hopper or tank scale	100 pounds–200 tons	45 kilograms–182 metric tons	hopper or tank	2 to 60 square feet	0.2 × 5.6 square meters
Hanging scale	5–200 pounds	2.2–91 kilograms	hook or suspended pan	—	—
Crane scale	1–100 tons	0.9–91 metric tons	hook	—	—
Monorail scale	50–5000 pounds	23–2268 kilograms	overhead track	2 × 8 overhead track	0.6 × 2.4 overhead track

WEIGHTING. 1. The artificial adjustment of measurements in order to account for factors which, in the normal use of the device, would otherwise be different from the conditions during measurement. For example, background noise measurements may be weighted by applying factors or by introducing networks to reduce measured values in inverse ratio to their interfering effects.

2. When several different determinations of the same physical quantity have been made, the more accurate of the measurements should be counted more heavily than the less accurate in estimating the most probable value of the quantity. Thus, if $q_1 \cdot q_2$, etc., are the individual values, the most probable value is

$$q = \frac{w_1 q_1 + w_2 q_2 + \cdots}{w_1 + w_2 + \cdots}$$

where w_1, w_2, etc., are weighting factors. The theory of the propagation of errors shows that the weighting factor for any individual determination is inversely proportional to the square of the uncertainty of that determination. The quantity q is called the weighted mean. 3. In statistical mechanics, certain states of a system are often more probable than others, especially when degeneracy is present. In computing average values, these states must be weighted with factors proportional to their probabilities, the computation then proceeding as in 2 above.

WEIGHTLESSNESS. 1. A condition in which no acceleration, whether of gravity or other force, can be detected by an observer within the system in question. Any object falling freely in a vacuum is weightless, thus an unaccelerated satellite orbiting the earth is weightless although gravity affects its orbit. Weightlessness can be produced within the atmosphere in aircraft flying a parabolic flight path. See also **Gravitation**.

2. A condition in which gravitational and other external forces acting on a body produce no stress, either internal or external, in the body. See also **Astronautics**.

An astronaut is said to be in a state of weightlessness when in orbit. Strictly speaking, the astronaut still has weight, for the Earth's gravity still acts on the person. Otherwise, the individual would fly off into outer space. However, when in free fall, the local effects of the gravitational field are eliminated for the astronaut. Objects that are released fall together with the astronaut and hence remain in the vicinity, unlike the situation on the ground. Therefore, the organs of the body respond as though the gravitational field were absent and this gives the sensation of weightlessness. Conversely, one senses the earth's gravity and feels weight because we are supported by the earth's surface.

WEIRS.   See **Flow Measurement**.

WELDING.   A method of joining metals by means of fusion or by solid-state processes. Metals having similar composition may be united in one homogeneous piece by fusing together the edges in contact, or by additional molten metal of the proper characteristics deposited where it will form a fused joint with each piece.

For special applications, welding techniques include electron beam welding and laser welding. In electron beam welding, a coalescence of metals is produced as the result of heat obtained from a concentrated beam composed primarily of high-velocity electrons impinging upon the surfaces to be joined. In laser-beam welding, the heat is obtained by directing a beam from a laser onto the weld joint.

In any machine or structure there are numerous cases where permanent junctions between the component parts are required. The use of welding in assembling metal structures and machines is thus very extensive in present-day technology. Typical of such practice might be mentioned the welding of joints in pipe lines, eliminating flanges, couplings, and cumbersome fittings. Bridges and buildings of structural steel are fabricated by arc welding, replacing the older riveted connections. Shipbuilders use welding as the principal means of fabrication. Where weight saving is as important as strength, welding is especially suitable because of the elimination of the weight of rivets and bolts, and often of the flanges in which they are seated. Many complicated machine parts, which at one time could be produced only by casting, are now built up from structural steel shapes by welding. To the above illustrations, may be added the aircraft and rocket structures, where welding has become standard practice. Machinery frames and bases, tanks, and steel frames are welded. Even high-temperature and pressure vessels, such as boiler drums, are accepted if welded in conformance with standards procedures.

Automatic Welding.   Welding is well adapted to automation. Numerous robots have been designed specifically for welding applications. In the more sophisticated systems, machine vision may be used. See also **Machine Vision**. The automotive industry in recent years has invested heavily in automated welding, as portrayed by the illustrations and captions of Figs. 1, 2, and 3.

Most present preprogrammed, automated welding systems use a set path and fixed parameters. They are not flexible and may not be cost effective for many applications, particularly small batches. As reported by engineers at the American Welding Institute (Knoxville, Tennessee), "Overcoming these limitations is the goal of next-generation 'intelligent' welding systems." Work is proceeding on programs that automate the entire process, from planning the welding procedure to real-time weld monitoring and quality control.

Welding appears to be an ideal application for computerized expert systems because it involves compliance with a complex and interrelated system of codes, specifications, tests, and inspections to ensure that welds will not fail in service. Today, this assurance is provided by a large number

Fig. 1.   Automated welding system. Computer-controlled robots achieve consistent welds of all components in the unitized body structure. (*Chrysler Corporation.*)

Fig. 2.   Master computer control station monitors robots on robotgate welding lines in automobile assembly plant. (*Chrysler Corporation.*)

visual-sensor data obtained during robotic welding is being developed. The intelligent seam tracker can visually guide a welding robot operating in an environment where the clarity of the visual image is reduced by as much as 70 percent. Seam tracking using conventional image-processing techniques is computation intensive and ineffective when noise exceeds about 20 percent.

Fusion Welding. Any two clean metal surfaces should form a bond as strong as the bulk material if they are brought into intimate contact. The key to success is to achieve extremely clean surfaces and really intimate contact. With rare exceptions, metals are covered with a layer of oxide, or some other material, formed by the interaction of the metal and its environment. This layer must be removed, or a new surface provided by cutting or deforming the piece in such a way as to expose clean metal. Then, the newly exposed metal atoms can join together intimately. To a metallurgist, intimate means that the atoms at the interface between

of engineers, designers, and welders. Problem-solving, expert-system software can help reduce the need for costly human experts. More into the future are combinations of expert systems and neural networks. The latter excel at dealing with complex, poorly characterized problems where multiple sensor inputs must be evaluated concurrently—as in welding. Neural-network/sensor systems must be able to perform in the processing environment and provide data fast enough for intelligent controls to operate in real time. A neural-network–based seam tracker that analyzes

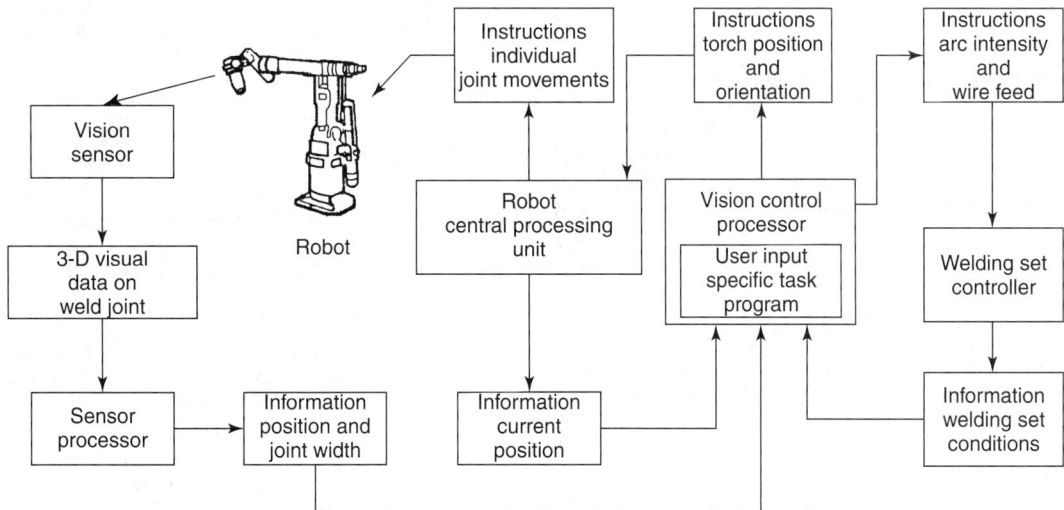

Fig. 3.   This vision-assisted robot system is capable of single-pass seam tracking during the welding process as well as real-time adaptive control of welding parameters and travel speeds. The system can perform a seam search to find the starting point of the weld. Then, while welding, it can measure joint gap, lateral seam position and standoff height. The robot controller uses the information supplied by the vision system to modify the robot's program path and alter welding parameters for varying conditions during welding. Advantages of the sophisticated system include: Compensates for poor part repeatability; imposes fewer limitations on product design; increases range of robotic arc welding applications; and permits welding on nonsymmetrical and multipass jobs. The combined welding torch and vision sensor contains a solid-state camera, laser pair, optics and associated video electronics mounted in a seven-axis drive unit. This compact drive unit rotates the camera about the centerline of the welding torch so the camera is always looking slightly ahead of the welding arc. The vision system operates on the principle of laser triangulation. It uses two 10 mW gallium arsenide lasers to project a stripe of structured infrared light onto the part being welded. An optical sensor views the laser stripe as it intersects the weld groove to feed back information on both the position of the weld groove and its geometry. At the wavelength which the optical sensor is viewing the laser stripe, the laser is approximately ten times the intensity of the welding arc, thus making it possible to seam track with or without the arc being present. (*GMF Robotics Corporation.*)

two welded surfaces are no further apart than the atoms within the materials. The principal methods of fusion welding are gas welding, arc welding, resistance welding, pressure welding, combination methods, such as atomic-hydrogen welding, as well as using more recently developed electron beam welding, laser welding, and plasma arcs.

Electron beam welding in vacuum chambers has made possible weld beads having a high ratio of depth to width, thus lessening the heat-affected zone. This process is used extensively for special alloys of titanium; aluminum, vanadium, and tungsten used on airplanes, helicopters, and space vehicles. In *precussion welding*, the source of heat is an electric arc obtained by discharging a capacitor; this technique has been found particularly useful for small-scale welding, as in electrical wiring. Other new developments in fusion welding include cored welding rods that carry their own flux in much the same way as rosin-core solder does; arcs shielded with an inert gas, usually helium or argon, to protect newly formed surfaces against atmosphere contaminants; and *submerged arc welding*, wherein an arc is sustained in the flux material, and metal is added to the gap.

Gas welding is usually accomplished by the oxyhydrogen or *oxyacetylene* methods, the latter being the most common, with *oxyhydrogen welding* being used primarily on aluminum. High pressure oxygen and acetylene are led through reducer valves to reduce the pressure to a torch where they are mixed and burned at the tip. The flame, either oxidizing, reducing, or neutral, is adjusted to the proper size and brought into contact with the work which has been previously prepared and clamped into position. The temperature of the work is raised until it melts and flows together. Additional metal necessary to accomplish the formation of a sound joint may be added as wire or rod as the welding progresses. In some cases, fluxes must be added to insure a sound weld, particularly with stainless steel or aluminum.

Arc welding is accomplished by utilizing the heat generated by an electric arc, either ac or dc, to fuse the metals together. A common method is to use a motor-generator set to supply dc at relatively low voltage. One electrode is fastened to the work, and the welding rod, usually coated with a flux, is made the other electrode, the arc being struck between the rod and the work. Metal from the rod melts off and is fused into the joint. Automatic machines for arc welding use coils of wire to feed into the arc, flux being supplied from a hopper as the work travels under the arc. This process is also known as *submerged arc welding* cited above. Another method of arc welding is *carbon arc welding* in which one electrode is a rod of carbon. It is generally used on automatic machines. The wandering of the arc is minimized by placing a magnetic field near the arc to stabilize it. Extra filler material is fed into the arc if needed, and it is often shielded by gas generated from a chemically-treated string or cord.

Atomic-hydrogen arc welding is a combination of gas welding and arc welding. In this case heat is supplied to work by a stream of hydrogen passing through an arc struck between two tungsten electrodes. As the hydrogen passes through the arc, it is dissociated into its atomic form. When it recombines to form molecular hydrogen, a large amount of heat is generated in a small space, making an intensely hot flame. The hydrogen blankets the weld as it is made and largely prevents loss of alloying materials. This method also has been applied to automatic machines, no flux being necessary. A large variety of alloy steels for exacting specifications are welded by this method.

Heliarc welding is a modification of the atomic hydrogen method for the welding of magnesium; here helium gas is blown along a single tungsten electrode, the work being the other terminal as in ordinary electric arc welding. The helium serves to protect the highly reactive magnesium from burning.

Electric-resistance welds may be divided into a number of different categories but all have the common characteristic, that the metal is heated by its own resistance to a semi-fused or fused state by the passage of very heavy currents for very short lengths of time and then welded by the application of pressure.

Solid-State Welding. Some combination of optimum temperature and deformation is required, but heat is not always necessary to remove the oxide film. This class of welding includes some "cold" welding methods; in vacuum welding, a clean cut is made and the surfaces to be joined are kept free from contamination in a vacuum chamber. When the two clean surfaces are brought into contact, a so-called diffusion bond is formed. In diffusion bonding without a vacuum, two surfaces to be welded are pressed together and deformed enough to bring them into reasonably good

contact; the temperature is held high enough, and for a long enough time, to eliminate the oxide already present at the interface, and the two pieces weld together. The oxide disappears either by dissolving into the joined surfaces, or by decomposing. Roll bonding of metal sheets is an example of welding that involves only deformation. It is accomplished by putting two sheets of material together and rolling them with considerable pressure and deformation, usually of the order of at least 50%. The oxide film at the interface between the two sheets is simply fragmented and mechanically dispersed, and the two new surfaces are clean enough to form a reliable bond. Oxides on the surface of a metal are relatively brittle; if the metal can be plastically deformed, the oxide film will rupture and a new metal surface will be exposed between the fragments.

Friction welding involves the application of considerable stress and deformation in the presence of relatively steep temperature gradients induced by rubbing the two surfaces together at high speeds. This is really a combination of fusion welding and deformation bonding. Explosive welding, in which one material is driven at high velocity onto the surface of another, creates large amounts of new surface area through extremely large, but highly localized deformation in the vicinity of the impact surface. Ultrasonic welding relies upon ultrasonic vibration in the plane of the interface to break up the oxide film and disperse it into the matrix.

Additional Reading

Bailey, N. et al.: "Welding Steels Without Hydrogen Cracking," 2nd Edition, ASM International, Materials Park, OH, 1993.

David, S.A., J.M. Vitek: "International Trends in Welding Science and Technology," ASM International, Materials Park, OH, 1993.

Davis, J.R., K. Ferjutz, N.D. Wheaton: "ASM Handbook: Welding, Brazing, and Soldering," ASM International, Materials Park, OH, 1994.

Gellerman, M., E. Francis: "Practical Gas Metal and Flux Cored ARC Welding," Prentice-Hall, Inc., Upper Saddle River, NJ, 1998.

Houldcroft, P.T.: "Welding Process Technology," Cambridge University Press, New York, NY, 2002.

Minnick, W.H.: "Gas Metal Arc Welding Handbook," 2nd Edition, Goodheart-Willcox Publisher, Tinley Park, IL, 1995.

Patterson, R.A., K.W. Mahin: "Weldability of Materials," ASM International, Materials Park, OH, 1990.

Staff: "Joining," in "Forecast Processes," *Advanced Materials and Processes*, 67 (January 1991).

Staff: "Joining Technology," in "Forecast '92," *Advanced Materials and Processes*, 59 (January 1992).

Staff: "Welding" in "Forecast '93," *Advanced Materials and Processes*, 27 (January 1993).

Staff: NCCER, "Welding," Prentice-Hall, Inc., Upper Saddle River, NJ, 1999.

William A., W.A. Bowditch, and K.E. Bowditch: "Welding Technology Fundamentals," Goodheart-Willcox Publisher, Tinley Park, IL, 1997.

WENTLETRAP (*Mollusca, Gasteropoda*). A marine mollusk with a white shell of elongated conical form, with many convex whorls bearing prominent ribs. Also called spiral staircases. The 200 species are widely distributed.

WESTERLIES. See **Winds and Air Movement**.

WEST INDIAN CORKWOOD. See **Balsa Tree**.

WEST NILE VIRUS. The West Nile (WN) virus has emerged in recent years in temperate regions of Europe and North America, presenting a threat to public, equine, and animal health. The most serious manifestation of WN virus infection is fatal encephalitis (inflammation of the brain) in humans and horses, as well as mortality in certain domestic and wild birds.

The West Nile virus is a member of the family Flaviviridae (genus *Flavivirus*). Serologically it is a member of the Japanese encephalitis virus complex that includes St. Louis encephalitis (SLE), Japanese encephalitis, Kunjin, and Murray Valley encephalitis viruses, as well as others (DeMadrid and Poterfield; and Calisher, et al.). All *Flaviviruses* share a common size (40–60 nm), symmetry (enveloped, icosahedral nucleocapsid), nucleic acid (positive-sense, single stranded RNA approximately 10,000–11,000 bases), and appearance in the electron microscope. See Figs. 1a and b.

History

The WN virus was first isolated from a febrile adult woman in the West Nile province of Uganda in 1937 (Hayes; and Hubalek and Halouzka). The ecology was characterized in Egypt in the 1950s. The virus became

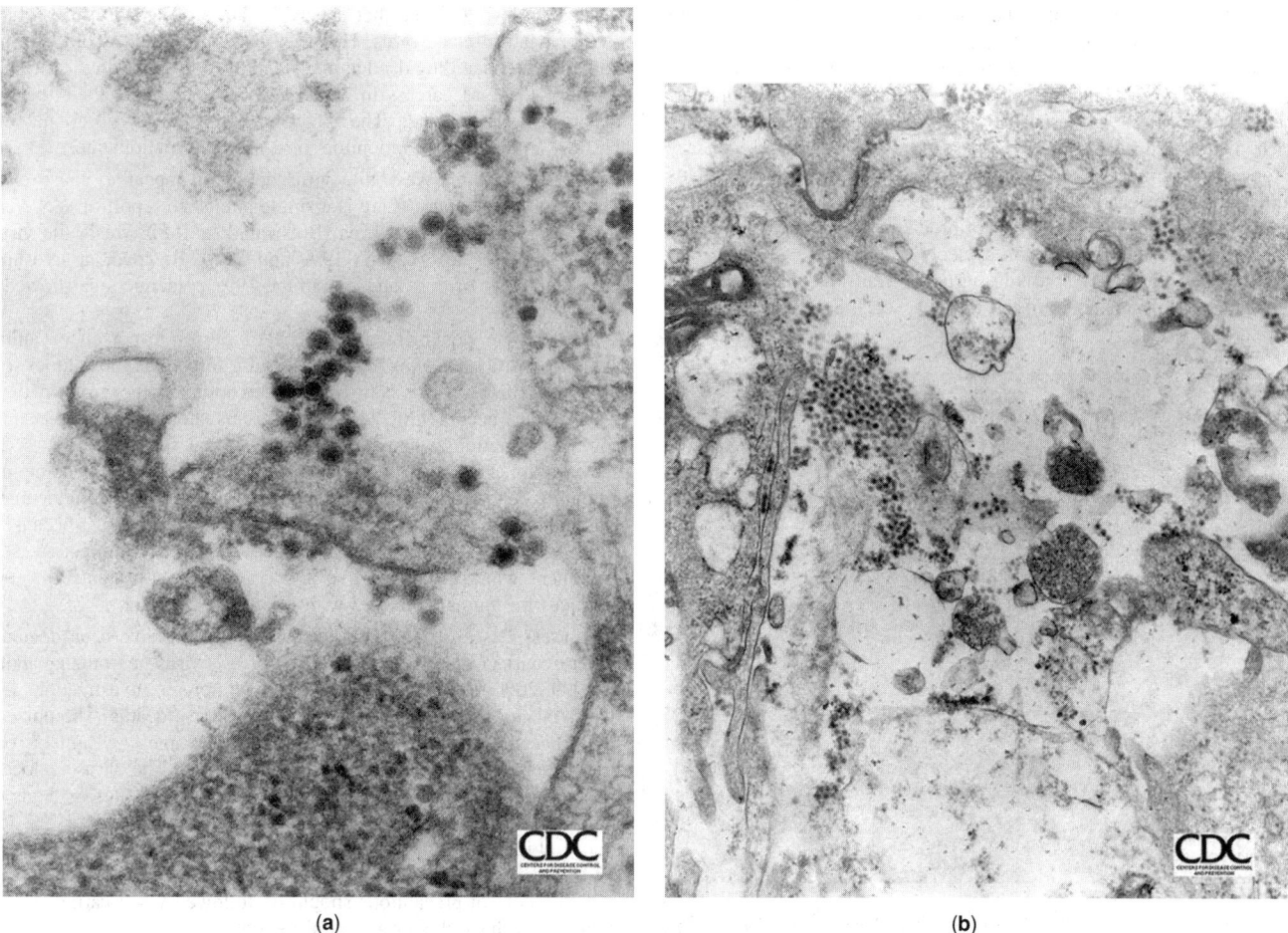

(a)   (b)

Fig. 1.   Scanned images are of West Nile virus isolated from brain tissue from a crow found in New York. The tissue was cultured in a Vero cell for a 3-day incubation period. The Vero cells were fixed in glutaraldehyde, dehydrated, placed in an Epon resin, thin sectioned, placed on a copper grid, and stained with uranyl acetate and lead citrate. The grids were then placed in the electron microscope and viewed. Total magnifications, for image 1 (**a**) 65,625x; and image (**b**) 171,250x. (Images courtesy of Bruce Cropp, Microbiologist, Division of Vector-Borne Infectious Diseases.)

recognized as a cause of severe human meningoencephalitis (inflammation of the spinal cord and brain) in elderly patients during an outbreak in Israel in 1957. The first recorded epidemics occurred in Israel during 1951–1954, and in 1957. Equine disease was first noted in Egypt and France in the early 1960s. The largest recorded epidemic caused by WN virus occurred in South Africa in 1974. A large human outbreak of WN encephalitis occurred in Israel in 2000. European epidemics of WN encephalitis have occurred in southern France in 1962, in southeastern Romania in 1996, and in south-central Russia in 1999 (Tsai, et al.; and Platanov, et al.). European equine outbreaks also have occurred in Italy in 1998 and in France in 2000.

In late summer 1999, the first domestically acquired human cases of West Nile (WN) encephalitis were documented in the U.S. (Staff: CDC 1999; Anderson, et al.; Briese, et al.; Jia, et al.; and Lanciotti, et al.). The discovery of virus-infected, overwintering mosquitoes during the winter of 1999–2000 predicted renewed virus activity for the following spring and launched early season vector-control and disease surveillance in New York City (NYC) and the surrounding areas (Staff: CDC 2000). These surveillance efforts were focused on identifying and documenting WN virus infections in birds, mosquitoes, and equines as sentinel animals that could predict the occurrence of human disease. By the end of the 2000 mosquito-borne pathogen transmission season, WN virus activity had been identified in a 12 state area from Vermont and New Hampshire in the north to North Carolina in the south. In 2000 there were 21 humans, 63 horses, 4,304 birds (78 species including 1999 data), and 480 mosquito pools (14 species) reported with WN virus (Staff CDC 2000). This annual human case incidence now ranks WN virus second only to LaCrosse encephalitis virus as the leading cause of reported human arboviral encephalitis in the U.S.

Through July 2001, the WN virus has been documented in Connecticut, Maryland, Massachusetts, New Hampshire, New Jersey, New York,

Pennsylvania, Rhode Island, Florida, Georgia, Virginia, Ohio, and the District of Columbia.

Although it is still not known when or how WN virus was introduced into North America, international travel of infected persons to New York, importation of infected birds or mosquitoes, or migration of infected birds are all possibilities. WN virus can infect a wide range of vertebrates; in humans it usually produces either asymptomatic infection or mild febrile disease, sometimes accompanied by rash, but it can cause severe and fatal infection in a small percentage of patients. In 1999 in New York, approximately 40% of laboratory-positive humans with encephalitis or meningitis had severe muscle weakness; 10% developed flaccid paralysis with electromyographic findings consistent with axonal neuropathy. The human case-fatality rate in the U.S. has been about 11%.

Unlike WN virus within its historical geographic range, or SLE virus in the Western Hemisphere, mortality in a wide variety of bird species has been a hallmark of WN virus in the U.S. The reasons for this are not known; however, public health officials were able to use bird mortality (particularly birds from the family Corvidae) to track effectively WN virus expansion in 2000. Early season field studies determined that areas with bird mortality due to WN virus infection were experiencing ongoing enzootic transmission. However, most birds survive WN virus infection as indicated by the high seroprevalence in numerous species of resident birds within the regions of greatest virus transmission. It is still not known to what degree migrating birds contribute to natural transmission cycles and dispersal of both viruses.

Entomology

Arthropod-borne viruses (termed "arboviruses") are viruses that are maintained in nature through biological transmission between susceptible vertebrate hosts by blood-feeding arthropods (mosquitoes, sand flies, ceratopogonids "no-see-ums", and ticks). Vertebrates can become infected when an infected arthropod bites them to take a blood meal. The term 'arbovirus' has no taxonomic significance.

The arboviral encephalitides are zoonotic, being maintained in complex life cycles involving a nonhuman primary vertebrate host and a primary arthropod vector. These cycles usually remain undetected until humans encroach on a natural focus, or the virus escapes this focus via a secondary vector or vertebrate host as the result of some ecologic change. Humans and domestic animals can develop clinical illness but usually are incidental or "dead-end" hosts because they do not produce significant viremia, and do not contribute to the transmission cycle.

In the United States, West Nile virus is transmitted by infected mosquitoes, primarily members of the *Culex* species. See Fig. 2.

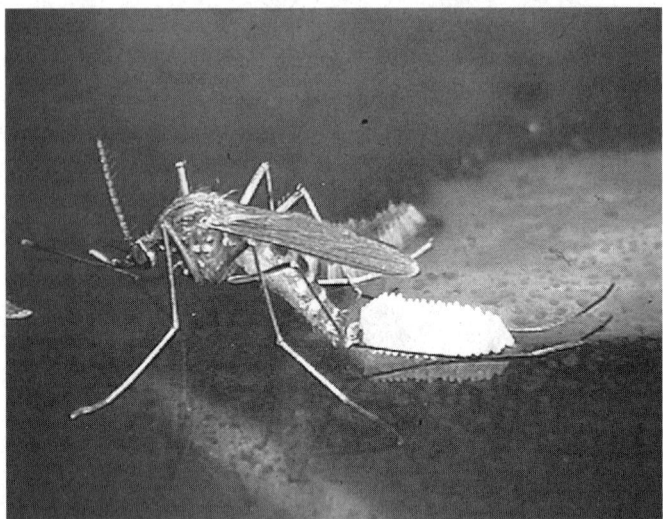

Fig. 2. *Culex* mosquito laying eggs.

Arboviral encephalitis can be prevented in two major ways: personal protective measures to reduce contact with mosquitoes and public health measures to reduce the population of infected mosquitoes in the environment. Personal protection measures include reducing time outdoors, particularly in early evening hours, wearing long pants and long sleeved shirts, and applying mosquito repellent to exposed skin areas. Public health measures include elimination of larval habitats or spraying of insecticides to kill juvenile (larvae) and adult mosquitoes. The combination of mosquito control methods selected for use in a control program depends on the type of mosquitoes to be controlled and the habitat structure. In emergency situations, wide area aerial spraying is used to quickly reduce the number of adult mosquitoes. In many states, aerial spraying may be available as a means to control nuisance mosquitoes. Such resources can be redirected to areas of virus activity when necessary.

Financing of aerial spraying costs during disease outbreaks is often provided by state or local emergency funds. Federal funding of emergency spraying is rare and almost always is associated with a natural disaster such as flood or hurricane.

Vertebrate Ecology

Transmission Cycle. West Nile (WN) virus is amplified during periods of adult mosquito blood-feeding by continuous transmission between mosquito vectors and bird reservoir hosts. Infectious mosquitoes carry virus particles in their salivary glands and infect susceptible bird species during blood-meal feeding. Competent bird reservoirs will sustain an infectious viremia for 1 to 4 days subsequent to exposure, after which these hosts develop life-long immunity. A sufficient number of vectors must feed on an infectious host to ensure that some survive the extrinsic incubation period (approximately 2 weeks, depending on temperature) to feed again on a susceptible reservoir host. People, horses, and most other mammals are not known to develop infectious-level viremias very often, and thus are probably "dead-end" or "incidental-hosts."

Birds. Through July 2001, more than 70 species of birds (mostly American crows) have tested positive for WN virus either by virus isolation or nucleic acid testing. Birds infected with WN virus can die or become ill.

There is no evidence that a person can get WN virus from handling live or dead infected birds. However, persons should avoid bare-handed contact when handling dead animals and use gloves or double plastic bags to place the bird carcass in a garbage can or contact their local health department for guidance. The U.S. Geological Survey's National Wildlife Health Center (NWHC) monitors bird mortality nationwide.

Dogs and Cats. West Nile virus does not appear to cause extensive illness in dogs or cats. There is a single published report of WNV isolated from a dog in southern Africa (Botswana) in 1982. West Nile virus was isolated from a dead cats in 1999 and 2000. However, a serosurvey in New York City of dogs and cats in the 1999 epidemic area showed a low infection rate.

There is no documented evidence of person-to-person or animal-to-person transmission of WN virus. Because infectious mosquitoes transmit the WN virus, dogs or cats could be exposed to the virus in the same way humans become infected. Veterinarians should take normal infection control precautions when caring for an animal suspected to have this or any viral infection. It is possible that dogs and cats could become infected by eating dead infected animals such as birds, but this is undocumented.

Nonetheless, there is no reason to destroy an animal just because it has been infected with WN virus. Full recovery from the infection is likely. Treatment would be supportive and consistent with standard veterinary practices for animals infected with a viral agent.

Horses. Cases of WN virus disease in horses have been documented, either by virus isolation or by detection of WN virus-neutralizing antibodies in 1999, 2000, and 2001. Results of epidemiologic investigations indicate that WN virus was responsible for several horse deaths. The horses most likely became infected with WN virus in the same way humans become infected, by the bite of infectious mosquitoes. The virus is located in the mosquito's salivary glands. When the mosquito bites or feeds on the animal, the virus is injected into the blood system of the horse. The virus then multiplies and may cause illness. Although there is no documented evidence of animal-to-person transmission of WN virus, normal veterinary infection control precautions should be followed when caring for a horse suspected to have this or any viral infection.

There is no documented evidence that WN virus is transmitted between horses. However, horses with suspected WN virus should be isolated from mosquito bites, if at all possible. Horses vaccinated against eastern equine encephalitis (EEE), western equine encephalitis (WEE), and Venezuelan equine encephalitis (VEE) are NOT protected against WN virus infection. A West Nile virus vaccine for horses was recently approved, but its effectiveness is unknown. It is not known how long a horse infected with WN virus will be infectious, but previously published data suggest that the virus is detectable in the blood for only a few days. Finally, there is no reason to destroy a horse just because it has been infected with WN virus. Data suggest that most horses recover from the infection. Treatment would be supportive and consistent with standard veterinary practices for animals infected with a viral agent.

Other Vertebrates. Through July 2001, The Centers for Disease Control and Prevention (CDC) has also received reports of WN virus infection in bats, chipmunks, raccoons, skunks, squirrels, and domestic rabbits.

Surveillance and Control

A special West Nile virus surveillance program has been initiated in 48 states, five cities, and the District of Columbia. See the Epidemic/Epizootic West Nile Virus in the United States: Revised Guidelines for Surveillance, Prevention, and Control, 2001: *http://www.cdc.gov/ncidod/dvbid/westnile/resources/wnv-guidelines-apr-2001.pdf* Data is being collected on a weekly basis and will be reported for the following five categories: wild birds, sentinel chicken flocks, human cases, veterinary cases, and mosquito surveillance. See the U.S. Geological Survey to view weekly maps and tables of data collected: *http://cindi.usgs.gov/hazard/event/west_nile/west_nile.html*.

Additional Reading

Anderson, J.F., T.G. Andreadis, and C.R. Vossbrinck, et al.: "Isolation of West Nile Virus from Mosquitoes, Crows, and a Cooper's Hawk in Connecticut," *Science* **286**, 2331–2333 (1999).

Briese, T., C. Huang, and L.J. Grady, et al.: "Identification of a Kunjin/West Nile-like Flavivirus in Brains of Patients with New York Encephalitis," *Lancet* **354**, 1261–1262 (1999).

Briese, T., W.G. Glass, and W.I. Lipkin: "Detection of West Nile Virus Sequences in Cerebrospinal Fluid," *Lancet* **355**, 1614–1615 (2000).

Brogdon, W.G., and J.C. McAllister: "Insecticide Resistance and Vector Control," *Emerg. Infect. Dis.* **4**, 605–613 (1998).

Calisher, C.H., N. Karabatsos, J.M. Dalrymple, et al.: "Antigenic Relationships Between Flaviviruses as Determined by Cross-neutralization Tests with Polyclonal Antisera," *J. Gen. Virol.* **70**, 37–43 (1989).

De Madrid, A.T., J.S. Porterfield: "The Flaviviruses (group B Arboviruses): A Cross-neutralization Study," *J. Gen. Virol.* **23**, 91–96 (1974).

Gubler, D.J., G.L. Campbell, L. Petersen, et al.: "West Nile Virus in the United States: Guidelines for Detection, Prevention and Control," *Viral Immunol.* **13**, 469–475 (2000).

Hayes, C.G.: "West Nile fever," Monath T.P "The Arboviruses: Epidemiology and Ecology," CRC Press, LLC., Boca Raton, FL, 1989, pp. 59–88.

Holden, P., D. Muth, and R.B. Shriner: "Arbovirus Hemagglutinin-inhibition in Avian Sera: Inactivation with Protamine Sulfate," *Am. J. Epidemiol.* **84**, 67–73 (1966).

Hubalek, Z. and J. Halouzka: "West Nile fever—a Reemerging Mosquito-borne Viral Disease in Europe," *Emerg. Infect. Dis.* **5**, 643–650 (1999).

Jia, X.Y., T. Briese, I. Jordan, et al.: "Genetic Analysis of West Nile New York 1999 Encephalitis Virus," *Lancet* **354** (1999).

Johnson, A.J., D.A. Martin, N. Karabatsos, and J.T. Roehrig: "Detection of Anti-arboviral Immunoglobulin G by Using a Monoclonal Antibody-based Capture Enzyme-linked Immunosorbent Assay," *J. Clin. Microbiol.* **38**, 1827–1831 (2000).

Lanciotti, R.S., J.T. Roehrig, V. Deubel, et al.: "Origin of the West Nile Virus Responsible for an Outbreak of Encephalitis in the Northeastern United States," *Science* **286**, 2333–2337 (1999).

Lanciotti, R.S., A.J. Kerst, R.S. Nasci, et al.: "Rapid Detection of West Nile Virus from Human Clinical Specimens, Field-collected Mosquitoes, and Avian Samples by a TaqMan Reverse Transcriptase-PCR Assay," *J. Clin. Microbiol.* **38**, 4066–4071 (2000).

Martin, D.A., T. Brown, A.J. Johnson, et al.: "Standardization of Immunoglobulin M Capture Enzyme-linked Immunosorbent Assays for Routine Diagnosis of Arboviral Infections," *J. Clin. Microbiol.* **38**, 1823–1826 (2000).

Monath, T.P., R.R. Nystrom, R.E. Bailey, et al.: "Immunoglobulin M Antibody Capture Enzyme-linked Immunosorbent Assay for Diagnosis of St. Louis Encephalitis," *J. Clin. Microbiol.* **20**, 784–790 (1984).

Platanov, A.E., G.A. Shipulin, O.Y. Shipulina, et al.: "Outbreak of West Nile Virus Infection, Volgograd Region, Russia, 1999," *Emerg. Infect. Dis.* **7**, 128–132 (2001).

Rose, R.I.: "Pesticides and Public Health: Integrated Methods of Mosquito Management", *Emerg. Infect. Dis.* **7**, 17–23 (2001).

Shieh, W.J., J. Guarner, M. Layton, et al.: "The Role of Pathology in an Investigation of an Outbreak of West Nile Encephalitis in New York, 1999," *Emerg. Infect. Dis.* **6**, 370–372 (2000).

Staff: American Mosquito Control Association, *American Mosquito Control Association partnership strategy document: Environmental Protection Agency pesticide environmental stewardship program.* AMCA, P.O. Box 586, Milltown, NJ, 1997. (*www.mosquito.org/PESPAMCA.htm*)

Staff: New Jersey Mosquito Control Association, *New Jersey Mosquito Control Association partnership strategy document: Environmental Protection Agency Pesticide Environmental Stewardship Program,* Cook College, New Brunswick, NJ, 1997. (*www-rci. rutgers.edu/~insects/psd.html*)

Staff: Florida Coordinating Committee Mosquito Control, *Florida mosquito control: The state mission as defined by mosquito controllers, regulators, and environmental managers,* University of Florida, Gainesville, FL, 1998. (*www.ifas.ufl.edu/~vero-web/whitep/whitep.htm*)

Staff: CDC, "Outbreak of West Nile-like Viral Encephalitis—New York, 1999," *MMWR Morb. Mortal. Wkly. Rep.* **48**, 845–849 (1999).

Staff: CDC, "Update: West Nile-like Viral Encephalitis—New York, 1999," *MMWR Morb. Mortal. Wkly. Rep.* **48**, 890–892 (1999).

Staff: CDC, Update: "Surveillance for West Nile Virus in Overwintering Mosquitoes-New York, 2000," *MMWR Morb. Mortal. Wkly. Rep.* **49**, 178–179 (2000).

Staff: CDC, Notice to Readers: Update: "West Nile Virus Isolated from Mosquitoes-New York, 2000," *MMWR Morb. Mortal. Wkly. Rep.* **49**, 211 (2000).

Staff: CDC, "West Nile Virus Activity—New York and New Jersey, 2000," *MMWR Morb. Mortal. Wkly. Rep.* **49**, 640–642 (2000).

Staff: CDC, Update: "West Nile Virus Activity—Northeastern United States, January–August 7, 2000," *MMWR Morb. Mortal. Wkly. Rep.* **49**, 714–718 (2000).

Staff: CDC, Update: "West Nile Virus Activity—Eastern United States, 2000," *MMWR Morb. Mortal. Wkly. Rep.* **49**, 1044–1047 (2000).

Staff: CDC, "Serosurveys for West Nile Virus Infection—New York and Connecticut Counties, 2000," *MMWR Morb. Mortal. Wkly. Rep.* **50**, 37–39 (2001).

Staff: Department of Health and Human Services, Public Health Service, Centers for Disease Control and Prevention. (*www.cdc.gov/ncidod/dvbid/arbor/arboguid.htm*).

Steele, K.E., M.J. Linn, R.J. Schoepp, et al.: "Pathology of Fatal West Nile Virus Infections in Native and Exotic Birds During the 1999 Outbreak in New York City, New York," *Vet Pathol.* **37**, 208–224 (2000).

Tsai, T.F., F. Popovici, C. Cernescu, G.L. Campbell, and N.I. Nedelcu: "West Nile Encephalitis Epidemic in Southeastern Romania," *Lancet* **352**, 767–771 (1998).

Work, T.H., H.S. Hurlbut, and R.M. Taylor: "Indigenous Wild Birds of the Nile Delta as Potential West Nile Virus Circulating Reservoirs," *Am. J. Trop. Med. Hyg.* **4**, 872–888 (1956).

Web References

APHIS West Nile Virus site: *http://www.aphis.usda.gov/oa/wnv/*

Centers for Disease Control and Prevention: *http://www.cdc.gov/ncidod/dvbid/*

Links to State and Local Government Sites: *http://www.cdc.gov/ncidod/dvbid/west-nile/city_states.htm*

National Institute of Allergy and Infectious Diseases (NIAID): *http://www.niaid.nih.gov/default.htm*

National Institutes of Health (NIH): *http://www.nih.gov/*

National Wildlife Health Center (NWHC): *http://www.nwhc.usgs.gov/*

Pesticides and Mosquito Control: *http://www.epa.gov/pesticides/factsheets/skeeters.htm*

Summary of West Nile Virus in the United States, 1999: *http://www.aphis.usda.gov/vs/ep/WNV/summary.html*

USGS West Nile Virus site: *http://www.usgs.gov/west_nile_virus.html*

West Nile Virus Maps, 2000: *http://nationalatlas.gov/virusmap.html*

WEST-WIND DRIFT. In the north and south Atlantic and Pacific oceans, the warm, poleward bound surface current produced by the prevailing westerlies. The best known of the west-wind drifts is the North Atlantic Drift, an eastward movement of warm water arising from the Gulf Stream and responsible for the warmth associated with Western Europe.

WETLANDS. The term *wetlands* is self-defining—namely, wetlands are areas of land that, unless disturbed, are *wet* (bordering on water saturation) all or part of a given time span. In general usage, wetlands and riparian areas are intermediate between land and water. As reported by the United States Department of Agriculture (USDA), wetland scientists have developed more than 50 different definitions of wetlands. In fact, defining wetlands has been controversial in the context of U.S. policies because of the implications for landowners who want to use and develop these areas and environmentalists who want to preserve them.

Since 1977, the U.S. federal government has used a three-part definition involving hydric soils, hydrophytic vegetation, and hydrology. According to the U.S. Army Corps of Engineers, wetlands are "Areas that are inundated or saturated by surface or groundwater at a frequency and duration sufficient to support, and that under normal circumstances do support, a prevalence of vegetation typically adapted for life in saturated soil conditions."

The phrase "under normal circumstances" has been interpreted to mean that an area with wetland hydrology and soils remains a wetland, even when adapted vegetation has been removed to make areas suitable for farming. While this definition has been accepted generally, criteria for delineating wetlands in the field on the basis of evidence of the characteristics in the three-part definition have not been agreed to so readily.

Wetland science incorporates numerous hydrologic and geologic variations, and the science today lacks a crystal-clear nomenclature. Thus, as of early 1994, the evaluation of wetlands is not an exact science. Serious readers interested in this topic may find the glossary (Table 2) at the end of this article helpful.

Examples of wetlands range from the obvious, such as cattail marshes, peat bogs, and tidal sloughs, to prairie potholes and alluvial flood plains that are wet during only part of the year and may rarely have standing water. As the 1993 flooding of the Mississippi and Missouri basins has shown, when some of these types of wetlands are wet, they can be *very wet*.

The ecological and hydrologic functions and values of wetlands depend on a wide array of factors, including morphology, vegetation, and the nature of adjacent areas. Wetness alone is an inadequate index of the importance of wetlands, but is getting increasing scrutiny as the wetland debate becomes more contentious.

Natural Values of Wetlands

Most, but not all, wetlands perform four very useful natural functions:

1. They deter small and major floods that result from abnormal precipitation in regions that feed water to great rivers. Wetlands either at or near the source of excessive precipitation act as sponges for storage of water and thus reduce the runoff that causes streams and rivers to swell way beyond their normal channels (banks), causing floods.

2. They provide sanctuaries for certain species of animals who survive best in wetland areas.
3. Particularly along the seacoasts, wetlands provide spawning and nursery areas for certain important species of fish and crustaceans, and, when wetlands are transformed to "dry" lands, this impacts adversely on fisheries.
4. Certain types of wetlands, such as cypress swamps, serve as water purifiers.

Land Values of Wetlands

Simply because they are land (and not considering their natural beneficial roles in the environment), wetlands have much economic value, particularly when they can be drained and filled in for agricultural production and, when located near large metropolitan areas, they can provide the necessary space for expansion. These are the two principal factors that have caused drastic reduction of wetland areas in the advanced industrial nations of North America and Europe since the early 1800s. Converted wetlands comprise some of the largest and most fertile agricultural lands and, commercially, have permitted urban communities to expand. For example, the airports of New York, Boston, and New Orleans have been constructed on transformed wetlands. In terms of so-called "urban sprawl," the natural advantages of wetlands have been negated by another natural phenomenon, namely the population "explosion."

Thus, the type of economic-environment conflict that has been present for many years in terms of adequate energy supplies also is present in the case of the wetlands in terms of land needs versus the environment. These kinds of dilemmas cannot be addressed by science and technology alone, but, in fact, seek sociological and societal solutions, tempered with whatever technical counsel that technology may provide.

Although a *few* naturalists over the past century have been aware of the fundamental natural values of wetlands, serious concerns over the large extent of wetlands conservation is relatively recent, extending back to the early–World War II era. Generally, society and some governments, including the United States, have shown an increasing awareness of the desirability at this juncture of providing special attention and caution, not only in terms of any future conversion of wetlands to "dry" land, but also the restoration of former wetlands.

1993 — A Reminder of Wetlands' Role in Flooding

The near-record long, late rains that swelled rivers in the upper midwestern United States and that tore into levees along more than 1000 miles (1609 km) of the Mississippi, Missouri, and Illinois rivers served as a vivid reminder of the flood-control function of the wetlands along these rivers. In the upper Midwest, cropland covers over half of the flood-prone land — that is, land that has more than a 1% chance of flooding in any year. In 1993, the rising rivers flooded cropland that once had been wetland. Although much of this land will return to crop production, land in areas where levees have been breached and ditches have been silted will require many months to dry out.

According to the U.S. Fish and Wildlife Service, the nine-state Midwest region had more than 57 million acres (23 mil hectares) of wetlands in 1780, but only 23 million acres (9.3 mil hectares) remained 200 years later. In Minnesota, North and South Dakota, Wisconsin, and Nebraska, about 40 to 60% of the wetlands have been developed, drained for cropland, or

otherwise converted. About 90% of the wetlands in Illinois, Iowa, Missouri, and Kansas have been converted. Shifts into agricultural uses accounted for the majority of conversions.

The USDA estimates that more than 29 million acres (11.7 mil hectares) of the 120 million acres (48.6 mil hectares) of cropland in the upper Midwest were originally wetlands that have been cleared and drained for crop production. Some hydrologists and ecologists believe that draining wetlands along rivers, combined with channeling and levee construction, contributes to higher flood levels and more powerful and destructive floods.

As early as 1987, a National Wetlands Policy Forum was convened. At that time, the forum suggested the following general policies:

● Establish a "no net loss" of wetlands.
● Provide incentives for private stewardship of wetlands.
● Initiate a wetland-restoration program.
● Commence regulatory reform, including improved wetlands delineation criteria and a consistent definition.

By 1990, considerable opposition to the aggressive field implementation of federal wetlands policies developed among farmers, developers, and small landholders. A review of federal interagency wetland guidelines was caught in the growing controversy and led to a major revision in 1991. These efforts, however, did not moderate a rapidly coalescing opposition to the regulation of wetlands. Property right interests targeted wetland regulation as an opening wedge in rolling back a wider array of regulation aimed at promoting the general welfare. Wetland regulation without compensation to landowners was portrayed as a "taking," proscribed under the Fifth Amendment to the U.S. Constitution.

Status

An inventory (1982) showed almost 48 million acres (19.4 mil hectares) of land with hydric soils in the nine Midwestern states that were affected by the 1993 flooding.

More than 1.4 million acres (0.6 mil hectares) of cropland on hydric soils in these states is still wet enough to be considered wetland, and about 200,000 acres (about 81,000 hectares) of this land shows some evidence of drainage. Most cropland on hydric soils was converted from wetlands through clearing and drainage. See also Table 1.

Environmental Issues

Although farmers have been concerned about losing some of their cultivatable land through a program to restore and revitalize the wetlands in some areas, and, although local land developers and city and town planners in various areas have resisted strong regulations over draining and filling in nearby wetland areas, the most publicized concern over the recent years has related to the effect of vanishing wetland on birds and other wildlife. This is a highly controversial topic and is reminiscent of nearly all efforts to match economic values with environmental values. Just a few locations where these considerations are quite serious are described here.

Major Wetlands in North America

Well covered in the literature is the plight of the wetlands in the Florida Everglades; the "Blackwater Country" of southern Georgia and northern Florida (including the Okefenokee Swamp and the Suwannee and St. Marys

TABLE 1. CROPLAND ON WETLAND SOILS IN UPPER MIDWESTERN UNITED STATES

| | Former Wetlands | Current Wetlands | | Total Current & Former Wetlands (Hydric Soils) | All Rural Land with Hydric Soils |
		Not Drained	Drained		
		Acres			
Illinois	7,239,600	176,500	78,000	7,494,100	8,723,300
Iowa	6,363,700	2,900	3,400	6,370,000	7,228,300
Minnesota	8,539,000	49,300	10,200	8,599,400	16,292,000
Missouri	3,669,000	2,400	0	3,671,400	4,982,400
Wisconsin	630,700	73,900	85,500	790,100	3,612,400
Kansas	523,500	7,600	0	531,100	736,500
Nebraska	607,300	35,100	1,300	643,700	1,486,700
North Dakota	715,100	543,900	10,900	1,269,900	2,547,200
South Dakota	469,800	303,900	18,200	791,900	2,215,300
Total	28,758,600	1,195,500	207,500	30,161,600	47,824,100

Note: Most recent comprehensive Survey (USDA, 1982).

ALLUVIUM

A general term for clay, silt, sand, gravel, or similar unconsolidated detrital material deposited during comparatively recent geologic time by a stream or other body of running water as a sorted or semisorted sediment in the bed of the stream, on its flood plain or delta, or as a cone or fan at the base of a mountain slope, especially a deposit of fine-grained structure (silt or silty clay) deposited during time of flooding.

ARROYO

A term applied in the arid and semiarid regions of the southwestern United States to the small, deep, flat-floored channel or gully of an ephemeral stream or of an intermittent stream, usually with vertical or steeply cut banks of unconsolidated material. It may be transformed to a temporary watercourse or short-lived torrent after heavy rainfall. Also called dry wash or wadi.

BACKWATER

Water that is retarded, backed up, or turned back in its course by an obstruction (such as a dam), an opposing current, or the movement of the tide (such as the water in a reservoir or the water obtained at high tide to be discharged at low tide).

Or a body of currentless or relatively stagnant water, parallel to a river and usually fed from it through a single channel at the lower end by the back flow of the river.

Or, generally, any tranquil body of water joined to a main stream, but little affected by its current, such as the water collected in side channels or flood-plain depressions after it overflowed the lowland.

Or a creek, arm of the sea, or series of connected lagoons, usually parallel to the coast and separated from the sea by a narrow strip of land, but communicating with it through barrel outlets.

BASIN AREA

For a given stream order, the total area, projected upon a horizontal plane, of a drainage basin bounded by the basin perimeter and contributing overland flow to the stream segment of order, including all tributaries of lower order. Sometimes called watershed area.

BAYOU

A term variously applied to many local water features in the Mississippi River basin and in the Gulf Coast region of the United States, especially in Louisiana. Its general meaning is a creek, large stream, minor river, or secondary watercourse that is tributary to another river or connecting with another body of water.

Or a sluggish and stagnant stream, characterized by a slow or imperceptible current, that follows a winding course through flat alluvial lowlands, coastal swamps, or marshes, or river deltas.

Or an effluent branch, especially sluggish or stagnant, of a main river or of a lake, as a distributary flowing through a delta or enclosing a low island. Also, the distributary channel that carries floodwater or affords a passage for tidal water through swamps or marshlands.

Or an intermittent, partly closed or disused watercourse, especially a lake or a sluggish stream formed on a river delta or in an abandoned river channel.

BAYOU LAKE

A lake or pool in an abandoned and partly closed channel or stream, as on the Mississippi River delta.

BOG

A waterlogged, spongy groundmass, acidic in nature, that supports primarily mosses. Decaying vegetable matter ultimately may develop into peat (hence a peat bog). Vegetation primarily is sphagnums, hedges, and heaths.

Bogs occur in damp, cool places. Cranberries grow well in bogs. If left undisturbed, bogs withhold the release of carbon dioxide to the atmosphere.

Sometimes associated with bogs are small shrubs, cotton grass, and stunted spruce.

BOREAL

Pertaining to the north, or located in northern regions. Or pertaining to the northern biotic zone, characterized by tundra and taiga and by dominant coniferous forests.

Boreal climate refers to a climatic zone that has a definite winter, experiences snow and a short summer that generally is hot, and is characterized by a large annual range of temperature. It includes large parts of North America and central Europe and Asia. This climate usually occurs between latitudes 60°N and 40°N.

BOTTOMLAND

Low-lying, level, usually highly fertile land, especially in the Mississippi River region of the United States and farther west, where the term signifies a grassy lowland formed by the deposition of alluvium along the margin of a watercourse.

Or an alluvial plain or a flood plain.

CANAL (Coast)

A long, narrow channel or arm of the sea connecting two larger stretches of water, usually extending far inland (sometimes between islands or between an island and the mainland) and approximately uniform in width. The Lynn canal in Alaska is an example.

Or a sluggish coastal stream, as may be found along the Atlantic coast of the United States.

CANAL (Stream)

An artificial watercourse of relatively uniform dimensions, cut through an inland area, and designed for navigation, drainage, or irrigation by connecting two or more bodies of water. A canal is larger than a ditch.

CREEK (Stream)

A term generally applied over most of the United States (except New England) and Canada and Australia to any natural stream of water, normally larger than a brook, but smaller than a river. Or a branch or tributary of a main river, a lowland water course of medium size, or a flowing rivulet. Also, a wide or short arm of a river, such as one filling a short ravine that joins the river.

CYPRESS DOME

A group of cypress trees growing in the middle of a small swamp. See Fig. 1. The trees purify the water by removing most of the nitrates and phosphates, binding them to mud and tree roots. Overloading the water with contaminants can cause the system to dysfunction.

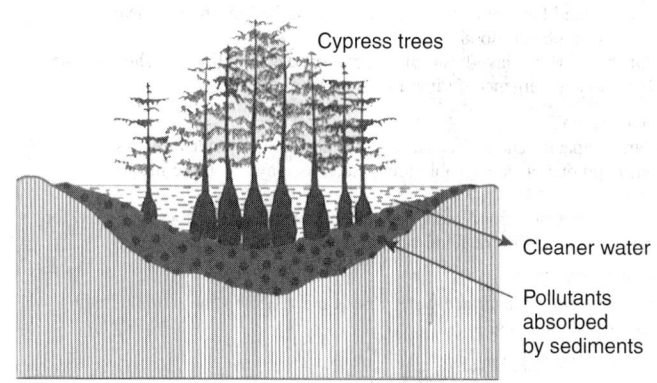

Fig. 1. Cypress swamp. Incoming polluted water is "treated" by the root system of the trees and discharged as less polluting. The pollutants are absorbed by sediments underlying the trees.

DELTA

The low, nearly flat, alluvial tract of land deposited at or near the mouth of a river, commonly forming a triangular or fan-shaped plain of considerable area enclosed and crossed by many distributaries of the main river. A delta sometimes extends beyond the general trend of the coast and results from the accumulation in a wider body of water (usually a sea or lake) of sediment supplied by a river in such quantities that it is not removed by tides, waves, and currents. Most deltas are partly subaerial and partly below water.

DELTAIC COASTAL PLAIN

A coastal plain composed of a series of more or less coalescing deltas. It consists initially of natural levee ridges separated by basins.

DELTA LEVEE LAKE

A lake on an advancing delta, formed between sandbars or natural levees deposited at the mouths of distributaries. An example is Lake Pontchartrain on the Mississippi River delta.

DELTA PLAIN

The level or nearly level surface composing the landward part of a large delta. Or strictly an alluvial plain characterized by repeated channel bifurcation and divergence, multiple distributary channels, and interdistributary flood basins.

DETRITUS

A collective term for loose rock and mineral matter that is worn off or removed directly by mechanical means, as by disintegration or abrasion, especially fragmental material (such as sand, silt, and clay) derived from older rocks and moved from its place of origin.

DISTRIBUTARY STREAM

An irregular, divergent stream flowing away from the main stream and not returning to it, as in a delta or on an alluvial plain. It may be produced by stream deposition choking the original channel.

Or one of the channels of a braided stream.

(continued)

TABLE 2. (*Continued*)

DRAINAGE BASIN
A region or area bounded peripherally by a drainage divide and occupied by a drainage system. In broader terms, the whole area or entire tract of country that gathers water originating as precipitation and contributes it ultimately to a particular stream channel or system of channels, lake reservoir, or other body of water.

ESTUARY
The seaward end of the widened funnel-shaped tidal mouth of a river valley, where freshwater mixes with and measurably dilutes seawater and where tidal effects are evident. A tidal river is a partially enclosed coastal body of water. Or a drowned river mouth formed by the subsidence of land near the coast or by the drowning of the lower portion of a nonglaciated valley due to a rise in sea level.

FLOOD PLAIN
The surface or strip of relatively smooth land adjacent to a river channel, constructed (or in the process of being constructed) by the present river in its existing regimen and covered with water when the river overflows its banks at times of high water. A river has one flood plain and may have one or more terraces representing abandoned flood plains.

Or any flat or nearly flat, usually dry lowland that borders a stream and that may be covered by its waters at flood stages.

Or the land beyond a stream channel, described by the perimeter of the maximum probable flood.

Or the part of a lake-basin plain between the shoreline and the shore cliff, subject to submergence during a high stage of the lake.

FLOODWAY
A large-capacity channel constructed to divert floodwater or excess streamflow from populous or damageable areas, such as a bypass route marked out by levees.

Or the part of a flood plain kept clear of encumbrances and reserved for emergency diversion of floodwaters.

GROUNDWATER
That part of the subsurface water that is the zone of saturation, including underground streams.

GROUNDWATER BASIN
A subsurface structure having the character of a basin with respect to the collection, retention, and outflow of water.

Or an aquifer or system of aquifers, whether or not basin shaped, that has reasonably well defined boundaries and more or less definite areas of recharge and discharge.

GROUNDWATER BARRIER
A natural or artificial obstacle, such as a dike or fault gouge, to the lateral movement of groundwater, not in the sense of a confining bed. It is characterized by a marked difference in the level of the groundwater on opposite sides.

HUMMOCK (Hammock)
A fertile area of deep, humus-rich soil, generally covered by hardwood vegetation and often rising slightly above a plain or swamp. An island of dense, tropical undergrowth, as found in the Florida Everglades. A rounded or conical knoll, mound, hillock, or other small elevation, generally of equidimensional shape and not ridgelike. A slight rise of ground above a level surface.

HYDRIC SOIL
A soil that in its undrained condition is saturated, flooded, or ponded long enough during the growing season to develop anaerobic conditions.

HYDROLOGY
The science that deals with continental water (both liquid and solid), its properties, circulation, and distribution, on and under the Earth's surface and in the atmosphere from the instant of its precipitation until it is returned to the atmosphere through evapotranspiration or is discharged into an ocean. See article on **Hydrology**.

KEY (Coastal)
A *cay*, particularly one of the coral islets or barrier islands off the southern coast of Florida.

LAGOON
A coastal lagoon is a shallow stretch of seawater, such as a sound, channel, bay, or salt-water lake, near or communicating with the sea and partly or completely separated from it by a low, narrow, elongate strip of land, such as a reef, barrier island, sandbank, or spit. Especially, it is the sheet of water between an offshore coral reef and the mainland. It often extends roughly parallel to the coast, and it may be stagnant.

Or a freshwater pond or lake near or communicating with a larger lake or a river. A stretch of freshwater cut off from a lake by a barrier, as in a depression behind a shore dune.

Or a body of water enclosed or nearly enclosed within an atoll.

The term also has been applied to other coastal features, such as an estuary, a slough, a bayou, a marsh, and a shallow pond or lake into which the sea flows. Lagoon also is used to describe any shallow, artificial pond or other water-filled excavation for the natural oxidation of sewage.

LANDLOCKED
Said of a body of water that is enclosed or nearly enclosed by land, such as a landlocked bay separated from the main body of water by a bar or a landlocked lake having no surface outlet.

MARSH
A water-saturated, poorly drained area, intermittently or permanently covered with water and having aquatic and grasslike vegetation.

A *coastal marsh* is a marsh bordering a seacoast, generally formed under the protection of a barrier beach or enclosed in the sheltered part of an estuary. Coastal marshes are subject to the daily cycle of ocean tides, which transport nutrients to a marsh and thus attract various forms of animal life. Coastal marshes are the habitat of fin and shell fishes that are harvested in the United States. The marsh often functions as a spawning locale and as a nursery for young fishes. Algae commonly is found in coastal salt marshes and serves as a nutrient for living forms.

MORPHOLOGY (River)
The study of the channel pattern and the channel geometry at several points along a river channel, including the network of tributaries within the drainage basin.

MUSKEG
A bog. Usually a sphagnum bog, frequently with tussocks of deep accumulations of organic material, growing in wet, poorly drained, boreal regions, often areas of permanent frost. Tamarack and black spruce commonly are associated with muskeg areas.

Or a term sometimes used in Michigan for *a bog lake*.

POND
A natural body of standing freshwater occupying a small surface depression, usually smaller than a lake and larger than a pool.

POOL
A small, natural body of standing water (usually freshwater), such as a stagnant body of water in a marsh, a transient puddle in a depression following a rain, or a still body of water in a cave.

Or a small, quiet, and rather deep reach of a stream, as between two rapids or where there is very little current.

Or a small or large body of impounded water, artificially confined above a dam or the closed gates of a lock.

PRAIRIE
An extensive tract of level to rolling, generally treeless grassland in the temperate latitudes of the interior of North America, especially in the Mississippi Valley region, characterized generally by a deep, fertile soil (suitable for wheat growing) and by a covering of tall, coarse grass and herbaceous plants.

Or a broad, low, wet, sandy, flat-bottomed, often water-covered, grass-grown tract or sink in the pinewoods of Florida.

PRAIRIE OR STREAM POTHOLE
A smooth, roughly circular, bowl-shaped, or cylindrical hollow that generally is deeper than it is wide, formed in the rocky bed of a stream. The feature is caused by the grinding action of stones or coarse sediment (sand, gravel, pebbles, boulders) that are whirled around and kept in motion by eddies or the force of a stream current in a particular location. Also, they may have been formed as the result of glacial action. Potholes frequently are found in seasonally flooded wetlands, notably in the northern plains of the United States. These features are removed, of course, when a wetland area is drained and filled.

Prairie potholes play an important role in the ecology of ducks. The holes serve as breeding oases for ducks. Nutrients (sometimes called "duck potatoes" and comprised of invertebrates and aquatic vegetation) fortify the ducks during periods of egg production. The holes are fed by melting snow and rain. Removal of the potholes causes the ducks to move elsewhere. See Fig. 2.

RIPARIAN AREA
The bank of a natural watercourse (as a river), but sometimes of a lake or tidewater. Legally, riparian rights are those of access to the water, including beaches.

RIVERINE BOTTOMLAND
Naturally occurring reservoirs, serving to trap and hold water during river flooding conditions. Entangled woody vegetation serves to trap sediment and thus functions to slow the flow of a river. Sometimes these areas are referred to as floodplain forests. Such bottomland in the northern flood plains of the United States often will feature cottonwood and silver maple trees, whereas tupelo and bald cypress are found along the slower-moving rivers of the south.

TABLE 2. (*Continued*)

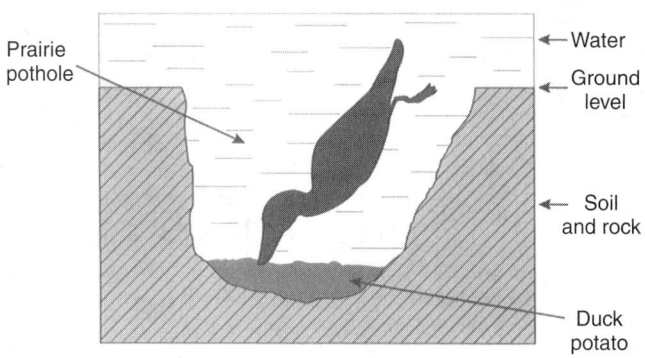

Fig. 2.   In some wetlands, pot holes form that hold water for fairly long time spans. At the base of the "prairie pothole," nutrients accumulate. These are sought out by canvasback ducks and are referred to as "duck potatoes."

SAVANNAH (Savanna)
An open, grassy, essentially treeless plain, especially as developed in tropical and subtropical regions. Usually there is a distinct wet and dry season. The trees that are found are drought-resistant.
   Or the tall grass characteristic of a savannah.
   Along the southeastern Atlantic Coast of the United States, the term is used for marshy alluvial flats that have occasional clumps of trees.

SLOUGH
A small marsh, especially a marshy tract lying in a swale or other local, shallow, and undrained depression on a piece of dry land, as on prairies of the midwestern United States. Sometimes, a dry depression that becomes marshy or filled with water.
   Or a large wetland, as a swamp, that occurs in the Everglades of Florida.
   Or a term used, especially in the Mississippi Valley, for a creek or sluggish body of water in a tidal flat, bottomland, or coastal marshland.
   Or a sluggish channel of water in which water flows slowly through low, swampy ground, as occurs along the Columbia River, or a section of an abandoned river channel that contains stagnant water and that occurs in a flood plain or delta.
   Also, an indefinite term indicating a small lake, a marshy or reedy pool or inlet, a bayou, a pond, or a small and narrow backwater. Or a place of deep mud, as a mudhole.

SPHAGNUM MOSS
A moss of the genus *Sphagnum*, often forming peat. Also referred to as *peat moss*.

STEPPE
An extensive, treeless grassland area in southeastern Europe and Asia developing in the semiarid mid-latitudes of that region. They are generally considered drier than the prairie that develops in the subhumid mid-latitudes of the United States.

SWALE
A slight depression, sometimes swampy, in the midst of generally level land.
   Or a shallow depression in an undulating ground moraine due to uneven glacial deposition.
   Or a long, narrow, generally shallow troughlike depression between two beach ridges that is aligned roughly parallel to the coastline.

SWAMP
A water-saturated area, intermittently or permanently covered with water, and having shrub- and tree-type vegetation essentially without peatlike accumulation. Where natural drainage is impeded, as by impervious soil or bedrock, or abnormal amounts of plant materials, a swamp may form. This usually occurs on level ground, but swamps also can exist on hillsides, mainly as the result of percolation of groundwater. Lake basins filled with vegetation and sedimentation ultimately may become swamps. Swamps may be formed on the flood plains of rivers as well as deltas. These are characteristic of the flat, poorly drained areas of the Atlantic Coastal Plain, such as the Great Dismal Swamp, which covers some 2000 square miles (5180 sq km) in Virginia and North Carolina and the Everglades of Florida. In northern North America, the accumulation of vegetation and sedimentation sometimes is referred to as muskeg. Some swamps ultimately become peat bogs. See also article on **Swamp**.

TAIGA
A swampy area of coniferous forest sometimes found lying between tundra and steppe regions.

TIDAL SWAMP
A swamp partly covered during the high tide by the backing up of a freshwater river.

TRIBUTARY
A stream feeding, joining, or flowing into a larger stream (at any point along its course) or into a lake. See also **Distributary System**.

TUNDRA
A treeless, level or gently undulating plain characteristic of arctic and subarctic regions. It usually has a marshy surface that supports a growth of mosses, lichens, and numerous low shrubs and is underlain by a dark, mucky soil and permafrost.

TUSSOCK
A dense tuft of grass or grasslike plants that usually form one of the many firm hummocks in the midst of a marshy or boggy area.

WETLAND PLANTS
Certain plants, such as cattails, have adapted to survival in the flooded conditions of a freshwater marsh. Cattails have a dense network of air spaces through which oxygen penetrates to the roots. Also, air holes are found in the flat leaves of water lilies. Because of its porous roots, the milfoil can grow even when it is totally submerged.

WADI
See article on **Wadi**.

rivers); the Mississippi River Basin, including the Missouri and Illinois rivers; the tidal of Minnesota, Manitoba, Ontario, and Quebec; the tidal salt marshes of the North American eastern seaboard; the Central Valley of California; the lower Savannah River in Georgia; the Sweetwater Marsh and the Tijuana River estuary near San Diego, California; the former Llano Seco Ranch along the Sacramento river in California; the San Jacinto River estuary near Houston, Texas; the mud flats in Gray's Harbor in northwestern Washington; the lower Klamath Basin in California; and so on.
   A few specific cases are described in the following paragraphs.

Tate's Hell. Located in the panhandle of Florida just east of the Apalachicola River and consisting of 128,000 acres (51,800 hectares) of *titi swamp*. Titi shrubs, some 10 feet (3 meters) high are intermixed with southern pines. Ecologists consider this swamp crucial to the productivity of the Apalachicola estuary, which is an important seafood harvesting area.

Pine Flatwoods. These cover millions of acres of the southeastern coastal plain, extending from the Carolinas to Texas. This is the habitat of numerous species of plants and animals, some of which are rare or endangered. From the standpoint of *biodiversity*, some ecologists consider

these areas as being among the most outstanding on Earth. The flatwoods are endangered for several reasons:

1. Pollution from urban areas;
2. Conversion to agriculture; and
3. Drying out because of anthropogenic disturbance of their upslope watersheds.

White Cedar Wetland Forests. These occur along the eastern coast of North Carolina (Dismal Swamp). Until the last 50 years, these wetlands had existed for many centuries and had provided a sanctuary for bear, deer, beaver, and other species. The cedar trees have provided a dense, cathedral-like canopy that slows the drying of the wetland carpet below. The areas sometimes are referred to as *cedar bogs*. As of the early 1990s, an estimate shows that about 90% of the cedars have been harvested for timber. Although most numerous in North Carolina, such wetlands once prevailed along the eastern and southern shore of the continent from Maine to Louisiana. Early settlers commonly harvested cranberries in the Cedar wetlands.
   For many years, no effort was made to alter the cedar harvesting practice in order to promote regrowth. Further, white cedars are extremely sensitive to urban pollutants that migrate to the swampy areas. Also, because white

cedars require up to 70 years to harvest economically, other species, such as the loblolly pine, which require only 40 years to harvest, have been planted in place of cedars. There are major differences in the ecosystems of the two species of trees.

Additional Reading

Batzer, D.P., R.B. Rader, and S.A. Wissinger: "Invertebrates in Freshwater Wetlands of North America: Ecology and Management," John Wiley & Sons, Inc., New York, NY, 1999.

Boyles-Sprenkel, C.: "Watershed Wars," *American Forests*, 17 (July/August 1992).

Brown, A.C. and A. McLachlan: "Ecology of Sandy Shores," Elsevier, Amsterdam, 1990.

Coniff, R. and M. Farlow: "Blackwater Country," *Nat'l. Geographic*, 34 (April 1992).

Giblett, R.J.: "Postmodern Wetlands: Culture, History, Ecology," Columbia University Press, New York, NY, 1997.

Goff, J.C. and B.P.J. Williams: "Fluid Flow in Sedimentary Basins and Aquifers," Blackwell Scientific, Palo Alto, CA, 1987.

Hamilton, D.P.: "Death of the Nile Delta?" *Science*, 1084 (November 23, 1990).

Hancock, J.: "The Birds of the Wetlands," Academic Press, Inc., San Diego, CA, 1999.

Johnson, R.: "New Life for the 'River of Grass'," *American Forests*, 38 (July/August 1992).

Kadlec, R.H. and R.L. Knight: "Treatment Wetlands: Theory and Implementation," Lewis Publishers, Boca Raton, FL, 1995.

Maitland, P. and N.C. Morgan: "Conservation Management of Freshwater Habitats: Lakes, Rivers and Wetlands," Chapman & Hall, New York, NY, 1997.

Mitchell, J.G., R. Gehman, and J. Richardson: "Our Disappearing Wetlands," *Nat'l. Geographic*, 3 (October 1992).

Mitsch, W.J. and J.G. Gosselink: "Wetlands," John Wiley & Sons, Inc., New York, NY, 2000.

Moore, P.D.: "Wetlands," Facts on File, Inc., New York, NY, 2000.

Parfit, M.: "Water," *Nat'l. Geographic*, Special Issue (November 1993).

Payne, N.F.: "Wildlife Habitat Management of Wetlands," Krieger Publishing Company, Melbourne, FL, 1998.

Rader, R.B. and D.P. Batzer: "Biomonitoring and Management of North American Freshwater Wetlands," John Wiley & Sons, Inc., New York, NY, 2000.

Richardson, J.L. and M.J. Vepraskas: "Wetlands Soils/ Hydric Soils," Lewis Publishers, Boca Raton, FL, 2000.

Seymour, R.J.: "Nearshore Sediment Transport," Perseus Publishing, Boulder, CO, 1989.

Smith, P.L. et al.: "Estuary Rehabilitation: The Green Bay Story," *Oceanus*, 12 (Fall 1988).

Staff: "Strategies for Wetlands Protection and Restoration," *Agricultural Outlook*, 32 (September 1993).

Trettin, C.C., M.F. Jurgensen, D.F. Grigal, and M.R. Gale: "Northern Forested Wetlands: Ecology and Management," Lewis Publishers, Boca Raton, FL, 1996.

Zedler, J.B.: "Handbook for Restoring Tidal Wetlands," CRC Press, LLC., Boca Raton, FL, 2000.

Web References

U.S. Fish & Wildlife Service: *http://www.fws.gov/*
USGS National Wetlands Research Center: *http://www.nwrc.nbs.gov/*
Wetlands Research & Technology Center: *http://www.wes.army.mil/el/wrtc/wrtc.html*

WETTING AGENT.   See Detergents.

WHALES, DOLPHINS, AND PORPOISES *(Mammalia, Cetacea).* The general organization of this interesting order of mammals is given in the table at end of this article. In general terms, whales are marine animals of completely aquatic habits. See Fig. 1.

Wide Range of Sea Mammal Physiology

The whales vary to a considerable degree in structure as represented by the various species. Whales 20 feet (6 meters) in length are among smaller species. Individuals of from 60 to 80 feet (18 to 24 meters) and even 90 feet (27 meters) have been reported among several of the larger species. The baleen or whalebone whales live on small marine animals, which are separated from the water by the sievelike whalebone fringes of the jaws. The toothed whales include some actively predacious species. Species of both orders were once widely sought for their oil and whalebone, but the latter has long been supplanted by manufactured products. Sperm oil is still valued as a fine lubricant and sperm whales are killed also for a waxy material, spermaceti, used in the manufacture of cosmetics. The peculiar substance called ambergris is formed in the intestine of the whale and is used to make the odor of perfumes more persistent.

Aside from their torpedolike fish shape, whales have many other internal and external structural adaptations that allow them to live, move, and

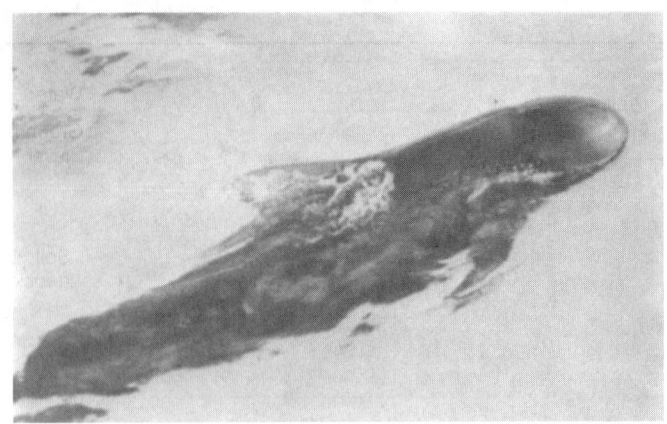

Fig. 1.   Whale. (*A.M. Winchester.*)

feed in the water. The skin is smooth and hairless, with the exception of a few single hairs or hair remains near the jaw, which may serve as tactile hairs. The fore and hind limbs appear in the embryo like those of any other mammal. However, the fore legs always develop into flat, more or less extended flukes. Whereas in a whale embryo that is 20 millimeters (0.8 inch) long one can see external indications of the hind feet, these appendages, indications and all, have completely disappeared by the time the embryo is 30 millimeters long. The blow hole (nostril) is located on the end of the snout in very young (4 to 5 millimeters; 0.2 to 0.3 inch long) whale embryos. (All other mammals have a similar location of the nostril). However, by the time the embryo has reached a length of 22 centimeters (8.6 inches), the blow hole has been shifted back to the upper part of the head, the same location as in adult whales. The location of the blow hole is probably related to the whale's weight distribution, which differs from that of land mammals. The sperm whales are the only members of this order which have their blow hole further up front.

The breast fins or flukes play a very minor role, or no role at all, in the whale's usual swimming motions; the tail provides all the diving power necessary for forward movement. Scientists have been able to study the tail movements of some whale species in the large American seaquariums, and, with the help of photographs, they have established that the tail is moved up and down on an exact vertical line. The real motor drive is produced almost exclusively by the broad. sweeping fluke, the sides of which are not supported by bones. The longer forward part of the tail provides the muscle power necessary for this diving motion. This part of the tail is a deep oval that is compressed at the sides, and it cuts through the water with very little resistance as the whale moves it up and down.

Speed of Movement. The speeds that whales can attain are astounding, at least insofar as baleen whales and dolphins are concerned. The slower right whales and gray whales can reach a peak speed of some 11 kilometers (7 miles) per hour; the humpback whale can reach a speed of up to 18 kilometers (11 miles) per hour. The usual speed of these three whales is generally from 3.5 to 5.5 kilometers (2 to 3.5 miles) per hour. Sperm whales can maintain a speed of 18 kilometers (11 miles) per hour over long distances, and they may travel as fast as 37 kilometers (23 miles) per hour over short periods of time. The large finback whales can maintain a general speed of between 22 and 26 kilometers (14 and 16 miles) per hour, while their peak speed is 50 kilometers (31 miles) per hour. The much smaller bottle-nosed dolphins and other dolphins are also able to reach similar speeds; scientists have measured a general velocity of 22 to 26 kilometers (14 to 16 miles) per hour for these animals.

Adaptation to Aquatic Environment. One of the greatest difficulties that a vertebrate land animal must overcome in its transition to life in and under water is the adaptation of its breathing to the very different requirements of aquatic living. Unlike fish, which breathe through gills, whales are forced to come to the surface of the water in order to breathe. When it reaches the surface, the whale expels the moist air from its lungs, under great pressure. When the pressure on the exhaled air is released, the air rapidly disperses and, as a result, is cooled to such an extent that the moisture it carries condenses and becomes visible as a white cloud. The cloud of steam, the "blast," can reach a height of 3 to 4 meters (10 to 13 feet) with right whales, a height of 2 meters (6.5 feet) with humpback whales, 4 to 6 meters (13 to 19.5 feet) with rorquals, 6 meters (19.5 feet) with blue whales, and 5 to 8 meters (16.5 to 26 feet) with sperm whales. Each whale species has a distinctive steam cloud, and a person acquainted with the types and shapes of clouds can easily distinguish the type of whale. Right whales have a double cloud. Finback whales release a cloud that is shaped somewhat like a pear, and the sperm whale's cloud is directed forward on a diagonal line.

Vertebrate animals are not really built for aquatic life; the salt content of their blood and body tissues is less than that of sea water. Thus, the excess salt accumulated from food and water must be eliminated from the body. Whales do not have salt excretory cells like those the bone fishes have on their gills. They are not able to excrete salt through their nasal glands like many sea birds. Whales also do not have perspiration glands, which help many other mammals give off excess salt. The kidneys are the whale's only salt-excreting organs. Whales have very large kidneys relative to those of land mammals; the kidneys of small whales are twice as large as those in land mammals of the same size.

Most whales are ocean-going animals; the few freshwater forms belong to the suborder of toothed whales. All baleen whales, as well as some species of toothed whales, have regular migrations. In 1931, scientists began to study those animals regularly, because of the economic importance of the whale industry; whales were tagged or marked so that scientists could plot their migrational routes and find out more about their way of life.

The Birth Process. Unlike most of the other larger mammals, whales are born tail first. The birth itself is usually a fairly smooth process, in spite of the size of the young, due to their smooth fish shape. Scientists have been able to observe the birth of some of the smaller whale species such as porpoises, dolphins, spotted dolphins, and bottle-nosed dolphins, in the larger seaquariums. The female bottle-nosed dolphin starts to swim more slowly when her labor pains begin. The other females of the group remain close to the pregnant female, presumably to protect her and to see that she does not get separated from the herd. The umbilical cord is very long, and it cannot be easily broken until the head of the whale calf is completely free; only then does the cord break off, near the calf's stomach. This prevents premature breathing motions by the calf. The mother pushes her young to the surface of the water immediately after birth; there it fills its lungs with air for the first time. The mother is often helped by several other adult members of the herd ("aunts") as she cares for the calf.

Blue whales and rorquals nurse their young for between five and seven months; most of the other whales nurse their young for almost a year. The nipples are situated in skin folds on both sides of the genital orifice; the young nurse underwater (a practice we are familiar with in hippopotamuses and sea cows). Whale milk is particularly rich in fats and protein; it is 40 to 50% fat (compared to between 2 and 17% in land mammals), and only 40 to 50% water (80 to 90% in land mammals). The protein content is twice that of land mammals, and the sugar content is 1 to 2% lower than that of land mammals, whose sugar content is 3 to 8%. This exceptionally nourishing milk helps the whale calf grow very rapidly; a young blue whale, for example, grows 9 meters (29.5 feet) in seven months, a daily growth of 4.5 centimeters (1.8 inches).

The females of some whale species can be re-impregnated immediately after the birth of their young. However, usually female blue whales, rorquals, and sperm whales do not become pregnant until after their young are weaned. For this reason these species generally give birth only one every two years. If we assume that the large whales have a life expectancy of 30 to 40 years, then a female can give birth to between 10 and 12 calves at the most during her lifetime. This is a rather low birth rate, adapted to the low natural death rate of these animals.

People have long hunted whales and used various parts of the whale for their needs. The early inhabitants of Alaska hunted whales in 1500 B.C. The first written report of whaling in North America is from A.D. 890. The most important part of the whale for man is the oil, which comes from the blubber layer in the subcutaneous connective tissue. Earlier, whale oil was used as lamp oil; today it is used principally in the manufacture of margarine. Whale oil also plays an important role as a raw material in the soap industry and in the production of linoleum and synthetic resins. The oil yield from whaling is only 2% of the world's fat production and only 5% of the production of animal fats. The meat of the larger baleen whales is eaten especially by the Japanese, but also in parts of Western

Europe. The meat of these whales is also used as dog food and, when it has been dried and crushed into meat powder, as cattle feed. People use whale bones to make glue and gelatin or manure (once the bones have been crushed). The baleen whale was formerly important as the supplier of the "whalebone" (baleen) which was a vital feature of the corset industry. Additional whaling products include vitamins and hormones from different internal organs, connective tissue fibers (used, for example, in stringing tennis rackets), ivory from the teeth of sperm whales, spermaceti, and ambergris.

Fig. 2.   Atlantic bottlenose dolphin.

Dolphins. These mammals are small-toothed whales that attain a length of about 8 feet (2.4 meters). There are many species of several characteristic forms. See Fig. 2. One of the more peculiar is the narwhal, *Monodon monoceros*, which has a single spirally twisted ivory tusk. The tusk is a single tooth, usually that of the left side. Its mate in the male and both of these teeth in the female are rudimentary. The narwhal has no other teeth except a few irregular rudiments. The tusk may attain a length of 8 feet (2.4 meters) and the animal may reach a length of from 12 to 16 feet (3.6 to 4.8 meters). These whales are dark gray and spotted in color. They appear in schools of from 15 to 20 individuals and can be described as gregarious and playful. They feed on crustaceans, cuttlefish, and small fishes. The purpose of the tusk is unknown. When in a playful mood, the animal seems to fence with it. They are not known to attack boats with the tusk. The tusk has been used as a source of good quality ivory, but the tusk does have a central cavity, which detracts from its value. The dolphin whales are not to be confused with dolphin fishes (*Osteichthyes*).

Several other species of dolphins live in large rivers of the Old and New World tropics, among them the Gangetic dolphin or susu of India and the Amazonian dolphin, *Inia geoffroyensis*, also called the inia or bouto, of South America. In most species of dolphins, the head is at least 12 to 15 inches (30 to 38 centimeters) broad, the beak about 6 inches (15 centimeters) long, and the blow-hole is crescent-shaped. Color is dark gray on the back and white underneath. Only one calf is produced at birth and the mother provides extremely attentive care.

Fig. 3.   Porpoise and young. (*American Museum of Natural History.*)

TABLE 1. WHALE ENDANGERMENT STATUS

Species	Status Listing U.S.	Status Listing IUCN	Population Estimate Pre-Exploitation/Present
Blue (*Balaenoptera musculus*)	Endangered	Endangered	228,000/14,000
Fin (*Balaenoptera physalus*)	Endangered	Vulnerable	548,000/120,000
Sei (*Balaenoptera borealis*)	Endangered	Not Listed	256,000/54,000
Bowhead (*Balaena mysticetus*)	Endangered	Endangered	30,000/7800
Sperm (*Physeter catodon or P. macrocephalus*)	Endangered	Not Listed	2,400,000/1,950,000
Northern Right (*Eubalaena glacialis*)	Endangered	Endangered	No Est./1000
Southern Right (*Eubalaena australis*)	Endangered	Vulnerable	100,000/3000
Humpback (*Megaptera novaeangliae*)	Endangered	Endangered	115,000/10,000
Gray (*Eschrichtius robustus*)	Endangered	Not Listed	20,000+/21,000
Bryde's (*Balaenoptera edeni*)	Not Listed	Not Listed	100,000/90,000
Minke (*Balaenoptera acutorostrata*)	Not Listed	Not Listed	140,000/725,000
Killer (*Orcinus orca*)	Not Listed	Not Listed	No Est./No Est.
Pygmy Right (*Caperea marginata*)	Not Listed	Not Known	No Est./No Est.
Narwhal (*Monodon monoceros*)	Not Listed	Not Known	No Est./35,000
Beluga (*Delphinapterus leucas*)	Not Listed	Not Known	No Est./50,000

Notes: U.S. = United States government estimate

IUCN = International Union for the Conservation of Nature and Natural Resources estimate

Endangered = Any species that is in danger of extinction throughout all or a significant portion of its range.

Vulnerable = Believed likely to move into the endangered category. Populations are decreasing because of overexploitation, extensive destruction of habitat, or other environmental disturbance.

The term grampus is a name meaning flat fish and is really applicable to almost any whale. It is usually given to Risso's dolphin which grows to about 15 feet (4.5 meters) in length, has a small rounded head, long tapering flippers, and feeds on squid and octopus. Grampus is sometimes also applied to a large dolphin, *Orcinus orca*, of vicious habits, and also known as the killer. These latter animals hunt in groups and are known to attack larger whales.

Porpoises. These are small animals related to the toothed whales. They reach a length of about 5 feet (1.5 meters). The muzzle is bluntly rounded, and most species have a dorsal fin. See Fig. 3. Porpoises are found chiefly in the northern oceans, although one species ranges from Japan to southern Africa and enters the rivers of China and India. The porpoise is usually smaller than the dolphin, but is much alike. Porpoise means "bottle-nose" as contrasted with the beak-type nose of the dolphin. Porpoises bear a single calf in the summer. Gestation period is about 1 year. The baby is about 3 feet (0.9 meter) in length. The life expectancy of a porpoise in captivity is about 30 years. A porpoise has between 80 and 100 teeth and feeds on fish and small marine animals. Color is black or dark gray above and white below. The flippers are black.

Endangered Whales. A status report for various species of whales is given in Table 1.

Additional Reading

Baskin, Y.: "Blue Whale Population May Be Increasing Off California," *Science*, 287 (April 16, 1992).

Beach, D.W. and M.T. Weinrich: "Watching the Whales," *Oceanus*, 84 (Spring 1989).

Booth, W.: "Unraveling the Dolphin Soap Opera," *Oceanus*, 76 (Spring 1989).

Braham, H.W.: "Eskimos, Yankees, and Bowheads," *Oceanus*, 54 (Spring 1989).

Brownell, R.L., Jr., K. Ralls, and W.F. Perrin: "The Plight of the 'Forgotten' Whales," *Oceanus*, 5 (Spring 1989).

Chapmasn, D.G.: "Whales (Antarctica)," *Oceanus*, 64 (Summer 1988).

Cousteau, J. and Y. Paccalet: "Whales," Henry N. Abrams, New York, NY, 1988.

Darling, J.C.: "Whales — An Era of Discovery," *Nat'l. Geographic*, 872 (December 1988).

Ellis, S.L.: "An Explanation for the Dolphin Die-Off," *Oceanus*, 79 (Spring 1989).

Ellis, S.L.: "Some Basics about the Whales: Ancient, Moustached, and Toothed," *Oceanus*, 26 (Spring 1989).

Epler, B.C.: "Whalers, Whales, and Tortoises," *Oceanus*, 30 (Summer 1987).

Fraker, M.A.: "A Rescue That Moved the World," *Oceanus*, 96 (Spring 1989).

Giddings, A.: "An Incredible Feasting of Whales — Alaska's Southeast," *Nat'l. Geographic*, 88 (January 1984).

Glass, K. and K. Englund: "Why the Japanese Are So Stubborn About Whaling," *Oceanus*, 45 (Spring 1989).

Golden, F.: "Fact and Fantasy (Whale Endangerment)," *Oceanus*, 3 (Spring 1989).

Harrison, R. and M.M. Bryden: "Whales, Dolphins and Porpoises," Facts on File, Inc., New York, NY, 1988.

Hoffman, R.J.: "The Marine Mammal Protection Act — A First of Its Kind Anywhere," *Oceanus*, 21 (Spring 1989).

Horgan, J.: "A Grave Tale: Do Whale Remains Help Life Spread on the Deep-Sea Floor?" *Sci. Amer.*, 18 (January 1990).

Katona, S.K.: "Getting to Know You: New Ways of Looking at Marine Mammals," *Oceanus*, 37 (Spring 1989).

Kawamura, A.: "Whaling (Japan)," *Oceanus*, 30 (Spring 1987).

Klinowska, M.: "How Brainy are Cetaceans (Whales)?" *Oceanus*, 19 (Spring 1989).

Leatherwood, S. and L.J. Hobbs: "Whales, Dolphins, and Porpoises of the Eastern North Pacific and Adjacent Arctic Waters; A Guide to Their Identification," Dover Publications, Inc., Mineola, NY, 1990.

MacLeish, W.H.: "The Gulf Stream: Encounters with the Blue God," Houghton Mifflin Company, Boston, MA, 1989.

Martin, A. and T. Martin: "Beluga Whales," Voyageur Press, Inc., Stillwater, MN, 1996.

Matem, B.R.: "Watching Habits and Habitats (Whales) from Earth Satellites," *Oceanus*, 14 (Spring 1989).

Natalie, R. and P. Goodall: "The Lost Whales of Tierra del Fuego," *Oceanus*, 89 (Spring 1989).

Nierenberg, W.A.: "Working with Whales," *Nat'l Geographic*, 886 (December 1988).

Polacheck, T.: "Harbor Porpoises and the Gillnet Fishery," *Oceanus*, 63 (Spring 1989).

Ridgway, S.H. and R. Harrison: "Handbook of Marine Mammals: River Dolphins and the Larger Toothed Whales," Academic Press, Inc., San Diego, CA, 1998.

Ryan, P.R.: "Buddha and the Whale," *Oceanus*, 52 (Spring 1989).

Scheffer, V.B.: "How Much Is a Whale's Life Worth, Anyway?" *Oceanus*, 109 (Spring 1989).

Sigurjónsson and Jóhann: "To Icelanders, Whaling is a Godsend," *Oceanus*, 29 (Spring 1989).

Simmonds, M.P. and J. Hutchinson: "The Conservation of Whales and Dolphins: Science and Practice," John Wiley & Sons, Inc., New York, NY, 1996.

Staff: National Geographic Society, "Whales: Dolphins and Porpoises," National Geographic Society, Washington, DC, 1995.

Stonehouse, B.: "A Visual Introduction to Whales, Dolphins, and Porpoises," Facts on File, Inc., New York, NY, 1998.

Swartz, S.L. and M.L. Jones: "Gray Whales," *Nat'l. Geographic*, 754 (June 1987).

Tyack, P.L. and L.S. Sayigh: "Those Dolphins Aren't Just Whistling in the Dark," *Oceanus*, 80 (Spring 1989).

Tyack, P.L.: "Let's Have Less Public Relations and More Ecology," *Oceanus*, 103 (Spring 1989).

Whitehead, H.: "The Unknown Giants," *Nat'l. Geographic*, 774 (December 1984).

Williams, H.: "Whale Nation," Harmony Books, New York, NY, 1988.

Würsig, B., M. W "ursig, and F. Cipriano: "Dolphins in Different Worlds," *Oceanus*, 71 (Spring 1989).

WHALE SHARK. See **Sharks.**

WHEAT (*Triticum sativum* and other species). *Gramineae*. Wheat is an annual plant producing the most valuable of cereal grains used by the white race. The plants grow either as summer annuals, seed being planted in the spring and the harvest gathered in the fall of the same year, or winter annuals, the seed then being planted in the fall and growing until stopped by cold weather, developing during that time an abundant root system which insures rapid growth in the springtime, the mature crop being ready for harvest in early summer.

When wheat seeds germinate a small primary root system is formed by the development of the radicle or seed root. This primary root system lasts but a short time, being soon replaced by a system of adventitious roots arising from the lowermost nodes of the stem and extending outward and downward to fill the soil with an extensive fibrous root system. The stem of the wheat plant is 2–4 feet (0.6–1.2 meters) tall and usually hollow, although some species have solid stems. See Fig. 1. The dried stems form wheat straw, frequently used in the manufacture of straw board. The leaves are of the ordinary grass type. The inflorescence is composed of very short-stemmed spikelets attached alternately on opposite sides of a zigzag axis. Each spikelet has 2–8 flowers. See Fig. 2. In certain varieties known as bearded wheats the lemma of the flower bears a long bristle or awn. Most species of wheat, particularly those grown in temperate climates, are close-pollinated. Durum wheat and primitive species are cross-pollinated. Evidence indicates, also, that wheats grown in hot dry climates are cross-pollinated. The fruit or grain of cultivated wheat varies somewhat according to the species. In many kinds of wheat the fruit separates readily from the surrounding floral bracts, while in a few kinds the lemma and palea tightly enwrap the grain. The grain bears at its apex a tuft of short hairs called the brush and has a distinct groove along the side which was against the palea. See Fig. 3. In section, a wheat grain shows several very distinct parts. Externally there is a layer several cells thick, called the pericarp, or ovary wall. Within this is a layer two cells thick, the tegmen, which is formed from the inner integument. The outer integument was absorbed during the development of the grain. Next is the nucellus, a single cell in thickness. These three layers constitute some 8% of the grain, and make up the substance which is called bran. Within these is the endosperm, which forms the bulk of the grain; the outermost layer of cells of the endosperm is the aleurone layer. In the basal portion of the seed next to the endosperm is the so-called germ, or embryo, from which the new plant may develop.

Classification and Varieties of Wheat

There are several ways of classifying wheats. Botanically they are separated according to the structure of the spikelet, the number of florets

Fig. 1. Sheaf of wheat. (*USDA photo.*)

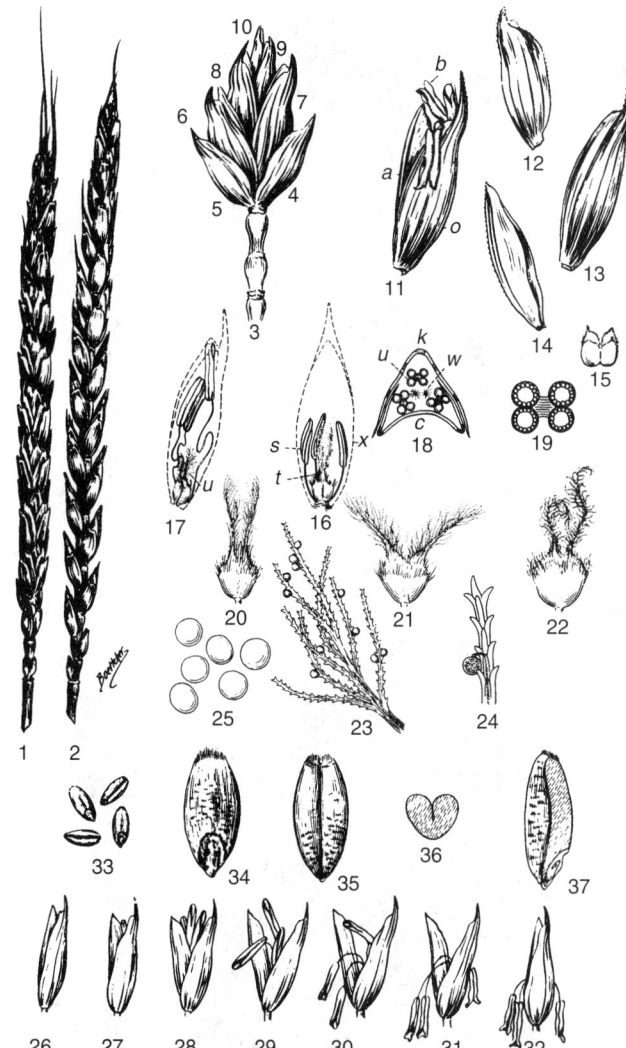

Fig. 2. Wheat inflorescence. (*USDA diagram.*)

(1) Spike, dorso-ventral view.
(2) Spike, lateral view.
(3) Spikelet, lateral view and subtending rachis.
(4) Upper glume.
(5) Lower glume.
(6) through (10) Florets.
(11) Floret, later view, opening in anthesis.
(12) Glume, lateral view.
(13) Lemma, lateral view.
(14) Palea, lateral view.
(15) Lodicules, which swell to open the glumes.
(16) Floret before anthesis, showing positions of stamens (s) and pistil (t)
(17) Floret at anthesis, showing position of pistil (u) and the elongating filaments of the stamens.
(18) Cross-section of floret: (c) palea, (k) lemma, (v) stamen, (w) stigma.
(19) Cross-section of anther.
(20) Pistil before anthesis.
(21) Pistil at anthesis.
(22) Pistil after fertilization.
(23) Portion of stigma (greatly enlarged), showing adhering pollen grains.
(24) Tip of stigma hair (greatly enlarged) penetrated by germinating pollen grain.
(25) Pollen grains (enormously enlarged).
(26) through (32) Florets during successive stages of blooming and anthesis. (Note: Time required for stages (26) to (31) is from 2 to 5 minutes; for total from [26] to [32] up to 40 minutes)
(33) Kernels (caryopses).
(34) Kernel, dorsal view.
(35) Kernel, ventral view.
(36) Kernel, cross-section.
(37) Kernel, lateral view, with one-half the kernel in longitudinal section.

and the nature of the parts of the flower. Among the kinds of wheat recognized in this classification are einkorn, spelt, emmer, durum, and common wheat. Again, if the palea and lemma adhere to the grain, the

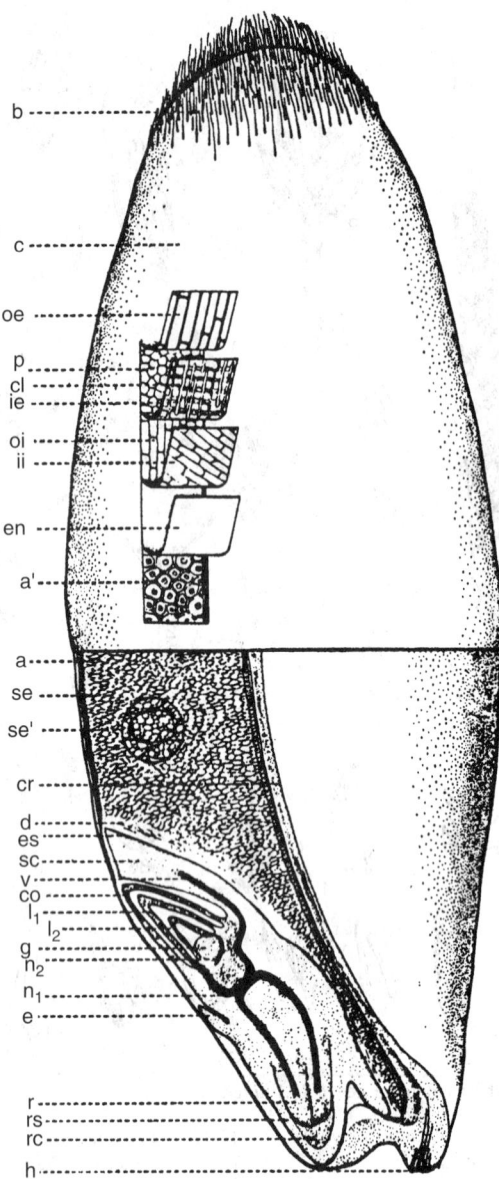

Fig. 3.   Kernel of wheat. (*USDA diagram by M.N. Pope.*) Note: The bran layer is comprised of the pericarp, testa, nucellar layer, and aleurone.

(b)	Brush.		

Pericarp

		Germ	
(c)	Cuticle.	(sc)	Scutellum.
(oe)	Outer epidermis,	(es)	Epithelium of scutellum.
(p)	Parenchymao		
(cl)	Cross layer.	(v)	Vascular bundle of scutellum.
(ie)	Inner epidermis.		
		(co)	Coleoptile.

Testa

		(l_1)	First foliage leaf.
(oi)	Outer integument of seed.	(l_2)	Second foliage leaf.
(ii)	Inner integument.	(g)	Growing point.
(en)	Epidermis of nucellus.	(n_2)	Second node.
		(n_1)	First node.

Endosperm

		(e)	Epiblast.
(a)(a′)	Aleurone.	(r)	Primary root.
(se)(se′)	Starch and gluten parenchyma.	(rs)	Root sheath or coleorhiza.
(d)	Crushed empty cells of endosperm.	(rc)	Root cap.
		(h)	Hilum.

wheat is classified as spelt wheat, while if the grain readily separates from these two parts it is naked wheat. If one turns to the grain itself there are hard wheats, in which the grain is horny and has a high protein content, and soft wheats with starchy grains. The nature of the soil in which the wheat is grown and the climate have considerable effect on the nature of the grain. Hard wheats are separated into hard spring wheat, hard winter wheat, and durum, the latter being especially rich in protein content.

Spring wheat is planted in the spring and harvested in the fall of the same year. *Winter wheat* is planted in the fall, growing for awhile and developing a good root system, after which growth ceases because of cold weather. Growth is resumed in the spring of the following year. The mature crop is harvested in early summer. See Fig. 4.

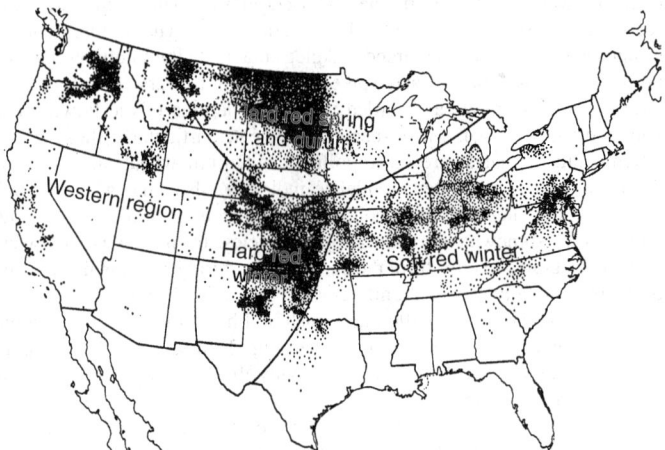

Fig. 4.   Regions of the United States where major classes of wheat are produced. The Western region produces white wheat mainly, but also produces some hard red winter, hard red spring, and soft red winter wheat. (*USDA.*)

Genetic Classification.   Of major importance to genetics and plant-breeding research is the following classification:

Diploid Group (*7 chromosomes*)
    Einkorn (*T. monococcum*)
Tetraploid Group (*14 chromosomes*)
    Emmer (*T. dicoccum*)
    Durum (*T. durum*)
    Persian (*T. persicum*)
    Poulard or Rivet (*T. turgidum*)
    Polish (*T. polonicum*)
Hexaploid Group (*21 chromosomes*)
    Common (*T. vulgare*)
    Club (*T. compactum*)
    Spelt (*T. spelta*)
    Shot (*T. sphaerococcum*)
    Macha (*T. macha*)
    Vavilovii (*T. vavilovi*)

High-yielding Varieties.   Sometimes referred to as HYVs, the high-yielding varieties of wheat (and rice) have formed the core of what became known as the *green revolution.* Well-deserved emphasis is given to the semidwarf varieties developed at the International Maize and Wheat Improvement Center (CIMMYT) in Mexico. Some of the current semidwarf HYVs, however, are the offspring of varieties developed from similar ancestors in other breeding programs. The relatively short and stiff stalk of the semidwarfs means that they respond to improved cultural practices through increased yields rather than through increased plant growth, which would also result in lodging (falling over of the plant). The semidwarf varieties in use, while considered by some to be revolutionary in their impact, are the product of a long developmental process.

Semidwarf wheats were noticed in Japan in the 1800s. Early in the twentieth century, several of these varieties found their way to Italy, where they were used to breed improved varieties that later found wide use. Japanese breeders also crossed their varieties with several American types, ultimately resulting in the release of a *Norin* variety in 1935. It was brought to the United States in 1946, again crossed with some American varieties, and taken to Mexico in the early 1950s. There the *Norin-Brevor* cross, in addition to some of the Italian varieties, was used by Norman Borlaug and his associates to develop the well-known Mexican varieties.

Mexican Varieties.   In 1946, S.C. Salmon, a U.S. Department of Agriculture USDA scientist acting as agricultural advisor to the occupation army in Japan, noticed *Norin 10* growing at the Morioka Branch Research Station in northern Honshu. The stems were short, about 23.5 inches (60 centimeters), but produced many full-sized heads. Dr. Salmon brought 16 varieties of this plant type to the United States. They were grown in a detention nursery for a year and then made available to breeders in seven locations. Although Norin 10 was not satisfactory for direct use in the United States, it was useful for breeding. Orville A. Vogel, a USDA scientist stationed at Washington State University, was the first to recognize its worth and to use it in a breeding program in 1949. Crossing *Norin 10* with U.S. varieties involved some problems, but a number of semidwarf lines were eventually developed. A *Norin* 10 X Brevor cross was to become particularly important.

In the interim, word about the short-strawed germ plasma had reached Norman Borlaug in Mexico. His breeding efforts had run into a yield plateau because of lodging under high levels of nitrogen fertilization. Introduction of the *Norin 10* genes led to the development of a number of Mexican dwarf and semidwarf bread wheat varieties. The genetic origins of some of these early hybrid varieties are shown in Fig. 5. International diffusion of these varieties began very quickly at the experimental level, and India and Pakistan were the first to be substantially involved.

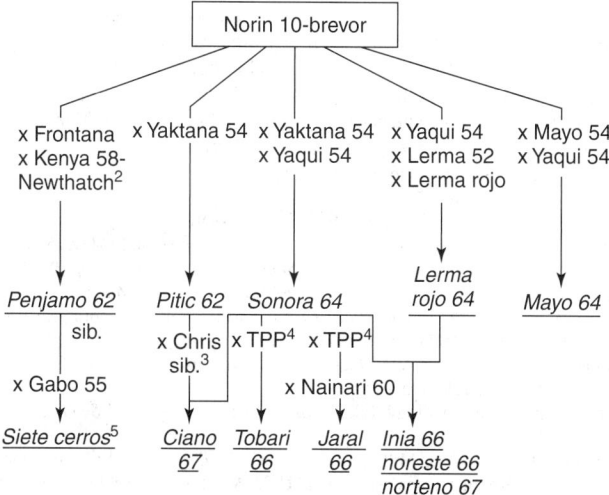

Notes:
1. This presentation of some of the more complex crosses is simplified for graphic purposes. For example, the parentage of *Lerma Rojo 64* would be more precisely written as: (*Yaqui 50 x Norin 10-Bevor*) *Lerma 52 Lerma Rojo*[2].
2. *Frontana x Kenya 58-Newthatch* was bred in Minnesota.
3. From Minnesota.
4. *Tezanos Pintos Precoz*, from Argentina.
5. Also known as cross 8156.

Fig. 5. Genealogy of early semidwarf wheat varieties emanating from work at CIMMYT (International Maize and Wheat Improvement Center). See note on diagram.

The Mexican varieties proved remarkably adapted to India and Pakistan for several reasons: (1) They had been bred in Mexico with alternate generations in different climatic and daylength regimes, primarily in order to get two generations each year. A valuable side effect of this system was to establish a good degree of insensitiveness to photoperiod. (2) Selection for disease resistance had also been practiced and the stocks introduced were found to show a remarkable level of resistance under the conditions in India and Pakistan. (3) The original stocks incorporated diversity. They had not been bred to pure line standards and there remained in them a reservoir of genetic potential that Indian wheat breeders were quick to exploit.

By the 1990s, the process of varietal change had gone through several stages in India. A large percentage of plantings in India, Pakistan, Afghanistan, and Nepal, among other less-developed countries, is planted to varieties of Mexican origin as just described.

Japanese Varieties. Japan had a long history in developing dwarf wheat. In 1873, Horace Capron, former U.S. Commissioner of Agriculture who headed an agricultural advisory group to Japan, wrote that "the

Japanese farmers have brought the art of dwarfing to perfection." He noted that "the wheat stalk seldom grows higher than 2 feet (0.6 meter), and often not more than 20 inches (51 centimeters)." The head was short, but heavy. The Japanese claimed that the straw had been shortened so that "no matter how much manure is used it will not grow longer, but rather the length of the wheat-head is increased." Capron noted that "on the richest soils and with the heaviest yields, the wheat stalks never fall down and lodge."

Probably unknown to Capron, some Japanese wheat varieties had already been introduced in France. The first introduction occurred in mid-1867 when the Société d'Acclimation of Paris received seed of a very productive early wheat "blé précoce," listed as *Haya Moughi*. The seeds were planted by a member of the Society, and a preliminary report was presented that fall. The stem or straw was short, and the plant flowered early. In the following years, other seeds were imported, and numerous reports of trials of the "blé précoce" appeared in the Bulletin of the Society. According to the description, the straw was very short, erect, and stiff; the plant was reported to flower 2 to 3 weeks ahead of all other spring varieties. The entry, however, noted that the variety was more of curious interest than of agricultural merit. "Blé Précoce du Japon" was sold commercially from 1882 to 1904 as a spring wheat and was used for experimental breeding work from 1930 to 1955. But it does not appear to have been involved in the parentage of any significant commercial varieties.

Two Japanese semidwarf varieties, however, did turn out to be of immense international interest in subsequent breeding programs, namely the *Akakomugi* and the *Shiro Daruma*.

Akakomugi (meaning red wheat) was often used as a cross-parent because of dwarfness and early maturity. It was mainly raised in southern Japan, but is no longer used commercially. *Akakomugi* played an important part in the breeding of Italian semidwarf varieties early in the 20th century. In 1917, *Shiro Daruma* was crossed with an American soft red winter variety, *Glassy Fultz*, and this produced *Fultz Daruma*. In turn, this variety was crossed with the American hard red winter variety Turkey Red at the Ehime Prefectural Agricultural Experiment Station in 1925. This was done in an effort to produce rust-resistant, short-stemmed, early-maturing varieties. The seeds of the first generation of the cross were transferred to the Konosu Experimental Farm of the National Agricultural Experiment Station and planted in 1926. Seed was subsequently sent to the Iwate Prefectural Agricultural Experiment Station in northeastern Japan.

A semidwarf selection developed from the seventh generation in 1932, *Tohoku No. 34*, was particularly promising. Following further testing, it was named *Norin* 10 and registered and released in 1935. The stem of *Norin 10* was particularly short, having a length of 20.5 to 21.25 inches (52 to 54 centimeters). *Norin 10*, in turn, was used in breeding programs in Japan, the United States, and Mexico. *Shiro Daruma* was also used at the Iwate Station to breed *Norin 1* in 1929 and *Norin 6* in 1932.

Italian Varieties. In 1911, seed from some of the short-straw early-maturing Japanese varieties was acquired by Ingegnoli, an Italian flower seed producer, during a trip to Japan. In turn, he provided the wheat seed to Nazareno Strampelli at the Royal Wheat Growing Experimental Station at Rieti. Strampelli started to use Japanese varieties in his breeding program in 1912. Strampelli was interested in developing wheat plants that would be both early ripening and resistant to lodging. Early ripening was desired to increase resistance to *blast* or *stretta* (wilting under hot wind stress) and rusts and to facilitate cropping. Resistance to lodging, obtained through shorter and thicker stems, was desired so that fertilizer applications could be increased.

Of the several Japanese varieties used by Strampelli, *Akakomugi* appeared to be the most important. This variety was crossed with *Wilhelmina Tarwe X Rieti* (a cross involving Dutch and Italian varieties originally made in 1906). This work was done in 1913. Among the well-known varieties to come out of this work were *Ardito* (1916), *Villa Glori* (1918), and *Mentana* (1918). *Ardito* was the first variety to gain wide use. It had a short straw (27.5 to 31.5 inches [70 to 80 centimeters]) and early-maturing characteristics. By 1926, this variety accounted for nearly all of the early-maturing varieties planted in Italy. Ardito was also grown elsewhere and became one of the progenitors of improved Argentine varieties and of the Russian winter variety, *Bezostaya*.

The second major variety, *Mentana*, attained international popularity because of its resistance to yellow rusts. Its genetic traits were bred into *Frontana* (Brazil) and *Kentana* (Mexico). *Mentana* was one of the three varieties that played a key role in the Mexican wheat-breeding program in the 1940s. By 1932, nearly a quarter of the early wheats planted in Italy were mainly *Mentana* and *Villa Glori*.

Italian varieties are now being grown in several of the less-developed countries in the Mediterranean region, notably Algeria, Morocco, and Turkey. *Strampelli* is one of the better known varieties. Italian varieties are also widely used in southeastern Europe. These varieties appear to have played an important role in the development of some of the early Mexican varieties.

Commercial Classifications of Wheats. Based upon the official grain standards of the United States, following are the principal commercial classes and subclasses of wheat that appear in worldwide trade. With exception of small percentages of durum wheat (*T. durum*) and small quantities of club wheat (*T. compactum*), all of the commercial wheats are common wheat (*T. vulgare*).

Classes and Subclasses

Hard Red Spring	**White**
Dark Northern Spring	Hard White
Northern Spring	Soft White
Red Spring	Western White
Hard Red Winter	**Durum**
Dark Hard Winter	Hard Amber Durum
Hard Winter	Amber Durum
Yellow Hard Winter	Durum
Soft Red Winter	**Red Durum**
Red Winter	**Mixed**
Western Red	

Official United States grain standards, when defining wheat, exclude einkorn, emmer, Polish wheat, poulard wheat, and spelt. All of these grains with exception of einkorn are grown in minor amounts in the United States and are used for feed.

In considering the commercial terminology of the classes and subclasses of wheat, one notes considerable inconsistency in selection of the titles. In most cases, the titles connote the color and physical characteristics of the kernel (red, yellow, amber, white; hard and soft); the season of the crop (spring or winter); the source of the wheat (northern, western); and wheat species in the cases of durum and club wheats. Kernels of hard wheats are described as "dark, hard, and vitreous," whereas the kernels of the soft wheats are described as soft and chalky.

Common wheat (*Triticum vulgare* Host); also *T. aestivum* L.) is also designated *bread wheat*. This wheat was first grown extensively in northern Europe.

1. Hard red spring wheat represents about 20% of total U.S. production and over 95% of Canadian production. This wheat produces a flour that is superior to other classes in breadmaking properties.
2. Hard red winter wheat represents nearly one-half the total U.S. wheat production. The kernels contain large amounts of high-strength gluten. The flour from this type is regarded as being almost equal to that produced from hard red spring wheat.
3. Soft red winter wheat represents about one-fifth total U.S. production. Flour made from this type of wheat is referred to as "weak flour" because doughs made from it lack elasticity and resulting breads tend to be heavy.
4. White wheat represents 6 to 8% of total U.S. production. This is a good quality wheat and particularly well suited for flours that are used in crackers, cakes, cookies, other pastries, and some breakfast foods.

Production of Wheat

Of the cereal crops, wheat ranks first in world tonnage and second in tonnage of cereal crops produced in the United States. Production of wheat is more widespread than the other major cereals, with the possible exception of maize (corn). Russia and former related countries account for about 21% of the total, followed by the United States (15%), China (11%), India (7%), and Canada (6%). Other major producing countries include France, Austria, Argentina, Italy, Pakistan, and Germany.

Wheat Flour

While a considerable quantity of wheat is used directly as food for domestic animals, the greater part is ground into flour. This milling of wheat is a very carefully controlled process. The first step is the thorough cleaning of the grain, removing therefrom any other substance. During this process the brush of the grain is removed. The cleaned grain is then moistened slightly in order to soften the outer layers so that they may be more easily removed subsequently. The moistened grain is then passed between iron rollers. The first of these are corrugated and break up the grains. Each successive pair of rollers grind the grain into finer and finer particles. Early in this grinding the coarse flakes of bran and the germ are removed. These are disposed of as such or ground up separately. The ground grain is passed through fine bolting silks, which insure a very even grade of fineness of the flour particles. Every precaution is taken in flour mills to prevent the accumulation of dust particles in the atmosphere, since these may form very dangerous explosive mixtures.

Flour is classified, according to the amount of the grain included in the final product, into graham flour, which contains the entire grain; whole wheat flour, which contains all the grain except about half of the bran; and straight bread flour, which results when all the bran is removed early in the grinding process. Wheat flour is used extensively in making breads, crackers, and pastries. Because it is so highly glutinous that pastes made from it will support their own weight, durum wheat is used in making spaghetti and macaroni. In making this product, a thick viscous paste or dough is prepared. This is forced, under great pressure, through holes in metal dies. Metal pins may project into the hole in the die, causing the dough which is forced through to merge as a hollow tube. This is macaroni. As they emerge, the tubes or strings are cut into suitable lengths and hung up to dry. Durum wheat alone has the necessary properties for making macaroni and similar substances. If any other wheat were used the product would break apart from its own weight. In addition to flour and macaroni, much wheat is used in making breakfast foods. The familiar puffed wheat results from heating wheat grains under pressure and suddenly releasing the pressure; the grain expands rapidly. Some wheat is used for making whiskey and certain varieties of beer.

Additional Reading

Abelson, P.H.: "Biotechnology in Third World," *Science*, 962 (May 25, 1990).

Abelson, P.H.: "Future of Agriculture," *Science*, 457 (August 3, 1990).

Beardsley, T.: "A Nitrogen Fix for Wheat," *Sci. Amer.*, 32 (March 1991).

Dover, M.J. and L.M. Talbot: "Feeding the Earth—An Agroecological Solution," *Technology Rev. (MIT)*, 26 (February 1988).

Gasser, C.S. and R.T. Fraley: "Genetically Engineering Plants for Crop Improvement," *Science*, 1293 (June 16, 1989).

Gibbons, A.: "Biotechnology Takes Root in the Third World," *Science*, 962 (May 25, 1990).

Harwood, J.: "U.S. Flour Milling on the Rise," *Food Review*, 34 (April–June 1991).

Holden, C.: "Russian Bugs Drafted in U.S. War (Wheat Aphids)," *Science*, 329 (October 20, 1989).

Lorenz, K.J. and K. Kulp: "Handbook of Cereal Science and Technology," 2nd Edition, Marcel Dekker, Inc., New York, NY, 2000.

Perkins, J.H.: "Geopolitics and the Green Revolution: Wheat, Genes, and the Cold War," Oxford University Press, Inc., New York, NY, 1997.

Pimental, D.: "Down on the Farm: Genetic Engineering Meets Ecology," *Technology Rev. (MIT)*, 23 (January 1987).

Reganold, J.P., R.I. Papendick, and J.F. Parr: "Sustainable Agriculture," *Sci. Amer.*, 112 (June 1990).

Rhoades, R.E. and L. Johnson: "The World's Food Supply at Risk," *Nat'l Geographic*, 74 (April 1991).

Satorre, E.H. and G.A. Slafer: "Wheat: Ecology and Physiology of Yield Determination," The Haworth Press, Inc., Binghamton, NY, 2000.

Schofield, J.D.: "Wheat Structure, Biochemistry and Functionality," The Royal Society of Chemistry," London, UK, 2000.

Smale, M.: "Farmers, Gene Banks and Crop Breeding: Economic Analyses of Diversity in Wheat, Maize, and Rice," Kluwer Academic Publishers, Norwell, MA, 1998.

Walsh, J.: "The Greening of the Green Revolution," *Science*, 26 (April 5, 1991).

WHEATGRASSES. See **Grasses**.

WHEAT MIDGE (*Insecta, Diptera*). A minute fly whose larva develops in the growing kernel of wheat. Introduced from Europe into Canada early in the nineteenth century, the species was troublesome in New York about the middle of the century but has not been serious since. Cultural methods are an adequate protection. They include crop rotation, fall plowing to bury and destroy the larvae, and the destruction of all debris from infested fields.

WHEATSTONE BRIDGE. See **Bridge Circuits (Electrical/Electronic)**.

WHEEL ORE. See **Bournonite**.

WHELK (*Mollusca, Gasteropoda*).   A moderately large marine mollusk with a spirally coiled shell. It is used as a bait in the cod fisheries and in Europe is eaten.

WHEY PROTEIN.   See **Protein**.

WHINSTONE.   A popular British term for basic, fine-grained igneous rocks belonging to the basaltic group and including diabase, dolorite, spidiorite, also greenstone and the lamprophyres. Synonym for trap or trap rock.

WHIPPOORWILL.   See **Nightjars and Nighthawks**.

WHIP SCORPION (*Arachnida, Pedipalpi*).   A large arthropod whose abdomen bears a whiplike posterior portion, although some species of the same order lack this terminal filament. The pedipalps are large and strong, either chelate or simple. The first pair of legs are modified as slender many-jointed sensory appendages. They are tropical animals, only a few species entering the southern part of the United States.

In the Southwest, many fears and superstitions are associated with these unpleasant looking animals, which are called vinegarroons. They discharge a disagreeable, sour-smelling secretion, but are entirely harmless to people.

WHISKER.   The fine, sharpened electrode forced into contact with the semiconductor material in a semiconductor diode or point-contact transistor.

WHISKEY.   A beverage of high alcoholic content made by distilling a fermented mash. There are four major producers of whiskey[1] in the world — Canada, Ireland, Scotland, and the United States.[2] Six principal operations are involved in the production of whiskey: (1) selection and preparation of raw materials; (2) fermenting; (3) distilling; (4) maturing and ageing; (5) blending (in most cases); and (6) containerizing (usually bottles). There are important variations in these operations from one country to the next and from one type of whiskey to the next.

Because distillation is a relatively complex procedure and essential to whiskey production and there are no immediately recognizable examples in nature to be copied — as in the case of natural fermentation of grape juice (see also **Wine**) — many centuries elapsed between the time when winemaking was pioneered and when the first alcoholic distillates appeared. Generally, it is believed that distillation was first used in connection with making brandies from wines, probably pioneered by monks during the Middle Ages. It is possible that the "discovery" of distillation came as the result of investigations in medicine and/or alchemy.

The question of whether Ireland or Scotland was the birthplace of whiskey has been debated for scores of years and remains inconclusive. Irish and Scotch whiskeys date back several centuries. The climate in both countries is favorable to whiskey production and is particularly suited to the production of barley, which both of these whiskeys use. Also, in both of these locations there were convenient sources of fuel (peat and coal) with which to fire the stills. It is interesting to note that as early as the 1770s there were over 1000 licensed distillers in Ireland. A contemporary report of the government at that time stated, in part, "(Distillers are) mostly men of little substance and no great honesty and well placed for defrauding the Revenue." A few years later, stills of low capacity were outlawed so that the government could get a better handle on revenue collection. It is to be noted, however, that in the early 1830s the government seized over 10,000 illicit stills. Through the mid- and late 1800s, because of much emigration from Ireland, the requirements for whiskey lessened and the number of operating skills became fewer so that by 1907 there were only 27 comparatively large distillers operating in Ireland. As of the late 1980s, there were four major distillers operating in Ireland for production of whiskeys. In addition to population diminishment, the imposing of higher and higher duties on the product caused a number of distillers to leave the field. Although Ireland exports whiskey, it in no way compares with the very high exports of Scotland.

The production of Scotch whiskey, dating back hundreds of years, has been an important factor in Scotland's export economy for many years,

exporting and local consumption supporting some 110 distillers operating there as of the early 1990s. *Highland malt whiskeys* are made in an area north of Dundee and Grennock; lowland malts are made south of these locations; Islay malts are from the Isle of Islay; and Campbelltown malts are from the town of that name, located in the Mull of Kintyre. Of the 110 distillers, nearly 100 of these produce malt whiskeys by the traditional pot still method, while the others produce grain whiskeys using the so-called Patent still method. Scotch whiskeys are exported to well over 150 countries worldwide — and possibly Scotch whiskey is the best known alcoholic drink throughout the world. The malt whiskies are heavy and strong and still remain most popular locally. It was not until the grain Scotch whiskeys were produced that they became popular in the British Isles south of Scotland, a popularity that rather rapidly extended to Europe, North America, and throughout the world. The malt whiskies are used for blending with the grain whiskies to provide an attractive balance.

The first whiskey stills operated in North America appeared as early as 1668 when Jean Talon established a brewery in the then small colony of Ville Marie, destined to become Montreal. As part of the brewery operation, a still was included for making spirits. Early Canadian settlers also prepared their own rum (in lieu of whiskey) from molasses shipped in from the West Indies. By the late 1770s, the grain farmers were expanding and, on occasion, would be faced with a surplus crop. This surplus was used for making distilled spirits. A high quality of the product was maintained and demand increased. The distinctive characteristics of Canadian whiskey established an excellent reputation and continued to increase demands so that, by 1860, Canada commenced to export whiskeys. As of the early 1980s, close to 25 distillers (most large-capacity) furnish Canadians as well as exporting to many parts of the world. Canadian whiskey, almost always regarded highly, has been responsible for furthering the awareness of Canada throughout the world. In recent years, the Canadian distilling industry has made a number of progressive moves and today operates some of the most modern facilities to be found anywhere.

As with the Canadians, the early settlers in Colonial America depended largely upon rum shipped in from Cuba and other West Indies islands. The Revolutionary War severed rum supplies and demand turned to whiskey. Earlier, farmers inland from the coast — in Ohio, Kentucky, Pennsylvania, etc., particularly in the mountainous areas — were making spirits from surplus grains. They found that a pack-horse could carry considerably more grain in the form of whiskey than of unprocessed grain — and the economic return was greater. For some years, in isolated regions particularly, whiskey was used as a medium of exchange. The Whiskey Rebellion, which occurred in the fall of 1793, was fought over tax collection in whiskey. Taxes had risen rapidly from just over 5¢ per gallon to over $10 per gallon.

Bourbon whiskey, made from corn (maize) and the most popular type in the United States, was first distilled in Kentucky in the mid-1830s. During the 1800s, numerous distilleries were established in Kentucky, Tennessee, southern Ohio, Indiana, and Illinois — all essentially located in the eastern portion of America's corn belt.

Raw Materials for Whiskey. What sugar in the grape is to wine so is starch in grain to whiskey. With grains, however, an additional step is required for enzymes to convert starch in the grains to sugar prior to fermentation.

Scotch Whisky. Malt Scotch whiskey is prepared from malted barley. As it is received, the barley is mechanically cleaned and stored until needed. At a controlled temperature, batches of barley are soaked in water (*burn water*) for 2 to 3 days until the grains become soft. The wetted grain is then spread over a large floor area and allowed to germinate. This may require from about a week up to nearly 2 weeks. Workers during this period turn the sprouting grains several times each day to prevent formation of molds. To prevent any damage to the grains, they use long-handled wooden shovels (*shiels*). During sprouting, the starch content of the grains slowly convert into sugar. At the proper time, the sprouting is halted by a kilning operation. The malted barley (*green malt*) is dried in a kiln that is fired with peat as a fuel. The characteristic aroma given off by the burning peat is called *peak reek*. The wet grains absorb a certain amount of the aromatic compounds from the smoke, and these account, in a large but indeterminate measure for the characteristic flavor of Scotch whiskey. Even in grain Scotch whiskies, some of the malt whiskey is added.

The dried malt is then transferred to large bins where it is slowly cooled and stabilized over a period of 3 to 4 weeks. The dried malt is then crushed to produce *grist*. In turn, the grist is transferred to a mash tun (usually a large cast iron vessel some 20 feet [6 meters] across and partially filled

---

[1] In Ireland and the United States, preferred spelling is whiskey (whiskeys); in Canada and Scotland, whisky (whiskies).

[2] Australia also produces substantial volumes of whiskey.

with water of the highest quality obtainable). In this operation, the sugar is dissolved to form *sweet liquor* or *wort*. The wort is drained off, and the residual (draff) is further processed for use as cattle feed. The wort becomes the starting material for the fermentation process.

Grain Scotch whiskey is made from a mixture of malted barley and unmalted cereals (usually barley and maize [corn]).

Irish Whiskey. The Irish fire the kilns for drying the green barley malt with coal instead of peat, an important difference because essentially no aromatic chemicals are picked up from coal smoke or combustion gases. The Irish do not make a straight malt whiskey, but rather use from 25 to 50% malted barley, mixed with from 50 to 75% of grist made up of unmalted cereals, barley, oats, wheat, and, on occasion, relatively small amounts of rye. Thus, Irish Whiskey is a grain whiskey essentially from the start.

Canadian Whiskey. The principle starting materials for Canadian Whiskey are cereal grains, normally maize (corn), rye, and barley malt. For Canadian Whiskey, the barley malt is not prepared by kilning as is done in Scotland and Ireland, but rather the ground grain meal is pressure-cooked, by which the starch is put into solution. It is then cooled and mixed with a predetermined small quantity of barley malt to commence conversion of the starch into sugar. Rather than screening off the grain in a mash tun, as is done by Scottish distillers, the resulting sweet mash is transferred directly to the fermenters.

In a separate operation, the Canadian distillers prepare a separate rye mash for fermentation and ultimately produce a rye distillate from it. The rye meal usually is not pressure-cooked. The formula for the rye mash may contain up to 90% rye grains and 10% barley malt. On the other hand, the rye content may be as low as about 50%, with the remaining grains, not including the barley malt, being maize (corn) or barley. This distillate enters into the manufacture of Canadian Whiskey later on in the process.

American Whiskey. In the United States, a number of whiskeys are offered, but bourbon made from corn (maize) leads by a wide margin over rye whiskeys. Ground meal is transferred to mash tubs or pressure cookers, where the starch is released, making it easier to convert into sugar. After cooling, barley malt is added, the enzymes of which convert the starch to sugar.

Fermentation Process

Scotch Whiskey. Previously mentioned sweet liquor or wort, after cooling, is transferred to the tun room where it is fermented. A tun is a large wooden vessel some 18 feet (5.4 meters) high and 12 feet (3.6 meters) across and is usually made from larch or Oregon pine. The vessel may hold from 2000 to 12,000 gallons. Yeast is added to commence conversion of sugar into raw spirits. During fermentation, the mass bubbles and froths. The tuns are equipped with *switchers* (wooden rotating arms that are turned near the top of the vessel to control foaming and thus keep the vessel from running over). Workers also agitate the fermenting mass periodically, using a long-handled perforated board (*rouse*). Fermentation requires about 48 hours, at which time the liquid in the tun contains about 10% alcohol, as well as some residual yeast and by-products of fermentation. The liquid is known as *wash* and is now ready for distillation. The tuns are thoroughly brushed and cleaned after each batch. The foregoing describes the process for fermenting used in making malt Scotch whiskey.

In the case of grain Scotch Whiskey, it is prepared much like the Canadian grain whiskey about to be described.

Canadian Whiskey. The sweet mash is transferred to fermenting vats, where a predetermined quantity of pure culture yeast is added. The culture is usually prepared by the distiller. Fermentation requires about 72 hours. The resulting liquid is termed "beer" and is now ready for distillation. At this point, the beer contains about 12% alcohol.

Irish Whiskey. The fermentation process is essentially the same as that used in Scotland.

American Whiskey. The previously mentioned malted mash is cooled to approximately room temperatures, after which it is transferred to the fermenting vessels. Yeast culture is added at this point. From this point the process is similar to that used by the Canadian distillers. The resulting liquid is also called beer.

Distillation

Distillation is a process in which a liquid is converted to a vapor and the vapor is then condensed and collected (*distillate*). The most volatile

component of a mixture (some distillers call these *foreshots*) tend to distill off first. However, some of the less volatile components also appear in small quantities in the early portion of the distillate. Similarly, toward the end of a simple or batch distillation, some of the low-volatile components (usually undesired) also tend to "come over" and contaminate the distillate. Some distillers call such components (*feints*). A skillful stillman, watching properly placed thermometers, can divert the foreshots at the commencement of a batch distillation and, toward the end of the distillation, can cut the heat input and divert the feints from commingling with the best portion of the distillate. In the simple or batch distillation of sugar solutions from fermenters, the principal purpose of which is to separate alcohol from water, the feints will contain undesirable fusel oil, aldehydes, which, if allowed to contaminate the distillate, would lower the quality, if not ruin it. However, the problem could be overcome by redistilling the contaminated distillate. See Fig. 1.

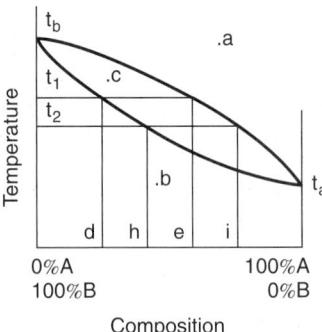

Fig. 1. Boiling-point diagram for a binary mixture, such as alcohol and water. Diagram depicts the boiling-point and equilibrium-composition relationships, at constant pressure, of all mixtures of liquid A (boiling point = t_a) and liquid B (boiling point = t_b). Liquid A is the more volatile of the 2 liquids. Note the 2 curves, the ends of which coincide. For all points above the top curve, such as point a, the mixture is entirely vapor. For all points below the bottom curve, such as point b, the mixture is completely liquid. For points between the 2 curves, such as point c, the system is partly liquid and partly vapor. If mixture of composition d is heated slowly, it will commence to boil at t_1. The first vapor produced will have a definite composition, represented by e on the diagram. When an appreciable amount of vapor has been formed, the composition of the liquid will no longer correspond to d, since the vapor is richer in the more volatile component than the liquid from which it was evolved. At a temperature, such as t_2, if the overall composition of the mixture is between i and h, there will be 2 phases in equilibrium, that is, a vapor phase of composition i and a liquid phase of composition h.

Over the last several decades, mainly since the great emergence of the chemical and petroleum industries after World War I, numerous improvements have been made in distillation equipment. Much of this equipment involves continuous distillation and the use of rectifying or fractionating columns in which very precise separations can be made. There are literally thousands of these modern distillation systems installed. Distilling improvements were relatively slow in being adopted by the whiskey and distillery industry because of its tight bonds to tradition and essential fears of change that would cause marked departures from the desired characteristics and properties of its products. In fact, some distillers emphasize their use of old and traditional methodologies in their advertising. Consequently, one finds today, notably in Scotland and possibly to a considerably lesser degree in other major whiskey-producing countries, examples of distillation technology that, from an engineering standpoint, appear quite antiquated. But these are equipments that produce whiskies of high quality and that have been extraordinarily successful.

Scotch Whisky Distillation. Malt whiskies made in Scotland are subject to two separate distilling operations. Two stills are placed side by side—a *wash still* and a spirits still. The wash from fermentation previously described goes to a wash still, which may be heated by a naked fire. A rummager inside the body of the still (essentially rotating arms that drag a copper mesh chain around the bottom of the still to keep particles from sticking) is used. Distillation is continued until the stillman determines that there is little if any recoverable alcohol remaining in the still. Meanwhile the distillate has been collected and is known as *low wines*.

The residue remaining (*pot ale*) usually is recharged into another batch for the wash still. The low wines, considering the type of distillation, will of course contain numerous impurities.

The low wines are charged into the adjacent *spirit still*, and this is where the skill of the stillman enters in terms of diverting the foreshots at the start and the feints toward the end of the distillation. The stillman decides when to switch the spirit still output to the spirit receiver—and when to stop the distillation. The foreshots and feints that have been collected are returned to the next batch of low wines. This entire process is referred to as the *pot still process*. It is interesting to note that, while all malt Scotch whiskey are made in this manner, the flavor and characteristics differ from one distiller to the next—even though water and other variable conditions may seem reasonably consistent. This is reminiscent of blast furnaces in the steel industry, which, even though identically designed and fed with the same raw materials, will differ some in operating characteristics.

The grain Scotch whiskeys are distilled in what is called a Patent still. An interesting point is that this column still was invented by Aeneas Coffey, an Irishman, in 1830, a century and a half ago and indeed ultramodern and ingenious for its time. See Fig. 2.

Irish Whiskey Distillation. Whereas Scottish distillers use two distillations, the Irish have traditionally used three distillations. The pot stills used in Ireland are considerably larger than those used in Scotland and may contain 30,000+ gallons. The Irish also introduced several innovations many years ago, including the use of taller still heads. They also use what is known as a *lyne arm* interposed between the still head and the condenser (*worm tub*). The lyne arm or lying-arm extends from the still head horizontally and is submerged in a water-filled trough to provide immediate cooling. The condensate from the lyne arm is returned to the still and thus the weaker (more volatile) spirits in the distillate are returned to the still for further rectification and concentration. In modern terms, this returned condensate would be termed reflux. The Irish call the return pipe a *foul pipe*. It will be recalled that the Scottish distillers use a wash still and a spirits still. The Irish interpose a second *low wines* still between the wash still and spirits still. The stillman thus is able to divide the distillate into strong and weak fractions, sending the stronger fractions forward and the weaker fractions backward in the process. Thus, a form of rectification is accomplished in a manually controlled intermittent fashion and accomplishes within the limitations of the equipment and the skills of the stillman, rectification that is much more easily (and probably more

effectively) accomplished in the modern continuous rectification systems. A continuous distillation system is illustrated in the entry on **Rum**.

The distillation of Canadian and American whiskeys have followed advancements in what might be termed North American distillation technology. Current installations range from designs of 50 to 60 years of age of installations made within the last few years. The actual distillation setups vary somewhat between distillers and ultimate objectives in terms of final product characteristics.

Ageing, Maturing, and Blending

There are many parallels in the manner in which the young whiskey is matured regardless of where it is produced. Some of the secondary constituents of whiskey, require evaporation for removal. Thus, the casks used for storage must be somewhat porous. Distillers frequently prefer oak wood casks, particularly casks that have previously been used for storing sherry. From an economic standpoint, it is interesting to note that as much as 15% of the young whiskey may evaporate. A recent estimate has indicated that up to 10 million gallons of Scotch whiskey may be lost per year through evaporation during storage. Temperature and humidity must fall within reasonable range for satisfactory storage. Scottish, Irish, and Canadian climates are particularly well suited in this respect. In Scotland, the malt whiskeys may be left to mature for 15 years, and improvement is possible beyond that age. However, most authorities agree that nearly any whiskey will take on a "woodiness" characteristic if stored for over 20 years in a cask. Grain whiskeys require less time because more of the undesirable constituents are removed during rectification. In Britain by law Scotch whiskey cannot be sold until it has matured a minimum of 3 years. Better brands require considerably longer, with the average Scotch whiskey being from 6 to 7 years old. When age is stated on a blended whiskey, the age must be that of the youngest whiskey in the bottle. Whiskey does not improve or deteriorate in the bottle. Most authorities agree that so-called "bottle character" of a whiskey is a myth. Whiskey is not overly sensitive to temperature after it is bottled. The practice of ageing in Ireland essentially is the same as that followed in Scotland.

The large majority of Scotch whiskey is blended, a practice that commenced in the mid-1800s. The availability of blended whiskeys greatly increased the popularity of Scotch whiskey throughout the world. A blend may contain from 20 to 40 different "single" whiskeys. Blending of Scotch is strictly an art, requiring much skill and judgment. Once a given "single" whiskey has been deemed sufficiently mature, 1/10 gallon (duty free) is

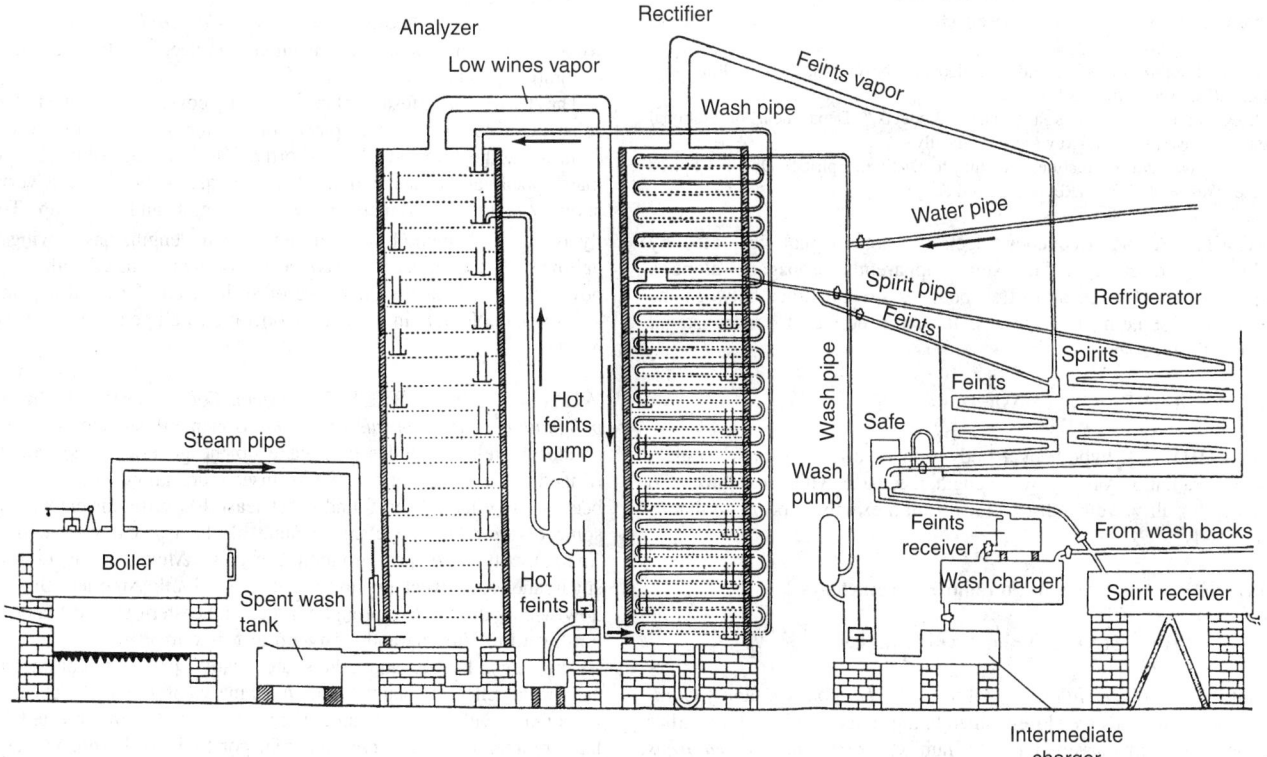

Fig. 2. Sectional view of distilling apparatus developed by Aeneas Coffee in Ireland in 1830. (*Adapted from early patent sketch.*)

drawn from the cask. The blender has such samples from many casks to judge almost on a continuing basis because the inventories of casks for maturing is so high. The blenders are not tasters, but depend essentially upon aroma, seeking out any trace of woodiness that may be derived from an imperfect cask; or hardness sometimes due to improper wood used in the cask; or any alien flavors that may have crept in due to contamination. The objective of the blender is not only that of maintaining high quality, but also to blend a product with qualities and characteristics that is consistent from year to year in the marketplace. Caramel coloring is added to Scotch and other whiskeys to maintain color uniformity.

It has only been within the last few decades that Irish whiskeys have been marketed as blends, although blends have been available since the mid-1800s on a localized basis. Blends have become much more popular in recent years.

In practice, most Canadian whiskey is aged in wood for a minimum of 3 years, but it can be sold legally at an age of 2 years. The average Canadian whiskey ranges from 4 to 6 years, with premium brands available up to 8 and 10 years and more. It will be recalled from a previous description of the raw materials used in Canadian whiskeys that a separate rye mash is prepared. The rye distillate is blended with that from other grains and barley malt. The distinctive characteristics of Canadian whiskeys are due to a large degree to the rye distillate. Canadian whiskeys are matured in charred oak barrels. The rye spirits may be blended with the grain whiskey prior to barreling, or they may be allowed to mature separately as "single" whiskeys and blended prior to bottling. Canadians at one time regarded their product essentially as rye whiskey and was frequently labeled so, but during the past several years, the word rye has been dropped by many distillers, even for products for local consumption, and the phrase Canadian Whiskey by itself is commonly used.

In the United States ageing is effected in charred new white oak barrels in climate-controlled warehouses. Most of the whiskey is made to achieve a maximum maturity within 4 to 6 years. Blends are very common in the United States, but there are several straight Bourbon whiskeys offered, as well as smaller number of rye whiskeys. Often, a blended whiskey will contain from 30 to 40% aged whiskey that has been blended with neutral spirits (60 to 70%). Grain neutral spirits are of approximately 190° proof and in terms of taste, color, and aroma are neutral.

Australian whiskeys are made somewhat along the pattern of Scotch whiskeys but do not exhibit the same characteristics as a true Scotch. Australian whiskey has a distinctive character of its own.

Additional Reading

Staff: "Annual Statistical Review," Distilled Spirits Institute, Washington, DC (Issued annually).

Staff: "Methods of Analysis," section on alcoholic beverages, Assoc. Offic. Agr. Chemists (Revised periodically).

Staff: "Regulations, Distilleries and Their Products," Department of National Revenue, Ottawa, Canada (Revised periodically).

Staff: "State Laws and Regulations Relating to Distilled Spirits," Distilled Spirits Institute, Washington, DC (Revised periodically).

WHISTLER. A radiofrequency electromagnetic signal generated by some lightning discharges. This signal apparently propagates along a geomagnetic line of force and often bounces several times between the Northern and Southern Hemispheres. Its name derives from the sound heard on radio receivers.

WHITEBEAM TREES. See **Ash Trees**.

WHITE BODY. A hypothetical body whose surface absorbs no electromagnetic radiation of any wavelength, i.e., one which exhibits zero absorptivity for all wavelengths; an idealization exactly opposite to that of the black body.

WHITE DEW. See **Precipitation and Hydrometeors**.

WHITE DWARF. See **Giant and Dwarf Stars**.

WHITEFISHES (Osteichthyes). Of the order Isospondyli, family Coregonidae, the whitefishes are closely allied to the salmon and trout. In earlier years, they were all classified in one family. Coreogonus clupeaformis (lake whitefish) is the largest of the species. This fish attains an average weight of about 4 pounds (1.8 kilograms) although records indicate a catch of one weighing 26 pounds (12 kilograms). The Coregonus artedii (shallow-water cisco) also is called the lake herring (although not a herring). The C. kiyi (chub of deep water whitefish) prefers waters as deep as 600 feet (180 meters). The whitefishes are found in the Great Lakes of America and have in the past supported significant commercial fisheries. They are an exceptionally good food fish. Unfortunately, overfishing, coupled with the disastrous damage rendered by the parasitic sea lamprey, have severely reduced the population of whitefishes in the Great Lakes and notably Lake Erie. See also **Lampreys (Agnatha)**.

WHITEFLY (Insecta, Homoptera). A minute sucking insect related to the aphid, phylloxeran, and scale insects. They are important economic pests, mainly on citrus, certain vegetables, and glasshouse plants. Some of the major species are described below.

Citrus blackfly (Aleurocanthus woglumi, Ashby). This insect occurs in Mexico and the West Indies. The nymphs are black with a white fringe. The adults have a dark brown body and dark blue wings.

Cloudy-winged citrus whitefly (Dialeurodes citrifolii, Morgan). This pest occurs in the citrus region of Florida. Eggs are black.

Common citrus whitefly (Dialeurodes citri, Ashmead). This insect once occurred widely in California before being brought under control. Eggs are pale yellow.

Orange spiny whitefly (Aleurocanthus spiniferus). This insect causes serious damage to citrus in southeast Asia. The adult is about $\frac{1}{16}$ inch (1.5 millimeters) long, yellowish orange, and coated with a waxy powder. The adults are active in good weather, but when it is cloudy or rainy, they do not move about. The larva of the fly is very tiny, oval, and brownish or black. It has a cottony wax-encrusted fringe about its body. The insects feed on leaves of citrus. They secrete honeydew on which a sooty mold grows. Usually, they are found on the underside of new leaves.

Woolly citrus whitefly (Aleurothrixus floccosus, Maskell). A pest of citrus regions of Florida. Brown, sausage-shaped eggs.

The habits and life cycles of the citrus whiteflies are all quite similar. All stages can be found throughout the year. Breeding, however, is confined to the warmer months. The tiny eggs, less than $\frac{1}{100}$ inch (0.25 millimeter) long, are laid near a short stalk and on the underside of a leaf. Eggs hatch after 4 days to 2 weeks, producing pale-yellow nymphs. They are miniature six-legged crawlers. Shortly thereafter, the nymphs begin sucking sap from the leaves. They then undergo two molts, after which the adults emerge. The adults are 4-winged and about $\frac{1}{12}$ inch (2 millimeters) long. A very fine white powder covers body and wings. In some areas, such as Florida, as many as 3 generations per year will be produced.

Greenhouse or *Glasshouse whitefly* (Trialeurodes vaporariorm, Westwood). These insects damage cucumber, lettuce, tomato, and a number of decorative plants.

The leaves of infested plants become covered with a sticky, glazy substance, making an ideal place for damaging fungus to grow. Unless controlled, the affected plants shortly wilt and die. Adults and nymphs may appear at the same time. The pale-green nymphs are very small, about $\frac{1}{30}$ inch (1 millimeter or less) in length and suck sap. The adult fly is about $\frac{1}{16}$ inch (1 to 2 millimeters) in length, has 4 wings, and a yellow body that appears to have been thoroughly dusted with a fine white powder. Sometimes the underside of an infested plant will appear almost snow white. Notably in glasshouse operations, all generations of the insect overlap.

WHITE-FRINGED BEETLE (Insecta, Coleoptera). The beetle (Graphognathus or Pantamorus leucoloma, Boheman) is one of a large number of closely related beetles that make up the generic group called white-fringed beetles. See Fig. 1. These insects are known as general feeders because they have been found on at least 400 different plants. However, some of their favorite plants include blackberry, cabbage, collard, corn (maize), cotton, cowpea, groundnut (peanut), Mexican clover, strawberry, sugarcane, and sweet potato. A native of South America, this insect is now widely distributed throughout the southeastern United States, ranging northward to Virginia and westward to Missouri.

The eggs, larvae, and adults are easily spread in commerce. Eggs can be attached to plants, farm machinery, or other objects that come in contact with the soil, and in hay harvested from infested fields in late summer or fall. Larvae are transported in soil, sod, nursery stock, and root crops. The adults attach themselves to seed cotton, hay, farm machinery, and tools. To prevent the spread of these beetles, both federal and state quarantines have been established. Any of these insects found outside the aforementioned areas should be reported immediately to the U.S. Department of Agriculture, Plant Protection Division, Hyattsville, Maryland 20782.

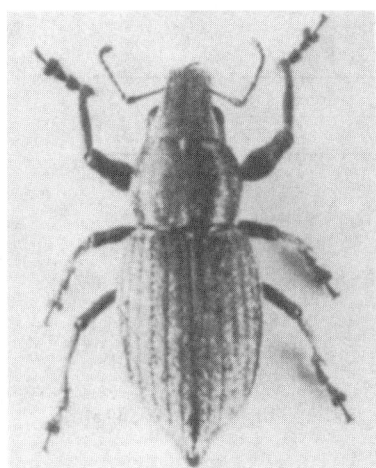

Fig. 1.   Adult white-fringed beetle. (*USDA*.)

As many as 300 grayish and long-headed beetles, ranging up to 1/2 inch (12 millimeters) in length and with pale white striping on the sides, may be found on one plant, usually feeding on the margins of leaves and milling about on the soil just under the plant. In the spring, large numbers of legless white grubs, ranging up to $\frac{1}{2}$ inch (12 millimeters) in length, will attack the lower stems and taproots of plants or even the planted seeds. An invasion like this will cause the plant to wilt and die very soon.

Cultural controls include crop rotation. Plants not preferred by these beetles include pasture grasses, small grains, and maize (corn). Corn should be planted in solid stands and not interplanted with a legume. Susceptible crops should not be planted in the same field more often than once every 3 to 4 years. Organic matter should be added to the soil. An economical method is to plow under a winter cover crop. This will help to reduce larval damage.

The beetles damage plants growing in well-drained, sandy loam more than those in heavy clay soils. A very dry summer can slow down development of the beetles. A heavy, week-long rain can kill many small larvae.

Larvae prefer root crops and plants with taproots to plants with fibrous roots. They may completely sever the main root. They sometimes feed on roots of young peach, pecan, tung, and willow trees, and on parts of dead plants in the soil. The larvae live in the soil. They rarely cause uniform damage over a large continuous area; sometimes they fully destroy plants in one part of a field and leave plants elsewhere in the field practically untouched. Beetles that feed on groundnut (peanut), cocklebur, ragweed, soybean, and strawberry lay large numbers of eggs; those that feed on cowpea, lespedeza, blackberry, and tobacco lay few eggs.

WHITEOUT.   See **Atmospheric Optical Phenomena**.

WHITE SQUALL.   See **Fronts and Storms**.

WHITE-TAILED DEER.   See **Deer**.

WHITETHROAT.   See **Warbler**.

WHITETIP SHARK.   See **Sharks**.

WHITING.   See **Codfishes**.

WIDGEON.   See **Waterfowl**.

WIEN BRIDGE.   See **Bridge Circuits (Electrical/Electronic)**.

WIEN BRIDGE OSCILLATOR.   A phase-shift, feedback oscillator that employs a Wien bridge circuit as the frequency-determining element.

WIEN LAWS.   From a study of the spectral energy distribution of thermal radiation, W. Wien, in 1896, arrived at three laws relating to the radiation from a black body.

1.   The wavelength λ_m of the spectral distribution, for which the radiation had greatest intensity, is inversely proportional to the absolute temperature T of the black body:

$$\lambda_m T = A$$

Thus as the temperature rises, the "peak" of the distribution curve is displaced or shifted toward the short-wavelength end of the spectrum. This is commonly called Wien's "displacement law." The value of the "displacement constant" is 2.8978 meter-degree absolute.

2.   The emissive power of a black body within the maximum intensity wavelength interval $d\lambda_{\max}$ varies as the fifth power of the absolute temperature

$$dE_\lambda {(\max)} = BT^5 d\lambda$$

3.   Wien's third law expresses the spectral energy distribution of the radiation from the black body at temperature T, as follows:

$$E_\lambda d\lambda = c_1 \lambda^{-5} e^{c2/\lambda T} d\lambda$$

in which E_λ is the emissive power within the wavelength interval $d\lambda$ and c_1 and c_2 are constants to be empirically determined.

The first and second laws are in accord with thermodynamic theory and with Planck's equation, and also agree very accurately with experiment. If $ch/k > \lambda T$ in the Planck radiation law, its denominator is very large compared to unity and the result is identical with Wien's third law. It thus holds for small values of λT but disagrees with experiment at long wavelengths. See also **Planck Radiation Formula**.

WIEN, WILHEM (1864–1928).   Wien was a German physicist who received his doctorate from the University of Berlin. He is known for his work on radiation. In 1896, he arrived at three laws relating to the radiation from a black body, now known as Wien Laws. Wien received the Nobel Prize for Physics in 1911 for his discoveries regarding the laws governing the radiation of heat. In addition, Wien made significant contributions in the study of cathode rays, X rays, and canal rays.

See also **Bridge Circuits (Electrical/Electronic)**; **Wien Bridge Oscillator**; and **Wien Laws**.

J. M. I.

WIGNER FORCE.   Short-range nuclear force of nonexchange type postulated phenomenologically as part of the interaction between nucleons. Postulated exchange forces are Bartlett, Heisenberg, and Majorana forces.

WIGNER NUCLIDES.   A special case of mirror nuclides. Pairs of odd-mass number isobars for which the atomic number and the neutron number differ by one, and in which the numbers of protons and neutrons are so related that each member of the pair would be transformed into the other by exchanging all neutrons for protons and vice versa.

WILCOXON'S TEST.   A test of the difference in means of two sets of observations, based on their rank order and therefore distribution-free.

WILD BOAR.   See **Swine**.

WILDEBEEST.   See **Antelope**.

WILLEMITE.   The mineral willemite is a zinc silicate, Zn_2SiO_4, occurring in hexagonal prisms, as masses or scattered grain. It is a brittle mineral with conchoidal fracture; hardness, 5.5; specific gravity, 3.9–4.2; subvitreous luster; usually some shade of yellow, yellowish-green, green, or reddish-brown, but may be colorless, white, or blue to nearly black; transparent to opaque. Much willemite is strongly fluorescent in yellow or yellowish-green hues. Willemite occurs associated with other zinc materials in Belgium, Algeria, Zaire, South West Africa, and Greenland. In the United States, except for three occurrences, one in Colorado, one in New Mexico, and one in Utah, Sussex County, New Jersey, is the only locality in the United States for willemite and is the only one in which that mineral is found in quantity. Here it is found associated with zincite and franklinite, forming an important ore of zinc. It was named by the French mineralogist, Michel Lévy, in honor of King William the First of the Netherlands.

WILLIWAW. See **Fronts and Storms**.

WILLOW TREES. Members of the family *Salicaceae* (willow family), these trees are of numerous species, many of which are found in Europe, Asia, and in several areas of the Americas. The willow usually is a medium-to-large tree, attaining a height of about 30 to 50 feet (9 to 15 meters), and a trunk diameter widely ranging from 3 to 30 inches (7.6 to 76 centimeters). As shown in Table 1, outstanding specimens attain much greater dimensions. The leaf of the willow is long, narrow, and swings gracefully in a breeze. The pendulous weeping willow (*Babylonica*) has

TABLE 1. RECORD WILLOW TREES IN THE UNITED STATES[1]

Specimen	Circumference[2]		Height		Spread		Location
	Inches	Centimeters	Feet	Meters	Feet	Meters	
Arroyo willow (1975) (*Salix lasiolepis*)	43	109	27	8.2	20	6.1	Oregon
Autumn willow (1985) (*Salix serissima*)	35	89	48	14.6	44	13.4	Michigan
Bebb willow (1991) (*Salix bebbiana*)	101	257	23	7	27	8.2	Idaho
Black willow (1995) (*Salix nigra*)	400	1016	76	23.2	92	28	Michigan
Bonpland willow (1999) (*Salix bonplandiana*)	169	429	63	19.2	74	22.6	Arizona
Coastal plain willow (1995) (*Salix caroliniana*)	106	269	52	15.8	56	17.1	North Carolina
Crack willow (1994) (*Salix fragilis*)	310	787	116	35.4	131	39.9	Michigan
Crack willow (1986) (*Salix fragilis*)	305	775	122	37.2	124	37.8	Michigan
Florida willow (1993) (*Salix floridana*)	10	25	20	6.1	23	7	Florida
Goodding willow (1993) (*Salix gooddingii*)	354	899	45	13.7	89	27.1	New Mexico
Hinds willow (1986) (*Salix hindsiana*)	58	147	50	15.2	32	9.8	Oregon
Hooker willow (1975) (*Salix hookerana*)	51	130	32	9.8	27	8.2	Oregon
Mackenzie willow (1999) (*Salix mackenzieana*)	17	43	30	9.1	27	8.2	Washington
Meadow willow (1976) (*Salix petiolaris*)	13	33	34	10.4	18	5.5	Michigan
Pacific willow (1999) (*Salix lasiadra*)	162	411	60	18.3	80	24.4	California
Peachleaf willow (1988) (*Salix amygdaloides*)	417	1059	58	17.7	82	25	Wisconsin
Purple-osier willow (1972) (*Salix purpurea*)	15	38	37	11.3	49	14.9	Michigan
Pussy willow (1983) (*Salix discolor*)	54	137	47	14.3	33	10.1	Michigan
Pussy willow (1991) (*Salix discolor*)	74	188	25	7.6	48	14.6	Rhode Island
Sandbar willow (1984) (*Salix exigua*)	69	175	36	11	46	14	Virginia
Scouler willow (1993) (*Salix scoulerana*)	168	427	40	12.2	50	15.2	Oregon
Scouler willow (1993) (*Salix scoulerana*)	144	366	64	19.5	47	14.3	Washington
Shining willow (1985) (*Salix lucida*)	130	330	74	22.6	81	24.7	Michigan
Silky willow (1991) (*Salix sericea*)	40	102	48	14.6	36	11	Virginia
Sitka willow (1999) (*Salix sitchensis*)	32	81	35	10.7	37	11.3	Washington
Tracy willow (1975) (*Salix tracyi*)	36	91	20	6.1	15	4.6	Oregon
Weeping willow (1990) (*Salix babylonica*)	309	785	117	35.7	116	35.4	Michigan
Weeping willow (1991) (*Salix babylonica*)	344	874	86	26.2	93	28.3	Michigan
White willow (1985) (*Salix alba*)	316	803	118	36	131	39.9	Michigan
White willow (1991) (*Salix alba*)	301	765	133	40.5	142	43.3	Michigan
Yellow willow (1985) (*Salix lutea*)	25	64	24	7.3	16	4.9	Idaho
Yewleaf willow (1996) (*Salix taxifolia*)	71	180	33	10.1	31	9.4	Arizona

[1]From the "National Register of Big Trees," American Forests (by permission).
[2]At 4.5 feet (1.4 meters).

long branchlets and long leaves that droop. It is native to Asia and is widely used there as an ornamental tree. The Thurlow weeping willow (*Salix elegantissima*) is quite similar to the pendulous weeping willow. The fruit of willow trees is a capsule, which breaks open when ripe, releasing pollen. Male and female flowers are found on separate trees. The roots are long and pliable.

Five species of willows can be classified as weeping willows, including the Pendula from China, the golden weeping willow, and the Japanese willow (*Salix urbaniana*). One of the principal American willows is the black willow.

The wood of willow trees generally is open-grained, hard, and has a brown-to-yellow color. The weight is about 30 pounds per cubic foot (480 kilograms per cubic meter). Willow wood is used in the United Kingdom for making cricket bats. It is also used for making artificial limbs and other products where toughness and nonshrinkage is particularly desired. The wood is high in glucoside content.

WILLY-WILLY. See **Fronts and Storms**.

WILSON'S DISEASE. A rare disorder, also known as *hepatolenticular degeneration*, which occurs only in one person per million population. This is an inherited disorder which involves a defect in copper transport and storage within the body. An excess of copper is deposited in most body tissues. Symptoms include episodes of jaundice, variceal bleeding, and hemolysis. At one time, a viable treatment was unknown, but in recent years, penicillamine therapy is used. D-Penicillamine acts as a chelating agent to bind copper and promote its excretion in the urine.

WILSON'S THEOREM. See **Number Theory**.

WIND AND AIR VELOCITY MEASUREMENTS. Wind velocity, direction, and pressure force are measured as important elements of weather observation and meteorology. Air velocity and other characteristics also are measured for industrial purposes, such as in the design and operation of ventilation, heating, and air-conditioning systems.

Anemometer. This term refers to a general class of instruments for measuring the speed (or force) of the wind in the atmosphere or of an air stream flowing in equipment and structures. Air speed, for example, is extremely important in wind tunnel research studies.

The *cup-type anemometer* operating on the principle shown in Fig. 1(a) and the vane-type instrument shown in Fig. 1(b) have been, for years, the traditional approaches to wind and air speed measurement. The cup-type instrument usually consists of from three to four hemispherical or conical cups mounted with their diametrical planes vertical and distributed symmetrically about the vertical axis of rotation. The rate of rotation of the cups, a measure of wind speed, is determined indirectly by gearing a mechanical or electrical counter to the shaft. The contact-type anemometer actuates an electrical contact at a rate depending upon the wind speed. The number of contacts occurring during a given time is measured. In gusty winds, the cup anemometer tends to register too high. In the *bridled-cup anemometer*, wind speed is measured by determining the drag exerted by the wind on a solid body. This instrument may consist of an array of cups about a vertical axis of rotation, the free rotation of which is restricted by a suitable spring restraining force. By proper adjustment of the force constant of the spring, it is possible to obtain an angular displacement that is proportional to wind velocity. The *pendulum anemometer* was invented about 1660 and consists of a plate that is free to swing about a horizontal axis in its own plane above its center of gravity. The angular deflection of the plate is a function of wind speed. The instrument is not used for station measurements because of the false readings that result when the frequency of the wind gusts and the natural frequency of the winging plate coincide. However, a version of this principle, the *normal-plate anemometer*, can be used for turbulence studies. The plate, restrained by a stiff spring, is held perpendicular to the wind, and its wind-activated motion is measured electrically. *Propeller-type anemometers*, of the type shown in Fig. 2, will determine both direction and speed of the wind.

Wind speed also can be measured from dynamic wind pressures with a *pressure-tube anemometer*. Wind blowing into a tube develops a greater pressure than the static pressure, while wind blowing across a tube develops a pressure less than the static pressure. The pressure difference is proportional to the square of the wind speed and is measured by a

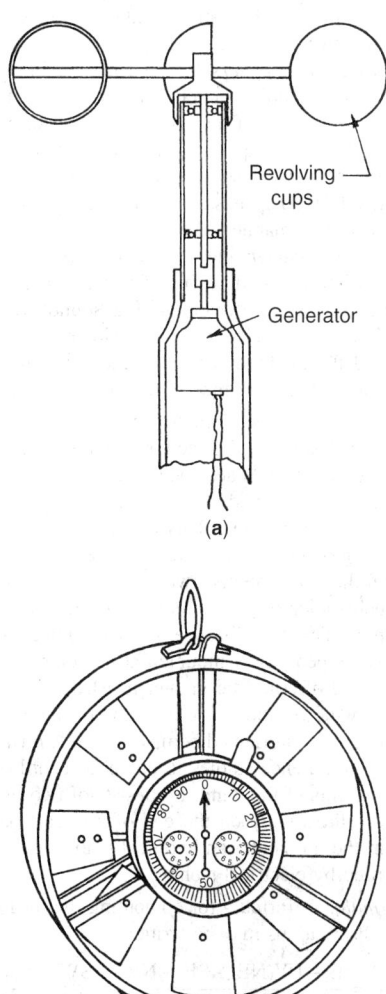

(a)

(b)

Fig. 1.   Anemometers: (**a**) cup-type; (**b**) vane-type.

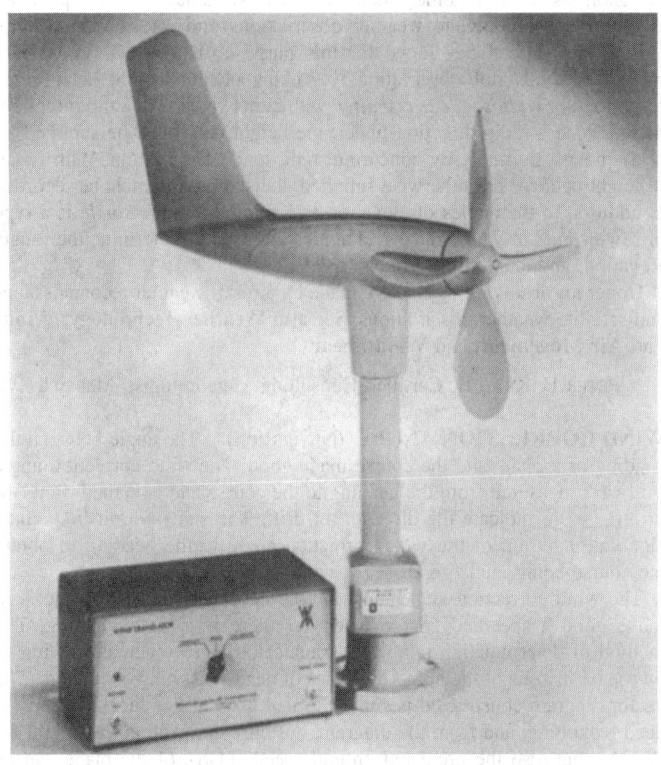

Fig. 2.   Wind speed and direction sensing instrument. (*Skyvane*.)

manometer. In connection with relative wind measurement for aircraft, see also **Airspeed Indicator**.

So-called *hot-wire anemometers* have been used in which the rate of heat transferred from an object at elevated temperature (such as a hot wire) to the air is a measure of air speed. Wind speed is determined by measuring either the current required to maintain the hot wire at a constant temperature, or the resistance variation of the hot wire while a constant current is maintained. Wind speeds as low as a few centimeters per second can be determined in this manner.

The *sonic-type anemometer* measures wind speed by means of the properties of wind-borne sound waves. The instrument operates on the principle that the propagation velocity of a sound wave in a moving medium is equal to the velocity of sound with respect to the medium, plus the velocity of the medium. The instrument has the advantages of a very short time-constant and absence of moving parts.

Wind Direction Indicators. Some of the anemometers just described have the ability to measure wind direction as well as velocity. There are, however, additional instruments designed strictly as direction indicators. A *wind cone* (or wind sleeve; wind sock) is a tapered fabric sleeve, shaped like a truncated cone and pivoted at its larger end on a standard, for the purpose of indicating wind direction. Because the air enters the fixed ends, the small end of the cone points away from the wind. Often seen at local airports, such indicators are simple and because of their size can be observed from considerable distance. A wind vane consists basically of an asymmetrically-shaped object mounted at its center of gravity about a vertical axis. The end offering the greater resistance to the motion of air moves to the downwind position. The direction of the wind is determined by reference to an attached oriented compass rose. A *bivane* is a sensitive wind vane used in turbulence studies to obtain a record of the horizontal and vertical components of the wind. It consists of two lightweight airfoil sections mounted orthogonally on the end of a counterbalanced rod that is free to rotate in the horizontal and vertical planes. The positions of the rod may be recorded by electrical techniques.

Lambert's Formula. A formula for computing the mean wind-direction from a series of observations may be written:

$$\tan \alpha = \frac{E - [W(NE + SE - NW - SW)\cos 45°]}{N - [S(NE + NW - Se - SW)\cos 45°]}$$

where α is the mean wind-direction, and each point of the compass replaced by the number of observations of wind from that direction.

High-Altitude Wind Measurements. Determination of wind speed and direction as well as other meteorological variables in the upper air is required for complete weather observations and forecasting. Several systems have been developed for this purpose. The *rawin system* is a method of winds-aloft observation by tracking a balloon-borne radar target, responder, or *radiosonde transmitter* with either radar or a radio direction-finder. With a radio direction-finder, the height data must be supplied by other means, normally by concurrent radiosonde observation. With radar, if height data are not otherwise supplied, the slant range must be recorded in addition to the angles of elevation and azimuth. *Radar-sonde* is a type of rawinsonde in which radar techniques are used to determine the range, elevation, and azimuth of a radar target.

Upper air observations also are made by aircraft weather reconnaissance and satellite weather observations. See also **Weather Technology**; **Winds and Air Movement**; and **Wind Shear**.

PETER E. KRAGHT, Certified Consulting Meteorologist, Mabank, TX.

WIND CORRECTION ANGLE (Navigation). The angle between the heading of a plane and the course made good. The wind correction angle is always measured from the heading to the course and is named right (+) or left (−) to indicate the direction of drift. The signs permit immediate algebraic addition of the wind correction angle to the heading to obtain the course being made good.

The wind correction angle must always be determined if a plane is to make good a specified course whenever there is a wind blowing. The method of determination is always graphical, either by actual drawing or by use of any one of the numerous types of dead-reckoning computers. The vector diagram constructed is similar to that used in graphical methods of dead-reckoning, and from the diagram, not only the wind correction angle (WCA), but also the predicted ground speed (PGS), of the plane can be determined.

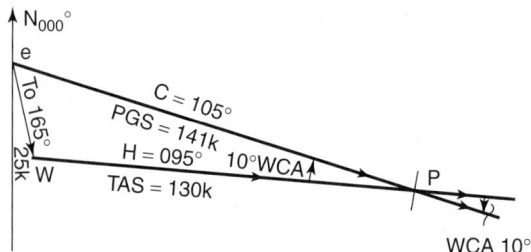

Fig. 1.   Velocity-vector diagram for determining wind correction angle and predicted ground speed.

The procedure may best be described by considering a specific situation. See Fig. 1. A plane is to proceed to a point on a rhumbline; bearing 105°; distance 180 miles (290 kilometers). The plane is to cruise at 130 knots; the wind is from 345°; speed 25 knots. The pilot wishes to know the wind correction angle, the heading of the plane, and the estimated time of arrival (ETA). The velocity-vector diagram is shown in the figure, which is constructed and labeled in accordance with the standard procedure of the U.S. Air Force. From a point e, a vector ew is drawn, representing the velocity with which the wind is moving the plane. From e, a line is drawn in the direction of the course to be made good. Then, from w, an arc is struck, using either dividers or a scale, with length equal to the true air speed of the plane (TAS). This arc cuts the course direction line in the point p. This completes the velocity-vector diagram, with ep representing the course to be made good and PGS; wp the heading and TAS; and the angle at p, measured from wp toward ep, representing WCA. Measurement with scale and protractor on the completed diagram gives PGS = 141 knots and WCA = 10° Right. Since, by definition, H + WCA = C, we have, in this case, H + 10° = 105° or H = 095°. Finally, with PGS = 141 knots, and the distance to be made good 180 miles, we have an expected elapsed time of 1 hour and 17 minutes (1^h17^m) for the flight. This interval added to the time of departure will give the ETA.

Throughout flight, the navigator must keep careful watch with the driftmeter, and if the drift is not equal to the predetermined wind correction angle, the wind must be determined and a new heading computed to make good the desired course. This procedure is illustrated under **Air Plot**.

See also **Course**; **Dead Reckoning**; and **Navigation**.

WIND EROSION.   See Soil.

WINDHOVER.   See Falcon.

WIND POWER. As with most other alternative energy sources that were aggressively researched during the energy crisis of the early 1970s, the interest in wind power has waned considerably in recent years. A number of new design concepts were produced during the energy-awareness period and one day may receive considerable attention. Directly attributable to energy from the sun, the winds arise from atmospheric thermal differences.

The Macroscopic View. In reviewing the macroscopic availability of total wind energy available on the Earth, Gustavson (1979) starts with the solar flux at the Earth's distance from the sun—at a value of 1400 watts per square meter which, when averaged over the Earth's surface, amounts to a total intercepted flux of 1.8×10^{17} W (350 watts per square meter). This, then, is the upper limit on all solar-energy-based processes. A number of authorities essentially agree that only about 2% of this gross amount of solar energy reaching the Earth results in gross global wind energy. Thus, the gross wind energy for the Earth becomes 3.6×10^{15} W, an upper limit on the theoretical potential of wind energy technology. For utilization of wind energy in terms of concepts proposed to date, only the near-surface winds would be tapped. Ellsaesser (1969) estimated that 35% of total global wind energy is dissipated within one kilometer (0.62 mile) of the earth's surface. It was further estimated (Brunt, 1939) that about 90% of this 35% component is dissipated close to or essentially adjacent to the "rough" surface of the earth where there is interference with a smooth flow of wind currents. Applying the foregoing considerations, the near-earth wind energy available over the earth approximates 1.3×10^{15} W.

The atmosphere is a great heat engine fueled by the sun. As in all such engines, only a fraction of the energy supplied can be transformed into work. The remainder must be rejected from the system, in this case by the loss of heat by long wave radiation to space from the earth and its atmosphere. The winds and ocean currents represent the work done by the engine and may be thought of as its flywheel. The windmill or aerogenerator converts the motion of translation of the winds of this heat engine into a rotation capable of driving an alternator. Since the earth-atmosphere system itself disposes of the rejected heat, wind power systems do not have to contend with the waste heat that presents problems in other kinds of power plants.

Based upon estimates (Putnam, 1948), there is 20 times more potential for windpower than for water power. An estimate of the annual average wind power available over the continental United States and coastal waters is given in Fig. 1. The estimates for the mountainous western half of the country and for the Appalachians are subject to considerable uncertainty. This is true because the values given are based upon measurements of standard weather stations, usually located at or near airports, the locations of which are usually selected so as to avoid high winds. For example, the map shows the Columbia River Gorge area lying east of Portland, Oregon as having only 100 watts per square meter per year of wind power, although actual wind measurements in the gorge show clearly a power level several times greater than this. Shoreline winds along the Oregon coast are also stronger than shown.

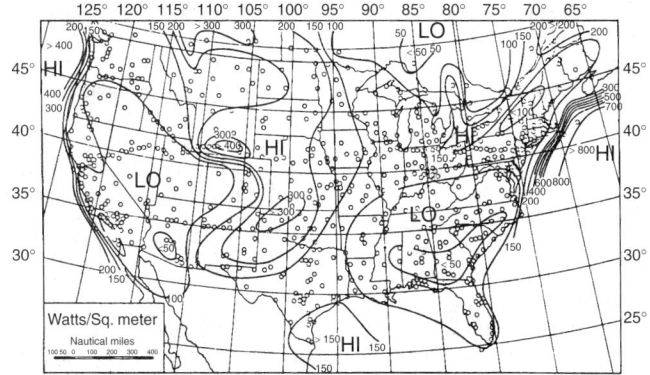

Fig. 1.   Available wind power in the United States (annual average). Markings on contour lines are watts/square meter.

Importance of Wind Speed. A very significant factor pertaining to the Earth's winds as a source of electrical energy is that the power produced is proportional to the cube of wind speed. The result is readily derived from the kinetic energy of wind. The mass of air m with speed V and density r flowing per unit time through area A swept by the blades of a conventional horizontal-axis wind turbine is rAV. Thus, the kinetic energy of this mass of air is given by:

$$\text{Kinetic energy} = \tfrac{1}{2} \, mV^2 = \tfrac{1}{2}\rho AV^3$$

As shown from the theoretical analysis by Betz (1919), the maximum fraction of this kinetic energy that can be extracted from the wind is $\frac{16}{27}$ or 0.593. Thus, the theoretical maximum energy output of a wind turbine is given by:

$$\text{Theoretical maximum power output} = 0.297 \; \rho AV^3$$

The process of energy conversion leads to power reductions that vary with the type of wind turbine and aerogenerator and which are roughly one-third of the theoretical maximum output. Hence:

$$\text{Available power output} = \left(\tfrac{2}{3}\right) 0.297 \; \rho AV^3 \simeq 0.2 \; \rho AV^3$$

If the diameter of the blades of the rotor system is D, then the foregoing equation becomes:

$$\text{Available power output} \simeq 0.05 \; \pi \rho D^2 V^3$$

The power available for a given wind speed is thus proportional to the square of the rotor diameter. Additional quantities requiring definition are:

$$\text{Power coefficient, } C_p = \frac{\text{Power Output of Wind Turbine}}{\tfrac{1}{2}\rho AV^3}$$

$$\text{Overall power coefficient, } C_{op} = \frac{\text{Power Output at Generator}}{\tfrac{1}{2}\rho AV^3}$$

The coefficient, C_{op}, thus includes the inefficiencies of the transmission and generator.

The term *windmill* was used for centuries to describe an assembly of equipment used for extracting power from the wind. More modern terms include *wind turbine generators* and *aerogenerators*.

Since the power generated by the wind is proportional to the cube of the wind speed, it is obviously important to locate areas of persistently high winds. For example, a 10 meter/second (22 miles/hour) wind will produce 8 times as much power as a 5 meter/second (11 miles/hour) wind. Surveys to locate suitable wind power sites have been conducted in the United States, France, and Great Britain. Wind measurements, especially in rough terrain, must be taken with great care. Misreading of wind power potential has resulted in certain past failures of wind power systems. See also **Winds and Air Movement**.

Velocity and Power Duration Curves. The single most valuable piece of information that can be obtained about a potential wind power site is its power duration curve, which is conveniently obtained from its velocity duration curve, shown in Fig. 2. The horizontal axis gives the number of hours in the year (8,760) and the vertical axis is wind velocity. Thus, the graph gives the number of hours in the year during which the indicated wind velocity was exceeded. A power duration curve is similar because power is proportional to the cube of wind speed. See Fig. 3. The power duration curve thus gives the number of hours per year that the power output exceeds the indicated values, which are obtained in part by cubing the wind speed. This particular curve is for a site with relatively little time with low wind during the year. During the interval ge, the wind is too light to produce significant power. At the wind speed corresponding to g (say 3 meters/second or 7 miles/hour), appreciable power is being generated. The point fg is called the "cut-in" point. With higher wind speeds, the power output is greater and at c the aerogenerator is operating at its rated capacity.

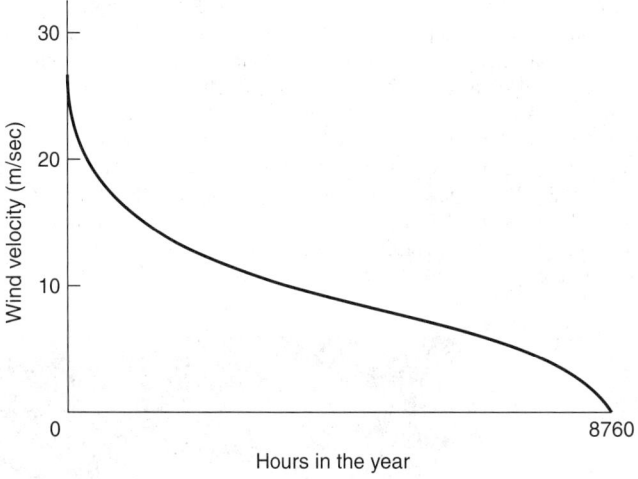

Fig. 2.   Wind velocity-duration curve.

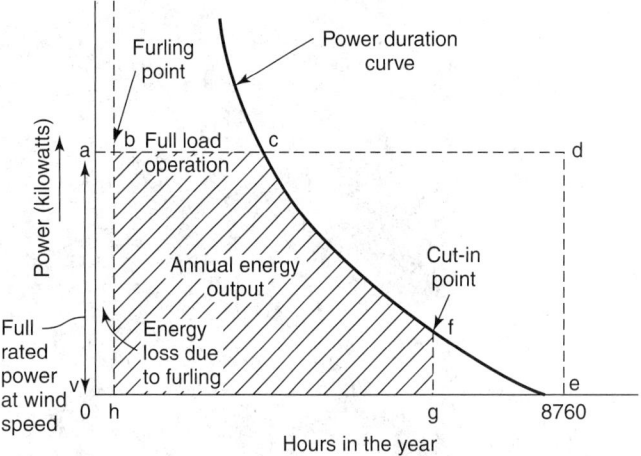

Fig. 3.   Wind power-duration curve.

At greater wind speeds, the output is generally held constant at this value (for component cost optimization reasons with conventional components) for full-load operation by adjusting the pitch of the blades, or by some other method. At some much higher wind speed, b, called the "furling point," perhaps about 27 meters/second (60 miles/hour), it is advisable to shut down the plant to avoid damage.

In the diagram, the hatched area bcfgh under the power duration curve represents the actual annual output of energy to the same scale as the rectangle adeo represents the annual output if the plant were running at full-rated power throughout an entire year. The ratio of area bcfgh to area adeo is the annual plant load factor and multiplication of this by 8,760 gives the specific output in kilowatt hours per year per kilowatt. Thus, the specific output is the equivalent number of hours of full-load operation.

Milestones and Wind Power

It is estimated that wind power has been utilized by people for at least 15 centuries, but that more scientific and engineering information concerning the utilization of this source of power has been accumulated during the past few decades than during the early centuries. However, some fundamental principles have been known for at least a few hundred years, as reflected by the refinements of earlier designs (so-called Dutch windmills and the small windmills, estimated at over 6 million units, used at one time throughout the farmlands of North America and in other areas of the world). Wind power became essentially obsolete during most of the present century because of the convenience and relatively low cost of electric power, which was brought to the hinterlands via very ambitious programs of rural electrification. It was not until many countries slowly awakened to the prospects of energy shortages and high energy costs that wind power, as a component of solar power, created new interest.

Some progress did occur nevertheless during the second quarter of this century when wind power, in general, had descended almost to the status of a curiosity. For example, Darrieus, built an aerogenerator at Bourget, France in 1929 that had a tower 20 meters (66 feet) high, with blades of the same diameter. A direct-current generator rated at 0.015 megawatts at wind speed of 6 meters/second (13.4 miles/hour) was wind-powered. During the late 1950s, another, larger unit rated at 0.8 megawatt and for operation at higher wind speeds was also built in France.

A large aerogenerator was installed at Yalta (near the Black Sea) in the former U.S.S.R. in 1931. The tower was 23 meters (75 feet) high, with blades 30.5 meters (100 feet) in diameter were used. An alternating current generator (rating not precisely known) was designed for operation at 11 meters/second (24.6 miles/hour).

In 1920, in Germany, an aerogenerator designed by Kumme and consisting of six blades was built. This unit included a generator on the ground and a long, vertical, flexible shaft topped by a bevel gear to transmit torque to the ground. A bit later, an aerogenerator was proposed by Honnef of Berlin, but was not built. This design involved 5 wind turbines, each nearly 40 meters (250 feet) in diameter, supported on a single large tower, 305 meters (1000 feet) high. See Fig. 4. The inventor rated the potential capacity of his design at 50 megawatts. Also, in Germany, several aerogenerators were designed by U. Hütter (Univ. of Stuttgart), built and constructed during the early 1960s. Capacities up to 0.1 megawatt and fiberglass blades were part of the design. These machines were reported to have good operating characteristics and resistance to fatigue. There remains much interest in Hütter's concepts.

In Great Britain, an aerogenerator using a pneumatic transmission system and designed by French engineer, J. Andreau, was built in the mid-1950s for the Enfield Cables organization. The tower was 30 meters (100 feet) high; the blades were 24 meters (80 feet) in diameter; and the unit had a rated capacity of 0.1 megawatt at 13.4 meters/second (30 miles/hours). The unit operated successfully for a number of years in Algeria.

In Denmark, during World War II, a total of 18 aerogenerators were constructed in the 70 to 90 kilowatt range. These supplied direct-current power to the dc power grids. Another aerogenerator, designed to produce 0.2 megawatt at speed of 15 meters/second (34 miles/hour) was constructed at Gedser. For this unit, the tower was 26 meters (85 feet) high; the blades (3 in number) were 24 meters (79) feet in diameter; and the electric generator located on a horizontally rotating platform atop the tower produced 380 volts ac. Although the unit was shut down in 1968, it was still standing as of the mid-1970s.

In the United States, the largest wind turbine ever built, as of that time, was the often-cited Smith-Putman generator, located on Grandpa's Knob near Rutland, Vermont. It was built in the 1940s and was in operation over test-run intervals in the period 1941–1945, after which it was continuously operated as a routine generating station on the lines of Central Vermont Public Service Corporation. It was ultimately shut down because of loss of a blade.

In 1974, the National Aeronautics and Space Administration (NASA) constructed an experimental wind turbine generator near Sandusky, Ohio. The tower was 30 meters (100 ft) high; the rotor was 38 meters (125 ft) in diameter. Planned output was 100 kW at 460 V, 3-phase, 60 Hz alternating current. Before being shut down, the unit gained considerable experience in designing and operating large aerogenerators.

Early in the revived interest in aerogenerators during the 1970s in the United States, ERDA (Energy Research and Development Administration)

Fig. 4.   Wind turbines arranged as a wind farm.

concentrated its attention on larger machines. However, in mid-1976, a program was established at the Rocky Flats laboratory (near Denver, Colorado) to evaluate the performance of many types of small wind machines. As pointed out at the time, whereas the large-wind-machine program is one of sequential development, the small-scale program is intended to emphasize parallel development of competing machines. Also, later in the 1970s, U.S. energy officials expressed an increasing interest in the wind-farm concept which involves lower-capacity machines.

Wind Power for Cargo Ships. Also, during the energy crisis period of the 1970s, much interest had been developed in wind-powered ships. Japan, Germany, and the United States researched the potentials of wind energy and a few experimental craft were built. There had been a lull in sailing craft research, except for pleasure vessels, since the very early 1900s. The final demise of large sailing ships is marked by some authorities as occurring at the end of World War I. They were forced into retirement by the advantages of low-cost fossil fuels in providing steady power that improved the reliability of scheduling arrivals and departures. Authorities estimated that between 20 and 30% of the operating costs of modern powered vessels goes into fuel. The some 25,000 commercial vessels of more than 1000 deadweight tons used worldwide in international trade consume between 5 and 8% of the total oil consumption in the world (communist countries excepted). Sail power is not only attractive from the standpoint of fuel savings, but also because the energy source is constantly renewable and provokes no significant environmental problems. It has been estimated (Bergeson, 1979) that a reduction of 10–20% in consumption of bunker oil by ships could reduce cargo transportation costs (1979 $) by as much as $5.5 billion per year. Another factor favorable to wind power for ships is the large bank of knowledge in this area. As of the early 1900s, sailing ships had been brought to an advanced state of development—and this knowledge is retrievable. To this is added prior experience, which has greatly enhanced the technology of land-based wind machines. Wind-powered ships of the future also can take advantage of advances in materials (as for sails), of hull antifouling chemicals, of auxiliary electrical, mechanical, and control equipment—not to mention major improvements in weather forecasting and ship-to-shore communications, which can be invaluable in plotting the most wind-efficient courses to follow.

Much pioneering research of the recent period in sailing ship technology was carried out at the Institut für Schiffbau in Hamburg, Germany. In wind tunnels and towing tanks, the performance of various hull configurations, including 1902 as well as recently designed hulls (as designed by Wilhelm Prölss of the Institute) were tested. As pointed out by Bergeson (1979), speed was measured at various angles of course with varying wind velocities and sea states, and actual log book data for the 1902 rig were correlated with the tank and wind tunnel test data. As a result, the propulsive force of the 1902 vessel was improved twofold. Further, the mean voyage speed was increased by 58% under sail alone, and up to 82% with the use of modest auxiliary power during part of the voyage time. Polar diagrams, which show ship speed under varying wind velocities, sea states, and headings, clearly summarized the gains in performance. In 1974, the U.S. Maritime Administration funded a study at the University of Michigan on the "Feasibility of Sailing Ships for the American Merchant Marine." Using information obtained from the German research and assuming a modern square rig of the same aerodynamic configuration as that designed by Prölss, the project postulated performance characteristics and made estimates of construction and operating costs of cargo sailing ships of 15,000, 30,000, and 45,000 deadweight tons. The report found no technical barriers to the development of such ships.

There are several schools of thought as regard the rigging (or arrangement of wind collectors) to be used on future wind-powered cargo vessels. Naval architects and marine engineers have looked back nearly a century or more to the most successful of the sailing ships whether they be square-rigged or of schooner configurations. Even in borrowing from older, tested designs, taking modern technology into consideration, the new designs are strikingly different. In some designs, rotating masts, fully equipped with sensors and automatic controls have been envisioned. See Fig. 5. Some authorities have a preference for a catamaran design. For example, Bergeson (1979) gives three advantages of a catamaran: (1) about half of the cargo of a catamaran can be considered as highly effective ballast—operating on a huge lever arm equal in length to the distance between the hull centerlines and thus creating enormous righting moment and impressive sail-carrying ability, as compared with the craft's low hull resistance, which results in impressive speeds for these craft; (2) when

beached, a catamaran gains immense stability with both bows grounded, whereas a trimaran, for example, can lurch violently, creating large shock loads; and (3) the catamaran is safer and more practical in larger sizes designed to carry heavy cargoes and water ballast. Bergeson envisions that a catamaran might be about 67 meters (220 feet) long with a deadweight capacity equivalent to that of 54-meter (180-foot) single-hull design. It is envisioned that such a catamaran with a light load would have a maximum potential speed of 25 knots in winds of 25 to 30 knots. Probably over the course of a year, the catamaran would achieve an average speed in excess of 12 knots.

Aerogenerator Configurations

Horizontal-Axis Rotors. Three well-established types are shown in Fig. 6. The modern propeller when used in a wind turbine utilizes variable-pitch in order to regulate rate of shaft rotation to achieve maximum efficiency. Other types include the flexible sailwing rotor and a design consisting of a circular array of 48 thin and narrow blades stressed in tension by a surrounding circular hoop, which serves as a rim drive to rotate at high speed and without step-up gearing the shaft of a generator. Various factors that cause dynamic loading of the rotors are shown in Fig. 7. The blades are attached to the hub in a flexible manner, permitting them to move to and fro in the direction of the wind in a vibration mode known as *flapping*. As a group, the blades of the horizontal-axis rotors are driven by lifting action. As a result, the speed ratio (ratio of blade tip speed to wind speed) is high, reaching a value as large as 12. The efficiency of a wind turbine at these large speed ratios is high.

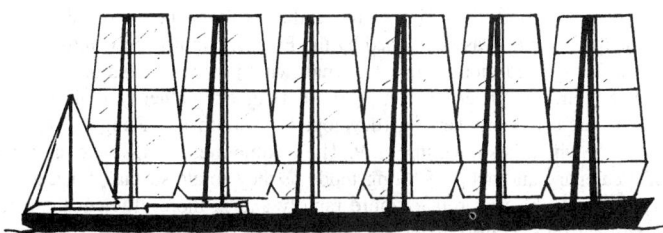

Fig. 5. Bulk cargo sailing ship of 45,000 deadweight tons. This design resulted from University of Michigan study. Tripod masts were used instead of rotating masts as used in the earlier *Dynaship* designed by Pröiss of Germany. Rotating masts can be used to set sails at most effective angle. In the tripod design, sails are trimmed by mechanically "bracing" the yards about a pivot on the leading side of the mast. Furling or unfurling sails is accomplished by rotating a vertical rod (jack stay) set between the yards and forward of the mast. (*After Bergesson.*)

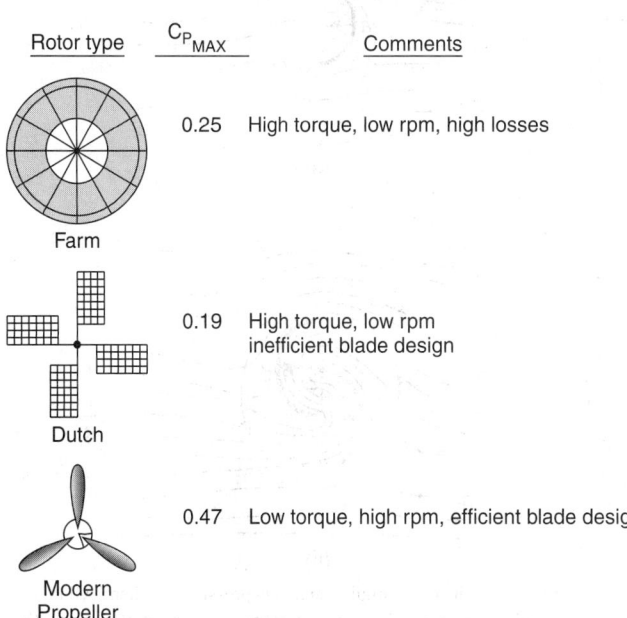

Fig. 6. Power extractable from three types of horizontal-axis rotors. (*From Savino.*)

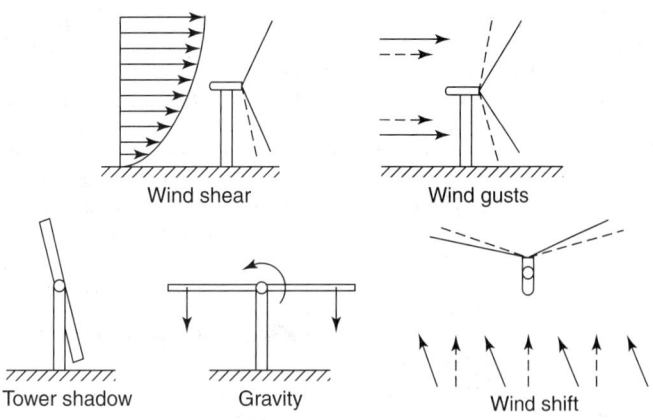

Fig. 7.   Various factors that cause dynamic loading of rotors.

Vertical-Axis Rotors. The best known vertical-axis rotor is made up of two identical hemicylinders with their axes vertical. The patterns of air flow around and air pressure on two such hemicylinders are shown in Fig. 8. As described by Klemin (1925), when the configuration is as shown in Fig. 8(a), there is a reduced pressure behind hemicylinder a as it moves against the wind, reducing the torque. If, however, there is an air passage between the hemicylinders, as in Fig. 8(b), the air pressure behind hemicylinder a is increased rather than reduced, with the result that the torque is about three times greater than without the air passage. This vertical-axis rotor was developed by the Finnish engineer Savonius (1931) and is being used increasingly for small wind-power installations.

The French engineer, Darrieus (1931), designed another type of vertical-axis rotor, a modern version (developed by South and Rangi, 1971 and 1972) of which is shown in Fig. 9. The flexible metal strips take the form of a catenary and act in a lifting mode as they rotate so that, for a given wind speed, the unit rotates more rapidly and is more efficient than the Savonius rotor. Unfortunately, the Darrieus rotor is not self-starting even in high winds.

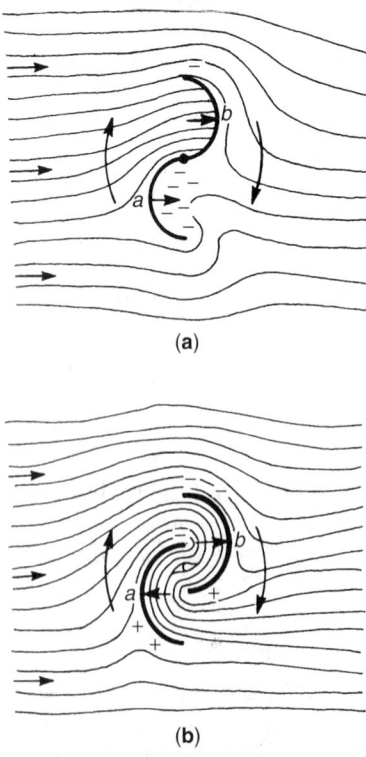

(a)

(b)

Fig. 8.   Patterns of air flow around and air pressure on hemicylinders. **(a)** Reduced pressure behind hemicylinder as it moves against the wind; **(b)** air passage between hemicylinders increases pressure behind hemicylinder with the result that torque is about three times greater than without air passage.

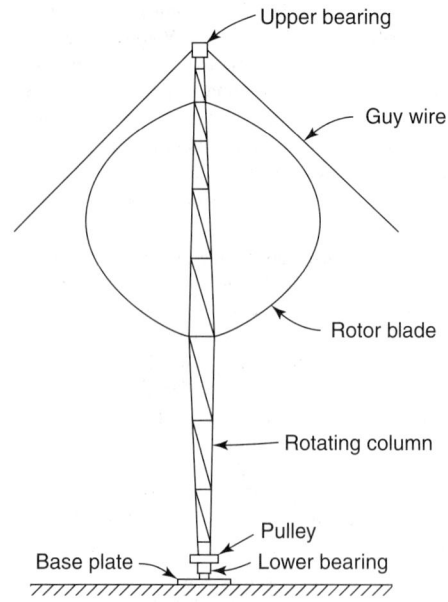

Fig. 9.   Vertical-axis rotor developed by South and Rangi.

The performance characteristics of various rotor configurations (according to Wilson and Lissaman, 1974) are shown in Fig. 10. This analysis suggests that the two-blade type is the most efficient type now available. Other studies (General Electric, 1975) and Kaman Aerospace (1975) reached the same conclusion.

Transmissions. The rate of rotation of large wind turbine generators operating at their rated capacity can be controlled by varying the pitch of the rotor blades. Because optimum generator output requires much greater rates of rotation (about 1800 revolutions per minute), it is necessary to increase greatly the lower rotor rate of turning. Among transmission options are mechanical systems involving fixed ratio gears, belts, and chains singly or in combination; or hydraulic systems involving fluid pumps and motors. Fixed-ratio gears are often preferred for top-mounted equipment because of their high efficiency and low system risk. For bottom-mounted equipment, which requires a right-angle drive, transmission costs can be reduced substantially by using large-diameter bearings with a ring gear mounted on the hub to serve as a transmission to increase rotor speed to generator speed. This approach has considerable flexibility.

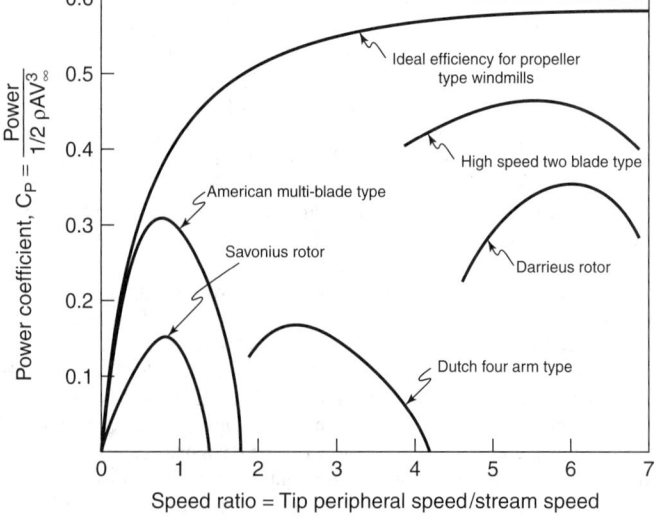

Fig. 10.   Performance characteristics of various rotor configurations.

Generators. Either constant- or variable-speed generators may be used, by variable-speed units are costly. Among constant-speed generator options are synchronous, induction, and permanent-magnet units. Like other components of aerogenerator systems, the transmission and generator technology is subject to continuing research and improvement.

Preferred Site Characteristics. Considerable attention has been given to those factors which effect aerogenerator capacity and efficiency. These are well summarized by Savino (1974): (1) a site should have a high annual wind speed; (2) there should be no tall obstructions for a mile or two upwind; (3) the top of a smooth, well-rounded hill with gentle slope lying on a flat plain or a site located on an island in a lake or sea is usually an excellent site; (4) an open plain or an open shoreline may be a good location; (5) a mountain gap which produces wind funneling is good. Some of these conditions are shown in diagrams of Fig. 11.

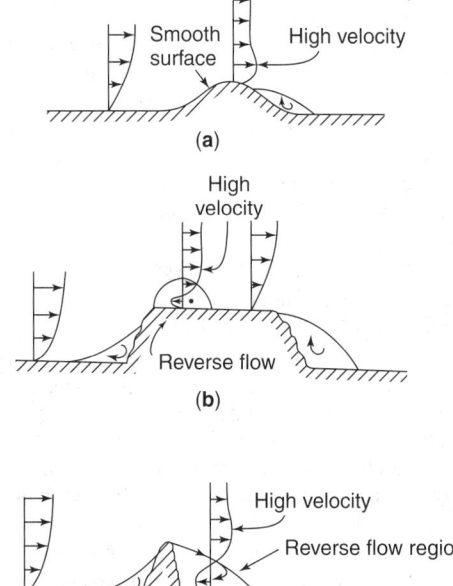

Fig. 11. Factors affecting wind-power locations: (**a**) well-rounded hill or ridge makes a favorable site; (**b**) hill or ridge with abrupt sides makes an unfavorable site; (**c**) sharp peak may possibly be suitable.

Ecological signs of a good wind site include: (1) flagging of trees, in which the branches stream downwind, as frequently found along shorelines where there is a prevailing on-shore wind; (2) throwing of trees, in which the main trunk is permanently bent downwind; (3) wind clipping, which causes abnormally low trees with tops of uniform height; (4) tree or bush carpets, in which vegetation never grows taller than a low scrubby bush. Mountainous coastal terrain provides some of the very best wind-power sites.

Aerogenerator Farms. The wind is a low density source of power compared with water. For instance, the energy developed is the product of the density of the air and the cube of wind speed. The density of air near sea level is about one-thousandth that of water—thus about 1,000 times more air than water must pass through a turbine at the same speed to generate equal amounts of power. This means that wind turbines must be larger than water turbines to achieve equal power output.

If wind power is to be considered as more than a source of electrical energy for isolated homes and farms, it becomes apparent that arrays or "farms" of wind turbines will be needed to generate enough power to justify feeding it into existing networks. A small farm consisting of 16 vertical rotor units of the Savonius type is shown in Fig. 12. The units are supported in a vertical position by an inexpensive system of guy wires. The only compression members are the vertical shafts of the rotors which are stiffened by the two hemicylinders of the Savonius rotors.

Wind Energy Storage. This is a problem common to most solar energy systems. A logical possibility is pump storage of water, similar to that currently used in conventional hydroelectric power plants. The use of wind to compress air and store it in caverns or preferably aquifers has been studied. The use of a very strong flywheel made up of radially oriented rods of high tensile strength has been explored by Rabenhorst. Other possibilities include use of the generated electricity to produce hydrogen and possibly other fuels.

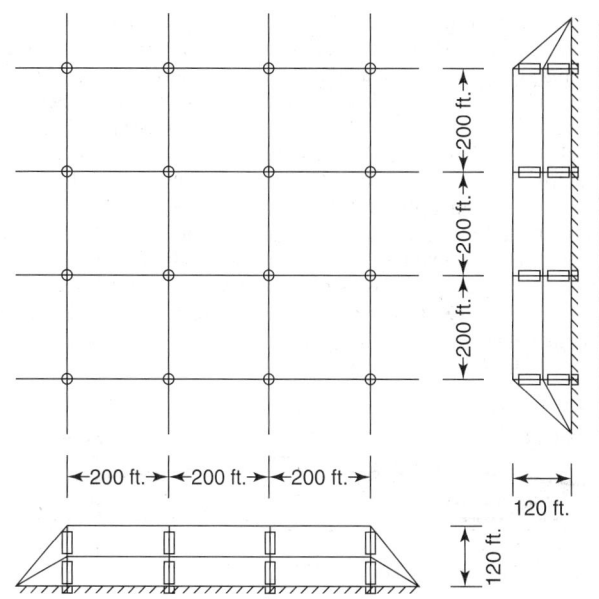

Fig. 12. Layout of small aerogenerator farm. (Note: 120 feet = 36 meters; 200 feet = 61 meters.)

Additional Reading

Bergerson, L.: "Sail Power for the World's Cargo Ships," *Technol. Rev. (MIT)*, **81**, 5, 22–36 (1979).

Cole, T.: "Planesail: New Era of Wings at Sea," *Pop. Mech.*, 13 (June 1989).

Cole, T.: "Blade Cuts Windpower Costs," *Pop. Mech.*, 17 (July 1990).

Dane, A.: "Landlubber's Sailwing," *Pop. Mech.*, 18 (September 1990).

Dane, A.: "Wind Beneath Its Wings," *Pop. Mech.*, 15 (April 1991).

Davis, G.R.: "Energy for Planet Earth," *Sci. Amer.*, 54 (September 1990).

Frank, D.: "Blowing in the Wind," *Pop. Mech.*, 41 (August 1991).

Hills, R.L.: "Power from Wind: A History of Windmill Technology," Press Syndicate of the University of Cambridge, New York, NY, 1996.

Kahn, R.D.: "Harvesting the Wind," *Technology Review (MIT)*, 56–61 (November 1984).

Klemin. A." "The Savonias Wing Rotor," *Mechanical Engineering*, **47**, 11, 911–912 (1925).

Miller, P. and R.H. Ressmeyer: "Our Electric Future," *Nat'l. Geographic*, 80 (August 1991).

Mills, A.: "Sunlight: The Energy Source of the Future," *Univ. of Wales Review*, 39 (Autumn 1988).

Moretti, P.M. and L.F. Divone: "Modern Windmills," *Sci. Amer.*, 110–118 (June 1986).

Patel, M.R.: "Wind and Solar Power Systems," CRC Press, LLC., Boca Raton, FL, 1999.

Putman, P.C.: "Power from the Wind," Van Nostrand Reinhold, New York, NY, 1948.

Staff: "Revised Solar Energy Budget," *Science*, 1403 (March 23, 1990).

Vosburgh, P.N.: "Commercial Applications of Wind Power," John Wiley & Sons, Inc., New York, NY, 1983.

Weinberg, C.J. and R.H. Williams: "Energy from the Sun," *Sci. Amer.*, 146 (September 1990).

Wilson, R.E. and P.B.S. Lissman: "Applied Aerodynamics of Wind Power Machines," Oregom State University, Corvallis, Oregon, 1974.

WIND ROSE. See **Winds and Air Movement**.

WINDS AND AIR MOVEMENT. Air motion relative to the surface of the Earth is termed wind. Since only slight horizontal variations in the density of the air can occur ordinarily, and the total volume of the atmosphere is substantially constant, it is evident that a wind is necessarily a circulation of the atmosphere, i.e., any movement of the air in one direction must be offset by a return current elsewhere. All such movements are the result of thermal energy due to the heat of the Sun.

Although the winds are caused primarily by differences of temperature, they are more directly related to differences of atmospheric pressure, through the continuous interaction of meteorological highs and lows. Further, circulatory modification comes from the rotation of the earth, surface friction, frontal movements, and ground surface characteristics.

Where averaged over long periods, local and small-scale irregularities in the motions of the atmosphere disappear, and a generalized pattern of winds is manifest. There are five generalized latitudinal wind belts in each hemisphere, which correspond to the four latitudinal pressure belts in each hemisphere: (1) the equatorial trough, or doldrums, is a narrow low pressure belt extending north and south from the equator, where winds are light and variable, and where frequent calms occur; (2) the trade wind belt, with winds remarkably constant in both direction and velocity, extends from the doldrum belt to (3) a narrow high-pressure belt of calm air movement at about 30° or 35°, known as the horse latitudes; (4) from the horse latitudes and extending toward the poles to 60° or 70° north and south are the westerlies, highly variable winds in both velocity and direction; and (5) at 60° or 70° is a narrow low-pressure belt known as the subpolar low; between this and the high-pressure belt at the 90° pole (the polar highs) blow the cold, sometimes violent winds of the polar easterlies. See Figs. 1 and 2.

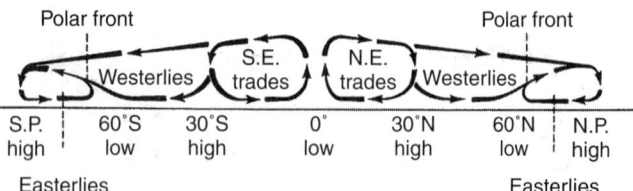

Fig. 1. Vertical cross section through the troposphere from the equator to the poles, showing air movements at and above the Earth's surface.

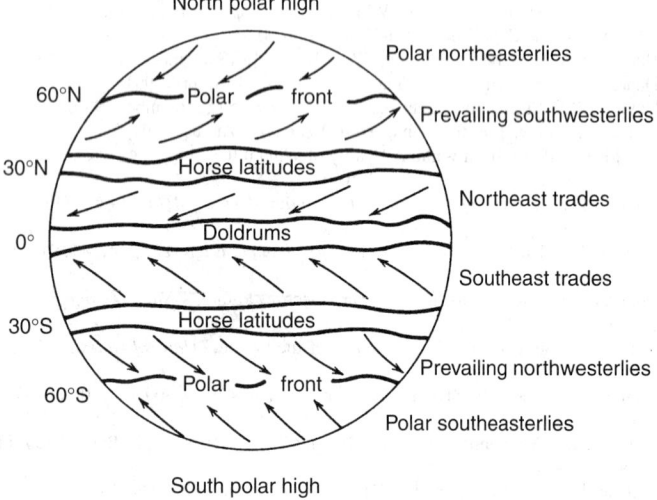

Fig. 2. Generalized map of the pressure and wind belts of the Earth.

Because pressure decreases more rapidly with altitude when air is cold, surface-wind characteristics disappear slowly with altitude; and at considerable height, the circulation is generally westerly.

Two well-developed, semi-permanent cyclones are present in the Northern Hemisphere, one located over the North Atlantic near Iceland, known as the Icelandic Low, and the other in the Aleutian area of Alaska, known as the Aleutian Low. In the Southern Hemisphere, there is a more-or-less continuous belt of low pressure at the corresponding latitude. There is considerable change in the general circulation of the atmosphere from winter to summer, the effects of which are known as monsoon effects. Another major seasonal change is the migration of the equatorial and subtropical circulations (doldrums, horse latitudes, and trade winds) north and south with the sun. This causes weakening of the westerlies during summer and strengthening during winter; weakening of the Icelandic and Aleutian lows during summer and weakening of the polar circulations in summer.

The winds may be roughly classified into three types: (1) winds that are, in a sense, permanent, including the easterly trade winds of the tropics, and the westerlies prevailing in the temperate zones; (2) seasonal winds, such as the monsoons and the smaller scale diurnal mountain and valley winds, and land and sea breezes; and (3) local winds and storms, which temporarily interrupt the more general air movements prevailing at the

time. Among these are the great cyclones and anticyclones characteristic of North American climate, ocean cyclones (including hurricanes or typhoons), thunderstorms, tornadoes, waterspouts (or ocean tornadoes), foehns (chinooks), and the cold fall winds, mistrals, and blizzards.

Winds may also be classified by their origin and movement:

1. *Gradient winds* blow in accordance with the existing pressure gradient, centrifugal force and Coriolis force. *Cyclonic winds* blow counterclockwise about regions of relatively low pressure in the Northern Hemisphere and clockwise in the Southern Hemisphere. *Anticyclonic winds* blow clockwise about regions of relatively high pressure in the Northern Hemisphere and counterclockwise in the Southern Hemisphere.

2. *Geostrophic winds* blow in accordance with the pressure gradient, but only where the pressure gradient is balanced by the Coriolis force. They blow, therefore, in straight or nearly straight lines over the earth, and are not possible at the equator because there is no Coriolis force present.

3. *Cyclostrophic winds* blow cylonically in both hemispheres in wind systems where the pressure gradient is balanced by centrifugal force in the absence of the Coriolis force. They occur near the equator as hurricanes and other local, less intense vortices.

4. *Antitriptic winds*, of small scale and short durations, blow, in general, along the pressure gradient. *Heliotropic winds*, land and sea breezes, and mountain winds, are of this type.

5. The narrow jet stream moves at high speed in an irregular easterly course through the upper air of the Temperate Zone.

Wind velocity increases, on the average, with altitude in the upper troposphere and decreases in the lower stratosphere. At higher elevations in the stratosphere, conditions of absolute calm may vary with wind velocities of 250 miles (402 kilometers) per hour. Winds-aloft observations, which determine such conditions, are commonly made by means of small pilot balloons, carrying radiosondes (see also **Weather Technology**). Winds of high velocity, changing in direction according to season, have been observed at heights of nearly 100 miles (161 kilometers).

Wind Equations. Many factors influence the movement of air in any quasi-horizontal plane at or above the surface of the Earth. In the friction layer, frictional forces play an important role. Convergence and divergence also play their part in determining air flow. Isallobaric fields have some influence. But, in general, three major forces play the main roles in establishing the winds. They are:

1. The Coriolis or deflective force due to the Earth's rotation, which is given by $2\omega V \rho \sin\phi$.
2. The pressure gradient of the atmosphere, which is given by dp/dx.
3. Centrifugal force due to curvature of path of air parcel, which is given by V^2/ρ.

If air movement is unaccelerated, these three forces are the only factors involved in determining the wind. When other forces are active, accelerations and decelerations alter equilibrium conditions among the three principal forces. However, equilibrium conditions hold, except under the excessive influence of one or more of the minor forces, and winds are generally determined by the balance among the three main forces.

There are several cases possible for establishing a balance of forces among the three. The following are for the Northern Hemisphere.

Case 1. If the parcel of air is moving in a field of pressure that is higher at the center of the field than at the exterior, the pressure gradient is directed outward from the center. In this case, the flow is clockwise and the defective or Coriolis force balances the pressure gradient and the centrifugal force:

$$\frac{dp}{dx} + \frac{V^2\rho}{r} = 2\omega\rho V \sin\phi$$

$$\begin{pmatrix} \text{pressure} \\ \text{gradient} \end{pmatrix} \quad \begin{pmatrix} \text{centrifugal} \\ \text{force} \end{pmatrix} \quad \begin{pmatrix} \text{Coriolis} \\ \text{force} \end{pmatrix}$$

where
dp/dx = pressure gradient
V = velocity of the parcel
ρ(rho) = density of the air
ω(omega) = angular velocity of the earth
r = radius of curvature of the percel's path
ϕ(phi) = latitude

Solving this equation, we obtain

$$V = r\omega\sin\phi \pm \sqrt{(r\omega\sin\phi)^2 - \frac{r}{\rho}\frac{dp}{dx}}$$

But $V = 0$ when $(dp/dx) = 0$, and we have

$$V = r\omega\sin\phi - \sqrt{(r\omega\sin\phi)^2 - \frac{r}{\rho}\frac{dp}{dx}}$$

It is not possible to give meaning in this case to a negative value for the radical; therefore, the quantity under the radical must always be positive. This places a limiting value in wind velocity in a clockwise or anticyclonic system of winds, which is given by the value of zero for the quantity under the radical. We then have

$$V_{max} = r\omega\sin\phi$$

For a given latitude, the quantity $\omega\sin\phi$ becomes fixed and V_{max}/r is constant. At the equator, the quantity $(\sin\phi)$ becomes zero and the velocity also becomes zero. Anticyclonic winds at the equator, therefore, are not possible.

For a given pressure gradient, the velocity of an anticyclonic wind increases toward the center of the system until the critical value of velocity is reached.

Case 2. If the parcel of air is moving in a field of pressure such that the pressure is less at the center than the exterior, the pressure gradient is directed inward toward the center. In this case, the flow is counterclockwise and the pressure gradient balances the deflective force and the centrifugal force:

$$\frac{dp}{dx} + \frac{V^2\rho}{r} + = 2\omega\rho V\sin\phi$$

The solution to this equation is

$$V\pm = \sqrt{(r\omega\sin\phi)^2 + \frac{r}{\rho}\frac{dp}{dx}} - r\omega\sin\phi$$

Again, when $(dp/dx) = 0$, $V = 0$ and the $+$ sign before the radical is necessary:

$$V = \sqrt{(r\omega\sin\phi)^2 + \frac{r}{\rho} - \frac{dp}{dx}} - r\omega\sin\phi$$

The quantity under the radical is always positive, so there can be no limiting value on the velocity. At the equator, where $\sin\phi$ is zero, the pressure gradient is still present under the radical to give the velocity a positive value. Winds blowing under these conditions are known as cyclonic winds and include both tropical cyclones and wave cyclones as the main types.

Case 3. If the flow of air is in a straight line, the centrifugal-force term drops out and the balance of forces is between the pressure gradient and the deflective force only:

$$\frac{dp}{dx} = 2\omega\rho V\sin\phi$$

$$V = \frac{dp/dx}{2\omega\rho V\sin\phi}$$

This wind is known as the geostrophic wind in contrast to Cases 1 and 2, which are gradient wind cases. Geostrophic winds increase southward for a given pressure gradient and, theoretically, become infinitely great at the equator, where the denominator becomes zero. They are directly proportional to the pressure gradient at any given latitude.

Case 4. At the equator, where the deflective force drops out entirely, a balance between the pressure gradient and centrifugal force is possible:

$$\frac{\rho V^2}{r} = \frac{dp}{dx}$$

$$V = \sqrt{\frac{r}{p}\frac{dp}{dx}}$$

Winds of this type are cyclostrophic winds and blow cyclonically or counterclockwise about a center of low pressure. Hurricanes are the best example of cyclostrophic winds, and even though they develop and migrate

from the equatorial zone, the deflective force plays no significant role until the center is considerably removed from low latitudes.

In the Southern Hemisphere, wind relations are the mirror image of the Northern Hemisphere, i.e., they blow counterclockwise about centers of high pressure, clockwise about centers of low pressure, and with high pressure to the left.

Specific Winds and Wind-Related Terms

Anabatic Wind. An upslope wind; usually applied only when the wind is blowing up a hill or mountain as the result of local surface heating, and apart from the effects of the larger-scale circulation; opposite of katabatic wind. The most common type is the valley wind.

Antitrade Winds. A deep layer of westerly winds frequently present above the surface trade winds of the tropics. They comprise the equatorward side of the mid-latitude westerlies but are found at upper levels rather than at the surface. They are best developed in the winter hemisphere and above the eastern extremities of the subtropical highs. The antitrades are dynamical in origin, and constitute an essential part of the atmosphere's primary circulation.

Beaufort Wind Scale. System of estimating and reporting wind speeds, invented in the early nineteenth century by Admiral Beaufort of the British Navy. It was originally based on the effects of various wind speeds on the amount of canvas that a full-rigged frigate of the period could carry, but has since been modified and modernized. See accompanying table. In its present form, for international meteorological use, it equates (a) Beaufort force, i.e., a number denoting the wind speed according to the Beaufort wind scale; (b) wind speed; (c) descriptive term; and (d) visible effects upon land objects or sea surface. See Table 1.

TABLE 1. BEAUFORT WIND SCALE

Code Number	Description	Wind Velocity	
		Miles/Hour	Kilometers/Hour
0	calm	0–1	0–1.6
1	light air	1–3	1.6–4.8
2	light breeze	4–7	6.4–11.3
3	gentle breeze	8–12	12.9–19.3
4	moderate breeze	13–18	20.9–29
5	fresh breeze	19–24	30.6–38.6
6	strong breeze	25–31	40.2–49.9
7	moderate gale	32–38	51.5–61.1
8	fresh gale	39–46	62.8–74.0
9	strong gale	47–54	76.5–86.9
10	whole gale	55–63	88.4–101.4
11	storm	64–75	103–120.7
12	hurricane	over 75	over 120.7

Buys Ballot's Law. Professor Buys Ballot at Utrecht in 1857 stated this law: "Standing with back to wind, low pressure is to left and high pressure to right in Northern Hemisphere, with the reverse being true in the Southern Hemisphere." This law was one of the earliest statements of the principles of winds now accepted as common meteorological knowledge.

Breeze. In general, any light or moderate wind. A land breeze is a coastal breeze blowing from land to sea, caused by temperature difference when the sea surface is warmer than the adjacent land. It usually blows by night, thus alternating with a sea breeze, which blows from sea to land on relatively calm, sunny, summer days when the sea surface is colder than the adjacent land. A lake breeze blows from the surface of a large lake onto the shores during the afternoon. Like land and sea breezes, it is caused by a difference in surface temperature of land and water.

Calm. The absence of apparent motion of the air. In the Beaufort wind scale, a condition of calm is reported when smoke is observed to rise vertically, or if the surface of the sea is smooth and mirrorlike. In United States weather observing practice, the wind is reported as calm if it is determined to have a speed of less than one mile per hour (or one knot).

Chinook. The name given to the foehn on the eastern side of the Rocky Mountains (U.S.). The chinook wind generally blows from the southwest, but its direction may be modified by the topography. When it sets in after a spell of intense cold, the temperature may rise by 20 to 40 °F (11 to 22 °C) in 15 minutes; a rise from 11 to 42 °F (6.2 to 23.5 °C) in 3 minutes has been recorded. After the first rise, the temperature may fluctuate violently as the station comes alternately under the influence of patches of warm and cold air. The chinook brings relief from the cold of winter, but its most important effect is the melting and evaporation of the snow; a foot of snow may disappear in a few hours. In other mountain regions, the foehn has a variety of local names, among which are "zonda," "puelche," "aspre," "sky sweeper," and "northwester."

Cyclostrophic Wind. Winds that blow as a result of a pressure gradient and centrifugal force, but in the absence of Coriolis force. They are, of necessity, cyclonic, and are restricted to equatorial zones—the only place where Coriolis force is zero, or nearly zero. The cyclostrophic component of a wind is the difference between the gradient wind and the geostrophic wind. Hurricanes are largely cyclostrophic winds, until they travel north or south sufficiently to be affected by Coriolis force.

Doldrums. A nautical term for the equatorial trough, especially with reference to the light and variable nature of the winds.

Duststorm (or Duster; Black Blizzard). An unusual, frequently severe weather condition characterized by strong winds and dust-filled air over an extensive area. A duststorm usually arrives suddenly in the form of an advancing dust wall, which may be miles long and several thousand feet high. Ahead of the dust wall, the air is very hot and the wind is light. Visibility is usually only a few yards (meters), and is sometimes reduced to nightlike darkness. Prerequisite to a duststorm is a period of drought over an area of normally arable land, thus providing the very fine particles of dust that distinguish a duststorm from the more common sandstorm of desert regions. The sand particles carried by the strong winds of a sandstorm are mostly confined to the lowest 10 feet (3 meters) and rarely rise more than 50 feet (15 meters) above the ground.

Easterlies. Any winds with components from the east, usually applied to broad currents or patterns of persistent easterly winds. (1) The *equatorial easterlies* refer to the trade winds, in the summer hemisphere, when they are very deep and, generally not topped by upper westerlies. In the winter hemisphere, these easterlies are restricted to a narrow belt along the equator. (2) The *tropical easterlies* refer to the trade winds when they are shallow, exhibit a strong vertical shear, and give way to the upper westerlies, which then govern the course of cloudiness and weather. The tropical easterlies occupy the poleward margin of the tropics in summer and can cover most of the tropical belt in winter. (3) The *polar easterlies* are a rather shallow and diffuse body of easterly winds located poleward of the subpolar low-pressure belt, and appreciable in the Northern Hemisphere, in the mean, only north of the Aleutian low and the Icelandic low. (4) *Krakatoa winds*, formerly called overtrades, is the name given to a layer of easterly winds over the tropics, which tops the antitrade winds. The name is derived from the observed behavior of the volcanic dust carried around the world after the great eruption of Krakatoa in 1883.

Ekman Spiral. An idealized mathematical description of the wind distribution in the planetary boundary layer, within which the Earth's surface has an appreciable effect on the air motion. The model is simplified by assuming that within this layer the eddy viscosity and density are constant, the motion is horizontal and steady, the isobars are straight and parallel, and the geostrophic wind is constant with height.

Fall Wind. A strong, cold, downslope wind. A fall wind differs from a foehn in that the air is initially cold enough so that it remains relatively cold despite adiabatic warming upon descent. It is a larger-scale phenomenon than the gravity wind (see *Katabatic Wind* described shortly), in that a fall wind requires a prior accumulation of cold air at high elevation.

A *bora* is a fall wind whose source is so cold that, when the air reaches the lowlands or coast, the dynamic warming is insufficient to raise the air temperature to the normal level for the region; hence it appears as a cold wind. It is very stormy and squally, the squalls sometimes reaching 100 miles (161 kilometers) per hour or more.

Foehn. A warm dry wind on the lee side of a mountain range, the warmth and dryness of the air being due to adiabatic compression upon descending the slopes. The foehn is characteristic of nearly all mountain areas. It is always associated with a strong wind blowing over a mountain ridge or chain and is found on the lee side where the air, after passing the crest of the mountains, flows down to lower levels. The exact nature of the foehn varies widely and depends upon local topography and the nature of the winds blowing across the terrain. Clouds and precipitation usually are present on the windward side of mountains when foehn conditions are present on the lee side.

Free Atmosphere. That portion of the Earth's atmosphere, above the planetary boundary layer, in which the effect of the Earth's surface friction on the air motion is negligible, and in which the air is usually treated (dynamically) as an ideal gas. The base of the free atmosphere is usually taken as the geostrophic wind level.

Friction Layer. The lower layer of air, below the "free atmosphere," where the friction with the Earth's surface affects its flow. Depending upon varying conditions, its thickness is usually from 1,500 to 3,000 feet (457 to 914 meters). The name is commonly used to include the surface and planetary boundary layers. The effects of friction in creating varying wind conditions are discussed under **Atmosphere (Earth)**.

Gale. A wind whose velocity ranges from 32 to 63 miles (51 to 101 kilometers) per hour.

Geostrophic Equilibrium. A state of motion of an inviscid fluid in which the horizontal Coriolis force exactly balances the horizontal pressure force at all points of the field. The assumption of geostrophic equilibrium in certain contexts of the equations of motion, but not in others, is known as quasi-geostrophic equilibrium (or approximation). This compromise arose from the fact that the horizontal divergence of the geostrophic wind grossly fails to estimate the real divergence, which is better represented by the field of vertical motion.

Geostrophic Wind. That horizontal wind velocity for which the Coriolis acceleration exactly balances the horizontal pressure force. The geostrophic wind is directed along the contour lines on a constant-pressure surface (i.e., a surface along which the atmospheric pressure is everywhere equal at a given instant), or along the isobars in a geopotential surface (i.e., a surface along which a parcel of air could move without undergoing changes in its potential energy), with low elevations (or low pressure) to the left in the Northern Hemisphere and to the right in the Southern Hemisphere.

Geostrophic Wind Level. The lowest level at which the wind becomes geostrophic in the theory of the Ekman spiral. In practice, this level is observed to be between 0.75 and 1 mile (1.2 and 1.6 kilometers), and it is assumed that this marks the upper limit of frictional influence on the Earth's surface. The geostrophic wind level may be considered to be the top of the Ekman layer and the base of the free atmosphere.

Gradient Wind. Any horizontal wind velocity tangent, at the point in question, to the contour line of a constant-pressure surface (i.e., a surface along which the atmospheric pressure is everywhere equal at a given instant) or to the isobar of a geopotential surface (i.e., a surface along which a parcel of air could move without undergoing changes in its potential energy). At such points where the wind is gradient, the Coriolis acceleration and the centripetal acceleration together exactly balance the horizontal pressure force.

Gust. A sudden brief increase in the speed of the wind. A gust is more transient in character than a squall and is followed by a lull or slackening in wind speed. With respect to aircraft turbulence, a gust is a sharp change in wind speed relative to the aircraft; or a sudden increase in air speed due to fluctuations in the air flow, resulting in increased structural stresses upon the aircraft. See also **Gust Front**.

An airplane flying horizontally into an upward gust of air experiences a sudden change of direction of relative wind. This induces a temporary excess of lifting power due to (a) an increase in the relative angle of attack and therefore a larger lift coefficient and (b) the greater, momentary, relative wind of the wing which thereby accelerates the airplane upward. The effect on the occupants is not unlike that of a surface vehicle passing over a bump in the roadway, hence the term "air bump." If the gust is downward, a downward acceleration is produced by the sudden decrease of angle of attack. This is sometimes referred to as an "air pocket" because the occupants feel as if the airplane had entered a region in which there was no air to sustain the lift. Actually, of course, there is practically no difference in the density of air in or out of the regions of the so-called air "bumps" and "pockets."

Gusts, vertical or horizontal, may be experienced particularly during line squalls and thunderstorms. Smaller gusts may be experienced due to convection currents, as when crossing the boundary between land and water, or between green fields and a desert, or above city buildings. Obstructional interferences may also cause gusts, such as updrafts or downdrafts, when approaching or leaving mountainous areas. See also **Gust Front**; and **Wind Shear**.

The relative intensities and the altitudes at which gusts have been encountered are listed in Table 2.

TABLE 2. CHARACTERISTICS OF GUSTS

Factor	Convection Currents	Line Squalls and Thunderstorms	Obstructions
Velocity of gust, knots	3 to 20	15 to 75	3 to 20
Altitude limit, feet meters	15,000 4572	50,000 15,240	Depends upon type of obstruction and prevailing wind.

High intensity gusts may be experienced over areas of a large conflagration, or explosion of a nuclear bomb.

Gustiness factor is a measure of the intensity of gusts given by the ratio of the total range of wind speed between gusts and the intermediate periods of lighter wind to the mean wind speed, averaged over both gusts and lulls.

Heliotropic Wind. A subtle, diurnal component of the wind velocity leading to diurnal shift of the wind or turning of the wind with the sun, produced by the east-to-west progression of daytime surface heating.

Isallobaric Wind. The wind velocity whose Coriolis force exactly balances a locally accelerating geostrophic wind. Mathematically, the isallobaric wind V is defined in terms of the local accelerations but approximated by the allobaric gradient as follows:

$$V_{is} = k \times \frac{1}{f} \frac{\partial V_g}{\partial t} = -\frac{\alpha}{f_2} \nabla_H \frac{\partial p}{\partial t}$$

where k is the vertical unit vector, f the Coriolis parameter, V_g the geostrophic wind, α the specific volume, ∇_H the horizontal deloperator, and p the pressure.

Jet Stream. See also **Jet Streams**.

Katabatic Wind. Any wind blowing down an incline; the opposite of anabatic wind. If the wind is warm, it is called a *foehn*; if cold, it may be a *fall wind*, such as the *bora*. The term is often taken as synonymous with *gravity wind*, whose direction down the slope of an incline is caused by greater air density near the slope than at the same levels some distance horizontally from the slope. Gravity wind is usually referred to when the density difference is produced by surface cooling along the incline, as in the case of a mountain wind. Fall winds, on the other hand, are considered to be larger-scale phenomena, such as the motion of cold air from an elevated interior toward an adjacent warm sea coast.

Local Winds. Winds that differ, over a small area, from those appropriate to the general pressure distribution, or which possess some other peculiarity. Local winds may be classified into four main groups: (1) winds upon which the effect of local topography is to intensify the general geostrophic flow, generally by being forced through a narrow gap; (2) antitriptic winds, including mountain and valley winds, fall winds, and foehn; (3) a number of instability winds caused by local heating or overrunning by cold air, such as several varieties of dust storms, and winds accompanying thunderstorms; (4) winds that are strong because of strong pressure gradients or uninterrupted flow over a level surface, or both, including sand- and snow-bearing winds, and some winds off the sea.

Logarithmic Velocity Profile. The variation of the mean wind speed with height in the surface boundary layer, derived with the following assumptions: (1) the mean motion is one-dimensional; (2) the Coriolis force can be neglected; (3) the shearing stress and pressure gradient are independent of height; (4) the pressure force can be neglected with respect to the viscous force; and (5) the mixing length l depends only on the fluid and the distance z from the boundary, $l = kz$.

Meridional Wind. The wind or wind, component along the local meridian, as distinguished from the zonal wind (along the local parallel of latitude). In a horizontal coordinate system fixed locally with the x-axis directed eastward and the y-axis northward, the meridional wind is positive if from the south, and negative if from the north (the zonal wind is positive if from the west, and negative if from the east).

Mistral. A north wind that blows down the Rhone Valley south of Valence, France and into the Gulf of Lions. It is strong, squally, cold, and dry; the combined result of the basic circulation, a fall wind, and jet-effect wind. It blows from the north or northwest in the Rhone Delta, where it is strongest, from northwest in Provence and from the northeast in the valley of the Durance below Sisteron.

Monsoon. A name for seasonal winds (derived from Arabic *mausim*, a season). It was first applied to the winds over the Arabian Sea, which blow for six months from northeast and for six months from southwest, but it has been extended to similar winds in other parts of the world. Even in Europe, the prevailing west-to-northwest winds of summer have been called the "European monsoon." The primary cause is the much greater annual variation of temperature over large land areas compared with neighboring ocean surfaces, causing an excess of pressure over the continents in winter and a deficit in summer; but other factors, such as the relief features of the land, have a considerable effect. Thus, monsoons are strongest in Asia, the largest land mass, but also occur on the coasts of tropical regions wherever the planetary circulation is not strong enough to inhibit them. Monsoon winds have been described in Spain, Australia, Africa, the United States, and Chile. In India, the term is popularly applied chiefly to the southwest monsoon and, by extension, to the accompanying rains.

Mountain and Valley Winds. A system of diurnal winds along the axis of a valley, blowing uphill and upvalley by day (anabatic wind), and downhill and downvalley by night (katabatic wind); they prevail mostly in calm, clear weather. The upvalley component, or valley wind, usually sets in about one-half hour before sunrise and continues until about one-half hour before sunset. On southerly slopes, it may reach 14 miles (22.5 kilometers) per hour; on northerly slopes, it is barely noticeable. The downvalley component, or mountain wind, is due to nocturnal cooling, and blows at night. It is somewhat weaker than the valley wind, reaching up to 9 miles (14.5 kilometers) per hour.

Relative Wind. Velocity of air with reference to a body in it; usually determined by measurements made at such a distance from the body that the disturbing effect of the body upon the air is negligible. The airspeed indicator, for example, located in an airplane, measures the speed of the relative wind. Equations dealing with aerodynamic forces are based upon the velocity of the relative wind.

Squall. See also **Fronts and Storms**.

Streamlines. Lines drawn everywhere tangent to wind vectors. They show the instantaneous flow-pattern of the air at a given time only and do not indicate trajectories of air parcels. Streamlines are used for two-dimensional flows (horizontal plane), although actually they are a three-dimensional function. The vertical velocity is usually considered insignificant in comparison with the horizontal flow. Streamline charts are useful and quickly-prepared forecast tools for obtaining wind forecasts at a given flight altitude. Trajectory charts, however, are more useful for projecting atmospheric phenomena.

Thermal Wind. The mean wind-shear vector in geostrophic balance with the gradient of mean temperature of a layer bounded by two isobaric surfaces.

Trade Winds. Commonly called Trades, the wind system occupying most of the tropics, which blows from the subtropical highs toward the equatorial trough; a major component of the general circulation of the atmosphere. The winds are northeasterly in the Northern Hemisphere and southeasterly in the Southern Hemisphere; hence, they are known as the northeast trades and southeast trades, respectively.

The trade winds are best developed on the eastern and equatorial sides of the great subtropical highs, especially over the Atlantic Ocean; in the Pacific Ocean, they are properly developed only in the eastern half. They are characterized by great constancy of direction and, to a lesser degree, of speed; the trades are the most consistent wind system on the earth.

Veering Wind. Any clockwise change in wind direction is known as veering of the wind, as opposed to backing of the wind, a counter-clockwise change.

Westerlies. Also known as Circumpolar Whirl, Countertrades, or Zonal winds, in general, any winds with components from the west. Specifically, the dominant west-to-east motion of the atmosphere, centered over the middle latitudes of both hemispheres. The equatorward boundary is fairly well-defined by the subtropical high-pressure belt; the poleward boundary is quite diffuse and variable. The *equatorial westerlies* are westerly winds occasionally found in the equatorial trough and separated from the mid-latitude westerlies by the broad belt of easterly trade winds.

Wind Rose. Any one of a class of diagrams designed to show the distribution of wind direction experienced at a given location over a considerable period; it thus shows the prevailing wind direction. The most common form consists of a circle from which 8 or 16 lines emanate, one for each compass point. The length of each line is proportional to the frequency of wind from that direction; and the frequency of calm conditions is entered in the center. Many variations exist. Some indicate the range of wind speeds from each direction; some relate wind directions to other weather occurrences.

Wind Shear. The local variation of the wind vector or any of its components in a given direction. The wind shear at a point is said to be cyclonic or anticyclonic according to the sense of rotation from the wind vector to the shear vector at that point.

See also **Atmosphere (Earth)**; **Atmosphere-Ocean Interface**; **Jet Streams**; and entries listed under **Meteorology**. For references see the lists at the ends of the entries on **Climate**; and **Meteorology**.

PETER E. KRAGHT, Certified Consulting Meteorologist, Mabank, TX.

WIND SHEAR. A change in wind direction and/or speed over a relatively short distance. For example, a west wind of 30 knots that changes to a northwest wind of 10 knots in 50 meters contains a shear of 45° and 20 knots in that short distance—0.9° and 0.4 knot per meter. Most winds blow parallel to the Earth's surface and shear is mostly in the horizontal direction. However, in cumulus clouds, showers, and thunderstorms, gust fronts and some cold fronts, shear may also involve the vertical direction, that is, up and downdrafts. Mathematically, shear in its simplest form is

$$\Delta V / \Delta D$$

where ΔV = change in wind vector
ΔD = distance over which the change occurs

Small-scale wind shear is present everywhere in the atmosphere. In some limited meteorological conditions, shear can become quite large. These conditions generally are restricted to various types of fronts, inversions, thunderstorms, showers, and gusty winds. See Fig. 1.

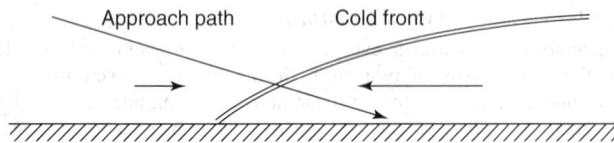

Fig. 1.   Schematic illustration (not to scale) of shear of the longitudinal wind component on an approach to an airport from a slight tailwind to a strong headwind.

Wind shear has its greatest impact on aircraft, especially on approach and climb-out to and from airports when the airspeed is nearer critical values than at cruising speeds. Wind shear below 500 meters (1500 feet) is regarded as low-level wind shear. Insofar as aircraft behavior is concerned, shear behaves as if it were in three components:

1. Shear of the longitudinal wind component, that is, the tailwind or headwind
2. Shear of the lateral wind component, that is, the cross wind
3. Shear of the vertical wind component, that is, the downdraft and updraft

Shear of the longitudinal wind component causes acceleration or deceleration of the airflow past the plane and affects the aerodynamically generated lift. If the operating airspeed is well above the stall value, the effect of sheer is seldom, if ever, of sufficient magnitude to create an emergency. However, on an approach or climb-out when the airspeed margin is often small, intense deceleration of the airspeed in shear can induce a borderline aerodynamical performance by the plane.

Shear of the lateral wind component induces an unwanted drift from the planned flight track. Shear of the vertical wind causes the plane to gain or lose altitude in an unplanned manner, which on approach may cause a plane to come too near the earth too soon.

Low-level wind shear is present primarily in and near thunderstorms and showers, in gust fronts, in cold and warm fronts, in sea breeze fronts, in whirlwinds, and gusty winds. Shear of a lesser magnitude is always present near the ground because of the frictional retardation of the wind across the earth.

See also **Fronts and Storms**; **Gust Front**; and other terms listed in entry on **Meteorology**.

PETER E. KRAGHT, Certified Consulting Meteorologist, Mabank, TX.

WIND TUNNEL. A device that provides an airstream of known and steady conditions, in which models requiring aerodynamic study are tested. The essential elements of a wind tunnel are: (1) a drive system consisting of either a compressor for continuous operation, or a tank of compressed air for intermittent operation; (2) a test section in which models are held and their orientation is changed; (3) instrumentation capable of reading force, pressure, and other effects produced by the model; and (4) an air efflux system consisting of free exit to the atmosphere or to a vacuum tank, or a tunnel returning the air to the compressor. There are many wind tunnels throughout the world, ranging in test section size from 1×1 inch to 30 feet $\times 60$ feet $(9.1 \times 18.2$ meters) with speeds from 50 to 7,000 miles $(81 - 11,265$ km) per hour and higher. See Fig. 1. See also **Aerodynamics and Aerostatics**.

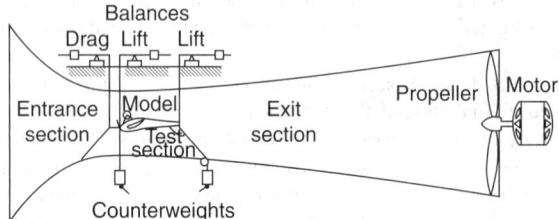

Fig. 1.   Simple wind tunnel of the open-circuit and closed-throat type. In some cases, the use of computer simulation has reduced the need for wind tunnel facilities.

With very marked improvements in mathematical modeling and computer simulation, some of the research that formerly required wind tunnels now can be handled by a computer workstation furnished with appropriate software. An example of the new technology is given in article on "Predicting Propeller Blade Loads without Testing," by Donham, Dupcak, and Conner in *Aerospace Engineering*, 4–9 (April 1987).

Computer simulation, however, does not suffice in all engineering test situations. In October 1990, for example, the largest wind tunnel facility in the world was dedicated by the Centre Scientifique et Technique du Batiment (CSTM) in Nantes, France. Known as the Jules Verne wind tunnel, the facility provides 54,000 square feet $(5000 \ m^2)$ for testing large and small machines and apparatus that are subject to high winds when in use. See also **Weather Technology**.

WINGED FRUIT.   See **Fruit**.

WINGNUT TREES.   See **Hickory and Wingnut Trees**.

WINTER EGG.   A form of egg produced by some invertebrates, usually in the fall. It has a thick protective shell and in species with a complex life cycle, such as the aphids, it is produced by the sexual generation. Such eggs usually pass through the winter before hatching, but similar eggs may be produced at the beginning of other unfavorable seasons, as periods of drought.

WIRE DRAWING.   The term "wire drawing" has two separate and distinct meanings. First, it is descriptive of the action whereby a rod is reduced to a wire by being pulled forcibly through a round die which reduces its diameter and increases its length (see Fig. 1). Secondly, wire drawing refers to a similar action occurring in fluid flow through a small aperture. The usage in the second case is derived from the first. Wire drawing as applied to a fluid is usually employed in discussions of the flow of steam or gas through the valves of engines. When the valves are partially open, the small area through which a fluid flows causes a reduction of pressure which is known as wire drawing. The description below refers entirely to the production of wires.

To make a wire, the metal is first prepared as a rod by rolling. It is then treated to give it a surface coating which will act as a lubricant during its passage through the die. The method of applying the lubricant gives rise to the wet and dry process of wire drawing. In the dry process the wire is dipped in an emulsion of hydrated lime and dried in baking ovens. Then just before it is passed into the die it passes through a greasing operation. The combination of lime film and grease provides the lubricant between the rod and the die during the drawing process. In the wet process the rod

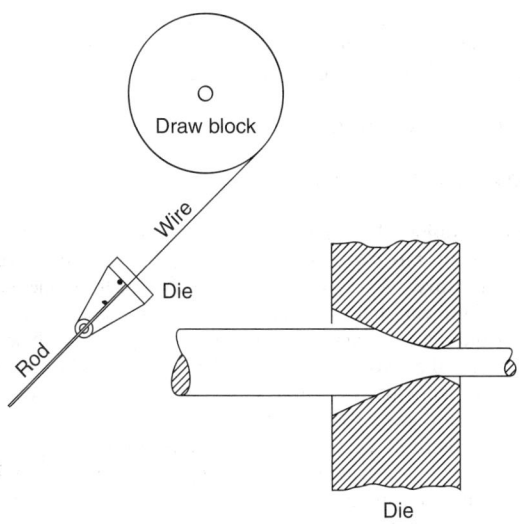

Fig. 1. Wire drawing machine.

is first dipped in a solution of copper sulfate, which results in a thin coating of copper in the wire. It is then dipped in a vat containing a fermented mixture made from meal and yeast. Upon being withdrawn from this vat, it is fed wet to the die. The liquor, together with the thin copper deposited in the salt bath, lubricates the rod in the wet drawing process.

Drawing consists of pointing the rod and threading it through the die. It is then gripped by pinchers and drawn a short distance by hand power applied through a draw bar. Then the end is clamped in a vise on a revolving drum called the drawblock. This drum is revolved slowly by power applied through a geared shaft, and in so revolving continues to pull the wire through the die until the rod has been completely reduced. When the stock is passed through the die it is elongated and reduced in cross section. The elongation is known as draft, and may amount to from 10 to 40 times the length of the original rod. Sometimes more than one draft is necessary to finish the wire. In such cases it is removed from the block and started through another smaller die.

The dies are the most essential part of wire drawing. They are made of chilled cast iron, steel, cemented carbides, or diamond. The hard cemented carbide type dies are widely used. Diamond dies are used only for fine wire. A typical cross section of a die is shown in the figure. The hole through the die is tapered, and the part where the reduction in wire size occurs is reamed to a very smooth finish. Two tapers are employed, one in which most of the reduction takes place, the other one of smaller taper, in order to ease the wire down to the exact size at exit from the die. The initial part of the tapered hole is not machined, as it serves only as a region for application of the lubricant to the rod. A plan of a draw bench which is the production unit in wire drawing, is also given.

WIREWORM (*Insecta, Coleoptera*). Slender hard-bodied larvae of the click-beetles. They live under bark and in logs or in the ground. Among the latter species are some important crop pests which damage the roots of grass and grains, and a few other crops including potatoes and cotton. The corn wireworm is one of the most widely known, often destroying entire fields.

The hard bodies of these worms render them immune from repellent substances that can be used in the soil, but crop rotation including clover, soy beans, buckwheat, flax, or some other species not subject to attack, is an effective method of avoiding serious loss.

WISHART'S DISTRIBUTION. The joint distribution of variances and covariances in samples from a multivariate normal population. It may be regarded as a generalization to p dimensions of the chi-square distribution, which is effectively that of the sample variance in univariate statistics.

WITCHES' BROOMS. The malformations known as witches' brooms are formed in many different kinds of woody plants. Many of them are caused by the presence of *Exoascus*, an ascomycete, which grows parasitically in the tissues of the plant. A bud infected by *Exoascus* is stimulated to rapid growth. So are the lateral buds of this shoot. The result is a bushlike mass of branches. The leaves of these infected branches are

commonly dwarfed and fall earlier than the other leaves of the plant. The stems contain a much greater amount of parenchymatous tissue than is found in the normal branches. No reproductive parts are found on the brooms.

Witches' brooms are formed on fir trees; here they are caused by *Aecidium elatinum*, one of the rusts. A rust, *Gymnosporangium*, also causes the formation of witches' brooms on white cedars. Hackberry trees are often conspicuously covered with witches' brooms. In this plant they are caused by a mite, *Phytoptus*, which attacks the plant.

WITCH HAZEL. A small shrub or tree (*Hamamelis virginiana L.*) and considered a most unusual native plant. It was called *Oe-eh-nah-kwe-ha-he* (spotted stick) by the Onondaga Indians of New York, who used witch-hazel extract as a natural healer. The elixir was prepared by boiling the leaves of the plant. They also used the twigs to prepare a liniment for back pains. It has been reported that the early American pioneers used a snuff prepared from powdering dry witch-hazel leaves for stopping nosebleed. The fresh leaves of the witch hazel and twigs are still used for preparing an after-shave astringent and as an additive in some rubbing alcohols. The branches of the witch-hazel are fibrous and are well suited for making brooms. The witch hazel tree has the appearance of a twisted shrub whose branches zig-zag in all directions. Generally, the tree does not exceed ten feet (3 meters) in height. The tree habitually grows in the shade of tall trees in dense northeastern forests. Delicately scented flowers do not appear until fall when the shade of taller trees is removed. Flowers appear at an inopportune time for pollination—hence the tree has adapted itself to self-pollination. Also the tree's black seeds are ejected and thus do not require wind or any other natural assistance for distribution. The seeds may be "exploded" to a distance of nearly 50 feet (15 meters).

The record witch hazel tree growing in the United States is located in Bedford, Virginia. As compiled by the American Forestry Association, this specimen has a circumference (at $4\frac{1}{2}$ feet; 1.4 meter above ground level) of 52 in. (132 cm), a height of 35 feet (10.7 meters), and a spread of 30 feet (9.1 meters).

WITCH OF AGNESI. Let *OBA* be a circle with diameter $OA = 2a$ and let *LK* be a tangent to the circle at *A*. From *O* draw any line intersecting the circle at *B* and the tangent *LK* at *C*. From *B* draw a line parallel to *LK* and from *C*, a line perpendicular to *LK*. The intersection of these two lines is *P* and its locus is the plane curve called the witch. Its equation in rectangular coordinates is

$$x^2 y + 4a^2 y = 8a^3$$

and, if ϕ is a parameter, $x = 2a \cot \phi$, $y = 2a \sin^2 \phi$. The curve is symmetric to the *Y* axis. Its asymptote is $y = 0$. See Fig. 1.

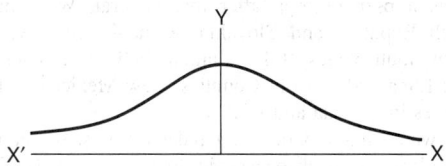

Fig. 1. Witch of Agnesi.

WITHERITE. The mineral witherite is barium carbonate, $BaCO_3$, crystallizing in the orthorhombic system. It is interesting to note that at 811 °C it changes to the hexagonal system, and at 982 °C it appears to become isometric. It has a rather imperfect prismatic cleavage; uneven fracture; hardness 3–3.7; specific gravity, 4.29; luster, vitreous to resinous; color, white to yellowish or grayish; streak, white; transparent to translucent. Witherite is found in veins, and often is associated with galena, as at Alston Moor, Cumberland, England. Associated with barite at Freiberg, Saxony, and at Lexington, Kentucky. Named in honor of Dr. William Withering, an English botanist.

WITTIG REACTION. This reaction provides an excellent method for the conversion of a carbonyl compound to an olefin:

$$(C_6H_5)_3 \xrightarrow{CH_3Br}$$

Triphenylphosphine

$$(C_6H_5)_3\overset{+}{P}CH_3(Br^-) \xrightarrow[-C_6H_6-LiBr]{C_6H_5Li}$$

Methyltriphenyl-
phosphonium bromide

$$(C_6H_5)_3P = CH_2 \longleftrightarrow (C_6H_5)_3\overset{+}{P}-\overset{-}{C}H_2$$

Wittig reagent

$$(C_6H_5)_2C = O$$
$$(C_6H_5)_3P \rightharpoondown CH_2 \longrightarrow (C_6H_5)_3P + CH_2$$
$$\qquad\qquad\qquad\qquad\qquad \underset{O}{\|} \quad \underset{C(C_6H_5)_3}{\|}$$
$$O \rightharpoondown \dot{C}(C_6H_5)_2$$

The reagent is unstable and so is generated in the presence of the carbonyl compound by dehydrohalogenation of the alkyltriphenylphosphonium bromide with phenyllithium in dry ether in a nitrogen atmosphere. There are various modifications, such as the phosphonate, in which diethylbenzylphosphonate, cinnamaldehyde, and sodium methoxide yield 1,4-diphenylbutadiene.

Dr. Georg Wittig, University of Heidelberg, was awarded the 1979 Nobel Prize for chemistry for his work in organic synthesis. Dr. Herbert C. Brown of Purdue University also participated in the joint award, but for separate work in organic synthesis.

More details on the early development of the Wittig reaction, dating back to the 1940s, is given in *Science*, **207**, 42–44 (1980).

WOLF. See **Canines**.

WOLFRAMITE. The mineral wolframite, tungstate of iron and manganese, is an isomorphous mixture of tungstate of iron, $FeWo_4$, and tungstate of manganese, $MnWO_4$, the amounts being variable. The pure iron tungstate is called ferberite and the manganese tungstate, hübnerite. It has been proposed that the name ferberite be applied to mixtures of not less than 80% $FeWO_4$ and not more than 20% $MnWO_4$, and that the term hübnerite be given to mixtures of not less than 80% $MnWO_4$ but not more than 20% $FeWO_4$. Wolframite would thus include the minerals of intermediate composition, and its formula would be written $(Fe^{2+}, Mn)WO_4$. Wolframite is monoclinic, usually appearing in tabular, columnar or bladed crystals, sometimes quite large; also may be massive. Its hardness is 4–4.5; specific gravity, 7.371; color, gray, reddish-brown, brown or black; streak, reddish-brown to black; luster, submetallic; opaque, occasionally magnetic.

Wolframite is found associated with apatite, cassiterite, quartz, and fluorite; in granites and pegmatites. Often with scheelite, $CaWO_4$, and sometimes as a pseudomorph after that mineral. Wolframite is found in the Czech Republic and Slovakia, Rumania, Saxony, Cornwall in England, New South Wales, Bolivia; and in the United States at Trumbull, Connecticut; Luna and Lincoln Counties, New Mexico. It also occurs in small quantities in Nevada and Utah.

Ferberite, which has monoclinic tabular crystals, sometimes massive, resembles wolframite, and occurs in Spain and in Boulder County, Colorado. Superb crystals of Ferberite in association with apatite and arsenopyrite crystals occur at Panasqueira, Portugal.

Hübnerite crystals are monoclinic, often long fibrous or bladed, may be massive; resembles wolframite and is found in Peru, the Black Hills, South Dakota; San Juan County, Colorado; White Pine County, Nevada; and Lemhi County, Idaho.

WOLLASTONITE. The mineral wollastonite is calcium metasilicate, $CaSiO_3$, which is found as tabular or short prismatic triclinic crystals. This mineral has a hardness of 4.5–5; specific gravity, 2.87–3.09; color, white to gray, rarely green to colorless; luster, vitreous to pearly; transparent to translucent. Wollastonite in the main is formed by the action of contact metamorphic processes on limestones at relatively high temperature (600°C+). Its common associates, diopside, vesuvianite, garnet and epidote, suggest this origin. Some of the more important localities are in the copper mines of Rumania, in the lavas of Monte Somma and Vesuvius, in Finland and Mexico. In the United States it is found in

Essex and Lewis Counties, New York; Keweenaw County, Michigan; and Riverside County, California. Wollastonite was named in honor of the English chemist, William Hyde Wollaston.

WOLVERINE. See **Mustelines**.

WOMBAT. See **Marsupialia**.

WOOD (or Timber; Lumber). The word wood is frequently applied to fuel, which, in this case, means forest products cut to a size suitable for burning in stoves. Again, especially in the plural form, the word refers to a stand of growing trees, especially if the stand covers a considerable area. The word may also apply to finished products such as boards, joists, and beams, all called wood. In botanical language, the word *wood* means: that part of the stem, trunk, or branches, which is composed of the water-conducting xylem and associated fibers, especially in perennial plants in which these cells form a considerable mass, as in trees, shrubs and vines.

The word timber is also applied to standing trees, especially in the phrase standing timber, meaning trees large enough to be commercially valuable. Another common phrase is timber-land. Timber also means prepared forests products, especially any large beams, or coarse products. Lumber has a similar meaning, but is used in a rather broader sense, applying generally to any forest products such as boards, planks, beams, etc.

Practically all plants yielding wood of any value are gymnosperms or dicotyledonous angiosperms. Among monocotyledons only certain bamboos and palms have any appreciable use, and that largely restricted to the regions in which these plants grow.

It is worthy of note that forests have a very great value to man in addition to their yields of wood. For a forest is a very important factor in conserving water, preventing it from rapid running-off, and so tending greatly to reduce the possibility of disastrous floods. This retention of water is due to several factors; one because the roots themselves tend to form hollows in which water stands and settles, to move slowly through the ground, another because the increased supply of humus accumulating in the ground under the trees acts as a sponge, absorbing and retaining water. Vast amounts of water are absorbed by trees and given off to the air by transpiration.

In North America, there are several types of forests, distinguished mainly by the kinds of trees most numerous in each type. In eastern North America, three main regions are recognized. There is the Northern Forest, occupying the northernmost tier of states from Minnesota east, and reaching downward in the mountains to Virginia and Tennessee. The dominant trees of this forest are coniferous, with white pine the most valuable species. Other coniferous species are hemlock, fir, and spruce. Hardwoods such as oaks, maples and birches are important trees of this northern forest. The southern forest occupies the region bordering the Atlantic and Gulf of Mexico and extending from the southeastern corner of Virginia into the eastern part of Texas. This also is a forest dominated by coniferous trees, mostly species of pine. Hardwood trees of this forest include sweet gum, ash, and tupelo. Between these two forests and extending west into the prairie region is the hardwood forest, in which are found oaks, black walnut, hickory, and basswood. In the western half of the country there are two regions, one, the Pacific forest, occupying the states bordering the Pacific ocean, and composed largely of Douglas fir, redwoods and several species of pine, all trees frequently attaining tremendous size. Occupying higher elevations in the western mountain states in the Rocky Mountain forest, containing such trees as Douglas fir, Engleman spruce, several pines, aspen, red cedar, and many others. Including the southernmost tip of Florida and the southern borders of Louisiana and Texas is a forest of an entirely different type, the tropical forest. This is characterized by such trees as magnolia, and live oak with an undergrowth of palmetto, which occupies vast regions of land in Central and South America, as well as the West Indies. These tropical forests, or rain forests, occur through the lowland areas of the tropical regions of South America and Africa, especially along the great rivers. Their luxuriant vegetation is correlated with rainfall, as indeed are the location and development of the temperate hardwood forests and northern coniferous forests in both hemispheres.

Woods are divided into two main classes, softwoods and hardwoods, distinguished not so much by the nature of the wood as by the trees. Softwoods are those obtained from coniferous trees such as pines, spruces, hemlocks and firs; hardwoods come from deciduous trees, and include such trees as oaks, ashes, maples, basswood, poplars, gums, as well as many tropical trees. It will be seen that the relative hardness of the wood is not an indication of its classification, since many hard pines, which are softwoods, are much harder than the soft wood of the bass, the poplar and the tulip tree, all of which are classed as hardwoods. Softwoods are much more extensively used by man, being employed widely in building construction. Hardwoods are used for furniture, interior finish, and for products demanding special wood structure.

The structure of wood is a very important factor in determining its ultimate use. The woody tissue of any plant is composed entirely of cells or vessels of various kinds, the majority of them being elongated and thick-walled. In gymnosperms, from which come the softwoods, these cells are tracheids, elongated cells in the walls of which there are many pits. These tracheids vary in size and in the thickness of their walls; those which are formed during the season of most active growth, in the spring, being larger and having thinner walls than those formed later. The hollow center, or lumen, of the spring-formed wood cells is much larger than that of the summer wood. As a result of this variation in cell size and structure, the wood of the tree is made up of distinct concentric layers of cells, often very sharply distinct. These are the annual rings. By counting them one may obtain an accurate knowledge of the age of the trees. In addition to the tracheids there are in the gymnosperm wood other cells which extend in narrow radial bands outward towards the surface of the stem. These are the ray cells which collectively form the rays of the wood. In softwoods the rays are not conspicuous; in many they are very minute. In hardwoods the cellular structure of the wood is more complex. Tracheids are present and resemble those of the softwood in appearance. In addition there are many vessels. These are composed of many cells arranged in vertical rows extending considerable distances along the stem. It is characteristic of vessels that there is no cross wall separating the component cells. Pits are found in the lateral walls of the vessels. In hardwoods, there are also many fibers, slender elongate cells with thick walls, in which there are only a few small pits. Compared with those of softwoods, the rays of hardwoods are very large and often form a conspicuous feature of the wood. All these cells in both hard and soft woods are formed by the cambium cells. When first formed, all wood cells contain protoplasm. Very early in their development, the protoplasm of the vessels, fibers and tracheids is lost, the cell becoming void of any living contents. Only the ray cells, and any parenchyma cells that may remain, retain their protoplasm. Therefore, the greater part of the woody tissue of the tree is dead. For some time this wood functions, it being the region through which water and dissolved mineral matter passes up through the stem. Wood thus functioning is called sapwood. It is usually pale in color. As the stem increases in size, changes occur in the cells, and there is less and less ascent of sap. Their walls frequently become much darker in color; often the lumina of the cells becomes filled up with various solid substances. When this condition exists the wood is known as heartwood. The separation between sapwood and heartwood is usually quite distinct; it does not necessarily coincide with an annual ring, but may form a very irregular region. The appearance of the heartwood differs greatly in different woods; in some it is never distinct from the sapwood, in others it forms a small mass in the center of the stem, while in others it includes almost the entire stem. In color heartwood ranges from white through yellows, reds, greens, browns, and even black in different trees; often it is streaked and mottled with different shades of color.

The first step in the preparation of wood is cutting it into suitable dimensions. Usually this means that the felled trunks or boles, known as logs, are cut into lengths varying from 2 to 3 feet to 16 feet (0.6 or 0.9 meter to 3.8 meters) or more, depending on the use to which the wood will be put. These logs are then sawed into boards, planks, or larger pieces. The manner in which a log is sawed is often a very important factor in the value of the product, since many woods are valuable for their grain, which differs according to the way the wood is cut. The word grain, applied to wood, has many meanings. Commonly it refers to the appearance of the wood when finished. This is determined by the nature of the cells composing the wood and their arrangement. In many woods, especially the softwoods, the cells are all parallel so that the wood splits easily. Such wood is said to have a straight grain. In other woods, the cells are oriented in various directions. Such woods split with great difficulty, and are called cross-grained woods. Wood composed mostly of small thick-walled cells, with few vessels, is called fine-grained wood, in contrast to that having many vessels, usually of large diameter, known as coarse-grained wood. The rays of the wood often form an important feature of the grain, since when they are large

they may appear in the form of distinct flakes or streaks in the surface of the wood. The size and appearance of these flakes are greatly changed by the method of sawing the wood.

The commonest way of sawing is called plain sawing. In this, successive boards or planks are cut from the log, beginning at one side and cutting successive pieces through the center of the opposite side. In those woods in which grain is of little value, especially those with very small rays, this method of cutting is entirely satisfactory, for it is quickly and easily done. In other woods, especially those in which the rays are large and form a conspicuous feature of the finished wood, it is not so satisfactory, since variation results. This variation is largely due to the nature of the rays, which extend directly outward from the center to the surface of the wood, and have little thickness compared to their other two dimensions. Consequently the first cuts will be across the rays, which will then appear as narrow linear streaks in the wood. When boards near the center are cut, the ray will be nearly parallel to the cut and so appear as an irregular broad patch on the surface, adding greatly to the beauty of the wood. Methods of cutting have been devised which increase the number of boards that show these radial cuts. These methods are known as quarter-sawing. There are many of these, all having as their object the cutting of the greatest number of radial cuts most economically. Another important feature of the grain of wood, which can be changed by the method of cutting, is found in the annual growth increment, which forms the annual rings of the cross section. In sawed woods these annual layers appear as lines or streaks, often elaborately twisting.

Another important step in the preparation of wood is drying or seasoning. This may be done either before or after the log is cut into boards. During this process much of the water present in the wood, both in the lumina of the cells and in the cell walls, themselves, is lost. As a consequence there is a certain amount of shrinkage of the wood, which is greater tangentially than in the other dimensions. As a result, the wood on drying tends to check or crack longitudinally, especially in large pieces of wood. For most purposes, undried wood, known as green wood, is entirely unsatisfactory; it will twist, warp, and crack as it dries; it cannot easily be glued, or finished, and it is more subject to attacks of fungi and boring insects. There are two methods of seasoning woods. The older method was to pile the wood in such a manner as to secure the maximum exposure to air, separating each piece from the next by narrow strips of wood, or by other means, and leaving the piled-up wood to dry slowly. This is called air-drying or air-seasoning. To dry wood in this way requires a month or more for boards an inch thick and several years for very large sticks, especially if the wood be a hardwood containing much water. Naturally the climate greatly affects the time required, wet seasons causing very slow drying. Recently much of the wood used, especially that used in making furniture, interior finish, shingles, etc., has been dried artificially in large rooms heated artificially. By this method it is possible to dry wood in a few days instead of months or years. The product is also much drier and better fitted for use. This method is known as kiln-drying or kiln-seasoning. Properly prepared wood has the following properties that give it its special value. It possesses great strength and incompressibility, yet is flexible, giving without breaking, and elastic, recovering after bending. Many woods have special properties that particularly fit them for special uses.

The uses made of wood are so numerous that only those of the greatest importance can be named. Immense quantities of wood, both hard and soft, are used in the pulp and paper industry. In construction work other large quantities are used, and in a variety of ways. Poorer grades of lumber are much used in making forms in which concrete work is poured. Scaffoldings and other framework also use quantities of the cheaper kinds of lumber. Wood is still a most important material for the construction of dwellings and other buildings. For the framework and for the rough finish of the walls and floors, the poorer kinds of wood are mostly used. For the outer parts, such as shingles and clap-boards, and for the interior finish materials and floors, better grades, often of special kinds of wood, are used. Shingles, for example, are mostly cut from white or red cedar, which is very resistant to the weather. Floors are made from hard pine, from maple or oak, and a few other woods which do not splinter easily and which resist heavy wear favorably. Nearly all the woods used in construction work are softwoods, which are light in weight, durable and easy to work.

For cabinet work and furniture an entirely different group of woods is used. Most of these are hardwoods, selected for their fine grains, or for their color, or because fashion and the whims of fancy happen to give them popularity. Many of the woods used for this purpose are tropical woods difficult to obtain and so of great value. In early days furniture was made from solid pieces; later it became the custom to use thin pieces of wood glued to a background of a different wood. These thin pieces were called veneer. There were several reasons for using veneer. One of course was economy, since an expensive block of wood would yield many more pieces of veneer than it would boards. Another reason was found in the fact that it was often difficult to fashion things from the hard dense woods used. They would split or crack easily, whereas when glued to a more easily handled more suitable woods they became usable. Finally, and very important, was the fact that often veneers could be obtained showing a beautiful grain which could rarely be obtained otherwise. So veneering became important.

There are three ways of cutting veneer wood. They may be sawed like any board: this method is wasteful since much material is wasted in the cut, as sawdust; it cannot produce very thin pieces, and it does not give the best grain in many cases. An advantage is found in the high polish that may be given veneers cut by this method. A second method of cutting veneers is slicing. In this method the short piece of wood, called a bolt, from which the veneers are to be cut, is moved up and down against a heavy stationary blade. This method has an advantage over sawing in that it allows much thinner veneers to be cut. As in sawed veneers, sliced veneers are limited to the size of the bolt from which they are cut, which is frequently a factor against them. The third method and the one used in cutting nearly all the veneers in the United States, is rotary-cutting. In this method the bolt is revolved against the knife, the veneer coming off in a thin sheet of any width desired. In order that the wood may be more easily cut and the veneer handled better, the bolt from which it is to be cut is usually soaked in water or steamed for some time before cutting, by both the slicing and the rotary method. After cutting, veneers are pressed flat and dried, in which condition they may be held indefinitely. They are commonly glued to some hardwood when they are used.

Another way in which veneers are used is in the making of plywood. This is made by gluing together three or more layers of veneer, the grain of each layer being at right angles to the one above it. Plywood is very resistant to blows and not easily cracked or broken. For which reason it is much used in making boxes, table tops, crates, and panels designed for various purposes.

In addition to these major uses of wood, there are many others. Cooperage is one of these. There are two kinds, slack and tight, each producing a barrel or container suited for a special purpose. Slack cooperage produces barrels, kegs, tubs, etc., which are used as containers for vegetables, for nails, and for many other things. Quantities of wood are used in the manufacture of charcoal, which is extensively used in making steel, explosives, and carbon dioxide gas, and also as a filter in the manufacture of sugar. Charcoal is made by allowing a closely packed pile of wood to burn with insufficient air, as a result of which the volatile materials are driven off and the carbon of the wood left. Millions of cords of wood are used for fuel each year in the United States; much of this is waste wood, which would be of little value for any other purpose. Large quantities of light softwoods, especially basswood, are used in making excelsior; in this the wood is first scored and then stripped off in thin pieces by knives. Excelsior is largely used to pack around glassware and dishes, and as a stuffing for mattresses and upholstery of the cheaper grades. Other uses of wood are for railroad ties, poles, posts, and piling, all of which call for woods which may be treated in such a way as to resist rotting and the attacks of various animals.

The following woods have properties that particularly fit them for special uses. *Lignum-vitae*, the heaviest of all woods, is a dark, greenish-brown, fine-grained wood with an oily appearance. It is obtained from *Guaiacum officinale*, a small tree native in tropical America. The wood of another species, *Guaiacum sanctum*, is also used. Both trees have compound leaves and showy flowers borne in clusters in the axils of the leaves. The fine-grained wood with its much-crossed fibers is used in making pulley sheaves, and as bushing around the propeller shafts of steamships, as well as for bowling balls and mallet heads. The wood contains a resin, guaiacum, which was formerly highly esteemed as a valuable medicine used in the treatment of social diseases. Today it is of slight importance as a drug. Rosewood is obtained from *Dalbergia nigra* and other species, natives of South America. It is hard close-grained wood with a pleasant fragrance. It is much used as a cabinet wood. Sandalwood is obtained from trees native in the East Indian region. It is a firm-textured wood of dull yellow color, which darkens with age. Like rosewood, sandalwood

has a characteristic aromatic odor. The tree, *Santalum album*, is interesting because it is a parasite on the roots of other plants. Satinwood is obtained from *Chloroxylon swietenia*, and other species, natives of India and Ceylon. The wood has a golden-yellow color and a brilliant luster, which makes it a valuable cabinet wood. Several species of walnut, particularly *Juglans regia* of Europe and Asia and *Juglans nigra* of North America, yield dark brown woods of great value. They are much used in furniture making, and in interior finish in houses, and also because, when once dried, they are not subject to any changes through swelling or shrinkage. They are particularly favored as material from which to make gun and rifle stocks. Walnut is capable of taking a very high polish. Another dark heavy wood which is much used in furniture making is ebony, the wood of *Diospyros Ebenum*, and other species. The sapwood of the tree is soft and creamy white and of little value; the heartwood is very dark, often black, and hard. There are many other less well known woods, which are frequently used in furniture making.

See also **Tree**.

WOODCHUCK. See **Squirrels and Other Sciuromorphs**.

WOODCOCK. See **Waders, Shorebirds, and Gulls**.

WOOD LOUSE (*Crustacea, Isopoda*). Small terrestrial crustaceans of oval form, somewhat flattened. They live under bark and debris near the surface of the ground, in decaying vegetation, and in other sufficiently moist situations and are common in gardens. Also called sowbugs. These forms constitute the families *Porcellionidae* and *Oniscidae*, and in the closely related family *Armadillididae* are found the pill bugs. The latter are able to roll themselves into almost perfect spheres when disturbed. The numerous species vary in color from gray to brownish and blackish.

WOODPECKERS AND TOUCANS (*Aves, Piciformes*). In addition to woodpeckers and toucans, the *Piciformes* include the barbets, jacamars, and honey-guides. These birds are found in all parts of the world except the Australian region. Most species nest in holes in trees. In the woodpecker, the beak is adapted for chipping wood, and the feet are formed for gripping the bark of trees. The tail is composed of stiff feathers and is used as a brace against the surface to which the bird clings. See Fig. 1. Woodpeckers excavate deep holes in trees as nests and deposit white eggs, the shells of which are like translucent china. They also dig into decaying wood for the insects contained in it. Capture of insects is facilitated by the sharp barbed tongue and sticky saliva. Both the barbed tongue and the stiff tail are lacking in a few genera.

Fig. 1. Downy woodpecker. (*Cruickshank.*)

Woodpeckers are well represented by numerous species, ranging from the diminutive piculets of South America and Asia to the great ivory-billed species, which attain a length of 18 inches (46 centimeters). North America has more than a score of species. The woodpecker's neck is so strong and its head so thick that it can withstand the vibrations set up for its chipping actions that no other bird could withstand. The woodpecker can beat against a tree at the rate of 20 pulses per second for nearly an hour at a time. The skull appears immune to this hard beating action. The cranium has been found to include a spongy shock-resistant tissue located between it and the beak. The tongue is about five times as long as the beak and is kept curled inside the mouth. Young birds cannot readily hold their heads up high because their heads are so heavy.

The wryneck is a bird related to the woodpeckers, but having soft tail feathers. The name stems from the curious habit of turning and extending the head displayed by the European species. The few known members of the group are found in Europe, Asia, and Africa. The flickers are moderately large woodpeckers of North America which differ from the other woodpeckers in their feeding habits. Flickers eat many insects on the surface of the ground and are particularly fond of ants, which they catch on their long sticky tongue. Three species are recognized: the common or yellow-shafted flicker, *Colaptes auratus*; the red-shifter, *C. cafer*; and the gilded flicker, *C. chrysoides*, of the southwest. Other species of the same genus occur in South America. The sapsucker is a North American woodpecker that cuts rows of small holes around the trunks of trees and visits them for the sap that exudes and for the insects that are attracted by it. The rows of holes are so regularly positioned as to be conspicuous. These birds are said to damage some trees seriously, causing the death of the tree in a few instances. The yellow-bellied sapsucker, *Sphyrapicus varius*, is the common eastern species. It is represented in the western United States by a variety, the rednaped sapsucker, and by two other species, the red-breasted and the Williamson sapsuckers, which occur in the far West. The jacamars are South American birds related to the woodpeckers. About the size of a robin, the jacamar is a slender bird with a chest of golden bronze and fiery red color. The beak is long and pointed and designed to catch flying insects. The feet are relatively weak and small. The jacamars drill cavities into tree branches for nesting. There are about 14 species in all.

Barbets are birds of several species intermediate between the woodpeckers and the toucans. They occur in the tropical regions of both Old and New Worlds, and some species are brilliantly colored. The name was formerly applied to other members of the order found in tropical South America, which are now known as puffbirds. The bill of the barbet incorporates bristles, which assist in catching insects. The coppersmith is a small Indian bird, *Xantholaema haematocephala*, a barbet with green coloring above and yellow-and-green markings below.

The toucan is a bird of South and Central America, characterized by an enormous beak. See Fig. 2. Most species are brilliantly colored and of large size. The beak, several times as large as the head, is of very light construction, with serrated edges used in cutting up fruit. The beak is covered with a thin horny shell and contains a fine bony reticulum. Toucans are also remarkable for their peculiar habit of tossing bits of food into the air and catching them in their mouth to be swallowed.

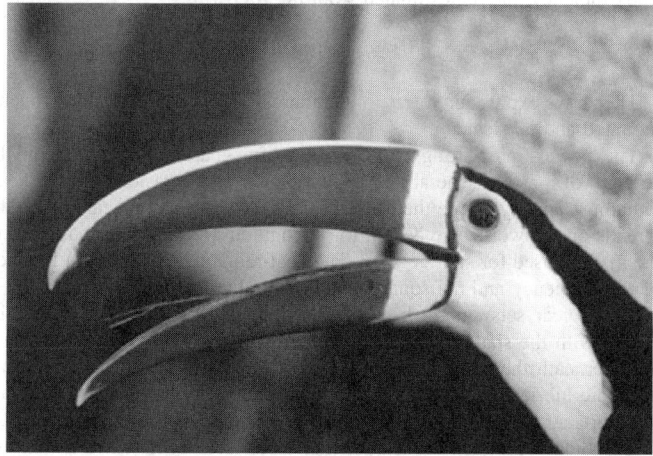

Fig. 2.   Toucan. (*A.M. Winchester.*)

Honey guides are birds of several African and Oriental species. They lay their eggs in the nests of other birds and are named from their reputed habit of leading the way to nests of bees.
   See also **Piciformes**.

WOOL.   The natural, highly crimped fiber from sheep, wool is one of the oldest fibers from the standpoint of use in textiles. Minute scales on the surface of the fibers allow them to interlock and are responsible for the ability of the fiber to *felt*, a phenomenon responsible for felt cloth and mill-finished worsteds. Crimpiness in wool is due to the open formation of the scales. Fine merino wool has 24 crimps per inch (~10 per centimeter). Luster of the fiber depends upon the size and smoothness of the scales. The basic wool protein, *keratin*, comprises molecular chains that are linked with sulfur. When sulfur is fed to sheep in areas deficient of the element, the quality of the wood improves. Wool fibers that fall below 3 inches (7.5 centimeters) in length are known as *clothing wool*; fibers 3−7 inches (7.5−17.8 centimeters) long are referred to as *combing wools*. The wool-fiber diameter ranges from 0.0025 to 0.005 inch (0.06−0.13 millimeter). See also **Fibers**; and **Goats and Sheep**.

WOOLLY BEAR (*Insecta, Lepidoptera*).   The densely hairy caterpillar of a moth of the family *Arctiidae*. Most hairy caterpillars belong to this family and most species of the family have hairy larvae.

WORD.   An ordered set of characters that is the normal unit in which information may be stored, transmitted, or operated upon within a given computer. For example, the set of characters 11598 is a word that may give a command for a machine element to move to a point 11.598 millimeters from a specified zero. A *computer word* is specifically defined as a sequence of bits or characters treated as a unit and capable of being stored in one computer location. (*IEEE.*)

WORK.   In the strict physical sense, work is performed only when a force is exerted on a body while the body moves at the same time in such a way that the force has a component in the direction of motion. In the simplest case where a constant force is applied in the same direction as the motion, it may be stated that work equals force multiplied by distance, or

$$W = Fs$$

An example could be the raising of mass m from elevation h_1 to elevation h_2 in a constant gravitational field where the acceleration of gravity is g. The work performed would be

$$W = mg(h_2 - h_1)$$

To generalize for more complex situations, the amount of work done during motion from point a to point b can be expressed by:

$$W = \int_a^b F \cos\theta \, ds$$

where F is the total force exerted and θ is the angle between the direction of F and the direction of the elemental displacement ds. In the cgs system, the unit of work is the dyne-centimeter or erg; in the mks system, the newton-meter or joule; and in the English system, the unit of work is the foot-pound.

In rotational motion, the foregoing definition can be exactly applied, but it is often convenient to express the force as a torque and the motion as an angular displacement. Thus, the work done will be:

$$W = \int_a^b \tau \cos\theta \, d\omega$$

where, in this case, θ is always the angle between the torque τ expressed as a vector quantity and the elemental angular motion $d\omega$, also expressed as a vector. The units of work performed in angular motion will, of course, be the same as in linear motion. It should be noted that the definition of work does not involve a time element. Power is defined as the rate at which work is performed.

WORK (Virtual).   When a system of particles subject to constraints is in equilibrium under the action of a set of impressed external forces, the total work done by these forces when the particles undergo small displacements compatible with the constraints (so-called virtual displacements) is zero. Analytically put,

$$\sum_{i=0}^{n} \mathbf{F}_{ie} \cdot \delta\mathbf{r}_i = 0$$

where the impressed external force on the ith particle of the system is $\mathbf{F}_{ie}$ and its virtual displacement is $\delta\mathbf{r}_i$. The sum is taken over all the particles.

WORK FUNCTION (Electronic). The energy (usually measured in electron-volts) needed to remove an electron from the Fermi level in a metal to a point an infinite distance away outside the surface. The work function is important in the theory of thermionic emission. In that case, as for example, that of an electron escaping from the heated, negatively charged filament of a vacuum tube, the work function may be called the thermionic work function. Photoelectric emission has a corresponding work function.

WORK HARDENING. Single crystals of pure metals show rapid plastic deformation at first, but this is often followed by a considerable increase in the shear strength. It is thought that this is due to the motion of dislocations along two different slip planes, which intersect and hence impede each other.

WORLD GEOGRAPHIC REFERENCE SYSTEM (GEOREF). A geographic reference system for the world, used in the Air Force for aircraft position reports and target designation, and for the control and direction of air units engaged in air defense, air–sea rescue, and tactical air operations.

WORM. A word without exact scientific limitations. Applied to creeping animals of the invertebrate phyla with long slender bodies, but also to some flatworms with broad thin bodies and only inaccurately to worm-like forms such as some of the insects. Scientifically it embraces six phyla: *Platyhelminthes* or flatworms, *Nemertea* or ribbon worms, *Nematoda* or roundworms, *Acanthocephala* or spinyheaded worms, *Nematomorpha* or hairworms, and *Annelida* or segmented worms.

WORM GEARING. Screw gearing or worm gearing is used to transmit power between shafts with perpendicular, non-intersecting axes. The worm is usually of cylindrical form, and resembles a screw; a section through the worm thread shows that the teeth are straight-sided and analogous to those of an involute rack. Worms are cut on a lathe or a thread milling machine and are often ground and polished after cutting and hardening to obtain surface precision and finish.

The worm wheel is essentially a helical gear with a face curved to fit a portion of the worm periphery. See Fig. 1. The tooth from and shape are obtained by cutting the wheel with a special form cutter known as a *hob*, which is essentially a replica of the worm, furnished with longitudinal flutes to provide cutting edges. In cutting the worm wheel teeth, the hob and the wheel blank are rotated at a speed ratio exactly that of the finished set; the hob is properly located with respect to the plane of the wheel and fed in radially until the teeth have been cut to full depth. This cutting action

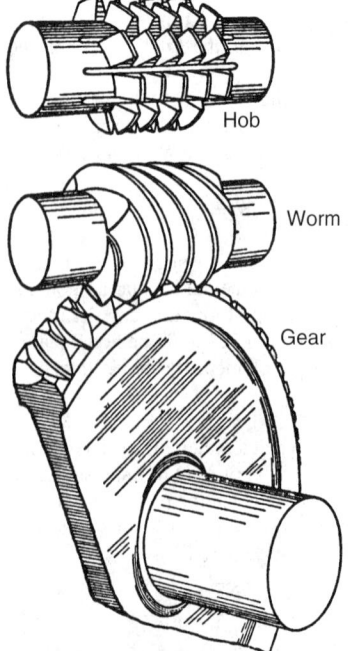

Fig. 1.   Worm gearing and hob.

generates worm wheel teeth that are of involute form at the midplane of the wheel, and are conjugate to the hob and consequently to the worm.

Tooth measurement in worm gearing is generally based on circular pitch, although diametral pitch gearing is manufactured and stocked by gear manufacturers. Circular pitch is measured in the diametral plane of the wheel and in a plane passing through the axis of the worm. If D_g represents the pitch diameter of the wheel, P_c the circular pitch, and N_g the number of teeth in the wheel, then

$$D_g = \frac{N_g P_c}{\pi}$$

The lead L of the worm is the distance that a thread advances in one turn, or the distance that a point on the pitch circle of the worm wheel will advance during one revolution of the worm. If N_w represents the number of threads or "starts" in the worm, then:

$$L = N_w P_c$$

A triple-threaded worm has a lead equal to three times the pitch; in a single-threaded worm the lead and pitch are alike.

The velocity ratio R of a worm gear set depends upon the lead of the worm and the pitch diameter of the wheel, or,

$$\frac{r/\min \text{Worm}}{r/\min \text{Wheel}} = \frac{N_g}{N_w}$$

Unlike most gearing, the velocity ratio is independent of the pitch diameter of one of the elements — the worm. The worm pitch diameter can therefore be selected to suit a particular center distance, or to make use of a stock hob and thereby dispense with the cost of a special cutter.

The lead angle H of the worm threads is the angle between a line tangent to the thread helix at the pitch line and a plane perpendicular to the axis of the worm. It is found from the following, where D_w represents the pitch diameter of the worm:

$$\tan H = \frac{P_c N_w}{\pi D_w}$$

The tooth pressure angle is measured in a plane passing through the axis of the worm, and is equal to one-half the thread profile angle A. Pressure angles of $14\frac{1}{2}°$ are commonly used for single- and double-threaded worms, and $20°$ for triple- and for quadruple-threaded worms. However, in many modern worm gear reducer sets, pressure angles as high as $30°$ are employed.

The nature of the tooth engagement in worm gearing causes greater sliding action between the surfaces in contact than in the case of spur gearing. The amount of this sliding action varies with the helix angle, and affects the efficiency of the gearing although it contributes to the smoothness of the drive. Efficiency depends not only on the material of the worm and worm wheel, the amount and character of the lubricant, the velocity of rubbing, but also upon the size of the helix angle of the worm. The efficiency may be estimated from the following:

$$E, \% = 100 - \frac{R}{2}$$

The above expression refers to commercial worm gear reducers properly mounted and lubricated.

The power-transmitting capacity of worm gearing is based upon the strength of the teeth, the ability to resist wear and abrasion, and the heat-radiating capacity. Since the teeth on the gear are usually weaker than the threads on the worm, the design for strength and for wear is analogous to the method used for spur gearing. The heat-radiating capacity is a function of the efficiency and the square of the pitch diameter of the gear.

Worm gear sets should be carefully aligned in the axial plane of the worm, with the shaft axes at $90°$. If the set is arranged so that the worm is underneath the wheel, the former may be run in an oil bath to insure adequate lubrication. Installations should preferably be enclosed to retain the lubricant and to prevent the admission of dust or foreign matter.

Globoidal worm gearing is a form of worm gearing in which a worm of "hour-glass" shape envelops a gear with straight-sided teeth. The capacity of this form of gearing is considerably greater than that of the conventional cylindrical type, and its efficiency is somewhat higher.

WRASSES (*Osteichthyes*). Of the order *Percomorphi*, family *Labridae*, there are over 600 species of wrasses, all carnivorous. There is a wide variety of sizes, from 3 inches (7.5 centimeters) in length (pencil varieties) up to 10 feet in length and weighing several hundred pounds (*Cheilinus*). They are frequently of a brilliant color. Wrasses prefer tropical reefs or open temperate marine waters. Some of the larger wrasses are considered good food items. They are noted for their canine teeth and protactile mouth. Many wrasses are considered to be of vicious temperament. They do not regularly travel in schools. The sand burrowing habit is a rather distinctive characteristic of the wrasses. They are active swimmers in daytime hours, but at night like to bury themselves completely under sand. A number of species of the smaller wrasses have been identified as cleaner fishes, that is, fishes that remove ectoparasites from other larger fishes, usually carnivorous. It has been learned that the larger fishes tend *not* to devour the known cleaners because of the services which they render. A commonly occurring labrid is the *Tautoga onitis*, which attains a length of about 3 feet (0.9 meter) and frequents the American Atlantic coastal waters. The *Lachnolaimus maximus* (pudding wife) with a length of about 18 inches (45 centimeters) and the *Halichoeres radiatus* (pearly razorfish) with a length of about 15 inches (37.5 centimeters) also prefer these Atlantic waters. Of about 12 inches (30 centimeters) in length and found in the eastern Atlantic from the Mediterranean northward to Norway is the *Labrus ossifagus* (cuckoo wrasse). Also among the wrasses are *Gomphosus varius* and *G. tricolor* (birdfishes) which are found in Indo-Pacific waters. The long-jawed wrasse (*Epibulus insidiator*), equipped with an extremely elongated lower jaw, is known for its different color phases and is found in Indo-Pacific waters.

WREN (*Aves, Passeriformes*). A small bird, related to the larger thrashers. Distinguished by its long curved beak and often sharply erected tail. Several species habitually build their nests about dwellings or in bird houses. From this habit the widely distributed and vociferous little house wren, *Troglodytes troglodytes*, has become widely known and the Bewick wren, *Thryomanes bewicki*, a more musical species, has introduced itself to many residents of the eastern states. The Carolina wren, *Thryothorus ludovicianus*, also frequents human habitations. North America is the home of a dozen species and Europe and Asia also have representatives of the group, but it attains its greatest diversity in South America.

WRIST. The slender part of the forearm at its attachment with the hand, especially the region containing the group of small carpal bones between the radius and ulna of the arm and the metacarpals of the hand. See also **Ganglion**.

WRONSKIAN. Let $y_1, y_2, \ldots, y_n$ be functions of x, each having non-vanishing derivatives $y_i, y_i, \ldots, y^{n-1}$, then the Wronskian of the function is the determinant

$$W = \begin{vmatrix} y_1 & y_2 & \cdots & y_n \\ y_1' & y_2' & \cdots & y_n' \\ \cdots & \cdots & \cdots & \cdots \\ y_1^{(n-1)} & y_2^{(n-1)} & \cdots & y_n^{(n-1)} \end{vmatrix}$$

If $W = 0$, the n functions are linearly dependent; if $W \neq 0$, they are linearly independent. When the y_i are solutions of a linear differential equation of nth order, evaluation of the Wronskian is a simple means for deciding whether or not the n functions give the complete solution of the differential equation.

See also **Determinant**.

WROUGHT IRON. A ferrous material aggregated from a solidifying mass of pasty particles of highly refined metallic iron, with which, and without subsequent fusion, is included a minutely and uniformly distributed quantity of slag. This definition of wrought iron indicates that it is a material made of two components; one, iron of a high degree of purity, the other, slag (chiefly silicate of iron). In the finished product the slag is distributed through the iron in threads and fibers, of which there is an enormous number. The slag imparts to the wrought iron a fibrous structure, quite different from the crystalline structure of cast metals. Wrought iron has made for itself a name as a metal which has resistance to corrosion, and which is exceptionally suitable for structural purposes where the structure is subject to shock. Wrought iron also can be readily worked, forged, machined, welded, galvanized, etc. Among the many applications of wrought iron might be mentioned tubes, pipes, and tanks.

WRYBILL. See **Waders, Shorebirds, and Gulls**.

WRYNECK. See **Woodpeckers and Toucans**.

WULFENITE. The mineral wulfenite is lead molybdate corresponding to the formula $PbMoO_4$, analyses showing that a part of the lead may be replaced by calcium. Wulfenite crystallizes in the tetragonal system usually in thin tabular forms, but is also found massive. It is a brittle mineral; hardness, 2.75–3; specific gravity, 6.5–7; luster, adamantine to resinous; color, yellowish to green or red, may be whitish or grayish; transparent to translucent. Wulfenite is a secondary mineral found in association with other lead minerals such as galena, and pyromorphite. It is believed to have been formed, at least in part, by the action of waters containing molybdenum salts on cerussite, anglesite, and pyromorphite.

Especially important localities are in the former Yugoslav Republics, the Czech Republic and Slovakia, Morocco, Zaire, New South Wales and Mexico. In the United States it has been found in Phoenixville, Pennsylvania, and in the Organ Mountains, New Mexico; Yuma County, Arizona; Box Elder and Salt Lake Counties, Utah; and in Clark and Eureka Counties, Nevada. Wulfenite was named in honor of F.X. von Wülfen, an Austrian mineralogist of the eighteenth century.

WÜRTZ-FITTIG-FRANKLAND REACTION. Sodium metal was used by Würtz as reagent for the preparation of paraffin hydrocarbons by treating alkyl iodide in ethereal solution, thus:

C_2H_5I	Na		C_2H_5	NaI
C_2H_5I	Na	(Ether)	C_2H_5	NaI
Ethyl iodide	Sodium		Normal-butane	Sodium iodide

The method has been applied to the preparation of paraffin hydrocarbons as high in the series as hexacontane $C_{60}H_{122}$. The alkyl radicals may be the same or different in the iodide or iodides taken.

Sodium metal was also used similarly by Fittig as reagent for the preparation of hydrocarbons by treating aryl bromide or iodide in the presence of dry ether, thus:

C_6H_5Br	Na		C_6H_5	NaBr
C_6H_5Br	Na	(Ether)	C_6H_5	NaBr
Phenyl iodide	Sodium		Biphenyl	Sodium bromide

When alkyl iodide and aryl bromide are taken, the hydrocarbon is of the mixed alkyl-aryl type, thus:

CH_3I	Na		CH_3	NaI
C_6H_5Br	Na	(Ether)	C_6H_5	NaBr
Methyl iodide	Sodium		Toluene	Sodium iodide
Phenyl bromide				Sodium bromide

CH_3I	Na		C_6H_4 $\diagup$ CH_3 (1)	NaI
C_6H_4 $\diagup$ CH_3 (1) $\diagdown$ Br (4)	Na	(Ether)	$\diagdown$ CH_3 (4)	NaBr
Methyl iodide	Sodium		Para-xylene	Sodium iodide
Para-bromo-toluene				Sodium bromide

The method has been applied to the preparation of substituted benzene hydrocarbons containing as many as four alkyl-groups (durene, $C_6H_2(CH_3)_4(1,2,4,5)$ and isodurene, $C_6H_2(CH_3)_4(1,2,3,5)$).

Frankland introduced the use of zinc instead of sodium to accomplish similar reactions. See also **Fittig Reaction**; and **Organic Chemistry**.

WURTZITE. A mineral zinc sulfide, (Zn, Fe)S, similar to sphalerite. Crystallizes in the hexagonal system. Hardness, 3.5–4; specific gravity, 3.98; color, brownish-black with resinous luster. Named after Adolphe Würtz, France.

X

XALSTOCITE. A pink to rose-pink variety of grossular garnet. Sometimes also called landerite or rosolite.

XANTHAN GUM. A very high-molecular-weight polysaccharide produced by pure culture fermentation of glucose by *Xanthamonas campestris*. The substance is readily soluble in hot or cold water, imparting a high viscosity at low concentrations. The solutions are pseudoplastic, with viscosity decreasing rapidly as shear rate increases. Heat, acid, and salt have little effect on the stability of its solutions. The substance is compatible with most other hydrocolloids, including starch. Xanthan gum undergoes a unique gel reaction with locust bean gum to produce a synergistic increase in viscosity. The gum is used as a thickening, suspending, emulsifying, and stabilizing agent in foods and has a number of important nonfood uses as well.

In 1974, the Northern Regional Research Center (Peoria, Illinois) of the U.S. Department of Agriculture and the Kelco Company were joint recipients of the Institute of Food Technologists award for the development and commercialization of xanthan gum. As early as 1956, researchers discovered unusual water-thickening abilities of a substance produced by the bacterium *Xanthamonas campestris*. They chemically identified the substance and named it *Polysaccharide B-1459* after the culture number. A food additive regulation for its use (when it was renamed xanthan gum) was issued in 1959. The process was brought into commercial production in 1964.

Glucose from starch is fermented by the aforementioned bacteria to produce xanthan gum, which is recovered by precipitation with isopropyl alcohol, then washed, dried, and milled. Among several applications, xanthan gum is used in pourable salad dressings for its emulsifying properties; in frozen foods for its freeze-thaw stabilizing effect; in juice drinks for its suspending properties; and in creamed cottage cheese for its stabilizing properties, pseudoplasticity, and mouthfeel. The gum makes a stable cream dressing that clings to the curd, but shears easily and thus does not have a gummy texture. The gum is also used as a stabilizer for frozen desserts and as a suspending agent in liquid feed supplements for cattle feeding and in milk replacers for calves. It also has been found that xanthan gum makes the proteins in nonwheat breads more extensible and thus produces better breads.

For references see the list in the entry on **Colloidal Systems**. Other gums and mucilages are described in the entry on **Gums and Mucilages**.

XANTHOPHYLLS. See **Carotenoids**.

X-BAND. A frequency band used in radar extending approximately from 5.2 to 10.9 kilomegacycles per second.

X-CHROMOSOME. One of the sex-determining chromosomes. It apparently carries genes for femaleness (XX) as opposed to male (XY) and, when possessing a fragile long arm, causes mental retardation. The X-chromosome can occur in single or double dose without deleterious effect. According to the Lyon hypothesis, a single X-chromosome is sufficient for normal development, and a Barr body may represent the second X-chromosome in condensed and inactive form. The inactivation of the X-chromosome appears to occur at random — so that paternal or maternal X-chromosomes have an equal chance of being inactivated. However, once inactivation has begun, early in embryonic life, the same X-chromosome will be inactivated in the descendants of each cell.

XENIA. That phenomenon, in which two plants, having been crossed, the characters of the male parent appear at once in the seed. Ordinarily the characters of neither parent are recognizable so early. Xenia is particularly well shown in corn plants, in which there is a conspicuous endosperm, the outermost layer of which is the aleurone layer. This endosperm results from the fusion of a sperm nucleus with two polar nuclei in the process of fertilization. The endosperm therefore is triploid. In corn it is frequently distinctly colored, with yellow dominating over white. It is possible therefore to pollinate a white-grained corn with a yellow parent, all the grains resulting from the cross being yellow since the gene which produces the yellow color is in every cell of the triploid endosperm. Red or purple colors often occur in corn, usually in the aleurone layer, and are dominant over colorless aleurone, so again crosses may be made showing directly the effect of genes from the male parent on the endosperm of the seed.

In some plants the influence of the male element is more extensive, involving tissues which are of female origin. This phenomenon is known as metaxenia. See also **Plant Breeding**.

XENOBLAST. A term proposed by Becke in 1903 for metamorphic crystals with undeveloped crystal faces.

XENOCRYST. A term proposed by Sollas in 1894 for crystals, usually corroded, which are foreign to the magma from which the igneous rock in which they occur has crystallized.

XENOLITH. A fragment, large or small, of a foreign rock included in an igneous mass. The term is derived from the Greek, meaning stranger and stone. Xenoliths, both large and small, are best displayed at the contacts or margins of batholiths.

XENOMORPHIC. The texture or fabric of an igneous rock having or characterized by crystals not bounded by their own crystal faces and which have their form impressed upon them by preexisting adjacent mineral crystals.

XENON. Chemical element, symbol Xe, at. no. 54, at. wt. 131.30, periodic table group 18 (inert or noble gases), mp $-112\,°\text{C}$, bp $-107.1 \pm 2.5\,°\text{C}$, density 3.5 g/cm^3 (liquid at $-109\,°\text{C}$). Specific gravity compared with air is 4.561. Xenon was found by Travers and Ramsay in 1898 when they were experimenting with liquefied air. Solid xenon has a face-centered cubic crystal structure. At standard conditions, xenon is a colorless, odorless gas and does not form stable compounds with any other element. Due to its low valence forces, xenon does not form diatomic molecules, except in discharge tubes. It does form compounds under highly favorable conditions, as excitation in discharge tubes, or pressure in the presence of a powerful dipole. Xenon forms a hydrate much more readily than argon, at a pressure slightly above 1 atmosphere pressure at $0\,°\text{C}$. The element also forms additional compounds with a number of organic substances, such as $\text{Xe} \cdot 2\text{C}_6\text{H}_5\text{OH}$ with phenol, which has a dissociation pressure of 1 atmosphere at $4\,°\text{C}$. See also **Chemical Elements**.

In 1962, the compound xenon platinum hexafluoride was synthesized by Bartlett. Later in the same year, Classen confirmed the synthesis and prepared the first binary compound of an inert gas, xenon tetrafluoride, a stable crystalline compound, mp about $90\,°\text{C}$. The compound was prepared by heating a 5:1 mixture of fluorine and xenon to $400\,°\text{C}$, then cooling it rapidly to room temperature. Since the original research, additional xylene-fluroine compounds have been reported, including XeF_2, XeF_4, XeF_6, XeF_6, XeSiF_6, XeO_2F_2, and Na_4XeO_6, as well as the hydrate. By heating xenon and fluorine above 10 atmospheres (up to 170 atmospheres), at $250\,°\text{C}$, Weinstock, Weaver, and Krop obtained XeF_5 and XeF_6 in an equilibrium mixture. D.F. Smith, S.M. Williamson, and C.W. Koch have reported the preparation of XeO_3 from the hexafluoride and tetrafluoride. XeO_3 is a white, crystalline, explosive compound.

Xenon also forms compounds, possibly clathrates, with certain substances in nonstoichiometric proportions. Crystalline compounds with benzene or hydroquinone, formed under 40 atmospheres pressure, contain about 26% xenon by weight. Alkaline hydrolysis of XeF_6 produces salts of octavalent xenon. No persistent divalent or tetravalent compounds are found in aqueous solution, but the former is intermediate in hydrolysis of the fluorides, and the latter in reactions of XeO_3 with XeF_2, H_2O_2, and various organics.

Xenon occurs in the atmosphere to the extent of approximately 0.00087%, making it the least abundant of the rare of noble gases in the atmosphere. In terms of abundance, xenon does not appear on lists of elements in the earth's crust because it does not exist in stable compounds under normal conditions. However, xenon because of its limited solubility in H_2O, is found in seawater to the extent of approximately 950 pounds per cubic mile (103 kilograms per cubic kilometer). Commercial xenon is derived from air by liquefaction and fractional distillation. There are nine natural isotopes ^{124}Xe, ^{126}Xe, ^{128}Xe, through ^{132}Xe, ^{134}Xe, and ^{136}Xe, and seven radioactive isotopes ^{123}Xe, ^{125}Xe, ^{127}Xe, ^{133}Xe, ^{135}Xe, ^{137}Xe and ^{138}Xe, all with relatively short half-lives, the longest ^{127}Xe with a half-life of about 36 days. See also Radioactivity. First ionization potential, 12.127 eV; second, 21.1 eV; third, 32.0 eV. Van der Waals radius 2.20 Å Electronic configuration $1s^2 2s^2 2p^6 3s^2 3p^6 3d^{10} 4s^2 4p^6 4d^{10} 5s^2 5p^6$.

Xenon is one of the elements of interstellar matter that is found in some meteorites. As reported by Lewis and Anders (1983), at least three types of xenon are present in carbonaceous chondrites. Two are abundant but controversial; the third is rare, but easy to explain. In 1964, Reynolds and Turner (University of California at Berkeley) examined the C2 chondrite Renazzo and were seeking xenon 129 (from radioactive decay of iodine 129). A controlled, step-heating process was used. In addition to finding xenon-129, the investigators found that in the fractions released between 600 and 1100 °C, the heavy isotopes of Xe ranging in mass from 131 through 136 were present. They found that these isotopes were enriched by as much as 6% with respect to primordial xenon. This enrichment increased from isotope 131 to isotope 136, as it does in the xenon formed by the fission of uranium and other heavy elements. Reynolds and Turner thus suggested that the new xenon component came from the fission of some extinct heavy element that had once been present in the meteorite. It was also noted that xenon 124 and 126 were also enriched in the meteorite, but inasmuch as these isotopes do not form by fission, their enrichment was not fully explained. The two sets of xenon components have been named H (heavy) and L (light) and, although apparently of different origins, they have proved inseparable in meteorites. See also **Meteoroids and Meteorites**.

Xenon also has been detected in some of the planetary atmospheres, such as that of Mars. Check various entries on planets in this book.

The gas finds principal application in special electronic devices and lamps. Xenon, in a vacuum tube, produces a beautiful blue glow when excited by an electrical discharge. Xenon lamps have been developed which provide a constant light (described as sunlight-plus-north-sky light) even when there are significant voltage changes. Thus, the lamps do not require voltage regulators. For a given wattage, xenon lamps have been found to deliver a greater light output. For example, an 800-watt xenon lamp will produce 2,000 lumens as compared with a 1,000-watt incandescent lamp that produces only about 200 lumens. Xenon also has found application in certain lasers. It has been found that xenon produces mild anesthesia, but cannot be used for surgery because the quantity required would cause asphyxiation. Xenon is used in bubble chambers, probes, and other applications where its high molecular weight is of advantage in the atomic energy field. Potentially, xenon is of interest as a gas for ion engines. The perxenates have been used as oxidizing agents in analytical chemistry. Xenon 133 and 135 are produced by neutron irradiation in air-cooled nuclear reactors. Xenon 133 has been found useful as a radioisotope in various studies.

Additional Reading

Anders, E.: "Noble Gases in Meteorites Evidence for Presolar Matter and Superheavy Elements," *Proceedings of the Royal Society of London*, Series A, Vol. 374, No. 1757, 207–238 (February 4, 1981).
Anderson, D.L.: "Composition of the Earth," *Science*, 367 (January 20, 1989).
Birgeneau, R.J. and P.M. Horn: "Two-dimensional Rare Gas Solids," *Science*, **232**, 329–336 (1986).

Greenwood, N.N. and A. Earnshaw: "Chemistry of the Elements," 2nd Edition, Butterworth-Heinemann, Inc., Woburn, MA, 1997.
Krebs, R.E.: "The History and Use of Our Earth's Chemical Elements: A Reference Guide," Greenwood Publishing Group, Inc., Westport, CT, 1998.
Lagowski, J.J.: "MacMillan Encyclopedia of Chemistry," Vol. 1, Macmillan Library Reference, New York, NY, 1997.
Lewis, R.S. and E. Anders: "Interstellar Matter in Meteorites," *Sci. Amer.*, **249**(2), 66–77 (August 1983).
Lide, D.R.: "CRC Handbook of Chemistry and Physics 2000-2001," 81st Edition, CRC Press, LLC., Boca Raton, FL, 2000.
Parker, P.: "McGraw-Hill Encyclopedia of Chemistry," 2nd Edition, The McGraw-Hill Companies, Inc., New York, NY, 1993.
Stwertka, A. and E. Stwertka: "A Guide to the Elements," Oxford University Press, Inc., New York, NY, 1998.

XENOPSYLLA CHEOPSIS. The species of Indian rat flea which is found in tropical regions. The flea is parasitic to humans and other animals and transmits bubonic plague and tapeworms.

XENOTOPIC. The fabric of a crystalline sedimentary rock in which the majority of the constituent crystals are anhedral. Fabric found in evaporites, chemically deposited cement, and recrystallized limestone or dolomite.

XEROPHYTES. These are plants that are adapted to grow in regions in which there is a decided lack of water, as in deserts. Plants growing in such localities have become modified in various ways that enable them to survive. Since the greatest problem is the lack of water, xerophytes must be formed as to avoid excessive loss of water. There are several factors that reduce water loss. In many of these plants the leaves are greatly reduced in size or are completely lost, as in most Cacti and many Euphorbias. Most of the water lost to the plant passes through the stomata of the leaves and stem epidermis. In many xerophytes the stomata are sunk deep in small pits. The number of stomata is often reduced and also their size. As a further protection against loss of water the epidermis may be covered with a very thick cuticle.

The sap of xerophytic plants has properties which make it hold water tenaciously, tending to reduce loss. Many xerophytes have a fleshy habit, so that greater space is available for storage of water. Sometimes the root is tremendously swollen, in others the stem becomes the enlarged part, while in many it is the leaves. Often the leaves form dense tufts which cause great reduction of free surfaces.

Xerophytes are extremely tolerant of prolonged drought. Some may be kept free from water for several years, yet revive and grow when supplied with water. Most of these plants cannot stand too much water, quickly rotting under such conditions. Many species of xerophytes are covered with thorns. This is commonly assumed to protect them against grazing animals.

Many xerophytes have extensive shallow root systems, which spread through a wide area, to take advantage of the occasional rains in the desert habitats. Others obtain water by sending roots very deeply into the soil. Creosote bushes, for example, may have roots several hundred feet (in excess of 120 meters) long.

XEROSIS. Abnormal dryness of the skin.

XEROSTOMIA. Insufficient flow or production of saliva, causing dry mouth.

XI PARTICLE. A hyperson with a rest-mass energy of about 1318.4 MeV, an isospin quantum number $\frac{1}{2}$, an angular momentum spin quantum number $\frac{1}{2}$, and a strangeness quantum number 2. Symbol, Ξ.

XIPHOSURA. The horseshoe or king crabs, a class of the phylum *Arthropoda* containing only a single genus, *Limulus*, with five species, one found along the Atlantic coast of North America from Florida to Nova Scotia and the other four on the eastern coast of Asia. See accompanying figure. They are animals of ancient lineage, showing no important change from fossils of the Triassic.

King crabs are distinguished by the following characters: (1) The cephalothorax is covered by a continuous arched plate of horseshoe shape. (2) Six segments of the abdomen are fused to form a continuous piece, hinged to the cephalothorax. (3) The abdomen bears a long caudal spine, the telson. (4) The appendages of the cephalothorax include six pairs associated with the mouth, five of them chelate. The five posterior pairs have the basal segments formed for crushing food. (5) The appendages of the abdomen are six pairs of broad plates bearing gills and used in swimming. (6) There are two large compound eyes and two smaller median eyes.

These animals burrow in sand and mud near the shore, feeding on small animals. They come to shore to deposit their eggs in sand. They are of little economic importance but have been used as food for pigs and domestic fowls, and to a limited extent as fertilizer. (See Fig. 1)

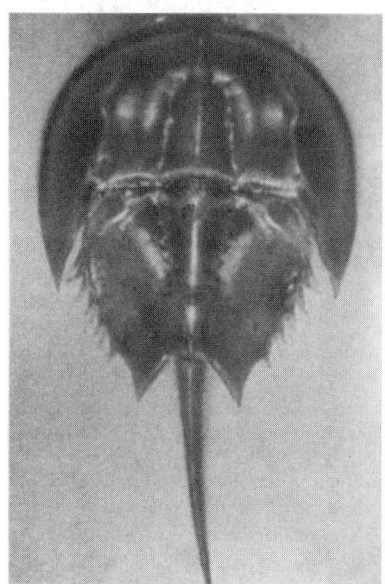

Fig. 1.   Xiphosura, King or horseshoe crab. (*A.M. Winchester.*)

XMM-NEWTON (The High-Throughput X-Ray Spectroscopy Mission). XMM-Newton is the European Space Agency's (ESA) second X-ray observatory. It was launched by an Ariane V rocket from Kourou, French Guiana on December 10, 1999, into an eccentric orbit with a perigee of 7000 km, an apogee of 107000 km and an inclination of $-40°$. This orbit allows long uninterrupted observations of up to 42 hours duration. With a projected lifetime of 10 years, XMM-Newton is allowing astronomers to conduct sensitive X-ray spectroscopic observations of a wide variety of cosmic sources. The first XMM-Newton announcement of opportunity (AO-1) attracted the attention of an astonishing $\sim$2000 professional astronomers worldwide, which amounts to approximately 25% of the world's astronomical community. The amount of proposals received was seven times higher than could be accommodated. XMM-Newton is the largest scientific satellite ever built in Europe with a mass of nearly 4 tons and a length of nearly 10 m (32.8 feet). See Fig. 1. Mission operations are conducted from the European Space Operations Center (ESOC) in Darmstadt, (Germany) and Science Operations from ESA's Vilspa tracking station in Villafranca (Spain). Ground stations in Kourou (French Guyana) and Perth (Australia) are used to provide almost continuous contact with XMM-Newton.

The limiting sensitivity for spectroscopic investigations with XMM-Newton is around 10^{-15} erg cm^{-2} s^{-1} and sources as faint as a few times 10^{-16} erg cm^{-2} s^{-1} can be detected. XMM-Newton uses grazing incidence Wolter I optics with a focal length of 7.5 m ($24\frac{1}{2}$ feet) to provide an effective aperture of 4500 cm^2 at 1 keV (12.4 Å) and 1000 cm^2 at 10 keV (1.24 Å). Each telescope module consists of 58 closely packed mirror shells with a maximum diameter of 70 cm ($27\frac{1}{2}$ inches) and a length of 60 cm ($23\frac{1}{2}$ inches). The angular resolution across the full energy range is 15 arc seconds half energy width, and the X-ray field of view is $\sim$30 arc minutes. A coaligned Optical Monitor (OM) provides simultaneous coverage of the wavelength range 1600–6000 Å. As well as broad band spectroscopy between 0.1 and 15 keV (0.8–120 Å) using the European Photon Imaging Counter (EPIC), a Reflection Grating Spectrometer (RGS) allows medium resolution spectroscopy with resolving powers between 100 and 700 over the wavelength band 5–35 Å (350–2500 eV).

EPIC consists of X-ray imaging cameras located behind each of the three telescope modules of XMM-Newton. The detectors are cooled CCD devices (2 MOS-CCD cameras, 1 pn-CCD camera), operating in a photon-counting mode to provide simultaneous imaging and nondispersive spectroscopy (spectral resolving power $\sim$7 at 0.2 keV to $\sim$70 at 15 keV) for every field that XMM-Newton observes. See Fig. 2.

The mission's aim of providing a medium-resolution spectroscopic capability is achieved by means of the two RGS. The grating arrays are placed directly behind two of the three mirror modules, in front of the EPIC MOS cameras, and each intercepts $\sim$50% of the converging beams. A position and energy sensitive readout at the spectroscopic (secondary) focus is performed by strip arrays of nine MOS CCDs for each RGS. The energy resolution of these CCDs is used to separate the overlapping diffraction orders -1 and -2 and to reject background arising from diffuse X-rays as well as from particle radiation or internal detector effects.

Fig. 1.   Artist's impression of XMM-Newton (*ESA/Ducros*).

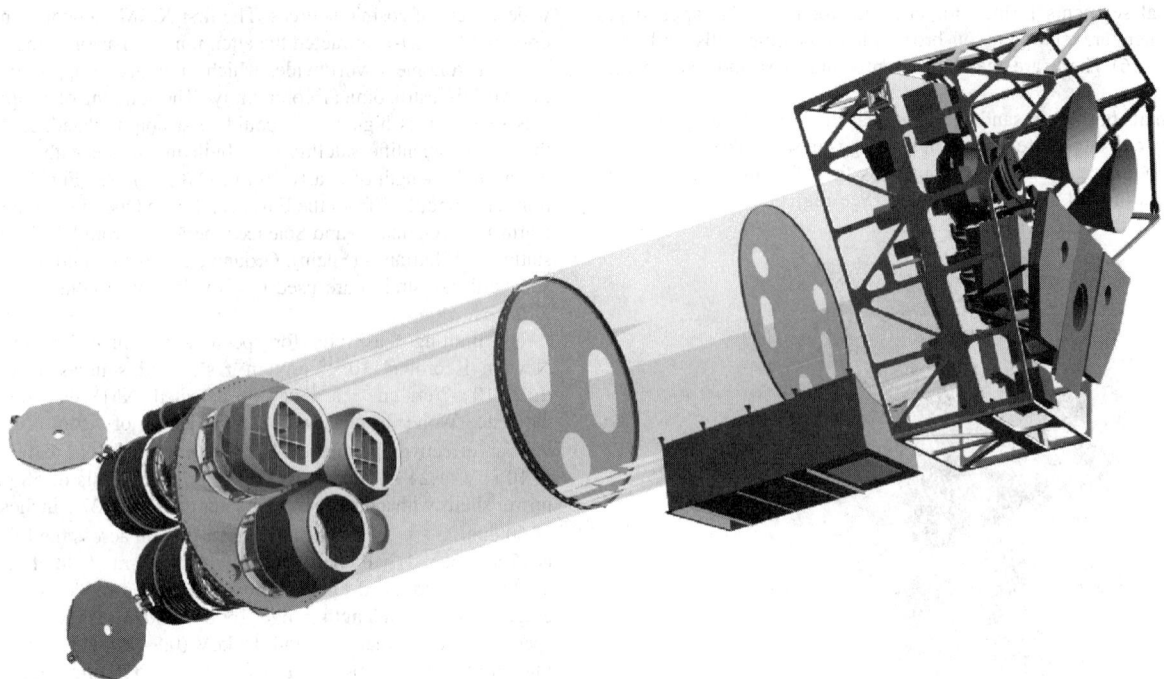

Fig. 2. If the XMM spacecraft were made of glass, one could gain a view of its payload. The X-ray telescopes, two with Reflection Grating Arrays, are visible at the lower left. At the right end of the assembly, the focal instruments are shown: The EPIC MOS cameras with their radiators (black/green "horns"), the radiator of the EPIC p-n camera (violet) and those of the (light blue) RGS receivers (in orange). The black box at the bottom of the bus is the outgassing device. (*Dornier Satellitensysteme GmbH.*)

The OM enables XMM-Newton to provide simultaneous coverage of the telescope field in the optical and ultraviolet wavebands The 30 cm Cassegrain telescope covers a 17 by 17 arcminute field with an angular resolution of ~1 arc second through the use of a photon-counting detectors. This instrument is also equipped with a standard set of U, B and V filters as well as specific UV wavelength band filters and 2 grisms for low-resolution dispersive spectroscopy over the full wavelength band.

The first science data from XMM-Newton fully demonstrate the capabilities of the spacecraft's telescopes and science instruments. The spacecraft viewed three regions of the sky: part of the Large Magellanic Cloud, the Hickson Cluster Group 16 and the star HR 1099. Results from these observations were officially presented at a press conference in 2000 February. These targets were chosen because they all present a variety of X-ray extended and point sources and are scientifically very interesting regions. XMM-Newton is now routinely making the observations proposed by the scientists involved in the development of the instruments and ground segment.

See also **X-ray Astronomy**.

Web References

European Space Agency XMM-Newton: *http://sci.esa.int/home/xmm-newton/index.cfm*

High Energy Astrophysics Science Archive Research Center HEASARC: *http://heasarc.gsfc.nasa.gov/*

F. JANSEN, ESA, Directorate of Scientific Programme, ESTEC, Noordwijk, The Netherlands.

X-NETWORK WINDOW. By way of software, a portion of a cathode-ray tube (CRT) can be singled out for detailed viewing upon command of the viewer. In this scheme, one or more windows can be designated as "dedicated" for use on an X-network, making it possible for several stations along the network to obtain instant information that may be of interest to several operators, as in a manufacturing or processing network.

X-RAY. In 1895, W. Roentgen of Würzburg, Germany discovered x-rays accidentally while experimenting with a Crookes tube. Roentgen observed the fluorescence of a barium platinocyanide screen that happened to lie near the tube and traced the effect to something that emanated from the spot where the cathode rays struck the tube wall. Putting the tube in a pasteboard box made no difference; so it was not light or ultraviolet radiation that caused the fluorescence. Investigation rapidly followed and

Fig. 1. Industrial-type x-ray photo of a Polaroid camera.

relatively soon, x-rays were being used by surgeons to examine the bones of living people. X-ray photos are also used in industry. See Fig. 1.

It was soon found that x-rays arise wherever cathode rays encounter solids; that "targets" of high atomic weight yield more copious X-rays; and that the greater the speed of the cathode particles, the more penetrating, or the "harder" the X-rays are. Special tubes were designed for producing x-rays. The earlier tubes were of the Crookes type, depending on the conduction of ionized gas. Those most used now are thermionic, of the Coolidge type, with a hot-wire cathode operating in a high vacuum; a construction which permits the passage of very high-speed electrons under voltage control, the quantity of them, and the intensity of the resulting X-rays, being regulated by the filament temperature.

Early experiments indicated that x-rays are something essentially different from light, an erroneous conclusion based upon the failure to observe regular reflection, refraction, or diffraction. We now know that this failure was due to the extremely short wavelength of the rays, the range of which extends from the extreme ultraviolet into the gamma-ray region, that is, from 10^{-7} to 10^{-9} centimeter. It is also known that X-rays are produced: (1) when electrons, accelerated in a vacuum, strike a target and lose kinetic energy in passing through the strong electric fields surrounding the target nuclei, thus giving rise to bremsstrahlung and resulting in a continuous x-ray spectrum; and (2) by the transitions of atoms from higher energy states to K, L, ... energy states, thus giving rise to characteristic x-rays. The term x-rays is not used to refer to the characteristic radiation from an element of atomic number Z less than 10, since the wavelengths of such radiation exceed those in the x-ray range. However, every element has its characteristic X-ray spectrum, when used as a target, although according to the Duane and Hunt law the radiation also depends on the accelerating voltage.

There are three principal means of detecting x-rays: the fluorescent effect, the photographic effect, and the ionizing effect. The only method at first available for distinguishing radiation of different wavelengths was to measure their penetration or their absorption coefficient in various substances. The discovery of the X-ray diffraction or grating effect of crystals, by von Laue, Friedrich, and Knipping, in 1912, made it possible to analyze the rays and measure their wavelengths very much as light is studied with the spectroscope. When X-rays of given wavelength are incident upon a crystal turned in various directions, the layers of atoms, at certain angles of incidence, reflect wave trains in phase with each other which, if caught on a photographic plate, produce a "Laue pattern." While the matter is not as simple as in the case of light incident on a diffraction grating, it is nevertheless possible to interpret such patterns in somewhat the same way as a line spectrum, and to deduce the wavelength from it. A unit convenient for expressing X-ray wavelengths is the "X-unit," which is 10^{-11} centimeter or 0.001 angstrom. See also **X-Ray Analysis**.

Additional Reading

Michette, A.G. and A. Pfauntsch: "X-Rays: The First Hundred Years," John Wiley & Sons, Inc., New York, NY, 1996.
See also list of references at the end of X-Ray Analysis.

X-RAY ANALYSIS. X-rays occupy that portion of the electromagnetic spectrum between 0.01 and 100 angstroms (Å). Their range of approximate quantum energy is from 2×10^{-6} to 2×10^{-10} erg, or from 106 to 100 eV. Important X-ray analytical methods are based upon: (1) fluorescence; (2) emission; (3) absorption; and (4) diffraction. These methods are used qualitatively and quantitatively to determine the element content of complex mixtures and to determine exactly the atomic arrangement and spacings of crystalline materials. See also **Ion Microprobe Mass Analyzer**.

Source of X-rays. X-rays are emitted by atoms that are bombarded with energetic electrons. This results from two separate effects: (1) deceleration of high-speed electrons as they pass through matter, and (2) ionization of individual atoms which abruptly stop the electrons. The first effect results in a continuous-type spectrum; the second effect results in characteristic line spectra.

Continuous Spectrum. The bulk of X-radiation arising from electron bombardment is the continuous spectrum. If an individual electron is abruptly decelerated, but not necessarily stopped, in passing through or near the electric field of a target atom, the electron will lose some energy DE, which appears as an X-ray photon of frequency $v = \Delta E/h$, where h is Planck's constant. An electron may experience several such decelerations before it is finally stopped, emitting x-ray photons of widely different energy and wavelength. A few electrons will be stopped in a single process, losing their entire energy and emitting an X-ray photon having the exact energy of the incident electron.

X-Ray Spectral Lines. These result when the incident electrons knock orbital electrons out of an atom. If an ejected electron is from one of the inner orbits of the atom (Fig. 1), an electron from an outer shell will fall to the inner orbit to fill the vacancy. The decrease in potential energy of this electron in approaching the nucleus results in the emission of an X-ray

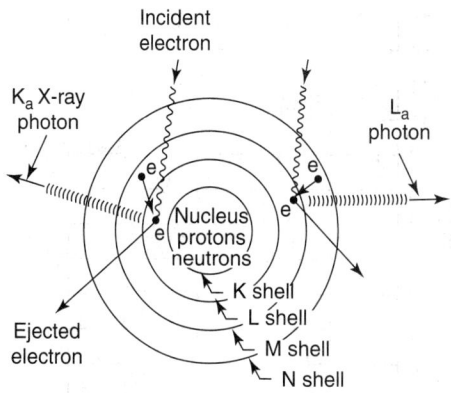

Fig. 1.   Origin of X-ray spectra due to electron bombardment.

photon having an energy exactly equal to that lost by the electron. The wavelength λ for such photons is related to ΔE by $\lambda = ch/\Delta E$, where c is the velocity of electromagnetic energy and h is Planck's constant. Because the energy of orbital electrons is quantized, the X-ray photons can have only certain definite wavelengths which are characteristic of the atom. This situation is somewhat analogous to the more familiar ultraviolet and visible-emission spectra of materials, the difference being that the optical spectra are the result of electron transitions between energy levels of just the outermost electrons of the atoms.

X-rays resulting from an electron transition filling an electron vacancy in the innermost shell of an atom are known as K x-rays or K lines; those from the L shell are known as L lines, and so on.

Generation of X-rays. An important component in an X-ray analytical device is an x-radiation generator. A high-vacuum Coolidge type tube, wherein electrons are emitted from a heated tungsten filament and accelerated by a high voltage to an anode (target) is a common source of x-rays. See Fig. 2.

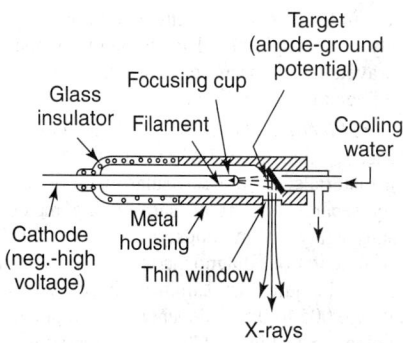

Fig. 2.   High-voltage, high-vacuum X-ray tube.

A wide variety of tubes is available. All high-power (high-current) commercial tubes employ a water-cooled anode. Tubes of this type have been built with ratings up to 10 kilowatts.

Detection of X-rays. Detectors include (1) Geiger-Mueller tube, (2) ionization chambers, (3) scintillation counters, (4) proportional counter, (5) electron-multiplier tubes, and (6) nondispersive detectors using cooled lithium-drifted Si detectors. See Fig. 3.

X-ray Crystallography. X-rays penetrating below the surface of crystalline materials are scattered by the individual parallel layers of atoms; each atomic layer acts as a new, although weak, source of X-rays. To be reinforced in a given direction at an angle θ (Fig. 4), the spacing d between crystal planes must be rigorously related to the wavelength of the radiation. At a given angle, X-rays of one definite wavelength will be constructively reinforced. These variables are related by Bragg's law.

$$n\lambda = 2d \sin \theta$$

where n is an integer. Note that θ is measured relative to the crystal face rather than to the perpendicular.

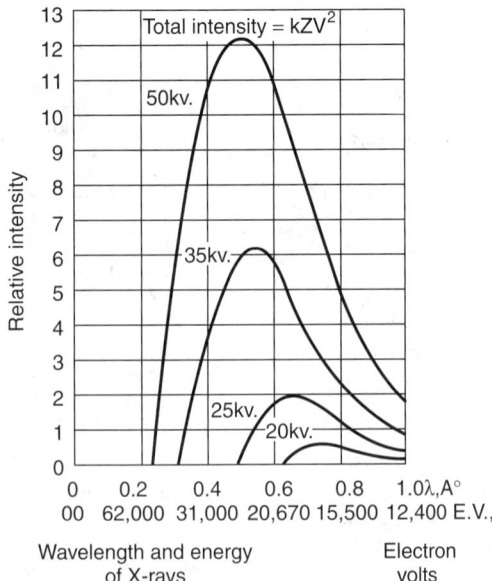

Fig. 3. Continuous X-ray spectrum of tungsten ($Z = 74$) at various tube voltages.

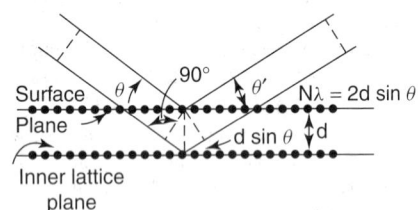

Fig. 4. Reflection of X-rays from internal crystal plane.

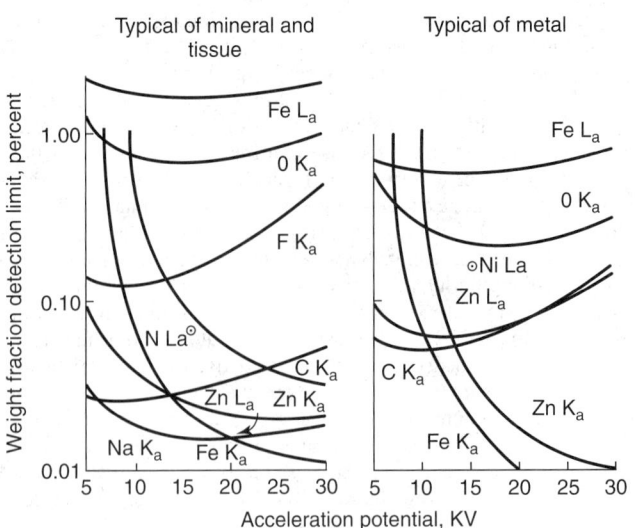

Fig. 5. Typical weight-fraction detecting limits of electron microprobe X-ray analyzer.

In addition to fulfilling this wavelength requirement, the energy will be diffracted only at an angle equal to the angle of the incident x-rays, independent of wavelength; otherwise, destructive interference is possible, and the "reflected" energy is negligible.

X-Ray Analyzer, Electron-Microprobe. Advantages of this analytical instrument include: (1) analysis can be confined to very small (microsamples) amounts of materials; (2) the particular material to be analyzed need not be physically separated from its surrounding materials, as is often required with many analytical methods; and (3) through the development of associated instrumentation, diagnostic techniques, and information displays, the method can be quite fast. Limits of detection in solid solution are from approximately 0.005 to 0.5%, depending upon the elements and sample matrixes involved. See Fig. 5. Concentrations as low as 10^{-16} gram may be measured.

Mainly used for metallurgical studies, nonmetallics also may be analyzed when samples are properly prepared. Biological applications include tooth and bone samples, cytochemical problems and staining techniques, physiochemical problems, and studies in pathology. Relative weight-fraction-detection limits for most elements in biological specimens are in the general range of 0.01 to 0.10%. Electronics industry applications include studies of diffusion phenomena, electrical-contact surfaces, interfaces on transistors, and microcircuitry analysis.

As shown by Fig. 6, electrons from an electron gun are directed to the sample through an electron optical system. Once the electron beam strikes the sample, a number of signal sources are activated, including (1) high-energy backscattered electrons, (2) low-energy secondary electrons, (3) cathodoluminescence, and (4) x-rays. Some heat also is generated within the sample. Volume d_3 of the specimen is that *volume from which X-rays are emitted.*

The X-rays produced may be detected nondispersively by a proportional counter whose output may be separated as a function of energy by a pulse-height analysis system into the various wavelength components. Better detection sensitivities, however, can be obtained through the use of a fully focusing diffracting-crystal spectrometer in conjunction with a proportional detector and the necessary pulse-height analyzer. As shown by Fig. 7, the

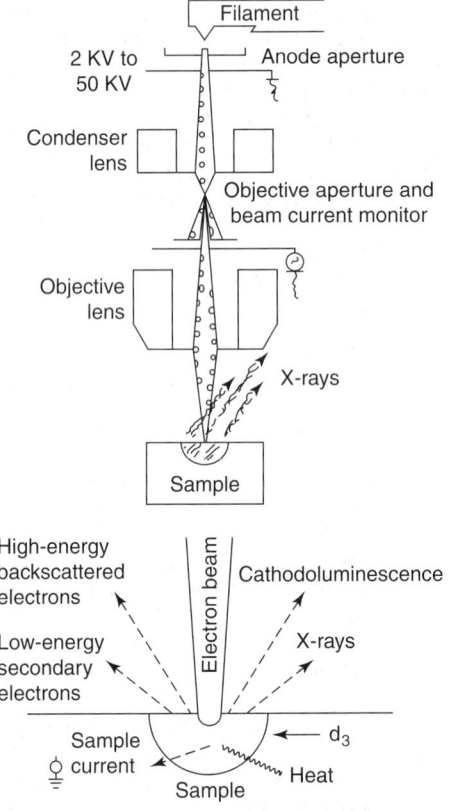

Fig. 6. Electron gun and probe-forming lens system of an integrated electron probe.

necessary condition for fully focusing optics is to have the x-ray source, the crystal, and the detector slit all placed on a common circle. This geometry requires that the diffracting-crystal planes be bent to the diameter of the Rowland circle.

Spherical aberration at the detector slit is minimized by further grinding the crystal surface to fit the radius of the Rowland circle. With the resultant Johansson optics, the crystal radius is fixed, and the 2θ range of the spectrometer is scanned by moving the crystal radially away from the source and, at the same time, rotating it into the detector to achieve a true focus throughout the spectrometer range. X-rays particularly of a wavelength greater than 2 Å, and electrons are highly absorbed in an air atmosphere. Thus, the spectrometer must be enclosed in a vacuum of the order of 10^{-5} torr. The present wavelength range of interest extends from approximately 1 to 100 Å. Diffracting crystals to cover this range

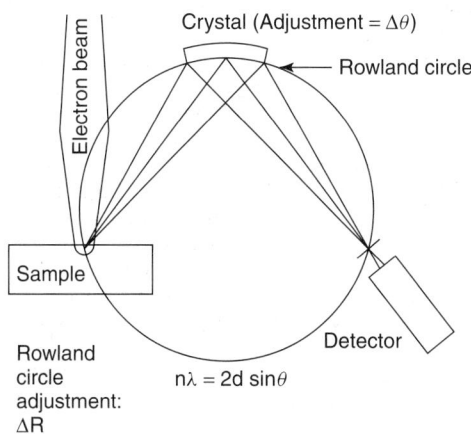

Fig. 7.   Geometry of a fully focusing diffracting-crystal spectrometer.

must provide broad wavelength coverage, high diffraction efficiencies for high peak-count intensities, good resolution, and good resulting peak-to-background ratios. Crystals that meet these objectives include lithium fluoride, ammonium dihydrogen phosphate (ADP), ethylenediamine *d*-tartrate. (EDT), quartz, and sodium chloride.

Detectors. Of the three commonly used X-ray detectors—(1) Geiger counter, (2) scintillation counter, and (3) proportional counter—the latter is used most frequently for electron-probe microanalysis. In the wavelengths from 1 to 10 Å, sealed proportional counters may be used. For longer-wavelength analysis—in the range from 10 to 93 Å—the thinnest possible detector window is required to limit spectral attenuation. Nitrocellulose windows have proved successful. Nondispersive detection systems using cooled Li-drifted Si are also applicable.

An *optical microscope* is required in the system to provide the analyst with a means of reference to identify various sample areas for analysis. Sample stages may hold single or multiple samples and are provided with means for moving the sample in x, y, and z planes without breaking the system vacuum. After the point of interest is located on the specimen, the data may be read out in a number of ways: (1) quantitative and seimquantitative information may be obtained by processing the x-ray detector signal through a rate meter to a strip-chart or X–Y recorder; (2) scaler systems also provide direct readout of quantitative data integration; and (3) for operational convenience, a data translator and typewriter or teletype printout system may be connected directly to record digital-counter information as hard copy.

X-ray Fluorescence Analysis. One of several types of spectrochemical techniques now used for laboratory analysis. The method is nondestructive. The characteristic X-ray spectrum of each element bears a simple direct relationship to the atomic number. The relation of the wavelength λ to the atomic number Z is

$$\frac{1}{\lambda} a Z^2 \quad \text{(Mosley's law)}$$

Since the X-ray spectral lines come from the inner electrons of the atoms, the lines are not related to the chemical properties of the elements or to the compounds in which they may reside. Because the characteristics of the X-ray spectra are associated with energies released through transitions of electrons within the inner shells of the atom, the spectra are simple. Most practical X-ray fluorescence analysis involves the detection of radiation release through electron transitions from outer shells to the K shell (K spectra), outer shells to the L shell (L spectra) and, in very few cases, from outer shells to the M shell (M spectra).

The simplest form of energy source available for commercial instrumentation is that obtained from an X-ray tube. For samples containing predominantly low-atomic-number elements, as in cement raw mix, the most efficient excitation is accomplished by using an X-ray tube target material of relatively low atomic number, such as chromium. Elements having higher atomic numbers are most effectively excited by high-atomic-number targets, such as tungsten or platinum. An optimum target material is rhodium for the analysis of a broad range of elements. The X-ray tube irradiates the sample, which in turn emits characteristic fluorescent radiation of its atoms.

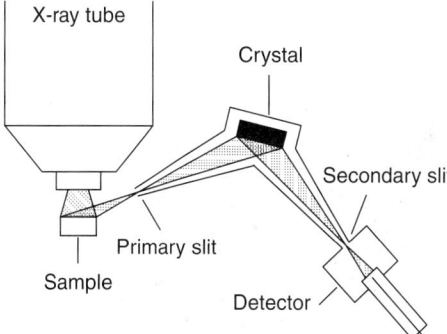

Fig. 8.   Optical path for one monochromator used in an industrial X-ray fluorescence analyzer.

Once X-ray fluorescence is produced from the sample by means of an X-ray tube, appropriate components of the instrument (Fig. 8) separate this radiation into its characteristic wavelengths, detect the energy emitted from each excited atom, and produce a signal that is representative of the number of atoms (concentration) of the elements in the sample. Typical excitation conditions (x-ray tube) are 50 kilovolts, 35 milliamperes. Bragg's law of X-ray diffraction is satisfied by the condition $N\lambda = 2d \sin q$, where λ is the wavelength and θ is the angle of incidence and diffraction of X-rays from a crystal whose lattice spacing is defined by d and N, the order of harmonic of the diffraction. For almost all fluorescence analysis, $N = 1$. The usable x-ray spectrum normally extends from 0.1 to about 20 Å. A helium or vacuum path is required for x-ray analysis of elements with atomic numbers lower than $Z \cong 24$ (wavelengths longer than 2.3 Å).

Instrumentation can provide for simultaneous elemental analysis using fixed, preselected X-ray detection channels and scanners or a goniometer to provide one or more channels that can be tuned to a wide wavelength coverage. Up to 30 monochromator positions are possible. Typical crystal materials covering the practical wavelength range of 1 to 20 Å are lithium fluoride, silicon oxide, sodium chloride, EDT, and ADP. Optimum analytical data for the elements of interest are obtained by using fully focusing Johansson curved and ground crystals.

An optical diagram of a Johansson curved-crystal spectrometer is given in Fig. 9. Each spectrometer of an x-ray quantometer may be equipped with optimum crystal-detector combinations for specific determinations in a wide variety of matrixes, including steel, aluminum, copper-base materials, ores, cement, and slags—in both liquid and solid states.

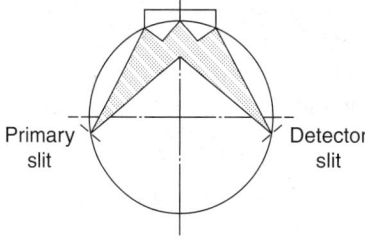

Fig. 9.   Optical diagram of Johansson curved-crystal spectrometer.

The diffracted X-radiation is detected by Geiger, proportional, or semiproportional detectors. See Fig. 10. The detector of each monochromator generates pulses, which are a measure of the intensity of radiation of each wavelength. The pulses are filtered through a discriminator in order to avoid undesired interferences. Pulses shrinking due to an increase of frequency of pulses is automatically compensated. Collected pulses are transferred to a computer for processing and output. See Fig. 11.

The transition from laboratory to automated instrument to achieve high-speed continuous analysis of dry or wet materials primarily involves the sample-handling and presentation hardware.

Limits of detectability for the desired elemental analyses vary depending upon the matrix, elements, methods of sample preparation, and quality of instrumentation applied. Generally, these are on the order of 1 to 100 parts per million. The limit of detectability, however, is only one criterion in evaluating methods of analysis. The time of analysis is important,

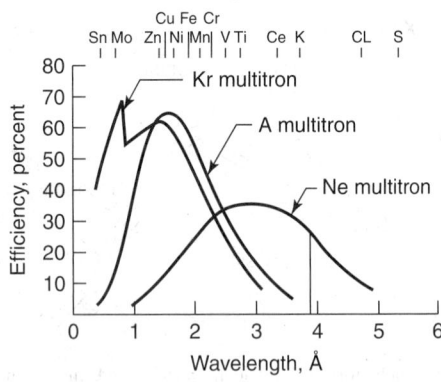

Fig. 10.   Relative efficiency of X-radiation detectors.

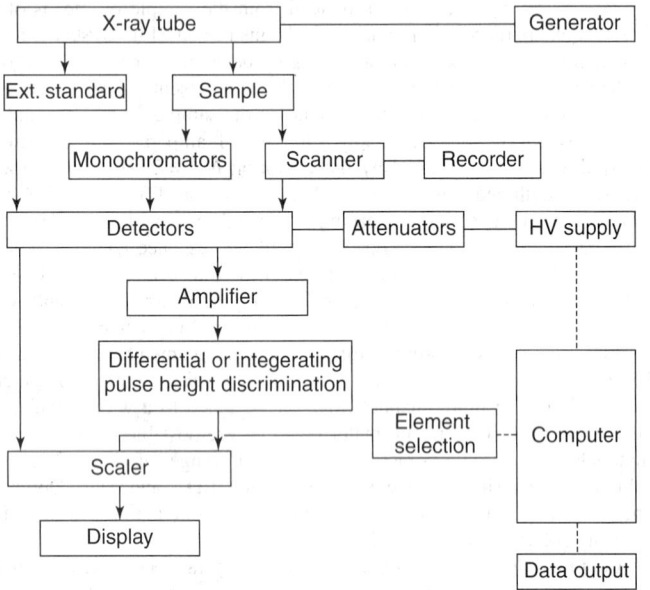

Fig. 11.   System configuration of X-ray fluorescence analyzer.

particularly in production and process control laboratories. In multi-element spectrometers, it is possible to perform as many as 30 simultaneous elemental determinations in from 20 to 120 seconds, depending upon the material being analyzed.

Additional Reading

Dunitz, J.D.D.: "X-Ray Analysis and the Structure of Organic Molecules," John Wiley & Sons, Inc., New York, NY, 1995.

Gilfrich, J.V., et al.: "Advances in X-Ray Analysis," Perseus Publishing, Boulder, CO, 1998.

Hammond, C.: "The Basics of Crystallography and Diffraction," Oxford University Press, Inc., New York, NY, 1998.

Janssens, K.H., F. Adams, and A. Rindby: "Microscopic X-Ray Fluorescence Analysis," John Wiley & Sons, New York, NY, 2000.

Jenkins, R., D.K. Smith, and V.E. Buhrke: "A Practical Guide for the Preparation of Specimens for X-Ray Fluorescence and X-Ray Diffraction Analysis," John Wiley & Sons, Inc., New York, NY, 1997.

Lifshin, E.: "X-Ray Characterization of Materials," John Wiley & Sons, Inc., New York, NY, 1999.

Suryanarayana, C. and M.G. Norton: "X-Ray Diffraction," Perseus Publishing, Boulder, CO, 1998.

Warwick, T. and D. Atwood: "X-Ray Microscopy," American Institute of Physics, College Park, MD, 2000.

W.G. SHEQUEN, P. E.

X-RAY ASTRONOMY.   The X-rays detected by X-ray astronomers, like those put to use in industry, medicine and laboratory research, must be produced by high-energy particles. It is not surprising, then, that an X-ray image of the cosmos can look very different from an optical image.

X-ray images reveal hot spots in the universe—regions where particles have been energized or heated to temperatures of millions of degrees by strong magnetic fields, violent explosions, or intense gravitational fields. X-ray astronomy has shown that these extreme conditions exist in an astonishing variety of places, ranging from the bizarre warped space around black holes to the vast spaces between galaxies.

Because the Earth's atmosphere absorbs X-rays, X-ray observatories must be placed high above the Earth's surface. This means that the ultra-precise mirrors and detectors, together with the sophisticated electronics that conveys the information back to Earth must be able to withstand the rigors of a rocket launch, and operate in the hostile environment of space.

The first hint that cosmic X-rays exist came in 1949, when a team of scientists led by Herbert Friedman at the Naval Research Laboratory in Washington, DC used detectors aboard a rocket to detect X-rays coming from the sun. Solar X rays come from the hot outer layers of the sun's atmosphere called the corona.

Thirteen years later, in 1962, a team led by Riccardo Giacconi of American Science and Engineering (AS&E) in Cambridge, Mass. using a greatly improved detector, discovered X-rays coming from a source beyond the solar system in the constellation of Scorpius. This source was named Scorpius X-1. They also discovered that the X ray sky is never dark, but is illuminated with a smooth background glow.

In contrast to the sun, which radiates only one-millionth of its energy in X rays, Scorpius X-1 radiates over 99% of its energy as X rays. This showed that the AS&E team had discovered an entirely new type of object. Within a year these results were confirmed by the AS&E group and by the Naval Research Laboratory group. A new field of astronomy had been born.

By 1970 about thirty cosmic X-ray sources had been discovered. Some were associated with the remnants of supernova explosions, one was associated with a giant galaxy called M87, and another with 3C273, an extremely distant, energetic type of source called a quasar. Most of the sources were X-ray stars like Scorpius X-1, whose nature at this time was still unknown.

In the early 1970s, NASA's Uhuru X-ray satellite, equipped with a relatively simple instrument, a sensitive X-ray detector similar to a Geiger counter attached to a viewing pipe to locate the source, made some remarkable discoveries. Uhuru, which was built and operated by the AS&E team, detected over 400 X-ray sources. Uhuru's observations helped scientists solve the mystery of the X-ray stars. They showed that they were neutron stars or black holes pulling matter from nearby companion stars. As the matter falls onto the neutron star or black hole, it is heated to tens, even hundreds of millions of degrees and becomes a bright X-ray source. See also **Galaxy**.

A black hole does not have a surface in the normal sense of the word. See also **Black Hole**. It is more like a whirlpool in space with a critical range of influence. The critical distance from a black hole is called the event horizon. Anything that is pulled inside the event horizon, matter, visible light, X rays, is doomed. It cannot escape the black hole and must fall inward. X ray observations allow scientists to observe matter just a fraction of a second before it passes beyond the event horizon. Since the discoveries of Uhuru, one of the primary objectives of X-ray astronomy has been to learn more about these bizarre objects.

Uhuru also detected X rays from over a dozen supernova remnants, from a few galaxies and from gigantic systems containing thousands of galaxies. These discoveries emphasized the need for an X-ray telescope that could make images like an optical telescope.

This was a difficult undertaking, because X-rays do not reflect off mirrors the same way that visible light does. The energetic X-ray photons penetrate into the mirror and are absorbed. They reflect only if they strike the mirror at grazing angles, in much the same way that a pebble skips across a pond. X-ray mirrors must be precisely shaped and aligned nearly parallel to incoming X-rays. The mirrors look more like barrels than the familiar dish shape of optical telescopes.

NASA's Einstein Observatory, launched in 1978, was the first large X-ray telescope with mirrors capable of making images of sources outside the solar system. It imaged, shock waves from exploded stars, detected X-rays from normal stars, white dwarfs, neutron stars, black holes, galaxies, and clusters of galaxies containing over a thousand galaxies. In all, Einstein located accurately over 7,000 X-ray sources. The observations of the hot gas in galaxy clusters made possible a new way to study the mysterious dark matter that surrounds many galaxies. See also **Einstein Observatory (HEAO-2)**.

The Roentgensatellite or ROSAT, a joint venture between Germany, the United Kingdom, and the United States, carried an even larger X-ray telescope into orbit in 1990. It expanded the number of known X-ray sources to more than 60,000 and has proved to be especially valuable for investigating galaxy clusters and the coronas of stars. See also **ROSAT (Roentgen Satellite)**.

In the 1990s, several X-ray satellites were launched, culminating with NASA's Chandra X-ray Observatory and the European Space Agency's XMM-Newton, both in 1999.

In 1993, the Advanced Satellite for Cosmology and Astrophysics (ASCA) was launched. ASCA, was Japan's fourth cosmic X-ray astronomy mission, and the second for which the United States is providing part of the scientific payload. ASCA was the first satellite to use a CCD X-ray camera in conjunction with its X-ray mirrors. This camera, which operates much like a camcorder or digital camera, allowed astronomers to make images with more information about the energy of the individual X-rays. ASCA operated successfully till July 15, 2000 when it was transferred into a safe-hold mode. The satellite re-entered on March 2, 2001 after seven and half years of scientific observations.

ASCA was followed by the Rossi X-ray Timing Explorer (RXTE), launched December 30, 1995. The RXTE was a special purpose NASA satellite that has the unique capability to study rapid time variability in the emission of cosmic X-ray sources over a wide band of X-ray energies, and has made valuable contributions to the understanding of neutron stars and black holes. See also **Rossi X-ray Timing Explorer (RXTE)**.

The BeppoSAX satellite is a major program of the Italian Space Agency with participation of the Netherlands Agency for Aereospace Programs. It was launched on April 30 1996 from Cape Canaveral and it is still operational. It is the first X-ray mission with a scientific payload covering more than three decades of energy from (0.1 to 300 keV) with a relatively large effective area, medium energy resolution and imaging capabilities in the range of 0.1–10 keV. BeppoSAX has proved to be especially useful for X-ray imaging the sources associated with gamma-ray bursts, and monitoring the X-ray afterglow of these puzzling sources. See also **BeppoSax (Satellite)**.

The XMM-Newton was launched from Kourou, French Guiana on December 10, 1999. XMM-Newton is a large Multi-Mirror X-ray astrophysics observatory developed by the European Space Agency (ESA) and the second cornerstone of the Horizon 2000 Science Programme. XMM-Newton carries three very advanced X-ray telescopes. They each contain 58 high-precision concentric mirrors, delicately nested to offer the largest collecting area possible to catch the elusive X-rays, and an optical monitor, the first flown on an X-ray observatory. This observatory will enable astronomers to conduct sensitive measurements of the X-ray spectrum of a wide variety of cosmic sources, and should be a powerful complement to the Chandra X-ray Observatory. See also **XMM-Newton (The High-Throughput X-Ray Spectroscopy Mission)**.

NASA's Chandra X-ray Observatory was launched by the Space Shuttle Columbia in July of 1999. It contains four sets of nested mirrors, which can produce images with 30 times more detail than any previous or existing X-ray telescope. It is the only X-ray telescope capable of producing X-ray images of the same quality as an optical telescope. The mirrors have been polished to a smoothness of a few atoms. If the surface of the state of Colorado were polished to same relative smoothness as Chandra's mirrors, Pike's Peak would be less than one inch tall!

Chandra's mirror assembly has four pairs of mirrors nested inside one another. The mirror surface is coated with iridium, a material more reflective than gold. Two sensitive, electronic X-ray cameras are used to collect the X rays focused by the mirrors. Scientists specify which camera will be at the focus for their observation. The High Resolution Camera uses an array of 69 million tiny glass tubes and a grid of electrically charged wires to determine the position and arrival time of each individual X ray and to reconstruct a high-resolution image of the celestial source of the X rays. The other camera is the Advanced CCD Imaging Spectrometer (ACIS) which works much like a camcorder or digital camera. It contains ten X-ray-sensitive CCD chips, each with over a million pixels, to record the position and energy, or color of the X rays. Even more detailed information about the energy of the incoming X rays can be obtained by inserting two screen-like instruments, called transmission gratings, into the path of the X rays. The gratings disperse the X rays into a high energy rainbow, containing thousands of distinct X-ray colors.

One of Chandra's early targets was the brilliant Orion star cluster, a cosmic birthing ground for stars. Stars in the Orion cluster were formed during the past few million years, so they are mere infants compared to our 4.5 billion year old sun. In one observation of the Trapezium region of the cluster, Chandra detected a thousand X-ray stars. Some are well known, massive, optically bright stars. Some are still embedded in the cloud of dust and gas from which they formed; they can be seen with an infrared, but not an optical telescope. A few are thought to be failed stars called brown dwarfs, because of their small mass.

Many young stars produce intense flares that make them a thousand times brighter in X rays than the sun. The cause of these flares is poorly understood, and scientists are hopeful that Chandra will help them to understand the bright, stormy youth of stars.

If the early stages of a star's development are wild, the end can be much more violent. When a massive star runs out of fuel, it undergoes a catastrophic explosion called a supernova. See also **Nova and Supernova**. The explosion, which is one of the most violent events in the universe, blows the star apart. The elements created by nuclear reactions inside the star over its lifetime — carbon, nitrogen, oxygen, and all the other elements heavier than helium — are spread into space.

The matter thrown off by the supernova creates a bubble of multimillion degree gas called a supernova remnant. This hot gas will expand and produce X-radiation for thousands of years. Chandra is the ideal instrument for studying these remnants. A team of scientists from MIT used Chandra to observe E0102-72, the remains of a star that exploded in a nearby galaxy. They found a ringlike structure that was thirty light years across and expanding at five million miles an hour. They determined that the ring was exceptionally rich in oxygen. It contained enough oxygen to mix with hydrogen and create more than a billion oceans. Thus confirmed the theory that massive stars provide most of the oxygen in the universe.

Chandra has also observed several giant black holes in the centers of galaxies, and come up with some surprises. The black hole in the center of the Milky Way galaxy, which has the mass of three million suns, is surprisingly weak in X rays, and the giant black hole in the center of the Andromeda Galaxy, which is surrounded by gas at an unexpectedly cool million degrees. See also **Andromeda**; and **Galaxy**. Current theories predict that the gas falling into the central black hole in Andromeda should be ten times or more hotter than is observed. Further Chandra observations of giant black holes should tell more about how a black hole swallows matter, and how space and time is warped by a black hole.

One of the most intriguing features of giant black holes is that they do not suck up all the matter that falls within their sphere of influence. Some of the matter falls into the black hole, and some explodes away from the black hole in high energy jets that move at near the speed of light.

Chandra has imaged a spectacular example of such a jet in Centaurus A, a nearby galaxy noted for its explosive activity. The image shows the nucleus, which is believed to harbor a giant black hole, and X-ray jets erupting from the center of the galaxy over a distance of 20,000 light years. The presence of bright X-ray jets means that electric fields are continually accelerating electrons to extremely high energies over enormous distances. Exactly how this happens is a major puzzle that Chandra may help to solve.

At the relatively nearby distance of eleven million light years from Earth, Centaurus A, has long been a favorite target of astronomers because it is the nearest example of a class of galaxies called active galaxies. Active galaxies are noted for their intense, bright cores and explosive activity, which are presumed to be due to a giant black hole in their center.

Active galaxies may be the solution to one of the oldest mysteries of X-ray astronomy, the X-ray background that was discovered in 1962. Many astronomers suspected that it was due to extremely distant active galaxies. They were unable to prove this, however, for no X-ray telescope until Chandra has had both the sharp vision and sensitivity to detect the individual sources.

Using Chandra, two teams of astronomers, from NASA's Goddard Space Flight Center, and Penn State University, have resolved over 70% of the background glow into individual sources, most of which are galaxies with a bright central core.

The nature of the remaining sources is not known. They could be galaxies so distant that the optical light gets absorbed by cool gas during its long journey across the universe. Or galaxies that are completely surrounded in dust. Or galaxies that have yet to form stars and are still just clouds of hot gas with a central giant black hole.

The giant clouds of hot gas in galaxy clusters pose another mystery. About 10% of all galaxies are part of huge clusters of hundreds or thousands of galaxies, X-ray observations show that the space between the galaxies is filled with a gas with temperatures in the range of 30 to 100 million degrees. There is enough gas in each galaxy cluster to make more than a hundred trillion stars.

The pressure of this hot gas is so great that it should have dispersed long ago. Since it has not, some additional force must be holding it together. The only known force that is strong enough is gravity. The galaxies cannot provide the necessary gravity, nor can the hot gas. An unseen form of matter, called dark matter, must be holding the cluster together. The amount of dark matter needed is greater than the galaxies and gas put together. If this percentage hold true for all galaxies and clusters of galaxies, then most of the matter in the universe is in some dark form.

The dark matter is one of the greatest riddles of modern astronomy. What is it? Possibilities include dark planetlike objects, white dwarfs, and black holes. Most scientists believe that it is in the form of some as yet undiscovered subatomic particle, called a WIMP, for Weakly Interacting Massive Particle, that does not give off light in any form, and can be detected only through its gravity.

With observations from Chandra and other telescopes, astronomers should soon know how much dark matter the universe contains. However, it may be a long time before they know what it is, and how it has shaped our universe. See also **Chandra X-Ray Observatory**; and **X-Ray**.

Additional Reading

Barcons, X. and A.C. Fabian: "X-Ray Background," Cambridge University Press, New York, NY, 1992.

Hoover, R.B.: "X-Ray Optics, Instruments and Missions II," SPIE International Society for Optical Engineering, Bellingham, WA, 1999.

Lewin, W.H., E.P. Van Den Heuvel, and J. Van Paradijs: "X-Ray Binaries, Vol. 26, Cambridge University Press, New York, NY, 1997.

Magill, F.N. and R. Smith: "The Solar System: Quantum Cosmology-X-Ray and Gamma-Ray Astronomy," Salem Press, Inc., Hackensack, NJ, 1998.

Seward, F.D.: "Exploring the X-Ray Universe," Cambridge University Press, New York, NY, 1995.

Tucker, W.H. and K. Tucker: "Revealing the Universe: The Making of the Chandra X-Ray Observatory," Harvard University Press, Cambridge, MA, 2001.

Web References

ASCA Guest Observer Facility: http://heasarc.gsfc.nasa.gov/docs/asca/

Cambridge X-Ray Astronomy Group: http://www-xray.ast.cam.ac.uk/

Chandra—X-Ray Astronomy Field Guide: http://chandra.harvard.edu/xray_sources/

High Energy Astrophysics Science Archive Research Center HEASARC: http://heasarc.gsfc.nasa.gov/

The RXTE Learning Center: http://heasarc.gsfc.nasa.gov/docs/xte/learning_center/

X-Ray Astronomy Gallery–Index, ROSAT images: http://wave.xray.mpe.mpg.de/rosat/mission/rosat

X-ray Astronomy: Neutron Stars http://imagine.gsfc.nasa.gov/docs/science/know_l1/neutron_stars.html

X-ray Astronomy: Supernovae and their remnants http://imagine.gsfc.nasa.gov/docs/science/know_l1/supernovae.html

WALLACE and KAREN TUCKER, X-ray Observatory Center, Harvard-Smithsonian Center for Astrophysics.

X-RAY ENERGY LEVELS. See **Energy Level**.

X-RAY SCAN AND OTHER MEDICAL IMAGERY. Introduction of computerized axial tomography in the form of the X-ray cat scanner (CAT) was a major event in the history of diagnostic medicine. The technology is also referred simply as computerized tomography (CT).

Tomography as a technique of X-ray photography by which a single selected plane is photographed, with the outline of structures in other planes eliminated, has been known for many years. X-rays lose energy in direct proportion to the density of the structure through which they pass. Thus dense structures of the body, such as bone, absorb much of the X-radiation and therefore in a traditional X-ray negative appear as light images—because less radiation has reached the film. The CAT scanner utilizes this same general principle (more or less energy absorbed, depending upon density of intervening tissue). But, instead of exposing a single X-ray film, or taking a series of single exposures from different angles and then comparing these in an attempt to create some semblance of the object in three dimensions, the CAT scanner indicates the exact amount of energy absorbed by the object from many different angles—so that the final image as seen by the radiologist is based upon a composite of contiguous one-millimeter cross sections. Film is replaced by electronic X-radiation detectors and the processing of so much information obtained

in such a short time span is processed by a computer. A rapid rate of calculation is needed to measure tissue density, requiring the solution of about one-half million equations every few minutes.

The CAT scanner has been likened to a large doughnut that houses x-ray tube and detectors. The doughnut is positioned around the part of the patient to be examined. As the patient lies on a table, the doughnut, housed in a large gantry, rotates, creating images as it turns. When the X-ray beam passes through the body, it strikes a detector in which calcium fluoride crystals scintillate (flash light). These scintillations are transmitted to a computer as electric signals. While the computer processes the information, registering it on a cathode ray tube, the doughnut rotates a few degrees and repeats the process until it has completed a 180-degree arc around the patient. The scanner forms a grid of readings with innumerable X-rays creating a matrix of nearly 100,000 intersections as they enter from various angles. The average amount of X-radiation absorbed by a patient during a CAT scan is estimated to be about one-fifth that absorbed in X-ray exposures in a typical executive physical given by many clinics and hospitals over 40 years of age.

The result is an X-ray image that gives the illusion of three dimensionality. Because the CAT beam is rotated around the body, it can image organs that overlap and are therefore obscured under conventional x-rays or radiograms. By using several hundred X-ray detectors to produce one exposure, the CAT is an order of magnitude more sensitive to slight gradations in density than radiographs, which frequently do not allow the practitioner to distinguish between tissues of approximately the same density.

Prior to the CAT scanner, normal X-ray photographs of the brain tended to be blurred beyond recognition by the skull. Since the first application of the CAT scanner to brain imaging in 1972, the instrument has come to be considered an indispensable neurophysiological tool. Prior to the availability of the CAT technique, an image of the brain's complex vascular system was obtainable only by means of an angiogram or arteriogram. Patients were given an injection of a dye directed to a specific site. In the diagnosis of some brain diseases, it was necessary to make a pneumoencephalogram ("air scan"), in which gas was injected into the lower spine and flowed upward to fill and outline brain cavities prior to x-ray procedures. These were both dangerous and painful procedures and required up to 2 or 3 days of hospitalization just for diagnostic purposes. The CAT can provide a detailed cross section of the brain in as brief a span as five minutes. Air scans can be eliminated and the need for angiograms has been sharply reduced. Rapid, usually accurate diagnosis of brain disorders and injuries has saved lives by reducing exploratory surgery and by greatly shortening delays in commencing treatment.

The image of an 80–90% underexposed medical radiograph can be increased to readable density and contrast by autoradiographic image intensification. The technique consists of combining the image silver of the radiograph with a radioactive compound, thiourea labeled with sulfur-35, and then making an autoradiograph from the activated negative.

Minimizing the X-ray dose received by patients during medical examinations and maximizing the quality of the radiographs are subjects of concern to the medical profession and the public. Some conflict is inherent in the two objectives because higher quality radiographs, i.e., those which convey more information to the physician, usually require higher exposure levels. Recent gains in quality or exposure reductions, or both, are due to developments, such as computer processing, electrostatic imaging systems, improvements in intensifying screens, and scatter rejection techniques.

Radiographs which are normally classified as "badly underexposed" actually contain most of the information which was intended to be recorded by the original exposure. Autoradiographic intensification effectively retrieves this information by increasing the image density and contrast to readable levels. The intensification occurs on an autoradiograph made from the underexposed film after the original image silver has been chemically combined with a radioactive isotope.

Other Photon Imaging Techniques

Since introduction of X-ray CAT scan technology, other photon imaging approaches have been developed. These concepts augment and, in some instances, are alternatives to the X-ray CAT scan. The more recent developments include positron emission tomography, K-edge dichromography, the use of synchrotron radiation, angiographic imaging, and nuclear magnetic resonance imaging. The latter topic is described in the article on **Nuclear Magnetic Resonance Spectroscopy (NMR)**. The recent use of synchrotron radiation is described under **Particles (Subatomic)**.

Positron Emission Tomography (PET). In PET, image construction is based on the location and intensity of gamma rays emitted in the region of a *neutron-poor* isotope. Planar images formed by computer PET result from the attenuation coefficients of the tissues that intercept a transmitted x-ray beam. Neutron-poor isotopes undergo radioactive decay by the process $P^+ \rightarrow N + e^+ + v$, where P^+ is a nuclear proton, N is a neutron, e^+ is a positron, and v is a neutrino. A neutron-poor isotope, such as ^{11}C will undergo beta decay, in which a proton becomes a neutron and a positron and a neutrino are ejected from the nucleus. Within a short distance, the positron encounters an electron, upon which the two annihilate each other and give rise to a pair of gamma-ray photons that depart at an angle of about 180°, each carrying an energy of 0.511 MeV.

The major isotopes used in PET are ^{15}O (half-life = 20 min), ^{13}N (half-life = 10 min), ^{11}C (half-life = 20 min), and ^{18}F (half-life = 110 min). Half-life spans are approximate. The foregoing isotopes are produced by a nearby cyclotron. They are either administered promptly, or are rapidly incorporated into appropriate molecules, such as metabolic substrates, substrate analogues, or drugs, which are then administered. Minicyclotrons for generating radionuclides are becoming available as of the late 1980s.

As pointed out by Ter-Pogossian and Brownell, the gamma photons derived from the decay of the isotopes within the patient's body are sensed by a circumferential array of collimated detectors, the circuitry of which is designed so that opposite members of the ring are coupled. A signal is recorded only when both members of the detector pair sense coincidental photons. By using a slight time difference in the activation of the detectors, one can locate the source of the photons on the basis of time-of-flight differences from an eccentrically positioned emitter. Data are fed into a computer, which generates the image based on location and source intensity. Tissue attenuation is taken into account. Spatial resolution is about 0.5 cm. Only minute amounts of tracers are needed. The radiation dose is small.

PET is particularly adapted to kinetic analysis of physiologic and biochemical events, including blood volume, blood flow, and consumption of oxygen and substrates. Some applications of PET include the following. (1) Identifying uptake of the glucose analogue ^{18}F-fluorodeoxyglucose in various regions of the brain. As pointed out by Mazziotta and Reivich, the laterality of neurologic responses to auditory stimulation has been correlated in this way. Visual stimuli of increasing complexity have been observed to produce symmetric increases in uptake in the primary and associative visual cortices, with a correlation of visual pathway abnormalities with neurologic findings. (2) Studies of Alzheimer's disease patients indicate dysfunction in temporal-parietal regions and in the structures near the third ventricle, as reported by de Leon and Friedland. (3) Studies of schizophrenic patients have suggested subtle local blood flow and metabolic changes. (4) Measurements of altered blood flow, oxygen utilization, and oxygen extraction fraction have been made in patients who have undergone a recent cerebral infarction, as noted by Baron. (5) A study of an adolescent who had cerebral vasculitis due to systemic lupus erythematosus showed local changes, when revealed by PET, that correspond with electroencephalographic findings. After remission, the PET abnormalities vanished. (6) Some patients with brain tumors have been examined by PET. In a very large percentage of cases, a positive correlation between tumor glycolysis and tumor histologic grade has been observed by PET (as noted by DeLaPaz). (7) Kuhl reports that in some patients with epilepsy, during interictal periods, the involved zones are hypometabolic. (8) In Duchenne's muscular dystrophy, biochemical abnormalities associated with the characteristic involvement of the posterolateral region of the left ventricle have been identified by PET. There are many other examples of the contributions of PET to patient diagnosis and biochemical research.

Digital Subtraction Angiography. Routine roentgenograms do not reveal vascular structures because blood vessels and surrounding soft tissues attenuate x-ray beams in the same way, thus not revealing any distinction between the two. Conventional arteriography requires high concentrations of iodine-containing contrast agents to be injected directly into an artery. The attenuation of the x-rays through Compton and photo-electric interactions with iodine provides the required contrast between vessels and surrounding tissues. An alternative technique known as digital subtraction angiography (DSA) requires considerably lower concentrations of contrast agents. DSA is now widely practiced by giving a peripheral or central intravenous injection of contrast materials. Most often, the agents will be administered by a pump-driven device. In this method, an x-ray

image is acquired on an area detector, such as a fluoroscope screen, or on a scanning line detector. Data then are digitized, amplified, and transferred to a computer.

An optimal approach to DSA angiography is that of using monochromatic x-ray beams at energy levels just above and just below the K-shell absorption edge of iodine. Image data above the K-edge include information arising from Compton and photoelectric interactions of x-ray beams with iodine and with atoms in the molecules of the soft tissues and bones of the body. The image recorded just below the K-edge includes virtually the same attenuation data except for additional information that arises from photoelectric absorption by K-shell electrons in iodine atoms. The logarithmic subtraction of the two images almost totally suppresses signals arising from soft tissue and bone, as observed by Rubenstein and Hughes.

Additional Reading

Baron, J.C., et al.: "Comparison Study of CT and Positron Emission Tomographic Data," *Amer. J. of Neurological Research*, 536 (April 1983).

Brownell, G.L., et al.: "Positron Tomography and Nuclear Magnetic Resonance Imaging," *Science*, **215**, 619–626 (1982).

DeLa Paz, R.L., et al.: "Positron Emission Tomographic Study of Suppression of Gray-Matter Glucose Utilization by Brain Tumors," *Amer. J. of Neurological Research*, 826, (April 1983).

DeLeon, M., A.E. George, and et al.: "Regional Correlation of PET and CT in Senile Dementia of the Alzheimer Type," *Amer. J. of Neurological Research*, 533, (April 1983).

Friedland, R.P., et al.: "Regional Cerebral Metabolic Alterations in Dementia of the Alzheimer Type: Positron Emission Tomography," *J. Comput. Assist. Tomogr.*, 590 (July 1983).

Kuhl, D.E., et al.: "Epileptic Patterns of Local Cerebral Metabolism and Perfusion in Human Determined by Emission Computed Tomography," *Ann Neurol.*, 348 (August 1980).

Mazziotta, J.C., et al.: "Local Cerebral Glucose Metabolic Response — Studies in Human Subjects with Positron CT," *Human Neurobiology*, 11 (February 1983).

Reivich, M., et al.: "Positron Emission Tomographic Studies," *Human Neurobiology*, 25 (February 1983).

Rubenstein, E., E.B. Hughes, L.E. Cambell, and et al.: "Synchrotron Radiation and Its Application to Digital Subtraction Angiography," *Conf. On Digital Radiography*, **314**, 42, Bellingham, WA, 1981.

Ter-Pogossian, M.M., M.E. Raichle, and B.E. Sobel: "Positron-Emission Tomography," *Sci. Amer.*, 171–178 (October 1980).

X-RAY STARS. See **Neutron Stars**.

X-RAY THICKNESS GAGE. See **Thickness Measurement and Gaging Systems**.

X-UNIT (or Xu). A unit used in expressing the wavelengths of x-rays or gamma-rays. It is about 10^{-11} centimeter of 10^{-3} Å. Accurately, 1 Xu = $1.00202 \pm 0.00003 \times 10^{-3}$ Å.

XYLEM. The cells composing the woody tissue of higher plants are xylem cells. They have undergone extreme modification during their development, becoming greatly elongated and having much-thickened walls. Xylem tissue includes tracheids, vessels, fibers, and parenchyma.

XYLENE. $C_6H_4(CH_3)_2$, formula weight 106.16. There are three xylenes, ortho-, meta-, and para-xylene. Sometimes referred to as dimethylbenzenes, the xylenes have the following key physical properties:

o-Xylene	mp −25 °C	bp 144 °C	sp gr 0.881
m-Xylene	−47.4 °C	139 °C	0.867
p-Xylene	13.2 °C	138.5 °C	0.861

All of these compounds are insoluble in H_2O, soluble in alcohol, and *o*-xylene and *m*-xylene are miscible in all proportions with ether; *p*-xylene is very soluble in ether.

The xylenes are very high-tonnage industrial chemicals and are raw materials or intermediate materials for numerous synthetic fibers, resins, and plastics. A large amount of *p*-xylene goes into polyester fiber production, while substantial quantities of *o*-xylene are consumed by the manufacture of phthalic anhydride. The prime source of xylenes are petroleum refinery reformate streams in conjunction with benzene and toluene extraction. The xylenes occur mixed in these streams.

When naphtha or naphthenic gasoline fractions are catalytically reformed, they usually yield a C_8 aromatics stream that is comprised of mixed xylenes and ethylbenzene. It is possible to separate the ethylbenzene and *o*-xylene by fractionation. It is uneconomic to separate the *m*- and *p*-xylenes in this manner because of the closeness of their boiling points. To accomplish the separation, a Werner-type complex for selective absorption of *p*-xylene from the feed mixture may be used. Or, because of the widely different freezing points of the two xylene isomers, a process of fractional crystallization may be used. To boost the *p*-xylene yield, the filtrate from the crystallization step can be catalytically isomerized.

XYLITOL.   See **Sweeteners**.

X-Y RECORDER.   An instrument used where it is desired to plot the relationship between two variables, $y = f(x)$, instead of plotting each variable separately as a function of time. An X–Y recorder closely resembles and functions like a single-pen recorder except that the chart (Y-axis) is moved in response to changes in a variable instead of at a uniform time rate. The chart may be driven by a separate servoactuated measuring element similar to that used in positioning the recorder pen, or by a self-synchronous motor in a remote-control system. In most applications of the X–Y recorder, a square-shaped graph sheet is used. Thus, the fullscale chart travel (Y-axis) is made equal to the fullscale pen travel (X-axis). The addition of a second servoactuated measuring element and recording pen to an X–Y recorder yields an X–X_1–Y, or an X–Y–Y_1 recorder in which the relationships among three associated variables can be plotted.

Y

YAG AND YIG. Synthetic yttrium aluminum and yttrium iron garnets, respectively. These materials were developed in the mid-1960s. They are pressed and sintered polycrystalline ceramics and are made by a solid-state reaction of Y_2O_3 with iron oxide or aluminum oxide. Garnets operate in microwave bandpass (filters) circulators and isolators in telephone, radar, and space-communication networks. The original electronic use led to the development of single-crystal yttrium aluminum garnets which approach the brilliance and hardness of diamond. Yttrium oxide is the base for neodymium-doped laser crystals. See also **Neodymium**; **Rare-Earth Elements and Metals**; and **Yttrium**.

YAGI ANTENNA. A type of directional antenna used on some types of radar and radio equipment consisting of an array of elemental, single-wire dipole antennas and reflectors.

YAM. See **Sweet Potato**.

YANG, FRANKLIN CHEN-NING (1922–). Yang is an American physicist born in China. As a child, Chen Ning Yang read Benjamin Franklin's autobiography and was so inspired by Franklin's life that he took the name of "Franklin" as his first name. He received is Ph.D. from the University of Chicago, 1948. Chen-ning Yang was a member of the Institute for Advanced Study at Princeton, New Jersey from 1949 to 1955 and a professor of physics there from 1955 to 1965. In 1965 he was appointed Albert Einstein Professor of Physics of the State University of New York at Stony Brook and director of the Institute of Theoretical Physics at Stony Brook.

In 1957 Yang shared the Nobel Prize in Physics with American physicist Tsung-Dao Lee. They were the first scientists of Chinese birth to win a Nobel Prize. They were also among the youngest men ever to receive a Nobel award. They obtained the shortest time interval ever between a discovery and the award of the Nobel Prize. Their research refuting the law of parity, stated that at the subatomic level, nature does not distinguish between left-and right-handed configurations: if a nuclear reaction or decay occurs in nature, then so does its mirror image and with equal frequency. He is also known for his researches in statistical mechanics and particle physics.

In 1980, he was given the Rumford Medal of the American Academy of Arts and Sciences. Yang helped to advance science by his membership on the boards of Rockefeller University, the American Association for the Advancement of Science, the Salk Institute for Biological Studies in San Diego, and Ben-Gurion University in Israel.

See also **Mathematics (State of the Art Reviews)**.

J. M. I.

YAPOCK. See **Marsupialia**.

YARDANG. A long, irregular, sharp-crested, undercut ridge between two round-bottom troughs, carved on a plateau or unsheltered plain in a desert region by wind erosion. Consisting of soft, but coherent deposits, such as clay-laden sand, it lies in the direction of the dominant wind.

YAW (Aircraft). This is the angular displacement about the normal axis of an aircraft. The normal axis of an airplane in horizontal flight is vertical. Intentional yaw is produced principally by the rubber. Yaw may be produced by the unsymmetric thrust forces when one engine quits in a multi-engine design. Yaw is produced by the usual correction of providing aileron rolling moments to counter the torque of propeller-engine combinations.

The yawing moment may be expressed mathematically as follows:

$$N = qSCC_N$$

where C_N is the yawing moment coefficient, N is the yawing moment, S is the wing area, and C is the mean aerodynamic chord of the wing.

YAWS *(Frambesia; Pian)*. A contagious disease caused by a spirochetal organism *Treponema pertenue* which is a close relative of the spirochete of syphilis. The disease is unevenly distributed throughout the tropics, being essentially a disease of hot, humid countries. It is not a venereal disease, and congenital transmission does not occur. Infection is by direct contact with an infective lesion.

After initially manifesting a small papule at the site of inoculation, the lesion enlarges to form a granulomatous papilloma, which usually ulcerates and contains infective treponemes. The lesion may regress and latent periods intervene, but relapses eventuate, infective lesions reappearing and lasting some 3 to 5 years.

Diagnosis is by microscopical examination of the lesion exudate. Humoral immunity develops soon after the initial infection and produces antibodies cross-reacting with other treponemes. They only partially protect the patient against reinfection.

It is to be noted that, as rural yaws dies out, men from the rural areas who visit cities are likely to acquire venereal syphilis. They then return to their villages with secondary syphilitic sores and from them infect children via the skin. The result is syphilis, not yaws, in the children.

Treatment of the disease relies, as in venereal syphilis, upon penicillin or tetracycline and mass campaigns using these drugs have had some success, but are hampered by the nomadic habits of many of the tribes in which the disease occurs.

R. C. V.

YAZOO STREAM. A stream (tributary) that flows parallel to the main stream for a considerable distance before joining it at a deferred junction. In particular, a stream forced to flow along the base of a natural levee formed by the main stream. Examples include the Yazoo River in western Mississippi, joining the Mississippi River at Vicksburg, Mississippi.

Y-CHROMOSOME. One of the sex-determining chromosomes. It apparently carries genes of maleness (XY), as opposed to female (XX). Since males possess X- and Y-chromosomes, spermatozoa may carry either an X or Y. Since females possess only X-chromosomes, all ova carry an X-chromosome. Thus the sex of offspring must depend upon the constitution of the spermatozoon. If it carries an X-chromosome fertilization will result in a female zygote, while a Y-bearing spermatozoon will give rise to a male. Since males produce two types of gametes with regard to their sex chromosome constitution, they are referred to as *heterogametic*, in contrast to the female, which produces one type of gamete and is called homogametic. This system prevails in mammals, but not in birds, for which the opposite situation holds.

Y-CONNECTION. Three-phase ac equipment is wound with three wires whose currents differ 120° electrically in phase. The windings can be connected either in Y or delta. In balanced electrical condition, voltages and currents are the same in all coils. In the Y-connections one end of all three coils is connected in a common joint, and leads from each of the other ends constitute the three-phase line. The Y-connection is preferred for alternators because of the usefulness of the neutral point, and because the line voltage is $\sqrt{3}$ times the phase voltage. The ability to bring out a neutral point and ground it either through resistance or reactance, is advantageous because it aids in working out protection and selectivity of control of parallel alternators.

YEAST CELLS. See **Ascomycetes (Sac Fungi; Fungi).**

YEASTS AND MOLDS. These are very important plant organisms that make both positive and negative contributions to mammalian life processes. Their plus and minus values are particularly noted in connection with the production and storage of food products.

Taxonomy. Plants that lack true roots, stems, or leaves, and that are without highly-organized conducting systems are called *simple plants*, and they make up the phylum *Thallophyta*. The two subdivisions of *Thallophyta* are *algae* and *fungi*. Algae have chlorophyll; fungi do not. Fungi utilize carbohydrates that are synthesized by green plants. Fungi are classified as *parasites* or *saprophytes*, the latter obtaining food from nonliving organic material. The science of fungi is *mycology*. See also **Fungus**.

The fungi include: *Slime molds* (*Myxomycetes*); *algal fungi* (*Phycomycetes*); *sac fungi* (*Ascomycetes*); and *club fungi* (*Basidiomycetes*); among others. See also **Ascomycetes**; and **Basidiomycetes**. The yeasts and most of the molds with which food products are associated in some way are of the *Ascomycetes* variety.

The varied interface between the *Ascomycetes* and foods ranges from their very negative, parasitic habit on certain crops (exemplified by Chestnut blight fungus; or ergot on grains) and their causative involvement in certain foodborne diseases (see also **Foodborne Diseases**) to their positive use in connection with the production (notably by fermentation) of major food products, such as bread and wine.

The term *yeast* is used to describe a relatively small number of *Ascomycetes* fungi (a few hundred—Lodder et al. (1970) classified 349 fermentative species, of which relatively few are used industrially), as compared with the vast number of other fungi (several thousand) that have been identified.

The term *mold* does not have a clear cut mycological definition. While the yeasts have a preponderantly positive value in food production and utilization, the molds make some positive contributions (some cheeses, etc.), but they participate in many more negative ways (leaf molds on plants; blue-green molds on fruits; machinery mold; causes of some foodborne diseases; etc.). Some authorities identify all or most mold as *saprophytes*, as previously defined. This identification, however, is not fully satisfactory in terms of the present loose ways in which the word mold is used in the professional literature. Molds do play an important role in the *biodegradation* of unwanted substances.

Yeasts

The importance of the economically useful yeasts can be attributed to two main factors: (1) *fermentation* — the transformation of simple sugars and other organic chemicals to other, more desirable chemicals; and (2) *respiratory* (*oxidative*) *metabolism* — the great capacity of some yeasts for a protein synthesis during growth in richly aerated media containing a wide variety of carbonaceous and nitrogenous nutrients. Thus, yeasts serve in many ways: (1) As living cells, they are biocatalysts in the production of bread, wine, beer, distilled beverages, among other important food products. (2) As dried, nonfermentative whole cells or hydrolyzed cell matter, yeasts contribute nutrition and flavor to human diets and animal rations. (3) As producers of vitamins and other biochemicals, yeasts are a rich source of enzymes, coenzymes, nucleic acids, nucleotides, sterols, and metabolic intermediates. See also **Coenzymes**; **Enzyme**; and **Enzyme Preparations**. (4) As a versatile biochemical tool, yeasts aid research studies in nutrition, enzymology, and molecular biology.

Background. It is estimated that the arts of making wine, leavened bread, and beer were practiced more than 4000 years ago. The phenomena for producing these foods were attributed to "yeast" at an early age. In many languages, the word for yeast describes the visible effects of fermentation, as observed in the expansion of bread dough and the accumulation of froth or barm on the surface of fermenting juices and mashes. Historically, it has been reported that yeast cells were first seen in a droplet of beer mounted on a slide in a crude microscope used by van Leeuwenhoek in 1680. He found globular bodies, but was not aware that they were living forms. For nearly two centuries the theory of spontaneous generation dominated thought and research on the causes of fermentation and disease. In 1818, Erxleben described beer yeast as

a living vegetable matter responsible for fermentation. In the following twenty years, yeasts were shown to reproduce by budding and, in 1837, Meyen named yeast *Saccharomyces*, or "sugar fungus." By 1839, Schwann observed "endospores" in yeast cells, later named ascospores by Reess. As early as 1857, Pasteur proved the biological nature of fermentation and later, in 1876, Pasteur demonstrated that yeast can shift its metabolism from a fermentative to an oxidative pathway when subjected to aeration. This shift, then named the Pasteur effect, is especially characteristic of bakers' yeast (*Saccharomyces cerevisiae*) and is applied in the large-scale production of yeasts.

Botanically, yeasts form a heterogeneous group of saprophytic forms of life occurring naturally on the surface of fruits, in honey, exudates of trees, and in soil. They are disseminated by airborne dusts, insects, and animals. Typically, industrial yeasts are generally oval, microscopic, unicellular organisms. In addition to lacking chlorophyll, they also lack locomotion. They reproduce vegetatively by budding and sexually by spore formation (ascospores) within the mother cell or ascus. These properties place them in the family *Endomycetacea* of the class *Ascomycetes*, as previously mentioned. Among the most important industrial yeasts are:

Saccharomyces cerevisiae (alcoholic beverages and bread)
S. cerevisiae var. *ellipsoideus*; *S. bayanus*; and *S. beticus* (wines)
S. uvarium (formerly called *S. carlsbergensis*) (beer, ale, etc.)
Kluyveromyces fragilis (formerly called *S. fragilis*) (whey disposal)

Food and feed yeast production employ several molds in the family *Cryptococcaceae*: *Candida utilis*, *C. tropicalis*, and *C. japonica*, which are cultivated on plant wastes (wood sugars, molasses, stillage), and *C. lipolytica*, which converts hydrocarbons to yeast protein.

Properties of Yeasts. The cell structure of *S. cerevisiae*, as observed in the optical microscope, reveals a rigid cell wall, a colorless, granular cytoplasm, and one or more vacuoles. Dimensions of a typical bakers' yeast are about 4 to 6 by 7 to 10 micrometers. Electron microscopy of ultrathin sections of a yeast cell show the microstructures, including: birth and bud scars on the cell wall, plasmalemma (cytoplasmic membrane), nucleus, mitochondria, vacuoles, fat globules, cytoplasmic matrix and volution or polyphosphate bodies. The cell walls of bakers' yeast contain 30–35% glucan (yeast cellulose), 30% mannan (yeast gum), which is bound to protein (about 7%), 1–2% chitin, 8–13% lipid material, plus inorganic components, largely phosphates.

Gross chemical composition of compressed bakers' yeast is approximately 70% moisture. The dry matter is made up of 55% protein (N × 6.25), 6% ash, 1.5% fat, and the remainder mostly polysaccharides, including about 15% glycogen and 8% trehalose.

Food yeast, molasses-grown, is dried to about 5% moisture and has the same chemical composition as bakers' yeast. In terms of micrograms per gram of yeast, the vitamin content is: 165 thiamine; 100 riboflavin; 590 niacin; 20 pyridoxine; 13 folacin; 100 pantothenic acid; 0.6 biotin; 160 para-aminobenzoic acid; 2710 choline; and 3000 inositol. Yeast crude protein contains 80% amino acids; 12% nucleic acids; and 8% ammonia. The latter components lower the true protein content to 40% of the dry cell weight.

Yeast protein is easily digested (87%) and provides amino acids essential to human nutrition. Most commercial yeasts show the following pattern of amino acids, among others, as percent of protein: 8.2% lysine; 5.5% valine; 7.9% leucine; 2.5% methionine; 4.5% phenylalanine; 1.2% tryptophan; 1.6% cystine; 4% histidine; 5% tyrosine; and 5% arginine. The usual therapeutic dose of dried yeast is 40 grams/day, which supplies significant daily needs of thiamine, riboflavin, niacin, pyridoxine, and general protein.

The ash content of food yeasts ranges from 6 to 8% (dry basis), consisting mainly of calcium, phosphorus, and potassium. Contained in quantities of less than 1% are magnesium, sulfur, and sodium. At the microgram level are included iron, copper, lead, manganese, and iodine.

Triglycerides, lecithin, and ergosterol are the main constituents of yeast lipid (fat). Oleic and palmitic acids predominate in yeast fat. These resemble the composition of common vegetable fats. Ergosterol, the precursor of calciferol (vitamin D_2) varies from 1 to 3% of yeast dry matter.

Metabolic Activity. This is generally associated with the familiar alcoholic fermentation in which theoretically 100 parts of glucose are converted to 51.1 parts of ethyl alcohol (ethanol), 48.9 parts of carbon dioxide (CO_2), and heat. In addition, however, the anaerobic reaction also yields minor byproducts in small amounts—mainly glycerol, succinic acid, higher alcohols (fusel oil), 2,3-butanediol, and traces of acetaldehyde, acetic acid, and lactic acid. Fusel oil is a mixture of alcohols, including *n*-propyl, *n*-butyl, isobutyl, amyl, and isoamyl alcohols.

Respiratory activity of oxidative dissimilation is characteristic of many species of yeasts. During aerobic growth, sugar is oxidized to carbon dioxide and water, with release of large amounts of energy (about 680 kcal when complete oxidation occurs). Aerobiosis produces a variety of byproducts, some in unusually high concentration, such as acetic acid, succinic acid, zymonic acid, polyhydric alcohols (glycerol, erythritol, etc), extracellular lipids, carotenoid pigments in shades of red and yellow, black pigment (melanin), and capsular polysaccharides (phosphomannan).

Production. Well over 85,000 tons (76,500 metric tons) of yeast dry matter are produced in the United States alone each year. About 75% of this is in the form of bakers' yeast, the remaining 25% represents about equal amounts of food yeast and feed yeast. This production issues from four types of manufacture: (1) Bakers' yeast is grown batchwise in aerated molasses solutions. (2) *Candida utilis* is obtained from wood pulp mill spent liquid. (3) *K. fragilis* is grown batchwise in cottage cheese whey. (4) Dried yeast is recovered as spent beer yeast. Worldwide production of all types of food and feed yeast is estimated at more than 450,000 dry tons (405,000 metric tons) per year.

The process for growing bakers' yeast is a model system for the propagation of microorganisms. The process commences with a laboratory culture of a pure strain of *S. cerevisiae.* Seed yeast is developed in successively larger volumes of nutrient solutions, beginning with a Pasteur flask and ending in a fermenting tank containing as much as 40,000 gallons (1514 hectoliters) of sterilized and diluted molasses maintained at 30 °C. During the highly aerated growth period, minerals are added, pH is adjusted to 4.5, and diluted molasses is continuously fed in proportion to the increase in cell mass. Under ideal conditions, yeast cells may double in number every 2.5 hours, converting more than half (56.7%) of the sugar supplied to cell components. Biosynthesis of cell matter requires an equal amount of oxygen. To produce 100 weight units of yeast dry matter with 50% protein content requires about 400 weight units of molasses, 25 weight units of aqua ammonia, 15 weight units of ammonium sulfate, and 7 weight units of monobasic ammonium phosphate. For each 100 pounds of dry yeast, 75,000 cubic feet of air are required; for 100 kilograms of dry yeast, 4,683 cubic meters of air are needed.

Fermentation Processes. The biochemistry of alcoholic fermentation involves a series of internal enzyme-mediated oxidation-reduction reactions in which glucose is degraded via the Embden-Meyerhof-Parnas pathway. See also **Carbohydrates**; and **Glycolysis**. Some typical reactions performed by yeasts are listed in the Table 1.

Post-Processing Spoilage of Food by Yeasts. Microbiological spoilage of food is a competitive process occurring among yeasts, bacteria, and molds. Yeasts normally play a small role in spoilage, because they constitute only a small portion of the initial population, because they grow slowly in comparison with most bacteria, and because their growth may be limited by metabolic substances produced by bacteria. Evidence does not show food poisoning as caused by the presence of spoilage yeasts in foods. The byproducts of metabolism are not considered toxic and, while there are a few yeasts that may be considered pathogenic, they are not known to be responsible for foodborne infections or intoxications (Walker, 1977; Peppler, 1977). The metabolism of yeasts can result in the development of unnatural flavors and odors and changes in pH because of the utilization of organic acids important in fermented foods. Many yeasts are capable of utilizing lactic, acetic, and citric acids, which are essential for production of flavor and for preservation of some foods. Decreased concentration of these acids causes an increased pH and produces conditions that favor growth of spoilage bacteria. Normally, several factors interact to establish conditions that favor growth and subsequent spoilage of foods by yeasts. Examples of food spoilage by yeasts are given in the lower half of Table 1.

Although yeasts are abundant in nature, notably on leaves and in the soil, they do not compete with bacteria and molds as sources of major problems. But, in assessing the potential for spoilage by yeasts, one must consider those factors which are favorable or unfavorable to the growth and multiplication of yeast populations. Availability of oxygen is important. No yeast is known that can grow under strictly anaerobic conditions; thus, all require oxygen to be present in some proportion. Temperature is also an important factor. Yeasts are not heat resistant in the sense that bacterial endospores are heat resistant. Most yeasts cannot withstand temperatures above 65–70 °C. The majority of yeast species have an optimum temperature for growth between 20 and 30 °C. Even though yeasts are very resistant to low temperatures and can survive frozen storage, they normally are not a major problem in spoilage of frozen or refrigerated foods. In refrigerated foods, psychrotrophic bacteria rapidly outnumber the yeasts, the latter constituting a minor part of the initial population. However, where antibiotics or ionizing radiation have been used to reduce or inhibit bacterial growth, then yeast population may predominate. If the definition of a *psychrophilic* organism is one that has an optimum temperature below 20 °C, then there are a number of strains of psychrophilic yeasts.

The survival and growth of all organisms is dependent on the presence of some water in the environment. Yeasts generally require more moisture for growth than molds, but less than bacteria. Systems containing high concentrations of sugar or salt have low a_w values or high osmotic pressure (Walker, 1977). Organisms growing in such systems are usually referred to as being *osmophilic.* A number of investigators have not been able to find any yeast that is clearly osmophilic. Windisch (1969) suggested the term osmotolerant to describe some yeasts. It is important to note that yeasts that normally will not grow in a high-sugar environment may appear, if the substance has been exposed to the air sufficiently to absorb water and dilute spots or edges of the material, thus favoring yeast growth. Species such as *Saccharomyces rouxii, S. rouxii* var. *polymorphus,* and *S. mellis* (Lodder, 1952) are capable of spoiling high-sugar foods, such as honey, maple sugar, sugar cane syrups, molasses, fruit syrups, candy, crystallized fruits, jams, jellies, and dried fruits. It is not uncommon for dried figs and prunes to be covered with a white coating of yeast during storage. The foods have a typical fermented odor and may contain gas pockets. This coating normally is a mixture of sugar and yeasts, with the principal yeast being *Schizosaccharomyces octosporus.*

Molds

The most critical factor in mold growth is the availability of sufficient moisture. Molds are widely distributed throughout nature. Chemicals for the destruction of molds, including those in the soil, are mentioned in the entry on **Antimicrobial Agents (Foods).** In addition to food spoilage, various molds can be destructive of various materials, notably of substances prepared from animal skins, such as leather, book covers, shoes, and materials prepared from vegetable substances, such as paper and wood products. In warm, humid areas, mold (mildew) grows on wooden structures, including painted surfaces unless these have been chemically treated. In the case of foods, molds cause innumerable problems during the full cycle of food production.

Storage temperature of food is less important than the presence of moisture, since fungi can grow and produce toxins over a wider span of temperatures than can any other microorganisms. Most species are able to grow at an a_w of 0.8 to 0.88; while xerophilic types can grow at an a_w of 0.65 to 0.75. A relative humidity of from 70 to 90% establishes suitable moisture equilibrium for initiation of mold growth and toxin formation in numerous food products. See also **Activity Coefficient**.

Unfortunately for the would-be consumer of food, fungal contamination is not always immediately apparent and thus may not seem to be cause either for the rejection of food by humans or for the rejection of feedstuffs by livestock. Concentrations of mold organisms are not always easy to see or to smell. Further, the relative stability of most mycotoxins to heat precludes the use of cooking as a detoxifying procedure. Among the most aggressive of the molds are species of *Candida, Aspergillus, Rhizopus,* and *Mucor.* Several of the fungi are not pathogenic to healthy humans, but may be virulent pathogens in debilitated persons, or those treated with broad-spectrum antibacterial drugs, or immunosuppressive measures. An example of this type of organism is *Cryptococcus neoformans* (Davis, 1973). Many of the fungi produce disease through infection rather than through mycotoxin production, but this usually requires predisposing factors. See also **Foodborne Diseases.**

Milner and Geddes pointed out in 1946 that the aflatoxigenic organism *Aspergillus flavus* is a common soil fungus throughout the world, and aflatoxin-contaminated food has been noted in many countries for a number of years (Wogan, 1966). Crops, such as groundnuts (peanuts), in intimate contact with soils, are likely to contact the necessary mold inoculum for aflatoxin production, resulting in toxin formation upon storage in air.

Several investigators have shown that toxic strains of *Fusarium, Cladosporium, Penicillium,* and *Mucor* can sporulate and grow at temperatures

TABLE 1. REACTIONS IN WHICH YEASTS PARTICIPATE[1]

Type of Reaction	Number of Known Reactions	Examples of Reactions
Reduction	156	Diacetyl to acetoin (in beer); cinnamic aldehyde to cinnamic alcohol.
Decarboxylation	21	Malic acid to lactic acid (in wines); amino acids to amines (histamine and tyramine accumulate in soft cheeses because of surface growth of *Torulopsis candida* and *Debaryomyces kloeckera*).
Deamination	17	Examples of deamination and decarboxylation include conversion of amino acids to fusel oil (leucine to isoamyl alcohol, isoleucine to amyl alcohol, and phenylalanine to phenyl ethanol). Fusel oil formation is a normal function of all yeast fermentations (in alcoholic beverages, levels range from trace to 2200 parts per million). Deamination: Glutamic acid to gamma-OH-butyric acid (*S. cerevisiae*).
Oxidation	14	Acids, alcohols, sugars, hydrocarbons. Also stepwise: alcohol to aldehyde or acid (sake yeast).
Esterification	10	Ethyl acetate (*Hansenula anomala*).
Condensation	9	Acetaldehyde to acetoin; acetaldehyde to pyruvic acid to alpha-acetolactic acid.
Hydrolysis	5	Starch hydrolysis: By *Endomycopsis fibuligera*; artichoke starch (inulin) by *Kluyveromyces fragilis*.
Amination	1	

Examples of Lesser-known Properties and Characteristics of Yeasts

Type of Reaction	Yeasts Involved	Examples
Lipolysis	*Candida lipolytica; C. rugosa; Torulopsis sphaerica*	Mainly in butter, margarine, and cheese.
Proteolysis	*C. lipolytica; T. sphaerica*	Especially on soft cheese surfaces.
Pectinolysis	*Saccharomyces kluyveri; Kluyveromyces fragilis; Hansenula anomala*	Softening of olives and cherries in brines, followed by formation of gas pockets in fruit. Strains of wine yeasts contain polygalacturonases which, during fermentation of grape juice, participate in the solubilization of pectin.
Acid formation	Species of Brettanomyces, Hansenula, Pichia, Saccharomyces	As a contaminant in wines, Brettanomyces spp. Forms a higher concentration of volatile acids (also isobutyric and isovaleric acids) than *S. cerevisiae*. Pichia species and other yeasts are responsible for acetic acid production in brines of domestic green olives; not lactobacilli, as assumed for years (Vaughn et al., 1976).
Pigmentation	*Rhodotorula glutinis.* Sporobolomyces spp.	Carotenoids (pink, red); *Rhodotorula glutinis* causes pink sauerkraut and discolors the surface of high-moisture cheeses. See also **Cabbage**.
Esterification	Species of Hansenula, Kluyveromyces, Brettanomyces	Ethyl acetate and ethyl lactate in cottage cheese and shredded Mozzarella cheese. See also **Milk and Dairy Products**.
Turbidity formation	In wine: *Saccharomyces bailii, S. chevalieri*, Brettanomyces spp. In beer: *S. diastaticus, S. bayanus*	In soft drinks; in wine: *S. bailli, S. chevalieri*; in beer: *S. diastaticus, S. bayannus*.

[1] Based, in part, upon Peppler et al. (1977).

well below 0 °C. It also has been shown that toxin formation can be associated with overwintering of grain in the field. In some varieties of barley, such as Siri and Mala, the incidence of *Aspergillus* and *Penicillium* species is related solely to temperature and moisture storage conditions and is not correlated to the percentage of unripe grains. Low humidities tend to predispose hosts to invasion by molds by causing them to lose turgor. Mold is the cause of serious problems in connection with rice. The predominant molds present in wild rice during fermentation curing are *Mucor*, aflatoxigenic and nontoxigenic *Aspergillus*, *Penicillium*, and *Rhizopus* species. Most of the B$_1$ aflatoxin is in the hulls and the level of the toxin can be reduced by parching, which serves to reduce, but not fully destroy molds.

Contamination of soybeans with *Aspergillus flavus* is found in approximately 50% of commercial samples. Fortunately, the incidence of aflatoxin from this route is quite low. Moisture at the time of maturity, development of the seed in a closed pod, and binding of zinc by phytic acid are suggested as reasons for resistance of soybeans to aflatoxin production.

The most active sugar contaminants are molds, particularly *Aspergillus*. Penicillium growth on cheddar cheese results in a pH gradient, increasing to a pH of 8 at the surface. Mycotoxins diffuse through foods and are not removed when the surface molds are removed. Therefore, molds of unknown variety or origin should not be ingested. On the other hand, many of the common molds found as contaminants of Western foods are used in Asian fermented foods and beverages, principally as sources of flavor.

Degradation of Aflatoxin. Numerous researchers have investigated a variety of chemical and physical means for degrading aflatoxin. These include the use of irradiation, heat, acids and bases, oxidizing agents, bisulfite, and biological agents. The practical application of ultraviolet radiation to date has not proved successful. Aflatoxin is quite heat stable. Numerous investigators have concluded that although heat can degrade aflatoxins, it is not an effective and economically feasible means for inactivating these toxins when present in foods or feeds.

Among oxidizers used have been sodium hypochlorite, potassium permanganate, and hydrogen peroxide. Bisulfite is commonly used in the wet milling of corn (maize) and in processing wines, fruit juices, jams, and dried fruits.

Machinery Mold. This mold generally refers to the buildup of the organism *Geotrichum candidum* on food-contact factory equipment in processing plants. The term *dairy mold* is used in the processing of milk to identify this mold.

Since passage of the original Federal Food and Drugs Act of 1906 (United States), the presence of machinery mold in any food processing plant is a violation of regulations. Antimicrobial agents have made it possible to keep machinery clear of this mold. The mold can be particularly bothersome in tomato and pineapple processing plants. See also **Antimicrobial Agents**.

Additional Reading

Adams, A., D.E. Gottschling, C. Kaiser, and T. Stearns: "Methods in Yeast Genetics, 1997," Cold Spring Harbor Laboratory Press, Cold Spring Harbor, NY, 1999.

Barnett, J.A., D. Yarrow, and R.W. Payne: "Yeasts" Characteristics and Identification," Cambridge University Press, New York, NY, 2000.

Davis, B.D., et al.: "Microbiology," Harper and Row, Hagerstown, Maryland, 1973.

Deak, T. and L.R. Beuchat: "Handbook of Food Spoilage Yeasts," CRC Press, LLC., Boca Raton, FL, 1996.

Fantes, P. and J. Beggs: "The Yeast Nucleus," Oxford University Press, Inc., New York, NY, 2000.

Jones, E.W., et al.: "The Molecular and Cellular Biology of the Yeast Saccharomyces: Gene Expression," Vol. 2, Cold Spring Harbor Laboratory Press, Cold Spring Harbor, NY, 1999.

Kurtzman, C.P. and J.W. Fell: "Yeasts: A Taxonomic Study," Elsevier Science, New York, NY, 1998.

Lodder, J., C.P. Kurtzman, and J.W. Fell: "The Yeasts, A Taxonomic Study," 4th Edition, Elsevier Science, New York, NY, 1998.

Milner, M. and W.F. Geddes: "Grain Storage Studies. III. The Relation between Moisture Content, Mold Growth, and Respiration of Soybeans," *Cereal Chem.*, **23**, 225 (1946).

Peppler, H.J.: "Yeast Properties Adversely Affecting Food Fermentations," *Food Technol.*, **34**, 2, 62–65 (1977).

Pringle, J.R., J.R. Broach, and E.W. Jones: "Molecular and Cellular Biology of the Yeast Saccharomyces," Vol. 3, Cold Spring Harbor Laboratory Press, Cold Spring Harbor, NY, 2000.

Rose, A.H. and J.S. Harrison: "The Yeasts," Vol. 5, Academic Press, Inc., San Diego, CA, 1997.

Walker, H.W.: "Spoilage of Food by Yeasts," *Food Technol.*, **31**, 2, 57–61 (1977).

Walker, G.M.: "Yeast Physiology and Biotechnology," John Wiley & Sons, Inc., New York, NY, 1998.

Windisch, S.: "Studies on Osmotolerant Yeasts," in "Yeasts" (A. Kocva-Kratochiflova, editor), Vyadavatel'stvo Slovenskej Akadmine Vied Bratislava, 1969.

Wogan, G.N.: "Chemical Nature and Biological Effects of the Aflatoxins," *Bacteriol. Rev.*, **30**, 460 (1966).

YELLOW FEVER. An endemic or epidemic disease caused by the *yellow fever virus*, which is a togavirus (Group B arbovirus, or flavovirus). Humans are the usual reservoir, but, in jungle areas, nonhuman primates may also carry the virus. The vector for transmitting the virus between humans is the mosquito, *Aedes aegypti*. Transmission between nonhuman primates is by an arthropod vector. Twelve days after biting an infected patient, the mosquito is then able to infect a nonimmune person. During the 18th and 19th Centuries, yellow fever was seen in North America and Europe, having been transmitted by mosquitoes breeding in the open water tanks of sailing vessels. In recent years, the mosquito has been essentially eliminated in the Caribbean, the southern United States, and Central America. Because the disease remains endemic in South America and sub-Saharan Africa, persons traveling to and through these areas require immunization with an effective yellow fever virus vaccine.

The U.S. Centers for Disease Control (Atlanta, Georgia) report that the last indigenous case of yellow fever was seen in the United States in 1911 and the last imported case was in 1924. However, devastating epidemics have occurred in Ethiopia (1960, 1962) and in Senegal (1965) causing many thousands of deaths.

Yellow fever is characterized by severe, fulminant hepatitis, accompanied by nephritis and hemorrhages. The incubation period is from 3 to 6 days. The symptoms in most cases are mild and consist of fever, headache, and perhaps jaundice. A severe case, however, is characterized by a sudden onset, acute prostration, fever, slow pulse rate, generalized aches and pains, and jaundice. The kidneys are particularly involved and large amounts of albumin in the urine is a feature. Vomiting may be described as uncontrollable. The tongue becomes bright red and the conjunctivae congested. Facial edema and flushing occur — to be replaced by a dusky pallor. The gums become swollen and there is a marked bleeding tendency with black vomit, melena, and ecchymoses. If the disease is left untreated, failure of the kidneys, with complete suppression of their function, or excessive bleeding from the gastrointestinal tract, usually occurs, and in such cases the mortality rate may be as high as 60%.

Death, when it occurs, comes usually within 6 to 7 days and rarely after ten days.

Treatment is essentially supportive. Prevention is by mosquito eradication and administration of vaccine. The 17D vaccine strain used causes few complications. In some recipients (about 5%), headache, myalgia, and low-grade fever may persist for 5–10 days after inoculation. Encephalitis is a very rare complication. It is estimated that over 34 million doses of 17D vaccine had been administered in the United States prior to 1980, as a result of which only two cases of encephalitis were reported. The French Dakar strain of vaccine is much less favored by most physicians because records indicate that it may cause meningoencephalitis in about 0.5% of the inoculations. Contraindications of yellow fever vaccine are hypersensitivity to eggs and pregnancy.

R.C. VICKERY, M.D., D.Sc., Ph.D., Blanton/Dade City, FL.

YELLOWFIN MENHADEN. See **Menhaden**.

YELLOWFIN TUNA. See **Tunas**.

YELLOW-GREEN ALGAE. A group of algae corresponding to the division *Chrysophyta*, the greenish-yellow to golden-brown color derived from chromatophores of that range of pigmentation. These algae usually have a cell wall composed of overlapping halves.

YELLOW-JACKET *(Insecta, Hymenoptera)*. A small black and yellow wasp that builds its nest of paper, either in a hole in the ground or under some object near the ground. The use of partly decayed wood in making the paper gives these nests a brownish color. Several species are known. The insect is found in Europe, Asia, and North America. It is an aggressive scavenger and possesses a vicious sting.

YELLOW LEGS. See **Waders, Shorebirds, and Gulls**.

YELLOW PERCH. See **Perches and Darters**.

YELLOWTAIL. See **Carangids**.

YELLOWTHROAT. See **Warbler**.

YERSINIOSIS. See **Foodborne Diseases**.

YEW TREES. Members of the family *Taxaceae* (yew family), these evergreen trees and shrubs have needlelike leaves, are very hardy and do well in most soils. Among conifers, the yews are distinctive in that a given tree is either male or female. Yews are considered among the best plants for hedges, having a very dense and evenly-distributed foliage. One disadvantage is that the yew grows more slowly than other plants suitable for hedges, such as the thujas or Lawson cypress. Hedges should not be overemphasized, however, because yews grow into medium-size trees as indicated by the accompanying table. Yews fall into four genera as indicated by the following list of principal species:

Chinese plum yew	*Cepholotaxus fortunei*
European or common yew	*Taxus baccata*
Japanese yew	*Taxus cuspidata*
Plum-fruited yew	*Podocarpus anidinus*
	(See entry on **Podocarps**.)
Prince Albert's yew	*Saxegothaea conspicua*

The yews grow abundantly in the British Isles, Europe, and parts of North America. The tree has been a favorite for hedges and gardens for many years, particularly in Britain and Europe. Possibly the oldest living tree in Europe is a yew at Fortingall in the Scottish Highlands. It is estimated to be about 1,500 years old. The common yew is found in the United States and Canada along the Pacific coast, southeastern Alaska, in the Sierra Nevada Mountains, through the northern states and in the Rocky Mountains. Often, the yew is found as a small tree or shrub. The Pacific yew or Oregon yew (*Taxus brevifolia*) can be considered a small forest tree, occurring on the Pacific slopes. It usually ranges in height from 20 to 40 feet (6 to 12 meters), but in some cases may reach as high as 70 feet (21 meters). The Savin or Florida yew (*Taxus foridana*) is a species local to western Florida. See Fig. 1.

Fig. 1.   Florida torreya (*Torreya taxifolia*) located in North Carolina. (*North Carolina Department of Natural and Economic Resources.*)

TABLE 1. RECORD YEW TREES IN THE UNITED STATES[1]

Specimen	Circumference[2]		Height		Spread		Location
	Inches	Centimeters	Feet	Meters	Feet	Meters	
Florida yew (1986) (*Taxus floridana*)	25	64	20	6.1	26	7.9	Florida
Pacific yew (1989) (*Taxus brevifolia*)	180	457	54	16.5	30	9.1	Washington

[1]From the "National Register of Big Trees," American Forests (by permission).
[2]At 4.5 feet (1.4 meters).

The outer bark of most yews is thin with many breaks. It peels away easily. The fruit is pulplike, bright red, showy, and full of seeds. The leaves are olive green, pointed, and leather-like to the touch. The wood is hard and durable, with a red coloration. In the past, it has found limited use for archer bows, canoe paddles, and some forms of cabinet work, when available.

The plum yews are set apart from the other yews in that the fruit is a nut, which is enclosed in a fleshy container that looks something like a plum—hence the name. Their leaves are also much longer, more widely spaced and of a lighter-green color. The plum yews are relatively small, open, and irregularly shaped trees. The Chinese plum yew is used extensively in gardens and landscaping, particularly in Europe. The smaller Japanese plum yew is also a favorite of some gardeners. As one authority points out, it is difficult to visualize the fruits of these yews as "cones."

Torreya california, the California nutmeg tree, is also a plum yew. It was named after the botanist whose name is also associated with the Torrey pine. The tree is much sharper in texture than the Chinese plum yew, but the fruit is similar. The tree is a native of California woodlands.

A record specimen of the *Torreya taxifolia* (Florida torreya) is shown in the accompanying figure.

Record yews growing in the United States are shown in Table 1.

YIELD POINT.   The minimum unit stress at which a structural material will deform without a significant increase in the load is called the yield point. Some materials do not have a yield point and in others it is not a well-defined value. Consequently, in these cases it has become common practice to use a quantity called the yield strength. The yield strength is the unit stress corresponding to a specific amount of permanent unit deformation.

YOKE.   1. The yoke of a motor or generator is the supporting and magnetic structure back of the poles, i.e., it is the part of the machine through which the flux passes in going between poles at the ends away from the armature. 2. The yoke of a magnetic-deflection cathode ray tube is the frame upon which is wound the horizontal and vertical deflecting coils.

YOLK GLAND.   A portion of the reproductive system of the female by which yolk is secreted. In some animals called the vitellarium.

YOLK SAC.   An accessory embryonic membrane formed in the vertebrates as an enveloping structure around the yolk of the egg. It is connected with the mid gut of the embryo and serves for the absorption of nourishment during embryonic life, and in some species, notably the fishes, after the individual has become active. The wall of the structure is composed of the same germ layers that form the gut, a lining endoderm, and a covering of splanchnic mesoderm. In the latter blood and blood spaces develop at an early period, later forming a network of vitelline vessels from which blood flows into the body of the embryo by way of a pair of large omphalomesenteric veins. Branches of the arterial system of the body extend into this plexus, completing a cycle for the transportation of the absorbed food to the developing body.

The yolk sac persists even in the mammals, where yolk is usually not present. In these forms it serves for the absorption of materials from the surrounding uterus during early development and is involved in the development of the circulatory system. It soon becomes a vestige, however, as its functions are taken over by other membranes.

YOSEMITE.   A deeply U-shaped portion of a glacial valley, with sheer walls, hanging troughs, and a wide almost level floor, encountered in the Sierra Nevada Mountains of California. These formations resemble the Yosemite Valley, California.

YOUDEN SQUARE.   If in a balanced incomplete block design the number of blocks is equal to the number of treatments, it is possible to arrange the blocks as columns of a rectangular array in such a way that each treatment occurs just once in every row. The corresponding experimental design is known as a Youden square (cf. **Latin Square**) and differences between rows as well as differences between columns can be eliminated from the estimate of error.

YOUNG-HELMHOLTZ THEORY.   The original theory of color vision. It assumes the existence in the eye of three mechanisms each sensitive to one of the three primary colors, red, green and blue. It accounts for some of the observed phenomena of color vision, but not for others. The theory is convenient for many purposes but has been modified by later students.

YOUNG INTERFERENCE EXPERIMENT.   In 1801 Thomas Young made the epochal discovery of the interference of light waves, by means of an experiment which has become classic. Light from a narrow slit L falls on a plate in which are two parallel slits, A, B, very close together, so that from the further side of the latter there emerge two exactly similar wave trains. See Fig. 1. These overlap in the region beyond AB and produce interference. If a screen F is placed at some distance from AB, alternate bright and dark bands or fringes appear on it, parallel to the two slits. If a translucent screen is used (or in the case of white light, a plate of colored glass acting as a light filter), these bands may be viewed by means of a magnifier beyond it at E, or better, a low-power micrometer eyepiece, with which the width of the band-interval can be measured.

It is easy to show from the elementary theory of interference that if $AB = s$, if the distance from AB to F is x, and if the wavelength of the light is λ, the distance on the screen between any two consecutive dark bands or any two consecutive bright bands is $b = x\lambda/s$. Therefore if b, s, and x are measured, we have at once a means of determining the wavelength: $l = bs/x$.

Other devices have proved more satisfactory than the pair of slits, such as the biprism, Fresnel's mirrors, or Lloyd's mirror; each of which produces a double virtual image of the slit L to serve as the two wave-train sources A and B.

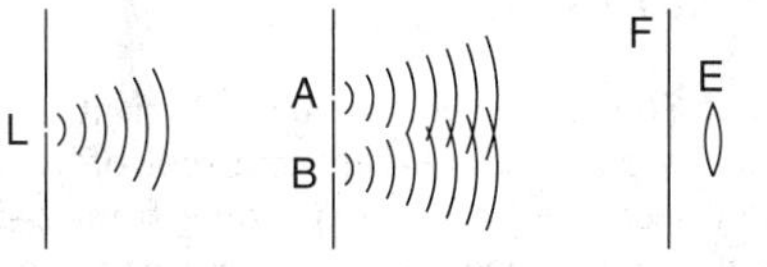

Fig. 1.   Light from single source L gives rise to two wave trains at A and B, which produce interference fringes on screen F.

YOUNG'S MODULUS.   A modulus of elasticity in tension or compression, involving a change of length. See also **Elasticity**.

YOUNG, THOMAS (1773–1829).   Young was an English inventor and scientist. He earned his medical degree in 1800 and began practicing medicine. In 1801, however, he became a professor of natural philosophy at the Royal Institution and became recognized for his published lectures.

Young began researching doing experiments with light and his work supported the wave theory by his introduction of the principle of interference, which explained the bright and dark bands seen when light passes through two narrow slits. Young is also recognized for his later work in which he worked with Egyptian Hieroglyphics and showed that the markings were not just pictorial but phonetic sounds.

See also **Young-Helmholtz Theory**; **Young Interference Experiment**; and **Young's Modulus**.

J. M. I.

YOUTH (Geological).   From a topographical standpoint, the youth stage is the first stage of the cycle of erosion in the topographic development of a landscape or region, during which stage the original surface or structure remains the dominant feature of the relief. The stage is typified by a few, small and widely spaced young streams; by broad, flat-top interstream divides and upland surfaces, only modified in a minor way by erosion; by partially developed or poorly integrated drainage systems, the latter having numerous swamps and shallow lakes; and by rapid and progressive increase of local relief, with sharp landforms, steep and irregular slopes, and a surface considerably above sea level.

From a coastal standpoint, the youth stage is that phase of development of a shore, shoreline, or coast that is typified by an ungraded profile of equilibrium. In the case of a shoreline of submergence, an irregular or crenulate outline, vigorous wave action, formation of sea cliffs, and associated erosional forms will be present, as well as a steep offshore profile, and the presence of bays, promontories, offshore islands, spits, bars, and other fairly minor irregularities. In the case of a shoreline of emergence, there usually will be a straight and simple outline, larger waves breaking well offshore, small waves coming to land to produce a nip or low cliff, and the formation of barrier beaches, lagoons, and marshes.

From the standpoint of a stream, the youth stage is the first stage in the development of a stream, a stream that has just commenced its work of erosion and is increasing steadily both in vigor and efficiency, enabled to erode its channel and not yet having reached a graded condition. The stream in this stage is typified by an ability to carry a load greater than what it is carrying; by forming a deep, narrow, steep-walled, V-shaped valley, gorge, or canyon with a steep and irregular gradient and rocky outcrops; by numerous waterfalls, rapids, and lakes; by a swift current and clear water; by a few, short, straight tributaries; by an ungraded bed; and by the absence of flood plains.

YTTERBIUM.   Chemical element symbol Yb, at. no. 70, at. wt. 173.04, thirteenth in the Lanthanide Series in the periodic table, mp 819 °C, bp 1196 °C, density 6.966 g/cm^3 (20 °C). Elemental ytterbium has a face-centered cubic crystal structure at 25 °C. The pure metallic ytterbium is silver-gray in color and is stable in moist or dry air up to 200 °C, after which oxidation occurs. The metal is readily dissolved by dilute and concentrated mineral acids. The metal dissolves in liquid NH_3 to yield a dark-blue color. There are seven natural isotopes ^{168}Yb, ^{170}Yb through ^{174}Yb, and ^{176}Yb. Ten artificially-produced isotopes have been identified. Ytterbium is one of the least abundant elements of the rare-earth group and 53rd among all elements occurring in the earth's crust. The element was first identified by J.D.G. Marignac in 1878. Electronic configuration

$$1s^2 2s^2 2p^6 3s^2 3p^6 3d^{10} 4s^2 4p^6 4d^{10} 4f^{13} 5s^2 5p^6 5d^1 6s^2.$$

Ionic radius Yb^{2+} 1.00 Å; Yb^{+3} 0.88 Å. Metallic radius 1.940 Å. First ionization potential 6.25 eV; second 12.18 eV. Other important physical properties of ytterbium are given under **Rare-Earth Elements and Metals**.

The principal sources of ytterbium are euxenite, gadolinite, monazite, and xenotime, the latter being the most important. Ytterbium is separated from a mixture of yttrium and the heavy Lanthanides by using the sodium amalgam reduction technique. Ytterbium metal is obtained by heating a mixture of lanthanum metal and ytterbium oxide under high vacuum. The ytterbium sublimes and is collected on condenser plates whereas the lanthanum is oxidized to the sesquioxide.

To date, the major uses of ytterbium have been in applied and fundamental research. The element and its compounds have been used in magnetic "bubble" domain devices (ytterbium orthoferrite), in phosphors to convert infrared to visible light, in lasers, and radioisotope ^{169}Yb has found application in portable industrial and medical radiographic units.

See references listed at ends of entries on **Chemical Elements**; and **Rare-Earth Elements and Metals**.

Additional Reading

Gschneidner, K.A., Jr., B.J. Beaudry, and J. Capellen: "Rare Earth Metals," pp. 720–732 in "Metals Handbook, 10th edition, Vol. 2, Properties and Selection: Nonferrous Alloys and Special Purpose Materials," ASM International, Metals Park, OH, 1990.

Lide, D.R.: "CRC Handbook of Chemistry and Physics 2000-2001," 81st Edition, CRC Press, LLC., Boca Raton, FL, 2000.

NOTE: This entry was revised and updated by K.A. GSCHNEIDNER, Jr., Director, and B. EVANS, Assistant Chemist, Rare-earth Information

Center, Institute for Physical Research and Technology, Iowa State University, Ames, IA.

YTTRIUM. Chemical element symbol Y, at. no. 39, at. wt. 88.905, periodic table group 3, mp 1522 °C, bp 3,338 °C, density 4.469 g/cm³ (20 °C). Most of the properties of yttrium are similar to those of the heavy rare-earth elements, falling between gadolinium and erbium. Elemental yttrium has a close-packed hexagonal crystal structure at 25 °C. The pure metallic yttrium is silver-gray in color, retaining a luster in air up to about 400 °C, above which it oxidizes to Y_2O_3. The metal is dissolved by most mineral acids, but is relatively inert in a 1:1 mixture of concentrated HNO_3 and 48% hydrofluoric acid. Yttrium is capable of working common metallurgical fabrication procedures. The metal is immiscible with liquid or solid uranium metal. It has a low thermal-neutron-absorption cross section and a low acute-toxicity rating. The natural isotope of yttrium is ^{89}Y. Fourteen artificial isotopes have been identified. See also **Radioactivity**. In terms of abundance, yttrium is present on the average of 33 ppm in the earth's crust and potentially is as plentiful as cobalt. The element was first identified by Fredrich Wohler in 1828. Electronic configuration $1s^2 2s^2 2p^6 3s^2 3p^6 3d^{10} 4s^2 4p^6 4d^1 5s^2$. Ionic radius $Y^{3}+$ 0.900 Å. Metallic radius 1.801 Å. First ionization potential 6.38 eV; second 12.24 eV. Other important physical properties of yttrium are given under **Chemical Elements**; and **Rare-Earth Elements and Metals**.

Residues from uranium mining operations in Canada have been a major source of yttrium. Xenotime (YPO_4) found in Malaysia is another source, as well as the ion-adsorption clay minerals in China. Some apatite deposits are unusually rich in yttrium and it also is found in gadolinite, euxenite, and samarskite.

In recovering yttrium, mixed rare-earth minerals or wastes are dissolved in HNO_3 or H_2SO_4. Liquid-liquid organic ion-exchange solvent extraction cells then separate a pure yttrium fraction, usually precipitated as an oxalate and then calcined to the oxide. The metal is obtained by metallothermic reduction, using calcium mixed with YCl_3 or YF_3 in a sealed retort at a temperature in excess of 1,550 °C. Alloys of yttrium and cobalt, or yttrium and magnesium have been deposited out of a molten electrolyte BaF_2-LiF-YF_3.

The major uses of yttrium are as phosphors in color television, computer monitors, trichromatic fluorescent lights, x-ray intensifying screens and temperature sensors. Yttrium oxide stabilized zirconium oxide is used as oxygen sensors for automobile engines and in iron making, wear-resistant and corrosion-resistant cutting tools, abrasives, high temperature refractories for continuous casing nozzles, jet engine coatings, simulated diamond gemstones (i.e. cubic zirconia), and as a solid electrolyte in solid oxide fuel cells (SOFC). Yttrium aluminum garnets are used as laser crystals for industrial cutting and welding, medical and dental surgery, temperature and distance sensing, photochemistry, photoluminescence, digital communication and non-linear optics. $YBa_2Cu_3O_{7-x}$ is an important high temperature ceramic superconductor with a 90 K superconducting transition temperature. It is primarily used in the form of thin films for superconducting electronics (such as SQUIDS, MRI and NMR coils, wireless communication subsystems and digital instruments) and as superconducting wires and tapes for power transmission lines, motors and generators, transformers, current limiters, magnetic separation, research magnet systems and current leads. Yttrium-iron garnets find use as microwave bandpass filters, circulators, and isolators in electronic and communications circuitry. Y_2O_3 is used as a metallurgical dispersion hardening agent in nickel-based superalloys made by powder metallurgy techniques. Yttrium metal also is specified in several cobalt-base superalloys in which it improves hot corrosion (sulfidation) resistance at high temperatures. When used in iron-chromium-aluminum alloys, yttrium improves workability and adds resistance to sag when the alloys are used as electrical-heating elements. The metal also is applied as cladding to rotating turbine engine parts to obtain superior oxidation resistance. Yttrium also has been used in permanent magnets, YCo_5, and shows great promise. These magnets are second only to $PrCo_5$ as the most powerful permanent magnet materials developed to date, far exceeding alnico and other more conventional materials.

Chemistry and Compounds. Yttrium hydroxide, $Y(OH)_3$ is precipitated by NH_4OH from solutions of yttrium salts. It differs in properties from the lanthanide hydroxides, both structurally and chemically, in its ability to absorb atmospheric CO_2. Yttrium oxide, Y_2O_3, is obtained by heating the hydroxide or oxy-acid salts; it forms mixed oxides when heated with other oxides, such as Fe_2O_3 and TiO_2. Yttrium also forms a peroxide,

Y_4O_9, obtained in hydrated form by treatment of yttrium solutions with hydrogen peroxide.

All four halides are known. The fluoride, YF_3, is readily formed by action of a fluoride on an yttrium nitrate solution. There is an oxyfluoride, YOF, formed by high-temperature, low-pressure heating of the mixed oxide and fluoride. Complex fluorides, containing $[YF_6]^{3-}$ exist; the cryolite minerals are of this composition. The group, $-YF_4$ is found in double salts formed by YF_3 and the alkali fluorides. The chloride, YCl_3, is formed as a hydrate by action of HCl solution upon the hydroxide. It may be dehydrated by slow heating; rapid heating gives the oxychloride, YOCl. The bromide, YBr_3, is formed as a hydrate by action of HBr upon the hydroxide. Its dehydration requires heating under vacuum. The iodide, YI_3 is best prepared from the anhydrous chloride, by reacting with HI and I_2.

Both normal and mixed carbonates are known. The former is precipitated, as $Y_2(CO_3)_3 \cdot 3H_2O$ from Y^{3+} solutions by alkali metal carbonates, which in excess dissolve the precipitate to form a soluble hydrated double carbonate. The oxycarbonate is also a double molecule, $3Y_2(CO_3)_3 \cdot 2Y(OH)_3$, formed by action of CO_2, upon the hydroxide.

The nitrate, $Y(NO_3)_3$ exists as a number of hydrates. The hexahydrate formed by action of HNO_3 upon the hydroxide, is dehydrated at 100 °C to give the trihydrate and the anhydrous salt. However, other hydrates are known, as well as double nitrates (especially with the lanthanide elements) and oxynitrates, of the general formula, $xY_2O_3 \cdot yN_2O_5 \cdot zH_2O$.

Yttrium hydroxide forms a hydrated sulfate with H_2SO_4, which is dehydrated on heating. It forms double sulfates with alkali and ammonium sulfates.

The carbide, formed from the oxide and carbon in the electric furnace, appears to have the composition, YC_2. It yields acetylene and other hydrocarbons upon hydrolysis.

See references listed at ends of entries on **Chemical Elements**; and **Rare-Earth Elements and Metals**.

Additional Reading

Gschneidner, K.A., Jr., B.J. Beaudry, and J. Capellen: "Rare Earth Metals," pp. 720–732 in "Metals Handbook, 10th edition, Vol. 2, Properties and Selection: Nonferrous Alloys and Special Purpose Materials," ASM International, Metals Park, Ohio (1990).

Gschneidner, K.A., Jr.: "Physical Properties of the Rare Earth Elements," pp. 4–112 to 4–121 in "CRC Handbook of Chemistry and Physics," 77th Edition, CRC Press, Boca Raton, Florida (1996–1997).

Hammond, C.R.: "The Elements," pp. 4–1 to 4–34 in "CRC Handbook of Chemistry and Physics," 77th Edition, CRC Press, Boca Raton, Florida (1996–1997).

NOTE: This entry was revised and updated by K.A. GSCHNEIDNER, Jr., Director, and B. EVANS, Assistant Chemist, Rare-earth Information Center, Institute for Physical Research and Technology, Iowa State University, Ames, IA.

YUCCA. See **Palm Trees**.

YUCCA BORER (*Insecta, Lepidoptera*). A giant skipper, *Megathymus yuccae*, whose larva bores in the stem and root of yucca plants. Several species of these insects are known, expanding from $2-3\frac{1}{2}$ inches (5 to 9 centimeters) in the adult stage. While yucca seems to be the prevailing food plant, some species are known to attack agave. They are limited in distribution to the southern states and south into Central America with the exception of one that has been taken in central Colorado and western Nebraska.

YUCCA MOTH (*Insecta, Lepidoptera*). A small moth, also commonly called the pronuba moth from the name of its genus. It lives in the flowers of yucca, depositing its eggs in the ovary of the plant and then pollinating the flower. The developing larva eats the seeds, but since many more are formed than it is capable of consuming, the exchange is beneficial to the plant. Several species of these moths are known. All belong to the same genus, whose early name, *Pronuba*, has been supplanted by another, *Tegeticula*. Yuccas are also frequented by moths of the genus *Prodoxus*, which are of no service in pollinating the flowers. They are called false yucca moths.

YUKAWA, HIDEKI (1907–1981). Yukawa was a Japanese physicist who did research in the field of quantum physics. He graduated from Kyoto Univesity. He is best known for predicting "heavy quantum" for a particle heavier than an electron. He is best known for his insight into mesons providing an explaination of the force responsible for binding together the atomic nucleus. For this insight and his research work, Yukawa won the Nobel Prize in Physics in 1949.

Yukawa's other awards include the Imperial Prize of the Japan Academy in 1940 and the Lomonosov Gold Medal of the USSR Presidium of the Academy of Sciences in 1964.

See also **Yukawa Potential**.

J. M. I.

YUKAWA POTENTIAL. A potential function of the form $V = (V_0/r)(e^{-r/b})$. It is used to describe the meson field about a nucleon. The Yukawa potential is employed rather frequently as one shape of nuclear potential well that can be used in attempts to fit theory with experimental results, for example, in high-energy scattering. It is characterized by (1) infinite strength at $r = 0$, (2) an exponential tail extending with appreciable strength to larger r rather than a Coulomb potential. See also **Nuclear Forces**; and **Particles (Subatomic)**.

Z

ZEBRA. See **Horses, Asses, and Zebras**.

ZEBRA FISH (*Brachydanio rerio*).   A popular, easy-to-breed aquarium fish highlighted by its four blue longitudinal stripes. The species reaches a length of 4.5 cm and is distributed in central India. The zebra fish, as of late 1990, was under consideration by molecular biologists as an organism that may one day be the vertebrate *Drosophila* as a model for studies of embryology and developmental biology.

ZEBRA TIME.   The same as mean time at the Greenwich meridian. The system is used in communication and for synchronized reckonings. Sometimes called Z time. The hour 2400 Zebra time is equivalent to 0300 (Moscow standard time); 0800 (Manila standard time); 0900 (Tokyo standard time); 1000 (Sydney standard time); 1400 (Hawaiian standard time); 1600 (Pacific standard time); 1700 (Mountain standard time); 1800 (Central standard time); 1900 (Eastern standard time).

ZEDOARY.   A perennial herb (*Curcuma zedoaria* Rosc.) of the family *Zingiberaceae*. The dried product prepared from the rhizomes is used as a spice and in the formulation of liqueurs. The bark extract has been used in the manufacture of bitters. There is no extensive use for the essential oil, which is obtained by steam distillation of the rhizomes.

The plant, native to India, is cultivated for edible purposes. The rhizomes are thick and tuberlike and are edible, as are the very large leaves. Flavoring products prepared from zedoary are described as having a warm and camphoraceous odor and a warm, spicy, and slightly bitter taste.

ZEEMAN EFFECT.   An effect of a moderately intense magnetic field upon the structure of the spectrum lines of a gas when subjected to its influence. The phenomenon, sought unsuccessfully by Faraday and finally observed by Zeeman in 1896, consists in the splitting up of each line into two or more components. In the simpler cases, when the source is viewed at right angles to the field, there are three components, of which the middle one has the same frequency as the unmodified line. This component is plane-polarized to vibrate parallel with the field, while the two side components vibrate at right angles to the field. When the source is viewed in the direction of the field, there are only two components, displaced in opposite directions, and circularly polarized in opposite senses. (See also **Polarized Light**.) These phenomena constitute the so-called "normal" Zeeman effect.

With most lines, however, an anomalous Zeeman effect is observed and the number of components is greater, in some cases reaching twelve or fifteen. They are symmetrically arranged and symmetrically polarized. The displacements, as in the simpler case, are proportional to the magnetic field intensity H, and are always expressible, in wave numbers, as rational multiples of the displacement in the normal effect, which is $4.67 \times 10^{-5} H$ (reciprocal centimeter), a quantity known as the "Lorentz unit." The Zeeman effects observed in sun spots give valuable information as to the magnetic conditions in those areas.

Closely related to the Zeeman effect are two others, the Paschen-Back effect, produced by very strong magnetic fields, and the Back-Goudsmit effect, observed with the spectra of elements having a nuclear magnetic moment, such as bismuth. See also **Electron Theory**; **Paschen-Back Effect**; and **Stark Effect**.

ZEEMAN, ERIK CHRISTOPHER (1925–   ).   Zeeman was an English mathematician. His doctoral work was in pure mathematics and he received his Ph.D. in 1954 for a thesis on knots and all the algebra you need to actually prove the existence of knots. He did research in topology, which is a type of geometry that examines the properties of shapes in many dimensions. His best known work was in catastrophe theory. His work has consequences for a broad range of fields from weather to psychiatry. Zeeman also made contributions in the development of the chaos theory.

Zeeman is not only known for this mathematical contribution but also for being an excellent educator. He spent much of his life teaching mathematics both to adults and children. He built the mathematics department at University of Warwick in England and made it an internationally known research center. He gave many mathematical talks on radio and televison to reach a broader audience feeling that many gifted were amongst the educationally deprived population. He was honored by Queen Elizabeth II knighting him in 1991 for his role in advancing mathematics education. See also **Chemical Elements**; **Electron Theory**; and **Zeeman Effect**.

J. M. I.

ZEIN.   See **Starch**.

ZEISBERG CONCENTRATOR.   A nitric acid concentrator, consisting of a packed tower, into which weak nitric acid vapors are introduced. Some steam is admitted into the bottom, and strong nitric acid vapors are discharged from the top to a condenser.

ZEISEL REACTION.   The demethylation of an organic compound by treatment with hydriodic acid, which leaves a hydroxy group in place of the methoxy group, and forms methyl iodide, which may readily be determined quantitatively, as in the study of the amount of methoxy groups present.

ZELLWEGER'S SYNDROME.   A rare familial condition in which are found severe hypotonia, abnormalities of the central nervous system, hepatic (liver) cirrhosis, and malformation of the skeleton. Persons with this malady seldom live beyond age 6 months. For persons who survive to an older age, there is frequently mental retardation and display of severe mitochondrial abnormalities. Hanson et al. (1980) have described the syndrome as a unique "experiment of nature" in which the role of the mitochondria in bile acid synthesis could be assessed in vivo.

ZENER CURRENT.   The current through an insulator in a very intense electric field, sufficient to excite an electron directly from the valence band to the conduction band.

ZENER DIODE.   A type of silicon diode that acts like a rectifier until the applied voltage reaches the avalanche breakdown voltage (zener voltage). At this point, the diode becomes conducting. The voltage drop across the diode remains essentially constant and independent of current. The diode is used in voltage-limiting circuits and power supplies. Zener diodes are now used in precise reference signals. A potentiometric instrument can be no more accurate than the calibrated (standard) source of emf against which the unknown potential is compared. Standard cells and a variety of batteries calibrated against standard cells have been of prime use, but in recent years, batteries such as the mercury cell are used directly as a standard when accuracy requirements permit. Since the 1960s, the zener diode used in an appropriate circuit has become most useful by offering low cost, very long life, small size, and accuracies equivalent or better than standard cells, particularly over long period. A typical circuit using zener diodes is shown in Fig. 1. The resistors R_1, R_2, R_3 and the dynamic resistance of *CR1* constitute a balanced Wheatstone bridge which tends to eliminate variations in V_{supply} from the nodes A and B. *CR1* and *CR2* are in a conventional cascade configuration because R_2 is inherently large compared to R_1 and R_3. *CR2* is also a temperature-compensated zener diode, and the resulting standard emf can be held to within 0.02% over a normal working range of supply voltage and ambient temperature.

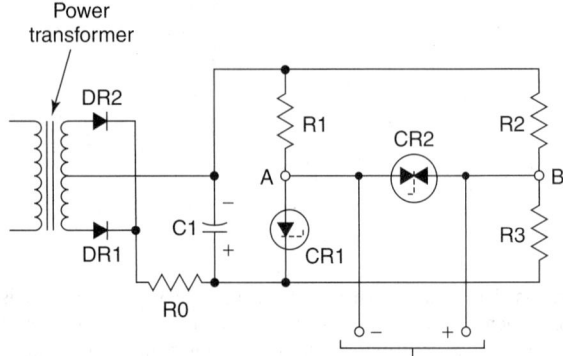

Fig. 1.   Bridge-type circuit in which zener diode is used as constant-voltage source.

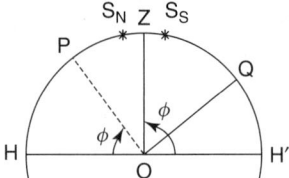

Fig. 1.   Principal parameters of zenith telescope.

ZENER VOLTAGE.   1. The field required to excite the Zener current, of the order of 1 volt per unit cell, or 10^7 volts/centimeter. 2. The voltage associated with that portion of the reverse volt-ampere characteristic of a semiconductor, wherein the voltage remains substantially constant over an appreciable range of current values.

ZENITH.   The point on the celestial sphere directly overhead is the *observer's zenith*. The *astronomical zenith* is defined as the point where the plumb line extended up from the surface of the Earth will intersect the celestial sphere. Owing to the fact that the plumb line may be affected by local gravitational effects, such as large mountains in the vicinity, the *geographical zenith* is the point where a line perpendicular to the surface of a smooth Earth would intersect the celestial sphere. The angular distance between the astronomical and geographical zenith is the station error of the point on the surface of the Earth. Because of the fact that the Earth is an oblate spheroid rather than a perfect sphere, neither the plumb line nor a perpendicular to the surface of the Earth will pass through the geometrical center of the Earth unless the observer is either at one of the poles or on the equator. The *geocentric zenith* is defined as the point where a line extended from the center of the Earth through the observer will intersect the celestial sphere. The angular distance between the astronomical and geocentric zenith is the reduction of latitude for the observer. See also **Celestial Sphere and Astronomical Triangle**.

ZENITH TELESCOPE.   This instrument, as the name implies, is designed for use at or very close to the zenith. A telescope is mounted in the same manner as the meridian circle, i.e., in the altazimuth form, with the azimuth so fixed that the instrument is always in the plane of the meridian. In place of the accurate circles for measuring altitude that are to be found on the meridian circle, this instrument carries a very accurate level so adjusted that the bubble is in the center of the tube when the telescope is pointing at the zenith. The reticle of the instrument carries, in addition to the set of wires parallel to the meridian, as in the meridian circle, two wires parallel to the axis of rotation, i.e., perpendicular to the meridian. One of these wires is fixed, and when the instrument is in proper adjustment, with the level bubble in the center of the tube, the fixed wire is in the plane of the prime vertical, i.e., passes through the zenith. The other wire may be moved by means of a fine screw, which is parallel to the meridian. The head of this screw is accurately calibrated so that the distance of the movable wire from the zenith wire may be determined in seconds of arc.

The zenith telescope is used primarily for the accurate determination of terrestial latitude by what is commonly known as Talcott's method. In Fig. 1, we have a representation of the celestial sphere drawn in the plane of the observer's meridian $HPZQH'$. HOH' is the plane of the horizon, Z is the astronomical zenith, P is the pole of rotation, and Q is the direction of the equator. HP is the altitude of the pole, and hence, by definition, is the astronomic latitude of O. Inspection of the figure shows that QZ, the declination of the zenith, is also equal to ϕ.

For the purpose of determining QZ, a pair of stars, S_s and S_n, are selected that have approximately the same right ascension and hence will cross the meridian within a short time of each other. The declinations of the stars are so selected that one will pass just south of the zenith and the other just north. Such pairs of stars, known as Talcott pairs, have been selected and their declinations accurately determined by various observatories. Of course, for the selection of the proper pairs to be used for a particular station, the latitude of that station must be approximately determined in advance by any of the methods discussed under latitude.

At the time that one of the stars is approaching the meridian, the observer watches in the eyepiece of the zenith telescope, and when the star appears, he sets the movable wire upon the star and keeps it there until the star has crossed the meridian. The reading of the screw gives the distance of the star north or south of the zenith, either S_nZ or S_sZ, depending upon which star is observed first. The second star is then observed, and the other zenith distance determined in the same manner. From the value of the measured zenith distances and the declinations of the stars, the value of the declination of the zenith, and hence the latitude, can be accurately determined.

The great advantage of this method is that both stars are observed close to the zenith, so that the correction for astronomical refraction is very small in either case. Furthermore, since one star passes north and the other south of the zenith, the refraction corrections practically neutralize each other. The measurement of the zenith distance by means of the screw can be very accurately made. The disadvantages of the method are that it requires the use of the accurately adjusted fixed instrument, and the declinations of the stars must be very accurately known.

Observations of this character are constantly being made at various stations all over the world for the purpose of determining the variation of latitude. See also **Celestial Sphere and Astronomical Triangle**; and **Telescope (Astronomical-Optical)**.

ZEOLITE GROUP.   To the zeolite group of minerals belong a number of hydrous silicates of aluminum which also ordinarily contain sodium or calcium, but rarely they may carry barium, strontium, magnesium, and potassium. These minerals are not related crystal-lographically as they occur in the isometric, orthorhombic, hex-agonal, and monoclinic systems, but they are all characterized by the presence of water, up to 10 or 20%, which is easily released with the application of heat. They are all rather soft minerals, hardness, 3.5–5.5; of low specific gravity, 2.0–2.5, and they will decompose readily upon treatment with acid, most of them yielding a gelatinous mass.

The easy fusion, together with the rapid expulsion of water, is responsible for the name of this interesting group; it is derived from the Greek words to boil and a stone, hence zeolite, "a boiling stone." The zeolites are secondary minerals, usually found filling fissures and cavities in the more basic igneous rocks as basalt, and gabbro, but occasionally in the more acidic types as granite or in gneisses. The following members of the zeolite group are described under **Analcime**; **Chabazite**; **Harmotome**; **Heulandite**; **Natrolite**; **Phillipsite**; **Scolecite**; and **Stilbite**.

ZEOLITE SOFTENING.   See **Feedwater (Boiler)**.

ZEOLYTIC CATALYST.   See **Petroleum**.

ZEOMORPHS.   See **John Dories**.

ZEPPELIN.   See **Dirigibles and Airships**.

ZEREVITINOV DETERMINATION.   A method of analysis of organic compounds containing active hydrogen atoms, as in hydroxy, carboxy, or imino groups, by reaction with methyl magnesium halide to yield methane quantitatively: $ROH + CH_3MgX \rightarrow CH_4 + ROMgX$.

ZERO.   A number symbolized by 0. If a is any other number, its properties are: $a \pm 0 = a$; $a \times 0 = 0$; $a^0 = 1$; $0/a = 0$, if $a \neq 0$. Division by zero is prohibited, for this operation is undefined. Certain other operations involving zero give ambiguous results known as indeterminate forms. They can be interpreted by the methods of differential calculus.

A zero matrix is usually called *null*. For the zero of a function, see also **Function**.

ZERO-BASED CONFORMITY.   See **Conformity**.

ZERO-BASED LINEARITY.   See **Linearity**.

ZERO-BEAT RECEPTION.   See **Homodyne Reception**.

ZERO ERROR.   The error of a device operating under the specified conditions of use when the input is at the lower range-value. The term *zero shift* is often used to represent a change or drift in zero error with time.

ZERO GRAVITY.   The absence of gravitational effects—either as a result of nullifying gravitational attraction, or counterbalancing gravitational attraction with centrifugal force. A zero-gravity switch actuates when a condition of weightlessness occurs. See also **Weightlessness**.

ZERO-ORDER REACTION.   A reaction that has a constant rate, as in the case of certain gases reacting on the surface of a solid when it is almost entirely covered. Under such conditions, the reaction rate may be independent of pressure, and thus, the process is kinetically of zero order.

ZERO POINT ENERGY.   The kinetic energy remaining in a substance at the absolute zero point of temperature. According to quantum mechanics, a simple harmonic oscillator does not have a stationary state of zero kinetic energy. The ground state has still one half quantum, hv, of energy, and the motion corresponding thereto. This agrees with the uncertainty principle, which does not permit the oscillator particle to be absolutely at rest exactly at the origin. In solids the zero-point energy is distributed in the normal modes of lattice vibration, and may be an appreciable term in the binding energy of the crystal, especially in hydrogen, helium, rare gases, etc. The motion may be observed in x-ray diffraction, but does not contribute to electronic resistivity.

ZERO POINT ENTROPY.   According to the third law of thermodynamics (see also **Thermodynamics**), the entropy of a system in equilibrium at the absolute zero must be zero. Such systems, as for instance a glass, which can have finite entropy at absolute zero are not in thermodynamic equilibrium.

ZEROTH LAW.   See **Thermodynamics**.

ZIGZAG RIDGE.   A continuous ridge that trends first in one direction, then in another, each alternating part being roughly parallel. Such a condition is formed by the converging of two ridges in one direction, each ridge converging with a different ridge in the opposite direction. Zigzag ridges are found in the Appalachian Mountains, where they resulted from truncation of plunging folds.

ZINC.   Chemical element, symbol Zn, at. no. 30, at. wt. 65.38, periodic table group 12, mp 419.58 °C, bp 907 °C, density 7.1 g/cm³. Elemental zinc has a close-packed hexagonal crystal structure. There are five stable isotopes ^{64}Zn, ^{66}Zn through ^{68}Zn, and ^{70}Zn. Six radioactive isotopes have been identified ^{62}Zn, ^{63}Zn, ^{65}Zn, ^{69}Zn, ^{71}Zn, and ^{72}Zn. With exception of ^{65}Zn which has a half-life of 245 days, the half-lives of the other isotopes are measured in minutes and hours.

Zinc is a bluish-white metal, malleable and ductile at 150 °C, but at 180 °C it changes rapidly so that at 205 °C it may be easily powdered; remains lustrous in dry air but is slightly tarnished in moist air or in water; burns upon heating to vaporization with a bluish flame, forming zinc oxide; soluble in acids—slowly when pure but rapidly on contact with copper or platinum; soluble in alkalies. Discovery prehistoric.

Zinc ranks 27th in order of abundance of the chemical elements in the earth's crust, an estimated 0.004% content of igneous rocks on an average basis. It is estimated that a cubic mile of seawater contains about 48 tons of zinc. First ionization potential 9.391 eV; second, 17.89 eV. Oxidation potential $Zn \rightarrow Zn^{2+} + 2e^-$, 0.762 V; $Zn + 4OH^- \rightarrow ZnO22^- + 2H_2O + 2e^-$, 1.216 V. Electron configuration $1s^2 2s^2 2p^6 3s^2 3p^{10} 4s^2$. Ionic radius Zn^{+2} 0.75 Å. Metallic radius 1.3324₅ Å. Other physical properties of zinc are described under **Chemical Elements**.

Zinc and lead usually occur together in nature as sulfides. Earlier separation processes involved the fine grinding of the combined sulfides and then treating the particles with chemical reagents to cause one sulfide to be preferentially wetted and thus the two sulfides separated by the froth flotation process. In a first stage, the lead sulfide is floated while the zinc sulfide sinks to the bottom of the tank. In the second stage, the process is reversed and the zinc sulfide is floated. Gangue and other nonmetals collect at the bottom of the tank. The separated sulfides are dewatered to a 6–7% moisture content and are referred to as the zinc concentrate and the lead concentrate.

A major zinc ore is ZnS (sphalerite) which frequently occurs with the major lead ore PbS (galena). The lead-zinc ores usually contain recoverable quantities of copper, silver, antimony, and bismuth as well. Major deposits of this type are worked in Australia, the United States, Canada, Mexico, Peru, the former Yugoslav Republics, and the former Soviet Union. Two other important zinc ores are $ZnCO_3$ (smithsonite) and iron-zinc-manganese oxide (franklinite). Several of these minerals are described under separate alphabetical entries.

Extractive Metallurgy of Zinc[1]

As of the early 1980s, zinc production throughout the world is based almost exclusively on two processes:

1. Electrolytic Process—in which oxidized zinc concentrates are leached in sulfuric acid and then electrolyzed to plate SHG (special high-grade zinc metal) on the cathode and to regenerate the acid on the anode.
2. ISP (Imperial Smelting Process)—a combined lead-zinc process in which oxidized concentrates are reduced with coke in a shaft furnace and the zinc vapor collected in a lead splash condenser.

The older retort and electrothermic processes have largely been replaced because of environmental and economic reasons.

The ISP evolved to fill a very special niche in nonferrous metallurgy because of its capability of treating lead-zinc concentrates which may also contain appreciable amounts of copper. The concentrate is normally oxidized in a sintering machine to produce a feed for the blast furnace where the zinc oxide is reduced with coke. Some effort has been underway to develop a hot briquetting operation to produce a suitable feed without sintering. Other efforts to improve the economic competitiveness of the process include air preheat and the use of an oxygen-enriched blast to reduce coke consumption.

Although the electrolytic zinc process can trace its industrial history back over 60 years, only during the last 15 to 20 years has it become the industry standard, commanding a large portion of the free world's zinc capacity. While the electrolytic zinc process has been varied to meet the demands of the particular feed, the flow sheet always contains the basic steps of roasting, leaching, solution purification, and electrolysis. The zinc concentrate is oxidized with air to produce acid-soluble zinc oxide (calcine), and sulfur dioxide-containing off gas suitable for conversion to acid, as well as byproduct steam. The Vieille-Montagne/Lurgi fluidized bed roaster is the industry standard.

The roaster product is leached with spent electrolyte (sulfuric acid) under near-neutral conditions to dissolve most of the zinc, copper, and cadmium, but little of the iron. The leach residue solids are releached in hot, strong acid to dissolve more zinc, since it attacks the otherwise insoluble zinc ferrites. The iron which is also dissolved in this second leach is then precipitated as jarosite, goethite, or hematite. The development of these iron precipitation techniques permitted the use of the hot, strong acid leach and an increase in zinc extraction from about 87% to greater than 95%. Simultaneously, the hot acid leach frequently generates a leach residue rich enough in lead and silver to provide significant byproduct value, as well as increased recovery of cadmium and copper.

The neutral solution is purified to remove impurities more noble than zinc, e.g., cadmium, copper, cobalt, nickel, arsenic, antimony, and germanium. The purification is accomplished by cementation in two or more steps with the addition of zinc dust. Generally, at least one cementation step is conducted at high temperature with arsenic, antimony, or copper-arsenic added. Cadmium is usually recovered in the metallic

---

[1] Information for this topic furnished by C.O. Bounds, St. Joe Minerals Corporation, Monaca, Pennsylvania.

state and copper, nickel, and cobalt are recovered as sludges if present in sufficient quantities.

Zinc is extracted from the purified solution in cells using lead/silver alloy anodes and aluminum cathodes at a current density of 38–60 amperes/square foot (400–650 amperes/square meter). The product is normally SHG zinc, particularly if strontium carbonate is added and/or lead/silver anodes of greater than 0.596 silver content are used. After deposition of 24 to 72 hours, the cathodes are removed from the cells and the zinc is stripped by automatic machines in modern plants, melted, and cast for market. The move to automated handling of large cathodes was a major factor in lowering the overall labor requirement in producing zinc.

The following technical developments have led to the acceptance of the electrolytic process:

1. Adoption of high-capacity, low-labor fluidized-bed roasters
2. Adoption of continuous schemes for leaching and solution purification, allowing more automation and lower operating costs
3. Improved raw material utilization via hot acid leaching and iron precipitation as previously described
4. Construction of higher-capacity plants with larger equipment
5. Adoption of mechanized cathode handling/stripping with dramatically lower labor demands
6. Improved byproduct recovery via the hot acid leach.

With exception of leach residue (jarosite, goethite, etc.) disposal, the process is environmentally sound.

The electrolytic process is inherently a somewhat energy inefficient (being based on electrical rather than directly on fossil energy) and capital-intensive operation. A return to pyrometallurgical smelting is conceivable if the environmental concerns are addressed and a noncoke (or at least low-coke) process is developed. New hydrometallurgical developments include the near-commercial Sherritt-Gordon pressure leach to eliminate roasting and the generation of sulfur-dioxide-rich gases, and laboratory experiments with chloride-based leaching/electrolysis.

Production. Principal producers of slab zinc include the former U.S.S.R., Japan, Canada, the United States, Germany (West), Australia, Belgium, France, Poland, Italy, Spain, Mexico, and China. The first four countries mentioned account for well over half of the production. Countries with limited production include Finland, the Netherlands, the former Yugoslav Republics, Bulgaria, and North Korea.

Uses of Zinc

Some concept of the major uses for zinc can be gleaned from Table 1. Slab zinc is available in three grades, as specified by the American Society for Testing and Materials (Table 2).

Zinc Coating. Constituting the largest single use of the metal, zinc coating is accomplished mainly by dipping the product in molten zinc or by electroplating (electrogalvanizing). Hot-dip galvanizing employs chiefly the less pure grades of zinc. The life of galvanized material is proportional to the thickness of the coating, and recent developments in both electrogalvanizing and hot-dip galvanizing have been toward application of heavy coatings that will withstand deformation without peeling.

Because of its relatively high electropotential (position in the emf series of metals), zinc can provide electrolytic protection against corrosion of several common metals, notably products made of iron and steel. Advantage of this characteristic also is taken in the use of zinc as the anode material for a number of types of batteries, power packs, and fuel cells. In

TABLE 1. CONSUMPTION OF SLAB ZINC IN THE UNITED STATES

Applications of Zinc	Percent of Total Tonnage Consumed	
Galvanizing		40.8
sheet and strip	26.5	
wire and wire rope	2.3	
tube and pipe	4.1	
fittings (tube and pipe)	0.6	
tanks and containers	0.3	
structural shapes	0.3	
fasteners	0.5	
pole-line hardware	0.4	
fencing, wire cloth and netting	1.5	
other	2.4	
In Brass and Bronze		13.8
sheet, strip, and plate	6.6	
rod and wire	5.0	
tube	0.6	
casting and billets	0.2	
copper-base ingots	0.6	
other copper-base products	0.8	
In Zinc-base Alloys		28.0
die casting alloy	27.4	
dies and rod alloy	0.1	
slush and sand casting alloy	0.5	
Other Uses		17.4
rolled zinc	2.2	
zinc oxide	3.5	
other applications	11.7	
Total		100.0

Source: U.S. Bureau of Mines.

providing a protective coating over ferrous metals, the attack of corrosive materials on the zinc produces a relatively inert reaction-product film, which deters destruction of the underlying zinc and base metal. When the coating is broken, as may result from mechanical means such as scratching, abrading, etc., the zinc, having a higher electropotential than the ferrous metals, slowly is expended in furnishing the required protective current. Thus, serious corrosion is delayed for a long period, providing long useful life to the zinc-coated products except in the most adverse cases, or where a poor selection of construction materials was initially made.

The most widely used form of zinc coating is effected by hot-dip galvanizing in which steel sheets, coils, structurals, hardware, wire, and other forms are dipped in a bath of molten zinc. The zinc readily adheres to a previously cleaned (pickled and thoroughly washed) iron or steel surface. The thickness of the coating is controlled by manipulating process temperatures, time the underlying metals are in contact with the molten bath, and mechanical means used. Products that are commonly hot-dip galvanized include roofing, siding, transmission towers, highway guard-rails, light poles, culverts, and fencing. Other forms of zinc coating include flame-spraying or metallizing, flake galvanizing, sherardizing or cementation, plasma arc spraying, vacuum metallizing, and also painting with materials that contain zinc pigments.

Electrogalvanizing makes it possible to apply a zinc coating to such products as steel strip, wire, conduits, hardware, etc. in a high-speed fashion through electroplating. One-side zinc coating of steel sheet for

TABLE 2. GRADES OF SLAB ZINC

Grade of Zinc	Composition (Percent)			
	Lead (maximum)	Iron (maximum)	Cadmium (maximum)	Zinc (minimum by difference)
Special High Grade	0.003	0.003	0.003	99.990
High Grade	0.03	0.02	0.02	99.90
Prime Western	1.4	0.05	0.20	98.0

Notes: When specified for use in manufacture of rolled zinc or brass, aluminum is held to 0.005% maximum.
Tin in Special High Grade zinc is held to 0.001% maximum.
Aluminum in Prime Western zinc is held to 0.05% maximum.
Source: American Society for Testing and Materials.

automotive applications by electrodeposition is a notable innovation in recent years. One advantage of electroplating is that the products to be coated are not subject to thermal conditions that may alter dimensions and shape. Electrodeposits range in thickness from 0.00015 inch (0.004 millimeter) to 0.001 inch (0.25 millimeter). Sherardizing is accomplished by heating the ferrous materials to be coated at a temperature of about 350 °C in contact with zinc dust in a closed vessel. The resulting coating consists of iron-zinc alloys. It has been found that sherardized coatings match the corrosion resistance of electrogalvanized or hot-dip coatings of equivalent thickness.

Die Castings. The major use of zinc as a structural material is in alloys for pressure die casting. Development of the modern zinc die-casting alloys was directly related to use of Special High Grade zinc, with the addition of particular alloying constituents held within close limits and control of impurities. For die castings it is essential to use this extra pure grade of zinc to ensure extremely low iron, lead, cadmium, and tin contents. It is also necessary to limit these same impurities in the metals added to make the desired zinc alloy composition. Only by such control can zinc die castings be produced that are stable in dimensions and properties.

The impurities lead, cadmium, and tin, if present in castings in amounts greater than the established maximums (0.005% lead; 0.004% cadmium; 0.003% tin), cause subsurface network corrosion. These limits are close to critical values. Iron is held to 0.10% maximum to prevent excessive skimming losses and machining problems.

Zinc alloys are low in cost of metal per casting, are easy to die cast, are cast at low temperatures, have greater strength than all other die-casting metals except the copper alloys, lend themselves to casting within close dimensional limits, permit the thinnest sections yet produced, and are machined at minimum cost. Their resistance to surface corrosion is adequate in a wide range of applications. Prolonged contact with moisture results in formation of white corrosion products, but surface treatments can be applied that largely prevent formation of such products.

Limiting service conditions for standard zinc-base die-casting alloys are as follows: At temperatures slightly above 95 °C (200 °F) their tensile strength is reduced 30% and their hardness 40%. At subzero temperatures, some embrittlement occurs, but impact strength is still in the same range as that of aluminum and magnesium die-casting alloys at normal service temperatures. At room temperature, impact strength of zinc die castings is much higher than that of aluminum or magnesium die castings or iron sand castings.

All die castings have at least a light flash at the die parting, and those requiring movable cores will have some flash around the cores. Flash is also formed around ejector pins at the points at which they make contact with the casting.

Although it often is cheaper to cut threads than to cast them, for many pieces cast threads usually are more economical. Male threads usually are made with a parting parallel to the axis; this leaves a flash at the parting. The flash can be removed with a shaving tool in some instances, but in others a chasing operation is necessary.

Zinc die castings are invariably cast within quite close dimensional limits, but some machining is commonly required in addition to removal of flash, even though it may consist only of such simple operations as punching, drilling reaming, or tapping of holes. Zinc die castings can be soldered or welded, but ordinarily neither technique is used except for special applications or repair.

Many of the finishes applied to other types of metal products can also be applied to zinc die castings, although some differences in formulation as well as occasional differences in method of application may be desirable. The types of finishes applicable to zinc die castings include: mechanical finishes (buffed, polished, brushed, and tumbled); electrodeposited finishes (copper, nickel, chromium, brass, silver, and black nickel); chemical finishes (chromate, phosphate, molybdate and black nickel); and organic finishes (enamel, lacquer, paint and varnish, and plastic finishes). Electrodeposited coatings of virtually any metal capable of electrodeposition can be applied to zinc die castings.

The automotive industry uses by far the largest number of zinc alloy die castings. Some of the most important mechanical parts made include carburetors, bodies for fuel pumps, windshield-wiper parts, speedometer frames, grilles, horns, heaters, and parts for hydraulic brakes. The electrical industry probably uses a larger diversity of die castings than the automotive industry. Such parts are used in washing machines, oil burners, stokers, motor housings, vacuum cleaners, electric clocks, and kitchen equipment

and utensils. Zinc die castings are used in business machines—typewriters, recording machines, picture projectors, vending machines, accounting machines, cash registers, cameras, slicing machines, garbage disposers, gasoline pumps, hoists, and drink mixers. Building hardware, padlocks, toys, and novelties also consume a substantial percentage of the total production of zinc-base die castings.

It is interesting to note that Gutenberg (1440s) used metal-based inks in printing the early Bibles. Zinc was used, along with copper and lead. These inks replaced those made of soot, which were prone to evanescence.

Much more detail on zinc metal products and applications will be found in the Horvick (1979) reference listed. Zinc alloys with copper are described under **Copper**.

Chemistry and Compounds. In virtually all of its compounds zinc exhibits the +2 oxidation state, although compounds of zinc(I) have been reported in the gaseous phase. Zinc is readily oxidized in the presence of hydroxide ions, e.g., by H_2O, this behavior being attributed to the stability of the $Zn(OH)_4{}^{2-}$ ion. Like other transition group 2 elements zinc has a marked tendency to form covalent structures, e.g., ZnO and ZnS.

Zinc oxide, formed by oxidation of the metal, dissolves in acids to yield Zn^{2+} ions, and in alkalies to form $Zr(OH)_4{}^{2-}$ ions. It reacts slowly with moist CO_2 to form the oxycarbonate, $5ZnO \cdot 2CO_2 \cdot 4H_2O$. Addition of an alkali metal hydroxide to a solution of a zinc salt does not precipitate the hydroxide, producing instead hydroxyzincate salt or a precipitate of flocculant zinc oxide. However, the hydroxide can be obtained from a sodium zincate solution on dilution and standing. Zinc hydroxide is amphiprotic, yielding Zn^{2+} ions with acids and $Zn(OH)_4{}^{2-}$ ions with (excess of) alkali hydroxides. Zinc peroxide is produced by treating a zinc chloride solution with sodium peroxide at a pH of 9.5. It is unstable, decomposing slowly on standing. Zinc oxide is used extensively in rubber, paints, and chemicals. Expanding uses include exterior latex paints, particularly alkyd-modified latex, the photocopy paper field, and as a substitute for mercury in mildew and fungus prevention formulations.

Unlike the other zinc halides, zinc fluoride, ZnF_2, is only slightly soluble in cold water. The anhydrous halides are prepared by direct union of the elements. In solution, zinc chloride, bromide, and iodide exhibit anomalous conductance properties attributed to undissociated molecules and complex ions. On heating these solutions, halogen acids, HX, are evolved, leaving oxyhalides in the fused residue. The zinc halides readily form double salts with halides of elements of main groups 1, 2, and sometimes 3 and 4. See also **Bromine**.

Zinc oxycarbonate is formed by the reaction of suspensions of the oxide or hydroxide with CO_2; to produce the normal carbonate, a very rapid stream of carbon dioxide and, usually, a somewhat higher pH is required.

Zinc also forms both nitrates and oxynitrates (in various hydrated forms) but the oxynitrate is less stable. It is formed by heating the hexanitrate, or treating the nitrate solution with NH_3.

Zinc forms a wide variety of other salts, many by reaction with the acids, though some can only be obtained by fusing the oxides together. The salts include arsenates (ortho, pyro, and meta), the borate, bromate, chlorate, chlorite, various chromates, cyanide, iodate, various periodates, permanganate, phosphates (ortho, pyro, meta, various double phosphates), the selenate, selenites, various silicates, fluosilicate, sulfate, sulfite, and thiocyanate.

Zinc sulfide, selenide, and telluride are more pronouncedly covalent than the oxide. They can be prepared from the elements, or in the case of first two, by the action of H_2S or hydrogen selenide upon zinc solutions. Zinc nitride, Zn_3N_2, prepared from zinc dust and NH_3, hydrolyzes readily to NH_3 and the oxide. Two zinc phosphides are known; Zn_3P_2, formed by heating the elements, yields phosphine with acids; ZnP_2 and $ZnHP$ have also been prepared.

One of the features of the chemistry of zinc is the fact that it is among the elements having a large number of complex compounds, mostly with coordination numbers of four and tetrahedral, but some with coordination numbers of 6, such as those of ethylenediamine which contain the ion $[Zn(en)_3]_{2+}$. The large number of halogen double salts have already been cited. The marked donor ability of oxygen toward zinc is evident from the number of basic salts, the existence of the zincates (containing $ZnO_2{}^{2-}$ and also $Zn(OH)_4{}^{2-}$, the latter in strong alkali), and the formation of such chelate complexes as acetylacetonates and dioxalato complexes. Sulfur is a better donor than oxygen, so that addition compounds of the type $(R_2S)_2ZnX_2$ are formed from dialkyl sulfides and zinc halides, while thiourea forms chelate complexes containing $[Zn(th)_2]^{2+}$. The ready reactions with ammonia, as with amines, give large numbers of complexes; those with ammonia include diammines, triammines and tetrammines, containing $[Zn(NH_3)_2]^{2+}$, $[Zn(NH_3)_3]^{2+}$ and $[Zn(NH_3)_4]^{2+}$ respectively.

Prominent among the carbon donor complexes are the cyanides, principally compounds of $[Zn(CN)_4]^{2-}$, although $[Zn(CN)_3]^-$ is also known. Other carbon donor complexes are the triethyl and tetraethyl complexes, which are readily electrolyzed.

Biological Aspects of Zinc. See also **Zinc (In Biological Systems)**.

Additional Reading

Bauccio, M.L.: "ASM Engineered Materials Reference Book," 2nd Edition, ASM International, Materials Park, OH, 1994.

Brooks, C.R.: "Heat Treatment, Structure and Properties of Nonferrous Alloys," ASM International, Materials Park, OH, 1989.

Craig, B.D. and D.S. Anderson: "Handbook of Corrosion Data," 2nd Edition, ASM International, Materials Park, OH, 1995.

Davis, J.R.: "Metals Handbook," 2nd Edition, ASM International, Materials Park, OH, 1998.

Dutrizac, J.: "Lead-Zinc 2000," The Minerals, Metals & Materials Society, Warrendale, PA, 2000.

Greenwood, N.N. and A. Earnshaw: "Chemistry of the Elements," 2nd Edition, Butterworth-Heinemann, Inc., Woburn, MA, 1997.

Guruswamy, S.: Staff, International Lead Zinc Research Organization, Marcel Dekker, Inc., New York, NY, 1999.

Hewitt, K. and T. Wall: "Zinc Industry," Woodhead Publishing, Ltd., Cambridge, UK, 2000.

Horvick, E.W.: "Properties of Pure Zinc," Metals Handbook, 9th Edition, Vol. 2, ASM International, Metals Park, OH, 1979.

Kubel, E.J., Jr.: "Expanding Horizons for Zinc-Aluminum Alloys," *Advanced Materials & Processes*, **51** (July 1987).

Lalo, J.: "Pennsylvania's Dead Mountain (Zinc Smelter)," *Amer. Forests*, **54** (March–April 1988).

Lide, D.R.: "CRC Handbook of Chemistry and Physics 2000-2001," 81st Edition, CRC Press, LLC., Boca Raton, FL, 2000.

Loffler, H.: "Structure and Structure Development of Al-Zn Alloys," John Wiley & Sons, Inc., New York, NY, 1995.

Parker, P.: "McGraw-Hill Encyclopedia of Chemistry," 2nd Edition, The McGraw-Hill Companies, Inc., New York, NY, 1993.

Porter, F.C.: "Zinc Handbook: Properties, Processing, and Use in Design," Marcel Dekker, Inc., New York, NY, 1991.

Sousa, L.J.: "The Changing World of Metals," *Advanced Materials & Processes*, 27 (September 1988).

Staff: "Properties and Selection: Nonferrous Alloys and Special-Purpose Materials," Vol. 2, 10th Edition, ASM International, Materials Park, OH, 1991.

Staff: "Alloy Phase Diagrams Handbook," Vol. 3, ASM International, Materials Park, OH, 1992.

Staff: "Nonferrous Metals–Nickel, Cobalt, Lead, Tin, Zinc, Cadmium, Precious, Reactive, Refactory Metals and Alloys," Vol. 4, American Society for Testing & Materials, West Conshohocken, PA, 1997.

Vander Voort, G.: "Atlas of Time-Temperature Diagrams for Nonferrous Alloys," ASM International, Materials Park, OH, 1991.

Zhang, X.G.: "Corrosion and Electrochemistry of Zinc," Kluwer Academic Publishers, Norwell, MA, 1996.

Web References

International Lead Zinc Research Organization: *http://www.ilzro.org/*
The Minerals, Metals & Materials Society: *http://www.tms.org/* Tables

ZINC (In Biological Systems). Zinc was one of the first of the trace elements known to be essential for both plants and animals, and yet problems of zinc nutrition are still of pressing importance. Evidence of zinc deficiency in crops is being recognized in new areas and the use of zinc in fertilizers has increased steadily in recent years. A dry, cracked condition of the skin of pigs (*parakeratosis*) caused by a zinc-deficiency has been a problem to pork producers. Diseases and syndromes that have been attributed to zinc deficiency in the human diet include loss of appetite, loss of sense of taste, and delayed healing of burns and other wounds. It is interesting to note that application of zinc-containing ointments to promote healing is an old practice in human medicine. Laboratory animals deficient of zinc may be subject to serious reproductive problems, including infertility of males, failure of conception or implantation of the embryo, difficult births, and deformed offspring. The extent to which zinc deficiency is a primary cause of reproductive problems in farm animals and humans is not thoroughly understood, and research continues.

Patients of both sexes with sickle cell anemia have been reported to be zinc deficient. Administration of zinc to males with this disease has been shown to improve the hypogonadism and the short stature associated with this deficiency. Administration of small dosages of zinc sulfate is part of the therapy in treating certain types of sickle cell anemia. See also **Anemias**.

Zinc deficiency in crops is frequently observed where fields have been graded to smooth them so that irrigation water can be applied more uniformly. Where the topsoil is cut away from small areas of these fields, such crops as corn (maize) and beans may be stunted and many leaves will be white instead of the usual green. If zinc fertilizers, supplying as little as 10 pounds of zinc per acre (11.2 kilograms per hectare) are applied, bumper crops may be grown on these soils. Citrus trees are commonly fertilized with zinc.

When zinc fertilizers are used on soils deficient in zinc, crop production may be increased even though the zinc concentration in the plant tissues and especially in the seed show no increase. With higher levels of zinc fertilization, the zinc concentration in plants may increase. Some evidence shows that the value of food and feed crops as sources of dietary zinc can be improved by using zinc fertilizers at rates exceeding those required for optimal plant growth. However, very high rates of zinc fertilization can depress crop yields.

The zinc contained in plants is not fully utilized by animals. Diets high in calcium and phosphorus have been associated with poor digestibility of dietary zinc. Diets with large amounts of soy protein are particularly likely to require extra zinc fortification for livestock. Meat is an important source of zinc for human diets. Where supplementation of zinc is indicated, zinc sulfate, zinc oxide, and zinc carbonate are commonly used.

Zinc in Metabolism. Zinc was first recognized as a trace element — then referred to as a growth factor for Aspergillus niger — by Raulin (1869). Evidence for a specific biochemical role of zinc was first obtained by Keilin and Mann in 1940, when the metal was shown to be a stoichiometric component of bovine carbonic anhydrases. The findings of many other zinc-containing enzymes during the interim have indicated a diverse biological role for zinc.

The high affinity of zinc for nitrogeneous and sulfur-containing ligands seems chiefly responsible for the occurrence of zinc in a wide variety of biological compounds, such as proteins, amino acids, nucleic acids, and porphyrins. Operationally, the enzymes affected by zinc can be considered in two groups: (1) zinc metalloenzymes; and (2) zinc metal-enzyme complexes. Zinc metalloenzymes incorporate zinc so firmly in the protein matrix that they can generally be considered as an entity. Under reasonably mild conditions, the metal and protein moiety are isolated together and exhibit an integral stoichiometric relationship. On the same basis, a strict correlation is preserved between metal and enzyme activity, allowing the inferential identification of a specific biological function of zinc in vivo. Zinc metal-enzyme complexes, in contrast, comprise enzymes which are activated in vitro by the addition of zinc ions. The loose association and the relative lack of metal ion specificity render it difficult in many cases to assign specific biological significance to zinc in vivo.

In zinc metalloenzymes, zinc is a selective stoichiometric constituent and is essential for catalytic activity. It is frequently present in numerical correspondence with the number of active enzymatic sites, coenzyme binding sites, or enzyme subunits. Removal of zinc results in loss of activity. Inhibition by metal complexing agents is a characteristic feature of zinc metalloenzymes. However, no direct relationship holds between the inhibitory effectiveness of these agents and their affinity for ionic zinc. Although zinc is the only constituent of zinc metalloenzymes in vivo, it can be replaced by other metals in vitro, such as cobalt, nickel, iron, manganese, cadmium, mercury, and lead, as in the case of carboxy-peptidases.

Zinc is a ubiquitous component of animal and plant tissue. In vertebrates, most organs, including pancreas, contain 20–30 micrograms of zinc per gram of wet tissue. Liver, voluntary muscle, and bone hold about double this amount. Zinc contents ranging from 100 to 1000 micrograms/gram weight have been measured in islet tissues of certain teleost fishes. Correlation between zinc content and insulin storage suggest a parallelism, but evidence is wanting for zinc-insulin complexes in vivo. The highest zinc content determined among mammals is found in the *tapetium lucidum cellulosum* of adult for seals — up to 150,000 micrograms/gram. Human blood contains 7–8 micrograms zinc/milliliter. About 12% of this is present in serum, 3% in leukocytes, and 85% in erythrocytes. In these compartments, zinc occurs as part of zinc proteins and zinc metalloenzymes. In erythrocytes, zinc is correlated to carbonic anhydrase activity.

In 1991, B.C. Cunningham, M.J. Mulkerrin, and J.A. Wells (Genetech) reported that size-exclusion chromatography and sedimentation equilibrium studies demonstrated that zinc ion (Zn^{2+}) induced the dimerization of human growth hormone (HGH). Scatchard analysis of $^{65}An^{2+}$ binding to HGH shows that two ZN^{2+} ions are associated per dimer of HGH in a cooperative fashion. Cobalt (II) can substitute for Zn^{2+} in the hormone dimer and gives a visible spectrum characteristic of cobalt coordinated in a tetrahedral fashion by oxygen- and nitrogen-containing ligands. Human growth hormone is synthesized and secreted into storage granules before its release from the anterior pituitary.

In 1991, R.E. Klevit (University of Washington) reported on the recognition of DNA by Cys$_2$, His$_2$ zinc fingers. The "zinc finger" protein motif was so named because of the tandemly repeating pattern observed in the amino acid sequence of the transcription factor TFIIA. Klevit points out, "According to the original hypothesis, each 30-residue sequence is an independently folded unit that binds a zinc ion and is responsible for sequence-specific DNA binding."

Underwood (1977) reported that high levels of dietary zinc interfere with the normal absorption and metabolism of several minerals. Earlier, Stewart and Magee (1964) reported that experience with laboratory animals indicated that 7500 parts per million (ppm) zinc in a purified diet lowered the concentrations of calcium and phosphorus in bone. Other investigators showed that animals fed the same level of zinc were anemic and had deformed, fragile erythrocytes. The livers contained reduced amounts of ferritin and lower concentrations of iron in the ferritin. Whanger and Weswig (1971) showed that in rats fed 2000 and 4000 ppm zinc, liver copper concentration was decreased. The most sensitive responses to zinc have been obtained with animals receiving minimally required or suboptimal levels of copper Campbell and Mills, 1974; and Murthy et al., 1974). Based upon these earlier findings, Hamilton et al. (1979) investigated possible zinc interference with copper, iron, and manganese in young Japanese quail (*Coturnix coturnix japonica*). The objective of the investigators was to identify the minimal level of excess dietary zinc that would produce physiological and metabolic deviations from normal and to define some of the most sensitive zinc-mineral interactions. Because other workers had found a sensitive zinc-copper antagonism, the Hamilton team studied the effects of supplemental zinc at copper levels of marginal deficiency. Data from the study showed that adequacy of copper intake is important when supplemental zinc is consumed either as a dietary supplement or in foods fortified with zinc. It was reported that results of the findings may be important for the general human population, whose dietary intake of many minerals, including copper, does not usually exceed the requirement or may be marginally deficient (Milne et al., 1978; Harland et al., 1978; Klevay, 1978). See also **Copper (In Biological Systems)**.

Zinc Poisoning: Soluble zinc salts are usually the etiologic agent in cases of zinc poisoning. The poisonous nature of zinc may be described as astringent, corrosive, and emetic. Sources of food poisoning include zinc-coated galvanized containers, pots, cans, tubs, and acids that convert zinc into soluble salts. Products that may be so involved include lemonade, cooked apples, mashed potatoes, spinach, chicken and tomatoes, and fruit punch. Zinc in combination with various organic and inorganic radicals (dimethyl dithiocarbamate, arsenate, etc.) can be quite poisonous.

Additional Reading

Campbell, J.K. and C.F. Mills: "Effects of Dietary Cadmium and Zinc on Rats Maintained on Diets Low in Copper," *Proc. Nutr. Soc.*, **33**(1), 15A (Abstract) (1974).

Considine, D.M. and G.D. Considine: "Foods and Food Production Encyclopedia," Van Nostrand Reinhold, New York, NY, 1983

Cunningham, B.C., M.G. Mulkerrin, and J.A. Wells: "Dimerization of Human Growth Hormone by Zinc," *Science*, 545 (August 2, 1991).

Gray, P.: "The Encyclopedia of the Biological Sciences," 2nd Edition, Krieger Publishing Company, Melbourne, FL, 1981.

Hamilton R.P., et al.: "Zinc Interference with Copper, Iron and Manganese in Young Japanese Quail," *J. Food Sci.*, **44**(3), 738–741 (1979).

Harland, B., L. Prosky, and J. Vanderveen: "Nutritional Adequacy of Current Levels of Zn, et al., in the American Food Supply for Adults, Infants, and Toddlers," in "Trace Element Metabolism in Man and Animals" (M. Kirchgessner, editor), p. 311, Institut fur Ernahrungsphysiologie, Technische Universitat Munchen, Freising-Weihenstephan, Germany, 1978.

Herrmann, B.W.A.: "Copper, Silver, Gold, Zinc, Cadmium, and Mercury," Vol. 5, Thieme Medical Publishers, Inc., New York, NY, 1999.

Klevay, L.M.: "Dietary Copper and Copper Requirements in Man," in "Trace Element Metabolism Man and Animals" (M. Kirchgessner, editor), p. 307, Institut fur Ernahrungsphysiologie Technische Universitat Munchen, Freising-Weihenstephan, Germany, 1978.

Klevit, R.E.: "Recognition of DNA by Cys$_2$, His$_2$ Zinc Fingers," *Science*, 1367 (September 20, 1991).

Lewis, R.J. and N.I. Sax: "Sax'x Dangerous Properties of Industrial Materials," 10th Edition, John Wiley & Sons, Inc., New York, NY, 1999.

Mills, C.F.: "Zinc in Human Biology," Springer-Verlag, Inc., New York, NY, 1989.

Milne, D.B., et al.: "Dietary Intakes of Copper, Zinic, and Manganese by Military Personnel," *Fed. Proc.*, **37**, 894 (Abstract) (1978).

Murthy, L., et al.: "Interrelationships of Zinc and Copper Nurtriture in the Rat," *J. Nutrition*, **104**, 1458 (1974).

Prasad, A.S.: "Biochemistry of Zinc," Kluwer Academic Publishers, Norwell, MA, 1993.

Rainsford, K.D., et al.: "Copper and Zinc in Inflammatory and Degenerative Diseases," Kluwer Academic Publishers, Norwell, MA, 1998.

Robson, A.D.: "Zinc in Soils and Plants," Kluwer Academic Publishers, Norwell, MA, 1993.

Sarkar, B.: "Genetic Response to Metals," Marcel Dekker, Inc., New York, NY, 1991.

Schollmerich, J. and J.D. Kruse-Jarres: "Zinc and Diseases of the Digestive Tract,"Kluwer Academic Publishers, Norwell, MA, 1997.

Stewartm, A.K. and A.C. Magee: "Effect of Zinc Toxicity on Calcium, Phosphorus and Magnesium Metabolism of Young Rats," *J. Nutrition*, **82**, 287 (1964).

Underwood, E.J. and W. Mertz: "Trace Elements in Human and Animal Nutrition," 5th Edition, Academic Press, Inc., San Diego, CA, 1990.

Whanger, P.D. and P.H. Weswig: "Effect of Supplementary Zinc on the Intracellular Distribution of Hepatic Copper in Rats," *J. Nutrition*, **101**, 1093 (1971).

ZINC BLENDE. See **Sphalerite Blende**.

ZINC CARBONATE. See **Smithsonite**.

ZINCITE. This mineral, (Zn, Mn)O, is an ore of zinc and occurs in considerable quantities at Franklin Furnace, New Jersey, where it is associated with willemite and franklinite. Its hexagonal crystals are rare, as it usually occurs massive, foliated, or in coarse to fine grains. When the crystals are observable, it reveals a perfect cleavage parallel to the base of the prism. The fracture is conchoidal. The mineral has a hardness of 4; specific gravity, 5.684; luster, subadamantine to vitreous; orange-yellow streak; color, red to orange-yellow; translucent to opaque. Zincite also has been found in Poland, Tuscany, Spain, Saxony, and Tasmania.

ZINC OXIDE PIGMENT. See **Paints and Coatings**.

ZINC SULFIDE. See **Sphalerite Blende**.

ZINGG PEBBLE CLASSIFICATION. In 1905, Theodor Zingg (Switzerland) proposed a classification of pebble shapes, based upon the graphical representation of the diameter ratio of intermediate (width) to maximum (length) plotted against the diameter ratio of minimum (thickness) to intermediate dimensions. Four classes are specified: spheroid; disk; blade; and rod.

ZIPF'S LAW. In statistics, a law that seems to be observed in various kinds of empirical data, connecting the size of aggregates with their rank order, according to size. For example, if the towns in a given country are ranked according to population x, beginning with the largest, it is often found that the rank r is connected with x by a relation of the type $r = k/x^p$ where k and p are constants. A special case occurs when $p = 1$, in which event the product of r and x is a constant. Laws of a similar but rather more complicated kind have been studied in such different fields as the distribution of businesses, of authors by number of publications, and of fragmentation in bombs.

ZIRCON. This mineral is zirconium silicate, $ZrSiO_4$, and is the chief ore of zirconium. Zircon occurs in square tetragonal prisms, although sometimes it assumes pyramidal or irregular forms. The mineral may be found in some beach and river placer deposits, but generally it is associated as an accessory mineral in siliceous igneous rocks, crystalline limestones, schists, and gneisses—and in sedimentary rocks derived from the foregoing. Zircon is without good cleavage; is brittle, with a conchoidal fracture; hardness, 7.5; specific gravity, 3.7–4.7; luster, adamantine, brilliant; color, green, yellow-green, golden-yellow, red, red-brown, brown, and blue. The name zircon derives from an Arabic word *zarqun*, meaning vermilion, or perhaps from the old Persian *zargun*, meaning golden-colored.

Zircon occurs in the Ural Mountains; Trentino, Monte Somma, and Vesuvius, Italy; Arendal, Norway; Ceylon, India; Thailand; at the Kimberley mines, Republic of South Africa; Madagascar; and in Canada in Renfrew

County, Ontario, and Grenville, Quebec. In the United States, zircon is found at Litchfield, Maine; Chesterfield, Massachusetts; in Essex, Orange, and St. Lawrence Counties, New York; Henderson County, North Carolina; the Pikes Peak district of Colorado; and Llano County, Texas.

Gem quality crystals from Ceylon (Sri Lanka) have been known for many years. They range from colorless to brownish orange, yellow, dark red, from light reddish-violet. Heat treated zircons provide a beautiful stone of light blue color. Colorless stones are used as a diamond substitute.

ZIRCONIUM. Chemical element symbol Zr, at. no. 40, at. wt. 91.22, periodic table group 4, mp $1,853\,^{\circ}C$, bp $4,376\,^{\circ}C$, density 6.44 g/cm^3, 6.47 g/cm^3 (single crystal). Metallic zirconium is allotropic. Up to about $863\,^{\circ}C$, the alpha phase (hexagonal close-packed) is stable; above this temperature, the metal assumes the beta phase (body-centered cubic). The most common impurity, oxygen, tends to stabilize the alpha phase.

Zirconium metal exhibits passivity in air due to the formation of adherent coatings of oxide or nitride. Even without the coating, it is resistant to the action of weak acids and acid salts, but dissolves in HCl (warm) or H_2SO_4 slowly, and more rapidly if F is present, forming compounds of ZrO^{2+} ions, or fluorozirconates in the last case.

Crystalline zirconium of high purity is a white, soft, ductile, and malleable metal, but that of 99% purity, when obtained at high temperature, is hard and brittle. Pure zirconium has a combination of properties which make it a valuable structural material for nuclear reactors. In addition to low neutron capture, zirconium has good strength at high temperatures, corrosion resistance to high velocity coolants, avoidance of formation of high activity isotopes, and resistance to mechanical damage from neutron radiation. Amorphous zirconium is a bluish-black powder. At about $500\,^{\circ}C$ zirconium burns in air; heated in hydrogen forms hydride; heated in nitrogen a nitride; and heated in chlorine the tetrachloride. On the laboratory scale, zirconium metal may be produced by the reduction of the chloride, oxide, or potassium zirconium fluoride with sodium metal.

Zirconium was first identified by Klaproth in 1789. The first crude powder was made by Berzelius in 1824 by reducing potassium fluorozirconate with potassium. A sample with a purity of 98% was not produced until the 1950s. There are five natural isotopes ^{90}Zr through ^{92}Zr, ^{94}Zr, and ^{96}Zr. Six radioactive isotopes have been identified ^{87}Zr through ^{89}Zr, ^{93}Zr, ^{95}Zr, and ^{97}Zr. With exception of 93Zr with a half-life of 1.1×10^6 years, the half-lives of the other isotopes are expressed in minutes, hours, or days. Zirconium is ranked 19th in abundance of the chemical elements occurring in the earth's crust with an estimated average content of zirconium in igneous rocks of 0.026%.

First ionization potential 6.95 eV; second, 13.97 eV; third, 24.00 eV; fourth, 33.83 eV. Oxidation potentials $Zr + 2H_2O \rightarrow ZrO_2 + 4H^+ + 4e^-$, 1.43 V; $Zr + 4OH^- \rightarrow ZrO(OH)_2 + H_2O + 4e^-$, 2.32 V. Electronic configuration $1s^2 2s^2 2p^6 3s^2 3p^6 3d^{10} 4s^2 4p^6 4d^2 5s^2$. Ionic radius Zr^{+4} 0.80 Å. Metallic radius 1.5895 Å. Other important physical properties of zirconium are given under **Chemical Elements**.

The most important ore for production of zirconium metal is zircon $ZrSiO_4$, which occurs in several regions in the form of a beach sand, often mixed with silica, ilmenite, and rutile. A floating-dredge technique is used in the mining operation. Early phases of beneficiation often take place on the dredge. Nearly all of the silica is separated by means of spiral concentrators, with the ilmenite and rutile removed by magnetic and electrostatic separators. The purest concentrates are used for metal production; others for refractories. Direct chlorination of the ore is the most modern method of extraction. In a simple reaction, water-soluble zirconium tetrachloride is yielded. Liquid-liquid extraction in several stages is required for the removal of hafnium. With this process, zirconium containing less than 50 ppm hafnium can be produced. Ammonium thiocyanate is used to complex the zirconium while hafnium is extracted by a methyl ethyl ketone solvent. In a similar system, HNO_3 is used in the aqueous phase and tributyl phosphate as the solvent. After separation, the two metals are precipitated as their sulfates or hydroxides. Calcination yields a pure ZrO_2. For production of pure metal, the pure ZrO_2 is chlorinated to $ZrCl_4$. Sometimes this is sublimed for additional purification. The zirconium tetrachloride in the gaseous phase is reacted with molten magnesium, forming zirconium metal and magnesium chloride. There are several minor variations of these processes. For example, sodium may replace magnesium.

Uses. The most important application of zirconium is in the formulation of the base metal in an alloy of 98% zirconium, 1.5% tin, 0.35%

iron-chromium-nickel, and 0.15% oxygen. This alloy is widely used in water-cooled nuclear reactors because of its excellent corrosion resistance up to about $350\,^{\circ}$ in H_2O, and its low neutron cross section. Currently, about 90% of the zirconium produced is used for this application. The excellent corrosion resistance of zirconium to both strong acids and alkalis, particularly its resistance to strong caustic solutions at all concentrations and temperatures, is attracting increasing attention for application in chemical processing equipment. See also **Nuclear Power Technology**.

Chemistry and Compounds. Due to its $4d^2 5s^2$ electron configuration, zirconium forms tetravalent compounds readily, although the Zr^{4+} ion does not exist as such in aqueous solution, except at very low pH values, the common cation being hydrated ZrO^{2+} (or $Zr(OH)_2^{2+}$). Many of the tetravalent compounds are partly covalent. There are also less stable Zr(III) compounds. The remarkably close similarity in chemical properties to those of hafnium is due to the identical outer electron configuration ($5d^2 6s^2$ for Hf) and the almost identical ionic radii (Hf^{4+} is 0.86 Å) this relatively low value for Hf^{4+} being due to the lanthanide contraction.

Zirconium oxide, ZrO_2 is widely known, both as a mineral, baddeleyite, and as an industrial product obtained from zircon, $ZiSO_4$. Moreover, the precipitate obtained by action of alkali hydroxides upon solutions of tetravalent zirconium is a hydrated oxide. The latter is readily soluble in acids to form oxysalts, which are usually formulated in terms of the ZrO^{2+} ion, without including its water of hydration, e.g., as $ZrO(H_2PO_4)_2$. The hydrated ZrO^{2+} ion is not amphiprotic; it does not dissolve in alkali hydroxides. While it does react on alkali carbonate fusions, the compounds formed have been shown to be mixed oxides rather than zirconates.

All four of the halides of zirconium and the common halogens are known and are solids at ordinary temperatures. They are readily hydrolyzed, to form hydrated oxyhalides such as $ZrOCl_2 \cdot 8H_2O$ or $ZrOBr_2 \cdot 8H_2O$. The tetraiodide yields both $ZrOI_2 \cdot 8H_2O$ and $ZrI(OH)_3 \cdot 3H_2O$, and this last composition probably represents best the structure of all these compounds. Like titanium tetrahalides, $ZrCl_4$ and $ZrBr_4$ act as Lewis Acids to form adducts, though of lesser stability, with H_2O and oxygen-function compounds, such as alcohols, ethers, and carboxy compounds generally. These include chelates formed with such compounds as 1,2-dihydroxy-benzene and acetylacetone, in which oxygen atoms act as electron donors. Zirconium also forms very stable complexes with $POCl_3$.

The trihalides of zirconium, like the dihalides of titanium, are extremely strong reducing agents, reacting even with H_2O.

Zirconium nitride, Zr_3N_4, is made by ammoniating the tetrachloride to yield $Zr(NH_3)4Cl_4$, which yields the nitride on heating. The nitride, like the boride and carbide, are alloy-like in character, with high fusing points, extreme hardness, and subject to considerable variation in composition. Thus Zr_3N_4 may vary in composition to ZrN without material change in its properties.

Unlike titanium, zirconium forms a few normal tetravalent salts, such as a tetranitrate and a tetrasulfate, as well as its more common basic salts. However, the normal salts readily undergo hydrolysis to form the basic salts.

Like titanium, zirconium forms halogen complexes, the most stable of which is the hexafluoride, ZrF_6^{2-}, as well as $ZrCl_6^{2-}$, and $ZrBr_6^{2-}$, which are less stable, except in concentrated solutions of the hydrogen halides. Zirconium also forms stable peroxy complexes, containing

$$\begin{array}{c} \diagdown \\ \diagup \end{array} Zr - O - O -$$

Unlike titanium, zirconium forms a heptafluoride ion, ZrF_7^{3-}, which is quite stable.

Additional Reading

Amato, I.: "Exploring the New Material World (Zirconium)," *Science*, 644 (May 3, 1991).

Anderson, D.L.: "Composition of the Earth," *Science*, 367 (January 20, 1989).

Bauccio, M.L.: "ASM Engineered Materials Reference Book," 2nd Edition, ASM International, Materials Park, OH, 1994.

Brooks, C.R.: "Heat Treaement, Structure and Properties of Nonferrous Alloys," ASM International, Materials Park, OH, 1989.

Craig, B.D.: "Handbook of Corrosion Data," 2nd Edition, ASM International, Materials Park, OH, 1995.

Elvers, B., et al.: "Water to Zirconium and Zirconium Compounds," Vol. 28, John Wiley & Sons, Inc., New York, NY, 1996.

Greenwood, N.N. and A. Earnshaw: "Chemistry of the Elements," Butterworth-Heinemann, Inc., Woburn, MA, 1997.

Krebs, R.E.: "The History and Use of Our Earth's Chemical Elements: A Reference Guide," Greenwood Publishing Group, Inc., Westport, CT, 1998.

Lewis, R.J. and N.I. Sax: "Sax's Dangerous Properties of Industrial Materials," 10th Edition, John Wiley & Sons, Inc., New York, NY, 1999.

Lide, D.R.: "CRC Handbook of Chemistry & Physics 2000-2001," 81st Edition, CRC Press, LLC., Boca Raton, FL, 2000.

Ondik, H.M. and H.F. McMurdie: "Phase Diagrams for Zirconium and Zirconia Systems," The American Ceramic Society, Westerville, OH, 1998.

Parker, P.: "McGraw-Hill Encyclopedia of Chemistry," 2nd Edition, The McGraw-Hill Companies, New York, NY, 1993.

Staff: "Properties and Selection: Nonferrous Alloys and Special-Purpose Materials," ASM International, Materials Park, OH, 1991.

Staff: "Alloy Phase Diagrams Handbook," ASM International, Materials Park, OH, 1992.

Vander Voort, G., Editor: "Atlas of Time-Temperature Diagrams for Nonferrous Alloys," ASM International, Materials Park, OH, 1991.

ZIRCONIUM SILICATE.   See **Zircon.**

ZODIAC.   A belt on the celestial sphere that extends for an angular distance of 9° from either side of the ecliptic. The zodiac is divided into 12 sections, each 30° long, which are known as the signs of the zodiac. These signs indicate the position of the sun for each month in the year and are named for the zodiacal constellations that occupied the signs about 2,000 years ago. Due to precession, the sign of Aries has moved back into the constellation of Pisces, so that the signs and constellation names no longer agree.

Since the planets all lie relatively close to the ecliptic, their paths along the celestial sphere will lie in the zodiac. It is a common practice with makers of almanacs to indicate the positions of the planets by the sign of the zodiac in which they are to be found. See also **Celestial Sphere and Astronomical Triangle.**

ZODIACAL LIGHT.   A faint glow that extends along the ecliptic or zodiac from the vicinity of the sun. The zodiacal light is so faint that it is completely masked by moonlight. It may best be observed in the western sky in the spring after the sunset twilight has completely disappeared, in the eastern sky in the fall just before the morning twilight appears, or during a total solar eclipse. It decreases in intensity with distance from the sun, but on very dark and clear nights, it has been followed completely around the ecliptic. In fact, the work of van Rhijn at the Mount Wilson Observatory indicates that the illumination is not confined to the ecliptic, but presumably covers the entire sky, being responsible for about 60% of the total skylight on a moonless night. There is a slightly increased illumination of the zodiacal light on the ecliptic directly opposite the sun known as the gegenschein.

Photographic observations of the spectrum of the zodiacal light indicate that it is composed of reflected sunlight and is essentially an extension of the F corona. The amount of material necessary to produce the intensity of the light is amazingly small. Calculation indicates that the zodiacal light could be accounted for if, inside the orbit of the Earth, there were particles 1 millimeter in diameter of the reflecting power of the moon, and each one 5 miles from its neighbors.

Observational studies indicate that the material producing the zodiacal light is located in a lens-shaped volume of space centered on the sun and extending well beyond the orbit of the Earth. Each individual particle, unless small enough to be held away from the sun by radiation pressure, must be moving about the sun in its individual orbit. The intensification of the reflected light at the point directly opposite the sun might be explained either by a concentration of the reflecting particles in this region, or by the fact that directly opposite the sun the particles would be in full phase.

Measurements over a 4-year period from the satellite OSO-5 have failed to show any temporal variations in the surface brightness and polarization of zodiacal light. These observations have tended to cast doubt on prior observations, from rockets and balloons, to the effect that variations in the light do exist, as for example that the brightness of the south pole exceeded that of the north pole by about 15% (at time of observation). Satellite observations have not indicated any north-south asymmetry. The situation is reviewed in more detail by Sparrow and Ney in *Science*, **181** (4098), 438–440 (1973).

ZOISITE.   This mineral is a hydrous aluminum silicate corresponding to the formula $Ca_2Al_3(Si_3O_{12})(OH)$, crystallizing in the orthorhombic system. Clinozoisite (monoclinic) is its isomorphous counterpart. Zoisite occurs as prismatic crystals, usually deeply striated vertically, and as compact or columnar masses. Perfect prismatic cleavage; brittle; uneven to conchoidal fracture; hardness, 6.5–7; specific gravity, 3.355; luster, vitreous to pearly; transparent to translucent; color, grayish white, green, pink (the manganese-rich variety, thulite), and blue to purple (tanzanite).

Zoisite occurs in crystalline schists which are products of regional metamorphism of basic igneous rocks rich in plagioclase, the calcium-rich feldspar; also in argillaceous calcareous sandstones, thulite from quartz veins, pegmatites, and metamorphosed impure limestones and dolomites.

Tanzanite is the blue to purple variety of zoisite and represents a recent discovery of this heretofore unknown variety from Tanzania. It occurs here as excellent transparent crystals from which fine gems have been cut.

ZOLLINGER-ELLISON SYNDROME.   A tumor involving the D-cells of the islets of Langerhans is called the D-Cell adenoma. This tumor causes a condition known as the Zollinger-Ellison syndrome. In this syndrome, the D-cells secrete an excessive amount of gastrin. This, in turn, stimulates the parietal cells of the stomach, to produce an increased volume of gastric juices. The gastric hyperactivity can lead to severe peptic ulcer sometimes accompanied by excessive diarrhea. The D-cell adenoma may be either benign or malignant, but requires removal to relieve the symptoms. The presence of a D-cell adenoma also causes an increase in insulin secretion by the beta cells.

ZONAL ABERRATION.   Spherical or monochromatic aberration of a lens of wide aperture due to the fact that the refracting power is different for different zones concentric at the axis.

ZONAL FLOW.   See **Atmosphere (Earth).**

ZONE.   A variously defined term, depending upon scientific field of use. A geographical zone is an area or region of latitudinal character, typically set off from surrounding areas by some distinctive characteristic. For example, the torrid zone, the two temperate zones, and the two frigid zones of the Earth are belts divided with respect to latitude and temperature. A geological zone is a regular or irregular belt, or a layer, band, or strip of earth materials, disposed horizontally, vertically, concentrically, or otherwise, thus providing a distinction from surrounding parts other than nature of content. Specific types of geological zones are a "zone of saturation," a "structural zone," or a "zone of fracture." An ecological zone is a biogeographic region typified by uniform climatic conditions, fauna, and flora; or a zone characterized by the dominance of a particular organism. A biostratigraphic zone is a belt of strata identified paleontologically, irrespective of thickness or lithology. Stratigraphic zone also is a generally used term for a stratigraphic unit of any kind, especially any rock stratum or body of strata which it is useful to recognize as a unit because of its being characterized by some unifying property, attribute, or content. Thus, such terms as "coal-bearing zone," "marine zone," "calcareous zone," "tar zone," etc.

ZONE PLATE.   A curious piece of diffraction apparatus, useful in illustrating certain aspects of the Huygens' principle. It consists of a transparent plate on which is a pattern composed of a very large number of concentric, opaque, circular rings with open spaces between them. They might be made by drawing on glass a system of circles whose radii are proportional to the square roots of the consecutive integers, and then covering every other annular space between these circles with black paint. Actually the rings must be so finely spaced that the design is first drawn to a large scale on white paper and a much reduced photograph of it then made on a glass slide. There should be two or three hundred of the rings. When finished, they look much like a set of Newton's rings.

A plate thus prepared, when placed across a beam of light, acts like a converging lens with its optical center at the center of the ring system. It will produce a faint, inverted, real image of a bright object, such as a candle. If there are N opaque rings and the radius of the outermost is R, then the focal length (see also **Mirrors and Lenses**) for light of wave-length l is $l = R^2/N\lambda$.

ZONE TIME.   The idea incorporated in zone time was suggested by Sandford Fleming in 1878 for use in establishing standard time. Zone time is used at present by all branches of air and sea navigation of the United States, and many other nations. The surface of the earth is divided into 24 time zones, each covering 15° of longitude and each centered on a meridian some integral multiple of 15° (1 hour) from Greenwich, England. Each zone is given a number, known as the zone designation, which is the number of hours that the central meridian of the zone is west or east of Greenwich. The zone designations are marked plus (+) for west longitude and minus (−) for east. The zero zone, centered on the meridian of Greenwich, is split; the −0 zone extending from longitude 7°.5 East to 0; and the +0 zone extending from 0° to 7°.5 West. The same is true for the zone centered on 180°, where we have +12 extending from 172°.5 West to 180°; and −12 from 180° to 172°.5 East.

Times on the ship or plane are always expressed on a 24-hour system, with midnight as 00 hours, noon as 12 hours, and midnight as 24 hours the same date, or 00 hours the following date. For example, a ship in longitude 78°52′ East would express an afternoon time of $03^h 43^m 52^s$ as $(-5) 15^h 43^m 52^s$ ZT. Such a system avoids the use of the A.M. and P.M. designations, which so frequently lead to serious confusion. The 24-hour system of naming time is in use by all branches of the armed forces of the United States, by many European countries, and by some business houses in America.

For ordinary working purposes, when time is not needed closer than the nearest minute, time is always expressed in four digits. For example, a ship in longitude 92°W would express 1 minute after midnight as (+6) 0001 ZT; a ship in longitude 130°E would express 6:34 P.M. as (−9) 1834 ZT. Since the zone designation corresponds to the longitude of the central meridian of the zone from Greenwich, Greenwich time can immediately be determined from zone time by the algebraic addition of the zone designation; e.g., if the zone time is (+4) 0946, Greenwich time is 1346. In transforming from zone time to Greenwich civil time, the civil calendar date must be carefully considered. If the ship's time and date is January 24 (+12) 1846 ZT, the Greenwich civil date and time is January 24, 3046 GCT, which should be expressed as January 25, 0646 GCT. Next, consider another ship, which may be only a few miles to the westward at this same instant. Since the Greenwich time is the same, we have the zone time for this ship as January 25 (−12) 1846 ZT.

ZOOCHLORELLA.   A green plant cell (alga) living in the tissues of various animals, including flatworms, coelenterates, sponges, and one-celled species. The association is a true symbiosis (animal association). Yellow or brown symbionts of the same habits are called zooxanthellae.

ZOOECIUM.   The body wall of the individual is colonies of Bryozoa. The zooecium acts as a sheath into which the polypide, consisting of the alimentary tract with the tentacles, can be retracted when the animal is disturbed.

ZOOGEOGRAPHY (also called Chorology).   The distribution of animal faunas on the surface of the earth. According to the prevailing characteristics of their faunas, several zoogeographical regions are recognized as principal divisions. These are the Palaearctic region, including all of Europe, Asia south to the Himalayas, and Africa above Lat. 20°N. Some biologists include Greenland and Labrador in the same region, but they are usually regarded as part of the Nearctic region with the rest of North America. Central and South America and the West Indies are in the Neotropical region. The Ethiopian region includes Africa and Madagascar. The Australian region includes Australia, Tasmania, New Zealand, and many of the Pacific islands to the north and east. The Oriental region includes all of Asia south of the Palaearctic, with the large adjoining East Indian islands and some of the smaller Pacific islands to the east. Various subdivisions have been recognized by some writers.

Later studies of animal distribution have attempted to correlate it with climatic conditions. Merriam's classification of the North American fauna included three principal regions, boreal, austral, and tropical. In the first the Arctic zone is limited to high altitudes and latitudes where the mean normal temperature of the six hottest weeks is 50°F at the lowest boundary. The Hudsonian zone ranges from this limit to a mean of 57°F, and the Canadian zone from this to 64°F. The transition zone of the Austral has an upper limit, expressed as the sum of mean daily temperatures above 43°F, of 10,000. The Upper Austral extends from this limit to 11,500, and the Lower Austral to 18,000. Tropical regions have a total of 26,000; in North America they occur only in southern Florida.

Still other attempts to classify the distribution of organisms in North America in relation to climatic conditions have resulted in the isophanes and phenomeridians of Hopkins' maps and in biotic areas based on rainfall, temperatures, and floral characteristics. In modern ecology these studies have become intricately associated with more detailed analyses of the relations of organisms to the environment, but for the more general needs of taxonomists and specialists in other fields of biology the older terms of distribution are still useful.

ZOOLOGY.   A branch of biology concerned with the animal kingdom and its members, both as individuals and as classes; as well as with aspects of animal life.

ZOOPLANKTON.   See **Ocean**; and **Plankton**.

ZOOSPORANGIUM.   A zoosporangium is a cell in which motile spores called zoospores are formed.

ZOOSPORE.   A zoospore is a motile spore or reproductive cell that develops into a new plant. It is formed from the protoplast of the cell, usually by repeated divisions of the protoplast. Zoospores are asexual.

ZORAPTERA.   An order of insects without a common name, made up of a few rare species found in the Oriental region, Africa, Central America, and the southern United States. They live in small colonies and resemble termites in appearance, although they differ conspicuously in the form of wings. As in the latter insects, some members of the colony are winged and some wingless.

ZORILLE.   See **Cats**.

ZSIGMONDY REAGENT.   A reagent for colloids which is a red colloidal solution of metallic gold obtained by reducing auric chloride by formaldehyde in the presence of an alkali. When mixed with sodium chloride, this reagent becomes blue because of an agglomeration of the particles of gold, but this color change is prevented by the presence of an adequate amount of certain other colloids. They can be classified according to the amount required to prevent the color change.

ZWITTERION.   An ion carrying charges of opposite sign, which thus constitutes an electrically neutral molecule with a dipole moment; looking like a positive ion at one end and a negative ion at the other. Most aliphatic amino acids form such dipolar ions, hence react with both strong acids and strong bases.

ZWORYKIN, VLADIMIR KOSMA (1889–1982).   Zworykin was an American physicist and electronic engineer born in Mourom, Russia. He attended the Petrograd Institute of Technology receiving his degree in electrical engineering. With his professor, Boris Rosing, he worked on cathode ray tubes and thought about their use for displaying images. He came to the United States in 1919 and began working for Westinghouse Electric Corporation in 1920 on radio research. He became an American citizen in 1924 and received his Ph.D. in Physics from University of Pittsburgh in 1926.

Zworykin is best known for inventing the iconoscope, or television pickup tube, and also for developing the kinescope, or television receiving tube. In 1929 he demonstrated an all-electric television system for the Institute of Radio Engineers at Rochester, New York. He was invited to join RCA to do commercial development of his invention. Zworykin's research came out of the lab and resulted in the invention of television.

Zworykin's contributions also include work on helping to develop the electron microscope. He was always particularly interested in the application of engineering and electronics in medicine and made important contributions.

Zworykin received many honors throughout his life some of which are, Rumford Medal of the American Academy of Arts and Sciences in 1941, the Medal of Honor of the Institute of Electrican and Electronic Engineers in 1951, and the Edison Medal of the American Institute of Electrical Engineers in 1953.

See also **Iconoscope**; and **Television (TV)**.

J. M. I.

ZYGOMATIC ARCH.   A narrow bridge of bone extending, in the mammalian skull, from before the ear to the cheekbone, below the temple. It is composed of a long process of the temporal bone squamosal extending forward and joining a short posterior process of the malar jugal bone.

ZYGOSPORE.   A spore formed by the union of two gametes of similar size (isogametes).

ZYGOTE.   The cell that is formed from the union of two sex cells or gametes. In many of the algae the two fusing cells are identical in size. In many cases these two cells are actively motile, as is the zygote which they form by their union. In other cases the zygote is nonmotile from the beginning. This is always true in the oögamous species of green algae and in all higher plants.

The zygote has at first a thin cell wall. Soon this thickens and becomes differentiated into layers. It then becomes a spore — a zygospore if formed from gametes of similar size, an oöspore if formed from gametes of unequal size (eggs and sperm). In the algae and fungi the zygote is often a stage in which the plant survives long periods of unfavorable conditions, such as drought or cold. In the higher plants the zygote fails to develop a thick wall and, by a series of cell divisions, becomes the sporophyte; in the seed plants, it becomes the embryo of the seed. See also **Cell (Biology)**.

ZYMOGENIC CELLS.   Those cells of the stomach that secrete pepsin.

ZYMOLYTIC REACTION.   A chemical reaction catalyzed by an enzyme, especially a reaction involving bond rupture or splitting, usually a hydrolysis.

INDEX

Aa, 1
AAAS (American Association for the Advancement of Science), 1
Aardvark, 1
Aardwolf, 1, 1881
Abaca, 1
Abacus, 585–586
Abalone, 1, 2383
Abalone poisoning, 1469
Abbe condenser, 1
Abbe number, 1, 1129
Abbe refractometer, 2965, 2967
Abbe sine condition, 2, 875
Abdomen, 2
Abel equation, 2
Abelian group, 2
Aberration of light, 2
Aberration (optical), 2
Ablate, 2
Ablating material, 2–3
Ablating nose cone, 3
Ablation (geomorphology/glaciology/meteorite/spacecraft), 3
Abney effect, 3
Abney mounting, of reflection grating, 2963
Abrasion, 3
Abrasion pH, 3
Abrasive ceramics, 710
ABS (acrylonitrile-butadiene-styrene) resins, 3–5
Absarokite, 5
Abscess, 5–6
Abscissa, 945
Abscission, 6
Absinthe, 232
Absolute, 6
Absolute altimeter, 6
Absolute altitude, 6
Absolute ceiling, 673
Absolute constant, 926
Absolute convergence, 945
Absolute coordinate system, 6
Absolute delay, 6
Absolute encoder, 1298
Absolute humidity, 6, 1853
Absolute magnitude (M), 6
Absolute manometer, 6
Absolute motion, 6
Absolute permeability, 2681
Absolute pressure, 6
Absolute reaction rate theory, 757
Absolute space-time, 6
Absolute system of units, 6–7
Absolute temperature scale, 3458
Absolute vacuum, 7
Absolute zero, 7
Absorbance, 7
Absorber, 7
Absorptimetry, 7
Absorption (energy/physiology), 8
Absorption band, 7
Absorption coefficient, 7–8
Absorption curve, 8
Absorption discontinuity, 8
Absorption dynamometer, 1156
Absorption edge, 8
Absorption flame photometry, 324
Absorption process, 8–9

Absorption spectrum, 9
Absorption tower, 8
Absorptivity (optical), 9
Abundance ratio, 9
Abyssal animals, 9
Abyssal hills, 9
Abyssal plain, 9
Abyssal rocks, 9
Abyssal zone, 9, 1208
Acacia gum, 1698
Acacia trees, 9–10
Acanthocephala (thorny-headed worms), 10
Acanthus, 10
Acaricide, 10, 2361
Acarina, 10
Accelerated flight (airplane), 10
Acceleration, 10
Acceleration density, 2945
Acceleration due to gravity, 10–11, 1682
Accelerator mass spectrometry, dating technique, 2928
Accelerometer, 11–14
Acceptable quality level, 2901
Accidental coincidence, 863
Accommodation (ocular), 14
Accommodation coefficient, 14
Accretion (geology), 14, 67
Accumulation (of snow in glaciers), 98
Accumulator (hydraulic), 14
Accuracy, 14–15
    chemical analysis, 160
Accutane, 24
ACELA express high-speed train, 2941–2943
Acesulfame-K, 3401
Acetaldehyde, 15, 88
Acetal group, 15
Acetal resins, 15
Acetanilid, 153
    antipyretic, 208
Acetate dyes, 1154
Acetate fibers, 1379
Acetates, as food antimicrobial agent, 204
Acetic acid, 15–16, 2685
    as food antimicrobial agent, 204
Acetoacetic ester condensation, 16
Acetone, 16–17, 2685
Acetophenetidin, 153
N-Acetyl-p-aminophenol, 153
Acetyl chloride, 767
Acetylcholine, 698
Acetylcholinesterase, 775–776
Acetylene, 17
Acetylene black, 621
Acetylene series, 17, 103
Acetylenic hydrocarbons, 103
Acetylsalicylic acid. See Aspirin
Achene, 17
Achernar (α Eridani), 17
Achilles tendon, 17
Achlorhydria, 17
Acholia, 422
Achondrites, 17
Achromat
    compound lens, 17
    optical microscopes, 2345
Achromatic, 17–18
Achromatic antenna, 185

Achromatic lenses, 779
Achromatopsia, 18
Acid anhydrides, nomenclature and examples, 2585
Acid-base metabolism, 18
Acid-base regulation (blood), 18
Acid deposition, 19, 20
Acid dyes, 1154
Acid halides, nomenclature and examples, 2586
Acidic solvent, 18
Acidification, 22
Acidimetry, 18
Acidity, 18, 2709
Acid number, 18
Acidosis, 18–19, 2142
Acidosis-prone/resistant diabetes, 1066
Acid-peptic diseases, 3555–3556
Acid radicals, 41
Acid rain, 19–21, 2804
Acids, 22–23
Acid salt, 3058
Acid sulfite pulp bleaching processes, 2881
Acidulant fermentation industry, 442–444
Acidulants (food), 23–24
Acler actiluatics, 440–441
Aclinic line, 24, 2027
Acmite-aegerine, 24
Acne, 1055
Acne rosacea, 24
Acne vulgaris, 24
    and androgens, 166
Acoela, 25
Acoelomata, 25
Acoustical absorptivity, 7
Acoustic capacitance, 606
Acoustic cavitation, 31
Acoustic emission, from bones, 507
Acoustic interferometer, 1969
Acoustic mode, 25
Acoustic power transmission ratio, 2837
Acoustic principle of similarity, 25
Acoustic resistance, 2978
Acoustics, 25–32
Acoustic scattering, 3109–3110
Acoustic scintillation, 25
Acoustic waves, 25
Acquired Immune Deficiency Syndrome (AIDS), 33–35
    AIDS-related eye disorders, 68
    impact on blood supply, 387
    and tuberculosis, 3534, 3535
Acquisition, 35
Acromegaly, 35, 2756
Acrylic acid, 35
Acrylic casting, 650
Acrylic fibers, 1379
Acrylic plastics, 35–36
Acrylonitrile, 36, 2685
Acrylonitrile-butadiene rubbers, 1220
Acrylonitrile-butadiene-styrene (ABS) resins, 3–5
ACTH. See Adrenocorticotropic hormone
Actin, 37, 931
Actinide contraction, 37
Actinide series, 37
Actinium, 37, 38, 39
Actinium series, 736, 2924
Actinolite, 38

Actinometer, 38–39
Actinometry, 39
Actinomycosis, 39, 1055, 1532
Actinomyosin, 37
Actinon, 39
Action, 39
Action current, 39
Action potential, 39–40
Action spectrum, 40
Activated alumina, 396
    as adsorbent, 51
Activated carbon
    adsorption using, 51
    decolorizing agent, 1032
Activated complex, 757
Activated silica, as adsorbent, 51
Activated sludge process, 52–53
Activation, 40
Activation energy, 40, 757
Activator, 40, 804
    for enzymes, 1311
Active center, 40
Active deposit, 40
Active immunity, 209
Active margins, 2528–2529
Active mass, 40
Active matrix LCDs, 2145, 2148–2152
Active matrix thin-film electroluminescent displays,
    1259–1260
Active network, 2444
Active optics (telescopes), 3444
Active site (catalyst), 652
Active transducer, 3510
Active transport, in cells, 679
Activity
    catalyst, 652
    radioactivity, 40
Activity coefficient, 40
Activity series, 40–41
Actuated mirror array, 2333
Actuating-error signal, 3164
Actuator, 41
    for robots, 3011
Acute bacterial arthritis, 236–237
Acute medical situation, 41
Acute myocardial infarction, 2027
Acute nephritic syndrome, 2064
Acyclic hydrocarbons, nomenclature and examples,
    2583–2584
Acyclovir, 210
Acyl, 41
    analytical determination, 161
Acylation, 41
    amino acids, 137
Adamatine compound, 41
Adaptation (ecology), 41
Adaptation luminance, 41
Adaptive control, 41–43, 931
Adaptive control system, 43
Adaptive pets, 247
Adcock antenna, 185–186
A/D converter. *See* Analog-to-digital converter
Addax, 43, 183, 184
Adder, 43–44, 3195–3196
Addison's disease, 44, 50
Addition, 44
Addition compounds, 895
Addition polymerization, 2377
Additive color process, 44
Additive compounds, 895
Additive number theory, 2506–2507
Additives (food), 44–45

Add-or-subtract circuit, 890
Addra, 183, 184
Address (computer system), 45
Address register, 2970
Adenine, 45
Adenine arabinoside, 209–210
Adenocarcinoma, 45
Adenoma, 45
Adenosine, 45
Adenosine diphosphate (ADP), 46, 2717
Adenosine phosphates, 46
Adenosine triphosphate (ATP), 46, 2717
Adenosis, 1628
Adenylic acid, 46
Adhara (ε Canis Majoris), 46
Adhesion (physics/work of), 46
Adhesives, 46–47
Adiabatic atmosphere, 47, 306, 1781
Adiabatic compressibility, 896
Adiabatic equivalent temperature, 1323
Adiabatic flame temperature, 880–881
Adiabatic process, 47–48
    in atmosphere, 307–308
Adiabatic wall, 48
Adiathermal, 1080
Adiposogenital syndrome, 2755
A-display, 48
Adjacency matrix, 2304
Adjutant (bird), 3358
Adjuvants, 198, 209
Adobe, 48
Adolescence, 48–49
ADP (Adenosine diphosphate), 46, 2717
Adrenal cortical hormones, 1788
Adrenal corticosteroids, 3355–3356
Adrenal glands, 49–50, 1302, 2061
Adrenaline, 49–50, 101
Adrenocorticotropic hormone (ACTH), 36–37, 50,
    699
    therapy applications, 2755
Adrenoleukodystrophy, and Addison's disease, 44
Adsorption (heat of), 52
Adsorption indicators, 50
Adsorption isotherm, 51
Adsorption process, 50–52
Adult acne, 1055
Adult respiratory distress syndrome, 302
Advanced glycation end products (AGE), 428
Advection, 52
    in atmosphere, 312
Advection fog, 1458
Adventitious buds, 52
Adventure games, 254
Adventurine, 2905
Aegerine, 24
Aeolian deposits, 1315
Aeolian tones, 52
Aeration, 52–53
Aerenchyma, 53
Aerial photography, 2726–2727
Aerobe, 53
Aerobic bacteria, 351
Aeroduct, 53
Aerodynamic chord, 777
Aerodynamic circulation, 60–62
Aerodynamic compressibility, 57–60
Aerodynamic heating, 3395
Aerodynamics, 53–62
    rotary-wing, 1745–1749
    supersonic, 62, 3391–3396
Aeroelasticity, 62
Aerogel, 62

Aerogel dust collectors, *Stardust* mission,
    3333–3335
Aerogenerator farms, 3773
Aerolite, 62
Aerology, 62, 2327
Aeronomy, 62
Aero-otitis media, 1723
Aeropause, 62
Aerosol, 62, 866, 868
    air pollutants, 2800–2801
Aerostatics, 53–62
Aerothermodynamic border, 63
Aestivation, 63
Affine tensor, 63
Aflatoxicosis, 1470
African freshwater butterfly fish, 1454
African hemorrhagic fever (Lassa fever), 2104–2105
Africanized bees, 1785
    beewolf against, 408
African trypanosomiasis, 63
Afterbirth, 63
Afterbody, 63
Afterburner, 63
Afterburning, 63, 80
Aftercooling, 63
Afterglow, 63, 320
    BeppoSAX observations, 414–415
After-image, 63–64
Aftershock, 1183
Agama, 64–65
Agamids, 64–65
Agar-agar, 65, 1698
Agarics, 65–66, 381, 1533
Agate, 66
Agave, 66
Age (geologic time), 1596
Agent Orange, 66
Age-related macular degeneration, 66–67, 2984
Agglomerate, 67
Agglomeration, 67
Agglutination, 67
Agglutinin, 67
Aggradation, 67
Aggregate, 67
Aggregate fruit, 1519
Aggressin, 67
Aging
    biochemical theory of, 427–428
    geriatrics, 1611–1612
    gerontology, 1614–1619
    of skin, 3180
Agitation, 803
Agonic line, 2028
Agouti, 3024
Agranulocytosis, 67
Agreement (coefficient of), 67–68
Agricultural waste, 3707
Agriculture
    effect of climate on, 809–810
    weather forecasting for, 3734
Agrostology, 68
Agulhas Current, 68
AIBO (Sony), 247
AIChE (American Institute of Chemical Engineers),
    68
AIDS. *See* Acquired Immune Deficiency Syndrome
AIDS-related eye disorders, 68
Ailanthus, 2231
Aileron, 68
Air, 69
    properties of, 53
Air compressors, 69–71

Aircraft icing, 71, 2841
Aircraft instrumentation, 75–76
Airflow, 71
Airfoils, 55–56, 71
Air gap, 71–72
Airglow, 72, 326–328
Air lift pump, 72
Air light, 72
Air lock, 72
Air mass, 72
Air-mass showers, 2842
Air movement, 3773–3778
Air parcel, 312
Airpath, 2562
Airplane, 72–84
    icing, 71, 2841
    instrumentation, 75–76
    optical fiber applications, 2568–2569
    propulsion methods, 77–81
Air plot, 84–85
Air pollution, 2798–2805
    and acid rain, 19–21
    indoor, 2938
Air preheater, 85
Air-puff test (tonometer), 3504
Airships, 1126
Airspeed indicator, 85–86
Air treatment, 2804–2805
Air velocity measurements, 3767–3768
Airy's experiment, on light aberration, 2
Aitken dust counter, 86
Aitken nuclei, 86
Akremite, 1343
Alabandite, 86
Alabaster, 86
Alanine, 133–137
Albacore, 3536–3537
Albatross, 490, 2684–2685, 2851
Albedo, 86, 504, 824
Albertite, 86
Albinism, 86–87
Albite, 1362
Albumin, 87
Alcoholate, 87
Alcohols, 87–88, 2581
    nomenclature and examples, 2585–2586
Alcoholysis, 88
Alcyonaria, 88, 189
Aldebaran (α Tauri), 88
Aldehydes, 88–89, 2581
    nomenclature and examples, 2586
Alder fly, 89, 2445
Alder trees, 89
Aldol condensation, 89
Aldose, 2059
Aldosterone, 49
    and sodium-potassium regulation,
        2823
ALDOX process, 88
Ale, 404
Aleurites, 1336
Aleurone grains, 89
Aleutian low, 311
Alewife, 89
Alexandrite, 89
Alfalfa, 90, 2118
    fungus diseases, 1530
Alfalfa mite, 2361
Alfalfa webworm, 3738
Alfisols, 3211
Alford slotted tubular antenna, 186
Alfvén, Hannes Olof Gösta, 90

Alfvén wave, 90
Algae, 90–95
    as protein source, 2864–2865
Algebra, 95
Algebraic equations, 95–96
Algebraic geometry, 1598
Algebraic number theory, 2506
Algicide, 96
Algin, 96
Alginic acid, 1698
ALGOL, 892
Algol Paradox, 424–425
Algol (β Persei), 96, 2682
    as visual binary, 3661
Algonkian (Proterozoic) era, 96, 2869
Algorithm, 96–97
Algorithmic language, 2082
Aliasing, 905
Aliasing error (data acquisition system), 97–98
Alicyclic compounds, 895
Alicyclic hydrocarbons, nomenclature and examples,
    2584
Alienation (coefficient of), 98
Alignment chart, 98
Alimentary tract, 2, 98
Alimentation, 98
Aliphatic alcohols, 87
Aliphatic carcinogens, 629
Aliphatic compounds, 98, 895
Alkali, 98
Alkali metals, 98
Alkaline battery, 393
Alkaline earths, 98
Alkaline proteases, 654
Alkalinity, 18, 2709
Alkalinity boosters, 1062
Alkali rocks, 98
Alkalizers (food), 23–24
Alkaloids, 98–102
    as alleopathic substances, 103
    in chemotherapy, 597
Alkalosis, 102
Alkane, 102, 2581
Alkene, 102, 2581
Alkyd resins, 102
Alkyl, 102
Alkylating agents, 596
Alkylation, 102–103, 2591
    catalytic, 654
Alkyne, 17, 103, 2581
Allanite, 103
Allantois, 103, 1346
Allegheny orogeny, 103
Allele, 103
Allen's rule, of ecological adaptation, 41
Alleomorph, 103
Alleopathic substance, 103
Allergenicity, 1586
Allergy, 103–105
Alligation, 105
Alligator gars, 3423
Alligators, 984–986
Allium, 105
Allobar, 105
Allochromatic, 105
Allochromy, 105
Allochthonous rocks, 105
Allogenic (ecology/geology), 105
Allograft, 819–820
Allogyric birefringence, 106
Allomerism, 106
Allomorphism, 106

Allotropes, 739
Allotype, 106
Allowable defects, 2901
Allowed band, 3231–3232
Allowed transition, 3516
Alloys, 106–108
Alloy steels, 2019
All-pass network, 2444
Allspice, 109
Alluvial fan, 109
Alluvium, 109
Allyl chloride, 767
Allyl ester resins, 109
Almach, 166
Almagest, 109–110
Almanac (astronomical), 110
Almandine, 1570
Almaz space station, 3265–3266
Almond tree, 3034–3035
Aloe, 110
Alopecia (hair loss), 110
Alpaca, 593, 594
Alpha Centauri, 110
Alpha chamber, 110
Alpha Crucis, 110
Alpha cutoff, 110
Alpha decay, 110–111
Alpha emitter, 110–111
Alphanumeric code, 854
Alphanumeric instruction, 1957
Alpha particle, 110
    binding energy, 425
Alpha ray emission, 2921
Alpha rhythm, 1257
Alpheratz (α Andromedae), 111
Alpine (ecology), 111
Alpine orogeny, 111
Altair (α Aquilae), 111, 220
Altazimuth, 111
Alteration, minerals, 2356
Alternating current, 111–113
Alternating current drives, 3144
Alternating current motors, 2397, 3138
Alternating-current series motor, 2397
Alternating gradient focusing, 113
Alternating gradient synchrotron, 2650
Alternating group, 113
Alternating series, 113
Alternation of generations, 113–114
Alternator, 114
Altimetry, 114–116
Altitude, 116
Altocumulus clouds, 826
Altostratus clouds, 826
Alum, 116
Alumel thermocouples, 3468
Alumina, 396
Aluminides, 125
Aluminizing, 1744
Aluminum, 116–120
    anodization, 178
    effect on steel, 2018
Aluminum alloys and engineered materials, 108,
    116, 120–125
Aluminum casting alloys, 124, 125
Aluminum composites, 119, 121
Aluminum electroplating, 125
Aluminum foil, 116
Aluminum hydroxide, antacid, 182
Aluminum-lithium alloys, 123
Aluminum oxide moisture sensor, 1885
Aluminum recycling, 117

Aluminum wrought alloys, 124
Alunite, 126
Alzheimer's disease, 126–128, 428, 531
Amalgam, 128
Amalthea, 2051
    orbital characteristics, 2760
Amanita poisoning, 1533
Amantadine hydrochloride, 209
Amaranthus, 128
Amatol, 144, 1343
Amatoxins, 1469
Amazon stone, 1362
Amber, 129
Ambergris, 129
Ambient compensation, 129
Ambient conditions, 129–130
Ambient protection, 129
Amblygonite, 130
Amblyopia, 130
Ambush bug, 130
Amebiasis (amebic dysentery), 130
Ameiva, 130
Amenorrhea, 131
Ament, 664
American antelope, 2860
American Association for the Advancement of
    Science (AAAS), 1
American Institute of Chemical Engineers (AIChE),
    68
American jackal, 603
American Meteorological Society (AMS), 152
American Society of Materials (ASM), 271
American Standard screws, alignment chart, 98
Americium, 37, 131
Amethyst, 131, 2905
Amici prism, 2849
Amicron, 131
Amides, 131–132, 2581
    nomenclature and examples, 2586
Amido group, 131
Amikacin, 196
Amination, 132
Amines, 132–133, 2581
    nomenclature and examples, 2586
Amino acid geochronology, 2928
Amino acid industry, 442–443
Amino acids, 133–138, 1829. *See also* Protein
    and genetic code, 1589
    neurotransmitters, 698
    racemization, 2928
Aminobenzene, 174
Aminoglycosides, 195
Aminopyrine, 153
Amino resins, 139
Amitosis, 139
Ammines, 139
Ammonia, 139–143
    catalysts for production, 654
Ammonium chloride, 143
Ammonium compounds, 143
Ammonium hydroxide, 143–144
Ammonium nitrate, 144, 1343
Ammonium perchlorate, rocket propellant,
    3015–3016
Ammonium phosphates, 144–145
Ammonium sulfate, 145
Ammonolysis, 132, 2592
Ammoxidation, catalytic, 654
Amnesia, 145
Amniocentesis, 145, 1294
Amnion, 145, 1346
Amoeba, 145

Amoeboid movement, 145
Amoebulae, 145
Amolytic fermentation, 1363
Amor asteroids, 275
Amorphous, 145–146
Amorphous alloys, 108
Amorphous carbon, 616
Amorphous mineral, 146
Amorphous silicon, 3134
    thin-film transistor, 2154–2155
Amosite, 146
Amount units, 3568
Amoxicillin, 194, 195
Ampere (unit), 3569
Ampere-hour (unit), 3569
Ampere per meter (unit), 3569
Ampere's law, 146
Ampere's rule, 146
Ampere's theorem, 146
Amperometer, 146
Amperometry, 146
Amphetamine, 146
Amphibia, 146
    fossils, 1478–1479
Amphibole, 146–147
Amphibole asbestos, 264
Amphibolite, 147
Amphidromic point, 147
Amphineura, 147
Amphioxus, 147
Amphipoda, 147
Amphiprotic, 147
Amphoteric ion, 1995
Ampicillin, 194, 195
Amplifier, 147–152
Amplifying-up converter, 151
Amplitude, 152
Amplitude comparison, 152
Amplitude discriminator, 152
Amplitude distortion, 1135
Amplitude hologram, 1777
Amplitude separation, 152
Ampoule (ampule), 152
Ampulla, 152
AMS (American Meteorological Society), 152
Amsler Grid, 152–153
Amygdala, 692
Amygdaloid, 153
Amylase, 655
Anabolism, 378
Anaclinal, 153
Anacoustic zone, 153
Anaerobe, 153
Anaerobic bacteria, 351
Anaerobic cellulitis, 1569–1570
Analcime, 153
Anal feelers, 153
Analgesics, 153
Anal infection, 5
Analog computer, 154, 903
Analog input, 154–155
Analog multiplexer, 155–156
Analog output, 156
Analog switch, 156
Analog-to-digital converter, 156–157
    integrating-ramp, 1965–1966
    parallel, 2634
    parallel-serial, 2635
    ramp, 2945
    successive-approximation, 3366–3367
    voltage-to-frequency, 3686
Analogy (dynamics/physiology), 157

Analysis
    chemical, 157–160
    organic chemical, 161–162
Analysis of covariance, 161
Analysis of variance, 161
Analytical calculi, 248
Analytical geometry, 1597–1598
Analytical instrumentation
    energy-matter interactions utilized, 158–159
    targets of, 160
Analytic continuation, 162
Analytic function, 162
Analytic geometry, 1009
Analytic number theory, 2506
Analyzer (optics/reaction-product/reagent-tape), 162
Anamnia, 162
Anamorphism, 162
Ananke, orbital characteristics, 2760
Anaphase, 677–678
Anaphylactic shock, 105
Anaphylaxis, 162
Anaplasia, 595
Ana-position, 162
Anastigmat, 162
Anatase, 162
Anatexis, 162
Anatomy, 163
Anchor ice, 1895
Anchor ring (torus), 163
Anchoveta, 164
Anchovy, 163–164
Ancillary statistic, 164
Andalusite, 164
AND circuit, 164–165
Anderson bridge, 541
Andesine, 1362
Andesite, 165
Andradite, 165
Androgenesis, 165
Androgens, 165–166, 3355
    in bile, 422
Andromeda constellation, 166
Andromeda Galaxy, 166, 1549
Androsterone, 165
5-β-Androsterone, 165
Anechoic chamber, 30
Anemias, 166–168, 2009
Anemometer, 3767–3768
Anemophilous plants (air-pollinated), 2797
Anemotaxis, 169
Aneroid, 169
Anestrus, 169
Aneurysm, 169, 232
    grafts for, 235
ANFO explosive (ammonium nitrate-fuel oil), 144,
    1343
Angel (radar echoes), 169, 482
Angelfishes, 169–170
Angel shark, 170
Angina pectoris, 2026–2027
Angiography, 170
Angioma, 170
Angiosperms, 170–171
Angle
    of lag, 2079
    mathematics/arrival/departure, 171
    of repose, 46, 172, 2976
    slip, 172
Anglerfishes, 171–172
Anglesite, 172
Angle-wing, 172
Angoumois grain moth, 1665

Angstrom (unit), 3568, 3569
Angstrom, Anders Jonas, 172
Ångstrom compensation pyrheliometer, 172
Ångstrom pyrgeometer, 172
Angular acceleration, 173
Angular distribution (particle), 172–173
Angular magnification, 173
Angular measurement (eccentricity correction), 173
Angular momentum, 173
Angular velocity, 173, 3613–3614
Angus cattle, 667
Anhedral, 173
Anhingas, 2671, 2672
Anhydrite, 173–174
Ani, 477
Aniline, 132, 174, 346
Animal anatomy, 163
Animal breeding, 1762
Animalcule, 174
Animal feedstuffs. See Feedstuffs
Animal pigmentation, 2747
Animal unit, 174
Animal-unit month, 174
Anion, 174, 1255–1256
Anionic reaction mechanism, 2588
Anisakiasis, 1470
Anise, 174–175
Anisodesmic structure, 175
Anisotropic medium, 175
Ankerite, 1138
Ankylosing spondylitis, 3313
Annabergite, 175
Annatto food colors, 175
Annealing, 175–176
    full, 1528
    glass, 1633
    steel, 2019–2020
Annelida, 177
Annihilation, 177
Annona, 177
Annual, 177
Annual ring, 177–178
Annuloaortic ectasia, 235
Annulus, 178, 797
Anoa, 520
Anode, 178, 1256
Anode sheath, 178
Anodic oxidation, 178
Anodic protection, 963
Anodize, 178
Anoles, 1900
Anolis, 178
Anomalodesmacea, 178
Anomalous dispersion, 178
Anomalous propagation, 178
Anomaly, oceanographic, 178
Anoplura, 178
Anorexia, 178–179
Anorexia nervosa, 179
Anorthite, 1362
Anorthoclase feldspar, 1362
Anorthosite, 179
Anosmia, 2558
ANOVA, 161
Anovulation, 1923
Anoxaemia, 1892
Anoxemia, 179, 2026
Anseriformes, 179–180
Ant, 180–182, 1885
Antacids, 182
Antarctic convergence, 182

Antarctic front, 1508
Antarctic Ice Sheet, 2787, 2791–2792
Antarctic krill, 990–991
Antarctic waters, 182
Antares (α Scorpii), 182–183
Anteater (aardvark), 1, 1210–1211, 2387
Antecedent stream, 183
Antelope, 43, 183–184, 2860
Antenna
    communications, 184–189
    temperature/zoology, 189
Antenna pattern, 2917
Anterograde amnesia, 145
Anther, 189
Antheridium, 189
Antherinid smelts, 3175
Anthesis, 189
Anthocyanins, 189, 872, 874, 2747
Anthophyllite, 189
Anthoxanthins, 2747
Anthozoa, 189–190
Anthracene, 190
Anthracite coal, 821
Anthracitization, 820
Anthraquinone, 190
Anthrax, 190–191, 355, 1053
Anthraxolite, 191
Anthropogenic, 191
Anthropoids, 191–193
Anthroxanthins, 872, 874
Anti-AIDS drugs, 33–34
Antiallatotropins, 1955
Antiarrhythmic agents, 229–230
Antibiotic industry, 441
Antibiotic resistance markers, 1586
Antibiotics, 193–198
    in chemotherapy, 597
    as food antimicrobial agent, 205
Antibody, 198–199, 3592
Anticaking agents, 199–200
Anticathode, 200
Anticline, 200
    as petroleum trap, 2688
Anticlinorium, 200
Anticoagulants, 200–201
Anticoincidence circuit, 201
Anticoincidence counter, 201
Anticorona ring, 319
Anticorrosive pigment, 2617
Anticyclone, 312
Antidote, 202
Antiferroelectric, 202
Antiferromagnetism, 202
Antifoaming agent, 1042
Antigen, 202, 1908
    clinical use, 198–199
Antigray-hair factor, 2628
Antihistamine, 202–203
Antihunting circuit, 203
Antikaons, 2321
Antilles current, 203
Antilock braking systems, 341
Antilogarithm, 2184–2185
Antimalarial drugs, 2233–2235
Antimatter, 203
Antimatter rockets, 3017–3018
Antimetabolites, 203, 596, 1020
Antimicrobial agents (foods), 203–205
Antimony, 205–207
Antinodes (loops), 207
Antioxidant, 207–208
Antiparallel vectors, 208

Antiparticles, 208, 2645
Antipodal cells, 208
Antipode, 208
Antiproton, 208
Antipyretic, 208
Antipyrine, 208
Antiredeposition agents, 1062
Antiresonance (parallel impedance), 208
Antisolar point, 208
Anti-stats, 1062–1063
Anti-Stokes bands, 2944
Antistokes lines, 208–209
Antisymmetric, 209
Antitoxin, 199, 209
Anti-twilight arch, 3552
Antiviral drugs, 209–210
Antlerite, 210
Ant lion, 210, 409–410, 2445
Ant-loving cricket, 210
Antonoff rule, 210
Antozonite, 1452
Anura, 210
Anus, 98, 210
Anvil (ear), 325
Aorta, 210, 489
Aortitis, 235
Aoudad, 1649, 1650
Apastron, 2677
Apatite, 210
Ape, 191
APECED (Autoimmune Polyendocrinopathy-Candidiasis-Ectodermal Dystrophy), 602
Aperiodic antenna, 186
Apertometer, 210
Aperture, 210
Aperture card, 210
Aphanite, 210
Aphasia, 211
Aphelion, 211
Aphid, 182, 211–212
Aphis-lion, 212, 410
Aphytic, 212
Apical growth, 212
Apids, 1782
API gravity, 2685, 3298
Aplacophora, 212
Aplanatic lens system, 212
Aplanatic points, 212
Aplastic anemia, 167
Aplite, 212
Apochromat, optical microscopes, 2345
Apodeme, 212
Apodiformes, 212
Apoenzyme, 857
Apogee, 2677
    satellites, 3065
Apollo asteroids, 275–276
Apollo Program, 3021–3022, 3265
Aponeurosis, 212
Apophyllite, 213
Apophysis, 213
Apostilb (unit), 3569
Appalachian Revolution, 103
Apparent horizon, 1786
Apparent molecular quantity, 213
Apparent sidereal time, 3493
Apparent solar time, 3492–3493
Appendage, 213
Appendicitis, 213
Appendicularian, 213
Appendix, 672

Appendix vermiformis, 213
Applanation tonometer, 3504
Apple bud moth, 566
Apple curculio, 1006
Apple flea-beetle, 1429
Apple fruitworm, 1522
Apple leaf hopper, 2114
Apple leaf roller, 2114
Apple leaf skeletonizer, 214
Apple maggot, 1520, 2217–2218
Apple redbug, 214
Apple tree, 3030–3031, 3034
    abscission, 6
    fungus diseases, 1530
Apple tree borer, 512
Apple-tree tent caterpillar, 1198
Approximate calculation, 214
Apricot tree, 3035–3036
Apterygota, 214
Aquaculture, 214–219
Aquamarine, 219, 417
Aqua regia, 219, 2458
Aquarius (the water bearer), 219
Aqueduct, 219–220
Aqueduct of Sylvius, 220
Aqueous humor (eye), 1636
Aquifer remediation, 3720
Aquila (the eagle), 220
Aquoion, 1995
Arachidic acid, 220
Arachnida, 220
Arago, Francois, 220
Aragonite, 220–221
Arago point, 320, 349
ARALL aluminum laminates, 125
Arapaima, 510
Araucarius, 221
Arbitrary constant, 926
Arbor, 221
Arborvitae, 221
Arc
    electrical, 221
    mathematics, 223
Arc back, 221
Arc cutting, 221
Arc furnace, 221
Arch, 221–222
Archaebacteria, 354, 675
Archean (Archeozoic), 222
Arched bridge, 546, 547, 550
Archegoniates, 222
Archegonium, 222
Archeocyte, 222
Archeophytic, 222
Archeozoic (Archean), 222
Archiannelida, 222
Archimedes principle, 568
Archimedes spiral, 223
Arching, 46
Architectural acoustics, 31
Arc lamp, 223
Arc shooting, 223
Arctic air, 72
Arctic circle, 223
Arctic front, 1508
Arctic haze, 2841
Arctic waters, 223
Arctic Zone, 223
Arcturus (α Bootes), 223
Arc welding, 3746
Area, 223
Area rule (for zero-lift drag), 223

Area sampling, 223, 3060
Arenaceous, 223
Arenaviruses, 224
Arête, 224
Argalis, 1649, 1650
Argand diagram, 224
Argental mercury, 128
Argentite, 224
Argillaceous, 224
Argillite, 224–225
Arginine, 133–137
Arginine-urea cycle, 3588
Argon, 225
Argonauta, 225
Argo Navis, 632, 2885
Argument, 225
Argyria, 225
Aridisols, 3211
Ariel, 3584–3585
    orbital characteristics, 2761
Aries (the ram), 225
Aril, 225
Aristotle, 225
Arithmetic, 225
Arithmetic instructions, 1109
Arithmetic mean, 225–226
Arithmetic progression, 2859–2860
Arkose, 226
Arm, 226
Armadillos, 1210, 1211
Armature, 226
Armature reaction, 226
Army cutworm, 1011, 1012
Armyworm, 226–227
Aroids, 227
Aromatic alcohols, 87
Aromatic amine carcinogens, 629
Aromatic carboxylic acids, 625–626
Aromatic compounds, 227, 895, 2581
    chlorinated, 767
Aromatic hydrocarbons, nomenclature and examples,
    2584–2585
Aromatic nitro group, analytical determination,
    161
Aromatization, catalytic, 653–654
ARPANET, 1971, 1972, 2442
Arrhenius, Svante, 227–228
Arrhenius equation (chemical kinetics), 757
Arrhenius-Guzman equation, 227
Arrhenius-Ostwald acid-base theory, 22
Arrhenius viscosity equations, 228
Arrhythmias, cardiac, 228–230
Arrow worm, 230, 718
Arroyo, 230
Arsenic, 230–231
Arsenopyrite, 232
Arsine, 230
Artemisia, 232
Arteries, 232–236
Arteriography, 599
Arterioles, 233
Arteriosclerosis, 233
Arteriosclerotic dementia, 128
Arthritis
    infectious, 236–237
    osteoarthritis, 2606–2607
    rheumatoid, 2993–2994
    and spondylarthropathies, 3313
Arthritis-dermatitis syndrome, 1661
Arthropoda, 237
    apodeme, 212
Arthropod appendage, 213

Arthroscopy, 237–238
Artichoke, 894
Artichoke stem maggot, 2218
Articulated model, 904, 905
Articulation (communications), 238
Artificial antenna, 1152
Artificial heart, 1731–1732
Artificial horizon, 238
    aircraft, 238–239
Artificial insemination, 239
Artificial intelligence, 239–244
    artificial life, 245–247
    automated reasoning, 247–249
    case-based reasoning, 250–251
    and control systems, 944
    expert systems, 251–252
    fuzzy reasoning, 252–254
    game playing systems, 254–255
    genetic algorithms and evolutionary computing,
        255–256
    machine learning, 257–258
    machine vision, 258–260
    natural language processing, 261
    neural networks, 262–263
    robotics, 263–264
Artificial language, 2082
Artificial life, 245–247
Artificial line, 264
Artificial replacement hip joint, 1771
Artificial respiration (CPR), 631
Artificial vision, 2209–2210
Artiodactyla, 264, 265
Arum, 227
Aryl, 102
Arylamides, 132
Asafetida, 264
Asbestos, 264
Asbestosis, 2781
Ascaris, 264
Ascending node, 1060
Ascent of sap, 264–265
Ascidiacea, 265–266
Ascidian, 265
Ascites, 266
Ascomycetes, 266–267, 1529–1530
Ascorbic acid (vitamin C), 267–268
Aseptic meningitis, 2122, 2315
Asexual reproduction, 268–269
Ash trees, 269–271
ASM (American Society of Materials), 271
Asparagine, 133–137
Asparagus, 271
Asparagus beetle, 271
Aspartame, 3400
Aspartic acid, 133–137
Aspen, 2814, 2815
Aspergillosis, 271, 356
Aspergillus, 266
Asphalt, 47, 271
Asphaltic oils, 3418
Aspheric surface, 271
Asphyxia, 271
Aspirin, 153, 271–272
    in anticoagulant therapy, 201
    antipyretic, 208
    rheumatoid arthritis therapy, 2994
Aspirin-induced asthma, 555
Assassin bug, 272, 410, 2073
Assembler, 272–273, 3516
Assembly system, 2852
Asses, 1798–1799
Associated colloids, 868

Associated compounds, 895
Associated liquid, 2166
Association
    chemical, 273
    coefficient of, 273–274, 961
    in ecology, 274
Associative pattern processing, in machine vision, 2212
Astatic, 274
Astatine, 274
A-station, 274–275
Asterism, 275
Asteroid
    Galileo mission flyby, 1555–1556
    mathematics, 279
    space objects, 275–279
Asthenosphere, 279, 1164
Astigmatic focus, 280
Astigmatic keratotomy (AK), 280
Astigmatism, 280, 3650
Aston, Francis William, 280
Aston dark space, 659
Aston whole number rule, 280
Astrobiology, 280–292
Astrobleme, 277, 278, 292–293
Astrocytes, 699–700
Astroglia, 699
Astrographic telescope, 293
Astrolabe, 293
Astrometry, 293
Astronautics, 293–296
Astronavigation, 296
Astron machine, 296
Astronomical clock, 296–297
Astronomical triangle, 674–675
Astronomical unit, 297
Astronomy, 297–299, 300. *See also* Star
Asymmetric top, 299
Asymmetry (chemical), 299–300
Asymptote, 300
Asymptotic relative efficiency (ARE), 300
Asymptotic series, 300
Asynchronous, 300
Atacamite, 300
Atavism, 300
Ataxia, 301
Ataxic, 301
Atelectasis, 302
Athermal transformation, 302
Atherogenesis, 232, 233
Atherosclerosis, 169, 233, 2026
    brain injury, 531
    of retinal artery, 2984
Athlete's foot, 1054–1055
Athodyd, 302
Atlantic cod, 855
Atlantic pompano, 609
Atlantic Suite, 302
Atlantic-type margins, 2526, 2527–2528
Atlantic weakfish, 984
Atlantique high-speed train, 2940–2941
Atlantis Space Shuttle, 3254–3257
Atmolysis, 302
Atmometer, 2842
Atmosphere
    Earth, 302–316
    Jupiter, 2044–2047
    Mars, 2263–2266
    Mercury, 2320
    Neptune, 2435–2436
    ocean interface, 316–318
    Saturn, 3089–3090

Uranus, 3581
    Venus, 3619–3621
Atmosphere, standard (unit), 3568–3569
Atmospheric boil, 3463
Atmospheric boundary layer, 2758
Atmospheric breaking, 318
Atmospheric convergence, 314
Atmospheric corrosion, 963
Atmospheric divergence, 314
Atmospheric duct, 318
Atmospheric electric field, 318
Atmospheric electricity, 318
Atmospheric instability, 308
Atmospheric Integrated Research Monitoring
        Network (AIRMoN), 21
Atmospheric interference (spherics), 318
Atmospheric inversion, 309
Atmospheric optical phenomena, 318–320
Atmospheric physics, 2327, 2745
Atmospheric pressure, 320–321
Atmospheric residue, 2701
Atmospheric scintillation, 3117
Atmospheric shimmer, 3463
Atmospheric stability, 308
Atmospheric tides, 315–316
Atmospheric turbulence, 321
Atoll, 321
Atom, 321–322
    Rutherford model, 2644
Atomic bomb, 2242–2243
Atomic clocks, 322, 818, 3493
Atomic disintegration, 322
Atomic energy, 322
Atomic energy levels, 322
Atomic frequency, 322
Atomic heat, 323
Atomic heat of formation, 323
Atomic-hydrogen arc welding, 3746
Atomic mass (atomic weight), 323
Atomic mass unit, unified (unit), 3569
Atomic number, 2644–2645
Atomic percent, 323
Atomic physics, 2644
Atomic plane, 323
Atomic radius, 743–750
Atomic refraction, 2964
Atomic species, 323
Atomic spectra, 324
Atomic spectroscopy, 324
Atomic structure, 739–743
Atomic time, 3493
Atomization, 324
    for spray drying, 3214–3315
Atom switch, 3133
ATP (adenosine triphosphate), 46, 2717
Atrial fibrillation, 1382–1383
Atrium, 324
Atrophy, 324
Atropine, 99, 100
Attention deficit-hyperactivity disorder (ADHD),
        1885–1886
Attenuation, 324
    sideband, 325
Attenuation coefficient (α), 324–325
Attenuation constant, 325, 1032
Attenuation factor, 325
Attenuator, 325
Attitude, 325
Atto (prefix for units), 3568
Attractive force, 1474
Attrition (geology), 325
Attrition mills, 325

Audibility, 325
Audiogram, 325
Audiometer, 325
Auditory localization, 325
Auditory organs, 325
Augen-gneiss, 323
Auger effect, 325–326
Augite, 326
Augmentation, 326
Auk, 726, 3695
Aureole (geology), 326
Auric (prefix for +3 gold salt), 326
Auriga (the charioteer), 326
Auroch, 520
Aurora, 326–328
Aurora australis, 328
Aurora borealis, 328
Auroral zone, 328
Aurus (prefix for +1 gold salt), 326
Auscultation, 328
Austenite, 328, 2016
Austenitic stainless steels, 785
Autecology, 1208
Authigenous (authigenic), 328
Autochthonous, 328
Autoclastic, 328
Autoclave (digester), 1101
Autocollimator, 328
Autoconvection gradient, 309, 328
Autoconvective lapse rate, 309, 328
Autocorrelation, 328
Autodyne detection, 1060
Autoerythrocyte purpura, 2886
Autogamy, 328
Autogenous, 328
Autogiro, 1751
Autograft, 820
Autoimmune
        Polyendocrinopathy-Candidiasis-Ectodermal
        Dystrophy (APECED), 602
Autoimmune skin diseases, 1055
Autoimmunity, 199
Autoinfection, 1922
Autointoxication, 329
Autoionization (preionization), 329
Autolysis, 329
Automated lamellar keratoplasty (ALK), 329
Automated reasoning, 247–249
Automatic burette, 569
Automatic focusing cameras, 2727–2730
Automatic pilot, 329
Automatic welding, 3744–3745
Automation, 329–337
Automotive acoustics, 31–32
Automotive aerodynamics, 57–60
Automotive electronics, 337–343
Autonomic nervous system, 343
Autoregression, 343–344
Autosyn, 344
Autotomy, 344
Autotransfusion, 487
Autozooid, 344
Autumnal equinox, 344, 807, 1323
Autunite, 344
Auxiliary storage, 3358
Auxins, 2766
Auxometer (auxiometer), 344
Auxospore, 344
Avahi, 2119
Avalanche (electronics/geology), 345
Average, 225, 344
Average deviation, 344

Average outgoing quality level, 2901
Avocado, 2107
Avocado leaf roller, 2114
Avogadro constant, 344
Avogadro law, 344–345
Avulsion, 345
Awantibo, 2119
Axes (aircraft), 345
Axial-flow compressors, 70
    in jet aircraft engines, 80
Axial magnification, 345
Axial organ, 345
Axil, 345
Axinite, 345
Axiom, 345
Axis (instantaneous/optic/optical), 345
Axis of rotation (fixed), 345
Axolotl, 345
Axon, 345, 695–697
Axoplasm, 696
Ayus, 3054–3055
Azeotropic distillation, 1133
Azeotropic system, 345–346
Azides, 346
Azidothymidine (AZT) (zidovudine), 33, 34
Azimexon, anti-AIDS drug, 34
Azimuth (astronomy/navigation/surveying), 346
Azimuth marker, 346
Azines, 346
Azo compounds, 346, 2581
Azores high, 311
AZT (azidothymidine), 33, 34
Azurite, 346–347
Azusa, 347

Babbler, 2655
Babesiosis, 349
Babinet compensator, 892
Babinet point, 320, 349
Babingtonite, 349
Baboons, 2385–2386
Bacillary dysentery, 1467
Bacilli bacteria, 350
Bacillus cereus gastroenteritis, 1467
Back-Goudsmit effect, 349
Background light, of sky, 1545
Background noise, 2466
Backlash, 349–350
Backscattering, 350
Back-swimmer, 350
Bacteria, 350–354
    as antigens, 202
    increased virulence, 352
Bacterial diseases, 352–353, 354–366
Bacterial endocarditis, 1301
Bacterial meningitis, 2315
Bacterial vaginosis, 3151–3152
Bactericidal agent, 193
Bacteriophage, 366, 3643
Bacteriostatic agent, 193
Badger, 2408
Bad lands, 366
Baeyer-Villiger oxidation, 2590
Baffle, 366–367
Bagasse, 367
Bagoong, 1651
Bag-worm, 367
Bahiagrass, 1676–1677
Bailey beads, 367
Bailey's mimosa tree, 10
Bainite, 367
Bakery products

staling, 3322–3323
sweeteners in, 3403
Balance (mechanical), 367–368
Balance coil, 367
Balanced amplifier, 147
Balanced detector, 367
Balanced draft, 1142
Balanced line, 367
Balanced modulator, 367
Balanced oscillator, 367
Balanced steel, 2018
Balancer set, 368
Balancing (mechanics), 368
Balanoglossus, 368
Baldpate, 3716
Ballast, 368
Ballastic body, 368
Ballastic camera, 368
Ballastic condition, 368
Ballastic density, 368
Ballastic measurement, 368
Ballastic missle, 368
Ballastic pendulum, 368
Ballastics, 368
Ballastic temperature, 369
Ballastic trajectory, 369
Balling scale, 3298
Ballistics standard artillery atmosphere, 306
Ballistocardiography, 1254–1255
Ball lightning, 369
Ball mills, 371
Balloon, 369–370
Balmer series, 324
Balsam of Gilead, 1400
Balsa tree, 371, 956
Balun, 371
Bamboo, 371
Banach space, 2140
Banana plant, 371–372
    fungus diseases, 1530
Band edge energy, 372
Banded krait, 3194
Band-elimination filter, 1392
Bandicoot, 2285
Bandpass filter, 1392
Bandspread, 372
Band theory, 3231–3232
Bandwidth, 372, 1543
    effective, 372–373
    optical fiber systems, 2566
Bank (aircraft), 373
Banner cloud, 828
Banteng, 520, 521
Banyan tree, 2400
Baobab tree, 373
Bar (unit), 3568, 3569
Baraboo, 373
Barberry shrub, 373
Barbet, 2745, 3782, 3783
Barbital, 373
Barbiturates, 373
Barchan, 373, 1152
Bar chart, 373, 374
Bar coding, 2212–2213
Bardeen and Brattain theory, 373
Bardet-Biedl syndrome, 374
Bardhan-Sengupta synthesis, 374
Bareberry plant louse, 211
Barff process, 374
Barite, 374
Barium, 374–375
Bark, 375

Bark beetle, 375
Barkhausen effect, 375
Bark louse, 375
Barkometer scale, 3298
Barley, 375–376
    fungus diseases, 1530
Barlow rule, 376
Barn (unit), 3568, 3569
Barnacle, 376, 412, 799, 988
Barnacle scale, 3204
Baroclinic atmosphere, 307
Baroclinic instability, 308
Baroclinity, 376
Barograph, 377
Barometer, 376–377
Barometric pressure, 320
Barometric wave, 314
Baroswitch, 377
Barotropic atmosphere, 307
Barotropic disturbance, 377
Barotropic instability, 308
Barotropic model, 377
Barotropic wave, 314
Barotropy, 376, 377
Barracuda, 377, 378
Barranco, 377
Barred spiral galaxies, 1547, 1549
Barrel (unit), 3569
Barrier beach, 378
Barrier layer, 378
Barrier layer cell, 378
Barrier reef, 378
Barringer Crater, 293, 2325
Barycentric parallax, 378
Baryon number, 926
Baryons, 378, 1364, 2321
Barytocalcite, 378
Basal cell carcinoma, 599, 3181
Basal cell nevus syndrome, 378
Basal cells, 600
Basal conglomerate, 378
Basal metabolism, 378–379
Basal plane, 379
Basalt, 379
    pillow lavas, 2748
Base level, 379–380
Base line, 380
Base-loaded antenna, 186
Base pair, 1829
Base period, 380
Base point, 380
Base pressure, 380
Bases, 22–23
BASIC, 892
Basic dyes, 1154
Basic oxygen process, for steelmaking, 2012–2014
Basic rock, 380
Basic salt, 380, 3058
Basic solvent, 380
Basidiomycetes, 380–381, 1530
Basil, 2360
Basilar membrane, 381
Basilisk (lizard), 381
Basin, 381
Basket star, 982, 1201
Basommatophora, 381
Bass (acoustics), 382
Bass (fish), 381–382, 477, 3383
Bass compensator, 382
Basswood trees, 382–383
Bast fibers, 383
Bastnasite, 383

Batagur, 383
Batch digester, 1101
Batch polymerization, 2378
Batch variation, 2901
Bathochrome, 383
Batholith, 383
Bath sponge (gourd), 1005
Bathyal zone, 1208
Bathymetric chart, 383
Bathypelagic zone, 383
Bathythermograph, 383
Bats, 383–388
    sonar, 387–388
Batten disease, 388–389
Battery, 389–395, 2835–2836
Bat tick, 395
Baud (unit), 395, 3569
Baudot code, 395
Baumé scale, 3298
Bauxite, 116, 395–396
Bay, 2107
Bayberry shrubs and trees, 396–397
Bayesian networks, 243
Bayou, 397
B cells, 1909, 3592
Bdelloidea, 397
Bdellonemertea, 397
B-display, 397
Beach erosion, 1876–1877
Beach-fleas, 147
Beach nourishment, 1876
Beach rock (coquina), 952
Beacon delay, 397
Beacon skipping, 397
Beam
    composite, 397
    structural, 397–398
Beam angle, 398
Beam bridge, 544
Beam-rider guidance, 1697
Beam width, 398
Bean, 398–399, 2118
    fungus diseases, 1530–1531
Bean thrip, 3482
Bean weevil, 399, 864
Bearing, 399
Bearing, navigational, 399
Bearing angle, 399
Bearing modulus, 399
Bears, 400–402
Beat, 401–402
Beat-frequency oscillator, 3162
Beat note, 402
Beaver, 402–403
Beccafico, 2655
Becke line, 403
Becke test, 403
Beckmann, Ernst, 403
Beckmann method, 403
Beckmann thermometer, 1499
Becquerel, Antoine Henri, 403
Becquerel effect, 403
Bedbug, 403, 1054
Bedding, 403
Bedrock, 403
Bee, 1147, 1885
    Africanized, 408, 1785
Beech trees, 403–404
Bee-eater, 952, 2071
Beef cattle, 667–669
Bee fly, 404
Beefmaster cattle, 668–669

Beef tapeworm, 3417
Bee louse, 404, 1123
Bee-martin, 2071, 2655
Bee-moth, 404
Beer, 404–406
Beer's law, 406
Beeswax, 1782
Beet, 406–407
Beet aphis, 211
Beet armyworm, 227
Beetles, 407, 864, 1145
Beet webworm, 3738
Beewolf, 407–408
Beggiatoa, 408
Beira, 183–184
Beisa, 183–184
Bel (unit), 3569
Belladonna, 408, 3214
Bellatrix (χ Orionis), 408
Bell-birds, 727
Bell crank, 408
Bellows, 408–409
Bellows gas meter, 1443
Bell's palsy, 409
Belt filter, 1396
Beluga, 3364
Bemporad formula, 409
Bénard convection cells, 409
Bench mark, 409
Bending deformation, 1042
Bending moment, 409
Bends, 581
Beneficial insects, 409–411
Beneficiation, 912
Benham top, 412
Benign, 412
Benign neoplasm, 595, 600
Bent, 412
Benthos, 412, 2550–2551
Bentonite, 412
Bent rod bacteria, 350–351
Benzaldehyde, 412
Benzedrine, 146
Benzene, 227, 412–413, 2685
Benzene cycloaddition, 2590
Benzenoids, 227
Benzidine reaction, 413
Benzine, 413
Benzoic acid, 413
    as food antimicrobial agent, 204
Benzoin, 413–414, 2978
Benzol, 412
Benzoyl chloride, 767
Benzyl benzoate, 414
Benzyl chloride, 767–768
Benzylpenicillin, 194
Benzyne, 414
BeppoSAX, 414–415
Bergamot oil, 415
Bergmann's rule, of ecological adaptation, 41
Bergschrund, 799
Berkelium, 37, 415
Berm (beach/geology/structural), 416
Bermudagrass, 1677–1678
Bermuda high, 311
Bernoulli, Daniel, 416
Bernoulli equation, 416
Bernoulli law, 416
Bernoulli method, 416
Bernoulli number, 416–417
Bernoulli polynomial, 416–417
Bernoulli theorem, 417

Berry, 417, 1517
Berthelot, Pierre Eugene Marcellin, 417
Berthelot equation, 417
Berthollide compounds, 731, 895
Beryl, 417
Beryllium, 417–419
Beryllium copper alloys, 108
Bessel function, 419, 1585
Besselian star numbers, 419
Besselian year, 1383
Beta-carotene, 634
Beta Centauri, 419
Beta Crucis, 419
Betacyanins, 872
Beta decay, 419–420
Beta distribution, 2669
Beta function, 420
Beta-glucans, 1089
Betalaines, 420, 872
Beta network, 2188
Beta radiation, 2921
Beta-ray chemical analyzers, 420
Beta rhythm, 1257
Betaxanthins, 872
Betelgeuse (α Orionis), 420
    Hubble Space Telescope image, 1818
Beusite, 420
Bevatron, 2650
Bevel gearing, 420–421
Bharal, 1649
Bias (statistics), 421
Bias cell, 421
Biceps, 421
Bichirs, 421
Biconical antenna, 186
Bicuspid, 421
Bicyclic alkanes, 2581
Bielids (meteor shower), 421
Biennial, 421
Bifidobacteria, 353–354
Bifocal contact lenses, 929
Big Bang theory, 969–970, 1545–1546, 3571–3572
Big bedbug, 2073
Big Dipper (Ursa Major), 3589
Bile, 421–423, 2171, 2173
Bile acids, 3356
Bile pigments, 421
Bile salts, 421
Biliary colic, 864
Biliary tract, 2
Biliary tract diseases, 1561–1562
Bilirubin, 422
Billfishes, 423
Bimetal thermometer, 423–424
Bimolecular reaction, 757
Bimorph cell, 424
Binary arithmetic, 1933
Binary asteroids, 276–277
Binary code, 854, 855
Binary-coded decimal (BCD), 424
Binary compounds, 895
Binary counter, 974
Binary digit, 1119
Binary erasure channel, 1930–1932
Binary granite, 424
Binary notation, 1107
Binary number, 424
Binary point, 2783, 2818
Binary stars, 424–425, 2744
    visual, 3661
Binary state, 425
Binder coke, 850

Binding energy, 323, 425–426
Binocular, 426–427
Binocular vision, 427
Binodals, 427
Binomial distribution, 427
Binomial series, 427
Bioastronautics, 427
Bioavailability, 1092
Biochemical engineering, 443–444
Biochemical imaging, 2739
Biochemical theory of aging, 427–428
Biochemistry, 429
Bioclimatology, 429
Biodynamics, 429
Bioengineered foods, 439–440
Bioherm, 429
Biological clock, 2748
Biological energy transfer, 429
Biological equilibrium, 429
Biological level, 429
Biological pest control, 1954–1955
Biological rhythms, 429
Biological systems, effects of minerals on. *See* Minerals
Biological therapy, 600
Biology, 429–430
Bioluminescence, 430, 2191
Biomass, 3707
    oceanic, 2544, 2549
Biome, 430–433
Bionics, 434
Biopak, 434
Biopharmaceutical industry, 444–449
Bioprocess engineering, 434–457
Biopsy, 459, 600
Bioremediation, 3720
Biosatellite, 459
Bioseparations, 455, 456–457
Bios II, 459
Bios IIa, 2628
Biosolids, 3713
Biosphere, 459, 1208
Biostratigraphy, 459
Biostrome, 459
Biotechnology, 434–457
    in microgravity, 1688
Biotic potential, 459
Biotin, 459–460
Biotite, 460
Bipartite curve, 2608
Biphenyls, 460
Bipolar coordinate, 460–461
Bipolar (manic-depressive) illness, 2243
Bipolar transistors, 156
Biprism, 461
Biquadratic equation, 461
Bi-quartz, 461
Biquinary code, 854–855
Biramous appendage, 461
Birch aphis, 211
Birch trees, 461–462
Bird louse, 462, 2189, 2235
Bird of paradise, 462–463
Birds, 463–472
    migration, 465, 467, 470–471
    pollination by, 2797
    West Nile virus, 3748
Bird's-nest fungi, 381
Bird songs, 471–472
Bird vocalism, 471–472
Birth process, 473
Biserial correlation, 473

Bishop-bird, 2655
Bishop ring, 319
Bismaleimide polymers, 473
Bismuth, 473–474
Bismuth glance, 474
Bismuthinite, 474
Bison, 474–475, 520
Bisulfite pulp bleaching processes, 2881
Bit, 475–476
Bit (unit), 3569
Biting lice, 462, 2189–2190, 2235
Bits per second (unit), 3569
Bittacus, 2308
Bitter chocolate, 578
Bitterling, 476
Bittern, 476, 793
Bitter patterns, 476
Bitter taste, 1419
Bitumen, 476
Bitumen adhesives, 46–47
Bituminous coal, 821
Bituminous fermentation, 820
Bituminous sands, 2687, 3418
Bivalves, 476, 2380–2383
Bivariate distribution, 1135
Bixin, 175
Bjerknes, Vilhelm Frimann Koren, 476–477
Black aphis, 211
Blackberry, 417, 3036
Blackbird, 477, 3483
Black body, 477
Black box, 477
Blackbuck, 183–184
Black-capped titmouse, 761
Blackcock, 1695
Black Death (bubonic plague), 563
Black diamond (carbonado), 621
Blackfish, 477
Black-fly, 477–478
Blackhaws, 1221
Black hole, 478
    Hubble Space Telescope observations, 1807, 1809
Blackhorse, 478
Black locust tree, 10
Black onyx, 720
Blackout, 478–479
Black parlatoria scale, 3204
Black pine, 2751
Black powder, 1343
Black scale, 3204
Black sea bass, 477
Black smoker, 2532–2533, 2554
Black ulua, 609
Bladder. *See* Urinary bladder
Bladder cancer, 2064
Bladdernut tree or shrub, 479
Bladder worm, 479
Blagden law, 1499
Blast furnace, 2010–2012
Blastocoele, 479
Blastocyst, 823
Blastomere, 479
Blastomycosis, 356, 479, 1531
Blastula, 479
Blast vane, 2041
Bleaching agents, 479–480, 1062
    in papermaking, 2880–2881
Bleak, 480
Bleeder resistance, 480
Blennies, 480, 1488
Blepharitis, 480–481, 1346–1347
Blepharoplasty, 481

Blepharospasm, 481
Blexbok, 183–184
Blind-fish, 481–482
Blindness, 3653
Blind worm, 482
Blister beetle, 482
Blob, 482
Bloch function, 482
Bloch wall, 482, 1138
Block chain, 719
Blocking capacitor, 482
Block switching, 156
Blood, 482–488
    acid-base regulation, 18
Blood-brain barrier, 489
Blood loss, 166–167
Blood poisoning (septicemia), 3136
Blood pressure, 489–490
    hypertension, 1888–1889
Blood recycling, 487
Bloodstone, 490
Blood substitutes, 487
Blood transfusion, 486–487
Blood worm, 490
Bloom, 490
Blower, 490
Blow-fly, 490
Blowhole, 490
Blue (butterfly), 490, 576
Blue asbestos, 984
Blueberry, 417
Bluebird, 490
Bluebottle, 490
Bluegill, 490, 3383–3384
Blue glow, 490
Blue-green algae, 90–91
Blue-sky (Linke) scale, 318, 1013, 2141
Bluestems, 1678
Blue supergiant stars, 1621
Bluethroat, 490
Bluff body, 490, 1440
Blushing, 491
Boa constrictor, 3191–3192
Boar, 3404–3405
Bobcat, 664, 666
Bobolink, 491
Bob-white, 2895
Bock beer, 404
Bode's relation, 491
Body cavity, 163
Body-centered structure, 491
Bodying agents (foods), 491–492
Body wall, 163
Boehmite, 120, 395
Bog, 492
Bog lake, 492
Bog-mosses, 562–563
Bohr, Niels Henrik David, 492
Bohr magneton, 492
Bohr theory of atomic spectra, 492–493
Boids, 245, 246, 3191–3192
Boiler (steam generator), 493–500
    feedwater, 1358–1361
Boiling, 500–501
Boiling curve, 501
Boiling point, 501–502
Boiling point constant, 502
Boiling point elevation, 502
Boiling-water nuclear reactors, 2481–2484
Boils, 502, 1053
Boise de rose, 502

Boise-Einstein statistics, 1340
Bole, 502
Bolide, 502, 2324
Boll weevil, 502
Bollworm, 502
Bolometer, 503
Bolson, 503
Boltzmann, Ludwig, 503
Boltzmann's distribution law, 503–504
Boltzmann transport equation, 504, 3518–3519
Bombadier beetle, 504
Bombardment, 504
Bond albedo, 86, 504
Bond energy, 748, 751–752
Bone, 504–509, 584
Bone conduction, 510
Bone densitometry, 509
Bone-derived growth factors, 506
Bone fractures, 506–507
Bone marrow, 505
    transplantation, 509
Bone morphogenetic protein, 506
Bone reconstruction, 506
Bonner Durchmusterung, 510
Bonnet form, of mean value theorem, 2307
Bontebok, 183, 184
Bony fishes, 1403, 1404–1406, 3423
Bony-tail, 510
Bony tongues, 510
Boobies, 2672
Book louse, 511, 962
Boolean algebra, 95, 511
Boole's inequality, 511
Boom, 511
Booster (electrical), 511
Booster amplifier, 147
Booster generator, 511
Booster transformer, 511
Boostrap amplifier, 147
Bootes (the herdsman), 511
Booth's lemniscate, 2118
Bootstrap, 511
Boracite, 511
Borax, 511, 864
Bordoni peak, 511
Bore, tidal, 513, 1330, 2534
Borehole, 2691
Borer, 511–513
Boriding, 1744
Boring (soil), 513
Born, Max, 513
Born approximation, 513
Bornite, 513
Born-Oppenheimer approximation, 513–514
Boron, 514–516
    effect on steel, 2018
Borosilicate glasses, 1630
Bose, Satyendra Nath, 516
Bose-Einstein statistics, 516
Bosons, 516
Boss, 516
Boston Harbor project, 3722
Bostonite, 516
Boswellia tree, 516–517
Botany, 517–518
Bot fly, 518
Bottom-fermentation yeast, 405
Bottom note (flavors), 1419
Bottom quark, 2647
Bottoms, 1131
Botulism, 352, 356–357, 1461, 1466
    in infants, and honey, 1782

Bougainvillea, 518
Bouger and Lambert law, 518
Bouguer's halo, 319
Boulangerite, 518
Boulder, 518
Boulder clay, 805
Boulder train, 1325
Boundary conditions, 519
Boundary layer, 519
    flowing fluids, 1446–1447
Boundary-value problem, 519
Bourdon tube, 519
Bournonite, 519
Bovine mastitis, 2293
Bovines, 519–521
Bovine spongiform encephalopathy (BSE), 521–526
Bowel, 1102–1103
Bowerbird, 527
Bowfin, 527
Bowfins, 3423
Bowman's capsule, 1341
Bowman's membrane, 959
Bow's notation, 527–528
Box annealing, steel, 2020
Box elder, 2253
Box trees and shrubs, 528
Boxwood, 528
Boyle, Robert, 528
Boyle-Charles law, 528
Boyle's law, 528
B.P. (before present), 528
Brachial, 528
Brachiopoda, 528
Brachistochrone, 2, 528–529, 1015
Brachyblast, 529
Brachycephalic, 529
Bracket fungi, 529
Bract, 529
Bradycardia, 228
Bradycardia-tachycardia syndrome, 228
Bragg, William Lawrence Sir, 529
Bragg angle, 529
Bragg equation, 994
Bragg's curve, 529
Bragg's law, 529
Bragg spectrometer, 529
Bragg's rule, 529
Brahe, Tycho, 529–530
Brahman bull, 668
Braid group, 2302
Braille system, 530
Brain, 689–693
    metabolic rate, 714
Brain death, 689, 1030–1031
Brain disorders, 530–531, 704
Brainerd diarrhea, 357
Brain imaging, 703–704
Brain injury, 531–533
BrainMap project, 704
Brain stem, 690
Brain tumors, 599
Braking ellipses, 533
Brambling, 568, 2655
Branch (computer), 533
Branch cut, 533
Branched alkanes, 2581
Branching, 533
Branching process, 473
Branch instruction, 1957
Branch point (mathematics), 533
Brass, 108, 109, 948
    season cracking, 3121

Brassica, 533
Bravais-Miller indices, 533
Brayton cycle, for solar power, 3225
Brazilian holly, 3114
Brazil-nut tree, 533
Brazilwood, 534
Brazing, 534–535
Bread, staling, 3322–3323
Breadfruit tree, 2400–2401
Breakdown voltage, 535
Breakpoint instruction, 1957
Breast, 535–536
Breast abscess, 5
Breast cancer, 597–598
Breccia, 536
Breed, 536
Breeder reactors, 2480, 2493–2495
Bremsstrahlung, 536, 2933
Brewer's malt, 2236
Brewster angle, 536
Brewster law, 537
Brewster point, 320, 349
Brick, 537, 710
Brickwork, 537
Bridge (structural), 543–550
Bridge amplifier, 147, 537–538
Bridge circuits, 538–543
Bridged hydrocarbon ring systems, 2585
Bridged-T bridge, 542
Bridge truss, 3532
Bridging gain, 551
Bridging loss, 551
Briggs logarithms, 2184
Brightness, 551
Brightness temperature, 1213
Bright segment arch, 3552
Bright spots, 1969
Brilliance, 551
Brillouin effect, 551
Brillouin zone, 551
Brill-Zinsser disease, 3002
Brine-fly, 551
Brinell hardness, 1714
Brine shrimp, 551
Bristlemouths, 551
British thermal unit (unit), 3569
British yellowhammer, 568
Brittle feather ore, 2034
Brittle fracture, 551–552
Brittle star, 1201, 2561
Brix scale, 3298
Broadband antenna, 186
Broadbills, 552–553, 2655
Broadening of spectral lines, 553
Broca, Pierre Paul, 553
Brochantite, 553
Brockets, 1039, 1040
Bromegrass, 1678
Bromelain, 654
Bromination, 2591
Bromine, 553–555
Bronchial asthma, 555–556
Bronchiectasis, 557
Bronchitis, 557
Bronchodilators, 556
Bronchoscopy, 557
Brontide, 557–558
Bronze, 108, 109, 948
Bronze Age, 558
Bronze birch borer, 512
Bronzed cutworm, 1011
Brookite, 558

Brooklyn Bridge, 544
Broomcorn, 558
Brown algae, 90, 93–94, 96
Brownian motion, 558–559
Brown mite, 2361
Brown pine, 2751
Brown rot, 266
Brown-tail moth, 559
Browntop millet, 2351
Brucellosis, 352, 357, 559
Brucite, 559
Brunton compass, 559–560
Brush (electrical machinery), 560
Brushless motors, 3138–3140
Brush turkey, 2308
Bruxism, 560
Bruxomania, 560
Bryonia, 560
Bryophyllum, 560
Bryophytes, 560–563
B-station, 563
Bubble chamber, 563
Bubble pressure, 1456
Bubo, 563
Bubonic plague, 352, 563–564
Buckeye trees, 564
Buckley gage, 564
Buckminsterfullerene, 616
Buckthorn shrubs and trees, 564, 565
Buckwheat, 564–565
Bud, 565
Budan theorem, 565
Bud aphis, 211
Budde effect, 566
Budding, 269, 566, 1663
Budgerigar, 2643
Bud moth, 566
Bud nematodes, 2431
Buerger's disease, 235
Buffalo, 520
    bison contrasted, 474
Buffalo carpet-moth, 566, 1056
Buffalo gnat, 477
Buffalograss, 1678
Buffer (chemical), 566–567
Buffer amplifier, 148
Buffer storage, 3358
Buffing, 567
Bufflehead, 3716
Bufotenine, 1712
Bug, 567
Buhrstone, 567
Bulb (botany), 567
Bulbil, 567
Bulblets, 567
Bulb nematodes, 2431
Bulbul, 567
Bulkhead, 567
Bulking agents (foods), 491–492
Bulk sampling, 3060
Bull, 567
Bullet train (Shinkansen), 2940–2941
Bullfinch, 567
Bullfrog, 567
Bullocksheart, 177
Bumblebee, 567
Bundle, 567–568
Bunting, 568, 1505
Buoy, 568
Buoyancy, 568–569
Burble, 569
Burden (instrument), 569

Burette, 569
Burgers vector, 569
Burn, 569–570
Burner, 571–574
Burn-in, 1031
Burning rate (r), 574
Burnishing, 574
Burrfish, 2816
Burrowing, 574
Burrowing nematodes, 2431
Bursa, 574
Bursitis, 574
Burst, 574
Bush-cricket, 574
Bushing, 574
Bush-pig, 3405
Bustard, 2939
Bustard-quail, 2939
Butadiene, 574–575
1,4-Butanediol, 87
Butcher bird, 3160
Butterfish, 575
Butterfly, 575–576, 657, 2120
Butterfly planetary nebula, 1814
Butterfly valve, 576
Butternut, 3696
Butylated hydroxyanisole (BHA), 207–208
Butylated hydroxytoluene (BHT), 207–208
Butyl rubber, 1220
Butyric fermentation, 1363
Buzz, 576
Buzzard, 1717
Bypass capacitor, 576
Bysmalith, 576
Byte, 576, 1108
Bytownite, 1362

CABARET system, 250
Cabbage, 533
Cabbage aphis, 211
Cabbage butterfly, 576
Cabbage root maggot, 2218
Cable (electrical), 577
Cableway, 577
Cabochon, 1585
Cacao tree, 577–578
Cacoa butter, 578
Cacomistles, 2911
Cactus, 578, 3367
CAD/CAM (computer-aided design/manufacturing), 333
CAD (computer-aided design), 334, 335
CADD (computer-aided design/drafting), 907
Caddis fly (or Caddice fly), 578
Caddis worm, 579
Cadelle, 1665
Cadmium, 579–580
Cadmium red line, 580
Cadmium selenide thin-film transistor, 2156–2157
Caecilians, 482
CAE (computer-aided engineering), 334, 907
Caecum, 672
Caesalpina tree, 580
Caffeine, 99, 100
Cage antenna, 186
Caimans, 985
Cairngorm stone (smokey quartz), 581, 2905
Caisson disease, 581
Calabar swellings, 1390
Calabash gourd, 1005
Calandra, 2655

Calandria, 581
Calaverite, 581
Calcalkalic rocks, 98
Calcarea, 581
Calcification, of bone, 505–506
Calcination, 581
Calcite, 581
Calcium, 581–583
    in biological systems, 583–585
    and bones, 505–506, 508
    in diet, 1087
    fertilizer requirements, 1372
    and hypertension, 1091
Calcium aginate, 1698
Calcium carbonate, antacid, 182
Calcium hypochlorite, 480
Calcium metabolism, 583–584
Calcium phosphate, anticaking agent, 199
Calco-uranite, 344
Calcrete, 1153
Calculator, 586–587, 903
    abacus, 585–586
Calculi, 587
Calculus, 587
Calculus of finite differences, 587
Calculus of natural deduction, 248
Calculus of residues, 587
Calculus of variations, 587, 3606
Caldera, 588
Calendar, 588
Calendering, of paper, 2632
Caleometer, 588
Calibration, 588
Caliche, 588, 1153
Calicivirus (Norwalk-like virus), 1461
Calico-back, 3356
Calico black bug, 1715
California pompano, 575
California vulture, 914
California Water Plan, 220
Californium, 37, 588–589
Californium-252, 589
Calla lilies, 227
Callisto, 1558, 2048–2049
    orbital characteristics, 2760
Call number, 589
Callus, 589
Calms of Cancer, 310
Calms of Capricorn, 310
Calorescence, 589
Caloric restriction, 1617–1618
Calorie (unit), 3569
Calorimetric dosimeter, 1141
Calorimetry, 589–590
Calorizing, 591
Calyx, 591
Cam, 591
Camber, 591–592
Cambium (plant), 592
Cambrian Period, 592–593
CAM (computer-aided manufacturing), 334–335
Camel, 593–594
Camel cricket, 593
Cameleopardalus, 593
Cameras, 2727–2732
    motion picture, 2732–2735
Campbell bridge, 542
Camphor, 594
Camptonite, 594
Campylobacter, 1461
Campylobacteriosis, 355, 357–358
Canada balsam, 594, 2977

Canadian, 594
Canadian whiskey, 3762
Canal (physiology), 594–595
Cananga, 595
Canard, 595
Canary, 595, 1505
Cancer (the crab), 595
Cancer and oncology, 595–600
    and cell cycle, 677
    and diet, 1090
    and occult blood, 488
Cancrinite, 601–602
Candela (unit), 3569
Candidiasis, 602
Candle-fly, 603
CANDU reactors, 2493
Canga, 603
Canines, 603–604
Canis Major (the great dog), 605
Canker worm, 605
Cannabis indica, 605
Cannel coal, 821
Cannizzaro method, 729–731
Cannizzaro's reaction, 89
Canonical, 605
Canonical ensembles, 1306–1307
Canonical time unit, 605
Canonical transformation, 605
Canopus (α Carinae), 605, 632
Canteloupe, 1005
Cantilever beam, 397–398
Cantilever bridge, 546, 547, 550
Canyon, 605
Capability ratio, 2897
Capacitance, 605–606
Capacitance-coupled circuit, 975
Capacitance transducer, 606
Capacitive load, 606
Capacitive proximity detectors, 2875
Capacitor, 606
Capacitor antenna, 186
Capacitor motor, 2399
Cap cloud, 828
Cape hunting dog, 604
Capelin, 606–607
Capella (α Aurigae), 607
Cape pigeon, 2684
Capercaillie, 1695
Capillarity, 607
Capillary, 607
Capillary electrophoresis, 1279
Capillary fringe, 607
Capillary pyrites, 2350
Capillary system (instrument), 607
Capric acid, 607
Capricornus (the sea-goat), 607
Caprimulgiformes, 607–608
Caproic acid, 608–609
Caprolactam, 145, 609
Capsule (fruit), 1519
Capsulotomy, 609
Capture release sampling, 3060
Capybara, 3023–3024
Caracal, 664, 666
Caracara, 609
Carangids, 609
Carapace, 609
Carapato, 609
Carat (unit of gem weight), 1585
Caraway, 3566
Carbamates, 609
Carbaminohemoglobin, 632

Carbanion, 609–610
Carbene, 610
Carbene addition, 2589
Carbenicillin, 194, 195
Carbide ceramic powders, 712
Carbohydrates, 610–615
    in fruit, 1519–1520
    industrial chemicals derived from, 453
    metabolism, 613–615
    transport in cells, 680
Carbon, 616–620, 623
Carbonaceous chondrite meteorites, 2326
Carbonado, 621
Carbon black, 621
Carbon-button microphone, 2342
Carbon cycle (nuclear), 621–622
Carbon dioxide, 622–623
    greenhouse gas, 813, 1639–1640, 1641
    on Mars, 2265–2266
Carbon fibers, 1379
Carbon Group, 623
Carbonitriding, 623, 1744
Carbonium ion, 624
Carbon monoxide, 624
    air pollutant, 2803–2804
Carbon pitch, 850
Carbon steels, 107, 108
Carbon suboxide, 624
Carbon tetrachloride, 624–625, 768
Carbonyls
    analytical determination, 161
    chlorinated, 766–767
Carbon-zinc battery, 393
Carboxyl group, analytical determination,
    161
Carboxylic acids, 625–626, 2581
    nomenclature and examples, 2585
Carbuncle (boil), 502
Carbuncle (geology), 626
Carburizing, 626, 1744
Carbynes, 616
Carcinogens, 626–630
Carcinoma, 595, 600
Cardamom, 630
Cardiac, 630
Cardiac arrhythmias, 228–230
Cardiac insufficiency, 915
Cardiac muscle, 2405
Cardiac transplantation, 1731
Cardinal, 568, 630–631, 1505
Cardinal number, 631
Cardiogenic shock, 3159
Cardioid, 631
Cardiologist, 630
Cardiopulmonary bypass, 1731
Cardiopulmonary resuscitation (CPR), 228, 631
Cardiovascular diseases, 232
Carey-Foster bridge, 541
Caribbean current, 631
Caribou, 1039, 1040
Caries, 631–632
Carina, 632
Cariology, 631–632
Carme, orbital characteristics, 2760
Carnallite, 632
Carnelian, 632, 720
Carnivora, 632–633
Carnot cycle, 633–634
    and absolute zero, 7
Carnotite, 634
Carnot theorem, 634
Caroba, 652

Carob-bean gum, 1698
Carol-roots, 3061
Carotenes, 634
Carotenoids, 634, 874
Carp, 634
    aquaculture, 216–217
Carpal tunnel syndrome, 1713
Carpenter-bee, 635
Carpenter-moth, 635
Carpet beetle, 566
Carpetgrass, 1678–1679
Carpopedal spasm, 2637
Carrageenan, 1698
Carrier (communications/food additive), 635
Carrier amplifier, 148, 635
Carrier-amplitude regulation, 635
Carrier current, 635
Carrier (food additive), 635
Carrier frequency, 635
Carrier phase, chromatography, 780
Carrier relaying, 635
Carrier Sense Multiple Access with Collision
    (CSMA/CD), 2181
Carrier suppression, 635
Carrier-to-noise ratio, 635
Carrion beetle, 636
Carrion flower, 3062
Carrot family, 3566
Cartesian coordinates, 636, 945–946
Cartesian tensor field, 3461
Cartilage, 504, 506, 636
    and osteoarthritis, 2606
Cartilage fishes, 1403, 1404
Carver, George Washington, 636
Caryopsis, 1519
Casale ammonia process, 140
Cascade, 636
Cascade amplifier, 148
Cascade controller, 42, 931
Cascade Range, 3678
Cascade shower, 636
Cascara, 636
Case-based reasoning, 242, 250–251
Case hardening, 636
Casein, 636
Caseinogen, 636
Cashew trees, 637–638
Cassegrainian focus, 3440
Cassegrain telescope, 638
Cassini, Gian Domenico, 638
Cassini division, 3086
Cassini mission to Saturn, 638–648
Cassiopeia (the chair), 648
Cassique, 2655
Cassiterite, 648
Cassowaries, 648–649, 2955
Castellanus clouds, 828
Cast ferrous alloys, 107–108
Casting, 649–650
    steel, 2014
Castor (α Geminorum), 650
Castor oil, 650
Casuarina tree, 651
Cat, 664–666
    West Nile virus, 3748
Catabolism, 378
Cataclastic, 651
Cataclysm, 651
Catalina ironwood, 1795
Catalpa trees, 651–652
Catalysis, 652–655, 758
Catalysts, 653–655

Catalysts, (continued)
  activator as, 40
Catalytic converter, 655–656
Catalytic cracking, 103, 653
Catalytic reforming, 653
Cataract, 656, 3651–3652
Catawberite, 656
Catbird, 2655
Cat briar, 3062
Catenary, 656, 1533–1534
Catenation compounds, 895
Caterpillar, 657
Catfish, 657–658
  aquaculture, 219
Catheter, 658
Cathetometer, 658, 890
Cathode, 658–659, 1256
Cathode dark space, 659
Cathode glow, 659
Cathode ray, 659
Cathode-ray tube, 659–664, 1129–1130
Cathodic protection, 963
Cathodoluminescence, 663, 2191
Cathodophosphorescence, 663
Cation, 664, 1255–1256
Cationic dyes, 1154
Cationic reaction mechanism, 2588
Catkin, 664, 1433
Catscratch disease, 355, 666–667
Cat's eye, 667, 2905
Catsharks, 3155, 3156–3157
Cattail, 667
Cattle, 519–520, 521, 667–669
Cattle grub, 518
Cattle lice, 2190
Cattle tick, 3487
Cat-whisker actuator, 2137
Cauchy, Augustin-Louis, 670
Cauchy conditions, 2643
Cauchy convergence test, 670
Cauchy distribution, 670
Cauchy integral formula, 670
Cauchy product, 2852
Cauchy-Riemann equation, 162, 670
Cauchy sequence, 2330
Cauchy theorem, 670
Cauda equina, 693
Cauldron-subsidence, 670
Cauliflower ear, 1724
Causality, 670
Caustic (optical), 670
Caustic chemical, 670
Caustic embrittlement, 1290
Cauterization, 670
Cavallas, 609
Cave, 670, 1324
Cave cricket, 593
Cavendish, Henry, 671
Cavitand, 671
Cavitation, 671–672
CCD. See Charge-coupled device
Cecum (Caecum), 98, 672, 1102–1103
Cedar tree, 672–673, 918, 2043
Ceiling (performance), 673
Celeriac, 673–674
Celery, 673–674, 3566
  fungus diseases, 1531
Celestial mechanics, 674
Celestial navigation, 296, 2422, 2423–2425
Celestial poles, 674
Celestial sphere, 674–675
Celestine, 675

Celestite, 675
Cell, 675–680
Cell cycle, 596, 676–677, 832–833
Cell division, 677–678
Cell membranes, 2375
Cellular automata, 245, 247
Cellular hypersensitivity, 1092
Cellular messengers, 2375
Cellular telephone, 3428, 3435
Cellulitis, 680
Cellulose, 680–681
  and diet, 1089
Cellulose adhesives, 46, 47
Cellulose ester plastics (organic), 681–683
Cellulosics applications, 683–684
Celom, 163
Celsius, Ander, 684
Celsius temperature scale, 684, 3458
Cement, 684–686
  as type of adhesive, 47
Cementation, 686
Cementite, 686, 2016
Cenophytic, 687
Cenozoic, 687
Centaurus (the centaur), 687
Center (instantaneous), 687
Center frequency, 635
Center gage, 687
Center of gravity, 687
Center of mass, 687, 2287
Center-of-mass system, 687
Center of oscillation, 687
Center of percussion, 687
Center of pressure (hydrostatic), 687
Center of symmetry, 687
Center punch, 687
Centi (prefix for units), 3568
Centrifugal casting, 649
Centigrade temperature scale, 684, 3458
Centipede, 237, 687–688, 761–762
Centipede bites, 1054
Central difference, 1095
Central force, 706
Central force motion, 294
Central limit theorem, 706
Central nervous system, 688–705
Central processing unit, 1106
  programmable controllers, 2853–2854
Centrechinoidea, 706
Centrifugal compressors, 70
  in jet aircraft engines, 80
  use in ammonia production, 140–141
Centrifugal fan, 1352, 1353
Centrifugal force, 706
Centrifugal pumps, 2885
Centrifugal still, 2376
Centrifuging, 706–707
Centriole, 677
Centripetal acceleration, 707
Centripetal force, 707
Centrode, 707–708
Centroid, 708
Centromere, 708
Centrosome, 708
Century plant, 66, 708
Cephalochordata, 708
Cephalopoda, 708–709, 2383–2384
Cephalosporins, 195
Cepheids, 709–710
  period luminosity law, 2680
Cepheus, 710
Ceramic-fiber reinforced metal-matrix composites, 711–712

Ceramic-matrix composites, 711
Ceramics, 710–712
Cereal grasses, 1680
  fungus diseases, 1531
Cerebellum, 690, 692
Cerebral angiography, 530
Cerebral embolism, 714–715
Cerebral ganglion, 713
Cerebral hemispheres, 693
Cerebral hemorrhage, 715
Cerebral malaria, 2010
Cerebral palsy, 713
Cerebral thrombosis, 714
Cerebral transient ischemia attacks, 714
Cerebrospinal fluid, 713–714
Cerebrovascular diseases, 232, 714–715
Cerebrum, 690
Čerenkov radiation, 715–716
Ceres, 274
Cereus, 578
Cerium, 716
Cerumen, 716
Cerussite, 716
Cervical cancer, 599, 1658
Cervical disk disease, 509
Cervix, 716
Cesarean section, 716–717
Cesium, 717
  alkali metal, 98
Cestoda, 717–718
Cetus (the whale), 718
CFC (chlorofluorocarbon), 62, 772
CGS system of units, 3568
Chabazite, 718
Chachalaca, 1006
Chadwick, James, 718
Chaetognatha, 718
Chaetopoda, 718
Chafer, 718
Chaffinch, 2655
Chaga's disease (South American Trypanosomiasis), 718–719
Chain, 719
Chain-balanced float, 3299
Chain block, 719–720
Chain-breaking antioxidant, 207
Chain radar beacon, 720
Chalazion, 1346, 3364
Chalcanthite, 720
Chalcedony, 720
Chalcid wasp, 411, 720
Chalcocite, 721
Chalcogenide glasses, 3132
Chalcopyrite, 721
Chalk, 721
Challenger Space Shuttle, 3249–3250, 3251
Chamber pressure (P_c), 721
Chamber volume (V_c), 721
Chameleon, 64, 65, 721
Chamois, 1649
Chamomile, 721
Champagne, 721–722
Chance coincidence, 863
Chandrasekhar, Subrahmanyan, 722
Chandra X-Ray Observatory, 722–724
Channel bass, 984
Channel carbon, 621
Channel frequency, 724
Channeling, 724
Channel precipitation, 1872
Channel protein, neuron membrane, 696
Channel tunnel, 3541–3545

Chaos theory, 2295–2299, 3590
    and weather forecasting, 3731
Chaparral, 724
Chaparral biome, 433
Chapman, Sidney, 724
Chapman region, 724
Chappius absorption bands, 7
Char, 3054
Characids, 724–725
Character (computer system), 725
Character coding, 1108
Character-controlled program generator, 2852
Characteristic equation, 725, 1320
Characteristic function, 725
Characteristic impedance, 725
Characteristic matrix, 725–726
Charadriiformes, 726
Charcoal, adsorption using, 50, 51
Charge conjugation, 726, 924
Charge-coupled devices (CCDs), 726
    for machine vision, 2210
    machine vision application, 258
    for telescopes, 3444
Charge-mass ratio, 726
Charge neutrality, 726
Charge number, 1273
Charles, Jacques Alexandre, 726
Charles law, 726–727
Charlock (wild mustard), 533
Charm, 2647
Charnockite, 727
Charolais cattle, 667
Charon, 2779
    orbital characteristics, 2761
Charpy test, 552
Charr, 3531
Chat (bird), 2656, 3697
Chatterer, 727
Chattermark, 727
Chebyshev, Pafnuty Lvovich, 727
Chebyshev polynomial, 1585, 2033
Check digit, 1119
Cheese antenna, 186
Cheese mite, 2362
Cheese skipper, 2362
CHEF system, 250
Cheilitis, 727
Chelates and chelation, 727–729
Chemical Abstracts Service (CAS) numbers, 767
Chemical adsorption, 51
Chemical affinity, 729
Chemical analysis, 157–160
    organic, 161–162
Chemical asymmetry, 299–300
Chemical composition, 729–731
Chemical compounds, 895–896
    number of, 759
    synthesis, 3408
Chemical derivative, 1052
Chemical elements, 731–752
Chemical equation, 752–753
Chemical equilibrium, 753–754
Chemical erosion, 1324
Chemical formula, 754–756
Chemical indicator, 1914
Chemical intermediate, 1969–1970
Chemical lasers, 2096
Chemical nomenclature, 2582
Chemical potential, 756, 3471
Chemical reaction rate, 756–759
Chemicals. See Chemical compounds
Chemical Substances Index, 2582, 2583

Chemical synthesis, 3408
Chemical transmitter, 696
Chemical trials, 446
Chemical vapor deposition (CVD)
        coatings, 2621
    for optical fiber manufacture, 2566
Chemiluminescence, 2191
Chemimechanical pulp, 2881
Chemisorption, 759
Chemithermo-mechanical pulp, 2881
Chemoreceptor, 759
Chemosphere, 759
Chemotherapy, 595–596, 600, 760
Cherimoya, 177
Cherry fruit fly, 1520–1521
Cherry fruitworm, 1522
Cherry tree, 3036–3037
Chert, 760
Chest cold, 557
Chestnut blight, 760, 1531
Chestnut blight fungus, 266
Chestnut trees, 760
Chevrotains, 3510
Chezy, Antione, 761
Chezy formula, 761
Chickadee, 761, 3499
Chicken, 2824–2832
    irradiation of meat, 2021
Chicken lice, 2190
Chicken mite, 2363
Chickenpox, 761
Chicory, 893, 894
Chicxulub cater, 2288
Chigger, 761
Chigger bites, 1054
Chilarity (quarks), 2652
Chilblain, 761
Children's Vaccine Initiative, 3594
Chile pine family, 221
Chile saltpeter, 3058, 3201
Chili current, 1853
Chill, 761
Chillblains, 1892
Chilopoda, 761–762
Chimachima, 609
Chimaeras, 762
Chimaeroids, 762
Chimango, 609
Chimney rock, 762
Chimpanzee, 191, 192–193
Chinaberry tree, 2231
China clay, 762
China grass, 2944
China stone, 762
China tree, 2231
China turpentine, 637
Chinch bug, 762, 763
Chinchilla, 3024
Chinese cedar, 2907
Chinese gooseberry, 2073
Chinook (game), 254
Chipmunk, 3319–3320
Chirality, 2029
Chiron, 276
Chirp, 762
Chiru, 1649
Chi-square, 762–763
Chital, 1039
Chitin, 763
Chitons, 147
Chive, 105, 2559
Chlamydia, 763, 3148–3149

Chloracne, 24
Chloral, 768
Chlorambucil, 596
Chloramphenicol, 197
Chlorargyrite, 763
Chlorella, 764
Chloride, biological aspects, 764–765
Chlorinated organics, 765–770
Chlorination, 2591
    water, 770, 3712
Chlorination-amination, 2589
Chlorine, 770–772
    biological aspects (of chloride), 764–765
    in seawater, 2523
    from volcanic eruptions, 3681
Chlorinity, 772
Chlorite, 772
Chlorite schist, 772
Chloritoid, 772
Chloroacetic acid, 768
Chloroacetylene, 768
Chlorobiphenyls, 768
Chlorofluorocarbon (CFC), 62, 772
Chloroform, 768
Chloronaphthalenes, 768
Chlorophylls, 773, 874, 2747
Chloroplast, 773
Chloroprene, 768
Chlorosis, 2009
Chlorostyrene, 769
Chlorpheniramine maleate, 203
Chlortetracycline, 196–197
Choke coil, 773–774
Choledocholithiasis, 1561
Cholelithiasis, 1561
Cholera, 355, 358, 774
    foodborne, 1461
Cholesky method, 774–775
Cholesteric liquid crystal, 2143, 2162
Cholesteric liquid crystal display, 2158
Cholesterin, 775
Cholesterol, 775, 3352
    in bile, 422
    and diet, 1088–1089
Cholic acid, 422
Choline, 775–776
Cholinesterase, 775–776
Chondrite, 776
Chondrite meteorites, 2326
Chondrodite, 776
Chondroitin sulfate, 506
Chondrostel, 776
Chonolith, 776
Chopper, 776
Chopper amplifier, 148
Chopper amplitude, 776–777
Chord, aircraft/mathematics, 777
Chordata, 777
Chordotonal organ, 325, 778
Chorea (Huntington's), 777–778
Chorea (Sydenham's), 778
Chorion, 778, 1291, 1346
Choroidal melanoma, 778
Chorology (zoogeography), 3818
Chough, 2656
Christoffel symbol, 778–779
Chroma (Munsell chroma), 779
Chromatic aberration, 779
Chromaticity, 779
Chromaticity diagram, 779
Chromaticness, 779
Chromatin, 677, 779

Chromatography, 779–784
Chromatophore, 784
Chromel, 786
Chromel thermocouples, 3468
Chrominance, 785
Chromite, 785
Chromium, 785–787
    biological aspects, 787
    effect on steel, 2018
Chromium plating, 786
Chromizing, 788, 1744
Chromogenic couplers, 788
Chromophore, 788
Chromophoric electrons, 788
Chromophytosis, 1055
Chromosome, 678, 1829
    and gene, 1588–1589
    X-chromosome, 3787
    Y-chromosome, 3799
Chromosphere, 3377
Chronic, 788
Chronic fatigue syndrome (CFS), 788–791
Chronic obstructive lung disease (COLD), 887
Chronic peptic esophagitis, 1327
Chronograph, 791
Chronolith, 3495
Chronometer, 792
Chronotron, 792
Chrysalis (chrysalid), 792
Chrysoberyl, 792
Chrysocolla, 792
Chrysophyta, 90
Chrysotile, 264, 792, 3137
Chuck, 792
Chuck-walla, 792
Chuck Will's widow, 2456
Chugging, 792
Chunnel (Eurotunnel), 3541–3545
Cicada, 792–793, 1782, 2183
Cichlids, 793
Ciconiiformes, 793–794
Cienega, 795
Cigar tree, 651
Ciguatera poisoning, 1468
Cilia, 794
Ciliophora, 794
Cilium, 794
CIM (computer-integrated manufacturing), 334, 335
Cinchona, 794–795
Cinder cones, 3678
Cinnabar, 795
Cinnamon, 2107
Cipolin, 795
Cipolleti weir, 1443
Circadian clock, 429, 795–796
    and jet-lag, 2035–2036
Circle, 796–797
Circuit breaker, 797
Circuit element, 1282
Circuits, fundamental (mathematical), 797–798
Circular curves, 799
Circular cylindrical coordinate, 1016
Circular distribution, 798
Circular mil (unit), 3569
Circulary polarized sound wave, 798
Circulating storage, 3358
Circulator (microwave), 798
Circulatory integral, 1963
Circulatory system, 163, 798–799, 1725–1732
Circumcision, 799
Circumhorizontal arc, 319
Circumzenithal arc, 319

Cirque, 799
Cirrhosis of liver, 2171
Cirripedia, 799
Cirrocumulus clouds, 826
Cirrostratus clouds, 826
Cirrus clouds, 799, 826
Cissoid of Diocles, 799–800
Cistron, 679, 1588
Citric acid, 800
Citriculture, 800
Citrine, 800, 2905
Citron, 801, 1005
Citrus blackfly, 3764
Citrus bud mite, 2361
Citrus mites, 2361
Citrus psylla, 800
Citrus rust mite, 2361
Citrus thrip, 3482
Citrus trees, 800–802
    fungus diseases, 1531
    nematodes affecting, 2431–2432
Citrus whitefly, 3764
Civet-cat, 3670
CIVEX process, 3576–3577
Civil time zones, 3493
Cladding waste, 2501
Clairaut, Alexis, 802
Clairaut equation, 802
Clam, 802, 2080, 2380–2381
Clapeyron-Clausius equation, 802–803
Clarain, 803
Clarifiers, 804
Clarifying agents, 803
Clarke belt, 803
Clarke cell, 395
Classical scattering (Thomson) cross section, 3109
Classifying (process), 803–804
Clastic rock, 805
Clathrate, 895, 2377
Claude ammonia process, 140
Clausius, Rudolf, 805
Clausius-Dickel column, 1100
Clausius equation, 805
Clausius equation of state, 805
Clausius law, 805
Claustrophobia, 805
Clavicle, 805
Clawfoot, 805
Clay, 805
    as adsorbent, 51
Clay ceramics, 710
Clean Air Act, and acid rain, 20, 21
Clean Air Status and Trends Network (CASTNET), 20–21
Clean rooms, 2336
Clear ice, 71
Clear icing, 1637
Clear lens extraction, 1985, 2964
Clear-winged moth, 805
Cleavage
    biology, 805
    minerals, 805, 2355
Clethra, 1734
Click beetle, 805
CLIMAP, 2522
Climate, 805–816
    effect of volcanic eruptions, 3682
    insect adaptability to, 1955–1956
    Milankovitch theory, 2348–2350
Climate change, 817
Climate forecasting, 812
Climax, ecological, 817, 1208

Climax association, 274
Climbing perch, 2077
Clinical trials, 446, 600
Clinker, 684
Clinometer, 817
Clinozoisite, 817
Clitoris, 1657
Clo (insulation), 817
Cloaca, 817
Clock, 817–818
Clonal selection theory, 819
Clone, 818–819
Cloning
    Dolly the sheep, 821–824
    mammals, 819–821
Closed curve, 1008
Closed-loop feedback control, 1356
Closed-loop gain, 1544
Closed-loop numerical control, 2507
Closed subroutine, 3366
Closed system, 824
Clostridium, 153
Clostridium perfringens, 1466–1467
Clot, 824
Clothes moth, 824
Cloud albedo, 824
Cloud chamber, 824
Cloud-detecting radar, 2842
Cloud feedback, 824
Cloud forcing, 824
Cloud-height indicator, 2842
Cloud point, 824
Clouds, 824–828
    precipitation processes in, 2840
Cloud seeding, 825
Clover, 2118
Cloverleaf antenna, 186
Clover mite, 2361
Clover root borer, 512
Clover seed midge, 2348
Clove tree, 828–829
Cloxacillin, 194
Clubfoot, 829
Cluster analysis, 829
Cluster headaches, 1720
Cluster sampling, 3060
CMV retinitis, 68
Cnidoblast, 829
Coacervation, 829, 868
Coagel, 829
Coagulation, 829
    water, 3712
Coagulation (Hofmeister series), 829
Coal, 829–841
Coal balls, 841
Coal conversion (clean coal) processes, 841–848
Coal gasification, 846–847
Coal hydrogenation, 1870
Coalification, 820
Coal liquefaction, 847–848
Coal mining, 833–836
Coal rank, 830, 831
Coal slurry pipelines, 837–839
Coal-tar analgesics, 153
Coal tar and derivatives, 848–850
Coastal Ocean Processes Program (CoOP), 2536
Coastal terrain hydrology, 1875–1877
Coastal waters, 2535–2536
Coastlines, 2535–2536
Coast redwood, 918, 2959–2962
Coatimundi, 2911, 2912–2913
Coating agents (foods), 850

Coatings, 2615–2622
  paper, 2632
Coaxial antenna, 186
Coaxial line, 577, 850
  for telephony, 3430
Cobalamin (vitamin B$_{12}$), 3664–3665
Cobalt, 850–852
  in biological systems, 853
  effect on steel, 2018
Cobalt bloom, 1326
Cobaltite, 853
Cobia, 854
COBOL, 892
Cobra, 3193–3195
Cocaine, 99–100
Cocci bacteria, 350
Coccidioidomycosis, 853, 1531
Coccolithophore, 2551–2552
Coccus, 854
Coccyx, 854
Cochlea, 325, 854, 1722
Cochran's theorem, 854
Cockatoo, 2643, 2876–2877
Cockchafer, 854
Cockle, 854, 2383
Cock-of-the-rock, 727, 2656
Cockroach, 854, 2603
Cocoa butter, 1412
Coconut oil, 1412
Coconut scale, 3204
Coddington eyepiece, 854
Code, 854–855
Codeine, 100, 153, 2393
Codfishes, 855–856
Codling moth, 856–857
Coefficient, 857
  absorption, 7–8
  accommodation, 14
  agreement, 67–68
  alienation, 98
  attenuation, 324–325
  collision, 866, 2983
  concentration, 913
  concordance, 913
  diffusion, 1100
  discharge, 1126–1127
  extinction, 1344
  internal conversion, 945
  luminous, 2192
  osmotic, 2605–2606
  performance (heat engines), 1734
  point-biserial correlation, 473
  reliability, 2975
  restitution, 866, 2983
  selectivity, 1996
  spreading, 3315
  variation, 3606
Coefficients
  emission (Einstein's), 1296
  trichromatic, 3527
Coel, 857–858
Coelacanths, 857, 3423
Coelenterata, 857
Coelom, 2, 857
Coelomata, 857
Coelomoduct, 857
Coelostat, 857
Coenchyme, 857
Coencytic, 857
Coenzyme A, 857–858
Coenzyme Q, 860
Coenzymes, 857–860

Coffeetree, 2056
Coffee tree and coffee, 860–862
Cogeneration (electricity and thermal energy), 493, 862–863
Cognitive science, 239
Cog project, 262
Coherence, 863
Coherence length, 3385
Coherency, 863
Coherent, 863
Coherent carrier, 863
Coherent echo, 863
Coherent scattering, 3109
Cohesion, 46
Cohesion pressure, 2846
Coil, 863
Coil-burning boilers, 572–574
Coil chain, 719
Coinage metals, 2323–2324
Coincidence, 863
Coking, 848–849
Cokite, 863
Col (saddle point), 3053
Colchicine, 100, 153
Cold (common cold), 887–888
Cold-cathode ionization gage, 863
Cold-flow test, 864
Cold front, 1508
Cold fusion, 1540
Cold-pressor test, 863
Cold-related injuries, 1892
Cold seeps, 2530–2534
Cold short, 863
Cold wall, 863
Cold-worked metal, 863–864
  annealing, 175–176
Cold-worked steels, 2019
Cole, 533
Colemanite, 864
Coleopter, 864
Coleoptera, 864
Coleoptile, 864
Colic, 864
Coliformes, 864
Colitis, 864–865
Collaboration, 1116
Collagen, 865–866
Collar, 866
Collar cell, 866
Collateral circulation, 866
Collateral series, 738
Collator, 866
Collectors, 803–804
Collembola, 866
Colles' fracture, 866
Colleterial glands, 866
Colligative property, 866, 3233
Collimator, 866
Collision, 866
Collision broadening, 866
Collision coefficient, 866, 2983
Collision frequency, 866
Collision parameter, 866
Collision rate, 866
Collodion, 2723
Colloid system, 866–870
  coacervation, 829
Colob, 2386
Colocynth, 1005
Cologarithm, 2184–2185
Colon, 2, 98, 870
Colon cancer, 597

Colony, 870
Colony-stimulating factor, 2204
Color, 870–871, 873–874. See also Pigmentation
  animals, 871
  minerals, 2354–2355
  nonspectral/plants, 874
Colorado potato beetle, 407, 871
Colorants (foods), 175, 871–872
Color blindness, 872–873, 3650
Color centers, 873
Colored glasses, 1630
Colorimetry, 873–874, 3302
Color index, 873–874
Color look-up table, 909
Color mixture curve, 874
Color pigment, 2617
Color raster technology, 908–909
Color saturation, 874
Color supertwisted-nematic LCDs, 2147–2148
Color temperature, 1213
Color temperature (stellar), 874
Color vision deficiency, 2983–2984
Colostrum, 874
Colpitts oscillator, 2604, 2746
Columba (constellation), 874
Columbia Space Shuttle, 3021, 3247–3249
Columbiformes, 874–875
Columbite, 875
Column (structural), 875
Columnar structure, 875
Column-binary code, 855
Coma (optics), 876
Coma (physiology), 875–876
  and brain death, 1030
Comagmatic, 875
Combination, 876
Combination frequencies, 876
Combinatorial explosion problem, in search, 242, 248
Combined sewers, 3713
Comb jelly, 2552
Combustion, 876–882
Comet, 882–886
  Hubble Space Telescope observations, 1808
  Stardust mission to study, 3328–3329
Comet Giacobini-Zinner, 883–884
Comet Hale-Bopp, 1809
Comet Halley, 882, 884, 885, 1709–1711
Comet Kohoutek, 882, 2073–2074
Comet Machholz, 886
Comet Shoemaker-Levy 9, 885–886, 1804
Comet Wild-2, *Stardust* mission to study, 3330
Command control, 886
Command destruct, 886
Command guidance, 887
Commensalism, 274, 3406–3407
Commiphora tree, 887
Commissure, 887
Common-base amplifier, 148
Common business-oriented language, 2082
Common cold, 887–888
Common-emitter amplifier, 148
Common hop, 2401–2402
Common ion effect, 888
Common machine language, 2082
Common-mode rejection ratio, 888–889
Common-mode voltage, 889
Communal entropy, 889
Communicable diseases, 929
Communication network, 2442
Communications cables, 577
Communications satellites, 3062–3073

Communication theory, 331
Commutation relations, 889
Commutative law, 889
Commutator (mathematics), 889
Compandor, 890
Companion cell, 890
Comparative anatomy, 163
Comparative biology, 890
Comparative genomics, 1843
Comparator, 890
Comparator amplifier, 890
Compass, 890–892
    Brunton compass, 559–560
    pelorus (dumb compass), 2673
    Sun compass, 3383
Compass north, 2471
Compensation theorem (network), 892
Compensator, 892
    use with microscopes, 2347
Compiler, 892, 1110, 3516
Complement, 892
Complemental male, 892
Complementarity principle, 892
Complement system, antibodies, 199
Complete integral, 1286
Complex compounds, 895
Complexity theory, 2299–2300
Complex salt, 3058
Complex variable, 892–893
Component, 893
Component analysis, 893
Component therapy (blood), 486
Composite course, 976
Composite family, 893–894
Composite materials, 895
Compositeness (quarks), 2652
Composite volcano, 3678
Compositional rule of inference, 253
Compound (chemical), 895–896
Compound distribution, 896
Compound nucleus, 896
Compressed air illness, 581
Compressibility factor, 896
Compression
    gas, 69–71, 896–897
    signal, 897
    structural, 897–898
Compression (ellipticity), 1286
Compression (flattening of Earth), 1418
Compressors, 69–71
Compton, Arthur Holly, 898
Compton effect, 898
Compton Gamma-Ray Observatory (CGRO), 414,
    898–903
Compton rule, 903
Compton scattering, 3109
Computation, 2295
Computational neuroscience, 704
Computed axial tomography (CAT), 530
Computed tomography (CT), 170
    nondestructive testing by, 2468
Computer, 903–904
    automation application, 332–333
Computer addresses, 45
Computer animation, 904–907
Computer-assisted ultrasonic microscopy, 2468
Computer code, 855
Computer error procedures, 907
Computer graphics, 907–911
Computer numerical control, 335
Computer operating system, 911
Computer processing, 1273

Computer programming, 1109–1110
Computer vision syndrome, 911–912
Computer word, 1108
Computing efficiency, 912
Concentration
    chemical/process, 912
    statistics, 912–913
Concentration cell corrosion, 962
Conch, 913, 2383
Conchoid of Nicomedes, 913
Conchology, 913
Concordance, 913
Concrete, 913
    prestressed, 2846
Concrete aggregate, 67
Concrete arch bridge, 546
Concretion, 914
Concurrent forces, 1474
Concussion, 506
    in sports, 532
Condensate, 914
Condensation, in atmosphere, 2839
Condensation compounds, 895
Condensation curve, 501–502
Condensation nucleus, 914
Condensation polymerization, 2593
Condensation rate, 914
Condensation shock wave, 914
Condensation trail, 914, 2840
Condensation-type hygrometer, 1884
Condensed film, 1391
Condenser microphone, 2342
Condensers, 1741
Conditional convergence, 945
Conditioning agents, 200
Condor, 914, 3693
Conduction heat transfer, 1739
Conductivity, 914
Conductivity meter, 1262–1263
Condyle, 914
Cone, 914–915
Cone-in-cone structure, 915
Confabulation, 145
Confidence interval, 915
Confidence level, 915
Confined-area pollution, 2799
Confluent, 915
Confocal conic, 2895
Confocal quadric, 2895
Conformal map, 915
Conformal representation, 915
Conformity, 915
Confused flour beetle, 1664
Congenital chloridorrhea, 1080
Congestive heart failure, 915–917
Conglomerate, 917
Congruent, 917
Conical antenna, 186
Conical ball mill, 371
Conical coordinate, 917
Conical cornea (keratoconus), 2057
Conical projections, 917
Conical surface, 914, 917
Conicoids, 2895
Conic section, 917–918
Conidia, 918
Coniferous forest biome, 433
Conifers, 918–920
Conjugate, 920
Conjugate acids and bases, 22–23
Conjugate directions (at a point P on a surface), 920
Conjugate elements of a group, 920

Conjugate numbers, 920
Conjugate point (mathematics), 920
Conjugate solutions, 920
Conjugate space, 2140
Conjugate vaccines, 3593
Conjugation, 920
Conjunction, astronomical, 920–921
Conjunctiva, 921–922
Conjunctival tumors, 922
Conjunctivitis, 921, 922
Connate water, 922–923
Connecting rod, 923
Connective tissue, 923
Connectivity, 924
Consanguinity, 924
Consecutive reactions, 757–758
Consequent streams, 924
Conservation laws
    nonlinear hyperbolic, 2300–2301
    and symmetry, 923–926
Conservative system, 926
Consistent statistic, 926
Consolute liquids, 926
Consolute temperature, 926
Consonance, 926
Constant, 926
Constantan thermocouples, 3468
Constant-current transformer, 926
Constant-deviation prism, 2849
Constant-level balloon, 926
Constant pressure specific heat, 3300
Constant volume specific heat, 3300
Constellations, 298, 926–928
Constipation, 928, 1090
    in elderly, 1612
Constituent, 928
Constraint, 928
Constraint-logic programming, 249
Constructional alloy steels, 2019
Consumer's risk, 2901
Contact angle, 928
Contact-deficiency diarrhea, 1080
Contact gettering, 1620
Contact lenses, 929
Contact-modulated amplifier, 148
Contact potential difference, 929
Contact tonometer, 3504
Contagion, 929–930
Continental air, 72
Continental crust, 1164
Continental drift, 1182, 2525, 3739–3740
Continental margins, 2526–2529
Continental rise, 930, 2526
Continental shelves, 930, 2526
Continental slopes, 930, 2526
Contingency coefficient, 930, 961
Contingency table, 930
Continual seizure, 312
Continuity equation, 930
Continuity of state, 930
Continuous frame, 3004
Continuous function, 930
Continuous group (topological group), 3505
Continuous information source, 1929
    capacity, 1932
Continuously variable transmission (Ford Motor
    Co.), 340
Continuous polymerization, 2378
Continuous time series, 3494
Continuous variable, 930
Continuous-wave radar, 930
Contour, 930–931

Contour integral, 1963
Contractile proteins, 931, 2405, 2865
Contractile vacuole, 931
Contractility, 931
Contrail, 914
Contrast (statistics), 931
Control action, 931–933
Control chart, 933–934
Control counter, 974
Control instructions, 1109
Controlled environment, 1310
Controlled oxidative fermentation, 1363
Controlled system, 2851
Controller
    automatic, 934–935
    cascade, 42, 931
    floating, 1432–1433
    hydraulic, 1854–1857
    integral, 934
    multiposition, 2403–2404
    on-off, 2404
    pneumatic, 2780–2781
    programmable, 2852–2857
    proportional, 934
    proportional-derivative, 934
    proportional-integral, 934
    proportional-integral-derivative, 934
    ratio, 2954
    reverse acting, 2988
    sampling, 3059–3060
    self-operated, 3128–3129
    self-tuning, 41–42
    smart, 42
    two-position, 2404
Control limits, 2898
Control system, 935–938
    architecture, 938–944
Control valve, 3601–3602
    cavitation in, 671–672
Convection, 945
    in atmosphere, 312
    chaos theory application, 2297
Convection heat transfer, 1739–1740
Convective atmosphere, 47
Convective cloud, 824
Convergence, 945
Convergent adaptation, 41
Convergent integral, 1963
Convergent series, 3137
Conversion, 945
Conversion coefficient, 1971
Conversion gain ratio, 945
Conversion ratio, 945
Conversion transducer, 3510–3511
Convict fish, 3397
Convolution, 945
Convolutional coding, 1933
Convulsion, 945
Convulsive seizures, 532
Cooling curve, 945
Cooling degree days, 1043
Coordinate-covalent bond, 2376
Coordinates, generalized, 945
Coordinate system, 945–996
Coordination compounds, 895, 946
Coordination number, 946
Coordination polyhedra, 946
Cootamundra wattle tree, 10
Cooter, 946, 3463
Coots, 2939
Copal, 946, 2977
Copalite (copaline), 946

Copepoda, 946
Coplanar forces, 1474
Copolymerization, 2377
Copper, 946–950, 2356
    in biological systems, 951–952
    effect on steel, 2018
Copper Age, 951
Copper alloys, 108
Copper glance, 721
Copperhead, 3197
Copper loss, 952
Copper pyrites, 721
Copper thermocouples, 3468
Copulation, 952
Coquina, 952
Coraciiformes, 952, 2071–2072
Coral, 88, 189, 412, 857, 953
    ocean harvesting, 2555
Coral reef, 429, 953–955
Coral tree, 955
Corbina, 984
Cordierite, 955
Cordless telephone, 3428
Core (magnetic), 955
Core (Earth), 1166
Core loss, 955
Core sampler, 955
Coriander, 3566
Coring, 955
Coriolis, Gustave-Gaspard de, 956
Coriolis effect, 955–956
Coriolis flowmeter, 1441
Coriolis force, 310
Cork, 956
Cork cambium, 375, 592
Cork oak, 956
Corkwood tree, 956
Corm, 956
Cormidium, 956
Cormorants, 2671, 2672–2673
Corn (maize), 956–958
Corncrake, 2939
Corn cutworm, 1011
Cornea, 959
Corneagen cell, 959
Corneal abrasion, 959
Corneal diseases, 3652
Corneal dystrophy, 959–960
Corneal topography (videokeratography), 960
Corneal transplantation, 3652
Corn earworm, 960
Corner cube prism, 2849
Corner reflector (optics), 960
Cornfield ant, 182
Corn oil, 968, 1412
Corn-root aphis, 211
Corn rootworm, 960, 3738
Corn sap beetle, 1146
Corn smut, 1530
Corn starch, 968
Corn starch maltodextrin, 492
Corn syrup, 3399
Cornu-double prism, 2849
Cornu-Jellet prism, 2849
Cornu spiral, 960
Corollary, 960
Corona (atmospheric optical phenomenon), 319
Corona (Sun), 3377–3378
    Ulsses mission to study, 3565
Corona Australis (southern crown), 960
Corona Borealis (northern crown), 960
Corona discharge (Saint Elmo's fire), 3054

Coronal holes, 3378
Coronary artery spasms, 2026–2027
Coronatae, 961
Coronaviruses, and colds, 887
Coronograph, 960, 3381
Corpus callosum, 961
Corpuscle, 961
Corpuscular cosmic rays, 961
Corpuscular theory of light, 961
Corrasion, 961
Corrasion-corrosion, 962
Correction to vacuum, 961
Correlation, 961
Correlation detection, 961
Correlation tracking and ranging (COTAR), 961, 3509
Correlation tracking and triangulation (COTAT), 961–962, 3509
Correlation tracking system, 962
Correlative spectroscopy, 2477
Correlogram, 962
Correspondence principle, 962
Corrodentia, 962
Corrosion, 962–965
    monitoring, 964–965
Corrosion embrittlement, 965
Corrosion fatigue, 962, 965
Corrosion inhibitors, 963, 1062
Corsite, 965
Cortex, 965
    kidney, 2061
Corti, organ of, 965, 1722
Corticosterone, 49
Corticotropin, 699
Cortisol (17-hydroxycorticosterone), 49
Cortisone (17-hydroxy-dehydroxycorticosterone), 49
Cortlandtite, 1796
Corundum, 396, 965
Corvus (the raven/crow), 965
Corydalis, 965–966
Cosecant function, 3527–3528
Cosecant-squared antenna, 186
Coset, 966
Cosine emission law, 966
Cosine function, 3527–3528
Cosmic background radiation, 969
Cosmic rays, 966–967
    corpuscular, 961
Cosmological Principle, 969
Cosmology, 968–971
Cosmotron, 2650
Costal, 971
Cotangent function, 3527–3528
Cotinga, 727, 2656
Cotter, 971
Cotton, 971–972, 1382, 2237
Cotton boll, 972
Cotton bollworm, 502, 960
Cottonmouth, 3197
Cottonseed oil, 1412
    source of proteins, 2864
Cotton stainer, 972
Cottontail, 2909
Cottonwood, 2814, 2815
Cotton wool spots, 68
Cottony cushion scale, 3204
Coucals, 1004
Coudé focus (telescope mount), 972, 3440
Couette flow, 972, 1445
Coulomb (unit), 3568, 3569
Coulomb, Charles, 972–973

Coulomb barrier, 2475
Coulomb collision, 973
Coulomb damping, 973
Coulomb degeneracy, 973
Coulomb energy, 973
Coulomb excitation, 973
Coulomb force, as exchange force, 1340
Coulomb law, 973, 1282
Coulometer, 973
Coumarin, 200, 201
Countdown, 973
Counter, 973–974
    computer system, 974
Counterbalancing, 974
Counterbore, 974
Counter-electromotive force, 1268
Countersink, 974
Country rock, 974
Couple, 974–975
Coupled circuit, 975
Coupled oscillator, 975
Coupled reaction, 131
Coupling
    chemical/mechanical, 975
    physics, 975–976
Coupling reaction, 2592
Coupon, 976
Courlan, 2939
Course, 976
Course, composite, 976
Covalent bond, 751, 2376
Covalent compounds, 895
Covalent radius, 745–746, 748–750
Covariance, 976
Covellite, 976
Coverage diagram, 2917
Coversine function, 3528
Cowbird, 976
Cowper's gland, 976
Cowpox, 3596
Cowry, 976
Cox, Allen V., 976
Cox model, 2303
Coxsackie virus, 976–977
Coyote, 603
Coypu, 3024
CPT symmetry, 2647
Crab, 412, 977, 988, 989, 1031, 3788
Crab apple tree, 3041
Crab-eating cabezon, 3119
Crabs (pubic lice), 2671
Cracking process, 977–978, 2591
Cramér-Rao inequality, 978
Cramer rule, 978
Cranberry, 417
Crane fly, 978, 2115
Cranes, 2939
Crankshaft, 978
Crappie, 3383
Crater (southern constellation), 978
Craton, 978–979
Crawfish, 990
Crayfish, 990, 1031
Creeper, 979
Creep (geology/metals), 979
Creosote, 849–850
Crepe (natural rubber), 3045–3046
Crescent moon, 2387
Cress, 533
Crest cloud, 828
Cretaceous mass extinctions, 2287–2288
Cretaceous Period, 979–980

Creutzfeldt-Jakob disease, 531, 980–981
    variant, 522–526
Crevasse, 981
Crick, Francis Harry Compton, 982, 3728
Cricket, 981–982, 2603
Crigler-Najjar's disease, 422–423, 1561
Crinoidea, 982
Crippled leapfrog test, 982
Critical composition, 982
Critical concentration, 982
Critical density, 982
Critical frequency, 982
Critical humidity, 1853
Critical inductance, 1915
Critical level of escape, 982
Critically damped, 1026
Critical mach number, 982
Critical mass, 983, 2480
Critical opalescence, 983
Critical point, 983
Critical potential, 983
Critical pressure, 983
Critical reactor, 983
Critical region, 983
Critical region (statistics), 983
Critical resolved shear stress, 983–984
Critical solution temperature, 984
Critical speed, 984
Critical temperature, 984
Critical velocity of flow, 984
Critical volume, 984, 3687
Croakers, 984
Crocidolite, 984
Crocodiles, 984–986
Crocoite, 986
Crohn's disease, 865
Cromwell current, 986–987
Crookes dark space, 659
Crookes tube, 726, 987
Crop, 987. *See also* specific crops
    pollination, 411
Crop (poultry), 2831
Crossbedding, 987
Crossbill, 1505, 2656
Cross correlation detection, 961
Crosscorrelation detection, 987
Cross-eyes (strabismus), 3358–3359, 3650
Crossing, of plants, 2765
Cross-matching, of blood, 486
Cross modulation, 987
Cross-pollination, 2797
Cross section, 987
Cross-slip, 999
Cross-staff, 987
Crosstalk, 987, 3429
Cross-translator, 3516
Croton bug, 987
Crow, 988
Crown of thorns, 1335
Crucible, 988
Crude distillation, 2699–2701
Crude fiber, 1089
Crude oil pipelines, 2703
Crunode, 988
Crush breccia, 651
Crustaceans, 237, 799, 988
    edible, 988–991
Crust (Earth), 1162–1164
Crux, 991
Cryogenic fluid pump, 993
Cryogenic freezing, 1499
Cryogenic gyroscope, 1702

Cryogenic materials, 991
Cryogenic propellant, 991
Cryogenics, 991–992
Cryohydrate, 993
Cryohydric point, 993
Cryolite, 116, 993
Cryology, 993
Cryometer, 993
Cryopump, 993
Cryoscope, 993
Cryoscopic constant, 993
Cryosphere, 993
Cryosurgery, 600
Cryotron, 993
Cryptococcal meningitis, 2315
Cryptococcosis, 993
Cryptocrystalline, 993–994, 2353
Cryptocrystalline quartz, 2905
Cryptomonads, 90
Crystal, 994–1000
    face-centered, 1000
    holocrystalline, 1776
    holohedral, 1780
    homometric pairs/isomorphous, 1001
    ionic, 1998
    lattice energy, 2106
    manufacture in microgravity, 2341
    mixed, 1003
    solid-state physics, 3230–3233
    twinning, 3552
Crystal class, 2353
Crystal detector, 1000
Crystal diode, 1120
Crystal field theory, 1000–1001
Crystal form, 2354
Crystal growth, 997
Crystal habit, 1001
Crystallization, 1002, 2353
Crystalloblastic, 1003
Crystallogram, 1003
Crystallographic indices, 2353–2354
Crystallography, 994
Crystal oscillator, 1003, 2604
Crystal phases, 1003
Crystal pickup, 1003
Crystal slip, 999
Crystal structure, 2353–2354
CS-85, anti-AIDS drug, 34
Ctenophora, 2552
Cube, 1003
Cubical parabola, 2633
Cubic equation, 1003–1004
Cuckoo, 1004
Cuckoo-rollers, 952
Cuckoo wasp, 1004
Cuckoo wrasse, 3785
Cuculiformes, 1004
Cucumber, 1005
Cucumber beetle, 407
Cucumber gall, 2430
Cucurbitaceae, 1004–1005
Cumene, 2685
Cummingtonite, 1005
Cumulants, 1005
Cumulative excitation, 1005–1006
Cumulative probability function, 1502
Cumulonimbus clouds, 828
Cumulose, 1006
Cumulus clouds, 828
Cupola, 2011
Cupola structure, 1006
Cuprite, 1006

Curassow, 1006, 1563
Curculio, 864, 1006
Curettage, 1006
Curie (unit), 3568, 3569
Curie, Marie, 1006
Curie, Pierre, 1006
Curie point (Curie temperature), 1007
Curie-Weiss law, 1007
Curium, 37, 1007
Curl, 1007
Currant, 417
Currant aphis, 211
Currant borer, 512
Currant fruit fly, 1521
Currant stem girdler, 3102
Current amplification, 1007
Current attenuation, 1007
Current balance, 1007–1008
Current correction angle (navigation), 1008
Current density, 1008
Current regulators, 2837
Current transformer, 1239
Curtius reaction, 1043
Curvature, 1008
Curvature of field (optics), 1008
Curvature of lens (total), 1008
Curve, 1008
    higher plane/plane/space, 1009
    smoothing, 3188
Curved rod bacteria, 350–351
Curve fitting, 1008–1009
Curve of concentration, 913
Curve plotter, 2776
Curvilinear orthogonal coordinates, 1009
Cushing, Harvey Williams, 1009
Cushing's disease, 2756
Cushing's syndrome, 49, 50, 2756
Cusp, 1009–1010
Cuspate foreland, 1009
Custard-apple, 177
CUSUM (cumulative sum) chart, 1010
Cutaneous abscess, 6
Cutaneous anthrax, 190, 355
Cuticle, 1010
Cuticula, 1010
Cutlassfishes, 1010
Cutoff, 1010
Cutoff frequency, 1010
Cut set, 1010
Cut set (fundamental/oriented), 1010
Cut set matrix, 1010
Cuttlefish, 708, 1010
Cutworm, 1010–1012
Cyanamides, 1012
Cyanic acid and related compounds, 1012
Cyaniding, 626
Cyanobacteria, 280, 351
Cyanogen, 1012–1013
Cyanohydrins, 1013
Cyanometer, 1013
Cyanometry, 318, 1013
Cyanosis, 1013
Cybernetics, 331, 1013
Cybotaxis, 1013
Cycad, 1013–1014
Cyclamate, 3401
Cycle (mechanical), 1014
Cycle (unit), 3569
Cycle efficiency, 1014
Cycle of stress, 1014
Cycle oil, 977, 2701
Cycle per second (unit), 3569

Cyclic AMP (adenosine monophosphate), 697,
    2375
Cyclic code, 1014
Cyclic curve, 1014
Cyclizine hydrochloride, 203
Cycloalkanes, nomenclature and examples, 2584
Cycloconverter, 1455
Cyclohexane, 2685
Cyclohexanol, 88
Cyclohexanol-cyclohexanone (KA oil), 1014–1015
Cyclohexatriene, 412
Cycloid, 1015
Cyclone, 312
Cyclone furnace, 573
Cyclone wave, 314
Cyclonic, 1015
Cyclophon, 1015
Cyclostomata, 1015
Cyclostrophic flow, 313, 1662
Cyclostrophic winds, 3774
Cyclotrimethylenetrinitramine, 1343
Cyclotron, 1015
Cyesis (pregnancy), 2844
Cygnus (the swan), 1015–1016
Cygnus X-1, 478
Cylinder, 1016
Cylindrical cam, 591
Cylindrical coordinate, 1016
Cyme, 1434
Cymophane, 792
Cypress trees, 1016–1018
Cyproheptadine hydrochloride, 203
Cyst, 1018
Cysteine, 133–137
Cyst-forming nematodes, 2431
Cystic fibrosis, 1018–1019
Cystine, 133–137
Cystine arabinoside, antimetabolite, 203
Cytochromes, 1019–1020
Cytogenic gland, 1020
Cytokinesis, 805
Cytology, 1020, 1587
Cytoplasm, 2375
Cytotoxic chemicals, 1020
Czochralksi method, 997

Dab, 1418
Dabchick, 2783
Dacite, 1021
Dacryocystisis, 1021
Dacryocystorhinostomy, 1318
Daddy longlegs, 978, 1717
Daguerre, Jacques, 1021
Daguerreotype, 2723
Daily wear contact lenses, 929
Dairy cattle, 669
Daisy, 893
D'Alambert, Jean -Le-Rond, 1021
D'Alambert principle, 157, 1021
D'Alambert's test, 670
Dallisgrass, 1679
Dalton, John, 1021
Daltonide compounds, 731
Dalton law, 1021
Dam, 1021–1025
Damalisks, 183, 184
Damascus steel, 2016–2017
Daminozide, 2767
Damping, 1026
Damsel bug, 410
Damsel fly, 2556
Danaite, 232

Danburite, 1026
Dandelion, 893, 894
Dandruff, 3122
Daniel cell, 1564
Darcy (unit), 3569
Dark adaptation, 1026
Dark current, 1026
Dark discharge, 1127
Dark-field illumination, 1026
Dark matter, 971, 1546
Darlington pair, 1026
Darters (birds), 2671, 2672
Darters (fish), 2677–2678
Darwin, Charles, 1026. See also Evolution
Darwinism, 1338
Dassies, 1892
Database, 1026–1027
Database management systems, 1027
Data Definition Language (DDL), 1027
Data link, 1027
Data Manipulation Language (DML), 1027
Data movement instructions, 1109
Date palm scale, 3204
Date-stone beetle, 1146
Datolite, 1028
Datum, 1028
Daughter element, 1032, 2920
Davydov splitting, 1028
Daw, 988
Dawn redwood, 2962
Day (unit), 3569
Daylight Saving Time, 3493
Dead band (instrument), 1028
Deadbeat escapement, 2673
Dead center, 1028
Deadly nightshade, 408
Dead men's fingers, 1028
Dead reckoning, 84, 1028–1030, 2422
Deafness, 1724–1725
Dearation, 1030
Death (brain), 689, 1030–1031
Death adder, 3195
Death hormone theory, 1615
Death process, 473
Death's head moth, 1031
Death watch, 1031
Debridement, 1031
De Broglie, Louis-Victor, 1031
De Broglie wavelength, 1031
Debug, 1031
Debye-Falkenhagen effect, 1031
Debye-Hückel limiting law, 1031
Debye-Sears effect, 1031
Debye theory of specific heat, 1031
Deca (prefix for units), 3568
Decaffeinated coffee, 861–862
Decalescence, 1031, 2956
Decanoic acid, 607
Decanting, 804
Decapoda, 1031–1032
Decarboxylation, 137
Decarboxylation coenzymes, 859
Decarburization, 1032
Decay constant, 325, 1032, 2920
Decay product, 1032
Decay time, 1032
Decca, 1032
DECCA navigation, 2427
Decemet's membrane, 959
Deci (prefix for units), 3568
Deciduous forest biome, 433
Deciduous plants, 1032

Decimal point, 2783, 2818
Decision function, 1032
Declination, 1032
Decoder, 1032
Decolorizing agent, 1032
Decomposition (chemical), 1032–1033
Decomposition voltage, 1033
Decompression sickness, 1033
Decorative coatings, 2616
Decoupling filter, 1033, 1392–1393
Decrement gage, 1033
Decrepitation, 1033
Dedifferentiation, 1033
Dedolomitization, 2931
De Donder's fundamental inequality, 729
Deep-bed filter, 1395
Deep Blue, 239, 242, 254
Deep earthquake, 1183
Deep-ocean drilling, 2543
Deep scattering layer, 1033
Deepsea anglerfishes, 171, 172
Deep-sea hot springs, 2530–2534, 2544, 2549
Deep Space Network, 189, 1033–1037, 1038
    artificial intelligence applications, 244
Deer, 1037–1040
Deer-antelopes, 183, 184
Deerfly, 1040, 1124
Deer tick, 2202
Defervescence, 1040
Deficiency diseases, 1040–1041
Deflagrating explosive, 1342
Deflation, 1041
Deflection (structural), 1041
Deflection angle, 1041
Deflection of the vertical, 1041
Deflection yoke, 1041
Deflector, 1041
Defoaming agents, 1041–1042
Deforestation, 814, 1042
Deformation (continuous, mathematics/materials),
    1042
Deformation bands, 1042
Degasification, 1042
Degassed water, 3711
Degeneracy, 1043
Degenerate electron gas, 1043
Degenerate oscillating system, 2603
Degenerate state, 1043
Degradation, 1033
    chemical/energy/nuclear, 1043
Degree Celsius (unit), 3569
Degree days, 1043
Degree (electrical/geometry/thermal), 1043
Degree Fahrenheit (unit), 3569
Degree Rankine (unit), 3569
Degrees of freedom, 1493
    robots, 3006
    statistics, 1043
Dehiscence, 1043
Dehumidification, 1043–1044
Dehydration, 1140–1151, 2063
    chemical, 1044
    physiological, 1044–1045
Dehydrochlorination, 2591
11-Dehydrocorticosterone, 49
Dehydrocyclization, 653
Dehydroepiandrosterone (DHEA), 1616
Dehydrogenation, 1045, 2592, 2609
    catalytic, 654
Dehydroisoandrosterone, 165
Deimos, 276, 277, 2274, 2276
    orbital characteristics, 2760

Deionization potential, 1045
Deionization time, 1045
Del (differential operator), 1045
Delay, 6, 1045
Delay circuit, 1045
Delayed coincidence, 863
Delayed coking, 977
Delayed neutrons, 2448
Delay equalizer, 1045
Delay line, 31, 1045
Delbrück scattering, 3109
Deliquescence, 1045–1046
Delta, 1046
    hydrology of, 1877
Delta connection, 1046
Delta function, 1046, 1124
Delta rhythm, 1257
Demagnetizing field, 1046
Demal solution, 1046
Dementias, 126–128, 1611
Demineralization, boiler feedwater, 1360
Demodulator, 1046, 1060
Demospongiae, 1047
DENDRAL expert system, 240, 251
Dendrite, 695–696, 1047
Dengue fever, 1047–1050
Dengue hemorrhagic fever, 1047–1050
Dense (set theory), 1050
Dense-media separation, 804
Density, 1050
Density currents, in ocean, 2524
Density meters, 3300
Dental caries, 631–632
Dentistry, 631–632
Dentition, 631, 1050–1051
Deodorizing, vegetable oils, 3609
Deoxidizing agent, 1051
Deoxyribonucleic acid. See DNA
Departure (navigation), 1051
Dephlegmation, 1131
Depolarization, 1051
Depressants, 804
Depressed pole, 1051
Depression (low), 311
Depth of field, 1051–1052
Derivative (chemical/mathematics), 1052
Derivative action gain, 1544
Derivative control, 931
Dermaptera, 1052
Dermatitis, 105, 1052–1056
Dermatomyositis, 2410
Dermatosis, 1052–1056
Dermestid, 1056
Dermolith, 1056
Dermoptera, 1056
Desalination, 1056–1059
Desalter, 2699
Desargues, Girard, 1060
Descartes, Rene, 1060
Descending node, 1060
Describing function, 1060
Descriptive geometry, 1060, 1598
Desert biome, 431–432
Desert saw viper, 1217
Desert winds, 432
Design of experiments, 1060
Desmine, 3356
Desorption, 1060
Desoxycorticosterone, 49
Detachable link chain, 719
Detection (radio), 1060
Detector, 1060

Detergents, 1060–1063
Determinant, 1064
Determinate structure, 1064
Detonating explosive, 1343
Detritus, 1064
Deuteric, 1064
Deuterium, 1064, 1864
Deuterium-tritium (D-T) reaction, 1536
Deuteron, 1064
Deuteron Compton effect, 898
Deutoplasm, 1064
Developmental anatomy, 163
Deviation, 1065
Deviation distortion, 1065
Devilfish, 708
Devil's darning needle, 1143
Devitrification, 1065
Devonian, 1065
Devonite, 1065
Dew, 2840
Dewar flask, 1065
Dewberry, 417, 3036
Dew cell, 2842
Dew claw, 1065
Dewlap, 1065
Dew point, 1065, 2839
Dextrin, 612
Dextro form, of stereoisomers, 137, 2028
Dextrorotary, 300
Dextrose, 3399
Dezincification, 1065
Dhole, 604
Diabase, 1137
Diabetes insipidus, 1065–1066
Diabetes mellitus, 1066–1072
Diabetic glomerulopathy, 2065
Diabetic retinopathy, 1073–1074
Diabimids, 1583
Diacetone alcohol, 16
Diagenesis, 1074
Diagnostic routine, 3043
Diagnostics (computer system), 1074
Diagonal (structural), 1074
Diagonal regression, 2970
Diallage, 1074
Dialysis, 1074
Diamagnetic materials, 2636
Diamagnetism, 1074
Diamond, 616, 1074–1076, 2356
    in meteorites, 2326
    ocean deposits, 2553
Diamond anvil high pressure cell, 1076–1079
Diamond cutting, 1585
Diamond films, 1076
Diamond pyramid hardness, 1715
Diaphragm, 1079
Diaphragm (respiratory system), 2982
Diarrhea, 1079–1080
Diarrheagenic *Escherichia coli*, 358
Diaspore, 1080
Diastem, 1080
Diastole, 1080
Diastolic pressure, 489, 1888
Diastrophism, 1080
Diathermal wall, 1080
Diathesis, 1080
Diatom, 90, 1080–1081
Diatomaceous earth, 1081. *See also* Fuller's earth
    decolorizing agent, 1032
Diatomic gases, 739
Diatomite, 1081
Diatreme, 1081

Diazo compounds, 346, 2581
Diazonium salts, 2581
Dibatag, 183, 184
Dichlorobenzenes, 768
Dichotomy, 1081
Dichroism, 1081
Dichroite, 955
Dichromatism, 1081
Dickcissel, 1505, 2656
Dicotyledons, 1081
Dicroscopic eyepiece, 1081
2,'3'-Dideoxyadenosine, anti-AIDS drug, 34
2,'3'-Dideoxycytidine, anti-AIDS drug, 34
Didymium glass, 1081
Die, 1081
Die casting, 1081–1082
Dielectric antenna, 186
Dielectric constant, 1082
Dielectric heating, 1082
Dielectric theory, 1082–1083
Diels-Alder cycloaddition, 2588
Diencephalon, 1083
Diesel cycle, 1014
Diesel engine, 1083–1086, 1970–1971
Die sinking, 1087
Diet, 1087–1094. See also Minerals
    and obesity, 2519
Diethanolamine, 1330
Diethylene glycol, 1333
Diethyleneglycol-bis-(allylcarbonate), 109
Diethyl ether, 1331
Diethyl pyrocarbonate, as food antimicrobial agent,
    205
Difference, 1094–1095
Difference equation, 1095
Difference machine, 1104
Difference of latitude, 1095
Difference of longitude, 1095
Differential
    mathematics, 1097
    vehicle, 1097–1098
Differential analyzer, 1095
Differential correction, 1095
Differential distillation, 1131
Differential equation, 1095–1096
Differential geometry, 1598
Differential leveling, 1096–1097
Differential manometer, 1097
Differential-mode voltage, 1097
Differential-producing flowmeters, 1436–1437
Differential pulley, 2208
Differential scanning calorimetry (DSC), 589, 590
Differential thermal analysis (DTA), 589
Differentiating network, 2444
Differentiating solvent, 3233
Differentiation (cells), 821
Differentiation (geology/mathematics/numerical),
    1098
Differentiation under the integral sign, 1098
Diffraction, 1098–1099
Diffraction bands, 1098
Diffraction grating, 1099
Diffuse characteristics, 1099
Diffuse mixing of organic solutions, microgravity
    experiment, 2340
Diffuse nebula, 2429
Diffuser, 1099–1100
Diffuse sky radiation, 1100
Diffusion, 1100
    in cells, 679–680
    facilitated, 680
Diffusion analysis, 1100

Diffusion coefficient, 1100
Diffusion current, 1100–1101
Diffusion equation, 2293, 2643
Diffusion equation (Einstein), 1101
Diffusion layer, 1101
Diffusion potential, 1101
Difluence, 1101
Digester, 1101
Digesters, in pulp making, 2882
Digestive system
    human, 163, 1101–1103
    other life forms, 1103
    ruminants, 1103–1104
Digit (computer system), 1119
Digital computer, 903
Digital computer systems, 1104–1116
Digital image, 258
Digitalis, 916, 1117
Digital micromirror device, 2332–2333
Digital multiplexer, 1117
Digital output, 1117
Digital radiography, 2468
Digital sound reproduction, 29–30
Digital subtraction angiography, 3797
Digital tachometer, 1298
Digital-to-analog converter, 156, 1117–1119
Digitate drainage, 1118
Digitizer, 908
Digitoxin, 916–917
Digoxin, 916–917
Digraph, 1119
Dihydric alcohols, 87
Dik-diks, 183, 184
Dike, 2077
Dike (dyke), 1119
Dilatancy, 1119
Dilatation, 1119
Dilation, 1119
Dilation number, 1119
Dilatometer, 1119
Dilaudid, 2393
Dill, 3566
Dilutional hyponatremia, 2062
Dilution refrigeration system, 2967–2968
Diluvium, 1119
Dimenhydrinate, 203
Dimensional analysis, 1119
Dimorphism, 1120
Dingo, 604
Dingy cutworm, 1011
Dinoflagellata, 1120
Dinoflagellate poisoning, 1468–1469
Dinosaurs, 1485
Dinostratus Quadratix, 2895
Diode, 1120, 1278
    in liquid crystal displays, 2149
Diode logic, 1120–1121
Diode transistor logic, 1121
Dioecious organisms, 1082
Dione, 3097
    orbital characteristics, 2761
Dionin, 2393
Diophantine approximation, 2507
Diophantine equation, 1121
Diopside, 1121
Dioptase, 1121
Dioptric system, 1121
Diorite, 1121
Dioxin, 1121–1122
Dip equator, 24
Diphenylhydramine hydrochloride, 203
Diphtheria, 352, 358–359, 1053, 1122

Diphyllobothriasis, 1470
Diploblastic, 1123
Diploid, 1123
Diplopoda, 1123
Dip needle, 1123
Dipnoids, 2200
Dipole, 1123
Dipole antenna, 186, 1123
Dipole moment, 1123
Dipole radiation, 1123
Dipper (bird), 2656
Dipper dredge, 1145
Diproton, 1123
Dip tank, 2618
Diptera, 1123–1124
Dirac, Paul Adrien Maurice, 1124
Dirac delta function, 1124
Direct-coupled amplifier, 148, 1124
Direct-current circuits, 1125
Direct current drives, 3143–3144
Direct current motors, 3138
Direct-current series motor, 2397
Direct-current shunt motor, 2397
Direct-distance dialing, 3436
Direct dyes, 1154
Direction (mathematics), 1126
Directional antenna, 187
Directional hydrophone, 1879
Directional luminous transmittance, 3518
Direction cosine, 1125–1126
Direct runoff, 1872
Direct sampling, 3060
Direct view storage tube, 908
Direct-vision prism, 2849
Dirichlet, Peter, 1126
Dirigibles, 1126
Dirofilariasis (heartworm disease), 1732
Disaccharides, 610
Discharge, 1234
    coefficient of, 1126–1127
    gaseous, 1127
Discone antenna, 187, 1127
Discontinuity, 1127
Discovery Program (NASA), 3327
Discovery Space Shuttle, 3250–3254
Discrete information source, 1928–1929
    capacity, 1930–1932
Discrete time series, 3494
Discrete variate, 1127
Discriminant, 1127
Discriminant function, 1127
Discriminator, 1127
Discriminatory analysis, 1127
Disease, 1922
Diseases (crop), 1127–1128
Disintegration (colloidal/nuclear), 1128
Disjoint subgraph, 3365
Diskette, 1128
Disk mill, 325
Disk storage, 1128–1129
Dislocation, 1129
    crystals, 997–999
Dislocation (bone), 506–507
Disorder pressure, 1129
Dispersal gettering, 1620
Disperse dyes, 1154
Dispersion, 869
Dispersion matrix, 1129
Dispersion (radiation/statistics), 1129
Dispersive power, 1129
Displacement current, 1129
Displacement measurement, 2817–2819

Displacement series, 40
Display technologies
    electroluminescent, 1258–1260
    field emission, 1384–1386
    flat panel, 1418
    inorganic light-emitting diode, 1941–1942
    interactive devices, 908
    liquid crystal, 2142–2162
    microdisplays, 2330–2335
    other, 1129–1130
    plasma panels, 2767–2771
    vacuum fluorescent, 3600–3601
Disposable contact lenses, 929
Disseminated gonococcal disease, 1661
Dissipation, 1131
Dissociating solvent, 3233
Dissociation, 1131
Dissolved gas-impeller agitation, 803
Dissonance, 1131
Distal, 1131
Distance marker, 1131
Distance modulus, 1131, 2678
Distillate, 1131
Distillation, 1131–1134
    crude petroleum, 2699–2701
    for desalination, 1058
Distillation, whiskey, 3762–3763
Distiller's malt, 2236
Distortion, acoustic/electromagnetic/optics, 1135
Distortion measure, 1929
Distributed amplifier, 148
Distributed capacitance, 605
Distributed capacitance (coil), 1135
Distributed control systems, 940–944
Distributed numerical control, 2507
Distribution (statistical), 1135
Distribution curve, 1135
Distribution-free methods, 1135
Distribution function, 1135
Distributive law, 1135–1136
Distributor-less ignition, 339
Ditch millet, 2351
Ditran, 1712
Diuresis, 1136
Diuretics, 1136
Divergence (mathematics), 1136
Divergence loss, 1136
Divergent series, 3137
Divers, 2187
Diversity reception, 1136
Diverticulitis, 1136–1137, 1612
Diverticulosis, 1136–1137
    and diet, 1090
Division, 1137
D-modules, 2299
DNA, 1589–1590, 2375
    bacteria, 351–352
    in chromosomes, 678
    histones associated with, 1772
    and human genome, 1829
    Recombinant, 1591–1592, 2375
DNA forensics, 1843
DNA replication, 678–679, 1592–1593
DNA technology, 434–457
DNA transcription, 1830
DNA virus, 679, 1588, 3643
Dobson fly, 965, 2445
Dobson spectrophotometer, 1137
Dobson unit, 1137
Dodders, 2636
Dodecanoic acid, 2108

Dodo, 874, 2746, 2747
Dog, 603
    rabies, 2910
    West Nile virus, 3748
Dogfish, 527, 3135
Dogwood shrubs and trees, 1137, 1138
Doherty amplifier, 148
Doldrums, 310, 1321
Dolerite, 1137
Dolly (cloned sheep), 821–824
Dolomite, 1138
Dolphins, 3754–3756
    marine fish, 1138
Domain, 1138, 2254, 3146
Domain structure, 1139
Dome, 1139
Dominance (ecology), 1139
Dominant epistasis, 1319
Domite, 1139
Donepezil, 127
Dopamine, 698
Doping, in semiconductors, 3130–3131
Doppler broadening, 553, 1139
Doppler effect, 1139–1140
Doppler-effect flowmeter, 1441
Doppler frequency, 1140
Doppler navigation system, 2428
Doppler radar, 1140
Doppler shift, 1140
Dorab, 1140–1141
Dormancy, seeds, 3124
Dormouse, 3022
Dosimeter, 1141
Dot cycle, 395
Double-beta decay, 2921
Double bond, 751
Double-dabble, 1141
Double decomposition, 1033
Double precision, 2844
Double salt, 3058
Double star, 1141
    binary stars, 424–425, 3661
    physical, 2744
Double-stream amplifier, 148
Doublet, 1141
Doublet lens, 1141
Douc, 2386
Dough rate of reaction, 2215
Douglas fir, 918, 1400, 1402–1403
Dove, 874, 2746
Dove prism, 2849
Down's syndrome, 1141–1142
Downwash, 56
Downwelling, 1142
Doxycycline, 197
Dracontic month, 1142
Draco (the dragon), 1142
Draft, 1142
Draft fan, 1353
Draft tube
    baffle crystallizer, 1002
    hydraulic turbine, 1860
Drag, 56–57
Drag effect, 1143, 1256
Drag fold, 1143
Drag line, 1143
Dragonets, 1143
Dragon fly, 1143, 2556
Dragon's blood, 2977
Drainage systems, 1143–1144
Drake, 2829–2830
Drallrohr Trocknung technique, 1151

Drawbridge, 550
Drawing, 1144
Dredge, 1145
Dreikanter, 1145, 1325
Drewite, 953, 1145
Driedfruit beetle, 1145–1146
Dried-fruit insects, 1145–1147
Dried-fruit moth, 1147
Dried-prune moth, 1147
Drift (geology/instrument), 1147
Drift (glacial deposits), 1627
Drift pin, 1147
Drilling mud, 2691
Drip irrigation, 2025–2026
Drizzle, 2840, 2841
Drogue, 1147
Drogue recovery, 1147
Drone bee, 1783
Drongo, 2656
Drop, 1147–1148
Drop attack seizure, 3125
Drop forging, 1475
Drosometer, 2842
Drosophila, 1148. *See also* Fruit fly
Dross, 1148
Droxtal, 1895
Drug discovery, 456
Drug-induced diabetes, 1072
Drum (fish), 984
Drumlin, 1148, 1627
Drumskin action, 1148
Drupe, 1518
Druse, 1148
Dry-adiabatic atmosphere, 47
Dry-adiabatic lapse rate, 309
Dry-adiabatic processes, 307
Dry aliabat, 1148
Dry-bulb measurements, 1882–1883
Dry-bulb temperature, 2878–2879
Dry eye, 1148
Dry fruits, 1519
Drying, 1148–1151
    paper, 2632
Dry-reed relay, 1151–1152
Dry rhinitis, 2995
D^2-statistic, 1152
Dubin-Johnson syndrome, 423, 1561
Dubin-Sprintz/Johnson disease, 423
DuBois-Reymond form, of mean value theorem, 2307
Duboscq colorimeter, 873
Duchenne's dystrophy, 2406
Duck, 180, 2829, 3715
Duckbill, 2387
Ducted-fan engine, 1152
Ductile fracture, 1152
Ductile iron, 107
Ductility, 1152
Dugong, 3119
Dulong and Petit law of specific heats, 1152
Dumas method for vapor pressure, 1152
Dumb compass (pelorus), 2673
Dummy, 1152
Dummy antenna, 187, 1152
Dumortierite, 1152
Dune, 1152
Dunnock, 2656
Duodenum, 2, 98
Duplexer, 1152
Duranickel, 108
Durbin-Watson statistic, 1153
Duricrust, 1153

Dusky raisin moth, 1147
Dust devil, 1516–1517
Dutch elm disease, 1287–1288
Duty cycle, 1153
Dwarfism, 1319, 2756
Dwarf stars, 1621–1622
Dwarf tapeworm, 3417
Dyadic, 1153
Dyes, for textiles, 1153–1155
Dynamical analogies, 1155
Dynamical friction, 1155
Dynamical mean sun, 1155
Dynamical parallax, 1155–1156
Dynamic balance, 1156
Dynamic characteristics, 1156
Dynamic consolidation, 125
Dynamic gain, 1544
Dynamic nonhierarchical network routing, 3436
Dynamic pressure, 57
Dynamic random access memory (DRAM),
    2311–2314
Dynamic relocation, 2859
Dynamic response, 2983
Dynamics, 1156
Dynamic scattering mode liquid crystal display, 2158
Dynamite, 1343
Dynamo, 1156
Dynamometer, 8, 1156–1157
Dyne (unit), 3569
Dynode, 1157, 1276
Dysbarism, 1157
Dyslexia, 1157
Dysmenorrhea, 1658
Dysphagia, 1326
Dyspnea, 1158
Dysprosium, 1158
Dystetic mixture, 1158
Dystrophic lake, 2138

e (transcendental number), 1331
Eagle, 1159–1160
Eagle mounting, of reflection grating, 2963
Ear, 325, 1720–1725
    and hearing, 1720–1725
Ear abscess, 5
Eardrum, 1721
    punctured, 1724
Earless monitors, 2178
Earth, 1160–1180. See also Moon
    atmosphere, 302–316
    effective radius of, 1213
    flattening, 1418
    orbital characteristics, 2759
    physical characteristics, 2760
    rate correction, 1181
    wind-Earth friction, 313–314
Earth axis, 1180
Earth current, 1180–1181
Earth dams, 1024
Earth inductor, 2228
Earthlight, 1181
Earth Observing System (EOS), 1181
Earth Observing System Data and Information
    System (EOSDIS), 1181
Earth point, 1181
Earthquakes, 1177, 1181–1195, 1183–1184
Earthquake swarm, 1183
Earth Radiation Budget Experiment (ERBE), 1181
Earth-rate unit (eru), 1181
Earth Science Enterprise, 1181, 3080–3081
Earthshine, 1181
Earth-star, 381

Earth tectonics, 1181–1195
Earth tide, 1198
Earthwork, 1198
Earthworm, 177, 574, 1198
Ear tick, 3487
Earwigs, 1052
Easterlies, 309
Easterly wave, 314
Eastern red cedar, 2043
Eastern tent caterpillar, 1198–1199
East Greenland current, 1199
Eberbach hardness, 1715
Ebert ion counter, 1199
Ebola hemorrhagic fever, 1199–1200
Ebola virus, 1391–1392
Ebonometer, 1200
Ebullioscopic constant, 502, 1200
Ebullism, 1200
Ebullition, 500–501
Eccentric, 1200
Eccentricity correction, in angular measurement,
    173
Ecdysis, 2399
Echelette, 1200
Echelle grating, 1200
Echelon, 1099, 1200–1201
Echinococcosis, 1854
Echinodermata, 1201
Echinoidea, 1201
Echo, 1201
Echocardiography, 170, 1201
Echo intensity, 1201
Echo-planar imaging, 2479
Echo power, 1201
Echo pulse, 1201
Echo sounder, 1201–1202
Echo suppressor, 1202
Eclipse
    lunar, 1206, 2388
    solar, 1202–1206, 3377–3378
Eclipse year, 1207
Eclipsing binary, 424, 1207
Ecliptic, 1207
Ecliptic system of coordinates, 1207
Eclogite, 1207
Ecological niche, 1208
Ecological pathology, 472
Ecological system, 1207
Ecology, 1207–1209
Ecosphere, 305
Ecotone, 1209
Ectoderm, 1291, 1614
Ectopic, 1210
Ectoprocta, 1210
Eczema, 104
Eddy, 1210
Eddy current, 1210
Eddy-current nondestructive testing, 2469
Eddy current probe, 14
Eddy-current transducer, 1210
Eddy viscosity, 1210
Edema, 1210
    eyelid, 1347
Edentata, 1210–1211
Edge, 1211
Edge dislocation, 998–999
Edge effect, 1211
Edgels, 258
Edge sequence, 1211
Edge tones, 1211

Edgeworth series, 1211
Edible snail, 2383
Edison, Thomas Alva, 1212
E-display, 1212
EDTA (ethylenediaminetriacetic acid), 727
Eel grass, 1212
Eels, 1212–1213
Eelworm, 1213
Effective atmosphere, 1213, 2576
Effective capacitance, 606
Effective data address, 1913
Effective exhaust velocity, 1213
Effective neutron cycle time, 1213
Effective propagation velocity, 1213
Effective radius of the Earth, 1213
Effective (RMS) electromotive force, 1268
Effective sound pressure, 1213
Effective temperature, 1213
Effective terrestrial radiation, 1213
Efficiency, 1214
    machine, 2209
Efflorescence, 1046, 1214
Effluent stream, 1214
Effusion, 1214
Effusive, 1214
Egg plant, 3214
Egret, 793, 1764
Eicosanoic acid, 220
Eigenfunction, 1214
Eigenvalue, 1214–1215
Eikonal equation, 1215
Einstein, Albert, 1215
Einsteinium, 37, 1215
Einstein Observatory (HEAO-2), 1215–1217
Einstein's diffusion equation, 1101
Einstein's emission coefficients, 1296
Einstein's heat equation, 1734
Einstein shift, 1217
Eja, 1217
Ekman layer, 1217
Ekman spiral, 1217, 2524
Elara, orbital characteristics, 2760
Elasmobranchs, 1403
Elastic cartilage, 636
Elastic collision, 866
Elastic constants and moduli, 1217
Elastic curve, 1217
Elastic fluid, 1444
Elasticity, 1217–1219
Elastic limit, 1219
Elastic scattering, 3109
Elastomers, 1219–1220
    bridge bearings, 544
E-layer, 1220
Elbow flowmeter, 1437
El Chicóon Volcano, 3676
Elder trees, 1220–1221
Electret, 1221
Electret microphone, 31
Electrical conductivity, 1221–1223
Electrical distance, 1223
Electrical ground fault circuit interrupters,
    1223–1224
Electrical instruments, 1224–1229
Electrical insulation, 1957–1958
Electrically erasable programmable read-only
    memory (EEPROM), 2312
Electrically programmable read-only memory
    (EPROM), 2312
Electrical power quality, 1229–1231
Electrical resistance, 2978
Electrical resonance, 2980

Electrical steels, 2019
Electric-arc lamp, 223
Electric cars, 1231–1233
Electric charge, conservation of, 925
Electric circuits, 1233–1234
Electric clock, 1234
Electric current units, 3568
Electric dipole, 1123, 1234
Electric discharge, 1234
Electric eel, 1700
Electric feedback, 1234
Electric field, 1234, 1384
Electric field strength, 1234
Electric field vector, 1234
Electric flux density, 1234
Electricity, 1235
Electric length, 1235
Electric motor, 2396–2399
Electric potential, 1235
Electric power, 1235
    chaos theory application, 2296–2298
    measurement, 1235–1239
    production and distribution, 1239–1251
Electric Power Research Institute (EPRI), 1240,
    1252
Electric power transmission, 1244–1250
Electric-resistance welding, 3746
Electric shock, 1252–1253
Electroacoustics, 30–31
Electroacoustic transducers, 976
Electroactive polymers, 3229
Electrocapillarity, 1254
Electrocardiography, 1254–1255
Electrochemical machining (ECM), 1255
Electrochemistry, 1255–1257
Electrochromic displays, 1130
Electrode, 1256, 1257
Electrodeless discharge, 1127
Electrodeposition, coatings, 2618
Electrodessication, 600
Electrodialysis, 1257
    for desalination, 1057
Electrodynamics, 1257, 2646
Electroencephalogram, 1257–1258
Electroendosmosis, 1258
Electrofocusing, 1280
Electroforming, 1258
Electrogalvanizing, 1565
Electrokinetic effects, 1258, 1394
Electrokinetic (zeta) potential, 1258
Electrokinetic transducer, 1258
Electroluminescence, 2191, 2593
Electroluminescent displays, 1258–1260
Electrolysis, 1256
    hydrogen fuel from, 1869
Electrolysis-type chemical analyzer, 1261
Electrolytes, 1223
    and dehydration, 1044
Electrolytic conductivity and resistivity
    measurements, 1261–1263
Electrolytic hygrometer, 1884–1885
Electrolytic transducer, 1263
Electromagnet, 1263–1264
Electromagnetic field, 1264, 1388
Electromagnetic force, 2645
Electromagnetic induction, 1264
Electromagnetic phenomena, 1264–1267
Electromagnetic pulse (EMP), 2243
Electromagnetic radiation, 1265, 1267
Electromagnetic spectrum, 1265, 1267
Electromagnetism, 1263
Electromechanical displays, 1130

Electrometer, 1267–1268
Electromotive force, 1268
Electromotive series, 40
Electromyography, 1268
Electron, 1268–1269, 2644
    binding energy, 425
    theory of, 1276–1277
    transport in solids, 3229, 3231–3232
Electron affinity, 1269
Electron avalanche, 1269
Electron beam, 1269
Electron beam lithography, 1269–1271
    coatings, 2619
Electron beam vacuum-evaporation process,
    1271–1272
Electron device, 1272
Electron diffraction, 1272
Electronegativity, 1272–1273
Electron emission, 1273
Electroneutrality, 1273
Electron gas, 1273
Electron gun, 1273
Electronically-controlled birefringence liquid crystal
    display, 2157–2158
Electronic balances, 3741–3742
Electronic data processing, 1273
Electronic gaging, 3475–3476
Electronic imagery, 2735–2736
Electronic nose, 2558
Electronic rectifiers, 2957
Electronics, 1273–1274
Electronic shunt, 2836
Electronics industry, 1274
Electron image tube, 663
Electron lens, 1274
Electron microscope, 1274–1276
Electron multiplication, 1276
Electron (photoelectron) spectroscopy, 1276
Electron spin, 2644
Electron spin resonance, 2476
Electron tube, 1277–1278
Electron volt, 1278, 3569
Electroosmosis, 1278
Electrophile, 2504
Electrophilic reaction, 1278
Electrophoresis, 1278–1279
Electrophoretic imaging displays, 1130
Electrophorus, 1280
Electroplating, 1280
Electropolishing, 1280–1281
Electrorheological fluids, 2166–2167
Electroscope, 1281
Electrosol, 1281
Electrostatic engine, 1995
Electrostatic generator, 1281
Electrostatic lens, 1281
Electrostatic microphone, 2342
Electrostatic precipitator, 1281
Electrostatics, 1281–1282
Electrostatic spraying, coatings, 2618
Electrostriction, 1282
Electrovalence, 750
Electrovalent compounds, 895
Electroviscous effect, 1282
Electrum, 1282
Element 100, 739
Element 104, 738
Element 105, 738
Element 106, 738–739
Element 107, 739
Element (graph), 1282
Elementary number theory, 2505–2506

Elements, 731–752. *See also* specific elements
    periodic table, 2678–2680
Elephant, 1282–1285
Elephant fish, 762
Elephant tree, 1285
Elevated pole, 1051
Elevator dredge, 1145
ELISA (enzyme-linked immunosorbent assay),
    testing blood supply for AIDS, 487
Elk, 1039, 1040
Ellipse, 1285
Ellipsoid, 1286
Ellipsoidal coordinate, 1286, 2896
Elliptical galaxies, 1547–1548
Elliptically polarized sound wave, 1286
Elliptical polarization, 1286
Elliptical system, 1286
Elliptic cylindrical coordinate, 1286
Elliptic geometry, 1597
Elliptic integral, 1286
Ellipticity (e), 1286
Ellipticity ratio, 1286
Elliptic partial differential equations, 2643
Elm trees, 1286–1288
El Niño, 317–318, 813, 1288–1290, 2082
Elongation, 1290
Eloxal process, 178
Elution, 1290
Elutriation, 1290
Eluvium, 1290
Embden-Meyerhof pathway, 613, 1647
Embolism, 232
Embrittlement, 1290
Embryo, 1290–1295
Embryo cloning, 819
Embryology, 823–824, 1295–1296
Embryonic fission, 1296
Embryo transfer, 823
Emerald, 417, 1296
Emery, 965
Emesis, 1296
Emetic, 1296
Emetine, 100–101
Emission (photoelectric), 1296
Emission coefficients (Einstein), 1296
Emission inventory, 1642
Emissive microdisplays, 2333–2334
Emissive power, 1296
Emissivity, 1296
Emittance, 1296
EMP (electromagnetic pulse), 2243
Emphysema, 1296–1297
Empress tree, 651
Empyema, 1297
Emu, 1297, 2955
Emulsion, 866, 868–869
Emulsion polymerization, 2378
Enamel, 2618
Enantiomers, 137
Enantiotropy, 1297
Enargite, 1297–1298
Enceladus, 3096–3097
    and *Cassini* mission, 648
    orbital characteristics, 2761
Encephalitis, 1298
Encephalon, 689–690
Encoder
    computer system, 1298
    electromechanical, 1298–1301
Encyclopedic knowledge, 261
Endangered species, 1301
    birds, 472

Endeavor Space Shuttle, 3257
End-effectors, robots, 3011
Endemic, 1301
Endive, 894
End moraine (terminal moraine), 3462
Endocarditis, 1301–1302
Endocarp, fruit, 1517
Endocrine glands, 1628
Endocrine system, 163, 1302–1303
Endoderm, 1291, 1614
Endogenic, 1303
Endogenous, 1303
Endometriosis, 1658
Endometrium, 1656
Endomorphism, 1303
Endorphins, 699
Endoscope, 1303
Endosmosis, 1303
Endothelium, 1303
Endothermic compounds, 895
End-point, of chemical reactions, 758–759
End-stage renal disease, 2066
Energine, 1306
Energy, 1303–1305
    conservation of, 924–925
    measurement, 1235–1239
Energy eigenstate, 3339
Energy level, 322, 1306
    selection rules, 3126
Energy state terms, 1306
Energy-transferring network, 2442
Enfleurage, 1421
Enhanced oil recovery methods, 2695–2696, 2698
Enkephalons, 699
Enrichment, 1306
Ensemble, 1306–1307, 3471
Ensign fly, 1307
Enstatite, 1307
Entemophilous plants (insect-pollinated), 2797
Enteric cavity, 1307
Enteric fever, 352
Enteritis, 1307
Enteron, 1307
Enterprise Space Shuttle, 3246
Enthalpy, 1307–1308, 3470
Enthalpy of solution, 1308
Entisols, 3211
Entner-Doudoroff pathway, 613
Entomology, 1308
Entomophagous parasite, 411
Entrance slit, 1308
Entrenched meander, 1308–1309
Entropy, 1309, 1322, 3470
    and energy, 1304–1305
    and information theory, 1928, 1929
Entry corridor, 1309
Enucleation, 1309
Envelope (mathematics), 1309
Envelope demodulator, 1046
Environment, 1309
Environment (controlled), 1310
Environmental lapse rate, 309
Environmental Protection Agency (EPA), 1310
    and acid rain, 20
Enzootic, 1310
Enzymatic catalysts, 652
Enzyme preparations, 1314–1315
Enzymes, 1310–1314
    in detergents, 1062
    growth of industrial, 449–451
    in neuron membrane, 696–697
Eocene, 1315

Eolian deposits, 1315
Eon (geologic time), 1596
Eophytic, 1315
Eosinophilia, 555
Eosinophils, 1315
Ephedrine, 101
Ephemeris second, 2306
Ephemeris time, 1315, 3493
Ephemeroptera, 1315–1316
Epicardium, 1316
Epicenter, 1185
Epichlorohydrin, 769, 1316
Epicontinental (marginal) seas, 1316
Epicyclic gear train, 1316
Epicycloid, 1316
Epidemic, 1462
Epidemic parotitis (mumps), 2404
Epidemiology, 1316–1317, 1462–1463
Epidermis, 1317
Epidiorite, 1317
Epidote, 1317, 2905
Epidural abscess, 6
Epigenetic, 1317
Epiglottitis, 1317
Epilepsy, 531, 1317–1318
    seizures, 3125–3126
Epinephrine, 49–50, 101
Epipelagic zone, 1318
Epiphora, 1318
Epiphyseal cartilage, 1318
Epiphysis, 1318
Epiphytes, 1318–1319
Epistasis, 1319
Epistaxis, 1319
Epitaxy, 1319
Epithelium, 1319
Epithermal neutrons, 2448
Epizootic, 1319
Epoch (geologic time), 1596
Epoxides, as food antimicrobial agent, 205
Epoxies, 47, 895
Epoxy resins, 1319–1320
Eppley pyrheliometer, 1320
Epsomite, 1320
Equal-area map, 1320
Equal-energy source, 1320
Equalizer (network), 1320
Equation, 1320
Equation of state, 1320–1321
Equation of time, 1321
Equator, 1321
Equatorial air, 72
Equatorial calms, 310
Equatorial coordinates (astronomy), 1321
Equatorial countercurrent, 1321
Equatorial stars, 927–928
Equatorial telescope, 1321
Equatorial trough, 310
Equatorial vortex, 312
Equatorial wave, 314
Equilibrium, 1321–1322, 3469
Equilibrium diagram, 1322–1323
Equilibrium distillation, 1131
Equilibrium states, 3340
Equilibrium vapor pressure, 3605
Equinoctial colure, 1323
Equinox, 1323
Equivalence principle, 1323, 1683
Equivalence theorem, 1323
Equivalent barotropic atmosphere, 307
Equivalent binary digits, 1119
Equivalent circuit, 1323

Equivalent electrons, 1323
Equivalent network, 2444
Equivalent potential temperature, 1323
Equivalent temperature, 1323
Equivocation, 1323
Era (geologic time), 1596
Erbium, 1323–1324
Erbium-doped optical fibers, 2568
Erg (unit), 3569
Ergodicity, 1324
Ergot, 267, 1324, 1531
Ergotamine, 1324
Eridanus, 1324
European red mite, 2362
Erosion, 1324–1325
Erratic (geology), 1325
Error, 1325–1326
Error, computer system, 1326
Error band, 1325–1326
Error correction codes, 1932–1933
Error detection, 1108
Error function, 1326
Error signal, 3164
Eruptive, 1326
Erysipelas (St. Anthony's fire), 1053, 1326
Erythema, 104, 1052, 1326
Erythrine, 1326
Erythrite, 1326
Erythrocytes, 482–483
Erythromycin, 197
Erythropoiesis, 167
Erythropoietin, 167
Escarpment, 1326
Eschar, 570
Escherichia coli
    diarrheagenic, 358
    use in biotechnology, 455
Escherichia coli 0157:H7 infection, 355, 359, 1461
Esker, 1326
Esophagus, 98, 1102, 1326–1327
    age-related disorders, 1611
Esophagus cancer, 1327
Esotropia, 1327
Essences, 1419–1427
Essential amino acids, 133, 134
Essential hypertension, 1888
Essential hypotension, 1892
Esterification, 1327, 2593
Esters, 1327–1328, 2581
    nomenclature and examples, 2586
Estimation, 1328
Estrogens, 3355
    and osteoporosis, 509
Estuary, 1328–1330
    hydrology of, 1877
Eta, 2321
Ethanal, 15
Ethane, 1330
Ethanoic acid, 15
Ethanolamines, 132, 1330–1331
Ethephon, 2767
Ether (substance postulated to fill space), 2971
Ethers, 1331, 2581
    chlorinated, 767
    nomenclature and examples, 2586
Ethology, 1331
Ethyl alcohol, 87, 1331–1332
Ethylbenzene, 2685
Ethyl cellulose, 1332
Ethyl chloride, 769
Ethylene, 1332–1333, 2685
    plant growth effects, 2767

Ethylenediaminetriacetic acid, 727
Ethylene dichloride, 769, 2685
Ethylene glycol, 87, 1333, 2685
Ethylene oxide, 1333–1334, 2685
   as food antimicrobial agent, 205
Ethylene-propylene elastomers, 1220
Ethylene-vinyl acetate (EVA) copolymers, 1334
Ethyl methacrylate plastics, 35
Ethyne, 17
Etidronate, 509
Etiolation, 1334
Etiology, 1334
Etymol, 3229
Eucalyptus trees, 1334–1335
Euclase, 1335
Euclid, 1335
Euclidean geometry, 1597
Euclidean space, 1335
Euclid's algorithm, 96
Eudiometer, 1335
Euglenophyta, 90
Eukaryotes, 675
Eulachon, 3188
Euler, Leonhard, 1335
Euler angle, 1335
Euler formula, 875
Eulerian coordinates, 946, 1335
Euler-Mascheroni constant, 1335
Euler's first integral, 420
Eunucleated host cell, 819
Euonymous scale, 3204
Euphorbiaceae, 1335–1336
Euphotic zone, 1208, 1336
Europa, 1558, 2050–2051
   orbital characteristics, 2760
European corn borer, 512, 1336
European fieldfare, 3483
European fruit scale, 3204
European Superlaser, 2096
Europium, 1336
Eurotunnel ("channel"), 3541–3545
Eustachian tube, 325, 1336
Eutaxic, 1336
Eutectic, 1336
Eutectoid, 1337
Eutrophic lake, 2138
Evacuated-tube solar collectors, 3218
Evaporation, 1337–1338
   boiler feedwater, 1360
   clouds, 825–826
Evaporative crystallizer, 1002
Evaporite, 1338
Evapotranspirometer, 2842
Evection, 1338
Even-even nuclei, 1338
Event, 1338
Evergreen bagworm, 367
Evolute, 1338
Evolution, 1338–1339
   and population senescence, 1614
   strategies, 255, 256
Evolutionary biology, 1338
Evolutionary computing, 242, 255–256, 263
Exacerbation, 1340
Exact differential, 1097
Excess-three code, 855
Exchange (particle), 1340
Exchange degeneracy, 1340
Exchange energy, 1340
Exchange force, 1340
Excimer lasers, 2096
Excitation, 1340

Excitation curve, 1340
Excited atom, 322
Excited state, 1340
Exciter, 1340
Exciting current (transformer), 1340
Excitonic matter, 3229
Excitron, 2957
Exclusion chromatography, 784
Exclusive OR circuit, 1340–1341
Excretion, 1341
Excretory system, 2, 163, 1341
Executive routine, 3043
Exit slit, 1341
Exocrine glands, 1628
Exogenetic, 1341
Exosmosis, 1341
Exosphere, 1341
Exothermic compounds, 895
Exotic nuclei, 2646, 2923
Exotropia, 1341–1342
Expanded film, 1391
Expanded polystyrene, 1457
Expander (signal), 1342
Expansion joint, 1342
Expansion turbines, 1573, 1576–1577
Expectation-Maximization algorithm, 262
Expected value, 1342
Experimental design, 1060
Expert systems, 42, 251–252
Explosive, 1342–1343
Exponent, 1344
Exponential distribution, 1344
Exponential smoothing, 1344
Exposure meters, 2730–2731
Expression (mechanical), 1344
Exsecant function, 3528
Exstrophy, 2062
Extended-wear contact lenses, 929
Extender pigment, 2617
Extensiometer, 1344
Extensive air shower, 967
Extensive sampling, 3060
Exterior ballistics, 368
External ear, 1721
External storage, 3358
Exteroceptors, 3136
Extinction, 1301. *See also* Mass extinction
Extinction coefficient, 1344
Extracapsular cataract extraction (ECCE), 656, 1345, 2615
Extraction (liquid-liquid), 1345
Extractive distillation, 1133
Extractive metallurgy, 1345
Extraembryonic membranes, 1346
Extranuclear inheritance, 1762
Extraordinary index, 1346
Extraordinary ray, 1346
Extrapolation, 1979
Extrapolation number, 2403
Extrasolar planets, 2759, 2761–2762
Extraterrestrial life, 1346
Extratropical cyclone, 1508, 1509
Extremely high pressure research, 3230
Extruding, steel, 2014
Extrusion, 1346
Extrusive, 1346
Exudative diarrhea, 1080
Eye (vertebrate), 1347
   itchy, 2032
   and vision, 3647–3660
   visual function, 3661
Eye bar, 1346

Eyelid, 1346–1347
Eyepiece, 1347
Eye tumors, 3653
Eye wall resection, 778

Fabric (geology), 1349
Fabry interferometer, 1969
Face-centered crystal, 1000
Faceted shading, 910
Facies, 1349
Facies, stratigraphic, 1349
Facilitated diffusion, 680
Facing, 1349
Facioscapulohumeral dystrophy, 2406–2407
Facsimile transmission, 1349
Factor, 1349–1350
Factor analysis, 1350
Factorial, 1350
Factorial experiment, 1350
Factor of safety, 1350
Factor theorem, 96
Faculae, 1350
Facultative anaerobe, 153, 351
Fading (communications), 1350–1351
Fagot, 1351
Fahlore, 3464
Fahrenheit temperature scale, 1351, 3458
Failure (structural), 1351
Fainting (syncope), 3408
Fairing, 1351
Falcon, 1351–1352
Falconiformes, 1351–1352
Fall armyworm, 226
Fallopian tubes, 2, 1290–1291
   blocked, 1923
   infection, 1657–1658
Fallout, radioactive, 1352
Fall webworm, 1199
False canines, 604
False chameleon, 1900–1902
False cirrus clouds, 828
False cleavage, 1352
False club tail lizards, 2176
Faltung, 945
Familial hypercholesterolemia, 233, 2171
Fan, 1352–1353
Fanglomerate, 1354
Fanned-beam antenna, 187
Fanning beam, 1354
Farad (unit), 605, 3568, 3569
Faraday, Michael, 1354
Faraday dark space, 1354
Faraday disk machine, 1354
Faraday effect, 1265, 2229
Faraday law of electromagnetic induction, 1354
Far-infrared region, 1935
Far point of the eye, 1354
Farsightedness (hyperopia), 1887–1888, 3649–3650
Fascia, 1354
Fast breeder reactors, 3576
Fastness, 1155
Fast neutrons, 2447
Fathom curve (isobath), 2027
Fatigue (metals), 1354
Fats, 1412
   and diet, 1088
   and hypertension, 1091–1092
   industrial chemicals derived from, 454
Fat substitutes, 491–492
Fatty acids, 625
   chlorinated, 767

Fault breccia, 536
Faults, 1185–1186
Fauna, 1355
F-display, 1355
Featherbacks, 1355
Feathering (aircraft), 1355
Feather key, 2060
Feather mite, 2363
Feather star, 982
Feature extraction, 258
Febrile, 1355
Fechner colors, 1355
Fechner fraction, 1355
Feedback, 1355–1356
    in automation, 331
Feedback amplifier, 148, 1355, 1356
Feedback control, 931, 935, 1356
Feedback oscillator, 1356
Feedback signal, 3164
Feeder
    gravimetric, 1356–1357
    parts/vibratory, 1357
    volumetric, 1357–1358
Feedforward control, 931, 1358
Feedstuffs
    antibiotics in, 197
    genetically engineered, 1586
Feedwater (boiler), 1358–1361
Feldspar, 1361–1362
    in ceramics, 710
Feldspathoids, 2434
Felsite, 1362
Female infertility, 1923
Female sex hormones, 1656, 1793, 2755
Femic, 1362
Femto (prefix for units), 3568
Fence lizard, 3403
Fennel, 3566
Fergusonite, 1362
Fermat, Pierre de, 1362
Fermat principle, 1362–1363
Fermat's last theorem, 1363, 3465
Fermat's spiral, 2633
Fermentation, 1331, 1363, 2591
    whiskey, 3762
Fermentation industry, 440–441
Fermi, 1364
Fermi, Enrico, 1364
Fermi-Dirac statistics, 1340, 1364
Fermi energy, 3385
Fermi level, 3232
Fermions, 516, 1364
Fermi plot, 420
Fermi resonance, 1364
Fermi selection rules, 1364
Fermi surface, 1364–1366
Fermium, 37, 1366
Ferns, 1366–1368
Ferreed, 1368
Ferrel law, 312
Ferric ammonium alum, 116
Ferricrete, 1153
Ferrielectric, 202
Ferrimagnetism, 1368
Ferrite, 1368, 2016
Ferrite circulator, 798
Ferritic stainless steels, 785
Ferroelectric effect, 1368–1369
Ferroelectric liquid crystal display, 2157
Ferroelectric materials, 1369
Ferromagnetic materials, 2225, 2636
Ferromagnetic minerals, 2355

Ferromagnetism, 1369
Fertile material, 1369
Fertilizer, 1369–1373
Fescue, 1679
Fetch, 316
Fetus, 1291, 1292–1293
Fever, 1374
Fever blisters, 1053, 3150
Fiber, and diet, 1089
Fiber glass, 1375–1377
Fiber optics. See Optical fiber systems
Fiber optic thermometers, 2919–2920
Fiber-reinforced composites, 1377–1378
Fibers, 1378
    generic designations and definitions of synthetic,
        1379–1382
Fibonacci numbers, 1382
Fibril, 1382
Fibrillation, 1382–1383
Fibroadenoma, 536
Fibroblast growth factor (FGF), 705
Fibrocartilage, 636
Fibroma, 1383
Fibroneurosarcoma, 599
Fibrosarcoma, 599
Fibrovascular bundle, 567–568
Fictitious, 1383
Fictitious year, 1383
Fidelity (communications), 1383–1384
Fiducial inference, 1384
Fiducial mark, 1384
Field, 1384
Field (mathematics), 1386
Fieldbus Messaging System, 2252
Field-effect transistors, 156
Field-effect transistor switch (FET), 1384
Field emission displays, 1384–1386
Field equations, 1387
Fieldfare, 2656
Field intensity, 1386
Field of view, 1386–1387
Field stop, 1387
Field strength, 1387
Field theory, 1387–1389
Fig scale, 3204
Fig trees, 2401
Figure of merit, 1389
Fig wasp, 720, 1147
Filament, 1389
Filariasis, 1389–1390
Filar micrometer, 1390–1391
File, 1391
Filled-system thermometer, 3473
Fillet, 1391
Filling fibers, 1378
Film (structure), 1391
Film boiling, 501
Film-compensated supertwisted-nematic LCD, 2147
Filoviruses, 1391–1392
Filter (communications system), 1392–1393
Filterable viruses, 3642
Filter-thickener, 1395
Filtration, 1393–1396
    water, 3712
Finch, 1397, 1505
Fineness ratio, 1397
Fine structure, 1397
Finger lake, 1397
Finger millet, 2350–2351
Finite element analysis, 1397–1399
    automation application, 334

Fiord, 1399
Fire ants, 181–182
Fireball, 1399
Fire brat, 1399
Fire-brick, 1399–1400
Fire bug, 1715
Fireclay, 1400
Firecrest, 3697
Fire damp, 2328
Fire eye, 2656
Firefly, 1400
Firetail, 2959
Firewall, 1400
Firming agents (foods), 1400
First-degree burn, 570
First in, first out (FIFO), 2907
First law of thermodynamics, 3470
First-order reaction, 757
First-order system, 1400
First quarter moon, 2387
Fir trees, 1400–1402
Fischer, Edmond H., 1403
Fischer-Tropsch synthesis, 844
Fish, 1403–1410
    fossils, 1479–1480
Fishbone antenna, 187
Fisher's z distribution, 1403
Fish farming, 214
Fish fly, 1411
Fish louse, 946, 1411
Fish meals, oils, and protein concentrates, 1411,
    2685
Fish tapeworm, 3417
Fission
    biology, 1411
    nuclear, 2474–2475
Fissure in ano, 1411
Fistula, 1411–1412, 2062
Fittig reaction, 1412
Fix, navigational, 1412–1413
Fixed bed, 1412
Fixed oils, 1412
Fixed-point arithmetic, 1412
Fixed-point arithmetic (computer system), 1412
Fixed-point system, 2783
Fizeau, Hippolyte, 1413
Fizeau experiment, 1413
Flagellates, 90, 1413
Flagellum, 1413
Flame cutting, 1413
Flame hardening, 1413–1414
Flame photometry, 1414
Flame retardant ABS, 4–5
Flame-retarding agents, 1414
Flame spectrometry, 1414
Flame tree, 2783
Flaming arc, 221
Flamingo, 1414–1415, 2708
Flange, 1415
Flares (Sun), 3380
Flare stars, 1415
Flash distillation, 1131
Flash flooding, 1510
Flashing (thermal), 1415
Flash point, 1415
Flash spectrum, 3377
Flatfishes, 1415–1418
Flatheaded apple tree borer, 512
Flat lizards, 2176
Flat panel display technology, 1418
    for television, 3456–3457
Flat-plate area (equivalent), 1418

Flat-plate solar collectors, 3216–3218
Flattening of the Earth, 1418
Flat-thin cathode ray tubes, 1129–1130
Flatulence, 1419
    and beans, 3236
Flatworms, 25, 2774, 3545–3546
Flavonoids, 1419
Flavor enhancer, 1426–1427
Flavor potentiator, 1427
Flavors, 1419–1427
Flax, 383, 1382, 1428
F layer (magnetosphere), 2229
Flea, 1428–1429
Flea-beetle, 407, 1428
Flea bites, 1054
Fleming, Sir Alexander, 1429
Flesh-eating bacteria, 359–360
Flexible manufacturing systems (FMS), 330, 334,
    335–336
Flexure, 1429–1430
Flight data recorder, 1430
Flight envelope, 1430–1431
Flight path, 1431
Flint, 1431
Flinty crush-rock, 1431
Flip-flop, 1431–1432
Flipper, 1432
Floaters (in eye), 1432
Floating amplifier, 148, 635, 1432
Floating controller, 1432–1433
Floating-point arithmetic, 1433
Floating-point system, 2783
Flocculi, 1433
Floccus clouds, 828
Flooding irrigation, 2022
Flood plain, 1433
Floods, 1176
Floor beam, 1433
Floppy lungs, 1297
Florican, 2939
Florida current, 1433
Florida holly, 3114
Florida red scale, 3204
Flotation, 803
Flounder, 1415–1416, 1418
Flour mite, 2362
Flowage (rock), 1433
Flow chart, 1433
Flowcharting, 1110
Flow diagram, 1433
Flower, 1433–1435
Flower fly, 411
Flower pecker, 2656
Flow injection analysis, 781
Flow measurement (liquids and gases), 1435–1444
Flow nozzles, 1436
Flow stress, 1444
Flow structure, 1444
Fluctuation, 1444
Fluctuation noise, 1444
Fluff louse, 2190
Fluid, 1444–1447
Fluid catalytic cracking, 977
Fluid flow, 1444–1447
Fluid friction, 1448
Fluidic amplifier, 149
Fluidic flowmeter, 1440
Fluidics, 1448–1449
Fluidity, 1445
Fluidization, 1449
Fluidized-bed combustion, chaos theory application,
    2297–2298

Fluidized fixed bed, 1412
Fluid parcel, 1449
Fluid resistance, 2978
Fluke, 1449, 3524–3525
Flume, 1449
Fluorescein angiography, 1449–1450
Fluorescence, 1450, 2191, 2711
    minerals, 2355
Fluorescence microscope, 2347
Fluorescence yield, 1450
Fluorescent lamps, 1900–1901
Fluorescent screen, 1450
Fluorescent whitening agents, 1062
Fluoridation, 1451
    and caries, 631
Fluoride, removal from water, 3713
Fluorine, 1450–1452
Fluorite, 1452
Fluoroantimonic (magic) acid, 23
Fluorocarbon, 1453
Fluorocarbon fibers, 1379–1380
Fluorocarbon polymers, 1453
Fluoroelastomers, 1220
Fluorometers, 1453
Fluoroplastics, 1453–1454
Fluorosis, 993
5-Fluorouracil, 203, 596, 600
Flushing, 46
Flutter, 1454
Flutter rate, 1454
Fluvial, 1454
Flux (luminous/physics/slag/solder), 1454
Flux-density threshold, 3482
Flux-gate magnetometer, 2229
Flux guide, 1454
Fluxmeter, 2228
Flux refraction, 1454
Fly, 1123–1124, 1454
Fly-back, 1454
Flying dragons, 65
Flying fishes, 1454
Flying hatchet fish, 725, 1454
Flying lemur, 1056
Flying spot scanning, 664
Flying squid, 3119
Flywheel, 1454–1455
F-number, lens, 210
Foam, 866, 1455–1456
Foamed ABS, 5
Foamed plastics, 1457
Focal collimator, 1457
Focal length, 1457
Focal point, 1457
Focal power, 2361
Focal seizure, 3126
Focometer, 1457
Focusing, 1457
Focusing collision, 1457
Fog, 1457–1458
Fog bow, 319
Fog tracks, 1458, 2840
Fokker-Planck equation, 1458
Folding (aliasing) error, 97
Folic acid, 1458–1459
Folic acid coenzymes, 859–860
Folium of Descartes, 1460
Follicle, 1460
Follicle cell, 1460
Fomalhaut (α Piscis Austrini), 1460
Fontanelle, 1460
Food
    antimicrobial agents, 203–205

bodying and bulking agents, 491–492
carbohydrates in, 615
carrier (food additive), 635
chelates used in, 728–729
coating agents, 850
colorants, 175
extrusion, 1346
firming agents, 1400
foodborne diseases, 1460–1471
fortification with amino acids, 133
hydrogen peroxide application, 1871
intermediate-moisture, 1970
irradiated, 2020–2021
staling, 3322–3323
Food acidulants, 23–24
Food additives, 44–45
Food alkalizers, 23–24
Food allergies, 104, 1092
Food anaphylaxis, 1092
Food begging, birds, 471
Food chain, 1471
Food colorants, 175, 871–872
Food crop. *See* Crop
Food Guide Pyramid, 1087
Foot, 1473
    trench foot, 3525
Foot-and-mouth disease, 1471–1473
Footcandle (unit), 3569
Footing, 1473
Footlambert (unit), 3569
Forage grasses, 1676–1680
Foramen, 1473
Forbe's disease, 2410
Forbidden band, 3231–3232
Forbidden line, 1473
Forbush decrease, 1473
Force, 1473–1474
Force-balance transducer, 1474
Ford Motor Company, automotive electronics
    evolution, 337–343
Foreshock, 1185, 1190
Forest tent caterpillar, 1199
Forging, 1474–1475
    steel, 2014
Forktail, 2656
Formaldehyde, 88, 1476–1477, 2685
Formal solution, 2365
Formation (ecology/geology), 1477
Form factor, 1477
Formic acid, 1477
FORMTOOL system, 250
Fornax galaxy, 1813
Fort Bragg fever, 2122
FORTRAN, 892, 1477
Forward breakover voltage, 3172
Foscarnet, anti-AIDS drug, 34
FOS process, 2013
Fossil ice, 1895
Fossils, 1477–1486
    invertebrate, 1987–1990
Foucault, Jean, 1487
Foucault pendulum, 1487
Foucault rotating mirror, 1487
Foundations, 1487
Four-address instruction, 1957
Four-color map theorem, 1487, 1488
Four-cycle engine, 1487–1488
Fourdrinier machine, 2630–2631
Four-eyed fishes, 1488
Fourier, Jean Baptiste Joseph, 1488
Fourier analyzer, 3304–3305
Fourier series, 1488

Fourier transform, 1488–1489
    in chemical analysis, 160
    and musical sound, 28–29
Fourier transform mass spectrometry, 2290
Fovea, 1489
Foveal vision, 1489
Fowl (chickens), 2824–2832
Fowl tick, 3487
Fox, 603–604
Foxglove, 1117, 3119
Foxtail, 1679
Foxtail millet, 2351
Fractal geometry, 1489–1491
    and chaos theory, 2298–2299
Fraction, 1491
Fractional distillation, 1131
Fractional quantized Hall effect, 1709
Fractography, 1491
Fracture, minerals, 2355
Fracture stress, 1492
Fractus clouds, 828
Frame (sampling theory), 1492
Framed structure, 1492
Frames, in expert systems, 252
Francis turbine, 1860
Francium, 1492
    alkali metal, 98
Frankincense, 517, 2978
Franklin, Benjamin, 1492
Franklin antenna, 187
Franklinite, 1492–1493
Fraunhofer diffraction, 1098–1099
Fraunhofer hologram, 1778
Fraunhofer lines, 9, 1493
Fraunhofer region, 1493
Frazil crystals, 1895
Freedom, degrees of, 1043, 1493, 3006
Free-electron lasers, 2094, 2096
Free electron theory of metals, 1493
Free energy, 1493
Free energy change, 1493–1494
Free fall, 1494
Free gyro, 1494
Free machining, 1494
Free-machining steels, 108
Free molecule flow, 1494
Free radical, 1494
    and aging, 428
Free radical addition, 2589
Free-radical scavengers, 207
Free rotation, 1494
Free space, 1494
Free surface, 1494
Free turbulent flow, 1494–1495
Free vector, 63
Free volume, 1495
Freeze-concentrating, 1495
Freeze-drying, 1495–1498
Freeze-preserving, 1498–1499
Freezing curve, 1499
Freezing index, 1043
Freezing point, 1499
Freezing-point depression, 1499
Freezing season, 1043
Frenkel defect, 1500
Frequency, 1500
    alternating currents, 111
Frequency band, 1500
Frequency changers, 2836
Frequency demodulator, 1046
Frequency deviation, 1500
Frequency distortion, 1135

Frequency distribution, 1500
Frequency divider, 1500
Frequency-division multiplex, 1500
Frequency doubler, 1500
Frequency drift, 1500
Frequency function, 1502
Frequency measurement, 1501–1502
Frequency monitor, 1502
Frequency polygon, 1502
Frequency response, 1502–1503
Frequency-shift keying, 1503
Frequency swing, 1503
Frequency synthesizer, 3162, 3163–3164
Frequency translation, 1503
Freshwater environment, 1207
Fresnel, Augustin Jean, 1503–1504
Fresnel-Aragon law, 1503
Fresnel diffraction, 1098, 1504
Fresnel hologram, 1778
Fresnel lens, 3241
Fresnel mirror, 1504
Fresnel region, 1504
Fresnel zone, 1504
Friability, 1504
Friction, 1504
    wind-Earth, 313–314
Frictional electricity, 1504
Friction damping, 973
Friction layer, 2758
Friction welding, 3746
Friedel-Crafts reaction, 1504–1505, 2590
Friedreich's ataxia, 301, 3118
Frigate-birds, 2671, 2672
Frilled lizard, 65
Fringillidae, 1505
Fritillary, 1505
Frog, 146, 210, 1505–1507
Frogfishes, 172
Frog hopper, 1505
Frogmouths, 608
Fröhlich's syndrome, 2755
Frontal bone, 1507
Frontal fog, 1458
Frontal inversion, 309
Frontal lobes, 693
Frontal type dementia, 127
Frontal wave, 314
Front focal length, 1507
Frontogenesis, 1508
Fronts, 1507–1517
Frost, 2840–2841
Frostbite, 1892
Frosted scale, 3204
Frost point, 2839
Frothers, 803
Froude, William, 1517
Froude number, 1517
Frozen-in-field, 1517
Fructose, 611–612, 3399
Fruit, 1517–1520
Fruit fly, 1148, 1520–1521
    homeobox, 2375
Fruit webworm, 3738
Fruitworm, 1522
Frustrum, of cone, 915
Fuch's dystrophy, 959
Fuel, 1522
Fuel cells, 1522–1528, 2835
Fuel rods, nuclear reactor, 2483
Fugacity, 1528
Fugu, 2880
Fulgurite, 1528

Full annealing, 1528
Full-bodied flavors, 1419
Fullerenes, 616, 618–620
Fuller's earth, 51, 1528. *See also* Diatomaceous earth
    decolorizing agent, 1032
Full linear group, 2125
Full moon, 2387
Fulminating, 1528
Fumarole, 1528
Fume, 1528
Function, 1528–1529
Functional analysis, 1529
Functional anatomy, 163
Functional genomics, 1842
Functional group analysis, 161–162
Functional isomers, 755
Functional programming languages, 2859
Functional reserves, 1529
Fundus, 1529
Fungal arthritis, 237
Fungal meningitis, 2315
Fungus, 1529–1533
Fungus gnat, 1533
Funicular polygons and catenaries, 1533–1534
Funiculus, 1534
Fur, 1534
Furan and related compounds, 1534–1535
Furfuraldehyde, 1535
Furnace carbon, 621
Furrow irrigation, 2022, 2023
Fuse (electric), 1535
Fusee, 3315
Fusion
    heat of, 1535
    nuclear, 1535
    phase change, 1535–1536
Fusion power, 1536–1541
Fusion welding, 3745–3746
Fuzzy logic, 253, 1115
Fuzzy reasoning, 252–254
Fuzzy set, 252–253

GABA. *See* Gamma-aminobutyric acid
Gabbro, 1543
Gabor hologram, 1777
Gadget (code name for first atomic bomb), 2242
Gadolinium, 1543
Gadwall, 3716
Gage, 1543
Gage line, 1543
Gahnite, 1543
Gain (magnitude ratio/transmission), 1544
Gain bandwidth product, 1543
Gait disorders, of elderly, 1612
Gal (unit), 3569
Galactose, 3399
Galactosemia, 1544
Galago, 2119
Galaxiids, 1544
Galaxy, 968, 1544–1552
    formation, 3572–3573
    Hubble Space Telescope images, 1806–1808, 1811–1812
    Hubble Space Telescope images of colliding, 1810, 1813
    Origins Program studies formation, 2597
Galaxy clusters, 1552
Galena, 1553
Galilean telescope, 1553
Galilean transformation, 1553, 2972
Galileo, Galilei, 1553–1554

Galileo mission to Jupiter, 275, 1554–1560, 2044
    Venus flyby, 3627–3628
Galileo number, 1561
Gall (botany), 1562
Gall aphis, 211
Gallbladder, 2, 421
    age-related disorders, 1611
    diseases, 1561–1562
Gallery forest, 432
Gall gnat, 1563
Galliformes, 1563
Gallium, 1563–1564
Gallium arsenide, 230–231
Gallium arsenide power sources, 3132–3133
Gallon (unit), 3569
Gallows telephone (Bell's first telephone), 3423
Gallstones, 421, 1561–1562
    and cirrhosis of liver, 2171
    in elderly, 1612
Gall wasp, 1564
Galvanic cell, 1564–1565
Galvanic corrosion, 962
Galvanizing, 1565
Galvanometer, 1565
Game of Life, 245
Game playing systems, 242, 254–255
Gamete, 1565–1566
Gametophyte, 1566
Gamma-aminobutyric acid (GABA), 698
    and epilepsy, 531
Gamma bursts, 1567
Gamma distribution, 2670
Gamma function, 1566
Gamma radiation, 1566, 2921
Gamma-ray astronomy, 1566–1567
Gamma-ray spectroscopy, 1567–1568
Gamma space, 1568
Gamow, George, 1568
Gamow-Teller selection rules, 1568
Ganglion, 1569
Gangliosides, 1569
Gangrene, 1053, 1569–1570
Gangue, 1570
Ganister rock, 1570
Gannets, 2671, 2672
Gantry, 1570
Ganymede, 1558, 1559, 2049–2050
    orbital characteristics, 2760
Gap, 1570
Gar, 1571
Gardener-bird, 527
Garden webworm, 3738
Garlic, 105, 2559
Garnet, 1570–1571
Garnierite, 1571
Gas, 1571
    flow measurement, 1435–1444
    ionized, 1999–2000
    kinetic theory, 323n, 2070–2071
    removal of dissolved from water, 3713
Gas absorption, 8–9
Gasahol, 1331
Gas analyzers
    combustion type, 1571–1572
    thermal-conductivity type, 1572–1573
Gas burners, 571
Gas calorimeter, 590
Gas chromatography (GC), 781–784
    with mass spectrometry, 2290
Gas compression, 69–71, 896–897
Gas constant, 1578
Gas discharge, 1578

Gas electrode battery, 394
Gaseous diffusion, for uranium isotope separation, 3578
Gaseous discharge, 1234
Gaseous film, 1391
Gas gangrene, 352, 1569
Gas lasers, 2093
Gas laws, 2677
    ideal, 1578, 1897–1898
Gas oil, 2701
Gasoline, 2701
Gaspra, 275, 301
Gas Research Institute (GRI), 1578
Gas scrubbing, 1578
Gasteropoda, 1578
Gastric ulcers, 3556
Gastrin, 699
Gastritis, 1578–1579
Gastroenteritis, 1307, 1579
Gastrointestinal cancer, 597
Gastrotricha, 1579
Gastrula, 1579
Gas turbines, 1573–1577, 1971
Gas welding, 3746
Gate (computer system), 1579
Gate-array with SRAM, 2314
Gate circuit, 1579
Gate-turnoff switch, 1579
Gate-turnoff thyristors, 3143
Gating, 1579
Gaucher's disease, 1579–1581, 2141
Gauge theories, 1581
Gaur, 520
Gauss (unit), 3569
Gauss, Carl Friedrich, 1581
Gauss error function, 1326
Gaussian (normal) distribution, 2470, 2898
Gauss law, 1282
Gauss-Markoff theorem, 1581
Gaussmeter, 2228
Gauss principle of least constraint (motion), 2396
Gauss theorem, 1581–1582
Gavials, 984, 985, 986
Gaviiformes, 1582
Gayal, 520
Gay-Lussac law (Charles law), 727
Gazelle, 183, 184
Gazelle-goat, 1649
G-display, 1582
Gear pumps, 2885
Gears, 2209
Gear train, 1582, 2209
Geaster, 381
Geckos, 1582–1583
GEE navigation, 2428
Geese, 180, 2828–2829, 3716
Gegenschein, 1583–1584
Geiger, Johannes Wilhelm Hans, 1584
Geiger counter (Geiger-Müller counter), 1584
Gel, 866, 869
Gelatins, 865–866
    film emulsions, 2723–2725
Gel permeation chromatography, 784
Gemini Program, 3021
Gemini (the twins), 1584
Gemsbock, 183, 184
Gem stones, 1584–1585
GenAID expert system, 252
Generalized coordinates, 1585
Generalized gangliosidosis, 2142
Generalized transmission function, 1585
General precession, 1585

General-purpose register, 2970
General relativity, 2972–2973
Generating function, 1585
Generations, alternation of, 113–114
Generative calculi, 248
Generic name, of chemicals, 2582
Genes, 679. *See also* Mutation
    defined, 1588–1590
    instability, 1591
Gene science, 1587–1593
Genet, 3670
Genetic algorithms, 242, 255–256
Genetic code, 676–677, 1589, 1830
Genetic diseases, 1592
Genetic engineering, 1585–1587, 1590, 1918–1919
    pollution cleanup applications, 3704–3705
Genetic expression, 679
Genetic programming, 256
Genetic recombination, 1588
Genetics, 1587–1593
    and senescence, 1618–1619
Genghis robot, 245, 247
Genital herpes, 3150–3151
Genital warts, 3150
Genomics, 439–440, 1842
Genotype, 1594
Gentamicin, 196
Geocentric coordinates (astronomy), 1594
Geocentric parallax, 1594
Geocentric zenith, 3810
Geochemical dating methods, 2931
Geochemistry, 1171
Geochronology, 1181
Geocorona, 303
Geocronite, 1594
Geode, 1594
Geodesic, 1594
Geodesy, 1169–1171
Geodetic astronomy, 1171
Geodetic coordinates, 1594
Geodetic datum, 1594
Geodimeter, 1594
Geoduck, 1594
Geogale, 2379
Geographical zenith, 3810
Geographic coordinates, 1594
Geography, 1594
Geoid, 1169–1170, 1595
Geologic time scale, 1595–1596
Geology, 1595–1596. *See also* numerous specific
    Minerals
    accretion, 14
    attrition, 325
    berm, 416
    carbuncle, 626
    cleavage, 805
    creep, 979
    differentiation, 1098
    drift, 1147
    erosion, 1324–1325
    erratic, 1325
    fabric, 1349
    formation, 1477
    gradient, 1662
    incompetent, 1913
    intrusion, 1986
    isocline, 2027
    joint, 2041
    platform, 2772
    structural, 3364
    vein, 3613
Geomagnetic anomalies, 1169

Geomagnetic equator, 1597
Geomagnetic field, 1166–1167, 1168
Geomagnetic latitude, 1597
Geomagnetic pole, 1597
Geometrical acoustics, 25
Geometrical optics, 1597, 2126
Geometric chord, 777
Geometric distortion, 1597
Geometric isomerism, 2028
Geometric progression, 2860
Geometry, 1597–1598
Geomorphology, 1598, 2745
    ablation, 3
Geophones, for petroleum prospecting, 2690
Geopotential, 1598
Geopotential height, 312–313
Geopotential surface, 313
Geoscience, 1598
Geosphere, 1598
Geostationary-Earth orbit satellites, 3064–3065,
    3067–3069
Geostrophic flow, 313, 1662
Geostrophic winds, 3774
Geosynchronous orbit satellites, 3066
Geosyncline, 1598
Geotectocline, 1598
Geothermal energy, 1598–1610
Geothermal gradient, 3467
Geotropism, 1610, 2767
Gephyrea, 1610–1611
Gerbil, 3022
Gerenuk, 183, 184
Geriatrics, 1611–1612
Germ, 1612
Germanium, 1612–1613
    Carbon Group member, 623
German measles (rubella), 1295, 3046
Germicide, 1613
Germination, 3124
Germ layer, 1613–1614
Germ plasm, 1614
Gerontology, 1614–1619
Gersdorffite, 1620
Gestation, 1620, 2844
Gettering, 1620
Geumals, 1039, 1040
Geyser, 1598, 1620–1621
Geysers (geothermal energy project), 1601–1602
Geyserute, 1621
Gherkin, 1005
Ghost image, 1621
Ghost sharks, 762
Giant beluga, 3364
Giant cell arteritis, 235, 715
Giantism, 35, 2756
Giant molecular clouds, 3573
Giant sequoia, 918, 1622–1623, 2959
Giant stars, 1621–1622, 2449
Giant tortoise, 2053
Gibber, 1145
Gibberellic acid, 1623–1624
Gibberellins, 1623–1624, 2766–2767
Gibbon, 191
Gibbous moon, 2387
Gibbs division surface, 1624
Gibbs-Duhem equation, 1624
Gibbs free energy, 1493
Gibbs free energy change, 1308
Gibbs function, 3470
Gibbs-Helmholtz equation, 1308, 1624
Gibbsite, 395
Gibbs-Konovalov theorems, 1624

Gibbs paradox, 1624
Gibbs surface tension formula, 3396–3397
Gibhead taper key, 2059
Giga (prefix for units), 3568
Gila monster, 2177–2178
Gilbert (unit), 3569
Gilbert, William, 1624
Gilbert's disease, 423, 1561
Gill, 1624–1625, 2982
Gilsonite, 1625
Gimbal, 1625
Gin, 1625
Ginger, 1625
Gingko, 529
Gingko tree, 2231
Gingko wood, 528
Gini mean difference, 1625
Ginseng, 1625
Giraffe, 1626
Girder, 397, 1626–1627
Girdle-tailed lizards, 2174, 2176
Gizzard, 1627, 2831
Gizzard shad, 1627
G-jitter, 2340
Glacial deposits, 1627
Glacier, 1627–1628
    ablation, 3
Glacier ice, 1895
Glaciology, 1628
Glan air-spaced prism, 2850
Glancing angle, 1628
Gland, 1628
Glanders, 359
Glan Thompson prism, 2850
Glare, 1904–1905
Glasgow Coma Scale, 876
Glass, 1628–1635
    annealing, 176
    in ceramics, 710, 711
    manufacture in microgravity, 2341
Glass blocks, 1634
Glass blowing, 1632
Glass-ceramics, 1630
Glass fibers, 1380
Glasshouse whitefly, 3764
Glass snake, 1636
Glassy cutworm, 1012
Glauberite, 1636
Glaucoma, 1636–1637
Glauconite, 1637
Glaucophane, 1637
Glaze ice, 1637, 2841
Glial fibrillary acidic protein, 699–700
Glide plane, 1637
Glider (artificial life), 246, 247
Glimmschicht method, 659
Global Atmospheric Research Program (GARP),
    806
Global change, 1637
Global Change Research Program, 1177, 1180,
    1181
Global Orbiting Navigation Satellite System
    (GLONASS), 3074
Global Positioning System (GPS), 3069,
    3073–3075
Global telephone networks, 3437–3439
Global temperature, 1637–1639
Global thresholding, in machine vision, 258
Global warming, 814, 817, 1639–1645
Global Weather Experiment, 806, 811–812
Globar (lamp), 1646
Globe artichoke, 894

Globe lightning, 369
Globular clusters, 1552
    giant stars in, 1621–1622
    Hubble Space Telescope images, 1810, 1811
Globulins, 1646
Gloger's rule, of ecological adaptation, 41
Glomerate, 1646
Glomerulonephritis, 2063–2064
Glomerulus, 1341, 2061
GLONASS (Global Orbiting Navigation Satellite
    System), 3074
Glory ring, 319
Glossitis, 1646
Glossmeter, 1646
Glovebox Flight Program, 1689
Glow worm, 1646
Glucagon, and diabetes, 1067, 1069
Glucocorticoids, 49
Glucose, 610–611, 3399
    metabolism, 613
Glucose isomerase, 449–450, 654
Glues, 46, 47
Glutamic acid, 133–137
    neurotransmitter, 698
Glutamine, 133–137
Glycation, and aging, 428
Glycerol, 87, 1646
Glycine, 133–137
    neurotransmitter, 698
Glycogen, 612
Glycol, 87, 1646
Glycolysis, 1646–1647
Glycosides, 1647
Glycyrrhizins, 3402
Glyoxylate shunt pathway, 613
Glyoxylic acid cycle, 613
Glyptolith, 1145
Gnat, 1123–1124, 1647
Gnatcatcher, 1647, 3698
Gneiss, 1647
Gnomonic projection, 1647
Gnu, 183, 184
Goal-Question software metric, 3209
Goat, 1647–1651
Goat-gazelle, 183, 184
Goat lice, 2189–2190
Goatsucker, 608, 2456
Go-away bird, 3545
Gobies, 1651
Goddard, Robert H., 1651–1652
GOES (geostationary operational environment
    satellites), 3736
GOES-NEXT program, 3736
Goethite, 1652
Goiter, 1993, 3485
Golay pneumatic cell, 1652
Gold, 1652–1654, 2356
Golden beryl, 417, 792
Golden-brown algae, 90
Golden-eye, 180, 212, 1654, 2445
Golden Ratio, 1382
Golden Rectangle, 1382, 1383
Goldfinch, 595
Goldfish, 634
Goldman kinetic perimeter test (for tunnel vision),
    2677
Gold number, 1654
Goldschmidt reduction process, 1654–1655
Golgi, Camillo, 1655
Gonadotropic hormones, 1793, 2755
Gonads, 1302, 1655–1659, 2755
Gondwanaland, 2525

Goniometer, 1659–1660
Gonococcemia, 1660–1661
Go/no-go detector, 1660
Gonorrhea, 352, 1660–1661, 3149
Gooseberry, 417, 2073
Gooseberry fruitworm, 1522
Goosefishes, 171–172
Gopher, 3319, 3320
Goral, 1649
Gorilla, 191–192
Goshawk, 1351–1352, 1717
Gossan, 1661
Gouge, 1661
Gourami, 2077
Gouraud shading, 910
Gourds, 1005
Gout, 1661–1662, 2886
Government Performance and Results Act, 20
Governor, 1662
GPS (Global Positioning System), 3069, 3073–3075
Grab bucket, 1662
Grackle, 1662
Grade (engineering), 1662
Graded bedding, 1662
Gradient
    geology, 1662
    mathematics, 1663
Gradient current, 1662
Gradient flow, 1662
Gradient winds, 3774
Grafting and budding, 1663
Graft-*versus*-host disease, 509
Graham law, 1663
Grain (unit), 3569
Grain alcohol, 1331
Grain boundary, 1663
Grain size, 1663–1664
Grain-storage insects, 1664–1665
Gram (unit), 3569
Grama grasses, 1679
Gram-atom, 1665
Gram-Charlier series, 665
Gram-equivalent, 1665
Grammar, 261
Gram-mole, 2365
Gram-molecular weight, 1665
Gram-negative bacteria, 351
Gram-positive bacteria, 351
Grana, 773
Granary weevil, 1664
Grand canonical ensemble, 1307
Grand conjunction, 921
Grand mal epilepsy, 3125
Grandry's corpuscle, 961
Granite, 1665–1666
Granitoid, 1666
Granulated salt, 3205
Granulation (Sun), 3376
Granulite, 1666
Granulocytes, 484
Grape curculio, 1006
Grapefruit, 801
Grape-leaf folder, 1666
Grape-leaf skeletonizer, 1666
Grape phylloxera, 1666
Grapes and wines, 417, 1666–1673
Grapevine flea-beetle, 1429
Grape-vine leaf hopper, 2114
Graph, 1673–1674
Graph component, 1674
Graphic tablet, 908

Graphite, 616, 1674–1675
    in steel, 2016
Graph rank, 1675
GRAS (generally regarded as safe), 1675
Grashof number, 1675
Grasses, 1675–1680
Grasshopper, 1680–1681, 2183, 2603
Grassland biome, 432–433
Grassquit, 2656
Graticule, 1681–1682
Grating, 1682
Grating light valve, 2333
Grauwacke, 1690
Gravel, 1682
Graveldiver, 480
Grave's disease, 3485–3486
Gravimetric feeder, 1356–1357
Gravimetry, 1171
Gravitation, 1682–1686
Gravitational collapse, 1684
Gravitational lensing, 1686, 3446–3447
    quasar seen through, 2906
Gravitational tide, 315
Gravitational wave antennas, 1685
Gravitational waves, 1684–1685
Gravity, 1687–1690
    acceleration due to, 10–11
    materials processing in microgravity, 2340–2341
Gravity wave, 314
Gray body, 1690
Gray-body radiator, 1936
Gray code, 255, 855
Gray copper ore, 3464
Gray field slug, 3187
Gray garden slug, 3187
Gray iron, 107
Grayling, 1690
Gray matter, 695, 1690
Gray scale, 1690
Graywacke (Grauwacke), 1690
Grease, 1690
Great attractor, 1546
Great Barrier Reef, 378, 953–954
Great-circle course, 1690
Great Dark Spot (Neptune), 2436
Greater apes, 191
Great laurel, 1735
Great Plague, 563–564
Great Red Spot (Jupiter), 2043, 2045–2046
Great White Spot (Saturn), 3099
Grebe, 1691, 2783
Green algae, 90, 91–93, 764
Greenback, 3531
Green flash, 1691
Green function, 1691
Greenhouse effect, 306, 814
    and global warming, 1639–1642
Greenhouse gases, 1639, 1641–1642
    climate effects, 813
Greenhouse slug, 3187
Greenhouse whitefly, 3764
Green June beetle, 2042
Greenockite, 1691
Green revolution, 1691–1692
Greenstick bone fractures, 505
Greenstone, 1692
Greenwich Mean Time, 3493
Gregariousness, 1692
Gregorian calendar, 588, 3493
Gregorian focus, 3440
Gregorian telescope, 1692
Gregory formula, 1692

Greisen, 1692
Grey (greige) goods, 1155
Gribble, 1692
Grid sampling, 3060
Griffith crack theory, 1692–1693
Grignard reaction, 1693, 2590
Grillage, 1693
Grimacing syndrome, 3464
Grindle, 527
Grippers, robots, 3011
Grit, 1693
Grosbeak, 1505
Gross anatomy, 163
Gross thrust, 1693
Grossularite, 1570
Gross weight, 1693
Ground (electrical), 1693–1694
Ground beetle, 407, 410
Ground-effect machine, 1693
Ground fault circuit interrupters, 1223–1224
Ground ice, 1895
Ground inversion, 309
Ground moraine, 1694
Groundnut (peanut), 2669
Ground pearl, 1694
Ground-state nuclear disintegration energy, 1128
Ground support equipment (GSE), 1694
Groundwater, 1694
Groundwater contamination, 3720
Ground wave, 1694
Ground wave, radio, 2934
Group, 1694
Group A Streptococcal (GAS) disease, 359–360
Group B Streptococcal (GAS) disease, 360
Group technology, 335
Group velocity (wave train), 1695
Grouse, 1695
Growth, 1695–1696
Growth curve, 1696
Growth hormone, 1792–1793
Growth hormone, and aging, 1615–1616
Grub, 511–512, 1696
Grub beetle, 407
Gruiformes, 1696
Grunerite, 1005
Grunts, 1696
Grus (the crane), 1696
Grysboks, 183, 184
Guanaco, 593, 594
Guanarito virus, 224
Guanidine, 1696
Guard cells, 1780–1781, 3357
Guar gum, 1698
Guava trees, 1696
Guayule, 3046
Guernsey cattle, 669
Guest-host liquid crystal display, 2158
Guiac gum, 1698
Guiana current, 1696–1697
Guidance, 1697
Guillemot, 3695
Guinea fowl, 2707, 2830
Guinea pig, 3023
Guinea worm, 1697
Guitarfish, 3177, 3178
Gulf Coast tick, 3487
Gulf minkfish, 984
Gulf Stream, 1697
Gulf Stream countercurrent, 1697
Gulf white butterfly, 576
Gullet, 2831
Gulls, 726, 3695–3696

Gum adhesives, 46, 47
Gumbo (okra), 2237
Gumbo (type of till), 3492
Gum resins, 2978
Gums and mucilages, 1697–1699
    fat substitute, 492
Gum trees, 1334–1335
Gunnels (fish), 480
Gunpowder, 1343
Guppy, 3671
Gurnards, 1699
Gusset plate, 1699
Gustatory receptors, 1425
Gust front, 1699–1700
Gutenberg, Beno, 1700
Gutta percha, 1700
Guttation, 1700
Gymnosperms, 1700
Gymnospore, 1700
Gymnotid eels, 1700
Gynandromorph, 1700
Gynecological cancers, 599
Gynecology, 1700
Gynecomastia, 536
Gypsum, 1700–1701
Gypsum plaster, 684
Gypsy moth, 1702
Gyre, 2544
Gyromagnetic ratio, 1702
Gyroscope, 1702–1703

Haber ammonia process, 140
Habit, 1705
Habit plane, 1705
HACCP system, for control of foodborne diseases,
    1470
Hachure, 1705
Hackberry trees, 1705
Hackly, 1705
Haddock, 856
Hadrons, 1705, 2321, 2647
Haemophilus influenzae Serotype b (Hib) disease,
    360–361
Hafnium, 1705–1706
Hagfishes, 1015, 1706
Haidinger fringes, 1706
Hail, 2841
Hair, 1706
Hair (ear), 1721
Hairstreak, 1706
Hairtails, 1010
Hairworm, 1706
Hairy fungus beetle, 1145
Hake, 1706–1707
Half-adder, 1707
Half-cell, 1707
Half-life (radioactive element), 2920
Half-life, biological, 1707
Half-period zone, 1504
Half-silvered surface, 1707
Half-step (semitone), 3135
Half-thickness (absorber), 1707
Halfwidth of a spectral line, 1707
Halibut, 1415, 1417
Halides, 1707
Halite (rock salt), 1707–1708
Hall effect, 1223, 1708–1709
Hall-effect magnetometer, 2229
Hälleflinta, 1709
Halley, Edmond, 1709
Halley's comet, 882, 884, 885, 1709–1711

Hall-Heroult electrolytic process, 116, 119
Hallucinogens, 1711–1712
Haloenzyme, 857
Halogenated compounds, nomenclature and
    examples, 2586
Halogen group, 1712
Halos, 318–319
Halter monitoring, 1255
Hamiltonian, 1712
Hammer (ear), 325
Hammerhead (Birds), 793–794
Hammett acidity function, 23
Hamming neighborhood, 255
Ham mite, 2362
Hamster, 3022
Hand, 1712–1713
Handedness (right- and left-), 1713
Hand hydrometer, 3298
Hanging pan balance, 3742
Hanging valley, 1713–1714
Hangnest, 2656
Hankel function, 419
Hansen's disease (leprosy), 361, 2121
Hantzch-Widman name, for heterocyclic compounds,
    2583
Hanuman, 2386
Hard automation, 334
Hard contact lenses, 929
Hardenability of steel, 1714
Hardening, 1714
    case, 636
    flame, 1413–1414
    precipitation, 2844
    work, 3784
Hard facing, 1349, 1714
Hard ferrite, 1368
Hardness, 1714–1715
    minerals, 2355
Hardpan, 588, 1715
Hardware relocation, 2859
Harelip, 1715
Hares, 2909–2910
Harlequin cabbage bug, 1715, 3356
Harmattan, 432
Harmonic, 1715–1716
Harmonic analysis, 1716
Harmonic conversion transducer, 3510–3511
Harmonic motion, 1716
Harmonic operation, 1716
Harmonic progression, 2860
Harmonic synthesizer, 1716
Harmotome, 1716
Harpy, 1160
Harrier, 1160, 1717
Hartebeest, 183, 184
Hartley, 1716
Hartley absorption bands, 7
Hartley oscillator, 2604
Hartley principles (information transmission),
    1716–1717
Hartmann test, 1717
Hartree-Fock approximation, 1717
Harvestman, 220, 1717
Harvey, William, 1717
Hashish, 605, 2255
Hastelloy, 108
Hatchet fishes, 725, 1454, 1717
Hausdorff space, 3505
Haverhill fever, 2954
Haversine function, 3528
Hawk, 1351–1352, 1717
Hawking, Stephen William, 1718

Hawk moth, 805, 1718
Hay bridge, 542
Hay fever, 103–104
Hazardous waste incineration, 3700–3701
Haze, 2841
Hazelnut shrubs, 1718–1719
Haze meter, 3518
H-display, 1719
HDL (high-density lipoprotein), 234
Head (zoology), 1720
Headache, 1719–1720
Header, 1720
Heading, 1720
Head-mounted displays, 2330
Heads-up vehicle speed display, 343
HEAO-2 (Einstein Observatory), 1215–1217
Hearing, 1720–1725
    audibility, 325
Hearing devices, 1725
Hearing loss, 1724–1725
Heart, 799, 1725–1732
Heart block, 228
Heartburn, 1326
Heart disease, 234–235
    ischemic, 2026–2027
Heart-lung transplantation, 1731
Heartworm disease (dirofilariasis), 1732
Heat, 1732–1734
    mechanical equivalent of, 2307–2308
Heat balance
    atmosphere, 305–306
    planet/process, 1734
Heat capacity, 1734. *See also* Specific heat
    and enthalpy, 1307
    quantum theory of, 2905
Heat capacity mapping, via satellite, 3078
Heat engine, 1734
Heat equation (Einstein), 1734
Heat exchangers, 1741–1743
Heathcock, 1695
Heather shrubs and trees, 1734–1735
Heating degree days, 1043
Heating value, selected fuels, 1522
Heat of adsorption, 52
Heat of combustion, 879–880
    combustible gases, 1572
Heat of decomposition, 1033
Heat of formation, 1033
    atomic, 323
Heat of fusion, 1535
Heat of reaction, 1308
Heat of sublimation, 3366
Heat of vaporization, 3604
Heat pump, 1736–1737
    with solar collector, 3218
Heat-resisting steels, 2019
Heat sink, 1738
Heat stress, exhaustion, and stroke, 1738
Heat transfer, 1738–1743
Heat-treated steels, 2019–2020
Heat treating, 1743–1744
Heaviside bridge, 542–543
Heaviside layer, 1220
Heavy ion, 2090
Heavy-ion fusion, 1540
Heavy water, 1744
Heavy water nuclear reactors, 2492–2493
Hecto (prefix for units), 3568
Hedgehog, 2379
Hedge sparrow, 3697
HEDTA (*N*-hydroxyethylenediaminetriacetic acid),
    727

Heel, 1744
Heel fly, 518
Heilgenschein, 2840
Heine formula, 2117
Heisenberg, Werner Karl, 1745
Heisenberg force, 1744
Heisenberg representation, 1744–1745
Heisenberg uncertainty principle, 3567
Heliarc welding, 3746
Helical antenna, 187
Helical gearing, 1745
Helicobacter infection, 355, 361
    and ulcers, 3556
Helicopters, 1745–1752
Heliopause, 1752
Heliostat, 857, 1752
Heliotrope, 490
Heliotropic winds, 3774
Helium, 1752–1753
    in Sun, 3378
Helix, 1753
Hellbender, 1754
Hellgrammite, 1754
Hellige turbidimeter, 3546
Helmholtz, Hermann von, 1754
Helmholtz equation, 1754
Helmholtz free energy, 1493
Helmholtz function, 3470
Helmholtz instability, 308
Helmholtz resonator, 1754
Helmholtz theorem, 1754
Helminthology, 1754
Helper T cells, 3592
Hematite, 1754
Hematology, 1754
Hematoma, 531–532, 1754
Hematopoiesis, 509
Hematuria, 1754, 2062
Heme, 1756
Hemerythrin, 484
Hemicelluloses, 1089
Hemichordata, 1754
Hemicolloid, 1754
Hemifacial spasm, 481
Hemihedrity, 1754
Hemimetabola, 1754
Hemimorphite, 1754–1755
Hemiplegia, 1755
Hemipodes, 2939
Hemiptera, 1755
Hemispherical scale (insect), 3204
Hemitropic, 1755
Hemlock trees, 1755–1756
Hemoccult card, 488
Hemochromatosis, 2172
Hemocyanin, 484, 1756
Hemodialysis, 2066–2067
Hemoglobin, 484, 1756
Hemoglobinuria, 1756–1757
Hemolymph, 1757
Hemolytic anemias, 167–168
Hemolytic disease of the newborn, 486
Hemolytic jaundice, 167
Hemophilia, 484, 486, 1757
Hemoptysis, 1757
HemoQuant, 488
Hemorrhage, 232, 1757
Hemorrhagic fevers, 224, 3639–3642
    African (Lassa fever), 2104–2105
    Dengue, 1047–1050
    Ebola, 1199–1200
    Korean, 2074

Marburg, 2254–2255
    viral, 3639–3642
Hemorrhagic telangiectasia, 2886
Hemorrhoids, 235
Hemp, 383, 1757–1758, 2255
Hemp seed oil, 1412
Henna shrub, 1758
Henry (unit), 1758, 3568, 3569
Henry, Joseph, 1758
Heparin, 200, 1758
    mini-dose, 201
Hepatitis, 2172
    foodborne, 1468
Hepatitis B, 3151
Heptatojugular reflux, 916
Herbicide, 1758–1760
    pyridine compounds, 2888
Hercules (constellation), 1760
Hereditary mechanics, 1760
Heredity, 1760–1762
Hereford cattle, 668
Herkimer diamonds, 191
Hermaphrodite, 1762
Hermaphroditism, 1762–1763
Hermite equation, 1763
Hermite polynomials, 1585
Hernandulcin, 3402
Hernia, 1763
Heroin, 101, 2393, 2394
Heron, 793, 1764
Herpes simplex virus, 1053
Herpes simplex virus diseases, 1764
    and AIDS-related eye disorders, 68
Herpetology, 1764
Herpolhode, 1764
Herring, 1764–1766
Herringbone gear, 1745
Herschel, F. W., 1767
Hertz (unit), 3568, 3569
Hertz, Gustav Ludwig, 1767
Herzberg absorption bands, 7
Hesperidium, 1517
Hessian, 1767
Hessian fly, 1767
Hessite, 1767
Heterocyclic compounds, 895
    chlorinated, 767
    nomenclature and examples, 2587
Heterodyne, 1767
Heterodyne conversion transducer, 3511
Heterodyne detection, 1060
Heterodyne meter, 1501
Heterogamy, 1767
Heterogeneous catalysts, 652
Heteroion, 1995
Heteromorphosis, 1767
Heteropolyacids, 1767
Heteropolymerization, 2377
Heterosphere, 1767
Heterospory, 1767
Heterozygous, 1767
Heulandite, 1767–1768
Heuristic routine, 3043
Heuristic search, 242
    automated reasoning, 248
Hevelian halo, 319
Hexactinellida, 1768
cis-9-Hexadecanoic acid, 2626
Hexadecimal numbers, 1768
Hexamine, 132, 1768
Hexanoic acid, 608
Hexapoda, 1768

Hexose monophosphate oxidative pathway, 613
Heydweiler bridge, 541
Hiatus hernia, 1327
Hibernating spacecraft, 1768
Hickory trees, 1768–1770
Hiding pigment, 2617
Hierarchical planning, 243
Higgs field, 2652
High-alloy irons, 107
High-alloy steels, 107–108
High-definition television (HDTV), 3453–3456
High-energy physics, 2644
High fidelity, 1770
High fructose corn syrup, 450
High-head hydroelectric power, 1858–1859
High-intensity discharge lamps, 1902–1903
High-level programming languages, 2858
High-level radioactive waste, 2501
High-lift devices, 56
High-pass filter, 1392
High-pressure/temperature boilers, 496
High-resolution scanning transmission electron
    microscope, 1276
Highs, 310–311
High-speed railways, 2939–2943
High-strength low-alloy steels, 108, 2019
High technology, 1770
High-temperature, high-strength, iron-based alloys,
    108
High-temperature gas-cooled nuclear reactor,
    2488–2492
High-temperature superconductors, 3384, 3387
High-Throughput X-Ray Spectroscopy Mission
    (XMM-Newton), 3789–3790
High vacuum, 1770
Hildebrand rule, 1770
Hill myna, 1662
Himalia, orbital characteristics, 2760
Hip, 1770–1771
Hippocampus, 692
Hippopotamus, 1771
Hip replacement, 1771
Hiran (high-precision short-range navigation), 3160
Hirsutism, 1771
Hirudinea, 1771–1772
Histamine, 1772
    and asthma, 555
    neurotransmitter, 698
Histamine poisoning, 1092
Histidine, 133–137
Histogram, 1772, 2898
Histology, 1772
Histones, 679, 1772
Histoplasmosis, 361, 1531, 1772–1774
Histosols, 3211
Hittorf, Johann Wilhelm, 1774
Hittorf dark space, 659
Hittorf principle, 1774
HIV (Human Immunodeficiency Virus), 33–34, 68
HIV-1, 33
HIV-2, 33
Hives (urticaria), 104, 3589–3590
H lines, 1774
Hoarfrost, 2841
Hoatzin, 1563, 1774, 2879
Hob, 1774–1775
Hock (hough), 1775
Hodgkin, Alan Lloyd, 1775
Hodgkin's disease, 1775
Hodograph, 1775
Hoffman reaction, 1043
Hogback, 1775

Hohlraum, 1775
Hoist, 1775–1776
Holding current, 3172
Holes (in semiconductors), 3130, 3231
Hollerith card, 96, 332, 586, 1776
Hollyhock, 2237
Holly trees and shrubs, 1776–1777
Holmium, 1777
Holocrystalline, 1777
Hologram, 1777–1778
Holography, 1777–1780
Holohedral crystal, 1780
Holothuroidea, 1780
Holotype, 1780
Homeobox, 2375
Homeostasis, 1780–1781
Homeotype, 1781
Homing guidance, 1697
Homocentric rays, 1781
Homocline, 1781
Homocyclic compounds, 895
Homodyne reception, 1781
Homogeneous (mathematics), 1781
Homogeneous atmosphere, 47, 1781
Homogeneous catalysts, 652
Homogenizing, 1781
Homoiotherapy, 1781–1782
Homologous series, 1782
Homology, 1782
Homopause, 1782
Homopolar bond, 2376
Homoptera, 1782
Honey, 1782, 3399
Honeybees, 1782–1785
Honey buzzard, 1160
Honey creeper, 2656
Honeydew, 1786
Honey eater, 2656
Honey guides (birds), 2745, 3782, 3783
Honey pecker, 2656
Honosphere, 1786
Hoodoo, 1786
Hookworm, 1054, 1981
Hoopies, 952
Hoopoe, 2071
Hop, 2401–2402
Hopfield absorption bands, 7
Hophornbeams, 1795
Hordeolum (stye), 1346, 3364–3365
Horizon (astronomical/celestial/geographic), 1786
Horizontal coordinate system (astronomy), 1786
Horizontal filters, 1395, 1396
Horizontal parallax, 2634
Hormones, 1302, 1786–1794, 1787–1794
    in bile, 422
    and carbohydrate metabolism, 615
    in chemotherapy, 597
    homeostasic regulation, 1780
    for pest control, 1955
Horn (electromagnetic), 1796
Horn (substance), 1797
Horn antenna, 187
Hornbeam trees, 1794–1795
Hornbill, 952, 2071, 2072
Hornblende, 1795–1796
Hornblendite, 1796
Horned toad, 1796
Hornet, 1796–1797, 3698
Hornfels, 1797
Horn fly, 1797
Horn-tail, 1797, 3102
Hornworm, 1797

Horologium (constellation), 1797
Horse, 1798–1800
    West Nile virus, 3748
Horse-antelopes, 183, 184
Horse chestnut trees, 564
Horseflesh ore, 513
Horsefly, 1124, 1797–1798
Horse latitudes, 310
Horse lice, 2190
Horsepower (unit), 3569
Horse silver, 763
Horshoe crab, 237
Host, 1800
Hot-dip galvanizing, 1565
Hot dry rock (geothermal energy), 1599, 1609–1610
Hot lime zeolite softening, 1360
Hot shortness, 1800
Hotwell, 1800
Hot working, 1800
Hough (hock), 1775
Hour (unit), 3569
Hour angle, 1800
Housefly, 1800
H-theorem (kinetic theory), 2070
Huallaga River, 3058
Hubara, 2939
Hubbing, 1775
Hubble, Edwin Powell, 1800–1801
Hubble Deep Field, 3572
Hubble's law, 968, 1546, 2959
Hubble Space Telescope, 1801–1825, 3448–3449.
    See also individual planets for images
Hue, 1825
Huggins absorption bands, 7
Huia, 2656
Human anatomy, 163
Human fungus diseases, 1531–1533
Human genome, 1829
Human Genome Project, 1593, 1825–1852
    major events in, 1827–1829
Human von Willebrand Factor, 485
Humboldt (Peru) current, 1853
Humectants, 1853
Hume-Rothery rules, 1853
Humidity, 6, 1853, 1882–1885, 3301
    psychometric chart, 2878–2879
Hummingbirds, 212, 3403–3404
Humphries equation, 3301
Hunchback (scoliosis), 3118
Hunting (rotating mechanisms), 1853
Huntington's chorea, 777–778
Hurricane, 1510, 1511–1515
Hurwitz criterion, for system stability, 3321
Huygens, Christian, 1854
Huygens eyepiece, 1347
Huygens' principle, 1854, 3729
Huygens' tractrix, 3510
Huygens' wavelets, 1854
Hyacinth, 2135
Hyades, 1854
Hyaline cartilage, 636
Hyalite, 2560, 2561
Hybrid, 1854
Hybrid ferromagnetic-semiconductor structures, 3134
Hybridization, 2764–2765
Hybridoma, 448
Hybrid vehicles, 1232
Hybrid vigor, 667
Hydantoin reaction, 138
Hydatid disease, 1854

Hydatogenesis, 1854
Hydra, 1881
    budding in, 269
Hydrate, 1854
Hydrated alumina, 396
Hydrated ion, 1995
Hydrated lime, 2136
Hydra (the serpent), 1854
Hydration, 2592
Hydraulic accumulator, 14
Hydraulic amplifier, 149
Hydraulic controller, 1854–1857
Hydraulic dredge, 1145
Hydraulic grade line, 1873
Hydraulic radius, 1856
Hydraulics, 1856
Hydraulic turbines, 1859–1860
Hydraulic uplift, 3575
Hydrazine, 1856–1857
Hydrazino group, analytical determination, 161
Hydrazoic acid, 1857
Hydrazones, 1857
Hydride, 1857–1858
Hydroboration-oxidation, 2589
Hydroboration reaction, 515
Hydrocarbons
    air pollutants, 2803
    nomenclature and examples, 2583–2585
Hydrocephalus, 1858
Hydrochloric acid, 1858
Hydrocracking, 653–654, 977
Hydrodynamics, 1445
Hydroelectric power, 1858–1863
Hydrofluoric acid, 1863–1864
    alkylation unit, 103
Hydrofluorocarbons, greenhouse gas, 1641
Hydrofoil, 1864
Hydroformylation, 2593
Hydrogen, 1864–1865
    alkali metal, 98
    catalysts for production, 654
    as fuel, 1865–1870
    radio emission, 2933
Hydrogenation, 1870, 2589, 2591
    catalytic, 654
    vegetable oils, 3609
Hydrogen bomb, 1865
Hydrogen bond, 273, 751
Hydrogen cyanide, 1870–1871
Hydrogen embrittlement, 1290
Hydrogenolysis, 2592
Hydrogen peroxide, 1871–1872
    bleaching agent, 480
    as food antimicrobial agent, 205
Hydrogen scale, 1872
Hydrogen sulfide, 1872
Hydrograph, 1872
Hydroid, 857, 1872
Hydroisoandrosterone, 165
Hydrokinetics, 1856, 1872–1873
Hydrolases, 1314
Hydrologic cycle, 1873–1875
Hydrology, 1873–1878
Hydrolymph, 1878
Hydrolysis, 1878–1879, 2592, 3233
Hydromagnetic equations, 1879
Hydromagnetics, 2228
Hydromedusa, 2308
Hydrometeorology, 2327
Hydrometeors, 2838–2843
Hydronium ion, 1879
Hydrophilic, 1879

Hydrophobic, 1879
Hydrophone, 1879
Hydrophytes, 1879
Hydroponics, 1879–1880
Hydrosphere, 1880
Hydrostatic equilibrium, 1880
Hydrostatic pressure, 1880, 2846
Hydrostatics, 1880
Hydrotreating, 654, 1880–1881
17-Hydroxycorticosterone (cortisol), 49
17-Hydroxy-dehydroxycorticosterone (cortisone), 49
17-Hydroxy-desoxycorticosterone, 49
Hydroxylamine, 1881
Hydroxyl group, analytical determination, 161
Hydroxyl radical, 23
Hydroxyproline, 133–137
Hydroxyzine hydrochloride, 203
Hydroxyzine pamoate, 203
Hydrozoa, 1881
Hyena, 1881–1882
Hygrometry, 1882–1885
Hygroscopic, 1885
Hygrothermograph, 2842
HyL process, 2011
Hymenoptera, 1885
Hypabyssal, 1885
Hyperactivity, in children, 1885–1886
Hyperbola, 1886
Hyperbolic function, 1886–1887
Hyperbolic geometry, 1597
Hyperbolic navigation, 2426–2427
Hyperbolic spiral, 1887
Hyperboloid, 1887
Hypercharge, 2322
Hyperconjugation, 1887
Hyperemia, 104
Hypereutectic alloy, 1887
Hyperfiltration, 2310
Hyperfine structure, 1887
Hypergeometric distribution, 1887
Hyperion, 3097–3098
    orbital characteristics, 2761
Hyperkalemia, 2063
Hyperlipidemia, 234
Hyperons, 1887
Hyperopia (farsightedness), 1887–1888, 3649–3650
Hyperparabolic partial differential equations, 2643
Hyperparathyroidism, 2636–2637
Hyperprolactinemia, 1923
Hypersensitive carotid sinus reflex, 228
Hypersonic flow, 1888
Hypersthene, 1888
Hypertension, 490, 1888–1889
    and diet, 1090–1092
    ocular, 2556
Hypertensive encephalopathy, 715
Hypertensive retinopathy, 2984
Hyperthecosis, 166
Hyperthyroidism, 3485–3486
Hypertrophy, 536
Hyperuricemia, 2886
Hyperventilation, 632, 1890
Hyperventilation syndrome, 1890
Hyploid gearing, 421
Hypobaric (controlled-atmosphere) systems, 1890
Hypocapnia, 1890
Hypochlorination, 2591
Hypochlorites, 479–480, 1890–1891
Hypochlorous acid, 1890
Hypocycloid, 1891
Hypodermis, 1891
Hypoeutectic alloy, 1891

Hypofluorite, 1891
Hypogene, 1891
Hypoglycemia, 1066
Hypoiodites, 1891
Hypoiodous acid, 1891
Hypomelanosis, 87
Hyponitrites, 1891
Hyponitrous acid, 1891
Hypoparathyroidism, 2637
Hypophosphates, 1891
Hypophosphites, 1891
Hypophosphoric acid, 1891
Hypophosphorous acid, 1891
Hypoplasia, 1891
Hypoplastic anemia, 2886
Hypoprothrombinemia, 1891
Hyposulfites, 1891–1892
Hyposulfurous acid, 1891–1892
HYPO system, 250
Hypotension, 1892
    and chronic fatigue syndrome, 789
Hypotenuse (Pythagorean) theorem, 2893
Hypothalamus, 692, 1302
Hypothermia, 1892
    and brain death, 1030
Hypothesis, 1892
Hypothyroidism, 3486
Hypoventilation, 632, 1892
Hypovolemic shock, 3159
Hypoxaemia, 1892
Hypoxia, 1892
Hyraxes, 1892
HYSOLAR project, 3226
Hysterectomy, 1892
Hysteresis, 1892–1893
    instrument, 1893
Hysteresis distortion, 1893
Hysteresis heater, 1893
Hysteretic error, 1893

Iapetus, 3097
    and Cassini mission, 648
    orbital characteristics, 2760
Iatrogenic, 1895
Ibex, 1649, 1650
Ibis, 793, 794, 1895
Ice, 1895
Iceberg, 2544
Ice crystal haze, 2841
Ice fishes, 1896–1897
Ice frost, 1897
Ice island, 1897
Icelandic low, 311
Iceland spar, 581
Ice pellets, 2841
Ice point, 1897
Ice-rafting, 1897
Ichneumon, 1897
Ichthyology, 1897
Ichthyosarco toxism, 1468
Ichthysauria, 1481
Iconoscope, 1897
Ideal gas law, 1578, 1897–1898
Ideal liquid, 2164
Ideal system, 1898
Ideal transducer, 3511
Idiopathic, 1898
Idiopathic hyperbilirubinemia, 423, 1561
I-display, 1898
Idoxuridine, 209
IgE-mediated asthma, 555

Igneous rock, 1898, 2356
Ignition temperature, 880
Ignitors (boilers), 571
Ignitron, 1898, 2957
Iguanids, 1898–1900
Ileum, 2, 98, 1102
Ilizarov procedure, 506
Illium G, 108
Illumination, 1900–1907
Ilmenite, 1907
Image antenna, 187
Image processing, 258
Imagery, 2722–2739
Imbricate structure, 1907
Imides, 1907
Imines, 132, 2581
    nomenclature and examples, 2586
Imino acids, 136, 137
Imino compounds, 1908
Immersion foot (trench foot), 1892
Immune system, 2375, 3592–3593
Immune system and immunology, 1908–1910
    and blood, 484
Immunogenicity, 198
Immunoglobulins, 1908–1910
Immunological tolerance, 199
Impact, 1910–1911
Impact testing, 1911
Impalla, 183, 184
Impedance, 1911
Impedance cardiography, 1254
Impedance-coupled amplifier, 149
Impedance matching, 1911
Impeded harmonic motion, 1716
Impeller, 1911
Imperative programming languages, 2858–2859
Imperfections, in solids, 911
Impetigo, 1053, 1911–1912
Impingement carbon, 621
Implantable contact lenses, 2964
Implosion, 1912
Impotence, 1912
Impounding reservoir, 1912
Impressed current, 1912
Impressed force, 1474
Impulse, 1912
Inactivated vaccines, 3593
Inbreeding, 1912
Incandescence, 2191
Inceptisols, 3211
Inch (unit), 3569
Inch of mercury (unit), 3569
Inch of water (unit), 3569–3570
Incisor, 1912
Inclination, 1912
Inclined plane, 2208
Inclined-tube manometer, 2244
Inclusion, 1912–1913
Inclusion compounds, 895
Inco, 108
Incoherent scattering, 3109
Incompetent (geology), 1913
Incomplete beta function, 420
Incompressible fluid, 1444
Inconel, 108
Incontinence, 1611, 2062
Incremental encoder, 1298–1299
Incremental inductance, 1915
Incubation period, 1462, 1913
Independent linearity, 2139
Indeterminacy principle, 3567
Indeterminate form, 1913

Indeterminate structure, 1913
Index (exponent), 1344
Index number, 1913
Index register, 1913, 2970
Index table, 1913–1914
Indian meal moth, 1147, 1665
Indian Ocean water, 1914
Indica rice, 2999–3000
Indicators
    adsorption, 50
    airspeed, 85–86
    chemical, 1914
        rate-of-climb, 2954
        true-airspeed, 85, 86
        turn-and-bank, 3549
Indifferent states, 1914
Indifferent system, 1914
Indigo, 1914
Indirect sampling, 3060
Indium, 1914–1915
Indoor air pollution, 2938
Indri, 2119
Inductance, 1915
Inductance-coupled circuit, 975
Induction
    electric/magnetic, 1915–1916
    mathematics, 1916
Induction field, 1916
Induction forces, 1916
Induction motors, 2398
Induction motor variable-speed drives, 3144–3145
Inductive-bridge transducers, for position
        measurement, 2818
Inductive interference, 1916
Inductive plates, for position measurement, 2818
Inductive proximity detectors, 2875
Induration, 1916
Indus (constellation), 1916
Industrial biotechnology, 1916–1919
Industrial enzymes, 449–451
Industrial scales, 3743
Inelastic scattering, 3109
Inequality, 1920
Inert gases, 1920
Inertia, 1920
    moments and products of, 1920
Inertial coordinates, 946
Inertial flow, in atmosphere, 313
Inertial guidance systems, 1920–1922
Infant botulism, 1466
Infection, 1922
Infection, foodborne, 1461–1468
Infection-associated fever, 1374
Infectious arthritis, 236–237
Infectious eczemoid dermatitis, 1053
Infectious mononucleosis, 1922
Inference, compositional rule of, 253
Inferior conjunction, 921
Infertility, 1922–1923
Infestation, 1922
Infinity, 1923
Inflammation, 1923–1924
Inflammatory bowel diseases, 864–865
Inflationary universe, 970–971
Influence line, 1924
Influent, 1924
Influenza, 1924–1927
Information, 1927–1928
Information extraction, 261
Information theory, 1928–1934
Infrared astronomy, 1934–1935, 1940
Infrared films, 2726–2727

Infrared imagery, 1936
Infrared process analyzer, 1936–1937
Infrared radiation, 1935–1940
Infrared radiation sources, 1940
Infrared spectroscopy, 1936
Infrasonic frequency, 1940–1941
Infusion, 1941
Ingestion, 1941
Ingot, 1941
Inguinal canal, 595
Inhalation anthrax, 191, 355
Inheritance, 1760–1762. See also Chromosome;
        Genes
    extranuclear, 1762
Iniomous fishes, 1941
Initial condition, 1095
Inner bremsstrahlung, 536
Inner compounds, 895
Inner ear, 1722
Inner salt, 3058
Inorganic compounds, 895
Inorganic light-emitting diode displays, 1941–1942
Inorganic synthesis, 3408
Inositol, 1942
Inositol triphosphate, 2375
Input/output devices, 1942
Insect, 237, 1942–1948
    beneficial, 409–411
Insecticide, 10, 1948–1952
    technology, 1952–1956
Insectivorous plants, 1956
Insect parasites, 411
Insemination, 1956
Insertion loss, 1956
Insolation, 1956
Instability, 1957
Instability line (fronts), 1509–1510
Instantaneous center, 687
Intrinsic images, 258
Instruction, 1957
Instruction address register, 1109
Instruction code, 855
Instruction counter, 974, 1957
Instruction register, 2970
Instrument, 1957
Instrumental variable, 1957
Instrumentation, 1957
Instrument burden, 569
Instrument transformers, 1239
Insulated gate transistor, 3143
Insulation
    electric, 1957–1958
    thermal, 1958–1963
Insulators, 3231
Insulin, and diabetes, 1067, 1068–1070
Insulin-dependent diabetes, 1066
Insulin resistance, 1069
Integral, 1963
    line/surface/volume, 1964
    volume, 196
Integral control, 931
Integral controller, 934
Integral equation, 1963–1964
Integral-furnace boiler, 496
Integral transform, 1964
Integrated circuit (IC), 1964–1965, 2335
Integrated Services Digital Network (ISDN), 3429,
        3437–3439
Integrating-ramp A/D converter, 1965–1966
Integration, 1966
    numerical, 1966–1967
Integration by parts, 1966

Integument, 1967
Integumentary system, 1967
Intelligent artifacts, 239
Intelligent telephone networks, 3438
Intellipath, 243
Intensity, 1967
Intensive sampling, 3060
Interactive computer graphics, 907
Interatomic potential, 1967
Intercept (mathematics), 1967
Interconnections (electronics), 1967–1968
Interdigital transducer, 31
Interface, 1968
Interface vehicles, 1864
Interfacial tension, 1968
Interference
    signal, 1968
    wave, 1968–1969
Interference (wave). See Wave interference
Interference filter, 1968
Interference microscope, 2346–2347
Interferogram, 1968
Interferometer, 1969
    angle of arrival measurement, 171
    optical telescopes, 3444–3446
    in radio astronomy, 2932–2933
Interferons, 600, 2204
    against viruses, 3646
Intergranular corrosion, 962
Interior ballistics, 368
Interleukin, 2204
Intermediate (chemical), 1969–1970
Intermediate-access storage, 3358
Intermediate boson, 516
Intermediate-frequency amplifier, 149
Intermediate-infrared region, 1935
Intermediate level transuranic waste, 2501
Intermediate-moisture foods, 1970
Intermediate neutrons, 2447
Intermetallic compound, 106, 896, 1970
Internal combustion engine, 1970–1971
Internal conversion, 1971
Internal-conversion coefficient, 945
Internal energy, 1307, 1971
Internal friction, 1504
Internally stored program, 2852
Internal pressure, 2846
Internal salt, 3058
International date line, 1971
International pollution, 2799
International Space Station (ISS), 3265, 3275–3293
    remote sensing application, 3078
International System of Units (Si), 3568
International temperature scale, 3459
Internet, 1971–1979
Interoceptors, 3136
Interphase, 677
Interpolation, 1979
Interpreter, 1110, 1979, 3516
Interpreter code, 2875
Interpretive routine, 3043
Interquartile range, 1979
Interrogation, 1980
Interrupt (computer system), 1980
Interstellar reddening, 1980
Interstellat dust, Stardust mission to study, 3329
Interstice, 1980
Interstitial compounds, 896
Intertial flow, 1662
Intervalometer, 1980
Intestinal nematodes, 1980–1984
Intestinal protozoa, 1984

Intra-abdominal abscess, 5
Intracapsular cataract extraction (ICCE), 1984–1985
Intracorneal lens implants, 2964
Intracranial abscess, 5–6
Intraocular lenses (IOLs), 656
Intrastromal corneal ring segments (ICRS), 1985–1986, 2964
Intrinsic contact potential, 929
Intrinsic factor, 168
Intrinsic proteins, 696
Intrinsic safety, 1986
Intron, 1588
Intrusion (geology), 1986
Intubation, 1986
Intussusception, 1986
Inulin, 612
Invariance principle, 1986
Invariance theorem, 1986
Invariant, 1986
Inverse, 1986
Inverse kinematics, 905
Inverse Laplace transform, 2087
Inverse matrix, 1986
Inverse plasticity, 1119
Inverse sampling, 3060
Inverse trigonometric function, 1986–1987
Invertebrate paleontology, 1987–1990
Inverted voltmeter, 151
Inverter, 1990–1991, 2837
Invert sugar, 3399
Investigational new drug, 446
Investment casting, 649, 1991
In vitro, 1991
In-vitro fertilization, 1991
In vivo, 1991
Involute, 1991
Io, 1558, 1559, 2051
    orbital characteristics, 2760
Iodine, 1991–1993
    in biological systems, 1993–1995
Iodine value, 3609
Iolite, 955
Ion, 1255–1256, 1995
Ion column, 1995
Ion concentration, 1995
Ion density, 1995
Ion engine, 1995
Ion exchange, for desalination, 1058
Ion exchange chromatography, 784
Ion-exchange resins, 1995–1998
Ionic bond, 751
Ionic charge, 1998
Ionic compounds, 896
Ionic crystal, 1998
Ionic equilibrium, 1998
Ionic migration, 1998
Ionic mobility, 1998
Ionic potential, 1998
Ionic radius, 744–745, 746
Ionic solids, 1995
Ionic strength, 1998
Ion implantation, 1998–1999
    coatings, 2621
    semiconductors, 3131
Ionization, 1999
Ionization chamber, 1999
Ionization potential, 1999
Ionized atom, 322
Ionized gases, 1999–2000
Ionizing energy, 2000
Ionizing particle, 2000

Ionizing radiation, 2921, 2924
Ion microprobe mass analyzer, 2000–2001
Ionomers, 2001
Ionosphere, 304, 2001–2002
    interference from meteorite trails, 2326
Ionospheric sounder, 2002
Ion pair, 2002
Ion plating, 1999
Iontophoresis, 1070
Iridescence, 2002
Iridium, 2002
Iridium satellite, 3072–3073
Iris diaphragm (microwave/optics), 2003
Irish whiskey, 3762, 3763
Iritis, 2003, 3653
Irminger current, 2003
Iron, 2003–2007
    in biological systems, 2007–2010
    in diet, 1087
    ocean deposits, 2553–2554
    removal from water, 3713
Iron Age, 2007
Iron carbide, 686
Iron-chromium alloys, 2019
Iron-deficiency anemia, 166–167, 2009
Ironmaking, 2010–2011
Iron metals, alloys and steels, 2010–2020
    annealing, 175
    hardenability, 1714
Iron meteorites, 2325–2326
Iron mica, 460
Iron overload, 2009
Iron oxide minerals
    goethite, 1652
    hematite, 1754
    limonite, 2138
    magnetite, 2226–2227
Iron thermocouples, 3468
Irradiated foods, 2020–2021
Irrational number, 2022
Irregular galaxies, 1549–1550, 1551
Irreversibility, and kinetic theory, 2070
Irreversible process, 2022, 2988
Irrigation, 2022–2026
Irritable bowl syndrome, 615, 1079
Irrotational field, 1964
Isallobars, 311–312
Isarithm, 2030
Ischemic heart disease, 2026–2027
Isentropic, 2027
Isentropic change, 2027
Isentropic mixing, 2027
Isentropic process, 47
Isentropic surface, 2027
Island Model Evolutionary Algorithm, 256
Island volcanoes, 3680
Islets of Langerhans, 1067, 1302
Isobaric equivalent temperature, 1323
Isobars, 736, 737, 2027, 2031
Isobath, 2027
Isochore, 2027
Isochrone, 2027
Isochronous pendulum, 2673
Isocline (geodesy/geology), 2027
Isodesmic structure, 2027
Isodiaphere, 2027
Isodimorphism, 2028
Isoelectric point, amino acids, 133–134, 137
Isoelectronic, 2028
Isogamy, 2028
Isogeotherm, 2028

Isogonal map, 915
Isogonic line, 2028
Isogram, 2028
Isohaline, 2028
Isolated amplifier, 149
Isoleucine, 133–137
Isoline, 2028
Isomagnetic, 2028
Isomalt, 491, 3400
Isomer, 755, 2028
    nuclear, 2029
Isomerases, 1314
Isomerism, 2028–2029
Isomerization, 2029, 2591
    catalytic, 653
Isometry, 2029
Isomorphism, 2029
Isonitriles, 2581
Isopach, 2029
Isopleth, 2029–2030
Isopropyl alcohol, 2030, 2685
Isoproternol, 50
Isoptera (termites), 2030
Isospin, 2904
Isostasy, 3739
Isostatic forging, 1475
Isotachophoresis, 1279
Isotherm, 3458
Isothermal, 2030
Isothermal annealing, steel, 2020
Isotone, 2031
Isotope, 736, 737, 2031–2032
    use in radiometric analysis, 2923–2924
Isotope dilution analysis, 2925
Isotope separation (uranium), 3578–3579
Isotopic medium, 2032
Isotropic antenna, 187
Isotropic phase, 2142
Itch mite, 2362
Itchy eyes, 2032
Iterated logarithm, law of, 2032
Iterative methods (for solving equations), 2032

Jacamar, 2745, 3782
Jacana, 2033, 3695, 3696
Jacaranda, 651–652
Jackal, 603, 604
Jackdaw, 988, 1662
Jack-in-the-pulpit, 227
Jack mackerel, 609
Jacks, 609
Jacobian, 2033
Jacobi elliptic function, 2033
Jacobi polynomial, 2033
Jacquard loom, 96
Jade, 2033–2034
Jadeite, 2033, 2034
Jaguar, 664, 665–666
Jamaica pepper, 109
Jamesonite, 2034
Jams, 3402
J antenna, 187
Janus, orbital characteristics, 2761
Japanese beetle, 407, 2034
Japanese millet, 2351
Japanese umbrella pine, 2751
Japanese wax scale, 3204
Japonica rice, 2999, 3000
Jarosite, 2034
Jasper, 760, 2034
Jaspilite, 2034

JATO (jet-assisted take-off), 2034
Jaundice, 2035
Jay, 2035
J-display, 2035
Jejunum, 2, 98, 1102, 2035
Jellies (sweetener), 3402
Jellyfish, 857, 3119
Jerboa, 3022
Jersey cattle, 669
Jet, 2035
Jetavator, 2035
Jet engine, 2035
Jet-lag, 2035–2036
Jet nozzle, 2036
Jet propulsion, 78–81
Jet Propulsion Laboratory (JPL), 2036–2040
Jet pumps, 2885
Jet streams, 2040–2041
Jet thrust, 2041
Jet vane, 2041
Jewel tetra, 725
Jewfish, 381
Jigger (chigger), 761
Jigging, 804
Jimson weed, 3214
John Dories (marine fishes), 2041
Johnsongrass, 1679
Johnson-Matthey trap, 1085
Johnson noise, 2466
Joint (anatomy/geology), 2041
Jointworm, 2041
Joist, 397
Josephson, Brian David, 2041
Josephson junctions, 2041, 3386
Joule (unit), 3570
Joule, James Prescott, 2041
Joule law, 2041–2042
Joule per Kelvin (unit), 3570
Joule-Thomson coefficient, 2042
Joule-Thomson effect, 2042
Joule-Thomson expansion, 991
Joule-Thomson inversion temperature, 2042
Jovian, 2042
    forming Jupiter-like planets, 3574
Joystick, 908
J particle, 2876
Judas tree (redbud), 2958
JUDGE system, 250
Julian calendar, 588
Jump, 2042
Jump discontinuity, 1127
Jumping plant lice, 1782
Junco, 2042
Junction diode, 1120
Junction transistor, 3514–3515
June beetle, 2042
June bug, 1315, 2042
June solstice, 3375
Jungle-cat, 664, 666
Jungle fowl, 2042–2043
Juniper scale, 3204
Juniper trees, 2043
Juno, 274
Jupiter, 2042, 2043–2052
    comet Shoemaker-Levy 9 collision, 885–886,
        1804
    Galileo mission to, 275, 1554–1560, 2044
    Hubble Space Telescope images, 1804, 1819,
        2044
    orbital characteristics, 2759
    physical characteristics, 2760
    ring, 1558, 1560, 2047–2048, 2049

satellites, 2048–2051
Ulysses mission flyby, 2051–2052, 3564–3566
Voyager missions to, 3689–3692
Jupiter rocket, 3021
Jurara, 2053
Jurassic Period, 2053
Jurin virus, 224
Jury problem, 2053
Jute, 383

Kaffir, 3235
Kagu, 2939
Kame, 2055
Kanamycin, 196
Kangaroo, 2284, 2285
Kaolinite, 2055
Kaons, 2321–2322
Kaposi's sarcoma, 68, 2783
Karaya gum, 1698
Kármán momentum integral, 1445
Kármán vortices, 2055
Karst, 1877
Katamorphism, 162
Katmai Range, 3678
Katoptric system, 2055
Katydid, 2055
Kauri gum, 2977
K-display, 2055
Keewatin, 222
Kelvin (unit), 3570
Kelvin, William Thomson, 1st Baron,
    2055–2056
Kelvin bridge, 538
Kelvin law, 2055
Kelvin temperature scale, 3458
Kendall's τ (rank correlation), 2946
Kennelly-Heaviside layer, 1220, 2001
Kennison nozzle, 1444
Kentucky coffeetree, 2056
Kepler, Johanners, 2056
Kepler's laws of planetary motion, 294–295,
    2056–2057
    and gravitation, 1682
Keratin, 2057
Keratitis, 2057, 3652
Keratoconus, 2057–2058
Keratohyalin, 2057
Keratopathy, 3652
Keratosis, 595, 600
Kerma, 2058
Kernel, 2058
Kernicterus, 2058
Kernite, 864, 2058
Kerogen, 2687
Kerr cells, 2058–2059
Kerr effect, 1265, 2058–2059
Kestrel, 1351
Ketones, 2059, 2581
    nomenclature and examples, 2586
Ketonic acids, 2059
Ketose, 2059
Ketosis, 2142
Ketosis-prone/resistant diabetes, 1066
Kettle holes, 2055
Key (machine), 2059–2060
Keyframing, 904, 905
Keyhole urchins, 1201
Keypunch, 2060
Khamsin, 432
Khapra beetle, 2060
Kidney, 2, 1341, 2060–2067
    and acid-base regulation, 18

and sodium-potassium regulation, 2823
    transplantation, 2067
Kidney cancer, 2064
Kidney infection, 2064
Kidney stones, 2065
KidSat, 2068
Kilauea Volcano, 3677
Killed steel, 2017
Killer bees. *See* Africanized bees
Killing frost, 2840
Kiln, 581
Kilo (prefix for units), 3568
Kilogram (unit), 3570
Kimberlite, 1075, 2068, 2677
Kinematic chain, 2068
Kinematics, 2068–2069
Kinetic energy, 1304, 2069
Kinetic polarimetry, 2785
Kinetic potential, 2079
Kinetics (chemical reactions), 756–759
Kinetics (fluid flow), 1445
Kinetics (mechanics), 2069–2070
Kinetic theory, 323n, 2070–2071
Kinetins, 2767
Kinetochore, 677
Kingbird, 2071
Kingfishers, 952, 2071–2072
Kinglet, 3697
King William pine, 2751
Kinins, 2072
Kinkajou, 2911, 2913
Kirchhoff, Gustav Robert, 2072
Kirchhoff equations, 805
Kirchhoff laws of networks, 157, 1233, 2072
Kiroombo, 2071
Kissing bug, 2072–2073
Kite (bird), 1160
Kiwi (bird), 2073, 2955
Kiwifruit, 2073
Kjeldahl method, 162
Klein-Gordon equation, 2904
Klinefelter's syndrome, 1656
Klipspringers, 183, 184
Klystron, 2348
Klystron amplifier, 149
Klystron oscillator, 3162
Knockout studies, 1843
Knot (unit), 3570
Knot theory, 2302
Knowledge base, 261
Knowledge capture, 243
Knowledge reasoning, 243
Knowledge representation, 243
Knudsen diffusion, 1100
Knudsen flow, 2073
Knudsen gage, 2073
Koala, 2284, 2285
Kob, 183, 184
Kobego, 1056
Kodochrome process, 2725
Koehler illumination, microscopes, 2344
Kohlrausch law, 2073
Kohoutek (comet), 882, 2073–2074
Kolmogoroff-Smirnoff test, 2074
Konjac flour, 492
Köppen climate classification system, 806, 808
Korean hemorrhagic fever, 2074
Korrigum, 183, 184
Korsakoff's psychosis, 145
Korteweg-de Vries equation, 2294
Kossel, Albrecht, 2074
Kouprey, 520

Kraft method, 1043
Krait, 3194
Krakatau, 3676
Krebs citric acid cycle, 613
Krill, 990–991, 2794
Krypton, 2074–2075
k-statistics, 2075
Kudzu, 2075
Kuiper belt, 1808, 3328
Kullenberg corer, 955, 956
Kumquat, 802
Kurtosis, 2075, 2898
Kuru, 522, 531, 980
Kwashiorker, 1040, 2863
Kyanite, 2075

Labaria, 2077
Labeled atom, 322
Labiatae (mint family), 2360
Laboratory balances, 3741
Labrador current, 2077
Labradorite, 1362
Labyrinth fishes, 2077
Laccolith, 2077–2078
Lacewing, 212, 1654, 2445
Lachrymator, 2078
Lacquer, 2618, 2977–2978
Lacrimal system, 2078
Lactic acid, 613, 2078
    and oxygen debt, 2613
Lactic acidosis, 2078
Lactic fermentation, 1363
Lactose, 612, 3399
Lactose intolerance, 1092
Lady beetle, 407, 410, 2078–2079
Lady bug, 2078
Lag (angle of), 2079
Lagering, 406
Lagoons, for sludge treatment, 3715
Lagrange coefficients, 2079
Lagrange formula for interpolation, 2079
Lagrange-Helmholtz equation, 1754
Lagrangian coordinates, 945, 946, 1585, 2079
Lagrangian function, 2079
Lagrangian point, 2079
Laguerre differential equation, 2079
Laguerre polynomials, 1585, 2079
Lake fly, 2305
Lake Nyos disaster, 3683
Lakes (bodies of water), 1174
    limnology, 2138
Lakes (dyes), 872
Lamarckism, 1338
Lambda hyperons, 2322
Lambda particle, 2079
Lambert (unit), 3570
Lambert-Beer law, 873
Lambert projection, 917, 2079–2080
Lambert's cosine law, 2080
Lamb shift, 2080
Lamé equation, 2080
Lamella (botany/zoology), 2080
Lamellar corpuscle, 961
Lamellibranchiata, 2080
Laminar boundary layer, 519, 1447
Laminar flow, 54–55, 1445, 2080–2081
Laminates, 1377
Lammergeier, 3693
Lampblack, 621
Lamprey, 1015, 2081
Lancelets, 147, 708, 777

Landing gear, 2081
Landing skid, 2081
Landsat program, 3075–3078
Landvaults, 3721–3722
Langevin ion, 2090
Langley, Samuel Pierpont, 2081
Langmuir, Irving, 2081–2082
Langmuir-Blodgett films, 2366–2374
Language (computer), 2082
Langur, 2386
La Niña, 2082–2084
    and global temperature, 1638
Lantern fishes, 1941
Lantern fly, 2084
Lanthanide contraction, 2084
Lanthanide series, 2084
Lanthanum, 2084–2085
Lapis lazuli, 2109
Laplace development, 1064
Laplace equation, 162, 2085, 2293
Laplace theorem, 2085
Laplace transform, 2085–2089
Laplacian, 2089
Lapping, 2089
Laproscopic cholecystectomy, 1561–1562
Lapse rate, 308–309, 2089
Larch trees, 919, 2089–2090
Large intestine, 2, 98, 1102
    age-related disorders, 1611
Large ion, 1995, 2090
Large Magellanic Cloud, 1550, 1551, 1552
Large numbers, law of, 2108
Large radio telescope, 188, 189
Large-scale integrated circuits, 2185–2186
Lark, 2090
Larmor's theorem, 2090
Larva, 2090
Laryngitis, 2090
Larynx, 2090–2091, 3672
Laser, 2091–2102
    for eye surgery, 2103–2104
    fiber optics applications, 2566–2567
    fusion power application, 1539
    with mass spectrometry, 2290–2291
Laser chemistry, 2719
Laser Doppler flowmeter, 2100
Laser femtochemistry, 2719–2720
Laser glasses, 1630
Laser gyroscope, 2100
Laser in-situ keratomileusis (LASIK), 280, 2102–2103, 2964
Laser metrology, 2100
Laser recording, 2102
Laser spectroscopy, 2098
Laser thermal keratoplasty (LTK), 2104, 2964
Lassa fever (African hemorrhagic fever), 2104–2105
Lassa virus, 224
Last in, first out (LIFO), 2907
Last quarter moon, 2388
Latent heat, 2105
Latent rabies, 2911
Lateral, 2105
Laterite, 2105
Latex, 2105, 2616
Latham's brush turkey, 2399
Latin square, 2105–2106
Latitude, 2106
    geomagnetic, 1597
Lattice (mathematics), 2107
Lattice cell, 2353
Lattice compounds, 2106
Lattice constant, 2106

Lattice designs, 2106
Lattice dimensions, 2106
Lattice energy, crystal, 2106
Lattice sampling, 3060
Lattice water, 1854
Latus rectum (parabola), 2633
Lauan tree, 2231
Laughing gas, 632
Laughing jackass, 2071
Launch window, 2107
Laurel, 2107–2108
Laurel family, 2107–2108
Laurentian, 222
Lauric acid, 2108
Lauter tun, 406
Lava, 2108
Lava dome volcano, 3678
Lavoisier, Antoine, 2108
Law of large numbers, 2108
Law of mass action, 1322
Lawrencium, 37, 2108
Lawson criteria (fusion power), 1538–1539
Lawsonite, 2109
Lazulite, 2109
Lazurite, 2109
Lazy eye (amblyopia), 130
Lazy H antenna, 187
L-band, 2109
LC oscillator, 3162
L-display, 2109
LDL (low-density lipoprotein), 234
Lead, 2109–2111
    biological effects, 2111–2112
    Carbon Group member, 623
Lead-acid battery, 389–390
Lead-acid battery, for electric cars, 1231
Leadcable borer, 1146
Lead (painter's) colic, 864
Lead glasses, 1630
Lead poisoning, 2112
Lead screw, 2112
Lead zirconate titanate (PZT), for piezoelectric
    accelerometers, 11
Leaf, 2112–2114
Leaf aphis, 211
Leaf filter, 1395
Leafhopper, 1782, 2114
Leaf insect, 2114
Leaf miner, 864, 2114
Leaf nematodes, 2431
Leaf protein concentrates, 2864
Leaf roller, 2114
Leaf sewer, 2114
Leaf tyer, 2114
Leakage current, 2114
Leakage reactance, 2115
Leaky-pipe antenna, 187
Leapfrogging, 2115
Leapfrog test, 2115
Least energy principle, 2115
Least significant bit, 2979
Least significant digit, 2979
Least squares, 2115
Leather bating enzymes, 654
Leather-jacket, 2115
Leathery turtle, 2201
Leavening agents, 2115–2116
Le Châtelier's principle, 2116
Lechwes, 183, 184
Leclanche battery, 393
Lee, Tsung-Dao, 2116
Leeches, 177, 1771–1772

Leek, 105, 2116, 2559
Leeway, 2116
Left-handedness (system), 1713
Left-handed screw, 1713
Left-sided congestive heart failure, 916
Legendre differential equation, 2116–2117
Legendre polynomial, 1585, 2033
Legionellosis, 361, 2116
Legionnaires' disease, 355, 361, 2117
Legume, 1518
Leguminosae, 2117–2118
Leibniz rule, 1098
Leishmaniasis, 2118
Lemma, 2118
Lemming, 3022
Lemniscate of Bernoulli, 2118
Lemon tree, 800, 801
    nematodes affecting, 2432
Lemur, 2119
Lemuroids, 2119
Lenard effect, 2119
Length of a curve, 2119
Length-preserving maps, 2029
Length units, 3568
Lens antenna, 187
Lenses. See also Mirrors
    achromat, 17
    aperture, 210
    aplanatic system, 212
    contact lenses, 929
    curvature (total), 1008
    doublet, 1141
    electron, 1274
    electrostatic, 1281
    intraocular, 1985
    optical center, 2562
    principal planes, 2848
    thin-lens relationships, 3479
    variable-focus, 3605
Lenticels, 375, 2119
Lenticular clouds, 828
Lenz' law, 2119
Leo (the lion), 2119
Leonids (meteor shower), 2120
Leopard, 664, 665, 666
Leopard-cat, 664, 666
Lepidolite, 2120
Lepidoptera, 2120–2121
Leprosy, 361, 2121
Leptite, 1666
Leptokurtic curve, 2898
Lepton number, 926
Leptons, 2121–2122, 2648
Leptospirosis, 362, 2122
Lepus (the hare), 2122
Lesch-Nyhan syndrome, 2122
Lesion, 2122
Lesser apes, 191
Lesser grain borer, 1665
Lettuce, 893, 894
Lettuce aphis, 211
Leucine, 133–137
Leucite, 2122
Leucocyte-inhibiting factor, 2204
Leukemias, 2122–2123
Leukocytosis, 2123
Leukopenia, 2123–2124
Level, surveyor's, 2124
Leveling, 1154
Leveling solvent, 3233
Level of escape, 2124
Level width (excitation energy), 2124

Lever, 2207
Levo form, of stereoisomers, 137, 2028
Levorotary, 300
Levorphanol, 2393
Lewis acids and bases, 22
Lewis salt, 22
Leyden jar, 606
L'Hospital rule, 2124
Libra (the scales), 2124
Librations, 2124
Lice, 178, 1054, 2671
Lichen, 2124–2125
Licorice, 2125
Lidar, 2099, 2125
Lie group, 2125, 2301
Life, 3574
    artificial (by artificial intelligence), 245–247
    extraterrestrial, 1346
Life sciences, 2125
Ligand, 2125–2126
Ligases, 1314
Ligature, 2126
Light, 2126–2129
    aberration of, 2
Light amplifier, 149
Light beer, 404
Light-coupled switch, 2130
Light curve (astronomical), 2130
Light-emitting diodes (LEDs), 2191
    fiber optics applications, 2567–2568
    inorganic displays, 1941–1942
    organic, 2593–2596
Lightguides, 2563
Lighting engineering, 1905
Light-ion beams, 1539–1540
Lightning, 1515, 2130–2132
Lightning recorder, 3155
Light pen, 908
Light pillar, 319
Light pollution, 2133–2135
Light sources, 1900, 1904–1906
    optical microscopes, 2344
    for photoelectric object detectors, 2872–2874
Light time, 2135
Light-water reactors, 2480–2481
Light-year, 2135
Lignin, 1089, 2135
Lignite coal, 821
Lignum vitae, 2135
Likelihood, 2135
Liliaceae, 2135–2136
Lily, 2135
Limacon, 2136
Limb darkening, 2136
Limb-girdle dystrophy, 2407
Lime and limestone, 2136–2137
    addition to brick, 537
    use in animal feed, 583
Limestone terrain hydrology, 1877
Lime tree, 802
Lime (basswood) trees, 382
Limit, 2137
Limited-area pollution, 2799
Limit switch, 2137
    magnetostrictive, 2875
Limnology, 2138
Limonite, 2138
Limonoids, 801
Limpet, 2138
Limpkin, 2939
Linalool, 502
Linden trees, 382

Line (mathematics), 2140–2141
Linear, 2138
Linear accelerators, 2650
Linear alkanes, 2581
Linear amplifier, 149
Linear block codes, 1933
Linear energy transfer, 2138–2139
Linear hypothesis, 2139
Linear inequalities, 2139
Linearity, 2139
Linearity control, 2139
Linearly independent vectors, 2139
Linear magnification, 2140
Linear-passive network, 2444
Linear power amplifier, 149
Linear programming, 2140, 2300
Linear shading, 910
Linear space (vector space), 2140
Linear stepper motors, 3350–3351
Linear synchronous motor, 2943
Linear systems, 2140
Linear topological space, 2140
Line hydrophone, 1879
Line integral, 1964
    vectors, 3608
Line of Apsides, 2141
Line of nodes, 1207, 2141
Line of position, 2141
Line of striction, of ruled surface, 3047
Line printer, 2141
Line-reversal pyrometer, 2141
Line sampling, 3060
Line scanners, 2210
Line width, 2141
Linke scale, 318, 1013, 2141
Linoleic acid, 2141
Linolenic acid, 2141
Linolic acid, 2141
Linsang, 3670
Linseed oil, 1428
Lion, 664–665
Liouville equation, 2141
Liouville-Neumann series, 2141
Lipidoses, 2141–2142
Lipids, 2142
Lipoma, 2142
Lipoproteins, 234
Lippmann hologram, 1777
Liquation, 2142
Liquefied natural gas (LNG), 2415–2416
Liquid, 2164–2168
    flow measurement, 1435–1444
Liquid-ceramic process, 712
Liquid-column chromatography (LC), 784
Liquid crystal display technology, 659, 2142–2162
    calculators, 586
Liquid-crystal-on-silicon devices, 2158–2159
    microdisplays, 2331–2332
Liquid crystal polymers, 2162
Liquid crystals, 2162–2163
Liquid-expanded film, 1391
Liquid helium, 1753
Liquid-immersion freezing, 1498
Liquid-in-glass thermometer, 2163–2164
Liquid junction, 2164
Liquid lasers, 2096
Liquid-liquid extraction, 1345
Liquid mixing, 2167
Liquid-phase projectors (loudspeakers), 2189
Liquid-propellant rocket engine, 2164
Liquidus curve, 2168
    alloy phase diagrams, 107

Lisp program, 256
Lissajous, Jules Antoine, 2168
Lissajous technique, cathode-ray tubes, 661
Listerial meningitis, 2315
Listeriosis, 362, 2168
    foodborne, 1470
Litchi tree, 2168
Liter (unit), 3570
Lithifaction, 2168
Lithium, 2168–2170
    alkali metal, 98
    biological effects, 2170
    and bipolar illness, 2243
Lithium batteries, 394
    for electric cars, 1231
    for spacecraft, 3238
Lithofacies, 1349
Lithology, 2170
Lithometeor, 2170
Lithosphere, 1162, 2170
Litmus paper, 2125
Little Dipper (Ursa Minor), 3589
Littoral, 2170
Littrow prism, 2849
Lituus, 2170
Live, attenuated vaccines, 3593
Liver, 2, 2170–2174
    transplantation, 2173–2174
Liver abscess, 5
Liverworts, 560–562
Livestock louse, 2189–2190
Lizard, 2174–2178
Lizard fishes, 1941
Llamas, 593, 594
Loaches, 2179
Load factor, 1350, 2179
Loading coil, 2179
Load matching, 2179
Load-measuring cells, 3742
Loa loa, 1390
Lobachevskian geometry, 1597
Lobectomy, 2179
Lobe pattern, 2917
Lobe pumps, 2885
Lobster, 988, 989–990, 1031
Local apparent time, 2179
Local area networks (LANs), 2179–2182
    automation application, 334
    optical fiber applications, 2570
Local astronomical time, 2182
Local Group (galaxy cluster), 1552
Localized spin, 1369
Local lunar time, 2182
Local meridian, 2321
Local sidereal time (LST), 2182
Local storage, 2311, 2312
Location counter, 974
Lockjaw, 352, 3463–3464
Locomotion, 2182–2183, 2406
    in fish, 1407
Locust, 1680, 2183, 2603
Locust (seventeen year), 2183
Locust bean gum, 1698
Locust borer, 635
Locust trees, 2183–2184
LODAR navigation, 2428
Lodestone, 2355
Loess, 1315, 1325, 2184
Log, navigational, 2184
Logarithm, 2184–2185
Logarithmic amplifier, 149
Logarithmic antenna, 187

Logarithmic chart, 2185
Logarithmic decrement, 2185
Logarithmic function, 2185
Logarithmic spiral, 2185
Logic (computer system), 2185
Logical calculi, 247
Logical design, 2185
Logical element, 2185
Logical instructions, 1109
Logical model, 2363
Logical operation, 2185
Logic-based programming languages, 2859
Logic diagram (computer system), 2185–2186
Logic-level signals, 2853
Logic programming languages, 249
Logistello (game), 254
Logistic curve, 2186–2187
Logistic equation, 1096
Logit, 2187
Log mean temperature difference (LMTD), 2187
Log-periodic antenna, 187
Logwood tree, 2187
Lolly ice, 1895
Lonar Crater, 293
London dipole theory, 2187
Longitude, 2187
Long-period comets, 883, 3328
Long-range accuracy (LORAC), 2187
Long-range navigation (LORAN), 2187
Long range order, 2187
Longwall mining, 835–866
Looming, 2187
Loon, 2187
Loop (computer system/mathematics), 2188
Loop antenna, 187–188, 2187
Loop diuretics, 1136
Loop gain, 1544, 2188
Loop gain, in feedback, 1356
Loop-Vee antenna, 188
LORAN navigation, 2427
Lorentz, Hendrik Antoon, 2188
Lorentz covariant, 1388
Lorentz frame, 2188
Lorentz invariance, 2188
Lorentz transformation, 2188
Lorenz force equation, 1265
Lorin tube, 2945
Lorises, 2119, 2188
Lorisoids, 2188
Loss (transmission), 2188
Losser, 2188
Loss factor, 2188
Loss function, 2188
Loss-in-weight feeder, 1357
Loss matrix, 2188
Lost wax casting, 649
Lottery sampling, 3060
Loudness level, 2188–2189
Loudspeaker, 2189
Louse, 2189–2190
Louse-Borne epidemic, 3001–3002
Lovegrasses, 1679
Love-lies bleeding, 128
Low-alloy steels, 107, 108
Low-Earth orbit satellites, 3065, 3071–3073
Lower control limit, 934, 2898
Lower Curie point, 1007
Lower consolute temperature, 926
Lower Ordovician, 594
Lower specification limit, 2899
Low-field electron mobility, 3132
Low-head hydroelectric power, 1858–1859

Lowitz arc, 319
Low-level programming languages, 2858
Low-level transuranic waste, 2501
Low-pass filter, 1392
Low-pressure storms, 1510
Lowry-Brønsted acid-base theory, 22
Lows, 311
Low vacuum, 2190
Low-voltage air circuit breakers, 797
LSD-25, 701, 1712
L-section, 2190
Lubricant, 2190
    and friction, 1504
Luciferase, 430
Luciferin, 430
Ludwig's angina, 680
Luft balanced condenser microphone system,
    1937–1938
Lumbar disk disease, 509
Lumber, 3780–3782
Lumen (unit), 3568, 3570
Lumen per square foot (unit), 3570
Lumen per square meter (unit), 3570
Lumen per watt (unit), 3570
Lumen second (unit), 3570
Luminance, 551
Luminescence, 2191
Luminol, 430
Luminosity function, 2191–2192
Luminous coefficient, 2192
Luminous density, 1050
Luminous emissivity, 1296
Luminous emittance, 1296
Luminous fishes, 1409
Luminous intensity units, 3568
Luminous reflectance, 2963
Lumped capacitance, 605
Lumpy jaw (cattle condition), 39
Lunar atmospheric tide, 316, 2192
Lunar-based telescopes, 3448
Lunar eclipse, 1206, 2388
Lunar Prospector mission, 2192–2200
Lunar rocks, 2358, 2391
Lung, 2982
Lung abscess, 5
Lung cancer, 598–599
Lungfishes, 2200, 3423
Lunisolar precession, 1585
Lupine aphis, 211
Lupus (the wolf), 2200
Lurgi process, 844
Luster, 2200
Lustre, minerals, 2355
Lutein, 2200
Lutetium, 2200–2201
Luth, 2201
Lutong, 2386
Lux (unit), 3568, 3570
Lyases, 1314
Lyceum, 225
Lycopene, 2747
Lycopsida, 2201
Lyell, Sir Charles, 2201
Lyman-alpha radiation, 2201
Lyme arthritis, 237
Lyme disease, 355, 2201–2203
Lymph, 2203
Lymphatic filariasis, 1389–1390
Lymph glands (nodes), 2, 600
Lymph nodes, 2203, 3592
Lymphocytes, 484
Lymphocytic choriomeningitis (LCM), 2203–2204

Lymphocytic choriomeningitis virus, 224
Lymphogranuloma venereum, 2204
Lymphokines, 2204
Lymphoma, 2204
Lymphosarcoma, 2204–2205
Lynx, 664, 666
Lyocell fibers, 1380
Lyophilic aerosol, 868
Lyophilic sol, 868
Lyotopic series, 820
Lyotropic liquid crystal, 2162
Lyra (the harp), 2205
Lyre birds, 2205, 2655
Lyrids (meteor shower), 2205
Lysine, 133–137
Lysis (bacteriology/physiology), 2205
Lysithea, orbital characteristics, 2760
Lysosomes, 2205

Maar, 2207
Macadamia tree, 2207
Macaques, 2386
Macassar butter, 1412
Macaw, 2643, 2876, 2878
Macdonals seamount, 3678
Mace, 225
Mach, Ernst, 2207
Mac Hack (game), 254
Mach angle, 2207
Machine, 2207
    simple, 2207–2209
Machine forging, 1475
Machine key, 2059–2060
Machine language, 2082, 2207
Machine language code, 855
Machine learning, 243, 257–258
Machinery molds, 3802–3803
Machine translation systems, 261
Machine vision, 258–260, 2209–2213
    proximity detection application, 2875
    with robots, 3008
Machine word, 2213
Machmeter, 85–86
Mach number, 85–86, 3391
Mach principle, 2213
Machupo virus, 224
Mackerels, 2213–2214
Maclaurin series, 2214
Maclaurin trisectrix, 3529
Macleod equation, 2214
Macroassembler, 2214
Macro instruction, 1957
Macrolibraries, 2214
Macrolides, 197
Macromolecular electronics, 3229
Macromolecular science, 2214
Macrophage, 484, 2705
Macrophage-activating factor, 2204
Macroscopic, 2215
Macrostatements, 2214
Macula, 2215
    age-related degeneration, 66–67
Macular degeneration, age-related, 66–67, 2984
Mad cow disease, 521
Madonna tree, 1734
Mafic, 2215
Magellanic Clouds, 1549–1550, 1551, 1552
Magellan mission to Venus, 2215–2217, 3626–3627
Maggot, 2217–2218
Magic acid, 23
Magma, 2218

Magnesite, 2218
Magnesium, 2218–2221
    anodization, 178
    in biological systems, 2221–2222
    fertilizer requirements, 1372
    and hypertension, 1091
Magnesium alloys, 108
Magnesium hydroxide, antacid, 182
Magnesium trisilicate, antacid, 182
Magnetically levitated vehicles (maglev), 2943
Magnetic amplifier, 149
Magnetic circuit, 2225
Magnetic cooling, 991–992
Magnetic dipole, 1123
Magnetic domains, 2224–2225
Magnetic double refraction, 2223
Magnetic equator, 24
Magnetic field, 1384, 2223, 2224–2225
    galaxies, 1546–1547
Magnetic field intensity, 2223
Magnetic films, 3478–3479
Magnetic flowmeters (magmeters), 1438–1439
Magnetic flux, 2225
Magnetic inclinometer, 1123
Magnetic induction, 1915–1916, 2223
Magnetic ink character recognition (MICR),
    2223–2224
Magnetic iron ore, 2227
Magnetic K-indices, 2224
Magnetic lunar daily variation (L), 2224
Magnetic materials, 2225–2226
Magnetic moment, 2224
Magnetic moment (particle), 2224
Magnetic north, 2471
Magnetic-particle NDT methods, 2469
Magnetic petroleum surveys, 2690
Magnetic pressure, 2224–2225
Magnetic proximity detectors, 2874–2875
Magnetic pyrites, 2892
Magnetic quantum number, 2904
Magnetic refrigeration, 2969
Magnetic refrigerator, 992
Magnetic resonance imaging (MRI), 2476–2479
Magnetic storm, 2225
Magnetic tape storage (computer), 2224
Magnetism, 2224–2226
Magnetite, 2226–2227
Magnetizing current, 1340
Magneto, 2227
Magnetoelastic transducers, 30
Magnetoelectric, 2227
Magneto-fluid dynamics, 1445
Magnetohydrodynamic generator, 1243, 2227–2228
Magnetohydrodynamics (MHD), 2228
Magnetoionic theory, 2228
Magnetometer, 2228–2229
Magneto-optical rotation, 2229
Magnetoresistance, 1223
Magnetosphere, 2229
    Jupiter, 2051
    Mercury, 2320–2321
    Neptune, 2439–2440
    Saturn, 3092–3094
    Uranus, 3585–3586
Magnetostriction, 2229
Magnetostrictive delay line, 1045, 2229
Magnetostrictive limit switch, 2875
Magnetotactic bacteria, 1169
Magnetotrail, 1169
Magnetron, 2347–2348
Magnetron amplifier, 149–150
Magnifying power, 2229–2230

Magnolia trees, 2230, 2231
Magot (monkey), 2386
Magpie, 2230
Mahalanobis' generalized distance, 1152
Mahogany trees, 2230–2231
Maidenhair tree, 2231
Maillard reaction, 138
Main memory, 2311, 2313
Main storage, 3358
Maize. *See* Corn
Majoram, 2360
Maksutov-Bouwers telescope, 2231–2232
Malachite, 2232
Malacology, 913
Malaria, 2232–2235
Malaria vaccines, 2235
Malaysian prawn, aquaculture, 215–216
Maleic anhydride, 2685
Maleic hydrazide, 2767
Male infertility, 1923
Maleo, 2235, 2308, 2399
Male pattern baldness, 110
Male sex hormones, 1656, 1793, 2755
Malignant cancer, 595, 600
Malignant hypertension, 3118
Malignant hyperthermia, 2410
Malleable iron, 107
Mallee fowl, 2399
Mallophaga, 2235
Mallow family, 2237
Malt, 2235–2237
Malta fever, 559
Malt beverages, 404–406
Maltitol, 3400
Maltodextrin, 3399
Maltose, 612
Malus cosine-squared law, 2237
Malvaceae (mallow family), 2237
Mambras, 3194
Mammalia, 2237–2238
    cloning, 819–821
Mammary gland, 535, 2238
Mammatus, 828
Mammography, 598
Manakin, 727, 2238
Manatee, 3119
Mandelbrot set, 1491
Mangabey, 2386
Manganese, 2238–2240
    in biological systems, 2241
    effect on steel, 2018
    removal from water, 3713
Manganese nodules, 2239, 2554–2555
Manganite, 2241
Mangel, 407
Mange mite, 2362
Mango tree, 637
Mangrove tree, 2241–2242
Manhattan Bridge, 544–545
Manhattan Project, 2242–2243
Manic-depressive (bipolar) illness, 2243
Manifold (topology), 3505
Manilla paper, 1
Manometer, 2244, 2845–2846
    absolute, 6
    differential, 1097
    micromanometer, 2341
Mantis, 410, 2244, 2603
Mantis fly, 2244
Mantispas, 2244
Mantle (Earth), 1164–1165, 1167–1168
Mantoux test, for TB, 3534–3535

Manucode, 2656

Manufacturing automation, 330–331

Manufacturing cells, 334

Manufacturing Message Specification (MMS), 2244–2252

Manufacturing Planning and Control Systems, 334

Many-body force, 2252

Many-to-few matrix, 1032

Maple trees, 2252–2254

MAP (Manufacturing Automation Protocol), 334

Map-matching guidance, 2254

Mapping (mathematics), 2254

Map projections, 2254

Maracaibo wood, 528

Marangoni effect, 1456

Marasmus, 1040

Marble bone disease, 509

Marburg-Ebola virus disease, 2254

Marburg hemorrhagic fever, 2254–2255

Marburg virus, 1391–1392

Marcasite, 2255

March equinox, 3629

Marching problem, 2255

Marconi, Guglielmo, 2255

Marconi antenna, 188

Marconi-Franklin antenna, 188

Mare, 2255

Margarines, 3613

Marginal seas, 1316

Mariculture, 214

Marijuana, 605, 1758, 2255–2256

Marine environment, 1207

Mariotte's law (Boyle's law), 528

Maritime air, 72

Markov, Andrei Andreyvich, 2256

Markov autoregression scheme, 343

Markov chain, 2256

Markov process, 2256, 2257

Marlin, 423

Marmoset, 2257, 2385

Marmot, 3319, 3320

Mars, 2257–2283

    Hubble Space Telescope images, 1820, 1821

    orbital characteristics, 2759

    Pathfinder mission to, 2657–2666

    physical characteristics, 2760

    satellites, 2273–2274

    Viking missions to, 2258, 2276–2283, 3633–3638

Marsh-antelopes, 183, 184

Marsh gas, 2328

Mars Observer, 297

Mars Polar Lander, 247

Marsupialia, 2284–2285

Marsupial mole, 2379

Marten, 2408

Martensitic stainless steels, 785

Martian rocks, 2358

Martin, 2286

Martingale, 2286

Mascon, 2286

Maser, 2286–2287

Mask (computer system), 2287

Masonry dams, 1024–1025

Mass, 2287

    center of, 687, 2287

    conservation of, 925

    origin of: Higgs field, 2652

Mass defect, 425, 2287

Mass-energy equivalence, 2287

Mass excess, 2287

Mass extinction, 2287–2289

    and asteroid collisions, 277–278

    and comet collisions, 885

    iridium as clue to, 2002

    in ocean, 2552

Mass flowmeters, 1440–1441

Massive pulmonary embolism, 236

Mass-luminosity relation, 2289

Mass memory, 2311

Mass number, 2289

Mass spectrometry, 2289–2292

Mass units, 3568

Mastic, 637

Masticatory substances, 2292–2293

Mastitis, 535, 2293

Mastoiditis, 1723

Material-transfer network, 2442

Mathematical lumping, 1398

Mathematical model, 2363

Mathematical physics, 2293

Mathematical symbols, 2293

Mathematics, 2293

    state of the art reviews, 2293–2303

Matrix

    adjacency, 2304

    inversion, 2305

    mathematics, 2304–2305

    triangular, 2305

Matrix (translator), 3516

Matrix tree theorem, 2304

Matter-antimatter symmetry, 2903

Mauve, 872

Maven (game), 254

Mavis, 2656, 3483

Maximum, 2305

Maximum boiling point, 502

Maximum freezing point, 1499

Maxwell (unit), 3570

Maxwell, James Clerk, 2305

Maxwell bridge, 541–542

Maxwell commutator bridge, 540

Maxwellian fluid, 1445

Maxwell's equations, 1264–1265

Maxwell's law of reciprocal deflections, 1041

Maxwell-Wagner effect, 1083

May beetle, 2042

Mayer, Maria Goeppert, 2305

Mayfly, 1315, 2305–2306

McArdle's disease, 2410

McCabe software complexity metric, 3208

Mcleod gage, 2306

McMath-Pierce Solar Telescope, 3381

M-display, 2306

Mead, 1782

Meadow-brown, 2306

Meadowlark, 2090, 2306

Mealybug, 1782, 2306

Mealyworms, 1664–1665

Mean, 225, 2898

Mean deviation, 2306

Mean life, 2306

Mean sea level, 2306

Mean solar day, 2306

Mean solar second, 2306

Mean solar time, 3492–3493

Mean square error, 2306

Mean value theorems, 2306–2307

Measles, 2307

Measured signal, 3164

Measurement

    air velocity, 3767–3768

    angular (eccentricity correction), 173

    ballastic, 368

    displacement, 2817–2819

    electric power, 1235–1239

    electrolytic conductivity and resistivity, 1261–1263

    frequency, 1501–1502

    humidity, 1882–1883

    liquid and gas flow, 1435–1444

    position, 2817–2819

    speed, 3609, 3613–3615

    thickness, 3475–3476

    torsional vibration, 3507

    wind velocity, 3767–3768

Measure of location, 2307

Measuring worm, 2307

Mechanical balance, 367–368

Mechanical equilibrium, 1321

Mechanical equivalent of heat, 2307–2308

Mechanical erosion, 1324–1325

Mechanical hysteresis, 349

Mechanical pulp, 2881

Mechanical resistance, 2978

Mechanics, 2308

Mechanization, 333

Mechlorethamine, 596

Mecoptera, 2308

Median, 2898

Mediastinitis, 5

Mediators, 199

MEDIATOR system, 250

Medical imagery, 3796–3797

Medical imaging, 2739

Mediterranean flour moth, 1665

Mediterranean fruit fly, 1521

Mediterranean water, 2308

Medium-alloy steels, 108

Medium-Earth orbit satellites, 3065, 3069–3071

Medium-head hydroelectric power, 1858–1859

Medlar tree, 3041

Medulla oblongata, 691–692

Medusa (or hydromedusa), 2308

Medusoid, 2308

Meerkat, 3670, 3671

Meerwein-Ponndorf-Verley reduction, 2590

Mega (prefix for units), 3568

Megaloblastic anemias, 167

Megaloblasts, 483–484

Megapode, 2308

Megasporangium, 2908

Meibomian glands, 3365

Meissner effect, 2308, 3384–3385

Mel, 2308

Melaleuca tree, 2308

Melamine, 132, 2308–2309

Melamine amino resins, 139

Melanin, 600

Melanocytes, 600

Melanoma, 600, 3181

Melatonin, 1302, 1616, 2749

Melioidosis, 362–363, 2309

Melon aphis, 211

Melt spinning, 3311

    polyester fibers, 2809

Membrana propria, 1319

Membrane separations technology, 2309–2310

Memory, 702–703

    and hippocampus, 692

Memory, electronic, 2310–2314

Memory hierarchy, 2310

Mendel, Gregor Johann, 2314

Mendelevium, 37, 2314

Mendeleyev, Dimitri, 2314

Mendelian genetics, 1760–1761
Menhaden, 1411, 2314–2315
Ménière's syndrome, 1724
Meningitis, 352, 363, 714, 2315
    brain injury, 532
Meningococcal infection, 355, 363
Meniscus, 2315–2316
Menopause, 1659
Meperidine, 2393
Mercaptans, 2316
6-Mercaptopurine, antimetabolite, 203
Mercator projection, 2316
Mercator sailing, 2316
Merchant grain beetle, 1145
Mercurial diuretics, 1136
Mercurian silver, 128
Mercury
    orbital characteristics, 2759
    physical characteristics, 2760
Mercury (element), 2316–2318
Mercury (planet), 2318–2321
Mercury-arc rectifier, 2957
Mercury-arc tube, 221
Mercury barometer, 377
Mercury fulminate, 1343
Merganser, 180, 3715
Meridian, 2321
Meridian circle, 2321
Meridional cell, 309
Merlin, 1351
Meromorphic function, 2321
Merozoites, 269
Merrill parauque, 2456
Mesa, 2321
Mesh method of analysis, circuits, 1125
Mesocarp, fruit, 1517
Mesoderm, 1291, 1614
Mesodesmic structure, 2321
Mesomorphic materials, 2143
Mesonephric duct, 1341
Mesons, 426, 1364, 2321–2322, 2646
Mesophase, 2142
Mesosphere, 304, 759, 2322
Mesozoic, 2322
Mesquite tree, 2322
Messenger RNA, 679, 1590, 1829–1830
Metachrosis, 2322
Metal casting, 649
Metal halide lamps, 1902
Metal hydrides, 1867–1868
Metal-insulator-metal (MIM) liquid crystal displays,
    2149–2150
Metallic glasses, 108
Metallic pigment, 2617
Metallic radius, 745, 747–748
Metallobiomolecules, 2323
Metallography, 2323
Metalloid, 2323
Metalloproteins, 2323
Metallothioneins, 2323
Metallurgy, 2323
    extractive, 1346
    phase diagram, 2706
Metal-mark, 2323
Metal-matrix composites, 123
Metal-organic chemical vapor deposition (MOCVD),
    coatings, 2621
Metal-organic compounds, 896
Metal oxide semiconductors, 3132
Metals, 2323–2324, 3231
    creep, 979
    fatigue, 1354

free electron theory of, 1493
    hardening, 1714
Metamere, 2324
Metamorphic rocks, 2324, 2356
Metamorphosis, 2324
    pupa, 2885
Metaphase, 596, 677
Meta position, on aromatics, 2585
Metastable-helium magnetometer, 2229
Metastable nuclei, 2324
Metastable state, 2324
Metastable system, 2324
Metastasis, 595, 600
Metatarsus adductus, 829
Meteoritic (meteor burst) communications, 2327
Meteoroids and meteorites, 2324–2327
    ablation, 3
Meteorological extremes, 1175
Meteorology, 2327–2328
    mesoscale and microscale, 3732
    radar application, 3735–3736
Meteor shower, 2328
    Bielids, 421
    Leonids, 2120
    Lyrids, 2205
    Perseids, 2682
    radiant point, 2916
Meteor trail, 1995
Meter (unit), 3570
Methadone, 2393
Methane, 98, 2328
    greenhouse gas, 1641
Methanogens, 2328
Methanol, 2328–2330, 2685
    catalysts for production, 654
Methanol diesel engine, 1086
Methicillin, 194, 195
Methionine, 133–137
Method of components, 2328
Methotrexate, 595, 596
    antimetabolite, 203
Methyl alcohol (methanol), 654, 2328–2330, 2685
Methyl chloride, 769
Methylene chloride, 769
Methylene radical, 610
Methyl methacrylate plastics, 35–36
Methylphenethylamine, 146
Metopon, 2393
Metric space, 2330
Metric topology, 2330
Metrology, 2330
Metronic cycle, 2330
Mexican bean beetle, 407
Mexican fruit fly, 1521
Mexican oregano, 2579
Mho (unit), 3570
Mica, 2330
Mice. See Mouse
Michelson-Morley experiment, 2971
Micri (prefix for units), 3568
Microchip, 3132
Microcline feldspar, 1362
Microclusters, 3134
Microcrystalline, 2353
Microcrystalline cellulose, 491
Microdisplays, 2330–2335
Microelectromechanical systems, for microdisplays,
    2332
Microelectronics, 2335–2340
Microfilm, 2340
Microfiltration, 2310
Microform, 2340

Microgravity, 1687–1690
    and calcium metabolism, 584
    and materials processing, 2340–2341
Micro instruction, 1957
Micromanometer, 2341
Micromechanical methods, 158
Micrometeorite, 2341
Micrometer, 2341
Micrometer (unit), 3570
Micron (unit), 3570
Micronutrients, 1372, 1373
Microphage, 2705
Microphone, 2341–2342
    infrared detector, 1937–1938
Microprocessor, 2342–2343
Microprogram, 2343
Microscope
    electron, 1274–1276
    scanning tunneling, 3106–3108
    ultramicroscope, 3557
Microscope (traditional-optical), 2343–2347
    resolving power, 2979–2980
Microscopic anatomy, 163
Microscopic reversibility, 368
Microspacecraft, 3245
Microstepping, 3349
Microwave acoustics, 30–31
Microwave amplifier, 150
Microwave circulator, 798
Microwave ferrite, 1368
Microwave freeze-drying, 1498
Microwave radiation, 2347
Microwave region, 2347
Microwave telephony links, 3430–3431
Microwave tubes, 2347–2348
Micturation, 2062
Midbrain, 692
Midcourse guidance, 1697
Middle ear, 325, 1721–1722
Middle note (flavors), 1419
Midge, 1123–1124, 2348
Mid-ocean ridges, 2529–2530
Mie scattering, 3109
Migraine headache, 1719
Migration-inhibition factor, 2204
Mil (unit), 3570
Milankovitch, Milutin, 2348–2350
Milankovitch theory, 2349–2350
Mile, nautical (unit), 3570
Mile, statute (unit), 3570
Milkfish, 2350
    aquaculture, 216
Milkweed, 2350, 3367
Milky quartz, 2905
Milky Way Galaxy, 968, 1544, 1547
    Hubble Space Telescope images, 1815, 1818
    and Local Group, 1552
    and vast galaxy drift, 3606–3607
Miller indices, 995, 2350, 2354
Millerite, 2350
Millet, 1679, 2350–2351
Milli (prefix for units), 3568
Millikan, Robert, 2351–2352
Millimicrosecond, 2411
Milling machine, adapative control, 42–43
Millipede, 237, 1123, 2352
Mimas, 3097
    orbital characteristics, 2761
Mimetite, 2352
Mimicry, 2352
Mimosa tree, 9
Mind-body problem, 239

Mineralocorticoids, 49
Mineralology, 2352–2359. *See also* specific
    Minerals
Minerals (effect on biological systems)
    calcium, 583–585
    chloride, 764–765
    chromium, 787
    cobalt, 853
    copper, 951–952
    iodine, 1993–1995
    iron, 2007–2010
    lead, 2111–2112
    lithium, 2170
    magnesium, 2221–2222
    manganese, 2241
    nickel, 2455–2456
    phosphorus, 2716–2717
    potassium, 2822–2823
    sodium, 2822–2823
    sulfur, 3372–3374
    zinc, 3814–3815
Mini-dose heparin, 201
Minimax (saddle point), 3053
Minimax method of estimation, 2360
Minimill concept, 2014
Minimum, 2305
Minimum boiling point, 502
Minimum freezing point, 1499
Minimum ionization, 1999
Minivet, 3160
Mink, 2407–2408
Minkowski space, 3364
Minor (of a matrix), 2360
Mint family (labiatae), 2360
Minute, time (unit), 3570
Miracidium, 2090
Miraculin, 3402
Mirage, 319–320
Miranda, 2439, 3585
    orbital characteristics, 2761
Mira (omicron Ceti), 2360
Mirror nuclides, 2360
Mirrors, 2360–2361. *See also* Lenses
    Foucault rotating, 1487
    Fresnel, 1504
    off-axis parabolic, 2556
    optical telescopes, 3446
    sign convention, 3164
Mir space station, 3268–3274, 3277
    remote sensing application, 3078
    Space Shuttle cooperative program, 3274–3275
Miscibility, 2361
Misoprostol, 3556
Mispickel, 232
Missilry, 2361
Mississippian Period, 2361
Missouri minnow, 634
Missouri sucker, 478
Mist, 2841
Mite, 10, 220, 2361–2363
Miter gearing, 420
Mithan, 520, 521
Mitochondrial myopathy, 2010
Mitosis, 596, 676
Mitotic apparatus, 677
Mixed crystal, 1003
Mixed ether, 1331
Mixed ketones, 2059
Mixed salt, 3058
Mixed sampling, 3060
Mixer, 2363
Mixing, 402

Mizar, 2363
Mnemonic operation code, 855
Mobile telephone, 3428, 3435
Mock moon/moon ring, 319
Mock sun/sun ring, 319
Modacrylic fibers, 1380
Mode, 2898
Model (scientific), 2363
Model atmosphere, 307
Modem, 3427, 3429
Moderation (neutron degradation), 1043
Moderator, 2363, 2480
Modulated amplifier, 150
Modulation, 2363–2364
Modulus, 2364
Modulus of elasticity, 2364
Modulus of rupture, 2364
Mohair, 1648, 2364
Mohorovicic, Andrija, 2364
Mohorovicic Discontinuity, 1164, 2364
Mohr-Westphal balance, 1050
Moh's hardness scale, 1714
Moire pattern, 2364
Moist-adiabatic lapse rate, 309
Moist-adiabatic processes, 307–308
Moisture-retaining agents, 1853
Molal concentration, 2365
Molar concentration, 2365
Molar heat, 2365
Molar refraction, 2964
Molds, 3800–3803
Mole (animal), 578
Mole (stoichiometry), 2365
Mole (unit), 3570
Mole cricket, 2365
Molecular beam, 2374–2375
Molecular biology, 2375–2376
Molecular clock, 2376
Molecular compounds, 896
Molecular distillation, 2376
Molecular electronics, 2365–2374
Molecular magnets, 2226
Molecular medicine, 1843
Molecular radio emission lines, 2933
Molecular sieve, 2376
Molecular sieve dehydration, 1151
Molecular spectra, 2377
Molecular streaming, 3360
Molecule, 2376–2378
Mole fraction, 2378–2379
Moles (animal), 2378–2379
Mole volume, 2379
Mollaret's meningitis, 2315
Mollier chart, 2379
Mollisols, 3211
Mollusca, 2379–2380
Molluscum contagiosum, 2380
Mollusks, 2380–2384
Moloch, 64–65
Molt, 465
Molybdenite, 2384
Molybdenum
    in biological systems, 2384
    effect on steel, 2018
    and nitrogen fixation, 2464
Moment, 2384–2385
Moment of inertia, 2384
Momentum, 2385
    conservation of, 925
Monal, 2707
Monarch butterfly, 575
Monatomic gases, 739

Monazite, 2385
Monel, 108
Monellin, 3402
Mongoloidism, 1141
Mongoose, 3670
Moniliasis, 602
Monitor, 2178
Monitoring amplifier, 150
Monkey bread, 373
Monkey puzzle tree, 221
Monkeys, 2385–2386
Monoceros (constellation), 2386
Monochlorobenzene, 769
Monochromatic, 2386
Monochromatic emissivity, 1296
Monochromator, 2386
Monoclonal antibody, 448
Monocoque, 2386
Monocyclic alkanes, 2581
Monocytes, 484
Monoecious plants, 2386
Monoethanolamine, 1330
Monolayer, 2386
Monomer, 2386
Monomode optical fibers, 2565
Monomolecular layer, 2386
Monomolecular reaction, 757
Monopole antenna, 188
Monorail, 2941
Monosaccharides, 610
Monosaturated vegetable oils, 3609
Monosodium glutamate, 442–443, 1427
Monotremeta, 2387
Monte Carlo method, 2299, 2387
Montreal Protocol, 2790
Moon, 2387–2392
    color analysis of images, 2737–2738
    geology, 2193–2194
    Lunar Prospector mission, 2192–2200
    orbital and physical characteristics, 2760, 2761
Moon's type (Braille), 530
Moose, 1039, 1040
Moraine, 1325
Morel, 267
Morera theorem, 2393
Morganite, 417
Mormon cricket, 2393
Mormyrids, 2393
Morphine, 100, 153, 2393–2394
Morphing, 904
Mortise gear, 421
Morton's neuralgia, 2444
Morula, 823
Mosaic vision, 2394
Mosquito, 1123, 2394–2395
    and malaria, 2232–2235
Mosquitofish, 3671
    aquaculture, 216
Mössbauer, Rudolf Ludwig, 2395
Mössbauer effect, 2395
Mosses, 559, 562–563
Moth, 657, 2120, 2395–2396
Mother Carey's chicken, 2684
Moth-miller, 603
Motion
    absolute, 6
    Brownian, 558–559
    Gauss principle of least constraint, 2396
    harmonic, 1716
    of planets, 2056–2057, 2762–2763
    reducing, 2959
    retrograde, 2985

Motion capture, 906–907
Motion control systems, 335–336
Motion picture cameras, 2732–2735
Motmot, 952, 2071
Motor, electric, 2396–2399
Motor-booster, 511
Motor cortex, 693
Mouflon, 1649, 1650
Moult, 2399
Mound birds, 1563, 2399–2400
Mountain, 2400
Mountain ash tree, 269
Mountain soils, 3211
Mountain waves, 315
Mount Etna, 3675–3676
Mount Pelée, 3676
Mount Pinatubo, 3676
Mount Saint Helens, 3682–3683
Mount Unzen, 3676
Mount Vesuvius, 3676
Mourning cloak, 2400
Mouse (animal), 2405, 3022
Mouse (computer input device), 907–908
Mousebird, 864, 2400
Mouse-deer, 1040, 3510
Mouth, 98
Movable span bridge, 550
Moving average, 2400
Moving-coil microphone, 2342
Moving phase, chromatography, 780
MRP-II (Manufacturing Resources Planning), 334
MRP-I (Materials Requirement Planning), 334
Mucilages, 1697–1699
Mucormycosis, 2400
Mucus, 2400
Mud dauber, 2400
Mudfish, 527
Mueller bridge, 538
Mugwort, 232
Mulberry family, 2400–2402
Mule, 1799
Muller's glass, 2560
Mullets, 2402
Multi-infarct dementia, 128
Multimode optical fibers, 2565
Multinomial (polynomial), 2811
Multiphase fluid flow, 2167–2168
Multiphase sampling, 3060–3061
Multiphoton resonance ionization mass spectrometry, 2291
Multiple-address instruction, 1957
Multiple alleles, 103
Multiple fruit, 1519
Multiple integral, 1963, 2402
Multiple interferometer determination of trajectories (MIDOT), 2402
Multiple object phase tracking and ranging (MOPTAR), 2402
Multiple scattering, 3109
Multiple sclerosis, 2402–2403
Multiplet, 2403
Multiple-tuned antenna, 188
Multiplexed twisted-nematic LCDs, 2146
Multiplexers, 3434–3435
Multiplication, 2403
Multiplicity, 2403
Multiposition controller, 2403–2404
Multiprocessing, 1112
Multiprogramming, 2404
Multivariate analysis, 2404
Mumps, 2404, 2642
Munia, 2656

Munsell chroma, 779
Muntjac, 1039, 1040
Muon, 2405
Muonium, 2405
Murine, 2405
Murine endemic, 3001
Murray bridge, 538
Musa antenna, 188, 2405
Muscle, 2405–2406
Muscovite, 2406
Muscular dystrophy, 2406–2407
Muscular system, 163
Mushroom, 65–66, 380–381, 1533
Mushroom-alcohol intolerance, 1469–1470
Mushroom maggot, 2218
Mushroom mite, 2362
Mushroom poisoning, 1469–1470
Musical instruments, 27–28
Musical sounds, 25–30
Musk-deer, 1039
Muskeg, 2407
Muskellunge, 2748
Muskmelon, 1005
Muskox, 1649
Muskrat, 3022
Mussel, 802, 2080, 2382
Mustard, 533
Mustelines, 2407–2408
Mutagens, 2408, 2409
Mutation, 679, 1590–1592, 2408
    and aging, 428, 1615
    point, 1591, 2375–2376
Mutualism, 3406
Myalgia, 2408
Myasthenia gravis, 2408–2409, 2409–2410
Myasthenic laryngitis, 2091
MYCIN expert system, 251, 252
Myclonic type seizure, 3126
Mycosis fungoides, 3181
Myelin, 696
Myelin sheath, 696
Myeloma, 509, 2409
Mylonite, 1431
Myna, 2409
Myocardial anoxia, 2026
Myocardial infarction, 2026, 2027
Myofibrils, 1382, 2405
Myoinositol, 1942
Myopathy, 2409–2410
Myopia (nearsightedness), 280, 2410, 3649–3650
Myosin, 931
Myotonic dystrophy, 2407
Myristic acid, 2410
Myrmecophile, 2410
Myrrh, 887, 2978
Myxedema, 3486

Nadir, 2411
Naevus (port-wine stain), 170
Nafcillin, 194
Naiad, 1754, 2411
Nalorphine, 2394
Naloxone, 101, 699
NAND circuit, 2411
Nannoplankton, 2411
Nano (prefix for units), 2411, 3568
Nanosecond (nsec), 2411
Nansen bottle, 2411
Naphtha, 2701
Naphthalene, 849
Napiergrass, 1679

Napoleonite, 965
Narcotics, 2393–2394
NASA Astrobiology Institute, 280, 289–292
NASA Astrobiology Roadmap, 281–292
Nasolacrimal, 2411–2412
Natalgrass, 1679
National Acid Precipitation Assessment Program (NAPAP), 21, 2804
National Atmospheric Deposition Program (NADP), 20–21
National Bureau of Standards (NBS), 2412
National Center for Atmospheric Research (NCAR), 2412
National Institute of Standards and Technology (NIST), 2458
National Research Council (NRC), 2412
National Surface Water Survey, 19
Natroalunite, 126
Natrolite, 2412
Natural bridge, 1324
Natural coordinates, 946, 2412
Natural frequency (mathematics), 2412
Natural gas, 2412–2421, 2687
Natural gas burners, 571
Natural language processing, 243, 261
Natural language understanding, 249
Natural logarithm, 2185
Natural rubber, 3044–3046
Natural selection (Darwinism), 1338
Nautical mile, 2421
Nautical mile (unit), 3570
Nautical twilight, 2421
Nautilus, 708
Navel orangeworm, 1147
Navier-Stokes equations, 2421
Navigation, 2421–2428
    automotive electronics for, 342
Navigation satellites, 3073–3075
Navigation sensors, 2422–2423
Navigator's stars, 2428
N-display, 2429
Neap tide, 3492
Nearest-neighbor mesh, 1113
Near-infrared region, 1935
Nearsightedness (myopia), 280, 2410, 3649–3650
Nebula, 2429–2430
    Hubble Space Telescope images, 1814, 1815
Necrosis, 2430
Necrotizing fascitis (flesh-eating bacteria), 359–360
Néel temperature, 202, 2430
Negative adsorption, 52
Negative azeotropy, 345, 1133
Negative ion, 1995
Negative kinetic energy (quantum mechanics), 2069
Negative resistance amplifier, 151
Negatron, 2122, 2430
Negro bug, 2430
Neighborhood of a point, 2430
Neil's parabola, 2633
Nekton, 2430, 2550
Nematic liquid crystal, 2143, 2162
Nematodes, 2430–2432
Nemertea, 2432
Nemertine worms, 25
Neo-Darwinism, 1338–1339
Neodymium, 2432–2433
Neohemocyte, 487
Neon, 2433
Neonatal, 2433
Neoplasm, 595, 2433–2434
Neoprene, 1220
Neosugar, 3402

Neo-Synephrine, 101
Neper (unit), 3570
Nepheline, 2434
Nephelometry, 2434
Nepheloscope, 2434
Nephoscope, 2434, 2843
Nephrite, 2033, 2034
Nephrolithiasis (kidney stones), 2065
Nephrology, 2060
Nephrometer, 2842–2843
Nephron, 2060, 2063–2064
Nephrotic syndrome, 2064
Neptune, 2434–2440
    Hubble Space Telescope images, 1821
    orbital characteristics, 2759
    physical characteristics, 2760
    ring system, 2436–2437
Neptunium, 37, 2441–2442
Neptunium series, 2925
Nereid, 2438
    orbital characteristics, 2761
Nernst, Walther Hermann, 2442
Nernst calorimeter, 590
Nernst effect, 2442
Nernst glower infrared source, 1940
Nernst heat theorem, 2442
Nernst principle of superposition, 3391
Nernst-Thompson rule, 2442
Nerve growth factor (NGF), 705
Nervous system, 163, 688–705
Nervous system disorders, 704, 530–531
Nessler tube, 873
Net, 2442
Nettle, 2360
Nettle hair, 2442
Network, 2442–2444
    optical fiber applications, 2569–2570
    programmable controller communications,
        2855–2856
Network databases, 2443
Network synthesis, 2444
Neuralgia, 2444
Neural networks, 242, 262–263, 1115
    nervous system as, 704
Neurilemma, 345, 696
Neuritis, 2444–2445
Neurofibrils, 1382
Neurofibromatosis, 2444
Neurogenic shock, 3159
Neuroglia, 699
Neurohormones, 698
Neuroleptic malignant syndrome, 2445
Neuromodulators, 698
Neuromuscular junctions, 694–695
Neuron, 688–689, 695–698
Neuronal ceroid lipofuscinosis, 388
Neuropeptides, 699, 700
Neuroptera, 2445
Neurosecretory cells, 2755
Neurotransmitters, 688, 698–700
Neutral, 2445
Neutral atom, 322
Neutraling power, of lens, 1507
Neutralization, 22
Neutral mutation, 2376
Neutrino, 2445–2446
    Solar, 3380–3381
Neutrino astronomy, 2446
Neutron, 2447–2448, 2645
    binding energy, 425
Neutron activation analysis, 2449, 2925
Neutron interferometer, 1685–1686

Neutron sources, 2448
    for nuclear reactors, 2483
Neutron stars, 2449
Neutrosphere, 2449
Nevado del Ruiz Volcano, 3676–3677
Newcastle disease, 2449–2450
New moon, 2387
Newt, 146, 2450
Newton (unit), 3568, 3570
Newton, Sir Isaac, 2450
Newton-Cotes formula, 2450
Newtonian fluid, 1444, 2992
Newtonian focus telescope, 3440
Newtonian telescope, 2450
Newton rings, 2450
Newton's formula for interpolation, 2450
Newton's laws of dynamics, 2450–2451
New Zealand brown midfish, 1544
New Zealand rimu, 2783
Neyman-Pearson theory, 2451
Niacin, 2451–2452
Nichrome, 786
Nickel, 2452–2455
    in biological systems, 2455–2456
    effect on steel, 2018
Nickel alloys, 108
Nickel-cadmium batteries, 393
    for electric cars, 1231
    for spacecraft, 3238
Nickel-hydrogen batteries, for spacecraft, 3238
Nickeline, 2456
Nickel-iron battery, for electric cars, 1231
Nicol prism, 162, 2849
Nicotinamide adenine dinucleotide (NAD), 2451
Nicotinamide adenine dinucleotide phosphate
    (NADP), 2451
Nicotine, 101
Nicotinic acid coenzymes, 858
Niemann-Pick disease, 2141–2142
Night blindness, 3650
Nighthawks, 2456
Nightingale, 2456
Nightjars, 608, 2456
Night lizards, 2176
Night vision, 2129
Nimbostratus clouds, 826
Nimbus clouds, 826
Ninhydrin reaction, 138
Niobium, 2456–2458
    effect on steel, 2018
NIST (National Institute of Standards and
    Technology), 2458
Nit (unit), 3570
Niter, 2458
Nitrates, as food antimicrobial agent, 204–205
Nitration, 2458, 2592
Nitric acid, 2458–2459
Nitride ceramic powders, 712
Nitriding, 1744, 2459
Nitriles, 132, 2581
    nomenclature and examples, 2586
Nitrites, as food antimicrobial agent, 204–205
Nitrobenzene, 346
Nitrochlorobenzene, 769–770
Nitro compounds, 2459–2460, 2581
    nomenclature and examples, 2586
Nitrocotton, 1343
Nitrogen, 2460–2464
    determination by Kjeldahl method, 162
    fertilizer requirements, 1370–1371
Nitrogen fixation, 140, 2463–2464
Nitrogen Group, 2464

Nitrogen mustard, 595, 596
Nitrogen oxides, air pollutant, 2802–2803
Nitrogen runoff, 1370
Nitroglycerin, 1343
    as heart medicine, 2026
Nitrohydrochloric acid, 219
Nitroso compounds, 2459–2460, 2581
    nomenclature and examples, 2586
Nitrous oxide (laughing gas), 632
NMDA receptors, 696
Nobel, Alfred Bernhard, 2464
Nobelium, 37, 2464–2465
Nobel Prizes, 2465
Noble gases, 1920
Nocardiosis, 363–364
Noctiluca, 2465
Noctilucent clouds, 828
Nocturnal radiation, 1213
Nodal lines, 2465
Nodal method of analysis, circuits, 1125
Nodal point, 147, 2465
Node, 2465–2466
Nodes, line of, 2141
Nodes of Ranvier, 696
Noise, 3339–3340
    statistics, 2467
    white and pink, 2466
Noise generator, 2466–2467
Noise suppressor (electrical noise), 3396
Noisemakers, 2655
Nonalcoholic beer, 404, 406
Non-aqueous dispersions, 2616–2617
Noncircuit element, 1282
Nondestructive testing, 2467–2469
Nonessential amino acids, 133, 134
Nonessential singular points, 2796
Non-Euclidean geometry, 1597
Nonferrous alloys, 108
Nonheterocyclic nitrogen compounds, 2586–2587
Nonheterocyclic sulfur compounds, 2587
Non-insulin-dependent diabetes, 1066
Nonlinear amplifier, 150
Nonlinear hyperbolic conservation laws, 2300–2301
Nonlinearity, 2139
Nonlinear resonance, 2981
Nonmelanoma skin cancer, 600
Non-Newtonian fluid, 1445, 2992
Nonnutritive sweeteners, 3400–3403
Nonpareil, 568
Nonpolar compounds, 896
Nonpolar solvent, 3233
Nonsinusoidal wave, 2470
Nonsteroidal anti-inflammatory drugs, 269
    for ulcers, 3556
Nonstoichiometric compounds, 731, 896
Nontransuranic low-level waste, 2501
Nonvolatile, 2474, 2681
Noradrenaline, 49–50, 698
Norbixin, 175
NOR circuit, 2470
Norem alloys, 851
Norepinephrine, 49–50, 698
Norfolk Island pine, 2751
Norin 10, 1691
Normal atom, 322
Normal concentration, 2470
Normal (Gaussian) distribution, 2470, 2898
Normal equivalent deviate, 2470
Normal (geometry/principal, to a curve at *P*),
    2470
Normalize (mathematics), 2470
Normalizing, steel, 2020

Normed linear space, 2140
North, 2471
North American high, 311
North Atlantic waters, 2471
Northbound node, 1060
Northern fowl mite, 2363
Northern hornworm, 1797
Northern lights, 328
Northern miller, 603
Northern pickerel, 2748
North magnetic pole, 1168–1169
North Pacific waters, 2471
Norwalk-like virus, 1461
Norwalk virus, 1461, 2471
No-see-'em, 477
Nose fly, 2471
NOT circuit, 2472
Notochord, 2472
Nova and supernova, 2472–2473, 3388–3390
    Hubble Space Telescope images, 1806, 1816
    neutron stars, 2449
    nucleosynthesis in, 734
    Supernova 1987A, 1806, 1816, 2472, 3388–3390
    Supernova 1993J, 3388–3390
NP-complete problems, 2299–2300
n-Type semiconductors, 3130
Nuclear binding energy, 425
Nuclear disintegration, 1128
Nuclear disintegration energy, 1128, 2895
Nuclear energy, 322
Nuclear equation of state, 2648
Nuclear fission, 2474–2475
Nuclear force, 2475–2476, 2645–2646
    as exchange force, 1340
Nuclear fusion, 1535
Nuclear induction, 1916
Nuclear isomer, 2029
Nuclear magnetic moment, 2476
Nuclear magnetic resonance (NMR), 170,
    2476–2479
Nuclear magnetometer, 2229
Nuclear magneton, 2479
Nuclear molecules, 2376
Nuclear Overhauser effect, 2477
Nuclear polyhedrosis virus (NPV), 1955
Nuclear potential, 2479
Nuclear power technology, 2479–2503
Nuclear radius, 2476
Nuclear spin, 2503, 3311
Nuclear structure, 2503
Nuclear transfer, in cloning, 821
Nuclear transmutation, 2503
Nucleate boiling, 501
    departure from, in boilers, 493–494
Nucleic acids, 2503–2504
Nucleonics, 2504
Nucleons, 2504
Nucleophile, 2504
Nucleophilic reaction, 2504
Nucleoproteins, 2503–2504
Nucleosynthesis, 734, 3573
Nucleotides, 1829
Nucleus, 2505
    Rutherford model, 2644
Nuclide, 736, 737, 2031, 2920
Null, 2505
Null vector, 2505
Number theory, 2505–2507
Numerical aperture, 209
Numerical control, 335, 2507–2509
Numerical integration, 1966–1967
Numeric code, 855

Numeric data types, 1107
Nusselt number, 2511
Nut, 1518–1519
Nutating-disk flowmeter, 1441–1442
Nutation, 2511
Nutcracker, 2656
Nuthatch, 2511
Nutmeg tree, 2511
Nut oil, 1412
Nutrition, 1087–1094. See also Minerals
Nutritional equivalency, 1092
Nutritive sweeteners, 3398–3400
Nux vomica tree, 2511
Nylon casting, 650
Nylon fibers, 1380
Nymph, 2511
Nyquist, Harry, 2511
Nyquist criterion, for system stability, 3321
Nyquist frequency, 2511
Nyquist rate (signaling), 2511
Nystagmus, 3650

Oak scale, 3204
Oak trees, 2513–2517
Oak wilt, 2513
Oarfish, 2517
Oat bran, 491–492
Oates bridge, 541
Oatgrass, 1679
Oats, 2517–2519
Oat smut, 1530
Oberon, 3583
    orbital characteristics, 2761
Obesity, 2519–2520
    in adolescence, 49
Object (real), 2520
Object detectors, 2871–2875
Objective prism, 2520
Object language, 2082
Object program, 1110, 2852
Obligatory anaerobes, 153
Obsequent stream valleys, 924
Observer's zenith, 3810
Obsidian hydration rate, 2928
Obsidian volcanic glass, 2520
Obstetrics, 2520
Occipital lobes, 693
Occluded front, 1509
Occlusion adsorption, 52
Occultation, 2520
Occult blood, 488
Occult hydrocephalus, 128
Occupational asthma, 555
Ocean, 2520–2537
    imaging, 2736, 2738–2739
    minerals in, 2357–2358
    pollution, 3722, 3723–3725
Ocean-atmosphere interface, 316–318
Ocean-atmosphere specific satellites, 3078
Ocean cataracts, 2524
Ocean current energy conversion, 2544, 2547–2548
Ocean currents, 2524
Oceanic biomass, 2544, 2549
Oceanic crust, 1164
Ocean perch, 2958
Ocean research vessels, 2538–2544
Ocean resources
    energy, 2544–2549
    living, 2550–2552
    mineral, 2553–2555
Ocean thermal energy conversion, 2544–2546

Ocean trenches, 2521
Ocean wave energy conversion, 2544, 2546–2547
Ocean waves, 316–317, 2534–2535
Ocelot, 664, 666
Octadecenoic acid, 2557
Octal, 2556
Octal digit, 1119
Octane number, 2701
Octave, 2556
Octopus, 708, 2384
Ocular accommodation, 14
Ocular dystrophy, 2407
Ocular hypertension, 2556
Oculocutaneous albinism, 87
Oculopharyngeal dystrophy, 2407
Odd-even rule of nuclear stability, 2556
Odonata, 2556
Odors
    classification, 2558
    deodorizing vegetable oils, 3609
    and flavor, 1419, 1421
    and pollination, 2798
    removal from water, 3712
Oersted (unit), 3570
Off-axis parabolic mirror, 2556
Office automation, 330
Offline, 2556
Offshore barrier, 378
Offshore oil and gas, 2554, 2692–2693
Ogive, 2556
Ohm (unit), 3568, 3570
Ohm, George Simon, 2556–2557
Ohmic contact, 2557
Ohm's law, 2557
Oilbirds, 608
Oil burners, 571
Oil-eating bacteria, 353
Oilless power breakers, 797
Oil of bitter almonds, 412
Oil of lavender, 2360
Oil of Patchouli, 2360
Oil power breakers, 797
Oils
    fixed, 1412
    industrial chemicals derived from, 454
Oil sands, 2687, 3419
Oil shale, 2557, 2687
Oil tanker spills, 3725
Okapi, 1626
Okra, 2237
Oldhams coupling, 975
Oldoinyo Lengai Volcano, 3677
Oleander aphis, 211
Olefin fibers, 1380–1381
Oleic acid, 2557–2558
Oleoresins, 1422, 2977
Olfactory system, 2558
Oligocene, 2559
Oligosaccharides, 610
Oligotrophic lake, 2138
Olingo, 2911–2912
Olive fruit fly, 1521
Olive oil, 2559
Olive tree, 2559
Olivine, 2559
OMEGA navigation, 2428
Omnidirectional antenna, 188
OMNI navigation system, 2428
Onchocerciasis, 1390
Oncogene, 595
Oncology. See Cancer and oncology
One-to-one mapping, 2254

Onion, 105, 2559, 2560
    irradiated, 2021
Onion maggot, 2218
Onion thrip, 3482
Online, 2559
On-off controller, 2404
Onychophora, 2560
Oocyte, 821
Oölite, 2560
Oology, 2560
Ooort cloud, 3328
Ooze, 2560
Opacity, 2560
Opah, 2560
Opal, 2560–2561
Opaque plasma, 2561
Open delta, 2561
Open-loop feedback control, 1356
Open-loop numerical control, 2507
open subroutine, 3366
Opera glass, 426, 3439
Operand, 2561
Operating characteristic, 2561
Operating ratio, 912
Operating system, 2561
Operational amplifier, 150
Operational helicopters, 1750
Operator (mathematics), 2561
Operator algebra, 2301–2302
Ophiasis, 110
Ophiolites, 2530
Ophitic texture, 2561
Ophiuroidea, 2561
Ophthalmia neonatorum, 922
Ophthalmology, 2561
Ophthalmoscope, 2577
Opiates, 153
Opium, 2815–2816
Oppenheimer, J. Robert, 2561
Opposition, 921
Opposum, 2284–2285
Opsonization, 199
Optical aberration, 2
Optical absorptivity, 9
Optical air mass (m), 2562
Optical anomaly, 2562
Optical antipodes, 2562
Optical axis, 336
Optical center, lens, 2562
Optical character recognition (OCR), 2562
Optical computer, 2129
Optical emission spectrochemical analysis,
    2562–2563
Optical encoder, 1298
Optical fibers, 2565–2567
Optical fiber systems, 2563–2575
    telephone cables, 3431–3432
Optical glass, 2575
Optical gratings, for position measurement, 2818
Optical haze, 3463
Optical images, graphical construction, 2575–2576
Optical lever, 2576
Optically effective atmosphere, 2576
Optical mode, 2576
Optical path, 2576
Optical phase conjunction, 2127–2128
Optical pumping, 2576
Optical pyrometer, 2576, 2919
Optical rotation, 300
Optical turbulence, 2576
Optical waveguide, 2574
Optic neuritis, 2576–2577

Optics, 2577
Optimum magnification, 2577
Optometry, 2577
Oral cavity, 2577
Orange maggot, 2218
Orange spiny whitefly, 3764
Orange tree, 801–802
    nematodes affecting, 2431
Orangutan, 191, 193
Orbit, 2577–2578
    aphelion, 211
    artificial satellites, 3064–3070
    perihelion, 2677
Orbital quantum number, 2904
Orchardgrass, 1679
Orchid, 2578
    saprophytic, 3061
OR circuit, 2578–2579
Order (of function), 2579
Order of magnitude, 2579
Ordinary ray, 2579
Ordinary-wave component, 2579
Ordinate, 945
Ordovician, 2579
Ore block, 2579
Orebody, 2579
Oregano, 2579
Ores, 2579, 2580
Organ, 2579
Organic chemistry, 2579–2593
Organic compounds, 896
Organic electroluminescence, 2593
Organic farming, 1370–1371
Organic light-emitting diodes (OLEDs), 2593–2595
    for microdisplays, 2334
Organic synthesis, 3408
Organ level, 429
Organ of Corti, 965, 1722
Organogeny, 2596
Organometallic compounds, 896
Organ-pipe coral, 87
Organ-system level, 429
Organ transplantation. See Transplantation
Oribis, 183–184
Ordinary differential equation, 1095–1096
Oriental black citrus aphid, 212
Oriental fruit fly, 1521
Oriental fruit moth, 2596
Oriented adsorption, 51–52
Oriented element, 1282
Orifice, 2596–2597
Orifice plates, 1436
Origins Program (NASA), 2597–2601
Oriole, 2601
Orion (the hunter), 2601
Ormer, 1
Ornithine cycle, 3588
Ornithology, 2601
Orogeny, 2601
Orographic lifting, 2840
Orpiment, 2601
Orthoclase feldspar, 1361
Orthogonal antennas, 2602
Orthogonal function, 2602
Orthogonal polynomials, 2602
Orthogonal squares, 2602
Orthographic projection, 2602
Ortho-hydrogen, 1864
Orthokeratology, 2603
Orthomorphic map, 915
Orthonormal set, 2602
Orthopnea, 916

Orthopneic cough, 916
Ortho position, on aromatics, 2585
Orthoptera, 2603
Orthoscopic system, 2603
Ortho-state, 2603
Orthostatic hypotension, 1892
Orthotomic system, 2603
Ortolan, 568
Oryx, 183, 184
Oscillating doublet antenna, 188
Oscillating-piston flowmeter, 1442
Oscillation, 2603
Oscillator, 2603–2604, 3162
Oscillatory flowmeters, 1440
Oscillograph, 2604
Oscillometry, 2604
Oscilloscope, 111, 1501, 2604
Osculating orbit, 2604
Osculating plane, 2604–2605
Osmium, 2605
Osmotic coefficient, 2605–2606
Osmotic diarrhea, 614, 1079
Osmotic pressure, 2606
Osteoarthritis, 2606–2607
Osteoblast, 505
Osteoclast-activating factor, 2204
Osteology, 2607
Osteomalacia, 508–509
Osteomyelitis, 507
Osteopetrosis, 509
Osteoporosis, 507–508
Ostrich, 2607–2608, 2954–2955
Otitis externa, 1723
Otitis media, 1723
Otosclerosis, 1725
OtterSkunk, 2408
Otto cycle, 1014
Otto cycle engine, 1084
Ouistiti, 2257
Outcrop, 2608
Outer bremsstrahlung, 536
Output line, 2608
Output signal, 3164
Ouzel, 477
Ouzel (ousel), 2608
Oval of Cassini, 2608
Ovarian tumors, 1657
Ovaries, 1302, 1656–1657
    and pituitary, 2755
Oven bird, 2656, 3697
Overdamped, 1026
Overfold, 2608
Overlap, 2608
Overrange, 2608
Over-the-horizon backscatter radar (OTH-B),
    2914
Overthrust, 2608
Overvoltage, 2608
Oviparous reproduction, 2608
Ovipositor, 2608
Ovoviviparous reproduction, 2608
Ovule, 1435
Owen bridge, 541
Owens process (glass blowing), 1632
Owl, 2608–2609, 3361–3362
Oxacillin, 194
Oxalic acid, 2608
Ox bow lake, 2608
Ox-goats, 1649
Oxidation, 2589, 2591, 2608
    amino acids, 138
    catalytic, 654

Oxidation number, 2610
Oxidation potential, 2610
Oxidative cleavage, 2589
Oxidative phosphorylation, 1019, 2717
Oxide ceramic powders, 712
Oxidizing agents, 2608
Oxidoreduction, 1314
Oxo process, 88–89, 2610
Ox-pecker, 2656
Ox warble, 518
Oxychlorination, catalytic, 654
Oxygen, 2610–2613
    determination by Unterzaucher method, 162
Oxygen debt, 2613
Oxygen Group, 2613
Oxygen lance, 1413
Oxytocin, 2613
Oyster, 412, 2080, 2383
    pearls from, 2669
Oyster poisoning, 1469
Oyster scale, 3204
Ozocerite (ozokerite), 2613
Ozone, climate effects, 813
Ozone data collection network (CASTNET), 20–21
Ozone hole, 2789
Ozone layer, 304, 2613
Ozonosphere, 2613

Pacemakers, 229
Pacific cod, 856
Pacific equatorial water, 2615
Pacific high, 311
Pacific mite, 2362
Pacific pines, 221
Pacific Suite, 302
Pacific-type margins, 2526
Pacinian corpuscle, 961
Packed erythrocytes, 486
Packet, 2442–2443
Pack ice, 1895
Packing
    for absorption towers, 8
    distillation, 1133–1134
Packing fraction, 425
Pacoemulsification, 2615
Pad (attenuator), 325
Paddlefishes, 2615
Paddle-wheel flowmeters, 1440
Painter's colic, 864
Paints, 2615–2622
Pair annihilation (matter-antimatter), 203, 208
Pair creation (matter-antimatter), 203, 208
Pair production, 2622
Paleobotany, 2622
Paleocene, 2622
Paleoclimatology, 815
Paleontology, 2622. See also Fossils
    invertebrate, 1987–1990
Paleophytic, 2622
Paleoseismology, 1189
Paleotectonics, 1192
Paleozoic, 2622
Pale western cutworm, 1012
Palladium, 2622–2623
Pallas, 274
Palmitic acid, 2626
Palmitoleic acid, 2626
Palm-kernel oil, 1412
Palm oil, 1412
Palm trees, 2623–2626
Palynology, 2626

Pancreas, 2, 1302, 2626–2627
    and diabetes, 1067
    and nonendrocine hormone sources, 1793–1794
    transplantation, 1071, 2627
Pancreas cancer, 2627
Pancreatitis, 2626–2627
Pandas, 2627–2628
Pangea, 2525
Panicle, 1433
Panther, 665
Pantograph, 2628
Pantothenic acid, 857–858, 2628–2629
Papain, 654
    for brewing, 406
Papataci fever, 2709
Papaw (papaya tree), 2629
Papaya tree, 2629
    irradiated fruit, 2021
Papermaking and finishing, 2629–2633
Paper nautilus, 225
Paper recycling, 3705
Paper wasp, 2633
Papillary ducts, 2061
Pap test, 599, 1657
Parabens, as food antimicrobial agent, 204
Parabola, 2633
Parabola (cubical/semicubical), 2633
Parabolic antenna, 188
Parabolic coordinate, 2633
Parabolic cylindrical coordinate, 1016, 2633
Parabolic partial differential equations, 2643
Parabolic spiral, 2633
Paraboloid, 2633–2634
Paraboloidal coordinate, 2634
Paradise fish, 2077
Paraffins, 102
    chlorinated, 766, 768
Paraformaldehyde, 1476
Parafoveal vision, 1489, 2634
Para-hydrogen, 1864
Parakeet, 2643, 2876
Paraldehyde, 15
Parallax (astronomy/instrument), 2634
Parallax error, 2634
Parallel A/D converter, 2634
Parallel adder, 43
Parallel circles, 3396
Parallel circuit, 1234
Parallel computing, 1111–1112
Paralleled-resonator filter, 1393
Parallelepiped, 2635
Parallel forces, 1474
Parallel impedance (antiresonance), 208
Parallelogram, 2635
Parallel sailing, 2635
Parallel-serial A/D converter, 2635
Parallel surfaces, 2635
Parallel-T bridge, 540–541
Parallel transmission, 2635
Paralysis, 2635–2636
Paralytic shellfish poisoning, 1468–1469
Paramagnetic amplifier, 150–151
Paramagnetic materials, 2636
Paramagnetism, 2636
Parameter, 2636
Parameter (statistics), 2636
Parametric acoustic array, 31
Paraphase amplifier, 151
Paraplegia, 2895
Para position, on aromatics, 2585
Para rubber plant, 1336
Paraselenae, 319

Paraselenic circle, 319
Parasite, 2636
    anaerobes, 153
Parasitic bumblebee, 567
Parasitic plants, 2636
Para-state, 2636
Parastillation, 1133
Parasympathetic autonomic nervous system, 343
Parathyroid glands, 1302, 2636–2637
Parathyroid hormones, 1788
Paratyphoid fever, 352
Paresthesia, 2637
Pareto chart, 2898
Parhelia, 319
Parhelic circle, 319
Parietal lobes, 693
Parity bit, 475
Parity check, 1933
Parkinson's disease, 1612, 2637–2642
Parotitis, 2642
Parrot, 2643, 2876–2878
Parrot fever (psittacosis), 2878
Parrotfishes, 2642–2643
Parr turbidimeter, 3546
Parsec, 2643
Parshall flumes, 1443–1444, 1449
Parsitism, 3407
Parsnip, 3566
Parsnip louse, 211
Parsnip webworm, 3738
Parthogenesis, 2643
Partial differential equation, 1095–1096, 2643
Partially balanced incomplete blocks, 2643–2644
Partial-order planning, 243
Partial pressure, 1882, 2644, 2846
Particle accelerators, 2648–2651
Particles (subatomic), 2644–2652
Particulate air pollutants, 2800–2801
Partition, 2654
Partition function, 2654
Partons, 2646
Partridge, 2654
Parvoviral enteritis, 2654
Pascal (unit), 3568, 3570
Pascal, Blaise, 2654
Pascal triangle, 2654–2655
Paschen, law of, 2655
Paschen-Back effect, 2655
Paschen-Runge mounting, of reflection grating, 2963
Pasiphae, orbital characteristics, 2760
Passeriformes, 2655–2657
Passive immunization, 209
Passive matrix LCDs, 2145–2148
Passive network, 2657
Passive solar, 3219
Passive submergence, 1876
Passive transducer, 3511
Passivity, 2657
Pasteur, Louis, 2657
Pasteur effect, 613
Patch (computer program), 2657
Patch-clamp technique, neuron research, 688, 697–698
Patenting, steel, 2019
Patent log, 2657
Path, 2657
Pathfinder mission to Mars, 2657–2666
Pathogenesis, 2666
Pathogenic, 2666
Pathological anatomy, 163
Pathology, 2666
Path trajectory, 3510

Patina, 2666
Pattern recognition, 2666–2667
  in machine vision, 2210–2213
Patterson function, 997
Pauli, Wolfgang Ernst, 2667
Pauli exclusion principle, 425–426, 740, 2667, 2904
Pauling, Linus Carl, 2667
Pauraque, 608
Pawpaw, 2108
Payoff matrix, 2188
Pea, 2118
Pea aphis, 212
Peachblossom ore, 1326
Peach scale, 3205
Peach tree, 3037–3038
Peach-tree borer, 512, 2667
Peach-twig borer, 512–513, 2668
Peacock ore, 513
Pea family, 2117–2118
Peafowl, 2668
Peaking circuit, 2668
Peaking transformer, 2668–2669
Peanut, 2118
Peanut-groundnut, 2669
Pear bud moth, 566
Pearl, 2669
Pearlite, 2016
Pearl millet, 2351
Pearlstone (perlite), 2681
Pearly razorfish, 3785
Pear midge, 2348
Pear slug, 3102
Pearson distributions, 2669–2670
Pearson's ρ (rank correlation), 2946
Pear thrip, 3482
Peat, 821
Peat bog, 492
Peat coal, 821
Peat-mosses, 562–563
Pea weevil, 864, 2670
Pebble classification (Zingg), 3815
Pebble mills, 371
Pecan bud moth, 566
Peccaries, 3404, 3405
Peclet number, 2670
Pectinase, 655
Pectins, 1089, 2670
Peculiar galaxies, 1549–1551
Pediatrics, 2671
Pediculosis, 2671
Pedigree analysis, 1587
Pedigree method, 2764–2765
Pedology, 2671
Pegasus (the flying horse), 2671
Pegmatites, 2356–2357, 2671
Pelagic fish, 2551
Pelecaniformes, 2671–2672
Pele's hair, 2672
Pelicans, 2671, 2672–2673
Pellagra, 2451
Pellet mills, 67
Pelorus (dumb compass), 2673
Peltic, 224
Pelvic inflammatory disease, 3153
Pelvis, 2673
Pemphigus vulgaris, 2673
Pencil beam, 2673
Pencil-beam antenna, 188, 2673
Pencil fish, 725
D-Penicillamine, anti-AIDS drug, 34
Pendulum, 2673

Pendulum clock, 2673–2674
Pendulum magnetometer, 2228
Peneplain, 2674
Penetrant methods, 2468–2469
Penguin, 2674–2675, 2794
Penicillin G, 194
Penicillins, 193–195, 267, 441
Penicillin V, 194
Penicillium, 266–267
Penis, 1656
Penning discharge, 2676
Penning effect, 2676
Penning gage, 863
Pennsylvanian Period, 2676
Penrose process, 478
Penstock, 2676
Pentachlorophenol, 770
Pentadactyl appendage, 2676
Pentaerythritol tetranitrate (PETN), 1343
Pentlandite, 2676
Pentode, 1278
Pentosans, 612–613
Pentose phosphate cycle, 613
Pentriode amplifier, 151
Penumbra, 3379
Pepo, 1517–1518
Pepper, 2676, 3214
Peptic ulcers, 3555–3556
Peptone, 2676
Percent relative humidity, 1882–1883
Percent word articulation, 238
Perch, 2677–2678
Perched, 2676
Perchloric acid, 1890–1891
Percutaneous, 2677
Perfect differential, 1097
Perfect fluid, 1444
Perfect gas, 2677
Perfect-gas law, 1578, 2677
Perfect vacuum, 7
Perfluorocarbons, greenhouse gas, 1641
Periapical abscess, 6
Periapsis, 2677
Periastron, 2677
Pericarp, fruit, 1517
Pericynthian, 2677
Periderm, 2677
Peridot, 2559
Peridotite, 2677
Perifocus, 2677
Perigee, 2677
  satellites, 3065
Perihelion, 2677
Perimetry, 2677–2678
Period (geologic time), 1596
Period, alternating currents, 111
Periodic function, 2678
Periodic table, 2678–2680
Period luminosity law, 2680
Periodontitis, 2680
Peripheral equipment, 2680
Peripheral nervous system, 688–705
Peripheral vision, 2680
Perissodactyla, 2680
Peristaltic movement, 1102
Peritonitis, 2681
Peritonsillar abscess (quinsy), 5, 2706
Periwinkle, 2681
Perlite (pearlstone), 2681
Perlucidus, 828
Permafrost, 432, 2793–2794
Permanent memory, 2681

Permanent set, 3146
Permeability (magnetic), 2681
Permeameter, 2681
Permeation, 2681
Permian Period, 2681
Permutation, 2681
Permutation group, 2681–2682
Pernicious anemia, 167
Perovskite, 2682
Peroxide decomposers, 207
Peroxides, nomenclature and examples, 2586
Peroxy group, analytical determination, 161
Perseids (meteor shower), 2682
Perseus (constellation), 2682
Persimmon trees, 2682–2683
Personal communications network (PCN), 3428
Personal equation, 2683
Perspective, 2683
Perspiration, 2683
Perthite, 2683
Perturbation, 2683
  astronomy, 2683–2684
Perturbation energy, 2683
Pertussis (whooping cough), 352, 2684
Peru (Humboldt) current, 1853
Pes cavus (clawfoot), 805
Pesticide, 1948–1952
  pyridine compounds, 2888–2889
  technology, 1952–1956
Petal, 1434
Petalite, 2684
Petit mal seizure, 3125
Petrel, 2684–2685, 2851
Petrochemicals, 2685
Petrogenesis, 2685
Petroleum, 2685–2699
Petroleum drilling, 2691–2697
Petroleum exploration, 2689–2691
Petroleum fly, 2699
Petroleum refining, 2699–2705
Petroleum reserves, 2697
Petrology, 2705
Petzval condition, 2705
Petzval surface, 1008, 2705
pH, 2708–2709
  abrasion, 3
Phacoemulsification, 1345
Phagocyte, 2705
Phalanger, 2285
Phalanx (pl. phalanges), 2705
Phalarope, 3695, 3696
Phanerogamic parasites, 2636
Phanerozoic time scale, 1595
Phantom circuit, 2705
Pharmaceutical industry, 442
Pharyngitis, 2705–2706
Pharynx, 98, 1102, 2706
Phase contrast microscope, 2346
Phase diagram (metallurgy/statistics), 2706
Phase difference, 2706
Phase distortion, 1135
Phase hologram, 1777
Phase index of refraction, optical fibers, 2565
Phase inverter, 2706–2707
Phase rule, 2707
Phase-shaped antenna, 188
Phase-shifting circuit, 2707
Phase-shift oscillator, 2707
Phase velocity (wave velocity), 3730
Phasor voltage, 111–112
Pheasant, 1563, 2042, 2043, 2707

Phenacetin, 153
    antipyretic, 208
Phenacite, 2707
Phenazocine, 2393
Phenobarbital, 373
Phenoclast, 2707
Phenocryst, 2707
Phenol, 2581, 2685, 2707–2708
Phenolics, 2708
Phenoxymethylpenicillin, 194
Phenylalanine, 133–137
Phenylamine, 174
Phenylbutazone, 153
Phenylephrine hydrochloride, 101
Phenylformic acid, 413
Phenyl hydride, 412
Phenylketonuria, 2708
Pheromones, for pest control, 1954–1955
Philips ionization gage, 863
Phillipsite, 2709
Phlebitis, 235
Phlebotomus fever, 2709
Phlebotomy, 2709
Phloem, 592, 2709
Phlogopite, 2709–2710
Phobos, 276, 2273–2275
    orbital characteristics, 2760
Phoebe, 3098
    orbital characteristics, 2761
Phoenicopteri, 2710
Phoenix (constellation), 2710
Pholidota, 2710
Phon, 2710
Phon (unit), 3570
Phong shading, 910
Phonometer, 2710
Phonons, 25, 2710
Phoocardiography, 1255
Phosgenite, 2710
Phosphates, as food antimicrobial agent,
    205
Phospholipids, 2710
Phosphorescence, 2191, 2711–2712
Phosphoric acid, 2710–2711
Phosphoric acid fuel cell, 1523–1524
Phosphors, 2711–2712
    cathode-ray tubes, 661
Phosphorus, 2712–2715
    in biological systems, 2716–2717
    effect on steel, 2018
    fertilizer requirements, 1371–1372
Phosphorylation
    oxidative, 1019, 2717
    photosynthetic, 2717–2718
Phot, 2718
Phot (unit), 3570
Photochemistry, 2588, 2718–2720
Photodynamic therapy, 600
Photoelasticity, 2720
Photoelectric constant, 2720
Photoelectric effect, 2720–2722, 2740
Photoelectric object detectors, 2872–2874
Photoelectron spectroscopy, 1276
Photoemission, 2722
Photogrammetry, 2722
Photographic color film, 2621, 2725
Photography, 2722–2739
Photoionization, 2740
Photoisomerization, 2590
Photoelectric polarimetry, 2785
Photoluminescence, 2191
Photolysis, 2718–2720

Photometers, 2740–2741
Photomultipliers, 2722
Photon, 2741
Photon engine, 2741
Photoneutron, 2741
Photonics, 2741
Photonuclear reaction, 2741
Photoperiodism, 2741–2742
Photophone, 2564
Photopic vision, 1489
Photoreceptor, 2742, 3647
Photorefractive keratectomy (PRK), 2742,
    2965
Photosphere, 2742, 3376–3377
Photosynthesis, 615, 2742–2743
    in leaves, 2114
Photosynthetic bacteria, 351
Photosynthetic phosphorylation, 2717–2718
Phototheodolite, 2744
Photovoltaic cell, 2744, 3227–3228
Photovoltaic effect, 2721
Phreatic, 2744
Phrenic nerves, 2744
Phthalic acid, 2744
Phthalic anhydride, 2685, 2744
Phugoid oscillation, 2744
Phycomycetes, 1529
Phylloxeran (phylloxerid), 2744
Physical acoustics, 25
Physical double star, 2744
Physical meteorology, 2745
Physical model, 2363
Physical optics, 2126
Physiography, 2745. See also Geomorphology
Physiological acoustics, 25–26
Physiological optics, 2126
Physiology, 2745
Phytoalexins, 1955
Phytochemicals, 452–455
Phytochrome, 773
Phytohormone, 2766
Phytometer, 2824
Phytoplankton, 2550, 2551
Pi (π), 2745
Picard method of successive approximations,
    2745
Piciformes, 2745
Pickerel, 2748
Pick's disease, 127
Pickup, 2745–2746
Pico (prefix for units), 3568
Pictorial representation, 2746
Picturephone, 3428
Piddock, 2746
Pierce oscillator, 2746
Piezoelectric accelerometers, 11–14
Piezoelectric effect, 2746
Piezoelectricity, 1003
Piezoelectric microphone, 2342
Piezometer ring, 2746
Pig, 3404–3405
Pigeon, 874, 2746
    domesticated, 2830–2831
Pigeon lice, 2190
Pigeon-toes, 829
Pigmentation
    animals, 2747
    plants, 2747–2748
Pigments, 2616–2617
Pika, 2909
Pike, 2748
Pileus cloud, 828

Pillbox antenna, 188, 2748
Pill bug, 2748, 3236
Pillow lava, 2748
Pilot (boilers), 571
Pilotage (piloting), 2748–2749
Pilot balloon, 370
Pimento, 109
Pinché, 2257
Pineal gland, 1302, 2749
Pineapple, 2749–2751
Pineapple beetle, 1146
Pineapple scale, 3204
Pine flatlands, 3753
Pine leaf scale, 3204
Pine lizard, 3403
Pine trees, 919–920, 2751–2753
Pinguecula, 922
Pinhole image, 2754
Pink bollworm, 2754
Pink-eye (conjunctivitis), 922
Pink noise, 2466
Pino, 2754
Pint (unit), 3570
Pintle chain, 719
Pinworm, 1983–1984, 2754
Pions, 2321–2322, 2405, 2645–2646
Pipefishes, 2754
Piping crow, 2656
Pipit, 2656
Piranha, 724–725
Pirani gage, 2754, 3598
Pisces (the fishes), 2754
Piscis Austrinus (constellation), 2754
Pi section, 2754
PI-section filter, 1393
Pistachio tree, 637
Pistil, 1434–1435
Piston corer, 955–956
Pitch, 850
Pitchblende, 2937, 3575
Pitching moment, 2754
Pi theorem, 2754
Pit-making oak scale, 3204
Pitot tube, 1437, 2755
Pitta, 2656
Pitting corrosion, 962
Pituitary gland, 1302, 2755–2757
Pituitary hormones, 1788
Pityriasis, 1055
Pityriasis rosea, 1053
Pixel, 2757
pK, 2757
Placenta, 1291–1292
Plages, around sunspots, 3377
Plagioclase feldspar, 1362
Plaice, 1415, 1417–1418
Plain old telephone service (POTS), 3426
Plaiting fibers, 1378
Plait point, 2757
Planck, Max, 2757
Planck law, 2757
Planck radiation formula, 2757
Plane (geometry), 2757–2758
Plane curve, 1009
Plane geometry, 1009, 1597
Plane sailing, 2758
Plane table, 2758
Planetarium, 2758
Planetary boundary layer, 2758
Planetary circulation, 2758
Planetary nebula, 2429
Planetary precession, 1585

Planetary ring systems
  Jupiter, 1558, 1560, 2047–2049
  Neptune, 2436–2437
  Saturn, 3090–3092
  Uranus, 3581–3582
Planetesimals, 3574
Plane tree, 3405–3406
Planets, 2758–2762. *See also* specific Planets
  extrasolar, 2759, 2761–2762
  formation, 3574
  heat balance, 1734
  imagery, 2736–2737
  Kepler's laws of motion, 2056–2057
  motions, 2762–2763
  search for habitable, 281
  sphere of influence, 3307
Plankton, 2550, 2551
Planning, artificial intelligence applications, 243
Plantain, 371, 2763
Plant anatomy, 163
Plant breeding, 1762, 2763–2766
Plant cutter, 2656
Plant formation, 274
Plant growth modification and regulation,
  2766–2767
Plant hoppers, 1782
Plant hormones, 2766
Plant louse, 211, 1782
Plant pigmentation, 2747–2748
Plaque (teeth), 631
Plasma, 2771–2772
  blood, 485
Plasma displays, 1903–1904, 2767–2771
Plasma frequency, 2771
Plasma oscillations, 2771
Plasmatron, 2772
Plasmid pBR322, 455
Plasmids, 1591
Plaster, 1701
Plaster of Paris, 47
Plastic casting, 650
Plastic deformation, 2772
Plastics, 2772
Plastic waste recycling, 3702–3704
Plastids, 2772
Plateau, 2772
Plate freezers, 1498
Plate girder, 1626–1627
Plate glass, 1632
Platelets, 484–485
Plate tectonics, 1181–1195, 2525–2526
  and volcanism, 3679–3680
Platform (geology), 2772
Platform beach, 2772
Platform reef, 2772
Platinum, 2772–2774
Platinum group, 2772
Platinum thermocouples, 3468
Platyhelminthes, 2774
Platykurtic curve, 2898
Platypus, 2387
Playa, 2774–2775
Plecoptera, 2775
Plecostomus, 2775
Pleiades, 2775
Plenum, 2775
Pleural cavity, 2982
Pleurisy, 2775
Plexus, 2775
PL/I, 892
Pliocene, 2775–2776
Plit hydrophone, 1879

Plotter, curve, 2776
Plotting sheet, navigational, 2776
Plover, 3695
Plumbicon, 3453
Plum curculio, 1006, 2776–2777
Plume moth, 2777
Plumeria tree, 2777
Plume tree, 3038–3039
Plural scattering, 3109–3110
Pluto, 2778–2779
  orbital characteristics, 2759
  physical characteristics, 2760
Pluton, 1986
Plutonium, 37, 2777–2778
Pneumatic, 2779
Pneumatic amplifier, 151
Pneumatic controller, 2780–2781
Pneumatic gaging, 3475
Pneumatic-probe pyrometer, 2781
Pneumatic transmission, 3517–3518
Pneumatolysis, 2781
Pneumocystosis, 2781
Pneumoencephalography, 530
Pneumokonioses, 2781
Pneumonia, 352, 2781–2783
p-n Junction, 3130, 3131
Pocket antenna, 188
Podargues, 608
Podicipediformes, 2783
Podocarps, 2783
Poi, 3418
Poinciana, 580, 2783
Point, 2783
Point-biserial correlation coefficient, 473
Point defects, 3230
Point drift, 1147
Point mutation, 1591, 2375–2376
Point source, 2783
Point-source pollution, 2799
Poise (unit), 3570
Poiseuille flow, 1445
Poison hemlock, 3566
Poisoning, (catalyst), 653
Poisoning, foodborne, 1461, 1468–1470
Poison ivy, 637
Poison oak, 637
Poisson, Simeon Denis, 2784
Poisson distribution, 2783–2784
Poisson equation, 2784
Poisson's ratio, 2784
Poki (game), 254
Polar air, 72
Polar biosphere, 2794
Polar compounds, 896
Polar coordinates, 2784
Polar creep, 2524
Polar cryosphere, 2791–2792
Polar distance, 2784
Polar front, 1508
Polar front theory, 2784
Polar high, 310
Polar hydrosphere, 2791
Polarimeter, 2784
Polarimetry, 2784–2786
Polaris (α Ursae Minoris), 2785
Polariscope, 2785
Polarization, 2785–2786
Polarized light, 2786
Polarizing microscope, 2347
Polarizing prism, 2850
Polar lithosphere, 2794
Polarographic analyzers, 2786

Polar orbit satellites, 3065
Polar research, 2786–2795
Polar solvent, 3233
Polar stratospheric clouds, 2789
Polar valence, 750
Polar vortex, 309
Pole (mathematics), 2796
Polecat, 2407
Pole figure, 2796
Poliomyelitis, 2796–2797
  brain injury from, 531
Pollack, 856
Pollen, 1435
Pollination, 2797–2798
Pollucite, 2798
Pollution
  acid rain, 19–21
  air, 2798–2805
  indoor, 2938
  light, 2133–2135
  ocean, 3722–3725
  polar atmosphere, 2788–2789
  rivers, 3718–3719
  and wastes, 3698–3705
  water, 3718–3726
Pollux (β Geminorum), 2806
Polonium, 2806
Polyacrylate elastomers, 1220
Polyamide-imide resins, 2806–2807
Polyarylates, 2807
Polybasite, 2807
Polybenzimidazole fibers, 1381
Polybenzimidazoles, 2807
Polybutylene resins, 2807–2808
Polybutylene terephthalate polyesters, 2808
Polycarbonate, 2808
Polychlorinated biphenyls (PCBs), 460
Polyconic projection, 917
Polycrystalline silicon thin-film transistor,
  2155–2156
Polycyclic alkanes, 2581
Polycyclic aromatic hydrocarbons (PAH), 628
Polycyclo-hexylene-dimethylene terephthalate, 2808
Polycystic disease of kidney, 2066
Polycystic ovary syndrome, and androgens, 166
Polyembryony, 1296, 2808
Polyene macrolides, 197
Polyester fibers, 1381, 2808–2809
Polyether-etherketone (PEEK), 2809
Polyetherimide, 2809
Polyethylene, 2685, 2809–2810
Polygeneration, 862
Polygon, 2810
Polyhalite, 2810
Polyhedron, 2810–2811
Polymastia, 535
Polymerase chain reaction (PCR), 451, 456
Polymer-dispersed liquid crystal display, 2157
Polymerization, 2377–2378
Polymerization catalysts, 654
Polymer liquids, 2167
Polymers, 2581
Polymorphism, 2811
Polymyositis, 2410
Polymyxins, 197
Polynomial, 2811
Polyp, 857, 2811
Polyp (coelenterates), 2812
Polypeptide, 2811–2812
Polyphase motors, 2398–2399
Polypropylene, 2685, 2812
Polysaccharides, 610

Polystyrenes, 2812–2813
Polysulfone, 2813
Polytropic processes, 2813–2814
Polyunsaturated vegetable oils, 3609
Polyurethanes, 1220, 2814
Polyvinyl chloride, 2685
Polyvinylpyrrolidone (PVP), clarifying agent, 803
Pomace fly, 1148
Pome, 1518
Pomegranate tree, 2814
Pompe's disease, 2410
Pons, 692
Pontiac fever, 361, 2117
Pontoon bridge, 550
Poorwill, 2456
Poplar trees, 2814–2815
Poppy, 2815–2816
Poppy seed oil, 1412
Population (statistics), 2816
Population genetics, 1588
Populations, 1208
Population senescence, 1614
Porcelain enamal, 710
Porcupine, 3023
Porcupine fish, 2816
Porgies, 2816
Porifera, 2816, 2817
Porkfish, 1696
Pork tapeworm, 3417
Porosity, 2816
Porphyrin, 2816, 2817
Porphyry, 2816
Porpoises, 3754–3756
Porter beer, 404
Portland cement, 47, 684–686, 1701
Port-wine stain, 170
Positional notation, 2819
Position angle (stellar), 2819
Position measurement, 2817–2819
Position vector, 3607
Positive azeotropy, 345, 1133
Positive-displacement flowmeters, 1441–1444
Positive ion, 1995
Positive-negative acid-base theory, 22
Positron, 1277, 2819–2820
Positron emission tomography (PET), 3797
    for brain imaging, 3797
Positronium, 2820
Positrons, 2122
Post-exercise asthma, 555
Positive-motion cam, 591
Postoperative wound infections, 353
Postural hypotension, 1892
Potassium, 2820–2822
    alkali metal, 98
    in biological systems, 2822–2823
    fertilizer requirements, 1372
    and hypertension, 1091
    and kidney abnormalities, 2063
Potassium aginate, 1698
Potassium aluminum sulfate (alum), 116
Potassium silicates, 3165
Potassium-sparing agents, 1136
Potato
    fungus diseases, 1531
    irradiated, 2021
    nematodes affecting, 2432
Potato family (solanaceae), 3213–3214
Potato leaf hopper, 2114
Potato stalk borer, 513
Potato starch maltodextrin, 492
Potential energy, 1304, 2823–2824

Potential scattering, 3109
Potential temperature, 1323
Potential transformer, 1239
Potential well, 2475
Potentiometer, 2824
Potentiometric titration, 3501–3502
Potherb butterfly, 576
Pot hole, 2824
Potometer, 2824
Potoos, 608
Potter wasp, 2824
Potto, 2119
Pouched mole, 2285, 2379
Poultry, 2824–2832
Poultry lice, 2190
Poultry mite, 2363
Pound, Robert, 2832
Poundal (unit), 3570
Powder metallurgy, 2832–2834
    aluminum alloys, 125
Powdery mildew, 266
Power, 2834
Power (mean), 2834
Power amplifier, 151
Power cables, 577
Power conditioning, 1230
Power factor meter, 1238–1239
Power function, 2834
Power gain, 2834
Power loading, 2834
Power series, 2834–2835
Power sources, 2835–2837
Power spectrum, 3301
Power supplies, 2835–2837
Power-switching drives, 3142–3143
Power transmission ratio (acoustic), 2837
Poynting-Robertson effect, 2837
ppb (parts per billion), 2837
ppm (parts per million), 2837
Praesepe (Beehive cluster), 595
Prairie, 432
Prairie chicken, 1695
Prairie dog, 3319, 3320
Prandtl number, 2837
Praseodymium, 2837–2838
Pratt and Whitney key, 2059
Prawn, 1031, 2838
Praying mantis, 410–411, 2244
Preamplifier, 151, 2838
Preantenna, 2838
Precambrian time scale, 1595–1596
Precancerous, 600
Precession, 2349–2350, 2838
Precession (astronomical objects), 2838
    general, 1585
Precious coral, 87
Precipitable water vapor, 2838
Precipitation, 2838–2843
Precipitation hardening, 2844
Precision, 2844
Preclampsia, 585
Precursor, 2844
Prediction (statistics), 2844
Pre-excitation syndromes, 229
Pregnancy, 2844
    and sexually transmitted diseases, 3152–3153
Prehension, 2844
Prehnite, 2844
Preionization (autoionization), 329
Premenstrual syndrome, 1658
Presbyopia, 2844–2845
Prescription Drug User Fee Act, 447

Preserves, 3402
Press, 1344
    papermaking, 2631–2632
Pressure, 2845–2846
Pressure broadening, 553
Pressure cooker, 502
Pressure filter, 1395
Pressure gradient force, atmosphere, 310
Pressure sensing, optical fiber applications,
    2570–2571
Pressure-sensing microphone, 2341–2342
Pressure suit, 2846
Pressure wave, 2846
Pressurized groundwood pulp, 2881
Pressurized water nuclear reactors, 2484–2488
Prestressed concrete, 2846
Pretibial fever, 2122
Prevost law of exchanges, 2846
Prickleback, 480
Prickly pear, 578
Priestly, Joseph, 2846
Primary alcohols, 87
Primary amebic meningoencephalitis, 2847
Primary amides, 131–132
Primary amines, 132
Primary amino group, analytical determination, 161
Primary battery, 389
Primary body, 2847
Primary circulation, 2847
Primary colors, 2847
Primary element, 2847
Primary irritant dermatitis, 105
Primary radar, 2847
Primary wastewater treatment, 3713
Primates, 2847
Prime meridian, 2847
Princess feather, 128
Principal components (statistics), 2847–2848
Principal planes (lens), 2848
Principal quantum number, 2904
Principle direction, 1126
Printed circuits, 2848
Prion, 353
Priority in, first out (PIFO), 2907
Prism
    mathematics, 2848
    optics, 2848–2850
Prism binocular, 426–427
Probability, 2850–2851
Probability density function, 1502
Probability distribution, 2851
Probability sampling, 3061
Probable error, 2851
Probably Approximately Correct (PAC) analysis, 258
Problem-oriented languages, 892, 2082
Proboscis, 2851
Procedure-oriented language, 2082
Procellariiformes, 2851
Process, 2851
Process-activating protein, 2865
Process analyzers, 158
Process annealing, 175
Process control, 2851
Process lapse rate, 309
Process management, 939
Process monitoring, 939
Proctitis, 2851
Procyon (α Canis Minoris), 2851–2852
Producer's risk, 2901
Product, 2852
Product demodulator, 1046
Product modulator, 2852

Product-moment, 2852
Profibus, 2252
Profile, 2852
Progesterone, 3355
Progestins, 3355
Progestogens, 3355
Program, 2852
Program amplifier, 151
Program generator, 2852
Programmable controller, 2852–2857
Programmable read-only memory (PROM), 2312
Programming flowchart, 2857–2858
Programming language, 2858–2859
Program register, 2970
Program relocation, 2859
Program storage, 3358
Progression, 2859–2860
Progressive multifocal leukoencephalopathy, 531, 2860
Progressive systemic sclerosis, 3117
Projective geometry, 1598
Projectors, 2735
Project Oxygen, 3438
Prokaryotes, 675
Prolactin, 1923
Prolapse, 2860
Proline, 133–137
Prolog program, 244, 247, 249
Promethazine hydrochloride, 203
Promethium, 2860
Prominences, 3379–3380
Prompt neutrons, 2448, 2480
Pronghorn antelope, 2860
Prony brake, 1156
Proof planning, 248–249
Propagation, direction of, 2861
Propagation constant, 2860–2861
Propaldehyde, 88
Propane, 2860–2861
Propeller fan, 1353
Propeller pumps, 2885
Propellers, 77–78
Propene nitrile, 36
Propenoic acid, 35
Proper motion (star), 2861
Proper subgraph, 3365
Prophase, 677
Propionates, as food antimicrobial agent, 204
Propionic acid, as food antimicrobial agent, 204
Propithque, 2119
Propolis, 2861
Proportional control, 931
Proportional controller, 934
Proportional counter, 2861
Proportional-derivative control, 931
Proportional-derivative controller, 934
Proportional gain, 1544
Proportional-integral control, 931–932
Proportional-integral controller, 934
Proportional-integral-derivative control, 933
Proportional-integral-derivative controller, 934
Proportional limit, 2861
Proportional sampling, 3061
Proprioceptive stimulation, 2861
Propulsion, 77–81
Propylene oxide, as food antimicrobial agent, 205
Proso millet, 2351
PROSPECTOR expert system, 252
Prostaglandins, 2861
Prostate, 1655–1656
Prostate cancer, 599
Prostate gland, 2

Prostate-specific antigen (PSA), 599
Protactinium, 37, 2861–2862
Protective coatings, 2616
Protective colloids, 868
Protein, 1829–1830, 2862–2869
    and acid-base regulation, 18
    in diet, 1087
Protein-based adhesives, 46, 47
Protein-calorie malnutrition, 1040, 2863
Protein glycosylation, 1072–1073
Protein hydrolysate, 2869
Proteomics, 1842
Proterozoic (Algonkian) era, 2869
Protium, 2869
Proton, 2645, 2869–2870
    binding energy, 425
Proton Compton effect, 898
Proton-proton reaction, 2870
Protopectin, 2670
Protoplanets, 3574
PROTOS system, 250
Protozoa, 2870
Protractor, 2871
Proustite, 2871
Proverb (game), 254
Prowfish, 480
Proximity detectors, 2871–2875
Pruritus, 2875
Psammitic, 223
Pseduoesotropia, 1327
Pseudo-adiabatic processes, 307–308
Pseudocode, 855, 2875
Pseudoequivalent temperature, 1323
Pseudoinstruction, 1957
Pseudoleucite, 2122
Pseudomembraneous enterocolitis, 1080
Pseudomorph, 2356, 2875
Pseudoplastic fluid, 2992
Pseudopotential theory, 752
Pseudo (weak) salt, 3058
Pseudoscalar, 2875
Pseudoscorpion, 220, 2875–2876
Pseudovector, 2876
Psilocin, 1712
Psilocybin, 1712
Psilomelane, 2876
Psi particle, 2876
Psittaciformes, 2876–2878
Psittacosis (parrot fever), 2878
Psocids, 962
Psoriasis, 1055
Psoriatic arthropathy, 3313
Psychogenic skin eruptions, 343
Psychological acoustics, 26–30
Psychometric chart, 2878–2879
Psychometry, 1882–1885
Psychomotor epilepsy, 3126
Psychosomatic disturbances, 343
Ptarmigan, 2879
Pteropsida, 2879
Pterygota, 2879
Ptomaine, 2879
Ptosis, 481, 1347
p-Type semiconductors, 3130
Puberty, 48–49
Public switched telephone network (PSTN), 3426
Public Utilities Regulatory Policies Avct (PURPA), 863
Pudding wife, 3785
Pudus, 1039, 1040
Puff-ball, 380, 381
Puffbirds, 2745

Puffer, 2880
Puffer fish poisoning, 1469
Puffin, 726, 2684, 3695
Puku, 183, 184
Pulfrich refractometer, 2965
Pulley, 2207–2208
Pulmonary embolism, 236
Pulp production and processing, 2880–2883
Pulsar, 2883
Pulse, 2883
Pulse (botany), 2883
Pulse code modulation, 3437
Pulse generator, 2884, 2885, 3162
Pulsejet engine, 80, 2035, 2884
Pulseless disease, 235
Pulse radar, 2884
Pulse-width modulated drives, 3146
Pulverized-coal boilers, 572–574
Puma, 664, 666
Pumice, 2885
Pump, 2885
Pumped-storage plant, 1860–1861
    as hydraulic accumulator, 14
Pumping protein, neuron membrane, 696
Pumpkin, 1005
Punch drunk, 532
Punctuated equilibrium, 1339
Punkie, 2348, 2885
Pupa, 2885
Puppis (constellation), 2885
Pure program generator, 2852
PUREX process, 3576–3577
Purines, 2885–2886
Purkinje effect, 2886
Purple light, 2886
Purple martin, 3397
Purple scale, 3204
Purpura, 1757, 2886
Push-button telephone, 3427
Push-pull amplifier, 151, 2836
Pustule, 2886
Pycnogonida, 2886
Pycnometer, 2886
Pygmy antelopes, 183, 184
Pyocyanase, 193
Pyramid, 2886
Pyranometer, 39
Pyrargyrite, 2887
Pyrgeometer, 39, 172
Pyrheliometer, 38–39, 2887
Pyridine and derivatives, 2887–2891
Pyridoxine (vitamin B_6), 3663–3664
Pyrite, 2891
Pyroclastic, 2891
Pyroclastics, 1878
Pyrogenetic minerals, 2891
Pyrolusite, 2891–2892
Pyrolysis, 977
    of urban waste, 3707
Pyrometer, 2892
Pyrometric cones, 2892
Pyrometry, 2892
Pyromorphite, 2892
Pyron, 2892
Pyronema confluens, 266
Pyrope, 1570
Pyrophanite, 2892
Pyrophyllite, 2892
Pyroxene, 2892
Pyrrhotite, 2892
Pyrrole and related compounds, 2892–2893

Pyrrophyta, 90
Pyruvic acid, 613
Pythagorean scale, 2893
Pythagorean theorem, 2893

Q, 2895
Q fever, 3002
Quad, 2895
Quadrant, 2895
Quadrate bone, 2895
Quadratic equation, 2895
Quadratix of Dinostratus, 2895
Quadratix of Tschirnhausen, 2895
Quadrature, 921, 2895
Quadrature (astronomy), 2895
Quadric surface, 2895–2896
Quadrifolium, 3030
Quadrilateral, 2896
Quadriplegia, 2896
Quadruple point, 2896
Quadrupole, 2896
Quadrupole mass spectrometer, 2291
Quadrupole moment, 2896
Quadrupole radiation, 2896
Quagga, 1799
Quahog, 2382
Quail, 2654, 2830, 2896
Quaking aspen, 2815
Qualitative chemical analysis, 158
Quality control, statistical, 2896–2901
Quantic, 2901
Quantile, 2901
Quantitative chemical analysis, 158
Quantization, 2901
Quantization of signals, 2901
Quantized Hall effect, 1708–1709, 3230
Quantized pulse modulation, 2901
Quantum, 2901
Quantum chemistry, 2901
Quantum-effect devices, 3133
Quantum efficiency, 2902
Quantum electrodynamics, 2902
Quantum liquid, 2165
Quantum mechanics, 2902–2904
    and invariance theorem, 1986
Quantum number, 2904
Quantum number (isospin), 2904
Quantum statistics, 2904–2905
Quantum theory of heat capacity, 2905
Quantum theory of radiation, 2905
Quantum theory of spectra, 2905
Quaquaversal, 2905
Quarantine, 2905
Quarks, 1705, 2646–2648
    formation after Big Bang, 3573
Quart (unit), 3570
Quarter-wave antenna, 188
Quartz, 2905–2906
    in ceramics, 710
    crystal oscillator, 1003
    for piezoelectric accelerometers, 11
    piezoelectricity, 2746
Quartzite, 2906
Quartz porphyry, 2906
Quasars, 2906–2907, 2937
    black holes in, 478
    and galaxies, 1545
Quasicrystals, 996, 1000
    in aluminum alloys, 125
Quasi-ferroelectric, 202
Quasi-periodic oscillations, 2883

Quasi-stellar radio sources. See Quasars
Quassia, 2907
Quaternary, 2907
Quaternion, 2907
Queen bee, 1783–1784
Quenching, 2907
Quetzal, 2907, 2908, 3529, 3530
Queue (computer system), 2907
Queueing problem, 2908
Quevenne scale, 3298
Quicklime, 2136
Quicksand, 2908
Quicksilver. See Mercury
Quiescent cell, 821
Quill, 2908
Quillfish, 480
Quillworts, 2908
Quince curculio, 1006
Quince tree, 3039
Quinine, 101, 794
Quinsy (peritonsillar abscess), 5, 2706
Quintuple point, 2908
Quota sample, 2908
Quota sampling, 3061
Quotient, 2908

Rabbit fever, 352
Rabbitfish, 2909
Rabbits, 2909–2910
Rabies, 2910–2911
Raccoons, 2911–2913
Racemate, 137
Race runner, 2176, 2913
Racon (radar beacon), 2913, 2915
Rad (unit), 3568, 3570
Radappertization, 2021
Radar, 2913–2915
Radar astronomy. See Radio astronomy
Radar beacon, 2915
Radar imagery, 2738–2739
Radar meteorology, 3735–3736
Radar mile, 2915
Radarscope, 663–664
Radial disk cam, 591
Radial distribution function, 2915
Radial keratotomy (RK), 2965
Radial symmetry, 3407
Radial velocity (star), 2915–2916
Radian (unit), 3568, 3570
Radiance, 2916
Radiant, 2916
Radiant boiler, 496–497
Radiant emittance, 1296
Radiant energy thermometer, 2916
Radiant point (meteors), 2916
Radiant power, 2916
Radiating atom, 322
Radiation, 2916
    quantum theory of, 2904
Radiational cooling, 306, 2916
Radiation belt, 2916
Radiation burn, 570
Radiation fog, 1458
Radiation hardening (electronics), 2917
Radiation heat transfer, 1740
Radiation laws, 2917
Radiation medicine, 2917
Radiation pattern, 2917
Radiation pressure, 2917
Radiation resistance, 2917
Radiation sickness, 2917–2918
Radiation therapy, 600

Radiation thermometry, 2918–2920
Radiative correction, 2920
Radiator, 2920
Radiatus cloud, 828
Radical (mathematics), 2920
Radicidation, 2021
Radioactivation analysis, 2923–2925
Radioactive dating techniques, 2928–2930
Radioactive elements, 736, 738
Radioactive fallout, 1352
Radioactive gas, 2920
Radioactive wastes, 2500–2503
Radioactivity, 2920–2931
Radio astronomy, 2931–2933
    antenna temperature, 189
Radio aurora, 327
Radio baton, 29
Radiobiology, 2934
Radiocarbon dating, 2929–2930
Radio communication, 2934–2936
Radio duct, 2936
Radio emission, 2933
Radio frequency allocation, 2936
Radio galaxies, 1551
Radioglaciology, 2792
Radiographic nondestructive testing,
    2467–2468
Radioisotope heater units, for spacecraft, 3239
Radioisotope power system technology, for
    spacecraft, 3238–3239
Radioisotopes. See Isotopes
Radioisotope thermoelectric generators, for
    spacecraft, 3239–3240
Radio meteorology, 2327
Radiometer, 2937
Radiometer vacuum gage, 2073
Radiometric analysis methods, 2923–2924
Radiometry, 2937
Radio navigation, 2425–2426
Radionuclide medical imaging, 2927–2928
Radio pulsar, 2883
Radiosonde, 2937
Radiosonde balloon, 370
Radio stars, 2937
Radio telescope, 2932–2933, 2937
Radiotopography, 599
Radium, 2920, 2937–2938
Radius vector, 3607
Radix point, 380, 2938
Radome, 2914, 2938
Radon, 2938–2939
Radurization, 2021
Raffinose, 3399
Rafter, 397
Ragweed, 894
Rails (bird), 2939
Railways, high-speed, 2939–2943
Rain, 2841–2842
Rainbow, 319
Rain forest, 432, 2944
Rain gage, 2843
Raisin moth, 1146–1147
Raisins, 1666–1667
RALPH system, 244
Ram, 1650
Raman spectroscopy, 2944
Ramark, 2944
Ramark (radar beacon), 2915
Ram drag, 79
Ramie fibers, 383, 2944
Ramjet engine, 80–81, 2035, 2945
Ramp A/D converter, 2945

Ramp response, 2983
Ramsden circle, 2945
Ramsden eyepiece, 1347
Random access memory (RAM), 2311–2314
   integrated circuits, 2335
Random-access storage, 3358
Random coincidence, 863
Random error, 2945
Randomization, 2945
   experiments, 1060
Random noise, 2466
Random-noise generator, 2467, 3162
Random number, 2945
Random selection, 2945
Random variable, 2945
Random vibration, 2945
Random walk, 2945
Range, 2254
   instrument, 2945–2946
   probability, 2946
Range marker, 1131, 2946
Range marks, 2946
Ranikhet disease, 2449
Rank (mathematics), 2946–2947
Rank correlation, 961, 2946
Rankine cycle, for solar power, 3225
Rankine temperature scale, 3458
Raoult's law, 2947
Rape, 533
Rape-kale, 533
Rapeseed, source of proteins, 2864
Rapid eye movement (REM) sleep, 3185
Rapier-horned antelopes, 183, 184
Raptors, 1351–1352
Rare-earth elements and metals, 2947–2954
Rare gases, 192
Rash, 2954
Raspberry, 417, 3039–3040
Raspberry-cane borer, 2954
Raspberry-cane maggot, 2218
Raspberry fruitworm, 1522, 2954
Raspberry sawfly, 3102
Rasputin, survives cyanide poisoning due to
   achlorhydria, 17
Rasse, 3670
Raster, 2954
Raster scanning, 3450
Raster scan technique, cathode-ray tubes, 661
Rat, 2405, 3022–3023
Rat-bite fever, 2954
Rate gain, 1544
Ratel, 2408
Rate-of-climb indicator, 2954
Ratfish, 762
Ratio controller, 2954
Ratio detector, 2954
Ratio thermometers, 2918–2919
Ratites, 2954–2955
RATO (rocket-assisted take-off), 2955
Rattlesnake, 3196–3197
Rauli tree, 404
Raven, 2955
Rayleigh law, 2955
Rayleigh number, 2955
Rayleigh scattering, 2944, 3109
Rayon fibers, 1381
Rays, 3177–3178
Ray tracing, 910, 2576, 2955
RC oscillator, 3162
Reaction coordinate, 757
Reaction curve, 2955
Reaction engine (motor), 2955

Reaction-product analyzer, 162
Reactive distillation, 1133
Reactive dyes, 1154
Read-only memories (ROM), 2312
Read only memory (ROM), integrated circuits, 2335
Readout, 2955
Reagent-tape analyzer, 162
Realgar, 2955
Real image, 2520
Real linear group, 2125
Real liquid, 2164
Realms (zoogeographic), 433
Real orthogonal group, 2125
Real-time computing, 2956
Real-time network routing, 3436
Real unimodular group, 2125
Reamer temperature scale, 2956
Reboilers, 1741
Recalescence, 2956
Receptor protein, neuron membrane, 696
Receptors, 695
Recessive pistasis, 1319
Reciprocal, 2956
Reciprocal vector system, 2956
Reciprocating compressors, 69–70
Reciprocating pumps, 2885
Reciprocity theorem
      (acoustical/electric-network/electroacoustical),
      2956
Reclamation disease, 951
Recognition genes, frogs and toads, 1507
Recoil atom, 322
Recoil particle, 2956
Recombinant DNA, 1591–1592, 2375
Recombinant vector vaccines, 3594
Recombination, 2956
Recombination energy, 2956
Recombination lines, 2933
Reconnaissance satellites, 3075–3082
Reconstitution, 1044
Recrudescent, 2957
Rectifiable curve, 1008
Rectification distillation, 1131
Rectifier, 2836, 2957–2958
Rectifier demodulator, 1046
Rectifier diodes, 2957
Rectilinear chart, 2958
Rectum, 2, 98
Rectum cancer, 597
Recurrence time, 2958
Red algae, 90, 94–95
Red ants, 1054
Redbird. *See* Cardinal
Red bone marrow, 505
Redbud, 2958
Red bug (chigger), 761
Red cedar, 221
Red clay, 2958
Red dwarf stars, 3573
Red eye, 68
Redfish, 984, 2958–2959
Red flour beetle, 1664
Red giant stars, 1621, 2449
Red hardness, 2959
Red jungle fowl, 2042–2043
Red mud, 2959
Redpoll, 2656
Red shift, 1683, 2959
Red spider mite, 2361
Redstart, 2959, 3698
Redstone rocket, 3019
Red supergiant stars, 1621, 2449

Redtop grass, 1679
Reduced mass, 2959
Reducing motion, 2959
Reduction, 2589
Reduction potential, 2959
Redundancy (structural), 2959
Redwing, 3483
Redwood, 1623
   coast, 918, 2959–2962
   dawn, 2962
Reedbucks, 183, 184
Reed canarygrass, 1679
Reed fish, 421
Reentrant (computer system), 2962
Reentry vehicle, 2962
   ablating materials used on, 2
Reference accuracy, 14–15
Reference ellipsoid, 2962
Reference-input signal, 3164
Refiner mechanical pulp, 2881
Reflectance, 2963
Reflected code, 1014
Reflecting telescopes, 3439–3440
Reflection, 2963
Reflection grating, 2963
Reflection nebula, 2429
Reflective liquid crystal display, 2159
Reflective microdisplays, 2331–2333
Reflectivity, 2963
Reflectometer, 2963
Reflector, 2963
Reflex activity, 693
Reflex circuit amplifier, 151
Reflux, in distillation, 1131
Reforming, 653
Refracting telescopes, 3439, 3440
Refraction, 2963–2964
   astronomical, 2964
Refractive eye surgery, 2964–2965
Refractive index, 2965
   optical fibers, 2571
Refractivity, 2965
Refractometers, 2965–2966
Refractories, 710
Refractory brick, 537
Refrigeration, 2966–2969
   expansion turbine application, 1577
Refrigerator, 1734
Refutation calculi, 248
Regeneration (zoology), 2969–2970
Regent-bird, 527
Regional pollution, 2799
Regional tectonics, 1192
Region of escape, 1341
Register (computer), 2970
Register memory, 2311
Regression, 2970
Regulator genes, 1589
Regulator mutation, 2375–2376
Regulus (α Leonis), 2970
Reheater, 497–498
Rehydration, 1044
Reindeer, 1039, 1040
Reindeer moss, 2125
Relapsing fever, 2970
Relational database management systems (RDBMS),
   1027
Relative coordinates, 946, 2970–2971
Relative humidity, 1853, 1882, 1883
Relativity, 2970–2974
Relaxation behavior, 2975

Relaxation frequency, 2975
Relaxation method, 2975
Relaxation time, 1032, 2975
Relay, 156, 2975
Reliability, 2975
Reliability coefficient, 2975
Rem (unit), 3570
Remainder, 1094
Remainder theorem, 96
Remanence, 2975
Remoras, 2975
Remote control, 2975
Remote sensing, 3075–3081
Removable discontinuity, 1127
REM sleep, 3185
Renal colic, 864
Renal corpuscle, 1341
Renal cysts, 2065
Renal fascia, 2061
Renal stones, 2065
Rendering, computer animation, 905
Rene 41, 108
Renin-angiotensin-aldosterone system, 1888–1889
Rennet, 654
Repeatability, 2975–2976
    chemical analysis, 160
Replication, 678–679
Replication, experiments, 1060
Repose (angle of), 46, 172, 2976
Reppe chemistry, 17
Representative sample, 3061
Reproducibility, 2976
    chemical analysis, 160
Reproduction
    asexual, 268–269
    oviparous, 2608
    ovoviviparous, 2608
    viviparous, 3671
Reproductive organs, 2
Reproductive system, 163
    and alternation of generations, 113–114
Reptilia, 2976
    fossils, 1480–1488
Repulsive forces, 2976
Reserpine, 100
Residue (mathematics), 2976
Residue theorem, 670
Resilience, 2977
Resin adhesives, 46–47
Resins (natural), 2977–2978
Resistance, 2978
Resistance-capacitance filter, 1393
Resistance thermometer, 2978–2979
Resistance transducer, 2979
Resistive-wall amplifier, 2979
Resistivity, 2979
Resistor capacitor transistor (RCTL) logic, 2979
Resistor transistor logic, 2979
Resolution, 2979
Resolution (computer system), 2979
Resolvers, 2818
Resolving power
    microscope, 2979–2980
    telescope, 2980
Resonance, 2980–2981
Resonance bridge, 541
Resonance neutrons, 2447
Resonance radiation, 2981
Resonator, 2981
Resorption, 2981
Respirable dust, 841
Respiration (plants), 2981

Respiratory system, 163, 2982
Response (instrument), 2983
Resting frequency, 635
Restitution coefficient, 866, 2983
Restless leg syndrome, 1611
Rest mass, 2983
Restoring force, 1474
Retaining wall, 2983
Retentivity, 2983
Reticle, 2983
Reticule, 2983
Reticulum (constellation), 2983
Retina, 2983–2984
Retinal detachment, 2984
Retinal disorders, 3650–3651
Retinal vein occlusion, 2984
Retinitis, 3652
Retinitis pigmentosa, 2984–2985
Retinoblastoma, 2984, 2985
Retinopathy, 2984
Retractor, 2985
Retroflector, 2985
Retrograde amnesia, 145
Retrograde motion, 2985
Retroreflector, 2985
Retrorocket, 2985
Return signal, 3164
Revegetation, 2985–2987
Revelle, Roger R,, 2987
Reverberation, 2988
Reverberatory furnace, 2988
Reverse acting controller, 2988
Reverse battery cell, 389
Reverse osmosis, 2310
    for desalination, 1058
Reverse transcriptase, 1591
Reversibility, principle of, 2988
Reversible processes, 2988
Reversion of series, 2988
Revolution, 2988
Revolution per minute (unit), 3570
Reye's syndrome, 272, 2172, 2988
Reynaud's disease, 235
Reynolds number, 1446–1447, 2988–2989
Rhea, 2989, 3097
    orbital characteristics, 2761
Rhebok, 183, 184
Rhenium, 2989–2990
Rheology, 2991–2992
Rheopectic fluid, 2992
Rheostat, 2992
Rhesus monkey, 2386
Rheumatic diseases, 2992–2993
Rheumatic fever, 2993
Rheumatoid arthritis, 2993–2994
Rheumatoid spondylitis, 3313
Rh group, 486
Rhinitis, 2995
Rhinoceros, 2995–2996
Rhino horn, superstitions about, 2996
Rhinophyma, 1055
Rhinosporidiosis, 2996
Rhinoviruses, and colds, 887
Rhizoids, 2996
Rhizome (rootstock), 2996
Rhodium, 2996–2997
Rhodochrosite, 2997
Rhododendron, 1734, 1735
Rhodonite, 2997
Rhombic antenna, 188
Rho-theta system, 2997
Rhubarb, 2997

Rhubarb curculio, 1006
Rhumb line, 2997
Rhyolite, 2997–2998
Ribavirin, anti-AIDS drug, 34
Ribbon parachute, 2998
Ribbon velocity microphone, 2342
Riboflavin coenzymes, 858–859
Riboflavin (Vitamin B_2), 2998–2999
Ribonucleic acid. See RNA
Ribosomal RNA, 1590
Ribs, 2999
Rice, 2999–3001
    irradiated, 2021
    nematodes affecting, 2432
Rice bird, 2656
Ricefish, 481
Rice leaf beetle, 407
Rice weevil, 1664
Richardson number, 3001
Richter, Charles Francis, 3001
Richter scale, 1188
Ricin, 650
Rickets, 505, 508
Rickettsial diseases, 3001–3002
Ridge line, 311
Riebeckite, 3002–3003
Rieke diagram, 3003
Riemann, Georg Friedrich Bernhard, 3003
Riemannian geometry, 1597
Riemann-Papperitz equation, 3003
Riemann seta function, 3003
Riemann surface, 3003
Riesz, Frigyes, 3003
Rifle bird, 2656
Rift valley fever, 3003–3004
Rigel (β Orionis), 3004
Righi-Leduc effect, 3004
Right ascension, 3004
Right-handedness (system), 1713
Right-handed screw, 1713
Right-sided congestive heart failure, 916
Rigid body, 3004
Rigid frame, 3004
Rill, 3004
Rime ice, 71, 2841, 3004
Rimming steel, 2017–2018
Ring (mathematics), 3004
Ring around, 3004
Ring canal, 594–595
Ring galaxies, 1549, 1550
Ring nebula, 2205
Ring systems, planets. See Planetary ring systems
Ring-tailed cat, 2911–2912
Ringworm, 1054–1055
Riometer, 3004–3005
Ripple, 3005
Ripple mark, 3005
Ripple voltage, 3005
Rise time, 3005
Risk/benefit analysis, 3005
Ritchey-Chretien telescope, 3005
River blindness, 478
River pollution, 3718–3719
Rivers, 1174
RNA, 1589–1590
    synthesis in DNA replication, 678–679, 1830
RNA virus, 679, 1588, 3643
Robber fly, 3005
ROBBIE system, 250
Roberval principle, 3005–3006
Robin, 3006
Robitzsch actinograph, 3006

Robotics, 3006–3014
    artificial intelligence, 263–264
    automation application, 334, 336
Robotic toys, 247
Robotic undersea vehicles, 2541–2543
Robustness, 3014
Roche limit, 424
Roches moutonnées, 3014
Rock, 3014
Rock beauty, 170
Rock drumlin, 1148
Rocket engine, 3015
Rocket propellants, 3015–3018
Rocket propulsion, 295
Rocketry, 3018–3022
Rockey Mountain wood tick, 3487
Rock-fill dams, 1024
Rock flour, 3022
Rock-goats, 1649
Rock salt (halite), 1707–1708
Rockwell hardness, 1714–1715
Rocky Mountain spotted fever, 3002
Rod-cone dystrophy, 374
Rodentia, 3022–3024
Rod mills, 371
Rodrigues formula, 2117
Roentgen, 3024
Roentgen (unit), 3568, 3570
Roentgen-equivalent-man (rem), 3024
Roentgen-equivalent-physical (rep), 3024
Role-playing games, 254
Roll, 68
Roller, 2071
Roller chain, 719
Rollers (bird), 952
Rolle theorem, 3024
Rolling, steel, 2014–2015
Rolling drum, 67
Rolling friction, 3024
Roll out, 3024
Rontgen, Wilhelm Conrad, 3024
Roof truss, 3532
Rook, 2656
Root
    mathematics, 3025
    plant, 3025–3027
Root-knot nematodes, 2431
Root-lesion nematodes, 2431
Root locus, 3024–3025
Root louse, 212
Root maggot, 1123
Root-mean-square, 3025
Root-mean-square electromotive force,
    1268
Root pressure, 264
Rootstock (rhizome), 2996
Rope, 3027–3028
Ropemaking fibers, 383
Ropy fermentation, 1363
Rosacea (rose family), 3030–3041
Rosacea (skin condition), 1055
ROSAT (Roentgen Satellite), 3028–3029
Rose bay, 1735
Rose chafer, 3030
Rose curve, 3030
Rose family, 3030–3041
Rosemary
    as antioxidant, 208
    oil, 2360
Rose of Sharon, 2237
Rose quartz, 2905
Rose scale, 3204

Rosin, 47, 2977
Rossby number, 313
Rossby wave, 377
Rossi X-Ray Timing Explorer (RXTE), 414,
    3041–3042
Rosy aphis, 212
Rotameter, 1437
Rotary disk filter, 1395
Rotary drum filter, 1394–1395
Rotary engine, 1971
Rotary pumps, 2885
Rotary telephone, 3427
Rotary-wing aerodynamics, 1745–1749
Rotatable dipole displays, 1130
Rotating compressors, 70
Rotating flowmeters, 1442–1443
Rotating generators, 2835
Rotation, 3042–3043
Rotational flow, 1447
Rotation axis, 3042
Rotation group, 2125
Rotation-reflection axis, 3043
Rotatoria, 3043
Rotifers, 3043
Rotor's disease, 423
Rough-weaving fibers, 1378
Rounding, 3043
Round-mouthed eels, 1015
Round-up, 440, 443
Roundworms, 25, 1980–1981
Route sampling, 3061
Routh criterion, for system stability, 3321
Routine (computer system), 3043–3044
Rove beetle, 3044
Rowland mounting, of reflection grating,
    2963
Royal jelly, 1782, 3044
Rubber
    natural, 3044–3046
        synthetic elastomers, 1219–1220
Rubber adhesives, 46, 47
Rubber latex cement, 686
Rubber plant (para), 1336
Rubella (German measles), 1295, 3046
Rubiaceae, 3046
Rubidium, 3046–3047
    alkali metal, 98
Ruby, 965, 1585
Ruby laser, 2093
Ruff, 3695
Ruffe, 2677
Ruffled grouse, 2707
Ruh beer, 406
Rule-based expert system, 252
Ruled surface, 3047
Rum, 3047–3048
Ruminant digestive system, 1103–1104
Run chart, 2898
Runoff, 1872, 1873, 1875
    nitrogen fertilizers, 1370
Rupture disk, 3048
Russell-Saunders coupling, 976
Russell (solar) collector, 3220
Rust fungi, 1530
Rusting, 962
Rutabaga, 533
Ruthenium, 3048–3049
Rutherford, Ernest, 3049
Rutherford scattering, 3109
Rutilated quartz, 2905
Rutile, 162, 3049–3050
Rydberg constant, 3050

Rye, 3050–3051
Ryegrass, 1679

Sabalo, 3418
Sabre-horned antelopes, 183, 184
Saccharides, 610
Saccharimeter, 3053
Saccharin, 3400–3401
Sac fungi, 266
Sacramento perch, 3384
Sacrum, 3053
Saddle point, 3053
Safety (intrinsic), 1986
Safety (limit) switch, 2137
Safety valve, 3053
Saffir/Simpson storm scale, 1511
Safflower, 3053
Saffron, 3053
Sage, antioxidant, 208
Sagebrush, 3053
Sagittarius (the archer), 3053
Sagitta (the arrow), 3054
Saguaro, 578
Saibling, 3531
Saiga, 1649
Sailfish, 423
Sailings, 2422, 3054
Sailplane, 3054
Saint Augustine grass, 1679
Saint Elmo's fire, 3054
St. Anthony's fire (erysipelas), 1053
Saki, 2385
Salamander, 146, 3054
Sal ammoniac, 140, 143
Salicylate drugs, 153
Salicylic acid, 272, 3054
Salientia, 210
Saline gradient, energy from ocean, 2544, 2549
Salinometer, 3054
Salivary glands, 1102
Salk, Jonas Edward, 3054
Salmon, 3054–3057
    aquaculture, 219
Salmonella, 1461
Salmonellosis, 355, 364, 1463–1464
Salmonfly, 1315, 2305
Salpin, 777
Salpingitis, 3057
Salsif, 894
Salt, 3057–3058
Salt (sodium chloride), 44, 3205–3206
    and hypertension, 1091–1092
Saltation, 3058
Saltator, 1505
Salt bridge, 3058
Salt dome, 2688
Salting out, 1345
Saltpeter, 3058
Salt river, 3058
Salty taste, 1419
Salve bug, 3058
Salyut space station, 3265, 3268
Samarium, 3058–3059
Sample, 2899, 3059
Sample-and-hold amplifier, 151, 3059
Sampled data system, 3059
Sampling (statistics), 3060–3061
Sampling controller, 3059–3060
Sampling error, 3060
Sampling with replacement, 3061
Sandbox tree, 1336

Sand cricket, 3061
Sand dollar, 1201, 3061
Sandfly, 2885
Sandfly fever, 2709
Sandgrouse, 874, 2746
Sandpike, 2748
Sandpiper, 3695
Sandstone, 3061
Sandstone dike, 3061
Sandworm disease, 1054
Sanitary sewers, 3713
San Jose scale, 375, 3205
Santa Gertrudis cattle, 668
Sap, ascent of, 264–265
Sapanwood, 534
Sapodilla, 3061, 3062
Saponification, 3199
Sapphire, 965, 1585
Saprophytes, 3061
SARAH (search and rescue and homing), 3062
Sarcoma, 599, 3062
Sardine, 1766
Saros, 3062
Sarsaparilla, 3062
Sassaby, 183, 184
Sassafras, 2108
Satellites (astronomical body), 2760, 3062
Satellites (communications and navigation),
      3062–3075
   telephony application, 3431
Satellites (scientific and reconnaissance),
      3075–3082
Satellite solar energy collectors, 3228
Satellite transmission frequencies, 3066–3067
Satin-bird, 527
Saturated compounds, 896
Saturated hydrocarbons, 2583
Saturated salt dew point sensors, 1883–1884
Saturated vapor, 3082
Saturated vegetable oils, 3609
Saturation, 2839, 3082
Saturation-adiabatic lapse rate, 309
Saturation-adiabatic processes, 307–308
Saturation current, 3082
Saturation curve, 983
   binary mixtures, 427
Saturation vapor pressure, 3605
Saturn, 3082–3101
   Cassini mission to, 638–648
   Hubble Space Telescope images, 1805,
      1822–1824
   orbital characteristics, 2759
   physical characteristics, 2760
   rings, 3090–3092
   satellites, 3094–3098
   Voyager missions to, 3689–3692
Sauger, 2676
Saurine poisoning, 1469
Sauropterygians, 1481–1485
Savanilla, 3418
Savanna biome, 433
Savannah, 3102
Sawfish, 3177, 3178
Sawfly, 1885, 3100, 3102
Sawtoothed grain beetle, 1145
Saw-toothed grain beetle, 1664
S-band, 3102
Scabies, 1054
Scad, 3102
Scalar field, 3102
Scalar product, 3608
Scalar quantity, 3102

Scald, 570
Scald fish, 1417
Scale, 3103
Scale factor, 3103
Scale insects, 375, 1782, 3103–3105
Scale symmetry, 2647
Scaling circuit, 3105
Scallop, 2080, 2382–2383
Scandium, 3105–3106
Scanning antenna, 188
Scanning electron microscope, 1275–1276
Scanning microdisplays, 2334–2335
Scanning tunneling microscope, 3106–3108
Scapolite, 3108–3109
Scarab, 3109
Scarlet fever, 352, 3109
Scarlet tanager, 3414
Scatter diagram, 3109
Scattering, 3109–3120
Scatterometry, 3110–3113
Scavenging, 3113
Scheelite, 3113–3114
Schering bridge, 540
Schinus tree, 3114
Schist, 3114
Schistosomiasis, 3114
Schizophrenia, 3114–3115
Schläfli formula, 2117
Schlieren, 3115
Schmidt, Maarten, 3115
Schmidt objective, 3115–3116
Schmidt process, 3116
Schmitt trigger, 3116
Schönlein-Henoch purpura, 2886
Schooling, in fishes, 1410
Schottky defect, 3116
Schrödinger, Erwin, 3116
Schrödinger equation, 2903, 3116
Schrödinger wave equation, 2293
Schuchert, Charles, 3116
Schumann-Runge absorption bands, 7
Schwarzschild, Karl, 3117
Schwarzschild radius, 478, 3117
Schwarzschild telescope, 3117
Sciatic neuralgia, 2444
Scientific satellites, 3075–3082
Scintillation
   astronomical, 3117
   atmospheric, 3117
Scintillation counter, 3117
Scintillometer, 3117
Sciumorphs, 3319–3320
Scleral buckle, 3117
Sclerenchyma, 3117
Scleroderma, 3114–3115
Sclerometer, 3118
Scleroscope, 1715
Scolecite, 3118
Scoliosis, 3118
Scombroid fish poisoning, 1092
Scombroid poisoning, 1468
Scoria, 3118
Scorodite, 3118
Scorpion, 220, 237, 3118
Scorpion fly, 2308, 3118
Scorpius (the scorpion), 3118
Scotch whiskey, 3761–3762
Scoter, 180, 3716
Scotopic vision, 2634
Scrapie, 531, 980
Screamer, 180, 3118

Screw, 2208–2209
Screw compressors, 70–71
Screw dislocation, 999
Screw-horned antelopes, 183, 184
Screw pumps, 2885
Screw-worm, 3118–3119
Scrophulariaceae, 3119
Scrub fowl, 2399–2400
Sculpins, 3119
Scup, 2816
Scurfy scale, 3205
Scurvy, 267, 505
Scyphozoa, 3119
Sea anemone, 189, 857, 3119
Sea arrow, 3119
Sea butterfly, 3119
Sea caves, 1324
Sea-cliff, 1324
Sea cows, 3119
Sea cucumber, 1201, 1780
Sea elephant, 3120
Sea fan, 87, 3119
Sea farming, 214
Sea feather, 857, 3119
Sea-floor spreading, 2526
Sea hare, 3119
Seahorses, 3120
Sea lemon, 3120
Sea leopard, 3120
Sealers, 2618
Sea level, 2521–2523
Sea-level horizon, 1786
Sea lily, 982, 1201
Sea lions, 3120–3121
Seals, 3120–3121
Sea moths, 3121
Seamounts, 3121
Sea mouse, 3121
Sea-pen, 87
Sea ranching, 214
Search
   artificial intelligence applications, 242–243
   automated reasoning, 248
Sea robins, 3121
Sea slug, 3121
Sea smoke, 1458
Season cracking, 3121
Seasonings, 1425
Sea squirt, 265, 266, 412
SeaStar, 3080–3081
Sea-surface temperature, 2523
Sea trout, 984
Sea urchin, 706, 1201
Sea walnut, 2552
Seawater
   composition, 2523
   density, 2523, 3161
Seaweeds, 412, 3121
Sebaceous cyst, 3122
Sebaceous gland, 3122
Seborrhea, 3122
Seborrheic dermatitis, 1055
Secant function, 3527–3528
Second (unit), 3570
Secondary alcohols, 87
Secondary amides, 131–132
Secondary amines, 132–133
Secondary battery, 389
Secondary emission, 3122
Secondary flow, 1447
Secondary glaucoma, 1637
Secondary sexual characteristics, 1655, 3122

Secondary wastewater treatment, 3713
Second-degree burn, 570
Second law of thermodynamics, 3470
    and degradation of energy, 1044
Second-order reaction, 757
Second-order system, 3122
Second reproductive caste, 892
Secretary bird, 1160, 1351
Secretory diarrhea, 1080
Sectile, 3122
Section modulus, 3122–3123
Secular determinant, 3123
Secular parallax, 2634, 3123
Secular terms, 3123
Sedge, 3123
Sedimentary rocks, 2356
Sedimentation, 804
    water, 3712
Seebeck effect, 3123
Seed, 3123–3124
    dissemination, 1519
Seed-corn maggot, 2218
Seeding (gas conductivity), 3125
Seeing (astronomy), 3125
Seger cone, 3125
Segregation ice, 3413
Seiche, 1330, 3125
Seismic moment, 1188
Seismic petroleum surveys, 2691
Seismograph, 1185
Seismology, 1185, 1187–1188
    Sun, 3381
Seizure, 3125–3126
Selection rules
    energy levels, 3126
    nuclear, 3126–3127
Selectivity coefficient, 1996
Selenium, 3127–3128
Selenographic, 3128
Selenology, 3128
Self-alignment, 2143
Self-centering chuck, 792
Self-conjugate direction, 920
Self-conjugate subgroup, 920
Self-diagnostics, 3128
Self-energy (particle), 3128
Self-induced gas-impeller agitation, 803
Self-inductance, 3128
Self-operated controller, 3128–3129
Self-pollination, 2797
Self-resident translator, 3516
Self-tuning controllers, 41–42
Sellmeier equation, 3129
Selsyn, 3129
Semantic knowledge, 261
Semianthracite coal, 821
Semiapochromat, 2345
Semiarid region hydrology, 1877–1878
Semi-automated analytical apparatus, 158
Semibituminous coal, 821
Semicarbazones, 3129
Semichemical pulp, 2881
Semiconductor diode, 1120
Semiconductor lasers, 2093–2094
Semiconductor rectifiers, 2957–2958
Semiconductors, 3129–3134, 3231
Semicubical parabola, 2633
Semi-decision procedures, 248
Semidiameter correction (sextant), 3135
Semikilled steel, 2018
Semipermeable membrane, 3135
Semipolar bond, 2376

Semisynthetic penicillins, 195
Semitone (half-step), 3135
Senescence, 1614–1618. *See also* Aging
Senile purpura, 2886
Sensible temperature, 3135
Sensillae, 3135
Sensitive plant, 2118
Sensitivity (instrument), 3135
    chemical analysis, 160
Sensitized decomposition, 1033
Sensitometry, 3135
Sensor, 3135
    for automation, 331
Sensory organs, 3135–3136
Sensory receptor, 3136
Sepal, 1434
Sepiolite, 3136
Sepsis, 3136
Septarian structure, 3136
September equinox, 344
Septic bursitis, 574
Septicemia, 3136
Septic shock, 3159
Septic sore throat, 2706
Septum, 3136
Sequence, 3136–3137
Sequence tags, 1834
Sequential-access storage, 3358
Sequential address accessing, 2311
Sequential analysis, 3137
Sequential decoding, 1933
Sequoia (giant sequoia), 918, 1622–1623, 2959
Serandite, 3137
Serial adder, 43
Serial storage, 3358
Seriate fabric, 3137
Seriemas, 2939
Series, 3137
    alternating, 113
    asymptotic, 300
    binomial, 427
    collateral, 738
    displacement, 40
    electromotive, 40
    homologous, 1782
    Liouville-Neumann, 2141
    lyotopic, 820
    Maclaurin, 2214
    Taylor, 3420–3421
Series (radioactive)
    actinium, 736, 2924
    neptunium, 2925
    thorium, 736, 2923
    uranium, 2922
    uranium-radium, 736, 738
Series circuit, 1234
Series-fed vertical antenna, 188
Series peaking circuit, 2668
Serine, 133–137
Sernyl, 1712
Serosa, 1346
Serotonin, 698
Serous gland, 3137
Serow, 1649
Serpentine, 3137
Serpentine asbestos, 264
Serum sickness, 209
Serviceberry, 3041
Service ceiling, 673
Service routine, 3043–3044
Servo accelerometers, 12
Servo-amplifier, 151

Servomechanism, 3137–3138
Servomotors, 3138–3146
Servopower, 331–332
Sesame oil, 1412
Set (mathematics), 3146
Set (permanent), 1042, 3146
Settling time, 3146
Seventeen-year locust, 2183
Sex, 3146–3147
Sex-influenced inheritance, 3147
Sex-limited inheritance, 3147
Sex-linked inheritance, 3147–3148
Sextans (constellation), 3148
Sextant, 3148
    semidiameter correction, 3135
Sexual intercourse, 952
Sexually transmitted diseases, 3148–3153
Sexual selection, 3154
Seyfert galaxy, 1550–1551, 1552, 3154
Sferics (spherics), 3154–3155
Sferics fix, 3155
Sferics observation, 3155
Sferics receiver, 3155
Sha, 1651
Shackle, 3155
Shad, 1766
Shadbush, 3041
Shad fly, 1315, 2305
Shale, 3155
Shallot, 2559
Shallow-water cisco, 3764
Shama, 2656
Shannon formula, 3155
Shannon-Nyquist sampling theorem, 97
Shaped-beam antenna, 188
Shaped-beam technique, cathode-ray tubes, 661
Shape-memory alloys, 125
Shapley, Harlow, 3155
Shared time control, 3155
Shark repellants, 3156
Sharks, 3155–3157
Shatter breccia, 536
Shaula (λ Scorpii), 3157
Shear, 3157–3158
Shear center, 3158
Shear deformation, 1042
Shearwater, 2684, 2851
Shechtmanite, 996
Sheep, 1647–1651
    Dolly, the cloned sheep, 821–824
Sheep bot, 2471
Sheep lice, 2189–2190
Sheep tick, 3158
Shelduck, 180
Shelf fungi, 529
Shell, 3158
Shellac, 47, 2618, 3158
Shellfish, 3158
Shell molding, 3158
Sheppard's corrections, 3158–3159
Sherardizing, 3159
Shield (geology), 3159
Shielded cables, 577
Shield volcano, 3678
Shift register, 2970
Shigellosis, 352, 364–365, 1467
Shilo, 3173
Shingles, 1053
Shipworm, 574, 3159
Shittah tree, 9
Shock front, 1445
Shock syndrome, 3159

Shock tube, 3159
Shock wave, 3159–3160
Shock-wave lithotripsy, 1562
Shoebill, 793
Shooting star, 3160
Shoran (short-range navigation), 3160
SHORAN navigation system, 2428
Shorebirds, 726, 3695–3696
Shore effect, 3160
Short circuit, 3160
Short-grass plains, 432–433
Short-period comets, 883, 3328
Short-period errors, 2945
Short-range force, 3160
Shortwall mining, 834–835
Short-wave radio fadeout, 3380
Shotgun sequencing, 1834
Shoveller, 3716
Showers (rain), 2842
Shrews, 2378–2379, 2379
Shrike, 3160
Shrimp, 988, 990, 1031
Shround of Turin, 2931
Shunt (electrical), 3160–3161
Shunt-fed vertical antenna, 188
Shunt peaking circuit, 2668
Sialons, 123, 125
Siamang, 191
Siamese fighting fish, 2077
Siberian high, 311
Sickle cell anemia, 168, 1756
Sick sinus syndrome, 228
Sidereal day, 3161
Sidereal hour angle, 3161
Sidereal month, 3161
Sidereal period, 3161
Sidereal year, 3161
Siderite, 3161
Siderolites, 2326
Siderostat, 857
Sidewinder, 3197
Siemens (unit), 3568, 3571
Sifakas, 2119
Sigma (statistical quality ocntrol), 2899
Sigma hyperons, 2322
Sigma particle, 3161
$Sigma_T$ (seawater density), 3161
Sigmoid colon, 2, 870
Sigmoidoscopy, 3161
Signal, 3161
    conditioning, 3161–3162
    instrument, 3164
Signal analysis, 3304
Signal generator, 3162–3164
Signal level, 3164
Signal-to-noise ratio, 3164
Signal transducer, 3164
Sign convention, lens and mirror, 3164
Sign digit, 1119
Significance tests, 3164
Significant digits, 3164
Significant wave, 3164
Silage, 3164–3165
Silent chain, 719
Silica gel
    anticaking agent, 199–200
    decolorizing agent, 1032
Silica glass, 1629–1630
Silicates (soluble), 3165
Siliceous, 3165
Silicic, 3165
Silicification, 3165

Silicon, 3165–3172
    amorphous, 3134
    Carbon Group member, 623
    effect on steel, 2018
Silicon controlled rectifier (SCR), 3172
Silicon controlled switch (SCS), 3172–3173
Silicone elastomers, 1220
Siliconizing, 1744
Silicosis, 2781
Silicrete, 1153
Silk, 657, 1382
Silk cotton trees, 3173
Silkworm, 3173
Sill, 2077, 3173
Sillimanite, 3173
Siltstone, 3173
Silurian Period, 3173
Silver, 2356, 3173–3175
    sterling, 3352
Silver amalgam, 128
Silver-cell battery, 3175
Silver coulometer, 973
Silverfish, 3175, 3418
Silver glance, 224
Silver king, 3418
Silver oxide battery, 394
Silversides, 3175
Silver thaw, 2842
Silver wattle tree, 9
Silvery pout, 856
Silver-zinc battery, for spacecraft, 3237–3238
Simal, 3173
Simmonds-Sheehan disease, 324, 2756
Simoon, 432
Simple curve, 1008
Simplex algorithm, 2140
Simplex operation (communication system),
    3175
Simulator (computer system), 3175
Sine function, 3527–3528
Sine magnetometer, 2229
Single-cell protein, 2864
Single-crystal ferrite, 1368
Single-ended amplifier, 151, 3175
Single-ended push-pull amplifier, 151
Single lens reflex camera, 2727
Single-mode optical fibers, 2565
Single-phase motor, 2399
Single-photon-emission computed tomography
    (SPECT), for brain imaging, 3797
Single scattering, 3109
Singular, 3176
Singular point of a curve, 3176
Singular point of a function, 3176
Sink hole, 670
Sinking speed (aircraft), 3176–3177
Sinope, orbital characteristics, 2760
Sintering, 67, 3177
Sinus, 3177
Sinus bradycardia, 228
Siphon (zoology), 3177
Sirius (α Canis Major), 605, 3177
Sirlin-type ice, 3413
Siskin, 1505
SI units, 3568–3571
Skarn, 3177
Skates, 3177–3178
Skeletal system, 163, 3178–3180
Sketchpad system, 907
Skew curve, 1009
Skewness, 2899, 3180
Skimmer, 726

Skin, 3180
    artificial, 570
Skin allergies, 105
Skin cancer, 599, 3180–3182
Skin diseases, 1052–1056
Skin effect, 3182
Skin friction, 3182
Skinks, 3182–3183
Skip distance, 3183
Skip effect, 2936
Skipper, 2120, 3183
Skraup quinoline synthesis, 2590
Skua, 726
Skunk, 2408
Skunk cabbage, 227
Skutterudite, 3183
Sky color, 318, 2129
Sky cover, 3183–3184
Skyhook balloon, 3184
Skylab space station, 3020, 3265, 3266–3268
Skylight, 1100
Sky wave, radio, 2934
Slag, 3184
Slate, 3184
Slaty cleavage, 805
Slave antenna, 3184
Sleep, 3184–3186
Sleep disorders, in elderly, 1611
Sleeping sickness, 63
Sleepwalking, 3186
Sleet, 2842
Slew, 3186
Slickhead fishes, 3186
Slider, 3186, 3463
Slide rule, 3186
Sliding friction, 1155
Slipper shell, 2138
Slip rings, 3186
Slit, 3186
Slit lamp, 3186
Slope, 3186
Slot antenna, 188
Sloth, 1210, 1211
Slot-hole borer, 513
Slow ion, 2090
Slow lemur, 2119
Slow neutrons, 2447
Slow viruses, 3645
Slow-worm, 2177
Sludge, 3187, 3713–3715
Sludge (ice), 1895
Slug (animal), 1578, 3187
    plant pollination by, 2797
Slug (unit), 3570
Slug tuning, 3187
Sluice, 3187
Slurry explosive, 1343
Small darkling beetle, 1145
Small intestine, 2, 98, 1102
Small ion, 1995
Small Magellanic Cloud, 1550, 1552
Smallpox, 3187
Smart controller, 42
SMART system, 250
Smectic liquid crystal, 2143, 2162
Smekal-Raman effect, 2944
Smell. *See* Odors
Smell blindness, 2558
Smelting, 581, 3188
Smelts, 3054–3055, 3188
Smith chart (Rieke diagram), 3003
Smith diagram, 3188

Smith III bridge, 538
Smithsonite, 3188
Smoketree, 637
Smokey quartz, 581, 2905
Smoky Mountain haze, 2801
Smoothing, of curve, 3188
Smut fungi, 1530
Snail, 412, 1578, 2383, 3188
    plant pollination by, 2797
Snake fly, 3188
Snake lizards, 1583, 2176
Snakes, 3188–3197
Snappers, 3198
SNC meteorites, 2326
Snellen eye chart, 3198, 3661
Snipe, 726, 3695
Snipe fly, 3198
Snow, 2842
Snow gage, 2843
Soaps, 98, 3198–3221
Society (ecology), 3201
Soda-lime-silica glasses, 1629–1630
Sodalite, 3201
Soda nitre (sodium nitrate), 3058, 3201
Sodium, 3201–3205
    alkali metal, 98
    in biological systems, 2822–2823
    and hypertension, 1091
    and kidney abnormalities, 2063
Sodium aginate, 1698
Sodium benzoate, as food antimicrobial agent, 204
Sodium bicarbonate, antacid, 182
Sodium bromide, 480
Sodium chloride (salt), 44, 1091–1092, 3205–3206
Sodium chrome alum, 116
Sodium hypochlorite, 480
Sodium perborate, 480
Sodium silicates, 47, 3165
Sodium-sulfur battery, for electric cars, 1232
Sodium thiosulfate, 3206–3207
Sodium zeolite softening, 1360
Sod webworm, 3738
Sofar (sound fixing and ranging), 3207
Soft brown scale, 3205
Soft contact lenses, 929
Soft drinks, 3402–3403
Softeners, 1062–1063
Soft ferrite, 1368
Software, 3207
Software engineering, 3207–3208
Software metrics, 3208–3209
Software relocation, 2859
Soft X-ray lasers, 2096
Soil, 3209–3213
Soil Classification System (USDA), 3211
Soil conservation, 1325
Soil profile, 3210
Soil washing, 3701–3702
Sol, 866, 867–868, 3213
Solanaceae (potato family), 3213–3214
Solar apex, 3214
Solar atmospheric tide, 316
Solar cells, 2835, 3227–3228
Solar cooling, 3216
Solar eclipse, 1202–1206, 3377–3378
Solar energy, 3215–3228
    spacecraft applications, 3241–3242
Solar flares, 3380
Solar furnaces, 3220–3221, 3225–3226
Solar heating, 3215–3219
Solar keratitis, 3181
Solar One plant, 3222–3224

Solar parallax, 3228
Solar seismology, 3381
Solar system, 2758–2762. *See also* Sun; specific
    Planets
    formation, 3574
Solar telescopes, 3381–3382
Solar time, 3228–3229
Solar tower energy collector, 3222
Solder glasses, 1630
Soldierfish, 3319
Soldier fly, 1147
Sole, 1415, 1418
Solenoid (meteorology), 313
Solenoidal, 3229
Solenoidal flow, 930
Solenoid (electrical), 41, 3229
Solfatara, 3229
Sol-gel glass, 1634–1635
Solid (geometry), 3229
Solid angle, 3229
Solid oxide fuel cells, 1526–1527
Solid solutions, 106
Solid-state ignition, automotive, 337–339
Solid-state lasers, 2094–2095
Solid-state physics, 3229–3232. *See also* Crystal
Solid-state TV cameras, 3453
Solid-state variable-speed drives, 3141–3142
Solid-state welding, 3746
Solid superacids, 23
Solidus curve, 3232
    alloy phase diagrams, 107
Solion, 3232
Soliton, 2568
Solstice, 807, 3232, 3375
Solubility, 3232
Solubility curve, 3232
Solubility product, 3232–3233
Solubilization, 868
Solution polymerization, 2378
Solvent, 3233
Solvent extraction, 1345
Solvent naphtha, 849
Solvolysis, 3233
Somatic cell cloning, 819
Somatoplasm, 3233
Somatostatin, and diabetes, 1067
Somite, 3233
Somnambulism, 3186
Sonar, 3233–3235
Sone, 3235
Songbirds, 2655
Song-thrush, 2656, 3483
Sonic barrier, 3235
Sonic boom, 3235
Sonics, 3235
Sonic speed, 3235
Sonochemistry, 31, 3558
Sonoluminescence, 32
Sorbates, as food antimicrobial agent, 204
Sorbic acid, as food antimicrobial agent, 204
Sore throat, 352, 2706
Sorghum, 3235
Sorghum grasses, 558
Sorghum midge, 2348
Sorption, 3236
Soufrière Volcano, 3677–3678
Sound, 25
Sound production, 3671–3675
Source encoding, 1929, 1934
Source program, 2852, 3516
Soursop, 177
Sour taste, 1419

South African albatross, 490
South American paper tree, 637–638
South American Trypanosomiasis (Chaga's disease),
    718–719
South Atlantic Central Water, 3236
South Atlantic Current, 3236
Southbound node, 1060
South Equatorial Current, 3236
Southern armyworm, 1012
Southern cabbageworm, 576
Southern hornworm, 1797
Southern lights, 328
Southern Oscillation, 317, 812
Southern sennet, 377
Southernwood, 232
South Pacific Central Water, 3236
South Pacific Current, 3236
Sowbug, 3236, 3782
Soybean, 399, 3236–3237
Space (Minkowski), 3246
Space charge, 3237
Space charge limitation of currents, 3237
Spacecraft power systems, 3237–3242
Space curve, 1009
Space frame (space structure), 3243
Space Infrared Telescope Facility (SIRTF), 1934,
    3243–3244
Space medicine, 3244
Space microelectronics technology, 3244–3246
Space Shuttle, 3021, 3246–3265
    Mir space station cooperative program,
    3274–3275
Space Shuttle *Atlantis*, 3254–3257
Space Shuttle *Challenger*, 3249–3250, 3251
Space Shuttle *Columbia*, 3021, 3247–3249
Space Shuttle *Discovery*, 3250–3254
Space Shuttle Earth Observations Project, 3078
Space Shuttle *Endeavor*, 3257
Space Shuttle *Enterprise*, 3246
Space stations, 3246–3294
Space structure, 3243
Space-time, 3295
Space Transportation System, 3257, 3259
Space vehicle
    ablation, 3
    aluminum applications, 119
    guidance and control, 3295–3297
Space velocity (stellar), 3297
Spadix, 1434
Span (instrument), 2945–2946
Spandex fibers, 1381–1382
Spanish cedar, 2907
Spanish fly, 482
Spar, 3297
Spark chamber, 3297
Sparrow, 568, 1505, 3297
Spastic paralysis, 693, 713
Spawning, 3297–3298
Spearfish, 423
Special relativity, 2971–2972
Specific activity, 3298
Specific adsorption, 51
Specification limits, 2899
Specific gravity, 3298–3300
Specific heat, 1733, 3300
    Debye theory of, 1031
    Dulong and Petit law, 1152
    electronic, 3300
    Humphries equation, 3301
Specific humidity, 1853, 3301
Specific ionization, 1999
Specificity, catalysts, 653

Specific permeability, 2681
Specific refraction, 2963–2964
Specific surface, 3301
Specific volume, 3301
Spectinomycin, 197
Spectral analysis, 3301
Spectral centroid, 3301
Spectral characteristic, 3301
Spectral class, 3301
Spectral colors, 3301
Spectral emissivity, 1296
Spectral energy distribution, 3301–3302
Spectral function, 3302
Spectral line halfwidth, 1707
Spectral orders, 2563
Spectral reflectance, 2963
Spectral sensitivity, 3302
Spectrochemical analysis (visible), 3302
Spectrofluorometers, 1453
Spectroheliograph, 3302, 3381–3382
Spectro instruments, 3302
Spectrophotofluorometer, 3302
Spectrophotometry, 873
Spectropolarimetry, 2785
Spectroscope, 3302–3303
Spectroscopic binaries, 424, 3303
Spectroscopic parallax, 3303
Spectrum analysis, 3304–3306
Spectrum viewer, 3305
Specular density, 1050
Specular reflectance, 2963
Speech clipping, 3307
Speech recognition systems, 261
Speed, 3307, 3614
Speed-length ratio, naval aerodynamics,
    1517
Speed measurement, 3609, 3613–3615
Sperm, 189
Spermophile, 3319, 3320
Sperrylite, 3307
Spessartine, 1570–1571
SPF (Sun Protection Factor), 600
Sphagnum moss, 563
Sphalerite blende, 3307
Sphene, 3307
Sphenisciformes, 3307
Sphere, 3307
Sphere of influence (planet), 3307
Spherical aberration, 3307
Spherical harmonics, 3307–3308
Spherical polar coordinates, 3308
Spherical surface, 3308
Spherical trigonometry, 3308
Spherics, 318
Spherics (sferics), 3154–3155
Spheroid, 3308–3309
Spheroidal coordinate, 3309
Spheroidize annealing, steel, 2020
Spherometer, 3309
Sphygmomanometer, 3309
Spica (α Virginis), 3309
Spice bush, 414
Spices, 1425
Spicules, 3309, 3377
Spider, 220, 237, 3309–3310
Spider bites, 1054
Spikelet, 1433
Spillway, 3310
Spin (nuclear), 2503, 3311
Spinach, 3310
Spinach aphis, 212
Spinach leaf miner, 3310–3311

Spinal cord, 690, 693
Spin-dependent force, 3311
Spine, 3311
Spin-echo correlative spectroscopy, 2477
Spinel, 3311
Spinneret, 3311
Spin-orbit coupling, 976
Spin quantum number, 2904
Spin state, 3311
Spinthariscope, 2921
Spiny anteater, 2387
Spiny lobster, 990
Spiny-tailed lizard, 64
Spiracle, 3312
Spiral, 3312
Spiral bevel gear, 421
Spiral curve, 3312
Spiral galaxies, 1547, 1548–1549
Spiral layer, 1217
Spiral-shaped bacteria, 351
Spiral valve, 3312
Spirochete, 351
Spiro-compounds, 896
Spiro hydrocarbon ring systems, 2585
Spittle bug, 1505, 1782, 3312
Spleen, 2, 3312
Splenomegaly, 3312
Splines, 905
Spodosols, 3211
Spodumene, 3312–3313
Spondylarthropathies, 3313
Sponge, 412, 2816, 2817
Spontaneous symmetry breaking, 2646
Spool valves, 1855–1856
Spoonbill, 1895, 2615
Spore, 560, 3313
Sporophyll, 3313
Sporotrichosis, 365, 3313
Sporozoites, 269
Sports drinks, 1044
Spot (fish), 984
Spotfin, 527
Spotted cutworm, 1011–1012
Spotted garden slug, 3187
Spotted irishlord, 3119
Sprain, 3313
Sprat, 1765–1766
Spray drying, 3313–3315
Spray electrification, 2119
Spray-type agglomerator, 67
Spreading coefficient, 3315
Spreading decline of citrus, 2431–2432
Spring, 3315
Springbuck, 183, 184
Spring clock, 3315
Spring index, 3315
Spring rate, 3315
Spring-tails, 866
Spring tide, 3492
Sprinkler irrigation, 2023–2025
Sprite animation, 904
Spruce budworm, 3317
Spruce trees, 3315–3318
Sprue (tropical), 3318
Spur gearing, 3318
Spurge family, 1335–1336, 3367
Spurious correlation, 3318
Spurious radiation, 3318
Sputnik, 3021, 3265, 3318
Sputtering, 3318
Squab, 2830–2831
Squall, 1517

Squal line, 1509
Squamous cell carcinoma, 599, 600, 3181
Square (of a number), 3318
Square law demodulator, 1046
Square law modulator, 3318
Square-loop ferrite, 1368
Square of Pegasus, 2671
Square root, 3318
Square root (Cholesky) method, 774
Squaretail, 575
Square wave, 3318
Squash, 1005, 3318–3319
Squash aphis, 212
Squash borer, 805, 3319
Squash bug, 3319
Squash vine borer, 513
Squeezed light, 2128
Squid, 708, 2384
Squirrelfishes, 3319
Squirrels, 3319–3320
Squirting cucumber, 1005
Stability, 3320
    mechanical, 3320
    system, 3320–3322
Stabilization (ship), 3322
Stable fly, 3322
Stachyose, 3399
Stack, 3322
Stadimeter, 3322
Stage, 3322
Stagger-tuned amplifier, 151
Stagnation pressure (total head), 416
Stainless steels, 107–108, 785, 2019
Stains (coating), 2618
Stalactite, 670, 3322
Stalagmite, 670, 3322
Stalagmometer, 3322
Staling (bakery and food products), 3322–3323
Stall, 3323
Stamen, 1434
Standard atmosphere (unit), 114–115, 306–307,
    3568, 3569
Standard conditions, 3323
Standard deviation, 2899–2900, 3323
Standard error, 3323
Standard free energy increase, 3323
Standard refraction, 2964
Standards, 3568–3571
Standard signal generator, 3162–3163
Standard state, 3323
Standard Time, 3493
Standard unit (statistics), 3323
Standard volume, 3687
Standing-wave ratio, 3323
Stannite, 3323
Staphylococcal intoxication, 1464–1465
Staple, 2809
Star, 3324. See also Double star
    Besselian star numbers, 419
    color temperature, 874
    element formation in, 3573
    flare stars, 1415
    giant and dwarf, 1621–1622
    heat index, 3343
    interstellar reddening, 1980
    luminosity, 3343
    magnitude, 6, 3343–3344
    navigator's, 2428
    neutron, 2449
    parallax, 3344
    position angle, 2819
    proper motion, 2861

Star, (continued)
  radial velocity, 2915–2916
  radio, 2937
  shooting, 3160
  space velocity, 3297
  variable, 3605–3606
  velocity curve, 3615
Star birth, 3573
Star catalogues, 3323, 3325
  Bonner Durchmusterung, 510
Starch, 612, 3324–3327
  in chromatophore, 784
  industrial chemicals derived from, 453
Starch adhesives, 46, 47
Star clusters, Hubble Space Telescope images, 1810,
    1811, 1812
Star connection, 3327
Star death, 3573–3574
Stardust mission, 3327–3338
Starfish, 279, 1201
Stark effect, 3338–3339
Starling, 3339
Starling, Ernest H., 3339
Star topology, networks, 2179
Starvation, 3339
State, 3339
Static and noise, 3339–3340
Static gain, 1544
Static induction thyristors, 3143
Static random access memory (SRAM), 2311–2312
Statics, 3340
  graphical, 3340
Stationary front, 1508
Stationary state, 3339, 3340–3341
Station error, 1041
Statistic, 3341
Statistical distribution, 1135
Statistical equilibrium, 1322
Statistical mechanics, 3341
Statistical moment, 2384
Statistical quality control, 2896–2901
Statistical thermodynamics, 3471
Stator, 3341
Statute mile (unit), 3570
Staurolite, 3341
Staybolt, 3341
Steady state, 3341
Steam calorimeter, 590
Steam cycles, 1014
  diagram factors, 3341
Steam distillation, 1131
Steam engine, 3341–3342
Steam fog, 1458
Steam separation, in boilers, 494–495
Steam turbine, 3546–3548
Steam washing, 495–496
Stearates, 3342
Stearic acid, 3342
Steel arch bridge, 547
Steelmaking, 2012–2014
Steels, 107–108
Steelyard, 3342–3343
Steerable antenna, 188
Stefan-Boltzmann law, 1740
Steinboks, 183, 184
Stele, 3343
Stellar heat index, 3343
Stellar luminosity, 3343
Stellar magnitude, 6, 3343–3344
Stellar parallax, 3344
Stellar scintillation, 3117
Stem, 3344–3347

Stem nematodes, 2431
Step-down amplifier, 151
Stephanite, 3347
Stephan's Quartet (galaxies), 1548
Steppe, 3347
Stepped leader, lightning, 2130–2131
Stepper motors, 3347–3351
Step response time, 3351
Steptoe, 3351
Steradian, 3351
Steradian (unit), 3568, 3570
Stereo broadcasting, 3351–3352
Stereographic projection, 3352
Stereoisomerism, 755, 2028
Stereo-power, 3352
Stereoscope, 427, 3352
Stereospectrogram, 3352
Stereo system, 3352
Sterling silver, 3352
Stern-Gerlach experiment, 3352
Sternum, 2982
Steroids, 1655, 3352–3356
Sterol, 3352, 3355
Stethoscope, 3356
Stevioside, 3402
Stibine, 206
Stibnite, 206, 3356
Stick insects, 2603
Stiffness, 3356
Stigma, 3356
Stigmatic, 3356
Stilb (unit), 3570
Stilbite, 3356
Stilt, 726
Stimulus, nerve, 3356
Stingray, 3177, 3178
Stink bug, 3356
Stirling converters, for spacecraft, 3240–3241
Stirling cycle engine, 1971
Stirred-pot still, 2376
Stirrup (ear), 325
Stoat, 2407
Stochastic, 3356
Stochastic process, 3356–3357
Stoichiometric compounds, 731
Stoker, 574
Stokes (unit), 3570
Stokes bands, 2944
Stokes flow, 3357
Stokes law for viscosity, 3357
Stokes lines, 209
Stokes theorem, 3357
Stolon, 3357
Stomach, 2, 98, 1102
  age-related disorders, 1611
Stomate (stoma), 3357
Stone Age, 3357
Stone canal, 594
Stonechat, 2656
Stone flies, 2775
Stone groundwood pulp, 2881
Stone roller, 3357
Stones (calculi), 587
Stony-iron meteorites, 2326
Stony meteorites, 2326
Stopping power, 3357
Storage (computer), 3357–3358
Storage battery, 389
Storage register, 2970
Storax, 2978
Stork, 793, 794, 3358
Storms, 1507–1517

Storm surge, 1511, 1512
Storyboard, 904
Stout beer, 404
Strabismus (cross-eyes), 1327, 3358–3359, 3650
Straight edge, 3359
Strain, 1042
Strain dyadic, 1217, 1218
Strain energy, 3359
Strain gage, 3359–3360
Strain hardening exponent, 3360
Strain theory, 3360
Strangeness, 1887
Strange particles, 2321–2322
Strangler fig tree, 2401
Strategy games, 254
Strath, 416, 3360
Stratification, 3360
Stratigraphy, 3360
Stratocumulus clouds, 826
Stratosphere, 304, 759
Stratus clouds, 828
Strawberry, 417, 1734–1735, 3040–3041
  irradiated, 2021
Strawberry corn borer, 513
Strawberry leaf roller, 2114
Stray capacitance, 605
Stream flow, 1872
Stream function, 3360
Streaming (molecular), 3360
Streamline flow, 1447
Streamlining, 53–54
Streamtubes, 53
Street of eddies (Kármán vortices), 2055
Streptococcal toxic shock syndrome, 360
Streptococcus pneumoniae infection, 365
Streptomycin, 196, 441
Stress (structural), 3360–3361
Stress corrosion cracking, 962
Stress dyadic, 1217
Stress relieving, 3360
  steel, 2019
Stress-rupture test, 3360
Stress-strain curve, 3360
Striated muscle, 2405
Striation, 3361
Strigiformes, 3361–3362
Stringer, 543, 3362
String filter, 1395–1396
String theory, 2647–2648
Stripped atom, 322
Stripping, 3362
Stroboscope, 3362
Stroke, 714–715
Stromatolite, 3362–3363
Stromatolith, 3363
Strong force, 2645
Strongyloides, 1981–1982
Strontianite, 3363
Strontium, 3363–3364
Stroud bridge, 541
Structural genes, 1588
Structural genomics, 1842
Structural geology, 3364
Structural protein, 2865
  neuron membrane, 697
Strut, 3364
Strychnine, 100, 101–102
Stuffing box, 3364
Stump tree, 2056
Sturgeons, 3364
Stye (hordeolum), 3364–3365
Stylet, 3365

Stylolite, 3365
Stylotypite, 3464
Styrene, 2685
Styrene-butadiene rubbers, 1219
Styrene-maleic anhydride, 3365
Subacute sclerosing panencephalitis, 531, 3365
Subaerial delta, 109
Sub-antarctic Water, 182
Subarctic Pacific Water, 3365
Sub-bituminous coal, 821
Subcortical dementias, 127
Subduction, 2529
Subdwarf stars, 1621
Subgiant stars, 1621
Subgraph, 3365
Subirrigation, 2022–2023
Sublimation, 3365
    in atmosphere, 2839
    in free-drying, 1495–1496
Sublimation (heat of), 3366
Submarine geothermal aquifers, 2544, 2549
Submarine geothermal springs, 2530–2534, 2544, 2549
Submarine hydrothermal deposits, 2554
Submarine telephone cables, 3431
Submarine volcano, 3678
Submerged fermentation, 440
Submersibles, 2540–2541
Sub-multiplexing, 156
Subpolar high, 311
Subpolar low-pressure belt, 310, 311
Subroutine, 3366
Subsequent stream valleys, 924
Subsidence, 3366
Subsidence inversion, 309
Subsolar point, 3366
Substance P, 699
Substandard propagation, 2936
Substantive colors, 1154
Substitute natural gas, 2421
Substitution, 2589
Subsurface currents, 2524
Subsurface runoff, 1872
Subtraction, 3366
Subtractive color process, 3366
Subtropical high, 311
Subunit vaccines, 3593–3594
Subwavelength illumination, 2128
Succession (plant), 3366
Successive-approximation A/D converter, 3366–3367
Succulents, 3367
Suckers, 3367
Sucking lice, 2189–2190
Sucralose, 3402
Sucrose, 612, 3399–3400
Sudden infant death syndrome (SIDS), 3367–3368
Sufficient statistic, 3368
Sugar alcohols, 3400
Sugar apple, 177
Sugarcane, 3368–3369
Sugar degrees, 3053
Sugar maple, 2252–2253
Sugars. See Carbohydrates
Suine (swine), 3404–3405
Sulfhydryl group, analytical determination, 161
Sulfites, as food antimicrobial agent, 205
Sulfonamides, 132, 3369
Sulfonation, 2592
Sulfone polymers, 3369
Sulfur, 3370–3372
    in biological systems, 3372–3374

effect on steel, 2018
fertilizer requirements, 1372
in petroleum, 2685
Sulfur dioxide
    and acid rain, 21
    air pollutant, 2802
    as food antimicrobial agent, 205
Sulfur fibers, 1382
Sulfur hexafluoride, greenhouse gas, 1641
Sulfuric acid, 3374–3375
Sulfurous acid, 3375
Sum, 3375
Sumac trees, 637–638
Summer solstice, 807, 3232, 3375
Sumner line, 3375–3376
Sumptner principle, 3376
Sun, 3376–3383. See also terms beginning "Solar"
    absorption spectrum, 9
    and skin cancer, 3181–3182
Sun block, 3181
Sun compass, 3383
Sun cross, 319
Sundangrass, 1679
Sundial, 818, 3383
Sundial time, 3492
Sun dog, 319
Sunfishes, 3383–3384
Sunflower, 893, 894
Sunflower oil, 1412
Sungrazing comets, 883
Sunis, 183, 184
Sun pillar, 319
Sun spider, 220
Sunspot, 301, 3376, 3378–3379
    and weather, 3736–3738
Superacids, 22–23
Superadiabatic lapse rate, 309
Supercalendering, of paper, 2632
Supercharger, 3384
Superconducting quantum interference device (SQUID), 2479
Superconducting Super Collider, 2647, 2652
Superconducting turbine generator, 1242–1243
Superconductivity, 3384–3387
Superconductors, 992
Supercooling, 3388
Superelevation, 3388
Superfatting, 3198
Superficial cellulitis, 1326
Superfinishing, 3388
Superfluidity, 992, 3388
Supergiant stars, 1621, 2449
Superheater, 497–498, 1741
Superheavy elements, 738
Superheterodyne receiver spectrum analyzer, 3305–3306
Superimposed river valley, 3388
Superior air, 72
Superior conjunction, 921
Supermolecular electronics, 1965, 2365–2374, 3129
Supernova. See Nova and supernova
Supernova 1987A, 1806, 1816, 2472, 3388–3390
Supernova 1993J, 3388–3390
Superposition (law of, in stratigraphy/Nernst principle of/principle of), 3391
Supersaturated vapor, 3391
Supersaturation, 2839–2840, 3391
Supersonic aerodynamics, 62, 3391–3396
Supersonic flow, 1445
Superstandard propagation, 2936
Superstring theory, 2647–2648
Supersymmetry, 2647, 2652

Supertwisted-nematic LCDs, 2146–2147
Supervised learning, 257
Supervisory control, 933, 3395–3396
Superwaves (ocean waves), 316–317
Suppressor (electrical noise), 3396
Suppressor mechanisms, antibodies, 199
Suppuration, 3396
Supraventricular arrhythmias, 228
Surface, 3396
Surface-active biomaterials, 507
Surface coal mining, 834
Surface integral, 1964
    vectors, 3608
Surface inversion, 309
Surface of positive total curvature, 3408
Surface of revolution, 3396
Surface runoff, 1872
Surface tension, 3396–3397
    Antonoff rule, 210
    Gibbs formula, 3397
Surfactants, 1060–1061, 3198
Surge, 3397
Surge arresters, 2132
Surgeonfishes, 3397
Surge tank/vessel, 3397
Suricate, 3670
Survival analysis, 2302–2303
Suspension bridge, 544, 545
Suspension polymerization, 2378
Suture, 3397
Swallow, 3397
Swallow float (swallow plinger), 3397
Swallowtail, 575, 3397
Swamp, 3397
Swamp cypress, 1016, 1017–1018
Swan, 180, 3716
Sweepback (aircraft wing), 3398
Sweep circuit, 3398
Sweeteners, 3398–3403
Sweet flag, 227
Sweet fly, 411
Sweet Gale, 397
Sweetgum tree, 3403
Sweetlip, 1696
Sweet potato, 3403
Sweetsop, 177
Sweet taste, 1419
Swells, 316
Swift (bird), 212, 3403
Swift (lizard), 2913, 3403–3404
Swimmer's itch, 1054
Swine, 3404–3405
Switching equipment (telephone), 3435–3437
Sword-bearer, 3404–3405
Swordfish, 423
Sycamore tree, 3405–3406
Sydenham's chorea, 778
Syenite, 3406
Sylvanite, 3406
Sylvitea, 3406
Symbiosis, 3406–3407
Symbiotic environment, 1207
Symbolic code, 855
Symmetric, 3407
Symmetric function, 95, 3407
Symmetric group, 2682
Symmetric kernel, 2058
Symmetry
    axis of, 3407
    center of, 687, 3407
    and conservation laws, 923–926
    plane of, 3407

Symmetry (*continued*)
   zoology, 3407
Sympathetic autonomic nervous system, 343
Sympathetic chain, 693
Symphyla, 3407
Symphylid, 687
Synapse, 694–695, 3407
Synaptic vesicles, 696
Synchronous, 3408
Synchronous computer, 3408
Synchronous motors, 2397–2398
Synchros, 2818
Synchrotron, 2650–2651
Synchrotron emission, 2651–2652, 2933
Synclastic surface, 3408
Syncline, 3408
Synclinorium, 200
Syncope, 3408
Syndrome, 3408
Syneresis, 3408
Synergic curve, 3408
Synfuels Program, 844–845
Synodic period, 3408
Synsets, 261
Synthesis (chemical), 3408
Synthesis gas, 3409
   ammonia production from, 140, 141–142
Synthesized speech, 3673
Synthetic division, 96, 3409
Synthetic fibers, generic designations and definitions, 1379–1382
Synthetic methyl acetone, 16
Synthetic rubber, 1219
Syntons, 2861
Syphilis, 352, 3149–3150, 3409–3411
Syphilitic arthritis, 237
Syrinx, 3672
Syrphid flies, 411, 1124
Systematic name, of chemicals, 2582
Systemic lupus erythematosus, 3411–3412
System programs, 1110–1111
Systems engineering, 3412
Systolic pressure, 489, 1888
Syzygy, 3412

Tabby-cat, 664, 666
Taber ice, 3413
Tabes, 3413
Tableaux, 248
Tableland, 3413
Table reef, 2772
Tableting, 67
Tabling, 804
Tabular iceberg, 3413
TACAN navigation system, 2428
Tachometers, 3614–3615
Tachycardia, 228
Tachylyte (tachylite), 3413
Tacoma Narrows Bridge, 545–547
Taconite, 3413
Tacrine, 127
Tactical theorem proving, 248
Tactile organs, 3413–3414
Taffrail log, 2657
Tahr, 1649
Takayasu's aortitis, 235
Take-off weight, 1693
Takin, 1649
Talbot law, 3414
Talc, 3414
Talegallas, 2399

Talus, 3414
Tamarack, 2089
Tamarao, 520
Tamarau, 520
Tamarin, 2257
Tamm levels, 3414
Tampan, 609
Tampon-related toxic shock syndrome, 352, 365
Tanager, 3414
Tandem (cascade), 3414
Tandem mass spectrometer, 2290
Tangelo, 801
Tangent, 3414
Tangent function, 3527–3528
Tangential acceleration, 3414–3415
Tangential Doppler effect, 1140
Tangential force, 1474
Tanget arc (atmospheric optical phenomenon), 319
Tank circuit, 3415
Tannin, 3415
   clarifying agent, 803
Tanning pills, 3181
Tantalite, 3415
Tantalum, 3415–3417
Tantochrone, 1015, 3417
Taos, 3499
Taped delay line, 1045
Tapestry moth, 824
Tapeworm, 717–718, 3417
Taphonomy, 3417
Tapioca dextrins, 492
Tapir, 3417
Tar acids, 849
Tarantula, 3417–3418
Tar bases, 849
Taro, 3418
Tarpan, 1798
Tarpon, 3418
Tarragon, 232
Tar sands, 2687, 3418–3419
Tarsi, 3510
Tarsiers, 3419
Tarsioids, 3419
Tarsus, 3419
Tartaric acid, 3419
Tarui's disease, 2410
Tasmanian devil, 2284
Tasmanian Huon pine, 2751, 2783
Tasmanian wolf, 2284–2285
Taste, 1419
   removal from water, 3712
Taste bud, 3419–3420
Tau particle, 3420
Taurine, neurotransmitter, 698
Taurus (the bull), 3420
Tautomerism, 2028–2029
Taxonomy, 3420
Taylor, Brook, 3420
Taylor series, 3420–3421
Tayra, 2408
Tay-Sachs disease, 2142, 2984
T cells, 1909, 3592
T distribution, 3421
t distribution, 3421
Tea, 3421
Teak tree, 3421
Teardrop balloon, 3421
Tears, 3421–3422
Technetium, 3422
Technology Demonstration Converter, 3241
Tectogenesis, 2601

Tectonism, 1080
Tegu, 2176–2177
Teju, 3422
Tektite, 3422
Telecentric system, 3422
Telecommunications, 3423
Telegraphy equation, 2293
Teleoperation, 3422–3423
Teleosteica, 3423
Telephony, 3423–3438
   queueing problem, 2908
Telephotometer, 3518
Telephoto photographs, 1349
Telescope
   astrographic, 293
   astronomical-optical, 3439–3449
   Cassegrain, 638
   equatorial, 1321
   Galilean, 1553
   Hubble Space Telescope, 1801–1825
   Maksutov-Bouwers, 2231–2232
   Newtonian, 2450
   radio, 2932–2933, 2937
   resolving power, 2980
   Ritchey-Chretien, 3005
   Schwarzschild, 3117
   seeing, 3125
   solar, 3381–3382
   Space Infrared Telescope Facility (SIRTF), 1934, 3243–3244
   zenith, 3810
Telescopium (constellation), 3449
Television, 3449–3457
Telluric lines, 3457
Tellurium, 3457–3458
Tellurometer, 1594
Telophase, 678
Temper, 3458
Temperature, 3458–3459
   and basal metabolism, 379
   human, 3459
Temperature inversion, 309
Temperature sensing, optical fiber applications, 2571
Temperature transfer standard, 3459
Temperature units, 3568
Tempering, glass, 1633
Temporal (bone of skull), 3459
Temporal lobes, 693
Tendon, 3459
Tenorite, 3459
Tenrec, 2378–2379
Tensile strength, 3459
Tensimeter, 3459
Tensiometer, 3459
Tension, 3459. *See also* Surface tension
   interfacial, 1968
Tension member, 3459
Tension test, 3459–3460
Tensor, 3460
Tensor contraction, 3460–3461
Tensor differentiation, 3461
Tensor field, 3461
   Cartesian, 3461
Tent caterpillar, 3738
Teosinte (ancestor of corn), 957
Tephra, 3461
Tera (prefix for units), 3568
Terbium, 3461–3462
Terephthalic acid, 3462
Terminal ballistics, 368
Terminal-based linearity, 2139
Terminal button, synapse, 696

Terminal moraine (end moraine), 3462
Termite, 2030, 3462–3463
Termitophile, 3463
Tern, 726
Ternary critical point, 983
Terrace, 1324
Terrapin, 3462
Terrestrial coordinates, 1594
Terrestrial environment, 1207
Terrestrial meridian, 2321
Terrestrial Planet Finder, 281
Terrestrial scintillation, 3117, 3463
Tertiary alcohols, 87
Tertiary amides, 131–132
Tertiary amines, 132, 133
Tertiary Period, 3463
Tesla (unit), 3568, 3571
Tesla coil, 3463
Testes, 1302, 1655–1656
    and pituitary, 2755
Testosterone, 165–166
Test pattern, 3463
Testudinata, 3463
Tetanus, 352, 3463–3464
Tetany, 2637
Tethys, 3097
    orbital characteristics, 2761
Tetrachoric correlation, 3464
Tetracyclines, 195–197
Tetradecanoic acid, 2410
Tetradic, 3464
Tetradymite, 3464
Tetrahedrite, 3464
Tetraodon poisoning, 1469
Tetrode, 1278
Texas leaf-cutting ant, 181
Textile dyes, 1153–1155
Thalamus, 690, 692
Thalassemia major, 2009–2010
Thallium, 3464–3465
Thaumatin, 3402
Thawing season, 1043
Theodolite, 3465
Theorem, 3465
Theorems of constraint/moderation, 2116
Theory of equations, 3465–3466
Therm (unit), 3571
Thermal black, 621
Thermal conductivity, 3466–3467
Thermal cracking, 977
Thermal diffusion, 1100
Thermal diffusivity, 3467
Thermal efficiency, 3467
Thermal equator, 3467
Thermal gradient (geothermal gradient), 3467
Thermal insulation, 1958–1963
Thermal jet engine, 3467
Thermal mass flowmeter, 1441
Thermal neutrons, 2448
Thermal noise, 2466
Thermal radiation, 3467
Thermal stability, 3467
Thermal tide, 315–316
Thermionic conversion, 3467
Thermionic emission, 3467
Thermionic regulation and switching devices, 2837
Thermistor, 3467–3468
Thermochemistry, 3468
Thermocline, 3468
Thermocouple, 3468–3469
Thermodynamic equilibrium, 1321–1322
Thermodynamic potentials, 3470

Thermodynamics, 3469–3471
    atmosphere, 306–307
Thermodynamic temperature units, 3568
Thermoelectric cells, 2835
Thermoelectric cooling, 3471–3472
Thermography, 3472–3474
Thermogravimetric analysis (TGA), 3473
Thermogravitational column, 1100
Thermoluminescence dating, 2928
Thermomechanical pulp, 2881
Thermometer, 3473
Thermometer (filled-system), 3473
Thermometric function, 3458
Thermometric titration, 3502
Thermophone, 3473
Thermoplastic resin adhesives, 46, 47
Thermoplastic resins, 2772
Thermosetting resin adhesives, 46, 47
Thermosetting resins, 2772
Thermosphere, 759
Thermotropic atmosphere, 307
Theta function, 3473
Theta wave, 1257
Thevenin theorem, 3473–3474
Thiamides, 132
Thiamine (Vitamin B_1), 3474
Thiazides, 1136
Thickeners, 804
Thickness measurement and gaging systems,
    3475–3476
Thief ant, 182
Thin-film electroluminescent displays, 2334
Thin-film fuel cells, 1527
Thin films, 3476–3479
Thin-film transistors, in liquid crystal displays,
    2150–2152, 2154–2157
Thin-layer chromatography (TLC), 780
Thin-lens relationships, 3479
Thiocyanic acid, 3479
Thioethers, 3479
6-Thioguanine, antimetabolite, 203
Thiokol rubbers, 1220
Thiophene, 3479–3480
Thiourea, 3480
Third-degree burn, 570
Third law of thermodynamics, 3470–3471
Third quarter moon, 2388
Thirteen-year locust, 2183
Thistle butterfly, 576
Thixotropic fluid, 1445, 2992
Thomas-Fermi differential equation, 3480
Thomson, J. J., 3480
Thomson cross section, 3109
Thomson law, 3480
Thomson parabola method, 3480
Thomson principle, 3480
Thomson scattering, 3109
Thorax, 3480
Thorianite, 3480
Thorite, 3480
Thorium, 37, 3480–3481
Thorium series, 736, 2923
Thorn apple, 3214
Thorny-headed worms (acanthocephala), 10
Thoroughbred race horses, 1799
Thousand-legged worm (millipede), 2352
Thrasher, 3481
Thread (yarn), 3028
Three-address instruction, 1957
Three-body problem, 3481–3482
Three-dimensional computed tomography (CT), 170
Three-dimensional computer animation, 904

Three-dimensional graphics, 909–911
Three-phase equilibrium, 3482
Threonine, 133–137
Threshold decoding, 1933
Threshold detector, 890, 3482
Threshold illuminance, 3482
Thrip, 3482
Throat, 3483
Throat breccia, 1878
Throat cancer, 2090–2091
Thrombin, 700
Thrombus, 232, 235–236, 531
Throttle, 3483
Throughput, 3483
Throwing power, 3483
Thrush, 602, 3483
Thrush (blackbird), 477
Thrust, 3483
Thulium, 3483
Thunder, 1515
Thunderstorm Project, 302
Thunderstorms, 1508, 1515
Thyme, 2360
Thymus gland, 3483–3484
Thyratron, 2957
Thyristor, 3484
    for servomotors, 3142–3143
Thyroid gland, 3484–3486
Thyroid hormones, 1788
Thyroiditis, 3486
Thyrotropic hormone, 1792
Tick, 10, 220, 3487
    Lyme disease from, 2201–2203
Tick bites, 1054
Tidal bore, 513, 1330, 2534
Tidal energy, 2544, 3488–3491
Tidal waves. See Tsunami
Tide gage, 3492
Tides, 3492
Tiger, 664, 665
Tiger beetle, 3492
Tiger's eye, 667, 2905
Tilapia, aquaculture, 217, 218
Till, 3492
Till (glacial), 1627
Tilleul trees, 382
Tillite, 917
Tilting-pan filter, 1396
Tiltmeter, 3492
Tilt table, 3492
Timber, 3780–3782
Timber dams, 1024
Time, 3492–3494
    rise, 3005
Time constant, 3494
Time delay, 3494
Time demodulation, 3494
Time discriminator, 3494
Time-division multiplexing, 3434, 3494
Time equation, 1321
Time-frequency duality, 1717
Time gate, 3494
Time modulation, 3494
Time pattern, 3494
Time-resolved photoacoustic calorimetry, 2720
Time response, 2983
Time series, 3494
Time sharing, 3494–3495
Time signal, 3495
Time slicing, 3495
Time-stratigraphic unit, 3495
Time units, 3568

Time zones, 3493
Timing belt, 3495
Timothy grass, 1680
Tin, 3495–3497
    Carbon Group member, 623
Tinamiformes, 3498
Tinamous, 3498–3499
Tincalconite, 2058
Tincture, 3499
Tinea, 1054
Tinea pedis, 1054–1055
Tinnitus, 1724
Tin stone (cassiterite), 648
Tintometer, 3499
Tissue, 675–676, 3499
Tissue culture, 821
Tissue level, 429
Tit, 3499
Titan, 3094–3096
    and *Cassini* mission, 638, 648
    orbital characteristics, 2760
Titania, 3583
    orbital characteristics, 2761
Titanite, 3499
Titanium, 3499–3500
    effect on steel, 2018
Titanium alloys, 108
Titanium carbide, 1744
Titanium dioxide, 3501
Titmice, 3499
Titration
    potentiometric, 3501–3502
    thermometric, 3502
Titration coulometer, 973
Toad bug, 3502
Toads, 146, 210, 1505–1507
Toadstool, 65, 380
Tobacco, 3213, 3214
Tobacco budworm, 960
Tobacco hornworm, 1797
Tobacco worm, 3502
Tobramycin, 196
Todies, 952, 2071
Todorokites, 3503
Toggle (flip-flop), 1431
Tokamak fusion reactor, 1537
Token bus, 2181–2182
Token ring, 2181–2182
Toluene, 2685, 3503
Tomaraw, 520
Tomato, 417, 3213, 3214
Tomato fruitworm, 960
Tomato hornworm, 1797
Tomato worm, 3503
Tombolo, 3503
Tone control, 3503
Tongue, 3503
Tonic seizure, 3463
Tonometer, 1636, 3504
Tonsillitis, 352
Ton (tonne) (unit of weight), 3571
Toon (Chinese cedar), 2907
Tooth, 3504
Topaz, 3504–3505
Topcoat, 2618
Top-fermentation yeast, 405
Topi, 183–184
Topical chemotherapy, 600
Top-loaded vertical antenna, 188
Top note (flavors), 1419
Topocentric, 3505
Topographical mapping, 3505

Topological group, 3505
Topological space, 3505
Topology, 3505
Top pan weighing, 3742
Topped tar, 849
Topping, 849
Top quark, 2647
Torbernite, 3505
Tornado, 1514, 1515–1516
Toroidal coordinate, 3505–3506
Torque, 2384, 3506
Torque amplifier, 3506–3507
Torque converter, 3507
Torr, 3507
Torsional stress, 3507
Torsional type magnetostrictive delay line, 1045
Torsional vibration measurement, 3507
Torsion balance, 3507–3508
Tortoise, 3463, 3552
Tortoise shell, 3508
Torus (anchor ring), 163
Total head (stagnation pressure), 416
Toucan, 2745, 3782–3783
Touch, 3413–3414
Touchscreen, 908
Touch-tone telephone, 3427
Tourmaline, 3508
Toussaint's formula, in altimetry, 115
Towering, 2187
Towhee, 2656
Townsend avalanche, 3508
Townsend discharge, 1578
Toxemia, 3508
Toxic fish and shellfish, 1468–1469
Toxicology, 3508
Toxic shock-like syndrome, 352
Toxic shock syndrome, 352, 355, 365, 1658–1659
Toxoid, 3508, 3593
Toxoplasmosis, 3508
    and AIDS-related eye disorders, 68
Trace (mathematics), 3509
Tracer compounds, 896
Tracer techniques, using isotopes, 2923, 2925–2926
Trachea, 3509, 3672
Tracheostomy (tracheotomy), 3509
Trachoma, 365, 922, 3509
Trachyte, 3509
Tracing routine, 3044
Track, navigational, 3509
Trackball, 908
Tracking, 3509
Traction, 3509–3510
Tractrix, 656
Tractrix of Huygens, 3510
Trade-wind inversion, 309
Tragacanth gum, 1698
Tragopan, 2707, 3510
Tragulines, 3510
Trailing mahonia, 373
Trajectory, 510
Transactinide elements, 738
Transactinium earths, 3510
Transcendental, 3510
Transcription, 679
Transcriptomics, 1842
Transducer, 3510–3511
    electroacoustics, 30–31
Transferases, 1314
Transfer characteristic, 3511
Transference number (transport number), 3511
Transfer function, 3511
Transfer magnetoresistance, 1708

Transfer margins, 2526
Transfer orbit, 296
Transferrin, 2009
Transfer RNA, 1590
Transfinite number, 3511
Transflective liquid crystal display, 2159
Transform, 3511
Transformation (mathematics), 3511
Transformer, 3511–3513
Transformer (instrument), 3513
Transformer bridge, 543
Transformer-coupled amplifier, 152
Transgenic animals, 820–821
Transglycosylation, 613
Transient (electrical), 3513–3514
Transient voltage surge suppressor, 1229–1230
Transistor, 3514–3515
Transistor blocking oscillator, 2604
Transit
    astronomy, 3515–3516
    surveyor's, 3516
Transition, allowed, 3516
Transition effect, 3516
Transition loss, 3516
Transition probability, 3516
Transition temperature, 3516
Translation, 3516
Translation operation (geometry), 3516
Translator, 3516
    computer program, 3516–3517
Translucidus cloud, 828
Transmissible mink encephalopathy, 531
Transmissible spongiform encephalopathy (TSE), 522
Transmission (pneumatic), 3517–3518
Transmission dynamometer, 1157
Transmission grating, 3517
Transmission level, 3517
Transmission limit, 3517
Transmission loss, 3517
Transmission mode, 3517
Transmissive microdisplays, 2331
Transmissometer, 3518
Transmittance, 3518
Transmittance (directional luminous), 3518
Transmittance meter, 3518
Transmittancy, 3518
Transmitter, 3518
Transmutation, 2503
Transonic, 3518
Transonic flow, 3518
Transpiration, 3518
Transplantation
    bone marrow, 509
    cardiac, 1731
    corneal, 3652
    heart-lung, 1731
    kidney, 2067
    liver, 2173–2174
    pancreas, 1071, 2627
Transponder, 3064
Transport equation, Boltzmann's, 504, 3518–3519
Transport number (transference number), 3511
Transport proteins, 2865
Transposons, 1834
Transuranium elements, 3519
Trap, 3519
Trapezoidal rule, 3519
Trapped ions, 3519
Trapping, 3519
Trap rocks, 2356
Traps, for petroleum, 2688–2689

Trauma, 3519
Traveling-wave amplifiers, 31, 152
Traveling-wave tube, 2347
Traverse, 3519
Traverse tables, 3519
Travertine, 3519
Trays, distillation, 1133–1134
Tree, 3519, 3520–3523
    annual ring, 177–178
Tree (mathematics), 3523
Tree architecture, 1113
Tree creeper, 2656
Tree cricket, 2183
Tree hopper, 1782, 3519, 3523
Tree-hyraxes, 1892
Tree of heaven, 2231
Tree path, 797
Tree pie, 2656
Tree toad, 3524
Trellis drainage, 3524
Trematoda, 1449, 3524–3525
Tremolite, 3525
Trench foot, 1892, 3525
Trench mouth, 2706
Trepanning, 3525
Trephine, 3525
Triangle, 3525
Triangulation, 3525–3526
Triangulum (the triangle), 3526
Triangulum Australe (constellation), 3526
Triassic Period, 3526–3527
Triboluminescence, 2191
Tricarboxylic acid cycle, 613
Trichinosis, 1470
Trichomoniasis, 1658, 3151
Trichroism, 3527
Trichromatic coefficients, 3527
Trickle irrigation, 2025
Trickling filter sludge treatment, 3715
Tridipole antenna, 188
Tridop, 3527
Tridymite, 3527
Triethanolamine, 1330
Triethylene glycol, 1333
Triethylenethiophosphoramide, 596
Trifoliate orange, 802
Trifolium, 3030
Trigeminal neuralgia, 2444
Trigone, 2062
Trigonometric curve, 3527
Trigonometric function, 3527–3528
Trigonometry, 3528–3529
Trihydric alcohols, 87
Trim (airplane), 3529
Trim cooler, 1741
Trimethylene glycol, 87
Trimmer capacitor, 3529
Trinitrotoluene (TNT), 1343
Triode, 1278
Tripelennamine hydrochloride, 203
Triphylite, 3529
Triple bond, 751
Triple point, 2707
Triple precision, 2844
Tripolite, 3529
Trisaccharides, 610
Tris-BP, 1414
Trisectrix of Maclaurin, 3529
Trismus (lockjaw), 352, 3463–3464
Trisomy 21 (Down's syndrome), 1141–1142
Tristimulus values, 3529
Tritium, 3529

Triton, 2438–2440
    orbital characteristics, 2761
Tritone paradox, 29
Trochoid, 3529
Trogon, 2907, 3529
Trogoniformes, 3529–3530
Trojan asteroids, 276
Trophallaxis, 3530
Trophic substances, 701
Trophosome, 3530
Tropical air, 72
Tropical cyclone, 1510–1512
Tropical depression, 1510
Tropical diseases, 3530
Tropical disturbance, 1510
Tropical fishes, 3531
Tropical forest biome, 432
Tropical month, 3531
Tropical showers, 1517
Tropical sprue, 3318
Tropical storm, 1510
Tropical ulcers, 3556
Tropical year, 3531
Tropic-birds, 2671
Tropism (zoology), 3531
Tropopause, 304, 3531
Troposphere, 304, 3531
Troupial, 2656
Trout, 3054, 3531
    aquaculture, 217–218
Trout-perch, 3531
Trout sturgeon, 172
True-airspeed indicator, 85–86
True north, 2471
Truffle, 267
Trumpeter, 2939
Truncation, 3531
Truncation error, 3531
Truncus arteriosus, 3532
Trunk, 3532
Trunks (telephony), 3434
Truss, 3532
Trypanosomiasis
    African, 63
    South American (Chaga's disease),
        718–719
Tryptophan, 133–137
Tschirnhausen Quadratix, 2895
T section, 3532
T-section filter, 1393
Tsetse fly, 3532–3533
Tsunami, 1185, 2535, 3533–3534
    and volcanic eruptions, 3682
Tuatara, 3534
Tubal pregnancy, 1658
Tube-noses, 2851
Tuberculosis, 352, 3534–3535
    foodborne, 1461
    of skin, 1053
Tuberculosis arthritis, 237
Tuberculosis meningitis, 3535
Tubers (solanaceae), 3213–3214
Tubfish, 3121
Tub transformer, 926
Tubules, 2060–2061
Tubuliform glands, 3536
Tucana (constellation), 3536
Tucotuco, 3024
Tuff, 3536
Tukon hardness, 1715
Tularemia, 3536
Tulip, 2135

Tumble-bug, 3536
Tumor, 600
Tumor necrosis factor, 600
Tuna, 3536–3538
Tundra, 3538
Tundra biome, 432
Tuned amplifier, 152
Tung, 3538
Tung oil, 1412, 3538
Tungsten, 3538–3540
    effect on steel, 2018
Tunguska event, 278, 293
Tunicate, 265, 777
Tuning fork, 3540
Tunnel engineering, 3540–3545
Tunnel vision, 2677
Tupelo trees, 3545
Tur, 1649
Turacos, 1004, 3545
Turbellaria, 3545–3546
Turbidimetry, 3546
Turbidity currents, 2525
Turbine
    expansion, 1573–1577
    gas, 1573–1577
    steam, 3546–3548
Turbine engine, 3546
Turbine flowmeters, 1439–1440
Turbocharged diesel engines, 1084–1085
Turboexpanders, 1576
Turbofan engine, 80, 3548
Turbojet engine, 79–80, 2035
Turboprop engines, 80
Turbot, 1415, 1416–1417
Turbulence, 3548
    atmospheric, 321
Turbulent boundary layer, 519, 1447
Turbulent flow, 54–55, 1445, 3548
Turing, Alan, 3548–3549
Turing machine, 586, 3549
Turing test, 239–240
Turkey, 2828, 3549
Turkey buzzard, 3693
Turkish boxwood, 528
Turmeric, 3549
Turn-and-bank indicator, 3549
Turner's syndrome, 3549
Turnover frequency (catalyst), 652–653
Turnstile antenna, 188, 3549
Turpentine, 2977
Turquoise, 3549
Turtle, 3463, 3549–3552
Tussock moth, 3552
Twaddle scale, 3298
Twilight, 320, 3552
Twine, 3028
Twin-lens camera, 2732
Twinning, crystals, 2356, 3552
Twins, 3552
Twisted-nematic LCDs, 2145–2146
Two-address instruction, 1957
Two-body problem, 3552–3553
Two-dimensional computer animation, 904
Two-drum Stirling boiler, 496
Two-out-of-five code, 855
Two-position controller, 2404
Tyndall, John, 3553
Tyndall effect, 3553
Type II restriction endonucleases, 446
Type SO/SBO galaxies, 1549
Typhoid fever, 352, 365, 3553–3554
    foodborne, 1461

Typhoon, 1510
Typhus, 3001–3002
Tyrant flycatcher, 2071
Tyrosine, 133–137
Tyuyamunite, 3554

Ubac, 3555
Udad, 1650
Udic, 3555
Ugrandite, 3555
Uintahite, 3555
Ulbricht sphere, 3555
Ulcer, 3555–3556
Ulcerative colitis, 864–865
Ulexite, 3556–3557
Ultimate analysis, 162
Ultimate period, 3557
Ultimate strength, 3557
Ultisols, 3211
Ultrabasic, 3557
Ultracold neutrons, 2448
Ultrafilter, 1395
Ultrafiltration, 2310
Ultrahigh-strength steels, 108
Ultramicroscope, 3557
Ultrasonic flowmeters, 1441
Ultrasonic nondestructive testing methods,
    2468–2469
Ultrasonics, 3557–3559
    MRI contrasted, 2479
Ultrasonic thickness gaging, 3476
Ultraviolet astronomy, 3559–3562
Ultraviolet catastrophe, in development of quantum
    theory, 2902
Ultraviolet films, 2727
Ultraviolet microscope, 2347
Ultraviolet photoelectron spectroscopy, 1276
Ultraviolet radiation (UV), 3562–3563
    and cancer, 600
Ultraviolet spectrometers, 3563–3564
Ultraviolet-visible spectropolarimeter, 3302
Ulysses mission, 2051–2052, 3564–3566
Umbel, 1433
Umbelliferae, 3566
Umbilical cord, 1291, 3566
Umbilical tower, 3566
Umbilicus, 3566
Umbra, 3379, 3566
Umbrella bird, 727
Umbriel, 3583–3584
    orbital characteristics, 2761
Umkehr effect, 3566–3567
Unavailable energy, 3567
Unbiased sample, 3061
Uncertainty principle, 742, 3567
Unconformity, geological, 3567
Underbreathing, 1892
Underdamped, 1026
Underwing, 3567
Undulant fever, 559
Ungulates, 519
Unification, in automated theorem proving, 248
Unified file, 866
Uniform flow, 1447
Unilayer, 2386
Unimodular group, 2125
Unimpeded harmonic motion, 1716
Uninterruptible power supply, 1230
Union dyeing, 1155
Unitary group, 2125
Unitary modular group, 2125

Unit normal, 2470
Unit pulse, 395
Unit ramp function, 2085–2086
Units and standards, 3568–3571
    absolute system, 6–7
Univalves, 2383
Univariate distribution, 1135
Universal chuck, 792
Universal motor, 2399
Universal-pressure boiler, 497
Universal Time, 3493
Universe, 3571–3574
Unsaturated compounds, 896
Unsaturated hydrocarbons, 2583
    analytical determination, 161
Unsaturated vegetable oils, 3609
Unstable angina pectoris, 2026–2027
Unsupervised learning, 257
Unsymmetrical bending, 3574–3575
Unterzaucher method, 162
Up-conversion amplifier, 151
Uplift, hydraulic, 3575
Upper air, 305
Upper air observations, 3575
Upper Atmosphere Research Satellite, 3080
Upper consolute temperature, 926
Upper control limit, 934, 2898
Upper Curie point, 1007
Upper specification limit, 2899
Upsetting (in forging), 1474–1475
Upsilon particle, 2647, 3575
Upslope fog, 1458
Urafic rocks, 1362
Uralite, 3575
Uraninite, 3575
Uranium, 37, 3575–3579
Uranium enrichment, 3578–3579
Uranium-radium series, 736, 738
Uranium series, 2922
Uranium spent-fuel reprocessing, 3576–3577
Uranus, 3580–3586
    Hubble Space Telescope images, 1824
    orbital characteristics, 2759
    physical characteristics, 2760
    ring system, 3581–3582
    satellites, 3582–3585
Urea, 3587–3589
Urea amino resins, 139
Uremia, 2066
Uremic syndrome, 2066
Ureter, 2, 2060–2062
Urethra, 2060–2062
Urinary bladder, 2060–2062
    cancer, 2064
Urinary tract, 2, 2060–2067
Urinary tract infections, 365
Urination, 2062
Urine, 3589
Urology, 2060, 3589
Ursa Major (the greater bear; Big Dipper), 3589
Ursa Minor (the lesser bear; Little Dipper),
    3589
Ursines, 400–402
Urticaria (hives), 104, 3589–3590
U.S. Canada Air Quality Agreement, 20
Uterine cancer, 599
Uterine tumors, 1658
Uterus, 2, 1657
    retrodisplacement, 1658
Utter chaos, 3590
U-tube manometer, 2244
UV. *See* Ultraviolet radiation

Uveitis, 3590, 3652–3653
Uxisols, 3211

Vaccination, 3591
Vaccines, 3591–3596
Vaccinia (cowpox), 3596
Vacuole, 3596
Vacuum, 3596–3600
Vacuum casting, 649
Vacuum fluorescent displays, 3600–3601
Vacuum fluorescent on silicon microdisplay,
    2334
Vacuum gages, 3598–3600
Vacuum induction melting, 649
Vacuum pumps, 3597–3598
Vacuum tube diode, 1120
Vagina, 1657
Valence, 750–752
Valency, 2376
Valine, 133–137
Valley fever (coccidioidomycosis), 853
Valve (control), 3601–3602
Valvular heart disease, 1728–1729
Vanadinite, 3602
Vanadium, 3602–3603
    effect on steel, 2018
Van Allen, James, 3603–3604
Van Allen radiation belts, 3604
Vancomycin, 197
Van der Waals, Johannes Diderik, 3604
Van der Waals adsorption, 51
Van der Waals equation, 3604
Van der Waals forces, 3604
Van der Waals radius, 744
Vane, 3604
Vanillin, 264, 399
Van Slyke reaction, 137
V antenna, 188, 3591
Van't Hoff, Jacobus Henricus, 3604
Van't Hoff equation, 3604
Van't Hoff law, 3604
Vapor, 3604
Vapor-absorption refrigerator, 2967
Vapor-compression refrigerator, 2966–2967
Vapor density, 3604
Vapor-filled thermometer, 3472–3473
Vaporization, 3604
Vaporization (heat of), 3605
Vapor lock, 3605
Vapor pressure, 3605
    Dumas method for, 1152
Vapor trail, 914
Var (unit), 3571
Variable (process), 3605
Variable-area flowmeters, 1437–1438
Variable-focus lens, 3605
Variable star, 3605–3606
Variance, 3606
Variance-covariance matrix, 1129
Variate difference method, 3606
Variation
    coefficient of/mathematics, 3606
    quality control concept, 2900
Variations, calculus of, 3606
Varicella-zoster virus, 1053
Varicose veins, 235
Variegated cutworm, 1011
Variegation, 86
Variola (smallpox), 3187
Variolite, 3606
Variometer, 3606
Varley bridge, 539–540

Varves, 3606
Vascular bundle, 567–568
Vascular headache, 1719
Vascular purpura, 2886
Vascular system, 163, 232–236
 plants, 3606
Vascular tissue, 3606
Vasculitis, asthma associated with, 555–556
Vas deferens, 1655, 2061–2062
Vasoactive intestinal peptide, 699
 bone resorption stimulation, 506
Vasopressin, 699
Vast galaxy drift, 3606–3607
Vat dyes, 1154
V-band, 3607
V center, 3607
Vector, 3607
Vector (for illness), 1462
Vector addition, 3607
Vector cardiography, 1254
Vector derivative, 3607–3608
Vector field, 3608
Vector integral, 3608
Vector multiplication, 3608
Vector processing, 1112
Vector space, 2140
Veery, 3483
Vega (α Lyrae), 3608–3609
Vegetable adhesives, 46–47
Vegetable oils (edible), 3609–3613
Vegetable-type fruits, 1519–1520
Vein (geology), 3613
Vein, vascular system, 232–236, 798–799
Veingerov single-sided microphone system, 1937
Vela (constellation), 3613
Velcimeter, 3613
Velocity curve, stellar, 3615
Velocity limiting control, 3615
Velocity measurement, 3613–3615
Velocity modulation, 3615
Velocity profile, 3615
Velocity-sensing microphone, 2341–2342
Velocity transducers, 14
Velogenic viscotropic Newcastle disease (VVND), 2450
Velvet ant, 3615
Vena contracta, 3615–3616
Venn, John, 3616
Venous thrombosis, 235–236
Ventifact, 3616
Ventilating fan, 1353
Ventricle (brain), 714, 3616
Ventricular fibrillation, 228–229, 1382–1383
Venturi tube, 1436–1437
Venus, 3616–3628, 3629, 3630
 Galileo mission flyby, 1554–1555, 3627–3628
 Magellan mission, 2215–2217, 3626–3627
 orbital characteristics, 2759
 physical characteristics, 2760
Venus flytrap, 1956
Venus' girdle, 3628–3629
Vernal equinox, 807, 1323, 3629
Vernalization, 3629
Vernier, 3629
Vernier engine, 3629
Versine function, 3528
Vertebra, 3179, 3629
Vertebrate appendage, 213
Vertebrates, 777
Vertex, 3629–3630
 of an angle, 171
Vertex matrix, 3630–3631

Vertical circle, 3631
Vertigo, 3631
Vertisols, 3211
Very Long Baseline Interferometry, 2933
Vesicant, 3631
Vesico-intestinal fistula, 2062
Vesico-vaginal fistula, 2062
Vesicular (petrology), 3631
Vesiculobullous, 2673
Vesta, 274
Vestibular, 3631
Vestibular neuronitis, 1724
Vestigial structures, 3631
Vesuvianite, 3631
Vetch, 3631
Vibration, 3631–3633
Vibration exciter, 3633
Vibration sensing, 11
Vibratory feeder, 1357
Vibrio parahaemolyticus infection, 365–366, 1467
Vibrio vulnificus infection, 366
Viburnums, 1220–1221
Vickers hardness, 1715
Vicuña, 593–594
Vidarabine, 209
Video-frequency amplifier, 152
Videokeratography (corneal topography), 960
Vidicon, 3452–3453, 3633
View camera, 2732
Vignetting effect, 3633
Vigorous cold front, 1510
Viking missions to Mars, 2258, 2276–2283, 3633–3638
Vincent's angina, 2706
Vinegar acid, 15
Vinegar eel, 3638
Vinegar fly, 1148
Vinyl chloride, 770, 2685
Vinyl cyanide, 36
Vinyl ester resins, 3638–3639
Vinyl polymerization, 2593
Violet plantain-eater, 3545
Viper, 3195–3197
Viper fishes, 3639
Viral hemorrhagic fevers, 3639–3642
Viral replication, 3643
Vireo, 3697–3698
Virga, 2842
Virgo (constellation), 3642
Virtual image, 2520
Virtual reality, 911
Virtual work, 3783
Viruses, 3642–3646
 for pest control, 1955
Viscacha, 3024
Viscera, 163, 3646
Viscoelasticity, 3646
Viscosity, 2991–2992, 3646–3647
 and aerodynamics, 54
 Arrhenius equations, 228
 eddy, 1210
 Stokes law for, 3357
Viscosity manometer, 1033
Viscous flow, 2080
Viscous fluid, 1444
Visibility, 3647
Vision, 3647–3660
Vissoirs, 222
Visual acuity, 3661
Visual binaries, 424, 3661
Visual field, 3661
Visual function (eye), 3661

Visual polarimetry, 2784–2785
Vitamin, 3661–3662
Vitamin A, 3662–3663
Vitamin B$_3$, 2628
Vitamin B$_{12}$ (cobalamin), 3664–3665
Vitamin B$_6$ (pyridoxine), 3663–3664
Vitamin B$_2$ (riboflavin), 2998–2999
Vitamin B$_1$ (thiamine), 3474
Vitamin C, 267–268
 and colds, 888
Vitamin D, 3666–3667
 and bone loss, 505, 508–509
Vitamin-D resistant rickets, 508–509
Vitamin E, 3667–3668
Vitamin H, 459
Vitamin K, 3669
Vitamins
 and carbohydrate metabolism, 615
 in diet, 1087–1088
Vitiligo, 87
Vitrectomy, 3651
Vitreous detachment, 3669–3670
Vitreous state, 3670
Vitrophyre, 3670
Viverrines, 3670–3671
Vivianite, 3670–3671
Viviparous reproduction, 3671
Viviparous topminnows, 3671
VLDL (very-low density lipoprotein), 234
Vocal cords, 2090
Vodka, 3671–3672
Voice and sound production, 3671–3675
Voice recognition system, automotive electronics, 343
Volatile, 3675
Volatile oils, 3675
Volatility product, 3675
Volcanic breccia, 536
Volcanic eruptions, 1178
Volcanic terrain hydrology, 1878
Volcanism
 Mars, 2270
 Mercury, 2319–2320
Volcano, 3675–3685
Vole, 3022
Volt (unit), 3568, 3571
Voltage amplifier, 152
Voltage clamp analysis, 698
Voltage divider, 3686
Voltage doubler, 3686
Voltage regulator, 2835–2836
Voltage-source inverter, 3145
Voltage-to-frequency A/D converter, 3686
Voltaic pile, 1255
Voltampere (unit), 3571
Volterra, Vito, 3686–3687
Volume
 critical, 984, 3687
 geometry/standard, 3687
Volume integral, 1964
 vectors, 3608
Volume-limiting amplifier, 152
Volumetric feeder, 1357–1358
Vomiting, 1296
Von Baeyer name, for alicyclic bridged compounds, 2583
Von Braun, Wernher, 3687–3688
Von Frisch, Karl, 3688
Von Hippel-Lindau disease (VHL), 3688
Von Klitzing, Klaus, 3688
Von Neumann, John, 3688–3689
von Recklinghausen's disease, 2444

Vortex flowmeter, 1440
Vortex laws, 3689
Voyager missions to Jupiter and Saturn, 3689–3692
    Neptune visit, 2434–2440
Vredefort Ring, 292–293
V/STOL craft, 1745, 1750–1752
VTOL craft, 1745, 1752
Vulcanization, 3044–3045
Vulture, 1351, 3692–3693
Vulva, 1657
Vulvitis, 1658

Wad, 3695
Wadden zee, 3695
Waders, 726, 3695–3696
Wadi, 3696
Wadsworth mounting, of reflection grating, 2963
Wagyu cattle, 668
Wake, 3696
Walking fish, 2077
Walkingstick, 3696
Wallaby, 2285
Wallace's eulipoa, 2399
Wallboard, 1701
Walleye, 2676–2677
Walnut curculio, 1006
Walnut husk fly, 1521
Walnut trees, 3696–3697
Walrus, 3120
Wapiti, 1039, 1040
Warble fly, 3697
Warbler, 3697–3698
Warfarin, 200, 201
Warm-bloodedness, 1781–1782
Warm front, 1508
Wart, 1053, 3698
Wart-hog, 3405
Wash boring, 513
Wasp, 1147, 1885, 3698
    as beneficial insect, 411
Waspaloy, 108
Waste-heat boiler, 1741
Wastes
    as energy sources, 3706–3707
    and pollution, 3698–3705
    radioactive, 2500–2503
Wastewater treatment, 3713–3715
Water, 3708–3715
    degassed, 3711
    greenhouse gas, 1639
    precipitable vapor, 2838
    role in colloid systems, 867
    in saturated air, 69
Water boatman, 3715
Waterbucks, 183, 184
Water chambers, for ship stabilization, 3322
Water-cooled arc lamp, 1906
Water erosion, 3213
Waterfall effect, 2119
Water flea, 3715
Waterfowl, 3715–3716
Water gas reaction, 624
Water hammer, 3716
Waterhouse-Friderichsen syndrome, 50
Water hyacinth, 3716
Water injection, for jet engines, 80
Water intoxication, 2062–2063
Waterjet cutting, 3716–3717
Water lilies, 3717
Water-loop heat pump, 1737
Water measurer, 3717–3718
Watermelon, 1005

Water penny, 3718
Water pheasant, 3695
Water plants, 1879
Water-pollinated plants, 2797
Water pollution, 3718–3726
    rivers, 3718–3719
Water resources, 3726–3727
Water scorpion, 3727
Watershed, 3727
Water softening, 3712–3713
Waterspout, 1514, 1516
Water strider, 3728
Water thrush, 3697
Water treatment, 3712
    boiler feedwater, 1358–1361
    wastewater, 3713–3715
Watson, James Dewey, 982, 3728
Watt, 1234
Watt (unit), 3568, 3571
Watthour (unit), 1234, 3571
Wattmeter, 1236–1239
Watt per meter Kelvin (unit), 3571
Watt per steradian (unit), 3571
Watt per steradian square meter (unit), 3571
Wave-cut beach, 1324
Wave-cut platform, 2772
Wave cyclone, 1508–1509
Wave equation, 3728
Waveguide, 3728
Wave interference, 1968–1969, 3728
    Young's experiment, 3805
Wavelength, 3728
Wavellite, 3728
Wave motion equation, 2293
Wave normal, 3728
Wave number, 3728
Wave period, 3728
Wave phase, 3728
Wave pole (wave staff), 3728–3729
Wave propagation (Huygens' principle), 1854, 3729
Waves and wave mechanics, 3729–3730
Wave staff (wave pole), 3728–3729
Wave train, 3730
    group velocity, 1695
Wave vector, 3730
Wave velocity (phase velocity), 3730
Wave velocity surface, 3730
Wax moth, 3730
Wax myrtle, 396
Waxplant, 2350
Waxwing, 3730
Weak force, 2645
Weak (pseudo) salt, 3058
Weasel, 2407
Weather alerts, 3734–3735
Weathering, 1324, 3731
Weather satellites, 3736
Weather technology, 3731–3737
Weaverbird, 3737
Weber (unit), 3568, 3571
Weber equation, 3737
Weber-Fechner law, 3737
Webworm, 3737–3738
Weddle rule, 3738
Wedge, 311
Weed control, 1758
Weeverfishes, 3738
Weevil, 864, 3738–3739
Wegener, Alfred Lothar, 3739–3740
Weibull distribution, 3740
Weierstrass function, 3741
Weighing, 3741–3744

Weight, 1682
Weighted mean, 225–226
Weighting, 3744
Weightlessness, 293–294, 1682, 1687, 3744, 3811
Weil's disease, 2122
Weirs, 1443
Weldalite, 123
Welding, 3744–3746
Wentletrap, 3746
Westerlies, 309
Western Union clock, 1234
West Nile virus, 3746–3748
Weston cell, 395
West-wind drift, 3749
Wet-bulb measurements, 1882–1883
Wet-bulb temperature, 2878–2879
Wet cell battery, 389
Wetlands, 3749–3754
Wet solar cells, 3227
Whales, 3754–3756
Wheat, 3756–3760
    irradiated, 2021
Wheatear, 2656
Wheat flour, 3760
Wheatgrasses, 1680
Wheat midge, 2348, 3760
Wheat rust, 1530
Wheatstem sawfly, 3102
Wheatstone bridge, 538, 1262
Wheel ore, 519
Whelk, 3761
Whinchat, 2656
Whinstone, 3761
Whippoorwill, 2456
Whipray, 3177, 3178
Whip scorpion, 220, 3761
Whiptail, 2176–2177
Whipworms, 1982–1983
Whisker, 3761
Whiskey, 3761–3764
Whistler, 3319, 3320, 3764
White body, 3764
White cedar, 221
White cedar wetland forests, 3753–3754
White cells, 484
White dwarf stars, 1621, 2449
Whitefish, 3764
Whitefly, 1782, 3764
White-fringed beetle, 3764–3765
White-man's-footsteps, 2763
White matter, 695
White noise, 2466
Whiteout (atmospheric optical phenomenon), 320
White peach scale, 3205
White point, 18
Whitethroat, 3697
White-throated poor-will, 608
Whiteware, 710
Whiting, 855–866, 984
Whooping cough (pertussis), 352, 2684
Wide-band amplifier, 152
Wiegand-effect switches, 2875
Wien, Wilhelm, 3765
Wien bridge, 541
Wien bridge oscillator, 3765
Wien laws, 3765
Wigner force, 3765
Wigner nuclides, 3765
Wilcoxon's test, 3765
Wild rice, 3000
Wildrye grasses, 1680
Willemite, 3765

Willow trees, 3766–3767
Wilm's tumor, 2064
Wilson cloud chamber, 824
Wilson's disease, 2172, 3767
Wind correction angle (navigation), 3768
Wind direction indicators, 3768
Wind erosion, 3213
Windhover, 1351
Windmilling, 1355
Wind power, 3768–3773
Winds and air movement, 3773–3778
Wind shear, 3778
Wind tunnel, 3778
Wind velocity measurements, 3767–3768
Wines, 1666–1673
Wing louse, 2190
Wingnut trees, 1768–1770
Winter egg, 3778
Winter forest, 432
Winterizing, vegetable oils, 3609
Winter solstice, 807, 3232
Wire drawing, 3778–3779
Wireless telephony, 3426, 3428–3429
Wireworm, 3779
Wisent, 520
Wishart's distribution, 3779
Witches' brooms, 3779
Witch hazel, 3779
Witch of Agnesi, 3779
Witherite, 3779
Wittig reaction, 3779–3780
Wobblestick actuator, 2137
Wolffish, 480
Wolff-Parkinson-White syndrome, 229
Wolf herring, 1140
Wolframite, 3780
Wollastonite, 3780
Wolverine, 2408
Wolves, 603
Wombat, 2285
Wood, 2880, 3780–3782
Wood borer, 864
Woodchuck, 3319, 3320
Woodcock, 3695
Woodhewer, 2656
Woodland biome, 433
Wood louse, 3782
Woodpeckers, 2745, 3782–3783
Wood pulp production and processing, 2880–2883
Woodruff key, 2059
Wood tick, 3487
Wood warblers, 3698
Wool, 1382, 3783
Woolly aphis, 212
Woolly bear, 3783
Wool-sorter's disease (inhalation anthrax), 191
Wooly citrus whitefly, 3764
Word, 3783
Work, 3783
    and first law of thermodynamics, 3470
    virtual, 3783
Worker bee, 1783–1784
Work function, 756
Work function, electronic, 3784
Work hardening, 3784
Working storage, 3358
Workplace pollution, 2799
    carcinogens, 627–631
Workstation clusters, 1116
World Geographic Reference System (GEOREF), 3784
Worldwide pollution, 2800

Worm (gear and gearing), 2209, 3784
Worm (grub), 1696
Worm lizards, 2178
Worm wheel, 2209
Wormwood oil, 232
Wort, 406
Wound tissue (callus), 589
Wrapper methods, 258
Wrasses, 3784–3785
Wren, 3785
Wrist, 3785
Wronskian, 3785
Wrought ferrous alloys, 108
Wrought iron, 3785
Wrybill, 3695
Wryneck, 3783
Wulfenite, 3785
Würtz-Fittig-Frankland reaction, 3785
Wurtzite, 3785

Xalstocite, 3787
Xanthan gum, 3787
Xanthophylls, 2747, 3787
Xantus becard, 727
X-band, 3787
X-chromosome, 3787
XCON expert system, 252
Xenia, 3787
Xenoblast, 3787
Xenocryst, 3787
Xenolith, 3787
Xenomorphic, 3787
Xenon, 3787–3788
Xenopsylla cheopsis, 3788
Xenotopic, 3788
Xerophytes, 3788
Xerosis, 3788
Xerostomia, 3788
X2000 high-speed train, 2940
Xi hyperons, 2322
Xi particle, 3788
Xiphosura, 3788–3789
XMM-Newton (High-Throughput X-Ray Spectroscopy Mission), 3789–3790
X-network window, 3790
X-ray, 3790–3791
X-ray analysis, 3791–3794
X-ray astronomy, 3794–3796
X-Ray CAT scanner, 3796
    for brain imaging, 703
X-ray computed tomography, 2468
X-ray fluorescence analysis, 393
X-ray lasers, 2096
X-ray photoelectron spectroscopy, 1276
X-ray scan and other medical imagery, 3796–3797
X-ray thickness gaging, 3476–3477
X-unit (Xu), 3797
Xylem, 592, 3797
Xylene, 2685, 3797–3798
Xylitol, 3400
X-Y recorder, 3798

YAC clones, 1834
YAG (yttrium aluminum garnets), 3799
Yagi antenna, 188–189, 3799
Yak, 520, 521
Yams, 3403
Yang, Franklin Chen-Ning, 3799
Yang-Mills equations, 2301
Yang-tao plant, 2073
Yardang, 3799

Yarn, 3028
Yaw (aircraft), 3799
Yaws (disease), 3799
Yazoo stream, 3799
Y-chromosome, 3799
Y-connection, 3799–3800
Yeasts, 3800–3803
    for brewing, 405
Yellowbird, 2656
Yellow bone marrow, 505
Yellow fever, 3803
Yellow-green algae, 90, 3803
Yellow gurnard, 3121
Yellow-jacket, 3803
Yellow nitidulid, 1146
Yellow-striped armyworm, 1012
Yellow-tail, 609
Yellow wagtail, 976
Yellow warbler, 595
Yersinia pestis, 563
Yersiniosis, 366
    foodborne, 1467–1468
Yew trees, 3803–3804
Yield point, 3804
YIG (yttrium iron garnets), 3799
Ylang-ylang oil, 595
Yoke, 1041, 3804
Yolk gland, 3804
Yolk sac, 1346, 3804–3805
Yosemite (U-shaped portion of glacial valley), 3805
Youden square, 3805
Young, Thomas, 3805
Young-Helmholtz theory, 3805
Young interference experiment, 3805
Young's modulus, 3805
Youth (geological), 3805
Ytterbium, 3805
Yttrium, 3806
Yttrium aluminum garnets (YAG), 3799
Yttrium iron garnets (YIG), 3799
Yucca borer, 3806
Yucca moth, 3806
Yukawa, Hideki, 3806–3807
Yukawa potential, 3807
Yule autoregression scheme, 343

Zebra bird, 2308
Zebra fish, 3809
Zebras, 1798, 1799–1800
Zebra time, 3809
Zebu, 520, 668
Zedoary, 3809
Zeeman, Erik Christopher, 3809
Zeeman effect, 741, 1277, 3809
Zeisberg concentrator, 3809
Zeisel reaction, 3809
Zelkova trees, 1705
Zellweger's syndrome, 3809
Zener current, 3809
Zener diode, 3809–3810
Zener voltage, 3810
Zenith, 3810
Zenith telescope, 3810
Zeolite group, 3810
    as adsorbents, 51, 52
    aluminum in, 120
Zeppelin, 1126
Zerevitinov determination, 3810
Zero, 3810–3811
Zero-based linearity, 2139

Zero-beat reception, 1781
Zero-emission vehicles, 1231
Zero error, 3811
Zero gravity, 3811
Zero-order reaction, 3811
Zero point energy, 3811
Zero point entropy, 3811
Zero-point internal energy, 1971
Zero shift, 3810, 3811
Zeroth law of thermodynamics, 3469–3470
Zeta potential, 1258
Zigzag ridge, 3811
Zinc, 3811–3814
    in batteries, 393–394
    in biological systems, 3814–3815
Zinc alloys, 108
Zinc blende (sphalerite blende), 3307
Zincite, 3815
Zinc oxide sunblock, 3181

Zinc spinel, 1543
Zinc spraying, 1565
Zingg pebble classification, 3815
Zipf's law, 3815
Zircon, 3815–3816
Zirconium, 3816
    effect on steel, 2018
Zodiac, 3817
Zodiacal light, 3817
Zoisite, 817, 3817
Zollinger-Ellison syndrome, 3817
Zonal aberration, 3817
Zonal flow, 309
Zona pelludica, 821
Zone, 3817
Zone bit, 476
Zone of capillarity, 607
Zone plate, 3817
Zone time, 3817–3818

Zoochlorella, 3818
Zooecium, 3818
Zoogeography (chorology), 3818
Zoology, 3818
Zoosporangium, 3818
Zoospore, 3818
Zoraptera, 3818
Zoster, 68
Zosterops, 2657
Zsigmondy reagent, 3818
Zucchini, 3319
Zwitterion, 1995, 3818
Zworykin, Vladimir Kosma, 3818
Zygomatic arch, 3818–3819
Zygospore, 3819
Zygote, 679, 1565, 3819
    in cloning, 819, 821
Zymogenic cells, 3819
Zymolytic reaction, 3819